P9-DMX-840

WORLD *of* BIOLOGY

WORLD *of* BIOLOGY

Kimberley A. McGrath, *Editor*

The Gale Group

DETROIT • SAN FRANCISCO • LONDON • BOSTON • WOODBRIDGE, CT

STAFF

Kimberley A. McGrath, *Editor*

Robyn V. Young, *Coordinating Editor (Images)*

Jacqueline L. Longe, *Contributing Editor*

Zoran Minderovic, *Associate Editor*

James Edwards, *Editorial Technical Consultant*

Margaret A. Chamberlain, *Permissions Specialist*
Shalice Shah-Caldwell, *Permissions Associate*

Mary Beth Trimper, *Production Director*
Evi Seoud, *Assistant Production Manager*
Cindy Range, *Production Assistant*

Cynthia D. Baldwin, *Product Design Manager*
Michelle DiMercurio, *Art Director*
Barbara Yarrow, *Graphic Services Manager*
Randy Bassett, *Image Database Supervisor*
Robert Duncan, *Imaging Specialist*

ISBN: 0-7876-3044-6
Printed in the United States of America
10 9 8 7 6 5 4 3 2 1

Library of Congress Cataloging-in-Publication Data
World of biology / Kimberley A. McGrath, editor.
    p.   cm.
  Includes bibliographical references (p.   ) and indexes.
   ISBN 0-7876-3044-6
   1. Biology—Encyclopedias. I. McGrath, Kimberley A.
QH302.5.W67 1999
570'.3—dc21
                           98-53855
                            CIP

CONTENTS

INTRODUCTION

Welcome to the *World of Biology*. We hope you will find this collection interesting and useful. The 1,034 individual entries in this volume offer an up-to-date overview of the dynamic field of biology and its allied disciplines. Ranging from acid rain, aging, and AIDS to zoology, zooplankton, and zygote, the entries explain in concise, detailed, and jargon-free language some of the most important topics, principles, and recent discoveries in biology. Brief biographies of the people who made those discoveries and shaped our understanding of the field also are included. Users will find here, for example, biographical sketches of almost everyone who has won a Nobel Prize in this century in a field related to biology.

Biology and its allied disciplines have changed at a breathtaking pace over the past fifty years. Those of us who lived through the molecular revolution that swept through biology over the past few decades can hardly believe how far we have come in such a short time. Experiments, procedures, and knowledge that once seemed completely beyond our reach have now become almost commonplace. Although we have known since 1953, for instance, that the deoxyribonucleic acid (DNA) in our chromosomes contains the genetic information controlling our growth, development, and metabolism, it seemed up until fairly recently that the mechanisms by which that information was stored and expressed must be so complex that we would surely never understand them in any detail. However, as this is being written in early 1999, the genetic material of some twenty species already has been completely sequenced. Molecular biologists confidently predict that within a year or two, all of the 100,000 or so genes and more than three billion nucleotide pairs that determine human heredity will have been spelled out and the information contained there will become available for biomedical technology.

While dramatic breakthroughs have occurred in genetics, progress in other biological fields has increased as well. Biochemists have mapped the chemical reactions that make life possible in exquisite detail. Anatomists and physiologists have studied the structures and processes that allow tissues and organs to carry out a multitude of functions. Endocrinologists have elucidated the amazing hormonal relationships that regulate growth, development, and functions of our bodies. Taxonomists have discovered and classified millions of different species and subspecies. Botanists and agronomists have uncovered the secrets of plant growth and have succeeded in transferring genes from cell to cell via genetic engineering, thus growing mature plants from single cells.

Ecologists now understand ecological processes much more fully than ever before. Even as we disrupt habitats and extirpate indigenous species, we are coming to recognize and appreciate the value of the ecological services that keep our world running. Whether human disturbances of the natural world will have catastrophic effects on global systems such as climate regulation, ocean circulation, and ecosystem productivity remain some of the most important questions faced by science. What we can do about these ecological problems may be our greatest challenge for the future.

Our rapidly expanding understanding of the world of biology already is finding practical applications that effect our lives in never-before imaginable ways. Before this century, medicine consisted mainly of amputation saws, bloodletting, morphine, and crude medicines that often did more harm than good. Modern advances have made life much safer and more comfortable than ever before in history. Terrible diseases such as small pox have been eradicated. Survival rates from colon cancer, childhood leukemias, heart attacks, and several other life-threatening diseases have increased dramatically. Installation of artificial joints and transplants of vital organs have become almost commonplace. Life expectancy in the United States has jumped from 47 to 76 years. Now biotechnology companies offer hope that in the not-too-distant future we may be able perform such miracles as curing cancer, growing new blood vessels to damaged hearts, creating synthetic organs from fetal stem cells, and replacing nonfunctioning neurons in our brains.

It may soon be possible to create designer drugs aimed at

a unique disease and condition. Understanding exactly which chemicals to which a pathogen is susceptible may determine which drugs or therapies will be most effective. Extracting immune cells from the body and culturing them in the laboratory makes it possible to produce enough specifically targeted killer cells to eliminate a specific pathogen or tumor. Already our detailed understanding of how certain chemicals affect cells has yielded potent medicines. Focusing on serotonin receptors in the brain, for example, led to the development of popular antidepressant drugs. Targeting histamine receptors in the stomach produced powerful antacid remedies. Will there be similar silver bullets to treat Alzheimer's disease or cancer?

Modern biomedical science now make giving birth much safer and more predictable. Where prenatal examinations once consisted mainly of listening to the baby's heartbeat and checking the mother's weight, it is now possible to screen for congenital problems such as cystic fibrosis, Down syndrome, hemophilia, and Huntington's disease. Surgery *in utero* to fix heart problems or other developmental abnormalities is now rare, but possible. Fertility drugs stimulate ovulation and increase the likelihood of conception. *In vitro* fertilization techniques make it possible for previously infertile couples to have children. However, these advances also raise ethical dilemmas. How much latitude should parents have in choosing characteristics for their babies. It is now possible to pick the sex of your baby and to screen for genetic illnesses. Will it someday be feasible to select for brains, beauty or athletic prowess too? In 1998, a Texas mother using fertility drugs gave birth to octuplets that are likely to suffer a lifetime of health problems. Is this fair to the babies? Is it acceptable to remove some fetuses? How should we choose who will live and who will die?

Genetic engineering techniques have already been used to replace cells making deficient products in our bodies. Children with a genetic immune deficiency (the kind made famous by the boy in the bubble with no natural immunity) have had blood cells removed from their bodies and treated with recombinant DNA technology to insert new genes in place of faulty ones. When these cells are transfused back into circulation, they replace defective cells and restore normal immune function. In some forms of leukemia, it is possible to destroy all the bone marrow with x rays and chemotherapy and then transfuse normal bone marrow back into the body if a close match can be found. Most of these therapies are expensive, however. How will we determine who gets treatment and how much society is willing to pay. Will these treatments be available only to young people with prospects for a long, productive life or will they be offered to everyone?

Among the difficult questions raised by this technology are how much would you want to know about your future and that of your children and who else should have access to this information? Suppose you or your children have a high probability of developing a fatal or debilitating condition later in life, would you want to know what lies ahead? If there is no treatment for your situation, would you rather not know what the future holds if there is nothing you can do about it? Should

insurance companies and employers have access to this information?

Gene analysis and recombinant DNA technology also have transformed law enforcement. DNA sequencing is increasingly used to identify criminals. Thanks to recently discovered amplification procedures, it is now possible to get enough nucleic acid for analysis from a single hair or a drop of blood. Although juries are sometimes reluctant to accept scientific evidence that they don't understand, it is possible, in some cases, to identify a suspect with a certainty of one in a million or one in a billion. It is now routine to keep a database of all sex offenders and violent criminals. In some countries, police have tested every male in a village in an effort to catch a rapist. How much testing of the general public would we accept? Should a genetic profile be a part of your permanent record? How would we protect privacy of innocent individuals and prevent misuse of such information?

Controversial applications of biotechnology also have been proposed for agriculture. Recombinant DNA techniques have been used to create plants that produce their own insecticide, tomatoes that don't rot on the shelf, frost-resistant strawberries, coffee with no caffeine, sheep that manufacture human hormones and clotting factors in their milk, "miracle" strains of grains that have higher yields and specific protein compositions. Cows treated with exogenous growth hormone produce higher milk yields. Until recently, growth hormone was too expensive to be used commercially, but genetic recombination techniques make it possible to inject milk cows routinely with synthetic hormone. Although Canada and many European countries have banned milk produced in this way, American officials have found no evidence that drinking milk from recombinant hormone-treated animals carries any additional risk. What do you think? Would you knowingly consume foods produced by genetic engineering or sterilized by nuclear radiation? Should such foods be labeled or would that just create fear and hysteria?

Ecologists have worries about genetic engineering and other modern technologies. It has been shown that altered genes can move from domestic crops to weedy relatives either by cross-breeding or transfer by insect or viral vectors. There is a possibility that we will produce super pests. Similarly, growing plants engineered to produce indigenous pesticides may simply increase the rate at which pests become resistant to controls. The use of subtherapeutic doses of antibiotics in animal feed may expose consumers to dangerous levels of these powerful chemicals and also is likely to select for resistant pathogens that will (or perhaps already do) infest our food supply. Chemicals once thought to be inert and innocuous, such as polychlorinated biphenyls, were released into our environment in large quantities. We now are finding that they have subtle but widespread effects on both human health and that of wildlife.

All of these examples demonstrate how biology is fraught with both promises and perils as we learn more and increase our power to transform the world around us. It becomes increasingly important for all citizens to have some scientific literacy and to understand some of the questions and problems that arise from our rapid progress in science and technology. We hope that the essays and articles contained in this book will help you understand some of these important issues and how they affect you and the world in which you live.

William P. Cunningham
Professor, Cell Biology
University of Minnesota
St. Paul, Minnesota

How to Use the Book

This first edition of *World of Biology* has been designed with ready reference in mind.

- **Entries are arranged alphabetically**, rather than by chronology or scientific field.
- **Bold-faced terms** direct reader to related entries.
- **Cross-references** at the end of entries alert the reader to related entries not specifically mentioned in the body of the text.
- A **Sources Consulted** section lists the most worthwhile print material we encountered in the compilation of this volume. It is there for the inspired reader who wants more information on the people and discoveries covered in this volume.
- The **Historical Chronology** includes more than 300 important events in the biological sciences spanning the period from 50,000 B.C. to 1999.

Special Thanks

In compiling this edition, we have been fortunate in being able to call upon the following people—our panel of advisors—who contributed to the accuracy of the information in this premier edition of *World of Biology*, and to them we would like to express sincere appreciation:

Nancy Bard
Librarian
Thomas Jefferson High School for Science and Technology
Alexandria, Virginia

Dr. Marie H. Bundy
Senior Scientist
Academy of Natural Sciences
Estuarine Research Center
St. Leonard, Maryland

Josephine Davies
Library Manager
Whitmore Library
Salt Lake City, Utah

Jennifer L. McGrath
Faculty Associate
West Noble High School
Ligonier, Indiana

Rupert Sheldrake
Biologist and Physiologist
London, England

A

ABIOTIC ENVIRONMENT

Ecologists quite commonly divide an **organism**'s surroundings, or environment, into two categories, its abiotic environment and its **biotic environment**. The abiotic environment is the non-living part. Included are all the physical elements of an organism's existence, especially the intersecting roles of the sun and solar **energy**, weather and climate, soil, and **water**. The basic dependence of biotic life on the abiotic environment is the need for solar energy to be transformed, by green **plants** through the processes of **photosynthesis**, into a form of energy useable by living organisms. Climate is widely considered a major determinant of life patterns on planet Earth. And, climate of course is driven by earth-sun relationships.

The quality of the abiotic environment as measured by how well it meets organisms' needs for survival and reproduction is a critical factor of life. Water, for example, can vary in terms of its spatial and seasonal availability, but also in terms of characteristics such as turbidity, alkalinity, pH levels, conductivity, level of dissolved oxygen, and the concentrations and combinations of various chemicals. Similar variances and spatial differences can be identified for each element of the physical or abiotic environment. Changes or perturbations in the abiotic environment can alter the conditions of organismic existence: drought, flood, fire, volcanic eruptions—all can change the conditions for life, sometimes drastically. Seemingly small, trivial, or subtle changes, such as variations or alterations in shade patterns, can also change ecological relationships. All types of ecological interactions vary in response to changes in the abiotic environment, not only those interactions between organisms and their non-living environment, but also the whole range of interactions among the organisms themselves. **Adaptation** and **speciation** are both measures of organismic response to changes in the abiotic environment.

Organisms also change the abiotic environment. In the course of the earth's history, organisms and their abiotic environment have co-evolved, each changing the other in the course of that **evolution**. Primitive organisms drastically changed the abiotic components of the earth, dramatically altering its receptivity to higher life forms, including, ultimately, the human **species**. Human activities (a biotic factor) are resulting in considerable changes in the earth's abiotic environments, including changes in global cycles: e.g., a wide variety of Earth's **ecosystem**s seem to be reacting, in multiple and varied ways, to increasing concentrations of **atmosphere** CO_2.

ACETIC ACID

Pure acetic acid (CH_3COOH) is a clear, colorless liquid that is found in vinegar. In fact, the name for the acid is derived from the Latin word for vinegar, acetum. When highly concentrated acetic acid is corrosive and can cause severe burns; in dilute form, it is useful in food preservation and a variety of other applications. Acetic acid is an organic acid because, like all **organic compound**s, it contains carbon. Like most acids, acetic acid **taste**s sour or tangy and has a pungent, biting odor. It is this quality that gives the vinegar used in salad dressings its characteristic tart flavor. Acetic acid constitutes about 5% of most table vinegars.

It is thought to have been first produced accidentally as a byproduct of wine production, because when fruit juices are fermented for too long they are converted to acetic acid. In commercial practice, manufacturers used to separate acetic acid from **water** by chilling the solution to solidify the acid into crystals. Thus the pure acid is known as ''glacial'' acetic acid. In the 700s, an Arabian alchemist named Jabir ibn Hayyan Geber (c. 721-815) prepared strong acetic acid by distilling vinegar. However, acetic acid was not produced in its pure form until Georg Ernst Stahl (1660-1734) isolated it in 1700. For more than a hundred years, scientists thought that acetic acid and other organic substances could be produced only from compounds that occur in living **organism**s. Then in the mid-1800s, chemists began synthesizing organic substances from chemicals. In 1845, German organic chemist Adolf Wilhelm Hermann Kolbe synthesized acetic acid from chemicals that do not occur in **plants** or **animals**. This work anticipated our modern distinction between organic chemistry, the study of compounds containing carbon, and biochemistry, the study of compounds and processes that occur in organisms.

About a century later, German-born American biochemist Konrad Emil Bloch discovered that acetic acid is the major precursor of **cholesterol**, a lipid found in nearly all human **tissue**s. Acetic acid is converted to cholesterol in the liver through a series of thirty-six chemical steps. Using radioactive tagging, Bloch determined which of cholesterol's carbon **atoms** come from acetic acid. His research laid the foundation for today's understanding of cholesterol **metabolism** and its relation to **heart** disease.

Acetic acid's importance is not limited to its role in biological processes. White vinegar is used as a household cleaner and in food processing as a preservative. Acetic acid is a valuable industrial chemical as well since it can be used as a solvent or converted into such useful compounds as acetic anhydride and acetate esters. Acetic acid is also important in the manufacture of synthetic fabrics and pharmaceutical products. It has also been used historically as a **fungi**de in certain agricultural application. However, in 1993, the U.S. EPA revoked this approval. Nonetheless, remains an important industrial chemical.

ACETYL COENZYME A · See Coenzyme

ACETYLCHOLINE

By the early 1900s, scientists had a reasonably clear idea of the **anatomy** of the **nervous system**. They knew that individual nerve **cells**—**neuron**s—formed the basis of that system. They also knew that nerve messages traveled in the form of minute electrical currents along the length of a neuron and then passed from the **axon** of one cell to the dendrites of a nearby cell.

One major problem remained, however. What was the mechanism by which the nerve message travels across the narrow gap—the **synapse**—between two adjacent neurons? The British neurologist Thomas R. Elliott (1877-1961) suggested an answer to that question as early as 1903. He proposed the idea that the nerve message is carried from one cell to another by means of a **chemical compound**. Elliott thought that adrenaline might be this chemical messenger, or neurotransmitter, as it is known today.

Nearly two decades passed before evidence relating to Elliott's hypothesis was obtained. Then, in 1921, the German-American pharmacologist, **Otto Loewi** (1873-1961), devised a method for testing the idea. Born in Frankfurt-am-Main, Germany, in 1873, Loewi received his medical degree from the University of Strasbourg in 1896 and then taught and did research in London, England, Vienna, Austria, and Graz, Austria. With the rise of Adolf Hitler (1889-1945), Loewi left Germany first for England and then, in 1940, the United States where he became a faculty member at the New York University College of Medicine.

In his 1921 experiment, Loewi found that when he stimulated the nerves attached to a frog's **heart**, they secreted at least two chemical substances. One substance he thought was adrenaline, while the second he named vagusstoffe, after the vagus nerve in the heart.

Soon news of Loewi's discovery reached other scientists in the field, among them the English physiologist Henry Dale (1875-1968). Dale completed his undergraduate work at Trinity College and then earned a medical degree from Cambridge in 1909. After a short academic career at St. Bartholomew's Hospital in London and at University College, London, Dale joined the Physiological Research Laboratories at the pharmaceutical firm of Burroughs Wellcome. Except for the war years, Dale remained at Burroughs Wellcome until 1960. He died in Cambridge on July 23, 1968.

While attending a conference in Heidelberg, Germany, in 1907, Dale became interested in the fungus ergot and the chemicals it secretes. By 1914, Dale had isolated a compound from ergot that produces effects on **organ**s similar to those produced by nerves. He called the compound *acetylcholine*. When Dale heard of Loewi's discovery of vagusstoffe seven years later, he suggested that it was identical to the acetylcholine (abbreviated ACh) he had discovered earlier. For their discoveries, Loewi and Dale shared the 1936 Nobel Prize for physiology or medicine.

Unraveling the exact mechanism by which acetylcholine carries messages across the synapse has occupied the energies of countless neurologists since the Loewi-Dale discovery. Some of the most important work has been done by the Australian physiologist, Sir **John Carew Eccles** (1903-1997), and the German-British physiologist, Bernard Katz (1911-). Eccles developed a method for inserting microelectrodes into adjacent cells and then studying the chemical and physical changes that occur when a neurotransmitter passes through the synapse. Katz discovered that **neurotransmitters** like acetylcholine are released in tiny packages of a few thousand **molecule**s each. He also characterized the release of these packages in resting and active neurons. For their work on neurotransmitters, Eccles and Katz each received a Nobel Prize for physiology or medicine in 1963 and 1970, respectively.

Acetylcholine, like other neurotransmitters, can be both excitatory or inhibitory, depending on the receptor to which they bind. For example, nicotinic acetylcholine receptors cause an excitatory response when stimulated, whereas muscarine acetylcholine receptors cause an inhibitory response when stimulated. There are still many unanswered questions as to how exactly acetylcholine elicits the transmission of a nerve impulse. It has been determined that acetylcholine binds to its target receptor and transfers its chemical signal. In order for the nerve impulse to stop, this binding must be reversible. Immediately after binding, the acetylcholine molecule is degraded by the **enzyme** acetylcholinesterase into acetate and choline, and the nerve impulse ends.

Acetylcholine can be inhibited by nerve gases and neurotoxins. They may work in one of two ways; either by inhibiting the action of acetylcholinesterase or by binding directly to the acetylcholine receptor. The effects of such neurotoxins could be uninhibited muscle stimulation, resulting in muscle spasms, or inhibiting muscle stimulation altogether. Use of these neurotoxins in the laboratory is helping scientists determine the exact mechanism by which acetylcholine causes a nerve impulse. Further study is needed before the final details about this neurotransmitter is known.

ACID AND BASE

Acid and base are terms used by chemists to categorize chemicals according to their **pH**. An acid is generally considered to be any material that gives up a hydrogen ion in solution, while a base is any material that creates a hydroxide ion in solution. Many of these acids and bases are familiar in everyday life. The vinegar we use in salad dressings gets its tart flavor from **acetic acid**, and one of the most common household drugs, aspirin, is a type of acid. **Protein**s, butter and oils, **fruits**, and berries all contain a number of natural organic acids. Bases, which feel slippery when dissolved in **water**, are used to make soap and other household products. When people get **heart**burn, they might take baking soda or an antacid tablet, which are both mild bases. Countless industrial processes use acids and bases as reactants or catalysts to make a variety of **consumer** goods.

People were probably aware of common acids and bases in prehistoric times, ever since they learned how to make wine. When wine turns sour, it changes to vinegar, or dilute acetic acid. In early times, people roasted limestone to obtain lime (calcium oxide), which is a base. Gradually scientists learned to formulate new acidic and basic substances. In the 700s, an Arabian alchemist named Geber (c. 721-815) prepared nitric acid and acetic acid, which he obtained by distilling vinegar. Some time before 1300, sulfuric acid was prepared, and alchemists created aqua regia, a mixture of sulfuric and nitric acids that is capable of dissolving gold, platinum, and many other materials. When strong acids became widely available during the Middle Ages, they launched an experimental revolution. For the first time alchemists were able to decompose substances without high **temperature**s and long waiting periods.

During the 1600s, alchemical methods of preparing acids were improved. Johann Rudolf Glauber (1604-1670), a German chemist, set up a small factory for making acids and salts, which are formed when acids and bases neutralize each other. Soon chemists became more interested in studying the properties of acids and bases and the neutralization reaction between the two substances. Dutch physician Franciscus Sylvius (1614-1672) diagnosed the human body in terms of its balance between acids and bases. Although Sylvius's ideas were simplistic, it is true that our health depends on a proper balance of acids and bases in our **cell**s and in body fluids such as **blood**.

During the 1660s, **Robert Boyle** discovered that certain plant extracts, such as litmus, can be used to distinguish acids from bases. Litmus paper turns red when dipped in an acid, blue when exposed to a base. Since then several other indicator substances have been found, which change color at different levels of acidity in a solution. Boyle went on to characterize acids, noting their sour or tart taste and their ability to corrode metals. Scientists speculated that acids were made of sharply pointed particles that literally pricked the tongue or scratched the metal. Neutralization, they theorized, occurs when an acid particle's spikes fit into a basic particle's pores.

During the 1700s, chemists attempted to describe the neutralization process in terms of the affinity, or degree of attraction, between acidic and basic particles. In 1791 German chemist Jeremias Benjamin Richter (1762-1807) demonstrated that a particular acid and base always neutralize each other in the same proportions. The idea that chemicals react in certain fixed proportions is called stoichiometry, upon which quantitative chemistry is based.

A new but erroneous definition of acids was developed by Antoine-Laurent Lavoisier in the 1770s. He mistakenly believed that all acids contain oxygen (which he named from the Greek words meaning "acid-producing") because he observed that acids are formed when oxygen compounds are dissolved in water. Lavoisier's theory was disproved in the early 1800s, when chemists began using the new tool of electricity to break compounds into elements. First, Humphry Davy demonstrated that hydrochloric acid (HCl) contains no oxygen. His finding was supported by Joseph Gay-Lussac, who proved that oxygen is not a component of prussic acid (now known as hydrocyanic acid, HCN).

Our modern understanding of acids began to take root in the 1830s, when German chemist **Justus von Liebig** defined an acid as a compound that contains hydrogen in a form that can be displaced by a metal. Bases, however, were understood only in terms of their ability to neutralize acids.

Then in the 1880s, Svante August Arrhenius proposed that when acids and bases dissolve in water, their **molecule**s break up into electrically charged particles called ion s (a term introduced by Michael Faraday). Acids produce positively charged hydrogen ions (H+), while bases produce negatively charged hydroxyl ions (OH-). Arrhenius's theory also explained, in very simple terms, what happens when an acid and base neutralize each other: the positive and negative ions unite to form water (H_2O), which is neutral.

The strength of a particular acid or base depends on its concentration of hydrogen ions, which is measured by the pH system on a scale of 1 (strongest acid) to 14 (strongest base). Strong bases are sometimes called alkalis. Because acidic and basic, or alkaline, solutions both conduct electricity, their strength can also be quantified by measuring their electrical conductivity.

Although Arrhenius's theory represented a giant step in our understanding of acids and bases, it had its limitations. What about solvents other than water, for example? And what about **ammonia**, which contains no oxygen but produces hydroxyl ions when dissolved in water? Another complication was the fate of the hydrogen ion in water. Instead of floating free, hydrogen ions combine with water molecules to produce a positively charged "hydronium" ion (H_3O^+).

In 1923 Arrhenius's concept was refined by Danish chemist Johannes Nicolaus Bronsted (1879-1947), who broadened our definition of acids and bases. The hydrogen ion, he pointed out, is a proton—a hydrogen atom without its electron, or negatively charged particle. So Bronsted defined acids as proton donors (they release hydrogen ions) and bases as proton acceptors (any substance that will combine with a loose proton). The same idea was proposed simultaneously by British chemist Thomas M. Lowry. The Bronsted-Lowry definition holds up no matter what the solvent is, and it explains why pure acids and dissolved acids behave differently.

The same year that Bronsted's work was published, American chemist Gilbert Newton Lewis suggested a slightly different way of looking at the new definition. Instead of donating protons, acids accept unattached pairs of electrons; conversely, instead of accepting protons, bases supply pairs of electrons. Under Lewis's definition, even substances that do not produce hydroxyl ions can be considered bases.

Despite all of these refinements, most common acids and bases behave just as Arrhenius described. Today these substances are used in refining oil and sugar and in manufacturing a great variety of products, including fertilizers, explosives, plastics, soap, paper, film, drugs, synthetic fabrics, dyes, solvents, and **pesticides**.

ACID RAIN

Acid rain, or acidic **precipitation**, first attracted scientific attention in the mid-1950s in Scandinavia, with studies focused on acidity and surface **water**s. Concern in the United States, Europe, and Canada began to rise in the 1960s when researchers noticed that fish **population**s in remote wilderness lakes were declining for unknown reasons and **forests** were showing significant **leaf** damage. Scientists speculated that sulfuric and nitric acid falling to the earth as acid precipitation might be the cause. These acids form high in the clouds when sulfur dioxide (SO_2) and **nitrogen** oxides (NO_x)—acid rain precursors emitted by coal-fired electric utilities and other fossil fuel-burning sources—react with water, oxygen, and sunlight. Sulfur dioxide (SO_2), the main source of acid rain, is a pungent toxic gas produced when sulfur laden coal is burned. The acids formed in the clouds are brought to the earth through rain, snow, or fog, and can also fall directly as particles or gases in a process called dry deposition. Rainfall is naturally slightly acid because it dissolves some of the **carbon dioxide** found in the **atmosphere** and produces weak carbonic acid [$CO_2 + H_2O = COOH^+ + OH^-$]. However, the acid rain caused by SO_2 and NO_x is considered a significant **pollution** problem.

As public concern about **air pollution** increased, Congress passed the 1970 Clean Air Act, which required every new power plant that burned **fossil fuels** to install sulfur controls. As a result, SO_2 emissions fell at a rate of 27% per year between 1970 and 1991, even though coal use nearly doubled. But passage of the 1970 law did not diminish public concerns about acid rain. Reports about dying forests and lakes that no longer supported fish continued to grab media and public attention. Canadian government officials stated that 42,000 lakes, mostly small ones, had died from cross border acid rain

caused by power **plants** in the midwestern United States. In Germany, foresters used the term *Waldsterben*, or forest **death**, to describe the decline of forests in parts of Germany where prevailing winds blew pollution across the border from coal burning plants in the Eastern bloc.

In the United States, dire predictions were made about acid rain's effects upon the Adirondacks and the Blue Ridge Mountains, where many lakes were found to have a **pH** of 5.0, which is too acidic for trout and other fish. By the time Ronald Reagan became president in 1981, pressure was mounting from Canada and the U.S. Congress to initiate a major research effort, the National Acid Precipitation Assessment Program, which cost $540 million over 10 years to sample 7,000 lakes and hundreds of woodlands. But even before its results were in, Congress passed the Clean Air Act Amendments of 1990. Title IV of that law called for a 10 million ton reduction in annual SO_2 emissions in the United States by the year 2010, an approximately 40% reduction in anthropogenic emissions from 1980 levels.

In 1991, the national assessment program concluded that "there is no evidence of a general or unusual decline of forests in the United States or Canada due to acid rain," and dangerous acidity was detected in only 4% of eastern lakes, not the 50% of lakes widely expected to show serious damage. However, episodic acidification could be possible in one out of every five lakes and streams in the United States, due to a low capacity to neutralize acids. In contrast to the national assessment, the National Surface Water Survey (NSWS) examined 1,000 lakes larger than ten acres and found that 75% of these lakes and 50% of the streams surveyed had been acidified by acid rain.

The national assessment also stated that **tree** mortality was found in 3% of eastern forests, with high altitude spruce trees suffering the worst effects. Acid rain precursors cause more than 50% of the visibility impairment or "haze" in the eastern United States, and 15% to 30% of haze in the West, the assessment also concluded. While acid rain precursors were being reduced in the United States and other countries, SO_2 emissions in Asia were expected to triple from 1990 levels by the year 2010 if the current coal and oil use trends continued.

Aside from acidifying water bodies, damaging forests, and contributing to haze, acid rain is linked to the corrosion of statues and monuments and to other material damages. Sulfate aerosols associated with SO_2 emissions are also linked to chronic bronchitis, asthma, and other respiratory health effects.

ACOUSTICS, PHYSIOLOGICAL

Physiological acoustics is the study of the transmission of sound and how it is heard by the human ear. Sound travels in waves, vibrations that cause compression and rarefaction of **molecule**s in the air. The **organ** of **hearing**, the ear, has three basic parts that collect and transmit these vibrations: the outer, middle, and inner ear. The outer ear is made of the pinna, the external part of the ear that can be seen, which acts to funnel sound through the ear canal toward the eardrum or tympanic **membrane**. The membrane is highly sensitive to vibrations and also protects the middle and inner ear. When the eardrum vibrates it sets up vibrations in the three tiny bones of the middle ear, the malleus, incus, and stapes, which are often called the hammer, anvil, and stirrup because of their resemblance to those objects. These bones amplify the sound. The stapes is connected to the oval window—the entrance to the inner ear which contains a spiral-shaped, fluid-filled chamber called the cochlea. When vibrations are transmitted from the stapes to the oval window, the fluid within the cochlea is put into motion. Tiny hairs that line the basilar membrane of the cochlea, a membrane that divides the cochlea lengthwise, move in accordance with the wave pattern. The hair **cell**s convert the mechanical **energy** of the wave form into nerve signals that reach the auditory nerve and then the **brain**. In the brain, sound is interpreted.

Early research into the physiology of hearing was conducted by **Hermann von Helmholtz**, a German physician who enjoyed the study of physics and made a close study of the function of both the eyes and ears. He theorized that the ear detected differences in pitch through the action of the cochlea, the snail-shaped organ of the inner ear. As a physicist he understood sound waves and their properties, such as pitch (the highness or lowness of a sound) and amplitude or loudness. He proposed that certain notes sounded pleasing together because their pitches had a mathematical relationship. However, the human ear can distinguish between two instruments playing the same pitch. He contended that the quality of a tone depended on the intensities of other pitches known as overtones which combine to give a sound a particular tone or timbre.

In 1857, Helmholtz proposed his resonance theory of hearing in which he suggested that the fibers along the basilar membrane of the cochlea were of different lengths and thus had their own natural vibration or frequency. When a sound of that same frequency entered the cochlea, that fiber would resonate and sense the sound. He also suggested that the cochlea's structure resonated at particular frequencies to enable both pitch and tone to be perceived.

Although many of Helmholtz's ideas were right, his grasp of what occurs inside the cochlea was incorrect. Many years later, **Georg von Békésy**, a Hungarian-American physicist, studied the cochlea by placing it in a fluid bath and thus could see in more detail what occurred. He also studied the cochlea indirectly by making mechanical models to observe what happened when the fluid in the cochlea begins to move.

For nearly a quarter of a century, Békésy worked for the Hungarian telephone system doing research on acoustics. He began research in physiological acoustics in 1923, first studying the eardrum, then the basilar membrane. He constructed a mechanical model of a cochlea, first made of a rubber membrane stretch over a metal frame and later one containing fluid. He found that vibrations transmitted to the fluid in the cochlea set up traveling waves in the basilar membrane. When the frequency (pitch) of the **stimulus** was increased, the section of sensed vibration moved toward the end of his model that was closest to the middle ear. When the frequency was decreased, the section of sensed vibration moved toward the inner ear.

Trees killed by acid rain in the Great Smokey Mountains. *(JLM Visuals. Reproduced by permission.)*

When he came to the United States in 1947, Békésy suggested a different theory of hearing to replace that of Helmholtz. The basilar membrane that separates the chambers of the cochlea is made up of about 24,000 fibers that stretch across its width. The fibers are progressively wider moving along the cochlea. Helmholtz thought that each fiber would have its own natural vibration and thus respond to sounds with that vibration. Békésy, using his artificial model to mimic the cochlea, found that sound waves passing through the fluid in the cochlea set up a wave in the membrane, and it is the shape of that wave that goes to the brain and is interpreted as sound. The hair cells along the wave transform the mechanical energy of the vibration into **nerve impulses** that can be sent to the brain and interpreted as sound. The wave travels from the stiffer basal part of the cochlea the more flexible upper part of the cochlea. Because of the shape of the cochlea, the resulting wave form is quite complex. Békésy likened the cochlea to a frequency analyzer, an electronic device that measures and interprets the frequency of waves. For his work on physiological acoustics, Békésy was awarded the Nobel Prize in Medicine or Physiology in 1961, the first time a physicist ever won in that category.

The understanding of the function of the inner ear, particularly the cochlea, has undergone a revolution in the last two decades. For example, scientists had believed that the cochlear tuning process was passive and mechanical. However, recent studies have shown that one group of cochlear hair cells have an active motion that enhances hearing. Research focused on the physiology of acoustics also laid the ground work for such advances as hearing aids and the cochlear implant, which involves surgically implanting electrodes in the cochlea to help stimulate the nerves involved in hearing. The implant helps people with hearing defects due to injury or loss of cochlear hair cells, which accounts for the most incurable forms of deafness.

ACTH · See Adrenocorticotropic hormone

ACTION POTENTIAL

An action potential is how **neurons**—or nerve **cells**—communicate. Neurons consist of four basic parts: the body (soma), dendrites, **axon**, and synaptic endings. Surrounding the body of each neuron is a semipermeable **membrane** that separates fluids inside from fluids outside the cell. These fluids are extremely different in ionic composition, causing the inside of the cell to be negative and the outside positive. In an undisturbed system, the inside is -70 millivolts (mV) in relation to outside. The membrane stores this voltage difference, called the **membrane potential**. The membrane is selectively permeable to ions, particularly potassium (K^+), which is more highly concentrated inside the cell, and sodium (Na^+), maintained at a higher concentration outside the cell.

Permeability of the membrane is influenced by three systems: (1) electrical gradient, in which ions move from high charge to low charge; (2) concentration gradient, in which ions move from areas of high concentration to lower concentration; and (3) sodium-potassium pumps, which pump ions through the membrane into or out of the cell. When the resting cell becomes stimulated, sodium ions travel down their electrochemical gradient into the cell, entering through voltage activated gates (special pores in the membrane), the membrane becomes more positive, moving toward zero mV (called depolarization), and reaches its peak value of about +40 or +50 mV, all within a fraction of a millisecond (ms), causing an action potential. Just as quickly, the cell returns to normal and the membrane returns to resting potential. The action potential moves like a wave from the soma along the axon to the synaptic (or terminal) endings. It is at the **synapse** that the message carried by the action potential is passed to the next cell by **neurotransmitters** released by the action potential. The next cell is similarly stimulated to produce its own action potential, and information is rapidly transmitted from cell to cell over rela-tively long distances within the **nervous system**. An action potential is an all-or-none response, which means that any **stimulus** which brings the membrane potential to the threshold at which Na^+ channels open—no matter how great or small this stimulus—will produce a full action potential.

ACTIVE TRANSPORT

Active transport is movement of **molecules** across a **cell membrane** or membrane of a cell **organelle**, from a region of low concentration to a region of high concentration. Since these molecules are being moved against a concentration gradient, cellular **energy** is required for active transport. Active transport allows a cell to maintain conditions different from the surrounding environment.

There are two main types of active transport: movement directly across the cell membrane with assistance from transport **proteins** and **endocytosis**, the "engulfing" of materials into a cell using the processes of **pinocytosis**, **phagocytosis**, or receptor-mediated endocytosis.

Transport proteins found within the phospholipid bilayer of the cell membrane can move substances directly across the cell membrane, molecule by molecule. The sodium-potassium pump, which is found in many cells and helps nerve cells to pass their signals in the form of electrical impulses, is a well-studied example of active transport using transport proteins. The transport proteins that are an essential part of the sodium-potassium pump maintain a higher concentration of potassium ions inside the cells compared to outside, and a higher concentration of sodium ions outside of cells compared to inside. In order to carry the ions across the cell membrane and against the concentration gradient, the transport proteins have very specific shapes, which only fit sodium and potassium ions. Because the transport of these ions is against the concentration gradient, it requires a significant amount of energy. It has been estimated that a full one third of the ATP used by a resting **animal** is used by the sodium-potassium pump.

Endocytosis is an infolding and then pinching in of the cell membrane so that materials are engulfed into a vacuole or vesicle within the cell. Pinocytosis is the process in which cells engulf liquids. The liquids may or may not contain dissolved materials. Phagocytosis is the process in which the materials that are taken into the cell are solid particles. With receptor-mediated endocytosis, the substances which are to be transported into the cell first bind to specific sites or receptor proteins on the outside of the cell. The substances can then be engulfed into the cell. As the materials are being carried into the cell, the cell membrane pinches in forming a vacuole or other vesicle. The materials can then be used inside the cell. Since all types of endocytosis use energy, they are considered active transport.

ACYCLOVIR · See Chicken pox

A boojum tree in Mexico during the dry season. The plant has adapted to seasonal changes in precipitation by restricting the growth of its foliage in the dry season. *(JLM Visuals. Reproduced by permission.)*

The long neck of the giraffe is an adaptation that allows it to survive competition for food resources. *(Photograph by Stan Osolinski, The Stock Market. Reproduced by permission.)*

ADAPTATION

In evolutionary terms, an **adaptation** is the change of a developmental, behavioral, physiological, or anatomical characteristic, resulting in increasing an **organism**'s chance of breeding and survival. Adaptation is the process whereby organisms undergo modification which enables them to function better in a given environment. Adaptations can be acquired during life, such as the development of more powerful muscles, or they can be inheritable, such as the length of limbs. All organisms have adapted in some way from their ancestors. **Natural selection** tends to establish adaptations in a **population**. Less-adapted organisms have a greater **tolerance** in a particular environment and are better suited to survive change or to colonize new areas, while highly adapted organisms are well-suited to living in their own environment but are less likely to survive change to that environment. A common example of a

highly adapted organism is a **flower**. The flowers of **plants** are adapted to make the most efficient use of the plant's resources by targeting pollinators. Many plants have flowers that attract only one **species** of insect. The plant needs only the characteristics to attract its target species of insect, not all species of insects. However, if the plant's target species of insect becomes extinct, the plant will follow the insect into **extinction**, unless the plant adapts to make use of another pollinator. The more highly adapted the plant is, the less likely it will become extinct.

In physiological terms, an adaptation is a change in an organism to react to conditions. This is a short-term change with a short-term benefit. An example of an adaptation of this type is the production of sweat to increase cooling on a hot day.

Another type of adaptation is sensory adaptation. If a receptor or **sense organ** is over-stimulated, its excitability is reduced. For example, continually applied pressure to an area of skin eventually causes the area to become numb to feeling and a considerably larger pressure has to be applied to the area subsequently to elicit a similar response.

Whether occurring within a span of minutes, during the organism's lifetime, or over many thousands or millions of years, adaptation serves to increase an organism's efficiency and ultimately its chances of survival.

ADAPTIVE RADIATION

With the process of **evolution**, all **organism**s have developed from an earlier set of ancestors. Adaptive radiation is the process whereby one organism can give rise to many different new forms. A primitive organism can leave offspring of a similar type which then adapt and evolve slightly differently due to the different pressures that they are placed under from different **habitat**s. For example, it is thought that at the start of the tertiary period there existed a type of relatively undifferentiated, simple mammal. All forms of subsequent **mammals**— both living and extinct—have evolved from this common ancestor. The different forms of mammal that have existed are produced because of the different environments that they have evolved to fit.

The classic example of adaptive radiation that can be easily followed and understood is on the Galapagos Islands. These islands were produced by volcanic activity and are some 600 mi (900 km) from the nearest land (Ecuador). All of the life that exists there has colonized the islands from the mainland. As a result, some **animal**s were unable to reach the islands and consequently some ecological **niche**s were left unfilled. As time has gone by, some of the organisms that made their way there have adapted to fill those niches. This effect is particularly noticeable in the finches of the islands. The mainland finches are all of one type, with short straight beaks for crushing seeds. On the Galapagos Islands there are thirteen distinct **species** which fall into six main groups. These groups can be readily recognized by their different beak structures. One group has a beak similar to the mainland species and feeds on nuts. Another has a long, slender beak which it uses to obtain nectar form the **flowers** of cactus **plants**. The remaining groups include parrot-like beaks, for obtaining buds and **fruits** and a similar group feeds on insects. Yet another group has a very slender beak which it uses to catch insects on the wing, and the final group has a beak which it uses to catch insects from within **tree**s. This last group does not behave like a woodpecker, holding a small stick in its beak which it uses to poke into the holes it bores into the tree with its beak. All of these finches are recognizably different both in their appearance and behavior. It is believed that they are all descended from the same ancestral stock, the mainland finch. The mainland finch is thought to have flown over to the islands, and in the absence of **competition**, evolved to take up the empty niches (niches which were filled with other species on the mainland).

See also Adaptation; Darwin, Charles; Natural selection

ADENOSINE TRIPHOSPHATE (ATP)

Adenosine triphosphate is an **organic compound** that serves as a source of **energy** for living **organism**s, and assists them in

A transmitted light micrograph of adenosine triphosphate. *(Photo Researchers, Inc. Reproduced by permission.)*

carrying out a wide variety of important functions. It is sometimes referred to as the **cell**'s currency of energy exchange. The compound consists of a substituted purine base (6-aminopurine) linked to a five carbon sugar (ribose) that is triply phosphorylated. In activities requiring energy assist, **hydrolysis** converts ATP to adenosine diphosphate (ADP) and phosphate by removing the terminal **phosphate group**. Because the total free energy of the products of this **chemical reaction** (ADP and phosphate) is less than that of ATP, energy is released. The liberated energy can be harnessed and used by living cells. Energy-requiring processes that make use of energy available in ATP include synthesis of **proteins**, **fats**, **polysaccharides**, and **nucleic acids**. ATP is also used to accomplish mechanical work by assisting muscle contraction in **animal**s or the movement of **cilia** and **flagella** in microscopic organisms. Many important physiological functions require ATP for osmotic work in which ions or metabolites are transported through **membrane**s against a concentration gradient. Nerve conduction, kidney function, and **secretion** of hydrochloric acid in the stomach are all examples of processes thus made possible. In **plants**, **photosynthesis** is assisted by the conversion of **light** energy into chemical energy in the form of ATP in a process called **photophosphorylation**. The ATP is used to drive energy-requiring synthesis of sugars and starch.

ADRENAL GLANDS

The **adrenal glands** are a pair of glands which are located on or very near the kidneys in most vertebrates. In humans, an adrenal gland perches on top of each kidney. The adrenal glands have two separate parts: the cortex and the medulla. Although the cortex and medulla are structurally close, they do not interact functionally. They each have their own unique characteristics and functions.

The adrenal cortex is the outer ''rind'' of the adrenal gland. It is made up of **tissue** that is similar to that found in the ovaries and testes. The cortex is responsible for producing three types of **hormones**: **sex hormones**, mineralocorticoids, and glucocorticoids. The most important sex hormones produced in the adrenal cortex are male sex hormones, or androgens. These hormones increase in both males and females at puberty, resulting in oily skin and hair in the armpits and pubic area. In males, androgens are also responsible for the development of the penis and testicles, and for increased muscle bulk. Mineralocorticoids are hormones which regulate the way the kidneys process sodium and potassium. Glucocorticoids affect the way the body processes, uses, and stores **carbohydrates**, **fats**, and **proteins**.

The core of the adrenal gland is called the adrenal medulla. It is made up of the same kind of tissue found in an area of the **nervous system**. The adrenal medulla produces two very important hormones called epinephrine and norepinephrine. These hormones are responsible for a mechanism called the "fight-or-flight" response.

When a person or **animal** is confronted with some type of danger, the adrenal glands pump out large quantities of norepinephrine and epinephrine (also called adrenaline). This adrenaline rush resets a variety of the body's systems, so that the frightened individual can respond quickly: **heart** rate increases, sugar is brought out of storage for a quick burst of **energy**, **blood** vessels which feed the muscles expand, bodily functions which are unnecessary for facing the danger at hand (such as **digestion** in the stomach and intestines) are temporarily shutdown, the bronchial tubes leading to the lungs increase in size so that more oxygen can be taken in.

A number of disorders can affect the adrenal glands. An overactive adrenal cortex is called Cushing syndrome, named after **Harvey Williams Cushing**. Individuals with this disorder are often obese, with abnormal hair growth, and high **blood pressure**. An underactive adrenal cortex is called Addison's disease; President John F. Kennedy suffered from this disorder. A tumor in the adrenal medulla (called a pheochromocytoma) can result in sudden spikes of blood pressure, with accompanying headache, sweats, anxiety, and nausea.

ADRENOCORTICOTROPIC HORMONE

Adrenocorticotropic **hormone**, also referred to as ACTH, is a chemical substance which is manufactured in the front (anterior) region of the **pituitary gland**. The pituitary is a small, kidney-bean sized gland located in the center of the **brain**. The pituitary receives feedback from other **organs** in the body about the body's chemical state. In response to this information, the pituitary releases more or less of its hormone products, including ACTH. These hormones travel via the **blood circulation** to various organs and **tissues** (target organs) throughout the body, where they exert their effects. Because the pituitary gland directs the function of so many other glands and tissues throughout the body, it is often referred to as the "master gland."

ACTH has a number of functions. It is responsible for causing the **adrenal gland** (a tiny gland perched on top of each kidney) to produce important hormones called **steroids**. Ste-

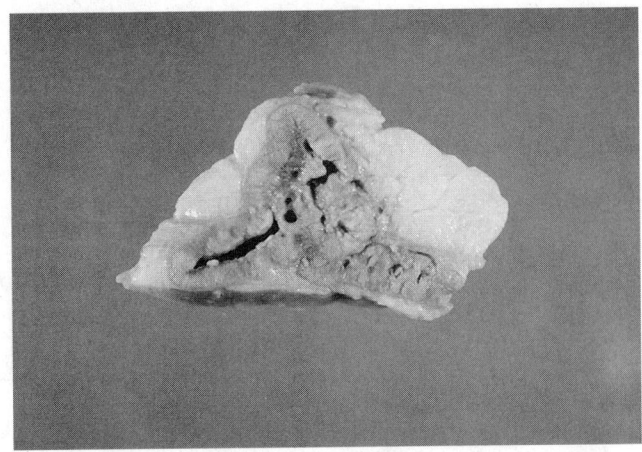

A cross section image of the human adrenal gland. *(Photograph by Martin M. Rotker, Photo Researchers, Inc. Reproduced by permission.)*

roids act throughout the body, affecting growth, sexual characteristics, response to various types of **stress**, immune function, and **inflammation**. ACTH also travels throughout the body to reach areas where fat is stored, encouraging the release of fatty acids back into the **blood**stream.

ACTH release is dependent on a feedback loop involving the adrenal gland, the **hypothalamus** (another gland located within the brain), and the anterior pituitary. When steroid hormone levels are low, this information is received by the hypothalamus. The hypothalamus responds by secreting a hormone called ACTH-releasing hormone, which travels directly to the anterior pituitary. ACTH-releasing hormone (as its name implies), encourages the anterior pituitary to release ACTH. This ACTH, then, reaches the adrenal glands and signals them to increase their release of steroid hormones. When steroid hormone levels increase above a particular level, the hypothalamus responds by shutting off production of ACTH-releasing hormone, which signals the anterior pituitary to stop producing ACTH. This elaborate feedback loop ensures that the body will have only the appropriate amount of steroid hormones available. In the event of the body being subjected to some type of stress requiring increased steroid levels, the hypothalamus can quickly respond by putting out a sudden quantity of ACTH-releasing hormone. This feedback loop allows the body's level of steroid hormone to increase by as much as 20 times within only minutes.

ACTH production can be affected by any process which affects the anterior pituitary gland. Many tumors of the anterior pituitary result in increased production of ACTH, causing increased levels of steroid hormone production by the adrenal glands. This causes a disease called Cushing's syndrome. Destruction of the anterior pituitary can occur due to tumor growth in the area or due to sudden bleeding into the pituitary gland (as occurs with Sheehan's syndrome). Loss of function within the anterior pituitary will result in decreased or absent production of ACTH, and dangerously low levels of steroid production by the adrenal glands.

ADRIAN, EDGAR DOUGLAS (1889-1977)
English physiologist

What physical changes occur within an organism's body when it sees, hears, smells, tastes, or feels some outside stimulus? That question has intrigued scientists for at least a century. By the early 1870s, some initial answers to the puzzle had begun to appear. Research showed that an electrical impulse causes heart muscle to contract in an "all-or-nothing" manner. That is, after stimulation, the muscle either responds in a specific manner independent of the stimulus's intensity and frequency or not at all. By the turn of the century, the all-or-nothing response was shown to be characteristic of all smooth muscle. This research also suggested that neurons (nerve cells) might behave similarly to muscle cells.

Confirmation of this view was provided over the next two decades by the work of a number of scientists, particularly by that of Edgar Douglas Adrian. Adrian was born in London, England, on November 30, 1889. He entered Trinity College, Cambridge, in 1908 and became a student of the physiologist Keith Lucas (1879-1916). Lucas had already completed some of the most critical research on the effect of electrical stimulation on muscle action.

Using sophisticated techniques of detection and analysis, Adrian was able to discover a number of facts about nerve transmission. He confirmed, first of all, that neurons, like muscle cells, respond in an all-or-nothing mode. He also showed that the electrical impulse traveling through a neuron does not change if the kind or the strength of the stimulus changes. In addition, he found that some sense organs eventually adapt to a stimulus that is applied steadily, while others do not.

Much of Adrian's research was inspired by or had significant impact on practical medical problems. For example, his early work on muscle and nerve cells was influenced by injuries incurred by soldiers during World War I. His later research on nerve transmission led to the development of the electroencephalogram and the treatment of deafness, paralysis, and other nerve disorders.

Adrian became a lecturer in physiology at Cambridge in 1919 and was promoted to professor in 1937. He left teaching in 1951 to become Master of Trinity College. From 1950 to 1955 he was president of the Royal Society. His two highest honors were creation as a hereditary baron of the realm by Queen Elizabeth II (1926-) in 1955 and his receipt of the Nobel Prize for physiology or medicine (shared with Charles Scott Sherrington) in 1932. Adrian died in London on August 4, 1977.

AEROBIC/ANAEROBIC

Until **Louis Pasteur** investigated the production of **alcohol** by **yeast**s, it was believed that life was only possible in the presence of air, i.e., under aerobic conditions. This is true for higher **organism**s, but many **bacteria** and some **protozoa** can grow only in the total absence of air, i.e., under anaerobic conditions, and are referred to as strict or obligate anaerobes.

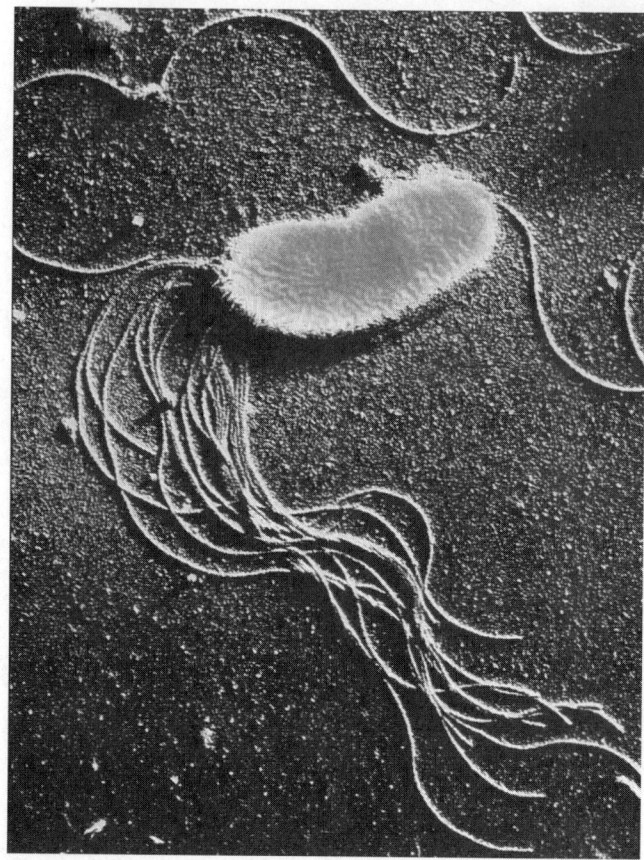

A scanning electron micrograph (SEM) of the aerobic soil bacterium *Psuedomonas fluorescens*. The bacterium uses its long, whip-like flagellae to propel itself through the water layer that surrounds soil particles. *(Photo Researchers, Inc. Reproduced by permission.)*

Some bacteria, **fungi**, and protozoa (facultative anaerobes) can grow in either the presence or absence of oxygen. Organisms that require oxygen are classified as strict or obligate aerobes. Some **microorganism**s, classified as microaerophiles, require low concentrations of oxygen and do not grow at atmospheric oxygen pressure or in the absence of oxygen.

Oxygen is required by aerobes for two purposes: (1) to serve as a terminal electron acceptor for theelectron transfer system, and (2) to participate in very small amounts in certain enzymatic reactions (e.g., in the oxidation of **hydrocarbon**s, the addition of molecular oxygen to the hydrocarbon **molecule** is required). In general, under aerobic conditions, reduced carbon materials such as **carbohydrate**s are metabolized to **carbon dioxide**, **water**, metabolic intermediates, **cell**ular materials, and **energy**. Many intermediate steps are involved in the overall reaction. Breakdown of some **organic compound**s also may result in the release of inorganic nutrient ions such as **nitrogen**, sulfur, and phosphorus.

Under anaerobic conditions, **metabolism** of organic compounds takes place much more slowly than when oxygen is plentiful. The products of anaerobic metabolism include a variety of partially oxidized compounds such as organic acids, alcohols, and methane gas. Anaerobic metabolism releases relatively little energy for the organisms involved. The end products still contain a significant amount of energy, so products such as alcohol and methane can be used as fuels. The most common anaerobic **heterotroph**ic reactions are that of **fermentation**, where an oxidized compound such as a sugar is converted to an alcohol or acid, carbon dioxide, and energy, and that of methane production, where hydrogen and carbon dioxide are converted to methane, water, and energy.

AGASSIZ, JEAN-LOUIS-RODOLPHE (1807-1873)

Swiss American naturalist

Agassiz was born in Motieren-Vuly, Switzerland, and grew up amidst the breathtaking beauty of the Swiss Alps. Agassiz's childhood was supervised by his minister father, who believed that supernatural powers created all natural wonders. Agassiz followed his family's wishes and pursued a degree in medicine. After attending the universities in Munich and Heidelberg, Germany, and Zurich, Switzerland, he eventually earned his Ph.D. in 1829.

Upon his graduation from the University of Munich, Agassiz published a monograph on the fish of Brazil that sparked the attention of the noted French anatomist **Georges Cuvier**. Although he possessed a strong interest in zoology, Agassiz went on to earn a medical degree. In 1832 he went to Paris to serve as an apprentice to Cuvier during that renowned scientist's last years.

Agassiz then accepted his first professional position as a professor of natural history at Neuchatel in Switzerland. For his first project, he published a five-volume work on fossil fish. This work helped establish his reputation as a naturalist and earned him the Wollaston Prize.

Agassiz then shifted his attention to the study of glaciers. Among many others, Agassiz was fascinated with the extreme heights of the Alps and the occasional sight of huge boulders that were thought to have been created by glacial movement. He spent his vacations in 1836 and 1837 exploring the glacial formations of Switzerland and compared them with the geology of England and central Europe.

The question of whether or not glaciers moved intrigued Agassiz, who discovered the answer in 1839 at a cabin that had been built on a glacier approximately ten years earlier. In one decade it had moved nearly 1 mi (1.6 km) down the glacier from its original site. In a unique experiment, Agassiz drove a straight line of stakes deeply into the ice across the glacierhill and then observed their movement. After moving, the stakes formed a U shape as middle stakes had moved more quickly than the side ones. Agassiz concluded that the center stakes moved faster since the glacier was held back at the edges by friction with the mountain wall.

This experiment demonstrated not only that the glacier moved, but that many thousands of years before massive ice blocks had probably moved across a great deal of the European land masses that now lacked the massive ice formations. The resulting conclusions led to the term Ice Age, which purported

Jean-Louis-Rodolphe-Agassiz. *(The Library of Congress. Reproduced by permission.)*

that glacial movement is responsible for modern geological configurations. One of the most significant developments that came out of his observations resulted when his discovery helped provide answers to studies pursued by such naturalists as **Charles Darwin** and Sir **Charles Lyell**. These two men concluded that glaciation was a primary mechanism in causing the geographical distribution and apparent similarities of flora and fauna that were otherwise inexplicably separated by land and water masses. Despite the evidence with which he was presented, Agassiz's background prevented him from agreeing with such conclusions, and he continued to believe that supernatural forces were responsible for the similarities.

AGGLUTINATION REACTION

Agglutination refers to the clumping of **cells**, such as **bacteria** or red **blood** cells, in the presence of an **antibody**. Because the clumping reaction occurs quickly and is easy to produce, agglutination is an important technique in diagnosis.

Two bacteriologists, Herbert Edward Durham and Max von Gruber, discovered specific agglutination in 1896. The clumping became known as Gruber-Durham reaction. Gruber introduced the term agglutinin for any substance that caused agglutination of cells. (The word comes from the Latin *agglutinare*, "to glue to.")

French physician Fernand Widal put Gruber and Durham's discovery to practical use later in 1896, using the reaction as the basis for a test for typhoid fever. Widal found that blood serum from a typhoid carrier caused a **culture** of typhoid bacteria to clump, whereas serum from a typhoid-free person did not. This Widal test was the first example of serum diagnosis.

Austrian physician **Karl Landsteiner** found another important practical application of the agglutination reaction in 1900. He was able to categorize human blood into four types, based on the clumping reaction of each type to blood serum. Landsteiner's agglutination tests made blood transfusion possible, since physicians could now avoid giving donor blood that would cause the recipient's blood to clump.

AGING

Aging is the natural effect of time and the environment on living **organism**s. These normal developmental changes include all which occur from embryonic and fetal development through senescence and **death**. One morbid aspect of aging is the increased vulnerability to death. The increased likelihood of death associated with aging makes it difficult to distinguish between natural aging and pathology (disease). Regardless of the difficulty, it is extraordinarily important to distinguish pathological changes from aging. For example, there is an age related increase in many types of **cancer**. Much of that cancer can be attributed to environmental causes and thus, while age related, is not aging. Lung cancer occurs disproportionately in mature people, however it results primarily from smoking and not aging. Alzheimer's disease, a progressive degenerative condition of the cerebral cortex, is associated with aging, however, as its name indicates, it is a pathological condition. Alzheimer's is not a result of normal aging.

Gerontology is the study of all aspects of aging. No single theory on how and why people age is able to account for all facets of aging. Consequently, some scientists study aging in **cell**s, others experiment with **animal**s such as the **fruit fly**, and still others carefully monitor elderly humans. Studies include the characterization of changes that occur in **cultured** *in vitro*. When a normal cell culture is first started, there is a period of time when the cells divide slowly. This is followed by a series of cell culture generations characterized by luxuriant and rapid growth before the cells become senescent and fail to grow. During growth *in vitro*, genetic changes can occur at any time that result in immortalized cell lines. If however, the changes that result in immortalization fail to occur and the cells remain normal and unaltered, aging occurs.

Life span is the maximum time that an individual may live under ideal circumstances. In humans, that figure is about 100 years. Life expectancy is calculated from the average of years lived by people in a particular **population**. While individuals may (and some do) live to be 100 in a developing country, the majority do not, and thus, life expectancy is relatively brief in developing countries. People in developed countries such as the United States have access to plentiful food and purified **water**. Further, they live in communities with excellent public health regulations, such as mandatory vaccinations for many communicable diseases, and have access to modern medicine. Consequently, life expectancy in developed countries is great-

er than it is in less developed countries. Further, because of changes in personal habits such as smoking cessation, attention to better diet, and more exercise, the already greater life expectancy found in developed countries is continuing to increase. Note, however, that there is no expectation that life span will increase even in developed countries. Biological changes intrinsic to aging are believed to have set a limit to the 100 years described above. The population of a developing nation may have a life expectancy of 50 years. A healthy 50-year-old person in that developing nation has aged no more than a healthy 50-year-old person living in a population with a life expectancy of 85 years.

Life span is **species** specific. Members of the same species have similar life expectancies. In most species, death occurs not long after the reproductive phase of life ends. This is obviously not the case for humans. However, there are some changes that occur in women with the onset of menopause when **estrogen** levels drop. Post-menopausal women produce less facial skin oil (which serves to delay wrinkling) and are at greater risk of developing osteoporosis (brittle bones). Men continue to produce comparable levels of facial oils and are thus less prone to early wrinkling. Osteoporosis occurs as calcium leaves bones and is used elsewhere; hence, sufficient calcium intake in older women is important, because bones which are brittle break more easily.

Protein cross-**linkage** and free radicals are also thought to contribute to aging. Faulty bonds can form in proteins with important structural and functional roles. Collagen makes up 25-30% of the human body's protein and provides support to **organ**s and elasticity to **blood** vessels. Cross-linkage in collagen **molecule**s alters the shape and function of the organs it supports and decreases elasticity. Free radicals are normal chemical byproducts from the body's use of oxygen, however they bind unsaturated **fats** into cell **membrane**s and cause cellular damage. Antioxidants, such as **vitamins** C and E, block free radicals and so are suggested to prolong life.

AGRICULTURE

Agriculture is the art and science of cultivating the soil, growing and harvesting crops, and raising livestock for human use. The word agriculture comes from the Latin words *ager* meaning a field and *cultura* meaning cultivation. Prehistoric peoples hunted, fished, or gathered their food, but by about 11,000 B.C., people began to domesticate and breed **plants** and **animals**.The early peoples noted which of the wild plants were edible or otherwise useful, saved the seeds, and replanted them in cleared land. Through time, cultivation of the most productive and hardiest plants yielded a stable strain. Young, wild animals were captured and those with the most useful traits, such as small horns or high milk yield were bred. The **evolution** from nomadic hunter-gatherers to cultivators allowed people to establish permanent villages because they had a close and reliable food supply. Fewer people were required to provide food and were freed to develop other technologies and services, such as building and crafts (e.g., pottery, weaving, or leatherwork).

The most important domesticated plants are cereals such as wheat, rice, barley, corn and rye; feed grains for animals such as soybeans, field corn, and sorghum; **fruits** and vegetables; and tobacco, coffee, and tea. The most important domesticated animals include meat animals, such as sheep, cattle, goats, and pigs; and poultry such as chicken, ducks, and turkeys. Other important domesticated plant and animal products include milk, cheese, **egg**s, nuts, and oils. Agricultural income is also produced from non-food crops (e.g., rubber, fiber plants that are made into clothing, rugs, curtains, ropes, and canvas; and oilseeds used in synthetic **chemical compound**s) and from raising animals for their pelts and hides and bees for their honey.

Farms vary in size according to the region and purpose of the farm. Commercial farming enterprises are usually conducted on large areas, such as single-crop (e.g., tea, rubber, sugarcane, or oranges) plantations, wheat farms, and Australian sheep farms, to produce cash crops or livestock that can be sold for profit. Individual subsistence farms, where most products are produced for a family's own use (though surplus products may be sold at a local market) and small-family mixed product farm operations are decreasing in number in developed countries but are numerous in developing countries. Nomadic herders are still important agricultural **producers** in the **grasslands** of sub-Saharan Africa, Afghanistan, and Mongolia.

The success of modern agriculture has resulted from soil **conservation** and improved methods of irrigation and drainage; the development of agricultural machinery which has increased farm efficiency and **productivity**; and agricultural chemistry, which has resulted in the production of herbicides and **pesticides** as well as the development of soil testing procedures that improve the use of fertilizers. Plant and livestock breeding in developed countries is now based on scientific principles and **genetics**. Improvements in packing, storage, processing, and transportation of agricultural products has increased the marketability of farm products.

Today, nearly 50% of the world's labor force is employed in agriculture, though the percentages ranged in the late 1980s from 64% of the economically active **population** in Africa to less than 4% in the United States and Canada.

AHLQUIST, JOHN • See Sibley, Charles Gald

AIDS

In the 1970s, many scientists assumed that infectious disease were no longer a major threat in developed countries, since they felt that they were winning the battle against polio, **smallpox**, and other serious infectious diseases. Many researchers began to focus on developing cures for non-infectious conditions such as **cancer** and **heart** disease. Their confidence was shattered in early 1981 by the advent of AIDS—Acquired Immune Deficiency Syndrome—a deadly infectious disease that suppresses the **immune system**, and was caused by **Human Im-**

mune Deficiency Virus (**HIV**), part of a group of **virus**es known as **retrovirus**es. The name AIDS was coined in 1982, and patients with AIDS die from opportunistic infections because HIV cripples the immune system. Opportunistic diseases have the opportunity to take hold because the immune system is severely impaired.

Following the discovery of AIDS, scientists tried to identify the virus that causes the disease. In 1983-4, two scientists and their teams reported isolating HIV, the virus that causes AIDS. One was Luc Montagnier (1932-) working at the Pasteur Institute in Paris and the other was Robert Gallo (1937-) at the National Cancer Institute in Bethesda. They both identified HIV as the cause of AIDS and showed the pathogen to be a retrovirus, meaning that its **genetic material** is **RNA**, instead of **DNA**. Following the discovery, a dispute ensued over who made the initial discovery, but today Gallo and Montagnier are credited as co-discoverers.

Inside its host **cell**, the HIV retrovirus uses an **enzyme** called reverse transcriptase to make a DNA copy of its genetic material. The single strand of DNA then replicates and, in double stranded form, integrates into the host cell's **chromosome** where it directs synthesis of more viral RNA. The viral RNA in turn directs the synthesis of **protein** capsids and both are assembled into HIV viruses. A large number of viruses emerge from the host cell before it dies. HIV destroys the immune system by invading **lymphocytes** and macrophages, replicating within them, killing them, and spreading to others.

Scientists believe that HIV originated in Africa and subsequently spread to Europe and the United States by way of the Caribbean. Since viruses exist that suppress the immune system in monkeys, scientists hypothesize that these viruses mutated to HIV in the bodies of humans that ate monkey meat, and subsequently caused AIDS. A fifteen year old male with skin lesions who died in 1969 is the first documented case of AIDS. Unable to determine the cause of **death** at the time, doctors froze some of his **tissue**s, and upon recent examination, the tissue was found to be infected with HIV. During the 1960s, doctors often listed leukemia as the cause of death in many AIDS patients. After several decades however, the incidence of AIDS was sufficiently widespread to recognize it as a specific disease. Epidemiologists, scientists that study the incidence and distribution of diseases and trace their causes, turned their attention to AIDS. James Curran, working with the Centers for Disease Control and Prevention (CDC) in Atlanta, Georgia, sparked an effort to track the occurrence of HIV. First spread through the homosexual community by male to male contact, HIV rapidly expanded through all **population**s. Presently, new HIV infections are increasing more rapidly among heterosexuals, with women accounting for approximately twenty percent of the AIDS cases. The worldwide AIDS epidemic is estimated to have killed more than 6.5 million people, and infected another 29 million. A new infection occurs about every fifteen seconds. HIV is not distributed equally throughout the world; most afflicted people live in developing countries. Africa has the largest number of cases, but the fastest rate of new infections is occurring in Southeast Asia and the Indian subcontinent. In the United States, although the disease is con-

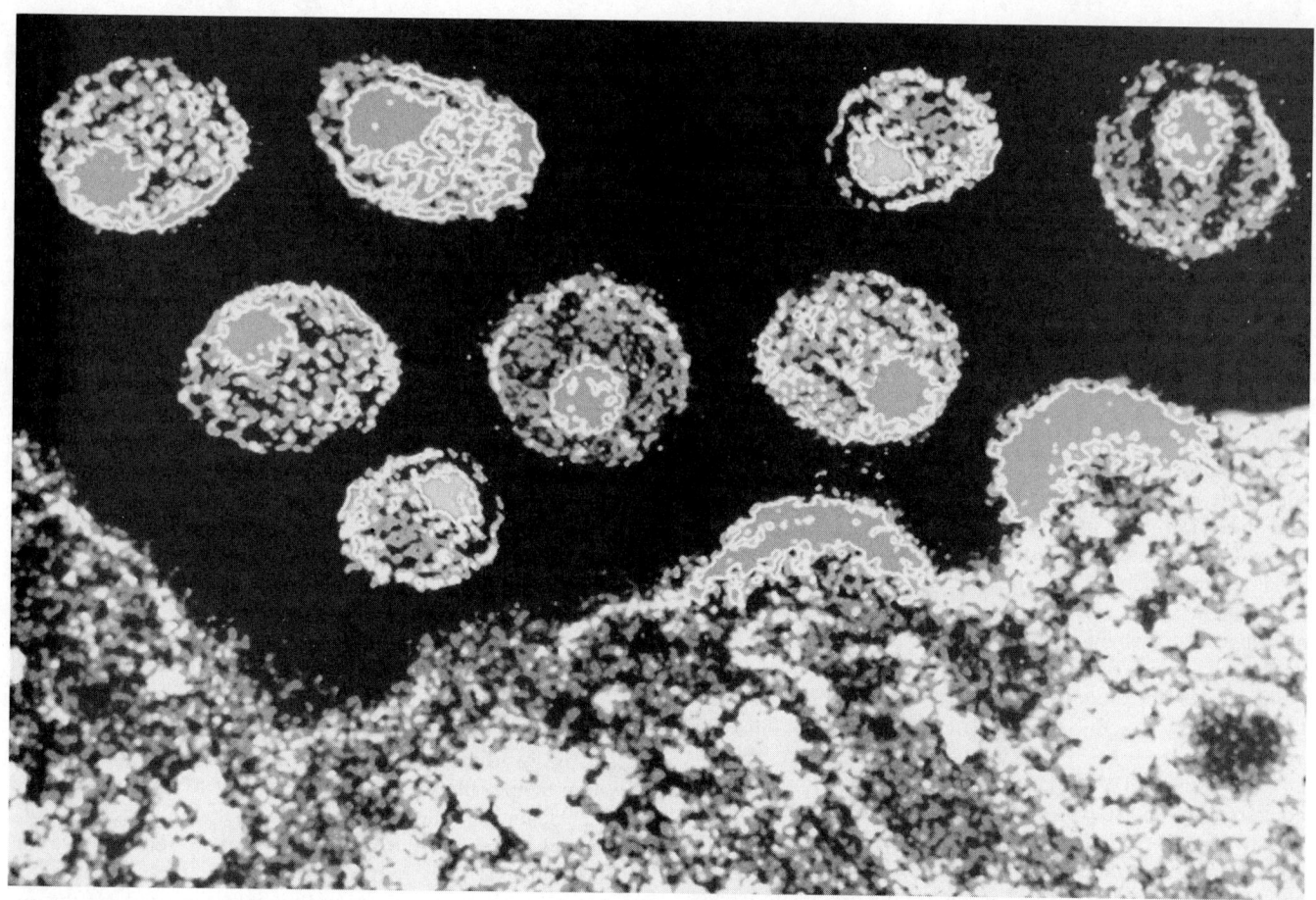

Mature HIV-1 viruses (above) and the lymphocyte from which they emerged (below). Two immature viruses can be seen budding on the surface of the lymphocyte (right of center). *(Photograph by Scott Camazir, Photo Researchers, Inc. Reproduced by permission.)*

centrated in large cities, it is also spreading to towns and rural areas. AIDS strikes people of all races, but statistically there is a higher prevalence rate among African-Americans and Hispanics in the United States due to the fact that they are minority populations. Once the leading cause of death among people aged 25-44 in the United States, AIDS is now second to accidents.

HIV is transmitted in bodily fluids. Its main means of transmission from an infected person is through sexual contact, specifically vaginal and anal intercourse, and oral to genital contact. Intravenous drug users that share needles are at high risk of AIDS, and an infected mother has a fifteen to twenty-five percent chance of passing H IV to her unborn child before and during **birth**, and also an increased risk of transmitting HIV through breastfeeding. Although rare in countries such as the United States where **blood** is screened for HIV, the virus can be transmitted by transfusions of infected blood or blood-clotting factors. Another consideration regarding HIV transmission is that a person who has had another sexually transmitted disease is more likely to contract AIDS.

Laboratories use a test for HIV-1 that is called ELISA (Enzyme-Linked-Immuno-Sorbent Assay). There is another type of HIV called HIV-2. First developed in 1985 by Robert

Gallo and his research team, ELISA is based on the fact that, even though the disease attacks the immune system, B cells begin to produce antibodies to fight the invasion within weeks or months of the infection. The test detects the presence of HIV-1 type antibodies, and reacts with a color change. Weaknesses of the test lie in the fact that it does not diagnose patients who are infectious but have not yet produced HIV-1 **antibody** or those that are infected with HIV-2. In addition, ELISA may give a false positive to persons suffering from a disease other than AIDS. Patients that test positive with ELISA are given a second more specialized test to confirm the presence of AIDS. Developed in 1996, this test detects HIV **antigen**s, proteins produced by the virus and can therefore identify HIV before the patient's body produces antibodies. In addition, separate tests for HIV-1 and HIV-2 have been developed.

When HIV is inside the body, the disorder goes through different phases, the last one being AIDS. During the earliest phase, the infected individual experiences general flu-like symptoms such as fever and headache within one to three weeks, and then remains relatively healthy while the viruses replicate and the immune system produces antibodies. This stage continues as long as the body keeps HIV in check. Pro-

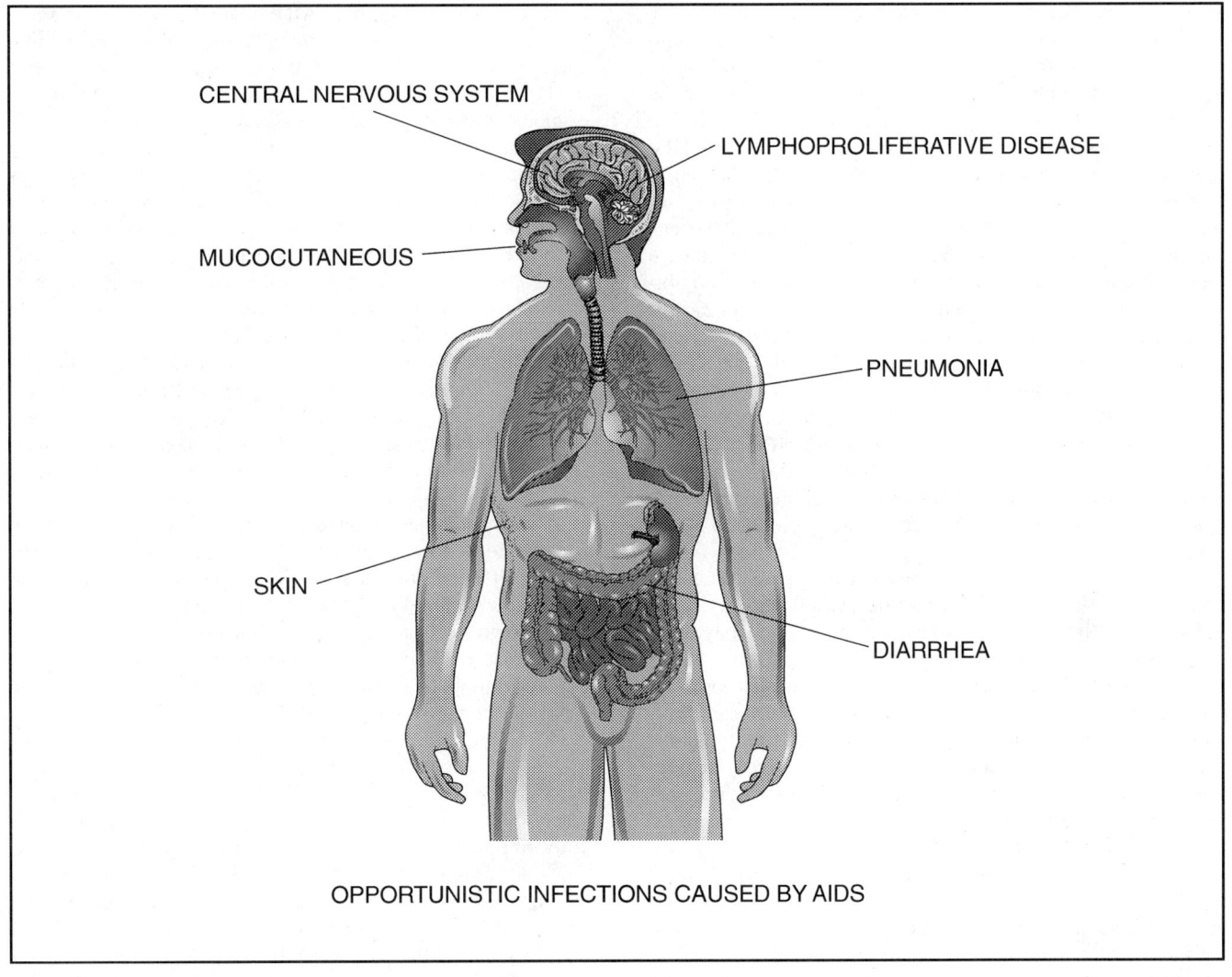

CENTRAL NERVOUS SYSTEM

LYMPHOPROLIFERATIVE DISEASE

MUCOCUTANEOUS

PNEUMONIA

SKIN

DIARRHEA

OPPORTUNISTIC INFECTIONS CAUSED BY AIDS

Because the immune system cells are destroyed by the AIDS virus, many different types of infections and cancers can develop, taking advantage of a person's weakened immune system. (Illustration by Electronic Illustrators Group.)

gression of the disease is monitored by the declining number of particular antibodies called CD4-T lymphocytes. HIV attacks these immune cells by attaching to their CD4 receptor site. The virus also attacks macrophages, the cells that pass the antigen to helper T lymphocytes. The progress of HIV can also be determined by the amount of HIV in the patient's blood. After several months to up to several years, the disease progresses to the next stage in which the CD4-T cell count declines and non-life-threatening symptoms appear. Such symptoms include swollen lymph glands, severe fatigue, night sweats, cough, and diarrhea. **Nervous system** impairment may involve **memory** loss, **depression**, and diminished ability to think and make judgments. Infections that take hold and persist include thrush (*Candida albicans*), a fungus that forms white spots and ulcers in the mouth and may spread to the vagina. Herpes simplex, a virus that causes sores in the mouth and anal area, may also become established. The immune system is critically impaired in the last phase, technically called AIDS. The CDC has

established a definition for the diagnosis of AIDS in which the CD4 T-cell count is below 200 cells per cubic mm of blood or an opportunistic disease has set in. The patient loses weight, and weakens due to diarrhea and coughing. AIDS patients die of rare diseases that a person with an intact immune system would fight off. Some of the opportunistic diseases that attack AIDS patients are pneumonia caused by *Pneumocystis carinii*, tuberculosis caused by mycobacterium, and toxoplasmic encephalitis, a disease of **brain** cells carried by cats and other **animal**s. Karposi's sarcoma, an unusual cancer of blood vessels that produces purplish-red lesions, as well as a form of cervical cancer in women that spreads to other tissues, are also opportunistic infections of AIDS patients.

Although progress has been made in the treatment of AIDS, a cure is yet to be found. In 1995, scientists developed a potent cocktail of drugs that help stop the progress of HIV. Among other substances, the cocktail combines zidovudine (AZT), didanosine (ddi), and a protease inhibitor. AZT and ddi are nucleosides that are building blocks of DNA. The enzyme, reverse transcriptase, mistakenly incorporates the drugs into the viral chain, thereby stopping **DNA synthesis**. Used alone, AZT works temporarily until HIV develops an immunity to the nucleoside. Proteases are enzymes that are needed by HIV to reproduce, and when protease inhibitors are administered, HIV replicates are no longer able to infect cells. In 1995, the Federal Drug Administration approved saquinavir as the first protease inhibitor to be used in combination with nucleoside drugs such as AZT, and then in 1996 they also approved the protease inhibitors ritonavir and indinavir to be used alone or in combination with nucleosides. The combination of drugs brings about a greater increase of antibodies and a greater decrease of HIV than either type of drug alone. Although patients improve on a regimen of mixed drugs, they are not cured due to the persistence of inactive virus left in the body. Researchers are looking for ways to flush out the remaining HIV. In the battle against AIDS, researchers are also attempting to develop a vaccine. In addition to the classic method of preparing a vaccine from weakened virus, they are trying to create a vaccine from a single virus protein.

In addition to treatment, the battle against AIDS includes preventing transmission of the disease. Infected individuals pass HIV-laden macrophages and T lymphocytes in their bodily fluids to others. Sexual behaviors and drug-related activities are the most important means of transmission. The virus gains entry into the bloodstream by way of small abrasions during sexual intercourse, or direct injection with an infected needle. A few common sense preventives include: abstain from sexual intercourse, or engage in a long-term monogamous relationship with a healthy partner. If you are uncertain that your partner has been free of HIV for five years, practice safe sex (use a condom) and avoid anal-rectal intercourse as well as oral sex. Do not start the use of drugs. If you are a drug user, however, always use a sterile needle, and avoid substances that may affect your judgment and self-control. Preventing HIV transmission among the peoples of the world requires intensive education and social programs to change sexual behaviors and drug dependence.

AIDS THERAPIES AND VACCINES

Since the first cases of acquired immunodeficiency syndrome (**AIDS**) and the related **human immunodeficiency virus (HIV)** became known in the early 1980s, scientists and researchers around the world have spent billions of dollars and thousands of hours to discover how the diseases are caused and what options are available to stop their progression. International conferences held in the mid-1980s allowed data to be shared on what exactly AIDS and HIV are, and to develop **blood** tests to diagnose them. Once the disease could be identified, however, the severity of the illness and the extent of the worldwide epidemic became a daunting challenge.

By the late 1980s, treatments for the **virus** became known throughout the medical community and the collection of horrific infections that prey on the severely weakened **im-** **mune system**s of those with AIDS could be managed to some extent. The **death** toll internationally was still extremely high, and remained so until the mid-1990s when the research collectively began to identify the best range of treatments for the 11 or more strains of HIV and AIDS. The 11th World AIDS conference held in 1996 in Vancouver, British Columbia, Canada, provided the turning point for treatment when several groups of researchers presented information on a combination of drugs that, when used together, could virtually erase any traces of the virus from the bloodstreams of those infected. A new blood test that could detect HIV much earlier than could previous tests was also introduced, which abruptly changed early treatment options, and established new ideas about how the disease works. Within a year, these HIV and AIDS treatments were being duplicated in medical offices and clinics around the world. By the late 1990s, HIV had moved from a progressively terminal disease to one that could be managed over the long term, at least for those who have access to and can afford these new treatment options.

The combined drugs therapy first introduced at the 1996 conference is now the standard of care for those with HIV. Eleven different drugs that fall into three categories make up the treatment quilt. Five nucleoside analog drugs, including the well-known AZT, are used in combination with four varieties of protease inhibitors and two non-nucleoside **reverse transcriptase** inhibitors to block the disease's progression. All three types of drugs are considered anti-retroviral treatments that work by interfering with the action of HIV reverse transcriptase inside infected **cell**s, thus ending the virus' replication process. The drugs attack HIV inside the body's immune cells, where it has already established an active infection. The mortality rate has since fallen sharply as more HIV-positive patients are prescribed this three-drug combination. The use of the improved blood test has also been expanded to measure the amount of HIV in the blood during drug treatment, which can help to pinpoint the most effective combination of drugs for each individual. In all cases, the goal of treatment is to keep the level of HIV in the body as low as possible, for as long as possible.

Many questions still remain concerning effective treatment of HIV and AIDS over the long term. The average cost for the three-drug combination treatment is $15,000 annually for each patient, which is exorbitantly expensive for those in developing countries (where the majority of HIV-positive patients live). While these new treatments can almost completely eliminate the virus in as few as two or more years, they do nothing to restore the functionality of the ravaged immune system. Patients may be just as susceptible to other illnesses that their immune systems cannot withstand, and even though the virus is not detected in the body, does not mean it is not present since HIV cells can mask themselves as almost healthy cells. A strict schedule and diet must be maintained for the drugs to be successful. The regimen is difficult for many patients to follow, either because of privacy constraints or the side effects some of the drugs can cause. Perhaps the most basic questions researchers are still struggling with are when to provide treatment, which drugs to begin with, how to identify when alterations are needed in the therapy, and which drugs to try next.

Because more than 16,000 new patients are diagnosed each day with AIDS worldwide, many researchers have also been studying how a vaccine can block HIV's entry into the immune cells. At least 25 experimental vaccines have been created since identification of the disease, but few have proved promising enough to complete the large-scale testing on human volunteers that is required to confirm the vaccine's success.

In June 1998 such a large-scale test began with 5,000 volunteers in 30 cities in the United States, and a smaller group in Thailand. Volunteers were given a series of shots that hopefully stimulate the immune system to resist the two most common strains of the AIDS virus. Previous smaller tests documented that 99.5% of the vaccinated volunteers produced strong levels of resistance in their immune cells, which then target and kill infections such as HIV. The trial was expected to last three years. Also in 1998, a group of researchers at the University of Michigan proposed to develop a vaccine that would prevent someone with HIV from passing it on to someone else. Such a vaccine would be given within the first three months after infection, when the virus is most contagious. Their theories resemble those used to create the Salk polio vaccine in the 1950s, which reduced the polio virus' symptoms and drastically reduced its ability to infect others, virtually eliminating the disease within a decade.

See also Adenosine Triphosphate (ATP); T cells

AIR POLLUTION

Air contamination is divided into two broad categories: primary and secondary. Primary **pollutants** are those released directly into the air. Some examples include dust, smoke, and a variety of toxic chemicals, such as lead, mercury, vinyl chloride and **carbon monoxide**. In contrast, secondary pollutants are created or modified into a deleterious form after being released into the air. A variety of chemical or photochemical reactions (catalyzed by **light**) produce a wide range of secondary pollutants especially in urban air. A prime example is the formation of **ozone** in smog. A complex series of **chemical reactions** involving volatile **organic compounds**, **nitrogen** oxides, sunlight, and molecular oxygen create highly reactive ozone **molecules** containing three oxygen **atoms**. Stratospheric ozone in the upper **atmosphere** provides an important shield against harmful ultraviolet radiation in sunlight. Ozone in ambient air (that surrounding us) is highly damaging to both living **organisms** and building materials.

The seven most important air pollutants regulated by the 1970 U.S. Clean Air Act included sulfur dioxide, particulates (dust, smoke, etc.), carbon monoxide, volatile organic compounds, nitrogen oxides, ozone, and lead. These pollutants were regarded as the greatest danger to human health. Because criteria were established to limit their emission, these materials are sometimes referred to as "criteria pollutants." Major revisions to the Clean Air Act in 1990 added another 189 volatile **chemical compounds** from more than 250 sources to the list of regulated air pollutants in the United States.

Smog over Los Angeles, California. *(Phototake NYC. Reproduced by permission).*

Some major pollutants are not directly toxic (poisonous) but have deleterious ecological effects. For example, excess nitrogen from fertilizer use and burning of **fossil fuels** is causing widespread damage to both aquatic and terrestrial **ecosystem**s by over fertilizing **plants** and favoring the growth of weedy **species**. Similarly, chlorofluorocarbons used as refrigerants and cleaning agents are generally chemically inert and non-toxic diffuse into the upper atmosphere where they catalyze and destroy the stratospheric ozone shield.

Air pollutants can travel surprisingly far and fast. About half of the fine reddish dust visible in Miami's air during the summer is blown across the Atlantic Ocean from the Sahara Desert. Radioactive fallout from an explosion at the Chernobyl nuclear reactor in the Ukraine was detected in Sweden within two days after its release and spread around the globe in less than a week. One of the best-known examples of long-range transport of air pollutants is **acid rain**. The acids of greatest concern in air are sulfuric and nitric acids, which are formed as secondary pollutants from sulfur dioxide and nitrogen oxides released by burning fossil fuels and industrial processes such as smelting ores. These acids can change the pH (a standard measure of the hydrogen ion concentration or acidity) of rain or snow from its normal, near neutral condition to an acidity similar to that of lemon juice. While this acidity is not directly dangerous to a humans, it damages building materials and can be lethal to sensitive aquatic organisms such as salamanders, frogs, and fish fry. Thousands of lakes in eastern Quebec, New England, and Scandinavia have been acidified to such an extent that they no longer support game fish **popula-**

tions. Acid **precipitation** has also been implicated in forest **death** in northern Europe, eastern North America, and other places where air currents carry urban industrial pollutants.

Because air pollution is so visible and obviously undesirable, most developed countries have had 50 years or more of regulations aimed at controlling this form of environmental degradation. In many cases, these regulations have had encouragingly positive effects. While urban air quality rarely matches that of pristine wilderness areas, air pollution in most of the more prosperous regions of North America, Western Europe, Japan, Australia, and New Zealand has generally been dramatically curtailed in recent years. In the United States, for example, the Environmental Protection Agency (EPA) reports that the number of days on which urban air is considered hazardous in the largest cities has decreased 93% over the past 20 years. Of the 97 metropolitan areas that failed to meed clean air standards in the 1980s, nearly half had reached compliance by the early 1990s. Perhaps the most striking success in controlling air pollution is urban lead. Banning of leaded gasoline in the United States in 1970 resulted in a 98% decrease in atmospheric concentrations of this toxic metal. Similarly, particulate materials have decreased in urban air nearly 80% since the passage of the U.S. Clean Air Act, while sulfur dioxides, carbon monoxide, and ozone are down by nearly one-third.

Unfortunately, the situation often is not so encouraging in other countries. The major metropolitan areas of other developing countries often have appalling levels of air pollution which rapid **population growth**, unregulated industrialization, lack of enforcement, and corrupt national and local politics only make worse. Mexico City, for example, is notorious for bad air. **Pollution** levels exceed World Health Organization (WHO) standard 350 days per year. More than half of all children in the city have lead levels in their **blood** sufficient to lower intelligence and retard development. The 130,000 industries and 2.5 million motor vehicles spew out more than 5,500 metric tons of air pollutants every day, which are trapped by mountains ringing the city. Most other Third World megacities (those with populations greater than 10 million people) have similar problems. Air quality in Cairo, Bangkok, Jakarta, Bombay, Calcutta, New Delhi, Shanghai, Beijing, Sao Paulo, and many lesser known urban areas regularly reach dangerous levels. Hopefully, as industrialization raises incomes in these developing countries, they will be able to take advantage of pollution control equipment and knowledge already available in already developed countries.

ALCHEMY

Alchemy is a precursor of modern chemistry. It represented the first systematic attempts of people to try to understand the workings of the universe without recourse to religion.

There were two main driving forces behind the "science" of alchemy—the search for the fountain of youth, whether it be a physical place or a potion capable of restoring and maintaining youth, and more famously a method of transforming base metals into gold. Alchemy as such achieved no-

toriety from its pre-eminence in fourteenth century Europe. It is true that stories of, and searches for, the two main tenets of alchemy had been around for many years previously, particularly in China and Egypt, but it was not until the middle ages that European nobles sponsored alchemists, thus making it a growth industry.

The search for a potion for eternal life was actually responsible for the premature **death**s of many practitioners of the art of alchemy. The consumption of compounds of antimony, mercury or **arsenic** was a common path in the search of everlasting life. Unfortunately these compounds are all poisonous and they very quickly had the opposite effect to that desired.

The transmutation of other metals into gold was attempted by the alchemists using a wide range of chemical and physical processes. The transmutation of gold is now possible, and it occurs as part of the **radioactive decay** of material and can be brought about by bombardment with a metal with a particle accelerator, unfortunately these materials are so radioactive as to be instantly deadly. It is interesting to note that if the ancient alchemists had been successful in transforming base metals, such as lead, into gold then the value of gold itself would have plummeted, leaving the inventor of the process no better off, financially.

Very few alchemists' names are now known to us due to their singular lack of success, but one who has been recorded in history is John Dee who was alchemist to Queen Elizabeth I of England, in the sixteenth century.

ALCOHOL

Alcohol is a term applied to a large group of hydroxyl derivatives of paraffin **hydrocarbon**s. They contain a hydroxyl (OH) group in place of one of the hydrocarbon's hydrogens. Examples of common alcohols include methanol, ethanol, isopropanol, butanol, vinyl, cetyl, amyl, and lauryl alcohol. Although many of the alcohols have important commercial or industrial applications, most biological interest is focused on ethanol, a product of biological **fermentation**. Ethanol is sometimes called ethyl alcohol, grain alcohol, or simply alcohol. It is the intoxicating agent found in wine, beer, and other fermented and distilled liquors and has the molecular formula C_2H_5OH. It has been produced for millennia, mostly by fermentation of fruit juices. Primitive humans discovered that the fermented juice could be stored in seal containers for extended periods of time, and provide a safe drink when other fluids were not readily available. Grapes and many other natural products can provide the sugars and starches needed for the production of alcohol beverages, although much of the present day industry depends on grains such as corn, wheat, rye, and barley. Ethanol production from grains involves conversion of grain starches to sugars that are then converted to ethanol with a complex of **enzyme**s called zymase. The ethanol content of fermented beverages is limited to about 12% by the inability of the biological fermenting system to tolerate higher concentrations of alcohol. Distillation processes are used to further concentrate ethanol in beverages such as brandy, whiskey, and liquors.

ALGAE

The algae are an extensive group of **plants**, which show a great diversity in their range of forms. Some taxonomists consider them to be a separate division in the plant **kingdom**, while others place them at a lower rank of **classification**, labeling them as a **class** within the division Thallophyceae.

All algae are primitive plants, and are taxonomically united by the presence of unicellular **organ**s of reproduction. The algae are comprised of representatives from two different groups of living **organism**, the **prokaryote**s (1,450 algal **species**) and the **eukaryote**s (23,500 algal species). The thallus (plant body) of algae can show a wide diversity of forms, ranging from unicellular to multi-**cell**ular, filamentous, flattened or ribbon-like. The smallest, diatoms (which range in size from 5 to 200 microns), are visible only through microscopic observation, but the largest, seaweeds, can be in excess of 150 ft (50m) long.

Algae are aquatic, either marine or freshwater, although some require only a little **water** to thrive. The classification of the algae is based on their structure, the presence of different photosynthetic pigments, the nature of the cell wall, and the arrangement and structure of the **flagella**. The more **evolutio**narily advanced forms of algae exhibit a relatively complex internal arrangement, although they can be readily distinguished from the so-called higher plants by their lack of vascular **tissue**.

There are fourteen main groups of algae. Diatoms, or Bacillariophyceae, have a wide economic importance. Many skin defoliants rely on diatoms to act as an abrasive. Diatoms are a group of microscopic algae that are enclosed by a silicaceous shell. The shell is composed of two parts, one overlapping the other in a lidded box arrangement. Diatoms reproduce by binary fission as well as sexually. They occur either singly or grouped into colonies, and are found abundantly in marine and freshwater plankton and benthos. With dinoflagellates and other algae, they are the primary **producers** in the food chain. Deposits of dead diatoms create both siliceous earths (more recent deposits) and oil reserves (older deposits).

The Charophyceae are more commonly known as the stoneworts. Instead of the silica of the diatoms, these organisms have **cellulose** cell walls encrusted with calcium carbonate. They are multi-cellular, filamentous, and are anchored to the bottom of fresh water ponds with rhizoids. They are unique among the algae because of their multi-cellular sex organs, called antheridia and oogonia. The Chlorophyceae, the so-called green algae, have the same photosynthetic pigments as higher plants. This is the largest and most diverse algal group. They also inhabit the widest range of environments. The aquatic species thrive in fresh or salt water for the aquatic species, while the "non-aquatic" prefer damp environments. The primitive Chlorophyceae are unicellular, single or colonial organisms, microscopic, motile or non-motile. The higher forms are multi-cellular, with either a filamentous or flattened thallus.

The Chrysophyceae are golden brown in color due to the carotenoid pigments associated with the chlorophyll. They differ from the other groups of algae in that their food storage

The cleanup of algal bloom washed up on an Adriatic beach in Italy. *(Greenpeace. Reproduced by permission.)*

product is oil rather than starch. The species encountered here are very diverse marine and fresh water algae with close evolutionary links to the **protozoa**. The Cryptophyceae are a small group of biflagellate unicellular organisms. They occur in both marine and fresh water **habitat**s. The Cyanophyceae (blue-green algae) are a primitive group of prokaryotic organisms. They can be microscopic as unicellular or colonial organisms, or they can be multi-cellular as filamentous forms. They are widely abundant in all forms of water and at extremes of **temperature**. Their color is variable from black to blue, via red, green, and purple, due to differing amounts of chlorophyll types. Some members of this group are responsible for toxin production. When conditions are right, whole bodies of water can be poisoned due to a bloom of these organisms depositing large quantities of toxin into the water. The Euglenophyceae are unicellular, with one to three flagellae. They are often found in sewage-polluted water and can occur in a colorless form.

The Phaeophceae are the brown algae, the seaweeds. They possess several types of chlorophyll, but the brown-colored fucoxanthin hides the expression of the other types.

They have a complex internal structure which, however, does not show vascularisation. They are virtually all marine benthic organisms and can vary in size from 0.2 in (1mm) to 150 ft (50 m). One group of algae bears the common name "fire algae," or Pyrrophyceae, due to their yellow and red colorization. They are mostly unicellular and have interlocking armored plates surrounding their cell wall. They are a major component of marine and fresh water plankton and require a vitamin supply. The Rhodophceae are another group of seaweeds, the red seaweeds. They are very diverse in form but relatively small compared to the brown seaweeds. They are usually benthic marine organisms. They can be used as a source of agar and are edible. The final group of algae are the Xanthophyceae, the yellow green algae. They are a very diverse group united by their chlorophyll pigments. If it were not for the pigments, their diversity of form would fit easily within all the other groups of algae.

Three recently discovered groups of algae, Prymnesiophyceae, Eustigmatophyceae, and Prochlorophyceae, were described in 1992.

See also Plankton

ALKALOID

Alkaloids are **nitrogen**-containing biochemicals, most of which are natural substances extracted from certain **species** of **plants**. The word alkaloid is derived from "like an alkali" (or "base"), when first discovered by chemists, these botanical chemicals were referred to as "vegetable alkalis." Alkaloids are usually extracted from plants using an **alcohol** solvent, such as ethanol. Alkaloids vary greatly in their chemical complexity and reactivity, and some show great physiological activity when ingested by **animal**s.

One familiar example of an alkaloid is nicotine, obtained from tobacco (*Nicotiana tabacum*), a herbaceous plant in the nightshade **family** (Solanaceae). In small doses, such as those obtained by smoking tobacco in cigarettes, nicotine is a highly addictive stimulant of the **nervous system**. In higher doses nicotine can be extremely toxic, and in the form of nicotine sulfate it is used as an insecticide.

Other alkaloids are used as medicines or as highly addictive, recreational drugs. These include codeine, heroin, morphine, and opium, addictive narcotics derived or manufactured from the sap of the opium poppy (*Papaver somniferum*). Cocaine is another natural alkaloid, derived from the foliage of coca shrubs (*Erythroxylon coca* and related species), and used as a stimulant and local anaesthetic. The alkaloids atropine and scopolamine are drugs derived from the belladonna plant (*Atropa belladonna*). A final example is quinine, a bitter-tasting, anti-malarial alkaloid extracted from cinchona **tree**s (*Cinchona ledgeriana* and related species).

ALLELE

Allele is the shortened, and more commonly used, form of the word alleleomorph. An allele is one of a series of possible forms of a given **gene**. Different alleles differ in their sequence of **DNA**. This difference can be as small as only one base pair change although it can be much larger. The difference must be such that the gene product, **RNA** or a **protein**, is different to the other form in an observable manner. Alleles are always at the same physical location (locus) on matching (homologous) **chromosome**s.

A **diploid organism** with two sets of identical alleles is termed **homozygous**, an organism with two different alleles is **heterozygous**. Several different forms of an allele can exist within a **population**, this is a phenomenon known as multiple allelism. Many genes occur in this manner, for example hair and eye color.

The first alleles studied in a scientific manner were those controlling varying characteristics of the pea. This work was undertaken by the Austrian monk, **Gregor Mendel**, in the late 1800s. Mendel studied a range of characters that were either present or absent, for example, smooth or wrinkled seed coats, tall or short **plant**s or axial or terminal **flower**s. Each of these characteristics is governed by a single gene and in the plants that Mendel used only the two alleles were present, providing a very useable model to study this system.

ALLOMETRIC GROWTH

Allometric growth characterizes probably all living beings. The term means differential growth and refers to developmental patterns of growth which are not uniform, that is, not all parts of the **organism** develop at the same rate. An example, known to all who look at babies and adults and note the differences therein, is head size. While it is true that an adult head is larger than a baby's head, it is in fact relatively (in relation to the baby) smaller. A late fetus has an enormous head in relation to its body length. Differential growth accounts for the changing proportion of head size. Perhaps an equally obvious instance of allometry is leg length in humans. The fetal and newborn baby have short, chubby, and ineffectual legs. The proportion of leg length to body length is low. About half of the adult height is accounted for by leg length. Obviously, legs accrue length disproportionately compared with body length.

The examples allometric growth of the head and legs of humans are from normal biology. There are, of course, pathologic examples. Acromegaly occurs in midlife with disproportionate enlargement (growth) of head, hands, and feet. Individuals with acromegaly have a burly appearance because of unequal growth of the jaw which lengthens and protrudes. Another example from pathology is achrondroplasic dwarfism. Here the relatively short arms and legs of the newborn persist while differential growth of the trunk permits a normal size head and trunk. The body is short but parts other than the limbs are of usual size. Another pathologic form of allometric growth is **cancer**, especially certain forms of leukemia. Instead of simply replacing white **blood cell**s with normal growth rate, the leukemic may produce white blood cells with a nonphysiologic and elevated rate.

It is difficult to discuss allometric growth without remembering Sir **D'Arcy Wentworth Thompson** of the University of Saint Andrews in Scotland. Thompson stated in simple terms: "An organism is so complex a thing, and growth so complex a phenomenon, that for growth to be so uniform and constant in all the parts as to keep the whole shape unchanged would indeed be an unlikely and an unusual circumstance. Rates vary, proportions change, and the whole configuration alters accordingly."

The fiddler crab, *Uca pugnax*, is a well studied example of allometric growth from **nature**. The **species** gets its name from the one enormous claw of the male. In one study, the weights of the great claw and total body weight were recorded in 400 crabs. The large claw was about 8% of the total body weight in young males. The proportion of the claw's weight increased to 38% in older males. Of course the body of the older male grew but its large claw grew even faster. Similar measurements were made of the claws of female fiddler crabs. The large claw remained at 8% of body weight regardless of how large the female became. The relative uniformity of growth in the female is an example of isometric growth in contrast to the allometric growth in males.

Thompson was not concerned exclusively with disproportionate growth within a species. He examined closely related species that differed significantly in shape. He showed that many such species, when mapped with Cartesian coordinates, have an orderly deformation of the coordinates when compared with the contrasting closely related species. This showed that while the organisms differed in form, the differences were harmonious, and supported the notion that the supposed relationship was in fact close. Regeneration was another instance where Thompson described allometric growth. In those forms which regenerate missing parts, such as the tail of Planaria, the rate of regeneration is a function of the amount of **tissue** lost.

Contemporary biologists do not simply record growth rate differences. Instead, they seek mechanisms that account for allometric growth. These include the function of specific **hormone**s, growth factors, and **oncogene**s.

ALLOPATRY AND SYMPATRY

Allopatry and sympatry are terms used in biogeography to describe the comparative distributions of **population**s and **species**. Species with sympatric distributions overlap in their geographical range to some degree. In contrast, allopatric species do not overlap in their distributions. **Evolution**ary biologists are particularly interested in these sorts of distributions, because they are believed to be one of the most important factors affecting the evolution of distinct, new populations and species.

Ecologists and evolutionary biologists define a population as "a group of individuals of the same species, which are capable of interbreeding with each other and producing fertile offspring." When populations of a particular species become physically isolated, they are effectively prevented from breeding with each other. Such an allopatric distribution might occur because the populations live on different islands, or because they inhabit opposite sides of a mountain chain. In either case, because of their physical isolation, the isolated, allopatric populations cannot easily breed with each other, if at all. Because the populations occur in **habitat**s characterized by different environmental conditions, they are exposed varying to regimes of **natural selection**. Consequently, over a long period of time the populations may evolve to become different in their physiology, **morphology**, behavior, and other attributes. (If one or more of the isolated the populations are small, their genetic characteristics will also change (i.e., evolve) because of the phenomenon of genetic drift.) Eventually, if the allopatric populations become different enough in their genetic attributes, they may no longer be able to interbreed successfully if they come into contact again (that is, if their distributions become sympatric). If this was the case, evolutionary biologists would consider them to have diverged sufficiently to be characterized as different species.

Allopatric distributions of populations are believed to be one of the most critical factors allowing the evolution of new species to occur. In contrast, populations that overlap to any significant degree will experience a relatively free exchange of genetic information. Even though partially sympatric populations may be experiencing quite different environmental conditions (at least at the extremes of their ranges), and therefore differing regimes of natural selection, the occasional interbreeding is sufficient to prevent them from diverging genetically to the degree that they can no longer successfully interbreed.

The mode of **speciation** just described, involving genetic divergence of allopatric populations, is believed to explain the occurrence of different, but closely related species in remote oceanic archipelagos. For example, the Galapagos Islands of the eastern, equatorial Pacific Ocean sustain about 13 species of so-called warbler finches (also known as Darwin's finches). These include six species of ground finches (in the **genus** *Geospiza*), six species of tree finches (genus *Camarhynchus*), and the warbler finch (*Certhidea inornata*). These are all believed to have originated from the long-ago colonization of one or several of the Galapagos Islands by a few, probably storm-blown finches from the South American mainland. Because there were no competitors for these small, seed-eating birds, the tiny founder population increased in size. Eventually populations occurred on each of the islands of the archipelago, where they were effectively isolated and could not interbreed with each other. These locally allopatric distributions allowed the isolated populations to diverge evolutionarily, and to eventually develop new species that occur nowhere else. Because there are few other kinds of landbirds on the Galapagos Islands, the original founder population of seed-eating finches was able to take advantage of various, empty ecological **niche**s. In this manner, species of warbler finches evolved with very heavy bills for feeding on large seeds (e.g., the large ground finch, *Geospiza magnirostris*), while medium-billed species specialize on smaller seeds (e.g., the small ground finch, *G. fuliginosa*), and thin-billed ones on small insects (e.g., the small insectivorous tree finch, *G. parvulus*. One species, the tool-using finch (*Camarhynchus pallidus*), has a specialized behavioral trait involving the use of a cactus spine to probe small tree-cavities so that insect **larva**e can be extracted. All of these distinct, unusual species are thought to have evolved because their originally homogeneous, founder population was able to diverge genetically in the isolated environments of the various islands of the Galapagos Archipelago.

ALTERATION OF GENERATION

Alteration of generation is a reproductive cycle and method of growth of certain **species**. It is found in both the **animal** and plant **kingdom**s although it is more common with certain plant species. It is characterized by the **haploid** phase (which reproduces sexually) alternating with the **diploid** phase (which reproduces asexually).

In mosses and vascular **plants**, there is a sexual and an asexual phase. The haploid (sexual) phase involves **gametophyte** (**gamete** or **sperm cell** plant) generation; the diploid (asexual) phase involves **sporophyte** (**spore** plant) generation.

The cycle begins with a sexually mature plant which is capable of producing **egg** and sperm cells. Both egg and sperm, as well as the parent plant, are haploid. The egg and sperm fuse to produce a diploid **zygote**. The zygote (a cell formed by the union of male and female gametes) does not immediately grow into a new gametophyte, instead it grows by **cell division** to produce a sporophyte. The sporophyte plant is a separate individual to the gametophyte, although in some cases it may remain dependent upon the gametophyte for nourishment, giving the appearance of a parasitic **organism**. The function of the sporophyte is to undergo **meiosis** to produce spores.

These haploid spores are powder-like in **nature** and are readily dispersed by the wind. Once they land on a substrate capable of supporting their growth they germinate and grow by **mitosis** (in which a cell divides into two so that the **nucleus** of each new cell has the full number of **chromosome**s) to produce a new gametophyte, and the whole process is repeated.

The sexually and asexually reproducing phases are often very different from each other and are easily recognized. In plants, certain phases consist of only a few partially parasitic cells reaching a maximum of several millimetres across, while other phases comprise the plant, shrub, or **tree**. One disadvantage of this system is the heavy reliance on **water**, which limits where the plant may grow. Within the thallophyte plants (i.e., **bacteria**, **algae**, **fungi**, and lichens) spores are not produced by the diploid generation, although clear differences are observable between the two generations.

The ascomycete fungi are haploid at all times other than when the gametes fuse to produce the zygote. As the first division of the zygote occurs, it simultaneously undergoes meiosis, immediately returning to the haploid phase.

In basidiomycete fungi, each cell usually contains two nuclei, a condition referred to as the dikaryotic state, which is generally held to be the equivalent of the diploid situation. Once a spore has germinated, it can join with the mycelium of another spore and a swapping of nuclei and subsequent migration occurs to produce the dikaryotic state. This can also occur with hyphae from the same spore.

Within the mosses and liverworts, the part of the plant that we are familiar with is the gametophyte. The sporophyte is a small capsule that is nutritionally dependent upon the larger plant. Within the flowering plants the gametophyte generation is vastly reduced in size. The male gametophyte is the pollen grain and the female gametophyte is the **embryo** sac that is contained within the ovule. The sporophyte generation is the plant itself.

If the sporophyte and gametophyte generation are markedly different in appearance the **life cycle** is termed heteromorphic, whereas if they are very similar the life cycle is said to be an isomorphic one.

The situation is very similar in animals such as jellyfish but one striking difference is that both phases are diploid in nature. In the case of the jellyfish, the asexual part of the reproductive life cycle is carried out in a planktonic phase, the polyp form, by an animal only one or two millimetres across. The corresponding sexually reproducing part of the life cycle is the medusa form, such as the Portugese man-of-war, which can have tentacles trailing for 10m (30 ft).

ALTMAN, SIDNEY (1939-)
American molecular biologist

In the early 1980s, Sidney Altman discovered that **ribonucleic acid (RNA)** molecules can act as enzymes. This disclosure, independently and concurrently made by Thomas R. Cech of the University of Colorado, broadened our understanding of the origins of life. Before this discovery, it was believed that all **enzymes** were made of **protein** and that primitive **cells**, therefore, used proteins to catalyze biochemical processes. Now it appears that RNA may have acted as a catalyst. Altman and Cech's work has not only had a ''conceptual influence on basic natural sciences,'' according to the Royal Swedish Academy of Sciences, but in addition, ''the discovery of catalytic RNA will probably provide a new tool for gene technology, with potential to create a new defense against viral infections.'' As a result of their findings, Altman and Cech were jointly awarded the 1989 Nobel Prize for chemistry.

Altman was born in Montreal, Quebec, on May 8, 1939, the second son of Victor Altman, an immigrant grocer, and Ray Arlin, who before her marriage worked in a textile mill. He attended West Hill High School in Montreal and the Massachusetts Institute of Technology, from which he graduated with a bachelor of science in physics in 1960. Between 1960 and 1962, he was a teaching assistant in the Department of Physics at Columbia University, while he waited for a suitable position in a lab. Around this time, Altman switched from physics to the newly emerging interdisciplinary field of molecular biology. He moved to the University of Colorado in Boulder in late 1962 to work as a research assistant. He was mainly preoccupied with studying the replication of the T4 bacteriophage, a substance that infects bacterial cells in much the same way as a virus infects human cells. Altman received his Ph.D. in biophysics in 1967.

After graduation, Altman briefly worked as a research assistant in molecular biology at Vanderbilt University before winning a grant from the Damon Runyon Memorial Foundation for Cancer Research. This permitted him to work as a research fellow in molecular biology at Harvard University. From 1967 to 1969, working under the biochemist Matthew Meselson, he continued his research into the genetic structure of the T4 bacteriophage. His receipt of the Anna Fuller Foundation Fellowship in 1969 enabled him to transfer to the Medical Research Council Laboratory of Molecular Biology in Cambridge, England, to work with molecular biologists Sydney Brenner and **Francis Crick**. The latter, in partnership with **James D. Watson**, discovered **DNA**'s double-helix structure in 1954.

Discovers Complexities in DNA Production

It was clear to scientists that genetic information is carried by DNA (deoxyribonucleic acid) into a cell's nucleus. In the cytoplasm (the substance inside the cell wall, surrounding the nucleus), the genetic code is copied into RNA. It is then converted into proteins, which are built of chains of amino acids. Altman originally intended to study the three-dimensional structure of **transfer RNA (tRNA)**, which is a small component of RNA that transfers **amino acids** onto a growing polypeptide chain as proteins are made. Much of the breakthrough work in this area had already been accomplished, however, so Altman decided to switch his attention to the transcription of tRNA from DNA. He found that the DNA from which tRNA is produced is not directly copied into tRNA but first undergoes an intermediary stage when it becomes a long strand of what is called "precursor RNA." This is composed of a strand of tRNA with additional genetic sequences at each end which are somehow later removed before it becomes tRNA.

While still working at the Medical Research Council in Cambridge, Altman studied the tRNA genes of the *Escherichia coli* (*E. coli*) bacterium, to which he added toxic chemicals. Subsequent mutations in the tRNA enabled him to isolate precursor tRNA from bacterial cells. Altman discovered that the additional sequences at each end of the strand of precursor tRNA were removed by something in the cells of the bacteria, probably an enzyme. The scientists found that the enzyme, named **ribonuclease** P (RNase P), would only cut off the extra sequences at a precise point.

When he returned the United States in 1971, Altman joined Yale University's biology department as an assistant professor. In 1972, he married Ann Korner. They have a son, Daniel, and a daughter, Leah. Altman was promoted to associate professor in 1975. In 1978, he published the results of an experiment carried out by one of his graduate students, Benjamin Stark. It demonstrated that RNase P was at least partially composed of RNA, which meant that RNA itself played an integral part in the activity of the enzyme. This finding was highly unorthodox, as it was then presumed that enzymes are made of protein, not nucleic acids.

Proves RNA is a Catalyst

In 1980, Altman attained a full professorship of biology at Yale. The following year, Cech at the University of Colora-

Sidney Altman. *(Photograph by Michael Marsland. Yale University Office of Public Affairs. Reproduced by permission.)*

do published independent results similar to Altman's. Cech discovered that the precursor RNA from the protozoan *Tetrahymena* were reduced to their final size as tRNA without the assistance of protein, and suggested that the precursor RNA catalyzed this itself. His findings lent weight to Altman's. Cech's use of the word "catalyst" to describe the action of the RNA was questioned, however, because rather than just speeding up a reaction, it used itself up in the process.

Three years later, by which time Altman had become chairman of Yale's biology department, his colleague, Cecilia Guerrier-Takada, was testing the catalytic activity of RNase P. She discovered catalysis even in the control experiments that used the RNA subunit of RNase P (the M1 RNA) but which contained no protein. Altman was able to prove that the M1 RNA demonstrated all the classical properties of a catalyst, especially as, unlike that studied by Cech, it remained unchanged by the reaction. This removed the last shadow of a doubt that RNA could act as an enzyme.

In 1984, Altman became a naturalized American, but retained his Canadian citizenship. From 1985 until 1989, as the dean of Yale College, Altman established a greater role for scientific education in all of Yale's curriculums. In 1989 he and Cech jointly received the Nobel Prize for chemistry for their discovery of RNA's catalytic ability. Their work put an end to the conundrum regarding proteins and nucleic acids which had long mystified scientists. They had been unable to discover which came first in the development of life, proteins or nucleic acids. Proteins catalyze biological reactions, whereas nucleic acids, such as RNA, transport the genetic codes that create the proteins. Altman and Cech proved that nucleic acids were the building blocks of life, acting as both codes and enzymes.

High hopes exist for the practical applications of their discovery, which was described by the Nobel Academy as one of "the two most important and outstanding discoveries in the biological sciences in the past 40 years," the other being Crick and Watson's discovery of DNA's double helix structure. If RNA enzymes are able to cut additional sequences of tRNA from a strand of precursor tRNA, doctors could possibly use RNA enzymes to cut infectious RNA from the genetic system of a person with an infectious viral disease. Research into this field is ongoing and, if fruitful, could contain the key to curing viral infections such as cancer and AIDS.

In addition to receiving the Nobel Prize, Altman was honored with the Rosentiel Award for Basic Biomedical Research in 1989. He is a member of the National Academy of Sciences and the American Society of Biological Chemists, of the Genetics Society of America, and is a fellow of the American Association for the Advancement of Science. He holds honorary degrees from the Université de Montréal, York University in Toronto, Connecticut University, McGill University, the University of Colorado, and the University of British Columbia. In 1991, he was selected to present the DeVane lecture series at Yale on the topic "Understanding Life in the Laboratory."

Altman has held a number of other part-time positions in addition to his full-time academic positions, including associate editor of *Cell* from 1983 to 1987, member of the Board of Directors of the Damon Runyon-Walter Winchell Fund for Cancer Research, member of the Board of Governors of the Weizmann Institute of Science, member of the Scientific Advisory Board of Bio-Méga, Inc., and Special Consultant to the Pathogenesis Corporation in Seattle.

ALTRUISM

Altruism is the behavior of an individual that benefits others, it can in a number of circumstances be detrimental to the individual but of benefit to the group as a whole. There are many examples of altruistic behavior in biology. Grazing **animal**s at risk from predators often employ members of the group to keep watch for predators. When a predator is sighted the individual sighting it lets out a warning to alert the other members of the herd. This warning is usually sufficient to save the lives of the majority of the animals involved but the individual giving the warning draws attention to itself, putting itself at greater risk.

Altruistic behavior is often more common amongst related individuals, the benefit that can be derived within a family group can, when viewed in one way be particularly great. Richard Dawkins argues that individuals are relatively unimportant and it is the survival of the **gene**s that is of paramount importance. An individual is merely a group of genes acting

together for convenience sake. Once a mating has taken place an individual will have 50% of the genes from each parent. If a parent were to sacrifice itself to save the life of a child of its own then 50% of the parents genes would survive. If the parent made the same sacrifice to save two of its children then two sets of 50% of its genes would survive (some of the same genes would be present in both individuals so it is not a case of 100% of the parents genes surviving). The closer the relationship the more genes they share, the more distant the relationship the smaller are the number of common genes. With the first example discussed the herd would be made of a number of individuals some with a close relationship to the animal giving the warning and some with very few genes in common. As a result of this the amount of risk involved in giving the warning is worth it. The increased level of risk to the individual is only slight but the potential benefit to common genes is large. This way of looking at altruism is known as the **selfish gene** hypothesis.

ALVEOLUS

An alveolus is a tiny air sac located within the lungs. The exchange of oxygen and **carbon dioxide** takes place within these sacs.

The basic structure of the **respiratory system** can be envisioned as an upside-down tree. Air is breathed into the trachea, which is the tree trunk, and thus the broadest part of the respiratory tree. The trachea divides into two major tree limbs, the right and left bronchi, each of which branches off into multiple smaller bronchi, which course through the **tissue** of the lung. Just as a tree's limbs branch off into ever smaller branches and twigs, so each bronchus divides into tubes of smaller and smaller diameter, finally ending in the terminal bronchioles. The air sacs of the lung, in which oxygen-carbon dioxide exchange actually takes place, are clustered at the ends of the bronchioles like the leaves of a tree at the ends of the smallest twig-like branches, and are called alveoli.

The alveoli are surrounded by tiny **blood** vessels, called capillaries. When air is inhaled (breathed into the lungs), it ultimately enters the alveoli. Because the alveoli are composed of only a single, thin layer of tissue, oxygen in the inhaled air can cross out of the alveoli and into the capillaries, where it binds with the **hemoglobin** found in red blood **cell**s. Blood containing oxygen is then carried throughout the body, for delivery to every type of tissue and **organ** system.

Carbon dioxide is one of the body's waste products. Carbon dioxide circulates through the body in the blood, until it reaches the alveolar capillaries. Carbon dioxide crosses out of the capillaries into the alveoli at the same time that oxygen is crossing out of the alveoli and into the capillaries. The carbon dioxide is then breathed out during exhalation.

It is interesting to note that, when comparing the alveoli of various **species**, the alveoli change in terms of both size and quantity. They are smallest but most numerous in **mammals**, intermediate in size and number in reptiles, and largest in size but smallest in quantity in amphibians. Humans continue to develop alveoli up until about the age of eight, when the human lung contains the adult number of approximately 300 million alveoli.

AMENSALISM

Amensalism refers to a relationship between two **species** in which one of the partners is inhibited, while the other is not affected in any significant way. Usually, the inhibited species is damaged by a chemical released by the other one into their shared environment.

One natural example of amensalism involves the growth of vegetation in the vicinity of breeding colonies of certain kinds of waterbirds. Beneath dense colonies of such species as cormorants and herons, the **trees** in which their nests are built, and most of the associated understorey vegetation, may be killed by the toxic excrement of the birds. The birds may eventually suffer a detriment from the loss of their nesting trees, but they receive no benefit (or harm) from the damage caused to the understorey **plants**.

Another natural example involves the black walnut (*Juglans cinerea*), a tree of forests of southeastern North America. Black walnut exudes a chemical known as juglone from its roots and foliage, which builds-up in its local environment. The juglone is quite toxic to most other species of plants, which therefore cannot grow beneath the canopy of a mature black walnut. The black walnut may receive a benefit in terms of their competitive relationship with larger plants, but there is no significant benefit from the damage caused to smaller plants of low abundance, such as mosses, ferns, and other low-growing vegetation.

Humans also have amensal relationships with numerous other species. In almost all such cases, the other species suffer a detriment as a result of one or more human activities. For example, **air pollution** caused by automobiles, electricity generating stations, or metal smelters often cause severe damage to lichens and plants in the affected area, whereas humans receive no direct benefit from this relationship. Another example is birds, **mammals**, and other wildlife that suffer **habitat** loss when forests are clear-cut to provide wood for industrial purposes. Although humans derive economic benefits from harvesting the timber, there are no particular benefits to people from the damage caused to habitat.

AMINO ACID

Amino acids, the building blocks of all **protein molecule**s, are **nitrogen**-containing **organic compound**s that consist of at least one acidic carboxyl group (COOH) and one amino group (NH$_2$). These two groups are attached to a carbon atom, which also carries a hydrogen atom, plus a side chain known as the ''R group.'' The R group varies from one amino acid to another and gives each amino acid its distinctive properties. Although relatively simple compounds, amino acids can vary widely and to date more than 80 different amino acids have been found in living **organism**s. Of these 80, 22 are considered the precursors of **animal** proteins.

A computer-generated model of glycine, an amino acid. *(Photo Researchers, Inc. Reproduced by permission.)*

The first few amino acids were discovered in the early 1800s. In 1806, the French chemist, Louis-Nicolas Vauquelin, isolated a compound in asparagus that proved to be the amino acid, asparagine. In 1812, William Hyde Wollaston found a substance in urine that he identified as a ''cystic oxide,'' and was later named cystine. And in 1820, another French chemist, Henri Braconnot, discovered the first two natural amino acids, glycine and leucine. Several other compounds were discovered toward the end of the century. In 1895, Sven Hedin isolated the compound arginine; in 1896, with the help of his colleague **Albrecht Kossel**, he discovered histidine. Three years later, in 1899, Edmund Dreschel identified another important amino acid, lysine. Though these scientists were able to determine that these were unique compounds, they were unsure of their exact significance. Scientists were also uncertain of the relationship between amino acids and protein molecules.

In 1899, the renowned German chemist, **Emil Fischer**, began investigating both questions. He synthesized many of the thirteen amino acids that were already known, and identified three more. He also showed how the various amino acids combined with each other inside the protein molecule—explaining, for instance, that the amino group of one amino acid is linked to the acidic carboxyl group of the next by a **peptide** bond. He suggested, as well, that the sequences and patterns formed by the various chains of amino acids helped establish the characteristics of different proteins.

In addition, Fischer developed a method for linking amino acids together, as they were in natural proteins, to form **polypeptide**s. In 1907, he managed to put together a synthetic protein molecule that contained eighteen amino acid units—a molecule so remarkably authentic that, as he demonstrated, digestive **enzyme**s attacked it just as they would a natural protein.

Although quite a lot was now known about the structure of amino acids, their nutritional significance had yet to be determined. Since the early 1800s, scientists such as Gerardus Mulder, François Magendie, and William Prout had established the nutritional importance of the proteins themselves.

But even here, with few exceptions (Magendie, for instance, had proven that gelatin had almost no nutritional value) the various proteins were considered roughly identical. So, most felt, were their amino acid units.

By the turn of the twentieth century, however, the situation began to change. In 1901, the British biochemist, **Frederick Gowland Hopkins**, not only discovered the amino acid tryptophan but later also showed that it played an important role in the diet. In one of his feeding experiments, Hopkins demonstrated that the protein in corn, zein—a protein that contains no tryptophan—could not sustain life in laboratory rats if used as the sole protein. Only when the tryptophan-rich protein casein was added to the diet did the rats once again begin to thrive. Hopkins's experiment suggested that, if proteins were not nutritionally identical (which seemed increasingly evident), perhaps it was the amino acids they contained that made the difference.

At roughly the name time, two Americans— **Thomas B. Osborne** and **Lafayette B. Mendel**—were reaching similar conclusions. Between 1909 and 1928, the two biochemists were investigating the proteins in a great variety of plant seeds. They found (among other things) that two amino acids in particular, tryptophan and lysine, were essential for normal growth in rats. Moreover, neither of the amino acids could be synthesized by the rats themselves, but had to be present in their diets.

In the 1930s, another American biochemist, **William Rose**, added the finishing touch to the amino acid story. In 1935, Rose isolated threonine, the last nutritionally important amino acid to be discovered, and, over the next decade or so, determined which amino acids could be synthesized by humans and certain **mammals**, and which had to be supplied by the diet. In humans, the dietary essential amino acids proved to be isoleucine, leucine, lysine, methionine, phenylalanine, threonine, tryptophan, valine, and, in growing children, histidine. Unless all these amino acids were attained through various protein foods, Rose explained, the body would not have the building blocks to form new protein molecules, and the growth and repair of body **cell**s would be impaired.

Nutritionists have since determined that all of these essential amino acids can be obtained from meats, **egg**s, milk, cheese and other foods derived from animals. Plant proteins, however, generally lack several of these amino acids and, unless supplemented, can lead to protein deficiency problems.

AMMONIA

Ammonia is a colorless gas composed of **nitrogen** and hydrogen with the formula NH_3. It is the simplest stable compound formed from these two elements. Ammonia is one of the most widely used compounds in the United States, and serves as a raw material for the production of many important compounds. A large amount of ammonia is used as a fertilizer. Its boiling point is about -27.4°F (-33.3°C) making it easy to liquefy at low **temperature**s. For this reason, it is commonly applied directly to the soil from portable tanks filled with the

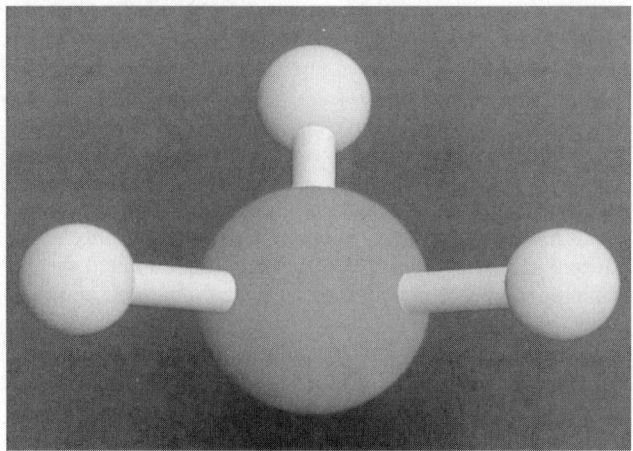

A computer-generated model of an ammonia molecule. *(Photo Researchers, Inc. Reproduced by permission.)*

liquefied ammonia. Ammonia is also used to produce salts such as ammonium nitrate and ammonium phosphate for use as commercial fertilizers. It is used in the manufacture of nylon and rayon and for scouring cotton, wool, and silk. Although nitrogen makes up approximately 79% of the gas in the **atmosphere**, **plants** are not able to assimilate gaseous nitrogen unless it is converted to ammonia or some other form. An industrial method for synthesizing ammonia known as the Haber-Bosch process involves the combining of elemental hydrogen and elemental nitrogen in a reaction that requires the use of a catalyst, high pressure (100-1,000 atmospheres), and temperatures of 750-1,200°F (400-650°C). Biological nitrogen fixation, which forms ammonia from atmospheric nitrogen and hydrogen supplied by the **organism**, is also an important source of ammonia for plant growth. Both free-living and symbiotic **bacteria** are known to carry out this reductive fixation process. An especially important example involves *Rhizobium* bacteria in root nodules of leguminous plants, and is the basis for the common agricultural practice of growing legumes in rotation with other crops to enrich fields with available nitrogen.

See also Nitrogen cycle

AMNIOCENTESIS

Amniocentesis is a procedure used to obtain amniotic fluid for prenatal diagnosis of a fetus. **Cells** naturally are exfoliated from the surface of the fetus and some of these cells survive for a time in the fluid surrounding the fetus in the amniotic cavity. Soluble biochemical material of clinical significance produced by the fetus may also accumulate in the amniotic fluid. The fluid can be analyzed for these substances directly. A local anesthetic is given and a hollow needle is inserted through the mother's abdominal wall into the amniotic cavity and a small sample of the fluid is withdrawn with a syringe attached to the needle. In order to insure the safety of the fetus, the procedure

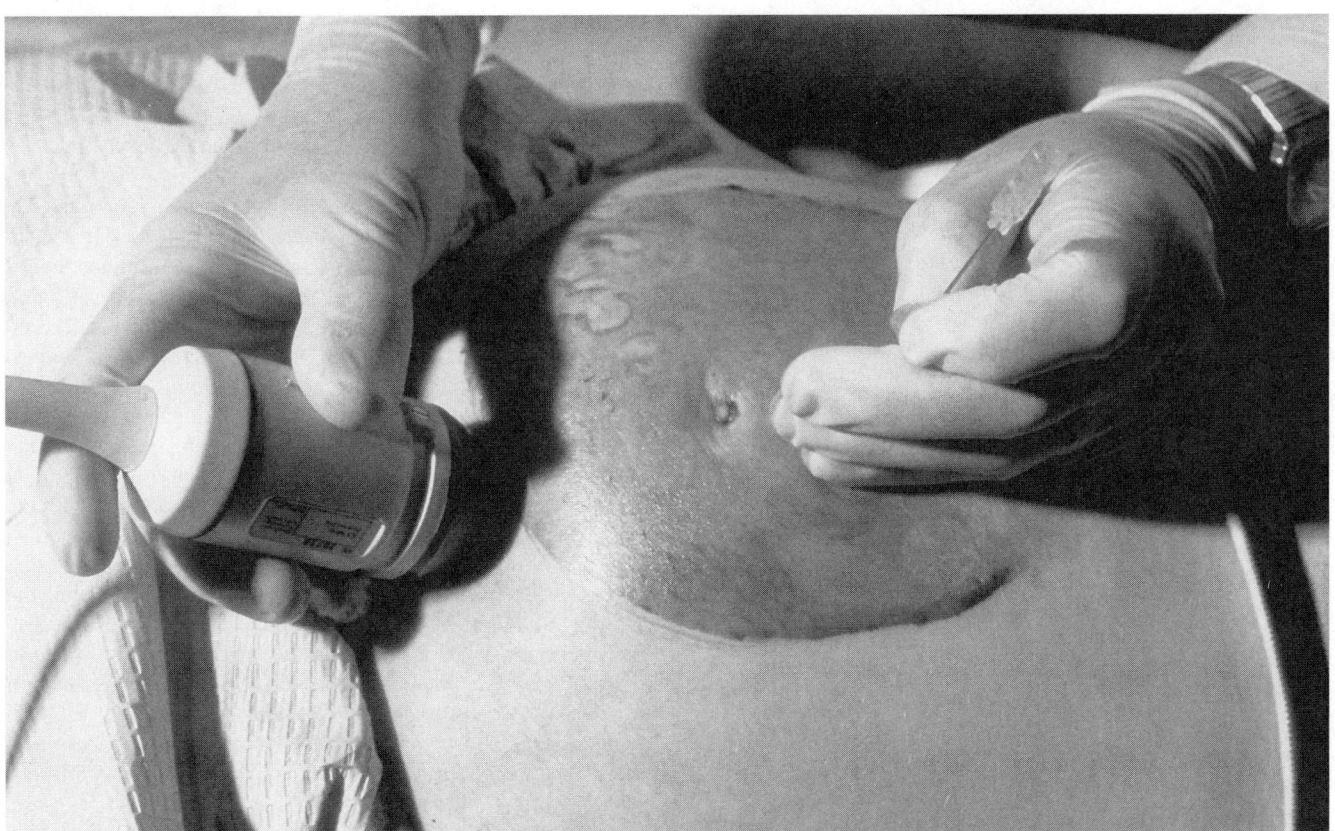

A physician uses an ultrasound monitor (left) to position the needle for insertion into the amnion when performing amniocentesis. *(Photograph by Will and Deni McIntyre, Photo Researchers, Inc. Reproduced by permission.)*

is monitored via an ultrasound scan. Viable cells in the fluid are then **culture**d (grown) *in vitro*. The **chromosome**s of the cultured cells can then be examined. Viewing the chromosomes under a **light microscope** will reveal if a normal **diploid** number of chromosomes are present or if extra or fewer chromosomes are present. Additionally, structural chromosomal aberrations can be detected.

Amniocentesis is an elective procedure that can detect the presence of many types of **gene**tic disorders, thus allowing doctors and prospective parents to make important decisions about early treatment and intervention. **Down syndrome** is a chromosomal disorder characterized by a diversity of physical abnormalities, mental retardation, and shortened life expectancy. It is by far the most common, nonhereditary, genetic **birth** defect, afflicting about one in every 1,000 babies. Since the risk of bearing a child with a nonhereditary genetic defect such as Down syndrome is directly related to a woman's age, amniocentesis is recommended for women who will be older than 35 on their due date. Thirty-five is the recommended age to begin amniocentesis because that is the age at which the risk of carrying a fetus with such a defect roughly equals the risk of miscarriage caused by the procedure—about 1 in 200.

Amniocentesis is ordinarily performed between the 14th and 16th week of pregnancy, with results usually available within three weeks. It is possible to perform amniocentesis as early as the 11th week but this is not usually recommended because there appears to be an increased risk of miscarriage when done at this time. The advantage of early amniocentesis is the extra time for decision making if a problem is detected. Potential treatment for the fetus can begin earlier. Elective abortions are safer and less controversial the earlier they are performed.

AMNION

The amnion is an extremely thin **tissue**, or **membrane**. It encloses a developing **embryo**, and is present within the **egg**s of all vertebrates which live on land. The presence of the amnion creates a cavity, called the amniotic cavity. The amniotic cavity is filled with amniotic fluid, in which the embryo floats. This fluid serves to cushion and protect the tiny embryo, acting as a kind of shock absorber. The presence of this fluid surrounding the embryo also prevents the thin, delicate tissues of the embryo from drying out, at the same time preventing the embryo from sticking to the inside of the egg shell. The amniotic fluid also helps maintain the embryo in an environment with a consistent **temperature**.

The amnion is sometimes popularly called "the bag of **water**s." At some time during a human **birth**, this bag of waters bursts.

AMPHIBIANS • See Herpetology

ANABOLISM • See Metabolism

ANAPHYLAXIS

As information about the body's immune responses accumulated in the late 1800s, researchers came to view this biological defense mechanism against invading **antigen**s as invariably protective. A few warning notes were sounded, however. **Edward Jenner**, the vaccination pioneer, observed in 1798 that patients given an inoculation a second time sometimes had violent reactions. In 1839 French physician **Francois Magendie** showed that rabbits who tolerated a first injection of **egg** albumin sometimes died when reinjected.

The first complete study and description of this dangerous immune response was produced by two Frenchmen, physiologist Charles Richet and physician Paul Portier (1866-1962). During a scientific cruise on the yacht of Prince Albert of Monaco, the prince suggested that Portier and Richet study the toxin produced by the tentacles of the Portuguese man-of-war, a jellyfish. They did this, showing that an extract of the tentacles was highly toxic. Back in France, the two continued their studies with extracts of toxin from sea anemone. Seeking to determine the toxic dose, they injected dogs with the venom. Dogs that survived were given time to recover and then reinjected. Richet had expected that the first exposure to the toxin would have created a certain amount of immunity in the dogs. Instead, to his surprise, the initial exposure had made the dogs hypersensitive. A second, much smaller dose of toxin quickly killed them. Since this was the opposite of protective prophylaxis, Richet in 1902 named this hypersensitive reaction *anaphylaxis*.

It soon became apparent that a wide range of substances could trigger anaphylactic reactions. This knowledge provided a valuable warning for physicians engaged in serum therapy; allowing them to check patients for possible sensitization before injecting potentially toxic amounts of serum. Understanding anaphylaxis, also called anaphylactic shock, also helped to explain the wider range of sensitization reactions involved in allergies. Although the majority of allergic reactions are irritating but rarely life threatening, people who develop anaphylaxis—for example, in reaction to insect bites and allergies to food and drugs—are in a potentially life-threatening situation. Usually, symptoms of anaphylaxis, such as hives, abdominal pain, and rapid pulse, arise with seconds or minutes of exposure to an allergen. Because of these important implications, Richet's discovery of anaphylaxis earned him the 1913 Nobel Prize for medicine.

ANATOMY

Anatomy is the study of the structure of **plant**s and **animal**s. This study can be divided into two fields, gross anatomy (or macroscopic anatomy) and microscopic anatomy. Gross anato-

my focuses on the study of structures which can be seen by the naked eye. Microscopic anatomy can be further subdivided into **histology**, which is the study of **tissue**s, and **cytology**, which is the study of **cell**s.

Study of gross anatomy began much earlier than that of microscopic anatomy, which could not develop until the invention of the **microscope**. A primitive study of anatomy began with the teachings of **Aristotle**, who argued that each **organ** had its own function, which could be determined by observing its structure. The first school of anatomy was found later in the third century B.C. by **Herophilius** of Chalcedon, who encouraged his students to dissect human bodies. Herophilus is credited both with determining that the **brain** controls the **nervous system** and with distinguishing between voluntary and involuntary nerves. Another Greek physician from the same century, **Erasistratus**, also performed human dissections and described theories relating the **arteries** with the **lungs**, the direction of circulation from the **veins** to the arteries, and described the functions of the trachea and epiglottis. The Greek physician Galen (c. 130-200 B.C.) correlated the works of Aristotle and other early physicians with his own experiments and animal dissections. He is credited with showing that arteries carry **blood** and not air, discovering the function of the kidneys, and adding to general knowledge of the nervous system. Unfortunately, his work was undisputed for years which hampered further discoveries in the field of anatomy for centuries.

The first post-classical medical schools in the West were founded in the Middle Ages, although researchers trusted the Greek medical texts before their own dissections. With the Renaissance came a renewed interest in anatomy. **Leonardo da Vinci** studied the muscular structure of humans and animals in his artwork. The first truly systematic studies of the human body began shortly thereafter, but were discouraged due to religious conflicts concerning the morality of dissecting human bodies. **Andreas Vesalius** was considered one of the founders of modern anatomy with his work *On the Structure of the Human Body* in 1543, only to be rewarded a death sentence during the Inquisition for his work. His advances lived on in his students, such as **Michael Servetus**, who discovered pulmonary circulation; and Realdo Columbus, who actually coined the term circulation.

The seventeenth century brought about many discoveries in gross anatomy with the founding of several medical schools and many advances in technology. Human dissections were common in medical schools and an understanding of human anatomy was well under way. The invention of the compound microscope at this time brought about the study of microscopic anatomy. **Marcello Malpighi** observed the movement of blood through capillaries and studied lung and gland structure. **Anthoni van Leewenhoek** helped discover **microorganism**s. **Jan Swammerdam** observed red blood cells and **Robert Hooke** coined the term cell. These discoveries led to the **cell theory** in the eighteenth and nineteenth centuries and served as a basis for the modern study of anatomy.

An underlying theme in the study of anatomy which has led to many of the modern theories is that structure is almost always related to function. For example, the structure of a bird's beak is related to the kind of food it eats, whether it picks berries or cracks open seeds. Another example is the structure of white blood cells. These cells have tiny projections on their surface, which helps them ''grab'' microorganisms, **virus**es, and cell particles which are then destroyed. The study of the structure of plants and animals, and how this structure is related to function, led to the branch of anatomy known as **comparative anatomy**.

Comparative anatomy is used to test **Charles Darwin**'s theory of **evolution**. Many studies in comparative anatomy have provided strong evidence for this theory. Structures in different organisms are compared, and organisms with more similar structures are considered to be more closely related evolutionarily. Comparative anatomy helps classify organisms into taxonomic categories and makes it possible to construct evolutionary trees. Structures that are studied are classified as either homologous or analogous. Homologous structures are the same in evolutionary origin, although presently are different in structure and function. Examples of homologous structures are the forelimbs in the wing of a bat, the fin of a porpoise, and the leg of a horse. Analogous structures have similar structure and function but different evolutionary origins, for example the eyes of vertebrates and octopuses. Another structure which an organism may have is a **vestigial organ**, which is underdeveloped and perhaps useless in an organism, but may be fully developed and functional in related organisms. An example of a vestigial organ is the human appendix.

The study of human anatomy has led to extensive knowledge about the many systems of the human body. The systems that are studied in human anatomy are: the **endocrine system**, which includes the **hormone**-releasing glands; the **reproductive system**, which includes the structures involved in the production of offspring; the **skeletal system**, which includes the bones of the body; the urinary system, or those structures concerned with the excretion of waste from the body; the **digestive system**, which consists of those organs which break down food into useable **energy** for the body; the **circulatory system**, which consists of the **heart** and **blood vessel**s; the **lymphatic system**, which helps defend the body against invasion by disease-causing agents; the **muscular** system, which includes the muscles of the body; and the nervous system, which controls the actions and reactions of the body. Developments in microscopic techniques have led to significant understanding of human anatomy on both a gross and a cellular level.

See also Botany

ANATOMY, COMPARATIVE

There are many forms of evidence for **evolution**. One of the strongest forms of evidence is comparative anatomy; comparing structural similarities of **organism**s to determine their evolutionary relationships. Organisms with similar anatomical features are assumed to be relatively closely related evolutio-

narily, and they are assumed to share a common ancestor. As a result of the study of evolutionary relationships, anatomical similarities and differences are important factors in determining and establishing **classification** of organisms.

Some organisms have anatomical structures that are very similar in embryological development and form, but very different in function. These are called homologous structures. Since these structures are so similar, they indicate an evolutionary relationship and a common ancestor of the **species** that possess them. A clear example of homologous structures is the forelimb of **mammals**. When examined closely, the forelimbs of humans, whales, dogs, and bats all are very similar in structure. Each possesses the same number of bones, arranged in almost the same way. While they have different external features and they function in different ways, the embryological development and anatomical similarities in form are striking. By comparing the **anatomy** of these organisms, scientists have determined that they share a common evolutionary ancestor and that in an evolutionary sense, they are relatively closely related.

Other organisms have anatomical structures that function in very similar ways, however, morphologically and developmentally these structures are very different. These are called analogous structures. Since these structures are so different, even though they have the same function, they do not indicate an evolutionary relationship nor that two species share a common ancestor. For example, the wings of a bird and dragonfly both serve the same function; they help the organism to fly. However, when comparing the anatomy of these wings, they are very different. The bird wing has bones inside and is covered with feathers, while the dragonfly wing is missing both these structures. They are analogous structures. Thus, by comparing the anatomy of these organisms, scientists have determined that birds and dragonflies do not share a common evolutionary ancestor, nor that, in an evolutionary sense, they are closely related. Analogous structures are evidence that these organisms evolved along separate lines.

Vestigial structures are anatomical features that are still present in an organism (although often reduced in size) even though they no longer serve a function. When comparing anatomy of two organisms, presence of a structure in one and a related, although vestigial structure in the other is evidence that the organisms share a common evolutionary ancestor and that, in an evolutionary sense, they are relatively closely related. Whales, which evolved from land mammals, have vestigial hind leg bones in their bodies. While they no longer use these bones in their marine **habitat**, they do indicate that whales share an evolutionary relationship with land mammals. Humans have more than 100 vestigial structures in their bodies.

Comparative anatomy is an important tool that helps determine evolutionary relationships between organisms and whether or not they share common ancestors. However, it is also important evidence for evolution. Anatomical similarities between organisms support the idea that these organisms evolved from a common ancestor. Thus, the fact that all vertebrates have four limbs and gill pouches at some part of their development indicates that evolutionary changes have occurred over time resulting in the diversity we have today.

ANFINSEN, CHRISTIAN BOEHMER
(1916-1995)
American biochemist

Biochemist Christian Boehmer Anfinsen is known for establishing that the structure of an **enzyme** is intimately related to its function. This discovery was a major contribution to the scientific understanding of the nature of enzymes. For this achievement, Anfinsen shared the 1972 Nobel Prize for Chemistry with the research team of **Stanford Moore** and **William Howard Stein**.

Anfinsen was born on March 26, 1916, in Monessen, Pennsylvania, a town located just outside of Pittsburgh. He was the child of Christian Anfinsen, an engineer and emigrant from Norway, and Sophie Rasmussen, who was also of Norwegian heritage. Anfinsen earned his B.A. from Swarthmore College in 1937. Subsequently, he attended the University of Pennsylvania, earning an M.S. in organic chemistry in 1939. After earning his master's degree, Anfinsen received a fellowship from the American Scandinavian Foundation to spend a year at the Carlsberg Laboratory in Copenhagen, Denmark. Upon his return in 1940, he entered Harvard University's Ph.D. program in biochemistry. His doctoral dissertation involved work with enzymes; he described various methodologies for discerning the enzymes present in the retina of the eye, and he earned his Ph.D in 1943.

After receiving his Ph.D., Anfinsen began teaching at Harvard Medical School, in the department of biological chemistry. From 1944 to 1946 he worked in the United States Office of Scientific Research and Development. He then worked in the biochemical division of the Medical Nobel Institute in Sweden under **Hugo Theorell**, as an American Cancer Society senior fellow, from 1947 to 1948. Harvard University promoted him to associate professor upon his return, but in 1950 he accepted a position as head of the National Institutes of Health's (NIH) National Heart Institute Laboratory of Cellular Physiology. He served in this position until 1962. Anfinsen returned to teaching at Harvard Medical School in 1962, but he returned to NIH a year later. This time he was named director of the Laboratory of Chemical Biology at the National Institute of Arthritis, Metabolism, and Digestive Diseases. He held this position until 1981; he spent a year at the Weizmann Institute of science and then in 1982 accepted an appointment as professor of biology at Johns Hopkins University, where he remained until his death. A former editor of *Advances in Protein Chemistry*, he was also on the editorial boards of the *Journal of Biological Chemistry*, *Biopolymers*, and the *Proceedings of the National Academy of Sciences*.

Anfinsen began his research concerning the structure and function of enzymes in the mid–1940s. Enzymes are a type of protein; specifically, they are what drives the many chemical reactions in the human body. All **proteins** are made up of smaller components called **peptide** chains, which are **amino acids** linked together. Amino acids are, in turn, a certain class of organic compounds. The enzymes take on a globular, three-dimensional, form as the amino acid chain folds over. The unfolded chain form of an enzyme is called the primary structure.

Once the chain folds over, it is said to be in the tertiary structure. From one set of amino acids for one particular enzyme there are 100 different possible ways in which these amino acids can link together. (Only certain amino acids can "fit" next to other amino acids.) However, only one configuration will result in an active enzyme. In general, Anfinsen's research concerned finding out how a particular set of amino acids knows to configure in a way that results in the active form of the enzyme.

Anfinsen chose to study the enzyme **ribonuclease** (RNase), which contains 124 amino acids and is responsible for breaking down the **ribonucleic acid (RNA)** found in food. This reaction enables the body to recycle the resultant smaller pieces. He felt that by determining how a particular enzyme assumes its particular active configuration, the structure and function of enzymes could be better understood. He reasoned that he could determine how an enzyme protein is built and when the enzyme becomes functional by observing it adding one amino acid at a time. He utilized techniques developed by Cambridge University's **Frederick Sanger** to conduct this research. Another research team headed by Stanford Moore and William Howard Stein was working simultaneously on the same enzyme as Anfinsen, ribonuclease; in 1960, using ribonuclease, Moore and Stein were the first to determine the exact amino acid sequence of an enzyme. However, Anfinsen remained more concerned with how the enzyme forms into its active configuration.

Anfinsen eventually changed his methodology of research during an opportunity to study abroad. While at the NIH, Anfinsen took yet another leave of absence when a Rockefeller Public Service Award allowed him to spend 1954 to 1955 at the Carlsberg Laboratory studying under the physical chemist Kai Linderstrøm-Lang. Anfinsen had been studying ribonuclease by building it up; Linderstrøm-Lang convinced him to start with the whole molecule and study it by stripping it down piece by piece. Anfinsen began with the whole ribonuclease molecule and then successively broke the various bonds of the molecule. The process is called denaturing the protein or, in other words, causing it to lose its functional capacity. By breaking certain key bonds, other bonds formed between the amino acids resulting in a random, inactive form of ribonuclease. By 1962, Anfinsen had confirmed that when this inactive form is placed into an environment that mimics the environment in which ribonuclease normally appears in the body, that inactive form would slowly revert to the active configuration on its own and thus regain its enzymatic activity. This discovery revealed the important fact that all the information for the assembly of the three-dimensional, active enzyme form was within the protein's own sequence of amino acids.

Receives Nobel Prize for Enzyme Research

For uncovering the connection between the primary and tertiary structure of enzymes, Anfinsen received half of the 1972 Nobel Prize for Chemistry. Moore and Stein were awarded the other half. In addition to his numerous journal articles on protein structure, enzyme function, and related matters, in

1959, Anfinsen published a book entitled *The Molecular Basis of Evolution.* After receiving the Nobel Prize, Anfinsen began focusing his research on the protein interferon, known for its key role as part of the body's immunity against both viruses and cancer. He succeeded in isolating and characterizing this important human protein.

Anfinsen's honors in addition to the Nobel Prize include The Rockefeller Foundation Public Service Award, a Guggenheim Fellowship, as well as honorary degrees from Georgetown University and New York Medical College and five other universities. Anfinsen was a member of the National Academy of Sciences, the American Society of Biological Chemists, and the Royal Danish Academy. As an opponent of biological weapons, he belonged to the Committee for Responsible Genetics. He married Florence Bernice Kenenger in 1941, and they had three children. Anfinsen and Kenenger divorced in 1978. In 1979, Anfinsen married Libby Esther Schulman Ely. He died on May 14, 1995, at Northwest Hospital Center in Randallstown, Maryland.

ANGIOSPERM • See Plant

ANIMAL

Animals are **organism**s in the **kingdom** Animalia. This is one of the five kingdoms (or major divisions) of organisms, the others being: Plantae (**plants**), Protists (**protozoa**ns), **Monera** (**bacteria**), and **Fungi** (fungi). Animals are eukaryotic organisms (i.e., the nuclear material of their **cell**s occurs within a **membrane**-bounded **nucleus**). They are composed of numerous cells, which do not have walls made of **cellulose**. In addition, animals can make voluntary, spontaneous movements, often in response to a sensory **perception**. Animals require a source of biologically fixed **energy** for their **nutrition**, such as the **biomass** of plants or other animals. The scientific study of animals is known as **zoology**.

The first animals were multicellular life forms that originated in late Precambrian times (4,600 million to 570 million years ago). However, little is known about these soft-bodied creatures or when they first evolved, because they did not preserve well as **fossil**s. The first definite fossils of animals date from about 640-670 million years ago, by which time phyla of animals already existed. Zoologists are still classifying animals into various groups, but 30-35 phyla are recognized. However, some of these are extinct and are only known from their ancient, fossil impressions.

The simplest of the living (or extant) animals are asymmetric or radially symmetric in shape, and include the Porifera (sponges) and Cnidaria (jellyfish and sea anemones). Bilaterally symmetric animals are somewhat more complex in their **anatomy**, and include the Platyhelminthes (flatworms) and Nematoda (nematodes). Animals known as coelomates have an enclosed body cavity and include the Mollusca (clams, snails, octopus, and squid), Annelida (segmented worms and leeches), Arthropoda (crustacea, insects, and spiders), Echinodermata (sea urchins and starfish), and Chordata (fish, amphibians, reptiles, birds, and **mammals**).

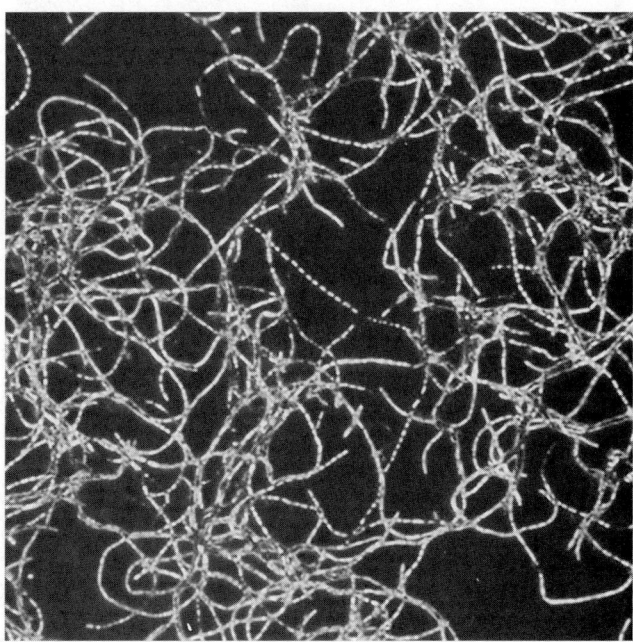

Anthrax bacterium, shown here chained together in long threads, has been used in biological weapons since World War II. *(Photo Researchers, Inc. Reproduced by permission.)*

Zoologists have given binomial (or scientific) names to about one million **species** of animals. These scientific names tell us both the **genus** (the first, capitalized word) and species (the second, lowercased word) of an organism. One example of this nomenclature is our own species, *Homo sapiens*). However, a much larger number of animal species has yet to be discovered (most of these occur in tropical **rain forest**s and the deepest parts of the oceans). In fact, some biologists estimate that there are several tens of millions of undiscovered species of animals, most of which are insects, particularly small species of tropical beetles.

See also Endangered species

ANTHRAX

The deadly outbreaks of anthrax in the nineteenth century baffled farmers who were raising their cattle in fields that had been cleared of any infected **animal**s. They did not understand the cause of the recurring outbreaks which killed farm animals as well as their handlers.

In humans, anthrax may manifest itself as malignant pustules (the cutaneous form of the disease), which could be contracted from the bristles of a shaving brush containing the **spore**s, or from exposure to hides or **blood** of infected animals. Anthrax pneumonia (inhalation anthrax), or ''wool sorter's disease,'' was prevalent among those working with hides and wool. An intestinal form of the disease resulted from eating anthrax-contaminated meat.

Rod-shaped **organism**s, *Bacillus anthracis*, were discovered in the blood of anthrax-infected animals as early as

1850 by French physician Casimir Joseph Davaine (1812-1882), who postulated that these organisms cause the disease. In 1876 **Robert Koch**, demonstrated that these rod-shaped bodies were present in the blood of infected animals, and absent from healthy ones, thus confirming Davaine's hypothesis.

During additional experimentation, Koch determined the **life cycle** of the bacillus. In presenting his findings, he demonstrated that the bacilli formed spores that were resistant to outside influences, including heat, and could remain for long periods in a state of suspended animation waiting to infect and kill the next victims. Therefore, under proper external conditions, the spores in the blood and **tissue**s of dead animals could become virulent and set off a recurrence of the deadly disease. This explained new outbreaks of anthrax in fields that had been cleared of infected animals. **Louis Pasteur** conducted his first experiments on anthrax with physicist Jules Joubert in 1877. They confirmed Koch's earlier experiments to show that dilution of *Bacillus anthracis* did not alter the virulence of the bacillus. He also discovered that earthworms, passing over the bodies of animals that had died of anthrax, were carrying the spores to the grass above and infecting healthy animals. Therefore, a vaccine was imperative for saving cattle, sheep, and horses from the reviving spores. Pasteur turned his attention to finding a vaccine for anthrax.

Pasteur, and his medical and veterinary colleagues Pierre-Paul-Emile Roux (1853-1933) and Charles-Edouard Chamberland (1851-1908), discovered that bacilli kept at 108–109°F (42–43°C) were reduced in virulence and lost their ability to make spores. This left the organism in an attenuated, or weakened, form. On May 5, 1881, the trio of Pasteur, Roux, and Chamberland answered a challenge by the Agricultural Society of Melun to give a public demonstration of their work. They inoculated 24 sheep, one goat, and six cows with a living attenuated vaccine, and on May 17 with a less attenuated **culture**. Another 24 sheep, one goat and four cows were not inoculated. On May 31, the animals were exposed to anthrax bacilli. Within two days, the uninoculated sheep and goat were dead; the four cows were sick. The inoculated animals were healthy and grazing in the fields.

In 1935 Mario Mazzucchi of Milan produced a vaccine by suspending anthrax spores in saponin. He claimed that his vaccine, unlike Pasteur's, was stable in virulence and protection. Four years later, Max Sterne in South Africa formulated a vaccine of avirulent uncapsulated sporing bacilli. It was reported to be even safer and to produce fewer side effects.

The first vaccine for humans was developed in 1948 by G.P. Gladstone of Oxford University. It is an alum precipitate of the **antigen** found in sterile filtrates of the bacillus, and it is used primarily to safeguard humans working with imported animal hairs, wools, hides, and bone meal.

Anthrax was also used in **biological warfare** during World War II, and may still be a threat as a weapon today. In 1979 an unusually severe outbreak of the inhalation form of anthrax resulted in over 1,000 fatalities around the Russian city of Sverdlovsk. Although information remains sketchy, it is thought that this outbreak may have resulted from an accident at a biological warfare facility.

In December of 1997, the Pentagon announced a plan to innoculate United States troops against anthrax as a precaution against biological warfare attacks, particularly from the Middle East and Asia. Fears that the civilian **population** of the United States and other countries may be targeted as well, has led to the formation and training of Hazardous Materials Teams, Hazmats, to contain such an outbreak. Worries that smallpox or other diseases may be used as biological weapons has increased concerns on this issue.

ANTIBIOTIC

Antibiotics are drugs such as **penicillin**, streptomycin, and erythromycin that are administered orally or by injection to rid the body of harmful **bacteria** that cause disease. Some microscopic bacteria that enter the human body through an opening or a wound quickly find abundant food and reproduce in great numbers, releasing **toxins** as they grow or when they die. The toxins can destroy human **cell**s or interfere with cell function, causing diseases like pneumonia or tuberculosis. Antibiotic drugs are derived from natural compounds that are antagonistic or harmful to bacteria. In **nature**, some **organism**s such as **fungi** and certain bacteria have evolved ways of defending themselves against harmful bacteria by producing their own toxins that destroy bacterial cells. Molds, such as those in the Penicillium **family** or Streptomyces griseus, have become a source of drugs for human use. These *anti-* (against) *biotic* (life) compounds usually work by either damaging the bacterial cell **membrane** or interfering with an important life function of bacteria.

The story of how antibiotic drugs came to be as accessible as the nearest pharmacy has a long history. The phenomenon of bacterial antagonism as it is known was observed as long ago as 1871, when the surgeon **Joseph Lister** first made note of it. In 1877, the French scientist **Louis Pasteur**, working with Jules François Jobert, discovered that when there were more common germs present first, something inhibited the growth of bacteria that caused **anthrax**, a fatal disease in farm **animal**s. In the 1880s German researchers found that rabbits gained protection against anthrax if they were administered bacteria from the streptococcus family. But research into bacterial antagonism as a cure for anthrax did not proceed because Pasteur had already developed a successful anthrax vaccine to prevent the disease.

A decade later, an Italian scientist, Dr. B. Gosio was investigating cases of pellagra, thought at the time to be caused by infected corn, but now known to be a deficiency disease. He studied **microorganism**s that attack corn, several of which belonged to the penicillium group. He **culture**d penicillium mold and found that the culture stopped the growth of anthrax germs in a test tube. However, Gosio did not have the resources to continue with this line of research. Several British scientists took up where Gosio had left off, producing a substance they called penicillic acid, which killed some bacteria—but it also killed experimental mice.

In 1928 **Alexander Fleming**, a Scottish doctor, made an important discovery that eventually led to the development of penicillin. In his crowded hospital laboratory, Fleming was ex-

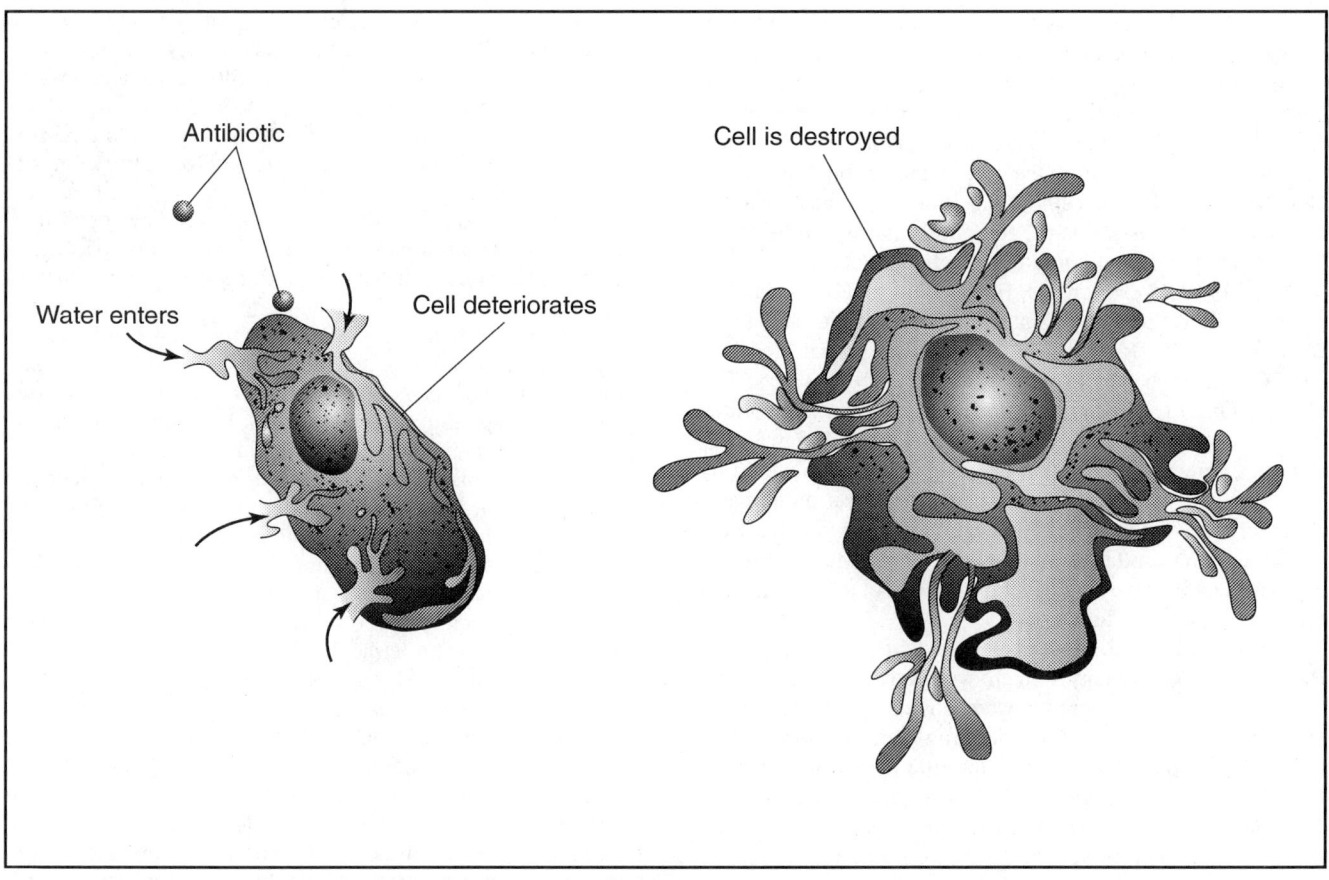

Antibiotic

Water enters

Cell deteriorates

Cell is destroyed

Different antibiotics destroy bacteria in different ways. Some short-circuit the processes by which bacteria receive energy. Others disturb the structure of the bacterial cell wall, as shown in the illustration above. *(Illustration by Electronic Illustrators Group.)*

amining the colorful patches of bacteria in each covered dish and noticed that a green mold had gotten into one dish. Fleming had seen molds growing in a laboratory dish before, knowing that mold **spore**s travel through the air and can easily land in a dish when left uncovered. Fleming noticed that the bacteria closest to the mold seemed to have disappeared or dissolved. He looked carefully at the mold with a trained eye and even photographed it. Fleming's colleague identified it as *Penicillium notatum*. Curious about how the bacteria in this dish were killed, Fleming used the greenish ''fluff'' on the dish to make a mixture that his laboratory workers called ''mold juice.'' Fleming named the juice penicillin and gave it to some laboratory mice. He found that the penicillin killed only the bacteria and not the cells in the mice, making Fleming's ''mold juice'' safer than any other known bacteria-killing substances. It was an incredible discovery. If this mold mixture could be made into a drug then someone with an infection could be cured of disease without harm.

Using penicillin as a drug sounded like a great idea, but Fleming found that when he grew mold in a dish it would kill bacteria for a while, then would lose its potency. He found also that the mold juice would not dissolve in common solutions, making it difficult to use. So although Fleming discovered penicillin and knew the value of it, he could not produce any-

thing useful to doctors. He continued to use the mold in his laboratory all the time to keep unwanted bacteria out of his culture dishes, and in 1929 Fleming wrote a report about his discovery. Other scientists read it, but it did not generate much excitement at the time.

Several years later one of Fleming's own students, Dr. C. G. Paine, cured several children of eye infections using a mixture like Fleming's mold juice. Dr. Paine found the mixture very hard to work with and very few people heard about how Fleming's penicillin could cure disease. In the mid-1930s Grieg Smith, an Australian bacteriologist, discovered that a group of soil organisms showed antibacterial activity. His paper was read with interest by scientists in the United States, steering research in a new direction.

In 1939 René Jules Dubos, a French-American microbiologist whose research interest was soil microbiology, made another important discovery. Working at the Rockefeller Institute for Medical Research in New York, Dubos isolated an antibacterial substance in the bacterium Bacillus brevis. He named the substance tyrothricin. With chemical analysis, he found tyrothricin to be a mixture of **protein**s consisting of relatively short **amino acid** chains. The compounds themselves did not prove to be effective agents against bacteria. However, Dubos's reports helped to revive interest in Fleming's earlier work on penicillin.

In 1939 Dubos's teacher at Rutgers University in New Jersey, Dr. Selman Abraham Waksman, began looking closely at the antibacterial properties of soil organisms. Waksman, also a soil microbiologist, had a large collection of streptomycetes fungi at his disposal at his Rutgers laboratory. He isolated antibacterial agents from the streptomycetes but found them all to be toxic to human cells. It was Waksman who coined the term antibiotic to describe the compound that would be harmful to bacteria without being toxic to human cells.

In 1940, two scientists, **Howard Florey** (1898-1968), a professor of pathology originally from Australia, and **Ernst Chain** (1906-1979), a German biochemist who had fled Hitler's Germany, began to experiment with penicillin at England's Oxford University. After many experiments, they found a way to get penicillin in its pure form and began toxicity tests on mice. Finding that the drug could be tolerated by mice with few side effects, they began testing it on humans. By this time, World War II had begun and the wounded were crowding into hospital, and Florey and Chain's team of workers rushed to finds ways of developing a useful form of penicillin to fight bacterial infection.

By 1942 penicillin was being made in large amounts by British companies. Many soldiers were saved from the infections that developed after they were wounded in battle. People called penicillin the new ''wonder drug'' for the thousands of cures it made possible, usurping the title for the sulfonamide drugs that had been developed a decade earlier. Penicillin cut the **death** rate from bacterial pneumonia from 60-80% down to 1-5%. Fleming received the 1945 Nobel Prize for Medicine as a result of the numerous uses found for penicillin.

Despite the effectiveness of penicillin, it was soon found that the drug worked only against Gram-positive bacteria. During the early 1940s Selman Waksman had been concentrating on Gram-negative bacteria. After numerous failures, Waksman succeeded in finding a nontoxic compound derived from *Streptomycetes griseus* mold which he named streptomycin. By January of 1944 he announced that this antibiotic could work against both Gram-positive and Gram-negative bacteria and was particularly effective against tuberculosis.

At the end of World War II, it became clear that the chemical structures of antibiotic compounds could be discovered. However, Ernst Chain, who figured prominently in the development of penicillin did not believe that the drug could ever be synthesized—that is, made in a factory without using molds or bacteria. But the great researcher was wrong. In postwar United States, drug companies hastily moved toward research in these areas. In fact, entering the race to produce antibiotics was the way many infant drug companies got their start, with researchers looking in nooks and crannies for all sorts of molds that could be potential drug sources. John Ehrlich and Quentin Bartz isolated another soil microbe in 1947, which the chemists of Parke Davis and Company found they could synthesize into an antibiotic. Thus the drug chloramphenicol, the first antibiotic that was antagonist toward a wide spectrum of bacteria, became the first of many synthetic drugs. Johnson and others developed bacitracin in 1945. Also in 1945, Benjamin Dugger, Y. Subbarow, and A. Dornbush discovered aure-

omycin, the first of the class of antibiotics known as tetracyclines. By 1948, Lederle Laboratories was producing aureomycin commercially. Working with Lechavalier, Waksman developed neomycin in 1949. A. C. Finlay and Gladys Lounsbury Hobby (who had also worked on penicillin) discovered another antibiotic, then marketed by Pfizer Drugs as terramycin in 1950. Some of these antibiotics, such as the cephalosporins and of course penicillin were developed from molds. Other antibiotics were developed from bacteria—chloramphenicol, erythromycin, and the tetracyclines such as aureomycin.

This enormous array of life-saving drugs can be classified into groups based on their chemistry. Included in the penicillin group are penicillin G, the most commonly used penicillin, ampicillin and amoxicillin. Penicillins are used to treat pneumonia, meningitis, strep infections, and sexually transmitted diseases. The cephalosporins, such as cephalothin and cephalexin, share many of their uses with penicillin. The aminoglycosides group includes streptomycin, used chiefly for Gram-negative bacterial infections like tuberculosis, and neomycin, which at one time was used to fight systemic infections and has now been replaced in many instances by kanamycin and gentamicin. The tetracylines, including tetracycline and chlortetracycline are broad-spectrum antibiotics that often cause side effects and thus are used in fewer cases. The macrolides include erythromycin, a drug that fights gram-positive bacteria and is often administered to patients that are allergic to penicillin. Bacitracin, an antibiotic often used topically as an ointment, belongs to the **polypeptide** group that is generally effective against Gram-negative bacteria. Sulfonamide drugs, such as sulfadiazine, are synthetic drugs used primarily in urinary tract infections often in conjunction with penicillin.

Today's antibiotics may be given as an injection with a needle, or in a pill that can be swallowed by the patient. Each antibiotic does something that prevents bacterial cells from growing or dividing normally. For example, penicillin works by preventing the cell wall of each new bacterial cell from forming. The antibiotic compound mimics similar compounds in the bacterial cell wall. If the antibiotic becomes part of the bacterial cell wall, it leaves a gap and thus the bacterium no longer has its cell contents encased and protected. The contents spill out and the cell dies. That is why Fleming saw bacterial cells that looked like they had dissolved.

Tiny, ubiquitous, and invisible to the eye, bacteria are capable of reproducing every 20 minutes. An antibiotic may be very effective in halting the reproduction of bacteria at first. However, there are always some bacteria which are naturally resistant to the drug. Soon the resistant strain is able to catch up and reproduce in strong numbers. The first observation of antibiotic resistance was noticed as early as the 1940s. Since that time, some authorities believe that antibiotics are often overused when they should be used only when necessary. Today in many quarters, antibiotics are routinely administered to farm animals to keep them from getting sick. Unfortunately, the antibiotics are then in the meat that people eat. It is also possible that antibiotic resistant bacteria can be consumed when eating meat. Excessive use of antibiotics in humans as well as in animals speeds up the development of antibiotic resistant bacteria.

Antibiotics had an enormous impact on the safety and quality of human life in the twentieth century. In 1924 President Calvin Coolidge's younger son had been playing tennis and got a blister. The blister became infected and the resulting **blood** poisoning killed him. At that time such small accidents could cause death. With a dose of an antibiotic he most probably would have been cured. Antibiotics do not work against **virus**es, only bacteria—and only some bacteria at that. Although antibiotics are not perfect, they changed forever how people are cured of disease and have lived up to their name, "wonder drugs."

ANTIBODY AND ANTIGEN

An antigen is any substance that provokes a response by the body's **immune system**. An antibody is a **protein** manufactured by the body to neutralize a specific invading antigen.

Before the 1880s, not much was known about the specific components and workings of the immune system. During the 1880s, researchers found that both **diphtheria** and tetanus were caused by a toxin produced by the disease bacillus and that the body responded by producing a neutralizer, or antitoxin. In 1890 **Emil von Behring** and **Shibasaburo Kitasato**, working in **Robert Koch**'s laboratory in Berlin, Germany, showed that guinea pigs became immune to diphtheria or tetanus **toxins** if they had been injected with serum from an already immune **animal**. From this, von Behring concluded that immunity was conferred by protective substances in the **blood**, which he called antitoxins, or antibodies, and that these substances were very specific, protecting only against one particular disease. In 1901, Von Behring won the first Nobel Prize in medicine for his work.

Other properties of antibodies were discovered in the 1890s. A German bacteriologist showed in 1894 that **cholera bacteria** were destroyed by antibodies (a process called *bacteriolysis*). In 1898 a young Belgian bacteriologist, **Jules Bordet**, found that when cholera serum was heated to 131° F (55° C), it retained its antibodies but lost its ability to destroy bacteria. He concluded that the heated serum lost a bactericidal substance, originally called alexine but which became known as complement. This explained an important element of immunity: an antibody combines with an antigen, and only after the antibody reacts with complement is the antigen made harmless. Bordet also explained that antigens can be detected by their reaction to specific antibodies, and by the fact that antibody-antigen complexes precipitate out of a solution when they react to complement. These immune reactions became the basis for countless numbers of diagnostic tests, including the famous Wassermann test for **syphilis**.

The specific nature of the antigen-antibody reaction was used by **Karl Landsteiner** in 1900 to make his very important discovery of the human **blood groups**. Landsteiner also showed that individual antibodies react to the specific chemical structure of individual antigens. German medical scientist **Paul Ehrlich** developed the "side-chain" theory of immunity around 1900 explaining that antibodies and antigens fit together in very specific molecular ways, like a key in a lock.

Although much was now known about how antibodies worked, researchers still didn't know what these substances actually were. In the 1930s **Arne Tiselius** developed sophisticated new methods of electrophoresis that allowed the isolation of antibodies in blood sera. By 1938 Tiselius had identified them as proteins of the gamma globulin portion of plasma. (Antibodies are now also called immunoglobulins.) In 1948 Astrid Fagraeus showed that antibodies are produced by plasma **cell**s in the bone marrow and lymph nodes.

The exact molecular nature of antibodies was difficult to discern, since the body produces about one million different antibodies, and they are all large **molecule**s. A pioneer in this field was **Linus Pauling**, who published his first paper on antibody structure in 1940. By 1962 both Rodney R. Porter (1917-1985) in England and **Gerald M. Edelman** (1929-) in the United States had worked out the basic molecular structure of antibodies. In 1968 researchers discovered that there are two types of **lymphocytes**, and that it is the B-lymphocytes that produce the antibody-forming plasma cells. In 1975 **Georges Kohler** (1913-) and **Cesar Milstein** at the Laboratory of **Molecular Biology** in Cambridge produced exactly similar (monoclonal) antibodies outside the body, by fusing a B-lymphocyte with a myeloma (**cancer**ous) cell. The resulting fused cell, like all cancer cells, divides indefinitely. Clones of the cell produce virtually unlimited amounts of identical antibodies, which can be used in the purification of **virus**es, bacteria, and **interferon**s; and as weapons against cancer.

By virtue of their ability to interact with protein molecules in blood and urine, these agents are also useful in diagnostic tests known as immunoassays. For example, home pregnancy tests are one type of immunoassay which use a monoclonal antibody (MAb) to react with hCG (Human **Chorion**ic Gonadotrophin), a **hormone** that is secreted during pregnancy.

ANTIBODY, MONOCLONAL

The **immune system** of vertebrates help keep the **animal** healthy by making millions of different **protein**s (immunoglobulins) called antibodies to disable **antigen**s (harmful foreign substances such as **toxins** or **bacteria**). Scientists have worked to develop a method to extract large amounts of specific antibodies from clones (exact copies) of a **cell** created by fusing two different natural cells. Those antibodies are called *monoclonal antibodies*.

Antibody research began in the 1930s when the American pathologist **Karl Landsteiner** found that animal antibodies counteract specific antigens and that all antibodies have similar structures. Research by the American biochemists Rodney R. Porter (1917-) and **Gerald M. Edelman** (1929-) during the 1950s determined antibody structure, and particularly the active areas of individual antibodies. For their work they received the 1972 Nobel prize in physiology or medicine.

By the 1960s, scientists who studied cells needed large amounts of specific antibodies for their research, but several problems prevented them from obtaining these antibodies. Animals can be injected with antigens so they will produce the desired antibodies, but it is difficult to extract them from among the many types produced. Attempts to reproduce various antibodies in an artificial environment encountered some complications. **Lymphocytes**, the type of cell that produces specific antibodies, are very difficult to grow in the laboratory; conversely, tumor cells reproduce easily and endlessly, but make only their own types of antibodies. A bone marrow tumor called a myeloma interested scientists because it begins from a single cell that produces a single antibody, then divides many times. The cells that divided do not contain antibodies and could, therefore, be crossed with lymphocytes to produce specific antibodies. These **hybrid** cells are called hybridoma, and they produce monoclonal antibodies.

One molecular biologist who needed pure antibodies for a study of myeloma **mutation**s was the Argentinean **Cesar Milstein** (1927-). After receiving a doctorate in biochemistry, specializing in **enzyme**s, from the University of Buenos Aires in 1957, he continued this study at the University of Cambridge in England. There he worked under the biochemist **Frederick Sanger** and earned another doctorate in 1961. Milstein had returned to Argentina, but political disturbances forced him to flee the country. He came back to Cambridge, where Sanger suggested that he work with antibodies.

In 1974, Milstein was working with **Georges Kohler** (1946-), a German postdoctoral student who had just received his doctorate from the University of Freiburg for work performed at the Institute for Immunology in Basel, Switzerland. To produce the needed antibodies, Milstein and Kohler first injected a mouse with a known antigen. After extracting the resulting lymphocytes from the mouse's **blood**, they fused one of them with a myeloma cell. The resulting hybrid produced the lymphocyte's specific antibody and reproduced endlessly. As Milstein soon realized, their technique for producing monoclonal antibodies could be used in many capacities. Milstein and Kohler shared part of the 1984 Nobel prize in physiology or medicine for their invention.

Today pure antibodies are made using the Milstein-Kohler technique and also through **genetic engineering**, which adds the **gene** for the desired antibody to bacteria that can produce it in large amounts. Monoclonal antibodies are instrumental in the performance of sensitive medical diagnostic tests such as: determining pregnancy with **chorion**ic gonadotropin; determining the **amino acid** content of substances; classifying antigens; purifying **hormone**s; and modifying infectious or toxic substances in the body. They are also important in **cancer** treatment because they can be tagged with radioisotopes to make images of tumors.

ANTICOAGULANT

Anticoagulants are substances that inhibit coagulation, or clotting, of the **blood**. They are used to keep blood for transfusions from clotting; to treat conditions involving dangerous blood clotting, such as cerebral thrombosis and coronary **heart** disease; and in situations where there is a risk of dangerous clotting, such as during some surgical operations, in particular the installation of artificial heart valves.

Donated blood tends to clot before it is absorbed into the recipient's **circulatory system**. Sodium phosphate was introduced as an anticoagulant to overcome this problem in 1869. In 1914 sodium citrate was shown to be an effective and harmless anticoagulant for donated blood; its use was of great value during World War I and had become standard by 1917.

For centuries, medical practitioners had used leeches to suck blood from patients. In 1884 J. B. Haycraft showed that blood flowed freely during this procedure because the leeches secreted an anticoagulant factor. A dry, powdered extract of leeches' heads called *hirudin* was introduced around 1900 and used in physiological experiments, but not in clinical practice.

The main anticoagulant used by patients is heparin, which is injected, and coumarin compounds, which are taken orally. Heparin was discovered by a medical student, Jay McLean (1890-1957), in 1916. McLean was studying at Johns Hopkins University under William Howell (1860-1945), who had been investigating blood coagulation for years. McLean took on a project of preparing pure samples of *cephalin*, a clotting substance obtained from **brain tissue**. Extracting compounds similar to cephalin from heart and liver tissue, McLean discovered that the liver extract did not cause blood to clot. McLean called the extract *heparphosphatid*.

After McLean left Johns Hopkins, Howell continued working on the liver extract, aided by Emmett Holt (1855-1924). They developed ways to extract an improved **water**-soluble anticoagulant from liver, which they named heparin in 1918. Howell continued his work on heparin during the 1920s, and **Charles Best** (1899-1978), David Scott, and Arthur Charles of the University of Toronto worked with heparin extracted from beef liver and developed practical methods of purifying and standardizing the drug. Clinical trials followed. The success of the artificial kidney in 1944 and the later heart-lung bypass procedure depended on the use of heparin to prevent fatal blood clotting. Heparin then came into standard use, although it does pose the risk of excessive bleeding.

Oral anticoagulants have their origins in an odd bleeding disorder of cattle that broke out in North Dakota and Canada in the 1920s. Cattle that ate hay made from spoiled sweet clover bled to **death**. F. W. Schofield, a Canadian veterinarian, traced the cause to clover in 1922, and a North Dakota veterinarian, L. M. Roderick, found in 1931 that the substance worked by reducing the activity of *prothrombin*, a clotting factor in the blood. Isolating the factor proved to be very difficult. In 1939, Karl Link, director of the experiment station run by the University of Wisconsin **agriculture** college, and his associate, H. A. Campbell, in 1939 finally accomplished the task. They discovered that the clotting factor was dicumarol, a coumarin compound that had been synthesized in an impure form in 1903.

In 1948 a very potent synthetic form of dicumarol was introduced as an extremely effective rodent killer, warfarin, named for the patent holder, Wisconsin Alumni Research Foundation. Although warfarin rapidly became very popular worldwide as a rat poison, medical practitioners hesitated to use it on patients because it was so toxic. This changed after a United States army recruit unsuccessfully attempted suicide in 1951 by taking massive doses of warfarin-based rat poison. Warfarin is now the most widely prescribed oral anticoagulant.

In addition to anticoagulants used for surgery and serious medical problems, aspirin also functions as an anticoagulant and has been recommended for reducing the possible formation of blood clots in healthy humans. Some studies have also suggested that red wine and grape juice have anticoagulant properties.

ANTIGEN • See Antibody and antigen

ANTIHISTAMINE

Antihistamines are drugs used to relieve the symptoms of allergies caused by histamine, an **amino acid** derivative found throughout the body. Histamine was first recognized and suggested as the cause of allergic reactions by **Henry Dale** (1875-1968) and Patrick Playfair Laidlaw (1881-1940) in 1910. By 1932 histamines were confirmed as causative agents in allergic response. Researchers then sought to find agents that could counteract the effects of histamines.

A Swiss-Italian pharmacologist, Daniele Bovet, of the Pasteur Institute in Paris, France, focused on this problem in 1936. Since histamine is extremely toxic except when introduced to the body by absorption through the intestine, Bovet reasoned that histamine must normally exist in the body in combination with a neutralizing substance. Only "free" histamine would produce allergic symptoms, so an antagonist (an substance that opposes or resists the action of another substance) to this free histamine had to be found. However, histamine has no naturally occurring antagonist. Because the two **hormone**s adrenaline and **acetylcholine** are structurally similar to histamine, Bovet investigated two groups of substances called sympatholitics and parasympatholitics, which block the effects of adrenaline and acetylcholine, hoping that some element of these compounds would also prove to be antagonistic to histamine. This approach proved fruitful, and in 1937 Bovet and his research student Anne-Marie Staub succeeded in synthesizing the first antihistamine, thymoxydiethylamine. This was too toxic to use in humans, however, so from 1937 to 1941 Bovet conducted thousands of experiments to produce a usable antihistamine. He succeeded with *pyrilamine*, which was introduced to the public in 1944.

Bovet's work laid the structural basis for safe, effective synthesis of antihistamines. Bernard N. Halpern, a French research biologist and physician, described the use of *phenbenzamine* in 1942. In 1943 a young lecturer at the University of Cincinnati named George Rieveschl developed diphenhydramine. Marketed to millions by Parke-Davis as *Benedryl*, this antihistamine made Rieveschl a wealthy man. Once developed, antihistamines became wildly popular, promoted by rival drug companies for relief of symptoms of the common cold as well as of allergies. Side effects of antihistamines include high **blood pressure** and sedation. Usually taken orally, antihistamine nasal sprays designed to work faster by going directly to the nasal passages have also been developed. Another function of these drugs was discovered by accident in 1947

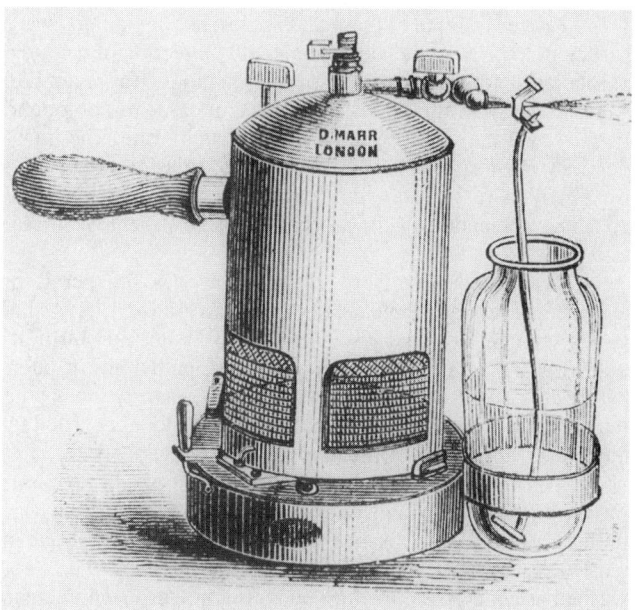

Invented by Joseph Lister in the 1860s, carbolic steam sprayers were used to create an antiseptic environment for surgery. *(National Library of Medicine. Reproduced by permission.)*

when an allergy patient took an antihistamine called *Dramamine* and unexpectedly found that, for the first time in years, she did not suffer from motion sickness when she rode a streetcar.

ANTISEPSIS

Antisepsis is the destruction or inhibition of growth of **microorganism**s on living **tissue**. Antiseptics are the substances that carry out antisepsis. Real understanding of the nature and use of antiseptics was not possible until the discovery of **bacteria**.

From the earliest times, physicians and healers were aware of the anti-infective and anti-spoilage properties of certain substances. Egyptian embalmers used resins, naphtha, and liquid pitch to decrease body decay, along with oils and spices. The ancient Greeks and Romans recognized the antiseptic properties of wine, oil, and vinegar. Balsam, an antiseptic of both southeast Asia and Peru, came to Europe in medieval times and remained in use in Europe through the 1800s. Turpentine was a favored antiseptic of the Middle Ages.

A thirteenth-century surgeon, Theodoric Borgognoni (1205-1298; Theodoric of Bologna), recommended dressings dipped in wine to ward off the development of pus in wounds. The eminent English physician Sir John Pringle (1707-1782) published a series of papers titled "Experiments Upon Septic and Antiseptic Substances (1750-52)," which contains one of the first uses of the word antiseptic (from the Greek *anti*, against, and *sepsis*, decay). Genevieve Charlotte d'Arconville introduced the use of chloride of mercury as an antiseptic in 1766. After iodine was discovered in 1811 by Bernard Courtois (1777-1838), it became popular as an antiseptic treatment for wounds.

However, none of these antiseptics, nor any others, were sufficient to prevent the almost inevitable infection of wounds, particularly following surgery. Amputations, for example, were common in the 1800s, especially in cases of compound fracture, yet they yielded a 40-45% mortality rate. The introduction of anesthesia in 1846 aggravated the problem, as it permitted more invasive and lengthy surgical operations, increasing tremendously the likelihood of postoperative infection.

Another deadly form of infection was puerperal, or childbed fever, a streptococcus infection of the uterus that struck women who had just given **birth**. As hospital birth attended by doctors became more common, epidemics of puerperal fever raced through maternity wards, sharply increasing maternal **death** rates. Most obstetricians were totally baffled by the causes and possible prevention of this fever.

The reason for this epidemic of infections in hospitals lies in the lack of knowledge about the existence of bacteria until **Louis Pasteur**'s discoveries. Physicians, and surgeons in particular, had no concern for cleanliness. They wore unwashed street clothes or filthy operating gowns, used unclean instruments, and did not wash their hands before examining or operating on patients—even going from a postmortem on an infected corpse directly to examination of a fresh patient. Many took pride in the accumulation of **blood** and pus on their medical garments.

Attempts to understand and stop puerperal fever brought some of the early advances in antisepsis. Dr. Charles White (1728-1813) of England in 1773 recommended antiseptic injection in some and strict cleanliness in all cases of childbirth. The Scottish physician Alexander Gordon (1752-1799) advocated hand and clothes washing for obstetricians in 1795. Dr. Oliver Wendell Holmes of the United States presented his conclusions about the spread of puerperal fever by unwashed doctors in 1843. The Hungarian doctor **Ignaz Semmelweis** made the same discovery in 1847. When he required his students to wash their hands in an antiseptic chloride solution before examining patients, maternal mortality rates plunged from a high of eighteen percent to a low of nearly one percent.

Unfortunately, while Semmelweis was correct about the transmission of infectious materials, he could not explain what those substances were. Pasteur pointed the way. In his studies of **fermentation**, Pasteur proved the existence of airborne microorganisms in the 1850s. **Joseph Lister** applied this new knowledge of bacteria to develop a successful system of antiseptic surgery, a tremendously important innovation that released surgery from the limitations imposed by the threat of infection.

Concerned about the high rate of infection after surgery, Lister—an English physician—studied wound healing, aided by microscopic analysis. After reading Pasteur's work, Lister concluded that microorganisms in the air caused the infection of wounds. Drawing on a report of the effects of carbolic acid on sewage bacteria, Lister developed an antiseptic system using the acid, both spraying the wound and surrounding areas to destroy infectious **organism**s and protecting the wound with multiple-layer dressings from new invasion by bacteria. He first used the method successfully in an operation on a compound fracture of the leg in 1865.

Lister's antiseptic method was not simple—it involved a series of six essential steps—but it was effective. Lister's published accounts of his successful surgical application of the technique *The Lancet* in 1867 and ignited controversy, especially since Pasteur's **germ theory** of disease was still in dispute. Nevertheless, Listerian antiseptic surgery gained adherents worldwide, especially in Germany, where they were applied somewhat successfully in treatment of soldiers during the Franco-Prussian War (1870-71). The United States was especially resistant to the practice of antisepsis. Widespread acceptance came in the 1890s after the German bacteriologist **Robert Koch** had effectively proven that germs cause disease.

Modifications of and improvements on Lister's techniques were soon developed. The carbolic spray that inundated the operating arena was abandoned in the 1880s in favor of cleanliness, sterilization, and topical antiseptics.

A final obstacle to surgical antisepsis was the human hands: while surgical instruments and dressings can be sterilized, the surgeons' and nurses' hands can only be washed with antiseptics. An American doctor, William Halsted, solved this problem in 1890. Halsted received his medical degree from Columbia University in 1877; he returned to the United States from two years of study in Europe as a convert to the Listerian method of antisepsis. After breaking an addiction brought on by his experiments with cocaine as an anesthetic, Halsted became chief of surgery at Johns Hopkins Medical School. There, he pioneered the use of rubber gloves in surgery to protect his head nurse, Caroline Hampton, from the antiseptic that was irritating her hands. Today, of course, sterile gloves are a requirement for surgical procedures.

Modern methods of preventing infection are very different from the techniques pioneered by Lister and others. **Antibiotics**, **penicillin**, and sulfa drugs fight infection internally, and aseptic methods such as sterilization prevent bacteria from existing in a given area. Nevertheless, external antiseptics continue to be important and are a lasting monument to Lister's vision.

AORTA

An elastic artery and the largest **blood** vessel in the vertebrate body, the aorta distributes blood from the **heart** to the **circulatory system**. Oxygenated blood is pumped into the aortic arch from the left ventricle. Several small **arteries** branch from the top of this arch to deliver blood to the head and upper torso. The aorta then curves sharply back to become the dorsal aorta, which runs parallel to the spinal column before it branches into many smaller arteries that supply blood to the rest of the body.

Like all other large blood vessels, the aorta is composed of three layers called tunicas. The interior layer, or tunica intima, is composed of a lining of very thin, flat endothelial **cell**s that sit on a thin connective **tissue** layer called the basal lamina. The middle layer, or tunica media, is the thickest and most unique part of the aorta. Composed of 40 to 70 thin sheets of elastin (a springy, flexible **protein**) interspersed with layers of spindle-shaped smooth muscle cells, the media makes the aorta limber and yet extremely strong. The elastic nature of the aorta is important in maintaining an even **blood pressure** and in keeping blood flowing smoothly through the vascular system. During contraction of the left ventricle (systole), the aorta stretches and stores blood. When the ventricle relaxes (diastole), the elastic wall of the aorta contracts and prevents blood pressure from falling to zero.

The outermost layer of the aorta, the tunica adventitia, is composed of a loose layer of collagen (protein) fibers that strengthen and help protect this important blood vessel. Nerves and small blood vessels (called *vasa vasorum* or vessels of vessels) run through this layer. These small blood vessels supply oxygen and **nutrients** to the muscle cells of the outer layers of the tunica media (which are too far from the lumen of the aorta to be reached by simple **diffusion**). The nerves regulate the contractions of the muscle cells and thus help control blood pressure. The adventitia also contains fibroblasts, which lay down the collagen and elastic fibers of the aorta wall as well as macrophages, **lymphocytes**, and other cells of the **immune system**, which locate and destroy damaged cells and infectious agents such as **bacteria** and **virus**es.

Because the aorta, especially the aortic arch, is subjected to very high blood pressure when the heart beats, its elastic walls often weaken and fail as a person ages. A weakened section of the aorta can balloon out into an aneurysm, which can suddenly burst. Or, in the case of a sharp blow, such as an automobile accident, the aorta can rip away from the heart. In either case the victim bleeds to **death** in a matter of minutes.

APGAR, VIRGINIA (1909-1974)
American physician

Within minutes of **birth**, virtually every child today receives an Apgar Score from a delivery nurse or midwife. This simple but crucial test devised in 1952 by Virginia Apgar evaluates infants immediately after birth and is able to identify babies that may be at risk during their first few minutes and hours of life. As an anesthesiologist who attended over 17,000 births, Apgar felt the need to act on her conviction that birth is the most hazardous time of life. Apgar's own life and achievements ranged far beyond her internationally recognized test of newborn health, resulting in the 1997 release of a twenty-cent U.S. postage stamp commemorating her many accomplishments.

Virginia Apgar was born in Westfield, New Jersey, to a very musical family. Her father, Charles Emory Apgar, a businessman and automobile salesman, was an amateur musician. Her mother, Helen May, shared the family's interest in music as did Virginia's brother. Apgar began studying the violin at six and soon was able to join in the family's living room concerts. Apgar eventually became a member of the local Amateur Chamber Music Players, performing with the Teaneck (N.J.) Symphony, and even learned how to build her own stringed instruments. Despite her love of music, Apgar set her sights on a career in medicine as early as high school. After graduating from Westfield High School in 1925, she entered

Mount Holyoke College, majoring in zoology and undertaking a rigorous premedical curriculum. During her college years, she demonstrated her abilities and versatility by working as both a librarian and waitress, while still having time to earn a letter in athletics, work as a reporter for the school paper, and play the violin in the school orchestra. After graduating in 1929, she entered Columbia University College of Physicians and Surgeons in New York and was awarded a degree in medicine in 1933.

Apgar First Encounters Gender Bias

As one of very few female medical students at Columbia during the early 1930s and one of the first women to graduate from its medical school, Apgar knew that her goal of becoming a surgeon would not be achieved easily in a male-dominated profession. Nonetheless, her record of excellence enabled her to become a surgical intern at Columbia Presbyterian Medical Center—only the fifth woman to be awarded that coveted internship. Yet after laboring for two years as an intern and performing many successful operations, she realized that the advice given to her by a professor of surgery, that a female surgeon would never have enough patients to make a living, was unfortunately true. Apgar herself summed up the situation realistically, saying that "even women won't go to a woman surgeon. Only the Lord can answer that one."

Switches to Anesthesiology

Although she reluctantly switched her medical specialty to anesthesiology, she embraced her new field with typical intelligence and energy. At this time, anesthesiology was a relatively new field, having been left by the doctors mostly to the attention of nurses. Apgar realized immediately how much in need of scientifically trained personnel was this significant part of surgery, and she set out to make anesthesiology a separate medical discipline. By 1937, she had become the fiftieth physician to be certified as an anesthesiologist in the United States. The following year she was appointed director of anesthesiology at the Columbia-Presbyterian Medical Center, becoming the first woman to head a department at that institution. It was mostly due to Apgar's hard work, excellent credentials, and growing national reputation that anesthesiology was established at Columbia-Presbyterian as an entirely new academic department. In 1949, when Columbia made her a full professor, she became the first full professor of anesthesiology ever.

Develops the Apgar Score System

As the attending anesthesiologist who assisted in the delivery of thousands of babies during these years, Apgar noted two post-delivery habits that she came to realize were sometimes detrimental to the health and survival of newborns. The first of these was the inclination of most medical staff to focus their immediate attention on the mother's condition and needs, leaving examination of the infant to be done later in the nursery. Second, she noticed that unless the newborn had suffered some obvious trauma during birth, its condition was assumed to be good. Many a time she realized that an infant had died from respiratory or circulatory complications that early treatment could have prevented. These endangered but seemingly normal babies were failing to receive the immediate medical care that could save their lives simply because no one knew enough to give their vital functions a quick, routine check.

Apgar then decided to bring her considerable research skills to this childbirth dilemma, and her careful study resulted in her publication of the Apgar Score System in 1952. Designed to assure that correctable problems are discovered immediately and addressed on the spot, her system enables the attending nurse or doctor to take a quick reading of a newborn's pulse, respiration, muscle tone, color, and reflexes. The Apgar System rates an infant from 0 to 2 in each of these five areas, with the test being performed from one minute to five minutes after birth. For each vital sign, 0 means no response, 1 is marginal, and 2 is normal or the best response. A newborn that is doing well would receive between 8 and 10, with lower scores suggesting that the baby might need some attention. A score of 4 or below would alert a physician to possible risk factors.

Apgar's quick and easy assessment soon was adopted worldwide, and medical schools adopted Apgar's own name as an acronym for teaching her test: Appearance (skin color), Pulse (heart rate), Grimace (reflex irritability), Activity (muscle tone), and Respiration (breathing). For her work in obstetrical anesthesia and for her development of the Apgar score, she is credited with laying the foundation for the new science called perinatology.

Becomes Director of Birth Defects Research

Although Apgar later published a book titled *Is My Baby All Right?*, she made no money from her test. She did, however, become well-known internationally, and in 1959, after 30 years at Columbia, teaching, administering, and assisting in the birth of some 20,000 babies, she became director of birth defects research for the March of Dimes Foundation. As the head of its division of clinical malformations, her work focused on birth defects and prenatal care.

Part of her responsibilities as director included the distribution of more than $5 million annually in research grants. She also realized that to be really effective, she herself would have to preach the gospel of good prenatal care. So Apgar, a woman of science who nonetheless was afraid of elevators, took to traveling a hundred thousand miles a year lecturing mothers-to-be about the benefits of seeking early prenatal care. She also warned of the dangers of drugs and radiation to the developing fetus. Today's awareness and emphasis on avoiding preventable birth defects owes much to the programs initiated by Apgar. She also proved to be more than a popular and visible proselytizer, as she doubled the annual March of Dimes income. Remaining with the Foundation until her death in 1974, she became vice-president and director of basic research in 1967. In 1973, she was elevated to senior vice-president in charge of medical affairs.

Although she had become internationally known and respected, Apgar remained a modest and unassuming individual. Despite never marrying—"I haven't found a man who can cook," —and living in an apartment in Tenafly, New Jersey, where she cared for her mother, she was by no means a one-dimensional person. She took up flying lessons when she was well over 50 and had as her goal being able to fly under New York City's George Washington Bridge. Described by many as a warm and compassionate person, she died in New York City of cirrhosis of the liver. The recipient of numerous awards and honors, Apgar is today silently acknowledged by the countless (and still-growing) numbers of individuals around the world who owe their lives and well-being to the physician who regarded birth as "the most hazardous time of life."

AQUINAS, ST. THOMAS (1225-1274)
Italian theologian

St. Thomas Aquinas was an Italian theologian and philosopher of the Dominican Order of the Catholic Church. He was educated or taught at Monte Cassino, the Universities of Naples and Paris, and the Theological School at Cologne. Aquinas wrote more than 60 works, some of which were massive in length. He is regarded as one of the greatest and most influential thinkers of the Catholic Church. He had an important influence on the intellectual awakening that occurred in western Europe during and after his lifetime. Previous to this time, from about 200–1200 C.E. (most of this period, from 500–1000 C.E., is now known as the Dark Ages), almost no important scientific advances had occurred in Europe or North Africa.

A central tenet of Aquinas' thinking and philosophy was that a omniscient (i.e., all-knowing), omnipotent (i.e., all-powerful) God had created the universe "out of nothing" (or *ex nihilo*. One of his major contributions to the development of scientific thought and methodology was advocacy of the idea that truth (or understanding) could be discovered through rational investigation of the natural world. As such, truth was not only the "word of God," as could be revealed by a literal interpretation of the Bible, which was commonly believed during those times. In other words, much could be learned about by the natural world by directly investigating organisms and phenomena. This so-called "rationalistic" approach to knowledge was crucial in the subsequent development of science as a way of understanding the natural world. It also fostered the development of a separation of scientific knowledge and speculation from the uncritical acceptance of the dogma of theology and the writings of such ancient natural philosophers as Aristotle.

ARBER, WERNER (1929-)
Swiss molecular biologist

Werner Arber's discovery of an **enzyme** that could cleave long strands of **deoxyribonucleic acid (DNA)** led to a revolution in genetics research, providing the foundation that led to techniques to separate and reassemble basic genetic material. **Gene splicing**, as it was called, proved invaluable for DNA sequencing and **gene mapping**, which focuses on genetic organization.

Arber received the 1978 Nobel Prize in physiology or medicine for his research in this area, sharing the prize with United States scientists and Daniel Nathans, who had also played an essential role in the development of gene splicing. The most controversial outcome of this research, however, was the eventual manipulation of DNA structures by geneticists, first in test tubes and then *in vivo*, or within a living organism. Arber warned his fellow scientists that such genetic research should be used carefully and conducted studies and participated in symposia on how to prevent the unintentional release of a genetically altered virus into the environment.

Werner Arber was born in Gränichen, Switzerland, on June 3, 1929. Educated in the Swiss public school system, he entered the Federal Institute of Technology in Zurich in 1949, where he focused on the natural sciences. Arber soon became exposed to experimental research and embarked on studies to isolate and characterize the radioactive isotope of chlorine. After graduation in 1953, he entered the University of Geneva as a graduate student, received an appointment as a research assistant in a laboratory, and studied biophysics. Werner became interested in bacterial **viruses** (**bacteriophages**) through the biophysicist Jean J. Weigle's studies of variations in these viruses, which aimed to show that a specific bacteriophage will only infect a specific host.

During his graduate studies, Arber assisted biophysicists at Geneva in developing high-level magnification techniques in electron microscopy to study bacteriophages. He completed his dissertation on deficiencies of a mutant strain of bacteriophage lambda and received his Ph.D. in 1958. Arber then went to the University of Southern California for further study and to refine his laboratory techniques in genetics and bacteriophage research. While in the United States, Arber also took the opportunity to visit several colleagues who were studying bacteriophages.

Arber returned to Switzerland to join the faculty at the University of Geneva in 1960. With support from the Swiss National Science Foundation, he embarked on studies of the molecular basis of bacteriophage restriction. Working with one of his graduate students, Daisy Dussoix, Arber found in 1962 that restriction was host-controlled and involved changes in the phage's DNA. In effect, the DNA of the invading phages is cut into component parts, although some phages survived the operation. This discovery set in motion a series of studies that jump-started genetics to become the new frontier in biomedical research.

Arber formulated a hypothesis presupposing that an endonuclease enzyme in the host severs the DNA of invading phages into component parts, while a methylase enzyme modifies the DNA of the host to make it invulnerable to its own endonuclease enzyme. Although he had yet to discover such an enzyme, Arber hypothesized that an endonuclease recognizes specific sequences of nucleotides, a fundamental building block of DNA and **RNA**, and cuts the DNA of the invading phages at the specific locations of these nucleotides. Arber called this two-enzyme theory a restriction-modification system. The theory received initial confirmation when Arber, with the biophysicist Urs Kühnlein, isolated phage mutants that

were inert to restriction and modification by mutation at specific nucleotide recognition sites. This discovery directly correlated Kühnlein's observation of DNA methylation with host-controlled modification in phages.

In 1965 Arber was appointed extraordinary professor of molecular genetics at the University of Geneva. He continued his research on restriction-modification and discovered, in 1968, the restriction endonuclease of *Escherichia coli* B, a common gut bacterium widely used in genetic studies. Although Arber's enzyme recognized specific nucleotide sequences, it cut the DNA at random spots and would later be known as a Type I restriction endonuclease. Since these Type I endonucleases severed the DNA at areas away from the recognition sites, they were unsuitable for studies of gene splicing. The second part of Arber's theory—that the endonuclease cut the invader's DNA at *specific* sites—was confirmed by the microbial geneticist **Hamilton Smith** and his colleagues, K. W. Wilcox and Thomas J. Kelley. Working at Johns Hopkins University, they identified what eventually came to be known as Type II, or specific, endonuclease. **Daniel Nathans**, also at Johns Hopkins, was a cancer researcher who first identified 11 cleaved fragments of a simian (monkey or ape) virus and eventually deduced the order in which individual fragments were replicated, showing that they began at a specific site and went in both directions around a circle, stopping approximately 180 degrees from where they started. Nathans and colleagues went on to isolate messenger RNA (mRNA), a type of RNA that is complementary to the protein-encoding segments of the host strand of DNA and communicates genetic information to proteins. They then began to map transcription sites (the origin and direction of each mRNA transcript during infection) by looking at different stages of infection and testing the RNA's ability to hybridize to the various ''restriction fragments'' due to their nucleotide sequences. This pioneering research led to a barrage of genetic studies aimed at mapping genetic codes, culminating in the international human genome project, which geneticists began in the late 1980s to develop a comprehensive road map of the human genetic system. Over the years, geneticists built upon the work of Arber, Smith, and Nathans to develop techniques to produce enough of a particular gene to study and then to artificially alter DNA through the transfer or insertion of genetic material.

Arber left the University of Geneva in 1970 and spent a year at the University of California at Berkeley as a visiting professor in the Department of Molecular Biology. Upon returning to Switzerland, he took an appointment as ordinary professor of molecular biology at the University of Basel and was reserved extensive modern facilities in the Biozentrum Research Institute, which was then under construction.

In 1978, Arber, Smith, and Nathans won the Nobel Prize for physiology or medicine, Arber being noted for his research showing that the host can alter DNA to prevent invasion by phages and other foreign genes through methylation (combining DNA with two carbons and three hydrogens), which cleaves the DNA. The cumulative efforts of these three scientists were an example of the growing emphasis on interdisciplinary communication and cooperation in scientific research as new discoveries in genetics were made simultaneously at many institutions throughout the world.

Arber's subsequent research has focused on genetic systems and their diversification. With the confirmation in the 1970s of **Barbara McClintock**'s theories of transposable genes that could "jump" to different strands of DNA during the early stages of meiosis (the process of cell division), Arber and other geneticists began to experiment with **gene** transplantation. Arber has theorized that genetic exchange through transposition may account for the diverse bacterial genetic codes that occur during evolution.

Investigations into recombinant DNA technology, however, also had controversial aspects. Studies of combining eukaryotic DNA (that is, DNA from an organism consisting of more than one cell) with bacterial or viral DNA in a molecule raised concerns about producing pathogens (a **microorganism** that can carry disease), especially since these pathogens could be cloned by copying the DNA molecules. Arber participated in discussions that led to a set of guidelines developed by the National Institutes of Health to conduct recombinant DNA research safely. As Arber points out in his introductory paper for the proceedings of the symposium, *Genetic Manipulation: Impact on Man and Society*, the initial risk was faced by the experimenters themselves, who were in direct contact with potential pathogens. What concerned the public, however, was the possibility of potential pathogens being accidentally introduced into the environment. Arber called for a realistic evaluation of the risks, saying that the guidelines had been designed to reduce the risks to a minimum.

Because of these precautions, geneticists have avoided serious mishaps and developed remarkable recombinant DNA studies, including a promising biomedical application known as gene therapy. **Gene therapy** begins with the splicing of DNA segments from various origins into a vector DNA molecule. Vectors act as "molecular delivery trucks," carrying genes to targeted cell types or organ systems, and are carefully chosen for their ability to colonize certain cell types and tissues in the body and become an inheritable trait. In gene therapy, disease-causing genes (specific nucleotide sequences) are replaced with normal sequences or additional genetic material is inserted to change the genes. This genetic material, for example, could carry a gene that expresses an inhibitor of a certain protein or hormone that causes a disease, such as arthritis. Investigators have also focused on gene therapies for cancer.

ARCHAEOBACTERIA • See Bacteria

ARCONVILLE, GENEVIEVE CHARLOTTE D' • See Antisepsis

ARISTOTLE (384 B.C.-322 B.C.)
Greek philosopher

Aristotle is counted among the greatest of philosophers for his inquiries into almost every branch of knowledge, including ethics, aesthetics, physics, and biology. In particular, his em-

Aristotle. *(The Library of Congress. Reproduced by permission.)*

pirical research into aspects of nature and history, including a lengthy classification of animals, was a truly new phenomenon in the Greek world.

The son of Nichomachus, a physician of Stageira in Thrace who served the Macedonian king Amyntas II, Aristotle went to Athens in 368 B.C. when he was about 17 years old to study at the philosopher Plato's Academy. He studied there for 20 years, but left after Plato died and went to Assos on the coast of Asia Minor to found a branch of the Academy. It was here that he married Pythias. Three years later he went to the island of Lesbos where he met Theoph rastus, who became his most celebrated disciple. In 343 B.C., King Philip of Macedon invited Aristotle to supervise the education of his 13 year old son, Alexander, who later became known as Alexander the Great. During his military campaigns in the East, Alexander is said to have sent biological specimens to Aristotle and to have provided funding for some of his research. When Alexander ascended the throne in 336 B.C., Aristotle returned to his native city Stageira where he stayed until returning to Athens in 335 B.C. to establish his own school, the Lyceum, where research was conducted and lectures were given. In 323 B.C. Alexander the Great died, and Athenians, who opposed Macedonian suzerainty, looked upon Aristotle suspiciously because of his former relationship of teacher to Alexander. Aristotle left Athens and went to Chalcis in illness in 322 B.C.

Aristotle's writings are grouped into exoteric works intended for the general public and pedagogical works that formed the basis of his lectures at he Lyceum. Only fragments of the exoteric works exist, but a large number of the pedagogical writings survive. They were first collected in an edition published by Andronicus of Rhodes in the first century B.C.

The writings fall into three main periods: the period of his relationship with Plato; the years when he taught at Assos and before he established the Lyceum; and the years when he headed the Lyceum. To the first period belong such works as the earliest parts of *On the Soul*, which is dominated by Plato's doctrine of Ideas and the soul's immortality, and parts of the *Physics*, which discusses nature, change, time, infinity, and the proof that an eternal Prime Mover exists. Aristotle's *On the Soul* briefly touches on lower activities of the soul, such as bodily motion and functions, that he later investigated more fully in four biological treatises. In his second period, Aristotle began to depart from Plato's teachings, publishing, for instance, *On Philosophy*, which contains Platonic influences but also criticizes Plato's theories. Aristotle's third period includes his major logical works, such as the *Prior Analytics* and *Posterior Analytics*, his works on ethics and politics, such as *Nichomachean Ethics* and *Politics*, and his biological treatises, including *History of Animals*, a lengthy classification of animals based on the similarities and differences in their parts, organs, and functions, and *On the Parts of Animals, On the Progression of Animals, On the Generation of Animals*.

While largely known today for his ethical and political philosophical writings, Aristotle is also recognized for his extensive empirical research into the natural world. He collected huge amounts of information about living creatures and systematically classified animal life. He emphasized the necessity for theories to rely on facts, noting, for instance, after giving a theory about the generation of bees, that "the facts have not been sufficiently ascertained" and stating that "credence must be given to the direct evidence of the senses more than to theories." Although some have credited Aristotle with originating the scientific method, Humanists and others have accused him of enslaving the human mind for two thousand years by establishing physical and biological laws and philosophical terminology that became dogma in western philosophy and Latin Christianity.

ARSENIC

Arsenic is the element with an atomic number of 33. Its atomic weight is 74.9216, and its chemical symbol is As. It usually occurs as a silver-gray brittle solid, but it has at least two other allotropic forms. One is a black amorphous solid, and another a yellow, crystalline solid. When heated, arsenic sublimes rather than melts, although at pressures of about 28 **atmospheres**, it can be made to melt at 1,502°F (817°C). Free arsenic is produced by reducing arsenic (III), As_2O_3, oxide with charcoal.

Arsenic is one of the four elements to become well known through the work of alchemists in the Middle Ages. The element derived its name from the Greek word *arsenikos*, for male. Some early scholars believed that metals, like **animals**, were either male or female. Two substances closely related to arsenic, *orpiment* and *realgar*, were generally regarded as "masculine." So when the new element arsenic was found, it was also given a "masculine" name. In fact, orpiment and realgar are sulfides of arsenic, a fact that was not clearly established until the early 1700s.

The great German scholar, Albertus Magnus, is sometimes given credit for having first isolated arsenic in the thirteenth century, and Schroeder is said to have prepared it in 1649. But there is no firm evidence to confirm these claims, and most scholars were unclear about the elemental **nature** of arsenic until the studies of J. F. Henckel (1679-1744) and Georg Brandt (1694-1768) in the 1720s and 1730s.

Compounds of arsenic were sometimes used by ancient peoples as dyes and pigments. Their most common use, however, was as poisons. Chinese farmers have used arsenic compounds for centuries to kill rats and insects that attacked their crops and food stores. Most cultures have been well aware of the poisonous quality of arsenic and its compounds. Greek and Roman records, for example, describe the high loss of life among slaves who were forced to work in orpiment and realgar mines.

Unfortunately, its toxic quality has made arsenic a favorite murder weapon—as well as a subject of inexhaustible public fascination. By 1836, its popularity as a tool of misconduct led British chemist James Marsh (1922-), to develop a test which detects traces of arsenic during autopsy. His Marsh test, used to convict criminals in many trials, created a sensation.

At one time, arsenic compounds, especially lead and calcium arsenate, were very popular as **pesticides**. Their use has been largely discontinued, however, because of the residue they leave in soils which remains poisonous to humans and animals. Some arsenic compounds have also been used as dyes and pigments, in fireworks, and in the tanning of leather. Because of the high toxicity of arsenic, these uses have been largely scaled back.

The element itself is used in alloys. One example is Babbitt metal, a soft alloy used in making bearings. Arsenic makes up about 3% of Babbitt metal. As an alloying metal, arsenic also helps lead and copper shot remain spherical. It is also used to strengthen metals that make up the grids in lead storage batteries, cable sheaths, and boiler tubes.

A very important role for arsenic today is as a "dopant" (additive) in germanium and silicon solid state products. Gallium arsenide is also widely used in solid state electronic devices.

ARTERIES

Arteries are elastic tubes which carry oxygen-rich (oxygenated) **blood** from the **heart** to all the other **organs** and **tissues** of the body. The body's largest artery is called the **aorta**. The aorta picks up blood from the heart which has already traveled through the lungs to become oxygenated. The aorta branches into many major arteries, which themselves branch into arterioles and then capillaries. The only artery in the body which carries unoxygenated blood is the pulmonary artery. This artery carries blood from the heart to the lungs to receive oxygen and to unload the waste product, **carbon dioxide**. All other arteries of the body carry blood which has already been to the lungs, to receive oxygen.

Arteries have a type of muscular layer within their walls which allows them to tighten up and grow smaller (constrict), or relax and grow larger (dilate). This allows the body to direct blood to tissues which have a greater need for oxygen at a particular time. This muscular layer also accounts for the fact that arteries play a major role in maintaining the **blood pressure**.

Complex interactions between the **nervous system** and chemicals produced in various organs throughout the body can affect blood pressure by causing the arteries to constrict or dilate. Constricted arteries result in a higher blood pressure; dilated arteries result in a lower blood pressure (think of a garden hose which spurts more strongly when slightly pinched, and flows more slowly when totally open).

Diseases of the arteries include high blood pressure, **arteriosclerosis** (in which plaques of fatty material clog the arteries), arterial thrombosis (in which blood clots block off normal blood flow through an artery), and arteritis (in which swelling and **inflammation** of the artery wall decreases blood flow). All of these diseases cause tissue damage by decreasing the appropriate amount of blood flow through a diseased artery. This results in too little oxygen being delivered to the tissues usually served by that artery.

ARTERIOSCLEROSIS

Arteriosclerosis, commonly called hardening of the **arteries**, is a disease in which the arteries become hard, inelastic, and thick, thus reducing **blood** flow. In atherosclerosis, a form of arteriosclerosis, fatty substances called plaque build up on the arteries' interior walls. Blocked coronary arteries can cause chest pain and, in severe cases, **heart** attack. Badly blocked arteries in the legs can cause **tissue death** requiring amputation. The kidneys can fail if their arteries are severely blocked, and when brittle arteriosclerotic **brain** arteries rupture, stroke occurs. Because of these serious effects, research into the causes, treatment, and prevention of arteriosclerosis has been intense.

Arteriosclerosis is a very ancient disease; 3500-year-old Egyptian mummies show signs of it. The hardened condition of the arteries was noted by Greek physician Galen in the second century A.D., but a full description of the disease was first made by Antonio Scarpa (1752-1832) in 1804. Scarpa noted the arteries' "rigidity and brittleness" as well as the fatty nature of their internal walls. An Alsatian pathologist, Johann Lobstein (1777-1835), refined Scarpa's description in 1833 and introduced the term arteriosclerosis. Modern attempts to understand the cause or origin of the disease began with the observations of William Gull (1816-1890) and Henry Sutton of England in 1872. German physiologist Rudolf Virchow suggested in the late 1800s that arteriosclerosis occurs when abnormally high levels of **cholesterol** and other **lipids** in the blood cause those lipids to deposit in the arterial walls. This hypothesis became generally accepted, although in recent years other ideas have been suggested as to how and why arteriosclerosis occurs.

Although the disease is widespread in industrialized nations, treatment methods were slow to develop. To prevent or reverse the process of arteriosclerosis, lifestyle changes are now recommended: a low-cholesterol diet, no smoking, weight loss, and exercise. Cholesterol-lowering medication may be prescribed. American biochemist Gladys Emerson (1903-1984) experimented with rhesus monkeys in the 1950s to learn about the effects of nutritional factors on the development of arteriosclerosis and found that a diet deficient in vitamin B_6 quickly produced hardened arteries. While smoking is also known to cause arteriosclerosis, studies have also shown that passive, or second-hand, smoke, which occurs when a non-smoker breathes in other people's tobacco smoke, can also cause the disease. In fact, second-hand smoke may contain a larger number of toxic substances, like **carbon monoxide**, nicotine, **ammonia**, and benzene, than the mainstream smoke which is actually inhaled by the smoker.

Since 1967, surgery has been used to treat severe cases of arteriosclerosis. In that year, Rene Favaloro performed the first coronary artery bypass surgery, in which a vein was grafted to an artery, bypassing the blocked portion of the artery. In 1964 Charles Dotter (1920-1985) and Melvin Judkins introduced transluminal angioplasty, in which a catheter is used to open blocked leg arteries. Andreas Gruentzig improved this technique by making the catheter a balloon that was inflated after being maneuvered into the blocked portion of an artery. Gruentzig carried out the first balloon angioplasty on the coronary artery of a patient in 1977. Since then, balloon angioplasty has come into wide use as a treatment for arteriosclerosis.

ARTHRITIS

Arthritis is a term for any of more than one hundred diseases that produce swelling in a joint, accompanied by pain and stiffness. The most common forms of arthritis are osteoarthritis (the degeneration of a joint) and rheumatoid arthritis ("the great crippler," **inflammation** of a joint that erodes bone and cartilage). Other forms include gout (caused by too much uric acid accumulating in the **blood** and most often affecting the big toe joint), ankylosing spondylitis (inflammation of spinal joints, mainly affecting young men), infectious arthritis (caused by invading **microorganism**s), and chronic Lyme arthritis (which appears in some people who contract Lyme disease). Lupus, an **autoimmune disease**, also has elements of arthritis, with painful and often swollen joints.

Neanderthal skeletons show signs of arthritis, as do Egyptian mummies. Ancient Greek and Roman physicians wrote detailed descriptions of arthritic conditions and methods of treatment. In fourteenth- and fifteenth-century Europe, gout became common among members of the upper classes, and an outbreak of rheumatoid arthritis swept through the masses of Europe during the Industrial Revolution. By the early nineteenth century, rheumatoid arthritis had been recognized as a distinct condition, separate from gout. Augustin Landre-Beauvais gave rheumatoid arthritis its first complete clinical description in 1800; in 1859 Alfred Garrod (1819-1907) distinguished gout by the presence of uric acid.

While the disease had been known for centuries, its cause remained unknown. Some thought arthritis was the result of an infectious disease, such as gonorrhea or tuberculosis.

In 1900 two physicians, Frederick J. Poynton (1869-1943) and A. Paine, discovered a **bacteria** in a group of children afflicted with rheumatism. They speculated that rheumatic arthritis could be the result of an immune reaction to an invading microorganism. In 1940 researchers found an *rheumatoid factor*, an **antibody**-like substance, in the blood of arthritis patients. Further study showed that rheumatic infections were caused by a group A streptococcus, so the rheumatoid factor was indeed an **immune system** response to that bacteria. Current research focuses on the relationship between specific genetically coded HLA **molecule**s (an element of the immune system) and the occurrence of various types of arthritis. For example, the HLA-B27 molecule is common in people with ankylosing spondylitis, and HLA-DR4 gene is associated with rheumatoid arthritis. In 1990, researchers discovered a faulty gene involved in hereditary osteoarthritis.

No cure exists for arthritis, so physicians have concentrated on alleviating the pain and crippling effects of the disease. The first really effective weapon against arthritic pain and inflammation was aspirin, introduced in 1899. Frenchman Jacques Forestier established the use of gold salts for treatment of arthritis in 1929, and interest in this approach revived in the 1960s. The most significant advance in arthritis treatment came with the discovery of **cortisone**. In the 1930s Edward Kendall, working at Minnesota's Mayo Clinic, isolated twenty-eight different **hormone**s, or corticoids, from the cortex of the **adrenal gland**. One of these, which had effects on laboratory **animal**s, Kendall called Compound E. He reported its discovery in 1936. At the same time, a Polish-Swiss biochemist, **Tadeus Reichstein**, also isolated the corticoids and Compound E (which he called F-a), and the American biochemist Joseph J. Pfiffner isolated the same compound.

Meanwhile, a colleague of Kendall's at the Mayo Clinic, Philip Hench, had become interested in rheumatoid arthritis. Hench theorized that hormones might relieve the symptoms of arthritis, and Kendall agreed that Compound E should be tried. In 1948 a young female patient of Hench's was desperately ill with arthritis. Lewis Sarrett of the Merck Drug Company in New Jersey had succeeded in synthesizing Compound E in usable amounts, and Hench secured enough of it to treat his patient. She made a speedy and dramatic recovery after a week of daily injections. Merck quickly produced 1,000 grams of Compound E, and Hench carried out further trials, with similar remarkable results. He reported his discovery in 1949. After the press enthusiastically heralded the new ''miracle'' drug as vitamin E, Compound E was renamed cortisone.

Corticosteroids like cortisone remain a potent weapon against arthritis, but they must be handled carefully because of their powerful side effects. A more radical treatment for joints severely deteriorated by arthritis is surgical joint replacement. This approach was pioneered by the English surgeon John Charnley, who developed a technique and prosthesis for total hip replacement in the 1960s.

Ongoing research into arthritis treatments are developing promising new therapeutic approaches. Researchers at the University of Pittsburgh are investigating a **gene therapy** approach which involves removing **cell**s from the afflicted joint

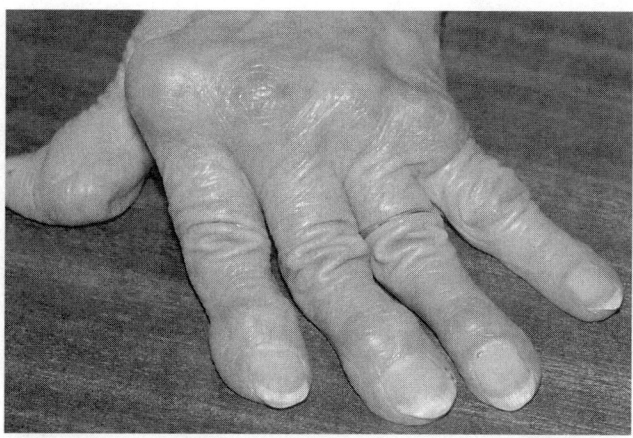

A close up of a hand deformed by rheumatoid arthritis. The knuckles are swollen and reddened and the fingers curve away from the thumb. The ends of the middle fingers are swollen with cartilage accretion. *(Photo Researchers, Inc. Reproduced by permission.)*

or area, modifying the cells to carry a gene that blocks inflammation, and then injecting the gene-bolstered cells into the joint area. Another area under investigation is nutrient therapy. For example, scientists at the University of Arizona have studied the use of two **nutrition**al supplements—glucosamine and chondroitin sulfate—to relive osteoarthritic pain and regenerate cartilage that has deteriorated. Considered to be the basic building blocks for joint cartilage, these substances are found in minute quantities in many foods and, interestingly, in shark cartilage, which contains chondroitin sulfate. Although many of these substances have been marketed to the general public as dietary supplements, like **vitamins**, more research is needed to confirm their beneficial effects.

ASEPSIS • See Antisepsis

ASTIGMATISM • See Eye disorders

ATMOSPHERE

Atmosphere refers to the mixture of gases and suspended particles surrounding a celestial object that has a gravitational field strong enough to prevent the gases and particles from escaping. The atmosphere of the earth is composed almost entirely of oxygen and **nitrogen** in their diatomic forms, i.e., as two **atoms** bound together by **chemical bond**s. Nitrogen gas (N_2) accounts for approximately 78% of the atmosphere, while oxygen (O_2) represents nearly 21%. In the remaining 1%, the inert noble gas argon makes up about 0.9% and **carbon dioxide** (CO_2) about 0.03%. The rest is composed of trace gases, including **carbon monoxide**, **water** vapor, hydrogen, methane, neon, helium, krypton, xenon, and **ozone**. The relative concentrations of these gases varies with their distance from the earth's surface.

Oxygen is required for the **respiration** of air-breathing **animal**s and to burn organic materials. Carbon dioxide, though

The earth's atmosphere. *(NASA. Reproduced by permission.)*

present in small amounts, maintains the heat balance of the earth and its atmosphere by strongly absorbing infrared (thermal) radiation. It is also necessary for **photosynthesis**. Increasing amounts of carbon dioxide in the atmosphere due to the combustion of **fossil fuels** has been linked to the **greenhouse effect** and global warming. Water vapor, present in variable amounts ranging from zero to 4% by volume, also absorbs infrared radiation and is an essential link in the **hydrologic cycle**.

Ozone, the triatomic form of oxygen (O_3) is concentrated in the atmospheric layer about 12-30 mi (19-48 km) above the earth's surface. Although present in concentrations of only about 0.001% by volume, ozone absorbs ultraviolet radiation so effectively that it can almost completely shield life on earth from harmful ultraviolet rays. In the 1970s, it was found that chloroflurocarbons (CFCs), used as refrigerants and propellants in aerosol dispensers, were being emitted into the atmosphere in large quantities. There was concern that these compounds, as they reacted with sunlight, could destroy atmospheric ozone and its protective effects. Therefore, manufacture of CFC propellants was banned in the United States in 1978. By the mid-1990s, manufacturers had phased out most of the production of CFCs to prevent atmospheric ozone depletion. However, ozone levels from others sources of **air pollution** are increasing at the earth's surface, which may cause crop damage as well as result in the formation of **acid rain**.

Another concern is the increasing levels of methane in the atmosphere produced from the intestines of grazing animals and from the cultivation of rice in rice paddies. Methane adds to the greenhouse effect, and reduces the volume of atmospheric hydroxyl ions, which aid in the ability of the atmosphere to cleanse itself of **pollutants**. Minute quantities of other gases, such as **ammonia**, hydrogen sulfide, and oxides of sulfur and nitrogen, may be temporary constituents of the at-

mosphere in the vicinity of volcanoes but are washed out by rain or snow. Oxides and other pollutants are added to the atmosphere through industrial operations and by automobiles and have become major sources of air pollution, including the damaging effects of acid rain.

The atmosphere is divided into distinct layers, beginning with the troposphere. It starts at the earth's surface and stretches 430 mi (700 km) to the top of the thin air of the exosphere, where satellites orbit the earth. The troposphere extends up about 10 mi (16 km) in tropical regions to a temperature of about -110°F (-79°C), and to about 6 mi (9.7 km) in temperate latitudes to a temperature of about -60°F (-51°C). Temperatures decrease upward in this layer at a rate of about 3°F per 1000 ft (5.5°C per 1000 m). The troposphere is the layer in which most clouds and weather occur; it contains 80% of the gases of the atmosphere.

The next layer is the stratosphere, with an upper boundary of about 30 mi (50 km). In the lower stratosphere, the temperature is almost constant, but within the ozone layer (at about 12-30 mi [19-48 km]), the temperature rises rapidly. At the top of the stratosphere, the temperature is about the same as the temperature at the surface of the earth. The layer from 30 50 mi (50-80 km), called the mesosphere, is characterized by a marked decrease in temperature as the altitude increases, reaching -184°F at the top. The next layer, called the ionosphere (also called the thermosphere because of high temperatures in this layer), extends to an altitude of 400 mi (640 km). At about 50 miles up within this layer, ultraviolet rays, x rays, and electrons from the sun ionize several layers of the atmosphere, causing them to conduct electricity and reflect radio waves of certain frequencies back to earth. Above this layer is the exosphere, which extends to about 6,000 mi (9,600 km), the outer limit of the atmosphere. Helium is the most abundant gas in the exosphere. The temperature is about 1,300°F (700°C) but may vary during sunspot activity.

The density of dry air at sea level is approximately 1/800 the density of water. At higher altitudes, the density decreases rapidly, being proportional to the pressure and inversely proportional to the temperature. Atmospheric pressure is measured by a barometer in torrs. Atmospheric pressure at sea level is 760 torrs (760 mm or 29.92 in of mercury). Measuring 3.5 miles into the air (5.6 km), atmospheric pressure is about 380 torrs (14.96 in), which means that about half of all air in the atmosphere lies below this level. Atmospheric pressure is approximately halved for each additional increase of 3.5 mi (5.6 km) in altitude. By 50 mi (80 km), air pressure is only 0.007 torr (0.00027 in).

ATOMIC THEORY

One of the points of dispute among early Greek philosophers was the ultimate **nature** of **matter**. The question was whether the characteristics of matter that we can observe with our five senses are a true representation of matter at its most basic level. Some philosophers thought that they were. Anaxagoras of Klazomenai (c. 498-428 B.C.), for example, taught that matter can be sub-divided without limit and that it retains its characteristics no matter how it is divided.

An alternative view was that of Leucippus of Miletus (about 490 B.C.) and his pupil, Democritus of Abdera (c. 460-370 B.C.). (The views of these scholars are preserved in a few fragments of their writings and of commentaries on their teachings. Some writers doubt that Leucippus even existed.) In any case, the ideas attributed to them are widely known. They thought that all matter consists of tiny, indivisible particles moving randomly about in a void (a vacuum). The particles were described as hard, with form and size, but no color, **taste**, or **smell**. They became known by the Greek word *atomos*, meaning indivisible. Democritus suggested that, from time to time, **atoms** collide and combine with each other by means of hook-and-eye attachments on their surfaces.

Perhaps the most effective popularizer of the atomic theory was the Roman poet and naturalist, Lucretius. In his poem, *De Rerum Natura*, ("On the Nature of Things"), Lucretius states that only two realities exist, solid, everlasting particles and the void. This atomistic philosophy was in **competition** with other ideas about the fundamental nature of matter. **Aristotle**, for example, rejected Democritus' ideas because he could not accept the concept of a vacuum nor the idea that particles could move about on their own.

In addition, debates between atomists and anti-atomists quickly developed religious overtones. As the natural philosophy of Aristotle was adopted by and incorporated into early Christian theology, anti-atomism became acceptable and "correct," atomism, heretical. In fact, one objective of Lucretius' poem was to provide a materialistic explanation of the world designed to counteract religious superstition rampant at the time.

In spite of official disapproval, the idea of fundamental particles held a strong appeal for at least some philosophers

down through the ages. The French philosopher Pierre Gassendi (1592-1655) was especially influential in reviving and promoting the idea of atomism. **Robert Boyle** and Isaac Newton were both enthusiastic supporters of the theory.

Credit for the first modern atomic theory goes to the English chemist, John Dalton. In his 1808 book, *A New System of Chemical Philosophy*, Dalton outlined five fundamental postulates about atoms: 1. All matter consists of tiny, indivisible particles, which Dalton called atoms. 2. All atoms of a particular element are exactly alike, but atoms of different elements are different. 3. All atoms are unchangeable. 4. Atoms of elements combine to form "compound atoms" (i.e., **molecule**s) of compounds. 5. In **chemical reaction**s, atoms are neither created nor destroyed, but are only rearranged.

A key distinguishing feature of Dalton's theory was his emphasis on the weights of atoms. He argued that every atom had a specific weight that could be determined by experimental analysis. Although the specific details of Dalton's proposed mechanism for determining atomic weights were flawed, his proposal stimulated other chemists to begin research on atomic weights.

Dalton's theory was widely accepted because it explained so many existing experimental observations and because it was so fruitful in suggesting new lines of research. But the theory proved to be wrong in many of its particulars. For example, in 1897, the English physicist Joseph J. Thomson showed that particles even smaller than the atom—electrons—could be extracted from atoms. Atoms could not, therefore, be indivisible. The discovery of radioactivity at about the same time showed that at least some atoms are not unchangeable but, instead, spontaneously decay into other kinds of atoms.

By 1913, the main features of the modern atomic theory had been worked out. The work of Ernest Rutherford, Niels Bohr, and others suggested that an atom consists of a central core, the **nucleus**, surrounded by one or more electrons, arranged in **energy** levels each of which can hold some specific number of electrons. Bohr's atomic model marked the beginning of a new approach in constructing atomic theory. His work, along with that of Erwin Schrödinger, Louis Victor de Broglie, Werner Karl Heisenberg, Paul Adrien Maurice Dirac, and others showed that atoms could be understood and represented better through mathematics than through physical models. Instead of drawing pictures that show the location and movement of particles within the atoms, modern scientists tend to write mathematical equations that describe the behavior of observed atomic phenomena. But the nature of the physical entities these equations describe is no longer clear.

ATOMS

Atoms are the smallest particles of **matter** that have distinct physical and chemical properties. Each different type of atom makes up an element which is characterized by an atomic weight and an atomic symbol. Since the **atomic theory** was first proposed in the early nineteenth century, scientists have discovered a number of subatomic particles.

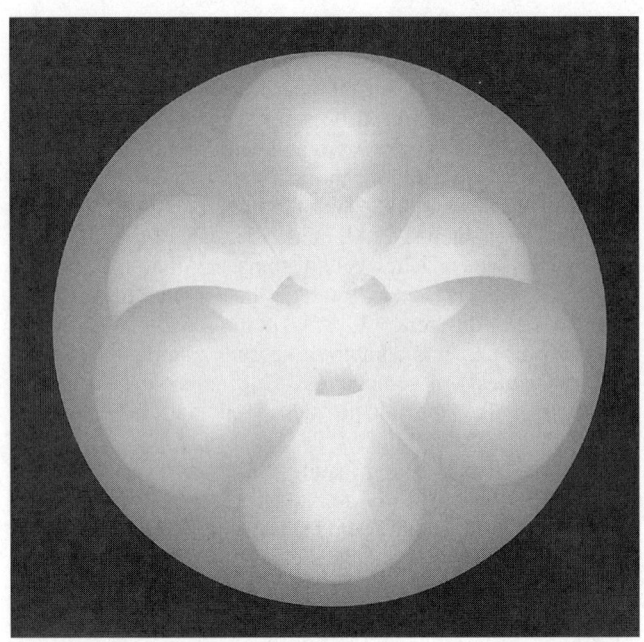

A computer generated model of a neon atom. The nucleus, at center, is too small to be seen at this scale and is represented by the flash of light. Surrounding the nucleus are the atom's electron orbitals: 1s (small sphere), 2s (large sphere), and 2p (lobed). *(Photo Researchers, Inc. Reproduced by permission.)*

The development of the atomic theory traces its history to early human civilizations. To these people, change was a concept to ponder. Ancient Greek philosophers tried to explain the causes of changes in their environment, typically, chemical changes. This led them to propose a variety to ideas about the nature of matter. By 400 B.C., it was believed that all matter was made up of four elements including earth, fire, air, and **water**. At around this time, Democritus proposed the idea of matter being made up of small indivisible particles. He called these particles *atomos*, or atoms. While Democritus may have suggested the theory of atoms, the Greeks had no experimental method for testing his theory.

This experimental method was suggested by **Robert Boyle** in the seventeenth century. At this time, he advanced the idea that matter existed as elements which could not be broken down further. Scientists built on Boyle's ideas, and in the early nineteenth century, John Dalton proposed the atomic theory.

Dalton's theory had four primary postulates. First, he suggested that all elements are made up of tiny particles called atoms. Second, all atoms of the same element are identical. Atoms of different elements are different in some fundamental way. Third, **chemical compound**s are formed when atoms from different elements combine with each other. Finally, **chemical reaction**s involve the reorganization of the way atoms are bound. Atoms themselves do not change.

Using Dalton's theory, scientists investigated the atom more closely. They wanted to determine the structure of these atoms. The first subatomic particle was discovered by J. J. Thomson (1856-1940) in the late nineteenth century. Using a cathode ray tube, he discovered negatively charged particles called electrons. Around this same time, scientists began to find that certain atoms produced radioactivity. In 1911, Ernest Rutherford (1871-1937) proposed the idea that atoms had a **nucleus** which the electrons orbited around. This led to the discovery of positively charged protons and neutral particles called neutrons.

Over time, scientists developed a chart known as the periodic table of elements to list all known elements. Atoms on this chart are symbolized by abbreviations called the atomic symbol. For example, oxygen atoms are denoted by the letter O. Each atom also has a unique mass denoted by its atomic weight. The atomic number is also distinct to each type of atom denoting the number of protons in their nucleus.

While atoms generally contain the same number of protons as neutrons, this is not always the case. Atoms which have more or less neutrons than protons are known as isotopes. For example, carbon atoms can have 12, 13, or 14 neutrons. When a nucleus as too many neutrons, as in the case of carbon-14, it is unstable and gives off radiation which can be measured. Radioactive isotopes have found many useful applications in biology. Scientists have used them in **radioactive dating** to determine the age of **fossil**s. They have also used them as tracer atoms to follow a chemical as it goes through metabolic processes in an **organism**. This has made them an important tool in medicine.

AUDUBON, JOHN JAMES (1785-1851)
American artist and naturalist

John James Audubon, the most renowned artist and naturalist in nineteenth century America, left a legacy of keenly observant writings as well as a portfolio of exquisitely rendered paintings of the birds of North America.

Born April 26, 1785, Audubon was the illegitimate son of a French naval captain and a domestic servant girl from Santo Domingo (now Haiti). Audubon spent his childhood on his father's plantation in Santo Domingo and most of his late teens on the family estate in Mill Grove, Pennsylvania—a move intended to prevent him from being conscripted into the Napoleonic army.

Audubon's early pursuits centered around natural history, and he was continuously collecting and drawing plants, insects, and birds. He developed a habit of keeping meticulous field notes of his observations at a young age. It was at Mill Grove that Audubon, in order to learn more of the movements and habits of birds, tied bits of colored string onto the legs of several Eastern Phoebes and so proved that these birds returned to the same nesting sites the following year. Audubon was the first to use banding to study the movement of birds.

While at Mill Grove, Audubon began courting their neighbor's eldest daughter, Lucy Bakewell, and they were married in 1808. They made their first home in Louisville, Kentucky, where Audubon tried being a storekeeper. He could not stand staying inside, and so he spent most of his time afield to "supply fowl for the table," thus dooming the store to failure. In 1810, Audubon met by chance Alexander Wilson, who is considered the father of American ornithology. Wilson had finished much of his nine volume *American Ornithology* at this time, and it is believed that his work inspired Audubon to embark on his monumental task of painting the birds of North America.

The task that Audubon undertook was to become *The Birds of America*. Because Audubon decided to depict each species of bird life-size, thus rendering each on a 36.5 in by 26.5 in (93 cm by 67 cm) page, this was the largest book ever published up until that time. He was able to draw even larger birds such as the whooping crane life-size by depicting them with their heads bent to the ground. Audubon pioneered the use of fresh models instead of stuffed museum skins for his paintings. He would shoot birds and wire them into life-like poses to obtain the most accurate drawings possible. Even though his name is affixed to a modern conservation organization, the National Audubon Society, it must be remembered that little thought was given to the conservation of birds in the early nineteenth century. It was not uncommon for Audubon to shoot a dozen or more individuals of a species to get what he considered the perfect one for painting.

Audubon solicited subscribers for his *The Birds of America* to finance the printing and hand-coloring of the plates. The project took nearly 20 years to complete, but the resulting double elephant folio, as it is known, was truly a work of art, as well as the ornithological triumph of the time.

Later in his life, Audubon worked on a book of the **mammals** of North America with his two sons, but failing health forced him to let them complete the work. He died in 1851, leaving behind a remarkable collection of artwork that depicted the natural world he loved so much.

John James Audubon. *(Archive Photos, Inc. Reproduced by permission.)*

AUSTRALOPITHECUS

The oldest known **hominids** are the Australopithecus found in Africa, where their **fossil** record extends from four to approximately one million years ago. Four Australopithecus **species** are recognized in the literature: *A. afarensis, A. africanus, A. robutus*, and *A. boisei*. Two more species are recent discoveries and not yet fully published: *A. anamensis* and *A. aethiopicus*. All species are clearly bipedal and thus hominids.

Australopithecine evolutionary origins remain vague, because of a sparse African fossil record between eight and four million years ago. A possible ancestor to some or all Australopithecus is the *Ardipithecus ramidus*, the oldest known hominid that is dated to 4.4 million years ago. In the nineteenth century Thomas Huxley and **Charles Darwin** postulated that people originated in Africa, since it was the home of their closest living relatives, the chimpanzee and the gorilla. The relationships of the Australopithecus species to each other and to the earliest widely recognized species of *Homo, H. habilis*, remain controversial. The continuing hominid fossil discoveries in Africa are slowly revealing the basic morphological and behavioral course of early **human evolution**, and confirm the prediction of Huxley and Darwin. So far, the australopithecines and earliest *Homo* have been found only in Africa, between Taung, South Africa, and Hadar, Ethiopia.

Known Australopithecus sites are concentrated in two areas from southern and eastern Africa. In South Africa, the sites are all caves containing a mixed stone and fossil matrix. Origin of these fossils included transport to the caves by **carnivore**s, and thus yield little or no evidence for hominid behavior. Since 1959 most Australopithecus research is in the Rift Valley, including Olduvai Gorge, from northern Tanzania to central Ethiopia in Eastern Africa. In this region, the Australopithecus sites are along ancient streams and lakes. Many of the Eastern African sites are found in stratigraphic position to volcanic layers whose age can be determined. One caution should be made: both the australopithecines and early Homo surely ranged widely within tropical and subtropical Africa. The known fossil concentrations reflect the occurrence of good to excellent bone preservation, and the australopithecines surely traveled into areas where conditions for fossil preservation did not exist.

AUTOIMMUNE DISEASE

Autoimmune diseases are complicated and often dangerous illnesses that affect the **endocrine system**, which produces the body's **cell**-regulating **hormones**. More than 60 different illnesses are considered autoimmune diseases, because each one causes the **immune system** to turn against different parts of the

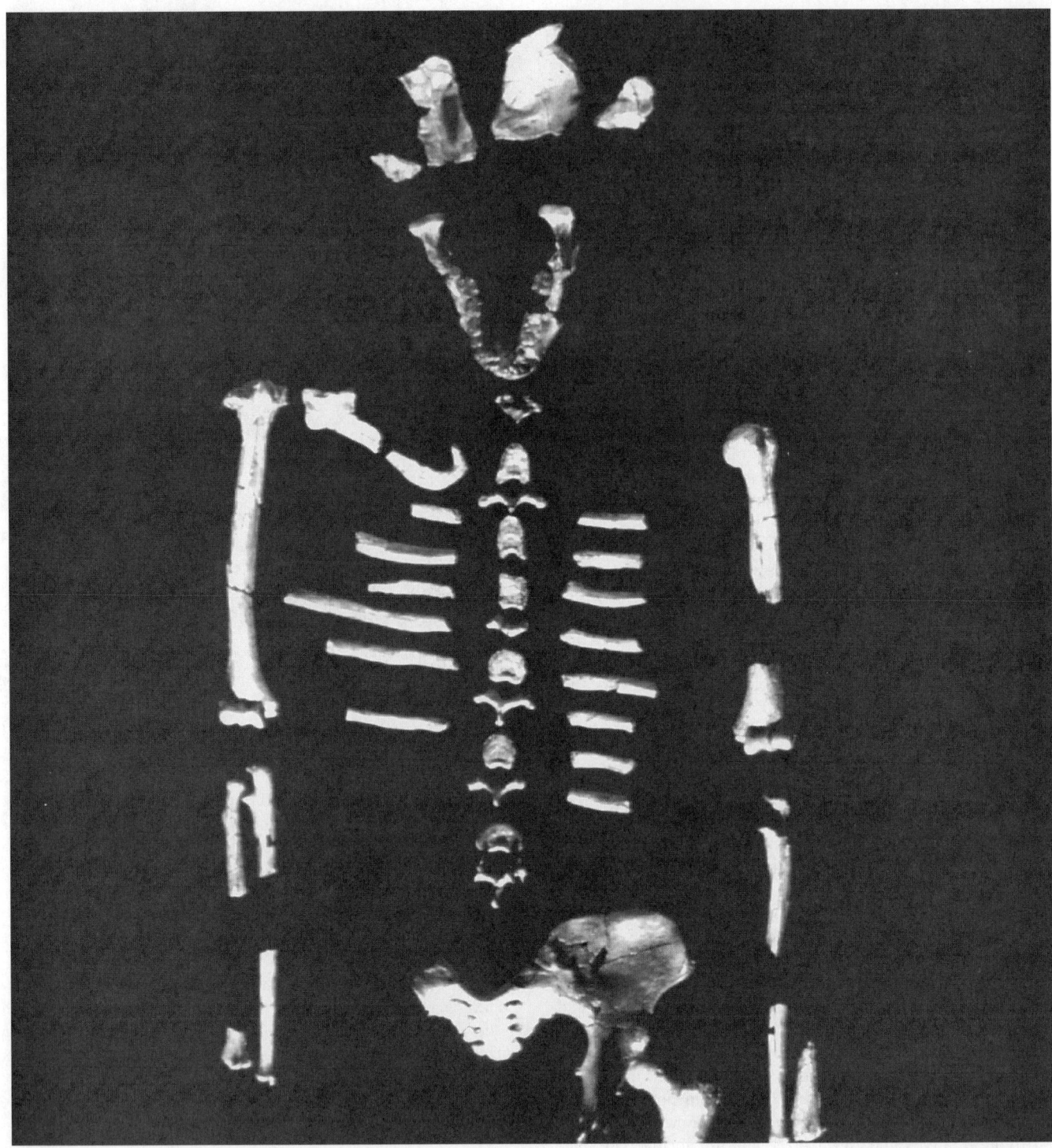

One of the most complete early hominid fossils is this *Australopitehcus afarensis* specimen commonly known as "Lucy," which was found by Donald Johanson in the Afar region of Ethiopia. *(Photo Researchers, Inc. Reproduced by permission.)*

body. In each case, the body sends signals through the hormones to produce anti-antibodies that attack the person's own **tissue**. Examples include rheumatoid **arthritis**, lupus, multiple sclerosis, juvenile diabetes, myasthenia gravis, Grave's disease, scleroderma, and chronic vasculitis.

A clear cause of these diseases has not been found. Some diseases may be caused by **heredity**, in which one parent may have rheumatoid arthritis, the child may have lupus and a cousin has multiple sclerosis. Others may be caused by a **virus** or **bacteria** that irreversibly alters the immune system. Or, it may be a combination of both when someone with a genetic predisposition to an autoimmune disease gets it only after a viral illness. Still others may be caused by an environmental hazard, such as recurrent exposure to toxic substances.

Women are much more likely to develop autoimmune diseases than men, although the reason why is not clear. Experts believe it may be related to the hormonal swings of pregnancy, nursing, and menopause that women experience over their lifetimes, as these shifts can sometimes exacerbate, calm or even eliminate the illness in some sufferers. But there is no consistent pattern that can clarify or confirm this suspicion.

These diseases often can go undiagnosed for years because clear parameters and effects are not known for many of the diseases, or the symptoms may be mild. Because of this, statistics on their prevalence are sketchy at best. Roughly 2.5 million Americans suffer from rheumatoid arthritis, which affects the lining of the joints and potentially the lungs and **heart**, and can cause pain, swelling, and deformation. At least 75% are women. Treatment usually includes pain medication and the use of corticosteroids, which can have various short and long-term side effects. Some are helped by joint fusion surgery, depending on the joint, or hip replacements.

More than 80% of the 130,000 people with lupus are women, which attacks several different areas of the body. Many with the disease develop rashes on their noses and cheeks, and become extremely sensitive to sunlight. It also attacks the joints and other **organ**s, particularly the kidneys, central **nervous system,** or heart. Lupus is one of the more difficult autoimmune diseases to treat because outbreaks are not consistent in their timing or duration, and their severity can range from mild to life-threatening. Most patients are treated with **steroids** such as the drug prednisone, but prolonged use can lead to osteoporosis, cataracts, weight gain, and increased chance of strokes and hardening of the **arteries.**

Multiple sclerosis affects almost 200,000 Americans, which damages the myelin coating on the nerves in the **brain** and **spinal cord.** The disease causes vision disorders, numbness, and tremors, and often leads to paralysis. Accurate estimates are not available for those with other autoimmune diseases such as myasthenia gravis, which attacks the connections between nerves and muscles, thus weakening the **muscular system**; Grave's disease, which stimulates the **thyroid gland** to produce excessive hormones and leads to weight loss, lethargy and insomnia; or scleroderma, deforming skin tissues, arteries, joint, and organs. Research completed in 1997 linked scleroderma with male fetal cells circulating in the **blood**stream of middle-aged women who had given **birth** to male children decades ago. While a normal immune system usually clears these fetal cells from the mother's body, an altered system may miss them or may try to destroy all the cells around the fetal ones, in a massive attack that leads to scleroderma.

Thus far those with autoimmune diseases have been treated with a variety of pain medications and steroids to counteract their altered endocrine system. However, researchers are trying to develop a treatment that corrects the defective immune system. One possibility is a T-cell vaccine that covers the self-destructive immune cells to prevent them from assaulting other cells, although this is still in the testing stage in 1998 and not likely to be available for several years. Other scientists are trying to develop medications that would in essence create a fake target for the defective system to attack, rather than the body. In the meantime, one successful new treatment option is to try to regulate the immune system rather than the hormones. Using a set of aggressive immune-suppressing drugs, some sufferers are finding at least temporary relief from the variety of effects caused by autoimmune diseases.

AUTONOMIC NERVOUS SYSTEM • See Nervous system

AUTOTROPH

Autotrophs are **organism**s that are capable of making their own food. They are "self feeders." These organisms make their own food by converting relatively simple inorganic **nutrients** into more complex, **energy**-rich, organic forms. Thus, the autotrophs do not need any outside source of organic material. There are two types of autotrophs, divided according to the processes by which they make their food. Photoautotrophs use the process of **photosynthesis**, while chemoautotrophs use the process of chemosynthesis. Photoautotrophs are far more common, and examples include green **plants**, **algae**, and some **bacteria**. This type of autotroph uses photosynthesis to convert the inorganic chemicals, **carbon dioxide** and **water**, into the organic sugar glucose, using sunlight as its source of energy. Glucose is the "food" produced by these photoautotrophs. Chemoautotrophs differ from photoautotrophs because they use chemicals from inorganic **chemical reaction**s, rather than sunlight, as their source of energy to produce organic materials. Certain types of bacteria are chemoautotrophs. For example, there are chemoautotrophic bacteria at deep sea hydrothermal vents where there is no sunlight. These bacteria support the entire food web at these great ocean depths through chemosynthesis, since photosynthesis cannot occur due to the lack of sunlight. In a food web, the autotrophs are the **producers**. They are the base of the food web, and all other organisms ultimately depend upon them for their energy and organic material. Autotrophs are consumed by other organisms, the **heterotroph**s, passing along organic nutrients and energy. Thus, without the autotrophs, other organisms would not be able to obtain the food or energy needed to survive.

AVERY, OSWALD THEODORE (ca. 1877-1955)
Canadian American biologist and bacteriologist

Oswald Theodore Avery was one of the founding fathers of immunochemistry (the study of the chemical aspects of immunology) and a major contributor to the scientific evolution of microbiology. His studies of the pneumococcus virus (causing acute pneumonia) led to further classification of the virus into many distinct types and the eventual identification of the chemical differences among various pneumococci viral strains. His work on capsular **polysaccharides** and their role in determining immunological specificity and virulence in pneumococci led directly to the development of diagnostic tests to demonstrate circulating antibody. These studies also contributed to the development of therapeutic sera used to treat the pneumonia virus. Among his most original contributions to immunology was the identification of complex carbohydrates as playing an important role in many immunological processes. Avery's greatest impact on science, however, was his discovery that **deoxyribonucleic acid (DNA)** is the molecular basis for passing on genetic information in biological self-replication. This discovery forced geneticists of that time to re-evaluate their emphasis on the **protein** as the major means of transmitting hereditary information. This new focus on DNA led to **James Watson** and **Francis Crick**'s model of DNA in 1952 and an eventual revolution in understanding the mechanisms of heredity at the molecular level.

Avery was born on October 21, 1877 (one source says 1887), in Halifax, Nova Scotia, to Joseph Francis and Elizabeth Crowdy Avery. His father was a native of England and a clergyman in the Baptist church, with which Avery was to maintain a lifelong affiliation. In 1887 the Avery family immigrated to the United States and settled in New York City, where Avery was to spend nearly sixty-one years of his life. A private man, he guarded his personal life, even from his colleagues, and seldom spoke of his past. He believed that research should be the primary basis of evaluation for a scientific life, extending his disregard for personal matters to the point that he once refused to include details of a colleague's personal life in an obituary. Avery's argument was that knowledge of matters outside of the laboratory have no bearing on the understanding of a scientist's accomplishments. As a result, Avery, who never married, managed to keep his own personal affairs out of the public eye.

Avery graduated with a B.A. degree from Colgate University in 1900 and received his M.D. degree from Columbia University's College of Physicians and Surgeons in 1904. He then went into the clinical practice of general surgery for three years but soon turned to research and became associate director of the bacteriology division at the Hoagland Laboratory in Brooklyn. Although his time at the laboratory enabled him to study species of bacteria and their relationship to infectious diseases and was a precursor to his interest in immunology, much of his work was spent carrying out what he considered to be routine investigations. Eventually, Rufus Cole, director of the Rockefeller Institute Hospital, became acquainted with Avery's research, which included work of general bacteriological interest, such as determining the optimum and limiting hydrogen-ion concentration for pneumococcus growth, developing a simple and rapid method for differentiating human and bovine *streptococcus hemolyticus*, and studying bacterial nutrition. Impressed with Avery's analytical capabilities, Cole asked Avery to join the Institute Hospital in 1913. Avery spent the remainder of his career there.

Research Focuses on Pneumonia Virus

At the Institute, Avery teamed up with A. Raymond Dochez in the study of the pneumococci (pneumonia) viruses, an area that was to take up a large part of his research efforts over the next several decades. Although Dochez eventually was to leave the institute, he and Avery maintained a lifelong scientific collaboration. During their early time together at the Rockefeller Institute, the two scientists further classified types of pneumococci found in patients and carriers, an effort which led to a better understanding of pneumococcus lung infection and of the causes, incidence, and distribution of lobar pneumonia. During the course of these immunological **classification** studies, Avery and Dochez discovered specific soluble substances of pneumococcus during growth in a cultured medium. Their subsequent identification of these substances in the blood and urine of lobar pneumonia patients showed that the substances were the result of a true metabolic process and not merely a result of disintegration during cell death.

Avery was convinced that the soluble specific substances present in pneumococci were somehow related to the immunological specificity of bacteria. In 1922, working with Michael Heidelberger and others at Rockefeller, Avery began to focus his studies on the chemical nature of these substances and eventually identified polysaccharides (complex carbohydrates) as the soluble specific substances of pneumococcus. As a result, Avery and colleagues were the first to show that carbohydrates were involved in immune reactions. His laboratory at Rockefeller went on to demonstrate that these substances, which come from the cell wall (specifically the capsular envelopes of the bacteria), can be differentiated into several different serological types by virtue of the various chemical compositions depending on the type of pneumococcus. For example, the polysaccharide in type 1 pneumococci is nitrogen-containing and partly composed of galacturonic acid. Both types 2 and 3 pneumococci contain nitrogen-free **carbohydrates** as their soluble substances, but the carbohydrates in type 2 are made up mainly of glucose and those of type 3 are composed of aldobionic acid units. Avery and Heidelberger went on to show that these various chemical substances account for bacterial specificity. This work opened up a new era in biochemical research, particularly in establishing the immunologic identity of the **cell**.

In addition to clarifying and systemizing efforts in bacteriology and immunology, Avery's work laid the foundation for modern immunological investigations in the area of antigens (parts of proteins and carbohydrates) as essential molecular markers that stimulate and, in large part, determine the success of immunological responses. Avery and his colleagues had found that specific anti-infection antibodies worked by neutralizing the bacterial capsular polysaccharide's ability to interfere with **phagocytosis** (the production of immune cells that recognize and attack foreign material). Eventually, Avery's discoveries led scientists to develop immunizations that worked by preventing an antigenic response from the capsular material. Avery also oversaw studies that showed similar immunological responses in *Klebsiella pneumonia* and *Hemophilus influenza*. These studies resulted in highly specific diagnostic tests and preparation of immunizing antigens and therapeutic sera. The culmination of Avery's work in this area was a paper he coauthored with Colin Munro Mac Leod and Maclyn Mc Carty in 1944 entitled "Studies on the Chemical Nature of the Substance Inducing Transformation of Pneumococcal Types. Induction of Transformation by a Desoxyribonucleic Fraction Isolated from Pneumococcus Type III." In their article, which appeared in the *Journal of Experimental Medicine,* the scientists provided conclusive data that **DNA** is the molecular basis for transmitting genetic information in biological self-replication.

Identifies DNA as the Basis of Heredity

In 1931 Avery's focus turned to "transformation" in **bacteria**, building on the studies of microbiologist **Frederick Griffith** showing that viruses could transfer virulence. In 1928, Griffith first showed that heat-killed virulent pneumococci could make a nonvirulent strain become virulent (produce disease). In 1932 Griffith stunned the scientific world when he announced that he had manipulated immunological specificity in pneumococci. At the time, Avery was on leave suffering from Grave's disease. He initially denounced Griffith's claim and cited inadequate experimental controls. But in 1931, after returning to work, Avery began to study transmissible hereditary changes in immunological specificity, which were confirmed by several scientists. His subsequent investigations produced one of the great milestones in biology.

In 1933 Avery's associate, James Alloway, had isolated a crude solution of the transforming agent. Immediately, the laboratory's focus turned on purifying this material. Working with type 3 capsulated pneumococcus, Avery eventually succeeded in isolating a highly purified solution of the transforming agent that could pass on the capsular polysaccharides' hereditary information to noncapsulated strains. As a result, the noncapsulated strains could now produce capsular polysaccharides, a trait continued in following generations. The substance responsible for the transfer of genetic information was DNA. These studies also were the first to alter hereditary material for treatment purposes.

Avery, however, remained cautious about the implications of the discovery, suspecting that yet another chemical component of DNA could be responsible for the phenomenon. But further work by McCarty and Moses Kunitz confirmed the findings. While some scientists, such as Peter Brian Medawar, hailed Avery's discovery as the first step out of the "dark ages" of **genetics**, others refused to give up the long-held notion that the protein was the basis of physical inheritance. The subsequent modeling of the DNA molecule by James Watson and Francis Crick led to an understanding of how DNA replicates, and demonstration of DNA's presence in all animals produced clear evidence of its essential role in heredity.

Avery also continued to work on other antigenic aspects of carbohydrates and the immune system. He was the first to create antibody-based treatments that were successful in protecting laboratory animals from infection, essentially by removing the protective capsular coat of the virulent cell. Collaborating with Dochez, he immunologically classified he-

molytic (destructive to blood cells) streptococcus and identified many of the specific antigens at work. These efforts revealed that hemolytic streptococcus had many serological types. Eventually hemolytic streptococcus was identified as the infectious agent in scarlet and acute rheumatic fever and hemorrhagic nephritis (kidney disease). Avery's work was the foundation for the eventual discovery of effective antibiotics for hemolytic streptococcus.

Despite the fact that Avery guarded his personal life, some information is known about his interests outside of science. A musician, he played cornet with the New York Conservatory of Music Orchestra and organized his own band. He also painted water colors. An independent Republican, he was a commissioned captain in the U.S. Army Medical Corps during World War I, assigned to the Institute for Medical Research. He served on various advisory committees during World War II, including the U.S. Army Board for the Study and Control of Epidemic Disease.

A highly reserved individual, Avery preferred to be remembered by his scientific accomplishments. He was fondly remembered by many of his colleagues and former students and clearly recognized for his efforts in helping to solve the puzzle of **heredity**. His honors were many, including several honorary degrees, the Paul Ehrlich Gold Medal, and the Copley Medal of the Royal Society of London. He also was a member of the National Academy of Sciences and foreign member of the Royal Society of London. He continued to conduct research in laboratories at the Rockefeller Institute Hospital for several years after his retirement. Eventually, he moved to Nashville, Tennessee, in 1947. He died there on February 20, 1955.

AXELROD, JULIUS (1912-)
American biochemist and pharmacologist

Julius Axelrod is a biochemist and pharmacologist whose discoveries relating to the role of **neurotransmitters** in the sympathetic **nervous system** earned him the Nobel Prize in physiology or medicine in 1970, together with **Ulf Euler** of Sweden and Sir Bernard Katz of Great Britain. As Axelrod himself has said, he was a late starter as a distinguished scientist, due to both the humble circumstances of his birth and his coming of age in the Great Depression of the 1930s. He only began real scientific research in 1946, and earned his Ph.D. in 1955. From then on he compensated for lost time and became the first chief of the pharmacology section of the National Institute of Mental Health, a branch of the prestigious National Institutes of Health.

Axelrod was born on May 30, 1912, in a tenement house in New York City, the son of Isadore Axelrod, a maker of flower baskets for merchants and grocers, and Molly Leichtling Axelrod. His parents had immigrated to the United States from Polish Galicia in the early years of the century, met and married in New York, and settled in the heavily Jewish area of the Lower East Side of Manhattan. Julius Axelrod attended public elementary and high schools near his home but later recalled that he got his real education in the neighborhood public library, reading voraciously through several books a week, everything from pulp novels to Upton Sinclair and Leo Tolstoy. He studied for a year at New York University, but when his money ran out he transferred to the tuition-free City College of New York, from which he graduated in 1933 with majors in biology and chemistry. He later claimed that he did most of his studying on the long subway rides between his home and the uptown Manhattan campus of City College.

Rejected for Medical School, Axelrod Pursues Research Career

Axelrod applied to several medical schools but was not admitted to any. It has been widely reported, in the *New York Times,* for example, that he failed to get into medical school because of quotas for Jewish applicants. It was difficult to find any work in New York in the depths of the Depression, and Axelrod was fortunate to find employment in 1933 as a laboratory assistant at the New York University Medical School at $25 per month. In 1935 he took a position as chemist at the Laboratory of Industrial Hygiene, a nonprofit organization set up by the New York City Department of Public Health to test vitamin supplements added to foods. He married Sally Taub on August 30, 1938, and they eventually had two sons, Paul Mark and Alfred Nathan. Axelrod took night courses and received an M.A. in chemistry from New York University in 1941. In the early 1940s he lost the sight of one eye in a laboratory accident.

Axelrod later speculated that he might have remained at the Laboratory of Industrial Hygiene for the rest of his working life. The work, he said, was moderately interesting, and the pay adequate. However, in 1946, quite by chance, he received the opportunity to do some real scientific research and found it exciting. The laboratory received a small grant to study the problem of why some persons taking large quantities of acetanilide, a non-aspirin pain-relieving drug, developed methemoglobinemia, the failure of hemoglobin to bind oxygen for delivery throughout the body. Axelrod, who had little experience in such work, consulted Dr. Bernard B. Brodie of Goldwater Memorial Hospital of New York. Brodie was intrigued with the problem and worked closely with Axelrod in finding its solution. He also found Axelrod a place among the research staff at New York University. The two men soon discovered that the body metabolizes acetanilide into a substance with an analgesic effect, and another substance that causes methemoglobinemia. They recommended that the beneficial metabolic product be administered directly, without the use of acetanilide. Related analgesics were investigated in the same manner.

In 1949, Axelrod, Brodie, and several other researchers at Goldwater Hospital were invited to join the National Heart Institute of the National Institutes of Health in Bethesda, Maryland. There Axelrod studied the physiology of caffeine absorption and then turned to the sympathomimetic amines, drugs which mimic the actions of the body's sympathetic **nervous system** in stimulating the body to prepare for strenuous activity. He studied such compounds as amphetamine, mesca-

line, and ephedrine and discovered a new group of **enzymes** which allowed these drugs to metabolize in the body. By the mid 1950s, Axelrod decided that he needed a doctorate to advance in his career at the National Institutes of Health. He took a year off to prepare for comprehensive examinations at George Washington University in the District of Columbia, submitted research work he had already done to satisfy the thesis requirements, and received a Ph.D. in pharmacology in 1955, at the age of forty-three. He was then offered the opportunity to create a section in pharmacology within the Laboratory of Clinical Sciences at the National Institute of Mental Health, another branch of the National Institutes of Health. He became chief of the section in pharmacology and held that position until his retirement in 1984.

Receives Nobel Prize for Research on Neurotransmitters

In 1957 Axelrod began the research which eventually led to the Nobel Prize. He and his colleagues and students studied the manner in which neurotransmitters, the chemicals which transmit signals from one nerve ending to another across the very small spaces between them, operate in the human body. In the 1940s the Swedish scientist Ulf von Euler had discovered that noradrenaline, or norepinephrine, was the neurotransmitter of the sympathetic nervous system. Axelrod was concerned with the way in which noradrenaline was rapidly deactivated in order to make way for the transmission of later nerve signals. He discovered that this was accomplished in two basic ways. First, he found a new enzyme, which he named catechol-O-methyltransferase (COMT). COMT was essential to the metabolism, and hence the deactivation, of noradrenaline. Second, through a series of experiments on cats, he determined that noradrenaline was reabsorbed by the nerves and stored to be reused later. These seemingly esoteric discoveries in fact had enormous implications for medical science. Axelrod demonstrated that psychoactive drugs such as antidepressants, amphetamines, and cocaine achieved their effects by inhibiting the normal deactivation or reabsorption of noradrenaline and other neurotransmitters, thus prolonging their impact upon the nervous system or the **brain**. His experiments also pointed the way to many new discoveries in the rapidly growing field of neurobiological research and the chemical treatment of mental and neurological diseases. The 1978 Nobel Prize in physiology or medicine, shared with Ulf von Euler and **Bernard Katz**, crowned his achievements in this area.

In his later years, Axelrod has worked in many areas of biochemical and pharmacological research, notably in the study of **hormones**. Especially important to the advancement of medical science was his development of many new experimental techniques which could be widely applied in the work of other researchers. He also had a great impact through his training of and assistance to a long line of visiting researchers and postdoctoral students at the National Institutes of Health. He continued his own research at the National Institute of Mental Health following his formal retirement in 1984. Early in 1993 Axelrod had the unusual experience of having his own life saved through a scientific discovery he had made many years before. At the age of eighty, he suffered a massive **heart** attack. The cardiologists at Georgetown University Medical Center soon determined that several of his coronary **arteries** were almost completely blocked by blood clots and that he must have immediate triple coronary-artery bypass surgery. The complication was that his blood pressure had fallen so dangerously low that he might not survive the operation. The solution to this crisis was to inject a synthetic form of noradrenaline to stimulate the contractions of his heart and thus raise his blood pressure to a more acceptable level. Axelrod survived the operation and within two months was back at work and attending conferences in foreign countries.

AXON

The axon, also called the **nerve** fiber, is the part of the nerve **cell** (**neuron**) which transmits nervous impulses. The axon leaves the nerve cell body (or soma) at an area called the axon hillock. Little offshoots from the axon are called axon collaterals, and may connect with other neurons. The area where the axon terminates, connecting with other nerve cells or with muscle cells, is called the synapse.

The nerve impulse originates in the soma, runs down the axon (possibly branching off to run down any axon collaterals), and ultimately reaches the end of the axon. At this synapse, the nerve impulse may be passed on to another neuron's soma, or to a muscle cell.

An axon may be quite short, or it may extend for as long as three feet throughout the nervous system. The thickness of axons also varies. Thicker axons tend to carry nerve impulses more quickly.

Many axons throughout the **brain** and body are covered by a specialized type of cell. In the brain and spinal cord, these cells are called oligodendrocytes. In the rest of the body, this wrapping function is performed by cells called Schwann cells. Oligodendrocytes or Schwann cells may be wrapped many times around the axon, so that the axon is covered with a thick layer of the oligodendrocyte's or Schwann cell's fatty membrane. This fatty white layer is referred to as the axon's myelin sheath, and greatly increases the speed of nerve conduction along the axon. Myelinated axons appear white, while unmyelinated axons appear gray. This is why some areas of the brain and spinal cord are referred to as containing white matter, while other areas are referred to as containing gray matter. Tiny interruptions in the myelin sheath are called nodes of Ranvier. These gaps are also considered to contribute to the speed of nerve conduction.

Problems which may affect axons include injury and disease. When an axon in the body is injured, but its soma remains intact, it is sometimes possible for the soma to grow a new axon. This may take up to two years. When an axon in the brain or spinal cord is injured, or when the neuron's soma is damaged or destroyed, no new axon will develop. Multiple sclerosis is a disease which destroys the myelin sheath along axons, slowing nerve impulses which travel along those axons.

B

BACON, ROGER (ca. 1214-ca. 1294)
English Philosopher

Roger Bacon, also called *doctor mirabilis*, was a natural philosopher and Franciscan monk who was regarded as one of the most controversial figures of the thirteenth century. His revolutionary ideas anticipated the intellectual and scientific revolution of the seventeenth century; in other words, he was ahead of his time. Bacon primarily devoted his life to learning languages, mathematics, optics (which was called perspectives) and experimental science. Many science historians consider Bacon's experiments as his greatest scientific accomplishment.

During Bacon's lifetime, university education was dominated by Christian theology, and classical authorities, such as, for example, **Aristotle**, were accepted only insofar as their ideas could be reconciled with Christian doctrine. Medieval thinkers, who generally dealt with purely speculative issues, paid little attention to the natural sciences, basically relying on known authorities. While Bacon, faithful to the spirit of the times, believed that all knowledge ultimately comes from God, he strongly objected to purely speculative thinking, maintaining that progress in science was impossible without experimentation. In other words, while fully accepting the wisdom found in the Scriptures, Bacon urged philosophers to, so to speak, read the *Book of Nature*.

Born into a wealthy family in Ilchester, England, Bacon was educated at Oxford. As a student, Bacon was profoundly influenced by Adam Marsh, the mathematician, and Robert Grosseteste, Chancellor of Oxford. During the 1240s, he was in Paris, where he taught in the Faculty of Arts. Bacon had little regard for his Parisian peers. A notable exception, however, was Peter of Maricourt, whose experimental work with mirrors included attempts to produce combustion at a distance. Around 1250, Bacon entered the Franciscan Order and returned to Oxford to teach. By 1257, because of his unconventional ideas, Bacon managed to arouse the suspicion and hostility of his religious superiors. Consequently, he was forbidden to teach or

publish his doctrines. Looking for support, Bacon wrote a letter to Guy de Foulques, an open- minded statesman who was interested in his work. In response, De Foulques, who was elected Pope in 1265, naming himself Clement IV, secretly instructed Bacon to write out his ideas. Among the works produced at the Pope's behest was *Opus maius*, considered his most important book. Unfortunately, Clement IV died in 1268, leaving Bacon essentially unprotected. In 1277, Bacon wrote a book in defense of astrology, provoking the anger of his superiors, who took a dim view of his researches in astrology, **alchemy**, magic, and perhaps even necromancy. Imprisoned in 1278, he was released in 1292.

In his *Opus maius*, Bacon presents a comprehensive world view, essentially covering all the areas of intellectual enquiry. Maintaining that the Scriptures contain all of knowledge, he urged scholars read them in Hebrew and Greek, the original languages, instead of relying on translations. While revering the Scriptures as the absolute source of knowledge, Bacon acknowledged the validity of scientific knowledge. The paradigm for scientific knowledge, according to Bacon, was mathematics, toward which, he believed, the human mind has a natural inclination. Exemplifying his interest in natural science, Bacon's book also contains a detailed discussion of optics. Since medieval scientists did not divide natural science into separate disciplines, Bacon's discussion of optics also includes a description of the anatomy of the eye.

While never doubting that **nature** was created by God, Bacon, did not hesitate to formulate his own general theoretical framework for natural science. Essentially, he believed that natural phenomena occur as a result of *seminal reasons*, creative forces which determine the development of the universe. Originally developed by the Stoic philosophers in Greece, the doctrine of seminal reasons was accepted by St. Augustine, also influencing other medieval thinkers, including St. Thomas Aquinas. In some ways futuristic, anticipating the triumph of empirical science in the distant future, Bacon's philosophy also has profound roots in classical philosophy and Christian theology. He is buried at Oxford, in the Franciscan Church.

Roger Bacon. *(The Library of Congress. Reproduced by permission.)*

BACTERIA

Bacteria are tiny **organism**s that are larger than **virus**es but can be seen only with the aid of a **microscope**. Although small, they are large in number and variety. Bacteria flourish in every possible **habitat** on Earth—in the air, soil, and **water**, as well as on and within the bodies of living things. Scientists have even discovered subsurface **population**s of bacteria that exist as deep as 1.7 mi (2.8 km) below the earth's surface and in locations with **temperature**s that reach 108.4°F (75°C). People often associate bacteria with disease, but bacteria also perform useful functions. They are part of chemical cycles during which they release essential elements such as carbon and **nitrogen** for recycling. They decompose dead and decaying organic matter, help **animal**s digest food and produce chemicals such as ethyl **alcohol**, butyl alcohol, **acetic acid**, and acetone. They aid in the production of food products such as cheese, butter, sauerkraut, coffee, wine, and cocoa. They aid in the manufacture of silk, cotton, and rubber. Scientists use **genetic engineering** on bacteria to produce medical substances such as human **insulin**, **interferon**, and other **protein**s. Bacteria synthesize certain **antibiotics**, and even clean up oil spills. Because of their size (diameters in the range of 1-5 um), bacteria were not classified or examined in detail until the nineteenth century. With the advent of better microscopes, many **species** of bacteria have been discovered.

As early as 1683, **Antoni van Leeuwenhoek**, a Dutch draper, observed bacteria from the human mouth using a single lens microscope that he designed. He was the first to discover and describe minute organisms, that we now know as bacteria. Leeuwenhhoek termed these organisms "animalcules" (little animals) since living things at that time were classified only as either **plants** or animals. Up until the mid- nineteenth century scientists considered bacteria to be animals all of one type. In 1857, a Swiss botanist named Karl von Naügeli suggested that bacteria have a class of their own within the plant **kingdom**. A century later, Otto Friedrich Muller, a Danish biologist, observed bacteria carefully and divided them into two groups according to shape: rod-shaped, known as *bacilli,* and spiral-shaped, called *spirilli.*

In the nineteenth century, much was learned about bacteria by French chemist and microbiologist, **Louis Pasteur**. In his early days studying **fermentation**, Pasteur had been asked to find out why some of the best of France's wines were spoiling. When he looked at wine under a microscope, he noticed that properly aged wine contained **yeast cell**s that cause the process of fermentation and produce alcohol, but in sour wine there was a proliferation of bacteria. Pasteur suggested that heating the wine gently at about 120°F (49°C) would kill the bacteria and not the yeast. He also suggested that greater cleanliness was need to eliminate bacteria, and inspired his student **Joseph Lister**, who went on to discover antiseptic surgery. A few years later the idea of using heat to kill **microorganism**s was applied to other perishable fluids, especially milk, and the process of pasteurization began. Pasteur identified two microbes that were killing silkworms, thereby saving the French silkworm industry from disaster.

Pasteur turned his attention to the link between diseases and microorganisms. He became interested in developing techniques for culturing and examining various disease-causing bacteria such as the bacteria that cause chicken **cholera** and **anthrax** and produced vaccines that successfully prevented them. He also created a vaccine for rabies. The Pasteur Institute was established for him and he headed this institution from 1888 until his death.

Inspired by the work of Pasteur, Ferdinand Julius Cohn, a German botanist, became increasingly interested in bacteria and in 1872 published three volumes in which he created a **classification** system based on their structure. He divided bacteria into four classes, Sphaerobacteria, Microbacteria, Desmobacteria and Spirobacteria, and provided a basis for bacterial classification. He is credited with coining the term *bacterium,* from the Latin, meaning little rod. Cohn was the first to demonstrate that the external form of bacteria was fixed: for example, rod-shaped bacilli can never become round or spirals. Cohn also discovered that some bacteria, such as *Bacillus subtilis,* form resistant **spore**s which make them difficult to kill with heat. Cohn's work was also instrumental in **Robert Koch**'s discovery of the relationship between specific bacteria and specific diseases.

In 1884 a discovery by **Hans Christian Gram**, a Danish bacteriologist, further advanced the study of bacteria. He developed a staining technique that distinguishes between two forms of bacteria, termed *gram-negative and* gram-positive that depends on the structure of the cell wall. This distinction is useful in identifying and classifying bacteria. It became more important years after its discovery when scientists realized that gram-negative bacteria, such as tuberculosis bacilli, resist taking up **penicillin** and other antibiotics in their cell walls, indicating diseases that cannot be treated with antibiotics.

Recently, with the invention of the powerful electron microscope, bacterial cells have been studied in great detail. They occur in three basic shapes, the round cocci, rod-like bacilli and helical spirilla and spirochetes. Scientists consider them *prokaryotes,*—organisms in which the **genetic material** is not enclosed in a definite nuclear **membrane**. Although bacteria lack most **organelle**s found in eukaryotic cells, they do have a type of **ribosome** that operates during **protein synthesis**. The genetic material in bacteria is **DNA** which is formed in a circular arrangement. Additional genes in smaller rings called plasmids also exist within bacterial cells. In 1997, researchers at the University of Wisconsin-Madison figured out the complete sequence of genes in *Escherichia coli*, an intestinal bacteria that has been studied by biologists for decades. *E. coli*'s DNA contains about 4,300 genes, and the sequencing is of great help to further biological research. The gene sequences of other bacteria are being worked out as well.

Bacteria are enclosed by a cell wall that contains a material called peptidoglycan, the basis of the Gram stain. Bacterial cell walls lack the **cellulose** found in plants. Some bacteria have another protective layer or capsule outside the cell wall for protection and adherence to substrate. Certain bacteria also have structures called pilli on the outside that helps them fasten to surfaces such as host membranes. Microscopy also shows that some bacteria move by hair-like projections called *flagella*. Although bacteria are unicellular, some form aggregates of cells called colonies. Bacteria reproduce by binary fission, and under unfavorable some bacteria form resistant spores called endospores.

Bacteria carry out different metabolic activities for **energy**. Some bacteria undergo **cellular respiration** as human cells do-using oxygen to help break down food for energy during the process of *aerobic respiration*. However, some bacteria produce their energy by means of fermentation in *anaerobic* conditions (without oxygen). Botulism and tetanus are diseases caused by **anaerobic** bacteria. Bacteria obtain their food in various ways. Certain bacteria are phototrophic, similar to green plants in that they use **light** as an energy source to make food from **carbon dioxide** (CO_2) and water and release oxygen. Other phototrophic bacteria use hydrogen sulfide gas instead of water, and release sulfur instead of oxygen. The smell of rotten **egg**s that you detect in some swamps is due to sulfur emission. Phototrophic bacteria contain pigments that are necessary to carry out food-making. Chemolithotrophs, on the other hand, are bacteria that do not get their energy from the sun. Instead, they oxidize inorganic compounds such as hydrogen, **ammonia**, or sulfur to obtain energy for the food-making process. These bacteria live near cracks in the earth where molten rock, magma, heats the sea water that seeps in. The heated water dissolves **minerals** from the magma, before it flows out to the sea, and these elements serve as the raw material for chemolithotrophs.

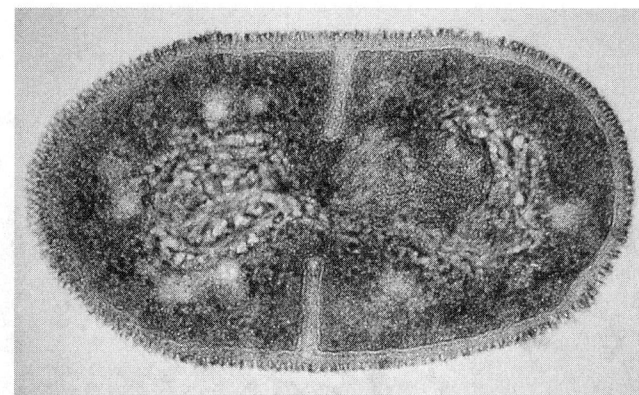

An electron micrograph of a dividing bacterium. Bacteria reproduce by means of binary fission, a type of asexual reproduction. *(JLM Visuals. Reproduced by permission.)*

Disease-causing bacteria are **parasite**s since they benefit at the expense of the host. Referred to as pathogens, these harmful bacteria release **toxins** or poisons that interfere with some function of the host's body. In a rare disease, a form of flesh-eating streptococcus called necrotizing fasciitis enters a wound in the skin of a patient and destroys the surrounding flesh at the rate of an inch an hour. Examples of more common infectious diseases caused by bacteria include strep throat, tuberculosis, typhoid fever, and tooth decay. Sexually transmitted diseases, such as **syphilis** and gonorrhea are also caused by bacteria. Bacterial diseases are transmitted from host to host through various means, such as direct contact, food, water, animal bites, contaminated objects, and droplets in the air. The body's **immune system** is capable of overcoming some bacterial diseases, but the control of bacterial diseases requires preventing transmission, using vaccines, and administering antibiotics. Antibiotics are medications that inhibit the synthesis of a functioning cell wall in bacteria, particularly in gram-positive species.

Most scientists believe that bacteria were the earliest life forms on Earth. Australian rocks that are 3.5 billion years old contain **prokaryote** microfossils that may be the first living things. Similar to today's bacteria, their DNA moved freely within their cells. In 1977, Carl Woese of the University of Illinois outlined the first **family** tree that involved two prokaryote domains. One domain, Archaea, microbes that usually inhabit extreme environments such as hot springs and salt ponds, are believed to be close to the group's original ancestor that were able to survive extreme conditions on the early Earth. The second domain, Bacteria, includes most of today's **eukaryote**s that differ from the Archeae in biochemical and physiological ways. The classification places all multicellular plants, animals and **fungi** in the Eukaryotic domain.

BACTERIOPHAGE

A bacteriophage, or *phage*, is a **virus** that infects a bacterial **cell**, taking over the host cell's **genetic material**, reproducing

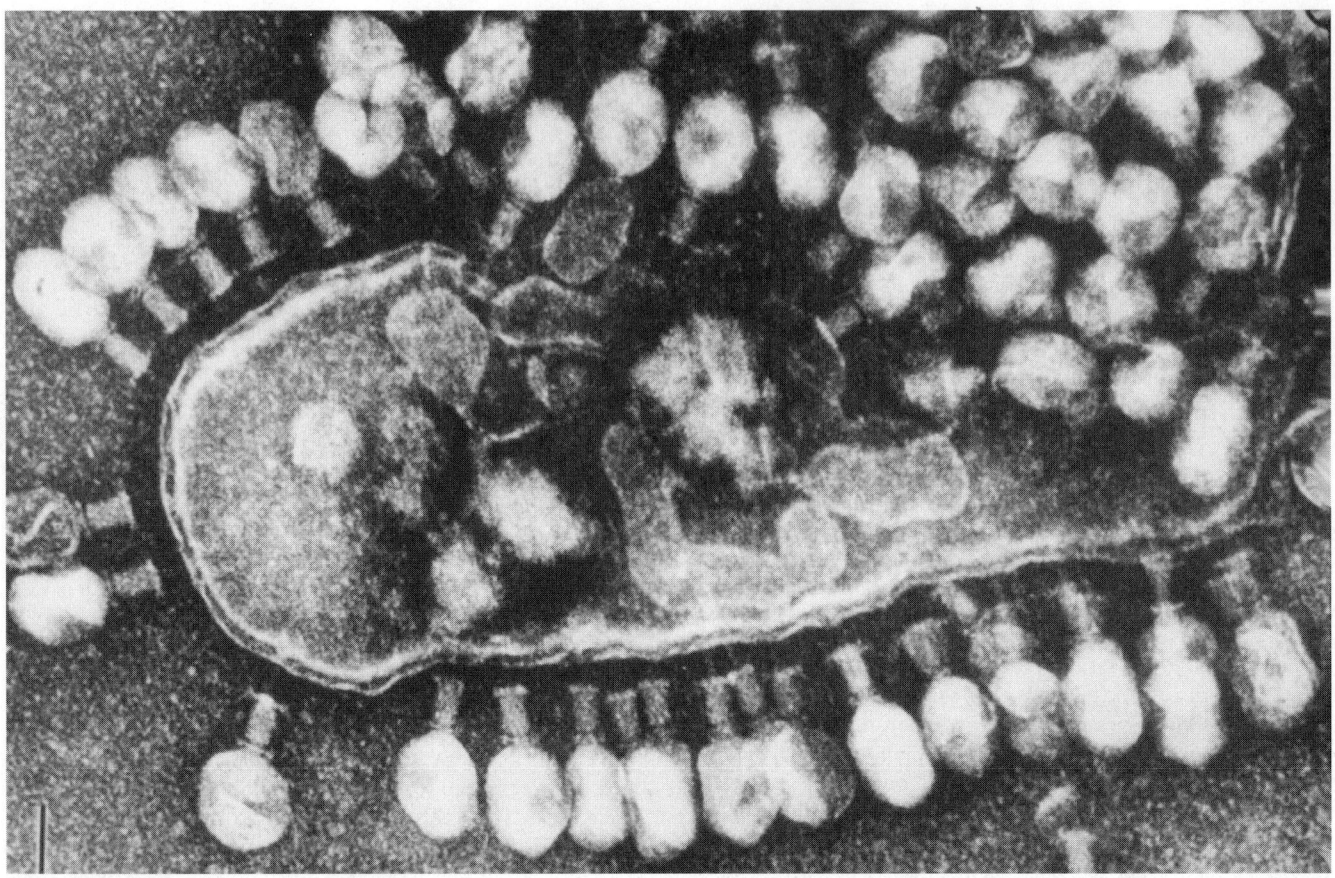

T4 bacteriophages attacking an *E. coli* bacterium. *(Photo Researchers, Inc. Reproduced by permission.)*

itself, and eventually destroying the bacterium. The word phage comes from the Greek word phagein which means to eat. Bacteriophages have two main components—a **protein** coat and a nucleic acid core of either **DNA** or **RNA**. Most DNA phages have double-stranded DNA, whereas phage RNA may be double or single-stranded. The electron **microscope** shows that phages vary in size and shape. Filamentous or threadlike phages, discovered in 1963, are among the smallest viruses known. Scientists have extensively studied the phages that infect *Escherichia coli* (*E.coli*), bacteria that are abundant in the human intestine. Some of these phages, such as the T4 phage consist of a capsid or head, often polyhedral in shape, that contains DNA, and an elongated tail consisting of a hollow core, a sheath around it, and six distal fibers attached to a base plate. When T4 attacks a bacterial cell, proteins at the end of the tail fibers and base plate attach to proteins located on the bacterial wall. Once the phage grabs hold, its DNA enters the bacterium while its protein coat is left outside.

Double stranded DNA phages reproduce in their host cells in two different ways: the lytic cycle and the lysogenic cycle. The lytic cycle kills the host bacterial cell. During the lytic cycle in *E.coli*, for example, the phage infects the bacterial cell, and the host cell commences to transcribe and translate the viral **gene**s. One of the first genes that it translates encodes an **enzyme** that chops up the *E.coli* DNA. The host now fol-

lows instructions solely from phage DNA which commands the host to synthesize phages. At the end of the lytic cycle, the phage directs the host cell to produce the enzyme, lysozyme, that digests the bacterial cell wall. As a result, **water** enters the cell by osmosis and the cell swells and bursts. The destroyed or lysed cell releases up to 200 phage particles ready to infect nearby cells. On the other hand, the lysogenic cycle does not kill the bacterial host cell. Instead, the phage DNA is incorporated into the host cell's **chromosome** where it is then called a prophage. Every time the host cell divides, it replicates the prophage DNA along with its own. As a result, the two daughter cells each contain a copy of the prophage, and the virus has reproduced without harming the host cell. Under certain conditions, however, the prophage can give rise to active phages that bring about the lytic cycle.

In 1915, the English bacteriologist Frederick Twort (1877-1950) first discovered bacteriophages. While attempting to grow *Staphylococcus aureus*, the bacteria that cause boils, he observed that some bacteria in his laboratory plates became transparent and died. He isolated the substance that was killing the bacteria and hypothesized that the agent was a virus. In 1917, the French-Canadian scientist Felix H. d'Hérelle independently discovered bacteriophages as well. The significance of this discovery was not appreciated, however, until about thirty years later when scientists conducted further bacteriophage research. One prominent scientist in the field was Salvador E. Luria (1912-1991), an Italian-American biologist especially interested in how X-rays cause **mutation**s in bacteriophages. He was also the first scientist to obtain clear images of a bacteriophage using an electron microscope. **Salvador Luria** emigrated to the United States from Italy and soon met **Max Delbruck** (1906-1981), a German-American molecular biologist. In the 1940s, Delbruck worked out the lytic mechanism by which some bacteriophages replicate. Together, Luria, Delbruck and the group of researchers that joined them studied the genetic changes that occur when viruses infect bacteria. Until 1952, scientists did not know which part of the virus, the protein or the DNA, carried the information regarding viral **replication**. It was then that Hershey (1908-), a United States biochemist, working with Martha Chase at the Carnegie Laboratory of **Genetics**, performed a series of experiments using bacteriophages. These experiments proved DNA to be the **molecule** that transmits the genetic information. (In 1953, the Watson and Crick model of DNA explained how DNA encodes information and replicates.) For their discoveries concerning the structure and replication of viruses, Luria, Delbruck, and Hershey shared the Nobel Prize for Physiology or Medicine in 1969. In 1952, two American biologists, **Norton Zinder** and Joshua Lederman at the University of Wisconsin, discovered that a phage can incorporate its genes into the bacterial chromosome. The phage genes are then transmitted from one generation to the next when the bacterium reproduces. In 1980, the English biochemist, **Frederick Sanger**, won a Nobel prize for determining the **nucleotide** sequence in DNA using bacteriophages.

In the last several decades, scientists have used phages for research. One use of bacteriophages is in **genetic engineering**, manipulating genetic molecules for practical uses. During genetic engineering, scientists combine genes from different sources and transfer the recombinant DNA into cells where it is expressed and replicated. Researchers often use *E. coli* as a host because they can grow it easily and they know a lot about it. One way to transfer the recombinant DNA to cells uses phages. Employing **restriction enzyme**s to break into the phage's DNA, scientists splice foreign DNA into the viral DNA. The recombinant phage then infects the bacterial host. Scientists use this technique to create new medical products such as vaccines. In addition, bacteriophages provide information about genetic defects, human development, and disease. One geneticist has developed a technique using bacteriophages to manipulate genes in mice, while others are using phages to infect and kill disease-causing bacteria in mice. In addition, microbiologists found a filamentous bacteriophage that transmits the gene that encodes the toxin for **cholera**, a severe intestinal disease that kills tens of thousands worldwide each year.

BAER, KARL ERNST VON (1792-1876)
Estonian biologist

One of ten children, Von Baer was born in Piep, Estonia, to parents descended from Prussian nobility. Due to the large size

Karl Ernst von Baer. *(The Library of Congress. Reproduced by permission.)*

of his family, he was sent to live with his childless uncle and aunt until the age of seven. Initially tutored at home, he later spent three years at a school for the children of nobility. Although his father and uncle encouraged him to pursue a military career, from 1810-1814, Von Baer attended the University of Dorpat in Vienna, Austria and obtained an MD degree. From 1814-1817, he studied comparative anatomy at Würzburg. In 1817, he was appointed prosector at the University of Königsberg, where in 1819, he accepted an appointment to teach **zoology** and anatomy and serve as chief at the new zoological museum that he organized. In 1828, he became a member of the St. Petersburg Academy of Sciences, where he taught zoology from 1829-30. He then returned to Königsberg until 1834. At that time, he became librarian at St. Petersburg Academy of Sciences, and conducted research in anatomy and zoology. In 1846, he was also appointed to the position of Professor of Comparative Anatomy and Physiology at the Medico-Chirugical Academy in St. Petersburg.

Von Baer also took part in other types of scientific projects. In 1837, he led a scientific expedition to Novaya Zemlya in Arctic Russia, and from 1851-6, studied the fisheries of lake Peipus and the Baltic and Caspian Seas. He served as inspector of fisheries for the empire from 1851-1852. He founded the St. Petersburg Society for Geography and Ethnography and the German Anthropological Society. Von Baer died in Dorpat, Estonia, on November 28, 1876.

Von Baer made significant contributions to the world of science. The first of Baer's most famous discoveries grew out of his work at Königsberg. For more than a century, scientists had attempted to determine the exact nature and location of the mammalian **egg**. In 1673, Regnier de Graaf had discovered follicles in the ovaries that he thought might be eggs. However, he later found structures even smaller than follicles in the uterus, raising doubts about the role of the follicles themselves. During his research at Königsberg, Baer discovered the mammalian egg by identifying a yellowish spot within the follicle visible only with a **microscope**. He developed this idea in his 1827 treatise, *De ovi mammalium et hominis genesi* (*On the Origin of the Mammalian and Human Ovum*).

Baer's second great accomplishment was his explanation of early embryonic development, a theory that he summarized in his two-volume textbook *Über die Entwicklunggeschichte der Thiere* (*On the Development of Animals*), 1828-1837. Here he set forth the theory that **embryos** of all animals begin as similar structures that are both simple and homogeneous and develop into complex heterogeneous forms. In fact, he said, it is impossible to distinguish among the early embryos of birds, reptiles, and **mammals** until their later stages of development. He set forth the germ-layer theory of development which stated that during the early stages of development, embryos ultimately form four (later shown to be three) distinct layers that eventually differentiate into specific organs or body structures. These are the ectoderm, mesoderm and endoderm. While observing embryos, Von Baer discovered the extraembryonic membranes, the **chorion**, **amnion**, and allantois and determined their functions. He also discovered the presence of a notochord in early vertebrate embryos. Although the notochord quickly changes into a spinal column, its early presence indicates the evolutionary connection between vertebrates and other organisms now classified together in the phylum *chordata*. For his work, Von Baer was awarded the Copley Medal of the Royal Society in 1876.

BALTIMORE, DAVID (1938-)
American microbiologist

David Baltimore won the Nobel Prize in physiology or medicine in 1975. He shared the award with the virologist **Renato Dulbecco** and the oncologist **Howard Temin** for the groundbreaking discovery that genetic information doesn't just travel from **DNA** (deoxyribonucleic acid, which contains genetic information) to **RNA** (ribonucleic acid, which communicates DNA information to **proteins**), although that concept had been at the heart of modern genetic theory. Rather, Temin and Baltimore independently discovered that some **viruses** could replicate their RNA into the DNA of healthy cells, causing tumors. This process is known as reverse transcription and is catalyzed by the **enzyme** reverse transcriptase. Its implications had a great effect on the study of cancer and the role of viruses in causing the disease. Born March 7, 1938, in New York City to Richard Baltimore and Gertrude Lipschitz, David was a gifted student of science. While still in high school, he attended a prestigious summer program for talented students of science at the Jackson Laboratory in Bar Harbor, Maine, where the focus was mammalian genetics. It was there that Baltimore decided to pursue a career in the research sciences and first met his future colleague, Howard Temin, also a student at the time.

Baltimore attended Swarthmore College in Pennsylvania and graduated in 1960 with high honors in chemistry. He started graduate work at the Massachusetts Institute of Technology (M.I.T.) but transferred after one year to the Rockefeller Institute, now called Rockefeller University, in New York. There he studied with Richard M. Franklin, who was a molecular biophysicist specializing in RNA viruses. Baltimore earned his Ph.D. in 1964 and returned to M.I.T. for a postdoctoral fellowship the following year.

Baltimore was interested in a specific group of RNA viruses, the picornaviruses, which include mengovirus and poliovirus, that do not have DNA but nevertheless seemed to reproduce in cells of complex organisms that carry their genetic information in DNA. Since 1964, Temin had been suggesting that RNA viruses could replicate themselves in DNA, but the scientific community disbelieved and even ridiculed him. Baltimore, however, persisted in looking for RNA or DNA enzymes in the genetic material of poliovirus to solve this riddle.

He continued his research in this area as a fellow at the Albert Einstein College of Medicine in the Bronx (1964–65) and as a research associate at the Salk Institute in California (1965–68). At the Salk Institute he met Renato Dulbecco, who had developed innovative techniques for examining animal viruses in the laboratory. He also met Alice Shih Huang, a postdoctoral fellow studying vesicular stomatitis virus.

Discovers Reverse Transcriptase

In 1968, Baltimore returned with Huang to Boston, where they were married. They have one daughter. Baltimore became an associate professor of microbiology at M.I.T. and continued to focus his research on poliovirus. By 1970, however, a number of scientists had suggested that all RNA viruses might not be alike. Baltimore's focus on poliovirus therefore might not reveal clues to the behavior of other viruses. Baltimore began to classify RNA viruses according to their varying replication strategies. It was during his work on this project that he discovered an enzyme that enabled an RNA virus to replicate its single strand of RNA and thus become compatible with the double-stranded DNA in a sample of Rauscher murine leukemia virus. The enzyme was later called **reverse transcriptase**.

Meanwhile, Temin had independently demonstrated the same thing, using a sample of Rous sarcoma virus. In 1970, both scientists made the initial announcements of reverse transcription within days of each other, at separate conferences. One month later they published an article detailing their findings in the journal *Nature*. The excitement of their news was instantaneous. Many scientists jumped to the conclusion that reverse transcription held the key to a cure for **cancer**. But Baltimore and Temin were more reserved in their response. They knew that their work did not establish a direct link between viruses and cancer. Their discovery did, however, quickly become a key to the study of cancer. Baltimore immediately began to be recognized for his achievement. In 1972 he was promoted to full professor at M.I.T. and in 1973 he was awarded a lifetime research professorship by the American Cancer Society.

While continuing to study reverse transcriptase, Baltimore and his colleagues at M.I.T. partially synthesized a mammalian **hemoglobin gene**. Other teams around the country were performing similar experiments at the same time, which raised the specter of genetic engineering. As a prominent figure in the scientific community, Baltimore became outspoken about the risks of genetic engineering. He was concerned that modern science—and biology in particular—might be misused. In 1975, he initiated a conference in which scientists attempted to design a self-regulatory system regarding experiments with recombinant DNA. In 1976, the National Institutes of Health established a committee to oversee federally funded experiments in the field of genetic engineering. After winning the Nobel Prize in 1975, Baltimore continued to be honored for his work. He was elected to the National Academy of Sciences and the American Academy of Arts and Sciences in 1974. In 1983 he became the director of the Whitehead Institute for Biomedical Research, where he remained until 1990. In that position, he made significant advances in the field of immunology and synthetic vaccine research. In 1990 he became President of Rockefeller University.

Implicated in Controversy over Falsified Data

Baltimore's career took a sudden turn in 1989 when it was revealed that the conclusions of a 1986 paper he had coauthored while still at M.I.T. were based on falsified data. A young scientist, Margaret O'Toole, confronted Thereza Imanishi-Kari, also a coauthor of the article and a supervising scientist in the M.I.T. lab, with suspicions of misconduct. Imanishi-Kari denied any wrongdoing. Baltimore stood by Imanishi-Kari, and O'Toole was subsequently demoted, later claiming that her career had been ruined because she had spoken out against her superiors.

The matter was taken up by a House subcommittee and the Office of Scientific Integrity, which eventually lent credence to O'Toole's suspicions. Baltimore retracted the article but it was too late. Though he was cleared of any wrongdoing and though Imanishi-Kari was not prosecuted, Baltimore's name had been attached to a major breach of scientific ethics. That Baltimore had earlier taken such a strong stand on the ethics of bioengineering was a particular irony not lost on the scientific community. In 1991, under pressure from the faculty, Baltimore resigned as President of Rockefeller University, though he remains a professor there and continues to do research.

BANTING, FREDERICK GRANT (1891-1941)
Canadian physician and medical researcher

Frederick Banting's principal achievement was the first isolation of the **hormone insulin** in 1921 and its successful use in treating diabetes. For this, Banting received the 1923 Nobel Prize for physiology or medicine along with John J. R. Macleod (1876-1935).

Frederick Banting was born in Alliston, Ontario. An average student, he graduated in 1916 from the University of Toronto medical school. After Medical Corps service in World War I, he completed his internship at Toronto's Hospital for Sick Children before beginning a medical practice in London, Ontario. Through an article on diabetes in a medical journal, Banting became interested in the disease, which he began to study at the University of Western Ontario.

By the early 1900s, scientists knew that the pancreas, an organ connected to the small intestine, was involved in diabetes. A **pancreas** hormone that reduced the blood glucose level was proposed in 1916 by the English physiologist Edward Sharpey-Schäfer. He thought the hormone was produced by cells called *islets of Langerhans* and called it *insuline*, from the Latin word for island. Various scientists tried unsuccessfully to isolate it.

In 1919, Moses Barron, a researcher at the University of Minnesota, showed that blockage of the duct connecting the two major parts of the pancreas caused shriveling of a second **cell** type, the acinar. It was Barron's article that inspired Banting, who thought that enzymes from the acinar cells digested the islet hormone. By tying off the pancreatic duct to destroy the acinar cells, he believed he could preserve the hormone and extract it from the islet cells.

At the suggestion of a colleague, Banting proposed his experiment to the head of the University of Toronto's Physiology Department, **John Macleod**, a noted expert on **carbohydrate** metabolism. At first Macleod rejected Banting's proposal, in part because he did not believe the islet hormone existed. Finally, however, Macleod supplied Banting with laboratory space, ten dogs for experimentation, and a medical student who was skilled in glucose measurement named **Charles Best** (1899-1978). Banting and Best began work in May 1921 while Macleod went on vacation to his native Scotland.

Banting and Best tied off the pancreatic ducts in some dogs so that the acinar cells would atrophy, then removed the pancreases to extract fluid from the islet cells. In the meantime, they removed the pancreases from other dogs to cause diabetes. These dogs were then injected with the islet cell fluid. After many attempts, the procedure was perfected. This required careful measurement of the glucose levels in the diabetic dogs before treatment to be sure they were diabetic and afterward to show that they maintained normal blood glucose levels.

When he returned from vacation in August 1921, Macleod learned of Banting and Best's success. He organized his entire laboratory to isolate and purify insulin from livestock for use in treating human diabetes. Banting and Best initially wanted to name their fluid "isletin," but Macleod insisted on Sharpey-Schäfey's term insuline, shortened to insulin. A major collaborator in the purification work was biochemist James Collip.

At the Hospital for Sick Children in January 1922, fourteen year-old Leonard Thompson became the first human to be successfully treated for diabetes using insulin.

Robert Bárány. *(UPI/Corbis-Bettmann. Reproduced by permission.)*

Banting's original theory was incorrect, as Banting and Best later discovered. The digestive enzymes in the acinar cells are inactive and insulin can be extracted from an intact pancreas. What led to their success was their precise method of glucose measurement.

Charles Best received his medical degree in 1925. Banting always insisted that both he and Best be credited for the discovery, and almost turned down his Nobel Prize because Best was not included. Best became head of the University of Toronto's physiology department in 1929 and director of the university's Banting and Best Department of Medical Research after Banting's death.

Frederick Banting also conducted research in cancer and heart disease. He again joined the Medical Corps during World War II, and died in a military air crash in Newfoundland in 1941.

BÁRÁNY, ROBERT (1876-1936)
Swedish physician

Robert Bárány made significant contributions to our understanding of the vestibular apparatus, part of the inner ear that plays an important role in maintaining balance. He devised tests to diagnose inner-ear disease and investigated the relationship between the vestibular and nervous systems. Because of his ground-breaking research in this area, he is credited with creating a new field of study, otoneurology. Bárány's achieve-

ments were recognized in 1914 with the Nobel Prize in physiology or medicine.

Bárány was born on April 22, 1876, in Rohonc (near Vienna), Austria-Hungary (now Austria), the eldest of six children. His father, Ignaz Bárány, was a bank official. His mother, Marie Hock Bárány, was the daughter of a well-known Prague scientist. A case of tuberculosis of the bone contracted when Bárány was young first awakened his interest in medicine.

Bárány received a doctor of medicine degree from the University of Vienna in 1900. He then spent two years studying internal medicine, neurology, and psychiatry at clinics in Frankfurt, Heidelberg, and Freiburg and returned to Vienna, where he received hospital surgical training. In 1903 he accepted a post at the Ear Clinic, also in Vienna, which was then directed by Adam Politzer, a leading figure in the history of otology (the study of the ear and its diseases).

It was the chance observation that clinic patients often became dizzy after having their ears irrigated that led Bárány to develop one test that still bears his name. The Bárány caloric test involves stimulating each of a patient's inner ears separately by syringing one with hot liquid and the other with cold. Normally, this results in rapid, involuntary movements of the eyeballs, termed nystagmus. Bárány demonstrated that the direction of the nystagmus is determined by the temperature of the water and the position of the head. He also showed that the absence or delay of nystagmus indicates a problem with the balance structures of the ear. The test was an eminently practical technique for diagnosis, since it could easily be performed at a patient's bedside.

Another diagnostic procedure introduced by Bárány was the chair test. The patient is turned in a rotating chair with a specially designed headrest that inclines the head slightly forward. Once again, any deviation from the normal pattern of nystagmus afterward indicates a problem. Yet another of Bárány's inventions during this period was the noise box, a much-used device that effectively isolates the hearing performance of one ear by creating a masking noise in the other.

At the start of World War I. Bárány, who was of Jewish descent, was dispatched by the army to the fortress of Przemysl on the border between Poland and Russia, where he served as a medical officer. While there, Bárány continued to study the connection between the vestibular apparatus and the nervous system. He also developed an improved surgical technique for dealing with fresh bullet wounds to the brain.

However, in April 1915, the Russians occupied Przemysl, and Bárány was transported along with other prisoners by cattle car to Merv in central Asia and Bárány came down with malaria. Bárány was placed him in charge of otolaryngology (the medical specialty concerned with the ear, nose, and throat) for both Russian natives and Austrian prisoners.

It was while he was still a prisoner of war that Bárány received the news that he had won the Nobel Prize. Thanks to the personal intervention of Prince Carl of Sweden, Bárány was released in 1916. He returned to Vienna that same year to be confronted by his colleagues' claim that he had inadequately cited their own contributions to his work. These accusations were investigated by the Nobel Prize Committee, which found them groundless. Nevertheless, the attacks prompted Bárány to accept a post as professor at the University of Uppsala in Sweden in 1917, where he rose to the position of chairman of the department of ear, nose, and throat medicine and remained there for the rest of his life.

While at Uppsala, Bárány studied the role of the part of the brain called the cerebellum in controlling body movement. He had previously devised another test for disturbances in cerebellar function, known as the pointing test, in which the patient points at a fixed object with the eyes alternately open and closed. Consistent errors while the eyes are closed indicates a **brain** lesion.

Bárány also developed a surgical technique for treating chronic sinusitis. For this, he was awarded the Jubilee Medal of the Swedish Society of Medicine in 1925. Among his numerous other awards were the Belgian Academy of Sciences Prize, the ERB Medal from the GermanNeurological Society, and the Guyot Prize from the University of Groningen in the Netherlands. He also received honorary degrees from several universities, including the University of Stockholm.

Bárány died in Uppsala on April 8, 1936, after a series of strokes.

BARR, MURRAY LLEWELLYN (1908-)
Canadian anatomist

Barr grew up on his father's farm in Canada. He was educated at the University of Western Ontario, and graduated in medicine in 1933. He spent nearly his entire career there—except for his term as a medical officer with the Royal Canadian Air Force during World War II.

Around 1949, when Barr was a professor at the university, he and graduate student Ewart G. Bertran were studying nerve cells, or **neurons**. Barr noticed that the nuclei of nerve cells in female cats contained a dense mass of chromatin (sex chromatin body) that males did not have.. Discovering the sex chromatin body, now called the Barr body, focused Barr's attention on human and medical genetics. Later studies by Barr and his co-workers found that these sex differences occur in the cells of most **mammals**.

In 1956, Tjio and Levan showed that there are forty-six **chromosomes** in human cells, twenty-three pairs. Female cells have a pair of sex chromosomes, known as XX, while male cells have only one **sex chromosome** in their XY sex configuration. By 1959, researchers identified Down, Turner, and Klinefelter syndromes as disease due to defects in chromosome number. Then, further studies of the Barr body led to the understanding of a process called inactivation. During **embryonic development**, one of the two X chromosomes is inactivated in each cell. The inactivated X condenses as a Barr body. The **embryo** cells continue to divide and form a clone of cells that have the same X chromosome active and the other inactive. British geneticist, Mary Lyons, showed that which X chromosome becomes inactive is a random occurrence. Inactivation helped scientists understand the mechanisms of sex-linked inheritance. Barr and his co-workers went on to develop a simple screening test for diseases involving sex chromosome anomalies. Their buccal smear technique using cells from the mouth replaced the practice of making skin biopsies.

Barr's research in human cytogenetics inspired the establishment of the first service facilities for medical cytogenetics. In the 1960s, health centers throughout Canada and elsewhere instituted similar facilities. He also had a major role in the growth of the University of Western Ontario. His later research, writing and teaching was devoted to neuroanatomy. Barr's neuroanatomy notes were published as a textbook, *The Human Nervous System: An Anatomical Viewpoint*, and is still one of the more popular books on the subject.

See also Heredity

BASES • See Acid and base

BASE PAIRING

DNA is formed of a double helically wound strand. The information found in a **molecule** of DNA is coded for by four different molecules arranged in particular orders. These information coding sections of DNA comprise four different organic bases. Adenine (A), guanine (G), thymine (T), and cytosine (C). In **RNA**, thymine is replaced with uracil (U). C, U, and T are members of a group of compounds known as pyrimidines, and A and G are purines. To hold together the **double helix** of DNA a purine produces hydrogen bonds with a pyrimidine. A always bond with T (or U in RNA) and G always bonds with C. The practical application of this is that if one has a single strand of DNA then it is possible to make the complimentary strand using this constant relationship of the pairing of the **nucleotide** bases. This allows **replication** of the DNA molecule in **cell division**, the production of RNA from DNA and the ability of the DNA molecule to carry conserved information is encoded in this pattern of base pairing. The sizes of the molecules and the arrangements of the **atoms** within them ensures that a purine never joins with a purine (the molecule produced would be too large to fit into the DNA chain) and a pyrimidine never joins with a pyrimidine (the molecule would be too small). Also when A links with T or U then two hydrogen bonds are produced whereas when G links with C three hydrogen bonds are produced, so an attempted mis pairing would produce a distorted molecule as well as being physically the wrong size for the DNA chain, this provides a self check mechanism when new DNA is manufactured.

BATESON, WILLIAM (1861-1926)
British geneticist

William Bateson was born in Whitby, Yorkshire, the son of a classical scholar. In 1883, he earned his bachelor of arts degree in natural science from Saint John's College, University of Cambridge. Although he had minimal training in physics, chemistry, and mathematics, throughout his career Bateson consistently surprised doubters with his outstanding abilities. He was a firm evolutionist and believed that the forms on Earth were descendants of a small number of ancestors. He made significant contributions to the science of genetics.

By 1894, **Charles Darwin**'s concept of continuous change had gained wide acceptance as an evolutionary theory. Darwin asserted that changes in species occur gradually, over a long period of time. Bateson, however, put forward the idea of discontinuous or abrupt change to explain the long process of evolution. According to Bateson, species do not develop gradually, but rather through abrupt jumps every once in a while. This controversial view was unacceptable to traditional biometric scientists, who were convinced that there was no break in evolutionary process.

Bateson was not deterred. Using sweet peas and poultry, he began experiments in hybridization, searching for answers to questions about the laws of **heredity**, such as: how are traits distributed among offspring and are there predictable patterns?

He was already convinced that traits were separate units, not products of blending. When **Gregor Mendel**'s work was rediscovered in 1900, Bateson had the support he needed. With **Reginald Crandall Punnett**, Bateson reinterpreted Mendel's experiments. They proved that Mendelian principles held for animals as well as plants—characteristics of inheritance were carried as individual "packets" of information, or **genes**, and these packets were passed on to offspring. Bateson's first major achievement was bringing Mendel's work to the attention of the world. It was no easy task, for most people had already accepted the "blending theories." Bateson translated Mendel's theories from German and promoted them with energy and enthusiasm. It soon became evident that a new science was emerging. Bateson named it **genetics**.

As the scientific community began accepting Mendelian genetics, Bateson and his partners continued to experiment. They were shocked to discover an apparent contradiction to Mendel's theories: some traits are consistently inherited together in what later became known as gene linkage.

Bateson did not realize that certain traits were linked because they were found on the same **chromosome**. Instead, Bateson tried to explain his observations by proposing his own vibration theory based on the physics of force and motion. This theory quickly faded as more evidence pointed to chromosomes as the only acceptable explanation for gene linkage.

BAYLISS, WILLIAM MADDOCK (1860-1924)
STARLING, ERNEST HENRY (1866-1922)
British physiologists

In 1902, William Bayliss and Ernest Starling collaborated in the discovery of secretin, which they named **hormone**. The men became brothers-in-law as well, when Bayliss married Starling's sister in 1893. Bayliss was also noted for work on Vaso-motor reflexes and treatment of surgical shock in World War I. Starling was also noted for studies of the **heart**, circulation, and kidneys.

Bayliss was born into a wealthy manufacturing family in Wolverhampton, Staffordshire, England. He first intended to become a physician and studied medicine at University College, London, but opted instead to study physiology at Oxford University. In 1888 he joined the faculty at University College.

Starling was born in London into a professional family—his father was a lawyer. He received his medical degree in 1889 from Guy's Hospital, London, where he became a lecturer in physiology. He met Bayliss in 1890 while doing research at University College and joined its faculty in 1899. That same year, Starling showed that food in the intestine triggers a nerve signal that causes some intestinal muscles to contract and others to relax. The action produces the wave pattern called peristalsis that moves food through the intestine. The discovery is sometimes called Starling's Law of the Intestine.

The Bayliss-Starling collaboration on hormones began in 1902 when they studied pancreatic secretion of digestive fluid as food leaves the stomach and enters the intestine. The Russian physiologist, **Ivan Petrovich Pavlov**, believed the secretion was controlled by nervous system signals. Hoping to prove this through experimentation, Starling and Bayliss cut the nerves of an animal's intestine, then injected food from its stomach into its intestine. The experiment disproved Pavlov's theory; the **pancreas** produced digestive fluids normally. Further investigation showed that the signal to the pancreas was chemical. The intestinal wall secreted a substance into the bloodstream that stimulated the pancreas. They named this substance secretin.

In 1905, they used the word hormone (from the Greek, meaning shock or impulse) to refer to secretin. Based on the earlier work of Edward Sharpey-Schäfer, Jokichi Takemine, and Jacob Abel, Bayliss and Starling applied the term to the entire class of chemicals that work in that manner.

BEADLE, GEORGE WELLS (1903-1989)
American geneticist

Early in his professional life, George Wells Beadle worked in the laboratory of **Thomas Hunt Morgan**, the geneticist who helped to revolutionize what we know about genetics—the inheritance of characteristics by the **deoxyribonucleic acid (DNA)** found in the chromosomes of cells. Beadle's innovative research on such diverse living things as corn, fruit flies, and bread mold helped to demystify the activities of genes, making it possible to reduce the inheritance of a particular characteristic to a series of steps needed for the manufacture of biochemicals, notably enzymes. For his work on the "**one gene-one enzyme**" concept, he shared the Nobel Prize for Physiology or Medicine with **Edward Lawrie Tatum** and **Joshua Lederberg** in 1958.

Beadle was born in Wahoo, Nebraska, on October 22, 1903, to Chauncey Elmer and Hattie Albro Beadle. He probably would have worked on the family farm if not for a high school science teacher who advised him to go on to college.

At the College of Agriculture at the University of Nebraska, Beadle gained an interest in genetics, especially that of corn. He received his undergraduate degree in biology in 1926, then left for Cornell University in New York where he earned his doctorate in genetics. During this time Beadle married Marion Cecile Hill. They would have one son, David.

In 1931 Beadle went to work in the genetics laboratory of Thomas Hunt Morgan at the California Institute of Technology (Caltech) in Pasadena, California. Morgan had pioneered genetics work on the **fruit fly**, *Drosophila melanogaster*. As Beadle studied inherited characteristics such as eye color, he began to think that genes might influence **heredity** by chemical means. When he left California for Paris in 1935, he continued this line of work with Boris Ephrussi at the Institut de Biologie Physico-Chimique. Carefully transplanting eye buds from the larvae of one type of mutant fruit fly to larvae of another, Beadle showed that eye color in the insects is not a quirk of nature but the result of a long chain of chemical reactions. For all the relative ease of working with fruit flies, however, Beadle sought a simpler organism and a simpler set of chemical reactions to study.

One Gene, One Enzyme

Several years later, Beadle found what he was looking for. When he returned from Paris in 1936 he briefly taught genetics at Harvard and then went on to Stanford University in California, where he remained from 1937 to 1946. As a professor of biology there, he began working with a red bread mold, *Neurospora crassa*. He would work with neurospora for seventeen years. In 1941 he began collaborating with Edward Tatum, and their work eventually won them—with Joshua Lederberg, who later worked with Tatum at Yale—the Nobel Prize.

Neurospora crassa, once the bane of bakers, became a boon for geneticists Beadle and Tatum. Not only does the mold have a short life cycle and grow on a basic sugar medium, but it reproduces both sexually and asexually. Also, the final cell division that produces its reproductive cells, known as ascospores, leaves them in a linear arrangement along the pod-like ascus (spore case), making the trail of inherited characteristics very clear to follow.

Taking a hint from fellow geneticist **Hermann Joseph Muller**, who in the mid–1920s had shown that the rate of mutation increases with exposure to x rays, Beadle and Tatum grew thousand of cultures of molds in which they had induced mutations. The wild strain of the mold can grow on a medium containing very few nutrients. With just some sugar sprinkled with a little biotin (a growth **vitamin**) and inorganic salts, a wild-type mold can synthesize all the proteins it needs to live. A mold with a mutation, however, loses the ability to make a particular compound it needs to grow, such as a specific **amino acid** (amino acids are the building blocks of proteins such as those used to construct DNA). Beadle expected that a missing amino acid would have to be supplied to the mold, but found to his surprise the mold was sometimes able to convert a similar compound to the necessary amino acid. Through a process of trial and error, Beadle was able to deduce the sequence of chemical steps involved in the work of conversion.

George Wells Beadle. *(The Library of Congress. Reproduced by permission.)*

Once Beadle had pieced together the pathways of chemical production, his ideas could be applied to other molds. One immediate application was to use his techniques to mass-produce the **antibiotic** penicillin. **Penicillin** and other antibiotics are derived from compounds produced naturally by certain molds, which use them as a defense against invading bacterial cells.

Beadle also crossed two different mutant strains of mold and found that the resulting hybrid could produce a particular amino acid that neither parent strain could produce alone. This was because one mutant lacked genetic coding for a certain enzyme (a protein that can encourage or inhibit chemical reactions), causing a breakdown in the chemical synthesis along one spot in the sequence, while the other mutant lacked different coding for an **enzyme** from another spot along the sequence. When crossed, the resulting mold could produce the missing amino acid because it had inherited both genetic patterns, one from each parent. Beadle concluded that specific genes (sequences of protein groups in DNA serving as functional units of inheritance) controlled each step in the sequence. Each gene held the information for the manufacture of a single enzyme, a concept that became known as "one gene, one enzyme."

Extended to other plants and animals, Beadle's theory could be used to explain all of genetic inheritance in terms of chemical reactions. Different genes control the different stages of chemical reactions. For example, cells must be able to produce the pigment that gives an animal's eyes their color. The production of pigment might occur in several steps, with enzymes used to hasten each chemical reaction. If the gene for any one of the enzymes is missing, the cells cannot produce the pigment.

The one gene-one enzyme concept caused a breakthrough in genetic research during the 1940s by shifting the study of genetics away from physical characteristics of organisms to the production of biochemicals. On the heels of this line of research, the compound deoxyribonucleic acid (DNA) was analyzed, and the mechanism of the genetic code was pieced together in the early 1950s. Beadle and Tatum parted ways when Tatum left for Yale University in 1945. Using the same mutation induction techniques on bacteria, Tatum worked along with Joshua Lederberg to show how genetic information can be transferred from one bacterium to another.

From Recognition with a Nobel Prize to the Corn Wars

Beadle became professor and chairman of the division of biology at Caltech in 1946 and stayed on until 1961. For his work in genetics he won the Lasker Award of the American Public Health Association in 1950. He and his first wife divorced, and he then married Muriel Barnett in 1953. With his second wife he wrote several books on genetics for a general audience. Recognition for years of work came in 1958 when Beadle, Tatum and Lederberg won the Nobel Prize. In that same year Beadle won the Albert Einstein Commemorative Award in Science, and in the following year he received the National Award from the American Cancer Society.

In the 1960s Beadle renewed his interest in the genetics of corn. He became a player in the ''corn wars,'' a debate among geneticists and archaeologists over the domestication of corn or maize in the Americas. Beadle contended that modern corn comes from a Mexican wild grass rather than a now-extinct species of maize. Beadle drew his conclusion from the corn remains that show that domestication occurred at the time of the Mayans and Aztecs.

In 1961 Beadle left California for Chicago, Illinois, where he became the sixth chancellor of the University of Chicago. He remained there until he retired in 1968. By then he had accumulated over thirty honorary degrees from many universities around the country and been awarded memberships into several prestigious academic societies. For their work in popularizing genetics, he and his wife Muriel won the Edison Award in 1967. In the late 1960s Beadle became director of the American Medical Association's Institute for Biomedical Research. He died on June 9, 1989, in Pomona, California, at age eighty-five from complications of Alzheimer's disease.

BEAUMONT, WILLIAM (1785-1853)
American surgeon

Beaumont, Connecticut-born and the son of a farmer, worked briefly as a school teacher, then studied medicine at St. Albans, Vermont. He received a license to practice medicine in time to serve as an assistant army surgeon during the War of 1812. Although he left the army in 1815 to start a medical practice in Plattsburgh, New York, he returned in 1820 and remained an Army surgeon, serving at various posts, until 1839.

It was at one of those army posts, Fort Mackinac in northern Michigan, that Beaumont met the patient that was to make both of them famous. The patient was a 19-year-old French Canadian trapper, Alexis St. Martin, who was accidentally shot on June 6, 1822, while visiting the Mackinac branch of the American Fur Company. The bullet wound tore a deep chunk out of the left side of St. Martin 's lower chest; and, although Beaumont was sent for immediately, everyone assumed that the young man would never survive. Miraculously, he did—although his wound needed to be rebandaged daily for a year—and in time St. Martin recovered virtually all his strength. (He lived to be 82, in fact.) However, St. Martin's bullet hole never fully closed. An inch-wide opening (called a fistula) remained through which Beaumont could put his finger all the way into the stomach.

About a year later, St. Martin needed a cathartic of rhubarb and sulphur and Beaumont decided to try administering the medicine through the hole in his patient's stomach. To the surgeon's surprise, the cathartic seemed to work exactly as it would have if it had been administered orally—and Beaumont promptly began planning other experiments as well.

Beaumont started by taking small chunks of food, tying them to a string, and inserting them directly into St. Martin's stomach. At varying intervals, he then pulled the food out—and was therefore able to observe, first hand, the results of **digestion**, hour by hour. Later, by using a hand lens, Beaumont began peering into his patient's stomach, and could actually see how the human stomach behaved at various stages of digestion and under varying circumstances. He was also able to extract and analyze samples of digestive juice.

Over the next few years, Beaumont conducted well over two hundred carefully detailed experiments and, in 1833, published his findings as *Experiments and Observations on the Gastric Juice and the Physiology of Digestion*. The book provided invaluable information on the digestive process and also suggested to other scientists (including **Claude Bernard**) that artificial fistulas might be a practical way to learn more about the body. A year after Beaumont's work was published, St. Martin, probably tiring of the scrutiny Beaumont had subjected him to, refused to cooperate with further studies and returned to Canada.

Although Beaumont resigned from the army in 1840, went into private practice in St. Louis, Missouri, and stayed out of the laboratory, his one classic work earned him lasting fame as one of America's more remarkable pioneer researchers.

Goslings following their mother. In a classic study of instinctive behavior, ethologist Konrad Lorenz showed that baby ducks and geese, which are observed to closely follow their mother on their early forays away from the nest, could also be induced to follow a substitute. The baby birds would form an attachment to whatever individual was present as they opened their eyes and moved about after hatching, regardless of that individual's species identity. *(Photograph by Robert J. Huffman, Field Mark Publications. Reproduced by permission.)*

BEHAVIORISM

Behaviorism is a highly influential academic branch of psychology that dominated the field between the two world wars. Behaviorism concerns itself with the use of strict experimental procedures to study observable behavior in response to environmental stimuli. It excludes ideas, **emotions** and inner mental experience in general. According to behaviorist theory, an individual simply responds to stimuli in their environment. These responses are observable and measurable and can therefore be predicted and controlled.

Behaviorism was developed in the early twentieth century by American psychologist John B. Watson. Watson has written: "Behaviorism claims that 'consciousness' is neither a definable nor a usable concept; that it is merely another word for the 'soul' of more ancient times. The old psychology is thus dominated by a subtle kind of religious philosophy" (*Behaviorism*, 1924). Watson sought to make the study of psychology scientific by using only objective procedures that produced tangible results.

Much of Watson's work was based on the experiments of **Ivan Petrovich Pavlov**, a Russian physiologist, who studied how **animal**s respond to certain stimuli and conditions. In Pav-

lov's famous experiment, he rang a bell as he fed some dogs. Under normal conditions, a dog will salivate whenever food is in its mouth. This is called an unconditioned response to an unconditioned stimulus. Pavlov built on this naturally occurring situation to see if the dogs could be taught to change their normal response. Pavlov devised an experiment so that each time the dogs heard the bell, a small amount of food was placed in the mouth. After several times, the dogs would begin to salivate whenever they heard the bell because they had learned that food would follow. The bell was a conditioned stimulus. Pavlov then removed the food, the unconditioned stimulus, and only rang the bell, the conditioned stimulus. The dogs continued to salivate (i.e. make the conditioned response) as if the food were still being presented. They had learned to respond by salivating to the sound of a bell ringing. This type of controlled response to a **stimulus** has been labeled "classical **conditioning**."

More recently, another psychologist, **B.F. Skinner** began testing Watson's theories in the laboratory. Skinner's studies led him to expand Watson's views of how individuals respond to their environment. Skinner believed that even as people respond to stimuli in their environment, they also operate on or change their environment to obtain certain results.

Although Skinner is not the originator of the theory of "operant conditioning," he has been the leading proponent. Operant conditioning differs from classical conditioning in that a reinforcement occurs only after the subject executes a predesignated behavioral act. No unconditioned stimulus is used. Instead, a spontaneous behavior is rewarded, or reinforced. In order to test this theory, Skinner invented the "Skinner Box" in which a rat or pigeon is put in an environment that requires the pressing of a lever to obtain food. At first the animal may press the lever infrequently and receive the food reinforcement. After a time, the animal begins to press the lever more often and therefore receive more reinforcement. The animal "operates" on its environment in order to receive a reward. In this way, animals can "learn" to behave in a certain way in order to receive a reward, or to avoid punishment. Thus, both Skinner and Watson would deny that the mind or feelings play any part in determining behavior. Instead, only our experience of consequences (rewards or punishments), determine our behavior.

A natural outgrowth of behaviorism is behavior therapy; a type of intervention which focuses on modifying observable behavior as a means to alleviate psychological suffering. Behavior therapy techniques emphasize symptoms of emotional distress. Emotional problems are considered the consequences of faulty acquired behavior patterns or the failure to learn effective responses to one's environment. The aim of behavior therapy, also known as behavior modification, is therefore to change behavior patterns. One of the most prominent behavior techniques is systematic desensitization or counter-conditioning which has been used successfully to treat phobias and fears. Patients are asked to imagine anxiety-producing situations or be presented with actual feared objects. Gradually, exposure to the feared object is increased and the patient learns to control their reaction. Often relaxation training is employed simultaneously in order to reduce anxiety further. The theoretic basis of this type of therapy is that once the appropriate overt expressions of emotions are learned, practiced and reinforced, the correlated subjective feelings will be felt.

Education is another field that has been influenced by the theories and concepts of Behaviorism. For example, programmed **learning** is based on Skinner's theory that learning can best be accomplished in small, incremental steps with immediate reinforcement for the learner. It is a self-paced, self-administered educational technique in which instruction is presented in a logical sequence. This technique can be applied through texts or computer-aided instruction programs. No matter what the medium, the concept of immediately reinforcing the correct response is emphasized. Behaviorism forces us to examine the issue of control in education. In the behaviorist's view, there is no alternative to control. It is simply a matter of who is to control. One does not grant the child "freedom" merely by leaving him alone. To refuse to use scientific control to shape human behavior is, for the behaviorist, a failure in responsibility.

BEHRING, EMIL VON (1854-1917)
German bacteriologist

Behring made major contributions to the understanding of the body's **immune system**, discovered the first successful treatment for tetanus, and came to be known as the "Children's Savior" for his success in conquering **diphtheria**.

Behring was born in Deutch-Eylau, Prussia (now Ilawa, Poland) to a family of 12 children. He studied at the University of Berlin, earning his medical degree in 1880. He served several years as a surgeon in the Prussian Army Medical Corps. It was then that he became interested in infection and how substances in the blood fight disease.

In 1889, Behring went to the University of Berlin to work in the laboratory of **Robert Koch**. Behring made some of his most important discoveries while working there with the Japanese bacteriologist **Shibasaburo Kitasato**. At the time, tetanus or lockjaw, a disease that causes muscle spasms, was widespread. Tetanus is brought about by **toxins** or poisons that are produced by **bacteria**. Behring and Kitasato injected the blood serum of an animal with tetanus into a healthy animal and found that the second animal developed an immunity to the disease. When serum from the immunized animal was injected into another animal, it produced immunity to tetanus as well. The two men studied how the blood produces substances that neutralize toxins. Behring called these substances antitoxins.

Diphtheria, a serious contagious bacterial disease that was a major killer of children, swept through Western Europe in the late 1800s. The death rate from diphtheria averaged thirty-five percent in general and ninety percent in cases involving the larynx. Behring suggested that the same principles of immunity that he and Kitasato discovered for tetanus be applied to diphtheria. In 1894, along with the French bacteriologist Pierre Roux, Behring developed a diphtheria antitoxin. Since then, researchers developed effective harmless forms of diphtheria toxin called toxoids that physicians use to immunize against diphtheria.

Behring became a professor at the University of Halle in 1894, and shortly after, at the University of Marburg. There he established what is known as the Behring Institute and continued one of his other research interests, the fight against tuberculosis. Behring was unable to find a tuberculosis vaccine, which was not discovered until 1924 with the work of French bacteriologists, Albert Calmette and Camille Guerin. Today, tuberculosis is under control due to the use of antibiotics.

Behring's vaccines helped to save the lives of millions of injured soldiers in World War I as well as countless others threatened by tetanus and diphtheria. For his work, Behring received the first Nobel Prize for physiology or medicine in 1901.

BELL, CHARLES (1774-1842)
Scottish anatomist

Charles Bell was a Scottish surgeon and anatomist who pioneered neurophysiological research. Bell's experimental work served as a catalyst to other researchers in neurology and led to several important discoveries. Bell is remembered today for giving his name to Bell's palsy after demonstrating that lesions on the seventh cranial nerve (facial nerve) can cause facial paralysis.

Born in Edinburgh, Charles Bell was the son of a clergyman and the younger brother of John Bell, an eminent surgeon and anatomist who first taught him **anatomy**. Charles Bell attended Edinburgh University and, after qualifying, became a surgeon at Edinburgh Royal Infirmary in 1799. In 1804, he moved to London where he lectured and became the owner of an anatomy school. Bell rose to prominence as a surgeon and in 1812 was appointed surgeon at Middlesex Hospital.

In 1821, Bell demonstrated that the seventh cranial nerve was a separate nerve. In 1824, he became Professor of Anatomy and Surgery at the Royal College of Surgeons. He helped found the Middlesex Hospital Medical School in 1828 and in that same year became the Principal of the Medical School at University College, London. In 1831 he was knighted and in 1836 he was appointed Professor of Surgery at the University of Edinburgh. He died on April 28, 1842 at Hallow, Worcestershire.

Bell executed meticulous dissection which played an important part in his discovery that facial paralysis or Bell's palsy occurs when one of the two facial muscles (right or left) become injured or inflamed causing paralysis to half of the face. This paralysis is usually temporary and is characterized by drooping of the eyelid and corner of the mouth. Bell also discovered the long thoracic nerve (Bell's nerve) which controls a muscle in the chest wall. Perhaps Bell's most important contribution to neurophysiology was his demonstration that nerves consist of separate fibers that are bound together; the fibers serve either sensory or motor functions but not both and transmit in one direction only. His ideas were set forth in his essays "Idea of a New Anatomy of the Brain" (1811) and "The Nervous System of the Human Body" (1830).

Bell sustained a life long interest in anatomy and was also a skilled draftsman. He enthusiastically taught anatomy to artists. He published *Essays on the Anatomy of Expression in Painting* (1806), which described the science of physiognomy. His interest in led him to treat gunshot wounds in the battle of Corunna. He was also instrumental in founding a hospital in Brussels following the battle of Waterloo.

See also Nervous system

BENACERRAF, BARUJ (1920-)
Venezuelan-American geneticist

Baruj Benacerraf was born in Caracas, Venezuela; his father was a wealthy textile merchant. Benacerraf grew up in France, then attended Columbia University in New York where he got his degree in 1942. He married and became a naturalized United States citizen in 1943. Benacerraf obtained a medical degree from the Virginia College of Medicine in 1945, and then served a tour of duty with the United States Army Medical Corps stationed in Nancy, France.

Emil von Behring. *(The Library of Congress. Reproduced by permission.)*

Benacerraf's experience with asthma as a child sparked his interest in the body's immune reactions. He researched immunological hypersensitivity with American chemist, Elvin Kabat at the College of Physician's and Surgeons at Columbia University in 1948. He went on to conduct research on immunity at the National Center for Scientific Research in Paris. Benacerraf returned to the United States in 1956, joining the faculty at the New York University School of Medicine. There, he began his Nobel Prize-winning research on **cells** involved in immune reactions. He continued these investigations at the National Institute of Allergy and Infectious Disease in Bethesda, Maryland from 1968 to 1970 and subsequently as chairman of the department of pathology at the Harvard University Medical School.

While attempting to produce uniform antibodies in test animals, Benacerraf noticed that some guinea pigs responded to antigens by producing antibodies, while others did not. He demonstrated that the animals' responses were determined by genes he called immune-response (IR) genes. Other researchers found similar genes in mice, rats, and monkeys. In 1969, Benacerraf confirmed that the IR genes were located within the MHC, (major histocompatibility complex). In the 1940s, American geneticist **George Snell** along with British researcher Peter Gorer had discovered a group of genes that later became known as the MHC. The NHC is the main system by which a mammal distinguishes between self and non-self and determines whether or not the body launches an **immune system response**. The identification of self vs. non-self by the immune system depends on the characteristics of surface **molecules** on cells. **Jean Baptiste Dausset**, a French immunologist and hematologist, discovered that humans carry the MHC as well as other animals, and in humans it is called the HLA (human leukocyte antigen) system. Benacerraf shared the 1980 Nobel Prize in physiology or medicine with Snell and Dausset for their work on immunological reactions. In 1980, Benacerraf became president and chief executive officer of the Dana Farber Cancer Center in Boston, Massachusetts.

BENDS • See Decompression sickness

BENEDEN, EDOUARD VAN (1846-1910)
Belgian cytologist

The son of zoologist Pierre-Joseph van Beneden (1809-1894), van Beneden was able to study many animal specimens at his father's small laboratory on the Belgian seaside. As a student, he began exploring **cell theory**; as a professor of zoology at the University of Liège, he studied how the eggs of an intestinal worm (called *Ascaris*) matured and were fertilized. In 1887, Beneden made two important contributions to the field of **cytology**. He discovered the amount of **chromosomes** in any body **cell** is constant and that the number of chromosomes characterizes a certain species. For example, humans always have 46 chromosomes.

Beneden's second important contribution was the discovery and naming of **meiosis**. He observed that when sex cells are formed prior to fertilization, the egg and sperm cell have only half the usual count of chromosomes. Chromosomes did not double as they did in regular cell division. Instead, the number of chromosomes in both the sperm and the egg cell halved. Therefore, fertilization created an embryo with one complete set of chromosomes from the sperm and the egg.

Beneden's findings were consistent with **Gregor Johann Mendel**'s heredity theories—every genetic factor exists in duplicate; one coming from the male and one from the female. **Theodor Boveri**, Hugo de Vries and others would make these concepts clearer in the coming years.

BERG, PAUL (1926-)
American biochemist

Paul Berg made one of the most fundamental technical contributions to the field of **genetics** in the twentieth century: he developed a technique for splicing together **deoxyribonucleic acid** (DNA)—the substance that carries the genetic information in living **cells** and **viruses** from generation to generation—from different types of organisms. His achievement gave scientists a priceless tool for studying the structure of viral chromosomes and the biochemical basis of human genetic diseases. It also let researchers turn simple organisms into chemical factories that churn out valuable medical drugs. In

1980 he was awarded the Nobel Prize in chemistry for pioneering this procedure, now referred to as recombinant DNA technology.

Today, the commercial application of Berg's work underlies a large and growing industry dedicated to manufacturing drugs and other chemicals. Moreover, the ability to recombine pieces of DNA and transfer them into cells is the basis of an important new medical approach to treating diseases by a technique called **gene therapy**.

Berg was born in Brooklyn, New York, on June 30, 1926, one of three sons of Harry Berg, a clothing manufacturer, and Sarah Brodsky, a homemaker. He attended public schools, including Abraham Lincoln High School, from which he graduated in 1943. In a 1980 interview reported in the *New York Times,* Berg credited a ''Mrs. Wolf,'' the woman who ran a science club after school, with inspiring him to become a researcher. He graduated from high school with a keen interest in microbiology and entered Pennsylvania State University, where he received a degree in biochemistry in 1948.

Before entering graduate school, Berg served in the United States Navy from 1943 to 1946. On September 13, 1947, he married Mildred Levy and they had one son, John Alexander. After completing his duty in the navy, Berg continued his study of biochemistry at Western Reserve University (now Case Western Reserve University) in Cleveland, Ohio, where he was a National Institutes of Health fellow from 1950 to 1952 and received his doctorate degree in 1952. He did postdoctoral training as an American Cancer Society research fellow, working with Herman Kalckar at the Institute of Cytophysiology in Copenhagen, Denmark, from 1952 to 1953. From 1953 to 1954 he worked with biochemist Arthur Kornberg at Washington University in St. Louis, Missouri, and held the position of scholar in cancer research from 1954 to 1957.

He became an assistant professor of microbiology at the University of Washington School of Medicine in 1956, where he taught and did research until 1959. Berg left St. Louis that year to accept the position of professor of biochemistry at Stanford University School of Medicine. Berg's background in biochemistry and microbiology shaped his research interests during graduate school and beyond, steering him first into studies of the molecular mechanisms underlying intracellular protein synthesis.

Experiments with Genetic Engineering

During the 1950s Berg tackled the problem of how **amino acids**, the building blocks of proteins, are linked together according to the template carried by a form of **RNA** (ribonucleic acid, the ''decoded'' form of DNA) called messenger RNA (mRNA). A current theory, unknown to Berg at the time, held that the amino acids did not directly interact with RNA but were linked together in a chain by special molecules called joiners, or adapters. In 1956 Berg demonstrated just such a molecule, which was specific to the amino acid methionine. Each amino acid has its own such joiners, which are now called transfer RNA (tRNA).

This discovery helped to stoke Berg's interest in the structure and function of genes, and fueled his ambition to combine genetic material from different species in order to study how these individual units of heredity worked. Berg reasoned that by recombining a gene from one species with the genes of another, he would be able to isolate and study the transferred gene in the absence of confounding interactions with its natural, neighboring genes in the original organism.

In the late 1960s, while at Stanford, he began studying genes of the monkey tumor virus SV40 as a model for understanding how mammalian genes work. By the 1970s, he had mapped out where on the DNA the various viral genes occurred, identified the specific sequences of nucleotides in the genes, and discovered how the SV40 genes affect the DNA of host organisms they infect. It was this work with SV40 genes that led directly to the development of recombinant DNA technology. While studying how genes controlled the production of specific proteins, Berg also was trying to understand how normal cells seemed spontaneously to become cancerous. He hypothesized that cells turned cancerous because of some unknown interaction between genes and cellular biochemistry.

In order to study these issues, he decided to combine the DNA of SV40, which was known to cause cancer in some animals, into the common intestinal bacterium *Escherichia coli.* He thought it might be possible to smuggle the SV40 DNA into the bacterium by inserting it into the DNA of a type of virus, called a bacteriophage, that naturally infects E. coli.

A DNA molecule is composed of subunits called **nucleotides**, each containing a sugar, a **phosphate group**, and one of four nitrogenous bases. Structurally, DNA resembles a twisted ladder, or helix. Two long chains of alternating sugar and phosphate groups twist about each other, forming the sides of the ladder. A base attaches to each sugar, and hydrogen bonding between the bases—the rungs of the ladder—connects the two strands. The order or sequence of the bases determines the **genetic code**; and because bases match up in a complementary way, the sequence on one strand determines the sequence on the other.

Berg began his experiment by cutting the SV40 DNA into pieces using so-called restriction enzymes, which had been discovered several years before by other researchers. These enzymes let him choose the exact sites to cut each strand of the double helix. Then, using another type of enzyme called terminal transferase, he added one base at a time to one side of the double-stranded molecule. Thus, he formed a chain that extended out from the double-stranded portion. Berg performed the same biochemical operation on the phage DNA, except he changed the sequence of bases in the reconstructed phage DNA so it would be complementary to—and therefore readily bind to—the reconstructed SV40 section of DNA extending from the double-stranded portion. Such complementary extended portions of DNA that bind to each other to make recombinant DNA molecules are called "sticky ends."

This new and powerful technique offered the means to put genes into rapidly multiplying cells, such as **bacteria**, which would then use the genes to make the corresponding protein. In effect, scientists would be able to make enormous amounts of particular genes they wanted to study, or use simple organisms like bacteria to grow large amounts of valuable

substances like human growth hormone, antibiotics, and insulin. Researchers also recognized that genetic engineering, as the technique was quickly dubbed, could be used to alter soil bacteria to give them the ability to "fix" nitrogen from the air, thus reducing the need for artificial fertilizers.

Questions the Ethics of Recombinant DNA Technology

Berg had planned to inject the monkey virus SV40-bacteriophage DNA hybrid molecule into E. coli. But he realized the potential danger of inserting a mammalian tumor gene into a bacterium that exists universally in the environment. Should the bacterium acquire and spread to other E. coli dangerous, pathogenic characteristics that threatened humans or other species, the results might be catastrophic. In his own case, he feared that adding the tumor-causing SV40 DNA into such a common bacterium would be equivalent to planting a ticking cancer time bomb in humans who might subsequently become infected by altered bacteria that escaped from the lab. Rather than continue his ground-breaking experiment, Berg voluntarily halted his work at this point, concerned that the tools of genetic engineering might be leading researchers to perform extremely dangerous experiments.

In addition to this unusual voluntary deferral of his own research, Berg led a group of ten of his colleagues from around the country in composing and signing a letter explaining their collective concerns. Published in the July 26, 1974, issue of the journal *Science,* the letter became known as the "Berg letter." It listed a series of recommendations supported by the Committee on Recombinant DNA Molecules Assembly of Life Sciences (of which Berg was chairman) of the National Academy of Sciences.

The Berg letter warned, "There is serious concern that some of these artificial recombinant DNA molecules could prove biologically hazardous." It cited as an example the fact that E. coli can exchange genetic material with other types of bacteria, some of which cause disease in humans. "Thus, new DNA elements introduced into E. coli might possibly become widely disseminated among human, bacterial, plant, or animal populations with unpredictable effects." The letter also noted certain recombinant DNA experiments that should not be conducted, such as recombining genes for antibiotic resistance or bacterial toxins into bacterial strains that did not at present carry them; linking all or segments of DNA from cancer-causing or other animal viruses into plasmids or other viral DNAs that could spread the DNA to other bacteria, animals or humans, "and thus possibly increase the incidence of cancer or other disease."

The letter also called for an international meeting of scientists from around the world "to further discuss appropriate ways to deal with the potential biohazards of recombinant DNA molecules." That meeting was held in Pacific Grove, California, on February 27, 1975, at Asilomar and brought together a hundred scientists from sixteen countries. For four days, Berg and his fellow scientists struggled to find a way to safely balance the potential hazards and inestimable benefits of the emerging field of genetic engineering. They agreed to collaborate on developing safeguards to prevent genetically engineered organisms designed only for laboratory study from being able to survive in humans. And they drew up professional standards to govern research in the new technology, which, though backed only by the force of moral persuasion, represented the convictions of many of the leading scientists in the field. These standards served as a blueprint for subsequent federal regulations, which were first published by the National Institutes of Health in June 1976. Today, many of the original regulations have been relaxed or eliminated, except in the cases of recombinant organisms that include extensive DNA regions from very pathogenic organisms. Berg continues to study genetic recombinants in mammalian cells and gene therapy. He is also doing research in molecular biology of HIV–1.

Nobel Prize Awarded for the Biochemistry of Nucleic Acids

The Nobel Award announcement by the Royal Swedish Academy of Sciences cited Berg ''for his fundamental studies of the biochemistry of nucleic acids with particular regard to recombinant DNA.'' But Berg's legacy also includes his principled actions in the name of responsible scientific inquiry.

Berg was named the Sam, Lula and Jack Willson Professor of Biochemistry at Stanford in 1970, and was chairman of the Department of Biochemistry there from 1969 to 1974. He was also director of the Beckman Center for Molecular and Genetic Medicine (1985), senior postdoctoral fellow of the National Science Foundation (1961–68), and nonresident fellow of the Salk Institute (1973–83). He was elected to the advisory board of the Jane Coffin Childs Foundation of Medical Research, serving from 1970–80. Other appointments include the chair of the scientific advisory committee of the Whitehead Institute (1984–90) and of the national advisory committee of the Human Genome Project (1990). He was editor of *Biochemistry and Biophysical Research Communications* (1959–68), and a trustee of Rockefeller University (1990–92). He is a member of the international advisory board, Basel Institute of Immunology.

Berg received many awards in addition to the Nobel Prize, among them the American Chemical Society's Eli Lilly Prize in biochemistry (1959); the V. D. Mattia Award of the Roche Institute of Molecular Biology (1972); the Albert Lasker Basic Medical Research Award (1980); and the National Medal of Science (1983). He is a fellow of the American Academy of Arts and Sciences, and a foreign member of the Japanese Biochemistry Society and the Académie des Sciences, France.

BERGSTRÖM, SUNE KARL (1916-)
Swedish biochemist

Sune Karl Bergström is best known for his research on **prostaglandins**. These substances, which were first discovered in the prostate gland and seminal vesicles, were found by Bergström and his colleagues to affect circulation, smooth muscle tissue, and general **metabolism** in ways that can be medically beneficial. Certain prostaglandins, for example, lower blood pressure, while others prevent the formation of ulcers on the stomach lining. For his research, Bergström shared the 1982 Nobel Prize in medicine or physiology with **John R. Vane** and **Bengt Samuelsson**.

Sune Bergström was born in Stockholm on January 10, 1916, to Sverker and Wera (Wistrand) Bergström. Upon completion of high school he went to work at the Karolinska Institute as an assistant to the biochemist Erik Jorpes. The young Bergström was assigned to do research on the biochemistry of fats and steroids. Jorpes was impressed enough with his assistant to sponsor a year-long research fellowship for Bergström in 1938 at the University of London. While there, Bergström focused his research on bile acid, a steroid produced by the liver which aids in the digestion of cholesterol and similar substances.

The following year Bergström received a British Council fellowship to continue his research in Edinburgh, but the fellowship was canceled after World War II broke out. In 1940 he received a Swedish-American Fellowship which allowed him to study for two years at Columbia University and to conduct research at the Squibb Institute for Medical Research in New Jersey. At Squibb, Bergström researched the steroid **cholesterol**, particularly its reaction to chemical combination with oxygen at room temperature, a process called auto-oxidation.

Bergström returned to Sweden in 1942, receiving doctorates in medicine and biochemistry from the Karolinska Institute two years later. He was appointed assistant in the biochemistry department of Karolinska's Medical Nobel Institute. While there, he continued experiments with auto-oxidation, working with linoleic acid, which is found in some vegetable oils. He discovered a particular **enzyme** was responsible for the oxidation of linoleic acid, and helped attempt to purify the enzyme while working with biochemist Hugo Theorell.

While attending a meeting of Karolinska's Physiological Society in 1945, Bergström met the physiologist Ulf von Euler, who discovered the hormone norepinephrine, and was doing research on prostaglandins. Scientists had observed in the 1930s that seminal fluid used in artificial insemination stimulated contraction and subsequent relaxation in the smooth muscles of the uterus. Von Euler isolated a substance from the seminal fluid of sheep and found it had the same effect in relaxing the smooth muscle of blood vessels. Impressed with Bergström's work on enzyme purification, von Euler gave him some of the extract for further purification.

Bergström began initial experiments but put his work on hold when in 1946 he was named a research fellow at the University of Basel. Returning from Switzerland in 1947, he was appointed professor of physiological chemistry at the University of Lund. His first task was to help revitalize the university's research facilities, which had fallen into disuse during the war. Afterwards, he resumed his research on prostaglandins, assisted by graduate students such as Bengt Samuelsson. Working with new large supplies of sheep seminal fluid, Bergström and his colleagues were able to isolate and purify two prostaglandins by 1957. Bergström was appointed professor of chemistry at Karolinska a year later, and brought his research on prostaglandins and his collaboration with Samuelsson with him. By 1962, six prostaglandins, identified as A through F, had been identified.

Bergström and Samuelsson then worked on determining how prostaglandins are formed. They discovered that prostaglandins are formed from common fatty acids, and further identified specific functions performed by each prostaglandin. Over the next few years, Bergström and Samuelsson surmised that certain prostaglandins could be used to treat high blood pressure, blocked arteries, and other circulatory problems by relaxing muscle tissue. These prostaglandins were also shown to prevent ulceration of the stomach lining and to protect against side effects of such drugs as aspirin, long known to irritate the stomach lining. Other prostaglandins could be used to raise blood pressure or stimulate uterine muscle by their contracting effect.

Bergström remained at Karolinska, serving as dean of its medical school from 1963 to 1966 and as rector of the institute from 1969 to 1977. He was chairman of the Nobel Foundation's Board of Directors from 1975 to 1987, and from 1977 to 1982 he served as chairman of the World Health Organization's Advisory Committee on Medical Research. He retired from teaching in 1981, choosing to devote his full time to research at Karolinska.

Bergström's memberships include the Royal Swedish Academy of Science (he served as its president from 1983 to 1985), the American Philosophical Society, and the American Academy of Arts and Sciences. Other awards given to Bergström besides the Nobel include the Albert Lasker Award in 1977, Oslo University's Anders Jahre Prize in Medicine in 1970, and Columbia's Louisa Gross Horwitz Prize in 1975.

BERIBERI

Beriberi, a disease caused by a thiamine deficiency, is most common in Far Eastern countries where boiled white rice makes up a large part of the daily diet. In other parts of the world, including the United States, the disease is seen today mostly in alcoholics, who often fail to nourish themselves properly. The name *beriberi* is Sinhalese for "I cannot"—an apt description of the patient in later stages of the disease who finds it difficult to perform even simple tasks.

Beriberi occurs in two forms. In the more commonly seen chronic form, the disease is characterized by *polyneuritis*, a generalized **inflammation** of nerves in the arms and legs. The polyneuritis may soon escalate to severe nerve damage, progressive paralysis of the legs, and a deterioration of muscles. In the more acute form of the disease, the beriberi causes fluid retention, which in turn causes swelling of **tissue**s, including the those around the **heart**. In time, potentially fatal heart problems tend to develop.

Before its cause was discovered, beriberi was almost impossible to treat and caused widespread suffering. In 1896 **Christiaan Eijkman**, a Dutch physician, found that he could give laboratory **animal**s beriberi by feeding them a diet restrict-

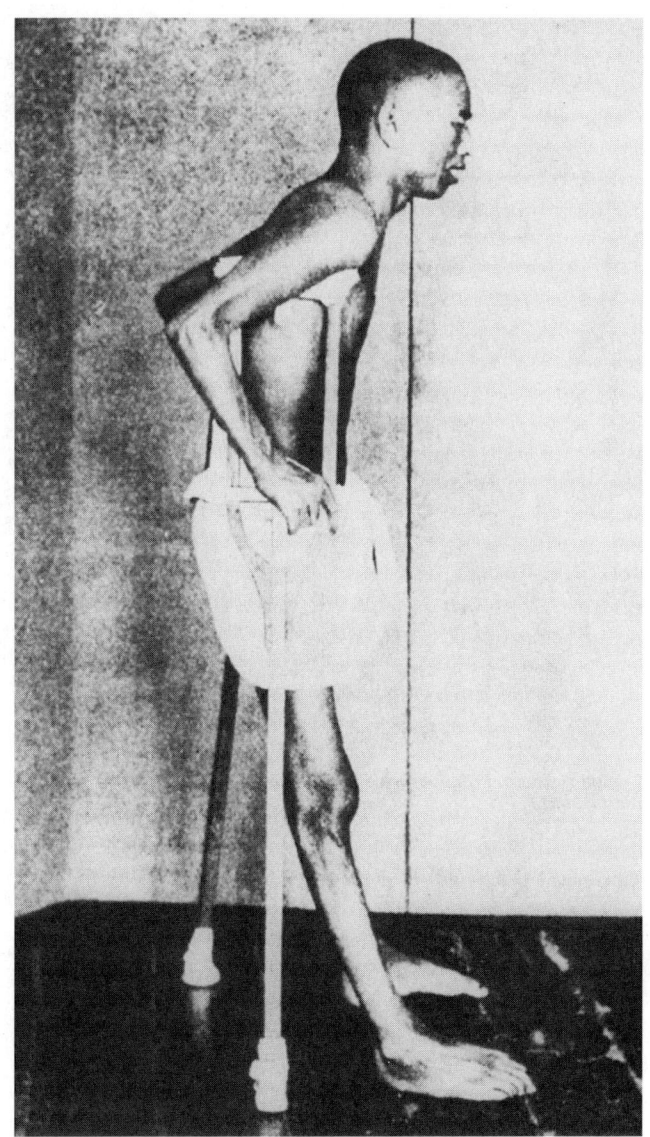

Side view of a beriberi victim. *(National Library of Medicine. Reproduced by permission.)*

ed to polished rice and that he could then cure them simply by switching their diet to unpolished rice. Although Eijkman thought a toxin in the rice might be the culprit, his colleague, Gerrit Grijns, correctly deduced that polished rice was lacking an essential substance somehow needed by the **nervous system**, and that this substance was present in the outer layers of rice and in other foods as well. Reports of Eijkman's and Grijns's work prompted a number of investigators to join the search for the elusive anti-beriberi factor, which in 1934 was finally isolated and identified as thiamine, the first member of the B family of **vitamins**.

Claude Bernard. *(The Library of Congress. Reproduced by permission.)*

BERNARD, CLAUDE (1813-1878)

French physiologist

Claude Bernard did not get off to a promising start. Born in 1813 in Saint-Julien, France, this son of poor vineyard workers attended a simple village school and originally dreamed of becoming a writer. At 21, he'd already written several plays and set off to Paris, France, to show them around. A well-known literary critic, however, strongly suggested he try a different career and, after some thought, Bernard took the man's advice. He entered the Faculty of Medicine, finished almost at the bottom of his class, but managed to obtain a medical degree in 1843. Four years later, the young man's fortune changed. He became an assistant to François Magendie, one of France's most prominent—and controversial—physiologists.

Unlike most of his contemporaries, Magendie firmly believed that researchers could study the body's reactions in the same way that they studied the reactions of inorganic material. Bernard agreed, learned a great deal about experimental physiology from his mentor, and then went on to design a number of experimental projects of his own. In 1855, when Magendie died, Bernard took his place as Professor of Experimental Medicine at the College de France. And, because Bernard tended to be more disciplined and organized than Magendie had been, he began attracting more and more scientific attention to the comparatively new discipline.

Many of Bernard's experiments centered around the digestive process. Inspired by William Beaumont—who had spent several years peering into the stomach of a patient with an accidentally caused opening (or fistula) in his side—Bernard decided to create artificial fistulas in live animals. Although his experiments infuriated antivivisectionists (including his own wife and daughters), Bernard made a number of discoveries. Among other things, he found that the stomach was not the sole digestive organ, as was then widely believed. While the stomach began the digestive process, much more **digestion** took place throughout the small intestine, Bernard reported. He also demonstrated the importance of the **pancreas**, whose secretions were clearly necessary to break down fat **molecules**, and later went on to identify the various nerves that control gastric secretion.

In 1857, Bernard isolated a starch-like substance in the liver of animals, a substance he named glycogen. Glycogen, Bernard showed, was a large molecule built up out of numerous tiny molecules of sugar taken from the blood stream. Its primary role was to serve as the body's reserve supply of **carbohydrates**. When the level of sugar in the blood became low, the stored glycogen broke down again into its components and released more simple sugars back into the blood stream. This continuing process, Bernard pointed out, indicated that the animal's body did not (as was then believed) merely break large molecules down into smaller ones, the way plants did. The animal's body could also take simple molecules and build them up into larger, more complex ones.

In 1851, Bernard devoted some of his attention to the portion of the nervous system which governs **blood circulation** (called the vasomotor system) and discovered that certain specific nerves governed the dilation and constriction of blood vessels. But why did the blood vessels need to keep widening and narrowing? Bernard theorized that, by doing so, the body was better able to control its distribution of heat. On hot days, he suggested, people looked flushed because the skin's blood vessels widened in order to release more excess heat from the body. On cold days, people looked pale because the skin's blood vessels narrowed in order to prevent body heat from escaping. While studying the vasomotor system, Bernard also discovered that the blood's red corpuscles carry oxygen from the lungs to body **tissues**.

Each of Bernard's findings convinced him that the body is constantly striving to maintain a stable, well-balanced internal environment, one that is not overly affected by outside influences. He therefore concluded that the body must be under the control of one strong and central regulating force. Bernard's theory, although widely accepted today, appeared quite radical in his own time, when most scientists believed that the body's various organs acted quite independently of each other.

Bernard's work brought him worldwide recognition. In 1865, he published the highly influential textbook, *An Introduction to the Study of Experimental Medicine*, which won him election to the prestigious French Academy in 1869. He even served in the French senate under Napoleon III (1808-1873) and, when he died in 1878, Bernard became the first scientist to be given a national funeral, an honor usually reserved for political and military leaders.

Best, Charles Herbert (1899-1978)

Canadian physiologist

Charles Herbert Best was most renowned as co-discoverer of **insulin** with **Frederick G. Banting**. Insulin, which is a **hormone** secreted by the **pancreas**, regulates the level of sugar in the **blood**. Its discovery in 1921 led to its use as a treatment for diabetes, which until that time had led swiftly to emaciation, coma, and death. Later in his career, Best assisted in the establishment of associations of diabetics to promote support groups and educational programs for their members. He also did important research on the nutrient choline and the blood anticoagulant heparin.

Best was born on February 27, 1899, in West Pembroke, Maine, a town near the border of the Canadian province of New Brunswick. His parents were Canadian citizens, both originally from Nova Scotia. Best was a direct descendant of Major William Best, who in 1749 was one of the founders of Halifax, Nova Scotia. Best's father, Herbert Huestes Best, was a country doctor whose practice straddled the U.S.-Canadian border. As a teenager, Best often accompanied his father on his rounds in a horse-drawn buggy. Best's mother was Luella Fisher Best.

After finishing high school, Best entered the University of Toronto in a liberal arts program. When World War I interrupted his education, he served as a sergeant in a regiment of the Canadian Tank Corps. He returned to Toronto in 1919 after the war to complete his education, but switched his course of study to physiology and biochemistry in preparation for a medical degree. Best played professional baseball in order to finance his education. He received his B.A. in 1921. In May of 1921, Best's physiology professor, John James Rickard Macleod, introduced him to Frederick Grant Banting, a 29-year-old orthopedic surgeon from London, Ontario. Best had worked as a research assistant for Macleod and planned to begin studying for master's degree under him in the fall. Banting would be using Macleod's lab during the intervening summer to do experiments to find out the function of the pancreas in preventing diabetes, and he needed an assistant to help with analyses of blood chemistry. Another of Macleod's students was also interested in the job, so he and Best flipped a coin. Best won. On May 17, 1921, the day after he completed his examinations for his undergraduate degree, Best began working with Banting. It was a collaboration that would set the course of his career.

Discovers Treatment for Diabetes

Experiments done 30 years earlier had shown that when a dog's pancreas was removed by surgery the animal developed the symptoms of **diabetes mellitus**: it would grow insatiably thirsty, begin excreting large amounts of sugar in its urine, and then become listless, go into a coma, and die. Banting's idea was that the pancreas must secrete something in addition to its digestive enzymes in order to prevent this process. He was convinced that the crucial substance would be found in groups of cells on the pancreas called the islets of Langerhans. These cells could be isolated by tying off a dog's pancreatic

ducts; the rest of the pancreas would atrophy after several weeks, but the islets of Langerhans would remain intact. An extract could then be made from the cells and injected into a diabetic dog. If Banting's idea was right, such an extract would relieve the symptoms of diabetes.

The way he originally planned the work, Banting would do the surgery, removing the pancreas from some dogs to make them diabetic and tying off the pancreatic ducts in others to isolate the islet cells. Best would do blood and urine tests on the dogs. As the research progressed, however, Best learned to do some surgery too. Best, for his part, had a personal interest in diabetes. His father's sister, who had lived with the Best family in West Pembroke, had died in a diabetic coma in 1918.

Banting and Best had expected to spend only eight weeks on their study. But it was July 30 before they were ready to prepare the extract. On that day, Banting removed the shriveled pancreas from a dog whose ducts had been tied. He and Best prepared an extract from it by chopping the pancreas into small pieces, grinding it in a chilled mortar with salt water, and filtering the mixture through cheesecloth. A blood sample from the diabetic dog showed its blood sugar level to be 0.2. Banting and Best injected some of their extract into the dog. An hour later its blood sugar level had dropped to 0.12. After another injection it registered 0.11. This dog died the next day, presumably from an infection. But Banting and Best were encouraged by the result and tested their extract on more diabetic dogs. They called the extract "isletin."

During the following months Banting and Best performed additional experiments to confirm and explain their results. With an injection of their extract they could revive a diabetic dog from its coma and prevent its imminent death. They found ways of obtaining the extract more easily and in larger quantities from the pancreases of fetal calves obtained from a local slaughterhouse. Macleod, who had been vacationing at his home in Scotland during the summer, returned in September and made suggestions for further studies. He also hired James Bertram Collip, a Ph.D. biochemist, to help purify the active component of the extract. Best continued with the work, but also began his M.A. program at the University of Toronto. That fall Banting and Best wrote their first paper describing the experiments with dogs, titled "The Internal Secretion of the Pancreas." It was accepted for publication in the February 1922 issue of the *Journal of Laboratory and Clinical Medicine.*

By the time the paper was published, however, Banting and Best had already treated a human diabetes patient with the extract. They had also begun to call their extract by the now familiar name of insulin, at the suggestion of Macleod. The word "insulin" is based on the Latin word for island. The first patient to receive insulin was 14-year-old Leonard Thompson, who was so weak after two years of suffering from diabetes that he had been admitted to Toronto General Hospital. Thompson's weight was down to 65 pounds, and his doctors expected him to live for only a few more weeks. Before administering insulin to the boy, Banting and Best performed a perfunctory clinical trial: they injected each other with their extract. Since there seemed to be no side effects other than soreness around the injection, in January 1922 they went ahead and treated the boy. After an initial problem with impurities in the insulin was solved, his condition began to improve. He regained his energy and put on weight. Thompson lived another 11 years, dying in 1935 from pneumonia contracted after a motorcycle accident. This success, a literal pulling back of a diabetic child from the brink of the grave, was repeated again and again in the next months, as insulin became a standard treatment for diabetes.

Nobel Committee Leaves Best Out

The 1923 Nobel Prize in physiology and medicine for that year was awarded to Banting and Macleod for the discovery of insulin. Banting was furious. In his opinion, Macleod had done little more than provide laboratory space, whereas Best had shared the work of research. Best was in Boston the day the news arrived, giving an address to medical students at Harvard. Banting immediately sent Best a telegram stating that he would share both the credit for the discovery and the Nobel Prize cash award with Best. Macleod, who considered the work a collaboration, divided his portion of the prize with Collip.

Best continued his studies, receiving his M.A. in 1922 and his M.D. in 1925, while also working on a commercial process for producing insulin. At the same time he received the M.D., Best was also awarded the Ellen Mickle Fellowhip for highest standing in the medical course. During the years Best was doing insulin research he had been courting Margaret Mahon, writing her love letters that also included details about the experiments on dogs. She was so well versed in the work that she helped Banting and Best write their first paper about it. Best married Margaret Mahon in 1924, and later they had two sons. In 1926 the couple sailed to England, where Best spent two years doing postgraduate research in the laboratory of Sir Henry Dale in London. This research led Best to the discovery of histaminase, an anti-allergic enzyme. He received his doctorate from the University of London in 1928.

Before the degree was awarded, however, Best had returned to the University of Toronto in 1927 to head the department of physiological hygiene, a post he held until 1941. In 1929, when Macleod retired, Best was also made chair of the department of physiology. He was just 30 years old at the time. He remained in that position until 1965.

Best's study of insulin led him to a related avenue of research. He had noticed that the laboratory dogs whose pancreases had been removed to render them diabetic developed fatty livers, similar to cirrhosis of the liver in alcoholics. Best and his colleagues found that feeding such dogs lecithin prevented this change in the liver. In the 1930s they isolated choline as the active nutritional component of lecithin, a component found in the **cells** of many plants and animals, and did studies on the role of choline in metabolism. In the 1930s Best also became interested in heparin, which had just been discovered. He recognized that heparin could be an important **anticoagulant** drug for preventing blood clotting and went to work purifying it for human use. With the outbreak of World War II, Best continued his research interest in blood. He established

the Canadian project for supplying dried blood serum to the wounded overseas and personally worked collecting blood from volunteers. This project was a predecessor to the blood transfusion service of the Red Cross. In 1941 Best was appointed director of the medical research unit of the Canadian Navy. In this capacity he coordinated studies to find ways to enhance night vision and to remedy motion sickness.

A Friend to Diabetics

In 1941, Frederick Banting was killed in a airplane crash en route to a wartime mission. After Banting's death, Best took over his directorship of the Banting and Best department of medical research at the University of Toronto. Best also worked to organize associations of diabetics that provided support groups and educational programs for their members, including summer camps for diabetic children. He was president of the American Diabetes Association from 1948 to 1949 and remained honorary president thereafter. He was also honorary president of the International Diabetes Foundation. In 1953 the University of Toronto named a new building for medical research the Best Institute. The same year, Best became the first president of the International Union of Physiological Sciences.

Best retired from the University of Toronto in 1965. In 1966 friends of Best purchased Best's parents' clapboard house in West Pembroke, Maine and gave it to the American Diabetes Association. Later the home was proposed to the U.S. National Trust for Historic Preservation as a cultural landmark and turned into a museum. Best spent his retirement years traveling around the world with his wife, who was a historian and a botanist, visiting friends and colleagues.

Best received scores of medals, awards, and honorary degrees and was praised by the Pope, the Queen of England, and other heads of state. He wrote numerous scientific articles, and was co-author of a widely used physiology textbook. In March of 1978 one of Best's sons died of a heart attack. Hours after hearing the news, Best himself collapsed from a ruptured blood vessel in his abdomen. He died several days later, on March 31, 1978, at Toronto General Hospital.

BIOACCUMULATION

Bioaccumulation is the build up of toxic chemical **pollutants** in the **tissue**s of **organisms**. Toxic contaminants often cannot be easily metabolized or excreted, and therefore are bioaccumulated at high levels within an organism. Bioaccumulation of toxic chemicals can result in **biomagnification**, an increase in the contaminant at higher levels of the food web.

Chemical pollutants that are bioaccumulated come from many sources. **Pesticides** are an example of a contaminant that bioaccumulates in organisms. Rain can wash freshly sprayed pesticides into creeks, where it will eventually make its way to rivers, estuaries and the ocean. Other major sources of toxic contaminants include atmospheric deposition with rainfall (initially from industrial smokestacks and automobile emissions), and both legal and illegal discharges to **water**. For example, industrial discharges can be major sources of toxic contaminants. Sludge from sewage treatment **plants** has been a source in the past.

Once a toxic pollutant is in the water or soil, it can easily enter the food web. For example, in the water the pollutants absorb or stick to small particles, including **phytoplankton**, which are microscopic **algae**. Since there is so little pollutant stuck to each phytoplankton, the pollutant does not do very much damage at this level of the food web. However, a small **animal** such as a **zooplankton** might then consume the particle. One zooplankton that has eaten ten phytoplankton would have ten times the pollutant level as the phytoplankton. Since the zooplankton may be slow to metabolize or excrete the pollutant, the pollutant may build up or bioaccumulate within the organism. A small fish might then eat ten zooplankton. The fish would have 100 times the level of toxic pollutant as the phytoplankton. This would continue throughout the food web until very high levels of contaminants have biomagnified in the top predator. While the amount of pollutant might have been small enough not to do any damage in the lowest levels of the food web, the biomagnified amount might do serious damage to organisms higher in the food web.

One of the classic examples of bioaccumulation that resulted in biomagnification occurred with the insecticide **DDT** and predatory birds such as bald eagles, osprey, peregrine falcons and brown pelicans. DDT is an insecticide that was sprayed in the United States prior to 1972 to help control mosquitoes and other insects. Rain washed it into creeks, where it eventually found its way into lakes and the ocean. The toxic pollutant bioaccumulated within each organism and then biomagnified through the food web to very high levels in predatory birds that consumed fish. Levels of DDT were so high that the birds' **egg** shells were abnormally thin. As a result, the adult birds broke the shells of their unhatched offspring and the baby birds died. The **population** of these birds plummeted. DDT was finally banned in the United States in 1972, and since that time there have been dramatic increases in the populations of many predatory birds.

Human health is also at risk because of bioaccumulation and biomagnification of toxic contaminants. When humans eat organisms that are relatively high in the food web, they can get high doses of some harmful chemicals. For example, marine fish such as swordfish, shark, and tuna often have bioaccumulated levels of mercury, and bluefish and striped bass sometimes have high concentrations of PCBs. The federal government and some states have issued advisories against eating too much of certain types of fish because of bioaccumulated and biomagnified levels of toxic pollutants.

BIOASSAY

A bioassay is a quantitative test of the adverse or toxic effects of a chemical on a living **organism**. Toxic effects may include both lethality (mortality) and subacute effects such as changes in growth, development, reproduction, pharmacokinetic responses, pathology, biochemistry, physiology, and behavior. Effects are expressed by quantifiable criteria such as number of organisms killed, percent **egg** hatchability, changes in length and weight, percent **enzyme** inhibition, number of skeletal abnormalities, or tumor incidence.

When measuring the toxicity of a chemical in a bioassay, the objective is to estimate as precisely as possible the range of chemical doses or concentrations that produce some selected, readily observable, quantifiable results in groups of the same test **species** under controlled laboratory conditions. The results of the exposure are plotted on a graph that relates the dose or concentration of the test chemical to the percentage of organisms in test groups exhibiting the defined response. This correlation is called a dose-response or concentration-response relationship. This correlation is based on the assumptions that effects observed are a result of exposure to the known chemical and that the production and severity of a response are functions of the dose or concentrations of the chemical. Generally, within limits, the greater the dose or concentration of the test chemical, the more severe the response. However, at doses or concentrations below some minimum (threshold) value, no measurable adverse response will be elicited, while at all doses or concentrations above some maximum value, most or all of the test organisms will be adversely affected.

A step-wise tier testing approach is usually used in bioassay testing in the laboratory, progressing from simple single species short-term tests to more complex longer-term tests that may include multispecies and **ecosystem** testing. Test organisms are exposed to various concentrations or doses of the test material. The criteria for effects (e.g., mortality, growth, reproduction, etc.), which are established before testing, are then evaluated by comparing the chemically exposed (treated) organisms with untreated (control) organisms. Control organisms may include three types: (1) untreated (negative control), (2) treated with any solvent or carrier materials used to deliver the test chemical (vehicle control), and (3) treated with materials known to produce a defined effect on the test organisms (positive or reference control).

Criteria for selection of species to use in a bioassay may include: (1) species that are widely available and abundant, (2) species that are indigenous to or representative of the ecosystem that may receive the impact of the test chemical, (3) species that are recreationally, commercially, or ecologically important, (4) species that can be maintained in the laboratory so that chronic toxicity testing can be conducted, and (5) species that are well-understood in term of physiology, **genetics**, and behavior so that results of testing may be more easily interpreted.

Data from bioassays have a variety of uses, including industrial decisions on product development, manufacture, and commercialization, registration of products to satisfy regulatory requirements, permitting for the discharge of municipal and industrial wastes, environmental hazard evaluations, and prosecution and defense of chemical-related activities in environmental litigation.

BIODEGRADABLE

A substance that can biodegrade is one that can decay over time, or be absorbed by the environment by natural or biochemical means. Most organic wastes and paper, for example, can be broken down by **bacteria** and sunlight into basic elements of **nature**, and thus recycled back into the **ecosystem**. How quickly this process occurs depends on how much of the substance is available and where it is deposited. A piece of paper left in a field, for example, will biodegrade at a much higher rate than one compacted into a landfill with several pounds of other wastes.

Some substances do not naturally biodegrade in the environment, which primarily include those created through synthetic means. Most plastics fall into this category.

See also Waste disposal

BIODIVERSITY

Biodiversity is the total richness of biological variation. Usually the scope of biodiversity is considered to range from the genetic variation of individual **organism**s within and among **population**s of a **species** to different species occurring together in ecological communities. Some definitions of biodiversity also include the spatial patterns and temporal dynamics of populations and communities on the landscape. The geographical scales at which biodiversity can be considered range from local to regional, state or provincial, national, continental, and ultimately to global.

Biodiversity at all scales is severely threatened by human activities, making it one of the most important aspects of the global environmental crisis. Humans have already caused permanent losses of biodiversity through the **extinction** of many species and the loss of distinctive, natural communities. Ecologists predict that unless there are substantial changes in the ways that humans affect **ecosystem**s, there will be much larger losses of biodiversity in the near future.

Species richness of the biosphere

About 1.7 million of Earth's species have been identified and designated with a scientific name. About 6% of the identified species live in boreal or polar latitudes, 59% in the temperate zones, and the remaining 35% in the tropics. However, knowledge of Earth's species is highly incomplete, especially for tropical countries. According to some estimates there could be as many as 30-50 million species on Earth, 90% of them occurring in tropical ecosystems. Tropical ecosystems are richer in species than are those at higher latitudes.

Most of the described species on Earth are **invertebrates**, especially insects, and most of the insects are beetles (Coleoptera). The famous scientist J.B.S. Haldane was once asked by a theologian to briefly explain what his knowledge of biology told of God's purpose. Haldane reputedly said that God has "an inordinate fondness of beetles," reflecting the fact these insects are so much more abundant than any other creatures on Earth. Some biologists believe that beetles account for most of the undescribed tropical insects.

The suggestion of enormous numbers of undescribed insects in tropical **forests** has mostly emerged from the work of

Terry Erwin. This entomologist performed experiments in which tropical-forest canopies were treated with an insecticide, and the subsequent "rain" of dead arthropods was collected using ground-level sampling devices. This innovative sampling procedure indicated that: (1) a large fraction of the insect species of tropical rainforests is unknown to science; (2) most insect species are confined to a single type of tropical forest, or even to particular **tree** species that are themselves of local distribution; and (3) most species of tropical-forest insects have little ability to disperse very far. Erwin's studies of tropical rainforest in Amazonia found that beetles accounted for most of the insect species and that most of the beetles are narrowly endemic, that is, of a local distribution and found nowhere else. The tree *Luehea seemanii,* for example, had more than 1,100 species of beetle in its canopy of which 15% were specific to that plant. The emerging conclusion from this and other descriptive research is that there is an enormous abundance of undescribed species of insects and other invertebrates living in tropical forests.

Compared with invertebrates, the numbers of species, that is, the species richness, of other groups of tropical-forest organisms is better known. Although it is very difficult to do so, the numbers of species of vascular **plants** have been described for a few tropical forests. For example: a plot of only 0.0004 sq mi (0.1 ha) in a moist forest in Ecuador had 365 species of vascular plants; there were 98 species of large trees in 0.006 sq mi (1.5 ha) of forest in Sarawak, Malaysia; there were 90 tree species in 0.0032 sq mi (0.8 ha) of forest in Papua New Guinea; 742 woody species occurred in 0.012 sq mi (3 ha) of forest in Sarawak with 50% of the species recorded as single individuals; and more than 300 species of woody plants were discovered on a 50-ha area of forest in Panama. These tropical forests are much richer than temperate forests which typically support fewer than 12-15 species of trees. The Great Smokies of the eastern United States have some of the richest temperate forests in the world, and they typically contain 30-35 species, far fewer than in tropical rainforests.

It is extraordinarily difficult to determine the numbers of birds in tropical forests because the dense foliage and darkness of the understory make it inconvenient to see small **animal**s, even if they are brightly colored. As a result, very few studies have been made of the birds of tropical rainforests. However, one study in Peru discovered 245 resident and 74 transient species in 0.4 sq mi (97 ha) of Amazonian forest. Another study of rainforest in French Guiana recorded 239 species of birds, and another found 151 species in a forest in Sumatra. In comparison, temperate forests in North America typically support only 15-30 species of birds.

Almost no systematic surveys have been made of all of the species of tropical ecosystems. In one case, a 42 sq mi (108 sq km) reserve of dry forest in Costa Rica was estimated to support about 700 plant species, 400 vertebrate species, and 13,000 species of insects, including 3,140 species of moths and butterflies.

Why is biodiversity important?

Biodiversity is valuable for the following classes of reasons:

(1) Intrinsic Value. Biodiversity has its own intrinsic values, regardless of its worth in terms of human needs. Because of this intrinsic merit, there are ethical considerations to any degradation of biodiversity. For example, do humans have the ''right'' to diminish or exterminate elements of biodiversity, all of which are unique and irretrievable? Is the human existence itself diminished by losses of biodiversity? Ethical issues cannot be resolved through science, but enlightened persons would mourn any loss of species, or of natural, ecological communities.

(2) Direct Utilitarian Value. Humans have an absolute requirement for the products of other species. Because of this need, wild and domesticated species and their communities are exploited in many ways to provide food, materials, **energy**, and other goods and services. This fact can be illustrated in many ways. In the United States, for example, about one-quarter of prescription drugs have active ingredients obtained from higher plants, and these uses contributed about $14 billion per year to the U.S. economy, and $40 billion per year worldwide. Potentially, harvests of biodiversity can be conducted in ways that foster renewal. Unfortunately, potentially renewable biodiversity resources are often harvested too intensively or inadequate attention is paid to regeneration, so the resource is consequently degraded or becomes extinct.

(3) Provision of Ecological Services. Biodiversity provides many ecological services that are directly and indirectly important to human welfare. Examples of these services include biological **productivity**, nutrient cycling, cleaning of **water** and air, control of erosion, provision of atmospheric oxygen, removal of **carbon dioxide**, and other functions related to the integrity of ecosystems. According to the biologist Peter Raven: ''Biodiversity keeps the planet habitable and ecosystems functional.'' There are many cases of the discovery of bio-products useful to humans as food, medicine, or for other purposes through research on previously unexploited plants and animals. Consider the case of the rosy periwinkle (*Catharantus roseus*), a small plant native to the tropical island of Madagascar. During an extensive screening of wild plants for anti-**cancer** chemicals, an extract of rosy periwinkle was observed to inhibit the growth of cancerous **cell**s. The active biochemicals are several **alkaloid**s in foliage of the plant, probably used to deter **herbivore**s. These natural substances are now used to prepare the drugs vincristine and vinblastine, which can be successfully used to treat childhood leukemia and a lymphatic cancer known as **Hodgkin's disease**. In this case, a species of wild plant known only to a few botanists has proven to be of great benefit to humans by treating previously incurable diseases in the process sustaining a large pharmaceutical economy. There is a tremendous undiscovered wealth of other biological products useful to humans in unexplored biodiversity.

Biodiversity and extinction

Extinction refers to the loss of some species or other taxonomic unit (e.g., subspecies, **genus**, **family**, etc.; each is known as a taxon) occurring over all of its range on Earth. (Extirpation refers to a more-local disappearance, with the taxon still surviving elsewhere.) The extinction of any species is an irrevocable loss of part of the biological richness of Earth, the only place in the universe known to support living creatures. Extinction can be a natural occurrence caused by unpredictable catastrophes, chronic environmental stresses, or ecological interactions such as **competition**, disease, or **predation**. However, there have been dramatic increases in extinction rates since humans have become Earth's dominant large animal and the perpetrators of global environmental changes.

Extinction has always occurred naturally. Almost all species that have ever lived on Earth have become extinct. Perhaps they could not cope with changes occurring in their environment such as climate changes or in the intensity of predation or disease. Alternatively, many extinctions may have occurred simultaneously as a result of unpredictable catastrophes. From the geological record it is known that species, families, and even phyla have appeared and disappeared over time. For example, numerous phyla of invertebrates proliferated during an evolutionary radiation at the beginning of the Cambrian era about 570 million years ago, but most of these are now extinct. The 15-20 extinct phyla from that period are known from the Burgess Shale of British Columbia, and they represent unique experiments in invertebrate form and function. Similarly, entire divisions of plants have appeared, radiated, and disappeared, such as the seed ferns Pteridospermales, the cycad-like Cycadeoidea, and woody plants known as Cordaites. Of the twelve **order**s within the class Reptilia, only three survive today: crocodilians, turtles, and snakes/ lizards. Clearly, the **fossil** record displays a great deal of evidence of natural extinctions.

Overall, the geological record suggests that there have been long periods of time characterized by uniform rates of extinction but punctuated by about nine catastrophic episodes of mass extinction. The most intense extinction event occurred at the end of the Permian period some 245 million years ago when 54% of marine families, 84% of genera, and 96% of species are estimated to have become extinct.

Another famous, apparently synchronous extinction of vertebrate animals occurred about 65 million years ago at the end of the **Cretaceous** period. The most renowned extinctions were of the last of the reptilian dinosaurs and pterosaurs, but many plants and invertebrates also became extinct at the same time. In total, perhaps 76% of species and 47% of genera became extinct in the end-of-Cretaceous crisis. One hypothesis to explain the cause of this mass extinction involves a meteorite impacting the Earth, causing great quantities of fine dust to be spewed into the **atmosphere**, and resulting in a climatic deterioration that most large animals could not tolerate. However, some scientists believe that the extinctions of the last dinosaurs were more gradual.

More recently, humans have been responsible for most of Earth's extinctions. These extinctions are occurring so quickly that they represent a modern mass extinction of similar intensity to those documented in the geological record. Recent extinctions caused by humans include well-known cases such as the dodo, passenger pigeon, and great auk. Many other high-profile species have been taken to the brink of extinction, including the plains bison, whooping crane, ivory-billed woodpecker, right whale, and other marine **mammals**. These losses have been caused by insatiable overhunting and intense disturbance or conversion of natural **habitat**s.

Beyond these well-known and tragic cases involving large animals, Earth's biodiversity is experiencing an even larger loss. This ruin is mostly being caused by extensive conversions of tropical ecosystems, particularly rainforests, to agricultural habitats that sustain few of the original species. As was described previously, tropical ecosystems have very large numbers of species, most of which have restricted distributions. The conversion of tropical forests to habitats unsuitable for specialized, native species inevitably causes the loss of most of the locally endemic biota. This is a great tragedy, and the lost species will never again occur.

The most important human influences causing the extinction or endangerment of species are: (1) excessive exploitation, (2) effects of introduced predators, competitors, and diseases, and (3) habitat disturbance and conversion. These stressors can result in small and fragmented populations which experience the deleterious effects of **inbreeding** and population instability and then decline further, ultimately to extirpation or extinction. The increased rate of extinction and endangerment of biodiversity during the past several centuries is best documented for vertebrates because, as noted previously, most invertebrate species, particularly insects, have not yet been described by scientists. During the last four centuries there have been more than 700 known extinctions globally, including about 100 species of mammals and 160 species of birds, all because of human influences.

A much larger number of species is facing imminent extinction; they are endangered. For example, more than one thousand species of birds are considered to be threatened with extinction. Of this total, 46% live on isolated oceanic islands, a situation in which species are especially vulnerable to extinction caused by stresses associated with human activities. Birds of tropical forests account for 43% of the threatened bird species, wetland species for 21%, grassland and savannah species 19%, and other habitats 17%. Only 1.5% of the threatened species are North American, 4.2% are European and Russian, 33% Central and South American, 18% African, 30% Asian, and 14% from Australasia and the Pacific.

Protection of threatened biodiversity

Biodiversity can be protected in ecological reserves. These are protected areas established for the preservation of natural values, usually the known habitat of **endangered species**, threatened ecological communities, or representative examples of widespread communities. In the early 1990s there were about 7,000 protected areas globally with an area of 651 million hectares. Of this total, about 2,400 sites comprising 379 million ha were fully protected and could be considered to be true ecological reserves.

Ideally, the design of a national system of ecological reserves would provide for the longer-term protection of all native species and their natural communities including terrestrial, freshwater, and marine systems. So far, however, no country

has implemented a comprehensive system of ecological reserves to fully protect its natural biodiversity. Moreover, in many cases existing reserves are relatively small and are threatened by environmental change and other stressors such as illegal poaching of animals and plants and sometimes tourism.

The World Conservation Union, World Resources Institute, and United Nations Environment Program are three important agencies whose mandates center on the conservation and protection of the world's biodiversity. These have developed the *Global Biodiversity Strategy,* an international program to help protect biodiversity. The broad objectives are to: (1) preserve biodiversity; (2) maintain Earth's ecological processes and life-support systems; and (3) ensure that **natural resources** will be sustainable used by humans. The *Global Biodiversity Strategy* is a mechanism by which countries and peoples can initiate meaningful actions to protect biodiversity to benefit present and future generations and for intrinsic reasons as well. Because it only began in the late 1970s, it is too early to evaluate the success of this program. However, the existence of this comprehensive international effort is encouraging as is the participation of most of Earth's countries, representing all stages of socioeconomic development.

Important progress is being made, and the progressive worldwide development of activities intended to identify, conserve, and preserve biodiversity will hopefully come to be regarded as an ecological "success story."

BIOLOGICAL COMMUNITY

In biology, the term **species** refers to all **organism**s of the same kind that are potentially capable, under natural conditions, of breeding and producing fertile offspring. The members of a species living in a given area at the same time constitute a **population**. All the populations living and interacting within a particular geographic area make up a biological (or biotic) community. The living organisms in a community together with their non-living or **abiotic environment** make up an **ecosystem**. In theory, an ecosystem (and the biological community that forms its living component) can be as small as a few mosquito **larva**e living in a rain puddle or as large as prairie stretching across thousands of kilometers. A very large, general biotic community such as the Boreal Forest is called a **biome**. It often is difficult, however, to define where one community or ecosystem stops and another starts. Organisms may spend part of their lives in one area and part in another. **Water, nutrients**, sediment, and other abiotic factors are carried from place to place by geologic forces and migrating organisms. While it might seem that a lake and the dry land surrounding it, for instance, are distinctly different in their environmental conditions and **biological communities**, there can be a great deal of exchange of materials and organisms from one to the other. Insects fall into the lake and are eaten by fish. Amphibians leave the lake to hunt on shore. Soil erodes from the land and fertilizes the water. Water evaporated from the lake surface falls back on the land as rain that nourishes plant life. Every biological community requires a more or less constant influx of **energy** to maintain living processes.

Several important ecological categories and processes characterize every biological community. **Productivity** describes the amount of **biomass** produced by green **plants** as they capture sunlight and create new **organic compound**s. A tropical rainforest or a Midwest corn field can have very high rates of productivity, while **desert**s and arctic **tundra** tend to be very unproductive. **Trophic level**s describe the methods used by members of the biological community in obtaining food. Primary **producers** are green plants that depend on **photosynthesis** for their nourishment. Primary **consumer**s are the **herbivore**s that eat plants. Secondary consumers are the **carnivore**s who feed on herbivores. Top carnivores are large, fierce **animal**s who occupy the highest level on the food chain or food web. Nobody eats the top carnivores except the **scavenger**s (like vultures and hyenas) and decomposers (like **fungi** and **bacteria**) that consume dead organisms and recycle their bodies back into the abiotic component of the ecosystem. Because of the second law of thermodynamics, a majority of the energy in each trophic level is unavailable to organisms in the next higher level. This means that each successive trophic level generally has far fewer members than the prey on which they feed. While there might be thousands of primary producers in a particular community, there might be only a few top predators.

Abundance is an expression of the total number of organisms in a biological community, while diversity is a measure of the number of different species in that community. The arctic tundra of Alaska has vast clouds of insects, enormous flocks of migratory birds, and great herds of a few species of **mammal**s during the brief summer growing season. Thus, it has a high abundance but very little diversity. The tropical rainforest, on the other hand, might have several thousand different **tree** species and an even larger number of insect species in only a few hectares, but there may be only a few individuals representing each of those species in that area. Thus, the forest could have extremely high diversity but low abundance of any particular species. Complexity is a description of the variety of ecological processes or the number of ecological **niche**s (ways of making a living) within a biological community. The tropical rainforest is likely to be highly complex, while the arctic tundra has relatively low complexity.

Biological communities generally undergo a series of developmental changes over time known as **succession**. The first species to colonize a newly exposed land surface, for example, are known as pioneers. Organisms such as lichens, grasses, and weedy flowering plants with a high tolerance for harsh conditions tend to fall in this category. Over time, the pioneers trap sediments, build soil, and retain moisture. They provide shelter and create conditions that allow other species like shrubs and small trees to take root and flourish. Larger plants accumulate soil faster than do **pioneer species**. They also provide shade, shelter, higher humidity, protection from sun and wind, and living space for organisms that could not survive on open ground. Eventually these successional processes result in a community very different from the one first established by the original pioneers, most of whom are forced to move on to other newly disturbed land. It was once thought that every area would have a climax community such as an oak forest or prairie grassland determined by climate, topography, and mineral composition. Given enough time and freedom from disturbance, it was believed, every community would inevitably progress to its climax state. It is now recognized, however, that some ecosystems experience continuous disturbance. Certain biological assemblages such as conifer **forests** that we once thought were stable **climax communities**, we now recognize as chance associations in an ever changing mosaic of regularly disturbed and constantly changing landscapes.

Many biological communities are relatively stable over long periods of time and are able to withstand many kinds of disturbance and change. An oak forest, for example, tends to remain an oak forest because the species that make it up have self-perpetuating mechanisms. When a tree falls, others grow up to replace it. We call the ability to repair damage and resist change resilience. For many years there has been an on-going debate between theoretical and field ecologists about whether complexity and diversity in a biological community increase resilience. Theoretical models suggest that a population of a few very hardy, weedy species, such as dandelions and box elder bugs, might be more resistant to change than a more highly specialized and more diverse community such as a tropical forest. Recent empirical evidence suggests that in at least some communities, such as prairies, higher diversity does impart greater resistance to change and a better ability to repair damage after **stress** or disturbance.

See also Ecological succession; Ecology

BIOLOGICAL WARFARE

The United Nations has defined biological warfare as the use of any living **organism**s (e.g. **bacteria**) or infective material derived from them, in order to cause disease or **death** in humans, **animal**s, or **plants**. The effectiveness of these organisms depends on their ability to multiply in the person, animal or plant attacked.

While the United Nations is a post-World War II institution, biological warfare is considerably older. Biological weapons have been used throughout world history: crops and livestock have been destroyed, **water** supplies have been contaminated, and humans have been exposed to lethal diseases.

Some historians have suggested that the second plague pandemic in the Middle Ages was a result of a rather crude form of biological warfare. During the fourteenth century, the city of Kaffa, or Caffa (now Feodosiya, Ukraine), was under siege by Tatars, whose own **population** was suffering from an outbreak of the plague. As a weapon, the attacking Tatar forces catapulted their cadavers into the city which subsequently initiated an epidemic in Kaffa. Refugees from Kaffa, which was then ruled by Genoa, moved to major ports, such as Venice, Genoa, and Constantinople, thereby spreading the disease over a large area. In another instance, a smallpox epidemic broke out among Native Americans in the Ohio River Valley after they were deliberately exposed to smallpox during French-

Indian War (1754-1767). At the suggestion of Sir Jeffrey Amherst, Captain Ecuyer distributed used blankets from the smallpox hospital among Native American tribes with the intention of spreading the disease. However, the cause of an epidemic is difficult to pin down, as modern epidemiologists have pointed out that other factors, such as initial contact with the Europeans, may have triggered the epidemic among the Native Americans. In the case of the plague pandemic, it is likely that other factors, such as poor sanitation and the extraordinary effectiveness of fleas and rats as pathogen-carriers, contributed to the spread of the disease.

The creation of modern biological weapons is the direct result of advances in the field of microbiology. For example, scientists have learned how to isolate strains of bacteria, enabling the mass production of pure bacterial colonies. Because of remarkable progress in microbiology, biological weapons are easily and cheaply produced, and easily dispersed. In addition, they are extremely potent.

Biological weapons are classified as: **virus**es, bacteria, rickettsia, and biological **toxins**. Of particular concern are genetically altered **microorganism**s, whose effect can be made to be group-specific. In other words, persons with particular traits are susceptible to these microorganisms.

There is evidence of biological weapon research and use during both World Wars. For example, during World War I, Germany developed a biological warfare program based on the **anthrax** bacillus (*Bacillus anthracis*) and a strain of pseudomonas (*Burkholderia mallei*), the causative agent of glanders, to infect livestock. During World War II, Japan conducted extensive biological weapon research in occupied Manchuria, China, where prisoners were infected with a variety of pathogens, including *Neisseria meningitis*, *Bacillus anthracis*, *Shigella spp*, and *Yersinia pestis*. It has been estimated that over 10,000 prisoners died as a result of either infection or execution following infection. In addition, the water supply and some food items were contaminated by biological agents, and an estimated 15 million potentially plague-infected fleas were released from aircraft, affecting many Chinese cities. However, as the Japanese military found out, biological weapons have fundamental disadvantages: they are unpredictable and difficult to control. After infectious agents were let loose in China by the Japanese, approximately 10,000 illnesses and 1,700 deaths occurred among Japanese troops.

Other countries encouraged biological weapon research with a view to determining the pathogenesis of specific bacteria. For example, prisoners in Nazi concentration camps were infected with pathogens, such as hepatitis A, *Plasmodia* spp., and two types of Rickettsia bacteria. These prisoners were used as guinea-pigs by Nazi physicians who attempted to develop vaccines and antibacterial drugs. During the Vietnam War, the U.S. government set up Operation Whitecoat, a research project which used conscientious objectors who nevertheless wanted to serve their country. For example, some of these young men were exposed to the causative pathogen of tularemia.

The first diplomatic effort to limit biological warfare was the Geneva Protocol for the Prohibition of the Use in War of Asphyxiating, Poisonous or Other Gases, and of Bacteriological Methods of Warfare. This treaty, ratified in 1925, prohibited the use of biological weapons. Unfortunately, the treaty failed to prohibit biological warfare research, nor did it ban the production and possession of biological weapons. Consequently, many countries developed extensive defensive biological weapon research. In the United States, several research facilities were constructed to develop antisera, vaccines, and various equipment for protection against a possible biological attack. From 1949–1968, in covert experiments, scientists used nonpathogenic agents to study aerosol use over large areas, as well as the effects of solar irradiation and climate on the pathogens' viability. Interestingly, several mysterious outbreaks of diseases occurred after some of these experiments, and, in 1977, senate hearings were held to investigate the outbreaks. The United States also had developed biological pathogens specifically for crops.

During the 1960s, there was worldwide concern regarding of biological warfare; the inadequacy of the 1925 Geneva Protocol was brought to light. In 1969, The United States began dismantling its offensive biological weapons program. In 1972, a treaty that took the Geneva Protocol a few steps further was proposed. This was the Convention on the Prohibition of the Development Production, and the Stockpiling of Bacteriological (Biological) and Toxin Weapons and on Their Destruction (BWC). The BWC prohibits development, possession, storing and stockpiling biological weapons, as well as devices used to disperse pathogens. In addition, it forbids transfer of pathogens or expertise to other countries. The treaty went into effect in March, 1995, and was signed by more than 100 countries. However, several signatories have since participated in using biological weapons, and some nonmilitary groups have used biological weapons as well. In 1978, the Soviet Union used ricin, a lethal toxin from castor beans, to assassinate the Bulgarian defector, Georgi Markov, and several others. During the Persian Gulf War (1991) thousands of U.S. troops were inoculated against several diseases because of the perceived threat of Iraqi biological warfare. In 1984, the Rajneeshee cult contaminated salad bars in several Oregon restaurants and; eleven years later, the cult Aum Shinriko released sarin gas in the Tokyo subway system. In 1996, a man from Ohio was able to obtain **bubonic plague culture**s through the mail!

Biological warfare is complex and the effects are extremely variable. Most importantly, only a small amount of a pathogen can be lethal to a large area or population. The continual threat of biological warfare challenges national security and questions the popular notions of warfare. Thus, even during peacetime, there is a perceived need for the development of detector devices, protective equipment, and decontamination procedures. In 1996, the U.S. Congress enacted the Defense Against Weapons of Mass Destruction Act to enhance disaster preparedness policies and to call upon the cooperation between civil and military institutions. As of the summer of 1997, the fire, police, rescue, and hospital emergency departments in more than 100 US cities have been designated to receive special training to respond to incidents involving the use of weapons of mass destruction.

BIOLUMINESCENCE

Bioluminescence is the production of **light** by living **organisms**. Some single-**cell**ed organisms (**bacteria** and **protista**) as well as many multicellular **animal**s and **fungi** demonstrate bioluminescence.

Bioluminescence in nature

Marine environments support a number of bioluminescence organisms including **species** of bacteria, dinoflagellates, jellyfish, coral, shrimp, and fish. On any given night one can see the luminescent sparkle produced by the single-celled dinoflagellates when **water** is disturbed by a ship's bow or a swimmer's motions. Many multicellular marine organisms have specialized light emitting **organ**s that project light in a particular direction or convey a unique shape to the light. The anglerfish has a light-emitting organ that projects from its head which serves as a bait to attract smaller prey fish. The light emitted from this organ in the anglerfish is actually produced by bacteria, living in a symbiotic relationship in which both the fish and bacteria profit from their shared existence.

Bioluminescent organisms in the terrestrial environment include species of fungi and insects. The most familiar of these is the firefly, which can often been seen glowing during the warm summer months. In some instances organisms use bioluminescence to communicate, such as in fireflies, which use light to attract members of the opposite sex. Certain reef fish use light produced from organs under their eyes to illuminate the interior of crevices and caves. This not only helps the fish to navigate but also allows it to locate prey. Organisms that are unpalatable or dangerous, such as jellyfish, use bioluminescence as a signal to warn off attacks by predators.

Biochemical mechanism

Light is produced by most bioluminescent organisms when a chemical called luciferin reacts with oxygen to produce light and oxyluciferin. The reaction between luciferin and oxygen is catalyzed by the **enzyme** luciferase. Luciferases, like luciferins, usually have different chemical structures in different organisms.

In addition to luciferin, oxygen, and luciferase other **molecule**s (called cofactors) must be present for the bioluminescent reaction to proceed. Cofactors are molecules required by an enzyme (in this case luciferase) to perform its catalytic function. Common cofactors required for bioluminescent reactions are calcium and ATP, a molecule used to store and release **energy** that is found in all organisms.

The terms luciferin and luciferase were first introduced in 1885. The French scientist Raphael Dubois obtained two different extracts from bioluminescent clams and beetles. When Dubois mixed these extracts they produced light. He also found that if one of these extracts was first heated no light would be produced upon mixing. Heating the other extract had no effect on the reaction, so Dubois concluded that there are at least two components to the reaction. Dubois hypothesized that the heat resistant chemical undergoes a chemical change during the reaction, and called this compound luciferin. The heat sensitive chemical, Dubois concluded, was an enzyme which he called luciferase.

Fireflies have a bioluminescent organ in their abdomen that they use to attract males. Enzymes within the organ react with oxygen to produce light. The insect controls the flashes by regulating the flow of oxygen. *(The Stock Market. Reproduced by permission.)*

Bioluminescence as a research tool

The two basic components needed to produce a bioluminescent reaction, luciferin and luciferase, can be isolated from the organisms that produce them. When they are mixed in the presence of oxygen and the appropriate cofactors these components will produce light with an intensity dependent on the quantity of luciferin and luciferase added as well as the oxygen and cofactor concentrations.

Scientists have used isolated luciferin and luciferase to determine the concentrations of important biological molecules such as ATP and Calcium. After adding a known amount of luciferin and luciferase to a **blood** or **tissue** sample, the cofactor concentrations may be determined from the intensity of the light emitted. Scientists have also found numerous other uses for the bioluminescent reaction such as using it to quantify specific molecules which do not directly participate in the bioluminescence reaction. To do this, scientists have attached luciferase to antibodies, which are molecules produced by the **immune system** that bind to specific molecules called **antigen**s. The **antibody**-luciferase complex is added to a sample where it binds to the molecule which is to be quantified. Following washing to remove unbound antibodies, the molecule of interest can be quantified indirectly by adding luciferin and measuring the light emitted. Methods used to quantify particular compounds in biological samples such as the ones described here are called assays.

BIOMAGNIFICATION

In most parts of the **ecosystem**, elements introduced at high levels as **pollution** or as a result of a region's geology can be degraded over time into harmless substances by the air, **water** and by biological decomposers. Most synthetic chemicals created by humans however, such as **DDT** as well as some radioactive materials and some mercury compounds, are not diluted or broken down by natural processes. In these cases, their concentrations can intensify as they are transferred up the food chain, or biomagnify.

While these contaminants may be present in the water or air at extremely low levels, their amount and intensity increase dramatically as they move up the food chain. For example, plankton and **invertebrates** absorb **toxins** as they filter the water for food. Forage fish such as alewife and smelt, which eat large amounts of **phytoplankton** and **zooplankton**, consume toxins from each one. When the trout, salmon, snapping turtles and herring gulls consume large amounts of forage fish, the levels and intensity of these contaminants again intensify. The **species** at the top of this food chain, such as eagles and other birds and humans, thus receive the highest concentration of these chemicals each time they consume the fish.

Because people have a varied diet, contaminants do not biomagnify as quickly as those in other species at the top of the food chain that rely on fish or other species for their entire diet. The concentration of some chemicals, which are stored in fatty **tissue**s and thus are not excreted, can be millions of times greater than that found in the water or air itself. Many of these chemicals persist in the environment from eight weeks to decades.

See also Bioaccumulation; Chemical compound; Pesticides; Pollution

BIOMASS

Biomass consists of living **organism**s, or parts of living organisms, as well as waste products and incompletely decomposed remains of living organisms. The term is quite encompassing and includes **plants**, referred to as phytomass, microbes, and **animal** material, or zoomass. Biomass density is a distinguishing feature of ecological systems and is usually presented as the amount of dry biomass per unit area. To insure a uniform basis for comparison, biomass samples are dried at 221°F (105°C) until they reach a constant weight.

In most settings, phytomass is by far the most important component. A square yard (0.84 m²) of the planet's land area has, on average, about 18-22 pounds (8.4-10 kg) of phytomass, although values may vary widely depending on the type of **biome**. Tropical **rain forest**s contain four or five times the average while **desert** biomes may have a value near zero. The global average for non-plant biomass is approximately 1% of the total. **Organic compound**s typically constitute about 95% by weight of biomass, and inorganic compounds account for the remaining 5%. An exception occurs in **species** that incorporate large amounts of inorganic elements such as silicon or calcium, in which case the inorganic portion may be several times higher.

Photosynthesis is the principle agent for biomass production. **Light energy** is used by chlorophyll containing green plants to remove (or fix) **carbon dioxide** from the **atmosphere** and convert it to energy rich organic compounds or biomass. It has been estimated that on the face of the earth approximately 200 billion tons of carbon dioxide are converted to biomass each year. **Carbohydrate**s are usually the primary constituent of biomass, and **cellulose** is the single most important component. Starches are also important and predominate in storage

organs such as tubers and rhizomes. Sugars reach high levels in **fruits** and in plants such as sugar cane and sugar beet. Lignin is a very significant non-carbohydrate constituent of woody plant biomass.

See also Productivity

BIOME

A biome is a large geographical region characterized by particular kinds of **plants** and **animals** and maintained by a distinct climate and soil conditions. There are eight generally recognized biomes throughout the world. The six terrestrial biomes (those found on land) are the **rain forest**, **desert**, **grasslands**, temperate deciduous forest, **taiga**, and **tundra**. The two aquatic biomes are the freshwater and marine biomes.

Rain forests are most commonly found in tropical regions close to the equator in areas of very heavy rainfall, at least 150 in (381 cm) per year, and with relatively constant warm **temperature**s. They are often recognized as one of the most biologically diverse **habitat**s on Earth. Deserts are the driest biome with only an average annual rainfall of 10 in (25 cm). **Organism**s living in the desert must have special **adaptation**s to survive in these dry conditions. Grasslands are the prairies. They can be found in both temperate and tropical climates; however, there is not enough rainfall in grasslands for many large **tree**s to survive. Rainfall in temperate grasslands (those lying between 25 and 65 degrees latitude) ranges from 10-30 in (25-75 cm) per year. At tropical and subtropical latitudes, annual **precipitation** averages 24-59 in (60-150 cm). As a result, grasslands are dominated by grasses. The temperate deciduous forest is dominated by deciduous trees, those that lose their leaves in autumn. Examples include maple, oak, beech, birch, and hickory. This biome is generally warm in the summer and cold in the winter. The taiga has extremely cold winters, with a relatively short growing season. The taiga is dominated by **forests** with conifer trees, such as pine, fir, and spruce. The tundra is the coldest of all of the terrestrial biomes. It has such a short growing season that the subsoil never thaws. As a result, no trees or shrubs can survive. The vegetation in the tundra is dominated by mosses, lichens, and other plants that do not grow very large and can reproduce very quickly during the short growing season. The freshwater biome includes creeks, rivers, ponds, and lakes. The marine biome includes all of the areas of salt **water** in the world, in particular, the oceans. The marine biome is the largest of all of the biomes, covering approximately 71% of Earth.

As one increases in latitude from the equator, the terrestrial biomes change in a relatively predictable pattern. Closest to the equator, the rain forest is most common. As one moves to higher latitudes the deciduous forest becomes more common. At still higher latitudes the taiga is most prominent due to the decreases in temperature. Finally, closest to the poles, in the coldest regions, one finds tundra. A similar, generalized pattern is observed going up a very tall mountain from low to high altitudes. The rain forest is dominant at the warmer bottom of the mountain, and at the highest altitudes a tundra-like community is found. Neither desert nor grasslands are included in this pattern because the predominant factor limiting their distribution is precipitation rather than temperature.

The terrestrial biomes are not necessarily completely distinct from each other. Two neighboring biomes blend together at their edges. For example, at the northern edge of the temperate deciduous forest biome in North America, both deciduous trees and conifers are present, which are characteristic of the taiga biome. A little farther north, the conifers will be dominant.

While biomes have recognizable types of vegetation, **species** will vary with location. For example, the vegetation in a North American and African desert will be similar and have similar adaptations. However, there will be different species in these two locations. For example, each location might have different species of cactus-like, succulent plants.

BIOREMEDIATION

Bioremediation is a type of biotechnology in which living **organism**s or ecological processes are utilized to deal with some environmental problem. The most common use of bioremediation is to metabolically break down or otherwise remove toxic chemicals before or after they have been discharged into the environment. In such uses, bioremediation takes advantage of the fact that certain **microorganism**s can utilize such toxic chemicals as metabolic substrates, in the process rendering them into simpler, less toxic compounds. Bioremediation is a relatively new and actively developing technology.

In general, bioremediation methodologies focus on: (1) enhancing the abundance of certain **species** or groups of microorganisms that can metabolize toxic chemicals (this is also known as *bioaugmentation*) and/or (2) optimizing environmental conditions for the actions of these organisms (also known as *biostimulation*). Bioaugmentation may involve the deliberate addition of strains or species of microorganisms that are specifically effective at treating particular toxic chemicals, but are not indigenous to or abundant in the treatment area. Biostimulation usually involves **fertilization**, aeration, or irrigation in order to decrease the importance of environmental factors in limiting the activity of microorganisms. Biostimulation focuses on rapidly increasing the abundance of naturally occurring, ubiquitous microorganisms that are capable of dealing with certain types of environmental problems.

Bioremediation of spilled hydrocarbons

Accidental spills of petroleum or other **hydrocarbon**s on land and **water** are regrettable but frequent occurrences. Such spills can range in size from a few gallons that may be spilled during refueling to enormous spillages of millions of tons as occurred to both the sea and land during the Gulf War of 1991. Once spilled, petroleum and its various refined products can be persistent environmental contaminants. However, these organic chemicals can also be metabolized by microorganisms which in the process transform them into simpler compounds, ultimately to **carbon dioxide**, water, and other inorganic chemicals.

Numerous attempts have been made to increase the rates by which microorganisms break down spilled hydrocarbons. In some cases, specially prepared concentrates of **bacteria** that are highly efficient at metabolizing hydrocarbons have been ''seeded'' into spill areas in an attempt to increase the rate of degradation of the spill residues. Although this technique has sometimes been effective, it commonly is not. This occurs because the indigenous microbial communities of soils and aquatic sediments contain many species of bacteria and **fungi** that are capable of utilizing hydrocarbons as a metabolic substrate. After a spill the occurrence of large concentrations of hydrocarbons in soil or sediment stimulates rapid growth of those microorganisms. Consequently, seeding of microorganisms that are metabolically specific to hydrocarbons does not always make much of a difference to the overall rate of degradation.

More important, however, is the fact that the environmental conditions under which spill residues occur are almost always highly sub-optimal for their degradation by microorganisms. Most commonly, the rate of microbial breakdown of spilled hydrocarbons is limited by the availability of oxygen or of certain **nutrients** such as nitrate and phosphate. Therefore, the microbial breakdown of spilled hydrocarbons on land can be greatly enhanced by occasionally tilling the soil to keep conditions aerated and by fertilizing with **nitrogen** and phosphorus while keeping conditions moist but not wet. Therefore, bioremediation systems for dealing with soils contaminated by spilled gasoline or petroleum can be based on simple tillage and fertilization.

Similarly, petroleum refineries may utilize a bioremediation process called landfarming in which oily wastes are spread onto land which is then tilled and fertilized until microbes reduce the residue concentrations to an acceptable level.

After some petroleum spills, more innovative approaches may prove to be useful. For example, it is difficult to fertilize aquatic **habitat**s, because the nutrients simply wash away and are therefore not effective for very long. In the case of the *Exxon Valdez* spill in Alaska in 1989, research demonstrated that nutrients could be applied to soiled beaches as an oleophilic (that is, oil-seeking), nitrogen and phosphorus-containing fertilizer. Because of its oleophilic nature, the fertilizer adhered to the petroleum residues and was able to significantly enhance the rate of oil degradation by the naturally occurring **community** of microorganisms. This treatment was applied to about 73 m (118 km) of oiled beach and proved to be successful in speeding up the process of degradation of the residues by increasing the rate of oxidation by about 50%. No attempts were made in this case to ''seed'' the microbial community with species that are specifically adapted to metabolizing hydrocarbons. It was believed that hydrocarbon-specific microbes were naturally present in the beach sediment and that their activity and that of species with broader substrate **tolerance**s only had to be enhanced by making the ecological conditions more favorable, that is, by fertilizing.

Bioremediation of metal pollution

Metals are common **pollutants** of water and land when they are emitted by many industrial, agricultural, and domestic sources. In some situations, organisms or ecological processes can be successfully utilized to concentrate metals that are dispersed in the environment, especially in water. The metals can then be removed from the system by harvesting the organisms. For example, metal polluted waste waters can be treated by encouraging the vigorous growth of certain types of **algae**, fungi, or vascular **plants**, usually by fertilizing thewater within some sort of constructed lagoon. This bioremediation system works because the growing plants and microorganisms absorb metals from the water (acting as so-called biosorbents), and thereby reduce their concentrations to a more tolerable range. The plants can then be harvested to remove the metals from the bioremediation system. In some cases, the plant **biomass** may even be processed to yield metal products of economic value.

Bioremediation of acidification

In some situations, artificial **wetlands** can be engineered to treat acidic waters associated with coal mining or other sources of acidity. Coal mining disturbs soil and fractures rocks and exposes large quantities of pyritic sulfur to atmospheric oxygen. Under such conditions, certain species of bacteria oxidize the sulfide of the mineral pyrites to sulfate generating large quantities of acidity in the process which is known as *acid mine drainage*. The resulting acidity is often treated by adding large quantities of acid-neutralizing chemicals such as lime or limestone. However, it has also been recently demonstrated that natural, acid-consuming, ecological processes operate in wetlands. These processes can be taken advantage of in constructed wetlands to decrease much of the initial acidity of acid mine drainages and thereby reduce the costs of conventional treatments with acid-neutralizing chemicals. The microbial processes that consume acidity are various, but they include: (1) the chemical reduction of sulfate to sulfide at the oxygen-poor interface between the sediment and the water column and around plant roots, (2) the reduction of ferric iron to ferrous in the same anoxic microhabitats, as well as (3) the primary productivity of **phytoplankton**, which also consumes some acidity.

A less intensive type of bioremediation can be used to mitigate some of the deleterious ecological effects associated with the acidification of surface waters, such as lakes and ponds. In almost any fresh waters, fertilization with phosphate will greatly increase the primary productivity of algae and vascular plants. In acidic waters, this process can be taken advantage of to reduce the acidity somewhat, but the most important ecological benefit occurs through enhancement of the habitat of certain aquatic **animals**. Ducks and muskrat, for example, can breed very successfully in fertilized acidic lakes, because their habitat is improved through the vigorous growth of vegetation and of aquatic insects and crustaceans. However, the productive but still acidic habitat remains toxic to fish. In this case, manipulation of the **ecosystem** by fertilization mitigates some but not all of the negative effects of acidification.

Bioremediation of sewage

Sewage represents a very complex mixture of wastes, usually dominated by fecal materials but also containing toxic chemicals that have been dumped into the disposal system by industries and home owners. Many advanced sewage-treatment technologies utilize microbial processes to both oxidize the organic **matter** associated with fecal wastes and to decrease the concentrations of soluble compounds or ions of metals, **pesticides**, and other toxic chemicals. The latter effect, decreasing the aqueous concentrations of toxic chemicals, is accomplished by a combination of chemical adsorption as well as microbial biodegradation of complex chemicals into their simpler, inorganic constituents. Microbial processes are relied upon in many sewage treatment systems including activated sludges, aerated lagoons, **anaerobic digestion**, trickling filters, waste stabilization ponds, composting, and disposal on land.

BIORHYTHMS

Biorhythms are natural, rhythmic cycles controlled by our "biological clock." Present from **birth** until **death**, biorhythms—or biological rhythms—it is believed occur independently of our will, health, **stress**, or other external influences and do not vary from person-to-person.

Four biorhythmic cycles have thus far been identified: physical biorhythms cycle every 23 days, influencing our coordination, **energy**, strength, endurance, initiative, and resistance to illness. Emotional biorhythms cycle in 28 days, affecting relationships, moods, sensuality, and feelings. Intellectual biorhythms cycle in 33 days, influencing cognitive, critical thinking, and problem-solving abilities; reasoning; **learning**; **memory**; and alertness. Intuition biorhythms cycle over 38 days, making us more or less receptive to "sixth sense" **perception**s such as **instinct** and hunches. Each biorhythm cycle has exactly the same capacity—from 100% to minus 100%. During the positive swing, energy levels and effectiveness in that particular cycle are high and we may find ourselves having really "good days" or handling problem situations smoothly. During the negative cycle, the converse is true, and we often have "bad days" or find problems difficult to handle. As a cycle rises from negative to positive, it crosses a mid-point. This day is called a "neutral day." As it crosses the midpoint moving from positive to negative, this is a "day of crisis." On both of these days, capabilities could be either extremely high or extremely low. Biorhythmic cycles are calculated mathematically based on date, month, and year of birth. Those who study biorhythms suggest important tasks relative to each particular cycle be planned for the positive days and avoided during the negative days, and that extra care be taken during the negative swing to avoid accidents. Therefore, knowing one's biorhythm cycle may be helpful in planning critical tasks and understanding otherwise unexplained highs and lows.

While the existence of biorhythms has not been proven scientifically, empirical evidence strongly suggests their influence is universal. Astute observers—probably from the beginning of time—were aware of cyclic rhythms in humans. In a book entitled *Is This Your Day?* published in 1973, George S. Thommen delves into the history of biorhythm research, referring to studies by Dr. Hermann Swoboda, professor of psychology at the University of Vienna, volumes published in the early 1900s by Dr. Wilhelm Fliess of Berlin, and other writings. Even **Hippocrates** apparently noted cyclic rhythms in his patients, taking them into account when planning treatment.

BIOSPHERE

The biosphere is the space on or near the Earth's surface which contains and supports living **organism**s. It is typically subdivided into the lithosphere, **atmosphere**, and hydrosphere. The lithosphere is the Earth's surrounding layer composed of solid substance such as soil and rock, the atmosphere is the surrounding gaseous envelope, and the hydrosphere refers to liquid environments such as lakes and oceans which lie between the lithosphere and atmosphere. The biosphere's creation and continuous existence results from chemical, biological, and physical processes. To study these processes a multi disciplinary effort has been employed by scientists from fields such as chemistry, biology, geology, and **ecology**.

History

The term biosphere was first used by Austrian geologist Eduard Suess (1831-1914) in 1875 to describe the space on Earth which contains life. The concept introduced by Suess had little impact on the scientific community until it was resurrected by Russian scientist Vladimir Vernadsky (1863-1945) in 1926 with the publication of *La biosphere*. In this work, Vernadsky extensively developed the modern concepts which recognize the interplay between geology, chemistry, and biology in biospheric processes.

Requirements for life

For organisms to live, appropriate environmental conditions (e.g., **temperature**, moisture, etc.) must exist, and the organisms must be supplied with **energy** and **nutrients**. In a biosphere such as the Earth's where no external nutrient supply exists, nutrients contained in dead organisms or waste products from living **cell**s must be transformed back into compounds which living **matter** can reutilize for the biosphere to continue to support life. Mineral sources of nutrients are also important.

Energy is needed for the functions which organisms perform, such as growth, movement, waste removal and reproduction. It is the only requirement for life which is supplied from a source outside the biosphere. This energy, in the form of **light** or solar radiation received from the Sun, is captured and stored by **plants** in a process called **photosynthesis**. Photosynthesis refers to the light induced **chemical reaction** between **carbon dioxide** and **water** which produces oxygen and large carbon compounds called organic **molecule**s. Energy is stored in the **chemical bond**s of organic molecules and can be released in the process of **respiration**; the enzymatic chemical reaction between organic molecules and oxygen to form carbon dioxide,water and energy. Organic molecules may also be used for growth since they are the major component of most

The Beni Biosphere Reserve in Bolivia. *(Conservation International. Reproduced by permission.)*

tissues. Plants and some **microorganism**s are the only organisms which can form organic molecules by photosynthesis. Heterotopic organisms like humans rely on plants for their energy needs.

The major elements or chemical building blocks which comprise all living organisms are carbon, oxygen, **nitrogen**, phosphorus and sulfur. Organisms are able to acquire these elements only if they occur in useable chemical forms termed nutrients. In a process called the nutrient cycle the elements are transformed from one chemical form to another and back to the original form. The different chemical forms in which carbon occurs illustrate this process. Carbon occurs in the gaseous molecule carbon dioxide or in the organic molecules which compose living and dead organisms. Gaseous carbon dioxide is transformed to solid **organic compound**s (simple sugars) by the process of photosynthesis as mentioned previously. When organisms grow they deplete the atmosphere of carbon dioxide. If this process were to continue without carbon dioxide being resupplied at the same rate that it is consumed eventually the atmosphere would no longer contain carbon dioxide and organisms could no longer use the energy supplied by the Sun to sustain life. Fortunately, carbon dioxide is returned to the atmosphere at the same rate that it is consumed when organisms respire their own stores of organic molecules, when microorganisms respire tissue from organisms which have died in a process known as **decomposition**, or when wildfires occur.

Biosphere evolution

During the Earth's long history, life forms have drastically altered the chemical composition of the biosphere and at the same time the biosphere's chemical composition has influenced which life forms inhabit the Earth. In the past, the rate that nutrients were transformed from one chemical form to another was not always the same as the transformation back to the original chemical form. This has resulted in a change in the relative concentrations of chemicals in the biosphere. For example, when life first evolved approximately 3.8 billion years ago, the atmospheric carbon dioxide concentration was much greater than what we find today and there was virtually no free oxygen. The decrease in carbon dioxide and increase in atmospheric oxygen which occurred over time was due to photosynthesis occurring at a faster rate than respiration. The carbon which was present in the atmosphere as carbon dioxide now lies in **fossil** fuel deposits and limestone rock.

Scientists believe that the increase in atmospheric oxygen concentration influenced the **evolution** of life. It was not until oxygen reached high concentrations such as we find on the Earth today that multicellular organisms like ourselves could have evolved. We require high oxygen concentrations to accommodate our high respiration rates and would not be able to survive had the biosphere not been altered by the organisms which came before us.

Current developments

Most research concerning the biosphere is being done to determine the effects which human activities have on our environment. **Pollution**, fertilizer application, land use, and fuel consumption affect nutrient cycles and may damage components of the biosphere such as the **ozone** layer which protects us from ultraviolet radiation. Fertilizer application increases the amount of nitrogen, phosphorus and other nutrients which organisms can use for growth. These excess nutrients damage lakes as demonstrated by large algal blooms and fish kills. Fuel consumption and land clearing increases carbon dioxide levels in the atmosphere and may cause global warming as a result of carbon dioxide's excellent ability to trap heat.

Recent interest in long term manned space operations has spawned research into the development of artificial biospheres. Extended missions will require that nutrients be cycled in a space no larger than a building. The Biosphere 2 project which received a great deal of press in the early 1990s should provide insights into the feasibility of such biospheres as well as the function and evolution of our own.

BIOSYNTHESIS

Biosynthesis is a process by which **organism**s create biologically important, complex **molecule**s from simpler ones. This process involves **chemical reaction**s which generally require an input of **energy**. Thus these reactions are called endergonic.

Within an organism biosynthesis, or anabolism, is a part of the **cell**'s overall **metabolism**. It is used to produce a variety of materials. To make energy storing molecules, **monosaccharides** are biosynthetically polymerized into **polysaccharides**. For example, the process of gluconeogenesis converts a simple molecule, pyruvate, into glucose. This glucose may be further reacted through glycogenesis to produce glycogen. Another example of biosynthesis is **protein synthesis**. In this reaction large polymers are created from smaller **amino acid**s. These **protein**s play a variety of essential roles in cell function. Certain amino acids are also produced biosynthetically. In addition to energy storage and protein synthesis, biosynthesis is also used in **DNA synthesis**. This process involves the production of **nucleic acids** from **nucleotide**s.

To drive biosynthetic reactions **enzyme**s are often required. These molecules act as catalysts to speed up reactions, and they do this by reducing the activation energy needed to drive reactions. In addition, cofactors such as **vitamins** and **coenzyme**s may aid in catalytic activity.

Biosynthesis has become an important new industrial technology. Various products are made through biosynthetic processes including **antibiotics**, drugs, and food products. These processes typically involve isolating a desired **microorganism**, fueling its growth in a large container, and refining and isolating the final product. However, our knowledge about biosynthetic processes remains incomplete and new advances are discovered frequently. In the future, these advances will be used by industry to produce more useful products.

BIOTIC ENVIRONMENT

The biotic environment of an **organism** refers to the living part of its environment, to the other organisms that help set the conditions for reproduction, and for growth and survival. The most obvious dependence of organisms on the biotic component of their environment is nutrient intake, **nutrients** that provide the **energy** needed for life processes. Green **plants** take their life-energy directly from the **abiotic environment**, but most other forms of life on earth must ingest other organisms to obtain nutrients and energy. **Ecosystem**s are often described in terms of their abiotic (non-living) and biotic (living) components.

Organisms as part of the environment of other organisms are often characterized in **ecology** by a typology of relational impacts. Typical is the following set of categories offered by **Eugene Odum**: neutral (where neither **population** affects the other); **competition** (both direct inhibition of one **species** by the other and indirect inhibition when a common resource is in short supply); **amensalism** (in which one population is inhibited, the other not affected); **predation** (in which one species preys on another by killing it); parasitism (in which one species takes its nutrients from the bodies of a host species, which are not usually killed outright though they may die eventually); **commensalism** (in which one population benefits and the other is not affected); protocooperation (in which both benefit but the relationship is not obligatory) and **mutualism** (in which interaction is favorable to both and obligatory).

Another way of thinking about the biotic environment is that it is comprised of the totality of the organisms in a particular area, ranging in scale from a small biotic **community** to the total **biomass** on the face of the earth, a mass of living things often called the **biosphere**.

BIOTIN

A member of the vitamin B family, biotin is an important **coenzyme** involved in the **metabolism** of both **carbohydrate**s and **protein**s and in the synthesis of fatty acids. It is required for **cell** growth and for the utilization of the other B-complex **vitamins**. It also may be involved in preventing hair loss. Like many vitamins, biotin was ''discovered'' several times—and given several different names as well.

In the l920s, for instance, various researchers isolated a growth factor for **yeast** that some named *bios*, and others called vitamin H. In 1927, biochemist M.A. Boas was the first to demonstrate a requirement for this compound in **animal**s. Rats who were fed a diet high in raw **egg** whites, Boas reported, soon developed severe skin rashes, lost their fur, and became paralysed—a syndrome known as *egg-white injury*. (We now know that egg whites contain a protein, avidin, that—unless destroyed by heat—keeps biotin from being absorbed by the body.) But Boas found a substance in liver that he soon discovered could cure this injury—a substance he named ''protective factor x.''

Finally, in 1940, **Vincent Du Vigneaud**—an American biochemist working for a leading pharmaceutical company—realized that biotin, a relatively little known compound, was identical both to vitamin H and to ''protective factor x.'' Intrigued by this discovery, Du Vigneaud went on to work out the coenzyme's complicated two-ring structure so that his firm's chemists could synthesize it.

Biotin is now known to be present in virtually every food, and is sometimes referred to as Vitamin B6. Moreover, the body can also synthesize it from intestinal **bacteria**. A biotin deficiency, therefore, is extremely rare and is seen only in infants born with a genetic disorder—and in people who, for one reason or another, eat large quantities of raw eggs.

BIRDS • See Ornithology

BIRTH

Birth, or **parturition**, in **mammals** is the process in which a fully developed fetus is expelled from the mother's uterus by the force of strong, rhythmic muscle contractions. The birth of live offspring is a reproductive feature shared by mammals, some fishes, and selected **invertebrates** (such as scorpions), as well as some reptiles and amphibians. **Animal**s who give birth to live offspring are called viviparous (meaning live birth).

In contrast to viviparous animals, other animals give birth to **egg**s; these animals are called **oviparous** (meaning egg

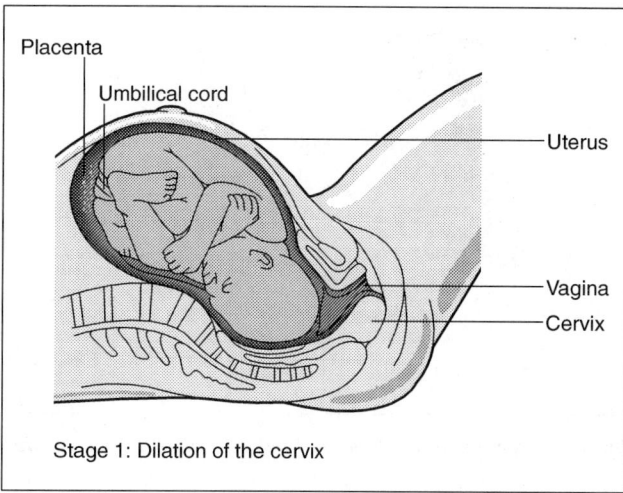

Stage 1: Dilation of the cervix

Stage 1: Dilation of the cervix. (Illustration by Hans & Cassady, Inc.)

birth). Some oviparous **species**, such as birds, retain their eggs inside their bodies for long periods of time; in these animals, the eggs are laid at an advanced stage of development. Other animals, such as frogs, give birth to less developed eggs, which undergo development outside the mother's body.

In both viviparous animals and oviparous animals, **fertilization** of the mother's egg with the father's **sperm** takes place inside the mother's body. One of the advantages to giving birth to live young is that the mother protects the fetus inside her body as it develops. The developing fetus derives **nutrients** from the mother's body, and so is assured of receiving all the nourishment it needs to complete development.

The length of time between fertilization and birth in viviparous animals is called the gestation period. The length of the gestation period varies according to species. The gestation period of mice is 21 days, of rabbits is 30-36 days, and of dogs and cats is 60 days. The largest mammal, the baleen whale, has a gestation period of 12 months—only three months longer than the gestation period of humans. Elephants have one of the longest gestation periods of all animals, 22 months.

Some viviparous animals such as humans, horses, and cows give birth to only one offspring at a time, although occasionally these animals produce twins or triplets. Other animals give birth to many offspring at a time. Usually, the multiple offspring in a litter are each derived from a separate egg, but the armadillo gives birth to four identical offspring that are derived from the same fertilized egg.

At the end of the gestation period, the mother's uterus begins to contract rhythmically, a process called labor. The initiation of labor leading up to birth is the result of a number of **hormone**s, notably oxytocin.

Shortly after fertilization the hormone progesterone increases and is maintained at high levels in the mother's **blood**stream. The high levels of progesterone prevent the uterus from contracting. The progesterone prepare the lining of the uterus (the endonestrium) for its supporting role in nurturing the developing fetus, and helps form the **placenta**. Maternal

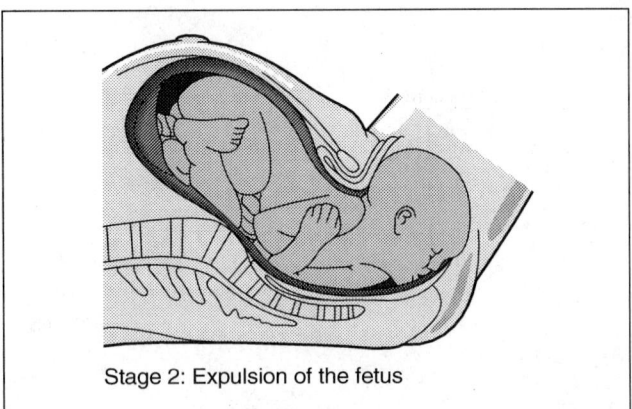

Stage 2: Expulsion of the fetus

Stage 2: Expulsion of the fetus. *(Illustration by Hans & Cassady, Inc.)*

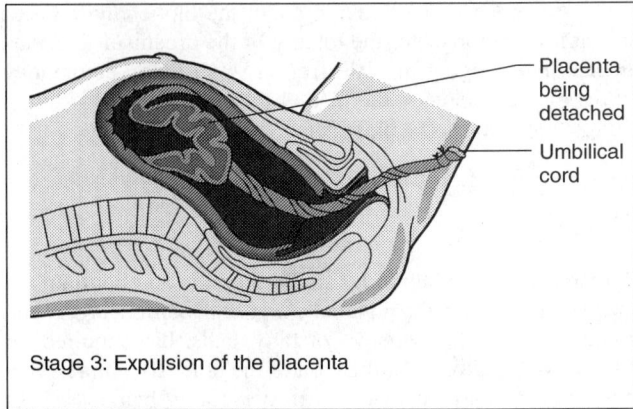

Placenta being detached

Umbilical cord

Stage 3: Expulsion of the placenta

Stage 3: Expulsion of the placenta. *(Illustration by Hans & Cassady, Inc.)*

progesterone levels begin to drop during the last weeks of gestation, while the levels of **estrogen** begin to rise. When progesterone levels drop to very low levels and estrogen levels are high, the uterus begins to contract.

Oxytocin is a hormone released from the **pituitary gland** in the **brain**, which stimulates uterine contractions and also controls the production of milk in the mammary glands of the breast (a process called lactation). Synthetic oxytocin is sometimes given to women in labor to induce labor.

The mechanism that prompts the **secretion** of oxytocin from the pituitary during labor is thought to be initiated by the pressure of the fetus's head against the cervix, the opening of the uterus. As the fetus's head presses against the cervix, the uterus stretches, and relays a message along nerves to the pituitary, which responds by releasing oxytocin. The more the uterus stretches, the more oxytocin is released.

Fetal hormones are also thought to play a role in initiating labor. At the end of gestation, the fetal **adrenal glands** secrete steroid hormones called cortico **steroids**, which cause the hormone-like substances known as **prostaglandins**. Prostaglandins contribute to the contraction of the uterus during labor.

Labor culminating in birth in humans begins with the rhythmic contractions of the uterus which dilate the cervix. This causes the fetus to move down the birth canal, and to be expelled together with the placenta—which had supplied the developing fetus with nutrients from the mother. These events can take anywhere from 48 hours to less than one hour but more usually the entire birth process takes about 16 hours.

In order for the fetus to leave the uterus and to enter the birth canal, it must pass through the cervix, the opening of the uterus. The cervix is normally tightly closed, and is sealed with a plug of mucus during gestation to protect the fetus from invading **microorganism**s. During the first stages of labor, the contractions of the uterus dilate the cervix which widens to about 4 in (10 cm), which can accommodate the passage of the fetal head.

In the last weeks of pregnancy, before labor begins, the uterus undergoes irregular contractions, which serve to exercise the muscles of the uterus and may even dilate the cervix; it's not unusual for a woman to go into active labor with a cer-

vix that's already one or two centimeters dilated. During the last weeks of pregnancy, the cervix also thins out (or effaces) which makes dilation easier.

In preparation for birth, the fetus moves further down into the mother's pelvis. When labor begins, the fetus is usually positioned with its head engaged with the top of the cervix. This engagement is called "lightening" or "dropping." When labor begins, the contractions loosen the mucus plug in the cervix, which causes small capillaries in the cervix to break, and the mucus and blood are discharged from the vagina. This discharge is sometimes called "bloody show" and signals the onset of labor.

Another sign that may signal the beginning of labor is the rupturing of the amniotic sac. In the uterus the fetus is encased in a **membrane** (the amniotic sac) and literally floats in amniotic fluid. When uterine contractions begin, this sac ruptures and the amniotic fluid can leak from the uterus. Not all women experience an abrupt rupturing of the amniotic sac; in some the amniotic fluid gradually leaks out as labor progresses. Once the amniotic sac has ruptured, or the amniotic fluid begins to leak, labor usually progresses more rapidly.

During the first stage of labor, the cervix dilates about 0.5-0.6 in (1.2-1.5 cm) an hour. The uterine contractions are about 5-30 minutes apart, and last for 15-40 seconds. The end of the first stage of labor is associated with the strongest uterine contractions. Contractions are two to five minutes apart, and last for 45-60 seconds. The cervix opens rapidly at this point. This period of labor, sometimes called transition, is usually the most difficult for the mother. The contractions are very strong and close together, and nausea and vomiting are common. After the cervix has dilated to its full width of 4 in (10 cm), the contractions slow down somewhat to about three to five minutes apart. The fetus is then ready to be born, and the second stage of labor begins.

During the second stage, lasting about one to two hours, the mother uses her abdominal muscles to push the fetus through and out of the birth canal. The pushing is actually a **reflex** action, but if a woman can help the reflex by actively using her muscles, birth goes much faster. As the fetus moves down the birth canal to the vaginal opening, the head begins to appear. The appearance of the head at the opening of the vagina is called crowning. After the head is delivered, first one shoulder is delivered, then the other. The rest of the body follows.

After the baby is born, the umbilical cord that has attached the fetus to the placenta is clamped. The clamping cuts off the circulation of the cord, which eventually stops pulsing due to the interruption of its blood supply. The baby now must breathe air through its own lungs.

Before delivery, the placenta separates from the wall of the uterus. Since the placenta contains many blood vessels, its separation from the wall of the uterus causes bleeding. This bleeding, if not excessive, is normal. After the placenta separates from the uterine wall, it moves into the birth canal and is expelled from the vagina. The uterus continues to contract even after the placenta is delivered, and it is thought that these contractions serve to control bleeding.

BISHOP, J. MICHAEL (1936-)
American molecular biologist

For work in **cancer** research, J. Michael Bishop shared the 1989 Nobel Prize in physiology or medicine with **Harold Varmus**. He and Varmus found that cancer genes (**oncogenes**) could be derived from normal cell **genes** which had not been inherently cancer-causing, as was previously thought; they stopped normal functioning and became cancerous under certain conditions. Their discovery the cellular origin of retroviral oncogenes set in motion a great amount of research on factors that govern the normal growth of cells.

John Michael Bishop was born February 22, 1936, to John and Carrie Grey Bishop. The family, which included another son and daughter, lived in York, Pennsylvania, where John Bishop was a Lutheran minister. Bishop's early schooling was almost entirely devoid of science. He entered Gettysburg College in 1953 as a premedical student and graduated with a chemistry degree in 1957. He attended Harvard Medical School but took several detours while he was there, first to work in the pathology department of the Massachusetts General Hospital and later to work with the virologist Elmer Pfefferkorn. He obtained his medical degree in 1962 and spent the required amount of time as an intern and resident at Massachusetts General, but his interest had finally focused on investigating the molecular biology of **viruses**. He worked for three years at the National Institutes of Health as a postdoctoral fellow, learning to do fundamental research. After a year of study in Germany with Gebhard Koch, Bishop took a teaching position in 1968 at the University of California at San Francisco. He was eventually appointed professor in the department of microbiology and immunology, as well as the director of the G. W. Hooper Research Foundation of the University of California Medical Center.

Bishop studied the genetic component of cancer. Many theories of cancer causation already existed when Bishop began his investigations, and new discoveries relevant to the field were made frequently. Robert Huebner and George Todaro had postulated that cancer genes (oncogenes) might lay hidden in cells, the result of viral infection many generations ago, waiting for particular environmental stresses to set them off. **Peyton Rous** had identified a sarcoma virus that caused tumors in chickens. G. Steven Martin found an **oncogene**, named *src* on the Rous sarcoma virus. **Howard Temin** identified the sarcoma virus discovered by Rous as a **retrovirus**—one that could somehow copy its own RNA information into the DNA of the host cell (which is reverse of the usual process of DNA to RNA reproduction). Temin also participated in David Baltimore's discovery of the **enzyme** called **reverse transcriptase** which accomplished that copying.

Bishop, Varmus, and their colleagues Deborah Spector and Dominique Stehelin conducted a search for *src* oncogenes in different species and found *src*- like genes just about everywhere, apparently as Huebner and Todaro had predicted. They also found that these genes were not inherently oncogenes but functioned as a regular part of the cellular machinery, performing work for the cell until their normal functioning was somehow changed. Bishop and his colleagues called these genes proto-oncogenes. Retroviruses apparently picked up these normal cellular genes and instigated changes that caused them to become cancerous, although retroviruses were only one possible cause of the transformation; some chemical carcinogens may also convert proto-oncogenes to oncogenes. In a review in *Science*, Bishop stated: "The proliferation of cells is governed by an elaborate circuitry that reaches from the surface of the cell to the nucleus. The products of proto-oncogenes may represent some of the junction boxes in that circuitry.... What we now know of oncogenes allows us to view their actions as 'short circuits' at the corresponding junction boxes."

For this discovery Bishop and Varmus received the 1989 Nobel Prize. Controversy erupted when Stehelin demanded a share of the prize for the work he had done with the two laureates, but the awarding committee remained firm. Stehelin, as well as Spector, had contributed important experiments, but the committee believed that the fundamental intellectual creativity belonged to Bishop and Varmus.

A strong proponent of basic research, in 1993 Bishop coauthored a paper in *Science* which sharply criticized the government's role in the field. The article mentioned "inadequate funding..., flawed governmental oversight of science, confusion about the goals of federally supported research, and deficiencies in science education," and offered a set of guidelines for solving these problems.

Bishop won the Gairdner Foundation International Award and the Armand Hammer Cancer Prize in 1984, the American Cancer Society National Medal of Honor in 1985, and the American College of Physicians Award in 1987.

BLACK, DAVIDSON (1884-1934)
Canadian anatomist and paleoanthropologist

Davidson Black was born in Toronto, Canada, in 1884. Black showed an early interest in biology and natural history and pursued a career in science despite a long family tradition in law. In 1903, he entered the medical school of the University of Toronto, graduating in 1909 with M.D. and M.A. degrees. Upon graduating from medical school, Black accepted a faculty appointment at Case Western Reserve University in Cleveland, Ohio. His primary responsibilities were in **anatomy**, but Black found an outlet for his interests in anthropology by helping in the expansion of a museum of comparative anthropology and anatomy. His studies in anthropology continued on sabbatical leave in 1914. In Manchester, England, Black studied anthropology under noted paleoanthropologist Grafton Elliot Smith.

In the early twentieth century, the scientific community was divided over the question of whether Asia or Africa was the birthplace of humanity. Black argued in favor of Asia, basing his position on the notion that evolutionary patterns and climate are closely interrelated, and the fact that geological evidence showed that China's climate had been suitable for the survival of ancient humans. Black's conviction was supported by a German physician's collection of mammal **fossils**.

In 1920, Black accepted a post as an anatomist at Peking Union Medical College. He welcomed the opportunity to go to China and the prospect of setting aside time to hunt fossils. Immersed in his duties at the college, Black laid aside his work in anthropology. However, in 1926, when Austrian paleontologist Otto Zdansky announced the discovery of two fossil hominid teeth at Zhoukoutian's Dragon Bone Hill site, Black successfully persuaded the Rockefeller Foundation to support a large-scale excavation at the site and again he was drawn into paleoanthropology.

Shortly before the end of the 1927 field season, Birger Böhlin, a Swedish paleontologist, found a beautifully preserved left lower molar. Based on this and the two earlier excavated teeth (which he had not seen), Black boldly proposed a new hominid **genus**, which he named *Sinanthropus pekinensis*, or ''Chinese man of Peking.'' Basing a new genus on such scanty evidence invited criticism, and his scientific colleagues refused to legitimate his claims for Peking Man. To gain support, Black traveled worldwide and allowed examination of the tooth, which he carried with him on a watch chain.

It was 1929 before additional finds of importance were discovered at Zhoukoutian. Chinese paleontologist W. C. Pei found a nearly complete skull of Peking Man partially embedded in a cave. After Black spent four months working to free the skull from the stone that encased it, he separated the bones, made casts of each, and reassembled the skull. Estimates of the brain capacity of *Sinanthropus pekinensis* placed it within the human range.

Based on the 1929 find, excavations were broadened and continued for nearly ten years. Black oversaw the work, kept detailed records, made casts and drawings, and photographed the finds from his Beijing laboratory.

Black died in 1934 of a heart attack. Franz Weidenreich (1873-1948), a German anatomist and anthropologist, carried on his work. Weidenreich studied the fossil materials extensively and published his findings between 1936 and 1943. During World War II, the Peking Man fossils were lost and never recovered. However, thanks to Weidenreich's scholarly pursuits, Black's work has been preserved.

BLACK, JAMES (1924-)
English pharmacologist

James Whyte Black was born on June 14, 1924, in Uddingston, Scotland, to a working-class family. His father was a Scottish coal miner who worked his way up to mining engineer. Black was the youngest of four sons. One of his older brothers studied medicine and Black soon followed in his footsteps. At age fifteen, he won a residential scholarship to St. Andrew's University, where he received his medical degree in 1946. He remained as an assistant lecturer from 1946 to 1947 before traveling to Malaysia to serve as a senior lecturer in physiology at the University of Malaya from 1947 to 1950. He returned to Scotland in 1950 and lectured in physiology at Glasgow Veterinary School until 1958. During this time he began research on the mechanism of increase in gastric secretions caused by the body's production of histamine. This research formed the basis for his later work on blocking histamine receptors (chemical groups in plasma membrane or **cell** interior that have an affinity for a specific chemical or compound, in this case histamine) to reduce gastric secretions. During his time in Glasgow, Black also became familiar with the alpha and beta adrenergic receptors, which are responsible for regulating heart beat.

Black joined Imperial Chemical Industries in 1958. There he sought better ways of treating angina pectoris, a painful disease caused by insufficient oxygenation of the heart. The painful episodes suffered by angina patients are caused by increased heart rate, which increases the heart's requirement for oxygen. Black's research led him to theorize that a drug that would neutralize the effects of the **hormones adrenaline** and noradrenaline, which mediate heart rate, would relieve the symptoms of angina.

The existence of receptors for these hormones had been understood since 1948, when the biochemist Raymond P. Ahlquist first described their action. Black developed a chemically similar but nonfunctional version of the active hormones that would block one of these receptors, the beta receptor. His first studies were with analogs of isoprenaline, a compound similar to noradrenaline. One of these analogs, known as propanolol or the trade name Inderol, had the desired effect. It constricted heart muscle, stopping angina attacks.

In 1964 Black joined the British subsidiary of Smith Kline & French Laboratories. There he worked on new approaches to treating intestinal ulcers. Black knew from his earlier studies that histamine stimulated the secretion of excess acid that causes ulcers. The antihistamines in use at that time inhibited muscle contractions but not acid secretion. Black attacked the problem using the same strategy that worked in the development of the angina treatment—he sought a chemical that would inhibit histamine receptors, blocking the action of the hormone. Many thousands of compounds were tested. Finally in 1972 a partial histamine receptor antagonist was found, guanylhistamine. Unfortunately, it had serious side effects and clinical tests were halted in 1974. After further modification to the chemical structure, Black's group introduced cimetidine, now known as Tagamet (registered trademark), a successful ulcer drug.

Black himself left Smith Kline in 1973. He spent four years as head of the department of pharmacology at University College in London. Then in 1978 he returned to industry, accepting a post as director of therapeutic research at the Wellcome Research Laboratories in Kent. He remained there until 1984, when the lure of academia led him to King's College of Medicine and Dentistry, where he remains today.

In 1988 Black was honored with the Nobel Prize in physiology or medicine, an award he shared with George Hitchings and Gertrude Elion, pharmaceutical researchers from Burroughs Wellcome in the United States. It is unusual for the prize to go to pharmacologists, and the award was a recognition of a truly outstanding contribution to medicine.

Black's success in designing new medicines may be attributed in part to the rational method he employed. Instead of randomly searching for chemicals with a physiological effect, he sought to understand the underlying biological processes and designed drugs that mimic life processes. To test his drugs, he designed "**bioassays**" that tested how well his drugs would work in the body.

BLASTULA

A blastula is an early **embryo** that has been formed by cleavage, or subdivision, of a fertilized **egg**. The fertilized egg, also known as a **zygote**, undergoes **cell division** that partitions the zygote into smaller and smaller **cell**s. Blastula formation is characteristic of only certain **species**. A classic example of a species that forms a blastula is the common leopard frog of North America, *Rana pipiens*. Chickens do not have a blastula stage due to the extreme amount of yolk in their eggs. Microscopic chick embryo cells cannot cleave the yolk mass. Instead of a blastula, a thin layer of cells forms on the surface of the yolk known as the blastoderm. Humans form a blastocyst which is not comparable to the amphibian blastula. The blastocyst of humans (and most other **mammals**) contains an inner cell mass which is comparable to the blastula. Blastula cells are not considered to have undergone **differentiation**. This means that the cells are not yet specialized as they are in an adult body. The onset of differentiation ordinarily occurs with gastrula formation which follows the blastula stage.

BLOCH, KONRAD (1912-)

German American biochemist

Konrad Bloch's investigations of the complex processes by which animal cells produce **cholesterol** have helped to increase our understanding of the biochemistry of living organisms. His research established the vital importance of cholesterol in animal cells and helped lay the groundwork for further research into treatment of various common diseases. For his contributions to the study of the metabolism of cholesterol, he was awarded the 1964 Nobel prize for Physiology or Medicine.

Konrad Emil Bloch was born on January 21, 1912 in the German town of Neisse (now Nysa, Poland) to Frederich

(Fritz) D. Bloch and Hedwig Bloch. Sources list his mother's maiden name variously as Steiner, Steimer, or Striemer. After receiving his early education in local schools, Bloch attended the Technische Hochschule (technical university) in Munich from 1930 to 1934, studying chemistry and chemical engineering. He earned the equivalent of a B.S. in chemical engineering in 1934, the year after Adolf Hitler became chancellor of Germany. As Bloch was Jewish, he moved to Switzerland after graduating and lived there until 1936.

While in Switzerland, he conducted his first published biochemical research. He worked at the Swiss Research Institute in Davos, where he performed experiments involving the biochemistry of **phospholipids** in tubercle bacilli, the **bacteria** that causes tuberculosis.

In 1936, Bloch emigrated from Switzerland to the United States; he would become a naturalized citizen in 1944. With financial help provided by the Wallerstein Foundation, he earned his Ph.D. in biochemistry in 1938 at the College of Physicians and Surgeons at Columbia University, and then joined the Columbia faculty. Bloch also accepted a position at Columbia on a research team led by Rudolf Schoenheimer. With his associate David Rittenberg, Schoenheimer had developed a method of using radioisotopes (radioactive forms of atoms) as tracers to chart the path of particular molecules in cells and living organisms. This method was especially useful in studying the biochemistry of cholesterol.

Cholesterol, which is found in all animal cells, contains 27 carbon atoms in each **molecule**. It plays an essential role in the cell's functioning; it stabilizes **cell** membrane structures and is the biochemical "parent" of cortisone and some sex hormones. It is both ingested in the diet and manufactured by liver and intestinal cells. Before Bloch's research, scientists knew little about cholesterol, although there was speculation about a connection between the amount of cholesterol and other fats in the diet and arteriosclerosis (a buildup of cholesterol and lipid deposits inside the arteries).

While on Schoenheimer's research team, Bloch learned about the use of radioisotopes. He also developed, as he put it, a "lasting interest in intermediary **metabolism** and the problems of biosynthesis." Intermediary metabolism is the study of the biochemical breakdown of glucose and fat molecules and the creation of energy within the cell, which in turn fuels other biochemical processes within the cell.

Conducts Research on Cholesterol

After Schoenheimer died in 1941, Rittenberg and Bloch continued to conduct research on the biosynthesis of cholesterol. In experiments with rats, they "tagged" **acetic acid**, a 2-carbon compound, with radioactive carbon and hydrogen isotopes. From their research, they learned that acetate is a major component of cholesterol. This was the beginning of Bloch's work in an area that was to occupy him for many years—the investigation of the complex pattern of steps in the biosynthesis of cholesterol.

Bloch stayed at Columbia until 1946, when he moved to the University of Chicago to take a position as assistant professor of biochemistry. He stayed at Chicago until 1953, becoming an associate professor in 1948 and a full professor in 1950. After a year as a Guggenheim Fellow at the Institute of Organic Chemistry in Zurich, Switzerland, he returned to the United States in 1954 to take a position as Higgins Professor of Biochemistry in the Department of Chemistry at Harvard University. Throughout this period he continued his research into the origin of all 27 carbon atoms in the cholesterol molecule. Using a mutated form of bread mold fungus, Bloch and his associates grew the fungus on a culture that contained acetate marked with radioisotopes. They eventually discovered that the two-carbon molecule of acetate is the origin of all carbon atoms in cholesterol. Bloch's research explained the significance of acetic acid as a building block of cholesterol, and showed that cholesterol is an essential component of all body cells. In fact, Bloch discovered that all steroid-related substances in the human body are derived from cholesterol.

The transformation of acetate into cholesterol takes 36 separate steps. One of those steps involves the conversion of acetate molecules into squalene, a hydrocarbon found plentifully in the livers of sharks. Bloch's research plans involved injecting radioactive acetic acid into dogfish, a type of shark, removing squalene from their livers, and determining if squalene played an intermediate role in the biosynthesis of cholesterol. Accordingly, Bloch traveled to Bermuda to obtain live dogfish from marine biologists. Unfortunately, the dogfish died in captivity, so Bloch returned to Chicago empty-handed. Undaunted, he injected radioactive acetate into rats' livers, and was able to obtain squalene from this source instead. Working with Robert G. Langdon, Bloch succeeded in showing that squalene is one of the steps in the biosynthetic conversion of acetate into cholesterol.

Bloch and his colleagues discovered many of the other steps in the process of converting acetate into cholesterol. Feodor Lynen, a scientist at the University of Munich with whom he shared the Nobel Prize, had discovered that the chemically active form of acetate is acetyl coenzyme A. Other researchers, including Bloch, found that acetyl coenzyme A is converted to mevalonic acid. Both Lynen and Bloch, while conducting research separately, discovered that mevalonic acid is converted into chemically active isoprene, a type of hydrocarbon. This in turn is transformed into squalene, squalene is converted into anosterol, and then, eventually, cholesterol is produced.

Awarded Nobel Prize for Cholesterol Research

In 1964, Bloch and his colleague **Feodor Lynen**, who had independently performed related research, were awarded the Nobel Prize for Physiology or Medicine "for their discoveries concerning the mechanisms and regulation of cholesterol and fatty acid metabolism." In presenting the award, Swedish biochemist Sune Bergström commented, "The importance of the work of Bloch and Lynen lies in the fact that we now know the reactions that have to be studied in relation to inherited and other factors. We can now predict that through further research in this field... we can expect to be able to do individual specific therapy against the diseases that in the developed countries are the most common cause of death." The same year, Block was honored with the Fritzsche Award from the American Chemi-

cal Society and the Distinguished Service Award from the University of Chicago School of Medicine. He also received the Centennial Science Award from the University of Notre Dame in Indiana and the Cardano Medal from the Lombardy Academy of Sciences the following year.

Bloch continued to conduct research into the biosynthesis of cholesterol and other substances, including glutathione, a substance used in protein metabolism. He also studied the metabolism of olefinic fatty acids. His research determined that these compounds are synthesized in two different ways: one comes into play only in aerobic organisms and requires molecular oxygen, while the other method is used only by anaerobic organisms. Bloch's findings from this research directed him toward the area of comparative and evolutionary biochemistry.

Bloch's work is significant because it contributed to creating "an outline for the chemistry of life," as E.P. Kennedy and F.M. Westheimer of Harvard wrote in *Science.* Moreover, his contributions to an understanding of the biosynthesis of cholesterol have contributed to efforts to comprehend the human body's regulation of cholesterol levels in blood and tissue. His work was recognized by several awards other than those mentioned above, including a medal from the Societe de Chimie Biologique in 1958 and the William Lloyd Evans Award from Ohio State University in 1968.

Bloch served as an editor of the *Journal of Biological Chemistry,* chaired the section on metabolism and research of the National Research Council's Committee on Growth, and was a member of the biochemistry study section of the United States Public Health Service. Bloch has also been a member of several scientific societies, including the National Academy of Sciences, to which he was elected in 1956, the American Academy of Arts and Sciences, and the American Society of Biological Chemists, in addition to the American Philosophical Society.

Bloch and his wife, the former Lore Teutsch, met in Munich and married in the United States in 1941. They have two children, Peter and Susan. Bloch is known for his extreme modesty; when he was awarded the Nobel Prize, the *New York Times* reported that he refused to have his picture taken in front of a sign that read, "Hooray for Dr. Bloch!" He enjoys skiing and tennis, as well as music.

BLOOD

While the importance of blood to life was understood by primitive peoples, they did not know that blood carries oxygen and food throughout the body, moves waste products, and fights disease. Early societies often ascribed mystical qualities to blood. The Greeks considered blood to be both one of the four essential humors and a carrier of the vital humor blood. For centuries, medical practice based on the findings of the Greek physician Galen advocated bloodletting, or bleeding, because the condition and amount of blood in the body was believed to profoundly affect health.

Nevertheless, the structure and function of blood remained unknown, although it had long been observed that blood standing in a container outside the body settled out into several components: a thick, dark red mass and a pale yellowish fluid (plasma) separated by a thin white layer. **William Harvey**'s revolutionary description of the **circulatory system** in 1628 focused investigative interest on blood. Some researchers, notably Antoine-Laurent Lavoisier and Richard Lower (1631-1691), studied **respiration**; Lower demonstrated that contact with air in the lungs turned dark venous blood into bright red arterial blood. Other researchers turned to the newly developed **microscope** to study the structure of blood. Dutch naturalist **Jan Swammerdam** discovered the red blood corpuscle in frog's blood in 1658 and observed human blood ''particles'' in 1662; his findings, however, were not published until 1737. **Marcello Malpighi**, who discovered the capillaries in 1661, also observed red blood **cell**s in **animal**s. It was the Dutch draper and devoted microscope hobbyist Antoni van Leeuwenhoek who first saw and described human red blood corpuscles in 1673-1674. This, plus his microscopic discoveries of **protozoa**, **bacteria**, and **sperm**atozoa, brought Leeuwenhoek world fame.

White blood cells, also called leukocytes, were first observed by Lancisi in 1740. William Hewson (1739-1774), a student of the famed English anatomist and surgeon **John Hunter** (1728-1793), studied the white cells in detail some 30 years later, managing to distinguish two types. **Paul Ehrlich** developed a staining method in 1875 that revealed several distinct kinds of leukocytes. In 1883, **Elie Metchnikoff**, a Russian-born microbiologist working in France, observed simple animal life under a microscope and noticed that certain cells moved immediately to damaged areas. He was soon able to show that white blood corpuscles in higher animals and human beings also attacked and ingested invading bacteria. Metchnikoff called these cells phagocytes.

The third formed objects in the blood, platelets, were first described by Alfred Donné (1801-1901) in 1842 and first studied in detail by Canadian physician William Osler (1849-1919), who reported his findings in 1873-1874. The role of platelets in blood clotting was explained by Giulio Bizzozero (1846-1901) in 1882. The origin of all types of blood cells was shown by Newmann in 1868 to be from parent cells in the bone marrow.

Methods of measuring blood elements were developed in the 1800s, an idea promoted during the 1840s by Gabriel Andral (1797-1876), who introduced the term anemia into medicine. François Magendie suggested a way to count red blood cells that was expanded on and used by Karl Vierordt in 1852 to make an actual count. The next year, Hermann Welcher found a way to count white corpuscles. Georges Hayem (1841-1933) was the first to accurately count platelets. The first practical hemoglobinometer—a blood ''counting chamber''—was invented by William Richard Gowers (1845-1915), an English neurologist, in 1878. These counting methods became increasingly important as knowledge about the medical implications of changes in blood composition grew.

While the red blood cells had been discovered in the 1600s, their function was not understood until the mid-1800s. In the early 1850s, two Germans, the physician Otto Funke

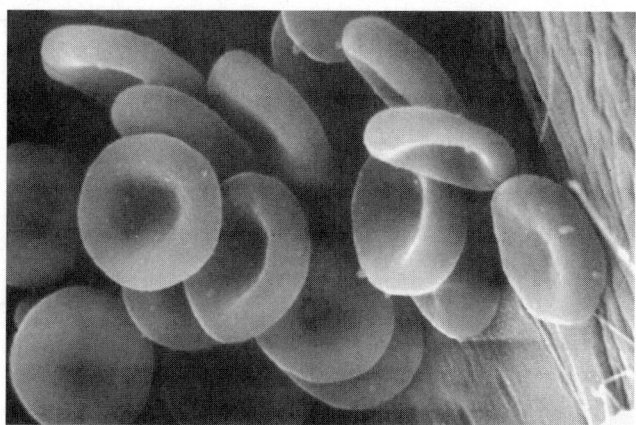

Red blood cells. *(Photograph by Dennis Kunkel, Phototake NYC. Reproduced by permission.)*

(1828-1879) and the chemist **Justus von Liebig**, showed that oxygen combined with a substance, as yet unnamed, in the red blood cell. The German physiological chemist Ernst Hoppe-Seyler isolated and crystallized the substance (a **protein**) in 1862 and named it **hemoglobin**. The studies of these men showed that it was hemoglobin that carried oxygen to the body's **tissue**s, giving it up at the cells. Another German chemist, **Hans Fischer**, worked out the chemical structure of hemin, the active portion of the hemoglobin **molecule**, and was able to synthesize it in 1929. While researching the causes of anemia, American surgeon George Whipple (1878-1976) discovered that iron, stored in the liver, was essential for the regeneration of hemoglobin; this 1925 discovery made it possible to treat cases of severe anemia (lack of red blood cells) with iron supplements. English biochemist Max Perutz finally worked out the precise atomic arrangement of hemoglobin—which contains 12,000 **atoms** per molecule—in 1960.

The study of blood cells was advanced in the twentieth century by Dr. **Florence Rena Sabin** of Johns Hopkins Medical School. Around 1920, she perfected a method of staining living cells so their movements and reactions could be studied. Sabin then used this technique to study the formation of red and white blood cells, which she actually saw occur in the blood vessels of a chick **embryo**.

Scientists have long known that hemoglobin carries oxygen from the lungs out to the body, then carries **carbon dioxide** back to the lungs to be exhaled. In 1996, cardiologist Jonathan Stamler and biochemist Joseph Bonaventura showed that hemoglobin plays another role as well, as part of a nitric acid cycle that also aids oxygen delivery.

BLOOD CIRCULATION

The **heart** and **blood** have been recognized since earliest times as vital elements of life, but the way in which they function was not understood until **William Harvey**'s discovery of the mechanics of the **circulatory system** in the seventeenth century.

Various Greek physicians had distinguished **arteries** from **veins**, but each—Alcmaeon (born c. 535 B.C.) and Praxagoras (born c. 340 B.C.)—thought that arteries carried air, since arteries are usually empty in corpses (the source for most of these investigations). The Alexandrian anatomist Herophilus observed the arterial pulse, which he associated with the heartbeat, and held that arteries contain blood, not air. His fellow Alexandrian, **Erasistratus**, came close to a concept of the circulatory system, maintaining that both arteries and veins originated from the heart and divided repeatedly throughout the body, ending in fine capillaries. Erasistratus saw the heart as a pump, but he repeated the error of regarding arteries as conduits for air.

All these early ideas about the circulatory system were superseded by the medical system described by Galen, a schema that remained firmly in place until early modern times. According to Galen, the veins originated in the liver, where blood was formed, and flowed from there to all parts of the body, where it was consumed. Some blood was carried to the heart, where it passed from one side to the other through minute pores in the septum. In the left side of the heart, Galen maintained, the blood mixed with air, creating a vital spirit that was carried throughout the body by the arteries.

In the centuries after Galen, Christian, Jewish, and Muslim authorities firmly opposed postmortem dissection, so Galen's beliefs about **anatomy** and physiology stood virtually unchallenged for nearly fourteen centuries. One exception was the Muslim physician Ibn an-Nafīs (c. 1205-1288), who accurately believed the right and left ventricles to be separate, and described movement of blood between them via the lungs. His work, however, passed into obscurity and was not rediscovered in the West until 1924.

The new scientific spirit that began stirring in the sixteenth century gave rise to questions about Galen's circulatory theory. **Leonardo da Vinci** (1452-1519) accurately drew cardiac valves. **Andreas Vesalius** rejected the idea of a permeable cardiac septum. In 1553, the Spanish theologian and physician **Michael Servetus** (1511-1553) published a theological treatise entitled *Christianismi restitutio* (*The Restoration of Christianity*) that, in describing the introduction and transmission of the divine spirit through the body, also described the "lesser circulation" of the blood—the passage of the blood from the left side of the heart through the pulmonary artery to the lungs and then back through the pulmonary vein to the right side of the heart. The theological views that Servetus espoused in his treatise resulted in his being burned at the stake as a heretic; almost all copies of his work were destroyed and his discovery of pulmonary circulation remained unknown until 1694. Although Servetus's idea about pulmonary circulation was correct, it was theoretical. An Italian anatomist, Realdo Colombo (1516?-1559), also described the "lesser circulation," his work being based on anatomical observations gained through vivisection, autopsy, dissection, and surgery. Colombo published his findings in 1559 in *De re anatomica*, and this work became widely known.

Two more discoveries paved the way for Harvey. The Italian physician-botanist Andrea Cesalpino (1519-1603)

coined the term *circulation*, described the heart's valves in 1571, and emphasized the outflow of arterial blood and inflow of venous blood from and to the heart. Hieronymous Fabricius (1537-1619), who taught Harvey at Padua, discovered the valves of the veins and published his findings in *De venarum ostiolis* in 1603.

The man who finally unraveled the mysterious mechanisms of blood circulation was the English physician William Harvey. A doctor with a flourishing practice, Harvey was also an assiduous medical researcher. For his studies of the heart and blood vessels, he vivisected and dissected many **species** of **animal**s as well as cadavers. In 1628, he published his ideas on the circulatory system in *De motu cordis et sanguinis* (*On the Motion of the Heart and Blood*), which transformed medicine.

Harvey reached his revolutionary conclusions through sound reasoning based on detailed observation. He found that the heart acted as a pump, forcing blood through the arteries by its contractions. He noted that the valves of both the heart and the veins allowed blood to flow in one direction only. Harvey then calculated the amount of blood propelled out of the heart with every beat and found that the volume of blood pumped out of the heart every hour amounted to three times the normal weight of a man. The same blood, he reasoned, must be continuously moving through the body in "a kind of circular motion"—that is, circulating.

Harvey's theory, while it generated some controversy, was generally accepted during his lifetime. The final proof of the theory came in 1661 when **Marcello Malpighi** used the newly invented **microscope** to discover the capillaries, the previously invisible network of extremely fine blood vessels linking arteries and veins. Harvey's findings ended the era of Galenic and Greek medicine and ushered in the era of modern experimental physiology.

BLOOD CLOT DISSOLVING AGENT

Over half of the cardiovascular-related **death**s in the United States are caused by **heart** attacks. When oxygenated **blood** cannot flow to an area of the heart muscle, a heart attack may occur. Heart attacks or coronary infarctions can result from a blocking of a blood vessel due to an accumulation of fatty substances on the inside wall of an artery (**arteriosclerosis**). They are also caused by blood clots which form in the heart (a *thrombus*), or by clots that form in other areas of the body and circulate to the heart (an *embolus*).

Many strategies have been developed to prevent the deposits of **cholesterol**, fibrin (a clotting material), and cellular debris which combine to form a thrombus. Such **anticoagulant**s as aspirin, warfarin, and heparin thin the blood and help to prevent initial clot formation. Until recently, however, very few therapies were able to break down previously existing clots. Scientists have now discovered several substances that can reduce the thrombus after it has already formed.

The process of dissolving blood clots is called thrombolysis. Thrombolytics, or clot-dissolving, agents have been shown to reduce heart muscle damage and heart attack deaths if administered within a few hours after symptoms appear. There are three groups of thrombolytic agents. The first group includes **enzyme**s, like plasmin, that act directly upon the fibrin strands within the thrombus. Certain agents that increase the amount of plasma activator comprise the second group. Finally, the third group contains the plasminogen activators which include streptokinase, urokinase, and **tissue** plasminogen activator. All these drugs digest the fibrin clot by stimulating production of the enzyme plasmin. Plasmin is produced when another substance, called plasminogen, is activated. The conversion of plasminogen to plasmin is controlled by certain enzymes known as plasminogen activators.

Streptokinase, produced with streptococci **bacteria**, has been used since about 1960. Although streptokinase is the least expensive activator, some negative side effects such as immune responses may accompany its use.

Urokinase is found naturally in humans, especially in the urine. Thus, no negative immune response is associated with its use. This therapy is usually administered in small doses and combined with other drugs because it is difficult to purify and rather expensive.

Tissue plasminogen activator (t-PA) is unique drug for dissolving clots in that it activates only fibrin-bound plasminogen and thus is directed more to target the clot site. t-PA in human blood is produced in very small amounts by vascular endothelial cells. Since about 1980, when t-PA was first purified from human uterine tissue, it has enjoyed widespread use among American physicians. In 1996, the U.S. Food and Drug Administration approved t-PA for the emergency treatment of stroke. In a 5-year clinical trial, investigators found that t-PA was an effective emergency treatment for acute ischemic (local anemia due to obstruction of the blood supply) stroke, although there is a risk of bleeding. Currently, t-PA is the most expensive thrombolytic agent on the market. A single treatment may cost $2,200 compared to $200 for a single treatment of streptokinase. In an effort to make the drug more affordable, studies are under way usingcloning technology to recombine the **gene**s that encode human t-PA.

Each of these thrombolytic agents shows great promise in reducing the severity of heart attack and will undoubtedly save many lives. However, it is essential to determine the most efficient means of production and the best dosages for these drugs. Researchers also continue to look at which are the best drugs to use. For example, a 1998 study has shown that the dual therapy of streptokinase and the anti-clotting agent hirudin reduces short-term risk of another heart attack or death as compared to those patients given streptokinase and heparin or t-PA therapy and either heparin or hirudin..

BLOOD GROUPS

The development of **blood** transfusion, a potentially lifesaving technique, was hampered by a serious problem: many times the patient would suffer an often fatal "transfusion reaction," apparently to the donated blood itself. The cause of such reactions was unknown.

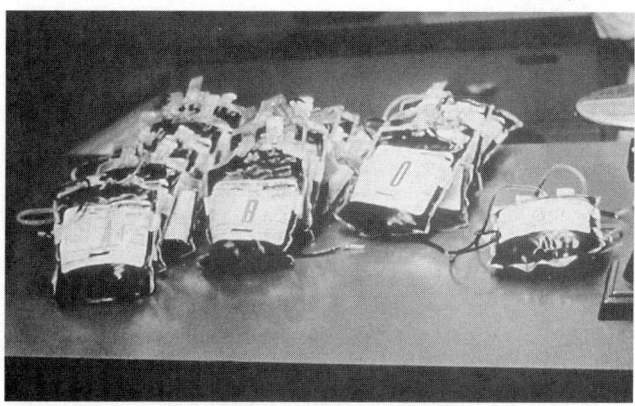

Bags of blood at a donation center, arranged by blood group. Before Karl Landsteiner discovered the existence of these distinct types, transfusion were often unsuccessful. *(National Library of Medicine. Reproduced by permission.)*

In the late 1800s, several researchers noted that when blood **cell**s from one **animal** or person were mixed with cells from another, agglutination occurred—the cells stuck together in clumps. An Austrian physician named **Karl Landsteiner** showed what caused this clumping in 1900. Mixing different samples of human red blood cells and human sera (blood without the cells), Landsteiner observed that one mixture would clump, while another different mixture would not. He determined that the different reactions were caused by differing **antigen**s in the red blood cells reacting to antibodies in the sera. In 1901, Landsteiner published a paper describing a method of categorizing human blood into three groups, or types: A, B, and C (later called O). In 1902, two of Landsteiner's assistants, Alfred von Decastello and Adriano Sturli, found a fourth blood type, AB. Depending on the antibodies and antigens present in each type of blood, serum from one type can cause clumping in another type. By identifying the blood type of both donor and recipient in a potential transfusion, it was now possible to avoid mixing bloods that would precipitate agglutination. Landsteiner's discovery paved the way to safe blood transfusion, and the physician (now an American citizen) received the Nobel Prize in medicine or physiology for his work in 1930.

Landsteiner and his associates, Alexander S. Wiener (1907-1976) and Philip Levine (1900-), discovered another important blood group, the Rh system, or **Rhesus factor**, in 1940. People with the Rh factor present in their blood are termed Rh positive; people without it are Rh negative. An Rh negative person exposed to Rh positive blood forms antibodies against the Rh factor. Once this has occurred, any future transfusion of Rh positive blood can cause a severe, possibly fatal, transfusion reaction. A fetus carried by an Rh-negative woman who has developed Rh antibodies (perhaps by previously carrying an Rh-positive baby) can have its red blood cells attacked by these antibodies, a condition called *erythroblastosis fetalis*, which may result in **brain** damage or **death**. The Landsteiner group's identification of the Rh factor made it possible to avoid Rh-incompatible transfusions and to diagnose and thus correct fetal Rh incompatibilities via a blood transfusion in the womb.

In 1910, Emil von Dungern and Ludwik Hirszfeld (1884-1954) demonstrated that blood groups are inherited, a concept confirmed by B. A. Bernstein in 1924. This, coupled with the discovery of many more blood groups—Landsteiner and his group found M, N, and P in 1927, for example—stimulated interest in the use of blood types in studying **heredity**. Soon, blood typing was being used in court cases to establish or disprove paternity. In the mid-twentieth century, American immunochemist William C. Boyd (1903-) pioneered the study of blood types as distributed throughout the human race, tracing the **evolution** and **migration** of the **species** and demonstrating that there is no genetically "pure" race, as all blood types are present in all human **populations**.

The most common blood types in the United States are O+ (O Rh-positive), found in 38 percent of the population, and A+, found in 34 percent of people. In an emergency, anyone can receive type O red blood cells. For this reason, people with type O blood are often called "universal donors."

Today an estimated 14 million units of blood are donated by about 8 million donors each year. These units are transfused into as many as 4 million patients annually, according to the American Association of Blood Banks. Among those who benefit from blood transfusions are accident victims, patients undergoing surgery, and people receiving treatment for diseases such as **cancer** and sickle cell disease.

BLOOD PRESSURE

One of **anatomy**'s great mysteries was solved when **William Harvey** explained the mechanics of **blood circulation** in 1628. Remaining to be investigated was blood pressure, the force of the **blood** pressing against the inner walls of the blood vessels. Blood pressure is higher during cardiac *systole*, when the **heart**'s ventricles contract, pumping blood into the **arteries**. Blood pressure is lower during cardiac *diastole*, when the ventricles relax.

Harvey had noted that when an artery was severed, blood was ejected in spurts. **Stephen Hales**, an English clergyman with an avid interest in physiology, became intrigued: how great was the pressure the arterial blood appeared to be under? Hales began exploring the mysteries of blood pressure and its measurement while a student at Cambridge University in the early 1700s and continued around 1712-13 at his parish in Teddington, England. He conducted a series of experiments on domestic **animals**, beginning with dogs. Using a glass tube connected via a brass pipe to an animal's blood vessel, Hales was able to measure blood pressure by the height to which the blood rose in the tube, most dramatically 9 ft. 6 in. (2.75 m 15.2 cm) in a white mare. Hales reported these first-ever blood pressure measurements in his *Haemastaticks* of 1733.

Hales continued his blood pressure investigations by studying the rate of blood flow through the capillaries, successfully demonstrating that resistance to flow in those small vessels could vary greatly. This indirectly proved that capillaries could dilate and contract. Hales also showed that blood pressure varied according to the size of the animal, and linked blood pressure measurement with rate of flow and peripheral resistance.

In the 1770s, Italian physiologist **Lazzaro Spallanzani** studied blood circulation and demonstrated that the arterial pulse is caused by the pressure exerted against the vessels' walls by the blood being pumped through.

Hales's method of measuring blood pressure was improved in 1828 by French physician Jean Leonard Marie Poiseuille (1797-1869). In place of Hales's glass tube, Poiseuille used a mercury manometer, a U-shaped tube containing mercury and calibrated in millimeters. Using his device, Poiseuille showed that blood pressure rises and falls as a person breathes out and in, and that blood pressure is the same in vessels both near the heart and far away. Today blood pressure is still expressed in millimeters (mm) of mercury (HG).

Blood pressure measurement became an important diagnostic tool after 1896, when Scipione Riva-Rocci (1863-1937) invented the sphygmomanometer, the prototype of today's standard blood pressure measuring instrument. The sphygmomanometer uses a flexible band around the arm that is inflated until arterial flow stops. Air is released from the band; when the pulse reappears, the pressure reading is taken (systolic pressure, while the ventricles are contracting). A Russian physician Nikolai Korotkoff added the use of the stethoscope to listen to the sound of both the maximum (systolic) and minimum (diastolic) pressure in the elbow's brachial artery. In a blood pressure measurement, the systolic number is placed over the diastolic number; for example, "120 over 80," written as 120/80 mm Hg.

Blood pressure can now be easily and accurately measured and is an important indicator of cardiac and circulatory health or distress. A blood pressure reading less than 140/90 mm Hg is considered normal, and a blood pressure below 120/80 mm Hg is even better. High blood pressure, also called hypertension, is an important risk factor for heart and kidney disease and stroke. About a quarter of all American adults now suffer from hypertension. Maintaining a healthy weight, being more physically active, choosing foods that are low in salt and sodium, and moderating any use of alcoholic beverages are steps that can help prevent this condition.

BLOOD TYPES

Until 1900, no one understood why **blood** could not be transfused from one person to another without occasional difficulties. Viennese pathologist **Karl Landsteiner** was the first to discover that not everyone's blood is exactly alike, and actually can be divided into four primary groups or types: A, B, AB, and O.

The various blood types are identified according to the various **proteins**, or **antigens**, on the surface of blood **cells**, and the antibodies to other antigens. Antibodies react to antigens as part of the body's defense mechanism to protect the blood supply, by combining with them chemically. One person with a blood antigen type A, for example, will have antibodies to blood type B, and vice versa. Some people have both A and B antigens (thus the AB blood type) and do not have antibodies for either group, or the O group. Those with this type of blood are called the "universal recipient," because they can receive blood from A, B or O donors. Approximately half of the world's **population** has antibodies to both A and B antigens, and make up the O group. They are called the "universal donors" because they can provide blood to people with all other blood types.

Since blood with one blood type's antigen does not carry antibodies to that particular antigen, blood of the same type can be easily transferred and mixed. Transfusing blood from one person to another is a common transplant today, but it is essential that the correct blood type be given to the recipient. If he or she has antibodies to the transfused blood type or antigens, an immune reaction occurs and the red cells in the blood clump together.

BLOOD VESSEL • See Blood circulation

BLUMBERG, BARUCH SAMUEL
(1925-   )
American research physician

When Baruch Samuel Blumberg was notified on October 14, 1976, that he was a co-winner of the Nobel Prize for physiology or medicine, he made a humorous and low-key comment to the *New York Times:* "I'm especially pleased that someone from Philadelphia won. It's appropriate in the Bicentennial year and makes up in part for the Phillies not making it to the World Series." But there was nothing low-key about the research Blumberg had done to win the prize. In 1963 he had discovered a **protein** in the **blood** of Australian Aborigines, the so-called Australia antigen, which he determined to be part of the hepatitis B **virus**. This discovery has led to the introduction of blood screening programs as well as a successful vaccine against this disease, which has a mortality rate of up to 15 percent.

Blumberg was born on July 28, 1925, in New York City, one of three children of Meyer Blumberg, a lawyer, and Ida Simonoff. After graduating in 1943 from Far Rockaway High School in Far Rockaway, New York, Blumberg enlisted in the Navy. He was assigned to study physics at Union College in Schenectady, where he earned a B.S. in 1946, and then enrolled at Columbia University graduate school in physics and mathematics. But Blumberg had become more and more interested in medical and biochemical matters, and partly at his father's urging he entered Columbia's College of Physicians and Surgeons in 1947. Four years later he earned his M.D. and completed his internship and residency at Bellevue and Presbyterian hospitals in New York. It was during this period that he met Jean Liebesman, another medical student, whom he married in 1954; they would have four children. Blumberg won a fellowship to Balliol College, Oxford University, in 1955, working toward a Ph.D. in biochemistry. His specific field of interest was hyaluronic acid, one of the major constitu-

ents of connective tissue, synovial fluid and the vitreous humor of the eyes. By 1957 he had earned his doctorate and was also hard at work on research which would later win him the Nobel Prize.

Seeks Clues to Variation in Disease Susceptibility

As a medical student working in Surinam (then Dutch Guiana), South America, Blumberg had become interested in the manner in which various ethnic groups respond to disease and infection. He began to ask himself a very simple question: why do some people get sick while others do not? It was this question that increasingly guided his work, even while at Oxford. Epidemiologists had already speculated that an answer to this question might lie in the blood, and more specifically in the variations of genetically reproduced proteins in the blood. To study such polymorphisms would necessitate a large variety of blood samples from around the world. Blumberg, on his return from England, took the perfect job for such research as chief of the geographic medicine and **genetics** section of the National Institutes of Health (NIH). From 1957 to 1964, his travels took him from Alaska, to Africa, the Pacific, South America, Europe and Australia. Often he journeyed to remote areas accompanied only by his blood drawing and testing equipment. It was during this time as well that Blumberg became interested in anthropology.

Soon Blumberg and former Balliol colleague Anthony C. Allison were studying blood samples from patients who had received multiple transfusions, such as hemophiliacs, focusing on the antigen/antibody connection. An antigen is the substance that causes the body to produce a chemical defense, or antibody, against a foreign substance. Their reasoning was that people who had received numerous transfusions might prove to be excellent test cases, producing antibodies other than those they had inherited. The serum of such patients would therefore provide a wide variety of antibody responses once they were tested against other serum samples. Blumberg hypothesized that antibodies created in the serum of hemophiliacs and other transfusion donors would react with unknown antigens in the homogeneous serum of donors from disparate geographic areas.

In 1963, the serum of a New York hemophiliac reacted with that of an Australian Aborigine, and Blumberg labeled the detected antigen the Australia antigen, Au. Initially he and other researchers thought Au, an antigen rare in North America but prevalent in Asia and Africa, might be an indicator of leukemia, because it appeared in many patients suffering from that disease. Later research dealt with groups of patients with Down's syndrome, who also show a high incidence of the antigen.

In 1964 Blumberg left NIH for the Institute of Cancer Research of the Fox Chase Cancer Center in Philadelphia, where he accepted the position of associate director of clinical research. He continued his researches on the Australia antigen, and in 1966 he discovered the link between Au and hepatitis B. A Down's patient who had previously tested negative for Au suddenly tested positive, and soon developed hepatitis, as did another with a sudden positive test for the antigen. Researchers in Japan and New York began a long series of controlled experiments which finally established the connection between hepatitis B and Au. That same year, the Australia antigen was identified as part of the B virus itself and was renamed HBsAg (hepatitis B virus antigen).

Heralds New Era in Hepatitis Research

The first practical result of Blumberg's discovery of HBsAg was a blood test which he and others developed to detect and screen out hepatitis B carriers — of which there are approximately one hundred million worldwide, and perhaps one million in the United States — from blood donors, thereby securing a safe blood supply. As early as 1969, such screening was underway at blood banks worldwide. After the American Association of Blood Banks ordered all of its members to use the hepatitis test in 1971, the incidence of hepatitis after transfusions dropped by 25 percent. In the 1970s, Blumberg, along with Irving Millman, developed a vaccine from the sera of patients with HBsAg which prevents hepatitis B infection. Since becoming commercially available in 1982, it has been widely and successfully used, especially among high-risk professionals such as healthcare workers. Another spin-off from Blumberg's work is research indicating that chronic infection with hepatitis B virus may be a precursor of **cancer** of the liver, the most common form of cancer in males in parts of Asia, India and Africa. The discovery of a vaccine against the disease may therefore reduce the risk of primary liver cancer. Mass vaccinations of newborns have been undertaken in some Asian and African nations to that effect.

Blumberg shared the Nobel Prize for physiology or medicine in 1976 with Dr. **D. Carleton Gajdusek** of the National Institute for Neurological Diseases, "for their discoveries concerning new mechanisms for the origin and dissemination of infectious diseases." They shared the $160,000 stipend equally. This is only one of a plethora of awards and honors Blumberg has won. Others include the Eppger Prize from the University of Freiburg (1973), the Distinguished Achievement Award in Modern Medicine (1975), the Gairdner Foundation International Award (1975), the Governor's Award in the Sciences from the Commonwealth of Pennsylvania (1988), and the Gold Medal Award from the Canadian Liver Foundation (1990).

In 1977 Blumberg became a professor of medicine and anthropology at the University of Pennsylvania; soon thereafter he was named vice president of population oncology at the Fox Chase Institute in Philadelphia. He has continued his researches in antigen systems as well as his studies in a wide range of other fields, including virology, physics, history, anthropology and philosophy. With the advent of the **AIDS** epidemic, Blumberg's antigen/antibody research has taken on new importance. After a long and distinguished career at Fox Chase, Blumberg returned to Oxford as master of Balliol College, becoming, at 64, the first scientist and first American ever to hold that prestigious chair.

Blumberg, known to friends, family and colleagues as Barry, is an avid movie-goer and reader. He also plays squash and enjoys running, hiking, swimming, and canoeing, in addition to his hobbies of carpentry and photography.

BONNET, CHARLES (1720-1793)
Swiss naturalist

Charles Bonnet was born in Geneva, Switzerland, to a wealthy family. He initially studied law, but a strong interest in insects led him to the field of natural history, to which he eventually devoted himself. His important observations have helped scientists better understand the process of insect **metamorphosis**. Many of Bonnet's observations are contained in his *Traité d'insectologie* (1745), which remains his highest regarded work in the field of insect biology.

Bonnet's observations on parthenogenetic reproduction, which is procreation without fertilization from sperm, are considered of special importance. He studied the reproduction of aphids and had determined that a number of females delivered live offspring even though their eggs had never been fertilized.

There was a debate over whether this reproductive phenomenon signified offspring that were "pre-formed" prior to their delivery, or offspring that were epigenetically developed, meaning that they formed during gestation. Bonnet was convinced that parthenogenetic reproduction involved pre-formed offspring.

From insect metamorphosis and parthenogenetic reproduction, Bonnet turned to the study of botany, but his research was cut short by failing eyesight. Even blindness did not still Bonnet's keen intellect; he simply applied himself to a more accessible discipline—philosophy.

See also Parthenogenesis

BORDET, JULES (1870-1961)
Belgian bacteriologist

Jules Bordet made a number of pioneering discoveries in immunology. He was the author of *Traité de l'Immunité dans les Maladies Infectieuses* (*Treatise on Immunity in Infectious Diseases*; 2nd ed., 1939) and numerous other medical publications.

Born in Soignies, Belgium, the second son of a schoolteacher, Bordet received his medical degree from the University of Brussels in 1892. In 1894, a Belgian government scholarship enabled Bordet to work in Elie Metchnikoff's laboratory at the Pasteur Institute in Paris, where he remained until 1901. In 1899, Bordet married Marthe Levoz; they had two daughters and a son.

At Metchnikoff's laboratory, Bordet investigated *bacteriolysis*, the phenomenon of **cholera** bacteria dying when injected into immunized animals (discovered in 1894 by Pfeiffer and Issaeff). Bordet found that serum contains two substances: a *preventive substance* (a specific **antibody**) and a *bactericidal substance*. When he heated serum to 55°C, the antibodies remained, but the bactericidal substance vanished, and the serum lost its ability to destroy **bacteria**. Bordet called this substance *alexin*, but Ehrlich renamed it *complement*. Bordet concluded that antibodies had to react with complement in order to be able to kill bacteria.

Bordet went on to show that the clumping and destruction of transfused red **blood** cells are also caused by antigen-antibody-complement reactions. He extended the concept of antigenic specificity—that an antigen can be identified by exposing it to blood serum containing the specific antibody to which it reacts. He also showed that when antibody-antigen complexes react with complement, they precipitate out of a solution. All these immune reactions were then used by numerous researchers as the basis for a myriad of diagnostic tests.

Bordet's findings made him famous, and in 1901, he moved to Brussels to direct its new bacteriological institute, which was renamed the Pasteur Institute of Brussels in 1903. Here, he demonstrated the complement-fixation reaction—the binding of complement to an antibody-antigen complex, which causes red blood cells or bacteria to clump. This, too, proved to be a valuable diagnostic tool. August von Wassermann (1866-1925) used it for his well-known test for syphilis.

In the course of his immune researches, assisted by his brother-in-law Octave Gengou, Bordet unexpectedly isolated the bacterium that causes whooping cough and developed a vaccine for the disease. Bordet's studies of this bacterium also revealed, in 1910, the phenomenon of antigenic variation in bacteria—the fact that disease organisms can change their antigens and therefore become resistant to antibodies.

From 1907 to 1935, Bordet was professor of bacteriology at the University of Brussels, in addition to being director of the Pasteur Institute. When he finally retired from the Institute in 1940, his son, Paul, succeeded him. For his discoveries relating to immunity, Bordet received the 1919 Nobel Prize for physiology or medicine. Bordet also studied blood coagulation and *bacteriophages* (viruses that destroy bacteria). He died in Ixelles, Belgium, in 1961.

Jules Bordet. *(The Library of Congress. Reproduced by permission.)*

BOTANY

Botany is the study of **plants** and includes plant **classification**, structure, physiology, and economic importance within human society. In the strictest sense, botany is the study of those **organism**s containing chlorophyll, however for historical reasons, it generally includes the study of non-chlorophyll-containing organisms such as the **fungi**.

Historically, the formal study of botany can be traced to the ancient Greeks and Romans. For them, botany was merely a part of the study of **nature** and natural history, rather than a distinct subject in its own right. Botany as a formal and separate subject did not exist until the mid-seventeenth century. The word botany comes from the Greek word *botane*, meaning plant or pasture.

At its inception, botany consisted of a purely comparative physical system in which plants were grouped together by appearance. Early in botany's history, the doctrine of signatures was used to classify plants. This stated that plants or plant parts that resembled areas of the human body could be used to treat disease in those areas. For example, a walnut with its wrinkled appearance was considered useful in the treatment of **brain** diseases. Increasingly, usage was considered along with physiology when classifying plants. It was not until the the mid-eighteenth century when **Carl Linnaeus** was compiling the

framework for modern classification of plant and **animal species** that the modern era of botany truly began. With the advent of the **microscope**, botany expanded to include a thorough investigation of **plant anatomy** and **cell** structure. At this point, botany was still an important subject for the clergy and very few professional botanists existed.

Many early botanists were merely collectors and recorders, and up until the invention of the printing press in the fifteenth century, much that was written about plants was merely copied from earlier studies of natural history. Mostly passed from the Greeks and Romans, plant information included the diagrams which became increasingly stylized and less accurate. With the invention of the printing press, herbals (documents on plants) with new and more accurate illustrations became available. New and original research became more common, and with the advent of the Renaissance, the production of botanical illustrations became more scientifically accurate.

Modern botany now makes use of a whole range of tools that have revolutionized the subject. These range from the molecular tools of **deoxyribonucleic acid** (**DNA**) investigation to micropropagation and a more holistic approach to the subject. Modern botany is just as interested in the range of organisms within a **habitat** and their interactions as with single plants. **Ecology** is an ever present underlying principle. There is much basic botanical research that remains to be accomplished. Many new plant species wait for discovery; many plants need to be further investigated for possible uses; molecular work needs to be done on the vast majority of plants; there is still much to be done on the **taxonomy** and evolutionary relationships between plants. In 1998, botanists at Kew Gardens in England produced a new classification of the plant world, which moved a large number of plant families into different **order**s. This re-evaluation of the taxonomy of the plants was based on an investigation of the DNA of the species examined. In total, three genes were examined from each **family**, and the similarity between these genes was considered. The more similar the gene sequence was the closer related were the plant families.

BOVERI, THEODOR (1862-1915)
German cytologist

Theodor Boveri was born on October 12, 1862, in Bamberg, Germany. He began his scientific career in 1881 as a student at the University of Munich. His original intent upon entering the university had been to study history and philosophy, but he soon decided to change his direction and concentrate on the natural sciences. He graduated in 1885, and in 1893, he became a professor of zoology and comparative anatomy at the University of Würzburg, where he remained for the rest of his career.

One of Boveri's contemporaries, **Edouard van Beneden**, was making exciting discoveries about **chromosomes** in the eggs of a species of roundworm during the 1880s. He found that the chromosome number is constant for each species, and that the chromosome number is reduced by half as reproductive cells are formed—a process we now know as **meiosis**. Van Beneden suggested that chromosomes represent continuous elements of **heredity**, but his reports were inconclusive.

It was Boveri, with his excellent powers of observation and interpretation, who proved that chromosomes were independent entities. He emphasized that they were organized structures. When Boveri first began his work, it was not yet known whether each chromosome contained factors responsible for the total development or whether each chromosome differed from others in being responsible for only particular hereditary features. Boveri's discoveries made it clear that certain chromosomes were responsible for certain characteristics. According to Edmund Wilson, a distinguished American cytologist, Boveri's theory provided the working basis for nearly all cytological interpretations of genetic phenomena.

Around 1887, Boveri and Van Beneden independently discovered a small structure that connects the chromosomes during cell division. Boveri called it the centrosome. He went on to demonstrate that it provided the division centers for the dividing egg **cell** and all its offspring.

Boveri paid a price for his outstanding scientific abilities. He was prone to bouts of depression and suffered numerous physical breakdowns. His health got progressively worse following the onset of World War I, and he died at the age of 53 in 1915.

BOVET, DANIEL (1907-1992)
Swiss Italian pharmacologist

Daniel Bovet had the distinction of making basic contributions in at least three distinct areas of pharmacology, the science of drugs. His research made possible the commercial development of sulfa drugs, **antihistamines**, and muscle relaxants. For his accomplishments in pharmacology he was awarded the Nobel Prize in physiology or medicine in 1957.

Bovet was born on March 23, 1907, in Neuchatel, Switzerland, the only son among the four children of Pierre Bovet and Amy Babut Bovet. Pierre Bovet was a professor of experimental education at the University of Geneva and the founder of the Institut J. J. Rousseau. His son later recalled in *Time* that he and his sisters were "guinea pigs" for testing his father's educational theories. Daniel Bovet received his primary and secondary school education in Neuchatel, then studied biology at the University of Geneva, from which he received his *license* in 1927. He did his graduate study in physiology and zoology at the same institution and earned his doctor of science degree in 1929.

Bovet went to Paris in 1929 to become an assistant in the Laboratory of Therapeutic Chemistry at the Pasteur Institute, working under the direction of Ernest Fourneau. In his 1965 article "Role of the Scientist in Modern Society," Bovet declared that being Fourneau's "pupil and collaborator... for nearly twenty years... was the greatest good fortune of my life." Bovet succeeded Fourneau as director of the Laboratory of Therapeutic Chemistry in 1939. It was there that he met Filomena Nitti, a fellow researcher and the daughter of Francesco Saverio Nitti, a former prime minister of Italy who had been driven into exile to Paris following Benito Mussolini's rise to power. Bovet and Filomena Bovet-Nitti, wife and collaborator (as she thereafter identified herself), were married in 1938 and she became his collaborator in nearly all of his research, as well as the coauthor of many of his scientific books and articles. They had two daughters and one son, Danièle Bovet, who became a professor of information science at the University of Rome.

Researches Sulfa Drugs and Antihistamines

In the early 1930s, the German scientist **Gerhard Domagk** discovered that Prontosil, a dye product, effectively combated streptococcal infections. Prontosil was a complex chemical, however, and expensive to produce. Bovet and his colleagues at the Pasteur Institute reasoned that the therapeutic action of the substance was probably due to some part of the drug's molecule that was only released when the molecule broke down in the body. After months of work and many experiments, they discovered that the active therapeutic agent was sulfanilamide. This product was much cheaper to produce than Prontosil and was soon being manufactured in quantity, becoming the first of the so-called "wonder drugs." Over the next several years Bovet and his associates went on to synthesize many other sulfanamide derivatives that together formed the group of sulfa drugs that were to save millions of lives during World War II and afterward. Domagk was awarded the

Nobel Prize in physiology or medicine in 1939 for his discovery of the therapeutic action of Prontosil, but it was the work of Bovet and his team that had made sulfa drugs a practical reality.

In 1937 Bovet turned his attention to **histamine**, a **hormone** that occurs naturally in all body tissues. When an irritant is introduced, an overproduction of free histamine can occur in some localized area of the body. The free histamine in turn causes swelling or an allergic reaction that often leads to severe discomfort, damage to body tissues, or—in extreme cases—to fatal shock. Bovet was struck by the fact that there was no natural product in the human body that would counteract the negative effects of free histamine; he believed that what was needed was an artificial substance which would block them. Bovet and his assistants soon synthesized the first antihistamine, although it had too many problems to be a viable commercial product. Between 1937 and 1941, Bovet and others performed some three thousand experiments to find a practical substitute. Eventually several were developed, including Bovet's own discovery, pyrilamine. These were the first of the many antihistamines now used in modern medicine.

Shifts Focus to Muscle Relaxants and Mental Illness

In 1947 Bovet and his family left Paris for Rome, where he was to organize and direct the Laboratory of Therapeutic Chemistry at the Istituto Superiore di Sanità. He also became an Italian citizen. It was about this time that he began to study the muscle relaxant properties of curare, the poison certain South American Indians had long used on their arrows. A chemically pure form of curare had been produced earlier and was used to relax body muscles before surgery, thus allowing the surgeon to use much smaller doses of potentially dangerous anesthetics. However, the effects of the curare itself were very unpredictable, and it was also expensive. Bovet set himself the task of finding a synthetic form of the drug that would have the advantages of predictability and low cost. During eight years of work he produced over four hundred synthetic forms of curare, including gallamine and succinylcholine, the latter becoming widely used. During his research on curare, Bovet spent some time with the Indians of South America to learn how they produced and used the drug. He later remarked humorously that he had done so out of a spirit of adventure; curare was only the pretext.

Bovet left Rome in 1964 to become professor of pharmacology at the University of Sassari on the Italian island of Sardinia. He returned to Rome as director of the Laboratory of Psychobiology and Psychopharmacology of the Italian national research council in 1969. He became professor of psychobiology at the University of Rome in 1971 and remained there as an honorary professor following his retirement in 1982. The positions in Rome reflected still another shift in the focus of his research, indicating an interest in the complex area of mental illness and its treatment through the use of chemicals.

As early as 1957, *Time* had reported Bovet's belief that the key to mental illness lay in chemistry. His studies centered

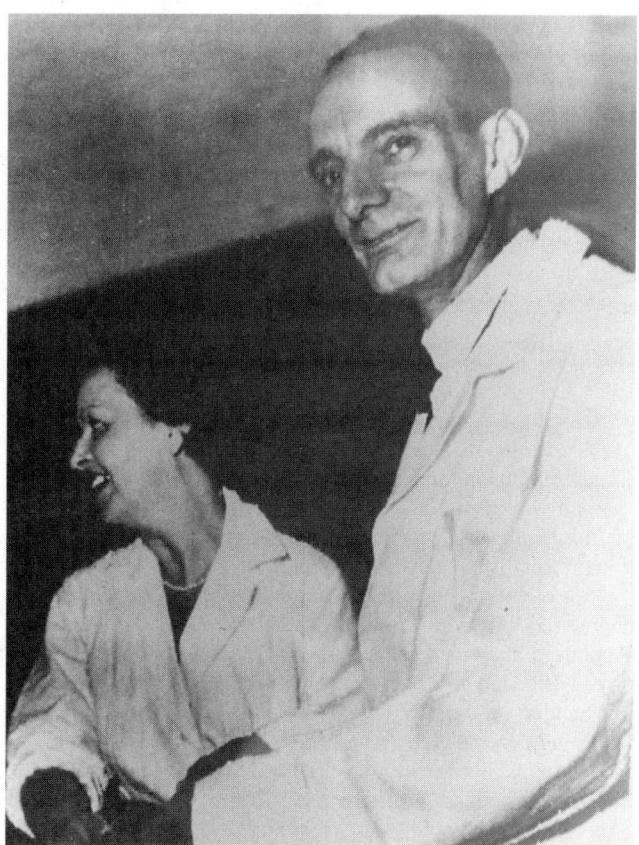

Daniel Bovet and his wife, Filomena, shown making a curare injection on a rabbit in his laboratory. *(The Library of Congress. Reproduced by permission.)*

on the effect of various chemical compounds on the central nervous system of the human body. While his work did not produce the kind of dramatic practical breakthroughs that he had achieved in sulfa drugs, antihistamines, and muscle relaxants, he did contribute much important basic research to this field.

Frequently collaborating with his wife, Bovet produced several books and over four hundred articles in the course of his professional life. Before 1947 most of his writings were in French; afterward many appeared in Italian and some in English. However, even this large output does not fully reveal the breadth of his intellectual interests. He was concerned with the impact of scientific discovery on political, social, and economic affairs and with the equally strong impact of those affairs on science. He illustrated this in ''Role of the Scientist in Modern Society.'' ''Unfortunately, in our century,'' he wrote, ''two-thirds of the global population are illiterate and walk barefooted, ten to fifteen percent suffer from hunger, thirty-three percent to forty percent do not have an adequate diet, seventy percent are not provided with sufficient water supply, and eighty percent lack adequate hygienic conveniences. Even the best drugs are ineffective for people living in very poor hygienic conditions.'' Science, he concluded, could not solve all of the world's problems. Personally, Bovet was a humble, enthusiastic man who single-mindedly pursued his quest for scientific progress without personal gain in mind. As *Time* noted when he received the Nobel Prize in 1957, Bovet had never taken out a patent in his own name and never made any money from his scientific discoveries. He was the recipient of numerous international awards in addition to the Nobel Prize. He died of cancer in Rome on April 8, 1992.

BOYER, HERBERT WAYNE (1936-)
Molecular geneticist

In 1973, Herbert Boyer was part of the scientific team that first described the complete process of **gene splicing**, which is a basic technique of **genetic engineering** (recombinant DNA). Gene splicing involves isolating **DNA**, cutting out a piece of it at known locations with an **enzyme**, then inserting the fragment into another individual's genetic material, where it functions normally.

Boyer was born in Pittsburgh and received a bachelor's degree in 1958 from St. Vincent College. At the University of Pittsburgh he earned an M.S. in 1960 and a Ph.D. in bacteriology in 1963. In 1966 he joined the biochemistry and biophysics faculty at the University of California, San Francisco, where he continues his research.

Boyer performed his work with Stanley Cohen from the Stanford School of Medicine and other colleagues from both Stanford and University of California, San Francisco. The scientists began by isolating a plasmid (circular DNA) from the bacteria *E. coli* that contains **genes** for an antibiotic resistance factor. They next constructed a new plasmid in the laboratory by cutting that plasmid with restriction endonucleases (enzymes) and joining it with fragments of other plasmids.

After inserting the engineered plasmid into *E. coli* bacteria, the scientists demonstrated that it possessed the DNA nucleotide sequences and genetic functions of both original plasmid fragments. They recognized that the method allowed bacterial plasmids to replicate even though sequences from completely different types of **cells** had been spliced into them.

Boyer and his colleagues demonstrated this by cloning DNA from one **bacteria** species to another and also cloning animal genes in *E. coli*.

Boyer is a co-founder of the genetic engineering firm Genentech, Inc. and a member National Academy of Sciences. His many honors include the Albert and Mary Lasker Basic Medical Research Award in 1980, the National Medal of Technology in 1989, and the National Medal of Science in 1990.

BOYER, PAUL D. (1918-)
American biochemist

Paul Boyer, Professor Emeritus of Biochemistry at the University of California, Los Angeles, has devoted his entire professional career to the study of **enzymes**. His efforts as an enzymologist focused particularly on the exploration of oxidative phosphorylation, the process by which organisms convert energy acquired from food into the energy currency of living cells, **adenosine triphosphate (ATP)**. He won the Nobel Prize in 1997 for his work on the mechanism of ATP formation. ATP is formed from adenosine diphosphate (ADP) and inorganic phosphate (Pi) in cellular organelles called **mitochondria**.

The pathway that synthesizes ATP is highly complex and has puzzled scientists for decades. Boyer developed a mechanistic model of how the various subunits of the enzyme ATP synthase cooperate like gears, levers and ratchets to produce ATP. The synthase enzyme was first identified in mitochondria in 1960. It was known that a series of mitochondial enzymes causes the breakdown of energy-rich compounds, using the released energy to pump hydrogen ions (protons) across an internal mitochondrial membrane, leaving the interior portion of the mitochondria with a shortage. British biochemist **Peter D. Mitchell** hypothesized that the hydrogen ions, as they flowed back through the membrane to the interior of the mitochondria, caused ATP synthesis.

By 1970, ATP synthase was known to consist of three sets of **protein** assemblies: a wheel-like structure in the mitochondrial membrane, a rod fixed to the wheel's hub on one end, and a cylinder that surrounds the other end of the rod and extends into the interior of the mitochondria. Several labs demonstrated that ATP is synthesized at three sites on the cylinder, and that the rod is needed to turn on catalytic activity at these sites. The underlying mechanism remained puzzling. It remained for Boyer to suggest a viable solution to the puzzle. He proposed that hydrogen ions (protons) cause the wheel to spin as they return through the mitochondrial membrane back to the center of the **organelle**, just as moving water causes a water wheel to turn. The rod spins, as well, because it is attached to the wheel, and causes the other end to rotate within the fixed cylinder. The rotation slightly changes the structure of the three active sites in the cylinder enabling them to capture ADP and phosphate, the building blocks of ATP, produce a molecule of ATP, and release it.

Boyer has described his theory as follows: "The membrane-bound ATP synthase of animals, plants, and microorganisms is a highly conserved enzyme with unusual subunit stoichiometry and properties. In the binding change mechanism for the synthase developed by our laboratory, translocation of protons is regarded as driving conformational changes that promote release of a tightly bound ATP at one catalytic site and the tight binding of ADP and Pi at another catalytic site. An interesting speculation is that catalysis is accompanied by a rotational movement of catalytic subunits relative to a noncatalytic core."

Boyer received his Ph.D. from the University of Wisconsin, and served on the faculty of the University of Minnesota for 17 years before joining the UCLA faculty in 1963.

BOYLE, ROBERT (1627-1691)
English physicist and chemist

For centuries people believed that everything was made of just three or four substances, which were mistakenly called elements. It was Robert Boyle who first set science on the right track and asserted the true nature of elements and compounds.

Boyle's father was an Englishman who made his fortune in Ireland and became a successful landowner there. Boyle was his seventh son and the youngest of fourteen children. By the time Boyle was born, his father had become an earl and was one of the wealthiest men in the country. Like his father, Boyle was an industrious worker; before he entered the prestigious Eton school at the age of eight, he was already speaking Greek and Latin. His passion for reading and learning continued to grow, and Boyle proved to be a gifted student with an excellent memory.

At an early age, Boyle and his brother went to Europe with a tutor to study French, mathematics, and many other subjects. For six years, they lived in Switzerland and traveled extensively through France and Italy, where Boyle learned of Galileo's experiments on the effects of gravity and other physical laws. When Boyle was 13, he witnessed a sudden, violent thunderstorm that changed his whole outlook on life. He developed a religious faith which he felt did not contradict scientific beliefs but instead reinforced his admiration for the creator of such a complex universe. From then on, he was a devout Christian, and he learned several ancient languages, including Hebrew and Aramaic, so that he could read the Bible in its original texts.

A civil war between England and Ireland demanded Boyle's return from continental Europe, and he reached home in 1645. By this time, Boyle had become interested in performing experiments in order to understand the way things work. Previously, people believed in making up a theory and then judging how well the facts fit the theory. Boyle, however, agreed with the philosopher Francis Bacon (1561-1626) that facts should be observed first, and then a theory should be developed to explain them. Boyle and other scholars interested in experimentation began meeting regularly in London, England, to discuss these new ideas. At first, they called themselves the "Invisible College," but when King Charles II (1630-1685) was restored to the throne in 1663, he granted a charter to the group of scientists; the group thus became known as the Royal Society.

When Boyle moved to Oxford in 1654, he met many more scientists and became interested in chemistry, because of its relation to medicine. During his 14 years at Oxford, Boyle contributed greatly to scientific philosophy in the fields of physics and chemistry. He set up an elaborate research laboratory and hired skilled assistants to conduct experiments. Unlike most scientists of his day, Boyle believed in meticulously recording his experiments and publishing them so that others could repeat his tests and confirm the data. He is credited with pioneering the modern scientific method and today this practice is universal in the research world.

In 1661 Boyle published his most famous work, *The Sceptical Chymist*, which revolutionized scientific thought and formed the basis of modern chemistry. In this work, Boyle defined an element as the simplest form of matter, one that cannot be broken down into any simpler form or changed into a different substance. Boyle's ideas contradicted beliefs held

Robert Boyle. *(The Library of Congress. Reproduced by permission.)*

ever since the ancient Greeks proposed that all things are made of only four elements—air, earth, fire, and water—which could be changed, or *transmuted*, into other substances. In another version of this idea, only three substances existed in nature (salt, sulfur, and mercury). But according to Boyle, none of these substances were true elements. Boyle argued that elements could be identified only by scientific experimentation. He also pointed out that a compound will usually have chemical properties that are very different from its parent elements.

Boyle's concept of an element arose from his experiments with gases, and he was the first scientist to succeed in collecting hydrogen in a device now called a *pneumatic trough*. In 1660 Boyle discovered a fundamental law of physics that helps explain the behavior of gases. When a gas is pressurized, Boyle found that the amount of space it takes up is related to the amount of pressure being exerted on it, as long as the gas's temperature doesn't change. For example, if the pressure on a given quantity of gas is doubled, the gas's volume is cut in half; if pressure is tripled, volume is reduced to one-third. (This relationship is called *inversely proportional*.) Boyle's law, along with a similar law that explains the effects of temperature, allows chemists today to calculate the volume of gasses under any pressure or temperature conditions. Boyle also realized that if air could be compressed, it must be composed of tiny particles separated by space. It was this conclusion that led Boyle to envision a universe composed of numerous tiny particles, and, in doing so, he anticipated the modern concept of atomic theory.

Vacuums were poorly understood but of much scientific interest to Boyle and his assistant, **Robert Hooke**. Air pumps were used in early laboratories to create a vacuum inside cylinders. Robert Hooke built an improved air pump based on German engineer, Otto von Guerricke's air pump design. Together, Boyle and Hooke developed a better vaccum. This new vacuum had better placement of the pumps valves and a preferable method of cranking its piston and supporting the air pump's cylinder. Boyle also proved for the first time that all objects, no matter how light or heavy, fall through a vacuum at the same speed. This showed, as Galileo had predicted, that the force of gravity is uniform. In another experiment, Boyle demonstrated that the sound of a clock ticking could not be heard in a vacuum, proving that sound waves depend on air for their transmission. Boyle showed, however, that electrical attraction could be felt through a vacuum.

Boyle was also interested in the nature of color, and he accurately described how the absorption and reflection of light produces the appearance of black and white, studying the changes in color that occur in certain plant extracts, such as *litmus*. He discovered that these substances, now called *indicators*, can be used to distinguish acids from bases. Boyle went on to develop tests for identifying other substances, such as copper, silver, and sulfur, via chemical reaction. He not only coined the term *analysis* in its modern sense, but also encouraged generations of chemists to determine the composition of substances through meticulous experimentation. In the late 1660s Boyle became the first scientist to study the phenomenon of **bioluminescence**, showing that certain **bacteria** and other **organisms** will glow in the dark if supplied with air. Boyle also found that water begins to expand just before it freezes. Throughout this period, Boyle and his staff published immense amounts of information for use by other scholars and scientists.

In 1668 Boyle returned to London to live with his favorite sister, Lady Ranelagh. In 1680 he invented the first match by coating a piece of coarse paper with phosphorus. He produced a flame by drawing a sulfur-tipped wooden splint through a fold in the paper. Also in 1680, he was elected president of the Royal Society, but he declined the honor, believing that the oath of office would conflict with his strict religious beliefs, and he continued to refuse all titles and other honorary positions. In his later years, Boyle wrote about medicine and diseases and devoted greater effort to promoting Christian ideals; in his will he left money for a series of lectures to defend Christianity from atheists and other "notorious infidels."

See also Acids and bases; Electromagnetism

BRAIN

With its hundreds of billions of interconnectednerve **cell**s, or **neuron**s, continuously transmitting electrochemical signals to each other, the human brain is the most complex object yet discovered. Fortunately, our understanding of the brain has been greatly aided in recent years by various technologies that allow scientists to "see" the brain, or "map" its electrochemical activity, without the surgery or autopsy required in the past.

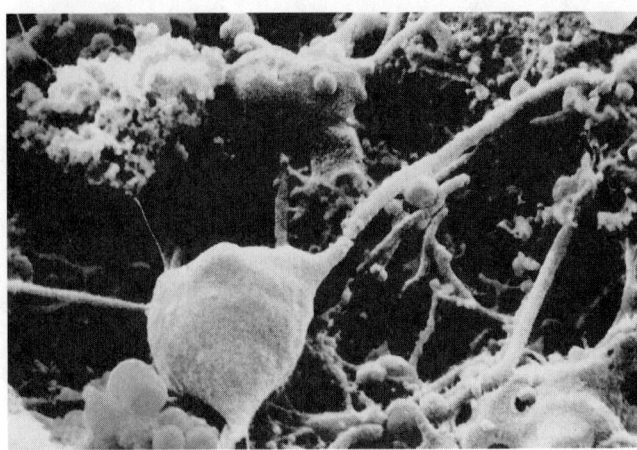

A scanning electron micrograph (SEM) of brain cells. The large gray cells with long thin branches are neurons. The glial cells are specialized structures which support and protect neurons. *(Photo Researchers, Inc. Reproduced by permission.)*

In higher **animals**, the brain and **spinal cord** constitute the central **nervous system**. Not all animals have brains, though all vertebrates do. The brain of the adult human being weighs between three and four pounds (1.35–1.8 kg). The human brain is protected by the cranium and three membranes, called meninges; the space between the two inner meninges is filled with cerebrospinal fluid. The fundamental building block of the brain in all animals is the nerve cell. This cell conducts electrical and chemical signals through an **axon** extending from the cell body. Shorter branches, called dendrites, conduct signals back to the cell body. A fatty substance called a **myelin sheath** covers the axons, protecting them from other **nerve impulses** nearby. Between any two nerve cells is an infinitesimal gap, called a **synapse**, across which electrochemical signals are transmitted in the continuously active brain.

Although astoundingly complex, the brain has been described in simple terms as having evolved from the overdevelopment of one end of the nervous system, becoming generally larger and more complex in vertebrates, especially **mammals**. In human beings, the brain begins developing in the four week old **embryo** starting as bulges at one end of the neural tube and forming into the hindbrain, midbrain, and forebrain. These three divisions are shared by all vertebrate brains, whether of sharks, cats, or humans.

The most primitive part of the brain from an evolutionary standpoint, the hindbrain (or brainstem) consists of the *medulla*, the *pons*, and the *cerebellum* (or "little brain"). These structures regulate several autonomic functions not consciously controlled but necessary for survival. For instance, the portion of the hindbrain at the top of the spinal cord regulates breathing, the heartbeat, and the diameter of **blood** vessels. All of the nerves running between the spinal cord and the brain pass through the medulla, an important junction for controlling deliberate movement. When signals for deliberate movement pass through the medulla they cross over from one side of body's movements on the right side, and vice versa. Besides its role in deliberate movement, the medulla contains sets of

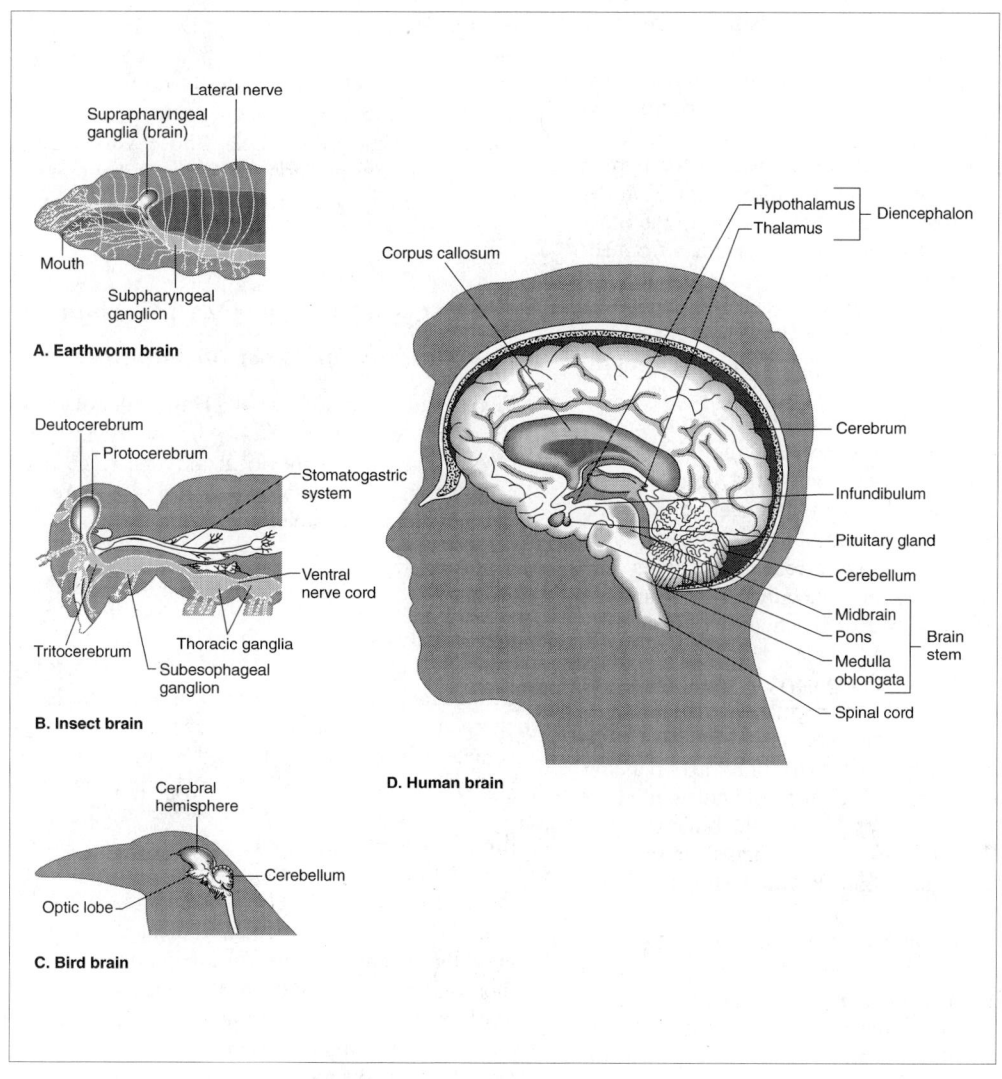

A. Earthworm brain

Lateral nerve

Suprapharyngeal ganglia (brain)

Mouth

Subpharyngeal ganglion

B. Insect brain

Deutocerebrum

Protocerebrum

Stomatogastric system

Ventral nerve cord

Tritocerebrum

Thoracic ganglia

Subesophageal ganglion

C. Bird brain

Cerebral hemisphere

Cerebellum

Optic lobe

D. Human brain

Corpus callosum

Hypothalamus

Thalamus

Diencephalon

Cerebrum

Infundibulum

Pituitary gland

Cerebellum

Midbrain

Pons

Medulla oblongata

Brain stem

Spinal cord

A comparison of the brains of an earthworm, an insect, a bird, and a human. *(Illustration by Hans & Cassady, Inc.)*

nerves for **organ**s in the chest and abdomen, for head and shoulder movements, and for such functions as salivating, swallowing, tasting, **hearing**, and maintaining equilibrium. Just above the medulla at the brainstem's top, the pons bridges the lower brainstem and the midbrain. It connects the **cerebellum** and cerebral cortex, the brain's largest, most recently evolved part. In the pons originate many facial and head nerves that regulate some eyeball movements and facial expressions. Working with nerves from the medulla, nerves from the pons also control breathing and maintain equilibrium. The hindbrain's largest part and the second largest part of the human brain the cerebellum conveys signals for movement from the cerebral cortex to the spinal cord, and from there to the muscles. Simultaneously, the cerebellum receives signals from activated muscles and joints and compares these against the movement signals from the cerebral cortex, thereby enabling adjustments to be made. The cerebellum alone does not initiate movement, but reroutes and refines signals for movement received from the cortex.

The midbrain, located between the hindbrain and forebrain, serves mainly to relay both sensory and motor nerve impulses passing between the pons and spinal cord and the thalamus and cerebral cortex. In many animals the midbrain is important in processing visual and olfactory information.

The forebrain comprises the **cerebrum**, with its thin outer cortex, and the limbic system. The human cerebrum consists of left and right hemispheres, each divided into four lobes. The occipital lobe receives and analyzes visual information. The temporal lobe deals with **memory**, hearing, and some language functions. The frontal lobe regulates movement and houses Broca's area, which handles language production. The parietal lobe deals with sensations. The brain's two hemispheres communicate through a thick nerve bundle called the *corpus callosum*, which enables one side of the brain to transmit what it learns to the other side. Nerve tracts from the right and left sides of the cerebellum connect with the opposite cerebral hemisphere.

The limbic system, which is also called the "emotional brain," forms a loose circuit of linked structures throughout the brain. Among the limbic system's structures are the olfactory bulb, by which odors are received; the amygdala, which receives input from both the olfactory system and cerebral cortex; the **hypothalamus**, which controls hunger, thirst, sex, and rage, as well as other vital drives and **emotions**; the **pituitary gland**, which secretes at least six **hormone**s vital for growth and other functions; the hippocampus, which plays an important role in the memory's consolidation of recently acquired information; and the thalamus, the "great relay station" located at the forebrain's base. The thalamus consists of two oval nerve masses that sort information about **sight**, sound, **taste**, and **touch**. Together with other sensory systems of the brain, neurons from the thalamus make up the reticular activating system. This system appears to filter means of maintaining consciousness. Overall, the limbic system is responsible for the basic drives, emotions, and involuntary behavior that are crucial for an animal's survival, among them, fear, anger, pain, pleasure, and sexual stimulation.

Brain disorders such as **schizophrenia**, dementia, Alzheimer's Disease, and alcoholism are a leading cause of **death** in the United States. They are also the nation's most common cause of social, economic, and psychological disability. For those reasons, the federal government is a primary sponsor of brain research, which can lead to drugs and other medical advances for dealing with brain disorders. In 1990, President George Bush proclaimed the Decade of the Brain. It was launched with a symposium sponsored by the National Institute of Mental Health and the Institute of Medicine and was attended by the world's leading neuroscientists. These are scientists from many disciplines whose work is unveiling the intricate structures and complex process that make up the human brain, offering hope for cures to the brain's many diseases.

BROCA, PIERRE PAUL (1824-1880)
French anthropologist and anatomist

Pierre Paul Broca, the son of a Huguenot doctor, was born near Bordeaux, France, in 1824. After studying mathematics and physical science at the local university, he entered medical school at the University of Paris in 1841. He received his M.D. in 1849. Though trained as a pathologist, anatomist, and surgeon, Broca's interests were not limited to the medical profession. His versatility and tireless dedication to science permitted him to make significant contributions to other fields, most notably to anthropology.

The application of his expertise in anatomy outside the field of medicine began in 1847 as a member of a commission charged with reporting on archaeological excavations of a cemetery. The project permitted Broca to combine his anatomical and mathematical skills with his interests in anthropology.

The discovery in 1856 of Neanderthal Man once again drew Broca into anthropology. Controversy surrounded the interpretation of Neanderthal. It was clearly a human skull, but more primitive and apelike than a modern skull and the soil stratum in which it was found indicated a very early date. Neanderthal's implications for evolutionary theory demanded thorough examination of the evidence to determine decisively whether it was simply a congenitally deformed **Homo sapiens** or a primitive human form. Both as an early supporter of **Charles Darwin** and as an expert in human anatomy, Broca supported the latter view. Broca's view eventually prevailed, though not until the discovery of the much more primitive Java Man (then known as *Pithecanthropus*, but later *Homo erectus*).

Broca is best known for his role in the discovery of specialized functions in different areas of the brain. In 1861, he was able to show, using post-mortem analysis of patients who had lost the ability to speak, that such loss was associated with damage to a specific area of the **brain**. The area, located toward the front of the brain's left hemisphere, became known as Broca's convolution. Aside from its importance to the understanding of human physiology, Broca's findings addressed questions concerning the evolution of language.

All animals living in groups communicate with one another. Non-human primates have the most complex communication system other than human language. They use a wide range of gestures, facial expressions, postures, and vocalizations, but are limited in the variety of expressions and are unable to generate new signals under changing circumstances. Humans alone possess the capacity for language rather than relying on a body language vocabulary. Language permits humans to generate an infinite number of messages and ultimately allows the transmission of information—the learned and shared patterns of behavior characteristic of human social groups, which anthropologists call culture—from generation to generation. The development of language spurred human evolution by permitting new ways of social interaction, organization, and thought.

Given the importance assigned to human speech in human evolution, scientists began to look for the physical preconditions of speech. The fact that apes have the minimal parts necessary for speech indicated that the shape and arrangement of the vocal apparatus was insufficient for the development of speech. The vocalizations produced by other animals are involuntary and incapable of conscious alteration. However, human speech requires codifying thought and transmitting it in patterned strings of sound. The area of the brain isolated by Broca sends the code to another part of the brain that controls the muscles of the face, jaw, tongue, palate, and larynx, setting the speech apparatus in motion. This area and a companion area that controls the understanding of language, known as Wernicke's area, are detectable in early fossil skulls of the genus *Homo*. The brain of *Homo* was evolving toward the use of language, although the vocal chamber was still inadequate to articulate speech. Broca discovered one piece in the puzzle of human communication and speech, which permits the transmission of culture.

Equally important, Broca contributed to the development of physical anthropology, one of the four subfields of anthropology. Craniology, the scientific measurement of the skull, was a major focus of physical anthropology during this period. Mistakenly considering contemporary human groups as if they were living **fossils**, anthropologists became interested in the nature of human variability and attempted to explain the varying levels of technological development observed worldwide by looking for a correspondence between cultural level and physical characteristics. Broca furthered these studies by inventing at least twenty-seven instruments for making measurements of the human body, and by developing standardized techniques of measurement.

Broca's many contributions to anthropology helped to establish its firm scientific foundation at a time when the study of nature was considered a somewhat sinister science.

BROOM, ROBERT (1866-1951)
Scottish physician and paleontologist

Born in poverty in Scotland, Robert Broom went on to become a respected physician and renowned paleontologist who made important **fossil** discoveries as to the possible ancestors of human beings. Broom received his M.D. from the University of Glasgow, but it was during a trip to Australia in 1892 that he found his true calling as a paleontologist. After moving to

South Africa in 1897, Broom joined the faculty at the University of Stellenbosch. However, he lost this position at the conservative religious institution because of his belief in the theory of evolution. Although he the worked as a physician, he devoted all his spare time to fossil hunting, conducting excavations staffed largely by volunteers who were attracted to his roguish personality. (Broom always wore a dark formal suit even while hunting for fossils, stripping naked when he got hot.)

Broom did his most notable work after reaching the age of 68 when he left medical practice to work at the Transvaal Museum in Pretoria, South Africa. Over the next few years, he made a succession of spectacular discoveries. Broom was convinced that an ape-like **mammal** (the Taung child) discovered by Raymond Dart in 1924 and named *Australopithecine,* was a hominid that walked erect like **homo sapiens,** the human beings of today. Intent on finding an adult *Australopithecine* to confirm Dart's findings, Broom found a fragmentary skull in the mid-1930s of just such an adult at Sterkfontein, which was eventually classified into the **genus** *Australopithecus africanus*. In 1947, Broom found a complete adult specimen estimated to be approximately 2.5 million years old. Broom's discovery, which remains the world's most complete specimen of *Australopithecus africanus,* was called ''Mrs. Ples.'' The name stems from Broom's initial description of the hominid as *Plesianthropus,*meaning almost man. In 1995, however, Mrs. Ples was renamed Mr. Ples after a reanalysis determined its sex was male.

In 1937, a young school-boy volunteering with Broom's expedition, Gert Terblanche, found the first specimen of *Australopithecus robustus* at Kromdraai, South Africa. This was Broom's most famous discovery. Estimated to have lived in southern Africa between 1.2–2 million years ago, this stockier form of *Australopithecine* had massive jaws and muscles that allowed it to chew the tough fruits and plants indigenous at that time. Broom also performed excavations at Swartkrans in 1948, unearthing fossils that were eventually connected to *Homo erectus,* one of the first hominids to migrate from Africa and adapt to both temperate and arctic climates.

Broom spent the remainder of his career exploring sites throughout South Africa and interpreting the early hominid remains that he had discovered. His book *Finding the Missing Link* was published in 1950, the year before he died. According to one account, Broom died only moments after writing the last lines of a monograph on the *Australopithecines*. His final words were, ''Now that's finished...and so am I.'' Mr. Ples and many of Broom's other findings are housed at the Transvaal Museum.

BROWN, MICHAEL S. (1941-)
American geneticist

Michael S. Brown, a genetics professor and director of the Center for Genetic Diseases at the University of Texas Southwestern Medical School, is one of America's foremost experts on cholesterol **metabolism** in the human body. In the 1970s,

Brown and Joseph Goldstein investigated familial hypercholesterolemia, a dangerous inherited disorder which causes elevated levels of **cholesterol** in the **blood**. Their research led them to the discovery of a **protein** in the **membranes** of a **cell**, called the LDL receptor, which plays a central role in the body's ability to lower cholesterol levels. For this discovery and their subsequent research on the LDL receptor, Brown and Goldstein shared the 1985 Nobel Prize in physiology or medicine.

Brown was born in New York City on April 13, 1941, to Harvey and Evelyn Katz Brown. He attended the University of Pennsylvania as an undergraduate, receiving his bachelor's degree in 1962. Following his graduation, Brown enrolled in the medical school at the University of Pennsylvania, where he was awarded the Frederick Packard Prize in Internal Medicine for his research. He earned his M.D. in 1966 and served as an intern and a resident at Massachusetts General Hospital in Boston. It was during his residency that he met **Joseph Goldstein**, his future research partner, who was also on the staff at Massachusetts General.

In 1968, Brown was made a clinical associate at the National Institutes of Health (NIH) in Bethesda, Maryland. He was assigned to the biochemistry lab, where he worked with Earl Stadtman, head of the laboratory for the National Heart, Lung, and Blood Institute. While at NIH, Brown focused his research on gastroenterology, particularly on the role of enzymes in digestive chemistry. In 1971, while studying a particular enzyme involved in the production of cholesterol, Brown was offered a position as an assistant professor at the University of Texas Southwestern Medical School in Dallas. He accepted, and Goldstein, who had also served at NIH in Bethesda, joined the Texas Southwestern faculty a year later. At this time the two began a collaboration which was to distinguish them as pioneers in genetics.

In Dallas during the 1970s, Brown and Goldstein examined skin samples from people who suffered from hypercholesterolemia, specifically those rare patients whose condition was homozygous, meaning that they had not just one defective gene but two. In these cases, patients often exhibited extremely high levels of low-density lipoprotein, LDL, even during childhood. LDL carries cholesterol to the cells, and in excessive quantities can clog arteries and encourage heart disease. Brown and Goldstein discovered that the cells of these patients were missing a crucial protein, called a receptor, which binds to LDL and regulates its level in the body. Without the protein, the body can not break down LDL, and it accumulates in the blood. Brown and Goldstein's breakthrough was the discovery and isolation of this LDL receptor protein.

Brown and Goldstein not only identified the LDL receptor, they also located the gene responsible for its production. By sequencing and **cloning** the **gene**, they were able to localize the **gene mutations** responsible for familial hypercholesterolemia, as well as other inherited conditions involving cholesterol metabolism. Their findings also led to possible drug therapies for people with cholesterol disorders. By administering a combination of drugs which would inhibit the liver's ability to synthesize cholesterol, Brown and Goldstein

increased their patients' need for cholesterol from outside sources. The patients' bodies subsequently produced more LDL receptors, and their cholesterol levels fell sharply. They also found that a liver transplant can correct genetic deficiencies in the production or expression of LDL receptors. In later research, Brown and Goldstein engineered a mouse which, because of its abnormally high numbers of LDL receptors, could eat a high-fat diet and yet show no significant rise in LDL.

In a series of experiments, Brown and Goldstein were ultimately able to define and analyze each step in the path of cholesterol through the body, from production to dissolution. They also demonstrated a mechanism by which a low-fat diet and regular exercise can decrease cholesterol levels. Brown and Goldstein's work had significant implications not only for genetic defects, but also for nutrition and fitness. In addition, the team's research methods contributed to a greater understanding of cell receptors in general, serving as a model for research on over 20 other receptors.

In addition to the Nobel Prize, Brown has received several honorary degrees and a number of awards for his research, including the Pfizer Award from the American Chemical Society in 1976, the Albert Lasker Medical Research Award in 1985, and the National Medal of Science in 1988. He has been a member of the National Academy of Sciences since 1980. He was appointed Paul J. Thomas Professor of Genetics and director of the Center for Genetic Diseases at the University of Texas Southwestern Medical School, positions he has held since 1977.

Brown balances his scientific and medical careers, continuing to makes rounds at the hospital.

BROWN, ROBERT (1773-1858)
Scottish naturalist

Robert Brown got his intellectual honesty and solid character from his father, an Episcopalian clergyman. But unlike his father, he did not develop a calling for religion. Instead, Brown completed college in Aberdeen, Scotland, then joined the army. Brown's journals during his five years as an assistant surgeon in the army show his determination to master details and his far-reaching curiosity.

In 1798, Brown was introduced to English botanist Joseph Banks (1743-1820), who was clearly impressed with Brown's talents. It may have been Banks' influence that got Brown the position of naturalist aboard the *Investigator* in 1801 to survey the plant life along the unfamiliar coasts of Australia.

When Brown returned to England in 1805, he brought with him over 4,000 species of plants and set about the arduous task of reporting on each species. The two systems of plant **classification** then in use were not suitable for the unusual Australian plant varieties, so Brown developed a modified system, making use of the *natural system* for the first time in England.

Brown concentrated on the classification of plants until 1827 when he made an unusual discovery. Using a microscope, he began observing grains of orchid pollen suspended in fluid. To his astonishment he saw that particles within the grain were moving. After ruling out the fluid and its gradual evaporation as possible causes for the movement, he proposed that the particle itself might be "alive." He tested fresh pollen from a variety of other plants—all with the same result. When he expanded his experiment to test powdered glass, coal, rocks and metals, he found the same movement. All minuscule particles that could be suspended in water exhibited what is now called Brownian motion.

Despite his experimentation, Brown could not explain the movement, and ultimately he left it to others for interpretation. (Eventually Brownian motion was linked to important theories in kinetic energy. This phenomenon was a visible manifestation of the idea that water was composed of particles.)

In 1831, Brown published a pamphlet on his observations of orchid pollen. The pamphlet contains a remarkably casual passage pinpointing a major revelation. Brown reported that the cells in orchid leaves contain a "single circular areola"—only one in each cell—generally more opaque than the cell membrane. He also noticed that this areola appeared in the planttissue as well. He dubbed it the cell *nucleus*, Latin for "little nut." It had been observed earlier, but Brown was the first to name it and to recognize it as a general feature in all living cells.

BROWNIAN MOTION • See Brown, Robert

BRUCE, DAVID (1855-1931)
Scottish microbiologist

David Bruce is noted for his work in parasitology, especially for his discovery of the cause of brucellosis and sleeping sickness. Born in Melbourne, Australia, to Scottish immigrants, Bruce and his parents returned to Scotland when he was five years old. Although Bruce longed to become a professional athlete, he was stricken with pneumonia at age 17. Bruce studied natural history and medicine at the University of Edinburgh, and after graduation, he found a job working with a doctor. He later met Mary Elizabeth Steele, whom he married in 1883. The couple subsequently began a lifelong partnership in medical science.

After joining the Army Medical Service, David and Mary Bruce were assigned to Malta in 1884, where Bruce began a study of an often-fatal disease suffered by English soldiers assigned to the Maltese garrison. The disease, known as *Malta*, *Mediterranean*, or *undulating fever*, caused chills, sweats, and weakness. Using a microscope, Bruce described the cause as a "micrococcus" growing in the spleens of patients. Eventually, the organism was isolated by Danish scientist Bernhard L. F. Bang (1848-1932).

In 1905, a scientific team headed by Bruce found that the soldiers were contracting the disease by drinking the milk of infected goats. Goats' milk was thus eliminated from the

David Bruce. *(The Library of Congress. Reproduced by permission.)*

soldiers' diets, and the disease vanished. Soon, physicians were calling the disease *brucellosis* in honor of Bruce. However, the fight against the disease was not yet over. Almost twenty years passed before physician Alice Catherine Evans discovered that brucellosis was often transmitted by the milk of cows as well as goats, leading to a drive to pasteurize all milk products and ultimately a decline in the disease's occurrence in humans.

After leaving Malta in 1889, Bruce was stationed in Africa. He conducted research in Zululand and Uganda on *nagana*, a common disease affecting domestic animals. He found that the infected tsetse flies could transmit the disease to humans. In 1903, after directing a hospital during the Boer War, Bruce was named director of the Royal Society's Sleeping Sickness Commission. With Aldo Castellani (1877-1971), Bruce and his colleagues isolated and described the microorganism that caused the disease, a worm-like parasite called a trypanosome. Bruce was then able to prove that the tsetse fly was the transmitter.

Bruce was knighted in 1908. By 1914, the Bruces had returned to England, where David served as commandant of the Royal Army Medical College. He directed scientific research during World War I and worked on tetanus antitoxins. He died in 1931, just four days after his wife's death. Before dying, David Bruce asked that any account of his work should acknowledge his wife's assistance and support.

BRUCELLOSIS

Brucellosis is a disease caused by **bacteria** in the **genus** *Brucella*. The disease infects **animal**s such as swine, cattle, and sheep; humans can become infected indirectly through contact with infected animals or by drinking *Brucella*-contaminated milk. In the United States, most domestic animals are vaccinated against the bacteria, but brucellosis remains a risk with imported animal products.

Characteristics of *Brucella*

Brucella are rod-shaped bacteria that lack a capsule around their **cell membrane**s. Unlike most bacteria, *Brucella* cause infection by actually entering host cells. As the bacteria cross the host cell membrane, they are engulfed by host cell **vacuoles** called phagosomes. The presence of *Brucella* within host cell phagosomes initiates a characteristic immune response, in which infected cells begin to stick together and form aggregations called granulomas.

Brucella species

Three **species** of *Brucella* cause brucellosis in humans: *Brucella melitensis*, which infects goats; *B. abortis*, which infects cattle and, if the animal is pregnant, causes the spontaneous abortion of the fetus; and *B. suis*, which infects pigs. In animals, brucellosis is a self-limiting disease, and usually no treatment is necessary for the resolution of the disease. However, for a period of time from a few days to several weeks, infected animals may continue to excrete brucella into their urine and milk. Under warm, moist conditions, the bacteria may survive for months in soil, milk, and even seawater.

Because the bacteria are so hardy, humans may become infected with *Brucella* by direct contact with the bacteria. Handling or cleaning up after infected animals may put a person with contact with the bacteria. *Brucella* are extremely efficient in crossing the human skin barrier through cuts or breaks in the skin.

Symptoms and treatment of brucellosis

The incubation period of *Brucella*—the time from exposure to the bacteria to the start of symptoms—is typically about three weeks. The primary complaints are weakness and fatigue. An infected person may also experience muscle aches, fever, and chills.

The course of the disease reflects the location of the *Brucella*bacteria within the human host. Soon after the *Brucella* are introduced into the bloodstream, the bacteria seek out the nearest lymph nodes and invade the lymph node cells. From the initial lymph node, the *Brucella* spread out to other **organ** targets, including the spleen, bone marrow, and liver. Inside these organs, the infected cells form granulomas.

Diagnosing brucellosis involves culturing the blood, liver, or bone marrow for *Brucella* **organism**s. A positive **culture** alone does not signify brucellosis, since persons who have been treated for the disease may continue to harbor *Brucella*bacteria for several months. Confirmation of brucellosis, therefore, includes a culture positive for *Brucella*bacteria as well as evidence of the characteristic symptoms and a history of possible contact with infected milk or other animal products.

In humans, brucellosis caused by *B. abortus* is a mild disease that resolves itself without treatment. Brucellosis caused by *B. melitensis* and *B. suis*, however, is chronic and severe. Brucellosis is treated with administration of an antibiotic that penetrates host cells to destroy the invasive bacteria.

Prevention

Since the invention of an animal vaccine for brucellosis in the 1970s, the disease has become somewhat rare in the United States. Yet the vaccine cannot prevent all incidence of brucellosis. In 1989, the Centers for Disease Control reported only 95 total cases in the United States. Most of these were reported in persons who worked in the meat processing industry. Brucellosis remains a risk for those who work in close contact with animals, including veterinarians, farmers, and dairy workers.

Brucellosis also remains a risk when animal products from foreign countries are imported into the United States. Outbreaks of brucellosis have been linked to unpasteurized feta and goat cheeses from the Mediterranean region and Europe. In the 1960s, brucellosis was linked to bongo drums imported from Africa: drums made with infected animal skins can harbor *Brucella*bacteria, which can be transmitted to humans through cuts and scrapes in the human skin surface.

In the United States, preventive measures include a rigorous vaccination program that involves all animals in the meat processing industry. On an individual level, people can avoid the disease by not eating animal products imported from other countries. If this is not possible or desirable, make sure that imported cheeses have been made with pasteurized milk. If the package does not indicate pasteurization, do not eat the cheese.

BUBONIC PLAGUE

The bubonic plague is a highly infectious and fearsome disease that attacks the lungs and lymph nodes. It is also called the Black Death or black plague. The bubonic plague is caused by Pasteurella pestis, a **bacteria** which resides within infected fleas and rats. Victims of the bubonic plague develop early symptoms, such as shivering, vomiting, headache, intolerance to **light**, back and limb pain, and a white coating on the tongue. Eventually, they develop black egg-sized swellings (buboes) filled with **blood** and pus under the armpits and in the groin. As the disease progresses, internal bleeding leads to black patches on the skin, and the victim may die in three to five days. Invasion of the lungs by the bacterium causes an equally fatal form of the plague called pneumonic plague, which can be transmitted from person to person by air droplets and saliva.

Historical records document outbreaks of the plague as early as 430 B.C., when an epidemic struck Athens, Greece; but the most notorious bubonic plague epidemic began in Europe around 1346, reportedly when a ship of sick and dying sailors arrived at the Black Sea port of Caffa. This plague lasted four years and killed about one-third of the **population** of Europe, or approximately 20 million people. For hundreds of years

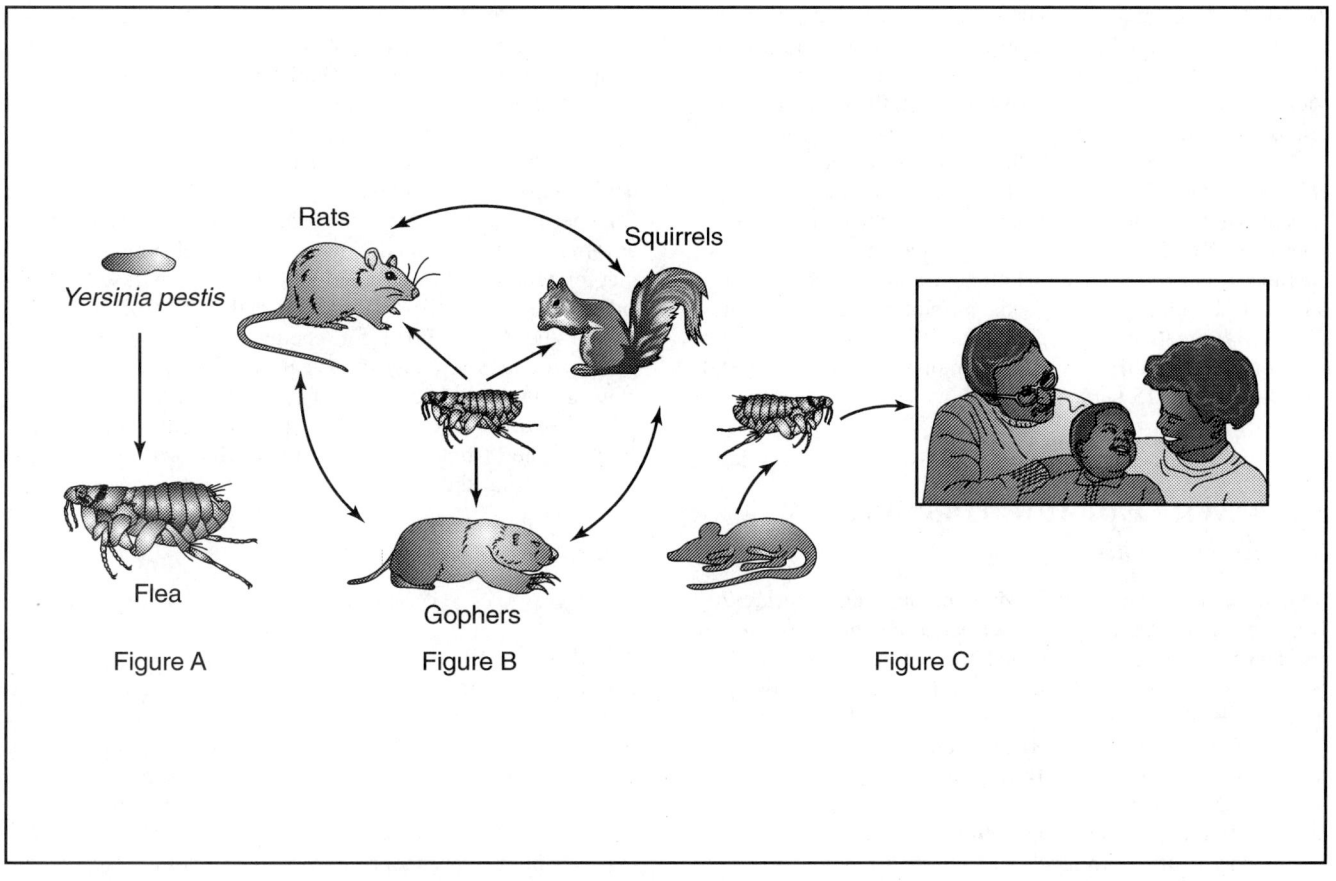

Plague is a serious infectious disease transmitted by the bites of rat fleas. There are three major forms of plague: bubonic, pneumonic, and septicemic. As illustrated above, fleas carry the bacterium *Yersinia pestis*. When a flea bites an infected rodent, it becomes a vector and then passes the plague bacteria when it bites a human. *(Illustration by Electronic Illustrators Group.)*

after, epidemics of bubonic plague would sweep across the world killing millions more. The disease was so lethal that some victims supposedly would go to bed healthy and die in their sleep. When the disease raged through a town, the population would die so quickly that it was nearly impossible to bury the dead.

Other serious epidemics arose from the fourteenth to seventeenth centuries, including the Great Plague of London in 1665. The incidence of plague began to decline after that. However, major epidemics still broke out occasionally; and, once **germ theory**, developed largely by **Louis Pasteur**, had provided a basis for understanding communicable disease around the mid-1800s, a major effort could be initiated to find its cause and cure. At the end of the nineteenth century, several major researchers made important discoveries that led to the control of bubonic plague.

Japanese-born scientist **Shibasaburo Kitasato** had studied in Germany where he was the first to grow a pure **culture** of the tetanus **virus**. He also developed, with **Emil von Behring**, the theory of antitoxin immunity. When Kitasato returned to Japan from abroad in 1894, he went to Hong Kong where a major outbreak of plague was in progress, and was able to isolate its cause—a bacterium called *Pasteurella pestis*. Around

the same time a Swiss bacteriologist, Alexandre Yersin (1863-1943), independently announced the same finding.

German bacteriologist Heinrich Hermann **Robert Koch**, who was the first to study bacteria in solid media and went on to identify the bacterial causes of **anthrax**, tuberculosis and **cholera**, turned his attention to bubonic plague in 1897. He determined that the disease was transmitted by a flea that infested rats; the germ lived in the stomach of the flea and the bloodstream of the rat and infected humans through the bite of either creature.

Today, **immunization** and **antibiotics** control the disease. However, occasional outbreaks of bubonic plague still occur in parts of the world where overcrowding and poor sanitation make the control of its pest carriers difficult. An outbreak of bubonic plague in Beed, India, in 1994 is believed to have resulted from an earthquake the year before that devastated the region, causing rodents to migrate and possibly enter grain stores. Although none of the 90 reported cases were fatal, public health officials believe that this outbreak may have set the stage for virulent outbreak of pneumonic plague in Surat, India. The theory is that some of the people who developed bubonic plague in Beed progressed to the pneumonic plague stage and then passed the sickness on to people in Surat. More than 6,000 cases of pneumonic plague and 56 deaths from the disease were reported in Surat in 1994. While some investigators thought the outbreak may have been caused by different bacteria or viruses and not the plague at all, the incident points to the problems caused by overcrowding and poor sanitation.

While plague vaccines have been developed, the most effective approach to prevent the plague continues to be improved sanitary conditions, including control of rat populations and immediate isolation of plague patients when they are diagnosed. For travelers to countries that still suffer from outbreaks of the plague, prophylaxis antibiotics are recommended. Early treatment with several antibiotics, including streptomycin, significantly reduces mortality from 60 to 100 percent to 10 to 15 percent.

BUCHNER, EDUARD (1860-1917)

German biochemist

Eduard Buchner was born in Munich, Germany on May 20, 1860, the same year **Louis Pasteur** performed his work on **fermentation**. Buchner's father died when Eduard was 12. His brother Hans, who became a famous bacteriologist, oversaw the boy's education. Following Army service, Eduard entered Munich Technical University to study chemistry. He was forced to withdraw because of financial hardship, and went to work in canning factories for four years. While there, he developed an interest in the fermentation process in which yeast breaks down sugar into alcohol and carbon dioxide.

When his brother Hans was once again able to provide support, Eduard resumed his studies in 1884 and soon received a scholarship. Buchner studied chemistry under Adolf von Baeyer and as well as botany at Munich. Under his brother's guidance, he began a systematic study of alcoholic fermentation which led to Eduard's first published paper in 1885. In his paper, Buchner revealed that fermentation could occur in the presence of oxygen, a conclusion contrary to the current prevailing view held by Louis Pasteur. In 1888 Eduard received his doctorate in chemistry and became Baeyer's assistant. With Baeyer's funding, Buchner set up a laboratory to continue research into fermentation.

By 1893 Buchner's was fully involved in seeking the active agent of fermentation. At that time there were two competing theories on fermentation. The vitalist theory, held by Louis Pasteur, held that living cells contained a "vital substance," an unidentified yet necessary component of living cells that was responsible for fermentation. The vitalists believed chemicals alone could not produce fermentation. The other view, the mechanistic theory, stated that yeast, continually decomposing in a liquid, set up chemical stresses that broke down sugar molecules making fermentation a complex, but otherwise normal, chemical reaction.

With the encouragement of his brother Hans, Eduard sought the active agent in fermentation. He obtained pure samples of the inner fluid of yeast cells by pulverizing yeast with a mixture of sand and diatomaceous earth, then squeezing the mixture through a canvas filter. This process avoided the destructive method of using solvents and high temperatures which had foiled previous investigations. Buchner and his assistant assumed the collected fluid was incapable of producing fermentation because the yeast cells were dead. However, when they attempted to preserve the fluid in concentrated sucrose (sugar), they were startled to observe carbon dioxide being released, a sign that fermentation was taking place. Buchner hypothesized that the fermentation was caused by an **enzyme** which he named zymase. His astonishing findings—that fermentation was the result of chemical processes both inside and outside cells—were published in 1897. Buchner received the Nobel Prize for chemistry in 1907. His work demonstrated that many chemical processes which occur in cells are under the regulation of enzymes, leading the way for other researchers to study the processes of biochemistry. Buchner died from wounds sustained while serving at a field hospital during the first World War.

BUDS AND BUDDING

A bud is a swelling on a plant stem, consisting of numerous overlapping, immature (i.e., not-yet developed), primordial leaves or floral structures (such as petals). The growth of plant buds occurs through the action of a region of rapidly dividing and differentiating **tissue**, known as an apical meristem.

A plant bud is a complex, compound structure, whose detailed **anatomy** is revealed by careful dissection and examination using a **microscope**. The smallest, least developed, **leaf** or floral primordia occur in the center of a bud, while the older, larger, more developed ones occur on the outside and curve up and over the younger ones, enveloping them in a sheath-like manner. In the center, beneath the youngest primordia, is the meristematic tissue. The dormant buds of many woody **plants** are protected by several tough, overlapping structures known as bud scales. When used in reference to botanical structures of this type (i.e., plant buds), the term "budding" describes the developmental process by which buds expand due to tissue growth and **differentiation**, and eventually develop into mature leaves or floral structures.

The buds of many plants have episodic growth, meaning they are dormant for long periods of time, only growing when environmental circumstances are favorable. In climates that are strongly seasonal, either because of cold winters or extended drought, buds are stimulated to grow (or "break") when the **temperature** warms in the springtime, or after rains occur. In such cases, the dormancy of buds is broken by the occurrence of specific concentrations of certain **plant hormones**, known as auxins (the amounts of the auxins are regulated by biochemical mechanisms).

In the stems of many plants, a dilute stream of auxins moves down the shoot from the bud at its apex, and this serves to inhibit the growth of buds lower on the stem (this is known as apical dominance). However, if the terminal bud on the shoot is injured (for example, by being browsed by a deer, or trimmed by a gardener), this stream is interrupted. This biochemical change results in previously dormant, lateral buds lower on the shoot breaking their auxin-enforced dormancy. This allows new, lateral shoots to develop and grow upwards, to replace the terminal bud-shoot that was previously growing (and dominant).

An alternative definition of ''budding'' refers to a kind of non-**sexual reproduction**, in which a new ''individual'' develops as a direct growth from a body tissue of the parent. The new individual may become detached from its parent, to grow as a discrete **organism**. However, the new individual is genetically identical to its parent (unlike the case in sexual reproduction). **Yeast**s are single-celled **fungi** that reproduce by budding new cells from the parent ones. In plants, **asexual reproduction** by budding can occur in various ways, such as the development of runners (i.e., above-ground stems, such as those of the strawberry), rhizomes (underground stems that grow outward from the parent, such as those of poplars), plantlets developed at the edges of leaves (as occurs in **species** of mother plant, or *Kalanchoe*), or bulbils produced in leaf axils (such as those of the tiger lily). Some **animal**s also reproduce asexually by budding, as is the case of many species of hydra, corals, and sea anemones.

BUFFON, GEORGES-LOUIS LECLERC, COMTE DE (1707-1788)

French naturalist

Georges-Louis Leclerc, Comte de Buffon was born to an aristrocratic family in Montbard, France. His affluent background allowed him to travel extensively and pursue a number of fields before he developed a passionate interest in natural history. After studying at the Jesuit College in Dijon, France, Buffon obtained a law degree in 1726. The intellectual life of Dijon was active but not oriented toward science, so Buffon went off to Angers, a city in northwestern France, to study medicine, mathematics, botany, and astronomy. The threat of a duel forced him to leave Angers in 1730, but he seized the opportunity to travel through France, England, and Italy. While he was traveling, Buffon's mother died and left him a sizable fortune.

Buffon had been so impressed with the upsurge of science in England that he dedicated the next couple of years to scientific endeavors. His first project, at the request of the French navy, was to write about the tensile strength of timber so that the government could improve the construction of war vessels. Next, he undertook a study of probability theory, *Mémoire sur le jeu du franc-carreau*, a project that contributed to his election to the Royal Society in 1730 and his admission to the Académie Royale des Sciences in 1734.

Buffon began to take an interest in botany and forestry. He wrote numerous dissertations and translated several works into French, including Stephen Hales' works on plants, *Vegetable Statiks*, and Isaac Newton's work on calculus. By this time, his work in the sciences began to elevate his standing, and he was advanced and transferred from the mechanical to the botanical section of the Académie Royale.

Nevertheless, Buffon's interest in natural history remained casual until he was appointed to the very prestigious

Georges-Louis Leclerc, Comte de Buffon. *(The Library of Congress. Reproduced by permission.)*

position of keeper of the Jardin du Roi, the French botanical gardens. This opportunity enabled him, for the next fifty years, to spend summers at the estate and return to Paris for the winters. During this time, he published forty-four volumes of his *Historie Naturelle* (Natural History), famous as the first modern work that attempted to treat nature as a whole. It was essentially the first encyclopedia on natural history to encompass both plant and animal kingdoms. Assisted by several eminent naturalists of the time, Buffon organized the often-confusing wealth of material into a coherent form. Moreover, in the work, he included suggestions on how the Earth might have originated, and he challenged the then-popular belief that the Earth was only 6,000 years old. Besides proposing that the Earth might be much older, he also suggested that the fact that animals retain parts that serve no known purpose to them is evidence that animals have evolved.

Buffon's popularity increased dramatically due to this work, and he remained a well-known scientific figure until his death in 1788. His prestige earned him an invitation to become a member of many academic societies, including those in Berlin, Germany, and St. Petersburg, Russia. Members of the aristocracy bestowed gifts upon Buffon and King Louis XV made him a count, commissioning a famous sculptor to create a bust of him.

See also Evolutionary theory

BURNET, FRANK MACFARLANE (1899-1985)
MEDAWAR, PETER BRIAN (1915-1987)
Australian and
English biologists

Working half a world apart, these two men made a highly significant contribution to immunology. Although Burnet developed the theoretical foundation of immunology, Medawar succeeded in proving it.

Born in Traralgon, a country town in Victoria, Australia, MacFarlane Burnet attended Geelong College, Victoria, and earned his medical degree from Melbourne University in 1923. He received his Ph.D. from the University of London in 1927. Burnet was associated with Melbourne Hospital for his entire career and also directed the University of Melbourne's Walter and Eliza Hall Institute for Medical Research. He conducted research on immunology throughout his career, even after retirement, and wrote scientific books for the general public.

On his return to Australia from England in 1928, Burnet investigated the deaths of twelve children following **diphtheria** vaccination. He found that the vaccine had been contaminated by staphylococcus **bacteria**. This focused Burnet's interest on the ways in which the body defends itself against infection. He began to study animal **virus**es and contributed to the basic understanding of **bacteriophages** (bacteria that attack viruses). In 1932, Burnet developed a laboratory technique for cultivating viruses; they could not grow outside living **cells**, and **mammal** cells were difficult to maintain in lab containers. Burnet's method of growing viruses in chick embryos became standard for the next twenty years.

Burnet noted that the chick embryos were unable to resist virus infection and produced no antibodies against viruses. This led Burnet to speculate that animals do not produce antibodies to substances they encounter very early in life, assuming such substances to be present at birth. That, Burnet thought, would explain how the body is able to distinguish between "self" cells and "antiself," or foreign, cells. Burnet attempted to prove his theory by inducing immunological tolerance in chick embryos with injections of antigens, but his efforts failed.

His idea, however, was picked up by P. B. Medawar at Oxford University. Medawar was born in Rio de Janeiro, Brazil, in 1915 to British parents, but his father was a native of Lebanon. At the age of four, Medawar moved with his family to England. He studied zoology at Oxford, graduating in 1939. From 1938 to 1947, he was a Fellow at Magdalen and St. John's Colleges, Oxford. In 1947, he became professor of zoology, first at Birmingham University and then at University College, London. He completed his career as director of the National Institute for Medical Research. Besides his research activities, Medawar became well-known for his books about the philosophy of science. During World War II, Medawar worked in a burns unit and became involved with the problem of skin-graft rejection. Patients with severe burns could not accept grafts from donors. Medawar used the fact that a second graft from the same donor was rejected more rapidly than the

first to prove that rejection of transplants was an immune reaction against antigens on the foreign tissue. Medawar's continued work in this field established transplantation immunobiology as an important area of specialized study.

When Burnet suggested that immunological tolerance could be induced by exposing embryos to antigens, Medawar decided to test the theory. He inoculated mouse embryos with tissue cells from mice of a different strain. The embryos did not reject the foreign tissue, and when the embryos matured, they accepted skin grafts from those other strains of mice as well. By 1953, Medawar had proven Burnet's theory of acquired immunological tolerance, which had important implications for tissue transplantation. The two men received the Nobel Prize for physiology or medicine in 1960 in recognition of their discovery.

Burnet married Edith Linda Druce in 1928; the pair had one son and two daughters. Burnet died in 1985. Medawar married Jean Shinglewood Taylor, with whom he had two sons and two daughters. Medawar died in 1987.

BUTENANDT, ADOLF FRIEDRICH (1903-1995)
German biochemist

Friedrich Butenandt devoted most of his career to isolating and synthesizing male and female **hormones**. He received the 1939 Nobel Prize in chemistry for demonstrating that the sex hormones were **steroids** and that the hormones are related to **cholesterol** and bile acids.

Butenandt was born to middle-class parents in what is now Wesermunde. In 1925, he graduated from the University of Marburg. His education continued at the University of Göttingen where he received his Ph.D. in 1927, and joined the faculty in 1931. In 1936, Butenandt became director of the Kaiser Wilhelm (later Max Planck) Institute for Biochemistry.

Working separately in 1929, Butenandt and the American biochemist Edward Doisy isolated the female hormone oestrone, which regulates the fertility-menstrual cycle. Butenandt showed that another **estrogen** hormone called estriol could be converted into oestrone by removing water. Butenandt's group continued their hormone research and in 1934, they isolated another female hormone, progesterone, which is important in pregnancy. Though other scientists also isolated progesterone the same year, Butenandt's isolation of the male hormone androsterone (which controls male fertility) was a sole accomplishment. The structure he deduced was verified experimentally when Swiss biochemist Leopold Ruzicka (1887-1976) synthesized it in 1934.

Butenandt isolated non-human hormones including ecdysone, the steroid hormone that transforms a caterpillar into a pupa and then into a butterfly. In 1959, Butenandt and a colleague were the first to isolate a **pheromone** or sex attractant, bombykol, from the scent gland of the silk moth *Bombyx mori*. Pheromones are hormones which perform important sexual functions in many organisms, especially insects. Released (usually by a female) in incredibly minute quantities, a pheromone can attract potential mates to the releasing individual from great distances.

In 1960 he became president of the Max Planck Society for the Advancement of Science. Butenandt's many honors have included membership in the New York Academy of Sciences, the Academy of Sciences at Göttingen, the Japanese Biochemical Society, the Deutsche Akademie der Naturforscher Leopoldina, Halle, and the Austrian Academy of Sciences; as well as the Grand Cross for Federal Services with Star in 1959 and six honorary doctorates.

C

CALORIE

A calorie is a unit of potential **energy** contained by a substance, which can be liberated when the material is oxidized, usually by combustion in the presence of oxygen. A calorie is defined as "the energy required to raise the **temperature** of one gram of pure **water** by one degree Centigrade under standard conditions." The standard conditions involve an atmospheric pressure of one **atmosphere**, and a temperature change from -4.1 to -2.3°F (15.5-16.5°C).

The calorie just defined is sometimes referred to as a gram-calorie or small calorie (short-form: cal). This is done to distinguish it from the calorie (or Cal) used as a measure of energy by dieticians, also known as a large calorie, or kilocalorie (kcal). A calorie is equal to 1,000 calories.

Energy can also be measured in other units, which can be converted among each other. One calorie is equivalent to 3.968 British thermal units (btu). A calorie is also equivalent to 4.187 joules (also known as an International Table calorie). The joule is the unit of energy that is most commonly used by scientists.

Data concerning the calorific energy contained by organic materials are usually obtained by completely oxidizing a known quantity of a substance by igniting it in a device known as a pressure-bomb. The amount of energy released is determined by measuring the increase in temperature of a known quantity of water contained within the pressure-bomb. Pure **carbohydrate**s have a calorific content of about 4,600 cal/g (4.6 Cal/g), while **protein**s contain about 4,800 cal/g, and **fats** about 6,000-9,000 cal/g.

CALVIN, MELVIN (1911-1997)
American biochemist

The winner of the 1961 Nobel Prize in Chemistry, Calvin unraveled most of the mystery surrounding **photosynthesis**, the process in which plants convert sunlight into food. The son of Russian immigrants, Calvin earned his Ph.D. from the University of Minnesota in 1935. His interest in photosynthesis began when he left for England and started his postdoctoral studies with British chemist Michael Polanyi (1891-1976) at the University of Manchester. Calvin's work there focused on organic **compounds**, such as chlorophyll and **hemoglobin**, that are composed of **atoms** of metal.

When he returned to America in 1937, Calvin began teaching at the University of California-Berkeley. During World War II, Calvin worked on the Manhattan Project, helping develop the atomic bomb. The simple method of extracting oxygen from the air, which Calvin designed for industrial purposes, also proved beneficial to the war effort.

After the war, while directing one of the University of California's laboratories, Calvin began his brilliant research on photosynthesis. Up to this time, no one understood the exact nature of the **molecules** involved in photosynthesis or the series of biochemical reactions that took place in plant cells during the process. It was generally known that plants combine carbon dioxide with water to form **carbohydrates**, the building blocks of all food, and that oxygen is released as a by-product. The process was so complex, however, that it had never been duplicated in a test tube.

Calvin confirmed that a *light reaction* involving chlorophyll, the plant's green pigment, instantly captures the Sun's energy, and is followed by a *dark reaction,* so-called because it can take place without sunlight, which begins to build carbohydrate molecules. To trace the path of carbon through the detailed reaction sequence, Calvin "tagged" his carbon dioxide with a radioactive isotope, carbon-14. Working with green algal **cells**, he interrupted the photosynthetic process at different stages by plunging the cells into alcohol. Then, using a new laboratory technique called paper **chromatography**, he analyzed the cells and the chemicals they produced. Calvin identified ten intermediate products that had been created within a few seconds. This cycle of reactions is now called the Calvin cycle.

Melvin Calvin. *(The Library of Congress. Reproduced by permission.)*

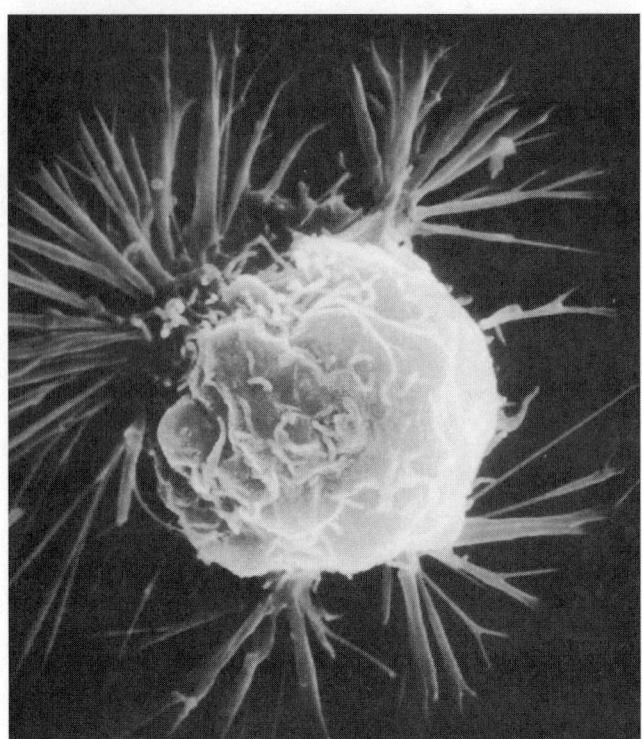

A breast cancer cell. *(Phototake NYC. Reproduced by permission.)*

Melvin Calvin was the recipient of several honors and awards, including the National Medal of Science in 1989, the Priestley Medal of the American Chemical Society in 1978, the Davy Medal of the Royal Society and the John Ericsson Award in Renewable Energy. He was on the Science Advisory Committees for both Presidents Kennedy and Johnson and in 1954 he became a member of the National Academy of Sciences.

See also Blood; Chlorophyll function and structure; Respiration

CANCER

Cancer may be as old as humankind. During the fifteenth century, what might now be considered a cancer-like growth was referred to as a *scirrus*, or scar. Environmental substances have long been associated with the disease. In 1775 Sir Percivall Pott connected frequent occurrences of scrotal cancer among chimney sweeps with their continual exposure to flue dust. Until the 1700s, Europeans treated cancer with crude methods like cauterization and **blood**letting. Although cancer research has been continuous for centuries, the most important conclusions have been drawn in the twentieth century.

Cancer is a group of many diseases in which certain **cell**s within the body lose their ability to regulate **cell division**. The cancerous cell multiplies uncontrollably, causing other normal cells to be crowded out and destroyed. If this growth takes place in a vital **organ**, malfunctions and **death** can result. The causes of cancer are diverse and the cure rates associated with different types of cancer vary. Scientists have long believed that cancer is linked to changes in the **genetic material** of the cell. It has been noted that the **chromosome**s of cancerous cells show abnormalities that may include deletions or **translocation**s of certain **gene**s. Many agents known to cause cancer (**carcinogens**) are also proven mutagens, or substances that cause atypical genetic changes. Ultraviolet radiation, x rays, **virus**es, and some chemicals are carcinogens that cause genetic **mutation**s.

These initial observations triggered the search for other causes of cancer. In 1950 scientists demonstrated that **nucleic acids** were important hereditary chemicals. Frank L. Horsfall, Jr., a clinician and virologist, was interested in finding the causes of cancer. He knew that once cancerous cells appeared, these cells would produce more cells with similar cancerous properties. In cell **culture**, the cancerous alteration was passed on even when the carcinogen was removed. This clue lead Horsfall to believe that the changes in the cell must have been a result of changes in DNA. He also saw similarities in **animal** cancers and changes in virus-infected **bacterial** cells. Horsfall's discovery that all cancer is attributed to changes in the nucleic acid of the cell provided a unifying concept for studying cancer.

Nearly 50 years before Horsfall's observations, a physician named **Peyton Rous** began research in pathology at what is now Rockefeller University. Rous was interested in the

COMMON PATHOGENS AND THE CANCERS ASSOCIATED WITH THEM	
Causative Agent	**Type Of Cancer**
Viruses	
Papillomaviruses	Cancer of the cervix
Hepatitis B virus	Liver cancer
Hepatitis C virus	Liver cancer
Epstein-Barr virus	Burkitt's lymphoma
	Hodgkin's lymphoma, Non-Hodgkin's lymphoma, Gastric cancers
	Cancers of the upper pharynx
Human immunodeficiency virus (HIV)	Kaposi's sarcoma Lymphoma
Bacteria	
Helicobacter pylori	Stomach cancer Lymphomas

(Table by Standley Publishing.)

physiology of cancer within **mammals** and birds and discovered the first virus-induced cancer. This connective **tissue** cancer in chickens, called Rous sarcoma, causes enlargement of the liver and is fatal. Rous's initial experiment included grafting sarcoma tumor cells from diseased hens into healthy hens. He noted that the healthy hens soon contracted the disease. Even filtered fluid extracts from diseased hens were contagious. Rous hypothesized that a virus may be the cause of the cancer. Unfortunately, other scientists were not able to duplicate Rous's experiments using different bird **species** and his work was ignored for several decades. Soon after Horsfall and other researchers, including Sarah Stewart and Bernice Eddy, established a better understanding of viruses and cancer, Rous's work was finally recognized when he received the Nobel Prize in 1966. Rous's viral theory of cancer engendered an entirely new perspective. It became accepted that some cancers are infectious, that is, able to spread from individual to individual via certain viruses. Today more than 24 oncogenic (cancer-causing) viruses are known. Several of these, including the Epstein-Barr and hepatitis viruses, cause other diseases in addition to cancer.

Because of our better understanding of the causes of cancer, various therapies have recently been introduced in an effort to improve cure rates for many cancers. **Chemotherapy** employs numerous chemical medications to attack and kill cancerous cells. Unfortunately, many of these compounds are toxic to the human body and cause severe side effects. One particularly inventive type of chemotherapy uses **light** sensitive compounds to deliver the active agent directly to the cancerous tumor. This is referred to as photodynamic chemotherapy. Radiotherapy uses radiation to shrink cancerous growths prior to or in place of surgical removal. Although radiation is site directed from an exterior source, or in some cases of breast cancer an implanted device, it also kills normal cells in massive numbers. Radiation therapy is frequently used in combination with chemotherapy. The side effects of both chemotherapy and radiotherapy, which include severe nausea and loss of hair, are well documented. Immunotherapy is yet another useful cancer treatment. This type of treatment is relatively new and appears to have great potential. Immunotherapy uses several different methods to bolster the patient's own **immune system**. The body is then better able to recognize and destroy the cancerous cells. Finally, herbal therapy has experienced a renewed interest in the medical community. Several herbs have been shown to contain cancer-inhibiting compounds.

The twentieth century's final decade heralded enormous advances in **genetics** research, including the ability to pinpoint genes and the role they play in disease. Scientists have identified mutations in three primary types of genes that can cause cancer. Mutations in tumor suppressor genes inhibit their ability to prevent the reproduction of damaged cells and tumor growth. **Oncogene**s, a mutated form of protooncogenes (before oncogenes), produce a dominant **protein** that encourages cell growth. **DNA** repair genes correct genetic errors; when they cannot perform their function due to a mutation, cancer occurs. Perhaps the most frequent genetic mutation associated with cancer occurs in the p53 gene and its encoded protein. Located on the short arm of the human chromosome 17, this tumor suppressor gene is often lost or mutated in tumors, leading to abnormal cell proliferation.

A growing number of genes have been identified as playing a major role in many types of cancer. A gene on chromosome 3 is thought to be a tumor suppressor gene associated with up to two-thirds of kidney cancer cases. A gene on the upper part of chromosome 2 causes colon cancer by fostering mutations in other genes. A mutation in a gene on chromosome 9 causes basal cell carcinoma, the most common type of skin cancer. A dysfunction in the p16 gene prevents it from producing a protein, leading to benign pituitary tumors, which can still cause vision problems and disrupt **hormone** balances.

While it is still uncertain whether many of these genetic mutations are inherited, the result of environmental influences, or a combination of both, researchers are closing in on many familial cancer-causing genetic mutations, including forms of colon and breast cancer. In 1995 the identification of the ''breast cancer genes'' BRCA1 and BRCA2 represented the first ''susceptibility'' genes found for a prevalent cancer. BRCA1, found on chromosome 17, is responsible for a prevalent form of hereditary breast cancer. BRCA2 is found on chromosome 13 and appears to cause as many cases of breast cancer as BRCA-1, including rare forms of male breast cancer. Originally thought to raise a woman's chances of getting breast cancer by 85%, follow-up research has indicated that women with either one of these mutations has an increased risk of closer to 50%. Nevertheless, these findings, as well as similar findings for other cancers, certainly will lead to tests that can identify individuals with genes that may predispose them to cancer. Understanding the genetic causes of cancer will also lead to **gene therapy** as a viable approach to curing and helping to prevent cancer.

Even with the confirmation of a genetic link to cancer, it will be many more decades before we fully understand this disease. Some forms of cancer are extremely rare and deadly, while others are common and may boast a cure rate of 90% or better. Every day potential carcinogens are discovered and novel therapies tested. Undoubtedly, much more information has yet to be learned before this feared disease is conquered.

CAPILLARY • See Blood circulation

CARBOHYDRATE

Carbohydrates are compounds consisting of carbon, hydrogen, and oxygen and are important sources of dietary **energy**. For most of the people in the world, carbohydrates—chiefly in the form of cereal grains and potatoes (or other root vegetables)—serve as the major source of energy. Even in the wealthier countries, such as the United States, carbohydrates provide 45-50% of the daily **calorie** count. In poorer countries, these easily digested and relatively inexpensive **nutrients** often provide well over 80% of every day's calories.

Chemically, carbohydrates are naturally occurring compounds composed of carbon, hydrogen and oxygen (CHO). In many cases the hydrogen-to-oxygen ratio of carbohydrates is the same 2-1 ratio as that of **water**. (Originally, chemists believed that all members of the class could be described as ''carbon hydrates,'' which is how the name originated. Although this assumption was soon disproved, the name remained.)

The carbohydrate group includes sugars, starches, **cellulose** and a number of other chemically-related substances. For the most part, these various carbohydrates are produced by green **plants**. In the process known as **photosynthesis**, countless varieties of plants are able to synthesize a simple sugar (glucose, mostly) using the **light** energy absorbed by the chlorophyll in their leaves. They also utilize water from the soil and **carbon dioxide** from the air. Typically, the simple sugar they produce is used partly to form the more complex carbohydrate cellulose, which makes up the plant's supporting framework, and partly to provide energy for its own metabolic needs. The rest, however, is stored away for later use in the form of seeds, roots, or **fruits.**

Interestingly, the digestive and metabolic processes in **animal**s and humans work almost in reverse fashion. When a fruit is eaten, for instance, the complex carbohydrates are broken down in the digestive tract to simpler glucose units. The glucose is then used primarily to produce energy in a process which involves oxidation and the **excretion** of carbon dioxide and water as waste products. (In the mid-1800s, **Justus von Liebig**, the famed German chemist, was one of the first to maintain that the body derived energy from the oxidation of foods recently eaten, and also declared that it was carbohydrates and **fats** that served to fuel the oxidation and not carbon and hydrogen as Antoine-Laurent Lavoisier had thought.)

Carbohydrates are usually divided into three main categories. The first category, the **monosaccharides**, are simple sugars that consist of a single carbohydrate unit that cannot be broken down into any simpler substances. The three most common sugars in this group are glucose (or dextrose), the most frequently seen sugar in fruits and vegetables (and, in **digestion**, the form of carbohydrate to which all others are eventually converted); fructose, associated with glucose in honey and in many fruits and vegetables; and galactose, derived from the more complex milk sugar, lactose. Each of these simple but nutritionally important sugars is a hexose, which means it contains *six* carbon **atoms**, twelve hydrogen atoms and six oxygen atoms. All three require virtually no digestion but are readily absorbed into the **blood**stream from the intestine.

Slightly more complex sugars are the disaccharides which contain two hexose units. The three most nutritionally important of these are sucrose (ordinary table sugar), maltose (derived from starch) and lactose, which is formed in the mammary glands and the only sugar not found in plants. All these sugars are split into the more easily absorbed monosaccharides in the digestive tract by specific **enzyme**s. If needed for future energy use, glucose units are typically squeezed together into larger, more slowly absorbed units and stored as **polysaccharides**, whose **molecule**s often contain a hundred times the number of glucose units as do the simple sugars. These highly complex carbohydrates include dextrin, starch, cellulose and glycogen. More efficient and more stable than the simple sugars, they are much easier to store. On the other hand, most of them need to be broken down by the digestive tract's enzymes before they can be absorbed. Some of them—cellulose, for instance—are almost impossible for humans to digest, but this indigestibility is useful since the colon needs a certain amount of bulk, or roughage, to perform at its best.

Glycogen is the form in which most of the body's excess glucose is stored. Both the liver and muscle are able to store glycogen, with muscle glycogen used primarily to fuel muscle contractions and liver glycogen used, when necessary, to replenish the bloodstream's dwindling supply of glucose. Glycogen was named by **Claude Bernard**, the French physiologist who, in 1856, discovered a starch-like substance in the liver of **mammals**. The substance, he later showed, was not only built out of glucose taken from the blood but could be broken down again into sugar whenever it was needed. In 1891, another physiologist, the German Karl von Voit, demonstrated that mammals could make glycogen even when fed more complex sugars than glucose. In 1919, Otto Meyerhof was able to show that, in working muscles, glycogen was converted into **lactic acid**. It wasn't until the 1930s, however, that two Czech-American biochemists, Carl Cori and Gerty Cori, were able to detail the complicated process by which glycogen, stored in the liver and muscle, was broken down in the body and resynthesized. Building on the work of the Coris, Fritz Lipmann, a few years later, was able to clarify even further the way carbohydrates could be converted into the forms of chemical energy most usable by the body.

The chemical structure of the various sugars was worked out in great detail by **Emil Fischer**, the noted German biochemist, who began his Nobel-prize winning work in 1884. Fischer not only was able to synthesize glucose and 30 more sugars, he also showed that the shape of these molecules was even more important than their chemical composition.

In the last 100 years much has been learned about carbohydrates' role in diet and health. We now know that not all carbohydrates are created equal. A 1988 University of Toronto study of 65,000 women showed that those with diets high in carbohydrates from white bread, potatoes, white rice, and pasta had more than twice the risk for Type 2 diabetes compared to those who ate a diet rich in high-fiber carbohydrates like whole wheat bread and whole grain pasta. The overall conclusion was that a high-carbohydrate diet that is low in fiber is not as healthy. According to the author of the study, the type of carbohydrate is more important than the total amount consumed. Low-fiber carbohydrates behave like sugar because they are digested fast and quickly drive up blood sugar levels and such rapid rises can contribute to diabetes. On the other hand, high-fiber carbohydrates like whole grains are digested more slowly and blood sugar levels rise more gradually and safely.

In addition to research on the dietary aspects of carbohydrates, there are efforts underway to create improved carbohydrate molecules and to identify new applications. For example, advances have been made with genetically altered carbohydrate-rich crops. A new type of potato has been cultivated which is designed to absorb less fat when fried. Other improved carbohydrate technologies include a new corn fiber by-product; a natural sweetener produced by fermenting glucose; a modified cellulose **polymer** which may have utility in low-fat foods; and an edible film that could be used as a coating for fruits and vegetables to prevent spoilage.

CARBON CYCLE

Carbon is an abundant element on Earth, particularly in its crust, surface **water**s, **atmosphere**, and biota. Carbon occurs in many different chemical combinations, including calcium carbonate ($CaCO_3$), **carbon dioxide** (CO_2), methane (CH_4), and a huge diversity of **organic compound**s (including **hydrocarbon**s and biochemicals). The storage of carbon in the various compartments of Earth and its **biosphere**, and the many transfers occurring among them, is referred to as the carbon cycle.

The most abundant mineral forms of carbon in crustal rocks and soil are limestone ($CaCO_3$) and dolomite ($CaMgCO_3$). These mostly occur in sedimentary rocks, which were formed in ancient marine environments through biological influences that resulted in the **precipitation** of limestone and dolomite from ions of calcium (Ca^{2+}), magnesium (Mg^{2+}), and bicarbonate (HCO_3^-) dissolved in water. The amount of carbon stored in sedimentary rocks has not yet been accurately estimated, but is thought to be much larger than that occurring in any other compartments of the carbon cycle.

Carbon also occurs in spaces within sedimentary crustal rocks in the form of hydrocarbons (i.e., compounds only containing carbon and hydrogen), such as coal, petroleum, and natural gas (collectively, these are known as ''**fossil fuels**''). These hydrocarbons are extremely important, non-renewable **natural resources** used as sources of **energy** and for the manufacturing of plastics and other materials. Wherever **fossil** fuel deposits can be accessed by mining and drilling technology, they are being rapidly used up. Other organic compounds of carbon also occur within crustal rocks (these may contain additional elements, such as oxygen, sulfur, and **nitrogen**), although in much smaller amounts than the hydrocarbons. All of these various kinds of organic carbon are derived from the partially decomposed **biomass** of ancient **plants** and other **organism**s, which became buried deep beneath marine sediment and were transformed extremely slowly, under conditions of intense pressure and heat and no oxygen, into their present forms.

Carbon occurs in the atmosphere mostly in the form of carbon dioxide at a concentration of about 360 parts per million (ppm), as well as methane at about 1-2 ppm. The concentrations of both of these gaseous compounds are increasing in the atmosphere. The concentration of CO_2 is increasing because of emissions associated with the combustion of fossil fuels, deforestation, the manufacturing of cement, and other sources related to human activity. Around 1850 the atmospheric concentration of CO_2 was about 280 ppm, compared with 360 ppm in 1998, an increase of 29%. During the same time, the concentration of CH_4 increased from about 0.7–1.7 ppm, mostly because of emissions associated with fossil-fuel mining, municipal landfills, agricultural livestock, and rice paddies.

Carbon also occurs in aquatic systems, usually in the form of the dissolved ion bicarbonate (CO_3^-). In acidic freshwater, however, it mostly occurs as dissolved CO_2.

Ecologically and biologically, carbon is crucial because it is the most abundant element in organisms, accounting for about 50% of typical, dry biomass. Carbon enters the ecological food web when photosynthetic organisms, such as plants (these, along with **algae** and photosynthetic algae, are referred to as **autotroph**s by ecologists), absorb CO_2 through tiny pores in their foliage, and fix it into simple sugars through the biochemical process known as **photosynthesis**. The plants then utilize the fixed energy to support their **respiration** (including the metabolic synthesis of an extremely wide variety of biochemicals) and to achieve growth and reproduction. Some of the organic **matter** of plants and other photosynthetic organisms (such as algae and blue-green **bacteria**) is consumed by **animal**s and other non-photosynthetic organisms (these are known as **heterotroph**s) and passed through ecological food webs. All organisms also release CO_2 to the atmosphere as a by-product of their respiratory **metabolism**.

CO_2 is also the most common emission associated with the **decomposition** of dead organic matter. However, if decomposition occurs under conditions in which oxygen (O_2) is not available, both CO_2 and CH_4 are emitted. Because the process of decomposition is relatively inefficient when it occurs under oxygen-deficient (or **anaerobic**) conditions, dead organic matter often accumulates in **wetlands** such as swamps and bogs, eventually forming peat. Under suitable geological circumstances (that is, involving anaerobic conditions, deep burial, and high pressure and high **temperature**) the peat may be very slowly transformed into carbon-rich fossil fuels.

Atmospheric CO_2 also dissolves into the waters of the oceans and lakes, where it occurs mostly as the ion CO_3^-, which is taken up by photosynthetic algae and blue-green bacteria and fixed into organic matter. These autotrophic organisms are the base of the marine food web. Oceanic CO_2 and CO_3^- are also utilized by many kinds of marine animals to manufacture their shells of $CaCO_3$ (calcium carbonate). This relatively insoluble mineral slowly accumulates in sediment, and may eventually become transformed into limestone or chalk.

Overall, the quantity of CO_2 taken up from the atmosphere by the global biota is rather similar to the amount released through respiration and decomposition. Consequently, the cycling of this nutrient can be viewed as a steady-state system. In modern times, however, various kinds of emissions associated with human activity (these are referred to as anthropogenic emissions) have significantly upset the atmospheric carbon balance. Global emissions of CO_2 and CH_4 are now larger than the uptake of these gases, an imbalance that has led to their increasing concentrations in the atmosphere. This change in atmospheric chemistry may be important in intensifying Earth's **greenhouse effect**, possibly leading to global warming and other climate changes.

CARBON DIOXIDE

Carbon dioxide (CO_2) is a compound consisting of one carbon atom and two oxygen **atoms**. It was the first gas to be distinguished from ordinary air, perhaps because it is so intimately connected with the cycles of plant and **animal** life. When we breathe air or when we burn wood and other fuels, carbon dioxide is released; when **plants** store **energy** in the form of food, they use up carbon dioxide. Early scientists were able to observe the effects of carbon dioxide long before they knew exactly what it was.

Around 1630, Flemish scientist Jan van Helmont discovered that certain vapors differed from air, which was then thought to be a single substance or element. Van Helmont coined the term "gas" to describe these vapors and collected the gas given off by burning wood, calling it gas sylvestre. Today we know this gas to be carbon dioxide, and van Helmont is credited with its discovery. He also recognized that carbon dioxide was produced by the **fermentation** of wine and from other natural processes. Before long, other scientists began to notice similarities between the processes of breathing (**respiration**) and burning (combustion), both of which use up oxygen and give off carbon dioxide. For example, a candle flame will eventually be extinguished when enclosed in a jar with a limited supply of air, as will the life of a bird or small animal.

Then in 1756, Joseph Black proved that carbon dioxide, which he called fixed air, is present in the **atmosphere** and that

it combines with other chemicals to form compounds. Black also identified carbon dioxide in exhaled breath, determined that the gas is heavier than air, and characterized its chemical behavior as that of a weak acid. The pioneering work of van Helmont and Black soon led to the discovery of other gases by Henry Cavendish, Antoine-Laurent Lavoisier, Carl Wilhelm Scheele, and other chemists. As a result, scientists began to realize that gases must be weighed and accounted for in the analysis of **chemical compound**s, just like solids and liquids.

The first practical use for carbon dioxide was invented by **Joseph Priestley**, an English chemist, in the mid-1700s. Priestley had duplicated Black's experiments using a gas produced by fermenting grain and showed that it had the same properties as Black's fixed air, or carbon dioxide. When he dissolved the gas in **water**, he found that it created a refreshing drink with a slightly tart flavor. This was the first artificially carbonated water (also called soda water or seltzer). Carbon dioxide is still used today to make colas and other soft drinks. In addition to supplying bubbles and zest, the gas acts as a preservative.

The early study of carbon dioxide also gave rise to the expression "to be a guinea pig," meaning to subject oneself to an experiment. In 1783, French physicist Pierre Laplace used a guinea pig to demonstrate quantitatively that oxygen from the air is used to burn carbon stored in the body and produce carbon dioxide in exhaled breath. Around the same time, chemists began drawing the connection between carbon dioxide and plant life. Like animals, plants "breathe," using up oxygen and releasing carbon dioxide. But plants also have the unique ability to store energy in the form of **carbohydrate**s, our primary source of food. This energy-storing process, called **photosynthesis**, is essentially the reverse of respiration. It uses up carbon dioxide and releases oxygen in a complex series of reactions that also require sunlight and chlorophyll (the green substance that gives plants their color). In the 1770s, Dutch physiologist **Jan Ingen Housz** established the principles of photosynthesis, which helped explain the age-old superstition that plants "purify" air during the day and "poison" it at night.

Since these early discoveries, chemists have learned much more about carbon dioxide. English chemist John Dalton guessed in 1803 that the **molecule** contains one carbon atom and two oxygen atoms (CO_2); this was later proved to be true. The decay of all organic materials produces carbon dioxide very slowly, and the Earth's atmosphere contains a small amount of the gas (about 0.033%). In our solar system, the planets of Venus and Mars have atmospheres very rich in carbon dioxide. The gas also exists in ocean water, where it plays a vital role in marine plant photosynthesis.

In modern life, carbon dioxide has many practical applications. For example, fire extinguishers use CO_2 to control electrical and oil fires, which cannot be put out with water. Because carbon dioxide is heavier than air, it spreads into a blanket and smothers the flames. Carbon dioxide is also a very effective refrigerant. In its solid form, known as "dry ice," it is used to chill perishable food during transport. Many industrial processes are also cooled by carbon dioxide, which allows faster production rates. For these commercial purposes, most carbon dioxide is obtained from natural gas wells, fermentation of organic material, and combustion of **fossil fuels**.

Recently, carbon dioxide has received negative attention as a "greenhouse" gas. When it accumulates in the upper atmosphere, it traps the Earth's heat, which could eventually cause global warming. Since the beginning of the industrial revolution in the mid-1800s, factories and power plants have significantly increased the amount of carbon dioxide in the atmosphere by burning coal and other fossil fuels. This effect was first predicted by Svante August Arrhenius, a Swedish physicist, in the 1880s. Then in 1938, British physicist G. S. Callendar suggested that higher CO_2 levels had caused the warmer **temperature**s observed in America and Europe since Arrhenius's day. Modern scientists have confirmed these views and identified other causes of increasing carbon dioxide levels, such as the clearing of the world's **forests**. Because **tree**s extract CO_2 from the air, their depletion has contributed to upsetting the delicate balance of gases in the atmosphere.

In very rare circumstances, carbon dioxide can endanger life. In 1986, a huge cloud of the gas exploded from Lake Nyos, a volcanic lake in northwestern Cameroon, and quickly suffocated more than 1,700 people and 8,000 animals. Today, scientists are attempting to control this phenomenon by slowly pumping the gas up from the bottom of the lake.

Despite these issues, industrial production of carbon dioxide remains strong in the 1990s. In fact, according to a 1997 chemical market study, commercial use of carbon dioxide is increasing with the worldwide market estimated to be $2 billion. The market has enjoyed steady price increases since 1995, when for the first time in many years, the demand exceeded the supply. Key areas of usage include beverage carbonation, frozen food and dry carbon dioxide for shipping. Alternate applications that are anticipated to keep demand high include the use of carbon dioxide in the production of precipitated calcium carbonate for alkaline paper, in making fire extinguishers, as an additive for wastewater treatment, and as an accelerant for greenhouse plant growth.

CARBON FIXATION

Carbon fixation refers to the chemical transformation of simple, inorganic compounds of carbon into more complex forms of organic **matter**. Examples of the simple compounds include **carbon dioxide** (CO_2) and bicarbonate (HCO_3^-), while the more complex forms include calcite ($CaCO_3$) and the organic matter of **organisms**.

In **ecosystem**s, carbon is fixed by **autotroph**s, which are organisms that utilize an external **energy** source to drive the synthesis of CO_2 and **water** (H_2O) into simple sugars. Usually, sunlight is the energy source for the fixation reactions, which is referred to as **photosynthesis**, and the organisms as photoautotrophs. Examples of photoautotrophs include **plants**, **algae**, and blue-green algae. Expressed simply, the photosynthetic reaction is:

Pure carbon dioxide gas can be poured because it is heavier than air. *(Photo Researchers, Inc. Reproduced by permission.)*

$$\text{light}$$
$$(1)\ CO_2 + 2H_2O \rightarrow CH_2O + 2O_2 + H_2O.$$

In reaction 1, the term CH_2O refers to a **carbohydrate**, which is a primary product of photosynthesis. The molecular oxygen (O_2) is a "waste" product of the photosynthetic reaction and is usually released by the autotroph. Through the many complex reactions of **metabolism**, the carbohydrate can then be used to synthesize the additional biochemicals needed by autotrophs for survival, such as complex carbohydrates, **protein**s, and **lipids**.

Some autotrophs are non-photosynthetic, meaning they utilize energy sources other than sunlight to drive their carbon-fixation reactions. These so-called chemoautotrophs use the stored energy of certain chemicals [usually sulfides such as hydrogen sulfide (H_2S) or iron sulfide (FeS_2)] to drive chemosynthesis. Chemoautotrophs are the basis of ecosystems that are independent of solar radiation.

In addition to photosynthetic fixations of CO_2, inorganic forms of carbon can be fixed by other biotic reactions occurring in aquatic ecosystems. One of these involves the fixation of dissolved carbon dioxide by a series of simple reactions, as follows:

$$(2)\ CO_2 + H_2O \rightarrow (H_2CO_3)$$
$$(3)\ H_2CO_3 \rightarrow (H^+ + HCO_3^-)$$
$$(4)\ HCO_3^- \rightarrow (H^+ + CO_3^{-2})$$
$$(5)\ Ca^{+2} + 2HCO_3^- \rightarrow (CaCO_3 + CO_2 + H_2O)$$

In reaction 2, dissolved carbon dioxide combines with water to form carbonic acid (a weak acid). In reaction 3, the carbonic acid dissociates to form hydrogen ion and bicarbonate. In reaction 4, the bicarbonate dissociates to hydrogen ion and carbonate. In reaction 5, calcium ion combines with bicarbonate to form calcite plus carbon dioxide and water. Reaction 5 commonly occurs within the bodies of certain aquatic organisms, which utilize calcite to construct their shells (invertebrate **animal**s) or bones (**vertebra**tes). The calcite is an insoluble mineral, which upon **death** of the organisms sinks to the floor of the body of water and accumulates in the sediment. Through extremely slow geological processes occurring in deep sediment, the calcite may eventually metamorphose into rocks, such as limestone or chalk.

CARBON MONOXIDE

Carbon monoxide, an odorless, poisonous gas, can be generated by gas furnaces and **water** heaters, ranges, space heaters, or wood stoves if they are malfunctioning or not vented properly. Cars, portable generators, and gas-powered gardening equipment also generate carbon monoxide (CO) and can cause problems if they are operated in enclosed areas or attached garages. Once inhaled, carbon monoxide inhibits the **blood**'s ability to carry oxygen by replacing oxygen in the red blood **cell**s which prevents the oxygen supply from reaching the **organ**s in the body. This oxygen deprivation can cause varying amounts of damage depending on the level of exposure. Low level exposure can cause flu-like symptoms including shortness of breath, mild headaches, fatigue, and nausea. Higher level exposure may cause dizziness, mental confusion, severe headaches, nausea, and fainting. Prolonged high level exposure can cause **death**. According to the U.S. **Consumer** Product Safety Commission, more than 2,500 people will die and 100,000 will be seriously injured by carbon monoxide over the next 10 years.

In the 1300s, Arnold of Villanova (1235-1311), a Spanish scientist, noticed that poisonous fumes were produced when wood was burned without adequate ventilation. Wood and other fuels containing carbon—such as coal, oil, and gasoline—produce carbon monoxide when they are burned without enough oxygen. (If enough oxygen were present, **carbon dioxide**, or CO_2, would be formed instead..) Although Arnold of Villanova did not understand the chemistry of carbon monoxide formation, he was the first to call attention to its harmful effects.

During the 1600s and 1700s, scientists began to learn more about the nature of gases. In the 1770s, oxygen and many other gases were discovered, and chemists determined what part gases play in the process of combustion, or burning. During this period of discovery, French chemist Joseph Marie François de Lassone (1717-1788) and British chemist **Joseph Priestley** independently prepared carbon monoxide gas in the laboratory for the first time. Soon afterward, in 1800, British chemist William Cumberland Cruikshank (1745-1800) determined its chemical composition.

However, carbon monoxide's poisonous effects were not well understood until the 1850s, when **Claude Bernard**, a French physiologist, explained how the gas acts in the human body. Bernard showed that when we breathe carbon monoxide, the gas prevents our blood from carrying oxygen through the body. It does this by combining with the blood's oxygen-carrying substance, called **hemoglobin**. Because carbon monoxide's affinity for hemoglobin is much stronger than that of oxygen, the bloodstream immediately absorbs carbon monoxide, and the body becomes starved of its life-supporting oxygen.

During the 1800s, chemists first attempted to turn gases into liquids. The first gas was liquefied in 1823 but it was not until 1877 that Louis Paul Cailletet, a French physicist, was able to liquefy carbon monoxide. To accomplish this, Cailletet simultaneously compressed and cooled a sample of carbon monoxide and then allowed it to expand. As it expanded, the gas cooled radically and condensed into a small amount of liquid.

Normally, carbon monoxide is present in the air in extremely minute amounts—not enough to pose a threat. However, in large cities, unhealthy amounts of carbon monoxide can build up, especially during heavy rush-hour traffic. Cars that are old give off the most carbon monoxide. Modern cars are equipped with catalytic converters that chemically change carbon monoxide into carbon dioxide. Despite its dangers, carbon monoxide is useful to industry as a fuel gas, which provides heat for manufacturing processes, and as a raw material for making chemicals and purifying such metals as iron and nickel. Over the last few decades technology has been developed to detect carbon monoxide for industrial applications. For example, the chemical industry has used a number of electronic gas sensors for analytical applications. Early industrial sensors involved a dual chambered sensor which oxidized CO and compared the heat of oxidation from the test chamber to a reference chamber. This type of oxidation requires a special platinum oxide catalyst and a heat source to burn the carbon monoxide. These systems were unacceptable for home use due their complexity of operation, expense, and lack of sensitivity. However, in the last decade or so, home carbon monoxide detectors have been made possible by improvements in advanced gas sensing technology. Other key factors have also contributed to the increased popularity of CO detectors. One is the rise in the use of other home safety appliances, such as smoke alarms. Another is the increased awareness of the dangers of carbon monoxide. Today, relatively inexpensive CO detectors can be purchased for as little as $30 to $80. In fact, many cities are now requiring that at least one smoke detector be installed in every home, apartment, and hotel.

CARCINOGENS

Cancer is a mysterious group of diseases which often arouses fearful images of **death**. There are over one hundred different types of cancer, which can be distinguished by the type of **cell** or **organ** which is affected, the treatment plan employed, and the cause of the cancer. Any substance or agent that can cause cancer is called a carcinogen. One of the first people to suspect that certain substances in the environment can cause cancer was Sir Percivall Pott, who in 1775 published a paper on cancer occurrence in chimney sweeps. It has since been discovered that benzo(a)pyrene, a chemical found in soot, is a potent carcinogen. Ionizing radiation was first suspected of being carcinogenic around the turn of the twentieth century, when physicians who developed the use of x rays and radium in medicine showed a high incidence of skin cancer. Later, in 1915, scientists in Japan documented carcinogens when they noticed that rabbits developed tumors when tar was applied to the inside of their ears. By 1930, Ernest Kennaway isolated polycyclic **hydrocarbon**s and proved that they were the carcinogenic agent.

Today, the media rarely misses an opportunity to report on newly discovered carcinogenic substances. Sometimes it seems as if *everything* causes cancer, but very few things are proven carcinogens. The two main categories of carcinogens are genetic and environmental. Specific environmental factors include tobacco, **alcohol**, diet, infection, sexual practices, occupation, geophysical phenomena, **pollution**, medications, food additives, and industrial products. Tobacco and diet together account for almost two-thirds of all cancer-related deaths. **Stress** and emotional factors should also be listed as elements that may contribute toward the development of some cancers.

Cigarette smoking is clearly the single most preventable cause of illness and premature death in the United States. The United States Surgeon General estimates that 30% of all cancer deaths are directly attributable to tobacco use. The carcinogens in tobacco include nicotine, tar, **carbon monoxide**, **ammonia**, formaldehyde, **phenol**s, creosote, anthracene, pyrene, hydrocyanic acid, **arsenic**, and lead. Tar is the particulate **matter** derived from burning **organic compound**s and is the leading cancer-causing chemical in tobacco smoke. Researchers have also identified 4-(N-nitrosomethylamino)-1-(3-pyridyl)-1butanone (NNK), as a carcinogen that is formed in the production, curing, and **aging** of tobacco. Studies in **ani-mal**s have shown that it can cause benign and malignant tumors, as well as other forms of cancer. Pipe and cigar smoke contain essentially the same array of poisons although the relative amounts may vary. In 1997, scientists produced the first chemical evidence that an increased risk for lung cancer is associated with passive, second-hand, smoke inhalation. The investigators found the metabolite of a tobacco carcinogen in the urine of non-smokers who have been exposed to second-hand smoke. Smokeless tobacco, including chewing tobacco and snuff, are not alternatives to avoid cancer. These products contain large quantities of cancer-causing chemicals called n-nitrosamines. Lip, mouth, tongue, and throat cancers have been positively linked to the use of smokeless tobacco products. Furthermore, the risk of these types of cancers are increased when tobacco is used with alcohol. This phenomenon is known as synergism. While marijuana does not contain nicotine or tobacco, it does contain tar and other carcinogens.

Diet and **nutrition** have been recognized as major factors that influence the development of many cancers. The National Cancer Institute recommends a low fat, high fiber diet with ad-equate allowances of vitamin A and vitamin C as a specific way to reduce the risk of developing colon cancer. Cruciferous vegetables like cabbage and brussels sprouts may also help reduce the chances of developing some stomach and colon cancers. Charred foods and alcohol contain carcinogens and should be avoided or consumed in moderation.

Infection may also lead to cancer. This usually occurs when a **virus**, **bacteria**, or **parasite** is contracted. Only about 10 percent of all cancers are believed to be caused by these **organism**s. **Retrovirus**es such as the herpes virus and the virus that causes **AIDS** have been implicated in certain types of cancer. According to the **oncogene** theory of virus-mediated cancer, when a virus infects a cell, it may enter the genes of the host cell that control growth and division. The infected cell may begin to divide uncontrollably, resulting in the formation of a tumor. Changes in a single oncogene rarely lead to malignant cancer. Rather, it usually requires many of the more than 100 oncogenes working in concert to initiate this chain of events. Parasitic and bacterial-induced cancer may be due to the chronic irritation, either internal or external, caused by these organisms over long periods of time. It also appears that other co-factors need to be present before many of these cancers can fully develop.

Changes in the body resulting from sexual intercourse, pregnancy, and childbirth are obviously in a different class of carcinogens than those produced by exposure to chemicals. They are, however, considered environmental since they are not controlled solely by one's own genes. This is not to suggest that childbirth causes cancer. In fact, pregnancy and childbirth may actually help prevent cancer of the uterus, ovary, and breast. However, frequent sexual intercourse with large numbers of partners has been positively linked to an increased risk of cervical cancer. Researchers think that the primary carcinogenic agent in this example may actually be an unknown virus.

The percentage of all cancers that can be attributed to work-related influences varies from about 4 percent to as high as 15 percent. Since 1971, the International Agency for Research on Cancer (IARC) has been categorizing carcinogens and occupations associated with high cancer rates. Some chemicals used in shoe, tire, and furniture manufacturing, as well as nickel refining, diesel fuel, and dry cleaning have been identified as ''anticipated'' carcinogens. Arsenic, asbestos, benzene, benzidine, chromium, 2-Naphthylene, oils, and vinyl chloride show occupational exposures causally associated with cancer in humans. Cadmium, **DDT**, and formaldehyde are other common chemicals that are suspected to cause cancer in humans.

Asbestos is perhaps the most familiar carcinogen in this category. This material, made of small silicate fibers, has been used in construction since the 1800s. Asbestos only becomes a cancer risk when the fibers are set free and inhaled during **decomposition** or renovation. The fibers irritate the body's alveoli and can lead to lung cancer. Asbestos exposure is now carefully regulated by the Occupational Safety and Health Administration (OSHA).

One of the most infamous carcinogens is PCB or polycholrinated biphenyls. These man-made chemicals encompass

209 individual compounds with various toxicities. PCBs were once widely used as coolants and lubricants before their manufacture was halted in October 1977 due to their accumulations in the environment and resultant health hazards, including their role as carcinogens that can cause cancer. However, PCBs are still widespread throughout the environment, including contaminated **water**, sediment, and fish. Another group of contaminants that exist worldwide are polychlorinated dibenzo-para-dioxins, or simply dioxins. Dioxins are typically by-products of **chemical reaction**s used for other purposes, like the production of some herbicides. A working group for IARC has determined that one dioxin, 2,3,7,8-tetrachlorodibenzo-para-dioxin (TCDD) is carcinogenic to humans, increasing risk of lung cancer and all cancers combined in those exposed to it. IARC continues to investigate the other dioxins. Benzene, a clear colorless liquid used widely as a solvent and reactant, has also been linked to leukemia, which is a type of **blood** cancer. Radiation refers to **energy** that is sent through space. It may be in the form of waves such as ultraviolet **light**, x rays, and microwaves, or in the form of charged atomic particles including electrons, protons, and neutrons. Radiation can alter genes within **chromosome**s. Skin cancer in early x-ray workers and radium workers, bone sarcoma in luminous dial painters, and lung cancers in miners of radioactive ores have all been studied. Ironically, some forms of radiation are used to diagnose and treat cancer.

Radiation from the sun is an increasing concern and may be linked to a depletion of the protective **ozone** layer in the Earth's upper **atmosphere**. The sun is the chief cause of non-melanoma skin cancer. The amount of ultraviolet-beta radiation from the sun varies with location, altitude, sky cover, and the time of year. Exposure levels can be reduced by using sun screen products and monitoring prolonged outdoor activities—especially between 11:00 a.m. and 1:00 p.m. during the late summer months.

In recent years, the approval of irradiation of food has raised growing concerns about its possible association with cancer. Food irradiation uses ionizing energy to kill bacteria, mold, and insects and can also prolong the food's shelf life. The major concern is that the process forms free radicals, or unidentified radiolytic products, which have been associated with the development of cancer. Those against irradiated foods argue that they contain high levels of carcinogens, including nitrosamines and formaldehyde. However, more than 39 countries, including the United States. approve the controlled use of irradiation for one or more food items, including meats, poultry, seafood, grains, vegetables, nuts, and spices.

Historically, substances have been tested for carcinogenicity by using them on human or animal test subjects. In the late 1970s, however, Bruce Ames of the University of California at Berkeley developed a carcinogenicity test that uses bacteria instead of human or animal subjects. The Ames test yields results in hours or days instead of the years required for human or animal tests, and it has a high rate of consistency with human tests. While this test is not accurate enough to give permanent proof of a substance's carcinogenicity, it does allow for speedy screening of new or previously unsuspected substances. The most difficult aspect of testing substances as potential carcinogens is determining what doses are actually harmful. For example, the Ames test on animals is performed with maximum tolerated doses (MTD), and some scientists question whether testing with MTD accurately reflet a substance's potential carcinogenic effect on humans.

Many environmental carcinogens are avoidable, and early diagnosis and treatment increases the recovery rate for all cancers. Genetically predisposed cancers are more difficult to regulate. They are also less common than many environmentally induced cancers. A growing area of interest are investigations of carcinogens like NNK and their ability to alter genes and cause cancer.

Research conducted by the World Health Organization (WHO) has shown that industrialized countries have high cancer rates compared with countries that have little industry. Despite the fact that industrialized countries make up only one-fifth of the world's **population**, one-half of all the world's cancers are found in people living in these countries. While many people fear the overall increase of chemical and physical carcinogens in the environment, it is especially important to prioritize risk factors of the highest magnitude. For example, smokers should be less concerned about food additives and more concerned about when they will stop smoking.

CARNIVORE

Carnivores include those **animal**s that obtain their **nutrition** by consuming other animals. It comes from a combination of Latin words literally meaning "flesh devourers." Carnivores are at the top levels of every food chain - primary carnivores feed on **herbivore**s, secondary carnivores feed on primary carnivores, and so on. Due to the Second Law of Thermodynamics, the number of **trophic level**s is typically limited to a maximum of four or five. There is a general correlation that the higher up on the food chain you go, the number of **organism**s decreases but their individual size increases. Herbivores typically have large or complex stomachs and long intestines to allow for enough time to adequately digest their high-**cellulose** diet. Because their food is similar in chemical composition to their own body chemistry, carnivores don't have to go through an extensive **digestion** process and thus have simple stomachs and short intestines. Another interesting contrast is that herbivores tend to be continuous feeders and carnivores tend to eat periodic meals. Carnivores thus have more leisure time in which to relax. During times of food gathering, however, carnivores are generally more active and more aggressive than their herbivore counterparts.

Animals within the **Order** Carnivora include the flesh-eating **mammal**s such as dogs, cats, wolves, bears, and weasels. These animals have carnassial teeth which are adapted for cutting and shearing their food. They are distributed worldwide except for Australia and Antarctica. Other carnivores which are not within this specific Order include a wide-range of animals such as blue-gill sunfish, large-mouth bass, sharks, and other fish; owls, hawks, and other birds; seals, sea otters,

Carnivores include those animals that obtain their nutrition by consuming other animals. Animals within the order Carnivora include the flesh-eating mammals such as the coyote pictured above. *(Photograph by Robert J. Huffman. Field Mark Publications. Reproduced by permission.)*

killer whales, and other marine mammals; and alligators, humans and other terrestrial vertebrates. Animals that ingest both plant and animal **matter** are called **omnivore**s.

CARR, ARCHIE FAIRLY (1909-1987)
American biologist

Archie Carr was born in Mobile, Alabama, in 1909. As a young person he was fascinated with wild **animals** and natural history, and this love of **nature** eventually developed into a career. Carr obtained his Ph.D. from the University of Florida in 1937, and he was a member of the teaching faculty there from 1937. From 1937–1943, he was also an associate of the Museum of Comparative **Zoology** of Harvard University, mostly conducting work in **taxonomy**. From 1945–1949, he took a leave of absence from the University of Florida to teach biology (in fluent Spanish) at the Escuela Agricola Panamericana (Pan-American Agricultural College) in Honduras. Carr was also a renowned naturalist, an author of numerous scientific works as well as popular literature, and an effective advocate of the **conservation** of sea turtles and other wildlife.

Archie Carr's research, and passion, largely involved marine turtles. For most of his career he was the world's foremost authority on that group of animals. Most of his research

A carnivorous pitcher plant, Isle Royale National Park, Michigan. *(Photograph by Robert J. Huffman. Field Mark Publications. Reproduced by permission.)*

was conducted in the Gulf of Mexico, the Caribbean Sea, northeastern South America, and the Pacific coast of Central America, but he also visited Africa and the South Pacific. During his fieldwork, Carr realized that all species of marine turtles were becoming endangered, mostly because of excessive harvesting of adults for their meat, and of eggs as a source of local food. Carr became an effective advocate of the conservation of marine turtles, particularly the green sea turtle and the olive ridley turtle. His actions, and those he inspired, have had a positive impact on the survival of sea turtles, and on marine conservation more generally.

Archie Carr published 10 books and more than 120 scientific papers and magazine articles during his career. Some of his better-known works include *Handbook of Turtles: The Turtles of the United States, Canada, and Baja California* (1952; Cornell University Press; this book won the Daniel Giraud Elliott Medal of the National Academy of Sciences), *High Jungles and Low* (1953; University of Florida Press), *The Windward Road* (1956, Knopf; a chapter in this book won an O. Henry Award for best short story of 1956), *The Reptiles* (1963; Life Nature Library), *Ulendo: Travels of a Naturalist in and out of Africa* (1964; Knopf), and *The Sea Turtle, So Excellent a Fish* (1986; University of Texas Press). Other awards won by Carr include the John Burroughs Medal of the American Museum of Natural History for nature writing.

In honor of Carr's contributions to the conservation of marine ecosystems, and of sea turtles in particular, the Archie Carr Center for Sea Turtle Research has been established at the University of Florida. The major aim of this institute is to continue Carr's mission of studying the biology and **ecology** of marine turtles, to advance the conservation of these animals by making scientific information readily available, and to provide training for graduate students and other highly qualified scientists. In addition, the Archie Carr National Wildlife Refuge is located near Melbourne, Florida.

CARREL, ALEXIS (1873-1944)
French medical scientist

Alexis Carrel was an innovative surgeon whose experiments with the transplantation and repair of body **organs** led to advances in the field of surgery and the art of **tissue** culture. An original and creative thinker, Carrel was the first to develop a successful technique for suturing **blood** vessels together. For his work with blood-vessel suturing and the transplantation of organs in **animal**s, he received the 1912 Nobel Prize in medicine and physiology. Carrel's work with tissue culture also contributed significantly to the understanding of **virus**es and the preparation of vaccines. A member of the Rockefeller Institute for Medical Research for thirty-three years, Carrel was the first scientist working in the United States to receive the Nobel Prize in medicine and physiology.

Carrel was born on June 28, 1873, in Sainte-Foy-les-Lyon, a suburb of Lyons, France. He was the oldest of three children, two boys and a girl, in a Roman Catholic family. His mother, Anne-Marie Ricard, was the daughter of a linen merchant. His father, Alexis Carrel Billiard, was a textile manufacturer. Carrel dropped his baptismal names, Marie Joseph Auguste, and became known as Alexis Carrel upon his father's death when the boy was five years old. As a child, Carrel attended Jesuit schools. Before studying medicine, he earned two baccalaureate degrees, one in letters (1889) and one in science (1890). In 1891, Carrel began medical studies at the University of Lyons. For the next nine years, Carrel gained both academic knowledge and practical experience working in local

hospitals. He served one year as an army surgeon with the Alpine Chasseurs, France's mountain troops. He also studied under Leo Testut, a famous anatomist. As an apprentice in Testut's laboratory, Carrel showed great talent at dissection and surgery. In 1900, he received his medical degree but continued on at the University of Lyons teaching medicine and conducting experiments in the hope of eventually receiving a permanent faculty position there.

Early Success with Blood Vessel Sutures

In 1894, the president of France bled to death after being fatally wounded by an assassin in Lyons. If doctors had known how to repair his damaged artery, his life may have been saved, but such surgical repair of blood vessels had never been done successfully. It is said that this tragic event captured Carrel's attention and prompted him to try and find a way to sew severed blood vessels back together. Carrel first taught himself how to sew with a small needle and very fine silk thread. He practiced on paper until he was satisfied with his expertise, then developed steps to reduce the risk of infection and maintain the flow of blood through the repaired vessels. Through his careful choice of materials and long practice at various techniques, Carrel found a way to suture blood vessels. He first published a description of his success in a French medical journal in 1902.

Despite Carrel's growing reputation as a surgeon, he failed to acquire a faculty position at the university. His colleagues seemed indifferent to his research, and Carrel, in turn, was critical of the French medical establishment. The final split between Carrel and his peers came when Carrel wrote a positive account of a miracle he apparently witnessed at Lourdes, a small town famous since 1858 for its Roman Catholic shrine and often visited by religious pilgrims. In his article, Carrel suggested that there may be medical cures that cannot be explained by science alone, and that further investigation into supernatural phenomena such as miracles was required. This conclusion pleased neither the scientists nor the churchmen of the day.

In June, 1904, Carrel left France for the French-speaking city of Montreal, Canada; an encounter with French missionaries who had worked in Canada had sparked Carrel's interest in that country several years earlier. Shortly after his arrival, Carrel accepted an assistantship in physiology from the Hull Physiology Laboratory of the University of Chicago, where he remained from 1904 to 1906. The university provided him with an opportunity to continue the experiments he had begun in France.

Blood transfusion and organ transplantation seemed within reach to Carrel, now that he had mastered the ability to suture blood vessels. In experiments with dogs, he performed successful kidney transplants. His bold investigations began to attract attention not only from other medical scientists but from the public as well. His work was reviewed in both medical journals and popular newspapers such as the *New York Herald*. In the era of Ford, Edison, and the Wright Brothers, the public was easily able to imagine how work in a scientific laboratory could lead to major changes in daily life. Human organ transplantation and other revolutions in surgery did not seem far off.

Begins Lifetime Career at Rockefeller Institute

In 1906, the opportunity to work in a world-class laboratory came to Carrel. The new Rockefeller Institute for Medical Research (now named Rockefeller University) in New York City offered him a position. Devoted entirely to medical research, rather than teaching or patient care, the Rockefeller Institute was the first institution of its kind in the United States. Carrel would remain at the institute until 1939. At the Rockefeller Institute, Carrel continued to improve his methods of blood-vessel surgery. He knew that mastering those techniques would allow for great advances in the treatment of disorders of the **circulatory system** and wounds. It also made direct blood transfusions possible at a time when scientists did not know how to prevent blood from clotting. Without this knowledge, blood could not be stored or transported. In the *Journal of the American Medical Association* in 1910, Carrel described connecting an artery from the arm of a father to the leg of an infant in order to treat the infant's intestinal bleeding. Although the experiment was a success, the discovery of **anticoagulants** soon made such direct transfer unnecessary. For his pioneering efforts, Carrel won the Nobel Prize in 1912.

Carrel's success with tissue cultures through animal experiments led him to wonder whether human tissues and even whole organs, might be kept alive artificially in the laboratory. If so, lab-raised organs might eventually be used as substitutes for diseased parts of the body. The art of keeping **cells** and tissue alive, and even growing, outside of the body is known as tissue culture. Successfully culturing tissue requires great technical skill. Carrel was particularly interested in perfusion—a procedure of artificially pumping blood through an organ to keep it viable. Carrel's work with tissue culture contributed greatly to the understanding of normal and abnormal cell life. His techniques helped lay the groundwork for the study of viruses and the preparation of vaccines for polio, measles, and other diseases. Carrel's discoveries, in turn, built upon the successes of, among others, Ross G. Harrison, a contemporary anatomist at Yale who worked with frog tissue cultures and transplants.

One of Carrel's experiments in tissue culture became the subject of a sensationalized news story and was viewed as a monstrosity by the public. In 1912, Carrel took tissue from the heart of a chicken embryo to demonstrate that warm-blooded cells could be kept alive in the lab. This tissue, which was inaccurately depicted as a growing, throbbing chicken **heart** by some newspapers, was kept alive for thirty-four years—outliving Carrel himself—before it was deliberately terminated. The *World Telegram,* a New York newspaper, annually marked the so-called chicken heart's ''birthday'' each January.

Though working in the United States, Carrel had not bought a house there, and did not become a U.S. citizen. Rather, he spent each summer in France, and on December 26, 1913, Carrel married Anne-Marie Laure (Gourlez de la Motte) de Meyrie, a widow with one son, in a ceremony in Brittany. They had met at Lourdes, where Carrel made an annual pilgrimage each August. Eventually, the couple bought some property on the island of Saint Gildas off the coast of Brittany, and lived in a stone house there. They had no children together.

Alexis Carrel. *(The Library of Congress. Reproduced by permission.)*

When World War I began, Carrel was in France. The French government called him to service with the army, assigning him to run a special hospital near the front lines for the study and prompt treatment of severely infected wounds. There, Madame Carrel, his wife of less than one year and a trained surgical nurse, assisted him. In collaboration with biochemist Henry D. Dakin, Carrel developed an elaborate method of cleansing deep wounds to prevent infection. The method was especially effective in preventing gangrene, and was credited with saving thousands of lives and limbs. The Carrel-Dakin method, however, was too complicated for widespread use, and has since been replaced by the use of antibiotic drugs.

After an honorable discharge in 1919, Carrel returned to the Rockefeller Institute in New York City. He resumed his work in tissue culture, and began an investigation into the causes of **cancer**. In one experiment, he built a huge mouse colony to test his theories about the relationship between **nutrition** and cancer. But the experiment produced inconclusive results, and the Institute ceased funding it after 1933. Nevertheless, Carrel's tissue culture research was successful enough to earn him the Nordhoff-Jung Cancer Prize in 1931 for his contribution to the study of malignant tumors.

Artificial Heart Collaboration with Charles A. Lindbergh

In the early 1930s, Carrel returned again to the challenge of keeping organs alive outside the body. With the engineering expertise of aviator Charles A. Lindbergh, Carrel designed a special sterilizing glass pump that could be used to circulate nutrient fluid around large organs kept in the lab. This perfusion pump, a so-called artificial **heart**, was germ-free and was successful in keeping animal organs alive for several days or weeks, but this was not considered long enough for practical application in surgery. Still, the experiment laid the groundwork for future developments in heart-lung machines and other devices. To describe the use of the perfusion pump, Carrel and Lindbergh jointly published *The Culture of Organs* in 1938. Lindbergh was a frequent sight at the Rockefeller Institute for several years, and the Lindberghs and the Carrels became close friends socially. They appear together on the July 1, 1935, cover of *Time* magazine with their "mechanical heart."

Carrel's mystical bent, publicly revealed after his visit to Lourdes as a young man, was displayed again in 1935. That year Carrel published *Man, the Unknown,* a work written upon the recommendation of a loose-knit group of intellectuals that he often dined with at the Century Club. In *Man, the Unknown,* Carrel posed highly philosophical questions about mankind, and theorized that mankind could reach perfection through selective reproduction and the leadership of an intellectual aristocracy. The book, a worldwide best-seller and translated into nineteen languages, brought Carrel international attention. Carrel's speculations about the need for a council of superior individuals to guide the future of mankind was seen by many as anti-democratic. Others thought that it was inappropriate for a renowned scientist to lecture on fields outside his own.

Unfortunately, one of those who disliked Carrel's habit of discussing issues outside the realm of medicine was the new director of the Rockefeller Institute. Herbert S. Gasser had replaced Carrel's friend and mentor, Simon Flexner, in 1935. Suddenly Carrel found himself approaching the mandatory age of retirement with a director who had no desire to bend the rules and keep him aboard. On July 1, 1939, Carrel retired. His laboratories and the Division of Experimental Surgery were closed.

Carrel's retirement coincided with the beginning of World War II in September, 1939. Carrel and his wife were in France at the time and Carrel immediately approached the French Ministry of Public Health and offered to organize a field laboratory, much like the one he had run during World War I. When the government was slow to respond, Carrel grew frustrated. In May, 1940, he returned to New York alone. As his steamship was crossing the Atlantic, Hitler invaded France.

Creates New Scientific Institute in Occupied Paris

Carrel made the difficult return to war-torn Europe as soon as he was able, arriving in France via Spain in February, 1941. Paris was under the control of the Vichy government, a puppet administration installed by the German military command. Although Carrel declined to serve as director of public health in the Vichy government, he stayed in Paris to direct the Foundation for the Study of Human Problems. The Foundation, supported by the Vichy government and the German military command, brought young scientists, physicians, lawyers,

and engineers together to study economics, political science, and nutrition. When the Allied forces reoccupied France in August, 1944, the newly restored French government immediately suspended Carrel from his directorship of the Foundation and accused him of collaborating with the Germans. Mercifully, perhaps, a serious heart attack forestalled any further prosecution. Attended by French and American physicians, and nursed by his wife, Carrel died of heart failure in Paris on November 5, 1944. After his death, his body was buried in St. Yves chapel near his home on the island of Saint Gildas, Cotes-du-Nord.

Carrel's reputation remains that of a brilliant, yet temperamental man. His motivations for his involvement with the Nazi-dominated Vichy government remain the subject of debate. Yet there is no question that his achievements ushered in a new era in medical science. His pioneering techniques paved the way for successful organ transplants and modern heart surgery, including grafting procedures and bypasses.

CARSON, RACHEL (1907-1964)
American biologist

Rachel Carson was a university-trained biologist, a longtime United States government employee, and a best-selling author of such books as *Edge of the Sea, The Sea Around Us* (a National Book Award winner), and *Silent Spring*. Her book on the dangers of misusing **pesticides**, *Silent Spring*, has become a classic of environmental literature and resulted in her recognition as the fountainhead of modern environmentalism. *Silent Spring* was reissued in a twenty-fifth anniversary edition in 1987, and remains standard reading for anyone concerned about environmental issues.

Carson grew up in the Pennsylvania countryside and reportedly developed an early interest in **nature** from her mother and from exploring the woods and fields around her home. She was first an English major in college, but a required course in biology rekindled that early interest in the subject and she graduated in 1928 from Pennsylvania College for Women with a degree in **zoology** and went on to earn a master's degree at Johns Hopkins University. After the publication of *Silent Spring*, she was often criticized for being a "popular science writer" rather than a trained biologist, making it obvious that her critics were unaware of her university work, including a master's thesis entitled "The Development of the Pronephros During the Embryonic and Early Larval Life of the Catfish (*Ictalurus punctatus*)."

Summer work also included biological studies at Woods Hole Marine Biological Laboratory in Massachusetts, where she became more interested in the life of the sea. After doing a stint as a part-time scriptwriter for the Bureau of Fisheries, she was hired full-time as a junior aquatic biologist. When she resigned from the United States Fish and Wildlife Service in 1952 to devote her time to her writing, she was biologist and chief editor there. First, as a biologist and writer with the Bureau and then as a free-lance writer and biologist, she successfully combined professionally the two great loves of her life, biology and writing.

Often described as "a book about death which exalts life," *Silent Spring* is the work on which Carson's position as the modern catalyst of a renewed environmental movement rests. The book begins with a shocking fable of one composite town's "silent spring" after pesticides have decimated insects and the birds that feed upon them. The main part of the book is a massive documentation of the effects of organic pesticides on all kinds of life, including birds and humans. The final sections are quite restrained, drawing a hopeful picture of the future, if feasible alternatives to the use of pesticides—such as biological controls—are used in conjunction with and as a partial replacement of chemical sprays.

Carson was quite conservative throughout the book, being careful to limit examples to those that could be verified and defended. In fact, there was very little new in the book; it was all available earlier in a variety of scientific publications. But her science background allowed her to judge the credibility of the facts she uncovered and provided sufficient knowledge to synthesize a large amount of data. Her literary skills made that data accessible to the general public.

Silent Spring was not a polemic against all use of pesticides but a reasoned argument that potential hazards be carefully and sufficiently considered before any such chemical was approved for use. Many people date modern concern with environmental issues from her argument in this book that "future generations are unlikely to condone our lack of prudent concern for the integrity of the natural world that supports all life." It is not an accident that her book is dedicated to Albert Schweitzer, because she wrote it from a shared philosophy of reverence for life.

Carson provided an early outline of the potential of using biological controls in place of chemicals, or in concert with smaller doses of chemicals, an approach now called integrated pest management. She worried that too many specialists were concerned only about the effectiveness of chemicals in destroying pests and "the overall picture" was being lost, in fact not valued or even sought. She pointed out the false safety of assuming that products considered individually were safe, when in concert, or synergistically, they could lead to human health problems.

Her holistic approach was one of the real, and unusual, strengths of the book. Prior to the publication of *Silent Spring*, she even refused to appear on a National Audubon Society panel on pesticides because such an appearance could provide a forum for only part of the picture and she wanted her material to first appear "as a whole." She did allow it to be partially serialized in *The New Yorker*, but articles in that magazine are long and detailed.

The book was criticized early and often, and often viciously and unfairly. One chemical company, reacting to that pre-publication serialization, tried to get Houghton Mifflin not to publish the book, citing Carson as one of the "sinister influences" trying to reduce the use of agricultural chemicals so that United States food supplies would dwindle to the level of a developing nation. The chemical industry apparently united against Carson, distributing critical reviews and threatening to withdraw magazine advertisements from journals deemed

Rachel Carson. *(The Library of Congress. Reproduced by permission.)*

friendly to her. Words and phrases used in the attacks included "ignorant," "biased," "sensational," "unfounded," "distorted," "not written by a scientist," "littered with crass assumptions and gross misinterpretations," to name but a few.

Some balanced reviews were also published, most noteworthy one by Cornell University ecologist LaMont Cole in *Scientific American*. Cole identified errors in her book, but finished by saying "errors of fact are so infrequent, trivial and irrelevant to the main theme that it would be ungallant to dwell on them," and went on to suggest that the book be widely read in the hopes that it "may help us toward a much needed reappraisal of current policies and practices." That was the spirit in which Carson wrote *Silent Spring* and reappraisals and new policies were indeed the result of the myriad of reassessments and studies spawned by its publication. To its credit, it did not take the science community long to recognize her credibility; the President's Science Advisory Committee issued a 1963 report that the journal *Science* suggested "adds up to a fairly thorough-going vindication of Rachel Carson's *Silent Spring* thesis."

While it is important to recognize the importance of *Silent Spring* as a landmark in the environmental movement, one should not neglect the significance of her other work, especially her three books on oceans and marine life and the impact of her writing on people's awareness of one of earth's great natural **ecosystem**s.

Under the Sea Wind (1941) was Carson's attempt ''to make the sea and its life as vivid a reality [for her readers] as it has become for me.'' And readers are given vivid narratives about the shore, including vegetation and birds, on the open sea, especially by tracing the movements of the mackerel, and on the sea bottom, again by focusing on an example, this time the eel. *The Sea Around Us* (1951) continues Carson's treatment of marine biology, adding an account of the history and development of the sea and its physical features such as islands and tides. She also includes human perceptions of and relationships with the sea. *The Edge of the Sea* (1955) was written as a popular guide to beaches and sea shores, but focusing on rocky shores, sand beaches, and coral and mangrove coasts, it complemented the physical descriptions in *The Sea Around Us* with biological data.

Carson was a careful and thorough scientist, an inspiring author, and a pioneering environmentalist. Her groundbreaking book, and the controversy it generated, was the catalyst for much more serious and detailed looks at environmental issues, including increased governmental investigation that led to creation of the Environmental Protection Agency (EPA). Her work will remain a hallmark in the increasing awareness modern people are gaining of how humans interact with and impact the environment in which they live and on which they depend.

Carver, George Washington (ca. 1865-1943)

African American agricultural chemist

George Washington Carver, born in slavery and orphaned in infancy, rose to national and international fame as an agricultural scientist. Carver grew up and was educated in the northern states and later became a faculty member at the all-black Tuskegee Institute in Alabama, working in the forefront of the infant discipline of ''scientific **agriculture**.'' Carver devised and promoted scores of uses for peanuts and sweet potatoes, and he had a significant effect on the diversification of southern agricultural practices. His testimony in 1921 before the House Ways and Means Committee achieved a tariff to protect the U.S. peanut industry and was the beginning of his identity as the peanut wizard. He also worked with **hybrid** cotton, conducted experiments in crop rotation and restoration of soil fertility, and developed useful products from Alabama red clay. Carver was a widely talented man who became an almost mythical American folk hero. He was deeply religious, explaining his wide-ranging interests as attempts to understand the work of the ''Great Creator.''

Carver was born near the end of the Civil War in Newton County, Missouri. His birth date is uncertain, although historian Linda O. McMurry suggests that he was likely born in the spring of 1865. His mother, Mary, was owned by Moses and Susan Carver, who were successful farmowners in the state. His father is believed to be a slave on a nearby plantation, and he was killed in an accident soon after Carver was born. His mother disappeared following a kidnapping by bushwhackers, and Carver and his brother were brought up by Moses and Susan. Carver was a frail and sickly child, and because of his weak health, he helped with the lighter tasks on the farm. He quickly mastered various household tasks, including cooking, laundering, canning, crocheting, needlework, as well as learning the alphabet and music. He also spent considerable time indulging his deep curiosity about **nature**, building a pond for his frog collection and keeping a little plant nursery in the woods. His talent with **plants** made him the neighborhood ''plant doctor.'' At the nearby Locust Grove Church, Carver heard a variety of Methodist, Baptist, Campbellite, and Presbyterian circuit preachers, and acquired a nondenominational faith.

Pursues an Education

In 1877 Carver left home for the county seat of Neosho, to attend a school for blacks. This was the beginning of a long journey through three states in pursuit of basic education. In these years he supported himself with odd jobs working for, and living with, various families along the way. In Neosho, he lived with a black couple, Andrew and Mariah Watkins, and helped with chores. He learned herbalism from Mariah, and he quickly recognized that his knowledge outstripped that of his teacher. In the late 1870s he hitched a ride to Fort Scott, Kansas, and moved in with the family of a blacksmith. Shortly thereafter he moved to Olathe, Kansas, where he made his home with another black couple, Ben and Lucy Seymour. He entered school, helped Lucy with her laundry business, and taught a class at the Methodist church. In the summer of 1880 he followed the Seymours to Minneapolis, where he established a laundry business and spent four years attending school. In 1884 he moved to Kansas City, acquired a typewriter, and took a job as a clerk at the Union Depot. His thirst for education continued, and he was accepted, by mail, into a small college in Highland, Kansas, only to be told when he arrived there that the college did not accept blacks. He stayed in Highland for a while, then moved to Ness County as a homesteader. On the frontier, Carver built a sod house, farmed, took his first art lessons, played accordion for local dances, and joined the literary society.

Around 1890 he sold his homestead and moved to Winterset, Iowa, where his talents and industry impressed a white couple, Dr. and Mrs. John Milholland. They persuaded him to enter Simpson College, a small Methodist College in Indianola. Carver quickly made friends on campus. He had intended to pursue art, but his art teacher, Etta Budd, encouraged him to consider a career in **botany** and suggested he enroll at the agricultural college at Ames, where her own father was a faculty member. The idea appealed to Carver; agriculture would allow him to be of service, and in 1891 he left for Ames and the Iowa State College of Agriculture and Mechanic Arts.

At Iowa, Carver was a popular student and active in a variety of campus affairs. During this time his painting entitled *Yucca and Cactus* was exhibited in Cedar Rapids and selected as an Iowa representative for the World's Columbian Exposition in Chicago in 1893. For his wide-ranging abilities, Carver was affectionately called "doctor" by the other students. His academic record was excellent, and his skills in raising, cross-fertilizing, and grafting plants were recognized by his professors. His bachelor's degree thesis, "Plants as Modified by Man," described the positive aspects of hybridization. He stayed on at Iowa for graduate work and was appointed an assistant in botany. Freed at last from odd jobs, Carver could now devote himself to greenhouse studies and teaching.

The Tuskegee Years

Carver received his master's degree from Iowa in 1896 and accepted a position at Tuskegee Institute, Alabama, at the invitation of its president, Booker T. Washington. He was to spend forty-seven years at Tuskegee, living most of that time in Rockefeller Hall, a dormitory occupied by students. The early years were difficult for Carver as he had numerous responsibilities at the institute. Besides heading the agriculture department, Carver was also director of the newly established Agricultural Experiment Station. Additionally, he managed the school's two farms, taught classes, and served on committees and councils. Despite all this, in 1910, unhappy with the number of agriculture graduates, Washington removed Carver from his charge of the Agriculture Department. Carver submitted his resignation, following which, Washington made him director of a new research department and "consulting chemist."

Experimental work was more to Carver's liking. From the beginning Carver had worked on a number of projects to help improve the lot of poor southern farmers. He analyzed **water**, feed, and soil. He experimented with paints that could be made with clay. He worked with organic fertilizers. He demonstrated uses for cheap and locally available materials, such as swamp muck. He searched for new, cheap foodstuffs to supplement the farmers' diets. In addition to human food items, he developed stock feeds, cosmetics, dyes, stains, medicines and ink from peanuts and sweet potatoes. In his agricultural bulletins, he offered elementary information to uneducated farmers.

In 1916 Carver received two prestigious invitations: to serve on the advisory board of the National Agricultural Society and to become a fellow of the Royal Society for the Arts in London. In 1919, under Tuskegee president Robert Russa Moton, he received his first salary increase in twenty years. He had become increasingly popular as a lecturer and his testimony before the House Ways and Means Committee in 1921 thrust him into the national limelight. In 1923 he was awarded the Spingarn Medal from the NAACP (National Association for the Advancement of Colored People) for his contributions to agricultural chemistry and for his lectures to religious, educational, and farming audiences that had "increased interracial knowledge and respect." Other honors included an honorary doctorate from Simpson College in 1928.

Becomes a Spokesman for "Chemurgy"

In the mid–1930s, the word *chemurgy* was coined to mean putting chemistry to work in industry for the farmer. Carver became a spokesman for chemurgy, just as he had been

George Washington Carver. *(The Library of Congress. Reproduced by pemission.)*

for the peanut industry and the "New South." In 1937 Carver met the industrialist Henry Ford at a chemurgy conference Ford had sponsored. A long friendship developed between the two men and when, in 1940, Carver established a foundation to continue and preserve his work, the Carver Museum in Tuskegee was dedicated by Ford. The museum contained seventy-one of Carver's pictures as well as handicrafts, case studies, and results of his research.

Carver received numerous awards and honors for his contributions to the field of scientific agriculture. Noteworthy among these were the Roosevelt Medal, which he received in 1939, an honorary doctorate from the University of Rochester, and the first award for "outstanding service to the welfare of the South" from the Catholic Conference of the South. In 1942 Ford erected a Carver memorial cabin in Greenfield Village, Michigan, and established a nutritional laboratory in Carver's honor in Dearborn, Michigan. Carver also received an honorary doctorate from Selma University, a fellowship from the Thomas A. Edison Institute, and was a member of Kappa Delta Pi, an honorary education society.

Carver's health had begun to fail in the 1930s. When he died on January 5, 1943, Tuskegee Institute was flooded with letters of sympathy from many people. Carver was buried in the Tuskegee Institute cemetery near the grave of Booker T. Washington. On January 9, 1943, President Franklin D. Roosevelt paid tribute to Carver in an address before Congress, and on July 14, 1943, Roosevelt signed legislation making Carver's Missouri birthplace a national monument.

CATABOLISM • See Metabolism

CATASTROPHISM

In the late seventeenth century, Anglican Archbishop **James Ussher** developed a biblical chronology that established the date of the Earth's creation at 9 a.m., October 26, 4004 B.C. His calculations were taken seriously for 200 years and promoted debates on the Earth's origin.

One of the most prevalent schools of thought that followed was catastrophism, which held that the Earth was formed by supernatural forces according to the account found in the Bible in the book of Genesis. Scientists who adhered to this belief system were called Neptunists.

Evidence that Earth was much older began to accumulate, and in 1785, **James Hutton** in a presentation to the Royal Society of Edinburgh challenged catastrophism. Hutton proposed that Earth was in a continuous but gradual process of change, constantly decaying, renewing, and repairing itself. He added that Earth had no semblance of a beginning and no prospect of an end. His ideas led to a school of thought, known as uniformitarianism, and scientists who adhered to it became known as Plutonists.

Uniformitarianism was sharply criticized by the Neptunists, led by Abraham Gottlob Werner, who believed that a great globe-engulfing sea had laid down all the rocky layers of Earth. They held that as the **water**s subsided, dry land emerged and sand and gravel washed down from the mountains into valleys and lowlands.

The debate continued to rage into the next decade and even the next century as scientists tried to align their observations and discoveries with catastrophism. For example, noted scientist **Georges Cuvier**, who was classifying and cataloging **fossil** remains, was pressured to accept the theory of catastrophism even though it did not coincide with his discoveries. In 1819, William Buckland attempted to resolve these differences by suggesting that there was no real need to reconcile geologic observations with the biblical account of creation because the Earth preceded the creation story.

As recently as the early 1900s, diluvialism creation theory, which held to the tenets of catastrophism, was popular for a couple of decades. It was based on the belief that one catastrophic event, the Mosaid flood (Noah's Ark), caused many of the Earth's surface features. It was abandoned, however, when scientists found evidence showing the dissimilarity of surface material, which refuted the theory of a worldwide flood.

A modern variant of catastrophism was born in 1980, when Luis Alvarez proposed that dinosaurs met their end 65 million years ago as the result of a huge asteroid impact. Eleven years later, the discovery of a gigantic crater off Mexico's Yucatan peninsula gave this theory a boost. However, many scientists remain unconvinced, noting that their interpretation of the fossil record doesn't support the idea of a cataclysmic **extinction**.

CECH, THOMAS R. (1947-)
American biochemist

The work of Thomas R. Cech has revolutionized the way in which scientists look at **RNA** and at **proteins**. Up to the time of Cech's discoveries in 1981 and 1982, it had been thought that genetic coding, stored in the **DNA** of the **nucleus**, was imprinted or transcribed onto RNA **molecule**s. These RNA molecules, it was believed, helped transfer the coding onto proteins produced in the **ribosome**s. The DNA/RNA nexus was thus the information center of the **cell**, while protein molecules in the form of **enzyme**s were the workhorses, catalyzing the thousands of vital **chemical reaction**s that occur in the cell. Conventional wisdom held that the two functions were separate—that there was a delicate division of labor. Cech and his colleagues at the University of Colorado established, however, that this picture of how RNA functions was incorrect; they proved that in the absence of other enzymes RNA acts as its own catalyst. It was a discovery that reverberated throughout the scientific community, leading not only to new technologies in RNA engineering but also to a revised view of the **evolution** of life. Cech shared the 1989 Nobel Prize for Chemistry with Sidney Altman at Yale University for their work regarding the role of RNA in cell reactions.

Cech was born in Chicago on December 8, 1947, to Robert Franklin Cech, a physician, and Annette Marie Cerveny Cech. As he recalls in an autobiographical sketch for *Les Prix Nobel,* he grew up in "the safe streets and good schools" of Iowa City, Iowa. His father had a deep and abiding interest in physics as well as medicine, and from an early age Cech took an avid interest in science, collecting rocks and **minerals** and speculating about how they had been formed. In junior high school he was already conferring with geology professors from the nearby university. He went to Grinnell College in 1966; at first attracted to physical chemistry, he soon concentrated on biological chemistry, graduating with a chemistry degree in 1970.

It was at Grinnell that he met Carol Lynn Martinson, who was a fellow chemistry student. They married in 1970 and went together to the University of California at Berkeley for graduate studies. His thesis advisor there was John Hearst who, Cech recalled in *Les Prix Nobel,* "had an enthusiasm for **chromosome** structure and function that proved infectious." Both Cech and his wife were awarded their Ph.D. degrees in 1975, and they moved to the east coast for postdoctoral positions—Cech at the Massachusetts Institute of Technology (MIT) under Mary Lou Pardue, and his wife at Harvard. At MIT Cech focussed on the DNA structures of the mouse **genome**, strengthening his knowledge of biology at the same time.

Concentrates on Gene Expression

In 1978, both Cech and his wife were offered positions at the University of Colorado in Boulder; he was appointed assistant professor in chemistry. By this time, Cech had decided that he would like to investigate more specific **genetic material**. He was particularly interested in what enables the DNA molecule to instruct the body to produce the various parts of itself—a process known as **gene expression**. Cech set out to discover the proteins that govern the DNA **transcription** process onto RNA, and in order to do this he decided to use **nucleic acids** from a single-cell **protozoa**, *Tetrahymena thermophila*. He chose *Tetrahymena* because it rapidly reproduced genetic material and because it had a structure which allowed for the easy extraction of DNA.

By the late 1970s much research had already been done on DNA and its transcription partner, RNA. It had been determined that there were three types of RNA: messenger RNA, which relays the transcription of the DNA structure by attaching itself to the ribosome where protein synthesis occurs; ribosomal RNA, which imparts the messenger's structure within the ribosome; and transfer RNA, which helps to establish **amino acids** in the proper order in the protein chain as it is being built. Just prior to the time Cech began his work, it was discovered that DNA and final-product RNA (after copying or transcription) actually differed. In 1977 Phillip A. Sharp and others discovered that portions of seemingly noncoded DNA were snipped out of the RNA and the chain was spliced back together where these intervening segments had been removed. These noncoded sections of DNA were called introns.

Cech and his coworkers were not initially interested in such introns, but they soon became fascinated with their function and the splicing mechanism itself. In an effort to understand how these so-called nonsense sequences, or introns, were removed from the transcribed RNA, Cech and his colleague Arthur Zaug decided to investigate the pre-ribosomal RNA of the *Tetrahymena,* just as it underwent transcription. In order to do this, they first isolated unspliced RNA and then added some *Tetrahymena* nuclei extract. Their assumption was that the catalytic agent or enzyme would be present in such an extract. They also added small molecules of salts and **nucleotides** for **energy**, varying the amounts of each in subsequent experiments, even excluding one or more of the additives. But the experiment took a different turn than they expected.

Research on Catalytic RNA Wins Nobel Prize

What Cech and Zaug discovered was that RNA splicing would occur even without the nucleic material being present. This was a development they did not understand at first; it was a long-held scientific belief that proteins in the form of enzymes had to be present for catalysis to occur. But here was a situation in which RNA appeared to be its own catalytic motivator. At first they suspected that their experiment had been contaminated. Cech did further experiments involving recombinant DNA in which there could be no possibility of the presence of splicing enzymes, and these had the same result: the RNA spliced out its own intron. Further discoveries in Cech's laboratory into the nature of the intron led to his belief that the intron itself was the catalytic agent of RNA splicing, and he decided that this was a sort of RNA enzyme which they called the ribozyme.

Cech's findings of 1982 met with heated debate in the scientific community, for it upset many beliefs about the nature of enzymes. Cech's ribozyme was in fact not a true en-

zyme, for thus far he had shown it only to work upon itself and to be changed in the reaction; true enzymes catalyze repeatedly and come out of the reaction unchanged. Other critics argued that this was a freak bit of RNA on a strange microorganism and that it would not be found in other organisms. They were soon proved wrong, however, when scientists around the world began discovering other RNA enzymes. In 1984, Sidney Altman proved that RNA carries out enzyme-like activities on substances other than itself.

The discovery of catalytic RNA has had profound results. In the medical field alone RNA enzymology may lead to cures of viral infections. By using these rybozymes as gene scissors, the RNA molecule can be cut at certain points, destroying the RNA molecules that cause infections or genetic disorders. In life sciences, the discovery of catalytic RNA has also changed conventional wisdom. The old debate about whether proteins or nucleic acids were the first bit of life form seems to have been solved. If RNA can act as a catalyst and a genetic template to create proteins as well as itself, then it is rather certain that RNA was first in the chain of life.

Cech and Altman won the Nobel Prize for Chemistry in 1989 for their independent discoveries of catalytic RNA. Cech has also been awarded the Passano Foundation Young Scientist Award and the Harrison Howe Award in 1984; the Pfizer Award in Enzyme Chemistry in 1985; the U. S. Steel Award in Molecular Biology; and the V. D. Mattia Award in 1987. In 1988, he won the Newcombe-Cleveland Award, the Heineken Prize, the Gairdner Foundation International Award, the Louisa Gross Horwitz Prize, and the Albert Lasker Basic Medical Research Award; he was presented with the Bonfils-Stanton Award for Science in 1990.

Cech was made full professor in the department of chemistry at the University of Colorado in 1983. He and his wife have two daughters. In the midst of his busy career in research, Cech still finds time for skiing and backpacking.

CELL

A cell is the basic unit of life—the smallest part of living **organism**s capable of surviving independently and reproducing itself. Every living organism is made of cells, but the size and shape of the cells varies according to their function. All cells, however, have the same basic structure. The **protoplasm**—the living substance that makes up the cell—consists of two main parts: the **nucleus** and the **cytoplasm**.

The nucleus is a round body near the center of the cell that contains the hereditary material determining the cell's structure and activities. This material, called **chromatin**, ordinarily looks like a ball of tangled string, but begins to turn into rodlike **chromosome**s during **cell division**. These chromosomes contain packages of **DNA (deoxyribonucleic acid)** information, which essentially acts as a blueprint for the human body. The cell nucleus also contains nucleoli, which contain **protein**s and **RNA (ribonucleic acid)**.

The cytoplasm is responsible for keeping the cell alive. It is made up mostly of **water**, but contains tiny **organelle**s that

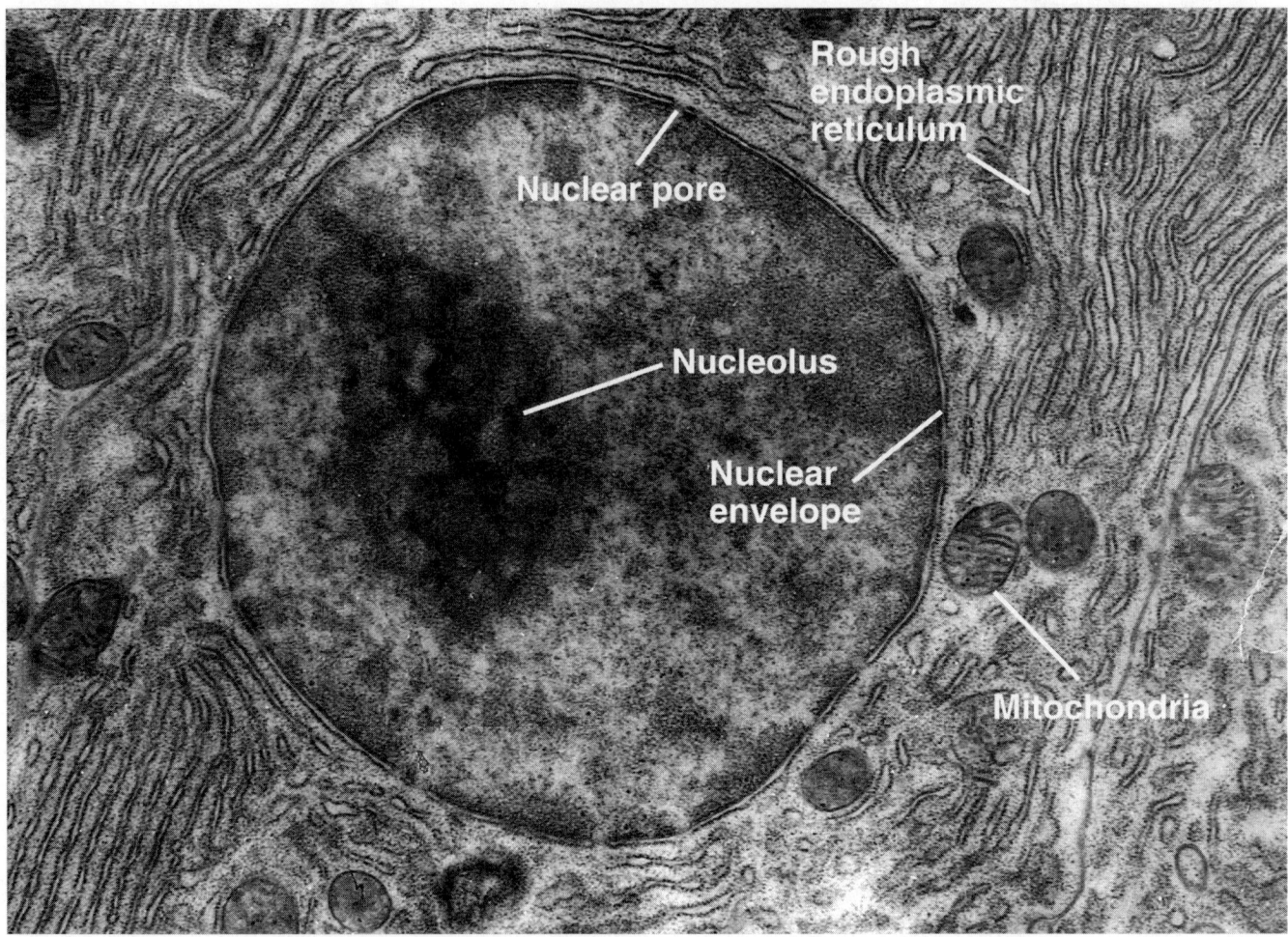

A living animal cell. *(Photo Researchers, Inc. Reproduced by permission.)*

have a particular function in sustaining the life of the cell. For instance, **mitochondria** produce the cell's power, **lysosome**s break down dead cells, **cilia** help move material outside the cell, the **endoplasmic reticulum** manufactures proteins for cell growth, **centriole**s assist in cell reproduction, and the **Golgi apparatus** makes proteins and **carbohydrate**s.

In 1665, an English scientist named **Robert Hooke** observed a thin slice of cork under his homemade **microscope**. He noticed that the cork was composed of neat holes enclosed by walls. Hooke called these formations *cells*, meaning "little boxes."

Over the next 150 years, improvements in microscopes allowed scientists to better study these cells and a few even observed the presence of a cell nucleus. But it was **Robert Brown**, a Scottish botanist, who was the first to show that the cell nucleus occurred in all plant cells. During his investigations into plant **fertilization**, Brown began noticing small bodies that always seemed to show up inside the cells. In 1831, Brown pointed out that the cell *nucleus*, as he called it, seemed to be fundamental in creating plant **tissue**.

Brown's discovery paved the way for the **cell theory** of **Matthias Schleiden** and **Theodor Schwann**. In 1838, Matthias

Schleiden knew that this nucleus (Schleiden called it a cytoblast) must play a role in the formation of cells. He also showed, by examining plant tissue under the microscope, that the cell was the basic unit of plant life. The following year, Theodor Schwann expanded Schleiden's findings to include all **animal**s as well as **plant**s. In addition, Schwann was convinced that cells were the smallest unit of individual life in the body. **Jan Evangelista Purkinjě** had also observed the characteristics of cell theory two years before Schwann, but his observations went unnoticed. In 1839, Purkinjě described the contents of animal **embryo**s using the term protoplasm in its scientific sense. To Purkinjě, the term meant "first formed," but eventually it took on a more general meaning: the living material inside a cell.

By 1882, Eduard Strasburger (1844-1912) had recognized that protoplasm had important components. He invented the terms *nucleoplasm*, to describe the protoplasm within the cell nucleus, and *cytoplasm*, for the protoplasm outside the nucleus.

From 1929 to 1948, **Albert Claude** studied the cell, using a method he developed for separating cell parts of different size, shape, and density with a **centrifuge**. He was the first to recognize that the mitochondrion was the cell's **energy** center. In 1942, he discovered the endoplasmic reticulum, ultimately identifying its role in the formation and transport of proteins and **fats**. Also during the 1940s, **George Emil Palade** made other important advances in understanding cell functioning. For instance, he showed that mitochondria produce a chemical called **adenosine triphosphate**, which provides energy to the cells. Then in 1949, **Christian René de Duve** discovered lysosomes. A few years later, his research team discovered another organelle called the peroxisome, found in plants and the livers and kidneys of **mammals**. Peroxisomes transform cellular material into hydrogen peroxide or glucose. In 1974, these three men shared the Nobel Prize in physiology or medicine for their work on the structure and function of the cell.

Scientists continue to seek out new organelles. In the early 1990s, biochemist Leonard Rome first identified an organelle, called a vault, that dots most cells by the thousands and is three times as large as a **ribosome**. The function of vaults is still uncertain, although they may serve to transport **molecule**s from the nucleus, where they are made, to the rest of the cell.

CELL DIVISION

Cell division is the process by which an **organism** grows or replaces damaged **tissue**. The growth of a fertilized **embryo** is accomplished through the division and **differentiation** of **cell**s, and while some cells, such as skin cells, divide almost continuously after **birth**, other highly specialized cells, such as some **neuron**s, do not and cannot be replaced after disease or injury. The two forms of cell division, **mitosis** and **meiosis**, are the biological mechanism by which the principles of **heredity** and **evolutionary theory** are realized.

German botanist Wilhelm Hofmeister (1824-1877) first examined the process of cell division in 1847 and showed that the **nucleus** did not truly disappear. He appeared to have been very close to discovering **chromosome**s, but that honor was left to **Walther Flemming** some 30 years later.

Prior to Flemming's work, little headway had been made in the field of **cytology** since Hofmeister's discovery, due in large part to the lack of effective **cell staining** techniques and poor **microscope**s. Advances in synthetic dyes allowed scientists to better study cells. By using the new dyes, Flemming was able to observe and correctly identify the stages of cell division.

Flemming found material within the cell nucleus that readily absorbed dye. He named the material **chromatin**, and by observing it at different phases, he could trace the action of cell division. As the process began, the chromatin arranged itself into short, thread-like objects, which **Wilhelm von Waldeyer-Hartz** later termed chromosomes. The chromosomes then doubled and pulled apart, each half migrating to opposite ends of the cell. In the final stage, the cell divided, leaving two daughter cells with equal amounts of chromatin.

By the end of 1879, Flemming had investigated all the stages of mitosis—a name which he derived from the Greek word for "thread"—and identified them in a variety of tissues. His terms for each stage (wreath, star, equatorial plate, and nuclear barrel) have since been replaced by today's familiar technical phrases (prophase, metaphase, early anaphase, and telophase).

Unfamiliar with the work of **Gregor Mendel**, Flemming was unaware of the genetic significance of his findings. However, with the rediscovery of Mendel by **Hugo de Vries** at the turn of the century, Flemming's work has since provided the physical basis of Mendel's theories of **inheritance** and his *Zellsubstanz, Kern, und Zeltlteilung* (1882; *Cytoplasm, Nucleus, and Cell Division*) is considered a classic text by cytologists.

In 1887, Flemming began concentrating on cell division in the **sperm**atozoa. Although he detected that differences existed as to how sex cells divided, Flemming failed to identify the process later termed *meiosis* by **Edouard van Beneden**. Beneden, a Belgian cytologist, began experimenting with bivalves in 1887. He observed that when sex cells are formed before **fertilization**, the **egg** and sperm cell possess only half the usual count of chromosomes. From this he surmised that in the creation of sex cells, chromosomes do not double as they do in mitosis. Instead, through the process of meiosis, the chromosomal pairs split, leaving sperm and egg cells with half the **genetic material** of the parent cell. Fertilization, therefore, result in an embryo with one set of chromosomes each from the mother and the father.

Like Flemming's works, Beneden's findings were consistent with Mendel's inheritance theories—every genetic factor existed in duplicate, one coming from the male and one from the female. Meiosis also provided the mechanism by which vast combinations of traits could be achieved within a **species** and upon which **natural selection** could work.

Scientists have since learned that, when cells divide, one copy of each chromosome is pulled to each daughter cell along **protein** guide wires called microtubules. In **cancer**, this process may goes awry, with daughter cells receiving more or less chromosomes than they should. Researchers suspected that the centrosome, part of the cell that helps form and organize microtubules, might be involved. In the late 1990s, Jeffrey Salisbury and his colleagues showed that cancer cells contained excess amounts of a centrosome protein, suggesting the presence of extra centrosomes that could recruit more protein guide wires. This remains an active area of research.

Another topic of current interest is the role of telomeres—pieces of **DNA** that protect the ends of chromosomes, but which gradually wear away as the cell divides repeatedly. There does seem to be a causal link between the wearing away of telomeres and the decline in a cell's ability to divide. However, scientists are still debating whether this may have significant implications for **aging** of the body as a whole.

CELL STAINING

The development of microscopy and **cell**-staining techniques made possible impressive advances in our understanding of

microorganisms. That fact should hardly be surprising. The **microscope** allows one to see structures too small to be visible with the naked eye; individual cells, for example. But those small structures are usually colorless and transparent, making it difficult to distinguish specific features contained within them. Thus, it became important to find ways of bringing out these features so that they could be studied in more detail. The father of microscopy, **Antoni van Leeuwenhoek**, found that dyeing muscle fibers with saffron allowed him to see the detailed structure of the fibers more easily.

From van Leeuwenhoek's time to the mid-nineteenth century, scientists searched for natural products to use in dyeing cells. They were only moderately successful until the discovery of the first synthetic dye by William Perkin in 1856. Then the new technology developed by Perkin unleashed a flood of colored compounds to the scientific **community** for potential use as stains.

One of the first scientists to explore these new dyes was the German anatomist **Walther Flemming**. Flemming used a number of different dyes to stain cells and observed structures that seemed to absorb these dyes strongly. Among these structures were bodies that he eventually called **chromatin**, after the Greek word for "color."

During the late 1800s, the art of cell staining became much more sophisticated. The German biologist Carl Weigert (1845-1904) observed, for example, that different kinds of **bacteria** are stained by different dyes. That discovery turned out to be a powerful influence on Weigert's cousin, the famous bacteriologist **Paul Ehrlich**. Ehrlich wrote his final college thesis on the techniques of cell staining. During his residency at the Charité Hospital in Berlin, Germany, Ehrlich discovered techniques for identifying **blood** disorders based on the way that cells absorb dyes. He also developed methods for staining white blood cells and mast cells.

Ehrlich's work on stains eventually led him in a quite different direction. He became convinced that staining might not be a neutral process in which cells simply absorb dye. There might, he thought, be some actual physical or chemical interaction between cell and dye that could result in the **death** of the cell. Thus, he began the search for a dye that would kill harmful bacteria in the process of staining.

Over a period of more than a decade, Ehrlich worked toward this objective. He finally found a dye called *trypan red* that kills trypanosomes, the **protozoa** that cause sleeping sickness. In 1907, he found an even more important bactericide called Salvarsan. This **arsenic**-containing compound proved to be a powerful agent in the treatment of **syphilis**.

Perhaps the most famous stain discovery was made in 1884 by the Danish bacteriologist **Hans Christian Joachim Gram**. Gram modified a traditional staining technique by adding iodine and then an **alcohol** wash to stained cells. He found that some cells retained the original dyes in this procedure, while others lost their color. The former were designated *gram positive* cells and the latter, *gram negative* cells. The procedure has since become a standard method for classifying bacteria.

Another approach to staining was suggested by the Italian histologist **Camillo Golgi** in the 1870s. Instead of using or-

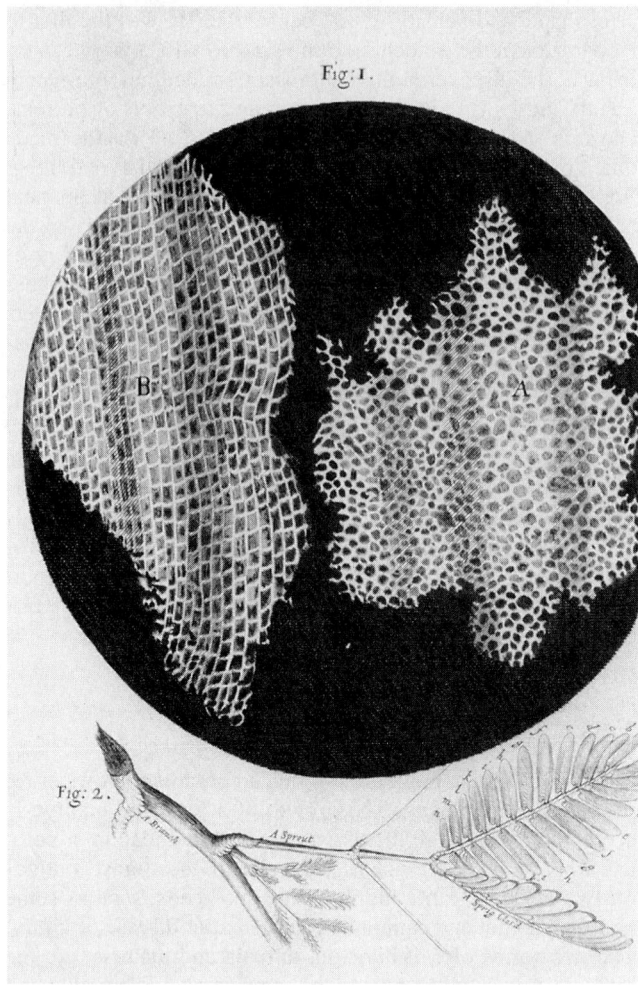

An illustration from Robert Hooke's *Micrographia*, in which he suggested that living things were composed of minute structures called "cells." *(The Library of Congress. Reproduced by permission.)*

ganic dyes, Golgi experimented with the use of silver salts as stains. He found that cell structures not visible with organic dyes were now easily seen. One of the first structures he discovered is now known as the Golgi body.

Yet another technique is immunoperoxidase staining, in which cells are washed with antibodies to various cell parts or cell chemicals. The antibodies are tagged with a dye that can be seen under a microscope, allowing observers to see certain components of the cell.

Today dozens of stains, both organic and inorganic, have been developed for specific uses. The names of many—Borrel's methylene blue, Ehrlich's triacid stain, Renault's eosin, Lugol's iodine, and Van Gieson's stain, for example-continue to memorialize their inventors.

CELL THEORY

The definition of cell theory has changed over the years. At first, it meant the specific microstructure of **cell**s. But over many years, several botanists and scientists contributed to and expanded the definition. Today cell theory explains the general principles of construction for all living things. Current cell theory consists of six statements: 1. All living **matter** is made up of cells. 2. All cells come from preexisting cells (most from **cell division**, but some from fusion of **egg** and **sperm**). 3. A cell is the basic unit of life. 4. Each cell is encased in a **plasma membrane** (a thin skin that separates it from the environment and other cells). 5. All cells have strong biochemical similarities. 6. Most cells are microscopic in size.

Before the 1800s, the basic units of life were thought to be fibers and vessels—people believed that living **organism**s just could not be broken down any more than that. Several scientists had made observations that pointed toward the existence of cells: **Robert Hooke**, who coined the term *cell*, had observed remnants of the tiny structures in cork and **Robert Brown** had even discovered the cell **nucleus** in **plants**. **Jan Purkinjě** also demonstrated that animal **tissue**s contained cells as well. Yet despite these findings, the connection between cells and life still was not recognized.

In 1838, however, **Matthias Schleiden** began incorporating some of these previous observations into his own research on plant **fertilization**. Using improved microscopic techniques, he was able to closely examine plant tissue and make some striking observations of his own—namely, that the cell is the foundation of the plant **kingdom**. Schleiden was first to recognize the importance of cells as fundamental units of life in plants, and his accurate conclusions about plant cells and cell activity marked the beginning of plant **cytology**.

The following year, **Theodor Schwann** adopted the term "cell theory" to describe his more elaborate observations regarding cells. Schwann extended Schleiden's cell theory of plants to include **animals**. What Schwann had observed was that, not only are all plant tissues made from cells, but all animal tissues are made from cells as well. Schwann also suggested that eggs are actually cells and that all life starts as a single cell. **Rudolf Virchow** neatly summarized the theories of Schleiden and Schwann with the phrase *Omnis cellula e cellula*, meaning "all cells from cells." Virchow correctly proposed that all cells originate from other cells and demonstrated that even diseased tissue comes from normal cells through the process of division. This assumption was essentially the foundation of cellular pathology.

Although both Schleiden and Schwann accurately described some key elements of cell theory, they also mistakenly believed that cells just erupted from the nucleus of a parent cell—free-cell formation theory. By 1861, **Rudolph Albert von Kölliker** had joined the search for answers to the growing number of questions about cells. Kölliker was particularly interested in the cell nucleus and believed it was the key to transmitting hereditary factors. He viewed the egg as a single cell, correctly stating that it continuously produces daughter cells. Kölliker was the first to interpret the developing **embryo** in terms of cell theory, thereby opposing Schwann's doctrine of free-cell formation. The observations from these and other scientists helped establish cell theory as an important biological principle.

Although the six statements of modern cell theory are now accepted as generally correct, there are some significant exceptions. For example, **virus**es are not composed of cells, but they still contain **genetic material** and can reproduce in a host cell. Also, **mitochondria** and **chloroplast**s are considered just parts of cells, but they also contain genetic material and can reproduce in a cell.

CELLULOSE

Cellulose is a complex **carbohydrate** that is a primary constituent of plant **cell** walls. It occurs in both primitive and highly evolved **plants**. Cellulose comprises about 20-30% of primary cell walls and about 40-90% of secondary cell walls. It is the world's most abundant **organic compound**. The biological function of cellulose is thought to be skeletal, providing shape and strength to the cell wall. Cellulose differs from other **polysaccharides** found in plants by consisting of molecular chains that are very long, by containing only one repeating glucose subunit, and by occurring naturally in a crystalline state. The glucose subunits are connected by beta **linkage**s. Most **mammal**s, including man, do not have **enzyme**s capable of promoting the **hydrolysis** of this beta linkage, so cellulose passes through the digestive tract unchanged. However, **microorganism**s found in the digestive tracts of herbivorous **animal**s (especially in ruminants (cud-chewing animals)) can break down cellulose into products that can be absorbed and used as a food source. In the environment, brown rot **fungi** are able to degrade cellulose.

Of many widely utilized natural substances, cellulose, especially cellulose fiber from cotton and wood, is one of the most important commercial raw materials for a large variety of chemical products. Cotton, flax, jute, and ramie fibers are comprised of nearly pure cellulose, while wood contains only about 42% cellulose. Since cellulose is insoluble in **water**, it can be readily separated from other constituents of plants. Cellulose can be produced from wood through the sulfite and sulfate pulping processes. Cellulose is treated with sodium hydroxide and exposed to carbon disulfide fumes to form sodium xanthate, an unstable ester. This solution, called viscose, can be forced through fine holes or slits into an acidic solution to form threads or sheets of rayon or cellophane.

Cellulose acetate esters are spun into fine filaments for the manufacture of some fabrics and are also used as photographic film, as a substitute for glass, in the manufacture of safety glass, and as a molding material. Cellulose ethers are used in paper sizings, adhesives, soaps, and synthetic resins. In mixtures of nitric and sulfuric acids, cellulose can form cellulose nitrates, which are flammable and explosive compounds. These compounds are used in various lacquers, plastics, medicines, and artificial leather. Guncotton, an explosive, is a type of cellulose nitrate.

Research has also been conducted to investigate the use of lignocellulosic **biomass**, which consists of cellulose, hemicellulose, lignin, and ash, to produce liquid fuels.

CENTRAL NERVOUS SYSTEM • See Nervous system

CENTRIFUGE

A centrifuge is a spinning device used to separate materials of different densities. Centrifuges are used industrially to separate impurities from products such as milk, lubricating oils, beer, and wine. They are also used in scientific research as a separation tool. In genetic research, for example, high-speed centrifuges are used to separate viruses, small cell components, and even individual protein and nucleic acid molecules.

Centrifuges work on the principle of centrifugal force and its counterpart, centripetal force. The relationship between these two forces can be demonstrated by considering the analogy of a person swinging a liquid-filled bucket on a rope. As the bucket is swung in a circle, it tends to pull outward, but the rope keeps it from flying away. The pull on the rope away from that person is called centrifugal force, and the tension of the rope that keeps the bucket from flying away is called centripetal force. When mixtures of materials are spun in this fashion, the heavier particles are driven outward, to the bottom of the container, and the lighter particles move less, staying on top.

The centrifuge was originally developed as a way to separate cream from milk. For centuries, farmers had simply allowed gravity to pull apart milk and cream; however, this process was extremely slow and inefficient. In 1877, a Swedish inventor, Carl Gustaf Patrik de Laval, introduced a device which could rapidly separate the two components. Laval invented a steam engine powered device which could spin a container of milk at 4,000 revolutions per minute in such a way that the heavier milk would sink to the bottom and the lighter cream would rise to the top. Over the next several years engineers made improvements in Laval's basic design and it began to be used in other areas of industry and science.

In 1923 a Swedish chemist, **Theodor Svedburg**, developed an faster centrifuge that could a force 100,000 times greater than normal gravity. By 1936, Svedburg had perfected an ultracentrifuge that spun at 120,000 times per minute and created a centrifugal force equal to 525,000 times that of normal gravity.

Using this state of the art technology, other scientists perfected centrifugation techniques used in biological sciences. Early pioneers in the field of biological centrifugation include **Albert Claude**, George Emil Palade, and Belgian biochemist Christian Rene de Duve. Claude was an expert in the use of the centrifuge, and Palade and de Duve developed advanced techniques to separate cell components. By applying differential centrifugation to ruptured cells, de Duve was able to collect different fractions of cells by centrifugation. Using these techniques he discovered two kinds of cell **organelles**: the **lysosome** and the peroxisome. The three men shared a shared the 1974 Nobel Prize for their work.

CENTRIOLE

A pair of interesting and somewhat mysterious cellular **organelles** called centrioles is found near the **nucleus** of most **animal** cells. Each centriole is a hollow, cylindrical structure about 0.66 ft (0.2 m) in diameter and 1.31 ft (0.4 m) long made of microtubules—very thin, stiff rods composed of the **protein** tubulin. Nine groups of fused, triplet microtubules make up the wall of each centriole. Each triplet set is angled inward like the blades of a turbine and adjacent rows are linked periodically by short protein bridges. In the center of some centrioles, faint radial spokes can be seen to point inward toward a hollow central structure. The two centrioles in each pair usually sit at right angles and are separated by a narrow space.

Centrioles appear to be autonomous organelles (new centrioles always arise from pre-existing ones). In each cell cycle, the pair of centrioles separates and a new centriole begins to grow perpendicular to, and some space from, the original one. During **cell division**, the centrioles serve as the organizing centers for the mitotic spindle—the bundle of microtubules that attach to **chromosome**s and separate the groups that will become daughter nuclei. How the growth and organization of the new centrioles is controlled is not known, but if the parent centrioles are removed or damaged, new centrioles are not formed.

Centrioles also are related in both structure and function to basal bodies—the structures that anchor **cilia** and **flagella** at the periphery of many cells. Basal bodies reproduce by a similar mechanism to that of the centrioles, and they play a comparable role in organizing the microtubules that make up the central skeleton of these organelles. In some cases, a precursor cell that has no cilia or flagella will develop them during **differentiation**. To do this, the centrioles will undergo repeated replication and the daughter centrioles will move to the cell surface to become basal bodies and organize cilia or flagella.

Plant cells generally lack centrioles, but a fuzzy structure called the centriolar-equivalent or centrasome seems to carry out the same functions as the animal centriole.

CEREBELLUM

The cerebellum, which is Latin for "little **brain**," is located at the back of the head. In the human brain it is the second-largest part, occupying a place partially tucked under the forebrain's cerebral hemispheres. In birds, it is the largest part, in relative terms, and processes the constant flow of **nerve impulses** between the brain and body that are necessary for flight. It is extensively folded, giving it an appearance of irregular pleats, and possesses right and left hemispheres that are connected with the **spinal cord** and forebrain. Each of the cerebellum's hemispheres connect with spinal cord nerves on the same side of the body, but with the opposite cerebral hemisphere.

The cerebellum's specialized function in the human brain is to maintain posture and balance, and to carry out coordinated movement, by processing signals that are transmitted from the cerebral cortex's motor area to the spinal cord and then to muscle groups, creating movements. The cerebellum also receives muscle and joint signals. It compares these with the cortex's signals, and makes adjustments as necessary to achieve the coordinated movement intended. Some evidence exists that the cerebellum can store a sequence of instructions for movements that are repeated frequently, and for repetitious skilled movements that are learned by rote. In some studies of the brain's responses to language-related tasks, researchers were surprised to find that, as tasks became more complex, several sites in the cerebellum were activated along with areas of the forebrain that process many types of information. The finding was a surprise because the cerebellum's functions are associated with movement.

Another important function played by the cerebellum is its role in the reticular activating system, a widespread network of nerve **cell**s that are the means by which humans maintain consciousness. The reticular activating system is also involved in the brain's ability to focus attention, blocking out some distractions that originate both within and outside the body.

CEREBRUM

The **cerebrum**, the largest part of the human **brain**, accounts for approximately 5/6 of its mass. Often regarded as the brain's thinking center, the cerebrum is divided into left and right cerebral hemispheres whose outermost layer — the cerebral cortex — consist of grayish **cell** bodies that make up most of the brain's "gray **matter**." The hemispheres communicate through the *corpus callosum*, or "hard body," which has a tough consistency in contrast with the brain's generally soft **tissue**. Below the convoluted cortex are nerves that lead to other parts of the brain and to the **spinal cord**. The cortex's convolutions are either *sulci*, the lines that demarcate the convolutions, or *gyri*, the brain tissue that forms ridges between the sulci. The sulci and gyri are similar in most brains and the most prominent ones have been mapped and named. For instance, the *central sulcus* and the *lateral sulcus* found in each cerebral hemisphere are very prominent. The cerebral hemispheres are also marked off into regions called lobes: the frontal, parietal, temporal, and occipital lobes. Each lobe is associated with different functions, including the auditory, visual, sensory, and motor areas. One major area of the human brain is called Broca's convolution, an area associated with language. It is named after the French surgeon, **Pierre Paul Broca**, who in 1861 discovered through postmortem studies that patients with *aphasia*—the inability to speak or understand speech—had physical damage to the same area of the brain.

While all vertebrates possess a cerebrum, it does not have the same importance in all of them as it does in human beings. Studies have shown, for instance, that a frog's behavior changes very little even after its cerebrum is removed. It can catch flies and its sexual functioning is unimpaired. Cats, on the other hand, while still able to purr, swallow, and move

Ernest Boris Chain. *(The Library of Congress. Reproduced by permission.)*

to avoid pain after their cerebrum has been removed, become sluggish and robotlike in their movements. Monkeys become severely paralyzed and can barely distinguish **light** and dark. A human whose cerebrum was removed would become totally blind, almost completely paralyzed, and would soon die, even though able to breathe and swallow.

CHAIN, ERNST BORIS (1906-1979)
German biochemist

Chain is renowned for his role in the discovery of **penicillin**, a drug that has saved millions of lives and was the first of the antibiotic "wonder drugs." The son of a wealthy chemist, Chain was born in Berlin in 1906. He earned his Ph.D. in chemistry in 1930 from the Friedrich-Wilhelm University in Berlin. When Adolf Hitler (1889-1945) came to power in early 1933, Chain immigrated to England, where he worked and studied under Frederick Gowland Hopkins at Cambridge University.

At the recommendation of Hopkins, **Howard Florey** invited Chain to join his pathology laboratory at Oxford to pursue studies of antimicrobial agents. While Chain conducted a literature search on lysozyme, an antibacterial **enzyme**, he came across a paper by **Alexander Fleming** published in 1929 describing his work with penicillin found in molds. Chain and

Florey decided to continue the work, which Fleming had abandoned shortly after his discovery.

Chain conducted the first chemical assay of penicillin. Florey and Chain concluded that penicillin was nontoxic yet effective in destroying a wide range of **bacteria**, and began conducting clinical trials in humans. The results were so successful that penicillin was quickly put into mass production to treat the infections of wounded soldiers during World War II. In 1945 Chain, Florey, and Fleming were awarded the Nobel Prize in Medicine for discovering penicillin.

Chain went on to discover penicillinase, an enzyme that causes the destruction of penicillin in the body. After World War II, Chain became scientific director of a health institute in Rome, but returned to England in 1961, where he headed a new laboratory at the University of London.

CHAPPARAL

Chaparral is a geographically widespread **ecosystem** (or **biome**) occurring in warm-temperate environments with a ''Mediterranean'' climatic regime, characterized by mild, cool, moist winters and hot, dry summers. Chaparral is a shrub-dominated ecosystem, particularly by **species** with evergreen, waxy, thick (or sclerophyllous), drought-resistant foliage. The shrub canopy is 3-13 ft (1-4 m) tall, sometimes with intertwining branches among closely-spaced **plants**. Periodic wildfires during the dry season are an important environmental influence on the structure of the chaparral ecosystem.

Chapparal vegetation is most widespread in coastal California and northwestern Mexico, in the vicinity of the Mediterranean Sea in southern Europe, and in smaller areas in southern Australia, Chile, and South Africa. Although different species of plants dominate the chaparral biome in each of these widely separated regions, they are all of a rather similar growth form, because they have evolved in response to similar regimes of **natural selection** (this is known as **convergent evolution**).

More than 100 species of shrubs occur in chaparral communities in California (plus many other kinds of plants and **animals**). The most widely distributed species of shrub is chamise (*Adenostoma fasciculatum*). On drier sites it is commonly joined by species of manzanita (*Arctostaphylos* spp.) and buckbrush (*Ceanothus* spp.). On moister sites, associated species include scrub oak (*Quercus dumosa*), chaparral holly (*Heteromeles arbutifolia*), chaparral cherry (*Prunus ilicifolia*), mountain mahogany (*Cercocarpus betuloides*), redberry and coffeeberry (*Rhamnus* spp.), silk-tassel bush (*Garrya* spp.), lemonadeberry and sugarbush (*Rhus* spp.), and laurel sumac (*Malosma laurina*).

CHARGAFF'S RULES

Chargaff's rules are a series of statements that refer to the composition of the **nucleotide** bases in **DNA**. They were suggested by Erwin Chargaff in the late 1940s.

Prior to 1944, scientists did not know what material in the **cell** was the **genetic material**. At this time, biologist **Oswald**

Theodore Avery and his collaborators showed that DNA was the principal component in microbial transformation. This prompted Chargaff, in 1947, to further investigate the composition of DNA.

DNA is a **polymer** made up of nucleotide **monomer**s. The nucleotides are composed of a **phosphate group**, a pentose sugar, and a **nitrogen** containing base. There are four specific bases found in DNA including two purines, adenine and guanine, and two pyrimidines, thymine and cytosine.

Chargaff found a number of properties of DNA that were consistent throughout most **organism**s. First, he found that there was not an equal distribution of bases within an organism. Also, the ratio of bases vary from **species** to species. He found an interesting regularity to the ratio of bases. The number of adenine components equaled the thymine bases and the number of guanines were equal to the cytosines.

These observations became the basis for Chargaff's rules, also known as the **base pairing** rules. These include the following statements. Within an organism, the number of purine bases are equal to the pyrimidine bases. The number of adenines and thyamines are equal. Also, the number of cytosines and guanines are equal.

While Chargaff's rules were supported by evidence, the reasons for these relationships among nucleotides remained unknown until the discovery of the DNA **double helix** structure. These rules were a substantial clue used by **James D. Watson** and **Francis Crick** to determine the structure of DNA.

CHEMICAL BOND

The term *chemical bond* refers to any force of attraction between two **atoms** or ions. Today, chemists recognize the existence of at least four types of chemical bonds: ionic (electrovalent), covalent, metallic, and hydrogen.

The concept of chemical bonds goes back—at least in its simplest and most general form—to the ancient Greeks. Philosophers who thought of **matter** as being composed of individual particles often considered the possibility that those particles might join, or bond with, each other in some way. As early as 100 B.C. Asklepiades of Prusa introduced the concept of ''clusters'' of atoms, somewhat similar to those that make up a **molecule**.

In 1789 the English chemist William Higgins speculated about the way in which the fundamental particles of matter might combine with each other. When John Dalton proposed his **atomic theory** around 1803, he specifically hypothesized that the atoms of elements would combine with each other to form ''compound atoms.'' No modern theory of bonding was possible, however, until the concept of a molecule was introduced by Amedeo Avogadro and then clarified by Stanislao Cannizarro and until the formulas of compounds could be agreed upon. Those steps were finally taken in the mid-1800s. Shortly thereafter, in 1858, Friedrich Kekulé attempted to describe the way in which atoms might combine with each other in forming **organic compound**s.

Kekulé hypothesized that a carbon atom was able to join with, or bond to, four other atoms. He also suggested that carbon atoms could bond with each other endlessly. The drawings he made to depict the bonding of atoms to each other nicely conveyed his ideas, but they were too clumsy to be used by chemists. They were quickly replaced by a system of chemical symbols joined by short dashed lines, introduced at about the same time by A. S. Couper.

The bonding in organic compounds was further clarified in 1874 by Jacobus van't Hoff. Van't Hoff emphasized the importance of thinking of molecules in three-dimensional terms. He suggested that the bonds in a carbon atom are directed to the four corners of a tetrahedron, with the carbon atom at its center. By placing the four other atoms at the corners of the tetrahedron, van 't Hoff was able to explain the existence of two or more forms of a compound that had identical molecular formulas, but different optical properties (optical **isomers**). The French chemist Joseph Le Bel proposed a similar theory at about the same time.

An important first step in explaining precisely how a bond forms was offered in 1904 by the German chemist Richard Abegg. Abegg came to the conclusion that the electronic structure of the atoms of inert gases—a complete outer shell of eight electrons—constituted a stable configuration. He suggested that atoms with more or less than eight electrons in their outer shell would gain or lose electrons in order to attain that stable configuration. This suggestion essentially describes the process now known as *ionic bonding*.

Abegg's theory of bonding was extended by Gilbert Newton Lewis, Walther Kossel, and Irving Langmuir in the late 1910s. Those investigators pointed out a second way in which atoms might bond with each other. Rather than losing or gaining electrons, they said, two atoms might share electrons with each other. In each case, the contribution of one electron from each atom would make up a shared pair that would provide a stable configuration for both atoms. Langmuir coined the term *covalent bond* for this kind of arrangement.

Another possibility was also evident. Both electrons in the shared pair might be contributed by only one of the atoms. This kind of bond explained the existence of coordination complexes first studied in detail by Alfred Werner and was called, therefore, a *coordinate covalent bond*.

The theory of chemical bonding attained its highest development in the work of **Linus Pauling** in the late 1930s. Before Pauling's work, most discussion of chemical bonds had assumed that the electrons in an atom are stationary. Pauling knew that they were actually in motion, and he developed a new explanation of bonding based on that notion. He first used the wave mechanical theory developed by Louis Victor de Broglie to write equations for the motion of electrons in an atom. Then he showed how the wave patterns of two electrons from adjacent atoms might overlap to produce a new pattern with less **energy** than that of the two original electrons. The lower energy of the combined pattern provided an explanation for the bond formed between the two atoms. Pauling also showed that the combined pattern is usually not one that resulted in a purely ionic or purely **covalent bond**, but a **hybrid** of the two.

In 1994 scientists for the first time successfully studied diradicals, the molecular **species** hypothesized to be archetyp-

al of chemical bond **transformations** in many classes of reactions. This research used femtosecond laser techniques with mass spectrometry to freezing the diradicals in time so they could be observed. These studies contributed to scientific knowledge about chemical bonds because their results established diradicals as intermediate agents in the formation of certain bonds.

CHEMICAL COMPOUND

A chemical compound is a substance created when two or more different elements are connected by **chemical bond**s or valence forces. Compounds and elements are both distinct substances which is one of the two basic categories of **matter**. The other category includes mixtures of substances.

Compounds have several defining characteristics. First, they must consist of different elements. Combinations of the same **atoms** form **molecule**s, not compounds. For example, when two atoms of oxygen combine they form a molecule of oxygen, not a compound. Therefore, while every compound is a molecule, not every molecule is a compound. A compound is created through a **chemical reaction** that creates electrostatic bonds between its atoms. Second, when elements combine they do so in definite, unchanging ratios. A **water** molecule, for example, always consists of two hydrogen atoms and one oxygen atom. The proportions of individual elements in a compound are defined by the principal of stoichiometry, which is the set of rules that which governs chemical bonding. Third, a compound has different physical and chemical properties than its parent atoms. When elements are combined, they lose their individual properties. For instance, the highly reactive metallic element sodium and the poisonous gas chlorine form a crystalline compound, sodium chloride, also known as common table salt. Unlike its parent atoms, sodium chloride can be safely ingested. In addition, compounds can not be broken apart by physical or mechanical means; a chemical reaction is required to sever the chemical bonds which hold the compound together.

A chemical compound is one of the two general categories of matter. The other category includes mixtures of substances. In contrast to compounds, mixtures do not have fixed proportions. They contain different substances in varying ratios, and in mixtures, each substance retains its original properties. Furthermore, mixtures can be prepared and separated mechanically; no chemical reaction is necessary for their formation.

The composition of chemical compounds can be expressed using standard chemical symbols and notation. These represent the structures of the reactants and the resulting compounds. The three dimensional structure of compounds is described using a notation that is designated by the rules of a special branch of chemistry known as stereochemistry. It is also important to note that the same elements can form more than one compound depending on their structural arrangement. For example, three carbon atoms, four hydrogen atoms, and one oxygen atom can bond together in different formations to create either propanol (which has an hydroxyl group (OH⁻) bonded to one of the end carbons) or isopropanol (where the hydroxyl group is connected to the middle carbon.) When more than one compound can be formed from the same atoms, the compounds are called **isomers** of one another. Isomers have different chemical and physical properties even though they are composed of identical atoms.

A compound is named according to its constituent atoms. Simple compounds consisting of only two atoms are named by combining the names of both elements and changing the ending of the second element's name to "ide." For example, the compound made from silver and chlorine is called silver chloride. For more complex compounds, numbers are used to indicate how many atoms are involved. For example, **carbon dioxide** (CO_2) has two oxygen atoms attached to one carbon atom, and phosphorous trioxide (P_2O_3) has three oxygen atoms connected to two phosphorous atoms.

CHEMICAL REACTION

A chemical reaction is a transformation resulting from the interaction between two **atoms** or **molecules**. Reactions involve the breakage and reformation of the **chemical bond**s which hold atoms together and can therefore cause changes in the structure or composition of substances.

A reaction may involve two or more reactants that combine to form a reaction product; it may involve a compound that breaks down to create individual reactants. In biological systems, chemical reactions are responsible for processes such as **metabolism**, **respiration**, and **digestion**.

There are four basic categories of chemical reactions. These are known as combination, **decomposition**, single replacement, and double replacement reactions. Specific types of reactions in these categories include oxidation, reduction, ionization, combustion, **polymer**ization, **hydrolysis**, condensation, and rearrangement reactions. Regardless of the type, reactions always involve an exchange of **energy**. If the energy requirements of the reaction product are less than that of the reactants, excess energy will be released in the form of heat and the reaction is called exothermic. If the energy of the end product is greater than the reactants, additional heat energy must be added to the system to make the reaction proceed and the reaction is called endothermic. Some reactions are reversible, meaning they have the ability to proceed in the opposite direction. The products of some reactions are unstable and can revert back to their original components.

Reactions cause chemical changes in substances as opposed to physical changes. An example of a physical change is the transformation of **water** from a solid to a liquid to a gas as the **temperature** increases. While these are dramatic physical changes, the water molecule has the same chemical properties in all three forms because its atomic bonds are unchanged. On the other hand, chemical changes cause a compound to have much different properties than those of the reactants. For example, the silvery metal sodium (Na) can be reacted with the poisonous, greenish yellow gas chlorine (Cl_2) to form common

table salt (sodium chloride, NaCl) which has properties unlike either of its parents. It is also important to note that chemical reactions are limited to changes that occur between atoms and should not be confused with nuclear reactions which are changes that occur inside the **nucleus** of the atom.

Reactions are written as equations with chemical symbols indicating the reactants on one side and the reaction products on the other. Arrows are used to show which direction the reaction proceeds and additional notation indicates if special conditions are required to drive the reaction. The two sides of the reaction equation must be in complete balance according to the laws of conservation of mass and energy because atoms can neither be created nor destroyed in a reaction. If there are twelve atoms present in the reactants at the beginning of the reaction, there must also be twelve atoms in the reaction products. The rate at which the reaction occurs varies depending on the energy of the system and is measured by tracking the concentration of the reactants or reaction products. The study of reaction rates and the factors that affect them is called kinetics.

See also Chemical compound

CHEMOTHERAPY

Chemotherapy is the treatment of a disease or condition with chemicals that have a specific effect on its cause, such as a **microorganism** or **cancer cell**. The first modern therapeutic chemical was derived from a synthetic dye. The sulfonamide drugs developed in the 1930s, **penicillin** and other **antibiotics** of the 1940s, **hormone**s in the 1950s, and more recent drugs that interfere with cancer cell **metabolism** and reproduction have all been part of the chemotherapeutic arsenal.

For thousands of years, medical practitioners have used **plants** and other substances to treat symptoms of disease. Modern chemotherapy began with the German physician **Paul Ehrlich**, who as early as 1905 began looking for specific chemicals to be "magic bullets" seeking out and destroying infectious **organism**s within the body without harming healthy **tissue**s. In 1910, Ehrlich discovered that an **arsenic** compound he had named Salvarsan was successful in treating the sexually transmitted disease syphillis. Other scientists experimented with drugs that killed **protozoa**.

The first drug to treat a widespread bacteria was developed in the mid-1930s by the German physician-chemist **Gerhard Domagk**. In 1932, he discovered that a dye named prontosil killed streptococcus bacteria, and it was quickly used medically on both streptococcus and staphylococcus. One of the first patients cured with it was Domagk's own daughter. In 1936, the Swiss biochemist Daniele Bovet, working at the Pasteur Institute in Paris, showed that only a part of prontosil was active—a sulfonamide radical that had long been known to chemists. Since it was much less expensive to produce, sulfonamide soon became the basis for several widely used "sulfa drugs" that revolutionized the treatment of formerly fatal diseases. These included pneumonia, meningitis, and puerperal ("childbed") fever. For his work, Domagk received the 1939 Nobel prize in physiology or medicine. Though largely replaced by antibiotics, sulfa drugs are still used against urinary tract infections, Hanson disease (leprosy), and malaria, and for burn treatment.

At the same time, the next breakthrough in chemotherapy—penicillin—was in the wings. In 1928, the British bacteriologist **Alexander Fleming** noticed that a mold on an uncovered laboratory dish of staphylococcus destroyed the bacteria. He identified the mold as Penicillium notatum, which was related to ordinary bread mold. Fleming named the mold's active substance penicillin, but was unable to isolate it.

In 1939 the American microbiologist René Jules Dubos (1901-1982) isolated from a soil microorganism an antibacterial substance that he named tyrothricin. This led to wide interest in penicillin, which was isolated in 1941 by two biochemists at Oxford University, **Howard Florey** and **Ernst Chain**.

The term antibiotic was coined by American microbiologist **Selman Abraham Waksman**, who discovered the first antibiotic that was effective on gram-negative bacteria. Isolating it from a Streptomyces fungus that he had studied for decades, Waksman named his antibiotic streptomycin. Though streptomycin had strong side effects, it paved the way for the discovery of other antibiotics.

The first of the tetracyclines was discovered in 1948 by the American botanist Benjamin Minge Duggar. Working with Streptomyces aureofaciens at the Lederle division of the American Cyanamid Co., Duggar discovered chlortetracycline (Aureomycin).

The first effective chemotherapeutic agent against **viruses** was *acyclovir*, produced in the early 1950s by the American biochemists **George Hitchings** and **Gertrude Bell Elion** for the treatment of herpes. Today's antiviral drugs are being used to inhibit the reproductive cycle of both **DNA** and **RNA** viruses. For example, two drugs are used against the **influenza** A virus, amantadine and rimantadine. And the **AIDS** treatment drug AZT inhibits the reproduction of the **human immunodeficiency virus (HIV)**.

Scientists began trying various **chemical compound**s for use as cancer treatments as early as the mid-nineteenth century. But the first effective treatments were the **sex hormones**, first used in 1945—**estrogen**s for prostate cancer and both estrogens and androgens to treat breast cancer. In 1946, the American scientist Cornelius Rhoads developed the first drug especially for cancer treatment. It was an alkylating compound, derived from the chemical warfare agent **nitrogen** mustard, which binds with chemical groups in the cell's DNA, keeping it from reproducing. Alkylating compounds are still important in cancer treatment.

In the next twenty years scientists developed a series of useful antineoplastic (anti-cancer) drugs, and, in 1954, the forerunner of the National Cancer Institute was established in Bethesda, MD. Leading the research efforts were the so-called ''4-H Club'' of cancer chemotherapy: the Americans **Charles Huggins**, who worked with hormones; **George Hitchings**, purines and pyrimidines to interfere with cell metabolism; Charles Heidelberger, fluorinated compounds; and British sci-

entist Alexander Haddow (1907-), who worked with various substances. The first widely used drug was 6-Mercaptopurine, synthesized by Elion and Hitchings in 1952.

Chemotherapy is used alone, in combination, and along with radiation and/or surgery, with varying success rates, depending on the type of cancer and whether it is localized or has spread to other parts of the body. They are also used after treatment to keep the cancer from recurring (adjuvant therapy). Since many of the drugs have severe side effects, their value must always be weighed against the serious short-and long-term effects, particularly in children, whose bodies are still growing and developing.

In addition to the male and female sex hormones androgen, estrogen, and progestins, scientists also use the hormone somatostatin, which inhibits production of growth hormone and growth factors. They also use substances that inhibit the action of the body's own hormones. An example is tamoxifen, used against breast cancer. Normally the body's own estrogen causes growth of breast tissues, including the cancer. Tamoxifen binds to cell receptors instead, causing reduction of tissue and cancer cell size.

Forms of the B-vitamin folic acid were found to be useful in disrupting cancer cell metabolism by the American scientist Seymour Farber (1912-) in 1948. Today they are used on leukemia, breast cancer, and other cancers.

Plant **alkaloid**s have long been used as medicines, such as colchicine from the autumn crocus. Cancer therapy drugs include vincristine and vinblastine, derived from the pink periwinkle by American Irving Johnson (1925-). They prevent **mitosis** (division) in cancer cells. VP-16 and VM-16 are derived from the roots and rhizomes of the may apple or mandrake plant, and are used to treat various cancers. Taxol, which is derived from the bark of several **species** of yew **tree**s, was discovered in 1978 and has been approved by the U.S. Food and Drug Administration for treatment of ovarian and breast cancer.

Another class of naturally occurring substances are anthracyclines, which scientists consider to be extremely useful against breast, lung, thyroid, stomach, and other cancers.

Certain antibiotics are also effective against cancer cells by binding to DNA and inhibiting RNA and **protein synthesis**. Actinomycin D, derived from Streptomyces, was discovered by Selman Waksman and first used in 1965 by American researcher Seymour Farber. It is now used against cancer of female reproductive **organ**s, **brain** tumors, and other cancers.

A form of the metal platinum called cisplatin stops cancer cells' division and disrupts their growth pattern. Newer treatments that are biological or based on **protein**s or **genetic material** and can target specific cells are also being developed. Monoclonal antibodies are genetically engineered copies of proteins used by the **immune system** to fight disease. Rituximab was the first monoclonal **antibody** approved for use in cancer, and more are under development. **Interferon**s are proteins released by cells when invaded by a virus. Interferons serve to alert the body's immune system of an impending attack, thus causing the production of other proteins that fight off disease. Interferons are being studied for treating a number of cancers, including a form of skin cancer called multiple myeloma. A third group of drugs are called anti-sense drugs, which affect specific genes within cells. Made of genetic material that binds with and neutralizes messenger-RNA, anti-sense drugs halt the production of proteins within the cancer cell.

Genetically engineered cancer vaccines are also being tested against several virus-related cancers, including liver, cervix, nose and throat, kidney, lung, and prostate cancers. The primary goal of genetically engineered vaccines are to trigger the body's immune system to produce more cells that will react to and kill cancer cells. One approach involves isolating white **blood** cells that will kill cancer and then to find certain **antigen**s, or proteins, that can be taken from these cells and injected into the patient to spur on the immune system. A ''vaccine gene gun'' has also been developed to inject DNA directly into the tumor cell. An RNA cancer vaccine is also being tested. Unlike most vaccines, which have been primarily tailored for specific patients and cancers, the RNA cancer vaccine is designed to treat a broad number of cancers in many patients.

In addition to drugs for treating cancer, scientists are also conducting studies of agents to stave off the side effects of chemotherapy. Anti-emetics are drugs used to reduce chemotherapy-induced nausea and vomiting. With the increasing use of chemotherapy to treat cancer, these therapies have become essential in the treatment of chancer. Anti-emetics include the phenothiazines, benzodiazepines, butryphenones, anticholinergics and **antihistamine**s, and the cannabinoids, which include marijuana. Non-pharmacological treatments for nausea and vomiting include behavior and relaxation therapy.

As research into cancer treatment continues, new cancer-fighting drugs will continue to become part of the medical armamentarium. Many of these drugs will come from the burgeoning biotechnology industry and promise to have fewer side effects than chemotherapy and radiation.

CHI-SQUARE MODEL

A chi-square model is a statistical method used to analyze the results of certain types of experiments. It involves a comparison of the deviation of real observations from expected values, and helps determine whether a result is significant or not. It has been a particularly useful model in the development of new drugs.

The chi-square function was first proposed in 1900 by a British statistician, **Karl Pearson**. He developed a formula which takes in consideration the squares of the deviations of each observation from the expected value and weighs them accordingly. By using this distribution, he was able to show whether or not the results from a certain data set matched the expected results from the total **population**. In this way he developed a measure of the ''goodness of fit'' for a data set. He also developed a method for comparing the relationship between two data parameters.

Many biological experiments involve enumerative data in which subjects from a population are counted and classified.

These type of experiments involve a reasonable degree of approximation and they are defined as multinomial experiments. For example, if a mouse were put in a maze and made to go through one of three doors depending on the **stimulus**, its response could be classified into one of three values. To determine whether the mouse's choice of door were random or a result of the specific stimuli given, a chi-square model is typically employed.

Chi square models are known as non-parametric statistical tests. This means that the data sets they measure do not necessarily follow a normal distribution about a mean. For an experiment to be valid by a chi square model it typically has the following restrictions. First, it consists of a set amount of identical trials. Second, the outcome of each trial falls into one of a set number of specific classes. Next, the probability that an observation falls in a certain class does not change over the course of the experiment. And finally, the trials are independent.

Since chi-square models can show whether there is a relationship between two data parameters, they have become important tools in the development of new drugs. In this type of research, a test and a control group of people are given a different treatments. Their conditions are monitored for a specified amount of time and data is collected. The incidences of recovery in the test and control groups are compared using a chi-square model to determine whether the treatment is truly effective or a result of random chance. It should be noted that while they are a powerful statistical tool, chi-square models never give a definitive answer and they must be interpreted by the experimenter.

CHICKEN POX

Chicken pox is a fairly common contagious disease of childhood caused by the varicella **virus**. Its primary symptom is a skin rash, which may be accompanied by fever, headache and muscle aches. The rash usually begins with red blotches on the back or the chest which then erupt into vesicles with a characteristic teardrop shape that enlarge and fill with clear liquid. As the vesicles dry up and form scabs in one part of the body, new vesicles erupt in another area of the body. (Chicken pox got its name because people thought the rash looked like chick peas lying on the skin.)

The time between exposure to the virus and the appearance of the symptoms (incubation period) is fairly long, ranging from fourteen to twenty-one days. It is communicable from a few days before the onset of symptoms until all the vesicles have crusted over, giving chicken pox (along with the measles) the highest rate of communicability. It is thought to be spread by infected droplets coming from the nose or throat. There are often epidemics of chicken pox, especially in winter and early spring. The virus which causes chicken pox is the same virus that causes shingles (a skin disorder usually found only in adults) as discovered by **Dr. Thomas Weller** while a researcher at Children's Hospital in Boston.

The general treatment for chicken pox is to keep the child clean and comfortable for the course of the disease which may last two weeks. Topical medications and cold compresses are often applied to prevent itching, as scratching the vesicles may cause scarring or secondary infection by strep or staph **bacteria**. In some severe cases, **antihistamine**s or an antiviral drug such as acyclovir may be recommended to shorten the course and severity of the disease. Usually a child who has chicken pox once has lifetime immunity, but there are examples of recurrence in adulthood. In March 1995, a vaccine for chicken pox was approved for use in the United States. The varicella vaccine is called a live vaccine because it contains live varicella zoster virus (VZV), which is responsible for the chicken pox rash. The vaccine is especially effective against the symptoms of chicken pox, including preventing them from progressing to a severe stage. Studies are still under way to determine how effective the vaccine is in actually preventing the disease.

CHIMERA

A **chimera** is an **organism** that contains **cell population**s derived from more than one fertilized **egg**. The term is derived from the Greek and refers to a mythical being with the body of a goat, the tail of a serpent, and the head of a lion with nostrils belching forth fire. Chimerae may be structured of **cells** from one **species** or of several species. Chimeric mice have been produced experimentally. Fusion of early **embryo**s is one method. Embryos whose parents differed in coat color and one or more biochemical markers are chosen for fusion. Expression of both coat colors (in patches of fur on the newborn) or biochemical markers indicate success in the operation. Embryos at the 8 to 12 cell stage are frequently chosen for fusion because the cells are sticky and the experimental combinations tend to form a single embryo. The experimental embryo is placed in a surrogate female for development to **birth**. A chimera formed in this manner is tetraparental, i.e., it has four parents (two for each of the embryos fused). A hexaparental mouse produced by Professor Clement Markert adorned a cover of the journal *Science* several years ago.

Does a **cancer** cell have the potentiality for giving rise to any kind of cell other than more cancer cells? The answer to that question was sought in an experiment with chimeras. It was judged that the interior of a early mouse embryo would have powerful biochemical substances that would guide the normal **differentiation** of the progeny of a malignant mouse cell. Cells from a mouse cancer, a teratocarcinoma, were introduced with a micropipette (a glass tube like a microscopic hypodermic syringe) into the interior of an early mouse embryo by Ralph Brinster in 1974. The interior of the embryo regulated the malignant cells and they gave rise to normal cell progeny that expressed the appropriate fur color after birth. The mouse chimerae refuted the notion that cancer cells can give rise only to other malignant cells. The demonstrated production of normal cell progeny must give hope that eventually in at least some kinds of cancer, patients will be treated to induce their cancers to give rise to normal cells. This would eliminate or minimize toxic **chemotherapy** and damaging radiation. This

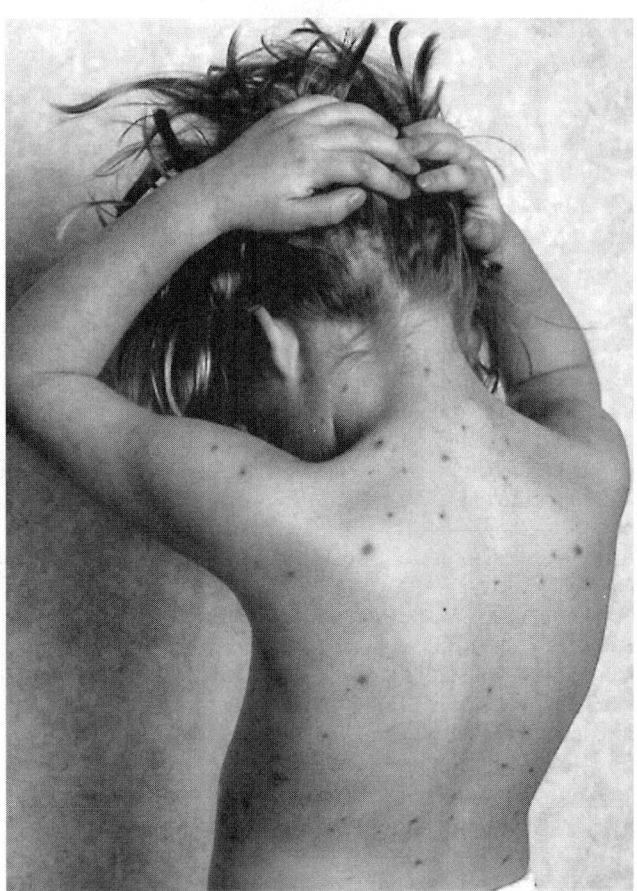

A young girl with chicken pox on her back and neck. *(Photograph by Jim Selby, Photo Researchers, Inc. Reproduced by permission.)*

is an example of an interesting biological procedure, the production of chimerae, being used to answer a fundamental question in pathology.

CHLOROPHYLL FUNCTION AND STRUCTURE

Chlorophyll comes from the Greek words meaning ''green **leaf**.'' Most plant **cells** contain yellow, orange, and red pigments, but these are obscured by the green pigment known as chlorophyll. It is in autumn, when woody **plants** stop producing chlorophyll, that the other colors become visible.

Chlorophyll was first isolated by French chemists in 1817. Pierre-Joseph Pelletier and his research partner Joseph-Bienaimé Caventou (1795-1877) extracted chlorophyll from green plants, but were unable to find any immediate use for it. Pelletier and Caventou were more interested in substances that could be used as medicines and later discovered such drugs as brucine, quinine, and strychnine.

About fifty years later, scientists learned of chlorophyll's role in plant life. Although Dutch scientist **Jan Ingen Housz** had described the process of **photosynthesis** in the late

1700s, no one knew exactly how green plants converted sunlight to food. In 1865 German botanist Julius von Sachs (1832-1897) discovered that chlorophyll *catalyzes*, or promotes, photosynthetic reactions in the presence of **light**. Sachs showed that chlorophyll is confined to certain specialized structures inside the cell, which he termed **chloroplast**s, and all the essential processes of photosynthesis are carried out within these structures. When a leaf is exposed to light, it is in the chloroplasts that the grains of starch, or **carbohydrate**s, first appear.

Just after the turn of the century, Russian scientist **Mikhail Tswett** developed the laboratory technique of **chromatography**, which separates individual chemicals by color so that they can be identified. Tsvett used chromatography to isolate different types of chlorophyll, but few other scientists knew about the technique. It was popularized a few years later by **Richard Willstätter**, who found that only two major types of chlorophyll exist in land plants: the *blue-green*, or ''a'' type, and the *yellow-green*, or ''b'' type. Scientists now know that there are three more forms of chlorophyll, the ''c,'' ''d,'' and ''e'' types.

Willstätter also discovered that chlorophyll's structure is very similar to that of **hemoglobin**, the red pigment in **blood**. Continuing Willstätter's work, German chemist **Hans Fischer** examined the subtle differences between chlorophyll and hemoglobin **molecule**s in the 1930s and worked out chlorophyll's complete structure. His discovery that chlorophyll contains magnesium was the first indication of the chemical's importance as a plant nutrient (the main difference between chlorophyll and hemoglobin is the fact that chlorophyll carries magnesium while hemoglobin carries iron). During the 1950s American biochemist **Melvin Calvin** confirmed that a *light reaction* involving chlorophyll instantly captures the sunlight and converts it into chemical **energy**. Robert Burns Woodward, an American chemist, synthesized the complex chlorophyll molecule in 1960.

CHLOROPLAST

Chloroplasts are plastid **organelle**s found in the **cell**s of **plants** and **algae**. They contain chlorophyll and carotenoid pigments, and an elaborate system of internal **membrane**s called thylakoids. They are active in **photosynthesis**, the all-important process that converts **light** into chemical **energy**. In algae, chloroplasts exist in a wide variety of sizes and shapes. Depending on the algal **species**, they may be simple disks or elaborate ribbons twisted into a spiral. Plant chloroplasts are usually disk-shaped and measure between 4 and 6 micrometers in diameter. They are especially abundant in **leaf** mesophyll **tissue** where a single cell may contain as many as 50. Specialized chloroplasts lacking grana, but not chlorophyll, are found in the bundle-sheath cells of plants with the C_4 or **Hatch-Slack photosynthetic pathway** for carbon fixation.

The internal structure of chloroplasts is very complex, with a structure that facilitates the capture of light photons and channels the energy of the captured photons into **chemical compound**s such as **adenosine triphosphate (ATP)**. ATP is

A transmission electron micrograph (TEM) image of a chloroplast from a tobacco leaf (*Nicotiana tabacum*). The stacks of flattened membranes that can be seen within the chloroplast are grana. The membranes that run between the stacks are stroma. The faint white patches within the chloroplast are nucleoids, where chloroplast DNA is stored. *(Photograph by Dr. Jeremy Burgess, Photo Researchers, Inc. Reproduced by permission.)*

formed from adenosine diphosphate and phosphate in a process called **photophosphorylation**. The energy stored in ATP is used to synthesize sugars within the chloroplast. A plasma envelope or membrane surrounds the interior of the chloroplast. Within the structure are chlorophyll-rich areas called grana formed from stacks of disk-like membranous structures called thylakoids. The grana are embedded in the stroma or ground substance of the plastid and are connected by stroma thylakoids that run between them. The light-capturing chlorophyll and carotenoid pigments are embedded, and precisely oriented, within the thylakoid membranes.

When chloroplasts have been in bright light for a time, they often contain starch grains formed from sugars produced by the rapidly photosynthesizing organelle. These temporary storage products soon disappear when the plant is placed in the dark, and the insoluble starch is converted back to soluble sugar and distributed throughout the plant.

Chloroplasts contain **DNA**, **RNA**, and **ribosome**s and are able to synthesize a number of **protein**s and **lipid**s needed for their structure and activity. The **replication** of chloroplasts requires the parent cell, and appears to involve a cooperation between chromosomal and chloroplast DNA. Chloroplast ribosomes are only about two-thirds as large as the ribosomes of plants and other **eukaryote**s. In this respect, they closely resemble the ribosomes of **bacteria**. They are also sensitive to bacteria-inhibiting **antibiotics**. Unlike eukaryote DNA but reminiscent of bacteria, chloroplast DNA generally has a circular rather than a linear structure, is not associated with **histones**, and is not bounded by a membrane. The similarities between chloroplasts and bacteria has led to speculation that chloroplasts originated as photosynthetic bacteria that found shelter in larger **heterotroph**ic cells and provided the larger cells with a built-in energy source. These larger cells became the precursors of eukaryotic cells when they acquired **mitochondria**.

See also Chlorophyll function and structure; Prokaryote

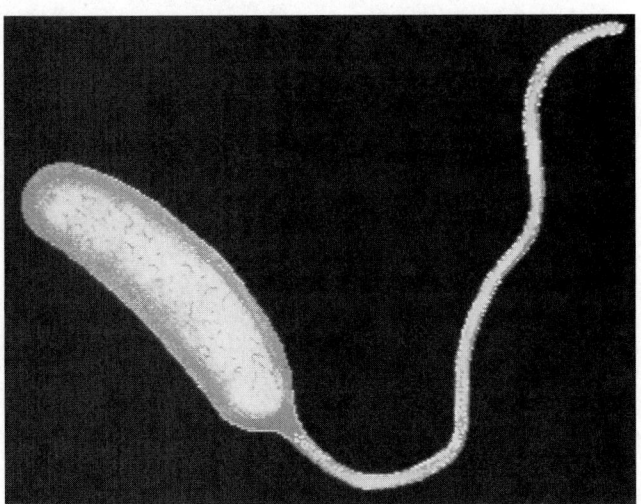

A transmission electron micrograph (TEM) image of *Vibrio cholerae* bacterium. *(Photograph by T. McCarthy, Custom Medical Stock Photo. Reproduced by permission.)*

CHOLERA

Cholera epidemics are bred in areas where poor sanitary conditions exist, particularly crowded urban settings. This often fatal intestinal disease is spread by the comma-shaped *Vibrio cholerae* **bacteria**. A highly contagious disease, cholera epidemics have been reported since antiquity. Despite advances in medicine, cholera outbreaks still occur today. In January 1998, 13,440 cholera cases were reported in the Democratic Republic of the Congo, Africa, with 778 **death**s. Such outbreaks are not uncommon and serve as grim reminders that technology has not yet conquered this dangerous disease.

Cholera victims initially suffer from diarrhea, vomiting, and cramps. Eventually, dehydration and cyanosis set in as the face takes on a bluish tint and the extremities become dark and cold. The disease is spread primarily by unwashed hands, unwashed **fruits**, and sewage-contaminated **water** supplies. Following an incubation period of 24 to 72 hours and the onset of diarrhea, victims are treated with large quantities of fluids, as well as the administration of glucose. Although less effective during severe outbreaks when health facilities are overcrowded with victims, rehydration therapy has, nevertheless, saved countless lives.

The cholera **organism** was isolated by **Robert Koch**, a German physician who led a government-supported scientific expedition into Egypt where an epidemic was underway in 1883. The epidemic ended before Koch had completed his research; he took his group to India, the sites of another cholera epidemic, to continue his investigation. In 1884, Koch cited the bacterium *Vibrio cholerae* as the causative agent of cholera.

Travelers are advised not to eat raw fish or unwashed foods when they visit countries where cholera outbreaks continue. Failure to treat the disease often results in death. The current cholera vaccine is not recommended to Americans traveling abroad, since the brief coverage it offers does not off-set its side effects. In fact, travelers have been warned to avoid vaccination in unsanitary conditions in developing countries. Researchers continue to work on an oral vaccine that could prove to be effective.

CHOLESTEROL

Cholesterol is one of the most common **steroids** and is found in almost all **animal** body **tissue**s, particularly the **nervous system**, liver, kidneys, and skin. It forms part of **cell membrane**s. Cholesterol is synthesized in the liver and other **organ**s, and the body uses it to produce other steroids. Its buildup in **arteries** has long been associated with increased risk of **heart** disease. The body uses cholesterol as the basis for synthesizing vitamin D, **cortisone**, **adrenal gland hormone**s, and **sex hormone**s.

During the late 18th century, cholesterol from gallstones became one of the first of the steroids to be studied and crystallized. In 1816, the French chemist Michel Eugène Chevreul named it "cholesterine," from the Greek words for bile and solid. In 1859, the French chemist Pierre Eugène Marcelin Berthelot showed it was an **alcohol** and the name was changed to cholesterol. In the 1840s it was found to be part of many animal tissues, including the arteries. In the 1880s, scientists determined cholesterol's formula, $C_{27}H_{46}O$, and isolated a similar substance—ergosterol—in rye infected with the fungus ergot.

The German chemists **Heinrich Otto Wieland** and Adolf Windaus demonstrated in 1912 the similarity of bile acids and cholesterol by converting each to cholanic acid. Wieland additionally demonstrated how the bile acids convert fatty substances such as cholesterol into **water**-soluble substances that can be easily digested. From his studies of the plant substance phytosterol, which is similar to the bile acids, he proposed the structure of cholesterol. For their work, Windaus received the 1927 Nobel prize in chemistry and Windaus was awarded the 1928 Nobel prize in chemistry.

After the German biochemist **Adolf Friedrich Johann Butenandt** isolated the sex hormones, the Swiss biochemist **Leopold Ruzicka**, showed their relation to cholesterol by partially synthesizing androsterone.

The first partial synthesis of cholesterol, in 1935 by the German chemist Otto Paul Diels, demonstrated cholesterol's steroid structure. In 1950, Diels's contribution earned him the 1950 Nobel prize in chemistry.

In the early 1940s, the American biochemist **Konrad Emil Bloch** used the then-new radioisotope tools deuterium (hydrogen-2) and carbon-14 to demonstrate that **acetic acid** ($C_2H_4O_2$) is cholesterol's originating **molecule**. In 1951, the German biochemist **Feodor Lynen** showed how acetic acid's 2-carbon fragment worked with a chemical called **coenzyme** A to build up the cholesterol chain. Lynen and Bloch shared the 1964 Nobel prize in physiology or medicine as a result of their research.

In all, over 30 different processes are required to turn acetic acid into ever-longer chains, until the movement of a single hydrogen from one 15-carbon chain to another forms a 30-carbon chain called squalene. Squalene molecules then combine to form the ring compound langesterol, which loses methyl (CH_3) groups to finally become cholesterol. For discovering this and other research on cholesterol, the British chemist John Cornforth (1917-) received part of the 1975 Nobel prize in chemistry. The complete structure was determined in 1955 by the American chemist Robert Burns Woodward, who also received a Nobel prize in 1965 for chemistry.

Cholesterol exists either in free form or as the ester of a fatty acid and is carried in the **blood** by lipoproteins (substances composed of **lipids** and **proteins**). The body's cholesterol level is regulated by low-density lipoprotein (LDL)-receptors, which pass cholesterol and LDLs into the cell for use. When there are more LDL molecules than LDL-receptor molecules, the LDL accumulates in the blood. It is currently believed that a related substance, high-density lipoprotein (HDL), carries cholesterol out of the tissues.

For several decades, scientists have associated cholesterol with the buildup of damaging plaque in the arteries (atherosclerosis) which can cause heart attacks. Health authorities have urged people to eat foods very low in animal **fats**, including cholesterol, to improve their health. However, research published in 1992 associates low cholesterol levels (below 160 mg per 100 ml blood serum) with a variety of diseases. Several new drugs have been produced which lower the blood LDL cholesterol levels while increasing the blood HDL cholesterol, and reduce the risk of heart attack in patients with atherosclerosis, including Mevacor, Zocor, and Cholestin. Cholestin was marketed as a dietary supplement until the FDA ruled in 1998 that it was actually an unapproved drug, prohibiting its sale. Mevacor and Zocor are available with a doctor's prescription. Research continues in order to understand cholesterol's precise role in human health and disease.

CHORDATES

Chordates are **animal**s in the **phylum** Chordata. At some stage of their development, all chordates have the following characters: a notochord, gill slits, a dorsal hollow nerve cord, and a post-anal tail. There are three sub-phyla of chordates: the Cephalochordata, Tunicata, and Vertebrata. The cephalochordates and tunicates are referred to as "invertebrate chordates," to distinguish them from the more complex, later evolved vertebrates.

The Cephalochordata consists of several groups of shallow-**water** marine **organism**s, including about 23 **species** of amphioxus and lancelets. These animals are tapered to a point a both ends, are filter-feeders, and can be as long as 2.8 in (70 mm). Although they are essentially **invertebrates** in their structure and function, these animals have a simple notochord and other characters of the chordates.

The Tunicata consists of about 1,250 species of marine organisms, including such familiar ones as sea-grapes and sea-peaches. Tunicates are filter-feeders, may be sessile or free-swimming, colonial or solitary, have a U-shaped gut, and are typically hermaphroditic (i.e., both sexes are represented in the same individual). The **larval** stage has a notochord and other characters of the chordates.

The Vertebrata consists of about 42,000 living species of fish, amphibians, reptiles, birds, and **mammals**. Vertebrates have a **brain**-case, vertebral column, lateral limbs, a skeleton of bone and/or cartilage, skin composed of surface epidermal and subsurface dermal layers, a **heart** and **blood cells**, a neural crest during embryonic development, and internal **organs** such as kidney, **pancreas**, and liver. The more complex (or more "highly evolved") vertebrates also have jaws, teeth, paired fins or limbs with an internal skeleton articulated to a limb girdle, lungs, and bony scales, feathers, or fur.

CHORION

In all vertebrates, the **chorion** is a membranous structure which completely encloses a developing **embryo**. After **birth**, this chorion is no longer needed, and is discarded.

In birds and reptiles, the chorion lies up against the interior of the **egg**shell, enclosing the entire contents of the egg. At one point, the chorion and another **membrane**, the allantois, unite to form the chorioallantoic membrane. Vessels pass through this area, delivering oxygen and **water** to the developing embryo, and allowing **carbon dioxide** to be released.

In **mammals**, the chorion is an important **tissue** which is produced very early in a pregnancy. Ultimately, it becomes part of the **placenta**. The placenta is an **organ** which connects the developing embryo with the mother's uterus, allowing the exchange of **nutrients**, oxygen, and waste between the embryo and its mother.

Soon after an egg is fertilized by a **sperm**, the resulting single-**cell**ed **zygote** divides repeatedly, to make up the multi-celled blastocyst. This blastocyst is composed of inner cells which will ultimately become the embryo, and outer cells which will ultimately form the chorion.

In mammals, the chorion is involved in the development of finger-like projections which push into the lining of the uterus. These chorionic villi will harbor **blood** vessels responsible for the delivery of oxygen and nutrients, and the removal of carbon dioxide and other wastes.

Human chorionic gonadotropin (hCG) is a **hormone** produced by the human chorion. This is the hormone which is measured to reveal the presence of a pregnancy in simple urine home pregnancy tests.

CHORIONIC VILLUS SAMPLING

Chorionic villus sampling (CVS) is a relatively new form of prenatal testing, completed earlier in gestation than the more traditional testing method, **amniocentesis**. Through CVS, the fetus's **gene**s can be examined and common chromosomal disorders and hereditary conditions, as well as the baby's sex, can be identified.

CVS can be performed in two ways. In transcervical testing, a thin tube is inserted through the cervix and into the uterus. Guided by ultrasound, the tube suctions microscopic projections called chorionic villi from within the **placenta**, which contain the same **genetic material** as does the fetus. These same fibers are extracted by transabdominal CVS, when a thin needle is inserted through the abdomen into the uterus. Again, ultrasound assists in identifying the exact location of the placenta.

The advantage of the CVS procedure is that it can be completed as early as 10 weeks in the pregnancy, but preferably closer to 12 or 13 weeks, and results are usually available within 48 hours. Amniocentesis is not done until the 15th to 19th week, and results can take up to two weeks. CVS was initially considered riskier because of increased chance of miscarriage or limb defects. With advances in procedure and training since its introduction in 1983, however, CVS is now considered as safe as amniocentesis and the preferred method for those women who need to obtain early information about the fetus.

See also Chorion; Chromosome; Down syndrome

CHROMATIN

Chromatin is a network of **deoxyribonucleic acid** (**DNA**) and nucleoproteins that constitutes a chromosome. Chromatin can only be found in a **cell** with a **nucleus** and is therefore not present in a prokaryotic cell. The DNA within a eukaryotic cell can be as long as 12 cm (4.7 in). Due to its length, the DNA must be arranged and organized in order to fit within the small area of a cell nucleus. To accomplish this task, the DNA is bound, through electrostatic forces, with nucleoproteins called **histones** and nonhistones. The assemblage of DNA with the nucleoproteins is called a nucleosome, which is the fundamental structural unit of chromatin and represents 1.8 turns of DNA wound around a core particle of another histone **protein**. It is the nucleosomes, along with the DNA material between nucleosomes (linker segment), that gives DNA the characteristic beads-on-a string appearance, with the nucleosomes representing the bead and the linker segment of DNA representing the string.

When chromatin is isolated, it appears to be composed as smooth fibers. While the highest level of chromatin organization is not well understood, scientists have found that chromatin fibers are divided into functional groups, called domains. The domains are grouped and arranged into loops called solenoids. In cells that are dividing, the solenoids are further condensed into chromatids; an identical pair of chromatids comprise the recognizable shape of a chromosome.

There are two types of chromatin: heterochromatin and euchromatin. Heterochromatin is chromatin in condensed form, is seen as dense patches and is **transcription**ally inactive while euchromatin is seen as delicate, thread-like structures that are abundant in active transcription cells.

When early scientists began staining cells, they noticed that granular material within the nucleus could become brightly colored when stained with a basic dye. The colored granular

structures were named chromatin, derived from the Greek word *khroma*, which means *color*.

See also Eukaryote

CHROMATOGRAPHY

The term chromatography was originally used about 100 years ago by a Russian botanist, **Mikhail S. Tswett** to describe the separation of bands of plant pigments (chlorophylls) extracted from green leaves. The process used petroleum ether on calcium carbonate packed in a vertical glass column. Though chromatography (from the Greek word for "color writing") was descriptive of colored bands, most modern chromatographic methods do not involve separation of colored compounds. Chromatography now describes the process of separating compounds and ions by a variety of matrices on large numbers and types of columns. The International Union of Pure and Applied Chemistry has defined chromatography as follows: "A method, used primarily for separation of the compounds of a sample, in which the components are distributed between two phases, one of which is stationary while the other moves. The stationary phase may be a solid, or a liquid supported on a solid, or a gel. The stationary phase may be packed in a column, spread as a layer, or distributed as a film, etc.; in this definition chromatographic bed is used as a general term to denote any of the different forms in which the stationary phase may be used. The mobile phase may be gaseous or liquid. Therefore, a chromatographic system consists of three components described as solute, solvent, and sorbent or more appropriately as the sample, mobile phase, and stationary phase."

The goal of chromatography is to separate mixtures of compounds into separate bands or peaks of individual compounds within a reasonable length of time. Separation is achieved when the solutes (or sample compounds) in the mobile phase demonstrate different affinities for the stationary solid phase, the mobile phase, or both, resulting in different retention times for the various sample compounds.

The simplest type of chromatography is column chromatography, in which a vertical tube is filled with a finely divided stationary phase. The mixture of materials to be separated is placed at the top of the column and is slowly washed down with a suitable mobile phase. Each type of material will move down the column at a different rate, depending on the its solubility and its tendency to be adsorbed. If the stationary phase used is a liquid adsorbed on a solid carrier, the process is called partition chromatography, since the mixture to be analyzed will be partitioned, or distributed, between the stationary liquid and a separate liquid mobile phase. If the stationary phase is solid, the separation process is called adsorption chromatography.

In thin-layer chromatography (TLC), the stationary phase is a thin layer on a glass plate or plastic film. The thin layer may be an adsorbent such as silica gel or alumina, which is made into a slurry, placed in a layer on the glass plate, and then dried. The sample mixture is dissolved in a volatile solvent. A small portion of this solution is placed on the thin layer. The solvent evaporates, leaving the mixture to be separated on the plate in the form of a small spot. The plate is placed upright in a jar, and a suitable developing solvent is added to the bottom. The jar is closed so that the atmosphere of the jar becomes completely saturated with the vapor of the solvent. The solvent rises up the plate by capillary action. When it has risen 4-6 inches (10-15 cm) in about 10-20 minutes, the plate is removed and dried. Separation of compounds is determined by examination under ultraviolet **light** or by spraying with a reagent that colors the various compounds. Paper chromatography, in which **water** adsorbed on paper acts as the stationary phase and an organic liquid is used as the mobile phase, is similar to TLC.

Gas chromatography (GC) usually uses a liquid on a solid support as the stationary phase and an inert gas such as **nitrogen**, hydrogen, helium, or argon, as the mobile phase. The stationary phase is contained in a narrow, coiled column from 4-15 ft. (1.5-5 m) in length. The mixture to be separated is injected with the mobile gas phase into the column, which is heated so that the mixture is vaporized. The different substances in the mixture pass through the column at different rates. Upon leaving the column, the substances pass through a detector, which gives a signal to a recording device. The resulting gas chromatogram shows a series of peaks, each of which is characteristic of a particular substance.

High-performance liquid chromatography (sometimes called high-pressure liquid chromatography or HPLC) is a more developed type of column chromatography. The particles that hold the stationary phase in the column are very small (0.004 inches or 0.01 mm) and uniform in size, provide a large surface area for the sample substances in the mobile liquid phase. The large pressure drop created in a column filled with such small particles is overcome by using a high-pressure pump to push the mobile liquid phase through the column and to the detector. The advantages of HPLC are high resolution and sensitivity. HPLC and GC are the two most commonly used separation techniques in analytical laboratories.

Gel permeation chromatography is based on the filtering or sieving action of the stationary phase, which has pores of uniform size in the range of 20-30 nm. A substance dissolved in a mobile liquid phase while moving down a column will be excluded from the stationary phase if its size is greater than that of the pores. If its molecular size is smaller, it will become trapped. Intermediate-sized **molecules** will be trapped by some pores but not others. Separation is based on molecular size, with larger molecules separating out first and smaller molecules last. This type of chromatography is used to measure the molecular weight of **polymers**, **proteins**, and other biological substances of high molecular weight.

Chromatography is used for the separation of pure substances from complex mixtures and is important in the analysis of environmental contamination, foods, drugs, **blood**, petroleum products, and radioactive fission products.

CHROMOSOME

Chromosomes are thread-like bodies in the **cell nucleus** of all **plants** and **animals** that hold the **genes**—the blueprints of he-

redity. Each chromosome carries a single strand of **DNA** that threads together about 1,000 genes.

Not much was known about chromosomes prior to the 1880s, due to the lack of adequate **cell staining** techniques and poor **microscopes**. In 1879, however, **Walther Flemming**, using new synthetic dyes, was able to discern bodies in cells that had previously gone undetected. He noticed that some material scattered throughout the nucleus heavily absorbed the dye and coined the word *chromatin* to describe this dark, stainable substance. Upon further observation, he noted that when a cell divided into two daughter cells, the **chromatin** first doubled, then split lengthwise, leaving each daughter cell with the same amount of chromatin as the parent cell. By 1882, Flemming had identified all the stages of this process, a fundamental operation of **cell division** now termed **mitosis**. In 1887, **Edouard van Beneden** observed a different form of division in sex cells. Rather than doubling, the chromatin split in half, with each half being distributed to the daughter cells. This phenomena was later named **meiosis**.

In 1888, the German anatomist **Wilhelm von Waldeyer-Hartz** renamed Flemming's chromatin *chromosomes*, meaning ''colored bodies.'' However, the connection between chromosomes and heredity was not made until 1902, when **Walter S. Sutton** published a short scientific article concerning the newly discovered work of **Gregor Mendel**. Sutton proposed that the ''factors'' which Mendel could not identify, but believed controlled heredity, were indeed contained in the chromosomes. **Theodor Boveri** independently came to a very similar conclusion, and in 1903, their work became known as the *Chromosomal Theory of Inheritance*.

For many scientists, the theories of Mendel and Sutton provided a sufficient explanation for heredity and **evolution**. However, the American geneticist **Thomas Hunt Morgan** remained skeptical of their work because the conclusions were speculative, based on nothing more than observation, inference, and analogy. Morgan wanted to draw firm conclusions based on quantitative and analytical data and so set out to test their theories using the **fruit fly** as his subject. The results of his experiments and those of his assistant **Hermann Muller** contributed greatly to the understanding of chromosomes and their role in heredity. Morgan found that genes—the term **Wilhelm Johannsen** coined for Mendel's ''factors''—were located on chromosomes. For the first time, the association of one or more hereditary traits with specific chromosomes was clear. He also discovered that genes on the same chromosome were often inherited together—an occurrence known as *autosomal linkage*. (Autosome is the name given to all the chromosomes that are not **sex chromosomes**.) However, chromosome pairs would sometimes break apart and exchange pieces—a process known as **crossing over**. The findings of Morgan and his colleagues focused the attention of the scientific world on chromosomes and prompted further research in the area of **genetics**.

In 1997, artificial human chromosomes were created for the first time by Huntington Willard and his colleagues at Case Western Reserve University in Cleveland, Ohio. Such artificial chromosomes may help scientists understand better how natural chromosomes work. They may also one day be used as a vehicle for carrying DNA into patients receiving **gene therapy**.

CHROMOSOME MAPPING

Also known as cytogenetic mapping or **genetic mapping**, chromosome mapping is a technique used to locate particular characteristics (coded for by specific **genes**) on specific **chromosome**s and regions within those chromosomes. By knowing the location of genes on a chromosome, predictive statements can be made about the **inheritance** of their characteristics. Initially, the maps of chromosomes were produced by looking at the inheritance of mutant conditions. Presently, a variety of techniques, ranging from physical observation of chromosomes to breeding experiments where the relative frequencies of specific offspring are considered, are used to map chromosomes.

If two characteristics are considered (and hence their controlling genes), both of which are present in one parent, then the frequency in which they occur together in the offspring is used in constructing the map. For example, **Gregor Johann Mendel** studied the **flower** color and plant height of peas and found that all possible combinations were in existence. Particular heights were found just as frequently with white flowers as with other colored flowers; similarly, dwarf **plants** occurred just as frequently with the two flower types. Mendel concluded that the forms of the two genes were transmitted from parent to offspring independently of each other. Named the **Law of Independent Assortment** by Mendel and considered the most basic form of chromosome mapping, it shows that the genes are on two separate chromosomes.

When genes occur together more frequently than would be expected, they are linked. Present on the same chromosome and members of the same **linkage** group, the closer the linked genes are to each other on the chromosome the more likely they are to be transmitted together, because there is less likely to be a cross over event between them during **meiosis**.

High resolution can be achieved on chromosome maps by using three, rather than two, linked genes. This allows a double crossover between two linked genes (which would give a false frequency) to be overcome. A double crossover is more likely the further apart the genes are on the chromosome. The more breeding experiments that are conducted, the greater the detail of the genetic map. As a consequence of such experiments, detailed maps exist for the classic **organism**s of **genetics** such as the **fruit fly** *Drosophila melanogaster* and the ascomycete fungus *Neurospora crassa*. For practical reasons, this method of chromosome mapping is not used for human beings. The majority of loci (positions of genes on chromosomes) that are known for humans are based on mutant **alleles** (alternative forms of a gene), which produce a particular disease or syndrome. Most mapping of non-mutant forms has developed from the studies of extended families.

See also Human Genome Project

CILIA

Cilia are tiny, hairlike structures found on surfaces of many **animal** and **protozoa**n **cells** as well as in some lower **plants**.

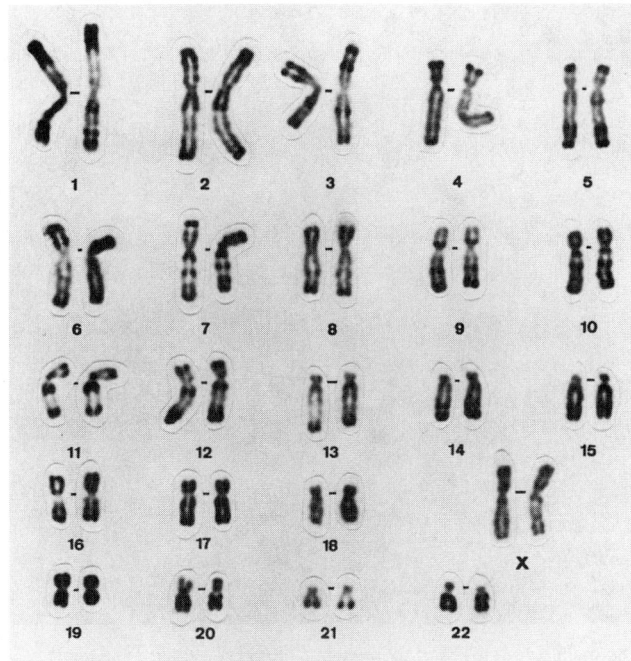

A human female karyotype. *(Phototake NYC. Reproduced by permission.)*

They are very similar in both structure and function to **flagella**. Typically about 0.82 ft (0.25 m) in diameter and 33 ft (10 m) long, cilia are enclosed within the cellular (plasma) **membrane** and have an interesting and complex interior structure called the **axon**eme made of microtubules and a large number of associated macromolecular **protein**s. The basic function of cilia is to move in a whip-like motion that provides motile force to allow single cells to swim through a fluid medium or to move liquids or particles across the cell surface. Protozoans, for instance, use cilia to swim, as well as to capture food and move it into their gullet. Many epithelial **tissue**s in higher animals have a carpet of cilia that moves mucus or particulate material across cell surfaces. For instance, respiratory passages are lined by cilia that push dust, pollen, and other contaminants up out of the lungs. Similarly, the uterine tubes in females have a ciliary carpet that moves ova down to the uterus for implantation.

The microtubules in the ciliary axoneme are generally arranged in a circle of nine sets of fused doublets arrayed in a circle around two individual central microtubules. The central pair is enclosed in a spiral protein sheath and the outer doublets are attached to this central sheath by radial spokes in a wagon wheel-like structure. Curving arms made of an important protein called dyein reach from one doublet pair to the next, while a linking protein called nexin forms temporary bonds between adjacent doublets. Dyein is one of a **family** of molecular motors that use **energy** from **adenosine triphosphate (ATP) molecule**s to change shape and move structures around within cells. In the case of cilia, the dyein molecules ''walk'' along microtubular surfaces and move one doublet set relative to the others. Coordinated attachment and release of many

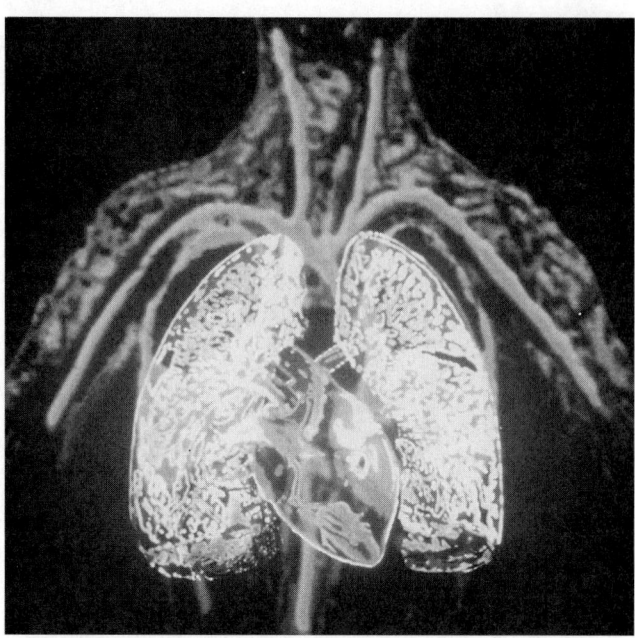

The main components of the human circulatory system. The heart (placed between the lungs) delivers blood to the lungs to pick up oxygen and circulates it throughout the body by means of a system of blood vessels. *(The Stock Market. Reproduced by permission.)*

dyein arms throughout the cilium causes bends to move progressively along the length of a cilium and create the complex motions that allow beating and recovery strokes for individual cilia. How these complex motions are coordinated between the many cilia that cover many cell types is not yet known, but it is thought that ion channels in the surface membrane probably play a role in this process.

Interestingly, the sensory **organ**s in higher animals often are derived from non-motile cilia. The **light**-sensitive rod and cone cells in the retina of the human eye, the chemical receptors of the olfactory cells in the human nose, and the sound sensitive hair cells in the human inner ear, for instance, all have highly modified, non-motile cilia as the basis of their sensory apparatus.

CIRCULATORY SYSTEM

Andrea Cesalpino, working in the fourteenth century, was one of the first scientists to use the term "circulation," and to discuss the concept of a closed circulatory system. **William Harvey** did a great deal of work which cemented the concept of the closed circulatory system, combining data he obtained from dissections, experiments, and calculations. Our current understanding of the circulatory system was born of Harvey's work.

The circulatory system is a complex network of **organ**s and tubes which serve to pump and carry **blood** throughout the body. The purpose of carrying blood throughout the body is two-fold: to deliver oxygen and other necessary substances to all of the **tissue**s and organs throughout the body; to remove waste products, including **carbon dioxide**, from all of the tissues and organs.

Some **species** of **organism**s have open circulatory systems, meaning that the blood is not entirely enclosed in a system of tubes. In these organisms (including some mollusks and arthropods), the blood leaves the blood vessels and flows directly into large spaces called sinuses. All vertebrate organisms (organisms with backbones) have closed circulatory systems, meaning that the blood never leaves the blood vessels. In fact, the circulatory systems of all of the vertebrates are remarkably similar.

The human circulatory system consists of a pump (the **heart**); a medium which is pumped through the vessels (the blood); a network of vessels to take blood from the heart to the tissues of the body (**arteries**, arterioles, and capillaries); and a network of vessels to return blood from the body's tissues back to the heart (**veins**, venules).

The right side of the heart processes unoxygenated blood coming from the tissues and organs of the body. This blood flows into the top of the right heart (right atrium), and is pumped into the bottom of the right heart (right ventricle). The right ventricle pumps this unoxygenated blood into the pulmonary artery, the only artery in the body which carries blood devoid of oxygen. The pulmonary artery carries blood to the lungs, where it loads up on oxygen and gets rid of carbon dioxide. The pulmonary vein (the only vein in the body carrying blood rich in oxygen) takes the blood to the top of the left heart (the left atrium), which pumps into the bottom of the left heart (the left ventricle). The left ventricle then pumps the blood into the body's largest artery, the **aorta**. The aorta has many major arteries which come off, and lead to all of the major organs and tissues of the body.

The arteries branch further into smaller vessels, called arterioles, which themselves branch into the tiny vessels called capillaries. Capillaries are only one-**cell** thick, allowing oxygen, **nutrients**, **hormone**s and other substances to diffuse across into the tissues and organs. Capillaries exist as networks of tiny tubes, which receive blood from arteries, and deliver blood to venules. Venules are small veins which collect the newly de-oxygenated blood, delivering it to the larger veins. Large veins from the lower body empty into the inferior **vena cava**; large veins from the upper body empty into the superior vena cava. Both venae cavae empty directly into the right atrium, where the cycle begins again.

The circulatory system can be broken down loosely into a number of important circuits. These include the pulmonary circuit (most important for oxygen delivery, and carbon dioxide removal); the hepatic portal circuit (which allows the blood to pick up various nutrients from the digestive organs, as well as waste products from the liver); the renal circuit (which plays an important role in regulating the quantity of water within the body, as well as removing various waste products); and the general body circuit (which includes muscle, bone, glands, and **brain**). The **lymphatic system**, which filters harmful substances out of the fluid surrounding body tissues, includes lymphatic fluid, lymph nodes, lymph vessels, and the thoracic duct. Lymphocytes, a type of white blood cell, are highly concentrated in lymphatic fluid. Originating in the bloodstream, lymphatic fluid bathes organs and tissues, and returns to the bloodstream after passing through lymphatic filters, which function as part of the body's defense system.

CITRIC ACID CYCLE • See Krebs, Hans Adolf

CLASS

The term Class refers to a taxonomic level within the hierarchical system of **classification** between **Phylum** and **Order**. **Organism**s are included in this taxon based upon several methods of scientific scrutiny. One area is comparative **morphology**, which examines shapes and sizes of structures and their developmental origins. A second method is comparative biochemistry, which compares segments of **amino acid**s in **protein**s and **nucleotide**s in nucleic acids to distinguish groups. A third method is comparative **cytology**, which looks at changes in the number, shapes, and sizes of **chromosome**s and their constituents. Examples of classes include: Mammalia (**mammals**), Aves (birds), Reptilia (reptiles), Amphibia (amphibians), Osteichthyes (bony fish), Chondrichthyes (cartilaginous fish), Insecta (insects), Gastropoda (snails, slugs, and their relatives), Cestoda (tapeworms), Liliopsida (monocots), Magniliopsida (dicots), and many more.

CLASSIFICATION

Classification is the categorization of living **organism**s into a hierarchical series of groups. Based on the work of **Carl Linnaeus** during the eighteenth century, classification was devised to organize scientific discussion of **plants** and **animals**. Its universal acceptance and use has done much to further science. One aim of modern classification is to show evolutionary relationships between organisms. All organisms are divided into one of five **kingdom**s and then into successively more exclusive groups. After the kingdom is the **phylum** (the term division is sometimes used instead of phylum for plants), then **class**, **order**, **family**, **genus**, and the smallest commonly used grouping is the **species**. Each grouping can be recognized by shared characters, the more characters in common the closer to the rank of species the organisms are. For example all animals are in the same kingdom, all animals with a backbone and notochord are in the phylum Chordata. All animals with hair that nourish their young by the production of milk are in the class Mammalia, those with opposable thumbs, binocular vision and well developed **brain**s are in the order **Primates**. Organisms with a prominent chin, small canine teeth, and the ability to use tools are in the family Hominidae. More specific, organisms with no eye ridges are in the genus *Homo*, and finally humans are the species *Homo sapiens*.

All groups at the same level, for example all families, are supposed to differ from each other by the same amount. Any newly discovered organism can be classified according to how similar it is to already classified organisms. The level of similarity with other organisms denotes at what point the new individual should enter the system. Various subgroups can be used in the system to denote minor differences—those differences that are not large enough to form a new group at that level of classification. For example, species can be divided up into subspecies.

The modern classification system is considered a natural system, representing genuine relationships between organisms. It is based on overall resemblances, unlike artificial classification systems that consider only one or two characters for a specific purpose. The more closely related organisms are the more features they will have in common so a natural system of classification will also reflect evolutionary relationships. It is a **phylogenetic** system. As more information becomes available, either from **fossil** evidence or new techniques such as cladistics and molecular and **DNA** studies, the classification system is modified accordingly.

Classifications are often debated within the scientific community, and inevitably, not all classifications are readily accepted. However, classification is a useful tool that encourages more effective communication. Classification is and always will be an artificial system trying to describe the real world. Sometimes the system breaks down and modifications have to be made.

See also Taxonomy

CLASSIFICATION, BIOLOGICAL

Biological classification is a formal system of identifying, naming, and grouping individual **organism**s. While various systems have existed since antiquity, the modern **classification** system was first developed by **Carl Linnaeus** in the eighteenth century. It is based on a binomial naming system and groupings of organisms based on important traits.

Methods for grouping organisms into formal systems have been around since antiquity. **Aristotle** established one of the first systems that identified **plants** and **animals** as distinct **kingdom**s. He even divided these kingdoms into more basic categories. Aristotle and many natural philosophers after him, followed an alphabetical ordering system. During the sixteenth century, the number of known plants and animals had increased to the point that a new organizational system was necessary. Gaspard Bauhin (1560-1624) introduced a two-part, or binomial, naming system and published a list of plants organized by similar characteristics. Bauhin's work was further developed by Joachim Jung (1587-1657) and **John Ray**. However, most of these sixteenth and seventeenth biologists still used many parts of the classification system proposed by Aristotle.

It was not until the eighteenth century that the modern classification system was developed. Carolus Linnaeus, a Swedish physician and botanist, is considered the founder of **taxonomy**, the branch of biology that concentrates on naming and classifying all forms of life. In 1735, he first proposed a binomial naming system in his work *The System of Nature*. According to this system, every organism was identified by a two part name. The first word in the name indicated the **genus** while the second was indicative of the **species**. The species were further grouped based on their physical similarities into a hierarchy of more general categories. During his lifetime, Linnaeus used his system to classify over 18,000 species of plants. Although Linnaeus sought to bring order in the diversi-

CLASSIFICATION HIERARCHY OF LIVING ORGANISMS

Common Name	Pink Carnation	Chanterelle Mushroom	*E. coli*	German Cockroach
Kingdom	Plantae	Fungi	Monera	Animalia
Phylum/ Division	Magnoliophyta	Basidomycota	Bacteria	Arthropoda
Class	Caryophyllidae	Hymenomycetes	Proteobacteria	Insecta
Order	Caryophyllales	Cantharellales	Cytophagales	Dictyoptera
Family	Caryophyllaceae	Cantharellaceae	Enterobacteriaceae	Blattellidae
Genus	*Dianthus*	*Cantharellus*	*Escherichia*	*Blattella*
Species	*Dianthus caryophyllus*	*Cantharellus cibarius*	*Escherichia coli*	*Blattella germanica*

Common Name	Albacore Tuna	Rose-ringed Parakeet	Siamese Cat	Humans
Kingdom	Animalia	Animalia	Animalia	Animalia
Phylum/ Division	Chordata	Chordata	Chordata	Chordata
Class	Actinopterygii	Aves	Mammalia	Mammalia
Order	Perciformes	Psittaciformes	Carnivora	Primates
Family	Scombridae	Psittacidae	Felidae	Hominidae
Genus	*Thunnus*	*Psittacula*	*Felis*	*Homo*
Species	*Thunnus alalunga*	*Psittacula krameri*	*Felis catus*	*Homo sapiens*

(Table by Standley Publishing.)

ty of life, he never believed that his groupings implied any relationship between organisms. Over the years, his system has been modified to reflect the evolutionary relationship between species.

The Linnaean system formalized the grouping of organisms in hierarchical levels. The most basic of these groups is the species. Next in order of complexity is genus. From here taxonomists group related genera into a **family**, related families into an **order**, related orders into a **class**, related classes into a **phylum** and related phyla into kingdoms. These seven categories can be divided further by using the prefixes sub- and super-. Each group is defined by specific characteristics that

its members share. For instance, animals with hair are grouped under the class **mammal**.

The groups of organisms at any taxonomic level are called **taxon**. For example, **Pinus** is the taxon for the genus of various pine **trees**. While early biologist developed taxonomy systems to list known animals, modern scientists have found it useful to have a system that reflects evolutionary pathways. The discipline of biology that deals with this kind of system is called **systematics**. There is a good deal of overlap between **taxonomy** and systematics.

According to the modern system, each taxon should reflect a common evolutionary background. If a single ancestor gives rise to all the species in a taxon, the taxon is said to be monophyletic. A taxon can also be polyphyletic if its members evolved from more than one type of ancestor. In an ideal taxonomy system, all the taxons would be monophyletic. However, this ideal has yet to be achieved.

The kingdom has traditionally been considered the highest level of classification for any organism. The two kingdom system, plant and animal, was the first one proposed. These kingdoms were distinguished by the way in which the organisms got food and whether they moved or not. So a stationary organism that produced food through chlorophyll was a plant and a mobile organism that ingests food was an animal. This system lasted well into the nineteenth century, but was found to be inadequate when **microorganism**s were discovered that could both produce their own food and move. In 1969, **Robert H. Whittaker** introduced the five kingdom system. He maintained the Plant and Animal kingdoms, but also included a group for **Fungi**, microorganisms (**Protista**) and **prokaryote**s (**Monera**). Recent comparisons of organisms on a molecular level have unveiled problems with the five kingdom system. In the future, additional kingdoms may be added.

CLAUDE, ALBERT (1898-1983)
Belgian American cell biologist

Biologist Albert Claude received the Nobel Prize in 1974 for his discoveries concerning the fine structure of the **cell**. His early work described the nature of **mitochondria** as the powerhouse of the cell, paving the way for much groundbreaking research by others. In addition, he demonstrated that the interior of cells was not merely an arbitrary mass of substances, but rather a highly organized space delineated by the net-like **endoplasmic reticulum**, a formation that he was the first to recognize.

Albert Claude was born in Longlier, Belgium (now Luxembourg), on August 24, 1898, to Florentin Joseph Claude and Marie-Glaudicine Wautriquant. He served in World War I, in which he won the Interallied Medal along with veteran status. The University of Liege admitted Claude under a special program designed for war veterans.

Claude earned an M.D. degree in 1928 under his continuing government scholarship and attended the Kaiser Wilhelm Institute in Berlin for further study. He relocated to the United States in 1929 to join the staff of the Rockefeller Institute in New York City, home to much of the great biomedical research and discoveries of the early twentieth century. There Claude studied the tumor agent of Rous sarcoma, a virus of chickens. Though Claude had not been invited to join the Institute, the director, Simon Flexner, one of the country's leading medical educators, approved his hiring.

In the laboratory of James B. Murphy, Claude began work on isolating the originating factor of the sarcoma, first discovered in 1911 at Rockefeller Institute by **Peyton Rous**, that was a type of soft-tissue **cancer** in chickens. Only recently had microbiologists first suggested that cancers might be caused by newly discovered agents known as viruses. But it

was not until 1932 that Rous' work was vindicated by the discovery of transmissible wild rabbit cancers that were proven to be viral in nature.

While pursuing the new field of virology, Claude developed a technique using a high-speed **centrifuge** to spin fractionated (broken-up) cells infected with viruses in an attempt to isolate their agents. Though his primitive machine was constructed from meat grinders and sieves, Claude was able to fractionate various components of cells that had never been separated before, paving the way for new understanding of their varying functions. Though he never succeeded in fully isolating the virus within the cell mixture (a development that came years later by other investigators), his discoveries nevertheless became crucial to the study of cell biology.

The Rous virus is among those now known as a ribonucleic acid (**RNA**) virus, that is, its genetic material is derived from RNA rather than the more common deoxyribonucleic acid (**DNA**). Claude found that it was not only virus-infected cells that showed a high RNA content, but also healthy cells. By the early 1940s Claude joined forces with biochemists George Hogeboom and Rollin Hotchkiss in an attempt to determine the origins of this cellular RNA.

Claude, Hogeboom, and Hotchkiss found a variety of different "granules" in the cells that they determined were mitochondria, which were first discovered in 1897. However, the purpose of these often abundant cell components, especially in the liver cells, still remained unknown. Claude found that the mitochondria were not the source of the cells' RNA, but they did harbor certain **enzymes** that seemed to be involved in the cells' energy **metabolism**, a process dimly understood at the time. Claude and his colleagues proved in 1945 that mitochondria are the "powerhouses" of all cells, from bacteria to liver, from plants to fungi to animals. The RNA, it turned out, was concentrated in other cell particles that fellow researcher **George Palade** discovered and called microsomes. Later renamed **ribosome**s, these particles were shown to be the centers of **protein** production in all cells of every type of living thing. In 1974 Claude, Palade, and a third researcher, Christian R. Duvé, shared the Nobel Prize for physiology or medicine.

By the early 1940s, Claude had significantly perfected ultracentrifugation (the process of separating cell particles) and was seeking other new technologies with which to probe the cell. In 1942 he became convinced that the newly developed electron microscope would be useful in furthering his studies and secured the use of the device at the Interchemical Corporation, home to the only electron microscope in New York City where it was used primarily for metallurgical purposes.

The cells that Claude and his associate, Keith Porter, observed under the microscope showed the presence of a "lacework" structure that was eventually proven to be the major structural feature of the interior of all but bacterial cells. This lace-work structure was also responsible in part for providing the shape of cells as well as the location for many granular cell components, including ribosomes. The discovery of this endoplasmic reticulum (derived from the Latin word for "fishnet") altered biologists' view of cells as simply bags of "stuff" to highly organized biological units.

Claude, who became a U.S. citizen at age 43, returned to his native Belgium in 1948 and for a time gave up active research to become an administrator at the Université Libre de Bruxelles, where he spent the next 20 years developing a significant cancer research center. During the same period he headed the Institut Jules Bordet, where he resumed research on the fine structure of cells.

In 1972 the Rockefeller University (formerly Institute) awarded Claude emeritus standing. Other honors accrued over the span of his career include the Medal of the Belgian Academy of Medicine, the Louisa G. Horowitz Prize of Columbia University, and the Paul Ehrlich and Ludwig Darmstaedter Prize of Frankfurt. In addition, Claude was a full member of the Belgian and French academies of science and an honorary member of the American Academy of Arts and Sciences. Other honors included the Order of the Palmes Académiques of France, the Grand Cordon of the Order of Léopold II, and the Prix Fonds National de la Recherche Scientifique from Belgium. Claude died in 1983.

CLIMAX COMMUNITY

The term ''climax community,'' or ''climax,'' is used by ecologists to describe a stable **ecosystem** that occurs at the end of a successional sequence. (**Succession** is the process of community recovery that occurs after an event of disturbance.) The climax is a theoretical ecosystem that is in an equilibrium with the environmental conditions occurring in the region. Terrestrial plant ecologists are particularly interested in the idea of climax ecosystems, but the theory is also relevant to **animal** and microbial communities, and to all environments.

Frederic Clements, an American ecologist, was influential in the initial development of the concept of climax ecosystems. Clements' most influential paper on this subject, published in 1916, suggested that there was only one true climax ecosystem for any given climatic region. He referred to this as the ''climatic climax,'' and theorised that it was the eventual end-point of all successional sequences. Clements believed this was the case regardless of whether the succession had started after fire, timber harvesting, or other kinds of disturbances, or from a pond or lake gradually filling in with sediment and organic debris, and even regardless of soil type. In eastern North America, for example, all areas climatically suitable to supporting forest would eventually become dominated by climax stands of old-growth sugar maple (*Acer saccharum*), beech (*Fagus grandifolia*), eastern hemlock (*Tsuga canadensis*), and other tolerant **species** (this means that they can reproduce themselves under deeply shaded, highly competitive conditions, and therefore eventually dominate stands).

The so-called ''monoclimax'' theory of Clements was criticised as being too simple by other ecologists. The British ecologist A.G. Tansley proposed a more realistic, ''polyclimax'' theory that accommodated the well-known influences of local soil type, topography, and disturbance history on community development during succession. Therefore, on drier sites in eastern North America, climax stands might be dominated by red oak (*Quercus rubra*) and other **tree**s that are tolerant of droughty conditions, and wet sites by silver maple (*Acer saccharinum*) and other species tolerant of flooding. Climax communities of sugar maple, beech, and eastern hemlock would only occur on intermediate, well-drained, moist sites (these are known as mesic conditions).

In the early 1950s, the American ecologist R.H. Whittaker suggested that landscapes actually support gradually varying climax communities, which are associated with continuous gradients of environmental change (as occurs, for example, with increasing altitude up a mountain). Therefore, according to Whittaker, ecological communities vary continuously, even in the climax (or old-growth) condition. Therefore, climax communities cannot be objectively divided into only one or several discrete types. This view of continuously varying community change is the one that most ecologists hold today.

In a practical sense, it is not possible for ecologists to identify the occurrence of an absolute climax ecosystem. However, the climax condition is suggested when there are relatively slow changes in the structure and function of old-growth ecosystems, compared with earlier, more dynamic stages of succession. Moreover, the old-growth ecosystem would be dominated by large, old individuals of the most **competition**-tolerant species occurring in the region (assuming they are tolerant of the local sites conditions). However, all ecosystems change over time, even in the climax (or old-growth) condition (therefore, the climax state cannot be regarded as being static in community composition). For example, even in old-growth forest, microsuccession is always occurring within stands, usually associated with the **death**s of individual trees or small groups of trees.

Moreover, it is common over many large regions that the frequency of stand-level disturbance events is rather short. For example, catastrophic wildfires, windstorms, or insect epidemics might occur every few decades or so. Under such conditions, an old-growth or climax community does not have an opportunity to develop before another disturbance initiates another succession.

CLONING

The phenomenon of identical twins has always attracted attention. After an **egg** is fertilized, it begins to divide repeatedly. If the egg completely separates during the two-**cell** stage, identical twins will result. Both individuals will have exactly the same combination of **gene**s (**genotype**) and each will have the same physical characteristics (**phenotype**). This is an example of how exact duplicates can naturally occur through **sexual reproduction**.

A clone is a group of genetically identical cells descended from a single common ancestor. Science has capitalized on the mechanisms of cellular reproduction to produce clones. Advances in biotechnology since the 1970s have enabled livestock breeders to clone virtually unlimited numbers of identical **animal**s from a single **embryo**, allowing the precise duplication of an animal with desired characteristics.

In 1979, veterinarian Steen Willadsen developed a way to divide sheep embryos in half at the two-cell stage, making

A pine forest in Wisconsin is an example of an ecological climax community. *(Photograph by Robert J. Huffman, Field Mark Publications. Reproduced by permission.)*

clones possible. In the next few years, several scientists, including Willadsen, J. P. Ozil, C. Polge, Stephen Voelkel, and R. A. Godke, made further strides in this area with both sheep and cattle embryos. Willadsen and Godke, working together, developed a simplified method of dividing and cloning sheep embryos in 1984.

In one cloning technique, dairy farmers trying to clone a cow with high milk-producing qualities begin by artificially inseminating this desired cow with the **sperm** from a prize bull. The resulting embryo, which contains the entire genetic instructions needed to form a complete calf, develops in the desired cow. After some time, the embryo divides into a mass of 32 identical cells. It is then carefully removed from the desired cow and meticulously separated into 32 separate cells. Next, the **nucleus** from each embryonic cell is removed microsurgically. The **genetic material** (genes) from each of the embryonic cells of the desired cow is then inserted into the space once occupied by the genetic material in unfertilized eggs of 32 carrier cows. Finally, each new embryo is transplanted into the 32 carrier cows, where it develops fully. After a normal pregnancy, each carrier cow gives **birth** to a calf that is genetically identical to the 31 other calves derived from the original 32 cell embryo. Each calf is a clone. The trait for increased

milk production has been cloned so that the farmer now has 32 high milk-producing cows instead of just one.

Cloning technology has enabled breeders to develop lines of cattle, sheep, and cotton **plants** that respectively produce more milk, wool, and cotton. These techniques have allowed us to increase **productivity** and quality control simultaneously. This work is also important to reproductive physiologists interested in how embryos develop and to researchers who require identical **organism**s for comparison during experimentation.

In another cloning technique, the animal is cloned from a cell taken from an adult rather than an embryo. Dolly, a sheep that was the first mammal produced this way, was born on July 5, 1996. This dramatic advance, achieved by **Ian Wilmut** and his colleagues at the Roslin Institute in Edinburgh, immediately caused a sensation within both the scientific community and the general public. In 1997, Roslin scientists announced two new additions to their high-tech flock: a pair of cloned lambs, named Molly and Polly, that carry a human gene in their cells. The goal was to create sheep whose **blood** would contain a substance that could be used to treat human patients with **hemophilia** B, a blood-clotting disorder.

Then in 1998, biologists James Robl and Steven Stice announced the creation of two other transgenic, or genetically

Dolly, the first mammal to be successfully cloned from an adult cell. *(AP/Wide World Photo. Reproduced by permission.)*

altered, cloned animals: a pair of calves known as George and Charlie. These scientists claim that their cloning method has a failure rate much lower than the one used with Dolly, whose birth occurred only after 270 failures, many of which ended in miscarriage or prenatal deformity.

Cloning is one area of **genetics** that is advancing very rapidly, and it is therefore not without controversy. For example, sometimes the deoxyribonucleic acid (**DNA**) in the **mitochondria** of the carrier animal egg interferes with the transplanted genetic material from the desired animal. This may result in life-threatening complications in the carrier animal and/or its offspring. And if this technology is ever applied to humans, who will decide which genes are "desired" and should be cloned? This is only one of many important questions that have arisen as a result of the amazing advances in genetic cloning.

COENZYME

While all **enzyme**s belong to the **protein** family, many of them are unable to participate in a catalytic reaction until they link with a nonprotein component, or coenzyme. This can be a metal ion— copper, iron, or manganese, for instance—or a moderately-sized **molecule** called a prosthetic group. Quite

often, though, coenzymes are composed wholly or partially of **vitamins**. Although some enzymes are attached very tightly to their coenzymes, others can be easily parted. In either case, the parting almost always deactivates both partners.

The first coenzyme was discovered by English biochemist Sir **Arthur Harden**. Toward the end of the nineteenth century, Harden began an intense study of the **fermentation** process, particularly **alcoholic** fermentation. Inspired by **Eduard Buchner**—who, in 1897 had discovered an active enzyme in **yeast** juice that he had named zymase—Harden used an extract of yeast in most of his studies.

In 1904, Harden, while working with William J. Young, made a surprising discovery. He'd already learned that boiling yeast juice appeared to destroy all its enzyme activity. However, he found that, when he added some of the boiled and presumably useless yeast juice to an active batch, the active yeast juice suddenly showed an increased capacity to ferment glucose. Some active principle, he reasoned, must have survived the boiling. To solve the chemical mystery, Harden used a filtration process called dialysis. He placed another batch of yeast juice in a semipermeable bag, then left the bag in a container of pure **water**. Before long, the juice's smaller molecules filtered through the bag's **membrane** and into the water, leaving the larger molecules behind in the bag. After further testing, Harden discovered that the yeast enzyme apparently consisted of two parts: a large-molecular part that could not survive boiling and was almost certainly a protein; and a small-molecular part that could survive boiling and was probably not a protein. (Harden called the nonprotein a coferment, but others soon began calling it a coenzyme.)

Several researchers quickly began studying the newly discovered component's chemical nature and, roughly 20 years later, Hans Euler-Chelpin, a German-Swedish chemist, was able to define its structure. (Harden and Euler-Chelpin shared the 1929 Nobel Prize in chemistry for their work.)

It soon became clear that virtually all coenzymes were composed of vitamins, particularly those in the water-soluble B family. For the most part, they functioned primarily in **energy** transfers and in the **metabolism** of **fats**, **carbohydrate**s, and proteins. During the 1930s Swedish biochemist Axel Theorell (1903-1982) and American biochemist Conrad Arnold Elvehjem (1901-1962) greatly furthered the understanding of vitamins through independently conducted research on oxidation enzymes and pellagra, respectively.

In 1947, a coenzyme particularly important in the metabolic process, Coenzyme A, or CoA, was discovered by **Fritz Lipmann**, a German-born American biochemist who received the 1953 Nobel Prize in medicine and physiology for the discovery. **Feodor Lynen** is also remembered for his research on CoA; because he succeeded in isolating acetylcoenzyme A (CoA combined with the two-carbon fragment first theorized by Lipmann), he, along with **Konrad Emil Bloch**, received the 1964 Nobel Prize for medicine and physiology.

Because most vitamins are inactive when first taken into the body, a two-step process must take place. The vitamin must be activated to its coenzyme form (with vitamins B_1, B_2, and B_6, for instance, this means the addition of a **phosphate**

group.) After that, the coenzyme must combine with its proper enzyme partner. Only then can the catalytic activity, for which both are programmed, be set in motion.

COGNITION

Cognition is the process of knowing. It encompasses all experiences of consciousness by which knowledge is acquired, as distinguished from experiences of feeling or willing. These experiences include the processes of perceiving, recognizing, conceiving, and reasoning. Cognition is also a branch of psychology. Cognitive psychologists study the physiological mechanisms that occur in the **brain** as a function of conscious knowing. Short and long-term **memory**, visual-spatial determination, problem solving, and language development are a few of the aspects cognitive psychologists refer to when talking about cognition. Cognitive science also entails studies of artificial intelligence, age-related changes in memory and **learning**, and the development of dementia (a condition marked by a decline in an individual's ability to think and learn).

One of the most basic cognitive functions is the ability to conceptualize, or group individual items together as instances of a single concept or category. Concepts provide the fundamental framework for thought, allowing people to relate most objects and events they encounter to preexisting categories. People learn concepts by building prototypes to which variations are added and by forming and testing hypotheses about which items belong to a particular category. Most thinking combines concepts in different forms. Examples include propositions (proposals or possibilities), mental models (visualizing the physical form an idea might take), schemas (diagrams or maps), scripts (scenarios), and images (physical models of the item). Other fundamental aspects of cognition are reasoning, the process by which people formulate arguments and arrive at conclusions, and problem solving—devising a useful representation of a problem and planning, executing, and evaluating a solution.

Memory is another cognitive function that is crucial to learning, communication, and even to one's sense of identity. Short-term memory provides the basis for one's working model of the world and makes possible most other mental functions; long-term memory stores information for longer periods of time. The three basic processes common to both short- and long-term memory are encoding, which deposits information in the memory; storage; and retrieval.

The cognitive function that most distinctively sets humans apart from other **animals** is the ability to communicate through language, which involves expressing propositions as sentences and understanding such expressions when we hear or read them. Language also enables the mind to communicate with itself.

Since the 1950s, cognitive psychology, which focuses on the relationship between cognitive processes and behavior, has occupied a central place in psychological research. The cognitive psychologist studies human **perception**s and the ways in which cognitive processes operate on them to produce

A young girl hones her cognitive skills by playing pick-up sticks. *(Photograph by Robert J. Huffman, Field Mark Publications. Reproduced by permission.)*

responses. Cognitive psychologists have studied such issues as the ways in which needs, motivations, and expectations affect perception; whether or not language exists in the mind prior to experience; curiosity and information seeking; personal constructs; and individual perceptual and cognitive styles.

The development of the modern computer has influenced current ways of thinking about cognition through simulation of cognitive processes and through the creation of information-processing models. Within these models, cognition is portrayed as a system receiving, representing, and manipulating information in various ways. The senses transmit information from outside stimuli to the brain, which perceives, interprets, and then responds to the information. The information may be stored in the memory or it may be acted on.

COHEN, STANLEY (1922-)
American biochemist

Stanley Cohen has devoted his career to the study of growth factors, **hormone**-like substances that are produced by individual cells rather than by organs. He shared the 1986 Nobel prize in physiology or medicine with the Italian-American biochemist **Rita Levi-Montalcini** (1909-) for their joint and individual growth factor discoveries.

Cohen was born in Brooklyn, New York, to Russian immigrant parents. His father was a tailor. Encouraged by his parents, Cohen concentrated on intellectual accomplishments and music during his teen years after recovering from polio. He received his bachelor's degree in 1943 from Brooklyn College, majoring in biology and chemistry, and his master's degree in zoology in 1945 from Oberlin College. His Ph.D. in 1948 from the University of Michigan included research on 5,000 earthworms that he hand-collected from the campus lawn.

Cohen began working in 1953 with Levi-Montalcini at Washington University in St. Louis. She had already discovered that mouse tumor extracts could stimulate **growth** of the fibrous axons that connect **brain** cells in chick embryos. Cohen purified the **protein**, which was named **nerve growth factor** (NGF), and determined its **amino acid** structure. He also developed antibodies that inhibit nerve growth and discovered NGF's presence in snake venom and in mouse saliva.

Earlier, while testing the mouse saliva extract on newborn mice, Cohen found that it speeded up development of their eyelids and corneas. Their eyes opened after one week instead of two, and their teeth also appeared earlier than normal. Cohen believed that it was a different substance than NGF, and named it epidermal growth factor (EGF). In 1962 Cohen, working at Vanderbilt University, isolated the factor. He later purified it and determined EGF's amino acid sequence in both mice, in 1972, and humans, in 1975. EGF has been used to grow skin for burn treatments and, because it is the same as one of the stomach hormones, to reduce stomach acid production.

Cohen's identification of EGF's receptor molecule and its method of binding was important in improving scientific understanding of the numerous growth factors that have since been discovered. His work also led him to discover the technique of DNA **cloning** in 1973 when he worked with **Herbert Boyer** from the University of California, sparking worldwide interest in **genetic engineering**. He is the recipient of several honors and awards including the National Medal of Science, the National Medical of Technology and the Albert Lasker Basic Medical Research Award.

COLBORN, THEODORA E. (1927-)
American environmentalist

Theodora Emily Decker Colborn, better known as Theo Colborn, is a leading proponent of the theory of endocrine disruption. This theory states that some synthetic chemicals interfere with the ways that **hormone**s work in humans and animals. Colborn argues that such disruption can have profound adverse effects, especially when a mother passes a contaminating agent to a growing fetus, and the contamination interferes with the hormonal signals used by the fetus to direct its **growth**. Colborn says the possible adverse effects, which in some cases are not apparent until adulthood, include impaired ability to reproduce, diminished intelligence, altered behavior, and reduced ability to resist disease. Colborn, along with journalist Dianne

Dumanoski and zoologist John Peterson Myers, presented her argument in a controversial 1996 book titled *Our Stolen Future: Are We Threatening Our Fertility, Intelligence, and Survival?—A Scientific Detective Story*.

Colborn was born on March 28, 1927, in Plainfield, New Jersey. Her parents were Theodore Decker and Margaret L. DeForge Decker. As a girl, Colborn was fascinated by water, spending many hours playing in the river by the farm where her family lived. This early love for the outdoors laid the groundwork for an enduring commitment to the environment. However, it was many years before Colborn found her calling as a professional environmental activist. Her bachelor's degree, earned from Rutgers University in 1947, was in pharmacy.

While studying at Rutgers, Colborn met Harry R. Colborn, and the couple married on January 20, 1949. Colborn and her new husband took over his father's drugstore in Newton, New Jersey. Over the years, they added two more stores to their business. At the same time, their family expanded with the births of their children, Harry, Christine, Susan, and Mark. By 1962, the demands of running three drugstores and raising four children had become so overwhelming that the couple felt they needed a change. They sold their New Jersey business and moved to Colorado, where they sought a simpler lifestyle in a sunnier climate.

Colborn and her husband owned pharmacies and raised sheep in western Colorado. It was during this period that Colborn first began to champion environmental causes. Her farm was located in a valley that was rich in coal. During the oil shortage of the 1970s, the coal began to be mined on a massive scale. Combined with the mining of other minerals, this led to significant damage of the local river, the Gunnison's North Fork. Colborn became active as a volunteer on western **water** issues. However, she felt hampered by the lack of official credentials. As she and her coauthors later wrote in *Our Stolen Future*, "Without a degree behind you, it was easy for opponents to dismiss you as a do-gooder, a 'little old lady in tennis shoes,' even though she was tall, middle-aged, and shod in cowboy boots."

In 1978, at age 51, Colborn entered the graduate program at Western State College of Colorado. For her master's degree in freshwater ecology, completed in 1981, she studied whether aquatic insects such as stone flies and mayflies could serve as indicators of river and stream health. For her doctoral work, Colborn moved to the University of Wisconsin at Madison, where she earned a Ph.D. degree in zoology in 1985. Her children were now grown, and her husband had died in 1983. Colborn was ready to embark upon a new stage of her life.

Promotes the Theory of Endocrine Disruption

In 1985, Colborn began a two-year stint with the Office of Technology Assessment of the U.S. Congress. As a congressional fellow and science analyst there, Colborn worked on studies of **air pollution** and water purification. Then in 1987, she joined the research team at the Conservation Foundation, a think tank in Washington, D.C. There, in the breakthrough assignment of her career, Colborn studied the environmental health of the Great Lakes. Her job involved sifting through hundreds of papers, trying to determine how well the Great Lakes were recovering from decades of acute pollution. At first, Colborn looked for a link between toxic chemicals in the lakes and **cancer** among people living in the region, but this proved to be a dead end. Yet she was still convinced that something was wrong. Colborn gradually came to believe that a disruption in hormones was the key to understanding the ill health effects seen in a long list of animals across the Great Lakes basin.

In 1988, Colborn accepted a position at the World Wildlife Fund in Washington, D.C., where she now serves as a senior scientist and director of the Wildlife and Contaminants Program. On sabbatical from 1990-1993, she served as a senior fellow of the W. Alton Jones Foundation, a private philanthropic trust. By this point, dozens of scientists around the world had begun collecting isolated pieces to the puzzle of hormonal disruption, but their work still had not been assembled into a single, coherent picture. In July 1991, Colborn helped bring together 21 key researchers from various disciplines for the Wingspread Conference in Racine, Wisconsin. Participants issued the Wingspread Statement, which warned that hormone-disrupting chemicals could jeopardize the future of humanity.

Not everyone agrees with Colborn's theory or her method of communicating it. For example, in a review of *Our Stolen Future* for *Scientific American*, Michael A. Kamrin wrote: ''The authors present a very selective segment of the data that have been gathered about chemicals that might affect hormonal functions. They carefully avoid evidence and interpretations that are not in accord with their thinking.'' Even the critics admit, however, that Colborn has been remarkably successful at raising public awareness of her theory. *Our Stolen Future* has been debated in publications ranging from *Environmental Science and Technology News* and *Science* to the *New York Times Book Review* and *Business Week*.

Colborn is in heavy demand as a speaker on environmental health issues. Her honors include the Women Leadership in the Environment Award from the United Nations Environment Programme in 1997 and the National Conservation Achievement Award in Science from the National Wildlife Federation in 1994. In her leisure time, Colborn enjoys bird-watching, a lifelong passion that has undoubtedly contributed to her choice of career path.

COLLIP, JAMES BERTRAM • See Insulin

COMMENSALISM

Commensalism (from Latin *com* meaning together and *mensa* meaning table) is a form of **symbiosis**, or close association between **organism**s of two or more **species**, in which one participant in the relationship benefits, and the relationship is neutral for the other participant. This type of symbiotic relationship is often symbolized as (+, 0). Commensalism is one of three

Theodora E. Colburn. *(World Wildlife Fund. Reproduced by permission of Ms. Colburn.)*

recognized categories of symbiotic relationships. The other two are **mutualism**, in which both participants benefit, and parasitism, in which one benefits and the other is harmed. Of the three types of symbiotic relationships, commensalism seems to be the least common. In addition, some scientists think it is unlikely that one species is completely unaffected by the relationship. Thus, examples of what at first appear to be commensalism may not prove to truly be commensalism upon more in-depth study. One example of commensalism occurs in the ocean. Certain types of barnacles attach to whales and obtain a free ride as the whale swims. Since barnacles rely on currents to bring them food that they can filter out of the **water**, movement by the whale greatly benefits the barnacles. Because of the whale's movements, the barnacles are always in a new environment with a new supply of food. The whale, however, does not appear to be helped or harmed by this relationship, and thus it is considered commensalism. There are other examples of commensalism throughout the world. Orchids, which are **plants** found in tropical areas, have such a commensal relationship with the **tree**s on which they live. The orchids benefit from the environment provided by these trees, while the trees do not appear to be helped or harmed. Cattle egrets are birds that live near grazing cattle. They seem to benefit from living with the cattle, because the cattle disrupt insects in the grasses, providing the birds with a readily available food source. There seems to be no cost or benefit to the cattle from this relationship. The pilotfish is a marine species that usually swims just in front of sharks. When the shark feeds, the pilotfish benefits by picking up scraps. There does not appear to be any effect, either positive or negative, to the shark. Some scientists even believe that commensal relationships exist between humans and a variety of species of **bacteria** living within them. For example, bacteria within human **digestive system**s benefit from the food that is eaten. However, in most cases the human neither benefits nor is harmed from this relationship. Occasionally, however, when these bacteria get into the wrong part of the human body, as is the case when *E. coli* is ingested, they can do serious harm. In such cases, because one participant is being harmed, the relationship would no longer be considered commensalism. It would be parasitism.

COMMUNITY

A community is made up of all of the **populations** of different **species** living in a specific environment. A community is only the living components of the environment. For example, all of the **algae**, **plants**, frogs, fish, and other **organism**s living in and around a pond make up a pond community. The most abundant organism in the community is known as the dominant member of that community.

The species living within a community interact together in numerous different ways. These interactions play an important role in shaping the size and structure of the community. Specifically, the interactions help determine what species are present in the community, and the size of each species' population. These interactions often result in **evolution** of the species involved and changes in the community. The interactions fall into three general categories: **competition**, **predation** and **symbiosis**.

When individuals of the same or different species share a limited resource, competition will result. Competition can have important effects on community structure. If the resource is severely limited, competition may result in **death** of some organisms, reducing population size. If competition is between individuals of different species, one species might eventually outcompete the other. The outcompeted species either would have to shift to another resource or would be excluded from that community. As a result of competition, each species has a **niche**, a specific job or role in a community, and in general, the niche of one species does not overlap with that of another species.

Predation occurs when one organism eats another live organism. As a result of predatory interactions, both the populations of the predator, the organism doing the eating, and the prey, the one being eaten, may evolve. The predator population would evolve better mechanisms to capture the prey. At the same time, the prey population might evolve better ways of escaping the predator. The numbers of both predators and prey will influence community structure. If there are many predators in a particular community, the numbers of prey will likely decrease, because so many are eaten. This could de-

crease the competition between prey species, resulting in increased species diversity within the community. Similarly, if there are only a few prey available, the numbers of predators will likely fall.

Symbiosis is a close association and interaction between organisms of two or more species within the same community. The interactions can be beneficial, cause harm or have no effect at all. Some symbiotic relationships have resulted in evolutionary changes in the organisms involved, and thus changes in the community.

Since there are so many members within a community, there are countless interactions. These interactions are very complex. As a result, it is often very difficult for scientists to figure out all of the interactions within a community and determine the exact causes of a particular community's structure.

COMPETITION

Competition is the situation that arises when two or more **organisms** require the same limited resource. The organisms involved can be of the same **species** (intra-specific competition) or from different species (inter-specific competition). Competition can occur over **nutrients**, **light**, space, a mate or any other item that an organism requires to successfully complete its **life cycle**.

Competition can be resolved in several ways, including the **death** of one of the competitors, the utilization of other resources by one or both of the competitors, the death of one species, the **migration** of individuals or of species, different temporal utilisation of the available resources, physiological changes which allow a different set of resources to be used, or both competitors living together but each suffering from a reduced **fitness** and viability.

When species compete for resources it is ultimately to the detriment of one or other of the species involved, sometimes both. It is for this reason that two different species rarely occupy identical ecological **niche**s. The closer the ecological niche the more fierce is the competition. This is known as the **competitive exclusion principle**, by the action of fierce competition two species will not inhabit the same area.

Competition is a powerful driving force in **evolution**, allowing only the most suitably adapted organism or species to survive when resources are limited. Competition ensures **survival of the fittest**. It must be remembered that an organism which is the fittest at one time and place may be out competed if the conditions change or if the organism moves to a new environment, competition is constantly occurring and any organism can be replaced by another if the ability to compete successfully is lost.

Competition can also occur at a biochemical level where two different **molecules** attempt to bind exclusively at the same site on a third molecule. An example of this form of competition is that seen between **carbon monoxide** and oxygen for the same binding site on a molecule of hemoglobin. With this example the hemoglobin binds preferentially and permanently to the carbon monoxide, reducing the ability of the **blood** to carry oxygen around the body.

See also Darwin, Charles; Malthus, Thomas Robert

COMPETITIVE EXCLUSION PRINCIPLE

In brief, the competitive exclusion principle states that two **species** with the same requirements cannot coexist in the same place. Though difficult to observe directly in **nature**, **competition** continues to be studied by biologists in the field and laboratory. One way in which this field work is done is to look at competition between two species and then remove one or the other; in a number of studies that removal resulted in a dramatic increase in the density of the other species, strong evidence of competition.

Competitive exclusion is also often labeled "Gause's Principle," from its formulation by the Russian microbiologist G.F. Gause. In a 1934 book titled *The Struggle for Existence*, Gause reported experiments on competition for common food in **protozoa**, experiments in which he tried to "approach the regularities which are characteristic for the **biosphere** as a whole." He concluded that "owing to its advantages, mainly a greater value of the coefficient of multiplication, one of the species in a mixed **population** drives out the other entirely." Gause derived his work from what have been called "the Lotka-Volterra equations," attempts to formulate an equation for the interactions between hosts and **parasites** (by Lotka) and similar attempts to present an equation covering the struggle for existence by Volterra.

But Garrett Hardin (in *Science*, April, 1960) traces the idea back to Joseph Grinnell in 1904: "Two species of approximately the same food habits are not likely to remain long enough evenly balanced in numbers in the same region. One will crowd out the others." Hardin also found similar thinking in Darwin's *Origin of Species*: "we can dimly see why the competition should be most severe between allied forms, which fill nearly the same place in the economy of nature." Hardin decided that the label "the competitive exclusion principle" was correctly descriptive of the concept.

Botanists especially have questioned competitive exclusion, because most green **plants** use similar resources. But, argument as to its validity is widespread in biology, with the arguments backed up with research. Some studies, e.g., of **bacteria**, **virus** populations, and mice, have tended to validate the concept. Other studies, e.g., of various insects and of silverside fish, have called the principle into question.

COMPUTED TOMOGRAPHY SCAN • See
Computerized axial tomography

COMPUTERIZED AXIAL TOMOGRAPHY

Since the invention of the x-ray machine, it has been apparent to physicians that an efficient and precise method for viewing

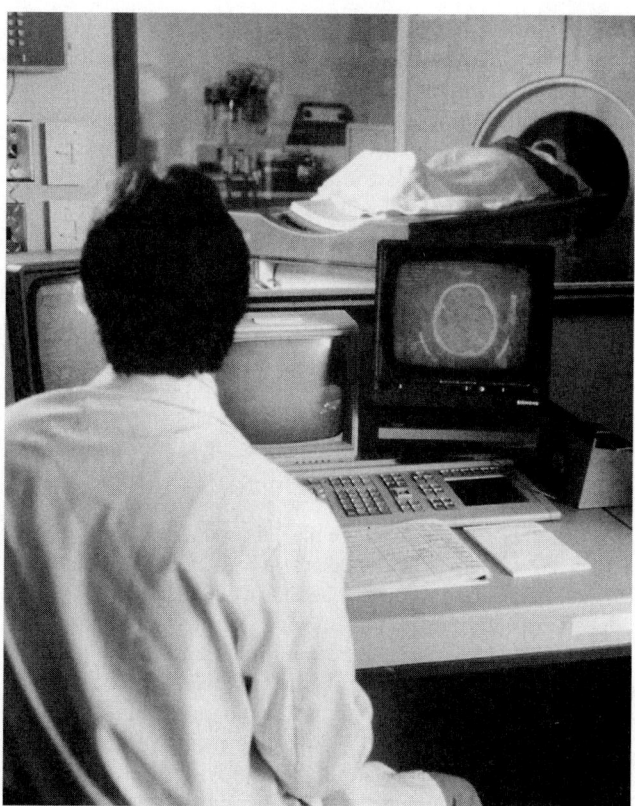

A computerized axial tomography (CT) scan in progress. CT scans use ionized radiation as do conventional x rays, but produce radiographic information of cross-sectional slices of the patient only. The typical CT scan consists of contiguous 10–millimeter-thick slices through the region requested. *(Custom Medical Stock Photo. Reproduced by permission.)*

patients' internal structures was essential. Fluoroscopes could outline bone structure and help locate foreign objects, but were not sensitive enough to show in detail organs such as the brain. During the late 1960s, **Allan Cormack**, an American, and **Godfrey Hounsfield**, an Englishman, independently developed a method called *computerized axial tomography* (CAT) scanning by which the internal structure of the body could be seen by assembling x-ray cross-sections, taken along a body axis, into a three-dimensional picture.

Cormack was the first to construct a tomographic device. His initial model used a thin beam of x-rays aimed at one section of the body but repeated from many different angles. This method allowed him to combine the many x-ray pictures into one complete view. Though Cormack published his findings and theories, he received little recognition, chiefly because he lacked a system to process the volumes of information that went into a single CAT scan.

The solution to this problem was the computer. Hounsfield began working on his own CAT scanner in 1967, using a system remarkably similar to Cormack's with one important difference: Hounsfield used computers to collate the x-ray data and create a tomographic picture. The prototype CAT scanner used gamma rays to view inside the body; they required very long exposures, and the first CAT scans (of a preserved human **brain**) took nine days to complete. This was far too long for patients to endure, so Hounsfield began to improve his design. Later models required nine hours, then nine minutes. His final apparatus could complete its x-ray scan in 4.5 minutes, with an additional 20 minutes for the computer to assemble the information. Hounsfield predicted that his scanner would not only provide the most accurate view of the body's internal structure, but that it could also identify areas of diseased tissue. The first CAT machine was installed at Atkinson Morley's Hospital in Wimbledon, England, in 1971, and Hounsfield's claims were tested on a human patient a year later. A woman, whose symptoms indicated the presence of a brain tumor, was scanned, and the results indeed showed a dark spot of diseased tissue on her brain. After several other equally successful experiments, Hounsfield patented the CAT scanner in 1972.

The design of most CAT scanners is relatively simple. A series of x-ray scanners are placed around the periphery of a tube-like cavity (large enough for a patient to insert his head, for example). Each x-ray scanner has a detector placed exactly opposite it, and both scanner and detector are built to move together so that, wherever the scanner moves, it is always aimed precisely at its detector. During the CAT scan, the x-ray scanners emit short pulses, usually lasting no more than a few milliseconds. Once this x-ray snapshot has been taken, the scanners and detectors rotate slightly (providing the same view but from a different angle) and another pulse is emitted. After all of the snapshots have been taken, a computer collates the information and constructs from it a complete picture. This can be a monumental task, since many CAT machines use up to 300 x-ray scanners taking 300 snapshots each, resulting in almost 90,000 x-ray "slices." Newer machines have also been designed to scan a patient's entire body. In the 1970s, a more advanced tomography technique was developed by Michael Phelps and Edward Hoffman, a pair of biophysicists from the UCLA School of Medicine. Their technique, called positron emission tomography (PET) uses radioactive tracers injected into a patient that emit positrons and gamma rays. The scanner "reads" the gamma rays in much the same way that x-rays are scanned in a CAT machine, but a much finer image can be obtained by tailoring the radioactive tracers used to the organs being viewed. For instance, certain tracers bond to **neurotransmitters** in the brain, so a detailed image of brain functions can be obtained.

CAT and PET scanners have significantly reduced the need for dangerous exploratory surgery, particularly on the brain. They also view soft tissue in unprecedented detail. However, tomographic technology is prohibitively expensive: the machines cost hospitals up to one million dollars each, and for this reason they have been criticized as expensive toys for wealthy establishments. However, CAT scanners are nearly one hundred times more efficient than x-ray machines, since they require less energy per view and use all of the information gathered, rather than the estimated one percent recorded by conventional x-ray machines. Cormack and Hounsfield shared a 1979 Nobel Prize for physiology or medicine for their work on computerized tomography. It is interesting to note that nei-

ther scientist held a doctoral degree at the time of their groundbreaking work. Research continues into various uses for CAT scans. One of the most recently developed uses for CAT scans is to help create a virtual three-dimensional reality via the computer that allows doctors to "fly" through areas of the body, in effect, doing a reconnaissance of diseased areas. The technique will also allow surgeons to "practice" procedures in a virtual reality setting.

CONDITIONING

Conditioning occurs when an **animal**'s behavior changes as a patterned response to a certain **stimulus**. The founder of classical conditioning was **Ivan Petrovich Pavlov**, who devoted more than 30 years to its study. Pavlov began his research career investigating **digestion**, winning a Nobel prize in 1904 for his work. Pavlov began his signature work on dogs in 1902. He restrained a hungry dog on a stand, and then rang a bell before giving it food. The hungry animal salivated freely in the presence of the food, as would be expected, however Pavlov's dog learned to associate the ringing of the bell with the imminent arrival of the meal and soon could be induced to salivate on hearing the sound alone. Pavlov called the food an unconditional stimulus, because the animal would salivate without fail, or unconditionally, when exposed to it. The ringing of the bell was thus a conditional stimulus, and the animal's **learning** process was called conditioning.

Pavlov repeated and refined such experiments for many years, using different stimuli and discovering, for example, how the conditioning could be reversed. His work was refined and extended by two American psychologists, Edward L. Thorndike and **B.F. Skinner**. Thorndike worked with cats, and did many conditioning experiments using a puzzle box. The cat was placed in a wooden box from which it could escape to its waiting food by doing a trick such as pressing a lever or pulling a loop of string. Thorndike observed that the cats would at first wander around the box purposelessly, finally letting themselves out by accident. After a few trials, the animals learned to do the trick and let themselves out of the box in only a few seconds. B.F. Skinner used a similar idea with his boxes. He placed laboratory rats in boxes that were rigged with a small lever. When the rat pressed the lever, the animal received a food pellet. Skinner also conditioned pigeons to peck at a light in order to receive food. Such conditioning is known as operant conditioning because the animal is rewarded (or punished) for performing a behavior. Once learned, the frequency of the behavior increases (or decreases) with the expectancy of the reward (or punishment).

Conditioning seems to explain more about how laboratory animals learn during laboratory experiments than how animals might act in the wild. Yet more sophisticated experiments in the 1990s have led to tantalizing insights into how the **brain** works in response to conditioning. So-called expectancy theory tries to explain what the brain believes about the immediate future. Medical researchers in the United States in the late 1980s studied such theories as the helpful effects

Ivan Pavlov (center, with beard) with his research staff and dog. *(Corbis Corporation (New York). Reproduced by permission.)*

of placebo drugs, and posited that the brain is organized to work in response to what it assumes will come next, as well as in response to what it actually senses. In the case of placebos, humans are conditioned by the sight of a doctor in a white coat, the prick of a needle, or the smell of antiseptic, for example, to expect to get relief from pain or illness. In many cases, a patient given a placebo improves just as dramatically as a patient given a proven drug because the patient is in some way conditioned to get well. Expectancy theory gained ground in the late 1990s, supported by fascinating research from around the world. Whereas the earliest conditioning experiments studied relatively simple and observable responses to stimuli, researchers in the late 1990s were able to study the brain's **chemical reaction**s and related responses in the immune, endocrine, and hormonal systems.

CONJUGATION

Conjugation is a process by which **genetic material** is transferred between two bacterial **cell**s that are temporarily joined. It is a type of genetic **recombination**, and is the bacterial version of sex.

During conjugation, two bacterial cells join through an appendage known as the sex pili. One cell is the "male" and

it is the **DNA** donor. The other is the "female" which is the DNA recipient. When they connect, a **cytoplasm**ic bridge is made between the respective cell **membrane**s. DNA in the form of a plasmid is then passed from one cell to the other through this bridge.

For conjugation to take place, the donor bacterial cell must have a resident F plasmid. This is a small, circular stretch of DNA that is separate from the bacterial **chromosome**. Plasmids are foreign pieces of genetic material that typically replicate independently. Unlike viral genetic material, they are generally beneficial to **bacteria**. The F plasmid, or fertility plasmid, contains about 25 genes that produce **proteins** related to the sex pili. At certain times, the F plasmid can become integrated into the bacterial chromosome. When this happens copies of some of the bacterial genes are transferred along with the F plasmid genes. This causes the recipient cell to have extra genes which can undergo recombination with the cell's own genes. When the cell divides via binary fission, a new colony of recombinant bacteria is formed.

CONSERVATION

The philosophy or policy that natural resources should be used cautiously and rationally so that they will remain available for future generations. Widespread and organized conservation movements, dedicated to preventing uncontrolled and irresponsible exploitation of **forests**, lands, wildlife, and **water** resources, first developed in the United States in the last decades of the nineteenth century. This was a time at which accelerating settlement and resource depletion made conservationist policies appealing both to a large portion of the public and to government leaders. Since then, international conservationist efforts, including work of the United Nations, have been responsible for monitoring natural resource use, setting up **nature** preserves, and controlling environmental destruction on both public and private lands around the world.

The name most often associated with the United States' early conservation movement is that of Gifford Pinchot, the first head of the U.S. Forest Service. A populist who fervently believed that the best use of nature was to improve the life of the common citizen, Pinchot brought scientific management methods to the Forest Service. He also brought a strongly utilitarian philosophy, which continues to prevail in the Forest Service. Beginning as an advisor to Theodore Roosevelt, himself an ardent conservationist, Pinchot had extensive influence in Washington and helped to steer conservation policies from the turn of the century to the 1940s. Pinchot had a number of important predecessors, however, in the development of American conservation. Among these was George Perkins Marsh, a Vermont forester and geographer whose 1864 publication, *Man and Nature*, is widely held as the wellspring of American environmental thought. Also influential was the work of John Wesley Powell, Clarence King, and other explorers and surveyors who, after the Civil War, set out across the continent to assess and catalog the country's physical and biological resources and their potential for development and settlement.

Conservation, as conceived by Pinchot, Powell, and Roosevelt was about using, not setting aside, natural resources. In their emphasis on wise resource use, these early conservationists were philosophically divided from the early preservationists, who argued that parts of the American wilderness should be preserved for their aesthetic value and for the survival of wildlife, not simply as a storehouse of useful commodities. Preservationists, led by the eloquent writer and champion of Yosemite Valley, John Muir, bitterly opposed the idea that the best vision for the nation's forests was that of an agricultural crop, developed to produce only useful **species** and products. Pinchot, however, insisted that "The object of [conservationist] forest policy is not to preserve the forests because they are beautiful...or because they are refuges for the wild creatures of the wilderness...but the making of prosperous homes...Every other consideration is secondary." Because of its more moderate and politically palatable stance, conservation became, by the turn of the century, the more popular position. By 1905 conservation had become a blanket term for nearly all defense of the environment; the earlier distinction was lost until it began to re-emerge in the 1960s as "environmentalists" began once again to object to conservation's anthropocentric (human-centered) emphasis. More recently deep ecologists and bioregionalists have likewise departed from mainstream conservation, arguing that other species have intrinsic rights to exist outside of human interests.

Several factors led conservationist ideas to develop and spread when they did. By the end of the nineteenth century European settlement had reached across the entire North American continent. The census of 1890 declared the American frontier closed, a blow to the American myth of the virgin continent. Even more important, loggers, miners, settlers, and livestock herders were laying waste to the nation's forests, **grasslands**, and mountains from New York to California. The accelerating, and often highly wasteful, commercial exploitation of natural resources went almost completely unchecked as political corruption and the economic power of timber and lumber barons made regulation impossible. At the same time, the disappearance of American wildlife was starkly obvious. Within a generation the legendary flocks of passenger pigeons disappeared entirely, many of them shot for pig feed while they roosted. Millions of bison were slaughtered by market hunters for their skins and tongues or by sportsmen shooting from passing trains. Natural landmarks were equally threatened—Niagara Falls nearly lost its water to hydropower development, and California's sequoia groves and Yosemite Valley were threatened by logging and grazing.

At the same time, post-Civil War scientific surveys were crossing the continent, identifying wildlife and forest resources. As a consequence of this data gathering, evidence became available to document the depletion of the continent's resources, which had long been assumed inexhaustible. Travellers and writers, including John Muir, Theodore Roosevelt, and Gifford Pinchot, had the opportunity to witness the alarming destruction and to raise public awareness and concern. Meanwhile an increasing proportion of the **population** had come to live in cities. These urbanites worked in occupations not directly dependent upon resource exploitation, and they were sympathetic to the idea of preserving public lands for recreational interests. From the beginning this urban population provided much of the support for the conservation movement.

As a scientific, humanistic, and progressive policy, conservation has led to a great variety of projects. The development of a professionally trained forest service to maintain national forests has limited the uncontrolled "**tree** mining" practiced by logging and railroad companies of the nineteenth century. Conservation-minded presidents and administrators have set aside millions of acres of public land for national forests, parks, and other uses for the benefit of the public. A corps of professionally trained game managers and wildlife managers has developed to maintain game birds, fish, and **mammals** for public recreation on federal lands. (For much of its history, federal game conservation has involved extensive predator elimination programs, however several decades of protest have led to more ecological approaches to game management in recent decades.) During the administration of Franklin D. Roosevelt, conservation projects included such economic development projects as the Tennessee Valley Authority (TVA), which dammed the Tennessee River for flood control and electricity generation. The Civilian Conservation Corps developed roads, built structures, and worked on erosion control projects for the public good. During this time the Soil Conservation Service was also set up to advise farmers in maintaining and developing their farmland.

At the same time, voluntary citizen conservation organizations have done extensive work to develop and maintain natural resources. The Izaak Walton League, Ducks Unlimited, and scores of local gun clubs and fishing groups have set up game sanctuaries, preserved **wetlands**, campaigned to control **water pollution**, and released young game birds and fish. Other organizations with less directly utilitarian objectives also worked in the name of conservation: the National Audubon Society, the Sierra Club, the Wilderness Society, the Nature Conservancy, and many other groups formed between 1895 and 1955 for the purpose of collective work and lobbying in defense of nature and wildlife.

An important aspect of conservation's growth has been the development of professional schools of forestry, game management, and wildlife management. When Gifford Pinchot began to study forestry, Yale had only meager resources and he gained the better part of his education at a French school of forest management in Nancy, France. Several decades later the Yale School of Forestry (financed largely by the wealthy Pinchot family) was able to produce such well-trained professionals as Aldo Leopold, who went on to develop the United States' first professional school of game management at the University of Wisconsin.

From the beginning, American conservation ideas, informed by the science of **ecology** and the practice of resource management on public lands, spread to other countries and regions. It is in recent decades, however, that the rhetoric of conservation has taken a prominent role in international development and affairs. The most visible international conservation organizations today are the United Nations Environment Program (UNEP), the Food and **Agriculture** Organization of the United Nations (FAO), and the World Wildlife Fund. In 1980 the International Union for the Conservation of Nature and Natural Resources (IUCN) published a document entitled the *World Conservation Strategy*, dedicated to helping individual states, and especially developing countries, plan for the maintenance and protection of soil, water, forests, and wildlife. A continuation and update of this theme appeared in 1987 with the publication of the UN World Commission on Environment and Development's paper, *Our Common Future*. The idea of sustainable development, a goal of ecologically balanced, conservation-oriented economic development, was introduced in this 1987 paper and has since become a dominant ideal in international development programs of the 1990s.

CONSERVATIVE AND DISPERSIVE REPLICATION • See Replication

CONSUMER

Consumers are sometimes synonymously known as **heterotroph**s ("other feeders"), although it is best to exclude decomposers from this group. Consumers include a diverse group of **organism**s that depend on **plants** and **animals** for their **nutrition**. Although most plants are **autotroph**s (primary pro-

ducers), the insectivorous plants and a few protistan **algae** are heterotrophic. Consumers cannot synthesize their own organic **matter**, and are grouped within the **food chain** as primary consumers (**herbivore**s), secondary consumers (primary **carnivore**s), tertiary consumers (secondary carnivores), and so on. An example of this type of food chain in the terrestrial environment might include rabbits, weasels, foxes, and coyotes. A similar example in an aquatic **ecosystem** might include **zooplankton**, blue-gill sunfish, large-mouthed bass, and great-blue heron. Humans are consumers who often eat both vegetable and animal matter, and are known as **omnivore**s. Detritivores are another consumer group comprised of organisms which feed on dead organic matter such as detritus.

CONTINENTAL DRIFT

Early scientists never seriously questioned the stability of the Earth's continents until 1620, when the famous philosopher Sir Francis Bacon (1561-1626) pointed out that the Atlantic coastlines of South America and Africa, if brought together, appeared to be a compatible fit. Likewise, the French naturalist **Georges de Buffon** suggested in 1750 that North America and Europe had once been joined because of the similarities of their present **plants** and **animals**.

For the next 300 years, scientists noticed other similarities between coastlines and **fossil** remains and attributed the similarities to a narrow land bridge that spanned the distance between the continents. It was thought that the land bridges had since subsided beneath the surface of the sea. It wasn't until the early 1900s when Alfred Lothar Wegener, a German geologist and climatologist, proposed a bold explanation for this phenomenon. In his 1915 book *The Origin of Continents and Oceans*, Wegener asserted that 200 million years ago, all the world's continents were part of one land mass, called *Pangaea*, or "all lands," which later drifted apart. His theory of continental drift set off a storm of scientific controversy that remained unsettled for decades.

Wegener speculated that this "supercontinent" broke up and slowly drifted apart to form the seven modern continents. Wegener supported his hypothesis with an impressive collection of data. He pointed to the similarities in ancient climatic conditions and fossil distributions on now widely separated continents, mapped similar rock types across oceans, and matched mountain chains on opposing shores.

Despite the popularity of Wegener's ideas in some scientific circles, geologists continued to debate them because he did not provide a compelling explanation for what caused the continents to move. Due to the lack of a driving mechanism, most geologists considered continental drift an unsupportable hypothesis. Unfortunately, a viable explanation did not appear until the early 1960s, about 25 years after Wegener's death.

After more than 20 years of studying the character of the seafloor, Harry Hammond Hess, an American geologist, introduced the concept of seafloor spreading in 1962. He argued that the ocean crust at mid-ocean ridges and rifts is young and that the seafloor is continually expanding and moving outward

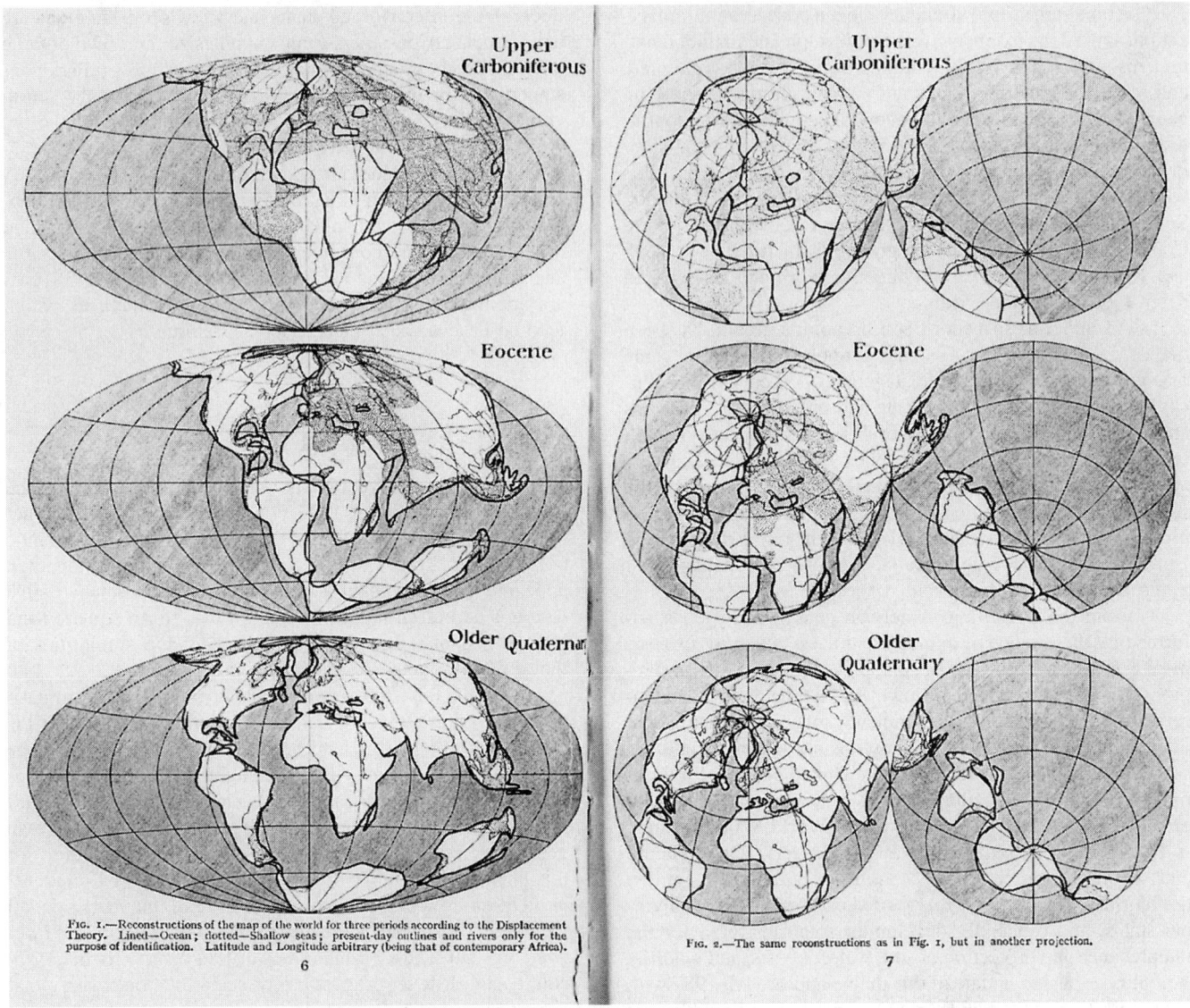

Fig. 1.—Reconstructions of the map of the world for three periods according to the Displacement Theory. Lined—Ocean; dotted—Shallow seas; present-day outlines and rivers only for the purpose of identification. Latitude and Longitude arbitrary (being that of contemporary Africa).
6

Fig. 2.—The same reconstructions as in Fig. 1, but in another projection.
7

An illustration from the English translation of Alfred Wegener's *The Origin of Continents and Oceans*, in which he proposed his theory of continental drift. *(The Library of Congress. Reproduced by permission.)*

from the ridge. This phenomenon is associated with volcanic activity that originates beneath the ocean bottom where molten rock periodically boils out of the ridge. With each underwater volcanic eruption, new seafloor forms and old seafloor ''spreads'', providing room for the new. Hess' paper synthesized many of the findings from the previous decade of marine geology research, and provided the necessary framework for further study and analysis.

This evidence also forms the basis of the new science of plate tectonics which builds upon the continental drift theory and is itself central to the new geology. Through studies at a variety of locations geologists have learned that the rigid exterior, or lithosphere, of the Earth consists of more than a dozen mobile plates floating along on the less rigid asthenosphere below. These lithospheric plates are much thicker

than the continental or oceanic crust, so as seafloor spreading pushes newly formed lithosphere away from the oceanic ridge, the continents passively drift across the surface of the Earth. In addition, other studies showed that as the lithospheric plates age, they grow denser and eventually sink, or subduct, into the asthenosphere at deep ocean trenches.

Despite the controversy over continental drift just a few decades ago, the vast majority of earth scientists now find plate tectonics theory thoroughly convincing in light of the evidence for seafloor spreading and plate subduction.

CONTINUOUS VARIATION

Continuous variation is the situation encountered when, for a given **phenotype**, all values between the two extremes are possible. Examples of this include such things as height and weight. Continuous variation within **organism**s does not display easily observable states, such as the color of eyes or hair, rather there is a steady gradation normally observed between the possible extremes.

Continuous variation is sometimes known as quantitative variation. It is common for the characteristic to have a mean value around which the **population** varies. Continuous variation is most common amongst those characteristics which are strongly affected by environmental factors and also amongst those characteristics which are under the control of many genes acting together. The larger the number of controlling genes the greater is the spread around the mean value.

Continuous variation is found within large populations. In smaller, more isolated communities it can be the case that many of the genes controlling a given character are represented only by one **allele** per gene (due to **inbreeding** and genetic drift). This gives the appearance of a characterisitc which is under the control of far fewer genes than is the actual case.

See also Discontinuous variation

COREY, ELIAS JAMES (1928-)
American chemist

Elias James Corey, a specialist in the synthesis of organic chemicals, developed many of the theories and methods that now define his field. Since his research career began in the 1950s, one of his goals has been to make the synthesis of chemicals more systematic, and he is best known for his logical approach to the creation of new substances, which he has named "retrosynthetic analysis." Corey's achievements in research, which resulted in far-reaching benefits for medicine and human health, were recognized in 1990 when he was awarded the Nobel Prize for Chemistry.

Corey was born July 12, 1928, to Fatina (Hasham) and Elias Corey in Methuen, Massachusetts. His father died eighteen months later. Corey had been named William at birth, but after his father's death his mother renamed him in his memory. Corey had three siblings, a brother and two sisters, and they were all raised by their mother, with the help of her sister and brother-in-law. His aunt and uncle actually moved in with them, and Corey still credits his mother's sister with being an important influence on his life. Corey enjoyed football and baseball as a child; he was also a good student, graduating from high school in 1945 and entering the Massachusetts Institute of Technology at the age of sixteen. He intended to study engineering but quickly developed an interest in chemistry. It was here that Corey began his research on organic synthesis, the manual formation of chemicals. He worked under John Sheehan on the organic synthesis of **penicillin** and received his Ph.D. in 1951.

After earning his doctorate, Corey accepted a position as instructor in chemistry at the University of Illinois at Champaign-Urbana. He continued his research on organic synthesis; in 1954 he became an assistant professor, and in 1955 he was named full professor of chemistry. In 1957, Corey received a Guggenheim fellowship. He left Illinois on sabbatical, and during this time he did research that laid the foundation for the rest of his career. He spent a portion of his sabbatical working on chemical synthesis with Robert B. Woodward at Harvard. The rest of the time he devoted solely to research in Europe, examining the problems of synthesizing **prostaglandins** (**hormone**-like substances with many different effects, found in the tissue of the body).

Develops Retrosynthetic Analysis

During the time Corey was a Guggenheim fellow, chemical synthesis was widely considered an intuitive process and often called an "art form." In a 1990 article in *Science,* Corey recalled his field as he found it in the 1950s: "Chemists approached each problem in an ad hoc way. Synthesis was taught by the presentation of a series of illustrative—and generally unrelated—examples of actual synthesis." It was while working with Woodward that Corey began his effort to systematize this intuitive process. Success in organic synthesis was often personal and difficult for the individual scientist to explain in full; Corey wanted the methods, as well as the results, to be both reproducible and teachable.

In 1959, Corey was offered a professorship of chemistry at Harvard, which he accepted. In his new position, he continued to search for what he has called the "deep logic" of chemical synthesis. His central innovation was to reverse the usual order of procedure by adding a planning process that began with the desired result, instead of the initial chemicals. Corey planned the process backwards from the **molecule** he wanted to synthesize, creating a chart or "tree" that included many possible compounds and reactions. This was retrosynthetic analysis, a formal system which eliminated much of the guesswork, as well as making it easier to use chemicals that were readily available or easy to synthesize. The system also made it possible to use computers for chemical synthesis, and Corey has been a pioneer in this application of artificial intelligence.

The actual results of Corey's work in chemical synthesis have been almost as important as his methodological innovations. In 1968, Corey and his colleagues were able to synthesize five different prostaglandins, which are involved in regulating many functions in the body including **blood pressure**, blood coagulation, and reproduction. Before this time only a small quantity of these substances was available, as they had to be extracted from the testes of Icelandic sheep. With scientists able to synthetically produce prostaglandins, their applications in medicine have increased profoundly. Eventually Corey's work on prostaglandins led to the development of what is now commonly known as the Corey lactone aldehyde. From this, prostaglandins of all three familial types can be derived.

Another result of Corey's work was the 1988 synthesis of ginkgolide B, a chemical naturally extracted from the ginkgo tree. This chemical is used for the treatment of asthma and circulatory problems in the elderly, and has grown to a market of over 500 million dollars a year. Besides these accomplishments, Corey also has improved or started over fifty new reactions. This has broadened the application of organic synthesis by increasing the tools available to the scientist.

Corey was awarded the 1990 Nobel Prize not for any specific scientific achievement but for his career as a whole. In conferring the award, the Nobel Prize committee said of him: "No other chemist has developed such a comprehensive and varied assortment of methods, often showing the simplicity of genius, which have become commonplace in organic synthesis laboratories." Corey continues his association with Harvard as head of an organic synthesis laboratory which operates with the help of graduate students. The "Corey research family" has contributed to the training of over one hundred-fifty university professors and an even greater number of scientists working in industry.

Corey has received eleven honorary degrees including doctorates from the University of Chicago in 1968 and Oxford University in 1982. He has received more than three dozen other awards from universities and scientific societies, including a 1971 Award for Creative Work in Synthetic Organic Chemistry, the Linus Pauling Award in 1973, and in 1988 the Robert Robinson Medal from the Royal Society of Chemistry. Corey was a member of the American Academy of Arts and Sciences from 1960 to 1968. He has been a member of the American Association for the Advancement of Science since 1966. He has served on the editorial board of several scientific journals and contributed over seven hundred articles for publication. He is co-author of a 1989 book, *The Logic of Chemical Synthesis.*

Corey was married in September 1961 to Claire Higham. He and his wife have three children (two sons and a daughter), and they reside in Cambridge, Massachusetts.

CORI, CARL F. (1896-1984)
CORI, GERTY THERESA RADNITZ (1896-1957)
Czech American biochemists

Carl and Gerty Cori were both born in Prague, Austria-Hungary (now the Czech Republic) in 1896. They entered the University of Prague's medical school at roughly the same time, each planning to become a physician. At some point during their student days, they met each other, fell in love, and teamed up to share laboratory research in biochemistry. They decided that they not only enjoyed working together but also preferred biochemical research to medical practice. But in 1920, after they married and obtained their medical degrees, only Carl was able to find a research position. It was in Vienna, Austria, and Gerty joined the staff of a hospital there, but they still longed to work together. In 1922, when Carl was offered a position as a biochemist at the New York Institute for the Study of Malignant Diseases in Buffalo, New York, he immigrated to the United States. Gerty followed him a few months later when a job opened up for her in the same institution.

At the New York Institute, the Coris did some research on the **metabolism** of abnormal growths but spent most of their time investigating the way normal healthy bodies utilize sugars and starches. The research team was particularly intrigued by two **hormones**—epinephrine and the recently discovered **insulin**—and their roles in **carbohydrate** metabolism.

In a series of papers published during the 1920s, the Coris provided the scientific world with a great deal of information about what happened to sugars after they were absorbed by experimental white rats. Among other things, they reported, normally about half the absorbed sugar (now called glucose) is converted to glycogen and stored in this form in the liver and muscles, with the rest either stored as fat or burned as fuel. The administration of insulin, however, not only decreases the amount of sugar stored in the liver, but increases its utilization elsewhere.

In 1932, the two scientists joined the faculty of Washington University in St. Louis, Missouri, where they were able to probe even more deeply into the mysteries of carbohydrate metabolism. The work of another biochemist, **Otto Meyerhof**, had already established the fact that, when muscles contract, the glycogen stored in them is somehow converted to **lactic acid**. The Coris wanted to find out exactly how the glycogen is broken down and how, after its conversion to lactic acid, it is then resynthesized into glucose. In their new laboratory, they were able to find the answers.

Using minced frog muscles to help them in their investigations, the Coris were soon able to isolate and identify a sugar and phosphate compound, previously unknown, which they named glucose-1-phosphate (often called Cori ester in their honor.) This discovery, plus their discovery of two new **enzymes**, helped them disprove a widely held belief—that the highly branched glycogen molecule breaks itself down to glucose molecules by adding water molecules at each of its many links. Although this breakdown process seemed simple and logical, the Coris pointed out it would also lead to a pronounced **energy** loss that would have impaired the eventual resynthesis. Instead of using water, glycogen—helped by one of the enzymes they discovered—adds inorganic phosphate at each of its links to form the newly-discovered phosphate-containing compound (which involved less energy loss) and then undergoes a long series of chemical changes before it is finally broken down. The Coris patiently detailed each of the changes and then went on to outline the resynthesis process—in time, actually managing to synthesize the glycogen in a test tube.

For their work on glycogen, the Coris shared (with the Argentinean **Bernardo Houssay**) the 1947 Nobel Prize in physiology or medicine. Gerty Cori was the third woman to be awarded a Nobel Prize in a scientific field. The other two were Marie Curie and her daughter, Irène Joliot-Curie. Other honors the couple shared included membership in the American Society of Biological Chemists, the National Academy of Sciences, the American Chemical Society, and the American Philosophical Society. They were joint recipients of the Midwest Award of the American Chemical Society in 1946 and the Squibb Award in Endocrinology in 1947.

CORMACK, ALLAN M. (1924-)
South African American physicist

Allan M. Cormack is a physicist whose theoretical analysis and experiments in the fields of nuclear and particle physics, computer tomography and math led to his invention of a mathematical technique for computer-assisted x-ray tomography. **Computerized axial tomography**, otherwise known as the CAT scan, is a process by which x rays can be concentrated on specific sections of the human body at a variety of angles. Once this information is analyzed by a computer, it is combined to reproduce images of internal structures previously unviewable by medical technology. It is considered the most revolutionary development in the field of radiography since the discovery of the x ray by **Wilhelm Conrad Röntgen** in 1895. Cormack was the first to analyze the possibility of such an examination of a biological system, in 1963 and 1964, and to develop the equations needed for computer-assisted x-ray reconstruction of pictures of the human brain and body. In 1979 Cormack was awarded the Nobel Prize for physiology or medicine, along with **Godfrey Hounsfield**, a British engineer who, independently of Cormack, developed the first commercially successful CAT scanning devices.

Allan MacLeod Cormack was born in Johannesburg, South Africa, on February 23, 1924, the son of George and Amelia (MacLeod) Cormack, a civil service engineer and a teacher respectively, who had emigrated from Scotland to South Africa prior to World War I. Young Cormack attended the Rondebosch Boys High School. At the University of Cape Town, South Africa, Cormack chose the field of engineering, but two years later he changed his major to physics, completing a baccalaureate of science in 1944.

He remained at the University of Cape Town, completing a Master of Science degree in the field of crystallography in 1945. During the years that followed, Cormack became a lecturer in physics at the University of Cape Town and pursued graduate studies in the field of theoretical physics for two years at Cambridge University in England. Working as a research student in the university's Cavendish Laboratory, he studied radioactive helium under the tutelage of Otto Robert Frisch. He also attended lectures on quantum physics given by Nobel Prize winner Paul Dirac.

In 1950 Cormack returned to South Africa from Cambridge to resume his position as a lecturer in physics at the University of Cape Town, where he would remain until 1956. During this period he was asked to serve a six-month service as resident medical physicist in the radiology department at the Groote Schuur Hospital in Cape Town, where he supervised the use of radioisotopes as well as the calibration of film badges used to measure hospital workers' exposure to radiation.

At Groote Schuur, Cormack witnessed first hand how radiation was being used in the diagnosis and treatment of **cancer** patients. Baffled by deficiencies in the technology used for such procedures, Cormack began a series of experiments and analyses, the results of which were two papers published separately between 1963 and 1964 in the *Journal of Applied Physics*. Cormack also conducted theoretical physics research in Boston on subatomic particles, following a 1956 Harvard University sabbatical as a Research Fellow, where he worked in the cyclotron laboratory under director Andreas Koehler.

Following a brief return to Cape Town in 1957, Cormack returned to the United States, accepting a post as assistant professor of physics at Tufts University in Medford, Massachusetts. Between 1956 and 1964, most of his research in connection with the development of computerized axial tomography was conducted on his own time. Neither of his two *Journal of Applied Physics* papers met with significant response, despite the fact that they proved the feasibility of his method for producing images of heretofore unviewable or barely viewable cross sections of the human body.

Cormak was naturalized as a citizen of the United States in 1966, and continued his academic career and his research in particle physics at Tufts. He was eventually promoted to associate and then full professor of physics, serving as chairman of the physics department from 1968 to 1976. Meanwhile, Hounsfield was independently coming to conclusions similar to Cormack's, and developed the first CAT scanner as early as 1972.

In 1979 Cormack and Hounsfield were awarded the Nobel Prize for physiology or medicine for their joint, though independent, development of CAT scan theory and technology. At the time, their selection as recipients of the prize was considered highly unusual. Unlike previous Nobel recipients, neither Cormack nor Hounsfield held a doctorate in medicine or science; further, their discovery was awarded the prize only after the Nobel Assembly vetoed the first choice of the selection committee; and, finally, it was highly unusual that the two men had never met or worked together, yet had worked on the same invention concurrently.

In 1980, Tufts appointed Cormack to university professor, its highest professorial rank, and awarded him with an honorary doctoral degree. In 1990, as one of several scientists receiving the National Medal of Science, Cormack was recognized by President George Bush. Cormack is a member of the National Academy of Science and the American Academy of Arts and Sciences, and is a fellow of the American Physical Society.

CORPUS LUTEUM

The corpus luteum is a temporary **endocrine gland** in the ovary necessary for the production of progesterone to prepare the **endometrium** (lining of the uterus) for implantation of a fertilized **egg** (blastocyst) and to sustain pregnancy. The corpus luteum maintains a healthy and **nutrition**al endometrial environment during embryonic development until the **placenta** is developed enough to produce progesterone in sufficient quantities to take over the maintenance of pregnancy. If the amount of progesterone secreted by the corpus luteum is insufficient or continues for too short a period, the developing fetus is unlikely to survive. This is called a luteal phase defect (LPD).

The corpus luteum—literally translated as yellow body—is actually the remnant of a follicle, a protective group of special **cell**s which surrounds each individual **oocyte** housed within the ovaries. The **metamorphosis** from follicle to corpus luteum takes place immediately following the ovulation stage of a woman's sexual cycle but begins about 14 days earlier during the follicular phase. In the follicular phase, luteinizing hormones (LH) and follicle stimulating hormones (FSH) released by the **pituitary gland** cause follicles, housed within the ovaries, to secrete **estrogen**. Several follicles begin to grow quickly, secreting increasing amounts of estrogen. One follicle, however, will grow more quickly than the rest, starving them of estrogen and maturing in approximately 14 days. It is from this follicle—the Graafian follicle—that the mature egg bursts forth and moves down the fallopian tube to the uterus. Once the egg is released, the empty follicle—previously containing follicular fluids—now fills with clotted **blood**. The follicle undergoes luteinization, changing dramatically in shape and color, and luteal cells begin to form. This transformation of cells is called **differentiation**. The follicle has now become the corpus luteum which consists of many blood vessels (70-80% of the blood which flows through the ovaries flows to the corpus luteum), and both small and large cells. Small cells contain receptors for LH, large cells do not. This means the large cells cannot be stimulated to secrete progesterone, even though they actually produce most of the progesterone secreted by the corpus luteum. The corpus luteum reaches peak performance within four days of development, continuing to grow for another four or five days. It stores **cholesterol**, secretes the steroid hormones estrogen and progesterone, and produces the **peptide** hormones oxytocin (which stimulates uterine muscle contractions and the production of breast milk), IGF-1 (a growth stimulator), and relaxin (which allows ligaments to relax during childbirth). Because estrogen causes uterine contractions, the amount of progesterone secreted by the corpus luteum must be great enough to completely override this effects of estrogen. The progesterone quiets contractions, allowing implantation of the blastocyst. Once implanted, the blastocyst begins to secrete a special hormone, **chorion**ic gonadotropin, a signal that it is now nestled in the endometrium and ready to grow. This signal stimulates the corpus luteum to continue its production and **secretion** of progesterone.

When an egg remains unfertilized, the corpus luteum regresses, stops producing hormones, shrivels up, and decomposes. This process takes approximately 12 to 16 days. The termination of luteal function and subsequent lowering of hormone levels once again triggers the release of LH and FSH, which trigger menstruation, marking the beginning of the next sexual cycle.

See also Menstrual cycle; Reproduction, human

CORTISONE

Cortisone is one of several steroid **hormones** secreted by the cortex (outer covering) of the **adrenal gland**. These hormones, called corticoids, are classified according to their functions, glucocorticoids controlling sugar **metabolism** and mineralocorticoids controlling the metabolism of **minerals** and **water**. The principal glucocorticoids are corticosterone and hydrocortisone (cortisol) and the principal mineralocorticoid is aldosterone. Cortisone is in both categories, because it quickly converts **protein** to the **carbohydrate** glucose and it helps regulate salt metabolism. Cortisone also helps the body withstand **stress**. It is used medically to reduce **inflammation**.

The adrenal cortex's production of cortisone is controlled by the hormone ACTH (**adrenocorticotropic hormone**), which is secreted by the **pituitary gland**. The pituitary, in turn, responds to corticotropin-releasing factor, a hormone-like substance produced by the portion of the **brain** called the **hypothalamus**.

Knowledge of cortisone is due primarily to three scientists, the Swiss chemist **Tadeus Reichstein** and the Americans **Edward Kendall**, a biochemist, and **Philip Showalter Hench**, a medical researcher. Kendall first began work on adrenal cortex hormones because an extract had been used successfully against Addison's disease, which is caused by adrenal gland dysfunction. The original hormone theory, developed by the British physiologists **William Bayliss** and **Ernest Starling**, held that each type of gland secreted only one hormone. But by the mid-1930s, Kendall and others believed that the adrenal gland produced many hormones. In 1936 Reichstein was the first to isolate what later was named cortisone. Kendall isolated a series of adrenal substances and converted the one he called Compound E into an active substance. He deduced that it was a steroid.

Hench and Kendall studied Compound E for possible use in treating **arthritis**. In 1948 and 1949, Hench and Kendall gave the name cortisone to Compound E, and the next year Hench and another colleague were the first to use it to successfully treat arthritis. For their work with cortisone and other adrenal hormones, Reichstein, Hench, and Kendall shared the 1950 Nobel prize in physiology or medicine.

Soon after cortisone's first successful medical use, treatments were discontinued. Researchers found that rheumatoid arthritis is not caused by hormone deficiency, and cortisone treatments have some serious side effects. These included edema (fluid retention), high stomach acidity, damage to bone, and abnormal metabolism of sodium, potassium, and **nitrogen**. Further experiments, however, yielded a refined product that reduced the side effects.

Cortisone (17-hydroxy-11-dehydrocorticosterone) has been synthesized by several different methods. It was originally derived from the bovine bile constituent deoxycholic acid in a costly 37-step process. A less expensive mass-production method was developed in 1948 by the American chemist Percy Julian, who was widely known for synthesizing chemicals from soybeans. Julian had already synthesized a substance that he called cortexolone, which was very similar to cortisone, except that it had one less oxygen atom. Julian added oxygen to his cortexolone, turning it into cortisone at a greatly reduced production cost.

The American chemist Carl Djerassi (1923-) in 1951 was the first to synthesize cortisone from raw plant materials—yams and sisal (diosgenin and hecogenin). That same year, the American chemist Robert Burns Woodward synthesized cortisone from orthotoluidine, a coal-tar derivative whose structure was one carbon ring with an attached methyl (CH_3) group. By adding various groups of **atoms**, Woodward transformed it into the basic steroid pattern of four rings and two methyl groups. Though twenty steps were involved, Woodward's process was considerably more streamlined than that for converting bovine bile.

Today cortisone is prescribed to reduce inflammation in allergy and in arthritis and other connective **tissue** diseases. It is also prescribed as a replacement hormone in Addison's disease, and for people whose **adrenal glands** have been removed. Other uses include **cancer** therapy, asthma treatment, and reduction of the body's immune response to prevent rejection of transplanted organs.

COURNAND, ANDRÉ F. (1895-1988)
French American physician

André F. Cournand shared the 1956 Nobel Prize in physiology or medicine with German surgeon **Werner Forssmann** and American physiologist **Dickinson Woodruff Richards, Jr.** for pioneering work in the field of cardiac and pulmonary physiology. Cournand helped develop the technique of cardiac catheterization, which permits **blood** samples to be obtained from the **heart** for determining cardiac abnormalities.

Cournand was born in Paris on September 24, 1895. His father, Jules Cournand, and his grandfather were both dentists. Cournand writes in his autobiography, *From Roots to Late Budding: The Intellectual Adventures of a Medical Scientist,* that his decision to study the sciences and medicine stemmed from his father's regrets of his own choice of dentistry over medicine. At age 15, young André began to accompany his parents to the salon of a physician friend where many internationally known scientists met and discussed issues of their day. Cournand's mother, Marguerite Weber Cournand, loved literature and learning and encouraged in her son a deep interest in philosophy and art, which Cournand maintained even while pursuing his medical studies and research.

In 1913, Cournand received his bachelor's degree from the University of Paris-Sorbonne, where he also began his medical studies in 1914. But in that year the first World War broke out, and many medical professors enlisted in the army. In the spring of 1915, Cournand decided to postpone his studies. In July of that year he joined a surgical unit that provided emergency care on the front lines. By 1916 he was trained as an auxiliary battalion surgeon and was serving in the trenches. He didn't return to medical school until 1919. After serving as an intern, he received his M.D. in 1930.

Begins Investigative Work in the United States

Cournand had decided to specialize in upper respiratory diseases and, delaying his entry into private practice, pursued further training in the United States. He joined a residency program at the Tuberculosis Service of the Columbia University College of Physicians and Surgeons at Bellevue Hospital in New York City. He stayed at Columbia for the remainder of his career, rising from his initial position as investigator to a full professor in 1951. He became a naturalized citizen of the United States in 1941.

At Bellevue Cournand began what would become a long collaboration with Dickinson W. Richards. Together, they investigated the theories of a Harvard physiologist, Lawrence J. Henderson, who had postulated that the heart, lungs, and **circu-**

latory system are a functional unit designed to transport respiratory gases from the atmosphere to the tissues in the body and back out again.

In order to study respiratory gases and their concentrations in the blood as it passed through the heart, samples of blood from the heart had to be obtained. At this time, there was no established technique for this task. Catheters—flexible tubes intended to introduce and remove fluids from organs—had been used for the past 100 years, but only in animal experiments. The safety of catheter use in humans was doubtful. But Cournand was aware that in 1929 a German scientist, Werner Forssmann, had dramatically demonstrated the safety of cardiac catheterization by performing it on himself. He had inserted a catheter into one of his arm veins and then threaded it into his right atrium. Cournand became convinced of the safety of catheterization after speaking with one of his professors in Paris who had also performed a type of catheterization on himself, and subsequently scores of others, without any problems.

The Bellevue team experimented on animals for four years, working to standardize the procedure and perfect the equipment they were convinced was necessary for their studies of the cardiac system. When at last cardiac catheterization was used to obtain a sample of mixed venous blood in humans, what could previously be only vaguely determined by clinical observation could be physiologically described. Cardiac catheterization not only allows for samples of mixed venous blood to be collected, but it also measures **blood pressure** in various parts of the cardiac circulatory system—the right atrium, the ventricles, and the arteries—and measures total blood flow and gas concentrations. In short, the functions of the heart and lungs can be fully specified through cardiac catheterization.

During World War II, Cournand led a team of physicians investigating the use of cardiac catheterization on patients suffering from severe circulatory shock resulting from traumatic injury. Obtaining physiological measurements of cardiac output in these patients helped identify the cause of shock—a fall in cardiac output and return. As a result of these findings, it was determined that the best treatment for shock was a total blood transfusion rather than simply replacing plasma, which had previously been used and was found to cause anemia.

After the war, Cournand applied the technique of cardiac catheterization to patients with heart and pulmonary diseases. The team continually worked to improve the technique and was able, at this time, to obtain simultaneous readings of blood pressure in the right ventricle and the pulmonary artery. This allowed for greater diagnostic accuracy of congenital defects as well as evaluations of treatment. Eventually these investigations led to increased understanding of acquired heart diseases and the relation between diseases of the lungs and cardiac function, thus opening up the field of pulmonary heart diseases.

Career Recognized

Cournand began to be recognized for his research in the mid–1940s, when he was invited to speak at and lead various conferences. In 1949 he won the Lasker Award, and in 1952 he was invited by the National Institutes of Health to screen grant applications for the Lung, Heart and Kidney Study Section. Cournand's increasing recognition culminated in the fall of 1956 when he was awarded the Nobel Prize. In 1958 he was elected to the National Academy of Science.

During his years of research, Cournand remained interested and involved in the arts. While still in Paris, he had become a follower of the modern art movement and was friends with such painters as Jacques Lipschitz and Robert Delaunay and such writers as Andre Breton. In 1924 he married Sibylle Blumer, a daughter of Jeanne Bucher, who was a prominent gallery owner in Paris. They were married until her death in 1959. In 1963 Cournand married Ruth Fabian, who died in 1973. He was married again, to Beatrice Bishop Berle, in 1975. He had four children, three daughters and an adopted son.

Cournand retired in 1964 and devoted the years until his death to the study of the social and ethical implications of modern science. He died on February 19, 1988, in Great Barrington, Massachusetts.

COUSTEAU, JACQUES (1910-1997)
French oceanographer

Jacques Cousteau was known worldwide through his television programs, feature-length films, and books, all of which have focused on the wonders and tragedies of the marine world. Through these films and publications, Cousteau helped demystify undersea life, documenting its remarkable variety, its interdependence, and its fragility. Through the Cousteau Society, which he founded in the 1970s, he led efforts to call attention to environmental problems, to reduce marine pollution, and perhaps most importantly, to bring lasting peace to the world.

Jacques-Yves Cousteau was born in St. André-de-Cubzac, France, on June 11, 1910, to Elizabeth Duranthon and Daniel Cousteau. At the time, the elder Cousteau worked as a legal adviser to American entrepreneur James Hyde, founder of the Equitable Life Assurance Society. In 1917, after a heated argument, Hyde fired Daniel and the family briefly fell on hard times, their problems compounded by the poor health of Jacques, who for the first seven years of his life suffered from chronic enteritis, a painful intestinal condition. In 1918, after the Treaty of Versailles, Daniel found work as legal adviser to Eugene Higgins, a wealthy New York expatriate. Higgins traveled extensively throughout Europe, with the Cousteau family in tow. Cousteau recorded few memories from his childhood; his earliest impressions, however, involve water and ships. His health greatly improved around this time, thanks in part to Higgins, who encouraged young Cousteau to learn how to swim.

In 1920 the Cousteaus accompanied Higgins to New York City. Here, Jacques attended Holy Name School in Manhattan, learning the intricacies of stickball and roller skating. He spent his summers at a camp on Vermont's Lake Harvey, where he first learned to dive underwater. At age thirteen, after

a trip south of the American border, he authored a hand-bound book he called "An Adventure in Mexico." That same year, he purchased a Pathé movie camera, filmed his cousin's marriage, and began making short melodramatic films.

During his teens, Cousteau was expelled from a French high school for "experimenting" on the school's windows with different-sized stones. As punishment, he was sent to a military-style academy near the French-German border, where he became a dedicated student. He graduated in 1929, unsure of which career path to follow. The military won out over film-making simply because it offered the opportunity for extended travel. After passing a rigorous entrance examination, he was accepted by the Ecole Navale, the French naval academy. His class embarked on a one-year world cruise, which he documented, filming everything and everyone—from Douglas Fairbanks, the famous actor, to the Sultan of Oman. After graduating second in his class in 1933, he was promoted to second lieutenant and sent to a naval base in Shanghai, China. His assigned duty was to survey and map the countryside, but in his free time he filmed the locals in China and Siberia.

In the mid–1930s, Cousteau returned to France and entered the aviation academy. Shortly before graduation, in 1936, he was involved in a near-fatal automobile accident that mangled his left forearm. His doctors recommended amputation but he steadfastly refused. Instead, he chose rehabilitation, using a regimen of his own design. He began taking daily swims around Le Mourillon Bay to rehabilitate his injured arm. He fell in love with goggle diving, marveling at the variety and beauty of undersea life. He later wrote in his book *The Silent World:* "One Sunday morning... I waded into the Mediterranean and looked into it through Fernez goggles.... I was astonished by what I saw in the shallow shingle at Le Mourillon, rocks covered with green, brown and silver forests of algae and fishes unknown to me, swimming in crystalline water.... Sometimes we are lucky enough to know that our lives have been changed, to discard the old, embrace the new, and run headlong down an immutable course. It happened to me at Le Mourillon on that summer's day, when my eyes were opened on the sea."

During his convalescence he met seventeen-year-old Simone Melchior, a wealthy high-school student who was living in Paris. After a one-year courtship, the couple married on July 12, 1937, and moved into a house near Le Mourillon Bay. The Cousteaus' first son, Jean-Michel, was born in March of 1938. A second son, Philippe, was born in December of 1939. Around this time, the new family's tranquil life on the edge of the sea was threatened by world events. In 1939 France began preparing for war, and Cousteau was promoted to gunnery officer aboard the *Dupleix*. The war was largely limited to ground action, however, and Germany quickly overran the ill-prepared French Army. Living in the unoccupied section of France enabled Cousteau to continue his experiments and allowed him to spend many hours with his family. In his free time, he experimented with underwater photography devices and tried to develop improved diving apparatuses. German patrols often questioned Cousteau about his use of diving and photographic equipment. Although he was able to convince authorities that the equipment was harmless, Cousteau was, in fact, using these devices on behalf of the French resistance movement. For his efforts, he was later awarded the Croix de Guerre with palm.

Undersea Work Leads to Development of the Aqualung

Although he loved diving, Cousteau regretted the limitations of goggle diving; he simply could not spend enough time under water. The standard helmet and heavy suit apparatus had similar limitations; the diver was helplessly tethered to the ship, and the heavy suit and helmet made Cousteau feel like "a cripple in an alien land," as cited in *Contemporary Authors New Revision Series*. A number of experiments with other diving equipment followed, but all the existing systems proved unsatisfactory. He designed his own "oxygen re-breathing outfit," which was less physically constrictive but which ultimately proved ineffective and dangerous. Also during this period he began his initial experiments with underwater filmmaking. Working with two colleagues, Philippe Taillez, a naval officer, and Frédéric Dumas, a renowned spearfisherman, Cousteau filmed his first underwater movie, *Sixty Feet Down*, in 1942. The eighteen-minute film reflects the technical limitations of underwater photography but was quite advanced for its time. Cousteau entered the film in the Cannes Film Festival, where it received critical praise and was purchased by a film distributor.

As pleased as he was with his initial efforts at underwater photography, Cousteau realized that he needed to spend more time underwater to accurately portray the ocean's mysteries. In 1937 he had begun a collaboration with Emile Gagnan, an engineer with a talent for solving technical problems. In 1942 Cousteau again turned to Gagnan for answers. The two spent approximately three weeks developing an automatic regulator that supplied compressed air on demand. This regulator, along with two tanks of compressed air, a mouthpiece, and hoses, was the prototype Aqualung, which Gagnan and Cousteau patented in 1943.

That summer, Cousteau, Talliez, and Dumas tested the Aqualung off the French Riviera, making as many as five hundred separate dives. This device was put to use on the group's next project, an exploration of the *Dalton,* a sunken British steamer. This expedition provided material for Cousteau's second movie, *Wreck.* The film deeply impressed French naval authorities, who recruited Cousteau to assist with the dangerous task of clearing mines from French harbors. When the war ended, Cousteau received a commission to continue his research as part of the Underwater Research Group, which included both Talliez and Dumas. With increased funding and ready access to scientists and engineers, the group expanded its research and developed a number of innovations, including an underwater sled.

In 1947 Cousteau, using the Aqualung, set a world's record for free diving, reaching a depth of 300 feet. The following year, Dumas broke the record with a 306-foot dive. The team developed and perfected many of the techniques of deep-sea diving, working out rigorous decompression schedules that

Jacques Cousteau. *(The Library of Congress. Reproduced by permission.)*

enabled the body to adjust to pressure changes. This physically demanding, dangerous work took its toll; one member of the research team was killed during underwater testing.

Begins World Adventures Aboard *Calypso*

On July 19, 1950, Cousteau bought *Calypso,* a converted U.S. minesweeper. On November 24, 1951, after undergoing significant renovations, *Calypso* sailed for the Red Sea. The *Calypso* Red Sea Expedition (1951–52) yielded numerous discoveries, including the identification of previously unknown **plant** and **animal** species and the discovery of volcanic basins beneath the Red Sea. In February of 1952, *Calypso* sailed toward Toulon. On the way home, the crew investigated an uncharted wreck near the southern coast of Grand Congloué and discovered a large Roman ship filled with treasures. The discovery helped spread Cousteau's fame in France. In 1953, with the publication of *The Silent World,* Cousteau achieved international notice. The book, drawn from Cousteau's daily logs, was written originally in English with the help of U.S. journalist James Dugan and later translated into French. Released in more than twenty languages, *The Silent World* eventually sold more than five million copies worldwide.

In 1953 Cousteau began collaborating with Harold Edgerton, a pioneer in high-speed photography, who had invented the strobe light and other photographic devices. Edgerton and his son, William, spent several summers aboard *Calypso,* outfitting the ship with an innovative camera that skimmed along the ocean floor, sending back blurry but intriguing photos of deep-sea creatures. The death of William Edgerton in an unrelated diving accident effectively ended the experiments, but Cousteau had already realized the limitations of such a method of exploring the ocean depths. Instead, he and his team began work on a small, easily maneuverable submarine, which he called the diving saucer, or DS–2. The sub has made more than one thousand dives and has been part of countless undersea discoveries.

In 1955 *Calypso* embarked on a 13,800-mile journey that was recorded by Cousteau for a film version of *The Silent World.* The ninety-minute film premiered at the 1956 Cannes International Film Festival, where it received the coveted Palme d'Or. The following year, the film won an Oscar from the American Academy of Motion Picture Arts and Sciences. In 1957, in part due to his film's success, Cousteau was named director of the Oceanographic Institute and Museum of Monaco. He filled the museum's aquariums with rare and unusual **species** garnered from his ocean expeditions.

Cousteau addressed the first World Oceanic Congress in 1959, an event that received widespread coverage and led to his appearance on the cover of *Time* magazine on March 28, 1960. The highly favorable story painted Cousteau as a poet of the deep. In April of 1961 he received the National Geographic Society's Gold Medal at a White House ceremony hosted by President John F. Kennedy. The medal's inscription reads: "To earthbound man he gave the key to the silent world."

Television Programs Bring Worldwide Recognition

After the White House ceremony, Cousteau appeared to be at the pinnacle of his career, but bigger things were still to come. During the early 1960s he and his crew participated in the Conshelf Saturation Dive program, which was intended to prove the feasibility of extended underwater living. The success of the first mission led to Conshelf II, a month-long project involving five divers. The Conshelf program and the DS–2 project provided material for the fifty-three-minute film *World without Sun,* which debuted in the United States in December of 1964.

Cousteau's first hour-long television special, "The World of Jacques-Yves Cousteau," was broadcast in April of 1966, with Orson Welles providing the narration. The program's high ratings and critical acclaim helped Cousteau land a lucrative contract with the American Broadcasting Company (ABC). The *Undersea World of Jacques Cousteau* premiered in 1968 and has since been rebroadcast in hundreds of coun-

tries. The program starred Cousteau and his sons, Philippe and Jean-Michel, and sea creatures from around the globe. The show ran for eight seasons, with the last episode airing in May of 1976. In 1977 the *Cousteau Odyssey* series premiered on the Public Broadcasting System. The new show reflected Cousteau's growing concern about environmental destruction and tended not to focus on specific animal species.

In the 1970s the Cousteau Society, a nonprofit environmental group that also focuses on peace issues, opened its doors in Bridgeport, Connecticut. By 1975 the society had more than 120,000 members and had opened branch offices in Los Angeles, New York, and Norfolk, Virginia. Eventually, Cousteau decided to make Norfolk the homebase for *Calypso.*

On June 28, 1979, Philippe Cousteau was killed when the seaplane he was piloting crashed on the Tagus River near Lisbon, Portugal. Philippe's death deeply affected Cousteau, who was to his death unable to talk about the accident or the loss of his son. Philippe was expected to eventually take command of his father's empire; instead, Jean-Michel was given increased responsibility for overseeing the Cousteau Society and his father's other ventures.

Finds New Outlet on Cable TV

In 1980 Cousteau signed a one-million-dollar contract with the National Office of Canadian Film to produce two programs on the greater St. Lawrence waterway. In 1984 the *Cousteau Amazon* series premiered on the Turner Broadcasting System. The four shows were enthusiastically reviewed, and called attention to the threatened native South American cultures, Amazon **rain forest**, and creatures who lived in one of the world's great rivers. The final show of the series, "Snowstorm in the Jungle," explored the frightening world of cocaine trafficking. In the mid–1980s "Cousteau/Mississippi: The Reluctant Ally" received an Emmy for outstanding informational special. In all, Cousteau's television programs have earned more than forty Emmy nominations.

In addition to his television programs, Cousteau continued to produce new inventions. The Sea Spider, a many-armed diagnostic device, was developed to analyze the biochemistry of the ocean's surface. In 1980 Cousteau and his team began work on the Turbosail, which uses high-tech wind sails to cut fuel consumption in large, ocean-going vessels. In spring of 1985 he launched a new wind ship, the *Alcyone,* which was outfitted with two 33-foot-high Turbosails.

In honor of his achievements, Cousteau received the Grand Croix dans l'Ordre National du Mérite from the French government in 1985. That same year, he also received the U.S. Presidential Medal of Freedom. In November of 1987 he was inducted into the Television Academy's Hall of Fame and later received the founder's award from the International Council of the National Academy of Television Arts and Sciences. In 1988 the National Geographic Society honored him with its Centennial Award for "special contributions to mankind throughout the years."

While some critics challenged his scientific credentials, Cousteau never claimed "expert" status in any discipline. His talents seemed to be more poetic than scientific; his films and books—which include the eight-volume "Undersea Discovery" series and the twenty-one-volume "Ocean World" encyclopedia series—have a lyrical quality that conveys the captain's great love of **nature**. This optimism was tempered by his concerns about the environment. He emphatically demonstrated, perhaps to a greater degree than any of his contemporaries, how the quality of both the land and sea is deteriorating and how such environmental destruction is irreversible.

Cousteau continued to speak publicly about environmental issues until he was well into his eighties, although he had given up diving in cold water. In the years before his death, he had been planning for the construction of the *Calypso 2* to replace the original *Calypso*, which had sunk in a Singapore shipyard in 1994. The $20 million vessel was to be powered by solar energy and include equipment for a television studio, marine laboratory, and satellite transmission facility. The oceanographer died of a heart attack on June 25, 1997, at his home in Paris. He had been suffering from a respiratory ailment for which he had been hospitalized for several months. He was 87.

COVALENT BOND

A covalent bond is a type of **chemical bond** that holds different **atoms** together by sharing electrons. Depending on how equally the electrons are shared between the atom, these bonds may be either polar (having an unequal distribution of charge) or nonpolar (having an equal distribution of charge) in nature. In their most stable form, the outer electron shells of an atom are evenly filled. If there are too many or too few electrons to evenly fill these shells, the atom will try to combine with other atoms to reach a lower **energy** state that is more stable. By sharing electrons, atoms can achieve greater stability.

Atoms can have varying degrees of attraction for its outer shell, or valence, electrons. This measure of attraction is called electronegativity. An atom with high electronegativity is strongly attracted to its electrons and will try to combine with another atom that has a lower electronegativity. The magnitude of the difference between the electronegativity values of the atom will determine how the atoms share electrons. If the difference is large, the atoms tend to form **ionic bond**s when they combine. If the difference is small, they tend to form covalent bonds. The degree of covalency varies depending on the exact difference.

If the electronegativity difference is less than 0.5 units, the atoms tend to share outer shell electrons equally. When this occurs, the bond is called a nonpolar covalent bond. **Chemical compound**s with this type of bond are electrostatically neutral and do not tend to form charged ions. Unlike ionic compounds, they do not conduct electricity. Nonpolar covalently bonded **molecule**s tend to be gases, low boiling liquids, or low melting solids. Examples include the noble gases.

When the difference in electronegativity is between 0.5 and 1.7 units, the electrons are not shared equally between the atoms. The electron balance is shifted somewhat toward the more electronegative atom because its attraction for the elec-

trons is stronger. When this situation occurs, the bond is called a polar covalent bond. Because the electron sharing is unequal, the molecules have a polar orientation; in other words, the electrons are not evenly distributed across the bond. Therefore, polar covalent bonds form what is known as a dipole, and the molecule itself is said to be polar. An example of a polar molecule is water (H_2O). In molecules made of three or more atoms, the polar bonds may be arranged symmetrically around the central atom. If this happens, the dipoles negate each other and make the entire molecule nonpolar. Examples of molecules with this type of bond symmetry include **carbon dioxide** (featuring two oxygen atoms balanced on either side of the central carbon atom) and carbon tetrachloride (where four chlorine atoms are uniformly spaced around the carbon atom in the middle). Methane has a similar configuration with its four hydrogens around its central carbon.

In the covalent bonds described above, each atom in the molecule contributes one or more electrons. Another type of covalent bond occurs when both electrons from the shared pair come from the only one of the atoms. These bonds are called coordinate covalent bonds and examples include the ammonium ion NH_4 and the sulfuric acid molecule H_2SO_4.

CREATIONISM • See Human evolution

CRETACEOUS

The Cretaceous, or Cretaceous period, refers to an interval of geologic time in the Mesozoic era spanning from approximately 144 million to 66 million years ago. It was during this time period that dinosaurs developed, flourished, and became extinct, and flowering **plants** and modern insects first made their appearance. It is often divided into early and late Cretaceous epochs. The Cretaceous was preceded by the Jurassic period and followed by the Tertiary period.

The Cretaceous was a time of great change in the Earth's landmass as continents separated and migrated on the globe's surface. At the start of the Cretaceous, South America and Africa were still connected and Antarctica was attached to Australia. North America had broken from Africa but was still attached to western Europe. India was drifting northward toward the Eurasian continent. By the end of the Cretaceous, South America had split from Africa and moved west, and North America had separated from western Europe opening the expanses of the Atlantic Ocean.

In early Cretaceous times seas began to encroach on the continental areas and sedimentary rocks including limestone, and marine clays, were deposited over large areas. The advancing seas reached their greatest extent in the late Cretaceous epoch and then began to recede. Global **temperature**s were warmer than today, but during the late Cretaceous extensive deposition of chalk (calcium carbonate) in deep ocean **water**s is thought to have lowered atmospheric concentrations of **carbon dioxide**. Reduced levels of atmospheric carbon dioxide diminished its role as a "greenhouse" gas, and this in turn led to a cooling of the Earth's surface.

Flowering plants (Angiosperms) became dominant on land, and insects increased in abundance with them. Reptiles were dominant land **animals** until the end of the Cretaceous. **Mammals** and birds were not important. Flying reptiles called pterosaurs were abundant. At the end of the Cretaceous, dinosaurs suddenly became extinct and other reptile groups were decimated. Many hypotheses have been advanced to explain the disappearance of dinosaurs and a dramatic decline in the abundance of other reptiles. One widely accepted theory postulates a catastrophic collision of an asteroid or comet with the Earth that created a mammoth dust cloud. The cloud engulfed the entire **atmosphere**, greatly decreasing the amount of sunlight reaching the Earth's surface, and reducing or eliminating **photosynthesis**. Plants, **herbivores** and animals dependent on them could not survive, and the biology of the Earth changed dramatically. This theory is supported by the very abrupt and widespread scope of **extinction**s occurring at the end of the Cretaceous period. Although other theories have also been proposed to account for these catastrophic events, including alterations of **habitat** due to the movement of continents, and the appearance of **egg**-eating reptiles or mammals, none has received the same support as the catastrophic collision hypothesis.

See also Continental drift; Evolutionary change, rate of; Mass extinction hypothesis

CRICK, FRANCIS HARRY COMPTON (1916-　)
British molecular biologist

Francis Crick worked closely with **James Watson** and together they were able to deduce the structure of the deoxyribonucleic acid (**DNA**) molecule. This was very important because it showed that DNA, not **protein** as previously believed, was the actual carrier of genetic instructions for the **cell**. The Watson and Crick model reveals valuable information about the organization and operation of life itself. Their discovery is generally considered to be the most important scientific breakthrough of this century. Their model, originally constructed in 1953, showed the DNA **molecule** as consisting of two chains, each wrapped around the same axis like a spiral staircase. These helical chains are made of alternating units of phosphate and sugar molecules. Each side of the chain is connected by four base pairs which include adenine, thymine, guanine, and cytosine. Every base is attached at one end to a sugar molecule. The opposite end of the base molecule will only bind to its complementary base. For example, adenine is always chemically bound to thymine, and cytosine is always bound to guanine. This consistent, complementary pairing of the bases revealed how DNA was able to make exact copies of itself and thus pass on hereditary information.

Crick was born in 1916 in Northampton, England. After graduating with a degree in physics from University College, London, he developed radar systems and magnetic mines for the British military during World War II. In 1947, he worked at Strangeways Research Laboratory by day and studied biology in the evenings. He later moved to the Cavendish Laboratory at Cambridge University. It was there that he first met James Watson and began work on the structure of DNA. Although Watson had to initially persuade Crick to work with DNA, it wasn't long before Crick enthusiastically embraced the project. Crick eventually became consumed with this mission and even named his house The Golden Helix after their working hypothesis.

Crick gained his Ph.D. from Caius College, Cambridge, in 1953. That same year several other Cambridge scientists were awarded the Nobel Prize in other areas of research. Nearly 10 years later, Francis Crick shared the Nobel Prize in physiology and medicine with James Watson and **Maurice Hugh Frederick Wilkins** for their work with DNA.

With fellow workers at Cambridge University, Crick later studied the structure and function of the **genetic code**—the sequence of **nitrogen** bases in DNA that directs the joining of amino acids to build protein molecules. Crick is credited with developing the term "codon" as it applies to the set of three bases that code for one specific **amino acid**. These codons are used as "signs" to guide protein synthesis within the cell. As a result of his later work with Drs. Barnett, Brenner, and Watts-Tobin, Crick was able to formulate a set of general properties of the genetic code. He has also used the common *Escherichia coli* **bacteriophage** to study genetic **mutation** mechanisms. In 1977, his distinguished status in the scientific community earned him a professorship at the famous Salk Institute for Biological Studies in San Diego, California. He continues to play an active role in several areas of ongoing research today.

CROSS-FERTILIZATION

Cross-fertilization (also known as allogamy or exogamy) is the fusion of two sex **cells** (**gamete**s) derived from two separate individuals of the same **species**. Examples are the mating between male and female in **animals** or the cross **pollination** of **plants**. The advantage of cross-fertilization is that it promotes the production of new genetic combinations within a **population**. This in turn leads to greater variation within the population, which gives it greater adaptive potential. As a mechanism of producing greater variation cross-fertilization is consequently a driving force of **evolution**.

There are a number of disadvantages to cross-fertilization, the two most major are the fact that two individuals of the same species need to be involved, and that desirable characteristics arising de novo in one individual can be rapidly diluted and eventually lost to the population. Due to the production of new genetic combinations mentioned above cross-fertilization is still a desirable trait. It allows **organisms** to evolve more rapidly and consequently aids in the spread of organisms and species into more and diverse **habitat**s.

CROSSING OVER

Crossing over is a method of exchange of **genetic material** during **meiosis**. When the homologous chromosomes line up the **gene**s on adjacent arms are aligned in sequence. During the **replication** of genetic material the **enzyme** responsible moves onto the homologous arm from the adjacent **chromosome**, the replication continues but the genetic material of the second chromosome has now been incorporated onto the original chromosome. This site of crossing over is visible as a chiasmata, where one arm forms a bridge over to the adjacent chromosome. This is an important method of rearranging the genetic material within an individual to produce new combinations of genetic material within the **gamete**s.

In some **organism**s, e.g. *Drosophilla* and certain **fungi**, crossing over has also been shown to occur during **mitosis**. Normally crossing over is equal between the chromosomes but occasionally it can be unequal which can give rise eventually to alterations in the chromosome number by loss of whole chromosomes. Crossing over alters the frequency with which genes on the same chromosome are transmitted together, the farther apart on the chromosome the greater is the chance of a chiasmata occurring between them. As a consequence of this, the phenomenon of crossing over has proved very important in the construction of genetic maps. The longer the chromosome the greater the chance of several chiasmata occuring, consequently the more varied genetically will be the resultant gamete. Crossing over can be observed by studying two genes which have previously been shown to be linked (i.e. on the same chromosome); a small percentage of the offspring from matings will show **independent assortment** as if the genes were on separate chromosomes. The greater the number of offspring showing non-linked transmission the farther apart the genes are on the chromosome, although it should be borne in mind that a double cross over event will reverse the effect of a single cross over for two widely separated genes.

It was originally thought that crossing over was a rare event and it is now understood that it is a normal and vital part of meiosis. Due to the rearranging of genetic material crossing over is an important part in the production of new characteristics and arrangements of characters, and as such it is an important building block of **evolution**.

CRYOBIOLOGY

Cryobiology is the science of freezing biological fluids, **cell**s, and **tissue**s. It is an extension of cryogenics, which is the study of the properties of **matter** at very low **temperature**s. Cryobiological techniques have application in genetic research, livestock breeding, **infertility** treatment, and **organ** transplantation. There are even those who believe that humans can be frozen and placed in suspended animation, although this concept is presently rejected by mainstream scientists.

The terms cryobiology and cryogenics are derived from the Greek *kryos*, meaning icy cold. Temperatures used in cryogenics range from -148°F (-100°C) to near absolute zero

Francis Harry Compton Crick. *(The Library of Congress. Repoduced by permission.)*

-459°F (-273°C). Such ultra-low temperatures can be achieved by the use of super-cooled gases. The study of these gases dates back to 1877, when Swiss physicist Raoul Pictet and French engineer Louis Cailletet first learned how to liquify oxygen. Although they worked independently and used different methods, both men discovered that oxygen could be liquefied at -297.67°F (-183.15°C). Soon after, other researchers liquefied **nitrogen** at -321.07°F (-196.15°C). Other breakthroughs in cryogenics included James Dewar's invention of the vacuum flask in 1898. Dewar's double-walled vacuum storage vessel allowed liquefied gases to be more readily studied. In the last 100 years, a variety of other methods for insulating super-cooled fluids have been developed.

In the twentieth century, scientists began applying cryogenic techniques to biological systems. They explored methods for treating **blood**, semen, tissue, and organs with ultra-low temperatures. In the last few decades, this research has resulted in advances in genetic research, livestock breeding, infertility treatment, and organ transplantation.

In the area of genetic research, cryobiology has provided an inexpensive way to freeze and store the embryos of different strains of research laboratory **animal**s, such as mice. Maintaining a breeding colony of research animals can be expensive, and cryogenic storage of embryos can reduce cost by 75%. When the animals are needed, they can be thawed and brought to life in a few months.

In **agriculture**, cryopreservation allows livestock breeders to mass produce the embryos of genetically desirable cattle. For example, hundreds of embryos can be harvested from a single prize dairy cow and frozen for later implantation in other mothers. Using similar techniques, pigs that are too fat to reproduce on their own can be artificially implanted with previously frozen embryos. In addition, cryobiologists are examining the possibility of increasing buffalo herds by freezing bison embryos and later implanting them into cows to give **birth**.

Cryobiology has met with great success in the treatment of human infertility. The use of frozen **sperm**, **egg**s, and embryos increases the success rate of fertility treatments because it allows doctors to obtain a large number of samples which can be stored for future **fertilization**. Techniques for freezing sperm were relatively easy to develop, and in 1953, the first baby fertilized with previously frozen sperm was born. The process for freezing embryos is much more complicated, however. It involves removing **water** from the cells and replacing it with an organic antifreeze that prevents the formation of ice crystals that can cause cells to burst. Advances over the last few decades have made this technique highly successful, and in 1984, the first baby was born from a previously frozen embryo. Freezing eggs is an even more difficult challenge because the fragile **membrane structure** that surrounds the eggs make them difficult to freeze without causing severe damage. However, scientists working at Reproductive Biology Associates in Atlanta, Georgia, have successfully frozen eggs using a chemical solution similar to the ovaries' natural fluids. They have also learned to collect eggs at a certain point in the **hormone** cycle to increase the eggs' chances of surviving the freeze-thaw process. Although still experimental, their technique has been used to freeze eggs, which were later thawed and used to impregnate a woman. In 1997, the first birth resulting from frozen eggs was recorded.

Organ storage is another important area of cryobiological research. Using conventional methods, organs can only be stored for very short periods of time. For example, a kidney can be kept alive for only three days, a liver for no more than 36 hours, and **heart**s and lungs for no more than six hours. If these organs could be frozen, storage times would be lengthened almost indefinitely. Although researchers have made great advances, they have not yet perfected the process of freezing and reviving organs. The problem they face is that the formation of ice crystals can damage fragile tissue. However, researchers at South Africa's H. F. Verwoerd Hospital have devised a way around this problem using a cryopreservant liquid that protects the organs during the freezing process. Boris

Rubinsky, another important researcher in the field, has discovered an antifreeze **protein** that has been successfully used to freeze and revive rat livers. Rubinsky's proteins are derived from fish living in the Arctic that have evolved to survive in very cold water. These proteins alter the structure of ice crystals so that they are less damaging to cells. Another experimental new technique, called vitrification, is used to cool organs so quickly that their **molecule**s do not have time to form damaging ice crystals. Continued success in these areas may one day lead to reliable methods of freezing and storing organs for future transplant.

Suspended animation is one application of cryobiology that has long been a topic of science fiction stories. The theory is that people dying of incurable diseases can be frozen before or just after **death** and thawed out in the future when medical science has found a cure for their affliction. While mainstream medical experts scoff at this notion, there are several companies that specialize in freezing people after death. Trans Time and CryoSpan are two California-based companies offering cryogenic suspension services. They freeze and maintain bodies in liquid-nitrogen baths at -320°F (-195°C).

See also Embryo and embryonic development

CT SCAN • See Computerized axial tomography

CULTURE

A **culture** is a single specie of **microorganism** that is isolated and grown under controlled conditions. The German bacteriologist **Robert Koch** first developed culturing techniques in the late 1870s. Following Koch's initial discovery, medical scientists quickly sought to identify other pathogens. Today **bacteria** cultures are used as basic tools in science and medicine.

The ability to separate bacteria is important because microorganisms exist as mixed **population**s. In order to study individual **species**, it is necessary to first isolate them. This isolation can be accomplished by introducing individual bacterial **cell**s onto a culture medium containing the necessary elements for microbial growth. The medium also provides conditions favorable for growth of the desired species. These conditions may involve **pH**, osmotic pressure, atmospheric oxygen, and moisture content. Culture media may be liquids (known broths) or solids. Before the culture can be grown, the media must be sterilized to prevent growth of unwanted species. This sterilization process is typically done through exposure to high **temperature**s. Some tools like the metal loop used to introduce bacteria to the media, may be sterilized by exposure to a flame. The media itself may be sterilized by treatment with steam-generated heat through a process known as autoclaving.

To grow the culture, a number of the cells of the microorganism must be introduced to the sterilized media. This process is known as inoculation and is typically done by exposing an inoculating loop to the desired strain and then placing the loop in contact with the sterilized surface. A few of the cells will be transferred to the growth media and under the proper conditions, that species will begin to grow and form an pure colony. Cells in the colony can reproduce as often as every 20 minutes and under the ideal conditions this rate of **cell division** could result in the production of 500,000 new cells after six hours. Such rapid growth rates help to explain the rapid development of disease, food spoilage, decay, and the speed at which certain chemical processes used in industry take place. Once the culture has been grown, a variety of observation methods can be used to record the strain's characteristics and chart its growth.

CUSHING, HARVEY WILLIAMS (1869-1939)

American neurosurgeon

Harvey Williams Cushing was born to Betsy and Henry Cushing on April 8, 1869 in Cleveland, Ohio. The son and grandson of physicians, Cushing's father specialized in diseases of women and in medical law. Cushing carried on the family tradition and attended Yale College and Harvard Medical School, earning his M.D. from Harvard in 1895. He then served as a resident at Massachusetts General Hospital and the Johns Hopkins Hospital. He traveled in Europe for one year (1900-1901) where he studied under two important physiologists, **Theodor Kocher** (1841-1917) in Bern, Switzerland, and **Charles Scott Sherrington** in London, England.

After his return from Europe, Cushing began to concentrate on the specialty for which he was to become famous: neurosurgery. In the early 1900s, the techniques of brain surgery were poorly developed and survival rates from such procedures were close to zero. Stimulated by his work in Bern and London, Cushing began to attack, one at a time, the specific problems that made neurosurgery so dangerous. For example, he developed silver clips that could be used to control bleeding during surgery. He was also successful in finding ways to cauterize **blood** vessels in the **brain** with electrical current. The use of these and other techniques made possible surgical procedures that could not be attempted earlier. As a result, survival rates rose dramatically.

Cushing's influence extended far beyond his own work. He established the Hunterian Laboratory at Johns Hopkins in 1905 and, later, the Laboratory of Surgical Research at Harvard. During his four decades at Harvard and Yale, he personally trained a large portion of the neurosurgeons who were to lead the field for generations to come.

Cushing's name is memorialized in medical terminology as a result of his research in another field. Beginning in 1908, he studied the **pituitary gland** and its malignancies. Eventually he was able to prove that tumors in the pituitary can result in a disorder characterized by wasting, obesity of the face and trunk, atrophy of the skin, and accumulation of fluids in the body. That condition is now known as Cushing's disease.

Harvey Williams Cushing. *(The Library of Congress. Reproduced by permission.)*

Cushing was also interested in the history of medicine and, in 1925, wrote a biography of William Osler (1849-1919) that earned him the 1926 Pulitzer Prize.

Cushing left Johns Hopkins in 1911 to become Professor of Surgery at the Harvard Medical School and Chief Surgeon at the Peter Bent Brigham Hospital in Boston, Massachusetts. He moved from Harvard to Yale in 1933, becoming Sterling Professor of Neurosurgery. He retired in 1937 and died in New Haven, Connecticut, on October 7, 1939.

CUVIER, GEORGES (1769-1832)

French naturalist

Georges Léopold Chrétien Frédéric Dagobert, Baron Cuvier was a French naturalist who is known as the founder of modern comparative **anatomy** and as the founder of the field of **paleontology**. He was born in 1769 in Montébeliard, near Basel. Although a French town, Cuvier's birthplace at that time belonged to the Duchy of Württemberg. Cuvier was an academically inclined young man, and, because his family lived in near-poverty, he accepted the offer to study for free at the Karlsschule in Stuttgart, Germany. He graduated at eighteen, returned home, and then found employment as a tutor in Normandy. While working in Normandy, he familiarized himself with the marine creatures he found on the beach, which he dissected and drew in detail, and while doing so, he referred to Aristotle's ideas of comparing different animal structures, **Carl Linnaeus**'s *System of Nature* and Buffon's *Natural Histo-*

ry of Animals. His marine animal drawings, which were rather impressive, came to the attention of Geoffroy Saint-Hilaire, and eventually led to Cuvier's appointment as assistant professor of comparative anatomy at the Museum of Natural History in Paris. Under Napoleon's regime, Cuvier became inspector General in the Department of Education and contributed to significant education reform in France. After Napoleon's fall, Cuvier retained his position and became an accepted authority in science and education, and earned several promotions which included a professorship at the Collège de France and permanent secretary for the Academy of Sciences. Cuvier died in 1832 of cholera, during the first major epidemic of that disease in Europe.

Prior to Cuvier, anatomists such as Louis Daubenton, Johann Friedrich Blumenbach, and Petrus Camper posited the human being as the fundamental form to which all other living creatures were compared. Cuvier, however, decided to create an objective system of comparative anatomy based on observation. His initial field of research were marine animals, particularly molluscs, worms, and various fishes. Later, he extended his investigations to vertebrates in general. The conceptual framework of Cuvier's research was a systematic method of comparative anatomy. According to Cuvier, living beings exhibit certain distinctive anatomic features which enable the scientist to place an individual specimen in the larger context of a general anatomic system. For example, one can make significant generalizations by observing individual features such as dental structure, foot structure, skull shape, etc. Cuvier's comparative research, which expanded from the study of vertebrates to include the entire **animal** kingdom, was presented in his work *The Animal Kingdom, Distributed According to Its Organization* (1817). While Cuvier's work did not contribute any new facts to the science of anatomy, his method earned him high praise and esteem in the scientific community.

An important element of Cuvier's methodology is his correlation theory, which posits the functional interdependence of particular **organs** within an in individual **organism**. For example, as Cuvier observed, carnivorous animals possess certain distinctive features which clearly separate them from, say, **herbivore**s. These features include sharp teeth, a certain jaw structure, a **digestive system** adapted to meat, acute eyesight, sharp claws, powerful and swift locomotion, etc.

In Paris, which is in a calcareous area, Cuvier applied his comparative method to study **fossil**s. In his carefully organized excavations, particular attention was paid to the specific location, position, and placement of the discovered fossils. In addition, using his correlation theory, he developed a reconstruction method which enabled researchers to identify incomplete skeletons. Furthermore, in order to validate a particular hypothesis concerning the identity of an incomplete skeleton, Cuvier would compare the extinct animal to its closest living relative, in an effort to complete the puzzle. These investigations were described in his seminal *Investigations on Fossile Bones* (1812), establishing Cuvier as the founder of modern paleontology. Using his comparative method, with particular emphasis on dentition and bone structure Cuvier was able to demonstrate that the two types of elephant, Indian and African, classified as examples of one **species**, in reality constituted two distinct species. In fact, Cuvier found that the extinct mammoth is closer to the Indian elephant than the two existing elephant species are to each other! Extending his research on elephants to Pachydermata in general, Cuvier studied both existing and extinct forms, identifying several new genera, including *Palaeotherium* and *Dinotherium*. In addition, he provided the first scientific description of the American giant sloth and named the pterodactyl.

Cuvier, like many of his colleagues, puzzled over the seemingly mysterious fact that animal forms changed through history. However, unlike some his colleagues, who approached the issue with extreme circumspection, Cuvier decided that species do not change. "The immutability of species," wrote Nordenskiöld, "is to Cuvier's mind an absolute fact." In order to explain why certain species were extinct and why fossils of some extinct creatures were unrecognizable from modern creatures, Cuvier invoked the **catastrophism** theory, which posits that a "new" species appear after the **extinction**, due to a violent upheaval (such as an earthquake) of its "old" counterpart. Thus, for example, Cuvier denied the existence of human fossils, asserting that, for example, lion fossils and lions in their present form represent two distinct species. Realizing the absurdity of the idea that species emerged out of nothing following a catastrophe, Cuvier attempted the explain the continuity of life by positing a type of near-extinction, which would allow the survival of small populations of a particular species, positing, as Cassirer has remarked, an evolution by analogy, whereby a particular species would be replaced by its new analogue, which to his mind seemed more reasonable than the notion of gradual **evolution**.

Cuvier's **classification** of animals is considered not only the most significant improvement of Linnaeus's system but essentially the foundation of all later classifications. Dividing the animal **kingdom** into four major phyla (mollusca, radiata, articulata, and vertebrata), Cuvier postulated a distinct "ground plan" for each group. According to Cuvier, the ground plan, determined the physiological and anatomical identity of a particular **phylum**. Naturally, he stated that species emerging from the same ground plan share many features, asserting, however, that there can be no comparisons across phyla. However, despite Cuvier's philosophical opposition to the idea of gradual, evolution, his systematic conception of animal life, reflected in his formal classification scheme, provided an intellectual framework, which, when the time came, fruitfully accommodated theories evolving from the evolutionary paradigm in biology. In other words, the intellectual power of Cuvier's system was not diminished by the fundamental paradigm change in nineteenth-century biology.

Cuvier's views of classification and evolution were vigorously opposed by several of his prominent contemporaries, who found his systematic philosophy, particularly his adamant insistence of four ground-plans, dogmatic. For example, Geoffroy Saint-Hilaire, who engaged Cuvier in a lengthy polemic, maintained that, because life manifests itself on the basis of a fundamental, indivisible impulse, Cuvier's claim that creatures emerging from different ground plans cannot be compared does not reflect the true nature of the animal world. Accused by his critics for speculative dogmatism, Cuvier nevertheless, as Cassirer has written, defended his views on the basis of empirical research. As scholars have observed, the polemic between Cuvier and Saint-Hilaire was never resolved owing to the fact that both antagonist defended points of view, which, while seemingly opposed, contributed, as complementary views, to the progress of biology.

Cuvier, whose schooling in Stuttgart introduced him to the tradition of German scientific and philosophical thought, retained, throughout a lifetime of painstaking experimental work, an interest in the philosophical foundations of science. For example, while rejecting Geoffroy Saint-Hilaire's speculations about the nature of life, Cuvier declared that life cannot be reduced to chemical processes, a speculative proposition which any proponent of **vitalism** would accept without hesitation. In addition, this almost fanatical proponent of empirical research never subscribed to the dogmatic materialism that seemed to arrive in the wake of the triumphal march of empirical science. Cuvier, who, as Nordenskiöld remarked, may have followed Immanuel Kant's theory of knowledge, denied that science could totally grasp reality as it is, approached life, particularly soul-life as a mystery, and defined materialism as ''an arbitrary hypothesis''—''so much the more so as philosophy cannot offer any direct proof of the true existence of matter.''

See also Anatomy, comparative

CYANOBACTERIA

Cyanobacteria, commonly known as blue-green **algae**, are members of the Kingdom **Monera**. Members of this **kingdom** lack a nuclear region clearly defined by a **membrane** or envelope and are thus known as **prokaryote**s. Members of the blue-green algae are among the oldest **fossil**s ever found, dating back to over three billion years.

Cyanobacteria are photosynthetic unicellular or colonial algae. They contain chlorophyll *a* and other accessory pigments, but these are not located in distinct **chloroplast**s. Instead, they are distributed throughout the **cell** on flat **organelle**s called thylakoids. Accessory pigments in this group include beta-carotenes, xanthophylls, and phycobilins. One type of phycobilin is phycocyanin, which gives the characteristic bluish green color, and hence their common name. Their cell wall is composed of mucopeptides. Many **species** also have a gelatinous sheath around their cells. Some species live as single-celled individuals like the **genus** *Sphaerocystis*, while others grow in colonies like the genera *Anabaena* and *Oscillatoria*. **Sexual reproduction** is rare. Most blue-green algae simply divide asexually by binary fission, or divide by fragmentation of the filaments if they are colonial.

Cyanobacteria are a successful group because they grow in a diversity of **habitat**s. Some species grow on moist **tree** bark and soil. A few genera like *Nostoc* develop a mutualistic relationship with their fungal symbionts and grow as lichens. Many species of blue-green algae grow in freshwater and ma-

rine habitats. Some genera such as *Aphanizomenon* and *Anabaena* are able to produce **ammonia** from **nitrogen** gas using the **enzyme** nitrogenase, a process known as nitrogen fixation. This process is sometimes done in specialized enlarged cells called heterocysts. Other species form the characteristic coloration seen in the upper spray regions of intertidal zones. Those species that live in shallow coastal areas can form dense mounds known as stromatolites. Some species proliferate in the near-boiling temperatures of thermal hot springs, while others tolerate the frigid temperatures found in Antarctic lakes. More commonly, many species of blue-green algae form dense mats floating on the surface of eutrophic lakes in the summer months. They are able to float on the surface using specialized gas **vacuoles**. Such species are often considered to be indicators of **eutrophication**. *Spirulina* is an example of blue-green algae that commonly grows in dense blooms in African lakes. Filter-feeding fish such as tilapia thrive on consuming this alga. Humans have also found them to be a valued health food because they are high in **protein**, **vitamins**, and linolenic acid. Aquaculturists in Mexico, Israel, and California have learned to cultivate *Spirulina*. Some species of blue-green algae produce a toxin that can inhibit filter-feeding **zooplankton** such as *Daphnia*. These **toxins** consist of various types of **alkaloid**s, **polypeptide**s, pteridines, and lipopolysaccharides, which can be lethal to cattle at high concentrations. There have even been reported cases of humans becoming sick after drinking this **water**. The bad taste in residential drinking water can often be attributed to the chemicals geosmin and methyl-isoborneol (MIB) released by some blue-green algae and **fungi**. These toxins are released when the cells die and decay.

CYBERNETICS

Cybernetics is a term that was originated by American mathematician Norbert Wiener (1894-1964) in the late 1940s. Based on common relationships between humans and machines, cybernetics is the study and analysis of control and communication systems. As Wiener explains in his 1948 book, *Cybernetics: or Control and Communication in the Animal and the Machine*, any machine that is ''intelligent'' must be able to modify its behavior in response to feedback from the environment.

This theory has particular relevance to the field of computer science. Within modern research, considerable attention is focused on creating computers that emulate the workings of the human mind, thus improving their performance. The goal of this research is the production of computers operating on a neural network. During the late 1990s work has progressed to the point that a neural network can be run, but, unfortunately, it is generally a computer software simulation that is run on a conventional computer. The eventual aim, and the continuing area of research in this field, is the production of a neural computer. With a neural computer, the architecture of the **brain** is reproduced. This system is brought about by transistors and resistors acting as **neuron**s, **axon**s, and dendrites. By 1998 a neural network had been produced on an integrated circuit, which contained 1024 artificial neurons. The advantage of these neural computers is that they are able to grow and adapt. They can learn from past experience and recognize patterns, allowing them to operate intuitively, at a faster rate, and in a predictive manner.

Another potential use of cybernetics is one much loved by science fiction authors, the replacement of ailing body parts with artificial structures and systems. If a structure, such as an **organ**, can take care of its own functioning, then it need not be plugged into the human **nervous system**, which is a very difficult operation. If the artificial organ can sense the environment around itself and act accordingly, it need only be attached to the appropriate part of the body for its correct functioning. An even more ambitious future for the cybernetics industry is the production of a fully autonomous life form, something akin to the robots often featured in popular science fiction offerings. Such an artificial life form with **learning** and deductive powers would be able to operate in areas that are inhospitable to human life. This could include long-term space travel or areas of high radioactivity.

See also Nerve impulses; Nervous system

CYCLIC AMP (CYCLIC ADENOSINE MONOPHOSPHATE)

Cyclic AMP was the first chemical to be isolated and identified as a second messenger—a substance produced by a **cell** when a **hormone** or other chemical (the first messenger) binds to specific receptor sites on the cell surface. The second messenger, in turn, influences the cell's processes. The discovery was a major step toward understanding how both hormones and cells function.

Until 1956, scientists thought that hormones performed their action on cells directly. Then the American physician and pharmacologist **Earl Sutherland** showed that hormones stimulate production by the cell of cyclic AMP (adenosine-3',5'-phosphoric acid), which affects the cell's activity. He received the 1971 Nobel Prize in physiology or medicine for his work.

Sutherland was born and raised in Burlingame, Kansas. He graduated from Washburn University in Topeka and received his M.D. from Washington University in St. Louis, Missouri. From there, Sutherland went on to conduct his research at Washington University and Case Western Reserve University, and was a faculty member at Vanderbilt University and the University of Miami.

Sutherland investigated the active and inactive forms of the **enzyme** phosphorylase, whose active form begins the breakdown of glycogen stored in the liver and muscle into glucose. He also co-discovered the source of the hormone glucagon in the pancreatic islet of Langerhans α cells, and investigated how glucagon and another hormone, epinephrine, cause the liver's release of glucose.

Sutherland and a colleague observed that inactive phosphorylase became active in the presence of an unrefined liver extract, but not when impurities were filtered from it. Next they added epinephrine and glucagon to the material removed during refinement. The result was a substance that in turn activated phosphorylase. Sutherland's research group and another group of scientists independently isolated cyclic AMP as the activating substance, which Sutherland designated a "second messenger."

It was soon demonstrated that cyclic AMP also worked as a second messenger with other hormones and in many other **tissue**s in different **species**. Further, it could either stimulate or inhibit cell activity. Subsequent research has discovered the second-messenger function of other chemicals, as well, including cyclic GMP (guanosine 3',5'-cyclic monophosphate), discovered by Sutherland.

Cyclic AMP has been shown to be an important link in many processes, such as **memory** and hormonal response. In 1998, scientists at the University of California at San Francisco discovered a possible link between a person's cyclic AMP levels and tolerance to **alcohol**—a factor which may prove important in determining the genetic predisposition to alcoholism. Further studies are necessary before this theory of cyclic AMP involvement is proven.

CYCLOSPORIN

With the advent of **organ** transplantation, it became clear that surgical techniques were far ahead of immunological control and protection of the body. Live organs could be transplanted from one body to another, but the drugs necessary to prevent rejection of the foreign **tissue** debilitated the patient's entire **immune system**. Frequently patients could not survive these severe infections, and mortality rates for transplantation were discouragingly high.

The discovery of cyclosporin brought about a major shift in the success of transplantation. Cyclosporin is an immunosuppressant whose action specifically inhibits graft rejection while allowing the bulk of the immune system to function normally and fight general infection. **Jean-François Borel**, a microbiologist working for Sandoz Laboratories in Switzerland, discovered cyclosporin in 1969 when he was vacationing in Norway. Sandoz employees were encouraged to pick up samples of naturally occurring **organism**s for analysis in the laboratory. When Borel visited Hardanger Vidda, a desolate highland plateau in southern Norway, he gathered some soil samples and brought them back to Sandoz for testing.

The laboratory isolated cyclosporin from a fungus (*Tolypocladium inflatum*) within the soil. Sandoz was primarily involved in **antibiotics** research, and the purpose of the first series of assays was solely to determine the substance's potential as an antibiotic. The tests yielded little of interest as far as antibiotics were concerned, but did show that cyclosporin had distinct immunosuppressive capabilities. Borel had received his Ph.D. in immunogenetics, and he wanted to learn more about cyclosporin.

Borel ran a second series of tests and found that cyclosporin acts specifically to inhibit T-lymphocyte activity, that part of the immune system that especially detects and attacks foreign invaders. The selectivity of cyclosporin is its salient feature. Other immunosuppressant agents had been used in transplantation and grafting, but these drugs were so indiscriminate in their paralysis of the immune system that the body became far too vulnerable to numerous types of infection; even those infections that were normally well controlled by an intact immune system would flourish. In addition, earlier immunosuppressant drugs were highly toxic.

Cyclosporin has the advantage of selectively interfering with only one specific subpopulation of the immune system—specifically, the T-**lymphocytes**. It does not actually destroy the T-lymphocytes, but fends them off, and it acts at an early stage in the **life cycle** of the T-lymphocyte, inhibiting its action by blocking the intercellular message carried by interleukin-2.

It looked as if Borel had discovered a superior drug for transplantation. But his employer, Sandoz, was not sufficiently impressed by the findings he reported in 1972. The estimated costs for production and testing of the drug were astronomical, and the transplantation field was only in its infancy, so the potential demand for cyclosporin was questionable. Sandoz was unwilling to put the necessary money and energy behind the drug for further exploration.

Two immunologists recognized the importance of Borel's discovery, however, when Borel presented his results to the British Society of Immunologists in 1976. David White and Sir Roy Calne asked Borel for samples of cyclosporin and began their own clinical studies of transplantation in rats. The results were remarkable: rejection was almost nonexistent, and the survival rate was far better than for other immunosuppressants. In mid-1977, White and Calne informed Borel of their findings and requested more samples of cyclosporin to continue their clinical trials, this time on dogs. Borel, in an effort to revive Sandoz's interest in cyclosporin, asked White and Calne to present their findings to Sandoz. The pharmaceutical company agreed that the drug looked much more promising now that there was evidence of its effectiveness. Sandoz decided to put money into research and development of cyclosporin, and White and Calne returned to England to undertake experiments at breakneck speed in order to present their results in time for the annual International Transplantation Congress in 1978.

The success of cyclosporin suffered an apparent setback in 1979 when further studies showed it to be nephrotoxic (poisonous to the kidneys) and to induce lymphomas (tumors). These side effects proved to be a matter of dosage, however. The practice at the time was to administer as much drug as the body could handle, just below toxicity. Researchers later opted to regulate the drug on the lower end of the scale—that is, only enough cyclosporin to prevent rejection. With the newly decreased dosage, the lymphoma was eliminated and nephrotoxicity was reduced.

Later research by Thomas Starzl (1926-) in Colorado indicated that cyclosporin worked most effectively when administered in tandem with **steroids**. In 1983, when the U.S. Food and Drug Administration approved cyclosporin for use in all transplant patients, it was stipulated that the drug could be used only in conjunction with steroids.

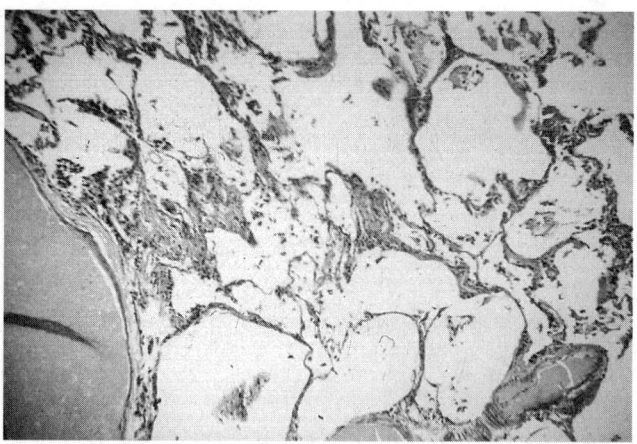

A sample of lung tissue containing mucous. High levels of mucous inside the lungs are a symptom of cystic fibrois. *(Custom Medical Stock Photo. Reproduced by permission.)*

Cyclosporin must be taken indefinitely by transplant recipients, and its long-term damage to the kidneys remains a serious danger. Although cyclosporin is not a wonder drug, it is one of the most potent and specific immunosuppressants available; it is effective in treating infectious complications, and is associated with a lower mortality rate. It is commonly used in kidney, **heart** and lung, liver and **pancreas**, and bone marrow transplants, although its effects are not confined to transplantation. Cyclosporin is also used to treat viral and fungal infections and autoimmune disorders like **arthritis**, and to promote healing of wounds, composite grafts, and allografts. One of the side-effects of cyclosporin has been excessive hair growth, or hypertrichosis. As a result, scientists have also studied cyclosporin as an agent to stimulate hair growth.

Cystic fibrosis

Almost everyone has experienced the congestion that accompanies a bad cold. The thick mucus that forms in the nose and throat makes breathing difficult. Luckily, this congestion dissipates in a few days as the body fights off the cold.

However, individuals afflicted with cystic fibrosis must constantly cope with the mucus which accumulates in the lungs, **pancreas**, and intestine. Cystic fibrosis, or mucoviscidosis, is an inherited disease that affects the exocrine glands. It is expressed only in the homozygous state without X chromosomal **linkage**. This genetic disease is often referred to as autosomal recessive. This means that children born to parents who each carry one recessive gene for the disease have a 25% chance of inheriting both copies of the defective gene—and with them the disease. This common disorder mainly effects about 1 out of every 2,000 Caucasians of European descent, and it is the leading fatal genetic disease in the United States. The disease is less common in African Americans and very rare in Asians and Native Americans.

Cystic fibrosis was not classified as a separate disease until 1938. In 1989, scientists at the University of Michigan

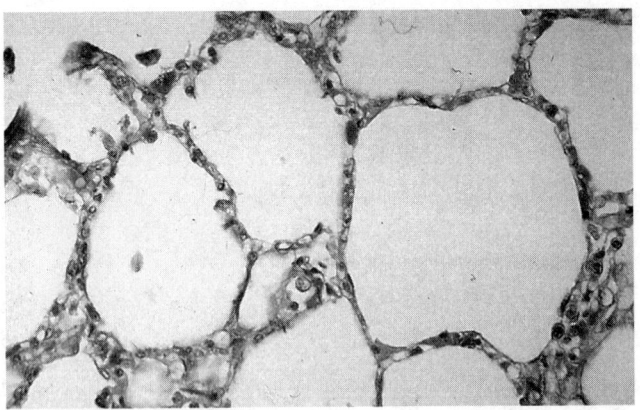

A sample of normal lung tissue. *(Custom Medical Stock Photo. Reproduced by permission.)*

and the Hospital for Sick Children in Toronto, Ontario, Canada, announced that they had identified the defective gene that causes cystic fibrosis. The gene is found on **chromosome** 7, and over 600 **mutation**s have been identified. Depending on the mutation, the symptoms and pathologies can range from mild to extremely severe. In 1992, scientists at the Cystic Fibrosis Center at the University of North Carolina were able to breed mice that showed human symptoms of cystic fibrosis. This newest achievement provided researchers with a model with which to study the disease.

Cystic fibrosis can be fatal if the mucus blocks the lungs. Patients may suffer from pneumonia caused by bacterial infections. Other serious complications include respiratory failure, diabetes, enlarged **heart**, liver cirrhosis, intestinal blockage, pancreatic disfunction, sodium deficiency, and sterility.

Abdominal cramps, malnutrition, growth retardation, and coughing are all symptoms associated with cystic fibrosis. However, the increased salinity of sweat is the most useful test to diagnose the disease. It is difficult to predict when any of these symptoms will appear or how severe they will be. While the disease used to be fatal to nearly all children who developed it, more than 50% of cystic fibrosis patients now live longer than thirty years.

Treatments are available for cystic fibrosis, but there is no cure. Often, **antihistamine**s and decongestants are prescribed to open air passages. Cough suppressants are avoided since coughing helps to loosen the mucus in the trachea and lungs. **Antibiotics** help to treat pneumonia, and studies of the over-the-counter anti-inflammatory drug ibuprofen indicate that it can slow pulmonary decline. Physical therapy and surgery have also been used.

In 1990, researchers used laboratory **cell** cultures to correct the genetic defect that causes cystic fibrosis. They knew that the accumulation of mucus was caused by a blockage of the cell channel used to regulate the flow of sodium and chloride ions (Na^+ and Cl^-) in and out of the cell. The proper balance of these ions maintains the **water** balance of the mucus outside the cell. This equilibrium is controlled by a specific **protein** produced by two copies of a single gene. If both genes are defective, cystic fibrosis occurs.

Scientists are currently investigating **gene therapy** as an approach to treating cystic fibrosis. The process involves inserting a copy of the normal cystic fibrosis gene into a **virus** that has been altered not to cause disease. The virus is then inserted into lung cells removed from a cystic fibrosis patient. Researchers have found that the normal gene reaches the **DNA** of the lung cells and begins producing the correct protein, leading to decreased mucus **secretion**. A unique approach to apply this technology to human use is the development of an aerosol spray. However, since viruses can cause problems in cystic fibrosis patients like inflamed lungs and swollen nostrils which make breathing more difficult, the spray approach uses liposomes (a sphere spontaneously formed when fat **molecule**s are in solution) to act as the gene carrier, or vector. The liposomes are coated with the healthy cystic fibrosis genes and then sprayed in the nostrils. Scientists have found that some of the genes do make it to cell nuclei in the lungs. Research continues into the effectiveness of this approach.

In 1996, scientists discovered that cystic fibrosis patients have a genetic defect that hinders the proper absorbtion of salt into lung epithelial cells. In turn, the resultant excess salt content inhibits a naturally occurring antimicrobial agent produced by these epithelial cells. With this natural defense system disabled, the **immune system** compensates by mounting a strong attack on **bacteria**. Unfortunately, this immune response is so overwhelming that it causes **inflammation** of the lung's minute branching airways, which further exacerbates the formation of mucus.

Rapid discoveries concerning the genetic factors in cystic fibrosis have led to research into new therapeutic approaches. Researchers at the National Institutes of Health, for example, have developed a "gene-assist" drug called CPX that helps promote chloride ion transport.With the development of new genetic testing technologies, physicians are able to test for a wider variety of the most common cystic fibrosis mutations known. As a result, a National Institutes of Health consensus panel has recommended that genetic testing should be offered to couples expecting a baby or planning to become pregnant, especially if they have a family history of cystic fibrosis. Many scientific and ethical concerns are associated with genetic testing of cystic fibrosis. Genetic testing sensitivity for cystic fibrosis can vary greatly, which carries the risk of not identifying a carrier or fetus with the disease. In addition, genetic counseling should be offered to couples at risk, but there is a shortage of genetic counselors. Finally, information about a person's genetic predisposition to cystic fibrosis and other diseases could lead to discrimination by employers and insurance companies.

CYTOCHROME

Since the mid-1800s it was known that oxygen was carried by the red **blood** corpuscles from the lungs to the **cell**s. Exactly what happened to the oxygen from there was still not known as late as the early 1920s.

In 1924, a Russian-born British parasitologist/ biochemist named David Keilin (1887-1963) discovered the missing element. Keilin was engaged in a study of the horse botfly. Using a microspectroscope, he noticed a four-banded absorption spectrum in the botfly's muscles. When he shook the cell suspension with air, the four bands disappeared, then reappeared soon afterward. Keilin reasoned that a pigment, or substance, in the cells absorbed the oxygen, which was what caused the spectrum to disappear as the cells mixed with air. Keilin called this respiratory **enzyme** *cytochrome* and suggested in his first paper on the subject in 1925 that it also acted as a catalyst for the combination of oxygen with hydrogen within the cells. While an earlier spectroscopic observer, C.A. MacMunn, had reported in 1884 a four-banded spectrum linked to the **respiration** process, Keilin established the existence of cytochrome and its function.

The German biochemist **Otto Warburg** furthered knowledge about cytochrome in the late 1920s. Noting that **carbon monoxide** molecules attach themselves to cytochrome as they do to **hemoglobin**, Warburg found that the respiratory enzyme iron oxygenase was a **protein** with iron-containing groups. It was the iron, Warburg concluded, that activated the oxygen transfer within the cells. For this discovery, Warburg was awarded the 1931 Nobel Prize for physiology or medicine.

It is now known that there are several forms of cytochrome in the body, most notably, cytochrome c. The different forms of cytochrome are involved in many important metabolic processes, such as electron transport systems and ATP production.

CYTOKINESIS • See Cell division

CYTOLOGY

Cytology is the branch of biology that studies **cell**s, the building blocks of life. The name for this science is translated from *kytos*, the Greek term for "cavity." Cytology's roots travel back to 1665, when British botanist **Robert Hooke**, examining a cross-section of cork, gave the spaces the name "cells," meaning "little rooms" or "cavities."

Cytology's birth as a science occurred in 1839 with the first accurately conceived **cell theory**. This theory maintains that all **organism**s—**plant**s and **animal**s alike — are comprised of one or more like units called cells. Each of these units individually contains all the properties of life, and is the cornerstone of virtually all living organisms. Further, cell theory states that hereditary traits are passed on from generation to generation via **cell division**. Cell division generally has a regular, timed cyclical period during which the cell grows, divides or dies. Virtually all cells perform biochemical functions, generating and transmitting **energy**, and storing genetic data carried down to further generations of cells. Cytology differs from its cousin, pathology, in that cytology concentrates on the structure, function and biochemistry of normal and abnormal living cells. Pathology pursues changes in cells caused by decay and **death**.

Cells can vary dramatically in size and shape from organism to organism. While plant and animals cell diameters generally average between 10-30 micrometers (0.00036-0.00108 inches), sizes can range from a few thousand atomic diameters for single-celled **microorganism**s, all the way up to 20 in (50 cm) diameters for the monocellular ostrich **egg**. Cell structures also differ between advanced single-celled and multicellular organisms (plants and animals) and more primitive prokaryotic cells (e.g., **bacteria**). Plant cells are the most representative of a prototypical cell, as they have a **nucleus**, cell **membrane** and cell wall. Animal cells, on the other hand, lack a formalized cell wall, although they contain the former two. **Prokaryote** cells (e.g., bacteria) are unique in that they lack a nucleus and possess no membrane-enclosed **organelle**s. Exceptions to the cell theory include syncytial organisms (e.g., certain slime molds and microscopic flatworms) without cellular partitions; however, they are derived secondarily from organisms with cells via the breakdown of cellular membranes. Finally, the number of cells within an organism can range from one for organisms like an amoeba, to 100 trillion cells for a human being.

Cytology has greatly benefitted from the electron **microscope**, which reveals internal and external cell dynamics too small to be monitored by traditional optical microscopes. Also, fluorescence or contrast microscopy with more traditional visual observation equipment enables the cell substance to be revealed when a specific cell material is stained with a **chemical compound** to illuminate specific structures within the cells. For example, basic dyes (e.g., hematoxylin) illuminates the nucleus, while acidic dyes (e.g., eosin) stain the **cytoplasm** (the cellular material within the membrane (excluding the nucleus). Finally, newer techniques including radioactive isotopes and high-speed **centrifuge**s have helped advance cytology.

Cytological techniques are beneficial in identifying the characteristics of certain hereditary human diseases, as well as in plant and animal breeding to help determine the chromosonal structure to help design and evaluate breeding experiments. A far more controversial discussion deals with the role of cytology as it relates to **cloning**.

Over time, cytology's prominence as a separate science has diminished, integrating into other disciplines to create a more comprehensive biological-chemical approach. Associated disciplines include cytogenetics (study of behavior of **chromosome**s and genes relating to **heredity**) and cytochemistry (study of chemical contents of cells and **tissue**s).

CYTOPLASM

Cytoplasm (also called **protoplasm**) is the fluid or semi-solid medium enclosed by the outer cellular or **plasma membrane** and outside the cell **nucleus**. Basically an aqueous solution rich in **protein**s, **carbohydrate**s, **lipid**s, **nucleic acid**s, and salts, the cytoplasm is the environment within which all the reactions of cellular **metabolism** take place. In many cells, the cytoplasm is in constant motion, transporting materials from one place to another. One of the best places to see this is in slime molds, **protozoa**, and certain algal **species**, where an ordinary light **microscope** shows streams of cytoplasm coursing through **cell** interiors.

One of the most interesting features of cytoplasm is its ability to change from a liquid (or sol) state to semisolid (or gel) state. This transition is brought about by polymerization of microtubules, microfilaments and other elements of the protoplasm **cytoskeleton**. Contraction of these cytoskeletal elements generates hydrostatic pressure that causes streaming of the liquid phase of the cytoplasm. Movement of molecular motors such as dynein along the elements of the cytoskeleton also helps move materials through the cytoplasm.

In addition to dissolved compounds, the cytoplasm also contains a wide variety of **organelle**s. Among the prominent **membrane**-bound cytoplasmic organelles are **mitochondria**, **chloroplast**s, the **endoplasmic reticulum**, and the **Golgi apparatus**. Some macromolecular structures, such as **ribosome**s, **centriole**s, microtubules, and microfilaments, could also be considered cytoplasmic organelles.

CYTOSKELETON

Eukaroytic **cell**s (those with a **membrane**-enclosed **nucleus**) have a great variety of complex shapes and are able to carry out coordinated and directed movement due to a complex, dynamic, internal scaffolding of **protein**s called the cytoskeleton. Many different types of proteins make up this three-dimensional structure, which usually extends throughout the cell interior. The main components of this apparatus are microtubules (stiff, hollow rods about 25 nanometers in diameter made of tubulin), microfilaments (thin, flexible, double-stranded helical **polymer**s around 5 nanometers in diameter made of globular actin **molecule**s), and intermediate filaments (tough, strong filaments 10-11 nanometers in diameter composed of a family of insoluble proteins including keratin, vimentin, desmin, neurofilament proteins, and nuclear lamins). The cytoskeleton is connected by linker proteins to both the **plasma membrane** enclosing the cell as well as to **organelle**s within both the nucleus and the **cytoplasm**. Other proteins known as molecular motors (such as dynein and myosin) are able to ratchet along the cytoskeleton to change cell shape as well as to move components around within the cell.

Some of the movements generated by elements of the cytoskeleton can be seen in the light **microscope**. For example, localized contraction of the actin network around the cell periphery creates hydrostatic pressure that results in cytoplasmic streaming in large cells such as amoeba. Similarly, microtubules attached to **chromosomes** during **cell division** slide past others anchored to centrisomes and **centriole**s creating the movement that separates **genetic material** during **mitosis** and **meiosis**. The importance of these structures can be demonstrated with chemical or physical treatments that disrupt them. Both cell organization and **metabolism** are disturbed by these treatments.

See also Cilia; Ectoplasm; Flagella

D

DA VINCI, LEONARDO (1452-1519)
Italian artist and scientist

Leonardo da Vinci has been called one of the world's few universal geniuses because of his knowledge and abilities in so many different areas of intellectual and artistic pursuit. While perhaps best known as the artist who created the paintings, the *Mona Lisa* and *The Last Supper*, he also was an accomplished sculptor, engineer, mechanic, inventor and architect. His detailed drawings of the human body linked art with science to provide a means for investigation into the human form and **anatomy.**

Leonardo was born in 1452 in Vinci, Italy, near the larger city of Florence. He began his formal artistic studies when he was 15, as an apprentice for a local artist named Andrea del Verrocchio. He studied painting, mechanical arts and sculpture, all of which served him well when he started his first job in 1482 as artist and engineer in residence for the duke of Milan, Italy. During his 17 years there, he became well-known for his painting abilities, as well as his designs for artillery, fortresses, canal locks and other mechanical needs. During this time he completed six paintings, including *The Last Supper*, painted on a wall in a Milan monastery. When the duke was forced out of Milan by the French in 1499, Leonardo returned to Florence.

While in Florence, Leonardo continued his artistic and engineering work. He painted one wall of the new city hall while Michelangelo worked on another; however, because he tried to use a new technique that didn't work, his portion was never completed. While working on that project he painted the *Mona Lisa*, probably the most famous painting in the world. At the same time his interest in scientific areas greatly expanded, and he dissected human and animal corpses to identify the form and function of each body part. His detailed drawings of the human body are considered the first accurate portrayals of the human anatomy.

Leonardo painted more than 17 paintings over his lifetime, and started several sculptures. His best legacy, however, are the prolific workbooks he wrote and sketched in constantly from his earliest years. He focused on four primary themes—the science of painting, architecture, mechanics, and human anatomy—as well as adding notes about **botany**, geology and hydrology. The greatness of his artistic and intellectual abilities are evident throughout the 31 volumes. For example, he created plans for a helicopter, airplane, parachute, war tank and machine gun, all of which were not invented until hundreds of years later. The drawing of human proportions called *Vitruvian man* is almost as famous as his paintings. Unlike most texts, the illustrations provide the primary information and the words further explain the drawings. He also wrote the text backwards, so that the page can only be read by another person when held up to a mirror. The best explanation for this is that Leonardo was left-handed, as it was not his intention to keep the notebooks private.

Leonardo spent the last part of his life living as a guest of Pope Leo X at the Vatican Palace, from 1513 to 1515, as did several of the prominent artists of the time. While there, he completed a series of drawings entitled The Deluge, which portrayed the world's destruction by a flood. These drawings combine the two elements that were the focus of Leonardo's life: the forces of life and nature. In 1515 he accepted an invitation to live and work at the palace of the French king, Francis I, where he virtually stopped all painting to focus on scientific topics. He lived in France until his death on May 2, 1519.

DALE, HENRY HALLETT (1875-1968)
English physiologist

Henry Hallett Dale was a British physiologist who devoted his scientific career to the study of how chemicals in the body regulate physiological functions. In 1936 Dale and German pharmacologist **Otto Loewi** were jointly awarded the Nobel Prize in physiology or medicine for research demonstrating that nerve **cells** communicate with one another primarily by the ex-

Leonardo da Vinci. (The Library of Congress. Reproduced by permission.)

change of chemical transmitters. In addition to his scientific work, Dale was a prominent figure in science and medicine in England at critical junctures in that nation's history. He was knighted in 1932.

Born June 9, 1875, in London, Henry Hallett Dale was the second son of seven children born to Charles Dale, a London businessman, and his wife, Frances Hallett Dale. After graduating from Tollington Park College, London, and the Leys School, Cambridge, Dale entered Trinity College at Cambridge University in 1894. His academic skills gained him first honors in the natural sciences and the Coutts-Trotter studentship at Trinity College.

Dale left Cambridge in 1900 to finish his clinical work in medicine at St. Bartholomew's Hospital in London. He received his bachelor's degree in 1903, and his medical doctorate in 1909. During this time, he also was awarded the George Henry Lewes studentship, which allowed him to pursue further physiological research. Later, Dale also received the Sharpey studentship in physiology at University College, London. Dale used these opportunities for research from 1902 to 1904, studying with **Ernest Henry Starling** and **William Maddock Bayliss** at University College. Starling and Bayliss identified

secretin—a substance secreted by the small intestine—as the first hormone, and Dale collaborated with the pair in further studies on the impact of secretin on cells in the **pancreas**. Dale's work with Starling and Bayliss instilled in him the idea that physiological functions could be affected by such chemicals as **hormones**. It was also in this laboratory that Dale first met Otto Loewi, who at the time was visiting University College from Germany. Dale and Loewi would go on to become lifelong friends, collaborators, and co-recipients of the 1936 Nobel Prize.

In 1904 Dale spent three months working in the laboratory of the chemist **Paul Ehrlich** in Germany. Members of Ehrlich's laboratory were studying the relationship between the chemical structure of biological molecules and their effect on immunological responses, research that would garner for Ehrlich the 1908 Nobel Prize in physiology or medicine. As did the experience at Starling's laboratory in London, Ehrlich's research introduced Dale to the potential impact that chemicals can have on mediating biological and physiological processes.

After Dale returned to Starling's London laboratory, he was recommended to chemical manufacturer Henry Wellcome for a position with London's Wellcome Physiological Research Laboratories, a commercial laboratory. Established in the 1890s to produce an antitoxin for diphtheria, the laboratories, by the first decade of the 1900s, had begun to promote and pursue basic scientific research.

Once Dale had settled at Wellcome, the company suggested that he consider examining the therapeutic properties of ergot, a fungus being used by obstetricians to induce and promote labor. For the next decade, Dale devoted his research efforts to studying the properties of the drug, leading to the accidental discovery of the phenomenon of adrenaline (or epinephrine) reversal, in which the normally excitatory effects of these drugs are neutralized.

Dale's research on the effects of ergot also introduced him to ongoing efforts to study the central nervous system. Dale showed, with the chemist George Barger, that epinephrine is one chemical in a class of such chemicals that has "sympathomimetic" properties.

Dale's accomplishments drew the attention of Henry Wellcome, and Dale was promoted in 1906 to the directorship of the Wellcome Laboratories. Dale began studies of the chemicals that operate in the posterior pituitary lobe of the **brain**.

Dale resigned from the Wellcome Laboratories in 1914, and joined the scientific staff of the Medical Research Committee; after 1920 this group came to be known as the Medical Research Council. During World War I, Dale joined the war effort by engaging in physiological studies of shock, dysentery, gangrene, and the effects of inadequate diet.

After the war, the Medical Research Council evolved to become the National Institute for Medical Research, and Dale served as the organization's first director from 1928 until 1942. His research efforts during the 1920s continued the work he began during the war—studying how histamine contributes to the swelling of tissue after traumatic shock. Dale demonstrated that histamine leads to the loss of plasma fluid into the tissues and produces swelling. This could lead to more serious problems, including decreased blood circulation, shock, and death.

Dale's study of histamine also contributed to his subsequent work on the **nervous system**. Histamine, like the neurotransmitter acetylcholine, dilates vascular tissue in the human body.

In 1927 Dale collaborated with H. W. Dudley to isolate acetylcholine from the spleen of an ox and a horse. Having isolated the crucial compound, Dale sought to understand how and where acetylcholine plays its role in vasodilatation, or the widening of the cavities of blood vessels. Over the next decade, Dale worked with colleagues at the National Institute for Medical Research and concluded that acetylcholine serves as a neurotransmitter and that this is the chemical mediator involved in the transmission of nerve impulses. Dale's findings disproved the proposition of **John Carew Eccles** and other neurophysiologists who maintained that nerve cells communicate with one another via an electrical mechanism. Dale demonstrated that a chemical process and not an electrical one was the underlying mechanism for nerve transmission. A similar conclusion had been reached by Otto Loewi: As early as 1921 Loewi suggested that a chemical mediator was responsible for the conduction of nerve impulses; it would be Dale who would identify the mediator.

For their work, Dale and Loewi were jointly awarded the 1936 Nobel Prize in physiology or medicine. During the 1930s, Dale continued collaborative research with G. L. Brown, W. Feldberg, J. H. Gaddum, and M. Vogt at the National Institute for Medical Research. Their efforts produced more evidence that acetylcholine is a neurotransmitter involved in nerve impulses.

During World War II, Dale served as chair of the Scientific Advisory Committee to the War Cabinet. Having been elected a fellow of the Royal Society in 1914, he served as secretary from 1925 to 1935, and as president from 1940 to 1945. His many other public affiliations included serving as president of various organizations, such as the Royal Institution of Great Britain during the mid–1940s, the British Association for the Advancement of Science in 1947, the Royal Society of Medicine from 1948 to 1950, and the British Council during the 1950s.

Other distinctions bestowed upon Dale include receiving the Copley Medal from the Royal Society in 1937 and being knighted with the Grand Cross Order of the British Empire in 1943. He also garnered the Order of Merit in 1944. Since 1959 the Society for Endocrinology has awarded the Dale Medal for the kind of excellence in research exemplified by Dale; and since 1961 the Wellcome Trust he chaired from 1938 until 1960 has endowed the Henry Dale professorship with the Royal Society.

In later years Dale worked with Thorvald Madsen of Copenhagen directing an international campaign to standardize drugs and vaccines. The 1925 conference of the Health Organization of the League of Nations adopted such standards for insulin and pituitary products largely because of Dale's efforts. He repeated these efforts to see into law the Therapeutic Sub-

stances Act in England. His other political activities included promoting both the peaceful use of nuclear energy and the value of scientific research. He died on July 23, 1968, after a brief illness.

DAM, HENRIK (1895-1976)
Danish biochemist

Henrik Dam is best known for his discovery of **vitamin K**, which gives blood the ability to clot, or coagulate. The discovery of vitamin K dramatically reduced the number of deaths by bleeding during surgery, and for the discovery Dam received the 1943 Nobel Prize in medicine and physiology. (**Edward A. Doisy**, the American biochemist who isolated and synthetically produced vitamin K, shared this prize with Dam.)

Carl Peter Henrik Dam was born in Copenhagen, Denmark, on February 21, 1895. His interest in science was shaped at least in part by his background. His father, Emil Dam, was a pharmaceutical chemist who wrote a history of pharmacies in Denmark. His mother, Emilie Peterson Dam, was a schoolteacher. He attended the Polytechnic Institute in Copenhagen, from which he received his master of science degree in 1920. He was associated with the Royal School of Agriculture and Veterinary Medicine in Copenhagen for the next three years, after which he spent five years as an assistant at the University of Copenhagen's physiological laboratory. He became assistant professor of biochemistry in 1928 and associate professor in 1929 (a post he held until 1941).

During these years Dam studied microchemistry under Fritz Pregl in Austria (1925) at the University of Graz, and collaborated with biochemist Rudolf Schoenheimer in Freiburg, Germany (on a Rockefeller Fellowship) from 1932 to 1933. He was awarded a doctorate in biochemistry by the University of Copenhagen in 1934. Afterwards, he worked with the Swiss chemist Paul Xarrer at the University of Zurich in 1935. Dam specialized in nutrition, which became his area of expertise.

Experiments with the Diet of Hens

It was while Dam was studying in Copenhagen that he became interested in what would become the vitamin K factor. In the late 1920s he began experimenting with hens in an attempt to discover how the animals synthesized **cholesterol**. Providing them with a synthetic diet, Dam discovered that they developed internal bleeding in the form of hemorrhages under the skin—lesions similar to those found in the disease scurvy. He added lemon juice to the diet (citrus fruits, high in vitamin C, had been found by the eighteenth century Scottish physician James Lind to cure scurvy in sailors), but the supplement did little to reverse the hens' condition.

After experimenting with a variety of food additives, Dam came to the conclusion that some vitamin must exist to give blood the ability to clot—and that this vitamin was what was missing from his synthetic hen diet. He made his findings known in 1934, naming the vitamin ''K'' from the German word *Koagulation*. Dam's continued research, along with the work of Doisy and other biochemists, led to the isolation of vitamin K and its synthetic production.

Dam's discovery proved vitally important in two areas: in surgical procedures and in treatment of newborn babies. Prior to surgery, patients are given vitamin K to assist in clotting the blood and reduce the risk of death by hemorrhage. Newborns are born deficient in vitamin K. Normally, beneficial bacteria that exist in the environment enter the intestinal tracts of infants and induce production of vitamin K. Modern hospitals are disinfected to such an extreme, however, that they kill these good bacteria along with the harmful ones. Mothers are injected with vitamin K shortly before giving birth to ensure that adequate amounts of the vitamin will be in the newborn's system.

Accepts Nobel Prize in New York

Dam's discovery led not only to the Nobel Prize but also the Christian Bohr Award in Denmark in 1939. Dam came to the United States in 1940 for a series of lectures in the U.S. and Canada under the auspices of the American-Scandinavian Foundation. During his visit Nazi Germany invaded Denmark. Dam chose not to return to his native country and accepted a position as senior research associate at the University of Rochester's Strong Memorial Hospital. Because of the war, the Nobel Prize Committee decided to present the awards in New York in 1943. The Nobel recipients of that year, including Dam, were the first to be awarded their prize in the United States. In 1945, Dam became an associate member of the Rockefeller Institute for Medical Research.

After Denmark was liberated, Dam returned in 1946 to accept the position of head of the biology department at the Polytechnic Institute (the position had been awarded to him in absentia in 1941). He returned to the U.S. in 1949 for a three-month lecture tour, this time to discuss vitamin E. In 1956, he was named head of the Danish Public Research Institute. He was a member of numerous organizations including the American Institute of Nutrition, the Society for Experimental Biology and Medicine, the Royal Danish Academy of Science, the Société Chimique of Zurich, and the American Botanical Society. During his career he published more than one hundred articles in scientific journals on vitamin K, vitamin E, cholesterol, and a variety of other topics. Dam married Inger Olsen in 1925. His primary form of recreation was travel. After he returned to Denmark, he pointedly criticized the American hospital system, saying it was hurt by too much emphasis on the business of running hospitals. He died in Copenhagen at the age of eighty-one on April 17, 1976. At his request, news of his death was delayed by one week to allow for private services.

DANIELL, JOHN FREDERIC (1790-1845)
English chemist

John Frederic Daniell was a scientist and inventor whose widespread interests reflected the relatively unified nature of the science of his day.

His contributions to technology started with the introduction of improvements to the sugar refining industry when he was a young man. Later, he brought improvements in the lighting industry to Europe and America by producing a new gas from a distillate of resin in turpentine.

In 1820, he made a contribution to meteorology by inventing the dew-point hygrometer, an instrument designed to measure relative humidity. It consisted of two thin glass bulbs suspended from a base and connected with a glass tube. One bulb contained liquid ether and a thermometer. As the other bulb was slowly cooled and reheated, dew would appear and disappear at the end with the thermometer The mean temperature of this action was taken as the dew point. A description of this was published in 1820.

He was rewarded by horticulturalists with a medal for his suggestion that the humidity and temperature in hothouses should be regulated. In 1836 he invented a new battery, the Daniell cell. Unlike the recently invented zinc-copper voltaic battery, its current did not decline rapidly. By introducing a barrier between the zinc and copper, he was able to stop the formation of hydrogen, which was impairing battery function.

Daniell was active as a teacher, writer and illustrator, and made himself known as a social philosopher. His early death occurred while attending a Royal Society meeting in London in 1845.

DART, RAYMOND A. (1893-1988)
Australian anatomist and anthropologist

A doctor and surgeon by training, Raymond A. Dart was drawn into the field of anthropology early in his career, abandoning plans to become a medical missionary in China. Shortly after beginning his academic career in South Africa, and partly out of necessity, Dart was to discover the first fossils of **Australopithecus** *africanus*, or "southern ape of Africa," forging the modern era of paleoanthropology.

Raymond Arthur Dart was born to Samuel Dart, a general store operator, and the former Eliza Anne Brimblecombe, on February 4, 1893, the fifth of nine children. Devout Baptists and pioneers in the settlement of Queensland, Australia, Dart's parents raised him in the Brisbane suburb of Toowong, later moving the family to a bush farm in Blenheim, where the future scientist spent his youth milking cows and hiking to school. In 1911, scholarship in hand, he entered Brisbane's newly founded University of Queensland, where he earned both bachelor's and master's degrees in biology. Another scholarship sent him in 1914 to St. Andrew's College at the University of Sydney, where, before the completion of his second year of medical studies, he was appointed as a tutor in biology and granted membership on the college staff. The year 1914 also saw the outbreak of World War I in Europe as well as the British Association's meeting in Sydney. Dart attended the gathering, which brought in noted scientists such as Grafton Elliot Smith and W. J. Sollas. The conference signalled a turning point in Dart's career as he became intrigued by the announcement of the discovery of Australia's first human fossil find, the Talgai Skull from Queensland, unearthed by Antarctic geologic explorer T. Edgeworth David. The description was presented by James T. Wilson, head of the university's department of anatomy. Soon thereafter Dart became an assistant to Wilson on neurological research and came to regard him as a mentor.

Dart's intentions to conduct more in-depth neurological research were hampered by his own schoolwork and administrative roles as demonstrator in anatomy and acting vice principal of St. Andrew's College. After receiving his bachelor's degree in medicine and a master's degree in surgery in August of 1917, he enlisted in the Australian Army Medical Corps, finishing his service in France as a captain. Upon his release from the military in 1919, he was immediately appointed to the post of senior demonstrator in anatomy at University College, London, by Grafton Elliot Smith. In 1920, at Elliot Smith's recommendation, the Rockefeller Foundation awarded Dart one of its first two foreign fellowships, allowing him the opportunity to teach at Washington University in St. Louis, and study at the Wood's Hole marine research station in Massachusetts.

By 1922, Dart had rejoined the University's faculty as a lecturer in **histology** and **embryology**. At the insistence of Elliot Smith, however, Dart applied for the vacant anatomy chair at the newly established School of Medicine at the University of Witwatersrand in Johannesburg, South Africa, a place of which he had never heard. "The very idea revolted me; I turned it down flat instantly," he wrote in recollections for the *Journal of Human Evolution.*

Acquires Rare Fossil

After some reflection, Dart changed his mind and arrived in South Africa in January of 1923. Dart found the School of Medicine in dire need of equipment, facilities, and a collection of bones with which to create a proper anatomy museum. To acquire the latter, Dart encouraged his students to search for **fossil** bones during holidays. In the summer of 1924, a fossilized baboon skull was brought to Dart's attention by Josephine Salmons, one of his student demonstrators, who had secured the skull from E. G. Izod of Rand Mines Limited. The fossil had been detected while mining a sheet of limestone at Taungs (now Taung) in the Bechuanaland Protectorate, where other fossil baboon skulls had been discovered as early as 1920. His interest piqued, Dart asked his colleague, geology professor R. B. Young, to look for similar specimens since Young was going to investigate the lime deposit at Taungs. By November 28, 1924, Dart had in his hands a fossil skull that would change the face of paleoanthropology.

Young sent back two crates full of bones, one of which held a face and skull still embedded in matrix, along with the internal cast of a cranium found by one of the quarry workers. A quick examination of this endocast brought to light a startling revelation, derived from Dart's earlier neurological studies with Elliot Smith. He had learned that convolutions toward the back of the primate brain cause two fissures which are farther apart in humans than in apes. This, Elliot Smith attributed to the evolutionary expansion of the cerebrum. Yet, the cast Dart then held in his hand exhibited fissures that were farther apart than those displayed by any primate he had ever seen.

Dart freed the face and skull from its rock casing shortly before Christmas of 1924. The face was nearly complete, exhibiting cranial and mandibular features of humanoid rather than anthropoid characteristics. The shape of the jaw and alignment of the teeth also resembled those of humans, as did the set of emerging molars which Dart likened to those of a six-year old child or slightly younger ape. Because of the near completeness of the specimen, Dart was able to measure the position of the foramen magnum—the opening at the base of the skull through which the spinal cord enters the cranial cavity. In his preliminary account of the specimen, published in *Nature* on February 7, 1925, he noted that the foramen magnum's "relatively forward situation" suggested "an attitude appreciably more erect than that of the modern anthropoids.... The specimen is of importance because it exhibits an extinct race of apes intermediate between living anthropoids and man." Dart believed he had found the "missing link," and that his discovery might bare out English naturalist Charles Darwin's earlier revelation that man's origins were linked to Africa. He named the creature *Australopithecus africanus*, the "southern ape of Africa."

Findings Are Subject of Controversy

Dart's views were immediately met with derision by the general public and adamant disagreement by many of his own colleagues, including Elliot Smith and Arthur Keith, who immediately categorized the find within the fossil family of modern gorillas and chimps. His opponents proposed a lack of evidence in morphology and geologic age, and were appalled that the discovery was made in Africa. During that era, Asia was seen as the cradle of humankind. Yet, Dart did have his defenders, particularly in the person of Scottish anthropologist Robert Broom, whose defense of Dart has been compared to English biologist Thomas Henry Huxley's staunch support of Darwin. Broom's successive series of Taung-like fossil discoveries in Sterkfontein in 1936, and Kromdraii in 1938, would turn the tide of evidence in favor of Dart's South African ape-man.

Despite the controversy that surrounded him, Dart was elected president of the Anthropological Section of the South African Association for the Advancement of Science in 1925, became dean of Witwatersand's School of Medicine in 1926, and was appointed vice president of the Anthropology Section of the British Association in 1929. The following year, with a growing belief in man's African origins, he set out on a series of archeological and anthropological expeditions. His joint venture with the Italian African Scientific Expedition of 1930 gave him his first glimpse of the gorilla in its natural habitat, precluding his sponsorship of the University of Witwatersrand's Gorilla Research Unit in the late 1950s. Dart's investigation of the Auni-Khomani groups of Southern Bushmen in 1936 was considered the most complete physical study conducted up to that time.

The year 1945 proved another turning point in Dart's scientific career. Again, baboon fossils had been found in South Africa, this time at Makapansgat in the northern Transvaal. Just as the baboons had foreshadowed the appearance of

Australopithecus africanus at Taung, as well as Broom's Sterkfontein, so did they at Makapansgat. The subsequent excavation of the site during April of 1947, turned up some three dozen australopithecine fossils, a number of fossilized baboon skulls, and thousands of animal bone fragments. Fractures in many of the baboon skulls found at Makapansgat, Sterkfontein, and Taungs, as well as those found in six australopithecine skulls, led Dart to the conclusion that his ape men had inflicted these mortal blows using bone weapons such as an antelope's upper arm bone, evident in abundance around the Makapansgat site. He further concluded that theirs was a culture adept in the manufacturing of tools and weapons from bones, teeth, and horns, and thus named it the *osteodontokeratic culture*. While this theory was ultimately rejected by the scientific community, Dart's observations helped create a new field of science called taphonomy, concerning the environmental circumstances that act upon bones after death.

Divorced from his first wife, Dora Tyree, in 1934, Dart married again on November 28, 1936. His second wife, Marjorie Gordon Frew, was chief librarian of the Witwatersrand Medical Library. They had two children, Diana Elizabeth and Galen Alexander. Dart retired from the chair of anatomy in 1958. From 1966 to 1986, he spent half of each year in Johannesburg and the other half in Philadelphia, where he had been appointed United Steelworkers of America Professor of Anthropology in the Avery Postgraduate Institute of the Institutes for the Achievement of Human Potential.

On the advent of his eightieth birthday, Dart remarked upon his achievements in the *Journal of Human Evolution*. ''To open closed doors, to find lost things, or to shed light where gloom enshrouded understanding, these are but types of the privileges common to all intelligent individuals.'' Dart died on November 22, 1988.

DARWIN, CHARLES (1809-1882)
English naturalist

One of the most influential scientists of the nineteenth century, Darwin is best known for establishing the theory of organic **evolution** by natural selection.

Darwin was born in Shrewsbury, England, the son of a respected physician. He was the grandson of the poet-physician **Erasmus Darwin** and the porcelain manufacturer Josiah Wedgwood (1730-1795). At age 16, Darwin entered the University of Edinburgh in Scotland with the expectation of becoming a physician. While attending medical classes, however, he was unable to watch actual surgical procedures (often done without anesthesia), and so he was pressured by his family to consider the ministry. Darwin transferred to Christ's College at Cambridge three years later to study theology, but he discovered that he had no religious aspirations, either.

One positive outcome for Darwin during his years at Cambridge was having the opportunity to meet Adam Sedgwick (1785-1873), a professor who interested him in geology. More importantly, he befriended John Stevens Henslow (1796-1861), a botany professor who drew out Darwin's pas-

sionate interest in natural history and helped him gain confidence. After graduating from Cambridge in 1831, Henslow invited Darwin to join the crew on a government survey ship, the H.M.S. *Beagle*, as an unpaid naturalist on a five-year voyage to South America and the South Pacific Islands. Despite his father's reservations, Darwin accepted the offer without hesitation and set sail in December 1831.

While in Brazil, Darwin found his first **fossil**, the skull of an extinct giant sloth. For the next three years, Darwin made geological and biological observations, took records, and collected specimens of every kind as the ship cruised back and forth along the coasts of South America. Darwin had begun to notice evidence that animals and plants had undergone evolutionary changes. In some areas, **species** had become extinct, like the gigantic fossil armadillos in South America. Yet Darwin noticed similar but not identical armadillos in other areas nearby. He was perplexed over the fact that existing species had demonstrated characteristics similar to those of extinct species. He also found slightly similar, though clearly different, species located in a variety of places around the world, but also completely lacking in other parts of the world. Moreover, Darwin was intrigued that the flora and fauna of oceanic islands were likely to resemble the same animal and plant species found on the neighboring continents. He thought it peculiar that islands with the same geological and physical features could be home to completely different animal species.

Four years after having set sail, Darwin landed in the Galápagos Islands, where he would make the most significant observations of the expedition. Darwin noticed that there were around 14 different types of finch birds on different islands of the Galápagos. Each type of finch appeared to have adapted completely to the island on which it lived. Moreover, some with sharper, finer beaks fed on insects and were more suited to stabbing their prey, while others ate seeds and had more powerful, parrot-like bills for breaking the shells. Another curiosity were the giant tortoises that appeared similar but possessed many distinctive features. The local island inhabitants could tell at sight from which island any of the giant creatures had come. Darwin began to ask if all of this biological diversity was arbitrary or whether a pattern of meaning could be discerned. Then a possible explanation began to emerge; he realized that species had to be mutable and diverged instead of fixed in form according to their original ancestry. A common ancestor could explain the similarities, but Darwin began to guess that each species could have given rise to new ones.

Upon returning to Britain in October 1836, Darwin's ideas came into focus and he began to synthesize a theory to explain his premonition. He began by asserting that if species had transformed, the issue of diversity was satisfied, but a whole new range of questions emerged. He asked why the bones of a human's arms and legs are similar in general to those of a dog and a horse. He questioned why the embryos of lizards and rabbits are similar, while their adult forms are different. He noticed that many animals, including humans, have functionless organs (e.g., the appendix) and wondered why many different organisms behave in similar ways. Darwin concluded that many of these questions were probably answerable, but only if species were related by descent from common ancestors.

About 50 years earlier, scientists had begun to suggest that there might be a common plan, or a connectivity among animals, that manifested itself in similarities between invertebrates and vertebrates. However, ideas of evolution were cautious and tentative until Darwin published his groundbreaking study *The Origin of Species by Means of Natural Selection* in 1859. Calling his **evolutionary theory** the process of *natural selection*, Darwin asserted that the living organisms that best fit their environment were more likely to survive and pass their characteristics on to their offspring. For example, the white fur of the polar bear blends in with the snowy environment, strongly contrasting with the brown and black fur of bears living in the forest. These different traits among similar animals represent genetic adaptations to different environments.

Darwin's observations were so convincing that he succeeded in persuading most of the scientific community of the real possibility of natural selection and evolution. Toward the end of the 1800s, however, as people increasingly accepted his ideas, they also recognized that Darwin lacked an explanation of how variations were produced or passed on. Without knowing how such variations occurred, natural selection could achieve nothing, asserted Darwin's critics. This rift in Darwin's theory lasted until his ideas merged with those of **Gregor Mendel**. Mendel proposed in 1866 that the **gene** was the basic unit of **heredity**. Although his work was not formally acknowledged until the early 1900s, Mendel demonstrated that genes are the molecular blueprints that are passed on to succeeding generations. With the knowledge of genes as the basic units of heredity, evolutionists (known as neo-Darwinists) were satisfied with the explanation that natural selection involves the evolution of not only physical and behavioral traits, but also the genes that serve as blueprints for those traits.

Darwin initially concentrated on animals and plants, but the ascent or evolution of humankind became the central focus of his book *The Descent of Man* (1871). Many people were repulsed at the suggestion that human beings could somehow be related to earlier, non-human life forms. In addition, many objections to the book were based on its implicit rejection of man's miraculous creation. Such criticism has lately reemerged in the United States in the form of creationism, which holds that evolution is nothing more than atheist speculation.

Darwin's important contributions to the biological sciences remain invaluable. Today, despite many modifications to evolutionary theory, it still plays a critical role in modern biology.

See also Cuvier, Georges; Genetic code

DARWIN, ERASMUS (1731-1802)

English physician

Erasmus Darwin was an English physician who had a significant influence on the development of theories of **evolution**. In particular, he helped to initiate the idea that traits developed by an **organism** during its life could somehow be passed on to its offspring. This idea was explained in Darwin's book

Charles Darwin. *(The Library of Congress. Reproduced by permission.)*

Zoonomia, or The Laws of Organic Life, which was published in 1794-1796. The idea was later expanded upon as "the inheritance of acquired traits" by the French naturalist Chevalier de Lamarck (1744-1829), who has become much more strongly identified with the idea than Erasmus Darwin.

However, few modern biologists believe that acquired traits can, in fact, be inherited by organisms. It is well known that individual organisms may display variable anatomical, biochemical, or behavioral traits as they develop through life (that is, their expressed **phenotype**). For example, a plant well supplied with nutrients, moisture, and light will be larger and more robust than if that same plant did not experience those relatively beneficial conditions. Such variable developmental possibilities are now known to be due to differing expressions of the fixed genetic potential of individual organisms (or their **genotype**; this refers to the specific qualities of their genetic material, or **DNA** [deoxyribonucleic acid]). Modern biologists refer to this variable expression of the genetic potential of organisms, as influenced by the environmental regime they encounter during development, as "phenotypic plasticity." In the times of Erasmus Darwin, Lamarck, and even **Charles Darwin**, it was thought that these variably expressed (or acquired) traits could somehow become fixed into the genetic make-up of an organism and its offspring, but this is now known to not be possible.

Erasmus Darwin was also the grandfather of **Charles Darwin**, one of the most famous naturalists of all time. Charles

Darwin is best known for his theory of the role of natural selection in driving evolutionary change, published in 1859 in his famous book, *On the Origin of Species*.

DAUSSET, JEAN (1916-)
French immunologist

Jean Dausset was born in Toulouse, France, and moved with his family to Paris at age eleven. After having earned an undergraduate degree in mathematics, he enrolled in the University of Paris medical school. When World War II broke out in 1939, Dausset joined the French medical corps. After France fell to the Germans in 1940, Dausset went to North Africa to fight with the Free French army. Before leaving Paris, Dausset gave all his identification papers to a Jewish colleague, who was therefore able to survive the Nazi occupation.

After the war ended, Dausset completed his studies at the university, receiving his medical degree in 1945. From 1946 to 1958 Dausset was director of laboratories at the French National Blood Transfusion Center, although he left for two years (1948-49) to study hematology under a fellowship at Harvard Medical School.

Dausset's work at the **blood** center, and earlier experiences during the war with transfusions, drew him into studies of abnormal transfusion reactions. In 1952 he discovered an antigen on certain people's white blood cells. By 1958, when he joined the medical faculty at the University of Paris, he had found more variants of the **antigen**. In 1965, Dausset suggested that these and other newly discovered antigens were all part of a single set of linked **gene**s, which constituted the human MHC (major histocompatibility complex). He called this the human leucocyte antigen group (HLA) and, in 1967, showed that tissue transplants are more successful when donor and recipient have matching HLA types. In 1967 Dausset was the first researcher to investigate possible links between an individual's HLA types and the person's risk for disease. For all of these findings about HLA, Dausset shared the 1980 Nobel Prize for physiology or medicine with George Snell and Baruj Benacerraf. Dausset had become head biologist of Paris's city hospital system in 1963 and also headed the Institute for Research into Diseases of the Blood. In 1968 he assumed directorship of the French National Institute for Scientific Research and became a professor of medicine at the University of Paris and, since 1978, the College of France. He remained at the university through the 1980s, continuing his studies of HLA. He has since founded the Human Polymorphism Study Center in Paris and has helped organize a collaboration to map the human **genome**. As his personal motto he chose "*Vouloir pour valoir*," or, roughly translated, "To achieve a lofty goal, you must aspire to it."

DAVIDSON, NORMAN R. (1916-)
American organic chemist

One of the hallmarks of Norman R. Davidson's career has been the tendency to find new and interesting projects in areas relatively unrelated to those in which he has previously worked. Davidson received his doctoral degree shortly before the beginning of World War II and, thus, some of his earliest research was related to military programs. After returning to an academic career at the end of the war, he began a series of studies on the kinetics of ultrafast reactions. In order to study these reactions, Davidson found it necessary to develop new techniques and instruments that could detect the presence of various chemical species over time periods of less than 10^{-5} second.

By the 1960s, Davidson had found a new topic of interest, the structure and function of **deoxyribonucleic acid (DNA)**. Again, Davidson developed methods of studying this important molecule that had previously been unavailable to chemists. Many of those techniques remain in wide use today. In recent years, Davidson has turned to yet another field of study, the structure and organization of **gene**s. In addition to his active research career, Davidson has been active in a variety of professional organizations and, in his eighth decade, remains Executive Officer of the Division of Biology at the California Institute of Technology.

Norman Ralph Davidson was born on April 5, 1916, in Chicago, Illinois. He attended the University of Chicago, from which he earned his B.S. degree in 1937 and his Ph.D. in chemistry in 1941. He also studied at Oxford University, which awarded him a B.Sc. degree in 1939. Davidson also holds an honorary degree, a Doctor of Science degree from the University of Chicago, awarded in 1992.

Ultrafast Reactions and War Research

Davidson's first major line of research focused on ultrafast reactions. In chemistry, it is often not difficult to write chemical equations that summarize the overall changes that take place during a reaction. However, the reaction itself generally tends to be much more complex, with a number of stages occurring between initial contact of products and final formation of products. The problem in studying these reaction stages is that they often occur very rapidly, often in much less than a thousandth of a second.

Davidson developed a variety of techniques for discovering what happens during these rapid reactions. One such technique is known as flash **photolysis**. In flash photolysis, reactants are exposed to very brief, high intensity pulses of radiant energy. This energy excites reactants in the mixture, and they give off characteristic spectral patterns that provide clues to their identity.

Davidson's work in this field was interrupted in 1942 by World War II. He joined the Manhattan Project and was assigned to research on the production of transuranium elements at the Division of War Research at Columbia University and, later, at the University of Chicago. He was also appointed Instructor of Chemistry at the Illinois Institute of Technology in 1942.

Joins Faculty at California Institute of Technology

At the war's conclusion, Davidson began working at the Radio Corporation of America (RCA), where he became interested in the development of an early electron microscope. After only a year at RCA, he accepted an offer to work at the California Institute of Technology (Caltech) as Instructor of Chemistry. Davidson has remained at Caltech ever since, eventually becoming full professor and then Norman W. Chandler Professor of Chemistry and Biology. In 1986, he was made Emeritus Professor in the Chandler chair. Since 1990, Davidson has also served as Executive Officer of the Division of Biology at Caltech and Senior Science Consultant at the Amgen Corporation, a pioneer in the industrial applications of genetic engineering technology.

Research on DNA and Genes

One of Davidson's earliest lines of research at Caltech involved studies of the DNA molecule. He claims that this research was motivated by the realization that DNA was not "just another **polymer**, but that it was an informational **molecule**." His research was part of a widespread attempt on the part of molecular biologists to find out how genetic information is stored and transmitted in chemical molecules such as DNA and **ribonucleic acid (RNA)**.

Among the critical breakthroughs made by Davidson was the process of DNA renaturation. Most beginning chemistry students are familiar with the process of **protein** denaturation, in which heat, exposure to chemicals, or other factors cause a protein molecule to lose its three-dimensional shape. A similar process occurs with DNA. Davidson discovered that DNA molecules once denatured, will renature spontaneously, that is, they will recover their original shape. The process by which renaturation occurs is a somewhat complex one whose elements Davidson was eventually able to decipher. An understanding of DNA renaturation has important applications in a number of fields, such as determining base sequences in a DNA molecule, comparing DNA from various organisms, and identifying RNA from particular **species**.

Davidson next turned to an analysis of the chemical composition of **gene**s. He decoded the gene structure for ribosomal RNA, as well as transfer RNA genes for *E. coli* plasmid F14 and for *B. subtilis*, as well as for a number of other cells. The motivation for this research, he said, is to discover what chemical mechanisms are involved in the expression of various genes at various stages of development.

In recent years, Davidson and his colleagues have turned to the study of the chemical structure of molecules involved in nerve transmission. One series of studies, for example, is designed to find out how changes in the chemical structure of molecules affects the ability of the potassium ion to flow through channels in neural membranes.

Davidson has been awarded the Peter R. Debye Award in Physical Chemistry in 1971, the Dickson Prize for Science in 1985, the Robert A. Welch Award in Chemistry in 1989, and the National Medal of Science in 1997. He was elected to the National Academy of Sciences in 1960 and to the American Academy of Arts and Sciences in 1984.

Norman R. Davidson. *(California Institute of Technology. Reproduced by permission.)*

DAVIS, MARGARET B. (1931-)
American paleoecologist

Margaret B. Davis is a distinguished paleoecologist noted for her analysis of ancient pollen to determine trends in **plant** growth and migration. Her work challenged the prevailing scientific idea that plant and animal communities tend to be stable, moving intact to new locations as the climate changes. By studying pollen from ancient plants (in a discipline known as **palynology**), she reconstructed past plant communities and showed how they change in response to variations in climate or other environmental influences. She found that associations between plant and animal communities are more fluid than once believed and that change is the order of the day for ecosystems.

Born in Boston, Massachusetts, on October 23, 1931, Margaret Bryan Davis grew up in the Boston area and graduated from Radcliffe College in 1953 with an A.B. degree in biology. While at Radcliffe, she took courses from noted paleobotanist Elso Borghoorn and was intrigued by the vegetational history of the late-Quaternary period, some ten thousand years ago. She believed that to best understand and interpret the history of ancient plant life it was more important to understand the physiology and ecology of flora, rather than just the stratigraphic interpretation of pollen records.

Davis was awarded a Fulbright fellowship and studied from 1953 to 1954 at the University of Copenhagen with palynologist Johannes Iversen of the Danish Geological Survey. Her work took her to Greenland where she recorded plant pollen deposited during the interglacial period. In 1954 she published her first paper, "Interglacial Pollen Spectra from Greenland," in a Danish geological journal. Combining both field work and paleoecology in Denmark, she returned to Boston and obtained a Ph.D. in biology from Harvard University in 1957. Davis was convinced that the growing new field of palynology was the best path to follow in tracing the history of ancient vegetation. Over the following four years, Davis continued her studies as a National Science Foundation postdoctoral fellow, first at Harvard, then at the California Institute of Technology where she concentrated in geology. In 1960, she became a research fellow at Yale University; her objective there was to clarify the relationship between pollen in lake sediments and vegetation composition, as a means to enhance the precision of pollen records for describing past vegetation.

In 1961, Davis accepted a position as a research associate in the University of Michigan's department of **botany**. She remained in Ann Arbor, Michigan, for the next twelve years and concluded her stay as professor of **zoology** before returning to Yale to serve as professor of biology. She remained at Yale for three years, then moved in 1976 to the University of Minnesota where she is Regents' Professor in the department of ecology, evolution and biology.

During Davis's stay at Michigan, she attracted international attention with a paper on the theory of pollen analysis, published in 1963 in the *American Journal of Science*. For years scientists had assumed that **fossil** pollen produced by trees tens of thousands of years ago could provide a clear picture of plant life during the period. However, since some species of **trees** produce more pollen than others, they had suggested that a correction factor ranging from 4:1 to 35:1 be used to equalize the difference. Davis' research on lake bed sediments showed that these correction factors were erroneous, and these factors could range as much as 24,000:1.

Davis' findings cast doubt on many theories of ancient plant life, and opened the door to a new scientific method of understanding the history and **ecology** of plants that lived on earth thousands of years ago. "These methods allow us a new approach to some long-standing environmental questions," Davis said in a University of Minnesota press release. "We can begin to distinguish between the impacts of humans and the natural process of change that has always occurred. For example, we can examine the response of **ecosystems** to past climatic change and use our findings to predict responses to future climatic warming." Davis has warned that the climatic changes predicted in response to the build-up of greenhouse gases will be at least one order of magnitude more rapid than climatic changes in the past, and will tax the ability of biotic systems to change and disperse from one area to another.

Davis also compiled maps for eastern North America depicting the **migration** of various **species** of **trees** during the past fourteen thousand years. Her maps indicate that the temperate forest trees moved at different rates and from different directions.

In 1982, Davis was elected to the National Academy of Sciences. She has also served as president of the American Quaternary Association and the Ecological Society of America. Davis is the author of more than sixty-five scientific publications and the recipient of numerous honors, including the Eminent Ecologist Award, bestowed by the Ecological Society of America. In 1993, she was also awarded the Nevada Medal and a cash prize of five thousand dollars in recognition for her work in "unlocking the history of environmental change and using it to understand present and future shifts in plant and animal communities."

DDT (Dichlorodiphenyl-trichloroacetic acid)

Dichlorodiphenyl-trichloroacetic acid (DDT) is a chlorinated **hydrocarbon** that has been widely used as an insecticide. DDT is virtually insoluble in **water**, but it is freely soluble in the fat of **organisms**. DDT is also persistent in the environment. The combination of persistence and lipid solubility means that DDT biomagnifies, occurring in organisms in preference to the non-living environment, especially in predators at the top of ecological food webs. Environmental contamination by DDT and related chemicals is a widespread problem, including the occurrence of residues in wildlife, in drinking water, and in humans. Ecological damages have included the poisoning of wildlife, especially predators.

DDT and other chlorinated hydrocarbons

Chlorinated hydrocarbons are a diverse group of synthetic compounds of carbon, hydrogen, and chlorine, used as **pesticides** and for other purposes. DDT is a particular chlorinated hydrocarbon with the formula 2,2-bis-(*p*-chlorophenyl)-1,1,1-trichloroethane.

The insecticidal relatives of DDT include DDD, aldrin, dieldrin, heptachlor, and methoxychlor. DDE is a related non-insecticidal chemical but an important, persistent, metabolic-breakdown product of DDT and DDD that is accumulated in organisms. Residues of DDT and its relatives are persistent in the environment, for example, having a typical half-life of 5-10 years in soil.

A global contamination with DDT and related chlorinated hydrocarbons has resulted from the combination of their persistence and a tendency to be widely dispersed with wind-blown dusts. In addition, their selective partitioning into **fats** and **lipids** causes these chemicals to bioaccumulate. Persistence, coupled with **bioaccumulation**, results in the largest concentrations of these chemicals occurring in predators near or at the top of ecological food webs.

Uses of DDT

DDT was first synthesized in 1874. Its insecticidal qualities were discovered in 1939 by **Paul Muller**, a Swiss scientist who won a Nobel Prize in medicine in 1948 for his research on the uses of DDT. The first important use of DDT was for the control of insect vectors of human diseases during and following World War II. At about that time the use of DDT to control pests in agricultural and forestry also began.

The peak production of DDT was in 1970 when 385.9 million lb (175 million kg) was manufactured globally. The greatest use of DDT in the United States was 79.4 million lb (36 million kg) in 1959, but the maximum annual production was 198.5 million lb (90 million kg) in 1964, most of which was exported. Because of the discovery of a widespread environmental contamination with DDT and its breakdown products and associated ecological damages, most industrialized countries banned its use in the early 1970s. Use of DDT continued elsewhere, however, mostly for control of insect vectors of human and livestock diseases in less developed tropical countries. Largely because of the evolution of resistance to DDT by many pest insects, its effectiveness for these purposes has decreased. Some previously well-controlled diseases such as malaria have even become more common in a number of countries (reduced effectiveness of some of the prophylactic pharmaceuticals used to threat malaria is also importance in the resurgence of this disease). Ultimately, the remaining uses of DDT will be curtailed and it will be replaced by other insecticides, largely because of its increasing ineffectiveness.

Until its use was widely discontinued because of its nontarget, ecological damages, DDT was widely used to kill insect pests of crops in **agriculture** and forestry and to control some human diseases that have specific insect vectors. The use of DDT for most of these pest-control purposes was generally effective. To give an indication of the effectiveness of DDT in killing insect pests, it will be sufficient to briefly describe its use to reduce the incidence of some diseases of humans.

In various parts of the world, **species** of insects and ticks are crucial as vectors in the transmission of disease-causing pathogens of humans, livestock, and wild **animals**. Malaria, for example, is a debilitating disease caused by the **protozoan**-*Plasmodium* and spread to people by mosquitoes, *Anopheles* spp. Yellow fever and related viral diseases such as encephalitis are spread by other species of mosquitoes. The incidence of these and some other important diseases can be greatly reduced by the use of insecticides to reduce the abundance of their arthropod vectors. In the case of mosquitoes, this can be accomplished by applying DDT or another suitable insecticide to the aquatic breeding **habitat,** or by applying a persistent insecticide to the walls and ceilings of houses which serve as resting places for these insects. In other cases, infestations of body **parasites** such as the human louse can be treated by dusting people with DDT.

The use of DDT has been especially important in reducing the incidence of malaria which has always been an important disease in warmer areas of the world. Malaria is a remarkably widespread disease, affecting more than 5% of the world's population each year during the 1950s. For example, in the mid-1930s an epidemic in Sri Lanka affected one-half of the population, and 80,000 people died as a result. In Africa, an estimated two to five million children died of malaria each year during the early 1960s.

Clutch of mallard eggs contaminated by DDT. The accumulation of DDT in many birds causes reproductive difficulties. Eggs have thinner shells that break easily, and some eggs may not hatch at all. *(National Geographic Image Collection. Reproduced by permission.)*

The use of DDT and some other insecticides resulted in large decreases in the incidence of malaria by greatly reducing the abundance of the mosquito vectors. India, for example, had about 100 million cases of malaria per year and 0.75 million deaths between 1933 and 1935. In 1966, however, this was reduced to only 0.15 million cases and 1,500 deaths, mostly through the use of DDT. Similarly, Sri Lanka had 2.9 million cases of malaria in 1934 and 2.8 million in 1946, but because of the effective use of DDT and other insecticides there were only 17 cases in 1963. During a vigorous campaign to control malaria in the tropics in 1962, about 130 million lb (59 million kg) of DDT was used, as was 7.9 million lb (3.6 million kg) of dieldrin and one million lb (0.45 million kg) of lindane. These insecticides were mostly sprayed inside of homes and on other resting habitat of mosquitoes rather than in their aquatic breeding habitat. More recently, however, malaria has resurged in some tropical countries, largely because of the development of insecticide resistance by mosquitoes and a decreasing effectiveness of the pharmaceuticals used to treat the actual disease.

Environmental effects of the use of DDT

As is the case of many actions of environmental management, there have been both benefits and costs associated with the use of DDT. Moreover, depending on socio-economic and ecological perspectives, there are large differences in the **perception**s by people of these benefits and costs. The controversy over the use of DDT and other insecticides can be illustrated by quoting two famous persons. After the successful use of DDT to prevent a potentially deadly plague of typhus among Allied troops in Naples during World War II, Winston Churchill praised the chemical as "that miraculous DDT powder." In stark contrast, Rachael Carson referred to DDT as the "elixir of death" in her ground-breaking book *Silent Spring*, the first public chronicle of the ecological damages caused by the use of persistent insecticides, especially DDT.

DDT was the first insecticide to which large numbers of insect pests developed resistance. This is an evolutionary process occurring because of the selection for resistant individuals that occurs when large populations are exposed to a toxic pesticide. Resistant individuals are rare in unsprayed populations, but after spraying they become dominant because the insecticide does not kill them and they survive to reproduce and pass along their genetically based **tolerance**. More than 450 insects and mites now have populations that are resistant to at least one insecticide. Resistance is most common in the flies (Diptera), with 156 resistant species, including 51 resistant species of malaria-carrying mosquito, 34 of which are resistant to DDT.

As mentioned previously, the ecological effects of DDT are profoundly influenced by certain of its physical/chemical properties. First, DDT is persistent in the environment because it is not readily degraded to other chemicals by **microorganism**s, sunlight, or heat. Moreover, DDE is the primary breakdown product of DDT, being produced by enzymatic **metabolism** in organisms or by inorganic de-chlorination reactions in alkaline environments. The persistences of DDE and DDT are similar, and once released into the environment these chemicals are present for many years.

Another important characteristic of DDT is its insolubility in water, which means that it cannot be "diluted" into this ubiquitous solvent, so abundant in Earth's environments and in organisms. In contrast, DDT is highly soluble in **fats and oils** (together known as **lipids**), a characteristic shared with other chlorinated hydrocarbons. In **ecosystem**s, most lipids occur in the **tissue**s of living organisms. Therefore, DDT has a strong affinity for organisms because of its lipid solubility, and it tends to biomagnify tremendously. Furthermore, top predators have especially large concentrations of DDT in their fat, a phenomenon known as food-web accumulation. In ecosystems, DDT and related chlorinated hydrocarbons occur in very small concentration in water and air. Concentrations in soil may be larger because of the presence of organic **matter** containing some lipids. Larger concentrations occur in organisms, but the residues in **plants** are smaller than in **herbivore**s, and the largest concentrations occur in predators at the top of the food web such as humans, predatory birds, and marine **mammals**. For example, DDT residues were studied in an estuary on Long Island where DDT had been sprayed onto salt marshes to kill mosquitoes. The largest concentrations of DDT occurred in fish-eating birds such as ring-billed gull (76 ppm), and double-crested cormorant, red-breasted merganser, and herring gull (range of 19-26 ppm).

Lake Kariba, Zimbabwe, is a tropical example of food-web bioconcentration of DDT. Although Zimbabwe banned DDT use in agriculture in 1982, it is still used to control mosquitoes and tsetse fly (a vector of diseases of cattle and other large mammals). The concentration of DDT in water of Lake Kariba was extremely small, less than 0.002 ppb, but larger in sediment of the lake (0.4 ppm). **Algae** contained 2.5 ppm, and a filter-feeding mussel contained 10 ppm in its lipids. Herbivorous fish contained 2 ppm, while a bottom-feeding species of fish contained 6 ppm. The tigerfish and cormorant (a bird) feed on small fish, and these contained 5 ppm and 10 ppm, respectively. The top predator in Lake Kariba is the Nile crocodile, and it contained 34 ppm. Lake Kariba exhibits a typical pattern for DDT and related chlorinated hydrocarbons-a large bioconcentration from water, and to a lesser degree from sediment, as well as a food-web magnification from herbivores to top predators.

Global contamination with DDT

Another environmental feature of DDT is its distribution everywhere in the **biosphere** in at least trace concentrations. This global contamination with DDT and related chlorinated hydrocarbons such as PCBs occurs because they enter into the atmospheric cycle and thereby become very widely distributed. This results from: (1) a slow evaporation of DDT from sprayed surfaces; (2) off-target drift of DDT when it is sprayed; and (3) entrainment by strong winds of DDT-contaminated dust into the **atmosphere**.

This ubiquitous contamination can be illustrated by the concentrations of DDT in animals in Antarctica, very far from places where it has been used. DDT concentrations of 5 ppm occur in fat of the southern polar skua, compared with less than 1 ppm in birds lower in the food web of the Southern Ocean such as the southern fulmar and species of penguin.

Much larger concentrations of DDT and other chlorinated hydrocarbons occur in predators closer to places where the chemicals have been manufactured and used. The concentration of DDT in seals off the California coast was as much as 158 ppm in fat during the late 1960s. In the Baltic Sea of Europe residues in seals were up to 150 ppm, and off eastern Canada as much as 35 ppm occurred in seals and up to 520 ppm in porpoises.

Large residues of DDT also occur in predatory birds. Concentrations as large as 356 ppm (average of 12 ppm) occurred in bald eagles from the United States, and up to 460 ppm in western grebes, and 131 ppm in herring gulls. White-tailed eagles in the Baltic Sea have had enormous residues-as much as 36,000 ppm of DDT and 17,000 ppm PCBs in fat, and **egg**s with up to 1900 ppm DDT and 2600 ppm PCBs.

Ecological damages

Some poisonings were directly caused by exposure to sprays of DDT. There were numerous cases of dying or dead birds being found after spraying of DDT, for example, after its use in residential areas to kill the beetle vectors of Dutch elm disease in North America. Spray rates for this purpose were large, about 1.5 –2.8 lb (0.7–1.4 kg) of DDT per **tree**, and resulted in residues in earthworms of 33-164 ppm. Birds that fed on DDT-laced **invertebrates** had intense exposures to DDT, and many were killed.

Sometimes, detailed investigations were needed to link declines of bird populations to the use of organochlorines. One such example occurred at Clear Lake, California, an important waterbody for recreation. Because of complaints about the nuisance of a great abundance of a non-biting aquatic insects called midges, Clear Lake was treated in 1949 with DDD at 1 kg/ha. Prior research had shown that this dose of DDD would achieve control of the midges but have no immediate effect on fish. Unfortunately, the unexpected happened. After another application of DDD in 1954, 100 western grebes were found dead as were many intoxicated birds. Eventually, the breeding population of these birds on Clear Lake decreased from about 2,000 to none by 1960. The catastrophic decline of grebes was linked to DDD when an analysis of fat of dead birds found residues as large as 1,600 ppm. Fish were also heavily contaminated. The deaths of birds on Clear Lake was one of the first well documented examples of a substantial mortality of wildlife caused by organochlorine insecticides.

Damages to birds also occurred in places remote from sprayed areas. This was especially true of raptorial (that is, predatory) birds, such as falcons, eagles, and owls. These are top predators, and they food-web accumulate chlorinated hydrocarbons to large concentrations. Declines of some species began in the early 1950s, and there were extirpations of some breeding populations. Prominent examples of predatory birds that suffered population declines from exposure to DDT and other organochlorines include the bald eagle, golden eagle, peregrine falcon, prairie falcon, osprey, brown pelican, double-crested cormorant, and European sparrowhawk, among others.

Of course, birds and other wildlife were not only exposed to DDT. Depending on circumstances, there could also be significant exposures to other chlorinated hydrocarbons, including DDD, aldrin, dieldrin, heptachlor, and PCBs. Scientists have investigated the relative importance of these chemicals in causing the declines of predatory birds. In Britain, the declines of raptors did not occur until dieldrin came into common use, and this insecticide may have been the primary cause of the damages. However, in North America DDT use was more common, and it was probably the most important cause of the bird declines.

The damage to birds was mainly caused by the effects of chlorinated hydrocarbons on reproduction and not by direct toxicity to adults. Demonstrated effects of these chemicals on reproduction include: (1) a decrease in clutch size (i.e., the number of eggs laid); (2) the production of a thin eggshell which might break under the incubating parent; (3) deaths of **embryo**s, unhatched chicks, and nestlings; and (4) pathological parental behavior. All of these effects could decrease the numbers of young successfully raised. The reproductive pathology of chlorinated hydrocarbons caused bird populations to decrease because of inadequate recruitment.

This syndrome can be illustrated by the circumstances of the peregrine falcon, a charismatic predator whose decline attracted much attention and concern. Decreased reproductive success and declining populations of peregrines were first noticed in the early 1950s. In 1970, a North American census reported almost no successful reproduction by the eastern population of peregrines, while the arctic population was declining in abundance. Only a local population in the Queen Charlotte Islands of western Canada had normal breeding success and a stable population. This latter population is non-migratory, inhabiting a region where pesticides are not used and feeding largely on non-migratory seabirds. In contrast, the eastern peregrines bred where chlorinated-hydrocarbon pesticides were widely used, and its prey was generally contaminated. Although the arctic peregrines breed in a region where pesticides are not used, these birds winter in sprayed areas in Central and South America where their food is contaminated, and their prey of migratory ducks on the breeding grounds is also contaminated. Large residues of DDT and other organochlorines were common in peregrinefalcons (except for the Queen Charlottes). Associated with those residues were eggshells thinner than the pre-DDT condition by 15-20% and a generally impaired reproduction.

In 1975, another North American survey found a virtual extirpation of the eastern peregrines, while the arctic population had declined further and was clearly in trouble. By 1985 there were only 450 pairs of arctic peregrines, compared with the former abundance of 5,000-8,000. However, as with other raptors that suffered from the effects of chlorinated hydrocarbons, a recovery of peregrine populations has begun since DDT use was banned in North America and most of Europe in the early 1970s. In 1985, arctic populations were stable or increasing compared with 1975 as were some southern populations, although they remained small. This recovery has been enhanced by a captive-breeding and release program over much of the former range of the eastern population of peregrine falcons.

It is still too soon to tell for certain, but there are encouraging signs that many of the severe effects of DDT and other chlorinated hydrocarbons on wildlife are becoming less severe. Hopefully, in the future these toxic damages will not be important.

DEATH

Death is the cessation of life. It involves a complete change in an **organism** and occurs on various levels, including somatic death, **organ** death, cellular death, and **organelle** death.

Somatic death refers to the death of a whole organism. It is typically preceded by the death of organs, **cells**, and cellular organelles. In **animal**s, it is characterized by the cessation of life functions such as **respiration**, **heart**beat, movement, and **brain** activity. However, it is difficult to determine the exact point of somatic death because it is similar to other conditions such as fainting or a coma.

A variety of changes occur after somatic death. One of the first changes is algor mortis. In this process the body cools, eventually reaching equilibrium with the surrounding **temper-**

ature. Another change is *rigor mortis* in which the skeletal muscles stiffen and become rigid. It typically begins five to ten hours of death and disappears after three or four days. *Livor mortis* is the condition in which the underside of the body takes on a reddish-blue color. This staining is the result of settling **blood** due to gravity. Since circulation ceases, blood clots begin to form right after death and cells begin to die (autolysis) in large numbers. Putrefaction is one of the final processes that occurs after death. This is the **decomposition** of the body by **microorganism**s such as **bacteria** and **fungi**. This process results in gases that give the body a greenish discoloration.

After somatic death, the body's organs die. However, the organs die at different rates. For example, the **heart** can survive for about 15 minutes after death. The kidneys and liver can last even longer, up to 45 minutes. This allows doctors to remove certain organs for transplant.

Cellular death occurs on a wide scale after a body dies, however, it also occurs on a smaller scale due to injury or disease. This process, known as necrosis, typically results in an inflammatory response. During the process of necrosis, a variety of cellular changes occur involving the cell's organelles, including the **plasma membrane**, **mitochondria**, **endoplasmic reticulum**, **lysosome**s, and the **nucleus**. The first noticeable change is a massive swelling of the cell. This is a result of an influx of ions caused because the cell loses its ability to actively transport them out of its plasma membrane. The nucleus also undergoes a similar change due to the lysis of its **membrane**. This process called karyolysis results in the leakage of **ribonucleic acid** (RNA) into the **cytoplasm** through nuclear pores. The lysosomes then release **enzyme**s, which are responsible for the degradation of cell components. After a cell dies, some of its organelles continue to function. For example the mitochondria may continue to function for over an hour. The endoplasmic reticulum may also continue to function, producing polyribosomes that synthesize **proteins**. The lysosomes continue enzyme activity for nearly two days after the death of a cell.

During a cell's **life cycle** autophagocytosis continually occurs. This is a process in which the organelles are naturally degraded. During autophagocytosis, sections of a cell's cytoplasm are enclosed in a phospholipid membrane that contains hydrolytic enzymes. Within these **vacuoles**, the contents (including any organelle) are broken down and recycled.

While the biological processes involved in death are generally understood, there is no single definition for what encompasses death. This is because the definition of death is a philosophical concept which is different for different cultures and has changed over time. Until the seventeenth century, western societies considered death to be the moment when the soul left the body. The determination was made when life signs such as breathing and **blood circulation** were not noticeable. As our understanding of life processes improved, society became concerned that people may have been declared dead before they actually were. During the eighteenth and nineteenth centuries, concern was so great that caskets were fitted with life signals so a person could send a signal if they were still alive before being buried.

Advances in technology have made the definition of death even more complex. In the past, death was assumed to occur when breathing and circulation had stopped. During these times, the brain would stop functioning fairly soon after the heart or lungs stopped. Today, modern medical devices have made it possible to maintain respiration and circulation even though spontaneous function has ceased. Thus, the concept of brain death was developed. In this definition, death is considered the point when there is an irreversible loss of brain activity. Even this definition has been challenged because it is possible that part of the brain could die even though lower-brain functions continue. As medical technology continues to improve, the debate on the definition of death will undoubtedly continue.

DECOMPOSITION

Decomposition is the process by which dead **organism**s and their wastes are broken down into an inorganic form usable by **plants** and other **autotroph**ic organisms. Decomposition is decay and rotting. The organisms that carry out decomposition are called decomposers, and they are primarily **bacteria** and **fungi**.

Because decomposers are not able to make their own food, they are considered **heterotroph**s. They must obtain all of their **nutrients** and **energy** from the food they consume.

Decomposition plays a major role in the cycling of nutrients through the food web. Organic nutrients are bound up in a living organism or within an organism's waste. Through the process of decomposition, they are released in an inorganic form that is usable by plants and other autotrophs. In effect, decomposition is responsible for recycling nutrients so that they can be reused. Some of the major nutrients that are recycled include **nitrogen**, phosphorus, carbon and oxygen. In addition, nutrients that are needed in smaller amounts, such as calcium, sulfur, iron and potassium, are also recycled through the process of decomposition. Without decomposition, an **ecosystem** would not be self-sustaining; it would be required to obtain the nutrients necessary for food production from outside the ecosystem.

In addition to cycling nutrients through the food web, decomposers are an important source of food for other **consumer**s. For example, some deposit-feeding organisms living in the muddy ocean bottom will feed by ingesting sediments. While the sediments themselves are later expelled, the bacteria of decay that live on the sediment particles are digested. In addition, other decomposers, such as fungi (e.g., mushrooms), are an important part of the diet of many organisms. Thus, decomposers have more than one role in an ecosystem's food web; they cycle nutrients and are a source of food for other organisms.

DECOMPRESSION SICKNESS

Just as people often feel pressure when flying in airplanes, professional and recreational divers are well aware of the drastic change in pressure as they descend into the **water** and return

Decomposition is the process by which dead organisms and their wastes are broken down into an inorganic form usable by plants and other autotrophic organisms. Because decomposition plays a major role in the cycling of nutrients through the food web, without it an ecosystem would not be self-sustaining. *(Photograph by Robert J. Huffman, Field Mark Publications. Reproduced by permission.)*

to the surface. Airplanes are pressurized, however, and the passengers usually experience only mild changes in air pressure as compared to divers, who may experience pressure changes many times greater than normal atmospheric pressure.

Unfortunately for divers, ascending too rapidly from depths greater than about 30 ft (9 m) can result in *decompression sickness*, also called dysbarism, caisson disease, the bends, and compressed air illness. When a diver remains at depth for sufficiently long periods of time, gases, especially **nitrogen**, accumulate in the body **tissue**s. If the diver surfaces too quickly, these nitrogen gases expand and form bubbles in various parts of the body. Mild cases involve only tingling of the skin, dizziness, or slight disorientation. When bubbles develop in nerves or joints, pain may occur. When they form in the **spinal cord** or **brain**, the diver can become blind, paralyzed, or even die. If the bubbles form in the lungs, the diver may choke, first experiencing rapid shallow breathing and ultimately death.

One of the earlier records of diving decompression problems appeared on a plaque at the entrance to Old Port Royal, Jamaica, West Indies in 1862. The Geological Society of Jamaica plaque states that Lucas Barret—age 25—died while conducting underwater research. Since then, diving has

become a much safer sport, especially after John Scott Haldane (1860-1936) developed a method for avoiding decompression sickness in 1907. He introduced the idea of stage decompression, in which the diver stops at prescribed depths for specific intervals of time during his or her ascent. As the diver waits at each level, the gases in the body tissues are allowed to adjust to the new pressure without forming hazardous bubbles.

Today, in deep commercial diving, this step-wise ascent has been replaced by the use of the decompression chamber. The diver is quickly brought to the surface before bubbles can form and then placed in the chamber. The air in the chamber is then compressed to the pressure that the diver experienced at depth. In some situations, divers actually enter the decompression chamber at depth and then are brought to the surface when it is convenient. In either case, over the next few hours or days — depending on the depth — the pressure is incrementally decreased and the diver's tissues slowly de-gas.

Divers are also susceptible to other gas-related illnesses. During rapid ascents, if air is unable to escape the lungs they may overexpand, resulting in an air embolism, which can be fatal. A combination of oxygen concentration and depth can also be deadly. For example, when breathing air (20% oxygen), toxic effects can occur below a depth of 132 ft (40 m), and in as little as 33 ft (10 m) with pure oxygen. Finally, nitrogen not only causes decompression sickness, it also has a narcotic effect that increases with depth. Nitrogen narcosis seriously impairs many divers in as little as 100 ft (30 m) of water. To combat decompression sickness, oxygen toxicity, and nitrogen narcosis, professional divers and to a lesser degree, recreational divers sometimes breath mixed gases rather than air. These consist of various mixtures of nitrogen, oxygen, and on occasion, helium, designed to decrease the side effects of atmospheric concentrations of nitrogen and oxygen.

DEHYDRATION SYNTHESIS

Dehydration synthesis is a **chemical reaction** in which one or more **molecule**s of **water** are removed from the reactants to form a new product. These reactions can occur when one of the reactants has a hydroxyl group (OH) that can be cleaved, thus forming the negatively charged hydroxide ion (OH $^-$). The other reactant must have a hydrogen atom which can be cleaved to yield a hydrogen, or hydronium, ion (H+). In solution, these ions are free to combine and form a water molecule. The respective reactants are then able to form a **chemical bond** that creates a new compound.

Because of the ready availability of their hydroxyl groups, many **alcohol**s readily participate in dehydration reactions. When an alcohol loses its hydroxyl group it can combine with other electron-loving **species** to form a new reaction product. If the alcohol loses the hydroxyl group in addition to one of its own hydrogen **atoms**, a new carbon-carbon bond can be formed within the molecule. When this occurs the alcohol is transformed into an unsaturated compound or into an ether. The nature of the reaction depends on the class of the alcohol. For example, ethyl alcohol, or ethanol, undergoes dehydration to form ethylene or diethyl ether. Certain **polymer**s are also formed by this type of reaction. Dehydration synthesis is essentially the opposite of **hydrolysis** which is the process of breaking down compounds by dissolving in water.

See also Protein synthesis

DEISENHOFER, JOHANN (1943-)

German biochemist and biophysicist

Johann Deisenhofer is a biochemist and biophysicist whose career has been devoted to analyzing the composition of molecular structures. An expert in the use of X-ray technology to analyze the structure of crystals, he became part of a team of scientists in the 1980s who were studying **photosynthesis**—the process by which plants convert sunlight into chemical energy. In 1988, he shared the Nobel Prize for Chemistry with Robert Huber and Hartmut Michel, awarded for their work in mapping the chemical reaction at the center of photosynthesis.

Deisenhofer was born September 30, 1943, in Zusamaltheim, Bavaria, approximately fifty miles from Munich, Germany. He was the only son of Johann and Thekla Magg Deisenhofer; his parents were both farmers and they expected him to take over the family farm, as was the tradition. It was clear from an early age, however, that Deisenhofer was not interested in agriculture, and his parents sent him away to school in 1956. Over the next seven years, Deisenhofer attended three different schools, graduating from the Holbein Gymnasium in 1963. He then took the *Abitur,* an examination German students must take in order to qualify for university. He passed the exam and was awarded a scholarship. He then spent eighteen months in the military, as was required for young German men, before enrolling at the Technical University of Munich to study physics. His interest in physics had been developed through reading popular works on the subject, and he had an early passion for astronomy. Deisenhofer soon found himself doing an increasing amount of work in solid-state physics, which concerns the structures of condensed matter or solids. He secured a position in the laboratory of Klaus Dransfeld, and there he narrowed his interests further to biophysics, the application of the principles of the physical sciences to the study of biological occurrences. In 1971, Deisenhofer published his first scientific paper and received his diploma, roughly equal to a master's degree. He then began work on his Ph.D. in biochemistry at the Max Planck Institute in Munich under the direction of Robert Huber. Here, Deisenhofer began using a technique known as X-ray crystallography, which had first been demonstrated by Max Laue in 1912.

A crystal is a solid characterized by a very ordered internal atomic structure. The structural base of any crystal is called a lattice, which is defined by M.F.C. Ladd and R.A. Palmer in *Structure Determination by X-ray Crystallography* as ''a regular, infinite arrangement of points in which every point has the same environment as any other point.'' Crystallography, the study of crystals, is considered a field of the physical sciences, and X-ray crystallography is the study of crystals using radiation of known length. When X rays hit crystals, they are scattered by electrons. Knowing the wavelength of the X rays used, and measuring the intensities of the scattered X rays, the crystallographer is able to determine first the specific electron structure of the crystal and then its atomic structure.

Deisenhofer finished work for his Ph.D. in 1974. He chose to remain in Huber's laboratory and continue his work with X-ray crystallography, first on a postdoctoral basis, and later as a staff scientist. At the same time, he was developing computer software to be used in the mapping of crystals. While working on his doctorate, Deisenhofer had embarked on a collaborative effort with Wolfgang Steigemann; they studied crystallographic refinement of the structure of Bovine Pancreatic Trypsin Inhibitor, and their findings were published in *Acta Crystallographica* in 1975.

Collaboration Leads to Nobel

In 1979, Hartmut Michel joined Huber's laboratory. He had been studying photosynthesis for several years and was trying to develop a method for a detailed analysis of the **molecules** essential to this reaction. Photosynthesis is a very complicated process, about which much is still not known. The photosynthetic reaction center, which is a membrane **protein**, is considered a key to understanding the process, since it is here the electron receives the energy which drives the reaction. In 1981, Michel discovered a way to crystallize the photosynthetic reaction center from the purple bacterium *Rhodopseudomonas viridis.* Once Michel had developed this technique, he turned to Huber for help in analyzing it. Huber directed Michel to Deisenhofer, and a four-year collaboration began.

Deisenhofer, with Kunio Miki and Otto Epp, used his X-ray crystallography techniques to determine the position of over 10,000 atoms in the molecule. They produced the first three-dimensional analysis of a membrane protein. *New Scientist* magazine, as quoted in *Nobel Prize Winners Supplement 1987–1991,* called the combined efforts ''the most important advance in the understanding of photosynthesis for twenty years.'' The Royal Swedish Academy of Sciences awarded the 1988 Nobel Prize for Chemistry jointly to Huber, Michel, and Deisenhofer for this work. Their findings opened the possibility of creating artificial reaction centers, but the scientists were credited with more than an increase in knowledge of photosynthesis. Their findings will aid efforts to increase the scientific understanding of other functions, such as respiration, nerve impulses, hormone action, and the introduction of nutrients to cells. Deisenhofer and Michel were also recipients of the 1986 Biological Physics Prize of the American Physical Society and the 1988 Otto-Bayer Prize.

In 1987, Deisenhofer accepted the Virginia and Edward Linthicum Distinguished Chair in Biomolecular Science at the University of Texas Southwestern Medical Center at Dallas; his goal there is to establish a major center for X-ray crystallography. He has continued his research interests in the areas of protein crystallography, macromolecules, and crystallographic software. Deisenhofer has been awarded the Knight Commander's Cross of the Order of Merit of the Federal Republic of Germany, as well as the Bavarian Order of Merit. He is a fellow of the American Association for the Advancement of Science and a member of the American Crystallographic Association, the German Biophysical Society, and Academia Europa. In 1993, Deisenhofer, with James R. Norris of the Argonne National Laboratory, published a two-volume book called *The Photosynthetic Reaction Center,* based on work that grew out of Diesenhofer's collaboration with Michel.

Deisenhofer was married in 1989 to a fellow scientist, Kirsten Fischer Lindahl. He enjoys music, history, skiing, swimming, and chess in his free time. After Diesenhofer won the Nobel Prize, Dr. Kern Wildenthal, president of the Southwestern Medical School, described him to the *New York Times* as ''very shy'' and a man whose ''life was his work.'' Wildenthal further observed that the scientist is ''quiet, peaceful and calm. But beneath that exterior, he is scientifically fearless.''

DELBRÜCK, MAX (1906-1981)
German American molecular biologist

Max Delbrück, the youngest of seven children, was born in Berlin, Germany, on September 6, 1906. In 1924 he began uni-

Max Delbrück. *(The Library of Congress. Reproduced by permission.)*

versity studies in astronomy and astrophysics and in 1930 received a Ph.D. in physics from the University of Göttingen. As a post-doctoral student, he traveled to Copenhagen, Denmark, where he worked with Niels Bohr, the theorist who proposed the model of the atom. After further work in Zurich, Switzerland, and Berlin, Delbrück left Germany for the United States in 1937, following Hitler's takeover.

In the United States, Delbrück studied biology and genetics at the California Institute of Technology under a Rockefeller Foundation grant and, while studying the genetics of fruit flies, became interested in the genetics of **bacteriophages**, large viruses that infect bacteria. Along with Emory Ellis, a biologist, Delbrück developed experimental methods to investigate bacteriophages and mathematical systems to analyze the results of the experiments. They published their results in 1939. In 1940, Delbrück met **Salvador Luria**, an Italian-American biologist conducting bacteriophage research at Columbia University. The two found that they shared interests and thus began a collaboration of research focused on mutations in bacteria that produced resistance to bacteriophages. Together, Delbrück and Luria published their work in 1943. They presented the first evidence that bacterial heredity is controlled by **genes**, and in doing so, overturned prevailing ideas about how genetic traits are acquired. Their work began the sciences of bacterial genetics and **molecular biology**.

Delbrück's work with Luria and Alfred Hershey, a microbiologist, standardized bacteriophage research. In 1946 Delbrück and Hershey independently demonstrated that genetic material from different **viruses** could be combined to form a virus different from either. They called this phenomenon genetic recombination, and their groundbreaking work paved the way for other geneticists to make further discoveries about bacteriophages, culminating with the discovery that **deoxyribonucleic acid**, or DNA, is the genetic material. By the early 1950s Delbrück received a letter from his friend James Watson, detailing the **double helix** structure of DNA. Delbrück's casual manner and wit provided a relaxed atmosphere, and thus his laboratory became a meeting place for many molecular biologists working on problems in genetics. His contributions were recognized in 1969 when he, Hershey, and Luria shared the Nobel Prize in physiology or medicine for work in replication and genetic structure of viruses.

DENATURATION

Denaturation describes physical changes in the three dimensional structure of **protein**s. These changes occur when the proteins are exposed to disruptive chemical or physical forces. These structural changes not only affect the physical form of the proteins but also their chemical functionality. One common example of a denatured protein is **egg** albumin which changes from a gelatinous liquid to a solid when heated.

Proteins are large **molecule**s composed of **amino acid**s which are arranged in a variety of complex structures. The primary protein structure is the simple linear sequence of amino acids within the protein. The secondary structure is determined by the configuration of certain subgroups of amino acids within the primary structure. These subgroups can be configured in three different ways depending on the type of protein. The alpha helix form can be thought of as a spiral staircase. It is found in structural proteins, such as the keratins in hair and skin. The beta-pleated sheet form is, as the name implies, a flat sheet shape, and it is found in silk fiber proteins. The third type of secondary structure is the random coil which does not have a defined shape and which is found in proteins such as collagen. The random coil structure links together the alpha helices and beta sheets so that proteins may contain all three secondary structures.

Proteins also adopt a tertiary structure which is achieved by looping and folding the chain over itself. This folded structure occurs because certain portions of the molecules have an affinity for **water** and other portions do not. Therefore, proteins will fold or bend into shapes so that the water loving or hydrophilic groups are on the outside of the molecule and the hydrophobic groups are buried in the internal parts of the molecule where they are shielded from the water. This folding process causes the proteins to take on complex helical patterns which result in stable structures and give the protein specific chemical and physical properties. For example, the way **enzyme**s are folded gives rise to specific catalytic areas, causing different enzymes to have different capacities for catalyzing **chemical reaction**s. Similarly, the structure of **hemoglobin** is responsible for that molecule's ability to carry oxygen.

When a protein is denatured, the molecule's tertiary structure is corrupted. This disruption affects the molecule's secondary (helical) structure without altering its primary structure. In other words, denaturation does not break any of the primary **chemical bond**s that link one amino acid to another but it changes the way the protein folds in upon itself. Denaturation occurs when proteins are exposed to strong acids or bases, high concentrations of inorganic salts, or organic solvents such as **alcohol**. In addition, heat or irradiation can cause denaturation. When the three-dimensional structure of the protein is disrupted, the molecule's biological activity is affected. Therefore, enzymes do not have the same catalytic function when they are denatured. Some proteins may be renatured by exposing the denatured protein to a solution that approximates normal physiological conditions.

Denaturation can have many detrimental side effects. In biological systems, denatured proteins can result in illness or even **death**. In fact protein denaturation is linked to many diseases such as prion encephalopathies, Alzheimer's disease and dementias. Denaturation can also have negative effects certain industrial processes. For example, in dairy processing certain milk proteins are denatured as a result of separation techniques and heat treatments. These treatments change the way the proteins, **minerals** and ions in the milk interact and may affect the milk's **nutrition**al content. Research has shown that whey protein denaturation can be reduced by the addition of small amounts of nonfat dry milk. Lowering the calcium ion concentration and reducing the **pH** of the milk protein concentrate also reduces the degree of denaturation. These findings will hopefully result in improved dairy processing techniques that enhance product stability.

Not all denaturing processes are harmful. Certain denaturing processes are beneficial. For example, albumin, a protein found in eggs is easily denatured. When albumin is denatured it forms a gelatinous solid which is capable of absorbing foreign **matter**. This property makes denatured albumin suitable for certain important industrial applications such as sugar refining and the manufacture of adhesives, varnishes, and inks.

Denaturation also has another, completely unrelated, meaning. It is the term that describes the process used to make ethanol undrinkable. In the United States drinking alcohol is heavily taxed but alcohol used industrially as a solvent is not. To prevent people from drinking the cheaper industrial grades of alcohol, trace amounts of contaminants are added which would induce vomiting or make the alcohol taste so terrible that no one would drink it. In the past, materials such as methanol, camphor, aldehol, amyl alcohol, gasoline, isopropanol, terpineol, benzene, castor oil, acetone, nicotine, acid, kerosene, and diethyl phthalate have been used as denaturants.

DENITRIFICATION

Denitrification is the conversion of nitrates to gaseous **nitrogen**. It is carried out by several genera of free-living soil **bacteria**. Denitrifying bacteria are especially active in **water**logged, **anaerobic** soils where they tend to deplete soil nitrates, forming free atmospheric nitrogen. Organic **matter** in the soil tends to promote denitrification when oxygen from the nitrate ion is used to oxidize the organics in the absence of atmospheric oxygen. Denitrification, thus, supports active **anaerobic** growth of the denitrifying **organism**s. In agricultural soils denitrifying bacteria can reduce soil fertility and lessen agricultural **productivity** by removing growth promoting nitrate from the soil before **plants** are able to utilize it. In swampy soils, denitrifiers may reduce the amount of soil nitrogen by as much as 50%. Some of the microbial forms involved in denitrification include *Thiobacillus denitrificans*, *Micrococcus denitrificans*, and **species** of *Serratia*, *Pseudomonas*, and *Achromobacter*. Denitrification has its positive side, however. The process helps to reduce the amount of nitrogen salts that might otherwise accumulate in oceans and lakes. Nitrates are highly soluble and readily leached from soil to aquifers, streams, and lakes, where they might accumulate to toxic levels. Converting this over supply of nitrate to the nontoxic nitrogen gas removes the threat to plants and **animal**s that might otherwise be harmed.

See also Nitrogen cycle

DEOXYRIBONUCLEIC ACID • see DNA

DEPRESSION

Depression is a psychoneurotic disorder characterized by lingering sadness, inactivity, and difficulty in thinking and concentration. A significant increase or decrease in appetite and time spent sleeping, feelings of dejection and hopelessness, and sometimes suicidal tendencies may also be present. It is one of the most common psychiatric conditions encountered, and affects up to 25% of women and 12% of men. Depression differs from grief, bereavement, or mourning, which are appropriate emotional responses to the loss of loved persons or objects.

Depression has many forms and is very responsive to treatment. Dysthymia, or minor depression, is the presence of a depressed mood for most of the day for two years with no more than two months' freedom from symptoms. Bipolar disorder (manic-depressive disorder) is characterized by recurrent episodes of mania and major depression. Manic symptoms consist of feelings of inflated self-esteem or grandiosity, a decreased need for sleep, unusual loquacity, an unconnected flow of ideas, distractibility, or excessive involvement in pleasurable activities that have a high potential for painful consequences, such as buying sprees or sexual indiscretions. Cyclothymia is a chronic mood disturbance and is a milder form of bipolar disorder.

Chemically speaking, depression is apparently caused by reduced quantities or reduced activity of the monoamine **neurotransmitters** serotonin and norepinephrine within the **brain**. Neurotransmitters are chemical agents released by **neuron**s (nerve **cell**s) to stimulate neighboring neurons, thus allowing electrical impulses to be passed from one cell to the next throughout the **nervous system**. They transmit signals between nerve cells in the brain.

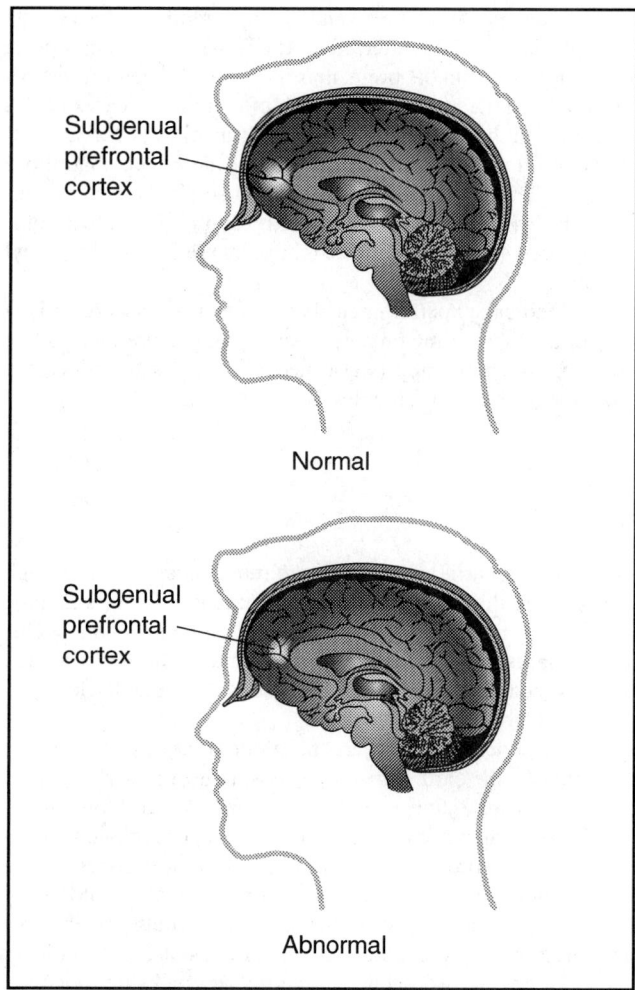

Subgenual prefrontal cortex

Normal

Subgenual prefrontal cortex

Abnormal

Recent scientific research has indicated that the size of the subgenual prefrontal cortex of the brain (located behind the bridge of the nose) may be a determining factor in hereditary depressive disorders. (Illustration by Electronic Illustrators Group.)

Introduced in the late 1950s, antidepressant drugs have been used most widely in the management of major mental depression. All antidepressants accomplish their task by inhibiting the body's re-absorption of these neurotransmitters, thus allowing them to accumulate and remain in contact longer with their receptors in nerve cells. These changes are important in elevating mood and relieving depression.

Antidepressants are typically one of three chemical types: a tricyclic antidepressant (so called because its **molecule**s are composed of three rings), a monoamine oxidase (MAO) inhibitor, or a serotonin reuptake inhibitor. The tricyclic antidepressants act by inhibiting the inactivation of norepinephrine and serotonin within the brain.

The MAOs apparently achieve their effect by interfering with the action of monoamine oxidase, an **enzyme** that causes the breakdown of norepinephrine, serotonin, and dopamine within nerve cells.

In the 1980s a new type of antidepressant called a serotonin reuptake inhibitor proved markedly successful. Its chemical name is fluoxetine, and it apparently achieves its therapeutic effect by interfering solely with the reabsorption of serotonin within the brain, thus allowing that neurotransmitter to accumulate there. Fluoxetine often relieves cases of depression that have failed to respond to tricyclics or MAOs, and it also produces fewer and less serious side effects than those drugs. It had thus become the single most widely used antidepressant by the end of the twentieth century. The most commonly used serotonin reuptake inhibitors are Prozac, Paxil and Zoloft.

Medical experts agree that antidepressants are only a part of the therapeutic process when treating depression. Some form of psychotherapy is also needed in order to reduce the incidence for chronic recidivism of the illness.

DESERT

A desert is an arid land area where more **water** is lost through evaporation than is gained from **precipitation**. Deserts include the familiar hot, dry desert of rock and sand that is almost barren of **plants**, the semiarid deserts of scattered **trees**, scrub, and grasses, coastal deserts, and the deserts on the polar ice caps of the Antarctic and Greenland.

Most desert regions are the result of large-scale climatic patterns. As the earth turns on its axis, large air swirls are produced. Hot air rising over the equator flows northward and southward. The air currents cool in the upper regions and descend as high pressure areas in two subtropical zones. North and south of these zones are two more areas of ascending air and low pressures. Still farther north and south are the two polar regions of descending air. As air rises, it cools and loses its moisture. As it descends, it warms and picks up moisture, drying out the land. This downward movement of warm air masses over the earth have produced two belts of deserts. The belt in the northern hemisphere is along the Tropic of Cancer and includes the Gobi Desert in China, the Sahara Desert in North Africa, the deserts of southwestern North America, and the Arabian and Iranian deserts in the Middle East. The belt in the southern hemisphere is along the Tropic of Capricorn and includes the Patagonia Desert in Argentina, the Kalihari Desert of southern Africa, and the Great Victoria and Great Sandy Deserts of Australia.

Coastal deserts are formed when cold waters move from the Arctic and Antarctic regions toward the equator and come into contact with the edges of continents. The cold waters are augmented by upwellings of cold water from ocean depths. As the air currents cool as they move across cold water, they carry fog and mist, but little rain. These types of currents result in coastal deserts in southern California, Baja California, southwest Africa, and Chile.

Mountain ranges also influence the formation of deserts by creating rain shadows. As moisture-laden air currents flow upward over windward slopes, they cool and lose their moisture. Dry air descending over the leeward slopes evaporates moisture from the soil, resulting in deserts. The Great Basin Desert was formed from a rain shadow produced by the Sierra

Nevada mountains. Desert areas also form in the interior of continents when prevailing winds are far from large bodies of water and have lost much of their moisture.

Desert plants have evolved methods to conserve and efficiently use available water. Some **flower**ing desert plants are ephemeral and live for only a few days. Their seeds or bulbs can lie dormant in the soil for years, until a heavy rain enables them to germinate, grow, and bloom. Woody desert plants can either have long **root system**s to reach deep water sources or spreading shallow roots to take up moisture from dew or occasional rains. Most desert plants have small or rolled leaves to reduce the surface area from which transpiration of water can take place, while others drop their leaves during dry periods. Often leaves have a waxy coating that prevents water loss. Many desert plants are succulents, which store water in leaves, stems, and roots. Thorns and spines of the cactus are used to protect a plant's water supply from **animal**s.

Desert animals have also developed protective mechanisms to allow them to survive in the desert environment. Most desert animals and insects are small, so they can remain in cool underground burrows or hide under vegetation during the day and feed at night when it is cooler. Desert amphibians are capable of dormancy during dry periods, but when it rains, they mature rapidly, mate, and lay **egg**s. Many birds and rodents reproduce only during or following periods of winter rain that stimulate vegetative growth. Some desert rodents (e.g., the North American kangaroo rat and the African gerbil) have large ears with little fur to allow them to sweat and cool down. They also require very little water. The desert camel can survive nine days on water stored in its stomach. Many larger desert animals have broad hooves or feet to allow them to move over soft sand. Desert reptiles such as the horned toad can control their metabolic heat production by varying their rate of **heart**beat and the rate of body **metabolism**. Some snakes have developed a sideways shuffle that allows them to move across soft sand. Deserts are difficult places for humans to live, but people do live in some deserts, such as the Aborigines in Australia and the Tuaregs in the Sahara.

Desert soils are usually naturally fertile since little water is available to leach **nutrients**. Crops can be grown on desert lands with irrigation, but evaporation of the irrigation water can result in the accumulation of salts on the soil surface, making the soil unsuitable for further crop production. Burning, deforestation, and overgrazing of lands on the semiarid edges of deserts are enabling deserts to encroach on the nearby arable lands in a process called desertification. Desertification in combination with shifts in global atmospheric circulation has resulted in the southern boundary of the Sahara Desert advancing 600 mi (1,000 km) southward. A desertification study conducted for the United Nations in 1984 determined that 35% of the land surface of the earth was threatened by desertification processes.

DIABETES MELLITUS

Diabetes is a metabolic disease caused by the body's inability to use the **hormone insulin** to effectively convert **carbohy-**

drates into the simple sugar glucose that **cell**s store and use to perform vital functions. Without glucose to fuel their activity, the cells use fat instead, producing ketones as a waste product. Ketones build up in **blood** and disrupt **brain** functions. Common signs of diabetes are excessive thirst, urination, and fatigue. The disease can also cause vision loss, decreased blood supply to hands and feet, pain, and skin infections. If left untreated diabetes can induce coma and cause **death**.

There are two main types of diabetes. *Juvenile* diabetes (also called Type I) occurs when the **pancreas**, a gland attached to the small intestine, fails to produce enough insulin; as a result, it is also referred to as insulin dependent diabetes mellitus (IDDM). *Maturity-onset* diabetes, or Type II non-insulin dependent diabetes mellitus (NIDDM), occurs when the body produces insulin but cannot use it efficiently. Juvenile diabetes is usually controlled by doses of insulin and a strict diet. Maturity-onset diabetes, which is often accompanied by obesity, is usually controlled by diet alone.

Diabetes often runs in families. In the United States about ten percent of the caucasian **population** suffers from diabetes, and it is even more common among African-American, Mexican-American, and certain Native American groups.

The symptoms of diabetes were identified 3,500 years ago in Egypt and were also known in ancient India, China, Japan, and Rome. The Persian physician Avicenna (980-1037) described the disease and its consequences. Thomas Willis (1621-1675), an English epidemiologist, was the first modern western physician to discover that the urine of diabetics tasted sweet. In 1815, the French chemist Michel Eugène Chevreul discovered that the sweetness came from ''grape sugar'' or glucose. The disease's formal name *diabetes*, meaning a siphon or running through, and *mellitus*, relating to sweetness or honey, was first used in 1860.

Injury to the **pancreas** was linked to diabetic symptoms by several scientists from the seventeenth to the nineteenth centuries. The existence of a pancreas **hormone** to reduce the blood glucose level was first proposed in 1916 by the English physiologist Edward Sharpey-Schäfer. Insulin was isolated in 1921 by the Canadians **Frederick Banting** and **Charles Best** (1899-1978), and in 1922 they used it for the first time to successfully treat fourteen-year-old Leonard Thompson of Toronto.

Diabetes is an autosomal dominant disease, but its expression is also thought to be influenced by other conditions, such as **aging**. Research on Type I diabetes shows that dominant genes either protect against the disease or increase susceptibility to it. Advances in molecular **genetics** have led to large-scale studies to identify the genes responsible for diabetes. The American Diabetes Association has established a national database that contains information and **genetic material** from families with Type II diabetes that will help investigators conduct genetic **linkage** studies to locate the specific genes involved. Scientists have already established that a **gene** on **chromosome** 7 is linked to Type II diabetes. When mutated, the gene produces a faulty **enzyme** that is unable to stimulate the pancreas to produce insulin.

Scientists are developing tests to accurately predict whether someone will develop diabetes or not by observing

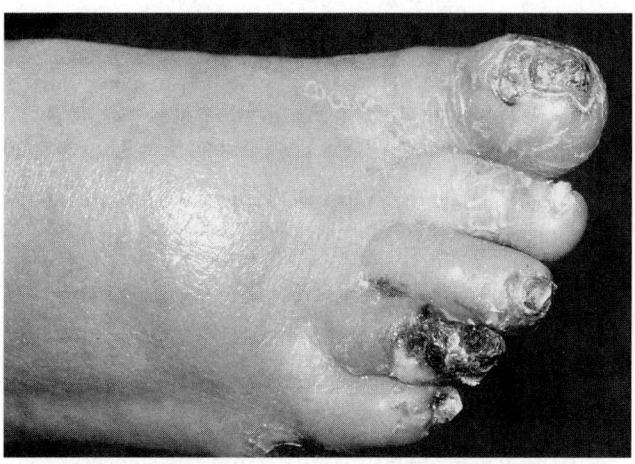

Persons with diabetes often suffer from foot ulcers, as shown above. *(Custom Medical Stock Photo. Reproduced by permission.)*

whether the **immune system** attacks the pancreas cells that make insulin. The research may also lead to a vaccine against diabetes and drugs to keep the immune system from identifying the pancreas as an enemy, similar to drugs that keep the body from rejecting transplanted **organ**s. New therapies for diabetes are continually under development. For example, studies of a combination drug therapy using metformin and troglitazone has been shown to significantly lower blood glucose levels in Type II diabetes patients by reducing glucose **secretion** by the liver (metformin) and enhancing the body's use of insulin (troglitazone). Another area under investigation is the use of islet **cell** transplantation, a promising approach for replacing whole organ transplantation by extracting and replacing only those cellular components that are needed to restore normal function. Islet cell transplantation could be a key to the successful replacement of the insulin-producing islets of Langerhans.

The seventh leading cause of death in the United States, diabetes remains a major health problem. Approximately 10.3 million people in the United States were diagnosed with diabetes in 1997, representing a six-fold increase over four decades. By the early 1990s, the costs associated with treating and caring for diabetes patients was estimated at over $90 billion a year.

DIAPHRAGM

Diaphragm is term used in physiology to describe several large muscle groups which are found only in humans and other **mammals**. These muscles are structural components that separate two adjacent regions of the body. Diaphragms in the human body include the pelvic diaphragm, the urogenital diaphragm, and the thoraco-abdominal diaphragm. The latter one, which is commonly referred to as ''the'' diaphragm, is the muscle that separates the chest cavity (which contains the **heart** and lungs), from the abdomen (which contains the stomach, intestines, and other **organ**s.)

The term comes from the Greek word "diaphragma," meaning "barrier." The thoraco-abdominal diaphragm, which is located in the midriff, consists of muscle and **membrane tissue** and is attached to the rib cage. It has three openings for the esophagus, **aorta**, and **vena cava**. The muscle has a slightly concave structure and it arches over the liver on the right and the stomach on the left. When we breath in, a nerve center in the **brain** stem triggers the diaphragm to contract and pull downward. This motion also causes the ribs to move outward, which helps to expand the chest cavity. The resultant decrease in pressure causes air to be drawn into the lungs. When we breath out, we relax the muscle and it snaps back up into its previous position, contracting the chest and forcing air out of the lungs. As the diaphragm moves up and down it also stimulates the stomach and liver and thus aids in **digestion.**

The diaphragm plays a critical role in the process of **respiration** and any interference with its free movement can prevent the proper intake of air into the lungs. Therefore, diseases which affect this muscle, such as poliomyelitis, seriously endangers life. Proper diaphragm function can also be impaired by developmental defects, hernia, injury, physical displacement, and infection. Minor perturbation of the diaphragm can cause a case of the hiccups, which occur when the diaphragm contracts with a series of quick spasms. These spasms make the diaphragm tighten suddenly, causing the lungs to draw in quick gulps of air. As this air rushes in, the vocal cords automatically close. As the air rushes by the vocal cords vibrate, creating the characteristic hiccup sound. Hiccups can be result from any irritation to the diaphragm that causes it to contract rapidly, such eating or drinking too fast. It can also be caused by muscle spasms that happen when the muscle is over exerted. After a few moments the diaphragm will usually relax and the hiccups will cease. Sometimes breaking the breathing cycle by taking a big drink of **water**, holding your breath, or breathing into a paper bag for a few minutes can also help. These tricks increase the amount of **carbon dioxide** in the lungs, which tends to slow or stop the muscle contractions.

DICHLORODIPHENYL-
TRICHLOROETHANE • See DDT
(Dichlorodiphenyl-trichloroacetic acid)

DICK, GEORGE (1881-1967)
American bacteriologist and pathologist

George Dick was an American bacteriologist and pathologist who, along with his wife, Dr. Gladys Rowena Henry Dick, isolated and identified the causative bacterium for scarlet fever and developed the toxin used for immunization. In addition, they devised the Dick method for preventing scarlet fever by toxin-nontoxin injection and developed the Dick skin test to determine the susceptibility to scarlet fever. The Dicks were nominated for the Nobel Prize in Medicine 1925 for their contributions and discoveries regarding scarlet fever, but no prize was awarded that year.

Dick was born on October 14, 1881 in Fort Wayne, Indiana, and received his medical degree from Rush Medical College in 1905. In 1910, he became pathology instructor at the University of Chicago were he met his future wife, Gladys. The Dicks married in 1914, and during the same year, George worked at the John R. McCormick Institute for Infectious Diseases. Dick studied scarlet fever among enlisted men during World War I while he served in the Army Medical Corps. After the war, from 1918 to 1933, Dick was professor of clinical medicine at the University of Chicago, and from 1933 to 1845 he served as professor of pathology and chairman of the Department of Medicine.

The Dicks identified the causative bacterium for scarlet fever in 1923 and attempted to produce scarlet fever in volunteers by swabbing their throats with organisms obtained from the throats of patients who had the disease. Surprisingly, while some of these volunteers developed tonsillitis, no one caught scarlet fever. In 1924, the Dicks isolated a strain from a nurse whose finger happened to be infected with surgical scarlet fever. This specific strain was swabbed onto the throats of ten volunteers. Of the ten subjects, two developed scarlet fever. From this experiment, they determined that the characteristic scarlet fever rash was actually caused by a toxin produced by a particular strain of the hemolytic streptococcus bacteria. Dick's discovery assured the medical community that the cause of scarlet fever was understood and could be brought under control, and further assurance was offered when he and his wife developed a skin test to determine the susceptibility or immunity to the disease. This skin text was also later used to predict if a pregnant woman would be susceptible to puerperal infection during childbirth. As antibiotics were discovered a couple decades later, their work became outdated.

George and Gladys Dick adopted two children in 1930, Roger Henry Dick and Rowena Henry Dick. After George retired, the Dicks moved to Palo Alto, California, where he died on October 14, 1967.

DIFFERENTIATION

When visiting the meat counter of a supermarket, one does not need a **microscope** to see that liver and steak are different. They are different because their constituent **cell**s are different. Microscopic observation confirms this notion. Liver cells differ from muscle cells in **morphology** (structure) and this difference is a reflection of physiological activities and biochemical functions that are ultimately under the control of **genes**. The differences that can be seen grossly (as at the meat counter), observed in the microscope, and detected by biochemical and molecular procedures together comprise what is known as differentiation.

Differentiation results from selective gene action of a **genome** (the entire genetic complement of an **organism**) held in common by all cells has been a tenet of modern genetic biology. Certainly the recent **cloning** of Dolly and other **mammals** supports this concept. It is the business of a cell to produce all of the **proteins** and **enzymes** held in common by most cells.

The commonly produced gene products are sometimes referred to as housekeeping proteins. However, the adult fly, frog or human are comprised of a great diversity of differentiated cells. The differentiated cells produce, in addition to the housekeeping gene products, **tissue**-specific proteins. A unique portion of the genome of differentiated cells is activated and this accounts for differentiation. While all other cells have these gene sequences, they are silent except in the specific cell type under consideration. Thus, liver differs from muscle not because of its genome, nor because of the activity of housekeeping genes, but by the activation of tissue-specific genes which convey cell specificity. Gene **regulation** that permits differentiation is the result of promoters and enhancers (which occur in **DNA** on either side of specific genes) and regulatory proteins which bind to the promoters and enhancers and which in turn enhance or inhibit **gene expression**.

Differentiation is associated with **embryology**. The undifferentiated cells of a **zygote**, **morula**, and **blastula** give rise to progressively more differentiated cells until the adult forms which is a mosaic of many highly differentiated cells. Ordinarily, differentiated cells have lost the competence to give rise any other kind of cell. Therefore, muscle never gives rise to liver and vise versa. Moreover, contained within the mosaic of terminally differentiated cells are a number of stem cells. A stem cell is a less than fully differentiated cell that has retained the competence to give rise to another stem cell and a cell that will become fully differentiated. Consider the skin—it would rapidly be lost because of abrasion and the wear and tear of use were it not for replacement by cells from the basal layer of epidermis. Differentiation of the skin stem cell progeny gives rise to post-mitotic keratinized protective cells. **Blood** is a tissue type that must be continually replaced. It is not surprising, therefore, to note that new blood cells develop from stem cells, which like skin stem cells, at division give rise to both more stem cells and cells which will differentiate as blood.

A final note about differentiation. **Cancer** is believed by many to not be a malignancy of fully differentiated adult cells, most of which have lost their potential for **cell division**. Rather, cancer is viewed as a disease of stem cells. Stem cells appear less than fully differentiated. Cancer cells appear less than fully differentiated. Stem cells are characterized by their mitotic (cell division) potential. Cancer cells are well known for their mitotic potential. Many fully differentiated cells are post-mitotic, i.e., they no longer retain the competence to divide. These ideas have given rise to the notion that, under appropriate circumstances, perhaps cancer cells could be induced to differentiate. A non-proliferating terminally differentiated cell would result which, by definition, is a cure. Differentiation is occurring in normal cells continuously. It is neither painful nor toxic. Perhaps someday, some cancer will be treated by differentiation therapy and if so, the therapy may be neither painful nor toxic. All-trans retinoic acid, a form of vitamin A, has induced remission of acute promyelocytic leukemia. While this is not a cure, it provides hope that differentiation therapy may have a place in the treatment of some cancer.

DIFFUSION

Diffusion is a process in which the random motion of **molecules** or other particles results in a net movement from a region of high concentration to a region of lower concentration. Familiar examples include the spread of tobacco smoke throughout the still air of an auditorium, or the dissemination of floral perfumes from a bouquet to all parts of the motionless air of a room. The rate of flow of the diffusing substance is proportional to the concentration gradient for a given direction of diffusion. Thus, if the concentration of the diffusing substance is very high at the source, and is diffusing in a direction where little or none is found, the diffusion rate will be maximized. Several substances may diffuse more or less independently and simultaneously within a space. Thus, in the examples mentioned earlier, tobacco smoke may diffuse throughout a room at the same time as the floral scent spreads. Because lightweight molecules have higher average speeds than heavy molecules at the same **temperature**, they also tend to diffuse more rapidly. Molecules of the same weight move more rapidly at higher temperatures increasing the rate of diffusion as the temperature rises.

DIGESTION

Digestion is the physiological process by which food is broken down, mechanically and chemically, into particles small enough to pass through the walls of the intestinal tract and into the **blood**. Once in the bloodstream, these tiny particles can then be distributed throughout the body and used for nourishment. The breaking-down process takes place in almost all parts of the digestive tract (also called the alimentary canal), beginning at the mouth and ending, some 15 ft (4.5 m) later, in the anus.

Specifically, the digestive process starts as soon as food, taken into the mouth, begins to be chewed into smaller pieces. While being chewed, the food is mixed with saliva that contains the **enzyme** ptyalin, the first of many enzymes that will help convert complex and indigestible food **molecules** into smaller and easier-to-absorb ones. While food is in the mouth, the activity of ptyalin is already at work, converting some of the complex starches into simple sugars.

After chewing, the food is swallowed, and it passes through the esophagus and into the stomach, where the breaking-down process goes into high gear. A strong churning motion causes the food to be thoroughly mixed by the stomach's potent digestive juices, which contain both hydrochloric acid and the enzyme pepsin. The foods are all dissolved into a thick liquid called chyme but, while the **protein** foods are partly digested, the other **nutrients** are basically unchanged. Then, in the small intestine, the digestive process is completed by a combination of pancreatic juices (containing the enzymes trypsin, amylase and lipase), intestinal juices, and bile. (The bile, stored in the gall bladder, works primarily on the digestion and absorption of fat.) Thoroughly digested, the nutrient molecules can now by absorbed by blood and lymph vessels in the small intestine's walls and carried into the circulation. Indigestible food particles pass into the large intestine, where some **water** and **minerals** are absorbed and bacterial action turns the rest into feces, which are eventually eliminated as waste products.

The digestive process is clearly a highly complicated one and, until fairly recently, only vaguely understood. In ancient times, for instance, the early philosophers could see that various foods entered the body and, while some of the food remained in the system presumably to provide nourishment, the rest emerged later in a completely changed form. How did they believe all this was accomplished? The Hippocratic philosophers, in the fourth and fifth centuries B.C., concluded that food was converted through body heat first into liquid form, then into what they called the ''four humors''—blood, phlegm, yellow bile and black bile—which could be absorbed into the body. Eventually, however, the theory changed.

A few centuries later, around 280 B.C., the philosopher **Erasistratus** pointed out that, after food was eaten, a great deal of activity soon took place in the stomach and intestines. To him, therefore, digestion was almost certainly a mechanical process. Somehow food was ground down in the stomach to liquid form, then carried to the liver where it was transformed into blood. The famous second century physician Galen, although he agreed that the liver was the major **organ** of digestion, decided that the Hippocratics were right: **animal** heat must be the guiding force behind digestion. Galen's views prevailed for hundreds of years.

In the seventeenth century, however, scientists began taking another look at the human body. A number of them, such as Franciscus Sylvius, a Dutch physician, believed that most bodily processes could be explained in purely chemical terms. Sylvius studied digestive juices, including saliva, and correctly concluded that digestion was a form of **fermentation**.

Other scientists of the time tended to see the human body as a kind of machine. To Giovanni Borelli, an Italian mathematician and a contemporary of Sylvius, the stomach was simply a binding device that could mechanically break food down into tiny particles. Erasistratus had been right all along, he concluded.

During the eighteenth century, several scientists made tremendous progress in settling the debate over whether digestion was a mechanical or chemical process. Through his experiments on animal **tissue**s, the Swiss physiologist Albrecht von Heller (1708-1777) discovered that bile was the key element in the body's digestion of **fats**. The debate was essentially settled by the French physiologist **René de Réaumur** in 1752. De Reaumur fed a hawk two small metal cylinders filled with meat and covered at their ends by gauze. Unable to digest the cylinders, the hawk eventually regurgitated them, and de Reaumur deduced that the meat could only have been dissolved by chemicals present in internal fluids.

Before long, a number of eighteenth-century scientists were able to study the role of digestive juices. The Italian physiologist **Lazzaro Spallanzani** (1729-1799) performed experiments similar to de Reaumur's and coined the term ''gastric juices.'' Spallanzani discovered that these juices helped

inhibit and prevent the putrefaction of food within the body. He also helped to further define the important role that saliva plays in the digestive process. Additionally, in 1822, the German chemist **Leopold Gmelin** (1788-1853) continued to study the chemical processes involved in food digestion by investigating stomach and pancreatic juices. And in 1842, William Prout proved that the most potent acid in gastric juices was actually hydrochloric acid. The discoveries made by these scientists validated the proposition that chemical processes governed the body's **digestive system**.

A particularly notable contribution to the understanding of digestion was made by the American army surgeon **William Beaumont**. Beaumont, who joined the army in 1812, was sent a few years later to a frontier post in Michigan. While there, he treated a young French-Canadian, Alexis St. Martin, who had been accidentally shot in the side. Although St. Martin recovered, his bullet wound never fully closed. He retained an inch-wide opening (or fistula) in his side that led to his stomach. Through this opening, Beaumont could not only observe changes in the stomach under varying conditions, but could remove samples of gastric juices. Beginning in 1825, then, the army surgeon conducted over two hundred experiments and, by so doing, provided the medical world with a great deal of previously unknown information about gastric physiology and the digestive process in a living human being.

Inspired by Beaumont's work, the French physiologist **Claude Bernard** began, in the mid-1800s, to create artificial fistulas in laboratory animals. Through these openings, Bernard made a number of important discoveries, among them that the small intestine, rather than the stomach, was the major site of digestion and that pancreatic juices were important digestive agents, particularly where fat molecules were concerned.

A few years later, in the 1870s, Willy Kuhne's research clarified the role of the intestines in the absorption of digested foods. And, toward the end of the nineteenth century, the Russian physiologist **Ivan Petrovich Pavlov**, by experimenting on living dogs, worked out the nervous mechanism that controls the **secretion** of gastric juices by the digestive glands.

Because digestion is such a complex process, there are many things that can go wrong. Today 60-70 million people a year in the United States are affected by digestive diseases, according to the National Institute of Diabetes and Digestive and Kidney Diseases. Among the most prevalent digestive diseases are abdominal wall hernias, constipation, gallstones, indigestion, hemorrhoids, diarrhea, irritable bowel syndrome, lactose intolerance, and peptic ulcers.

DIGESTIVE SYSTEM

The digestive system is that system which exists within all **organism**s to allow nutritive particles to be taken in, broken down, and either used by the body or removed from the body as waste. Digestive systems range from the very simple, primitive systems of the once-**cell**ed organisms, to the complex, multi-**organ** systems used by vertebrates (organisms with backbones).

Single-celled organisms engulf food particles with their outer **membrane**. The food particles are brought inside the cell;

hence this type of **digestion** is referred to as intracellular. Food particles are ultimately engulfed by bags of potent chemicals within the cell (called digestive **vacuoles**) which break down the particles into usable components. Waste products may be re-packaged into little bags, and passed back out through the membrane(interestingly enough, this type of digestion is used by the **immune system**'s white **blood** cells in **mammals**; white blood cells protect the body from infection by engulfing and digesting foreign invaders such as **virus**es and **bacteria**).

More complex organisms may have a mouth which leads to a large, open body cavity. Sponges, for example, carry **water** into this cavity, where any tiny food particles are extracted and distributed throughout the cells of the sponge for digestion. The water and any waste is then sent back out through the mouth.

Flatworms have a blind gut. This is a tube within which digestion takes place. It is equipped with a mouth opening, but is otherwise totally sealed off. Food enters the mouth and is partially digested by chemicals released into the gut. Because this type of digestion occurs within the cavity of the gut, and not within a cell, it is referred to as extracellular digestion. Once the food has been broken down through this extracellular digestion, the smaller bits can be absorbed by the cells which line the gut. Waste products are passed back out through the mouth.

As organisms become more complex, their digestive systems become increasingly complex. Higher up the evolutionary ladder, the blind sac eventually leads to a separate opening for the passage of waste products, the primitive anus. This is evident in various worms called nematodes. In higher **animal**s, outpouchings from the digestive tube become specialized organs such as the liver, **pancreas**, and gallbladder.

The basic schemata for vertebrate (organisms with backbones) alimentary canals or digestive tracts involves structures which are responsible for receiving food, conducting and storing food, breaking food down and absorbing its useful **nutrients**, and absorbing water and eliminating wastes (defecating).

The mouth is the opening responsible for receiving food. Within the mouth, food is often broken down into smaller bits by the mechanical actions of chewing with teeth and by the chemical action of substances found within saliva. Saliva helps to lubricate food, allowing it to pass into the esophagus.

The esophagus serves to conduct the food to its next destination. In some animals (certain birds and reptiles), an outpouching from the esophagus called the crop is used for storage of food. It may also contain small rocks, which help to grind tough bits of food, such as seeds.

The stomach is a muscular pouch which works to mix the food, usually with a highly acidic combination of chemicals which will further breakdown the food. The stomach sends food into the small intestine, where a variety of alkaline chemicals produced by the liver and pancreas continue the process of digestion. **Enzyme**s and bacteria along the border of the small intestine aid in the digestive process. As food becomes broken down into its component nutrients, these nutrients are absorbed into the bloodstream.

The components of food which are not absorbed continue passing down the large intestine, which works to absorb

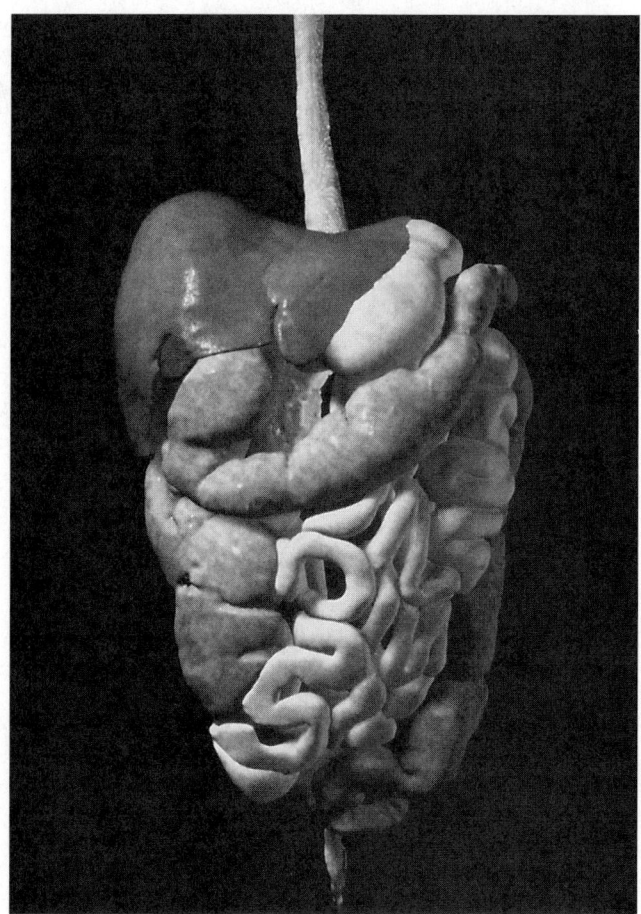

A clay model of the human digestive system. *(Custom Medical Stock Photo. Reproduced by permission.)*

water from this mass. Ultimately, the mass of waste (feces), appropriately dehydrated by the absorption of water from it, arrives at the rectum, where feces are stored. When a sufficient quantity has accumulated, the feces are expelled through the anus.

DIGITALIS

An extremely valuable drug in the treatment of **heart** disease, digitalis is derived from the leaves of the foxglove plant (*Digitalis purpurea*), a popular garden **flower**. Herb doctors and "old wives" had used foxglove for centuries, but the plant's efficacy was unknown to the physicians of early modern times.

In 1775, the eminent English physician William Withering (1741-1799) began studying digitalis. Withering's keen interest in **botany** led him to collect plant specimens, as did his love for one of his patients (whom he married), a flower painter. Withering noted that old country women used foxglove to treat dropsy (edema), an accumulation of fluids caused by a failing heart. Willing to consider these "old wives' tales," Withering embarked on a detailed study of digitalis. He determined the most effective treatment form—a powder made

Digitalis purpurea. (Photo Researchers, Inc. Reproduced by permission.)

from dried leaves picked just before the plant blossomed—and, of critical importance, the correct dosage for different cardiac conditions. Equally important, Withering established clear standards for when to discontinue administration of the drug, which can be toxic.

Withering published a monograph on his findings, *Account of the Foxglove*, in 1785, but he had spread knowledge about the "new" drug before then, because digitalis became part of the Edinburgh *Pharmacopoeia* in 1783. In spite of Withering's clear warnings about overdoses of digitalis, the drug was commonly prescribed in dangerously large doses for a host of medical conditions. Finally, in the early twentieth century, investigators clarified the effect digitalis had on the heart and the correct circumstances for the drug's use.

The active principles of digitalis eluded researchers until the mid-1800s. E. Homolle and Theodore Ouevenne won a cash prize in 1841 from the Société de Pharmacie in Paris when they isolated digitalin. Oscar Schmiedeberg (1838-1921) isolated the highly potent digitoxin in 1875. The English chemist Sydney Smith obtained digoxin from woolly foxglove (*Digitalis lanata*) in 1930.

When used correctly digitalis increases the circulatory power of the heart without increasing the heart rate. It remains a widely prescribed heart medicine today, especially for heart failure.

DIPHTHERIA

Diphtheria is a serious infectious disease of the **respiratory system** caused by the bacterium *Corynebacterium diphtheriae*. Now a rare disease, it ravaged the world in the late nineteenth century and was a major killer of children. Symptoms included sore throat, fever, and body aches. The disease causes yellowish-gray mucous **membrane**s to form and spread on the inner throat and nose. In some cases, the membrane can impede breathing, requiring a life-saving tracheotomy to open the windpipe. The infection can cause fatal damage if it spreads to the **heart** or kidneys. The disease is spread from person to person by coughing and sneezing and is more common among poor **population**s living in crowded conditions. The diphtheria bacterium produces **toxins** carried throughout the body by the **blood**stream. These can be neutralized with injections of antitoxin. **Antibiotics**, such as erythromycin, are also used to treat the disease.

Diphtheria received its name in 1826 from French physician Pierre-Fidèle Bretonneau (1778-1862), who was the first to study its symptoms. He took the name from the Greek word for parchment to describe the membranes that form in the throat. Bretonneau was the first physician to perform a tracheotomy on a diphtheria patient, a four-year-old girl. He made an opening from her neck into her windpipe to enable her to breathe. The surgery was successful and saved the girl's life.

In the late 1800s, deadly diphtheria epidemics raged through Europe and the United States, and the scientific community rallied to find the cause and cure. Much of the research took place in the laboratories of microbiologists **Louis Pasteur** of France and Robert Koch of Germany. The rod-shaped bacterium responsible for diphtheria was first identified by two German bacteriologists working in Koch's laboratory—Edwin Klebs (1834-1913), who isolated the **bacteria** in 1883, and Friedrich August Löffler (1852-1915), who proved it was the causative **organism** by inducing diphtheria in laboratory **animal**s with **culture**s of the bacteria.

Shortly after, French scientist Pierre-Paul-Emile Roux (1853-1933) and Swiss bacteriologist Alexandre Yersin (1863-1943), working at the Pasteur Institute in Paris, showed that a toxin produced by the bacillus was the actual cause of the disease. Roux, who was also known for his research on the **anthrax** and rabies **virus**es, made the first analysis of the toxin's chemical properties.

Once the cause of diphtheria had been identified, German bacteriologist **Emil von Behring** and Japanese scientist **Shibasaburo Kitasato**, who had been conducting research on the human body's ability to produce antitoxins to neutralize poisons produced by invading bacteria, were quickly able to develop a method to vaccinate individuals against diphtheria. They did this by injecting laboratory animals with sublethal doses of diphtheria toxin. Then they introduced blood serum of the animals containing antitoxin into humans to immunize them.

Behring and Kitasato's method was refined in the following decades. German microbiologist **Paul Ehrlich** developed standard dosages for their diphtheria antitoxin; and, in 1891, New York City pathologist Anna Wessels Williams isolated a stronger and more effective strain of diphtheria antitoxin. In 1894, Pierre Roux developed a diphtheria antitoxin serum using horses. He used this serum to successfully treat more than 300 cases of diphtheria. In 1913, Behring announced the development of a longer-lasting vaccine for diphtheria which was a mixture of toxin and antitoxin. The same year, Hungarian scientist and pediatrician **Bela Schick** introduced a simple, reliable test to determine whether a person is susceptible to diphtheria. The Schick test involves injecting a small amount of toxin under the skin. If a red, swollen rash appears around the injection, the person is susceptible and should be immunized.

The incidence of diphtheria has declined dramatically since the large-scale introduction of diphtheria vaccine. The vaccine, which is injected into the forearm, contains diphtheria toxoid. This is a form of diphtheria toxin chemically treated to be nonpoisonous. The toxoid enables the body to build up antibodies to the disease so that, if the person becomes exposed to the germ, the disease will not develop.

One widely used vaccine for diphtheria is DPT, which also includes vaccines for whooping cough and tetanus. Three injections of this vaccine are recommended for children every few months after **birth**, followed by boosters at ages eighteen months, at four to six years, and then every ten years.

The research that led to the vaccine for diphtheria also created the scientific basis for the understanding of the body's **immune system**. Furthermore, research into the nature of diphtheria and other bacterial diseases at the turn of the century paved the way for the development of numerous antibiotics.

Despite the advances against diphtheria, diphtheria outbreaks still occur. These outbreaks can be large and spread rapidly. According to the National Coalition for Adult **Immunization**, one out of every ten people who gets diphtheria will die from it.

DIPLOID

The diploid **chromosome** complement of a **species**, designated 2n, consists of the combined **haploid** chromosome sets, each designated 1n, contributed by the male and female parents at **fertilization**. The diploid complement of chromosomes consists of pairs of each chromosome. The haploid **gamete** (sex) **cell**s have only one member of each chromosome pair. In most frogs of the **genus** *Rana*, the diploid number is 26 and the haploid number is 13. In humans, the diploid chromosome number is 46 and the haploid number is thus 23. Most body cells, also known as somatic cells, are diploid.

The diploid number of body cells is maintained by **mitosis** which is nuclear division. Mitosis gives rise to two identical sets of chromosomes which are distributed to daughter cells at **cell division**. Diploid cells have a precise amount of **DNA** which is characteristic of the species. **Meiosis** is the process by which diploid germ cells give rise to haploid cells of the mature **sperm** and mature **egg**s. Haploid cells contain pre-

cisely half the quantity of DNA as that of diploid cells. Fertilization of the egg with a sperm restores the diploid chromosome complement and the diploid amount of DNA in the **zygote**. Subsequent **embryo**nic and adult cells are diploid and are derived by mitosis from the zygote.

There are exceptions to the general rule that body cells are diploid. One exception are cells of the regenerating liver which may contain exactly twice the diploid amount of DNA. **Cancer** cells are another exception. Cancer cells commonly have an irregular complement of chromosomes which is known as aneuploidy. Cancer cells may have a diploid number of chromosomes but one or more chromosomes may be structurally abnormal.

DIRECTIONAL SELECTION

Directional selection is a type of **natural selection** that emphasizes one type of **mutation** that eventually phenotypically manifests itself within the **population**. The darkened wing color of the peppered moth (*Biston betularia*) located in industrial areas is an example of directional selection. The darker wings provided appropriate camouflage from predators as the moths dwelled on soot-covered **tree**s.

Directional selection is commonly instigated by the human population on other **species** for a particular use. A plant breeder may only choose to grow seed from a plant with a particular characteristic or set of characteristics. Eventually, successive breeding of the plant will alter the population and the desired characteristic will become the common **phenotype**. Dog breeding is an example of directional selection applied by humans.

See also Selective pressures

DISCONTINUOUS VARIATION

Discontinuous variation is variation within a **population** of a characteristic that falls into two or more discrete classes. Classic examples include such things as eye color in **animal**s and the tall and short pea **phenotype**s used by Austrian botanist **Gregor Johann Mendel**. Characteristics that display discontinuous variation are present in one state or another; there is no blending or merging of the different forms possible. Unlike **continuous variation**, discontinuous variation is displayed by characteristics that are usually controlled by only one or two **gene**s and that have little or no environmental component in their expression. For example, height in humans is a characteristic that shows continuous variation. The maximum height an individual can grow to is governed by the genes controlling the system as well as the intake of food. Thus, an undernourished individual is unlikely to achieve maximum height potential. With eye color, a person's eyes will be a certain color irrespective of any external factors operating. This type of variation is sometimes known as qualitative variation. Discontinuous variation can be produced by a single **gene mutation** or it can be evidence of a polymorphic gene system.

DISRUPTIVE SELECTION

Disruptive selection is a special form of directional selection. By choosing two phenotypic extremes within a **population**, successive breeding can be carried out to produce two discontinuous strains. For example, if a plant breeder breeds the longest and the shortest seeds within a population, then after several generations two different strains would be produced: one bearing long seeds and the other bearing short seeds. The strains would be morphologically dissimilar in respect of their seed lengths, but they would be the same in all non-linked effects.

Disruptive selection forces the **organism** to produce two different forms. When it occurs under natural conditions, disruptive selection can be the first step on the production of two different **species**. An example of this occurrence could be seen if a river drove its path through a population of **plants**. The plants cannot move so they have to adapt or die out under the new conditions. Within the population, some individuals may be better suited to **water**logged conditions and will thrive at the riverbed. Other individuals living the furthest from the water may be better suited to dryer conditions. The alteration of the river's course forces a wedge within the population, producing two rapidly diverging forms of the plant species.

DNA (DEOXYRIBONUCLEIC ACID)

The modern science of **genetics** can be traced to the research of a Moravian monk, **Gregor Mendel**, in the mid-1800s. Mendel was able to develop a series of laws that described mathematically the way hereditary characteristics are passed along from one generation to the next. These laws could best be understood by assuming that hereditary characteristics are contained in discrete ''packages'' in an **organism**. These ''packages'' were called by a variety of names such as unit character, gemmule, biophere, pangene, and eventually, **gene**.

By whatever name these ''packages'' were called, however, the term had no concrete referent. It was simply used to refer to some abstract concept that was needed in order to understand the laws of **heredity**.

The story of genetics during the twentieth century is, in one sense, an effort to discover more specifically what a gene is. An important breakthrough came in the early 1900s with the work of the American geneticist, **Thomas Hunt Morgan**. Working with fruit flies, Morgan was able to show that genes are somehow associated with the **chromosome**s that occur in the nuclei of **cell**s. By 1912, Hunt's colleague, A. H. Stirdivant, was able to construct the first chromosome ''map'' showing the relative positions of different genes on a chromosome. The gene now had a concrete, physical referent: it was a portion of a chromosome.

During the 1920s and 1930s, a small group of scientists began to look for an even more specific description of the gene. They wanted to find out what kind of chemical **molecule** a gene was made of.

Most researchers assumed that genes were some kind of **protein** molecule. Protein molecules are large and complex.

They can occur in an almost infinite variety of structures. This quality is just what one would expect for a class of molecules that must be able to carry the enormous variety of genetic traits of which we know.

A smaller group of researchers looked to a second **family** of compounds as potential candidates as the molecules of heredity. These were the **nucleic acids**. The nucleic acids were first discovered in 1869 by the Swiss physician Johann Miescher. Miescher originally called these compounds *nuclein* because they were first obtained from the nuclei of cells. One of Miescher's students, Richard Altmann, later suggested a new name for the compounds, a name that better reflected their chemical nature: nucleic acids.

Nucleic acids seemed unlikely candidates as molecules of heredity in the 1930s. What was then known about their structure suggested that they were too simple to carry the vast array of complex information needed in a molecule of heredity. Each nucleic acid molecule consists of a long chain of alternating sugar and phosphate fragments to which are attached some sequence of four of five different **nitrogen** bases: adenine, cytosine, guanine, uracil and thymine. (The exact bases found in a molecule depend slightly on the type of nucleic acid.)

It was not clear how this relatively simple structure could assume enough different conformations to ''code'' for hundreds of thousands of genetic traits. In comparison, a single protein molecule contains various arrangements of twenty fundamental units (**amino acid**s) making it a much better candidate as a carrier of genetic information.

Yet, experimental evidence began to point to a possible role for nucleic acids in the transmission of hereditary characteristics. That evidence implicated a specific sub-family of the nucleic acids known as the deoxyribosenucleic acids, or DNA. DNA is characterized by the presence of the sugar deoxyribose in the sugar-phosphate backbone of the molecule and by the presence of adenine, cytosine, guanine, and thymine, but not uracil.

As far back as the 1890s, the German geneticist **Albrecht Kossel** obtained results that pointed to the role of DNA in heredity. In fact, historian John Gribbin has suggested that the evidence was so clear that it ''ought to have been enough alone to show that the hereditary information...*must* be carried by the DNA.'' Yet, somehow, Kossel himself did not see this point, nor did most of his colleagues for half a century.

As more and more experiments showed the connection between DNA and genetics, a small group of researchers in the 1940s and 1950s began to ask how a DNA molecule could code for genetic information. The two who finally resolved this question were a somewhat unusual pair, **James Watson**, a 24-year old American trained in genetics, and **Francis Crick**, a 36-year old Englishman, trained in physics and self-taught in chemistry. The two met at the Cavendish Laboratories of Cambridge University in 1951 and became instant friends. They were united by a common passionate belief that the structure of DNA held the key to understanding how genetic information is stored in a cell and how it is transmitted from one cell to its daughter cells.

In one sense, the challenge facing Watson and Crick was a relatively simple one. A great deal was already known about

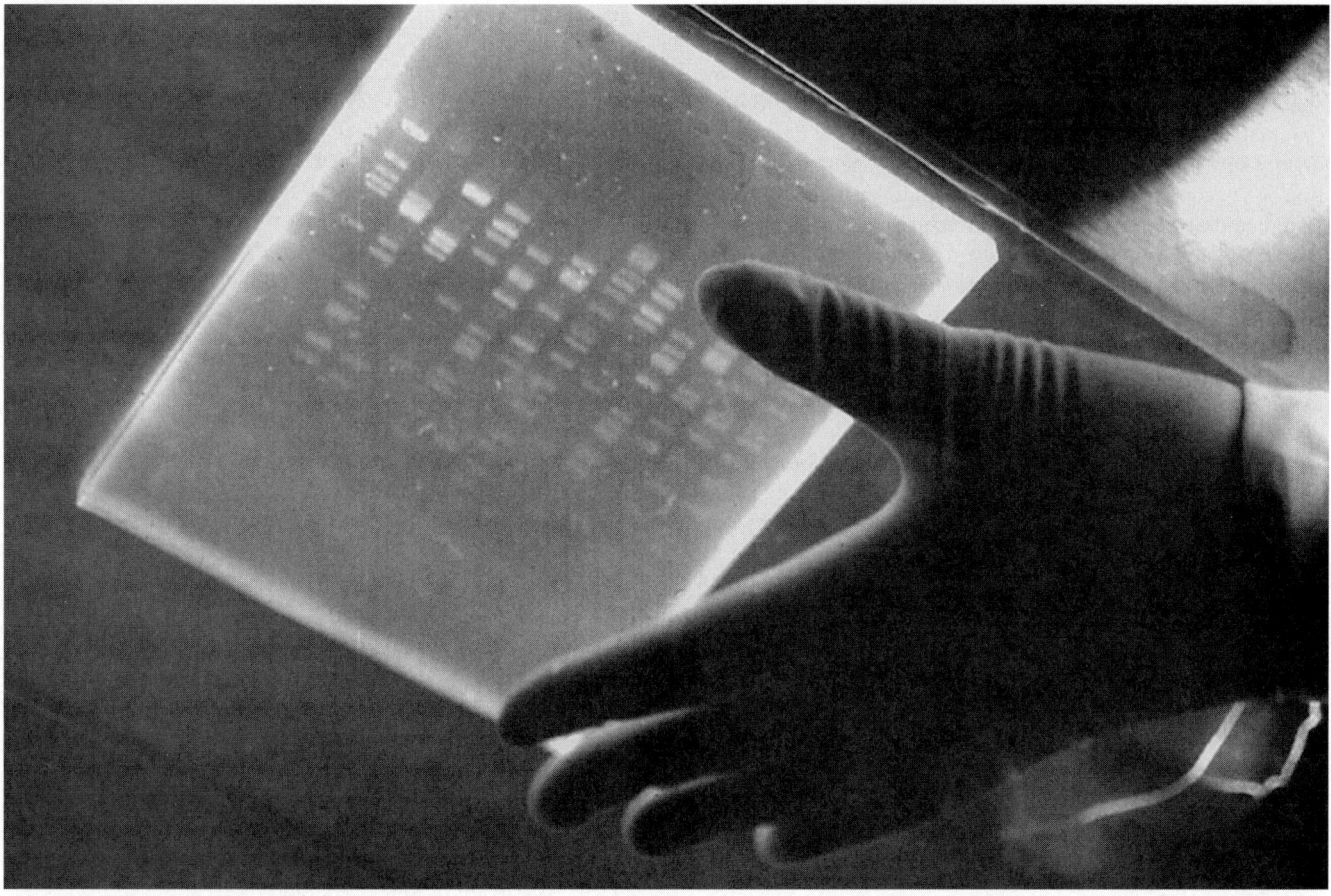

A DNA blueprint obtained by electrophoresis. DNA fragments are placed on top of a gel surrounded by a solution that conducts electricity. *(Photo Researchers, Inc. Reproduced by permission.)*

the DNA molecule. Few new discoveries were needed, but those few discoveries were crucial to solving the DNA-heredity puzzle. Primarily the question was one of molecular architecture. How were the various parts of a DNA molecule oriented in space such that the molecule could "hold" genetic information?

The key to answering that question lay in a technique known as x-ray crystallography. When x-rays are directed at a crystal of some material, such as DNA, they are reflected and refracted by **atoms** that make up the crystal. The refraction pattern thus produced consists of a collection of spots and arcs. A skilled observer can determine from the refraction pattern the arrangement of atoms in the crystal.

The technique is actually a good deal more complex than described here. For one thing, obtaining good x-ray patterns from crystals is often difficult. Also, interpreting x-ray patterns—especially for complex molecules like DNA—can be extremely difficult.

Watson and Crick were fortunate in having access to some of the best x-ray diffraction patterns that then existed. These "photographs" were the result of work being done by Maurice Wilkins and Rosalind Elsie Franklin at King's College in London. Although Wilkins and Franklin were also working on the structure of DNA, they did not recognize the information their photographs contained. Indeed, it was only when Watson accidentally saw one of Franklin's photographs that he suddenly saw the solution to the DNA puzzle.

Racing back to Cambridge after seeing this photograph, Watson convinced Crick to make an all-out attack on the DNA problem. They worked continuously for almost a week. Their approach was to construct tinker-toy-like models of the DNA molecule, shifting atoms around into various positions. They were looking for an arrangement that would give the kind of x-ray photograph that Watson had seen in Franklin's laboratory.

Finally, on March 7, 1953, they had the answer. They built a model consisting of two helices (corkscrew-like spirals), wrapped around each other. Each helix consisted of a backbone of alternating sugar and **phosphate group**s. To each sugar was attached one of the four nitrogen bases, adenine, cytosine, guanine, or thymine. The sugar-phosphate backbone formed the outside of the DNA molecule, with the nitrogen bases tucked inside. Each nitrogen base on one strand of the molecule faced another nitrogen base on the opposite strand of the molecule. The base pairs were not arranged at random, however, but in such a way that each adenine was paired with a thymine, and each cytosine with a guanine.

The Watson-Crick model was a remarkable achievement, for which the two scientists won the 1954 Nobel Prize in chemistry. In the first place, the molecule had exactly the shape and dimensions needed to produce an x-ray photograph like that of Franklin's. Furthermore, Watson and Crick immediately saw how the molecule could "carry" genetic information. The sequence of nitrogen bases along the molecule, they said, could act as a **genetic code**. A sequence, such as A-T-T-C-G-C-T...etc., might tell a cell to make one kind of protein (such as that for red hair), while another sequence, such as G-C-T-C-T-C-G...etc., might code for a different kind of protein (such as that for blonde hair). Watson and Crick themselves contributed to the deciphering of this genetic code although that process was long and difficult and involved the efforts of dozens of researchers over the next decade.

Watson and Crick had also considered, even before their March 7th discovery, what the role of DNA might be in the manufacture of proteins in a cell. The sequence that they outlined was that DNA in the **nucleus** of a cell might act as a template for the formation of a second type of nucleic acid, **RNA (ribonucleic acid)**. RNA would then leave the nucleus, emigrate to the **cytoplasm** and then itself act as a template for the production of protein. That theory, now known as the *Central Dogma*, has since been largely confirmed and has become a critical guiding principal of much research in **molecular biology**.

Scientists continue to advance their understanding of DNA. Even before the Watson-Crick discovery, they knew that DNA molecules could exist in two configurations, known as the "A" form and the "B" form. After the Watson-Crick discovery, two other forms, known as the "C" and "D" configurations were also discovered. All four of these forms of DNA are right-handed double helices that differ from each other in relatively modest ways.

In 1979, however, a fifth form of DNA known as the "Z" form was discovered by Alexander Rich and his colleagues at the Massachusetts Institute of Technology. The "Z" form was given its name partly because of its "zig-zag" shape and partly because it is so different from the more common A and B forms. Although Z-DNA was first recognized in synthetic DNA prepared in the laboratory, it has since been found in natural cells whose environment is unusual in some respect or another. The presence of certain types of proteins in the nucleus, for example, can cause DNA to shift from the B to the Z conformation. The significance and role of this most recently discovered form of DNA remains a subject of research among molecular biologists.

Another subject of research has been **genetic mapping**, determining the DNA sequence of an organism's complete set of genes, or **genome**. A group of scientists are working on sequencing the human genome, a collaboration known as the **Human Genome Project**. In 1997, important steps were taken towards completing this project. A major breakthrough occurred at the University of Wisconsin at Madison by Dr. Frederick Blattner. Blattner and his team of scientists have successfully mapped the entire genome of the bacterium *Escherichia coli*. Every base pair along the entire set of *E. coli*

genes is now known. This tiny organism has over four million base pairs in its DNA sequence, comprising over four thousand genes. Scientists have identified the functions of almost two-thirds of the genes. The process of **replication** of these genes has also been studied extensively. *Transcription*, the process of creating a strand of RNA from a DNA template, has been shown to be a discontinuous process in *E. coli*. As the RNA strand is being formed, there are periods of pausing that occur. This is important in determining how the DNA intermediates are formed during replication, ensuring a correct copy of each of the organism's genes. These data about *E. coli* can be used to discover new genes and also to compare the genome of various organisms, including humans.

DNA REPLICATION

DNA, short for **deoxyribonucleic acid**, is a double-stranded, helical **molecule** that forms the molecular basis for **heredity**. For DNA replication to occur, this molecule must first unwind, or "unzip," itself to allow the information-encoding bases to become accessible. The **base pairing** within DNA is of a complementary **nature** and, consequently, when the molecule unzips, due to the action of **enzyme**s, two strands are temporarily produced, each of which acts as a template. A **replication** fork is first made—the DNA molecule separates at a small region and then the enzyme DNA **polymer**ase adds complementary **nucleotide**s to each side of the freshly separated strands. The DNA polymerase adds nucleotides only to one end of the DNA. As a result of this, one strand (the leading strand) is replicated continuously, while the other strand (the lagging strand) is replicated discontinuously, in short bursts. Each of these small sections is finally joined to its neighbor by the action of another enzyme, DNA ligase, to give a complete strand. This whole process gives rise to two completely new and identical daughter strands of DNA.

This method of replication, known as the semi-conservative hypothesis, was proposed from the outset of the discovery, with the description of the structure of DNA by biochemists **James D. Watson** and **Francis Harry Compton Crick** in 1953. In 1957, biochemist **Arthur Kornberg** first produced new DNA from the constituent parts and a parent strand, forming synthetic but not biologically active molecules of DNA outside the **cell**. However, not until the work of Matthew Meselson and Franklin W. Stahl in the late 1950s was the semi-conservative hypothesis conclusively proven true.

In the semi-conservative method, two strands of the parent molecule unwind and each becomes a template for the synthesis of the complementary strand of the daughter molecule. A competing hypothesis, which would eventually be disproved, was the conservative hypothesis, which states that no unzipping occurs and that a new DNA molecule is formed alongside the original parent molecule. Consequently, of the two molecules of DNA produced after a round of replication, one of them is the intact parent molecule. By using radioactively labeled **nitrogen** to produce new DNA over several generations of cell replication by a bacillus **species**, all of the

DNA in the daughter cells contained labeled nitrogen. The bacilli were then placed in media containing unlabeled nitrogen. After a further round of DNA replication the DNA was examined and it was found to contain equal amounts of labeled and unlabeled nitrogen. In the second generation two types of DNA were found—half was identical to the DNA from the first generation and the remaining half was found to consist of entirely unlabeled nitrogen. These results are consistent with the zip fastener model of the semi-conservative hypothesis, but not at all consistent with the conservative hypothesis. Thus it was shown that DNA replication proceeds via the semi-conservative replication method.

See also Genetic code; Genetic mutation; Mutation

DNA SEQUENCING • See DNA (deoxyribonucleic acid)

DNA SYNTHESIS

Deoxyribonucleic acid (DNA) synthesis is a process by which strands of **nucleic acids** are created. In a **cell**, DNA synthesis takes place in a process known as **replication**. Using **genetic engineering** and **enzyme** chemistry, scientists have also developed man-made methods for synthesizing **DNA**.

The DNA **molecule** was discovered in 1951 by Francis Crick, James Watson, and Maurice Wilkins. In 1953, Watson and Crick used x-ray crystallography data from **Rosalind Franklin** to show that the structure of DNA is a **double helix**. For this work, Watson, Crick, and Wilkins received the Nobel Prize for physiology or medicine in 1962.

To understand how DNA is synthesized, it is important to understand its structure. DNA is a long chain **polymer** made up of chemical units called **nucleotides**. It is the **genetic material** in most living **organisms** that carries information related to **protein synthesis**. Typically, DNA exists as two chains of nucleotides that are chemically linked following **base pairing** rules. Each nucleotide is made up of a deoxyribose sugar molecule, a **phosphate group**, and one of four **nitrogen** containing bases. The bases include the purines adenine (A) and guanine (G), and the pyrimidines thymine (T) and cytosine (C). In DNA, adenine generally links with thymine and guanine with cytosine. The chains are arranged in a double helical structure that is similar to a twisted ladder or spiral staircase. The sugar portion of the molecule makes up the sides of the ladder and the bases compose the rungs. The phosphate group holds the whole structure together by connecting the sugars. The order in which the nucleotides are linked is known as the sequence that is determined by DNA sequencing.

In a eukaryotic cell, DNA is synthesized prior to **cell division** by a process called replication. At the start of replication the two strands of DNA are separated by various enzymes. Each strand then serves as a template for producing a new strand. Replication is catalyzed by an enzyme known as DNA polymerase. This molecule brings complementary nucleotides to each of the DNA strands. The nucleotides connect to form new DNA strands, which are exact copies of the original strand known as daughter strands. Since each daughter strand contains half of the parent DNA molecule, this process is known as semiconservative replication. The process of replication is important because it provides a method for cells to transfer an exact duplicate of their genetic material from one generation of cell to the next.

After the nature of DNA was determined, scientists began to examine the cellular genes. When a certain gene was isolated, it became desirable to synthesize copies of that molecule. One of the first ways in which a large amount of a specific DNA was synthesized was though genetic engineering. Genetic engineering begins by combining a gene of interest with a **bacterial** plasmid. A plasmid is a small stretch of DNA that is found in many bacteria. The resulting **hybrid** DNA is called recombinant DNA. This new recombinant DNA plasmid is then injected into a bacterial cell. The cell is then cloned by allowing it to grow and multiply in a **culture**. As the cells multiply so do the copies of the inserted gene. When the bacteria has multiplied enough, the multiple copies of the inserted gene can then be isolated. This method of DNA synthesis can produce billions of copies of a gene in a couple of weeks.

In 1985, researchers developed a new process for synthesizing DNA called polymerase chain reaction (PCR). This method is much faster than previous known methods producing billions of copies of a DNA strand in just a few hours. It begins by putting a small section of double stranded DNA in a solution containing DNA polymerase, nucleotides, and primers. The solution is heated to separate the DNA strands. When it is cooled, the polymerase creates a copy of each strand. The process is repeated every five minutes in an automated machine until the desired amount of DNA is produced.

See also DNA replication

DNA TECHNOLOGY

DNA technology is a broad area that covers any use of **deoxyribonucleic acid** (**DNA**) modification for purposes that benefit humans. This includes relatively straightforward routines, such as the production of genetic maps using **restriction enzymes**, all the way through to highly controversial procedures, such as modification of the **genome** itself to produce desirable effects.

Technology utilizing DNA can be used for purely investigative reasons. This includes such techniques as genetic fingerprinting in all of its guises. This suite of techniques allows researchers to produce genetic maps of **chromosomes** based on restriction enzyme digests. One of the better-known applications of genetic fingerprinting is in the investigation of parentage of an individual. Another is in the analysis of a crime scene and the consequent detection of the culprits. In both of these cases the same basic idea is utilized. DNA from a sample (either from a child or from a crime scene) is digested to produce characteristic bands that are visualized under UV **light** after the dyed and digested DNA has been subjected to an electric current. The pattern of bands can then be compared to the presumed parents' DNA banding, and, if matching bands are present, the parentage can be confirmed. With criminal investigations, the DNA sample is compared to that of a suspect, and, if there is a perfect match of bands, the DNA is said to belong to that individual.

DNA technology that deals with actual modifications of the genome is far more controversial than the above cases (which are themselves not free from controversy). Any **organism** that has DNA can have this DNA modified in one of two ways. The first and simplest involves a modification for that individual. This approach can be used to treat some human diseases. For example, where specific genes are missing and the appropriate **protein**s are not being manufactured, it is possible in some cases to add the relevant DNA to **cell**s, and this DNA is then expressed to provide the missing protein. Because the corrected DNA cannot be passed from a parent to a child, it is called a non-heritable change. The second and more complex approach involves heritable modifications of the DNA, which can then be passed on to offspring. Heritable changes to human genomes are ethically problematical and, consequently, this approach is more commonly applied to other organisms.

Heritable changes may be carried out on nonhuman organisms for a number reasons, the commonest being to mark **population**s. For example, many **bacteria** have genes for antibiotic resistance inserted so the populations can be studied. Genes from other **species** can also be added; for example, bacteria, mice, and **plants** have all had luminescent genes from jelly fish added to their genomes for monitoring purposes. Another common reason for adding genes to a foreign organism is for the manufacture of alien products. Some cows have been modified so that they can produce human **insulin** in their milk. Pigs have been modified to overcome a number of transplantation problems so that some limited transplantation of **organ**s can be carried out from pigs to humans.

DNA technology is a new area of research that still retains enormous controversy. This situation will likely continue for many years as public debate continues and advancing technology constantly out paces the requisite legislation.

See also Gene therapy; Genetic engineering; Human Genome Project

DNA TRANSLATION

DNA translation is the formation of a **protein molecule** using the coded instructions present within the **deoxyribonucleic acid (DNA)** molecule. The first stage of the process is called **transcription** and involves the production of a molecule of single-stranded **ribonucleic acid** called messenger **RNA** (mRNA) from the DNA template. This molecule is transferred from the **nucleus** to the **ribosome**s, which are the site of the **protein synthesis**. Translation starts at one end of the mRNA and proceeds much like the reading of a magnetic tape in a cassette recorder. As the mRNA passes through the ribosome, every

codon (set of three bases) is read, and each codon codes for a specific **amino acid**. These amino acids are brought together and joined to give a chain of amino acids, a **polypeptide**. The amino acids are brought to the ribosomes by another molecule of RNA, transfer RNA (tRNA). As the amino acids are joined to each other, hydrogen bonds start to form between amino acids within the chain, giving a three dimensional structure to the protein. As the mRNA emerges from the first ribosome, it is quickly attached to a second so the whole process of translation commences again before the first molecule of protein is produced. As soon as the mRNA leaves the first ribosome, that ribosome is ready to start reading another molecule of mRNA. These two processes allow for a fast production of the **gene** product.

Within the triplet code of the codons, certain sequences code for a start signal that tells the ribosome when to start manufacturing the polypeptide chain. Similarly, there are also stop signals that tell the ribosome when to stop manufacturing the chain of amino acids. These stop signals are sometimes referred to as nonsense codons because they do not code for any amino acids and the mRNA basically falls off the ribosome because there is nothing to add to it.

Before the mRNA is read and translated, certain pieces of the message are removed or excised. These sections of DNA are the nonsense coding sequences that interrupt the DNA/RNA and carry no useful information for the manufacture of proteins. They are called **intron**s. Only the coding parts of the DNA, called the **exon**s, are read and translated. This again speeds up the process of the production of the gene product.

Translation is a fast and efficient process that can yield a large amount of protein in a very short time. The whole process can be brought to a halt in some instances by a build up of the product; in other words, the process can be self-limiting.

See also Chromosome

DOBZHANSKY, THEODOSIUS (1900-1975)

Russian American biologist

Theodosius Dobzhansky's work in **genetics** and evolutionary biology ensures his recognition as an influential scientist of the twentieth century. The recipient of the 1964 Medal of Science, his most significant contributions were in the development of the modern theory of **evolution**. His first book on **evolutionary theory**, *Genetics and the Origin of Species,* is considered by many scientists to be the most important book on the subject to be written in the twentieth century. In it, Dobzhansky was able to link the work of Austrian botanist **Gregor Mendel** and English naturalist **Charles Darwin**. He accomplished this by gathering empirical evidence that supported Mendel's mathematical theories of inherited traits, while extending critical evolutionary issues far beyond the mathematical model.

A common thread of biological evolution runs through his enormous output of written work—nearly six hundred publications, including twelve books—spanning biological, philosophical, and humanistic disciplines. Dobzhansky's gifts as an original thinker and theoretical synthesizer, combined with his research skills in the field and laboratory, were central to the rapid advances that were made in the study of population genetics during his time, and pivotal to his evolutionary concepts.

Dobzhansky was born on January 25, 1900, in Nemirov, Russia (now Ukraine). He was the only child of Sophia Voinarsky and Grigory Dobzhansky, a high school mathematics teacher. Until he was nine years old, Dobzhansky was educated at home by his mother and a German governess. In 1910, the family moved to Kiev, where he entered the second year of an eight-year Russian gymnasium (high school) program. His interest in the natural sciences surfaced at this time, and he spent considerable time collecting butterflies. However, when he was fifteen, he was persuaded by a young entomologist, Victor Luchnik, that to be a success, he would have to specialize. Dobzhansky chose to study the genus *Coccinella*, or ladybird beetles. (He was to continue this study throughout his early career; they were the subject of his first published work in 1918.) As a young man, he read Charles Darwin's *On the Origin of Species*, which also had a great influence on his decision to become a biologist.

He was graduated from the University of Kiev in 1921, and took a position as an instructor of **zoology** at the Polytechnic Institute in Kiev. His friendship with a young professor of botany, Gregory Levitsky, was the stimulus to a beginning interest in genetics. Literature on new genetic discoveries in Germany, England, and the United States was just beginning to be disseminated in Russia, and Dobzhansky was able to study these works. He remained at Kiev until 1924, when he became an instructor in genetics at the University of Leningrad. It was during this time that he began to study the genetics of *Drosophila* fruit flies Drosophila.

In 1927, on a two-year fellowship from the International Board of the Rockefeller Foundation, he came to Columbia University to work with geneticist Thomas Hunt Morgan. When Morgan left Columbia for the California Institute of Technology in 1928, Dobzhansky went with him. He remained there as professor of genetics until he returned to Columbia University in 1940 as professor of zoology. He was to stay at Columbia until 1962, when he became professor at the Rockefeller Institute in New York City. In 1971, he returned to California to become adjunct professor in the department of genetics at the University of California at Davis. He remained at Davis until his death.

Early Years of Genetic Research

Dobzhansky brought his skills as an anatomist and field naturalist to Morgan's laboratory at Columbia, but what he needed to learn was *Drosophila* genetics. Even though he had conducted research at Leningrad on the pleiotropic effects (the observation that each gene acts on each part of the body) of mutant *Drosophila* genes, he believed that current research had surpassed his knowledge. It was at this point that he came under the guidance of American geneticist Alfred H. Sturtevant. Their collaboration, at first one of student and teacher,

Theodosius Dobzhansky. *(The Library of Congress. Reproduced by permission.)*

and later one of equals, was to continue for many years, although their association was later marked by professional disagreement, and, at the end, personal bitterness. Dobzhansky's research on population genetics of *Drosophila* in the early thirties involved the physiological, developmental, and genetic causes of hybrid sterility in this genus. This led him, in 1935, to formulate the genetic concept that what sets species apart is reproductive isolation. It also defines speciation as a slowly changing evolutionary process. Dobzhansky coined the term "isolating mechanisms" to describe events that interfere with gene exchange between species.

Beginning in 1932, Dobzhansky began to extrapolate his genetic studies toward a theory of natural evolution. In 1933, he published a major study in the *American Naturalist* on geographical variation and evolution in ladybird beetles. Although important because it could be understood by both geneticists and systematists, it pointed to a major problem that remained to be solved before evolution could be studied by means of modern genetics: genetic analysis of species differences in ladybird beetles and fruit flies was next to impossible; therefore, no genetic experiments could be conducted. The study of evolution would have to wait for a more cooperative organism.

It was during this period that *Drosophila pseudoobscura* entered the picture. It was this fruit fly, determined to be a separate species of the American form found in the Pacific Northwest, that would be the basis of Dobzhansky's evolutionary research for the rest of his scientific career. The study and reconstruction of geographical chromosome arrangements in *D. pseudoobscura* was crucial, because this research gave him the tools to reconstruct the evolutionary history of many species. His study of chromosomal variability led to a series of writings on hybrid sterility in *D. pseudoobscura* that began in 1933. Through his research, Dobzhansky was able to determine that the sterility factor was connected in some way with the autosomes, or non-sex chromosomes. This discovery was of great importance toward the development of a theory of evolution in wild populations.

Toward A Modern Evolutionary Theory

Although previous groundwork on *D. pseudoobscura* had been done by other geneticists in the United States and Europe, it was Dobzhansky's published accounts of his experiments that led to widespread *D. pseudoobscura* collection and subsequent development of detailed chromosome maps by many other groups. The group headed by Dobzhansky and Sturtevant at the California Institute of Technology was now recognized as the center of *D. pseudoobscura* Drosophila pseudoobscura research. From 1936 through 1976, Dobzhansky, in collaboration with Sturtevant and others, produced a body of work that was eventually published under the title "Genetics of Natural Populations." This series of forty-three papers—a synthesis of genetic theory backed up with empirical evidence—is a major contribution to modern evolutionary genetics. In 1937, Dobzhansky wrote the seminal *Genetics and the Origin of Species,* building on the outcome of thirty-five years of genetic laboratory research; the theoretical framework of English geneticists Ronald A. Fisher and John Burdon Sanderson Haldane and American geneticist Sewall Wright; and his own experimental work.

In the early 1940s, Dobzhansky's experiments on speciation with *D. willistoni* augmented his studies of *D. pseudoobscura.* Collection of this group of fruit flies took him to Colombia, South America. His work with this species involved the relationships between ecological and genetic diversity, as well as a study of physiological adaptation. He published papers on his conclusions in 1957 and 1959.

Ecological and Philosophical Contributions

Dobzhansky's work in South America extended to other areas of biological inquiry, such as ecology and systematics. In 1943, he made the first of many visits to Brazil, where he compared population structure and evolutionary process in the constant environment of a tropical rain forest to the changing seasons of a temperate climate. His study of the diversity of species in tropical forests led him to develop a hypothesis ac-

counting for the causes of this diversity. In the early 1960s, he also published several papers discussing the innate reproductive capacity of fruit fly populations. Although Dobzhansky conducted only brief research into human genetics, he had a strong interest in human evolution and wrote several books on the subject. His *Mankind Evolving,* published in 1962, used the integration of Mendelism, or mathematical theory of inherited traits, and Darwinism, the theory of evolution by natural selection, as a basis of understanding human nature. Through a synthesis of anthropology, evolutionary theory, genetics, and sociology, Dobzhansky discusses the two-dimensional aspects of human nature—the biological and the cultural—and their interrelatedness as a force in human diversity and race.

His concern for humanity led him to speak out against racial bigotry. After World War II, he wrote several publications criticizing the eugenic movement, which was concerned with improving the hereditary qualities of the population. His first major book about race was *Heredity, Race, and Society,* published in 1946. This was followed that same year by an exposé of Soviet biologist Trofim Denisovich Lysenko's suppression of genetics in the Soviet Union. Dobzhansky also explored the evolutionary roots of religion in articles published in the 1960s and 1970s, and published *The Biology of Ultimate Concern* in 1967. Also during that time, in an effort to show how evolutionary biology raises new philosophical problems and enlightens old ones, he wrote several essays on various philosophical subjects.

The Rewards of a Distinguished Career

Dobzhansky was a man possessed with tremendous energy and discipline. He was a world traveler and fluent in six languages. He enjoyed the outdoors and was always eager to get out in the field, whether for work or for relaxation. His enthusiastic personality made him a successful teacher, and under his tutelage more than thirty students obtained their Ph.D. degrees.

He received numerous honors and awards throughout his academic and scientific career. He was elected to the U. S. National Academy of Sciences, the American Academy of Arts and Sciences, and the American Philosophical Society, as well as many scientific societies in other countries. He served as president of six professional societies, and was a member or honorary member of many more, both in the United States and abroad. He received, among others, the Daniel Giraud Elliot Medal in 1946, the Kimber Genetics Award in 1958, the Darwin Medal in 1959, the Pierre Lecomte du Nouy Award in 1963, the National Medal of Science in 1964, and capped off his honors with the Gold Medal Award for Distinguished Achievement in Science, given by the American Museum of Natural History in 1969.

Dobzhansky married Natalia (Natasha) Petrovna Sivertzev, also a geneticist, on August 8, 1924. She died in 1969. Their only child, Sophia, became an anthropologist. Dobzhansky was diagnosed as suffering from chronic lymphatic leukemia in 1968. Surprisingly, the disease did not affect his productivity during the remaining years of his life, and he continued to work until the day before his death. He died of heart failure on December 18, 1975.

DOCHEZ, ALPHONSE RAYMOND (1882-1964)
American bacteriologist

Born in San Francisco of Belgian parentage, Alphonse Raymond Dochez was educated at Johns Hopkins University, where he acquired a B.A. in 1903 and an MD in 1907. He began his career as a researcher at the newly established Rockefeller Institute for Medical Research. There, working with **Oswald T. Avery**, the pair made notable discoveries about *pneumococci*, the **bacteria** that cause the most common type of bacterial pneumonia. In studying the **blood** of pneumonia patients, they found not only pneumococcus bacteria, but also specific **antibodies** formed in the blood to fight the bacteria and help the body recover from the disease. Dochez and Avery were then able to isolate these specific antibodies and use them to develop an antiserum that could be inoculated into patients to treat pneumonia.

When World War I interrupted this research, Dochez, now a major in the Medical Corps, worked on respiratory diseases among the military on the European front. After his return to the United States in 1919, he went back to Johns Hopkins and began work on connecting streptococcus bacteria with another infectious disease, scarlet fever. He found a direct causal link between a type of streptococcus and the disease, something many other physicians and bacteriologists had disputed. Dochez, now on the staff of the College of Physicians and Surgeons of Columbia University in New York City, then developed an antiserum that he tested on laboratory animals. He obtained a British patent for the antiserum in 1924 and a U.S. patent in 1926. However, another research team, George Dick and Gladys Dick of Evanston, Illinois, were working on a diagnostic test and an antiserum for scarlet fever at the same time. Their U.S. patent had preceded his, so with a court ruling in their favor, Dochez concluded his work in scarlet fever.

Later in the 1920s, Dochez focused his research on the most common upper respiratory infection—the common cold. His research on animals showed that viruses and not bacteria are the microscopic culprits that cause colds. Although he tried isolating cold-producing viruses, he was unable to—research techniques of that era were not adequate. (Now scientists have found almost 200 cold-producing viruses—and there is still no one effective treatment).

Throughout his long career at Columbia University, Dochez always reasoned out a problem systematically and gave great thought to experimental design. Those who worked on his laboratory staff always felt included and lauded for their work, and their names were included on scientific papers he presented. Working with Dochez meant being part of the "Tea Club," where scientific discussions abounded as well as commentary on world events. His inner circle referred to him as "Doh"—a man who enjoyed fashionable clothes and an evening at the Opera. Although he retired from Columbia in 1949, he continued an active interest in medical science and was studying causes of cancer at the time of his death in 1964 at the age of 82.

DOHERTY, PETER C. (1940-)
Australian immunologist and virologist

Peter C. Doherty was born to a piano teacher and a government employee in Brisbane, Australia, on October 15, 1940. A fine student, he earned his bachelor's degree from the University of Queens a year early. He went on to pursue a veterinary (agricultural) career, obtaining his B.V.Sc. from Queensland in 1962, and an M.V.Sc. in 1966.

Later, though, Doherty's career took an unexpected turn. By reading the books of Sir **Frank Macfarlane Burnet**, a countryman who had won the 1960 Nobel in Medicine for his work on immune tolerance, Doherty became interested the **immune system**. Doherty went to the University of Edinburgh, Scotland, earning a Ph.D. there in 1970 for his studies of viral infection of sheep brains. In 1972 he returned to Australia as a research fellow at the John Curtin School of Medical Research, Australian National University, Canberra—setting the stage for his seminal discovery.

At the Curtin school, space was at a premium. Doherty had his own laboratory for a project studying killer **T cells**—white **blood** cells that destroy virus-infected cells in the body by recognizing viral **proteins** on the **cells'** surfaces. In the nearby laboratory of Robert Blanden, which was working on a similar project, space was so short that they needed to export a worker. That worker, **Rolf Zinkernagel**, moved into Doherty's lab; together the two began studying how killer T cells know which cells to attack.

At the time, immunologists were very interested in a group of **genes** collectively called the major histocompatibility complex, or MHC. These genes, clustered together in the **DNA** sequence, encode a series of proteins called the MHC antigens, which determine whether a transplanted organ will be accepted or rejected by a recipient. If the MHC genes of the donor and the recipient match, the organ survives; if they do not, the organ is attacked by the recipient's immune system and dies.

A number of researchers had guessed that the rejection of MHC-mismatched organs was essentially the same process as the killing of virus-infected cells by killer T cells. Doherty and Zinkernagel demonstrated that this was true, and that the MHC **antigens** were necessary for killer T cells to tell friend from foe. When they investigated further, they found something unexpected. Most immunologists had expected that when virus-infected cells and killer cells were poorly MHC matched, the immune cells' killing response would be strongest, much as in badly matched transplants. But the opposite was true. In order to get proper T cell killing of the virus-infected cells, Doherty and Zinkernagel discovered, the cells' MHC regions had to match.

The two had discovered that T cells—indeed, the immune response in general—can only recognize viral proteins when they are displayed in the context of properly matched MHC antigens. The immune system, which had evolved to recognize "self" from "other" did not react most strongly to "other," but to a third state, "altered self." This discovery finally put transplant rejection into biological context: the body does not purposely reject mismatched organs because they are different, it rejects them because it mistakenly identifies the mismatched MHC antigens as "self" antigens that have been altered by interaction with viral proteins. The finding also opened the way to better methods for heading off transplant rejection, for creating vaccines, and for further unraveling the workings of immunity; vulnerability to certain infections; and autoimmune disease, where the body mistakenly attacks its own tissues.

While immunologists recognized that Doherty's and Zinkernagel's 1973-1974 discovery was important, its true significance took time to sink in. The successful collaborators went their separate ways, with Doherty heading for the Wistar Institute in Philadelphia, as an associate professor in 1975. He returned to the Curtin school in 1982 as professor and head of the Department of Experimental Pathology, leaving once again in 1988 to be chairman of the Department of Immunology at St. Jude Children's Research Hospital, Memphis, Tennessee. At St. Jude's, in addition to heading the hospital's immunology research program, he has maintained his own laboratory work on viruses that cause cancer as well as the Epstein-Barr virus, which causes severe infections in the compromized immune systems of **AIDS** patients.

Along the way, international recognition of Doherty's work grew. In 1983 he was given the Paul Ehrlich Prize in Germany; in 1986, he gained the International Award of the Gairdner Foundation of Canada; and in 1995 he received the Albert Lasker Medical Research Award, often a prelude to a Nobel. Finally, in 1996, for his collaborative work with Zinkernagel, the two were awarded the 1996 Nobel Prize in Physiology or Medicine.

Edward A. Doisy. *(The Library of Congress. Reproduced by permission.)*

DOISY, EDWARD A. (1893-1986)
American biochemist

Edward Adelbert Doisy was an acclaimed biochemist whose contributions to research involved studying how chemical substances affected the body. In addition to research on **antibiotics**, **insulin**, and female **hormones**, he is noted for his successful isolation of **vitamin** K, a substance that encourages **blood** clotting. Because he was able to synthesize this substance, many thousands of lives are saved each year. For this research, Doisy shared the 1943 Nobel Prize in medicine or physiology with Danish scientist **Henrik Dam**.

Doisy, one of two children, was born November 13, 1893, in Hume, Illinois, to Edward Perez Doisy, a traveling salesman, and Ada (Alley) Doisy. Doisy received his baccalaureate degree in 1914 from the University of Illinois at Champaign and then obtained his master's in 1916. The advent of World War I interrupted his schooling for two years, during which time he served in the Army. After the war, Doisy received his Ph.D. from Harvard University Medical School in 1920. Beginning in 1919 he rapidly rose through the academic ranks, achieving the position of associate professor of biochemistry in the Washington University School of Medicine, St. Louis. He left this position in 1923 to go to the St. Louis University School of Medicine, and a year later he was ap-

pointed to the chair of biochemistry, where he engaged in research and teaching. He also was named the biochemist for St. Mary's Hospital. Doisy held these positions until his retirement in 1965.

For 12 years—from 1922 until 1934—Doisy worked with biologist Edgar Allen to study the ovarian systems of rats and mice. During this time he participated in research that isolated the first crystalline of a female steroidal hormone, now called oestrone. He later isolated two other related products, oestriol and oestradiol–17β. When Doisy administered these in tiny quantities to female mice or rats whose ovaries had been removed, the creatures acted as if they still had ovaries. Many women have benefitted from this research, as these compounds and their derivatives have been used to treat several hormonally-related problems, including menopausal symptoms.

Doisy, in 1936, turned from this line of research to trying to isolate an antihemorrhagic factor that had been identified by Danish researcher Henrik Dam. Dam had discovered a chemical in the blood of chicks that decreased hemorrhaging; he called this substance *Koagulations Vitamine*, or vitamin K. Using Dam's work as a springboard, Doisy and his co-workers spent three years researching this new vitamin. They discovered that the vitamin had two distinct forms, called K1 and K2, and successfully isolated each— K1 from alfalfa, K2 (which differs in a side chain) from rotten fish. Alter Doisy had isolated these two compounds he successfully determined their structures, and was able to synthesize the extremely delicate vitamin K1.

Synthesizing vitamin K enabled large quantities of it to be produced relatively inexpensively. It has since been used to treat hemorrhages that would previously have been fatal, especially in newborns and other individuals who lack natural defenses; it is estimated that the use of vitamin K saves almost 5,000 lives each year in the United States alone. For these research advances, Doisy shared the 1943 Nobel Prize for medicine or physiology with Dam. Some of this research was funded by the University of St. Louis and some of the funds were contributed by the pharmaceutical manufacturer Parke-Davis and Co.—a financial arrangement that Doisy saw as a model for future industry-university research relations.

Over the course of his career, most of Doisy's research focused on how various chemical substances worked in the human body. In addition to vitamin K, his team studying the effects of certain antibiotics, sodium, potassium, chloride, and phosphorus. He also developed a high-potency form of insulin, for use in treating diabetes.

Doisy was made St. Louis University's distinguished service professor in 1951, and later was named emeritus professor of biochemistry. As a sign of his contributions, the university's department of biochemistry was named in his honor in 1965, and he was made its emeritus director. Because of his prominence and his loyalties to the University, there are numerous plaques and buildings bearing his name.

Doisy's contributions to the field of biochemistry are recognized by the numerous honorary awards he held and the scientific societies to which he belonged. He was member of the League of Nations Committee on the Standardization of Sex Hormones from 1932 to 1935, and in 1938 was elected to the National Academy of Sciences. In 1941 he was honored with the Willard Gibbs Medal of the American Chemical Society, which is perhaps the highest distinction in chemical science. He served as both the vice president and then president, from 1943 to 1945, of the American Society of Biological Chemists, and was the 29th president of the Endocrine Society in 1949. Doisy died October 23, 1986.

DOMAGK, GERHARD (1895-1964)
German biochemist

Gerhard Domagk discovered the first synthetic drug that could be used to battle the effects of many bacterial diseases. He was born in Lagow, Brandenburg, Germany (now Poland), on October 30, 1895. Domagk began his studies at the University of Kiel but abruptly stopped at the outbreak of World War I during which he served in the military and was wounded in action. He returned to school to study medicine and was awarded a medical degree in 1921.

After his schooling, Domagk began working for I.G. Farbenindustrie, a large company that manufactured industrial dyes. Because of his medical training, he did research on dyes with an eye toward their medical applications. One newly

Gerhard Domagk. (The Library of Congress. Reproduced by permission.)

manufactured dye, called Prontosil Red, was of particular interest to Domagk. In 1932, Domagk found that when he injected dye into mice infected with *Streptococcus* **bacteria**, it cured the animals of the usually fatal effects and seemed to have few side effects. More dramatically, when his daughter Hildegarde contracted a serious *Streptococcus* infection after pricking herself with a knitting needle infected with a virulent bacteria in the laboratory, Domagk, in desperation, administered large doses of Prontosil to her, judging the dosage based only on his experiments with mice. In 1935, the story of her recovery spread like wildfire all over the world. Prontosil was later used by Franklin D. Roosevelt's son who was dying of an infection.

Domagk and many others quickly followed up on his discovery. The same year, a French researcher, Daniele Bovet, discovered that it was just one portion of the Prontosil molecule that was effective against bacteria—a sulfonamide called sulfanilamide. This portion of the molecule blocks coenzyme action in bacteria and kills them. In England it was found that the chemical was effective in controlling bacterial meningitis, pneumonia and gonorrhea. In rapid order, other sulfonamide drugs, dubbed "sulfa" drugs, were developed which saved many lives and gave the world community hope of curing infectious disease during the 1930s. These drugs included sulfanilamide, sulfapyridine, sulfathiazole, and sulfadiazine.

In 1939 Domagk was recognized for his contribution with a Nobel Prize in medicine and physiology. He initially accepted the award but later was forced to refuse it because Adolf Hitler coerced German scientists into refusing awards by threatening them with arrest and even jailing. After Hitler's death and the end of World War II, Domagk accepted his medal and went on to do further research in the area of chemotherapy and its application to tuberculosis and **cancer**.

See also Ehrlich, Paul

DOMINANCE HIERARCHY

Research by biologists on social organization in **animal** groups led to the frequent observation that some individuals have preferential rights over others. The result is a dominance hierarchy in a **population**. Dominant animals have first choice of desired resources (often including space and **territory**) and of mates. The dominance may be accepted, without open conflict, by subordinate animals, or it may be contested, often repeatedly. Oftentimes, the subordinates to a dominant individual may then dominate other individuals lower in a hierarchy. Much of the early research was done on domestic fowl (leading to the term "pecking order" becoming a vernacular term in wide usage) but dominance hierarchies occur in a wide range of animal **species**, including humans. Dominance hierarchies have been documented in most primate species, in social bees, in the sea anemone, in wallabies and in coyotes, wolves and wild dogs. Dominance hierarchies have been observed in hedge sparrows, mole rats, freshwater crabs, greylag geese, ravens, and various species of fish, to give some indication of the range of **organism**s exhibiting such behavior.

Dominant individuals do not destroy subordinates to establish Hierarchical status relationships can be maintained by the dominant or by the subordinate individuals. Dominance relationships can also change according to context, with different individuals assuming dominance in different situations.

Hormone output is one important factor related to the outcome of initial encounters and to the establishment and maintenance of dominance hierarchies. A frequently noted correlation is that between the amount of male sex hormone and rank position. Other hormones, e.g., **estrogen** and epinephrine, can also affect dominance relationships. Sex and gender; territorial familiarity; breeding season; age, maturity and experience; seniority in the group; size and strength; and **heredity**, can all influence an individual's ranking and status. The size of the group and whether it is confined (such as in a zoo, a chicken coop or a fenced field) can also be contributing factors.

Research on different species has revealed all kinds of dominance hierarchies. For example, a fifteen-year, on-going study of an Asian colobine monkey concluded that "the dispersal pattern of langur males seems to reflect a long-lasting effort to achieve reproduction. That goal was tied to the necessity to rise through the ranks of a male band to the top of its dominance hierarchy, because only the highest ranking males can replace those males already holding a harem." A seven-year study of brown hyenas of the Central Kalahari **desert** in Africa showed a "separate, linear dominance hierarchy within

each sex." In another study of an ant in Australia, it was found that "worker's age regulates the linear dominance hierarchy."

Study of dominance hierarchies help biologists to better understand reproductive success and stability in wild animals (in trout, for example), and thus aid in conserving diversity and forestalling **extinction**. Or it can be economically practical. The long-term research on dominance behavior and territoriality among chickens has allowed farmers to increase the **egg** production of their hens. One group of researchers was able to develop a "winner" and a "loser" strain of chickens in as little as four generations.

Extrapolation from non-human species to human behavior is difficult, but many studies of dominance relations among human groups have been completed. A few illustrations illustrate its importance. Studies have demonstrated that young children in play and study groups tend to agree as to which are the dominant members. Dominant facial appearance has been correlated with the attainment of rank in the military. And, sociologists have studied the formation of dominance hierarchies in committees and task groups.

DOMINANCE, LAW OF

There are three different laws of dominance within biology: one in **genetics**, one in plant **ecology**, and one in **animal** behavior.

The genetics law of dominance states that when a dominant and recessive form of a **gene** come together, the dominant form masks the recessive form. Even though the recessive **allele** (or member of the gene pair) is still present, it is not visible.

Within plant ecology, the law of dominance refers to **succession**. At each stage of succession, a particular **species** has a greater impact on the immediate environment than any other. This is said to be the dominant species. The law of dominance states that the stages of succession proceed towards one final type of plant cover; for example, oaks in an oak woodland.

In animal behavior, the law of dominance says that the individual animal that is capable of winning aggressive encounters with other members of the group becomes the dominant animal in the group. This greater prominence within the group leads to many advantages for the dominant animal, such as greater choice of mates and food.

See also Dominance and recessiveness; Dominance hierarchy

DOMINANCE AND RECESSIVENESS

An **allele** is a member that makes up a **gene** pair. Different alleles of the same gene can have different levels of phenotypic expression (outward appearance). Because genes are usually present in two copies (**sex chromosome**s being the exception in **diploid organism**s like humans), this difference in expression can be important to the **phenotype** of the organism. Some alleles mask the effect of all other forms of the gene present. These alleles are said to be dominant. A dominant gene has the same effect when it is present as a single copy (heterozygous) as when it is present as two copies (homozygous). The allele not expressed in the heterozygous form is said to be recessive to the masking allele, the dominant allele. For example, with the pea **plants** studied by Austrian botanist **Gregor Johann Mendel**, the allele for tall plants was dominant to the allele for dwarf plants. When a gene is said to be dominant or recessive without any qualification, then it is assumed that effect is with reference to the wild type allele. People often refer to the characteristic without acknowledging that it is caused by an allele; this is implied. For example, one may state that, in humans, blue eyes are recessive to all other eye colors when what we actually mean is that the allele for blue eyes is recessive to all other eye color alleles.

An effect called co-dominance can also occur where there is a blending of the effects of the two alleles. Many genes for characteristics that are not for discrete states (e.g. presence versus absence) can be co-dominant. Examples of this include such characteristics as **blood types** with the ABO **blood** system.

In certain instances, there is no dominance shown at all, and when two differing parental types cross there is simply an intermediate blending of the two forms. For example, snap dragons with white **flower**s when mated with those with red flowers give progeny with pink flowers. There are also situations between dominance and no dominance, which are known as partial dominance. With these situations there is an intermediate form produced in the first generation, but it is markedly more similar to one of the parents than the other.

For the effects of a recessive allele to be observable it has to be present in the homozygous form. In other words, there have to be two copies of the same allele present. Recessive alleles can be hidden in one generation but they may reappear in the next depending upon the breeding that has occurred. Just because the allele cannot be directly observed does not mean it is not present. **Punnet squares** can be used to calculate the frequency of different phenotypes and **geno-type**s (genetic constitution) in particular breeding experiments, and from observing the results of these the genetic makeup of the parental types can be inferred. It is also possible to perform crosses with a known genotype and then infer the unknown genotype from the resulting progeny.

Alleles do not act in an all or nothing manner. All degrees of blending are possible; dominant and recessive are only part of the overall picture.

See also Dominance, Law of; Gene expression

DOUBLE HELIX

The double helix refers to **DNA**'s ''spiral staircase'' structure, consisting of two right-handed helical polynucleotide chains coiled around a central axis. **Gene**s, which are specific regions of DNA, contain the instructions for synthesizing every **protein**. Because life cannot exist without proteins, the discovery

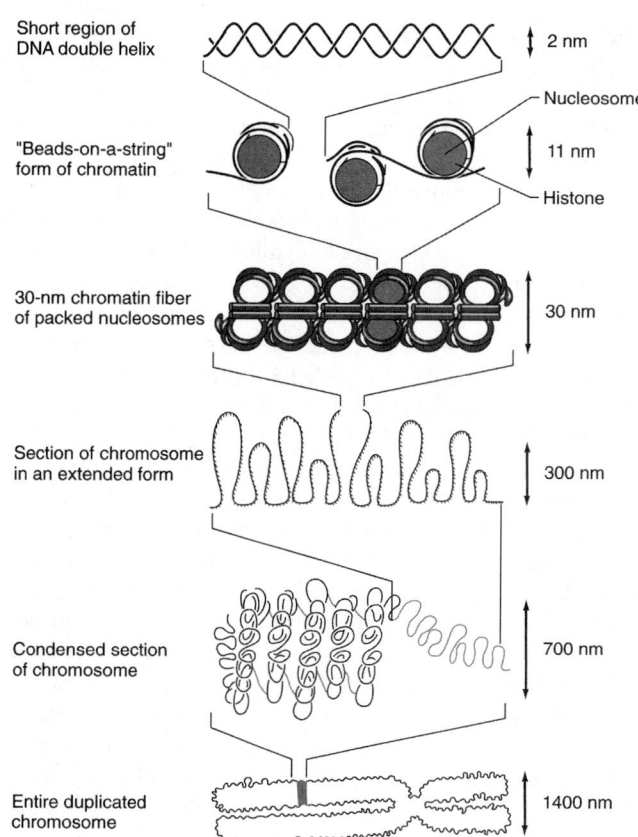

Short region of DNA double helix — 2 nm

Nucleosome

"Beads-on-a-string" form of chromatin — 11 nm

Histone

30-nm chromatin fiber of packed nucleosomes — 30 nm

Section of chromosome in an extended form — 300 nm

Condensed section of chromosome — 700 nm

Entire duplicated chromosome — 1400 nm

The structure of the DNA molecule. (Illustration by Hans & Cassady, Inc.)

of DNA's structure unveiled the secret of life: **protein synthesis**. In fact, the ''central dogma'' of **molecular biology** is that DNA is used to build **ribonucleic acid (RNA)**, which is used to build proteins, which in turn play a role in building DNA and RNA.

The discovery of the double-helix molecular structure of deoxyribonucleic acid (DNA) in 1953, one of the major scientific events of the twentieth century, and some would say in the history of biology, marked the culmination of an intense search involving many scientists. But ultimately, credit for the discovery, and the 1962 Nobel Prize in the Physiology or Medicine category, went to **James Dewey Watson**, who at the time of the discovery was an American postdoctoral student from Indiana University; and **Francis Harry Compton Crick**, a researcher at the Cavendish Laboratory in Cambridge University, England. Their work, conducted at Cavendish Laboratory, significantly impacted the emerging field of molecular biology.

Prior to Watson and Crick's discovery, it had long been known that DNA contained four kinds of **nucleotide**s, which are the building blocks of **nucleic acids**, such as DNA and RNA. A nucleotide contains a five-carbon sugar called deoxyribose, a **phosphate group**, and one of four **nitrogen**-containing bases: adenine (A), guanine (G), thymine (T), and cytosine (C). Thymine and cytosine are smaller, single-ringed structures called pyrimidines; adenine and guanine are larger, double-ringed structures called purines. Watson and Crick drew upon this and other scientific knowledge in concluding that DNA's structure possessed two nucleotide strands twisted into a double helix, with bases arranged in pairs such as A T, T A, G C, C G. Along the entire length of DNA, the double-ringed adenine and guanine nucleotide bases were probably paired with the single-ringed thymine and cytosine bases. Using paper cutouts of the nucleotides, Watson and Crick shuffled and reshuffled combinations. Later, they used wires and metal to create their model of the twisting nucleotide strands that form the double-helix structure. According to Watson and Crick's model, the diameter of the double helix measures 2.0 nanometers (nm). Each turn of the helix is 3.4 nm long, with 10 bases in each chain making up a turn.

Before Watson and Crick's discovery, no one knew how hereditary material was duplicated prior to **cell division**. Using their model, it is now understood that **enzyme**s can cause a region of a DNA **molecule** to "unwind" one nucleotide strand from the other, exposing bases that are then available to become paired up with free nucleotides stockpiled in **cell**s. A half-old, half-new DNA strand is created in a process that is called "semiconservative **replication**." When free nucleotides pair up with exposed bases, they follow a base-pairing rule which requires that A always pairs with T, and G always with C. This rule is constant in DNA for all living things, but the order in which one base follows another in a nucleotide strand differs from **species** to species. Thus, Watson and Crick's double-helix model accounts for both the sameness and the immense variety of life.

It is fair to say that Watson and Crick's discovery of the double helix would not have been possible without significant prior discoveries. In his 1968 book, *The Double Helix, A Personal Account of the Discovery of the Structure of DNA*, Watson wrote that the "race" to unveil the mystery of DNA was chiefly "a matter of five people": Maurice Wilkins, **Rosalind Franklin**, Linus Pauling, Crick, and Watson. Wilkins, an Irish biophysicist who shared the 1962 Nobel Prize in Physiology or Medicine with Crick and Watson, extracted DNA gel fibers and analyzed them using X-ray diffraction. The diffraction showed a helical molecular structure, and Crick and Watson used that information in constructing their double-helix model. Franklin, working in Wilkins' laboratory, between 1950 and 1953, produced improved X-ray data using purified DNA samples, and through her work confirmed that each helix turn is 3.4 nm. Although her work suggested DNA might have a helix structure, she did not postulate a definite model. Pauling, an American chemist and twice Nobel laureate, in 1951 discovered the three-dimensional shape of the protein collagen. Pauling discovered that each collagen **polypeptide** or **amino acid** chain twists helically, and that the helical shape is held by hydrogen bonds. With Pauling's discovery, scientists worldwide began "racing" to discover the structure of other biological molecules, including the DNA molecule.

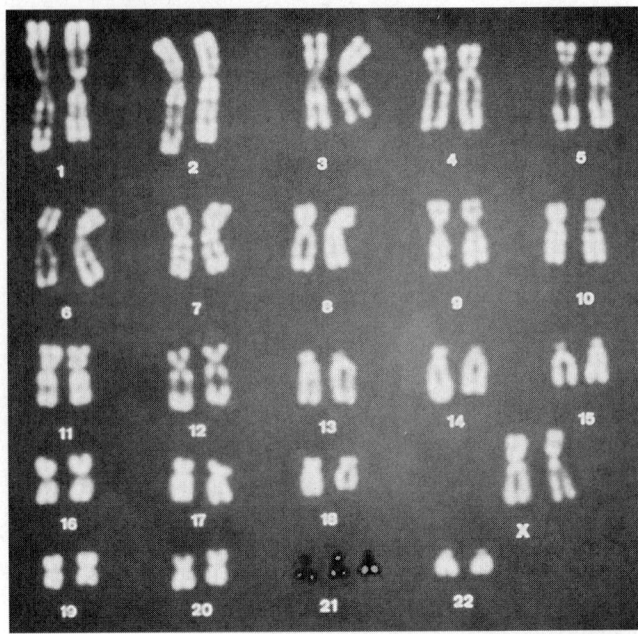

A false-color karyotype indicating Down syndrome. Down syndrome is a congenital disorder resulting from trisomy (3 chromosomes instead of 2 in pair 21). This abnormal number of chromosomes usually results from faulty cell division in the egg of the mother or the sperm of the father. *(Phototake NYC. Reproduced by permission.)*

DOWN SYNDROME

Records about the existence of Down syndrome date back to the Saxons, from whom we have anecdotal reports of the disorder. The first medical report of the condition was not published, however, until 1866. In that year, the British physician, J. Langdon Down (1828-1896) wrote a paper dealing with "Ethnic classification of idiots." In this paper, Down described children with the disorder as "representative[s] of the great mongolian race: when placed side by side," he continued, "it is difficult to believe that the specimens compared are not children of [Mongols]." Down based his comparison on the presence of vertical folds of skin above a patient's eyelids, giving him or her an Oriental appearance. Edouard Séguin's (1812-1880) text on *Idiocy and Its Treatment by Physiological Means*, published in the same year, also carried a detailed discussion of the disorder, then known as mongolism.

Credit for the most complete early description of the disorder is usually given to two British physicians, John Fraser and Arthur Mitchell. Fraser and Mitchell discussed 62 cases of the condition, known to them as kalmuc idiocy, and suggested that it occurred as the result of bad health conditions during pregnancy. They found no evidence to suggest that the disorder was hereditary.

Over the next half century, however, the condition did indeed prove to be hereditary. In 1932, the Dutch physician P. J. Waardenberg suspected that it occurred as a result of a chromosomal abnormality. That hypothesis was confirmed in 1959 by the French pediatrician, Jérôme Lejeune. Lejeune and his colleagues found that Down's patients have 47 **chromosome**s instead of the usual 46. The 47th chromosome is a third copy of chromosome 21. Because of this characteristic, Down syndrome is also known as trisomy 21. Although 95% of all Down syndrome cases have trisomy 21, there are two other major types of the syndrome. Approximately 4% of Down syndrome patients have **translocation**, in which the extra chromosome 21 has broken off and attached to another chromosome. In Mosaicism, which accounts for 1% of the cases, only certain **cell**s have trisomy 21.

Scientists believe that Down syndrome is a contiguous **gene** syndrome, meaning that most of the Down syndrome phenotypic features are unlikely to develop from a single chromosomal region. Chromosome 21, itself, contains nearly 800 genes. Although scientists have yet to isolate all the chromosomes and genes involved in Down syndrome, they are working on mapping of regions which play a role in determining specific syndrome traits. New methods for karyotyping, including one that uses spectrally classified chromosomes, are being used for identifying chromosomal abnormalities.

Down syndrome results in retarded physical and mental growth and is, in fact, the most common identifiable form of mental retardation. Patients tend to develop Alzheimer's disease-like dementia and die by the age of 35. In recent years, research and educational organizations have been very successful in helping the general public understand more about the nature and diversity of Down syndrome patients and to see that many can lead long and productive lives.

Charles R. Drew. *(AP/Wide World Photo. Reproduced by permission.)*

DREW, CHARLES R. (1904-1950)
African American American surgeon and blood researcher

Charles R. Drew was a renowned surgeon, teacher, and researcher. He was responsible for founding two of the world's largest **blood** banks. Because of his research into the storage and shipment of blood plasma—blood without **cells**—he is credited with saving the lives of hundreds of Britains during World War II. He was director of the first American Red Cross effort to collect and bank blood on a large scale. In 1942 a year after he was made a diplomat of surgery by the American Board of Surgery at Johns Hopkins University, he became the first African American surgeon to serve as an examiner on the board.

Charles Richard Drew was the eldest of five children. He was born on June 3, 1904, in Washington, DC, to Richard T. Drew, a carpet layer, and Nora (Burrell) Drew, a school teacher and graduate of Miner Teachers College. As a student, Drew excelled in academics and sports, winning four swimming medals by the age of eight. In 1922 he graduated from Paul Laurence Dunbar High School, where he received the James E. Walker Memorial Medal in his junior and senior years for his athletic performance in several sports, including football, basketball, baseball, and track.

Drew attended Amherst College in Western Massachusetts on an athletic scholarship. He would be one of sixteen black students to graduate from Amherst during the years 1920 to 1929. He served as captain of the track team; he was enormously popular and was awarded several honors, including the Thomas W. Ashley Memorial Trophy for being the football team's most valuable player.

Although Drew was a gifted athlete, he worked hard in school to keep high grades. By the time he graduated in 1926, he had decided to apply to medical school. However, his funds were severely limited. Before he could go to medical school, he had to work for a couple of years. He accepted a job at Morgan State College in Baltimore, Maryland, as a professor of chemistry and biology, as well as director of the college's sports program. During the next two years, he paid off his undergraduate loans and put some money aside for medical school.

In 1928 he was finally able to apply to medical school. However, African Americans who wished to become doctors at that time did not have many opportunities. There were two colleges open to them. Drew applied to Howard University and was rejected because he did not have enough credits in English. Harvard University accepted him for the following year, but he did not want to wait so he applied to and was immediately accepted to McGill University in Montreal, Canada.

Embarks on Research in Blood

At McGill, Drew continued to excel in sports and academics. In 1930 he won the annual prize in neuroanatomy and was elected to Alpha Phi Omega, the school's honorary medical society. During this time, under the influence of Dr. John Beattie, a visiting professor from England, Drew began his research in blood transfusions. The four different types of blood—A, B, AB, and O—had recently been discovered. Subsequently, doctors knew what type of blood they were giving to patients and were avoiding the negative effects of mixing incompatible blood types. However, because whole blood was highly perishable, the problem of having the appropriate **blood type** readily available still existed. In 1930 when Drew and Beattie began their research, blood could only be stored for seven days before it began to spoil.

In 1933 Drew graduated from McGill with his Medical Degree and Master of Surgery degree. He interned at the Royal Victoria Hospital and finished his residency at Montreal General. During this time, he continued researching with Beattie. Because of his father's death in 1934, Drew decided to return to Washington, D.C., to take care of his family. In 1935 he accepted a position to teach pathology at Howard University Medical School. The next year he obtained a one-year residency at Freedmen's Hospital in Washington, D.C.

Develops Process to Preserve Plasma

In 1938 having accepted a two-year Rockefeller Fellowship, Drew continued his work in blood at Columbia University-Presbyterian Hospital in New York. Under the auspices of the Department of Surgery, he worked with Dr. John Scudder and Dr. E. H. L. Corwin on the problem of blood storage. Drew began to study the use of plasma as a substitute for whole blood. Because red blood cells contain the substance that determines blood type, their absence in plasma means that a match between donor and recipient is not necessary, which makes it ideal for emergencies. In 1939, while supervising a blood bank at Columbia Medical Center, Drew developed a method to process and preserve blood plasma so that it could be stored and shipped to great distances. (Dehydrated plasma could be reconstituted by adding water just before the transfusion.)

Drew graduated from Columbia University in 1940, with a Doctor of Science degree; he was the first African American to receive this degree. In his dissertation, "Banked Blood: A Study in Blood Preservation," Drew showed that liquid plasma lasted longer than whole blood. He was asked to be the medical supervisor on the "Blood for Britain" campaign, launched by the Blood Transfusion Betterment Association. At the height of World War II, Nazi warplanes were bombing British cities regularly and there was a desperate shortage of blood to treat the wounded. In order to meet the huge demand for plasma, Drew initiated the use of "bloodmobiles"—trucks equipped with refrigerators. The Red Cross has continued to use them during blood drives. In 1941 after the success of "Blood for Britain," Drew became director of the American Red Cross Blood Bank in New York. He was asked to organize a massive blood drive for the U.S. Army and Navy, consisting of 100,000 donors. However, when the military issued a directive to the Red Cross that blood be typed according to the race of the donor, and that African American donors be refused, Drew was incensed. He denounced the policy as un-

scientific, stating that there was no evidence to support the claim that blood type differed according to race. His statements were later confirmed by other scientists, and the government eventually allowed African American volunteers to donate blood, although it was still segregated. Ironically, in 1977 the American Red Cross headquarters in Washington, D.C., was renamed the Charles R. Drew Blood Center.

Drew was asked to resign from the project. He returned to Washington, D.C., and resumed teaching. In 1941 he was made professor of surgery at Howard University, where he had been rejected 13 years earlier, and chief surgeon at Freedmen's Hospital. In 1943 he became the first black surgeon to serve as an examiner on the American Board of Surgery. He was an inspiration and role model to his students and received numerous honorary degrees and awards during this period of his life, including the National Association for the Advancement of Colored People (NAACP) Spingarn Medal in 1944. He wrote numerous articles on blood for various scientific journals, and in 1946 was elected Fellow to the International College of Surgeons.

In 1939 Drew married (Minnie) Lenore Robbins, and they had four children. Drew continued teaching in Washington, DC; and, during the summer of 1949, as a consultant to the Surgeon General, he travelled with a team of four physicians, assessing hospital facilities throughout Occupied Europe. On March 31, 1950, after performing several operations, Drew allowed his colleagues and some of his students to talk him into attending a medical meeting being held at Tuskegee Institute as part of its Founder's Day celebrations. When Drew dozed off while driving near Burlington, North Carolina, his car overturned, and he was killed.

Despite his untimely death at the age of 45, Drew left behind a legacy of life-saving techniques. Additionally, many of his students rose to prominence in the medical field. In 1976 Drew's portrait was unveiled at the Clinical Center of the National Institutes of Health, making him the first African American to join its gallery of scientists. Four years later, his life was honored with a postage stamp, issued as part of the U.S. Postal Service's "Great Americans" series.

DRIESCH, HANS ADOLPH EDUARD (1867-1941)
German biologist and philosopher

Hans Driesch was born in Bad Kreuzsnach, Germany. He studied at Freiburg with August Weismann and then at Jena where he received his Ph.D. with Ernst Heinrich Haeckel in 1889. Driesch's work is especially remembered because it was a critically important antecedent to the animal **cloning** experiments of the late twentieth century. Prior to Driesch, another German biologist **Wilhelm Roux**, showed that a half embryo develops when one cell of a two **cell** frog **embryo** is killed. This suggested that the embryo is a mosaic and that cells develop independently. Further, it implied that the genetic material contained in the nucleus is qualitatively unequally divided. Thus, embryonic cell division would resulting in differential distribution of the genetic material. Driesch, using sea urchin **eggs**, obtained diametrically different results. He separated embryonic cells (blastomeres) at the two cell stage and found that each of the cells cleaved (divided), formed ciliated blastulae, and subsequently developed into perfect but miniature pleuteus larvae. The isolated cells were viewed as a ''harmonious equipotential system'' meaning that they ordinarily developed harmoniously, when naturally together, to give rise to a normal pleuteus larva but when separated, they evidenced that they were equipotential, i.e., each could give rise to an entire pleuteus larva. Using modern terminology, one would say that the genetic material **(DNA)** was not divided but it replicated with cell division in the two cell stage embryo.

How long in development does complete genetic potential persist, i.e., how long do embryonic cells remain totipotent, with replication of the entire DNA **genome**, during development? This was the question raised by the pioneer animal cloners Robert Briggs and Thomas J. King of Philadelphia when they reported the first successful animals produced by cloning from embryonic nuclei. More recently, **Ian Wilmut** in Scotland, and others, have cloned mammals from adult nuclear donors. Obviously, totipotentiality exists in at least some cells of the adult.

Of course Driesch did not know what would happen many years later. The fact that he got whole organisms from half embryos caused him to seriously doubt the prevailing notion that life could be explained in terms of physics and chemistry. He could not envision a mechanism that could divide and reconstitute itself into two whole mechanisms. He believed this was impossible following the principles of physics and chemistry. Because of this, he adopted a vitalistic view of development which proposed that the principles of physics and chemistry are inadequate to explain living phenomena and therefore, there must be a vital force or ''entelechy'' which guides development.

Driesch abandoned experimental embryology altogether and spent his later years in philosophy. His liberal views, desire for peace, and lack of sympathy for the Nazi regime resulted in his forced retirement in 1933. While retired, he continued his philosophical studies until his death in 1941.

DRIVES • See Instinct

DUBOIS, EUGÈNE (1858-1940)
Dutch anatomist and paleoanthropologist

An anatomy assistant under the Dutch morphologist **Max Furbringer**, Eugène Dubois knew little of the science of paleoanthropology—the study of the **fossil** remains of humankind's ancestors—when, in 1887, he decided flatly that he would devote his scientific efforts to the search for **Charles Darwin**'s proposed ''missing link.'' His unearthing of Java man, near the village of Trinil on Java, was the first fossil discovery of *Homo erectus*, a direct ancestor of modern man, and the first deliberate search for man's fossil ancestors. Despite

Hans Adolph Eduard Driesch. *(UPI/Corbis-Bettman. Reproduced by permission.)*

the controversy surrounding his discovery, Dubois worked at his research on the growth relationship between brain and body size and conducted investigations into the climate of the geologic past.

On January 28, 1858, Marie Eugène Francoise Thomas Dubois was born into the family of Jean Joseph Balthasar Dubois and Maria Catharina Floriberta Agnes Roebroeck at Eijsden in the province of Limburg, Netherlands. He was cast into a world that teetered on the brink of enormous scientific enlightenment and heresy. His birth had been preceded two years earlier by the discovery of primitive human fossils in Germany's Neander Valley, the bones of which were still confounding scientists up to the time of Dubois's own discovery. Then, on October 26, 1859, several months before Dubois's second birthday, Darwin's *On the Origins of Species* was released upon an unsuspecting public.

Dubois proved to be something of a naturalist from an early age, learning the names of local flora and collecting herbs for his father's pharmacy. His father encouraged the younger Dubois, lending Eugène his equipment with which to perform chemistry experiments. Dubois's path toward a career in the natural sciences was further plotted while attending the state high school in Roermond, where his interest in human origins began. In 1877, Dubois attended the University of Amsterdam. Initially a medical student, his interest in the natural sciences proved too strong a force to ignore and Dubois devoted himself to their study.

As his talents became recognized among the faculty, he was appointed assistant to Furbringer in 1881, thus guiding his career, somewhat reluctantly, into anatomy. According to Bert Theunissen in *Eugène Dubois and the Ape-Man from Java*, ''Dubois's main achievement in this study was to establish that the thyroid cartilage, part of the larynx, was homologous to the fourth and fifth branchial arches.'' In 1886, Dubois became a lecturer in anatomy at Amsterdam University, but he was never really content with his work. After a falling-out with Furbringer over his work on the larynx, Dubois decided on an alternative route.

On the Path That Leads to Java

Dubois was certainly familiar with the discovery of Neanderthal man shortly before his birth, and his growing interest in human ancestry was further incited by his talks on the subject with the Dutch botanist **Hugo De Vries**. Dubois studied **Charles Darwin**, **Charles Lyell**, and **Ernst Haeckel**, who had proposed the name *Pithecanthropus alalus* Pithecanthropus alalus for Darwin's so-called ''missing link.'' In 1876 and 1877, Dubois tried his hand at excavation near previously discovered Neolithic flint mines in his own country, but found nothing of prehistoric value.

The Neanderthal fossils remained the subject of scientific scrutiny decades after their discovery, and while Dubois acknowledged their primitiveness, he believed they were closely related to modern man and not deserving of a new taxon. He determined that still older forms would be found but not in prehistoric Europe, where the cold climate could not have supported the evolutionary progress of humankind's forebears. Therefore, the tropics of the Old World, inhabited by man's closest primate relative, the ape, must have been the jumping off point for transitionary man. And while Darwin had suggested Africa as the birthplace of humankind based on humans' close anatomical relationship to chimpanzees and gorillas, Lyell and Alfred Russel Wallace had opted in part for the East Indies on zoogeographical grounds. Dubois's determination to find the intermediate species between man and ape sent him packing for the Dutch East Indies on October 29, 1887.

Because the Dutch government refused his request to financially support such a venture, Dubois secured passage to the East Indies by enlisting in the Royal Dutch East Indies Army as a surgeon. He was assigned to a post on Sumatra, but his medical duties interfered with his plans to search out fossil man. A transfer to the Sumatran village of Pajakombo in May 1888 led to the discovery of animal fossils in the Lida Adjer Cave. This proved noteworthy for Dubois, as he had earlier posited the theory that, since human fossil remains up to that point had been discovered in caves, he would then have his greatest chance of success searching there. While later excavations in Sumatra turned up little, his finds at Lida Adjer se-

cured a government subsidy for further research on Java, where a fossil human skull had recently been discovered in Wadjak Cave. This discovery excited in Dubois the notion that yet older fossils could be found there.

In June 1890, Dubois began his work on Java near the Wadjak discovery site. By September, his crew had unearthed a skull similar to the first Wadjak find, followed by fragments of the skeleton. A year later, Dubois's crew was concentrating much of its efforts near Trinil, a site along the Solo River that proved rich in mammal fossils. By August of that year they had uncovered the fossilized molar of a primate, followed in October by a skullcap that appeared not quite human, yet not quite ape. These, Dubois believed, belonged to a chimpanzee with decidedly human traits. In May of 1892, more proof of the apes' human-like tendencies came to light when a nearly complete left thigh bone was discovered at Trinil. By Dubois's estimate, the femur had been found approximately twelve meters from where the earlier fossil had been found, suggesting to him that all three fossils belonged to the same individual. More than that, the structure of the thigh bone bore a striking resemblance to a human femur, testifying to the fact that his ''chimpanzee'' had walked erect. The thigh bone and later recalculation of the skull capacity left no doubt in Dubois's mind that he had discovered the ''missing link''. He named the specimen *Pithecanthropus erectus,* Pithecanthropus erectus the ''ape-man who walked upright.''

Faces Criticism Over *Pithecanthropus* Interpretation

The discovery also confirmed for Dubois two theories popular in the scientific circles of the time. One, that upright posture was ''the first step on the road to becoming human,'' and that the East Indies, not Africa, was the cradle of humankind. From the study of the fossils, estimated at roughly half a million years old, Dubois also began to develop a saltationist theory concerning man's evolutionary progress, which emphasized a more rapid, punctuated rate of evolution than Darwin's gradualist theory presumed. In August 1894, Dubois's findings along with his theories were published in the thirty-nine page treatise *Pithecanthropus erectus, eine menschenaehnliche Uebergangsform.* The publication met with immediate criticism, mainly in Dubois's assertion that the molar, skull cap and thigh bone belonged to the same individual. There seemed no argument, even from his fiercest opponents, that the femur was human-like if not altogether human, but the structure of the skull cap, according to the French and Germans, was too decidedly gibbon-like for any serious discussion of its relationship to the femur. While he was derided for his ape-man theory, he was applauded both for his discovery of a new species of gibbon and for further dating man's existence to the Pliocene epoch some five million years earlier.

British criticism was altogether different, regarding as a mistake Dubois's neglect of modern human and the earlier Neanderthal skulls in the initial anatomical comparisons of his ape-man. The British anatomist Daniel J. Cunningham concluded that the skull was unquestionably human and furthermore the skull capacity fit a gradualist interpretation, wherein

Pithecanthropus formed the bottom rung of man's climb to humanity, followed by the intermediate Neanderthals. Dubois and his Java man returned to the Netherlands in August of 1895 to confront the critics. On September 21, the remains of *Pithecanthropus erectus* Pithecanthropus erectus were displayed before attendants of the Third International Congress of Zoology held in Leiden. Included among the group of prominent scientists were Dubois's American defender, Othniel C. Marsh, and the German pathologist, Rudolf Virchow, who felt that Dubois's arguments were wildly speculative at best and attributed the femur to a specialized species of gibbon. Dubois stood his ground on the interpretations, adding to them more comprehensive information than had been published in his treatise a year earlier, including his comparative analysis of *Pithecanthropus* and Neanderthal crania.

Filling the Evolutionary Gaps

Dubois spent the next several years traveling with his ape-man, defending his interpretations of the remains, and studying those of other fossil primates and men. In 1897, he and his wife, Anna Geertruida Lojenga, whom he married in 1886, settled in Haarlem, west of Amsterdam. There he became curator of the Teylers Museum, a position he would hold until his death, and was awarded an honorary doctorate in botany and zoology from the University of Amsterdam. As the 1890s came to a close, some of Dubois's more influential critics had begun to shift their opinions of *Pithecanthropus* in Dubois's favor. Even the prominent British anatomist Arthur Keith, who had at once been skeptical of Dubois's interpretations, came to agree with Dubois's assigned age for the creature and its genealogical position on the human family tree. But after 1900, regardless of this acceptance, Dubois would have no further word on the subject for nearly 25 years.

From 1900, Dubois kept himself busy on a series of studies that examined in mammals the evolution of brain size relative to body size. His earlier calculations concluded that brain size doubled with every evolutionary progression, but since these calculations did not work with his assignment of *Pithecanthropus erectus* Pithecanthropus erectus to an intermediate form between man and ape, he had to revise his methods. This method proved self-serving, in that Dubois augmented *Pithecanthropus'* body weight to fit the scheme and support his saltationist theory of evolution. Dubois's theory later led to the misconception that he had consigned his ape-man to the lower status of gibbon, just as Virchow had suggested in 1895. But Dubois never doubted or intended to change its stature in the annals of evolutionary history. As he wrote in 1932, as quoted in *Natural History,* his reinterpretation suggested that *Pithecanthropus* was "a gigantic genus allied to the gibbons, however superior to the gibbons on account of its exceedingly large brain volume and... its faculty of assuming an erect attitude and gait.... I still believe, now more firmly than ever, that the *Pithecanthropus* of Trinil is the real 'missing link.'"

Throughout the 1920s and 1930s, fossils similar to those of *Pithecanthropus erectus* Pithecanthropus erectus were coming to light in the East, beginning in 1929 with the partial skull of Peking (Beijing) man in China. The continuing acceptance of Dubois's claims for Java man led the Dutch anatomist Ralph von Koenigswald to Java, where he eventually unearthed the fossil remnants of nearly 40 *Pithecanthropus*-like individuals. This new evidence proved that Dubois's ape-man was not an intermediate species between man and ape, but rather a direct ancestor to modern man. These fossils, including Dubois's original *Pithecanthropus erectus,* were later reclassified as *Homo erectus.* Having never changed his mind on the matter, Dubois died at his estate in central Limburg on December 16, 1940.

DULBECCO, RENATO (1914-)
Italian American virologist

Renato Dulbecco was a pioneer in the field of virology, the study of **viruses**. Dulbecco developed the plaque assay technique which allowed scientists to quantify the number of viral units in a laboratory **culture**, thus making possible most of the later major discoveries in virology. For his work in the study of viruses that could cause **cancer** in animals and humans, Dulbecco shared the 1975 Nobel Prize in Medicine or Physiology with microbiologist **David Baltimore** and oncologist **Howard Temin**.

Dulbecco was born in Catanzaro, Italy, on February 22, 1914, the son of Leonardo Dulbecco, a civil engineer, and Maria Virdia Dulbecco. During World War I he lived with his mother and siblings in Turin and Cuneo after his father was called into military service. After the war, the family relocated to Imperia, where Dulbecco received his primary and secondary education. His interest in physics led him to build an electronic seismograph, one of the earliest of its kind. He entered the University of Turin in 1930 at the age of 16 to study medicine. By the end of his first year of study, his interests turned to biology and he went to work as a laboratory assistant for Giuseppe Levi, a professor of anatomy and an expert on nerve tissue, where he learned histology and the techniques of cell culture. His fellow students included microbiologist Salvador Edward Luria and neurologist Rita Levi-Montalcini, both of whom were to be Nobel Prize winners and were to influence Dulbecco's scientific career.

Dulbecco received his doctorate of medicine in 1936 and was drafted into the Italian army as a physician. He was discharged in 1938 but was recalled in 1939 at the outset of World War II. After Italy, led by dictator Benito Mussolini, became a belligerent in 1940, Dulbecco served in France and then in Russia. A serious wound in Russia in 1942 hospitalized him for several months, after which he went home. Following the fall of Mussolini's government, Dulbecco went into hiding in a small village near Turin and became a physician to the local partisan units resisting the German occupation. After the end of the war in 1945, he was elected a city councilor of Turin but soon gave up the position to return to scientific study and research at the University of Turin. In 1946 Luria invited Dulbecco to join his research group at the University of Indiana at Bloomington. Dulbecco and Levi-Montalcini both immigrated to the United States the following year. He became an American citizen in 1953.

At Indiana, Dulbecco experimented with **bacteriophage**, viruses that invade and kill bacteria **cells**. His principal discovery at this time was that bacteriophage previously rendered inactive by exposure to ultraviolet light could be reactivated by exposure to white light of short wavelength. This work attracted the attention of **Max Delbrück**, a German-born physicist-turned-microbiologist. In 1949, Delbrück invited Dulbecco to join him at the California Institute of Technology (Caltech) in Pasadena. Dulbecco became a research fellow and later a professor of biology at Caltech, where he remained until 1963.

In the early 1950s, Dulbecco developed a method for determining the number of units of a given virus in a culture of animal **cell** tissue. This method, called the plaque assay technique, enabled the researcher to count the viral units in a culture by examining the number of plaques, or clear spots, in the culture, where the viruses had killed the host cells. This method was the basis for many of the later important advances made in animal virology. One spectacular practical result of the use of the plaque assay technique was the development of physician **Albert Sabin**'s polio vaccine, developed from a living virus, used to prevent poliomyelitis, a paralyzing and sometimes lethal disease. This vaccine eventually superseded the vaccine produced earlier by physician Jonas Salk, which was made with a virus killed by formaldehyde.

In the late 1950s Dulbecco's interest shifted to the study of animal viruses that could cause cancerous tumors. His research over the next 20 years was devoted to an investigation of the precise manner in which particular viruses could transform host cells in such ways that the cell was either killed or multiplied indefinitely (that is, became cancerous). While working on the polyoma virus, which causes tumors in mice, Dulbecco and his colleagues discovered that the virus's **DNA (deoxyribonucleic acid)** combined with the DNA of the host cell and remained there as a provirus (a virus that is integrated with a cell's genetic material and that can be transmitted without causing disintegration when the cell reproduces) which controlled the genetic mechanism of the cell. In a process called cell transformation, the virus could induce a cancer-like state, causing the cell to multiply endlessly in a tissue culture environment in the laboratory. In an animal body, the same process of cell transformation and subsequent cell multiplication led to the growth of cancerous tumors.

In 1963 Dulbecco left Caltech to become one of the original fellows of the Salk Institute, a research organization founded by Salk in La Jolla, California. There Dulbecco continued his research on animal tumor viruses.

In 1972 Dulbecco moved to London to become assistant (later deputy) director of research at the Imperial Cancer Research Fund. He was by then involved in the study of breast cancer in human beings. While in London, Dulbecco, Baltimore, and Temin were jointly awarded the Nobel Prize in medicine or physiology for their work on tumor virology. In his Nobel Prize lecture Dulbecco made a strong plea for the governments of the world to ban or otherwise remove cancer-causing substances from the environment.

Dulbecco returned to southern California in 1977 to become a distinguished research professor at the Salk Institute.

He became president of the institute in 1982 and held that position until his retirement in 1992. In addition, during the late 1970s Dulbecco taught at the University of California in San Diego.

DUVÉ, CHRISTIAN DE (1917-)
English Belgian biochemist and cell biologist

Christian René de Duvé's ground-breaking studies of cellular structure and function earned him the 1974 Nobel Prize in physiology or medicine (shared with **Albert Claude** and **George Palade**). His discovery of the two key cellular organelles— **lysosomes** and peroxisomes— earned him an honor from the Swedish Academy. This work, along with that of his fellow recipients, established the field of cell biology. De Duvé introduced techniques that have enabled other scientists to better study cellular **anatomy** and physiology and his research has also been of great value in helping clarify the causes of and treatments for a number of diseases.

Christian René de Duvé was born on October 2, 1917 in England after his parents, Alphonse and Madeleine (Pungs) de Duvé, fled Belgium after the German army invaded it during World War I. De Duvé returned with his parents to Belgium in 1920, where they settled in Antwerp. (De Duvé later became a Belgian citizen.) In 1934, intending to become a physician, de Duvé entered the medical school of the Catholic University of Louvain.

De Duvé joined J. P. Bouckaert's group, where he studied physiology, concentrating on the hormone **insulin** and its effects on uptake of the sugar glucose. De Duvé's experiences in Bouckaert's laboratory convinced him to pursue a research career when he graduated with an M.D. in 1941. During World War II De Duvé spent time in a prison camp, but managed to escape and returned to Louvain to resume his investigations of insulin. Before obtaining his Ph.D. from the Catholic University of Louvain in 1945, de Duvé published several works, including a 400-page book on glucose, insulin, and **diabetes**. The dissertation topic for his *Agrégé de l'Enseignement Supérieur* was also insulin. De Duvé then obtained an M.Sc. degree in chemistry in 1946.

After graduation, de Duvé studied with Hugo Theorell at the Medical Nobel Institute in Stockholm for 18 months, then spent six months with Carl Ferdinand Cori, Gerty Cori, and Earl Sutherland at Washington University School of Medicine in St. Louis. Thus, in his early postdoctoral years he worked closely with no less than four future Nobel Prize winners. De Duvé returned to Louvain in 1947 to take up a faculty post at his alma mater teaching physiological chemistry at the medical school. In 1951, de Duvé was appointed full professor of biochemistry. As he began his faculty career, de Duvé's research was targeted at unraveling the mechanism of action of the anti-diabetic **hormone**, insulin and his early experiments opened new avenues of research.

As a consequence of investigating how insulin works in the human body, de Duvé and his students also studied the **enzymes** involved in **carbohydrate metabolism** in the liver. Duvé separated liver cell components by spinning them in a centrifuge, a machine that rotates at high speed. De Duvé assumed that particular enzymes are associated with particular parts of the **cell**. These parts, called cellular organelles (little organs) can be seen in the microscope as variously shaped and sized grains and particles within the body of cells. It had long been recognized that there existed several discrete types of these organelles, though little was known about their structures or functions at the time.

Using a technique called differential centrifugation, developed some years earlier by fellow-Belgian Albert Claude at the Rockefeller Institute for Medical Research, in which cells are ground up slightly by hand prior to being spun to separate various components, de Duvé got better separation of liver cell organelles, and was able to isolate certain enzymes to certain cell fractions. One of his first findings was that his target enzyme, glucose–6-phosphatase, associated with the cellular organelles, microsomes, were the site of key cellular metabolic events. This was the first time a particular enzyme had been clearly associated with a particular organelle.

De Duvé and his students also applied the differential centrifugation technique to the enzyme acid phosphatase, which in cells acts to remove phosphate groups from sugar molecules under acidic conditions. He and his students observed that the cell fraction initially showed a lower level of enzyme than expected, but when allowed to sit in the refrigerator for several days, the enzyme activity increased to expected levels. This phenomenon became known as enzyme latency.

De Duvé found an organelle devoted to cellular **digestion**. With this research, de Duvé identified lysosomes and elucidated their pivotal role in cellular digestive and metabolic processes. Later research in de Duvé's laboratory showed that lysosomes play critical roles in a number of disease processes as well.

De Duvé eventually uncovered more associations between enzymes and organelles. His research on monoamine oxidase showed that the enzyme was associated with a separate cellular organelle, the peroxisome. Further investigation led to more discoveries about this previously unknown organelle. It was discovered that peroxisomes contain enzymes that use oxygen to break up certain types of molecules. They are vital to neutralizing many toxic substances, such as alcohol, and play key roles in sugar metabolism.

Using the technique that he had used in these early experiments, de Duvé pioneered its use to answer questions of both basic biological interest and immense medical application. His group discovered that certain diseases result from cells' inability to properly digest their own waste products. Disorders of glycogen storage, including Tay Sachs disease, result from malfunctioning lysosomal enzymes.

In 1962 de Duvé joined the Rockefeller Institute (now Rockefeller University) while keeping his appointment at Louvain. Working with research groups at both institutions, he has studied inflammatory diseases such as arthritis and arteriosclerosis, genetic diseases, immune dysfunctions, tropical maladies, and cancers, leading to the creation of new drugs used in combatting some of these conditions. In 1971 de Duvé formed the International Institute of Cellular and Molecular Pathology, affiliated with the University at Louvain. Research at the institute focuses on incorporating the findings from basic cellular research into practical applications.

De Duvé helped found the American Society for Cell Biology. He has received awards and honors from many countries, including more than a dozen honorary degrees. In 1974, de Duvé, along with Albert Claude and George Palade, both also of the Rockefeller Institute, received the Nobel Prize in physiology or medicine, and were credited with creating the discipline of scientific investigation that became known as cell biology. De Duvé was elected a foreign associate of the United States National Academy of Sciences in 1975, and has been acclaimed by Belgian, French, and British biochemical societies. He has also served as a member of numerous prestigious biomedical and health-related organizations around the globe.

E

ECCLES, JOHN C. (1903-1997)
Australian neurophysiologist

John Carew Eccles was a neurophysiologist whose research explained how nerve cells communicate with one another. He demonstrated that when a nerve **cell** is stimulated it releases a chemical that binds to the membrane of neighboring cells and activates them in turn. He further demonstrated that by the same mechanism a nerve cell can also inhibit the electrical activity of nearby nerve cells. For this research, Eccles shared the 1963 Nobel Prize for Physiology or Medicine with **Alan Lloyd Hodgkin** and **Andrew Huxley**.

Born on January 27, 1903, in Melbourne, Australia, Eccles was the son of William James and Mary Carew Eccles. Both of his parents were teachers, and they taught him at home until he entered Melbourne High School in 1915. In 1919, Eccles began medical studies at Melbourne University, where he participated in athletics and graduated in 1925 with the highest academic honors. Eccles's academic excellence was rewarded with a Rhodes Scholarship, which allowed him to pursue a graduate degree in England at Oxford University. In September 1925, Eccles began studies at Magdalen College, Oxford. As he had done at Melbourne, Eccles excelled academically, receiving high honors for science and being named a Christopher Welch Scholar. In 1927, he received appointment as a junior research fellow at Exeter College, Oxford.

Embarks on Neurological Research

Even before leaving Melbourne for Oxford, Eccles had decided that he wanted to study the **brain** and the **nervous system**, and he was determined to work on these subjects with **Charles Scott Sherrington**. Sherrington, who would win the Nobel Prize in 1932, was then the world's leading neurophysiologist; his research had virtually founded the field of cellular neurophysiology. The following year, after becoming a junior fellow, Eccles realized his goal and became one of Sherrington's research assistants. Although Sherrington was then nearly seventy years old, Eccles collaborated with him on some of his most important research. Together, they studied the factors responsible for inhibiting a **neuron**, or a nerve cell. They also explored what they termed the ''motor unit''—a nerve cell which coordinates the actions of many muscle fibers. Sherrington and Eccles conducted their research without the benefit of the electronic devices that would later be developed to measure a nerve cell's electrical activity. For this work on neural excitation and inhibition, Eccles was awarded his doctorate in 1929.

Eccles remained at Exeter after receiving his doctorate, serving as a Staines Medical Fellow from 1932 to 1934. During this period, he also held posts at Magdalen College as tutor and demonstrator in physiology. The research that Eccles had begun in Sherrington's laboratory continued, but instead of describing the process of neural inhibition, Eccles became increasingly interested in explaining the process that underlies inhibition. He and other neurophysiologists believed that the transmission of electrical impulses was responsible for neural inhibition. Bernhard Katz and Paul Fatt later demonstrated, however, that it was a chemical mechanism and not a wholly electrical phenomenon which was primarily responsible for inhibiting nerve cells.

Returns to Australia

In 1937, Eccles returned to Australia to assume the directorship of the Kanematsu Memorial Institute for Pathology in Sydney. During the late 1930s and early 1940s, the Kanematsu Institute, under his guidance, became an important center for the study of neurophysiology. With Katz, Stephen Kuffler, and others, he undertook research on the activity of nerve and muscle cells in cats and frogs, studying how nerve cells communicate with muscle or motor cells. His team proposed that the binding of a chemical (now known to be the neurotransmitter **acetylcholine**) by the muscle cell led to a depolarization, or a loss of electrical charge, in the muscle cell. This depolarization, Eccles believed, occurred because charged ions in the muscle cell were released into the exterior of the cell when the chemical substance released by the nerve cell was bound to the muscle cell.

During World War II, Eccles served as a medical consultant to the Australian army, where he studied vision, hearing, and other medical problems faced by pilots. Returning to full-time research and teaching in 1944, Eccles became professor of physiology at the University of Otago in Dunedin, New Zealand. At Otago, Eccles continued the research that had been interrupted by the war, but now he attempted to describe in greater detail the neural transmission event, using very fine electrodes made of glass. This research continued into the early 1950s, and it convinced Eccles that transmission from nerve cell to nerve cell or nerve cell to muscle cell occurred by a chemical mechanism, not an electrical mechanism as he had thought earlier.

In 1952, Eccles left Otago for the Australian National University in Canberra. Here, along with Fatt and J. S. Coombs, he studied the inhibitory process in postsynaptic cells, which are the nerve or muscle cells that are affected by nerve cells. They were able to establish that whether nerve and muscle cells were excited or inhibited was controlled by pores in the membrane of the cells, through which ions could enter or leave. By the late 1950s and early 1960s, Eccles had turned his attention to higher neural processes, pursuing research on neural pathways and the cellular organization of the brain.

Begins a Second Career in the United States

In 1966, Eccles turned sixty-three and university policy at the Australian National University required him to retire. Wanting to continue his research career, he accepted an invitation from the American Medical Association to become the director of its Institute for Biomedical Research in Chicago. He left that institution in 1968 to become professor of physiology and medicine and the Buswell Research Fellow at the State University of New York in Buffalo. The university constructed a laboratory for him where he could continue his research on transmission in nerves. Even at a late stage in his career, Eccles's work suggested important relationships between the excitation and inhibition of nerves and the storing and processing of information by the brain.

In 1975, he retired from SUNY with the title of Professor Emeritus, subsequently moving to Switzerland. During the final period of his career, Eccles focused on a variety of fundamental problems relating to consciousness and identity, conducting research in areas where physiology, psychology, and philosophy intersect. He died at his home in Contra, Switzerland.

Eccles received a considerable number of scientific distinctions. His memberships included the Royal Society of London, the Royal Society of New Zealand, and the American Academy of Arts and Sciences. He was awarded the Gotch Memorial Prize in 1927, and the Rolleston Memorial Prize in 1932. The Royal College of Physicians presented him with their Baly Medal in 1961, the Royal Society gave him their Royal Medal in 1962, and the German Academy awarded him the Cothenius Medal in 1963. Also in 1963, he shared the Nobel Prize for Physiology and Medicine with Alan Hodgkin and Andrew Huxley. He was knighted in 1958.

In 1928, Eccles married Irene Frances Miller of New Zealand. The marriage, which eventually ended in divorce in 1968, produced four sons and five daughters. One of their daughters, Rosamond, earned her doctorate and participated with her father in his research. After his divorce from Irene Eccles, Eccles married the Czech neurophysiologist Helena Tabořiková in 1968. Dr. Tabořiková also collaborated with Eccles in his scientific research.

ECHOLOCATION

Echolocation is a physiological process that some **animals** use to gain information about their environment. By emitting and detecting its own sounds reflected from objects, animals can avoid obstacles, find food and communicate with others. Echolocation used by many types of bats, some whales and dolphins, as well as other **mammals** and birds.

The sounds transmitted by animals using echolocation are of a very high frequency, 1,000-200,000 hertz. These sounds are often beyond the audible range of humans, which is 20-20,000 hertz. The sounds of bats and whales can only be heard using high frequency sound detection systems.

The way echolocation works is the animal emits a high-frequency sound and then waits for it to bounce off an object. Some of the sound **energy** is absorbed by the object, some is transmitted and re-radiated on the other side of the object, and some is reflected or echoed back to the emitter. By interpreting returning echoes, animals can accurately identify the direction, distance, velocity and some aspects of the size of the object in their path. The bat can quickly identify and zero in on a mosquito, as well as avoid **tree**s and other large objects while flying in total darkness. Likewise, a dolphin can detect a school of fish over 100 yd (91 m) away.

ECOLOGY

Ecology is the study of life and its relation to the environment. An ecologist attempts to understand how **plants** and **animals** depend upon their physical settings and upon one another in order to live. By emphasizing this dependence, ecologists assert that man cannot view **nature** as separate and detached; any change man makes in his environment affects all the **organisms** in it. The word ecology is derived from the Greek *oikos*, meaning "home, household, or place to live," and *logos*, meaning "study of." **Ernst Haeckel**, a German biologist and philosopher, coined the term ecology in 1866 after recognizing the importance of studying the environment as a separate scientific field. Before then, ecological issues were pursued under the study of natural history; a student of the great outdoors was called a naturalist. Later natural history was subdivided into special disciplines, such as geology, **zoology**, and **botany**. Even today, the ecologist must draw upon many scientific disciplines, including biology and geology.

In its earliest form, ecology was suggested in the 1700s as the concept of the economy of nature in which the Earth's

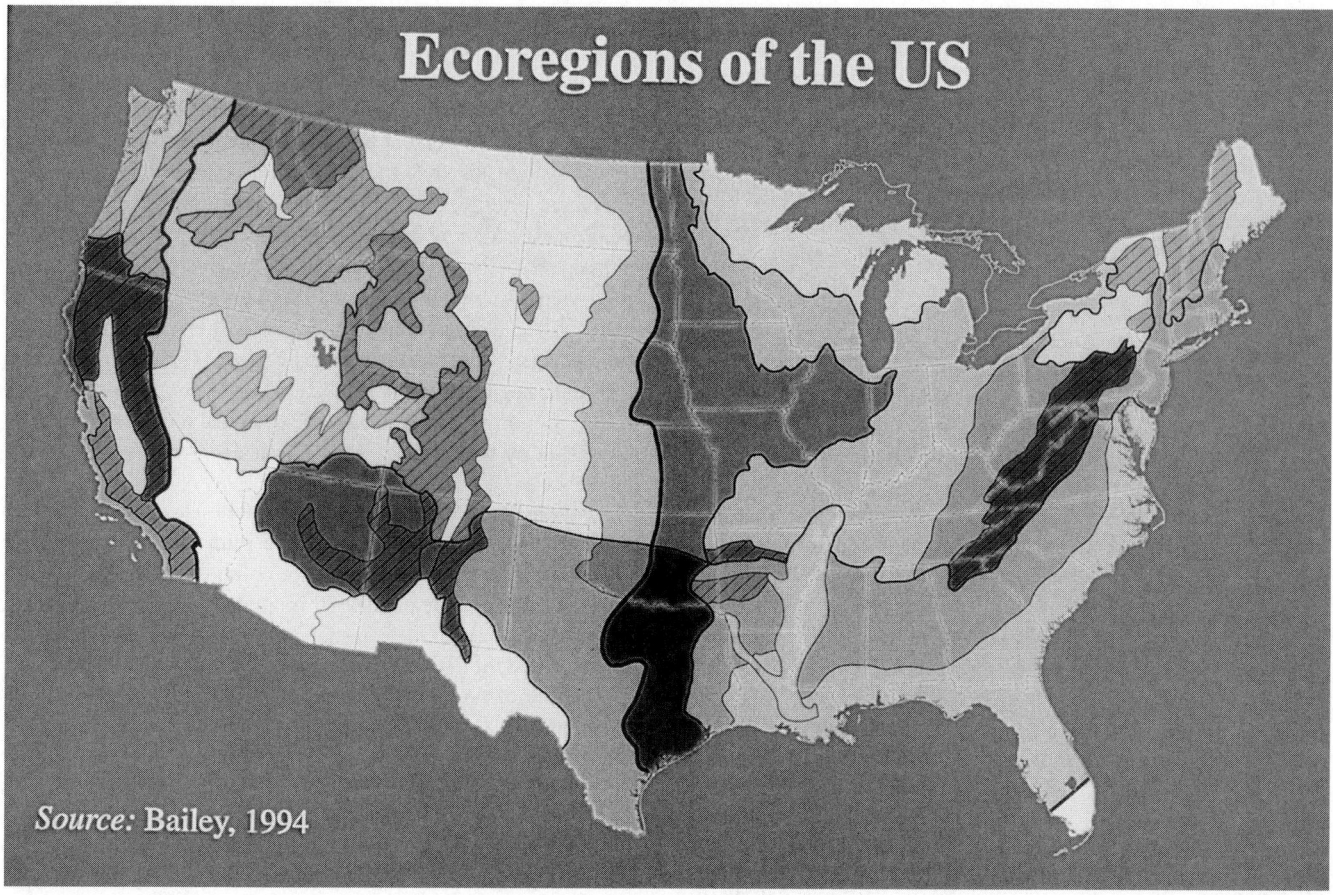

Ecoregions of the US

Source: Bailey, 1994

Map of the ecological regions of the coterminous United States. *(U.S. Forest Service. Reproduced by permission.)*

interrelated organisms were seen as harmoniously balanced. When one creature created large amounts of offspring that perished or were eaten by another, this apparent imbalance actually served to perpetuate the **species** of another type of creature. In the 1800s, **Charles Darwin**, in his book *The Origin of Species by Means of Natural Selection*, examined how living organisms in nature adapt to the conditions of life and their environment and the struggle for existence (**natural selection**). However, instead of emphasizing **ecosystem**s, Darwin concentrated on an organism's ability to genetically reproduce more abundantly within certain environments and among other competing living organisms. Even though the studies of ecology and **evolutionary theory** are completely separate, the ecological understanding of where and how an organism lives is rooted in an understanding of how it has evolved.

Ecology is a relatively young science, but four basic principles have earned wide acceptance. The first tenet states that life patterns reflect the patterns of the physical environment. Unlike man, who can create livable conditions for his species nearly anywhere on Earth, each plant and animal flourishes only when certain physical conditions are present. Closely related is the second principle, which applies to biotic communities: living creatures of an area (its biota) tend to group themselves into loosely organized units known as com-

munities, that become natural homes of each member species. For example, oak and hickory **trees** are generally found together in **forests**.

The third principle of ecology states that an orderly, predictable sequence of development takes place in any area. From barren soil to a self-sustaining **community**, this sequence is called *ecological succession*. The fourth principle asserts that a community and its environment constitute an ecosystem. Every ecosystem consists of organisms that draw vital materials from their surroundings and transfer materials to it. The inhabitants of an ecosystem are classified as **producers**, **consumer**s, and decomposers. Through these four principles, ecologists investigate the interactions of organisms in various kinds of environments to learn how nature establishes orderly patterns among a great variety of living things.

In recent years scientists have begun to recognize the value of **biodiversity** to the environment. Biodiversity refers to the complex and great variety of animals and plant life on Earth, or in any ecosystems. These ecosystem members interact with one another in a great number of ways, not all of which are understood or recognized by man. Therefore, man's action that might harm or destroy a particular species could well have effects on other species that were not anticipated. Human development—roads, buildings, and other structures—

threaten particular species and hence biodiversity. At present, species are going extinct about 100 times faster than the natural rate. **Conservation**s organizations are encouraging governments to protect at least 10% to 12% of their total land area to aid in the preservation of biodiversity.

Also in recent years, climitalogists and other scientists have begun to see that man's actions are influencing his environment and, potentially, his climate. Through the release of man-made gases like **carbon dioxide**, heat is being trapped within the **atmosphere**, resulting in a slight warming that, if it continues, could have great reprecussions in the future. One recent comprehensive study used tree rings (which show **temperature** variations as well as **precipitation** variations), ice core samples (where trapped air can give clues about past environments), and coral records to trace climate patterns over the last 600 years. Researchers combined this evidence with temperature readings, which have been available for only about the last 150 years, and with historical records, and concluded that the 20th century has been the warmest century in the last 600 years. Moreover, they concluded that the warmest years in all of that period were 1990, 1995, and 1997. Such changes are a clear example of one species affect on the Earth, one that could eventually have an impact on all other plant and animal species on the planet.

ECOSYSTEM

The notion of ecosystem (or ecological system) refers to indeterminate ecological assemblages, consisting of communities of **organism**s and their environment. Ecosystems can vary greatly in size. Small ecosystems can be considered to occur in tidal pools, in a back yard, or in the rumen of an individual cow. Larger ecosystems might encompass lakes or stands of **forests**. Landscape-scale ecosystems comprise larger regions, and may include diverse terrestrial and aquatic communities. Ultimately, all of Earth's life and its physical environment could be considered to represent an entire ecosystem, known as the **biosphere**.

Often, ecologists develop functional boundaries for ecosystems, depending on the particular needs of their work. Depending on the specific interests of an ecologist, an ecosystem might be delineated as the shoreline vegetation around a lake, or perhaps the entire waterbody, or maybe the lake plus its terrestrial watershed. Because all of these units consist of organisms and their environment, they can properly be considered to be ecosystems.

Through biological **productivity** and related processes, ecosystems take sources of diffuse **energy** and simple inorganic materials, and create relatively focused combinations of these, occurring as the **biomass** of **plants**, **animal**s, and **microorganism**s. Solar electromagnetic energy, captured by the chlorophyll of green plants, is the source of diffuse energy most commonly utilized by ecosystems. The most important of the simple inorganic materials are **carbon dioxide**, **water**, and ions or small **molecule**s containing **nitrogen**, phosphorus, potassium, calcium, magnesium, sulfur, and some other **nutrients**.

Because diffuse energy and simple materials are being ordered into much more highly structured forms such as biochemicals and biomass, ecosystems (and life more generally) represent rare islands in which negative **entropy** is accumulating within the universe. One of the fundamental characteristics of ecosystems is that they must have access to an external source of energy to drive the biological and ecological processes that produce these localized accumulations of negative entropy. This is in accordance with the Second Law of Thermodynamics, which states that spontaneous **transformations** of energy can only occur if there is an increase in entropy of the universe; consequently, energy must be put into a system to create negative entropy. Virtually all ecosystems (and life itself) rely on inputs of solar energy to drive the physiological processes by which biomass is synthesized from simple molecules.

To carry out their various functions, ecosystems also need access to materials-the nutrients referred to above. Unlike energy, which can only flow through an ecosystem, nutrients can be utilized repeatedly. Through **biogeochemical cycles**, nutrients are recycled from dead biomass, through inorganic forms, back into living organisms, and so on.

One of the greatest challenges facing humans and their civilization is to develop an understanding of the fundamentals of ecosystem organization-how they function and how they are structured. This knowledge is absolutely necessary if humans are to design systems that allow a sustainable utilization of the products and services of ecosystems. Humans are sustained by ecosystems, and there is no tangible alternative to this relationship.

ECOTONE

Ecotones are the transition zones between two distinct communities or **ecosystem**s. These edge **habitat**s typically have a greater diversity of **species** than the neighboring communities. Many species of **plants** and **animal**s require a heterogeneous mixture of habitat, and thus flourish in these areas. This is known by ecologists as the "edge effect." The species diversity tends to increase with greater distinction between the two communities. For example, ecologists often find a higher number of species in the ecotone between a mature forest and the adjoining grassland than between two types of **forests**. Wildlife biologists recognize the importance of ecotones and often promote these edge habitats through selective cutting or prescribed burning. For example, ruffed grouse require both dense and open forests as well as areas with low-growing herbaceous plants. Other examples of ecotones include riparian habitats along stream banks, lake shores, marsh edges, ocean beaches, and forest meadows. All of these areas are subject to the negative impact of human intrusion such as dams on rivers, marshes drained for housing developments, roads along the edges of forests, and oil spills along coast lines.

The region between the mountain ecosystem of Pike's Peak (background, left) and Rampart's Range, Colorado (foreground) is known as an ecotone. *(U.S. Geological Survey. Reproduced by permission.)*

ECTODERM

Cell division of a fertilized **egg** results in the formation of a **population** of presumably equivalent **cell**s. The initially large egg is cleaved into smaller embryonic cells. Cleavage is followed by gastrulation. Gastrulation is the beginning of the process of cell specialization. Gastrulation is an extraordinarily critical stage in development because the cells of the former **blastula** are rearranged by folding and cell migration to construct an **embryo** of three layers. The outer cell layer of an amphibian gastrula and the superficial cell layer of a bird, reptile and mammalian primitive streak stage is referred to as ectoderm. Ectoderm, and the other cell layers, known as **mesoderm** and **endoderm**, comprise the three primary germ layers of vertebrate embryos.

Embryonic ectoderm differentiates into a diversity of adult cells, **tissue**s and **organ**s. These include the integument, the **nervous system**, and external **sense organ**s. The ectodermal integument derivatives consist of epidermis, nails, hair, tooth enamel, and glands associated with the skin, such as the mammary, sweat and sebaceous glands. The ectodermal central nervous system is comprised of the **brain** and **spinal cord**. Ectoderm also gives rise to spinal nerves, cranial nerves, the autonomic nervous system, and the adrenal medulla. The nervous system contains **neuron**s which are the essential conduct-

ing cells. The neurons are supported and maintained by ectodermal neuroglia cells. External sense organs of ectodermal origin include the nose, eyes, and ears. The mucous **membrane**s of the oral cavity and anus are ectodermal in origin.

ECTOPLASM

The **ectoplasm** is a clear region of the **cytoplasm** just under the **cell**ular or **plasma membrane** from which all cellular **organelle**s and **macromolecule**s visible in the **light microscope** have been excluded. This is most easily seen in amoeba and other **protozoa**ns that lack a thick cell wall and are thin enough to observe in a live mount. The origin of the ectoplasm is the sol-gel transformation responsible for cytoplasmic streaming. Elements of the **cytoskeleton** such as microtubules and microfilaments **polymer**ize along the cell periphery and change a portion of the cytoplasm into the gel state. Contraction of this gel creates hydrostatic pressure that causes streaming of the soluble regions of the cytoplasm (also known as the **endoplasm**).This hydrostatic pressure also causes some of the aqueous fluid of the cytoplasm to squeeze through the gel and to accumulate under the plasma membrane. As the fluid passes through the tight network of filaments in the gel, all macromolecules and organelles are filtered out leaving only the clear, watery solution known as ectoplasm.

This phenomenon is not of much physiological importance. Probably the most important reason for retaining the distinction between ectoplasm and endoplasm is that the later is the location of the **endoplasmic reticulum**, an important membranous network responsible for **protein** and lipid synthesis, **metabolism** of many toxic compounds, and a host of other vital processes.

EDELMAN, GERALD M. (1929-)
American biochemist

Gerald M. Edelman and his associate Rodney Porter received the 1972 Nobel Prize in physiology or medicine for their discoveries concerning the chemical structure of antibodies. Edelman used these discoveries to draw conclusions not only about the **immune system** but about the nature of consciousness as well.

Born in New York City on July 1, 1929, to Edward Edelman, a physician, and Anna Freedman Edelman, Gerald Maurice Edelman attended New York City public schools through high school. After graduating, he entered Ursinus College, in Collegeville, Pennsylvania, where he received his B.S. in chemistry in 1950. Four years later, he earned an M.D. degree from the University of Pennsylvania's Medical School, spending a year as medical house officer at Massachusetts General Hospital.

In 1955 Edelman joined the United States Army Medical Corps, practicing general medicine while stationed at a hospital in Paris. Following his 1957 discharge from the Army, Edelman returned to New York City to take a position at Rockefeller University studying under Henry Kunkel. Kunkel, with whom Edelman would conduct his Ph.D. research, was examining the unique flexibility of antibodies at the time.

In 1967, while a debate raged between two schools of scientists to explain antibody synthesis, Edelman and his associate, Joseph Gally proposed a radical theory that would later be confirmed as essentially correct. It depended on the vast diversity that can come from chance in a system as complex as the living **organism**. Each time a **cell** divided, they theorized, tiny errors in the transcription—or reading of the code—could occur, yielding slightly different proteins upon each misreading. Edelman and Gally proposed that the human body turns the advantage of this variability in immunoglobulins to its own ends. Many strains of antigens when introduced into the body modify the shape of the various immunoglobulins in order to prevent the recurrence of disease.

Edelman's doctoral thesis investigated several methods of splitting immunoglobulin molecules, and, after receiving his Ph.D. in 1960 he remained at Rockefeller as a faculty member, continuing his research.

In 1961 Edelman and his colleague, M.D. Poulik succeeded in splitting IgG—one of the most studied varieties of immunoglobulin in the blood—into two components by using a method known as "reductive cleavage." The technique allowed them to divide IgG into what are known as light and heavy chains. Data from their experiments and from those of the Czech researcher, Frantisek Franek, established the intricate nature of the antibody's "active sight." The sight occurs at the folding of the two chains which forms a unique pocket to trap the antigen. Porter, who was the first to split an immunoglobulin, combined these findings with his, and, in 1962, announced that the basic structure of IgG had been determined.

In 1965, Porter and Edelman began studying the **amino acid** sequence in subsections of different myeloma, cancers of the immunoglobulin-producing cells. The project, completed in 1969, determined the order of all 1,300 amino acids present in the protein, the longest sequence determined at that time.

By the end of the 1970s, the principle Edelman and Poulik uncovered led him to conceive a radical theory of how the **brain** works.

Rather than an incoming sensory signal triggering a predetermined pathway through the **nervous system**, Edelman theorized that it leads to a selection from among several choices. Edelman envisioned the nervous system as a fluid system based on three interrelated stages of functioning.

In the formation of the nervous system, cells receiving signals from others surrounding them fan out like spreading ivy—not to predetermined locations, but rather to regions determined by the concert of these local signals. The signals regulate the ultimate position of each cell by controlling the production of a cellular glue in the form of cell-adhesion **molecules** and anchoring neighboring groups of cells together. Once established, these cellular connections are fixed, but the exact pattern is different for each individual.

The second feature of Edelman's theory allows for an individual response to any incoming signal. While the vast complexity of these connections allows for some of the variability in the brain, it is in the third feature of the theory that Edelman made the connection to immunology. The neural networks are linked to each other in layers. An incoming signal passes through and between these sheets in a specific pathway. The pathway, in this theory, ultimately determines what the brain experiences, but just as the immune system modifies itself with each new incoming **virus**, Edelman theorized that the brain modifies itself in response to each new incoming signal. In this way, Edelman sees all the systems of the body being guided in one unified process, a process that depends on organization but that accommodates the world's natural randomness.

Dr. Edelman has received honorary degrees from a number of universities, including the University of Pennsylvania, Ursinus College, Williams College, and others. Besides his Nobel Prize, his other academic awards include the Spenser Morris Award, the Eli Lilly Prize of the American Chemical Society, Albert Einstein Commemorative Award, California Institute of Technology's Buchman Memorial Award, and the Rabbi Shai Schaknai Memorial Prize.

A member of many academic organizations, including New York and National Academy of Sciences, American Society of Cell Biologists, Genetics Society, American Academy of Arts and Sciences, and the American Philosophical Society, Dr. Edelman is also one of the few international members of the Academy of Sciences, Institute of France. In 1974 he became a Vincent Astor Distinguished Professor, serving on the board of governors of the Weizmann Institute of Science and is also a trustee of the Salk Institute for Biological Studies.

EGAS MONIZ, ANTONIO (1874-1955)

Portuguese neurologist

Egas Moniz was born in Avança, Portugal, on November 29, 1874. He received his early education from his uncle, an abbot, and later entered the University of Coimbra in 1891 where he pursued a degree in mathematics. He eventually changed his mind, however, and entered the medical degree program. He received his M.D. from Coimbra in 1899.

For much of his life, Egas Moniz divided his time between political action and medical research. The first decade of the twentieth century was a period of revolutionary upheaval in Portugal and Egas Moniz was active in the Republican movement that led to the overthrow of the monarchy in 1910. He went on to serve as a deputy in the new parliament, ambassador to Spain, foreign minister, and Portuguese delegate to the 1918 Paris Peace Conference. He retired from politics in 1919 after becoming involved in a duel over a political disagreement.

Egas Moniz's scientific research focused on neurology, especially pertaining to the **brain**. His first major contribution was the development of a technique for studying the brain. Previously, the use of X-rays in studying the brain had met with little success. However, Egas Moniz developed a technique in which he injected solutions into the brain that are opaque to X-rays. With this approach, X-rays could be used to identify the precise location and size of brain tumors and brain injuries. This technique of cerebral angiography is still widely used today.

Later in life, Egas Moniz began to explore brain surgery and its possible use in treating mental illness. In 1935, he attended a conference in which he learned about the experimental removal of the prefrontal lobe of the brain of two monkeys. After this surgery, symptoms of anxiety and frustration could no longer be induced in the monkeys, although the animals had also lost the ability to learn.

Despite the fact that scientists knew nothing about the function of the prefrontal lobes, Egas Moniz saw a possible application of the monkey experiment to human mental disorders. He proposed the use of a surgical procedure for mental patients in which the prefrontal lobes were severed from the rest of the brain, a process now known as prefrontal lobotomy.

Because of a serious case of gout, Egas Moniz was unable to carry out this surgery himself. A colleague, Pedro de Almedia Lima (1903-1983), performed the actual operations under Egas Moniz's direction. Of the first 20 operations performed, seven patients were said to be cured of their disorder, eight experienced some improvement, and five were unchanged. For his development of this technique, Egas Moniz was awarded a share of the 1949 Nobel Prize for physiology or medicine.

Prefrontal lobotomy has occupied a controversial place in medicine. In the late 1940s and early 1950s, lobotomies became popular in the United States for the treatment of a variety of mental disorders. By one estimate, an average of 5,000 operations were performed annually between 1949 and 1952. Opposition to the procedure grew in the 1960s, however, as it became clear that lobotomies often turned humans into "vegetables." Lobotomies eventually fell into disfavor as other methods for treating mental disorders became available. Much more sophisticated versions of Egas Moniz's original procedure have been developed and currently are in use for the treatment of highly specialized conditions, such as intractable pain.

EGG

Sex **cell**s of most **animal**s are either eggs in females (ova) or **sperm** in males. Eggs have the potential, when fertilized with the sperm of a male, to divide, form an **embryo**, and develop into an **organism**. In contrast to sperm, which are small, mobile, and produced in great numbers, eggs tend to be nonmotile, larger, and produced in limited numbers.

Eggs are cells. That means that they contain a **nucleus** and **cytoplasm** and all of the other common cell structures. They differ from body cells, known as somatic cells, by the fact that they must eventually become **haploid** (having only one set of **chromosome**s) in preparation for fusion with the haploid sperm nucleus. The ovary is made up of **diploid** (having two sets of chromosomes) cells. The egg-forming ovarian cells are known as oogonia and are in fact stem cells that give rise to mature female **gamete**s (sex cells). These diploid cells divide by ordinary **mitosis** to form more diploid oogonia. An egg forms when one of these cells doubles its **DNA** in preparation for meiotic maturation and forms a primary **oocyte**. The oocyte then begins a period of growth. During the growth period, the oocyte retains its double quantum of DNA. Ovulation, the release of the oocyte, is controlled by the pituitary follicle stimulating **hormone** (FSH) and luteinizing hormone (LH). The first meiotic maturation division is completed at ovulation forming a secondary oocyte and a polar body (a nonfunctioning cell containing a nucleus but very little cytoplasm), each with the diploid quantum of DNA. The second meiotic division occurs (in many animals, including humans) at the time of **fertilization** with the extrusion of the haploid second polar body which leaves the mature egg (ovum) with a haploid chromosome complement. Fertilization of the mature egg with a haploid sperm results in restoration of the diploid chromosome number and the beginning of embryonic development.

The larger size of eggs, compared with sperm, is attributed to the fact that eggs must contain all of the material necessary for the onset of growth and development. A fertilized egg cannot feed, and thus sufficient **nutrition**al supplies must be stored to enable **cell division** to proceed. The built-in reserves of the egg must last until the developing organism becomes free-living or obtains parental nourishment. The care, or lack thereof, of parents correlates with egg size and abundance. For example, marine **invertebrates** tend to produce thousands of eggs which are, for the most part, tiny. For example, sea urchins do not nurture their young, and accordingly, they produce vast numbers of **larvae** that are highly vulnerable to **predation**. The sea urchin egg has a limited store of **nutrients** associated with its relatively small size and it rapidly develops into a free-living larva. The diameter of the sea urchin egg is about the same as for mammalian eggs which is approximately 0.004 ins (0.1 mm). Despite its relatively small size, the sea urchin egg has a volume about 10,000 times larger than that of its sperm. Fish and amphibia have eggs much larger than sea urchins. The common leopard frog, *Rana pipiens*, has an egg about 0.04 ins (1 mm) in diameter. Obviously, the roughly 3,000 eggs produced by the female frog are far fewer than those of the sea urchin. The frog ovum has sufficient nutrients to last the 10-14 days required to develop to a feeding larva (tadpole). Reptiles and birds have large eggs. The size of a chicken egg is well known. The largest egg known is that of the ostrich which may weigh 3 lb (1.4 kg). Fertilization of the chicken egg is internal. Sperm in the female reproductive tract encounter the huge egg (the yolk) just after ovulation. The newly fertilized egg continues down the reproductive tract gaining first an albumen layer and then a shell **membrane** and shell. The egg is then laid. No feeding is required of the developing chick because of the enormous stores of food. The egg at the time of laying already contains an embryo. The embryo, known as a blastodisc, develops after fertilization and is incubated during transit of the reproductive track. Twenty one days after laying, a fully developed chick emerges that can move and feed independently. In this instance, the large size of the ovum is associated with advanced development at hatching. The human egg is extraordinarily tiny compared to the bird egg. The human egg is ovulated from the ovary, it is received by the upper end of the Fallopian (uterine) tube, and it begins its transit to the uterus. Fertilization occurs in the upper end of the tube and development begins prior to the embryo entering the uterus (the womb). The embryo implants in the **endometrium** of the uterus shortly after its arrival in the womb and **birth** occurs about nine months later.

EHRLICH, PAUL (1854-1915)

German bacteriologist

Through his comprehensive study of the effects of chemicals in the human body, Ehrlich fathered the fields of **chemotherapy** (the treatment of disease with chemical agents) and hematology (the study of **blood**). He also made important contributions to the understanding of immunity and discovered Salvarsan, the first effective treatment for **syphilis**.

Ehrlich was born on March 14, 1854, in Strehlen, Silesia (then part of Germany), to a prosperous Jewish family. He was the son of Ismar Ehrlich and his wife, Rosa, the aunt of bacteriologist Karl Weigert. Ehrlich's interest in biology and chemistry led him to study medicine. He attended universities in Breslau, Strasbourg, Frieberg-im-Briesgau, and Leipzig, earning his medical degree in 1878. Ehrlich was fascinated by the reactions of cells and tissues to dyes. Using aniline dyes, for example, Ehrlich investigated white blood cells. In the pro-

Paul Ehrlich. *(The Library of Congress. Reproduced by permission.)*

cess, he developed new ways of staining cells for research, including the methylene blue stain for **bacteria**. Heinrich Koch used this stain when he discovered the bacillus that causes tuberculosis.

In 1890, Ehrlich became a professor at the University of Berlin, where he worked with **Emil von Behring** and **Shibasaburo Kitasato** on the study of immunity, or the body's own defense against disease. The group searched for a substance that would give immunity against **diphtheria** using antitoxins. Antitoxins are antibodies produced by the body's immune system to fight poisons invading the body. Ehrlich worked on the chemical aspects of the study and, in 1892, the group announced the development of a diphtheria antitoxin for medical use. Ehrlich also pioneered the production of large quantities of the antitoxin using horses. He shared the 1908 Nobel Prize in Physiology or Medicine with Soviet biologist **Elie Metchnikoff** for his work on immunity and serum therapy.

In 1894, Ehrlich was made director of a new institute for serum research in Frankfurt, where he studied the concepts of *active* and *passive* immunity and developed his ''side-chain'' theory of immunity to explain how antitoxins work at the cellular level in response to toxins. Ehrlich also continued his study of blood using staining techniques. Realizing that stains colored bacteria but not surrounding cells, he looked for a way to combine the stain with a substance that could kill the bacteria. This, he reasoned, could be a ''magic bullet'' in the fight against bacterial diseases. He also identified dyes, such as trypan red, that had the ability to destroy microorganisms on their own.

Ehrlich began working with organic compounds containing arsenic because he felt its properties were similar to those of the nitrogen atoms that gave trypan red its effectiveness. He studied literally hundreds of arsenic compounds and, by 1907, he had reached number 606, which he put aside because it was not effective against trypanosomes. However, two years later, Ehrlich's assistant, Sahachiro Hata (1872-1938), discovered that the compound number 606 was effective against the dread disease syphilis. Caused by a **microorganism** called a *spirochete*, syphilis meant a slow and painful death for thousands of people. In 1910, Ehrlich announced that chemical 606, which he called Salvarsan, could cure syphilis.

For several years, Ehrlich suffered personal and professional attacks because of his work with syphilis. Some felt the disease was a just punishment for sinful sexual behavior and attacked Ehrlich for searching for a cure. The administration of the drug was also complicated, even risky at first, and when a few patients died because doctors administering the drug failed to follow Ehrlich's instructions, Ehrlich was accused of fraud. The attacks finally ceased in 1914, when the German parliament at last endorsed his cure as authentic.

Ehrlich was married in 1883 to Hedwig Pinkus. The couple had two daughters. Unfortunately, the strain surrounding Ehrlich's controversial efforts to cure syphilis took its toll on his health and he suffered a series of strokes during his last year, which led to his death in Bad Homburg, Germany, in 1915.

See also See also Antibody and antigen

EIJKMAN, CHRISTIAAN (1858-1930)
Dutch physician

Born in the Netherlands in 1858, Eijkman received his medical degree from the University of Amsterdam in 1883, then went to Germany to study under the famous bacteriologist, Robert Koch. Encouraged by Koch, in 1887 Eijkman joined a commission sent to the Dutch East Indies (now Indonesia) to investigate **beriberi**—and began the work that was to make him famous.

At the time, beriberi was a widely prevalent disease, characterized by polyneuritis, the kind of nerve damage that causes numbness, paralysis and, in many cases, death. Because Louis Pasteur's germ theory of disease had already led to so many successful cures, physicians now assumed that all diseases must be caused by **microorganism**s. The scientific commission sent to investigate beriberi, therefore, was primarily searching for its causative **organism**—an organism they failed to find. Disappointed, most of the group returned home in 1887, but Eijkman remained behind to serve as director of a new bacteriology lab set up in a medical school constructed for native doctors. It was there that, around 1890, Eijkman helped solve the problem of beriberi, at least partly by accident.

When a group of laboratory chickens suddenly developed a strange disease—one with symptoms that resembled polyneuritis—Eijkman promptly commandeered the chickens and once again tried to find the causative germ, without success. Moreover, he was unable to transfer the disease from sick chickens to healthy ones. And then, to add to his frustration, the disease vanished as suddenly as it had started.

Fortunately Eijkman refused to give up. He stubbornly continued to delve into every aspect of the peculiar vanishing disease. Before long, he learned that, for a brief period of time, one of the cooks had been feeding the lab chickens boiled rice from the hospital's own stores. A second cook, however, decided it was wrong to feed rice meant for people to mere chickens, and switched back to cheaper unpolished rice. Oddly enough, Eijkman learned that the chickens had developed their illness while eating the "better" polished rice.

To determine whether the polished rice was actually responsible for causing the sickness, Eijkman began feeding it to other chickens which quickly developed the beriberi-like illness. And even more intriguing, Eijkman could then cure this new illness simply by switching the sick chickens back to the unpolished rice. Eijkman, therefore, became the first researcher to pinpoint a dietary-deficiency disease. At first, he didn't fully understand the meaning of his findings, assuming that there must be a toxin in rice grains that could be neutralized by something in the hulls. But others would quickly clarify his results.

A younger colleague, Gerrit Grijns, took over the nutrition studies when an illness compelled Eijkman to go home in 1896, and in 1901 he proposed that beriberi was caused, not by germs, but by the lack of some natural substance present in rice hulls and other foods (this substance turned out to be thiamine). Over the next decade, a number of investigators—most notably, England's **Frederick Gowland Hopkins**—came to similar conclusions about a number of diseases and a new era in medicine was underway. Eijkman, whose work served as the basis for the modern theory of **vitamins**, shared the Nobel Prize in physiology or medicine with Hopkins in 1929.

EINTHOVEN, WILLEM (1860-1927)
Dutch physiologist

Although trained in medicine, Willem Einthoven was always very much interested in physics, and his greatest contributions to science involve the application of physical principles to the development of new instruments and techniques in physiological studies. One such instrument, the string galvanometer, made possible the first valid and reliable electrocardiogram, thereby providing physicians with one of their most valuable tools for the study of cardiovascular disorders. For his invention of the string galvanometer, Einthoven was awarded the Nobel Prize for Physiology or Medicine in 1924.

Einthoven was born on May 21, 1860, in Semarang, Java, in what was then the Dutch East Indies and is now Indonesia. His father, Jacob Einthoven, was a physician in Semarang. When Jacob died in 1866, his wife, Louise M. M. C. de Vogel, returned to her native Holland with her six children, Willem included. The family settled in Utrecht, where young

Willem attended local grammar and high schools. Upon graduation from high school in 1879, he enrolled in the medical program at the University of Utrecht. Six years later, Einthoven received his Ph.D. in medicine, having written his doctoral thesis on the use of color differentiation techniques in spectroscopic analysis. He was immediately offered an appointment as professor of physiology at the University of Leiden, a job he actually began after passing his final state medical examinations on February 24, 1886. Einthoven would remain at his post at the University of Leiden for the next forty-two years until his death in 1927. Also in 1886, Einthoven was married to Frédérique Jeanne Louise de Vogel, a cousin, with whom he would father four children: a son and three daughters. Perhaps the most significant feature of Einthoven's career is the way he made use of his interest in—and knowledge of—physics in his study of physiological problems. The research for which he is best known involved the detection of the association between electrical currents and the beating human heart. Physiologists in the 1880s knew that each contraction of the heart muscle is accompanied by electrical changes in the body, but no precise quantitative data existed for this phenomenon. At that time, the only equipment available to measure electrical charges in the body was not sensitive enough to detect the minute changes in potential difference—the amount of energy released—associated with a heartbeat. The most commonly used device, a capillary electrometer, made use of the rise and fall of a thin column of mercury in a glass tube. Unfortunately, the measurement process of such an instrument took place too slowly to determine actual changes in potential difference resulting from muscular contractions.

Around 1903, Einthoven invented an improved method for measuring such changes: the string galvanometer. Einthoven's new instrument consisted of a very thin quartz wire suspended in a magnetic field. An electric current, even one as small as those associated with muscular contraction, caused a deflection of the wire. By focusing a moving picture camera on the wire, Einthoven could obtain a visual record of the movement of the wire as it was displaced by electrical currents from the heart.

As a result of his research, Einthoven was able to detect and identify a number of different kinds of electrical waves associated with a beating heart, waves that he originally labeled as P, Q, R, S, and T waves. He was eventually able to show that some of these waves result from contractions and electrical changes in the atria and others from contractions and electrical changes in the ventricles of the **heart**.

Einthoven published a complete description of his string galvanometer in 1909 as *Die Konstruktion des Sitengalvanometers*. In this work, he outlined a method for using the galvanometer to record heart action using three combinations of electrode placement: right hand to left hand, right hand to left foot, and left hand to left foot. Such arrangements of the electrodes could be used, he showed, to locate the position of the heart and to detect any abnormalities in its function.

In *Nobel Prize Winners*, Einthoven's biographer points out that the invention of the string electrode "revolutionized the study of heart disease." For his accomplishment, Eintho-

ven was awarded the Nobel Prize for Physiology or Medicine in 1924. Interestingly enough, the basic principles of electrocardiography, while first developed by Einthoven, were also derived independently a short time later by the English physicians Sir Thomas Lewis, Sir William Ogler, and James Herrick.

Einthoven continued to refine, develop, and extend the applications of his string galvanometer throughout the rest of his career. For example, later in his life he modified the device so that it could be used to receive long-distance radio telegraph signals and to measure changes in electric potential in nerves. He was also very popular as a lecturer and made a number of trips to Europe and the United States to talk about his work. Among the many honors Einthoven received was his election as an honorary member of the Physiological Society in 1924, and his induction into England's prestigious Royal Society two years later in 1926.

Einthoven died in Leiden, Netherlands, on September 28, 1927. His obituary in the periodical *Nature* spoke of the "grace, beauty, and simplicity of his character." Although he left few students or disciples behind, Einthoven's impact on the development of electrocardiography was profound.

ELEMENT

Elements are a form of **matter** that can not be broken down by ordinary chemical or physical means. Each element is represented on the periodic table of elements by its own symbol. Currently, 112 elements are known to exist although only 92 of these occur in **nature**. The other 20 are known as transuranium elements and have only been produced artificially. While elements were once thought to be irreducible, they have since been found to be composed of subatomic particles.

The idea that matter was made up of elements was first proposed by ancient Greek philosophers. Many of these early scholars believed that nature was not fundamentally complex, but was instead made up of a small number of basic materials. These ideas eventually led to a proposal by Empedocles (c. 490-430 B.C.) in the fifth century B.C. that all matter was based on four elements: earth, air, fire, and **water**. He imagined that every substance was formed by blending various proportions of two or more of these elements. His concept was adopted and further developed by a variety of other scholars including **Aristotle**.

Over the next 2,000 years, our concept of the composition of matter evolved slowly. Most of the study of chemistry was dominated by pseudoscientific alchemists who focused their efforts on turning cheap metals into gold. However, some new discoveries were made during this time. For example, in the eighth century A.D. the Arabic philospher Jābir ibn Haiyān (c. 721-815) discovered mercury and sulfur. He suggested that these substances were the only true elements.

As more and more "fundamental" materials were discovered, the concept of elements became confused. Some scholars tried to maintain the Greek ideas that matter was composed of only a small number of elements. However, these new discoveries made this notion impossible. In 1661, **Robert Boyle** proposed the modern definition for chemical elements. In his theory, any substance that could not be broken down into simpler substances was an element. As this theory was accepted, the Greek system of four elements was discarded.

With Boyle's definition of elements in hand, scientists began to compile lists of elements. In 1789, Antoine-Laurent Lavoisier (1743-1794) published the first modern textbook of chemistry, *Elementary Treatise on Chemistry*. This book contained a list of 33 known elements. Many of these substances, such as sulfur and phosphorous, continue to be considered elements. Others such as chaux and silice were found to be compounds. Two of the materials on Lavoisier's list, lumiere (**light**) and calorique (heat) were shown to be forms of **energy**.

Over time it became apparent to scientists that certain elements had similar properties and may be organized in groups. The first chemist to formally recognize that the properties of elements followed certain patterns was Johann Dobereiner. He developed a system that grouped elements in triads, but applications of this approach were severely limited. Other scientists proposed alternate systems to describe the periodic nature of the properties of elements. Eventually two scientists, Dmitri Mendeleev and Julius Lothar Meyer, independently developed the modern periodic table of elements. It was first published in 1872.

The periodic table of elements lists substances by their atomic symbol. Each element is made up of a single type of atom that has a distinct atomic weight and atomic number. The atomic number is the number of protons in the atom's **nucleus**. All elements can be classified as either metals or nonmetals, although sometimes a separate class called metalloids is used. Elements that have similar chemical and physical properties are grouped together on the periodic table. These groups represent families of elements such as noble gases, halogens, alkaline earth metals, and alkali metals.

It is estimated that 25 of the known elements are critical to the existence of life. In fact, four of these (oxygen, carbon, **nitrogen**, and hydrogen) make up 96% of all living matter. Oxygen is a primary component of all types of biomolecules, including **polysaccharides**, **nucleic acids**, and **proteins**. It makes up about 65% of human body weight. Carbon is also found in biomolecules, constituting 18.5% of human body weight. Nitrogen is an important component in proteins. Other elements such as phosphorous, calcium, potassium, sulfur, sodium, chlorine, and magnesium make up a large portion of the remaining 4%.

Trace elements are substances that **organisms** require in tiny amounts. Iron is one trace element that is required by nearly all forms of life. Certain trace elements are required only by certain organisms. For example, iodine is a required element in vertebrates. Other significant trace elements include boron, chromium, cobalt, copper, fluorine, iodine, manganese, molybdenum, selenium, silicon, tin, vanadium, and zinc.

See also Atomic theory

ELION, GERTRUDE BELLE (1918-)
American biochemist

Gertrude Belle Elion's innovative approach to drug discovery furthered the understanding of cellular metabolism and led to the development of medications for leukemia, gout, herpes, malaria, and the rejection of transplanted organs. Azidothymidine (AZT), the first drug approved for the treatment of **AIDS**, came out of her laboratory shortly after her 1983 retirement. One of the few women who has held a top post at a major pharmaceutical company, Elion worked at Wellcome Research Laboratories for nearly five decades. Her work, with colleague **George H. Hitchings**, was recognized with the Nobel Prize for physiology or medicine in 1988. Her Nobel award was notable for several reasons: few winners have been women, few have lacked the Ph.D., and few have been industrial researchers.

Elion was born on January 23, 1918, in New York City, the first of two children, a daughter and a son, of Robert Elion and Bertha Cohen. Robert, a dentist, immigrated to the United States from Lithuania as a small boy. Bertha came to the United States from Russia at the age of 14. Elion, an excellent student who was accelerated two years by her teachers, graduated from high school at the height of the Great Depression. As a senior in high school, she had witnessed the painful death of her grandfather from stomach cancer and vowed to become a cancer researcher. She was able to attend college only because several New York City schools, including Hunter College, offered free tuition to students with good grades. In college, she majored in chemistry because that seemed the best route to her goal.

In 1937 Elion graduated Phi Beta Kappa from Hunter College with a B.A. at the age of 19. Despite her outstanding academic record, Elion's early efforts to find a job as a chemist failed. One laboratory after another told her that they had never employed a woman chemist. Her self-confidence shaken, Elion began secretarial school. That lasted only six weeks, until she landed a one-semester stint teaching biochemistry to nurses and then took a position in a friend's laboratory. With the money she earned from these jobs, Elion began graduate school. To afford tuition, she continued to live with her parents and to work as a substitute science teacher in the public schools. In 1941, she graduated summa cum laude from New York University with a M.S. degree in chemistry.

Upon her graduation, Elion again faced difficulties finding work appropriate to her experience and abilities. The only job available to her was as a quality control chemist in a food laboratory, checking the color of mayonnaise and the acidity of pickles for the Quaker Maid Company. After a year and a half, she was finally offered a job as a research chemist at Johnson & Johnson. Unfortunately, her division closed six months after she arrived. The company offered Elion a new job testing the tensile strength of sutures, but she declined.

Seeks Opportunity at Wellcome Research Laboratories

As it did for many women of her generation, the start of World War II ushered in a new era of opportunity for Elion.

As men left their jobs to fight the war, women were encouraged to join the workforce. "It was only when men weren't available that women were invited into the lab," Elion told the *Washington Post.*

For Elion, the war created an opening in the research lab of biochemist George Herbert Hitchings at Wellcome Research Laboratories in Tuckahoe, NY, a subsidiary of Burroughs Wellcome Company, a British firm. When they met, Elion was 26 years old and Hitchings was 39. Their working relationship began on June 14, 1944, and lasted for the rest of their careers. Each time Hitchings was promoted, Elion filled the spot he had just vacated, until she became head of the Department of Experimental Therapy in 1967, where she was to remain until her retirement 16 years later. Hitchings became vice president for research. Over the years, they have written many scientific papers together.

Settled in her job and thrilled by the breakthroughs occurring in the field of biochemistry, Elion took steps to earn a Ph.D., the so-called "union card" that all serious scientists are expected to have as evidence that they are capable of doing independent research. Only one school offered night classes in chemistry, the Brooklyn Polytechnic Institute (now Polytechnic University), so that's where Elion enrolled. Attending classes meant taking the train from Tuckahoe into Grand Central Station and transferring to the subway to Brooklyn. Although the hour-and-a-half commute each way was exhausting, Elion persevered for two years, until the school accused her of not being a serious student and pressed her to attend full-time. Forced to choose between school and her job, Elion had no choice but to continue working. Her relinquishment of the Ph.D. haunted her, until her lab developed its first successful drug, 6-mercaptopurine (6MP).

In the 1940s, Elion and Hitchings employed a novel approach to fighting the agents of disease. By studying the biochemistry of **cancer** cells, and of harmful **bacteria** and **virus**es, they hoped to understand the differences between the metabolism of those **cell**s and normal cells. In particular, they wondered whether there were differences in how the disease-causing cells used nucleic acids, the chemicals involved in the replication of **DNA**, to stay alive and to grow. Any dissimilarities discovered might serve as a target point for a drug that could destroy the abnormal cells without harming healthy, normal cells. By disrupting one crucial link in a cell's biochemistry, the cell itself would be damaged. In this manner, cancers and harmful bacteria might be eradicated.

Elion's work focused on purines, one of two main categories of nucleic acids. Their strategy, for which Elion and Hitchings would be honored by the Nobel Prize 40 years later, steered a radical middle course between chemists who randomly screened compounds to find effective drugs and scientists who engaged in basic cellular research without a thought of drug therapy. The difficulties of such an approach were immense. Very little was known about nucleic acid biosynthesis. Discovery of the double helical structure of DNA still lay ahead, and many of the instruments and methods that make molecular biology possible had not yet been invented. But Elion and her colleagues persisted with the tools at hand and

their own ingenuity. By observing the microbiological results of various experiments, they could make knowledgeable deductions about the biochemistry involved. To the same ends, they worked with various species of lab animals and examined varying responses. Still, the lack of advanced instrumentation and computerization made for slow and tedious work. Elion told *Scientific American,* "if we were starting now, we would probably do what we did in 10 years."

Discovers Drug That Fights Leukemia

By 1951, as a senior research chemist, Elion discovered the first effective compound against childhood leukemia. The compound, 6-mercaptopurine (6MP) (trade name Purinethol), interfered with the synthesis of leukemia cells. In clinical trials run by the Sloan-Kettering Institute (now the Memorial Sloan-Kettering Cancer Center), it increased life expectancy from a few months to a year. The compound was approved by the Food and Drug Administration (F.D.A.) in 1953. Eventually 6MP, used in combination with other drugs and radiation treatment, made leukemia one of the most curable of cancers.

In the next two decades, the potency of 6MP prompted Elion and other scientists to look for more uses for the drug. Robert Schwartz, at Tufts Medical School in Boston, and Roy Calne, at Harvard Medical School, successfully used 6MP to suppress the immune systems in dogs with transplanted kidneys. Motivated by Schwartz and Calne's work, Elion and Hitchings began searching for other immunosuppressants. They carefully studied the drug's course of action in the body, an endeavor known as pharmacokinetics. This additional work with 6MP led to the discovery of the derivative azathioprine (Imuran), that prevents rejection of transplanted human kidneys and treats rheumatoid arthritis. Other experiments in Elion's lab intended to improve 6MP's effectiveness led to the discovery of allopurinol (Zyloprim) for gout, a disease in which excess uric acid builds up in the joints. Allopurinol was approved by the F.D.A. in 1966. In the 1950s, Elion and Hitchings's lab also discovered pyrimethamine (Daraprim and Fansidar) a treatment for malaria, and trimethoprim (Bactrim and Septra) for urinary and respiratory tract infections. Trimethoprim is also used to treat Pneumocystis carinii pneumonia, the leading killer of people with AIDS.

Launches Antiviral Program

In 1968, Elion heard that a compound called adenine arabinoside appeared to have an effect against DNA viruses. This compound was similar in structure to a chemical in her own lab, 2,6-diaminopurine. Although her own lab was not equipped to screen antiviral compounds, she immediately began synthesizing new compounds to send to a Wellcome Research lab in Britain for testing. In 1969, she received notice by telegram that one of the compounds was effective against herpes simplex viruses. Further derivatives of that compound yielded acyclovir (Zovirax), an effective drug against herpes, shingles, and chicken pox. An exhibit of the success of acyclovir, presented in 1978 at the Interscience Conference on Microbial Agents and Chemotherapy, demonstrated to other scientists that it was possible to find drugs that exploited the differences between viral and cellular enzymes. Acyclovir (Zovirax), approved by the F.D.A. in 1982, became one of Burroughs Wellcome's most profitable drugs. In 1984 at Wellcome Research Laboratories, researchers trained by Elion and Hitchings developed azidothymidine (AZT), the first drug used to treat AIDS.

Although Elion retired in 1983, she continued at Wellcome Research Laboratories as scientist emeritus and keeps an office there as a consultant. She also accepted a position as a research professor of medicine and pharmacology at Duke University, where she works with a third-year medical student each year on a research project. Since her retirement, Elion has served as president of the American Association for Cancer Research and as a member of the National Cancer Advisory Board, among other positions. Hitchings, who retired in 1975, also remains active at Wellcome Research Laboratories.

In 1988, Elion and Hitchings shared the Nobel Prize for physiology or medicine with Sir **James Black**, a British biochemist. Although Elion had been honored for her work before, beginning with the prestigious Garvan Medal of the American Chemical Society in 1968, a host of tributes followed the Nobel Prize. She received a number of honorary doctorates and was elected to the National Inventors' Hall of Fame, the National Academy of Sciences, and the National Women's Hall of Fame. Elion maintained that it was important to keep such awards in perspective. ''The Nobel Prize is fine, but the drugs I've developed are rewards in themselves,'' she told the *New York Times Magazine*.

Elion never married although she was engaged once. Sadly, her fiance died of an illness. After that, Elion dismissed thoughts of marriage. She is close to her brother's children and grandchildren, however, and on the trip to Stockholm to receive the Nobel Prize, she brought with her 11 family members. Elion has said that she never found it necessary to have women role models. ''I never considered that I was a woman and then a scientist,'' Elion told the *Washington Post*. ''My role models didn't have to be women—they could be scientists.'' Her interests are photography, travel, and music, especially opera. Although her home is in North Carolina, she still keeps her subscription to the Metropolitan Opera in New York.

EMBRYO AND EMBRYONIC DEVELOPMENT

An embryo is a stage in the development of an **organism**. The term can be applied to **plants**, invertebrate and vertebrate **animals**. Rather than consider development of the literally millions of **species** of organisms, here will be considered the embryo of the common North American leopard frog, *Rana pipiens*. The species has been studied extensively because it is indeed common, it is relatively easy to induce ovulation in females, *in vitro* **fertilization** was first demonstrated in the species, and developmental progress can be monitored easily in a glass dish because the embryo is neither within a shell (as in reptiles and birds) nor within the body of the mother as in

mammals. Development is arbitrarily divided into an embryonic period which occurs prior to hatching from the jelly **membrane**s that enclose the embryo, and **larval** development which is the period of the free-living feeding tadpole.

Embryonic development ordinarily is considered to begin with the formation of a the fertilized **egg** known as the **zygote**. In the case of North American leopard frogs, zygote formation occurs in breeding ponds or **wetlands** in the early spring. The frogs overwinter in the cold **water** of northern lakes. When the ice melts and the days grow longer, the frogs leave the cold lakes and seek shallow ephemeral bodies of water. Water has a high specific heat which means that lakes resist a change in **temperature**. The shallow water of breeding ponds warms readily which is essential for both ovulation and embryo development. Encountering warmth permits the release of ovarian eggs, a process known as ovulation. Male frogs clasp mature female frogs and this may encourage egg release. The clasping male releases **sperm** as the eggs are extruded. A female may release as many as 3,000 eggs which, when fertilized, results in a potential of 3,000 embryos. The processes leading to fertilization can be accomplished in the laboratory. Laboratory frogs are ordinarily kept cold until embryo formation is required. Female frogs brought to laboratory temperature (18° C or about 65° F) are treated with an injection of pituitary extract and progesterone. Ovulation occurs within 48 hours. The release of sperm (spermiation) is induced with an injection of human **chorion**ic gonadotropin. A zygote is formed when freshly ovulated ova are combined with freshly released sperm. The zygote cleaves (divides) into two **cell**s within 3.5 hours. It is exquisitely beautiful to observe the yolky mass of the zygote dividing in the process of creating a new individual. First there are two blastomeres (cells), then four, then eight, and so on until a ball of cells forms known as a **morula**. With continued division, the morula gives rise to the **blastula**. The frog blastula is also a ball of cells but it is hollow. The mature blastula has about 3,000 cells and they form within the first 24 hours in the laboratory.

Blastula cells vary in size depending upon the amount of yolky granules they contain. Very yolky cells are large; cells with minimal yolk are tiny but clearly visible in the **microscope**. Regardless of size, the cells are considered to be developmentally equivalent. That means that they have not begun the process of specialization known as **differentiation**. Classic grafting experiments early in this century demonstrated the lack of specialization by grafting small **population**s of blastula cells to a new site within an embryo with no effect on development. While 3,000 cells are present at the definitive blastula stage, there has been no net growth, i.e., no increase in mass. A frog blastula has the same diameter as a zygote. Cleavage gives rise to an increasing number of cells which simply partitions the former zygote into ever smaller compartments (cells).

Gastrulation, which follows the blastula stage, is a time of much developmental change. Many living cells have the competence to migrate within an organism and the first great **migration**s are observed at gastrulation. Some cells on the surface migrate to the interior. In the process of migration, the former population of equivalent and unspecialized cells becomes a structured sphere with a complicated interior with the beginnings of differentiation. The three primary germ layers may be detected at this time; the external **ectoderm**, the internal **endoderm**, and the intermediate **mesoderm**. The rearrangement of cells that resulted from migration forms an area of invagination where cells move to the interior. The site of movement inwards is known as the blastopore. The blastopore will eventually become the posterior opening of the **digestive system** and the positioning of the blastopore within the gastrula permits identification of an anterior and posterior axis as well as left and right sides.

The onset of differentiation has occurred with gastrulation. Cells begin their specialization and this limits their competence to differentiate all cell types. Thus, when a graft is made which exchanges populations of cells from one area to another, the grafts develop in their new locations as if they had not been moved. An example: when cells destined to form neural structures are grafted to a skin forming area, they do not form skin but continue on their pathway to differentiate neural cells.

Specific **tissue** types form during embryo development. A portion of the ectoderm on the dorsal side of the embryo rolls up into a tube which is the beginnings of the central **nervous system**. The anterior portion of the tube becomes the **brain** and the posterior portion becomes the **spinal cord**. Some mesoderm cells become specialized to form muscle and the muscle functions well before the onset of feeding. This can be observed by muscular activity (bending and twisting of the embryo body) in the as yet unhatched embryo in its jelly membranes. Another form of muscle is obvious. Careful examination of the throat region of the embryo reveals a regular pulsation. The pulsation is the beating of the **heart** muscle which by this process begins to circulate embryonic **blood** cells. Embryonic gills are exposed on either side of the head. The structure of the gills is so delicate that blood cells, with a dissecting microscope, can be seen surging in synchrony with the beating heart. Embryonic development is similarly proceeding with the **excretory system** and the digestive system. Within six days in the laboratory, all embryonic systems are well underway. Hatching occurs at about this time, which marks the end of the embryonic period and the beginning of larval development. Larvae feed in about a week after fertilization. Embryonic development has been characterized by differentiation of **organ** systems but no increase in mass. Feeding begins and the tadpoles grows enormously in size compared with its origin. Feeding and growing will continue until the larval tadpole begins its **metamorphosis** into a juvenile frog. From zygote to adult is **epigenesis**; understanding how **gene**tic information (**DNA**) is used to ultimately yield an adult is central to developmental biology today.

EMBRYO TRANSFER

Developments in reproductive technology are occurring at a rapid rate in agricultural **animal** science as well as in human

biology. *In vitro* **fertilization**, **embryo culture**, preservation of embryos by freezing (cryopreservation) and **cloning** technology yield embryos that are produced outside of the female **reproductive system**. Embryo transfer permits continued survival of embryos by insertion into the female reproductive system.

Obviously, there would be no need for embryo transfer if mammalian embryos could be cultured to maturity in the laboratory. While **cell** culture *in vitro* has made remarkable strides, embryos can be sustained in culture for only a few days. Thus, their survival is dependent upon transfer to the hospitable and nurturing environment of the uterus of a foster mother. While embryo transfer may seem to be high technology, it actually got its start well over a century ago. The uterus of one variety of rabbit (a Belgian hare doe) was used for the nourishment, growth and complete fetal development, of another variety of rabbit (an Angora) in 1890 by Walter Heape of Cambridge, England. Heape referred to the Belgian hare doe as a "foster-mother." Since that time, foster-mothering of this type has been extended not only to sheep, pigs, goats, cattle, horses and humans but to more exotic **species** such as the water buffalo, mink, llama, antelope, baboon and to pet species such as the cat and dog. There are economic and humane factors in some embryo transfer. From time to time, it becomes important to ship cattle to distant sites on the globe. Cattle are heavy and require special care in shipping. Further, the welfare of cattle in transit is now closely monitored. To circumvent the costs of shipment and possible problems in the welfare of the animals, it is now becoming practice to ship frozen embryos to the new herd site, place the thawed embryos in the reproductive systems of host animals, and await the **birth** of a new herd that arrived safely in small, frozen, packages.

Transfer can be accomplished by direct non-surgical insertion of the embryo into the female reproductive tract (via the vagina and cervical opening of the uterus) or surgically. A common procedure is to utilize an injection catheter that in some ways functions as a hypodermic syringe. The injection apparatus consists of a long thin hollow tube which contains a plunger. At the distal end of the tube is connected a plastic straw containing the embryo and a small drop of culture medium. The apparatus is inserted into the uterus, via the vagina, and the embryo is released within the uterus by gentle pressure on the plunger. Alternatively, surgical embryo transfer is an option in some cases. The upper end of the Fallopian (ovarian) tube is exposed and the embryo is placed within that site. Or, a small incision may be made in the wall of the uterus for insertion of the embryo. In cattle, the female recipient must have been in estrus (sexual receptivity, or heat) at the same time as the mother of the embryo donor. Whether surgical or non-surgical procedures are used, hundreds of thousands of transferred cattle embryos have been reared to birth which witnesses to the efficacy of the procedure. And, the availability of human *in vitro* fertilization clinics attests to the usefulness of the procedure in certain kinds of fertility problems.

EMBRYOLOGY

Embryology is the study of the development of **organisms**. This is as true of **plants** as it is of **animals**. Seed formation pro-

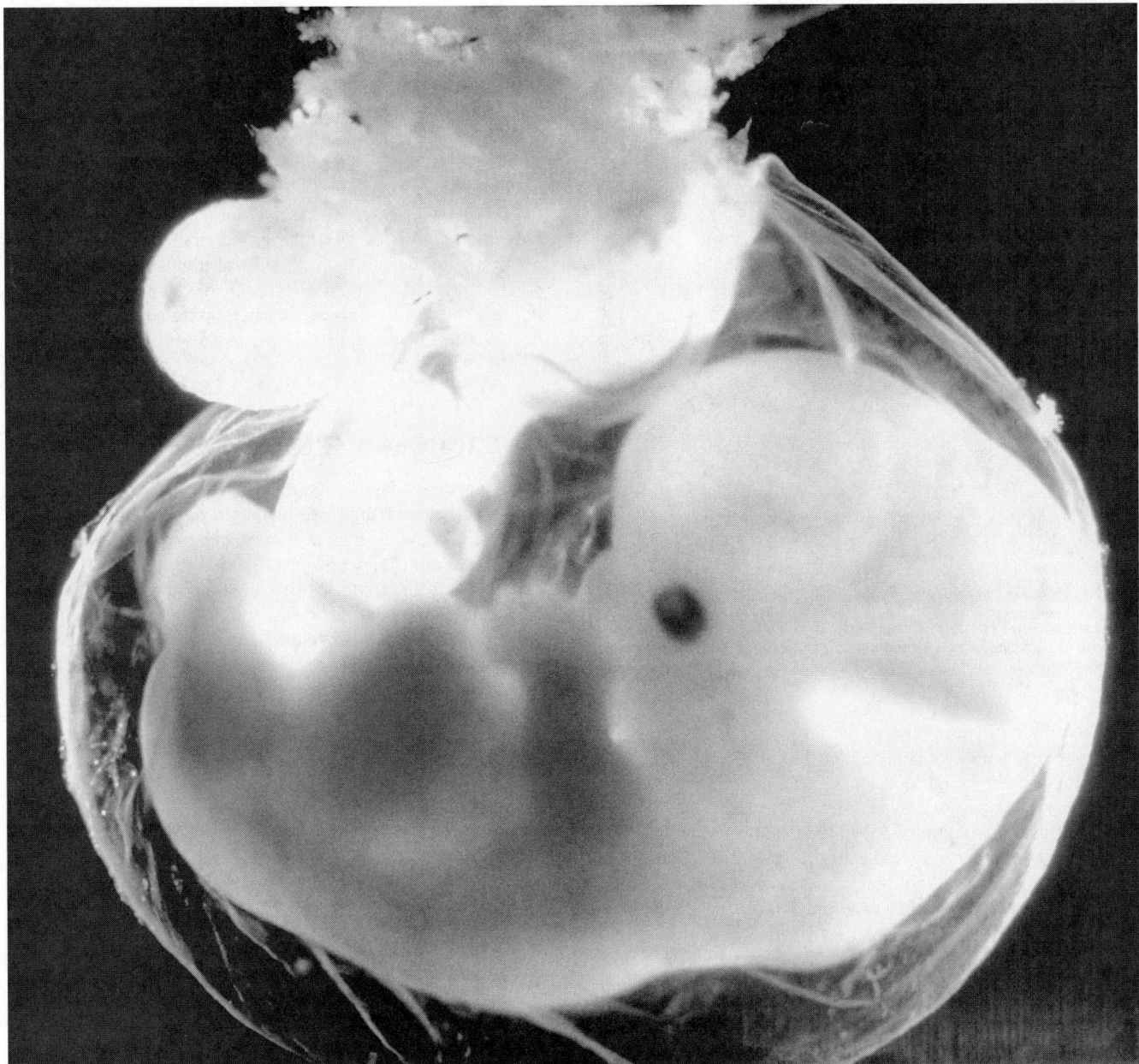

A human embryo at 5-6 weeks of development. *(Photo Researchers, Inc. Reproduced by permission.)*

ceeds following **fertilization** in higher plants. The seed consists of the **embryo**, the seed coat and another part sometimes called the endosperm. While plants are extraordinarily important for survival of animal life, animal embryology is described here. The dictionary definition limits the meaning of the term "embryo" to developing animals that are unhatched or not yet born. Human embryos are defined as developing humans during the first eight weeks after conception. The reason that many embryologists have difficulty with this terminology is that it is purely arbitrary. It would be difficult indeed, if not impossible, to discriminate a human embryo nearing the end of the eighth week from a developing human during the ninth

week after conception. There are no morphological events that distinguish a pre-hatching frog tadpole from a post-hatching tadpole (hatching never occurs synchronously in an **egg** mass—there are always those that hatch early and those **larva**e which are dilatory). What do embryologists study if they deign not to be limited by dictionary definitions? The answer is that they consider development from a **zygote** to a multicellular organism. In the particular case of humans, development does not even stop at **birth**. Note that teeth continue to develop and sex glands with sexual **differentiation** mature long after birth. For a number of years, many embryologists have referred to their discipline as developmental biology to escape from the

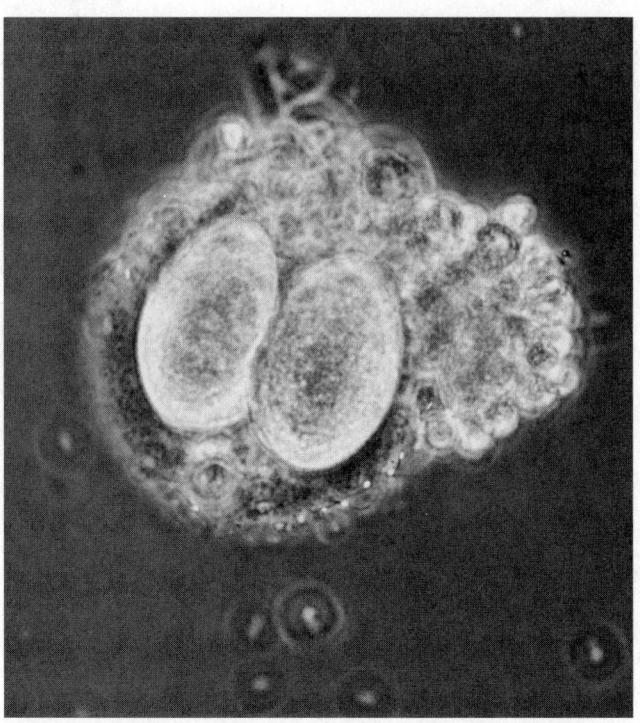

A human two-cell embryo 24 hours after fertilization. *(Photograph by Richard G. Rawlins, Ph.D., Custom Medical Stock Photo. Reproduced by permission.)*

need to confine their studies to earlier stages. Embryology in the modern sense is the study of the life history of an animal and human embryology considers developmental aspects of life as a whole and not just the first eight weeks.

While modern embryology, which seeks to know developmental mechanisms in molecular terms dates from after World War II, embryology has its roots over two millennia ago. The Greek philosopher **Aristotle** scrutinized developing chick embryos. This was a good start because he looked. He argued correctly that development of an embryo was not simply growth from a minute preformed organism derived from either the egg or **sperm**, but was a process of form acquisition from a formless precursor. While it is recognized that the zygote is not a simple **cell**, it certainly does not contain within its **plasma membrane** a perfect and preformed organism. The embryo becomes progressively more complex and this correct view of Aristotle is now known as **epigenesis**.

Little changed in embryology for almost 2,000 years. About a century ago, careful observations were made of a number of developing organisms. By this time, there was a **cell theory** and good **microscope**s were available. Next came a causal analysis. For instance, it was known that the dorsal **ectoderm** of all vertebrate embryos rolls up into a tube to form the central **nervous system**. What factors control the very regular appearance of the nervous system and subsequent differentiation into the various parts of the **brain** and the **spinal cord**? It was hypothesized that the underlying chordamesoderm cells of the gastrula signaled the ectoderm to become neural. The signal was referred to as induction. Other embryonic **organ**s

also seemed to appear as a result of induction. Chemical embryology sought to characterize the nature of inducing signals. Now, modern molecular embryology seeks to examine on the level of the **gene** what controls differentiation of specific **tissue** and cell typed of a developing organism.

There are practical considerations that drive some embryologists. The causes of developmental abnormalities (congenital malformations) in humans becomes more understandable with a consideration of embryology. The human embryo is extraordinarily vulnerable to drugs, **virus**es, and radiation during the first several months of development when many critical organ systems are developing. Understanding this aspect of embryology can help, to some extent, to reduce the toll of human suffering.

ENDANGERED SPECIES

Some **species** adapt to natural changes in their environment such as climate change and increased **competition** from other species. The two to four million **plants** and **animal**s that exist today have adjusted to meet these challenges, out of the estimated 500 million species that have existed since life began on Earth. Because of increased human activities around the world that destroy other **habitat**s, most **extinction** (**death** of a species in the wild) has occurred over the past 200 years. With the accelerated destruction and **pollution** of the world's tropical **rain forest**s, in particular, at least 500,000 insect, plant, and animal species are expected to disappear over the last two decades of the twentieth century.

Before a species becomes extinct, it is **class**ified by the International Union for the Conservation of **Nature** and Natural Resources (IUCN) as either endangered, critically endangered, threatened, or rare. An endangered species faces immediate danger of extinction even with human action, while a critically endangered species is one that will not survive without human intervention and protection. Threatened species may still be abundant in their own habitat but their **population** is rapidly declining; a rare species is considered at risk because of low overall population numbers.

Most species become threatened and endangered because of a combination of factors: their habitat is either disturbed or eliminated (most often as a result of human activities); commercial and sport hunting are not managed to prevent population declines; competing species overtake the primary food sources; or the species is killed to protect livestock and crops. More than two-thirds of the species listed as endangered by the IUCN and the United States, through its Endangered Species Act, are included because their habitats were overtaken by human activities, and particularly land development. Large animals and birds such as the lion, Bengal tiger, elephant, and California condor are at increased risk because they require large areas of habitat for survival and breed in low numbers. Thus, the chances that they can adjust to human encroachment of their native habitat are low.

Various international and national laws have been passed to try to reverse the extinction of hundreds of plants and animals. In the United States, the primary law governing species protection and rehabilitation is the Endangered Species Act, first passed as the Endangered Species Conservation Act in 1966 and revised to its present name in 1973. The act commits the federal government, in cooperation with state and local governments, to develop recovery plans and other methods to protect and restore endangered and threatened plants and animals. Individuals and groups can sue to halt an action that is considered to violate the act, and federal agencies must not initiate projects that would jeopardize endangered species or modify habitats critical to their survival. This last requirement has been frequently debated in subsequent revisions of the act, and in 1998, the U.S. Congress is again revisiting the act's provisions.

One of the first species placed on the nation's endangered list was its national symbol, the bald eagle. In May 1998, the bird was delisted, as was the gray wolf, American peregrine falcon, tidewater goby fish, and almost one dozen other species. Since the act was passed, less than 1% of listed species have become extinct. As of mid-1998, 1,135 species, including 466 animals and 669 plants, are included on the endangered species list.

See also Ecosystem

ENDERS, JOHN F. (1897-1985)

American virologist

John F. Enders' research on viruses and his advances in tissue culture enabled microbiologists **Albert Sabin** and **Jonas Salk** to develop their vaccines against polio, a major crippler of children in the first half of the twentieth century. His work also served as a catalyst in the development of vaccines against measles, mumps and chicken pox. As a result of this work, Enders was awarded the 1954 Nobel Prize in medicine or physiology.

John Franklin Enders was born February 10, 1897, in West Hartford, Connecticut. His parents were John Enders, a wealthy banker, and Harriet Whitmore Enders. Entering Yale in 1914, Enders left during his junior year to enlist in the U.S. Naval Reserve Flying Corp following America's entry into World War I in 1917. After serving as a flight instructor and rising to the rank of lieutenant, he returned to Yale, graduating in 1920. After a brief venture as a real estate agent, Enders entered Harvard in 1922 as a graduate student in English literature. His plans were sidetracked in his second year when, after seeing a roommate perform scientific experiments, he changed his major to medicine. He enrolled in Harvard Medical School, where he studied under the noted microbiologist and author **Hans Zinsser**. Zinsser's influence led Enders to the study of microbiology, the field in which he received his Ph.D. in 1930. His dissertation was on **anaphylaxis**, an allergic condition that can develop after a foreign protein enters the body. Enders became an assistant at Harvard's Department of Bacteriology in 1929, eventually rising to assistant professor in 1935, and associate professor in 1942.

Following the Japanese attack on Pearl Harbor, Enders came to the service of his country again, this time as a member

Aruba Island rattlesnake. It was once widespread in rocky habitats on Aruba, but most of its natural habitat has been destroyed by various human influences. *(Photograph by Robert J. Huffman, Field Mark Publications. Reproduced by permission.)*

of the Armed Forces Epidemiology Board. Serving as a consultant to the Department of War, he helped develop diagnostic tests and immunizations for a variety of diseases. Enders continued to work with the military after the war, offering his counsel to the U.S. Army's Civilian Commission on Virus and Rickettsial Disease, and the Secretary of Defense's Research and Development Board. Enders left his position at Harvard in 1946 to set up the Infectious Diseases Laboratory at Boston Children's Hospital, believing this would give him greater freedom to conduct his research. Once at the hospital, he began to concentrate on studying those viruses affecting his young patients. By 1948 he had two assistants, Frederick Robbins and Thomas Weller, who, like him, were graduates of Harvard Medical School. Although Enders and his colleagues did their research primarily on measles, mumps, and chicken pox, their lab was partially funded by the National Foundation for Infantile Paralysis, an organization set up to help the victims of polio and find a vaccine or cure for the disease. Infantile paralysis, a **virus** affecting the **brain** and **nervous system** was, at that time, a much-feared disease with no known prevention or cure. Although it could strike anyone, children were its primary victims during the periodic epidemics which swept through communities. The disease often crippled and, in severe cases, killed those afflicted.

During an experiment on chicken pox, Weller produced too many cultures of human embryonic **tissue**. So as not to let them go to waste, Enders suggested putting polio viruses in the cultures. To their surprise, the virus began growing in the test tubes. The publication of these results in a 1949 *Science* magazine article caused major excitement in the medical community. Previous experiments in the 1930s had indicated that the polio virus could only grow in nervous system tissues. As a result, researchers had to import monkeys in large numbers from India, infect them with polio, then kill the animals and remove the virus from their nervous system. This was extremely expensive and time-consuming, as a single monkey could provide only two or three virus samples, and it was difficult to keep the animals alive and in good health during transport to the laboratories.

The use of nervous system tissue created another problem for those working on a vaccine. Tissue from that system often stimulate allergic reactions in the brain—sometimes fatally—when injected into another body, and there was always the danger some tissue might remain in the vaccine serum after the virus had been harvested from the culture. The discovery that the polio virus could grow outside the nervous system provided a revolutionary breakthrough in the search for a vaccine. As many as 20 specimens could be taken from a single monkey, enabling the virus to be cultivated in far larger quantities. Since no nervous system tissue had to be used, there was no danger of an allergic reaction through inadvertent transmission of the tissue. In addition, the technique of cultivating the virus and studying its effects also represented a new development in viral research. Enders and his assistants placed parts of the tissues around the inside walls of the test tubes, then closed the tubes and placed the cultures in a horizontal position within a revolving drum. Because this method made it easier to observe reaction within the culture, Enders was able to discover a means of distinguishing between the different viruses in human **cell**s. In the case of polio, the virus killed the cell, whereas the measles virus made the cells fuse together and grow larger.

Although Enders had as good an opportunity as anyone to develop a vaccine against polio, he refused suggestions by Robbins and Weller to take that road. The exact reason for his refusal is unclear, although it may have been that Enders was reluctant to submit himself to the restrictions which the National Foundation might have placed on his research. But because his breakthrough made it possible to develop a vaccine against polio, Enders, Robbins and Weller were awarded the Nobel Prize for medicine or physiology in 1954. Interestingly enough, Enders originally opposed Salk's proposal to vaccinate against polio by injecting killed viruses into an uninfected person to produce immunity. He feared that this would actually weaken the immunity of the general population by interfering with the way the disease developed. In spite of their disagreements, Salk expressed gratitude to Enders by stating that he could not have developed his vaccine without the help of Enders' discoveries.

Enders' work in the field of immunology did not stop with his polio research. Even before he won the Nobel Prize, he was working on a vaccine against measles, again winning the acclaim of the medical world when he announced the cre-

ation of a successful vaccine against this disease in 1957. Utilizing the same techniques he had developed researching polio, he created a weakened measles virus which produced the necessary antibodies to prevent infection. Other researchers used Enders' methodology to develop vaccines against German measles and chicken pox.

In spite of his accomplishments and hard work, Enders' progress in academia was slow for many years. Still an assistant professor when he won the Nobel Prize, he did not become a full professor until two years later. This may have resulted in his dislike for university life—he once said that he preferred practical research to the "arid scholarship" of academia. But by the mid-fifties, Enders began receiving his due recognition. He was given the Kyle Award from the U.S. Public Health Service in 1955 and, in 1962, became a university professor at Harvard, the highest honor the school could grant. Enders received the Presidential Medal of Freedom in 1963, the same year he was awarded the American Medical Association's Science Achievement Award, making him one of the few non-physicians to receive this honor.

Enders married his first wife, Sarah Bennett, in 1927, and in 1943 she passed away. They had two children, John Enders II and Sarah Steffian. He married again in 1951 to Carolyn Keane. Affectionately known as "The Chief" to students and colleagues, Enders took a special interest in those he taught, keeping on the walls of his lab portraits of those who became scientists. When speaking to visitors, he was able to identify each student's philosophy and personality. Enders wrote some 190 published papers between 1929 and 1970. Towards the end of his life, he sought to apply his knowledge of immunology to the fight against **AIDS**, especially in trying to halt the progress of the disease during its incubation period in the human body. Enders died September 8, 1985, of heart failure, while at his summer home in Waterford, Connecticut.

ENDOCRINE GLAND

Endocrine glands are collections of **tissue** which manufacture and release chemical messengers called **hormone**s. Hormones are chemicals which affect the functioning of some distant organ or tissue.

Unlike exocrine glands (such as sweat glands and salivary glands), endocrine glands do not have ducts (ducts are tube-like passageways through which substances produced in an exocrine gland are transported). Instead, endocrine glands pass their products directly into the **blood**stream. Endocrine hormones are thus transported to the **organ** or tissue where they will have their effect: the target organ or target **cell**. This target organ or cell may be at a considerable distance form the endocrine gland from which the hormone originates. Endocrine glands may be well-defined structures (such as the **thyroid gland**), or may be loosely-associated collections of cells (existing within the gastrointestinal tract).

The functioning of endocrine glands is regulated in several ways. Positive feedback refers to a situation in which a hormone circulating in the bloodstream passes through a particular gland, signaling that gland to produce more of its product. Negative feedback refers to a situation in which a hormone circulating in the bloodstream signals a particular gland to turn off production of its hormones.

The major endocrine glands of the human body include the thyroid, pituitary, **hypothalamus**, ovaries, testicles, and parts of both the **pancreas** and the **adrenal glands**. During pregnancy, the **placenta** plays an important role as an endocrine organ. The variety of endocrine glands are vital to the development of the fetus, the development of secondary sexual characteristics, growth, the appropriate **regulation** of the body's chemical milieu, the transport and utilization of **nutrients** (especially sugars and **fats**), the body's response to various **stress**es, the regulation of basic bodily functions such as **heart** rate and **blood pressure**, and even the regulation of mood. In the **animal kingdom**, endocrine activity is responsible for the utilization of nutrients, growth, molting, **metamorphosis**, coloration, reproductive capacity, and sexual behavior.

ENDOCRINE SYSTEM

Often called the ductless glands, the nine distinct glands of the human endocrine system secrete **hormone**s or ''chemical messengers'' directly into the **blood** stream, without the ducts found in the exocrine system. The endocrine system also includes numerous specialized **tissue**s, including stomach **cell**s that secrete hormones. The hormones, secreted in minute amounts and controlled by feedback mechanisms, travel throughout the body to target **organ**s, often remote from the secreting gland. They cause no changes in tissues they pass through but cause significant changes in the target organs that have receptors for the hormones. Working in concert with the **nervous system**, the hormones released by the endocrine system regulate growth, ovulation, milk production, sexual development, and many other physiological and behavioral functions. Thus, the endocrine system is the body's chemical coordinating system, initiating and deactivating bodily changes as needed. The hormones secreted by each gland or tissue degrade rapidly once their messages have been delivered. Through feedback mechanisms, the **endocrine gland**s are told to stop their **secretion**s, thereby maintaining a delicate balance of hormone initiation and deactivation. If the system malfunctions and too much or too little of a hormone is secreted, the result can be gigantism, dwarfism, goiter, and other abnormalities or diseases.

Of the human body's nine endocrine glands, the pituitary influences so many other glands that it has been called the ''master gland.'' Situated at the base of the **brain**, just below the **hypothalamus**, the tiny pituitary, about the size of a pea, works with the hypothalamus as part of a direct link between the endocrine system and the nervous system. In fact, the pituitary's posterior lobe is actually part of the brain, and the anterior or front lobe responds to secretions from the brain. The hypothalamus, which monitors internal organs and emotional states, produces the hormones oxytocin and antidiuretic hormone, which are stored in the cells of the pituitary's poste-

rior lobe, which then releases the hormones into the blood. The hypothalamus also produces substances called releasing factors that control hormonal secretions from the important hormones, including growth hormone that is essential for normal growth, thyroid stimulating hormone, and follicle-stimulating hormone and luteinizing hormone, both of which play a role in stimulating the production of **gamete**s and **sex hormones**. Because the hypothalamus and pituitary work so closely to integrate such a variety of the activities within the human body, they are also referred to as the neuroendocrine control center.

The thyroid is an endocrine gland that produces hormones containing iodine, which is necessary to maintain thyroid hormone levels in the blood. If thyroid secretions are not sufficient, a condition called hypothyroidism can result. In adults, hypothyroidism produces dry skin, sluggish behavior, and an inability to tolerate cold. In children, it can lead to mental retardation and stunted growth. If thyroid secretions are in excess, hyperthyroidism can result. Symptoms of this condition include nervousness and agitation, weight loss even with normal food consumption, profuse sweating, increases in metabolic rates, **heart** rate, and blood flow. This condition is known as Graves' disease. The four pea sized parathyroid glands found in the thyroid's tissue secrete a hormone called parathormone that helps regulate the blood's calcium levels.

The gonads, which are called testes in males and ovaries in females, are the primary reproductive organs. These endocrine glands, which control reproductive function, not only produce gametes but also secrete **estrogen**s, progesterone, and androgens. The two **adrenal glands** found in humans, located above each kidney, have an outer adrenal cortex that secretes hormones called glucocorticoids. These help maintain the blood's glucose level and suppress tissue **inflammation** resulting from injury or infection. The inner part of the adrenal glands, called the adrenal medulla, secrete epinephrine and norepinephrine, which help regulate **blood circulation** and **carbohydrate metabolism**. During times of excitement or **stress**, the adrenal medulla also helps the body prepare its defenses. Its hormonal secretions cause the heart to beat faster, the lung's airways to dilate, and other bodily changes known as the ''fight or flight'' response. The pancreatic islets, another part of the endocrine system, are clustered throughout the **pancreas**, an exocrine gland that secretes digestive **enzyme**s. Among the hormones secreted by the approximately two million pancreatic islets is **insulin**, which stimulates the liver, muscle, and other cells to take up glucose after a meal, when blood glucose levels are high. The pineal gland, a **light** sensitive organ that evolved from a third eye that vertebrates had on the top of their head until approximately 240 millon years ago, secretes the hormone melatonin. This hormone, which is secreted when light is absent, plays a role in the development of gonads and in reproductive cycles. The thymus gland cells whose job is to fend off disease.

Recently, intense controversy has arisen over **pesticides** and other chemicals that some scientists believe could be creating widespread adverse effects on human health and wildlife by acting as hormone mimics and thereby disrupting the normal functioning of endocrine systems. The 1996 book, *Our*

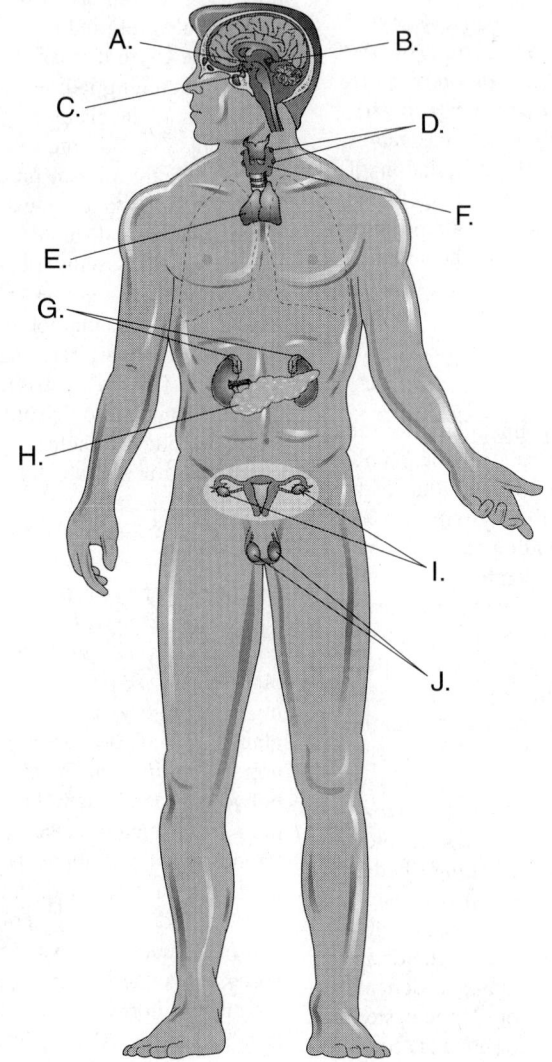

The human endocrine system: A. Hypothalamus. B. Pineal gland. C. Pituitary gland. D. Parathyroid gland. E. Thymus. F. Thyroid gland. G. Adrenal glands. H. Pancreas. I. Ovaries (female). J. Testes (male). *(Illustration by Electronic Illustrators Group.)*

Stolen Future, by Theo Colborn, Dianne Dumanoski, and John Myers, sets forth the concerns about "hormone havoc" that have become a major environmental issue of the 1990s.

ENDOCYTOSIS

Endocytosis (from Greek *endon*, meaning within and *kytos*, meaning vessel) is the process by which materials are engulfed into a **cell**. There is an infolding of the cell **membrane**, which then pinches around substances, forming a vacuole or vesicle, and the materials are transported into the **cytoplasm** of the cell. Since endocytosis requires **energy** some scientists consider it a form of **active transport**. There are three types of endocytosis: **pinocytosis**, **phagocytosis**, and receptor-mediated endocytosis.

Pinocytosis is the process in which cells engulf liquids. The liquids may or may not contain dissolved materials. Pinocytosis can occur in many types of cells in multicellular **organisms**. For example, as a human **egg** cell matures in the ovary, it is surrounded by other cells. These cells pass **nutrients** to the egg cell, which engulfs them using pinocytosis. Pinocytosis has also been observed in white **blood** cells (macrophages and leukocytes), kidney cells, epithelial cells of the intestine, and plant root cells.

Phagocytosis is the process by which solid particles are engulfed into the cell. Some unicellular organisms such as amoeba feed by phagocytosis. Their pseudopods move around solid food particles, which are engulfed into a vacuole or vesicle in the cytoplasm of the organism. The vacuole or vesicle can then fuse with a **lysosome**, containing digestive **enzymes**, so that the food may be broken down and used by the organism. Phagocytosis can occur in multicellular organisms as well. For example, some types of white blood cells in humans engulf invading **bacteria** and other foreign particles and destroy them. Thus, these phagocytic cells are an important part of our **immune system**.

With receptor-mediated endocytosis the substances which are to be transported into the cell first must bind to specific sites or receptor **proteins** on the outside of the cell. The substances are then engulfed into the cell. For example, in **animals**, **cholesterol** enters cells through receptor-mediated endocytosis. In order for cholesterol **molecules** to enter the cell, they must first bind with receptors on the cell membrane. Once they bind, a vesicle forms, carrying the cholesterol into the cell.

Once materials are engulfed into the cell and digested or used, wastes can be removed by the process of exocytosis. In this process, which is almost the opposite of endocytosis, the vacuole or vesicle joins the cell membrane and releases substances to the outside of the cell.

ENDODERM

Endoderm (also called entoderm) is one of the three original layers of **tissue** which appear during early formation of the

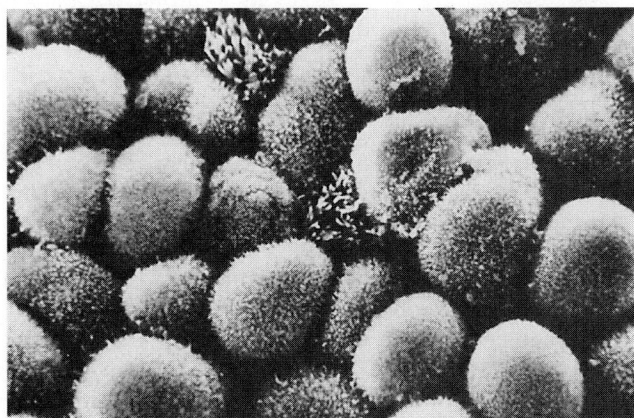

A micrograph image of the internal human uterine wall showing the mucosa, or endometrium. *(Photograph by Professors P.M. Motta and S. Makabe, Custom Medical Stock Photo. Reproduced by permission.)*

embryo (the embryo is the offspring, developing within an **egg**, seed, or the body of its mother). The three layers of embryonic tissue are collectively referred to as germ tissue, meaning that they exist during the **germination** or developmental phase of an **organism**. These germ layers of tissue include an outer layer called **ectoderm**, a middle layer called **mesoderm**, and an inner layer called endoderm. Endoderm ultimately becomes the delicate lining (**epithelium**) of such structures as the pharynx, respiratory tract (excluding the nose), gastrointestinal tract (excluding the mouth and anus), bladder, and urethra. Furthermore, most of the glands associated with these structures are invested with endodermally-derived epithelium. These include the liver, **pancreas**, gall bladder, thyroid, parathyroid, and thymus glands.

ENDOMETRIUM

The uterus (womb) is the **organ** in which an early **embryo** implants and develops until **birth**. The body of the uterus is known as the corpus and the opening of the uterus into the vagina is known as the cervix. The thick walled uterus is structured of smooth muscle (the myometrium). The endometrium is the lining of the uterine corpus and its thickness and structure vary with the stages of the **menstrual cycle**. The endometrium is comprised of the *stratum functionale*, which is lost and subsequently replaced during each menstrual cycle, and the *stratum basale* which is retained after the menstrual period and which supplies **cells** for the renewal of the endometrium. Uterine glands are found deep within the *stratum basale* and it is cells from these glands that provide for repopulation of the *stratum functionale*. The endometrium is well supplied with **blood** vessels. Sloughed *stratum functionale* causes blood to be released from ruptured vessels and the mix of sloughed cells and blood comprises menstrual fluid.

Ordinarily, **tissues** remain in place throughout life. Sometimes, however, endometrial cells wander away from the uterine corpus and establish colonies elsewhere. This pathological condition is known as endometriosis.

ENDOPLASM

Endoplasm is the liquid material in the center of **cell**s outside the **nucleus**. It is the fluid matrix within which all the **cytoplasmic organelle**s are found. Essentially synonymous with the term **protoplasm**, this rich, colloidal mixture of **proteins**, **carbohydrate**s, **lipids**, and **nucleic acids** is the very stuff of life; it is the medium in which the myrid **chemical reaction**s required for life occur. In large cells such as amoebae and giant algal cells, the endoplasm can be seen to flow back and forth through the cell interior by cytoplasmic streaming caused by contraction of the **cytoskeleton**. This motion probably helps distribute **nutrients** and newly synthesized products throughout the cell. The endoplasm is not, however, a simple, continuous space. It is divided into many individual compartments by the membranous bags and tubules of the **endoplasmic reticulum** and other organelles. Within these separate compartments, mutually incompatible enzymatic reactions can take place simultaneously while raw materials and products can be concentrated and stored until they are needed.

Most biologists now use the words protoplasm, endoplasm, and cytoplasm interchangeably. Perhaps the most important reason to retain the term endoplasm is that it helps to explain the origins of the name endoplasmic reticulum for the predominant **membrane** compartments found in this region. Cell biologists once made a distinction between the endoplasm and the ectoplasm because they looked so different under the **light microscope**. We now know that the main differences between these regions is that the peripheral mat of cytoskeletal elements just under the **plasma membrane** of some cells filters out larger organelles and elements of the cytoplasm such as **mitochondria**. Thus the ectoplasm appears to be optically clear because it contains only small, soluble **molecule**s while the endoplasm is seen, even in the light microscope, to be packed with a wide variety of **macromolecule**s and membranous organelles.

ENDOPLASMIC RETICULUM

A major component of the internal **membrane** system of eukaroytic **cell**s (those with nuclei), the endoplasmic reticulum (or ER) is an enclosed network of membranous tubules, vesicles, and flattened cisternae (bags) extending throughout the **cytoplasm** of most cells. Comprising as much as 60% of all cellular membrane, the ER is generally divided into two major types. The rough endoplasmic reticulum (RER) is generally made up of extensive membrane sheets, the cytoplasmic surfaces of which are studded with **ribosome**s that carry out synthesis of **protein**s to be embedded in membranes, sequestered in endomembrane compartments or exported out of the cell. The smooth endoplasmic reticulum (SER) is composed of a branching network of membranous tubules. **Enzyme**s on or within the SER are responsible for detoxifying poisonous chemicals, synthesizing **lipids** and **steroids**, and assembly of glycogen. The SER also is an major storage site for calcium, which is released to stimulate muscle contraction, enzyme action, membrane fusion, assembly of the **cytoskeleton**, and a host of other important reactions. Without these membrane systems, eukaryotic cells could not carry out their many different functions.

Both RER and SER are interconnected via direct contacts and mobile vesicles to other membranous components such as the **Golgi apparatus**, **lysosome**s, the nuclear membrane, and the **plasma membrane**. Coordinated cooperation between these compartments allow the numerous, complex activities of cellular **metabolism** to occur without interference between incompatible reactions. The intracellular traffic between these compartments is both rapid and complex and is regulated by an array of signal and docking proteins on various membrane surfaces. The motile force for this vesicular transport is provided by molecular motors attached to the cell cytoskeleton.

ENERGY

Energy is a fundamental physical element, which can be defined simply as: the capacity of a body or system to accomplish ''work.'' Work is defined as: the result of a force being applied over some distance. As such, work can be accomplished in various ways, as is illustrated by the following examples: (a) a baseball bat strikes a ball, causing it to fly though the air; (b) an engine transforms energy of gasoline into motion of a driveshaft, allowing an automobile to be driven along a highway; (c) heat from a stove is absorbed by **water** in a kettle, causing it to boil, and (d) the pigment chlorophyll absorbs sunlight, converting its energy into a form that **plants** and **algae** can use to manufacture simple sugars through **photosynthesis**. In each of these cases of ''work,'' energy has been transformed from one state to another, and a measurable outcome has been achieved.

Energy can exist in several states, each of which is fundamentally different from the others. Under suitable conditions, however, any of these forms of energy can be converted into the others, through various physical or chemical **transformations**. The three categories of energy are known as: electromagnetic, kinetic, and potential, each of which can also exist in various states.

Electromagnetic energy (or electromagnetic radiation) is associated with photons. These are infinitesimally small entities that travel through space at a constant speed of 9.8×10^8 ft/sec (3×10^8 m/sec; this is known as the speed of **light**). Electromagnetic energy exists as a continuous spectrum of wavelengths, which are known as: gamma, X-ray, ultraviolet, visible light, infrared, microwave, and radio (ordered from the shortest to longest wavelengths). The human eye can perceive electromagnetic energy having wavelengths between about 0.4-0.7 micrometers, a range referred to as visible radiation or ''light.''

Kinetic energy is associated with motion, and can be divided into two types: mechanical and thermal. Mechanical kinetic energy is associated with objects in motion, and is determined by the mass of the object and its speed at travel.

Mechanical energy is typified by a deer running through a meadow, water flowing in a stream, and a planet travelling through space. Thermal kinetic energy is associated with the rate at vibration of **atoms** or **molecule**s. Such vibrations are frozen at 460° F (-273° C; this is known as "absolute zero"), but they occur increasingly vigorously at higher **temperature**s. Thermal energy is sometimes referred to as "heat."

Potential energy is the stored ability to perform work. For work to actually be performed, potential energy must be transformed into either electromagnetic or kinetic energy. The various types of potential energy are associated with gravity, chemicals, compressed gases, electrical potential, magnetism, and the sub-atomic organization of **matter**, as is described below:

(1) Gravitational potential energy results from attractive forces existing between objects (that is, gravity). For example, water stored at any height above sea level contains gravitational potential energy, which is converted into kinetic energy if a pathway allows the water to flow down-hill.

(2) Chemical potential energy is stored in the bonds between atoms within molecules. This energy is liberated during exothermic reactions (i.e., reactions that have a net release of thermal energy). For example: (a) chemical potential energy in the molecular bonds of sulphide **minerals** (such as iron sulphide, FeS_2) is released when the sulphides are oxidized by burning or by the **metabolism** of certain **bacteria**; (b) **hydrocarbon**s store energy in inter-atomic bonds of their hydrogen and carbon atoms, which is liberated when coal, gasoline, or natural gas are combusted; (c) biochemicals produced metabolically by **organism**s store potential energy in their inter-atomic bonds: **carbohydrate**s contain about 16.8 kilojoules/gram, **protein**s about 21.0 kJ/g, and **lipids** or **fats** about 38.5 kJ/g.

(3) Electrical potential energy results from differences in quantities of electrons (subatomic, negatively charged particles), which will flow through a conducting material (such as a metal) from locales where they occur in a high density to those with less (i.e., along a gradient of electrical potential energy).

(4) The potential energy of compressed gases can be released to do work when the gases are allowed to expand and match the pressure of the surrounding **atmosphere**.

(5) Nuclear potential energy is associated with the extremely strong binding forces existing within atoms, and is by far the densest form of energy. Fission nuclear reactions involve the splitting of isotopes of certain heavy atoms, such as uranium-235 and plutonium-239, to generate smaller atoms plus enormous quantities of energy (essentially, this represents the conversion of matter into energy). Fission reactions occur in nuclear explosions, and under controlled conditions in nuclear reactors. Fusion reactions involve the combining of certain light elements, such as hydrogen, to form heavier atoms under conditions extremely high temperature and pressure, while liberating huge amounts of energy. Fusion reactions involving hydrogen occur in stars and in hydrogen bombs.

Although energy can exist in various forms, all of them can be measured in the same or equivalent units. The internationally accepted unit is the joule (J), which is defined as the energy required to accelerate 1 kg of mass at 1 m/s^2 (1 m per second per second) for a distance of 1 metre. A **calorie** (or gram-calorie, abbreviation cal) is another unit of energy, equivalent to 4.184 J, and equal to the amount of energy required to raise the temperature of one gram of pure water by one degree Centigrade (specifically, from 15° C to 16° C). Note that the dietician's "calorie" is equivalent to 1000 calories.

All transformations of energy must behave according to certain physical principles, known as the **laws of thermodynamics**. These are universal principles, meaning they are always true, regardless of the circumstances. The First Law of Thermodynamics states that energy can undergo transformations among its various states, but is never created or destroyed; consequently, the energy content of the universe always remains constant. The Second Law of Thermodynamics states that energy transformations can occur spontaneously only if there is an increase in the **entropy** of the universe. Entropy is a physical attribute related to disorder, or the degree of randomness in the distributions of matter and energy. As the randomness (i.e., disorder) increases, so does entropy.

ENTOMOLOGY

Entomology is the study of insects. As a subdiscipline of the biological sciences, it focuses on the life, history, **morphology**, physiology, **genetics**, reproduction, development and **ecology** of this most abundant and diverse taxonomic group of terrestrial **organism**s. The number of **species** of insects is estimated in the range of 5-10 million; the biology and ecology of most of these species are virtually unknown as only about 1 million species have been described by scientists.

Insects are most abundant and diverse in the tropics. They occupy nearly every conceivable terrestrial **habitat** and many freshwater ones. In contrast, the marine **ecosystem**s possess only a few species. Due to their small size insects can inhabit a wide variety of microhabitats unavailable to larger **animal**s. Many insects are phytophagous, feeding and living on various parts of **plants** such as seeds, **flower**s and stems. Some of these species, such as aphids, may transmit plant diseases by their feeding activities. Other species are soil dwellers, decomposers, pollen and nectar gatherers of flowering plants, symbionts in the dens of vertebrates, or predators of other insects and other small **invertebrates**. Some species, such as mosquitoes and black flies, cause irritation, **blood** loss and transmit disease organisms as they feed on the blood of animals. This tremendous diversity in lifestyles and feeding habits makes it quite improbable to find a terrestrial animal or plant that does not interact in some way with an insect.

Various important ecosystem functions are performed by insects. Many insects, such as soil-dwellers, carrion-feeders, and woodborers, aid in the **decomposition** process and nutrient cycling. Insects are consumed by fish, birds, and other vertebrates and thus serve as an important food source in both aquatic and terrestrial food webs. In some **cultures**, insects are even an important **protein** source for humans. Many insects are entomophagous predators, preying on other insects, and hence, are important regulators of **population** levels of many herbivorous insects. Many flowering plants are dependent upon pollinating bees, flies, and butterflies for their continued survival.

Although most insects perform some vital ecosystem services that indirectly benefit humans, a minority of insect species adversely affect agricultural crops, livestock and poultry production, stored products, wood and lumber materials, and human health. Consequently, most entomologists focus their study on the interactions of these insects with humans and our products and activities. Most crops are damaged to various degrees by plant-feeding insects. Insects may feed or oviposit into stems, leaves, roots, flowers or **fruits** greatly limiting the crop yield. Thus, agricultural entomologists devise methods to limit insect damage by using a variety of techniques such as **pesticides**, adjusting crop practices (e.g., crop rotation), and encouraging natural predators and diseases of the pest insects (i.e., biological control). The compatible use of a variety of such pest suppression techniques in an ecological sensitive manner that is economically feasible to **agriculture** is termed integrated pest management (IPM). This philosophy is a cornerstone principle of entomologists involved in controlling pestiferous insects.

Medical entomologists study insects that directly impact human health, e.g., bloodsucking flies or stinging wasps, and those that serve as vectors for human disease organisms. Mosquitoes, lice, fleas and other blood-sucking insects transmit some of the most prominent infectious diseases in the world such as malaria, yellow fever, dengue fever, plague, and typhus.

Other entomologists investigate the physiology, development, genetics, predators, diseases, and behavior of pest insects to discover new ways of controlling insect populations. For example, investigations of insect development have led to the use of specific chemicals that disrupt the **hormone**s guiding **metamorphosis**. Also, behavioral studies have discovered the widespread use of insect-produced and plant-produced chemicals orienting insects to mates, food or other habitat needs. Entomologists use these behavior- modifying chemicals to monitor or even suppress pest populations. Many universities and government agencies house and maintain insect museums so entomologists can conduct systematic studies of species and other higher categories of insects.

Insects are also important in environmental studies. For example, aquatic insects are used as key indicators of the effects of **pollution** in streams and other aquatic systems while other insects, such as butterflies, are used as ecological indicators of changing terrestrial habitats. Other entomologists focus on **conservation** efforts to protect certain insect species that are endangered of **extinction**. These biological studies of insects contribute to advances in biology, especially to the subdisciplines of **ethology**, hormonal physiology, chemical **ecology**, population ecology, **evolution**, and genetics. For example, enormous strides in genetics have resulted from the study of the *Drosophila* **fruit fly**. Consequently, the field of entomology is interdisciplinary in scope as the study of insects often involves basic and applied sciences within the agricultural, environmental, and biomedical fields.

ENTROPY

Nicolas-Léonard Sadi Carnot first suggested that all **energy** will eventually break down into an unusable form. This idea was seized by William Thomson (Lord Kelvin), who reintroduced it several years after Carnot's death. Thomson explained that this unusable form is waste heat, and that it dissipates into the environment. While this was to be an important concept, it was not viewed as such until the German physicist Rudolf Clausius discussed it in his 1850 essay. It was not until 1865 that Clausius finally named the concept *entropy*, from the Greek word for "transformation."

Entropy, as described by Clausius, was the ratio of the amount of heat in a system to that system's absolute **temperature**. He assumed that the system was closed—that is, that no heat could escape to the outside world, nor could heat enter from without. In such a system, the internal energy could, and would, be converted into heat, but the heat could not be reconverted into usable energy. Thus, all of the energy would eventually be "used up," filling the system with waste heat.

Clausius proposed that since the amount of heat within a closed system could only increase, so would the ratio of entropy. While such a system could not be created in the laboratory, the concept could be applied to theoretical physics. Since the time of its origination, the concept of entropy has been adopted as the second law of thermodynamics—that all energy is eventually transformed into unusable heat. The first law of thermodynamics states that heat can move from a hot substance to a cool substance, but never the reverse.

Modern theorists have explored the consequences of entropy with disturbing results. It is commonly accepted that the universe as a whole is a closed energy system, with no environment outside the universe for energy to move to or from. If the concept of entropy is applied to this assumption, then it must be true that all of the energy within the universe will eventually be converted into unusable heat. The temperature would be uniform at a few degrees above absolute zero; no heat would flow, no energy would be exchanged. The finite amount of heat contained inside the universe would be used up.

This theory is known as the heat-death of the universe. Though it is based on two strong assumptions (entropy and the idea that the universe is a closed system), it is not widely accepted by cosmologists. The more popular theory is that different areas of the universe are governed by different laws of physics, and that entropy does not necessarily apply to the entire universe.

Another modern interpretation of entropy is the idea that it represents not the breakdown of energy, but the move from order to disorder. These theorists claim that all things—**matter**, energy, and thought—are moving inexorably toward a state of chaos.

ENZYMATIC ENGINEERING

Enzymatic engineering includes a wide variety of techniques related to the manipulation of **molecule**s to create new enzymes, or **protein**s, that have useful characteristics. These techniques range from the modification of existing enzymes to the creation of totally new enzymes. Typically, this involves changing the protein on a genetic level by mutating a **gene**. Other ways that modifications to enzymes are made are by **chemical reaction**s and synthesis.

Enzymes are relatively large, complex molecules that catalyze biochemical reactions. These include such diverse reactions as the **digestion** of sugars, **DNA replication**, or the combating of diseases. For this reason, the creation of improved enzymes has been an important field of study for the last fifty years. A variety of steps are involved in the creation of modified enzymes. These entail a determination of the molecular structure, function and modifications to improve the effectiveness.

The key to creating new and improved enzymes and proteins is determining the relationship between the structure and function of the molecule. The structure of an enzyme can be described on four levels. The primary structure is related to the sequence of **amino acid**s that are bonded together to form the molecule. The amino acids that make up the enzyme are determined by the **DNA** sequence of the gene that codes for the enzyme. When enough amino acids are present, the molecule folds into secondary structures known as structural motifs. These structural motifs are further organized into a three dimensional, or tertiary structure. In many enzymes, multiple tertiary structures are combined to give the overall quaternary structure of the molecule. Scientists have found that the specific structural organization of amino acids in an enzyme are directly related to the function of that enzyme.

Structural information about enzymes is obtained by a variety of methods. The amino acid sequence is typically found by using DNA sequencing. In this method, the pattern of **nucleotide**s is determined for the gene that encodes the enzyme. This information is then compared to the **genetic code** to obtain the amino acid sequence. Other structural information can also be found by using spectroscopy, **chromatography**, magnetic resonance, and x-ray crystallography.

To investigate the function of an enzyme, it is helpful to be able to create large quantities of an enzyme. This is done by **cloning** a desired gene. **Polymer**ase chain reaction (PCR) which is a method of making a large number of copies of a gene is particularly helpful at this point. The cloned genes are then incorporated into a biological vector such as a bacterial plasmid or phage. Colonies of **bacteria** that have the specific gene are grown. Typically, these bacteria will express the foreign gene and produce the desired enzyme. This enzyme is then isolated for further study.

To create new enzymes, the DNA sequence of the gene can be modified. Modifications include such things as deleting, inserting, or substituting different nucleotides. The resulting gene will code for a slightly modified enzyme which can then be studied for improved stability and functionality.

One of the major goals in enzymatic engineering is to be able to determine enzyme structure and functionality based on the amino acid sequence. With improvements in computer technology and our understanding of basic molecular interactions, this may someday be a reality.

ENZYME

Enzymes are complex **protein**s that act as catalysts for the countless biochemical reactions that keep humans, **animal**s, **plant**s, and **microorganism**s alive. Constituents of every living **cell**, enzymes have relatively large **molecule**s that contain one or more **amino acid** chains. The sequence of amino acids within the chains and the distinctive way each chain folds into its own characteristic three-dimensional shape help determine the enzyme's particular activity. In order for them to act, many enzymes also need to be attached to a nonprotein substance called a **coenzyme**. In most cases, these coenzymes are composed wholly or partially of **vitamins**, especially those in the **water**-soluble B family. The typical animal cell (roughly onebillionth the size of a drop of water) contains about three thousand different enzymes, almost all programmed to perform specific **chemical reaction**s necessary for **metabolism**.

For example, in the digestive tract certain enzymes are involved in breaking down oversized fat, **carbohydrate**, and protein molecules into smaller and easier-to-absorb molecules; a different set of enzymes assists in moving these molecules into the **blood**stream; then other enzymes utilize some of these molecules in the **biosynthesis** of new cellular structures.

Enzymes have important industrial and commercial uses as well. Since ancient times, people have observed enzymes at work fermenting their wine and beer, turning their sour milk into cheese, and causing their bread dough to rise. However, these reactions were generally considered as **fermentation**s of some mysterious kind and only vaguely understood. Then, in the early 1800s, biochemists began taking a closer look at the "ferments" causing some of these reactions.

In 1833, French chemist Anselme Payen separated a substance from an extract of malt that, he realized, seemed capable of speeding up the conversion of starch to sugar. Payen called the substance diastase, the first enzyme to be isolated and prepared in concentrated form. Three years later, German physiologist Theodor Schwann prepared an extract containing **tissue**s from an animal's stomach, mixed it with hydrochloric acid, and demonstrated that the animal tissue extract greatly increased the acid's meat-dissolving properties. Schwann then isolated the substance in the extract that appeared to provide the added potency, and named the substance pepsin from the Greek word meaning "to digest." Pepsin proved to be a vitally important enzyme and was the first such to be prepared from animal tissues.

In 1876, another German physiologist, Wilhelm Kühne (1837-1900), pointed out that certain ferments—such as pepsin and the trypsin he himself had recently isolated from pancreatic juice—should be given a separate name. He suggested the name enzyme, meaning "in **yeast**," because these substances merely resembled the more important ferments found in living cells, notably yeast, which were then believed to be governed by so-called "vital forces." In 1896, however, **Eduard Buchner** (who isolated the enzyme zymase from "dead" yeast cells) proved that the ferments in yeast were no different from those in digestive juice and from then on all were called enzymes, though scientists had yet to prove their protein structure because of their fragility.

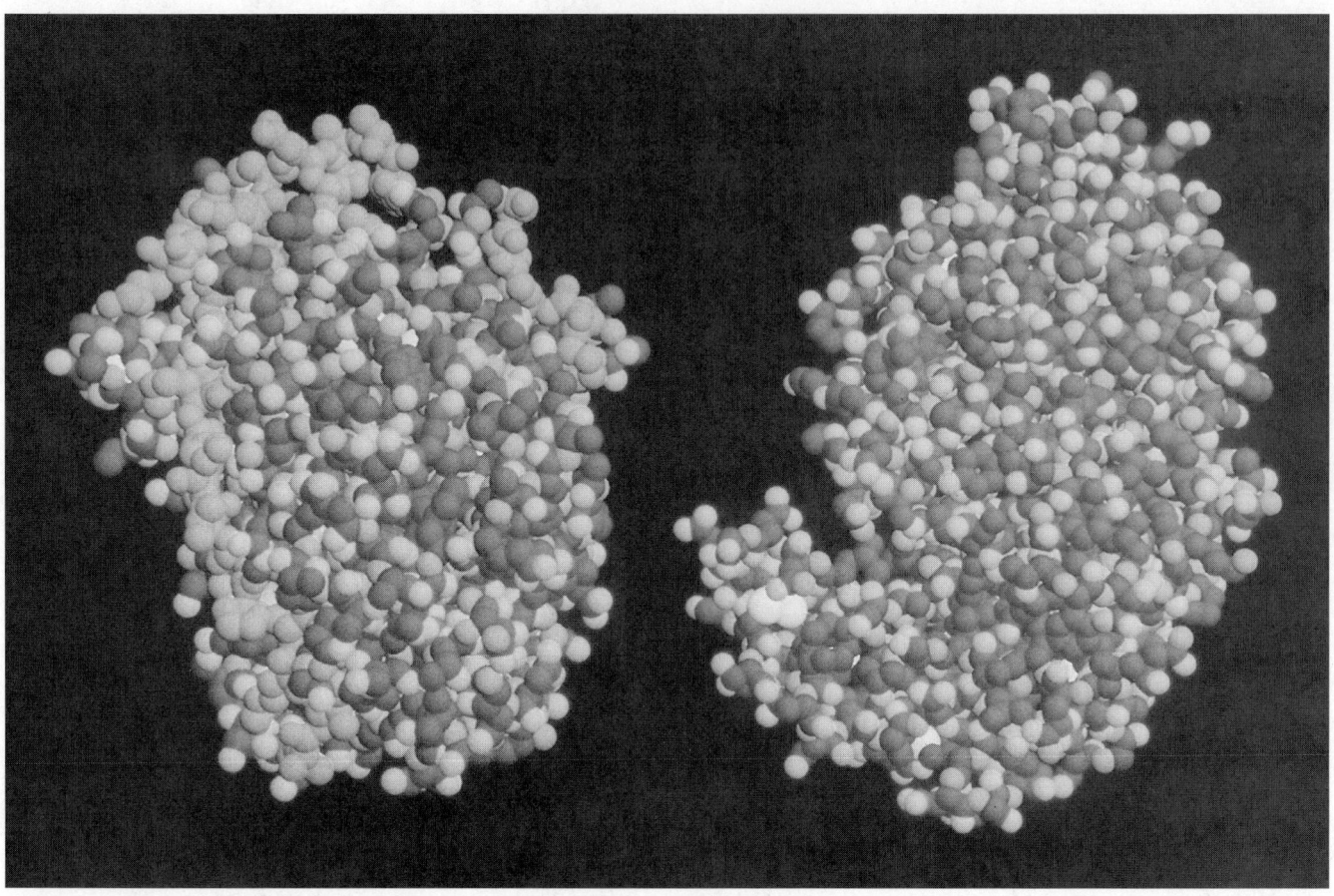

Computer-generated models of pepsinogen (left), a pre-enzyme found in the stomach, and pepsin, a digestive enzyme. In the presence of increased acidity, persinogen cleaves a segment, transforming it into pepsin. *(Photo Researchers, Inc. Reproduced by permission.)*

In the 1920s, the highly respected German chemist **Richard Willstätter** declared that enzymes were not proteins, and his views dominated the scientific world for many years, despite the pioneering protein **classification** research of Ernst Hoppe-Selyer beginning in the 1870s. In fact, in 1926, when American biochemist James Sumner isolated an enzyme in pure form (the enzyme urease) and proved it was a protein, his work was hotly challenged. Another ten years passed before the corroborative research of **John Northrop** settled the issue.

When Payen named the first enzyme diastase, he started the tradition of ending the names of most enzymes with the suffix ''ase.'' Today, the suffix is usually also added to the names of classes of enzymes, and these names typically indicate the reaction that the enzyme catalyzes. For example, the name transferase indicates that the enzymes in this general classification all act to help transfer chemical groups from one molecule to another. More specifically, the transaminase members of this classification transfer amino groups, while the transmethylase members transfer methyl groups.

Whatever their classification, most enzymes tend to act on only one kind of substance (called a substrate) and trigger only one kind of reaction. Because enzymes are relatively large molecules, only a section of the enzyme actually reacts

with the substrate, a section called the active site. Most researchers believe that a particular substrate can only fit into the active site of a particular enzyme. This lock and key hypothesis was first suggested by **Emil Fischer** in 1894. (In related research begun in 1913, Leonor Michaelis [1875-1949] was able to define the rates of reactions with the Michaelis-Menten equation, which supported the theory of such unions.) Because a perfect fit between active site and substrate is so important, any change in either could obviously impede the catalytic reaction.

More serious problems can result when a particular enzyme is actually missing. In a number of hereditary human diseases—such as **phenylketonuria (PKU)** and galactosemia—geneticists have discovered that the affected individuals are born missing certain specific enzymes. Some of these enzyme-deficiency diseases can now be effectively treated and many researchers are concentrating on the search for more of these disorders, which may ultimately revolutionize the practice of medicine.

Another current research focus is in the development of ''artificial enzymes.'' These are enzymes, synthesized in a lab, which mimic the action of naturally occurring enzymes. These artificial enzymes can catalyze reactions more specifically and efficiently than their natural counterparts. This is accomplished through combining known binding sites with different catalytic groups. The new enzyme is then studied to determine its ability to carry out selective reactions in an efficient manner. Artificial enzymes have been designed which catalyze such reactions as selective nucleic acid cleavage, chemical synthesis, and cell response control. Through the creation of such enzymes, researchers hope to both gain a better understanding of enzyme mechanics as well as discover practical applications for these enzymes, such as medicinal use.

EPIGENESIS

Epigenesis is development from an **embryo** with the sequential appearance of **tissue**s and **organ**s. The tissues and organs differentiate from a **population** of equivalent (undifferentiated) embryonic **cell**s which in turn descended from a single cell, the **zygote**. The concept holds that a relatively formless **egg** gives rise to the highly structured adult **organism**. Epigenesis is to be distinguished from preformation, which is the notion that organisms are fully formed but minuscule in the sex cells. Curiously, some preformationists were animalculists who thought they saw a fully formed human in the head of a **sperm** cell. Others, known as ovists, believed that eggs contain the miniature human. To be logical, the fully formed ''homunculus'' that existed in either the egg or sperm must have had within it sex cells with even more tiny formed humans. And those tiny humans must have had all future generations encased in their sex cells as ever more tiny ''homunculi.'' Obviously, this was nonsense. Both animalculists and ovists were completely in error. With the development of the **light microscope**, it became obvious that egg **cytoplasm** was devoid of tissues and organs and, of course, the head of a sperm was comprised almost entirely of the **nucleus**. Actually, early observers were epigenetic. **Aristotle** and **William Harvey** looked at the developing chick. They did not see a chicken in the egg. Rather, they saw a developing embryo that gradually became more complex with time. These early observers clearly reported the epigenetic development of chick embryos.

EPILEPSY

Epilepsy is a disease of abnormal **brain** activity that causes recurrent seizures. Seizures are sudden, brief episodes of altered states of consciousness, often accompanied by spasmodic or unusual motor activity. For centuries, their frightening and violent quality made objective analysis of their nature difficult.

In ancient Greece, the cause of epilepsy was usually thought to be supernatural—the Greeks felt it to be an affliction of the gods. Indeed, the name for the disease comes from the Greek *epilepsis*, which means ''a taking hold'' in the sense that a great power might take over someone's body. Because epilepsy was believed to be a ''sacred disease,'' its interpretation was put into the hands of religious adepts who administered an assortment of ineffective remedies such as tortoise

blood or camel hair. Their practices were attacked as fraudulent in the fifth century by the Hippocratic school of thought and Greek physicians who were convinced that epilepsy had a natural cause, although these physicians had no idea of what that cause might be. The idea that epilepsy was a supernatural phenomenon persisted into the Middle Ages and became linked with demonic possession cured only by exorcism. Even during the intellectual enlightenment of the Renaissance period that followed, epilepsy remained a mystery. A scientific understanding of epilepsy would not occur until the nineteenth century, when the disease came to the attention of physicians who studied the brain and **nervous system**. Their realization that epilepsy had a neurological cause opened the door for its improved treatment—both social and medical.

In the 1860s John Hughlings Jackson (1835-1911), an English neurologist, made one of the first important studies of epilepsy. He concluded from his careful experiments and observations that epileptic seizures began at specific locations somewhere on the cerebral cortexes of its victims. He also clarified the symptoms of epilepsy. In particular, he characterized a type of seizure that typically involves only one side of the body and does not involve a loss of consciousness. This type of epilepsy has come to be known as *Jacksonian* epilepsy.

With the invention of the electroencephalograph (*EEG*) by Hans Berger (1873-1941) in the 1920s, it became possible to observe the electrical activity of the brain directly while a seizure occurred. Subsequent observations made it clear that a seizure resembled an ''avalanche'' of brain activity in which an electrical impulse beginning in a given spot on the cortex spreads uncontrollably throughout an entire hemisphere or even the entire cortex. In time, medications providing effective relief for the symptoms of epilepsy were developed, and as neurologists studied the life histories of their patients, the causes of their seizures (including serious childhood illness and brain damage resulting from injuries, or tumors) began to be better understood.

According to international surveys, approximately one adult in 200 adults suffers from recurrent epilepsy. When infants who have had seizures due to fevers, which is usually a temporary condition, and adults who have only had one seizure are factored in, the figure rises to approximately one in 80. One of the earliest successful treatments for epilepsy were bromides, compounds formed by combining bromine with another element or radical. Barbiturates and phenobarbitone have also been used to treat epilepsy, with phenobarbitone being the most widely prescribed epilepsy medication by the 1970s. New anti-epileptic drugs are constantly undergoing development and testing.

In the past dozen years, computerized axial tomography (CAT) and **magnetic resonance imaging (MRI)** have given physicians the ability to view detailed images of the brain and locate precisely where certain types of seizures begin, paving the way for surgery to offer epileptics a complete cure. Because of the risks involved, brain surgery to treat epilepsy is used primarily for patients who do not respond to medications. Surgery works best when the abnormal electrical discharge that is the source of epilepsy originates in one specific area of the brain. For instance, it appears that certain types of epilepsy are caused when an episode of high fever and convulsions during infancy leaves a small area of scar **tissue** in the brain, which then triggers periodic seizures. Using accurate scanning of the brain and delicate surgery, specialists can sometimes remove the offending speck of tissue, giving the patient a new life free from debilitating seizures. Another approach to controlling epilepsy, called vagus nerve stimulation, is undergoing testing. The approach uses a device similar to a **heart** pacemaker to deliver low-level, intermittent electrical pulses to the vagus nerve in the neck, which seems to control seizures by traveling to certain brain centers like the temporal lobe.

Research has shown that there is no known cause in nearly 60% of epilepsy cases. Known causes included head trauma, cerebrovascular disease, brain tumors, developmental disorders, infections, and poisoning. Epilepsy may also be inherited, although such cases are rare. In recent years, scientists have identified some of the **gene**s responsible for inherited epilepsy. A gene on **chromosome** 21 is responsible for progressive myoclonus epilepsy, which causes nerve degeneration and progressive dementia. It is believed that the disease results from a **mutation** blocking the production of the **enzyme** cystatin B. A gene on the X chromosome has also been identified as causing a rare form of epilepsy and mental retardation only in women, even though the gene is passed on by the father.

EPITHELIUM

Cells form **tissue**s and tissues form **organ**s. The several types of tissues in humans include nerve, muscle, **blood**, connective, reproductive and epithelial. Epithelial tissues are comprised of continuous sheets of cells that cover the body, line the digestive tract and other body cavities, and form glands. Epithelial cells have little intercellular material and because of the close proximity between cells, have strong adhesion. Protection is one function of some epithelia. For example, the epithelium covering the body protects from fluid loss, the entry of pathogens (microbes), and from damage by abrasion and other forms of wear and tear. The multilayered skin epithelium, known as a stratified squamous epithelium, is constantly replacing itself from its basal layer. The loss of surface cells is known as exfoliation and this can be demonstrated by briskly rubbing the skin in a darkened room with bright sun entering a window. The gut from the stomach to the beginning of the rectum is lined with a single layer of columnar shaped cells. The columnar epithelium functions both in absorption (of **nutrients** and **water**) and in **secretion** (of **enzyme**s by the stomach). Gut cells are subject to abrasion and thus there is rapid cell replacement. The epithelium of the renal glomeruli and capsule are structured of squamous epithelia that is one cell thick and function in **excretion**. Some epithelium is ciliated such as in the **respiratory system**. The **cilia** clean the air passageways and are essential for survival. Epithelial cells are anatomically separated from the underlying connective tissue by a basement **membrane**.

Because the bulk of the body mass is not epithelial, it may come as a surprise to learn that perhaps 85% of all human

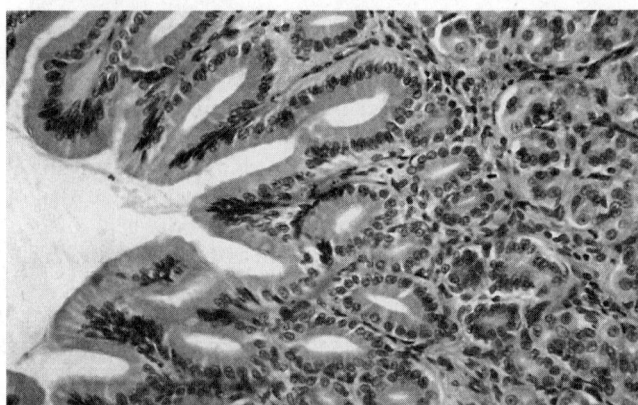

Epithelial cells in human stomach. *(Custom Medical Stock Photo. Reproduced by permission.)*

cancer is of epithelial origin (such cancer is referred to as carcinoma). It may be less surprising when one remembers that the epithelium is the tissue that has contact the environment. Ultraviolet radiation does not affect muscle, nerves or tissue types other than the skin epithelium where is causes melanoma and other skin cancers. Smoking is unlikely to cause cancer of the **nervous system** or of the **reproductive system**. The first tissue type that the carcinogen-bearing smoke encounters is the respiratory epithelium.

A tumor of epithelium that does not have the capacity to transgress (invade) the basement membrane is benign. Malignant epithelium by definition has the competence to invade the basement membrane and to gain access to connective tissue and ultimately the vascular system where dissemination (metastasis) occurs. Metastasis therefore requires that invading epithelial cells be motile and that they release enzymes that digest the basement membrane and components of the extracellular matrix.

ERASISTRATUS   (304 B.C.-250 B.C.)
Greek physician and anatomist

Erasistratus, considered the father of physiology, was born on the island of Chios in ancient Greece. His father and brother were doctors, and his mother was the sister of a doctor. He studied medicine in Athens and then, around 280 B.C., enrolled in the University of Cos, a center of the medical school of Praxagoras. Erasistratus then moved to Alexandria, Egypt, where he taught and practiced medicine, continuing the work of Herophilus. In his later years, he retired from medical practice and joined the Alexandrian museum, where he devoted himself to research.

Although Erasistratus wrote extensively in a number of medical fields, none of his works survive. He is best known for his observations based on his numerous dissections of human cadavers (and, it was rumored, his vivisection of criminals, a practice allowed by the Ptolemy rulers). Erasistratus accurately described the structure of the **brain**, including the cavities and membranes, and made a distinction between its *cerebrum* and *cerebellum* (larger and smaller parts). Contrary to popular belief at the time, he viewed the brain, not the **heart**, as the seat of intelligence. By comparing the brains of humans and other **animal**s, Erasistratus rightly concluded that a greater number of brain convolutions resulted in greater intelligence. He also accurately described the structure and function of the gastric (stomach) muscles and observed the difference between motor and sensory nerves. Erasistratus promoted hygiene, diet, and exercise in medical care.

In his understanding of the heart and blood vessels, Erasistratus came very close to working out the circulation of the blood (not actually discovered until **William Harvey** in the seventeenth century), but he made some crucial errors. Erasistratus understood that the heart served as a pump, thereby dilating the arteries, and he found and explained the functioning of the heart valves. He theorized that the arteries and veins both spread from the heart, dividing finally into extremely fine capillaries that were invisible to the eye. However, he believed that the liver formed blood and carried it to the right side of the heart, which pumped it into the lungs and from there to the rest of the body's organs. He also believed that *pneuma*, a vital spirit, was drawn in through the lungs to the left side of the heart, which then pumped the pneuma through the arteries to the rest of the body. The nerves, according to Erasistratus, carried another form of pneuma, animal spirit.

After Erasistratus, anatomical research through dissection ended, due to the pressure of public opinion. Egyptians believed in the need of an intact body for the afterlife—hence mummification. Scientific anatomical studies were not resumed until the thirteenth century.

ERLANGER, JOSEPH (1874-1965)

American physiologist

Joseph Erlanger was an American physiologist whose pioneering work with his collaborator, **Herbert Spencer Gasser**, helped to advance the field of neurophysiology. For their work on "the highly differentiated functions of single nerve fibers" Erlanger and Gasser shared the 1944 Nobel Prize in medicine or physiology. The awarding of the Nobel Prize to Erlanger and Gasser also recognized their roles in developing the most basic tool in modern neurophysiology: the amplifier with cathode-ray oscilloscope. The prize culminated for Erlanger a distinguished career in medical education and physiological research.

Erlanger was born on January 5, 1874 in San Francisco, California, the sixth of seven children, to Herman Erlanger and Sarah Galinger, both immigrants from Southern Germany.

In 1889, Joseph Erlanger entered the classical Latin curriculum at the San Francisco Boys' High School. After graduating in 1891, he began studies in the College of Chemistry at the University of California at Berkeley, receiving a bachelor's degree in 1895. At Berkeley Erlanger performed his first research—studying the development of newt **egg**s. He then enrolled at the Johns Hopkins University School of Medicine in

Joseph Erlanger. *(The Library of Congress. Reproduced by permission.)*

Baltimore and earned a medical degree in 1899, graduating second in his class. This distinction allowed Erlanger to work as an intern in internal medicine for William Osler, the renowned physician and teacher.

In Baltimore, in the summer of 1896, he worked in the histology laboratory of Lewellys Barker, studying the location of horn cells in the spinal cord of rabbits; the following summer, he undertook a project to study the digestive process of dogs. This study led to Erlanger's first published paper in 1901, and to his appointment as assistant professor of physiology at Johns Hopkins by William H. Howell, one of America's most important physiologists and head of the department. He was later promoted to associate professor of physiology.

In 1904, at Johns Hopkins, Erlanger designed and constructed a sphygmomanometer—a device that measures **blood** pressure. Erlanger improved on previous designs by making it sturdier and easier to use. Later that year, he used the device to find a correlation between **blood pressure** and orthostatic albuminuria, wherein **protein**s appear in the urine when a patient stands. His last few years at Johns Hopkins were spent studying electrical conduction in the **heart**, particularly the activity between the auricles and the ventricles that is responsible for the consistent beating of the heart. Using a clamp of his own design, he was able to determine that a conduction blockage, or heart block, in the bundle of His, a connection between the auricles and ventricles, was responsible for the reduced pulse and fainting spells associated with Stokes-Adams syndrome.

In 1906, Erlanger left Johns Hopkins and moved to the University of Wisconsin, where he became the first professor of physiology at the university's medical school. The following year Erlanger left Wisconsin for the Washington University School of Medicine in St. Louis, where he worked for the remainder of his career, serving as professor of physiology and department chairman.

In 1917, the United States' entry into World War I presented him with the opportunity to return to the laboratory and to his research on cardiovascular physiology. He participated with other physiologists in the study of wound shock and helped to develop therapeutic solutions that were used by the U.S.Army in Europe. He also continued the work that he had begun at Johns Hopkins, studying the sounds of Korotkoff, the sound one hears in an artery when measuring blood pressure with a stethoscope.

In the early 1920s, Erlanger turned to neurophysiology. The arrival at Washington University of Herbert Spencer Gasser, a student of Erlanger's from Wisconsin and a fellow Johns Hopkins graduate, spurred this change. Erlanger and Gasser would collaborate at Washington University until Gasser's departure in 1931 for the Cornell Medical College. Understanding how nerves transmit electrical impulses preoccupied Erlanger and Gasser during the 1920s. The difficulty in studying nerves was that the electrical impulses were too weak and too brief to measure them accurately. In 1920, one of Gasser's former classmates, H. Sidney Newcomer, developed a device that would amplify nerve impulses by some 100,000 times, allowing physiologists to measure and study the subtle changes that occur during nerve transmission. A year later, Erlanger and Gasser, based on advances made at the Western Electric Company, constructed a cathode-ray oscilloscope that could record the nerve impulse. The cathode-ray oscilloscope with amplifier was a technological breakthrough that permitted neurophysiologists to overcome the barrier posed by the subtlety and brevity of nerve activity. Erlanger and Gasser went on to study the details of nerve transmission. Their most significant contribution derived from these researches was their conclusion that larger nerve fibers conducted electrical impulses faster than smaller ones. Also, they demonstrated that different nerve fibers can have different functions.

Erlanger's wholly American education, consisting of a full-time research effort, represented a new generation of American physiologists. For his scientific efforts, Erlanger was elected a member of the National Academy of Sciences, the Association of American Physicians, the American Philosophical Society, and the American Physiological Society. He also received honorary degrees from universities of California, Michigan, Pennsylvania, Wisconsin, and Johns Hopkins University, Washington University, and the Free University of Brussels. His highest honor came when he shared, with Gasser, the 1944 Nobel Prize for physiology or medicine. After his retirement in 1946, Erlanger continued to work part-time performing research and helping graduate students in their work. Erlanger died of heart failure on December 5, 1965, one month before his 92nd birthday.

ERYTHROCYTE

Erythrocytes, also known as red **blood** cells, contain the pigment **hemoglobin** which has the remarkable capacity to combine with and release oxygen. Human red blood cells contain a 33% solution of hemoglobin. Oxygen is transported to living **tissue**s of the body as oxyhemoglobin in red blood cells. The human red blood **cell** is a biconcave disc with an average of about 0.0003 inch (7.5 cm) in diameter. Erythrocytes are the most common cell type in blood (with an average of about 5,500,000 per ml in men and 5,000,000 per ml in women). Newborn babies have an even greater number of erythrocytes with as many as 7,000,000 per ml. Red blood cells are suspended in plasma which is the straw colored liquid part of the blood. The characteristic red color of blood is due to the erythrocytes. Human, and most mammalian erythrocytes, have nuclei while they develop in the bone marrow. The nuclei and some **cytoplasm**ic structures are lost as the red blood cell matures. The life span of an erythrocyte is about 120 days. Old cells are removed from the circulation by the spleen and bone marrow. The old cells are constantly being replaced by fresh new red blood cells.

The pathological condition of having too few erythrocytes, or erythrocytes containing too little hemoglobin, is known as anemia. Anemia can be caused by blood loss or by other conditions. Too many red blood cells is referred to as polycythemia and may occur as an **adaptation** to living in mountains to compensate for reduced oxygen in the air.

The mature erythrocytes of many lower **animal**s, such as many fish, amphibians, and birds, contain nuclei in contrast to humans. Frog erythrocyte nuclei are virtually inert. In a sense, that means that frog erythrocytes are physiologically quite similar to erythrocytes of higher animals. While it is true that **mammals** do not have mature nucleated red blood cells, the nuclei of comparable cells in frogs and other lower animals behave as if they were not there. Curiously, while frogs and fish tend to have small bodies, their erythrocytes are huge compared to humans.

ESTROGEN

Estrogen is one of the **sex hormones** produced by the ovaries in females and the testes in males. They are known as female **hormone**s because they are found in greater amounts in females than in males. It plays a dominant role in the female **reproductive system**. It influences growth, puberty, and regulates the reproductive cycle of menstruation and pregnancy. In addition, it affects many other body parts such as the bones, skin, **arteries**, the **brain** etc.

During puberty, estrogens stimulate linear growth and skeletal maturation. They cause behavioral changes and promote the development of both primary and secondary sexual characteristics. Primary female sexual characteristics include the internal and external genitalia; secondary sexual characteristics include breast development and the female pattern of fat deposition. Estrogen is not just one hormone, but a group of hormones of varying degree of activity. The three most important hormones of the estrogen group are estrone, estriol, and estradiol. Estradiol is the most potent and the most abundant estrogen hormone.

In premenopausal women, estrogen is produced at high levels. It regulates the growth of the women's secondary sexual characteristics; stimulates the growth of the uterine lining during the **menstrual cycle** and helps maintain the lining during pregnancy. However, after menopause the production of estrogen tapers off. This lack of estrogen may affect many of the body **tissue**s including the reproductive system, urinary tract, **heart**, **blood** vessels, bones, breast, skin and parts of the brain. Some women will have no signs or symptom of decreased estrogen, while others may have many symptoms such as hot flashes, disturbed sleep, mood changes, vaginal dryness, and urinary problems. Estrogen replacement therapy (ERT), where estrogen is administered therapeutically, either alone or in combination with the hormone progestin may yield many short- and long-term benefits to the menopausal women. In the short term, it will relieve symptoms associated with menopause such as hot flashes, mild **depression**, and changes in vaginal tissue. In the long-term, it could help increase calcium absorption and thus increase bone mineral density. Loss of estrogen due to menopause has been shown to contribute to the development of osteoporosis (weakening of the bones). Estrogen therapy may prolong life by preventing heart disease, probably by increasing ''good'' **cholesterol**, HDL. One of the risks of ERT is the threat of endometrial **cancer** (cancer of the uterus). However, in combination with another hormone progestin, there appears to be no significant increase in the risk of uterine cancer. Estrogen therapy may be administered either through pills that are taken orally or can be given via a transdermal patch. However, before making the decision to take ERT, each woman should discuss the risks and benefits with her own personal physician.

Estrogens are currently being used in the treatment of certain types of breast cancer in men and women and certain kinds of prostate cancer in men. This hormone has been found to have many beneficial effects on brain **cell**s, and brain chemicals. In the laboratory, it has been observed that brain cells form better connections in the presence of estrogen. Hence, the use of estrogen to treat Alzheimer's disease is being investigated in many clinical studies that are being carried out all over the country.

ETHOLOGY

The study of **animal** behavior under natural conditions is the science of **ethology**. It emphasizes the objective investigation of entire patterns of behavior in regards to function and evolutionary **adaptation** in an ecological context. Ethology encompasses all aspects of animal behavior, such as communication, foraging, **courtship**, sensory **perception**, **learning**, and orientation.

Although humans have been observing animal behavior for millennia, ethology has emerged has a relatively new sci-

ence in the latter part of the twentieth century. The principle founders of ethology were the three co-winners of the 1973 Nobel Prize in Physiology and Medicine: **Karl von Frisch**; **Konrad Lorenz**; and **Nikolaas Tinbergen**. Each scientist made great contributions to the study of the behaviors of particular animals and to the methodology of studying their behavior. Karl von Frisch conducted extensive studies of the sensory world of the honey bee, while Lorenz and Tinbergen made landmark contributions to the study of vertebrate behavior. Lorenz demonstrated the importance of descriptive work in behavior studies and identified several inborn or innate components of behavior. His most celebrated studies concern **imprinting**; a phenomenon of young animals fixating on particular objects or individuals during a brief critical time. Tinbergen developed novel experiments to investigate animal behavior and is especially noted for his study of key stimuli involved in communication.

Ethology is an integrated discipline encompassing neurobiology, **ecology**, and **evolution**. For example, ethologists would examine sensory and **nervous system**s of an **organism** for their role in communication among the **population** and ask of its adaptation and evolutionary significance. Several other disciplines, such as physiology, psychology, **cell**ular biology, **genetics**, and population biology, also interact with ethology. Ethological studies may focus on the individual organism and determine the specific processes of sensory and motor nervous systems. Examining the sensory receptors on a moth antennae or the neurological mechanisms of how a cricket sings typify this approach. At another level, this discipline focuses on the behavior among a population of a **species** such as sexual, territorial, or foraging behavior and how these enable the population to survive and adapt to their changing environment. This population approach to ethology has now been defined by Harvard biologist, **Edward O. Wilson**, as **sociobiology**, dealing with all aspects of sexual and social behavior, and behavioral ecology. Mechanisms of communication among nestmates in an ant colony or ranking the social status of individual **primates** would constitute a sociobiological study.

See also Behaviorism

EUBACTERIA

The eubacteria are the largest and most diverse **taxon**omic group of **bacteria**. Some regard this as an artificial assemblage, merely a group of convenience rather than a natural grouping. The eubacteria are all easily stained, rod-shaped or spherical bacteria. They are generally unicellular, but a small number of multicellular forms do occur. They can be motile or non- motile and the motile forms are frequently characterised by the presence of numerous **flagella**e. Many of the ecologically important bacteria responsible for the fixation of **nitrogen**, such as Azotobacter and Rhizobium, are found in this group.

The **cell** walls of all of these **species** are relatively thick and unchanging, thus shape is generally constant within groups found in the eubacteria. Thick cell walls are an evolutionary **adaptation** that allows survival in extreme situations where thinner walled bacteria would dry out. Some of the bacteria are gram positive whilst others are gram negative. One commonality that can be found within the group is that they all reproduce by transverse binary fission, although not all bacteria that reproduce in this manner are members of this group.

EUGENICS

Eugenics is the study of improving the human race by selective breeding. Its rationale is to remove bad or deleterious **gene**s from the **population**, increasing the overall **fitness** of humanity as a result. Campaigns to stop the criminal, the poor, the handicapped, and the mentally ill from passing on their genes were supported in the past by such people as British feminist Marie Stopes and Irish playwright George Bernard Shaw. In the United States in the early 1900s, enforced sterilization was carried out on those deemed unfit to reproduce.

One problem with practical eugenics is who decides what is a desirable characteristic and what is not. Both before and during the Second World War there was a eugenics program in place in Nazi Germany. Initially this was carried out on inmates of mental institutions, but very quickly the reasons for sterilization became more and more arbitrary. After the Second World War there was a backlash against eugenics and anthropologists such as Franz Boas and Ruth Benedict championed the opposite view, *tabula rasa* (blank slate). This view favored the theory that humans are born with an empty mind, which is filled by experience. American anthropologist Margaret Mead carried out work in Samoa that confirmed these ideas, demonstrating that people are programmed by their environment, not by their genes, and that providing a decent environment produces people who behave decently towards each other. Modern work suggests that it is a subtle interaction of both the genetic make-up and the environmental conditions that shape an individual.

Eugenics happens to a minor degree in modern society, most notably when a couple with **family** histories of genetic disorders decide not to have children or to terminate a pregnancy, based on genetic screening. In 1994 China passed restrictions on marriages which involved individuals with certain disabilities and diseases.

There is evidence that the practice of eugenics could never be truly effective. When one calculates the frequency of deleterious **allele**s in the population, it is found humans all carry at least 1% of alleles which if present in homozygous form would prove fatal. When scientists predict the effects that might be achieved by preventing all individuals possessing a given allele from breeding, it is found that the effect would be minimal. One problem associated with this prediction is the difficulty in detecting certain alleles when they are present in the heterozygous form.

EUKARYOTE

Eukaryote is the **cell** type found in **plants**, **animals**, **fungi** and protists. The cells of **organism**s in each of these four biological kingdoms have discrete **membrane**-bound nuclei and **organelle**s. For this reason, the cells and the organisms that contain them are commonly referred to as eukaryotes, or cells and organisms with true or genuine nuclei. Eukaryotes are distinguished from prokaryotic organisms such as **bacteria** that lack a discrete **nucleus**. The eukaryotic nucleus contains **DNA**, the carrier of genetic information in cells, associated with **protein** and arranged into structures called **chromosome**s that typically divide in a highly organized process called **mitosis**. The nuclear membrane or envelope serves to separate the two phases of **protein synthesis: transcription** (the assembly of an **RNA** messenger **molecule** complementary to a strand of DNA) in the nucleus, and **translation** (protein synthesis based on the messenger RNA pattern) in the **cytoplasm**. This separation gives eukaryotic cells greater control over development by determining which proteins are produced. Cells with identical DNA may develop differently based on which DNA successfully sends RNA transcripts to the cytoplasm. As a result, **differentiation** of **tissue**s and organs is possible even though the cells of various tissues and organs contain identical DNA.

See also Prokaryote

EUPHENICS

The term **euphenics** was coined in the early 1960s as an alternative or counterpart to the word **eugenics**. Euphenics means specifically intervention in human development at the molecular and **cell**ular level. The term has been used to refer to treatment of genetic diseases.

The first time the term euphenics was used in print was in a 1963 article in the British journal *Nature* by **Joshua Lederberg**, who won the Nobel prize in medicine in 1958. It appeared sporadically in several other British journals thereafter. Judging from its absence from the indices of many **genetics** and **molecular biology** textbooks, euphenics was still not a term in wide use by the late 1990s. However, as genetic testing and **genetic engineering** advanced sometimes startlingly in their capabilities, debate over their ethical use intensified.

''Eugenics'' was already freighted with such negative associations that the need for a different term was clear. After World War II, when the Nazis' horrific eugenics program was revealed to the rest of the world, eugenics essentially dropped out of scientific vocabulary. For example, the American Eugenics Society became the Society for the Study of Social Biology. As eugenics was strongly associated with racism, euthanasia, genocide, and forced sterilization, scientists lacked a suitable term to denote some of their work on human genetics. The 1990s saw even more of the human **genetic code** uncovered. Genetic testing had grown so sophisticated that couples known as carriers of genetic abnormalities, Tay-Sachs disease for example, could have their eight-cell *in vitro* **embryo**s tested before implantation. The **Human Genome Project**, begun in 1990, intensified the speed of new genetic discoveries, and sparked ever firier debate over the ethics of work in this area. Euphenics is thus a badly needed neutral term that allows scientists to discuss new genetic testing and treatments for genetic disease without invoking the heated response that "eugenics" engenders.

EUTROPHICATION

The process of heightened biological **productivity** in a body of **water** is call eutrophication. The major factors controlling eutrophication in a body of water, whether large, small, warm, cold, fast-moving, or quiescent, are nutrient input and rates of primary production. Not all lakes experience eutrophication. Warmth and **light** increase eutrophication, (which in Greek means "well nourished") if nutrient input is high enough. Cold dark lakes may be high in **nutrients**, but if rates of primary production are low, eutrophication does not occur. Lakes with factors that limit plant growth are called oligotrophic. Lakes with intermediate levels of biological productivity are called mesotrophic.

Many lakes around developed areas experience cultural eutrophication, or an accelerated rate of plant growth, because additional nitrates and phosphates (which encourage plant growth) flow into the lakes from human activities. Fertilizers, soil erosion and **animal** wastes may run off from agricultural lands, while detergents, sewage wastes, fertilizers, and construction wastes are contributed from urban areas. These nutrients stimulate the excessive growth of green **plants**, including **algae**. Eventually these plants die and fall to the bottom of the lake, where decomposer **organism**s use the available oxygen to consume the decaying plants. With accelerated plant growth and subsequent **death**, these decomposers consume greater amounts of available oxygen in the water; other **species** such as fish and mollusks thus are affected. The water also becomes less clear as heightened levels of chlorophyll are released from the decaying plants. Native species may eventually be replaced by those tolerant of **pollution** and lower oxygen levels, such as worms and carp.

While at least one-third of the mid-sized or larger lakes in the United States have suffered from cultural eutrophication at one time or another over the past 40 years, Lake Erie is the most publicized example of excessive eutrophication. Called a "dead" lake in the 1960s, the smallest and shallowest of the five Great Lakes was inundated with nutrients from heavily developed agricultural and urban lands surrounding it for most of the twentieth century. As a result, plant and algae growth choked out most other species living in the lake, and left the beaches unusable due to the smell of decaying algae that washed up on the shores. New pollution controls for sewage treatment plants and agricultural methods by Canada and the United States led to drastic reductions in the amount of nutrients entering the lake. Almost 40 years later, while still not totally free of **pollutants** and nutrients, Lake Erie is a biologically thriving lake, and recreational swimming, fishing and boating are strong components of the region's economy and aesthetic benefits.

See also Chlorophyll; Decomposition; Nitrogen; Phosphate group; Water pollution

EVANS, ALICE CATHERINE (1881-1975)
American microbiologist

Evans was born in Neath, Pennsylvania, on January 29, 1881. After completing her primary education and studying for one year at the Susquehanna Institute in Tonawanda, she began teaching in a local elementary school. Four years later, she decided to pursue a college education and entered Cornell University for a two-year nature study course. She became so interested in science that she decided to remain at Cornell for a bachelor of science degree in agriculture.

With the support of her bacteriology professor at Cornell, Evans received a scholarship to the University of Wisconsin and, in 1910, received a master's degree from that institution. She often thought about studying for a doctoral degree, but became so involved in her own research that she never attained that goal.

After graduation from Wisconsin, Evans accepted a research job with the Department of Agriculture, working first in Wisconsin and later in the Dairy Division Laboratories in Washington, D.C. One of her early projects was the study of bacteria in dairy products. She was intrigued to discover a close similarity between two types of **bacteria**, *Bacillus abortus* and *Micrococcus melitensis*. The former caused spontaneous abortions in cows and was not thought to be transmitted to humans. The latter organism caused a disease in humans originally known as Malta fever, but latter referred to as undulating fever.

Evans' research was of significance for two reasons. First, spontaneous abortion among cows was a serious problem for the dairy industry. The development of a vaccine for *B. abortus* eventually proved to be of enormous benefit to the industry. Second, the similarity of *B. abortus* to *M. melitensis* raised the possibility that, popular opinion to the contrary, harmful bacteria might be transferred from cows to humans.

In 1918, Evans discovered the first confirmed case in which **brucellosis**—a common disease in cows—was transmitted to humans. The agent responsible was yet a third microorganism, *Brucella suis*. Evans was able to demonstrate that this serious disease in cows could be transmitted to humans. Bacteriologists were at first dubious of Evans's work. Some rejected her findings on irrelevant grounds: that she was a woman or that she had no Ph.D., but others wondered how the transmission of brucellosis from cow to human had escaped scientists' notice for so long.

Gradually the answer to that question became clear. First, Evans began to collect more and more reports of brucellosis in humans from countries around the world. Apparently the disease was more common than anyone had imagined. Second, she discovered that the disease can occur in two forms: *acute*, in which the symptoms occur quickly and are easy to recognize, and *chronic*, in which symptoms develop slowly and over many years. It soon became obvious that chronic brucellosis was relatively common among families exposed to raw milk, but that it was typically diagnosed as another condition.

Evans herself developed chronic brucellosis as a result of her research. She experienced a number of devastating attacks of the disease, but survived them all. She lived a long and productive life, serving in 1928 as the first female president of the American Society for Microbiology. She died in Alexandria, Virginia, on September 5, 1975, at the age of 94. As a result of her work, vaccination of cows and pasteurization of milk are now routine and are responsible for dramatic declines in both bovine and human diseases.

EVOLUTION

Evolution is the gradual, cumulative change over time of the characteristics of groups of **organisms**, in a heritable manner. Eventually, these minute changes add up to produce an individual that is markedly different from its distant ancestors, but almost indistinguishable from its most immediate ancestors. These changes are brought about by the organism's response to the environment, and, over the entire course of the history of life on earth, evolution has given rise to all the different forms of life that have ever existed.

Evolution does not work on the individual unit of life; changes are too small and too slow to be effective at that level. In fact, evolution works at the **population** level—in other words, among groups of organisms that are capable of successfully breeding with each other. With organisms that do not breed with other individuals, the rate of **evolutionary change** is slower than it is among **outbreeding** organisms.

Evolution leads to increasing complexity and, eventually, to the production of new **species**, which can survive or become extinct depending upon their reaction to the environment and its continuing changes. Evidence for evolution comes from the **fossil** record, **genetics**, and comparative studies.

The mechanism behind evolution is **natural selection**. Small, individual changes that arise by chance can confer an advantage to those possessing them; this group then has better success at breeding, and their **gene**s are consequently spread further throughout the population. The theory of evolution is now widely accepted, but when it was first put forward in the nineteenth century by English naturalist **Charles Darwin** there was much opposition, particularly from religious quarters. The opposing theory to evolution is the theory of special creation, which states that each type of species was created in the form in which it currently exists, and that no two species are related, by descent, to any other. Most scientists now accept the theory of evolution, although it must be stressed that this is still a theory. The theory of evolution fits all of the available evidence, which is a large amount, and to overthrow this theory, a large amount of supposition and work would have to be disproved. It is true that there are many gaps in scientific knowledge, such as the missing link, which is the common ancestor for both apes and humans, but as time goes on these gaps in our knowledge become smaller.

Evolution does not proceed at a constant rate; at times, a gradual change occurs that allows for a good reconstruction of the process form the fossil record. This is known as phyletic **gradualism**. The other method of evolution, which can leave gaps in the fossil record is the quicker and more explosive form, called **punctuated equilibrium**.

See also Creationism; Survival of the fittest; related essays on Evolution

EVOLUTION, CONVERGENT

In ecologically similar **habitat**s it can sometimes happen that apparently similar **organism**s arise. These organisms may be entirely unrelated, their similarities coming about as a result of the same **selective pressures** being applied to each set of organisms. These selective pressures may produce similar structures in unrelated **species**, for example the wings of all flying **animals** are very similar because the same laws of aerodynamics apply governing what makes an efficient shape for a wing, irrespective of the animal involved or the physical location. With species of **plants** which share the same pollinators it can be seen that analogous structures and methods of attracting the pollinating species to the plant are similar, this is because the plants are operating under the same set of selective pressures.

Convergent evolution creates problems for those using **evolution**ary patterns to try to answer questions relating to **taxonomy**, in particular to those studying cladistics. It can provide evidence of false relationships and incorrect evolutionary pathways.

Convergent evolution is sometimes called convergence.

See also Evolution, divergent

EVOLUTION, DIVERGENT

Divergent evolution is also known as divergence. This is the opposite of convergent evolution.

Starting from a point at which there is one basic form of **organism**, then when different **selective pressures** are placed on that one organism type, that one form can produce a wide variety of new types. If only one structure on the organism is considered, these changes can either add to the original function of the structure or they can change it completely. Divergent evolution is one of the first steps in **speciation**, the production of a new **species**. Divergence can be seen when looking at any group of related organisms; the differences are produced from the different selective pressures under which the life forms live. Any **genus** of **plants** or **animals** can show divergent evolution, for example, the diversity of floral types in the orchids. The greater the number of differences present, the greater the divergence. Scientists speculate that the greater the divergence, the greater the length of time since the organisms in question shared a common ancestor.

Divergent evolution can occur quite easily in nature. If a freely interbreeding **population** on an island is separated by the appearance of a barrier to breeding, such as the presence of a new river, then, over time, the organisms may start to diverge and show differences. If the opposite ends of the island have different pressures acting upon them this pressure will produce divergent evolution. When the organisms were capable of freely interbreeding the differences caused by the different ends of the island would be averaged out over the whole population and not be discernible.

EVOLUTION, EVIDENCE OF

Evidence of evolution can be observed in a number of different ways, including distribution of **species** (both geographically and through time), comparative anatomy, **taxonomy, embryology, cell** biology, and **paleontology**.

English naturalist **Charles Darwin** formulated the theory of organic **evolution** through **natural selection** in his ground breaking publication *The Origin of Species by Means of Natural Selection,* published in 1859. One of the first pieces of evidence that started the young scientist thinking along evolutionary lines was given to him on his journeys aboard the *HMS Beagle* as a naturalist. Prior to the work of Darwin, most people accepted the biblical account of creation where all **animal**s and **plants** were brought into the world fully formed and immutable. Darwin made extensive collections of the plants and animals that he came across wherever the ship stopped, and very soon he started to notice patterns within the **organism**s he studied.

There were similarities between organisms collected from widely differing areas. As well as the similarities, there were also striking differences. For example, **mammals** are present on all of the major landmasses, but one does not find the same mammals even in similar **habitat**s. One explanation of this is that in the past when the landmasses were joined, mammals spread over all of the available land. Subsequently this land moved apart, and the animals became isolated. As time passed, random variation within the **population**s was acted upon by natural selection. This process is known as **adaptive radiation**—from the same basic stock, many different forms have evolved. Each environment is slightly different, and slightly different forms are better suited to survive there. An example of this, which is seen at a formative stage, is the case of the finches on the Galapagos Islands. All of the Galapagos finches bear similarities to the mainland finches, but each species has evolved to fill a particular **niche**, which is not already filled by an animal on the islands even though there are species filling these ecological openings on the mainland.

If it is true that widely separated groups of organisms have ancestors in common, then logic dictates that they would have certain basic structures in common as well. The more structures they have in common then the more closely related they must be. The study of evolutionary relationships based on commonalties and structural differences is termed comparative anatomy. What scientists look for are structures that may serve entirely different functions but are basically similar. Such homologous structures suggest a common ancestor. A classic example of this is the pentadactyl limb, which in suitably modified forms can be seen in all mammals. A greater modified version of this can also be seen amongst birds. This limb has been used by different groups for slightly different purposes and so provides an example of divergent evolution.

These evolutionary relationships are reflected in taxonomy. Taxonomy is an artificial, hierarchical system showing relationships between species. Each level progressed within the taxonomic system denotes a greater degree of relatedness for the organism in that group to the level above.

In embryology, the developing fetus is studied, and similarities with other organisms are observed. For example, annelids and molluscs are very dissimilar as adults. If, however, the **embryo** of the ragworm and the whelk are studied, one sees that for much of their development they are remarkably similar. Even the **larva**e of these two species are very much alike. This adds evidence to a past recent common ancestor. It is not however true that a developing organism replays its evolutionary stages as an embryo; there are some similarities with the more conserved regions, but embryonic development is subjected to evolutionary pressures as much as other areas of the **life cycle**.

Cell biology is an area where many similarities can be seen between organisms. Many structures and pathways within the cell are vital for the continuance of life. The more important and basic to the whole structure of life a pathway or **organelle** is, the more likely it is to be observed. For example, the **DNA** code is the same in virtually all living organisms as are such structures as **mitochondria**. These are virtually ubiquitous throughout known life. It is inconceivable that each of these things can have arisen separately for each species of living organism. The conclusion that must be drawn is that they all arose from the same basic source many millennia ago. The examples visible today have survived, and the organisms carrying these processes on have adapted, yielding the diversity of forms seen today.

Perhaps the most persuasive argument in favor of evolution is the **fossil** record. Paleontology (the study of fossils) provides a clear record that many species no longer exist. By such techniques as **carbon dating** and studying the placement of fossils within the ground, an age can be given for fossils. By placing fossils together based on their ages, a gradual change in form can be surmised which can be followed and extrapolated to the species that exists today. It must be stressed that the fossil record is very incomplete, and many intermediate species are missing, but the weight of evidence from those that do exist favors the theories of evolution and natural selection.

Evolution is a theory. As such it must have evidence to support it, and, if new evidence appears to overthrow it, then a new theory must be formulated. The amount and quality of evidence in favor of evolution is overwhelming. Any evidence requiring a totally new theory would have to be staggering in its scope and strength. The new evidence that has been forthcoming supports the theory of evolution and merely fine tunes scientific understanding of the mechanisms involved.

See also Creationism; Survival of the fittest; related essays on Evolution

EVOLUTION, HUMAN • See Human Evolution

EVOLUTION, PARALLEL

Parallel evolution is a system whereby different **populations** can, even though they are in widely different areas, follow the same course of **evolution** to produce the same **organisms** at the end of the process. As evolution is based on random changes of **genetic material** interacting with the environment, the idea of two identical **species** evolving in two separate areas is clearly remote. Parallel evolution is more commonly observed between structures. For example, a large number of organisms from different evolutionary backgrounds that now live in the same environment have the same structures. The sea is an example of just such an environment, and when one studies the **animals** present within the sea, there are many similarities. Fish and **mammals** living in the sea both use their tails for propulsion. This is actually an example of convergent evolution. Examining fish alone, it can be seen that in different areas of the world they have independently moved into very deep **water habitats**. Due to the initial starting stock being similar, and the evolutionary pressures the same, **adaptations** such as loss of eyes and coloration have occurred independently in many locations. This is parallel evolution. Similar examples can be found among all living organisms, for example, among **plants** that develop the same types of structures to attract pollinators and indeed among the pollinators themselves. Without knowing something of the history of the organisms under study, observers have difficulty distinguishing between parallel and convergent evolution.

See also Natural selection; Survival of the fittest; related essays on Evolution

EVOLUTIONARY CHANGE, RATE OF

The rate of evolutionary change is the speed at which new **species** arise. It is a long process that is not observable within a human lifetime. **Evolutionary** change can only be observed by examining **fossils** and species that are related to each other. The rate of change is governed by the generational turnover of the species under examination, fast lived species are capable of changing more quickly than those that have a long life span and reproduce less often.

Even fast-lived species such as **bacteria**, some of which have generation times which can be measured in minutes, cannot be said to have shown evolutionary change in the time that humans have been observing them. Some **virus**es such as HIV are often reported as being present in new forms, but they are still recognizable as HIV. Over sufficient time, there will be enough variation and divergence to say that a new disease has arisen, but again this has never been shown to occur within the time span that humans have been capable of observing.

One of the basic building blocks of evolution is variability. The raw material of variability is a change in **DNA**, known as **mutation**. For a mutation to have any effect, it has to be passed on to the next generation. Differences present within **populations** in relation to their environment are exploited. At times, the changes that are produced can occur relatively quickly; but at other times such as when the population is under stable conditions, there may be no factors selecting for any change, and the population characteristics remain constant.

One technique that has been used to examine the rate of evolutionary change is DNA analysis. To do this, the percentage similarity between samples of DNA from the **organisms** under study is examined. The greater the similarity, the more recently the organisms are believed to have diverged from a common stock. The information that is obtained in this manner agrees with information obtained from other sources such as the fossil record and comparative anatomy studies.

At certain times the rate of evolutionary change can be very rapid, leaving little fossil evidence of intermediate forms. This method of evolution is known as **punctuated equilibrium**. The other, more gradual approach, is known as phyletic **gradualism** (phyletic meaning evolutionary).

It must be born in mind that life has been present on this planet for some 3,000 million years, with a fossil record extending back for 1,000 million years. The time that has been available for these processes to work is vast by human reckoning.

Eutherian **mammals** have been in existence for 60 million years, and recognizable humans have been in existence for one million years. This gives some indication of the lengths of time necessary for changes to occur. The rate of evolutionary change is governed by the interaction of the rate of mutation present in the population and the rate of change in the environment.

See also Related essays on Evolution

EVOLUTIONARY THEORY

The concept of evolution first emerged in the early nineteenth century. Up until that time, the post-classical Western world had generally regarded the account of creation provided in the Biblical book of Genesis as literal fact. Thus, it was held that the world had been created relatively recently, that every **species** was created separately and distinctly, and that these species had remained unchanged over the centuries.

As the eighteenth-century scientific **community** began attempting to classify **plants** and **animals** systematically, however, the immense diversity and interrelatedness of living things cast doubt on this traditional model. The plant and animal **fossils** sought out by excavators were particularly at issue, since they implied that the history of earth extended back much further than previously thought, and that life developed only gradually and unevenly from simple to advanced **organisms**.

In 1809, the French botanist **Jean Baptiste de Lamarck** made an important attempt to formulate an explanation for this apparent complexity of organic life. Although he did not use the term ''evolution,'' Lamarck argued that species gradually progressed over time from simpler to more complex types. To account for this change, Lamarck proposed that organisms possess an innate drive toward perfection and an ability to adapt to their environment.

Lamarck further believed that acquired characteristics could be passed on from generation to generation. For example, the ancestors of the giraffe, in reaching high leaves to eat, would have stretched and elongated their necks; this trait was then passed on to their descendants. Although this belief in the transmission of acquired traits has since been discredited, Lamarck was correct in assuming that traits could somehow be inherited, and that this process could lead, over long periods of time, to significant evolutionary changes.

It was the English naturalist **Charles Darwin** who initiated the understanding of evolution (in Darwin's own terms, "descent with modification") held today. Darwin's theories were based on the geological and biological studies he conducted in the 1830s during travels in South America and the South Sea islands. Darwin's detailed observations of plants and animals provided a considerable body of evidence for evolutionary change. Fossils indicated that some species had become extinct, although clearly related species survived. Darwin also recognized that similar but not identical species were now found in different geographical regions. In the Galapagos Islands, furthermore, Darwin observed that the beaks of different types of finch birds were evidently adapted to the food supplies of the geologically distinct islands they populated. Similar **adaptation**s characterized the local species of giant tortoises.

Synthesizing his observations—ultimately presented to the world with the publication of *On the Origin of Species* in 1859—Darwin postulated that species were mutable and divergent rather than fixed in form. Darwin recognized that many species had common ancestors. Thus, lizards and rabbits are similar in their embryonic stage, and the bones in human arms and legs correspond to those in the limbs of dogs or horses. To explain these parallels, Darwin proposed that species had branched and evolved through **natural selection.**

Natural selection implies that, given the **competition** for limited resources such as food, those organisms best adapted to their specific environment are most likely to survive, reproduce, and transmit their traits to offspring. This process, occurring in conjunction with environmental changes, causes the advantageous traits of members of a particular species to predominate and the disadvantageous traits to be lost. Over thousands of years, certain forms of life accumulate enough changes for an apparently new and distinct species to emerge.

Darwin's theories, though bold and controversial, took place within the context of a dynamic nineteenth-century scientific and intellectual community. Thus, Darwin's *Origin of Species* was modeled on **Charles Lyell**'s similarly groundbreaking *Principles of Geology*, published in the 1830s. **Alfred Russel Wallace**, working from independent studies in the East Indies, formulated his own theory of natural evolution contemporaneous with Darwin's.

It was Wallace who coined the phrase "the **survival of the fittest**"—a phrase that does not signify that the strongest species will necessarily flourish, but rather that those species best "fitted" to their environments have the best chances of survival. In addition, prominent intellectuals such as Herbert Spencer (who popularized the term "evolution") and T. H.

Huxley (called "Darwin's Bulldog" for his promotion of Darwin's work) were vital in the process of debating, elaborating, and circulating Darwin's theories.

It was the presence of this community that allowed Darwin to publish his second important book, *The Descent of Man*, in 1871. This publication, which exposed Darwin to considerable religious opposition and public scorn, followed through on the implications of his earlier work by asserting that human beings had descended from, and were biologically related to, earlier life forms.

Even as the scientific community came to accept Darwin's principles of descent with modification and evolution through natural selection, they recognized that Darwin had failed to provide an adequate explanation of how biological variations were produced or passed on. Without this component, critics argued, the theory of natural selection accounted for very little.

This issue was essentially resolved by the Austrian botanist **Gregor Mendel**, who in the 1860s isolated the basic unit of **heredity** now known as the **gene**. Working in the garden of the monastery he had entered in 1843, Mendel discovered that differences among varieties of common garden peas (such as variations in shape, size, flowering, coloring, and seed characteristics) were due to paired units of heredity. Mendel was the first to discover that these molecular "blueprints," passed on from parents to offspring, determined which features a living organism will inherit. Mendel received credit for this contribution to evolutionary theory only in the early 1900s, when his work was confirmed by independent scientists such as **Hugo de Vries**. Mendel's discoveries, nonetheless, are now incorporated into our understanding of evolution.

The integration of Darwin's descent with modification and Mendel's work in **genetics** was substantially achieved by the Russian-American scientist **Theodosius Dobzhansky**, who published his influential book *Genetics and the Origin of Species* in 1937. Dobzhansky's experiments with fruit flies clarified the process of evolutionary adaptation by demonstrating the variability of genes and thus the possibility of rapid evolutionary change through genetic **mutation** and **recombination**. "Neo-Darwinism," as the synthesis of Darwinism and Mendelism genetics is often called, thus recognizes that evolution involves not only physical and behavioral traits, but also the genes that serve as the basis for those traits. This process of genetic evolution is known as genetic drift.

Continuous and substantial advances have been made in the study of evolution in the twentieth century. Evolutionary theory, in fact, has served as a point of reference for various specialists working within a complex network of scientific and mathematical disciplines that rely on increasingly sophisticated methodologies. **Walter S. Sutton, Theodor Boveri, Wilhelm Johannsen, Thomas Hunt Morgan,** and **Hermann Muller** have investigated the complex relationship between **chromosome**s, genes, and the laws of heredity. Biometricians such as **Ronald Fisher, John Haldane,** and **Reginald Crandall Punnett** have used mathematical and statistical techniques to analyze genetic changes, thereby establishing the field of **population genetics.** Julian Huxley, the grandson of T. H. Huxley, made important contributions to the field of **embryology**, among other areas. The paleonthologist George Simpson focused on the intercontinental **migration** patterns of ancient species. **James Watson and Francis Crick** introduced a model for **deoxyribonucleic acid** (**DNA**) to explain the chemical basis of genes, heredity, and evolution.

Today biologists continue to study the patterns, mechanisms, and pace of evolution, leading to some proposed changes in the original theory. For example, Darwin's model of evolution was based on **gradualism**, which assumes slow, steady rates of change. A newer model, called **punctuated equilibrium**, proposes that change may occur in relatively quick bursts, followed by longer periods of stasis. Rather than undermining evolution, such changes just underscore the vitality of evolution and the theory that explains it.

EXCRETION

Excretion is a process by which **animal**s remove **nitrogen** containing waste products from their bodies. Other forms of waste removal that are not covered under this definition of excretion include **blood**-gas transport, which involves the removal of **carbon dioxide** waste, and body-**temperature regulation**, which considers the removal of heat. These topics are examined in discussions of **respiration** and thermoregulation.

The basic physiological problem that excretion has evolved to overcome is the problem of maintaining a consistent internal environment. This process, known as **homeostasis**, involves the removal and gain of equal amounts of material. The most important compounds involved in excretion are **water** and nitrogen. Water regulation is crucial to maintaining osmotic pressures in the **cell**. When too much is present, a cell will burst. If too little is present, the **chemical reaction**s in the cell will not proceed correctly. Nitrogen waste is a result of the chemical breakdown of **protein**s. These wastes can be toxic to animals and must be removed.

A variety of excretory structures have evolved in the animal **kingdom**. In lower **order organism**s such as **protozoa** and sponges, the primary excretion structure is a contractile vacuole. Water enters the vacuole **membrane** and is incorporated into the cell. When the membrane is digested, the material inside becomes part of the **protoplasm**. If the material is unneeded, the membrane diffuses to the outer membrane of the cell and is expelled.

In most multicellular animals, the primary excretion structure is the transport **epithelium**. This is a collection of cells, or an opening in the body that leads directly to the surface. In most animal **excretory system**s, the transport epithelia are organized in tubular networks which have a large surface area. The simplest of these transport structures is the nephridia.

Flatworms have a simple excretory system called protonephridia. The protonephridium is a network of tubes that go throughout the body. At the end of each is a bulb which is surrounded by **cilia**. Wastes drain from these tube to the excretory ducts which then expel the waste to the surrounding via openings called nephridiopores. Worms have another type of tubular excretory system known as the metanephridium. Each segment of the worm has a pair of metanephridia which are soaked in the coelomic fluid. Nitrogenous wastes remain in the metanephridium and are eventually secreted. Insects have a different type of excretory structure called malpighain tubules. These tubules remove nitrogen waste from the hemolymph fluid. Eventually, the waste may be are eliminated as nearly dry **matter**.

In vertebrates, most have kidneys which contain segmented excretory tubules. Kidneys employ three basic steps in urine formation including filtration, reabsorption and **secretion**. When blood travels through the kidney, it is filtered. The crucial compounds such as salts, **amino acid**s, **carbohydrate**s, and a small amount of water are reabsorbed. The rest of the material is waste which is secreted. Since most animals can not tolerate a buildup of **ammonia**, the nitrogen waste is converted to urea. Urea and water are the major components of urine which is the final product of kidney processing.

EXCRETORY AND URINARY SYSTEMS

The excretory system in all types of **organism**s has one main function: to maintain **homeostasis** within a given organism. Homeostasis is that condition in which all internal systems and chemicals of that organism are in consistent balance. In order to maintain homeostasis, an excretory system has to be capable of three main functions: it must be able to rid the organisms of waste products; it must be able to keep both the fluid and the salt content of the organism within normal parameters; and it must keep the concentration of other substances in body fluids at normal levels.

The primary waste products of all organisms are **carbon dioxide**, **nitrogen**-based substances (created by the breakdown of **protein**s into **amino acid**s), and **water**. Carbon dioxide and some water **excretion** is performed by the **respiratory system**. Nitrogen and water are processed and released (excreted) by the excretory/**urinary system**.

Sponges and cnidarians are examples of very simple organisms which somehow manage to regulate fluid and waste without the benefit of any excretory structures. Even **invertebrate**s such as flatworms have simple excretory structures, called nephridial organs. These nephridial organs are simple tubes. Waste from body **cell**s crosses through the wall of these tubes via a process called **diffusion**. The waste is then conducted down the tubes with help from **cilia** (tiny hair-like projections which beat regularly, aiding in waste movement). The waste is expelled out of the organism through excretory pores (tiny openings to the outside).

As organisms become more complex in body structure, so too does their excretory system. Even earthworms have kidney-like structures which allow wastes to be removed from **blood**, along with appropriate quantities of water, salts and sugars. These organs also allow the earthworm's body to retain water, salt, and sugars as needed to ensure homeostasis.

Vertebrates have well-developed excretory systems, which include urinary systems for the collection, storage, and ultimate passage of nitrogen and salt-containing urine from the body. These systems vary according to the specific needs of the vertebrate organism. For example, **animal**s which live on dry land (especially **desert**-dwelling animals) need to have the opportunity to retain as much fluid as possible. Therefore, their excretory systems are designed to allow these organisms to retain appropriate quantities of fluid.

The human excretory system consists of the kidneys (which filter blood received from the renal artery, a branch off of the descending **aorta**); the ureters (a tube which leaves each kidney, to conduct urine into the bladder); the bladder (a distensible, balloon-like bag which stores urine); and the urethra (a tube which allows urine to leave the body). The human kidneys are complex organs made up of individual units called nephrons. Nephrons do the major work of filtration, removing nitrogen products (in the form of urea), and excess salts and water from the blood. When the body needs salts and water, the nephrons are capable of reabsorbing these substances. The salts and waters then re-enter the bloodstream.

Nephrons are complex structures, with many parts. These parts are sensitive to various **hormone**s which signal the cells within the nephrons to allow more or less water or salts to be released, based on the body's need for those substances. The major hormones responsible for adjusting the nephrons handling of water and salts are called anti-diuretic hormone (ADH) and aldosterone.

ADH is secreted by the **hypothalamus** gland in the **brain**. When the hypothalamus senses that the blood is getting to concentrated, it releases an increased amount of ADH. The ADH goes to the cells of the nephron, signalling them to retain more fluid within the body, releasing less as urine.

Aldosterone is released by the **adrenal glands**, little glands which sit atop each kidney. When the adrenal glands sense that the blood contains to little salt, they secrete aldosterone. Aldosterone directs the cells of the nephrons to hold onto more salt, allowing little to escape in the urine.

Kidney failure is a serious, life-threatening problem which has numerous causes, including infection, poisoning, tumors, shock, circulatory disorders, immune disorders. When the kidneys cannot work properly to maintain homeostasis, severe derangements of body chemistry result. Fluid may be retained, causing swelling (edema) throughout the body; the body becomes increasingly, dangerously acidic; and poisonous nitrogen waste-products accumulate. The end result of untreated kidney failure is coma and **death**. Treatment for this devastating disorder includes a procedure called dialysis. In dialysis, a patient's blood is sent through a tube into a complex system of plastic tubes, which act as an artificial kidney, filtering the blood and allowing urea and other wastes to be removed.

EXERGONIC AND ENDERGONIC REACTIONS

Exergonic refers to **chemical reaction**s that proceed spontaneously from reactants to products with the release of **energy**.

Endergonic reactions require energy input to proceed. Although the terms are often used rather loosely, they are precisely defined thermodynamic concepts based on changes in an entity called Gibbs free energy (G) accompanying reactions. Reactions in which -G decreases are exergonic, and those in which -G increases are endergonic. **Exergonic reaction**s often involve the breakdown of **organic compound**s found in food, whereas endergonic reactions frequently entail synthesis of complicated **molecule**s. Biological **metabolism** contains many examples of both types, and living **organism**s have developed elaborate techniques for coupling the two.

Although a negative -G indicates that energy must be added to the system before a reaction will occur, it tells us nothing about the rate at which it will progress. As is often the case, it may go very slowly if substantial activation energy is required to start the reaction. Living organisms have found a way around this problem by forming **protein** catalysts, called **enzyme**s, that effectively reduce the amount of activation energy needed, and allow the reaction to proceed at a satisfactory rate. Enzymes do not affect the free energy of the reaction, and will not enable reactions to proceed that are not energetically feasible.

By coupling exergonic and endergonic reactions, organisms are able to use the available energy in food they consume to construct complex proteins, **lipids**, **nucleic acids** and **carbohydrate**s needed for their growth and development. A well-known example involves coupling the formation of energy-rich **adenosine triphosphate (ATP)** from adenosine diphosphate (ADP) and phosphate (an endergonic reaction), with the transfer of hydrogen, removed from organic food materials, to oxygen (an exergonic reaction). The process is called oxidative **phosphorylation**. Energy stored in ATP may be used subsequently when the exergonic conversion of ATP back to ADP and phosphate is coupled with the endergonic synthesis of a needed **cell**ular component.

EXONS AND INTRONS

Exons and introns refer to specific **nucleotide** base sequences in the **genetic code** that are involved in producing **protein**s. Exons are the **DNA** bases that are transcribed into mRNA and eventually code for **amino acid**s in the proteins. Introns are DNA bases which are found between exons, but are not transcribed. **Gene**s which contain introns are known as interrupted genes.

The discovery of introns and exons occurred in 1977 independently by **Richard Roberts** and **Phillip Sharp**. Prior to this time, it was thought that **eukaryote** genes were strange because they contained more DNA than was transcribed into **RNA**. These men ran experiments which attempted to identify DNA from the resulting mRNA. It was assumed that the mRNA would have the same base sequence as the DNA from which it was transcribed. This was not the case however. They found stretches of DNA sequences that were not part of the mRNA. Further, these sequences were interspaced between coding sequences thereby interrupting the code. These data led to the description of exons, the coding DNA, and introns, the interrupting DNA. For their work, Roberts and Sharp shared in the 1993 Nobel Prize in physiology.

In most living systems, genes are made up of nucleic acid sequences which are translated into mRNA and then into proteins. In eukaryotic **cell**s, the interrupted genes have a sequence with four different regions including a regulatory region, exons, introns and a stop sequence. The proportion of interrupted genes varies with each **organism**. Simple organisms such as **yeast**s, have very few interrupted genes. In more complex organisms like **mammals**, almost every gene has an intron.

While each section is important, the exons are the sequences that actually code for proteins. The number of exons that code for a protein vary. Some proteins may have three or four exons but others can have 30 or 40. These differences are found in most **species**.

The introns are interspaced between the exons in an alternating fashion. Their nucleic acid sequence is highly variable and can be as short as a few dozens bases or as long as a few hundred. One thing that is constant however, is the sequences at the beginning and end of the intron. These capping sequences are of importance when the gene is transcribed into mRNA.

When a gene is transcribed into RNA, the entire sequence, including the introns, is copied. This primary transcript of RNA is further processed to produce the protein coding mRNA. In this process, the exons are spliced together by a series of **enzyme**s. First, the ends of exons are brought together. Then the introns are removed and the exons are chemically bonded.

Since introns do not get translated into proteins, scientists have tried to determine how and why they evolved. While theories about their **evolution** are complex, it is generally believed that on primeval Earth both introns and exons existed and were formed by the random combination of nucleotides. Exons helped code for proteins so they were incorporated into living organisms. Introns were also incorporated, perhaps randomly, and are thought to play a regulatory role in cell activity.

EXTINCTION

Extinction is the termination of an entire **species**. It occurs when a species can no longer reproduce at replacement levels. Extinction usually occurs as a result of major environmental or climactic changes. The affected species responds to the change in one of three ways: (1) by adapting and successfully continuing to prosper at current levels; (2) by not adapting and quickly perishing; or (3) by adapting in such a way as to become a distinctly different species. Another cause of extinction is the effect of humans on the environment. **Habitat** destruction, hunting and **pollution** have become significant factors in plant and **animal** species survival in the last few millennia. The introduction of new species may also serve as a major stress to indigenous **organism**s when the new organism competes for food and/or space.

Extinction is an ongoing feature of the Earth's everchanging **ecosystem**. Most of the plant and animal species ever to have lived on Earth are now extinct. The **fossil** record reveals the occurrence of a number of mass extinction events, each involving the destruction of large numbers of different species. The most well known mass extinction occurred at the end of the **Cretaceous** period, some 66,000,000 years ago. It was at this time that almost half of all known species disappeared, including the dinosaurs and marine animals. In order to determine the cause of this mass extinction, geologists study the sediment from a layer of rock that marks the boundary between Cretaceous and Tertiary time periods called the K-T boundary.

The asteroid theory was first postulated by American geologist Walter Alvarez. Together with his father, Nobel prize-winning physicist Luis Alvarez, Walter analyzed sediment he collected in the 1970s from the K-T layer near the town of Gubbio, Italy. The sample showed a high concentration of the element iridium, a substance rare on Earth but common in meteorites. Other samples of K-T boundary strata from around the world were also analyzed and excess iridium was found in these samples as well. Using the average thickness of the sediment as a guide, they calculated that a meteorite would require a diameter of about 6 mi (10 k) to produce that much iridium.

If a meteorite that size had hit Earth, the dust thrown up in to the air would have produced an enormous cloud of dust that would have encircled the Earth and blocked out sunlight for many months, possibly years. This climactic change would have severely depressed **photosynthesis**, resulting in the **death** of many **plants**, the subsequent death of **herbivore**s, and finally the death of their predators as well. This chain of events would have occurred so rapidly that there would have been no chance for **adaptation** to the new environment. A major problem with this theory, however, was that a 10-k meteorite would leave a very large crater, between 93-124 mi (150-200 k) in diameter. While Earth has many impact craters on its surface, few are even close to this size.

Because 65 million years had passed since the hypothetical impact, scientists decided to shift the search underground. A crater that old would almost certainly have been filled in by now. In 1992 an impact crater was discovered under the surface near the village of Chicxulub in Mexico's Yucatan Peninsula. When core samples were analyzed, they showed the crater to be about 112 mi (80 k) in diameter and 65 million years old.

The asteroid theory is widely accepted as the most probable explanation of the K-T iridium anomaly, but many geologists still debate whether the impact of this large meteorite was the single cause of the mass extinction of the dinosaurs and other life forms at that time.

The Great Ice Age that occurred during the Pleistocene era (1.6 million-10,000 years ago) also caused the destruction of many plants and animal species. This was a time of extremely diverse animal life, and **mammals** were dominant (in contrast to the reptiles of previous periods). In the late Pleistocene (50,000-10,000 years ago), several other extinction events occurred. These events wiped out several species known collectively as megafauna, including mammoths, mastodons, ground sloths, and giant beavers. The late Pleistocene extinction of megafauna did not occur all at once, nor was it of equal magnitude throughout the world. However, the continents of Africa, Australia, Asia and North America were all affected.

The causes of the extinction events of the late Pleistocene are still debated; however, the arrival of humans seems to have been a major factor. Indeed, the past 10,000 years have seen dramatic changes in the **biosphere**. The invention of **agriculture** and animal husbandry and the eventual spread of these practices throughout the world has allowed humans to utilize a large portion of the available productivity of the Earth. Calculations show that humans currently use approximately 40 percent of the **energy** produced by the Sun. Use of such an inordinately large proportion of the Earth's principal energy source by a single animal species is unique in the history of the planet.

Extinction remains a threat to many species of plants and animals, but the time to extinction is greatly accelerated due to the human factors of habitat destruction, pollution and over-harvesting. Today many countries, especially the United States, are passing laws to protect species that are threatened or in danger of extinction. Of the many natural causes of extinction, human activity is the single most effective and yet also the most manageable. Laws have been passed in many countries that protect wild areas from destruction, and individual species from hunting and harassment. The role of the zoological park has changed from one of simple exhibition to a more activist, wildlife **conservation**ist stand. Zoos are banding together worldwide to improve conditions for their inhabitants, and provide extensive public education in order to slow the extinction rate of species.

EYE DISORDERS

The eye is the delicate **organ** of **sight** with which we keep informed about the natural world around us. **Light** from an object first strikes the transparent covering of the eyeball, the cornea, then passes through the lens, a clear, flexible disk, which bends the light rays enough so that they focus on the back of the eye, the retina.

Some early discoveries about the physiology of vision were made by Thomas Young, a British physicist and physician. In 1793 he observed that the eye focuses on near or distant objects by changing the shape of the lens. This adjustment is called accommodation. Around the turn of the nineteenth century, Young discovered that color sensation is controlled by tiny structures on the retina which we now call cones, Charles Babbage (1792-1871), an eccentric English mathematician who is sometimes referred to as the grandfather of the modern computer, invented an ophthalmoscope with which to study the retina of the eye in 1847. He gave it to a doctor friend to test, but it was mislaid and forgotten about. Then in 1851 German physician and physicist **Hermann von Helmholtz** independently invented an ophthalmoscope for which he is usually credited. Helmholtz also invented an ophthalmometer for measuring the curvature of the eye, and he renewed interest and

American bison (*Bison bison*). The American bison was once distributed from Alaska and western Canada throughout the United States and into northern Mexico. With the beginning of the European settlement of America, however, the bison became a target of commercial hunters, and by 1889 there were fewer than 1,000 bison left in the United States. To combat imminent extinction of the plains bison, several closely guarded preserves were established and captive breeding programs were initiated; and today the numbers of plains bison have increased to more than 50,000. While this species is no longer considered threatened, its survival remains substantially dependent on the continuation of conservation activities. *(Superstock. Reproduced by permission.)*

expanded Young's theory about color vision. Each of these theories and inventions enabled physicians to study more closely eye disorders.

The following are fairly common vision problems that do not threaten vision and can often be corrected with corrective lenses—eyeglasses and contact lenses or with surgery. Hyperopia or farsightedness was first explained by Franciscus Donders (1818-1889), a Dutch physiologist who discovered in 1858 that the cause of farsightedness was a too-shallow eyeball that caused the image to focus behind the retina. To correct hyperopia, convex corrective lenses are used. In myopia or nearsightedness, the image is focused in front of the retina so that only near objects can be seen clearly. Concave lenses can be worn to permit focusing for objects far away.

The technique of radial keratotomy involves making incisions on the cornea to correct myopia surgically. It was first done in Japan in 1955 by T. Sato and later in 1979 by Saviatoslav Feodorov in Russia. Whereas radial keratotomy was once considered a risky procedure, in the 1990s the technique has improved. Follow-up interviews on hundreds of patients who underwent the procedure show that two-thirds of the patients were able to stop wearing corrective eyeglasses or contact lenses. A newer, "corneal sculpting" laser surgery that takes only about 30 seconds to complete has been approved to correct myopia. Called photoreactive keratectomy (PRK) the process uses an excimer laser to sculpt a minute area on the surface of the eye. Since the surgery removes only 5–10% of the cornea, the corneal dome's strength and integrity is maintained. This approach has also been used to treat hyperopia and astigmatism. Astigmatism is an eye condition that makes objects appear blurred. Thomas Young suffered from astigmatism and discovered that this condition is due to irregular

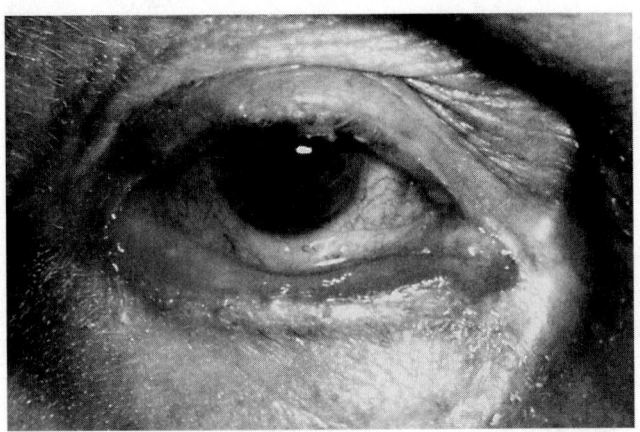

A close-up of the eye of an elderly person showing ectropion of the lower eyelid. Ectropion is an eye disorder in which the eyelid turns away from the eye. The most common type is senile ectropion (seen here), in which the droop of the eyelid is due to loss of tissue elasticity in old age and weakness in the muscles surrounding the eye. *(Photograph by Dr. P. Marazzi, Photo Researchers, Inc. Reproduced by permission.)*

curves in the cornea. In 1862 Donders found that astigmatism is caused by unevenness in the curvature of either the cornea or lens. The irregular curvature makes it impossible for light rays to focus on a single point. The condition may cause headaches or eye strain. It can be corrected with glasses and certain kinds of contact lenses. Presbyopia occurs when the lens loses it elasticity and it can no longer accommodate. The condition is usually associated with age and becomes evident after 40. Usually presbyopia is corrected by wearing reading glasses. Strabismus is a condition whereby one eye is not able to focus with the other resulting in a deviation toward or away from the nose. Sometimes in children the condition will self correct or surgery is recommended to "uncross" the eyes. Color blindness is the inability to distinguish between certain colors, most often red and green. It is a congenital condition that cannot be cured, but contact lenses to correct color blindness have been invented by Dr. Jay Schlanger.

Blindness is a complete or partial loss of vision. People born with the inability to see have congenital blindness. Other people who lose their vision through accident or disease have acquired blindness. Some of the following eye disorders can lead to blindness if untreated. Cataracts are spots on the lens of the eye which cut off the light that passes through. They are

sometimes attributed to age, although they may form in children or persons with diabetes or other conditions. Recent evidence shows that cataracts can be caused by exposure to ultraviolet light. If unchecked, cataracts cause the entire lens to becomes cloudy and blindness results. Glaucoma is a condition in which the eyeball hardens because fluid inside the eye does not drain properly, leading to an increase in pressure that can kill the optic nerve and cause blindness. Keratitis is an **inflammation** of the cornea of the eye. In some cases both the cornea and conjunctiva are inflamed. These two conditions represent a large portion of all eye diseases. Keratitis follows cataracts and glaucoma as a leading cause of blindness. The condition has many causes, including injury, infection, radiation, irritation by chemicals, or allergy. With the use of contact lenses, especially soft and extended-wear contacts, physicians are seeing an increase in the number of cases of ulcerative keratitis, probably because of a decrease in hygiene associated with wearing lenses around the clock without disinfecting them. Conjunctivitis is a fairly common, very contagious inflammation of the eyelid caused by infection and usually treated with drug therapy. Trachoma is an extremely infectious eye disease which inflames the lining of the eyelid and forms small ulcers on the cornea. It is more prevalent in developing countries than in the United States. Ophthalmia neonatorum affects the eyes of newborns, causing ulcers to form on the cornea, the swelling of eyelids and discharging of pus. It is often caused by gonorrhea passed on from mother to child and is preventable if drops are applied to the infant's eyes. Retinitis is an inflammation of the retina which may be caused by **bacterial** infection, injury of the eyeball or intense light. Degenerative retinopathy may be brought on by many conditions that change the pressure in the eye including advanced age, diabetes, high **blood pressure**, **arteriosclerosis**, anemia, and leukemia. Retinopathy is successfully treated with argon laser surgery. Cytomegalovirus (CMV) retinitis is a frequent eye infection found in **AIDS** patients and is treated with drug therapy.

The ubiquitous use of the computer, especially in the work environment, has also led to the development of computer vision syndrome (CVS). Recognized as a medical condition by the American Optometric Association, the syndrome affects people who use video display terminals for long periods of time. Although no long-term disability is associated with the condition, which usually abates within hours of leaving the computer, it does cause dry burning eyes, blurred vision, delayed focusing, altered color **perception**, and headaches.

F

FABRICI, GIROLAMO (1537-1619)
Italian surgeon and anatomist

Girolamo (also Geronimo) Fabrici (also Fabrizio or Fabricius) was born on May 20, 1537, in the town of Aquapendente, near Orvieto, Italy. He studied first humanities, then medicine, at the University of Padua. His teacher in anatomy and surgery was Gabriel Fallopius. After graduating with his M.D. in 1559, he taught anatomy privately and practiced surgery until 1565. In that year, he was appointed to replace Fallopius, who had died, at Padua. Fabricius held that post until 1613, when he retired because of ill health. He died in Bugazzi, near Padua, on May 21, 1619.

Fabricius made a number of advances in the field of anatomy. He studied the structure of the eye and of the larynx, the mechanics of respiration, and the movement of muscles. One of his most important accomplishments was his detailed description of the semilunar valves in veins. Although these valves had been observed earlier, Fabricius published the most complete description of their structure and function in his *De venarum ostiolis (On the Valves of the Veins)* in 1603. He was incorrect, however, in his explanation of the valves' function.

Fabricius is perhaps best known as the founder of the modern science of embryology. From 1600 until his death, he carried out important and original research on the late fetal stages of many different animals. In 1612, he published the first detailed description of the development of the chick embryo from the sixth day onward.

In addition to his scholarly work, Fabricius made other contributions. He attained considerable fame as a teacher. His most famous student was **William Harvey**, who studied with him from 1597 to 1602. In addition, Fabricius was instrumental in establishing the first permanent anatomical theater at the University of Padua.

FALOPIUS, GABRIELE (1523-1562)
Italian anatomist

Gabriele Fallopius was one of the most noteworthy Italian anatomists of the sixteenth century. His family lived in poverty and, as a young man, he served the Catholic Church. Fallopius studied at Ferrara and then at Pisa, and then had the opportunity to study anatomy in Padua, which at that time was considered the best place for anatomical study. Other areas of Europe were not as advanced in the biological sciences. For example, scientists in France thought that the work of Galen could not be improved upon, and the teaching of the natural sciences were being suppressed in Germany by the ongoing religious struggles. Fallopius was a student of **Andreas Vesalius** who, through his method and technique, laid the foundation for modern anatomy and is considered to be one of the most important scientists in history. As Castiglione has pointed out, according to the eminent medical historian, C. V. Daremberg, "Fallopius was a genius while Vesalius was only a scientist."

By the age of twenty-four, Fallopius became a professor in Ferrara, Italy. Several years after Vesalius's death, Fallopius taught at Padua, where he was entirely supported by the government and continued in his mentor's tradition of attention to detail. Fallopius became very well known as somewhat of a pioneer in his field and his lectures were attended by large audiences. In addition to his research, lecturing, and teaching, Fallopius was a physician and surgeon, and maintained an extensive medical practice. During his career Fallopius published *Observationes anatomicae*, which contained many descriptions of his anatomical research. The first edition of his book was published in Venice in 1561, and was followed by later editions published in Italy and several other countries. His collected works, *Opera omnia*, were published after his death, in Venice (1584), Frankfort (1600), and once again in Venice (1606). Fallopius made several important anatomical discoveries and improved upon many of his predecessor's findings. He performed an extensive study of the structures of the ear and was the first anatomist to describe the semicircular canals (*chorda tympani*). Fallopius was also the first anatomist to describe the circular folds of the small intestine and the inguinal band, later called Poupart's ligament. He corrected Vesalius's findings on the course of the cerebral arteries, and provided a more detailed description of the ocular muscles and cerebral nerves. Perhaps his best known discoveries are the structures of the male and female reproductive organs—he described the clitoris and what are now known as the Fallopian tubes, as well as the *arteria profunda* of the penis.

FAMILY

A level of **classification** in the taxonomic system, a **family** is a group of similar genera, and related families are grouped together in the same **order**. Classification is based on levels of similarities; **organism**s with a broad outline of similar characteristics are placed in the same family. Families can vary in size; for example, the Orchidaceae contains all members of the orchids, some 18,000 **species** contained in 750 genera.

Within **botany**, all family names end in -aceae. In **zoology**, the family names end in -idea. There is no specific fixed number of characteristics which must be similar to group organisms into a family. The division into this group is artificial and must be decided by a taxonomist who is an expert in the group under discussion. Over time proscribing characteristics have been defined for most families, so the presence of those characteristics instantly identifies the organism under question to belong to that family. A family is not necessarily as apparent as the similarities that are present at the generic level, or indeed at some of the higher levels of **taxonomy**.

See also Genus

FARSIGHTEDNESS (HYPEROPIA) • See Eye disorders

FATS AND OILS

Fats and oils are a broad class of **animal** and vegetable compounds which are used in products such as foods, cleansers, and lubricants. They are members of the lipid **family** and are **energy**-rich compounds that are basic components of the normal diet. Fats and oils have essentially the same chemical structure—a mixture of fatty acids combined with glycerol (a trihydroxy **alcohol**) and are similarly insoluble in **water**. However, while fats remain solid (or semisolid) at room **temperature**, most oils very quickly become liquid at increased temperatures. Unfortunately, this definition can lead to confusion with petroleum and essential oils. Therefore chemists have begun to classify both as fats. Animal fats and oils include butter, lard, tallow and fish oil. **Plants** provide a number of oils, such as cottonseed, peanut and corn oils.

The fundamental nature of natural fats and oils was determined almost two hundred years ago by the French chemist Michel Eugéne Chevreul. Back in 1811—when Chevreul first began his investigation of animal fats—the science of organic chemistry was still in its earliest stages. Chevreul began simply enough by analyzing a potassium soap that had been made from pig fat. After treating the soap with various chemicals, he found that it yielded a crystalline material with acid properties—the first fatty acid to be discovered.

During the next decade, Chevreul decomposed a variety of soaps, most of which had been made from different animal fats. Not surprisingly, he obtained a whole series of fatty acids from these soaps. After isolating and studying the fatty acids, he gave them names—such as butyric acid and stearic acid which are still used today. By 1816 he also established the fact that animal fats were composed not only of fatty acids, but also of glycerol, an alcohol that, in 1783, another chemist, Carl Scheele, had already found in both olive oil and in other vegetable and animal fats.

Since Chevreul's time, a great deal more has been discovered about the chemical structure of fats. For one thing, chemists have found that, in food, fats are primarily triglycerides, which are composed of one **molecule** of glycerol and three molecules of fatty acids. And, while the glycerol component tends to remain the same in virtually all fats, the fatty acids themselves come in a variety of types and configurations.

The typical fatty acid is simply a chain of carbon **atoms** with hydrogen atoms attached to them. The length of the chain can vary widely, however, most range from four carbon atoms to 24 or more. Most edible fats and oils have 16 to 18 carbon atoms and are considered long-chained. The degree of saturation can also vary, depending on the number of hydrogen atoms in the chain. The chain with the maximum number of hydrogen atoms, for instance, is considered saturated. Saturated fats usually come from animal sources, have very high melting points, and—most importantly—can often increase the body's **blood cholesterol** and, by so doing, possibly contribute to the onset of certain diseases such as atherosclerosis.

Most of the nutritionally important fatty acids can be synthesized by the body from other substances but a few must be supplied by the diet. These are the ones termed essential fatty acids (EFAs). Among EFAs, linoleic acid is considered the most valuable. The most abundant polyunsaturated fatty acid in nature, linoleic acid helps maintain the function of cellular and subcellular **membrane**s and is involved in the synthesis of **prostaglandins**, which are potent substances that appear to mimic or inhibit many **hormone**s. A deficiency of this essential fatty acid, seen most often in poorly-nourished infants and young children, causes dry, scaly skin and some delay in wound healing.

Fats have two main functions: they provide some of the raw material for synthesizing and repairing **tissue**s and they serve as a concentrated source of fuel energy. Fats, in fact, provide humans with roughly twice the energy, per unit weight, as do **carbohydrate**s and **protein**s —nine kilocalories per gram (Kcal/gram) versus about four Kcal/gram for either sucrose or most proteins. Fats are not only an important source of day-to-day energy, they can, if not immediately needed, be stored indefinitely as adipose tissue in case of future need.

Fats also help transport fat-soluble **vitamins** throughout the system; cushion and form protective pads around delicate organs, such as the **heart**, liver and kidneys; make up the layer of fat under the skin that helps insulate the body against too much heat loss; and even add to the palatability of other foods that might otherwise be inedible.

While normal amounts of fat in the diet are essential to good health, unnecessarily high amounts can lead to a number of problems. For instance, a certain amount of excess adipose tissue can be valuable during periods of illness, overactivity, or food shortages; however, too much adipose tissue can not only be unsightly but can overwork the heart and put added **stress** on other parts of the body. High levels of certain circulating fats may not only lead to atherosclerosis, but have been linked to other illnesses, including **cancer**.

How much fat in the diet is considered too much? In the past, nutritionists considered reasonable a diet that obtained 40 percent of its **calories** from fats. These days, however, they recommend that no more than 30 percent (and preferably even less) come from fat. In healthy adults, too, body fats typically should make up no more than 18-25 percent body weight in females and 15-20 percent in males.

New research on fats and oils has continued into the 1990s. For example, a 1997 report details attempts by genetic engineers to develop **enzyme**s which will allow plants to convert saturated fats and oils to healthier unsaturated materials. In addition to nutritional applications, advances have been made on oils for other applications. Because they are environmentally friendly, renewable resources, an increasing number of plant oils are being evaluated for use as industrial lubricants. Two examples which are anticipated to have a significant industrial impact are the well known soybean oil and the lesser known meadowfoam oil.

FERMENTATION

In the absence of the gas oxygen, certain living things are capable of breaking down **carbohydrates** (starches and sugars) to form **alcohol** and **carbon dioxide** gas. This process is known as anaerobic respiration or *fermentation*, and it has been used for centuries in the production of certain foods and beverages.

Throughout history, the process of fermentation was shrouded in mystery and superstition. Many thought that the process was spontaneous, just as they believed that life arose spontaneously from nonliving things. During the seventeenth century, English chemist **Robert Boyle** proclaimed that an understanding of the fermentation process would lead to an understanding of the cause of other phenomena like disease. His prediction came true two centuries later. In the mid 1800s, a French scientist, Charles Cagniard de Latour, first discovered that tiny living **cell**s called **yeast** can cause fermentation. This paved the way for the seminal experiments on yeast fermentation performed several years later by French scientist **Louis Pasteur**.

Pasteur was dedicated to analyzing what really happens during fermentation after being asked to find out why some of the greatest burgundy wine produced in France was spoiling. After observing wine under the **microscope**, Pasteur noted that wine normally contained yeast cells which were producing the desired alcohol but wine that had become sour contained **bacteria**, other **microorganisms** that produced **lactic acid** during fermentation, spoiling the wine. Pasteur's experiments showed that fermentation could take place only in the presence of living cells. His hypothesis countered the argument of a great chemist of the time, **Justus von Liebig**, who erroneously explained that fermentation was a **chemical reaction** caused by the **decomposition** of dead yeast cells. Pasteur's correct theory of fermentation led him to research in other related fields of microbiology which included his work on the cause and treatment of numerous diseases caused by microorganisms.

The fermentation process is caused by **enzyme**s, catalysts in chemical reactions similar to the digestive enzymes in the human body. Certain enzymes act on starch to break down the long chain-like **molecule**s into smaller units of sugar. Then other enzymes convert one kind of sugar molecule to another. Still other enzyme reactions break apart the sugar molecule (composed of carbon, hydrogen and oxygen) into ethyl alcohol and carbon dioxide gas.

The production of carbon dioxide and alcohol are incidental to the release of **energy** needed by **organism**s such as yeast to survive. But these metabolic by-products have been used in human enterprise for centuries. The yeast *Saccharomyces cerevisiae* is traditionally added to liquids derived from grains and **fruits** to brew beer and wine. The natural starches and sugars provide food for the yeast and during fermentation the desired alcohol is released. In China for thousands of years, traditional soy sauce or shoyu was brewed by adding the fungus *Aspergillus oryzae* to a mixture of boiled soybeans and wheat and allowing it to ferment for about a year.

In recent times, yeasts have been used to aid in the production of alternative energy sources. Yeasts are placed in large fermentation vats containing organic material. During fermentation the yeast convert the organic material into ethanol fuel. Researchers are working on developing yeast strains that will convert even larger organic **biomass**es into ethanol more efficiently.

The fermentation process is not limited to microorganisms. It takes place on occasion in **animal**s. If goldfish are placed in anaerobic conditions they can survive for a limited time without oxygen by carrying on fermentation provided that the **water** is cold enough to reduce their metabolic needs. In human beings, fermentation occurs in muscle **tissue** during periods of exercise. When energy requirements are high during strenuous exercise, oxygen cannot be delivered fast enough. To continue supplying the muscle with energy, sugar molecules in the tissue are converted to lactic acid. However, the buildup of lactic acid causes fatigue and even after exercise ceases, breathing continues to be heavy to get more oxygen back into the tissue.

FERTILIZATION

For many years, the process of fertilization was a mystery to biologists. It was not until the late nineteenth century, when

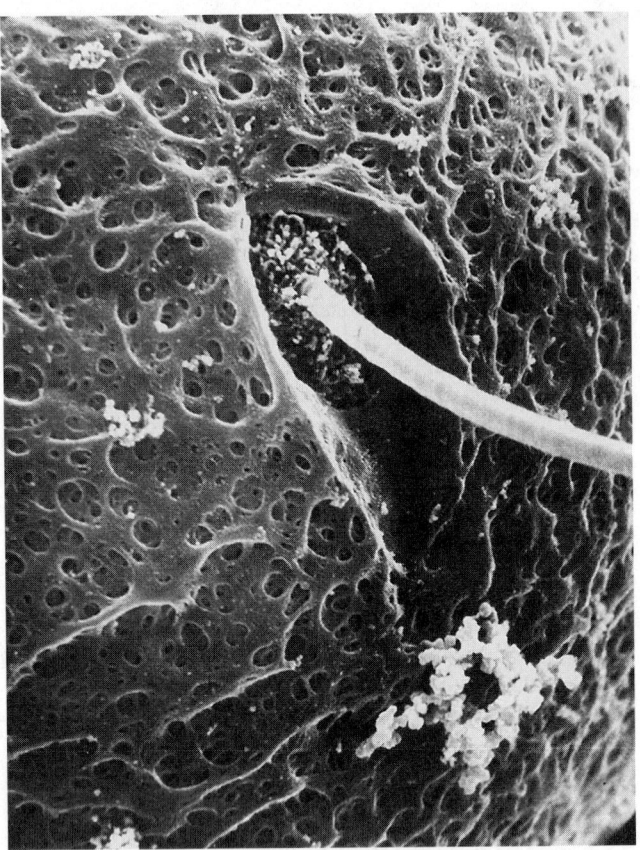

Sperm penetrating a hamster egg. *(David M. Phillips, Photo Researchers, Inc. Reproduced by permission.)*

the science of microscopy had begun to take off, that scientists began to observe and understand the chemical and biological processes that occurred within the human body, both before and after insemination.

It is now known that fertilization cannot occur until the (male) **sperm** and the (female) **egg**, or ovum, are brought together. In many aquatic and amphibian **species**, the eggs are released by the female, while the male deposits his sperm on or near them; however, in most mammalian species (including humans), the male uses a special **organ** to deposit the sperm safely within the female. The sperm then swims randomly until it dies or encounters an egg.

Once the sperm and egg meet, the fertilization process can begin. As the tip of the sperm touches the outer surface of the ovum, certain **chemical reaction**s are set into motion: the two bodies are fused together, while the contents of the sperm are allowed to pass into the egg, where it commingles with the contents therein; the surface of the egg becomes impermeable and unreceptive to further sperm, thus insuring fertilization by one sperm only. Once the egg is "activated" it begins to divide meiotically—first into two **cell**s, then four, then eight, and so on. These multiplying cells will eventually form the fetus. The rapidly developing egg is now called a **zygote**; the fertilization process has ended.

Although much of this process is common knowledge today, it comes in the wake of years of tedious research and much misunderstanding. The first scientist to shed light upon the mystery of fertilization was the German botanist Christian Sprengel. He pointed out that **plants** were unable to pollinate themselves and that **pollination** could not take place without the assistance of birds, insects, or weather to spread the plants' seed. When extrapolated to **animal**s, this seemed to indicate that two "ingredients" were necessary to create life—a male and a female.

The Italian physiologist **Lazzaro Spallanzani** was one of the first to conduct this sort of research using a **microscope**. Using animals, he deduced that it was necessary for the male seminal fluid to come in contact with the female ova for activation to occur. However, he also supported the popular view that the spermatozoa were **parasite**s living within the testicles of all male creatures—a view probably fueled by **Antoni van Leeuwenhoek's** recent discovery of microscopic **bacteria** in **water**.

It was not until the late 1800s that two European scientists, Hermann Fol and Wilhelm Hertwig, illustrated the importance of the sperm. Both students of **Ernst Haeckel**, Fol and Hertwig experimented using near-transparent sea urchins—creatures whose reproductive organs could be easily observed. In 1877, Fol observed that the sperm actually penetrates the egg, while Hertwig showed that only one sperm was necessary to fertilize an egg.

In addition to their experimental observations, Hertwig and Fol began to hypothesize about the nature of **heredity** and how it was passed from one generation to the next. This concept had been popularized in the mid-1800s by the German biologist **August Weissmann**. Weissmann located heredity within a variety of germ cells, which he called germ plasm. He explained that the germ plasm was distinguishable even in **embryo**s; as the **organism** matured, the germ plasm would be safely encased deep within its body. When the organism eventually reproduced, the germ plasm would be passed along during fertilization. In Weissmann's model, the organism itself is used simply as protection and transportation for the germ plasm. This concept was summed up by the English writer Samuel Butler as "a hen is only an egg's way of producing another egg." Man's understanding of heredity has since been greatly enhanced by the discovery of **DNA**.

Strangely, it is also possible for an egg to be activated without encountering a sperm. This phenomenon is called **parthenogenesis** ("virgin **birth**") and is not uncommon among certain species of marine life. During the early years of the twentieth century, scientists began to experiment with artificial parthenogenesis—that is, ways to chemically activate the egg. The defining work in artificial parthenogenesis was performed separately by the American scientists Jacques Loeb and **Ernest Just**. Using solutions of sea water to activate the eggs, they succeeded in raising parthenogenetic frogs and sea urchins to sexual maturity; these creatures were physically identical to fertilized creatures but were genetically inferior since they carried the hereditary code of just one parent. The work of Loeb and Just led other scientists to look more closely for the location of heredity and provided some of the keys to the secrets of **evolution**.

Fertilization can also occur through artificial insemination, in which semen is introduced into the vagina or uterus without sexual contact. This process was first developed for breeding cattle and horses. In humans, the process is sometimes used by couples who wish to have a child, but cannot because the husband is infertile or has a genetic disease. It may also be used by a woman who wants children but has no male partner. Typically, artificial insemination involves frozen human spermatozoa from an anonymous male donor. The semen is injected into the cervix with a syringe.

Another form of artificial insemination, called in vitro fertilization, involves mixing the sperm and ovum in a laboratory dish, where fertilization occurs. The fertilized egg is then transplanted into the woman's uterus. This method is used when a woman's fallopian tubes are blocked, severely damaged, or missing, keeping the sperm from reaching the ovum naturally. The first "test-tube baby" born as a result of this procedure was Mary Louise Brown, born in England in 1978. The procedure was used successfully in the United States for the first time in 1981. Since then, more then 45,000 American babies have been born as a result of this technique.

FIBIGER, JOHANNES (1867-1928)
Danish pathologist and bacteriologist

Johannes Fibiger was a Danish bacteriologist whose early work on childhood **diphtheria** and tuberculosis demonstrated the vital role medical research could play in controlling diseases that threatened public health. In 1926, Fibiger received the Nobel Prize in physiology or medicine for demonstrating how **cancer**-like **tissue**s could be induced experimentally in the laboratory.

Johannes Andreas Grib Fibiger was born on April 23, 1867, in the Danish village of Silkeborg. His father, Christian Fibiger, was a district physician; his mother, Elfride Muller, was a writer and the daughter of a Danish politician. When Fibiger was three, his father died and the family moved to Copenhagen, where he attended the University of Copenhagen at age sixteen and studied medicine, biology, and zoology. After earning his medical degree in 1890, he undertook several years of medical apprenticeship in various hospitals and with the Danish army. In 1891, he married Mathilde Fibiger, a distant cousin and physician's daughter, with whom he had two children.

It was while working as an assistant in a bacteriological laboratory at the University of Copenhagen that Fibiger was persuaded to undertake doctoral work on diphtheria, a virulent childhood disease that caused its victims to suffocate. Fibiger discovered better methods of growing diphtheria **bacteria** in the laboratory and demonstrated that there were two distinct forms of the bacillus, an important step in identifying carriers of the disease who frequently displayed no symptoms. At the turn of the century, diphtheria was a major public health problem, and epidemics were frequent in Denmark and throughout the rest of the developed world. Fibiger produced an experimental serum against the disease and carefully monitored the results of an inoculation program. In 1897, the International Medical Congress published his report, a model of its kind, which brought Fibiger international attention and confirmed the effectiveness of the serum. The young scientist had received his Ph.D. only two years earlier. Fibiger later came to regard his work on diphtheria as his highest scientific achievement.

In 1900, at age thirty-three, he joined the faculty of the Institute of Pathological Anatomy, one of a number of young professors hired by the University of Copenhagen. He was also appointed director of the institute and launched a successful program to construct a modern research facility for pathology and anatomy. Within its walls, Fibiger and another faculty member, C. O. Jenson, conducted research on tuberculosis in cattle and humans. Flying in the face of popular opinion, they demonstrated that humans could contract tuberculosis from infected cattle, especially by drinking their milk. Supported by the research of other investigators in Europe, these findings led to the passage of strict regulations governing the sale of raw milk, resulting in fewer adolescent deaths due to tuberculosis.

Seeks to Produce Cancer in the Laboratory

Fibiger's experiments on tubercular rats led him to the discovery for which he won the Nobel Prize. Performing a series of routine dissections in 1907, he discovered abscesses that appeared to be cancerous in the stomach lining of three wild rats. Microscopic examination revealed that the abscesses contained the **larvae** of a minute parasitic worm or nematode.

By the early 1900s, scientists had ample observational data suggesting that environmental irritants such as soot and harsh chemicals produced cancer in chimney sweeps and chemical workers. Many scientists thought that chronic irritation from mechanical or chemical agents was the basis of all cancer, but no one had yet succeeded in turning normal **cell**s into cancerous cells under laboratory conditions.

Working on the hypothesis that the parasites produced a chemical toxin that induced cancer of the stomach, Fibiger undertook an ambitious research program. He trapped and examined more than a thousand wild rats, feeding them worm larvae, and even injecting them with the parasite, all without result. Surmising that the larvae was not passed from rat to rat but through an intermediate host, he traced the parasite to a rare **species** of cockroach found near a Copenhagen sugar refinery. By feeding healthy rats a diet of white bread and cockroaches, Fibiger finally succeeded in producing stomach abscesses in more than a hundred **animal**s. For the first time, a researcher had induced what at the time was thought to be cancer in a laboratory setting. Fibiger reported his achievement in the *Journal of Cancer Research* and was awarded the 1926 Nobel Prize in medicine or physiology for his discovery of *Spiroptera carcinoma*, the parasitic worm that he thought had produced the cancer. Yet, in his acceptance speech, Fibiger expressed doubt that parasites played any great role in gastric cancer in humans.

Later investigators would find a number of weaknesses in Fibiger's research. Like most scientists of the period, Fibiger had not thought to check his findings against a control group of rats fed on a diet of only white bread. Nor was it easy to reproduce Fibiger's findings in other laboratories due to the lack of a standard strain of laboratory rats in the 1920s; Fibiger's animals had all been caught in the wild. Other investigators expressed doubt that the abscesses described by Fibiger were truly cancerous. There was some evidence that the abscesses might have been caused by a diet deficient in vitamin A. Nonetheless, the lasting effect of Fibiger's prize-winning discovery—later refuted by other researchers—was the great impetus it gave to other investigators to pursue laboratory research on the causes of cancer.

Fibiger abandoned parasitology after World War I to follow the work of two Japanese scientists who induced skin cancer in rabbits by painting their ears with coal tar. Conducting his own experiments by painting the backs of rats with the irritant, Fibiger reported two valuable insights: that cancer did not occur with the same frequency in all **species** or even within the same species, and that individual predisposition played an important role in susceptibility to cancer. At the time of his death, he was working with two colleagues on a vaccine for cancer, hoping to demonstrate that inoculating laboratory animals with matter drawn from malignant tumors would induce immunity to the disease.

During his long career as director of the Institute of Pathological Anatomy at the University of Copenhagen, Fibiger divided his time between research and teaching. He was a generous colleague who was widely respected for his meticulous laboratory methods. He published seventy-nine scientific papers and served as secretary and then president of the Danish Medical Society, and as president of the Danish Cancer Commission. He was co-editor and founder of *Acta Pathologica et Microbiologica Scandinavica*. In 1927, he was awarded the Nordhoff-Jung Cancer Prize.

On January 30, 1928, less than two years after delivering his Nobel Prize speech, Johannes Fibiger died in Copenhagen of a massive heart attack. He was sixty years old and had recently learned that he had colon cancer.

FINSEN, NIELS RYBERG (1860-1904)
Danish physician

Niels Ryberg Finsen, the father of phototherapy, received the Nobel Prize in 1903 for physiology or medicine after proving his radical theory that **light** rays could cure disease and save lives. Called "The Light Hunter" by an early biographer, Finsen was born in Thorshavn in the Faeroe Islands in the North Sea. His father, Hannes Steingrim Finsen—whose ancestry dates back to Icelandic Vikings of the tenth century—was a govenor in the islands; his mother—Johanne Fröman—was also born in Iceland.

Finsen attended school in the Faeroe Islands before being sent to prep school in Denmark from which he was expelled for "...small ability and total lack of energy." In 1876, his father sent him to Reykjavík school in Iceland where he was exposed to "...an absolutely unique system of teaching him to believe nothing but what he found out for himself." In

1891, he received his medical degree from the University of Copenhagen where he became an anatomy instructor. He married Ingeborg Balslev in 1892 (with whom he had four children) and left the university in 1893 to pursue his fascination with sunlight.

Finsen's first success was with smallpox. He observed that, although smallpox blisters covered the entire body, they became infected and caused scars only where the skin was exposed to sunlight—primarily on the hands and face. Through previous experiments, he hypothesized that the blue, violet, and ultraviolet light rays caused this irritation, and proposed keeping patients in rooms with red glass and red curtains to allow only infrared light in. While local doctors laughed at his crazy theory, two doctors in Bergen, Norway—Lindholm and Svendsen—put it into practice. Every patient they treated in "red rooms" recovered without a scar. Shortly thereafter, scientific confirmation came from Gothenburg, Sweden, that people with the deadly black smallpox recovered with the same treatment.

Finsen gained international recognition, yet something more bothered him. Contrary to physiological evidence which said sunlight was harmful, he believed it was somehow good. Astute observations and instinct drove him to pursue its therapeutic properties. By 1883, he was suffering from Pick's disease, a progressive and fatal illness. One day, while shivering through a period of sudden chill which often engulfed his ailing body, he watched a cat bathing in the sun on a roof outside his window. Repeatedly, as the shade moved over its body, the cat edged back into the sunlight. He pondered that cats were perfectly healthy creatures, seldom needing a doctor. On another occasion, he noticed a skating bug floating downstream. Each time the current drew it into the shadow of the bridge, it darted back into the sunlight. Because nature came alive when the sun shone—bringing birds into song, flowers into bloom, and bees buzzing about busily—he suspected sunlight played a larger role than just the warmth it provided. Even his own body seemed more vital and energetic after spending time in the sun.

Finsen began crude experiments—a famous one on his wife's ear lobe and another on the tail of a tadpole—which led to his next great discovery and the Nobel Prize. Contrary to his findings that ultraviolet rays aggravated smallpox blisters, he believed they had beneficial properties. He ultimately proved that high concentrations of these rays killed microbes. Laughed at again by the scientific and medical communities, he turned to the chief engineer of the local power plant, Winfred Hanson. Together they designed an electric arc lamp, gathering the ultraviolet rays through a series of lenses and focusing them on the huge sore on the face of a Danish engineer and friend of Hanson's by the name of Morgensen. For eight years, Morgensen had suffered with *lupus vulgaris*, a form of tuberculosis which caused oozing, disfiguring, and incurable open sores. No medical treatment had helped; however, after sitting under Finsen's clumsy lamp two hours a day for a month, the sore began to change. From November 1895 until March, 1896 he received treatment, which totally eradicated the disfiguring lesion. Soon thereafter, two manufacturers by the name of Jörgensen and Hagemann founded the Finsen Institute in Copenhagen. The scientific community could no longer deny Finsen's success, and four university professors agreed to serve on the institute's board.

Confined to a wheel chair during most of his final experiments, and too ill to attend the Nobel ceremony, Finsen's tenacious research earned him Knighthood to the Order of Danneborg, the Silver Cross, the Danish gold medal for merit, the Cameron Prize from the University of Edinburgh, and other honorary awards. Finsen died in his wife's arms at the age of forty-four.

FISCHER, EDMOND H. (1920-)
Chinese American biochemist

Edmond H. Fischer was the joint recipient with his longtime associate, **Edwin Krebs**, of the Nobel Prize in Physiology or Medicine in 1992 for discoveries dealing with reversible **protein** phosphorylation as a biological regulatory mechanism. Responsible for a wide range of basic processes, including **cell** growth and differentiation, regulation of **gene**s, and muscle contraction, protein phosphorylation is now the subject of one in every 20 papers published in biology journals. Application of Fischer and Krebs's work to medicine has elucidated mechanisms of diseases such as **cancer** and **diabetes**, and has yielded drugs that inhibit the body's rejection of transplanted **organ**s.

Edmond H. Fischer was born on April 6, 1920, in Shanghai, China. His father, Oscar Fischer, had come to China from Vienna, Austria, after earning degrees in business and law. Fischer's mother, Renée Tapernoux Fischer, was born in France. She had come to Shanghai with her family after first arriving in Hanoi, where her father was a journalist for a Swiss publication. In Shanghai, Fischer's grandfather founded the first French newspaper published in China and helped to establish a French language school that Fischer attended.

At the age of seven, Fischer was sent, along with two older brothers, to a Swiss boarding school near Lake Geneva After entering the School of Chemistry at the University of Geneva at the start of World War II, Fischer earned a degree in biology and another in chemistry. He received his doctorate at Geneva in 1947 and worked at the university on research until 1953. Fischer then moved to the United States where there were more educational and career opportunities in the field of biochemistry. His first position was at the California Institute of Technology, where he was given a postdoctoral fellowship.

Fischer came to Seattle to accept an offer of an assistant professorship from Hans Neurath, chair of the University of Washington's department of biochemistry. Thus began a long association with Edwin Krebs, who had worked in the laboratory of **Carl Ferdinand Cori** and **Gerty T. Cori** in St. Louis on the **enzyme** phosphorylase in the late 1940s. The Coris won the Nobel Prize in 1947 for their isolation of phosphorylase, showing its existence in active and inactive form. Fischer had worked on a plant version of the same enzyme while he was in Switzerland.

In the mid–1950s Fischer and Krebs set out to determine what controlled the protein's activity. Their experiments centered on muscle contraction. A resting muscle needs energy (stored as glycogen in the body) in order to contract, and phosphorylase frees glucose from glycogen for use by the muscle. Fischer and Krebs discovered that an enzyme they called protein kinase was responsible for adding a phosphate group from the compound ATP (adenosine triphosphate, the **cell**'s energy store) to phosphorylase, which activated the enzyme. In a reverse reaction, an enzyme called protein phosphatase removed the phosphate, turning phosphorylase off. Protein kinases are present in all cells and are critical for many phases of cell activity, including metabolism, respiration, protein synthesis, and response to stress.

By the 1970s biochemical research in the area that Fischer and Krebs opened up was so extensive that 5 % of papers in biology journals dealt with protein phosphorylation. Between 1 and 5 % of the **genetic code** may be concerned with protein kinases and phosphatases. Science has made connections that show the role of protein kinases in diseases, including cancer, diabetes, and muscular dystrophy. Fischer and Krebs have also been able to demonstrate in their research how the **immune system** is activated. They showed how a surface protein starts a chain reaction that recruits **lymphocytes** to fight infection.

In the field of organ transplants, drugs that influence phosphorylation prevent rejection of the transplants by the body's immune system. The drug cyclosporin has been developed and is widely used to prevent the rejection of liver, kidney, or pancreatic transplants in human beings. Cyclosporin and another drug, FK–506, inhibit protein phosphatase, thereby preventing the rejection of tissues in organ transplant operations. Irregular protein kinase activity can cause abnormal cell growth leading to tumors and cancer. Research scientists are expecting protein kinases and phosphatases to be the major drug targets of the 21st century.

Besides the Nobel Prize, Fischer won the Werner Medal from the Swiss Chemical Society in 1952 and was elected to the American Academy of Arts and Sciences in 1972. Fischer and his colleague, Krebs continue in the role of emeritus professors at the University of Washington.

FISCHER, EMIL (1852-1919)
German chemist

Emil Fischer was a professor of chemistry for forty years who also served as director of the German chemical industries during World War I. Fischer's research on important organic substances such as sugars, **enzyme**s, and **protein**s, built the foundation for modern biochemistry. He was the scientist who initially described the action of enzymes as a lock and key mechanism where the structure of an enzyme fits exactly into the molecule with which it reacts to ''unlock'' a biochemical reaction. In 1902 he received the Nobel Prize in chemistry for his laboratory synthesis of sugars and purine, a substance found naturally in all deoxyribonucleic acid (**DNA**). Fischer

Emil Fischer. *(The Library of Congress. Reproduced by permission.)*

was dedicated to academic research and was among the first scientists in the world to promote substantial industrial as well as governmental support for university laboratories.

Emil Hermann Fischer was born on October 9, 1852 in Euskirchen, Germany, near Bonn and Cologne. With five older sisters, he was the only son of Laurenz Fischer and Julie Poensgen Fischer. His father was a successful businessman who started as a grocer, then added a wool spinning mill and a brewery as he prospered. Fischer described his youth as happy in his unfinished autobiography, *Aus meinem Leben* (*Out of my Life*). Fischer was a brilliant student, graduating in 1869 at the top of his class from the Gymnasium (high school) of Bonn. After graduation, Fischer tried working in business with an uncle, but he was much more interested in building a laboratory. He entered the University of Bonn in the spring of 1871.

Follows a Master Chemist to a Lifelong Career

After less than a year at the University of Bonn, Fischer transferred to Strasbourg where he studied under the noted chemist, Adolf Baeyer. Fischer's creativity flourished in the academic atmosphere of Strasbourg; he especially noted in his autobiography the accessibility of his professors, and the opportunities to travel and visit other chemical laboratories. For his doctorate Fischer did research on fluorescein, a coal tar dye that shows a fine yellow-green fluorescence in solution, and is used to trace **water** through systems. Fischer's researches into coal, coal tar, and the synthesis of organic chemicals, did much to build the German dye industry. Dyes manufactured in Germany soon captured the world market.

Expands Research

Fischer received his doctoral degree in 1874 from Strasbourg, but he continued his research on coal tar dyes with a cousin, Otto Philipp Fischer, until 1878. Ultimately he acquired a number of patents for industrially useful chemicals. In 1875 Fischer was invited to follow Baeyer to the University of Munich where Fischer became associate professor of analytical chemistry in 1879. His researches included the discovery of a new compound, phenylhydrazine, a chemical he later used extensively in research on sugars. By 1878 he figured out the chemical formula for phenylhydrazine, and this discovery stimulated other researches leading to the development of such synthetic drugs as novocaine. In 1881 Fischer began investigations into a new field, purine chemistry (part of a group of **nucleic acids**), identifying three **amino acid**s and synthesizing many more. This research resulted in many more advances in the German drug industry.

Fischer left in 1882 to accept the position of professor of chemistry at the University of Erlangen, near Nuremberg. At Erlangen, Fischer continued his work on purines and began to study carbohydrates in 1884. His subsequent work with phenylhydrazine in an unventilated laboratory caused him to suffer the effects of phenylhydrazine poisoning which attacks the kidney, liver, and respiratory system. Fischer had, from an early age, periodically suffered from stomach disorders; the added contamination to phenylhydrazine made him extremely ill. Upon his recovery in 1885 he accepted a chair in Würzburg, where, he wrote in his autobiography, "gaiety and humor flourished." In 1888 Fischer married Agnes Gerlach. They had three sons before she died in 1895. While Fischer was at Würzburg he was honored with a Bavarian medal.

Berlin Brings World Recognition and a Nobel Prize

In 1892 Fischer accepted the position of professor in charge of the chemistry department at the University of Berlin, the most prestigious position for an academic chemist in Germany at that time. He was offered full freedom in the construction of a new building at the chemical institute of Berlin, and his subsequent design of a well-ventilated laboratory became a model for university laboratories all over the world. In addition, his teaching methods led to the formation of small groups of students involved in basic scientific research. With the help of his cooperative teams of students, and fellow researchers from many countries, he designed a careful plan for each research project. As the work progressed he always looked for deviations from the expected results. Each unusual occurrence was researched systematically to its conclusion. This strategy permanently influenced both graduate education in chemistry and the expectations of universities for research and publication from their professors worldwide.

Fischer's researches into sugar and purines had proven especially successful. He synthesized about one hundred and thirty purines, which included caffeine, theophylline (used in the preparation of the motion sickness drug Dramamine), and uric acid. In addition, after studying the three-dimensional shapes of sugar molecules, Fischer synthesized glucose as well as about thirty other sugars. By 1899 Fischer finished most of his work on sugars and purines and began research on proteins and enzymes in an effort to identify their chemical nature. Fischer was elected to membership in the Academy of Sciences, and, in 1902, he received the Nobel Prize "for his synthesis in the groups of sugars and purine," as quoted by Eduard Farber in *Nobel Prize Winners in Chemistry*. In 1909 he received the Helmholtz Medal for his work on sugar and protein chemistry.

Fischer believed in basic research. Determined to keep the preeminent position of world leader in chemical research for Germany, a position he did much to create, he gathered support from industry, government, and other scientists to establish a number of research institutes—the Kaiser Wilhelm Society for the Advancement of Sciences, the Kaiser Wilhelm Institute for Chemistry in Berlin-Dahlem, and the Kaiser Wilhelm Institute for Coal Research in Mulheim-Ruhr. Fischer was interested in research in every branch of chemistry. As director of the University of Berlin laboratories he started a radiochemistry laboratory where, years after his death, scientists Otto Hahn and Lise Meitner worked on research that led to the fission of uranium and the ultimate development of the atomic bomb.

Makes Major Contribution to German War Effort

World War I took Fischer away from most of his experimental investigations as he redirected his research concentrations toward the war effort. Besides being the leading chemist in Germany, he had long worked closely with industry and government. The British blockade would have brought the defeat of Germany by 1915 had Fischer and his colleagues not succeeded in using the resources they had to synthesize much of what they could no longer get on the world market. He led the development of synthetic saltpeter (potassium nitrate) and nitric acid, both used in the manufacture of explosives. As food became in short supply he coordinated research and production of synthetic fertilizers. Fischer directed research to replace diminishing supplies of camphor (used to stabilize gunpowder) and pyrites which supplied sulfur for explosives.

Before World War I scientists had enjoyed the freedom to travel and communicate with other scientists regardless of political differences and skirmishes between their respective countries. However, World War I brought a change. Scientists became national resources. Fischer ended his long friendship with British chemist, Sir William Ramsay, also a Nobel laureate. But research alone could not win the war, and not all of Fischer's projects were successful. It was obvious to Fischer that Germany would be defeated. In an effort to organize the rebuilding of chemical research and industry in Germany to gain back the leadership it had before the war, Fischer and a friend made plans to form the German Society for the Advancement of Chemical Instruction.

The war years were personally tragic for Fischer. He lost his two younger sons, which left him depressed, and he was suffering from **cancer**. Emil Fischer died in Berlin, July 15, 1919. Some reports say he died of cancer, most say it was sui-

Hans Fischer. *(The Library of Congress. Reproduced by permission.)*

cide. His remaining son, Hermann Otto Laurenz Fischer (1888–1960) went on to become a Professor of Biochemistry at the University of California in Berkeley. On October 9, 1952, Fischer's son dedicated the Emil Fischer Library at the University of California which is the repository of the collected works of Fischer, including the manuscript for his autobiography, research files, and Fischer's correspondences in World War I.

FISCHER, HANS (1881-1945)
German chemist

Hans Fischer was a medically-minded chemist who won the Nobel Prize for chemistry for his pioneering investigations into the chemical structure of pyrroles, molecular compounds which give the specific color to many important biological substances, including **blood**, bile, and the leaves of **plants**. Building on the foundations laid by his predecessors and colleagues, many of them from Fischer's homeland of Germany, he spearheaded a series of investigations lasting more than two decades that led to the synthesis of **hemoglobin**, bilirubin, and (more than 25 years after his death) **chlorophyll**. During the course of his investigations, Fischer developed and oversaw an extremely productive microanalytical approach to studying chemical compounds, especially the pigments that occur in nature. By overseeing specific laboratory procedures conducted simultaneously by several labs, Fischer was able to conduct more than 60,000 microanalyses of chemical substances. In 1930, he won the Nobel Prize, primarily for his work in elucidating the structure of and synthesizing the blood pigment hemin.

Fischer was born at Höchst am Main in Germany on July 27, 1881, to Eugen Fischer, a dye chemist, and Anna Herdegen Fischer. Through his father's work as laboratory director at the Kalle Dye works, Fischer developed an early interest in the chemical nature of pigments, or coloring matter. Interested in both chemistry and medicine, Fischer received his doctorate in chemistry in 1904 from the University of Marburgh and his M.D. in 1908 from the University of Munich. After working on chemical structure of peptides and sugars with **Emil Fischer** (no relation) in Berlin, Fischer went to the Physiological Institute in Munich, where he first began his study (under Freidrich von Müller) of pigments, an area that was to become the overriding focus of his scientific pursuits. Fischer's dual expertise in chemistry and medicine led him to become chair of medical chemistry at the University of Innsbruch in 1916. Although he published his first notable scientific paper (on the subject of bilirubin, or bile pigment) in 1915, his research efforts soon came to a standstill due to World War I and the following years of reconstruction after Germany's defeat. Fischer's ill health also impeded his research efforts. He contracted tuberculosis when he was 20 years of age and had a kidney removed in 1917 due to complications from the disease.

Lays Groundwork for Nobel Prize-winning Discovery

In 1921, Fischer's investigation of pigments began in earnest as he accepted an appointment as director of the Institute für Anoreganische Chemise at the Technische Hochschule, or Technical University, in Munich. It was there that Fischer would conduct his groundbreaking research into pyrrole chemistry for nearly a quarter of a century. Fischer immediately reinitiated his studies of bile pigments and organized a number of specialized laboratories to simultaneously conduct the specific tasks needed to determine their chemical structures. Using a process known as Gattermann aldehyde synthesis to systematically prepare the numerous compounds needed for pyrrole derivatives, Fischer organized teams of microanalysts, sometimes referred to as "Gattermann cooks." He also set up specific laboratories to work on individual segments of a chemical problem, such as making calorimetric determinations and developing X-ray diagrams. By segmenting the work, Fischer's laboratory turned out more new chemical compounds than any laboratory that had preceded it.

Fischer's first major advance was the discovery of porphyrin synthesis in 1926. Porphyrins are made up of pyrroles joined in a chemical ring and are the pigments that appear throughout nature. The accepted view in chemistry prior to Fischer's work was that a single basic porphyrin structure was the primary component for all pigments occurring in nature. Fischer began to unravel the fundamental chemical structure

of the porphin (the nucleic core of porphyrins), which had been proposed by W. Küster in 1912. This accomplishment led to the discovery of specific molecular structures of individual porphyrin groups that make up certain pigments. Specifically, Fischer had found that porphyrins are made up of four pyrrole nuclei bound by methane groups into a ring structure. This led to the creation of porphyrin in a laboratory setting. With the ability to synthesize porphyrin, Fischer and his colleagues were able to further determine thousands of specific porphyrin structures. In *Great Chemists,* Heinrich Wieland, an organic chemist, describes Fischer's attempt to synthesize porphyrins. "Fischer began to put the pyrrole segments together in mosaiclike arrangements and then to weld together, by brilliant synthetic procedures, the semimolecules of the pyrrometheenes produced in this manner." Fischer soon recognized that porphyrins differed primarily through the components that made up the rings. He also discovered that bilirubin was derived from hemin and identified it as a porphyrin.

In 1929, Fischer successfully synthesized hemin, showing that its ring had a center atom of iron. Fischer received the Nobel Prize in chemistry in 1930 for his synthesis of hemin, which is one of two components of hemoglobin, the red respiratory protein of erythrocytes (red blood **cell**s or corpuscles). During the Nobel Prize ceremonies, Fischer was also noted for his demonstration that hemin is related to chlorophyll, the light absorbing, green plant pigment. In 1944, Fischer finally worked out the chemical structure of and synthesized the pigment bilirubin, which he had first begun investigating during World War I. Over the years, Fischer's laboratory had synthesized approximately 130 porphyrins. He also conducted in-depth studies of the specific structure of chlorophyll and published 129 papers on the topic. He successfully identified chlorophyll's pyrrole rings, which had a center of magnesium rather than iron like hemin's pyrrole rings. The synthesis of chlorophyll, while based largely on Fischer's work, was not accomplished until 1960, 15 years after his death.

Obsession with Work Leads to Suicide

Fischer was a dedicated scientist who had few outside interests. He was also secretive and seldom discussed his work with other scientists outside of his laboratory. Fischer's lack of outside interests extended to politics as well. Although he privately expressed concern over the rise of dictator Adolf Hitler and Nazi Germany, he chose not to speak out publicly. In 1935, Fischer married Wiltrud Haufe. Despite being three decades older than his bride, Fischer was a happily married man and once confided to Wieland, who was a personal friend, that his wife had greatly enriched his life.

Despite Fischer's dedication to his work, which some colleagues called obsessive, he did enjoy taking long motoring vacations in his car. His other love was the outdoors. Although he constantly battled the debilitating effects of tuberculosis, Fischer was an expert skier, accomplished hiker, and an avid mountain climber, a passion he shared with his father until an accident claimed the elder Fischer's life. Germany's involvement in World War II added to Fischer's woes. Because of supply restrictions and frequent bombing raids made by Allied forces, his work was seriously restricted. When a bombing run destroyed Fischer's institute, the scientist gave in to despair. In 1945, Fischer committed suicide, despondent over what he viewed as the destruction of his life's work.

Although he was able to organize large scientific efforts and had an intuitive feel for the chemical structures involved in the field of pyrrole chemistry, Fischer was not noted for his ability to clearly write or lecture on such topics. Despite this fact, he published the definitive work on pyrrole chemistry in three volumes, *Die Chemie des Pyrrols,* which has remained a standard text on the subject. In addition to the Nobel Prize, Fischer received the Leibig Memorial Medal in 1929 and the Davy Medal in 1936. He also received an honorary degree from Harvard University in 1935.

FISH • See Ichthyology

FISHER, RONALD ALYMER (1890-1962)
English mathematician and biologist

Fisher made many important contributions to the field of statistics. Born in London, England in 1890, Fisher attended Caius College in Cambridge. He graduated in 1912, specializing in mathematics and theoretical physics. After graduation he spent an additional year in Cambridge studying statistical mechanics and the theory of errors, subjects in which he was particularly interested. Following the outbreak of World War I, Fisher attempted to enlist, but his poor eyesight exempted him from military duty. During the next six years he worked at several jobs, including investment broker, high school teacher, and farm laborer, but in 1919 he found his niche at the Rothamstead Experimental Station, where he became a one-man statistics department. It was Fisher's job to analyze a huge backlog of weather data accumulated over 60 years—a seemingly impossible task when faced with traditional statistical analysis methods.

Fisher originated the idea of using small samples for his random sampling distributions, a highly successful concept that provided a completely objective, yet valid analysis. In the following years he applied similar methods to determine exact distributions of other statistical functions, including regression coefficient and discriminate functions. Fisher also developed new methods for using statistics in experiments, which included the analysis of variance, co-variance, and multivariate analysis. In addition, he made improvements to several existing statistical functions. In 1925 Fisher released Statistical Methods for Research Workers, a book that revolutionized research methods in many fields, particularly **genetics** and **agriculture**.

While at Rothamstead, Fisher dabbled in his other passion—genetics. He bred poultry, mice, snails, and other creatures and published his findings in several papers that contributed to scientists' overall understanding of genetic dominance. His study of human **blood** types and the formation of a blood-grouping department at the Galton Laboratory in 1935 were of significant aid in clarifying the inheritance of Rhesus blood types among humans.

Ronald Alymer Fisher. *(The Library of Congress. Reproduced by permission.)*

Fisher left Rothamstead in 1933 to become professor of eugenics at University College in London. Ten years later he moved to Cambridge University as Balfour Professor of Genetics. In 1959 Fisher retired from his position and moved to Australia, although he continued to work in the Division of Mathematical Statistics of the Commonwealth Scientific and Industrial Research Organization. He died in Adelaide, Australia in 1962. Fisher's many contributions to data collection and interpretation methods symbolized the beginning of a new era in statistical analysis.

See also Blood group; Heredity; Mendelian laws of inheritance; Meteorology; Rhesus factor

FITNESS

Fitness is a term used in **genetics** and evolutionary biology, to indicate the contribution of the **genotype** (i.e., the specific genetic information encoded in the **DNA** of an **organism**) of a particular individual to the next generation of its **population**, in comparison to the contribution of other genotypes. (The

Flagella are tiny, hair-like appendages of the surface of many cell types responsible for movement or sense receptors. Here, sperm is moving over the surface of a uterus. *(Photo Researchers, Inc. Reproduced by permission.)*

term fitness is also sometimes used in reference to a particular **allele**, i.e., to one of two or more possible combinations of DNA comprising the information of a specific **gene**.)

All individual organisms eventually die. However, according to **evolutionary theory**, individuals can continue to influence the **evolution** of their population and **species**, if they have passed as much as possible of their genetic information on to succeeding generations, that is, by having a high fitness.

The fitness of an organism is related to the degree to which it is adapted to its environmental and social circumstances. An **adaptation** is a characteristic that enhances the probability of an individual passing its genes along to its descendents, who may then pass their genetic legacy on to their own progeny, and so on. Adaptations can involve traits that are anatomical, physiological, or behavioral, and they may be due to simple genetic attributes (such as a single gene), or be under the control of numerous genes. Often, adaptations also enhance an individual's chances of surviving for a relatively long time, although this is not a necessary condition for it to have a high level of fitness (which keys on the organism passing its genetic information on to the next generation, and not necessarily on its own longevity). In fact, in many species, individu-

als die as soon as they manage to reproduce, as is the case of annual **plants** and Pacific salmon.

The notion of fitness has also been extended to groups of closely related individuals, through the theory of **kin selection**. According to this idea, an individual can increase its own "inclusive fitness" by fostering the fitness of closely related individuals, with which it shares a high degree of genetic information (such as its parents, sisters and brothers, and even cousins and nieces). Kin selection has been used to explain the occurrence of so-called "altruistic" (or helping) behavior in many species. Examples include the tendency of younger progeny of Florida scrub jays (*Aphelocoma caerulescens*) to help in the rearing of their younger siblings, and of social insects such as ants, bees, and termites to cooperate in colonies. In all of these cases, the reproduction of selected individuals is supported by the work of other, closely related, non-breeding individuals, who have an evolutionary stake in the breeding success of the group, through their inclusive fitness.

It is important to understand that fitness and adaptations are not the only factors influencing the evolution of populations and species. In some cases, apparently unpredictable environmental catastrophes, such as a forest fire or hurricane, may result in the survival of some individuals rather than others. Over the longer term, however, fitness and adaptation are thought to be the major forces driving evolution.

FIXED ACTION PATTERN • See Instinct

FLAGELLA

Flagella are tiny, hair-like appendages on the surface of many **cell** types that provide either the motile force for movement or that serve as sensory receptors for information about the cell's external world. There are two very different kinds of flagella in the biological world. Bacterial flagella are very thin, solid rods made up of a chain of globular **molecules** of a single **protein** called flagellin. These rods, which have a helical shape, move only when spun by molecular motors in the cell **membrane**. They allow bacterial cells to swim through a liquid medium.

In contrast, the flagella of eukaryotic cells (those with a true **nucleus**) are much larger and more complex than those of bacteria. Typically about 0.82 ft (0.25 m) in diameter and 66-164.05 ft (20-50 m) in length, eukaryotic flagella usually are very similar in structure and function to **cilia**. The most obvious differences between cilia and flagella is that the former are relatively short and numerous, while the later are typically longer and more sparse. A typical flagellated cell usually only has two to eight flagella.

The internal structures of both cilia and flagella are generally the same. Each has an **axon**eme (a hollow bundle of microtubules and associated proteins) composed, generally, of a central pair of single tubules surrounded by a protein sheath, and a peripheral set of nine doublet microtubules. Each of the external doublets has a radial spoke linking it to the central pair. It also has two curving arms made of a protein called dynein. This is a molecular motor that "walks" along an adjacent doublet pair. Because the doublets are firmly anchored to the cell wall by a **centriole**, sliding of adjacent tubules causes a bend to propagate along the length of the flagellum. The three-dimensional bending causes flagella to move in a continuous helical spiral that propels a cell through a fluid medium. In contrast, cilia move in a pattern that resembles the arms of a person doing the breast stroke, with a rigid power stroke followed by a flexible recovery. The **energy** source for both ciliary and flagellar motion is **adenosine triphosphate (ATP)**, which powers the "walking" motion of the dynein arms in both **organelles**.

Most motile, single-celled **organisms**, such as **algae** and **protozoa**, have either cilia or flagella. In multicellular **animals**, **sperm** are among the most commonly flagellated cells. Interestingly, the flagella of insect sperm show many variations from the typical nine plus two microtubular arrangement of most cilia and flagella. In certain insect **species** the ratio can be nine plus zero, nine plus one, nine plus five or seven, nine plus dozens, nine plus nine plus two, nine plus nine plus hundreds, or even no organization at all other than a random bundle of doublets. Amazingly, all these variations appear to be motile.

Alexander Fleming. *(The Library of Congress. Reproduced by permission.)*

In many sensory organs, the actual sensory receptors are highly modified, non-motile flagella. Some examples are the outer segments of the rods and cones in the retina of the eye, the hair cells of the cochlea in the inner ear, and the olfactory cells of the nose. Apparently, the sensitive cellular membrane covering flagella makes a good receptor for a wide variety of information from the world around us.

FLEMING, ALEXANDER (1881-1955)
Scottish bacteriologist

With the experienced eye of a scientist, Alexander Fleming turned what appeared to be a spoiled experiment into the discovery of the first of the "wonder drugs," **penicillin**.

Fleming was born on August 6, 1881, to a farming family in Lochfield, Scotland. Following school, he worked as a shipping clerk in London and enlisted in the London Scottish Regiment. In 1901, he began his medical career, entering St. Mary's Hospital Medical School, where he was a prizewinning student. After graduation in 1906, he began working at that institution with Sir **Almroth Edward Wright**, a pathologist. From the start, Fleming was innovative and became one of the first to use Paul Ehrlich's arsenical compound, Salvarsan, to treat syphilis in Great Britain.

Wright and Fleming joined the Royal Army Medical Corps during World War I. They studied wounds and infection-causing bacteria at a hospital in Boulogne, France. At that time, antiseptics were used to treat bacterial infections, but Wright and Fleming showed that, especially in deep wounds, **bacteria** survive treatment by antiseptics while the protective white **blood** cells in the wound are destroyed. This creates an even worse situation in which infection can spread rapidly. Forever affected by the suffering he saw during the war, Fleming decided to focus his efforts on the search for safe antibacterial substances. He studied the antibacterial power of the body's own leukocytes contained in pus. In 1921, he discovered that a sample of his own nasal mucus destroyed **bacteria** in a petri dish. He isolated the compound responsible for the antibacterial action, which he called lysozyme, in saliva, blood, tears, pus, milk, and egg whites.

Fleming made his greatest discovery in 1928. While he was growing cultures of bacteria in petri dishes for experiments, he accidentally left certain dishes uncovered for several days. Fleming found a mold growing in the dishes and began to discard them, when he noticed, to his astonishment, that bacteria near the molds were being destroyed. He preserved the mold—a strain of Penicillium—and made a culture of it in a test tube for further investigation. He deduced an antibacterial compound was being produced by the mold, and named it penicillin. Through further study, Fleming found that penicillin was nontoxic in laboratory animals. He described his findings in research journals but was unable to purify and concentrate the substance. Little did he realize that the substance produced by his mold would save millions of lives during the twentieth century.

Fleming dropped his investigation of penicillin and his discovery remained unnoticed until 1940. It was then that Oxford University-based bacteriologists **Howard Florey** and **Ernst Chain** stumbled upon a paper by Fleming while researching antibacterial agents. They had better fortune than Fleming, for they were able to purify penicillin and test it on humans with outstanding results. During World War II, the drug was rushed into mass-production in England and the United States and saved thousands of injured soldiers from infections that would otherwise have been fatal.

Accolades began pouring in for Fleming. He was elected to fellowship in the Royal Society in 1943, knighted in 1944, and shared the Nobel Prize in physiology or medicine with Florey and Chain in 1945. Fleming continued working at St. Mary's Hospital until 1948, when he moved to the Wright-Fleming Institute. He died in London on March 11, 1955.

See also Antisepsis

FLEMMING, WALTHER (1843-1905)
German anatomist

Of Flemish descent, Flemming was born in Sachenberg, Germany, where his father worked as the director of an insane asylum. As a young man, Flemming showed great aptitude for literature and language, but chose instead to study medicine at several German universities. He served briefly as a hospital as-

sistant before becoming an assistant in the department of **zoology** at Wurzburg University. In 1872, he accepted a position at the University of Prague, but the rise of Czech nationalism among his colleagues and students compelled him to return to Germany. He eventually became professor of anatomy and director of the Anatomical Institute at the University of Kiel, where he remained until his retirement.

Prior to Flemming's work, little headway had been made in the field of **cytology** since **Matthias Schleiden** and **Theodor Schwann** had proposed their **cell** theory in 1839, due in large part to the lack of effective cell staining techniques and poor **microscope**s. However, during the 1850s, advances in synthetic dyes were made, allowing scientists such as Flemming to better study cells. By using the new dyes, Flemming was able to observe and correctly identify the stages of **cell division**, a process that he later named **mitosis**.

Flemming found material within the cell **nucleus** that readily absorbed dye. He named the material chromatin, and by observing it at different phases, he could trace the action of cell division. As the process began, the chromatin arranged itself into short, thread-like objects, which Wilhelm von Waldeyer-Hartz later termed **chromosome**s. The chromosomes then doubled and pulled apart, each half migrating to opposite ends of the cell. In the final stage, the cell divided, leaving two daughter cells with equal amounts of chromatin.

By the end of 1879, Flemming had investigated all the stages of mitosis—a term that he derived from the Greek word for "thread"—and identified them in a variety of **tissue**s. In 1887, he began concentrating on cell division in spermatozoa. Although he detected that differences existed in the way sex cells divided, Flemming failed to identify the process later termed **meiosis** by **Edouard van Beneden**.

Unfamiliar with the work of **Gregor Mendel**, Flemming was unaware of the genetic significance of his findings. However, with the rediscovery of Mendel by Hugo de Vries at the turn of the century, Flemming's work has since provided the physical basis for Mendel's theories of **inheritance**. Flemming's *Zellsubstanz, Kern, und Zeltlteilung* (1882; *Cytoplasm, Nucleus, and Cell Division*) is considered a classic text by cytologists.

FLOREY, HOWARD WALTER (1898-1968)
Australian biochemist

Born in Adelaide, Australia, Florey became a leading researcher of the disease process. With colleague **Ernst Chain**, he isolated **penicillin**, making possible its wide production and use in treating bacterial diseases.

The son of a boot maker, Florey showed no interest in learning the family business. Instead, his natural curiosity and scholastic ability led him to pursue medical research. Florey attended the University of Adelaide, and after earning his medical degree in 1921, he received a Rhodes Scholarship to study at Oxford University. He also studied at Cambridge University and in the United States as a Rockefeller Foundation traveling fellow before returning to Oxford to earn his Ph.D. in pathology and biochemistry.

Florey had been interested in antibacterial agents for years, and in 1930 he began studying a natural antibacterial substance called lysozyme which had been discovered by **Alexander Fleming** almost a decade earlier. Florey was the first to purify it and determine how it acted. This line of study was to lead to his best-known achievement.

After four years as professor at the University of Sheffield, Florey was appointed professor of pathology at Oxford in 1935. He consulted chemist **Frederick Gowland Hopkins** regarding a suitable person to lead the biochemistry work at Oxford. Hopkins recommended Chain. Beginning in 1938, Florey and Chain began to study antibacterial agents found in **bacteria** and molds. They decided first to study penicillin, described (but never isolated) by Fleming almost a decade earlier. By 1941, the two had produced concentrated penicillin and shown that it could successfully treat bacterial infections in laboratory **animal**s without toxic effects. Florey and Chain began clinical trials on nine humans, all with dramatically successful results. The mass bloodshed of World War II created a desperate need for medications that could bring relief to thousands of victims of injury and sickness, so efforts to produce penicillin in large quantities were begun. Florey went to the United States to encourage production of the drug. He even traveled to battlefields to investigate the effectiveness of penicillin on wounded soldiers. Soon the drug was in common use.

After the war, Florey continued antibiotic research and later concentrated on experimental pathology. He also contributed research on the biology of mucus secretions, electron microscopy, and circulatory and pulmonary illnesses. Florey remained at Oxford as a professor of pathology until 1962, when he became provost of Queen's College, Oxford, and served as president of the Royal Society from 1960 to 1965. For his work with penicillin, Florey shared the 1945 Nobel Prize in Medicine with Fleming and Chain. Florey was knighted in 1944 and in 1965 was named Baron Florey of Adelaide.

FLOWER

Virtually everywhere on earth flowers blossom on **plants** and **tree**s. They are essential for reproduction of vegetation, since they develop the seeds for further growth, and flowering plants are the nutrient source for almost all **species** of **animals**. Humans consume grains, **fruits** and vegetables, and the animals many people use for food, such as cattle, hogs and sheep, also live on flowering plants.

Any plant that produces some sort of flower, even a small colorless one, is a flowering plant. Thus, grasses, roses, apple trees, lilacs and oaks are all flowering plants. At least 200,000 kinds of flowers have been classified, from the microscopic water blossom to the three-foot-wide tropical plant called the Giant Rafflesia. A blossom's shape may be long and thin, such as grass, or flowers that resemble stars, balloons or even insects. Flowering plants also may have pleasant fragrances to attract the birds and insects that help to fertilize the

The center of a composite flower showing florets. Flowering plants are the dominant terrestrial plants in the world. *(Photograph by Robert J. Huffman, Field Mark Publications. Reproduced by permission.)*

plant, or they may have unpleasant odors to repel enemies. These and other variations have emerged over millions of years, as plants adapt to their environment.

Despite various shapes, sizes, and smells, each flowering plant includes some or all of the following parts: roots, stems, leaves, and flowers. The roots are underground and serve as anchors or support structures, and absorb water and dissolved **minerals** which nourish the plant. Stems may be partially underground, such as bulbs of onions or tulips or tubers of potatoes, but most of the stems grow upright above the earth. They support the rest of the plant and convey the food and **water** up and down the plant. Leaves spread out and absorb sunlight, which acts on their green **chlorophyll** to produce sugar and starch, which feed the plant. Leaves may be simple or compound. Simple leaves contain a single undivided blade, while compound leaves are divided into **leaf**lets.

Flowers are the reproductive parts of the plant. In the center of the flower is the pistil, which produces the seeds. At the flower's base is the ovary, where the ovules that will grow into seeds after **fertilization** are stored. Above the ovary is the style, a stalk that grows to the right height for the plant to receive pollen, or the fertilizing substance. Five arms at the style's top form the stigma, which open when the plant is ready to be fertilized.

The pollen is produced in the stamens, which encircle the pistil. A thin stalk called the filament holds up the anther, the small sac that comprises the top part of the stamen. Inside the anther is the yellowish, powdery pollen. The size and shape of these parts vary according to whether the plants pollinate via insects or the air. Around the stamens are the protective petals, which collectively are called the corolla. Their shape and color attract and provide a landing base for insects, and also protect the reproductive parts they surround. Holding the blossom are leaves called sepals (collectively called the calyx). Flowers that have all of these parts are termed perfect or complete, while those missing one or more parts are called incomplete or imperfect flowers. Either type may grow singly or in clusters.

The flowering plants of the world have been classified into approximately 300 families according to these flower parts. Thus, even though certain plants may live in different climates and soils and have vastly differing shapes, they may be part of the same **family** because of how their reproductive **organ**s look and operate. The lily family, for example, is found around the world, and includes plants that are pollinated by insects. They contain three sepals and petals which closely resemble one another, six stamens, one pistil, long sheathlike leaves with parallel veins, and fruit that contains many seeds within one capsule. Examples of the lily family include the tulip, hyacinth, trillium, lily-of-the-valley, onion, garlic, asparagus, crocus, yucca and aloe, and all the lily plants. The pea family includes trees such as the locust and rosewood; vines like the wisteria and bean; herbs that include clover, lupine, licorice and alfalfa; and the **protein**-rich seed plants of the peas, beans, lentils, peanuts, and soybeans.

Flower families are further divided into two large groups that have evolved separately for millions of years: monocots (having one sheath or casing) and dicots (having two casings). Monocots, or moncotyledons, have only one seed leaf, petals and sepals which resemble one another, simple leaves with parallel veins without teeth or serrated edges, and singular stems that are all or partially underground. Because they are mostly low plants, tree forms of monocots are rare. Examples of moncots include lilies, grasses, and orchids. Dicots, or dicotyledons, are a large and varied group that range from a violet to an oak tree. They all have two seed leaves, which means they send down roots and send up a stem, their petals are distinct from the sepals, the leaves are simple or compound and have a wide variety of shapes, and there are multiple branches from one or more stems. The leaves' veins also have a radiating network of veins, versus the parallel veins of a monocot. While almost all flowering plants in either group need soil to survive, tropical plants called epiphytes, or air plants, live on the trunks and branches of trees. They get their food from the air and dirt and decaying leaves that collect around their open roots.

FLU VACCINE • See Influenza

FOOD CHAIN

There is no waste produced in a functioning, thriving natural **ecosystem**. All **organism**s, dead and alive, are potential sources of **energy** and **nutrition** for other members of the environment. For example, a worm digests tiny soil **nutrients**; a robin eats the worm; a wild cat eats the robin and when the cat dies, it is then consumed by bacterial decomposers. This process of exchanging energy and nutrients among living organisms by feeding on each other is are called food chains.

The word "chain" is probably misleading, however, because it implies an orderly **linkage** of equal parts. Actual food chains are extraordinarily complex because there is no exact order specifying what creatures eats whom. It would be more accurate to consider each of the links as an energy carrier in a complex network of many interconnected food chains, called a food web. Another way to picture a food chain is in the shape of a pyramid. The bottommost layer of a food pyramid consists of the most abundant elements: **plants** that trap solar energy through **photosynthesis**. These plants are then eaten by herbivorous **animal**s, which are considered the primary **consumer**s at the second level of the pyramid. In turn, these primary consumers become food for other animals called secondary consumers at the third level of the pyramid. Each succeeding level consists of fewer, usually larger flesh-eating animals.

The consumers at the top of the pyramid do not represent the uppermost end of the food chain. When they die, they are eaten by tiny, microscopic organisms which serve as decomposers. When devoured by these **bacteria**, the nutrients are returned to the soil to be recycled into yet another food chain.

It may seem like the food chain consists of predator-like continuous killing and devouring of all living creatures. However, in the ecological sense, predators of the food chain can be **herbivore**s and feed off of plants, or they can be parasitic and continuously feed off live organisms. **Predation** is a more general term to explain this because it separates live-feeders from **scavenger**s and decomposers. An organism does not necessarily have to kill another organism to be a predator. For example, **parasite**s are organisms that derive nourishment by living in or upon the body of a host organism that remains alive.

FORESTS

Forests are a crucial component of the world's **ecology**, providing a wealth of biological diversity. Although **tree**s are the most visible component, within the woodland and forest **biome**s that cover about 30% of the world's ice-free land surface, there are other **animal** and plant **species** as well. Forest trees provide a huge number of products upon which human society depends, including fuel, fiber, paper, materials for construction and building, food, and medicines. They also play a vital role in many **ecosystem**s, such as maintaining the structure and fertility of forest soil, and preventing soil erosion and floods. In addition, forests provide the invaluable ecological function of absorbing **carbon dioxide** and releasing oxygen through **photosynthesis**. Through this process, forests control greenhouse gas levels, which help maintain global **temperatures**, and produce the **atmosphere** upon which all living things depend.

Trees began their **evolution** more than 400 million years ago. Botanists believe the first tree was a plant probably less than a meter tall with tiny stems and no true leaves. By 350 million years ago, huge Devonian forests of fern-like trees had grown and were home to a great variety of insects. By the time warm-**blood**ed **mammals** appeared, seed-bearing conifers had become dominant, succeeding earlier **spore**-bearing **plants**. Conifers in turn were overtaken by deciduous trees with broad leaves and **flower**s, which had spread worldwide by 60 million years ago.

Today, three types of forest trees occur in regions of varying distances from the equator. Evergreen broadleaf biomes or tropical **rain forest**s flourish in the tropical parts of Africa and southeastern Central America, where the annual rainfall is never less than 130 centimeters. Throughout the year, the trees sprout new leaves and shed old ones. Tropical forests contain more species of animals than any other biome.

In moist, temperate latitudes where winters are mild, the deciduous broadleafs dominate. Temperate deciduous forests occur in much of the eastern United States, extending northward into southeast Canada, as well as in central and northern Europe, China, the eastern region of the former Soviet Union, Korea, Japan, coastal Brazil, east Africa, the eastern coast of Australia, and most of New Zealand. In these forests, where rainfall is somewhat evenly distributed throughout the year, the seasons change and leaves fall from the tree branches, creating a rich soil that supports lush forest growth. Wolves, bears, foxes, and mountain lions once inhabited these forests; today the wildlife largely comprises squirrels, raccoons, opossums, rabbits, and rodents.

Evergreen coniferous forests are found mostly in mountainous regions of the temperate zone and at higher, colder latitudes of northern Europe, Asia, and North America. Also called boreal forests or **taiga**, which means "swamp forest," these forests occur where the winters are long, cold, and dry and summers are short. Bogs can be found in certain parts of the taiga, along with cold lakes and streams. In the taiga's far north, the dense forest becomes scattered, eventually giving way to **tundra**. Moose, elk, deer, grizzly bear, black bear, wolves, lynx, and wolverines can be found in the taiga, along with rabbits, hares, porcupines, and rodents. Besides in the boreal forests, evergreen conifers also grow in such biomes as montane coniferous forests, temperate rain forests, and pine barrens.

See also Acid rain, Biodiversity, Carbon dioxide, Evolution, Photosynthesis

Redwood multiple trunks in forest glade. *(Photograph by Robert J. Huffman, Field Mark Publications. Reproduced by permission.)*

FORSSMANN, WERNER (1904-1979)
German physician

Werner Forssmann, a surgeon and urologist, was relatively unknown in his native Germany when he won the Nobel Prize in 1956 for his work in **heart** catheterization. His groundbreaking experiment had been done almost three decades earlier, and when he received word of the award—after a morning of surgery during which he had operated on three patients with kidney disease—he commented, as quoted in *Mayo Clinic Proceedings,* "I feel like a village parson who has just learned that he has been made bishop."

Werner Theodor Otto Forssmann was born on August 29, 1904, in Berlin, the only child of Julius Forssmann, a lawyer employed by a life insurance company, and Emmy Hindenberg. Forssmann's father died in World War I while young Forssmann was still a student in the Askanische Gymnasium, a school emphasizing a humanistic approach to education. His mother worked as an office clerk and his grandmother took over the role of running the household. Forssmann's uncle, a doctor just outside of Berlin, became an influential force in his nephew's life, ultimately convincing Forssmann to pursue a career in medicine. In 1922, after graduating from the Gymnasium, Forssmann entered the Friedrich Wilhelm University in Berlin, passing the state examination in 1928. Forssmann's doctoral thesis on the effects of concentrated liver on pernicious anemia, a blood deficiency, marked the way for his later experiments. Together with a small group of fellow students, Forssmann experimented on himself, taking large doses of liver concentrate daily and demonstrating its healthful effects on **blood**. After receiving his doctor's diploma in early 1929 and being frustrated in his efforts to obtain a post as an internist, Forssmann worked for a short time in a private women's clinic in Spandau. Then, through family connections, he secured an internship at the August Viktoria Home in Eberswalde, a small town northeast of Berlin.

Seeks Cardiac Diagnostic Tool

Training as a surgeon, Forssmann nevertheless gave thought to an earlier passion of his, one inspired by a teacher he encountered in medical school: heart diagnosis. He was dissatisfied with the inaccuracy and uncertainty of diagnostic techniques such as percussion, auscultation, X ray, and even electrocardiography. He became convinced that there was an internal diagnostic method that would not involve major risks, trigger automatic reflex actions, or disturb the balance of pressure in the thorax. As early as the mid-nineteenth century, there had been a procedure known as cardiac catheterization in **animal** experiments. Doctors had performed the procedure in the late nineteenth century to determine blood pressure in the right and left chambers of the heart. Some of these procedures employed the use of a catheter inserted through the jugular vein of a horse. Forssmann believed that he could do this on humans through a vein at the elbow traditionally used for intravenous injections. His research on cadavers supported his idea, and by the summer of 1929 Forssmann approached his supervisor, Dr. Richard Schneider, with a plan to catheterize his own heart with a ureteric catheter. Schneider, however, would not allow such a dangerous experiment in his hospital.

Undaunted, Forssmann set out to convince a surgical nurse in his section of his experiment's feasibility so he could gain access to the sterilized instruments he needed. Eventually, the nurse agreed to aid him, even agreeing to be the first subject. Forssmann, however, had no intention of experimenting on anyone but himself initially. He gave himself a local anesthetic in the left elbow and then made an incision. Once he had opened his vein, he inserted the catheter about a foot up his arm and had the nurse accompany him to the X-ray lab. There, Forssmann stood behind a fluoroscope screen with a mirror placed so that he could see the image of the catheter, which he pushed up until it was in the right ventricle of his heart. Then he calmly ordered that photographs be made of this momentous achievement.

The results of this experiment were published in a short paper in the prestigious *Klinische Wochenschrift* and won Forssmann a position at the Charité Hospital in Berlin. But the reception to his article by other physicians was cool and his superior at the Charité did not approve of his unorthodox tech-

niques, so Forssmann was soon back in Eberswalde. He continued his experiments for the next two years, during which time he proved that the insertion of a catheter in the heart was painless and caused no damage to the blood vessels. He also pioneered techniques for measuring pressure inside the heart and for injecting opaque material for X-ray studies of the heart. Still, his work was reviled by most physicians, who called it unethical and considered his experiments stunts. By 1931, Forssmann, discouraged by the response to his work, gave up experimental medicine. He returned to the Charité Hospital in Berlin and soon moved on to the Mainz City Hospital. It was there, in 1932, that he met the woman who would become his wife, Dr. Elsbet Engel, a resident in internal medicine. Though their marriage was happy and fruitful (they would have six children together), it also necessitated another change of hospitals for Forssmann, for it was against the hospital's policy for a married couple to work together. Forssmann trained as a urologist in Berlin at the Rudolph Virchow Hospital, then took a position as a surgeon and urologist at the City Hospital of Dresden-Friedrichstadt for two years. Later, he became a senior surgeon at the Robert Koch Hospital in Berlin. His colleagues considered him a fine surgeon.

Fame Comes Late in Life

During World War II, Forssmann served as an army surgeon, surviving six years spent in Germany, Norway, Russia, and in a prisoner of war (POW) camp. Back in Germany after the War, he practiced as a country doctor in the Black Forest village of Wambach for three years before returning to the practice of urology in 1950 at Bad Kreuznach. It was only after the war that Forssmann discovered that others had continued working with his cardiac catheterization experiment of 1929. The most notable implementation was by two Americans, **Dickinson Woodruff Richards, Jr.**, and **André F. Cournand**, who developed it into a tool for diagnosis and research. In 1954, Forssmann received the Leibniz Medal from the German Academy of Science in Berlin, yet he was refused a professorship at the University of Mainz. He had resigned himself to being a little-known doctor in Bad Kreuznach when, on October 18, 1956, he was notified that he had won, along with Richards and Cournand, the Nobel Prize for Physiology or Medicine for his contribution to the knowledge of heart catheterization and pathological changes in the **circulatory system**.

The Nobel Prize finally earned Forssmann renown and respect; in *Clinical Cardiology,* H. W. Heiss called him "one of the great fathers of cardiology." In 1958, he became the chief of the surgical division of the Evangelical Hospital of Düsseldorf, and ten years later he was awarded the gold medal of the Society of Surgical Medicine of Ferrara. After he retired, Forssmann spent his time in the Black Forest, where he enjoyed the outdoors and nature. He died of a heart attack in Schopfheim, West Germany, on June 1, 1979.

Fossil

Fossils are the remains, traces or impressions of prehistoric **organism**s. There are wholly intact **animal**s which have been pre-

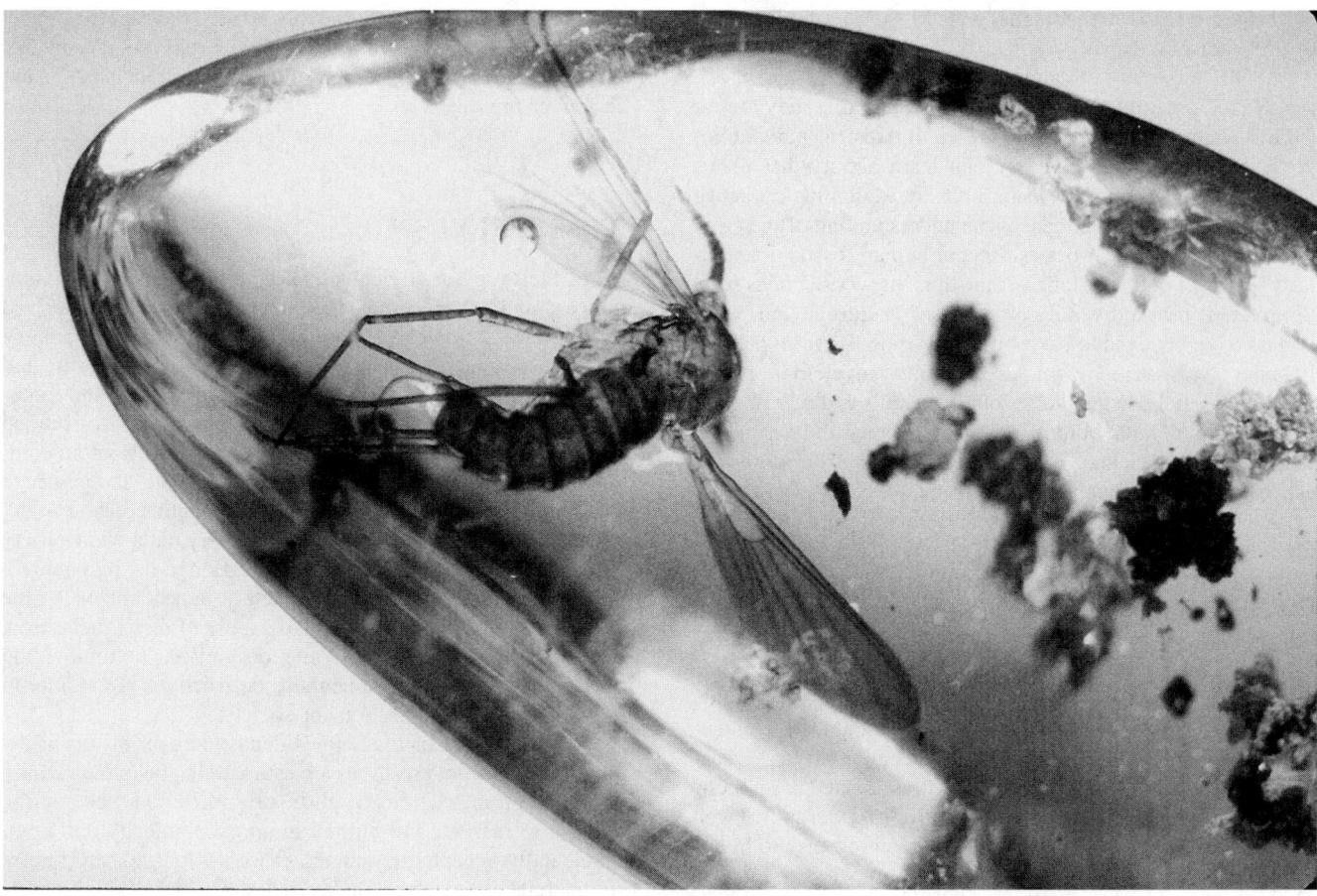

A fly in amber, approximately 35 million years old. *(JLM Visuals. Reproduced by permission.)*

served such as the wooly mammoth (discussed below) or the thousands of **species** of insects caught up in the resins of ancient pine **tree**s and preserved in the yellow-orange translucent fossil-bearing material amber. Prehistoric humans which fell into bogs have been preserved intact. Other fossils are produced by replacement of organic **matter** by inorganic matter. The inorganic replica is preserved in rock (petrified). The petrification process may be so subtle that microscopic examination, especially of petrified **plants**, reveals fine **cell**ular structure. Footprints and preserved fecal material, while not actual remains of preexisting animals, are considered to be fossils because they were made by living animals. Footprints reveal something of the weight of the animal, length of stride, and whether it was bipedal or if it moved on all four feet. The diet of prehistoric animals may be revealed, in part, by examination of fossil feces known as coprolites. Unfortunately, far from all preexisting organisms are known. Many plant and animals have soft bodies and it would be extremely unlikely if they would be preserved down through the ages. Thus, little is known of literally thousands of organisms that lived in the past. Another disappointment when looking at the fossil record is that the overwhelming majority of fossils are incomplete fragments of organisms. **Charles Robert Darwin** knew full well the inadequacy of the record of preexisting life—he referred

to fossils as a multivolume history of life, but unfortunately, there are only a few volumes remaining. And, here and there, only a chapter has been preserved, and often, many of the pages are blank. The Darwinian evaluation of the incompleteness of the fossil record of 1859 still holds.

An example of elegantly preserved fossils are the wooly mammoths. They were huge elephant-like animals found in many parts of the world including North America. They lived during the Pleistocene period which ended approximately 10,000 years ago. The reason that we know so much about these enormous fossils is that many of them lived in areas permanently frozen. Thus, when a mammoth would fall into, or be trapped, in an icy crevasse or meet some other fate that resulted in an icy tomb, the animal would be frozen and remained as such until it was discovered thousands of years later. The preservation was so good that the plants of the last meal could sometimes be identified and the muscle (meat) was occasionally still red thereby providing a meal for sled dogs. The Siberian mammoths had extremely long tusks which were so abundant that they were traded on the medieval markets. Mammoths are extinct now but they existed at the same time as prehistoric humans. Cave dwellers depicted masses of mammoths which witnesses to both the former abundance of these titanic fossil **mammals** as well as to the artistic ability of prehistoric humans.

Of course, fossil mammoths are of relatively recent origin. So too are some human fossils. The fact that they are far from ancient does not diminish their interest to paleontologists. As mentioned above, we can learn about what plants were in the diet of extinct mammoths by examining **digestive system** contents. This enhances the understanding of the environment of these great beasts. Recent human fossils are similarly interesting. For instance, some humans whose bones were discovered from a diversity of locations in Europe and North America are believed to have suffered during life from the malignancy (**cancer**) known as multiple myeloma. Paleopathologists believe that holes found in skulls and other bones, with sharply defined borders and which appear to be punched out, are strong evidence of multiple myeloma. The fossils are estimated to be as much as 5,000 years old. No one knows for certain what the contemporary risk factors are for this human malignancy but exposure to petroleum products and radiation have been suggested as possibly involved in causation. Paleontologists, and cancer epidemiologists as well, would suggest that other factors may be important to the causation of multiple myeloma due to the dearth of petroleum industry, diagnostic X-rays, nuclear power plants and nuclear explosions 5,000 years ago. However, multiple myeloma is thought to be not uncommon in fossil humans which suggests alternative risk factors. Other human cancers, generally associated with bone because of bone durability, are known from antiquity. It is well that this be remembered when it becomes all too easy to blame cancer in modern humans because of industry or the **stress** of modern life.

As interesting as frozen carcasses of ten millennia ago, or evidence of cancer in humans thousands of years ago are, it should be emphasized that they are not the primary focus of much of the research that has attended the study of fossils. Ancient fossils provide scholars with considerable insight into a diversity of subjects including climatology, geography, geology and **evolution**. Examples: Ancient plant and animal fossils gives information about the climate when they existed as living beings. Modern coral reefs occur in tropical or semitropical areas where the **water** remains warm throughout the year. It seems reasonable to suspect that sites of fossil coral reefs are places where the **temperature** remained warm. The presence of fossil corals, accompanied by other fossil **invertebrates** known to live in the sea, permits the mapping of ancient oceans. Ancient geography was indeed dissimilar to the geography of today. A vast ocean extended from the Gulf of Mexico to the Arctic Ocean thereby cleaving the North American continent into two great land masses. The geological study of fossils has shown that similar fossils are found in sedimentary rock throughout the world. It is also known that the oldest fossils are found deepest and newer fossils are found in more superficial layers. Any one particular site is unlikely to have all of the layers of sedimentary rock that are known. However, age can be estimated by the examination of the fossils in the particular layers that are present. Finally, even though it is incomplete, the fossil record is a testament to the fact of evolution. Some groups of animals evolved into many species, many of which still survive. In contrast, the fossil record contains many examples of organisms that flourished in the past but are no longer present.

FOSSIL FUELS

Fossil fuels are buried deposits of petroleum, coal, peat, natural gas, and other carbon-rich **organic compound**s derived from the dead bodies of **plants** and **animals** that lived many millions of years ago. Over long periods of time, pressure and heat generated by overlying sediments concentrate and modified these materials into valuable **energy** sources for human purposes. Fossil fuels currently provide about 90% of all commercial energy used in the world. They provide the power to move vehicles, heat living spaces, provide **light**, cook our food, transmit and process information, and carry out a wide variety of industrial processes. It is no exaggeration to say that modern industrial society is nearly completely dependent on (some would say addicted to) a continual supply of fossil fuels. How we will adapt as supplies become too limited, too remote, too expensive, or too environmentally destructive to continue to use is a paramount question for society.

The amount of fossil fuels deposited over history is astounding. Total coal reserves are estimated to be in the vicinity of 10 trillion metric tons. If all this resource could be dug up, shipped to market, and burned in an economically and environmentally acceptable manner, it would fuel all our current commercial energy uses for several thousand years. Petroleum (oil) deposits are thought to have originally amounted to some four trillion barrels (600 billion metric tons), about half of which has already been extracted and used to fuel industrial society. At current rates of use the proven oil reserves will be used up in about 40 years. World natural gas supplies are thought to be at least 10 quadrillion cubic feet or about as much as energy as the original oil supply. At current rates of use, known gas reserves should last at least sixty years. If we substitute gas for oil or coal, as some planners advocate, supplies will be used up much faster than at current rates. Some unconventional **hydrocarbon** sources such as oil shales and tar sands might represent an energy supply equal to or even surpassing the coal deposits on which we now depend.

In the United States, oil currently supplies about 40% of all commercial energy use, while coal contributes about 22%, and natural gas provide about 24%. Oil and its conversion products, such as gasoline, kerosene, diesel fuel, and jet fuel are the primary fuel for internal combustion engines because of the ease with which they can be stored, transported, and burned. Coal is burned primarily in power plants and other large, stationary industrial boilers. Methane (natural gas) is used primarily for space heating, cooking, **water** heating, and industrial processes. It is cleaner burning than either oil or coal, but is difficult to store or to ship to places not served by gas pipelines.

The use of fossil fuels as our major energy source has many adverse environmental effects. Coal mining often leaves

This scientist is using an argon ion laser to determine the flow of coal ash particles in an exhaust gas model. The research is being done to find ways of reducing emissions from coal-fired power plants. *(Photo Researchers, Inc. Reproduced by permission.)*

a devastated landscape of deep holes, decapitated mountain tops, toxic spoil piles, and rocky rubble. Acid drainage and toxic seepage from abandoned mines poisons thousands of miles of streams in the United States. Every year the 900 million tons of coal burned in the United States (mainly for electric power generation) releases 18 million tons of sulfur dioxide, five million tons of **nitrogen** oxides (the main components of **acid rain**), four million tons of **carbon monoxide** and unburned hydrocarbons, close to a trillion tons of **carbon dioxide**, and a substantial fraction of the toxic metals such as mercury, cadmium, thallium, and zinc into our air. Coal often contains uranium and thorium, and that most coal-fired power plants emit significant amounts of radioactivity—more, in fact, than a typical nuclear power plant under normal conditions. Oil wells generally aren't as destructive as coal mines, but exploration, drilling, infrastructure construction, **waste disposal**, and transport of oil to markets can be very disruptive to wild landscapes and wildlife. Massive oil spills such as the grounding of the *Exxon Valdez* n Prince William Sound, Alaska, in

1989 illustrate the risks of shipping large amounts of oil over great distances. Nitrogen oxides, unburned hydrocarbons, and other combustion byproducts produced by gasoline and diesel engines are the largest source of **air pollution** in many American cities.

One of the greatest concerns about our continued dependence on fossil fuels is the waste carbon dioxide produced by combustion. While carbon dioxide is a natural atmospheric component and is naturally absorbed and recycled by **photosynthesis** in green plants, we now burn so much coal, oil, and natural gas each year that the amount of carbon dioxide in the **atmosphere** is rapidly increasing. Because carbon dioxide is a greenhouse gas (it is transparent to visible light but absorbs long wavelength infrared radiation), it tends to trap heat in the lower atmosphere and increase average global **temperature**s. Climatic changes brought about by higher temperatures can result in heat waves, changes in rainfall patterns and growing seasons, rising ocean levels, and could increase the frequency and severity of storms. These potentially catastrophic effects of global climate change may limit our ability to continue to use fossil fuels as our major energy source. All of these considerations suggest that we urgently need to reduce our dependency on fossil fuels and turn to environmentally benign, renewable energy sources such as solar power, wind, **biomass**, and small-scale hydropower.

FOX, SIDNEY WALTER (1912-)
American biochemist

Born in Los Angeles, California, Fox graduated from the University of California in 1933 and earned his doctorate degree at the California Institute of Technology in 1940. Throughout his career, Fox taught at institutions in Florida and became associated with the National Aeronautics and Space Administration (NASA) after 1960. Fox pursued questions on the origins of life, but unlike other proponents of **evolutionary theory** who approached the question from a biochemical standpoint, Fox's approach was biological.

Fox asserted that a mixture of **amino acid**s subjected to intense heat (a scenario in which the earth formed with a steaming ocean and exposed rocks) becomes a **protein**-like **polymer**. Fox named these polymers "proteinoids," or "coacervates." When dissolved in **water**, they form tiny spheres that share some properties with cells, including a double **membrane** and the ability to carry on simple **chemical reactions**. Adding more proteinoids to water will cause the spheres to combine, grow, and divide. Fox speculated from this evidence that the earliest **cell**s were formed through this process. It has also been suggested that the formation of cells and **nucleic acids** may have paralleled each other and combined at some point, and that the advent of nucleic acids did not come about through proteinoids. Fox's theory is currently a subject of ongoing research throughout the scientific community.

FRACASTORO, GIROLAMO (ca. 1478-1553)
Italian physician and philosopher

Girolamo Fracastoro, famous for his insight into and literary works on natural philosophy, astronomy, and medicine, was born in Verona, Italy, the sixth of seven brothers in a well-respected family. His grandfather was physician to the reigning nobility of Verona; his mother—Camilla Mascarelli—is believed to have died while he was very young. Although his father's occupation is not known, it was he who introduced the young Fracastoro to literature and philosophy, tutoring him personally before sending him off to the Academy of Padua under the guardianship of an old family friend and teacher, Girolamo della Torre, and under whom Fracastoro would ultimately study medicine. In keeping with tradition, however, he first studied liberal arts—including literature, astronomy, mathematics, and philosophy—as well as geography, botany, and pharmacology. Fracastoro married Elena de Clavis around 1500 with whom he had a daughter and four sons. Two sons

died very young, and Fracastoro expressed his heartache in a beautiful poem written for them.

Upon receiving his medical degree in 1502, Fracastoro began teaching logic and **anatomy** at the academy where he met Copernicus, who entered in 1501 to study medicine. In 1508, the war between the Roman Emperor Maximilian I and Venice closed the university, and Fracastoro fled to live near the border of Veneto where he apparently practiced medicine. In 1509 he returned to Verona, practicing medicine and managing Incaffi, the estate on the shores of Lake Garda inherited from his father. During the ensuing years, he developed relationships with well-known philosophers, scientists, and influential figures such as bishop Gian Matteo Giberti of Verona, a great patron of the arts and sciences.

At the turn of the sixteenth century, **syphilis**, the then mysteriously transmitted, ravaging, and untreatable epidemic, was spreading wildly through a terrified population. Around 1510, Fracastoro began composing a long and beautifully written narrative poem on the disease. Considered his most famous work, the 1,300-verse epic entitled *Syphilis sive morbus Gallicus,* (the French disease), written in tones of Virgil, Ovid, and Dante, combined fact and fantasy. He presented his poem in two books to Cardinal Pietro Bembo— considered the best stylist of the age—in 1525 for Bembo's advice. Published in 1530 in three volumes and rapidly becoming popular, the first volume describes the horrors of the disease. The second, devoted to cures and preventions, culminates with a mythical tale of cause and cure. The third contains two mythical stories. The first story tells of Christopher Columbus's journey to the West Indies where the disease was rampant among the natives. Fracastoro depicts the natives as descendants of the lost city, Atlantis. As punishment for its wickedness, the gods plagued the city with the dreaded disease before plunging it into the depths of the ocean with violent earthquakes. It is in the West Indies that Columbus discovers the holy guaiacum tree, extractions from which could cure the disease. The second story tells of how a young shepherd named Sifilo blasphemed against the Sun god, Apollo. In his rage, Apollo afflicted the shepherd with a disease. Only after Sifilo appeased the god did Apollo and Juno provide the healing guaiacum tree. The disease is believed to have received its named from this tale. Fracastoro dedicated his volumes to Bembo, who claimed it was the "most precious gift he had ever received."

De contagione et contagiosis morbis et curatione, written by Fracastoro in 1546, contains the culmination of his philosophy on contagious diseases. Virtually centuries before the origins, causes, and transmission of diseases were understood scientifically, at a time when disease was perceived as punishment from God and influenced by natural phenomenon such as phases of the moon, Fracastoro began rejecting such theories. Because his medical educators believed in the "separation of the two realms of theology and science," he thought more scientifically, postulating that diseases were spread either by simple contact, by *fomites* (clothing, sheets, or other physical objects), or from a distance by *seminaria morbi* (seeds of contagion) which enter the body and rapidly multiply. Fracastoro has been called "the forerunner of the **germ theory** of infectious diseases," and credited with "prophetic intuition."

In 1545, Pope Paul III nominated Fracastoro as the physician to the Council of Trent, and around 1546 he became canon of Verona. He suffered a stroke on August 6, 1553, and died that same day, most likely in his villa at Incaffi. In 1555, his statue was erected near those of Pliny and Catullus in the Piazza dei Signori, Verona.

FRAENKEL-CONRAT, HEINZ (1910-)
German American biochemist

Born in the German city of Breslau (now Wroclaw, Poland), Heinz Fraenkel-Conrat studied medicine in his native country but left for Great Britain during Adolf Hitler's rise to power. After earning his Ph.D. in 1936 from the University of Edinburgh, he came to the United States and remained, becoming a citizen several years later.

Fraenkel-Conrat contributed much to the study of **viruses**, particles that attack and damage living **cells**. Viruses basically contain two parts: an inner core of **nucleic acid**, which allows them to pass on genetic material, and a protein coat. In 1955 Fraenkel-Conrat successfully separated the outer protein coat from the inner nucleic acid core of the tobacco **mosaic** virus or TMV. He then reassembled the two components to show that the viruses were still capable of infecting living cells. Continuing his work, Fraenkel-Conrat showed that the protein portion was relatively inert and that the nucleic acid was responsible for infecting **bacteria**. The nucleic acid in the tobacco mosaic virus is **RNA**, ribonucleic acid, and his discovery showed that RNA is capable of transmitting genetic information as is **DNA**, deoxyribonucleic acid, the nucleic acid found in some other viruses as well as in living things. Later Fraenkel-Conrat worked with Wendell Stanley to show the entire sequence of 158 amino acids that make up the tobacco mosaic virus protein.

Over his long career as a professor of virology and later molecular biology at the University of California at Berkeley, Fraenkel-Conrat has written a number of virology texts considered the most important in the field.

FRANKLIN, ROSALIND (1920-1958)
English physical chemist

Rosalind Franklin was born on July 25 in London, England, the daughter of an affluent Jewish family. At the age of 15, Franklin determined she would become a scientist—a daunting aspiration for a young girl in what was then a decidedly male-dominated field. Franklin's most famous work—done between 1951 and 1953—involved the use of a new invention known as x-ray crystallography. Using this technology, she produced the first clear photos of a **deoxyribonucleic acid (DNA)** molecule. These photographs ultimately led to the identification of DNA's **double helix** structure. She died of ovarian cancer at the age of 37.

As a youth, Franklin attended St. Paul's Girls' School, one of the few girls' schools in London at that time which taught chemistry and physics. Her father adamantly opposed her attending university, insisting she go into social work. However, she ultimately gained his approval to attend Newnham College at Cambridge University. She obtained a bachelor of arts degree in 1941 and then accepted a fellowship there, which she relinquished a year later to take her first position at the British Coal Utilization Research Association. There she focused on the structure of graphite and other carbons, using her analytical skills and chemistry background to differentiate between carbons that became graphite when heated and those that did not. She used this research as the subject for her doctoral studies.

After earning her doctorate in physical chemistry at Cambridge in 1945, Franklin travelled to Paris, working in the Laboratoire Central des Services Chimiques de L'Etat from 1947 until 1950. There she was introduced to and began using the newly-developed technology of x-ray diffraction known as x-ray crystallography with which atoms could be identified and mapped in any type of crystal.

Upon returning to England in 1951, she was hired as a research associate in the lab of John Randall at King's College, University of London. Randall gave her sole responsibility for researching DNA, a project which had been begun but on which no work had been done for some time. From 1951 to 1953 she studied the structure of DNA. Using a precise and painstaking analysis which she gained her noteriety, she determined—and was the first to report— that the sugar-phosphate backbone was on the outside rather than the inside of the DNA molecule while gleaning crucial evidence of the molecule's two- stranded, helical structure. J. D. Bernal, a noted peer, called her photographs of the DNA molecule "...among the most beautiful x-ray photographs of any substance ever taken."

At the height of her research, before she documented or published her findings, Randall shared her discoveries with colleagues at a routine seminar. Apparently some friction existed between Franklin and another colleague at the lab, **Maurice Wilkins**, who also had an interest in DNA but had been assigned by Randall to a different project. In several historical accounts, Wilkins is reported to have had difficulty accepting a woman as his peer. Regardless, he obtained one of Franklin's DNA photographs and, without permission from either her or Randall, showed it to her competitors at Cambridge—**James Watson** and **Francis Crick**. This team used her data, along with information gleaned from other scientists, to accurately define and describe the double helix structure of DNA. They quickly publishing their findings in *Nature*. Although a supporting article on Franklin's work was published in the same issue, she received little credit for her crucial role in solving the DNA riddle.

Following her work on DNA, Franklin began studying the structure of the polio virus and plant viruses. She showed that the tobacco mosaic **virus** was a hollow tube rather solid, as previously thought, and its **ribonucleic acid (RNA)** was contained within the protein, not inside the tube.

In 1962, several years after Franklin's untimely death, Wilkins, Watson, and Crick received the Nobel Prize for physiology or medicine for discovering the structure of DNA. No reference was made to Franklin's work, and debate still continues among the scientific and other community as to whether sexism may have played a significant role in denying Franklin due recognition for her important contribution to the discovery.

FRISCH, KARL VON (1886-1982)

Austrian zoologist

Karl von Frisch won the Nobel Prize in 1973 for his pioneering work in the field of animal physiology and behavior. Frisch was a leading researcher in the study of insect behavior, and his studies proved that fish have acute hearing and that bees communicate effectively through a ritual dance. Frisch's discoveries and subsequent Nobel Prize were also significant because this was the first major acknowledgement of advances made in the study of **ethology**.

Frisch was born in Vienna in 1886 into a family dedicated to science. His father, Anton Ritter von Frisch, was a physician, and his mother, Marie Exner, came from a long line of distinguished scientists and scholars. From his earliest years, Frisch was exposed to the natural world, in large part due to a country house that his family retreated to every summer. There, the young Frisch spent his time collecting various species of animals. "Even before I went to school," he wrote in his autobiography, *A Biologist Remembers,* "I had a little zoo in my room." But Frisch was not simply a collector; he was also a keen observer. "I discovered that miraculous worlds may reveal themselves to a patient observer where the casual passer-by sees nothing at all," he said in his autobiography. A few early observations—most notably that the sea animals he collected in an aquarium in his room waved their tentacles when he turned on the lights—piqued an interest in the sensory systems of **animals** that would last his lifetime.

By the time Frisch reached college age, it was clear that his interests were focused on **zoology**. Nevertheless, his father thought medicine a more practical field than zoology, and in 1905 Frisch enrolled as a student of medicine at the University of Vienna. Medical school, Frisch later wrote, proved invaluable in providing a background in **histology**, **anatomy** and human physiology. He studied with his uncle, Sigmund Exner, who was a renowned physiologist and lecturer at the university. Though Exner taught human physiology, he encouraged his nephew to pursue his interest in animals by aiding him in a research project on the position of pigments in the compound eyes of certain beetles, butterflies and crustaceans. According to Frisch, his uncle's openness toward the study of animals in a course limited to human physiology was unheard of at the time. Comparing the physiology of humans and animals would only later be seen as so invaluable that it was made into a separate discipline. In the middle of his third year as a medical student, Frisch found himself increasingly frustrated by the "medical character" of the curriculum. He finally decided to drop his medical studies to pursue the field of ethology, or the study of animal behavior. He transferred to the Zoological In-

stitute at the University of Munich, where he studied under **Richard von Hertwig**. He continued to cultivate the interests he had developed under his uncle's leadership, researching light perception and color changes in minnows. It was at this time that he discovered minnows had an area on the forehead filled with sensory cells—a "third, very primitive eye," he called it in *A Biologist Remembers.* This explained why blind minnows reacted to light by changing color in the same way as minnows with sight. Frisch wrote his doctoral thesis on this subject and received his degree in 1910.

Frisch also began to question the common assumption of the time that fish and all invertebrates were color blind. He successfully trained minnows to respond to colored objects, proving that they could perceive color. These findings, however, were not kindly received by members of the scientific community, and Frisch's most notable opponent was Karl von Hess, the director of Munich Eye Clinic. The debate arose partly because of the theoretical connection between Frisch and the views of the famous naturalist **Charles Darwin**. Frisch believed in Darwinism, which theorized that the survival of certain species of animals depended on the development of their senses. Frisch hypothesized that animal behavior, rather than simply being a fixed mechanism, had an "adaptive biological significance," assumptions that were still a source of disagreement among scientists at the time. Despite the arguments about his research, Frisch was offered a teaching job at the University of Munich in 1921.

While teaching at the University of Munich, Frisch continued to study color perception in animals on vacations spent at his family's summer home. Having proved that color-blindness in fish was a fallacy, he turned to prove the same for bees. He conjectured that the adaptive purpose of the bright coloration of flowers was to guide bees to nectar. The bees, in turn, aided the flowers through pollination. That bees would be color-blind seemed untenable to Frisch. To test his hypothesis, he used research strategies similar to the ones he had used with fish. He conditioned their behavior by placing drops of sugar water on squares of blue-colored cardboard. He then placed these blue squares among plain gray squares. Eventually, he placed blue squares without sugar water among the gray squares. He found that the bees continued to go to the blue squares for their food, proving that they could differentiate color.

In 1914 Frisch's research was interrupted by the outbreak of World War I. He was excused from military duty because of poor eyesight but accepted a plea from his brother, who was a physician, to volunteer at a Red Cross hospital in dire need of help. His background in medical school qualified him to establish a bacteriologic laboratory at the hospital, enabling rapid diagnosis of diseases such as cholera, dysentery and typhoid. While at the hospital, he met a nurse, Margarethe Mohr, whom he married in 1917. Eventually they had three daughters and a son.

Meanwhile his research on bees continued to deepen. During the war, he would take a few weeks' leave from the hospital every summer, returning to his country house to research the bees. As the war came to an end, his work at the hospital lessened and his students returned to the Zoological Institute. After a four-year hiatus, he began teaching again and in January 1919 became an assistant professor.

Eventually Frisch became interested in scout bees—those that left the hive to explore a region for food. He set out dishes of sugar water and observed their behavior. When the dish was empty a scout bee occasionally came to the dish. When the food dish was full the scout would return in a matter of minutes with a whole company of bees. "It was clear to me that the bee community possessed an excellent intelligence function," Frisch wrote in his autobiography, "but how it functioned I did not know."

In the spring of 1919 Frisch developed a glass cage in which he placed a single honeycomb that could be observed from all sides. Through continuous observation and experimentation, Frisch concluded that scout bees, who foraged for food for the whole honeycomb, conveyed this information to the other bees by performing a kind of dance on the honeycomb. This dance excited the forager bees, who then flew directly to the food. In retrospect, Frisch called his first discovery of the bees' dance "the most far-reaching observation of my life." It would be another 20 years before Frisch fully understood the complexity of this dance.

In the fall of 1921, Frisch was appointed professor of zoology and director of the Zoologic Institute at Rostock University and began investigating whether fish could hear. The physiology of fish indicated that they could not. They did not have any of the characteristics thought to be necessary for the sense of hearing, like ear lobes, auditory canals, middle ears, or a cochlea in the inner ear, which was thought to be the center of hearing in humans. Frisch used his proven methods of behavior conditioning to test hearing in fish. He whistled to blind catfish before feeding them. Eventually he whistled but did not feed them and the catfish continued to respond. The answer seemed simple—or, as one skeptical scientist put it, "There is no doubt. The fish comes when you whistle." Frisch eventually refined his early research in this area with the help of his students and discovered other facts that supported his initial findings.

In 1925 Frisch began working at the Zoological Institute of the University of Munich. However, during World War II, the Zoological Institute at the University was destroyed, and Frisch spent those years in his country home and at the University of Graz. In 1950 he returned to Munich to rebuild the Institute as its director. During this time, he wrote many books for the general public as well as for the scientific community. Frisch retired in 1958 and died in 1982.

About his life's work, Frisch wrote philosophically in *A Biologist Remembers:* "The layman may wonder why a biologist is content to devote 50 years of his life to the study of bees and minnows without ever branching out into research on, say, elephants or at any rate the lice of elephants or the fleas of moles. The answer to any such question must be that every single species of the animal kingdom challenges us with all, or nearly all, the mysteries of life."

This attitude was shared by the Nobel committee, who rewarded him with the prize in medicine and physiology in 1973. The prize, which Frisch shared with two other animal behaviorists, **Konrad Lorenz** and **Nikolaas Tinbergen**, was a departure for the Nobel Committee. Never before had there been such public recognition of the interactive study of animals and humans. In an article in *Science* magazine regarding the Nobel Prize, Frisch was praised for teaching the world that "human behavior [is not] something... outside nature" but something that is "subject to the principles that mold the biology, adaptability and the survival of other organisms."

See also Behaviorism

FRUIT FLY (*DROSOPHILA MELANOGASTER*)

This small insect, also known as the vinegar fly, was a favorite subject for laboratory studies in **genetics** beginning around 1900. Why did researchers choose to study *Drosophila melanogaster* instead of another type of insect or **animal**? Because of its small size, only 1/16 to 3/32 inch (1.5 to 2.5 millimeters), the fruit fly could be kept by the thousands in small containers, ensuring that researchers always had an abundant supply of experimental subjects. *Drosophila* didn't need much to eat; a bowl of fermenting or spoiled fruit could feed thousands of the tiny insects. And this particular **species** had simple, easy-to-study characteristics: bright red eyes, clear wings, shiny black abdomen with yellow bands, and only four **chromosome**s in each **cell**.

Most important, though, were its reproductive characteristics. *Drosophila melanogaster* has the most rapid reproductive rate of any dried-fruit insect. The whole maturation process takes only seven days: about 24 hours in the **egg**, three days as **larva**, and three days as a pupa. Sometimes, if the female keeps the mature eggs within her body, they may hatch within an hour after they are laid. Consequently, many generations of fruit flies can be bred in a short time for genetic study.

An adult female fruit fly may lay as many as 2,000 eggs in her lifetime. However, the average number of eggs during a female's life span is about 1,000. The life span of the female can vary, depending on the weather conditions. Females live about 39 days in warm weather or up to 70 days in cooler weather. But males don't seem to have the same weather sensitivity. They typically live about 41 days.

The characteristics of *Drosophila melanogaster* made the insect especially valuable to researchers like **Thomas Hunt Morgan**, who needed to study several generations of a particular species in order to observe the mechanisms of **heredity**.

Today fruit flies are still popular as experimental subjects. They are used to study everything from circadian rhythms and sex-determining **gene**s to alcoholism and cocaine addiction.

FRUITS

Fruits are ripened ovaries of flowering **plants**, and are developed from the ovary wall from fertilized ovules (or seeds). The size, shape, color, and other attributes of fruits are extremely variable, depending upon each particular plant **species**.

Fruits may be dry in texture, such as those of the sunflower *Helianthus annua*, and the key-like fruits (or samaras) of the sugar maple (*Acer saccharum*); or they may be thick and fleshy, like the tomato (*Lycopersicum esculentum*), the apple (*Malus pumila*), and the watermelon (*Citrullus vulgaris*).

Some fruits only contain a single seed, as is the case of the avocado (*Persea gratissima*), the cherry (*Prunus avium*), the mango (*Mangifera indica*), and the date palm (*Phoenix dactylifera*). These single-seeded, fleshy fruits are known as drupes. Other fruits contain numerous seeds, such as the orange (*Citrus sinensis*), broad bean (*Vicia faba*), and cucumber (*Cucumis sativa*). All of these edible, multi-seeded fruits are technically known as berries, although the bean is also known as a legume. Some fruits develop into an aggregate structure, consisting of the fruits of numerous **flower**s. Examples of aggregate fruits include the mulberry (*Morus nigra*), the pineapple (*Ananas sativus*), the blackberry (*Rubus fruticosus*), and the fruit-head of maize (or corn, *Zea mays*); consisting of numerous, dry fruits known as achenes.

Fruits that release their ripe seeds for dispersal into the environment are termed dehiscent. For example, the milkweed (*Asclepias syriaca*) has pod-like fruits that split open when ripe, releasing their seeds into the air. Fruits that do not release their seeds are called indehiscent. One example is a squash or gourd (such as the pumpkin; *Cucurbita pepo*). Indehiscent fruits must be eaten by an **animal** (which later defecates the seeds) or rot on the ground to disperse the seeds.

Many fruits are characterized by the mode in which their seeds are dispersed into the environment. Edible fruits that are fleshy and tasty are designed to attract animals, which then feed on the ripe fruits and eventually pass the undigested seeds some distance away from the parent plant. One example of this is the wild strawberry (*Fragaria vesca*). In comparison, the fruit of the coconut (*Cocos nucifera*) is designed for long-distance dispersal in the ocean, and has a fibrous, extremely tough, waterproof covering around its single, large seed. The light, conspicuously plumed, one-seeded fruits of the common dandelion (*Taraxacum officinale*) are designed for dispersal by the wind. The fruits of the burdock (*Arctium lappa*) are covered in tiny, recurved bristles, which readily stick the fruits to the fur of a passing mammal for dispersal

FUNGI

Fungi play an essential role in breaking down organic **matter** and thereby allowing **nutrients** to be recycled in **nature**. As such, they are important decomposers and without them living communities would become buried in their own waste. Some fungi, the saprobes, get their nutrients from nonliving organic matter, such as dead **plants** and **animal** wastes, clothing, paper, leather, and other materials. Others, the **parasite**s, get nutrients from the **tissue**s of living **organism**s. Both types of fungi obtain nutrients by secreting **enzyme**s from their **cell**s that break down large organic **molecule**s into smaller components. The fungi cells can then absorb the nutrients.

Although the term fungi invokes unpleasant images for some people, fungi are a source of **antibiotics**, **vitamins**, and industrial chemicals. **Yeast**, a kind of fungi, is used to ferment bread and **alcohol**ic beverages. Nevertheless, fungi also cause athlete's foot, yeast infections, food spoilation, wheat and corn diseases, and, perhaps most well known, the Irish potato famine of 1843-1847, which caused the **death** of 250,000 people in Ireland.

Fungi are classified as a division of simple plants, though, lacking **chlorophyll**, true stems, roots, and leaves, they cannot photosynthesize. The fungi body, called *mycelium*, is composed of threadlike filaments called *hyphae*. All fungi can reproduce asexually—by **cell division**, budding, fragmentation, or **spore**s—though some reproduce sexually. The number of existing fungi **species** may be more than a million.

The main groups of fungi are chytrids, water molds, zygosporangium-forming fungi, sac fungi, and club fungi. Chyrids live in muddy or aquatic **habitat**s and feed on decaying plants, though some live as parasites on living plants, animals, and other fungi. Water molds, distantly related to other fungi, play an important role as decomposers in aquatic habitats. Some, however, live as parasites on aquatic animals and terrestrial plants, including potato plants that can be destroyed by certain types of water molds. Zygosporangium-forming fungi also can be either saprobes—such as the well-known black bread mold—or parasites on insects, such as houseflies. Sac fungi, of which more than 30,000 species are known, include the yeast used to leaven bread and alcoholic beverages. But many of these fungi also cause diseases in plants. Club fungi, numbering more than 25,000 species, include mushrooms, stinkhorns, and puffballs. While some are edible, others produce deadly poisons.

FUNK, CASIMIR (1884-1967)
Polish biochemist

Casimir Funk was a Polish-born American biochemist who pioneered research on dietary requirements and **vitamin**s, especially vitamin B1 which is also known as thiamine. He is also recognized for his research into **animal** hormones, and the biochemical aspects of **diabetes**, **cancer** and ulcers.

Casimir was born in Warsaw on February 23, 1884, the son of a well-known dermatologist. He attended the University of Bern, Switzerland and in 1904 obtained his Ph.D. in organic chemistry at age 20. Prior to emigrating to the United States in 1915, he worked at the Pasteur Institute in Paris and the Lister Institute in London. In 1920, he became a naturalized American citizen. Funk returned to Warsaw in 1923 under the sponsorship of the Rockefeller Foundation to assume the role of Director of the Biochemistry Department of the State Institute of Hygiene. He left this position in 1927 due to political turmoil and went to Paris where he founded the Casa Biochemica, a privately funded research foundation. Funk abandoned this when France was invaded by Germany at the start World War II in 1939. Funk returned to the United States to work for the United States Vitamin Corporation and in 1940 became the president of the Funk Foundation for Medical Research. He retained this position until he died on November 20, 1967, in Albany, New York.

Funk is best known as a pioneer in the study of vitamins. He began his research while at the Lister Institute where he attempted to find a cure for **beriberi**. He isolated an amine (a organic compound of **nitrogen**) which was known to cure a beriberi-like disease in pigeons. He suggested that deficiency diseases such as beriberi, rickets and scurvy resulted from a dietary deficiency. Funk postulated (wrongly it was later proved) that all deficiency diseases could be cured by amines for which he coined the term vitamine. This was later shortened to vitamin in 1920 by Sir Jack Drummond. Funk continued to search unsuccessfully for a cure to beriberi in humans. Robert Williams is credited with having isolated vitamin B1 (thiamin) from rice polishings as a cure for beriberi in 1934. Funk analyzed the molecular structure of thiamin and synthesized it in 1936.

Funk performed extensive research into male sex **hormone**s and animal hormones. In 1929, he was able to extract the male sex hormone androsterone from human urine in a crude form. He contributed to the understanding of the biochemistry of cancer, ulcers and diabetes.

An American fly agaric (*Amanita muscaria formosa*). This mushroom is very common in all of North America, but is more slender, tinged with a salmonlike coloration, and somewhat more rare in the southern states. *(Photograph by Robert J. Huffman, Field Mark Publications. Reproduced by permission.)*

G

GAIA HYPOTHESIS

The word *Gaia* is an umbrella term that is derived from the Greek word meaning Goddess of the Earth. It has now come to symbolize "Earth-Mother," or "Living Earth," hypothesizing that Earth acts like a "superorganism," with all its biological and physical systems cooperating to keep it healthy.

The Gaia concept evolved from the work of a few noted scientists. Initially the hypothesis was formulated by the eighteenth-century Scottish geologist **James Hutton**, who was the first to use the term superorganism in reference to Earth. He asserted that it was essential to view Earth's systems as affecting and affected by a single **organism**, and pointed out that physiology would be the proper science to study this codependence in the planet's systems. Hutton became known as the father of geology after he published his *Theory of the Earth*, which pointed to volcanism as being the primary force shaping Earth.

In its modern form, the Gaia theory was put forward by James Lovelock whose book *Gaia: A New Look at Life on Earth* achieved considerable attention when it was published in 1979. Lovelock, an English chemist, suggested that Earth's **biosphere** acts as a single living system called Gaia. If left alone, he wrote, it can regulate itself.

Lovelock drew on his background as a chemist to explain his theory. He asserted that Earth provides a delicate balance between atmospheric **carbon dioxide** and oxygen, maintained by living organisms, and is responsible not only for creating a unique atmospheric chemical composition, but also for other environmental characteristics that make life possible. For example, he argued that it is no accident that the level of oxygen is kept remarkably constant in the **atmosphere** at 21%. Likewise, limestone has been stored deep under the oceans and has buried large amounts of carbon dioxide that would otherwise affect Earth's heat balance. There has been a suggestion that microscopic marine **algae** are using Earth's weather system to disperse themselves around the planet, by their creation of a gas, dimethyl sulphide, which ultimately creates particles on which **water** vapor can condense—a cloud.

Today, Lovelock suggests that the biosphere has the ability to create the environment that most favors its own stability, and he warns that by tampering with Earth's own environmental balancing mechanisms, we are placing ourselves and our planet at grave risk. He points to global warming and **ozone** depletion as indications of this risk.

Lovelock's arguments have been incorporated into opposing sides of Earth **conservation** movement. Environmentalists insist that man's destruction is upsetting Earth's ability to regulate itself, while industry representatives argue that Earth can continue to survive based on its history of self-preservation through **adaptation**.

GAJDUSEK, D. CARLETON (1923-)
American virologist

Gajdusek was born in Yonkers, New York, on September 9, 1923, the son of Hungarian immigrants. His parents provided a rich intellectual environment at home, and Gajdusek became interested in science at an early age. While still in high school, he spent summers working at the Boyce Thompson Institute for Plant Research in Yonkers. Gajdusek entered the University of Rochester in 1940 at the age of 16. He graduated with a bachelor of science degree in biophysics three years later. In 1946, he also received an M.D. from Harvard Medical School.

In the eight years following graduation from Harvard, Gajdusek completed residencies in Boston, Massachusetts, and New York City, New York; worked for two years at the California Institute of Technology with Linus Pauling; did research in virology at Harvard with John Enders; served in the United States Army at the Walter Reed Medical Center; and continued his studies in Iran at the Pasteur Institute. Finally, in 1954, he moved to the Walter and Eliza Hall Institute of Medical Research in Melbourne, Australia, to carry out additional research in virology.

During his tenure at the Hall Institute, Gajdusek learned about an unusual disease that infected members of the Fore tribe in New Guinea. The disease—called kuru—caused a slow, but ultimately fatal, degeneration of the **brain** and had never been studied by medical scientists. Gajdusek decided to travel to New Guinea, where he spent a year learning more about kuru.

Gajdusek spent much of the next decade trying to discover the causative agent for kuru. He felt sure that the disease was caused by a **virus**. An important clue was that the Fore, in a ritualistic practice, honored their dead by eating their brains. Gajdusek reasoned that this practice would be an ideal mechanism by which a viral infection could be transmitted from one person to another.

However, standard techniques for identifying viruses produced no results in this case. For a while, Gajdusek considered the possibility that kuru might be a hereditary disorder rather than an infectious disease.

In 1963 another possibility occurred to him. He realized that kuru was similar in some ways to scrapie, a neurological disorder that affects sheep. Scrapie begins to appear in sheep long after they have been infected. The incubation period can be many years. Scientists believed that scrapie was caused by an unusual type of virus that acts extremely slowly.

Gajdusek considered the possibility that kuru is also caused by a slow-acting virus. To test this hypothesis, he implanted pieces taken from the brains of kuru victims into apes. More than two years later, the disease began to appear in the apes. Evidence for the existence of a scrapie-like, slow-moving viral infection appeared to exist.

Gajdusek's success with his kuru research prompted him to attack other unexplained brain diseases. In 1971, he found that a slow-moving virus might also be responsible for Creutzfeldt-Jacob disease, a degenerative brain disorder that occurs throughout the world. For his work on slow-moving viruses, Gajdusek received a share of the 1976 Nobel Prize for Physiology or Medicine.

Some scientists now believe that the slow-moving virus responsible for scrapie, kuru, and Creutzfeldt-Jacob disease is actually a new type of infectious particle called a **prion**. First suggested in 1982 by American neurologist **Stanley Prusiner** (1942-), the prion is thought to be a naked piece of **protein** that has the ability to cause certain types of viral-like diseases.

In 1997, Gajdusek admitted in a Maryland circuit court that he had sexually molested a 17-year-old boy. The boy was one of more than 50 children he had brought back to the United States since the 1960s from the South Pacific. These children often lived with Gajdusek, who financed their educations. As a part of his plea bargain, Gajdusek received a one-year prison sentence and five years probation. While he has been sued in civil court for $2.2 million by one of the boys he took in, others he adopted rallied around Gajdusek and supported him during his trial. He retired from his position as chief of the Laboratory of Central Nervous Systems at the National Institutes of Health in 1997.

GALTON, FRANCIS (1822-1911)

English scientist

Francis Galton was born near Birmingham, England, in 1822. His impressive talents appeared early. At the age of three, he was already reading, and at four, he was studying Latin. His I.Q. at adulthood was estimated at 200. But as a young man of 22, a fresh graduate of Cambridge, Galton did not continue his training as a physician. His wealthy father had died and left Galton a large inheritance, so the young Galton was free to do whatever he fancied.

For a time he chose to travel, exploring virtually unknown parts of Sudan, Syria, and southwest Africa. Upon his return, Galton published two books about his travels. He was even recognized by the Geographical Society and the Royal Society with various medals and honors. He was also knighted in 1909.

Galton never held any academic or professional post. Instead, he did his experimenting at home or during his travels. His earliest researches had to do with meteorology. In his 1863 book *Meteorgraphica*, Galton presented the modern technique of weather-mapping and coined the term anticyclone to describe the high pressure systems that bring fair, calm weather. Because of these studies, Galton was instrumental in establishing the Meteorological Office and the National Physical Laboratory.

Galton's natural curiosity about **nature** and mankind soon led him to explore a new frontier: **heredity**. He felt that the study of heredity was not progressing as quickly or accurately as possible because of a lack in quantitative research. Galton strongly believed that virtually everything could be proven mathematically—everything was quantifiable. He even went so far as to develop a system to measure beauty and the effectiveness of prayer. These researches established him as the founder of the biometric school. (Followers of the biometric school use statistics to prove hypotheses in **genetics**.)

In 1859, **Charles Darwin**, who was Galton's relative, published *The Origin of Species*. In that book, Darwin asserted that certain characteristics from two different individuals blend together to create variations in their offspring. Darwin also believed that these characteristics are ''copied'' within the parents' bodies and carried to the reproductive organs, where they wait to pass on the traits to the offspring.

Galton, seeing that this theory could be tested statistically, set out to support Darwin's ''pangene'' theory. It made sense to him that these ''copied'' characteristics were probably floating around in the bloodstream. So he chose to transfuse or switch the blood of a purebred silver-gray rabbit with a common lop-eared rabbit. After breeding the silver-gray rabbits that had the blood of lop-eared rabbits, Galton found no difference in the rabbits' offspring. Silver-gray rabbits still produced silver-gray offspring. He thought this revealed the weakness of the inheritance aspect of Darwin's theory.

Several years later, Galton came up with his own theory to explain inheritance: the theory of ancestral heredity. According to Galton's theory, each parent contributes half of the traits to the offspring, each grandparent one-fourth, and so on.

With each generation, the traits become more diluted and the offspring begin exhibiting the average of race, not the average of the parents. Unfortunately, Galton's theory was still too similar to Darwin's "blending." It did not survive the rediscovery and eventual acceptance of **Gregor Mendel**'s work, which stressed particulate heredity—or the notion of traits being inherited individually rather than being blended. Nevertheless, Galton's theory turns out to be mathematically sound in that one-half of any individual's genes come from each parent, one-fourth from each grandparent, and so forth.

Galton pioneered studies of identical twins, whose differences can be attributed to environmental factors since they are genetically identical. He also studied talented families to determine how artistic, intellectual, or athletic skill might be inherited. Galton's heredity studies led him to believe that scientific breeding could be applied to human populations. Galton called this science eugenics. **Eugenics** entails "breeding in" desirable traits of the human population, such as talent and healthiness, and "breeding out" undesirable traits, such as stupidity and weakness. Galton suspected that getting such a concept accepted would be difficult, but he devoted significant time and energy to ensure that the topic was not forgotten. He even donated a large amount of his inheritance money to establish a Chair of Eugenics at University College in London.

Later in his life, Galton became interested in fingerprinting. He thought it might be a way to track differences in families, race, morals, and intellect. Although Galton never found any correlations to support this assumption, he did establish fingerprinting as an easy and almost infallible means of human identification. The fingerprinting methods he developed are essentially the same methods used today.

Among Galton's other books were *Hereditary Genius* (1869), *Inquiries into Human Faculty* (1883), *Natural Inheritance* (1889), and *Finger Prints* (1892). He was knighted in 1909, two years before his death in 1911.

GALVANI, LUIGI (1737-1798)
Italian anatomist

Luigi Galvani was an Italian anatomist who, through his discovery that the legs of frogs would move when touched by two different metals, essentially founded the study of current electricity, also contributing significantly to the field of animal physiology. He was born in Bologna, in the Papal States (now Italy) on September 9, 1737. Galvani received a medical degree from the University of Bologna in 1762, where he was named Professor of Anatomy and Gynecology in 1775. In 1797, when Bologna fell to the French, Galvani lost his professorship because he refused to become a sworn supporter of Napoleon. He died the following year, in poverty.

Scientists had been experimenting with static electricity for more than a century by the time Galvani conducted his frog leg experiments. Knowing that an electric spark could provoke movement in live muscle tissue, Galvani noticed, while dissecting a frog, that nerve action/muscle movement was induced by electrical phenomena. For example, in one

Francis Galton. *(The Library of Congress. Reproduced by permission.)*

experiment, Galvani used silver and brass rods, one touching the spinal cord, the other touching the foot. When the opposite ends of the rods made contact, the leg muscle contracted. Another experiment required the placement of the legs and spinal cord, respectively, upon brass and copper foil. A muscle contraction would occur when the ends of a metal rods would simultaneously touch both pieces of foil. In a third experiment, the legs and spinal cord were immersed in fluid. When the fluid came in contact with metal, a muscle contraction would result. A key element in these experiments was the fact that two different types of metal were needed to provoke a contraction and that electricity could be produced by chemical action. Electricity was definitely involved, but Galvani did not know whether the electricity was coming from the metal or from the muscle. Galvani, along with other scientists, referred to this phenomenon as "animal electricity" and eventually as "galvanism." The concept of galvanism was challenged by the Italian physicist Alessandro Volta (1745-1827), whose electrical experiments demonstrated that the electrical current originated in the metals.

Galvani's frog experiments may have encouraged assumptions that were eventually invalidated, but his contributions to physiological and electrical science initiated subsequent discoveries and spurring research in a variety of fields, including electric currents and electrotherapy. Processes and devices named after him include the galvanometer, which measures the current in a conductor, galvanic skin response a change in the skin's electrical conductivity due to any **stimulus**, and galvanization, the process whereby an electric current is used to apply a layer of zinc crystals to any other metal, usually iron and steel. The many terms derived from his name include *galvanotaxis*, which denotes movement in response to an electrical stimulus. In everyday language, the verb denotes the operation of providing an impetus for an action.

See also Biology and physics

GAMETE

A gamete is a sexual reproductive **cell** such as a **sperm** or an **egg** that must fuse with another gamete to produce a **zygote**, and eventually, a new **organism**. Gametes are typically **haploid** (1N), containing only half the number of the **chromosome**s needed to form a new **diploid** (2N) organism. During **fertilization**, two unlike gametes, each containing a single set of chromosomes, fuse to produce a diploid cell, or zygote, containing paired chromosomes. The zygote develops into a new diploid organism. Specialized cells of the mature organism undergo **meiosis** to form haploid gametes that begin the process of reproduction again. Although fusing gametes normally differ from each other genetically, they may be identical in form (isogamy) or they may differ in size and shape (heterogamy). An extreme case of heterogamy, where one gamete is large and nonmotile (the egg), and the other is small and motile (the sperm) is called oogamy. Oogamy is the type of gamete formation typical of **animals**, **plants**, and some forms of **fungi** and **algae**. Isogamy and heterogamy are found in many fungi, algae, and **protista**.

See also Life Cycle

GAMETOPHYTE

A gametophyte is a **haploid**, **gamete**-producing sexual phase in the **life cycle** of the **plants** that exhibit a phenomenon called the **alteration of generations**. The other distinct phase in the life cycle of such plants is called a **sporophyte**. The gametophyte with a single set of **chromosome**s gives rise to the sporophyte with a double set when two gametes fuse to form a **zygote**. The zygote develops into a sporophyte and produces haploid **spore**s by **meiosis** that develop into gametophytes. In the gametophyte phase, male and female **organ**s (gametangia) develop and create gametes, which unite in **fertilization** (syngamy). In some **algae** and **fungi** almost the entire life cycle is gametophyte and in others it is almost all sporophyte, and either form may exist independently of the other. In liverworts, hornworts, and mosses, the gametophyte phase of the life cycle dominates, and the sporophyte tends to be dependent on the gametophyte. In seed producing plants, on the other hand, the sporophyte is the dominant phase and the gametophytes exist only as **parasite**s on the sporophytes. In flowering plants, or angiosperms, the gametophyte portion of the life cycle consists of the microgametophyte or pollen grain that produces **sperm**, and the megagametophyte or **embryo** sac which produces the **egg**. Fusion of the sperm and egg gives rise to a zygote that develops into an embryo encased in the seed.

GANGLION

A ganglion is a cluster of **neuron cell** bodies which occur outside the central **nervous system** (the **brain** and **spinal cord**). Neurons are made up of the cell body, with dendrites to conduct **nerve impulses** towards it and an **axon** to conduct nerve impulses away from it. The cell body contains the usual **organelle**s, as well as the cell's **nucleus** (containing genetic information). When a group of neuron cell bodies exists within the central nervous system, it is called a nucleus; when such a group exists outside of the central nervous system (within, therefore, the peripheral nervous system), it is referred to as a ganglion.

In very primitive **organism**s, no central nervous system can be identified. No distinct brain structure exists. Instead, nervous information is totally relayed through the existence of ganglion bundles.

GARROD, ARCHIBALD (1857-1936)
English physician and chemist

Archibald Garrod was a physician whose innovative work in clinical medicine and chemistry led him to discover a new class of human disease based on hereditary factors. A pioneer in biochemistry, Garrod stressed the chemical uniqueness of each person. For his work on inborn errors of metabolism, Garrod was elected to the Royal Society and received a knighthood.

Archibald Edward Garrod was born in London on November 25, 1857, the fourth and youngest son of Sir Alfred Baring Garrod and Elisabeth Ann Colchester. Garrod's father, a distinguished professor of medicine at University College in London, was the first physician to note the presence of uric acid in patients suffering from gout. In later years, Garrod would cite his father's discovery as the first quantitative biochemical investigation performed on living humans.

As a child, Garrod demonstrated an early talent for illustration and a lasting interest in color. He studied physical geography at Marlborough and astronomy at Oxford, where he graduated with first-class honors in natural science. Deciding to follow in his father's footsteps, Garrod began the study of medicine at St. Bartholomew's Hospital in London. He received a number of scholarships and spent a year attending medical clinics in Vienna, resulting in the publication of a book on the laryngoscope, a device used to examine the interior of the throat. A tall, handsome man, Garrod became a skilled clinician whose reassuring manner enabled him to gather detailed medical histories from his patients. In 1884, Garrod joined the staff of St. Bartholomew's Hospital, but promotion was slow and for nearly three decades he had ample time to pursue his interest in chemistry and disease. He wrote a number of papers on joint disorders, his father's specialty, pointing out the difference between rheumatism and rheumatoid arthritis as diseases.

Work in Chemistry Leads to Discovery of Genetic Disease

Garrod's interest in joint disease led him to study the chemistry of pigments in urine. While working as a visiting physician at the Great Osmond Street Hospital for Sick Children, he examined a three-month old boy, Thomas P., whose urine was stained a deep reddish-brown. Garrod's diagnosis was alkaptonuria, which is caused by an abnormal build-up of homogentisic acid, or alkapton. In a normal person, the acid is broken down through a series of chemical reactions into carbon dioxide and water. But in rare cases, the metabolic process is interrupted and the acid is excreted in the urine, where it turns black upon contact with the air. According to the germ theory of disease, which had transformed the study of medicine in Garrod's time, alkaptonuria was thought to be a bacterial infection of the intestine. The disorder was almost always diagnosed in infancy, lasted throughout life and was thought to be contagious. Garrod's training in physical science, however, led him to investigate the disease as a series of chemical reactions. He reviewed 31 cases of alkaptonuria from his own practice and from the medical literature, and presented his findings to the Royal Medical and Chirurgical (Surgical) Society of London in 1899. Alkaptonuria, he noted, although rare, tended to appear among children of healthy parents. It was not contagious and seemed to be a harmless error in metabolism.

When a third child with alkaptonuria was born to the parents of Thomas P., Garrod suspected that something more than mere chance was involved. When he learned that Thomas P.'s parents were blood relations—their mothers were sisters—he inquired into the backgrounds of other families with one or more children with alkaptonuria. In every instance, their parents were also first cousins. It was while walking home from the hospital one afternoon that Garrod conceived of the possibility that alkaptonuria might be a disease caused by **heredity (genetics)**. **Gregor Mendel**'s work on the principles of heredity, newly discovered in England, offered a simple explanation. The mating of first cousins apparently created conditions under which a rare, recessive Mendelian factor (or **gene**) appeared in the offspring. Garrod's classic paper on alkaptonuria was published in *Lancet* in 1902.

Garrod went on to study other metabolic disorders, including the pigment disorders porphyria, the cause of George III's madness, and albinism. Like alkaptonurics, albinos tended to be children of parents who were first cousins. In a series of lectures delivered before the Royal College of Physicians in 1908, Garrod described such disorders as ''inborn errors of metabolism.'' In each instance, he claimed, a genetic factor caused a deficiency in a certain **enzyme** which led to a premature block in the chemistry of normal metabolism. In his book, *Inborn Errors of Metabolism* (1909), Garrod described an important new class of diseases which were genetic, not bacteriological in origin.

In recognition of his contributions to science, Garrod was made a fellow of the Royal Society in 1910, and was knighted in 1918. He spent World War I in the Army Medical Service on the island of Malta as consulting physician to the British forces in the Mediterranean. Two of his sons were killed in combat, a third died of the Spanish **influenza** following the armistice.

After the War, Garrod returned briefly to St. Bartholomew's, but was soon summoned to Oxford to become Regius Professor of Medicine. In his lectures, Garrod urged students to think of disease in terms of biochemistry. Clinicians, he argued, were uniquely placed to observe anomalies of nature which they could then investigate in the laboratory. In his later writings, Garrod hypothesized that there might be a molecular (genetic) basis for all variations in life functions, including physical appearance, susceptibility to disease, even behavior.

Garrod retired in 1927. He and his wife, Laura Elizabeth, whom he married in 1886, moved to Cambridge to be near their daughter, Dorothy, a noted archaeologist and teacher at Newnham College. Archibald Edward Garrod died at home on March 28, 1936. He was 78.

The significance of Garrod's contribution to the science of genetics was not appreciated in his lifetime; he was an elderly physician when most young geneticists were botanists and zoologists. It was not until the 1940s that Garrod's path-breaking work in human genetics was rediscovered and applied to gene theory.

GASSER, HERBERT SPENCER (1888-1963)
American neurophysiologist and physician

Herbert Gasser was born in Platteville, Wisconsin, on July 5, 1888. His mother, Jane Griswold, who descended from an early Connecticut family, was a teacher trained in Wisconsin's first State Normal School in Platteville. Gasser's father, Herman, was born in the Tyrol and came to the United States as a boy. Herman was a self-educated man who eventually qualified in medicine and became a country doctor.

After attending State Normal School, Gasser received two degrees in science at the University of Wisconsin, a bachelor's degree in zoology in 1910 and a master's in **anatomy** in 1911. However, Gasser's future interests were determined by a physiology course in the University's newly organized medical school. The young lecturer who emphasized the new spirit of research in medicine was **Joseph Erlanger**, the man with whom Gasser would share the Nobel Prize 33 years later. In 1915, he earned his medical degree from Johns Hopkins University, where he conducted research on **blood** coagulation in his spare time. After another year of research in Wisconsin, Gasser joined Erlanger at Washington University in St. Louis, in 1916.

Earlier scientists had provided painstaking microscopic slides of **neurons** and general theories of nerve networks in the body. Gasser's contributions made it possible to trace pathways while keeping the **nervous system** intact. Physiologists knew that impulses (action potentials) travel along nerves to convey sensation and to stimulate muscles, and that these impulses could be recorded by electrical instruments. A hypothesis existed that impulses moved faster along thick fibers than they did thin ones. Gasser's dramatic new method involved stimulating a given region of nerves and then reading the transmitted signal as it reached its destination, much like a physician tests a patient's knee jerk response with a rubber mallet. His problem was in finding recording devices capable of measuring, in fractions of a second, impulses that were small in quantity and short in duration. The available devices were inadequate. The string galvanometer and the capillary electrometer were slow and insensitive. The cathode-ray oscillograph, although quick, was insensitive to small currents.

The first breakthrough for Gasser came with the same vacuum tube amplifier that made radio possible. The three-stage amplifier had been brought to St. Louis by H. Sidney Newcomer, one of Gasser's classmates at Hopkins, who had built the device with the help of friends at the Western Electric Company. Nerve impulses could now be recorded, though the instrument's inertia caused distortions in timing the impulses. Their report describing this apparatus and experiments on nerves in the diaphragm appeared in 1921. This article was less important for its new knowledge about nerves than for its description of how sensations could at last be signalized.

A new technology, again from Western Electric, allowed Gasser and Joseph Erlanger to conduct the pioneering studies that eventually led to their Nobel Prize. It had been believed for over a decade that, should a means be discovered to test the Braun tube, the nerve impulse might accurately be recorded. But the tube, invented in 1897, used a cold-cathode technology, wherein the emission of electrons from the cathode's electrode is triggered by an outside force—this proved to be its downfall. Western Electric had, on the other hand, developed an oscillograph tube fitted with a hot cathode. This permitted the instrument to operate at a low voltage, which made it more sensitive to the small currents of the nerve action potentials. The instrument could record both the time elapsed between impulses and the change in nerve reactions. Though the tube was a breakthrough for Gasser and Erlanger, they still had to devise auxiliary apparatus to coordinate their induction shocks with the action potentials that were displayed on the screen. This work was reported in 1922.

Using the cathode-ray oscilloscope, Gasser and Erlanger almost immediately made two discoveries about the unexpected complexity they found in nerve trunks. In one, they determined that the sequence of events of nerve impulse transmission consists of two parts. There is an initial, large, rapid deviation in electric potential, called the spike, which ascends then descends during the actual transmission. The spike is followed by a sequence of small, slow potential changes, called the after-potential, that first has a negative and then a positive deviation.

In their other discovery, Gasser and Erlanger found that the composite action potential of a nerve has a range of velocities. They eventually identified three distinct patterns based on the length of spikes and their after potentials, and classified the fibers into three main groups. The fastest and thickest are A fibers, the intermediate size are B fibers, while the thinnest and slowest are C fibers. Their findings thus confirmed the hypothesis that thick fibers conduct impulses faster than thin ones.

Erlanger and Gasser next showed how these three types of fibers are distributed over the incoming and outgoing fibers of the spinal cord, the sensory and motor roots. The perception of pain is carried by the thin, slow fibers, while muscle sense and touch and muscle movement are conducted by the fast fibers. Gasser subsequently explored the excitability of nerve fibers in relation to after-potentials. He also continued to refine the oscilloscope, first using x-ray film and eventually a camera to record the impulses.

Gasser served as professor of pharmacology at Washington University from 1921-31. During a two-year leave of absence between 1923-25, he worked with Archibald V. Hill and Henry Hallett Dale in London, Walter Straub in Munich, and Louis Lapicque at the Sorbonne, on investigations involving muscle contractions and excitation of nerves. In 1931, Gasser became professor of physiology at Cornell University Medical College in New York City. In 1935, at age 47, Gasser became the second scientific director of the Rockefeller Institute for Medical Research, succeeding Simon Flexner. Gasser's medical training and his grasp of mathematical and physical sciences equipped him well to lead and to comprehend the expanding field of scientific medicine. His tenure bridged the economic depression of the 1930s, World War II, and the unsettling changes in the funding of scientific research after the war. Despite these trying times, Gasser, nevertheless, led the Institute's transition from its original emphasis on pathology and infectious diseases to a broader biological approach to human diseases. From 1936-57, he also served as editor of *The Journal of Experimental Medicine.*

During World War II, many Rockefeller Institute laboratories closed and their facilities and staff were organized to support war efforts. Gasser returned to work he had done on chemical warfare during the first world war, chairing a civilian committee on research development in that field. So it was a great surprise for Gasser when a cable arrived in 1944 from Stockholm, announcing that he had won a Nobel prize. Gasser retired from the Institute in 1953 and was succeeded by Detlev W. Bronk. With a change to emeritus status came the opportunity for Gasser to return to the laboratory. Instead of plunging into new areas of nerve physiology, he returned to unfinished work on differentiation of the thin C fibers. The introduction of electron microscopy helped him confirm many of his earlier findings. Gasser's scientific contributions were recognized by honorary degrees from twelve universities. He was elected to the National Academy of Sciences in 1934, the American Philosophical Society in 1937, and was a member of more than twenty other scientific societies in the United States, Europe, and South America. He received the Kober Medal in 1954, from the American Association of Physicians.

Following a second stroke, Gasser died in New York Hospital on May 11, 1963.

GENE

Genes are the physical units of **heredity** and are located along each **chromosome** in the **cell**s of the human body. For each physical trait—eye color, height, hair color—a person inherits two genes or two groups of genes, one from each parent. Because both of these genes cannot be expressed together, one usually overpowers the other. The more powerful gene is called the dominant gene and the weaker, the recessive gene.

All genes on the same chromosome are called linked genes because they are usually inherited together. Genes on the X and Y chromosomes are called sex-linked because the X and Y chromosomes determine sex.

Sometimes genes on the same chromosome are not inherited together. When sex cells divide to form an **egg** or **sperm** cell (a process called **meiosis**), each chromosome pairs off with a partner. As the chromosomes lie side by side, groups of genes from one chromosome may change places with groups of genes from the partner chromosome. This is called **crossing over** and thus explains how families inherit different combinations of linked traits.

An Austrian monk, **Gregor Mendel**, introduced the world to hereditary factors—genes—that determine all hereditary traits. During his experiments with pea **plants**, Mendel noticed that the plants inherited traits in a predictable way. It was as if the pea plants had a pair of factors responsible for each trait. Even though he never actually saw them, Mendel was convinced that tiny independent units determined how an individual would develop. Before Mendel's findings, traits were thought to be passed on through a blending of the mother and father's characteristics, much like a blending of two liquids.

When Mendel's laws of heredity were rediscovered in 1900, they became vitally important to biologists. Among other things, Mendel's laws established heredity as a combining of independent units, not a blending of two liquids. Wilhelm Johannsen, a strong supporter of Mendel's theories, coined the term gene to replace the variety of terms used to describe hereditary factors. His definition of the gene led him to distinguish between **genotype** (an **organism**'s genetic make-up) and **phenotype** (an organism's appearance).

In 1910, **Thomas Hunt Morgan** began to uncover the interesting relationship between genes and chromosomes. Morgan discovered that genes are located on chromosomes. He also found that the appearance of the characteristics associated with linked genes occur together in the offspring. In other words, genes located on the same chromosome are usually inherited together, or ''linked.'' But as Morgan and his colleagues worked with more and more characteristics simultaneously, they discovered that genes on the same chromosome were not always linked. Sometimes crossing over occurred; paired chromosomes would break apart and rejoin during meiosis, causing many different gene combinations. Morgan and his colleagues went on to develop and perfect these and other gene concepts, which laid the groundwork for all genetically based medical research.

Currently, almost 4,000 genetic diseases are known to affect humans. In recent years, scientists have been identifying genes associated with various diseases and conditions at a furious pace. For example, in 1995, scientists found genes linked to at least some cases of male **infertility, epilepsy, schizophrenia**, Alzheimer's disease, and breast **cancer**. In 1996, they

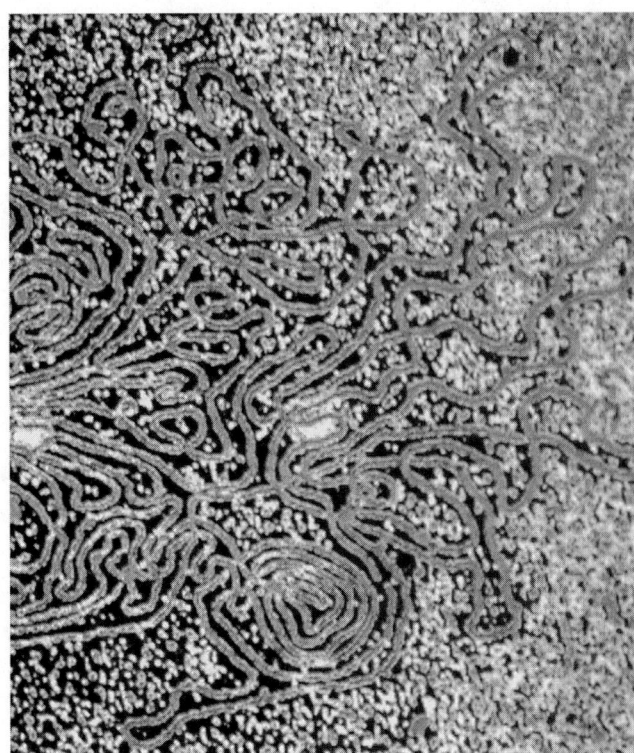

Strands of DNA. *(The Stock Market. Reproduced by permission.)*

found genes associated with some cases of retinitis pigmentosa (a progressive form of blindness) and basal cell carcinoma (the most common form of skin cancer), and in 1997, with heroin addiction, Parkinson's disease, fetal **alcohol** syndrome, glaucoma, and obsessive-compulsive disorder. Given enough time and effort, scientists may one day learn how to prevent or treat many such diseases.

GENE EXPRESSION

Genes, the biological unit of **inheritance**, are present in all living **organism**s. The majority of the time these genes are silent; they are not expressed. At certain times, however, genes are switched on to produce their particular product. This is called gene expression. Different genes have different triggers to activate them and then to subsequently switch them off. For some genes, such as developmental genes, expression occurs only once in a life time, even though the genes are present at all times within the body.

When **DNA** is folded to take up the least amount of room, the genes are inaccessible and it is impossible for the genes to be expressed. For gene expression, **enzyme**s are used to unfold the DNA to allow **transcription** and **translation**.

Gene expression can be switched on by external **stimulus**, and once the product has been manufactured, the genes are then deactivated. For example, in certain **species** of **bacteria** the enzyme B galactosidase is only produced when lactose is present in the growth medium. The enzyme is basically produced to order. Some genes, which are present in high copy number, are present as such because their expression is required in high quantities. This allows for a rapid expression and production of product. The current hypothesis of how the gene expression is controlled is called the **Jacob-Monod Hypothesis**.

See also Allele; Gene theory

GENE FLOW

Gene flow is the movement of **gene**s within a **population**. This movement of genes is the product of mating and gene exchange within populations. A genetically isolated population is known as a deme, gene flow can occur within demes or between demes when populations meet. On a limited scale gene flow can occur across **species** barriers. This occurrence is one of the arguments against releasing genetically modified **organism**s into the environment, particularly those that have been modified to include antibiotic resistance for screening purposes. Because of gene flow, these genes can spread through the wild population of that species and antibiotic resistance can be spread between species.

Gene flow is a method by which advantageous genes or **allele**s can be spread throughout a population, making them more common and more likely to remain within the population. If genes or alleles do not spread within a population, they are very quickly lost and, in evolutionary terms, they are a dead end. Gene flow creates new combinations of genes in individuals that can then be tested against the environment. If the new combination is more able to survive in the given environment, that gene and combination of genes is more likely to spread within the population.

See also Evolution; Natural selection

GENE FREQUENCY

Gene frequency is a measure of how common a particular **allele** is as a proportion of all possible alleles at a given locus (a physical location on a **chromosome**) within a **population**. This measure is also referred to as allelic frequency.

If a given allele exists in a population in two different forms ''A'' and ''a'', then the percentage of ''A'' taken as a measure of the overall occurrence of both ''A'' and ''a'' would be the gene frequency for ''A'' in that population. So, for example, if there are 30 individuals with a **genotype** ''AA,'' 50 with ''Aa,'' and 20 with ''aa,'' then it can be seen that ''A'' is present a total of 110 times (30 + 30 + 50). This is from a total possible occurrence of 200 (30 + 30 + 50 + 50 + 20 + 20) alleles at the locus under study. The gene frequency in this case for ''A'' is 110 / 200 = 0.55 or 55%.

Gene frequency can only be quoted for a population; the population can, however, be of any number of individuals. Gene frequency can, in certain circumstances, only be an estimate due to the hidden, recessive alleles that occur in populations.

GENE MUTATION

Gene mutation is a sudden change in the **DNA** making up an individual **gene**. Gene mutation is a fairly infrequent event, although its occurrence can be speeded up by the action of radiation or of certain chemicals (termed mutagens). Within a gene, the most common form of **mutation** is a single change in one of the bases on the DNA chain. Because of the redundancy of the **genetic code**, the resulting new codon may still code for the same **amino acid**. This is a neutral mutation. There is no observable effect on the product. With certain changes, a different amino acid is the result of an alteration in the genetic code. This new amino acid in the gene product may have a profound affect on the shape or functioning of the **polypeptide**; this is now a new **allele** of the original gene. In some instances, these mutations are in key components of the gene, and hence the polypeptide, and these forms are potentially nonfunctioning. Where a gene mutation is functional, it may produce a product slightly different to the original type. It is this difference in functionality that is then acted upon by **natural selection**. Mutation is the raw material of **evolution**. Any mutation, no matter how beneficial, will be short lived if it is unable to be passed on to the next generation. Consequently, the only gene mutations that are of any evolutionary consequence are those that are heritable, that is those which occur in the **gamete**s or the gamete precursors.

See also Heredity

GENE POOL

The **genetic material** that is contained within a **population** of sexually reproducing **organism**s is known as the gene pool. For a gene pool to exist, the genetic material must be interchangeable in some way between individuals. Therefore it is specific to **outbreeding** populations. A gene pool is of no fixed size. Small gene pools may die out quickly, whereas large ones are more stable.

A genetically isolated population, or ''deme,'' depends upon its gene pool for its evolutionary future. Within the population there is **gene flow**, a movement of the genes within the deme as a result of breeding. The overall **gene frequency** remains the same within the deme. This constancy is known as the genetic equilibrium.

Within a small population it is possible by chance for genes or **allele**s to be lost from the gene pool because the individual carrying them fails to reproduce. This natural variation in gene frequency within the gene pool is known as genetic drift. If adjacent populations were to meet, a small amount of breeding could possibly take place which would alter the composition of the gene pool. If the two populations were to start to freely interbreed, then they would become one enlarged gene pool.

GENE SPLICING

The first wave of genetic revolution involved study of the structure of the **deoxyribonucleic acid (DNA) molecule**. This new insight into the mysteries of **heredity** quickly led to another revolution. In the early 1970s, scientists discovered unique ways to manipulate genes. This technology, known as ligation, has generated scores of studies. The methods involve copying, dissecting, modifying, and even recombining portions of DNA from different sources. More importantly, these foreign genes can be made to become an integral part of the bacterium. They will replicate along with the original genes just as if they had always been a part of the bacterial **cell**. Thus, when the genetically altered bacteria replicate, the spliced genes are transmitted from one bacteria to the next. The foreign genes can also be functional within their bacterial host. They can be introduced to make their normal gene products. The substances secreted by altered bacteria can often be produced more efficiently than by conventional methods.

Scientists were eager to use this general knowledge to solve practical problems. Presently, gene splicing has many applications, especially in medicine and **agriculture**. Recombinant vaccines for hepatitis and therapeutics such as **insulin** have already been produced. A new type of tomato that has a longer shelf life was made using gene-splicing techniques. This variety of tomato is expected to save farmers and shippers millions of dollars each year. Efforts are also focused on **cloning** bacterial genes for **nitrogen** fixation into crop **plants**, thus eliminating the need for most fertilizers. Private gene-splicing companies are emerging to help meet the need for these new products, fostering the rapid growth of the biotechnology industry. Among the recently developed foods derived from new plant varieties created through this technology are an herbicide-tolerant soybean and a **virus**- and/or insect-protected tomato, potato, corn, squash, and papaya.

The stage for such growth was set in 1953, when **James Watson** and **Francis Crick** discovered the molecular structure of DNA. DNA ligases had been discovered and purified in five separate laboratories in 1967. These enzymatic compounds are important because they enable certain ''sticky ends'' of the broken DNA to be fused together. They can also be fused together with any other segment of DNA that has been cleaved by the same restriction enzyme. Hugh Smith, a molecular geneticist, described the first restriction endonuclease in 1970. This **enzyme** protects the bacteria cells against viral infection and is also able to break foreign DNA at specific sites. **Herbert Boyer**, a bacteriologist and molecular geneticist, is credited with isolating and describing a specific restriction endonuclease called EcoRI. This enzyme has proved to be especially useful in the development of cloning methods. Thus, the two main tools required for gene-splicing, the ''glue'' (DNA ligase) and the ''scissors'' (restriction endonuclease) were now available. It wasn't long before **Paul Berg**, a biochemist, and his colleagues were able to successfully recombine different DNA molecules in a test tube.

The recombinant DNA revolution gained momentum in the fall of 1973 at Stanford University and at the University of California at San Francisco (UCSF). Stanley Cohen (Stanford) and Herbert Boyer (UCSF) were the scientists credited with pioneering the techniques that now allow scientists to insert DNA from virtually any source into bacteria cells and de-

tect the expression of foreign genes in the bacteria. Cohen achieved the first successful transformation experiments in the common bacteria *Escherichia coli*. Cohen and Boyer later published a landmark report that detailed the mixing and recombining of DNA from two separate plasmids in *E. coli*. A plasmid is simply a small, circular extrachromosomal DNA molecule that replicates independently from the other bacterial DNA. Later experiments showed that it was possible to go one step further and combine DNA from an unrelated bacterium, *Staphylococcus aureus*, with *E. coli*. A year later they reported on the first successful cloning of **animal** genes in *E. coli*. In this experiment, specific genes from the African clawed toad, *Xenopus laevis*, were spliced into the *E. coli* bacterium.

In addition to the practical applications of ligation, scientists hope that recombinant DNA methods will help answer some of the basic questions of cell biology. They would also like to gain a better understanding of the processes that control human heredity and development.

GENE THEORY

Gene theory is the idea that **gene**s are the unit of **inheritance**. This theory has been modified many times over the years, as new evidence has become available. In the nineteenth century, **Gregor Mendel**, an Austrian monk, studied peas and how certain characteristics were transmitted from one generation to another. Mendel realized that the transmission was occurring through their **gamete**s. He found that the adult **plants** contained two copies of a factor controlling the observed characteristic, but that the gametes only had the one copy.

In 1900, Mendel's work was rediscovered and reinterpreted. By this time **chromosome**s had been discovered and **meiosis** described. In the early part of the twentieth century, **Thomas Hunt Morgan** showed the relationship between the number of **linkage** groups in an **organism** and the number of chromosomes. Each of these inheritable factors were called genes. **George Beadle** and **Edward Tatum** proposed that one gene was responsible for the production of one **enzyme** (the one gene one enzyme theory). In the mid-twentieth century, Oswald Avery showed that **DNA** was the **molecule** carrying the genes. Edwin Chargaff worked out the constant relationship between the bases within DNA. In 1953, **James Watson** and **Francis Crick** worked out the structure of DNA along with its method of **replication**.

Because of these discoveries, scientists have been able to cut DNA with **restriction enzyme**s and remove very precise sequences of DNA. This DNA can then be removed from one **cell** and placed into another. This is commonly carried out using **bacteria** as the recipient cell. This bacterium containing the new DNA can often be ''tricked'' into reading the DNA and making the product. This evidence is proof that genes are indeed the unit of inheritance and that the theory stating genes are responsible for the production of living organisms is largely correct. This is noted because many genes and their products do interact dramatically with the environment, and anything that we see is ultimately a product of the interaction of the organism's genes with its environment.

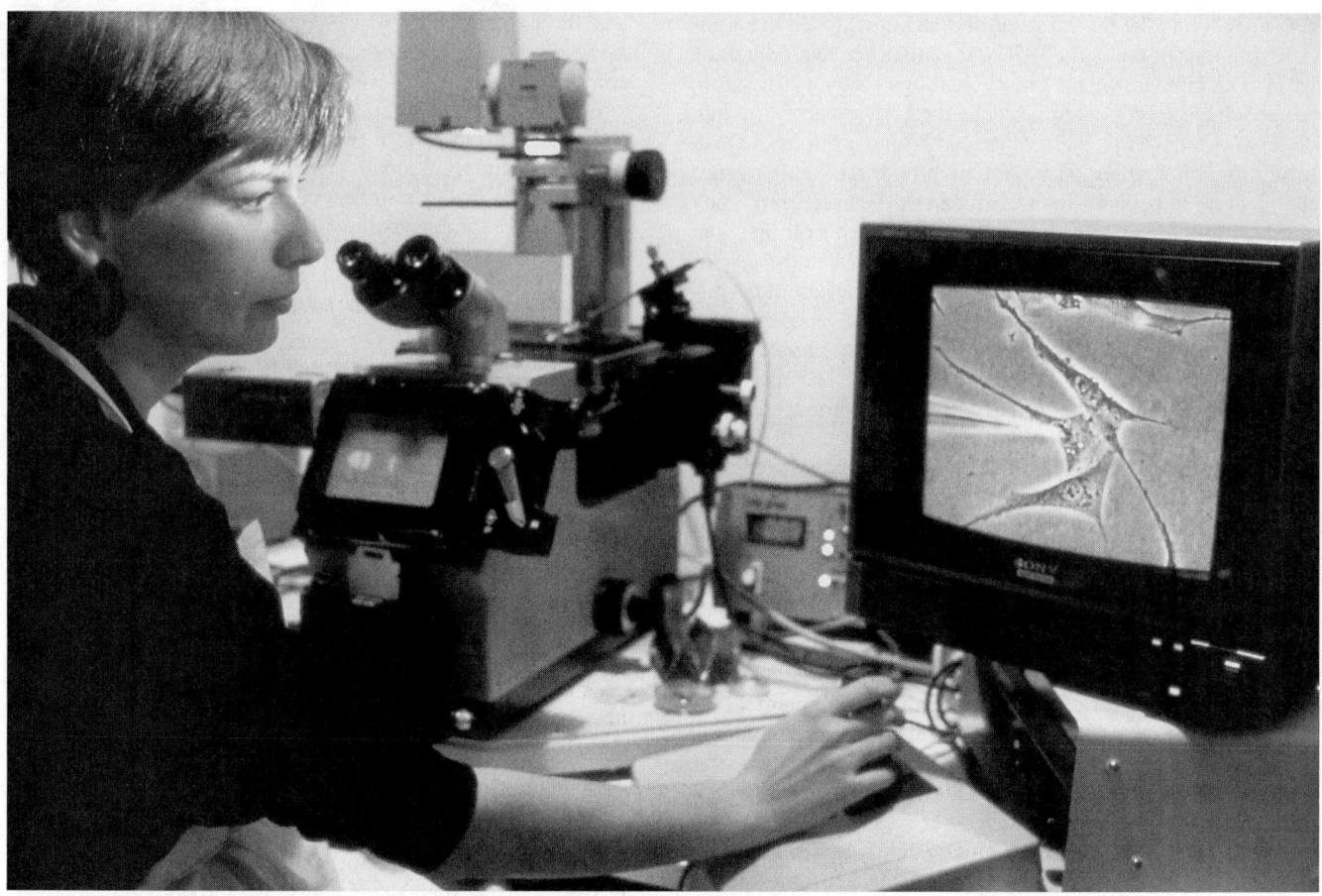

A scientist performing a microinjection of a corrective gene into a human T-lymphocyte cell (a white blood cell). At the left is the tip of the micropipette used for the injection and at the right the cell. *(Photo Researchers, Inc. Reproduced by permission.)*

GENE THERAPY

A **cell** is regarded as the basic unit of life. Located in each cell is a set of 46 **chromosome**s, 23 of which we receive from the mother and 23 from the father at conception. These chromosomes contain **gene**s that act as the blueprint for the production of specific **proteins**, which code for our appearance, different facets of our personality, our health, etc. Each gene is composed of **DNA**, which carries the hereditary information in an individual.

Each of us carries about half a dozen of defective genes. However, most of us do not suffer any harmful effects from our defective genes because we carry two copies of nearly all genes, one derived from the mother, and the other from the father. In most cases, the potentially harmful defective gene is ''recessive'' and the normal copy of the gene is enough to avoid all the symptoms of the disease. If, however, the disease-causing gene is dominant, then even if its counterpart is normal, it will still produce the disease. In addition, there are the X-chromosome linked genetic diseases. As males have only one copy of the X-chromosome, if the genes residing on the X are defective, there are no others available to compensate for it. Examples of such X-linked diseases are **hemophilia**, and Duchenne muscular dystrophy.

The discovery of the structure of the DNA by **James Watson** and **Francis Crick** in the 1950s proved to be the first major step in understanding genes and DNA. In the 1970s, scientists were successful in isolating specific genes from DNA. The 1980s proved to be a very remarkable decade in the field of **molecular biology**. Significant advances were made, several disease genes were isolated and mapped to specific chromosomes. DNA was synthesized outside of the cell and recombinant **DNA technology** was introduced. Until this point, our genetic make-up was entirely controlled by **nature**. Now however, the notion that we could go into the cell and manipulate the gene in any desired fashion did not seem like science fiction. The idea that science and technology could be used to manipulate genes was first planted in the minds of scientists.

This process of manipulating **genetic material** either to treat a disease or to change a physical characteristic is known as gene therapy. The concept of gene therapy is simple: introduce genes whose products can correct a defect and thus cure the disease, or slow the progression of a disease. It is not fundamentally different from other forms of medicine, in that instead of a therapeutic product such as a drug or a protein, genes are delivered to the body. The use of genes has many advantages. Because **gene expression** can be regulated, genes can be delivered in a directed manner to specific **tissue**s. The body itself can be used as the protein manufacturing facility. The constraints that apply to routine protein therapy, such as the biodegradability of proteins, its toxicity, its half-life, etc. can be overcome by gene delivery. Besides treating diseases and altering physical characteristics, gene therapy could also have applications in preventing our pre-disposition to certain familial, chronic diseases.

If an individual has a disease that is a result of inheriting two copies of a defective gene, then he could be a candidate for gene therapy. In its simplest form, what is done is that a new and properly functioning gene, (e.g. "ADA") is isolated from the normal cells. This gene, called the transfer gene, is removed by cutting DNA at specific locations, using a technique called **gene splicing**. A genetically altered **retrovirus** is used as the vehicle (vector) to deliver the new gene to the cells. During **cell division** when the cell is actively synthesizing DNA, **RNA** from the retrovirus is converted into DNA and incorporated into the DNA of the cell. The genetically altered cell now has a properly functioning ADA gene, which produces ADA within the cell. William French Anderson, Michael Blaise, and Ken Culver performed the first successful gene therapy experiment in 1990. They performed this experiment for treating Adenosine deaminase (ADA) deficiency, a severe combined immunodeficiency, also known as the "Boy in the bubble" disease. The results of the first gene therapy experiment were quite impressive.

However, the major limitation of gene therapy is the efficiency of gene delivery. The retrovirus infects the cells in an unpredictable manner. They insert the therapeutic gene at random positions. This newly inserted gene may actually interrupt a normal and essential gene sequence and thus could potentially harm the cell. Alternatively, the new gene could be inserted in dormant inactive stretches of DNA, and they may not be turned on frequently enough to produce enough of the therapeutic product. Nevertheless, new strategies are being researched and suggested and a new wave of enthusiasm for gene therapy is again being generated. Many ethical and legal issues arise when discussing gene therapy. People could make "designer children" by altering the genetic make-up before **birth**. Since it opens up the possibility of genetic manufacturing and gives us the ability to make a "perfect master race" etc.... the process must be carefully monitored.

See also Genetic engineering

GENETIC CODE

By the early 1950s, scientists knew that **gene**s were made of **deoxyribonucleic acid (DNA)** and that specific **protein**s were the products of specific genes. The exact link between DNA and proteins was less well understood, however. Since proteins are considered the language of life, researchers believed that the DNA **molecule**, with its four **nitrogen**ous bases, might be the code for this language. This is how the term genetic code originated.

Protein molecules are comprised of **amino acid**s. There are 20 biologically important amino acids. Only four different

bases—adenine (A), thymine (T), cytosine (C), and guanine (G)—are found in DNA. When each of these bases combines with a sugar and a phosphate molecule, a **nucleotide** unit is formed. How could only four different nucleotides code for 20 different amino acids? Scientists reasoned that if a single nucleotide coded one amino acid, only four amino acids could be provided for. If two nucleotides specified one amino acid, then there could be a maximum number of sixteen possible arrangements. George Gamow, a Noble Prize winner, demonstrated that at least three nucleotides in sequence were required to code for a single amino acid. This would provide for 64 possible combinations or codons—more than enough different "messages" to code for the 20 amino acids. Marshall Nirenberg, Har Khorana, and Robert Holley, three American biochemists, immediately adopted this three-nucleotide, or triplet codon, hypothesis as a foundation for their research. Each worked independently to discover how DNA influences **protein synthesis**. It was important to understand how the **genetic material** in a living **cell** directs the synthesis of proteins, because these proteins determine the structure and activities of the cell.

Marshall Nirenberg was born in New York City on April 10, 1927. He obtained his Ph.D. from the University of Michigan in 1957. He conducted research as head of the Laboratory of Biochemical **Genetics** at the National Institutes of Health; there he started his work with the genetic code. He wanted to solve the code by answering the following questions: 1) What was the ratio of the four bases in each triplet that coded for a particular amino acid? 2) What was the exact sequence of the bases in each triplet? and 3) Which triplet coded for which amino acid?

To simplify the task of identifying the triplet responsible for each amino acid, he used a man-made **ribonucleic acid (RNA) polymer**. Chemically, RNA is very similar to DNA, except that RNA is single-stranded and nonhelical and contains the base uracil (U) instead of thymine (T). Nirenberg first used a pure RNA polymer that contained only uracil (U). He found that only one amino acid, phenylalanine, was produced from this code, concluding that the code for phenylalanine must be the triplet UUU. Similarly, a pure cytosine (C) RNA polymer produced only the amino acid proline. Thus, the corresponding code must be CCC.

It wasn't long before a working dictionary of the RNA codes was established. Nirenberg began to notice that certain amino acids could be specified by more than one triplet. Statistical tests were later used to determine the amino acid produced from RNA made of a mixture of U and C by measuring the relative proportions of the resulting proteins. It was found that the code was redundant. A particular amino acid could be specified by more than one codon. Thus, the amino acid serine could be produced from any one of the combinations UCU, UCC, UCA, or UCG. Some triplets didn't specify any amino acid. These "nonsense" triplets signaled the beginning or end of synthesis.

After Nirenberg decoded 50 triplets, he worked on finding the exact orders of the three bases of the triplet. By painstakingly labeling one amino acid at a time with radioactive carbon-14, he passed the experimental material through a filter that retained only the cell's **ribosome**s and attached amino acids. The amount of radioactivity in the filter was measured to determine exactly which triplet code specified which amino acid. For example, AAC coded for asparagine, while CAA coded for glutamine. He obtained clear results for over sixty of the possible codons using this technique.

Har Khorana used this information to introduce new techniques of comparing DNA possessing a known structure with the RNA it would produce. Khorana was born in 1922 in Raipur, a village in Punjab, which is now part of West Pakistan. He studied for several years at universities throughout the world, joining the faculty at the University of Wisconsin, where he became interested in the genetic code, in 1960. He showed that separate nucleotide triplets do not overlap. He also helped determine the genetic code by re-creating each of the 64 possible triplets of DNA nucleotides that work in combinations as instructions for the protein-synthesizing ribosomes within the cell. His work proved that the genetic code is linear and consecutive and confirmed that three nucleotides make up one amino acid.

Khorana and his colleagues also determined the direction in which the code is read. In 1968, Khorana produced the first complete and functional synthetic gene. This achievement proved that it was possible to change genes and observe the results of those changes. This has subsequently been used as a valuable tool to study genetic disorders and the mechanisms of **cancer**ous cells. Artificial genes are now used to obtain large amounts of valuable proteins for human dietary and medical needs.

Robert Holley was born in Urbana, Illinois, on January 28, 1922. He received his Ph.D. from Cornell University in 1947 and became interested in the mechanics of proteins after working with **penicillin** in World War II. He discovered that small RNA molecules called transfer RNA (tRNA) existed and acted as acceptors for activated amino acids. Holley isolated specific tRNAs for three amino acids—alanine, tyrosine, and valine—in 1958.

As a direct result of Marshall Nirenberg, Har Khorana, and Robert Holley's work, the genetic code has been solved. The 20 amino acids are coded by 61 triplet codons; three additional codons do not code for any amino acids but direct the cell as to when it should cease protein synthesis. Each of these chemists shared the 1968 Nobel Prize in Physiology or Medicine for their achievements. Holley died in 1993.

GENETIC DRIFT · See Human evolution

GENETIC ELEMENTS, TRANSPOSABLE ·
See McClintock, Barbara

GENETIC ENGINEERING

Genetic engineering is the artificial altering of **gene**s. This usually takes the form of the insertion of alien genes into the **genome** of a recipient **organism**. Genetic engineering is also called recombinant **DNA technology**. This name is slightly more descriptive of the process, which is actually recombining the **DNA** of existing organisms into new organisms.

Before scientists had the ability to directly alter the genes of an organism, many attempts were made to change living organisms. At its simplest, these attempts involved selective breeding, whereby **plants** or **animals** with the desired characteristics were bred together to enhance those characteristics. The offspring not showing the required combinations were not used to breed again, whereas those showing them were bred again. Successive generations gave new combinations. Examples of this technique are the many different breeds of dogs that exist today. Selective breeding is a slow process, the exact time being dependent upon the length of the breeding cycle and the age at which sexual maturity of the organisms concerned is reached.

Cell fusion is a quicker way of altering the DNA. This technique, however, is suitable only for individual cells. Two selected cells under the mediation of viral attack will stick together and fuse. The nuclei of the two cells remain intact, resulting in a binucleate cell. The cell then starts to undergo **mitosis**. Not all of the **chromosome**s are capable of lining up properly, so the cell is usually incapable of developing into a living organism. However, it is still often capable of producing **protein**s and **polypeptide**s.

Genetic engineering is a more direct approach, resulting in quicker, more reliable, and more usable products. With this process, a desired piece of DNA, which may have been manufactured artificially or removed from another organism, is relocated into another cell. The cell which has had the DNA inserted now becomes able to manufacture the appropriate protein. The cells into which DNA is most frequently inserted are generally bacterial cells. Their rapid reproductive cycle (sometimes as fast as every 20 minutes) allows for a quick build of modified cells so that a large quantity of product can be effectively gathered.

Another scientific stride is the **cloning** of the gene. This procedure has been undertaken to allow bacteria to manufacture useful proteins. It is also used to mark bacteria so they can be rapidly screened. This is carried out most frequently with the addition of a gene for resistance to **antibiotics** in conjunction with the desired gene. By screening the resultant bacteria in antibiotic containing media, only the recombinants will survive. Fluorescent genes from jellyfish have also been added to organisms in this manner to show which are successful recombinants.

It is also possible to splice genes into higher organisms. This procedure has been undertaken with a number of plants and animals. Genes have been successfully inserted into cattle to produce human **blood**-clotting agents, and into various plants to make the fruit to last longer. Due to the contentious nature of such work in humans, very little research has been carried out in this area.

Genetic engineering in the form of recombinant DNA technology is a very powerful tool. It can produce great benefits, however, it poses great moral issues as well. Ethical committees from many disciplines are in place to oversee this sort of work and to consider the future of this technology.

GENETIC MAPPING

Hundreds of scientists are currently working together toward one common goal: to decipher the coded information in the human **genome**. A genome includes all the **DNA** within a **cell**'s **chromosome**s. The human genome consists of an estimated 50,000-100,000 **gene**s, plus a massive amount of DNA of unknown function. This monumental project, launched in 1990, is estimated to cost over three billion dollars and take more than 15 years to complete. It is by far the largest coordinated effort ever undertaken in the biological sciences. Named the **Human Genome Project**, this undertaking promises to revolutionize medicine and biology by creating a complete map of the location of genes on all 24 human chromosomes.

This has already been achieved for simple **virus**es with very short DNA sequences. In 1995, Craig Venter and his research team at the Institute for Genomic Research in Maryland announced that they had sequenced the entire genome for the *Haemophilus influenzae* bacterium. This bacterium has 1.8 million base pairs—the rungs of the DNA **double helix**—that make up its single circular chromosome.

By 1995, the Human Genome Project had already yielded an impressive amount of information about the human genome as well. Thanks to new research tools emerging from the project, the pace of gene discovery had nearly quadrupled. Examples of important discoveries already made by the project include the gene involved in **cystic fibrosis** and two genes involved in a hereditary form of colon **cancer**.

Coordinating the mapping of the entire human genome, with its three billion base pairs, requires successive approximations. First, the chromosome set and number of genes are identified. Then mapping begins by establishing which gene is linked to each chromosome, the order of the genes within a specific chromosome, and the distances between the genes on the same chromosome. Sometimes it is also possible to resolve the **nucleotide** sequence within a particular gene. Some physicians believe that this map will help diagnose, cure, and eventually prevent many diseases caused by faulty genes. Biologists hope to use the map to learn more about life itself. However, some opponents believe that the project's goal will reveal little about the genes' roles and suggest alternative research strategies.

Everyone does agree that this research would not be possible without the earlier contributions of several insightful biochemists. **Thomas Hunt Morgan**, Alfred H. Sturtevant, **Frederick Sanger**, and Edwin M. Southern helped develop the foundation for today's research projects.

Nearly one hundred years after **Gregor Johann Mendel**'s work with inherited traits, Thomas Hunt Morgan began studying **genetics** at Columbia University. He eventually founded what was to become the most important genetic laboratory of his day. Morgan used the **fruit fly** as a model for his experiments because it was easy to breed and maintain. Each tiny fly produced a new generation every two weeks. Morgan initially planned to use the flies to conduct breeding experiments similar to those Mendel had carried out using pea **plants** in the 1860s. Morgan noted that the eyes of the fruit fly are either red or white. During his studies of this inherited trait he bred a red-eyed female with a mutant white-eyed male fly. All the offspring had red eyes. This indicated that the gene for white eyes was recessive and was masked by the dominant red-eyed trait.

Surprisingly, when the offspring of the first generation-cross were mated, several white-eyed males appeared. However, among hundreds of flies, no white-eyed females were found. To explore this phenomenon Morgan mated the original white-eyed male with a red-eyed female from the first offspring generation. Their offspring included red- and white-eyed male and female flies. Morgan wondered why the white-eyed flies didn't show up in the first generation. He reasoned that the gene that determines eye color in the fruit fly must be carried on the same chromosome that determines the sex of the fly (the X chromosome). These experiments introduced the concept of **sex-linked traits**. They also helped to show that genes are located on chromosomes.

Subsequent crossbreeding resulted in additional unexpected results. The ratios of traits such as eye color, wing length, and leg length did not agree with Mendel's earlier work. Morgan proposed that the two alternative forms of a gene, called **alleles**, must occupy the same location on similar chromosomes. He correctly reasoned that during **meiosis**, when the chromosomes are copied, some pieces were breaking off and rejoining with the opposite chromosome. This explained the strange variations he observed in a few of the crossbred flies. This phenomenon is known today as **crossing over**.

With the discovery of crossovers, it became clear that genes must be positioned at particular spots, or loci, on the chromosomes. Furthermore, the alleles of any given gene must occupy corresponding loci on similar chromosomes. Morgan and his undergraduate assistant, Alfred Sturtevant, noticed that the genes for different traits were recombining at different rates. As Morgan's experiments had shown, these were fixed and predictable **recombination**s. It occurred to Sturtevant, however, that the percentage of recombination was probably related to the distance between genes on the same chromosome. Sturtevant postulated that genes are arranged in a linear series on chromosomes, like beads on a string. He also believed that genes located close together are less likely to separate than are genes located far apart. Sturtevant worked to determine the frequencies of recombination and plotted the sequences of the genes along the chromosome, reporting the relative distances between them. In 1913, Sturtevant began constructing chromosome maps using data from fruit fly crossover studies. By 1951, Sturtevant had completely mapped all four chromosomes of the fruit fly.

Several years later, Frederick Sanger was involved in similar work at Cambridge University. He was specifically interested in determining the exact structure of the **amino acid** chain of **protein molecule**s. In 1945, he discovered a chemical that could attach itself to only one end of an amino acid chain. By attaching this agent to an amino acid and then breaking the chain down with **enzyme**s, Sanger could tell which amino acid had been at the attached end by using a common technique called paper **chromatography** separation. By 1953, Sanger had found the exact order of the amino acids that comprise the **insulin** molecule. This paved the way for others to synthesize insulin as a treatment for diabetes. Sanger's long and very distinguished career did not stop here. In 1977, he went on to determine the entire sequence of DNA in a small virus. He is one of only three scientists to ever receive two Nobel Prizes in Chemistry—one for his insulin work and another for his viral work.

In addition to these breakthrough discoveries, a valuable lab tool was developed by Edwin Southern. The technique, known as Southern blotting, makes it possible to identify a DNA fragment by separating a mixture of fragments using electrophoresis, which allows the amino acids to be separated from proteins based on their different rates of **migration** in an electric field at a controlled **pH**. Southern then devised a way to transfer the DNA by blotting them on a special nitrocellulose sheet. A radioactive probe was then added to the DNA sequence. This probe is complementary to a specific DNA sequence and binds to the DNA fragment, which can later be visualized. This procedure makes it possible to pick out a specific fragment from a mixture containing millions of other fragments. Similar methods have recently been used to separate **RNA** and proteins and are whimsically referred to as Northern and Western blotting.

The knowledge that emerged from these studies has been characterized as the product of the "golden age" of genetics. Contemporary scientists are applying this knowledge to the human genome and other related projects in an attempt to better understand the mechanisms of life.

GENETIC MARKER

A genetic marker is an easily observed **gene**. When the gene is expressed, the phenotypic effects are very obvious. The gene is consequently used to identify an individual or a **cell** that carries it. It can also be used as a probe to mark a **nucleus, chromosome,** or locus.

Genetic markers can include the production of various easily screened chemicals, or more obvious effects such as colors or different structures. Genetic markers can also be used as an identifier for **genetic engineering**. After the marker is linked to a gene, its subsequent presence in cells indicates a successful genetic transformation. Markers of this **nature** can include antibiotic resistance, fluorescent genes, or other characteristics that the non-transformed **organism** is lacking.

Genetic markers are a powerful tool. Because of speed and cost-effectiveness, they are still used as an efficient method of screening organisms and producing initial maps of the genetic material of organisms, even when many more precise molecular tools are available.

GENETIC MATERIAL

Genetic material is the inheritable material of an **organism**. The genetic material of most life forms is comprised of **DNA**.

DNA is present in the **cell**s as **chromosome**s, which are found in the **nucleus** of **eukaryote**s. This is not the only location for genetic material. Some DNA is located in the **organelle**s of various organisms. These organelles are chiefly known as **mitochondria**, or in **plants** as **chloroplast**s. **Gene**s that are contained in organelles are said to show "extra nuclear **inheritance**." These types of genes show a slightly different pattern of inheritance compared to nuclear genes. Some organisms contain **RNA** as their genetic material. This type of inheritable material is restricted to only a small number of **virus**es. Organisms which have their genetic material dispersed throughout the cell are called **prokaryote**s. Some **bacteria** and all viruses are prokaryotes.

Genetic material must be able to replicate information about itself to pass on to future generations. In 1944, Oswald Avery showed that genetic material was carried in the nucleic acid of the cell. In 1953, **James Watson** and **Francis Crick** suggested the method by which DNA was able to replicate itself.

GENETIC SYSTEM

A genetic system is how the **genetic material** of a given **species** is organized and transmitted. It is the arrangement of the inheritable material within an **organism**, and how that material is transmitted to the next generation.

With all eukaryotic organisms, the genetic material is contained in specific structures. It is usually located in the **nucleus**, although some **DNA** can also be contained in various **cell organelle**s such as the **mitochondria**, or in the **chloroplast**s of **plants**. The majority of living organisms are **eukaryote**s.

Prokaryotes are a type of living organism that do not have their genetic material in a specific site. Instead the genetic material is spread throughout the cell. Some **bacteria** and all **virus**es are prokaryotes.

All living organisms, with the exception of some viruses, have DNA as their genetic material. Those viruses have **RNA** as their genetic material. Eukaryotic organisms have their DNA arranged in the form of **chromosome**s, whereas prokaryotes have threads or circular bands of DNA or RNA.

The method of transmission of the genetic material is initially similar for all organisms. First, in all organisms, the genetic material must be reproduced. Differences begin to occur in the transmission process. With lower organisms, a simple process of splitting, or binary fission, takes place, dividing the cell contents between the resulting new organisms. With higher organisms, more complex systems operate, ensuring a mix of genetic material from different individuals. This process involves the production of **gamete**s which fuse to produce a **zygote**. The exact details of how this process is carried out vary dramatically within the living world. The genetic system has the same function for all living organisms, however, the exact mechanics can differ, even between quite closely related species.

GENETIC VARIANCE

Genetic variance is the observable diversity of a trait in a **population** which can be attributed to **gene**tic heterogeneity. Genes can often exist in several different forms or **allele**s. Some of these alleles produce the same effect as others, but at certain times, there are alleles present in a population which have an observable effect on the phenotype of the **organism**.

Genetic variance is a measure of the level of dissimilarity between alleles within a given population. Some characteristics show no genetic variance. These are ones that are usually highly conserved due to their importance. One example of such a characteristic would be the **hemoglobin molecule** responsible for the transport of oxygen around the body in humans. Any alteration in the **genotype**, and hence the phenotype, of this molecule would be fatal. Other characteristics can show a wide genetic variance, such as eye and hair color. These characteristics are less important, hence a greater variability is seen within the population.

Genetic variance is a measure within a population. Consequently, the genetic variance present within a **species** may be greater than encountered in any breeding population. This would be due to barriers halting the intermixing of genes between all the populations making up a specific species.

GENETICS

Genetics is the scientific study of **heredity** and variation. It is concerned with the similarities and differences between individual **organism**s within and among **species**. The characteristics of an individual organism are largely the result of information transmitted from the parent to the offspring, via the **gamete**s (there is also an important environmental component). The understanding of how this occurs is the study of genetics.

Our growing knowledge of genetics is due to the work of many people, including **Gregor Mendel, Edwin Chargraff, Oswald Avery, Thomas Morgan, George Beadle, Edward Tatum, James Watson**, and **Francis Crick**. The discoveries made by them and others have shown that inheritable characteristics are present in organisms in the form of **gene**s. These genes are arranged on **chromosome**s made of **DNA**, except in the case of certain **virus**es that utilize **RNA** in place of DNA. Organisms that store chromosomes in their **nucleus** are known as **eukaryote**s. Organisms whose chromosomes are dispersed throughout the **cell** are known as **prokaryote**s. The laws and study of genetics apply to all living organisms, no matter how their genetic information is carried or arranged.

Since Mendel's first experiments with genetics in the nineteenth century, discoveries have been made at an exponential rate. Mendel, an Austrian monk, started his work with pea **plants** and studied the frequencies of certain easily observable characteristics such as the plant's height and **flower** color. He related the frequency of characteristics in the parental type to their occurrence and reoccurrence in the first and subsequent generations from carefully controlled breeding experiments.

The next step in genetics utilized the study of **bacteria** and then fruit flies. Using these organisms had two advantages: first, their small size allowed many experiments to be carried out in a small area; second, their fast generation time allowed many experiments to be conducted quickly. The study of genetics was then applied to human beings and other higher **mammals** by using family histories and direct observation. Currently the study of genetics can be carried out by directly looking at the genes themselves. Sequenced DNA can be produced from genes whose function is known. These sequences can then be tested in other organisms to see if they have the same capacity to produce the product. Genes can be moved around, transferred between species, and used to repair certain genetic abnormalities.

Even with the rapid advance of knowledge, there is still much that remains unknown. At the end of the twentieth century, scientists took the first steps in the **cloning** of mammals and in the sequencing of the entire human **genome**. There are many genes whose functions are still unknown, and many genes that are not mapped to specific areas of specific chromosomes. While there are more types of organisms alive than can possibly be studied, scientists hope that what is learned from the model systems they study is equally applicable to other related organisms. So far, all evidence points to this being the case.

In the future, genetics could be used to cure certain diseases, as many common ailments in humans are genetically linked, ranging from color blindness to life-threatening diseases like **cystic fibrosis**. Crops and farm **animal**s have already been genetically engineered to provide such items as tomatoes with a longer shelf life and cows that produce human **blood**-clotting agents in their milk. Due to the power of genetics, particularly the modern molecular approaches, much is misunderstood of this science. Furthermore, there are many ethical problems that need to be considered in this field. The advances in the study of genetics sometimes happen before legal statutes are in place to govern them. Legal controls are necessary, as they protect the work that is being undertaken, provide safeguards, and give confidence to the public.

GENOME

The **genome** is the full set of **gene**s carried by a **gamete**. It comprises one representative of each of the **chromosome** pairs of the adult **diploid** parent. In other words, a genome is a single set of genetic instructions. Not all of the **allele**s within a genome are expressed. Some are masked by the presence of dominant forms. A genomic library, or genomic bank, is a full genome stored as recombinant **DNA molecule**s.

The genomic formula is a mathematical expression of the number of genomes present in an individual **cell** or **organism**. One of the commonly encountered genomic formulae is the **haploid**, which is represented by n. This is sometimes called the basic number. The diploid form, which has two sets of a genome, is 2n; the triploid is 3n, and the tetraploid is 4n. Genetic abnormalities, where one chromosome is missing from the genome, can be represented in the same manner. For example, a diploid organism with one chromosome missing is a monosomic cell and is represented by 2n-1. A diploid with two chromosomes missing is termed a nullisomic and is represented by 2n-2. Additions of chromosomes can also occur and are represented in the same form; for example, 2n+1 is trisomic.

GENOME PROJECT • See Human Genome Project

GENOTYPE

Genotype is the genetic constitution of an **organism**. The genotype determines potential characteristics and limitations of the individual organism from **embryo** to adulthood. For organisms that reproduce sexually, genotype embodies the entire complex of genes inherited from both parents. Genotype can be demonstrated mathematically to show that individual offspring inherit different traits from their parents. Genes for certain traits can be recessive or dominant, which determines the likelihood of their expression as **phenotype** in a given individual. Phenotype is the observable features of an organism, such as size, skin or fur color, eye color, etc. Phenotype is determined by the interaction of genotype and the environment. Different individuals with the same genotype can have different phenotypes if raised in different environments.

GENUS

A **genus** (plural, genera) is a category in **biological classification**, referring to groups of **species** that are closely related phylogenetically (that is, through their evolutionary history) and share many elements of their **morphology**, biochemistry, and behavior. The genus is the principal taxonomic category between species and **family**. Related genera are separated from others in the same family on the basis of differences in their taxonomic characters. Although the degree of difference is not precisely defined by biologists, it is greater than that occurring between closely related species, and less than that between families.

All species are designated by a latinized, two-word name called a binomial. That of humans, for example, is *Homo sapiens*. The first word of the binomial refers to the genus of the species. In the particular case of *Homo*, only one living species occurs in the genus, which is referred to as monospecific. Most genera are polyspecific and contain more than one species. This is the case of the **plants** known as sedges (*Carex*), and the insects known as fruit flies (*Drosophila*), each of which has hundreds of species in the genus.

GEOGRAPHICAL ISOLATION

Geographical isolation occurs when members of one **species** living in a particular area become physically separated from

each other. The immediate result is the formation of two separate **populations** of the species. The long-term result can be **speciation**, the **evolution** of two separate species.

Organisms are geographically isolated by physical barriers, such as mountains, rivers, oceans, islands, **forests**, or **deserts**. These physical barriers prevent organisms from intermingling and interbreeding with one another. For example, part of a population of field mice might be separated from the rest of the population during a period of heavy rain, when a creek forms in the middle of the field from rainwater runoff. If the creek becomes permanent and turns into a river, the mice on one side would be geographically isolated from those on the other side. Organisms with greater mobility, such as birds and some types of fish, are less likely to be geographically isolated from other members of their species.

Geographical isolation plays a very important role in evolution, particularly in the development of new species. If members of a species are geographically isolated, the environments in which each new population lives and the subsequent **selective pressures** might change. Over time, each population will evolve to survive the particular environmental conditions in which it lives. If geographical isolation continues for a very long time (on an evolutionary scale), the species may evolve a great deal. If for some reason, the populations are no longer geographically isolated, that is, they come back together, the individuals in one population might not be able to reproduce with members of the other population. The geographical isolation would have resulted in reproductive isolation and the two populations would now be considered two separate species.

One of the most dramatic examples of geographical isolation occurred millions of years ago when Australia separated from other continents as a result of plate tectonics (the construction of Earth's crust). At that time most of the **mammals** on Earth were marsupials, the type of mammal in which the young develop in pouches. As a result of the plate separation, the mammals in Australia were geographically isolated from those on the other continents. Slowly, over millions of years, the environments and selective pressures of the continents changed and two separate species evolved. **Placental** mammals, those in which the young develop primarily within the mother, evolved to become dominant in most of the world, displacing most of the marsupials. However, since Australia was geographically isolated from the rest of the world, and environmental conditions and therefore selective pressures were different, placental mammals did not evolve and displace the marsupials. Today, most of the marsupial species that have survived are in Australia. Similarly, there are a very large number of native placental mammal species throughout the rest of the world and none in Australia. Thus, geographical isolation resulted in very different types of mammals in Australia than the rest of the world.

GEOLOGIC TIME

Geologic time describes the immense span of time—billions of years—revealed in the complex rock surface of the earth.

GEOLOGICAL TIME SCALE

Millions of Years	Era	Period	Epoch	Major Geological Events	Biological Life
0.01	Cenozoic	Quaternary	Holocene		Humans dominate earth
1.0	Cenozoic		Pleistocene	Extensions of ice caps in arctic and north temperate regions; continents are elevated	Earliest humans; extinction of large mammals
13	Cenozoic	Tertiary	Pliocene	Formation of mountains in northwestern North America; Alps and Himalaya mountains rise	
25	Cenozoic		Miocene	Most areas on earth become cooler and drier	Earliest hominids
36	Cenozoic		Oligocene	Development of flood plain and river deposits in North American Great Plains	
58	Cenozoic		Eocene	Jungles and forests are widespread; climates are consistently warm	Earliest grasses
65	Cenozoic		Paleocene	In North America, basins develop between Pacific coast and Rockies	Earliest large mammals; dinosaurs are extinct
135	Mesozoic	Upper Cretaceous		Development of Rocky Mountains; seas cover Atlantic and Gulf coastal plains	Dinosaurs dominate earth; earliest flowering plants
180	Mesozoic	Lower Jurassic		Climates are consistently mild; mountains develop along Pacific coastline	Earliest birds and small mammals
230	Mesozoic	Triassic		Deserts and dead seas develop in Eurasia and North America	Dinosaurs first appear
280	Paleozoic	Permian		Formation of Appalachian and Ural mountains completed; continents are elevated	About 96% of all plant and animal life on earth disappear at the end of this period
310	Paleozoic	Pennsylvanian		Southern continents covered by ice; Appalachian and Ural mountains begin to develop	Earliest reptiles
		Mississippian		Much of North American interior is covered by seas; mountain formation in southern North America and central Europe	Earliest winged insects
345	Paleozoic	Devonian		Mountain formation in northeastern North America	Earliest amphibians and and vascular plants
405	Paleozoic	Silurian		Dead seas in Michigan, New York, Ohio, southeastern Canada	Earliest insects and land plants
425	Paleozoic	Ordovician		Mountain formation in northeastern North America	Earliest corals
500	Paleozoic	Cambrian		Seas begin to appear in North America; climates are consistently mild	Earliest fish
600 to 4,500	Precambrian			Oldest dated rocks	Origin of life (primordial algae and bacteria)

(Table by Standley Publishing.)

Geochronology is the science of finding out how old rocks and minerals are. Absolute time and relative time are terms used to describe the age of rocks and events used by geologists. Radiometric age determination is a method used by geologists to determine the actual age, in years, of rocks and minerals. Knowledge of stratigraphy, the branch of geology that catalogues the earth's successions of rock layers, is essential to establish the relative ages of rock units. By finding which rock unit came first, the order of the events in Earth's history can be sorted out.

Before scientific methods were used to find out about geologic time, ideas about time and Earth history came from religious theories. The Hindu and Mayan religions believed in endlessly repeating cycles of time, each lasting for billions of years. Ideas in western culture about the age of the earth were just as precise—and just as wrong. In the 1650s, the Irish clergyman and scholar James Ussher (1581-1656) used the book of Genesis in the Bible to determine that Earth was created in 4004 B.C. For its time, Ussher's exercise was perfectly good science. He was basing his results on the only information that was considered relevant to the question of the earth's age. Within those assumptions, Ussher engaged in rational philosophical inquiry, and certainly did not intend to suppress earth science. Isaac Newton also speculated on the age of the earth, using the investigative techniques of the time—which would be considered archaic today.

As early as the eighteenth century scientists knew that the earth's lifetime was immense. But geologists were not able to measure the earth's history was until mass spectrometers became available in the 1950s. The mass spectrometer is an instrument used to separate different varieties of atoms from each other. Before that time, educated guesses had been made by comparing the rock record from different parts of the world and estimating how long it would take natural processes to form all the rocks on Earth.

Georges Louis Leclerc de Buffon (1707-1788), for example, calculated the earth to be 74,832 years old by figuring how long it would take the planet to cool down to the present temperature. Writing around 1770, he was among the first to suggest that Earth's history can be known about by observing the planet's current state.

James Hutton (1726-1797) did not propose a date for the formation of the earth, but is famous for the statement that the earth contains ''no vestige of a beginning—no prospect of an end.'' The German geologist Abraham Werner (1750-1817) invented the stratigraphic column, a diagram of layers within the earth. An original approach to geological history was suggested by the French zoologist and paleontologist **Georges Cuvier** (1769-1832), who observed that specific **fossil** animals occurred in specific rock layers, forming recognizable groups, or assemblages. William Smith (1769-1768) combined Werner's and Cuvier's approaches, using fossil assemblages to identify identical sequences of layers distant from each other, linking or correlating rocks which were once part of the same rock layer but had been separated by faulting or erosion.

In 1897, the physicist Lord Kelvin (1824-1907) developed a model for Earth history, which assumed that has been cooling steadily since its formation. Because he did not know that heat moves around in currents in the earth (convection), or that the earth generates its own heat from the decay of radioactive minerals buried inside it, he proposed that the earth was formed from 20-40 million years ago.

In the late eighteenth century, geologists began to name periods of geologic time. In the nineteenth century, geologists such as William Buckland (1784-1856), Adam Sedgwick (1785-1873), Henry de la Beche (1796-1855), and Roderick Murchison (1792-1871) identified widespread rock layers beneath continental Europe, the British isles, Russia, and America. They named periods of time after the places in which these rocks were first described. For instance, the Cambrian period was named for Cambria in Wales, and the Permian, for the Perm basin in Russia. The Mississippian and Pennsylvanian periods widely used by American geologists were named for the American states. By the mid-nineteenth century, most of the modern names of the periods of geologic time had been proposed; many of them are still in use.

Relative age determination

A rock layer may or may not contain evidence which reveals its age. Rock layers whose ages are defined by relationships with the dated rock units around it are examples of relative age determination. That relationship is found by observing the unknown rock layer's stratigraphic relationship with the rock layers whose ages are known. If the known rock layer is on top of the unknown layer, then the lower layer is probably the older of the two. That is based on the principle of superposition, derived from the writings of Nicolaus Steno (1638-1686), which states that when two rock layers are stacked one above the other, the lower one was almost always formed before the higher one. The undated rock can thus be integrated into a frame of reference.

Radiometric age determination

Every rock and **mineral** exists in the world as a mixture of elements, and every element exists as a population of **atoms**. One element's population of atoms will not all have the same number of neutrons, and so two or more kinds of the same element will have different atomic masses or atomic numbers. These different kinds of the same **chemical element** are called nuclides of that element. A nuclide of a radioactive element is known as a radionuclide.

The nucleus of every radioactive element spontaneously disintegrates over time. This process results in radiation, and is called radioactive decay. Losing high energy particles from their nuclei turns the atoms of a radioactive nuclide into the daughter product of that nuclide. A daughter product is either a different element altogether, or is a different nuclide of the same parent element. A daughter product may or may not be radioactive. If it is, it also decays to form its own daughter product. The last radioactive element in a series of these transformations will decay into a stable element, such as lead.

While there is no way to tell whether an individual atom will decay today or two billion years from today, the behavior of large numbers of the same kind of atom is so predictable that certain nuclides of elements are called radioactive clocks. The use of these radioactive clocks to calculate the age of a rock is referred to as radiometric age determination. First, an appropriate radioactive clock must be chosen. The sample must contain measurable quantities of the element to be tested for, and its radioactive clock must tell time for the appropriate interval of geologic time. Then, the amount of each nuclide present in the rock sample must be measured.

Each radioactive clock consists of a radioactive nuclide and its daughter product, which accumulate within the atomic framework of a mineral. These radioactive clocks decay at various rates, which govern their usefulness in particular cases. A three-billion-year-old rock needs to have its age determined by a radioactive clock that still has a measurable amount of the parent nuclide decaying into its daughter product after that long. The same radioactive clock would reveal nothing about a two-million-year-old rock, for the rock would not yet have accumulated enough of the daughter product to measure.

The time it takes for half of the parent nuclide to decay into the daughter product is called one half-life. The remaining population of the parent nuclide is halved again, and the population of daughter product doubled, with the passing of every succeeding half-life. The amount of parent nuclide measured in the sample is plotted on a graph of that radioactive clock's known half-life. The absolute age of the rock, within its margin of error, can then be read directly from the time axis of the graph.

When a rock is tested to determine its age, different minerals within the rock are tested using the same radioactive clock—similar to questioning different witnesses at a crime scene to determine if they saw the same event happen in the same way. Ages may be determined on the same sample by using different radioactive clocks. When the age of a rock is measured in two different ways, and the results are the same, the results are said to be concordant.

Discordant ages means the radioactive clock showed different absolute ages for a rock sample, or different ages for different minerals within the rock. A discordant age result means that at some time after the rock was formed, something happened to it which reset one of the radioactive clocks back to zero. For example, if a discordant result happens in the potassium-argon test, the rock may have been heated to a blocking **temperature** above which a mineral's atomic framework becomes active and wiggly enough to allow trapped gaseous argon-40 to escape.

Concordant ages mean that no complex sequence of events—deep burial, metamorphism, and mountain building, for example has happened that can be detected by the two methods of age determination that were used.

A form of radiometric dating is used to determine the ages of organic matter. A short-lived radioisotope, carbon-14, is accumulated by all living things on Earth. Upon the **organism's death**, the carbon-14 begins to change into carbon-12 at a known rate (its half-life is 5,730 years). By measuring how much of the carbon-14 is left in the remains, and plotting that amount on a graph showing how fast the carbon-14 leaves the body, the approximate date of the organism's death can be known.

When uranium atoms decay, they emit fast, heavy alpha particles. Inside a zircon crystal, these subatomic particles tear long trails of destruction through the zircon's crystal framework. The age of a zircon crystal can be estimated by counting the number of these trails. The rate at which the trails form has been found by determining the age of rocks containing zircon crystals, and noting how torn-up the zircon crystals become over time. This age determination technique is called fission-track dating. This technique has detected the world's oldest

rocks, between 3,800,000,000 and 3,900,000,000 years old, and yet older crystals, which suggest that the earth had some solid ground on it 4,200,000,000 years ago.

The age of the whole earth is deduced from the ages of other materials in the solar system, namely, meteorites. Meteorites are pieces formed from the cloud of dust and debris left behind by a supernova, the explosive death of a star. Through this cloud, the infant Earth spun, attracting more and more pieces of matter like a ball of wet mud rolled through a sandpile. The meteorites which fall to Earth today have orbited the Sun since that time, unchanged and undisturbed by the processes which have destroyed the earth's first rocks. Radiometric ages for meteorites fall between 4,450,000,000 and 4,550,000,000 years.

The radionuclide iodine-129 is formed in nature only inside stars. A piece of solid iodine-129 will almost entirely decay into the gas xenon-129 within a hundred million years. If this decay happens in open space, the xenon-129 gas will float off into space, blown by the solar wind. Alternatively, if the iodine-129 was stuck in a rock within a hundred million years of being formed in a star, then some very old rocks should contain xenon-129 gas. Both meteorites and the earth's oldest rocks contain xenon-129. That means the star that provided the material for the solar system died its cataclysmic death less than 4,650,000,000 years ago.

GERM THEORY

If you have ever suffered from a cold or needed an inoculation in order to go to school, then you know that everyday life can expose you to contagious disease. Many major health problems are caused by tiny **microorganisms** such as **bacteria** and **virus**es that are impossible to see without a powerful **microscope**. Only in the last two hundred years did scientists propose the idea that diseases are caused and transmitted by microorganisms. The connection between microorganisms and disease is generally called germ theory.

Since the time of the ancient Greeks, people had believed in the theory of **spontaneous generation**, the idea that living things can arise from nonliving substances like flies coming from rotting meat. Even a rational thinker like **Theophrastus** (372-287 B.C.), who is considered the father of **botany**, did not know that living things can come only from other living things.

In 1668, an Italian scientist, **Francesco Redi**, performed a series of now-famous experiments using maggots to prove that flies do not arise spontaneously from rotting meat. However, even with the proof of his experiments, people still believed that microscopic **organism**s could arise from nonliving **matter**. Later, **Lazzaro Spallanzani**, an Italian biologist, performed experiments in which he attempted to question the English scientist John Needham's earlier "proof" that microorganisms developed from spontaneous generation. Spallanzani used vessels containing a mixture of seeds and **water**, a kind of nutrient broth that would normally breed microorganisms that were originally attached to the plant matter.

He boiled the broth for 30-45 minutes and poured it into sterile jars, which were then sealed. He observed that no microorganisms developed. His results showed that Needham in his similar experiments had not boiled the solution sufficiently to kill the germs. Spallanzani had refuted the principle of spontaneous generation and established that bacteria were somewhat heat resistant, but that the higher **class** of **animal**cula (what we know as **protozoa**ns) were destroyed more easily. Years later, in the 1830s, the theory of spontaneous generation once again was back in fashion, and German physiologist Theodor Schwann, who is also given credit for the **cell theory**, once again repeated Spallanzani's experiment to set the argument to rest.

Early in the nineteenth century, improved microscopes made it possible to see bacteria. Friedrich Henle (1809-1885) declared in the 1840s that contagious substances are not only organic but are living organisms. He believed there were two types of disease: miasmas, which come from the environment, and contagions, which spread from person to person. Italian doctors had results to show that the deadly disease **cholera** was caused by bacteria. Scientists began to share information confirming this new idea that bacteria might cause disease. But it was not until later in the century when the work of several scientists led to the germ theory—the idea that different microbes such as bacteria or viruses cause a specific disease. Two great scientists that helped shape this theory were **Louis Pasteur** (1822-1895), a noted French chemist and microbiologist, and **Robert Koch** (1843-1910), a German doctor.

Louis Pasteur was born on December 27, 1822, at Dôle, in eastern France. In 1847, he received his doctorate from the École Normale in Paris. Pasteur became a serious, hardworking scientist who was trained as a chemist and did research on crystals and **fermentation**. However, he is best remembered for his dedicated work on microorganisms. While studying the process of fermentation of grape sugar by **yeast** to produce wine, Pasteur observed that bacteria were contaminating the ferment and turning it sour as the organisms produced **lactic acid**. When asked to find the cause of a disease killing silkworms and destroying the French silk industry, Pasteur discovered that microorganisms were responsible for the disease.

Pasteur experimented with **culture**s of many disease-causing bacteria. While injecting healthy chickens with the bacteria that cause cholera, Pasteur got unexpected results. Because the bacteria were old, they did not give the disease to the chickens but instead gave the animals immunity to the disease. Pasteur had discovered that weakened microbes make a good vaccine. He was successful in developing a similar vaccine for **anthrax**, a deadly disease of livestock that was prevalent at the time.

Pasteur also developed a treatment for rabies, a deadly disease contracted from the bite of an infected, rabid animal. He methodically worked on a rabies vaccine using fluid from the **spinal cord**s of infected rabbits, and in 1885, he saved a young boy from the dread disease in a celebrated case.

Working around the same time as Pasteur was a young German doctor named Robert Koch, a former student of Henle. Koch had heard of the early work done by Pasteur, and

he was interested in the new germ theory of contagious disease. Koch set up a laboratory in a small room next to his doctor's office, and there he began experimenting with many types of disease-causing bacteria. Koch, an amateur photographer, was the first person to take pictures of bacteria through a microscope. In 1876, Koch showed beyond a doubt that a specific kind of bacteria caused the disease anthrax. With his painstaking experiments to support him, Koch published a book a couple of years later in which he announced that specific diseases were caused by specific microbes.

In 1881, Koch worked with **tissue** taken from an ape that had died of tuberculosis, a disease which at the time claimed many lives. Koch isolated the rod-shaped bacteria that cause tuberculosis by growing them in culture dishes separate from any other germs. He inoculated healthy guinea pigs with the bacteria, and when the guinea pigs became sick, he found the tuberculosis bacteria growing in them. He then removed some bacteria from the guinea pigs and grew them in yet another culture. Then he infected a second group of healthy animals with this cultured bacteria. When those healthy animals contracted tuberculosis, then he could be sure that the same bacteria were responsible. This long procedure invented by Koch became the standard way in which disease-causing organisms could be identified. It is known as Koch's Postulates.

As the 1800s drew to a close, the scientific community worked toward isolating more disease-causing bacteria. Most scientists agreed that the organisms that caused many contagious diseases could be isolated and a vaccine for each produced. Pasteur and others knew that there must be microbes smaller than bacteria that they could not see, and indeed viruses were not visible until the invention of the electron microscope. The technique of injecting animals with deadly germs may seem cruel by today's standards, but during the nineteenth century, scientists were unable to maintain live **cell**s outside of animals. The scientists that proposed the germ theory contributed much to our knowledge of disease and immunity.

GERMINATION

Germination refers to the beginning of growth of a mature seed, to produce a small plant, or seedling.

All seeds require access to adequate moisture and oxygen and a suitable **temperature** regime before they can germinate. The seeds of some **species** must also be exposed to **light** before they will germinate, although others are inhibited by this condition. The seeds of many **plants** are capable of germinating as soon as they ripen. In most cases, such seeds are dispersed into the environment, where if they encountered suitable conditions, they quickly germinate to establish a seedling. In rare cases, such as the red mangrove (*Rhizophora mangle*), the seed actually germinates while still attached to the parent plant, an unusual **habitat** known as **viviparity**.

The seeds of many plants, however, must experience a resting or dormancy phase before they are ready to germinate. Dormancy can be related to a number of attributes of the seed, including a seedcoat that is impervious to **water** and/or oxygen, a hard seedcoat that prevents expansion of the **embryo**, the presence of chemicals that inhibit germination, or the need for some sort of pre-treatment, such as exposure to cold or dry conditions. Seeds with an impervious or hard seedcoat may eventually soften or otherwise weaken after exposure to moisture and decay in the environment. Alternatively, they may require passage through the gut of an **animal** (after feeding on the fruit in which the seed was enclosed), or sometimes to fire or repeated freeze-thaw cycles (this requirement of hard-coated seeds is known as scarification). Seeds containing anti-germination chemicals may break dormancy after these substances are washed away by rainwater. Seeds requiring exposure to cold temperature usually germinate in the springtime or early summer (this dormancy requirement is known as after-ripening, and it can often be hastened by keeping the seeds in a refrigerator or freezer for a few weeks or months).

The seeds of some plants can remain viable for a long time, even centuries. In such cases, the seeds persist in the ground (as a so-called seedbank) until conditions suitable for germination are encountered. Seeds of pin cherry (*Prunus pensylvanica*), red raspberry (*Rubus strigosus*), and arctic lupin (*Lupinus arcticus*) have persistent seedbanks of this kind, as an **adaptation** to recovery after rare events of disturbance of their **ecosystem**.

GESTALT

Defined simply, the word gestalt means any structure or configuration of physical, biological, or psychological phenomena so integrated as to constitute a functional unit with properties not derivable from its parts in **summation**. It also refers to the pattern or figure assumed by such a gestalt structure or system. Gestalt is the German term for pattern or shape, meaning the configuration of patterned relationships of parts to the whole.

Most people are familiar with the idea, if not with the details, of using gestalt therapy in psychology. Psychology, as the science of human and **animal** (especially primate) behavior, can be considered a branch of biology. Much of the research of the early Gestalt theorists focused on **perception**, on how the parts of an object or setting interact with one another and in the process of interacting produce a perceived whole distinct from the sum of its parts. Their work had many implications for perception, **learning**, and social psychology.

Gestalt concepts have been used, in biology, in many other ways beyond psychological studies of primate behavior, studies that have tried to better understand the behavior of monkeys, apes, and humans. In the field of comparative psychology, concepts developed in the study of **primates** are tested on other **species**. Gestalt theory has been applied, for example, to the study of how Rufous hummingbirds learn through spatial association.

The holistic, ordered conception of an **ecosystem** is one obvious gestalt pattern studied by biologists (though ecologists do not often use the term gestalt). Other studies have sought gestalt patterns through experiments that tested whether or not there is a substantial gestalt component to colony odor among one species of ants, and through explorations of the role of the postpharyngeal gland in another ant species as a gestalt **organ** for recognition of nestmates. Biologists have also examined animal motility to determine whether it reflects piecemeal assembly or exhibits gestalt patterns.

Biologists doing medical research have studied zinc in rats to find out whether an intake exceeding requirements for the mineral could make up for previously deficient intake; the result was a complex, gestalt-like, biphasic growth response pattern. Medical researchers have used gestalt patterns of craniofacial features of children with de Lange and other syndromes to distinguish mild from severe cases. The word syndrome itself refers to the pattern of symptoms in a disease or to the occurrence together of a number of characteristic symptoms. To help physicians in their fight against allergies, a visual gestalt has been applied to categorize pollen types.

In founding gestalt theory, Max Wertheimer was reacting to the blind atomism that prevailed in the psychology of his time. Today, in research work in psychology and behavioral biology, the study of organized patterns are still providing a complement to standard reductionist approaches.

GILBERT, WALTER (1932-)
American biophysicist and molecular biologist

Walter Gilbert developed a method for determining the sequence of bases (individual components) in DNA, for which he shared the 1980 Nobel Prize in chemistry with **Frederick Sanger** and **Paul Berg**. Gilbert was born in Boston, Massachusetts, to a psychologist mother and an economist father. Interested in science since childhood, he majored in chemistry and physics at Harvard and received his Ph.D. from Cambridge University in England in 1957 for work in mathematical physics.

While teaching at Harvard, Gilbert shifted to experimental biophysics through acquaintanceship with **James Watson**, co-discoverer of the **double helix** shape of **DNA**. They experimented with messenger **RNA**, the **nucleic acid** that carries **gene** information to a cell's **ribosomes** for **protein** production.

In 1965-66, Gilbert and Benno Muller-Hill explored what mechanisms a **cell** employs to turn genes ''on'' or ''off.'' Using a radioactive tracing technique, they isolated a protein which turns off production of lactose-digesting **enzymes** in the bacteria *E. coli*. This verified the model of cellular gene expression proposed by **François Jacob**, **André Lwoff**, and **Jacques Monod** a few years earlier.

This work led to Gilbert's sequencing of the DNA bases in and near the same genes. In 1975, at the suggestion of Andrei Mirzabekov, a Soviet scientist, Gilbert used the chemical dimethyl sulfate to fragment a strand of DNA at sites of the bases guanine and adenine. He labeled the segments with radionucleotides, cut them with restriction enzymes, which break DNA at known locations, and separated the fragments by electrophoresis. Gilbert and Allan Maxam developed a similar method for the other two DNA bases, thymine and cyto-

Alfred Goodman Gilman. *(AP/Wide World Photo, Inc. Reproduced by permission.)*

sine, using the chemical hydrazine. After identifying the base sequences in the fragments, Gilbert was able to reconstruct entire DNA strand sequences thus "reading the letters" in which genetic instructions were coded.

In the 1970s and 1980s, Gilbert found himself in the midst of social and medical controversy when he decided to use the information he had to further commercial biotechnology when he joined Biogen. In 1996 he became CEO of Net-Genics, a pharmaceutical biotechnology company.

See also Tiselius, Arne

GILMAN, ALFRED GOODMAN (1941-)
American biochemist and pharmacologist

Alfred Goodman Gilman was born in 1941 in New Haven, Connecticut, to Alfred Gilman Sr. and Mabel Schmidt Gilman. Gilman grew up in White Plains, New York, where his father was on the faculty of The College of Physicians and Surgeons of Columbia University and then later a founding chairman of

the Pharmacology Department at the new Albert Einstein College of Medicine. Visits to his father's laboratory peaked his interest in biology. Gilman was also able to observe intricate pharmacological experiments designed for medical students.

In 1955 Gilman was sent to the Taft School in Watertown, Connecticut, a prep school for boys. Gilman was not happy about being sent away, nor did he enjoy the rigid structure of the boys' school. From there, he went on to Yale where he majored in biochemistry. Gilman describes his first laboratory project, to test **Francis Crick**'s adapter hypothesis, as "wildly overambitious." The experience was rewarding for him though, because of the encouragement he received from his lab instructor Melvin Simpson.

After receiving his B.A. from Yale in 1962, Gilman worked for Burroughs Wellcome in New York and published his first papers. He knew he wanted to go into research when he entered a unique M.D.-Ph.D. program at Case Western Reserve University in the fall of 1962. Gilman conducted research on the **thyroid gland** and was also interested in studying **cell**s and **genetic**s. He earned an M.D. and Ph.D. in pharmacology in 1969. His interest in genetics led him to the Pharmacology Research Associate Training Program at the National Institute of General Medical Sciences, where he researched cyclic adenosine monophosphate (AMP), a genetic regulator that moderates hormone actions. His work with Nobel laureate **Earl Sutherland** was the beginning of his interest in cell communication.

In 1971 Gilman accepted a position as an assistant professor of pharmacology at the University of Virginia in Charlottesville. It was here that Gilman began his Nobel Prizewinning work. He and his colleagues knew about **Martin Rodbell**'s work at the National Institutes of Health with guanine **nucleotide**s (components of **deoxyribonucleic acid [DNA]** and **ribonucleic acid [RNA]**). Rodbell and his research associates at the NIH ascertained that the guanine nucleotides were somehow related to cell communication, but could not prove it. Gilman's research began where Rodbell left off, and in the late 1970s, he and his colleagues started looking for the chemicals that would confirm Rodbell's work. Gilman used genetically altered leukemia cells to detect the presence of G-**protein**s. He found that without the G-protein, the cells did not respond to outside stimulation the way a normal cell would. In 1980 they found the G-proteins, named because they bind to the guanine nucleotides.

G-proteins are instrumental in the fundamental workings of a cell. They allow us to see and smell by changing light and odors to chemical messages that travel to the brain. Understanding how G-proteins malfunction could lead to understanding serious diseases like **cholera** or **cancer**. Scientists have linked improperly working G-proteins to everything from alcoholism to diabetes. Pharmaceutical companies are developing drugs that would focus on G-proteins.

In 1979 Gilman was asked to chair the Department of Pharmacology at the University of Texas Southwestern Medical Center in Dallas. He eventually accepted the postion and his time at Southwestern was filled with many other awards. Among the awards are the Paul Edvard Poulson Award from the Norwegian Pharmacological Society in 1982 and the Gairdner Foundation International Award in 1984. In 1987 he shared the Richard Lounsbery Award from the National Academy of Sciences with Martin Rodbell in 1987, foreshadowing the Nobel. In 1989 he won the Albert Lasker Basic Medical Research Award. Finally, Gilman and Rodbell were awarded the 1994 Nobel Prize in Medicine for their collaborative work. Since his discovery, Gilman has been in the forefront of G-protein research. He predicts that eventually scientists will be able to map cell communication in a way that will allow scientists to predict how cells will respond to a variety of signals, leading the way to major advances in the treatment of disease.

GLOBAL WARMING • See Greenhouse effect

GMELIN, LEOPOLD • See Digestion

GOLDSTEIN, JOSEPH L. (1940-)
American molecular geneticist and physician

Joseph Leonard Goldstein, the only son of Isadore E. and Fannie (Albert) Goldstein, was born on April 18, 1940, in Sumter, South Carolina. He graduated from Washington & Lee University in 1962 with a B.S. degree in chemistry, and attended Southwestern Medical School of the University of Texas Health Science Center in Dallas. There, Donald Seldin, chairman of the Health Science Center's Department of Internal Medicine, offered him a future faculty appointment, provided he would specialize in **genetics** and then return to Dallas to establish a division of medical genetics there. He received his M.D. degree in 1966.

Goldstein's internship and residency at Massachusetts General Hospital brought him to **Michael Brown**, who had arrived from the University of Pennsylvania, having also obtained his M.D. degree in 1966. The two served in the same internship and residency program, and both were interested in research. After finishing their training in 1968, they joined the National Institutes of Health (NIH) in Bethesda, Maryland.

At the NIH biochemical genetics laboratory, Goldstein studied under the leadership of **Marshall Warren Nirenberg**, who was awarded the 1968 Nobel Prize in physiology or medicine for unraveling the way in which the genetic code determines the structure of **protein**s. Here, he learned about the excitement and efficiency of biology on a molecular level. At the same time, he worked under Dr. Donald S. Fredrickson, clinical director of the National Heart Institute, who was investigating people with hypercholesterolemia, or abnormally high cholesterol levels. In particular, Goldstein was interested in those patients with homozygous familial hypercholesterolemia. Familial hypercholesterolemia, identified as a genetically acquired disease by Carl Müller of Oslo, Norway, involved a genetic defect which caused a metabolic error resulting in high blood cholesterol levels and **heart** attacks. But it was Fredrickson and Avedis K. Khachadurian of the Ameri-

can University of Beirut, who identified two forms of the disease: a heterozygous form, involving a single defective **gene** found in one in 500 people; and a homozygous form, in which two defective genes are present and which strikes about one in a million. Blood cholesterol levels reach four to eight times the normal amount with symptoms of atherosclerosis, or hardening of the **arteries**, beginning in childhood. Nearly every sufferer from the homozygous form dies from a heart attack before the age of 30.

In 1972 Goldstein left the National Institutes of Health for Seattle under a two-year NIH fellowship in medical genetics. During this time, he worked with Arno G. Motulsky, an internationally recognized expert in the field of genetic aspects of heart disease, and devoted himself to a study investigating the frequency of various hereditary hyperlipidemias (diseases of high blood-fat levels) in a random sampling of heart attack survivors. The samples were taken from 885 patients (who survived three months or more) out of 1,166 coronary victims admitted in an 11 month period to 13 Seattle hospitals from 1970 to 1971. Studying 500 of those survivors and 2,520 members of their families revealed that 31% of the survivors had high blood-fat levels, either high cholesterol, high triglycerides, or a mixture of both. Eleven percent had an inherited combination of high cholesterol and high triglycerides. Goldstein and his associates defined this disease as familial combined hyperlipidemia. He knew that due to its complexity, combined hyperlipidemia would be an arduous area in which to begin research. Patients with homozygous hypercholesterolemia—having no normal genes at the area of the unknown defect—might be easier to study regarding gene functioning and cholesterol level.

Returning to the University of Texas Health Science Center in 1972 as head of the medical school's first division of medical genetics, and assistant professor in the department of internal medicine which was still directed by Donald Seldin, Goldstein addressed the task of identifying the fundamental genetic defect in familial hypercholesterolemia (FHC). Brown had joined the staff the previous year.

The idea of cell receptors was known, but it had never been studied in relationship to fat and cholesterol in the blood. Over 93% of the cholesterol in the human body is found inside cells. There, it participates in functions critical to cell development and cell membrane formation. Cholesterol also contributes to the essential production of **sex hormones**, corticosteroids and bile acids. The remaining 7% is dangerous, however, if it is not absorbed into the cells as it courses through the **circulatory system**, and sticks instead to the walls of blood vessels disrupting the flow of blood to the heart and **brain**.

Dietary cholesterol, found only in animal foods, is not necessary to the human body since the body produces its own cholesterol in the liver. If no cholesterol is available in the bloodstream, individual cells will produce their own. The human liver excretes that cholesterol which is not used by cells or deposited on artery walls. Cholesterol is fat-soluble, but attaches itself to water-soluble proteins, or lipoproteins, manufactured in the liver, as a means of moving through the bloodstream. The lipoproteins most favored by cholesterol are low-density ones, called LDLs, which are composed of much more fat than protein. Thus, high levels of LDLs are equated with the threat of heart disease.

Goldstein and Brown started their study by observing **tissue** cultures of the human skin cells known as fibroblasts, harvested from six FHC homozygotes, 16 FHC heterozygotes and 40 normal people. The cultured fibroblasts, like other animal cells, need cholesterol for the formation of the cell membrane. During this process, Goldstein and Brown were able to follow the manner in which the cells obtained cholesterol, and identify the process of cholesterol extraction from the lipoproteins in the serum of the culture medium, specifically LDLs. This discovery was made in 1973 with their demonstration of the presence of receptor molecules on the cells, which function to adhere LDLs and carry them into the cell. Goldstein and Brown noted that each individual cell normally has 250,000 receptors that bind low-density lipoproteins, and further located LDL receptors on circulating human blood cells as well as cell membranes from assorted animal tissues.

The cells of individuals with the heterozygous form of FHC have 40-50% of the LDL receptors that are typically present on normal cells. Cells of individuals with the homozygous form of FHC have no LDL receptors or a very small number. Cholesterol, manufactured by the liver and attached to LDLs, is passed into the blood, but is removed from the circulatory system rather slowly. Under normal circumstances an LDL molecule spends a day and a half in the bloodstream, but in FHC heterozygotes this length of time is extended to three days, and in FHC homozygotes to five days, providing increased opportunity for cholesterol to accumulate in the walls of the blood vessels.

Cholestyramine, a drug used to treat high cholesterol levels, had been synthesized over 20 years before Goldstein's and Brown's study, but had never been fully understood. Goldstein and Brown discovered that cholestyramine works by multiplying LDL receptors in the liver, which then converts cholesterol into bile acids and passes them into the intestines. However, in spite of this action, cholestyramine had only limited effect on levels of serum cholesterol. Goldstein and Brown determined the reason for this: The increased numbers of LDL receptors in the liver signaled the need for more cholesterol and the liver responded by increasing cholesterol production. This increase in cholesterol level then shut down the production of LDL receptors in the liver. These findings indicated the need for a drug to impede the liver's synthesis of cholesterol that could be administered in tandem with cholestyramine. In 1976 Akiro Endo, a Japanese scientist, isolated compactin, an anticholesterol **enzyme**, from **penicillin** mold, and in the same year Alfred W. Alberts of Merck, Sharp and Dohme research laboratories isolated a structurally similar enzyme, mevinolin, from a different mold. Goldstein and Brown combined mevinolin and cholestyramine in **animal** experiments with good results, and in 1987 the Food and Drug Administration approved mevinolin, now called lovastatin, for marketing. The FDA made the recommendation with the stipulation that the drug should be used only when diet and exercise proved inadequate in treating high cholesterol. Goldstein anticipated a lapse of five to ten years before use of the drug would affect the nation's coronary death rate.

For revolutionizing scientific knowledge about the regulation of cholesterol metabolism and the treatment of diseases caused by abnormally elevated cholesterol levels in the blood, Goldstein and Brown received the 1985 Nobel Prize in physiology or medicine. They also have received awards from the National Academy of Sciences, the American Chemical Society, the Roche Institute of Molecular Biology, the American Heart Association and the American Society for Human Genetics.

Goldstein's and Brown's research illuminating the activity of LDL receptors and their function in the management of cholesterol levels has had far-reaching effects. Not only has their work increased understanding of an important aspect of human physiology, but it has also had a practical impact on the prevention and treatment of heart disease. The National Institutes of Health, in part because of Goldstein's and Brown's work, recommended the lowering of fat intake in the U.S. diet.

GOLGI, CAMILLO (1843-1926)
Italian histologist

Golgi was born in Corteno, Italy, on July 7, 1843, the son of a physician. His home town was later renamed Corteno-Golgi in his honor. Golgi studied medicine at the University of Pavia, where he received his M.D. in 1865. After graduation, he worked briefly in a psychiatric clinic, but eventually decided to pursue a career in histological research.

Financial difficulties forced him in 1872 to accept a position as chief medical officer at the Hospital for the Chronically Ill in Abbiategrasso, Italy. No research facilities were available there, however, and he was able to continue his studies only by converting an unused kitchen into a laboratory. By 1875, Golgi had earned sufficient fame to receive an appointment as lecturer in histology at the University of Pavia. Four years later he was appointed Professor of Anatomy at the University of Siena, but he stayed only a year there before returning to Pavia as Professor of Histology. There he married Donna Lina Aletti, the niece of one of his former professors.

Golgi's earliest research involved the study of **neurons**, or nerve cells. Neurons present a number of problems for researchers that other **cells** do not. While most cells are compact and have a relatively fixed shape, neurons are commonly very long and thin with structures that are difficult to see clearly. In the 1860s, techniques used to stain and study non-nerve cells were well developed, but they were largely useless with neurons. As a result, a great deal of uncertainty surrounded the structure and function of neurons and neuron networks.

In 1873, Golgi found that silver salts could be used to dye neurons. The neurons turned black and stood out clearly from surrounding **tissue**. Golgi perfected his technique so that the addition of just the right amount of dye for just the right period of time would highlight one or another part of the neuron, a single complete neuron, or a group of neurons.

Golgi's new technique resolved some questions about the nervous system, but not all. He was able, for example, to confirm the view of **Wilhelm von Waldeyer-Hartz** that neurons are separated by narrow gaps—**synapses**—and are not physically connected to each other. He was unable to completely explain, however, the complex, overlapping network of dendrites.

While studying the **brain** of a barn owl in 1896, Golgi made a second important discovery. He found previously undetected bodies near the nuclear **membrane**. Those bodies, now known as *Golgi bodies* or *Golgi complexes*, seem to be involved in the manufacture of **proteins** and **carbohydrates**. For his research on the **nervous system**, Golgi was awarded a share of the 1906 Nobel Prize for physiology or medicine.

Between 1885 and 1893, Golgi was also involved in research on malaria. He made one especially interesting discovery in this field; namely, that all the malarial **parasites** in an **organism** reproduce at the same time, a time that corresponds to the recurrence of fever.

In addition to his scientific work, Golgi was active in Italian politics. He was elected a Senator in 1900 and served in a number of administrative posts at Pavia. He died in Pavia on January 21, 1926.

GOLGI APPARATUS

The Golgi apparatus is a collection of flattened **membrane** sacks called cisternae that carry out the processing, packaging, and sorting of a variety of cellular products in higher **plants** and **animals**. This important cellular **organelle** was named in honor of **Camillo Golgi**, a famous Italian neuroanatomist, who first described it in **brain** cells late in the nineteenth century. An individual Golgi apparatus is usually composed of four to eight cisternae, each a micron or less in diameter stacked on top of each other like pancakes. Many cisternal stacks interconnected by tubules and mobile transport vesicles make up a Golgi complex, which often is located near the **nucleus** in the center of the cell. In some animal cells, this complex can be huge, filling much of the **cytoplasm**ic space. In some plant cells, on the other hand, many small, apparently independent Golgi apparatuses are distributed throughout the cell interior.

Each Golgi stack has a distinct orientation. The *cis* or entry face is the site at which transport vesicles bringing newly synthesized products from the **endoplasmic reticulum** dock with and add their contents to the Golgi cisternae. A complex network of anastomosing membrane tubules attach to and cover the fenestrated cisternae on the *cis* face and serves as a docking site for transport vesicles. From the *cis* face a flow of vesicles carry transport and chaperone **proteins** back to the endoplasmic reticulum, while secretory products move on into the medial cisternae where further processing takes place. Finally, the products move to the *trans* or exit face where they undergo final processing, sorting, and packaging into vesicles that will carry them to the cell surface for **secretion** or to other cellular organelles for storage or use. Complex oligosaccharides are synthesized in the Golgi apparatus, and glycoproteins are assembled as materials move through the compartments of this organelle. A unique set of **enzymes** and chaperone proteins occur in each of the Golgi compartments to direct and carry out this complex set of reactions.

Jane Goodall with "Lulu," a baby chimpanzee. *(The Library of Congress. Reproduced by permission.)*

GONDWANALAND · See Continental drift

GOODALL, JANE (1934-)
English ethologist

Jane Goodall is known worldwide for her studies of the chimpanzees of the Gombe Stream Reserve in Tanzania, Africa. She is well respected within the scientific community for her ground-breaking field studies and is credited with the first recorded observation of chimps eating meat and using and making tools. Because of Goodall's discoveries, scientists have been forced to redefine the characteristics once considered as solely human traits. Goodall is now leading efforts to ensure that animals are treated humanely both in their wild habitats and in captivity.

Goodall was born in London, England, on April 3, 1934, to Mortimer Herbert Goodall, a businessperson and motor-racing enthusiast, and the former Margaret Myfanwe Joseph, who wrote novels under the name Vanne Morris Goodall. Along with her sister, Judy, Goodall was reared in London and

Bournemouth, England. Her fascination with animal behavior began in early childhood. In her leisure time, she observed native birds and **animals**, making extensive notes and sketches, and read widely in the literature of zoology and ethology. From an early age, she dreamed of traveling to Africa to observe exotic animals in their natural habitats.

Meets Leakey in Africa

Goodall attended the Uplands private school, receiving her school certificate in 1950 and a higher certificate in 1952. At age eighteen she left school and found employment as a secretary at Oxford University. In her spare time, she worked at a London-based documentary film company to finance a long-anticipated trip to Africa. At the invitation of a childhood friend, she visited South Kinangop, Kenya. Through other friends, she soon met the famed anthropologist **Louis Leakey**, then curator of the Coryndon Museum in Nairobi. Leakey hired her as a secretary and invited her to participate in an anthropological dig at the now famous Olduvai Gorge, a site rich in fossilized prehistoric remains of early ancestors of humans. In addition, Goodall was sent to study the vervet monkey, which lives on an island in Lake Victoria.

Leakey believed that a long-term study of the behavior of higher **primates** would yield important evolutionary information. He had a particular interest in the chimpanzee, the second most intelligent primate. Few studies of chimpanzees had been successful; either the size of the safari frightened the chimps, producing unnatural behaviors, or the observers spent too little time in the field to gain comprehensive knowledge. Leakey believed that Goodall had the proper temperament to endure long-term isolation in the wild. At his prompting, she agreed to attempt such a study. Many experts objected to Leakey's selection of Goodall because she had no formal scientific education and lacked even a general college degree.

While Leakey searched for financial support for the proposed Gombe Reserve project, Goodall returned to England to work on an animal documentary for Granada Television. On July 16, 1960, accompanied by her mother and an African cook, she returned to Africa and established a camp on the shore of Lake Tanganyika in the Gombe Stream Reserve. Her first attempts to observe closely a group of chimpanzees failed; she could get no nearer than 500 yd (457m) before the chimps fled. After finding another suitable group of chimpanzees to follow, she established a nonthreatening pattern of observation, appearing at the same time every morning on the high ground near a feeding area along the Kakaombe Stream valley. The chimpanzees soon tolerated her presence and, within a year, allowed her to move as close as 30 ft (9.1m) to their feeding area. After two years of seeing her every day, they showed no fear and often came to her in search of bananas.

Chimpanzee Research Yields Numerous Discoveries

Goodall used her newfound acceptance to establish what she termed the "banana club," a daily systematic feeding method she used to gain trust and to obtain a more thorough understanding of everyday chimpanzee behavior. Using this method, she became closely acquainted with more than half of the reserve's one hundred or more chimpanzees. She imitated their behaviors, spent time in the trees, and ate their foods. By remaining in almost constant contact with the chimps, she discovered a number of previously unobserved behaviors. She noted that chimps have a complex social system, complete with ritualized behaviors and primitive but discernible communication methods, including a primitive "language" system containing more than 20 individual sounds. She is credited with making the first recorded observations of chimpanzees eating meat and using and making tools. Tool making was previously thought to be an exclusively human trait, used, until her discovery, to distinguish man from animal. She also noted that chimpanzees throw stones as weapons, use touch and embraces to comfort one another, and develop long-term familial bonds. The male plays no active role in family life but is part of the group's social stratification. The chimpanzee "caste" system places the dominant males at the top. The lower castes often act obsequiously in their presence, trying to ingratiate themselves to avoid possible harm. The male's rank is often related to the intensity of his entrance performance at feedings and other gatherings.

Ethologists had long believed that chimps were exclusively vegetarian. Goodall witnessed chimps stalking, killing, and eating large insects, birds, and some bigger animals, including baby baboons and bushbacks (small antelopes). On one occasion, she recorded acts of cannibalism. In another instance, she observed chimps inserting blades of grass or leaves into termite hills to lure worker or soldier termites onto the blade. Sometimes, in true toolmaker fashion, they modified the grass to achieve a better fit. Then they used the grass as a long-handled spoon to eat the termites.

Finds Audience through Television and Books

In 1962 Baron Hugo van Lawick, a Dutch wildlife photographer, was sent to Africa by the National Geographic Society to film Goodall at work. The assignment ran longer than anticipated; Goodall and van Lawick were married on March 28, 1964. Their European honeymoon marked one of the rare occasions on which Goodall was absent from Gombe Stream. Her other trips abroad were necessary to fulfill residency requirements at Cambridge University, where she received a Ph.D. in ethology in 1965, becoming only the eighth person in the university's long history who was allowed to pursue a Ph.D. without first earning a baccalaureate degree. Her doctoral thesis, "Behavior of the Free-Ranging Chimpanzee," detailed her first five years of study at the Gombe Reserve.

Van Lawick's film, *Miss Goodall and the Wild Chimpanzees,* was first broadcast on American television on December 22, 1965. The film introduced the shy, attractive, unimposing yet determined Goodall to a wide audience. Goodall, van Lawick (along with their son, Hugo, born in 1967), and the chimpanzees soon became a staple of American and British public television. Through these programs, Goodall challenged scientists to redefine the long-held "differences" between humans and other primates.

Goodall's fieldwork led to the publication of numerous articles and five major books. She was known and respected first in scientific circles and, through the media, became a minor celebrity. *In the Shadow of Man,* her first major text, appeared in 1971. The book, essentially a field study of chimpanzees, effectively bridged the gap between scientific treatise and popular entertainment. Her vivid prose brought the chimps to life, although her tendency to attribute human behaviors and names to chimpanzees struck some critics as manipulative. Her writings reveal an animal world of social drama, comedy, and tragedy where distinct and varied personalities interact and sometimes clash.

Advocates Ethical Treatment of Animals

From 1970-75 Goodall held a visiting professorship in psychiatry at Stanford University. In 1973 she was appointed honorary visiting professor of Zoology at the University of Dar es Salaam in Tanzania, a position she still holds. Her marriage to van Lawick over, she wed Derek Bryceson, a former member of Parliament, in 1973. He has since died. Until recently, Goodall's life has revolved around Gombe Stream. But after attending a 1986 conference in Chicago that focused on the ethical treatment of chimpanzees, she began directing her energies more toward educating the public about the wild chimpanzee's endangered habitat and about the unethical treatment of chimpanzees that are used for scientific research.

To preserve the wild chimpanzee's environment, Goodall encourages African nations to develop nature-friendly tourism programs, a measure that makes wildlife into a profitable resource. She actively works with business and local governments to promote ecological responsibility. Her efforts on behalf of captive chimpanzees have taken her around the world on a number of lecture tours. She outlined her position strongly in her 1990 book *Through a Window:* "The more we learn of the true nature of non-human animals, especially those with complex brains and corresponding complex social behaviour, the more ethical concerns are raised regarding their use in the service of man—whether this be in entertainment, as 'pets,' for food, in research laboratories or any of the other uses to which we subject them. This concern is sharpened when the usage in question leads to intense physical or mental suffering—as is so often true with regard to vivisection."

Goodall's stance is that scientists must try harder to find alternatives to the use of animals in research. She has openly declared her opposition to militant animal rights groups who engage in violent or destructive demonstrations. Extremists on both sides of the issue, she believes, polarize thinking and make constructive dialogue nearly impossible. While she is reluctantly resigned to the continuation of animal research, she feels that young scientists must be educated to treat animals more compassionately. "By and large," she has written, "students are taught that it is ethically acceptable to perpetrate, in the name of science, what, from the point of view of animals, would certainly qualify as torture."

Goodall's efforts to educate people about the ethical treatment of animals extends to young children as well. Her 1989 book, *The Chimpanzee Family Book,* was written specifically for children, to convey a new, more humane view of wildlife. The book received the 1989 Unicef/Unesco Chil-

dren's Book-of-the-Year award, and Goodall used the prize money to have the text translated into Swahili. It has been distributed throughout Tanzania, Uganda, and Burundi to educate children who live in or near areas populated by chimpanzees. A French version has also been distributed in Burundi and Congo.

In recognition of her achievements, Goodall has received numerous honors and awards, including the Gold Medal of Conservation from the San Diego Zoological Society in 1974, the J. Paul Getty Wildlife Conservation Prize in 1984, the Schweitzer Medal of the Animal Welfare Institute in 1987, the National Geographic Society Centennial Award in 1988, and the Kyoto Prize in Basic Sciences in 1990. Many of Goodall's endeavors are conducted under the auspices of the Jane Goodall Institute for Wildlife Research, Education, and Conservation, a nonprofit organization located in Ridgefield, Connecticut.

GORGAS, WILLIAM CRAWFORD (1854-1920)

American physician and Surgeon General, U.S. Army

United States military man and physician, William Crawford Gorgas, first gained recognition for eradicating yellow fever and malaria in Havana, Cuba. Due to this amazing feat, in 1902, U.S. Surgeon General Dr. George M. Sternberg recommended Gorgas be placed in charge of solving the same problem in Panama. When the U.S. government assumed control of Panama from the French and took over construction of the Panama Canal, they were forced to abandon the project because of yellow fever. Gorgas was appointed Chief Sanitary Officer of the zone in 1904, receiving commission to implement mosquito-control methods along the entire canal zone. By late 1907, his efforts had totally eliminated the disease from the area, enabling continued construction and successful completion of the canal. Gorgas was elected to the Alabama Hall of Fame in 1953.

Gorgas was born near Mobile, Alabama, the first child of General Josiah Gorgas who, originally from Pennsylvania, was a Chief Ordnance Officer in the Confederate Army; and Amelia Gayle, the daughter of a governor of Alabama. Gorgas first attended Sewanee University, ultimately becoming a gold medal student. From 1876-79, he attended Bellevue Medical College in New York City where he studied medicine and accepted an internship. His dream of entering the military was fulfilled in June 1880 when he was commissioned as second lieutenant in the U.S. Army Medical Corps.

When Dr. Ronald Reed discovered that mosquitoes carried the dreaded malaria and yellow fever, not everyone believed his findings. Gorgas did, however, and it was he who implemented the huge mosquito control program in Havana, Cuba where, in the early 1900s, 500 people were dying each year from the diseases. By 1902, Gorgas' efforts had yellow fever entirely eradicated. When the United States assumed control of Panama, tuberculosis, **cholera**, smallpox, **bubonic** plague, and diphtheria were all troublesome; however, yellow fever and malaria alone had killed an estimated 10,000-20,000 people there between 1882 and 1888. When Gorgas was commissioned to take on the daunting task of eradicating those diseases, he found an abundance of open sewage drains and shallow trenches dug by the French and filled with water as protection from ferocious umbrella ants. These pools became stagnant and putrid in the hot climate—perfect breeding grounds for the disease-carrying mosquito.

Amazingly, Gorgas' work was greatly hampered by U.S. government officials in the zone who refused to accept the mosquito theory. They not only denied him authority to implement his plans and hindered shipment of supplies from the United States, they tried desperately to have him dismissed from the project—until, in 1905, several top officials died from yellow fever. Fortunately for the entire project, President Theodore Roosevelt knew of Gorgas' success in Havana and, encouraged by John F. Stevens, the newly appointed chief engineer of the canal project, gave Gorgas his full support. Stevens immediately assigned 4,000 workers to the sanitation effort and approved an unlimited budget for supplies.

The crews set to work, first covering the stagnant waterways with an oil and pesticide combination, then fumigating every house. Infected people were quarantined in screened-in tents, running water was supplied to the homes, streets were paved, and entirely new communities emerged. Within two years, 24,000 workers lived in a disease-free zone. When Gorgas wrote his book *Sanitation in Panama* in 1915, he said, "...it is now more than eight years, and not a case of yellow fever has originated in the Isthmus." He also wrote, with amazing perception, "No doubt the great centers of civilization will remain for centuries much as they are at present. The white settlers will go to the valleys of the Amazon and Congo, building up large agricultural communities which will supply the European and American centers...I believe that the peoples of that day will look back upon the sanitary work done at the Canal Zone as the first great demonstration that the white man could live as well in the tropics as in the temperate zone. I am inclined to think that at this time the sanitary phase of the work will be considered more important than the actual construction of the canal itself, as important to the world as this great waterway now is, and will be for generations to come."

GRADUALISM

There is currently an ongoing debate among **evolution**ary biologists about the rate at which evolution occurs. There are two prevailing theories. Some scientists believe that evolution occurs gradually in very small increments over millions and millions of years. This theory is called gradualism. Others believe that evolution does not occur continuously, but occurs in fits and starts. This theory, called **punctuated equilibrium**, proposes that there are relatively long periods of time during which no evolutionary changes are occurring; it is during relatively short time spans that there are many evolutionary changes in the **species** that inhabit the earth.

When **Charles Darwin** first proposed his ideas on evolution, he expressed the view that evolution and **speciation** oc-

curred when **organism**s changed slowly, but at a relatively constant rate through time. He thought that over millions of years, small, incremental variations added up until eventually individuals were distinct enough from their ancestors that they could be considered a new species. Although he did not call it such, Charles Darwin believed that evolution occurred via gradualism.

The **fossil** evidence that so clearly supports evolution, does not necessarily support gradualism; the idea that evolution occurred slowly and at a constant rate over very long periods of time. If this was the case, the fossil record would show many transitional forms, individuals with very slight changes over time that clearly link one species to the next. However, this is not the case. In general, the fossil record shows relatively few of these transitional individuals. There are far more fossil examples of somewhat larger jumps from one species to the next. While one species clearly evolved from the other, the fossil record seems to show that the jump occurred relatively quickly in an evolutionary sense, rather than slowly over a long period of time. The supporters of gradualism argue that the transitional forms are rare in the fossil record because they were less common, and thus less likely to be preserved. Their opponents (supporters of punctuated equilibrium) state that the transitional forms never really existed; evolution and speciation occurred in short, rapid spurts so that the transitional forms were not present at all.

Despite the great advances in evolutionary biology since the time of Darwin, it is still not entirely clear whether evolution occurs slowly and continuously as explained by the theory of gradualism or whether it occurs quickly as is explained by punctuated equilibrium. Evolutionary biologists need to delve further into the fossil record to determine the rate of evolution.

GRAM, HANS CHRISTIAN JOACHIM (1853-1938)
Danish physician

Hans Christian Joachim Gram was a Danish physician and bacteriologist who developed a method of staining **cells** for microscopic study. Gram was born in Copenhagen, Denmark, on September 13, 1853. He received a B.A. in the natural sciences from the Copenhagen Metropolitan School in 1871 and served as an assistant to the zoologist Japetus Steenstrup from 1873-1874. He subsequently became interested in medicine and earned a medical degree from the University of Copenhagen in 1878. Gram, who worked in several areas of science and medicine, earned a gold medal in 1882 for a study on human erythrocytes. The following year he received a doctoral degree for his work in this field.

After obtaining his degree, Gram pursued post-doctoral studies in Berlin, focusing on bacteriology and pharmacology and pursuing post-doctoral studies in Berlin. It was in Berlin, in 1884, that he published his work on staining **cells**, which became widely known as **Gram staining**.

At that time, the method of staining cells was not entirely new to scientific research and several methods were already being used. Gram borrowed from a procedure initially devised by **Paul Ehrlich**, who used alkaline aniline solutions to stain bacteria cells. Experimenting with pneumococci **bacteria**, Gram first applied Gentian violet, which stained the cells purple, and then washed the cells with Lugol's solution (iodine), which served as a mordant to fix the dye. He followed those steps by applying **alcohol**, which washed away any dye that was not permanently fixed. Gram found that some cells remained purple (Gram positive), while others stayed essentially unstained (Gram negative). Gram's method aided microscopic study of bacteria, as well as provided a means of differentiating and classifying bacteria cells. Several years later, the pathologist Carl Weigart improved upon Gram's method by adding another staining step, which consisted in dyeing the Gram negative cells with saffranine.

Gram remained in Berlin working as an assistant in a hospital until 1891, when he was appointed as a professor of pharmacology at the University of Copenhagen. In 1889, Gram married Louise I. C. Lohse, and in 1892, advanced to the position of Chief of Internal Medicine at the Royal Frederiks Hospital. Extremely active in the field of medical education, Gram also maintained a large internal medicine practice. From 1901-1921, Gram served as chairman of the Pharmacopoeia Commission, during which time he abolished the use of many useless and obsolete therapeutic treatments. In addition, he published a four-volume book on the importance of rational pharmacology in clinical science. After his retirement in 1923, he returned to an earlier interest: the history of medicine. During his career, Gram received several honors including the Danneborg Commander's Cross, the Golden Medal of Merit, and an honorary M.D. Gram died in Copenhagen, on November 14, 1938.

GRAM STAINING

In the second half of nineteenth century, scientists proved that specific bacterial **organism**s caused specific diseases, and the field of microbiology was on its way to becoming a distinct science. The **microscope** was also further developed during that time, and scientists were concerned with identifying and classifying **bacteria**. Most bacteria are difficult to see with a compound **microscope**, but can be seen when there is obvious contrast between the bacteria **cell**s and their surrounding medium. Various dyes are used to stain **cell**s so that they are more easily seen. As early as the late eighteenth century, scientists had developed some basic methods of staining cells to aid in their study and used natural substances such as saffron that stained some parts of a cell. The discovery of synthetic dyes in the mid-1800s enabled scientists to utilize more colors to stain cells.

In 1884, the Danish physician **Hans Christian Joachim Gram** further developed a method of staining bacteria originally developed by the German biologist **Paul Ehrlich**. Ehrlich used aniline **water** and gentian violet (a cationic dye) to stain cells, and the cell walls would appear purple after staining. Gram added a potassium triiodide solution, which acted as a

mordant for the gentian violet dye, and then poured ethanol over the cells to wash away the unfixed dye. Gram found that some of the cells remained purple, while others did not. Bacteria that remained purple were "positive" and those that did not remain purple were called "negative." A few years later, Carl Weigert, director of the Senckberg Foundation in Frankfurt, Germany, added another step to the staining method. Weigert followed Gram's procedure with a final staining using saffranine (an anionic dye), which subsequently stained the "negative" bacteria red. The Gram stain is still considered to be the definitive, differential test to determine the chemical make-up of a bacterium cell wall. On the basis of a cell's reaction to the Gram stain, bacteria are divided into two groups, Gram positive and Gram negative.

The distinguishing feature between Gram positive and negative bacteria is the difference in the structure of the cell walls. The cell wall of a Gram positive is a thick, single layer of a cross-linked polysaccharide that is easily stained by gentian violet, while the cell wall of a Gram negative bacteria consists of a thin layer of polysaccharide and is covered by a lipid layer that resists the gentian violet, but can be stained by saffranine. Many dyes, which are **organic compound**s, are positively charged and easily combine with the negatively charged, acidic polysaccharide wall. Other dyes are negatively charged and combine with **protein** based cell constituents.

The chemical make-up of the cell wall also determines the penetrability of the wall by various drugs. Knowing if a bacteria is Gram positive or negative determines what type of antibiotic is suitable for treatment, as some **antibiotics** act against Gram positive bacteria (i.e. **penicillin**), while others act against Gram negative bacteria (i.e. tetracycline or streptomycin). Another important consideration is the fact that some Gram negative bacteria release endotoxins which can be fatal. When pharmaceutical companies develop new antibacterial drugs, the Gram stain is the method by which scientists determine the effectiveness of the drug.

See also Microorganism

GRASSLANDS

Grasslands, which are **biome**s wet enough to avoid becoming **desert**s but too dry to support **forest**s, are usually flat or rolling regions that receive an average 9.8-39.4 in (25-100 cm) of rain a year.

Although large portions of the earth's grasslands have been turned into farmlands, a number of grasslands can still be found in various countries of the world. Because rains are seasonal, grassland **plants** have adapted to drought conditions. In lowland areas, roots may go as far as three meters (more than nine feet) below the surface to locate **water**. More often the grasses rely on a huge **root system**, and some grasses produce a network of subsurface stems, called rhizomes, that stay alive after the above-ground foliage has died. The complex food webs supported by grasslands contain larger **population**s of **animal**s than any other terrestrial biome. Examples include the huge herds of zebra and wildebeests found in African

Florida Everglades, USA. *(Photograph by Robert J. Huffman, Field Mark Publications. Reproduced by permission.)*

grasslands. Many annual grasses and legumes, including wheat, barley, onions, and peas, originated in the Mediterranean grassland region known as the Fertile Crescent (between the Nile Vally and the Mesopotamian flood plain, in what is what is now Turkey), extending eastward from Greece.

Because of differences in rainfall patterns, the grasslands of various regions have distinct characteristics, but three main grassland types can be found. Shortgrass prairies experience strong winds, light and infrequent rainfall, and rapid evaporation. In the United States, the North American Great Plains, a shortgrass prairie east of the Rocky Mountains, was plowed to grow wheat, and was overgrazed, making it vulnerable to the drought, strong winds, and poor farming practices during the 1930s that transformed the prairie into a Dust Bowl. The region's original population of some 60 million bison, at one time the dominant large **herbivore** of the North American grasslands, was hunted to near **extinction**.

Tallgrass prairie, which once extended west across North America from the temperate deciduous forests, were mostly turned into farmland, though some of these grasslands can still be found, for instance, in eastern Kansas. These wetter grasslands abound with daisies, sunflowers, and other **flower**s, as well as legumes.

South America and Australia vary in their vegetation, depending on rainfall. Where rainfall is lowest, rapidly growing grasses dominate; where rainfall is greater, acacia and other shrubs grow in patches; where rainfall is highest, savannas shift into tropical woodlands with tall, coarse grasses, shrubs, and low trees. Grasses can grow from three to 12 ft

high (0.9 2.7 m). These subtropical grasslands experience a rainy summer period followed by a dry winter. In winter, the grasses wither. Africa possesses the largest savannas, home to wildebeests, zebras, Cape buffalo, impalas, and other herds of large ungulates (hoofed animals), which are preyed upon by lions and cheetahs. Scavengers, including vultures, jackals, and hyenas consume the remains.

In South Africa and Zimbabwe can be found the Veldt or Veld, the grassy plateaus that support potato and maize crops and large herds of cattle, in addition to industrial and mining centers. Veldt elevations range from approximately 500 to 6,000 ft (150-1,830 m). Another well-known grassland is the Pampas in southeastern South America, an area of approximately 300,000 sq mi (777,000 sq km). The region extends from North Argentina into Uruguay. In the more humid East Pampa, corn and other crops are grown, while in the dry West Pampa livestock are raised. The Steppes are the temperate grasslands of Eurasia, extending from Southwest Siberia to the lower reaches of the Danube River. They include the wooded steppe, where rainfall is highest and deciduous **tree**s grow; the tillable steppe, with productive agricultural land; and the semi desert non tillable steppe.

In recent years, ecologists have become concerned about the damage to grasslands. Overgrazing and changing grasslands into croplands can change weather patterns by, for instance, affecting how much sunlight is reflected or absorbed in a region. Ecologists point out that grasslands provide valuable **ecosystem** services, in addition to the economic value they furnish through the meat, milk, wool, leather, and other products produced in grasslands. By processing large amounts of carbon as soil organic **matter**, grasslands help maintain the composition of the **atmosphere**. Tilled grasslands release large amounts of carbon, an issue of growing concern as scientists try to assess the potential global warming impacts from increased concentrations of **carbon dioxide** in the atmosphere.

GRAY, ASA (1810-1888)

American botanist

Asa Gray was an American botanist best known as an nineteenth century authority on botanical **taxonomy** and a pioneer in plant geography. He was a strong supporter of **Charles Darwin** and his theory of **natural selection**.

Gray was born on November 18, 1810 in Sauquoit, New York. He attended Fairfield Medical School where he studied medicine. Subsequent to graduation, he briefly practiced medicine (1831-1832), then taught science at a Utica, New York, high school while collecting plants and teaching himself **botany**. In 1934 he met John Torrey, a chemistry professor at the College of Physicians and Surgeons, New York. The two collaborated on Torrey and Gray's *Flora of North America* (1838-43); this was expanded and published as volume I of the *Synoptical Flora of North America* in 1878. Gray worked for a year in 1835 as librarian and curator of the New York Lyceum of Natural History. In 1836, he planned to join the United States Exploring Expedition as a botanist but he resigned before sailing due to frustration with delays. In 1838, Gray accepted a position at the University of Michigan, Ann Arbor Professor of Natural History and spent a year in Europe purchasing books for the university library, meeting botanists, and studying American plants in European herbaria. In 1842, he accepted a position as Professor of Natural History at Harvard University, specializing in botany, a post he retained for 31 years until his retirement in 1873. Harvard named its botanical garden and herbarium, which houses Gray's priceless collections of books and plants donated to the university, after him. Gray died in Cambridge, Massachusetts, on January 30, 1888.

Gray's career was highlighted by his relationship with Charles Darwin of whom he was a strong supporter. Using Darwin's theory of natural selection which Darwin shared with Gray in 1857 two years before the publication of Darwin's *On The Origin of the Species*, Gray explained the geographical distribution of flora occurring in Japan, northern and eastern America, and Europe and theorized that all were descendents of circumboreal flora carried southward by glaciation during the Pleistocene era.

Gray published several hundred collected works, the best known of which was his *Manual of the Botany of the Northern United States, from New England to Wisconsin and South to Ohio and Pennsylvania Inclusive* (1848), also known as *Gray's Manual*. This comprehensive text helped establish systematic classification of botany in the United States.

See also Anatomy; Taxonomy

GREENHOUSE EFFECT

The greenhouse effect, the cause of global warming, is an unprecedented, and possibly irreversible, environmental condition in which damaging human-produced gases build up and trap heat within the earth's protective atmospheric shield, called the **ozone** layer.

The threat of global warming was first recognized in 1896 by Swedish chemist Svante August Arrhenius, who suggested that the burning of **fossil fuels** might have a serious impact on the earth's **temperature**, but scientists at that time could not have predicted that it would become one of the world's most pressing environmental issue less than one hundred years later.

Arrhenius's early warning of the dangers of **carbon dioxide** was not taken seriously until 1938 when G. S. Callender, an English physicist, pointed to meteorological records showing the gradual warming since 1880 of North America and Northern Europe. Callender was the first scientist to gather data from several sources on the danger of increasing **carbon dioxide** levels. Even with this data, Callender's study did not garner significant support because many scientists believed that the excess carbon dioxide would be absorbed in the oceans, not in the **atmosphere**. Also, the warming trend was inexplicably replaced by a temporary temperature decline around 1940. Callender remained adamant, however, and noted in 1958 that the warming had resumed in 1942. Finally, in the 1960s, concern about atmospheric **pollution** was taken

seriously as the link between **air pollution** and the earth's temperature became abundantly clearer.

The earth's fragile atmosphere is comprised primarily of **nitrogen**, oxygen, and carbon dioxide. These three ingredients, along with methane, provide Earth with a protective blanket that regulates how much of the Sun's enormous heat reaches Earth, as well as how much heat leaves our atmosphere. For this reason, these gases are called greenhouse gases, and they appeared to be balanced until the advent of the industrial revolution.

Through this delicate system, the earth's atmosphere allows the visible and infrared wavelengths of radiation from the sun to reach the ground. Once it has hit the surface, this visible **light** is absorbed and reflected by the earth as infrared radiation that cannot be seen but which can be felt as heat. If not for the presence of atmospheric gases such as carbon dioxide, **water** vapor, and other greenhouse gases, the heat would escape out beyond the earth's atmosphere. Instead, it is absorbed by the greenhouse gases, and much of it is re-emitted down towards the surface, resulting in extra heat.

A great deal of the carbon dioxide that is released by industrialized societies is absorbed by **forests**, oceans, and the process of limestone deposition. However, these resources are either limited or depleting, and extra output of carbon dioxide collects in the atmosphere. Global warming from the build-up of greenhouse gases is also exacerbated by **ozone** depletion, which is allowing harmful radiation to reach the earth's surface.

Recently, computers have allowed scientists to develop models that estimate the consequences of global warming while suggesting ways to slow the process. Measurements now show that between 1957 and 1975, the amount of carbon dioxide in the air has increased from 312-326 parts per million, a jump of approximately 5%. These measurements were collected all around the world from the top of the highest mountain, Mauna Loa in Hawaii, to the South Pole, where air was gathered through airplane air-intake systems. Overall, carbon dioxide concentrations are up about 30% since preindustrial times. Methane is up 145%, and nitrous oxide is up 15%. A car emits about 20 lb (9 kg) of carbon dioxide for every gallon of gas it burns.

Most climatologists—scientists who study the earth's climate—now believe it is clear that this human-produced greenhouse effect is changing the earth's climate. One recent comprehensive study used **tree** rings (which show temperature variations as well as **precipitation** variations), ice core samples (where trapped air can give clues about past environments), and coral records to trace climate patterns over the last 600 years. Researchers combined this evidence with temperature readings, which have been available for only about the last 150 years, and with historical records, and concluded that the twentieth century has been the warmest century in the last 600 years. Moreover, they concluded that the warmest years in all of that period were 1990, 1995, and 1997.

Although it is difficult to precisely predict exactly how greenhouse effect processes will affect the earth's temperature in the future, one computer model estimates that by the year

2050 the temperature level could rise by as much as 4°F (2°C). This average increase would be unevenly distributed, ranging from as little as less than a degree at the equator to up to six degrees at higher latitudes. Although this may not sound like a large increase, it may be enough to affect the rate of glacial melting, raising sea levels, and to otherwise have an impact on the earth's climate. Scientists believe the global sea level has already risen by about 4-10 in (10-25 cm) over the past 100 years. By the year 2100, it is expected to rise an additional 20 in (50 cm), doubling the number of people in the way of storm surges. Warmer climates would affect growing seasons, shift crop zones, and could increase the risk of certain tropical diseases like malaria.

Valuable research continues as scientists try to compare today's atmospheric and air quality with much older historical data. Industries and governments are trying to establish programs that would help protect the environment. At the 1997 climate-change conference in Kyoto, Japan, delegates from 159 nations agreed to a pact that takes the first steps toward legal regulations that will reduce industrial gases—Europe by 8%, Japan by 6%, and the United States by 7%. As these same gases are often behind the engines and industries of every country's economies, conflicts will need to be resolved between industrial users and environmental advocates.

GRIFFITH, FREDERICK (ca. 1879-1941)
English microbiologist

Frederick Griffith's work with streptococci and pneumococci **bacteria** gave him an important place in the history of biology. However, the impact of his work on the science of **genetics** was even more crucial, although it is not clear whether Griffith himself ever realized his contributions to this field. His classic experiments, published in a single seminal paper in 1928, showed that some strains of bacteria could appropriate the disease-causing characteristics of other strains. Although interesting enough for the light it shed on the virulence of certain **organisms**, what he called the "transforming principle" was also the first clear evidence linking **DNA** to **heredity** in **cell**s.

The details of Griffith's life are not completely known, partly because he lived very quietly and reclusively and partly because the importance of his work was not appreciated until well after his death. He was born in 1879 (some sources say 1877 or 1881) in Hale, in Cheshire, England, and he attended Liverpool University. His one older brother, A. Stanley Griffith, was also a microbiologist. After his graduation from the University of Liverpool in 1901, the younger Griffith worked for the Liverpool Royal Infirmary, the Thompson Yates Laboratory, and the Royal Commission on Tuberculosis. In 1910 he began working for the government, in what would later be called the Ministry of Health, under the supervision of Arthur Eastwood. The facilities were primitive, but as a friend wrote of Griffith in *The Lancet:* "He could do more with a kerosene tin and a primus stove than most men could do with a palace."

Griffith researched many kinds of **microorganism**s, but his most important work dealt with pneumococcus, the bacte-

ria that can cause pneumonia. All types of these bacteria can theoretically cause disease, but some types (such as Type III) cause disease more readily than others (such as Type II). When Griffith began his work, he knew that the difference in virulence was due to a **polysaccharide** coating, or capsule, on the Type III organisms, which protected the bacteria from the host's immune system. The Type II pneumococcus lacked the "capsule" that protected Type III. Bacterial colonies with capsules are called (S) colonies, and ones without capsules are called (R) colonies. They look quite different and are easy to identify in culture.

Research on Bacteria is Precursor to Genetics

Griffith injected some mice with Type II pneumococcus alone and other mice with Type III pneumococcus that had been killed by heating. None of the mice developed pneumonia. When he injected mice with both live Type II and dead Type III pneumococcus, however, the mice not only developed pneumonia, but live Type III bacteria could be extracted from their blood. Somehow the Type II bacteria had made protective capsules for themselves, "transforming" themselves into Type III. They had apparently acquired the characteristics from the dead Type III bacteria.

After later researchers managed to obtain transformed bacteria in a test tube instead of a live animal, work in the area declined for awhile. It was not until 1944 that **Oswald Avery**, Colin Munro MacLeod, and Maclyn McCarty took up Griffith's experiments again and tried to explain his results. They extracted the active transforming principle from Type III (S) pneumococcus and showed preliminarily that it was DNA. In "Studies on the Chemical Nature of the Substance Inducing Transformation of Pneumococcal Types," they cautiously stated that if DNA actually proved to be the transforming principle, "... then nucleic acids of this type must be regarded not merely as structurally important but as functionally active in determining the biochemical activities and specific characteristics of pneumococcal cells."

DNA is a very long **molecule**, or **polymer**, made up of linked, individual units. There are only four of these units, or **nucleotide**s, however, scrambled in varying order along the length of the DNA. Biochemists of the time knew about **nucleic acids**, but they were certain that it was protein that caused **inheritance**; they were not inclined to suspect much of a hereditary role for a molecule (DNA) that seemed too simple for such a complex activity. Finally, in 1952, other researchers used radioactive labeling to prove that DNA was indeed the hereditary material that Griffith had first observed transforming bacteria. Griffith's work may thus be seen as pivotal in beginning the science of molecular biology.

Little is known about Griffith's private life except that he enjoyed skiing, walking, and vacations at his country cottage in Sussex. In the first Griffith Memorial Lecture, given in 1966, W. Hayes said, "Fred Griffith has been described as a shy and reticent man, whose quiet kindly manner, and his devotion to his job, made him a lovable personality to those few who got to know him." He published very little, but what he did was of a very high quality, and Hayes believed that this

"... must be ascribed to an innate humility and capacity for self-criticism, so that he offered to posterity only those products of his research which he judged to be new and important." Griffith and a longtime colleague at the Ministry of Health, the bacteriologist William M. Scott, were killed in 1941 during the bombing of London when a bomb blew up the building in which they worked.

GROSSETESTE, ROBERT (1168-1253)
British scholar and theologian

Robert Grosseteste was an unusual combination of scientist and theologian, acclaimed by modern theorists as "...the central figure in England in the intellectual movement during the first half of the thirteenth century." Grosseteste wrote prolifically on philosophy, astronomy, and science, and diligently encouraged the addition of a greater number of scientific subjects to the curriculum of the university of his time.

Information about Grosseteste's life until he reached his early fifties is virtually nonexistent. Even the exact year of his birth is uncertain. He was, however, born in Suffolk County, probably in a little town called Stow Langtoft. His family were not wealthy, and survival during the first half of his life was most likely difficult. He had a sister, Juetta (or Ivetta), who became a nun, and letters written by him indicate he may have had relatives who were clergymen. He was admired for his knowledge of the arts and literature; was well-versed in music, architecture, poetry, mathematics, astronomy, optics, and physics; understood Greek and Hebrew; and historic evidence ssuggests he was conversant in both medicine and law. Writings attributed to him from the year 1200 indicate he studied and lectured on theology at the Cathedral School at Oxford, and it seems likely he was the first chancellor when Oxford was declared a university in 1215.

In 1221 he left that post, holding several church-related positions over the next few years. His devout faith and adherence to Biblical theology gained him the position of lector (lecturer) to Oxford's newly established order of Franciscans, and in 1235 when he was somewhere in his late sixties, he was elected to the distinguished post of Bishop of Lincoln, the largest diocese in England. Almost immediately, he instituted strict reforms in his new diocese based on Biblical principles. So strong was his faith that he unhesitatingly spoke out—sometimes violently—against immoral and unethical conduct within the churches, adamantly berating the pope, cardinals, and bishops whenever he believed their power and influence deviated from Biblical standards. He even publicly denouncing the king for his misuse of power.

A fervent believer in the philosophy of **Aristotle** and a translator of Greek and Arabic scientific texts, Grosseteste was one of the first scholars to bring Aristotelian theory to Europe. He wrote commentaries of Aristotle's *Posterior Analytics* around 1228 and *Physics* between 1228 and 1231, delving into the theory of scientific reasoning and expressing theories on the systematic and scientific search for knowledge and truth. The foundation of Grosseteste's scientific reasoning was the

Bible in which God is often referred to as Light. In Genesis, the first book of the Bible, God used light to created the World. Grosseteste firmly believed that, to understand God, one must understand light. He also believed that experiments were essential to validate any theory. This led him to an intense investigation into light using Euclid's methods of geometry. Ultimately, he correctly determined that a rainbow was caused by the refraction of light passing through mist, and not—as previously believed—by reflection. Grosseteste wrote *De Luce*, a famous essay on his study of light; *De Natura Locorum*, in which he diagrams a glass of **water** refracting light; and his study of optics using lenses and mirrors is recorded in his *De Iride*, in which he wrote "...we may make things a very long distance off appear as if placed very close, and large near things appear very small..."

Until his death at around the age of eighty-three or eighty-five, Grosseteste was vital, active and worked harder than ever before in his already rich life. In his lifetime, Grosseteste was either loved or hated. His outspoken convictions made him unpopular in many circles. However, in 1253, his church honored him as being their greatest bishop ever, and several recommendations were made to the pope by the king, archbishops, and other clergy in 1280, 1286, 1288, 1307 to declare him a saint. Although modern historians still find him controversial, most agree he was a "...portentous person, a man of universal genius."

GUILLEMIN, ROGER (1924-   )
French American endocrinologist

Roger Guillemin is one of the founders of the field of neuroendocrinology, the study of the interaction between the central **nervous system** (such as the **brain**) and endocrine glands (such as the pituitary, thyroid, and pancreas). Guillemin focused his research on **hormones** produced by the brain, and their subsequent effect on body processes. He proved the correctness of a hypothesis first proposed by English anatomist Geoffrey W. Harris that the hypothalamus releases hormones to regulate the **pituitary gland**. For discoveries which led to an understanding of hypothalamic hormone productions of the brain, Guillemin and fellow endocrinologist **Andrew V. Schally** shared the 1977 Nobel prize for physiology or medicine with physicist **Rosalyn Sussman Yalow**. Guillemin and Schally were pioneers in isolating, identifying, and determining the chemical nature of such hormones as TRF (thyrotropin-releasing factor which regulates the **thyroid gland**), LRF (luteinizing-releasing factor which controls male and female reproductive functions), somatostatin (which regulates the production of growth hormones and **insulin**), and endorphins (which may be involved in the onset of mental illness). Guillemin's work led to scientific advances including an understanding of thyroid diseases, **infertility**, juvenile diabetes, and the physiology of the brain. According to Guillemin, the determination of the chemical structure of TRF marked an end to the pioneering era in neuroendocrinology and the beginning of a major new science.

Roger Charles Louis Guillemin was born on January 11, 1924, and raised in Dijon, France, the son of Raymond Guillemin, a machine toolmaker, and Blanche Rigollot Guillemin. He attended the University of Dijon where he received a Bachelor's degree in 1942, and then entered the University of Lyons medical school, graduating with a medical degree in 1949. However, Guillemin interrupted his studies during World War II in order to join the French underground during the Nazi occupation, becoming part of an operation helping refugees escape to Switzerland over the Jura Mountains. During and after the war Guillemin received three years of clinical training and briefly practiced medicine before joining a well-known Canadian physiologist, **Hans Selye**, as a research assistant. To work with Selye, Guillemin moved to the Institute of Experimental Medicine and Surgery at the University of Montreal in Canada. In 1950, he suffered a near-fatal attack of tubercular meningitis. After his recovery in 1951, Guillemin married Lucienne Jeanne Billard, who had been his nurse during his illness. They had six children, five daughters—Chantal, Claire, Helene, Elizabeth, and Cecile, and a son François.

Embarks on Career at Baylor University

Guillemin received his Ph.D. from the University of Montreal in 1953, and accepted an assistant professorship at Baylor University Medical School in Houston, Texas. His research involved endocrinology, the study of the hormones that circulate in the **blood**. The **endocrine system** is a hierarchical one in which hormones from the pituitary gland regulate other endocrine glands. It was thought that the head of the entire system was the **hypothalamus**, located at the base of the brain just above the pituitary gland. However, the way in which hypothalamic hormonal regulation occurred was unclear. The theory of regulation by nerve impulses was marred by the anatomical fact that there are few nerves that extend from the hypothalamus to the pituitary. Anatomist Geoffrey W. Harris theorized that hypothalamic regulation occurred by means of hormones, which are transported by the blood. Harris's experiments supported his hypothesis, proving altered pituitary function when the blood vessels were cut between the hypothalamus and the pituitary. The problem was that no one had yet been successful in isolating and identifying a hormone from the hypothalamus.

Begins Intense Research on Hypothalamic Hormones

Guillemin began an investigation to find the missing evidence, a task of extraordinary difficulty because very minute amounts of hypothalamic substances are involved. At Baylor, Guillemin worked together with Schally using a technique called mass spectroscopy and a new tool developed by physicists Solomon Berson and Rosalyn Sussman Yalow called radioimmunoassays (RIAs) which enabled scientists to isolate and identify the chemical structure of hormones. In the early 1960s, Guillemin considered continuing his research in France, and obtained a concurrent appointment at both Baylor and the Collège de France in Paris. However, he left the Collège de France in 1963, and was appointed director of the Laboratory for Neuroendocrinology at Baylor University. By this time he and Schally had ended their scientific cooperation and had become fiercely competitive in a race to identify hypothalamic hormones.

Guillemin worked with sheep hypothalami which he obtained from slaughter houses. Obtaining the specimens was a large-scale, difficult operation. Only very minute amounts of substance existed in each sheep hypothalamus and it had to be extracted very soon after death. Guillemin and Roger Burgus, a chemist who worked with Guillemin, reported that their laboratory collected about five million hypothalamic fragments from sheep brains, which involved handling about five hundred tons of brain tissue. Finally in 1968, Guillemin and his coworkers isolated the hypothalamic hormone that effects the release of thyrotropin. The following year Guillemin, as well as Schally, who had been working independently, revealed the structure of TRF (a hypothalamic hormone which today is called thyrotropin-releasing hormone or TRH). When TRH is secreted by the hypothalamus, it causes the pituitary gland to secrete another hormone that in turn causes the thyroid gland to secrete its own hormones. Shortly thereafter Guillemin and his colleagues isolated and determined the chemical structure of GRH (growth-releasing-hormone), a hypothalamic hormone that causes the pituitary to release gonadotropin which in turn influences the release of hormones in the testicles or ovaries. This discovery led to advancements in the medical treatment of infertility.

In 1970 Guillemin moved to the Salk Institute in La Jolla, California. There he isolated a third hypothalamic hormone which he named somatostatin. This hormone acts by inhibiting the release of growth hormone from the pituitary gland. In 1977 Guillemin and Schally were awarded the Nobel Prize for their research on hypothalamic hormones. Guillemin wrote on the importance of their discoveries in an autobiography, published in *Pioneers in Neuroendocrinology II*, stating that: "I consider the isolation and characterization of TRF as the major event in modern neuroendocrinology, the inflection point that separated confusion and a great deal of doubt from real knowledge. Modern neuroendocrinology was born of that event. Isolations of LRF, somatostatin, and the recent endorphins were all extensions (as there will be still more, I am sure) of that major event—the isolation of TRF, a novel molecule in hypothalamic extracts, with hypophysiotropic activity, the first so characterized.... The event was the vindication of 14 years of hard work."

Guillemin soon turned his attention to another class of substances, known as neuropeptides. Produced by the hypothalamus and other parts of the brain, neuropeptides act at the **synapses** of the nerves (the area where the nerve impulse passes from one neuron or nerve cell to another). One group of neuropeptides, for example, called endorphins, seem to affect moods and the perception of pain. Guillemin's recent research includes neurochemistry of the brain and growth factors.

Guillemin is known as an urbane conversationalist who is interested in the arts and enjoys painting. He and his wife have a collection of contemporary French and American paintings, pre-Columbian art objects, and artifacts from around the world. Guillemin is also a connoisseur of fine food and wine.

GULLSTRAND, ALLVAR (1862-1930)
Swedish ophthalmologist

Major contributions to our understanding of the human eye were made by Swedish ophthalmologist Allvar Gullstrand, particularly in the area of how the eye forms images. His mathematical approach to solving physiological problems had a great significance in the science of ophthalmology, and his discoveries won him the Nobel Prize for medicine or physiology in 1911. He also developed a number of devices, such as the slit lamp and the reflector ophthalmoscope, which became valuable tools in eye examinations and for the treatment of optical disorders. Gullstrand also served for many years as a member, and later as president, of the Nobel Committee responsible for awarding the prize for physics.

Gullstrand was born June 5, 1862, in Landskrona, Sweden, to Pehr Alfred Gullstrand and Sophia Korsell Gullstrand. His father, the city physician, influenced his decision against a career in engineering, in favor of one in medicine. After studying at universities in Uppsala, Stockholm, and Vienna, Gullstrand received his medical degree from Stockholm's Royal Caroline Institute in 1888. He earned his Ph.D. one year later through a dissertation on astigmatism, an eye defect involving faulty curvature of the optic lens. Utilizing his early training and natural aptitude in engineering, he formulated complex theories in optics, which considerably advanced knowledge in this field.

Increases Understanding of How the Eye Functions

During this time, Gullstrand began working as chief physician at the Stockholm Eye Clinic, and by 1892 he was both clinic director and lecturer at the Karolinska Institute. He left the University in 1894 to serve as a professor at the University of Uppsala, where his research in geometrical optics began to flourish. His studies in dioptrics of the eye, or the science of refracted light and its effect on the retina, helped clear up certain misconceptions regarding the way the eye functions. One such misunderstanding concerned the accommodation theory of optics, by which the eye adjusts its focus on objects near and far. In his *Handbook on Physiological Optics,* German biologist and physicist **Hermann von Helmholtz** postulated that the eyes react to the problems of focusing by altering the curvature of their lens. When the eye focused on a nearby object, the lens became more convex (curving-outward), while focusing on something farther away made the lens more concave (curving-inward). In his commentaries on the third edition of the *Handbook* (1908), which he reedited, Gullstrand demonstrated that Helmholtz's theory accounted for only two-thirds of the accommodation. The remaining one-third could be explained by what Gullstrand called "extracapsular accommodation," where the fibers behind the lens made the necessary adjustments. The concept of the human eye as an optical system was among Gullstrand's most important achievements.

Gullstrand was given an honorary degree from Uppsala University in 1907 for his advances in eye research. He invented two devices commonly used even today in eye examina-tions—the slit lamp and the ophthalmoscope (sometimes called the Gullstrand ophthalmoscope), in cooperation with the Zeiss Optical Works in Germany. The slit lamp, consisting of a light used in combination with a microscope, permits doctors to pinpoint the location of a foreign object or tumor in three dimensional space. The ophthalmoscope is a combined light and magnifying lens enabling doctors to look at the retina at the back of the eye, as well as the optic disk. Doctors use it in an inspection for eye defects, as well as **arteriosclerosis** and diabetes. Gullstrand also designed aspheric lenses for those patients whose lenses had been removed as a result of cataracts.

Earns Nobel Prize for Work in Optics

The Nobel Prize in medicine or physiology was awarded to Gullstrand in 1911 for his work on the refraction of light and formation of images in the eye. In his lecture to the Nobel Academy, Gullstrand noted that the laws concerning the formation of optical images had been completely unknown when he began studying the eye lens, and that much of what had been known at that time had since been proven false. A special chair in physical and physiological optics was established for him in 1914 at Uppsala, and he became a member of the Nobel Academy's Physics Committee, and later its president, serving until 1929. Gullstrand received honorary degrees from the University of Dublin and the University of Jena in Germany. He was also awarded the Björken Prize from the Uppsala Faculty of Medicine, the Swedish Medical Association's Centenary Gold Medal, and the Graefe Medal from the German Ophthalmological Society.

Gullstrand married the former Signe Christina Breitholtz in 1885. The couple had a daughter who died while still a young girl. After retiring from Uppsala University in 1927, Gullstrand died of a stroke on July 21, 1930.

GYMNOSPERM

A gymnosperm is a primitive type of plant characterized by the presence of naked seeds. The gymnosperms were an important group of **plants** during the carboniferous period, which lasted from approximately 345-280 million years ago. They coexisted and co-evolved with the dinosaurs and are now well represented in the **fossil** record. The gymnosperms were also responsible for the vast majority of the coal and oil deposits in the world today. While some **species** became extinct, the most numerous existing examples are the conifers. Other groups existed, but are represented by only a small number of species. These include the cycads, ginkgos, and the gnetum group.

The gymnosperms are distinguished from the angiosperms (the flowering plants) by the unprotected ovule produced on the surface of the megasporophylls (a type of modified **leaf** which produces **spore**s). The spore producing structures are usually arranged in cones. The **gametophyte** generation is vastly reduced and the **tree** form is the visible **sporophyte**. One limitation on the spread of this group has been its continued reliance on **water** for reproduction. In this situation, the male spore can only reach the female spore if it is drawn in by water.

H

HABITAT

The word habitat is often used by humans to designate a particular kind of local environment—a hardwood forest, a marsh, or a coral reef. In biology, it is used to designate that distinct environment inhabited by a particular **species**. Habitat is species specific. Writers often append the word 'natural' to habitat, meaning the habitat of an indigenous species, that place to which a species has long been adapted, that place with the conditions suitable for its survival. But habitat can also refer to places with conditions commensurate to the needs of an exotic, migrant species. Finding favorable habitat conditions, especially with an advantageous competitive situation, invaders can become established and even prosper. Habitat requirements of the world's **organism**s vary widely, and can be very different even for similar species. Changes in habitat can quickly alter the conditions for survival of a species.

Human impacts on the habitat of other species range widely. General examples of such impact include numerous alterations of weather and increasing influence on climate, widespread deforestation and **desert**ification, lowering the **water** table below the level where plant roots can reach moisture, and the **sedimentation**, salinization, acidification, and even desiccation of water bodies and streams. Seemingly subtle and small-scale changes, such as in the shade pattern for example, can mean drastically altered conditions for some species.

Degradation, fragmentation, or complete destruction of habitat specifically required for survival by a particular species is one of the major ways in which human activities threaten the existence of other organisms. This is particularly true for species that are very specialized. One example is the giant panda, which feeds only on certain species of bamboo, species that are being decimated by human developments. Destruction of the conditions essential for the survival of the bamboo eliminates the habitat necessary for the survival of the panda. Biologists who study the giant panda do not expect it to survive in the wild, only in zoos. Another example is the radical alteration of the habitat for anadromous fish resulting from the construction of dams on river systems.

HAECKEL, ERNST HEINRICH (1834-1919)

German biologist

Ernst Haeckel was a naturalist who spent much of his career examining the relationship between evolutionary development (phylogeny) and the development of the embryo (ontogeny). Although the pursuit of this relationship often led Haeckel to advance some rather extreme theories, the environment of controversy and debate that surrounded him produced a number of genuine advances in the field of biology.

Haeckel was born in Potsdam, Prussia (now Germany), in 1834. His father, a lawyer and government official, decided early on that young Ernst would be a physician. Haeckel himself displayed little interest in medicine, preferring the biological sciences; he chose, however, to honor his father's wishes, enrolling at the University of Berlin's college of medicine. While a student, he took part in an expedition to the North Sea to study tiny sea creatures and, though he continued his medical coursework, his passion for biology was reignited. He obtained his M.D. in 1857 but practiced medicine for just one year before accepting a position at the University of Jena as lecturer (and eventually professor) of **zoology**.

Haeckel subscribed to the philosophical school of thought known as monism, which viewed the world as being a unified whole, and this philosophy formed the basis for much of his work. For example, one of Haeckel's earliest theories claimed that tiny one-celled animals called Radiolaria were created from inorganic matter through a kind of spontaneous crystallization, thus bridging the gap between organic and inorganic **matter**.

It was in 1859, the year he read **Charles Darwin**'s *The Origin of Species*, that Haeckel turned his full attention toward the burgeoning science of **evolution**. He became a staunch advocate of Darwinism, the first such supporter in Germany and one of the first in Europe, using Darwin's findings as a springboard to launch his own hypotheses. He wrote a number of

Ernst Haeckel Heinrich. *(The Library of Congress. Reproduced by permission.)*

popular articles further describing organic and inorganic nature as being interconnected; these articles, though almost universally disputed, were widely read and soon earned Haeckel the reputation as an "expert" in his field.

Probably Haeckel's most famous theory was that of the relationship between embryonic and evolutionary development, a hypothesis first suggested by Karl von Beer. This theory, summarized in Haeckel's famous phrase "ontogeny recapitulates phylogeny," claimed that the stages an embryo went through during its development were representative of the stages of evolution; that is, that each embryonic stage represented the final stage of an evolutionary predecessor. This theory is based upon the knowledge that the earliest embryonic stages of many animals—such as turtles, chickens, and humans—resemble each other. With this theory as a starting point, Haeckel constructed a number of "ancestral trees" showing man's evolution from inorganic matter to lower animals to present day.

Although many of Haeckel's "unifying" theories were questionable, he was responsible for several important advances in his field. His suggestion that all life stemmed from a hypothetical two-layered creature called a Gastrea led to extensive research into the nature of many marine invertebrates, such as jellyfish, sponges, and medusae. His 1876 publication *Die Perigenesis der Plasidule* contained the first attempt to place heredity on a molecular level within the **cell** nucleus. He was the first to divide life into protozoan (one-celled) and

metazoan (multi-celled) categories, and he was the first scientist to use the word "ecology" to describe the relationship between living organisms and the environment that surrounds them. On the other hand, another of Haeckel's quotable phrases, "politics is applied biology," was later adopted by Nazi propagandists. In fact, his theories were used by the Nazis as a justification for racism, nationalism, and social darwinism.

Haeckel retired from the University of Jena in 1909. He died in Jena on August 8, 1919. Haeckel left behind the Phyletic Museum at Jena, which he founded, and the Ernst Haeckel Haus, containing the bulk of his books and research as well as a number of mementos from his many expeditions.

HALDANE, JOHN BURDON SANDERSON (1892-1964)

English Indian geneticist and physiologist

Haldane was born in Oxford, England, the son of well-known physiologist John Scott Haldane. As a boy, he participated in his father's experiments in human respiratory physiology, and as a student, he carried out breeding experiments on mice and guinea pigs. Haldane was educated at Eton and later attended New College, Oxford. Although he majored in mathematics, he excelled in classical studies and philosophy.

Haldane became an officer in the Black Watch regiment in France and the Middle East at the outset of World War I. After being wounded, he returned to study physiology at New College in 1919. He held a fellowship there until 1923, when he moved to Cambridge to work under English biochemist Frederick Gowland Hopkins for two years. After returning to London, Haldane accepted a position at University College as genetics chair and a part-time position at the John Innes Horticultural Institution at Merton. Haldane remained a Weldon professor of biometry at University College for twenty years. During much of this time, he was a member of the British Communist party and spoke on political issues. An outspoken Marxist during the 1930s, Haldane served temporarily as chairman of the editorial board of the *London Daily Worker*.

After serving in World War II, Haldane left the Communist party, disappointed with the fame accorded to Soviet biologist Trofim Denisovich Lysenko (1898-1976). In 1957, Haldane immigrated to India in objection to the Anglo-French invasion of Suez. There, he was appointed director of the Genetics and Biometry Laboratory in Orissa, which possessed exceptional research facilities. Haldane became a naturalized Indian citizen in 1961 and held positions at the Indian Statistical Institute in Calcutta. He remained in India until his death from cancer in 1964.

Haldane's eventful personal life is evident in his numerous advances in the field of **genetics** and **evolutionary theory**, which were augmented by his contributions to population genetics and population dynamics. Upon reading **Gregor Mendel**'s work in 1901, Haldane began a mathematical analysis of genetics and **evolution**. He later informally indulged his interest in genetics by studying the **laws of inheritance** using his sister's three hundred guinea pigs.

Through mathematical analysis of mutation rates, Haldane became convinced that natural selection and not mutation was the driving force behind evolution. As a theoretical geneticist he formulated a mathematical theory of natural selection during the 1920s. This theory, combined with similar theories of two other geneticists, set the framework for the neo-Darwinian interpretation of evolution. This approach correlated classical genetics and evolutionary phenomena by incorporating mathematical analyses of mutation rates, intensities in selection, and rates of evolutionary change.

In 1932, Haldane estimated for the first time the mutation rate of a human gene and predicted the effect of recurring harmful mutations on a population. In 1936, he examined the link between **hemophilia** and color blindness. Haldane is also responsible for introducing the concept of genetic load, which is the percentage of deleterious or lethal genes present in a species population. After his introduction to **enzyme** reactions in 1924, Haldane produced the first proof that these reactions obey the laws of **thermodynamics**.

The interest in physiology Haldane had harbored since participating in his father's research led him to study the physiology of breathing, particularly with respect to deep-sea diving and safety in mines. During World War II, Haldane worked with his father again, this time to improve gas masks. Haldane wrote several books that are considered scientific classics, including *Enzymes* (1930) and *The Causes of Evolution* (1932).

See also Respiration

HALES, STEPHEN (1677-1761)
English clergyman and physiologist

Professionally, Stephen Hales was a clergyman, serving as "perpetual curate" of Teddington, Middlesex, England, from 1709 until his death in 1761. Avocationally, Hales was a leading scientist of his time—the founder of plant physiology, a trailblazer in the study of **blood circulation** and **blood pressure** measurement, and a pioneer in public health.

Hales was born on September 17, 1677, in Kent, to an influential county family. Little is known about him until he began studies at Cambridge in 1696. At the university, Hales enthusiastically immersed himself in scientific studies and biological experiments while also pursuing his clerical degree. He received his bachelor of arts degree in 1702 and his master's degree in 1703. He was also ordained a deacon in 1703, and in 1709, he went to the "perpetual" post in Teddington.

During his early years at Teddington, Hales continued experiments that he had begun at Cambridge, achieving the first blood pressure measurements with a glass-tube manometer. He also investigated reflex actions in a decapitated frog. Hales then gave up his animal experiments, "being discouraged by the disagreeableness of anatomical dissections," and turned to the investigation of the movement of sap in plants, accidentally discovering that the force exerted by flowing sap would expand a bladder tied over a stem. From this, Hales realized he could use a glass tube, as in his animal work, to measure the force of the sap's flow.

Stephen Hales. *(The Library of Congress. Reproduced by permission.)*

Hales's study of plant transpiration led to his investigations of air, both "fixed" in varying substances and as given off or absorbed under different conditions. This, in turn, led to his invention of various measuring devices. Hales's work was tremendously important to later chemists. Another of his contributions was the investigation of ways to chemically dissolve kidney and bladder stones, in the course of which the cleric invented a surgical forceps.

An important aspect of Hales's career was the application of his findings to practical uses. He used his knowledge of air and breathing to invent ventilators to remove noxious air from hospitals, merchant and slave ships, and prisons. He adapted a gauge from his plant experiments to ocean-depth measurement. He worked on ways to distill fresh water from ocean water. He involved himself in social issues, working for passage of the 1736 Gin Act, and was active in the founding of the colony of Georgia, while also attending to his parish duties. Noted for cheerfulness and serenity, Hales died on January 4, 1761, after a brief illness. He was buried at Teddington, with a monument in his memory at Westminster Abbey.

HALLER, (VICTOR) ALBRECHT VON (1708-1777)
Swiss biologist

Haller was one of the great heroic figures of early biology. Born in Bern, Switzerland, he was not a healthy child, but he displayed prodigious intellectual talents at an early age. He wrote scholarly articles at the age of eight, and by the age of ten, he had completed a Greek dictionary.

Haller enrolled as a medical student at the University of Leyden and earned his degree at the age of 19. At Leyden, he studied under the famous Hermann Boerhaave (1668-1738). Haller began his own medical practice in 1729 at the age of 21 and continued in private practice until 1736. He was then appointed Professor of Anatomy, Botany, and Medicine at the newly created University of Göttingen. He served at Göttingen until 1753, when he returned to Bern. He spent the remaining twenty years of his life in research, writing, and government service. Haller died in Bern on December 17, 1777.

Haller displayed interests and talents in a wide range of fields, but he is probably best known for his work on nerves and muscles. When he began his research, little was understood about the structure and function of nerves or about their interaction with muscles. A popular theory of the time held that nerves are hollow tubes through which a spirit or fluid flows. Haller rejected this idea, however, since no one had ever been able to locate or identify such a spirit or fluid.

Instead, Haller concentrated on two specific and identifiable nerve-related phenomena: irritability and sensibility. By irritability, he meant the contraction of a muscle that occurs when a stimulus is applied to the muscle. Haller found that irritability increases when the stimulus is applied to the nerve connected to a muscle. He concluded that the stimulus was transmitted from the nerve to the muscle, thus clarifying for the first time the relationship of nerve to muscle.

In his study of sensibility, Haller found that ordinary tissue does not respond to stimuli, but that nerves do. He showed that stimuli applied to nerve endings travel through the body, into the spinal column, and eventually into the **brain**. By removing certain parts of the brain, he was then able to show how each part affects specific muscular actions.

Because of his pioneering research on the nervous system, Haller is often credited as the founder of the science of neurology. He published many scientific works, including the eight-volume *Elementa Physiologiae Corporis Humani* (*Elements of the Physiology of the Human Body*, 1757-66). He also published three philosophic romances and a well-known poem, "Die Alpen" ("The Alps," 1729).

HALLUCINOGENS

Hallucinogens are natural and synthetic substances that when ingested, significantly alter one's state of consciousness. Hallucinogenic compounds often cause people to envision or perceive random colors, patterns, events, and objects that do not exist. Many different types of substances are classified as hallucinogens, solely because of their capacity to produce such hallucinations.

Some users of hallucinogens have reported feeling mystical and insightful, while others are fearful, paranoid, and hysterical. Unlike such drugs as barbiturates and amphetamines, hallucinogens are not physically addictive; however, people can become psychologically dependent upon them. The real danger of hallucinogens is not their toxicity, but their unpredictability. People have had extremely varied reactions to these substances, especially to LSD (lysergic acid diethylamide), that it is virtually impossible to predict the effect a hallucinogen will have on any given individual.

When produced naturally, hallucinogens are formed in dozens of psychoactive **plants**, including the peyote cactus, various **species** of mushrooms, and the bark and seeds of several **tree**s and plants. These natural forms of "psychedelic" (mind-expanding) substances have been available for centuries. In Mexico, exotic hallucinogenic mushrooms called *Psylocybe mexicana*, which contain the **fungi** psilocybin and psilocin, have been used in religious rituals since the time of the Aztec civilization. In Europe, the fungus *Amanita muscaria* was thought to have been used by the Vikings. Amanita muscaria and its close relative, Amanita pantherina, are also found in the United States; their major psychoactive ingredients are ibotenic acid and muscimol.

Members of the Native American Church, an organization comprised of members of Indian tribes throughout North America, advocate the use of mescaline, a form of psychedelic drug found in the peyote cactus. Currently peyote is the only psychedelic agent that has been authorized by the federal government for limited use during Native American religious ceremonies.

There are a few less common natural hallucinogens. These include ololiuqui (morning glory) seeds, which are ingested by Central and South American Indians both as intoxicants and hallucinogens. Oloiuqui is used ritually as a way to communicate with the supernatural and was first written about by the Spanish explorer Hernández, who wrote "when the priests wanted to commune with their Gods, they ate ololiuqui seeds and a thousand visions and satanic hallucinations appeared to them." Harmine, another psychedelic agent that has been used for centuries, is obtained from the seeds of *Peganum harmala*, a plant found in the Middle East. The feeling of exhilaration brought about by this drug is sometimes followed by nausea, sedation, and sleep, but the psychic excitement involves visual distortions similar to those induced by LSD. Marijuana and hashish, two substances derived from the hemp plant, are also considered natural hallucinogens, although their potency is extremely low when compared to others. Marijuana, a green herb from the **flower** of the hemp plant, is considered a mild hallucinogen; hashish is marijuana in a more potent, concentrated form.

Yet even the most potent of these naturally occurring hallucinogens is hardly as powerful and unpredictable as a synthetic hallucinogen, such as LSD. LSD became well known in the 1960s when many people sought spiritual enlightenment through drug-induced states of consciousness. A form of LSD was first produced in 1938, when **Albert Hoffman**, a Swiss research chemist at Sandoz Laboratories, synthesized many important ergot **alkaloid**s, including hydergine, LSD-25, and psilocybin. Hoffman accidentally experienced the first "LSD high" when a drop of the material entered his **blood**stream through the skin of his fingertip. The psychoactive potency of the compound was regarded as extraordinary, and Hoffman could hardly recount his episode after it was over. He was the first to record LSD's unique ability to induce *synesthesia*, an overflow of one sensory ability into another. For example, a person experiencing synesthesia may hear colors and see sounds. The physical effects of hallucinogens are considered inconsequential when compared to their effects on the mind. **Death** from an overdose of hallucinogens is highly unlikely; however, deaths have resulted from accidents or suicides involving people under the influence of LSD. Other synthetic hallucinogens include phencyclidine (PCP) and ketamine (ketamine hydrochloride), which were both once used as anesthetics. Ecstasy, or MDMA (methylenedioxymethamphetamine), is a stimulant that also has mind-altering effects.

Ingesting some psychedelic mushrooms can induce profuse sweating, excessive salivation, decreased **heart** rate, increased **blood pressure**, and reduction in pupil size. In the case of illicit "designer drugs" like MDMA, illegal production has resulted in deaths because of the inadvertent creation of other substances. Although it has not been conclusively documented, hallucinogens are believed to affect serotonin, a neurotransmitter in the **brain**. Recently several hallucinogenic compounds have been found to resemble serotonin structurally, and the hypothesis has been advanced that at least some drug-induced hallucinations are due to alterations in the functioning of serotonin **neuron**s. It was demonstrated that LSD interfered with the transmitter action of serotonin.

HAPLOID

A **cell** is haploid (from Greek *haploos*, meaning single, and *ploion*, meaning vessel) if it contains only one set of **chromosome**s, as opposed to most body cells that contain two sets of chromosomes. A haploid cell is symbolized as "n."

In all cells of **organism**s, with the exception of the sex cells and the **gametophyte** generation in **plants**, chromosomes are arranged in pairs, called homologous pairs. The chromosomes in homologous pairs are the same size and have genes for the same traits. Each chromosome in the homologous pair comes from a different parent. A cell that has both chromosomes of each homologous pair is a **diploid** cell. A diploid cell is symbolized as "2n." In the human body, diploid cells have 46 chromosomes arranged in 23 homologous pairs.

A cell is haploid if it has only one chromosome from each homologous pair. The only cells in an organism that are haploid are the **gamete**s, the **sperm** and **eggs**. In addition, in plants, the gametophyte generation is also haploid. During the process of **meiosis**, a series of divisions of the **nucleus**, the homologous pairs of chromosomes in a diploid cell split apart, forming haploid sperm and egg cells. In the human body, haploid cells have 23 chromosomes, no longer arranged in pairs.

During **fertilization**, a haploid sperm cell joins together with a haploid egg cell. The homologous pairs of chromosomes are restored, and the resulting cell will no longer be haploid. It will be a diploid cell, with the normal number of chromosomes for a body cell. For example, in the human body,

Arthur Harden. *(The Library of Congress. Reproduced by permission.)*

a sperm with 23 chromosomes will join together with an egg containing 23 chromosomes. The resulting **zygote** will be diploid, with 46 chromosomes arranged in 23 homologous pairs.

HARDEN, ARTHUR (1865-1940)
English biochemist

Born in Manchester, England, Harden did his undergraduate work at a local school, Owens College, but went to the University of Erlangen in Germany for graduate studies. After obtaining his Ph.D. there in 1888, he returned home and spent ten years teaching at his alma mater. In 1898, he began studying fermentation simply because he thought it might help him differentiate between several varieties of **bacteria**. Although his original theory did not work out, Harden gradually became more and more interested in **fermentation** itself, the process by which a certain few agents seemed able to cause a number of organic compounds first to decompose, then to transform themselves into other compounds altogether.

The most commonly seen example of the process—and the one that especially intrigued chemists during the mid-to-late 1800s—was alcoholic fermentation, the conversion of glucose and other sugars into alcohol apparently through the action of yeast preparations. For a long time, scientists had assumed that living cells in yeast contained a "vital force" that enabled yeast to transform other compounds. But in 1897, a German biochemist, **Eduard Buchner**, proved that a **cell**-free juice extracted from yeast could also act as a fermenting agent.

Buchner named the active principle in his yeast juice zymase, which he assumed was an **enzyme**. It was Buchner's zymase-containing yeast juice that Harden used in his research studies for the next several years.

In 1904, working at the Jenner Institute of Preventive Medicine along with a student named William J. Young, Harden made an interesting discovery. He had already learned that boiling his yeast juice appeared to destroy its enzymatic activity, rendering it no longer able to ferment sugar. But during one particular experiment, he found that, if he added a little of the boiled and apparently "dead "yeast juice to an active batch, the addition seemed somehow to increase the yeast's fermentation powers. The boiled juice, then, must still contain some sort of active principle. But what could it be?

To find out, Harden decided to try *dialysis*, a filtering process often used in chemistry. Another batch of active yeast juice was poured into a semipermeable bag and the bag itself placed in a container of pure water. Gradually, some of the juice's smaller molecules began filtering out through the bag's membrane-like sides and into the water. The rest, containing the larger molecules, remained behind.

After repeatedly testing both sections of juice—separately and once again combined—Harden reported that zymase, the yeast's enzyme, apparently consisted of two parts: a large-molecular part that could not survive boiling and was almost certainly a protein ; and a small-molecular part that could survive boiling and was probably not a **protein**. Neither part could ferment sugar by itself. Although Harden called the non-protein factor a *coferment*, it was clearly the first example of a coenzyme, the non-protein molecule—usually composed of vitamin s—that many enzymes needed as an attachment before they could do their work.

Harden studied the fermentation process for several more decades, discovering, among other things, that inorganic phosphate groups played an important but temporary role in fermentation. It was this discovery that helped arouse interest in an important new branch of biochemistry, intermediary metabolism. This branch of biochemistry concentrates on the study of certain intermediate compounds that come to life, sometimes very briefly, during the course of many biochemical reactions. Harden's discovery also caused later researchers, such as Carl and Gerty Cori and Fritz Lipmann, to realize the vital importance of the phosphates to biochemistry.

Harden became a professor of biochemistry at the University of London in 1912. With Hans Euler-Chelpin, he shared the 1929 Nobel Prize in chemistry for his work in fermentation. He was the Joint Editor of *The Biochemical Journal* with William Bayliss from 1913 to 1938 and received the Davy Medal in 1935. In 1936, four years before his death, he was knighted.

HARTLINE, HALDAN KEFFER (1903-1983)

American physiologist and biophysicist

Haldan Keffer Hartline was born on December 22, 1903, in Bloomsburg, Pennsylvania, to Daniel Schollenberger Hartline

and Harriet Franklin Hartline. He attended college at Lafayette College in Easton, Pennsylvania, graduating with a B.S. in 1923. He went on to study retinal electrophysiology as a graduate student at Johns Hopkins University, obtaining his M.D. in 1927. Hartline spent the next two years at Johns Hopkins University as a National Research Council fellow in medical sciences.

Between 1929 and 1931, Hartline was a Johnson Traveling Research Scholar from the University of Pennsylvania to the universities of Leipzig and Munich. He travelled extensively in Germany during those years before returning to the United States, where he joined the Eldridge Reeves Johnson Research Foundation for Medical Physics as an assistant professor of biophysics at the University of Pennsylvania. Hartline married Mary Elizabeth Kraus on April 11, 1936. They had three sons: Daniel Keffer, Peter Haldan, and Frederick Flanders.

From the early days of his career, Hartline was fascinated by the **metabolism** of nerve **cells**, and he eventually focused his attention on the workings of individual cells in the retina of the eye. During the late 1920s and early 1930s, Hartline used recently-developed methods of fiber isolation to record the activity of single nerve fibers in the retina. He began by experimenting with *Limulus polyphemus,* the horseshoe crab. He chose this primitive creature because it possessed a feature that was ideal for his research: a compound eye with a long optic nerve and large individual photoreceptors. It seemed to Hartline that working with the horseshoe crab might allow him to record the electrical behavior of single nerve fibers. He succeeded in 1932, while working at the Eldridge Reeves Johnson Foundation. Hartline and Columbia University psychophysiologist Clarence H. Graham managed to isolate single nerve fibers from the optic nerve, placed electrodes on those single fibers, stimulated them with light, and recorded the nerve impulses that occurred. This was the first record of the activity of a single optic nerve fiber, and it proved to Hartline and Graham that their theories had been correct: information is relayed through individual optic nerve fibers by a series of uniform nerve impulses.

Hartline moved into another field of vision in 1938, when he began to study the vertebrate eye, using microdissection techniques to record the activity of individual fibers in the optic nerve of frogs. While recording the nerve impulses from the single nerve fibers lying behind the rods and cones of the eye, he found that the fibers making up the nerve did not all behave in the same way. Some were stimulated by steady light, others were stimulated by the light when it first hit the retina, and still others were stimulated only as the light was shut off. Hartline demonstrated that visual information begins to be differentiated in the retina and in the receptors themselves, as soon as the stimulation occurs, before the information can be conducted more deeply into the **central nervous system**. This research afforded new insights into the working of the retina. It also provided a new understanding of how the mechanisms of vision were integrated with, and how they affected, the nervous system as a whole. For this discovery, Hartline was awarded the Howard Crosby Warren Medal of the Society of Experimental Psychologists in 1948.

Hartline continued his teaching and research at the University of Pennsylvania, becoming professor of biophysics and chair of the department at Johns Hopkins in 1949. In 1953, Hartline joined the faculty of Rockefeller University in New York as professor of neurophysiology. There, Hartline began investigating the phenomenon of inhibition in the retina of the compound eye, using the horseshoe crab as a subject once again. He and his colleague, Floyd Ratliff, demonstrated the electrical response of nerve fibers and cells to light hitting the retina, and the mechanism by which this response allows the eye to differentiate shapes. He found that the receptor cells in the eye are interconnected in such a way that when one is stimulated, others nearby are depressed, thus sharpening the contrast in light patterns. In the 1960s, Hartline extended these studies to the dynamics of the receptors and their interactions, with a view to understanding visual phenomena such as motion detection. Hartline's findings eventually led to the development of a set of mathematical equations expressing the interrelationship of the receptor units of the compound eye; this information has been key to understanding brightness and contrast in the retinal image.

For his work on electrical activity on the cellular level within the eye, Hartline shared the 1967 Nobel Prize for physiology or medicine with the American biologist **George Wald** and the Swedish neurophysiologist **Ragnar Granit**. This was not the only award received by Hartline during this period; he also received the A. A. Michelson Award of Case Institute, 1964, and the Lighthouse Award in 1969. In addition to the Nobel Prize and the other awards and honors received during his lifetime, Hartline was also presented with a number of honorary degrees. He was awarded doctorates from Lafayette College in 1959, the University of Pennsylvania in 1971, Rockefeller University in 1976, the University of Maryland in 1978, and Syracuse University in 1979; an LL.D. from Johns Hopkins University in 1971; and an M.D. from the University of Freiburg in 1971.

Hartline was a member of many important scientific organizations, many of them elective. He was elected to the National Academy of Sciences in 1948, and to the American Academy of Arts and Sciences in 1957. Hartline also held memberships in the American Philosophical Society and the Biophysics Society, and in 1966 was elected a foreign member of the Royal Society, London. The Optical Society of America made him an honorary member, as did the Physiology Society (U.K.).

Hartline died on on March 17, 1983, in Fallston, Maryland.

HARVEY, WILLIAM (1578-1657)

English physician and biologist

William Harvey, the discoverer of **blood circulation**, was born in Folkestone, England on April 1, 1578. He went up to Cambridge in 1593, where he received his B.A. in 1597, after which he journeyed to Padua in Italy to study medicine. Padua was reputed to be one of the finest universities at the time and

William Harvey. *(The Library of Congress. Reproduced by permission.)*

had many famous scholars, including Galileo, while Harvey was a student. Harvey received his medical degree in 1602. Back in England, he became a Fellow of the Royal College of Physicians in 1607, and was appointed to the Saint Bartholomew Hospital, London where he served from 1609 until 1643. He attended the death of James I and he was the personal physician to Charles I. He became President of the Royal Society of Physicians in 1654. Obviously, Harvey was an extraordinarily distinguished physician. However, his publication on the circulation of blood in 1628, a notable example of experimental biology, and correct in its interpretations and conclusions, earned for him scorn and he was denounced by the major scholars of the day. It resulted in significant injury to his practice.

It is difficult in the modern world to understand the tenacious hold ancient scholars had on the world in the time of Harvey. Galen's (born in 129 in Pergamon which is now called Bergama and is located in Turkey) view of the function of the **heart** and vessels remained paramount and unchanged for 1,400 years. Galen held that the liver was the most important organ in the **blood**-vascular system. Galen believed that the liver was the site of blood formation and venous blood from the liver moved throughout the body. This blood ebbed and flowed as the tides, i.e., back and forth. He thought that some venous blood from the liver moved to the right auricle and to the right ventricle. Here it gave off noxious vapors which dissipated via the lungs. The remainder of the right ventricular blood moved from there to the left ventricle via tiny pores. Blood in the left ventricle was mixed with blood of the arteries and air from the lungs and it then went to the periphery of the body where, like venous blood, it ebbed and flowed in the system. Note, these beliefs of Galen were not harmonious with the notion that blood circulated.

Harvey, after carefully examination of animal and human hearts, wrote his famous *Exercitatio anatomica de Motu Cordis et Sanguinis in Animalibus* which challenged and corrected the assertions of his ancient predecessor. Harvey looked for pores in the septum between the ventricles. He looked, ''but, damn it, no such pores exist, nor can they be demonstrated'' (an English translation of Harvey who wrote in Latin). He monitored the beating heart, made rough calculations of the volume of blood leaving the heart per unit time and compared that with the total blood volume. He concluded that blood could not be produced fast enough to account for this volume, thus it must circulate. He examined valves in **veins**. He concluded that venous blood could move in only one direction and that was toward the heart. His conclusions were made from direct observation, often of living animals, and the observations led him to postulate that blood was in motion all of the time and that it circulated. As far as contemporary knowledge of circulation is concerned, the only information missed by Harvey pertained to the capillaries and they were too small to see without a microscope. Microscopes had not yet been invented.

Harvey was not correct in all of his assertions however. For instance, he believed that the heart was the source of body heat and that blood, after providing heat to the distal parts of the body, returned to the heart for warming. However, any contemporary student of heart and vessel function should read Harvey.

Harvey is less well known for his embryological studies. Regardless, his studies of chick **embryology** were the best of his time and for many years to come. He clearly was epigenetic in view (i.e., he did not believe in preformation of a minute being in either the egg or sperm) but described the gradual development and differentiation of an individual from an essentially homogenous **egg**. All embryologists to this day know the aphorism *ex ovo omnia* (all animals from eggs).

HATCH-SLACK PHOTOSYNTHETIC PATHWAY

In the 1960s, two Australian scientists, M.D. Hatch and C.R. Slack described a new pathway for carbon fixation in **photosynthesis**. The Hatch-Slack pathway describes a biochemical system in which carbon is first incorporated into the four-carbon **molecule**, oxaloacetic acid. Earlier, in the 1950s, an American scientist, **Melvin Calvin**, uncovered a chemical pathway for photosynthetic carbon fixation that came to be called the Calvin Cycle. In his scheme, carbon was first fixed into the three carbon compound, 3-phosphoglyceric acid (PGA). He was awarded the Nobel Prize in for this discovery in 1961. For

a time the Calvin Cycle was thought to be the only carbon fixation pathway. Hatch and Slack's surprising discovery proved otherwise. Their efforts were aided by the work of G.O. Burr and his colleagues at the Sugar Cane Research Institute in Hawaii who reported that PGA was not the first photosynthetic product in sugarcane **plants**, a **species** that is especially efficient in photosynthesis. The experimental procedure for discovery of early products involved exposing photosynthesizing plant material for a very brief time to **carbon dioxide** enriched with radioactive ^{14}C. The plant **cells** were then extracted and analyzed by **chromatography** to determine what chemicals had incorporated radioactivity carbon from the carbon dioxide. Burr's laboratory found that PGA was not the first compound to become radioactive, but it remained for Hatch and Slack to sort out details of the newly discovered pathway.

The two fixation mechanisms are sometimes called the C_3 and C_4 pathways based on the number of carbons in the initial product. Each pathway has its own unique catalyzing **enzyme** and acceptor molecule. The molecule that accepts the carbon in the C_4 pathway is a three-carbon compound called phosphoenolpyruvate (PEP) and the catalyzing enzyme is PEP carboxylase. The acceptor molecule for the C_3 pathway is a five-carbon sugar called Ribulose 1,5 bisphosphate (RuBP), which splits into two molecules of PGA after adding a carbon, and the enzyme is RuBP carboxylase. The two carboxylating enzymes use different forms of carbon dioxide as substrate. RuBP carboxylase incorporates carbon dioxide, whereas PEP carboxylase uses the hydrated bicarbonate ion form. The enzymes also differ in their cellular location. RuBP carboxylase is located in the **chloroplast**, and PEP carboxylase in the **cytoplasm**. Sugarcane, and other plants that first fix carbon by the Hatch-Slack pathway also use the Calvin cycle in a curious system that involves the uptake of carbon dioxide twice. The complex double fixation process is commonly found in tropical plants and other plants that are highly efficient in photosynthesis. This efficiency may result from PEP carboxylase's greater affinity for carbon dioxide and its ability to assist the less efficient enzyme.

HAWORTH, WALTER (1883-1950)
English chemist

Walter Haworth's earliest research was influenced by his contact with William Perkin at the University of Manchester and involved a study of terpenes, a class of **hydrocarbon**s often found in **plants**. The work for which he is best known, however, involves his studies of various **carbohydrate**s, including a number of important **monosaccharide**s, disaccharides, and **polysaccharide**s. Among his finest achievements was the determination of the molecular structure for glucose, perhaps the most important of all monosaccharides. The method he used for designating the formula of glucose and those of other carbohydrates is well known today to any student of organic chemistry as the Haworth formula. The 1937 Nobel Prize in chemistry was awarded to Haworth in recognition not only of his work on carbohydrates but also for his elucidation of the structure of **vitamin** C and the first artificial synthesis of this important compound.

Walter Norman Haworth was born in Chorley, Lancashire, England, on March 19, 1883. He was the fourth child and second son of Thomas and Hannah Haworth. Thomas Haworth, whose family enjoyed a distinguished reputation in business, was the manager of a linoleum factory and took it for granted that his son would follow him into that business. And, indeed, after completing school at the age of fourteen, young Haworth did take a job at the linoleum factory. He soon decided, however, that he had no interest in making his career in that kind of work. Instead, he had become fascinated with the chemical applications he saw all around him and had decided that he wanted a career in that field.

Becomes Interested in Terpenes

That road was made all the more difficult, however, when Haworth's parents withheld their approval and support for any additional education for their son. It was only through great personal effort and the aid of a private tutor that he was finally able in 1903 to pass the entrance examination at Manchester University. There he studied chemistry under the department chairperson, William Perkin, Jr., and became particularly interested in Perkin's own specialty, the chemistry of terpenes. Haworth received his degree in 1906, earning first-class honors in chemistry, and then stayed on at Manchester to work as Perkin's assistant.

In 1909 Haworth left Manchester to spend a year at the University of Göttingen studying with future (1910) Nobel Prize winner Otto Wallach, an expert on terpenes. In only one year, Haworth had earned his Ph.D. and was on his way back to Manchester. One year later, he had qualified for his second doctorate, a D.Sc. in organic chemistry. Over the next fifteen years, Haworth held posts at three institutions. He was senior demonstrator at the Imperial College of Science and Technology in London from 1911 to 1912, lecturer at United College in the University of St. Andrews from 1912 to 1920, and professor of organic chemistry at Armstrong (later King's) College in the University of Durham from 1920 to 1925. In the latter year he was appointed Mason Professor of Chemistry at the University of Birmingham, a post he held until his retirement in 1948.

Begins Work in Carbohydrate Chemistry

The most important period for Haworth in his pre-Birmingham days was his tenure at St. Andrews. It was there that he was introduced to the new field of carbohydrate chemistry by Thomas Purdie and James Colquhoun Irvine, two of England's foremost authorities in the field. In the early 1910s, scientists knew a fair amount about the chemical composition of the carbohydrates, but relatively little about their molecular structure. It was to the question of molecular structure that Haworth soon turned his attention at St. Andrews, and before long, he had abandoned his work on terpenes.

World War I interrupted Haworth's new line of research, however. For the duration of the war, the chemical laboratories at St. Andrews (like other such facilities) were completely given over to the manufacture of chemicals with military importance. At the war's conclusion, however, Haw-

orth returned to his work on carbohydrates. The first stages of that research were devoted to the monosaccharides, the simplest of the carbohydrates. Haworth developed a method by which he could determine the sequence of linkages within a molecule and was able to elucidate the detailed formulas for many compounds. Among the most important of these was glucose, which Haworth showed in 1926 to exist as a six-membered ring consisting of five carbon atoms and one oxygen atom. The convention he used to represent the glucose structure, showing the three-dimensional orientation of its components, has since become known as a Haworth formula or Haworth projection.

In his later work at Birmingham, Haworth took on more and more complex structures, eventually finding formulas for lactose and sucrose, two important disaccharides. He also took on yet another challenge—the determination of the structure for hexuronic acid. Hexuronic acid had been discovered in 1932 by Albert Szent-Györgyi in extracts taken from the adrenal gland, in cabbages, and in oranges. Szent-Györgyi suspected that his hexuronic acid might be identical to vitamin C, the antiscurvy agent, that had also been discovered recently.

In his own research, Haworth was able to elucidate the structure of this compound and then to synthesize it in his laboratory. That accomplishment was historic since it was the first time that a vitamin had been produced synthetically. Because of the compound's antiscurvy properties, Haworth suggested that it be renamed ascorbic acid ("not-scurvy" acid), a name by which it is now universally known. For his work both with carbohydrates and with vitamin C, Haworth was awarded a share of the 1937 Nobel Prize in chemistry with Paul Karrer.

Haworth's health failed him in 1938, but three years later he had recovered sufficiently to return to his research and other commitments. Included among those other commitments were a number of political and professional responsibilities. He served as chairperson of the British Chemical Panel for Atomic Energy during World War II. He also became dean of the faculty at Birmingham from 1943 to 1946 and served as president of the British Chemical Society from 1944 to 1946. At the same time, he continued an active program of research, concentrating on the most complex of all carbohydrates, the polysaccharides.

Haworth was married to Violet Chilton Dobbie in 1922. The couple had two sons. Haworth died at his home in Birmingham of a heart attack on March 19, 1950, his birthday. In addition to the Nobel Prize, he had been awarded the Longstaff Medal of the British Chemical Society in 1933, the Davy Medal in 1934, and the Royal Medal of the Royal Society in 1942. He was made a fellow of the Royal Society in 1928 and was knighted in 1948.

HAZEN, ELIZABETH (1885-1975)
American microbiologist

Hazen was born in Rich, Mississippi, on August 24, 1885, and raised by relatives in Lula, Mississippi, after the death of her parents. Hazen attended the public schools of Coahoma County, Mississippi, and earned a B.S. from the State College for Women, now Mississippi University for Women. She began teaching high school science and continued her own education during summers at the University of Tennessee and University of Virginia.

In 1916, Hazen began studying bacteriology at Columbia University, where she earned an M.A. the following year. World War I provided some opportunities for women scientists, and Hazen served in the Army diagnostic laboratories and subsequently in the facilities of a West Virginia hospital. Following the war, she returned to Columbia University to pursue a doctorate in microbiology, which she earned in 1927 at age 42.

After a four-year stint at Columbia University as an instructor, Hazen joined the Division of Laboratories and Research of the New York State Department of Health. She was assigned to special problems of bacterial diagnosis and spent the next few years researching bacterial diseases. She investigated an outbreak of **anthrax**, tracing it to a brush factory in Westchester County. Hazen discovered unknown sources of tularemia in New York and was the first to identify imported canned seafood that had spoiled as the cause of type E toxin deaths.

Her discoveries led Hazen to try to better understand mycotic (fungal) diseases. In 1944, she was given the responsibility of investigating such diseases, and she acquired **cultures** of fungi from local laboratories and specialized collections. Although Hazen was learning more about mycotic diseases, fungal infections continued to spread in epidemic proportions among school children in New York City. In addition to pneumonia, many other fungal diseases caused widespread ailments, such as moniliasis (thrush), a sore mouth condition that makes eating excruciatingly painful. Despite personal health problems and stressful working conditions, Hazen persevered.

In the mid-1940s, she teamed up with Rachel Brown (1898-1980), a chemist at the Albany laboratory who prepared extracts from the cultures sent by Hazen. In the fall of 1950, Hazen and Brown announced at a National Academy of Sciences meeting that they had successfully produced two antifungal agents from an antibiotic. This led to their development of Nystatin, the first fungicide safe for treating humans. Nystatin was immediately used nationwide, earning $135,000 in its first year.

Nystatin, which is still sold as a medication today under various trade names, turned out to be an extremely versatile substance. In addition to curing serious fungal infections of the skin and **digestive system**, it can also combat Dutch Elm disease in trees and even restore artwork damaged by water and mold. Remarkably, Hazen and Brown chose not to accept any royalties from the patent rights for Nystatin. Instead, they established a foundation to support advances in science. The donated royalties totaled more than $13 million by the time the patent expired. Hazen died on June 24, 1975.

HEARING

Hearing is one of a human being's five senses. In order to hear sound, the outer ear—the visible part of the ear called the pinna and the external ear canal—functions like a funnel, catching sound waves and directing them to the tympanum, or tympanic **membrane**, called the ear drum. The sound waves cause the tympanum to vibrate and these vibrations are transferred through three tiny bones in the middle ear which lies between the eardrum and the inner ear. The middle ear is lined with a membrane containing **blood** vessels and **secretion** glands, and is connected to the back of the nose by a tiny tube called the Eustachian tube. This tube balances air pressure within the ear with the air pressure in the **atmosphere**, which is why our ears "pop" as we drive or fly at different elevations. It also helps us maintain our sense of balance. The three bones in the middle ear—the malleus (hammer), the incus (anvil), and stapes (stirrups)—carry the tympanic vibrations to the inner ear which is encased by rock- hard bone and which contains a coiled **organ** called the cochlea (which means snail). The cochlea is a complicated, fluid-filled system containing thousands of microscopic receptors called hair **cells**. The bending of these hairs by the wave-like action of the fluid transforms the fluid waves into electrochemical impulses which are then transmitted to the **brain** via the hearing or auditory nerve which runs through a small, bony, internal auditory canal. In the brain, the auditory nerve divides into an extremely complex system called the central auditory **nervous system**. **Nerve impulses** are decoded in the hearing center of the brain into recognizable sounds such as music or speech. Normal hearing in both ears (binaural hearing) allows us—among other important functions—to pinpoint and localize a sound accurately, discern speech from a noisy background, and enjoy a melody. Exposure to extremely loud sound can irreversibly damage the fragile hair cell receptors causing permanent hearing impairment or loss, as can middle ear infections, **heredity**, some illnesses and **birth** defects, and the natural **aging** process.

HEART

The heart is a muscular pump which serves to circulate the blood throughout the body. The structure of the heart in vertebrates (organisms with backbones) is remarkable similar. All vertebrates have at least one atrium (a chamber which receives **blood** returning from circulation throughout the body) and at least one ventricle (a chamber which pumps blood into the arteries which circulate it throughout the body). As organisms become more complex, so do their hearts. Therefore, fish have only one atrium and one ventricle; amphibians have two atria and one ventricle; reptiles have two atria and two ventricles, although with an incomplete wall separating the two ventricles); birds and **mammals** have two atria and two ventricles, with complete separation between all four chambers.

In the human heart, the major job of the right side of the heart is to receive oxygen-poor (deoxygenated) blood, and then pump it back to the lungs to load up on oxygen. The major job of the left side of the heart is to receive oxygen-rich (oxygenated) blood, and then pump it out to all the other organs and tissues of the body.

The receiving chambers of the heart are called the atria (singular=atrium). The pumping chambers of the heart are called ventricles. Specialized valves are closed to prevent backflow from the ventricles to the atria during pumping. The valve operating between the right atrium and the right ventricle is called the tricuspid valve; the valve operating between the left atrium and the left ventricle is called the mitral valve. Valves also exist between each ventricle and the exiting arteries.

The basic flow of blood through the heart is as follows: deoxygenated blood is picked up throughout the body by **veins** which flow into the major veins, the venae cavae the vena cavae empty into the right atrium the right atrium empties into the right ventricle the right ventricle pumps blood through the pulmonary artery (the only artery in the body to carry deoxygenated blood) to the lungs, where the blood loads up on oxygen and gets rid of the waste product, carbon dioxide the pulmonary vein (the only vein in the body to carry oxygenated blood) carries blood to the left atrium the left atrium empties into the left ventricle the left ventricle pumps oxygenated blood into the body's largest artery, the aorta; numerous arteries branching off of the aorta deliver the oxygenated blood to all of the organs and tissues of the body

The heart is composed of specialized muscle tissue. It beats due to complex electrical networks which stimulate the carefully organized contraction (systole) and relaxation (diastole) of the muscle fibers. The functioning of the heart can be studied using a device which detects the electrical activity of the heart and graphs it on paper. This device is called an electrocardiogram (ECG or EKG). The beating of the heart, its strength, and its rate are all affected by various nerves and hormones. This allows the heart to respond to all kinds of stress, including exertion, emotional stress, and changes in the bodily environment such as infections.

A number of conditions and diseases can affect the heart. Because the heart is essential to life, these disorders are all potentially life-threatening. Such disorders include: infections or inflammations of the sac around the heart (pericarditis), the lining of the heart (endocarditis), or the cardiac muscle itself (myocarditis); electrical derangements within the heart, called arrhythmias or blockade; stiffening of and fatty blockages within the arteries which supply oxygen to the heart muscle (coronary artery disease, or atherosclerosis, which can lead to heart attack); long-term high blood pressure and stiffening in the arteries of the body, which cause the heart to have to pump more strongly against this pressure, possibly causing the heart to enlarge.

HELMHOLTZ, HERMANN VON (1821-1894)

German physiologist and physicist

Hermann Helmholtz was one of the few scientists to master two disciplines: medicine and physics. He conducted break-

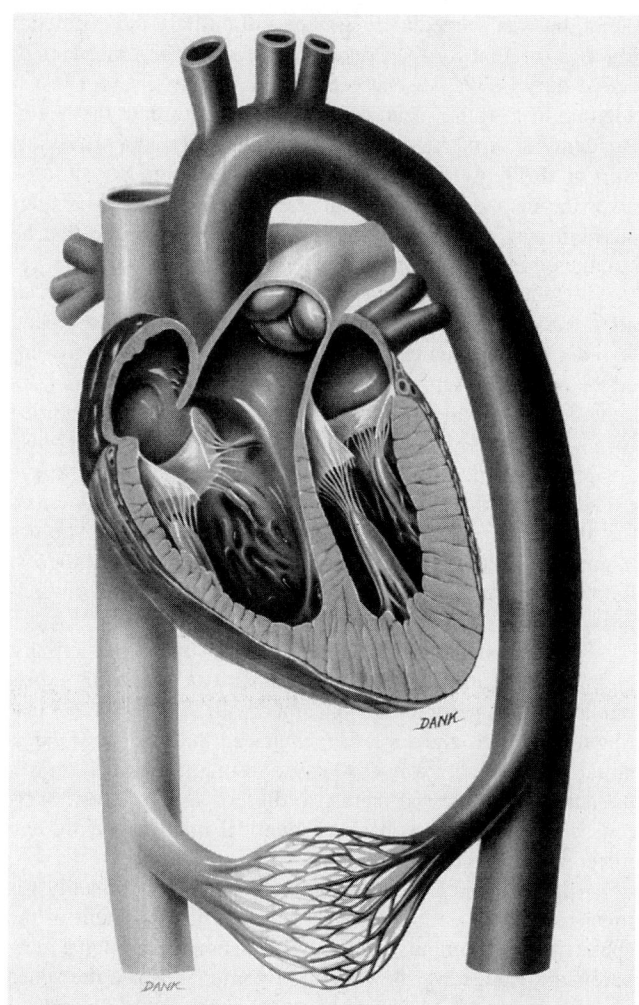

An anterior cross section of the human heart. *(Illustration by Leonard D. Dank, Custom Medical Stock Photo. Reproduced by permission.)*

through research on the nervous system, as well as the functions of the eye and ear. In physics, he is recognized (along with two other scientists) as the author of the concept of conservation of energy.

Helmholtz was born into a poor but scholarly family; his father was an instructor of philosophy and literature at a gymnasium in his hometown of Potsdam, Germany. At home, his father taught him Latin, Greek, French, Italian, Hebrew, and Arabic, as well as the philosophical ideas of Immanuel Kant and J. G. Fichte (who was a friend of the family). With this background, Helmholtz entered school with a wide perspective. Though he expressed an interest in the sciences, his father could not afford to send him to a university; instead, he was persuaded to study medicine, an area that would provide him with government aid. In return, Helmholtz was expected to use his medical skills for the good of the government—particularly in army hospitals.

Helmholtz entered the Friedrich Wilhelm Institute in Berlin in 1898, receiving his M.D. four years later. Upon graduation he was immediately assigned to military duty, practicing as a surgeon for the Prussian army. After several years of active duty he was discharged, free to pursue a career in academia. In 1848 he secured a position as lecturer at the Berlin Academy of Arts. Just a year later he was offered a professorship at the University of Konigsberg, teaching physiology. Over the next twenty-two years he moved to the universities at Bonn and Heidelberg, and it was during this time that he conducted his major works in the field of medicine.

Helmholtz began to study the human eye, a task that was all the more difficult for the lack of precise medical equipment. In order to better understand the function of the eye he invented the ophthalmoscope, a device used to observe the retina. Invented in 1851, the ophthalmoscope—in a slightly modified form—is still used by modern eye specialists. Helmholtz also designed a device used to measure the curvature of the eye called an ophthalmometer. Using these devices he advanced the theory of three-color vision first proposed by Thomas Young. This theory, now called the Young-Helmholtz theory, helps ophthalmologists to understand the nature of color blindness and other afflictions.

Intrigued by the inner workings of the sense organs, Helmholtz went on to study the human ear. Being an expert pianist, he was particularly concerned with the way the ear distinguished pitch and tone. He suggested that the inner ear is structured in such a way as to cause resonations at certain frequencies. This allowed the ear to discern similar tones, overtones, and timbres, such as an identical note played by two different instruments.

In 1852 Helmholtz conducted what was probably his most important work as a physician: the measurement of the speed of a nerve impulse. It had been assumed that such a measurement could never be obtained by science, since the speed was far too great for instruments to catch. Some physicians even used this as proof that living organisms were powered by an innate "vital force" rather than energy. Helmholtz disproved this by stimulating a frog's nerve first near a muscle and then farther away; when the stimulus was farther from the muscle, it contracted just a little slower. After a few simple calculations Helmholtz announced the impulse velocity within the nervous system to be about one-tenth the speed of sound.

After completing much of the work on sensory physiology that had interested him, Helmholtz found himself bored with medicine. In 1868 he decided to return to his first love—physical science. However, it was not until 1870 that he was offered the physics chair at the University of Berlin and only after it had been turned down by Gustav Kirchhoff. By that time, Helmholtz had already completed his groundbreaking research on energetics.

The concept of conservation of energy was introduced by Julius Mayer in 1842, but Helmholtz was unaware of Mayer's work. Helmholtz conducted his own research on energy, basing his theories upon his previous experience with muscles. It could be observed that animal heat was generated by muscle action, as well as chemical reactions within a working muscle. Helmholtz believed that this energy was derived from food and that food got its energy from the sun. He proposed

that energy could not be created spontaneously, nor could it vanish—it was either used or released as heat. This explanation was much clearer and more detailed than the one offered by Mayer, and Helmholtz is often considered the true originator of the concept of conservation of energy.

While this was undoubtedly Helmholtz's greatest legacy, he also began several projects that were later completed by other scientists. He advanced a number of hypotheses on electromagnetic radiation, speculating that it lay far into the invisible ranges of the spectrum. This line of research was later resumed, very successfully, by one of Helmholtz's students, Heinrich Rudolph Hertz, the discoverer of radio waves. Helmholtz's theories on electrolysis were also the basis for future work conducted by Svante August Arrhenius.

Helmholtz had been a sickly child; even throughout his adult life he was plagued by migraine headaches and dizzy spells. In 1894, shortly after a lecture tour of the United States, he fainted and fell, suffering a concussion. He never completely recovered, dying of complications several months later.

HELMONT, JOHANNES (JAN) BAPTISTA VAN (1579-1644)
Flemish physician, chemist, and physicist

Jan van Helmont lived and worked during a unique period of history—the dawn of the scientific revolution. As Europe began to shrug off the old-fashioned superstitions of the Middle Ages, scientists learned how to conduct experiments and make rational observations, rather than believing in religious phenomena to explain the world around them. However, the clash between science and faith eventually caused difficulty for van Helmont. He was interrogated by the notorious Spanish Inquisition for giving scientific explanations for supernatural events.

Born into a noble family, van Helmont became a medical doctor in 1599. He often treated people for free, refusing to profit from the misery of his fellow human beings. In 1609, after extensive travels and medical experience (including treating victims of **bubonic plague**), he turned down several attractive job offers and devoted himself to pure research on the principles of nature.

In many respects, van Helmont followed the teachings of Paracelsus, a Swiss physician of the early 1500s who pioneered the use of alchemy to prepare medicine instead of trying to make gold. Like Paracelsus, van Helmont was interested in alchemy and mysticism. He also believed in the fallacious theory of spontaneous generation—that **organisms** can spring to life from decaying materials such as dusty grain or old rags.

But van Helmont disagreed with Paracelsus' belief in the ancient Greek theory that all matter is composed of four elements (earth, air, fire, and water). Instead, van Helmont asserted that the basic element of the universe was water, an idea that went even further back to the Greek philosopher Thales, who lived around 600 B.C. Van Helmont, to prove his theory in a scientific manner, grew a willow tree in a tub for five years, giving it nothing but pure water and even protecting it from dust in the air. He weighed the tree and the soil carefully before starting the experiment and then again at the end. Although the soil lost practically no weight, the tree had gained more than 160 pounds (72 kg), which van Helmont attributed, mistakenly, to the water he had added. Because this experiment represented the first application of quantitative methods to a biological question, van Helmont is sometimes called the father of biochemistry.

Van Helmont also discovered gases as a class of substances and first coined the term *gas* (from the phonetic spelling of the Greek word *chaos*) to describe vapors that differed from ordinary air. He applied scientific techniques to study several gases, most notably *gas sylvestre*, which he produced from burning wood. This gas is now known to be carbon dioxide. Van Helmont realized that carbon dioxide was also produced by other chemical processes, such as the fermentation of wine and the reaction of acids with limestone and other carbonates. He also described carbon monoxide, chlorine gas, digestive gases, sulfur dioxide, and a ''vital gas'' in the blood, and he showed that a burning candle would use up air in an enclosed space. In these studies, van Helmont paved the way for more definitive chemical analysis of gases by **Joseph Priestley**, Antoine-Laurent Lavoisier, Joseph Black, and other scientists who had better apparatus to work with.

Another popular belief of van Helmont's day was *transmutation*, which held that one metal could be changed into another or destroyed altogether. Relying upon scientific methods, van Helmont refuted this idea by showing that dissolved metals could be recovered in their original quantity. His insight was a forerunner of physical laws regarding the indestructibility of matter. Van Helmont also studied the behavior of pendulums and recommended that they be used to measure time.

Much of van Helmont's work, however, still focused on medicine and health. He demonstrated that acid is the stomach's digestive agent, and he suspected that the substance was hydrochloric acid. Through ingenious observations, van Helmont identified many causes of asthma. He also studied the symptoms of bronchitis, tuberculosis, epilepsy, and hysteria, and he diagnosed illness by analyzing the specific gravity of urine. Van Helmont realized that fever is part of the body's natural healing process. Instead of the traditional treatment of bloodletting or purging, he prescribed remedies according to the specific disease, its cause, and the bodily organ being affected. In fact, van Helmont's medical research introduced the concept that diseases are caused by specific harmful agents, rather than by a general imbalance of the body's ''humors,'' or fluids.

Throughout his scientific career, van Helmont clung to his religious and mystical beliefs which, unfortunately, differed from those of the Catholic authorities. Spain had occupied the Flemish territories, and the Church still had the force of the law behind it. During the early 1600s, van Helmont was embroiled in a controversy over a pamphlet he had published on curative ointments. He was condemned for heresy by the General Inquisition of Spain, a special committee set up to punish those who contradicted the Catholic Church. Van Helmont was also denounced by the University of Louvain's med-

ical and theological faculties. Eventually, he was detained and interrogated, then kept under house arrest for years. Perhaps because of this experience, van Helmont published little of his work. On his deathbed, he gave his writings to his son to edit and publish.

See also Gases, existence of

HEMOGLOBIN

Hemoglobin is the **protein** responsible for red **blood** cell's ability to bind oxygen and **carbon dioxide**. It is distributed throughout the red cell **cytoplasm** and allows the **cells** to transport oxygen from the lungs to the **tissue**s where it exchanges the oxygen for carbon dioxide. The cells then carry the carbon dioxide from the blood stream to the lungs where it is released as part of the process of **respiration**.

Hemoglobin production occurs in bone marrow and requires iron, folic acid, and vitamin B12. Absorption of the latter requires a substance known as Intrinsic Factor, which is produced by the stomach. Hemoglobin actually consists of two parts: a helical protein chain, known as globin, and a carbon ring complex, called heme, which contains iron. This carbon ring, also called a porphyrin ring, is made up of many carbon **atoms** connected to four **nitrogen** atoms facing a central hole. The nitrogen atoms trap an iron atom in this hole where it bonds to an oxygen or carbon dioxide **molecule** or to part of the globin chain.

When hemoglobin production is reduced or when its oxygen transferring ability is impaired, anemia can result. Minor symptoms of anemia include pale skin, weakness, fatigue, and dizziness. Severe symptoms include breathing difficulty and abnormal **heart** rhythm. Iron deficiency anemia is the most common type and it results from either chronic blood loss, lack of iron in the diet, impaired absorption of iron from the intestine, or increased need for iron, as in pregnancy and **birth**. Pernicious anemia occurs when the Intrinsic Factor is lacking and the body can not process vitamin B12. Abnormal production of hemoglobin can cause a hereditary condition known as **sickle-cell anemia**.

HEMOPHILIA

Suffering an occasional cut, scratch, or bruise is a normal consequence of life. When there is damage to the skin and a **blood** vessel ruptures, bleeding occurs. The human body is then able to initiate a series of reactions which cause the bleeding to stop. First, platelet **cell**s in the blood move toward and attach to the site of the wound. The platelets are further held in place by strands of *fibrin*. The formation of the strands is the key event in a complex series of enzymatic reactions that are still somewhat of a mystery today. Without this cascade clotting process, people would be in danger of bleeding to **death** from very minor injuries.

However, the scenario described above could be fatal for a person afflicted with hemophilia. The term *hemophiliac,*

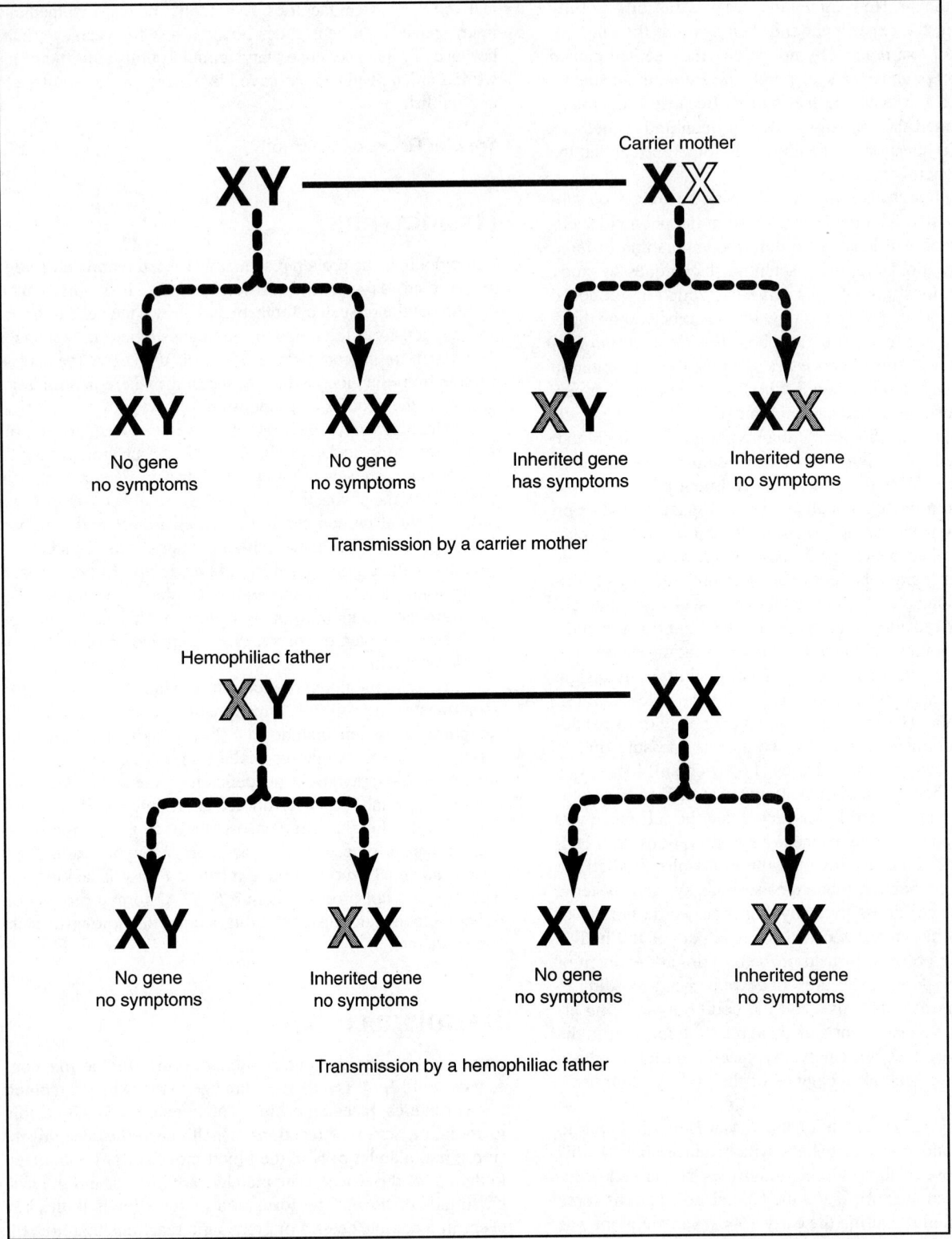

Hemophilia A and B are both caused by a genetic defect present on the X chromosome. Approximately 70% of people with hemophilia A or B inherited the disease, while the remaining 30% have hemophilia due to a spontaneous genetic mutation. *(Illustration by Electronic Illustrators Group.)*

coined by the German physician Johann Schönlein (1793-1864) in 1828, is made up of Greek and Latin terms which refer to "one who loves to hemorrhage or bleed." First described by the Islamic surgeon Abu al-Qasim in the tenth century, this genetic disease has existed for several hundred years and has directly influenced history. Queen Victoria (1819-1901) had several hemophilic sons that died before they had the opportunity to become King of England. As early as the nineteenth century, scientists suspected that hemophilia may be passed from parents to offspring, or inherited. They also noticed that generally only males showed the uncontrolled bleeding that is a major symptom of the disease.

Today we know that hemophilia is caused by a small defect in a single human **gene**. When this particular defect occurs, as in classic or type A hemophilia, the body lacks an important **protein** which helps to form fibrin. This protein is called factor VIII. About 85% of all hemophiliacs are missing the gene that instructs the body's cells how to produce factor VIII. Hemophilia B is a less common type of genetic disease caused by a deficiency in another necessary protein, called factor IX. In each type of bleeding disorder, proteins are either missing or deficient and thus fibrin is not able to form. Approximately 2 out of every 10,000 males are afflicted with either type A or type B hemophilia.

Hemophilia is a sex-linked trait because the genes for factor VIII and IX are located on the X **sex chromosome**. Female cells contain two X **chromosome**s and male cells contain one X chromosome and one Y chromosome. All males inherit one factor VIII gene on the X chromosome from their mothers. If this gene is defective, the male will be hemophilic. A female, in contrast, has two factor VIII genes, one inherited from each parent. If one gene is defective but the other is not, the female will not be affected by this disease. The normal gene on the second X chromosome protects her. She will, however, carry the defective gene and may pass this gene on to her children. Generally, carrier females will pass their defective gene on to half their daughters, who will be carriers, and to half their sons, who will be hemophilic. For example, Queen Victoria was a carrier, and the genetic profile of her offspring supports this fact. Only in a very rare situation would a female acquire two defective genes and be hemophilic.

Fortunately, hemophiliacs can be treated with transfusions of concentrated factor VIII protein. This has increased the life expectancy of some hemophiliacs. The protein concentrate can be prepared by combining volumes of blood donated from many humans with normal clotting blood. However, this method usually spreads viral diseases which were present in the original donated blood. Many hemophiliacs are chronically infected with these **virus**es which include the virus that causes **AIDS** and hepatitis. In fact, in the early to mid-1980s, over 90% of the people with severe hemophilia contracted the AIDS virus through contaminated human blood plasma. Only recently has a process been invented where certain viruses, including the AIDS virus, can be detected in the blood supply. Cow and pig blood have also been concentrated and used as a therapy, since these **animal**s have higher concentrations of factor VIII than humans. Transfusion reactions and other problems have caused a decline in the use of animal-derived factor VIII.

Philip Showalter Hench. *(The Library of Congress. Reproduced by permission.)*

In the early 1980s, a completely new way of making factor VIII was discovered. Scientists from several research companies have been able to make the factor VIII protein by isolating the normal gene and inserting this human gene into hamster cells. The hamster cells then produce large amounts of pure factor VIII protein. The protein is then harvested and used as a therapy for hemophiliacs. This process is called **genetic engineering**. In addition to eliminating many of the side effects associated with previous factor VIII therapies (for example, complications from bleeding and from transfused blood and blood products that may be contaminated with a virus), a **gene therapy** approach promises to provide a permanent programming of cells to make clotting factor. Studies in mice has shown the process to be effective, and clinical trials in humans are on the horizon. According to the World Federation of Hemophilia, the cure for hemophilia using gene therapy is imminent.

HENCH, PHILIP SHOWALTER (1896-1965)
American pathologist

Philip Showalter Hench, an American clinical pathologist, performed groundbreaking research in rheumatoid **arthritis**. His clinical tests of adrenal compound E, which Hench named cortisone, and of **ACTH**, which produces **cortisone** naturally by stimulating the adrenal cortex, offered the first hope for patients suffering from rheumatoid arthritis. Hench and his colleague, biochemist Edward C. Kendall, gained immediate worldwide attention when they filmed the miraculous recovery of arthritis patients—some of whom could barely walk—as they climbed stairs and even jogged in place. Although prolonged clinical trials showed that neither cortisone or ACTH was a viable long-term therapy for arthritis due to side effects such as high **blood pressure** and high glucose levels, Hench's efforts opened new vistas in medical research, particularly in the study of both hormones and rheumatoid arthritis. A meticulous researcher who methodically collected his clinical data before publishing his results, Hench shared the 1950 Nobel Prize in physiology or medicine with Kendall ''for their discoveries relating to the hormones of the adrenal cortex, their structure and biological effects.'' (Chemist **Tadeus Reichstein** also received a share of the prize for his independent work with the adrenal cortex and its **hormones**.)

Hench was born in Pittsburgh, Pennsylvania, on February 28, 1896. The son of Jacob Bixler Hench, a classical scholar and school administrator, and Clara John Showalter, Hench attended a private high school, Shadyside Academy, and then enrolled at the University of Pittsburgh in 1916. His education looked as though it would be interrupted when he enlisted in the U.S. Army Medical Corps. But Hench was transferred to the reserves so he could return to his studies, and he enrolled in Lafayette College in Easton, Pennsylvania. He received his B.A. from Lafayette in 1916 and enrolled at the University of Pittsburgh School of Medicine, where he received his M.D. in 1920.

After completing his internship at St. Francis Hospital in Pittsburgh, Hench became a fellow in medicine at the Mayo Foundation of the University of Minnesota. The bright young physician and scientist would spend his entire career at the Mayo Clinic, where, in 1926, he cofounded the Department of Rheumatic Disease, which was the first training program in rheumatology in the United States. Hench spent the 1927–28 academic year on sabbatical studying research medicine with Ludwig Aschoff, a leading rheumatic fever investigator, at Freiburg University. He also studied with clinician Freidrich von Müller in Munich. Hench completed his formal education in 1931 when he received a master of science degree in internal medicine from the University of Minnesota.

A physician first, Hench's research was clinically based. He began studying rheumatoid arthritis in 1923. Unlike osteoarthritis, a degenerative joint disease common in later life, rheumatoid arthritis is a chronic inflammatory disease of the joints often contracted at the relatively young age of 30 to 35. In advanced stages, rheumatoid arthritis could cause deformity due to bone and surrounding muscle atrophy. In 1929, Hench took note of a patient who had suffered from severe arthritis for more than four years. The patient had entered the Mayo Clinic suffering from jaundice, a disease caused by excess bilirubin, a liver product, in the bloodstream. Amazingly, the man's arthritis had abated and remained dormant for several weeks after his recovery from jaundice. Carefully collecting data, Hench waited until he had authenticated nine similar cases, among them patients who experienced remissions from painful fibrositis and sciatica, two other inflammatory conditions, before publishing his data in 1933.

Hench was convinced that these cases held a vital clue to a therapy for arthritis and set out to induce jaundice artificially. Hench's initial experiments used infusion or ingestion of bile to emulate jaundice's production of excess bile in the blood or the liver. Although these experiments failed, Hench's attention was soon drawn to another group of patients, women whose arthritis vanished during pregnancy. He also observed that some arthritic patients went into less complete remission after surgical operations, anesthesia, or severe fasting. Looking for a common physiological denominator, Hench, who enjoyed reading Sir Arthur Conan Doyle's novels of Sherlock Holmes, had a prime suspect—glandular hormones. Furthermore, the fact that both jaundice and pregnancy caused remission in almost the exact same manner led Hench to believe that his missing compound was not bilirubin or a female-only sex hormone.

Collaboration with Kendall Leads to Treatment for Arthritis

Fortunately, the Mayo Foundation's own Edward C. Kendall was a world renowned chemist in the field of steroids, a specific group of hormones. Kendall had isolated six steroids from the adrenal cortex, the outer part of the endocrine glands located atop the kidneys, which he alphabetized compound A through F. Hench's first try with compound A was a failure. Both Hench and Kendall then decided to try compound E. But at that time, in 1941, compound E was extremely difficult to synthesize and, as a result, costly. With both high (300°F [147°C]) and low (-100°F [-128°C]) temperatures needed to produce compound E, the delicate work took time and attention and the slightest mistake could result in a useless compound. It wasn't until more than two years after World War II that scientists from the pharmaceutical firm Merck & Co. had developed a process that could produce enough compound E for Hench to attempt his experiment. Still, the compound was expensive to produce. Hench recalled in an interview for an article in the *Saturday Evening Post* that he and his colleagues ''almost went into shock'' when a $1,000 bottle of compound E was dropped on a marble floor.

Hench's results with compound E were miraculous. The first patient, a 29-year-old woman, experienced total remission of symptoms after three injections over three days. Hench's results were quickly confirmed by five other researchers across the country. Hench and his colleagues received instant public notoriety as a result of their studies both with compound E, which Hench named cortisone, and with adrenocorticotropic hormone (ACTH), a hormone produced by the pituitary gland which spurs the body's natural production of cortisone through the adrenal cortex.

Unfortunately, Hench's miraculous ''cure'' for arthritis turned out to be short lived. Without the use of cortisone or ACTH, rheumatic symptoms returned; and long-term use of cortisone or ACTH causes several side effects, including high blood glucose and high blood pressure, as well as obesity associated with adrenal or pituitary gland tumors. Much to Hench's credit, he maintained his scientific cautiousness throughout the heady early days of the discovery, quickly recognizing the harmful side effects and outlining future directions in research of these hormones. Nevertheless, the studies of Hench and Kendall, along with those of Tadeus Reichstein, opened entirely new avenues of medical research; as a result, the three scientists were awarded the Nobel Prize for medicine or physiology in 1950.

Hench retired from the Mayo Foundation in 1957. In addition to the Nobel Prize, he was a recipient of the numerous awards, including the prestigious Lasker Award, which he also shared with Kendall. Hench married Mary Genevieve Kahler in 1927, and the couple had two sons, Philip Kahler and John Bixler, and two daughters, Mary Showalter and Susan Kahler. His hobbies included photography, tennis, opera, and Sherlock Holmes novels. He died from pneumonia on March 30, 1965, while vacationing with his wife in Ocho Rios, Jamaica. To honor him, Hench's alma mater, the University of Pittsburgh, presents the annual Hench Award to a distinguished university alumnus.

HEPATOPORTAL SYSTEM

The hepatoportal system is a unique arrangement of **blood** vessels leading to and from the liver. The vast majority of the **circulatory system** is arranged in such a way that blood containing oxygen is delivered by **arteries** and then capillaries to various **organ** systems. Oxygen then leaves the blood, feeding the organs. Oxygen-depleted blood is then picked up from these same capillaries, and carried through **veins** back to the **heart**, where it is recirculated to the lungs to pick up oxygen again.

In the case of the hepatoportal system, the blood leaving capillaries throughout the intestinal system is carried by the hepatic vein not to the heart for re-oxygenation, but to a second set of capillaries within the liver. This blood is rich in all types of **nutrients** which it has picked up from the intestines. The hepatic vein sends this nutrient-rich blood through a capillary system in the liver, as well as into tiny little spaces within the liver, called liver sinuses. Here, various nutrients which the liver is responsible for processing and storing are removed from the blood. **Toxins** and waste products are also removed by the liver, and are processed to make them less dangerous.

After passing through the hepatoportal system, the blood is sent into the inferior **vena cava**, which delivers it to the heart for re-oxygenation. Because blood arriving at the liver via the hepatic vein has already passed through the capillary beds of the intestines, it is mostly depleted of oxygen by the time it gets to the liver. Therefore, oxygenated blood is sent to the liver through the hepatic artery, a branch off of the **aorta**.

The major difference, then, between the hepatoportal system, and **blood circulation** through other organs, is the fact that the blood passes through two sets of capillaries prior to its return to the heart: that of the intestines, followed by that of the liver.

Major disease processes involving the hepatoportal system most commonly stem from liver damage and scarring, called cirrhosis. Cirrhosis can be due to a number of underlying diseases, although by far the most common etiology is damage from **alcohol** abuse. The scarring of cirrhosis blocks the outflow of blood through the hepatoportal system, resulting in high **blood pressure** within the hepatoportal system. This high blood pressure (called portal hypertension) is conveyed to nearby capillary systems, particularly that of the esophagus. As a result, severe, difficult-to-control, life-threatening bleeding (hemorrhaging) can take place. Other types of processes which can interfere with appropriate flow through the hepatoportal system include damage due to hepatitis (an infection and **inflammation** of the liver), tumors, congenital defects (defects which are present at **birth**), or worm infestation.

HERACLITUS (525-475)
Greek philosopher

The thought of Heraclitus—a Greek philosopher who lived probably within the span of approximately 525–475 B.C. (actual dates unknown)—can be accessed only through a few fragments of text preserved in a second-hand way from quotes and citations by contemporaries and followers. No copies have been preserved of his one known book. Of the approximately 125 identified fragments, fewer than 90 are accepted as fully verbatim citations. Very little is known either, of the life of Heraclitus.

Despite the deficiency of established information, Heraclitus remains among the most discussed and influential of the pre-Socratic Greek philosophers. His thoughts and statements are still read and widely quoted today, and some of those fragments have become well known even to people unfamiliar with his name, including for example, the aphorism that "one cannot step twice into the same river." The river persists but the flow of the waters is ever-changing.

For biologists, the Heraclitean fragments suggest attention to process, tension, and complexity. He can be considered an ancient precursor of **Charles Darwin**. In perhaps his most influential idea, Heraclitus broke with the notion that the natural world was fixed and stagnant, but taught instead that everything changes and evolves, that everything is in flux and nothing remains stagnant. He declared clearly, for example, the very modern notion that "the sun is overseer and sentinel of cycles, for determining the changes and the seasons which bring all things to birth."

His thought is also often summarized as a "doctrine of opposites," perhaps nowhere better demonstrated than in the following fragment: "Graspings: wholes and not wholes, convergent divergent, consonant dissonant, from all things one and from one thing all." Similar dialectics reemerge frequently in contemporary biology, e.g., in the debates about unity and diversity.

In some of his fragments, Heraclitus also underlined the kind of inquiry that has become central to the thinking of modern scientists and to the scientific method. His ideas were experientially based: ''Whatever comes from sight, hearing, learning from experience: this I prefer.'' No one can forget the simple statement that helps to illuminate centuries of scientific curiosity: ''Nature loves to hide.''

HERBIVORE

An herbivore (from Latin *herba*, meaning grass, and *vorare*, meaning to devour) is an **animal** that obtains its **nutrition** by consuming **plants** and **algae**, the primary **producers** (**autotroph**s). Since these **consumer**s do not eat meat, one can think of herbivores as vegetarians. Examples of herbivores are cows and horses which eat grasses, hay, and other plants; caterpillars, which consume leaves; and marine **zooplankton**, which consume **phytoplankton**. Each **ecosystem** has its own specific set of herbivores, and within each ecosystem there are often several different types of herbivores, each eating a different type of producer. For example, in a rainforest, one herbivore might eat only a certain type of leaves, while another eats only specific fruit, and a third eats only nuts. Yet another herbivore might eat all three.

Because herbivores are not able to make their own food, they are considered **heterotroph**s. They must obtain all **nutrients** and **energy** from the food they consume.

Herbivores are a very important link in food webs. While they do not make their own food, they do pass energy and nutrients along the food web between producers and higher **order** consumers (the **carnivore**s), thus linking these two levels. Because they are the first level of consumers in a food web, herbivores are often called primary or first-order consumers.

HEREDITY

All living things pass on traits from one generation to the next according to a systematic set of ''blueprints.'' These blueprints are contained in the long, thread-like **chromosome**s that lie inside the cell **nucleus** of all living things. On these chromosomes are genes that determine the hereditary traits of the offspring.

Egg and **sperm** cells, or sex **cells**, are specially formed to carry only one set of the 23 different chromosomes that are normally found in the human body. (Regular body cells have *two* sets of the 23 chromosomes.) When a mother's egg is fertilized by the father's sperm, the egg inherits one set of chromosomes from each parent, for a total of 46 chromosomes.

Some characteristics can only be inherited through genes and chromosomes: **blood** type, eye color, maleness or femaleness, etc. These are called hereditary traits. Most characteristics, however, are a result of both heredity and environment. For instance, a person can inherit a general body type, but environmental factors such as diet and exercise may change that body type.

The study of heredity—a science called **genetics**—started in the 1800s, when scientists first began trying to explain the existence of different **species** and variations within the same species. At that time, French biologist **Jean Baptiste de Lamarck** strongly believed that acquired characteristics would improve when routinely used over time. Those characteristics that were not used simply faded away. Lamarck also maintained that acquired characteristics were inherited from one generation to the next. In other words, Lamarck believed that if a giraffe continuously stretched its neck to reach for food, it would develop a longer neck. And the longer neck would be passed on to the next giraffe generation. Although his belief that acquired characteristics were inherited was incorrect, Lamarck was on the right track. He implied that traits can be inherited from generation to generation—that species undergo long-term evolutionary changes.

In 1859, **Charles Darwin** published his landmark book *The Origin of Species*, in which he outlined his theory of evolution through **natural selection**. Darwin believed that members of a particular species have slightly different characteristics. In the competition for space, food, and shelter, some of these characteristics would make a particular plant or **animal** better able to survive and produce offspring than others of its species. Therefore, these advantageous characteristics would persist in future generations, while those less advantageous ones would disappear as their carriers died out. After centuries or millennia of competition or natural selection, recent members of a species might be quite different from their ancestors. This theory gained advocates like the revered English physician Thomas Huxley (1825-1895), who, as ''Darwin's Bulldog,'' did more than anyone else to overcome opposition to Darwinian theory. But even with all the support, Darwin's theory still lacked an explanation for how the differences in species occurred.

Darwin, realizing that he needed to explain the mechanics of variation, asserted that tiny particles floating in an individual's bloodstream entered the eggs and sperm to determine hereditary characteristics. But **Francis Galton** proved him wrong with a simple blood transfusion experiment between two different types of rabbits. The transfusion didn't change the offspring of the rabbits as it should have if Darwin were correct.

In 1884, August Weissmann proposed that a special hereditary substance existed in the egg nucleus, which he termed ''germ plasm.'' His theories concerning the behavior of this substance—later identified as chromosomes—were eventually proved correct. However, he mistakenly believed that the germ plasm passed intact from generation to generation, unchanged by any environmental factors. Weissmann's theory, therefore, could not adequately account for the changes that occurred between generations and drove Darwin's theory of evolution.

It wasn't until 1900 that the second important theory concerning heredity was discovered, although it had been formulated some forty-five years earlier. **Gregor Mendel**, an Austrian monk, had begun experimenting with pea **plants** at about the same time that Darwin set forth his ideas on natural selection. Through his efforts, Mendel demonstrated that actual physical ''hereditary factors'' could be transmitted independently. Mendel ultimately established the basic laws of heredity—the missing key to Darwin's natural selection theory—and set the standard for the field of genetics. His revolutionary theories, however, were met with disinterest during his lifetime and remained largely unknown until 1900, when they were independently rediscovered by **Hugo de Vries**, Karl Correns (1864-1933), and Erich Tschermak (1871-1962).

De Vries took Mendel's theories further. Unlike the Austrian monk, he believed that variations, rather than arising from gradual or transitional steps, occurred in jumps he called **mutation**s. This formed the cornerstone of de Vries's mutation theory, which he proposed in 1901.

Despite these theories, no biological mechanism for heredity had yet been found. **Walther Flemming** had discovered chromosomes during the 1870s but, unaware of Mendel's work, did not understand their genetic significance. In 1903, a young graduate student, Walter S. Sutton, at last made the connection. He had observed that during **cell division** in regular cells, chromosomes were present in pairs. But in the cell division of reproductive cells, only one member of each pair entered a sperm or egg. The chromosomes became pairs again when the egg joined the sperm in the **fertilization** process. Sutton saw that this pairing, unpairing, and pairing again paralleled the movement of Mendel's "hereditary units." **Theodor Boveri** independently came to the same conclusion, and together their hypothesis came to be known as the chromosomal theory of **inheritance**.

By 1909, when Wilhelm Johannsen coined the term *gene* to describe the "hereditary units" on the chromosome, Mendelian theory and chromosomal theory had been widely accepted by scientists. American geneticist **Thomas Hunt Morgan**, however, remained unconvinced and set out to empirically prove or disprove Mendel's theory of inheritance. Following his many experiments with the **fruit fly**, Morgan was won over, convinced that genes were the trait-determiners and that they are arranged in a certain order on each chromosome. He also noticed that all the genes on the same chromosome were usually inherited together. Morgan referred to these as *linked genes*.

Further experiments showed that traits did not always follow Mendel's basic laws of heredity. Morgan showed that offspring don't always inherit all of the genes on a chromosome. He called this occurrence *crossing over*. By 1915, Morgan, along with Hermann Muller, Calvin Bridges (1889-1938), and Alfred Sturtevant (1891-1970), had fully developed the concepts of **linkage** and **crossing over**.

Yet some still refused to acknowledge the great strides made by biologists. The Ukrainian biologist Trofirm Denisovich Lysenko (1898-1976) gained control of Soviet biological research between 1928 and 1965, and, with the backing of Joseph Stalin (1879-1953), imposed his erroneous view that acquired characteristics could be inherited. Although his influence waned with the rise of Nikita Khrushchev (1894-1971), Lysenko severely damaged the Soviet Union's reputation in the international scientific community. His legacy would not be erased until the launching of *Sputnik I* in 1957.

By 1953, **James Watson** and **Francis Crick** had developed a model of **deoxyribonucleic acid** (**DNA**), the building

A mother with her five children. It is probable that each sibling inherited their dark hair from their parents. *(Photograph by T. McCarthy, Custom Medical Stock Photo. Reproduced by permission.)*

blocks of genes, thus deciphering the **genetic code** and providing a key to the chemical basis of heredity. In recent decades, most research on heredity has focused on the function of DNA, its regulatory processes, and its evolution.

HEROPHILUS (335 B.C.-280 B.C.)
Greek anatomist and physiologist

Sometimes called the father of **anatomy**, Herophilus was born in Chalcedon, Asia Minor. Little is known about his life; the date and place of his death have been completely lost, as have all his writings. Herophilus studied medicine under Praxagoras of Cos and then at Alexandria, where he later taught and practiced medicine. In Alexandria, Herophilus had the unique opportunity to practice human dissection, a research technique not allowed elsewhere. Herophilus even performed public dissections. His work was highly regarded, and the medical school he founded at Alexandria attracted scores of students.

Herophilus made many anatomical studies of the **brain**. He distinguished the *cerebrum* (larger portion) from the *cerebellum* (smaller portion), pronounced the brain to be the seat of intelligence, and identified several structures of the brain, several of which still carry his name. He discovered that the nerves originate in the brain, was the first to distinguish nerves from tendons, and noted the difference between *motor nerves* (those concerned with motion) and *sensory nerves* (those related to sensation). He traced the optic nerve and described the retina. He studied the liver extensively and described and named the duodenum, the first part of the small intestine.

Drawing a distinction between arteries and **veins**, Herophilus noted the arterial pulse and developed standards for its measurement and use in diagnosis. He thought that arterial pulsation was involuntary, rising from dilation and contraction of arteries due to impulses sent from the heart. He corrected the idea that arteries carry air rather than **blood**. Herophilus also wrote a treatise on midwifery and accurately described the ovaries, the uterus, and the tubes leading from the ovaries to the uterus (later named the Fallopian tubes). In the field of medical treatment, Herophilus sensibly recommended good diet and exercise, but was also an enthusiastic advocate of bleeding and frequent drug therapy.

Although his medical school languished after his death, Herophilus (and his younger successor Erasistratus) established the disciplines of anatomy and physiology, which did not significantly advance before Galen in the second century and then the early modern anatomists of the thirteenth century.

HERPETOLOGY

Herpetology is the scientific study of amphibians and reptiles. The term ''herpetology'' is derived from the Greek and refers to the study of creeping things. Birds and **mammals**, for the most part, have legs that lift their bodies above the surface of the ground. Amphibians (**class** Amphibia) and reptiles (class Reptilia), with the exception of crocodiles and lizards, generally have legs inadequate to elevate their bellies above the terrain, thus they creep.

Both Amphibia and Reptilia are within the **phylum** Chordata, which also includes several classes of fishes, reptiles, birds and mammals. Amphibia include the anurans, which are frogs and toads; the urodeles, which include salamanders and sirens; and the gymnophioma, which are peculiar worm-like legless caecilians. **Larva**l amphibians (tadpoles) respire with gills whereas adults breathe with lungs. Amphibian skin is ordinarily scaleless. Reptilia includes lizards, snakes, turtles and crocodiles. They have scaly skin and respire with lungs. Extinct reptiles are of great scientific and popular interest and include dinosaurs, pterosaurs and ichthyosaurs.

Some scientists are both herpetologists and ecologists. They study **habitat**, food, **population** movements, reproductive strategies, life expectancy, causes of **death** and a myriad of other ecological problems. Their studies have significance not only to survival of the **animal**s that they study but also to humans. Amphibians and reptiles manage their **metabolism** of xenobiotic (foreign to the body) toxic substances in much the same way as humans do, by metabolic change in the liver and other **organ**s that permits rapid **excretion**. It becomes a notable concern when amphibians and/or reptiles cannot survive in an altered environment. Amphibians in a number of countries have been reported to be found in diminishing numbers and many are anatomically abnormal. Because of their similarity in managing toxic substances, whatever is causing the population perturbations and anomalous **anatomy** in the lower creatures may be of equal concern to humans.

The studies of amphibians and reptiles relating to pathology and medicine is less well known than similar studies with higher **organism**s. Herpes **virus**es are now recognized as being microbial agents related to animal and human **cancer**. Burkitt's lymphoma and Kaposi's sarcoma are two human cancers with an established link with herpes viruses. The first

Alfred Day Hershey. *(The Library of Congress. Reproduced by permission.)*

cancer of any type known to be causally associated with a herpes virus was the Lucké renal adenocarcinoma of the leopard frog, *Rana pipiens*. Virologists working with the frog cancer can perform a multiplicity of experiments with the herpes virus and frog cancer. The frog experiments would be very difficult to perform on other animals, and would be precluded for ethical reasons from human experimentation.

The feasibility of vertebrate **cloning** was first demonstrated in the frog, *R. pipiens*, in Philadelphia in 1952, and later in the South African clawed toad, *Xenopus laevis*, in England in 1958. Prior to the frog experiments, it was generally thought that cloning was a ''fantastical'' dream. **Cloning** has since been achieved with sheep, cows, and other mammals.

As economic resources, turtle meat and crocodile (raised on farms for that purpose) hides have a significant role in the Louisiana economy. Further, many amphibians and reptiles are collected for scientific study. Only rodents exceed in number frogs used for biomedical research.

HERSHEY, ALFRED DAY (1908-   )
American virologist

Alfred Day Hershey was born on December 4, 1908, in Owosso, Michigan. He received his Ph.D. in bacteriology from Michigan State in 1934. He was appointed to the staff of the

Department of Bacteriology at Washington University in St. Louis, Missouri. There he worked under J.J. Bronfenbrenner, who had been studying **bacteriophages**. A bacteriophage, or simply phage, is a type of **virus** that infects **bacteria** cells. It consists of only **nucleic acid** and **protein**. Nucleic acids, notably **DNA** and **RNA**, contain genetic material inherited by every **cell**. The basic structure of a bacteriophage includes a head and a tail; the head is made of protein encasing a core of nucleic acid, and the tail is protein.

During the 1940s, other scientists working with bacteriophages included **Max Delbruck** at Vanderbilt University in Nashville, Tennessee, who was working on the life cycle of phages, and Salvador Luria of Columbia University. Together they showed that bacterial cells can undergo spontaneous mutation in order to resist destruction by phages. Delbruck, Luria, and Hershey formed the core of the Phage Group, a group of scientists dedicated to bacteriophage research, especially those phages that infect the bacillus *E. coli* strain B.

In 1946, both Hershey and Delbruck, working separately, found that different strains of phage will exchange genetic material if both have infected the same bacterial cell. Hershey called this phenomenon genetic recombination. Then in 1952, at Cold Spring Harbor Laboratory in New York with Martha Chase, a geneticist, the pair discovered how phages infect bacteria. First the phage attaches itself to the bacterial cell membrane by its tail, and then the nucleic acid core is injected into the bacterial cell. The phage DNA enters the bacterial cell and then directs the cell to produce more bacteriophages. This experiment helped to prove that DNA is the genetic material of bacteriophages as well as other organisms. Throughout the remainder of the 1950s and 1960s, Hershey investigated the nucleic acid of bacteriophages, establishing that bacteriophage DNA is single-stranded, unlike the double-stranded DNA that exists in higher life forms. For their work on the replication and genetic structure of viruses, the 1969 Nobel Prize for physiology or medicine was awarded to Hershey, Luria, and Delbruck.

Hershey married Harriet Davidson in 1945. They had one son. Hershey died in 1997.

HESS, WALTER RUDOLF (1881-1973)
Swiss physiologist

Walter Rudolf Hess was born in the Swiss town of Frauenfeld to Clemens and Gertrud (Fischer Saxon) Hess on March 17, 1881. He inherited a strong interest in science from his father, a physics teacher. After finishing high school, Hess began his college career, changing universities frequently and taking every opportunity to travel. He eventually received a medical degree from the University of Zurich in 1905, and took a hospital residency under the famous surgeon Dr. Konrad Brunner.

While working for Brunner, Hess designed an improved **blood** viscometer (to measure blood's thickness and consistency) and began thinking about research in earnest. He took a second residency in Zurich and specialized in ophthalmology (the physiology and diseases of the eye) under the mistaken

Walter Rudolf Hess. *(UPI/Corbis-Bettmann. Reproduced by permission.)*

impression that the discipline would allow him time to continue his circulatory system investigations. He indeed developed a successful ophthalmology practice with a good income, but it took up all of his time. In 1912 Hess gave up his practice and moved to the Institute of Physiology in Zurich. Eventually he was named chair of the Physiology department, and began traveling to conferences and meetings throughout Europe. The stresses inherent in administrative work and World War I cut into his research time again, but he still managed to publish two important monographs, *The Regulation of the Circulatory System* in 1930, and *The Regulation of Respiration* in 1931.

Hess brought an unusual variety of tools and skills to his research. He had learned the basic principles of physics from his father, he knew a great deal about optics and hand-eye coordination from his days as an ophthalmologist, and he was a skilled surgeon. These all proved useful when he began conducting brain research on experimental **animals**. Hess's work on the **circulatory** and **respiratory systems** had included investigations of their interrelationship with other parts of animal physiology, including how blood flow and breathing were affected by the **nervous system**. Gradually this led to research on the areas of the **brain** responsible for regulating internal organs.

Of particular interest to Hess was the diencephalon, which is located under the cerebellum and is thus very difficult to access without damaging the rest of the brain. Hess designed very small electrodes and a mechanical guidance system that could implant the electrodes in experimental animals (cats) with the least possible disruption of their normal behavior. He also designed a method of delivering electrical stimulus pulses swiftly and accurately. On at least one occasion there was a public outcry about the use of animals for experimentation. Hess was instrumental in convincing the activists that, if properly regulated and humanely conducted, animal experiments were important for human welfare.

Using the electrodes to stimulate different areas of the brain, Hess observed the results on other areas of bodily function, such as blood pressure, **respiration**, and body temperature. He recorded his observations not only on paper, but also on film, and maintained meticulous records of dissections and cell studies. He also compared the results of electrical stimulation with behaviors resulting from naturally occurring brain lesions. He found that the diencephalon, and particularly the hypothalamus, controlled many of the body's responses, such as fear and hunger, and he was able to map out some of these responses in detail. Partly due to the isolation imposed by World War II, and partly because his papers were written entirely in German, the outside world knew little of his work until he had accumulated about 25 years worth of experiments. This may have been fortunate, because, as he wrote in his sketch, "The vast number of experiments turned out to be decisive; for generalization concerning symptoms, syndromes, and localizations could be supported only by such a large body of data."

In 1949, Hess won a share of the Nobel Prize for physiology or medicine for his work in analyzing the function of the diencephalon, part of the interbrain, and its role in coordinating the body's internal organs; Portuguese neurosurgeon Antonio Egas Moniz shared the award for his work on white brain matter. Other recognitions he received included Switzerland's Marcel Benorst Prize in 1933 and the German Society for Circulation Research's Ludwig Medal in 1938.

Hess married the former Louise Sandmeyer in 1908; the couple had two children, Rudolf and Gertrud. He retired in 1951, although he continued his work and was instrumental in the establishment of an institute for brain research. He died in Locarno, Switzerland on August 12, 1973.

HETEROTROPH

Heterotrophs (from Greek *heteros*, meaning other or different and *trophos*, meaning feeder) are **organism**s that are not able to make their own food. They must ingest or absorb food produced by other organisms. Therefore, the heterotrophs rely on other organisms for their **nutrition**. Heterotrophic organisms include **animal**s, **fungi**, and some single-**cell**ed **protozoa** (e.g., ameba, paramecia) and **bacteria**. While **autotroph**s make their own food by converting inorganic **nutrients** into organic forms, heterotrophs cannot do this. Heterotrophs require most

nutrients in an already produced, organic form. They use these nutrients both as a source of **energy** and as building blocks to form cell and body parts. In a food web the heterotrophs are the **consumer**s. There are many different types of heterotrophs in a food web, depending on what they consume. If they ingest autotrophs (**producers**), they are known as **herbivore**s (primary consumers). Some heterotrophs eat other heterotrophs. These are the **carnivore**s (secondary or higher level consumers). Predators, which capture live food, and **scavenger**s, which eat already dead food, are two types of carnivores. **Omnivore**s are heterotrophs that eat both autotrophs and other heterotrophs. Decomposers, which break down organic material into an inorganic form usable by **plants**, are also examples of heterotrophs. One hypothesis about the evolution of life on Earth states that the first living cells were heterotrophs. These primitive organisms absorbed or ingested simple organic **molecule**s for use as energy and building blocks. When **competition** for these organic molecules increased, those organisms that could use alternative sources of energy, such as the sun or inorganic **chemical reaction**s, to make their own organic molecules were better able to survive and reproduce. Thus, according to this hypothesis, autotrophs evolved from heterotrophs.

HETEROZYGOTE SUPERIORITY

A heterozygote is a **diploid** (**nucleus** that contains two sets of **chromosome**s) **organism** with two different **allele**s (alternate form of a **gene**) at the corresponding loci. Sometimes the heterozygote is not phenotypically distinguishable because of the phenomenon of **dominance and recessiveness**. At other times, there is incomplete dominance and a blending of the two alleles occurs, or both alleles continue to manufacture their **protein**s. In certain circumstances, the heterozygous form can confer an advantage onto the organism containing it. The classic example of this is seen with humans and a **blood** disorder called sickle **cell** anemia. In sickle cell anemia, the individual is recessive homozygous and the **hemoglobin** in the blood is not as capable of carrying oxygen around the body as normal hemoglobin. The red blood cells are characteristically sickle shaped. The heterozygous form produces both normal and sickle blood cells. This is known as sickle trait. Sickle cell anemics are poorly adapted to their environment and many die very young. It might be thought that such a deleterious **mutation** would quickly be lost from the **population**. This would be the case were it not for the advantage that the heterozygote confers. In countries with a prevalence of malaria, sickle cell trait provides a slight immunity, thus ensuring the survival of the appropriate allele. This means that in malaria-infected areas the heterozygous form is more advantageous than either of the two homozygous forms. The heterozygous form, in this environment, has a selective advantage. This is heterozygote superiority.

Heterozygote superiority is only conferred in a particular environment, beyond that environment the heterozygote may be disadvantaged. This is the case with sickle cell anemia. In areas without malaria there is no advantage in the heterozygote. Indeed, there is a disadvantage due to the reduced capacity to carry oxygen.

HETEROZYGOUS AND HOMOZYGOUS

The term heterozygous refers to the situation where a specific **gene** is represented by two differing forms or alelles within the **diploid nucleus**. The opposite case within the **cell** is considered homozygous. A homozygote has both copies of the gene present in the same form, or as identical **alleles**. The alleles are present in the same loci, the same physical location on the **chromosomes**.

A heterozygote is capable of producing **gametes** of two different forms—one for each allele present. This is as a result of the homologous chromosomes moving to opposite ends of the dividing cell during **meiosis**. Conversely, the homozygote is only capable of producing the one type of gamete because there is only one type of allele present.

Phenotypically, the heterozygote and homozygote may be indistinguishable from each other. This effect is due to the phenomenon of **dominance and recessiveness**. If one allele is dominant, it will mask the observable effects of the recessive allele. It will only become clear that there are two different alleles present when the offspring are observed and the genetic make-up of the gametes is inferred. The dominant allele present in the homozygous form is phenotypically indistinguishable from the heterozygote. This form is referred to as homozygous dominant. The homozygous form of the recessive allele (homozygous recessive) has the **phenotype** of the recessive allele and it can be easily inferred from observation if it is known that that particular characteristic is recessive. There are cases where dominance is incomplete and the heterozygote shows a blending of the two alleles.

A heterozygote need not be a diploid as described above, polyploid heterozygotes can also exist. The recognizing characteristic of a heterozygote is that it does not breed true, or in other words, it produces two (for a diploid **organism**) or more (for a polyploid individual) forms of gametes. Polyploid homozygotes can also occur and are recognized by their ability to produce only the one type of gamete.

See also Genetics; Heterozygote superiority; Punnett square

HEVESY, GEORG VON (1885-1966)

Swedish radiochemist

Georg von Hevesy developed radioactive tracer analysis, a method widely used in chemistry and medicine. For this accomplishment, which had far-reaching consequences in physiology, biochemistry, and mineralogy, he was awarded the Nobel Prize in chemistry in 1943. Hevesy was also the co-discoverer of the **element** hafnium.

Georg Charles von Hevesy was born in Budapest, Hungary, on August 1, 1885, to Louis Bisicz and his wife, the former Baroness Eugenie Schosberger. The family, who was given a title by Emperor Franz Joseph I in 1895, first changed their name to Hevesy-Bisicz and then simplified it to Hevesy; Hevesy always used the ''von'' in German correspondence and publications. (Many sources refer to him as ''de Hevesy,''

while his first name often appears as ''George'' or ''György.'') Both sides of the family were well-to-do; facing no financial obstacles, Hevesy moved smoothly through the Piarist Gymnasium in Budapest, then studied physics and chemistry at the University of Budapest, and finally earned his doctorate at the University of Freiburg in 1908 with a thesis on the chemical behavior of sodium hydroxide in fused sodium metal.

Hevesy received his degree and became an assistant at the Eidgenössische Technische Hochschule in Zürich, commencing a career in which he knew or worked with nearly every major scientist of the first half of the twentieth century. He was acquainted with Albert Einstein, for example, in Zürich, where he continued his work with fused salts. After two years he moved on to work with Fritz Haber at the Technische Hochschule in Karlsruhe, but he realized that he lacked the research techniques for the electron-emission studies that Haber had set for him. Hevesy suggested that he join Ernest Rutherford's group at the University of Manchester in England, and Haber agreed. He received an honorary research fellowship and left in 1911.

Develops Radioactive Tracer Analysis

At Manchester Hevesy worked with, among others, Neils Bohr, Frederick Soddy, Henry Moseley, and Hans Geiger. His first project was the chemistry of the radioactive decay products of actinium, and the effort yielded the finding that successive alpha decay products differ in chemical valence in steps of two. This provided support for the proposal Soddy had recently put forward concerning the existence of alpha-decay—the ejection of an alpha particle, or helium nucleus, from the radioactive nucleus. It was here, however, that Hevesy began the research which was to occupy him for the rest of his life, and it was Rutherford who set him on the road to it. The Austrian government had given Rutherford a hundred kilos of radioactive lead whose activity was known to be that of ''radium-D,'' a decay product of radium. Hevesy was assigned the task of separating the radium-D from ''all that lead.'' Over many months, he tried every chemical separation he knew, with a uniform lack of success. We now know that radium-D is the radioactive isotope lead–210; it is, in other words, a form of lead with the same number of protons and electrons, and hence the same chemistry, as any other lead, but with a different number of neutrons in its nucleus. Separation of isotopes can be done only by painstaking physical methods, and chemical separation is impossible.

Having failed in the separation study, Hevesy then acted on the principle of the popular saying, ''If life hands you lemons, make lemonade.'' Since he could not separate it, he decided to use radium-D to trace the course of lead in chemical processes. Working with Friedrich Paneth at the Vienna Institute of Radium Research in 1913, he was able to conduct precise solubility studies of lead salts by mixing an insignificant mass of radium-D with a regular lead salt, then determining the amount dissolved not by the usual tedious gravimetric methods, but by simple measurement of the proportion of radioactivity found in solution. He was also able to demonstrate lead exchange between solid and solution, and the migration of lead atoms in the metal. He and Paneth showed that the electrochemical properties of radium-D were identical with those of lead, thereby adding to the growing evidence of the existence of isotopes.

In 1913 Hevesy returned to Hungary, where he served for a time as a lecturer at the University of Budapest before joining the Austro-Hungarian army during World War I. His post during the war was as technical supervisor at the state electrochemical copper plant, and he was able to continue his research on a limited basis. After the war he became a full professor at Budapest, continuing his lead tracer work, but the political situation in Hungary was disintegrating rapidly, and in 1920 he accepted an invitation to join Bohr's Institute for Theoretical Physics at the University of Copenhagen.

Participates in the Discovery of Hafnium

His first project there, carried out with Johannes Bro, was an attempt at isotopic separation by fractional distillation. They had limited success with metallic mercury, but obtained fairly pure isotopic samples of chlorine, whose two stable isotopes differ by about six percent in mass. Hevesy wished to learn X-ray spectroscopy, and in 1923 he turned for help to physicist Dirk Coster. The two of them set about finding element seventy-two, which Bohr's recent revision of the periodic table suggested should be a transition metal corresponding to zirconium. They found the anticipated spectral lines in extracts of zirconium ores, and were able to isolate and characterize the new element as its fluoride. They named it "hafnium" after the Latin name for Copenhagen.

In 1923 Hevesy also returned to his work with radioactive lead tracers, and for the first time he ventured into biology to study the uptake of lead in bean seedlings. This work was published in 1924, the year in which Hevesy married Pia Riis, who would bear him three daughters and a son. Two years later, he moved his new family to the University of Freiburg, where he developed X-ray fluorescence as an analytical tool, while expanding the university's X-ray spectroscopy program. As an administrator, however, Hevesy came into increasing contact with the new Nazi regime, and this caused him to return to Copenhagen in 1934.

Heavy water (water in which some of the hydrogen atoms are the heavier isotope hydrogen–2, or deuterium) had just become available, and in Copenhagen Hevesy was pleased to have this first non-toxic isotope available to study animal and human physiology. He and his colleagues quickly demonstrated the rapid exchange of internal and external **water** in goldfish, and measured the average turnover time of a water molecule in the human body (about thirteen days) and the approximate number of water **molecules** in the body (10^{27}).

In 1934, Irène and Frédéric Joliot-Curie succeeded in producing artificial radioisotopes by alpha particle bombardment. Hevesy seized the possibilities of this development by making radioactive phosphorus–32 from sulfur–32, a very large advance for the study of physiology. Here was an element central to all animal physiology, and a means of following its intake, circulation, exchange, and excretion. A number of discoveries followed from the use of this tracer, including

the dynamic exchange of serum and bone phosphate and the synthesis and distribution of **DNA** and **RNA**. Today, these discoveries form the foundation of our understanding of body chemistry.

Phosphorus was only the first element that Hevesy used or introduced into use as a radioactive tracer. Others included calcium–45, potassium–42, sodium–24, chlorine–38, and carbon–14. There is little of physiological importance that lies outside this list except nitrogen, oxygen, and sulfur, and it is clear just how fundamental his contributions to science have been. It was in recognition of these accomplishments that Hevesy was awarded the Nobel Prize in chemistry in 1943. Announced in 1944 and overshadowed by the closing battles of World War II, the award received little public notice.

At the time he received the Nobel Prize, Hevesy had moved again. He had left Copenhagen in 1943, moving away from the Nazis for a second time, and settled in Stockholm. He worked at the Institute for Organic Chemistry there for the remainder of his life, becoming a Swedish citizen in 1945. Much of his later research focused on physiology and medicine, particularly the study of **cancer**. He published over four hundred books and papers in the course of his career and received many awards and honors, including honorary doctorates from nearly a dozen universities and honorary memberships in many scientific societies. He was also a foreign member of the Royal Society. In addition to the Nobel Prize, he received the Cannizzaro Prize in 1929 from the Academy of Sciences in Rome, the Copley Medal of the Royal Society in 1950, and the Faraday Medal in 1959. He also received the Atoms for Peace Award in 1959. Hevesy died at a clinic in Freiburg on July 5, 1966, after a long illness.

HEYMANS, CORNEILLE JEAN-FRANÇOIS (1892-1968)
Belgian physiologist and pharmacologist

Born in Ghent, Belgium on March 28, 1892, Corneille Jean François Heymans was the eldest of six sons of Jan-Frans Heymans, a noted pharmacologist who founded the J. F. Heymans Institute of Pharmacology and Therapeutics at the University of Ghent. He and his father were to become a scientific team of considerable reputation—one of the few father-son scientific teams in history.

Heymans' career was delayed by four years of service as a field artillery officer in the Belgian Army during World War I. His performance won him the Belgian War Cross and the Order of the Crown of Leopold, among other decorations for valor.

After the war, Heymans received his medical degree from the University of Ghent in 1920. His father was his principal teacher and later would become his primary co-researcher in the experiments that ultimately led to the Nobel Prize in 1938. Had his father not died in 1932, he most likely would have shared the award with his son.

The year following his graduation from the university, Heymans married Berthe May, an ophthalmologist. The young couple studied abroad for several years, permitting Heymans to establish valuable contacts with some of the leading scientists of the day in his field, among them Eugène Gley at the Collège de France, Maurice Arthus at the University of Lausanne in Switzerland, Ernest H. Starling at University College in London, and Carl Wiggers at Western Reserve University's medical school in Cleveland, Ohio.

Heymans returned to the University of Ghent in 1922 to become a lecturer in pharmacodynamics, the study of the action of drugs in the body. He succeeded his father as professor of pharmacology and director of the Institute in 1930, but father and son continued to collaborate on many projects, including respiratory experiments that revealed previously unknown facts about how breathing is regulated in human beings and animals.

At that time, it had been well known for half a century that changes in **blood pressure** were associated with changes in the rate and the depth of breathing. The mechanism enforcing these changes in respiration was not known. It was believed, however, that alterations of breathing rates were the result of the direct action of blood pressure on the **brain**'s respiratory center, the medulla. It was assumed that the medulla was able to detect changes in the **blood** circulating through it and regulate the rate of breathing accordingly.

Another scientist, Heinrich E. Hering, however, had noted a reflex action in the carotid artery (two major arteries on each side of the neck) that appeared to influence the heart beat. Through a series of experiments originally intended to refute Hering's contention, Heymans instead demonstrated that the reflex in the artery also exerted control over breathing.

The effort to determine this fact involved what became known as the "isolated head" technique. The head of an anesthetized dog, attached to its body only by the vagus aortic nerves, was kept alive by the shared circulation of blood of a second anesthetized dog. The Heymans found that when they induced hypertension (increased blood pressure) in the isolated body of the first dog its medullary respiratory center was stimulated or inhibited appropriately. But when the aortic nerves were severed, all respiratory response to changes in the blood pressure ceased. This experiment enabled the Heymans team to demonstrate conclusively that the aortic nerves were the reflex mechanism's sole sensory pathway.

The experiment thus disproved the classical theory of the blood's direct action on the brain and provided the evidence for an alternative explanation. The Heymans later determined the sites at which changes in the blood were detected. They discovered that the reflex in the carotid artery contains pressure-sensitive areas, or presso-receptors, that can detect even slight changes in blood pressure. They also found small structures on the inside walls of the carotid artery and the aorta. These chemoreceptors responded to changes in the chemical composition of the blood. By making clear why certain drugs affected respiration and circulation, Heymans' discovery opened the way for improvements in the treatment of many diseases.

Heymans' colleagues appreciated the thoroughness and accuracy of his work, which he documented in over eight hundred articles and papers published during his career. Heymans also won great recognition as a gifted teacher.

Many scientific honors came to him. In addition to the Nobel Prize, he was awarded the Alvarenga Prize of the (Belgian) Académie Royale de Medécine, the Prix Quinquennal de Medécine of the Belgian government, the Pius XI Prize of the Pontificia Academia Scientiarum and the Monthyon Prize of the Institut de France. Heymans held sixteen honorary degrees and belonged to more than forty scientific and medical societies.

Throughout his career, he traveled widely both as a lecturer and a tourist. He lectured at several major American universities, including Harvard and the University of Chicago. He was fluent in many languages and conducted seminars in Montevideo, Chile, to help organize scientific exchange programs between that country and his own. He visited India on behalf of the World Health organization. During World War II, he helped organize relief efforts to provide food for Belgian children. In so doing, he made several trips to Berlin to obtain the cooperation of German officials in getting Red Cross food shipments into Belgium.

Heymans and his wife had four children: Marie-Henriette, Pierre, Joan and Berthe. In 1963, upon his retirement from the Heymans Institute, he was designated professor emeritus. He continued to visit the institute several times a week until his death following a stroke in Knokke, Belgium on July 18, 1968.

HIBERNATION

Hibernation, or winter dormancy, is the act of passing the cold season in a dormant physiological state of low **metabolism**. Hibernation is a common trait in many **animal**s living in **ecosystem**s having a prolonged cold period during the winter.

Many **invertebrates** that hibernate must be capable of withstanding extreme cold conditions. To prevent damage to their **cells** and **tissue**s caused by needle-like ice crystals, they may develop extremely high concentrations of dissolved substances (such as specialized **protein**s or sugars) in their plasma, which prevents freezing unless super-cold **temperature**s occur. However, many terrestrial invertebrates are capable of being frozen solid for months during hibernation, without apparent ill-effect. Two examples are the goldenrod gall fly (*Eurosta solidaginis*) and the yellowjacket wasp (*Vespula maculata*). A few **species** of vertebrate animals can also survive the freezing of much of their body **water**, including the wood frog (*Rana sylvatica*), chorus frog (*Pseudacris triseriata*), and hatchlings of the painted turtle (*Chrysemys picta marginata*).

Some species of **mammals** spend the winter in a deep sleep, which conserves **energy** associated with movement and other activities. Such species include the black bear (*Ursus americanus*) and raccoon (*Procyon lotor*). This condition is considered to represent an extreme lethargy, rather than a true hibernation, which would be accompanied by significantly reduced body temperature.

Many species of hibernating mammals lower their body temperature greatly during hibernation, achieving a condition known as hypothermia. Examples include the woodchuck

Archibald Vivian Hill. *(The Library of Congress. Reproduced by permission.)*

(*Marmota monax*), the golden-mantled ground squirrel (*Citellus lateralis*), and the little brown bat (*Myotis lucifugis*). Another is the poor-will (*Phalaenoptilus nuttallii*), the only bird known to hibernate. Hypothermia is an adaptive trait associated with hibernation, because of the large energy savings achieved by not keeping the body at the warm temperature required for normal activities during the active season (typically, the energy expenditure is less than about 2-6% of normal). The characteristics of the hypothermic condition during hibernation include a low body temperature, oxygen consumption as low as 1-5% of that occurring during normal basal metabolism, extremely slow breathing rate, a state of dormancy that is much more profound than deep sleep, and a reduced **heart** rate. The ability to arouse upon stimulation is usually maintained, although this response is slow and groggy, and requires the re-activation of the normal metabolic rate and functions. Once suitable environmental conditions occur in the springtime, the hibernating mammal increases its metabolic rate, and can resume its normal activities.

HILL, ARCHIBALD VIVIAN (1886-1977)
English physiologist

Hill was born in Bristol, England, on September 26, 1886. His father, a timber merchant, abandoned the family when Hill was

three, leaving his mother to educate the boy and his younger sister. After completing his primary education, Hill earned a scholarship to Trinity College, Cambridge, which he entered in 1905. At Trinity, Hill majored in mathematics and completed the usual three-year course in two years. In the process, however, he found that he was more interested in physiology than in mathematics.

After graduating in 1909 with a degree in natural sciences, Hill began research on frog muscle at the Cambridge Physiological Laboratory. At the time, muscle research was proceeding in a number of directions. Walter Fletcher (1873-1934) and Frederick Gowland Hopkins, for example, had earlier studied the chemical changes that occur in muscles during contraction, discovering the role of **lactic acid** in that process.

Hill, however, decided to forego the study of the chemical process to concentrate instead on the heat changes that occur during muscular activity. Scientists had already found that a small amount of heat is produced during muscular contraction. Hill's goal was to analyze that process in greater detail. The task was a challenging one. Only very small amounts of heat are produced during muscular activity and for only very short periods of time. To deal with these problems, Hill used a thermocouple to measure heat changes. A thermocouple is a very sensitive kind of thermometer that converts heat changes into electrical currents that are more easily read and recorded.

With the thermocouple, Hill was able to find that heat is formed twice during muscular activity, once during the contraction itself and once following the contraction. In the former case, heat is evolved rapidly, and in the second case, more slowly, but often in larger quantities. Hill also demonstrated that oxygen is consumed during the second phase rather than during the contraction. His techniques were so precise in this research that he was able to detect temperature changes as small as 0.003° C in a few hundredths of a second. For his accomplishments, Hill was awarded a share of the 1922 Nobel Prize for physiology or medicine.

After serving with distinction in World War I, Hill did research on anti-aircraft artillery at King's College, Cambridge, work for which he was knighted in 1918. He became Professor of Physiology at Manchester University in 1920 and then moved to University College, London, in 1923. From 1926 to 1951, he was professor at the Royal Society, serving also as Secretary of the organization from 1935 to 1946. He continued his research in physiology after his retirement in 1952.

Hill married Margaret Neville Keynes in 1913. They had two sons and two daughters. Hill died in Cambridge on June 3, 1977.

HIPPOCRATES OF COS (460 B.C.-377 B.C.)
Greek physician

Hippocrates was a famous Greek physician who is widely known as the "Father of Medicine." Although little is known about his life, a few facts are considered accurate.

Hippocrates was born on the Greek island of Cos around 460 B.C. to a family of physicians. Cos was the site of one of the great medical schools of ancient Greece, and Hippocrates taught there for many years. He also traveled widely, lecturing in Greece and probably throughout the ancient Mideast. He was well known in his lifetime and died around 377 B.C. in Larissa.

Hippocrates is considered the father of medicine because, through his school, he separated medical knowledge and practice from myth and superstition basing them instead on fact, observation, and clinical experience. Our knowledge of Hippocrates' methods and teachings comes from the *Corpus Hippocraticum*, the Hippocratic Collection. This is a series of about 60 books that seem to have been collected in the great Library of Alexandria after about 200 B.C. While Hippocrates may have written only a few and possibly none of these books, they are considered to be an expression of his medical teachings and philosophy, which became an important basis of Western medicine.

The Hippocratic approach to medicine, expressed in the books, emphasized that disease arose from natural causes, not from whims of the gods. Hippocrates insisted on careful observation of medical conditions; the books contain dozens of detailed clinical descriptions of diseases. He recommended as little interference as possible with the body's own ability to heal. Treatment focused on diet, rest, and cleanliness. He advanced the doctrine of the four humors, whereby disease was supposed to result from an imbalance among the body's four important fluids.

Hippocrates also emphasized a high ethical standard for physicians. The Hippocratic Oath is a statement of medical ethics. Developed over 2,000 years ago, it probably reflects the views of Hippocrates while not actually having been written by him. The oath pledges a physician to serve only the benefit of the patient, and to keep confidential anything he or she sees or hears in the course of treatment. Many medical students today still take a form of the Hippocratic Oath when they receive their medical degrees.

HISTOLOGY

Histology, also known as microscopic **anatomy**, deals with the study of **cell**s as they are constituted in the various **tissue** types. Single cells, isolated from the fabric of tissue, are generally difficult to classify. However, when cells are examined in their normal configuration and relationship to neighboring cells, the tissues formed by those cells are readily identified. A **microscope** is required because cells are microscopic and so too are the identifying characteristics of tissues.

While the body is structured of a bewildering array of **organ**s, the number of tissue types is limited. Tissue types include **epithelium**, connective tissue, muscle, nervous tissue, **blood** and reproductive tissue. These categories include more than one type of tissue. For instance, connective tissue encompasses bone, cartilage, tendons, ligament, and fibrous connective tissue. Another example: the catagory muscle includes skeletal, cardiac and smooth.

Hippocrates of Cos. *(The Library of Congress. Reproduced by permission.)*

The term microscopic anatomy is descriptive in that it indicates that tissue structure must be examined with a compound microscope. Because **light** does not penetrate thick masses of tissues and because tissue detail is revealed only by staining, it is necessary to process tissue in a somewhat elaborate manner for proper evaluation. A small fragment of tissue is cut and fixed (preserved). The fixative is chosen to maximize the retention of tissue structure. Dehydration follows fixation. This step removes **water** so that the tissue may be infiltrated with paraffin or a combination of paraffin and other **waxes** in preparation for sectioning. Frequently an intermediate step is required between dehydration and paraffin to enhance proper infiltration. The tissue in its paraffin block is now mounted on a microtome which is similar to a meat slicer but it can make slices of tissue routinely at 0.0001-0.0004 in thick. The slices of tissue are then mounted on a glass slide, the paraffin is removed, the tissue is stained, water is removed and a permanent mounting medium of the refractive index of glass is applied. Finally, a thin cover slip of glass is placed over the biopsy section. The microscopic slide is now ready for examination.

Histopathology uses knowledge of histology to evaluate tissues that may be malignant (**cancer**ous) or otherwise abnormal. A lump is biopsied by a physician. A trained pathologist, a medical doctor who specializes in diagnosis of disease, will examined the biopsy. The tissue may be either normal or it may display varying degrees of pathological change. The histological diagnosis by the pathologist determines if further treatment is required. Physicians, because of time constraints, have devised short cuts that permit rapid examination of tissues. However, there is no short cut for microscopic examination of the biopsied tissue. Despite the explosion of biochemistry and **molecular biology**, pathologists still rely on microscopic examination of suspected malignant tissue.

HITCHINGS, GEORGE H. (1905-)
American pharmacologist

George H. Hitchings is among the most prolific of modern pharmaceutical scientists. He worked at Burroughs Wellcome Company, a British pharmaceutical company with research facilities in the United States, for more than thirty years before his retirement in 1975. Hitchings produced many important pharmaceuticals for treating diseases such as **cancer**, gout, and malaria, and for preventing rejection of transplanted organs. His contributions were based on the premise that an understanding of what makes diseased **cells** different from normal cells makes it possible to exploit those differences to destroy cancer cells or foreign invaders such as **bacteria** or **virus**es with drugs. For his work in finding treatments for serious diseases, Hitchings and his long-time Burroughs Wellcome collaborator Gertrude Elion shared the 1988 Nobel Prize in physiology or medicine with British pharmaceutical scientist Sir James Black. It was the first time since 1957 that pharmaceutical scientists had been awarded the prize.

George Herbert Hitchings was born to George Herbert Hitchings, Sr., a naval architect, and Lillian H. Belle Hitchings on April 18, 1905, in Hoquiam, Washington, on the Olympic Peninsula. His father's death when he was twelve and his admiration for Louis Pasteur, a preeminent scientist-philanthropist who became his role model, aimed Hitchings toward a career in medicine. As the salutatorian of his high school class, Hitchings gave an address to the graduating class on the germ theory and Pasteur's life.

Hitchings attended the University of Washington, where he received a bachelor's degree in chemistry in 1927 and a master's degree in chemistry in 1928. He also showed a fondness for many scholarly subjects, studying the arts and history in college. He began his career in scientific research at an early age. "The Chemistry of the Waters of Argyle Lagoon," the first of his more than three hundred scientific publications, appeared in the publications of the Puget Sound Biological Station in 1928, when he had just entered graduate school. He continued his graduate work in biological chemistry at Harvard College, where he received his Ph.D. in 1933. Hitchings' doctoral dissertation concerned the **metabolism** of **nucleic acid**s, the chemicals that make up **DNA**, the carrier of genetic information. Hitchings did his work on nucleic acids before **James Watson** and **Francis Crick** discovered the structure of DNA, and at that time no one was interested in nucleic acids. Hitchings couldn't find a job. Finally, after working for nine years as a teaching fellow at Harvard (1933–39) and Western Reserve University (1939–42), he was hired by Burroughs Wellcome in 1942 and resumed his work on nucleic acids. He became vice president of research in 1967 and held the position until 1975, when he became scientist emeritus.

Pioneers Rational Drug Design

Until Hitchings and the pharmacologist **Gertrude Elion** came along, drug researchers sought new drugs by modifying natural products. The two pioneered a method that has come to be known as "rational" drug design. They reasoned that if they understood the differences between normal and diseased or infected cells, these differences could serve as a entry point to selectively kill diseased tissue without harming surrounding normal tissue. They implemented these ideas by investigating the chemical pathways of nucleic acid synthesis, which is crucial to cell metabolism. Hitchings synthesized chemicals similar in structure to natural nucleic acids, the purines and pyrimidines. These related compounds interfered with **DNA synthesis**. Because cancer cells divide quickly, the compounds are particularly disruptive to them, killing them as they try to divide. This form of chemotherapy is just one instance of the rational drug design that helped Hitchings accumulate eighty-five patents over his thirty-year career.

One compound in particular, 6-mercaptopurine (6MP), a purine analog synthesized in 1951, proved to be particularly effective. Working with scientists at Sloan-Kettering Institute, Hitchings and Elion perfected the drug, which was used to combat childhood leukemia. 6MP and thioguanine, also produced by Hitchings and Elion, are still used to treat acute leukemias.

In 1959 Hitchings discovered that 6MP inhibited production of antibodies in rabbits. A less toxic form called azathioprine, marketed under the trade name Imuran, was developed in 1957 to control rejection of transplanted organs and treat autoimmune diseases. In the nearly nine thousand kidney transplants performed each year, Imuran remains the drug most commonly used to prevent organ rejection. 6MP is broken down in the body by xanthine oxidase, the same enzyme that converts purines into uric acid, the cause of the painful joint disease gout. Further investigation of purine analogs led to the development of allopurinol in the 1960s. It blocks uric acid production by competing for xanthine oxidase, an enzyme that converts purines to uric acid. Hitchings was also active in the development of other drugs, including pyrimethamine, which is used to treat malaria, and trimethoprim, which is used to treat urinary tract infections and other bacterial infections.

Philanthropy has always been a part of Hitchings' life, and he has said that when he was baptized his father dedicated his life to the service of mankind. He served as president of the Burroughs Wellcome fund, a charitable organization, from 1971 to 1990, and continues to serve as its director. In addition, he has served as director of a dozen local chapters of philanthropic organizations.

Hitchings married Beverly Reimer in 1933. The couple had two children, Laramie Ruth and Thomas Eldridge. Beverly died in 1985. In 1989, Hitchings was remarried to Joyce Shaver.

Besides the Nobel Prize, Hitchings has received numerous awards, including the Gregor Mendel Medal from the Czechoslovakian Academy of Science in 1968 and the Albert Schweitzer International Prize for Medicine in 1989. He has been awarded eleven honorary degrees and has been a member of the National Academy of Sciences in 1977. In addition, he has traveled widely, lecturing in Africa, Asia, Europe and South America.

HIV • See Human immunodeficiency virus (HIV)

Ho, David Da-I (1952-)
Chinese-born American medical scientist

David Ho, one of the world's foremost researchers of **acquired immune deficiency syndrome (AIDS)**, made his greatest contribution to science when he demonstrated the power of combinations of new state-of-the art AIDS drugs. Ho's research showed that by administering these drugs to patients within weeks after they caught the **human immunodeficiency virus (HIV)**, the virus could be halted in its tracks. Ho's work is considered a major breakthrough in the battle against AIDS, offering increased hope that a cure and vaccine will eventually be found.

Ho, the first chief executive officer and scientific director of the Aaron Diamond AIDS Research Center, the largest AIDS research center in the world, was also the fourth scientist to identify the virus that causes AIDS. He was named *Time* magazine's Man of the Year in 1996. Ho attributes his success in AIDS research to his tenacity, instilled in him by his Chinese immigrant family. ''You always retain a bit of an underdog mentality,'' he explained in *Time* magazine. ''People get to this new world, and they want to carve out their place in it. The result is dedication and a higher level of work ethic.''

Ho was born in Taiching, Taiwan, the oldest child of Paul and Sonia Ho. At birth, he was given the name Da-I, meaning ''Great One,'' a name that conveyed the family's expectations for their oldest son. In Taiching, the family lived in a small four-room house with a ditch serving as outdoor toilet. To create a better life for his family, Paul Ho left Taiwan for America, sacrificing time spent with his wife and growing son. The family, now consisting of Sonia, Paul, Da-I, and his younger brother, reunited nine years later in Los Angeles. Paul Ho then began studying for his engineering degree at the University of Southern California. A devout Christian, he chose new American names from the Bible for his sons—David and Philip. A third son, Sidney, was born after the family's move to central Los Angeles, where they settled in a predominately African-American neighborhood.

As a child, Ho was inspired to excel academically by the example of his father and other relatives, all scientists and engineers. By the time he had reached age 10, he had already developed a keen interest in science, conducting amateur scientific experiments in the family garage. At first, though, David's new American schoolmates called him ''stupid'' because he was practically mute; he couldn't communicate because he spoke only Chinese. This experience, however, made him determined to excel, and in later years he would become upset if he didn't make As in every subject. Six months after

he started school, Ho began speaking English well enough to communicate with schoolmates after being enrolled in an English as second language program. He and his brothers also picked up English words by watching television comedies such as *The Three Stooges*.

After high school, Ho studied physics at the Massachusetts Institute of Technology and then the California Institute of Technology, where he graduated summa cum laude with a B.S. degree. Deciding that gene splicing was the most exciting area of research, he then attended medical school at Harvard, earning an M.D. degree in 1978. He started his residency in internal medicine in 1981 at Cedars-Sinai Medical Center, later serving for one year as chief resident.

Seeks Virus that Causes AIDS

While at Cedars-Sinai, Ho began seeing increasing numbers of homosexual men with a mysterious illness that targeted the body's defense mechanism—the **immune system**. The men began dying from infections that do not normally take hold in people with healthy immune systems, such as *Pneumocystis pneumonia*, a disease of the lungs, and toxoplasmosis, a disease that attacks the brain. At the time, doctors attributed the cause to everything from recreational drugs to allergic reactions to too many sex partners. Ho, like some other researchers, suspected a **virus**—and decided to specialize in research on the disease, which would later become known as AIDS.

In 1982, he started work at Massachusetts General Hospital in the infectious disease lab of Martin Hirsch, a virologist with a sharp interest in the mysterious disease. Ho now calls Hirsch his mentor. Ho wanted to be the first to isolate the virus, but that honor fell to Luc Motagnier of the Pasteur Institute in Paris, France. Ho was the fourth scientist to identify the virus, with Robert C. Gallo, a scientist at the National Cancer Institute and Jay Levy, who worked at the University of California Medical Center in San Francisco, being respectively second and third.

Yet Ho made important contributions in Hirsch's lab. ''While working in Hirsch's lab, Ho became expert at detecting HIV in places where few were able to find it,'' *Time* said in its 1996 article on the scientist. ''He was the first to show that it grows in long-lived immune cells called macrophages and among the first to isolate it in the nervous system and semen. Just as important, he showed that there isn't enough active virus in saliva for kissing to transmit the disease.'' In 1986, Ho moved back to Los Angeles to work in the infectious disease department of Cedars-Sinai Medical Center. He also became an assistant professor at the University of California, Los Angeles, School of Medicine. In 1989 he was approved for the post of associate professor.

By the time Ho joined Cedars-Sinai, scientists knew that HIV decimates the immune system by targeting **T cells**—cells that help keep disease from taking hold. They also had determined that HIV slips into T cells by attaching and unlocking CD4 receptor proteins, which are located on the surfaces of these cells. Some scientists then suggested that massive infusions of CD4 particles into the bodies of HIV patients might fool the virus, preventing it from latching on to the true receptors.

Ho and Robert Schooley of the University of Colorado Medical Center in Denver tried out this theory by testing soluble CD4 in two dozen AIDS patients. Unfortunately CD4 proved ineffective against the virus because some HIV viruses, termed "wild viruses," could tell which CD4 molecules were decoys and avoid them. But by embarking on this experiment, Ho and Schooley discovered that there were tens of thousands of infectious virus particles in their patients' bodies, much more than had ever been thought to exist in HIV patients.

Ho then decided to return to researching HIV in the earliest stages of infection to investigate the virus's mysteries. He and his team at UCLA found four homosexual men suffering from flu-like symptoms similar to an early HIV infection, and turned to a newly devised genetic tool—the PCR test—to find out how much virus was in his patients' **blood**. To many scientists' surprise, he discovered that there were millions of viral particles coursing through the patients' blood, even when patients showed no symptoms. And there were as many virus particles in these men with early stage disease as there were in those in the latest stages. His studies, like those conducted by Robert W. Coombs at the University of Seattle, demonstrated that the virus was always actively reproducing even in early stages, and never went into hiding as previously believed. His work also disproved earlier theories by revealing that HIV did direct damage to the immune system, rather than destroying it through an autoimmune reaction—a reaction to an invader substance in which the immune system attacks itself.

In 1990, Ho received a chance to embark on even more challenging research when philanthropist Irene Diamond named him the chief executive officer and scientific director of the new Aaron Diamond AIDS Research Center. "I said 'I don't want a star. I want a wonderful scientist,'" Diamond recalled telling people who criticized her then-unknown choice, as relayed in *Time* magazine. At the same time, Ho also became professor of medicine and microbiology at the New York University School of Medicine and co-director of the Center for AIDS Research.

Uses New AIDS Drugs to Battle the Illness

With the Diamond Research Center's funds, Ho was able to attract more than 50 world-class researchers to the center. As he began his research at the new center, Ho began to theorize that the virus load in AIDS patients might be much bigger than was apparent because the immune system could be killing off some of the viral particles. Eventually, he reasoned, the immune system becomes exhausted, letting the AIDS virus overtake the immune system's ability to combat it. Ho sought to determine whether the virus could be contained, and if so, whether the immune system could recover enough to restore health to HIV patients.

In 1994, drug researchers developed protease inhibitors and demonstrated for the first time that drugs could actually halt HIV's replication. Ho then embarked on a study with HIV-infected patients which indeed revealed that the immune system created billions of white blood cells as HIV replicated—evidence that the immune system works hard to combat the virus.

Then in a ground-breaking study in 1995, Ho shocked the AIDS research world by showing that a combination of three drugs, including protease inhibitors, halted the AIDS virus, allowing the body's immune system to reduce the number of viral particles in patients to undetectable levels—in patients in both early and late stages of the disease. By using several drugs in addition to the protease inhibitors, the scientists could drastically reduce the chances of HIV virus particles' mutating to survive and become resistant to the therapy.

The news astounded the research community, though scientists were careful to note that the therapy was not a cure for AIDS, and more research needed to be done to test the long-term effectiveness of the new drug combinations, called "drug cocktails." Scientists make two important cautions regarding the advances gained by Ho and his colleagues. First, even though HIV is apparently eliminated from the blood of patients who have been treated shortly after having contracted the virus, further experimentation is needed to determine whether it has been eliminated permanently from such body organs as the brain, lymph nodes, and testes. Second, the drugs cannot reverse the damage done to the bodies—particularly the immune systems—of patients who have suffered with AIDS for years. Nonetheless, Ho's work has provided both a scientific basis of hope for a cure and a direction for future research. "We basically have, for the first time, staggered the virus, and the new optimism comes from the fact that we now realize maybe, just maybe, the virus is not as invincible as we previously thought," Ho said in a 1997 CNN television interview.

Ho—who is married to Susan Kuo, an artist, and has three children, Kathryn, Jaclyn, and Jonathan—has found time during his research career to author or co-author 110 scientific articles. He has been awarded many scientific prizes, including the Scientific Award of the Chinese-American Medical Society, the Ernst Jung Prize in Medicine, and the New York City Mayor's Award for Excellence in science and technology. He and his family reside in a suburb of New York City.

See also AIDS therapies and vaccines

HOAGLAND, MAHLON BUSH (1921-)
American biochemist

Mahlon Hoagland is best known for discovering, in the 1950s, that **ribonucleic acid (RNA)** molecules in the cytoplasm retrieve specific **amino acids** and take them to the **ribosomes** for assembly into **proteins**.

Hoagland was born in Boston, the son of Hudson Hoagland (1899-1983), the American physiologist who co-founded the Worcester foundation for Experimental Biology with Gregory Pincus. Mahlon Hoagland received his M.D. in 1948 from Harvard Medical School, and served on the bacteriology department faculty from 1953 through 1967.

In the 1950s scientists had already determined that messenger RNA carried instructions for protein production from *codons*, triplets of base pairs on **deoxyribonucleic acid (DNA)**, from a **cell**'s nucleus into the cytoplasm. There, RNA-rich ribosomes somehow used the information to assemble **amino acids** into proteins. American biochemist **Paul Berg** had already determined that the amino acids were activated by combining with **adenosine triphosphate (ATP)**, but it was not known how the amino acids recognized the *anticodons*, or the complimentary RNA triplets that carried their code.

Hoagland and his associates accidentally discovered that the amino acids first attached themselves not to the RNA in the ribosomes, but to specific locations on small molecules of what had been called *soluble RNA*. Hoagland theorized that these molecules, whose name soon became transfer RNA, were complementary to the ribosomal RNA, where the amino acids were then attached. Although he was unaware of it at the time, Hoagland's discovery fit a theory of **Francis Crick**, the co-discoverer of the DNA **double helix**, that the amino acids attached to an adaptor RNA molecule that is complementary to the ribosomal RNA. Both theories were proven to be correct.

Transfer RNA was also discovered independently by Paul Berg and by Robert Holley.

Hoagland's other work has included determining the **cancer**-causing properties of beryllium, biosynthesizing coenzyme A (which is required for cell metabolism), and studies of liver regeneration, growth control, and amino acid activation of protein synthesis.

In 1967 he joined the biochemistry faculty at Dartmouth Medical School, and also became president of the Worcester Foundation for Experimental Biology. He is a member of the U.S. National Academy of Sciences.

HODGKIN, ALAN LLOYD (1914-)
English biophysicist

Alan Lloyd Hodgkin is best known for his work in defining the electrical and chemical characteristics of nerve impulses. Along with **Andrew F. Huxley** he performed experiments on the nerve fibers of squid and described the **nerve impulses** with a series of mathematical equations. For their research in this area, which resulted in the ionic theory of nerve impulses, the two men shared the 1963 Nobel Prize in physiology or medicine with **John C. Eccles.**

Hodgkin was born on February 5, 1914, in Banbury, Oxfordshire, England, to George L. and Mary Wilson Hodgkin. Hodgkin's father died in Baghdad during World War I, only a few years after his birth. Hodgkin was educated at the Downs School in Malvern and the Gresham School in Holt. In 1932, he entered Trinity College, Cambridge, where he first became interested in physiology. Hodgkin became a fellow at Trinity in 1936, serving as lecturer and later as assistant director of research at the physiological laboratory.

Hodgkin began studying the electrical properties of the nerve fibers in the shore crab while at Cambridge. He spent a year at the Rockefeller Institute in New York City between 1937 and 1938, and while there he met scientists who had developed new methods for studying nerve fibers. Hodgkin brought these ideas back to Cambridge, where with Andrew Huxley he devised an experiment to test an hypothesis about nerve impulses first proposed by German physiologist Julius Bernstein.

Establishes Relationship Between Resting and Acting Potentials

Bernstein had hypothesized that nerve **cells** possess a resting or unstimulated potential and an action or stimulated potential. During the resting potential, he believed, the nerve cell **membrane** had an unequal distribution of positively and negatively charged ions, with more negative ones on the inside. During resting potential, the membrane was permeable to the positively charged ions, but the negatively charged ions could not permeate the cell membrane. When the cell was stimulated, Bernstein argued, the membrane "gates" were temporarily opened, allowing ions to pass in both directions. By using the nerve cells of the shore crab, Hodgkin was able to establish that the resting potential was due to an outward movement of potassium ions; during the action potential the cell membrane's gates allowed in the more concentrated sodium ions. He also discovered that the action potential was usually much larger than the resting potential.

Some of the researchers Hodgkin had met in the United States were working with squid, whose nerve fibers are larger than those of most organisms. Hodgkin and Huxley were able to develop a method to study these fibers using microelectrodes, and they were able to confirm the results of their earlier experiment. Their progress, however, came to a halt during World War II, when Hodgkin worked on radar systems for aircraft for the Air Ministry. Hodgkin and Huxley were back in Cambridge in 1945, and they formed a small research group to pursue their pre-war investigations into nerve fibers.

In 1951, Hodgkin and his colleagues published the results of their research. They found that the membrane is permeable only to specific ions during the resting potential, because of the differing concentrations of potassium and sodium. The concentration of the positively charged sodium ions is greater on the outside of the membrane and the concentration of negative potassium ions higher on the inside during resting potential. During the action potential, the negative and positive ions travel through the membrane, so that the interior charge becomes positive and the exterior negative. This is followed by an equilibrium charge, then a return to the resting potential charge state. All this happens in milliseconds.

The work done by Hodgkin and Huxley which was most responsible for bringing them to the attention of the Nobel Prize committee was the development of a series of mathematical formulae they published in 1952. The purpose of these equations was to synthesize the experimental information then available about the electrical and chemical nature of nerve transmissions. Their goal was to analyze and predict each stage in the passage of the nerve cell membrane from resting to action potential. They were awarded the 1963 Nobel Prize in physiology or medicine, which they shared with John C. Eccles, an Australian who advanced the British team's findings by showing what happens to nerve impulses transmitted across the **synapses**, or intersections, between nerve cells.

Hodgkin was appointed Foulerton Research Professor of the Royal Society in 1952, and was awarded the Royal Medal in 1958. He was John Humphrey Plummer Professor of Biophysics at Cambridge from 1970 to 1981, president of the Marine Biological Association from 1966 to 1976, and a master of Trinity College.

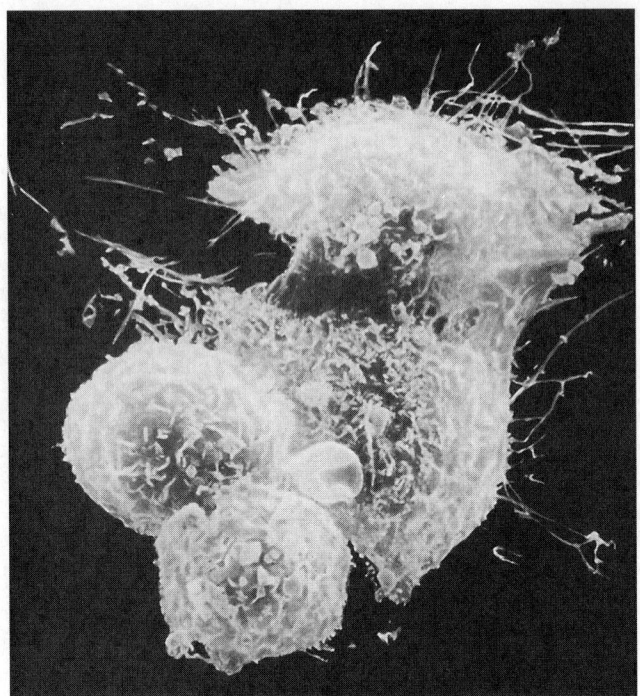

A scanning electron micrograph (SEM) image of dividing Hodgkin's cells from the pleural effusions (abnormal accumulations of fluid in the lungs) of a 55-year-old male. *(Photo Researchers, Inc. Reproduced by permission).*

Hodgkin has been married since 1944 to Marion Rous, the daughter of American Nobel Laureate Peyton Rous. The couple met during Hodgkin's year at the Rockefeller Institute in New York. They have four children.

HODGKIN'S DISEASE

Hodgkin's disease is a **cancer**ous enlargement of the lymph nodes, spleen, and other lymphoid **tissue**. Examples of the syndrome were first recorded by the Italian physician **Marcello Malpighi** in 1666, but it was an English physician, Thomas Hodgkin (1798-1866), who first described the disease in detail in 1832.

Born in London in 1798, Hodgkin received his medical degree in 1821 from the University of Edinburgh and served as lecturer on morbid **anatomy** at Guy's Hospital in London from 1825 to 1837. Hodgkin introduced the use of the newly invented stethoscope to Great Britain from the Continent and promoted the importance of postmortem examination. Passed over for an appointment as assistant physician at Guy's, Hodgkin, a Quaker, devoted increasing amounts of time in his later years to philanthropic and humanitarian concerns. He died of dysentery in Jaffa on a mission to Palestine in 1866.

Hodgkin's interest in postmortem investigations led to the presentation of a paper in 1832 titled "On Some Morbid Appearances of the Absorbent Glands and Spleen" describing a particular type of lymphoma characterized by swollen lymph tissue. The importance of this paper wasn't recognized until 1856, when the English doctor Samuel Wilks redescribed the condition and named it Hodgkin's disease.

Hodgkin's disease is distinguished from other conditions that cause lymphatic tissue swelling by the presence of giant, mostly multi-nuclear, **cell**s. These Sternberg-Reed cells were first recorded by pathologists Carl Sternberg (Germany) in 1898 and Dorothy Reed (U.S.) in 1902.

Hodgkin's disease occurs primarily between the ages of 15 and 35 and after the age of 55. Once almost certainly fatal, Hodgkin's disease can now, for the most part, be successfully treated, especially if it is discovered in the early stages. The standard therapy combines radiation and **chemotherapy**. Clinical trials are also being conducted using bone marrow transplants as an adjuvant therapy. The transplants are designed to replace bone marrow which may have been destroyed by chemotherapy, thus compromising the patient's **immune system**. Ironically, immune deficiency may be a risk factor for acquiring Hodgkin's disease. As a result, it is often found in patients with the acquired immunodeficiency syndrome, or **AIDS**.

HOLISM

Holism is usually defined as entities existing in **nature**, wholes, which are irreducible to the mere sum of their parts. Holism is often referred to as a theory or doctrine according to which a whole cannot be analyzed without residue into the sum of its parts or reduced to discrete elements. Wholes exist at every level: a **molecule** can be 'atomized' or it can be studied as an entity unto itself, as a whole. Molecular science is, then, more holistic than atomic science. Complete **reductionism** would allow only the atomic or even the sub-atomic level as worthy of analysis. Reductionism and holism are complementary, and both are needed to better understand the earth and its complexity.

In biology, **ecology** is the field most engaged with a holistic approach. Ecologists have developed a number of concepts, strategies and techniques to help their work to be holistic. Holism in ecology is generally defined as work beyond the molecular level. However, excluding any particular level depends on definitions of both holism and ecology as not relevant below the level of the **organism**, which is becoming less and less true as the various life sciences meld into a more unified science. Many examples can be cited of biological ecologists working to realize the mandate they have created through their own definitions of—and rationales for—ecology as a discipline. Much of their success derives from the units they work with, and the levels of complexity on which they focus. Many ecologists argue that ecology begins with attempts to understand organisms-in-relation. Often, that "in-relation' refers to inclusive conceptions like **species** and **niche**, and to aggregate units such as **population, community,** and **ecosystem**.

All ecological units are at least implicitly holistic, and all have been used to attempt a holistic account of ecological relationships. Niche is holistic because it is a concept developed in ecology to attempt measurement and understanding of what one ecologist labeled the ''n-dimensional hypervolume''—i.e., all of the dimensions of one particular species existence that can reasonably be measured or included in some way in a researcher's assessment of the multiple relationships of that organism with its surroundings. Community, in ecology, is an attempt to assess holistically, again within reasonable limits, the most significant interactions among the major species of organisms, each with its most crucial others, in a given space in a given time.

For most ecologists, the study of ecosystems is at the **heart** of ''the holistic approach'' in ecology. The ecosystem concept is intended to go one step beyond community and incorporate the interactions of the members of a given community with their surroundings, usually the **abiotic environment** in all its measurable dimensions. Ecologists employ holism broadly to grasp complexity at the level of the **biosphere**, and specifically in the restoration of individual lakes, to cite just two current areas of ecological interest. Ecologists claim, with some justification, that they can use the hierarchical organization of ecosystems to reconcile reductionism and holism.

In all of these concepts within a concept, ''reasonable'' and ''n-dimensional'' are used cautiously to limit the scope of inquiry: niche, community and ecosystem are holistic, but they don't measure or include everything, even in a bounded space in a limited time frame. And, these unit concepts do not exclude other levels as ''wholes,'' but instead are a reflection of those levels by which ecologists have traditionally defined their discipline.

Most biologists, whatever the scale at which they work, are committed to a big picture evolutionary view of their science. As **Theodosius Dobzhansky** stated, ''nothing in biology makes sense except in the light of **evolution**.'' Evolution is an inclusive, holistic tent that incorporates and integrates virtually all of biology. Many biologists insist on evolutionary biology as the most synthetic of the biological sciences, comprehending and unifying, as it does, all of biology from the molecular to the ecological. Evolution, in its incorporative, holistic scope, has at least since the early 1940s been described as ''the modern synthesis.''

The study of evolutionary change is central in ecology, as it is in all of the life sciences. Traditional debates about diversity and complexity and between reductionists and holists become manifest in the theory of evolution. The argument centers on emergence, the idea that at each level of complexity in the hierarchy of [both physical and biological] systems, new qualities emerge that, at lower levels, are not only absent, but are clearly meaningless.

HOLLEY, ROBERT WILLIAM (1922-1993)

American biochemist

Robert William Holley is best known for his 1962-1965 work in chemically isolating pure strands of transfer **ribonucleic acid** (tRNA), which retrieves specific amino acids for assem-

Robert William Holley. *(The Library of Congress. Reproduced by permission.)*

bly by the cell's **ribosomes** into proteins. He also determined the complete sequence of one RNA strand. For this work, he shared the 1968 Nobel Prize for physiology or medicine with Har Gobind Khorana and Marshall Nirenberg.

Holley was born in Urbana, Illinois, where his parents were teachers, and received his bachelor's degree from the University of Illinois. He obtained his Ph.D. from Cornell University and spent almost his entire career there, in 1964 becoming professor of biochemistry and **molecular biology**.

Working with yeast RNA, Holley used techniques developed for **proteins** by Frederick Sanger, including chemical fragmenting, cutting with restriction **enzymes**, and separating by ion-exchange chromatography and paper electrophoresis. He used snake venom for the first cuts on long strands. After deciphering the base sequences on the fragments, he reconstructed the sequence of the entire strand.

Following this, Holley studied the relationship between messenger RNA (mRNA), the first copy of a **DNA** instruction, and tRNA. He also partially identified the three-dimensional structure of RNA and showed that tRNA was a cloverleaf, rather than a helix as deoxyribonucleic acid (DNA) is.

HOMEOSTASIS

Living things are incredibly complex. They are constantly using and creating **energy** with countless cellular reactions. An **organism** must constantly maintain conditions which are favorable for these reactions to occur. The internal state of every living thing must be kept fairly constant. Homeostasis is the maintenance of internal conditions within certain boundaries.

The concept of homeostasis was developed by a French physiologist, **Claude Bernard**, in 1851. He was the first to recognize the presence and relative constancy of an organism's internal environment, which he called the *milieu intérieur*, while living in an ever-changing external environment, or *milieu extérieur*. An organism needed to maintain a dynamically constant internal environment in order to survive. Shortly after Bernard's pioneering work, an American physiologist, Walter Cannon (1871-1945), coined the term homeostasis, referring to this internal dynamic constancy.

An organism has many mechanisms and structures for maintaining homeostasis. Some of these mechanisms are under involuntary control from the autonomic **nervous system**; for example internal **temperature**, **blood pressure**, or the **digestion** of food. Motor pathways carry commands from the central nervous system to regulate the glands and nonskeletal muscles of the body, regulating the body's internal physiological condition. Some mechanisms for maintaining homeostasis are more directly controlled by the organism. For example, a lizard needs to maintain its internal temperature by modifying its external environment; moving from direct sunlight to shade as needed.

Homeostasis operates at many levels, including the molecular, cellular, organismic, and populational levels. Homeostasis at the molecular level involves a process known as feedback inhibition. This limits the amount of product produced in a **chemical reaction**. With feedback inhibition, the end product of a reaction has an inhibitory effect on the reactants, which causes the reaction to cease until the levels of the product fall below a certain level, at which time the reaction continues. This maintains a fairly constant level of product in a system.

At the cellular level, the **cells** themselves regulate their internal and external environment. One example is contact inhibition. Many cells will stop dividing if they become so numerous that they touch each other. A chemical messenger is passed from cell to cell, inhibiting further **cell division**. **Cancer** cells do not exhibit contact inhibition which may account for the uncontrolled growth of tumors.

At the organismal level, the autonomic **nervous system** in cooperation with other body systems help maintain homeostasis. An example of homeostasis in an organism is the **endocrine system**. **Hormone** levels help regulate the activities of many body systems. Hunger and thirst are also examples of a homeostasis mechanism in an organism. These sensations help regulate the amount of **nutrients** and **water** that are digested by an organism.

Finally, at the population level, homeostasis can also be maintained. Relationships between predatory **animal**s and their prey are examples of mechanisms of homeostasis. These relationships help control population size. If a prey animal becomes too numerous, so do their predators. The increase in **predation** will eventually decrease the prey population, which will in turn cause the predator population to decrease as well. The population size of both organisms are maintained in a delicate balance with each other. This balance, much like the balance maintained in the individual cells of an organism, is a kind of homeostasis.

HOMEOTIC MUTATION

Homeotic **mutation**s are changes in the genetic makeup of an **organism** which result in an incorrect positional placement of certain **organ**s or structures. Genetic mutations such as these can lead to the development of bizarre organisms that have misplaced structures on their bodies.

Homeotic mutations were first described by Edward Lewis and his colleagues during the 1950s. Lewis dedicated much of his academic career to investigating the genetic make-up of the **fruit fly**. One facet of his studies involved finding the cause for a strange mutation in a fruit fly that resulted in legs growing from the head in place of antennae. During his investigation, he found that specific **genes**, he called homeotic genes, were responsible for the development of each region of the fly's body. He also discovered that these genes were arranged on the **chromosome** in a linear sequence. They were even arranged in the order corresponding to the order of the body region that they control. For example, the homeotic genes for the head were located at the beginning of the chromosome while those for the posterior were located at the end. The importance of Lewis' work to biology was recognized in 1995, when he was awarded the Nobel prize in physiology.

Homeotic genes are regulatory genes that dictate the type and placement of specialized structures on an organism. These genes even determine how body parts are shaped. For example, human toes and fingers are both composed of bone and muscle, but the shape and details are different. Toes and fingers develop differently because of the action of the homeotic genes. The **protein**s that these genes code for are regulatory proteins that bind to **DNA**. This binding action effectively activates or inactivates certain structural genes, thereby controlling which ones are expressed. A genetic mutation in a homeotic gene can effect the development of the organism.

Genetic mutations involve any change in the **genetic code** of a **cell**. They are rare events that typically occur during **DNA replication**. They may also be caused by mutagenic agents or radioactivity. A variety of **point mutation**s can occur in which a single **nucleotide** in a gene is changed. Substitution mutations are those in which the wrong nucleotide is copied into the gene. Insertion mutations involve the addition of an extra nucleotide, and deletion mutations involve the deletion of a nucleotide from a gene. Any of these mutations can result in the expression of a malfunctioning protein. When a mutation occurs in a homeotic gene, the structural gene that it controls may be expressed inappropriately, creating a deformed organism.

In an organism like the fruit fly, a single homeotic gene can be responsible for the development of a whole structure. If a mutation occurs in this gene, a dramatic structural change can result. In more complex organisms, such as **mammals**, homeotic genes are typically found in multiple copies. For example, mice have four copies of their homeotic genes. These extra copies make it difficult to observe drastic changes as a result of mutations in homeotic genes. To some extent, the extra copies of the homeotic gene help compensate for the mutated one.

Since homeotic mutations lead to a change in body shape but are not necessarily lethal, it is thought that they may be responsible for evolutionary changes. This would occur if a change in body shape lead to a significant advantage for the organism. For example, a homeotic mutation could allow an **animal** to obtain food more effectively, or it could make the animal more attractive. This would lead to greater reproductive **fitness** of the animal and thus propagate the mutation to future generations.

HOMINIDS

Modern humans (*Homo sapiens sapiens*) and their known predecessors *Ardipithecus*, *Australopithecus*, *Homo habilis*, *Homo erectus*, *Homo sapiens neanderthal* are hominids. The key characteristic of hominids is upright posture and a bipedal mode of locomotion. Other hominid anatomical distinctions that developed through time are a larger **brain** capacity, opposable thumb and precision grip, shorter arms and longer legs, wider **birth** canal, and shorter big toes. In addition to these biological differences, the humans ancestors evolved culture. The cultural changes the joined biological factors in **human evolution** changes involve the way of life, use of the environment, and interpersonal relations. Specific highlights of cultural developments through time were the development of stone tool use, control of fire, formation of social groups, cooperative work, burial rites, and portable and nonportable art by the end of the Pleistocene era.

The biological changes serve to mark the hominids as distinct from the other **primates**. Hominids exhibit size reduction in their faces, particularly the jaws and teeth as well as having thicker enamel on the teeth. The size reduction includes losing the bony ridges over the eyes and sagittal crest on top of the head. Conversely, the brain becomes larger in comparison to the body.

The closest living relatives of hominids are the apes. According to the biochemical evidence, the last common ancestor occurred at least five million years before present. The oldest known hominid is *Ardipithecus ramidus* from Ethiopia, dated to 4.4 million years ago. Next, *Australopithecus* existed in Africa from approximately 4-1 million years ago. The *gracile australopithecines* probably became extinct before two million years ago, either through **evolution** into the **genus** *Homo* or through replacement. *Homo habilis* is known from circa 2.3 million years ago in association with stone tools. Approximately 1.8 million years ago *H. habilis* was replaced by or

evolved into *H. erectus*, which existed until 300,000 years ago. *H. erectus* was the first hominid to venture out of Africa and into Asia and Europe. Archaic *Homo sapiens* began appearing ca. 500,000 years before present. ***Homo sapiens neanderthalensis*** appeared in Europe, the Near East, and Africa about 200,000 years ago and persist until about 30,000 years ago. Modern humans (*H. sapiens sapiens*) first appear ca. 120,000 B.P. and are a result of the last 100,000-40,000 years. Controversy does exist for the **classification** of various hominid **fossil** finds and their ages of occurrence.

HOMO SAPIENS NEANDERTHALENSIS

Neanderthals fill a prominent yet ambiguous position in studies of human origins. In the nineteenth century their discovery and less than flattering description was responsible for the popular view of cave men. The original description emphasized the anatomical differences from modern humans, particularly as short and brutish appearing individuals. Today it is known that anatomical differences were more obvious in the skull, shoulder blade, and pelvis. Their short and muscular build with short limbs is similar to modern peoples adapted to cold climates. A large number of stone tools and weapons have been found, more advanced than those of *Homo erectus*. Neanderthals are associated with the Middle Paleolithic in Europe, Near East, and West Asia. Their sites are found between England and Afghanistan and the temporal range is approximately 200,000 to circa 30,000 years ago. Neanderthals are the first people known to have buried their dead.

For many years anthropologists considered the Neanderthals as a variety of *Homo sapiens* occurring between *Homo erectus* and modern humans (*Homo sapiens sapiens*). It is now known that Neanderthals were contemporary with fully-modern humans. Today controversy exists whether Neanderthals were part of the **gene pool** that give rise to modern humans, and therefore our ancestors. Some researchers believe that Neanderthals were an evolutionary divergent side-branch and played no ancestral role for modern humans. This problem is virtually impossible to solve from studying the **fossil** remains. A recent study extracted and isolated **mitochondrial DNA** from the bones of a Neanderthal. Next, the Neanderthal DNA sequences were compared to modern humans. The results indicate that Neanderthal mitochondrial DNA is much closer to modern human DNA than any other living **species**. Interpretation of this result is controversial. The differences between Neanderthal and modern DNA fall outside the range of variation established for modern humans. Taken literally, this result implies that Neanderthals are not our ancestors and had previously split off from the main line of **human evolution**. While the available data supports a claim that Neanderthals were a different species, it is not demonstrated by results from just one individual.

Researchers previously thought speech originated ca. 40,000 years ago with modern humans. A recent study argues that for thousands of years prior to this, Neanderthals had the ability to speak. This claim is based on the diameter of a nerve canal that connects the **brain** and tongue in Neanderthal skulls. This hypoglossal canal is roughly the same size as that in a modern human skull and implies Neanderthals had the necessary physical attributes for speech.

HOMOLOGY

Homology is a term used in **comparative anatomy** and evolutionary biology in reference to traits of **organisms** that have a common phylogenetic ancestry, but are now dissimilar in their structure, function, or behavior. Homology is based on the observation that there are basic, repeating patterns in the attributes of organisms, and that these are often due to a shared ancestry. However, depending on the environmental circumstances affecting the evolution of particular **species** within related groups (i.e., differences in **natural selection**), the attributes may eventually adopt differing forms. Nevertheless, their essential homology is retained, and this can often be discerned during earlier stages of development.

The study of homologies has been an important aspect of biology since the nineteenth century, and numerous examples have been identified in various groups of both closely and distantly related organisms. For example, the three tiny bones of the middle ear of humans (these are known as the malleus, incus, and stapes) are ultimately derived from certain jawbones of early ancestors of our vertebrate **phylogeny**. Similarly, the specific facial muscles which humans use to frown and smile are derived from muscles used by ancient fishes in our evolutionary lineage to pump **water** through their gills. Another case of homology is the wings of birds, which are derived from the forelimbs of their ancient, reptilian ancestors. A botanical example involves certain elements of the floral structures of higher **plants**, such as petals and bracts, which are derived from the modified leaves of their ancestors. In all of these examples of homologies, structures occurring in ancient ancestors became evolutionarily modified through natural selection into dissimilar, but homologous structures.

As such, homology is different from analogy. The latter term refers to a similarity of the attributes of organisms (i.e., of their structure, function, or behavior), occurring because of **convergent evolution**, rather than through a common ancestry. Examples of analogous structures are the wings of flying insects, birds, and bats, all of which are rather similar in their shape and function, but are derived from different anatomical elements. Another case involves the drought-adapted **morphology** of certain non-related groups of plants that live in **desert ecosystem**s, such as species in the cactus and spurge families (Cactaceae and Euphorbiaceae, respectively). These may have a similarly tubular or barrel-shaped stem, be armed with sharp spines, lack leaves, have a thick cuticle to conserve water, and have green stem **tissue**s capable of **photosynthesis**. However, these plants of different families look similar because of convergent evolution, and not because they are closely related.

HOOKE, ROBERT (1635-1703)

English physicist

Robert Hooke was one of a special breed of scientist whose intellect and ingenuity spanned many different disciplines. Like his contemporaries Isaac Newton (1642-1727) and Christiaan Huygens, Hooke worked in many fields, often with remarkable results.

Hooke was born in Britain, on the Isle of Wight in 1635. A sickly child who was stricken with smallpox at an early age, he was not expected to survive more than a few years. His persistent ill health forced him to remain indoors, where he found amusement in taking apart and reassembling mechanical devices. By his tenth birthday he had become adept at constructing intricate mechanical toys, including working boats and clocks.

After his father's death in 1648, Hooke was sent to London to attend boarding school, where the headmaster recognized his potential and placed him in a curriculum that included Latin, Greek, and mathematics. Hooke attended Oxford in 1653. Though he never completed his bachelor's degree, it was at Oxford that Hooke met some of Britain's greatest scientists, around whom the British Royal Society would later form. Among these was the physicist Robert Boyle, for whom Hooke served as a laboratory assistant. Under Boyle's tutelage Hooke constructed the precursor to the modern air pump, the first in a long line of ingenious scientific tools he would invent.

Using this new air pump, Boyle performed the research that would ultimately lead him to the finding known today as Boyle's law. (Boyle's Law states that there is an inverse relationship between the volume and pressure of an ideal gas, at constant temperature.) In fact, some scientists have suggested that Hooke himself was the author of this law, reasoning that Hooke may have been pressured to relinquish credit for his discovery to his instructor.

Around this same time many European inventors were vying to develop the first accurate device to determine longitude on a sailing ship. Already in use, the chronometer, essentially a modified clock, was unreliable since the pendulum used to regulate its motion was thrown off by the ship's rocking. Sometime near 1660 Hooke introduced a chronometer design based upon a spring rather than a pendulum. (Today Hooke's name is associated with the law that, in many situations, the force on an object is proportional to its displacement from its resting, or equilibrium, position.) Although his design was sound, he was unable to find investors to back him, and it was not until 1674 that Christian Huygens patented his own spring-driven chronometer. Hooke immediately claimed that Huygens' invention was a derivative of his own, beginning a dispute that remains unresolved to this day.

That was not the only confrontation Hooke had with one of his peers. Perhaps the most famous was his feud with Isaac Newton, which began in the early 1670s. Newton, then a young student, had submitted a paper on light and colors to the British Royal Society. Hooke reviewed the paper and quickly dismissed it. Newton published a second paper on light in

1675, introducing a theory describing light as an undulatory wave. Hooke's reply was that Newton had stolen this wave theory outright from his own earlier publication, *Micrographia.*

Hooke later made a similar claim to Newton's theory of gravitation. The verbal battles between these two scientists were very bitter, several times driving Newton to a nervous breakdown. While his true contribution to the canon of theoretical science is unclear, Hooke was unquestionably one of society's most productive inventors of scientific equipment. Among his list of accomplishments are the universal joint, the reflecting telescope, the compound **microscope**, the wheel barometer, the anemometer, the spring-driven wristwatch, the "cross-hairs" sight for telescopes, and new standards for microscopy. The bulk of his inventions were constructed during his term as Curator of Experiments for the British Royal Society, where he was commissioned to explore new avenues and create new devices.

Though mechanics was certainly his first love, Hooke turned to architecture after a great fire burned most of London in 1666. To help with the reconstruction of the city and to aid his colleague, English architect Christopher Wren (1632-1723), Hooke designed several prominent buildings, most of which still stand.

HOPKINS, FREDERICK GOWLAND (1861-1947)

English biochemist

Born in Sussex, England, Hopkins had a lonely and unhappy childhood. He was brought up by his widowed mother and an unmarried uncle who tended to ignore him. When Hopkins was seventeen, his uncle chose a career in insurance for him and for several years he dutifully gave in to his uncle's wishes. At the same time, however, he also took part-time courses in chemistry at the University of London, eventually getting his degree. In 1888, already twenty-seven years old, Hopkins received the small inheritance that finally enabled him to enter medical school at Guy's Hospital in London.

After getting his doctoral degree in 1894, Hopkins joined the staff of Guy's Hospital and taught for several years. In 1898, he was invited to teach physiology and **anatomy** at Cambridge University and it was at Cambridge—when Hopkins was well into his thirties—that his long, distinguished career really began.

Hopkins' early research was in uric acid and his studies of the effects of various diets on uric acid excretion first aroused his interest in proteins. In 1901, working with S.W. Cole, a student at Cambridge, Hopkins discovered tryptophan, an important **amino acid**, and was able to isolate it from protein. A few years later, he demonstrated that tryptophan, and certain other amino acids, could not be manufactured in the body from other nutrients but had to be supplied as such in the diet. (By so doing, he laid the foundation for the concept of the essential amino acid outlined by **William Rose** a generation later.)

After his work with tryptophan, Hopkins' primary interest became the study of diet and its effect on **metabolism**. At the time, nutritional science was in a fairly primitive stage. Most researchers confidently believed that a well-rounded diet consisted of the proper mixture of fats, **proteins, carbohydrates**, mineral salts, and water, and that the so-called diet-linked illnesses—such as beriberi or scurvy—were caused by some toxic substance in certain foodstuffs. Hopkins, studying the literature—including reports by **Christiaan Eijkman** that polished rice seemed to cause beriberi, while unpolished rice effected a cure—began to have serious doubts.

Hopkins had already noticed that his laboratory rats failed to grow on a diet of artificial nutrients, but grew rapidly when he added tiny amounts of cow's milk to their daily rations. He suspected, therefore, that normal food must contain substances missing from the pure fats, proteins and carbohydrates routinely used—for consistency—in nutritional studies. Terming these substances "accessory food factors," he pointed out that they appeared to be necessary for growth. His two papers on the subject, in 1906 and 1912, are considered the first explanations of the concept of **vitamins**.

In 1907, Hopkins and Sir Walter Fletcher conducted pioneering research in another area of biochemistry when they demonstrated that working muscles accumulate **lactic acid**. And in 1922, Hopkins isolated the tripeptide (triple-linked) enzyme, glutathione, from living tissue and demonstrated its importance to the utilization of oxygen by tissue cells.

For his pioneering work in vitamin research, Hopkins received the 1929 Nobel Prize in medicine or physiology (sharing the prize with Eijkman). He was knighted in 1925 and received numerous other awards, including the Royal Medal of the Royal Society of London in 1918 and the Copley Medal in 1926.

HOPPE-SELYER, ERNST FELIX (1825-1895)

German biochemist

Ernst Hoppe-Selyer was one of the leaders in making biochemistry (or physiological chemistry, as it was called then) a scientific field distinct from medical physiology. He performed the first study of the **nucleic acids**, gave the name hemoglobin to the red **blood** cells, and discovered the **enzyme** invertase.

Ernst Hoppe was born in Freiburg-an-der-Unstrut, Germany. His father, a minister, and his mother died when he was a child. After he was adopted by his brother-in-law, he added Selyer to his name. He received his medical degree from the University of Berlin in 1851, then combined a medical practice with scientific research. Hoppe-Selyer's interest shifted gradually from physiology to chemistry. After serving on the faculties of the Universities of Berlin and Tubingen, in 1872 he became professor of physiological chemistry at the University of Strasbourg (then part of Germany). He established the first independent biochemistry laboratory in 1877 and the first biochemical journal.

Hoppe-Selyer's first important discovery came in 1862, when he used the newly invented spectrograph to determine the structure of the red blood **cells**, which he called **hemoglobin**. He later showed how hemoglobin binds and releases oxygen and how carbon monoxide can take oxygen's place in the blood cell. He also demonstrated some of the chemical similarities of hemoglobin and chlorophyll.

Hoppe-Selyer began studying the nucleic acids after they were discovered in 1869 by one of his students, the Swiss biochemist **Johann Friedrich Miescher**. Hoppe-Selyer showed that nucleic acids were present in yeast, and his work was extended by his one-time assistant, Albrecht Kossel.

Hoppe-Selyer's other research included the discovery in 1871 of invertase, the enzyme that converts sucrose (table sugar) into the simpler sugars glucose and fructose. He helped determine that lecithin is composed of nitrogen, phosphorus, fat, and choline (one of the B **vitamins**). And he demonstrated that lecithin and the steroid **cholesterol** are found in every cell.

HORMONE

Hormones are chemicals produced by glands, **tissue**s, and **organ**s to control the function of a target organ or regulate the production of another hormone. The body produces a specific amount of a hormone in response to stimuli (signals) from inside and outside. The hormone balance helps keep the body functioning in ordinary or **stress**ful situations—a process called **homeostasis**.

Dozens of human hormones play important roles in growth, sex and reproduction, **digestion**, **blood** composition, and stress control. Other **animal**s and **plants** produce hormones as well. Several hormones may work as a team on the same organ or tissue. Their combined effect may be greater than the sum of their single effects (synergism).

Many hormones are produced by the **endocrine system**, a large group of ductless glands. They include the pituitary and pineal, at the base of the **brain**; the thyroid and parathyroid in the throat; and the **adrenal glands**, sex glands (ovaries and testes), thymus, and **pancreas** in the trunk. The **digestive system** and other tissues and organs also produce hormones.

Hormones are either **protein**s or lipid-like **steroids**. Some protein hormones are long chains (**polypeptide**s) ranging from three to over two hundred **amino acid**s. Glycoprotein hormones have both **carbohydrate** and **peptide** structures. Protein hormones are manufactured and then stored or circulated until they are needed. Steroids, including all sex and **adrenal gland** hormones, are synthesized as needed from the steroid **cholesterol**.

The first modern European scientists to be interested in body glands and their **secretion**s were sixteenth-and seventeenth-century anatomists such as **Andreas Vesalius** in Italy and Regnier de Graaf in the Netherlands. The concept of a hormone (from the Greek word meaning shock or impulse) was developed by the British biochemists **Ernest Starling** and William Bayliss after they isolated the digestive substance secretin in 1906. Scientists soon realized that the first hormone to actu-

ally be isolated and synthesized was the adrenal substance epinephrine, by the Japanese-American chemist Jokichi Takemine in 1901. His achievement was based on earlier work by the British physiologist Edward Sharpey-Schäfer and the American pharmacologist John Jacob Abel.

The next milestone was the isolation of the thyroid hormone thyroxine in 1914 by the American biochemist Edward Kendall. Too much or too little thyroxine can cause illness. One of the earliest known thyroid-gland disorders is called Graves's disease, characterized by the **thyroid gland**'s increased size and activity. It was first defined in 1835 by Irish scholar and physician Robert James Graves. Its cause is not known, but Graves's disease is linked to stress and may be hereditary.

In 1921, the Canadian physicians **Frederick Banting** and **Charles Best** isolated the pancreatic hormone **insulin**. It was soon used to control diabetes.

Another area of important exploration was with adrenal gland cortex hormones. The cortex (outer covering) of the adrenal glands produces several hormones. In the 1930s, the American biochemist Edward Kendall and the Swiss chemist **Tadeus Reichstein** independently isolated the first in a series of these hormones, **cortisone**, and established that it was a steroid. Cortisone was the first hormone to be used medically, by the American researcher Philip Hench in the 1940s to reduce **inflammation** in rheumatoid **arthritis** and other connective tissue diseases. Other corticoids (adrenal cortex steroid hormones) are hydrocortisone (corticol), corticosterone, and aldosterone. Cushing's disease, described in 1912 by the American neurosurgeon and physiologist Harvey Cushing (1869-1939), results from excessive production of cortisone, hydrocortisone, and corticosterone. It causes fat redistribution from the lower body to the trunk, facial puffiness, and diabetes.

In the 1930s, many scientists, including Adolf Butenandt in Germany, **Leopold Ruzicka** in Switzerland, and Percy Julian in the United States, began studying hormones produced by the male testes and female ovaries.

Hormones produced by the **pituitary gland** are human growth hormone (hGH), or somatotropin (hST). Growth hormone is necessary for body growth and development. The American biologist **Herbert Evans** experimented during the 1930s with pituitary extracts that increased growth in laboratory animals. In the 1940s, Evans and his colleague Choh Hao Li isolated it. In the 1960s and 1970s, Li and others independently synthesized it through **genetic engineering**.

Eventually, scientists learned to make some hormones in the laboratory. American biochemist **Vincent Du Vigneaud** synthesized the small (eight amino acid) pituitary hormone oxytocin, which generates milk production in the mammary glands and causes uterine contractions. This achievement led to the synthesis of many larger and more complex hormones for medical use.

No systems of the body operate completely independently of others. The Argentinean physiologist **Bernardo Houssay** showed how the pituitary gland affected the secretion of insulin and other hormones. For this work, he shared the 1947 Nobel prize in physiology or medicine. **Adrenocorti-**

cotropic hormone (ACTH) stimulates the adrenal gland cortex to secrete corticol (hydrocortisone), corticosterone, and aldosterone. ACTH's properties were first investigated in the 1930s by the Canadian James Collip, as well as Evans and Houssay. Li was one of several scientists to isolate and synthesize ACTH.

The body must have some way of regulating hormone levels and production. In the 1950s, the British anatomist Geoffrey Harris proposed that a part of the brain called the hypothalamus and the pituitary gland jointly control production of hormones. His theory was confirmed by the American endocrinologists Roger Guillemin and Andrew Victor Schally, as well as other scientists. Feedback or signals such as a hormone's blood level begins the process.

The increase in knowledge about hormones is reflected in studies and treatment of Addison's disease. In 1849, Thomas Addison, a British physician and pathologist, matched symptoms of progressive weakness, increased skin pigmentation, and weight loss to an atrophied condition of the adrenal gland. This was the first medical description of an adrenal gland disease. Addison 's disease symptoms are now known to be caused by low levels of two hormones, aldosterone and hydrocortisone (cortisol), and can be controlled with replacement hormones.

Today scientists are increasing their understanding of hormones' interaction with target tissue cells and how this information can be used in the treatment of illness. Researchers are also interested in the activity of hormone-like substances called growth factors, produced by individual cells throughout the body.

HOUNSFIELD, GODFREY N. (1919-   )
English biomedical engineer

Godfrey Newbold Hounsfield was born August 28, 1919, in Newark, England, the youngest of five children of a steel-industry engineer turned farmer. He graduated from London's City and Guilds College in 1938 after studying radio communication. When World War II erupted, Hounsfield volunteered for the Royal Air Force (RAF), where he studied and later lectured on the new and vital technology of radar at the RAF's Cranwell Radar School. After the war he resumed his education, and received a degree in electrical and mechanical engineering from Faraday House Electrical Engineering College in 1951. Upon graduation, Hounsfield joined Thorn EMI (Electrical and Musical Industries) Ltd., an employer he has remained with his entire professional life.

At Thorn EMI, Hounsfield worked on improving radar systems and then on computers. In 1959, a design team led by Hounsfield finished production of Britain's first large all-transistor computer, the EMIDEC 1100. Hounsfield moved on to work on high-capacity computer memory devices, and was granted a British patent in 1967 titled "Magnetic Films for Information Storage."

Hounsfield's work in this period included the problem of enabling computers to recognize patterns, thus allowing them to "read" letters and numbers. In 1967, he envisioned a medical diagnostic system in which an x-ray machine would image thin "slices" through the patient's body and a computer would process the slices into an accurate representation which would display the tissues, organs, and other structures in much greater detail than a single x ray could produce. Computers available in 1967 were not sophisticated enough to make such a machine practical, but Hounsfield continued to refine his idea and began working on a prototype scanner. He enlisted two radiologists, James Ambrose and Louis Kreel, who assisted him with their practical knowledge of radiology and also provided tissue samples and test animals for scans. The project attracted support from the British Department of Health and Social Services, and in 1971 a test machine was installed at Atkinson Morely's Hospital in Wimbledon. It was highly successful, and the first production model followed a year later. These original scanners were designed for imaging the brain, and were hailed by neurosurgeons as a great advance. Before the computerized axial tomography (CAT) scanner, doctors wanting a detailed brain x ray had to help their equipment see through the skull by such dangerous techniques as pumping chemicals or air into the brain. As head of EMI's Medical Systems section, Hounsfield continued to improve the device, working to lower the radiation exposure required, sharpen the images produced, and develop larger models which could image any part of the body, not just the head. This "whole body scanner" went on the market in 1975.

CAT scanners generated some resistance because of their expense: even the earliest models cost over $300,000, and improved versions several times as much. Despite this, the machines were so useful they quickly became standard equipment at larger hospitals around the world. The scanner won Hounsfield and his company more than thirty awards, including the MacRobert Award, Britain's highest honor for engineering. In 1979, Hounsfield's collection of scientific tributes was topped off with the Nobel Prize. That year's Nobel was shared with Allan M. Cormack, an American nuclear physicist who had separately developed the equations involved in reconstructing an image via computer. A surprising feature of the selection was that neither man had a degree in medicine or biology, or a doctorate in any field.

Hounsfield moved on to positions as chief staff scientist and then senior staff scientist for Thorn EMI. He continued to improve the CAT scanner, working to develop a version which could take an accurate "snapshot" of the heart between beats. He has also contributed to the next step in diagnostic technology, nuclear magnetic resonance imaging. In 1986, he became a consultant to Thorn EMI's Central Research Laboratories in Middlesex, near his longtime home in Twickenham.

HOUSSAY, BERNARDO A. (1887-1971)
Argentine physiologist

Bernardo Alberto Houssay was born in Buenos Aires, Argentina, on April 10, 1887; his parents had emigrated from France before his birth. His father was Albert Houssay, a lawyer who also taught literature at the National College of Buenos Aires, and his mother was the former Clara Laffont.

Houssay completed his secondary education at the Colegio Británico at the age of 14. Three years later he earned his degree in pharmaceutical chemistry from the University of Buenos Aires, receiving the highest honors in his class. He then enrolled in the school of medicine at the university and was granted his M.D. at the age of 23. Houssay's medical studies took somewhat longer to complete than might have been expected, given his previous academic record, because he simultaneously worked as a hospital pharmacist in order to help pay for his expenses.

Having completed his studies, Houssay was appointed provisional professor, and, in 1912, full professor of physiology at the university's school of veterinary science. In 1913, he became chief physician at Alvear Hospital as well as a laboratory director in the newly created National Public Health Laboratories. Houssay's 1919 return to the university as chair of physiology marked the beginning of his greatest impact in the field. It was at the university that he established and became director of the Institute of Physiology, a research center that was to attain worldwide distinction. At its peak, the Institute was home to 135 graduate students from every part of the world, extending Houssay's influence far beyond the borders of Argentina. In 1920 Houssay married María Angélica Catán, a chemist. All three of their children, Alberto Bernardo, Héctor Emilio José, and Raúl Horacio, earned medical degrees.

In spite of his many administrative responsibilities, Houssay continued to be very active in research throughout his life. He was intensely interested in every aspect of physiology, from the cardiovascular to the respiratory to the gastrointestinal systems. But his major accomplishments resulted from his studies of the endocrine system, studies that dated to research begun while he was still a medical student. That research received an important impetus in 1921 when Canadians Frederick Banting and Charles Best and Scottish physiologist John Macleod discovered the role of **insulin** in the development of **diabetes**.

From 1923 to 1937, Houssay studied the interaction between the **pancreas** and insulin, on the one hand, and the **pituitary gland** (then called the hypophysis) and its secretions, on the other. One of his first major discoveries was the role of the anterior lobe of the pituitary gland in the metabolism of carbohydrates. A more important discovery was that the oxidation of sugars in the body depends not simply on the presence or absence of insulin, but on a complex interaction between insulin and other **hormones**, such as prolactin and somatotropin, produced in the pituitary gland. For his unraveling of this process, Houssay received a share of the 1947 Nobel Prize for physiology.

The political turmoil that swept Argentina in the 1940s altered Houssay's career. During the uprisings of 1943, he signed a petition calling for the democratization of the Argentine government. As a result, he was dismissed from his post at the university. Two years later, the dismissal was voided, and Houssay returned to the university. He was there only briefly, however, before he was asked to retire, which he did in 1946. In the meantime, he and some colleagues had founded the independent Institute of Biology and Experimental Medicine in order to continue with their research. Even when Houssay was yet again reinstated to his old post at the university in 1955, he continued to serve as director of the Institute.

Houssay was a major leader of Argentine science for many years. He founded, assisted in the establishment of, or served as head of nearly every major scientific organization in the country between 1920 and 1970. He was honored not only by his own nation, but by scientific societies all over the world. He was given honorary doctorates by more than 25 universities and was elected to membership in scientific societies in Great Britain, Germany, France, Italy, Spain, and the United States. Houssay died in Buenos Aires on September 21, 1971.

HUBEL, DAVID H. (1926-)
Canadian American neurobiologist

Born February 27, 1926, in Windsor, Ontario, of American parents, Elsie M. Hunter Hubel (pronounced hyü-ble) and Jesse H. Hubel, David Hunter Hubel grew up in Montreal. From his father, who was a chemical engineer, Hubel developed an interest in science, especially chemistry and electronics.

From 1932 to 1944, Hubel attended the Strathcona Academy in Outremont, Ontario. He began his college studies at McGill University in 1944. Although he received his B.S. with honors in mathematics and physics, he decided to enter McGill University Medical School in 1947—a decision which he appears to have made almost on the spur of the moment, since he had not taken any college course in biology. He also worked summers at the Montreal Neurological Institute, where he began his studies of the nervous system. He received his medical degree in 1951 and spent the next four years studying clinical neurology, first at the Montreal Neurological Institute and then at Johns Hopkins University in Baltimore, Maryland.

In 1955, Hubel was drafted into the United States Army, which sent him to the Neurophysiology Division of the Walter Reed Army Institute of Research in Washington, D.C. At Walter Reed, Hubel discovered a stimulating group of physiologists who encouraged him to do original research for the first time in his life. Determined to study sleep, he developed a device, known as a tungsten microelectrode, to record the electrical impulses of nerve cells. He used this device on cats to measure the activity of nerve cells in sleep.

During his research on sleep, Hubel became more interested in the reactions of his subjects to the firing responses recorded by the microelectrodes during waking states. He had placed the microelectrodes in the visual cortex area of the brain for his sleep experiments, and he began to realize that it was possible to understand how the **brain** operates in the visual process. In reading the work of other scientists on this subject, Hubel discovered the research papers of Stephen Kuffler, who was then a leading figure in the neurophysiology of vision.

After his army service ended in 1958, Hubel went to Johns Hopkins University where he did further research on the surface of the brain, the gray matter of the cerebral cortex, in the laboratory of Vernon Mountcastle. But shortly afterwards he moved to the Wilmer Institute, also at Johns Hopkins, and joined Stephen Kuffler's research team. There he met **Torsten Wiesel**, and under the direction of Kuffler the two of them began to make discoveries about the relationship of the retina to the visual cortex as part of the general physiology of the brain.

In 1959, Hubel and Wiesel, along with the rest of Kuffler's research team, followed Kuffler to the Harvard Medical School in Boston. By 1964, Harvard had formed a new department of neurobiology, naming Kuffler as its chairman. Hubel became chairman of this department in 1967, and in 1968 he was named the George Packer Berry Professor of Physiology.

Much of the work done by Hubel and Wiesel, using microelectrodes and electronic equipment, centered around a section of the visual cortex in the brain known as area 17. The cells in this section of the visual cortex form several thin layers that are arranged in columns running through the cortex. Hubel and Weisel discovered that certain cells of area 17 in the brain respond to the stimulation of specific retinal **cells** in the eye. In particular, they found that cells in the cortex are specialized to respond to different types of stimulation. There are types of cortical cells that respond to light spots and others that respond specifically to the different angles of a tilted line. They discovered that some respond only to definite directions of movement, while others respond only to definite colors.

Hubel and Wiesel's research has made the visual cortex the most mapped-out section of the brain, and it has deepened the scientific understanding of how the visual system works. In addition, their work has led to practical ophthalmological applications for the treatment of congenital cataracts, as well as a condition occurring in childhood known as strabismus, where one eye is unable to focus with the other because of a muscle imbalance. Hubel and Wiesel discovered that at birth the visual cortex begins to develop its structures from the stimulation of the newborn's retina. The development of the brain is shaped by the activity of the eye, and the sooner childhood eye disorders are corrected, therefore, the better the chances of avoiding serious visual impairments in the future. Before their research, the customary medical practice had been to delay operating on these conditions, but today doctors recognize the importance of the early removal of cataracts and the prompt treatment of strabismus.

For their work on how the retinal image is read and interpreted by the cells of the visual cortex, Hubel and Wiesel shared the first half of the 1981 Nobel Prize for physiology or medicine. For his work on split-brain physiology, **Roger W. Sperry** won the second half.

Hubel has been married to Shirley Ruth Izzard Hubel since 1953, and they have three sons.

HUBER, ROBERT (1937-)
German biochemist

The study of **photosynthesis** —the ability of **plants, algae,** and **bacteria** to translate sunlight into energy to build various

David H. Hubel. *(Science Photo Library, Photo Researchers, Inc. Reproduced by permission).*

chemical compounds—has long intrigued scientists, yet it is only since the 1950s that this process has begun to be understood in any detail. The analytic work of Robert Huber has played a significant role in the development of this understanding, and his most important achievement was matching the structure of a photosynthesizing **protein** complex to its function. Huber's work in X-ray diffraction enabled him and a co-worker to map the atomic structure of a bacterial photosynthetic reaction center—the basic unit or heart of the photosynthetic process. Such a description has helped advance not only photosynthesis research, but also various medical investigations. For his work in "unraveling the full details of how [such a] protein is built up, revealing the structure of the molecule atom by atom," the Nobel committee awarded Huber and two other German co-researchers the 1988 Nobel Prize in chemistry.

Robert Huber was born on February 20, 1937 in Munich, Germany to Sebastian and Helen Kebinger Huber. His father was a bank clerk and the family had a hard time during World War II and the years following. In 1947 Huber entered the Humanistisches Karls-Gymnasium in Munich, a school with an emphasis on humanistic studies. In an autobiographical piece for *Les Prix Nobel,* Huber remembers the teaching of Latin and Greek as being "intense," but it was here he developed an interest in chemistry. Few chemistry classes were available, so he taught himself "by reading all the textbooks I could get." In 1956 he graduated from the gymnasium and entered the Technische Hochschule of Munich—later renamed the Technical University—to study chemistry. He graduated in 1960, and he married Christa Essig that same year. Various stipends and grants helped Huber and his growing family—they would have four children—through the years he spent as a graduate student in the crystallography laboratory of W. Hoppe.

Develops X-ray Crystallography Techniques

As a graduate student, Huber worked with a number of prominent chemists, including Walter Hieber in the field of inorganic chemistry, Ernst Fischer who studied organometallic chemistry, and F. Weygand in organic chemistry. But it was crystallography that won Huber's interest. Though his thesis work for his 1963 doctorate was done on the crystal structure of a diazo compound, it was crystallographic studies on the insect metamorphosis hormone ecdysone that set him on the path of X-ray crystallo- graphy. Working with Hoppe at both the Technical University and at the Physiological-Chemical Institute of the University of Munich, Huber was able to determine the molecular weight and steroid nature of ecdysone. He employed X-ray diffraction techniques (where an X-ray beam is shot at a crystallized substance) to determine the atomic structure of ecdysone by analyzing how the beam was dispersed by the crystal. Huber was so impressed by the results he attained that he decided to concentrate on crystallographic research.

After determining the structure of several organic compounds and developing some improvements in existing X-ray crystallography methods, in 1967 Huber and H. Formanek set out to elucidate the structure of erythrocruorin, an insect protein. Their results showed a marked similarity between erythrocruorin and mammalian proteins. Their work also suggested for the first time that there might be a universal globin fold—the globin fold being the manner in which the chain of amino acids constituting the protein folds upon itself, endowing the protein with a shape specific to its function. In 1968, Huber became a lecturer at the Technical University, and three years later he accepted a position as a director at the prestigious Max Planck Institute for Biochemistry at Martinsried near Munich. He maintained his affiliation with the Technical University as well, becoming a full professor there in 1976.

Throughout the 1970s, Huber and his co-workers refined and perfected the techniques of X-ray crystallography, elucidating the structures of various proteins in collaboration with both foreign and domestic laboratories. His work in enzyme inhibitors and immunoglobulins has been of particular interest to researchers developing technologies for drug and protein design. Huber's laboratory at Martinsried became internationally recognized for the high quality of its work, and for Huber's delight in undertaking projects others thought impossible.

Wins Nobel Prize for Work on Photosynthesis

In 1982 a fellow researcher at Martinsried named Hartmut Michel came to Huber with a monumental task: to eluci-

date the atomic structure of the protein complex that powers photosynthesis in the purple bacteria, *Rhodopseudomonas viridis*. Michel had managed to isolate and crystallize a protein complex known as a membrane-bound protein, which is situated on the outer membrane of the bacterium. These proteins, made up of a tangle of four protein subunits and molecules of chlorophyll, help transport energy across the walls of cells. Yet they had been extremely difficult to isolate, because of their intermediary position on the membrane wall. Many believed these proteins were impossible to isolate, but by 1982 Michel had grown crystals of this protein complex, which functions as a photosynthesis reaction center. The reaction center is the place where electrons—released by a photon-excited chlorophyll molecule—create an electrical charge difference that produces the energy to power the synthesis of chemical compounds such as sugar, **carbohydrates**, and other **nutrients**.

Huber agreed to take on the task of developing a structural analysis of the proteins Michel had crystallized. Working with German biochemist **Johann Deisenhofer** at Martinsried and several other biochemists, his team used their improved X-ray crystallographic techniques to determine the exact atomic structure of the reaction center. By 1985 they had mapped over 10,000 separate atoms, and their structural analysis confirmed predictions as to the path that electrons follow in the reaction center. Though there are significant differences in the process of photosynthesis in green plants and in bacteria, the three-dimensional atomic model that Deisenhofer and Huber developed has proved to be of immense importance in further photosynthesis research in general. It has also been vital in research into the part that membrane-bound proteins may play in diseases such as **cancer** and **diabetes**. The work of the three main researchers in this project—Huber, Michel and Deisenhofer—was recognized by a joint award of the Nobel Prize for chemistry in 1988.

Huber has also been instrumental in developing computer models, such as FILME, PROTEIN, FRODO, and MADNESS to help in determining atomic structures through X-ray crystallography. Besides the Nobel Prize, Huber's work has been recognized by the E. K. Frey Medal from the German Society for Surgery in 1972, and the Otto Warburg Medal from the German Society for Biological Chemistry in 1977. He has received the Emil von Behring Medal from the University of Marburg in 1982, and the Keilin Medal from the London Biochemical Society and the Richard Kuhn Medal from the Society of German Chemists, both in 1987, as well as the Sir Hans Krebs Medal in 1992. He has also received numerous honorary doctorates and memberships in foreign chemical and biochemical societies.

HUGGINS, CHARLES BRENTON (1901-1997)
Canadian-American physician

Charles Brenton Huggins was born in Halifax, Nova Scotia, Canada in 1901. He died in Chicago, Illinois in 1997. He received his undergraduate education from Acadia University where he graduated in 1920. He received his M.D. degree in 1924 from Harvard University after which he trained in surgery at the University of Michigan. He later became director of the Ben May Laboratory for Cancer Research at the University of Chicago from which he retired in 1969. Huggins was the first physician to use a chemical substance in the treatment of **cancer** and thus was the father of **chemotherapy**.

Huggins is remembered for his innovative treatment of prostate cancer and other malignancies that respond to **hormone** therapy. Prostate cancer is the most common internal cancer in the United States accounting for more than 40% of new cancer cases in men. African American males have the highest rate of prostate cancer in the world. Prostate cancer has a propensity for moving (metastasizing) to the lower spine where it can cause unrelenting pain. Obviously, any improvement in treatment of this feared and common malignancy would have a great impact.

The prostate is a sex gland that surrounds the neck of the bladder and provides most of the fluid that carries sperm to the exterior. It had long been known that castration in boys inhibits development of male sex glands including the prostate, and that castration in adult males caused regression of those sex glands. Castration removes the source of testosterone. Thus, it is not surprising that Huggins and his students focused on the effects of loss of male hormones and its effect on prostate cancer. In 1941, Huggins and his colleagues showed that most prostate cancer **cells** will respond to androgens (male sex hormone). Many of the cancer cells are androgen- dependent and will die if deprived of androgens. An obvious means of androgen ablation is surgical castration. This operation has a profound and immediate effect on prostate cancer. The level of the enzyme acid phosphatase, a marker for prostate cancer, drops concurrently with pain. Castration causes an immediate and precipitous drop in testosterone and thus loss in tumor mass. Huggins, at the same time, showed that exposure to the synthetic female estrogenic hormone, diethylstilbestrol (DES), has an effect similar to castration, i.e., tumor shrinking and abolition of pain occur. Chronic exposure to DES inhibits the pituitary gland from stimulating the testosterone-secreting cells of the testis. Treatment by DES (and selected other drugs) thus is a pharmacological androgen ablation therapy. Neither orchiectomy (castration) nor estrogens cures prostate cancer but both result in immediate shrinking of the tumor with significant eradication of pain.

It has often been thought that cancer cells do not follow physiological signals from the body, i.e., they are autonomous. Certainly the work of Huggins and his associates showed that this is not necessarily the case in all cancers. Some cancers, such a prostate, are now known to be sustained by hormone action. Deprivation of that hormone directly, by removing the source of that hormone, or by using a chemical that will interfere with hormone production have a common and desired effect which is relief to the patient.

The hormone therapy by Huggins was the beginning of cancer chemotherapy. Prior to Huggins, cancer was treated only with surgery or radiation. Huggins research opened the door for others to develop and test chemicals that have efficacy in cancer treatment. The work of Huggins was recognized by the Albert Lasker Award for Clinical Research in 1963. This was followed by the Nobel Prize in Physiology or Medicine in 1966 which he shared with Dr. **Peyton Rous**.

HUMAN CHORIONIC GONADOTROPIN

Human **chorion**ic gonadotropin (HCG) is a glycoprotein **hormone** produced by the extraembryonic **tissue** of the early human **embryo**. After **fertilization**, the human **zygote** undergoes cleavage followed by the formation of a blastocyst. The blastocyst is a hollow sphere constructed of an inner **cell** mass, which becomes the embryo proper, and a trophoblast, which is embryonic tissue that will contribute to the formation of the **placenta**. The portion of the trophoblast that is invasive into the maternal uterus is known as the syncytiotrophoblast. The syncytiotrophoblast is the tissue of origin of HCG. The hormone is produced early in pregnancy and increases in rate of production until about the tenth week of pregnancy. Thereafter it decreases. The function of HCG is to stimulate the production of progesterone by the **corpus luteum**. This assures a continual supply of ovarian progesterone until the placenta develops a supply of progesterone around seven weeks of gestation. Progesterone prepares the uterine lining, the **endometrium**, for implantation and maintainence of the embryo.

The presence of HCG in the urine of a woman is indicative of pregnancy. Actually, the test reveals the presence of trophoblast cells and does not in any way indicate the health of the fetus. Early on, there were mice (Aschheim-Zondek) and rabbit (Friedman) tests for the presence of HCG in urine. However, these were expensive. Later, the leopard frog, *Rana pipiens*, was shown to be much less expensive as a biological test **organism**. A male leopard frog will release living **sperm** in an hour after receiving an injection of morning urine containing HCG. Somewhat similarly, female African clawed toads, *Xenopus laevis*, will release **egg**s after receiving an injection of HCG-containing urine. These tests have now been replaced with even more sensitive clinical tests, one of which will reveal pregnancy prior to the first missed period.

Cryptorchidism is a condition where the testes do not descent into the scrotum of a newborn baby. This is a serious condition because abdominal testes are vulnerable to testicular **cancer** at a much higher incidence than normal testes. Further, abdominal testes are generally sterile. Some infants respond to HCG treatment of this condition. HCG enhances maturation of the external genitals and often causes the undescended testes to move into the scrotum.

HUMAN EVOLUTION

For well over a century, ever since **Charles Darwin**, Western scientific thought has stated that all of today's **species**, including man, have arisen by the modification of earlier, simpler forms of life. This means that the story of human evolution begins with a creature that most of us today would not consider human.

Illustration depicting the stages of human evolution. *(Custom Medical Stock Photo. Reproduced by permission.)*

Today's human beings, or *Homo sapiens sapiens*, belong to the *Hominid* family tree. *Hominid* means "human types" and describes early creatures that split off from the apes and took to walking upright, or on their hind legs. In the overall history of life on Earth, the human species is a very recent product of **evolution**. There are no human-like **fossils** older than 4 million years, which makes them only one-thousandth the age of life on Earth. The oldest and first ancestor of all known **hominids** was probably *Australopithecus afarensis*, named for a region in Ethiopia. What distinguished it from the *African pongids* (gorillas and chimps), from which it split, was that it was clearly built for two-legged walking. It was only about 3.5–4 ft (about 1m) tall, and it had a **brain** the size of a chimpanzee. By about 2.5 million years ago, it appears to have evolved into the slightly taller *Australopithecus africanus* with a slightly larger brain. Altogether, there were probably four main species of australopithecines.

From *Australopithecus* came the oldest known hominid to be given the Latin **genus** designation *Homo*, or "Man." This was *Homo habilis*, called "nimble," "capable," or "handy" man. Taller than its predecessors, it also had a bigger brain and, for the first time, used tools made of stone. It is possible that it also was a hunter. By about 1.5 million years ago, the hominid brain had increased in size to about half what it is today, and this difference made for a new **classification**, *Homo erectus*, or "upright man." This was the first hominid to use fire and hand axes and to substantially travel about. Early or archaic Homo sapiens, called "wise" or "intelligent" man, appeared about 300,000 to 400,000 years ago, and although it wore clothing and buried its dead, it still did not have a modern-size brain. It was only about 40,000 years ago

that *Homo sapiens sapiens*, or doubly "wise" man, appeared. This creature was anatomically indistinguishable from ourselves and had real language. Thus, in a relatively short period of time, modern humans turned from strictly hunting and gathering and took to domesticating **animal**s and then **plants**. Soon settlements turned into real cities and a civilization based on **agriculture** came to be.

Although the course of human evolution may have followed this seemingly logical, progressive road, its discovery and understanding was by no means easy or quick, nor did the discovery story follow the same route. Since about the middle of the nineteenth century, scientists have had to struggle to try to put together the puzzle of human evolution using the equivalent of randomly found pieces. But well before Charles Darwin offered a unifying theory into which we could try to fit all of these pieces, the foundations of mankind's knowledge about human evolution had been laid.

The Greek philosopher **Aristotle** was one of the first to speculate on the possibility that **organism**s had evolved from one another. However, by medieval times, such ideas were overwhelmed by the power generated by a literal interpretation of the first chapter of Genesis in the Bible. This interpretation said that humankind came about from a single, unique act of creation. During the Renaissance, the Italian artist **Leonardo da Vinci** performed comparative studies of men and beasts and wrote in his notes that "Man in fact differs from animals only in his specific [characteristics]," but his thoughts had little influence. Although the literal Biblical interpretation of human origins held sway even during the Scientific Revolution of the seventeenth century, it was during that time that a Frenchman, Isaac de la Peyrere (1594-1676), discovered what he believed were stone tools made by extremely ancient people. He published his findings in 1655, only to have his book burned. In 1700, the earliest recognition of a fossilized human part was given to a skull fragment found at Canstatt, near Stuttgart, Germany. Although it was described then as "ancient" in origin, this meant in the early eighteenth century that it was thought to be about 4,000 years old.

In the same century, however, the Swedish botanist **Carl Linnaeus** published his *Systema naturae*, a methodical classification of all living things. It was here that he invented the system of binomial nomenclature that is still used today. Nomenclature is Latin for "list of names," and binomial means "two names" in Latin. With this system, the first name designates an organism's genus—a group of organisms that are closely related. Each genus is made up of smaller groups different from each other, called species. This is the second name. Linnaeus classified apes with humans by including them in the genus *Homo*, but gave only modern man the name *Homo sapiens*. His contemporary, the French naturalist Georges Buffon, endorsed an evolutionary concept of man and the earth itself, arguing that everything in **nature** develops and changes slowly and continuously.

It was not until the nineteenth century, however, that anyone said that modern human beings were the result of an extremely long and slow evolutionary process. In 1809, the French naturalist **Jean Lamarck** was the first biologist to state

STAGES IN HUMAN EVOLUTION

Genus/Species	Approximate Period of Existence	Fossil Discovery
Ardipithecus ramidus	4.4 million years ago	Eastern and Southern Africa
Australopithecus		
A. anamensis	4 million years ago	Eastern Africa
A. afarensis	3.5 million years ago	Eastern Africa
A. africanus	3 million years ago	Southern Africa
A. aethiopicus	2.6 million years ago	Eastern Africa
A. boisei	2 million years ago	Eastern Africa
A. robustus	2 million years ago	Eastern Africa
Homo		
H. habilis	2.2 million years ago	Eastern and Southern Africa
H. erectus	1.3 million years ago	Asia, Europe, Indonesia, Eastern Africa
H. sapiens (archaic)	400,000 years ago	Asia, Eastern Africa, Europe
H. sapiens neanderthalensis	230,000 years ago	Europe, Middle East
H. sapiens sapiens (modern)	120,000 years ago	Africa, Europe, Middle East

(Table by Standley Publishing.)

boldly that humans evolved from four-footed animals. Although he was wrong when he argued that changes came about because acquired characteristics were passed on to offspring, he was the first real evolutionist.

In 1859, the view of man's history and his place on earth was changed forever by the publication of *On the Origin of Species by Means of Natural Selection*, written by the English naturalist Charles Darwin. In this revolutionary book, which stated that all living things had achieved their present form by slowly occurring natural processes, Darwin barely mentioned human evolution. However, in his 1871 book, *The Descent of Man*, he argued that man had descended from subhuman forms of life.

It was during the same decade in which Darwin's epochal first book was published that the physical evidence to back up his theory began to accumulate. The following, which describes these discoveries in the order in which they were made, can be better understood if read in light of the brief story of man's evolution outlined earlier, since it arranges these discoveries using today's knowledge and benefits from hindsight. Furthermore, today's physical anthropologists have at their disposal not only an extensive fossil record and an excellent knowledge of **comparative anatomy**, but they benefit from other disciplines and scientific areas such as geology and plate tectonics, molecular anthropology, radioactive isotope dating methods, and observations made of primate behavior. One amazing bit of new information gained from recent molecular **DNA** studies is that man and the chimpanzee share more than 99% of their **genetic material**.

Back in 1856, when human fossils were found in the Neander Valley near Dusseldorf, Germany, science had no methods for judging their age and could only say that they were very ancient individuals. This particular ''ancient'' skeleton was the first extinct human form ever recognized and is today classified as *Homo sapiens neanderthalensis*. It was re-

garded then as being halfway between apes and humans, and it is now known to have lived between 100,000 and 70,000 years ago. It is still debated whether Neanderthals evolved into fully modern people or were driven to **extinction** by an invasion of modern types from elsewhere.

In 1868, five skeletons were found in Cro-magnon caves in southwest France that were unquestionably modern, or *Homo sapiens sapiens*, and were given the name Cro-magnon man. The geological evidence at the site seemed to indicate that they were around 40,000 years old. In 1894, a Dutch paleontologist, Marie-**Eugene Dubois**, discovered the first-known fossil of *Homo erectus*, which came to be called Java man.

After the turn of the century, there was much discussion that a ''missing link''—or half-man, half-ape—probably existed in the distant past. In 1912, a British lawyer, Charles Dawson (1864-1916), announced the discovery near Piltdown Common, near Lewes, England, of skull pieces that showed its owner had a large, modern brain and human teeth set in the jaw of an ape. Although this turned out to be a hoax, many scientists embraced so-called Piltdown man as the missing link. This deliberate fraud confused the human evolution picture for a full 40 years until it was finally exposed.

In 1923, another variety of Dubois's *Homo erectus* was found in China by the Austrian paleontologist Otto Zdansky and was called Peking man (the complete skull of which was discovered in 1929 by Weng Chung Pei). The year 1924 became significant when **Raymond A. Dart**, an Australian anthropologist, discovered the first *Australopithecus* fossil in South Africa. The Scottish paleontologist **Robert Broom** supported Dart's theory that it was a primitive precursor of modern man, and he found a similar skeleton on his own in South Africa in 1936.

It was not until 1959 that the priority of African hominids over Asian hominids assumed a strong position. In that year, Mary Douglas Leakey (1913-1996), the wife of British anthropologist Louis S. B. Leakey (1903-1972), found skull fragments at Olduvai Gorge in Tanzania that proved to be a species of *Australopithecus*, called *Zinjanthropus*, that was 1,750,000 years old. It was also **Louis Leakey** who discovered in 1961 what he called *Homo habilis*. The skull of this creature held a larger brain than *Australopithecus* and was between 1.8 and 2 million years old.

In 1974, the American archeologist Donald C. Johanson (1943-   ) discovered a 4-million-year-old fossil whose scientific name is *Australopithecus afarensis*, but whose popular name became Lucy. Although her brain was only about one-third as large as today's human, the interesting thing about her was that she was completely bipedal. Many think that the sudden brain development that later occurred in hominids was the result of having their hands freed from walking and becoming available to use tools.

In recent years, new fossil discoveries and genetic evidence have fueled a debate concerning when and where *Homo sapiens sapiens* emerged. In 1988, researchers found numerous fossil fragments in a cave in Israel that suggest that anatomically modern humans lived there about 92,000 years ago. These results suggest that modern humans may have existed much longer than supposed, and they also support the theory that modern humans evolved first in Africa and then spread throughout the world. This notion, called the out-of-Africa model, says that Neanderthals found in Europe and elsewhere were a distinct and parallel human species that came to a dead end. This model is opposed by the multi-regional model, which argues that modern humans arose virtually simultaneously and independently in several different places in Africa, Europe, and Asia.

In 1997, Svante Pääbo of the University of Munich and his colleagues looked at a small stretch of DNA from a Neanderthal bone. They compared it with the DNA from more than 1,600 modern humans of various racial groups. The scientists saw no evidence of a relation. This suggests that Neanderthals were indeed a distinct species, which supports the out-of-Africa model. By Pääbo's calculations, Neanderthals and modern humans must have evolved separately for more than a half million years to have become so different. However, research is still needed to confirm this finding.

As we approach the end of the twentieth century, only one thing is certain: The mystery of human origins is a difficult and perhaps even more complicated problem than ever imagined.

See also Piltdown hoax

HUMAN GENOME PROJECT

The Human Genome Project (HGP) is an international effort to map out the human **chromosome**s, pinpointing the location and details of all the **gene**s and interconnecting segments. The term **genome** represents an individual's total **genetic material**, those chemical segments which control the activities of **cell**s and human **heredity**. The HGP, simply stated, aims to identify the **DNA** sequences that determine each **organism**'s characteristics.

The first stirrings of the HGP in the United States began in the mid-1980s. By 1990, a much more ambitious research program had taken form, including similar research projects in the United Kingdom, Denmark, France, Germany and Italy. The overall mapping and sequencing efforts are coordinated through an international body called the Human Genome Organization, or HUGO.

The Human Genome Project has numerous implications. On the one hand, further understanding and mapping of human **genetics** could lead to the development of new therapies for hitherto incurable diseases and new ways of manipulating DNA. There are over 3,000 known inherited diseases, and we've only identified the responsible gene for a mere 3% of them. On the other hand, it brings to mind all sorts of potential dangers and ethical questions. Some scientists and politicians have raised the concern that the ability to generate information has been pursued without adequate attention to its ethical implications.

One potential danger was signalled by the U.S. National Institute of Health's attempt to patent complementary DNA in 1991. While patenting is touted as a necessary financial incentive for research initiatives, it also restricts access to the information generated and to the use of those discoveries. That places a great deal of control in the hands of the private firm that funded the research, and may limit the spread of real benefits which result from the discovery. When the product being patented is human DNA, it also rekindles the debate over patenting life forms.

While patenting may restrict access to some information, other critics worry that the availability of genetic information may have serious privacy and equity effects. Most states do not have laws protecting citizens against the misuse of genetic information by, for instance, employers and insurers. In the absence of effective legal remedies, genetic testing may be used to bar people from employment or insurance coverage. Insurers may even make mandatory testing a requirement for coverage. Also, existing laws may not be adequate to protect peoples privacy: whereas there may be some protection from unwanted testing of an individual, genetic information about that same individual may be accurately predicted by tests on relatives. Privacy protection has not kept pace with genetic science.

Within the budget for the HGP in the United States is a small allocation (3% of its total) towards study of the ethical, legal and social implications (ELSI) of the project. The ELSI Working Group is charged with studying the issues of fairness, privacy, delivery of health care, and education. However it only has the power to make recommendations to policy makers in these areas.

Beyond these questions lie deeper ethical considerations. What are the implications and dangers of changing genetic structures? Who decides what are good genes and what are bad genes? The Human Genome Project has left those broader questions unanswered as it races towards the creation of a genetic road-map.

HUMAN GENES AND CORRESPONDING MEDICAL DISORDERS

Disorder	Gene Identified	Chromosome
Alzheimer's disease, type 3	AD3	14
Alzheimer's disease, type 4	AD4	1
Amyotrophic lateral sclerosis	SOD1	21
Breast cancer, type 1	BRCA1	17
Breast cancer, type 2	BRCA2	13
Colon cancer, nonpolyposis, type 1	MSH2	2
Colon cancer, nonpolyposis, type 2	MLH1	3
Cystic fibrosis	CFTR	7
Duchenne muscular dystrophy	DMD	X
Huntington's disease	HD	4
Juvenile onset diabetes	IDDM1	6
Malignant melanoma	CDKN2	9
Obesity	OBS	7
Polycystic kidney disease	PKD1	16
Severe combined immunodeficiency	ADA	20
Testis-determining factor	TDF	Y
Wilson's disease	ATP7B	13
X-linked mental retardation	FMR1	X

Source: National Center for Biotechnology Information, National Institutes of Health.
Reproduced by permission.

(Table by Standley Publishing.)

HUMAN IMMUNODEFICIENCY VIRUS (HIV)

The **human immunodeficiency virus (HIV)** belongs to a **family** of **virus**es known as the "**retrovirus**es." These viruses are known as **RNA** viruses because they have RNA as their basic **genetic material** instead of **DNA**. The retroviruses are unable to replicate outside of living host **cell**s, because they contain only RNA. However, they have an **enzyme** called "**reverse transcriptase**" that can make DNA from the RNA and allow them to integrate into the host cell **genome**. The retroviruses are composed of three subfamilies, two of which are pathogenic to humans. They are the oncarnovirus subfamily and the lentivirus subfamily. The human immunodeficiency virus, which belongs to the lentivirus subfamily, is further divided into two types based on the diseases they produce. The HIV-1 produces the acquired immunodeficiency syndrome (**AIDS**), while the HIV-2 produces a similar disease that is restricted to West Africa.

The genetic material of the HIV virus consists of two short strands of RNA about 9,200 **nucleotide**s long, enclosed in an outer lipid envelope. A viral glycoprotein (gp120) is displayed on the surface of the envelope. This **protein** recognizes and binds to the CD4 receptor on T-helper cells. The HIV genome contains a long terminal repeat (LTR) and the gag, pol, env, and tax/rex **gene**s. The LTR helps in the integration of the virus into the host cell DNA. The gag gene codes for the proteins that make up the outer core or capsid while the env gene codes for the envelope glycoprotein including the outer envelope glycoprotein (gp 120) and the transmembrane glycoprotein (gp141). The major proteins coded by the pol gene are the reverse transcriptase, protease, and the integrase. The tax/rex gene codes for certain factors that have a regulatory role.

The HIV infects cells that have the CD4 receptor **molecule** on their surface. In macrophages and cells lacking this molecule, an alternate receptor molecule (such as the Fc receptor, or the complement receptor site) may be used for entry of HIV. The immune cells such as the **blood** monocytes, macrophages, **T cells**, B cells, natural killer (NK) cells, dendritic cells, hematopoietic stem cells, etc are the primary targets of HIV infection.

After entering the body, the virus attaches itself by fusion to a cell with the appropriate CD4 receptor molecule. On gaining entry into the cell, the viral particle uncoats from its envelope and releases the RNA. The reverse transcriptase encoded by the pol gene, reverse transcribes the viral RNA into DNA, and the integrase enzyme (also coded by the pol gene) inserts the HIV proviral DNA into the genomic DNA of the host cell. The HIV provirus is replicated by the host cell and transcribed to produce new progeny RNA molecules. The infected host cells either release the new HIV virions by lysis, or the viruses can escape by surface budding. These go on to infect additional host cells.

sponse to develop. Therefore, the Center for Disease Control (CDC) recommends testing for HIV at six months after the last possible exposure to the virus (through unprotected sex or sharing needles).

HIV is primarily spread as a sexually transmitted disease. However, one can also acquire the virus through either intravenous drug use or transfusions. The virus can be present in a variety of body fluids and **secretion**s, but the presence of HIV in blood, and genital secretions, and to a lesser extent breast milk, is significant for the spread of HIV. In addition, HIV infection can be acquired as a congenital infection during **birth** or in infancy. Mothers with HIV infection can pass the virus either transplacentally at the time of delivery through the birth canal or through breast milk. The diagnosis of clinical AIDS often occurs because of the presence of rare diseases such as Kaposi's sarcoma, *pneumocystis* pneumonia, or other serious recurrent infections. The patient's lifestyle, and medical history could also provide clues. Laboratory diagnosis of the infection is based on measuring the antibodies to HIV using a test known as ELISA. Positive results are further confirmed with another test known as a Western Blot. Together, the two tests are more than 99.9% accurate.

No vaccines are currently available to prevent infection by HIV. However, scientists and researchers the world over are working on making a vaccine to HIV and have some interesting leads. The drugs used to treat HIV fall into three categories: the nucleosides, non-nucleosides, and the protease inhibitors. The nucleoside and non-nucleoside inhibit the reverse transcriptase enzyme while the third category of drugs inhibits the enzyme protease. These drugs are given in combinations of two or three to attack the HIV in different ways.

Educating the people and increasing the awareness about the virus and the infection is perhaps the best means of preventing the disease. The use of condoms during sexual contact can greatly reduce the risk of infection. By avoiding dangerous practices such as sharing syringes, and needles for intravenous drug use, direct inoculation of the virus into the blood stream could be prevented. Proper testing of blood and blood products will eliminate transmission of the virus by transfusion.

HUMUS

Humus is the more or less stable fraction of organic **matter** remaining in a soil after the major portions of added plant and **animal** residues have decomposed. No longer recognizable as living **tissues**, humus comprises large **molecules** (e.g., 2,000-300,000 g/mol) with variable structures and composition. Humus is characterized by dark-colored, complex, colloidal-sized, amorphous aromatic ring-type compounds, including polyphenols and polyquinones. Because of the chemical complexity of these compounds, humic materials are resistant to further microbial degradation.

Humus in a soil supplies **nutrients** as well as growth-promoting substances (e.g., **vitamins, amino acid**s, auxins, and gibberellins) that are necessary for plant growth. Humus also

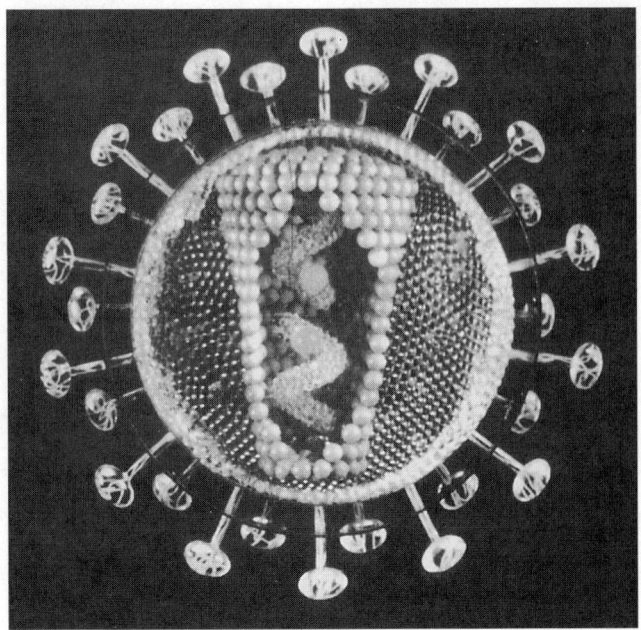

A three-dimensional model of the HIV virus. *(Corbis Corporation. Reproduced by permission.)*

The primary target of the HIV is the **immune system** itself, with a special affinity for CD4+ (T-helper) cells. Following infection, there is a "latent" phase during which the viral replication continues actively, accompanied with a progressive destruction of the CD4+ cells. During latency there are enough immune cells remaining to provide an immune response and fight infections. Eventually, when a significant number of T cells are destroyed, and the rate of production of the cells cannot match the rate of destruction, there is a loss of both cell-mediated and **humoral immunity**. This failure of the immune system leads to the appearance of clinical AIDS. The patients generally die of secondary causes such as Kaposi's sarcoma (a rare form of **cancer** that occurs in HIV-infected individuals) or **bacterial** and fungal infections.

Primary HIV infection may go undetected in more than half the cases, because the symptoms produced are mild and they subside quickly. This is followed by a clinical latent period, which could last on an average 8-11 years. The latency period varies from person to person and depends on a variety of factors including the person's health status and life style. In cases of acute HIV infection, the most common symptoms are fever, swelling of the lymph glands, a red, diffuse rash all over the body, sore throat or upper respiratory infection, muscle ache, diarrhea and headache. These symptoms subside in a couple of months. A symptomatic acute HIV infection is generally seen in people who acquire HIV infection through sexual transmission. Within three months of infection, the body mounts an immune response to the virus, and detectable levels of antibodies are seen. Both humoral and cell-mediated immune responses play a role. There is a decline in the viral counts and the levels of CD4+ T helper cells increase. In rare cases, it may take as long as six months for the immune re-

indirectly affects plant growth by affecting soil properties and processes. Humic substances reduce the plasticity, cohesion, and stickiness of clayey soils, making these soils easier to till. They also improve the **water**-holding capacity and infiltration rate of a soil by improving the aggregation (structure) of soil mineral particles. Increased aggregation of soil particles also increases the number and volume of macropores in a soil, resulting in better aeration and oxygen supply to plant roots.

Humic substances are negatively charged due to the dissociation of hydrogen ions from carboxylic (-COOH) or **phenol**ic (-OH) groups. Because of this charge, they can hold positively-charged nutrient cations (e.g., potassium, calcium, magnesium) in an easily exchangeable form so they can be used by **plants** but at the same time are prevented from leaching out of the soil profile by percolating water.

Traditionally humic substances are classified into three chemical groupings, based on solubility. Fulvic acid is the humic fraction that is lowest in molecular weight and lightest in color. Fulvic acid is soluble in both acid and alkaline solutions and is more susceptible to microbial degradation than other types of humic substances. Humic acid is medium in molecular weight and color, soluble in alkaline solutions but insoluble in acidic solutions, and is intermediate in resistance to microbial degradation. Humin is the highest in molecular weight, darkest in color, insoluble in both acid and alkaline solutions and is the most resistant to further microbial degradation. However, all three groups are relatively stable in soils. The half-life (time required to degrade half of the amount of a substance) of fulvic acid may range from 10-50 years, while the half life of humic acid is usually centuries.

HUNTER, JOHN (1728-1793)
Scottish physiologist and surgeon

John Hunter was a Scottish physiologist and surgeon who lived during the eighteenth century. Considered the father of modern surgical techniques, he is also well known for his large collection of anatomical specimens.

Hunter was born on February 13, 1728, in Long Calderwood, Lanarkshire. After working as a cabinet maker for a short while, he moved to London at the age of 20 to work with his brother, William, who was a renowned surgeon and obstetrician. Hunter started with no formal medical education; instead he helped his brother prepare anatomical specimens for his lectures. While working with these specimens, John conducted a detailed study of the lymphatic vessels and of the growth and structure of bone. He dissected a great many corpses and kept a large number of animal specimens as well, including such disparate animals as a whale and a bull. During the course of his work, Hunter improved upon an arterial injection embalming technique originally developed by **William Harvey**. As he grew more experienced, John began to conduct his own anatomical investigations. Meanwhile, he attended surgical classes at nearby London hospitals. In 1753 after only five years of study, he was appointed a Master of **Anatomy** of the Surgeon's Corporation. Three years later, he served as house surgeon at St. George's Hospital.

From 1759-1763, Hunter served the British army as a surgeon in France and Portugal. When he returned to London in 1763, he established a private practice. In the late 1760s, he accepted a senior surgical post at St. George's and was soon appointed physician extraordinary to George III. One of the most famous, or infamous, examples of his work during this period occurred in 1775. The body of the late Mrs. Martin van Butchell was brought to Hunter for embalming. It so happened that the woman's will stipulated that her husband would receive his inheritance only so long as her body remained above ground. Her husband had Hunter embalm the body in such a way that he could dress it and keep it in a glass storage case, thereby ensuring he met the conditions of his late wife's will.

In the late 1770s, Hunter was promoted to Deputy Surgeon to the Army, and during this time, he wrote an impressive number of papers. A few of his works include his *Treatise on Natural History of Human Teeth*, which was published in 1771. It was followed by the *Treatise on the Venereal Disease and Treatise on the Blood* in 1786, and finally *Inflammation and Gun-Shot Wounds*, which was published posthumously in 1794. In his last promotion, in 1790, Hunter achieved the title of Inspector General.

On October 16, 1793, during a meeting of the board of governors at St. George's Hospital, Hunter collapsed and died. He was originally buried at St. Martins-in-the-Fields, but later (on March 28, 1859) his remains were moved to Westminster Abbey. After his death, his collection of over 13,000 specimens was donated to the Royal College of Surgeons in 1795 and was later used as the basis for the Hunterian Museum.

HUTTON, JAMES (1726-1797)
Scottish physician

James Hutton, a Scottish physician and farmer, is considered by many to be the "father of geology." Hutton observed geological changes and theorized that the forces which were changing the landscape of his farm were the same forces that had changed the Earth's surface in the past. He built on this theory to form his "Principle of Uniformitariansm" in 1785.

This principle states that current geological processes, for example volcanic activity and erosion, are the same processes that were at work in the past, and will still be at work in the future. A summary of his theory is the phrase "the present is the key to the past." He watched these slow changes occurring on his own farm and theorized that over time, a stream could carve a valley, rain would erode rock, and sediment could accumulate and form new landforms. He realized that these forces must be acting very slowly, and therefore, the earth must be older than it was thought to be at the time (6,000 years). He published this theory in 1790 in his work *The Theory of the Earth*.

We now know his theory to be true. The Earth is approximately 4.6 billion years old, plenty of time for these slow processes to mold and shape the planet. The same forces are acting now that were acting in the past, even though the relative rates that these forces work to produce change may vary over time. When Hutton published his theories, however, they were not met with enthusiasm. Uniformitarianism went against both religious beliefs and the theory of catastrophism, the accepted theory of the time. Catastrophism states that the earth was formed not by slow processes, but by violent, worldwide disturbances such as earthquakes and floods. This theory is congruent with Biblical references and the theory of creationism. It was not until the nineteenth century that Sir **Charles Lyell**, in his 1830 work *Principles of Geology*, popularized the theory of uniformitarianism.

James Hutton was not only known for his uniformitarianism theory, but also for developing the concept of the rock cycle. This theory describes the interrelationships between igneous, sedimentary, and metamorphic rocks. The matter that makes up these rocks is neither created nor destroyed, but instead transformed from one rock type to another. He also suggested that the study of the earth be called "geophysiology." Hutton's theories about the earth as an entity which undergoes dynamic cycles such as these are considered by some to be the basis of the **Gaia hypothesis**, the concept of the "living earth."

HUXLEY, ANDREW FIELDING (1917-)
British physiologist

Sir Andrew Fielding Huxley shared the 1963 Nobel Prize for physiology or medicine with colleague **Alan Lloyd Hodgkin** for research which unlocked the secret of excitation and inhibition in nerve cells. Huxley theorized that the movement of sodium ions, and not just potassium ions, across the cell membrane was necessary to create an action potential. To test his hypothesis, in 1947, his colleagues Hodgkin and Bernard Katz manipulated the levels of sodium ions in the fluid outside the cell and discovered they could create or eliminate an action potential. This finding began the exciting road to discovering how **neurons** communicate.

Huxley was born in Hampstead, London. His father, Leonard Huxley, taught classics before becoming a writer, marrying Rosalind Bruce after his first wife died. Andrew, the youngest of their two sons, is half-brother to biologist Sir Julian Huxley and writer Aldous Huxley from his father's first marriage. His grandfather, Thomas Henry Huxley, was a writer and scientist whose work helped **Charles Darwin** establish his theory of **natural selection**.

Huxley attended University College School from 1925-1930 then Westminster School where became a King's Scholar. When he entered Trinity College, Cambridge on a scholarship in 1935, his interests lay in physics and engineering. However, a course in physiology excited him so much he transferred to medicine. He became a research assistant with Hodgkin at the Marine Biology Laboratory in Plymouth in 1939 studying electrical impulse transmission from neuron to neuron in squid giant **axons**. Earlier researchers, including Julius Bernstein, had already determined these impulses were action potentials and hypothesized (correctly) that a **cell**'s resting potential was determined by the distribution of ions within and outside the cell. They knew the larger, positively charged sodium ions were more highly concentrated outside the cell, and the smaller, positively charged potassium ions were more highly concentrated inside the cell. They also believed that the tiny pores in the **membrane** allowed only the smaller ions to move in or out of the cell, and assumed the cell's resting potential was maintained solely by the movement of potassium ions down the concentration gradient—from inside to outside the cell. Bernstein believed that action potentials caused a temporary "breakdown" of the membrane, permitting ions from both inside and outside the cell to come into contact, thus shifting the resting potential from negative to neutral.

Huxley and Hodgkin decided to test this hypothesis. Using microscopic electrodes to stimulate an action potential in a neuron of a squid giant axon, they discovered that, during an action potential, the voltage was much greater than neutral and that the inside of the cell became positively charged while the outside became negatively charged. This meant that the direction of voltage was actually reversed. They decided that special "gates" or "channels" must penetrate the membrane, opening when subjected to voltage stimulation to allow larger sodium ions to move through the membrane. In 1947, Sir Bernard Katz, Huxley, and Hodgkin tested this theory. By inserting an electrode into cell membranes of squid giant axons and using a special voltage clamp to maintain the membrane at a particular voltage level, they watched the flow of ions with the aid of radioactive isotopes. Using these observations, Huxley created a mathematical model of an action potential using only a hand-cranked adding machine. The remarkable nature of this feat is even more remarkable considering modern biochemical methods of studying gates, channels, sodium-potassium exchange pumps, and the movement of ions across cell membranes, did not exist until the 1980s.

Huxley's research was interrupted by World War II. In 1940, he began conducting operational research in gunnery for the Anti-Aircraft Command and, later, the Admiralty. In 1946 he assumed a research fellowship awarded to him in 1941 by Trinity College, Cambridge and continued his former research while teaching in the Department of Physiology. Around 1952, Huxley began investigating muscle contraction, studying the striation pattern of isolated muscle fibers. He developed a special interference light **microscope** which allowed the observance of living, unstained muscle tissue. In 1960, he was named head of the physiology department and Jodrell Professor of Physiology at University College, London, and was appointed Royal Society Research Professor at the University of London in 1969 where he remains as professor emeritus. He has honorary degrees from universities in several countries, became a Fellow of the Royal Society in 1955, received the Copley Medal of the Royal Society in 1972, and was knighted in 1974.

Huxley married Jocelyn Richenda Gammell Pease, daughter of the Honorable H.B. Pease and M.S. Pease, a geneticist. Huxley's wife is a Justice of the Peace and extremely active in public work. The couple have five daughters and one son.

HUXLEY, HUGH ESMOR (1924-)
English-born American molecular biologist

Hugh Huxley has devoted his career to answering the question of how muscles contract. Although his initial answer—the sliding filament theory—has since become part of every standard biology and physiology textbook, Huxley has relentlessly pursued the finer details of muscle contraction well into his seventies.

Huxley was born to a middle-class Welsh family in Birkenhead, Cavendish, England, in 1924. His father, Thomas, was an accountant for the post office. But both he and Huxley's mother, Olwen (Roberts), "were people of remarkable intellect, great readers, lovers of music and with great moral strength and power of judgment," Huxley wrote in an autobiographical essay in 1996. "They instilled in my sister and me the idea that if we worked hard and tried hard enough we would win scholarships to University, perhaps even to Cambridge." As a boy, Huxley developed an interest in physics, and tinkered with electric motors, shocking coils, and short-wave radio receivers. "Atomic physics, relativity and quantum theory—or what little I knew of it—were to me then subjects of magical interest, offering glimpses of the ultimate nature of reality." When not puttering with science projects, he took long bicycle rides in the countryside.

Huxley entered Cambridge University in 1941 as World War II intensified. After two years of physics, however, he began to feel "very restive at playing no direct part in, nor even being very near to, the great wartime events that were taking place," and joined the Royal Air Force as a radar officer in 1943. During his four-year tour of duty, Huxley participated in the development of several radar advancements, work which later resulted in his being made a member of the Order of the British Empire.

Enters Molecular Biology Field

The bombing of Hiroshima and Nagasaki in 1945 devastated Huxley, causing him to fundamentally question his idealistic view of nuclear physics. He later wrote, "There was dismay and disillusionment that the first practical consequences of all that beautiful work in atomic and nuclear physics had been the atom bomb, and my reluctant conclusion that I would never be able to enjoy working in that field without feelings of guilt." He considered switching his course of study to economics, but then elected to finish his physics training, returning to Cambridge in 1947. "I wanted to do scientific research involving physics," he wrote, "but far away from its wartime uses." Believing he could make a contribution to the life sciences by applying physics techniques to biology and medicine, Huxley secured a place as a student researcher at the renowned Cavendish Laboratory at Cambridge, where such scientists as **James Watson** and **Francis Crick** then conducted their work as part of a small, newly formed research group of molecular biologists. At Cavendish, Huxley worked on his thesis, using x-ray crystallography to analyze the structure of muscle tissue.

Andrew Fielding Huxley. *(The Library of Congress. Reproduced by permission.)*

Discovers Sliding Filament Structure

Upon receiving his Ph.D. in **molecular biology** in 1952, Huxley came to the Massachusetts Institute of Technology (MIT) to work with the newly invented electron **microscope**. Huxley soon realized that the images produced by x-ray crystallography and electron microcopy could provide much more information about the molecular structure of cells if they were analyzed together rather than separately. Huxley also became intrigued by the work of Jean Hanson, a fellow British researcher at MIT who had been using phased-contrast light microscopy to investigate the contraction of rabbit and insect muscle fibers. The two researchers combined their efforts and, as they produced increasingly improved images of their specimens, they slowly realized that muscle fibers were organized into bundles of overlapping filaments made from two **proteins**, actin and myosin. They published their findings in the journal *Nature* in 1953, but on the advice of their supervisor, did not include any of their ideas as to how the structure might work to make a muscle contract.

Later that year, Huxley met British physiologist **Andrew Huxley** (no relation), who was to win the 1963 Nobel Prize in medicine or physiology for his work on how nerves stimulate muscles to contract. Andrew Huxley, on a visit to MIT, expressed great interest in Hugh Huxley's theory of a sliding-filament structure, and said that he had been conducting research on the same idea. Their concept was that, in the presence of the biochemical **adenosine triphosphate (ATP)**, the myosin and actin fibers would slide past each other, effectively shortening the muscle. In 1954, by agreement, the two Huxleys published adjacent papers in *Nature* describing their views on the principle. Time, improved instruments, and countless experiments have proven them correct, but Hugh Huxley's contribution, which was first published in a smaller paper that appeared after the 1953 paper with Hanson and before the 1954 *Nature* piece, has been obscured. Perhaps because of confusion over their last name or Andrew Huxley's greater prominence as a Nobel Prize winner, some science historians have mistakenly attributed the sliding filament theory to only Andrew Huxley. "I think I'm credited with the first mention of the sliding mechanism in print in 1953," Hugh Huxley said in a 1997 interview with contributor Karl Leif Bates.

In 1956, Hugh Huxley returned to Great Britain as a Medical Research Council external staff member at University College, London. Four years later, he moved to the Medical Research Council Laboratory of Molecular Biology at Cambridge, an institution which had grown out of the original Cavendish research group. Huxley eventually became department director in 1977 and served in that capacity until 1987 when he accepted a professorship at Brandeis University and their Rosensteil Basic Medical Sciences Research Center. He also became the director in 1988, a post he held until 1994.

More Precise Investigations

In the decades following his discovery of the sliding filament structure, Huxley devoted his career to first providing definitive experimental proof of his theory and then to understanding the exact chemical and physical process of the sliding mechanism or cross-bridge function. In the late 1950s, he spent several years trying to generate more detailed images of the fine structures of the fibers. As a diversion from this often frustrating work, he applied his skills to the study of virus structures. In the process he developed new staining techniques that, when later applied to his muscle studies, resulted in his first significant images of the sliding mechanism. By the early 1970s, the efforts of Huxley and others had led in the widespread acceptance of the sliding filament theory. "In fact, several people asked me what I was going to work on next, now that the muscle problem was essentially solved, and were puzzled and disappointed when I said I would continue working on muscle because I did not think the evidence was really there yet," Huxley wrote in his autobiographical article.

Indeed, since that time Huxley has continued to probe the structure of the actin and myosin proteins themselves and how the chemical impulse from ATP is turned into motion by these large molecules. Determining how myosin and actin behave has been a persistent problem since it is very difficult to observe the molecules change their shape in microseconds. Huxley resorted to using more and more powerful x-ray beams for diffraction studies at the Cornell University and Argonne

National Lab synchrotrons. Also, myosin itself proved a difficult protein to map. Its full structure was not determined until 1993. "It looks like it was built to change shape, but we still don't know how it does," Huxley told Bates. "It's a fascinating problem. Very intractable. And the problem continues to be important," Huxley added, "because researchers over the years have identified proteins very similar to actin and myosin that are responsible for all kinds of motion in non-muscle cells, like cells that change their shape. It's a very basic problem, and it's about time we figured it out."

HYBRID

A hybrid is the offspring of a mating between two genetically dissimilar parents. This often occurs between two related **species** or varieties. When hybridization is attempted between two unrelated species, the genetic incompatibility is usually too great and the **embryo** fails to thrive or **fertilization** does not occur.

Oftentimes, hybrids are sterile, resulting from the failure of the **chromosome**s to pair up properly during **meiosis**. The more distantly related the **organism**s the more likely it is that the offspring will be sterile. Sterile **plants** are bred using vegetative means of propagation, while sterile **animal** hybrids must always be bred from the parental types.

Perhaps the most famous hybrid is the cross between a horse and a donkey: the mule. The mule is sterile because the horse has a **diploid** chromosome number of 62 and the donkey has a diploid chromosome number of 64. Subsequently, the hybrid mule has a diploid number of 63 chromosomes, which are unable to pair correctly during meiosis. The mule is a cross between a female horse and a male donkey. The cross between a male horse and a female donkey is a hinny.

HYDROCARBONS

Hydrocarbons are **chemical compound**s that are composed entirely of carbon and hydrogen **atoms**. The simplest hydrocarbon is methane (CH_4), which contains one carbon and four hydrogen atoms, and usually exists as a gas. The largest hydrocarbons can contain hundreds of carbon atoms and larger numbers of hydrogen atoms, and exist as solids.

Hydrocarbons occur naturally in petroleum (also known as crude oil), coal, and natural gas. These materials are known as "**fossil fuels**," because they are derived from ancient plant **biomass** that became buried deep in the ground, where **chemical reaction**s occurring in an oxygen-deficient environment and under conditions of extremely high pressure and **temperature** resulted in the formation of complex mixtures of hydrocarbons. Fossil fuels are mined from deposits in the environment and are used directly as fuels, or are processed in refineries into hydrocarbon fractions which are used as fuels (such as gasoline) or as industrial feedstocks for manufacturing lubricants, plastics, and other materials.

Hydrocarbons can be classified into three groups: aliphatics, alicyclics, and aromatics.

Aliphatic hydrocarbons are compounds in which the carbon atoms occur as an open chain. Saturated aliphatics (or alkanes) have a single bond between all of the adjacent carbon atoms, and they cannot contain any additional hydrogen atoms in their molecular structure (this also means that they cannot react with hydrogen gas). In contrast, unsaturated aliphatics (or paraffins) contain one or more double or triple bonds, and under suitable conditions will react with hydrogen to form saturated compounds. These differences can be illustrated by the following series of twocarbon aliphatic carbons: ethane, H_3CCH_3; ethylene, $H_2C=CH_2$; and acetylene, $HC(CH$. Unsaturated aliphatics are relatively unstable chemically, and for this reason do not occur naturally in petroleum. They are produced during industrial refining and by photochemical reactions in the environment after crude oil is spilled.

Alicyclic hydrocarbons have some or all of their carbon atoms arranged as a ring structure, which can be saturated or unsaturated.

Aromatic hydrocarbons contain one or more five- or six-carbon rings in their molecular structure. The simplest aromatic hydrocarbon is benzene, with a C_6H_6 ring structure.

HYDROLOGIC CYCLE

The hydrologic cycle, sometimes called the **water** cycle, is the movement of water between the air, land and bodies of water. The hydrologic cycle includes numerous physical, chemical and biological processes. Almost 98% of the water on Earth is in liquid form. The rest is either bound up in ice, in the soil, in living **organism**s, or in the **atmosphere**. The hydrologic cycle is balanced on Earth because any water moving out of an environment is balanced by water moving in.

The hydrologic cycle also shows how water moves from one state to another. Evaporation is the changing of water from a liquid state to a gaseous state. Because of the heat of the sun, water evaporates from oceans, lakes, ponds, rivers, streams and glaciers and enters the atmosphere as gaseous water vapor. A small amount of water evaporates from the soil surface. In addition, water enters the atmosphere through transpiration from **plants** and **respiration** by **animal**s and plants. Transpiration (sometimes called evapotranspiration) is evaporation of water from the leaves of plants. This is the primary way that plants move water and dissolved **nutrients**. Respiration is the process of releasing **energy** from food **molecule**s. Water is produced during respiration and some of this water is released into the atmosphere as a gas.

Once in the atmosphere, wind can carry the water vapor around. Under cool atmospheric conditions, the gaseous water vapor molecules can condense into a liquid or solid form. This is how clouds are formed. If the liquid or solid drops get large and heavy enough, they fall to the ground as **precipitation**. Rain, snow, sleet, and hail are all forms of precipitation.

Most of this precipitation falls into the oceans, which cover about 71% of the Earth. However, there are several fates of the water that falls on the land. Much of it seeps into the ground until it reaches a zone where all of the cracks and spaces between soil particles are completely saturated with water. The top of this saturation zone is the water table. The groundwater in this saturated zone seeps slowly through the ground until it is discharged directly from the ground as a spring or into existing lakes, rivers, estuaries or oceans. In some locations, humans drill wells into the ground below the water table and use this groundwater as a source of fresh water for drinking and other purposes. Plants, whose roots reach into the soil, take up some of the water that percolates into the ground for their biological processes.

The remainder of the water that falls on the land is acted on by gravity, forming runoff. Small trickles of runoff eventually join together to form creeks and rivers, which eventually open into ponds, lakes, estuaries and oceans. Water evaporates once again during its passage to these bodies of water, and continues to evaporate from them, continuing through the hydrologic cycle. Because water is always being moved through the hydrologic cycle, it is constantly being renewed. The amount of time that a water molecule stays in a particular part of the hydrologic cycle is referred to as its residence time.

Humans have a great influence on the hydrologic cycle. By damming rivers we prevent runoff and change evaporation rates. Changes in land use such as cutting down a forest or building impervious structures such as roads and parking lots reduce the amount of transpiration and increase the amount of runoff. In addition, chemical **pollutants** can enter runoff, surface water and groundwater. This can have major impacts on water quality and organisms that live in and use this water, including human life. Overuse of water can reduce supplies of surface water and groundwater and change groundwater flow. Atmospheric **pollution** resulting from global warming may increase evaporation rates, which will result in further, perhaps devastating changes in global climate.

HYDROLYSIS

Hydrolysis is a **chemical reaction** in which **water** reacts with another compound to split it apart. Appropriately, this process takes its name from the Greek expression meaning ''breaking up with water.'' Hydrolysis occurs because water forms negative and positive ions which can displace other chemical **species** in solution, thus causing changes in **chemical bond**s. Hydrolytic reactions may be accelerated by the addition of a catalyst, such as an **enzyme**, or by controlling other reaction factors such as **temperature** and **pH**. These reactions are important in many biochemical processes, particularly in enzyme catalyzed reactions such as those used in **digestion**. Hydrolysis reactions are also commonly used in the synthesis of industrial chemicals such as glycerine and **alcohol**.

In inorganic chemistry, hydrolysis is typified by the dissolution of salts made from weak acids and strong bases, such as sodium acetate. For example, when sodium acetate dissolves in water, it is separated into its component ions, (Na^+ and $CHCHOH^-$. In organic chemistry, hydrolytic reactions can split apart much larger **molecule** such as **proteins**. Protein hydrolysis is measured by evaluating differences in solubility,

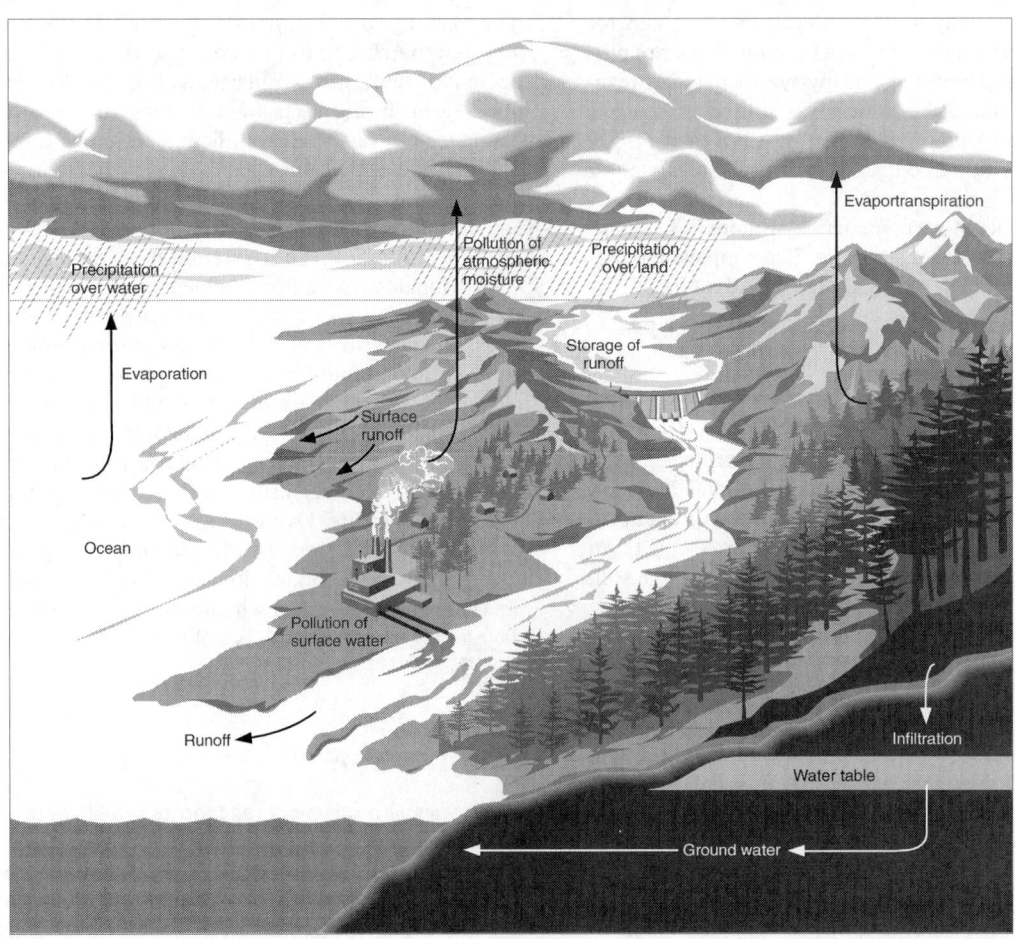

The hydrologic cycle on Earth. *(Illustration by Hans & Cassady, Inc.)*

optical rotation of **light**, and with spectrophotometric methods which measure how much light is measured at a given wavelength. While these methods can determine the rate of the hydrolysis, they do not necessarily identify what new compounds are formed in the reaction. Other analytical methods are used to characterize hydrolytic byproducts.

Strong acids or bases are typically used to aid in these reactions. In many biochemical processes, proteins, **fats**, oils, and **carbohydrate**s are broken down, which are hydrolytic reactions catalyzed by enzymes. These enzymes, respectively referred to as proteases, lipases, and amylases, are important in the process of digestion.

Hydrolysis is also useful industrially; for example, hydrolytic reactions are used to split fats into fatty acids and glycerol. This particular reaction, which is conducted under alkaline conditions, is also known as saponification and is part of the soap making process. Similarly, starches can be broken down into sugars by using strong acids or bases. Enzymes that are used to assist in this reaction are generally classified as amylases. There are a wide variety of specific amylases available including alpha ammylase, glucoamylase, and glucose isomerase. These amylases are used to break down starch into components that can be used in adhesives, sweeteners, and chemical feedstocks for other reaction processes. Another use of the hydrolysis reaction is to synthesize alcohols from olefins, for example, ethanol, CH_3COOH, from ethene, CH_2CH_2. This reaction requires a strong acid to drive it at an appropriate rate.

Chemical researchers continue to explore the mechanisms of hydrolysis. One area of research is focused on building **polymer**s which are easily hydrolyzed to make them more **biodegradable**. In the textile industry, the chemicals used to finish fabrics are evaluated to find ways to make them more resistant to hydrolysis. Also, hydrolysates made from collagen, corn, soy, wheat, and silk proteins are used in a variety of commercial products like cosmetics.

See also Digestion; Protein

HYMAN, LIBBIE HENRIETTA (1888-1969)

American invertebrate zoologist

Libbie Henrietta Hyman earned an international reputation for her monumental six-volume work on the classification of invertebrates. Although she considered her invertebrate treatise essentially a ''compilation'' of the literature, others have called it a remarkable synthetic work. Compiled by one independent woman with enormous knowledge of the field and a great facility for translating European languages, it represents a textbook of the invertebrate animal **kingdom** that whole academies might have attempted. Hyman's treatise consists of judicious analysis and integration of previously scattered information; it has had a lasting influence on scientific thinking about a number of invertebrate animal groups, and the only works that can be compared with hers are of composite authorship. Hyman also influenced the teaching of **zoology** classes nationwide with the publication of her laboratory manuals.

Hyman was born on December 6, 1888 in Des Moines, Iowa, the third of four children and the only daughter. Her parents were Jewish immigrants; her father, Joseph Hyman, came to the United States from Konin, Poland, at age fourteen, and her mother, Sabina Neumann, was born in Stettin, Germany. Hyman's childhood and youth were spent in Fort Dodge, Iowa, where her father kept an unsuccessful clothing store. Her home life was strict and without affection. Her father, twenty years older than her mother, worried about his declining fortunes and ignored his children, although he did have scholarly inclinations, keeping volumes of Dickens and Shakespeare, which Hyman read. In her brief autobiography, Hyman remembered her mother as being ''thoroughly infiltrated with the European worship of the male sex.'' Her mother required her to do ''endless housework'' caring for her brothers, whom Hyman believed were ''brought up in idleness and irresponsibility.''

From an early age, Hyman demonstrated an interest in nature. She learned the scientific names of **flowers** from a high-school botany book that belonged to her brothers, and she made collections of butterflies and moths. She remembered being initially puzzled by classification, until she suddenly realized that the flowers of a common cheeseweed were the same as the flowers of a hollyhock. In 1905, she graduated from Fort Dodge High School. She was class valedictorian but had failed to attract the attention of her science teachers. Although she passed the state examination for teaching in the country schools, she was too young to be appointed to a teaching position and so returned to high school during 1906 for advanced studies in science and German. When these classes ended, she took a factory job, pasting labels on oatmeal cereal boxes.

Attends the University of Chicago

On her way home from the factory one fall afternoon, she met Mary Crawford, a Radcliffe graduate and high school language teacher who was ''shocked'' to learn what she was doing. Crawford arranged for Hyman to attend the University of Chicago with scholarship money that was available to top students. ''To the best of my recollection,'' Hyman said, ''it had never occurred to me to go to college. I scarcely understood the purpose of college.'' At the university, she began a course in botany, but was discouraged by anti-semitic harassment from a laboratory assistant. Instead, she majored in zoology and graduated in 1910 with a B.S. degree. Professor Charles Manning Child, from whom she had taken a course during her senior year, encouraged her to enter the graduate program. As Child's graduate assistant, she directed laboratory work for courses in elementary zoology and comparative vertebrate **anatomy**.

Hyman was not free from family responsibilities, however. Her father had died in 1907; her possessive mother moved to Chicago with her brothers, and Hyman was again required to keep house for them and endure their continuing disapproval of her career.

Hyman received her Ph.D. in 1915, when she was twenty-six years old, for a dissertation entitled, ''An Analysis of the Process of Regeneration in Certain Microdrilous Oligochaetes.'' She then accepted an appointment as Child's research assistant, a position she held until he neared retirement. Her work in Child's laboratory consisted of conducting physiological experiments on lower **invertebrates**, including hydras and flatworms. It was during this time that Hyman realized that many of these common animals were misidentified because they had not been carefully studied taxonomically. She became a taxonomic specialist in these invertebrate groups. Hyman's interest in invertebrates had a strong aesthetic component; she confessed a deep fondness for "the soft delicate ones, the jellyfishes and corals and the beautiful microscopic **organisms**."

During her time as a laboratory assistant, helping Child direct his classes, Hyman had felt that a better student guide book was needed, and now she wrote one. *A Laboratory Manual for Elementary Zoology* was published in 1919 by the University of Chicago Press. The first printing quickly sold out, and in 1929 she wrote an expanded edition. She also published, in 1922, *A Laboratory Manual for Comparative Vertebrate Anatomy,* which also enjoyed brisk sales. The second edition of this manual was published in 1942 as *Comparative Vertebrate Anatomy.* She was never excited about vertebrates, however, and she refused to consider a third edition. (The third edition was published in 1979, the work of eleven contributors.)

Laboratory Manuals Assure Financial Independence

By 1930, Hyman had realized she could live on the royalties from the sale of her laboratory manuals, and she resigned her position in the zoology department, leaving Chicago in 1931 to tour western Europe for fifteen months. She never again worked for wages. When she returned from her travels, she settled near the American Museum of Natural History in New York City, where she lived modestly, close to the museum's "magnificent" library, determined to devote all of her time to writing a treatise on the invertebrates. In 1937, she was made an honorary research associate of the museum. Although unsalaried, she was given an office, where she placed food and water at the window for pigeons. The first volume of *The Invertebrates* appeared in 1940.

Hyman had always wanted to live in the country and indulge her interest in gardening. In 1941, she bought a house in Millwood, Westchester County, about thirty-five miles north of Times Square. She commuted to her work at the museum until 1952, when she sold the house and returned to New York City. Although she said that gardening and commuting had taken time away from her treatise, during those years of residence in the country she completed the second and third volumes, which were both published in 1951. At the museum, Hyman spent most of her time in the library. She read, made notes, digested information, composed in her head, and typed the first and only draft of her books on her manual typewriter. She also taught herself drawing, and her books contain her own illustrations. She apparently never had a secretary or an assistant. The fourth volume of the treatise was published in 1955, and the fifth in 1959.

Hyman loved music and regularly attended performances of the Metropolitan Opera and the New York Philhar-

monic. Her physical appearance had been altered by a bungled sinus operation in 1916, and to many she presented a brusque and formidable exterior, but she was not a recluse. She carried on a lively correspondence with scientists who sent her specimens or consulted her. She encouraged young scientists and contributed to charitable causes. She acquired a small, but valuable art collection, and made summer collecting trips to marine laboratories.

Receives Awards and Honors

Hyman's recognition began with publication of her first invertebrate volume. The University of Chicago awarded her an honorary doctor of science degree in 1941, and honorary degrees followed from other colleges. She received the Daniel Giraud Elliot Medal of the National Academy of Sciences in 1951, the Gold Medal of the Linnaean Society of London in 1960, and the American Museum presented her with its Gold Medal for Distinguished Achievement in Science in April 1969, a few months before she died.

Hyman served as president of the Society of Systematic Zoology in 1959, and she edited the society's journal, *Systematic Zoology,* from 1959–1963. She was vice president of the American Society of Zoologists in 1953 and a member of the National Academy of Sciences, as well as Phi Beta Kappa, Sigma Xi, the American Microscopical Society, the American Society of Naturalists, the Marine Biological Laboratory of Woods Hole, the American Society of Limnology and Oceanography, and the Society of Protozoologists. In addition to her books, she published 135 scientific papers between 1916 and 1966. Her early papers represent contributions to Child's physiological projects; her taxonomic and anatomical papers began to appear in 1925.

In the last decade of Hyman's life, her health was poor and her work on invertebrates had become more difficult. In 1967, at the age of seventy-eight and suffering from Parkinson's disease, she published the sixth volume of her treatise. She announced in its preface that this would be the last volume of *The Invertebrates* from her hands, although McGraw-Hill intended to continue the series with different authors. "I now retire from the field," Hyman wrote, "satisfied that I have accomplished my original purpose—to stimulate the study of invertebrates." She died on August 3, 1969.

HYPOTHALAMUS

Deep within the center of the **brain**, below the thalamus and above the pituitary, lies the essential link between the human body's endocrine and **nervous system**s, and the overall controller of the autonomic or automatic nervous system. Connected to the brain and **spinal cord**, the hypothalamus automatically monitors and adjusts the body's nervous system and its **metabolism**, which in turn impact eating, drinking, **hormone**s and sexual drives, **temperature** and emotional reactions. Thus, this small part of the brain controls each person's basic drives for hunger, thirst, sex, and **emotions**.

The hypothalamus is found in the limbic system of the brain, which is connected by nerve fibers with the mental or intellectual part of the brain, the cortex. Because the hypothalamus directs both emotions and physical changes, it is the link between subjective feelings of emotion based in the brain's cortex and the physiological changes throughout the rest of the body. For example, the hypothalamus controls **heart** rate and **blood** flow, breathing, and dilation of the pupils in the eyes. **Cell**s in the hypothalamus react to increased temperature caused by exercise or heat by sending signals to the nervous system to begin the body's cooling processes, including sweating and dilation of the blood vessels. Likewise, it signals the nerves to initiate shivering, and the blood to slow its flow when the body is cold. Each of these physical parameters can change drastically according to a person's emotions, an interaction balanced by the hypothalamus.

The balance of hormones in the body, or the **endocrine system**, is also controlled by the hypothalamus. It regulates this part of the body by sending messages to the main hormone gland, the pituitary, which controls other **endocrine gland**s. **Nerve impulses** from other parts of the brain influence the hypothalamus to send such messages. These hormones control the body's metabolism, responses to **stress**, fluid and sugar balance, and reproductivity. Thus sexual development and arousal are also controlled by the hypothalamus, which secretes a gonadotropin-releasing hormone, or GnRH, at the beginning of puberty. This hormone then stimulates other hormonal production and thus development of the body's sexual **organ**s. Recent research by Dr. **Theo Colborn** and other international researchers has found that many persistent chemicals found throughout the world may be impacting the hypothalamus, thus affecting some people's sexual, emotional and physical development.

ICHTHYOLOGY

Ichthyology is the scientific study of the multiplicity of aquatic **chordates** known as fish. It has been noted that fishes of various **class**es differ among themselves as much as, or more than, other chordate classes (amphibians, reptiles, birds, and **mammals**) differ between each other. Jawless fish, the Class Agnatha, are parasitic and include the lampreys and hagfishes. They are bereft of jaws and paired fins. Sharks, rays, and skates are fish with a cartilaginous skeleton and belong to the Class Chondrichthyes. The largest class of fishes, Class Osteichthyes, are the common fishes with bones and are well known to sportsmen and commercial fishermen and people who enjoy eating non-red meat.

The vertebrate kidney has evolved from a primitive pronephros through a mesonephros to the highly evolved kidney of humans, the metanephros. Not just humans but all mammals have a metanephric kidney. All amphibians have a mesonephric kidney. As stated above, fishes of various classes differ significantly. For instance, hagfishes retain the primitive pronephros; most bony fish develop a mesonephros.

Fish are harvested for commercial purposes. Thus, many millions of fish are taken each year for human consumption. It should not be surprising therefore to learn that many kinds of malignancies have been observed in these **animals**. The Registry of Tumors of Lower Animals, located at George Washington University, reported that in its collection, almost three-fourths of the tumors of chordates were of tumors of fish. Most of the fish tumors were of bony fish. This knowledge gleaned from examining fish helps in the evaluation of **pollution** of our fresh **water** and oceans. A reduced prevalence of fish tumors at a particular location suggests the efficacy of clean-up efforts; an increased incidence is a warning flag concerning environmental deterioration.

Fishes are vulnerable to infections by **virus**es as are higher vertebrates. A herpesvirus of the channel catfish has been isolated and its viral **DNA** has been characterized in detail. Other fish herpesviruses have been studied and they are thought to cause **cancer** in carp and salmon. Virus study in fish may lead to enhanced understanding how viruses cause disease in fish and in higher **organism**s.

IMMUNE SYSTEM

The immune system is the body's biological defense mechanism that protects against foreign invaders. Only in the last century have the components of that system and the ways in which they work been discovered, and more remains to be clarified.

Since ancient times medical observers had noticed that the body seemed to have powers to protect itself and resist disease. In particular, people who survived some infectious diseases did not suffer from those diseases again during their lifetime. This led to the practice of variolation in Asia, whereby people were injected with a mild case of smallpox to prevent the later development of a severe case of the disease. Lady Mary Wortley-Montague introduced variolation to Britain from the Ottoman Empire in 1720. The procedure was rather risky, however, because the injected person could develop an acute rather than mild case of smallpox, which could lead to an epidemic.

The true roots of immunology—or the study of the immune system—date from 1796 when an English physician, **Edward Jenner**, discovered a method of smallpox vaccination. He noted that dairy workers who contracted cowpox from milking infected cows were thereafter resistant to smallpox. In 1796 Jenner injected a young boy with material from a milkmaid who had an active case of cowpox. After the boy recovered from his own resulting cowpox, Jenner inoculated him with smallpox; the boy was immune. After Jenner published the results of this and other cases in 1798, the practice of Jennerian vaccination spread rapidly.

It was **Louis Pasteur** who established the cause of infectious diseases and the medical basis for immunization. First,

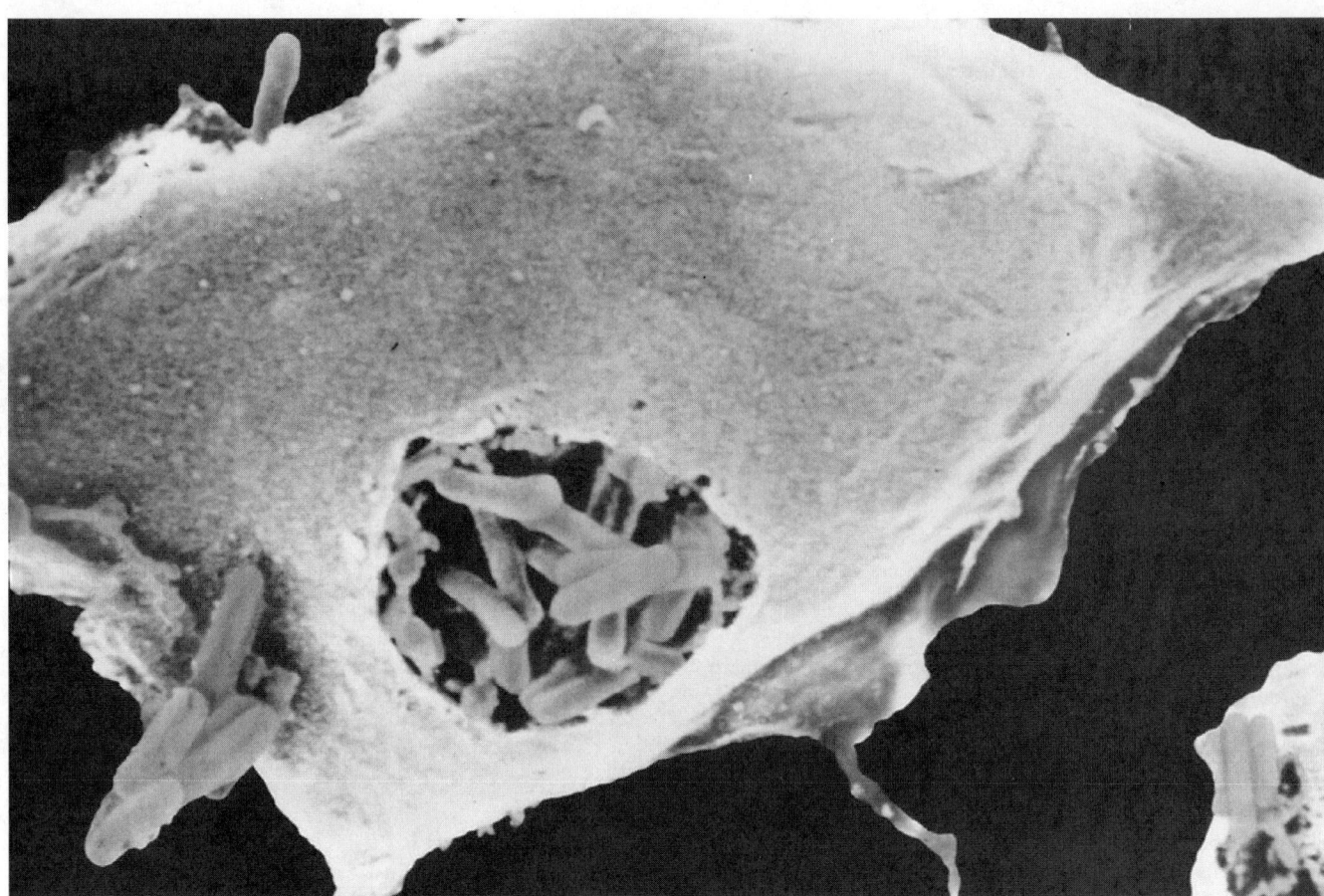

A scanning electron micrograph (SEM) image of a macrophage cell engulfing the small, rod-shaped bacteria that cause tuberculosis. The modern study of how the body combats infection began with Louis Pasteur's germ theory of disease. *(Photo Researchers, Inc. Reproduced by permission.)*

Pasteur formulated his **germ theory** of disease—the concept that disease is caused by communicable **microorganism**s. In 1880, Pasteur discovered that aged **culture**s of fowl **cholera bacteria** lost their power to induce disease in chickens but still conferred immunity to the disease when injected. He went on to use *attenuated* (weakened) cultures of **anthrax** and rabies to vaccinate against those diseases. The American scientists Theobald Smith (1859-1934) and Daniel Salmon (1850-1914) showed in 1886 that bacteria killed by heat could also confer immunity.

Why vaccination imparted immunity was not yet known. In 1888, Pierre-Paul-Emile Roux (1853-1933) and Alexandre Yersin (1863-1943) showed that **diphtheria** bacillus produced a toxin that the body responded to by producing an antitoxin. **Emil von Behring** and **Shibasaburo Kitasato** found a similar toxin-antitoxin reaction in tetanus in 1890. He discovered that small doses of tetanus or diphtheria toxin produced immunity, and that this immunity could be transferred from **animal** to animal via serum. Von Behring concluded that the immunity was conferred by substances in the **blood**, which he called antitoxins, or antibodies. In 1894, Richard Pfeiffer (1858-1945) found that antibodies killed cholera bacteria (bacterioloysis). Hans Buchner (1850-1902) in 1893 discovered

another important blood substance called complement (Buchner's term was *alexin*), and **Jules Bordet** in 1898 found that it enabled the antibodies to combine with **antigen**s (foreign substances) and destroy or eliminate them. It became clear that each **antibody** acted only against a specific antigen. **Karl Landsteiner** was able to use this specific antigen-antibody reaction to distinguish the different **blood groups**.

A new element was introduced into the growing body of information during the 1880s by the Russian microbiologist **Elie Metchnikoff.** He discovered **cell**-based immunity: white blood cells (leucocytes), which Metchnikoff called phagocytes, ingested and destroyed foreign particles. Considerable controversy flourished between the proponents of cell-based and blood-based immunity until 1903, when **Almroth Edward Wright** brought them together by showing that certain blood substances were necessary for phagocytes to function as bacteria destroyers. A unifying theory of immunity was posited by **Paul Ehrlich** in the 1890s; his ''side-chain'' theory explained that antigens and antibodies combine chemically in fixed ways, like a key fits into a lock. Until now, immune responses were seen as purely beneficial. In 1902, however, Charles Richet and Paul Portier demonstrated extreme immune reactions in test animals that had become sensitive to antigens by previous exposure. This phenomenon of hypersensitivity, called **anaphylaxis**, showed that immune responses could cause the body to damage itself. Hypersensitivity to antigens also explained *allergies*, a term coined by Pirquet in 1906.

By the early 1900s, immunology had become an established medical field with its own journals, first in Germany in 1909 and then in the United States in 1916 (the latter published by the world's first immunology society, founded in 1913).

Much more was learned about antibodies in the mid-twentieth century, including the fact that they are **protein**s of the gamma globulin portion of plasma and are produced by plasma cells; their molecular structure was also defined. An important advance in immunochemistry came in 1935 when Michael Heidelberger and Edward Kendall (1886-1972) developed a method to detect and measure amounts of different antigens and antibodies in serum. Immunobiology also advanced. **Frank Macfarlane Burnet** suggested that animals did not produce antibodies to substances they had encountered very early in life; **Peter Medawar** proved this idea in 1953 through experiments on mouse **embryo**s.

In 1957 Burnet put forth his clonal selection theory to explain the biology of immune responses. On meeting an antigen, an immunologically responsive cell (shown by C. S. Gowans [1923-] in the 1960s to be a lymphocyte) responds by multiplying and producing an identical set of plasma cells, which in turn manufacture the specific antibody for that antigen. Further cellular research has shown that there are two types of **lymphocytes** (nondescript lymph cells): B-lymphocytes, which secrete antibody, and T-lymphocytes, which regulate the B-lymphocytes and also either kill foreign substances directly (killer T cells) or stimulate macrophages to do so (helper T cells). Lymphocytes recognize antigens by characteristics on the surface of the antigen-carrying **molecule**s. Researchers in the 1980s uncovered many more intricate biological and chemical details of the immune system components and the ways in which they interact.

Knowledge about the immune system's role in rejection of transplanted **tissue** became extremely important as **organ** transplantation became surgically feasible. Peter Medawar's work in the 1940s showed that such rejection was an immune reaction to antigens on the foreign tissue. Donald Calne (1936-) showed in 1960 that *immunosuppressive* drugs—drugs that suppress immune responses—reduced transplant rejection. These drugs were first used on human patients in 1962. In the 1940s **George Snell** (1903-) discovered in mice a group of tissue-compatibility **gene**s, MHC that played an important role in controlling acceptance or resistance to tissue grafts. **Jean Dausset** found human MHC, a set of antigens to human leucocytes (white blood cells), called HLA. Matching of HLA in donor and recipient tissue is an important technique to predict compatibility in transplants. **Baruj Benacerraf** in 1969 showed that an animal's ability to respond to an antigen was controlled by genes in the MHC complex.

Exciting new discoveries in immunology are on the horizon. Researchers are investigating the relation of HLA to disease; certain types of HLA molecules may predispose people to particular diseases. This promises to lead to more effective

treatments and, in the long run, possible prevention. Autoimmune reaction—in which the body has an immune response to its own substances—may also be a cause of a number of diseases, like multiple sclerosis, and research proceeds on that front. Approaches to **cancer** treatment also involve the immune system. Some researchers, including Burnet, speculate that a failure of the immune system may be implicated in cancer. In the late 1960s, Ion Gresser (1928-) discovered that the protein **interferon** acts against cancerous tumors. After the development of genetically engineered interferon in the mid-1980s finally made the substance available in practical amounts, research into its use against cancer accelerated. The invention of monoclonal antibodies in the mid-1970s was a major breakthrough. Increasingly sophisticated knowledge about the workings of the immune system holds out the hope of finding an effective method to combat one of the most serious immune system disorders, **AIDS**.

Avenues of research to treat AIDS include a focus on supporting and strengthening the immune system. (However, much research has to be done in this area to determine whether strengthening the immune system is beneficial or whether it may cause an increase in the number of infected cells.) One area of interest is cytokines, proteins produced by the body that help the immune system cells communicate with each other and activate them to fight infection. Some individuals infected with the AIDS **virus** HIV (**human immunodeficiency virus**) have higher levels of certain cytokines and lower levels of others. A possible approach to controlling infection would be to boost deficient levels of cytokines while depressing levels of cytokines that may be too abundant. Other research has found that HIV may also turn the immune system against itself by producing antibodies against its own cells.

For many years it was believed that the immune system responded only to invading antigens and was not influenced by psychological events. However, building on research that began in the mid-1960s, scientists have determined that the immune system is also affected by a person's psychological health, or state of mind. This branch of research is referred to as pscyhoimmunology, or psychoneuroimmunology (the study of the relationship among psychology, neurology, and immunology). A complex network of nerves, **hormone**s, and neuropeptides appear to link the immune system and an individual's psyche. For example, extreme psychological **stress** has been shown to suppress the immune system and accelerate disease in people with HIV. (Short-term stress is believed to have certain benefits to the body.) Other psychosocial factors—such as a fixation on dying, clinical **depression**, a lack of purpose in life, inability to be assertive, and lack of a supportive network of friends and family—may also affect the immune system. Research into pscyhoimmunology focuses on treatments that can impact stress levels and other psychological factors.

Advances in immunological research indicate that the immune system may be made of more than 100 million highly specialized cells designed to combat specific antigens. While the task of identifying these cells and their functions may be daunting, headway is being made. By identifying these specific cells, researchers may be able to further advance another promising area of immunological research—the use of recombinant **DNA technology**, in which specific proteins can be mass produced. This approach has led to new cancer treatments that can stimulate the immune system by using synthetic versions of proteins released by interferons.

IMMUNITY, CELL-MEDIATED

The immune system is a network of **cell**s and **organ**s that work together to protect the body from infectious **organism**s. Many different types of organisms such as **bacteria**, **virus**es, **fungi**, and **parasite**s are capable of entering the human body and causing diseases. It is the immune system's job to recognize these agents as foreign and destroy them.

The immune system can respond to the presence of a foreign agent in one of two ways. It can either produce soluble **protein**s called antibodies, which can bind to the foreign agent and mark them for destruction by other cells. This type of response is called a humoral response or an **antibody** response. Alternately, the immune system can mount a cell-mediated immune response. This involves the production of special cells that can react with the foreign agent. The reacting cell can either destroy the foreign agents, or it can secrete chemical signals that will activate other cells to destroy the foreign agent.

During the 1960s, it was discovered that different types of cells mediate the two major classes of immune responses. The T **lymphocytes**, which are the main effectors of the cell-mediated response, mature in the thymus, thus the name T cell. The B cells, which develop in the adult bone marrow, are responsible for producing antibodies. There are several different types of T cells performing different functions. These diverse responses of the different T cells are collectively called the "cell-mediated immune responses."

There are several steps involved in the cell-mediated response. The pathogen (bacteria, virus, fungi, or a parasite), or foreign agent, enters the body through the **blood** stream, different **tissue**s, or the respiratory tract. Once inside the body, the foreign agents are carried to the spleen, lymph nodes, or the mucus-associated lymphoid tissue (MALT) where they will come in contact with specialized cells known as **antigen**-presenting cells (APC). When the foreign agent encounters the antigen-presenting cells, an immune response is triggered. These antigen presenting cells digest the engulfed material, and display it on their surface complexed with certain other proteins known as the Major Histocompatibility Class (MHC) class of proteins.

Next, the T cells must recognize the antigen. Specialized receptors found on some T cells are capable of recognizing the MHC-antigen complexes as foreign and binding to them. Each T cell has a different receptor in the cell **membrane** that is capable of binding a specific antigen. Once the T cell receptor binds to the antigen, it is stimulated to divide and produce large amounts of identical cells that are specific for that particular foreign antigen. The T lymphocytes also secrete various chemicals (cytokines) that can stimulate this proliferation. The cytokines are also capable of amplifying the immune defense functions that can eventually destroy and remove the antigen.

In **cell-mediated immunity**, a subclass of the T cells mature into cytotoxic T cells that can kill cells having the foreign antigen on their surface, such as virus-infected cells, bacterial-infected cells, and tumor cells. Another subclass of T cells called helper T cells activate the B cells to produce antibodies that can react with the original antigen. A third group of T cells called the suppressor T cells are responsible for regulating the immune response by turning it on only in response to an antigen and turning it off once the antigen has been removed.

Some of the B and T lymphocytes become "memory cells," that are capable of remembering the original antigen. If that same antigen enters the body again while the memory cells are present, the response against it will be rapid and heightened. This is the reason the body develops permanent immunity to an infectious disease after being exposed to it. This is also the principle behind **immunization**.

IMMUNITY, HUMORAL

One of the ways in which the **immune system** responds to pathogens is by producing soluble **protein**s called antibodies. This is known as the humoral response and involves the activation of a special set of **cell**s known as the B **lymphocytes**, because they originate in the bone marrow. The humoral immune response helps in the control and removal of pathogens such as **bacteria**, **virus**es, **fungi**, and **parasite**s before they enter host cells. The antibodies produced by the B cells are the mediators of this response.

The antibodies form a **family** of plasma proteins referred to as immunoglobulins. They perform two major functions. One of the functions of an **antibody** is to bind specifically to the **molecule**s of the foreign agent that triggered the immune response. A second function is to attract other cells and molecules to destroy the pathogen after the antibody molecule is bound to it.

When a foreign agent enters the body, it is engulfed by the **antigen** presenting cells or the B cells. The B cell that has a receptor (surface immunoglobulin) on its **membrane** that corresponds to the shape of the antigen binds to it and engulfs it. Within the B cell, the antigen-antibody pair is partially digested, bound to a special **class** of proteins called MHC-II, and then displayed on the surface of the B cell. The helper T cells recognize the pathogen bound to the MHC-II protein as foreign and becomes activated.

These stimulated T cells then release certain chemicals known as cytokines (or lymphokines) that act upon the primed B cells (B cells that have already seen the antigen). The B cells are induced to proliferate and produce several identical cells capable of producing the same antibody. The cytokines also signal the B cells to mature into antibody producing cells. The activated B cells first develop into lymphoblasts and then become plasma cells, which are essentially antibody producing factories. A subclass of B cells does not differentiate into plasma cells. Instead, they become memory cells that are capable of producing antibodies at a low rate. These cells remain in the immune system for a long time, so that the body can respond quickly if it encounters the same antigen again.

The antibody destroys the pathogen in three different ways. In neutralization, the antibodies bind to the bacteria or toxin and prevent it from binding and gaining entry to a host cell. Neutralization leads to a second process called opsonization. Once the antibody is bound to the pathogen, certain other cells called macrophages engulf these cells and destroy them. This process is called **phagocytosis**. Alternately, the immunoglobulin IgM or IgG can bind to the surface of the pathogen and activate a class of serum proteins called the complement, which can cause lysis of the cells bearing that particular antigen.

In the humoral immune response, each B cell produces a distinct antibody molecule. There are over a million different B lymphocytes in each individual, which are capable of recognizing a corresponding million different antigens. Since each antibody molecule is composed of two different proteins (the **light** chain and the heavy chain), it can bind two different antigens at the same time.

IMMUNIZATION

When a foreign agent (pathogen) enters the body, a protective system known as the **immune system**, consisting of a complex network of organs and **cell**s that can recognize the pathogen, mounts an immune response against it.

Any substance capable of generating an immune response is called an **antigen** or an immunogen. Antigens are not the foreign **bacteria** or **virus**es themselves; they are substances such as **toxins** or **enzyme**s that are produced by the **microorganism**. In a typical immune response, certain cells known as the antigen-presenting cells trap the antigen and present it to the immune cells (**lymphocytes**). The lymphocytes that have receptors specific for that antigen binds to it. The process of binding to the antigen activates the lymphocytes and they secrete a variety of cytokines that promotes the growth and maturation of other immune cells such as cytotoxic T lymphocytes. The cytokines also act on B cells stimulating them to divide and transform into **antibody** secreting cells. The foreign agent is then either killed by the cytotoxic T cells or neutralized by the antibodies.

The process of inducing an immune response is called **immunization**. It may be either natural, i.e. acquired after infection by a pathogen, or, the immunity may be artificially acquired with serum or vaccines.

In order to make vaccines for immunization, the **organism**, or the poisonous toxins of the microorganism that can cause diseases are weakened or killed. These vaccines are injected into the body or are taken orally. The body reacts to the presence of the vaccine (foreign agent) by making antibodies. This is known as active immunity. The antibodies accumulate and stay in the system for a very long time, sometimes for a lifetime. When antibodies from an actively immunized individual are transferred to a second non-immune subject, it is referred to as passive immunity. Active immunity is longer lasting than passive immunity because the **memory** cells remain in the body for a long time.

Immunizations are the most powerful and cost-effective way to prevent infectious disease in children. Because they have received antibodies from their mother's **blood**, babies are immune to many diseases when they are born. However, this immunity wears off during the first year of life. That is why immunization programs which help children build their own defense against disease should be started early and carried out faithfully.

Each year in the United States, as many as 50,000-70,000 adults die needlessly from vaccine-preventable diseases or their complications. Eight childhood diseases (measles, mumps, rubella, **diphtheria**, tetanus, pertussis, hemophilus **influenza**e type b, and polio) are preventable by immunization. These illnesses are serious, and their complications can be terrible. With the exception of tetanus, all the other diseases are contagious and could spread rapidly, resulting in epidemics. Vaccines could prevent the children from ever getting the diseases and they do not have any serious side effects. Hence, they are perhaps among the safest and most effective medicines. In addition, a child can also be vaccinated against hepatitis B and **chicken pox**. Vaccines are given at **birth**, and at 2,4,6, and 12-15 months of age. A child needs about 15 vaccines before the age of two to prevent the above-mentioned 10 dangerous diseases.

Vaccinations against flu (influenza), hepatitis A, and pneumococcal disease are also recommended for some adolescents and adults. The vaccines indicated for adults will vary depending on their lifestyle factors, occupation, chronic medical conditions, and travel plans.

IMPRINTING

Imprinting is a term used in **ethology** (study of **animal** behavior) to describe the development of a stable behavioral pattern during a brief period of juvenile life (known as the "sensitive phase") in a social **species**. It occurs as a result of a timely exposure to a particular **stimulus**. Imprinting is usually associated with the juvenile's developed recognition of its own species, or of particular individuals within its species (such as its parents). Ethologists have demonstrated that sexual (or species) imprinting and parental imprinting are separate events in the behavioral development of young birds, each with its own sensitive phase.

The concept of imprinting was first discovered by **Konrad Lorenz**, a German biologist and pioneer in ethology, who recorded the behavior in ducks and geese. Lorenz discovered that a chick will learn to follow the first conspicuous moving object it sees after hatching. Normally, this object would be the mother bird, but in various experiments, ducklings and goslings have imprinted on artificial models of birds, bright red balls, and even human beings.

Imprinting on its parents is a learned behavior, without which a young bird might not be fed sufficiently, or might wander away from its caregiver. Parental imprinting is an especially common learned behavior in precocial birds, which are born in a relatively advanced state of development and leave their nest soon after **birth**, but are still tended by one or both parents as they move about while foraging. Examples of precocial birds include ducks, geese, swans, grouse, chickens, pheasants, and shorebirds such as sandpipers and plovers.

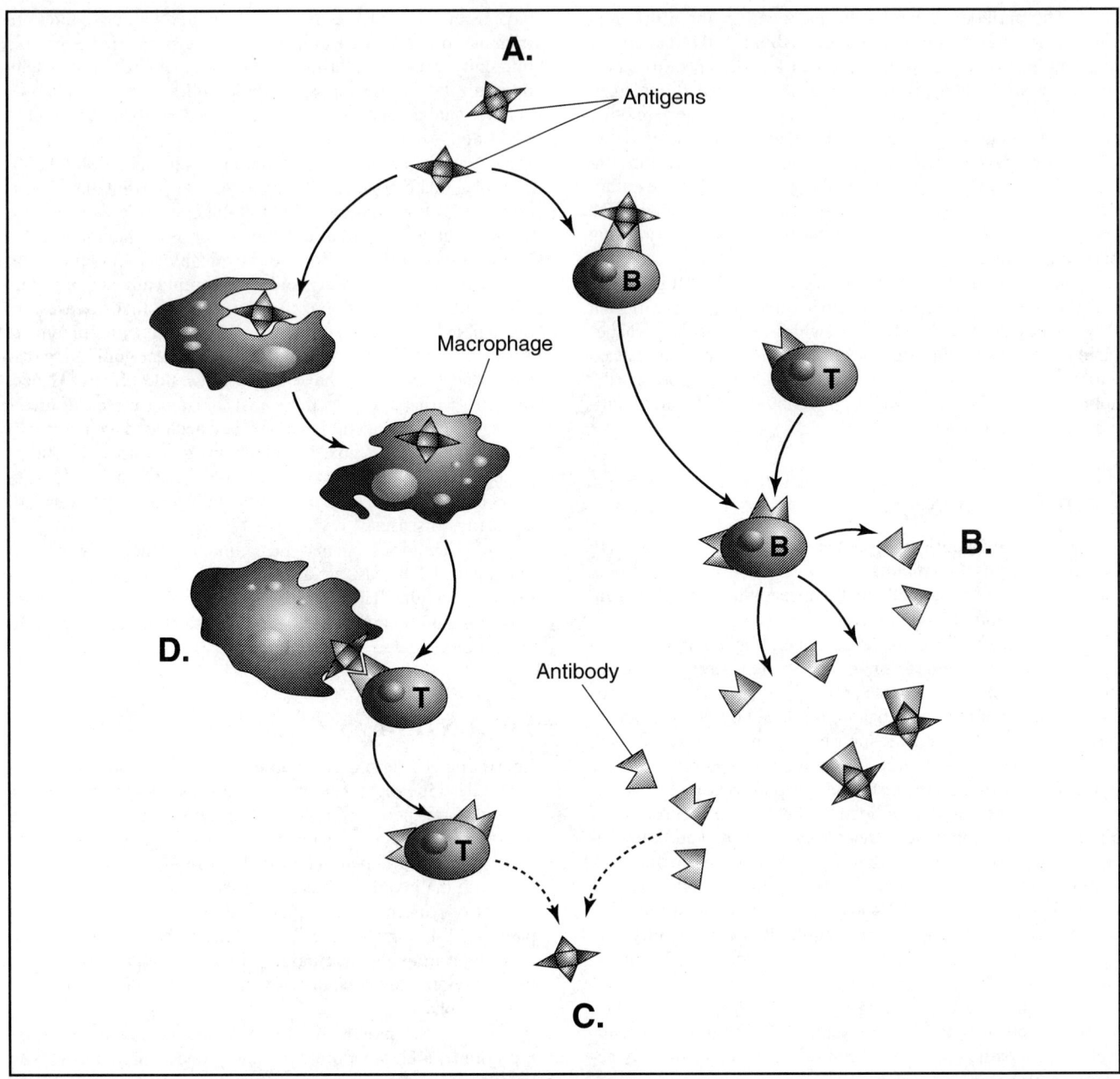

How vaccines used in immunizations work: A. Vaccines contain antigens (weakened or dead viruses, bacteria, and fungi that cause disease and infection). When introduced into the body, the antigens stimulate the immune system response by instructing B cells to produce antibodies, with assistance from T cells. B. The antibodies are produced to fight the weakened or dead viruses in the vaccine. C. The antibodies "practice" on the weakened viruses, preparing the immune system to destroy real and stronger viruses in the future. D. When new antigens enter the body, white blood cells called macrophages engulf them, process the information contained in the antigens, and send it to the T cells so that an immune system response can be mobilized. *(Illustration by Electronic Illustrators Group.)*

The effects of the imprinting process carry over into the adult life of the animal as well. It is crucial in the individual's ability to later recognize its own species, participate in successful breeding practices, and flock with the appropriate social group. Experimental studies by ethologists have shown that young birds that are only in contact with other species during this critical period of sexual imprinting will not seek to mate with individuals of their own species. Instead, they will try to mate with an individual of their "foster" species. Inappropriate sexual imprinting can be a problem when individuals of certain **endangered species** are being raised in captivity. For example, rare whooping cranes (*Grus americana*) raised by sandhill cranes (*G. canadensis*) may later have extreme difficulty in mating with individuals of their own species.

Imprinting in **mammals** is most thoroughly studied in birds, although it is believed to be especially important in the hoofed mammals, which tend to congregate in large herds in which a young animal could easily be separated from its mother. Other **organisms**, such as salmon, imprint on chemical cues in their environment, and use these to remember the location of their breeding grounds. Atlantic salmon (*Salmo salar*) are born in freshwater **habitat**s (cool rivers and streams), and later migrate to the ocean to grow into adults. When they are young, the salmon imprint on chemical cues in their natal habitat. When they are adults, the salmon use their chemical memory to find the same breeding river or stream (or at least its vicinity) in which they were born, in order to breed there themselves.

The existence of a critical period for imprinting is a genetically fixed trait in certain species of animals. **Learning**, however, is a circumstantial experience, being dependent on environmental conditions occurring at a particular time. Learning through imprinting is, therefore, an interesting case of a behavioral interaction between **inheritance** and opportunity.

INBREEDING

Inbreeding is the mating of closely related individuals that share common ancestry. Inbreeding is used in **animal** husbandry to retain desirable characteristics and eliminate undesirable ones. However, the mating of siblings in farm animals can only be continued for a few generations as viability and fertility are adversely affected. Inbreeding can cause harmful genes that are recessive in both parents to be manifest in the offspring. Linebreeding is a mechanism used by agricultural breeders to concentrate the genes of a certain ancestor in a strain of animals. This is accomplished by mating a female with her grandsire or uncle, for example. The probability of undesirable genes in the offspring is reduced.

There are some types of inbreeding which occur in **nature**. Self-**fertilization**, for example, occurs in bisexual **organisms** such as most flowering **plants**, some **protozoan**s, and **invertebrates**. It is the fusion of male and female **gamete**s produced by the same individual. As an evolutionary and reproductive mechanism, self-fertilization allows an isolated individual to create a local **population**. However, without a degree of variability among the individuals produced, possibilities for **adaptation** to environmental change are reduced.

Outbreeding is defined as mating individuals that are not related at all. Some cross-**species** outbreeding has occurred, particularly between the horse and the ass to produce the mule.

Crossbreeding has been practiced for a long time and refers to mating within the same species. The main function of crossbreeding is to produce offspring with carry the desirable traits of both parents.

See also Genetic trait; Heterozygous and homozygous; Hybrid

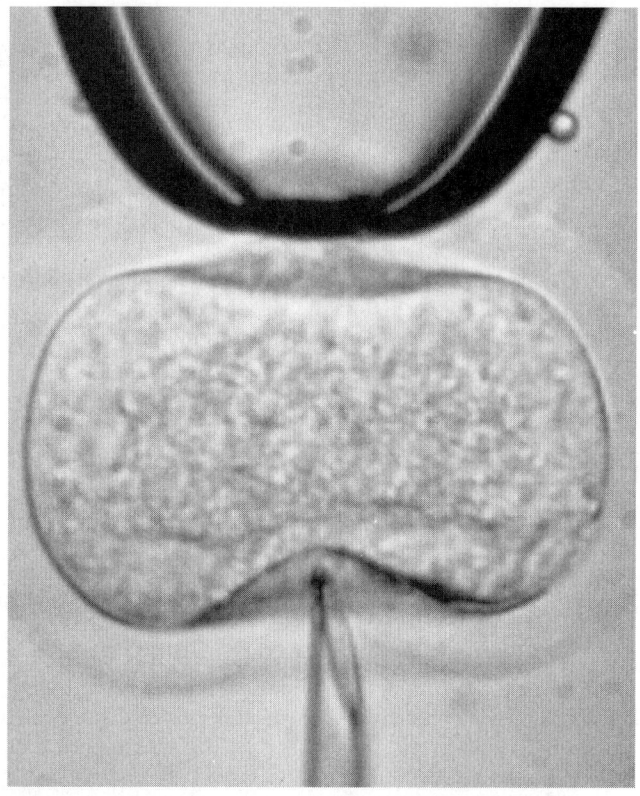

A microscopic image of a needle (bottom) injecting sperm cells directly into a human egg (center). The broad object at top is a pipette used to hold the ovum steady. *(Phototake NYC. Reproduced by permission.)*

INDEPENDENT ASSORTMENT

Originally formulated by the Austrian monk **Gregor Johann Mendel**, the law of independent assortment states that the distribution of alleles to gametes during **meiosis** is random. If one particular **allele** goes to one **gamete**, it has no influence on the likelihood of any other allele going to the same gamete.

The law of independent assortment holds true for genes on separate chromosomes, however, there are many cases where the law of independent assortment does not hold true. When the alleles are present on the same linkage group or **chromosome**, they are physically attached to each other and cannot show independent assortment. The further apart the alleles are the more likely that there is going to be a cross over event between them and they will be able to show independent assortment.

See also Genetics

INFERTILITY

Infertility refers to the inability of a couple to achieve a pregnancy after attempting to do so for at least one full year. Infertility can be primary or secondary. Primary infertility occurs when a couple has never been able to conceive; secondary infertility describes a couple who have previously conceived a pregnancy, but are unable to conceive again after a full year of trying.

About 40% of the time infertility is due to a problem with the male partner; 40% of the time the problem is with the female partner; 20% of the time there are fertility problems with both the man and the woman in the couple; and about 3-4% of the time, no cause for infertility can be found.

The male contribution to **fertilization** and the establishment of pregnancy is the **sperm**. The sperm are mixed into a fluid called semen, which is discharged from the penis during a process called ejaculation. The whip-like tail of the sperm allows the sperm motility; that is, it permits the sperm to essentially swim up the female reproductive tract, in search of the **egg** it will attempt to fertilize.

Male factor infertility can be caused by a number of different characteristics of the sperm. To check for these characteristics, a semen analysis is carried out, during which a sample of semen is obtained and examined under a **microscope**. The four most basic characteristics which are evaluated are: sperm count, motility, morphology, and volume.

The normal number of sperm present in just one milliliter (ml) of semen is over 20 million. An individual with only 5-20 million sperm per ml of semen is considered subfertile, and an individual with less than five million sperm per ml of semen is considered infertile.

Sperm are also examined to see how motile they are; better swimmers indicate a higher degree of fertility, as does longer duration of survival. Sperm are usually capable of fertilization for up to 48 hours after ejaculation.

Morphology refers to the structure of the sperm. Not all sperm within a specimen of semen will be perfectly normal. Some may be developmentally immature forms of sperm, some may have abnormalities of the head or tail. A normal semen sample will contain no more than 25% abnormal forms of sperm.

The total quantity of semen produced in a single ejaculation is important. The semen is made up of a number of different substances, and an abnormal proportion of these substances could affect the ability of the sperm to successfully fertilize an egg.

Because the female makes multiple contributions to fertilization, the establishment of pregnancy, and the maintenance and development of a pregnancy, exploration of female infertility factors can be quite complex. The female partner must be able to produce a normal egg, move that egg into the fallopian tube where eggs are normally fertilized, harbor the resulting **zygote** safely in the fallopian tube while the **cell**s divide to form the blastocyst, then move the blastocyst into the uterus, where it should implant itself into the uterine wall. The woman's body must then be capable of sustaining the pregnancy for nine months, during which time the blastocyst passes through multiple stages of development, including **embryo** and fetus.

Female infertility factors can include a wide variety of conditions. Causes are divided up into basic categories: peritoneal, ovulatory, cervical, and uterine factors.

Peritoneal factors are those factors which are due to any problems within the abdomen of the female partner (excluding those involving specifically the ovaries, fallopian tubes, or uterus). The most common peritoneal factors include pelvic adhesions (thick, fibrous scars due to past infections or surgeries) and endometriosis (the abnormal location of uterine **tissue** outside of the uterus, which bleeds on a monthly basis during the menstrual period, resulting in the formation of scar tissue in the surrounding area).

Ovulatory factors are factors which prevent the maturation and release of the egg from the ovary. These factors include a whole host of endocrine abnormalities, in which appropriate levels of the various **hormone**s which influence ovulation are not produced. Because many hormones produced by multiple **organ** systems interact to bring about normal ovulation, ovulation abnormalities can stem from problems with the ovaries, the **adrenal glands**, the **pituitary gland**, the **hypothalamus**, or the **thyroid gland**.

Cervical factors—the cervix is the opening from the vagina into the uterus through which the sperm must pass. Mucus produced by the cervix helps to transport the sperm into the uterus. When the cervix has been injured during a previous **birth**, due to surgery, or after an infection, the resulting scarring can result in a smaller than normal cervical opening which is difficult for sperm to enter. When the cervical mucus is not produced in an appropriate quantity, or when its composition is incorrect, it may hamper the sperm's travels. Furthermore, some women produce antibodies (immune cells), which are specifically directed to identify sperm as foreign invaders, and to kill those invaders.

Uterine/tubal factors include any factors that interfere with the normal structure and function of the uterus and the fallopian tubes. This can include tumors or abnormal growths within the uterus, chronic infection and **inflammation** of the uterus, and a variety of endocrine problems (problems with the **secretion** of certain hormones), which prevent the uterus from developing the thick lining necessary for implantation by a blastocyst. Tubal factors are often the result of previous infections which have left scar tissue. This scar tissue blocks the tubes, preventing the ovum from being fertilized by the sperm. Ectopic pregnancies occur when a pregnancy accidentally continues to develop within the tube past the blastocyst stage. In this situation, the tube will eventually rupture, requiring emergency surgery. Remaining scar tissue from such an ectopic pregnancy can cause subsequent infertility. Other problems with the fallopian tubes include rare conditions in which someone is born with defective or absent **cilia** (cilia are hair-like projections which line the fallopian tubes, and beat in a regular fashion in order to move the egg and then the blastocyst along). Without cilia, the egg may not move in an appropriate fashion through the fallopian tube.

INFLAMMATION

Inflammation refers to the body's response to a number of different types of injury or infection. The body will produce certain chemicals which have predictable effects on the surrounding **tissue**s. **Blood** vessels in the area will dilate (expand) and become leaky. The leaky vessels cause fluid to collect, resulting in swelling. The expansion of blood vessels near the surface produces an increase in blood flow to the injured area, leading to a reddening of the skin, and causing the area to feel hot to the **touch**. Some of the chemicals which accumulate will cause pain.

As early as the first century B.C., a Roman writer named Cornelius Celsus described what have become known as the four classic hallmarks of inflammation: tumor (swelling), dolor (pain), rubor (redness), and calor (heat). An eighteenth-century surgeon in Scotland, **John Hunter**, was the first to recognize that inflammation is not actually a disease, but rather the body's unfortunately detrimental response to some other injury or infection.

It is currently known that inflammation can also result from the body's mis-recognition of its own tissues as foreign. This means that the body's **immune system**, designed to protect the body from attack by foreign invaders such as **bacteria**, **virus**es, or **fungi**, accidentally identifies the body's own tissues as foreign, and sets off a cycle of inflammation. In the case of this type of autoimmune reaction, the inflammation serves no appropriate purpose, and instead causes damage to the body's own tissues and **organ**s.

Inflammation can be acute (quick and sudden in onset, and usually short in duration. Acute inflammation usually occurs in response to a sudden injury or exposure to some type of injurious agent (heat, radiation, or caustic chemicals). Chronic inflammation occurs when inflammation lasts for a relatively long duration. In this case, there may be an overgrowth of blood vessels and tissues in the area, resulting in a pattern of scarring. Autoimmune disorders such as rheumatoid **arthritis** follow this pattern. Some of the disability caused by such disorders, then, is due to the accumulation of scar tissue in an affected area of the body.

INFLUENZA

Influenza is an infectious disease caused by the influenza **virus**. The term *influenza* is used to describe influenza and other similar illnesses. The general symptoms of influenza, also known as the flu or grippe, are chills, headache, fever, weakness, and aching of the joints. The incubation period is short—between one and three days—and the first symptom is a fever that may reach 103°F (39.4°C). Cough and gastrointestinal discomfort may also accompany the disease. Usually, the viral infection runs its course in about a week. However, the virus weakens the **immune system**, making the human body subject to secondary infections such as bacterial pneumonia. Fatalities associated with influenza usually result from such secondary complications. The presence of viral pneumonia was probably the cause of many deaths during the great influenza epidemics of the past.

Historically, influenza has been known for centuries and may have been what **Hippocrates of Cos** described as the cause

A tranmission electron microscopy (TEM) image of influenza viruses budding from the surface of an infected cell. *(Photo Researchers, Inc. Reproduced by permission.)*

of an epidemic as early as 412 B.C. In the sixteenth century, John Keys (1510-1573), also known by his Latinized name, Johannes Caius, was a well-educated English physician who wrote a treatise on the "Sweatyng Sickness." In this curious book, he describes a highly-contagious illness in which death frequently occurred within a few days or even hours after symptoms began to show. The symptoms he described included fever, delirium, and labored breathing—not very different from influenza symptoms. At that time the cause was sometimes attributed to the English diet—but nobody really knew how the sweating sickness was transmitted. Treatment included bed rest and doing anything that would promote sweating.

Even at the beginning of the twentieth century, doctors had difficulty determining how influenza spread. They knew it was not spread by bacteria from unsanitary conditions, nor by insects. But what doctors did know was that the disease could be fatal. Between 1918 and 1919, influenza killed about 20 million people all over the globe. In the United States more than 550,000 people died of the flu—ten times more Americans than were killed in action during World War I. There were emergency hospitals set up to treat flu patients. Soldiers were made to gargle with salt **water** to prevent the disease.

Scientists and physicians searched for the cause of influenza but could not find one. Not until the 1930s with the invention of more powerful **microscope**s did scientists began to see pathogens much smaller than bacteria—viruses. During that decade the nature of viruses was discovered. Wendell Stanley, an American biochemist, prepared large quantities of viruses, and found that they could be crystallized. Viruses are very simple structures made of only **protein**s and **nucleic acids** which could be crystallized in much the same way as other nonliving chemicals. However, when a virus is inside a living **cell** it uses the cell's genetic machinery to make more copies of itself. Stanley had discovered that viruses are on the borderline between living and nonliving things because they grow only when inside living cells. During World War II, Stanley worked on culturing the influenza virus. Today, influenza is grouped with other viruses known as adenoviruses.

Because viruses are so tiny, vast numbers are found in human body fluids like mucus and saliva. Coughing or sneezing releases an aerosol spray filled with viruses. The pathogen easily travels from one person to another, making influenza very contagious. The viruses usually enter the upper respiratory tract and begin reproducing. The body's immune system, including antibodies and T **lymphocytes**, work to kill the viruses, but may be not be enough to stop the disease from running its course.

Scientists are still puzzled by the influenza outbreak that occurred around the time of World War I. Originating in the American Midwest, the epidemic spread worldwide, possibly carried abroad by American soldiers during World War I. Incredibly, lung **tissue** from an Army private who died from the flu was preserved in paraffin and stored in Washington, DC. In 1997, scientists used **polymer**ase chain reaction—which amplifies minute **genetic material**—to extract the virus. Genetic studies found that the flu contained a pig virus. The study further proved that pigs, which are susceptible to both bird and human viruses, can serve as a cauldron where new and dangerous viral strains can intermingle and form other strains. The finding also dispelled a long-time belief that the epidemic may have developed from an avian, or bird, virus.

In 1997 a three-year-old boy in Hong died from a virus previously thought to occur only in birds. Investigators believe that the transmission was direct from bird to human without the usual intermediary pig. Prior to the boy's infection, thousands of chickens in Hong Kong died from the virus. By the end of 1997, 15 additional cases had occurred in the city.

Research has shown that viruses have a phenomenal ability to change, or mutate, very quickly. A treatment that may have worked on one flu virus may not work on another. Each year different strains develop and are named for places where they first occur. The Asian flu caused a worldwide epidemic in 1957-1958 and the Hong Kong flu in 1968-1969. Fewer deaths resulted from the Hong Kong flu because **antibiotics** were used to treat secondary infections.

The best treatment for influenza is prevention; and flu vaccines (both killed and live virus) are recommended, especially for the very young and the very old. In addition the antiviral drug known as amantadine has been used to treat certain strains of influenza.

INGENHOUSZ, JAN (1730-1799)
Dutch physician and plant physiologist

While working at a London hospital, Ingenhousz became an expert in the technique of preventing smallpox by inoculation. Treatment at that time used a hazardous live virus instead of today's safer vaccine. In 1768, he traveled to Vienna, Austria, to inoculate Austria's royal family. His treatment was so successful that he was appointed court physician and awarded a lifetime income, which he used to finance independent research.

After returning to England in the late 1770s, Ingenhousz began the experiments that led to his discovery of **photosynthesis**, the process by which **plants** convert sunlight into food. During photosynthesis, plants absorb **carbon dioxide** from the

Jan Ingenhousz. *(The Library of Congress. Reproduced by permission.)*

air and release oxygen. **Respiration** is essentially the reverse process—oxygen is used up and carbon dioxide is released. Inspired by **Joseph Priestley**'s research on oxygen, Ingenhousz learned that only the green parts of plants can revitalize air that has been depleted of oxygen, and they do so only in the presence of sunlight. This was the first indication of light's role in plant life. He also discovered that only the visible light of the sun, not its heat, is necessary for photosynthesis. Ingenhousz's research additionally proved that respiration occurs in all parts of the plant, including its roots and flowers. He later discovered that plants produce much more oxygen than they use up, and, subsequently, that **animal** and plant life are naturally balanced in a system of mutual support.

Ingenhousz also broke new ground in physics and chemistry. For example, he improved phosphorus matches, invented a hydrogen-fueled lighter, and mixed an explosive propellant for firing pistols.

See also Chlorophyll function and structure

INHERITANCE • See Mendelian laws of inheritance

INSECTS • See Entomology

INSTINCT

An instinct is a stereotyped, **species**-typical behavior that appears fully functional the first time it is performed, without the need for **learning**. Such behaviors are usually triggered by a particular **stimulus** or cue, and are not readily modified by subsequent experience. For instance, a kangaroo rat instantly performs an automatic escape jump maneuver when it hears the sound of a striking rattlesnake, even if it has never encountered a snake before. Clearly, instinctive behaviors play an important role in survival, but our understanding of the forces that promote and guide their development in living **animal**s is in fact quite limited.

Classic examples of animal instinct

Researchers of animal behavior, ethologists, first named the stereotyped, species-typical behaviors exhibited in particular circumstances fixed action patterns, which were later called instincts. A cocoon-spinning spider ready to lay its **eggs** builds a silk cocoon in a particular way, first spinning a base plate, then the walls, laying its eggs within, and finally adding a lid to seal the top. The spider performs all these actions in a specific sequence, and, indeed, cannot spin its cocoon in any other way. If the spider is relocated after having spun the base plate, she will still make the walls, deposit the eggs (which promptly fall out the bottom), and spin the lid for the top. When ready to begin the next cocoon, if the spider is returned to her original base plate, she will nonetheless begin by spinning a new base plate over the first, as if it were not there.

Many fixed action patterns occur in association with a triggering **stimulus**, sometimes called a releaser. Baby gulls respond to the sight of their parent's bill by pecking it to obtain a tasty morsel of food. The releaser here is a bright red spot on the parent's bill; neither the shape nor the color of the adult's head have a significant influence on the response. When a female rat is sexually receptive, rubbing of her hindquarters (the releaser) results in a stereotypical posture known as lordosis, in which the front legs are flexed, lowering the torso, while the rump is raised and the tail is moved to one side (a fixed action pattern). A male rat who encounters a female in lordosis experiences another releaser and initiates copulation. Neither sequence requires any prior experience on the part of the animal.

The role of instinct in learning

Imprinting

In another classic study of instinctive behavior, ethologist **Konrad Lorenz** showed that baby ducks and geese, which are observed to closely follow their mother on their early forays away from the nest, could also be induced to follow a substitute. The baby birds would form an attachment to whatever individual was present as they opened their eyes and moved about after hatching, regardless of that individual's species identity. Young birds that had thus imprinted on Lorenz followed him everywhere as they matured, and as adults, these birds were observed to court humans, in preference to members of their own species.

These baby birds instinctively open their mouths in anticipation of food when an adult bird returns to the nest. Because of this instinct, the baby bird is able to compete for limited resources, increasing its chances for survival. *(The Stock Market. Reproduced by permission.)*

Lorenz concluded that **imprinting** represented a kind of preprogrammed learning, guided by a mechanism that under normal circumstances would not be corrupted by individuals of the wrong species. In the natural situation, imprinting would facilitate the babies' social attachment to their mother, which later allows them to recognize appropriate mating partners.

Critical periods

Bird song is a largely species-specific behavior performed by males in their efforts to establish and maintain their territories and to attract females. Many songbirds develop their mature songs through a process involving a critical period when, as a nestling, the bird hears the song of its father. The juvenile bird does not sing until the following spring, when it begins to match its immature song to the one it heard from its father during its critical period. If the nestling is prevented from hearing adult song during the critical period, it will never develop a species-typical song. Evidently, there is also a strongly instinctive aspect to what may be learned during the critical period; most birds cannot produce every song heard during that time, but appear to be selective toward songs that are produced by other members of their species.

Instincts can be exploited

Some animals have evolved the capacity to take advantage of the reliable, instinctive behavior of others. Avian brood parasites, including the North American cowbird and the European cuckoo, exploit the parental behavior of other birds and lay their eggs in the host's nest. The unwitting host feeds the interloper's hatchlings, which are often bigger than its own, and thus may represent a greater releaser of the powerfully instinctive feeding behavior of the parents. The adult brood parasite is literally parasitizing the parental behavior of the host bird, for it exerts no further parental investment in its offspring, leaving them instead in the care of the host.

Instinct and learning: a continuum

We use the term instinct to describe species-typical behavior that is seemingly performed without aid of prior experience, but what we seem to mean is that the animal moves and behaves as if mysterious and unknown forces were guiding it. Many people who study animal behavior argue that the term instinct is not ultimately helpful because it tells us little about the real mechanisms underlying behavior. The use of the term indicates only that the behavior is relatively closed to modification by experience and nothing more. Since **nervous system tissue**s are soft, delicate, and often very complex, understanding the operation of these structures in producing behavior presents a great challenge. This, combined with the role of experience in producing many superficially "instinctive" behaviors, makes things even more difficult.

Many behaviors held up as examples of instinct are shown to have an experiential component: for instance, as new gull chicks continue to peck at bill-like objects, the accuracy of their pecking improves and the kinds of bill-like objects they will peck at are increasingly restricted. Thus, the wide variety of behavioral patterns observed in living **organism**s surely represents a continuum, from those not much influenced by learning to those that are greatly influenced by it; a strict "**nature** versus nurture" dichotomy is probably too simplistic to describe any animal behavior.

The answer to the question "Under what conditions should a behavior be genetically closed, and when should a provision be made for learning?" seems to be related to the situation's predictability in nature. When it is crucial that the correct response to some occurrence be carried out the first time (like a kangaroo rat faced with a striking rattlesnake), **natural selection** should favor a fairly rigid, infallible program to underlie an appropriate response. The existence of a reliable relationship between some environmental cue and a biologically appropriate response permits the development of a releaser for triggering the "right" reaction the first time, whether to a predator, potential mates, or one's own offspring.

INSULIN

Insulin is a hormone produced by the pancreatic islet of Langerhans **cell**s that helps convert **carbohydrate**s to the simple sugar glucose and regulates glucose levels in the **blood**. Its other functions include regulating specific events in a cell's **life cycle**. Insulin is best known for treatment of the metabolic disease **diabetes mellitus**.

The importance of glucose as an indicator of diabetes was noted for thousands of years by Asian and Middle Eastern physicians. In seventeenth century Europe, a sweet flavored urine was listed as a symptom of diabetes. In 1815, the French chemist Michel Eugène Chevreul discovered that the sweetness came from "grape sugar" or glucose. Later in the nineteenth century scientists learned how the body manufactures, stores, and uses glucose.

Injury to the **pancreas**, a gland attached to the small intestine, was linked to diabetes beginning in the seventeenth century and confirmed by **animal** experiments in the late nineteenth century, particularly in the work done in 1889 by German physiologist Joseph von Mehring (1849-1908) and Russian pathologist Oscar Minkowski (1858-1931). The pancreas has two main cell types, *acinus* (found in the seventeenth century by the Dutch anatomist Regnier de Graaf) and islet of Langerhans (named for the German pathologist **Paul Langerhans** (1847-1888), who discovered it in 1869).

In 1905, English physiologists **Ernest Starling** and **William Bayliss** discovered **hormone**s—substances secreted by glands and carried in the blood to control cell activity elsewhere. Eleven years later, the English physiologist Edward Sharpey-Schäfer theorized that a hormone produced by the pancreas lowers the level of glucose in the blood. He called it *insuline*, from the Latin word for island, since he believed it originated in the islet (island) cells. By 1921 the Canadian researchers **Frederick Banting** and **Charles Best** had isolated a crude but usable extract from animal sources that worked as Sharpey-Schäfer had predicted. Insulin, as it was renamed, was used soon afterward to successfully treat a teenaged boy with diabetes. Banting shared the 1923 Nobel Prize in physiology or medicine for the accomplishment. Crystalline insulin, stable and pure enough to produce commercially, was first isolated in 1926 by the American pharmacologist John Jacob Abel.

In the 1950s, the British biochemist **Frederick Sanger** determined that insulin is composed of two chains, with 21 **amino acid**s in the A chain and 30 in the B chain. The two chains are connected by one double sulfur bond at the seventh amino acids and another linking the 20th A chain amino acid and the nineteenth B chain amino acid. Sanger received the 1958 Nobel Prize in chemistry for his work. Other scientists showed insulin's folded, or three-dimensional, structure.

In 1982 insulin became available as a genetically engineered product called Humulin that is identical with human insulin. Humulin is produced jointly by Genentech and Eli Lilly and Co. The A and B chains are produced separately in different strains of *E. coli* **bacteria**. Each chain is separated from the bacteria and purified. Finally the two chains are combined chemically and repurified.

Insulin performs its glucose-regulating function with another hormone called glucagon, which is produced by the islet of Langerhans cells (a fact discovered by the American researcher **Earl Sutherland**). Glucagon converts glycogen to glucose and raises blood sugar level. Together the two hormones ensure that the body stores and uses the correct level of glucose to meet its energy needs.

Insulin's other roles in body **metabolism** include helping muscle and fat cells take in and use glucose, inducing production of glycogen, encouraging storage of fat within cells by preventing its use as a fuel, and stimulating the movement of amino acids into cells so they can produce **protein** in various phases of the cell cycle.

Traditionally, diabetics who use insulin have had to inject themselves via a syringe. However, alternative approaches for insulin delivery, including nasal and oral methods, are being studied. Alternatives to syringe injections may help ensure patient compliance to insulin therapy by making insulin delivery easier and more convenient. For example, miniature computerized pumps, which fit in a pocket or handbag, deliver a steady dose of insulin through a flexible tube with a small needle inserted through the skin into fatty **tissue**. New technology is also under development to provide automatically controlled insulin delivery. One such device uses a glucose sensor implanted into the patient. The sensor responds to changes in blood glucose levels and then adjusts the rate of insulin release via a pump.

Commercial fermentation units like these are used to grow cultures of microorganisms for biological products like interferon. *(Photo Researchers, Inc. Reproduced by permission.)*

INTEGUMENTARY SYSTEM

The integumentary system consists of all of those **tissue**s which create the external wrapping of the body (including skin, hair, fur, nails, etc.), as well as glands of the skin (including sweat glands and oil glands). All of these structures are derived from the same primitive tissues, which differentiate to form the different components of the integumentary system.

The primary functions of integumentary structures include protection of the internal **organ**s of the body. The skin, in particular, serves as the body's largest organ, protecting it from physical, chemical, and infectious threats. The integumentary system also serves to help maintain appropriate body **temperature**, and prevents excess fluid loss. It guards the rest of the body against the threat of ultraviolet radiation, and is integral to the conversion of vitamin D. The integumentary system is involved in the acquisition of information from the external environment, including information on ambient temperature, pressure, and pain. Some fatty substances can be absorbed into the skin; other substances can be removed from the body by **excretion** through the sweat glands.

INTERFERON

While the middle years of the twentieth century saw the development of **antibiotics**, potent new weapons against bacterial diseases, no such chemical defenses against viral diseases had yet emerged other than a few anti-viral vaccines. This was and continues to be a significant gap, since more than half of the communicable diseases that affect human beings are caused by **ribonucleic acid** (**RNA**) **virus**es. The first step was to find out how the body protected itself against viruses; it was known that antibodies acted only against bacteria. That step was taken by Alick Isaacs, a Scottish virologist, in 1957.

Isaacs was born in Glasgow, Scotland, in 1921, to a Russian Jewish family. He studied medicine at Glasgow Universi-

ty but found he preferred research to the actual practice of medicine. Accordingly, he pursued graduate studies in bacteriology at Glasgow and secured fellowships to research **influenza** with eminent microbiologists Stuart Harris of England and **Frank MacFarlane Burnet** in Australia. Returning to England, Isaacs joined the World Influenza Centre at the National Institute for Medical Research in 1951. There, he carried out his continuing studies of viral influenza until his untimely **death** in 1967.

Early in his studies of influenza, Isaacs became interested in the viral interference phenomenon—the fact that an RNA virus in a **cell** inhibits the growth of any other viruses in that cell. He found that this interference seemed to be caused by something inside the cell. In 1957, while working with the visiting Swiss scientist Jean Lindenmann, Isaacs found that chick **embryo**s injected with influenza virus released very small amounts of a **protein** that destroyed the virus and also inhibited the growth of any other viruses in the embryos. Isaacs and Lindenmann named the interfering protein *interferon*.

Further research showed that interferon was produced within hours of a viral invasion (antibodies take several days to form), and that most living things, including **plants**, can make the protective protein. Interferon was seen as the cell's first line of defense against viral infections, and its discovery was expected to pave the way for successful treatment of viral diseases.

Because interferon was thought to be **species**-specific (meaning only human interferon will work in human beings) and the body produces it in only minute amounts, interferon research inched forward at a snail's pace.

Interest in interferon was revived in the late 1960s when Ion Gresser (1928-), an American researcher in Paris, discovered that the protein stopped or slowed the growth of tumors in mice and also stimulated the production of tumor-killing **lymphocytes**. Gresser and the Finnish virologist Kari Cantell then developed a way to make interferon in useful amounts from human **blood** cells. Monoclonal antibodies, first produced in 1975, made large-scale purification of interferon possible, and the mid-1980s saw the advent of genetically engineered interferon, the first example of which was produced from bacteria by Swiss scientist Charles Weissmann in 1980. Scientists now know that there are three major types of interferon: alpha, beta, and gamma. They have also learned that interferons are not species specific but can have activity in other species.

Research of interferon's ability to kill **cancer** cells is active. It has been used successfully against leukemia, osteogenic sarcoma (a bone cancer), and as a therapy for delaying disease recurrence and prolonging survival in patients with melanoma (skin cancer). Research also continues on the use of interferon to treat viral diseases like rabies, hepatitis, and herpes infections. In December 1997, researchers from the Duke University Medical Center also announced study results that indicate interferon may be a way to preserve donor livers longer prior to transplantation.

INTERPHASE

Interphase refers to that period of time between periods of mitotic division in the **cell**. Mitotic division (**mitosis**) is that time in a cell's development when it has duplicated its genetic information, and is actively involved in dividing into daughter cells. Because much attention has been paid to the fascinating process of mitosis, many people have mistakenly looked upon interphase as a quiet, resting phase existing between the more important, more active phases of **cell division**. In fact, interphase is more than just a "between" phase, as its name mistakenly suggests. The largest part of a cell's existence is spent in interphase. It is the period of time during which the cell's true functions are carried out, during which time any products of the cell are elaborated, and during which the cell actually has the opportunity to grow. While the interphase **nucleus** (the part of the cell containing **genetic material** in the form of **chromosomes**) exists in a form which is hard to study under a **microscope**, the rest of the cell (the **cytoplasm** and its **organelles**) are at their peak of activity. Later in interphase, the chromosomes in the nucleus will duplicate (replicate) themselves), in preparation for mitosis.

INTRODUCED SPECIES

Some **species** of **plants**, **animals**, and **microorganisms** have been spread by humans over much wider ranges than they occupied naturally. Some of these introductions have been deliberate and were intended to improve conditions for some human activity, for example, in **agriculture**, or to achieve aesthetics that were not naturally available in some place. Other introductions have been accidental, as when plants were introduced with soil transported as ships' ballast or insects were transported with timber or food. Most deliberate or accidental introductions have not proven to be successful, because the immigrant

species were unable to sustain themselves without the active intervention of humans. (In other words, the introduced species did not become naturalized.) However, some introduced species have become very troublesome pests, causing great economic damages or severe losses of indigenous species.

Deliberate introductions

The most common reason for deliberate introductions of species beyond their natural range has been to improve the prospects for agricultural productivity. Usually this is done to provide agricultural species of plants or animals that would not otherwise be available for cultivation. In fact, all of the most important species of agricultural plants and animals are much more widespread on Earth today than they were prior to their domestication and extensive cultivation by humans. Wheat (*Triticum aestivum*), for example, was originally native only to a small region of the Middle East, but it now occurs virtually anywhere that conditions are suitable for its cultivation. Corn or maize (*Zea mays*) originated in a small area in Central America, but this species is now cultivated on all of the habitable continents. Rice (*Oryza sativa*) is native to Southeast Asia, but it is now very widespread under cultivation. The domestic cow (*Bos taurus*) was native to Eurasia, but it now occurs worldwide. The turkey (*Meleagris gallopavo*) is native to North America, but it now occurs much more extensively. There are many other examples of plant and animal species that have been widely introduced because they are useful as agricultural crops.

Other species have been widely introduced because they are useful in improving soil fertility for agriculture or sometimes for forestry. For example, various species of **nitrogen**-fixing legumes such as clovers (*Trifolium* spp.) and alfalfa (*Medicago sativa*) have been extensively introduced from their native Eurasia to improve the fertility of agricultural soils in far-flung places. In other cases, species of earthworms (such as the European nightcrawler, *Lumbricus terrestris*) have been widely introduced because these animals help to humify organic **matter** and are thereby useful in aerating soil and improving its structural quality. There have also been introductions of beneficial microorganisms for similar reasons, as when mycorrhizal **fungi** are inoculated into soil or directly onto **tree** roots. When their roots are infected with a suitable root mycorrhiza, plants gain significant advantages in obtaining **nutrients**, especially phosphorus, from the soil in which they are growing.

In some cases, species of animals have been introduced to improve the prospects for hunting or fishing. For example, Eurasian gamebirds such as ring-necked pheasant (*Phasianus colchicus*) and gray or Hungarian partridge (*Perdix perdix*) have been widely introduced in North America, as have various species of deer in New Zealand, especially red deer (*Cervus elaphus*). Species of sportfish have also been widely introduced. For example, various species of Pacific salmon (*Oncorhynchus* spp.) and common carp (*Cyprinus carpio*) have been introduced to the Great Lakes to establish fisheries.

Species of plants and animals have also been widely introduced in order to gain aesthetic benefits. For example, whenever people of European cultures discovered and colonized new lands, they introduced many species with which they were familiar in their home countries but were initially absent in their new places of residence. Mostly, this was done to make the colonists feel more comfortable in their new homes. For example, parts of eastern North America, especially cities, have been widely planted with European trees such as Norway maple (*Acer platanoides*), linden (*Tilia cordata*), horse chestnut (*Aesculus hippocastanum*), Scots pine (*Pinus sylvestris*), Norway spruce (*Picea abies*), as well as with many species of shrubs and herbaceous plants. The European settlers also introduced some species of birds and other animals with which they were familiar and comfortable, such as the starling (*Sturnus vulgaris*), house sparrow (*Passer domesticus*), and pigeon or rock dove (*Columba livia*).

Accidental introductions

Humans have also accidentally introduced many species to novel locations, and where the **habitat** was suitable these species became naturalized. For example, when cargo ships do not have a full load of goods they must carry some other heavy material as ballast, which is important in maintaining stability of the vessel in rough seas. The early sailing ships often used soil as ballast, and after a trans-oceanic passage this soil was usually dumped near the port and replaced with goods to be transported elsewhere. In North America, many of the familiar European weeds and soil **invertebrates** probably arrived with solid ships' ballast, as is the case for water horehound (*Lycopus europaeus*), an early introduction to North America at the port of New York. Ships have used water as ballast since the late nineteenth century, and many aquatic species have become widely distributed by this source. This is how two major pests, the zebra mussel (*Dreissena polymorpha*) and spiny water flea (*Bythothrepes cederstroemii*), were introduced to the Great Lakes from European waters.

An important means by which many agricultural weeds became widely introduced is through the contamination of agricultural seedgrains with their seeds. This was especially important prior to the twentieth century when the technologies available for cleaning seeds intended for planting were not very efficient.

Introduced species as an environmental problem

In most places of the world, introduced species have caused important environmental degradations. There are so many examples of this phenomenon that in total they represent a critical component of the global environmental crisis. A few selected examples can be used to illustrate problems associated with introduced species.

Several European weeds are toxic to cattle if ingested in large quantities, and when these plants become abundant in pastures they represent a significant management problem and economic loss. Some examples of toxic introduced weeds of pastures in North America are common St. John's wort (*Hypericum perforatum*), ragwort (*Senecio jacobaea*), and common milkweed (*Asclepias syriaca*).

Some introduced species become extremely invasive, penetrating natural habitats and dominating them, to the exclu-

Zebra mussels were introduced to the Great Lakes from European waters by ships using water as ballast. *(Photograph by C. Childs, Custom Medical Stock Photo. Reproduced by permission.)*

sion of native species. Purple loosestrife (*Lythrum salicaria*), originally introduced in North America as a garden ornamental, is becoming extensively dominant in **wetlands**, causing major degradations of their value as habitat for other species of plants and animals. In Florida, several introduced species of shrubs and trees are similarly degrading habitats, as is the case of the bottlebrush tree (*Melaleuca quinquinerva*) and Australian oak (*Casuarina equisetifolia*). In Australia, the prickly pear cactus (*Opuntia* spp.) was imported from North America for use as an ornamental plant and as a living fence, but it became a serious weed of rangelands and other open habitats. The cactus has since been controlled by the deliberate introduction of a moth (*Cactoblastis cactorum*) whose **larvae** feed voraciously on its **tissue**s.

Some introduced insects have become troublesome pests in **forests**, as is the case of the gypsy moth (*Lymantria dispar*), introduced in 1869 to North America from Europe, and a defoliator of many tree species. Similarly, the introduced elm bark beetle (*Scolytus multistriatus*) has been a factor in the spread of Dutch elm disease, caused by an introduced fungus (*Ceratocystis ulmi*) that is deadly to North American species of elm trees (especially *Ulmus americana*). Another introduced fungus (*Endothia parasitica*) causes chestnut blight, a disease that has eliminated the once abundant American chestnut (*Castanea dentata*) as a canopy tree in deciduous forests of eastern North America.

Other introduced species have caused problems because they are widely feeding predators and **herbivore**s. Vulnerable animals in many places, especially isolated oceanic islands, have been decimated by introduced predators such as mongooses (**family** Viverridae), domestic cats (*Felis catus*), and domestic dogs (*Canis familiaris*), by **omnivore**s such as pigs (*Sus scrofa*) and rats (*Rattus* spp.), and by herbivores such as sheep (*Ovis aries*) and goats (*Capra hircus*). The recent deliberate introduction of the predatory Nile perch (*Lates niloticus*) to Lake Victoria, Africa's largest and the world's second largest lake, has recently caused a tragic mass **extinction** of native

An ochre sea star (*Pisaster ochraceus*), an invertebrate, on the California coast. *(Photograph by Robert J. Huffman, Field Mark Publications. Reproduced by permission.)*

fishes. Until recently, Lake Victoria supported an extremely diverse **community** of more than 400 species of fish, mostly cichlids (family Cichlidae), with 90% of those species occurring nowhere else. About one-half of the endemic cichlid species are now extinct in Lake Victoria because of **predation** by the Nile perch, although some species survive in aquaria, and a few are still in the lake.

Ecologically, it is reasonable to consider humans and their symbiotic associates (that is, the many species of plants, animals, and microorganisms with which humans have intimate, mutually beneficial, and essential relationships) as the ultimate in invasive species. Humans, in fact, are widely self-introduced.

INVERTEBRATES

Invertebrates are **animals** without backbones. This simple definition hides the tremendous diversity found within this group which includes **protozoa** (single-**cell**ed animals), corals, sponges, sea urchins, starfish, sand dollars, worms, snails, clams, spiders, crabs and insects. In fact, more than 98% of the nearly two million described **species** are invertebrates, ranging in size from less than a millimeter to several meters long. Invertebrates display a fascinating diversity of body forms, means of locomotion, and feeding habits.

Invertebrates are an essential part of every **ecosystem** on this planet. We could not function without them. They are responsible for the **decomposition** of organic waste, which allows the recycling of the chemicals in the ecosystem. Invertebrates also are involved with the **pollination** of **plants**, and are crucial as links in food chains where **herbivore**s convert the **energy** in plants into energy that is available to animals higher up the food web.

Most invertebrates live in **water** or have some stage of their life in water because the carditrais are relatively stable in water, especially sea water. The external layers of aquatic invertebrates are generally thin and are permeable to water, allowing the exchange of gas, although some have specialized respiratory structures on their body surface. Aquatic invertebrates feed by ingesting directly, by filter feeding, or actively capturing prey.

Some groups of invertebrates such as earthworms, insects, and spiders live on land. These invertebrates need to have special structures to deal with life on land. For example, because air is less buoyant than water, earthworms have strong muscles for crawling and burrowing while insects and spiders move by means several pairs of legs. Drying out on land is a problem so earthworms must secrete mucous to keep their bodies moist, while insects and spiders are waterproof and are physiologically adapted to conserve water.

IONIC BOND

An ionic bond is an electrostatic force that holds **atoms** together through electron transference. **Chemical compound**s held together with ionic bonds separate in solution to form individual ions.

In their most stable configuration, the outer electron shells of an atom are evenly filled. If there are too many or too few electrons to evenly fill these shells, the atom will tend to lose or gain electrons to reach a lower **energy** state which is more stable. An atom's tendency to gain or lose electrons is measured by its electronegativity. A high electronegativity value means the atom tends to hold on to its electrons, while a low value means the atom is electropositive and tends to give them up. Electronegative and electropositive atoms can combine to form stable **molecule**s. The magnitude of the difference between their electronegativity values determines the type of bond the atoms will form. If the difference is low, the atoms tend to share electrons more equally and form **covalent bond**s. If the difference is large (greater than 1.7 units), the atoms will form ionic bonds because some of the atoms will donate electrons and other atoms will receive electrons.

Inorganic compounds, such as salts, are held together by ionic bonds. In simple compounds, the electric charge is easily spread across the composite atoms. In more complex solids, the atoms form three-dimensional configurations in which anions might be shared by several adjacent cations. This leads to a network of closely packed atoms held together by powerful electric forces. Therefore, ionic solids tend to be strong, crystalline materials with definite patterns of cleavage and high melting points.

When ionic compounds are dissolved in **water**, they separate into their component ions. This separation process is called ionization. For example, when sodium chloride (NaCl), also known as table salt, is dissolved in water, it ionizes to release positive sodium ions (Na+) and negative chloride ions (Cl-). These ions allow the solution to conduct electricity. Therefore, ionic compounds are also called electrolytes. Covalent compounds like sugar are non-conductive and are called non-electrolytes.

When solutions of different ions are mixed, **chemical reaction**s may take place. Different combinations of ions can come together to form more stable compounds that are not soluble in water. These compounds are called precipitates because they fall out of solution. For example, when silver nitrate ($AgNO_3$) and sodium chloride (NaCl) are dissolved in water, four ions are formed: Ag+, NO_3 -, Na+, and Cl- . Because of their relative electronegativities, the silver ions and the chloride ions will bond together to form solid silver chloride (AgCl) which is not soluble in water.

See also Chemical bond

ISOMER

The term isomer (meaning same member) refers to two **molecule**s that have the same chemical formula, but which differ according to the spacial arrangement of their **atoms**. Molecules that are isomers have the same number and kind of atoms, but they differ in respect to the arrangement or configuration of their atoms. For example, dimethyl ether and ethyl **alcohol** have the formula C_2H_6O, but the two compounds have quite different properties. The term isomer was first used by the German chemist Jons Jakob Berzelius. Much of what is known about this class of compounds came as a result of work done by Friedrich Wohler and Justus Liebig in the early 1820s.

Isomerism can occur in one of several forms depending on the structure, position, and geometry of the **chemical bond**s. Chain isomerism exists among the alkanes, which consist of carbon atom chains. Because these chains may be straight or branched, they may yield different possible configurations. For example, normal butane ($CH_3CH_2CH_2CH_3$ is a straight chain, and isobutane $CH-(CH_3)_3$ is branched. Position isomerism occurs because not all hydrogen atoms in a molecule are identical. When these hydrogens are replaced by other elements, alternate isomeric structures can be created. Therefore, propanol (C_3H_7OH) occurs in two forms: one has the hydroxyl group attached to a terminal carbon atom and the other has it on the middle carbon. Functional group isomerism applies to compounds that cannot be fully identified solely by reference to their chemical formula. For example, the formula C_3H_8O may be an ether or an **alcohol**. The final type of isomerism, geometric isomerism, refers to molecules in which the atoms are attached in the same order, but which have different spatial relationships. One type of geometric isomerism, called cis-trans isomerism, refers to asymmetry across the carbon-carbon double bond. For example, the four carbon molecule known as 2-butene can exist as either the cis isomer or trans isomer. In the cis form, the two methyl groups are on the same side of the double bond. In the trans form, the two methyl groups

are on opposite sides. Each isomer is chemically identical but they have different physical properties, such as melting, boiling point, and density.

In 1848, the French scientist **Louis Pasteur** discovered that certain geometric isomers of tartaric acid behaved differently toward **light**. He manually separated the two isomers by handpicking differently shaped crystals from a mixture. Pasteur determined that these two forms, although identical in many respects, behaved differently toward polarized light. He found that one of these isomers would rotate a beam of polarized light counterclockwise. He called this form the levorotatory or (L) form. He found the other isomer would rotate the beam clockwise and he called this the dextrorotatory or (D) form. Interestingly, he found that only one of these forms could be used as a medium to **culture** molds.

Van't Hoff expanded on Pasteur's work by recognizing that the presence of an asymmetric center in a molecule would give rise to two isomeric molecules called enantiomers, one of which is the exact mirror image of the other. Enantiamers, also called optical isomers, are studied in the specialized branch of chemistry known as stereochemistry. These isomers are important in biology because on a molecular level nearly all biochemical processes involve the spatial recognition of one molecule by another. Such recognition serves as the means by which molecules interact and determines how complex biological structures are built.

ITAKURA, KEIICHI (1942-)
American molecular biologist

Keiichi Itakura was the first to use a bacterial **cell** to synthesize the mammalian hormone somatostatin, a **polypeptide** produced by the brain's hypothalamus. Somatostatin inhibits secretion of growth hormone (somatotropin), **insulin**, and glucagon, and is being studied for use in treating insulin-dependent diabetes and other diseases. He selected it because of its possible usefulness, its small size, and also because it was known to have low toxicity and was thus safe to work with.

Itakura was born in Tokyo, Japan, and received a B.S. in 1965 and a Ph.D. in 1970 from the Tokyo College of Pharmacy. After doing further research for the National Research Council of Canada and at the California Institute of Technology, in 1978 he joined the staff the City of Hope National Medical Center in California, where he is a senior research Biologist. He still performs research at Caltech and is also affiliated with Genentech, Inc., a company specializing in **genetic engineering**.

Itakura and his staff were able to synthesize the **gene** for somatostatin and have it expressed by transcribing into **RNA** and translating that RNA into a specifically-designed protein (sequence of amino acids). To begin the process, he took the known 14-**amino acid** sequence of somatostatin and assembled codons (three DNA bases) for each amino acid. Since several codons can express each amino acid, he selected those that where most likely to be expressed in the *E. coli* **bacteria** in which the synthesis took place. The gene was then fused to a gene in *E. coli* plasmid (circular **DNA** found in some bacteria). There, it was expressed as if it were a native *E. coli* gene, producing a precursor form of the **hormone**. Itakura then removed this and chemically converted it to the active form.

He used radioimmunoassay techniques to show that bacteria had produced the mammalian somatostatin. He also demonstrated that it was able to inhibit the release of growth hormone from rat pituitary cells.

IVANOVSKY, DMITRI IOSIFOVICH
(1864-1920)
Russian botanist

Ivanovsky, the son of a landowner, was born in Gdov, Russia. He attended the Gymnasium of Gdov and later graduated as a gold medalist from the Gymnasium of St. Petersburg in 1883.

At the University of St. Petersburg, he enrolled in the natural science department and studied under several prominent Russian scientists. While a student, he became interested in diseases that destroy tobacco plants. He graduated in 1888 after presenting his thesis "On Two Diseases of Tobacco Plants."

The following year, he was asked by the directors of the Department of Agriculture to study a new tobacco disease, called *tobacco mosaic*, that had afflicted plants in the Crimean region. He crushed the infected leaves, which were distinguished by their mosaic pattern, into sap and forced the material through a Chamberland bacterial filter that was known to remove all **bacteria**. Despite following this procedure, the sap, when brushed on the leaves of healthy plants, was still toxic enough to cause disease. Ivanovsky's 1892 report on the tobacco mosaic disease detailed what he maintained must be an agent smaller than bacteria. It was the first study in which factual evidence was offered concerning the existence of this new kind of infectious pathogen.

Ivanovsky's work was ignored by the scientific community, and he eventually abandoned his study of this pathogen without understanding the implications of his research. The Dutch botanist Martinus Willem Beijerinck repeated Ivanovsky's experiments with this new pathogenic source, giving it the name filterable virus in 1898.

See also Stanley, Wendell

J

JACOB, FRANÇOIS (1920-)
French biologist

François Jacob made several major contributions to the field of **genetics** through successful collaborations with other scientists at the famous Pasteur Institute in France. His most noted work involved the formulation of the Jacob-Monod operon model, which helps explain how genes are regulated. Jacob also studied messenger ribonucleic acid (mRNA), which serves as an intermediary between the **deoxyribonucleic acid (DNA)**, which carries the genetic code, and the ribosomes, where proteins are synthesized. He also demonstrated that **bacteria** follow the same general rules of **natural selection** and **evolution** as higher **organisms**. In recognition of their work in genetic control and viruses, Jacob and two other scientists at the Pasteur Institute, **Jacques Lucien Monod** and **André Lwoff**, shared the 1965 Nobel Prize for Physiology or Medicine.

Jacob was born on June 17, 1920, in Nancy, France, to Simon Jacob, a merchant, and the former Thérèse Franck. Jacob attended school at the Lycée Carnot in Paris before beginning his college education. He began his studies toward a medical degree at the University of Paris (Sorbonne), but was forced to cut his education short when the German Army invaded France during World War II in 1940. He escaped on one of the last boats to England and joined the Free French forces in London, serving as an officer and fighting with the Allies in northern Africa. During the war, Jacob was seriously wounded. His injuries impaired his hands and put an abrupt end to his hopes of becoming a surgeon. For his service to his country, he received the Croix de Guerre and the Companion of the Liberation, two of France's highest military honors.

Despite this physical setback, Jacob continued his education at the University of Paris. In his autobiography, *The Statue Within,* Jacob said he got the idea for his thesis from his place of work, the National Penicillin Center, where a minor **antibiotic** called tyrothricin was manufactured and commercialized. For his thesis, Jacob manufactured and evaluated

the drug. Nearing 30 years old, he earned his M.D. degree in 1947, the same year he married Lysiane "Lise" Bloch, a pianist. They would eventually have four children: Pierre, Odile, Laurent, and Henri.

With his professional future unsure, Jacob continued to work for a while at the National Penicillin Center. The tide turned when he and his wife had dinner with her cousins, including Herbert Marcovich, a biologist working in a genetics lab. Jacob recalled, "As Herbert spoke, I felt an excitement rising like a storm. If a man of my generation could still go into research without making himself ridiculous, then why not I?" He decided to become a biologist the next day.

Begins Career as Biologist in "The Attic"

Jacob joined the Pasteur Institute in 1950 as an assistant to André Lwoff. Lwoff's laboratory location and its cramped quarters earned it the name of "the attic." The year 1950 was an exciting one in Lwoff's lab. Lwoff had been working with lysogenic bacteria, which are destroyed (lysed) when attacked by bacteria-infecting **virus** particles called **bacteriophages**. The bacteriophages invade the bacterial cell, then multiply within it, eventually bursting the cell and releasing new bacteriophages. According to Lwoff's research, the bacteriophage first exists in the bacterial cell in a non-infectious phase called the prophage. He could stimulate the prophage to begin producing infective virus by adding ultraviolet light. These new findings helped to give Jacob the background he would need for his future research.

Jacob continued his education at the University of Paris during his first years at the Pasteur Institute, earning his bachelor of science in 1951 and studying toward his doctor of science degree, which he received in 1954. For his doctoral dissertation, Jacob reviewed the ability of certain radiations or **chemical compound**s to induce the prophage, and proposed possible mechanisms of immunity.

Once on staff in the lab, Jacob soon formed what would become a very fruitful collaboration with Élie Wollman, also

François Jacob. *(The Library of Congress. Reproduced by permission.)*

stationed in Lwoff's laboratory. In the summer of 1954 he and Wollman discovered what they termed erotic induction in the bacteria *Escherichia coli.* They later changed the name of the phenomenon to zygotic induction. In zygotic induction, the **chromosome** of a male bacterial cell carrying a prophage could be transferred to a female cell that was not carrying the prophage, but not vice versa. Zygotic induction showed that both the expression of the prophage and immunity was blocked in the latter instance by a variable present in the cytoplasm which surrounds the cell's nucleus.

In another experiment, he and Wollman mated male and female bacterial cells, separating them before they could complete conjugation. This also clipped the chromosome as it was moving from the male to the female. They found that the female accepted the chromosome bit by bit, in a certain order and at a constant speed, rather similar to sucking up a piece of spaghetti. Their study became known as the "spaghetti experiment," much to Wollman's annoyance.

In the book *Phage and the Origins of Molecular Biology,* Wollman explained that by following different genetic markers in the male, they could determine each gene's time of entry into the zygote and correctly infer its position on the DNA. Jacob and Wollman also used an electron microscope to photograph the conjugating bacteria and time the transmission of the genes. "With Élie Wollman, we had developed a tool that made possible genetic analysis of any function, any 'system,' " Jacob said in his autobiography. The two scientists also discovered and defined episomes, genetic strains which

automatically replicate as part of the development of chromosomes.

Jacob and Wollman also demonstrated that bacteria could mutate and adapt in response to drugs or bacteriophages. Evolution and natural selection worked in bacteria as well as in higher life forms. Jacob and Wollman summarized their research in the July 1956 issue of *Scientific American:* "There is little doubt that the basic features of genetic recombination must be similar whether they occur in bacteria or in man. It would be rather surprising if the study of sexual reproduction in bacteria did not lead to deeper understanding of the process of genetic recombination, which is so vital to the survival and evolution of higher organisms."

In 1956 Jacob accepted the title of laboratory director at the Pasteur Institute. Within two years Jacob began to work with Jacques Monod, who had left Lwoff's lab several years earlier to direct the department of cellular biochemistry at the Pasteur Institute. Arthur Pardée also often joined in the research. Jacob and Monod studied how an intestinal **enzyme** called galactosidase is activated to digest lactose, or milk sugar. Galactosidase is an inducible enzyme, that is, it is not formed unless a certain substrate—in this case lactose—is present. Inducible enzymes differ from constitutive enzymes which are continuously produced, whether or not the inducer is present. By pairing a normal inducible male bacteria with a constitutive female, they showed that inducible enzyme processes take precedence over constitutive enzyme synthesis. In the experiments conducted by Jacob and Monod, the inducer, lactose, served to inhibit the gene that was regulating the synthesis of galactosidase.

Afterward, Jacob realized that his work with Monod and his earlier work with Wollman on zygotic induction were related. In *The Statue Within,* he said, "In both cases, a gene governs the formation of a cytoplasmic product, of a repressor blocking the expression of other genes and so preventing either the synthesis of the galactosidase or the multiplication of the virus." Their chore then was to determine the location of the repressor, which appeared to be on the DNA.

Discovers mRNA, Genetic Control Mechanism

By the end of the decade, Jacob and Monod had discovered messenger RNA, one of the three types of **ribonucleic acid.** (The other two are ribosomal RNA and transfer RNA.) Each type of RNA has a specific function. Messenger RNA is the mediator between the DNA and ribosomes, passing along information about the correct sequence of amino acids needed to make up proteins. While their work continued, Jacob accepted a position as head of the Department of Cell Genetics at the Pasteur Institute.

In 1961 they explained the results of their research involving the mRNA and the now-famous Jacob-Monod operon model in the paper, "Genetic Regulatory Mechanisms in the Synthesis of Proteins," which appeared in the *Journal of Molecular Biology.* Molecular biologist Gunther S. Stent in *Science* described the paper "one of the monuments in the literature of molecular biology."

According to the Jacob-Monod operon model, a set of structural genes on the DNA carry the code that the messenger

RNA delivers to the ribosomes, which make **proteins**. Each set of structural genes has its own operator gene lying next to it. This operator gene is the switch that turns on or turns off its set of structural genes, and thus the oversees the synthesis of their proteins. Jacob and Monod called each grouping of an operator and its structural genes an **operon**. Besides the operator gene, a regulator gene is located on the same chromosome as the structural genes. In an inducible system, like the lactose operon (or lac operon as it is called), this regulator gene codes for a repressor protein. The repressor protein does one of two things. When no lactose is present, the repressor protein attaches to the operator and inactivates it, in turn, halting structural gene activity and protein synthesis. When lactose is present, however, the repressor protein binds to the regulator gene instead of the operator. By doing so, it frees up the operator and permits protein synthesis to occur. With a system such as this, a cell can adapt to changing environmental conditions, and produce the proteins it needs when it needs them.

A year after publication of this paper, Jacob won the Charles Leopold Mayer Prize of the French Academy of Sciences. In 1964, Collège de France also recognized his accomplishments by establishing a special chair in his honor. His greatest honor, however, came in 1965 when he, Lwoff, and Monod shared the Nobel Prize for Physiology or Medicine. The award recognized their contributions ''to our knowledge of the fundamental processes in living matter which form the bases for such phenomena as adaptation, reproduction and evolution.''

During his career, Jacob wrote numerous scientific publications, including the books *The Logic of Life: A History of Hereditary* and *The Possible and the Actual*. The latter, published in 1982, delves into the theory of evolution and the line that he believes must be drawn between the use of evolution as a scientific theory and as a myth.

Throughout his autobiography, *The Statue Within,* Jacob notes his drive to continually look ahead. He writes, ''I am bored with what has been done, and excited only by what is to do. Were I to frame a prayer, I would ask to be granted not so much the 'strength' as the 'desire' to do.''

JACOB-MONOD HYPOTHESIS

In 1961, **François Jacob** and **Jacques Monod**, two French biologists, publicized their two part theory that was later coined the Jacob-Monod hypothesis. They postulated that **ribosome**s were not manufactured anew each time a **protein** was made and that the ribosomes did not contain the template necessary for the manufacture of the chains of **amino acid**s and hence the proteins.

Jacob and Monod proposed that ribosomes are general structures which, when supplied with the appropriate instructions, can manufacture any proteins. They are merely assembly areas. Primary to the hypothesis was the idea that each **deoxyribonucleic acid** (**DNA**) cistron (a length of functional DNA) produced a short-lived piece of **ribonucleic acid** (**RNA**). The produced RNA would have the amino acid sequence coded within its **nucleotide** sequence and would then move from its production site within the **nucleus** to the main body of the **cell**. Once there, it would join into temporary association with a ribosome, and then the **ribosome** would read the message on the RNA and use the information contained to produce a protein. Because of its specialized role in the cell, this type of RNA was termed messenger RNA (mRNA). After the manufacture of several **molecule**s of protein, the mRNA template becomes worn out and the molecule disassociates. This allows the ribosome to accept another piece of mRNA and produce another protein.

The Jacob-Monod hypothesis accommodated the observed characteristics of **protein synthesis** and the arrangement of the cell. The ribosomes were known to be constant, unchanging structures within the **cytoplasm** and, yet, also capable of manufacturing all the proteins necessary for the functioning of the cell and the **organism** as a whole. The proposal that the ribosomes were factories that worked on the blueprint, which was provided by DNA through the intermediate mRNA, aligned with what was already understood about their function. The rapid and sometimes short-lived production of protein was also explained by this theory.

At the time the theory was introduced, mRNA was not known as a separate molecule, but before the end of 1961, its existence was independently proven by several scientists. The Jacob-Monod hypothesis is now accepted as the paradigm for the processes that occur within the cell as part of the **transcription** and **translation** process.

In 1965, Jacob and Monod, along with **Andre Lwoff**, were awarded the Nobel Prize in physiology or medicine for their contribution to **molecular biology**.

JAMES, WILLIAM (1842-1910)
American psychologist

William James was an esteemed scientist credited with bringing laboratory science to the study of psychology. Trained as a medical doctor and a physiologist, James insisted that mental processes—thought and consciousness—originated from physical processes, that is, the actions of nerves, muscles and glands. His monumental *Principles of Psychology* assimilated much earlier research, and grounded psychology as a natural science. In spite of his pivotal role in the development of American psychology, James left the field behind later in his life, and concentrated on philosophy.

William James was born January 11, 1842, eldest son of a theologian. It was a distinguished intellectual family. James's younger brother was the novelist Henry James. The James children were educated somewhat haphazardly, as the family traveled between New York, Boulogne, France, and Geneva. William James was multi-lingual at an early age, and in his father's company he was exposed to many religious and philosophical thinkers.

He began studying art formally at age 18, but quickly abandoned the pursuit and started taking science courses at Harvard. Eventually he entered Harvard Medical School. His studies were interrupted by a trip to the Amazon, which led to a breakdown in his health. He later studied with some of the foremost scientists in Germany, and finally completed his medical degree in 1869, when he was 27 years old. But his health was so poor that he could barely leave the house, and he spent his time reading. He sank into **depression** and phobia, and thought of taking his life. This awful period ended when James decided to believe in free will—or at least he willed himself to believe in free will. With this act of faith, he regained health and energy, and began teaching physiology and then psychology at Harvard.

James's approach to psychology was notably different from the prevailing theories of the time. He brought a more biological bent to psychology, insisting that thinking was part of an organism's survival apparatus. He wrote technical papers on dizziness in deaf-mutes and on the human experience of space and time. He created the first psychological laboratory in the United States, at Harvard, and enthusiastically supervised many student experiments. His background in medicine lent his psychology a much more empirical and physiological bent than it had before, and he is credited with formulating the framework from which modern experimental psychology sprang.

In 1890 he published *Principles of Psychology*, an exhaustive two-volume work which organized and synthesized virtually all the extant research on psychology. James later condensed this into a one-volume textbook, which became quite popular.

James lost interest in psychology after his great success with it, and went on to write seminal works of American philosophy, *Varieties of Religious Experience* and *Pragmatism*. In the last years of his life, he was hailed as America's greatest philosopher. His philosophical works are still studied, whereas his writings and research on psychology now seem antiquated. Nevertheless, he made an enormous contribution to the growth of this new science. Much modern research is indebted in spirit to James, and for this he is remembered.

JENNER, EDWARD (1749-1823)
English physician

Jenner was born in Berkeley, England, the third son and youngest of six children of Stephen Jenner, a clergyman of the Church of England. He was orphaned at age five and was raised by his older brother, also a clergyman. When Jenner was 13 years old, he was apprenticed to a surgeon. Then in 1770, he moved to London, England, to work with **John Hunter** (1728-1798), an eminent Scottish anatomist and surgeon who encouraged Jenner to be inquisitive and experimental in his approach to medicine. Jenner returned to Berkeley in 1773, and set up practice as a country doctor. His curiosity about natural phenomena and dedication to medicine ultimately earned him status as a pioneer of virology and immunology, as well as the founder of the practice of vaccination.

During and prior to Jenner's lifetime, smallpox was a common and often fatal disease worldwide. Many centuries

Edward Jenner. *(The Library of Congress. Reproduced by permission.)*

before Jenner's time, the Chinese had begun the practice of blowing flakes from smallpox scabs up the nostrils of healthy persons to confer immunity to the disease. By the seventeenth century, the Turks and Greeks had discovered that, when injected into the skin of healthy individuals, the serum from the smallpox pustule induced a mild case of the disease and subsequent immunity. This practice of inoculation reached England by the eighteenth century. However, it was quite risky as those who were inoculated frequently suffered a severe or fatal case of smallpox. Despite the risk, people willingly agreed to inoculation because of the widespread incidence of smallpox and the fear of suffering from terribly disfiguring pockmarks that resulted from the disease.

As a young physician, Jenner noted that dairy workers who had been exposed to cowpox, a disease like smallpox only milder, seemed immune to the more severe infection. He continually put forth his theory that cowpox could be used to prevent smallpox, but his contemporaries shunned his ideas. They maintained that they had seen smallpox victims who claimed to have had earlier cases of cowpox.

It became Jenner's task to transform a country superstition into an accepted medical practice. For up until the mid-1770s, the only documented cases of vaccinations using cowpox came from farmers such as Benjamin Jesty of Dorsetshire who vaccinated his family with cowpox using a darning needle.

After observing cases of cowpox and smallpox for a quarter century, Jenner took a step that could have branded him a criminal, just as easily as a hero. On May 14, 1796, he removed the fluid of a cowpox from dairymaid Sarah Nelmes, and inoculated James Phipps, an eight-year-old boy, who soon came down with cowpox. Six weeks later, he inoculated the boy with smallpox. The boy remained healthy. Jenner had proved his theory. He called his method *vaccination*, using the Latin word *vacca*, meaning cow, and *vaccinia*, meaning cowpox. He also introduced the word virus.

The publication of Jenner's *An Inquiry into the Causes and Effects of the Variolae Vaccinae* set off an enthusiastic demand for vaccination throughout Europe. Within 18 months, the number of deaths from smallpox had dropped by two-thirds in England after 12,000 people were vaccinated. By 1800, 100,000 people had been vaccinated worldwide. As the demand for the vaccine rapidly increased, Jenner discovered that he could take lymph from a smallpox pustule and dry it in a glass tube for use up to three months later. The vaccine could then be transported.

Jenner was honored and respected throughout Europe and the United States. At his request, Napoleon released several Englishmen who had been jailed in France in 1804 while France and Great Britain were at war. Across the Atlantic Ocean, Thomas Jefferson received the vaccine from Jenner and proceeded to vaccinate his family and neighbors at Monticello. However, in his native England, Jenner's medical colleagues refused to allow him entry into the College of Physicians in London, insisting that he first pass a test on the theories of Hippocrates and Galen. Jenner refused to bow to their demands, saying his accomplishments in conquering smallpox should have qualified him for election. He was never elected to the college.

Nearly two centuries after Jenner's experimental vaccination of young James, the World Health Organization (WHO) declared smallpox to be eradicated. However, when WHO announced its plan to destroy the last remaining stocks of the smallpox **virus** (which was used for research) on June 30, 1999, not everyone was pleased with the decision. Some scientists believe the stockpiled virus could still prove beneficial in terms of research to help fight other deadly viruses, including the **human immunodeficiency virus** (**HIV**), which causes **AIDS**.

See also Immune system

JERNE, NIELS K. (1911-1994)
Danish immunologist

Niels Kaj (sometimes transliterated Kai) Jerne was born on December 23, 1911, in London, England, to Danish parents Else Marie Lindberg and Hans Jessen Jerne. The family moved to the Netherlands at the beginning of World War I. Jerne earned his baccalaureate in Rotterdam in 1928 and studied physics for two years at the University of Leiden. Twelve years later, he entered the University of Copenhagen to study medicine, receiving his doctorate in 1951 at the age of 40. From 1943 until 1956 he worked at the Danish State Serum Institute, conducting research in immunology.

In 1955, Jerne traveled to the United States with noted molecular biologist **Max Delbrück** to become a research fellow at the California Institute of Technology at Pasadena. The two worked closely together, and it was not until his final two weeks at the Institute that Jerne completed work on his first major theory—on selective Max Delbrück formation. At this time, scientists believed that specific antibodies (molecules that defend the body from infection) do not exist until an antigen (any substance originating outside the body such as a **virus**) is introduced and acts as a template from which cells in the immune system create the appropriate antibody to eliminate it. Jerne's theory postulated instead that the **immune system** inherently contains all the specific antibodies it needs to fight specific antigens; the appropriate antibody, one of millions that are already present in the body, attaches to the antigen, thus neutralizing or destroying the antigen and its threat to the body.

In 1960, Jerne left his research in immunology to became chief medical officer with the World Health Organization in Geneva, Switzerland, where he oversaw the departments of biological standards and immunology. From 1960-62, he served on the faculty at the University of Geneva's biophysics department.

From 1962-66, Jerne was professor of microbiology at the University of Pittsburgh in Pennsylvania. During this period he developed a method, now known as the Jerne plaque assay, to count antibody-producing **cell**s by first mixing them with other cells containing antigen material, causing the cells to produce an antibody that combines with red **blood** cells. Once combined, the blood cells are then destroyed, leaving a substance called plaque surrounding the original antibody-producing cells, which can then be counted. Jerne became director of the Paul Ehrlich Institute, in Frankfurt, Germany, in 1966, and, in 1969, established the Basel Institute for Immunology in Switzerland, where he remained until taking emeritus status in 1980.

In 1971, Jerne unveiled his second major theory, which deals with how the immune system identifies and differentiates between self **molecule**s (belonging to its host) and nonself molecules (invaders). Noting that the immune system is specific to each individual, immunologists had concluded that the body's self-tolerance cannot be inherited and is therefore learned. Jerne postulated that such immune system "learning" occurs in the thymus, an organ in the upper chest cavity where the cells that recognize and attack antigens multiply, while those that could attack the body's own cells are suppressed. Over time, mutations among cells that recognize antigens increase the number of different antibodies the body has at hand, thereby increasing the immune system's arsenal against disease.

Jerne introduced what is considered his most significant work in 1974—the network theory—wherein he proposed that the immune system is a dynamic self-regulating network that activates itself when necessary and shuts down when not needed. At that time, scientists knew that the immune system contains two types of immune system cells, or lymphocytes: B cells, which produce antibodies, and T cells, which function as "helpers" to the B cells by killing foreign cells, or by regulating the B cells either by suppressing or stimulating their antibody producing activity. Further, antibody molecules produced by the B cells also contain antigen-like components (idiotypes) which can attract another antibody (anti-idiotype), allowing one antibody to recognize another antibody as well as an antigen. Jerne's theory expanded on this knowledge, speculating that a delicate balance of lymphocytes and antibodies and their idiotypes and anti-idiotypes exists in the immune system until an antigen is introduced. The antigen, he believed, replaces the anti-idiotype attached to the antibody. The immune system then senses the displacement and, in an attempt to find the anti-idiotype a "mate," produces more of the original antibody. This chain-reaction strengthens the body's immunity to the invading antigen.

Jerne shared the 1984 Nobel Prize for medicine or physiology with **Cesar Milstein** and **Georges J. F. Kohler** for his body of work that explained the function of the immune system

Jerne retired to southern France with his wife, Ursula Alexandra Kohl, whom he married in 1964; the couple had two sons. A citizen of both Denmark and Great Britain, Jerne received honorary degrees from American and European universities, was a foreign honorary member of the American Academy of Arts and Sciences, a member of the Royal Danish Academy of Sciences, and won, among other honors, the Marcel Benorst Prize in 1979 and the Paul Ehrlich Prize in 1982. Jerne died on October 7, 1994, at his home in Pont du Gard, southern France.

JOHANNSEN, WILHELM LUDWIG (1857-1927)
Danish biologist

Wilhelm Johannsen, whose family did not have the financial means to provide him a college education, served as an apprentice to a pharmacist in 1872 and taught himself chemistry and botany. Johannsen's interest in botany flourished when he became an assistant at the newly-formed Carlsberg laboratory. Here he was given the freedom to experiment and research, even after his resignation in 1887.

Johannsen's extensive research in botany attracted him to the plant experiments surrounding **heredity** and **natural selection**. So many opposing theories existed at that time to explain evolutionary change: **Charles Darwin**'s theory of small, continuous changes; **Francis Galton**'s theory of regression; not to mention the rediscovery of **Gregor Mendel**'s theory. Johannsen's work suggested that there might be common ground in these seemingly opposing viewpoints.

Johannsen experimented with bean **plants** to show that natural selection had no effect on pure line **species**. In other words, self-fertilizing bean plants didn't produce abnormally large or abnormally small seeds, despite Johannsen's attempts at "selecting" these characteristics. Johannsen knew that the offspring of a pure line were genetically identical and his pure line theory suggested that any differences among them must

be due to chance and environment—not natural selection. Johannsen was the first to attribute the origin of small, continuous differences to **mutation**. Johannsen's discoveries sent biologists to investigate mutations in search of other answers to evolution.

In 1909, there were as many terms to describe what biological component caused inheritance as there were heredity theories. Johannsen clarified the terminology by coining the term **gene** to describe a unit of inheritance. He felt it was a term that was not associated with any of the many theories that existed to explain evolution. And he didn't stop there. Johannsen came up with the terms **phenotype** and **genotype**. Phenotype refers to the observed traits—the actual appearance of an organism. Genotype is the actual genetic code inherited from parents that produces visible differences. The phenotype may change with time, but the genotype is fixed for life.

"JUMPING" GENE • See McClintock, Barbara

JUST, ERNEST EVERETT (1883-1941)
American biologist

Ernest Just was a pioneer in the study of human **egg** fertilization, artificial **parthenogenesis**, the relationship of **bacteria** to human **cell**s, and the effect of environment on **cell division**. A native of Charleston, South Carolina, Just was born the oldest of three children on August 14, 1883. His mother, Mary Matthews Just, opened a private religious school, where she taught her son. His father, Frazier Just, a dock worker and wharf builder, died in 1887, when Just was only four years old. Just learned early that he would have to work in the fields to help his mother support the family.

In 1900, having completed six years of education at the Industrial School in Orangeburg, South Carolina, Just worked his way to New York aboard a coastal steamer. He studied at the Kimball Union Academy in Meriden, New Hampshire, where he edited the school paper and led the debate club. He studied biology at Dartmouth College, where he enrolled in every available biology course and developed a specialty in cytology. In 1907, he graduated magna cum laude and Phi Beta Kappa.

Just began teaching **zoology** at Howard University in 1907. He was named head of the zoology department in 1912 and remained in that position the rest of his life. He married colleague Ethel Highwarden, joined the medical staff, and, as a means of upgrading the school's training of physicians, headed the physiology department. In 1915, he took a leave of absence from his teaching post to complete a doctorate in zoology at the University of Chicago.

From 1909-30, Just taught during the school year, then spent his summers immersed in productive research among Nobel scientists at the renowned marine lab at Woods Hole, Massachusetts. With his knowledge of French and German, he studied widely the nature of the cell, particularly the transfer of water within the cell, the effects of ultraviolet light on the number of **chromosome**s in a cell, and other variations in the reproductive cell caused by manipulation of physical and chemical conditions. His findings resulted in more than 60 published articles in scientific journals.

A precise researcher, Just examined the formation of egg and **sperm** cells in sea urchins and sandworms and perfected methods of identifying and standardizing cell samples. Departing from theories of his day, he used seawater to produce parthenogenesis, or spermless **fertilization**, in sea urchins. Supported by the National Research Council, the Carnegie Corporation, the Rosenwald Foundation, Julius Rosenwald, and the General Education Board, his studies determined the importance of cytoplasm and ectoplasm to fertilization and heredity. He also produced more current data on the function of the kidneys, **pancreas**, and liver, which proved crucial to the study of **cancer**. His findings, which fellow scientists valued highly, assisted the differentiation between living and non-living matter and between **animal**s and plants, determination of sex in unborn fetuses, and an understanding of evolution. He published two books, *The Biology of the Cell Surface* (1939) and *Basic Methods for Experiments on Eggs of Marine Animals* (1939), and edited *The Biological Bulletin, Protoplasma, General Cytology, The Journal of Morphology*, and *Physiological Zoology*. He served as vice president of the American Society of Zoologists and was named a fellow of the American Association for the Advancement of Science.

During the 1930s, Just grew bitter because of the scientific community's callous indifference to black scholarship. Unable to obtain a suitable post at a major university research center, he moved to Berlin to work at the Kaiser Wilhelm Institute. Just later worked in Russia, Italy, and France, where he lectured at the Sorbonne. For his work, Just became the first notable black to receive the NAACP's Spingarn Medal. Even more to his credit, he trained a cadre of research scientists who followed his meticulous laboratory method and furthered his inquiry into the nature of the cell. Just died of cancer in Washington, DC, on October 27, 1941, and was buried in the Lincoln Cemetery. In 1996, the U.S. Postal Service issued a stamp honoring Just as part of its Black Heritage series.

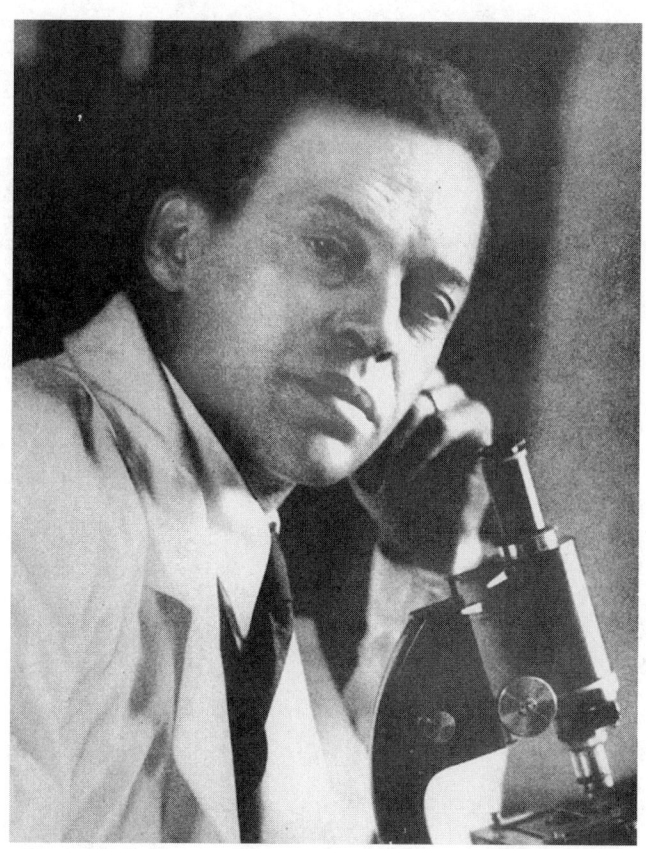

Ernest Everett Just. *(AP/Wide World Photos, Inc. Reproduced by permission.)*

K

KAMEN, MARTIN DAVID (1913-)
American chemist

Martin D. Kamen was responsible, with **Samuel Ruben**, for the discovery of carbon-14. His discovery of the long-lived radioactive isotope made possible much further research on **photosynthesis**, so this was of particular importance to biologists. Carbon-14 was also subsequently found to be extremely useful in dating **fossils**, so Kamen's work had broad implications for scientists in varied fields such as geology, **paleontology**, and archaeology.

Kamen was born August 27, 1913, the first-born son of Russian Jewish immigrant parents. The Kamens ran a photographic portrait studio in the Hyde Park area of Chicago. Kamen was a child prodigy on the violin, and through his high school years it was assumed that he would have a career in music. But balking at family pressure to perform, Kamen stopped studying music seriously, and entered the University of Chicago in 1930 intending to major in English. However, his family's fortune had slid considerably in the Great Depression, and Kamen's parents urged him to switch his major to chemistry. Though Kamen had felt no previous interest in science, he began the chemistry courses, because this was one of the few fields that offered any hope of future employment. Kamen finished his bachelor of science degree in three years, and then continued at the University of Chicago for his Ph.D., which he received in another three years.

In 1936, he set out for Berkeley, California to visit Edward O. Lawrence's famed Radiation Laboratory, home of the cyclotron. After working for six months for no pay, Kamen was hired on at the lab. He used the cyclotron to isolate a radioactive carbon isotope that could be used as a tracer to study the still mysterious process of photosynthesis. Kamen and his colleague Samuel Ruben first worked with Carbon-11, which had a half-life of only 21 minutes. The team worked furiously to perform plant and animal experiments with Carbon-11 before it decayed. Logistical problems sometimes led the team to use carrier pigeons to fly samples between labs. Because of its short life, the usefulness of Carbon-11 was limited. Apparently Edward Lawrence, head of the Radiation Laboratory, feared that he would lose grant funding if the lab could not come up with a better product. So Lawrence ordered Kamen to find something better: a long-lived carbon isotope. The scientists had an idea that such a thing existed, and Kamen's team worked practically around the clock for months to come up with it. Finally they found Carbon-14, which proved to have a half-life of 5,730 years.

Carbon-14 was one of the most important radioactive substances ever discovered. It is present in all living tissues, but it begins to decay once an organism dies. Because it decays at a uniform rate, it can be used to determine when something died, and thus how old it is. This process, called **carbon dating**, was invented by **Willard Libby** in 1946. Kamen's discovery also led to experiments that finally mapped the process of photosynthesis.

After his momentous discovery, Kamen worked on the Manhattan Project, developing equipment for the separation of uranium isotopes. But in July 1944, Kamen was ordered off the project. He had been tracked by the Federal Bureau of Investigation (FBI), and apparently was considered a security risk because of left-wing leanings and because he was outgoing and talkative. After this blow, Kamen for a time worked in a shipyard, the only job he could get. He later found employment at Washington University in St. Louis, where he did significant biomedical research. But he was still shadowed by the FBI, and in 1948 the House Un-American Activities Committee (HUAC) declared him a spy. Kamen battled the allegation for years. He eventually won a libel suit and forced the State Department to grant him a passport.

Kamen's later research covered photosynthesis, biological oxidation, **protein** structure, and several other areas. He made a lasting contribution to many branches of science with his discovery of Carbon-14. He carried out eminent scientific work under the cloud of political persecution, and finally succeeded in clearing his name.

Martin David Kamen. *(The Library of Congress. Reproduced by permission.)*

KARRER, PAUL (1889-1971)
Russian Swiss organic chemist

Paul Karrer's long and distinguished career in chemistry included the study of sugars and plant pigments, subjects that led him to the description and synthesis of vitamin A as well as several other **vitamins**. Karrer's work with vitamins helped to solve their chemical riddle, enabling physiologists to define the way in which the body utilizes them. In 1937 Karrer shared the Nobel Prize for chemistry for research that incorporated the vitamins A and B.

Paul Karrer was born in Moscow, Russia, on April 21, 1889, the son of Julie Lerch Karrer and Paul Karrer, a Swiss dentist who was practicing in Russia. At three years of age Karrer and his family returned to Switzerland, initially to Zürich, but later settling in the canton of Aargau, a region in the north of the country. Karrer was educated in schools in this canton, and it was while in secondary school that he began showing a passion for science. In 1908 he entered the University of Zürich, ultimately studying chemistry under Alfred Werner, whose work on the linkage of **atoms** in **molecules** won him the Nobel Prize in 1913. Karrer, after completing his doctoral dissertation on cobalt complexes in 1911 and earning his Ph.D., became Werner's lecture assistant. His attention

soon turned to organic arsenical compounds, and Karrer's first paper on the subject, published in 1912, caught the eye of **Paul Ehrlich**, a renowned chemist in Germany whose work at the turn of the twentieth century had helped to explain the action of poisons and how to neutralize their effects by antitoxins. Ehrlich subsequently invited Karrer to join him as a research assistant in Frankfurt-am-Main at the Georg Speyer-Haus, a research institute.

Directs Chemical Research in Germany and Switzerland

Karrer remained in Germany until the beginning of World War I when he was called back to Switzerland for national service. While serving in an artillery unit, he met and married Helene Frölich, the daughter of the director of a psychiatric clinic. They would have three sons, but only two of them, Jurg and Heinz, survived infancy. In 1915 Ehrlich died in Frankfurt, and Karrer accepted the position of researcher and director of the chemical research division of the Georg Speyer-Haus, returning to war-time Germany. Karrer stayed in Germany for the next three years, and during this time he focused more closely on plant product chemistry. Then in 1918 he returned to the University of Zürich, first taking a position as associate professor of organic chemistry, and with Werner's death in 1919, becoming a full professor of chemistry as well as director of the Chemical Institute. Karrer would remain at the University of Zürich for the rest of his career, acting as rector from 1950 to 1953.

Karrer, of necessity, had to split his time between administration, teaching, and research. With the latter, he turned his attention to the spatial or steric configuration of atoms in molecules of **amino acids**, **proteins**, and **peptides**. But by the late 1920s he had shifted his focus to the pigmentation of plants, and more specifically, to the anthocyanins, the blue and red colors of berries and flowers. Though these substances had been isolated by another researcher, Karrer—by splitting their **macromolecule**s with **enzyme**s—helped to clearly describe their chemical make-up. More importantly, it was these researches in plant pigments that eventually led him to his work on carotene.

Vitamin Research Leads to Nobel

From anthocyanin, Karrer moved on to crocin, the yellow pigment of flowers such as the crocus. In connection with yellow pigments, Karrer tackled the structure of carotenoids, the orange-to-yellow pigments found in foods such as carrots and sweet potatoes. **Richard Kuhn**, a German chemist, had managed to isolate beta-carotene at about this time, and he and Karrer became something of rivals in explaining beta-carotene's chemical constitution. By 1930 Karrer had solved the structure of the carotene molecule. It was a logical progression from the study of plant pigments to that of vitamins, for Karrer learned that the body actually synthesizes vitamin A from carotene. Thus, he was soon on the track of the chemical make-up of vitamins themselves. By 1931 Karrer had become the first scientist to describe the structure of a vitamin, successfully demonstrating that vitamin A is very similar to one half of the symmetrical carotene molecule. Up until the time of his discovery, scientists had thought vitamins to be some peculiar state of matter, perhaps a colloid—a dispersed solution in suspension. But Karrer managed to show that vitamin A, in specific, is made up of atoms of hydrogen, carbon, and oxygen in a regular ring-like formation.

Karrer carried on his vitamin research over the next decade, ultimately synthesizing vitamin A in the laboratory. He then went on to research the chemical structure of several B vitamins, riboflavin being the first that he actually synthesized. In 1937, for his work on carotenoids and flavins, he shared the Nobel Prize for chemistry with **Walter Haworth**, an Englishman who researched the make-up of **carbohydrate**s and vitamin C. Karrer, however, was not one to rest on his laurels, and the very next year he synthesized vitamin E, and soon after, vitamin K.

From vitamin research, Karrer turned to an investigation of the enzyme nicotinamide adenine dinucleotide (NAD) which is involved with the **energy** system of **cell**s. By 1942 he had contributed greatly to the understanding of both the NAD structure and function in cellular electron transfer. In his sixties, Karrer went back to earlier work, both in carotenes and in poisons—this time **alkaloid**s. In the former, he successfully synthesized all the carotenoids, some forty different compounds. His work in alkaloids helped to determine the structure of curare, a resinous extract from certain South American trees used by indigenous peoples for poison arrows. Its medicinal uses include general anesthesia and reduction of muscle spasms.

Apart from his research, Karrer was a tireless administrator and teacher, directing over 200 dissertations in his academic career. He was also a prolific writer, with over 1,000 publications to his credit, including the 1928 organic chemistry textbook, *Lehrbuch der organischen Chemie*. Aside from the Nobel, Karrer won numerous awards and prizes, among them the Marcel Benoist Prize from Switzerland in 1923, the Cannizzaro Prize from Italy in 1935, and the Officier de la Legion d'Honneur from France in 1954. Despite fame, wealth, and offers from universities around the world, Karrer stayed on in Zürich until his retirement in 1959, eschewing luxuries such as a car. He died on June 18, 1971, after a short illness.

KARYOTYPE

The full complement of **chromosome**s, arranged in a logical order, is known as a karyotype. Chromosomes of a **cell** are visible with a **microscope** only during **mitosis** (nuclear division). When a cell's nuclei are in **interphase** chromosomes are not ordinarily visible. **DNA replication** occurs during interphase. Nuclear division begins with mitotic prophase. Chromosomes first appear as long threads in prophase. However, they become condensed as mitosis proceeds. The chromosomes are double because of prior DNA replication. The two parts of the chromosome are known as chromatids, which are held together at the centromere. Metaphase follows prophase and here the chromosomes are maximally shortened and readily stain

(hence the name chromosome which means colored body). The nuclear **membrane** disappears at mitotic metaphase and the chromosomes can be prepared for karyotyping. First, a good metaphase spread must be prepared. Historically, this was done by literally squashing a fixed and stained cell in metaphase. With luck, the squashing separated individual chromosomes such that they could be photographed. Then the chromosomes were arranged in a logical order which was generally by size, from long chromosomes to short chromosomes.

Contemporary karyotyping involves first obtaining a **culture** of cells in a liquid medium. Many cells in mitosis can be obtained with culture. The cells are treated and stained which results in unambiguous identification of each chromosome pair. Formerly, short chromosomes were lumped together as were long chromosomes. Now, with the introduction of a diversity of stains which include some that fluoresce differentially in the ultraviolet, each chromosome pair may be clearly distinguished.

Some birth abnormalities (e.g., **Down syndrome**) and certain **cancer**s are characterized by specific chromosomal aberrations. Further, exposure to **teratogen**s, mutagens, and **carcinogens** may result in a non-specific karyotypic change from the normal.

KATZ, BERNARD (1911-)
German English physiologist

Katz was born in Leipzig, Germany, on March 26, 1911. He received his M.D. from the University of Leipzig in 1934, just as Adolf Hitler (1889-1945) was gaining power in Germany. Katz, who was Jewish, left his homeland for England where he pursued his post-graduate studies at the University of London. He eventually earned a Ph.D. in 1938 and a Sc.D. in 1934 from London.

During his years in London and (briefly, during World War II) Sydney, Australia, Katz worked with other scientists on the transmission of **nerve impulses** along a **neuron** and across the synaptic gap between neurons and muscles. With **Alan Hodgkin** (1914-) and **Andrew Huxley** (1917-), he found that the movement of a nerve impulse along a single neuron can be described in terms of the **diffusion** of potassium and sodium ions across the cell **membrane**. This diffusion of ions creates a small electric potential that corresponds to the movement of the electrical impulse in the neuron.

In the 1950s, Katz concentrated on the transfer of the nerve message across **synapses**, the spaces between neurons or between neurons and muscle cells. Earlier research by **Henry Dale** (1875-1968) and **Otto Loewi** (1873-1961) had indicated that the neural message is carried across the synaptic gap by means of a chemical substance, a **neurotransmitter**, later identified as **acetylcholine**.

Katz eventually found that molecules of acetylcholine are apparently "packaged" in tiny vesicles stored at the end of a neuron. Upon stimulation, the neuron releases packages of acetylcholine that travel across the synapse and stimulate the adjacent muscle cell or second neuron. For this discovery,

Katz was awarded a share of the 1970 Nobel Prize for physiology or medicine. He was also knighted by Queen Elizabeth II (1926-) in 1969.

KENDALL, EDWARD CALVIN (1886-1972)

American biochemist

Edward Kendall is known for two major contributions to biochemical knowledge. The first of these was his isolation of the thyroid **hormone** thyroxine. In addition, he isolated several steroid hormones produced by the cortex (outer covering) of the **adrenal gland**, one of which is **cortisone**, playing a major role in demonstrating its medical use. For his work with cortisone, Kendall shared part of the 1950 Nobel Prize in physiology or medicine with his colleague **Philip Hench**.

Throughout his career, Kendall often relied on intuition rather than strict laboratory procedure in performing his research. After retiring from the Mayo Clinic in 1951, he continued his research at Princeton University. Kendall was born in South Norwalk, Connecticut, where his father was a dentist. He received both his bachelor's degree (1908) and his doctorate (1910) from Columbia University. The theory of hormones had been developed in the early years of the century by the British physiologists **Ernest Starling** and **William Bayliss** (1860-1924), describing glandular secretions that control body functions. In 1910, Kendall began working to isolate a thyroid hormone, first at a pharmaceutical company (Parke Davis and Co.) and then at St. Luke's Hospital in New York. By 1913, he had made a much more pure **thyroid gland** extract than previously was available, showing its activity in both experimental dogs and human patients.

In 1914, he joined the staff of the Mayo Clinic, in Rochester, Minnesota, where the founders were interested in studying and treating thyroid diseases. By the end of the year, Kendall had isolated the crystalline form of the hormone, later named thyroxine, and shown its activity in successful treatment of patients with underactive thyroid glands. In studying how thyroxine affects oxidation processes in the body, he needed the **coenzyme** glutathione, a **peptide** formed from the **amino acid**s cysteine, glutamic acid, and glycine. Since it was not available in pure form, Kendall and his associates independently isolated and synthesized it.

During the 1930s, Kendall turned his attention to the hormones of the adrenal cortex after an adrenal extract had been successfully used to treat Addison's disease, which results from atrophy of the gland. In 1933, he isolated a crystalline substance, which he called ''the'' hormone, because it was then believed that the cortex secreted only a single hormone. However, further research showed that several substances were present in his crystals, none of them necessarily a hormone. The next year, Kendall and others independently suggested that the adrenal cortex secretes more than one hormone.

Over the next two years, Kendall isolated a series of crystalline substances, to which he gave alphabetical titles. Kendall's Compound E, later named cortisone, was independently isolated by three groups of scientists, but only Kendall converted it to a related compound, a diketone, which had already been demonstrated to be active. Kendall deduced that Compound E is a **steroid**. During World War II Kendall directed a program to synthesize and produce quantities of Compound E, because some medical authorities thought it might help prevent injury-related stress and surgical shock in wounded military personnel. Hench, who was working on treatment methods for rheumatoid **arthritis**, and Kendall determined to try using Compound E on arthritic patients. In 1948, after a supply of it was available, they used it as a treatment for the first time. In 1949, they named Compound E cortisone. Kendall's Compounds B and F were later identified by other scientists as the hormones corticosterone and cortisol (hydrocortisone).

KETTLEWELL, BERNARD (1907-1979)

English geneticist and entomologist

Bernard Kettlewell is best known for his research on industrial melanism, or the effects of industrial pollution on pigmentation in insects, particularly moths and butterflies. He specialized in entomological fieldwork and made a significant contribution to the Lepidoptera collection at the British Museum of Natural History. Kettlewell's work on industrial melanism at Oxford in the 1950s was considered to be the first rigorous scientific study to confirm **Charles Darwin**'s theory of **natural selection**.

Henry Bernard Davis Kettlewell was born in Howden, Yorkshire, on February 24, 1907, to Kate (Davis) and Henry Kettlewell, a member of the British Corn Exchange. He attended the prestigious Charterhouse public school from 1920 to 1924, spent a year studying in Paris, then enrolled at Caius College, Cambridge University, in 1926 to study medicine and **zoology**. He accepted appointments at several hospitals in England, including St. Bartholomew's in London and St. Luke's in Guildford, where he served for a short time as an anesthetist. After receiving his medical degree in 1935, Kettlewell established a general practice in the town of Cranleigh, Surrey. A year later he married Hazel Margaret Wiltshire, with whom he had two children. Kettlewell left his medical practice at the onset of World War II, when he was assigned to Woking War Hospital.

Following the war, the British government instituted the National Health Service program, which significantly changed the profession of medicine in England. Kettlewell left medicine at that time to pursue his lifelong hobby, entomological fieldwork, in a professional capacity. In 1949 he accepted a research appointment at the International Locust Control Center at Cape Town University in South Africa. While in Africa, Kettlewell pursued his passion for fieldwork by making scientific expeditions throughout the continent, including Mozambique, Zaire, the Knysna Forest, and the Kalahari Desert. He was awarded a Nuffield Research Fellowship at Oxford University and returned to England in 1952. Two years later he was appointed senior research officer in the zoology department at Oxford, where he was to perform his most celebrated experiments—those dealing with industrial melanism. Over a span of two decades, Kettlewell worked with a small research team in the laboratory of his good friend Edmund Brisco Ford to link the world of the field naturalist with that of professional biology.

Long before Kettlewell's research in England, animals indigenous to regions subject to industrial pollution were known to exhibit darker coloration as time passed. More than seventy species of moths, for example, had darkened in color since the onset of the Industrial Revolution in Britain. This phenomenon was often interpreted as evidence that the polluting chemicals themselves were darkening, or melanizing, the animals. Moreover, those scientists who suspected that the melanization was actually a result of natural selection could not demonstrate a mechanism by which the animals were being selected.

Kettlewell's work was significant because he discovered, and captured on film, the process of selection at work in the peppered moth, *Biston betularia*. The peppered moths—and all moths subject to industrial melanization—were active at night, resting motionless on tree trunks during the day. In nonindustrial regions, the peppered moth was usually white with black spots and was well camouflaged against the pale, lichen-covered surfaces of most tree trunks. In industrial areas, Kettlewell found that the soot from the factories killed the lichens and darkened the tree trunks. In this environment, the light-colored moth was no longer protected by its coloration and was subject to bird **predation**. The dark form of the peppered moth, quite rare before industrialization, became the new beneficiary of camouflage and flourished with its selective advantage. Kettlewell concluded that in polluted areas, twice as many black moths as pale moths survived, and that in unpolluted areas, the opposite was true. He strengthened his case by producing film footage of birds eating the conspicuous moths.

In his later research, Kettlewell developed more sophisticated experiments concerning the evolution of dominance in inherited characteristics and the creation of new **species**. The centerpiece of his work remained Lepidoptera, moths and butterflies, and he was one of the primary contributors of specimens to the British Museum of Natural History's Rothschild-Cockayne-Kettlewell collection. In addition to this contribution to evolutionary biology, Kettlewell's research also led to several advancements in the methods of **entomology**. For example, Kettlewell developed new techniques for radioactive labelling of insects for purposes of identification, through which he was once able to demonstrate that a moth captured in his yard had migrated from North Africa by determining that it had flown through a French nuclear test site.

Kettlewell's research was particularly embraced in the Communist Bloc, and he received the Soviet Union's Darwin Medal in 1959, as well as Czechoslovakia's Mendel Medal in 1965. He retired in 1974 and was awarded an honorary degree from Oxford University the following year. Renowned for his ability to turn enthusiasm for field research into progress in evolutionary science, Kettlewell was praised in *Antenna* by geneticist Cyril A. Clarke as "by far the best field-worker cum scientist of his generation." Kettlewell died in 1979.

KHORANA, HAR GOBIND (1922-)
Indian American biochemist

Har Gobind Khorana is best known for developing chemical methods to determine the **nucleotide** sequence of ribonucleic acid (**RNA**) and for deciphering the **genetic code**. For this he shared the 1968 Nobel Prize for physiology or medicine with **Marshall Nirenberg** and **Robert Holley**.

Khorana was born in the small village of Raipur, India, where his father was a clerk in the British colonial service. Encouraged by his parents to obtain an education, he received a master's degree in 1945 from the University of Punjab, Lahore (now part of Pakistan). He then went to the United Kingdom, where he earned a doctorate in 1948 from the University of Liverpool. After brief visits to Switzerland and India, he returned to the United Kingdom to do research at Cambridge University during 1950-52, becoming interested in the nucleic acid investigations there. After further study in Switzerland, Khorana took a research position with the British Columbia (Canada) Research Council in Vancouver. In 1960 he joined the faculty of the University of Wisconsin.

Khorana's genetic code work involved using chemical methods and the **enzymes** deoxyribonucleic acid (**DNA**) polymerase and RNA polymerase, which he called "beautifully precise copying machines," to produce long strands of **nucleic acids** with known base sequences. This allowed him to compare the information-bearing DNA with the corresponding transfer RNA (tRNA) that retrieved specific **amino acids** for assembly into **proteins**. Building on Nirenberg's work, Khorana precisely spelled out all sixty-four genetic code triplets (three-base information units) and related them to specific amino acids or terminating functions.

Two years after receiving the Nobel Prize for this work, he made pioneering progress in the construction of yeast and **bacteria** genes in the laboratory. Khorana also advanced knowledge of the way a **cell**'s protein synthesizing bodies recognize a triplet and correctly assemble the amino acids, as well as how RNA and amino acids recognize each other.

KIMURA, MOTOO (1924-)
Japanese biologist

One key to the theory of Darwinian **evolution** is the concept of **natural selection**. According to this concept, random **mutations** occur in the **gene pool** of any given **population**. Those mutations that increase the likelihood of survival among future generations survive. Those that do not, disappear. Mutations that have a selective advantage are, according to this theory, likely to increase in number in the given population. An alternative to this theory was suggested in 1968 by the Japanese geneticist Motoo Kimura. Kimura found that mutations with no apparent selective advantage can also increase in a population. His research focused on mutant **genes** that are not expressed phenotypically. Since these genes can not be found morphologically, they can only be detected by biochemical means. Kimura believed that the mutant genes he studied are neither more nor less advantageous to the **organism** than are the normal genes they replace. Therefore, evolutionary change depends not on natural selection but on the random drift of these neutral mutations. The Kimura theory of neutral evolution has not, as yet, been widely accepted by population geneticists. Some think that evolutionary effects of the mutant genes may still be found in future research.

Motoo Kimura was born in Okazaki, Japan, on November 13, 1924. After earning his master's degree at Kyoto University, he attended the University of Wisconsin, where he received his Ph.D. in 1956. After graduation, Kimura accepted an appointment at the Japanese National Institute of Genetics in Mishima. He became head of the Population Genetics Department at the Institute in 1964.

KINGDOM

The **classification** or **taxonomy** hierarchy can be credited to **Carl Linnaeus**, a Swedish botanist who lived during the eighteenth century. This classification hierarchy starts at its broadest level with kingdom, which are made up of phyla, which in turn are made up of **class**es, which are equally made up of **or**ders, and so on down to **species**. At the time of **Aristotle** in the fourth century B.C., two kingdoms of living forms were recognized-the Plantae and Animalia. Today, with more sophisticated techniques, most scientists recognize the five kingdom approach first proposed by R. H. Whittaker in 1969. The Kingdom **Monera** includes the prokaryotic (''before seed or kernels,'' meaning before the **nucleus**) **bacteria** and blue-green bacteria (formerly known as blue-green **algae**). This group is differentiated by the lack of **membrane**-bound nuclei. **Prokaryote**s also lack other cell **organelle**s such as **mitochondria**, **chloroplast**s, and other specialized structures. Bacteria function primarily as decomposers, and some groups are able to chemosynthesize. Blue-green bacteria are common photosynthetic **organism**s in aquatic **ecosystem**s, and can form dense mats on lakes during the summer. All other organisms are quite different from prokaryotes because they have the structures mentioned above, and are thus called eukaryotic (''true kernels,'' or true nucleus). The Kingdom **Fungi** consists of molds, **yeast**s, and mushrooms. These organisms function primarily as decomposers and obtain their food by secreting **en**zymes that break down organic **matter** in the **tissue**s of living or dead organisms and absorbing the released **nutrients**. Organisms in the Kingdom **Protista** include unicellular **protozo**ans and unicellular algae. This kingdom is alternately known as the Protoctista Kingdom. Members of the Kingdom Plantae include mosses, liverworts, ferns, macro-algae (green, brown, and red algae), and seed-bearing **plants** (conifers and flowering plants). Lastly, members of the Kingdom Animalia include the **invertebrates** (e.g., sponges, flatworms, nematodes, arthropods, etc.) and the vertebrates, which are included in the **Phylum** Chordata.

KITASATO, SHIBASABURO (1852-1931)
Japanese bacteriologist

Bacteriologist Shibasaburo Kitasato made several important contributions to the understanding of human disease and how the body fights off infection. He also discovered the bacterium that causes **bubonic plague**.

Born in Kumamoto, Japan, Kitasato, completed his medical studies at the University of Tokyo in 1883. Shortly thereafter, he traveled to Berlin to work in the laboratory of **Robert Koch**. Among his greatest accomplishments, Kitasato discovered a way of growing a pure **culture** of tetanus bacillus using anaerobic methods in 1889.

In the following year, Kitasato and German microbiologist **Emil von Behring** reported on the discovery of tetanus and **diphtheria** antitoxins. They found that animals injected with the microbes that cause tetanus or diptheria produced substances in their blood, called antitoxins, that neutralized the toxins produced by the microbes. Furthermore, these antitoxins could be injected into healthy animals, providing them with immunity to the microbes. This was a major finding in explaining the workings of the **immune system**. Kitasato went on to discover **anthrax** antitoxin as well.

In 1892, Kitasato returned to Tokyo and founded his own laboratory. Seven years later, the laboratory was taken over by the Japanese government, and Kitasato was appointed its director. When the laboratory was consolidated with the University of Tokyo, however, Kitasato resigned and founded the Kitasato Institute in 1914 at Shirogane, Tokyo.

During an outbreak of the bubonic plague in Hong Kong in 1894, Kitasato was sent by the Japanese government to research the disease. He isolated the bacterium that caused the plague. (Alexandre Yersin (1863-1943) independently announced the discovery of the organism at the same time). Four years later, Kitasato and his student Kigoshi Shiga were able to isolate and describe the organism that caused dysentery.

Kitasato was named the first president of the Japanese Medical Association in 1923 and was made a baron by the Emperor in 1924. He died in Japan in 1931. The institute he founded lives on as Kitasato University, established in 1962.

KLINEFELTER SYNDROME

Named in 1942 for Massachusetts General Hospital physician Dr. Harry Klinefelter, Klinefelter syndrome is a genetic disorder affecting males. People with this syndrome are born with at least one extra X **chromosome**.

Chromosomes are structures in the **nucleus** of every **cell** in the body. Chromosomes contain the genetic information necessary to direct the growth and functioning of all the cells and systems of the body. A normal person has a total of 46 chromosomes in each cell, two of which are responsible for determining that individual's sex. Females have two X chromosomes; males have one X and one Y chromosome.

In Klinefelter syndrome, an error very early in development results in an abnormal number and arrangement of chromosomes. Most commonly, a patient with Klinefelter syndrome will be born with 47 chromosomes in each cell. The extra chromosome is an X chromosome. Some Klinefelter patients have even more complex chromosomal errors, including the presence of 48, 49, or even 50 chromosomes, with all of the extras beings Xs.

Klinefelter syndrome is one of the most common chromosomal abnormalities. About one in every 1,000 infant boys is born with some variation of this disorder. While the cause of Klinefelter syndrome is now known, it has been noted that the disorder is seen more frequently among the children of older mothers.

The presence of more than one X chromosome in a male results in a significant delay in the onset of puberty. The penis and testicles tend to be smaller than normal, and **infertility** is common. The testicles may remain up in the abdomen, instead of descending into the scrotum as is normal. Body hair is decreased. Breast size is increased. Sexual drive is often below normal. Boys with Klinefelter syndrome tend to be tall and thin.

While it was once believed that boys with Klinefelter syndrome were uniformly mentally retarded, it has more recently become clear that the disorder can exist with no such retardation. However, children with Klinefelter syndrome frequently have difficulty with language, including **learning** to speak, read, and write. Some children have difficulty with social skills, and are noted to be more shy, anxious, or immature than their peers. Overly aggressive behavior has also been noted.

The greater the number of X chromosomes present, the greater the disability. Boys with multiple extra X chromosomes have distinctive facial features, deformities of bony structures, more disordered development of male features, and a tendency to be more severely intellectually retarded.

Other conditions accompany Klinefelter syndrome. Due to abnormal development of male features, almost 100% of all men with Klinefelter will be sterile (unable to produce a child). Certain lung disease and rare tumors are more commonly found among patients with Klinefelter syndrome. Furthermore, because men with Klinefelter syndrome have enlarged breasts, they have nearly the same chance of developing breast **cancer** as do women.

Some symptoms of Klinefelter syndrome are treatable. Regular injections of the male **hormone** testosterone at the onset of puberty can promote strength, facial hair growth, and a more muscular physique.

KLUG, AARON (1926-)
Lithuanian English molecular biologist

Aaron Klug made many breakthroughs that advanced the knowledge of the basic structures of **molecular biology**, but he is best known for his creation of the new technique of crystallographic electron microscopy which made possible not only his own scientific discoveries but those of many other scientists as well. For his development of this technique as well as

for his contributions to scientific knowledge, he was awarded the Nobel Prize in chemistry in 1982. He was also knighted as Sir Aaron Klug by Queen Elizabeth II in 1988.

Klug was born on August 11, 1926, in Zelvas, Lithuania, the son of Lazar Klug, a cattle dealer, and Bella Silin Klug. When he was two years old, he and his parents emigrated to Durban, South Africa, where members of his mother's family were already established. He was educated in the Durban public schools and, while attending Durban High School from 1937 to 1941, he developed an interest in science. Klug became especially interested in microbiology through reading Paul De Kruif's well-known book, *Microbe Hunters*, first published in 1926. He entered the University of Witwatersrand in Johannesburg in 1942 to take the premedical curriculum but extended his courses to include additional chemistry, mathematics, and physics before graduating with a B.S. degree in 1945. He then attended the University of Cape Town on a scholarship to take a master's degree in physics. There he first learned the techniques of x-ray crystallography, a method of determining the arrangement of atoms within a crystal. This is accomplished by studying the patterns formed on a photographic plate after a beam of x rays is deflected by the crystal. This methodology was to be basic to much of his later research. He married Liebe Bobrow in 1948 and eventually became the father of two sons, Adam and David Klug.

In 1949 Klug and his wife moved to Cambridge, England, where he had received a fellowship to study at the Cavendish Laboratory of Cambridge University. He hoped to work in the research group examining biological materials under the direction of Max Perutz and John Kendrew. When he found that there were no positions available in that group, he decided to study the molecular structure of steel and wrote a thesis on the changes that occur when molten steel solidifies, for which he received his Ph.D. in 1952. Still wishing to use his training to study biological materials, Klug in 1953 obtained a fellowship to work at Birkbeck College in London. Here he came under the influence of **Rosalind Franklin**, a reticent scientist who had pioneered x-ray crystallographic analysis. This technique had made a vital contribution to the discovery of the double-helical structure of deoxyribonucleic acid (**DNA**), an accomplishment that had won a Nobel Prize for **Francis Crick** and **James Watson**. Franklin introduced Klug to the x-ray study of the structure of viruses, an undertaking that was to occupy him for several years. They worked together to determine the structural nature of the tobacco mosaic **virus**, which attacks tobacco plants. After Franklin's death in 1958, Klug became the director of the Virus Structure Research Group at Birkbeck College.

Esteemed Colleagues Spur Success

In 1962 Klug returned to Cambridge to accept a position at the Laboratory of Molecular Biology recently established by the British Medical Research Council. Here he found himself stimulated by a large group of distinguished scientists, including Francis Crick, Max Perutz, and John Kendrew, all three of whom won Nobel Prizes in 1962. Over the next thirty years Klug himself became an increasingly important part of the research team, becoming joint head of the division of structural studies in 1978 and director of the entire laboratory in 1986.

Klug's most important contribution to scientific research was the development over many years of a technique which came to be known as crystallographic electron microscopy. X-ray crystallography had proved adequate for many biological discoveries such as the double-helical structure of DNA. However, many complex biological molecular structures were simply not available in crystal form suitable for x-ray diffraction. The obvious alternative was the use of powerful electron microscopes which could magnify an object up to a million times and reproduce it on a photographic plate, or micrograph. But the problem with electron micrographs was that they presented an essentially two-dimensional picture of a three-dimensional object. The micrographs did, however, contain in a confused form much of the information necessary for a three-dimensional reconstruction of the object, especially if the object was examined from different angles in successive micrographs.

Klug's new idea was essentially to combine the techniques of the electron microscope and x-ray crystallography by doing x-ray diffractions of the electron micrographs themselves, just as one would do with a crystal. Then the researcher could gradually put together a three-dimensional reconstruction of the object under consideration, filtering out extraneous specks on the photographic plates that confused the picture. The process was a complex one that required, among other things, very sophisticated mathematical analysis in the course of the work. Armed with the new research technique, Klug and his colleagues were able to reveal the structures and modes of operation of many basic biological materials such as viruses, animal **cell** walls, subcellular particles, **chromatin** from the **genetic material** of the cell **nucleus**, and various **proteins**. His accomplishments were rewarded in 1982 when he received the Nobel Prize. He has also been awarded the H. P. Heineken Prize from the Royal Netherlands Academy of Arts and Sciences, Columbia University's Louisa Gross Horwitz Prize, and a host of honorary degrees.

KOCH, HEINRICH HERMANN ROBERT (1843-1910)
German bacteriologist

Koch was born on December 11, 1843. As one of thirteen children born to a mining engineer and his wife, Koch spent his youth in the Harz Mountains in Clausthal, Germany. During his adolescent years, his father insisted he learn the shoemaker's trade, but when money became available for an academic career, Koch entered the University of Göttingen as a student of medicine and natural science at the age of 19. He graduated in 1866. After service as a surgeon in the Franco-Prussian War, Koch settled down as a country doctor in Wollstein, what is now Wolsztyn, Poland.

Working in a homemade laboratory that was separated from his examining room by a curtain, Koch began to study **microorganisms**. His microscopic studies led him to develop a technique by which he spread a liquid gelatin on glass slide plates to produce a transparent solid medium for the isolation of pure **cultures**. Considered one of his greatest achievements, the methods of plating pure cultures and preparing dried and heat-fixed smears of **bacteria** for staining proved to be major contributions to bacteriology. Koch's techniques are still used in the study of diseases.

Koch's work with bacteria led him to examine the causative agent of **anthrax**, a deadly disease of cattle and sheep. For years, farmers had been confused about the outbreaks of anthrax in fields where infected cattle had been removed years earlier. After isolating strains of the anthrax bacillus, Koch showed that, under certain conditions, the bacilli formed **spores** that could remain dormant for several years. These spores remained in infected fields and could develop into the disease-causing anthrax bacillus under favorable conditions. By the late nineteenth century, researchers had put forth the **germ theory** of disease, but no one had been able to prove that a single identifiable microorganism was responsible for a given disease. Koch's publication of his work with the anthrax bacillus proved the germ theory.

In 1883, Koch made great strides in public health through his work with **cholera**. In competition with French researcher **Louis Pasteur**, who had taken a team to Egypt, Koch, likewise, took a group of German scientists to Egypt in an attempt to win the race to isolate the causative agent. But the epidemic in Egypt ended before Koch's research was completed, and he subsequently went to India, where he was able to isolate the comma-shaped bacillus responsible for cholera, *Vibrio cholerae*, from samples of drinking water, food, and clothing.

Koch became internationally known for his advancement of the field of bacteriology. In 1885, he was named director of the new Institute of Hygiene in Berlin. Five years later, Koch published the Four Postulates on which modern bacteriology is built: 1) the organism must be present in every case of the disease; 2) it must be cultivated in a pure culture; 3) it must produce the disease in a susceptible animal upon inoculation; and 4) it must produce the same disease when healthy animals are inoculated.

Koch also contributed to the study of tuberculosis, a debilitating respiratory disease. In 1882, he was able to isolate *Mycobacterium tuberculosis*, the tiny tubercle bacillus that causes tuberculosis. It was Koch's search for a cure for tuberculosis, however, that would cause him temporary shame in the eyes of his fellow researchers. He thought he had found the cure, and in 1890, under pressure from the German government, he announced that he had ''at last hit upon a substance that has the power of preventing the growth of the tubercle bacillus not only in the test-tube, but in the body of an animal.'' Thousands of tubercular patients rushed to Berlin for treatment with the magic potion, tuberculin. Those inoculated eventually died of miliary tuberculosis, considered the worst form of the disease. Koch, who had always conducted his research in secrecy, was forced to reveal the method by which he obtained tuberculin. In his search for a cure, he had cultured tubercle bacilli on a glycerine broth, heat-killed them, and filtered off the liquid. He had hoped the result would be an antitoxin similar to the one developed for **diphtheria**. But rather than discover the cure for tuberculosis, Koch actually had discovered the substance used for diagnosis of the disease. Despite this disappointment, Koch received the Nobel Prize for physiology or medicine in 1905 for his work with tuberculosis.

Koch had married Emmy Fraats in 1866; she bore him one child. He later married Hedwig Freiberg in 1893. Koch died on May 27, 1910, in Baden-Baden, before he or any of his German colleagues could find a biological treatment for tuberculosis. In fact, it was his rivals in France who ultimately triumphed in the quest for a cure. Today, recognized for his pioneering work in the field of bacteriology, Koch is best remembered for the studies in which he established the germ theory of disease.

KOCHER, THEODOR (1841-1917)
Swiss surgeon

In 1870s Switzerland colloid goiter was a common ailment, usually marked by a huge glandular swelling on the front of the neck. In later years it would be understood that a simple iodine supplement to the diet could significantly reduce the disorder. But in the nineteenth century, surgical removal of the **thyroid gland** was the only known cure. However, in the absence of effective anesthetics and **antisepsis**, surgical attempts to remove a goiter meant almost certain death for the patient. This was the challenge faced by Swiss surgeon Theodor Kocher, who devoted his medical career to making thyroidectomy, or the removal of a thyroid gland, a relatively safe procedure by applying new notions of antisepsis. Kocher performed thousands of thyroidectomies in his career, and the post-operative research and data he collected helped amass new knowledge about the physiology of the thyroid gland and its related disorders. For his many contributions to medicine, and especially the treatment of goiter, Kocher received the Nobel Prize in medicine in 1909.

Emil Theodor Kocher was born August 25, 1841, the son of Jacob Alexander and Maria (Wermuth) Kocher, in Bern, Switzerland. His father was an engineer and his mother a descendant of the Moravian Brethren. She passed on to her son a deeply religious philosophy which would help him gain an empathetic understanding of his patients in years to come. Schooled in Berlin, Germany; London, England; Paris, France; and Vienna, Austria, Kocher received his M.D. from the University of Bern in 1869. That same year he married Marie Witschi-Courant—the couple would have three sons. Newly married and newly graduated from medical school, Kocher visited various European clinics, including one in Vienna, where he studied under the most famous European surgeon of the day, Theodor Billroth. In 1872 Kocher, who was only thirty-one years old at the time, was named professor of clinical surgery at Bern University, a post he would hold for the next forty-five years.

Kocher first gained recognition for developing a method for treating a dislocated shoulder, a technique now known by his name. Subsequently, he also created new methods or improvements in existing methods for operations upon the lungs, stomach, gall bladder, intestine, cranial nerves, and hernia. He also developed a special pair of surgical forceps, now known as "Kocher's forceps," instruments that were used for many years after his death. Despite his many successes and contributions that improved surgical procedures, Kocher was open to other suggestions and ideas. "It is an indication of his scientific objectivity that he was always ready to abandon any of his own techniques or gadgets in favor of improvements introduced by other surgeons," Theodore L. Sourkes has written in *Nobel Prize Winners in Medicine and Physiology*. The example Sourkes provides is Kocher's ready abandonment of his own style of operating on hernias in favor of another approach.

Kocher further contributed to medicine with his *Textbook of Operative Surgery* (the book was translated into several languages, including an English edition in 1895), his pioneering of ovariotomy and, especially, his application of the antiseptic techniques of the English researcher and doctor **Joseph Lister**.

Understanding the Thyroid Gland

Kocher himself credited his success with thyroidectomy operations in part to Lister's method of antisepsis. He said while receiving his Nobel Prize that it was because of Lister that one of the "most dangerous operations, the removal of the thyroid gland, so often appearing urgently necessary because of severe respiratory disturbances, could be performed without substantial danger." However, despite his mastery over the operation, Kocher himself considered the increased knowledge about the *physiological* function of the thyroid gland an even greater advancement in medical science. In 1883, at the congress of the German Surgical Society, Kocher reported that out of his first 100 thyroidectomies, 30 had resulted in a serious disorder. This ailment was apparently a result of the whole, rather than partial, removal of the goiter. The symptoms Kocher described were called operative myxedema, and were akin to naturally occurring myxedema. Patients suffering from myxedema usually reported weight gain, slowing of intellect and speech, hair loss, tongue thickening, and abnormal heart rates, as well as developing blood-related problems of anemia and altered white blood-cell counts. Kocher further related that myxedema symptoms were similar to problems experienced by patients suffering from sporadic cretinism and cachexia strumipriva, diseases that resulted in mental retardation and dwarfism. Because of Kocher's postulations, it was discovered that a lack of thyroid secretions was the cause of all these diseases. Kocher further pointed out that hypothyroidism can be traced not only to absence of the gland, whether congenital or surgical, but also to a goiter which has caused the gland to stop working. His descriptions of the thyroid disorders clarified and brought together a series of medical observations on this subject over the years.

Kocher's observations also opened the way for future treatment of thyroid disorders. Although initial attempts to rectify the condition by administering thyroid **hormone** were not particularly successful, researchers recognized the importance of iodine, and in 1914 the effective part of the hormone, thyroxin, was isolated for effective treatment. Meanwhile, Kocher helped perfect surgical technique for thyroidectomy, and his surgical mortality rates dropped by a great margin over the years.

Georges Köhler. *(UPI/Corbis-Bettmann. Reproduced by permission.)*

During his long surgical career Kocher performed more than 2,000 thyroidectomies. In time the need for the operation declined as iodine-deprived regions, like the "goiter belt" of the Great Lakes area in the United States and certain parts of Switzerland, incorporated supplements into their diets. Nevertheless, Kocher's contributions to combatting endemic goiter continue to be recognized in a world where nearly five percent of the population still continues to suffer this disorder.

Kocher died in Bern eight years after winning the Nobel. While placing a wreath on his tomb, American neurosurgeon **Harvey Cushing** said in a speech at the First International Neurological Congress in 1931, "From hard work and responsibility surgeons are prone to burn themselves out comparatively young, but Kocher had been blessed with an imperturbility of spirit or had cultivated these habits of self-control which enabled him to bear his professional labours, his years, and his honours with equal composure to the very end."

KÖHLER, GEORGES (1946-1995)

German immunologist

Born in Munich, in what was then occupied Germany, on 17 April 1946, Georges Jean Franz Köhler attended the Universi-

ty of Freiburg, where he obtained his Ph.D. in biology in 1974. From there he set off to Cambridge University in England, to work as a postdoctoral fellow for two years at the British Medical Research Council's laboratories. At Cambridge, Köhler worked under Dr. **César Milstein**, an Argentinean-born researcher with whom Köhler would eventually share the Nobel Prize. At the time, Milstein, who was Köhler's senior by 19 years, was a distinguished immunologist, and he actively encouraged Köhler in his research interests. Eventually, it was while working in the Cambridge laboratory that Köhler discovered the hybridoma technique.

Antibodies are produced by human plasma cells in response to any threatening and harmful bacterium, or tumor cell. The body forms a specific antibody against each antigen; and César Milstein has told the *New York Times* that the potential number of different antigens may reach "well over a million." Therefore, for researchers working to combat diseases like **cancer**, an understanding of how antibodies could be harnessed for a possible cure was of great interest. And although scientists knew the benefits of producing antibodies, until Köhler and Milstein published their findings, there was no known technique for maintaining the long-term **culture** of antibody-forming plasma cells.

Köhler's interest in the subject had been aroused years earlier, when he had become intrigued by the work of Dr. Michael Potter of the National Cancer Institute in Bethesda, Maryland. In 1962 Potter had induced myelomas, or plasma-cell tumors in mice, and others had discovered how to keep those tumors growing indefinitely in culture. Potter showed that plasma tumor cells were both immortal and able to create an unlimited number of identical antibodies. The only drawback was that there seemed no way to make the cells produce a certain *type* of antibody. Because of this, Köhler wanted to initiate a **cloning** experiment that would fuse plasma cells able to produce the desired antibodies with the "immortal" myeloma cells. With Milstein's blessing, Köhler began his experiment.

For several weeks after he made the hybrid cells, Köhler put off testing the outcome of his experiment for fear of failure. But disappointment turned to joy when Köhler discovered his test had been a success: Astoundingly, his hybrid cells were making pure antibodies against the test antigen. The result was dubbed "monoclonal antibodies." For his contribution to medical science, Köhler—who in 1977 had relocated to Switzerland to do research at the Basel Institute for Immunology—was awarded the Nobel Prize in medicine in 1984.

The implications of Köhler's discovery were immense. In the early 1980s Köhler's discovery had led scientists to identify various **lymphocytes**, or white blood cells. Among the kinds discovered were the T-4 lymphocytes, the cells destroyed by **AIDS**. Monoclonal antibodies have also improved tests for hepatitis B and streptococcal infections by providing guidance in selecting appropriate **antibiotics**, and they have aided in the research on thyroid disorders, lupus, rheumatoid **arthritis**, and inherited brain disorders. More significantly, Köhler's work has led to advances in research that can harness monoclonal antibodies into certain drugs and toxins that fight cancer, but would cause damage in their own right. Researchers are also using monoclonal antibodies to identify antigens specific to the surface of cancer cells so as to develop tests to detect the spread of cancerous cells in the body.

Despite the significance of the discovery, which has also resulted in vast amounts of research funds for many research laboratories, for Köhler and Milstein—who never patented their discovery—there was little remuneration. In fact, during the years following the discovery until they won the Nobel Prize, Köhler received only a single honorary doctorate. Following the award, however, he and Milstein, together with Michael Potter, were named winners of the Lasker Medical Research Award.

In 1985, Köhler moved back to his hometown of Freiburg, Germany, to assume the directorship of the Max Planck Institute for Immune Biology. He died on March 1, 1995.

See also Antibody, monoclonal

KÖLLIKER, RUDOLPH ALBERT VON (1817-1905)
Swiss anatomist and physiologist

Rudolph Albert von Kölliker was born in Zurich, Switzerland, on July 6, 1817, the son of a bank officer. After graduating from the Zurich gymnasium, he studied science and medicine at the University of Zurich. He earned his Ph.D. from Zurich in 1841 and his medical degree from Heidelberg a year later. For his Ph.D. thesis, Kölliker studied the origin of spermatozoa and first demonstrated that they are cells. Some years later, he was also able to demonstrate the cellular nature of ova.

Kölliker was a prolific worker and writer, publishing about 300 research papers in his lifetime on such diverse topics as **histology**, **embryology**, comparative anatomy, physiology, **zoology**, and **evolution**. In addition, he wrote the first modern textbook on histology and one of the earliest embryology texts. One of Kölliker's most important contributions was the development of **cell theory**. He helped to confirm the view that cells arise only from other cells and cannot be generated from noncellular material. He also advocated the view that **tissue** should be studied and understood as a mass of individual cells.

During the 1840s, Kölliker carried out extensive research on the **nervous system**. Probably his most important accomplishment was documenting the connection between nerve fibers and the cell bodies to which they are attached. This research formed the groundwork of the **neuron** theory, developed to its highest degree some years later by **Santiago Ramón y Cajal**.

In other research, Kölliker studied the effects of poisons such as curare and strychnine on the nervous and **muscular systems**. He also worked to advance **Charles Darwin**'s newly announced theory of evolution. He disagreed with Darwin, however, as to the mechanism by which evolutionary change occurs. He argued that rather than being a slow, regular process, evolutionary change is more likely to occur in short, dramatic spurts. This view presaged the discovery of **mutations** by **Hugo de Vries** a half century later.

Arthur Kornberg. *(The Library of Congress. Reproduced by permission.)*

Kölliker's first academic appointment was as Associate Professor of Physiology and Comparative Anatomy at Zurich in 1844. Three years later he took a similar position at the University of Würzburg in Bavaria, a post he held for the next 53 years. In addition to his teaching, Kölliker also founded in 1848 the zoological journal *Zeitschrift für Wissenschaftliche Zoologie*, which is still published today. Kölliker died in Würzburg on November 2, 1905.

See also Evolutionary theory

KORNBERG, ARTHUR (1918-)
American biochemist

Arthur Kornberg was the first to synthesize deoxyribonucleic acid (**DNA**) outside the **cell**. He also isolated and purified one of the **enzymes** necessary for successful synthesis. His results showed that a **chromosome** is composed of a continuous strand of DNA. For his success, he received half of the 1959 Nobel Prize in physiology or medicine.

Kornberg was born in Brooklyn, NY. He graduated from the City College of New York in 1937, and received his M.D. from the University of Rochester in 1941. After serving in the Coast Guard during World War II, he began his career in biochemical research. He headed the biochemistry department at Stanford University for many years.

After American biochemist **Severo Ochoa** synthesized ribonucleic acid (**RNA**) nucleotides in 1955, Kornberg began attempting to synthesize DNA. As a starting point, Kornberg isolated a pure form of the *E. coli* bacterial enzyme now known as DNA polymerase I, which plays a role in copying DNA within the cell. He then used as a synthesizing template a circular, single-stranded DNA from a **bacteriophage** (**virus** which parasitizes bacteria) called $\phi X174$, which was isolated by Robert Sinsheimer at the California Institute of Technology. In nature, $\phi X174$ reproduces inside *E. coli*, forming a complementary second DNA circle that acts as a template for its new copy. Kornberg, Sinsheimer, and Mehran Goulian (Stanford University) isolated this second DNA circle to use as their template. Then they added the enzyme and the four DNA nucleosides (base-sugar groups) with three-phosphate groups attached—ATP (adenosine triphosphate), GTP (guanosine triphosphate), CTP (cytidine triphosphate), and TTP (thymidine triphosphate). Finally, DNA nucleotides were formed. The scientists also showed that their synthetic DNA was biologically active. Since Sinsheimer's group had already found that even a single copying error would make the virus inactive, the activity proved that error-free DNA could be synthesized.

Later research by other scientists revealed details of the natural DNA copying process. It is now known that DNA polymerase I's main task is examining copied DNA for errors, removing incorrect nucleotides, and repairing damaged ones. Both DNA polymerase I and polymerase III—the main synthesizing enzyme in many species—are needed for this replication.

Kornberg's current research focus at Stanford has changed from **DNA replication** to inorganic polyphosphate (poly P), a linear **polymer** with many functions, including phosphate and energy reservoirs, buffering against alkali, bacteria capsule formation, and regulation of growth and development especially under stress or deprivation.

KOSSEL, ALBRECHT (1853-1927)

German biochemist

Albrecht Kossel isolated several major structural parts of the **nucleic acids** and discovered histidine, an essential **amino acid**. He received the 1910 Nobel Prize in physiology or medicine for his work with the nucleic acids and cellular **proteins**.

Kossel was born in Rostock, Germany, where his father was a merchant. Kossel's first scientific love was botany, but he was convinced by his father to study medicine and received his medical license in 1877. He next went to the University of Strasbourg (then part of Germany), where he became interested in physiological chemistry and worked with **Ernst Hoppe-Seyler** on the nucleic acids. After serving on university faculties in Berlin and Marburg, in 1901 he became head of the physiology department at the University of Heidelberg.

Between 1877 and 1881 he served as Ernst Hoppe-Seyler's assistant and began his studies of the nucleic acids, which had been discovered ten years earlier by one of Hoppe-Seyler's pupils, the Swiss biochemist Johann Miescher. Miescher had called his substance nuclein because it was found in

a cell's **nucleus**, and it was thought to be a phosphorus-rich **protein**. Kossel showed that nuclein was actually composed of a protein portion and a non-protein portion, which was the nucleic acid.

Over the next 20 years, Kossel and his own research team made important discoveries about the structure of both the nucleic acids and cellular proteins. He showed that nucleic acid is made up in part of purines and pyrimidines. He identified their structures, showing that a pyrimidine has a single six-sided ring, while a purine has a six-sided ring that shares one side with a five-sided ring. He isolated the two purines, adenine and guanine, which are now known to exist in both ribonucleic acid (**RNA**) and deoxyribonucleic acid (**DNA**). He also isolated the three pyrimidines— thymine and cytosine, found in DNA, and uracil, which RNA contains instead of thymine. Adenine is composed of carbon, hydrogen, and **nitrogen**; the others also contain oxygen. In addition, he identified **carbohydrates** in the nucleic acids and made preliminary predictions about the types of sugars involved. His work was continued by the American biochemist Phoebus Levene, who had once studied with him. In addition to these findings, Kossel was the first to isolate the protein histone—a component of **chromatin**, the structural material of the **chromosome**s that supports the DNA.

Kossel was always motivated to find the biological functions of the chemicals he studied and isolated. In a famous 1912 lecture, he expressed his conviction about the importance of nucleic acids and cellular proteins as the chemical basis for **genetics**. He published many papers detailing his work and held honorary doctorates at six universities. He was also a member of the Royal Swedish Academy of Sciences and the Royal Society of Sciences of Uppsala.

KRANZ ANATOMY

The leaves of green **plants** using the C_4 or Hatch-Slack pathway for photosynthetic carbon fixation almost invariably have a specialized internal arrangement of **cell**s surrounding the vascular bundles that is called Kranz anatomy. Kranz, the German word for "halo," or "wreath," refers to a ring of mesophyll cells just to the outside of another ring of large bundle-sheath cells, both of which encircle the vascular bundle. In transverse sections viewed under the **microscope**, the two cell layers give the appearance of a wreath surrounding each bundle. In addition to the unique "wreaths," other features that typify leaves with Kranz anatomy include small intercellular spaces, and frequent veins.

Kranz anatomy and the C_4 photosynthetic pathway are especially characteristic of tropical grasses such as sugar cane, where it was first discovered, and corn, although it has also been found in other plants. It is sometimes possible to distinguish C_4 plants by their dark green veins, which are a consequence of the chlorophyll rich "wreaths" surrounding the conductive **tissue**. The **chloroplast**s of mesophyll cells typically contain chlorophyll-bearing internal membranous structures called grana, while the large and conspicuous chloroplasts in bundle-sheath cells lack grana or have only a poorly developed type. During active **photosynthesis**, bundle-sheath chloroplasts tend to form larger and more numerous starch grains than the mesophyll chloroplasts.

Plants with Kranz anatomy and the C_4 photosynthetic pathway tend to be highly efficient in photosynthesis. They generally have higher maximum rates of photosynthesis, and become **light** saturated at higher light intensities enabling them to capture and store large amounts of light **energy** even in tropical areas. They are able to photosynthesize more effectively at higher **temperatures**, and at low **carbon dioxide** concentrations that severely limit photosynthesis in less efficient C_3 plants. The basis for the superior photosynthetic ability of C_4 plants with Kranz anatomy is based in part on cooperation between the C_3 and C_4 photosynthetic pathways, both of which occur in these plants, and a peculiar system that involves the double fixation of carbon dioxide. Carbon dioxide is first fixed in the mesophyll cells by the C_4 pathway with a three-carbon compound, phosphoenolpyruvate, as the acceptor **molecule**, and the four-carbon molecule, oxaloacetate, as the product. Oxaloacetate is quickly converted to malate using reducing power produced in mesophyll chloroplasts. The malate is transported to bundle-sheath cells where it releases carbon dioxide. The released carbon dioxide is quickly captured by the C_3 pathway, forming a three-carbon compound, 3-phosphoglycerate, and is then incorporated into sugars and starch.

See also Carbon dioxide fixation; Hatch-Slack photosynthetic pathway.

KREBS, EDWIN G. (1918-)

American biochemist

Edwin Gerhard Krebs was born to William Carl Krebs and Louisa Helena Stegeman Krebs in Lansing, Iowa, on June 6, 1918. He was the third of four children. His father, a Presbyterian minister, died while Krebs was in his first year of high school. In order to keep Krebs's two older brothers enrolled at the University of Illinois in Urbana, Louisa Krebs moved the family from Greenville, where Edwin Krebs grew up, to the university town.

In 1940, after completing his high-school and undergraduate work in Urbana, Krebs entered medical school at Washington University School of Medicine in St. Louis, Missouri. He had the opportunity to work under Arda A. Green, who was associated with **Carl Ferdinand Cori** and **Gerty T. Cori**. The Coris were a husband-and-wife team who had won the Nobel Prize in 1947 for research on **carbohydrate** metabolism and the **enzyme** phosphorylase. Krebs's later collaboration with Fischer at the University of Washington in Seattle had its beginning in the research conducted by the Coris.

After receiving his medical degree in 1943 and completing an eight-month residency in internal medicine at Barnes Hospital in St. Louis, Krebs became a medical officer in the navy, serving in that capacity until 1946. Due to the unavaila-

bility of a resident position, and on the advice of one of his professors, Krebs now began studying science. Because of his background in chemistry, Krebs chose to work in biochemistry and was accepted by the Coris as a postdoctoral fellow in their laboratory. For two years, while working for the Coris, Krebs studied the interaction of protamine (a basic **protein**) with rabbit muscle phosphorylase. This work seemed so rewarding to him that he decided to continue his efforts in the field of research, and when in 1948 he was invited by Hans Neurath to join the faculty as an assistant professor in the department of biochemistry at the University of Washington.

At this time Neurath's department greatly emphasized protein chemistry and enzymology (enzymes are proteins that act as catalysts in biochemical reactions). Work in the Coris' laboratory had established that the enzyme phosphorylase existed in active and inactive forms, but what controlled its activity was unknown. Combining his experience on mammalian skeletal muscle phosphorylase with **Edmond Fischer**'s experience with potato phosphorylase after Fischer joined the department, Krebs and Fischer teamed up to uncover the molecular mechanism by which phosphorylase makes **energy** available to a contracting muscle. What they discovered was reversible protein **phosphorylation**. An enzyme called protein kinase takes phosphate from **adenosine triphosphate** (ATP), the supplier of energy to cells, and adds it to inactive phosphorylase, changing the shape of the phosphorylase and consequently switching it on. Another enzyme, called protein phosphatase, reverses this process by removing the phosphate from phosphorylase, thus deactivating it. Protein kinases are present in all cells.

Once it became evident that reversible protein phosphorylation was a general process, the impact of Krebs and Fischer's work was immeasurable. Their collaboration opened the field of biochemical research and paved the way to much of the work done in the area of biotechnology and **genetic engineering**. Protein phosphorylation has even been posited as the basis of learning and memory. Medical applications have included development of the drug **cyclosporine**, which blocks the body's immune response by interfering with phosphorylation to prevent rejection of transplants. As important as what happens when the process functions normally is what happens when it goes awry: protein kinases are involved in almost 50 percent of **cancer**-causing oncogenes.

Recognition for Krebs's work came through various awards besides the Nobel Prize. In 1988 Krebs and Fischer shared the Passano Award for their research, and Krebs was one of four scientists to share the Lasker Award for Basic Medical Research in 1989. He was co-recipient of the Robert A. Welch Award in Chemistry in 1991, followed by the Nobel Prize in physiology or medicine a year later. Besides concentrating his research on protein phosphorylation, Krebs has investigated signal transduction and carbohydrate metabolism.

In 1968 Krebs had left the University of Washington to accept the position of founding chairman of the department of biological chemistry at the University of California in Davis. When he returned to Washington in 1977 he became chairman of the department of pharmacology. From 1977 until 1983, Krebs was associated with the Howard Hughes Medical Institute as well.

Hans Adolf Krebs. *(The Library of Congress. Reproduced by permission.)*

Krebs was married on March 10, 1945, to Virginia Deedy French, and they have three children, Sally, Robert, and Martha.

KREBS, HANS ADOLF (1900-1981)
German British biochemist

The son of a physician, Hans Krebs attended several German universities before receiving his medical degree from the University of Hamburg in 1925. Although he set up practice as an ear, nose and throat specialist (his father's occupation), he soon realized he preferred doing research and, a year later, became an assistant to the noted biochemist **Otto Warburg** at the Kaiser Wilhelm Institute in Berlin, Germany. While there, Krebs became interested in **amino acid**s, the building blocks of **protein**. He particularly wanted to know more about the then-unknown process by which the body, under certain circumstances, breaks down the amino acids instead of using them for constructive purposes.

In the course of several years of research, Krebs discovered that when amino acids were broken down (or degraded), their **nitrogen** atoms were the first to be stripped away. After this deamination process, the nitrogen atoms were excreted from the body in the form of urea, a major component in urine. By 1932, Krebs was able to describe several of the basic steps in urea formation and to discuss what happened to the remain-

der of the amino acids. His "urea cycle" won Krebs some fame but by then the Nazi movement was becoming more powerful and, like almost all of Germany's Jewish scientists, Krebs decided to leave the country.

In 1933, Krebs went to England, where he studied for a while at Cambridge University, working under **Frederick Gowland Hopkins**. In 1935, he joined the faculty of Sheffield University, where he remained until 1954, then moved on to Oxford University, finally retiring in 1967. While at Sheffield, Krebs concentrated much of his attention on **carbohydrate** metabolism and discovered the process for which he is best known: the citric acid cycle (also called the tricarboxylic acid cycle or, more simply, the **Krebs cycle**). Several other biochemists—in particular, **Otto Meyerhof** and **Carl and Gerty Cori**—had already shown that glycogen, the carbohydrate stored in the liver and muscles, was broken down to **lactic acid** by a process that required no oxygen and that released very little **energy**. Krebs was interested in discovering how the lactic acid was then broken down into **carbon dioxide** and **water**—and was now somehow able to release a comparatively great deal of energy.

From the work of **Albert Szent-Györgyi**, Krebs knew that several four-carbon compounds were able to increase the consumption of oxygen by cellular tissues. Krebs then located two six-carbon acids with similar oxygen-increasing powers, (one of them the citric acid that gave the cycle its original name). All these various compounds, Krebs felt, must be involved in the chain that led from carbohydrates to carbon dioxide, water and energy. And, by performing countless studies—most of them on pigeon breast muscles—by 1937, he was able to work out most of the long, complicated chain.

The chain is, in fact, a cycle—a series of chemical changes that begins when lactic acid (a three-carbon compound) is broken down into a mysterious two-carbon compound (determined by **Fritz Lipmann** to be acetyl **coenzyme** A). The two-carbon compound, when combined with the four-carbon oxaloacetic acid, becomes the six-carbon citric acid. The citric acid then undergoes a complicated series of chemical changes, during which hydrogen atoms occasionally break off (and combine with atmospheric oxygen to produce energy), and two-carbon fragments either attach themselves to the chain to regenerate certain compounds or break off (to be broken down into carbon dioxide and water, liberating more energy in the process).

The Krebs cycle, it soon became clear, not only helped explain the metabolic pathways followed by carbohydrates but by fats and proteins as well. The cycle, in fact, appears to play a fundamental role in virtually all cell **metabolism** and Krebs' discovery is therefore considered a major contribution to biochemistry. For his work, Krebs shared the 1953 Nobel Prize in physiology or medicine with Lipmann. He was also knighted by Queen Elizabeth II (1926-) in 1958.

KREBS CYCLE

The Krebs cycle is part of the process used by **cell**s to convert foodstuffs, such as **carbohydrate**s, into usable **energy**. The cycle was named for Sir **Hans Adolf Krebs** who first explained its operation in 1936. It is also referred to as the citric acid cycle or the tricarboxylic acid (TCA) cycle. The Krebs cycle has a key role in the **metabolism** of humans and **animal**s. Metabolism is the sum of all biochemical processes involved in life and is divided into two subcategories: anabolism and catabolism. Anabolism is the building up of large organic **molecule**s from simpler precursors, and catabolism is the breakdown of complex substances into simpler molecules. The Krebs cycle is primarily a catabolic process because it breaks down larger carboxylic acids into smaller units. This process involves oxidative reactions which release chemical free energy; some of this energy is lost as heat and the rest of it is conserved by coupling it with molecules capable storing energy.

Since the early 1900s, many biochemists, including Harden, Meyerhof, and Warburg, have contributed to the understanding of cellular metabolism. In 1928, Szent-Györgyi found that four different four-carbon acids could stimulate oxygen uptake in **tissue** samples. He theorized that these carboxylic acids must play a role in changing carbohydrates into energy. In 1937, the German-born British biochemist Hans Adolf Krebs (1900-1981) found that six-carbon acids, particularly citric acid, were also involved in this process. He eventually detailed the entire cycle through which foods are changed into energy.

The Krebs cycle consists of a series of six **chemical reaction**s which occur in a repeating loop. These reactions take place inside cellular **organelle**s known as **mitochondria**, which can be thought of as tiny energy "factories." These factories release hydrogen by breaking bonds between carbon **atoms** in the food molecules. After each step, the fuel molecule is shortened and it goes on to the next step in the process. The hydrogen released in this process is eventually transferred to another molecule known as **adenosine triphosphate (ATP)**.

ATP is an **organic compound** composed of adenine, the sugar ribose, and three **phosphate group**s. When ATP is broken down by subsequent **hydrolysis**, it yields adenosine diphosphate (ADP), inorganic phosphorus, and energy. The released energy is used to fuel almost all cellular functions. The Krebs cycle is only one portion of carbohydrate catabolism. It occurs between the other stages which are **glycolysis** and oxidative **phosphorylation**.

The first step of the Krebs cycle acts upon a material known as acetyl **coenzyme** A (acetyl CoA), which is a derivative of the vitamin called **pantothenic acid**. Acetyl CoA is formed from fatty acids or glucose prior to entering the cycle. Fuel molecules, such as carbohydrates, fats, and **protein**s, must be converted to acetyl CoA before they can be utilized in the the Krebs cycle. Acetyl CoA enters at the start of the cycle and reacts with oxaloacetic acid to form citric acid. The energy that is released in this reaction is stored in the form of ATP. Citric acid, a six carbon acid, then begins the next step in the cycle. It is broken down to form ketoglutaric acid, a 5 carbon acid. The third step breaks down ketoglutaric acid to yield succinic acid, which contains 4 carbons. The fourth step begins with succinic acid and converts it to oxaloacetic acid. Oxaloaetic acid then starts the cycle all over again by reacting with more acetyl CoA.

Each of the steps in the Krebs cycle is catalyzed by a specific **enzyme**. As carbons are removed from the carboxylic acids they are converted to **carbon dioxide**, which animals expire as a metabolic byproduct. The hydrogens which are removed are bound to a vitamin-derived carrier compound, nicotinamide adenine dinucleotide (NAD), which eventually stores the energy in the form of ATP.

In most higher **plants** and in certain **microorganism**s, such as the bacterium *Escherichia coli*, this process operates slightly differently and is called the glyoxylate cycle, because its primary intermediate is glyoxylic acid.

KROGH, AUGUST (1874-1949)
Danish physiologist

Schack August Steenberg Krogh (pronounced Krawg) was born on November 15, 1874, in Grenaa, Jutland. Throughout his life Krogh was active in both **zoology** and human physiology, accomplishing his major discoveries in the physiology of **respiration**.

Krogh attended school at the Gymnasium at Aarhus, and then went on to the University of Copenhagen in 1893. He first entered Copenhagen with the intention of studying physics and medicine, but under the influence of zoologist William Sorensen, he changed to the study of zoology and physiology. Sorensen had advised Krogh to attend the lectures of Christian Bohr, an expert in circulatory and respiratory physiology. After attending Bohr's lectures at Copenhagen, Krogh began to work in Bohr's laboratory in 1897, and after receiving his master of science degree in 1899, he became Bohr's laboratory assistant.

One of Krogh's earliest achievements was his invention of a microtonometer—an instrument that measures gas pressure in fluids—which he developed to help in his research with a marine organism named *Corethra*. As a student Krogh had done research on the larvae of *Corethra* to determine how its air bladders operated (he found that they worked like the diving tanks of submarines). Traveling in 1902 to Greenland, Krogh studied the amounts of oxygen and **carbon dioxide** dissolved in fresh and sea water. His research cast a new understanding on the role of the oceans in carbon dioxide regulation and at the same time he was able to improve his techniques for measuring gas pressures in fluids.

In 1903 Krogh received a Ph.D. in zoology from the University of Copenhagen, where, in his doctoral dissertation, he demonstrated the difference between the skin and lung respiration of the frog. Whereas the frog's skin respiration was constant and regular, Krogh found that the frog's lung respiration varied and was controlled by the autonomic system through the mechanism of the vagus nerve. Oxygen passed from the air sacs (alveoli) of the lung through a membrane to the capillaries and then to the **blood** stream where it formed carbon dioxide after it was used by the different tissues in the body. The process then reversed when the blood carried carbon dioxide to the alveoli of the lungs, where it was exhaled.

Krogh was married in 1905 to Marie Jorgensen, a physiologist who also worked in Bohr's laboratory. (The couple would eventually have three daughters and one son.) In 1906 the first of Krogh's papers to receive international recognition, a work which showed that **nitrogen** is not involved in animal **metabolism**, was awarded the Seegen Prize from the Vienna Academy of Sciences. In 1907 Krogh received further international attention at Heidelberg, Germany, when he discussed his findings on the diffusion of pulmonary gases at the International Congress of Physiology.

In 1908 Krogh made another trip to Greenland with his wife to study the Eskimo's meat-eating dietary habits and the effects it had on their respiration and metabolism. He was also given an associate professorship of zoophysiology at the University of Copenhagen that year. Two years later Krogh and his wife were given a laboratory at Ny Vestergade for physiological research. Krogh then became a full professor at Copenhagen in 1916.

From 1908 to 1912 Krogh was engaged in research to resolve the question of how oxygen was transferred in the lungs to the blood. Bohr and **John Burdon Sanderson Haldane**, along with other scientists, believed that the lung acted as a gland in the alveolar transfer of oxygen to the blood; in other words, the lung secreted the oxygen. Krogh, in 1912, convincingly delivered the fatal blow to the secretion theory by first showing that in fishes there is no secretion of oxygen into the air sacs, and then by demonstrating that the amount of oxygen in the blood always equalled the amount that should be provided by his diffusion theory.

It was not until 1916, however, that Krogh accomplished the work that would, in 1920, earn him the Nobel Prize for physiology or medicine. He showed that muscle tension was always slightly lower than the tensions in the capillaries, even when the muscle was at work. Noting that there were few open capillaries when a muscle was at rest, Krogh demonstrated that as soon as the muscle became active many capillaries began to open up. He was also able to show that blood did not enter the capillaries through the pressure of the blood vessels but from the relaxed tonus (partial contraction) of the active muscle. The relaxation of the muscle allowed the field of capillaries to open and the blood to flow in, thus providing more oxygen to the muscle, organ, or tissue.

Krogh's discoveries relating to gas exchanges in the lung and to the operation of the capillary system helped to develop medical techniques for breathing through the trachea. His work also improved surgical methods for open heart surgery, such as the procedure for reducing body temperature to below normal levels to slow down the rate of gaseous exchange.

In 1922 Krogh became interested in **insulin** (which had been discovered by **Frederick G. Banting** and **John James Rickard Macleod** the year before), partly because his own wife had diabetes. Besides being active in insulin research, Krogh helped to promote manufacturing facilities in Denmark for its production. Krogh also maintained his interest in zoology, writing about insects and becoming particularly attentive to theories about the way honey bees communicate.

Krogh died on September 13, 1949, in Copenhagen.

KUHN, RICHARD (1900-1967)
German chemist

Richard Kuhn was a Nobel Prize-winning organic chemist who devoted much of his life to studying the synthesis of **vitamins** and carotenoids, the fat-soluble yellow pigments that are found in **plants**. He researched the chemistry of **algae** sex cells and optical stereochemistry, and spent a great deal of time understanding **carbohydrate**s. He was determined to succeed in his work by uncovering the practical applications of substances in the fields of medicine and **agriculture**. Later in his career, Kuhn concentrated on studying how the body fights disease using **organic compound**s.

Kuhn was born in Vienna, Austria, on December 3, 1900, to Hofrat Richard Clemens, a hydraulics engineer, and Angelika (Rodler) Kuhn, an elementary school teacher. After spending almost ten years of his life at home under the educational guidance of his mother, Kuhn entered the Döbling Gymnasium, where he attended classes with future Nobel Prize-winning physicist Wolfgang Pauli. After graduating from the Gymnasium in 1917, he was drafted into the German (Austro-Hungarian) army and served until World War I ended in November of 1918.

Once Kuhn was discharged from the military, he entered the University of Vienna where a professor of medical chemistry, Ernst Ludwig, turned his interests towards chemistry. Just three semesters after entering the university, Kuhn transferred to the University of Munich, where he studied chemistry under noted scientist **Richard Willstätter**. Kuhn received his Ph.D. in 1922 for his thesis, "On the Specificity of Enzymes in Carbohydrate Metabolism." He worked briefly as Willstätter's assistant before leaving Munich in 1926 to join the Federal Institute of Technology at Zurich, a Swiss technological high school, where he spent three years as professor of general and analytical chemistry.

In 1929 Kuhn left Zurich and joined the University of Heidelberg's newly established Kaiser Wilhelm Institute for Medical Research (renamed the Max Planck Institute in 1950, with Kuhn's assistance) as both a professor of organic chemistry and director of the institute's chemistry department. Kuhn would turn down a number of other offers to spend the remainder of his career at the institute; he became its director in 1937.

Makes Advances in Research of Vitamin Production

Kuhn was particularly interested in how the chemistry of organic compounds was related to their function in biological systems. His early work concentrated on carotenoids. One such substance was carotene, the pigment found in carrots, whose chemical formula had been determined earlier by Willstätter at the University of Munich. After further research, Kuhn discovered that carotene was a precursor in the chemical production of vitamin A and that nature uses all kinds of chemical structures for biological actions. In addition, Kuhn and his colleagues discovered that carotenoids existed in numerous plants and animals and that vitamin A was an essential part of maintaining the body's mucous membranes.

At the time, Kuhn was just one of two scientists working with carotene; **Paul Karrer** at the University of Zurich was the other. The two men would remain fierce competitors throughout their careers. Through Kuhn's investigations, he found two distinct compounds in carotene: beta-carotene, which bends light, and alpha-carotene, which does not. Two years later Kuhn's work led to the discovery of a third form of carotene, called gamma-carotene. These compounds have exactly the same chemical formulas but different molecular structures; therefore, they are known as **isomers**.

In the 1930s, Kuhn's turned his attention to researching members of the water-soluble vitamin B group. Working with other scientists, he painstakingly isolated and crystallized a small amount of vitamin B_2 (riboflavin) from skim milk. By determining the structure of riboflavin, Kuhn was able to clearly explain the chemical composition and to eventually synthesize this compound. He also demonstrated that B_2 plays a primary role in respiratory **enzyme** action and provided the key to how vitamins function and what their applications are in living systems. For this work, Kuhn was offered the 1938 Nobel Prize. Then, in the late 1930s, Kuhn and three other coworkers determined both the chemical composition and molecular structure of adermin, now commonly referred to as vitamin B_6, which acts against skin disease and helps to regulate the metabolism of the **nervous system**.

Forced to Reject Nobel Prize

Although Kuhn was to be awarded the Nobel Prize in 1938, he did not actually receive it until the late 1940s. Due to the political climate in Germany and the fact that a Nazi concentration camp prisoner, Carl von Ossietzky, was honored with the Nobel Peace Prize in 1934, Hitler instituted a policy forbidding German citizens from accepting the award. As a result, Kuhn was forced to turn down the award and was not properly honored until after the war ended in 1945. He received his medal and certificate in 1949 at a special ceremony in Stockholm for his work with carotenoids and vitamins.

The 1940s saw Kuhn expand his research to include carbohydrates. Kuhn researched **alkaloid** glycosides, which appear in potatoes and tomatoes, and tried to unlock their pigments and biological structures. He also returned to researching milk, from which he extracted carbohydrates using **chromatography**. In doing so, he greatly improved the use of chromatography, which is the chemical separation of mixtures into their original form. After becoming a professor of biochemistry at the Max Planck Institute in 1950, Kuhn focused much of his effort on the study of organic substances that are instrumental in the body's resistance to infection. His investigations into a variety of "resistance" factors effective against **cholera** and **influenza** uncovered the molecular interaction between an organism and its attacker. He also went on to identify **pantothenic acid**, an important ingredient in **hemoglobin** formation and the release of energy from carbohydrates, and **para-aminobenzoic acid** (PABA), a compound that proved useful in the synthesis of anesthetics.

Known to have an upbeat and outgoing personality, Kuhn enjoyed such activities as billiards, chess and tennis. He was also a skilled violinist who frequently played with a chamber ensemble for public enjoyment. Kuhn met his wife, Daisy Hartmann, while he was a professor and she a student at the Federal Institute of Technology at Zurich. They would marry in 1928; the couple had four daughters and two sons together.

In addition to winning the Nobel Prize, Kuhn was awarded the Pasteur, Paterno and Goethe Prizes for his work. He was a member of numerous national and international scientific societies and received honorary degrees from a variety of institutions including the Munich Technical University, the University of Vienna, and the University of St. Maria, Brazil. A charter senate member of the Max Planck Society for the Advancement of Science, he later served as vice-president. Kuhn became the editor of the chemical journal, *Annalen der Chemie,* in 1948, and served as president of the German Chemical Society. He published more than seven hundred scientific papers and received over fifty distinctions. Shortly before his death, the University of Heidelberg gave its first commemorative medal struck in honor of a scientist. Kuhn died July 31, 1967, at age sixty-six, in Heidelberg, Germany.

L

LACTIC ACID

Lactic acid, or 2-Hydroxypropanoic acid, is one of a family of organic carboxylic acids produced by biological **organism**s. It is found in the **blood** and muscles of **animal**s, and is a constituent of some plant juices. It is an important acidic component of fermented food products such as yogurt, buttermilk, sauerkraut, green olives, and pickles, where its formation typically results from the activity of **bacteria**. Lactic acid bacteria may be members of any of several genera of rod- or sphere-shaped gram-positive bacteria that form lactic acid as a major product of **carbohydrate fermentation**. Two genera, Lactobacillus and Streptococcus, contain some of the most active lactic acid producers.

Lactic acid is produced in active muscle **tissue** from the break down of glycogen, a carbohydrate used as an **energy** source for muscle movement. The pathway consists of the breakdown of glycogen to glucose, which is then converted to lactic acid in a series of reactions coupled to formation of ATP from ADP and phosphate. The ATP is used directly to energize muscle movement. The lactic acid is picked up by the blood and delivered, in the form of lactate salts, to the liver where it is converted back to glycogen. In an active muscle, the removal of lactate from the muscle is slower than its production, and if the muscle remains active for an extended period lactate will accumulate and eventually cause muscle fatigue and cramps. Lactic acid is used in a variety of industrial processes including tanning leather and dyeing wool. It was first isolated in 1780 by the Swedish chemist, Carl Wilhelm Scheele.

See also Muscular system

LAMARCK, JEAN-BAPTISTE (1744-1829)
French botanist

Jean Baptiste Lamarck was born on August 1, 1744, in the village of Bazentin-le-Petit in northern France. He was the youngest of eleven children born to noble parents, but he lived a life that was hardly prosperous. His father, a military officer, expected his son to become a priest. He sent Lamarck, then eleven years old, to the Jesuit school at Amiens, France, where he remained until his father's death in 1760. At age sixteen, Lamarck left school and joined the army in search of adventure. He fought in the Seven Years' War and afterward spent five years at various French ports on the Mediterranean and eastern borders of France. This traveling introduced him to the plant species of many different French regions. In 1768, an illness forced Lamarck to leave the military, and after several years, he found a job in a Paris bank. During the next four years, he studied medicine and became increasingly interested in meteorology, chemistry, and shell collecting.

In 1778, Lamarck published *Flore française*, a meticulously compiled catalog of French flora. To identify each plant, Lamarck used a dichotomous key; that is, a systematic list of key characteristics. By comparing the plant's characteristics to the listed traits at each stage of identification, large groups of dissimilar plants could be quickly eliminated and the plant's identity easily determined. The book and Lamarck's method soon attracted the attention of noted biologist Georges Buffon, who nurtured Lamarck's interest in **botany** and in 1781 secured him the position of botanist to King Louis XVI. Lamarck continued to work at the Jardin du Roi until the French Revolution resulted in its dissolution.

When the institute was reopened as the National Museum of Natural History in 1793, Lamarck was made a professor of **zoology** and placed in charge of organizing the museum's collection of **animal**s and **fossil**s without backbones. During the early 1700s, Swedish botanist Carl Linnaeus had grouped these creatures into two general categories: insects and worms. Lamarck, however, did away with Linnaeus's categories. He named the entire group invertebrates and then set about classifying them according to their anatomic similarities. He differentiated eight-legged arachnids from six-legged insects, and echinoderms such as starfish and sea urchins from crabs, shrimp, and other crustaceans. He eventually published the re-

Jean-Baptiste Lamarck. *(The Library of Congress. Reproduced by permission.)*

During his lifetime, Lamarck's theories were largely ignored by the scientific community. At age 65, he lost his eyesight but continued writing with the help of his daughter, who took his dictation. Lamarck died in great poverty on December 28, 1829, and his family was forced to sell his papers and scientific collections to pay for his funeral. During the mid-1800s, critics of Charles Darwin briefly revived interest in Lamarck's work in hopes of debunking the English naturalist's theory of natural selection, an attempt that proved futile. Although Lamarck's belief in acquired characteristics was incorrect, he remains the first scientist to acknowledge the adaptability of organisms and develop a consistent evolutionary theory.

See also See also Heredity

LAMARCKISM

Chevalier de Lamarck was a French naturalist and invertebrate zoologist who lived from 1844-1829. He is best known for a theory of **evolution** developed in his book, *Philosophie zoologique*, published in 1809. This theory, known today as Lamarckism, is based on the so-called "**inheritance** of acquired traits," meaning that characteristics that an **organism** may develop during its lifetime are heritable, and can be passed on to its progeny.

The anatomical, biochemical, and behavioral characteristics that an individual organism displays as its develops through life is known as its **phenotype**. However, the phenotype that an individual actually develops is somewhat conditional, and is based on two key factors: (1) the fixed genetic potential of the organism (or its **genotype**; this refers to the specific qualities of its **genetic material**, or **DNA** ([**deoxyribonucleic acid**]); and (2) the environmental conditions which an organism experiences as it grows. For example, an individual plant (with a particular, fixed genotype) that is well supplied with **nutrients**, moisture, and **light** throughout its life will grow larger and will produce more seeds than if that same plant did not experience such beneficial conditions. Conditional developmental possibilities as these are now known to be due to differing expressions of the genetic potential of the individual (biologists refer to the variable expression of the **genome** of an organism, as influenced by environmental conditions encountered during its development, as "phenotypic plasticity."

However, at the time of Lamarck and other biologists of the late eighteenth and nineteenth centuries the mechanisms of inheritance were not known (this includes **Charles Darwin** and Alfred Russel Wallace, the co-discoverers of the theory of evolution by **natural selection**, first published in 1859). These scientists thought that the developmental contingencies of individual organisms (which they called "acquired traits") were not initially fixed genetically, but that they could somehow become incorporated into the genetic make-up of individuals, and thereby be passed along to their offspring, so that evolution could occur. For example, if the ancestors of giraffes has to stretch vigorously to reach their food of **tree** foliage high in the canopy, this physical act might somehow have caused the individual **animal**s to develop somewhat longer necks. This

sults of his efforts in his seven-volume work, *Natural History of Invertebrates* (1815-22), which is today considered his most important contribution to botany.

While creating his classification system, Lamarck, who had previously believed in the fixity of species, gradually developed an evolutionary theory to explain the differences between living animals and fossils. Having already rejected the idea of extinction, Lamarck proposed that species gradually changed over time. He also theorized that living creatures fit into a linear, hierarchical scheme that began with the simplest life form and progressed to the most complex—humans. To account for such a progression, he set forth four laws, which he published in his *Zoological Philosophy* in 1809. The laws stated that organisms possessed an innate drive toward perfection; that they could adapt to the environment; that spontaneous generation occurs frequently; and that acquired characteristics could be passed from one generation to the next.

Lamarck is best known for this last law, the *Theory of Inheritance of Acquired Characteristics*, although it has since been disproved. The most famous example he gave of this law concerned the giraffe. He hypothesized that the giraffe gradually evolved its long neck from a primitive antelope-like animal that had to keep stretching its neck to reach higher leaves. According to Lamarck, this stretching caused the primitive animal's neck to grow longer, and this acquired characteristic could be passed on to its offspring.

"acquired" trait somehow became fixed in the genetic complement of those individuals, to be passed on to their offspring, who then also had longer necks. Eventually, this presumed mechanism of evolution could have resulted in the appearance of the modern, extremely long-necked giraffe.

Modern biologists, however, have a good understanding of the biochemical nature of inheritance. They know that phenotypic plasticity is only a reflection of the variable, but strongly fixed genetic potential that exists in all individuals. Therefore, the idea of the inheritance of acquired traits is no longer influential in evolutionary science. Instead, biologists believe that evolution largely proceeds through the differential survival and reproduction of individuals whose genetic complement favors these characters in particular environments, compared with other, "less-fit" individuals of their **population**. If the phenotypic advantages of the "more-fit" individuals are due to genetically fixed traits, they will be passed on to their offspring. This results in genetic change at the population level, which is the definition of evolution. This is, essentially, the theory of evolution by natural selection, first proposed by Darwin and Wallace in 1859.

LANDSTEINER, KARL (1868-1943)
American immunologist

Karl Landsteiner was one of the first scientists to study the physical processes of immunity. He is best known for his identification and characterization of the human **blood** groups, A, B, and O, but his contributions spanned many areas of immunology, bacteriology and pathology over a prolific forty-year career. Landsteiner identified the agents responsible for immune reactions, examined the interaction of **antigens** and **antibodies**, and studied allergic reactions in experimental animals. He determined the viral cause of poliomyelitis with research that laid the foundation for the eventual development of a polio vaccine. He also discovered that some simple chemicals, when linked to proteins, produced an immune response. Near the end of his career in 1940, he and the immunologist Philip Levine discovered the **Rh factor**, which helped save the lives of many fetuses with mismatched Rh factor from their mothers. For his work identifying the human blood groups, he was awarded the Nobel Prize for medicine in 1930.

Born June 14, 1868, Landsteiner was the only child of Dr. Leopold Landsteiner, a famous Viennese journalist, and Fanny Hess Landsteiner. Leopold Landsteiner was the Paris correspondent for several German newspapers and the founder of the daily *Presse,* an influential liberal newspaper. The family lived in Baden bei Wien, an upper-middle-class suburb of Vienna. Karl was six years old when his father suffered a massive heart attack and died. Karl was placed under the guardianship of a family friend, but remained extremely close to his mother.

In 1885, when he was seventeen, Landsteiner passed the entrance examination for medical school at the University of Vienna, where early in his training he expressed enthusiasm for the study of chemistry. He took a year off from school at the age of twenty for his obligatory military service. When he was twenty-one, Landsteiner and his mother converted from Judaism to Catholicism and Karl was christened Karl Otto Landsteiner. Landsteiner graduated from medical school at the age of 23 and immediately began advanced studies in the field of organic chemistry, working in the research laboratory of his mentor, Ernst Ludwig. In Ludwig's laboratory Landsteiner's interest in chemistry blossomed into a passion for approaching medical problems through a chemist's eye.

For the next ten years, Landsteiner worked in a number of laboratories in Europe, studying under some of the most celebrated chemists of the day: Emil Fischer, a celebrated protein chemist who subsequently won the Nobel Prize for chemistry in 1902, in Wurzburg; Eugen von Bamberger in Munich; and Arthur Hantzsch and Roland Scholl in Zurich. Landsteiner published many journal articles with these famous scientists. The knowledge he gained about organic chemistry during these formative years guided him throughout his career. The nature of antibodies began to interest him while he was serving as an assistant to Max von Gruber in the Department of Hygiene at the University of Vienna from 1896 to 1897. During this time Landsteiner published his first article on the subject of bacteriology and serology, the study of blood. He had found a subject that was to occupy his entire scientific career.

Discovers Blood Types

Landsteiner moved to Vienna's Institute of Pathology in 1897, where he was hired to perform autopsies. He continued to study immunology and the mysteries of blood on his own time. In 1900, Landsteiner wrote a paper in which he described the agglutination of blood that occurs when one person's blood is brought into contact with that of another. He suggested that the phenomenon was not a pathology, as was the prevalent thought at the time, but was due to the unique nature of the individual's blood. In 1901, Landsteiner demonstrated that the blood serum of some people could clump the blood of others. From his observations he devised the idea of mutually incompatible blood groups. He placed blood types into three groups: A, B, and C (later referred to as O). Two of his colleagues subsequently added a fourth group, AB.

In 1907 the first successful transfusions were achieved by Dr. Reuben Ottenberg of Mt. Sinai Hospital, New York, guided by Landsteiner's work. Landsteiner's accomplishment saved many lives on the battlefields of World War I, where transfusion of compatible blood was first performed on a large scale. In 1902 Landsteiner was appointed as a full member of the Imperial Society of Physicians in Vienna. That same year he presented a lecture, together with Max Richter of the Vienna University Institute of Forensic Medicine, in which the two reported a new method of typing dried blood stains to help solve crimes in which blood stains are left at the scene.

In 1908 Landsteiner took charge of the department of pathology at the Wilhelmina Hospital in Vienna. His tenure at the hospital lasted twelve years, until March of 1920. During this time, Landsteiner was at the height of his career and produced fifty-two papers on serological immunity, thirty-three on bacteriology and six on pathological anatomy. He was among the first to dissociate antigens, which stimulate the production of immune responses known as antibodies, from the antibodies themselves. Landsteiner was also among the first to purify antibodies, and his purification techniques are still used today for some applications in immunology.

Landsteiner also collaborated with Ernest Finger, the head of Vienna's Clinic for Venereal Diseases and Dermatology. In 1905, Landsteiner and Finger successfully transferred the venereal disease syphilis from humans to apes. The result was that researchers had an animal model in which to study the disease. In 1906, Landsteiner and Viktor Mucha, a scientist from the Chemical Institute at Finger's clinic, developed the technique of dark-field microscopy to identify and study the microorganisms that cause syphilis.

Works Toward Polio Vaccine

One day in 1908 the body of a young polio victim was brought in for autopsy. Landsteiner took a portion of the boy's spinal column and injected it into the spinal canal of several species of experimental animals, including rabbits, guinea-pigs, mice and monkeys. Only the monkeys contracted the disease. Landsteiner reported the results of the experiment, conducted with Erwin Popper, an assistant at the Wilhelmina Hospital.

It had generally been accepted that polio was caused by a microorganism, but previous experiments by other researchers had failed to isolate a causative agent, which was presumed to be a bacterium. Because monkeys were hard to come by in Vienna, Landsteiner went to Paris to collaborate with a Romanian bacteriologist, Constantin Levaditi of the Pasteur Institute. Working together, the two were able to trace poliomyelitis to a virus, describe the manner of its transmission, time its incubation phase, and show how it could be neutralized in the laboratory when mixed with the serum of a convalescing patient. In 1912 Landsteiner said that the development of a vaccine against poliomyelitis might prove difficult but was certainly possible. The first successful intravenous polio vaccine, developed by Jonas Salk, wasn't administered until 1955.

Landsteiner kept a grueling work schedule that allowed little time for social activity. He was serving at a war hospital in 1916 when, at the age of 48, he married Leopoldine Helene Wlasto. Helene bore a son christened Ernst Karl on April 8, 1917. After the war, Landsteiner's Austria was in chaos, with extreme shortages of food and fuel. He accepted a position as chief dissector in a small Catholic hospital in The Hague, Netherlands. There, from 1919 to 1922, he performed routine laboratory tests on urine and blood. Nevertheless, he managed to publish twelve papers on different aspects of immunology. It was during this time that Landsteiner began working on the concept of haptens, small molecular weight chemicals such as fats or sugars, that determine the specificity of antigen-antibody reactions when combined with a protein "carrier." He combined haptens of known structure with well-characterized proteins such as albumin, and showed that small changes in the hapten could affect antibody production. He developed methods to show that it is possible to sensitize animals to chemicals

that cause contact dermatitis (inflammation of the skin) in humans, demonstrating that contact dermatitis is caused by an antigen-antibody reaction. This work launched Landsteiner into a study of the phenomenon of allergic reactions.

Post-War Europe Prompts Move to United States

In 1922, Landsteiner accepted a position at the Rockefeller Institute in New York. Throughout the 1920s Landsteiner worked on the problems of immunity and allergy. He discovered new blood groups: M, N and P, refining the work he had begun 20 years before. Soon after Landsteiner and his collaborator, Philip Levine, published the work in 1927, the types began to be used in paternity suits.

The Landsteiner family spent their summers in an isolated house on Nantucket that reminded Landsteiner of his Scheveningen home in the Netherlands. Landsteiner developed a profound dislike for his growing celebrity as the world's foremost authority on the mechanisms of immunity. He never got used to the noise and crowds of New York City, confessing to friends that he wished he could lock his family away when he was not home. Despite these problems, he became a United States citizen in 1929. Always shunning publicity, even avoiding offers to give public seminars, Landsteiner was stunned when he was besieged by reporters in 1930, upon the news that he had won the Nobel Prize.

In his Nobel lecture, Landsteiner gave an account of his work on individual differences in human blood, describing the differences in blood between different species and among individuals of the same species. This theory is accepted as fact today but was at odds with prevailing thought when Landsteiner began his work. In 1936 Landsteiner summed up his life's work in what was to become a medical classic: *Die Spezifität der serologischen Reaktionen,* which was later revised and published in English, under the title *The Specificity of Serological Reactions.*

Landsteiner officially retired in 1939, at the age of seventy-one, but went on working. With Levine and Alexander Wiener he discovered another blood factor, labeled the Rh factor, for Rhesus monkeys, in which the factor was first discovered. The Rh factor was shown to be responsible for the dreaded infant disease, erythroblastosis fetalis, which occurs when mother and fetus have incompatible blood types and the fetus is injured by the mother's antibodies. During his later years, Landsteiner formed a friendship with Linus Pauling, the American biochemist who won the Nobel Prize in chemistry in 1954. Their discussions led Pauling to apply his knowledge to immunology and to contribute a chapter to the revised edition of Landsteiner's book, *The Specificity of Serological Reactions.*

Landsteiner was said to worry incessantly and was overcome toward the end of his life with fear that the Nazis would take over the civilized world. He began to fear for his family's lives. Something of a scandal developed when he tried to prevent publication of his Jewish descent. Later his fear of fascism was surpassed by the discovery that Helene had a malignant thyroid tumor. On June 14, 1943, Landsteiner celebrated his seventy-fifth birthday with his wife, Helene, and his son, who had completed medical school and was a practicing physician. On June 24, Landsteiner had just sent off the final revision of the manuscript for his book, when he was seized by a coronary obstruction. He died two days later on Saturday, June 26, 1943. Helene died the same year on Christmas day. Upon his death, tributes were published around the world, but no mention of his death was published in his native Austria or Germany until 1947, after the war and the defeat of Nazism.

LANGERHANS, PAUL (1847-1888)
German physician

Paul Langerhans is known primarily for his microscopic anatomical research. Langerhans, who was from a family of physicians, was born in Berlin on July 25, 1847. He attended the Gymnasium zum Grauen Kloster, and studied medicine at the University of Jena, and later in Berlin, where he received his medical degree in 1869. As a student, Langerhans developed new **cell** staining techniques and, using gold chloride as a stain, was able to identify nerve endings in the Malpighian layer of the skin, which were later called Langerhans cells. Applying another one of his staining techniques, Langerhans studied the **pancreas** (mainly rabbits), and, during his research, noticed unique polygonal cells within the parenchyma of the pancreas. He subsequently used his pancreas research as the topic for his dissertation and was the first to provide a detailed description of microscopic pancreatic structures. Langerhans did not know the significance of these cells at the time of his research. In 1893, the French histologist, G. E. Laguesse, discovered that the polygonally-shaped cells were the endocrine cells of the pancreas that secreted insulin, to be distinguished from the exocrine cells, which secrete digestive enzymes. Laguesse named these endocrine cells the Islets of Langerhans.

While Langerhans was studying medicine in Berlin, he was a pupil of several well known pathologists, including Julius Cohnheim and Rudolf Virchow, with whom he shared an interest in anthropology. In 1870, after completing his medical studies, Langerhans went on an expedition to Egypt and Palestine with the geographer Heinrich Kiepert. There, Langerhans made skull measurements and conducted an extensive anthropological study of the Palestinian population.

He joined the army during the Franco-Prussian War and worked as a physician in a military hospital. In 1871, Langerhans was offered a position as prosector of pathology at the University of Freiburg im Breisgau, where he eventually became associate professor. During his career, Langerhans discovered and described many unknown microscopic structures. He also conducted research on the structure and innervation of the skin, discovering granular cells in the external Malpighian layer of skin, which were later named, Langerhans' layer, or *stratum granulosum*. He also examined cardiac muscle fibers and human accessory genital glands. Another fruitful area of Langerhans's research was the study of the macrophage system, an effort which eventually led to the concept of the reticuloendothelial system.

In 1874, Langerhans contracted tuberculosis and gave up his career. He traveled to Switzerland, Italy, and Germany

A cecropia moth larva (*Hyalophora cecropia*) in Royal Oak, Michigan. (*Photograph by Robert J. Huffman, Field Mark Publications. Reproduced by permission.*)

seeking cures, eventually settling on Madeira, an island off the coast of North Africa, where the mild climate helped stabilize his condition. At one point, he felt well enough to practice medicine in the capitol, Funchal, where he died in 1888 of a kidney infection. While practicing medicine on Madeira, Langerhans studied all aspects of tuberculosis and wrote several zoological and topographical papers, which his mentor, Virchow, presented to the Prussian Academy of Sciences. He also wrote a pamphlet about the health-enhancing qualities of Madeira's natural environment.

See also Endocrine gland; Immune system

LARVA

Larva refers to the stage in the life cycle of certain organisms between the egg and adult. These juveniles do not resemble the adult, but undergo a morphological transformation into the adult stage. This type of indirect development is termed **metamorphosis**. In several groups of sessile (attached) marine organisms, the larval stage functions in dispersal. Examples of these include the planula larva of sponges and cnidarians, the pluteus larvae of echinoderms, the nauplius and zooea larvae of arthropods, and the trochophore and veliger larvae of certain molluscs. These structures are typically propelled by **cilia** and may travel with water currents over long distances. The veliger larva of zebra mussels (*Dreissena polymorpha*) are suspected to have been transported from northern Europe to Lakes Erie and Huron in 1986 by tanker ships emptying their ballast water. These small molluscs have now invaded much of the Great Lakes watershed and cause hundreds of millions of dollars in damage each year by clogging water intake pipes. Some parasitic animals have intermediate larval stages that mature in vectors and intermediate hosts before entering humans and other vertebrate final hosts. Examples of these include the miracidia and cercaria larvae of trematode flukes. Many insects also have a larval stage, which are known in various groups by different names such as caterpillars (moths and butterflies), grubs (beetles), maggots (flies), and nymphs or naiads (aquatic insects). A tadpole stage of the frog is also considered to be a larval stage.

LAVERAN, ALPHONSE (1845-1922)
French biologist

Alphonse Laveran was a French army physician who took advantage of his period of service in Algeria to study malaria, a disease known since ancient times and common in tropical and subtropical areas. Using very primitive technology, he discovered and ultimately proved that malaria was caused by a minute animal **parasite**; he also suggested, though he did not himself prove, that the parasite was transmitted to human beings by some species of mosquito. He later went on to study other diseases caused by parasites. For the work he did in this field throughout his career, he was awarded the Nobel Prize in medicine in 1907.

Charles Louis Alphonse Laveran was born on June 18, 1845, into a military family in Paris. He was the second child and only son of Louis-Theodore Laveran, a career military physician, and Marie-Louise Anselme Guénard de la Tour Laveran. When Laveran was five years old, he went with his parents to Algeria, where his father was stationed. His father was his first teacher, and after the family returned to Paris in 1856, Laveran received his secondary education at the College Sainte-Barbe and the Lycée Louis-le-Grand. In 1863, he entered the military medical school at Strasbourg, which his father had also attended; Laveran graduated in 1867. He joined the military medical service following graduation and saw active duty during the Franco-Prussian War of 1870–1871. It was at this time that he first witnessed the ravages which diseases can cause in an army at war. In 1874, he won by competitive examination an appointment to a professorship earlier held by his father at the École du Val-de-Grace, a military medical school in Paris. This was a temporary appointment, and at its conclusion in 1878 he was sent to the military hospital at Bône (now Annaba) in Algeria.

Discovers the Malaria Protozoan

It was while at Bône that Laveran began a careful study of malaria, common in many parts of Algeria, in an effort to learn its cause. He set up a small laboratory and with the primitive, low-powered **microscope** available to him, he spent much time examining **blood** samples from malaria patients both living and deceased. His studies were briefly interrupted when he was transferred to Biskra, Algeria, where malaria was rare, but they were resumed when he moved on to Constantine, also in Algeria. There, on November 6, 1880, he first observed under the microscope circular and cylindrical bodies which had moving filaments, or flagella. This confirmed his earlier suspicion that malaria was caused by living animal cells, minute single-celled creatures called **protozoa**, which acted as parasites in the human body. The particular protozoan which Laveran had discovered to be the cause of malaria later came to be called plasmodium.

Laveran's discovery was presented to the Academy of Medicine in Paris on November 23, 1880. A second paper, based upon further research, was published by the Société Médicale des Hopitaux on December 24 of that year. In 1881, Laveran published a brief monograph, *Nature parasitaire des accidents de l'impaludisme,* which provided more details of his findings. Laveran's conclusions were not immediately accepted by other scientists studying malaria. His microscopic research proved difficult to replicate. Moreover, in the wake of the discoveries of the German scientist Robert Koch and others, it was widely assumed at the time that bacteria were the causes of most diseases, including malaria.

Laveran, however, continued his research, examining the blood of hundreds of malaria patients, both in Algeria and in Italy. By 1884, in a personal microscopic demonstration, he was able to persuade Louis Pasteur that his theory was correct. Other noted scientists such as William Osler were convinced during the course of the 1880s. Also in 1884, Laveran published a book, *Traité des fièvres palustres avec la description des microbes du paludisme,* which summarized all of his research on malaria. In this work, he revealed his suspicion that the malaria protozoa were nurtured and transmitted to human beings by some species of mosquito. It remained for the British physician, Ronald Ross, working in India in the late 1890s, to prove that the malaria parasite was indeed transmitted by the Anopheles mosquito.

Laveran returned to Paris from Algeria in 1883 and became professor of military hygiene at the École du Val-de-Grace in 1884. He married Sophie Marie Pidancet in 1885; they had no children. When his professorship came to an end in 1894, he was offered only temporary administrative positions at the military hospitals at Lille and at Nantes. Angry because he was not offered a post at a military laboratory where he could continue his research, he resigned from the military medical service in December of 1896 and accepted a position at the Pasteur Institute in Paris. There he pursued his research for the rest of his life.

Expands Research to Other Disease-Causing Parasites

Laveran's demonstration that protozoa, as well as bacteria, could be the causes of disease in both human beings and animals led many other researchers into the field. Laveran himself did much significant work on disease-causing parasites. He was especially concerned with the trypanosome family of protozoa, one of which is the cause of the disease trypanosomiasis, or African sleeping sickness, transmitted by the tsetse fly. He also studied the trypanosome responsible for another tropical disease, kala azar, or dumdum fever,. He was awarded the Nobel Prize in 1907 for his work on all disease-causing protozoa. He used half of the prize money to establish a laboratory for research on tropical diseases at the Pasteur Institute.

Laveran was honored with membership in the French Academy of Sciences in 1901. The French government made him a Commander of the Legion of Honor in 1912. During World War I he served on several committees concerned with preserving the health of French soldiers, and he served as president of the Academy of Medicine in 1920. He died after a short illness on May 18, 1922.

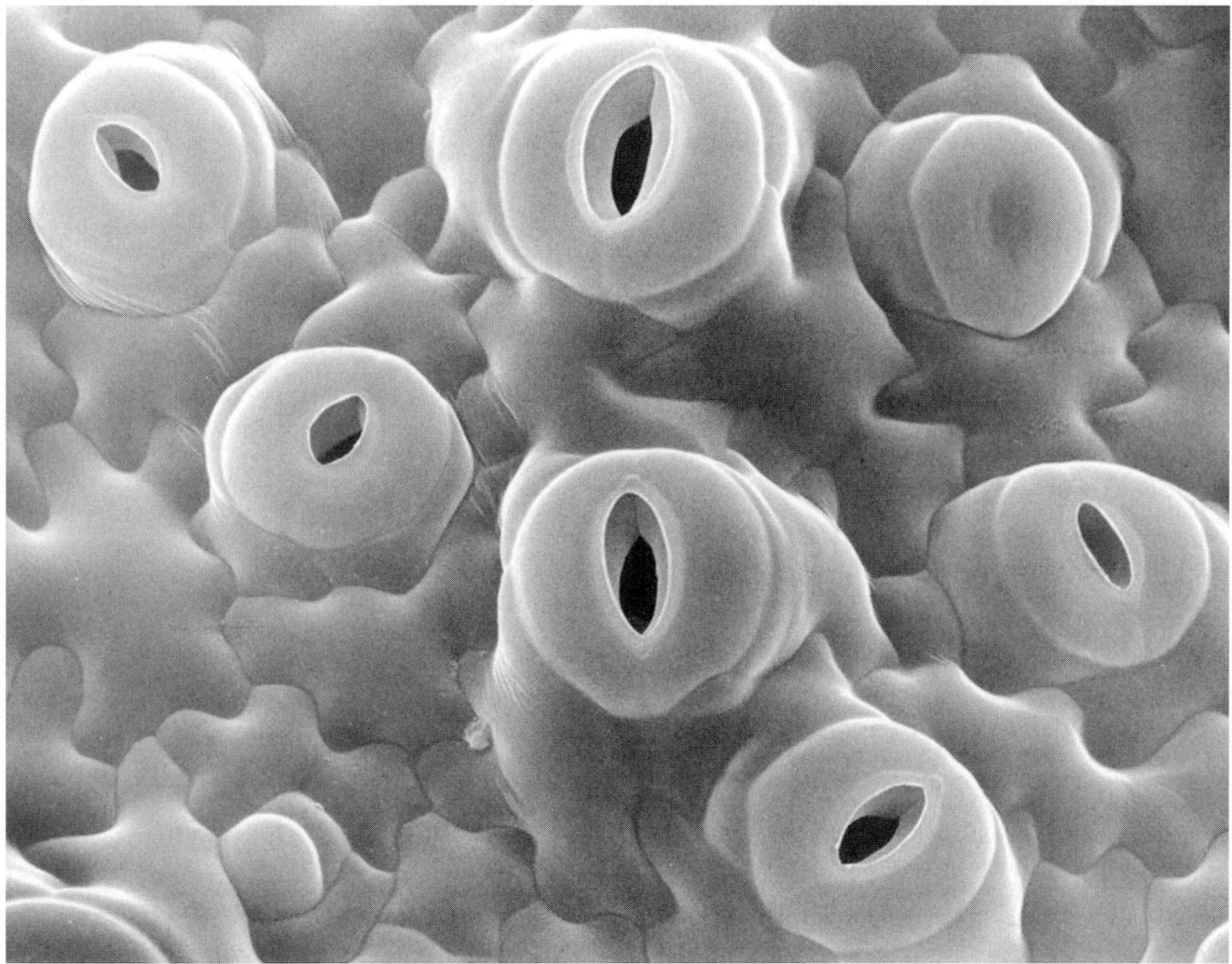

A scanning electron micrograph (SEM) of open stomata of the surface of a tobacco leaf (*Nicotiana tabacum*). Stomata are breathing pores scattered over the leaf surface, and sometimes stem, that regulate the exchange of gases between the leaf's interior and the atmosphere. Stomatal closure is a natural response to darkness or drought as a means of conserving water. Each pore is controlled by the turgor of two guard cells (sides seen beneath pore's opening) on either side of it. When they are full of water the pore is open; when they lose turgor, the pore closes. *(Photo by Dr. Jeremy Brugess, Photo Researchers, Inc. Reproduced by permission.)*

LAW OF INDEPENDENT ASSORTMENT •
See Independent assortment

LAW OF SEGREGATION • See Segregation

LEAF

The leaf is the principal **energy**-capturing and food-producing **organ** of **plants**. Leaves are attached to and supported by the plant stem, which also provides the leaf with **water** and inorganic **nutrients** from the soil. They usually have a flattened surface, called a blade (or lamina) which absorbs radiant energy from the sun, and a slender stalk called the petiole, which supports the blade and connects it to the stem. Leaves provide

nourishment to the plant by converting **light** energy into chemical energy used for growth and development. The conversion process is called **photosynthesis** and requires uptake of **carbon dioxide** gas from the **atmosphere** and the subsequent release of oxygen. Tiny pores, called stomata, in the surface of the leaf facilitate this exchange of gases. The stomata can open and close as needed to support rapid movement of gases into and out of the leaf. The extensive surface area of the leaf tends to result in extensive water loss into the atmosphere. To minimize this loss, and reduce the possibility of wilting, the leaf is covered with a tight outer layer of **cell**s called an epidermis. The epidermis is covered with a waxy coating called a cuticle. Because the epidermis and cuticle impede gas exchange, stomata are crucial for photosynthesis. When the stomata are open, they also allow water to escape to the atmosphere. To minimize water loss, stomata tend to close at night when photosynthesis is not possible, and open only during the day when rapid gas exchange is needed. Under unusually dry conditions when soil moisture is inadequate the plant may become water-stressed. The stomata may then close during the day the to prevent wilting. Photosynthesis is then impeded also, and plant growth is reduced.

The main light-absorbing compound in leaves is called chlorophyll. It is this pigment that gives plants their characteristic green color. It is found in **organelle**s called **chloroplast**s in the mesophyll cells in the interior of the leaf. With the aid of internal **enzyme** systems, chlorophyll transfers absorbed light energy to sugars, starch, and other energy storing carbon-based compounds. These compounds are then used to support plant growth and development. The oxygen produced in photosynthesis is released into the atmosphere where it replaces oxygen used by **animal** and plant **respiration** and by combustion. In the autumn chlorophyll is bleached and disappears from leaves. Leaves then change color to yellow, orange or red, as other pigments such as carotenoids and anthocyanins, previously masked by high concentrations of chlorophyll, are revealed.

Leaves originate in the apical (top) bud of the growing plant, together with the **tissue**s of the developing stem. They vary greatly in form, structure, and arrangement. These differences are useful in differentiating among plant **species**. Leaves of monocotyledons tend to have internal conducting tissue (veins) arranged in parallel patterns, whereas dicotylendons are more apt to have a net-like pattern. Leaves may be arranged on the stem in several different ways. Some plants have leaves set opposite one another on the stem, others have three or more leaves in a whorl at each node, and still others may have an alternating or spiral arrangement. Leaves of dicotyledons may be simple or compound. Simple leaves contain a single undivided blade, while compound leaves are divided into leaflets. Some leaves issue directly from the main stem (lacking a connecting support or petiole) and are said to be sessile. The nature of the margin of the leaf blade is highly variable. Some leaves have a smooth edge; others are toothed, and still others deeply lobed. In some species, leaves may be reduced to spines as in the Japanese barberry, or to the scale-like food containing structures such as those found in onion bulbs.

See also Carbon dioxide fixation; Plant anatomy

LEAKEY, LOUIS (1903-1972)

African English paleontologist and anthropologist

A pioneer in the field of paleoanthropology—the study of early humans and prehumans through both their fossilized remains and the cultural artifacts (mostly stone tools) they left behind—Louis Leakey helped change the prevailing view of humankind's origins. Along with other paleoanthropologists, he sought clues to, among other mysteries, how and when modern humans and apes split off from a common ancestor, and the identification of the point at which a creature appeared on the earth who can accurately be given the designation "human."

Louis Seymour Bazett Leakey was born on August 7, 1903, in Kabete, Kenya. His parents, Mary Bazett (d. 1948) and Harry Leakey (1868–1940) were Church of England missionaries at the Church Missionary Society, Kabete, Kenya. Louis spent his childhood in the mission, where he learned the Kikuyu language and customs (he later compiled a Kikuyu grammar book). As a child, while pursuing his interest in **ornithology**—the study of birds—he often found stone tools washed out of the soil by the heavy rains, which Leakey believed were of prehistoric origin. Stone tools were primary evidence of the presence of humans at a particular site, as toolmaking was believed at the time to be practiced only by humans and was, along with an erect posture, one of the chief characteristics used to differentiate humans from nonhumans. Scientists at the time, however, did not consider East Africa a likely site for finding evidence of early humans; the discovery of *Pithecanthropus* in Java in 1894 (the so-called Java Man, now considered to be an example of *Homo erectus*) had led scientists to assume that Asia was the continent from which human forms had spread.

The Search for Africa's Oldest Hominid

Shortly after the end of World War I, Leakey was sent to a public school in Weymouth, England, and later attended St. John's College, Cambridge. Suffering from severe headaches resulting from a sports injury, he took a year off from his studies and joined a **fossil**-hunting expedition to Tanganyika (now Tanzania). This experience, combined with his studies in anthropology at Cambridge (culminating in a degree in 1926), led Leakey to devote his time to the search for the origins of humanity, which he believed would be found in Africa. Anatomist and anthropologist Raymond A. Dart's discovery of early human remains in South Africa was the first concrete evidence that this view was correct. Leakey's next expedition was to northwest Kenya, near Lakes Nakuru and Naivasha, where he uncovered materials from the late Stone Age; at Kariandusi he discovered a 200,000-year-old hand ax.

In 1928 Leakey married Henrietta Wilfrida Avern, with whom he had two children: Priscilla, born in 1930, and Colin, born in 1933; the couple was divorced in the mid–1930s. In 1931 Leakey made his first trip to Olduvai Gorge—a 350-mile ravine in Tanzania—the site that was to be his richest source of human remains. He had been discouraged from excavating at Olduvai by Hans Reck, a German paleontologist who had fruitlessly sought evidence of prehistoric humans there. Leakey's first discoveries at that site consisted of both animal fossils, important in the attempts to date the particular stratum (or layer of earth) in which they were found, and, significantly, flint tools. These tools, dated to approximately one million years ago, were conclusive evidence of the presence of **hominid**s—a family of erect primate mammals that use only two feet for locomotion—in Africa at that early date; it was not until 1959, however, that the first fossilized hominid remains were found there.

In 1932, near Lake Victoria, Leakey found remains of *Homo sapiens* (modern man), the so-called Kanjera skulls (dated to 100,000 years ago) and Kanam jaw (dated to 500,000 years ago); Leakey's claims for the antiquity of this jaw made it a controversial find among other paleontologists, and Leakey hoped he would find other, independent, evidence for the existence of *Homo sapiens* from an even earlier period—the Lower Pleistocene.

In the mid–1930s, a short time after his divorce from Wilfrida, Leakey married his second wife, Mary Douglas Nicol; she was to make some of the most significant discoveries of Leakey's team's research. The couple eventually had three children: Philip, Jonathan, and Richard E. Leakey. During the 1930s, Leakey also became interested in the study of the Paleolithic period in Britain, both regarding human remains and geology, and he and Mary Leakey carried out excavations at Clacton in southeast England.

Until the end of the 1930s, Leakey concentrated on the discovery of stone tools as evidence of human habitation; after this period he devoted more time to the unearthing of human and prehuman fossils. His expeditions to Rusinga Island, at the mouth of the Kavirondo Gulf in Kenya, during the 1930s and early 1940s produced a large number of finds, especially of remains of Miocene apes. One of these apes, which Leakey named *Proconsul africanus,* Proconsul africanus had a jaw lacking in the so-called simian shelf that normally characterized the jaws of apes; this was evidence that *Proconsul* represented a stage in the progression from ancient apes to humans. In 1948 Mary Leakey found a nearly complete *Proconsul* skull, the first fossil ape skull ever unearthed; this was followed by the unearthing of several more *Proconsul* remains.

Louis Leakey began his first regular excavations at Olduvai Gorge in 1952; however, the Mau Mau (an anti-white secret society) uprising in Kenya in the early 1950s disrupted his paleontological work and induced him to write *Mau Mau and the Kikuyu,* in an effort to explain the rebellion from the perspective of a European with an insider's knowledge of the Kikuyu. A second work, *Defeating Mau Mau,* followed in 1954.

During the late 1950s, the Leakeys continued their work at Olduvai. In 1959, while Louis was recuperating from an illness, Mary Leakey found substantial fragments of a hominid skull that resembled the robust australopithecines—African hominids possessing small brains and near-human dentition—found in South Africa earlier in the century. Louis Leakey, who quickly reported the find to the journal *Nature,* suggested that this represented a new genus, which he named *Zinjanthropus boisei,* the genus name meaning ''East African man,'' and the species name commemorating Charles Boise, one of Leakey's benefactors. This species, now called *Australopithecus boisei,* was later believed by Leakey to have been an evolutionary dead end, existing contemporaneously with *Homo* rather than representing an earlier developmental stage.

In 1961, at Fort Ternan, Leakey's team located fragments of a jaw that Leakey believed were from a hitherto unknown genus and species of ape, one he designated as *Kenyapithecus wickeri,* and which he believed was a link between ancient apes and humans, dating from fourteen million years ago; it therefore represented the earliest hominid. In 1967, however, an older skull, one that had been found two

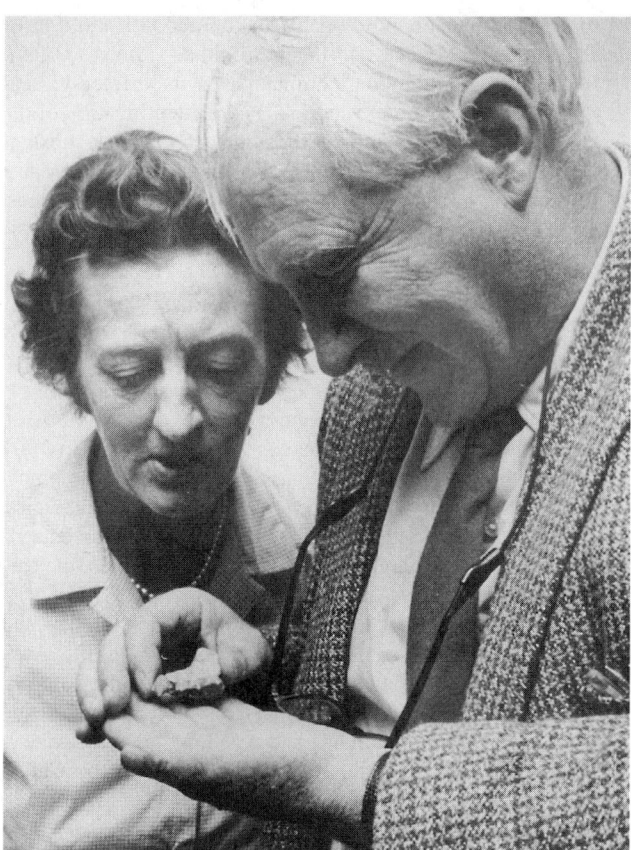

Dr. Louis Leakey and his wife, Mary, look at a fragment of bone taken from a 14 million year old hominid found in Kenya. *(AP/Wide World Photos. Reproduced by permission.)*

decades earlier on Rusinga Island and which Leakey had originally given the name *Ramapithecus africanus,* was found to have hominid-like lower dentition; he renamed it *Kenyapithecus africanus,* and Leakey believed it was an even earlier hominid than *Kenyapithecus wickeri.* Leakey's theories about the place of these Lower Miocene fossil apes in human evolution have been among his most widely disputed.

The Discovery of *Homo Habilis*

During the early 1960s, a member of Leakey's team found fragments of the hand, foot, and leg bones of two individuals, in a site near where *Zinjanthropus* had been found, but in a slightly lower and, apparently, slightly older layer. These bones appeared to be of a creature more like modern humans than *Zinjanthropus,* possibly a species of *Homo* that lived at approximately the same time, with a larger brain and the ability to walk fully upright. As a result of the newly developed potassium-argon dating method, it was discovered that the bed from which these bones had come was 1.75 million years old. The bones were, apparently, the evidence for which Leakey had been searching for years: skeletal remains of *Homo* from the Lower Pleistocene. Leakey designated the creature whose remains these were as *Homo habilis* (''man with ability''), a creature who walked upright and had dentition resembling that of modern humans, hands capable of toolmaking, and a large cranial capacity. Leakey saw this hominid as a direct ancestor of *Homo erectus* and modern humans. Not unexpectedly, Leakey was attacked by other scholars, as this identification of the fragments moved the origins of the genus *Homo* back substantially further in time. Some scholars felt that the new remains were those of australopithecines, if relatively advanced ones, rather than very early examples of *Homo*.

His Last Years

Health problems during the 1960s curtailed Leakey's field work; it was at this time that his Centre for Prehistory and Paleontology in Nairobi became the springboard for the careers of such paleontologists as Jane Goodall and Dian Fossey in the study of nonhuman primates. A request came in 1964 from the Israeli government for assistance with the technical as well as the fundraising aspects involved in the excavation of an early Pleistocene site at Ubeidiya. This produced evidence of human habitation dating back 700,000 years, the earliest such find outside Africa.

During the 1960s, others, including Mary Leakey and the Leakeys' son Richard, made significant finds in East Africa; Leakey turned his attention to the investigation of a problem that had intrigued him since his college days: the determination of when humans had reached the North American continent. Concentrating his investigation in the Calico Hills in the Mojave Desert, California, he sought evidence in the form of stone tools of the presence of early humans, as he had done in East Africa. The discovery of some pieces of chalcedony (translucent quartz) that resembled manufactured tools in sediment dated from 50,000 to 100,000 years old stirred an immediate controversy; at that time, scientists believed that humans had settled in North America approximately 20,000 years ago. Many archaeologists, including Mary Leakey, criticized Leakey's California methodology—and his interpretations of the finds—as scientifically unsound, but Leakey, still charismatic and persuasive, was successful in obtaining funding from the National Geographic Society and, later, several other sources. Human remains were not found in conjunction with the supposed stone tools, and many scientists have not accepted these "artifacts" as anything other than rocks.

Shortly before Louis Leakey's death, Richard Leakey showed his father a skull he had recently found near Lake Rudolf (now Lake Turkana) in Kenya. This skull, removed from a deposit dated to 2.9 million years ago, had a cranial capacity of approximately 800 cubic centimeters, putting it within the range of *Homo* and apparently vindicating Leakey's long-held belief in the extreme antiquity of that genus; it also appeared to substantiate Leakey's interpretation of the Kanam jaw. Leakey died of a heart attack in early October, 1972, in London.

A Controversial Career

Some scientists have questioned Leakey's interpretations of his discoveries. Other scholars have pointed out that two of the most important finds associated with him were actually made by Mary Leakey, but became widely known when they were interpreted and publicized by him; Leakey had even

encouraged criticism through his tendency to publicize his somewhat sensationalistic theories before they had been sufficiently tested. Critics have cited both his tendency toward hyperbole and his penchant for claiming that his finds were the "oldest," the "first," the "most significant"; in a 1965 *National Geographic* article, for example, Melvin M. Payne pointed out that Leakey, at a Washington, D.C., press conference, claimed that his discovery of *Homo habilis* had made all previous scholarship on early humans obsolete. Leakey has also been criticized for his eagerness to create new genera and species for new finds, rather than trying to fit them into existing categories. Leakey, however, recognized the value of publicity for the fundraising efforts necessary for his expeditions. He was known as an ambitious man, with a penchant for stubbornly adhering to his interpretations, and he used the force of his personality to communicate his various finds and the subsequent theories he devised to scholars and the general public.

Leakey's response to criticism was that scientists have trouble divesting themselves of their own theories in the light of new evidence. "Theories on prehistory and early man constantly change as new evidence comes to light," Leakey remarked, as quoted by Payne in *National Geographic*. "A single find such as *Homo habilis* can upset long-held—and reluctantly discarded—concepts. A paucity of human fossil material and the necessity for filling in blank spaces extending through hundreds of thousands of years all contribute to a divergence of interpretations. But this is all we have to work with; we must make the best of it within the limited range of our present knowledge and experience." Much of the controversy derives from the lack of consensus among scientists about what defines "human"; to what extent are toolmaking, dentition, cranial capacity, and an upright posture defining characteristics, as Leakey asserted?

Louis Leakey's significance revolves around the ways in which he changed views of early human development. He pushed back the date when the first humans appeared to a time earlier than had been believed on the basis of previous research. He showed that human evolution began in Africa rather than Asia, as had been maintained. In addition, he created research facilities in Africa and stimulated explorations in related fields, such as primatology (the study of primates). His work is notable as well for the sheer number of finds—not only of the remains of apes and humans, but also of the plant and animal species that comprised the ecosystems in which they lived. These finds of Leakey and his team filled numerous gaps in scientific knowledge of the evolution of human forms. They provided clues to the links between prehuman, apelike primates, and early humans, and demonstrated that human evolution may have followed more than one parallel path, one of which led to modern humans, rather than a single line, as earlier scientists had maintained.

LEAKEY, MARY (1913-1996)
English paleontologist and anthropologist

For many years Mary Leakey lived in the shadow of her husband, **Louis Leakey**, whose reputation, coupled with the prejudices of the time, led him to be credited with some of his wife's discoveries in the field of early human archaeology. Yet she established a substantial reputation in her own right and came to be recognized as one of the most important paleoanthropologists of the twentieth century. It was Mary Leakey who was responsible for some of the most important discoveries made by Louis Leakey's team. Although her close association with Louis Leakey's work on Paleolithic sites at Olduvai Gorge—a 350-mile ravine in Tanzania—led to her being considered a specialist in that particular area and period, she in fact worked on excavations dating from as early as the Miocene, (an era dating to approximately 18 million years ago) to those as recent as the Iron Age of a few thousand years ago.

Developing an Interest in Archaeology

Mary Leakey was born Mary Douglas Nicol on 6 February 1913, in London. Her mother was Cecilia Frere, the great-granddaughter of John Frere, who had discovered prehistoric stone tools at Hoxne, Suffolk, England, in 1797. Her father was Erskine Nicol, a painter who himself was the son of an artist, and who had a deep interest in Egyptian archaeology. When Mary was a child, her family made frequent trips to southwestern France, where her father took her to see the Upper Paleolithic cave paintings. She and her father became friends with Elie Peyrony, the curator of the local museum, and there she was exposed to the vast collection of flint tools dating from that period of human prehistory. She was also allowed to accompany Peyrony on his excavations, though the archaeological work was not conducted in what would now be considered a scientific way—artifacts were removed from the site without careful study of the place in the earth where each had been found, obscuring valuable data that could be used in dating the artifact and analyzing its context. On a later trip, in 1925, she was taken to Paleolithic caves by the Abbé Lémozi of France, parish priest of Cabrerets, who had written papers on cave art. After her father's death in 1926, Mary Nicol was taken to Stonehenge and Avebury in England, where she began to learn about the archaeological activity in that country and, after meeting the archaeologist Dorothy Liddell, to realize the possibility of archaeology as a career for a woman.

By 1930, Mary Nicol had undertaken coursework in geology and archaeology at the University of London and had participated in a few excavations in order to obtain field experience. One of her lecturers, R. E. M. Wheeler, offered her the opportunity to join his party excavating St. Albans, England, the ancient Roman site of Verulamium; although she only remained at that site for a few days, finding the work there poorly organized, she began her career in earnest shortly thereafter, excavating Neolithic (early Stone Age) sites in Henbury, Devon, where she worked between 1930 and 1934. Her main area of expertise was stone tools, and she was exceptionally skilled at making drawings of them. During the 1930s Mary met Louis Leakey, who was to become her husband. Leakey was by this time well known because of his finds of early human remains in East Africa; it was at Mary and Louis's first meeting that he asked her to help him with the illustrations for his 1934 book, *Adam's Ancestors: An Up-to-Date Outline of What Is Known about the Origin of Man.*

In 1934 Mary Nicol and Louis Leakey worked at an excavation in Clacton, England, where the skull of a **hominid**—a family of erect primate mammals that use only two feet for locomotion—had recently been found and where Louis was investigating Paleolithic geology as well as fauna and human remains. The excavation led to Mary Leakey's first publication, a 1937 report in the *Proceedings of the Prehistoric Society.*

Excavating at Olduvai Gorge

By this time, Louis Leakey had decided that Mary should join him on his next expedition to Olduvai Gorge in Tanganyika (now Tanzania), which he believed to be the most promising site for discovering early Paleolithic human remains. On the journey to Olduvai, Mary stopped briefly in South Africa, where she spent a few weeks with an archaeological team and learned more about the scientific approach to excavation, studying each find *in situ*— paying close attention to the details of the geological and faunal material surrounding each artifact. This knowledge was to assist her in her later work at Olduvai and elsewhere.

At Olduvai, among her earliest discoveries were fragments of a human skull; these were some of the first such remains found at the site, and it would be twenty years before any others would be found there. Mary Nicol and Louis Leakey returned to England. Leakey's divorce from his first wife was made final in the mid-1930s, and he and Mary Nicol were then married; the couple returned to Kenya in January of 1937. Over the next few years, the Leakeys excavated Neolithic and Iron Age sites at Hyrax Hill, Njoro River Cave, and the Naivasha Railway Rock Shelter, which yielded a large number of human remains and artifacts.

During World War II, the Leakeys began to excavate at Olorgasailie, southwest of Nairobi, but because of the complicated geology of that site, the dating of material found there was difficult. It did prove to be a rich source of material, however; in 1942, Mary Leakey uncovered hundreds, possibly thousands, of hand axes there. Her first major discovery in the field of pre-human fossils was that of most of the skull of a *Proconsul africanus* Proconsul africanus on Rusinga Island, in Lake Victoria, Kenya, in 1948. *Proconsul* was believed by some paleontologists to be a common ancestor of apes and humans, an animal whose descendants developed into two branches on the evolutionary tree: the *Pongidae* (great apes) and the *Hominidae* (who eventually evolved into true humans). *Proconsul* lived during the Miocene, approximately 18 million years ago. This was the first time a fossil ape skull had ever been found—only a small number have been found since—and the Leakeys hoped that this would be the ancestral hominid that paleontologists had sought for decades. The absence of a "simian shelf," a reinforcement of the jaw found in modern apes, is one of the features of *Proconsul* that led the Leakeys to infer that this was a direct ancestor of modern humans. *Proconsul* is now generally believed to be a species of *Dryopithecus,* closer to apes than to humans.

Discovering "Dear Boy": *Zinjanthropus*

Many of the finds at Olduvai were primitive stone hand axes, evidence of human habitation; it was not known, however, who had made them. Mary's concentration had been on the discovery of such tools, while Louis's goal had been to learn who had made them, in the hope that the date for the appearance of toolmaking hominids could be moved back to an earlier point. In 1959, Mary unearthed part of the jaw of an early hominid she designated *Zinjanthropus* (meaning "East African Man") and whom she referred to as "Dear Boy"; the early hominid is now considered to be a species of *Australopithecus*— apparently related to the two kinds of australopithecine found in South Africa, *Australopithecus africanus* and *Australopithecus robustus*— and given the species designation *boisei* in honor of Louis Leakey's sponsor Charles Boise. By means of potassium-argon dating, recently developed, it was determined that the fragment was 1.75 million years old, and this realization pushed back the date for the appearance of hominids in Africa. Despite the importance of this find, however, Louis Leakey was slightly disappointed, as he had hoped that the excavations would unearth not another australopithecine, but an example of *Homo* living at that early date. He was seeking evidence for his theory that more than one hominid form lived at Olduvai at the same time; these forms were the australopithecines, who eventually died out, and some early form of *Homo,* which survived—owing to toolmaking ability and larger cranial capacity—to evolve into *Homo erectus* and, eventually, the modern human. Leakey hoped that Mary Leakey's find would prove that *Homo* existed at that early level of Olduvai. The discovery he awaited did not come until the early 1960s, with the identification of a skull found by their son Jonathan Leakey that Louis designated as *Homo habilis* ("man with ability"). He believed this to be the true early human responsible for making the tools found at the site.

Working on Her Own

In her autobiography, *Disclosing the Past,* released in 1984, Mary Leakey reveals that her professional and personal relationship with Louis Leakey had begun to deteriorate by 1968. As she increasingly began to lead the Olduvai research on her own, and as she developed a reputation in her own right through her numerous publications of research results, she started believing that her husband felt threatened by her. Louis Leakey had been spending a vast amount of his time in fundraising and administrative matters, while Mary was able to concentrate on field work. As Louis began to seek recognition in new areas, most notably in excavations seeking evidence of early humans in California, Mary stepped up her work at Olduvai, and the breach between them widened. She became critical of his interpretations of his California finds, viewing them as evidence of a decline in his scientific rigor. During these years at Olduvai, Mary made numerous new discoveries, including the first *Homo erectus* pelvis to be found. Mary Leakey continued her work after Louis Leakey's death in 1972. From 1975 she concentrated on Laetoli, Tanzania, which was a site earlier than the oldest beds at Olduvai. She knew that the lava above the Laetoli beds was dated to 2.4 million years ago, and the beds themselves were therefore even older; in contrast, the oldest beds at Olduvai were two million years old. Potassium-argon dating has since shown the upper beds at Laetoli to

be approximately 3.5 million years old. In 1978, members of her team found two trails of hominid footprints in volcanic ash dated to approximately 3.5 million years ago; the form of the footprints gave evidence that these hominids walked upright, thus moving the date for the development of an upright posture back significantly earlier than previously believed. Mary Leakey considered these footprints to be among her most significant finds. She died in Nairobi, Kenya, on 8 December 1996.

In the late 1960s, Mary Leakey received an honorary doctorate from the University of the Witwatersrand in South Africa, an honor she accepted only after university officials had spoken out against apartheid. Among her other honorary degrees are a D.S.Sc. from Yale University and a D.Sc. from the University of Chicago. She received an honorary D.Litt. from Oxford University in 1981. She has also received the Gold Medal of the Society of Women Geographers.

Louis Leakey was sometimes faulted for being too quick to interpret the finds of his team and for his propensity for developing sensationalistic, publicity-attracting theories. Late in her career, Mary Leakey has been critical of the conclusions reached by her husband—as well as by some others—without adding her own interpretations to the mix. Instead, she remained more concerned with the act of discovery itself; she had written that it was more important for her to continue the task of uncovering early human remains to provide the pieces of the puzzle than it was to speculate and develop her own interpretations. Her legacy lies in the vast amount of material she and her team unearthed; she left it to future scholars to deduce its meaning.

LEARNING

Learning is the alteration of behavior as a result of experience. When an **organism** is observed to change its behavior, it is said to learn. Many theories have been formulated by psychologists to explain the process of learning. Early in the twentieth century, learning was primarily described through behaviorist principles that included associative, or conditioned response. Associative learning is the ability of an **animal** to connect a previously irrelevant **stimulus** with a particular response. One form of associative learning—classical **conditioning**—is based on the pairing of two stimuli. Through an association with an unconditioned stimulus, a conditioned stimulus eventually elicits a conditioned response, even when the unconditioned stimulus is absent. The earliest and most well-known documentation of associative learning was demonstrated by Ivan Pavlov, who conditioned dogs to salivate at the sound of a bell. In operant conditioning, a response is learned because it leads to a particular consequence (reinforcement), and it is strengthened each time it is reinforced. Without practice any learned behavior is likely to cease, however, repetition alone does not ensure learning; eventually it produces fatigue, boredom, and suppresses responses. Positive reinforcement strengthens a response if it is presented afterwards, while negative reinforcement strengthens it by being withheld. Generally, positive reinforcement is the most reliable and produce the best results.

Once the pattern of behavior has been established, it may be sustained by partial reinforcement, which is provided only after selected responses.

In contrast to classical and operant conditioning, which describe learning in terms of observable behavior, other theories focus on learning derived from motivation, **memory**, and **cognition**. Wolfgang Köhler, a founder of the **Gestalt** school of psychology, observed the importance of cognition in the learning process when he studied the behavior of chimpanzees. In his experimentation, Köhler concluded that insight was key in the problem-solving conducted by chimpanzees. The animals did not just stumble upon solutions through trial and error, but rather they demonstrated a holistic understanding of problems that they solved through moments of revelation. In the 1920s, Edward Tolman illustrated how learning can involve knowledge without observable performance. The performance of rats who negotiated the same maze on consecutive days without reward improved drastically after the introduction of a goal box with food, indicating that they had developed cognitive maps of the maze prior to the reward although it had not been observed in their behavior.

In the 1930s, Clark L. Hull and Kenneth W. Spence introduced the drive-reduction theory. Based on the tendency of an organism to maintain balance by adjusting physiological responses, the drive-reduction theory postulated that motivation is an intervening factor in times of imbalance. Imbalances create need, which in turn create drives; both encourage action in order to reduce the drive and meet the need. According to drive-reduction theory, the association of stimulus and response in classical and operant conditioning only results in learning if accompanied by drive reduction.

Perceptual learning theories postulate that an organism's readiness to learn is of primary importance to its survival, and this readiness depends largely on its perceptual skills. Perceptual skills are intimately involved in producing more effective responses to stimuli. In the laboratory, perceptual learning has been tested and measured by observing the effects of practice on perceptual abilities. Subjects are given various auditory, olfactory, and visual acuity tests. With practice, subjects improve their scores, indicating that perceptual abilities are not permanent but are modifiable by learning. In studies of animal behavior, the term perceptual learning is sometimes used to refer to those instances in which an animal learns to identify a complex set of stimuli that can be used to guide subsequent behavior. Examples of such perceptual learning include imitation and observational learning, song learning in birds, and **imprinting** in newborn birds and **mammals**. Imprinting occurs only during the first 30 or so hours of life. It is a form of learning in which a very young animal fixes its attention on the first object with which it has visual, auditory, or tactile experience and thereafter follows that object.

Observational learning, also known as modeling or imitation, proposes that learning occurs as a result of observation and consequence. Behavior is learned through imitation, however behavior that is rewarded is more readily imitated than behavior that is punished. Termed vicarious conditioning, this type of learning is present when there is attention to the behavior, retention and the ability to reproduce the behavior, and motivation for the learning to occur.

Current research on learning is highly influenced by computer technology, both in the areas of computer-assisted learning and in the attempt to further understand the neurological processes associated with learning by developing computer-based neural networks that simulate different types of learning.

LEDERBERG, JOSHUA (1925-)
American geneticist

Joshua Lederberg is a Nobel Prize-winning American geneticist whose pioneering work on genetic recombination in bacteria helped propel the field of molecular **genetics** into the forefront of biological and medical research. In 1946, Lederberg, working with **Edward Lawrie Tatum**, showed that **bacteria** may reproduce sexually, disproving the widely held theory that bacteria were asexual. The two scientists' discovery also substantiated that bacteria possess genetic systems comparable to those of higher **organism**s, thus providing a new repertoire for scientists to study the genetic basis of life.

Continuing with his work in bacteria, Lederberg also discovered the phenomena of genetic **conjugation** and transduction—or the transfer of either the entire complement of **chromosome**s or chromosome fragments, respectively—from **cell** to cell. In his work on conjugation and transduction, Lederberg became the first scientist to manipulate **genetic material**, which had far-reaching implications for subsequent efforts in **genetic engineering** and **gene therapy**. In addition to his laboratory research, Lederberg lectured widely on the complex relationship between science and society and served as a scientific adviser on biological warfare to the World Health Organization.

Lederberg was born in Montclair, New Jersey, on May 23, 1925. His father, Zwi Hirsch Lederberg, was a rabbi; his mother, Esther Goldenbaum, had emigrated from Palestine two years before Lederberg was born. Lederberg's parents moved to New York City, eventually settling in the Washington Heights district. Lederberg attended the city's public schools, where, as he wrote in the book *The Excitement and Fascination of Science: Reflections of Eminent Scientists,* he was a precocious youth whose inquiring mind was nurtured by "a cadre of devoted and sympathetic teachers." At Stuyvesant High School (which specialized in science education), Lederberg first encountered other youths who could compete with him intellectually. Through a program known as the American Institute Science Laboratory, Lederberg was given the opportunity to conduct original research in a laboratory after school hours and on weekends. Here he pursued his interest in biology, working in cytochemistry, or the chemistry of cells. A voracious reader, Lederberg was influenced early on by science-oriented writers such as Bernard Jaffe, Paul de Kruif, and H. G. Wells. For a Bar Mitzvah present he received Meyer Bodansky's *Introduction to Physiological Chemistry,* and on his sixteenth birthday, E. B. Wilson 's *The Cell in Development and Heredity.*

After graduating from high school in 1941, Lederberg entered Columbia University as a premedical student. He received a tuition scholarship from the Hayden Trust, which, coupled with his living at home and commuting to school, made it financially possible for him to attend college. Although his undergraduate studies focused on **zoology**, Lederberg also received a foundation in humanistic studies under Lionel Trilling, James Gutman, and others. Lederberg's work in zoology was fostered by H. Burr Steinbach, who helped Lederberg get space in a histology lab to pursue his own research. This early undergraduate research included an interest in the cytophysiology of mitosis in plants and the uses of genetic analysis in cell biology. In 1942, Lederberg met Francis Ryan, whose work in the biochemical genetics of *Neurospora* was Lederberg's first opportunity to see significant scientific research as it occurred. Lederberg graduated with a B.A. with honors in 1944 at the age of nineteen.

At the age of seventeen, Lederberg had enlisted in the United States Navy V–12 college training program, which featured a condensed pre-med and medical curriculum to produce medical officers for the armed services during World War II. While an undergraduate he also was assigned duty at the U.S. Naval Hospital at St. Albans in Long Island. He began his medical courses at the Columbia College of Physicians and Surgeons in 1944, but left after two years to study under Edward L. Tatum in the microbiology department at Yale University.

Embarks on Nobel Prize-winning Research

In spring of 1945 Ryan had suggested that Lederberg ask Tatum—who had made substantial contributions to biochemical genetics—if Lederberg could work in Tatum's lab at Yale. Lederberg was interested in natural **recombination**; and Tatum, working with George W. Beadle, had done pioneering investigations proving that the **DNA** (**deoxyribonucleic acid**) of *Neurospora* (a **genus** of **fungi**) played a fundamental role in many of the chemical reactions in *Neurospora* cells. While Lederberg helped Tatum continue his studies of *Neurospora,* the two proceeded to embark on a more tenuous line of research, studying *Escherichia coli* (a bacterium that lives in the gastrointestinal tract) for evidence of genetic inheritance. At the international Cold Spring Harbor Symposium of 1946, Lederberg and Tatum were graciously granted additional time to talk about their *E. Coli* research in addition to the *Neurospora* studies. The scientists' announcement that they had discovered sexual or genetic recombination in the bacterium was met with keen interest by an audience that included the leading molecular biologists and geneticists in the world. The prevailing theory among biologists of the time was that bacteria reproduced asexually by cells essentially splitting, creating two cells with a complete set of chromosomes (threadlike structures in the cell nucleus that carry genetic information). Lederberg and Tatum had found evidence that some strains of *E. coli* pass on hereditary material cell to cell. They found that a conjugation of two cells produced a cell that subsequently began dividing into offspring cells. These offspring showed that they inherited traits from each of the parent strains.

In *The Excitement and Fascination of Science,* Lederberg recalled the intense scrutiny this discovery came under at the Cold Spring Harbor Symposium. **André Lwoff** suggested that perhaps what they had found was a cross-feeding of nutrients between the cells. But in general at that meeting and a second one the following year, Lederberg found the giants of genetics, such as **Jacques Monod**, **Salvador E. Luria**, and Lwoff, to be supportive of and interested in his research. Lederberg also received requests for *E. coli* cultures by others who wanted to investigate his findings.

Lederberg's interest in basic research began to draw him further and further away from pursuing a medical career. In 1947 while at Yale he received an offer from the University of Wisconsin to become an assistant professor of genetics with a focus on the new field of microbial genetics. Although only two years away from receiving his M.D. degree, Lederberg viewed his return to medical school in *The Excitement and Fascination of Science* as a "grave (if not total) interruption of research at its most exciting stage." His prospective appointment at Wisconsin was met with some skepticism concerning his youth (he was only twenty-two) and his yet-to-be fully-accepted research. More troubling personally were references to his character and his Jewish heritage. But the strong support of senior colleagues at Wisconsin and Yale prevailed. Lederberg accepted the position at Wisconsin (receiving his Ph.D. degree from Yale in 1948) and spent a fruitful and satisfying decade there. He never regretted abandoning his medical training, although he noted his later honorary medical degrees from Tufts University and the University of Turin as being among his most valued.

Lederberg continued to make groundbreaking discoveries at Wisconsin and firmly established himself as one of the most promising young intellects in the burgeoning field of genetics. By perfecting a method to isolate mutant bacteria species using ultraviolet light, Lederberg was able to prove the long-held theory that genetic mutations occurred spontaneously. He found he could "mate" two strains of bacteria—one resistant to penicillin and the other to streptomycin—and produce a bacteria resistant to both antibiotics. He also found that he could manipulate a virus's virulence.

Working with graduate student Norton Zinder, Lederberg discovered genetic transduction, which involves the transfer only of hereditary fragments of information between cells as opposed to complete chromosomal replication (conjugation). Lederberg went on to breed unique strains of viruses. Although these strains promised to reveal much about the nature of viruses in hopes of one day controlling them, they also posed a clear threat in terms of creating harmful biochemical substances. At the time, the practical aspect of Lederberg's work was hard to evaluate. The Nobel Prize Committee, however recognized the significance of his contributions to genetics and, in 1958, awarded him the Nobel Prize in physiology or medicine for the bacterial and viral research that provided a new line of investigations of viral diseases and cancer. Lederberg shared the prize with Beadle and Tatum. Lederberg's work in genetics eventually proved to be one of the foundations of gene mapping, which eventually led to efforts to genetically treat disease and identify those at risk of developing certain diseases.

Addresses Role of Science in Society

A brilliant laboratory scientist and technician, Lederberg was also concerned with the role of science in society and the far-reaching effects of scientific discoveries, particularly in genetics. In a Pan American Health Organization/World Health Organization lecture in biomedical sciences called "Health in the World Tomorrow," Lederberg acknowledged concerns of the public, and even some scientists, over the new-found ability to tamper with the genetic code of life. But he was more concerned with the many ethical questions that would arise over the inevitable successes the advancing fields of microbiology and genetics were ushering in. Lederberg saw the biological revolution as "a philosophical one" that was to bring a "new depth of scientific understanding about the nature of life." He foresaw advancements in the treatment of cancer, organ transplants, and geriatric medicine as presenting a whole new set of ethical and social problems, such as the availability and allocation of expensive health-care resources.

Although Lederberg had a profound faith in science, he was not so confident of scientists' ability to rationally communicate the ramifications of their work. In *Man and His Future,* he lamented the "archaic clumsiness of our basic mechanisms of communication. Man's dilemma," he said, "is the discrepancy between the size of his population and the complexity of his institutions, on one hand, and his individual feebleness, measured as a data input rate of no more than fifty bits per second."

Lederberg was also interested in the study of biochemical life outside of earth and coined the term *exobiology* to refer to such studies. Along with physicist Dean B. Cowie, he expressed concern in *Science* over the possible contamination of biological life on other planets from microbes carried by human spacecraft. He was also a consultant to the U.S. Viking space missions to the planet Mars.

Lederberg's career included an appointment as chairman of the new genetics department at Stanford University in 1958. In 1978 he was appointed president of Rockefeller University. Working with his first wife, Esther Zimmer, a former student of Tatum's whom Lederberg married in 1946, Lederberg investigated the role of bacterial enzymes in sugar metabolism. He also discovered that penicillin's ability to kill bacteria was due to its preventing synthesis of the bacteria's cell walls. Among Lederberg's many honors were the Eli Lilly Award for outstanding work by a scientist under thirty-five years of age and the Alexander Hamilton Medal of Columbia University. After divorcing his first wife, Lederberg married Marguerite Stein Kirsch in 1968, with whom he had two children, a daughter and a son.

While Lederberg recognized the intense competition that sometimes arises among modern-era scientists, he described his own personal scientific dealings as congenial. "The shared interests of scientists in the pursuit of a universal truth," said Lederberg in *The Excitement and Fascination of Science,* "remain among the rare bonds that can transcend bitter personal, national, ethnic, and sectarian rivalries."

Antoni van Leeuwenhoek. *(The Library of Congress. Reproduced by permission.)*

LEEUWENHOEK, ANTONI VAN (1632-1723)
Dutch biologist and microscopist

Antoni van Leeuwenhoek is best remembered as the first person to study **bacteria** and "animalcules," or one-celled animals, now known as **protozoa**. Unlike his contemporaries **Robert Hooke** and **Marcello Malpighi**, Leeuwenhoek did not use the more advanced compound **microscope**; instead, he strove to manufacture magnifying lenses of unsurpassed power and clarity that would allow him to study the microcosm in far greater detail than any other scientist of his time.

Leeuwenhoek was born on October 24, 1632, in Delft, Holland. Although his family was relatively prosperous, he received little formal education. After completing grammar school in Delft, he moved to Amsterdam to work as a draper's apprentice. In 1654, he returned to Delft to establish his own shop, and he worked as a draper for the rest of his life. In addition to his business, Leeuwenhoek was appointed to several positions within the city government, which afforded him the financial security to spend a great deal of time and money in pursuit of his hobby—lens grinding. Lenses were an important tool in Leeuwenhoek's profession, since cloth merchants often used small lenses to inspect their products. His hobby soon turned to obsession, however, as he searched for more and more powerful lenses.

In 1671, Leeuwenhoek constructed his first simple microscope. It consisted of a tiny lens that he had ground by hand from a globule of glass and placed within a brass holder. To this, he had attached a series of pins designed to hold the specimen. It was the first of nearly six hundred lenses ranging from 50 to 500 times magnifications that he would grind during his lifetime. Through his microscope, Leeuwenhoek examined such substances as skin, hair, and his own blood. He studied the structure of ivory as well as the physical composition of the flea, discovering that fleas, too, harbored parasites.

Leeuwenhoek began writing to the British Royal Society in 1673. At first, the Society gave his letters little notice, thinking that such magnification from a single lens microscope could only be a hoax. However, in 1676, when he sent the Society the news that he had discovered tiny one-celled animals in rainwater, the interest of member scientists was piqued. Following Leeuwenhoek's specifications, they built microscopes of comparable magnitude and confirmed his findings. In 1680, the Society unanimously elected Leeuwenhoek as a member.

Until this time, Leeuwenhoek had been operating in an informational vacuum; he read only Dutch and, consequently, was unable to learn from the published works of Hooke and Malpighi (though he often gleaned what he could from the illustrations within their texts). As a member of the Society, he was finally able to interact with other scientists. In fact, the news of his discoveries spread worldwide, and he was often visited by royalty from England, Prussia, and Russia. The traffic through his laboratory was so persistent that he eventually allowed visitors by appointment only. Near the end of his life, he had reached near-legendary status and was often referred to by the local townsfolk as a magician.

Amid all the attention, Leeuwenhoek remained focused upon his scientific research. Specifically, he was interested in disproving the common belief in **spontaneous generation**, a theory proposing that certain inanimate objects could generate life. For example, it was believed that mold and maggots were created spontaneously from decaying food. He succeeded in disproving spontaneous generation in 1683, when he discovered bacteria cells. These tiny organisms were nearly beyond the resolving power of even Leeuwenhoek's remarkable equipment and would not be seen again for more than a century.

Leeuwenhoek created and improved upon new lenses for most of his long life. For the forty-three years that he was a member of the Royal Society, he wrote nearly 200 letters that described his progress. However, he never divulged the method by which he illuminated his specimens for viewing, and the nature of that illumination is still a mystery. Upon his death on August 30, 1723, Leeuwenhoek willed twenty-six of his microscopes—a few of which survive in museums—to the British Royal Society.

LELOIR, LUIS F. (1906-1987)
French Argentine biochemist

Luis F. Leloir began a career in medicine but found himself drawn to the relatively more tractable problems posed by bio-

chemistry. His early research involved investigations of fatty acids in the liver, which led to the discovery of antihypertensives. In a subsquent search for the "missing link" in the conversion of **carbohydrate**s in the body into **energy**, Leloir discovered a group of substances called sugar **nucleotide**s, which allowed him and others to determine the precise mechanism of energy conversion. Leloir also discovered glycogen, which is synthesized along with nucleotides and is the major store of energy in animal cells. Leloir's work with sugar nucleotides won him the Nobel Prize in chemistry in 1970.

Luis Federico Leloir was born on September 6, 1906, in Paris, France. Leloir's grandparents on both sides were immigrants to Argentina from France and Spain. When they moved to Argentina they invested in land, which turned to considerable profit as cattle and crops took on great importance in Argentine industry. This money would serve Leloir well later, as it would allow him to follow a career solely in scientific research at a time when such opportunities were very scarce in Argentina.

Leloir's parents, Federico and Hortensia Aguirre Leloir, were in France in 1906 only for a visit, and returned to their home in Buenos Aires, Argentina, when Leloir was two years old. Federico Leloir was educated as a lawyer, though he never practiced in that field. The younger Leloir grew up in a house filled with books on a variety of topics. Later in his life Leloir maintained there was no specific reason for his foray into the field of science, as it was clearly not a family tradition. He described himself as ill-suited to a career in music, sports, politics, or law, but acknowledged that he had a tremendous capacity for teamwork. Leloir completed his primary and secondary education in Buenos Aires, and then enrolled at the University of Buenos Aires to study medicine. He graduated with a medical degree in 1932, followed by employment in the hospital of the university as an intern. He found medicine somewhat limited in terms of the treatment options available at the time, and had no confidence in his own ability to diagnose and treat his patients.

He decided to try a position in research at the Institute of Physiology, still at the university, to help develop new options in treatment for physicians and to work on a Ph.D. degree. He worked under Bernardo Houssay, a Nobel winner in 1947 in the area of adrenal gland research, and consequently developed an interest in biochemistry. His doctoral thesis was written on the influence of the adrenal glands on carbohydrate metabolism, and his thesis won the annual prize of the faculty for best thesis. Leloir's relationship with Houssay would last the rest of Houssay's life, and Leloir described him as an intellectual inspiration.

In 1936 Leloir left Argentina to spend a year in Cambridge, England, conducting postdoctoral work in enzyme research at the Biochemical Laboratory of Cambridge University. He then returned to Buenos Aires and began research on breakdown of fatty acids in the liver. Eventually, his work led to a collaborative discovery of the peptide hypertensin, so named by this group because of its vasoconstrictive action. Vasoconstriction is the constriction of blood vessels, which causes high blood pressure or hypertension. Another group of scientists at Eli Lilly, a pharmaceutical company in Indianapolis, made a similar discovery around the same time, and named their peptide angiotensin. Both groups used these different names for the same peptide for several years, fighting over which it should be called, until finally a compromise was reached. Today the peptide is known as angiotensin. Later, in 1946, Leloir and his group issued a book based on their research findings in this area called *Renal Hypertension.*

Seeks Research Opportunities in the United States

Though averse to political involvement, Leloir was affected by the Argentine government's decision in 1943 to dismiss Houssay from his position at the university. Houssay had innocently signed a letter that was interpreted as antigovernment, and the country's politics at the time were in a state of upheaval. Many others at the university resigned their positions in support of Houssay, and Leloir decided it would be a good time for him to work abroad. He had married Amelie Zuherbuhler in 1943; together they left for the United States with no positions secured. After a short time in New York City, the Leloirs settled in St. Louis, Missouri, where Leloir worked as a research assistant in a biochemistry laboratory at Washington University. Later he moved to the Enzyme Research Laboratory of the College of Physicians and Surgeons of Columbia University in New York.

In 1945 Leloir returned to Argentina to work again under Houssay, who had been reinstated at the university. Leloir, though, had begun to hatch a plan to start a private research institute, and slowly began gathering the necessary team. Houssay was eventually removed from his post again, this time for being "over age," but Leloir had his team assembled, and finally received backing from Jaime Campomar. The owner of a textile firm, Campomar had expressed an interest to Houssay in sponsoring a research institute specifically in the area of biochemistry. Thus the Institute for Biochemical Investigations was begun.

The future of Leloir's institute was in question after Campomar died in 1957. In a bit of a last-ditch effort, Leloir applied to the National Institutes of Health in the United States for funding, and to his surprise obtained it. The institute continued to receive monies from the NIH for several years until rules for granting money to foreign applicants were changed. In 1958 the government of Argentina offered assistance as well, giving the institute a former girls' school for a new home. Further financial backing came a short while later after the formation of the Argentine National Research Council, and the institute became associated with the faculty of the University of Buenos Aires.

Scientists at this time were familiar with the idea that carbohydrates are broken down by the body into simpler sugars for energy. Beginning in the late 1940s, Leloir believed there was a "missing link" in the understanding of this process, and set out to find it. What he eventually discovered was a group of substances, now known as sugar nucleotides, that are responsible for the conversion into energy of sugars stored in the body. The discovery of these substances helped Leloir,

Luis F. Leloir. *(UPI/Corbis-Bettmann. Reproduced by permission.)*

among others, to determine specifically the process of carbohydrate conversion into energy. Leloir also found that a complex sugar called glycogen is synthesized with these sugar nucleotides, stored in the liver and muscles, and then broken down by the body into simpler glucose as energy is needed.

For his work with sugar nucleotides, Leloir was awarded the Nobel Prize in chemistry in 1970. He was only the third Argentine to receive the Nobel in any field, and the first in the area of chemistry. He instantly became a national hero, and was later honored as the subject on a postage stamp. Leloir was somewhat leery of the Nobel, telling *Newsweek* that his prize money would be spent on further research, "if I'm ever allowed to work again in the peace and quiet that I'm used to."

Leloir played a major role in the establishment of the Argentine Society for Biochemical Research as well as the Panamerican Association of Biochemical Sciences. Among his memberships were the National Academy of Sciences (United States) and the American Academy of Arts and Sciences. He was elected to membership in the Royal Society of London in 1972, and to the French Academy of Sciences in 1978. In addition to the Nobel, he received prizes and honors from universities all over the globe, including the Gairdner Foundation Award in 1966 and honorary degrees from the Universities of Paris, Granada (Spain), and Tucumán (Argentina). In 1971 he was keynote speaker at a biochemistry symposium held in his honor.

Leloir was known to be courteous and accessible. He has been given credit for performing major scientific research with limited funding. He often used homemade apparatus and gadgets, and encouraged inventions for use in his laboratory. In one instance Leloir constructed makeshift gutters out of waterproof cardboard to protect the library in his research laboratory from a leaky roof. Leloir died on December 2, 1987, in Buenos Aires, leaving his wife, one daughter, and several grandchildren.

LEUKOCYTES • See Blood

LEVI-MONTALCINI, RITA (1909-1989)
Italian American biochemist

Rita Levi-Montalcini revealed a fundamental process for **cell** growth and **differentiation** by discovering the **hormone**-like **protein** nerve growth factor (NGF). For this work, she received part of the 1986 Nobel prize in physiology or medicine.

Levi-Montalcini was born in Turin, Italy, where her father was an electrical engineer and mathematician and her mother an artist. Despite the objections of her father, who did not approve of education for women, she earned two medical degrees in 1936 and 1940 from the University of Turin, specializing in neurology and psychiatry. She was particularly interested in embryo nervous systems, inspired by an article on limb growth in chick embryos published in 1934 by the nerve development specialist Viktor Hamburger. During World War II Levi-Montalcini, who is Jewish, lived and worked underground to avoid the Italian government's anti-Semitic practices, developing a theory that many immature nerve cells are normally programmed to die. After Italy was liberated in 1944, she worked as a physician in a refugee camp. In 1947, Hamburger invited Levi-Montalcini to Washington University in St. Louis, Missouri, to pursue her nerve theory, which their research confirmed in 1949. Intending to stay there just one year, she remained for thirty years, becoming a United States citizen in 1956.

After 1962, she divided her research time between St. Louis and Rome. Levi-Montalcini's discovery of NGF began when she observed that mouse tumors grafted to chick **embryos** stimulated the development of embryo nerves. Furthermore, rapid growth occurred whether the tumors were in direct contact with the embryo or not. In searching for a chemical that accounted for the growth, she went to Brazil (with one of her research mice in her purse) to use the latest procedures for the then-new technique of **tissue** culture—the mixing of tumor slices with chick blood and embryo extract. During twenty-four hours of incubation, a dense halo of nerve axons grew near the tumor. Further research conducted in St. Louis with her assistant Stanley Cohen identified a substance she named nerve growth factor.

Her work showed how target cells produce NGF and determine the direction axons grow. It also showed that nerve cells die when antibodies block NGF. After 1977, when she retired from Washington University, she lived in Rome with her twin sister Paola Levi-Montalcini, a well-known painter, where she published her autobiography, *In Praise of Imperfection* in 1988, one year before she died.

LEWIS, EDWARD B. (1918-)
American developmental geneticist

Edward B. Lewis, sometimes called the father of developmental **genetics**, has dedicated a lifetime of research to the study of **gene** clusters responsible for early embryonic development. His tenacity resulted in important discoveries and led to formal recognition of his work. In 1995, Lewis was awarded the Nobel Prize in Physiology or Medicine for his groundbreaking genetic research. He shared the prize with two other scientists, Eric Wieschaus of Princeton University and Christiane Nüsslein-Volhard of the Max Planck Institute for Developmental Biology in Germany. Working independently of his co-recipients, Lewis studied "master control" gene clusters in fruit flies and subsequently discovered their corresponding human counterparts. Such a discovery promises to explain and eventually prevent congenital human malformations (about 40% of all human birth defects). It may also lead to improved in-vitro fertilization techniques, as well as a better understanding of substances harmful to early pregnancy.

Edward B. Lewis was born May 20, 1918, in Wilkes-Barre, Pennsylvania, to Edward B. Lewis and Laura (Histed) Lewis. His early years were spent trying to satiate his thirst for scientific knowledge in an environment that did not lend itself to learning. Books were not commonplace at home and as he remembered, "the high school library had nothing at all on genetics." Lewis found solace in playing the flute. He practiced daily, and during high school played with the local symphony orchestra. His musical abilities led to a scholarship at Bucknell University; however, Lewis transferred to the University of Minnesota, which offered course work in genetics. In 1939, Lewis received a B.A. degree in biostatistics from the University of Minnesota. He went on to earn a Ph.D. in genetics at the California Institute of Technology (Caltech) in 1942 and a M.S. in meteorology the following year. After serving as a weatherman in the Army during World War II, Lewis returned to Caltech to reestablish his affiliation with his alma mater.

Since the 1940s, Lewis has been a pioneer in the field of developmental genetics. The direction of his research was already set as a sophomore in high school: with the encouragement of a biology teacher, Lewis and a friend, Edward Novitski, purchased 100 fruit flies from Purdue University for one dollar. Lewis and Novitski let the flies breed, checking each day for any unusual new hatchlings. Their eagerness to learn something from a living specimen sparked careers in biology for both boys. In Lewis it created a lifelong obsession with the genetic workings of the fruit fly. In fact, it was a mutated fruit fly discovered by Novitski that led to Lewis's first postulations about the genetic factors causing mutations in the flies. Like Lewis, Novitski spent his professional life immersed in genetics research. Now retired, he resides in Eugene, Oregon.

Continuing his work with fruit fly specimens, Lewis was able to collect, crossbreed, and ultimately study an enormous amount of mutant flies. By mutating fly embryos so that the flies developed extra pairs of wings, Lewis was able to discern that it was not only the wings that were duplicated but the whole body segment that contained the wings. Because the fruit fly has only eight chromosomes (humans have 23 sets), Lewis was able to pinpoint the gene sequence responsible for the development and order of each fly-body segment. His findings were published in a 1978 *Nature* paper entitled "A Gene Complex Controlling Segmentation in Drosophila." Since then, geneticists have discovered that the gene sequences are almost identical for all other animal species as well.

Lewis has often received recognition for his contributions to developmental genetics. In 1981, he was honored with a Ph.D. from the University of Umeå in Sweden. He received the Thomas Hunt Morgan Medal from the Genetics Society of America in 1983. He was awarded the Canadian Gairdner Foundation International Award in 1987 and Israel's Wolf Prize in Medicine in 1989. In 1990, he received three separate awards: the Lewis S. Rosenstiel Award in basic medical research, the National Medal of Science, and an honorary membership in the Genetical Society in Great Britain. Lewis won the prestigious Albert Lasker Basic Medical Research Award in 1991, the Louisa Gross Horwitz Prize in 1992, and was given an honorary Doctor of Science degree from the University of Minnesota in 1993.

LEYDIG, FRANZ (1821-1908)

German zoologist

Franz Leydig is a zoologist recognized for his work in **comparative anatomy**. Franz was born in Rothenburg-ob-der-Tauber, Germany on May 21, 1821. He was the only son in a family of three children. His father, Melchior Leydig was a Catholic and minor public official while his mother, Margareta, was a protestant. The elder Leydig was an avid gardener and a beekeeper, and Franz shared father's interest in both religion and the natural sciences. When young Franz received a **microscope** as a present at age 12, it helped to cement science as his life's calling.

At the age of 19, Leydig began attending the Universities of Würzburg and Munich where he studied medicine. After receiving his doctorate from Wurzburg, he became an assistant professor and taught physiology, **histology** and developmental anatomy under professor **Rudolph von Kölliker**. In 1849 he qualified as a University lecturer. On a trip to Sardinia in the winter of 1850 he became fascinated with the diverse marine life there, and he decided to devote much of his future research efforts to these creatures.

In 1857 Leydig was made a full professor of **zoology** at the University of Tübingen where he published his text on **cell** morphology, *Lehrbuch der Histologie des Menschen und der Tiere*. This work described his own theories on cell behavior and structure as well as the work of Purkyne, Valentin, and Schwann. Leydig was also influenced by **Johannes Müller**'s

work on cellular doctrine and the role of glands. *Lehrbuch* is generally considered to be one of the best accounts of comparative microscopical anatomy of its time. It is also noteworthy for introducing one of Leydig's most important discoveries: that certain fishes and amphibians contain a large secretory cell which helps lubricate their skin. This cell was eventually named after Leydig.

Leydig is also credited with discovering a testosterone-producing cell in the testes which he described in his account of the male sex organs, *Zur Anatomie der Männlichen Geschlechtesorgane und Analdrüsen der Säugetiere*, published in 1850. While Leydig was first to describe the morphology, or the structure, of these cells, their mode of operation has only recently been described. In 1883 he also discovered a new type of cell that occurs in connective tissue and in blood vessels of crustaceans. In 1892 he discovered the gland of Leydig whose secretions aid in the sexual reproduction of certain vertebrates.

Later in his career Leydig joined various scientific societies, most notably the Royal Society of London, the Imperial Academy of Science, and the New York Academy of Sciences. He became professor of comparative anatomy at University of Bonn in 1875 and a professor emeritus in 1887. He retired to the town of his birth where he died on April 13, 1908. Leydig was survived by his wife Katharina Jaeger, but they had no children.

LI, CHOH HAI (1913-1987)

Chinese American biochemist

Choh Hai Li's lifelong research into the **pituitary gland** resulted in a series of **hormone** discoveries, including the isolation of human growth hormone (somatotropin), ACTH (adrenocorticotropic hormone), and melanocyte-stimulating hormone.

Li was born in Canton, China, and graduated from the University of Nanking in 1933. Two years later he emigrated to the United States, receiving his Ph.D. from the University of California, Berkeley, in 1938. He spent his entire career at that university, eventually becoming head of the Hormone Research Laboratory.

In 1921-22, the American biologist **Herbert Evans** (1882-1971) was one of several scientists studying the effect of pituitary gland extract on the **adrenal gland**. One such effect was to make animals grow very large (gigantism). During the 1940s, Li worked with Evans to isolate the growth hormone in cattle, then showed that it was a branched peptide molecule and partially determined its **amino acid** concentration. Other scientists used their method to isolate the growth hormones of different animal species. Next, working with Harold Papkoff (1925-), Li purified human growth hormone, using the then-new technique of ion exchange. By 1966, Li and his coworkers had determined the exact sequence of human growth hormone's 245 amino acids. Using **genetic engineering** methods, in 1970 Li was one of several scientists to independently synthesize the hormone, the largest one up to that time. Li's studies showed the great variation in growth hormone's size and composition from one species to another, and also showed that growth hormone is active only in its own species or a very closely related one.

Willard F. Libby. *(The Library of Congress. Reproduced by permission.)*

In 1944, Evans and Li were one of several scientific teams to produce a pure form of the hormone ACTH. This hormone stimulates the adrenal gland cortex to produce several hormones that control the body's glucose (sugar), water, and salt metabolism, and also affects numerous other body functions. In 1956 Li showed that ACTH has 39 amino acids, of which the first 13 are required for production of the glucose-controlling hormones. Li also isolated melanocyte-stimulating hormone (MSH), which controls the production of cell pigments. He showed how MSH's structure is similar to ACTH's and how the two hormones' functions are interrelated.

LIBBY, WILLARD F. (1908-1980)
American chemist

Chemist Willard F. Libby developed the radiocarbon dating technique used to determine the age of organic materials. With applications in numerous branches of science, including archaeology, geology, and geophysics, radiocarbon dating has been used to ascertain the ages of both ancient artifacts and geological events, such as the end of the Ice Age. In 1960, Libby received the Nobel Prize for his radiocarbon dating work. During World War II, Libby worked on the Manhattan Project to develop an atomic bomb and was a member of the Atomic Energy Commission for several years in the 1950s. An

outspoken scientist during the Cold War between the U.S. and the former Soviet Union, Libby advocated that every home have a fallout shelter in case of nuclear war. Opposed to bans against nuclear weapons testing, Libby was considered by some to be a pawn for a federal administration that wished to continue the arms race. Libby, however, was a strong proponent of the progress of science, which he believed resulted in more benefits than detriments for the human race.

Willard Frank Libby was born to Ora Edward and Eva May Libby on December 17, 1908, on a farm in Grand Valley, Colorado. In 1913, the family, which included Libby and his two brothers and two sisters, moved to an apple ranch north of San Francisco, California, near Sebastopol, where Libby received his grammar school education. A large boy who would eventually grow to be six-feet three-inches tall, Libby developed his legendary stamina while working on the farm. He played tackle for his high school football team and was called "Wild Bill," a nickname used by some throughout Libby's life. After graduating from high school in 1926, Libby enrolled at the University of California, Berkeley. He made money for college by building apple boxes, earning one cent for each box and sometimes $100 in a week. "I was the fastest box maker in Sonoma County," he told Theodore Berland, who interviewed Libby for his book *The Scientific Life.*

Although Libby was interested in English literature and history, he felt obligated to seek a more lucrative career and entered college to become a mining engineer. By his junior year, however, Libby became interested in chemistry, spurred on by the discussions of his boarding house roommates, who were graduate students in chemistry. Libby took on an heavy course load, focusing on mathematics, physics, and chemistry. After receiving his B.S. in chemistry in 1931, he entered graduate school at Berkeley and studied under the American physical chemist Gilbert Newton Lewis and Wendell Latimer, who were pioneering the physical chemistry field.

During graduate school, Libby built the United States' first Geiger-Muller tube for detecting radioactivity, which results from the spontaneous disintegration of an atom's nucleus. Libby refined the mechanism in order to detect minute amounts of radioactivity in elements not previously thought to be radioactive, including samarium. Libby continued to make his own Geiger counters throughout his life, claiming that they were much more sensitive than those manufactured for the open market.

Libby received his Ph.D. in 1933 and was appointed an instructor in chemistry at Berkeley. After the Japanese bombed Pearl Harbor in 1941, Libby, who was on a year sabbatical as a Guggenheim Fellow at Princeton University, joined a group of scientists in Chicago, Illinois, to work on the Manhattan Project, a government-sponsored effort to develop an atomic bomb. During this time, he worked with American chemist and physicist Harold Urey at Columbia University on gaseous diffusion techniques for the separation of uranium isotopes (isotopes are different forms of the same element having the same atomic number but a different number of protons). After the war, he accepted an appointment as a professor of chemistry at the University of Chicago and began to conduct research at the Institute of Nuclear Studies.

Embarks on Nobel Prize-winning Research

In 1939, scientists at New York University had sent radiation counters attached to balloons into the earth's upper atmosphere and discovered that neutron showers were created by cosmic rays hitting atoms. Further evidence indicated that these neutrons were absorbed by nitrogen, which then decayed into radioactive carbon–14. In addition, two of Libby's former students, Samuel Ruben and Martin Kamen, made radioactive carbon–14 in the laboratory for the first time. They used a cyclotron (a circular device that accelerates charged particles by means of an alternating electric field in a constant magnetic field) to bombard normal carbon–12 with neutrons, causing it to decay into carbon–14.

Intrigued by these discoveries, Libby hypothesized that radioactive carbon–14 in the atmosphere was oxidized to carbon dioxide. He further theorized that, since plants absorb carbon dioxide through photosynthesis, all plants should contain minute, measurable amounts of carbon–14. Finally, since all living organisms digest plant life (either directly or indirectly), all animals should also contain measurable amounts of carbon–14. In effect, all plants, animals, or carbon-containing products of life should be slightly radioactive.

Working with Aristide von Grosse, who had built a complicated device that separated different carbons by weight, and graduate student Ernest C. Anderson, Libby was successful in isolating radiocarbon in nature, specifically in methane produced by the decomposition of organic matter. Working on the assumption that carbon–14 was created at a constant rate and remained in a molecule until an organism's death, Libby thought that he should be able to determine how much time had elapsed since the organism's death by measuring the half-life of the remaining radiocarbon isotopes. (Half-life is a measurement of how long it takes a substance to lose half its radioactivity.) In the case of radiocarbon, Libby's former student Kamen had determined that carbon–14's half-life was 5,370 years. So, in approximately 5,000 years, half of the radiocarbon is gone; in another 5,000 years, half of the remaining radiocarbon decays, and so on. Using this mathematical calculation, Libby proposed that he could determine the age of organisms that had died as many as 30,000 years ago.

Since a diffusion column such as von Grosse's was extremely expensive to operate, Libby and Anderson decided to use a relatively inexpensive Geiger counter to build a device that was extremely sensitive to the radiation of a chosen sample. First, they eliminated 99% of the background radiation that occurs naturally in the environment with eight-inch thick iron walls to shield the counter. They then used a unique chemical process to burn the sample they were studying into pure carbon lampblack, which was then placed on the inner walls of a Geiger counter's sensing tube.

Libby first tested his device on tree samples, since their ages could be determined by counting their rings. Next, Libby gathered tree and plant specimens from around the world and discovered no significant differences in normal age-related radiocarbon distribution. When Libby first attempted to date historical artifacts, however, he found his device was several hundred years off. He soon realized that he needed to use at least several ounces of a material for accurate dating. From the Chicago Museum of Natural History, Libby and Anderson obtained a sample of a wooden funerary boat recovered from the tomb of the Egyptian King Sesostris III. The boat's age was 3,750 years; Libby's counter estimated it to be 3,261 years, only a 3.5% difference. Libby spent the next several years refining his technique and testing it on historically significant—and sometimes unusual—objects, such as prehistoric sloth dung from Chile, the parchment wrappings of the Dead Sea Scrolls, and charcoal from a campsite fire at Stonehenge, England. Libby saw his new dating technique as a way of combining the physical and historical sciences. For example, using wood samples from forests once buried by glaciers, Libby determined that the Ice Age had ended 10,000 to 11,000 years ago, 15,000 years later than geologists had previously believed. Moving on to man-made artifacts from North America and Europe (such as a primitive sandal from Oregon and charcoal specimens from various campsites), Libby dispelled the notion of an Old and New World, proving that the oldest dated human settlements around the world began in approximately the same era. For many years after Libby's discovery of radiocarbon dating, the journal *Science* published the results of dating studies by Libby and other scientists from around the world. In 1960, Libby was awarded the Nobel Prize in chemistry for his work in developing radiocarbon dating. In his acceptance speech, as quoted in *Nobel Prize Winners,* Libby noted that radiocarbon dating ''may indeed help roll back the pages of history and reveal to mankind something more about his ancestors, and in this way, perhaps about his future.'' Further progress in radiocarbon dating techniques extended its range to approximately 70,000 years.

In related work, Libby had shown in 1946 that cosmic rays produced tritium, or hydrogen–3, which is also weakly radioactive and has a half-life of 12 years. This radioactive form of hydrogen combines with oxygen to produce radioactive water. As a result, when the U.S. tested the Castle hydrogen bomb in 1954, Libby used the doubled amount of tritium in the atmosphere to date various sources of water, deduce the water-circulation patterns in the U.S., and determine the mixing of oceanic waters. He also used the method to date the ages of wine, since grapes absorb rain water.

Enters the Political Arena

In 1954, U.S. President Dwight D. Eisenhower appointed Libby to the Atomic Energy Commission (AEC). Although he continued to teach graduate students at Chicago, Libby drastically reduced his research efforts and plunged vigorously into his new duties. Previously a member of the commission's General Advisory Committee, which developed commission policy, Libby was already acquainted with the inner workings of the commission. He soon found himself embroiled in the nuclear fallout problem. Upon a recommendation by the Rand Corporation in 1953, Libby formed and directed Project Sunshine and became the first person to measure nuclear fallout in everything from dust, soil, and rain to human bone.

As a member of the AEC, Libby testified before the U.S. Congress and wrote articles about nuclear fallout. He noted that all humans are exposed to a certain amount of natural radiation in sources such as drinking water. He went on to point out that the combination of the body's natural radioactivity, cosmic radiation, and the natural radioactivity of the earth's surface was more hazardous than fallout resulting from nuclear testing. Libby believed, and most scientists concurred, that the effects of nuclear fallout on human genetics were minimal.

Many scientists, however, thought that Libby was merely a "yes man" for the federal administration. In reply, Libby often responded to what he considered misguided thinking. He once wrote to the French physician and author Albert Schweitzer, who had publicly declared that future generations would probably suffer from fallout, that he doubted whether Schweitzer was aware of the most recent data on the subject and that nuclear testing was necessary for the defense effort and the free world's survival. Even after Libby resigned from the AEC in 1959 to resume his scientific studies, he continued to argue with zeal about the necessity for nuclear testing. He also urged the nation's industrial community to employ isotopes in factories and on farms and was a member of the international Atoms for Peace project that supported nuclear energy production for non-military purposes. Libby's experiences in Washington, DC, convinced him that more scientists needed to be in positions of political power and not just advisers. As a result, he was pleased when U.S. President John F. Kennedy appointed Glenn T. Seaborg, a nuclear chemist, chair of the AEC in 1961.

Returns to Academia

After retiring from the AEC, Libby took a position in the chemistry department at the University of California, Los Angeles (UCLA), largely due to his first wife's desire to live in California again. Libby had married Leonor Lucinda Hickey in 1940, and the couple had twin daughters. When the family moved to their new home in Bel-Air, California, Libby proceeded to build his own fallout shelter, using sandbags and railroad ties, for approximately $30. During the Cold War, Libby believed that every home should have a fallout shelter in case of nuclear war and wrote a series of articles for the Associated Press news service proposing this necessity. According to Berland, Libby once complained, after hearing a physician say that perhaps it would have been better if scientists had never discovered the power of the atom, that the only way to stop such inevitable discoveries was to "kill all the scientists." He went on to tell Berland that physicians would then have to "go back to witchcraft" for treating people.

In 1962, Libby received a joint appointment as director of the Institute of Geophysics and Planetary Physics. He believed that the new frontier in science was outer space and said that the U.S. must support a large space exploration program in order to prevent the Soviets from controlling outer space, which would probably enable them to rule the world.

Libby and his wife Leonor divorced in 1966. Libby later married Leona Woods Marshall, a professor of environmental engineering at UCLA. He retired in 1976 and died in Los Angeles on September 8, 1980, from complications suffered during a bout of pneumonia.

LIEBIG, JUSTUS VON (1803-1873)
German chemist

Justus Von Liebig was a German chemist who made important contributions to the development of organic chemistry, and is considered the "father" of agricultural chemistry. His first academic position was gained in 1824, at the age of only 21 years, wh en he was appointed an Assistant Professor of Chemistry at the University of Giessen. Von Liebig is best known as the originator of the "**Law of the Minimum**" and the concept of radicals, and as the discoverer of chloroform. Von Liebig's most enduring influences were the demonstration of the importance of quantitative analysis in chemical research and teaching, and showing that chemistry was relevant to the needs of society, through agricultural production, medicine, and sanitation. Von Liebig was famous during his own time, and is now considered one of the most influential chemists of the nineteenth century.

Von Liebig's work in agricultural chemistry demonstrated that plants had a requirement for inorganic forms of nutrients in order to grow. As such, plants obtain **carbon dioxide** (and some **nitrogen**) from the atmosphere, and inorganic forms of nitrogen (ammonium and nitrate), phosphorus (phosphate), potassium, and other essential chemicals from the soil. Von Liebig developed the first inorganic fertilizers for use in agriculture (today, such fertilizers are the principal means by which nutrients are supplied to crops in intensively managed agricultural systems.

Von Liebig's "Law of the Minimum" developed from the observation that the productivity of agricultural crops could be increased by adding fertilizers (or nutrients). Von Liebig's law essentially states that agricultural (and ecological) productivity is limited by whichever nutrient is available in the least supply relative to the potential biological demand. If, for example, the productivity of a crop of maize (or corn; *Zea mays*) in a particular field is limited by the avail ability of inorganic nitrogen in the soil, then addition of this nutrient (usually in the form of ammonium) as a fertilizer will result in an increase in crop yield. In contrast, the addition of other potentially limiting nutrients, such as phosphate or potassium, would not have any effect on crop yield. However, if the need for ammonium was first satiated by intensive fertilization, then phosphate or potassium could, in turn, become limiting nutrients to the productivity of the corn crop.

Von Liebig was also influential in the development of the theory of radicals (or free radicals; these are an atom or group of atoms containing at least one unpaired electron and existing for a very brief period of time, before reacting to produce a stable molecule). Radicals are now a central idea in theoretical and applied chemistry.

Justus Von Liebig published books and more than 200 papers in scientific journals, including *Organic Chemistry and Its Application to Agriculture and Physiology* in 1840, and *Organic Chemistry in Its Application to Physiology and Pathology* in 1842. These books revolutionized food production.

LIFE CYCLE

Life cycle refers to the series of changes that the members of a **species** undergo as they pass from the beginning of a given developmental stage to the beginning of that same developmental stage in a subsequent generation.

In many simple **organism**s, including **bacteria** and various protists, the life cycle is completed within a single generation: an organism begins with the fission of an existing individual; the new organism grows to maturity; and it then splits into two new individuals, thus completing the cycle. In higher **animal**s, the life cycle is also complete in a single generation. The individual animal begins with the union of male and female sex **cell**s (**gamete**s); it grows to reproductive maturity; and it then produces gametes, at which point the cycle begins anew.

By contrast, in most **plants**, the life cycle is multigenerational. An individual plant begins with the **germination** of a **spore**, which grows into a gamete-producing organism (the **gametophyte**). The gametophyte reaches maturity and forms gametes, which, following **fertilization**, grow into a spore-producing organism (the **sporophyte**). Upon reaching reproductive maturity, the sporophyte produces spores, and the cycle starts again. This multigenerational life cycle is called alternation of generations; it occurs in some protists and **fungi** as well as in plants.

The life cycle of bacteria is termed haplontic. This term refers to the fact that it encompasses a single generation of organisms whose cells are **haploid** (i.e., contain one set of **chromosome**s). The term, ploidy refers to the number of chromosomes occurring in the **nucleus** of a cell. In normal somatic (body) cells, the chromosomes exist in pairs. This condition is called **diploid**y. During **meiosis** the cell produces gametes, or germ cells, each containing half the normal or somatic number of chromosomes. This condition is called haploidy. When two germ cells (e.g., **egg** and **sperm**) unite, the diploid condition is restored. Thus the one-generational life cycle of the higher animals is diplontic; it involves only organisms whose body cells are diploid (i.e., contain two sets of chromosomes). Organisms with diplontic cycles produce sex cells that are haploid, and each of these gametes must combine with another gamete in order to obtain the double set of chromosomes necessary to grow into a complete organism.

The life cycle typified by plants is known as diplohaplontic, because it includes both a diploid generation (the sporophyte) and a haploid generation (the gametophyte). This phenomenon is also called alternation of generations because the life cycle includes the alternation of a sexual phase and an asexual phase. The two phases, or generations are often physiologically distinct. In **algae**, fungi, mosses, ferns, and seed plants, alternation of generations is common; but is not always easy to observe, however, since one or the other of the generations is often very small, even microscopic. In the sexual phase, the organism is called a gametophyte, and it produces gametes, or sex cells. In the asexual phase, the organism is a sporophyte, and produces spores asexually. In terms of chromosomes, the gametophyte has a single (haploid) set, and the sporophyte has a double (diploid) set.

Among animals, many **invertebrates** have an alternation of sexual and asexual generations (e.g., **protozoa**ns, jellyfish, flatworms), but the alternation of haploid and diploid generations is unknown.

LIGHT

Light, also known as visible radiation, is electromagnetic radiation within the wavelength band that can be perceived by the human eye. This range is from about 380-780 nanometers (or nm; 1 nm = 10^{-9} m). Wavelengths shorter than the lower end of the visible range are known as ultraviolet, and longer ones are infrared.

The electromagnetic spectrum of visible radiation is commonly partitioned into a number of sub-ranges, which correspond to the "colors of the rainbow": red, orange, yellow, green, blue, and violet. These colors have the following wavelength ranges: red, 740-620 nm; orange, 620-585 nm; yellow, 585-575 nm; green, 575-500 nm; blue, 490-445 nm; violet, 445-390 nm.

Sometimes, the word "light" is also used in reference to the entire electromagnetic spectrum, or to components such as gamma, x ray, ultraviolet, infrared, microwave, or radio. In the physical sense, however, this is an inappropriate use of the word light, which should be restricted to wavelengths in the range of 380 to 780 nm.

LIMITING FACTOR

Limiting factors are environmental influences that prevent individual **organism**s, their **population**s, or multi-**species** communities from attaining as high a rate of **productivity** as they could potentially achieve under optimal conditions. Limiting factors can be a single environmental influence, or a complex of related ones. Oftentimes, limiting factors are divided into two categories: density-dependent and density-independent. Density-dependent factors that affect a population relate to population density, such as **predation** and the number of organisms per a given area. Density-independent factors are environmental factors affecting a population, such as **temperature** and oxygen supply.

Numerous environmental factors must be suitable if organisms are to persist and flourish. For example, temperature cannot be too hot or too cold, and **nutrients** must be available in appropriate amounts. The minimal requirement for a metabolically essential environmental factor is the least availability that will sustain organisms and ecological processes. On the other hand, the maximally acceptable levels are associated with toxicity and other biological damage that may be caused by too intense an exposure. Copper, for example, is an essential micronutrient, which is needed at some minimal level by all organisms if they are to live. At higher levels of exposure, however, copper is poisonous to those same organisms. The same is true, more or less, of all environmental factors. The minimum and maximum levels of most environmental factors bound a relatively broad range. Within this range, there is an optimal exposure, at which the factor exerts no significant constraint on biological productivity.

The "principle of limiting factors" is related to the controlling factors for ecological processes. It suggests that, at any particular time, the productivity of an **ecosystem** is limited by a single essential factor (the one that is present in the smallest supply relative to the potential biological demand. Such a limiting factor might be climatic, and could be due to extreme values of temperature, wind-speed, or moisture. Alternatively, the limiting factor could involve an insufficient supply of an essential nutrient (such as carbon, **nitrogen**, or phosphorus), or an intense exposure to a toxic chemical (such as sulfur dioxide, an important air pollutant). Interpreted in this manner, limiting factors can be viewed as a kind of ecological stress, which if alleviated will result in a higher rates of productivity and greater development of the individual, population, or **community**.

One of the best ways of determining the limiting roles of particular environmental factors is to perform experiments. Within limits, these can be done in the laboratory. However, the most convincing experiments investigating limiting factors are done in the field. For example, plant ecologists have studied the limitations posed to **tundra** vegetation by climatic factors (such as cool ambient temperature) by enclosing small areas of natural vegetation within greenhouses erected in the field. Commonly, the warmer climate within such field greenhouses allows the **plants** to grow significantly faster and larger. In other experimental studies, the limitations associated with certain nutrients, such as phosphate or nitrate, have been examined by **fertilization** trials in which nutrients are added to ecosystems alone or in combination. Similarly, the limiting influences of toxic factors have been studied by transplanting organisms between polluted and cleaner environments, such as away from (or in to) a local area polluted by sulfur dioxide. If experiments such as these do not result in a significant response to the manipulation of a particular environmental factor, then it was not the limiting factor.

One of the clearest examples of the application of the principle of limiting factors is in the study of an environmental problem known as **eutrophication**. Eutrophication occurs when lakes or other bodies of **water** have an extremely high rate of algal productivity, usually because they are being fertilized with a variety of nutrients in sewage and agricultural runoff. Severely eutrophic waters are often smelly, devoid of fish, and unsuitable for use as drinking water or for recreation. In most freshwater lakes, the productivity of **phytoplankton** (the community of unicellular **algae** that live in the water column) has been demonstrated to be limited by the availability of inorganic phosphorus, occurring in the form of the ion phosphate. Studies involving the manipulation of entire lakes have shown that the experimental fertilization with phosphate will usually cause a large increase in the productivity of phytoplankton. In contrast, if a lake is fertilized with another important nutrient, such as nitrate, ammonium, potassium, or inorganic carbon, there is no increase in productivity, indicating these are not limiting nutrients. However, if the lake water is first well fertilized with phosphate, its productivity will then respond to nitrate addition, indicating that the nitrate is the secondary limiting factor to the productivity of phytoplankton.

Knowledge of limiting factors has been utilized to control **pollution** associated with eutrophication. Once scientists discovered that phosphorus is the primary limiting nutrient for algal productivity, regional government authorities were advised to enact regulations to reduce the amount of phosphate inputs into surface waters. This has been done mostly by banning the use of household detergents containing large amounts of phosphorus (present as polyphosphates), and by treating human and livestock sewage to reduce its phosphorus content before the wastewater is discharged into bodies of water. These actions have been effective in decreasing the importance of eutrophication as an environmental problem in many areas.

LINKAGE

Linkage is a term used to describe the phenomenon of two or more non-allelomorphic **gene**s repeatedly occurring in the same **gamete** (a **haploid** reproductive **cell**). These genes and their appropriate **allele**s are passed from generation to generation as a joined unit. They do not obey the law of **independent assortment**.

Linkage occurs on the same **chromosome**, and in fact, one chromosome is one linkage group. When cross over occurs during **meiosis** (**cell division** that produces four haploid gametes), the linkage breaks down. **Crossing over** is the exchange of genes by chromosomes during meiosis that alters the genetic pattern contained within the chromosome. The further apart the alleles are on the chromosome the more likely a cross over event will occur between them and linkage will not be evident.

Linkage and its occasional breakdown during cross over provided was studied in order to produce the earliest form of genetic (or linkage) maps. Linkage maps were based on the percentages of cross overs between linked non-allelomorphic genes and showed the relative locations of genes within the chromosomes of an **organism**.

LINNAEUS, CARL (1707-1778)
Swedish botanist

Linnaeus decisively broke through centuries of confusion over how to revise the **classification** system that had been in place since antiquity. With few parallels in the history of science, Linnaeus's contribution to **botany** will remain intact perhaps as long as the first classification system.

Linnaeus was born on May 23, 1707, in Stenbrohult, Sweden. His father, a clergyman, maintained a small botanical garden on the parsonage grounds, where Linnaeus earned the nickname "little botanist." In 1716, Linnaeus entered a Latin school and began to formalize his interest in botany and the natural sciences. In 1727, he transferred to the University of Lund to study medicine, but he also undertook extensive botanical excursions. One year later, he went on to the University of Uppsala, which was considered a better school for medicine, but Linnaeus was disappointed to find that its facilities were no better than those at the University of Lund. Nevertheless, Uppsala did have something that made up for the shortcoming—a botanical garden containing rare foreign **plants**.

As his academic ideas started to mature along with his research, Linnaeus constructed a new theory of plant sexuality. In 1735, he published his *System Naturae*, and two years later, *Genera Planetarum*. He also moved briefly to Holland, where he received his M.D. In 1739, Linnaeus began to practice medicine, and two years later, he became the chair of botany, dietetics, and materia medica at the University of Uppsala. For the rest of his life, Linnaeus remained in this position, while his fame as a premier botanist spread throughout the world because of his influence in revising the 2,000-year-old system of classification.

The philosopher Aristotle had devised the first classification system over 2,000 years earlier, when he established the basic principles of dividing and subdividing plants and **animals**. At that time, only about a thousand species were known. Therefore, he grouped them into simple categories of animals with backbones and animals without backbones. Plants were divided into different categories that dealt more with size and appearance. By the sixteenth century, however, the system was proving to be less and less adequate as the body of knowledge of plants and animals grew. Modification came slowly, often marked with debate and controversy, succeeding only in revealing the complexity of the process. In 1753, Linnaeus published his *Species Planetarium*, in which he replaced the antiquated Aristotelian system with the principles of classification used today.

In creating his system, Linnaeus's primary consideration was the number of observable characteristics of the organism, specifically its **anatomy**, structures, and details of reproduction. Based on his observations, Linnaeus created a hierarchical system in which living things were grouped according to their similarities, with each succeeding level possessing a larger number of shared traits. He named these levels **class**, **order**, **genus**, and **species.**

Linnaeus also popularized binomial nomenclature, giving each living thing a Latin name consisting of its genus and species, which distinguished it from all other organisms. For example, the cougar received the scientific name *Felis concolor*, while the lion became *Panthera leo*. This system allowed scientists to communicate worldwide about organisms without having to understand different languages. Also, each type of organism can be fitted into the scheme in a logical and orderly manner, allowing for infinite expansion. The various hierarchical levels in the system provide as well a conceptual framework for understanding the relationships among different organisms or groups of organisms.

Linnaeus's desire to classify all living things often bordered on the compulsive; he believed his work to be divinely inspired and considered those who did not follow his system to be ''heretics.'' However, he was also a skilled and caring instructor who nurtured the interests of his many students, often sending them abroad to the Middle East, China, and the Pacific Islands for new specimens. In 1761, Linnaeus was given the noble title von Linné, and while the king of Spain

Carl Linnaeus. *(The Library of Congress. Reproduced by permission.)*

offered him generous compensation to settle in his country, Linnaeus remained in Sweden at Uppsala until his death after a stroke in 1778.

Today an international commission of scientists maintains the Linnaeus classification system and adheres to the rules for adopting scientific names when newly discovered species or subspecies need to be classified. Although the system depends on the judgments and opinions made by biologists, its concept and general organization are accepted by scientists throughout the world.

LIPIDS

Lipids belong to a broad class of animal- and plant-derived compounds which include **fats and oils**. They are insoluble in **water** because they are composed primarily of long chains of **hydrocarbon**s. Some lipids, such as fats, are solid or semisolid at room **temperature** while others, such as oils, are liquids. Examples of naturally derived lipids include butter, lard, tallow and fish oil from animals and cottonseed, peanut, soy, and corn oil from **plants**. Lipids are important dietary constituents and are present in many foods; in addition they are used in the production of many industrial products such as cosmetics, cleansers and lubricants.

Depending on their chemical composition, lipids are classified as hydrocarbons, simple lipids, or complex lipids. Hydrocarbons are the most basic form because they only contain carbon and hydrogen. Simple lipids consist of carbon, hydrogen and oxygen; complex lipids contain one or more additional elements, such as phosphorus, **nitrogen**, or sulfur. Examples of the latter include sterols and **phospholipids**.

The first scientific study of natural lipids is credited to the French chemist Michel Eugéne Chevreul. In 1811 Chevreul analyzed potassium soap made from pig fat and isolated a fatty material with acid properties. By working with soaps made from different animal fats, he found that a series of these fatty acids existed. He also found that the properties of these lipids depended on the structure of the hydrocarbon chain making up the fatty part of the **molecule**. By 1816 he had established that animal fats were composed not only of fatty acids but also of glycerol, an **alcohol** that was known to exist in olive oil and other vegetable and animal fats.

Scientists subsequently found that fatty acids can combine with glycerol to form triglycerides, one of the most common dietary fats. Such fats are classified as saturated or unsaturated depending on their chemical structure. The degree of saturation, also known as degree of hydrogenation, refers to the number of hydrogen **atoms** that are associated with the **chemical bond** between two carbon atoms. A carbon - carbon double is said to be unsaturated because the bond can react with additional hydrogen atoms. Lipids are unsaturated if they contain one or more of these double bonds. They are said to be saturated when all the double bonds have been fully hydrogenated, in other words when they have been ''filled up'' with hydrogen atoms.

Lipids aid in a number of biochemical functions in both plants and animals. For example, fatty lipid deposits below the skin help insulate **mammals** from cold temperatures. Various hydrocarbons and waxy lipids in animal skin, insect exoskeleton cuticle, and the waxy layers that coat plant leaves and fruit help prevent moisture loss. Complex lipids participate in hormonal responses in mammals, and certain fatty acid derivatives serve as sex attractants and growth regulators in insects. Furthermore, **cholesterol**, a complex lipid, is the basis for **sex hormones** of higher animals.

In humans, the liver can synthesize necessary lipids from other lipids that are ingested, or they can be made from **carbohydrate**s and **amino acid**s. Even with a fat free diet the body can still produce at least some of the lipids it requires to function. Still, some proportion of a healthy diet should include lipids, usually in the form of triglycerides. Essential dietary lipids include certain polyunsaturated fatty acids as well as the **vitamins** A, D, E, and K. However, diets high in fats are known to cause serious health problems such as **arteriosclerosis**, **heart** disease and **cancer**.

Industrially, lipids are processed from animal and plant sources on a large scale. They are isolated and purified by a number of techniques and may be bleached to remove undesirable color or odor. The purified lipids may then be reacted with other **organic compound**s to create materials with the desired functionality. Ultimately, these materials are used to make soaps, detergents, lubricants, thickeners, and a host of other raw materials which are subsequently used in cosmetics, foods, pharmaceuticals, and agricultural products.

New research on fats and oils has continued into the 1990s. A 1997 report details ttempts by genetic engineers to develop **enzyme**s which will allow plants to convert saturated fats and oils to healthier unsaturated materials. In addition to **nutrition**al applications, advances have been made on oils for other applications. Because they are environmentally friendly, renewable resources, an increasing number of plant oils are being evaluated for use as industrial lubricants. Two examples which are anticipated to have a significant industrial impact are the well known soybean oil and the lesser known meadowfoam oil.

LIPMANN, FRITZ ALBERT (1899-1986)
German American biochemist

Born in Königsberg, Germany, Fritz Lipmann earned his medical degree at the University of Berlin in 1922 and, five years later, received his Ph.D. there as well. For the next several years, Lipmann conducted research at Otto Meyerhof's laboratory in Heidelberg, Germany, and taught at the Kaiser Wilhelm Institute in Berlin. In 1932, when the Nazi movement in Germany made life increasingly uncomfortable, he accepted a position with the Carlsberg Foundation in Copenhagen, Denmark. In 1939, Lipmann immigrated to the United States, settling first at the Cornell Medical School and then, two years later, moving on to Harvard (1941-49) and then the staff of Massachusetts General Hospital (1949-57). In 1957, he became Professor of Biochemistry at the Rockefeller Institute in New York.

For a long time, most of Lipmann's research at these institutions centered around **carbohydrate metabolism**, especially the role played by phosphates. In 1937, Lipmann had discovered, almost by accident, that phosphates were somehow important to the metabolic process. He was not certain exactly what role they played, but believed it had something to do with the delivery of **energy** to the body's **cell**s. In 1941, he finally came up with some of the answers he had been seeking. He found a molecule that released low-energy phosphate—*adenosine monophosphate*—and discovered that, during the course of carbohydrate metabolism, the molecule picked up two energy-rich phosphate bonds, and became **adenosine triphosphate (ATP)**, a high-energy configuration that was able to release small traces of energy, when needed, to cells throughout the body.

In 1947, Lipmann made an even more important discovery. Working with pigeon liver extracts, he found a catalytically-active, heat-stable compound that appeared able to control the transfer of acetyl groups from one molecule to another. After isolating the compound and determining its structure (it was composed largely of pantothenic acid, or vitamin B2) he named it coenzyme A (CoA), with the ''A'' standing for acetylation. Lipmann speculated that CoA probably played an important role in the **Krebs cycle**, a complicated cycle of fatty acid, carbohydrate and **protein** oxidation.

Hans Krebs had already shown that **lactic acid** was broken down to carbon dioxide by way of a two-carbon compound that was part of his cycle. Lipmann believed that Krebs' two-carbon compound needed the help of CoA in order to enter the cycle and, by 1951, proved this to be the case. The two-carbon compound in the Krebs' cycle combined with CoA to form acetylcoenzyme A, a kind of super coenzyme that served as the hub of numerous biochemical reactions. For instance, in 1950, the coenzyme was found by Feodor Lynen (1911-1979) to play a key role in the metabolism of fats.

For his work on coenzyme A, Lipmann received the 1953 Nobel Prize in physiology or medicine, sharing the prize with Krebs. He also received several other honors including membership in the Faraday Society, the Danish Royal Academy of Sciences and the Royal Society of England.

LISTER, JOSEPH (1827-1912)

English surgeon

Identifying antiseptics and disinfectants has been challenging part of medicine for centuries. Joseph Lister, an English surgeon, contributed to a fundamental revolution in surgery with the introduction of his antiseptic method. At the time Lister was practicing medicine, the mortality rate for certain injuries and surgeries was extremely high due to infection. The mortality rate dropped drastically with the use of an antiseptic method, and when used in conjunction with the anesthetics that were available at the time, surgeons dared to perform more complicated surgical procedures.

Lister was born to a well known Quaker family at Upton, England. Lister studied medicine at University College, and received his medical degree in 1852. As a student, Lister had the opportunity to be a spectator at the first surgery performed with general anesthesia, performed by Robert Liston (1794-1847). He also studied **histology** under William Sharpey during which time, Lister wrote an important paper on **inflammation** where he discussed the susceptibility to disease of inflamed tissue. Lister was also interested in microscopic anatomy and physiology—perhaps because his father, Joseph Jackson Lister, was a microscopist. At one point, Lister wanted to become a surgeon and left England to study at Edinburgh University with James Syme (1799-1870), who was well known for his success with performing amputations and joint excisions. Syme, the first surgeon to adopt **antisepsis** and anesthesia, eventually became Lister's father-in-law.

As a surgeon, Lister was concerned with the high mortality rate of post-amputation patients and the high rate of gangrene after surgery. Applying the knowledge that **bacteria** caused disease, and drawing from **Louis Pasteur**'s work that proved the existence of airborne **microorganisms**, Lister concluded that airborne bacteria could cause infection in surgical wounds. Lister read about the affect of carbolic acid used on sewage bacteria in outhouses, cesspools, and stables in the nearby town of Carlisle, and developed an antiseptic system whereby he would spray carbolic acid in the operating room, and use it to sterilize the surgical instruments and his hands. In addition, he applied the acid in and around the wound and directly on the dressings. Lister first used this method in 1865 while treating a compound fracture of a leg, an injury that often claimed about 60% of patients, and where amputation of

Joseph Lister. *(Archive Photos, Inc. Reproduced by permission.)*

a limb was usually the only treatment. The procedure was successful. Lister published his antiseptic method in *The Lancet*, the highly respected English medical journal, in 1867. There was one problem: carbolic acid, especially the spray, was harmful to those who came in contact with it. However, Lister found milder antiseptics and also heat-sterilized the surgical instruments. At first, the medical community did not support Lister's theory, but eventually his antiseptic method gained recognition and was adopted as standard procedure for treating wounds and during surgery. Medics used Lister's antiseptic method, which proved to be very effective, during the Franco-Prussian War (1870-1871). In 1877, Lister became Professor of Surgery at King's College, London.

Lister received many honors and awards. A dedicated surgeon, he treated both inflicted and surgical wounds; he experimented with various antiseptics, developed absorbable sutures, and introduced a method of draining wounds. He was the first British surgeon to be elevated to the peerage (became a member of the House of Lords), and upon his death in 1912, his remains were interred in Westminster Abbey. When he died, it was said that Lister had saved more lives than all the wars in history had claimed.

LOEWI, OTTO (1873-1961)

German American pharmacologist and physiologist

Otto Loewi (pronounced *lō*–ee) was born in Frankfurt am Main, Germany, on June 3, 1873. On the advice of his parents, Loewi entered the University of Strasbourg to study medicine, and received his medical degree in 1896. After graduation, he briefly visited Italy, and then returned to Germany for more training in chemistry and experimental methods. During this period, he also worked in the tuberculosis and pneumonia wards at the City Hospital of Frankfurt, where he was discouraged from continuing with clinical medicine because of high death rates. Instead, he turned his attention to an academic career in scientific research, and in 1898 he joined the department of pharmacology at the University of Marburg, first with an assistantship and then as a lecturer.

By 1902, Loewi had published the results of his scientific research at Marburg. His work dealt with the functioning of the kidneys and the effects on these organs of substances that increase the production of urine, known as diuretics. In 1903, along with other researchers including **Henry Hallett Dale**, Loewi began to consider the chemical transmission of nerve impulses. The **hormone adrenaline** and the chemical muscarine had already been identified as possible nerve transmitters by several English physiologists. In 1905, Loewi followed Hans Meyer, under whom he had worked at Marburg, to the University of Vienna. The same year, he met Gulda Goldschmiedt, the daughter of a chemistry professor, in Switzerland, and he married her the following year. They would have four children.

At the University of Vienna, Loewi concentrated on the effects of adrenaline and noradrenaline on diabetes and blood pressure. He also studied the response of the **heart** to the stimulation of the vagus nerve, one of the main cranial nerves in the autonomic system. In 1909, he was appointed to the University of Graz as a professor of pharmacology, where he remained until the German occupation of Austria in 1938.

By 1921, fifteen years after the idea of chemical transmission of nerve impulses had first been proposed by the English physiologists, scientists had still not discovered definite evidence of the existence of a chemical transmitter within the nervous system. One night Loewi had a dream that would help; he dreamt the design of an experiment that would determine the existence of a chemical transmitter. He jotted down some notes from the dream, still half asleep, but when he awoke the next morning he could not read his scrawl. The next night at three o'clock the idea returned to him; this time he immediately went to his laboratory.

For this pathbreaking experiment, Loewi used two hearts from frogs. He removed the vagus nerve from the first heart, and he stimulated the same nerve in the second one. After stimulating the nerve in the second heart, he removed some fluid and injected it into the heart without the vagus nerve. He observed that the rate of this heart slowed as if the vagus nerve had been stimulated. Then he stimulated the heart with the vagus nerve so it would beat faster. He again removed fluid and injected it into the heart without the vagus nerve. Its rate increased as if it had been stimulated directly by the missing nerve.

Loewi had established the role chemicals play in the transmission of nerve impulses, but he was not sure at first what these chemicals were. He called one "vagus substance" and the other "accelerator substance." Over the next fifteen years, Loewi, along with his colleagues, published a number of papers on the results of his initial experiment. What he had called vagus substance was identified as acetylcholine in 1926; other transmitters were later identified. In 1936, Loewi identified adrenaline as one of the sympathetic nervous system transmitters and noradrenaline as the most important one. Henry Hallett Dale shared the 1936 Nobel Prize with Loewi for his discovery of chemical transmitters in the voluntary nervous system.

After the German occupation of Austria in 1938, Loewi was only allowed to leave because he turned over his Nobel Prize money to the Nazis. His family was also able to escape, and they joined him in New York City in 1940. He became a United States citizen in 1946. He spent the rest of his life writing articles, delivering lectures, and writing his memoirs. He died on December 25, 1961, at the age of 88 in New York City.

LORENZ, KONRAD (1903-1989)

Austrian zoologist and ethologist

Konrad Lorenz was born on November 7, 1903, in Vienna, Austria, as the younger of two sons born to Adolf Lorenz and his wife and assistant, Emma Lecher. Lorenz's love of **animal**s began outside of school, primarily at the family's summer home in Altenberg, Austria, and his interests became more grounded in science when he read about **Charles Darwin**'s evolutionary theory at the age of 10. In 1922, Lorenz began premedical training at Columbia University in New York but returned early to Austria to continue the program at the University of Vienna. Despite his medical studies, Lorenz found time to informally study animals. He also kept a detailed diary of the activities of his pet bird Jock, a jackdaw. In 1927, his career as an animal behaviorist was launched when an ornithological journal printed his jackdaw diary. During the following year, he received an M.D. degree from the University of Vienna and became an assistant to a professor at the anatomical institute there. His interests led him to study zoology at the University of Vienna, and in 1933, Lorenz earned his Ph.D. in that field.

Lorenz developed the theories for which he is best known during the years 1935 to 1938. He spent what he called his "goose summers" at the Altenberg home, concentrating on the behavior of greylag geese and confirming many hypotheses that he had formed while observing his pet birds. While working with the geese, Lorenz developed the concept of **imprinting**. Imprinting occurs in many species, most noticeably in geese and ducks, when—within a short, genetically set time frame—an animal will accept a foster mother in the place of its biological mother, even if that foster mother is a different species.

In addition, he and **Nikolaas Tinbergen**, future Nobel Prize cowinner, developed the concept of the innate releasing mechanism. Lorenz found that animals have instinctive behavior patterns, or fixed-action patterns, that remain dormant until a specific event triggers the animal to exhibit this behavior for the first time. The fixed-action pattern is a specific, ordered series of behaviors, such as the fighting and surrender postures used by many animals. He emphasized that these fixed-action patterns are not learned but are genetically programmed. The stimulus is called the "releaser," and the **nervous system** structure that responds to the stimulus and prompts the instinctive behavior is the innate releasing mechanism. Lorenz later devised a hydraulic model to explain an animal's motivation to perform fixed-action patterns.

While the research continued, Lorenz accepted an appointment in 1937 as lecturer in **comparative anatomy** and animal psychology at the University of Vienna. In 1940, he became professor of psychology at the University of Konigsberg in Germany but a year later answered the call to serve in the German Army. In 1944, Lorenz was captured by the Russians and sent to a prison camp. It was not until 1948 that he was released. Upon his return, Lorenz went back to the University of Vienna before accepting a small stipend from the Max Planck Society for the Advancement of Science to resume his studies at Altenberg. By 1952, Lorenz had published a popular book *King Solomon's Ring,* an account of animal behavior presented in easily understood terminology. Included in the book are many of his often-humorous experiences with his study subjects. The book also includes a collection of his illustrations.

In 1955, with the increased support of the Max Planck Society, Lorenz, ethologist Gustav Kramer, and physiologist Erich von Holst established and then codirected the Institute for Behavioral Physiology in Seewiesen, Bavaria, near Munich. During the ensuing years at Seewiesen, Lorenz again drew attention, this time for the analogies he drew between human and animal behavior—which many scientists felt were improper—and his continuing work on instinct. Following the deaths of codirectors von Holst and Kramer, Lorenz became the sole director of the Seewiesen Institute in 1961.

In 1966, Lorenz again faced some controversy with his book *On Aggression,* an example of his shift from solely studying animal behavior to including human social behavior. In the book, Lorenz describes aggression as "the fighting instinct in beast and man which is directed against members of same species." He writes that this instinct aids the survival of both the individual and the species, in the latter case by giving the stronger males the better mating opportunities and territories. The book goes on to state that animals will use rank, territory, or evolved instinctual behavior patterns to avoid actual violence and fatalities. Lorenz says only humans purposefully kill each other—a fact that he attributes to the development of artificial weapons outpacing the human evolution of killing inhibitions.

In 1973, Lorenz, Tinbergen, and Karl Frisch, who studied bee communication, jointly accepted the Nobel Prize for their behavioral research. In the same year, Lorenz retired

Konrad Lorenz. *(AP/Wide World Photos, Inc. Reproduced by permission.)*

from his position as director of the Seewiesen institute. He then returned to Altenberg where he continued writing and began directing the department of animal sociology at the Austrian Academy of Science. In addition, the Max Planck Society for the Promotion of Science set up a research station for him at his ancestral home in Altenberg.

In 1927, the same year his career-launching diary was published, Lorenz married childhood friend Margarethe "Gretl" Gebhardt, a gynecologist. They had two daughters, Agnes and Dagmar, and a son, Thomas. Lorenz died February 27, 1989, of kidney failure at his home in Altenburg, Austria, at age 85.

LUDWIG, KARL FRIEDRICH WILHELM (1816-1895)

German physiologist

Karl Friedrich Wilhelm Ludwig was one of the greatest researchers and teachers in the history of physiology. Born in Witzenhausen, Germany, he was the son of a former cavalry officer in the Napoleonic wars who became a civilian official at Hanau. After finishing his schooling at the Hanau gymnasium in 1834, Ludwig studied medicine at the University of Marburg, where he was expelled for dueling and political activities. He later returned to the university and earned his medical degree in 1840 after writing a dissertation on renal secretion. He then progressed through a series of anatomy and physiology professorships combined with research in Marburg in 1846; Zurich in 1849; Vienna, at the Josephinum (1855); and finally Leipzig (1865), where he remained until his death thirty years later.

Ludwig had based his investigation and teaching of physiology on chemical and physical laws, explaining all physiological processes on the basis of measurable and experimentally demonstrable phenomena—not on some speculative "vital force." He invented a number of devices and methods to carry out his scientific approach to physiology, many of them related to his interest in circulation and respiration. In 1847 he devised the kymograph to prove that blood is moved by mechanical forces, not the invisible "vital force." Ludwig's kymograph used a mercury manometer tube and a revolving drum to graphically record blood pressure variations and other vital signs. With later modifications, the kymograph became a standard tool for recording results of experiments.

In 1859 Ludwig and a student designed a mercury pump that separated and measured quantities of gases—oxygen and carbon dioxide—in the blood. Ludwig's *stromuhr*, or stream gauge, of 1867 measured the flow of blood. Ludwig also discovered that he could keep organs alive outside an animal by pumping blood or a saline solution through the excised part, a process called perfusion.

In addition to his important inventions, Ludwig produced a long list of major physiological observations and findings, including discoveries about salivary secretions, the mechanism of cardiac activity, **respiration**, **blood**, and **blood circulation**. Ludwig used his appointment in 1864 to the newly created chair of physiology at Leipzig to create a model teaching center for physiology, which became a world center for physiological study. Ludwig's textbooks of 1852 and 1856 (the first modern one on physiology), his hundreds of students, and his numerous scientific contributions have made him perhaps the most influential physiologist of the second half of the nineteenth century.

LURIA, SALVADOR EDWARD (1912-1991)

Italian American biologist

Salvador Luria was born on August 13, 1912, in Turin, Italy. In 1929, he entered medical school there and soon developed a technical facility for culturing **cell**s; that is, he was able to grow cells in artificial media inside culture dishes. He received a medical degree in 1935 and then served in the Italian Army for three years. After this service, he studied at the Curie Institute in Paris, where he became curious about **bacteriophage**s, **virus**es that attack **bacteria**. He was especially interested in how x-ray radiation affects bacteriophages by causing mutations. Luria emigrated to the United States in 1940 and accepted an appointment to Columbia University. He met Max Delbruck shortly thereafter, and the latter invited Luria to carry on research at Vanderbilt University in Nashville, where Delbruck was teaching.

Together, Luria and Delbruck planned a series of experiments to study bacteria that had developed resistance to bacte-riophages. They wanted to see if the resistance was the result of some action by bacteriophages on normal cells, or if the bacteria had become resistant because of mutation. Luria's ideas about random mutation were inspired by an unlikely source—watching people play slot machines at a country club dance during Luria's tenure at Indiana University in Bloomington. He observed how the payoff from a slot machine varied from just a few coins to a big pile of coins, a rather rare occurrence. Luria likened the bacteria to slot machines. Sometimes there were small clusters and sometimes large clusters, probably the descendants of a mutation. Luria's idea led to the development of the fluctuation test, which, with Delbruck's mathematical analysis, demonstrated that the bacterial resistance was an adaptive response caused by spontaneous, random mutation.

Luria and Delbruck began to collaborate with Alfred Hershey, a biologist also conducting bacteriophage research. The three became members of the "Phage Group," an informal assembly of scientists who worked exclusively with seven strains of bacteriophage so that their experimental results could be compared. Luria's research contributed a great deal to the understanding of the structure of viruses. This work was recognized in 1969, when Luria, Delbruck, and Hershey shared the Nobel Prize in physiology or medicine.

Luria had married Zella Hurwitz, a psychology professor, in 1945. They had one son. Luria died at home in Lexington, Massachusetts, after suffering a heart attack in 1991.

LWOFF, ANDRÉ MICHEL (1902-1994)

French microbiologist

André Michel Lwoff shared the 1965 Nobel Prize in physiology or medicine with **François Jacob** and **Jaques Monod** "for their discoveries concerning genetic control of **enzyme** and **virus** synthesis," and contributing to "our knowledge of the fundamental processes in living matter which form the bases for such phenomena as **adaption**, reproduction, and **evolution**." Lwoff's fascination with microscopic life and his highly analytical mind helped identify disease production in organisms, including its genetic and metabolic mechanisms of action.

Lwoff was born in Ainay-le-Château, a tiny town in Central France, to Russian immigrants of Jewish faith—Marie Siminovitch, a sculptor, and Salomon Lwoff, a psychiatrist and chief physician at a psychiatric hospital. He often accompanied his father to different hospitals and, on one such occasion, met his father's friend, Elie Metchnikoff, who allowed the young lad to peer at a typhoid bacillus under a microscope.

As Lwoff grew older, his interest in biology grew. His father, however, wanted him to enter the more lucrative profession of medicine. At the age of seventeen, he enrolled at the Sorbonne—the University of Paris—to study medicine and biology. He spent his summers at the Marine Biology Laboratory in Roscoff, Brittany studying eye pigment in small salt or freshwater parasites. In 1921, at the age of nineteen, he began working at the Pasteur Institute in Paris. In 1925, he married microbiologist Marguérite Bourdaleix, a co-worker at the Institute. The couple collaborated extensively in their professions over many years.

Lwoff received his M.D. in 1927 and D.Sc. in 1932 from the University of Paris. Receiving a Rockefeller Foundation fellowship, he studied with Otto Meyerhof in Heidelberg. He was appointed head of a laboratory at the Pasteur Institute in 1929 and became head of Microbial Physiology 1938. In 1936, a second Rockefeller grant took him to Cambridge, England where he and his wife identified growth factor X and its inhibiting properties in the growth of the *Haemophilus influenzae* bacterium. He became professor of microbiology at the Science Faculty in Paris in 1959.

Lwoff's important research into **microorganisms** include the **morphology** (formation of **tissue** and **organ**s) and nutritional needs of ciliates—single-celled animals covered with hair-like **cilia**; defining and understanding the role of growth factors, without which organisms cannot synthesize, grow, or multiply; establishing a microorganism classification system based on their energy sources and method of synthesis; and correctly classifying a bacterium **genus** ultimately named *Moraxella lwoffii* in his honor.

Almost a century earlier, in 1866, one **Gregor Mendel** became the first to suggest a law of inheritance in organisms, ultimately known as **genetics**. In the 1940s, the era of discovery of **deoxyribonucleic acid** (DNA) and the identification of the life cycle of **bacteriophage**s (virus particles which infect bacterial cells), Lwoff turned his attention to the genetic properties of viruses and bacteria. It was already understood that, when virus particles entered a bacterial cell, they first remained dormant but sometimes would multiply and caused lysis (cell death). When the bacteria cell died, the bacteriophages—or lysogenic bacteria—escaped. In 1950, Jacob and Monod encouraged Lwoff to investigate this phenomenon. He placed one lysogenic bacterium in culture and watched it divide through nineteen generations of "daughter cells." Each generation was lysogenic. His amazing observation proved that lysogeny (cell death) is a genetic phenomenon.

Lwoff also discovered a difference between lysogenic (bacteriophage) and noninfectious phage particles (which he named prophages). He found that ultraviolet light would cause the harmless prophages to multiply, become destructive, and destroy cells. He, Jacob, and Monod subsequently discovered how the prophage attached to the chromosome of the infected bacterial cell—where the genes normally reside—and acted like a bacterial gene. The prophage particle cannot multiply under normal conditions; however, the trio determined that ultraviolet light and other external factors change the cell's environment, stimulating phage duplication which ultimately destroys the bacterial cell. Based on their observations, they began to wonder if viruses may cause cancer. They accurately conjectured that dormant cancer-causing properties live in the protein covering of the virus, and that certain external factors stimulate those carcinogenic qualities into life, just as prophages become lysogenic when stimulated by ultraviolet light.

In 1968, Lwoff left the Pasteur Institute to assume directorship of the Cancer Research Institute at Villejuif, a town close to Paris. He was president of the International Association of Microbiological Societies, a member of scientific academies and societies in several countries, received honors from numerous organizations, and honorary degrees from prestigious universities such as Oxford and Harvard.

LYELL, CHARLES (1797-1875)
Scottish geologist

Charles Lyell was a scientist whose ideas were extremely important in the development of theories of geological and evolutionary change. His most influential textbook was *Principles of Geology*, published in 1833. Other well-known books written by Lyell are *Travels in North America, with Geological Observations* and *The Antiquity of Man*. Because of his great influence on the development of the principles of his discipline, Charles Lyell is sometimes referred to as a father of modern geology (along with another Scot, **James Hutton**).

Lyell's most important theory, which built upon earlier work of Hutton, was the so-called theory of uniformitarianism, which largely refuted the previously widely believed doctrine of **catastrophism**.

According to the theory of uniformitarianism, major geological forces that are observed today also occurred in the past, and likely throughout the history of Earth. Examples of such forces include volcanism, earthquakes, and erosion by wind, water, and gravity. Moreover, the theory states that these "existing causes" have been responsible for major changes that have occurred in the structure of the Earth during its geological history. A central element of the interpretation of the theory of uniformitarianism is the importance of time (over extremely long periods of time, even relatively slow-acting forces like erosion by wind and water can have an enormous influence on the character of Earth's surface.

Catastrophism is an earlier, very different doctrine that was widely believed by many scientists, but was largely been replaced by Lyell's (this kind of rapid change in scientific understanding is sometimes referred to as a "revolution"). According to catastrophism, major changes in Earth's structure, as implied by rapid changes in geological stratification and in fossil assemblages, were caused by sudden, violent, cataclysmic events, rather than by gradual, evolutionary and environmental changes. Before the influence of Lyell, most scientists and other people (at least, those of European culture) believed that Earth and its species only had a history of about 6,000 years, based on a literal interpretation of the Book of Genesis in the Bible. Moreover, according to that doctrine Earth and all life had been formed by God over a period of only six days. However, the observations of Lyell and other geologists of his time were reporting clearly contradictory evidence about the forces influencing the geological character of Earth and the evolution of its existing species. These observations of the real, natural world suggested that life was much more ancient than only a few thousand years, and that existing species appeared to have evolved from previous ones, which are now extinct. These ideas of Lyell and his colleagues were extremely influential on other scientists, including **Charles Darwin**, who is best known for his theory of the role of **natural selection** in driving evolutionary change, published in 1859 in his famous book, *On the Origin of Species.*

Another important concept championed by Lyell, also building upon previous work by James Hutton, was that older rocks were generally buried beneath younger ones. As such, careful excavation and studying of geological layers and the fossils they contain could be used to understand the geological and evolutionary history of Earth.

LYMPHATIC SYSTEM

The lymphatic system is complex network of thin vessels, capillaries, valves, ducts, nodes and **organ**s that runs throughout the body, helping protect and maintain the internal fluids system of the entire body by producing, filtering and conveying lymph and by producing various **blood cell**s.

The three main purposes of the lymphatic system are to drain fluid back into the bloodstream from the **tissue**s, filter lymph and fight infections.

This system, which includes the spleen, the thymus, lymph nodes and lymph ducts, is a major component of the **immune system**.

The lymphatic system branches through all parts of the body carrying lymph, a milky fluid made of chyle (intestinal fluid produced after **digestion**, containing **fats** and **protein**s). Lymph contains red blood cells and many white blood cells, especially **lymphocytes**. Lymphocytes are the cells that attack **bacteria** in the blood.

Lymph is produced by small bean-shaped "lymph nodes," soft nodules that are not usually externally visible or easily felt. They are located in clusters in various parts of the body, such as the neck, armpit and groin.

Lymph seeps outside the blood vessels in spaces in body tissues and is stored in the lymphatic system to flow back into the bloodstream.

Lymph nodes help prevent infection by filtering out foreign material, such as bacteria and **cancer** cells, and destroying **toxins** and germs. They also make lymphocytes that produce antibodies and attack cells infected with **virus**. Antibodies are proteins that detect invading foreign cells and help to kill them. This process is sometimes called the immune response.

When bacteria are detected in the lymph fluid, the lymph nodes swell as they produce and supply additional white blood cells to help fight off whatever pathogen has entered the body.

At this stage, they may be felt by infection sufferers. An extreme example is that of **Bubonic Plague** victims, who could be identified by the large bumps on their body called "buboes." These were swollen lymph nodes filled with puss.

There are two main types of lymphocytes: T lymphocytes and B lymphocytes. The B lymphocytes make antibodies. One kind of T lymphocyte assists in the production of antibodies, and the other fights infection directly. These killer T cells are being studied for their potential to fight cancer.

The lymphatic system runs parallel to the bloodstream. As the blood circulates, fluid seeps into the body tissues. This fluid carries **nutrition** to the cells and wastes back to the bloodstream.

The lymphatic system also returns **water**, fats, proteins and other substances to the blood.

The spleen filters the lymph to remove old red blood cells. These are destroyed and replaced by new red blood cells made in the bone marrow.

Think of the lymphatic system as a slow-flowing drainage network that collects interstitial fluid throughout the body, filters it, destroys foreign matter and germs, and returns it to the bloodstream.

LYMPHOCYTES

There are three different types of **cell**s in the **blood**, the red blood cells, white blood cells, and the platelets. All these cells are produced in the bone marrow. It is the white blood cells or the leukocytes that are involved in the defense systems of the body. There are five types of leukocytes. These are grouped into two distinct classes based on their **morphology**. The "polymorphonuclear granulocytes" have nuclei and specific granules inside the cells. The neutrophils, eosinophils, and basophils comprise this class of leukocytes. The cells that lack nuclei and have no cell-specific granules fall into the second category of leukocytes, known as "mononuclear agranulocytes". The monocytes and lymphocytes belong to this class.

The lymphocytes are the most abundant of the white blood cells. They are also the main functional cells of the **immune system**. There are two main types of lymphocytes, the T-lymphocytes, and the B-lymphocytes. The B-lymphocytes are produced in the bone marrow and they mature over there as well. During the process of development in the bone marrow, the B-cells develop unique **protein** structures (receptors) on the surface of their **membrane**. The precursor to the T lymphocyte leaves the bone marrow and migrates to the thymus. That is where they develop and differentiate. Hence the name, "T cell" for thymus-derived cell. Like the B cells, during the process of developing and differentiating in the thymus, the T-lymphocytes acquire protein structures on their surface called receptors. These receptors are capable of recognizing different structures present on the surface of pathogens. If a T lymphocyte is created that has a receptor capable of reacting with structures present on normal healthy cells of the body, then the immune system purges itself of such lymphocytes, by a process called clonal deletion. This process prevents the lymphocytes from attacking the body's own **tissue**s and causing **autoimmune disease**.

There are two main types of T cells, the T-4 lymphocytes (bearing CD4 as their accessory **molecule**) and the T-8 lymphocyte (having CD8 as its accessory molecule). The T lymphocytes play a role in **cell-mediated immunity** and in immunoregulation. The B-lymphocytes produce specific antibodies when activated, hence they are known as the **antibody** producing cells.

The protein structures or receptors present on the surface of both the T and the B lymphocytes enable them to recognize specific **antigen**s. The receptors have grooves that can fit into complementary grooves existing on the surface of the antigen molecules. It is like a lock fitting a key. When the lymphocyte receptor finds its complementary structure on the surface of an antigen and binds to it, the whole immunological cascade is initiated.

The T-4 lymphocytes release specific chemicals called cytokines that induce the production of millions of cells bearing the same receptor. This process known as clonal proliferation guarantees that there are adequate cells to mount a successful immune response. Some of the proliferated cells differentiate into CD8+ cytotoxic T lymphocytes (CTLs) that can kill the pathogen-infected cells, while others activate the B lymphocytes. The stimulated B lymphocytes start multiplying and produce clones of cells with identical B-cell receptors. The rapidly proliferating B cells start producing a soluble version of the receptor. They, later, differentiate into plasma cells capable of secreting these soluble receptors, which are called antibodies.

Some of the T cells and the B lymphocytes become circulating memory cells. These cells are capable of remembering the original antigen. If that antigen enters the body again, while the memory cells are present, the body can mount a very rapid and an intense immune response against the pathogen. This is the reason why the body develops a permanent immunity after an infectious disease and is the principle behind **immunization**. Some other cells known as the T8 suppressor cells play a role in regulating the immune response. That is to say, they ensure that the immune response is turned on only in response to an antigen and is turned off once the antigen has been removed.

LYNEN, FEODOR (1911-1979)
German biochemist

Feodor Felix Konrad Lynen was born in Munich, Germany, on April 6, 1911, the seventh of eight children, to Wilhelm and Frieda (Prym) Lynen. Lynen showed an early interest in his older brother's chemistry an eventually, enrolled in the Department of Chemistry at the University of Munich in 1930. There he studied with German chemist and Nobel laureate **Heinrich Wieland**, who was Lynen's principal teacher both as an undergraduate and graduate student. On February 12, 1937, Lynen received his doctorate degree. Three months later, on May 17, he married Wieland's daughter, Eva, with whom he would have five children: Peter, Annemarie, Susanne, Eva-Marie, and Heinrich.

Upon his graduation, Lynen stayed at the University of Munich in a postdoctoral research position. In 1942, he was appointed a lecturer, and eventually was made a full professor in 1953. A year later, he was named director of the newly established Max Planck Institute for Cell Chemistry. Throughout his years with the University, where he stayed until his death, Lynen supervised the research of nearly ninety students, many of whom reached leading positions in academia or industry.

In the first years after World War II, German scientists were spurned by their European and American colleagues.

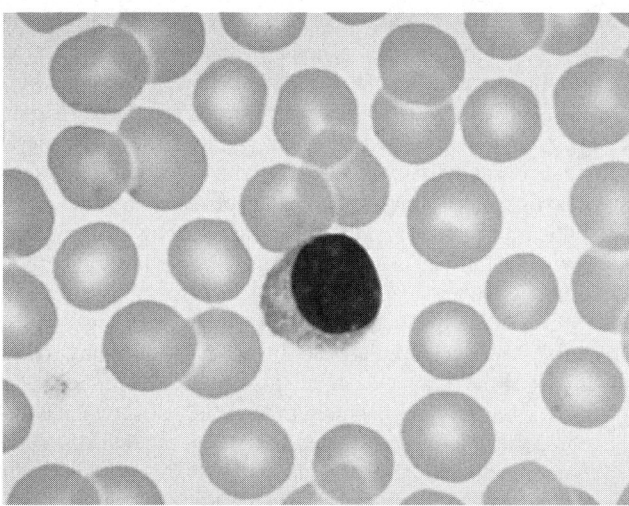

A lymphocyte cell. *(Photograph by Lester V. Bergman, Corbis Images. Reproduced by permission.)*

Only four German biochemists were invited to attend the First International Congress of Biochemistry held in Cambridge, England, in July of 1949. Lynen, one of the four, made an ideal good-will ambassador for Germany because of his good sense of humor and the fondness he had for parties. His cheery nature and solid research drew many foreign scientists to Munich. His magnetic personality was formally recognized years later when, in 1975, he was chosen to serve as president of the Alexander von Humboldt Foundation, an institution devoted to fostering relations between Germany and the international scientific community.

During the 1940s, Lynen began studying how the living **cell** changes simple chemical compounds into sterols and **lipids**, complex **molecule**s that the body needs to sustain life. The long sequence of steps and the roles various **enzyme**s and **vitamins** played in this complicated metabolic process were not well understood. After the war, Lynen began to publish his early findings. At the same time, he became aware of similar work being conducted in the United States by Bloch. Eventually, Lynen and Bloch began to correspond, sharing their preliminary discoveries with each other. By working in this manner, the scientists determined the sequence of thirty-six steps by which animal cells produce **cholesterol**.

One of the breakthroughs in the cholesterol synthesis work came in 1951 when Lynen published a paper describing the first step in the chain of reactions that resulted in the production of cholesterol. He had discovered that a compound known as acetyl-coenzyme A, which is formed when an acetate radical reacts with coenzyme A, was needed to begin the chemical chain reaction. For the first time, the chemical structure of acetyl-coenzyme A was described in accurate detail. By solving this complex biochemical problem, Lynen established his international reputation and created a new set of challenging biochemical problems. Determining the structure of acetyl-coenzyme A supplied Lynen with the discovery he needed to advance his research.

During his rehabilitation from a serious ski injury at the end of 1951, Lynen contemplated how the structure and action of acetyl-coenzyme A made it a likely participant in other biochemical processes. Upon his return to the lab, Lynen began investigating the role of acetyl-coenzyme A in the biosynthesis of fatty acids and discovered that, as with cholesterol, this substance was the necessary first step. Lynen also investigated the catabolism of fatty acids, the chemical reactions that produce energy when fatty acids in foods are burned up to form carbon dioxide and water.

In addition to elucidating the role of acetyl-coenzyme A, Lynen's research revealed the importance of many other chemicals in the body. One of the most significant of these was his work with the vitamin biotin. In the late 1950s, Lynen demonstrated that biotin was needed for the production of fat.

Lynen and Bloch shared the Nobel Prize in medicine or physiology in 1964, largely because the Nobel Committee recognized the medical importance of their work. Medical authorities knew that an accumulation of cholesterol in the walls of arteries and in blood contributed to diseases of the circulatory system, including arteriosclerosis, heart attacks, and strokes. In its tribute to Lynen and Bloch, the Nobel Committee noted that a more complete understanding of the metabolism of sterols and fatty acids promised to reveal the possible role of cholesterol in heart disease. Any future research into the link between cholesterol and heart disease, the Nobel committee observed, would have to be based on the findings of Lynen and Bloch.

In 1972, Lynen moved to the Max Planck Institute for Biochemistry, which had just recently been founded. Between 1974 and 1976, Lynen was acting director of the Institute. He continued to oversee a lab at the University of Munich, however-er.

At the end of his life, Lynen was a renowned scientist, and a proud Bavarian. The author of over three hundred scholarly pieces, Lynen was also praised as a hard-working man who expected much of himself and his students. Six weeks after an aneurism operation, Lynen died on August 6, 1979, at the age of 68.

LYSOSOME

Lysosomes are small membranous bags of digestive **enzyme**s found in the **cytoplasm** of all eukaryotic **cell**s (those with true nuclei). As the principle site of intracellular **digestion**, they contain a variety of enzymes capable of degrading **proteins**, **nucleic acids**, sugars, **lipids**, and most other ordinary cellular components. These enzymes hydrolyze (break down) their target compounds best under acidic conditions. Although lysosomes vary considerably in size even within a single cell, the normal range is usually 0.82-3.28 ft (0.25-1.0 m) or slightly smaller than the average mitochondrion.

The **membrane** enclosing lysosomes appears to be similar to that of other cellular **organelle**s, but it has several unique properties. First of all, hydrogen pumps in the membrane acidify the lysosomal interior to a pH of five, an optimal level for the activity of its internal enzymes. The membrane has docking sites on its exterior that allow both materials to be digested and the enzymes to carry out the job to be transferred into the lysosome from transport vesicles derived from the **Golgi apparatus**, the **endoplasmic reticulum**, or from **endocytosis** by the **plasma membrane**. The lysosomal membrane also has transport complexes that allow the final products of digestion such as **amino acid**s, simple sugars, salts, and nucleic acids to be exported back into the cytoplasm, where they can be either excreted or recycled by the cell into new cellular components. Finally, by mechanisms that are not yet fully understood, the lysosomal membrane is able to avoid digestion by the enzymes it contains even though it is composed of the same compounds that those enzymes routinely destroy.

See also Endomembrane system; Eukaryote

M

MACLEOD, JOHN JAMES RICKARD
(1876-1935)
Scottish physiologist

John James Rickard Macleod was born in Cluny, near Dunkeld, Scotland, on September 6, 1876, the son of the Reverend Robert Macleod. Soon after his birth, the family moved to Aberdeen. Macleod attended Aberdeen Grammar School and Aberdeen University. He went on to study medicine at Marischal College, where he graduated with honors in 1898. With an Anderson traveling scholarship, Macleod continued his education at Leipzig's Physiology Institute where he studied biochemistry for a year. In 1900 he returned to London to become a demonstrator in physiology at the London Hospital Medical College, and the following year became a biochemistry lecturer there. The same year he was named a Mackinnon research scholar by the Royal Society. During this period, Macleod published his experiments on intracranial circulation and caisson's disease. In 1902 he attended Cambridge University and obtained a diploma in public health. He married Mary Watson in 1903, and journeyed to America to become professor of physiology at Western Reserve University in Cleveland, where he stayed for 15 years. The same year he arrived at Western Reserve, his text, *Practical Physiology,* was published.

Macleod began his investigations into the human body's **carbohydrate** metabolism during his early years at Western Reserve, studying salt and urea metabolism. Studies of the breakdown of liver glycogen followed, and in 1913, he published *Diabetes: Its Physiological Pathology.* In 1918, Macleod went to the University of Toronto to become a professor of physiology and associate dean of the faculty of medicine. Macleod's major areas of interest at this time were the effects of oxygen excess and deprivation. He also studied **respiration** in **animal**s whose **brain**s had been removed or spines cut. During this period, he wrote *Physiology and Biochemistry in Modern Medicine.* The 1,000-page book went through seven editions and became a standard text in the field.

In 1921 Macleod returned to his work on carbohydrate metabolism, comparing the **blood**-sugar level in normal animals with that of animals with their pancreas removed. At the time it was known that diabetes was caused by the failure of the pancreas to secrete a substance that regulates sugar metabolism, causing an abnormally high concentration of glucose in the blood and an excretion of sugar into the urine. It was also understood that the unidentified substance sped the passage of sugar in the form of glucose through the body to be oxidized as a source of **energy** or converted the sugar into glycogen for storage for later use as glucose. Macleod appointed **Frederick Grant Banting**, a Canadian orthopedic surgeon, to specifically investigate the function of a cluster of **cell**s in the pancreas known as the islet of Langerhans. Macleod chose **Charles Herbert Best**, one of his senior medical students, to be Banting's chief laboratory assistant. Together the three men planned how they would separate the islet from the pancreas to isolate the substance secreted by the cell cluster. This substance, Macleod believed, was the sugar regulator they were seeking.

When Banting and Best tied the ducts of the pancreas so that it would atrophy, they were able to isolate a residue in the islet. They injected this extract into dogs and found that it did indeed lower blood glucose. The problem of how to obtain larger quantities of the extract, named **insulin**, remained. For help, Banting turned to James Bertram Collip, a young Canadian biochemist. Collip used pancreas glands bought from a butcher as a source of insulin. To demonstrate how safe the extract was, Banting and Best injected themselves with insulin. By January, 1922, they had begun clinical trials. A youngster named Leonard Thompson was the first diabetic to receive insulin injections. The results proved that the new treatment controlled the debilitating disease. Solving the final problem of making insulin therapy available to the general public, an American biochemist, John Jacob Abel, converted insulin into a crystalline form in 1926, so that it could be given in precise dosages. Today insulin is prepared from the pancreatic tissue of domestic animals.

The Nobel Prize committee acted with remarkable swiftness to recognize the achievement. Macleod and Banting were given the Nobel Prize in 1923, only a year after their discovery. Collip later noted that Macleod's outstanding position in the field of carbohydrate metabolism had made it appropriate and fortunate that the discovery of insulin had been made in his laboratory. Macleod, for his part, maintained that it was only through team work that insulin could be isolated. He shared his prize money with Collip, while Banting divided his with Best.

In 1928 Macleod returned to Scotland to become chairman of the physiology department at the University of Aberdeen. He continued his research into carbohydrate metabolism there and at the Rowett Institute, publishing numerous papers on insulin, experimental glycosuria—the presence of sugar in the urine—respiration and lactic acid metabolism. Arthritis forced him to discontinue his laboratory work, but he continued supervising the work of the physiology department. He died on March 16, 1935, at the age of 58.

MAGENDIE, FRANÇOIS (1783-1855)
French physiologist

François Magendie was born on October 6, 1783, the son of a surgeon who was known for his radical views and his active support of the French Revolution. The elder Magendie raised his two sons to be fiercely independent and to think for themselves—two traits that François never lost.

Apprenticed at sixteen to a surgeon friend of his father's, the young man began his formal medical studies a few years later, obtaining his medical degree in 1808 from the University of Paris. Although he was interested at first in **anatomy**, Magendie later switched to physiology and almost at once ran into trouble. At the time, most European scientists believed strongly in **vitalism**—the idea that biological processes were governed by "vital forces" that could not be explained in strictly scientific terms. Magendie disagreed. He was convinced that in the biologic sciences, just as in the physical sciences, facts were more important than theories—and even here, all facts had to be verified in the laboratory.

Magendie's strong opinions—and his desire to experiment in sometimes forbidden areas—won him numerous enemies and even, at times, the reputation of being a vivisectionist. Nevertheless, he established the idea of experimental physiology (an idea further popularized by his disciple, **Claude Bernard**) and made a number of important discoveries. Magendie, for instance, was interested in the action of various plant-derived drugs on the body. One of the drugs he studied was strychnine and, in 1809, he described in detail the effects of strychnine injections on animal subjects—and also proved that the poison reached the animal's **spinal cord** by the **blood**stream and not, as was then commonly believed, by the **lymphatic system**. Because of such experiments, Magendie was able to introduce into French medicine a variety of new drugs, including morphine, codeine, quinine and, of course, strychnine.

In 1815, post-revolutionary France was short of food, and Magendie was asked to serve as chairman of a special commission set up to investigate the nutritional value of various food extracts. Intrigued by the problem, Magendie continued his nutritional investigations long after the commission disbanded. Among other things, he found that mammals could not be kept alive by diets that lacked any **nitrogen**-containing foods (in other words, **protein**s). He found, as well, that not all these substances were equally life-sustaining—a few, like gelatin, had very little nutritional value. Magendie's findings pointed the way for later nutritional researchers, like **Frederick Gowland Hopkins**, **Thomas B. Osborne**, and **William Rose**, who were able to provide more definitive answers.

Magendie devoted much of his time to studies on the nervous system and, in 1822, published a classic paper in which he distinguished the separate motor and sensory roots of the spinal nerves. He showed that the ventral roots, those entering the spine in front of the dorsal roots, carry impulses to the muscles; and the dorsal roots, those entering the spine slightly behind the ventral roots, carry impulses to the **brain** from receptor **neuron**s.

Magendie also was the first person to describe cerebrospinal fluid and a foramen (natural hole) in the brain, now called the foramen of Magendie. He died on October 7, 1855.

MAGNETIC RESONANCE IMAGING (MRI)

Magnetic resonance imaging (MRI) is a medical technique which utilizes a magnetic field and the natural resonance of **atoms** to provide an image of human **tissue**. While the foundation for its development first took place in the late 1930s, it was not until the late 1960s that it was used by doctors to view the inner workings of the human body.

The development of MRI began in the early 1900s with discoveries made in nuclear magnetic resonance (NMR). During this time, scientists were just starting to develop theories about the structure of atoms and the nature of visible and ultraviolet **light**. NMR was discovered to be related to the magnetic properties of an atom's **nucleus**. It is a phenomenon in which atomic nuclei absorb and emit radio waves when placed in a large magnetic field. These properties were first demonstrated in 1924 by the Austrian physicist Wolfgang Pauli.

In 1938, the first instrument to utilize an atom's nuclear magnetic resonance for analysis was developed. This device was able to provide data related to the magnetic properties of certain substances. However, this crude instrument had two major drawbacks including its ability to only analyze gaseous materials and its inability to provide direct measurements. These limitations were overcome in 1945 when two groups led by two scientists Felix Bloch and Edward M. Purcell independently developed improved NMR devices. These new devices were more useful than the first NMR, providing researchers with the ability to collect data on many different types of systems. After some technological improvements scientists were able to use this technology to investigate biological tissues in the mid-1960s.

The application of NMR to medicine soon followed. Scientists discovered that different types of tissue gave differ-

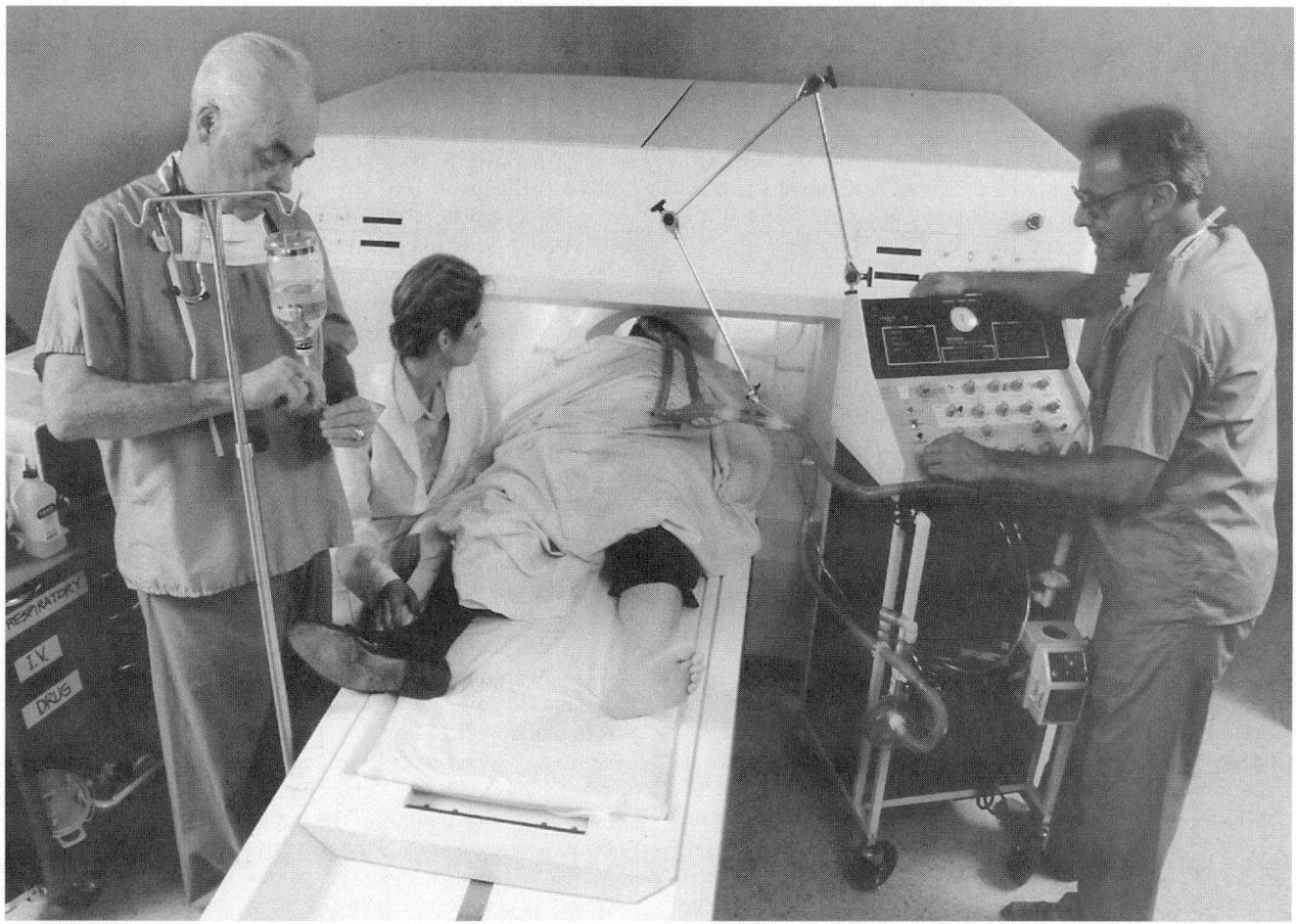

A patient undergoing a magnetic resonance imaging test. *(FPG International. Reproduced by permission.)*

ent magnetic signals. One of the first applications of NMR was using it to distinguish between normal and **cancer**ous tissue. This was done by Raymond Damadian in 1971. Later experiments showed that many different body tissues could be distinguished by NMR scans. In 1973, NMR data was integrated with computer calculations of tomography and the first magnetic resonance image (MRI) was produced. To test this new method, a device was designed to study a living mouse. While it took more than an hour to scan, an image of the mouse's internal **organ**s was obtained. Human imaging followed a few years later. The first commercial MRI scanners were sold around 1981. Since then, various technological improvements have been made which helped reduce the scanning time required and improve the resolution of the images. Most notable improvements have been made in the three-dimensional application of MRI.

Modern MRI devices are used frequently by doctors to produce images of the interior tissues of their patients. First the patient is put on a flat bed device which is surrounded by several coils which can produce a strong constant magnetic field. A known radio frequency (RF) signal is then applied to the system causing certain atoms within the patient to resonate.

When the RF signal is stopped, the atoms continue to resonate for a short time. Eventually, when the resonating atoms return to their natural state they emit their own RF signal which is received by detectors on the MRI. The signals are then processed through a computer and converted into a visual image of patient.

The signals that are emitted from the body are produced by protons within the body. In a typical MRI scan these signals come from hydrogen atoms in the body. The first magnetic resonance images were constructed solely on the concentration of protons within a given tissue. However, these images were fuzzy and did not have good resolution. When the relaxation time, which is the time it takes for the protons to emit their signals, was included in the calculations of the scan, MRI became much more useful for constructing an internal image of the body. In all body tissues, there are two types of relaxation times, T1 and T2, that can be detected. The different types of tissues exhibit different T1 and T2 values. For example the **brain** tissue has a different T1 and T2 value then **blood**. By using the three variables, proton density, T1 and T2 values, a clear image can be constructed.

MRI is now used for a variety of applications. It is used by far the most for creating images of the human brain. It is particularly useful for this purpose because the soft tissue emits a distinct signal making it easy to distinguish between healthy tissue and lesions. In addition to structural information, MRI also allows scientists to study brain function. This type of imaging is based on the fact that during brain activity the rate of blood flow changes. When the scans are taken with sufficient speed the blood can actually be seen moving through the organ. This has important consequences for studying the various parts of the brain. Another application for MRI is in sports medicine where it is used for muscular skeletal imaging. Using MRI, injuries to the ligaments and cartilage in the joints of the knees, wrists, and shoulder can be readily seen. This technique has eliminated the need for traditional invasive surgeries. A developing use for MRI is in tracking chemical components in the body. In these scans a person is injected with a compound containing **molecule**s such as Carbon 13 or Phosphorus 31. These atoms produce distinguishable signals so they can be easily tracked in the body.

MALPIGHI, MARCELLO (1628-1694)
Italian physiologist

In the second half of the seventeenth century, Marcello Malpighi used the newly invented **microscope** to make a number of important discoveries about living **tissue**s and structures, earning himself enduring recognition as a founder of scientific microscopy, **histology** (the study of tissues), **embryology**, and the science of **plant anatomy**.

Malpighi was born at Crevalcore, just outside Bologna, Italy, on March 10, 1628. The son of small landowners, Malpighi studied medicine and philosophy at the University of Bologna. While at Bologna, Malpighi was part of a small anatomical society headed by the teacher Bartolomeo Massari, in whose home the group met to conduct dissections and vivisections. Malpighi later married Massari's sister.

In 1655, Malpighi became a lecturer in logic at the University of Bologna; in 1656, he assumed the chair of theoretical medicine at the University of Pisa; in 1659, he returned to Bologna as lecturer in theoretical, then practical, medicine; from 1662 to 1666, he held the principal chair in medicine at the University of Messina; finally in 1666, he returned again to Bologna, where he remained for the rest of his teaching and research career. In 1691, at the age of sixty-three, Malpighi was called by his friend Pope Innocent XII to serve as the pontiff's personal physician. Reluctantly, Malpighi agreed and moved to Rome, where he died on November 29, 1694, in his room in the Quirinal Palace.

Early in his medical career, Malpighi became absorbed in using the microscope to study a wide range of living tissue—**animal**, insect, and plant. At the time, this was an entirely new field of scientific investigation. Malpighi soon made a profoundly important discovery. Microscopically examining a frog's lungs, he was able for the first time to describe the lung's structure accurately—thin air sacs surrounded by a network of tiny **blood** vessels. This explained how air (oxygen) is able to diffuse into the blood vessels, a key to understanding the process of **respiration**. It also provided the one missing piece of evidence to confirm **William Harvey**'s revolutionary theory of the **blood circulation**: Malpighi had discovered the capillaries, the microscopic connecting link between the **veins** and **arteries** that Harvey—with no microscope available—had only been able to postulate. Malpighi published his findings about the lungs in 1661.

Malpighi used the microscope to make an impressive number of other important observations, all "firsts." He observed a "host of red **atoms**" in the blood—the red blood corpuscles. He described the papillae of the tongue and skin—the receptors of the senses of taste and touch. He identified the rete mucosum, the Malpighian layer, of the skin. He found that the nerves and spinal column both consisted of bundles of fibers. He clearly described the structure of the kidney and suggested its function as a urine producer. He identified the spleen as an **organ**, not a gland; structures in both the kidney and spleen are named after him. He demonstrated that bile is secreted in the liver, not the gall bladder. In showing bile to be a uniform color, he disproved a 2,000-year-old idea that the bile was yellow and black. He described glandular adenopathy, a syndrome rediscovered by Thomas Hodgkin (1798-1866) and given that man's name 200 years later.

As if this catalog of human-tissue discovery weren't enough, Malpighi also conducted groundbreaking research in plant and insect microscopy. His extensive studies of the silkworm were the first full examination of insect structure. His detailed observations of chick embryos laid the foundation for microscopic embryology. His botanical investigations established the science of plant anatomy. The amazing variety of Malpighi's microscopic discoveries piqued the interest of countless other researchers and firmly established microscopy as a science.

MALTHUS, THOMAS ROBERT (1766-1834)
English economist

Thomas Robert Malthus was born in 1766 southwest of London in Surrey. Malthus had an enormous impact on scholars of biology, human **population**s and economics. He entered Jesus College at Cambridge in 1784 and graduated four years later at which time he was ordained in the Church of England. He served for a brief time as a curate in a parish not far from where he was born. He later became a Fellow of Jesus College and also a Fellow of the Royal Society. In 1805 he became a professor of history and political economy at the East India Company's college at Haileybury in Hertfordshire.

He is remembered not because of the holy orders that he took. Rather, his essays on population brought recognition to him worldwide. He was not optimistic with regard to the lot of working humans at the time. He believed that the means of subsistence would always be in short supply because of **human reproduction**. He thought that human reproduction would always outstrip the capacity of the land to produce food. Ultimately, it was the availability of food that would set population limits. Famine, disease, and infant mortality would take their toll because of inadequate food supplies. The views of Malthus were articulated at a time when it was hoped that the poor laws would encourage population growth and the increased population would insure national wealth. The dire views of Malthus were exactly opposite those of most other scholars at the time. He urged that a nation balance production with consumption. He suggested that public works would provide relief for the working poor. He published some of his ideas in his *Essay on the Principle of Population* in 1798. The essay was revised and enlarged for a total of six editions all of which carried his name except the first which was anonymous.

Charles Darwin acknowledged that he read Malthus. Darwin recognized that uncontrolled reproduction coupled with limited means of subsistence would create a struggle for existance with survival of only those best suited. Animals, as humans, vary in capabilities related to survival. The struggle, according to Darwin, selected those best fit for survival. Darwin had just returned from the voyage of the Beagle and he recognized that the Malthusian population concept applies to animals as well as humans. Thus, Malthus indirectly contributed to Darwin's theory of natural selection. Alfred Russell Wallace, whose views on natural selection were similar to Darwin, also acknowledged reading Malthus.

Malthus, who believed that public works were a means of minimizing economic distress of the poor, was said to have influenced the English economist John Maynard Keynes (1883-1946) who advocated governmental aid for full employment as a means of digging out of economic depression. And, certainly the administration of Franklin Delano Roosevelt (1882-1945) did in fact institute vast public projects to rescue the poor and to stimulate production.

Public concern for overpopulation and inadequate food supply is greater now than it was in the time of Malthus. Malthus did not provide easy answers for this problem. He was against contraception and against abortion. In lieu of disease and famine, he offered delayed marriage. The issues are more complicated now. The American Nobel Laureate Norman E. Borlaug (1914-), the father of the green revolution, has averted famine for much of the world, at least for the time being, by vastly increasing the food supply with modern agriculture. Here is an instance where the means of subsistence has increased at an incredible pace not foreseen by Malthus. Curiously, modern refrigerators have decreased food spoilage and waste. Hence, there has been a period when food supplies have been relatively abundant. Birth control by means of the pill, or with the condom, offer options not previously available for population control. Welfare laws are being changed in an effort to not encourage large families. Thomas Robert Malthus would be excited indeed to learn how human society will handle the problem of limited means of subsistence and the pressing burden of population growth over two centuries after his death.

MAMMALOGY

Mammalogy is the branch of zoology that deals with the study of mammals, which are any organisms in the class Mammalia.

Mammals are warm-blooded (or homoiothermic) vertebrates having mammary glands in the female, a four-chambered heart, a diaphragm, and hair or fur. Major subject areas in mammalogy include anatomy, physiology, behavior, ecology, evolution, and classification and systematics.

Mammals first evolved from an ancient group of reptiles known as the Therapsida, which were most abundant and diverse during the Triassic period (which began about 245 million years ago). The time of the evolution of first mammals is not known exactly, but probably occurred early in the Triassic. The first fossils of placental mammals are from the Jurassic period (about 210 million years ago), but mammals did not become prominent until the late Mesozoic (just before 290 million years ago). The Cenozoic era, during which numerous species of mammals evolved to fill ecological niches made vacant by the demise of the last of the great reptiles (including the dinosaurs), is sometimes known as the ''Age of Mammals.'' The Cenozoic era extends from 65 million years ago to the present time.

There are 18 orders of mammals, containing about 4,300 living species of extremely diverse biology in terms of size, shape, behavior, and other attributes. The most ''primitive'' (or evolutionarily ancient) order is the Monotremata (including only the platypus and several species of echidnas). These are the only egg-laying mammals. The order Marsupialia is characterized by embryos born in an early stage of development, followed by further growth in a specialized pouch called a marsupium. The Marsupialia includes about 260 species of kangaroos, wallabies, opossums, bandicoots, numbats, and marsupial ''mice'' and ''cats.'' All other species are placental mammals that bear their young in a much later stage of development. The placental mammals are divided into 16 orders, including the Insectivora (moles, shrews), Chiroptera (bats), Primates (lemurs, monkeys, apes, humans), Edentata (sloths, anteaters, armadillos), Lagomorpha (rabbits, hares, pikas), Rodentia (rats, mice, squirrels, gophers, beavers, porcupines), Cetacea (whales, dolphins, porpoises), Carnivora (cats, dogs, bears, raccoons, weasels, skunks, hyenas, seals), Proboscidea (elephants), Sirenia (dugongs, manatees), Perissodactyla (horses, zebras, tapirs, rhinoceroses), and Artiodactyla (pigs, hippopotamuses, camels, deer, giraffes, antelopes, cattle, sheep, goats, bison).

Mammalian anatomy changes remarkably during development from the relatively simple structures of the fertilized egg, to the much more complex adult form. Some important anatomical characteristics of adult mammals include a four-chambered heart, a diaphragm in the thoracic cavity, a high degree of separation of the breathing and alimentary (food) passages (this is done by the palate), a lower jaw composed of a single bone, teeth that are differentiated for various functions, a relatively large brain, and covering of the body with insulating hair or fur (some species, such as whales, have secondarily lost this characteristic).

Almost all mammals give birth to live young (exceptions are the platypus and echidnas). The marsupials are non-placental mammals, which means that their young are born at an extremely early stage of development, then crawl to the mother's abdominal pouch, where they continue their growth while suckling on a teat for nourishment. The placental mammals evolved after the marsupials. Placental mammals have a much longer gestation period, and their developing fetuses are nourished by **nutrients** and respiratory gases that pass through a specialized maternal **organ** known as a placenta (wastes also transfer across the placenta, from the fetus to the mother). After birth, all young mammals are nourished by a specialized, fat-rich substance known as milk, which is produced by the mother in a mammary gland). Depending on the species, the young may not be weaned (i.e., become independent of its mother's milk) for as long as several years.

A distinguishing element of mammalian physiology is homoiothermy, or metabolic activity that maintains the body **temperature** within a narrow, warm range that is optimized for muscular function and **enzyme** efficiency. (Birds and some other vertebrates are also warm-blooded.) Homoiothermy is a crucial physiological trait that allows most mammal species to have an extremely active lifestyle, and in some cases to live in cold environments.

Mammalian behavior is typically characterized by an great ability to change with experience. As they discover opportunities and dangers in their environment, mammals learn to seek or avoid these conditions. Other classes of **animal**s do this as well, but not to the same degree to which most mammals are capable.

Our own species, *Homo sapiens*, is mammalian, so technically the study of our anatomy, physiology, behavior, and evolution falls within the field of mammalogy. In fact, scientific knowledge of the human species far exceeds that of any other kind of mammal (or any species of any other group). This immense body of knowledge clearly shows many parallels between all aspects of the biology of *Homo sapiens* and that of other species of mammals (and other animals). In fact, a great deal of research on other species of mammals (such as rats, dogs, monkeys, and apes) is undertaken with the ultimate purpose of finding out more about the biology of humans. (This is because most kinds of experimental, invasive research is not considered to be ethically permissible with humans, but is allowed with other species.) Nevertheless, our species is clearly unique in many important respects, particularly with regards to intelligence, cultural evolution, and an extraordinary ability to adaptively manage social and ecological systems, often using advanced technologies. Humans are also unique in their ability to carelessly damage or even destroy those systems, even though each of us, and our species as a whole, ultimately depend on them.

MAMMALS

The more than 4,000 **species** of living mammals belong to the vertebrate **class** Mammalia. This diverse group of **animal**s has certain common features: all have four legs, bodies covered by hair, a high and constant body **temperature**, a muscular **diaphragm** used in **respiration**, a lower jaw consisting of a single bone, a left systemic aortic arch leaving the left ventricle of

the **heart**, and three bones in the middle ear. In addition, all female mammals have milk-producing glands. There are three living subclasses of mammals: the Monotremata (**egg**-laying mammals), the Marsupialia (pouched mammals), and the **Placenta**lia (placental mammals).

Mammals range in size from bats, some of which weigh less than 1 oz (28.4 g), to the blue whale, which weighs more than 200,000 lb (90,800 kg). Mammals are found in cold arctic climates, in hot **desert**s, and in every terrain in between. Marine mammals, such as whales and seals, spend most of their time in the ocean. While mammals are not as numerous and diverse as, for example, birds or insects, mammals have a tremendous impact on the environment, particularly due to the use of Earth's **natural resources** by one species of mammal: humans.

Species of mammals have developed varying **adaptation**s in response to the different environments in which they live. Mammals in cold climates have insulating layers-a thick coat of fur, or a thick layer of fat (blubber)-that help retain body heat and keep the animal's body temperature constant. Some mammals that live in deserts survive by special adaptations in their kidneys and sweat glands, as well as by their ability to avoid heat by behavioral means. Other adaptations for survival in extreme climates include **hibernation** (a state of winter dormancy) or estivation (summer dormancy). These responses make it possible for the animal to conserve **energy** when food supplies become scarce.

The care of the young (parental care) is notable among mammals. Born at an average of 10% of its mother's weight, mammalian young grow rapidly. The protection the young receive from one or both parents during the early stages of their lives enables mammals to maintain a strong survival rate in the animal **kingdom**.

The subclass Placentalia contains the majority of living mammals. The **embryo** of placentals develops in the mother's uterus, is nourished by **blood** from the placenta, and is retained until it reaches an advanced state of development. The Marsupialia are found in Australia and in North and South America. Their young develop inside the uterus of the mother, usually with a placenta connected to a yolk sac. Young marsupials are born in a very undeveloped state and are sheltered in a pouch (the marsupium) which contains the nipples of the milk glands. Kangaroos, wallabies, and most Australian mammals are marsupials, as is the opossum of the New World. The Monotremata of Australia include the duck-billed platypus and two species of spiny anteaters. Monotremes lay eggs, but have hair and secrete milk like other mammals.

MASS EXTINCTION HYPOTHESIS

Extinction is the permanent loss of a **species** or another **taxon** of **organism**s (such as a **genus, family, order,** or **phylum**), occurring over all of its range. Extinction has occurred naturally throughout the history of life on Earth. In fact, biologists believe that the species that live today are only a small fraction of the total number that have ever lived. For example, the remarkable **fossil** record of the Burgess Shale of British Columbia demonstrates that many phyla of **invertebrates** lived at the beginning of the Cambrian era, about 570-million years ago, during an evolutionary proliferation of the first complex (or metazoan) **animal**s. Most of these phyla are now extinct. Biologists interpret the 15-20 extinct phyla of the Burgess Shale as representing some fantastic evolutionary experiments in the form and function of invertebrates. Modern palaeontologists only know about these extinct creatures because of extraordinary circumstances that allowed the preservation of even soft-bodied species as fossils.

The rate of extinction has not been uniform during the history of life. The geological record shows that extremely long periods of time have been characterized by relatively slow and uniform rates of extinction. However, these relatively tranquil periods have been punctuated by some catastrophic episodes of so-called "mass extinction." There have been five particularly massive mass extinctions. These are: the Ordovician event, occurring 440 million years ago; the Devonian, 365 million years ago; the Permian, 245 million years ago; the Triassic, 210 million years ago; and the **Cretaceous**, 65 million years ago. The most catastrophic mass extinction occurred at the end of the Permian era, and resulted in the losses of about 54% of marine families, 84% of genera, and 96% of species.

Perhaps the most renowned mass extinction, however, was that occurring about 65-million years ago at the end of the Cretaceous era. This was the event that resulted in the extinctions of the last of the dinosaurs and pterosaurs, as well as many other species of **plants** and animals. About 76% of species and 47% of genera may have became extinct during this cataclysm.

The "mass extinction hypothesis" suggests that these catastrophes may have been caused by extremely rare, unpredictable impacts of planet Earth with large meteorites from outer space. This event could have caused enormous quantities of tiny, inorganic particulates to be ejected from Earth's surface into the **atmosphere**. When suspended in the atmosphere, the fine dusts can interfere with global climate in several ways, particularly: by reflecting sunlight back to space; and by absorbing incoming solar radiation in the upper atmosphere, thereby preventing its penetration to the surface. This could result in a severe cooling of the global climate, creating conditions inhospitable to many, even most of the existing species and **ecosystem**s. A mass extinction could be the result.

This theory has been used to explain the end-of-Cretaceous cataclysm that apparently killed off the last of the dinosaurs. It has been suggested that at that time Earth was impacted by a 10-km-wide meteorite, causing an enormous quantity of fine dust to be spewed into the atmosphere, resulting in a climatic deterioration that many species could not tolerate. Evidence in support of this idea includes the occurrence of a thin layer of rock aged at about 65-million years ago, which is unusually rich in the trace element iridium. Iridium occurs in relatively high concentrations in meteorites, and smaller concentrations in Earth's crust. Therefore, dusts originating from a gigantic meteoritic explosion could explain the dated iridium anomaly. In addition, geologists have recently discovered a giant meteorite-impact crater near Yucatan in eastern Mexico, aged at about the time of the end-of-Cretaceous event.

Markhors (*Capra falconeri*) are found in various mountain ranges in India, Pakistan, Afghanistan, Tajikistan, and Uzbekistan. These goats are considered endangered throughout their range. The primary cause of the markhor's decline is excessive hunting, primarily for its horns, but also for its meat and hide. *(Photograph by Robert J. Huffman, Field Mark Publications. Reproduced by permission.)*

Some scientists believe that all of the great mass extinctions of the distant geological record were caused by meteorite impacts, or perhaps by periods of intense volcanic activity, which could also have injected immense quantities of fine particulates into the atmosphere. Nevertheless, this theory is controversial among scientists.

It should also be noted that the modern species of planet Earth are now undergoing another event of mass extinction. This catastrophe is the best documented of the extinction catastrophes. It is largely being caused by the clearing of natural ecosystems by humans to develop the land for agricultural, agricultural, and urbanized uses. To a much lesser degree, this ongoing mass extinction is also being caused by excessive harvesting of some species as sources of food, hides, or other commodities. Many biologists are predicting that this modern catastrophe will turn out to be the most intense mass extinction ever experienced by life on Earth.

MATTER

The simplest definition of matter is that it is anything that has weight or fills space. Matter is all the things we can see and feel in the world. Rocks, giraffes, **water**, and cars all consist of matter, whether they are living or non-living. Air is also matter. Though it is invisible, its effects can be felt, as when the wind blows. And with careful measurements, its weight can be calculated. Matter is often broken down into three categories: solid, liquid, and gaseous. For biologists, such a basic description of matter often suffices. But as long ago as ancient Greece, scientists theorized about what matter is comprised of. And in terms of modern-day physics and chemistry, our understanding of matter has grown quite complex.

Many of the renowned Greek philosophers discussed matter. Empedocles (c. 495-435 B.C.) believed all matter was made of a combination of four elements: earth, air, fire, and water. The elements mixed through the interaction of love and hate. Democritus (c. 460-370 B.C.) held a less mystical view, that matter is composed of tiny indestructible units called **atoms**. Empedocles had the upper hand for about two thousand years.

A later proponent of **atomic theory** was the English chemist **Robert Boyle**. Boyle was in some sense an alchemist, who was interested in the transmutation of gold. But he was unusual for his time in that he was convinced of the value of experiments. He discovered that air was compressible, and this led him to theorize that air was composed of discrete particles. These, he thought, must be atoms. He published *The Skeptical Chemist* in 1661, which proposed for the first time that the four Greek elements should be replaced with what we think of as modern **chemical element**s: material substances that cannot be broken down further. Isaac Newton (1642-1727) also theorized that matter was composed of indivisible particles, which were atoms.

Scientists following Boyle and Newton developed their theories and experimental methods, leading to a much fuller understanding of matter. By the mid-nineteenth century, European scientists had a working knowledge of atoms. The Russian chemist Dimitri Ivanovich Mendeléev first published the periodic table of the elements in 1869, arranging the known chemical elements by atomic weight.

The next advance in scientists' understanding of matter came with twentieth century physics. Physicists began to explore the nature of atoms, and found that they were not indivisible but made up of smaller particles. And Albert Einstein (1879-1955) postulated that matter can be transformed into **energy**, and energy into matter. To the modern physicist, matter is not simply stuff with weight that fills space. Matter is made of particles that have many qualities, including spin, electrical charge, and gravitational attraction. For life scientists, this sophisticated description of matter is most useful in microbiology, where molecular processes in and between **cell**s can sometimes be studied using theories of force and motion derived from physics.

MATTER, CONSERVATION OF

Matter, by definition, has mass and takes up space. Mass is generally defined as a quantitative measure of inertia. Inertia is the resistance of matter to a change in its speed or position as a result of the application of a force. The greater the mass of a object (matter), the smaller the change produced by an applied force.

Conservation Laws in physics state that physical properties (i.e. measurable quantities such as mass) do not change. These laws of conservation govern matter, **energy**, momentum and electric charge. Each such law states that fundamental, physical, measurable quantities of matter, energy, momentum and electric charge remain constant with the passage of time. An important function of conservation laws is that they make it possible to predict visible behavior without having to consider the microscopic details of the course of a physical process or **chemical reaction**.

Conservation of matter, therefore, implies that matter can be neither created nor destroyed—i.e., processes that change the physical or chemical properties of matter (such as change of state, from a solid to a liquid to a gas) leave the total mass of matter unchanged.

Einstein's recent theory of relativity has caused the notion of mass and the law of conservation of matter to undergo a radical revision. Matter has lost its absoluteness. The mass of an object is now seen to be equivalent to energy, to be inter convertible with energy, and to increase at very high speeds near that of **light** (about 186,000 miles [299,460 km] per second). Matter, therefore is no longer understood to be constant, or unchangeable. In both chemical and nuclear reactions, some conversion between matter and energy occurs, so that the products of these reactions can have smaller or greater mass than the parts that caused the reaction. However, these inter conversions of matter and energy are too small to be detectable except in cases involving subatomic particles or speeds comparable to that of light. In these situations, conservation of matter is better defined as a special case of the more general law of conservation of mass-energy.

MAYER, JULIUS ROBERT (1814-1878)
German physicist and physiologist

Robert Mayer was a brilliant but unlucky man. He originated the concept of **conservation** of **energy**, now considered the first law of thermodynamics, and was the first person to determine the mechanical equivalent of heat. He was a physician by training and did not have a reputation as a scholar in the scientific community. Thus, it was up to later scientists to "rediscover" the landmark research Mayer had performed years earlier.

Mayer was born in Heilbronn, Germany, in 1814. He was the youngest son of an apothecary, and the only son who chose not to enter his father's profession. He enrolled at the University of Tubingen and, despite being expelled in 1837 for his participation in a secret society, received his medical degree in 1838. In 1840 he signed on as a ship's doctor for a vessel sailing to East Indies. It was on this ship, near Java, that Mayer began to notice the remarkable redness of the **blood** drawn from the men aboard. Mayer supposed that, because of the intense tropical heat, the men's bodies required less oxygen to maintain their temperature; thus, more unused oxygen remained in the bloodstream. This discovery fascinated Mayer, for it seemed to indicate that oxygen was the true and single source of **animal** heat. He further proposed that food energy could be the only source of oxidation within a living **organism**. Thus, energy within the body could not be created from nothing, nor could it be reduced to nothing—it could only be used or stored.

What Mayer had conceived was the concept of conservation of energy. He planned to publish his ideas immediately but found he lacked the schooling in physics necessary to explain them. He began to study the nature of work and energy, conducting a number of experiments on the subject. One such experiment used a horse-powered device to stir a container full of paper pulp. By measuring the increase in temperature of the pulp in relation to the amount of work performed by the horse, Mayer came up with a figure for the mechanical value of heat (the amount of heat required to raise the temperature of a certain amount of water by one degree celsius). This figure was remarkably close to that of the present day and was skewed only by the inaccurate values for specific heat used at that time. Mayer announced both his theory of conservation of energy and his figure for the mechanical equivalent of heat in a paper published in 1842. It received very little attention, but Mayer was determined to explore the subjects in even greater detail.

In 1845 he published another paper, this one applying energy conservation to the motion of the planets, the tides, the sun, magnetism, electricity, and the living world. In many of these cases, he identified solar energy as the source of most of the world's energy. In this way, Mayer explained the greater ramifications of the First Law of Thermodynamics to an extent no other scientist could match; still, he would never be given the credit he deserved. In 1847 James Joule announced his own independent research into the mechanical equivalent of heat; and, because of his international reputation, his work was given priority. That same year **Hermann von Helmholtz** published his work on the conservation of energy; he was also given credit over Mayer, whose research was still relatively unknown.

Two of Mayer's children died in 1848, which, coupled with the lack of respect paid him by the scientific community, drove him to despair. He attempted suicide in 1849, jumping from a third-story window. Mayer succeeded only in breaking his legs and laming himself. He was committed to a mental hospital in 1851 and remained there for nearly a decade.

While Mayer languished in the asylum, his work was finally beginning to gain status. Justus Liebig, who had originally published Mayer's papers, spread his theories during lectures. He also mistakenly announced, in 1858, that Mayer was dead. By the time he was released from the hospital's care, he had become a well-respected theoretical physicist. He was awarded the Copley Medal in 1871 from the Royal Society and was granted the privilege of adding ''von'' to his name—a very high German honor. He died of tuberculosis in 1878.

MAYR, ERNST (1904-   )
German American biologist

Considered one of the century's most important evolutionary biologists, Ernst Mayr has made major contributions to **ornithology**, **evolutionary theory**, and the history and philosophy of biology. He is best known for his work on **speciation**—how one species arises from another. In his more than sixty years in the United States, however, he has published hundreds of articles and more than a dozen books. Through these writings, he has not only clarified certain aspects of earlier scientific theories but proposed new theories which have changed the course of biological research.

Mayr was born on July 5, 1904, in Kempten, Germany, near the borders of Austria and Switzerland. He was one of three sons of Helene Pusinelli Mayr and Otto Mayr, who was a judge. As a boy, Mayr enjoyed bird watching. He received a broad education, including Latin and Greek. In 1923, he followed in the footsteps of several physicians in his family and began studying for a medical degree at the University of Greifswald. Within two years, however, he had become so enthralled with the evolutionary theories and the exploratory voyages of nineteenth-century British naturalist **Charles Darwin** that he switched from medicine to **zoology**. He moved to the University of Berlin, where he had once worked at the zoological museum during his summer vacations. In 1926, he received his Ph.D. in zoology, *summa cum laude* from the university. Soon afterward, he became the zoological museum's assistant curator.

Leads Expeditions to New Guinea and the Solomon Islands

While still working at the museum, Mayr went to Budapest in 1928 to attend a zoological conference. There he met Lionel Walter Rothschild, a British baron and well-known zoologist. Impressed by the young man, Lord Rothschild asked him to lead an ornithological expedition to Dutch New Guinea, in the southwest Pacific. Mayr jumped at the chance. New Guinea at that time was extremely inaccessible, but Mayr was eager to investigate the birds of several remote mountain ranges. The trip was not easy, and Mayr's party suffered a variety of illnesses and injuries. But Mayr was undaunted and he decided to remain in the region, making a second expedition sponsored by the University of Berlin to mountain ranges in the Mandated Territory of New Guinea. Then in 1929 and 1930, Mayr participated in a third expedition, the American Museum of Natural History's Whitney South Sea Expedition to the Solomon Islands. The experiences and insights crowded into these few years in the south Pacific were to stimulate Mayr's thinking about biology and the development of **species** for decades to come.

When the expedition to the Solomons was over, Mayr was invited to be a Whitney research associate in ornithology at the American Museum of Natural History in New York City. In 1932, he was named associate curator of the museum's bird collection and he decided to stay in the United States, eventually becoming an American citizen. Over the next de-

Ernst Mayr. *(AP/Wide World Photos, Inc. Reproduced by permission.)*

cade, Mayr worked at identifying and classifying bird species, studying their geographical distribution and relationships. These years resulted in two of his most influential books.

He published the first of these books in 1941, the *List of New Guinea Birds.* This book is much more complex than its title might imply; it explores the ways closely related species can be distinguished from one another and how variations can arise within a species. In a short article in *Science,* Stephen Jay Gould calls each species Mayr discusses in this book "a separate puzzle, a little exemplar of scientific methodology." In writing the *List of New Guinea Birds,* Gould observes, Mayr "sharpened his notion of species as fundamental units in **nature** and deepened his understanding of **evolution.**"

Proposes Theory of Geographic Speciation

While most biologists in the 1930s and 1940s accepted the broad premise of Darwin's theory about evolution—that species change and evolve through a process called **natural selection**, sometimes loosely called "the survival of the fittest"—there was little understanding of how the process worked. If the fittest members of an **animal** population were the ones that survived, where did those especially well adapted creatures come from in the first place? These questions were complicated during this period by the fact that there was no clear understanding of exactly what constituted a plant or animal **species**. Instead, there were two conflicting approaches: one school of thought tried to classify species by their shape and appearance, and another tried to identify species through their **genes**.

In December 1939, Mayr attended a lecture series at Yale University given by a well-known geneticist named Richard Goldschmidt, who argued that new species can arise through sudden genetic mutation. Goldschmidt believed that these changes could take place within a single generation, and Mayr was appalled by what he heard. As Fred Hapgood explained in *Science 84,* "What Mayr heard in those lectures seemed so wrong that he decided, in his own words, to 'eliminate' those ideas 'from the panorama of evolutionary controversies.'" Mayr was convinced that extremely long periods of time were required for the development of a new species, and he set out to demolish Goldschmidt's argument.

The result was *Systematics and the Origin of Species,* which he published in 1942. Based partly on what he had learned in the South Pacific, Mayr argued that geographic speciation is the basic process behind the formation of new species. This theory had been advanced more than a hundred years earlier—even before Darwin—but it had never taken hold. Mayr showed how the process works: when a few animals become separated from their original **population** and breed among themselves over a great many generations, they eventually change so much that they can no longer breed with their original group. For example, birds from the mainland who once settled on an island may look like their ancestors and will have many similar genetic traits, yet the two groups will not be able to interbreed. The island birds have then become a new species. This concept, which Mayr gradually elaborated, was to form the core of his thinking.

Mayr's ideas about speciation not only found general acceptance but won him great respect. *Systematics and the Origin of Species* has been called the "bible" of a generation of biologists. **E. O. Wilson** remembers how in this book Mayr offered him the "theoretical framework on which to hang facts and plan enterprises" as reported in *Science 84.* Wilson continued, "He gave **taxonomy** an evolutionary perspective. He got the show on the road."

Accepts Professorship at Harvard

Over the next several years, Mayr continued to expand and refine his ideas about speciation. In 1946, he founded the Society for the Study of Evolution, becoming its first secretary and later its president. In 1947, he founded the Society's journal, *Evolution,* and served as its first editor. In 1953, at the age of forty-nine, he was named Alexander Agassiz Professor of Zoology at Harvard. From 1961 to 1970, he was director of Harvard's Museum of Comparative Zoology, and during those years he brought about an important expansion of the museum, which is now a major center of biological research. By this time, Mayr was recognized as a leader of what has been called "the modern synthetic theory of evolution." In this context, he is often mentioned with three other eminent researchers in the field: George Gaylord Simpson, **Theodosius Dobzhansky**, and Julian Huxley. In 1963, Mayr published *Animal Species and Evolution,* in which he wrote of man's place in the **ecosystem**.

In a 1983 interview with Carol A. Johmann in *Omni,* Mayr discussed many of the concerns he has expressed throughout his career: "Man must realize that he is part of the ecosystem and that his own survival depends on not destroying that ecosystem." But Mayr remained pessimistic about the future of the human race. Implicit in his work is the firm belief that animal species, including the human one, do not improve toward some higher state as they evolve; they merely adapt to changing conditions. And though it might someday be possible to alter the human race by manipulating its genetic structure, Mayr found this idea not only morally offensive but futile, since even with scientific advances there can never be any agreement about which genes are worth manipulating.

Still, Mayr has said, all this does not mean that humankind cannot improve, but improvement must be through education and cultural advances. Writing in *Scientific American* in 1978, Mayr pointed out that it is cultural evolution, not genetic changes, which has permitted the human species to fly more powerfully than birds, bats, or insects. He explained that cultural evolution "is a uniquely human process by which man to some extent shapes and adapts to his environment.... Cultural evolution is a much more rapid process than biological evolution."

Life's Work Culminates in Balzan Prize

In 1975, Mayr retired from Harvard as emeritus professor of zoology. But he continued to work intensely and his interests continued to expand. He changed careers, as Stephen Jay Gould has observed; from a scientist he became a historian of science. He undertook to write not only about evolution, but also about the entire history of biology. In 1982, he published *The Growth of Biological Thought,* intended to be the first of two volumes. Then in 1991, at the age of eighty-seven, he published yet another carefully wrought discussion of evolution, *One Long Argument.* In that book, he wrote, "The basic theory of evolution has been confirmed so completely that modern biologists consider evolution simply a fact.... Where evolutionists today differ from Darwin is almost entirely on matters of emphasis. While Darwin was fully aware of the probabilistic nature of selection, the modern evolutionist emphasizes this even more. The modern evolutionist realizes how great a role chance plays in evolution."

Mayr married Margarete Simon on May 4, 1935, and they had two daughters. For many years, Mayr was a familiar figure around Harvard, a lean man with brown eyes and white hair. Many people noted that Mayr's natural assertiveness and strong writing style added to the weight of his arguments, helping to win over others. During an exceptionally long career, Mayr was awarded ten honorary degrees, including Doctor of Science degrees from both Oxford and Cambridge Universities and a Doctor of Philosophy degree from the University of Paris. In 1954, he was elected to the National Academy of Sciences. His numerous prizes include the Darwin-Wallace Medal in 1958, the Linnean Medal in 1977, the Gregor Mendel medal in 1980, and the Darwin Medal of the Royal Society in 1987. In 1983, he received the Balzan Prize, which has been called the equivalent of the Nobel Prize in the biological sciences. Commenting on Mayr's winning the Balzan Prize in the article in *Science,* Stephen Jay Gould called Mayr "our greatest living evolutionary biologist."

McCLINTOCK, BARBARA (1902-1992)
American geneticist

Barbara McClintock was born on June 16, 1902, in Connecticut. The daughter of a physician, she grew up in the Flatbush section of Brooklyn, New York—a rural area during that time. During high school, McClintock's interests turned from sandlot baseball to science. Despite her parents' misgivings, she persuaded them to let her study at Cornell University, an agricultural college in Ithaca, New York. McClintock was fascinated with a **genetics** course she took in her junior year. However, in the plant breeding department where most of the genetics courses were taught, women were not permitted as graduate students. So McClintock signed up in the **botany** department and became assistant to a **cytology** professor who encouraged her interest in plant genetics.

After McClintock completed her graduate work at Cornell, she stayed on as an instructor and researcher, associating herself with Rollins Emerson's famous maize (corn) genetics group. During this time, she discovered the function of each **chromosome** in corn. Cultivating her own crops on university grounds, she spent summers recording changes in each plant, then returned to the lab after the harvest. By observing **cell divisions** at various stages of a corn plant's development, she was able to tell what each chromosome did in the process. Between 1929 and 1931, she published nine papers on the subject of genetics.

In 1931, McClintock assigned graduate student Harriet Creighton a problem in genetic crossover in corn. With McClintock's guidance, Creighton was able to show that genetic information is transferred after chromosomes cross over during the formation of sex **cell**s. While on a visit to Cornell, **Thomas Hunt Morgan**, one of the pioneers in the study of genetics, urged McClintock and Creighton to publish their findings on genetic crossover in corn. They did, receiving favorable reviews for their work.

McClintock's major contribution to genetics came after she left Cornell, which at the time had no women above the level of professor outside the home economics department. It was difficult at first, because she had gained a reputation as being a difficult, independent loner. She worked briefly at the University of Missouri, but felt isolated from the administration and constrained by its bureaucratic rules (she often ignored building hours and would break into her lab on Sundays). In 1941, she found a permanent position at Cold Spring Harbor Laboratory on Long Island, a residential research site run by the Carnegie Institution of Washington, where she could conduct her experiments free of administrative interference.

Over the years, McClintock had developed an interest in the **mutation**s caused by x rays. She was so familiar with maize plants and their chromosomes that she instantly recognized anything unusual. One season, some of her corn seedlings had developed odd streaks and spots on the leaves. There appeared to be a pattern, but it was unlike any pattern made from the x rays. Something in the plant itself seemed to be controlling the appearance of the patterns. She studied more closely and found that many color patches showed up in opposite pairs. For instance, if part of a **leaf** contained more green streaks than usual, a nearby portion would contain fewer than usual.

To McClintock, these were not signs of disorder, but of a larger and different system of order. She set out to make that difference understandable. For six years, she researched her unusual observations, and in 1951, she presented her theory of genetic transposition (also called genetic mobility or gene jumping). According to her theory, **gene**s in a chromosome could actually move from one position to another within that same chromosome. McClintock saw transposition as a way that individual **organism**s regulate their development, and even their survival, in times of stress. McClintock's discovery, however, was met with derision. The majority of geneticists believed that genes were preprogrammed, and, like beads on a string, occupied a fixed place on chromosomes. Any changes, therefore, occurred only as a result of random accidents. When McClintock presented her findings at a symposium, her colleagues dismissed her as "an old bag who'd been hanging around Cold Spring Harbor for years."

By the 1970s, McClintock had expanded her theory, arguing that genes could not only change their positions on a chromosome, but that they could also serve different purposes in those different positions. Slowly, her ideas gained acceptance, as scientists began to witness transposition in other life forms. But it wasn't until 1983, some 30 years after her discovery, that McClintock received the Nobel Prize in physiology or medicine. She was also the recipient of the National Medal of Science (1970) and the first MacArthur Laureate Award (1981), a lifetime annual prize of $60,000.

When asked about the long delay in recognition for her discovery, McClintock observed, "If you know you're right, you don't care. You know that sooner or later, it will come out in the wash." McClintock died on September 2, 1992.

McCollum, Elmer Verner (1879-1967)

American biochemist

Elmer McCollum was born on a farm near Fort Scott, Kansas, where he spent his first seventeen years. Money was scarce and the rural community's single school was rarely in operation. But the young Kansan was bright, energetic, and determined to get an education. By moonlighting at numerous jobs, he worked his way, first through high school, then through the University of Kansas, graduating in 1903, and finally through Yale University, where he earned his doctorate in 1906.

Throughout his academic career, McCollum's first love was always organic chemistry. Shortly after graduation, however, he was offered a position at the University of Wisconsin as an instructor in biochemistry (then known as agricultural chemistry). With no better job offers in sight, he decided to accept—a decision that was to alter his life. In 1907, when McCollum arrived at the University's Agricultural Experiment Station, a research study was already in progress. The study was designed to examine the dietary effects of three widely-used grains on the health and reproductive capacity of dairy cattle. Because the three grains were chemically similar, the researchers expected similar results. To everyone's intense surprise, however, only one grain—corn—kept the cattle healthy and strong. McCollum immediately resolved to try some **nutritional** experiments of his own.

Before long, McCollum had set up own laboratory and established the country's first colony of albino laboratory rats devoted to nutritional research. With the help of a young biochemist-in-training named Marguerite Davis, McCullum began the studies that were to lead to the discovery of vitamin A. By 1913, McCollum was able to report that the laboratory rats failed to grow when fed diets in which lard or olive oil was the only source of fat. These same rats, however, quickly resumed normal growth when ether-soluble extracts of butter or eggs were added to the diet. He concluded that butterfat and egg yolks contained some "growth-promoting factor" missing in other fats—a factor he soon isolated and termed *fat-soluble A* (to distinguish it from a water-soluble factor previously discovered by **Christiaan Eijkman**), eventually to be named vitamin A.

In the years that followed, McCollum contributed to the discovery of other fat-soluble vitamins, such as vitamin D in 1922, and did important work in the field of trace **minerals**. In 1917, he went to Johns Hopkins University as professor of biochemistry, remaining there until 1944. In his later years, he lectured widely on nutritional topics, wrote several outstanding textbooks, and received numerous awards. Shortly before he died, at the age of 88, McCollum mused on his accomplishments and concluded: "I have had an exceptionally pleasant life and am thankful."

Mechnikov, Ilya • See Metchnikoff, Elie

Medawar, Peter Brian (1915-1987)

British biologist and writer

Sir Peter Brian Medawar shared the 1960 Nobel Prize in physiology or medicine with Australian Sir **Frank MacFarlane Burnet** for discovering that **tissue** transplantation failed because of **immune system** rejection and not because of **gene**tic differences, as previously believed. During five years of experiments, Medawar learned that "immunological tolerance" could be produced by grafting **cell**s from one **animal** into the fetus of another, permitting successful tissue grafts between the two subjects. As the first clear evidence that tissue rejection could be overcome, this discovery opened the door to transplant surgery and the development of complex immunosuppressant drugs which help prevent rejection of transplanted **organ**s.

Medawar, called "a great scientist, a man of great courage, and a great writer," was born in Rio de Janeiro, Brazil. His father, who was born in Lebanon, became a naturalized British subject while living in London, England as a paying guest in the home of the Dowling family. There he fell in love with, and ultimately married, their oldest daughter, Muriel. The couple moved to Brazil when he became an agent for a British dental supply company.

The young Medawar left Brazil for England in 1928 where he first attended Marlborough College and then Magdalen College, Oxford where he studied **zoology**. He received two scholarships in 1935 and became a Fellow of the college in 1938. In 1937, he married fellow undergraduate, Jean Shinglewood Taylor, the daughter of a Cambridge physician, who also became a scientist and who, for many years, was chairwoman of the Family Planning Association. Mrs. Medawar wrote in her book, *A Very Decided Preference: Life with Peter Medawar,* "I realized quite soon that he was extraordinary and exceptional. Before long, others recognized this, and by the time Peter began his life's work in immunology, his quality had become apparent." The Medawars had two sons and two daughters.

Medawar's early research into tissue **culture** and nerve regeneration led him to develop a "biological glue" with which severed nerves could be rejoined and which became widely used in skin grafts. His work in immunology began early in World War II when the Medical Research Council requested he investigate why skin taken from a donor would not permanently graft to that of a recipient. In 1944, he became assistant professor of Zoology at St. John College, Oxford and moved to the University of Birmingham in 1947, continuing his investigation into tissue rejection in collaboration with colleagues. For identifying what they called "actively acquired tolerance," (the ability to induce tolerance to foreign tissue by injecting grafts into a fetus), he was elected Fellow of the Royal Society of London. In 1951, he became professor of Zoology at University College in London and received the Royal Medal in 1959. In 1960 he introduced tissue typing which was used for the first time in 1962 for a human kidney transplant. From 1962 until 1971 he was director of London's National Institute for Medical Research which, under his administrative guidance, became world renowned in immunology.

Medawar was a Reith Lecturer for the British Broadcasting Corporation, spoke and lectured in many countries, and was a Foreign Member of the New York Academy of Sciences, the American Academy of Arts and Sciences, and the American Philosophical Society. Among many other distinguished awards and honors from around the World he was named by England's Queen Elizabeth II as Commander of the Order of the British Empire (OBE) in 1959, was knighted in 1965, and in 1981 was presented the most prestigious of all royal honors—the Order of Merit (OM).

Medawar also gained fame as a prolific and extremely talented writer. His energy, positive outlook, and zest for life endeared him to almost everyone he met. He has been called both a "paragon of rationalism" and a "paragon of humanism." His incredible scientific mind was matched only by his compassion and love of people and passion for life. In 1969, at the age of fifty-four, he suffered a massive and almost fatal stroke while reading the lesson at the annual service of the British Association for the Advancement of Science at Exeter

Cathedral. After long months in therapy, he could walk with a caliper and walking stick. Although his left arm was paralyzed and half the visual field in both eyes was lost, he continued courageously and enthusiastically to research, write, travel, lecture, and enjoy his family. A second stroke in 1980 left him with slurred speech and wheel-chair bound, and he suffered a third stroke in 1985 shortly after completing his famous autobiography, *The Thinking Radish.* Of Medawar's spirit, his wife wrote, "He came to be admired as much for his courage after these strokes as he had been for his intellect before them."

MEIOSIS

Meiosis, also known as reduction division, consists of two successive **cell divisions** in **diploid** cells. The two cell divisions are similar to **mitosis**, but differ in that the **chromosome**s are duplicated only once, not twice. The end result of meiosis is four daughter **cell**s, each of them **haploid**. Since meiosis only occurs in the sex **organ**s (gonads), the daughter cells are the **gametes** (**sperm**atozoa or ova), which contain hereditary material. By halving the number of chromosomes in the sex cells, meiosis assures that the fusion of maternal and paternal gametes at **fertilization** will result in offspring with the same chromosome number as the parents. In other words, meiosis compensates for chromosomes doubling at fertilization. The two successive nuclear divisions are termed as meiosis I and meiosis II. Each is further divided into four phases (prophase, metaphase, anaphase, and telophase) with an intermediate phase (**interphase**) preceding each nuclear division.

Events of meiosis

The events that take place during meiosis are similar in many ways to the process of mitosis, in which one cell divides to form two clones (exact copies) of itself. It is important to note that the purpose and final products of mitosis and meiosis are very different.

Meiosis I

Meiosis I is preceded by an interphase period in which the **DNA** replicates (makes an exact duplicate of itself), resulting in two exact copies of each chromosome that are firmly attached at one point, the centromere. Each copy is a sister chromatid, and the pair are still considered as only one chromosome. The first phase of meiosis I, prophase I, begins as the chromosomes come together in homologous pairs in a process known as synapsis. Homologous chromosomes, or homologues, consist of two chromosomes that carry genetic information for the same traits, although that information may hold different messages (e.g., when two chromosomes carry a message for eye color, but one codes for blue eyes while the other codes for brown). The fertilized **eggs** (**zygote**s) of all sexually reproducing **organism**s receive their chromosomes in pairs, one from the mother and one from the father. During synapsis, adjacent chromatids from homologous chromosomes "cross over" one another at random points and join at spots called chiasmata. These connections hold the pair together as a tetrad (a set of four chromatids, two from each homologue). At the chiasmata, the connected chromatids randomly exchange bits of genetic information so that each contains a mixture of maternal and paternal genes. This "shuffling" of the DNA produces a tetrad, in which each of the chromatids is different from the others, and a gamete that is different from others produced by the same parent. **Crossing over** does, in fact, explain why each person is a unique individual, different even from those in the immediate family. Prophase I is also marked by the appearance of spindle fibers (strands of microtubules) extending from the poles or ends of the cell as the nuclear **membrane** disappears. These spindle fibers attach to the chromosomes during metaphase I as the tetrads line up along the middle or equator of the cell. A spindle fiber from one pole attaches to one chromosome while a fiber from the opposite pole attaches to its homologue. Anaphase I is characterized by the separation of the homologues, as chromosomes are drawn to the opposite poles. The sister chromatids are still intact, but the homologous chromosomes are pulled apart at the chiasmata. Telophase I begins as the chromosomes reach the poles and a nuclear membrane forms around each set. Cytokinesis occurs as the **cytoplasm** and **organelle**s are divided in half and the one parent cell is split into two new daughter cells. Each daughter cell is now haploid (n), meaning it has half the number of chromosomes of the original parent cell (which is diploid-2n). These chromosomes in the daughter cells still exist as sister chromatids, but there is only one chromosome from each original homologous pair.

Meiosis II

The phases of meiosis II are similar to those of meiosis I, but there are some important differences. The time between the two nuclear divisions (interphase II) lacks **replication** of DNA (as in interphase I). As the two daughter cells produced in meiosis I enter meiosis II, their chromosomes are in the form of sister chromatids. No crossing over occurs in prophase II because there are no homologues to **synapse**. During metaphase II, the spindle fibers from the opposite poles attach to the sister chromatids (instead of the homologues as before). The chromatids are then pulled apart during anaphase II. As the centromeres separate, the two single chromosomes are drawn to the opposite poles. The end result of meiosis II is that by the end of telophase II, there are four haploid daughter cells (in the sperm or ova) with each chromosome now represented by a single copy. The distribution of chromatids during meiosis is a **matter** of chance, which results in the concept of the **law of independent assortment** in **genetics**.

Control of meiosis

The events of meiosis are controlled by a **protein** enzyme complex known collectively as maturation promoting factor (MPF). These **enzyme**s interact with one another and with cell organelles to cause the breakdown and reconstruction of the nuclear membrane, the formation of the spindle fibers, and the final division of the cell itself. MPF appears to work in a cycle, with the proteins slowly accumulating during inter-

phase, and then rapidly degrading during the later stages of meiosis. In effect, the rate of synthesis of these proteins controls the frequency and rate of meiosis in all sexually reproducing organisms from the simplest to the most complex.

Human gamete formation

Meiosis occurs in humans, giving rise to the haploid gametes, the sperm and egg cells. In males, the process of gamete production is known as **spermatogenesis**, where each dividing cell in the testes produces four functional sperm cells, all approximately the same size. Each is propelled by a primitive but highly efficient flagellum (tail). In contrast, in females, **oogenesis** produces only one surviving egg cell from each original parent cell. During cytokinesis, the cytoplasm and organelles are concentrated into only one of the four daughter cells—the one which will eventually become the female ovum or egg. The other three smaller cells, called polar bodies, die and are reabsorbed shortly after formation. The process of oogenesis may seem inefficient, but by donating all the cytoplasm and organelles to only one of the four gametes, the female increases the egg's chance for survival, should it become fertilized.

Mistakes during meiosis

The process of meiosis does not work perfectly every time, and mistakes in the formation of gametes are a major cause of genetic disease in humans. Under normal conditions, the four chromatids of a tetrad will separate completely, with one chromatid going into each of the four daughter cells. In a disorder known as nondisjunction, chromatids do not separate and one of the resulting gametes receives an extra copy of the same chromosome. The most common example of this mistake in meiosis is the genetic defect known as **Down syndrome**, in which a person receives an extra copy of chromosome 21 from one of the parents. Another fairly common form of nondisjunction occurs when the **sex chromosome**s (XX, XY) do not divide properly, resulting in individuals with Klinefelter syndrome or Turner syndrome. Other mistakes that can occur during meiosis include **translocation**, in which part of one chromosome becomes attached to another, and deletion, in which part of one chromosome is lost entirely. The severity of the effects of these disorders depends entirely on the size of the chromosome fragment involved and the genetic information contained in it. Modern technology can detect these genetic abnormalities early in the development of the fetus, but at present, little can be done to correct or even treat the diseases resulting from them.

MEMBRANE

Cell membranes or **plasma membrane**s surround cells, separating the **cytoplasm** and **organelle**s on the inside from the extracellular fluid on the outside. Several cell organelles (**mitochondria**, **endoplasmic reticulum**, and Golgi bodies) are also bounded by membranes. The membrane allows a cell or

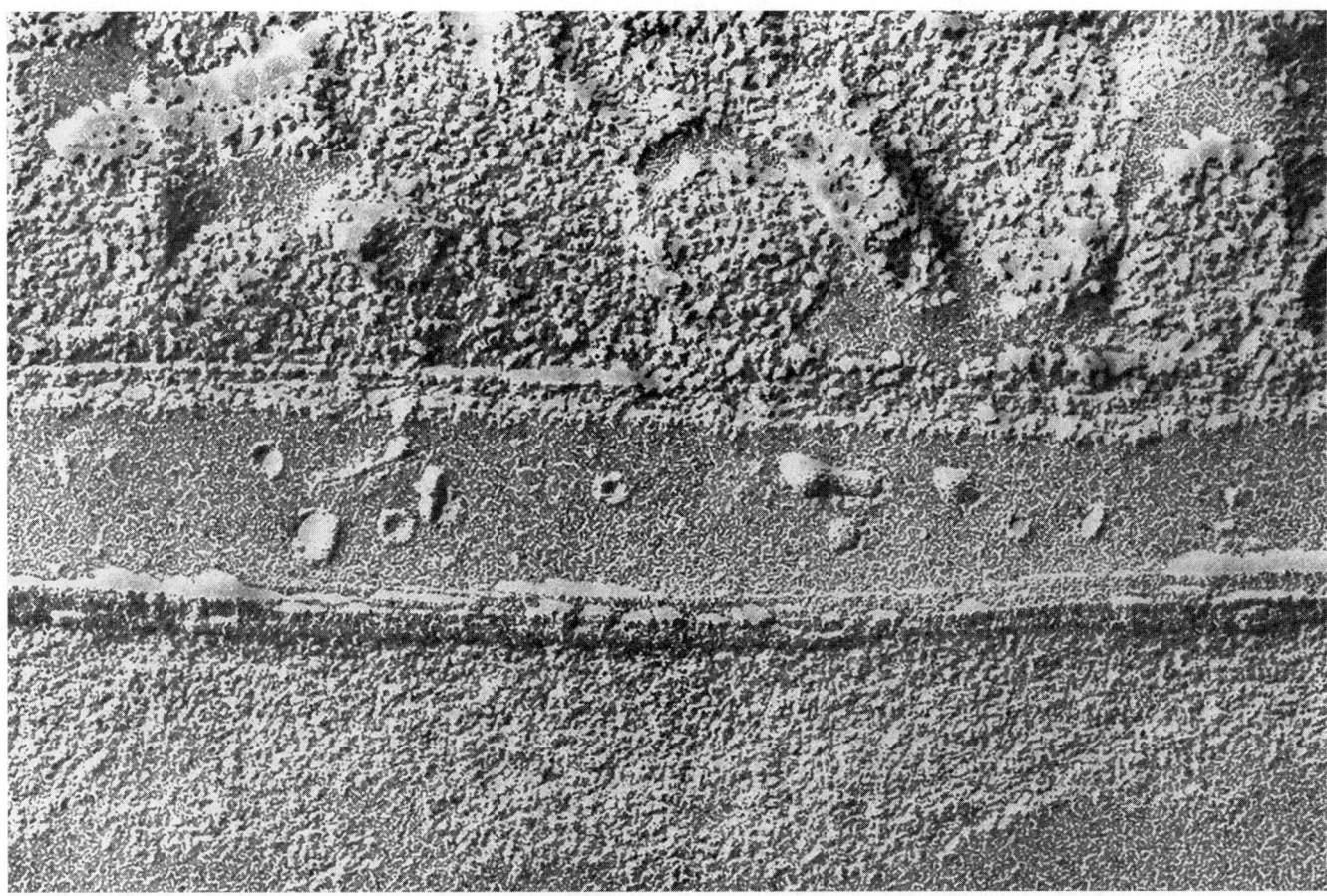

A freeze fracture image across the cell wall and membrane of a blue-green alga. *(Dr. Dennis Kunkel, Phototake NYC. Reproduced by permission.)*

organelle to maintain a constant internal environment, usually one that is quite different from the medium surrounding it. This is accomplished by the semipermeable nature of the membrane that regulates the passage of all substances going through it.

The detailed chemical composition of a membrane varies, depending on its location and the functions it performs. However, all membranes do have the same basic structure. The majority of the membrane is composed of two layers of phospholipid **molecule**s lined up side by side with their fatty acid "tails" facing inward. The outer edges of the membrane are hydrophilic (soluble in **water**), while the interior area is hydrophobic (insoluble in water). Because of this dual chemical nature of the phospholipid bilayer, the entire membrane surface is permeable to gases (such as oxygen and **carbon dioxide**), to small, uncharged polar molecules (such as water and **ammonia**), and to nonpolar molecules (such as **lipids**). However, the membrane is impermeable to charged molecules (such as ions and **protein**s) and to larger, uncharged polar molecules.

Embedded within and spanning the phospholipid bilayer are various transport proteins that serve as "gates," selectively allowing charged molecules and ions and larger molecules to pass through the membrane. These transport proteins channel molecules by a variety of methods, including facilitated **diffusion** (movement with the concentration gradient, using no ATP **energy**) and **active transport** (movement against the concentration gradient, using ATP energy).

The plasma membrane that forms the boundary of a cell has several other molecules in addition to the basic **membrane structure**. These include integral proteins, **cholesterol**, glycoproteins, and glycolipids. The phospholipid bilayer with its biochemical inclusions is known as the fluid mosaic model of membrane structure. Some membrane proteins serve as receptors for **hormone**s, transferring the signal to the interior of the cell (via G proteins) without allowing the "messenger" molecule to enter, thus protecting the integrity of the cell. Other **carbohydrate** molecules attached to the exterior of the plasma membrane act as "markers," identifying the cell as a particular type.

Cystic fibrosis—a fatal, hereditary disease characterized by a heavy mucus buildup in the lungs—is caused by a defective plasma membrane protein. In persons with cystic fibrosis this transport protein, known as the sodium-potassium pump, abnormally transports sodium ions across the membrane without carrying the chloride ions that usually accompany them. Research is currently underway to correct through **genetic engineering** the faulty gene that codes for the plasma membrane protein.

MEMBRANE POTENTIAL

Certain **cell**s within the **nervous system** are called "excitable cells." Surrounding each of these cells is a highly complex, semipermeable membrane, an insulating barrier which separates fluids inside the cell from fluids outside. In the quiet state, when the cell is resting or unstimulated, the inside of a **neuron**'s membrane is negatively charged—about -70 millivolts (mV)—and the outside is positively charged. This difference in voltage or charge is stored in the membrane and called the membrane potential because it provides a source of potential **energy** which can be used to generate and propagate electrical signals from one cell to the next through ion currents. The membrane is selectively permeable to ions, particularly potassium (K+), which is more highly concentrated inside the cell, and sodium (Na+), more highly concentrated outside the cell. When an event excites the cell, these ions change places through voltage-gated channels—or special pores—in the membrane. The inner and outer membrane charges reverse, triggering an **action potential**, an all-or-none response which travels down the cell's **axon** to the synaptic terminals where it causes the release of **neurotransmitters** into the **synapse**. These neurotransmitters, in turn, stimulate the adjoining cell. Neurons have a threshold **stimulus**, which means there is a minimum level of stimulation needed to trigger the exchange of K+ and Na+ ions across the membrane. They also need a refractory period, time during which the K+ and Na+ to return to their normal intra and extracellular levels, once again giving the membrane its potential to fire off another signal.

MEMBRANE STRUCTURE

Surrounding each **cell** within the **nervous system** is a highly complex lipid bilayer called a **membrane** formed by microscopic phospholipid **molecule**s. **Phospholipids** have a round, negatively charged "head" consisting of phosphate, choline, and glycerol; and two "tails" comprised of fatty acids. The head is hydrophilic (likes **water**), but the tails are hydrophobic (do not like water). Because intra and extracellular fluids are aqueous (watery), the phospholipids huddle together in a double layer, the heads of the outer layer pointing toward the extracellular fluid and the heads of the inner layer pointing toward the intracellular fluid. The tails of both layers point away from the fluids and meet to form a hydrophobic environment between the heads. Molecules of **cholesterol** dotted throughout the membrane help prevent phospholipids from packing together too tightly.

The bilipid layer is variably permeable—some substances can move through while others cannot. Substances move through the membrane by 1) simple **diffusion**, in which low molecular weight substances pass directly through the membrane down their concentration gradient (from areas of high to low concentration); 2) facilitated diffusion, in which minute channels in certain **protein**s allow charged molecules (ions) such as sodium, potassium, and hydrogen to flow in and out of the cell; and 3) **active transport**, in which sodium-potassium pumps operated by cellular **energy** (ATP) force ions across the membrane, often against their concentration gradient. The many different types of proteins interspersed throughout the membrane—some with a **carbohydrate** side chain—are essential for most of the membrane's functions: integral proteins run completely through the membrane, surface proteins are embedded in the outer layer only, inner proteins are in the inner layer only, and transmembrane proteins—which contain a tiny channel through which fluids can pass—also connect the intra and extracellular environments. Proteins join cells together to form cell groupings such as **tissue**, help keep the cell in shape by connecting to portions of the **cytoskeleton**, act as receptors for **neurotransmitters** and **hormone**s, participate in **enzyme** activity to aid in **metabolism** and other functions, and control the movement of molecules into and out of the cell.

MEMORY

Memory is the retention and retrieval in the human mind of past experiences. Memory involves both remembering and forgetting. Reasoning and intelligent behavior requires memory, since remembering is a prerequisite to reasoning. Also, the ability to solve problems or even to recognize that a problem exists depends on remembering.

Forgetting has many functions as well. The most important may be time orientation since memories tend to fade over time, and therefore new **learning** cannot occur without the loss or suppression of old patterns. Another function of forgetting is the relief from the anxiety of painful past experiences.

Many factors influence the retention of memories. Practice tends to build and maintain memory for a task or for any learned material. Practice has a cumulative effect on memory. It can lead to skillful performance on a musical instrument, the recitation of a poem, and even to reading and understanding the written word. Over a period of no practice what has been learned tends to be forgotten.

The biological importance of memory has been studied for a number of years. While the adaptive and survival reasons for remembering are obvious, forgetting also has an important biological function. For example, it might be helpful to speculate on what would happen if memories did not fade and disappear. Without forgetting, adaptive ability would suffer. Reasoning behavior and thoughts that were appropriate a decade ago may no longer be correct and may even lead to fatal consequences. Cases are recorded of people who forgot so little that their everyday lives were very confused. Such profound confusion can lead to extreme psychosis and an inability to distinguish between present time reality and past memory. Thus, forgetting seems to serve the survival of the individual and as well as the **species**.

Another biological reason memories fade and disappear may simply be that the human **brain** at our current stage of **evolution** has a finite capacity for storage. Older memories must simply be lost in order to make room for new ones. Much research has been done on the concept of short-term and long-term memory. Short-term memory is thought to hold a limited amount of information (about 5-9 items) for only a few seconds. After that, the information is either coded into a separate long-term system or lost. There is organic evidence to support this two-system theory. Persons who have suffered damage to an area of the brain called the hippocampus can retain short-term memory functions but are apparently unable to store any new information into long-term memory.

MENDEL, JOHANN GREGOR (1822-1884)

Austrian biologist

Born on July 22, 1822, in Heinzendorf, Austria (now the Czech Republic), Mendel was the son of a peasant farmer and the grandson of a gardener. As a child, Mendel benefited from the progressive education provided by the local vicar, and he eventually enrolled at the Philosophical Institute in Olmutz (now Olomouc). However, Mendel's worsening financial condition repeatedly forced him to suspend his studies, and in 1843, he entered the Augustinian monastery at Brünn (now Brno).

Although Mendel felt no personal vocation at the time, he believed that the monastery would provide him the best opportunity to pursue his education without the financial worries. He took the name Gregor and eventually was placed in charge of the monastery's experimental garden. In 1847, he was ordained as a priest. Four years later, he was sent to the University of Vienna to study **zoology**, **botany**, chemistry, and physics. Following his studies, he returned in 1854 to the monastery and also began teaching the natural sciences at the Brno Technical School.

From then until 1868, in his limited spare time, Mendel performed most of his now-famous **heredity** experiments. No one had yet been able to make any statistical analysis in breeding experiments, but Mendel's strong background in the natural sciences and his coursework in principles of combinatorial operations prompted him to try. Mendel worked mostly with pea **plants**, carefully selecting pure varieties that had been cultivated for several years under strictly controlled conditions. He crossed different plants until he produced seven easily distinguishable seed and plant variations (yellow vs. green seeds, wrinkled vs. smooth seeds, tall vs. short plant stems, etc.).

Mendel discovered that, while short plants produced only short offspring, tall plants produced both tall and short offspring. Since only about one-third of the tall plants produced other tall plants, Mendel concluded that there must be two types of tall plants: those that bred true and those that did not.

Mendel continued experimenting. He thought that, by crossing these different plants, he would find intermediate varieties of the offspring. In other words, if he crossed a tall plant with a short plant, the result would be a medium-sized plant. He soon found that this was not the case. Mendel crossed short plants with tall plants, planted the seeds from that union, then self-pollinated the plants from this second generation. He followed the results by counting and recording each generation.

Johann Gregor Mendel (1866), as published in the frontispiece to the 1930 edition of *Mendel's Principles of Heredity. (The Library of Congress. Reproduced by permission.)*

All of the offspring that sprouted from the short-tall cross were tall, but the offspring from the self-pollination of those tall plants gave him half tall plants (non-pure), one-quarter pure tall, and one-quarter pure short. Tallness, the more powerful characteristic (the one that shows up the most), was dubbed the dominant trait. Shortness, the weaker characteristic (the one that is frequently masked), was called the recessive trait. It did not seem to matter whether Mendel used male or female plants, the results were always the same. Mendel's quiet, methodical investigation took over eight years to complete and involved more than 30,000 plants.

The results of Mendel's initial plant breeding experiments formed the basis of his first law of heredity: the **law of segregation**. This law states that hereditary units (**genes**) are always in pairs, that genes in a pair separate during division of a **cell** (the **sperm** and **egg** each receive one member of the pair), and that each gene in a pair will be present in half the sperm or eggs.

Mendel's further experiments established a second law: the **law of independent assortment**. This law states that each pair of genes is inherited independently of all other pairs. However, it holds true only if the characteristics are located on different **chromosomes**. By sheer coincidence, Mendel had indeed selected such characteristics. But genes located on the same chromosome, as Thomas Hunt Morgan later discovered, are usually inherited together.

In all, Mendel uncovered the following basic laws of heredity: 1) heredity factors must exist; 2) two factors exist for each characteristic; 3) at the time of sex cell formation, heredity factors of a pair separate equally into the **gamete**s (the sperm or the egg); 4) gametes bear only one factor for each characteristic; 5) gametes join randomly no matter what factors they carry; 6) different hereditary factors sort independently when gametes are formed.

Mendel, however, never received acknowledgment during his lifetime for the important contribution he had made to the study of heredity. Although he carefully documented his experiments, presented his findings to the Brünn Society for the Study of Natural Science in 1865, and published *Experiments with Plant Hybrids* the following year, the scientific community was indifferent. Botanists, including Karl Wilhelm von Nägeli, to whom Mendel sent his work, were unaccustomed to statistical analysis. Also, scientists as a whole were hesitant to give credence to such novel theories regarding heredity from such an obscure man.

Mendel died in Brünn on January 6, 1884. Ironically, because of Mendel's refutation of the intermediacy theory that he himself had once posited, **Charles Darwin**'s evolutionary theory was greatly bolstered, for prior to Mendel natural selection was believed to be counteracted or compromised by repeated blending of gene characteristics throughout the hereditary cycle. Not until 1900, when Mendel's pioneering work was rediscovered by Hugo de Vries and others, did Mendel begin to receive his due in the annals of science.

Mendel, Lafayette Benedict
(1872-1935)
American physiological chemist

Lafayette Benedict Mendel was a chemist known for his contributions to the field of **nutrition**. He was born in New Delhi, New York on February 5, 1872 to Benedict Mendel, a local merchant, and Pauline Ullman, both of whom had both come to the United States from Germany. Mendel attended local schools until age 15 when he won a New York State scholarship. He attended Yale University as an undergraduate where he studied classics, economics, and the humanities. He graduated in 1891 and was awarded a fellowship that enabled him to take graduate studies at the Sheffield Scientific School. There he began the study of physiological chemistry under Russell H. Chittenden.

Although Mendel's interest in science came relatively late in his academic career, he still completed the requirements for his PhD in two years. His thesis was on the synthesis of a **protein** derived from hemp seed, a topic which he would return to in future research efforts. That same year began teaching at the Sheffield Laboratory of Physiological Chemistry. In 1897, after a year's break during which he traveled to Germany to study physiology and chemistry, he was appointed assistant professor. In 1903 Mendel was made a full professor of physiological chemistry. During this time he was known for the clarity and forcefulness of his lectures, as well the warmth of his relationships with his students.

Mendel was also recognized for the quality of his scientific work was in the field of nutrition. He worked with **Thomas B. Osborne** of the Connecticut Agricultural Experiment Station and together they authored more than 100 papers dealing with nutrition. Their first work involved ricin albumin, a poisonous protein derived from castor beans which is so potent that it is fatal to rabbits at only two one-thousands of a milligram per kilogram of body weight. In 1909, Mendel and Osborne began studying the nutritive effects of proteins. They devised experiments in which they fed rats on a controlled basis; they accurately measured the rats food intake and then recorded subsequent changes in weight of the animal. Using this method they discovered a variety of substances that were necessary for a healthy diet. For example, their studies with protein-free milk led them to discover **vitamin** A. Unfortunately for Mendel, McCollum at the University of Wisconsin had made the same discovery a few weeks earlier. Nonetheless, the pair are given credit for the discovery of other key **nutrients**, such as water soluble vitamin B. With annual grants from the Carnegie Institute of Washington, they also found that a proper diet must include lysine and tryptophan, two essential **amino acid**s. In addition, they established that the body can not produce certain nutrients on its own and it must receive these elements, or their precursors, thorough diet.

Even though Mendel was made a Sterling professor of physiological chemistry at Yale in 1921, he maintained his collaboration with Osborne until Osborne's retirement in 1928. Mendel continued to work with junior scientists until his death in New Haven, Connecticut on December 9, 1935. Today Mendel's work in identifying nutritional compounds is considered to be of inestimable value to modern agricultural science. In his lifetime Mendel helped develop nutrition into a well documented branch of biochemistry, and he established several methodologies which are still used today.

Mendelian laws of inheritance

The foundations of the modern science of **genetics** were laid by **Gregor Mendel**, an Austrian Monk, who carried out experiments on the **inheritance** of characters between generations. Mendel worked on inheritance in sweet-peas, and selected characters that bred true; that is, the characters did not blend into one another in the next generation. Characters chosen for study by Mendel included **flower** color (such as red versus white), plant height (tall versus dwarf), seed coat (smooth-coated seeds verses wrinkled seeds), pod length (long pods versus short pods), and so on. Mendal eventually formulated the three laws of genetics, known today as the Mendelian laws of inheritance. These are the **law of segregation**, the **law of independent assortment**, and the **law of dominance**. Mendel's work went unnoticed for nearly two decades after his **death** in 1887, but was eventually recognized widely by the scientific community.

In order to understand Mendel's three laws of inheritance, it is necessary to review the plant breeding experiments

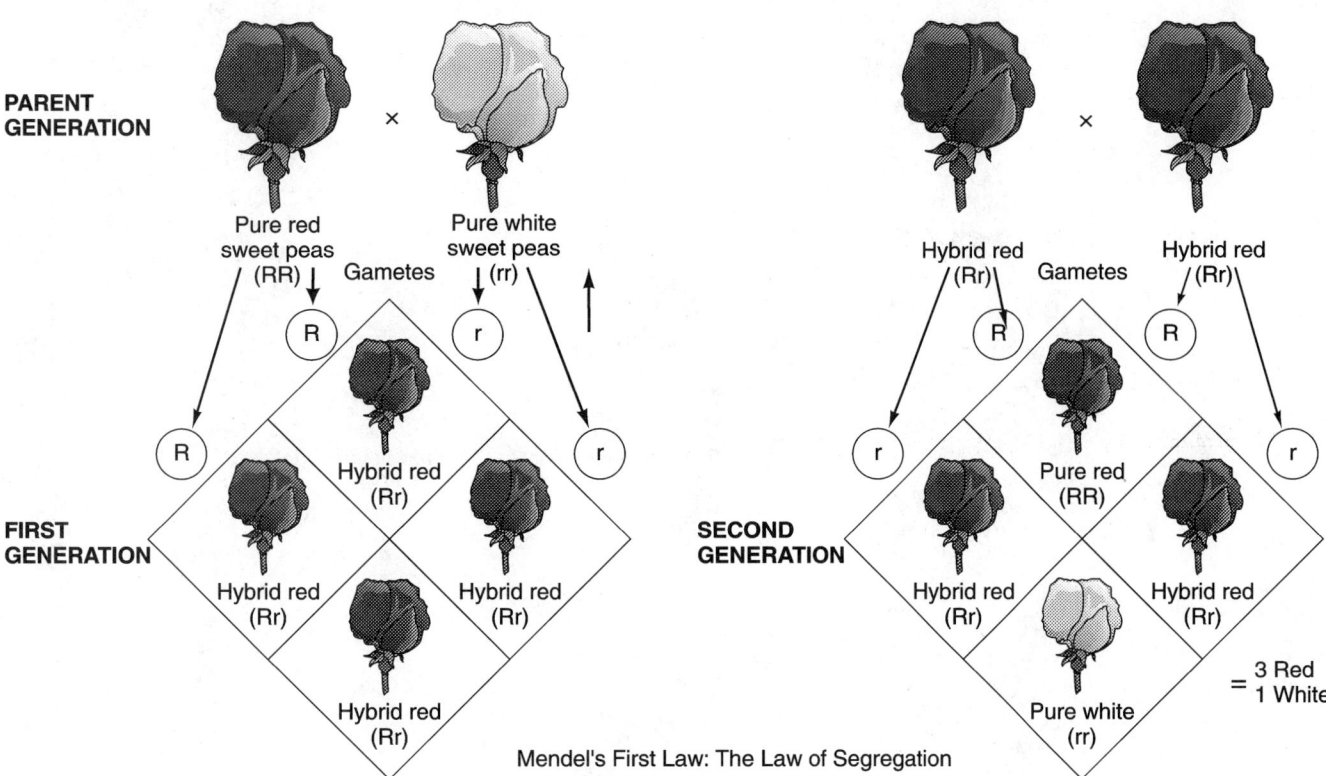

PARENT GENERATION

Pure red
sweet peas
(RR)   Gametes   Pure white
sweet peas
(rr)

R   r

FIRST
GENERATION

R   r

Hybrid red
(Rr)

Hybrid red
(Rr)   Hybrid red
(Rr)

Hybrid red
(Rr)

SECOND
GENERATION

Hybrid red
(Rr)   Gametes   Hybrid red
(Rr)

R   R

r   r

Pure red
(RR)

Hybrid red
(Rr)   Hybrid red
(Rr)

Pure white
(rr)

= 3 Red
1 White

Mendel's First Law: The Law of Segregation

(Illustration by Hans & Cassady, Inc.)

which inspired the laws. Two of the three laws involve dihybrid crosses where two different sets of traits are studied together. For example, if tall **plants** (controlled by two dominant **allele**s, TT), red-flowering (controlled by two dominant alleles, RR), sweet-peas are pollinated by dwarf (controlled by two recessive alleles, tt), white-flowering (controlled by two recessive alleles), firn plants the pollen and ova will contain either TR or tr, which represent only a single allele of each set of genes. When **fertilization** takes place, the resulting first filial generation (F1) will all have progeny of the same outward appearance (**phenotype**); they will all be tall, red-flowering (TtRr) sweet-peas. The progeny of the first filial generation (the F2 generation) are also all tall, red-flowering plants, because R, the allele for red and T, the allele for tall are dominant to r and t, which are known as recessive alleles. Since sweet-peas are self-pollinating and each plant now produces **gamete**s of four different **genotype**s, namely TR, Tr, tR, and tr. When the seeds resulting from the random combination of gametes with one of these genotypes, the second filial generation (F2) will have four different phenotypes: tall with red flowers, tall with white flowers, dwarf with red flowers, and dwarf with white flowers, in the proportion 9:3:3:1.

Mendel found that sweet peas with the same phenotype (purple flowers) had different genotypes which were only expressed in subsequent generations. Purple flowers were produced when the alleles contained at least one allele for purple flowers, which is dominant over the allele for white flowers. White flowers were produced when both alleles coded for white flowers which is the homozygous recessive allele.

During Mendel's time neither the existence of genes nor their structure and function were understood. Indeed, **chromosomes** remained unknown for several years after Mendel's death. Mendel's experimental plants had factors that occurred in pairs. These factors have only one member of a pair in the gametes (pollen and ova). Mendel described the three laws of inheritance that described the passage of genes from one generation to the next.

Mendel's three laws of inheritance:

• Mendel's first law of inheritance is the law of segregation. This states that genes segregate during gamete formation into their different alleles. The two members of a pair of alleles separate (segregate) into two different gametes, and exert their influence in the offspring as one of a new pair of alleles. Segregation is the result of the separate carriage of genes on **chromosomes**, which are not altered or blended by forming pairs. A gene for red flowers in the sweet-pea does not become diluted from having been paired with the gene for white flowers and is passed to subsequent generations unaltered in the gametes.

• The second Mendelian law of inheritance is the law of independent assortment which describes the chance distribution of alleles to the gametes (ova or spermatozoa). If an individual has two pairs of alleles, Aa and Bb, its gametes will contain equal numbers of the four possible combinations (AB, Ab, aB, ab), with one member from each pair. Independent assortment applies only to **genes**

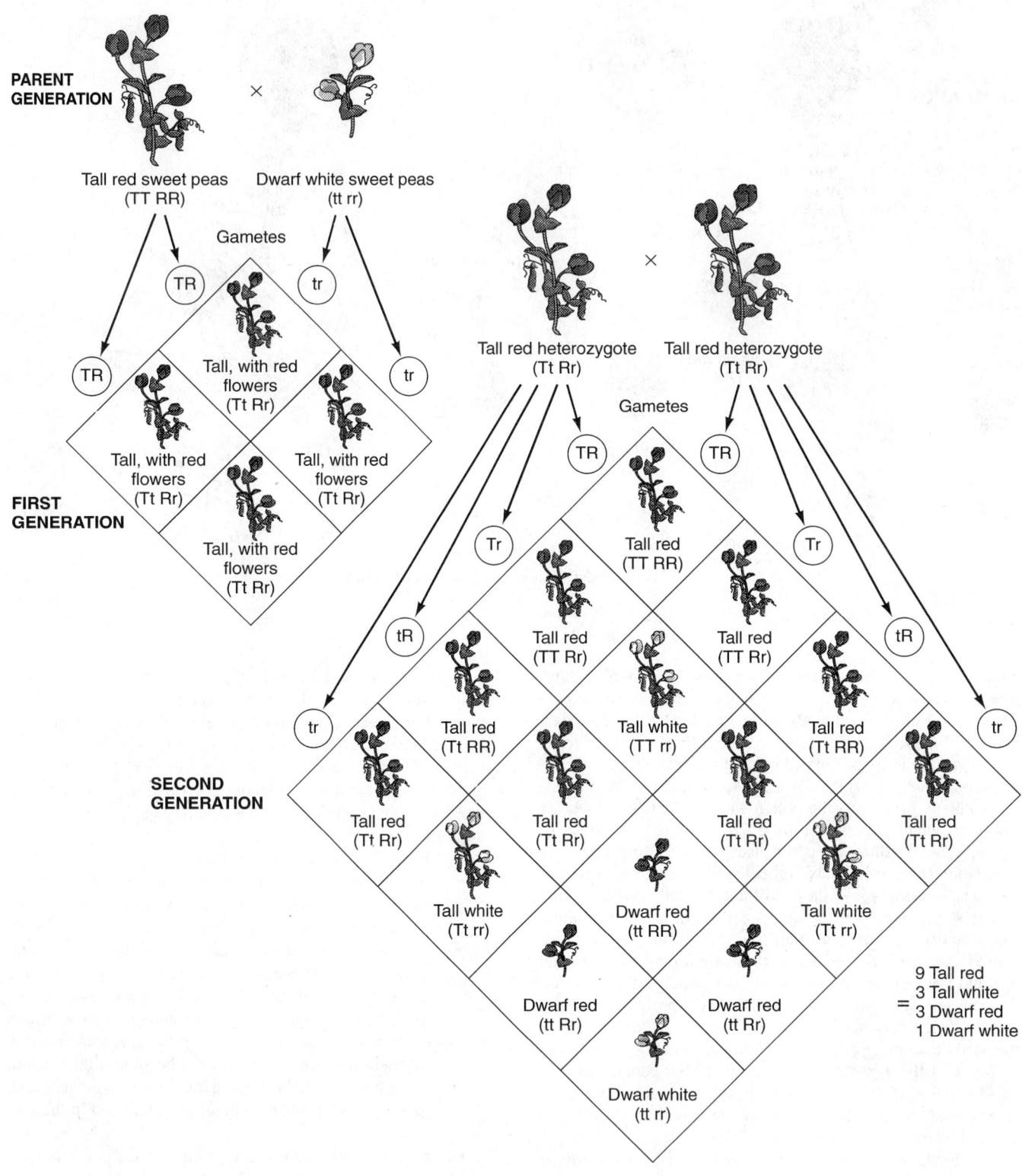

PARENT GENERATION

Tall red sweet peas
(TT RR)

Dwarf white sweet peas
(tt rr)

Gametes

TR

tr

TR

Tall, with red
flowers
(Tt Rr)

tr

FIRST GENERATION

Tall, with red
flowers
(Tt Rr)

Tall, with red
flowers
(Tt Rr)

Tall, with red
flowers
(Tt Rr)

Tall red heterozygote
(Tt Rr)

Tall red heterozygote
(Tt Rr)

Gametes

TR

TR

Tr

Tr

Tall red
(TT RR)

tR

tR

Tall red
(TT Rr)

Tall red
(TT Rr)

tr

tr

Tall red
(Tt RR)

Tall white
(TT rr)

Tall red
(Tt RR)

SECOND GENERATION

Tall red
(Tt Rr)

Tall red
(Tt Rr)

Tall red
(Tt Rr)

Tall red
(Tt Rr)

Tall white
(Tt rr)

Dwarf red
(tt RR)

Tall white
(Tt rr)

Dwarf red
(tt Rr)

Dwarf red
(tt Rr)

= 9 Tall red
3 Tall white
3 Dwarf red
1 Dwarf white

Dwarf white
(tt rr)

Mendel's Second Law: The Law of Independent Assortment

(Illustration by Hans & Cassady, Inc.)

lying on different chromosomes, and does not apply to linked genes on the same chromosomes. The F2 generation shows a ratio of nine tall red-flowers, three dwarf red-flowering, three tall white-flowering, and one dwarf, white-flowering. A monohybrid cross, which involves a single character, such as plant height produces 12 tall plants, and four dwarf plants, in a typical 3:1 ratio. Each pair of alleles making up a gene, whether controlling plant height or flower color, behaves though each pair were the only gene present for each pair segregates independently of other characters.

- The third Mendelian law of inheritance, the law of dominance, states that heredity factors (genes) work together as sets, usually as pairs of alleles. The total number of different alleles represents some of the variation available to species. Frequently, in heterozygotes (with one dominant and one recessive allele) only the dominant allele of the gene is expressed in the phenotype, the recessive allele being concealed and is not expressed.

MENSTRUAL CYCLE

The normal human menstrual cycle is 28 days, but no woman is always precisely regular, and cycles as short as 21 days or as long as 35 days are not abnormal. Menstruation is the shedding of the uterine lining, the **endometrium.**

The menstrual cycle begins with about five days of menstruation, followed by a proliferative phase during which ovulation occurs. The proliferative phase lasts to about the fourteenth day, and then a secretory phase begins.

At the end of menstruation, just at the beginning of the proliferative phase, the endometrium is thin, and the ovary is dormant. With the production of gonadotropic **hormone**s from the **pituitary gland,** an ovarian follicle (occasionally more than one) ripens in one of the ovaries. This ovarian follicle contains the ovum. **Cell**s surrounding the ovum multiply, and secrete an **estrogen**ic hormone, estradiol This hormone causes multiple changes to occur in the endometrium, so that it becomes thicker and more vascular.

At about mid-cycle ovulation occurs: The ovum is released from the follicle and received into the fallopian tube, down which it is carried to the uterus.

If the ovum is fertilized during the proliferative phase, it reaches the uterine cavity at a time when the endometrium is in the secretory phase, and the ovum embeds itself in the endometrium and starts its growth. If the ovum is not fertilized the endometrium breaks down and menstruation occurs. The disintegrating endometrium is shed, together with some **blood.** The endometrium contains plasmin, an **enzyme** that dissolves blood clots, so that the menstrual discharge is normally fluid. The total blood loss does not generally exceed 1.69 oz (50 ml).

See also Reproduction, human

MESODERM

Mesoderm is one of the three primary germ layers of a vertebrate **embryo.** A **zygote** is formed after **fertilization** which develops by **cell division** into a **blastula.** This is followed by gastrulation which is the process of **cell** movements and **translocation**s that results in the formation of a three layered embryo. The superficial layer is known as **ectoderm**, the interior layer is **endoderm**, and the middle germ layer is known as mesoderm. Through the process of **differentiation**, mesoderm gives rise to all the **organ**s between the ectoderm and endoderm. These include striated (voluntary) skeletal muscle, **heart** muscle, **blood**, spleen, adrenal cortex, kidneys, and the various kinds of connective **tissue**, e.g., cartilage, bone, smooth muscle (this includes the involuntary muscle of the uterus, the muscular layers of the gut, and urinary bladder), the connective tissue layers of the skin, and endothelial lining of blood vessels.

A fish is quite different from a frog, a lizard, a mouse or a human. Yet, the mesoderm gives rise to the same organs in each of these diverse forms. The kidneys of a frog (mesonephros) and the kidneys of a human (metanephros) are both derived from mesoderm as is the kidney of all vertebrates. The germ layer origin of differentiated tissues is strictly conserved which is why students of **embryology** still learn about the primary germ layers.

Clearly, the most of the mass of the adult body is of mesodermal origin. Malignancies of mesodermal origin are referred to as sarcomas. Prefixes to that term give a more precise definition of that which is malignant. For example, chondrosarcoma is a malignancy of cartilage and fibrosarcoma is a malignancy of fibrous connective tissue. While mesoderm is the dominant tissue of the body, it is vulnerable to fewer malignancies than epithelial cells.

MESSENGER RNA • See RNA (ribonucleic acid)

METABOLISM

The term *metabolism* refers to all the chemical changes that take place in the body's **tissue**s when the **cell**s are producing both **energy** and essential new organic materials. While these metabolic activities are many and varied, most of them fall into two broad categories: *anabolic* processes and *catabolic* ones. These two quite different processes take place constantly and simultaneously.

Anabolism is the cell's synthesizing or building-up phase. Through anabolic reactions, simple substances—usually the **molecule**s of glucose, fatty acids, and **amino acid**s that have been derived from foodstuffs—are combined in various ways to form more complex substances, such as the new cellular material needed for growth and tissue maintenance. By combining amino acids, for instance, the cells can form structural **protein**s and use them to repair or replace worn-out tissues. The cells can also form functional proteins, such as **en-zyme**s, antibodies, and **hormone**s.

Catabolism, the cell's degradative or breaking-down phase, is almost exactly the reverse. Through catabolic reactions, complex compounds—proteins, **fats**, and **carbohydrate**s—are broken down primarily to produce the energy needed for all metabolic activities. Not all the energy is used up at once: as complex nutrient molecules are broken down and oxidized—or burned—as fuel, only about 60% of the energy released is used for immediate needs. The rest is stored in the **chemical bond**s that link certain **atoms**, most notably those of the phosphate **adenosine triphosphate (ATP)**. When the body needs energy, enzymes break these chemical bonds and release the stored energy for use by the cells.

Enzymes help regulate most metabolic activities. For example, hormones secreted by special cells in the **pancreas** help determine whether metabolic reactions will be largely anabolic, as is usually the case soon after a meal is eaten, or catabolic, generally during periods when additional energy is needed. And thyroxine, a hormone secreted by the **thyroid gland**, is one of several hormones that help determine the rate at which these activities will occur—the metabolic rate or, roughly, the rate at which the body uses up energy in the course of its various metabolic reactions.

The metabolic rate is influenced by a number of factors, such as the individual's age, sex, level of activity, state of health, and, of course, the amount of the hormone thyroxine he or she secretes. Probably the most influential factors, though, are the **temperature** of the surrounding environment and the calorigenic effects of the foods most recently eaten. Because an individual's metabolic rate can give physicians a great deal of important information, it often needs to be measured. And since almost all the energy used by the body is eventually converted to heat, the metabolic rate is usually calculated by measuring the amount of heat loss an individual displays during resting (or basal) conditions. The person's basal metabolic rate (BMR) is then judged to be normal or abnormal by comparing it to standardized rates—that is, rates that reflect the average BMRs of healthy individuals of various ages taken under identical and standardized conditions.

Back in the 1830s, the German physiologist **Theodor Schwann** coined the word metabolism for the chemical changes that take place in living tissues. Fittingly enough, it was during Schwann's lifetime that many important metabolic concepts were first formulated. In 1828, for instance, Friedrich Wöhler discovered that urea, an organic product present in urine, could be manufactured in the laboratory from inorganic chemicals, a discovery that stimulated interest in studying the body's own chemicals. Later, Wöhler also showed that a chemical taken by mouth somehow had combined with other chemicals when it appeared in the urine, one of the first studies that demonstrated chemical changes definitely could—and did—occur inside the body.

In 1842, **Justus von Liebig** showed that **animal** heat stemmed almost entirely from the oxidation of recently consumed foods. Liebig, who also believed that not all foods provided equal amounts of heat, pioneered a number of studies aimed at determining the caloric values of different foods. In the 1850s, **Claude Bernard**, the noted French physiologist,

made an important contribution to the study of metabolism. For one thing, Bernard discovered a starch-like substance in the liver that he named glycogen. He went on to show that glycogen was composed of simple **blood** sugars stored in the liver and could be broken down into glucose when needed. This process, Bernard argued, indicated that the body's various mechanisms appeared to work together, in an integrated fashion, to maintain a constant and well-balanced inner environment.

In the 1890s, German physiologist Max Rubner (1854-1932) showed that the energy used and released by animals followed the same basic chemical principles that inanimate systems were already known to follow, and **Eduard Buchner**'s experiments with **yeast** proved that metabolic processes in **organism**s resulted strictly from **chemical reaction**s and were not energized by a "life force." In the years that followed, numerous scientists—among them **Hermann von Helmholtz**, German physiologist Eduard Pflüger (1829-1910), and **Arthur Harden**—were able to establish most of the important metabolic parameters.

During the twentieth century, therefore, many metabolic studies began to center around the concept of intermediary metabolism. The concept, introduced in the 1860s by Karl von Voit, held that chemical reactions were often much longer and more complicated than had previously been realized, and that it was vital for chemists to search for the intermediate steps that helped a chemical reaction get from one stage to another. Among those scientists working on intermediary metabolism were two German physiologists, Gustav Embden (1874-1933) and **Otto Meyerhof**, who, in 1933, were able to work out the complicated sequences of chemical reactions that are involved in the breakdown of glycogen. Four years later, **Hans Krebs** introduced a model of what he called the citric acid cycle, a central feature of one of the body's major catabolic pathways. And in the late 1940s and 1950s, German biochemist **Fritz Lipmann** made a number of valuable contributions to intermediary metabolism through his discovery of **coenzyme** A and by his emphasis on the vital role played by phosphates in various metabolic pathways.

One focus of recent research is the role of metabolism in the **aging** process. Studies have found that reducing **calorie** intake substantially, but not to the point of malnutrition, extends the life span of animals ranging from spiders and fleas to mice and monkeys. The increased longevity goes beyond the effects of simply avoiding diseases that are linked to obesity. Scientists are now trying to better understand this observation.

METAMORPHOSIS

A fertilized **egg** develops into an **embryo**. Sometimes embryos develop directly into adults. However, many **species** have an intermediate form between the embryo and the adult which is known as a **larva**. Often the larva bear little resemblance to the adult. Accordingly, for the adult to form, the larva must undergo significant anatomical and physiological changes. Those changes are referred to as metamorphosis.

The selective advantage of metamorphosis is relatively clear in some cases. Consider the sea urchin, a sessile (non-swimming) marine invertebrate related to the starfish. Sea urchins would have little opportunity to disperse if it were not that they have a larval form, the pluteus, which is **cilia**ted and which moves rapidly with sea currents. The pluteus larva bears essentially no resemblance to the adult form. Pluteus larvae permit dissemination of sea urchins which in turn prevents overcrowding of adults and allows for colonization of new sites. Tunicates are for the most part sessile also. They grow permanently attached to rocks or seaweeds. Some take the form of a sac with two openings which permits the flow of sea **water**, containing microscopic food, through the **animal**. It is a primitive **organism** indeed and its principal response to a disturbance is to eject sea water. The ejection of water has led to the common name ''sea squirts''. Tunicates would have essentially no means of dispersal were it not for a swimming larva which looks very much like a tadpole complete with a muscular tail. The tadpole bears essentially no resemblance to the adult tunicate. Most people have observed metamorphosis in certain insects such as the monarch butterfly. After a period of feeding as caterpillars, these form pupae which in turn eventually undergo metamorphosis to form adults butterflies. The endocrinology of insect metamorphosis has been studied extensively. It has been said that if the **life cycle** of marine animals and certain insects were not known, many if not all, larval forms would be judged unrelated to their adult form.

Frog (*Rana pipiens*) development is perhaps the best known instance of metamorphosis. What is quite obvious is that an aquatic swimming legless larva, the tadpole, gives rise to a terrestrial four-legged form, the frog. Metamorphosis is quite complex. **Respiration** of the larva is accomplished by gills contained within a gill chamber. Water containing oxygen flows over the delicate gills which permits oxygenation of red **blood cell**s. Lungs replace gills at metamorphosis. The lungs allow for oxygenation of red blood cells from the **atmosphere**. **Herbivore**s (plant eating animals) have long **digestive system**s to permit utilization of plant food. **Carnivore**s have short digestive systems related to the greater ease of digesting animal food. Frog tadpoles feed on plant food and it is not surprising that they have a long coiled gut that permits utilization of this diet. Frogs on the other hand eat animal food such as insects and worms. Metamorphosis is a time of remodeling the long tadpole gut to the short digestive system of a frog. Tadpoles, as do other aquatic organisms, excrete **nitrogen**ous waste (from **metabolism**) as **ammonia**. Frogs convert nitrogenous waste to the less toxic urea which is eliminated from the body in urine. Other metamorphic changes included drastic changes in the skin, eyes, **nervous system**, etc. The many complex **transformations** are under the control of thyroid **hormone**s thyroxine and triiodothyronine.

The rationale for metamorphosis in northern leopard frogs, *R. pipiens*, is unknown. However, they overwinter at the bottoms of selected lakes and rivers which do not freeze solid. The frogs leave the lakes in the spring to reproduce in ponds and **wetlands**. The advantage of the transient bodies of water is the lack of predator fish. Ponds and wetlands are bereft of

The transformation from larva to adult that a butterfly undergoes during pupation is an example of metamorphosis. *(Photograph by Nelson-Bohart & Associates, Phototake NYC. Reproduced by permission.)*

fish because these bodies of water dry by the end of summer. Thus, the period when an aquatic life is appropriate for larvae ends in late summer. Metamorphosis to the terrestrial form seems to be related to the loss of an aquatic **habitat**.

METCHNIKOFF, ELIE (1845-1916)
Russian microbiologist

An important early researcher in immunology, Elie Metchnikoff was born in Kharkov, Russia, on May 16, 1845. His father was an officer of the Imperial Guard. His mother was the daughter of a Jewish writer, and she encouraged her son's interest in natural sciences. After graduating from the University of Kharkov in 1864, Metchnikoff continued his studies in Germany, then returned to Russia and earned his Ph.D. from the University of St. Petersburg in 1867. Metchnikoff's early career was difficult. His first faculty positions, at the University of Odessa beginning in 1862 and at St. Petersburg in 1868, were marred by difficult working conditions and severe eyestrain. After his first wife died in 1873, Metchnikoff attempted suicide. Thanks to a happy and financially successful second marriage, Metchnikoff became financially independent and moved to Messina, Italy, in 1882 to devote himself to research. There, he made his great discovery.

In 1865, Metchnikoff had studied roundworms for the purpose of observing intracellular **digestion**. In Messina, he studied transparent starfish **larvae** and observed a similar process, whereby mobile **cell**s surrounded and engulfed invading foreign particles. He called these **bacteria**-eating cells phagocytes and devoted the next twenty-five years of his life to developing and promoting his concept of **phagocytosis**. Continuing his research, he showed that white **blood** corpuscles in higher **animal**s and humans are also phagocytes. Metchnikoff first published his ideas about phagocytes in 1883 and wrote a comprehensive book on immunity in 1901. This new concept met with serious objections: first, because it contradicted the prevalent idea that white blood cells aided rather than attacked bacteria, and second, because it seemed to conflict with findings that antibody substances in the blood were responsible for immune responses. Metchnikoff worked, wrote, and spoke vigorously in support of his ideas, which became accepted as a component of the **immune system** in the early 1900s. He shared the 1908 Nobel Prize for medicine or physiology with Paul Ehrlich for his work on white blood corpuscles.

Metchnikoff returned to Russia to head Odessa's Bacteriological Institute from 1886-1887. After his work came to the attention of **Louis Pasteur**, Metchnikoff settled in Paris in 1888 as a member and then director of the Pasteur Institute. Late in his life, he became interested in longevity, which he linked to bacteria in the intestinal tract. He believed that regularly ingesting **lactic-acid** bacilli—found in sour milk and yogurt—would increase a person's life span. Metchnikoff died in Paris of cardiac failure on July 16, 1916.

MEYERHOF, OTTO (1884-1951)

German American biochemist

The son of a merchant, Meyerhof received his medical degree from the University of Heidelberg in 1909. Although he was originally attracted to psychology, a meeting with Otto Warburg aroused his interest in cellular physiology and, in 1913, when he joined the faculty of Kiel, he began what was to be a lifelong investigation into the biochemistry of muscle.

At the time, it was already common knowledge that muscle contained glycogen and that—as **Frederick Gowland Hopkins** and his coworkers had shown a decade earlier—a working muscle accumulates **lactic acid**. With these facts as starting points, Meyerhof conducted a series of experiments and, in 1919, demonstrated how the glycogen-lactic acid cycle

works. In phase one, when the muscle begins to contract, glycogen is converted to lactic acid. Oxygen is not consumed, but an oxygen debt is built up. In phase two, when the muscle rests after work, molecular oxygen is then consumed to pay off the "debt" and to oxidize about one-fifth of the lactic acid. The **energy** yielded from the oxidation process makes it possible for the remaining four-fifths of lactic acid to be reconverted to glycogen.

Meyerhof's discovery elaborated the observation of **Archibald Vivian Hill**, in 1913, that heat was emitted and oxygen consumed only during a muscle's contraction and its recovery. Hill and Meyerhof shared the 1922 Nobel Prize in physiology or medicine for their work on the biochemistry of muscular action. Meyerhof's work also laid the groundwork for **Carl** and **Gerty Cori**'s more detailed explanation, a few years later, of the steps by which glycogen is converted to lactic acid (a process thereafter often known as the Embden-Meyerhof pathway, after Meyerhof and a coworker)

Like many other Jewish scientists, Meyerhof left Germany in 1938 after the Nazis rose to power. Unfortunately, the biochemist settled in Paris, France, and was therefore forced to flee a second time when the Germans invaded France in 1940. Meyerhof then came to the United States and became a professor of physiological chemistry at the University of Pennsylvania, in Philadelphia, remaining there until his death.

MICROSCOPE

Microscopes magnify images so that we can see things that otherwise would be invisible with the naked eye. The most common type, **light** microscope, uses glass lenses to focus light and create a high resolution image. Some light microscopes collect light reflected from the surface of solid objects. Using stereo (twin) eyepieces, they produce a three-dimensional view of surfaces but usually don't reveal any internal structure. This type of instrument usually is limited to enlargements between 10 and a few hundred times. Large plant or **animal** cells are generally visible in at this magnification but smaller structures are mostly indistinguishable.

The microscopes most commonly used in biology laboratories transmit light through a thin specimen held on a glass slide. A condenser lens between the light source and the specimen increases illumination brightness while one or more objective lenses collects light passing through the specimen and forms an image. The most common magnification ranges for these instruments is 4x, 10x, 40x, and 100x. The image usually is further magnified 10 or 15 times by an eyepiece. The objective lens most important in determining both ultimate magnification and resolution of the instrument. With a very good objective lens, the highest useful magnification for a light microscope is about 1,000 diameters, meaning that smallest space between objects that can be resolved is about 0.2 micrometers.

The most important limit to resolution in any optical system is caused by diffraction as light rays pass through the lenses. Late in the nineteenth century, the German physicist Ernst Abbe showed both theoretically and experimentally that best possible resolution in a light microscope is equal to about 0.6 times the wavelength of the illuminating light divided by a factor called the numerical aperture that combines the refractive index of the medium through which light passes and the diameter of the lens. The best possible numerical aperture of an oil immersion lens (one in which the specimen and the lens are immersed in oil) is about 1.4. The wave length of visible light is approximately 0.5 micrometer and thus the limit to resolution is about 0.2 micrometers.

Some very small biological specimens such as living **bacterial** or **protozoa**n cells can simply be suspended in a drop of **water** and observed directly with a light microscope. For most **tissue**s from higher **plants** or animals, however, elaborate preparation techniques are required to preserve specimens, stain them to make internal structures visible, and to make it possible to cut slices as thin as about 1 micrometer that are relatively transparent to light. The thinner the section is, the crisper the resulting image. Chemical fixatives such as formaldehyde are used to preserve tissue. Organic solvents such as **alcohol** are used to extract water and to allow an embedding medium such as paraffin wax to infiltrate the specimen. The most common stains in light microscopy are organic dyes such as hematoxlyn and eosin, which bind to **proteins** and **nucleic acids** to create distinctive colors that help identify cellular components.

A recent development that provides high resolution images with thick sections is called confocal light microscopy. In these instruments, a very narrow beam of light produced by a laser shining through tiny pin-hole aperture focuses at a very specific spot in the specimen. Fluorescent light re-emitted from special stains in the specimen at a wavelength different than that in the incident illumination is collected by a mirror and passed through a pin-hole that is confocal (that is at exactly the same focal length) with the laser aperture. The fluorescent light is collected by an extremely sensitive photodetector, amplified, and displayed on a video terminal. Because computer controls the position of the confocal apertures very exactly, the image is as clear as if it had been made with an extremely thin section. Digital enhancement of the electronic signal from the photodetector gives very good images of structures that would otherwise be invisible in a light microscope.

The resolution limits imposed by light optics are overcome by electron microscopes, which use a beam of very high **energy** electrons rather than light to create an image. Although roughly a thousand times larger and more expensive than an ordinary light microscope, these instruments also have approximately one thousand times better resolution. This means that magnifications of a million times or more are possible and that objects as small as a fraction of a nanometer can be resolved. Large **molecule**s are generally visualized easily in an electron microscope and in special cases even individual **atoms** can be seen.

Preparation of biological tissues in electron microscopy is difficult because the specimen must be able to withstand both very high vacuums and the intense bombardment of the electron beam. Biological samples are usually fixed (pre-

served) with glutaraldehyde or some other chemical that retains fine structure. They are then dehydrated and embedded in a very hard epoxy resin that can be sliced into sections about 20 nanometers thick by an extremely sharp diamond knife. The only stains useful in a TEM are heavy metals such as lead, tungsten, uranium, or gold. In some cases these stains can be coupled to antibodies or unique enzymatic reactions to give useful information about the location of specific cellular components. An alternative technique for sample preparation for the TEM involves freezing biological tissues in liquid **nitrogen**, fracturing them in a high vacuum system, and then making a very thin carbon-platinum replica of the fractured surface. This freeze-fracture technique has been useful for studying **membrane structure**.

In spite of the expense of confocal and electron microscopes, and the difficulties in preparing tissues for observation, a good deal of what we know about the internal organization of **cell**s has come about since the introduction of these instruments in the past few decades. An explosion of information in cell biology brought about in recent years by higher resolution instruments and better preparation techniques brought us a wealth of information about cellular structure and how living **organism**s function.

See also Cytology

MIGRATION

Migration refers to periodic or seasonal movements of **animal**s over a relatively long distance, from one **habitat** or climate to another. Migrations may be made by particular individuals, or by an entire **population** of a **species**. Migration is an extremely "expensive" undertaking for animals to make, in terms of the energetic costs of the extensive movements from one place to another. Migration can also be a risky venture, during which animals are exposed to predators, the possibility of becoming irretrievably lost, and other natural hazards. However, there are also substantial benefits to be gained by migrating. These are usually associated with having access to a seasonal abundance of food in certain kinds of habitats. It is because of the opportunity to exploit such abundant resources that many species of animals have adopted the extraordinary behavioral attribute of migration.

Individuals of some species undertake regular short-distance migrations, usually to take advantage of seasonally rich habitats at higher altitude in the mountains. These consistent, seasonal movements are referred to as "vertical" migrations. For example, in the Rocky Mountains, individual grizzly bears (*Ursus arctos*) commonly migrate to high-altitude **tundra** meadows in the early summer, where they feed on a lush growth of sedges and other **plants**. Later in the summer they migrate downward to feed in lower-altitude valleys and meadows, where they also spend the winter sleeping deeply in a den. Some birds of the mountains also undertake migrations of this sort, as is the case of Clark's nutcracker (*Nucifraga columbiana*).

A much larger number of animal species undertake longer-distance seasonal migrations. For example, caribou

Caribou in the Arctic National Wildlife refuge. Some caribou migrate over 600 miles (966 km) to spend the winter in forests. *(U.S. Fish and Wildlife Service. Reproduced by permission.)*

(*Rangifer tarandus*) living in the subarctic regions of Alaska and northern Canada spend the long winter in the coniferous boreal forest, where habitats are relatively protected from the cold and wind. They then migrate northward in the summer to graze in the open tundra, where there is a lush **productivity** of herbaceous plants to feed upon, a relatively windy landscape where the density of biting flies is somewhat less, and fewer wolves (*Canis lupus*) and other predators of the vulnerable caribou calves.

Some large **mammals** of temperate prairie and tropical savanna also undertake long-distance, seasonal migrations. The once-abundant buffalo (*Bison bison*) of North America once migrated in enormous herds from their summer range in the northern prairie, to their winter habitats further south. Wildebeest (*Connochaetes taurinus*) still have similarly impressive migrations in the savanna and grassland of eastern Africa.

Some species of fish are also long-distance migrants. Atlantic salmon (*Salmo salar*) breed in large rivers, but the juveniles migrate to the ocean, where they feed for several years until they become reproductively mature, after which they migrate back up their **birth** streams to breed. This migratory habit is known as anadromous, and it is shared by the various species of Pacific salmon (*Oncorhynchus* spp.), such as the coho salmon (*Oncorhynchus kisutch*). In contrast, American eels (*Anguilla rostrata*) are known as catadromous, because they breed in the ocean (in the Sargasso Sea), and the young elvers migrate up rivers and lakes to grow into adults.

Most of the species of birds that breed in the temperate, boreal, and tundra zones undertake long-distance migrations, from their northern breeding habitats to their wintering habitats in the subtropics or tropics of the southern parts of their

range. (Of course, in the Southern Hemisphere, these directions are reversed.) In North America, the many species of migratory birds include almost all of the flycatchers, hummingbirds, tanagers, thrushes, vireos, warblers, and most of the other songbirds that breed in the **forests**, tundra, and **wetlands** of North America. These birds are essentially tropical species that somehow discovered that more northerly **ecosystem**s supported a great abundance of insects, spiders, and other critical foods during the growing season of spring, summer, and autumn. They learned to migrate to those seasonally profitable habitats, where they spend their breeding season, which is the time that birds need access to the greatest abundance of insect foods. This is because of the enormous demands of their ravenous young, which in most cases do not feed themselves and must be nourished by the parents, even for some time after they have learned to fly. By undertaking long-distance migrations, these species of birds are able to successfully raise their large families, while avoiding the relatively intense **competition** for food that occurs in most tropical habitats.

During the past several decades, ecologists and birdwatchers have been reporting alarming declines in the populations of many species of migratory birds in North America. Most of the declining species breed in mature temperate and boreal forests. The reasons for the songbird declines are not totally understood, but the most important factors are thought to be the following: (1) extensive deforestation in the tropical wintering range; (2) disturbance of mature-forest habitat in the northern breeding range; (3) fragmentation of breeding habitats into "islands" too small to sustain populations over the longer term, and that are easily penetrated by forest-edge predators and nest **parasite**s; (4) loss of critical habitats used for staging and migration; and (5) effects of **pesticides** and other toxic chemicals. One of these migratory songbirds, the Bachman's warbler (*Vermivora bachmanii*), used to breed in mature hardwood forests in the southeastern United States. Unfortunately, it became extinct several decades ago. This tragic loss was probably caused by the loss of the critical wintering habitats of Bachman's warbler on the island of Cuba in the Caribbean. These tropical forests were destroyed and converted into sugar-cane plantations.

Not all of the avian migrants are songbirds (many species of sandpipers, plovers, raptors, and seabirds also undertake long-distance migrations. Some of them travel extraordinarily long distances. One example is the whimbrel (*Numenius phaeopus*), which breeds on the northern tundra, and then migrates over land and the ocean to spend the winter on the pampas of Argentina. However, the longest migrations of any animal are made by the arctic tern (*Sterna paradisaea*), some of whom breed in the High Arctic, at the limits of land on the northern tips of Ellesmere Island, Greenland, and Spitzbergen. Once their broods are raised, these terns migrate south to spend their winter foraging in Antarctic waters. Therefore, twice each year, they travel virtually the entire length of the globe!

Miller, Stanley Lloyd (1930-)
American biochemist

Born in Oakland, California, Miller earned a B.A. in Chemistry at University of California. At age 24, he earned his Ph.D. at the University of Chicago, where he worked under the Nobel prize winner Harold Urey. After earning his doctorate, Miller became a Jewett Fellow at the California Institute of Technology. He worked for five years as an instructor and later as an assistant professor in the Biochemistry Department at the Columbia College of Physicians and Surgeons. Beginning in 1960 Miller held positions at the University of California in San Diego as an assistant professor, associate professor and professor of Chemistry.

Miller is best known for his work on **spontaneous generation**, or the formation of life out of non-living matter. In the nineteenth century the theory of spontaneous generation had been proven false after centuries of speculation. People had concluded that life could spontaneously and inexplicably appear after observing such incidents as mushrooms seemingly appearing out of nowhere. But the development of the microscope and more detailed observations by **Louis Pasteur** laid to rest any possibility of forms of life materializing out of nothing. Rather, it was discovered that microscopic life, such as **bacteria** and **larvae** invisible to the naked eye, was the source for the visible life that purportedly had suddenly appeared. Pasteur demonstrated that life would cease to develop spontaneously in completely sterile conditions. He demonstrated that if a room were kept completely sterile, no life would form. Many scientists wondered if Pasteur increased the dimensions of his experiments he would achieve the same results. These

scientists particularly wondered if the four year timespan of Pasteur's experiment were increased to a billion years and if the size of his small laboratory were stretched to the size of the universe, the conditions would remain constant.

Miller began research to answer such questions. He attempted to recreate the conditions that could have made first life possible. While at the University of Chicago, Miller performed an experiment that was meant to recreate the conditions of primordial life—the time when only lifeless matter such as gas and rock existed. Earlier, **John Burdon Haldane**, a British biochemist, had suggested that the Earth's early **atmosphere** contained no free oxygen. Based on this, Miller surmised that the Earth's environment billions of years ago was similar to conditions found in the environment on the planet Jupiter. Miller's experimental conditions consisted mainly of an atmosphere of hydrogen, with strong admixtures of **ammonia**, methane, and sterilized **water**. He then applied an electrical discharge to represent the influence of electricity from lightning storms. Miller then analyzed the water and found under his microscope simple **organic compound**s and even a few simple **amino acid**s.

The results strongly suggested that amino acids, the components of **proteins**, could have been synthesized by lightning discharges in the oxygen-poor atmospheric conditions believed to have characterized the Earth in its earliest existence. Since Miller's initial experiments, other researchers have demonstrated that the fundamental building blocks of other kinds of **molecule**s basic to life could be created in a similar way.

Miller introduced an entirely new field of research with his pioneering experiment. Following this experiment, Miller and many other scientists have tried variations on his original research by modifying gas combinations and utilizing other types of **energy** sources. These subsequent tests have reconfirmed Miller's original results and provided a solid basis for further biochemical research and work on **nucleic acid**s by such scientists as **James Watson** and **Francis Crick**. As of this writing, Miller is a professor of Chemistry at the University of California, San Diego.

Milstein, César (1927-)
Argentine English biochemist

César Milstein was born on October 8, 1927, in the eastern Argentine city of Bahía Blanca, one of three sons of Lázaro and Máxima Milstein. He studied biochemistry at the National University of Buenos Aires from 1945 to 1952, graduating with a degree in chemistry. Heavily involved in opposing the policies of President Juan Peron and working part-time as a chemical analyst for a laboratory, Milstein barely managed to pass with poor grades. Nonetheless, he pursued graduate studies at the Instituto de Biología Química of the University of Buenos Aires and completed his doctoral dissertation on the chemistry of aldehyde dehydrogenase, an **alcohol** enzyme used as a catalyst, in 1957.

With a British Council scholarship, he continued his studies at Cambridge University from 1958 to 1961 under the

César Milstein. *(Science Photo Library, Photo Researchers, Inc. Reproduced by permission.)*

guidance of **Frederick Sanger**, a distinguished researcher in the field of **enzyme**s. Sanger had determined that an enzyme's functions depend on the arrangement of **amino acid**s inside it. In 1960 Milstein obtained a Ph.D. and joined the Department of Biochemistry at Cambridge, but in 1961, he decided to return to his native country to continue his investigations as head of a newly-created Department of **Molecular Biology** at the National Institute of Microbiology in Buenos Aires.

A military coup in 1962 had a profound impact on the state of research and on academic life in Argentina. Milstein resigned his position in protest of the government's dismissal of the Institute's director, Ignacio Pirosky. In 1963 he returned to work with Sanger in Great Britain. During the 1960s and much of the 1970s, Milstein concentrated on the study of antibodies, the **protein** organisms generated by the **immune system** to combat and deactivate antigens. Milstein's efforts were aimed at analyzing myeloma proteins, and then **DNA** and **RNA**. Myeloma, which are tumors in cells that produce antibodies, had been the subject of previous studies by **Rodney R. Porter**, **Frank MacFarlane Burnet**, and **Gerald M. Edelman**, among others.

Milstein's investigations in this field were fundamental for understanding how antibodies work. He searched for **mutation**s in laboratory **cell**s of myeloma but faced innumerable difficulties trying to find antigens to combine with their

antibodies. He and **Georges Köhler** produced a hybrid myeloma called hybridoma in 1974. This cell had the capacity to produce antibodies but kept growing like the cancerous cell from which it had originated. The production of monoclonal antibodies from these cells was one of the most relevant conclusions from Milstein and his colleague's research. The Milstein-Köhler paper was first published in 1975 and indicated the possibility of using monoclonal antibodies for testing antigens. The two scientists predicted that since it was possible to hybridize antibody-producing cells from different origins, such cells could be produced in massive **culture**s. They were, and the technique consisted of a fusion of antibodies with cells of the myeloma to produce cells that could perpetuate themselves, generating uniform and pure antibodies.

In 1983 Milstein assumed leadership of the Protein and Nucleic Acid Chemistry Division at the Medical Research Council's laboratory. In 1984 he shared the Nobel Prize with Köhler and **Niels K. Jerne** for developing the technique that had revolutionized many diagnostic procedures by producing exceptionally pure antibodies. Upon receiving the prize, Milstein heralded the beginning of what he called "a new era of immunobiochemistry," which included production of **molecule**s based on antibodies. He stated that his method was a by-product of basic research and a clear example of how an investment in research that was not initially considered commercially viable had "an enormous practical impact." By 1984 a thriving business was being done with monoclonal antibodies for diagnosis, and works on vaccines and **cancer** based on Milstein's breakthrough research were being rapidly developed.

In the early 1980s Milstein received a number of other scientific awards, including the Wolf Prize in Medicine from the Karl Wolf Foundation of Israel in 1980, the Royal Medal from the Royal Society of London in 1982, and the Dale Medal from the Society for Endocrinology in London in 1984. He is a member of numerous international scientific organizations, among them the U.S. National Academy of Sciences and the Royal College of Physicians in London. His hobbies include walking, outdoor cooking, and attending the theater. Milstein is married to biochemist Celia Prilleltensky; they have no children.

MIMICRY

Mimicry is a physical or behavioral resemblance of one **species** to another, or to an inanimate object. Although it is more common in **animal**s, it is also found in **plant**s. This arrangement benefits either the originator only, or both species. Mimicry was first reported in 1862 by the British naturalist Henry Bates while observing tropical forest butterflies in Brazil. He described a resemblance between two unrelated families of butterflies, where the one **family** that was edible (mimic) closely resembled another inedible family (model). This is now known as Batesian mimicry. This type of arrangement confers an advantage to the mimic when predators such as birds (known as dupes) avoid the edible along with the unpalatable

An Indian leaf butterfly sitting on leaf. The butterfly blends in with the leaf to avoid predators. *(The Stock Market. Reproduced by permission.)*

individuals. The original model species is argued to have evolved a warning coloration to advertise its distastefulness to potential predators. **Natural selection** favors palatable species that have evolved similar coloration. Thus, **predation** on both groups is diminished. A familiar example in North America is the close resemblance between non-toxic Viceroy butterflies and distasteful Monarch butterflies, which obtain their toxin from the milkweed plant that the **larval** stage ingests. Both Viceroy and Monarch butterflies have prominent orange-colored wings with white spots and black venation, and are even difficult for humans to distinguish. Another example of Batesian mimicry is the resemblance of harmless king snakes to the venomous coral snakes, which both have red, yellow, and black bands. The red and black bands adjoin in the former, whereas the red and yellow bands are next to each other in the latter species.

A second type of mimicry is known as Mullerian, originating with the work of the German zoologist Fritz Muller in 1879. This type of mimicry differs from Batesian in that the two unrelated species that closely resemble each other are both unpalatable. An example of this is the close resemblance of the yellow and black color banding in many different species of stinging wasps. This strategy of "pooling" their **adaptation**s reinforces the **learning** behavior of vertebrate predators such as birds to avoid all individuals that resemble these wasps. Recent research has suggested that the Viceroy-Monarch example may in fact be Mullerian because some birds find the Viceroy to be unpalatable. Another type of mimicry is orchids that possess floral structures that mimic female wasps to entice male wasps to land and attempt to mate, and in the process transfer pollen. The eye spots on butterfly and moth wings and the fake rattle behavior in some non-venomous snakes are further examples of mimicry that help protect these animals from predators. Mimicry differs from cryptic coloration (blending in with the environment) in that the bright coloration or distinctive behavior broadcasts an obvious message to would-be predators.

MINERALS

Minerals are the basic materials which make up the earth's crust. Most rocks are combinations of minerals. A mineral can be either a **chemical element** or a **chemical compound** and is defined as a natural, inorganic, crystalline solid.

Each mineral has its own characteristic chemical composition and can be identified by its properties. These properties, which include hardness, crystalline structure, luster, and color, are a result of the way the **atoms** in a particular mineral are held together.

Over 3,000 different minerals are known to exist on earth. Of these, only about 20 are considered common. These are the ''rock-forming'' minerals. Ten of these minerals alone make up 90% of the mass of the earth's crust. These ten minerals are quartz, orthoclase, plagioclase, muscovite, biotite, calcite, dolomite, halite, gypsum, and the ferromagnesian minerals. Minerals that are not rock-forming are known as ''accessory'' minerals.

Minerals can be divided into two main groups based on their chemical composition: silicate and non-silicate. The silicates contain the elements silicon (Si) and oxygen (O). These minerals constitute 96% of the earth's crust. Quartz is a silicate which contains only silicon and oxygen. Other silicates, such as feldspars, contain metals which combine with silicon and oxygen. Quartz and the feldspars alone make up over 50% of the earth's crust. The silicates also occur in a very specific crystal structure known as the silicon-oxygen tetrahedron. This is a four-sided structure which occurs when four oxygen atoms combine in a pyramid with one silicon atom in the center. These tetrahedra can then combine with each other to form a single chain (as in pyroxene), a double chain (like the amphibole minerals), or sheets (such as is found in the micas).

The non-silicate minerals do not contain the element silicon. These minerals can be subdivided into six groups; the carbonates, halides, native elements, oxides, sulfates, and sulfides. The carbonates contain a CO_3 group. The halides contain either the element chlorine (Cl) or fluorine (F) in combination with sodium (Na), potassium (K), or calcium (Ca). The native elements are pure elements which occur in **nature**, for example silver (Ag) or copper (Cu). The oxides contain oxygen in combination with any other element except silicon. The sulfates contain a SO_4 group, and the sulfides contain the element sulfur (S).

With so many different minerals one would think that identification would be problematic. Mineralogists, scientists who study minerals, have devised seven simple tests which help identify many minerals. These tests are color, luster, streak, cleavage and fracture, hardness, crystal shape, and density. Some minerals also have special properties which can be used for definitive identification.

The color of a mineral can be used for identification. While color alone is not conclusive evidence, it can be used in combination with the other tests to determine the identity of a mineral. Many minerals have a very specific characteristic color. Examples include: sulfur, which is yellow; or cinnabar, which is red.

The way in which **light** is reflected from the surface of a mineral determines its luster. Luster can be described as metallic or nonmetallic. If a mineral has a metallic luster, it reflects light much like polished metal. If a mineral does not have a metallic luster, it is said to be nonmetallic and can be described as glassy, waxy, pearly, brilliant, or dull.

The streak test for mineral identification involves rubbing the mineral on an unglazed ceramic tile (called a streak plate). The color of the mineral in powdered form is the color of the streak. The streak color may be different from the color of the mineral itself.

The way the atoms in a mineral combine to form its crystalline structure determines how it will appear when broken. Cleavage refers to a mineral's ability to split easily along certain flat surfaces, or planes. If a mineral fractures when broken, it breaks unevenly. Uneven fractures can be either fibrous (resembling strands of fibers) or conchoidal (resembling chipped glass).

Another useful test is the hardness test. Hardness refers to the ability of a mineral to resist scratches and is determined by the strength of the bonds in its atomic structure. The test is performed by scraping a mineral with different materials and noting which will produce a scratch on the mineral. The mineral is then compared with a scale developed in 1822 by a German mineralogist named Friedrich Mohs, appropriately named the ''Mohs hardness scale.'' This rates a mineral's hardness on a scale from one to ten, with ten being the hardest. Examples of materials on the Mohs scale include talc with a hardness of 1, a fingernail with a hardness of 2.5, a copper penny with a hardness of 3, glass with a hardness of 6, steel with a hardness of 6.5, quartz with a hardness of 7, and diamond with a hardness of 10. If a mineral can be scratched by a material found on the scale, its hardness must be less than that of the scratching material. If a mineral can scratch a material found on the scale, it must be harder than that material.

The last two tests which can be used to identify a mineral are crystal shape and density. A mineral is formed in a specific crystalline shape which is determined by how the atoms in a mineral combine. A mineral's density, or its mass divided by its volume, is determined by the kinds of atoms it has as well as how closely these atoms are packed together. These tests, combined with those described above, can be used to identify an unknown mineral sample.

Some minerals have special properties besides those described above. For example, a mineral might be magnetic. Some minerals display fluorescence or phosphorescence. Some are radioactive. Others may display double refraction, the bending of light shining through a crystal to form a double image. These specific characteristics can be used to further identify a mineral.

Minerals are important natural resources as well as items of beauty. Steel is made from the iron (Fe) mined from iron-rich minerals. Glass is made from silicon which is found in many minerals. Other metals that come from mineral ores include titanium (Ti) and molybdenum (Mo). Diamond is used for cutting and grinding instruments. Many technological advances have been made as a result of scientific study of the properties and characteristics of the minerals found in the Earth's crust.

MINIMUM, LAW OF THE

The Law of the Minimum was first proposed in 1840 by **Justus von Liebig**, a German chemist. He worked on agricultural crops and concluded that the level of individual chemicals in the manure fertilizer he added proportionally influenced growth of the crops. Each **plant** and **animal** requires certain kinds and amounts of **nutrients** (both macro- and micronutrients) for optimal growth and survival. If all chemicals are provided except for one key ingredient, the plant or animal will either stop growing or die. If that one chemical is added in excess, then the next essential element in limiting concentration will limit the growth, and so on. An analogous illustration would be a wooden barrel with staves where each stave is cut at a certain height. The level of **water** poured into this barrel would reach only that of the shortest stave, which would be equivalent to the limiting nutrient.

Phosphorus is the number one limiting nutrient in most lakes. Limnologists (aquatic ecologists) know that if they can limit the phosphorus input into a water body, then they can control algal growth in this system. People who fertilize their lawns often unknowingly add excess nutrients when only one element such as **nitrogen** may be limiting the growth. This accounts for much of the excess nutrient input into aquatic **ecosystem**s.

Not only is too little of essential chemicals important, but too much can also be detrimental. For example, copper is an essential trace element for **algae**, but it can be added in excess (known as ''blue-stoning'') to control algal blooms. Thus, the Law of **Tolerance**, which describes the optimum environmental conditions for **organism**s, is a more realistic ecological concept than the Law of the Minimum.

MIRSKY, ALFRED EZRA (1900-1974)
American biochemist and physiologist

In the 1940s, Alfred Ezra Mirsky, working with colleague Arthur Pollister at the Rockefeller Institute for Medical Research (now the Rockefeller University) in New York, became the first person to develop a method of isolating the **genetic material** in **animal** cells called **chromatin**. Chromatin is made up of deoxyribonucleic acid (**DNA**), special **protein**s called histones, and other proteins. These materials combine to form **chromosome**s at the stage of development of a single **cell** just before it divides into two ''sister'' cells.

Mirsky's discovery led to his further discovery that each cell in the entire body has exactly the same amount of DNA—except for the **sperm** cell and **egg** cell, which have half the amount of DNA contained in other cells. When the egg and sperm cells unite, they become a single cell containing the full complement of DNA. Mirsky's discoveries led the way to understanding the **nucleus** of the cell, opening the door to the fascinating field of molecular and cellular biology and **genetics**.

Mirsky was born in Flushing, New York on October 17, 1900 to Michael D. and Frida I. Mirsky. He attended high school at the Ethical Cultural School in New York City, gradu-ated from Harvard College, studied medicine for two years at Columbia University's College of Physicians and Surgeons, and received his doctorate from Cambridge University in England in 1926 writing his thesis on ''The Hemoglobin Molecule.'' In 1927, he joined the Rockefeller Institute, was named associate in 1929, associate member in 1940, member in 1949, and professor in 1954—the year he received an honorary M.D. from the University of Gothenburg. In 1965, he assumed the responsibility of librarian of the school, holding that position while actively pursuing his research until his retirement in 1971. He also founded the Christmas lectures for high school children which have been held on campus every year since 1959. Mirsky married Reba Paeff in 1926 with whom he had a daughter and a son. The year after her death in 1966, he married Sonia Wohl, associate librarian at the university.

In the 1930s, in collaboration with a fellow Rockefeller investigator, Mirsky began research into protein molecules. Between 1935 and 1936, during leave from Rockefeller and as a visiting faculty member at the California Institute of Technology, Mirsky continued these investigations, working with Nobel prizewinner **Linus C. Pauling**. Together, they determined the chemical conditions that cause the long, spiral chains which comprise a protein molecule to unwind (denaturation) and re-wind to assume their original shape.

Mirsky's research accomplishments include the discovery in the 1950s that ribonucleic acid (**RNA**) plays a role in forming proteins from microsomes. As an outgrowth of this work on protein **biosynthesis** within the cell, he and several collaborators determined that the contents of the cell nucleus have the ability to break down glucose and produce **adenosine triphosphate** (ATP), then use ATP to generate RNA and several other different proteins.

After several years of close association with the prestigious *Journal of General Physiology,*, Mirsky became editor-in-chief in 1951. Although he relinquished the position in 1961, he remained actively involved with the journal until his death. Mirsky's role in administrative matters at the Rockefeller Institute, particularly as chairman of the Faculty Committee on Education Policies and other committees, influenced greatly the restructuring of the institute to a graduate university.

MISSING LINK • See Human evolution

MITCHELL, PETER D. (1920-1992)
English biochemist

Peter D. Mitchell was awarded the 1978 Nobel Prize in chemistry for his chemiosmotic theory, which explained how organisms use and synthesize **energy**. In his Nobel Prize address, Mitchell honored his long association with Professor David Keilin of Cambridge University, whose work provided the takeoff point for Mitchell's discoveries. Keilin had discovered **cytochromes**—electron-carrier **protein**s that assist in energy transfer via a respiratory chain. Mitchell's revolutionary chemiosmotic hypothesis changed the way scientists view energy transformation, and though it was initially viewed as controversial, it eventually won almost universal acceptance.

In 1961, when Mitchell's idea was first introduced, it was greeted by some in the scientific community with skepticism: what he was proposing was radically different than the prevailing thought on energy conversion at that time, and those opposing his conclusions questioned the validity of his research. Also, although Mitchell viewed his small research staff and unconventional laboratory at Glynn House mansion in Cornwall as positive elements conducive to productive research, others viewed his unorthodox working environment with suspicion. Mitchell's chemiosmotic theory generated intense debate, but the positive result for science as a whole was the creation of much additional scientific experimentation and productivity attempting to prove or disprove his theory, and advancing the discipline of bioenergetics—the study of energy exchanges and transformations between living things and their environments—in the process.

Peter Dennis Mitchell was born in Mitcham, Surrey, England, on September 29, 1920, the son of Christopher Gibbs Mitchell, a civil servant, and Kate Beatrice Dorothy Taplin Mitchell. He received his secondary education at Queens College in Taunton, England, and was admitted to Jesus College at Cambridge University in 1939. A graduate student of James F. Danielli in the department of biochemistry at Cambridge, Mitchell earned his doctorate degree in 1951. He taught biochemistry at Cambridge from 1951 until 1955, when he left to develop a chemical biology unit at Edinburgh University. He remained there until 1963, when poor health caused him to look for a calmer working atmosphere.

Mitchell found a peaceful environment in an eighteenth-century manor house in Cornwall. The manor house, called Glynn House, was in disrepair, and was restored by Mitchell and converted to family living quarters and a research laboratory. Glynn Research Laboratories was organized and directed in 1964 by Mitchell and his colleague, Jennifer Moyle, whose background work was instrumental to Mitchell's development of the chemiosmotic hypothesis. By the time Mitchell received the Nobel Prize, the laboratory had grown to require a staff of six.

Development of the Chemiosmotic Theory

The intriguing question of how organisms take energy from their surroundings and transform it for use in specialized functions, such as movement and **respiration**, was thought to have been answered by a theory called chemical coupling. This theory postulated that energy was carried down the respiratory chain by an unknown high-energy intermediate compound formed during oxidation. The energy derived from the intermediate compound was thought to form a "universal energy currency" known as **adenosine triphosphate** (ATP).

The search was on to identify the energy-rich intermediary when Mitchell upset prevailing thought by proposing that the process was an electrical, not a chemical, one. He coined the term "proticity" to explain the process by which protons flow across **cell** membranes to synthesize ATP. Mitchell likened this process to the way electricity moves from a high con-

centration to a concentration low enough to power an electric appliance. Laboratory experiments crucial for the support of his chemiosmotic theory were successfully carried out during the 1960s by Mitchell and Moyle at Glynn Research Laboratories, as well as in other research labs throughout the world. These experiments included identifying the **membrane** protons that provide a link to the movement of other molecules across the cell membrane and showing that the membrane also serves to halt the movement of other molecules.

Recognition for his work on cell energy transfer culminated in Mitchell's receipt of the Nobel Prize in 1978. Later, Mitchell and his staff at the Glynn laboratory studied the biochemical actions involved in energy transfer within cells, seeking precise details of this complex process. Those contributions advanced scientific knowledge of how cells use, transform, and generate energy. Mitchell maintained that he was just one more link in science's intellectual and historical chain. He believed that the practice of science was a continuing process, whereby one scientist builds on the discoveries and knowledge of another, and was quick to give credit to those whose past work had advanced and made possible his own. In the December 15, 1978 issue of *Science,* Frank Harold quoted Mitchell as saying, "Science is not a game like golf, played in solitude, but a game like tennis in which one sends the ball into the opposing court and expects its return."

Many awards other than the Nobel Prize were presented to Mitchell. Among them were the CIBA Medal and Prize, Biochemical Society, England; the Warren Triennial Prize, Massachusetts General Hospital; the Louis and Bert Freedman Award of the New York Academy of Sciences; the Lewis S. Rosenstiel Award for Distinguished Work in Basic Medical Research of Brandeis University; and the Copley Medal of the Royal Society. He held memberships in various professional societies and received honorary degrees from universities in Berlin and Chicago, as well as from numerous British institutions. Although immersed in his work, Mitchell found time to participate in local affairs, respond to environmental issues, and restore medieval farmhouses. He and his wife, Mary Helen French, were married in 1958; they had six children: Jeremy, Daniel, Jason, Gideon, Julia, and Vanessa. Peter Mitchell died at Glynn House, Bodmin, on April 10, 1992.

MITOCHONDRIA

Mitochondria are double **membrane**-bound **cytoplasmic organelle**s (they are enclosed by two independent lipoprotein membranes) found in almost all higher plant and **animal** cells. Although they vary considerably in size and shape even within a single **cell** over time, mitochondria are usually bean-shaped or thread-like with an average diameter of about 0.5 nm and a length ranging from one to hundreds of meters. Mitochondria often move rapidly around the cell interior, constantly changing size or shape. There might be a few giant interconnected mitochondrial networks in a cell at one time, or they may break up into hundreds or thousands of individual mitochondria under different conditions.

Mitochondria serve as the principle **energy** source for all animal cells and for plant cells when **light** is absent. A complex set of **enzyme**s and cofactors located both in the interior of the mitochondrion and in its inner membrane carry out a series of reactions that break down energy-rich **organic compound**s. The energy released by these reactions is captured by a process called oxidative **phosphorylation** and used to generate **adenosine triphosphate (ATP)**, a stable **chemical compound** that serves as the currency for macromolecular synthesis, mechanical motion, transmembrane transport, and almost all other metabolic and physiological activities at the cellular level.

It is widely believed that mitochondria probably arose sometime far back in history as a result of a symbiotic relationship between a prokaryotic (bacterial-like) cell and a **nucleus**-containing proto-eukaryotic cell. Mitochondria have many similarities to free-living bacteria including their own **genetic material** and a bacterial-like **protein** synthesizing system. There are also many similarities between the enzymes that carry out oxidative phosphorylation in mitochondria and those that are responsible for **photosynthesis** in **chloroplast**s, suggesting that one of these systems might be evolutionary descendants of the other.

See also Boyer, Paul D.; Respiration, cellular

MITOSIS

Mitosis is the process during which two complete, identical sets of **chromosome**s are produced from one original set. This allows a **cell** to divide during another process called cytokinesis, thus creating two completely identical daughter cells.

In order for an **organism** to grow and develop, the organism's cells must be able to duplicate themselves. Three very basic events must take place to achieve this duplication: the deoxyribonucleic acid (**DNA**), which makes up the individual chromosomes within the cell's **nucleus** must be duplicated; the two sets of DNA must be packaged up into two separate nuclei; and the cell's **cytoplasm** must divide itself to create two separate cells, each complete with its own nucleus. The two new cells, products of the single original cell, are known as daughter cells.

During much of a cell's life, the DNA within the nucleus is not actually organized into the discrete units known as chromosomes. Instead, the DNA exists loosely within the nucleus, in a form called **chromatin**. Prior to the major events of mitosis, the DNA must replicate itself, so that each cell has twice as much DNA as previously. Mitosis is then ready to begin.

The first stage of mitosis is called prophase. During prophase, the DNA organizes or condenses itself into the specific units known as chromosomes. Chromosomes appear as double-stranded structures. Each strand is a replica of the other and is called a chromatid. The two chromatids of a chromosome are joined at a special region, the centromere. Structures called **centriole**s position themselves across from each other, at either end of the cell. The nuclear **membrane** then disappears.

During the stage of mitosis called metaphase, the chromosomes line themselves up along the midline of the cell. Fibers called spindles attach themselves to the centromere of each chromosome.

Mitosis in an onion root tip, magnified 330 times at 35mm. Mitosis is the process by which new cells, during asexual reproduction, receive identical genetic material to that of the parent cell. This continuous process is arbitrarily divided by scientists into stages called prophase, metaphase, anaphase and telophase. Here in anaphase the chromosomes separate towards opposite sides of the dividing cell; in telophase the physical process of division begins. *(Photograph by J.L. Carson, Custom Medical Stock Photo. Reproduced by permission.)*

During the third stage of mitosis, called anaphase, spindle fibers will pull the chromosomes apart at their centromere (remember that chromosomes have two complementary halves, similar to the two nonidentical but complementary halves of a zipper). One arm of each chromosome will migrate toward each centriole, pulled by the spindle fibers.

During the final stage of mitosis, telophase, the chromosomes decondense, becoming unorganized chromatin again. A nuclear membrane forms around each daughter set of chromosomes, and the spindle fibers disappear. Sometime during telophase, the cytoplasm and cytoplasmic membrane of the cell split into two (cytokinesis), each containing one set of chromosomes residing within its nucleus.

Mitosis always creates two completely identical cells from the original cell. In mitosis, the total amount of DNA doubles briefly, so that the subsequent daughter cells will ultimately have the exact amount of DNA initially present in the original cell. Mitosis is the process by which all of the cells of the body divide and therefore reproduce. The only cells of the body which do not duplicate through mitosis are the sex cells (**egg** and **sperm** cells). These cells undergo a slightly different type of **cell division** called **meiosis**, which allows each sex cell produced to contain half of its original amount of DNA, in anticipation of doubling it again when an egg and a sperm unite during the course of conception.

MOLECULAR BIOLOGY

Molecular biology is the field of study concerned with the structural and functional properties of biological systems. Specifically, it attempts to understand the function of **molecules** within a living system and how they interact. It includes such topics as biophysics, **genetics** and biochemistry. Researchers in this field have made many significant contributions to modern biology including discovery of the structure of **nucleic acids** and **protein**s, the mechanism for **heredity**, and the processes involved in **metabolism**. Recently, advances in **genetic engineering** technology have led to significant discoveries making molecular biology an active area of study for many new scientists.

Molecular biology is a relatively new area of study in the field of biology. While it traces its roots back to the 1850s when **Gregor Mendel** proposed his laws of **inheritance**, it actually began in the early 1940s. Around this time, scientists had already figured out that **chromosome**s contained the hereditary information of the **cell**. They also knew that chromosomes were composed of **DNA** and protein. In 1941, **George Beadle** and **Edward Tatum** established the relationship between **gene**s and **enzyme**s. They determined that each gene codes for a single protein. Later in 1944, **Oswald Avery** and his colleagues showed that DNA is the **genetic material**. This established a molecular basis for heredity and focused the study of molecular biology on determining the structure of DNA. This structure was determined in 1953 by **James Watson** and **Francis Crick**. With the structure in hand, scientists worked on new theories to explain how genes express themselves in living **organism**s.

The study of genetic material has been one of the primary focuses of molecular biology. These investigations have been focused on three main areas including: the composition, organization and structure of the chromosomes which contain DNA, the molecular reactions involved in **gene expression**, and the control of gene expression.

DNA has been found to be a nearly universal genetic material. All organisms use it except certain **virus**es that contain a similar molecule called ribonucleic acid (**RNA**). It is a long **polymer** made up of **monomer** units called **nucleotide**s. Nucleotides are composed of three parts including a **phosphate group**, a five carbon sugar, and a **nitrogen**-containing base. These nucleotides are arranged in a **double helix** structure which coils around a central axis. The nitrogen bases provide the code which is translated into proteins. Typically, a gene is made up of a sequence of these nucleotides that code for a specific protein. Additional stretches of code influence gene expression.

In **prokaryote**s, the genetic material is arranged in a single chromosome which contains a circular DNA molecule. This chromosome contains about 4000 genes and each nucleotide is important in gene expression. In **eukaryote**s, the genes are typically arranged on multiple chromosomes which are made up of linear DNA molecules. The number of chromosomes differs depending on the organism. Human cells contain 23 pairs of chromosomes, each encoding thousands of genes. In both prokaryotic and eukaryotic cells, the DNA is associated with chromosomal proteins.

Since DNA was determined to be the material responsible for heredity, a mechanism for its **replication** was needed. This mechanism was determined in 1958 by Matthew Meselson and Franklin Stahl. In an experiment, they showed that the replication of DNA followed a semiconservative path. This meant that the two strands of DNA separated and served as templates for two new strands. Each resulting strand was composed of half of the original ''parent'' strand and half of a new ''daughter'' strand.

After a model for the replication of DNA was worked out, scientists turned their attention to figuring out how nucleotide sequences were expressed in cells. This led to the discovery in the 1960s that the **genetic code** is translated into proteins by **ribosome**s. In this process, DNA is first transcribed into messenger RNA (mRNA). The mRNA then interacts with **organelle**s in the cell called ribosomes. In this process, known as **protein synthesis**, the nucleotides are translated into **amino acid**s and strung together to create proteins. During this process, three nucleotides are read at once by the ribosome. This triplet then results in the addition of a single amino acid to the growing protein chain.

While investigating a variety of problems in molecular biology, scientists have developed numerous concepts and techniques that have helped solve various biological and biochemical problems. One of these was the use of intact cells and cell-free systems. Since bacterial cultures were simple systems and relatively easy to propagate, they were used for much of the early work in molecular biology. Many of the significant discoveries were made with these cultures. Work with eukaryotic cells has only recently been possible. Molecular biologists also developed cell-free systems in which they could study processes such as **DNA synthesis** and protein synthesis.

After the structure of DNA was determined and the nature of the genetic code was known, scientists attempted to create maps of where genes were physically located on the chromosomes. At first these maps were crude, only showing relative positions of few genes. A major breakthrough came in 1977 when DNA sequencing became possible. This allowed scientists to find the order of the nucleotide bases on any gene. Work is currently ongoing to sequence the entire human **genome**. It is hoped that this information will provide invaluable clues in the diagnosis and treatment of genetic disorders.

A variety of new techniques which come under the heading of DNA biotechnology have recently been developed by molecular biologists. Recombinant DNA is one such technology that has had important implications to science, medicine, and industry. In this technique, a certain gene is isolated and then transferred to another organism. This host organism then expresses the gene thereby producing the desired protein. **Cloning** is another technique that has led to some exciting new developments. In this procedure, the genetic material from one organism is duplicated and a new organism is created. New biotechnologies such as these promise to revolutionize medicine.

MOLECULE

A molecule is defined as the smallest particle of an element or a compound that possesses the properties of the original substance. Some materials, such as helium, neon, and argon, tend to form monoatomic molecules. Most molecules, however, are made of two or more **atoms**. Many gases, such as oxygen, **nitrogen**, chlorine, and **carbon monoxide**, are diatomic (meaning they are composed of only two atoms). Triatomic molecules (which are composed of three atoms) include **ozone** (O_3) and **carbon dioxide** (CO_2). In complex molecules, such as **protein**s, the number of atoms can range up to hundreds or thousands. Molecules vary in size from less than 1 millimicron to more than 500 millimicrons and in weight from 4 units (for helium) to 40 million units (for the tobacco mosaic **virus**).

Molecules differ from **chemical compound**s because compounds consist of different elements. Combinations of the same atoms form molecules, not compounds. For example, when two atoms of oxygen combine they form a molecule of oxygen, not a compound. Therefore, while every compound is a molecule, not every molecule is a compound. Each molecule of a given substance contains the same number and kinds of atoms which are held together by **chemical bond**s. The two basic types of chemical bonds are covalent and ionic. A **covalent bond** holds atoms together by sharing electrons; an **ionic bond** holds them together through electron transference. The nature of these bonds determine the physical form of the molecule. Gases have weak attractive forces between molecules, whereas liquids and solids have stronger forces.

These bonds may be broken or rearranged in a number of different **chemical reaction**s which fall into four basic categories. These are known as combination, decomposition, single replacement, and double replacement reactions. Specific types of reactions in each of these categories include oxidation, reduction, ionization, combustion, **polymer**ization, **hydrolysis**, condensation, and rearrangement reactions.

In certain cases, atoms within a molecule may be arranged in different configurations to form molecules which contain the same number and same type of atoms, but which have different properties because of their spacial arrangement. Such molecules are known as **isomers**. Examples of isomers include ethyl **alcohol** and methyl ether which both contain one oxygen atom, two carbon atoms, and six hydrogen atoms. Other types of isomers include stereoisomers, geometric isomers, and optical isomers.

One important characterization of a molecule is its molecular weight. Molecular weight is the sum of the atomic weights of the atoms in a molecule. For convenience, the

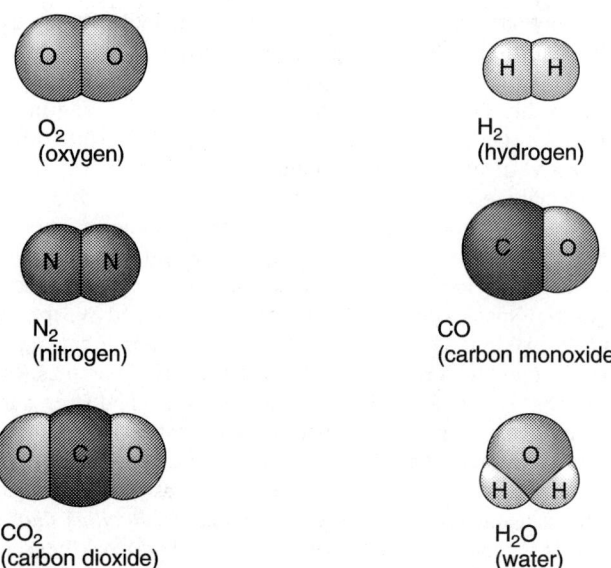

Molecules of various elements. *(Illustration by Hans & Cassady, Inc.)*

chemical formula is used to add up these values. For gases, a more accurate molecular weight can be determined by measuring the volume of a given weight and then calculating the weight of 22.4 L at 0°C and 760 mm of mercury.

MONERA

The Monera is the largest **kingdom** of prokaryotic **microorganisms**. **Prokaryote**s are single-celled **organism**s that lack an organized **nucleus**, i.e., one that is separated from the **protoplasm** by a **membrane**-like envelope. (Organisms in all other kingdoms have a cellular nucleus and are termed eukaryotic.) Monerans have their **genetic material** organized as a single strand of **DNA**, which occurs throughout their **cytoplasm**. They have no subcellular **organelle**s, such as **chloroplast**s, **mitochondria**, or **flagella**. Monerans were the first organisms to evolve, about 3.5 million years ago (the first **eukaryote**s evolved 2 billion years afterwards).

Bacteria and blue-green bacteria (or **cyanobacteria**) are the major groups in the Monera. They have rigid or semi-rigid cell walls, propagate by binary division of the **cell**, and do not undergo **mitosis** or **meiosis**. The cyanobacteria are photosynthetic, using chlorophyll dispersed within the cytoplasm as their primary **light**-capturing pigment.

Biologists have named about five thousand **species** of monerans. There are, however, large numbers of additional species that have not yet been described by microbiologists. Among the great diversity of monerans are species capable of exploiting a phenomenally wide range of ecological and metabolic opportunities. Some species are photosynthetic, utilizing the **energy** of sunlight to fix **carbon dioxide** and **water** into simple sugars. Others are chemosynthetic, being capable of using energy released by the oxidation of sulfide **minerals** to manufacture their organic **nutrition**. Yet other monerans can utilize virtually any organic material as a substrate in their **heterotrophic** nutrition, either in the presence of oxygen or under anaerobic conditions. Some species of monerans can tolerate remarkably extreme environments, for example, living in hot springs at **temperature**s as high as 172°F (78°C), while others are active at sub-zero temperatures as deep as 1,300 ft (400 m) in glacial ice.

In addition to free-living monerans, many species of bacteria live in **mutualism**s (i.e., mutually beneficial symbioses) with more-complex organisms. For example, many bacterial species live in the rumens of cows and sheep, and others live in the human gut, in both cases aiding in the **digestion** of complex organic foods. Other bacteria, known as *Rhizobium*, live in a mutualism with the roots of leguminous **plants** (such as peas and clovers), fixing **nitrogen** gas into a form (**ammonia**) that plants can utilize as a nutrient.

Many species of bacteria are **parasite**s of other organisms, causing various diseases. For example, *Bacillus thuringiensis* is a pathogen of moths, butterflies, and blackflies, and is used as a biological insecticide against certain pests in **agriculture** and forestry. Other bacteria cause diseases of humans, including various infections, bacterial pneumonia, **cholera**, **diphtheria**, gonorrhoea, Legionnaire's disease, leprosy, scarlet fever, **syphilis**, tetanus, tooth decay, tuberculosis, whooping cough, most kinds of food poisoning, and the "flesh-eating disease" caused by a virulent strain of *Streptococcus*.

MONOCLONAL ANTIBODY • See Antibody, monoclonal

MONOD, JACQUES LUCIEN (1910-1976)
French biologist

The structure of all living matter is determined by the composition of its deoxyribonucleic acid, or **DNA**, molecule; the discovery in the early 1950s that the **genetic code** carried by DNA is responsible for the shape of all the **proteins** that make up skin, eyes, hair—all the **tissues** of life—astounded the scientific community at the time. But how this master plan is carried out, and how its instructions are read and followed by the body, were facts discovered much later by French biologist Jacques Lucien Monod and a small cadre of scientists working with him. Monod and his colleagues postulated, and later demonstrated, the process by which messenger ribonucleic acid (mRNA) carries instructions for **protein synthesis** from the DNA in a cell's **nucleus** to its **cytoplasm**, where the instructions are carried out. Monod and two fellow researchers, **Francois Jacob** and **André Lwoff**, won the 1965 Nobel Prize for physiology or medicine.

Early Interest in Biology

Monod was born in Paris, on February 9, 1910, to Lucien Hector Monod, a painter and intellectual of Swiss Huguenot descent, and Charlotte Todd (MacGregor) Monod, a Scottish-American from Milwaukee, Wisconsin. At the age of seven, Monod moved with his family to Cannes in the south of France. His parents were very influential in his education, and Monod later credited his father for his own passionate interest in music, and, later, biology. The young Monod learned to play the cello at an early age, and even during the years he was doing research in **molecular biology**, he played in and directed a string quartet and a Bach choir. Although Monod later confessed a serious inclination towards a career in conducting, he also showed an early interest in biology, collecting beetles and tadpoles in the woods around his southern France home. His interest developed further, and he entered the College de Cannes from where he graduated in the summer of 1928. Monod went on to receive a B.S. from the Faculte des Sciences at the University of Paris, Sorbonne, in 1931. Although he stayed on at the university for further studies, Monod felt that the academic curriculum at the Sorbonne was deficient and did not reflect contemporary biological research. Therefore, it was through the personal contacts he developed during excursions to the nearby Roscoff marine biology station that Monod received his true scientific grounding.

While working at the Roscoff station, Monod met André Lwoff, with whom he would establish a life-long collaboration. Lwoff introduced Monod to the potentials of microbiology and microbial nutrition, and these became the focus of Monod's early research. Boris Ephrussi, another scientist working at Roscoff, opened Monod to the importance of physiological and biochemical **genetics**. And Louis Rapkine, also a Roscoff contemporary, impressed upon Monod the importance of learning the chemical and molecular aspects of living organisms.

Embarks on His Career

During the autumn of 1931 Monod took up a fellowship at the University of Strasbourg in the laboratory of Edouard Chatton, France's leading protistologist. Then, in October, 1932, he won a Commercy Scholarship that called him back to Paris to work at the Sorbonne once again. This time he was an assistant in the Laboratory of the Evolution of Organic Life, which was directed by the French biologist Maurice Caullery at the time. Moving to the **zoology** department in 1934, Monod became an assistant professor of zoology in less than a year. That summer, Monod also embarked on a natural history expedition to Greenland aboard the *Pourquoi pas?* This expedition was a great success and developed in Monod a life-long love for sailing. In 1936 Monod left for the United States with Ephrussi, where he spent time at the California Institute of Technology on a Rockefeller grant. His research centered on studying the **fruit fly** (*Drosophila melanogaster*) under the direction of **Thomas Hunt Morgan**, an American geneticist. Here Monod not only met with refreshingly new opinions, but he also got his first look at a new way of studying science—a research style based on collective effort and a free passage of critical discussion. This was in contrast to the rigid, sometimes sterile, attitude among the faculty at the Sorbonne. Returning to France, Monod completed his studies at the Institute of Physiochemical Biology. In this time he also worked with Georges

Teissier, a scientist at the Roscoff station, who influenced Monod's interest in the study of bacterial growth. This later became the subject of Monod's doctoral thesis at the Sorbonne.

This was a time of war in Europe, and despite a medical exemption from military service which allowed him to retain his academic position, Monod joined the French resistance movement. His Sorbonne laboratory became an underground meeting place and propaganda print shop. Thereafter Monod also joined the Franc-Tireurs Partisans and was captured by the Gestapo. He managed to escape and continued his underground resistance efforts. Monod is also credited with helping to organize the general strike that led to Paris' ultimate liberation, and he was honored with several military commendations for his efforts.

During this period Monod also continued his pursuit of music, forming a Bach choir, La Cantate, which he would direct until 1948. In 1938, he met his future wife, Odette Bruhl, an archeologist and orientalist, through the choir. In the postwar period Monod served as the laboratory director of Lwoff's Department, and he also became an officer in the Free France Forces. As a member of General de Lattre de Tassigny's staff, he met a number of American scientists. They provided Monod with several scientific journals in which he read articles about the spontaneous **mutations** of **bacteria**. Monod later recalled the influence these articles had on the course of his career. He noted that these journals in particular, lead him to the study of genetics and later, his research into the structure of DNA.

Studies Enzyme Induction

Monod's work comprised four separate but interrelated phases beginning with his practical education at the Sorbonne. In the early years of his education, he concentrated on the kinetic aspects of biological systems, discovering that the growth rate of bacteria could be described in a simple, quantitative way. The size of the colony was solely dependent on the food supply; the more sugar Monod gave the bacteria to feed on, the more they grew. Although there was a direct correlation between the amount of food Monod fed the bacteria and their rate of growth, he also observed that in some colonies of bacteria, growth spread over two phases, sometimes with a period of slow or no growth in between. Monod termed this phenomenon ''diauxy'' (double growth), and guessed that the bacteria had to employ different **enzyme**s to metabolize different kinds of sugars.

When Monod brought the finding to Lwoff's attention in the winter of 1940, Lwoff suggested that Monod investigate the possibility that he had discovered a form of ''enzyme adaptation,'' in which the latency period represents a hiatus during which the colony is switching between enzymes. In the previous decade, a similar phenomenon had been recorded by the Finnish scientist, Henning Karstroem while working with protein synthesis. Although the outbreak of war and a conflict with his director took Monod away from his lab at the Sorbonne, Lwoff offered him a position in his laboratory at the Pasteur Institute where Monod would remain until 1976. Here

Jacques Lucien Monod (left). *(The Library of Congress. Reproduced by permission.)*

he began working with Alice Audureau to investigate the genetic consequences of his kinetic findings, thus beginning the second phase of his work.

To explain his findings with bacteria, Monod shifted his focus to the study of enzyme induction. He theorized that certain colonies of bacteria spent time adapting and producing enzymes capable of processing new kinds of sugars. Although this slowed down the growth of the colony, Monod realized that it was a necessary process as the bacteria needed to adapt to varying environments and foods to survive. Therefore, in devising a mechanism that could be used to sense a change in the environment, and thereby enable the colony to take advantage of the new food, a valuable evolutionary step was taking place. In Darwinian terms, this colony of bacteria would now have a very good chance of surviving, by passing these changes on to future generations. Monod would summarize his research and views on relationship between the roles of random chance and **adaptation** in **evolution** in his 1970 book *Chance and Necessity.*

Between 1943 and 1945, working with Melvin Cohn, a specialist in immunology, Monod hit upon the theory that an ''inducer'' acted as an internal signal of the need to produce the required digestive enzyme. This hypothesis challenged the German biochemist Rudolf Schoenheimer's theory of the ''dynamic state'' of protein production, which stated that it was the mix of proteins that resulted in a large number of random combinations. Monod's theory, in contrast, projected a fairly stable and efficient process of protein production which seemed to be controlled by a master plan. In 1953, Monod and Cohn published their findings on the generalized theory of induction.

Discovers Role of Messenger Ribonucleic Acid (mRNA)

That year Monod also became the director of the department of cellular biology at the Pasteur Institute and began his collaboration with Francois Jacob and Jacob's team. In 1955, working with Jacob, he began the third phase of his work by investigating the relationship between the roles of **heredity** and environment in enzyme synthesis, that is, how the organism creates these vital elements in its metabolic pathway and how it knows when to create them.

It was this research that led Monod and Jacob to formulate their model of protein synthesis. They identified a gene cluster they called the operon, at the beginning of a strand of bacterial DNA. These genes, they postulated, send out messages signalling the beginning and end of the production of a specific protein in the cell, depending on what proteins are needed by the cell in its current environment. Within the operons, Monod and Jacob discovered two key genes, which they named the "operator" and "structural" genes. The scientists discovered that during protein synthesis, the operator **gene** sends the signal to begin building the protein. A large molecule then attaches itself to the structural gene to form a strand of messenger **RNA** (mRNA). In addition to the operon is the regulator gene, which codes for a repressor protein. The repressor protein either attaches to the operator gene and inactivates it, in turn, halting structural gene activity and protein synthesis; or the repressor protein binds to the regulator gene instead of the operator gene, thereby freeing the operator and permitting protein synthesis to occur. As a result of this process, the mRNA, when complete, acts as a template for the creation of a specific protein encoded by the DNA, carrying instructions for protein synthesis from the DNA in the cell's nucleus, to the **ribosome**s outside the nucleus, where proteins are manufactured. With such a system, a cell can adapt to changing environmental conditions, and produce the proteins it needs when it needs them.

Word of the importance of Monod's work began to spread, and in 1958 he was invited to become professor of biochemistry at the Sorbonne, a position he accepted conditional to his retaining his post at the Pasteur Institute. At the Sorbonne, Monod was the chair of chemistry of metabolism, but in April, 1966, his position was renamed the chair of molecular biology in recognition of his research in creating the new science. His Nobel prize in 1965 both increased his responsibilities and thrust him to the center of a growing limelight in the field of biochemistry.

Monod's life following the Nobel prize reveals a dramatic shift to the administrative side of scientific research. Elected to the College de France and named chair of molecular biology in 1967, Monod used his influence and fame to bolster the cause of the organized student movement against the academic establishment in France. In 1971 he was offered the directorship of the Pasteur Institute and, on April 15, 1971, he was named director general of the institute. At this time the in-

stitute was on the verge of financial collapse, and Monod set aside his research to devote all his efforts to modernize and revitalize the organization. As his administrative duties grew, his research activities rapidly slowed and finally stopped in 1972 after the death of his wife. Not long thereafter, Monod himself fell ill with aplastic anemia. Four years later, with his own death imminent, and having completed only the first phase of his intended sweeping changes at the institute, Monod returned to his home in Cannes. He died there on May 31, 1976.

His twin sons, Olivier and Philippe, followed him into scientific research, Olivier as a geologist and Philippe as physicist. Monod's list of awards and honors was impressive, and it included the Montyon Physiology Prize, the Louis Rapkine Medal, and the Charles Leopold Mayer Prize. He was made a Chevalier de l'Ordre des Palmes Academiques and later an officer in the Legion of Honor. He also received both the Croix de Guerre and the Bronze Star Medal.

MONOMER

A monomer (meaning "single member") is any **molecule** capable of bonding with others of the same kind to form a long chain. Monomers are the single repeating units of any **polymer** (which means "many members.") For example, styrene is the monomer from which polystyrene polymers are produced and vinyl chloride is the monomer of polyvinyl chloride (or PVC). Other common monomers include methyl methacrylate, adipic acid, and hexamethylenediamine.

Monomers are usually molecules of relatively low molecular weight and simple structure. For a monomer to form a polymer, it must be di-functional in other words it must have at least two reactive sites. Under the proper conditions monomers are highly reactive and they can chemically bond together through a variety of **chemical reaction**s. When reacted in this fashion, monomers form polymers which are used in such products as rubber, finishes, synthetic fibers, and plastics.

One of the most common applications of monomer chemistry is found in plastics. Here the monomers, which are microscopic chemical units in liquid form, combine with one another in a "head to tail" fashion to form long fibers. These long chains of connected monomers are called polymers. Polymers link together in a processes known as "crosslinking" to form rigid networks. To picture this network of polymers, think of a piece of rope which is like a single polymer strand. Two polymer strands which are crosslinked are similar to two parallel strands of rope that are connected with smaller pieces of rope to make a ladder. When several rope ladders are joined together they form a net. The net is stronger and holds it shape better than individual ropes. The same is true when multiple polymer strands interconnect. The average number of polymer chains, in other words the lengths of the resulting polymers, determines the average molecular weight of the polymer. These factors can be controlled by determining the reaction kinetics. This fact is important, because large variations in chain lengths can result in variations in properties. For example, in the case of polyethylene, longer polymer chains result in plastics that are harder and stronger but more difficult to shape. Although polymer chains are typically linear, they may also include side chains which may significantly modify their properties.

Monomers that occur naturally in biological systems can create biopolymers. For example, **plants** use a monomer based on sugar molecules to create the polymer known as **cellulose**.

MONOPHYLOGENY

Monophylogeny is a term used in the science of **systematics** to describe a group of **organism**s that are classified in the same **taxon** and share a common ancestor. The term monophylogeny comes from the Greek words *mono* meaning ''one,'' *phylon* meaning ''tribe'' and *genesis* meaning ''origin.'' Systematics is concerned with investigating relationships between groups of organisms based on their biological diversity and evolutionary history. It is related to the science of **taxonomy** which deals with naming and grouping of organisms depending on various physical characteristics.

Grouping organisms based on their relationship with each other has been an important part of biological studies since antiquity. **Aristotle** proposed one of the first systems although his groupings were mainly alphabetical. **Carl Linnaeus** was the first to suggest ordering organisms in a hierarchical arrangement. When **Charles Darwin** proposed his **evolutionary theory**, some scientists began grouping organisms by their evolutionary relationships. This led to the development of systematics.

Scientists who study systematics create genealogy diagrams, called phylogenic trees. These diagrams are designed to graphically depict evolutionary relationships. To construct them, **species** are grouped using a variety of data such as **fossil** records, comparative anatomy and molecular biological similarities. The ideal phylogenic tree would contain only taxons that are monophylogenic, that is, each related organism comes from a single ancestor. However, since much of the data that are currently available are conflicting, polyphylogenic groups are created. These constructs show organisms in a single taxon with two or more different evolutionary ancestors. As additional data are collected about specific species, scientists hope to develop a unified phylogenic tree that shows the complete evolutionary history of every organism.

See also Ontogeny and phylogeny; Polyphylogeny

MONOSACCHARIDES

Monosaccharides are the simplest form of a large and diverse group of **organic compound**s called **carbohydrates**. They are sweet, **water** soluble substances that contain more than one hydroxyl group (OH) and a carbonyl group (C=O). They are either polyhydroxy aldehydes or ketones depending on whether the carbonyl group is formed from a terminal carbon (aldoses), or not (ketoses). They are grouped by the number of carbon

atoms they contain; trioses have three, tetroses four, pentoses five, and hexoses six. The smallest monosaccharides are the two forms with a three carbon skeleton: glyceraldehyde and dihydroxyacetone. Pentoses and hexoses are most common, but monosaccharides may contain as many as nine carbons. They are named by combining a prefix that may designate the number of carbons in the **molecule** or some other descriptive feature, and a generic suffix -ose.

Important pentoses include xylose and arabinose, both of which are components of plant **cell** walls, and ribose and deoxyribose, which are found in ribonucleic acid (**RNA**) and deoxyribonucleic acid (**DNA**), respectively. Common hexoses include glucose, galactose, (both aldoses) and fructose (a ketose). Glucose is found in many **fruits** and is the sugar that circulates in the **blood** of higher **animal**s. It is also the basic sugar unit in important **polysaccharides** such as starch, glycogen and **cellulose** which may contain many thousands of glucose units. Fructose is the building block for the polysaccharide, inulin, found in the roots and tuber of dahlias and Jerusalem artichoke. Inulin serves as a commercial source for fructose, which because it is the sweetest of all monosaccharides, is valued as an ingredient in many soft drinks. Honey, composed of approximately 80% fructose and glucose, was almost the only source of sugar available in ancient times. Sucrose, or table sugar, which has replaced honey as the most used sweetener, is a disaccharide made up of glucose and fructose. Lactose, the disaccharide that makes up 2-8% of mammalian milk, is composed of galactose and glucose.

Cane sugar (sucrose) was introduced to Europe by the Moors about 700 A.D., although it had been used in parts of the Orient long before that. Cane production for European use was most prevalent in North Africa, the Canary Islands, and Madeira. Sugar beets were not exploited as a source of sucrose until about 1800. The development of sugar beets as a commercial crop was encouraged by Napoleon when the French were not able to import cane sugar because Britain controlled the seas. The use of sucrose has expanded enormously in the last hundred years. Today, the world production, almost entirely from sugar cane and sugar beets, is in excess of ten million tons annually.

When humans ingest sucrose it is hydrolyzed to glucose and fructose by sucrase **enzyme**s in the intestinal mucosa. The monosaccharides are then further metabolized to provide needed **energy**.

Important derivatives of monosaccharides include ascorbic acid (vitamin C), a derivative of glucose, and sorbitol and mannitol, sweetening agents derived from glucose and mannose, respectively.

MOORE, STANFORD (1913-1982)
American biochemist

Stanford Moore's work in **protein** chemistry greatly advanced understanding of the composition of enzymes, the complex proteins that serve as catalysts for countless biochemical processes. Moore's research focused on the relationship between

Stanford Moore. *(UPI/Corbis-Bettmann. Reproduced by permission.)*

the chemical structure of proteins, which are made up of strings of **amino acid**s, and their biological action. In 1972 he was awarded the Nobel Prize in chemistry with longtime collaborator **William Howard Stein** for providing the first complete decoding of the chemical composition of an **enzyme**, ribonuclease (RNase). This discovery provided scientists with insight into cell activity and function, which has had important implications for medical research.

Moore was born on September 4, 1913, in Chicago, Illinois, but spent most of his childhood in Nashville, Tennessee, where his father, John Howard Moore, was a professor at Vanderbilt University's School of Law. His mother was the former Ruth Fowler. In 1935 Moore earned a B.A. in chemistry from Vanderbilt. Moore continued his education in organic chemistry at the graduate school of the University of Wisconsin and completed the Ph.D in 1938. Though Moore considered attending medical school, he accepted a position in 1939 as a research assistant in the laboratory of German chemist Max Bergmann at the Rockefeller Institute for Medical Research (RIMR), later renamed the Rockefeller University.

Bergmann's research group focused on the structural chemistry of proteins. During his early years at RIMR, Moore questioned whether proteins actually had specific structures. The direction of research in Bergmann's laboratory was greatly influenced by the arrival of William Howard Stein. At Berg-

mann's suggestion, Moore and Stein began a long-lived and successful collaboration. With the exception of his wartime service, the years abroad, and a year at Vanderbilt University, Moore remained at RIMR all of his professional life. In 1952 he became a member of the institute, and his association with RIMR/Rockefeller University continued until his death in 1982.

From 1942 to 1945, Moore worked as a technical aide in the National Defense Research Committee of the Office of Scientific Research and Development (OSRD). As an administrative officer for university and industrial projects, Moore studied the action and effects of mustard gas and other chemical warfare agents. As a technical aide, Moore also coordinated academic and industrial studies on the actions of chemical agents. In 1944 he was appointed to the project coordination staff of the Chemical Warfare Service and continued to contribute to this research until the end of the war in 1945.

Though his initial investigation of **chromatography**—the process of separating the components of a solution—began in the late 1930s, the war interrupted this work. After the war and following the death of Max Bergmann in 1944, Moore returned to RIMR to resume work with William Howard Stein. This marked a productive period for the two men, leading to their work on chromatographic methods for separating amino acids, **peptide**s (compounds of two or more amino acids), and proteins. Moore's work in chromatography was influenced by the methods of paper chromatography developed by A. J. P. Martin and Richard L. M. Synge of England. However, limitations in these earlier methods prohibited the study of protein chemistry; new techniques in chromatography had to be developed so that amino acids could be separated. Moore and Stein utilized column chromatography, in which a column or tube is filled with material that separates the components of a solution. In 1948 they successfully separated amino acids by passing the solution through a column filled with potato starch. The process was time consuming, however, and presented inadequate separations for amino acid analysis. To facilitate the procedure, Moore and Stein replaced the filler material with a synthetic ion exchange resin, which separated components of a solution by electrical charge and size. In 1949 they successfully separated amino acids from blood and urine.

Study and Work Abroad

Further interruptions in the work at RIMR occurred in 1950, when Moore held the Francqui Chair at the University of Brussels (where he established a laboratory for amino acid analysis), and in 1951, when he was a visiting professor at the University of Cambridge (England) working with **Frederick Sanger** on the amino acid sequence of **insulin**.

In 1958 Moore and Stein contributed to the development of the automated amino acid analyzer. This instrument facilitated the complete amino acid analysis of a protein in twenty-four hours; previously employed procedures, such as chromatography, had required up to one week. The automated technique afforded researchers a tool with which to separate and study the large chemical sequences in protein molecules. This instrument is used worldwide for the study of proteins, enzymes, and **hormones** as well as the analysis of food. While Moore and Stein endeavored to determine the chemical composition of proteins, scientists concurrently worked to determine their three-dimensional structure.

Discovery of the Chemical Nature of Enzymes

By 1959 Moore and Stein had determined the amino acid sequence of pancreatic ribonuclease (RNase), a digestive enzyme that breaks down ribonucleic acid (**RNA**) so that its components can be reused. They discovered that ribonuclease is made up of a chain of 124 amino acids, which they identified and sequenced. This marked the first complete description of the chemical structure of an enzyme, a discovery that earned Moore and Stein the 1972 Nobel Prize in chemistry (an award shared with **Christian Boehmer Anfinsen** of the National Institutes of Health). Understanding protein structure is essential to understanding biological function, which opens the door to the treatment of disease. Moore's findings influenced research in neurochemistry and the study of such diseases as **sickle-cell anemia**. Scientists later discovered that related ribonucleases are present in nearly all human cells, which prompted studies in the fields of **cancer** and malaria research.

In 1969, Moore returned to Rockefeller University following one year spent at Vanderbilt University School of Medicine as a visiting professor of health sciences. Upon his return, the team of Moore and Stein resumed their studies of protein chemistry with an investigation of deoxyribonuclease, the enzyme that breaks down deoxyribonucleic acid (**DNA**).

In addition to scientific work, Moore also served as an editor of *The Journal of Biological Chemistry,* treasurer and president of the American Society of Biological Chemistry, and president of the Federation of American Societies for Experimental Biology. Moore received many honors, including membership in the National Academy of Sciences and receipt of the Richards Medal of the American Chemical Society and the Linderstro slash m-Lang Medal. He was also awarded honorary degrees by the University of Brussels, the University of Paris, and the University of Wisconsin.

Moore was diagnosed with amyotrophic lateral sclerosis (Lou Gehrig's disease) some time in the early 1980s. On August 23, 1982, he committed suicide in his home in New York City. Upon his death Moore, who never married, left his estate to Rockefeller University.

MORGAN, THOMAS HUNT (1866-1945)
American geneticist

Thomas Hunt Morgan was born on September 25, 1866, in Lexington, Kentucky. As a child growing up in rural Kentucky, he was surrounded by nature and wildlife. Perhaps that environment contributed to his intense interest in biology, for Morgan later majored in **zoology** at State College of Kentucky. After his graduation in 1886, he investigated chemistry and **morphology** (the study of **organism** development to better understand evolutionary relationships) at Johns Hopkins University, completing his doctorate in 1890. From his graduate days

on, Morgan believed that **heredity** was in some way central to understanding all biological phenomena—especially development and **evolution**. His persistence in trying to prove and develop heredity theories led to his winning the Nobel Prize for physiology or medicine in 1933.

In 1903, there were several attempts to explain variations in **plants** and **animal** species. One was **Charles Darwin**'s theory of **natural selection**, a process by which organisms best adapted to local environments leave more offspring that survive to spread their favorable traits throughout a **population**. But Morgan wondered how complex organisms such as humans could have evolved from such a process. To him, the theory seemed incomplete. Morgan viewed natural selection as a process that sorted out variations in an organism, not as one that created the variations. So what was it that determined whether a baby would be a boy or a girl, or whether it would have blue eyes or green eyes? The three widely known heredity theories of the time offered competing explanations: the Mendelian (or **gene**) theory, the **chromosome** theory, and the **mutation** theory.

Gregor Mendel, by cross-breeding pea plants, had first determined some of the rules of inheritable traits—those of **sex determination**, gene **linkage** (inheritance of characteristics together), and **mimicry**. Advocates of the chromosome theory maintained that genes located on chromosomes were responsible for specific inherited traits. Morgan was skeptical of the Mendelian and chromosome theories because the conclusions were speculative, based on nothing more than observation, inference, and analogy. Morgan wanted to be able to draw firm, rigorous, testable conclusions based on quantitative and analytical data. His strong belief in experimental analysis attracted Morgan to **Hugo de Vries**'s mutation theory. De Vries, a Dutch botanist, had physical evidence that large-scale variations in one generation could produce offspring that were of a different species than their parent plants. Morgan set out to test de Vries's theory in animals and also to disprove the other heredity theories.

His first experiments using the **fruit fly** (*Drosophila melanogaster*) were unsuccessful; Morgan was not able to duplicate the magnitude of mutations that de Vries had claimed for plants. Then in 1910, Morgan noticed a natural mutation in one of the male fruit flies: it had white eyes instead of red. He began breeding the white-eyed male to its red-eyed sisters and found that all of the offspring had red eyes. When Morgan bred those offspring, he found that they produced a second generation of both red- and white-eyed fruit flies. Morgan was fascinated to find that all of the white-eyed flies were male. He traced the unusual finding to a difference between male and female chromosomes. The white-eye gene of the fruit fly was located on the male sex chromosome. By studying future generations of fruit flies, Morgan found that genes were linearly arranged on chromosomes. His work with the fruit fly strongly backed Mendel's gene concept and, moreover, established that chromosomes definitely carried genetic traits. For the first time, the association of one or more hereditary characteristics with specific chromosomes was clear, thereby unifying Mendelian "trait" theory and chromosome theory.

That was only the first of Morgan's discoveries. Working with students **Hermann Muller, Alfred H. Sturtevant,** and

Calvin Bridges, Morgan went on to develop and perfect his concepts of linkage by explaining why, for instance, he occasionally found a white-eyed female in his studies. Morgan concluded that traits found on the same chromosome were not always inherited together. This genetic "mistake" was called **crossing over**, because one chromosome actually exchanged material with (or crossed over to) another chromosome. This process was an important source of genetic diversity. In 1915, Morgan, along with his students, published the culmination of his work, *The Mechanism of Mendelian Heredity.* These results provided the key to all further work in the area of **genetics** and laid the groundwork for all genetically-based research.

In 1904, Morgan had married Lilian Vaughan Sampson, who assisted in his research. They had one son and three daughters. Morgan died in 1945.

MORPHOLOGY

The term morphology as used by biologists refers to a study of the shape, structure, and size of **plants**, **animals**, and **microorganisms**, and of their parts. It is sometimes used synonymously with **anatomy**, although anatomy usually implies detailed study of either gross or microscopic structure. Interest in morphology, especially comparative morphology, was greatly stimulated when the concept of **evolution** was widely accepted. Prior to that time, similarities of form in different kinds of **organisms** was merely accepted as fact. **Charles Darwin**, on the other hand, regarded them as evidence that living organisms evolve by a series of steps from pre-existing forms. This view encouraged morphologists to look for similarities and differences in various forms of life in an effort to establish evolutionary relationships.

Early morphologists attempted to establish structural patterns within an organism, or between different types of organisms. Examples include comparisons of the bone pattern in the human foot, with those of the hand, or of the human arm with the wing of a bird. Stages of embryonic development were also compared. The aim of much of this work was to establish evolutionary relationships that could provide a natural **classification** system based on genetic similarities.

Developmental morphologists, who focus their attention on patterns of development as an organism grows from an ovum to seed to an adult, provide another approach to the study of form and structural relationships of living organisms. This approach is called developmental morphology, or morphogenesis. It attempts to furnish an understanding of the mechanisms responsible for development, and the ways in which adjacent parts of a developing organism influence the growth of neighboring structures.

The methods used by morphologist vary depending on the type of organism and the structural level of the study. Studies of gross structure rely on careful dissection and precise description of all parts. **Tissues** and **cells** are better studied microscopically, relying on histochemistry and autoradiography. Histochemistry makes use of dyes that differentially stain structural or molecular components in carefully prepared tissue samples. This technique allows the histochemist to identify acid or basic components, as well as the location of specific chemicals such as **protein**, **DNA**, or glycogen within the cell or tissue. It has also been possible to pinpoint the location of some **enzyme**s using this technique.

Autoradiography involves supplying the tissue with a radioactive substance and allowing time for the components to interact. The tissue is then prepared and placed in contact with a photographic film or emulsion. Radioactivity causes the silver grains in the film to darken, and allows the scientist to locate sites where the substance has been concentrated.

Modern techniques such as transmission and scanning electron microscopy have provided the contemporary morphologist with a basis for comparative studies of the fine structure of cells and tissues in different organisms. They have also enabled study of the smallest components of cells, and have contributed to an understanding of the basis for the contraction of muscle cells and the movement of **cilia** and **flagella** in microorganisms.

MORULA

The morula is an early stage in the development of an embryo. After the **sperm** and **egg** unite to form the **zygote** within the fallopian tube, the **cell**s of the zygote continue to replicate through the process known asmitosis. The cilia lining the fallopian tube continue to beat, moving the developing mass of cells along towards the uterus. When development reaches the 16-cell stage, it is referred to as a morula. This morula is the form which reaches the uterus. The morula is a solid ball of cells, with no fluid or hollow interior. The change in the interior of the ball of cells occurs during the next stage of development, called the blastocyst stage.

See also Embryo and embryonic development

MRNA • See RNA

MULLER, HERMANN JOSEPH (1890-1967)
American geneticist

Hermann Joseph Muller was the first to show that genetic **mutations** can be induced by exposing chromosomes to x rays. For this demonstration he was awarded the 1946 Nobel Prize in physiology or medicine. He also took up a crusade to improve the condition of the human **gene pool** by calling for a cessation of the unnecessary use of x rays in medicine and a halt to nuclear bomb testing in order to prevent further damage to the genetic makeup of the human population.

Hermann Joseph Muller was born on December 21, 1890, in New York City. His father was also named Hermann Joseph Muller; his mother was the former Frances Lyons. Muller's paternal grandfather had come to the United States from Germany after the revolution that had swept Europe in 1848. Muller's father had wanted to be a lawyer, but instead had taken up the family business of producing bronze art work. The elder Muller died when young Hermann was only nine.

Hermann attended Morris High School in the borough of the Bronx in New York City. When he founded its science club, perhaps the first of its kind in the country, it seemed obvious that science would be his calling. On a scholarship, Muller enrolled at Columbia University in 1907 and by his sophomore year decided that he would major in **genetics**. He received his bachelors degree in 1910 and then continued at both Cornell Medical School and Columbia for his master's degree, studying the transmission of nerve impulses.

An Early Interest in Fruit Fly Mutations

Two of Muller's classmates, **A. H. Sturtevant** and Calvin B. Bridges, were working at Columbia with **Thomas Hunt Morgan**, a zoologist who was performing ground-breaking work in genetics. In 1912, Muller joined this group. Together the four became something of a legend at Columbia, where the "fly room" buzzed with talk of **chromosome**s, **gene**s, **crossing over**, and mutations. The flies in this case were *Drosophila melanogaster*, fruit flies with a brief three-week breeding cycle making them ideal for genetic study. These fruit flies also have just four pairs of chromosomes, the dark-staining microscopic structures within the **nucleus** of each **cell**. Experiments were done to study mutations, abnormal traits that seem to arise spontaneously in the **fruit fly** population. The mutations were tracked in order to infer which part of each chromosome contained the gene responsible for a particular trait, such as eye color or wing shape.

Muller's doctoral thesis in 1916 was on "crossing over," a phenomenon discovered in 1909 by a Belgian scientist, F. A. Janssens, when he noticed that during the duplication and separation of like chromosomes, sometimes part of a chromosome would break off and reattach at a comparable place on the other chromosome. If two genes were far apart on a chromosome, then it would be more likely that a break could occur between them. Thus, a high frequency of crossing over observed between any two traits would mean a long distance between the genes, while a low frequency of crossing over between two traits would mean the genes were close together. The team used this information to "map" each chromosome in order to show how genes for each trait might be arranged along its length. The findings of the group were published in *The Mechanism of Mendelian Heredity,* written by the four in 1915.

In 1916, Muller took a teaching position at Rice Institute in Texas, where he did further research in genetics, especially mapping "modifier" genes, which seem to control the expression of other genes. Upon his return to Columbia two years later, Muller did some of his most important theoretical work. Realizing that genes on the chromosomes are self-replicating and are responsible for synthesizing the other components of cells, he theorized that all life must have started out with molecules that were able to self-replicate, which he likened to "naked genes." These molecules, he suggested, must have been something like **virus**es, a very astute hypothesis given the little that was known about viruses at the time.

Hermann Joseph Muller. *(The Library of Congress. Reproduced by permission.)*

In 1921, Muller returned to Texas, this time to the University of Texas in Austin, where he remained until 1932. Muller had grown impatient with waiting for mutations to happen on their own, so he began seeking methods of hastening rates of mutation. In 1919, he had discovered that higher temperatures increase the number of mutations, but not always in both chromosomes in a chromosome pair. He deduced that mutations must involve changes at the molecular or sub-molecular level. He struck on the idea of using x rays instead of heat to induce mutations, and by 1926 he was able to confirm that x rays greatly increased the mutation rate in *Drosophila*. He also concluded that most mutations are harmful to the organism, but are not passed on to future generations since the individual affected is unlikely to reproduce; nonetheless, he suggested, if the rate of harmful mutations were too become too high, a species might die out.

Muller reported his success in inducing mutations in a 1927 article in *Science* entitled "Artificial Transmutation of the Gene." The article gained him international status as an innovator and introduced other scientists to a technique for studying a large number of mutations at once. This led to the realization that mutations are actually chemical changes that can be artificially induced with any number of other chemicals. It also help spawn the infant study of radiation genetics.

Genetics and Politics Prove a Volatile Mix

In the early 1930s, personal problems led Muller to leave the United States. In 1923 he had married Jessie Mary Jacob, a mathematician. But the pressure of his work and a divorce from his wife led to a nervous breakdown. Muller left Texas in 1932 and moved to Berlin to work at the Kaiser Wilhelm Institute. There he spent a year as a Guggenheim fellow doing research on mutations and exploring the structure of the gene. However, Hitler was rising to power and Muller, being a strong supporter of socialism, left Germany.

Muller's next stop was the Soviet Union, where he stayed from 1933-1937. At the Academy of Sciences in both Leningrad and Moscow he studied radiation genetics, cytogenetics and gene structure. However, he soon became openly critical of Trofim Denisovich Lysenko's theories of genetics, which were dominant in the Soviet Union in the mid 1930s. Lysenko held a number of erroneous beliefs about how genetic traits are passed on to future generations. As Lysenko's hypotheses were compatible with certain aspects of Marxist theory, he was a favorite among those in political power. Disputing Lysenko proved dangerous, so in 1937 Muller was forced to leave the Soviet Union. Still following his socialist beliefs, Muller volunteered to fight in the Spanish Civil War.

The next year Muller got a job at the Institute of Animal Genetics in Edinburgh, Scotland, again working on radiation genetics. There he met Dorothy Kantorowitz, a German refugee. In 1939 they married and, as both of them were of part Jewish heritage, left for the safety of the United States in 1940. He continued his research at Amherst College and, starting in 1945, at Indiana University, where he was appointed professor of **zoology** and where he stayed until his death in 1967.

In 1946 Muller was awarded the Nobel Prize for medicine or physiology for his important work on mutations. Muller was also a member of the National Academy of Sciences and a fellow of the Royal Society. He used the opportunity of his world fame to campaign for many social concerns sparked by his interest in the genetic health of the human population. He spoke out against needless x rays in medicine and for safety in protecting people regularly exposed to x rays. In the 1950s he campaigned to outlaw nuclear bomb tests because he believed that nuclear fallout would cause mutations in future generations. Toward the end of his life, Muller believed that the human race should take action in order to keep healthy genes in the population. His idea came out of the belief that modern culture and technology suspend the process of **natural selection** and thus increase the number of mutations in human genes. He believed that there should be programs to promote eugenics, literally "good genes." He supported the idea of establishing **sperm** banks in which the sperm of exceptionally healthy and gifted men would be frozen as an "endowment" to be used for future generations. The concept of such massive intervention in the human gene pool and in the private lives of individuals was and remains highly controversial.

MÜLLER, JOHANNES PETER (1801-1858)
German physiologist

Johannes Peter Müller was a German physiologist known for his research in the fields of **anatomy**, **zoology** and embryology. He is considered by many to be the most significant life scientist and biological researcher in Germany during his era.

Müller was born in Coblenz, Germany, on July 14, 1801 to a well to do shoemaker who hailed from a family of winegrowers. He studied at the newly opened University of Bonn where he was awarded his medical degree in December 1822. He continued his studies for 18 months in Berlin where he successfully took the state medical exam in 1824. He returned to Bonn that year, and in 1826 he was appointed to the Chair of Physiology at the age of 24 and a full professorship at 29. In 1833, Müller took a professorship at Berlin University in anatomy and physiology where he remained until the end of his life. He died on April 28, 1858.

Müller began his work in **embryology** by entering a competition at Bonn to answer the question: Does the fetus breathe in its mothers womb? Müller proved that **respiration** occurred by studying the umbilical cord and observing that **blood** flowing to the fetus is brighter red than the blood flowing from the fetus to the placenta. Another of Müller's key achievements was the discovery that the senses respond to stimuli in a characteristic way regardless of the type of stimuli, which he proposed as the law of specific nerve energies in 1840. For example, the eye responds to stimuli with a sensation of light and the ear with a sensation of sound. He also investigated the eyes response to internal stimuli such as the imagination or organic malfunction. Müller's interest in zoology led to his work on the **classification** of marine creatures, specifically amphibians and reptiles. He also did much to advance the use of **microscope**s in the study of pathological tumors.

Müller's most notable work was the *Handbuch der Physiologie des Menschen* (1833-1840) which deals with human physiology including his work on the composition of blood, the origin of fibrin, secretion, the voice in the larynx of humans and animals, the lymph, sound in the tympanic cavity, the sympathetic nerve and various other components of the **nervous system**. The book also included a lengthy section on the soul in which Müller examines whether the soul is foreign to the body and only temporarily united with it or inherent in all matter.

MÜLLER, PAUL (1899-1965)
Swiss chemist

Paul Müller was an industrial chemist who discovered that dichlorodiphenyltrichloroethane (**DDT**) could be used as an insecticide. This was the first insecticide that could actually target insects; in small doses it was not toxic to humans and yet it was stable enough to remain effective over a period of months. When DDT was introduced in 1942, the effects it would have on the environment were not well understood. It was widely hailed, in particular for its ability to reduce the incidence of tropical diseases by reducing insect populations. For his work with DDT and the role his discovery played in the fight against diseases such as typhus and malaria, Müller was awarded the 1948 Nobel Prize in medicine or physiology.

Paul Hermann Müller was born in Olten, Switzerland, on January 12, 1899, to Gottlieb and Fanny Leypoldt Müller. His father was an official on the Swiss Federal Railway, and the family moved to Lenzburg and then to Basel, where Müller was educated until the age of seventeen. After finishing his secondary education, Müller worked for several years in a succession of jobs with local chemical companies. In 1919, he entered the University of Basel to study chemistry. He did his doctoral work under F. Fichter and H. Rupe, and his dissertation examined the chemical and electrochemical reactions of m-xylidine and some related compounds. Xylidines are used in the manufacture of dyes, and when Müller received his Ph.D. in 1925 he went to work in the dye division of the J. R. Geigy Corporation, a very large Swiss chemical company. Müller married Friedel Rügsegger in 1927; they had two sons and a daughter.

Müller initially conducted research on the natural products that could be derived from green **plants**, and the compounds he synthesized were used as pigments and tanning agents for leather. In 1935, he was assigned to develop an insecticide. At that time the only available insecticides were either expensive natural products or synthetics ineffective against insects; the only compounds that were both effective and inexpensive were the **arsenic** compounds, which were just as poisonous to human beings and other **mammal**s. Müller noticed that insects absorbed and processed chemicals much differently than the higher animals, and he postulated that for this reason there must be some material that was toxic to insects alone. After testing the biological effects of hundreds of different chemicals, in 1939 he discovered that the compound DDT met most of his design criteria. First synthesized in 1873 by German chemist Othmar Zeidler, who had not known of its insecticide potential, DDT could be sprayed as an emulsion with water or could be mixed with talcum or chalk powder and dusted on target areas. It was first used against the Colorado potato beetle in Switzerland in 1939; it was patented in 1940 and went on the market in 1942.

Müller had set out to find a specific compound that would be cheap, odorless, long-lasting, fast in killing insects, and safe for plants and animals. He almost managed it. DDT in short term application is so non-toxic to human beings that it can be applied directly on the skin without ill effect. It is cheap and easy to make, and it usually needs to be applied only once during a growing season, unlike biodegradable **pesticides** which must often be applied several times, in larger amounts and at much higher cost. Typhus and malaria are very severe, often fatal illnesses, which are carried by body lice and mosquitoes respectively; in the 1940s several potentially severe epidemics of these diseases were averted by dusting the area and the human population with DDT. The insecticide saved many lives during World War II and increased the effective-

ness of Allied forces. Soldiers fighting in both the Mediterranean and the tropics were dusted with DDT to kill lice, and entire islands were sprayed by air before invasions.

Despite these successes, environmentalists were concerned from the time DDT was introduced about the dangers of its indiscriminate use. DDT was so effective that all the insects in a dusted area were killed, even beneficial ones, eradicating the food source from many birds and other small creatures. Müller and other scientists were actually aware of these concerns, and as early as 1945 they had attempted to find some way to reduce DDT's toxicity to beneficial insects, but they were unsuccessful. Müller also believed that insecticides must be **biodegradable**.

Hailed as a miracle compound, DDT came into wide use, and the impact on beneficial insects was not the only problem. Because it was such a stable compound, DDT built up in the environment; this was a particular problem once it began to be used for agricultural purposes and applied over wide areas year after year. Higher animals, unharmed by individual small doses, began to accumulate large amounts of DDT in their tissues (called **bioaccumulation**). This had serious effects and several bird species, most notably the bald eagle, were almost wiped out because frequent exposure to the chemical caused the shells of their eggs to be thin and fragile. Many insects also developed resistances to DDT, and so larger and larger amounts of the compound needed to be applied yearly, increasing the rate of bioaccumulation. The substance was eventually banned in many countries; in 1972 it was banned in the United States.

In addition to the 1948 Nobel Prize in physiology or medicine, Müller received an honorary doctorate from the University of Thessalonica in Greece in recognition of DDT's impact on the Mediterranean region. He retired from Geigy in 1961, continuing his research in a home laboratory. He died on October 13, 1965.

MULLIS, KARY (1944-)
American biochemist

Kary Mullis is a biochemist who designed polymerase chain reaction (PCR), a fast and effective technique for reproducing specific **genes** or **DNA** fragments that is able to create billions of copies in a few hours. Mullis invented the technique in 1983 while working for Cetus, a California biotechnology firm. After convincing his colleagues of the importance of his idea, they eventually joined him in creating a method to apply it. They developed a machine which automated the process, controlling the chain reaction by varying the temperature. Widely available because it is now relatively inexpensive, PCR has revolutionized not only the biotechnology industry, but many other scientific fields, and it has important applications in law enforcement, as well as history. Mullis shared the 1993 Nobel Prize in chemistry with **Michael Smith** of the University of British Columbia, who also developed a method for manipulating **genetic material**.

Kary Banks Mullis was born in Lenoir, North Carolina, on December 28, 1944, the son of Cecil Banks Mullis and Ber-

Kary Mullis (left) shaking hands with Japanese Emperor Akihito upon receiving the Japan Prize for his achievements in science and technology. *(Photograph by Masaharu Hatano, Reuters/Corbis-Bettmann. Reproduced by permission.)*

nice Alberta (Barker) Fredericks. He grew up in Columbia, North Carolina, a small city in the foothills of the Blue Ridge Mountains, where his temperament and his curiosity about the world set him apart from others. As a high school student, for example, Mullis designed a rocket that carried a frog some 7,000 feet in the air before splitting open and allowing the live cargo to parachute safely back to earth. Even at a young age, Mullis was considered a maverick and nonconformist. He entered Georgia Institute of Technology in 1962 and studied chemistry. As an undergraduate, he created a laboratory for manufacturing poisons and explosives. He also invented an electronic device stimulated by brain waves that could control a light switch.

Upon graduation from Georgia Tech in 1966 with a B.S. degree in chemistry, Mullis entered the doctoral program in biochemistry at the University of California, Berkeley. In Berkeley at that time there was growing interest in hallucinogenic drugs; Mullis taught a controversial neurochemistry class on the subject. His thesis adviser, Joe Nielands, told *Omni* that as a graduate student Mullis was "very undisciplined and unruly; a free spirit." Yet at the age of twenty-four, he wrote a paper on the structure of the universe that was published by *Nature* magazine. He was awarded his Ph.D. in 1973, and he accepted a teaching position at the University of Kansas Medical School in Kansas City, where he stayed for four years. In 1977, he assumed a postdoctoral fellowship at the Universi-

ty of California, San Francisco. After two years there, discouraged by universities and uncertain what to do with his life, he left and took a job in a restaurant. One day his graduate advisor encountered him there and convinced him that he was wasting both his mind and his education waiting tables. In 1979, he accepted a position as a research scientist with a growing biotech firm, Cetus Corporation, in Emeryville, California, which was in the business of synthesizing chemicals used by other scientists in genetic **cloning**.

Invents PCR to Reproduce Small Sections of DNA

At Cetus, Mullis was bored by the routine demands of corporate life. He spent much of his time sunbathing on the roof and writing computer programs that would automatically respond to certain kinds of administrative requests. "I'd no real responsibilities for about two years," he told *Omni* magazine. "I was playing," he told *Parade Magazine.* "I think really good science doesn't come from hard work. The striking advances come from people on the fringes, being playful." He conceived of polymerase chain reaction (PCR) while driving out to his ranch in Mendocino county and thinking, as he describes it, somewhat "randomly" about ways to look at individual sections of the **genetic code**.

Reproducing deoxyribonucleic acid or DNA had long been an obstacle to anyone working in **molecular biology**. The most effective way to reproduce DNA was by cloning, but however much of a scientific advance this process represented, it was still cumbersome in certain respects. DNA strands are long and complicated, composed of many different **chromosomes**; the problem was that most **genetic engineering** projects were tasks that involved tiny fragments of the DNA molecule, almost infinitesimal sections of a single strand. Cloning works by inserting the DNA into bacteria and waiting while the reproducing bacteria creates copies of it. The cloning process is not only time-consuming, it replicates the whole strand, increasing the complexity. The revolutionary advantage of PCR is its selectivity. It is a process that reproduces specific genes on the DNA strand millions or billions of times, effectively allowing scientists to amplify or enlarge parts of the DNA molecule for further study.

Mullis remembers that it took a long time to convince his colleagues at Cetus of the importance of this discovery. "No one could see any reason why it wouldn't work," he told *Popular Science.* "But no one seemed particularly enthusiastic about it either." Once they had become convinced of its importance, however, PCR became the focus of intensive research at Cetus. Scientists there developed a commercial version of the process and a machine called the Thermal Cycler; with the addition of the chemical building blocks of DNA, called **nucleotide**s, and a biochemical catalyst called polymerase, the machine would perform the process automatically on a target piece of DNA. The machine is so economical that even a small laboratory can afford it, and the technique, as one microbiologist told *Time,* "can reproduce genetic material even more efficiently than nature."

The selectivity of the PCR process, as well as the fact that it is simple and economical, have profoundly changed the course of research in many fields. In an interview with *Omni,* Mullis remarked of PCR: "It's so widely used by molecular biologists that its future direction is the future of molecular biology itself." In the field of **genetics**, the process has been particularly important to the **Human Genome Project**—the massive effort to map human DNA. Nucleotide sequences that have already been mapped can now be filed in a computer, and PCR enables scientists to use these codes to rebuild the sequences, reproducing them in a Thermal Cycler. The ability of this process to reproduce specific genes, thus effectively enlarging them for easier study, has made it possible for virologists to develop extremely sensitive tests for acquired immunodeficiency syndrome (**AIDS**), capable of detecting the **virus** at early stages of infection. There are many other medical applications for PCR, and it has been particularly useful for diagnosing genetic predispositions to diseases such as **sickle cell anemia** and **cystic fibrosis**.

PCR has also revolutionized evolutionary biology, making it possible to examine the DNA of woolly mammoths and the remains of ancient humans found in bogs. PCR can also answer questions about more recent history; it has been used to identify the bones of Czar Nicholas II of Russia who was executed during the Bolshevik revolution, and scientists at the National Museum of Health and Medicine in Washington, DC, are preparing to use PCR to amplify DNA from the hair of Abraham Lincoln, as well blood stains and bone fragments, in an effort to determine whether he suffered from a disease called Marfan's syndrome. In law enforcement, PCR has made genetic "fingerprinting" more accurate and effective; it has been used to identify murder victims, and to overturn the sentences of men wrongly convicted of rape. Some have suggested the PCR can be used to create tags or markers for industrial and biotechnological products, including oil and other hazardous chemicals, to insure that they are used and disposed of in a safe manner.

Cetus awarded Mullis only ten-thousand dollars for developing the PCR patent. Frustrated both by the size of this award and the restrictions the company continued to place on his scientific research, Mullis left Cetus in 1986. He became director of molecular biology at Xytronyx, a San Diego research firm, but two years later, again frustrated with the routine of corporate research, he left to become a private biochemical research consultant. The Du Pont Corporation challenged Cetus for Mullis's patent for PCR, filing suit in the late 1980s; they argued that while working for them in the early 1970s, Nobel laureate **Har Gobind Khorana** had written a paper which anticipated the process. Although he had already left Cetus, Mullis agreed to testify on their behalf, and in February 1991 a federal jury decided against Du Pont. That same year, Cetus sold the process to Hoffman-LaRoche, Inc., for 300 million dollars, the most money ever paid for a patent. When he invents something again, Mullis told *Omni,* "I'm not going to hand it over to some company like Cetus without something saying it's mine. If anyone makes $300 million off it, I'm going to be part of that."

In 1990, Mullis received both the Preis Biochemische Analytik Award from the German Society of Clinical Chemis-try and the Allen Award from the American Society of Human Genetics; in 1991, he received the Gairdner Foundation Award and the National Biotech Award. In 1993, he was presented with the Japan Prize, in addition to the Nobel Prize. He is a member of the American Chemical Society. He has been married and divorced three times and is the father of three children. He works and lives in an apartment overlooking the Pacific Ocean in La Jolla, California.

MURPHY, WILLIAM P. (1892-1987)
American physician and pathologist

William P. Murphy won the 1934 Nobel Prize for physiology or medicine for his role in the discovery of liver as the successful dietary treatment for pernicious anemia, a deadly disorder in which bone marrow ceases to produce the fully mature red **blood** cells needed to carry oxygen to all parts of the body. Murphy's professional persistence following the discovery led to the simple, effective, and inexpensive treatment of the disease by intramuscular injection of a highly-concentrated liver extract. Murphy shared the Nobel Prize with **George Hoyt Whipple**, who had observed that a diet of liver, kidney, meat, and vegetables had a regenerative effect on the blood of dogs in which he had induced anemia; and George Richards Minot, who, building on Whipple's research, isolated liver as the effective dietary factor. Murphy and Minot collaborated on the highly successful study in which pernicious anemia patients were fed one-quarter to one-half pound of liver daily. Reputed for his diligence and dedication, Murphy assumed the painstaking, time-consuming responsibility of counting the microscopic reticulocytes (red blood cells) in the blood samples of pernicious anemia patients before and during the liver diet. The dramatic increase in reticulocytes in the samples following the patient's consumption of liver clearly identified the critical connection between liver ingestion and the production of mature red blood cells.

William Parry Murphy was born on February 6, 1892, in Stoughton, Wisconsin, to Congregational minister Thomas Francis Murphy and his wife, Rose Anna Parry. He attended public schools in Wisconsin and Oregon and received his B.A. in 1914 from the University of Oregon. Murphy taught high school math and physics for two years in Oregon before entering the University of Oregon Medical School in Portland, where he also worked in the anatomy department as a laboratory assistant. In 1918, he took a summer course at Rush Medical School in Chicago. He later received the William Stanislaus Murphy Fellowship award and entered Harvard Medical School in Boston, from which he graduated in 1922. He interned at Rhode Island Hospital, Providence, then returned to Boston to become an assistant resident physician at Peter Bent Brigham Hospital.

Two Men Battle Pernicious Anemia

In 1925, Minot had put pernicious anemia patients at Boston's Huntington Memorial Hospital on a liver-rich diet and observed their improvement. Wanting more evidence, he told no one of his experiment, not even the resident from Boston's Peter Bent Brigham Hospital whose collaboration he recruited. Minot was an attending physician at Brigham where Murphy, a hard-working resident with a keen interest in blood disorders, attracted his attention. Without saying why, Minot asked Murphy to feed liver to pernicious anemia patients at Brigham. Murphy followed Minot's instructions, and two independent surveys were underway at two different institutions.

Murphy encountered difficulties, however. Not only did he have to convince the hospital to obtain and prepare tender, palatable liver on a daily basis, he also had to convince patients to eat it every day. Murphy himself was a lover of liver—he ate it because he liked it. He virtually became a liver salesman to his patients and observed results identical to Minot's: he watched with excitement as life returned to patients who had been dying and, in fact, should have been dead.

Overcomes Obstacles to Therapy

The liver diet therapy for pernicious anemia presented certain problems. Patients found it difficult to eat half a pound of liver every day of their lives. Particularly troublesome was the question of how to feed it to patients so ill they could no longer eat. This problem was partially overcome by a suggestion from one of Murphy's female patients: Couldn't uncooked liver be pulverized and fed to patients in orange juice? Not too proud to listen to a suggestion from a nonprofessional, Murphy tried it. He and Minot literally force-fed their dying patients, pouring liquid liver into stomach tubes as long as the patient showed any sign of life. Within a week, patients who had been too ill to eat were sitting up asking for food.

Murphy, however, was not satisfied. The two doctors enlisted the expertise of Edwin J. Cohn, a physical chemistry professor at Harvard Medical School. Cohn chemically reduced large amounts of liver to a concentrated extract fifty to one hundred times more potent than the liver itself. Ingestion of three vials a day of this extract, which cost $17.00 a month, proved just as effective as the cheaper but less palatable liver diet, which cost approximately $5.50 a month. Still not satisfied, Murphy felt the cost of the extract was prohibitive for many people and continued to search for a less expensive method of administering it. He sought the help of Guy W. Clark of the Lederle Laboratories; soon they developed an extremely concentrated extract. Injected into the muscle only once a month, the extract provided the same therapeutic effect as the liver diet or the oral extract. The monthly cost of this injection was $1.20.

Medical professionals, however, virtually refused to believe the results of the carefully documented study, which Murphy and Minot presented at a medical meeting in Atlantic City in 1926. The treatment was entirely too simple. Pernicious anemia had been thought to be caused by some type of poison, and patients were treated with **arsenic**, blood transfusions, or removal of the spleen, the organ which breaks down red blood cells, all to no avail. But worldwide treatment by the liver diet soon convinced the skeptics. Murphy's lifesaving contribution to society was further advanced by Harvard physician William Castle, who, in 1948, isolated the active ingredient in liver which promoted the development of fully mature red blood cells in patients suffering from pernicious anemia. That factor, named cyanocobalamin for its high concentration of cobalt, is commonly called Vitamin B_{12}, which is now used universally via intramuscular injection for the lifesaving treatment of pernicious anemia.

In addition to working with Minot on the liver diet study, Murphy became Minot's partner in private practice in Boston. In 1924, he was appointed assistant in medicine at Harvard Medical School, promoted to associate in medicine at the Brigham Hospital in 1935, and became a senior associate in medicine and consultant in hematology there. Harvard and Brigham both granted him emeritus status in 1958, when he retired to a suburb of Boston. He married Pearl Harriet Adams in 1919; they had a son, William Murphy, Jr., and a daughter, Priscilla Adams. Murphy's honors include the Cameron Prize and Lectureship of the University of Edinburgh, the Bronze Medal of the American Medical Association, and the Gold Medal of the Massachusetts Humane Society. He died on October 9, 1987, in Brookline, Massachusetts.

MURRAY, JOSEPH E. (1919-)
American surgeon

Joseph Edward Murray was born 1919 in Massachusetts, the son of William Andrew Murray and Mary DePasquale Murray. He earned an A.B. in 1940 from Holy Cross College and then went on to Harvard University to earn his medical degree in 1943. His early work specialized in plastic surgery, in particular reconstructive surgery of the eye and hand. It was a training that would stand Murray in good stead with his later research, for one of the major problems plastic surgeons had to deal with was the rejection of skin grafts by the **immune system**. Murray and other plastic surgeons soon learned that grafts would take between identical twins.

In the late 1940s, Murray became drawn to the work of a team of doctors at Brigham Hospital who were studying end-stage renal disease, and one of the directions their researches was taking was transplantation. Research had been progressing over the past half century on kidney transplants in dogs, but there had never been a successful human transplant. These Harvard researchers, led by John Merrill and David Hume, had been doing experiments transplanting kidneys from cadavers onto the thigh of patients with kidney failure, grafting the third kidney to the femoral vessel of the recipient. One such thigh transplant functioned for about six months, enough time to allow the patient's own kidneys to heal and resume functioning. Kidney dialysis was also being perfected at this time, but Murray felt that it was only a temporary solution. He developed a surgical technique to connect the blood vessels of the donor kidney with those in the abdomen of the recipient, implanting the ureter directly into the urinary bladder.

The new procedure required the right patient; he or she would have to be one of a pair of identical twins with the other twin willing and able to donate a kidney, thus avoiding rejection by the immune system of the recipient. Such an opportunity came in 1954 when the Herrick brothers turned up at Brigham Hospital. The subsequent operation lasted five and one-half hours and was an immediate success. Richard Herrick lived another seven years on the transplanted kidney before dying of heart failure.

Murray continued to perform more successful operations on identical twins, including Edith Helm, who went on to have children and grandchildren, but the real problem now became how to suppress the immune reaction so that the operation would be more generally available. At first Murray and other researchers tried total body x rays and infusions of bone marrow from the donor to adapt the recipient's immune system. In most cases the transplants functioned for several weeks, but there were many failures. Finally in 1959, after a course of total body x rays, a non-identical twin survived a kidney transplant from his brother and went on to lead a normal life. Later in 1959 two Boston hematologists, William Dameshek and Robert Schwartz, demonstrated that the compound 6-mercaptopurine would prevent a host animal from rejecting a foreign **protein**. This was the opening Murray was looking for, and working with chemists and other researchers, Murray developed a drug regimen to suppress the immune system and thus allow an organ from a non-related donor to be accepted by the recipient's body. In 1962 Murray successfully completed the first organ transplant from a cadaver.

Murray's successes became known worldwide and inspired other surgeons to experiment with a variety of organ transplants. With the development of less toxic immune suppressants such as azathioprine, transplants became a growth industry with registries for organs documented worldwide. A related medical benefit was the increase in research into the rejection phenomenon, and thus into the functioning of the human immune system, research that has proved invaluable with the onset of Acquired Immunodeficiency Syndrome (**AIDS**).

After this work on renal transplants, Murray went back to his first love, plastic surgery, developing ways to repair inborn facial defects in children. He headed the plastic surgery divisions of Peter Bent Brigham Hospital from 1951–1986 and Children's Hospital Medical Center from 1972–1985, and he has also been a professor of surgery at Harvard Medical School since 1970. Murray was the recipient of the Gold Medal from the International Society of Surgeons in 1963. Four years after retiring from surgery, but not from administrative duties at Brigham Hospital, Murray was awarded the Nobel Prize for physiology or medicine along with **E. Donnall Thomas**, whose work in bone marrow transplants was closely related to Murray's research. By tackling the difficult problem of **organ** transplants, he provided a definitive solution to endstage renal disease as well as stimulating worldwide research into immunology. His work in craniofacial reconstruction as a plastic surgeon has not only mended and saved lives, but also enlarged the scope and diversity of plastic surgery.

MUSCULAR SYSTEM

The muscular system includes those **tissue**s of the body which, by virtue of being composed of contractile tissue, are able to

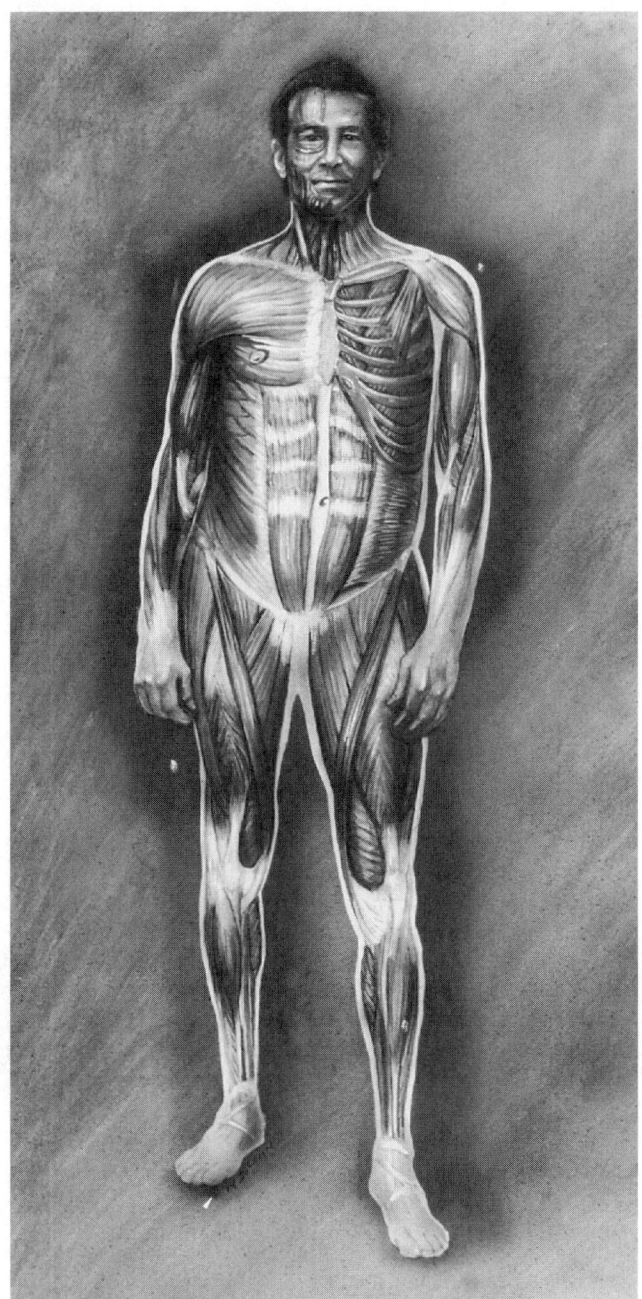

Muscle is a type of tissue made up of fibers that are able to contract and relax affording bodily movement. (Illustration by Delilah R. Cohn, Custom Medical Stock Photo. Reproduced by permission.)

cause some type of movement. The term "contractile" refers to the fact that muscles are made of complex muscle fibers, composed of myofibrils. These myofibrils are composed of filaments which overlap each other in a way that allows them to slide past each other, changing the degree of overlap. This allows the muscle to grow shorter or longer, resulting in movement.

Some muscles are under voluntary control, meaning that an individual can decide to move those muscles. Voluntary muscles are also known as skeletal muscles, because most of these muscles are attached to the bones of the body, and are responsible for moving the **skeletal system**. Voluntary muscles which are not directly attached to bones include those muscles around the mouth and the anus. While skeletal muscles are under voluntary control, they also may receive some automatic input from areas of the **brain**. This is certainly true of many of the muscles of **respiration** (breathing), which are automatically programmed to continue contracting and relaxing, so that an individual does not have to decide to take each breath. There are more than 650 different voluntary muscles in the human body.

A second type of muscle, called smooth muscle, is considered to be under involuntary control. Smooth muscle makes up the muscles of the intestine, the uterus, the **blood** vessels, and the eye. The intestine contracts and relaxes without an individual even being aware of its actions, allowing food to be churned up and moved along its length. An individual cannot exert control over these muscles; one cannot, for example, decide to speed up the amount of time it takes for a meal to travel down the length of intestine. Similarly, when a woman goes into labor, she cannot prevent her uterus from contracting simply by willing it to stop. Once the correct **hormone** environment is in place, the uterus will continue contracting entirely without voluntary control. An individual's pupillary muscles contract or relax due to their response to the presence or absence of **light**, and not due to an individual's desire to have dilated or contracted pupils.

A third type of muscle is cardiac (**heart**) muscle, which is responsible for the forceful contraction of the heart beat. This type of muscle is primarily involuntary, although individuals have been taught to manipulate heart rate through a process known as biofeedback. Cardiac muscle is amazingly strong and resilient, given its ability to beat continuously over an individual's lifetime. In contrast, skeletal muscle needs periods of rest, a luxury which cardiac muscles neither require nor can afford to take.

MUTAGENESIS

Mutagenesis is the induction of genetic change in a **cell** by the alteration of **DNA**. The change or alteration is heritable in cell progeny and is almost always detrimental. While most **mutation**s are indeed detrimental, it is well to remember that mutations provide the fabric of **evolution**. Further, many mutations have been economically valuable to humans. The Delicious apple is such a mutation.

Mutagenesis can result in alteration of an individual gene, a **point mutation**, or can result in aberrations in **chromosome** number and structure. If a mutagenic event occurs to a body cell, the cell may survive but it may show no observable effect, it may be lethal to the cell (which for the most part would also have no observable effect), it might be propagated as a useful variant (the example of the Delicious apple), or it may give rise to variant cells that produce either a benign tumor or a malignant **cancer**. If the mutation occurs in a ga-

mete (sex cell), the genetic alteration may be passed to subsequent generations. Leopard frogs, *Rana pipiens*, are spotted as the common name suggests. A dominant mutation occurred sometime in the past that caused some progeny of spotted frogs in Minnesota to have no spots. The mutation is known as "burnsi" and it behaves as a dominant Mendelian gene.

Mutagenesis can occur as the result of exposure to ionizing radiation and certain chemicals. Ionizing radiations include cosmic rays, x rays, and ultraviolet **light**. It is of interest to note that melanoma, caused almost exclusively by exposure to the ultraviolet radiation of sunshine, is the most rapidly increasing lethal cancer in the United States. The heightened prevalence of melanoma is presumably due to increased leisure time and greater exposure of skin to the mutagenic effects of ultraviolet. A number of chemicals have been identified as mutagenic. Curiously, many of the chemicals are not intrinsically the cause of mutagenesis. The chemicals must be metabolically changed by **enzyme**s to become activated as mutagenic chemicals.

Teratogenesis is the origin of birth abnormalities. Many **teratogens** are mutagens. Carcinogenesis concerns the causation of cancer. Many **carcinogens** are mutagens. One may wonder if survival is possible because of exposure to mutagenesis-causing radiation and chemicals. It is reassuring to know that the body has evolved mechanisms for DNA repair. One mechanism, excision repair, involves recognition and removal of damaged DNA by a repair endonuclease, synthesis of the missing DNA by a polymerase, and DNA ligase to seal the site of repair.

MUTATION

Mutation is an alteration in an **organism**'s **gene**s. The mutation is carried to the organism's offspring, so the offspring may look or act different from its ancestors. Mutations may be beneficial or harmful, or they may have no noticeable effect at all. Genes consist of long strings of **DNA** molecules. DNA takes the form of a spiral ladder, with rungs made of two interlocking substances called bases. The sequence of the rungs of the DNA ladder determines the **genetic code** of the organism. Any change or break in the DNA ladder alters the genetic code, and so makes a change in the next generation of the organism, whether it is a bacterium or a human being.

The discovery and naming of mutation is credited to a Dutch botanist, **Hugo De Vries**. He studied the evening primrose **flower**, and actually came up independently with the same laws of **genetics** that **Gregor Mendel** had published earlier. De Vries brought forth his primrose study only as a confirmation of Mendel's genetic theory. But De Vries had found something that Mendel hadn't mentioned, which was that some of the primroses came up with changes that had not existed in earlier generations of the plant. He called these changes mutations.

The idea of mutation was central to Darwin's theory of **natural selection**. Organisms mutated at random, but some of the mutations actually helped an organism survive. That mutated organism then had a better chance than others at surviving long enough to reproduce. So helpful mutations were passed on to succeeding generations, allowing a **species** to adapt to changing environments.

A mutated green frog with six legs. Dutch botanist Hugo de Vries first proposed in the mid-nineteenth century that the underlying reason for such abrupt morphological changes could be traced to changes in hereditary material passed from one generation to the next. *(JLM Visuals. Reproduced by permission.)*

The idea that mutations were entirely random was sacrosanct to **evolutionary theory**. A British geneticist, John Cairns, caused considerable controversy in 1989 by publishing his study of **bacteria** that seemed to be able to produce beneficial mutations when needed. He called this directed mutation. Cairns put bacteria that could not digest lactose into a petri dish, and fed them only lactose. He found that a larger than random percentage of the bacteria acquired the beneficial mutation that allowed them to digest the lactose. Scientists around the world attempted to reproduce, explain or gainsay Cairns' results. Ten years later, scientists seemed to agree that there might be a phenomenon known as ''hypermutation,'' where bacteria under stress can increase their mutation rate. Mutations spread through the bacteria **population** quickly in hypermutation, allowing the strain to evolve faster. Though most mutations are harmful to the individual, the faster **evolution** seems to help the population as a whole. So in Cairns' experiment, a flurry of mutations, possibly brought on by the stress of starvation, brought out the helpful lactose-tolerant trait, and individuals carrying this trait survived and flourished.

Hypermutation seemed more palatable to biologists than directed mutation, in which the bacteria somehow came up with only the required mutation. Probably much further research will be done in this area. In any case, the work of Cairns and others opened up the theory of evolution to new complexities and controversy.

See also Genetics

MUTUALISM

Mutualism (from Latin *mutuus* meaning lent or borrowed) is a form of **symbiosis**, or close association between **organisms** of two or more **species**, in which both participants in the relationship benefit. Because both participants are gaining something from the relationship, this type of symbiosis is often symbolized as (+, +). Mutualism is only one of three recognized categories of symbiotic relationships. The other two are **commensalism**, in which one participant benefits and the other is neither helped nor harmed, and parasitism, in which one benefits and the other is harmed. Some mutualistic relationships have existed over such long periods of time that the two species have evolved simultaneously to live with each other. In some cases, the association is so close that it is obligatory; one or both of the individuals in the relationship would not be able to survive without the other. For example, in some types of lichens, a mutualistic relationship between **algae** and **fungi** have evolved so closely together that they apparently cannot survive apart from one another. Lichens are an example of mutualism because the algae photosynthesize and provide **nutrients** to the fungus, while the fungus protects the algae from the often harsh environmental conditions where these organisms live, by providing moisture and **minerals**. Another example of a mutualistic symbiotic relationship is that of the **nitrogen**-fixing **bacteria** *Rhizobia* living in nodules on the roots of legumous **plants** such as clover, alfalfa, peas, and beans.

A cleaner shrimp participating in a symbiotic relationship with a zebra moray. *(JLM Visuals. Reproduced by permission.)*

These bacteria are able to take atmospheric nitrogen, a form that plants cannot use, and change it into a form that is usable by plants. The relationship is mutualistic because both organisms benefit; the plants gain the important nutrient nitrogen in a usable form, and the bacteria get a source of **energy**, usually glucose. There are numerous other examples of mutualism throughout the world. For example, large sharks often have small suckerfish known as remora living on them. These fish attach to the surface of the shark, while it is swimming. When the shark feeds, the remora detaches and picks up scraps. In return for the free ride and food, the remora cleans the shark of external **parasites**. Termites have symbiotic **protozoa** living within their digestive tract. The termites themselves cannot digest the wood they consume; they rely on the protozoa. Therefore, in a mutualistic relationship, the termites provide food for

the protozoa, and the termites get to use the end products of protozoan **digestion**.

MYELIN SHEATH

The myelin sheath is a fatty white covering which is wrapped around nerve fibers (**axon**s) throughout the body, **brain**, and **spinal cord**. This covering is composed of the **membrane**s of one of two different types of specialized **cell**s. In the central **nervous system** (the brain and spinal cord), the myelin sheath is composed of cells called oligodendrocytes. Throughout the rest of the body, Schwann cells create the myelin sheath. When oligodendrocytes or Schwann cells spiral around a nerve's axon, several layers of their membranes serve to insulate the axon. Because myelin has a white color, fibers which are myelinated are referred to as white matter. Unmyelinated fibers are referred to as gray matter. Small areas of interruption along the myelin sheath are called nodes of Ranvier. These nodes of Ranvier occur approximately every millimeter along the myelin sheath.

The function of this insulating layer of myelin around a nerve fiber is to increase the speed of a nerve impulse's conduction along the fiber. Without myelin, a nerve impulse travels at a rate of several meters per second along the nerve axon. **Nerve impulses** along a myelinated axon travel at approximately 100 meters per second.

Multiple sclerosis is a disease which results in the destruction of the myelin sheath. As the myelin sheath is stripped from axons, nerve impulses slow throughout the body, and movement becomes slow and difficult. Eventually, scar **tissue** may replace the myelin along various axons, resulting in further disability, and eventually in paralysis.

MYOPIA • See Eye disorders

N

NATHANS, DANIEL (1928-)
American molecular biologist

Nathans was born in 1928 in Delaware. He was the last of nine children born to Samuel and Sarah Nathans, Russian Jewish immigrants. Nathans received his B.A. from the University of Delaware in 1950 and his M.D. from Washington University in St. Louis in 1954. It was during the summer after his first year of medical school that Nathans had his initial exposure to laboratory work.

After medical school, Nathans completed a one-year internship at Columbia-Presbyterian Medical Center. After this, he spent two years (1955–57) at the National Cancer Institute as a clinical associate studying **protein synthesis**. In 1956, Nathans married Joanne Gomberg, with whom he had three sons. Returning to Columbia-Presbyterian, Nathans completed his residency in 1959. That same year Nathans won a United States Public Health Service grant to do biochemical research at Rockefeller University in New York with **Fritz Lipmann** and **Norton Zinder**. It was at this point that Nathans fully committed to work in the laboratory rather than in a clinical practice. In New York, Nathans continued his work on protein synthesis and began viral research, mostly related to host-controlled variations in **virus**es.

In 1962, Nathans began his long relationship with Johns Hopkins University as assistant professor of microbiology and director of **genetics**. He was elevated to associate professor in 1965 and full professor in 1967. He was named director of the molecular biology and genetics department in 1972 and Boury Professor of **Molecular Biology** and Genetics in 1976, positions he retained for many years.

In 1962, when Nathans first arrived at Johns Hopkins, **Werner Arber**, at Basel University in Switzerland, predicted the existence of an **enzyme** capable of cutting DNA at specific sites. **Deoxyribonucleic acid (DNA)** is assumed to be the source of autoreproduction in many viruses. An ability to cut or cleave the DNA into specific and predictable fragments was important to greatly improving our capabilities for researching and understanding viruses. The necessity of "specific" and "predictable" fragments relates to the need of the scientist to know the fragment he or she is studying is identical to the fragment any other scientist would get following the same laboratory procedure.

In 1968, Arber got halfway to his goal, finding an enzyme (type I) capable of cleaving DNA, but in seemingly random patterns. In 1969, **Hamilton O. Smith**, a colleague of Nathans at Johns Hopkins, wrote to Nathans (who was in Israel at the time) to tell him he had developed a type II enzyme. This enzyme, named Hind II, was capable of cleaving DNA into specific and predictable fragments.

At this time, Nathans was working on a simian virus (SV40) which causes tumors in monkeys. SV40 was particularly impervious to then-current methods of study, so Nathans immediately saw an application of Smith's tool. Nathans, with Kathleen Danna, used Hind II to cut SV40 into eleven pieces and show its method of **replication**. One technique they employed in this process was radioactive labeling. The combined efforts of Arber, Smith, and Nathans over a period of more than a decade led to their receipt of the Nobel Prize in physiology or medicine in 1978. Their inter-laboratory cooperation greatly advanced the potential for consistent DNA and **gene** research.

Nathans continued his work with Hind II and cleared the path for much of the work that has been done since in research on DNA function and structure (such as restrictions maps, used to define DNA structure). This early work has also led to the area of recombinant DNA research, which involves the process of joining two DNA fragments from separate sources into one **molecule**. Since this field of research was uncharted territory and carried some risks, including the creation of new pathogens, Nathans was among an early group of scientists who, in 1974, encouraged the publication of research guidelines and some self-imposed limits on DNA research. Despite the risks, recombinant DNA research has been put to good use in creating supplies of heretofore scarce enzymes and **hor-**

Daniel Nathans. *(UPI/Corbis-Bettmann. Reproduced by permission.)*

mones, including human-produced **insulin**. In the 1980s, Nathans's research continued to be linked closely to DNA and genetics. A good portion of his scientific work during this time related to the effect of growth factors on genes and gene regulation.

NATURAL SELECTION

Evolution and the theory of natural selection are the principle unifying concepts in biology. This is as true now in the age of **molecular biology** as it was a century ago. The unification stems from the fact that details of embryology, **morphology**, molecular biology, physiology, **paleontology**, **animal** distribution, etc. only become understandable when considered in the context of evolution and natural selection. Certainly evolution had been considered prior to the publication of **Charles Robert Darwin**'s the *The Origin of Species* in 1859. However, a mere consideration was no longer appropriate for biological scientists. Darwin's book was an extensive catalog containing a plethora of evidences that support the reality of evolution. His theory of natural selection provided the mechanism for evolution. The 1859 book was the beginning of modern biology.

Evolutionary theory explains the bewildering array of similarities in related biological groups. For instance, frogs, mice, and humans, while obviously different, share similar **organ** systems. A histological (**tissue**) section of a mouse liver is difficult to distinguish from that of a human and both share similarities with frog liver. The frog liver functions in detoxification of xenobiotic substances in much the same manner as that of the mouse and human, and of course, all other vertebrates. Striated (voluntary) muscle is essentially identical in all vertebrates. The similarities of **cell** biology, **anatomy**, organ systems and physiology are of course well known to both lay people and scientists. What accounts for this notable similarity? The frogs, mice, humans and other vertebrates descended from a common ancestral **species** in the distant past. Species with a more recent divergence from a common ancestor, such as humans and apes, share more similarities than creatures with a common ancestor in the remote past. Darwinian evolution seeks to explain these relationships by the mechanism of natural selection.

Natural selection rests on a number of obvious facts. First, in natural **population**s, reproduction has the potential to result in more individuals than can survive. Consider the frog again. A female northern leopard frog, *Rana pipiens*, has the capacity to produce about 3,000 offspring per year. What is true for frogs is true for all species, i.e., all species have the competence to produce offspring in greater numbers than can survive. This results in **competition** between individuals and not all individuals of the species survive. Second, scrutiny of any population of **organism**s will reveal that not all individuals are identical. Variation exists in form and function. Some of the variation is heritable and that means the variation will be passed from the parents to the next generation. Some variation will affect survival and reproduction. Some differences are advantageous and will enhance survival and reproduction; others will not. It is not difficult to see that, in time and with competition, the inherited differences that impart enhanced survival and reproduction will tend to become more common. The population changes (evolves) as the frequency of a particular variant changes.

The natural world is constantly changing which results in new **selective pressures**. Where previously a particular population was well adapted for survival and reproduction, another environment may result in a new variant having enhanced survival and reproduction potential. In other words, the environment changes and natural selection results in the evolution of a changed population.

As the environment changes, continued selection leads to a steady change in the structure and physiology of a species; the change will always be toward enhanced **adaptation** to the environment (which has sometimes been referred to as "**fitness**"). The surviving individuals of a population may become exquisitely fit. Darwin observed all of the above. All that was missing was an understanding of **heredity**. Modern **genetics** has augmented understanding of natural selection, which over a century after the publication of the *The Origin of Species*, remains central to the understanding of all aspects of biology.

NATURE

Nature is defined in the dictionary as the totality of physical reality, or the total system of spatiotemporal phenomena and events, or as real and objective existence. It has also been defined as the world of mind and **matter** external to the observer, reality as observed. In the vernacular, nature is often described as the out-of-doors, as natural scenery or a landscape. Many people still take their image of nature from the nineteenth century British poet Alfred, Lord Tennyson, who described "nature red in tooth and claw." Or they see it as a benign, domesticated place to camp and play.

Ralph Waldo Emerson, a nineteenth century American thinker, wrote one of the most influential commentaries on nature. He suggested that "strictly speaking, all that is separate from us [humans] must be ranked under this name, Nature." Nature is multi-faceted, incorporating as it does, all the material world of living **organism**s and natural objects and forces. As Emerson noted, "there is a property in the horizon [of nature] which no man has but he whose eye can integrate all the parts." "Nature," Emerson claimed, "is not only the material, but is also the process and the result." Emerson also cautions that, to appreciate nature, "the unity of Nature [and] the unity in variety" must both be apprehended. In appreciating Nature today, though, we should also remember Emerson's caution that "Nature is so pervaded with human life that there is something of humanity in all and in every particular."

Biology is primarily concerned with what might be called "living nature," the realm of living organisms. That interest is, of course, intertwined with recognition that organisms depend on, and draw their sustenance from non-living nature, the **abiotic environment** of the sun and the **minerals** and **water**s of the earth. The Latin root of the word nature is biological, meaning "to be born," and nature is still the mother of all organisms, including the human, the source from which all sustenance flows.

The purpose of biology is to understand how nature works, how the complex living systems of planet earth survive, evolve, and interact. Since biology is the science of living things, humans are active units in all its processes. The human body is natural, that is it is composed of the same elements and structural units found in other organisms, and it performs the same basic functions. The human starts life as a simple, undeveloped cellular form, is born, grows as do other living entities, and develops through similar processes.

There is much talk these days about "the social or cultural creation of nature" and even about "the end of nature." If nature is defined as the material world, that world of interest to scientists, including biologists, or is defined as equivalent with the universe, then nature as such is not humanly created and cannot end. This is true even if nature is confined to the material world of the planet earth. Nature is perceived in different ways by different cultures, or even by different individuals, but there is also a nature that in fact exists as a reality independent of what humans think about it. As long as the planet exists, nature exists, whatever happens to the human **species**.

How nature is defined, including what is natural, becomes critical as humans take over more and more of the surface, and the **productivity**, of the earth and its ecological systems. One example is provided by on-going debates over

Yellowstone Park. The debate centers on the degree to which the park should be left "natural." If the park is to be an accurate representation of nature—so that part of the human environment can be visited as a way to keep in touch with raw nature—then how much human impact and interference can be allowed? Is fire a natural component of the system, to be left alone, or should fires be controlled? Is the park too small an island of nature to function as a complete 'natural' **ecosystem**? For example, can a grizzly bear **population** be maintained in such a small island of nature? Should wolves have been reintroduced? Should the local elk population be managed or allowed to fluctuate in response to changes in vegetation and predators?

Humans cannot end nature, but human impacts on the natural world are increasing at a steep rate. Nature defined on an earthly scale has nowhere escaped human modification. Nature defined as synonymous with the universe has hardly been dented. But biologists are concerned mostly with the nature of living things on earth, and on this planet, scientists are documenting rather dramatic impacts. Research by biologists is documenting human reduction of the diversity of life on earth, and is recording an acceleration in the natural rate of species **extinction**. Humans are altering and destroying the **habitat**s of innumerable organisms, every day and on a large scale.

The natural world is dynamic and always changing. Nature evolves. Nature cannot be held in one place, at some static point that humans may consider ideal. But humans still need to consider carefully the changes they are making, intentionally and inadvertently, on a small scale and every day, and on a global scale that reaches far into the future. The fate of nature and the fate of the human species are indelibly intertwined.

NEHER, ERWIN (1944-)
German biophysicist

Erwin Neher was born in Landsberg, Germany, in 1944, the son of Franz Xavier Neher and Elisabeth Pfeiffer Neher. In 1967, he earned his master's degree from the University of Wisconsin under a Fulbright scholarship. He then went on to complete his doctorate at the Institute of Technology in Munich, Germany, in 1970.

While the existence of ion channels that transmit electrical charges was hypothesized as early as the 1950s, no one had been able to see these channels. As a doctoral student, Neher was drawn to the question of how electrically charged ions control such biological functions as the transmission of **nerve impulses**, the contraction of muscles, vision, and the process of conception. He realized that in order to get answers to these questions he would have to look for the ion channels.

It was in his doctoral thesis that Neher first developed the concept of the patch clamp technique as a way of discovering the ion channels. In 1974 he shared a laboratory space with **Bert Sakmann** at the Max Planck Institute in Göttingen. They both agreed that understanding the nature of ion channels was the most important problem in the biophysics of the cell membrane, and they set out to develop the techniques of patch clamping.

In 1976 Neher and Sakmann published their landmark paper on the use of glass recording electrodes with microscopic tips, called micropipettes, pressed against a **cell** membrane. With these devices, which they called patch clamp electrodes, they were able to electrically isolate a tiny patch of the cell **membrane** and to study the **proteins** in that area. They could then see how the individual proteins acted as channels or gates for specific ions, allowing certain ions to pass through the cell membrane one at a time, while preventing others from entering. Their work with patch clamps allowed them to remove a patch of the membrane and to enter the interior of the cell. They then were able to conduct various experiments to observe the intricate mechanism of ion channels. Several years later, Neher, Sakmann, and their colleagues refined the technique of patch clamping. Creating a better seal between the micropipette and the patch of cell membrane it pressed against was one of the refinements they sought. Without a tight seal there was interference by "noise" that overshadowed the smaller electrical currents.

Neher solved the problem of outside noise interference in 1980 when he was able to observe on his oscilloscope a marked drop in the noise level to almost zero. From this drop he was able to infer that he had produced a seal that was one hundred times better than previously attained. While other researchers had noticed an abatement of noise at times, Neher was the first to realize the significance of the drop in noise level.

Neher found that by using a light suction with a super clean pipette, he could create a high-resistance seal of 10-100 gigohms (a gigohm is a measure of electrical resistance equal to one billion ohms). He called this seal a "gigaseal." With the gigaseal, background noise could be decreased, and a number of new ways could be used to control cells for patch clamp experimentation. Patches from the cell could now be torn away from the membrane to act as a membrane coating over the mouth of the pipette, thus allowing for more exact measurement of electrical ion movement. A strong suction could force the pipette into the cell while still maintaining a tight seal for the cell as a whole.

In 1976 Neher returned to the Max Planck Institute in Göttingen. In 1978, he married microbiologist Eva-Maria Ruhr; they have five children. He became director of the membrane biophysics department at the Max Planck Institute in 1983, and in 1987 he was made an honorary professor.

In 1991 Neher and Sakmann won the Nobel Prize for proving the existence of ion channels. The Nobel Committee also praised the work of Neher and Sakmann for helping in research on **heart** disease, **epilepsy**, and disorders affecting the nervous and muscle systems. Patch clamp research has helped in the development of new drugs for these conditions.

NEISSER, ALBERT LUDWIG SIEGMUND (1855-1916)

German dermatologist

Albert Neisser was a German dermatologist who devoted his career to studying various skin diseases. Most notably, Neisser made important contributions to the study of venereal diseases and discovered the bacterium that causes gonorrhea. *Neisseria* is a **genus** of non-endospore forming diplococci which includes the **species** *N. gonorrhoeae*, the causative agent of gonorrhea and *N. meningitidis*, the causative agent of meningococcal meningitis.

Neisser was born on January 22, 1855, in Schweidnitz, Germany (now Swidnica, Poland). His father was a physician. Neisser, who lost his mother before he turned one, attended the Volksschule in Münsterberg, continuing his education at St. Maria Magdalena Humanistic Gymnasium in Breslau, where he was a classmate with **Paul Ehrlich**. He entered medical school in 1872 at Breslau, where, under the direction of Anton Biermer, wrote a thesis on the Echinococcus, and graduated in 1877. It is interesting to note that Neisser was not an outstanding student and needed to repeat some of his coursework.

Initially, Neisser wanted to enter the field of internal medicine, but since there were no openings in Biermer'a clinic, he decided to go into dermatology and became an assistant to Oskar Simon. In 1879, he discovered the gonococcus, which Neisser called micrococcus (it was later renamed by Ehrlich). At the time, bacteriology was a growing field of study. Neisser had use of the Zeiss microscope that had Abbe's condenser and oil immersion system, and he had learned how to make smear tests for bacterium identification, as well as **cell staining** techniques. Also in 1879, Neisser presented an important paper that clarified the etiology of venereal diseases. Later that year, he travelled to Norway, were he examined leprosy patients and brought back enough smears of the small thin rod bacilli to conduct further study. As a result of his work, Neisser identified that particular bacillus as the causative agent of leprosy. The etiology, diagnosis, and prophylaxis of leprosy intrigued Neisser during his entire career. In addition, he became very interested in and studied other diseases such as lupus, anthrax, glanders, actinomycosis, and psoriasis.

In 1880, he was named *Privatdozent* at Breslau, eventually assuming the chair of dermatology after Simon died. In 1881, Neisser married Toni Kaufmann, who became an invaluable assistant, actively participating in all his investigations and accompanying him on his research trips. In 1892, he opened a dermatology clinic and became very well known in his field. Several years later, Neisser began to study **syphilis**, and attempted to develop preventive inoculations. Drawing an analogy from the inoculations that Behring had used against diphtheria and tetanus, Neisser used infected serum to inoculate some young prostitutes and was subsequently accused of malicious conduct. When, in 1903, **Elie Metchnikoff** and **Wilhelm Roux** found that syphilis could be communicated to apes, Neisser journeyed to Java to obtain several apes, so that he could continue the research. Later, he studied syphilis among the Dutch sailor population on Java. While Neisser did not isolate and identify the syphilis spirochete (that discovery was made by Fritz Richard Schaudinn and Erich Hoffmann in 1905), he contributed valuable information about the transmission of the disease and helped develop a serological test for syphilis, named for August von Wassermann. His research of venereal diseases pushed him into the public health field where he was an advocate for public education and thought that better hygiene among prostitutes was far more effective than throwing them in jail. Neisser supported the idea of a central Board of Health and was an outspoken proponent of doctor/patient confidentiality.

In 1907, Neisser became a full time professor of dermatology at Breslau, eventually training many well known dermatologists. Neisser was very attached to his wife who, as stated earlier, was a partner in his work, and when she died in 1913, he was extremely saddened. His health rapidly declined over the next few years, and he died on July 30, 1916, in Breslau. His house in Breslau was made into a museum in 1920; in 1933, the house was confiscated by the Nazis and turned into a guest house. Fortunately, all of Neisser's papers were salvaged by a physician named Brock.

NEARSIGHTEDNESS • See Eye disorders

NEOTENY

Neoteny is a term that describes **sexual reproduction** by **larval organism**s. In theory, it could occur as the result of premature sexual maturation in larvae, or because of sexual maturity in an organism that retains larval characteristics. Neoteny occurs widely among tailed amphibians known as salamanders. Perhaps the best know example of neoteny is that of the Mexican axolotl, *Ambystoma mexicanum*, which was found near Mexico City (this salamander is rare now in natural **population**s). The axolotl retains a broad head, bushy external gills, teeth in both jaws and lacks eyelids. It has small legs and feet and it has a dorsal fin that runs from the back of the head to the tip of the tail. A ventral fin is found between its posterior limbs which extends to the tail tip. The axolotl is larger than average (up to 10 in [25 cm]) for salamanders. The retention of larval characters permits an aquatic life and reproduction in **water**.

It would be plausible to think of the axolotl as an fully mature aquatic **species** with no suspicion that they are in fact larval forms. However, in Paris in 1865, Mexican axolotls underwent spontaneous **metamorphosis** to the adult terrestrial form in an aquarium. Not long thereafter, it was discovered that thyroid extracts would routinely cause metamorphosis of axolotls. Metamorphosis caused loss of gills, tail fins, and other larval characteristics.

Axolotl is an Aztec word appropriate for *A. mexicanum*. However, the term can be used for any salamander that reproduces as a larva. All families of salamanders have some neotenic forms and they are relatively common in North America. For example, the blotched tiger salamander *Ambystoma tigrinum melanostictum*, occurs in Yellowstone and Grand Teton National Parks, and may reproduce either by neoteny or by reproduction as an adult. It is not known why some populations reproduce by neoteny and other populations do not.

NERVE GROWTH FACTOR

Nerve growth factor (NGF) is a **hormone**-like **protein** produced by target **cell**s throughout the body that controls growth of **axon**s (fibers) that connect embryonic nerve cells and is also required for their survival. NGF, discovered in the 1950s by the Italian-American biochemist **Rita Levi-Montalcini**, was the first of a number of growth factors that control cell development.

When NGF binds with receptors on the nerve cell and moves into the **nucleus**, the cell responds by growing an axon (a fiber for transmitting signals) toward the target cell and forms a **synapse** (connection) with it, building the body's **nervous system** network. When NGF is blocked by antibodies, the nerve cells die.

Levi-Montalcini first observed rapid nerve growth in chick embryos that were in direct or indirect contact with mouse tumors. Believing that a chemical was the cause, she used the then-new technique of **tissue culture**—combining tumor material with embryo **tissue** and **blood**—to produce nerve growth in the laboratory. The chemical, which she named nerve growth factor, was purified and its protein structure determined by Levi-Montalcini's colleague **Stanley Cohen** in 1953. The two scientists received the 1986 Nobel prize in physiology or medicine for this work. NGF's complete **amino acid** sequence was accomplished in 1971 by the American biochemists Ruth Hogue Angeletti and Ralph Bradshaw using a technique developed by the Italian biochemists Vincenzo Bocchini and Pietro Angeletti. Several groups of scientists in the United States have since sequenced the **gene** that codes for the **molecule**.

Scientists know that production of axons and development of NGF receptors are two separate functions, but many mysteries about the receptors must still be solved. For example, sympathetic nervous system cells in very early embryos do not need NGF and will not even respond to it. But more mature sympathetic **neuron**s and some **sensory neuron**s must have NGF to survive. Scientists are trying to discover how and when the cells begin expressing NGF receptors and responding to it. Many believe that electrical activity may be involved. Also, the composition of the receptor molecule is not yet known, but a protein produced by an **oncogene** (**cancer**-causing gene) called trk appears to be a component. NGF produced by **genetic engineering** is being studied for possible use in treating nerve cells damaged by Alzheimer's disease, Huntington's chorea, and **AIDS**-related peripheral neuropathy.

NERVE IMPULSE

A nerve impulse is a message which is transmitted along a nerve fiber or **axon**. It originates in the **neuron**'s soma, or **cell** body, passes into the axon hillock, and then speeds down along the axon until it reaches the axon terminal or synaptic knobs. The impulse must then make its way across a gap called the **synapse**, to stimulate another neuron's or muscle cell's soma.

The nerve impulse occurs due to changes in the electrical charge along the axon. Normally, at rest, the chemical makeup of the axon's **membrane** allows the inside of the membrane to be slightly more electrically negative compared to the slightly more positive outside of the membrane. This condition is maintained by the presence of various channels or pumps within the membrane which work to maintain a particular proportion of potassium and sodium **molecule**s on the inside and the outside of the neuronal membrane.

When a nerve is stimulated, however, this potassium-to-sodium proportion is changed, so that the inside of the axon's membrane becomes more positive than the outside. When this positive charge reaches a particular level, called the threshold, the charge is conducted down the axon. This is called the **action potential**.

A nerve impulse traveling down an axon moves at a rate of slightly greater than one mile per hour (about 0.5 meters per second). Some axons, however, are wrapped with layers of specialized cells. These cells have fatty membranes which line the outside of the axons, creating what is called a **myelin sheath**. Small interruption occurring about every millimeter along the myelin sheath are called nodes of Ranvier. These myelinated axons can transmit nerve impulses much more quickly, at a rate of nearly 300 miles per hour (about 130 meters per second). This is accomplished because the action potential skips over the myelinated area of the axon, jumping rapidly from node to node. This type of nerve impulse is called saltatory conduction or saltatory propagation.

The nerve impulse continues down the axon until it reaches the axon terminal, where the synaptic knobs are located. The electrical charge reaching this location causes the synaptic knobs to unload little packets of chemicals (**neurotransmitters**) into a gap (synapse) which exists between the axon terminal and the next neuron's soma. These neurotransmitters cross the synapse, heading toward the new neuron. The arrival of these chemicals at the next neuron's soma thus stimulates an action potential which runs down its axon.

See also Brain; Nervous system

NERVE PLEXUS

A nerve plexus is a complicated intertwining of many nerve fibers. Two nervous projections from the **spinal cord**, called the dorsal root (dorsal roughly means back) and the ventral root (ventral roughly means front) combine to form a single spinal nerve. This spinal nerve then divides into two separate branches, the dorsal branch (also called the dorsal ramus) and the ventral branch (also called the ventral ramus). A number of the ventral rami (rami is plural of ramus) combine together, exchanging nerve fibers, and creating a complex intertangled web of nerve fibers called a nerve plexus. The major nerve plexuses are the cervical plexus, the brachial plexus, and the lumbosacral plexus.

The cervical plexus provides nerves which serve the muscles and skin of the neck, upper shoulders, chest, and part of the head. The nerve which operates the **diaphragm** also emerges from the cervical plexus. The brachial plexus allows the lower shoulder and the entire arm to function. The lumbo-sacral plexus sends nerves to the skin and the muscles of the thigh and leg, including the body's largest nerve, the sciatic nerve.

NERVOUS SYSTEM

How are human functions such as movement, **learning**, thought, and reflexive action regulated by the body? For a very long time, scholars divided these processes into two large categories: those controlled by the **brain** and those controlled by the mind. According to this duality, intellectual processes such as critical thinking and **memory** were a function of the mind, an immaterial entity that was not subject to scientific study, but was to be understood in terms of philosophical analysis. The brain, on the other hand, was thought to be a concrete, physical entity responsible for phenomena such as muscular activity that could be investigated by scientific means. During the Middle Ages, for example, many scholars assigned the control of higher intellectual processes to an immortal and invisible soul and the control of muscular behavior to a tangible and mortal soul that could be studied scientifically.

Studies of the physical aspect of this duality have a very long history. For example, the Greek physicians **Erasistratus** and **Herophilus** discovered the existence of nerves and traced their connection through the central nervous system into the brain, thereby refuting the then widely held belief that the **heart** was the center of sensation. Both carried out detailed investigations of the brain and identified many of its regions. The great Roman physician Galen carried out a number of experiments on the nervous system and identified seven of the twelve cranial nerves. He also expanded on an earlier theory that nerves are hollow tubes through which ''**animal** spirits'' flow. Galen's work was among the last research on the nervous system for nearly 1,500 years.

When such research began again, it involved a new assumption; namely, that all forms of animal behavior—both intellectual and physical—are functions of the physical brain and, thus, subject to concrete, scientific analysis. Over time, it has become more and more clear that this assumption is valid and that all animal behaviors can be understood in terms of the nervous system and, on a more basic level, in terms of biochemical and biophysical changes that take place within this system. Today, there are probably no human or animal behaviors ascribed by scientists to the control of an intangible ''mind'' or ''soul.''

One landmark figure in early research on the nervous system was Thomas Willis (1621-1675). In 1664, Willis wrote the *Cerebri anatome*, a long article that summarized all that was then known about the brain and nervous system. Some historians claim that this article provided the first complete description of the nervous system. In that article, Willis also reported on his own research, describing the role of the **cerebellum** and the **cerebrum**, respectively, in the control of involuntary and voluntary activity.

At about the same time, the brilliant French scientist René Descartes applied his mechanistic philosophy to the analysis of animal behavior. He taught that muscular actions are controlled by a nervous system that consists of a complex arrangement of pumps and tubes. Messages are carried from the brain to the muscles by means of "vital spirits" that flow through these tubes, Descartes wrote. Although his theory was proved incorrect, Descartes did introduce an important new concept to the study of the nervous system: the reflex. The term reflex refers to any involuntary response that a body makes when exposed to a **stimulus**. The knee-jerk reaction that occurs when a physician tests your reflexes is perhaps the most familiar of these reflex actions.

Understanding reflex action became one of the major goals among researchers of the nervous system over the next century and a half. In 1811, the English surgeon Sir **Charles Bell** made an important discovery. He found that nerves leading into the spinal column consist of two different kinds. Some (the sensory nerves) carry messages from receptors in the eyes, ears, tongue, skin, and other places on the body, while others (motor nerves) transmit messages from the central nervous system to the muscles. Bell's discovery was confirmed a decade later by **François Magendie**, working with no knowledge of Bell's own research. The term reflex action was not actually introduced until 1833. It was first used by the English physiologist Marshall Hall (1790-1857), who did extensive studies on the phenomenon.

A detailed explanation of the reflex process was eventually provided by **Charles Scott Sherrington**, working with the "scratching" reflex in dogs. Sherrington was able to trace the path of a nerve message from regions on a dog's back, across sensory nerves, to the **spinal cord**, across **synapse**s, to motor **neuron**s, and then to muscle **cell**s in the dog's leg. Because of his extensive research on nerve processes such as this one, Sherrington has become known as the father of modern neurophysiology. Studies of reflex action have become an important part of psychological research also. The Russian physiologist **Ivan Petrovich Pavlov** carried out his now-famous research on conditioned reflexes in the 1880s and 1890s. He showed that, by exposing an animal to two stimuli at the same time (for example, food and a bell), he could train an animal to respond (by salivating) to the stimulus (the bell) that was otherwise unconnected with this response. Throughout much of the twentieth century, the American psychologist **B. F. Skinner** carried this type of research on conditioned responses to even higher levels of sophistication.

Nineteenth-century neurophysiologists also explored a number of other features of the nervous system. For example, the German physiologist **Johannes Müller** focused on the sensory **organ**s, especially the response of the eye to color and of the ear to sound. Some of his most important discoveries are summarized in his "doctrine of specific nerve energies." According to this doctrine, a sensory nerve can respond in only one way. For example, no matter how one stimulates the optic nerve, that stimulation will always be interpreted in the brain as **light**.

The exact mechanism by which messages travel along nerve fibers was also a topic of great interest. As early as 1771, the Italian anatomist **Luigi Galvani** had found that nerve mes-

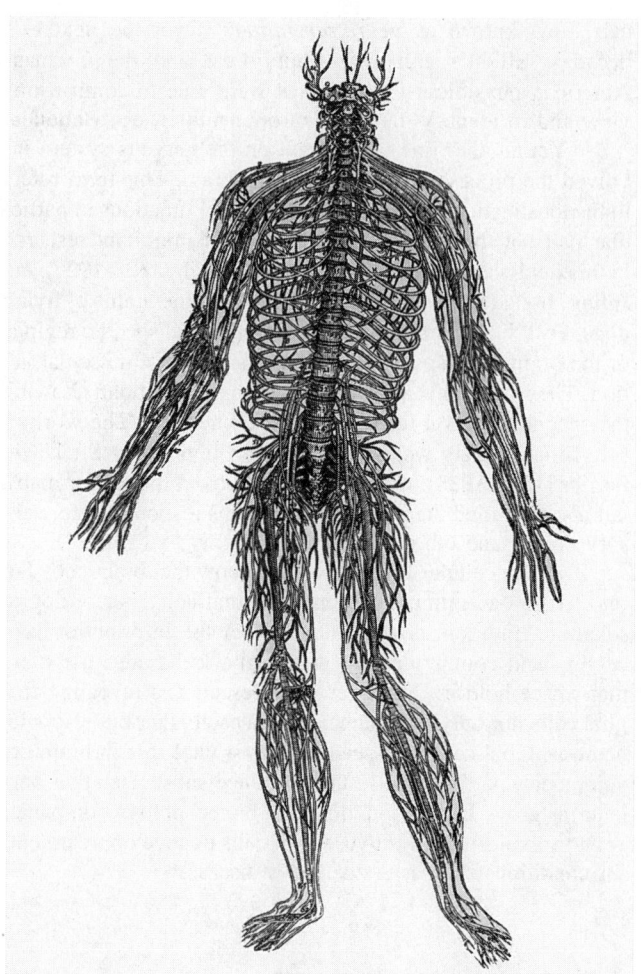

The human nervous system. *(Corbis Corporation. Reproduced by permission.)*

sages somehow have an electrical component. This line of research was further developed by Emil du Bois-Reymond (1818-1896), sometimes called the father of modern electrophysiology. The electrical nature of nerve transmission was put to use in the 1920s by **Joseph Erlanger** and **Herbert Gasser** in their adaptation of the cathode ray tube for the measurement of nerve messages. With their cathode ray tube, Erlanger and Gasser were able to greatly amplify the naturally small electrical currents passing along nerve cells and to study them in much better detail. This research allowed them to calculate, for example, the speed at which messages travel along a nerve.

A related problem concerning nerve transmission involved the mechanism by which a message travels from one cell to the next. In the 1890s, **Wilhelm von Waldeyer-Hartz** had suggested that adjacent nerve cells do not actually touch each other, but are separated by a small gap. This view was confirmed experimentally by **Camillo Golgi** shortly thereafter. How nerve messages are transmitted across these tiny gaps—synapses—soon became a topic of intense research. In 1903, the English physiologist Thomas R. Elliott (1877-1961) suggested that such messages are carried across the gap by chemicals, now known as *neurotransmitters*. Over the next two decades, Elliott's colleague **Henry Dale** and the German-American physiologist **Otto Loewi** were able to confirm this view and to identify the first neurotransmitter, **acetylcholine**.

Yet another line of research on the nervous system involved the process of "mapping" the brain. This term refers to the localization of certain specific bodily functions in particular parts of the brain. Some of the most important research in this field was carried out by Gustav Fritsch (1838-1927) and Julius Hitzig (1838-1907). By operating on the brains of living dogs, Fritsch and Hitzig were able to show that specific regions of the brain are responsible for specific kinds of muscular action. They were thus able to draw a map of the brain showing the specific location for each of these functions. The work of Fritsch and Hitzig was advanced and extended by Sir David Ferrier (1843-1928), who not only improved the earlier maps, but also modified them to include regions responsible for sensory organs and other functions.

At the cellular level, scientists know that brain cells fall into two broad groups: neurons and glial cells. According to scientific tradition, neurons do most of the information processing and communicating, while glial cells are little more than space-holders. However, new research is revealing that glial cells not only communicate with each other but also with neurons. Glial cells also seem to play a vital role in brain development, partly because they produce substances that help neurons grow. In addition, drug and biotechnology companies are now exploring ways to use glial cells to develop treatments for conditions such as Alzheimer's disease.

NEURILEMMAL SHEATH

Nerve fibers (**axon**s) are often covered by a special insulating layer called the **myelin sheath**. This covering is composed of the **membrane**s of specialized **cell**s called oligodendrocytes (in the **brain** and **spinal cord**, or central **nervous system**) and Schwann cells (throughout the rest of the body). In axons throughout the body, the outermost sheath formed by the presence of Schwann cells wrapping around the axon is called the neurilemmal sheath. This neurilemmal sheath is composed of the **nucleus** and **cytoplasm** of the Schwann cells. It is thus also called the sheath of Schwann. Because axons in the central nervous system are not insulated by Schwann cells, they have no neurilemmal sheath.

The neurilemmal sheath plays an important role in the potential regeneration of injured axons throughout the body. When an axon is cut or seriously damaged, causing it to be detached from its nucleus or soma, the axon will degenerate. Specialized **immune system** cells will clean up the resulting debris. The neurilemmal sheath, now an empty tube, is left behind. If the soma is uninjured, it may be able to begin to sprout a new axon. This new sprout will find its way into the empty neurilemmal sheath. When full recovery is possible, Schwann cells will myelinate the new axon within the old neurilemmal sheath.

NEURON

Neurons are nerve **cell**s (neurocytes), which, together with neuroglial cells, comprise the nervous **tissue** making up the **nervous system**. The neuron is the integral element of our five senses and of countless other physical, regulatory, and mental faculties, including **memory** and consciousness. A neuron consists of a nerve cell body (or soma), an elongated projection (**axon**), and short branching fibers (called dendrites). Neurons receive nerve signals (**action potential**s), integrate action potentials, and transmit these signals to other neurons or effector **organ**s, such as muscles and glands. The structure and function of neurons is essentially the same in all **animal**s, although the human nervous system is much more specialized and complicated than that of lower animals. Humans are born with a large, but finite, supply of neurons and those cells that are lost through **aging**, injury, or disease cannot be replaced.

Structure and function

Neurons exist in many shapes and sizes. Their structure, like that of other cells in the body or in **nature**, illustrates that structure often determines function. There are three basic structural and functional classifications of neurons.

Structural classification

The structural classification of a neurons depends upon the number of dendrites extending from the cell body. Multipolar neurons have several dendrites; the majority of neurons in the spinal chord and **brain** are multipolar. Bipolar neurons have only two processes: a single dendrite and an axon. Bipolar neurons are found in the **sense organ**s-and in the retina of the eye and in olfactory cells. Unipolar neurons lack dendrites and have a single axon, and are also **sensory neuron**s.

Nerve cell body

The nerve cell body contains a **nucleus**, a nucleolus, and **cytoplasm** containing the cell (such as **mitochondria**, **endoplasmic reticulum**, and so on). Unique to the nerve cell body are Nissl bodies, which are rough surfaced vesicles in the endoplasmic reticulum (cytoplasm located near the nucleus), and are involved with **protein synthesis**. Another characteristic structure of nerve cells are the neurofibrils, which are delicate threadlike structures that help to maintain the shape of the cell, and which transport substances between the cell body and the axon terminals. The **plasma membrane** around the cell separates the cytoplasm on the inside of the cell from the extracellular fluid on the outside. Cell **membrane**s of neurons contain electrically gated channels, which when properly stimulated allow electrically charged particles (such as sodium and potassium ions) to pass across the barrier. This ionic exchange is the basis for the flow, or action potential, of the nerve impulse.

Axons

An axon is a single smooth projection arising out of the nerve cell body at a raised area called the axon hillock. Axons conduct **nerve impulse**s away from the nerve cell body. Axons vary in length and diameter; some (such as those in the central nervous system) are very short, as small as 0.01 in (0.25 mm), while others (such as those in the peripheral nervous system) conduct impulses over long distances in the body, and can be 3 ft (1 m) long. The speed at which an impulse travels depends upon the diameter of the axon, with axons with large diameters (0.001 in/0.025 mm) conducting impulses more rapidly. Axons may have branches called axon collaterals. The main axon and its collaterals can split into smaller branches ending in small filaments called axon terminals. Axon terminals have knob-like swellings at the very end called synaptic knobs or end buttons. Each synaptic knob communicates with a dendrite or cell body of another neuron, the point of contact being a **synapse**. Under very high magnification, a very tiny space, the synaptic cleft or gap (about one millionth of an inch, or mm), can be detected between the synaptic knob and dendrite or cell body. Synaptic knobs contain hundreds of neurovesicles that contain a transmitter substance (or neurotransmitter). When a nerve impulse reaches the synaptic knob the neurotransmitter is ejected into the synaptic cleft and serves as a **stimulus** to the next adjacent neuron. The vast majority of all impulses transmitted occur at the synaptic gaps, although recent research indicates that chemical transmission can occur at other points along the axon. Many neurological diseases and psychiatric disorders result from a disturbance or alteration of synaptic activity. Drugs such as tranquilizers, anesthetics, nicotine, and caffeine target the synapse and can cause an alteration of impulse transmission.

Dendrites

Dendrites are so named because they resemble tiny **tree**s (the Greek word for tree is *dendron*). The main function of a dendrite is to receive and integrate signals from neighboring neurons and conduct these signals to the nerve cell body. Dendrites are branched extensions of the cell's cytoplasm and contain all the normal cytoplasmic structures. A nerve cell's dendrites can branch out quite extensively, thus increasing the total surface area of the neuron and making more room to receive incoming signals from other nerves. A single neuron can receive signals from hundreds of other nerves. The dendrites in sensory neurons have specialized cells called receptors, which convert stimuli into electrical signals. Under the **microscope**, dendrites appear hairy; the little hairs or projections are called spines, and each spine is the site of a synapse, which is the point of communication between neurons.

Glial cells

One cannot discuss the neuron without mentioning glial cells or neuroglia. It was once thought that these cells simply held everything together (*gloios* means glue, in Greek), but we now know that neuroglia are highly specialized cells. For example, neuroglia are responsible for physical support, protection against infection (through **phagocytosis**), and the connection of nerve cells to **blood** vessels. The Schwann cell (or neurolemmocyte) is a common type of glial cell found in peripheral nerve axons. Schwann cells wrap ''jelly roll style'' around the axon, forming a whitish phospholipid (fatty) protective and insulating cover known as the **myelin sheath**. The

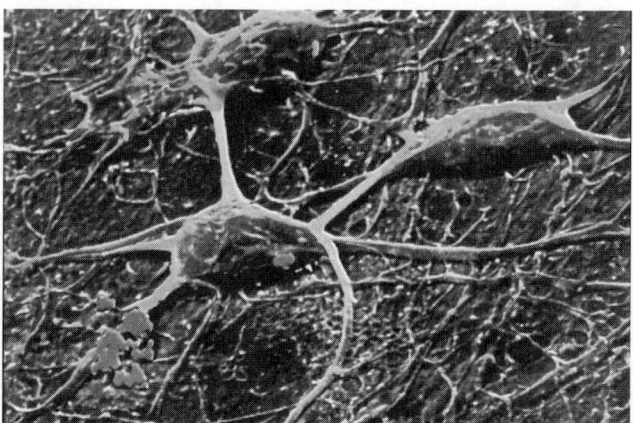

A scanning electron micrograph (SEM) of three neurons of the human cerebral cortex. *(Photo Researchers, Inc. Reproduced by permission.)*

myelin sheath of the axon of peripheral nerves has interruptions or exposed gaps known as the Nodes of Ranvier. Schwann cells also make up the neurolemma, a continuous sheath that covers both the myelin sheath and the axon at the Nodes of Ranvier. Action potentials traveling down the axon occur only at the Nodes of Ranvier, jumping rapidly from gap to gap (saltatory conduction), which conducts impulses significantly faster than in nonmyelinated nerves. The neurolemma is found only in peripheral nerve fibers and plays a crucial part in nerve fiber regeneration. Damaged axons will regenerate; damaged cell bodies will not. The myelinated sheaths of the axons of neuron in the brain and **spinal cord** (the central nervous system) are made from different glial cells (oligodendrocytes), which lack a neurolemma, so making the regeneration of their axons impossible. Multiple sclerosis is a serious demyelinating disease of the central nervous system. Not all axons are myelinated; the presence of myelin is one difference between white **matter** (which has myelinated axons) and gray matter (which does not).

Functional classification

Sensory neurons transduce physical stimuli, such as **smell**, **light**, or sound, into action potentials, which are then transmitted to the spinal cord or brain. Sensory neurons, which bring information into the central nervous system, are also referred to as afferent neurons. Motor neurons transmit nerve impulses away from the brain and spinal cord to muscles or glands and are also called efferent neurons. Interneurons transmit nerve impulses between sensory neurons and the motor neurons. Interneurons are responsible for receiving, relaying, integrating, and sending nerve impulses. Interneurons are found exclusively in the central nervous system and account for almost 99% of all the nerve cells in the body.

NEUROSCIENCE

Neuroscience is the study of the **nervous system** and its components. Neuroscientists may examine the nervous systems of humans and higher animals as well as simple multicellular nervous systems, or investigate nervous phenomenon at the cellular, **organelle**, or molecular level.

Neuroscience principally originated with three European scientists working at the end of the nineteenth and the beginning of the twentieth century. **Camillo Golgi**, an Italian physician, perfected a vital laboratory technique that first allowed scientists to trace the workings of the nervous system. Golgi completed a medical degree at the University of Padua in 1865, and then became a medical researcher at the University of Pavia. He was interested in cells and tissues, and experimented with ways to stain cells so they could be seen. Researchers before him had prepared cells with organic dyes, but Golgi found that staining with silver salts gave much clearer results. He became fascinated with nerve tissue, and using his staining process, he was the first to see in fine detail how this tissue was organized. He proved that the fibers of nerve cells did not meet completely, but left a gap, now known as a **synapse**. He devoted his life to mapping the structure of the nervous system. Golgi's work was furthered by a Spanish medical researcher, **Santiago Ramón y Cajal**. Ramón y Cajal, working at the University of Zaragoza, first improved on Golgi's staining method, then used it to discover the connection between the gray matter in the brain and the spinal cord. He shared the Nobel Prize for medicine with Golgi in 1906.

Golgi and Ramón y Cajal established the anatomy of the nervous system. The English neurologist **Charles Scott Sherrington** is credited with founding modern neuroscience with his work on the functioning of the nervous system. In other words, he brought the science from describing what the nervous system was to showing how it worked. His research explored the brain's ability to sense position and equilibrium, and the reflex actions of muscles.

Many other researchers continued to explore the workings of the nervous system. As laboratory imaging techniques progressed, neuroscientists were able to look at nerve cells at the molecular level. This allowed scientists to map the growth of nerve cells and nerve networks, and to study how individual cells process, store, and recall information. Working with living brains to explore nerve function was all but impossible until the late 1970s, when sophisticated brain imaging machines were first developed. Positron emission tomography (PET) revolutionized neuroscience by allowing scientists to produce pictures of a working brain. Since then, scientists and engineers have come up with even better brain imaging systems, such as functional magnetic resonance imaging (fMRI). Using fMRI, neuroscientists can detect increases in blood oxygenation during brain function, and this shows which areas of the brain are most active. Brain activity occurs very quickly—neurons can respond to stimulus within 10 milliseconds—and very sophisticated equipment is needed to capture such fleeting movements. So-called neuroimaging is one of the hottest fields in neuroscience, as neurologists and technicians work together to find new ways of recording nerve action. Researchers in the late 1990s explored ways to map the flux of sodium ions within the brain, giving a direct record of neural activity, or to measure the scattering of light by brain tissues with fiber-

optics. Both these techniques hope to give a more precise picture of which areas of the brain become active when a person thinks.

NEUROTRANSMITTERS

Neurotransmitters are special chemicals which help to pass **nerve impulses** on from one **neuron** to another. When a nerve impulse runs down the **axon** of a neuron, ultimately it arrives at the axon terminal, where the synaptic knobs are located. The synaptic knobs contain packets (vesicles) of specialized chemicals, the neurotransmitters. When the nerve impulse reaches the synaptic knobs, the vesicles spill out into a space located between the axon terminal and the next neuron's **cell** body or soma. This space is referred to as the **synapse**, or synaptic junction.

The neurotransmitters spread across the synaptic junction, until they reach the next neuron's soma, where they attach themselves to special receptors along the soma's **membrane**. This allows the nerve impulse to continue to be transmitted along this next neuron's axon.

Almost immediately, special **enzyme**s also arrive in the synaptic junction, to degrade any leftover neurotransmitters lingering there. This prevents the neurotransmitters from continuously stimulating the next neuron, even after the need for nerve impulse transmission has passed.

A number of neurotransmitters exist, including **acetylcholine**, norepinephrine, ephinephrine (also called adrenaline), serotonin, glycine, dopamine, histamine, enkephalins, and endorphins. Some of these neurotransmitters are considered to be excitatory, meaning that they cause the chemical and electrical changes necessary to send a nerve impulse down a particular neuron. Other neurotransmitters are considered to be inhibitory, meaning that they cause the chemical and electrical changes necessary to temporarily prevent a nerve impulse from traveling down a particular neuron.

Neurotransmitters are responsible for a large variety of bodily functions, including muscle movements; **learning** and **memory**; **regulation** of moods, **emotions**, and sleep; transmission of sensations such as pain; and regulation of such basic functions as **temperature** and **water** balance in the body. A number of illnesses and diseases can occur due to problems with the neurotransmitters. Curare (remember those poison-tipped arrows in the movies?) is a poison which acts by filling the neurotransmitter receptor sites, preventing the true neurotransmitters from entering. This results in total paralysis. In fact, this process has been mimicked with drugs which are given to surgical patients, to purposely induce paralysis, preventing any muscle movement during surgery. Organophosphate insecticides, as well as nerve gas substances used in warfare, cause poisoning by interfering with the enzymes which should clear neurotransmitters from the synaptic junctions. The neurotransmitters continue to stimulate the neurons, resulting in uncontrollable muscle twitching, and ultimately leading to paralysis. The presence of excess dopamine in the **brain** has been identified as the cause of Parkinson's disease.

The excess dopamine causes muscles to be over-stimulated, resulting in stiffness and uncontrollable movements. Too little of the neurotransmitters dopamine, serotonin, and norephinephrine has been identified as a cause for serious **depression**. Medications to treat depression interfere with the enzymes which normally degrade these neurotransmitters within the synaptic junction, allowing these neurotransmitters to continue stimulating the next neuron's soma.

NIACIN

Niacin, or nicotinic acid, is a member of the **water**-soluble vitamin B family and, for the most part, functions as part of two important **coenzymes**. Both **enzyme**s play vital roles in a number of metabolic pathways, in particular, those pathways concerned with **cellular respiration** (the process by which **tissue** cells "burn" **carbohydrate**s and **protein**s in order to release **energy**) and, to a lesser extent, those pathways involved in the synthesis of fatty acids and **steroids**.

A deficiency of niacin causes pellagra, a serious disease which has plagued mankind for centuries, in most cases striking people whose diet consists mainly of corn and cornmeal. Until fairly recently, pellagra was a major health problem in the United States, especially in poorer rural areas. In the 1920s, in fact, pellagra not only killed thousands of people but pellagra patients filled both hospitals and (because mental confusion was one of its symptoms) mental institutions as well.

Although no one knew the exact cause of the disease, by the beginning of the twentieth century, more and more researchers began to suspect that a dietary deficiency was responsible. The search for an "anti-pellagra factor" intensified in both Europe and the United States. In 1912, **Casimir Funk**, the Polish-born biochemist who coined the term *vitamine*, did manage to isolate the right factor—nicotinic acid—from rice polishings. Unfortunately, at the time Funk was actually hunting for the substance that would cure **beriberi**, another serious deficiency disorder. When he found that nicotinic acid had only a minimal effect on beriberi, he pushed the compound aside. As a result, in the years that followed, it was largely ignored.

However, in the 1930s, a number of researchers—among them Hans Euler-Chelpin, **Otto Warburg**, and **Arthur Harden**—began reporting that nicotinic acid appeared to be part of quite a few vital coenzymes. Perhaps, they suggested, the compound was a lot more important than was originally supposed.

Niacin, however, wasn't fully established as a vitamin until 1937 when a team of researchers—headed by an American biochemist, Conrad Arnold Elvehjem—administered 30 mg of nicotinic acid to a dog suffering from *blacktongue*, the canine equivalent of pellagra. The dog improved immediately and, with further doses, was soon completely cured. Other biochemical researchers quickly confirmed the fact that niacin was the "anti- pellagra" vitamin for humans and that adding foods high in niacin to the diet, such as meat, green vegetables, **yeast**, and most grains, dramatically cured the disease. More-

over, since tryptophan is converted by the body into niacin, adding milk and other tryptophan-rich foods to the diet worked equally well.

Very quickly, pellagra cases began declining. And in 1941—when breads and cereals routinely began to be fortified with the vitamin—pellagra ceased to be a problem in the United States (although it still crops up occasionally in other parts of the world).

NICHE

Four distinct stages of niche theory development in biological **ecology** can be identified: (1) Joseph Grinnell's original formulation of niche (in 1917 and 1928) as a spatial unit; (2) Charles Elton's formulation (in 1927) of niche as a functional unit; (3) Gause's (1934) **competitive exclusion principle**; and (4) E. Evelyn Hutchinson's concept of multidimensional niche in the 1950s.

Although Darwin understood the idea of niche and a few other biologists used the term earlier, Grinnell is credited with its formal development. To Grinnell, niche was a spatial unit that stood for the "concept of the ultimate distributional unit, within which each **species** is held by its structural and **instinc**tive limitations." His conception of niche was "preinteractive," that is, it referred to the entire area within which an **organism** could survive in the absence of other organisms. This is in contrast to the "post-interactive" niche, the actual place occupied by the organism in an environment after it has interacted with other organisms.

At about the same time, Charles Elton was developing the niche concept along somewhat different lines. Elton conceived of niche as a functional unit to describe the organism's "place in the **biotic environment**, relations to food and enemies." Although Elton presented niche as an organism's ecological position in a larger framework like a **community** or **ecosystem**, he then restricted its use to the food habits of an organism. Elton's niche is postinteractive.

Gause is credited with being the first investigator to perceive the connection between **natural selection**, **competition**, and niche and to see the interacting aspects of these concepts. Gause stated that "it is admitted that, as a result of competition, two similar species scarcely ever occupy similar niches, but displace each other in such a manner that each takes possession of certain peculiar kinds of food and modes of life in which it has an advantage over its competitor. Gause experimentally tested the general conclusions drawn from the Lotka-Volterra competitive equations, confirming and amplifying them. These conclusions are summarized in the "Competitive Exclusion Principle," which states that two species cannot coexist at the same locality if they have identical ecological requirements. Gause based the principle on an Eltonian definition of niche.

The Eltonian niche dominated ecological theory during the period 1930-1950 and began to be referred to as an organism's "occupation" or "profession." Hutchinson responded to this rather limited idea of niche by incorporating selected features from both Grinnell's and Elton's niche definitions and redefining niche as an "n-dimensional hypervolume," an abstract multidimensional space defining the environmental limits within which an organism is able to survive and reproduce. Hutchinson's "fundamental niche" is preinteractive, composed of "close to innumerable" dimensions, each corresponding to some requisite for a species. By setting the number of defining dimensions at "close to innumerable," Hutchinson attempted to illustrate the complexity of the systems within which organisms exist and interact. He depicted it by plotting each identifiably important environmental variable along an axis to show the points below which and above which the given organism could not survive.

Hutchinson's "realized niche" usually corresponds to a smaller hypervolume because competition and other interactions serve to restrict organisms from some parts of their fundamental or potential niche. Although most current works in niche theory use some variation of Hutchinson's multidimensional niche, both the Eltonian and the Hutchinson niches are still found in contemporary ecology and are still useful. Any application of niche, however, is only an approximation of reality, because niche dimensions are too numerous to be counted.

NICOLLE, CHARLES J. H. (1866-1936)
French bacteriologist

Born in 1866, in Rouen, France, Charles Jules Henri Nicolle was the son of physician Eugène Nicolle. Charles took his medical degree in 1893 in Paris, then returned to Rouen for a staff position in a hospital. Shortly thereafter, he married Alice Avice. Nicolle agreed in 1902 to assume the directorship of the Institute Pasteur in Tunis, Tunisia. Until his death in 1936, Nicolle lived and worked in Tunis with occasional lecturing in Paris.

Affiliated with the original Institute Pasteur (which was founded in Paris in 1888), the institute in Tunis was basically an organization in name only. Over the years to come, however, Nicolle improved a run-down antirabies vaccination unit into a leading center for the study of North African and tropical diseases. It was in Tunis where Nicolle accomplished his groundbreaking work on typhus. He became intrigued by the observation that an outbreak of typhus did not seem to take hold in hospital wards as it did among the general populace of the city. Although the contagion infected workers who admitted patients into the hospital, it did not affect other patients or attendants in the actual wards. Those who collected or laundered the dirty clothes of newly admitted patients typically came down with the disease.

Realizing that the washing, shaving, and providing of clean clothes to the new patient was possibly the key to the pattern of infection, Nicolle initiated a series of experiments in 1909 to confirm his suspicion of the arthropod-borne nature of typhus. He theorized that lice, which attached themselves to the bodies and clothes of human beings, transmitted the disease, so he began his investigation by infusing a chimpanzee with human blood infected with typhus, then transferred the chimpanzee's blood to a healthy macaque monkey. When the fever and rash of typhus was seen on the monkey, Nicolle placed twenty-nine human body lice obtained from healthy humans on the skin of the macaque. These lice were later placed on the skin of a number of healthy monkeys, which all contracted the disease.

For his research into the cause of typhus, Nicolle was awarded the 1928 Nobel Prize for physiology or medicine. Once Nicolle isolated the relationship between typhus and the louse, preventative measures were established to counter unsanitary conditions. The development of the insecticide **DDT** by **Paul Müller** in 1939 was the most effective prophylactic against typhus.

Nicolle is also responsible for other important contributions to the science of bacteriology. Stemming from his research into typhus was his recognition of a phenomenon known as "inapparent infection," a state in which a carrier of a disease exhibits no symptoms. This theoretical discovery suggested how diseases survived from one epidemic to another.

Nicolle, along with a variety of other colleagues over time, also researched African infantile leishmaniasis, which affected humans, and a related disease in dogs. Another significant discovery concerned the role of flies in the transmission of the blinding disease trachoma. For these and other works, Nicolle received the French Commander of the Legion of Honor and was named to the French Academy of Medicine. In 1932 he became a professor in the College de France.

NICOTINIC ACID • See Niacin

NIRENBERG, MARSHALL WARREN (1927-)
American biochemist

Marshall Warren Nirenberg is best known for deciphering the portion of **DNA** (**deoxyribonucleic acid**) that is responsible for the synthesis of the numerous **protein** molecules which form the basis of living **cell**s. His research has helped to unravel the DNA **genetic code**, aiding, for example, in the determination of which genes code for certain hereditary traits. For his contribution to the sciences of genetics and cell biochemistry, Nirenberg was awarded the 1968 Nobel Prize in physiology or medicine with **Robert W. Holley** and **Har Gobind Khorana**.

Nirenberg was born in New York City on April 10, 1927, and moved to Florida with his parents, Harry Edward and Minerva (Bykowsky) Nirenberg, when he was ten years old. He earned his B.S. in 1948 and his M.Sc. in biology in 1952 from the University of Florida. Nirenberg's interest in science extended beyond his formal studies. For two of his undergraduate years he worked as a teaching assistant in biology, and he also spent a brief period as a research assistant in the nutrition laboratory. In 1952, Nirenberg continued his graduate studies at the University of Michigan, this time in the field of biochemistry. Obtaining his Ph.D. in 1957, he wrote his dissertation on the uptake of hexose, a sugar **molecule**, by ascites tumor cells.

Shortly after earning his Ph.D., Nirenberg began his investigation into the inner workings of the genetic code as an American Cancer Society (ACS) fellow at the National Institutes of Health (NIH) in Bethesda, Maryland. Nirenberg continued his research at the NIH after the ACS fellowship ended in 1959, under another fellowship from the Public Health Service (PHS). In 1960, when the PHS fellowship ended, he joined the NIH staff permanently as a research scientist in biochemistry.

Nirenberg cracks the genetic code

After only a brief time conducting research at the NIH, Nirenberg made his mark in genetic research with the most important scientific breakthrough since **James D. Watson** and **Francis Crick** discovered the structure of DNA in 1953. Specifically, he discovered the process for unraveling the code of DNA. This process allows scientists to determine the genetic basis of particular hereditary traits. In August of 1961, Nirenberg announced his discovery during a routine presentation of a research paper at a meeting of the International Congress of Biochemistry in Moscow.

Nirenberg's research involved the genetic code sequences for **amino acids**. Amino acids are the building blocks of protein. They link together to form the numerous protein molecules present in the human body. Nirenberg discovered how to determine which sequences patterns code for which amino acids (there are about 20 known amino acids).

Nirenberg Honored with Nobel Prize

Nirenberg's discovery has led to a better understanding of genetically determined diseases and, more controversially, to further research into the controlling of hereditary traits, or genetic engineering. For his research, Nirenberg was awarded the 1968 Nobel Prize for physiology or medicine. He shared the honor with scientists Har Gobind Khorana and Robert W. Holley. After receiving the Nobel Prize, Nirenberg switched his research focus to other areas of biochemistry, including cellular control mechanisms and the cell differentiation process.

Since first being hired by the NIH in 1960, Nirenberg has served in different capacities. From 1962 until 1966 he was Head of the Section for Biochemical Genetics, National Heart Institute. Since 1966 he has been serving as the Chief of the Laboratory of Biochemical Genetics, National Heart, Lung and Blood Institute. Other honors bestowed upon Nirenberg, in addition to the Nobel Prize, include honorary membership in the Harvey Society, the **Molecular Biology** Award from the National Academy of Sciences (1962), National Medal of Science presented by President Lyndon B. Johnson (1965), and the Louisa Gross Horwitz Prize for Biochemistry (1968). Nirenberg also received numerous honorary degrees from distinguished universities, including the University of Michigan (1965), University of Chicago (1965), Yale University (1965),

Marshall Warren Nirenberg. *(The Library of Congress. Reproduced by permission.)*

University of Windsor (1966), George Washington University (1972), and the Weizmann Institute in Israel (1978). Nirenberg is a member of several professional societies, including the National Academy of Sciences, the Pontifical Academy of Sciences, the American Chemical Society, the Biophysical Society, and the Society for Developmental Biology.

Nirenberg married biochemist Perola Zaltzman in 1961. While described as being a reserved man who engages in little else besides scientific research, Nirenberg has been a strong advocate of government support for scientific research, believing this to be an important factor for the advancement of science.

NITRIFICATION

Nitrification is one process within the important **nitrogen cycle**. It is a biological process involving the breakdown of **ammonia** or organic **nitrogen** into nitrates and nitrites. This breakdown process is via oxidation, a chemical process that requires oxygen in this case. It is accomplished by two groups of chemosynthetic **bacteria** that utilize the **energy** yield in these conversion processes. The first step involves the oxidation of ammonia to nitrate, and is accomplished by *Nitrosomas* in the soil and *Nitrosococcus* in the marine environment. The second step involves the oxidation of the nitrate produced above into nitrites, releasing 18 kcal of energy. It is accomplished by *Nitrobacter* in the soil and *Nitrococcus* in salt **water**. Nitrification is usually considered a beneficial process. However, in certain situations it may have detrimental effects through production of excess nitrates, polluting streams and lakes. The reverse process of nitrification, occurring in anoxic (oxygen-deprived) environments, is called **denitrification** and is accomplished by other **species** of bacteria.

NITROGEN

According to the definition of an organic chemical, nitrogen need not apply; carbon is the essential ingredient. Yet nitrogen is found in all **proteins** and in many other important organic substances such as **protoplasm**, the living material in plant and **animal cells**. All life depends on the cycling of nitrogen from the **atmosphere** to the soil and into the **food chain**.

Although nitrogen was one of the earliest gases to be discovered, chemists did not understand its true **nature** for many years. The atmosphere contains 78 percent nitrogen by volume, but because nitrogen is chemically inactive, it does not support combustion or **respiration**. It is the oxygen in the air that takes part in these processes. Without nitrogen to dilute the oxygen, however, substances would burn and explode much more easily.

When scientists began to identify individual gases during the 1700s, they believed that a substance called phlogiston is released during combustion and absorbed by the surrounding air. In 1772, Daniel Rutherford was studying the portion of air (mainly nitrogen and some **carbon dioxide**) that does not support combustion. When Rutherford removed all of the carbon dioxide from his sample, the gas that was left over would still not support combustion. He called it "phlogisticated" air because he believed it had absorbed all of the phlogiston it could hold. Today this gas is known as nitrogen.

Around the same time, several other chemists, including Karl Wilhelm Scheele, **Joseph Priestley**, and Henry Cavendish, were also studying the portion of air that does not support combustion. However, Rutherford's report was the first to distinguish nitrogen from carbon dioxide. A few years later, Antoine-Laurent Lavoisier developed the modern theory of combustion, which explained the behavior of oxygen and other gases. Lavoisier recognized that nitrogen is an element, which he called "azote" (without life).

In 1790, nitrogen's modern name was introduced by French chemist Jean Antoine Claude Chaptal (1756-1832).

The name was chosen to indicate that the gas is a constituent of "nitre" (potassium nitrate, or saltpeter). Nitrogen (atomic number 7) has no color, odor, or taste. The gas is abundant, occurring "free" (as an element, N_2) in the atmosphere. However, in order to be useful to plant and animal life, free nitrogen must be combined with other elements into compounds, such as nitrates, which are found in fertile soils. These nitrogen compounds can be restored to depleted soils by the addition of manure and other natural fertilizers.

In the mid-1800s, Jean Boussingault showed that **plants** could also flourish in soil that had been treated with chemical fertilizers such as ammonium salts. Boussingault also found that certain plants called legumes (which include beans, peas, clover, and alfalfa) can replenish the soil's nitrogen by extracting, or assimilating, it from the air. Boussingault's agricultural research marked the beginning of our modern understanding of the nitrogen cycle.

Since then, scientists have learned that nitrogen is assimilated not by the plant itself, but by **bacteria** that live on the plant's roots. These nitrogen-fixing bacteria convert free nitrogen into **ammonia** (NH_3), which is then converted by nitrifying bacteria into compounds that can be used by plants to make proteins, which are essential **nutrients** for all life. When plants and animals die and decay, their nitrogen compounds are recycled to the soil. Some nitrogen is also recycled during thunderstorms, which produce nitrogen oxides that are carried to the earth by the rain.

Many types of nitrogenous fertilizers are used in **agriculture**. Natural fertilizers include garbage, sewage, fish scraps, meat-processing wastes, guano (seabird waste), and animal manure. During decay, bacteria decompose these organic materials and release ammonia, which then enters the nitrogen cycle in the soil. Farmers also use manufactured chemical fertilizers such as sodium and calcium nitrates. When an excess of nitrogen fertilizer is applied to the soil, however, it can contribute to **water pollution**. In addition to fertilizers, nitrogen is used to manufacture explosives such as TNT (trinitrotoluene) and nitroglycerin. These compounds are made by converting ammonia into nitric acid.

Until the turn of the twentieth century, most nitrogen compounds were produced from Chile saltpeter (sodium nitrate), but William Crookes predicted that the world would soon run short of this raw material. Scientists feared that, without fertilizers to increase crop yields, the growing world **population** would go hungry. This began the search for a way to "fix" free nitrogen from the air into usable compounds such as ammonia.

In the early 1900s, Norwegian physicist Kristian Olaf Bernhard Birkeland (1867-1917) and his co-worker Samuel Eyde (1866-1940) developed an electric-arc method of combining atmospheric nitrogen and oxygen into nitrogen oxides, but the process required large amounts of electricity. Another nitrogen-fixation method called the cyanamide process was invented around the same time by Adolf Frank (1834-1916) and Heinrich Caro (1834-1910), but it also used too much power to be practical. Then in 1913, German chemists Fritz Haber (1868-1934) and Carl Bosch (1874-1940) developed a com-

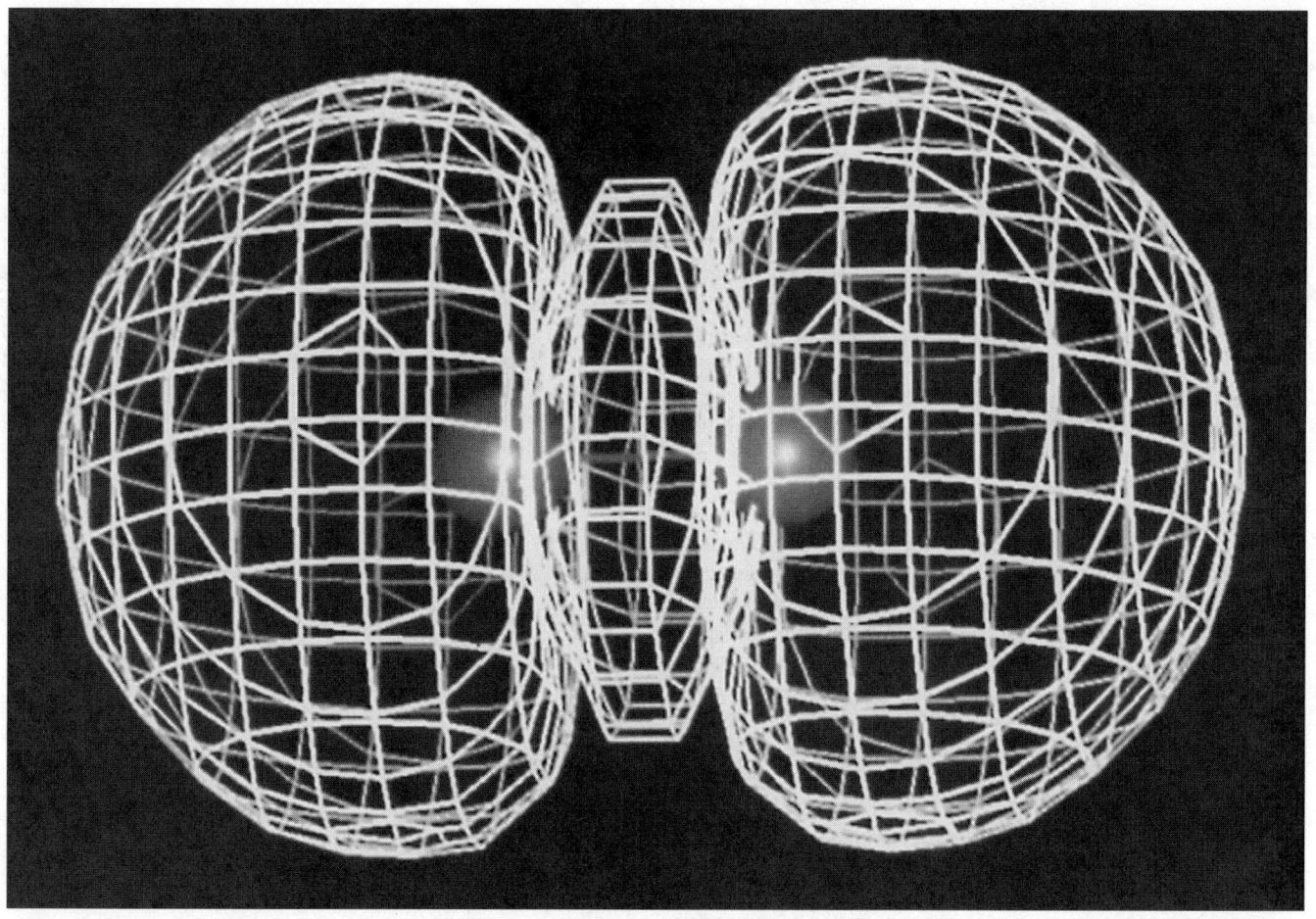

Computer graphics representation of a diatomic molecule of nitrogen. The dark spheres are the nitrogen atoms, and the white area between the atoms represents their strong covalent bonds. *(Photo Researchers, Inc. Reproduced by permission.)*

mercial process for synthesizing ammonia from atmospheric nitrogen. During World War I, Germany was able to continue manufacturing ammunition from synthetic ammonia after its supplies of raw materials had been cut off by Allied ships.

Nitrogen is one of the more difficult gases to liquefy. By the mid-1800s, most gases had been liquefied, but a few so-called permanent gases resisted all attempts at liquefaction. Then in 1877, French physicist Louis Paul Cailletet (1832-1913) succeeded in liquefying small amounts of nitrogen. Liquid gases were still a laboratory curiosity rather than a commercially available product, however.

In 1895, German engineer Carl von Linde (1842-1934) invented a continuous process for producing large quantities of liquid air, which is mainly nitrogen and oxygen. Linde's process immediately became a commercial success. In an improved version, it is still used today to produce liquid air, and liquid nitrogen has found many practical uses in research and industry. In biological research, for example, liquid nitrogen is used to freeze **blood** cells, **sperm**, **tissue**s, and even whole small **organism**s. When frozen, the **cell** stops its normal activities, allowing scientists to examine a "freeze-frame" of cell life. In industry, nitrogen is used for refrigeration, food pro-

cessing, and metal heat treating. The electronics industry uses nitrogen as a blanketing medium in the manufacture of components such as transistors. Nitrogen is also used to make the anesthetic "laughing gas," or nitrous oxide.

Although nitrogen is very valuable to humankind, the proliferation of fertilizers and **fossil fuels** today has created too much of a good thing. As of the late 1990s, human activities had doubled the natural rate at which nitrogen was made available on land. This new nitrogen in the system is due mainly to man-made fertilizers. Other major sources include the increased cultivation of legumes and other crops that harbor nitrogen-fixing bacteria and the burning of fossil fuels. Researchers think that this excess nitrogen may be decreasing biological diversity in some areas. In addition, nitrous oxide is building up in the atmosphere, where it can eat away at the **ozone** layer. Other nitrogen compounds contribute to smog and **acid rain**.

NITROGEN CYCLE

All **plants** and **animal**s need certain organic nitrogen compounds to live. During the 1700s, scientists knew that nitrogen is obtained by plants, which use it to make **proteins**. But no one was sure where plants got their nitrogen. One obvious possible source is air, which contains nearly 80% nitrogen. But in the late 1700s, H. B. de Saussure discovered that most plants cannot extract, or assimilate, nitrogen from the air. Instead, he concluded, they must absorb it somehow from the soil through their roots. Most people thought that nitrogen had to be restored to the soil by the addition of **humus**, manure, or other decaying organic **matter**. Farmers routinely added these fertilizers to their soil to supply nitrogen compounds in a form that plants could use.

Then in the mid-1800s, Jean Boussingault proved that plants could flourish without organic **fertilization** as long as other sources of nitrogen, such as nitrates or ammonium salts, were supplied. Eventually, these chemical fertilizers took the place of the farmer's smelly, germ-breeding manure pile and compost heap. Boussingault also showed that beans, peas, clover, and other legumes can replenish the soil's nitrogen by assimilating it from the air. But most other plants, Boussingault found, depend entirely on fertilizers for their nitrogen, because they cannot obtain it from the air as legumes do. Still, no one had figured out how leguminous plants assimilate nitrogen. Boussingault and his contemporaries believed that the process was strictly chemical.

Then in 1862, **Louis Pasteur** first suggested that **microorganisms** might be involved. During the 1880s, as a test of this concept, Pierre Berthelot grew plants in soil that had been sterilized. When he found that nitrogen is not assimilated in sterile soil, he concluded that microorganisms are concerned in the process.

Soon chemists were studying how **bacteria** and other microorganisms break down and recreate nitrogen compounds. By the late 1880s, scientists had learned that certain bacteria live in a *symbiotic*, or mutually beneficial, relationship with leguminous plants. Today we know that the process takes place through tiny hairs on the plant's roots, and it involves the production of nodules, or lumps, that contain colonies of bacteria. These "nitrogen-fixing" bacteria convert nitrogen from the air into **ammonia**. Once ammonia has been formed, it is converted by "nitrifying" bacteria into nitrates—compounds that can be used by plants to make proteins and other organic **nutrients**. Some bacteria that live free in the soil are also capable of assimilating nitrogen directly from the air, and in tropical regions, certain **algae** perform this function. Other types of nitrifying bacteria produce nitrate compounds from ammonium salts in the soil, rather than from atmospheric nitrogen. This process requires soil aeration and a neutral or alkaline soil environment. Still other types of bacteria "denitrify" soil by breaking down its nitrogen compounds and returning free nitrogen to the air.

Thus the idea of a nitrogen cycle in **nature** had gradually been discovered: Organic material decays and forms ammonia; bacteria use nitrogen from ammonia or from the air to make nitrate compounds; nitrates are absorbed by plants and used to make proteins; plants are eaten by people and other animals; and when plants and animals die and decay, their nitrogen compounds are recycled to the soil. More recently,

scientists have also discovered that, when thunderstorms discharge electricity into the **atmosphere**, nitrogen oxides are formed. These compounds are literally rained onto the earth in the form of nitric and nitrous acids, which replenish the earth's nitrogen content by forming nitrates and nitrites in the soil.

However, harmful amounts of **acid rain** can result when automobile exhaust, power plants, and other artificial sources overload the atmosphere with nitrous oxides. Acid rain falling back on earth, in turn, causes acidification of lakes and streams and contributes to damage of **trees** at high elevations. In addition, acid rain hastens the decay of building materials and paints. Nitrogen oxide emissions also contribute to smog and poor visibility. For all these reasons, they are controlled in many ways by different levels of government throughout the United States.

Nomura, Masayasu (1927-)
Asian American molecular biologist

Masayasu Nomura is the American molecular biologist who demonstrated that those **ribosomes** present in **bacteria** can be reduced to their molecular components of **ribonucleic acid** (**RNA**) and **proteins**. Four years later he further demonstrated that they can then be reunited to regenerate themselves.

Nomura was born on April 27, 1927, in Hyogo-ken, Japan. He married Junko Hamashima on February 10, 1957; they had two children—Keiko and Toshiyasu. After receiving his Ph.D. in microbiology at the University of Tokyo in 1957, Nomura went to the United States to work as a postdoctoral fellow in Sol Spiegelman 's laboratory. While at Spiegelman's lab, Nomura isolated a kind of RNA that receives information from a bacteriophage (a **virus** that infects bacteria) **genome**, then serves as an model for producing the proteins within the bacteriophage. This type of RNA later became known as messenger RNA (mRNA). He then briefly returned to Japan as an assistant professor at the Osaka University Institute of Protein Research before emigrating to the United States in 1963 to join the faculty of the University of Wisconsin's department of **genetics**, where he became a full professor in 1966.

Demonstrates the Reversibility of Protein Splitting in Ribosomes

By the early 1960s, the basic decoding of genetics had been clarified and the protein biosynthesis components identified, thanks to the efforts of such scientists as **Paul Zamecnik** and **Marshall Warren Nirenberg**. The term "ribosome "had been introduced in 1958 to describe the tiny **organs** that are present in all living cells, and that synthesize proteins; furthermore, it was known that ribosomes were the site where **amino acids** were assembled to form proteins, and that ribosomes were made up of two different subunits, one larger than the other, each with its own complicated structure consisting of RNA molecules and various protein **molecules**. But what was not known was how the molecular components were assembled into sophisticated ribosome structures, or how the assembled structures performed their functions.

Using the work of Zamecnik and Nirenberg as a springboard, Nomura discovered that by centrifuging bacterial ribosomes in heavy salt concentrations some ribosomal proteins would split off from the ribosomes, and it occurred to him that the situation might be reversible. Four years later, under very specific conditions, Nomura mixed those particles lacking protein with the split-off proteins; like magic, functionally active ribosomal particles were formed. The significance of this reconstitution proved that the necessary information for the proper construction of ribosomal particles was contained in their molecular components, rather than in some extraneous factor. It also opened the way for study of the molecular components of ribosomes. In 1968 Nomura and his colleagues actually reconstituted the small ribosome subunits from purified RNA and dissociated ribosomal proteins, and in 1970, as professor of genetics and biochemistry, he did the same with the larger ribosomal subunit. He and his team also did significant research into the ribosomal makeup of chromosomes, isolating a number of genes from *Escherichia coli* (also known as E coli) cells.

During the 1960s and 1970s, Nomura also showed how certain strains of enterobacteria, called colicins, have a tendency to kill other, related bacterial strains. Some colicins kill bacteria by splitting RNA in ribosomes, while others eliminate it by causing **deoxyribonucleic acid** (**DNA**) breakdown. This discovery initiated many modern studies of colicins.

Nomura shed much light on the processes of information transfer from genes to proteins. In 1970 he was appointed co-director of the Institute for Enzyme Research at the University of Wisconsin. He was elected to the National Academy of Sciences in 1978.

NORTHROP, JOHN HOWARD (1891-1987)

American biochemist

Born in Yonkers, New York, Northrop received his B.S. from Columbia University in 1912, his M.A. in 1913, and Ph.D. in 1915. His desire to achieve academically is not surprising since he came from a long lineage of illustrious academics, including Princeton University President Jonathan Edwards and philanthropist Frederick C. Havemeyer, who gave a chemical laboratory to Columbia University. His father, John Northrop, who was killed in a laboratory accident, was a member of the zoology department of Columbia University. His mother, Alice Belle Rich Northrop, taught botany at Hunter College.

Northrop himself is best known for his studies of **enzyme**s and his contribution to **virus** research. Northrop continued the work of James B. Sumner, who had isolated and crystallized a bean **protein** called urease in 1926. While many researchers dismissed Sumner's work as insignificant, Northrop drew upon it to conduct his own research.

In 1930 Northrop crystallized pepsin, a digestive substance that splits proteins and is found in gastric secretions. Two years later, he crystallized trypsin, and three years after that, he crystallized chymotrypsin, both of which are protein-splitting substances found in pancreatic secretions. Because of Northrop's ability to isolate and crystallize these substances, researchers today are familiar with the proteins known as enzymes—substances that are critical for **digestion**, **respiration**, and other processes that support life.

Northrop took his theory about isolating enzymes and applied it to the isolation of the virus. He joined forces with **Wendell Stanley** to crystallize the tobacco mosaic virus. The result of their work was the knowledge that viruses, until then identified neither as living nor nonliving organisms, were actually nucleoproteins, nonliving compounds that consist of proteins and **nucleic acids**. For their work in isolating the chemical nature of enzymes and viruses, Northrop and Stanley shared the 1946 Nobel Prize in chemistry.

In 1938 Northrop isolated the first bacterial virus, which also proved to be a nucleoprotein. He studied the intestinal bacteriophage *Staphylococcus*, isolating it to determine the presence of nucleic acid and protein in the virus. Today, it is known that nucleic acid is the primary part of the **molecule**. Northrop also introduced a purified diphtheria antitoxin in 1941. In crystalline form, he maintained it was at least forty times as effective as the more primitive antitoxin, and additionally it produced no side effects.

Northrop, who had been appointed to the staff of the Rockefeller Institute in 1916 after a stint as a captain in the Chemical War Service of the United States Army, remained a member of the Institute until his retirement in 1962. The impact of Northrop's research concerning the chemical nature of enzymes was not fully realized until subsequent research confirmed his findings nearly a decade after he received the Nobel Prize.

NUCLEIC ACIDS

Nucleic acids are complex **molecule**s that contain a **cell**'s genetic information and the instructions for carrying out cellular processes. The two nucleic acids, **ribonucleic acid** (**RNA**) and **deoxyribonucleic acid** (**DNA**), work together.

A molecule is made of phosphate-base-sugar **nucleotide** chains; its three-dimensional shape affects its genetic function. In humans and other higher **organism**s, DNA is shaped in a two-stranded helix (spiral) and further organized on structures called **chromosome**s. DNA in some **bacteria** is circular. Most RNA molecules are single-stranded and take various shapes, such as a cloverleaf.

Nucleic acids were discovered by the Swiss biochemist Johann Miescher (1844-1895). Born in Basel where his father was a physician, Miescher was an assistant to **Ernst Hoppe-Seyler** at the University of Strasbourg (then in France, but soon to become part of Germany) in 1859 when he isolated a cellular substance containing **nitrogen** and phosphorus. Thinking it was a phosphorus-rich nuclear **protein**, Miescher named it nuclein.

It was actually a protein plus nucleic acid, as the German biochemist **Albrecht Kossel** discovered in the 1880s. Kossel, another former assistant of Hoppe-Seyler, also isolated nucleic acids' two purines (adenine and guanine) and three pyrimidines (thymine, cytosine, and uracil), as well as **carbohydrate**s.

The American biochemist **Phoebus Levene**, who had once studied with Kossel, identified two nucleic acid sugars—ribose in 1909 and deoxyribose (meaning that it had less oxygen than ribose) in 1929. This meant that there were two nucleic acids, one named for each type of sugar. Levene also defined a nucleic acid's main unit as a phosphate-base-sugar nucleotide. The nucleotides' exact connection into a linear **polymer** chain was discovered in the 1940s by the British organic chemist **Alexander Todd**.

In 1951 the American **James Watson** and the British **Maurice Wilkins** and **Francis Crick** determined DNA's two-stranded helical shape. Adenine is always paired with thymine and guanine is always paired with the cytosine. In RNA, uracil replaces thymine. In the 1960s scientists discovered that three consecutive DNA or RNA bases (a codon) comprise the **genetic code** or instruction for production of a protein. The codons were matched to specific **amino acid**s in the 1960s, mainly by the American **Marshall Warren Nirenberg** who found that a gene may be one codon or many.

A gene is transcribed into messenger RNA (mRNA), which moves from the **nucleus** to structures in the **cytoplasm** called **ribosome**s. The American biochemist **Mahlon Bush Hoagland** discovered by accident in the 1950s that transfer RNA (tRNA) molecules already in the cytoplasm read the instructions and bring the required amino acids to a ribosome for assembly. Some proteins carry out cell functions while others control the operation of other genes. DNA is replicated (completely copied) when a cell prepares to divide, one ''parent'' strand and the ''child'' of the other strand going to each of the two new cells. The process is called semiconservative **replication**. Until the 1970s cellular RNA was thought to be only a passive carrier of DNA instructions. It is now known to perform several enzymatic functions within cells, including transcribing DNA into messenger RNA and making protein. In certain **virus**es called **retrovirus**es, RNA itself is the genetic information. This, and the increasing knowledge of RNA's dynamic role in DNA cells, has led some scientists to believe that RNA was the basis for the Earth's earliest life forms, an environment called the RNA World.

Since the 1970s nucleic acids' cellular processes have become the basis for **genetic engineering**, in which scientists add or remove genes in order to alter the characteristics or behavior of cells. Such techniques are used in **agriculture**, pharmaceutical and other chemical manufacturing, and medical treatments for **cancer** and other diseases. In the 1990s scientists are determining the precise sequence of DNA nucleotides for humans and other **species**. The information available when these **genome** projects are completed is expected to allow further advances in **genetics**-based medicine.

NUCLEOTIDE

A nucleotide is a single chemical unit which, when bonded with other nucleotides, forms **nucleic acids**. Nucleic acids such as **deoxyribonucleic acid (DNA)** and **ribonucleic acid (RNA)** are the basis for all life on Earth.

Chemically speaking, nucleotides are composed of three types of molecular groups including a sugar structure, a **phosphate group**, and a cyclic base. A sugar **molecule** is the primary structure for all nucleotides. In general, the sugars are composed of five carbon **atoms** with a number of hydroxy (-OH) groups attached. The sugars differ depending on the type of nucleotide, and can be either D-ribose or 2-deoxy-D-ribose. When incorporated into a nucleotide, the sugar molecule exists in a closed ring structure.

A key part of a nucleotide is a heterocyclic base that is covalently bound to the sugar at its first carbon. These bases are either pyrimidine or purine groups, and they form the basis for the nucleic acid code. Two types of purine bases are found including adenine and guanine. In DNA, two types of pyrimidine bases are present, thymine and cytosine. In RNA, the thymine base is absent and uracil is found instead.

A phosphate group makes up the final portion of a nucleotide. This group is derived from phosphoric acid and is covalently bonded to the sugar structure on the fifth carbon. Chemical **linkage**s between nucleotides are made possible by the presence of the phosphate group. In a nucleic acid **polymer**, the phosphate group from one nucleotide is bonded to the third carbon on another nucleotide. Multiple bonds are made in this way creating a sequence of bases which become an **organism**'s **genetic code**.

NUCLEUS

Intact **cell**s are comprised of a nucleus and **cytoplasm**. A nuclear envelope encloses **chromatin**, the nucleolus, and a matrix which fills the remaining space. The chromatin consists primarily of the genetic material **DNA** and histone **protein**s. The nucleolus is a small structure within the nucleus which is rich in ribosomal **RNA** and proteins. Nucleoli are associated with **protein synthesis** and enlarged nucleoli are observed in rapidly growing embryonic **tissue** (other than cleavage nuclei), cells actively engaged in protein synthesis, and malignant cells. The nuclear matrix contains mainly proteins.

The genetic instructions for an **organism** are encoded in nuclear DNA. Hence, when genetic replicates are desired, they are cloned by nuclear transplantation. Frogs were the first creatures produced by **cloning** and this was accomplished in Philadelphia in 1952. More recently, sheep (Dolly) and other creatures have been produced by cloning nuclei from adult **animal** donors. The cloning procedures for frogs or **mammals** consists of insertion of a nucleus into an **egg** that has been deprived of its own genetic material. The reconstituted egg, with a new nucleus, develops in accordance with the genetic instructions of the nuclear donor.

There are, of course, cells which do not contain the usual nuclear structures. Embryonic cleavage nuclei (cells forming a **blastula**) do not have a nucleolus. Obviously, they have the genetic competence to produce nucleoli when needed because gastrula and all later cells contain nucleoli. Mature red **blood** cells, **erythrocyte**s, in most mammals have lost their nuclei. Nuclear loss does not preclude the competence to carry oxygen.

NÜSSLEIN-VOLHARD, CHRISTIANE (1942-)
German genetic researcher

Christiane Nüsslein-Volhard was born on October 20, 1942, in Magdeburg, Germany. The daughter of Rolf Volhard, an architect, and Brigitte (Hass) Volhard, a musician and painter. And while few women of her generation chose scientific careers, Nüsslein-Volhard found that being female in a male-dominated field presented little in the way of an obstacle to her studies. She received degrees in biology, physics, and chemistry from Johann-Wolfgang-Goethe-University in 1964 and a diploma in biochemistry from Eberhard-Karls University in 1968. In 1973 she earned a Ph.D. in biology and **genetics** from the University of Tübingen. Nüsslein-Volhard was married for a short time as a young woman and never had any children. She decided to keep her husband's last name because it was already associated with her developing scientific career.

In the late 1970s Nüsslein-Volhard finished postdoctoral fellowships in Basel, Switzerland, and Freiburg, Germany, and accepted her first independent research position at the European **Molecular Biology** Laboratory (EMBL) in Heidelberg, Germany. She was joined there by **Eric F. Wieschaus** who was also finishing his training. Because of their common interest in *Drosophila*, or fruit flies, Nüsslein-Volhard and Wieschaus decided to work together to find out how a newly fertilized fruit fly egg develops into a fully segmented embryo.

Nüsslein-Volhard and Wieschaus chose the fruit fly because of its incredibly fast embryonic development. They began to pursue a strategy for isolating **gene**s responsible for the embryos' initial growth. This was a bold decision by two scientists just beginning their scientific careers. No one had done anything like this before, and it wasn't certain whether they would be able to actually isolate specific genes.

Their experiments involved feeding male fruit flies sugar water laced with **deoxyribonucleic acid (DNA)**-damaging chemicals. When the male fruit flies mated with females, the females often produced dead or mutated embryos. Nüsslein-Volhard and Wieschaus studied these embryos for over a year under a microscope which had two viewers, allowing them to examine an embryo at the same time. They were able to identify specific genes that basically told **cell**s what they were going to be—part of the head or the tail, for example. Some of these genes, when mutated, resulted in damage to the formation of the embryo's body plan.

Nüsslein-Volhard and Wieschaus published the results of their research in the English scientific journal *Nature* in 1980. They received a great deal of attention because their studies showed that there were a limited number of genes that control development and that they could be identified. This was significant because similar genes existed in higher **organism**s and humans and, importantly, these genes performed similar functions during development. Nüsslein-Volhard and Wieschaus's breakthrough research could help other scientists find genes that could explain birth defects in humans. Their research could also help improve in-vitro **fertilization** and lead to an understanding of what causes miscarriages.

In 1991 she and Wieschaus received the Albert Lasker Medical Research Award, which is considered second only to the Nobel. During this time Nüsslein-Volhard had begun new research at the Max Planck Institute in Tübingen, Germany, similar to the work she did on the fruit flies. This time she wanted to understand the basic patterns of development of the zebra fish. She chose zebra fish as her subject because most of the developmental research on vertebrates in the past was on mice, frogs, or chickens, which have many technical difficulties, one of which was that one couldn't see the embryos developing. Zebra fish seemed like the perfect organism to study because they are small, they breed quickly, and the embryos develop outside of the mother's body. The most important consideration, however, was the fact that zebra fish embryos are transparent, which would allow Nüsslein-Volhard a clear view of development as it was happening.

Despite her prize-winning research on fruit flies, she received skeptical feedback on her zebra fish work. Other scientists claimed it was risky and foolish. When she submitted papers about her laboratory's work for publication, one reviewer even asked her why she was bothering. Nüsslein-Volhard was not one to be stopped by criticism or to rest on her laurels. Even though her reputation was built on her fruit fly research, her love of new challenges pushed her to take on this risky new project and set her sights to the future.

On October 9, 1995, in the midst of criticism about her new research, Nüsslein-Volhard (the first German woman to win in this category), Wieschaus, and **Edward B. Lewis** of the California Institute of Technology won the Nobel Prize in Physiology or Medicine for their work on genetic development in *Drosophila*. Lewis had been analyzing genetic mutations in fruit flies since the forties and had published his results independently from Nüsslein-Volhard and Wieschaus.

NUTRIENTS

Nutrients are any chemicals required for life. Nutrients are of two basic types: (1) inorganic substances that are absorbed by **autotroph**ic **organism**s, such as **plants** and certain **microorganism**s, for use in their synthetic reactions and **metabolism**, and (2) **biomass** ingested by **animal**s and **heterotroph**ic microorganisms as sources of organic nourishment.

Plants absorb a wide range of mineral nutrients from their environment. They utilize these nutrients in photosynthetic reactions and other metabolic processes to manufacture all of the biochemicals that they require for growth and reproduction. Nutrients required by plants in relatively large quantities are known as macronutrients, and include compounds of the following elements: carbon, hydrogen, oxygen, **nitrogen**, phosphorus, potassium, calcium, magnesium, and sulfur. Carbon is required in the largest amounts, typically making up about one-half of the dry weight of plant **tissue**, while hydrogen, oxygen, nitrogen, and calcium typically occur in concentrations of one to several percent. Phosphorus, potassium, magnesium, and sulfur account for 0.1%-1% of dry plant tissues. Plants require micronutrients in much smaller amounts, such as the metals copper, iron, and zinc.

Animals must feed on plants or other animals to obtain virtually all of their nutrients. After it is ingested, the food is typically digested, which breaks it down into relatively simple biochemicals and inorganic chemicals, which are absorbed through the gut and used in the animal's metabolism. The most important **nutrition**al need of animals is for **energy** to support their **respiration** and growth (the fixed energy of plant or animal biomass is ingested for this purpose). There are also some micronutrients that animals require in small quantities, but they cannot synthesize themselves. These include biochemicals called **vitamins**, which must be obtained through their food. Some animals can also utilize mineral forms of certain nutrients, which may be obtained directly from the inorganic environment, without eating biomass. For example, many grazing **mammals** will utilize salt licks when they are available, because these animals crave sodium, which often is not present in sufficiently large concentrations in the plants that they eat.

Appropriate nutrition for all organisms is a **matter** of both quantity and balance. For good nutritional health, all essential inorganic and organic nutrients must be available, and they must be obtainable in an appropriate balance. A severe shortage of even a trace micronutrient can result in severe metabolic dysfunction, and even the **death** of organisms.

NUTRITION

Nutrition is the biological process by which an **organism**, such as a plant or **animal**, takes in and utilizes food. This food is converted to **energy** which is used to keep the organism alive. While **plants** and animals have distinctly different nutritional requirements, there are some commonalities.

Theories about nutrition have been posed throughout history. During the time of the ancient Greeks, Anaxagoras suggested that nutrition was a result of the fact that everything contained a small amount of everything else. For example, he believed that when food was eaten, the part of the food that was hair would become part of the hair while the part of it that was muscle would become muscle. Theories about plant nutrition were also advanced. For example, **Aristotle** believed that plants got all their nutritional needs from the soil. This theory was widely accepted until the seventeenth century when a Belgian physician named **Johannes Baptista van Helmont** ran an experiment to see whether that was true. He thought that if plants grew by absorbing soil, then the amount of soil that he started with would become less overtime as the plant grew. After five years he found that almost no soil was lost. He concluded that the main compound responsible for the increase in the plant's mass was the **water** that he gave the plant. In the eighteenth century, **Stephen Hales** postulated the notion that plants were also nourished by the air. Later, all of these scientists were found to be at least partially correct.

Plants grow mainly by accumulating water in their **cells**. In this case, water is a nutrient because it supplies the hydrogen and some oxygen that is incorporated into compounds for **photosynthesis**. Most of the water that enters a plant is lost by transpiration. The water that is left behind works as a solvent and helps maintain the form of soft **tissue**. The dry weight of a plant is about 95% organic material and 5% inorganic **minerals**. Carbon, oxygen and hydrogen are the most abundant elements found in the dry weight of a plant. **Nitrogen**, sulfur and phosphorus are also relatively abundant.

Since plants and other photosynthetic organisms can produce many of their own nutrition requirements they are known as **autotrophs**. This means they can readily transform inorganic compounds into biologically useful **organic compounds**. And while plants can produce much of their own nutritional needs such as starch, they still require at least seventeen essential **nutrients**.

To some extent the minerals in a plant reflect the mineral content of the soil they are growing in. Some nutrients are essential. Elements required in large amounts by plants are known as macronutrients. Nine have been identified. These include carbon, oxygen and hydrogen which are the major components of the plant's organic compounds. Nitrogen and phosphorus are important components of the plant's **genetic material**, the **nucleic acids**. Nitrogen is also a key component of **proteins**, **hormones**, and **coenzymes**. Phosphorus is found in the phospholipid bilayer of the cell **membranes**. Other macronutrients include potassium which helps maintain water balance and is involved in **protein synthesis**. Calcium is an important part of the formation of cell walls. It also helps activate some **enzymes** and is involved in the way plant cells respond to stimuli. Magnesium is a component of chlorophyll, the energy producing **organelle** in plant cells.

In addition to the macronutrients, plants also require smaller amounts of micronutrients to grow properly. Most of these function as cofactors in biological reactions. These include chlorine which is involved in photosynthesis and water balance control. Iron makes up part of the **cytochrome**s, the organelles which control electron transport. Other micronutrients include boron, manganese, zinc, copper, molybdenum, and nickle. Mineral deficiency can weaken or even kill a plant. Typically, when a plant is deficient in a nutrient, a variety of symptoms might result. For example, if a plant is deficient in magnesium, a key part of chlorophyll, the leaves turn yellow.

Plants take in their nutrient requirements in different ways. Some nutrients are absorbed from the soil by the roots. Others are absorbed through the air. **Carbon dioxide**, the most important, is taken in through the air. Some plants get nutrition by being predators. For example, mistletoe grows on **tree**s and supplements its nutrition by absorbing nutrients from the tree. Carnivorous plants have specialized leaves that can act as traps for insects.

Unlike plants, animals are **heterotroph**s which means they are unable to live on inorganic nutrients alone. They need a supply of organic compounds for energy and growth. There are different types of animals and they can be classified by their nutrition requirements. **Herbivore**s eat plants, **algae** and

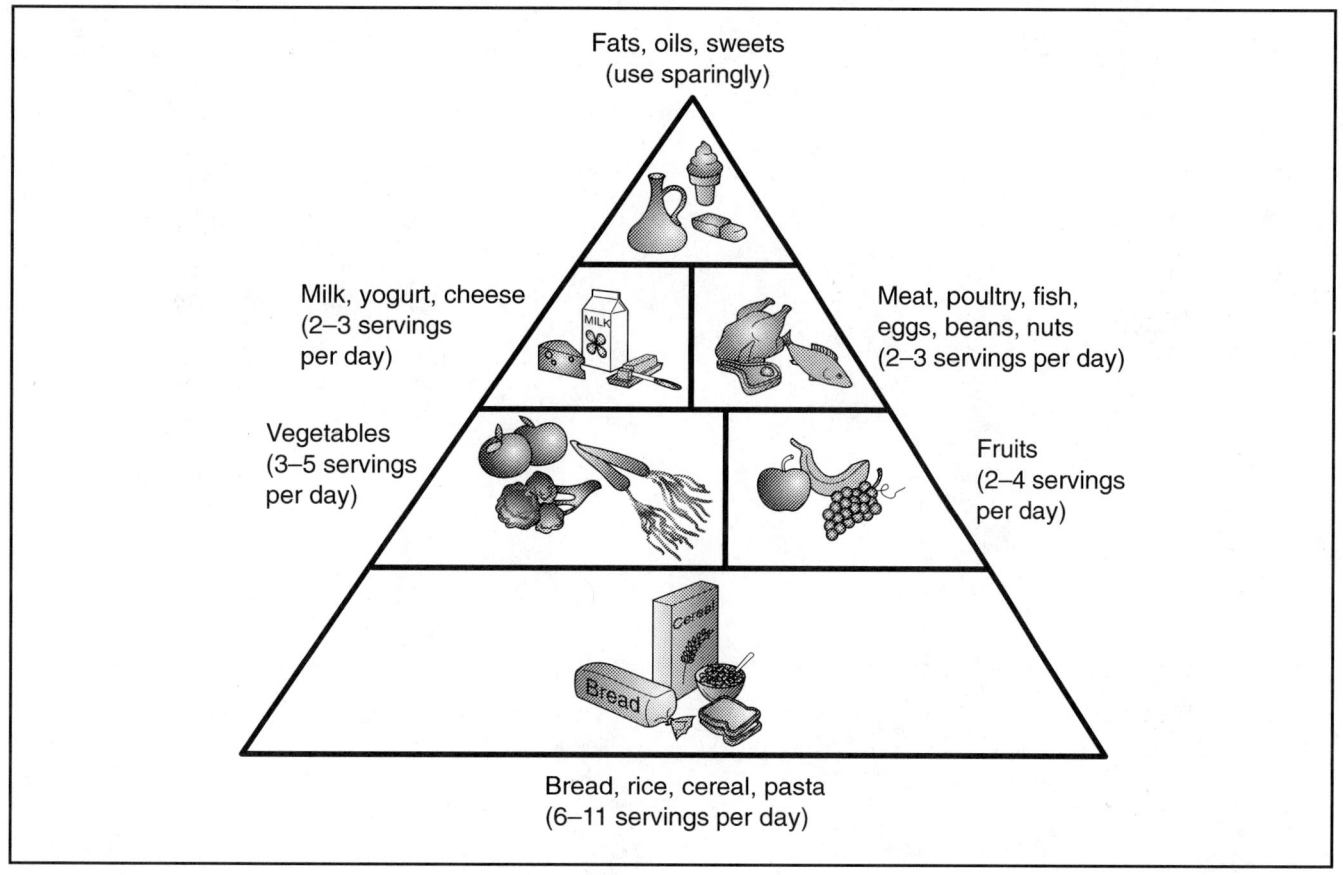

Suggested daily food servings. (Illustration by Electronic Illustrators Group.)

autotrophic **bacteria**. **Carnivore**s eat other animals. **Omnivore**s eat both plants and other animals. Animals are also categorized by the way in which they get there food. Suspension feeders sift food particles from the water. This includes animals such as clams and whales. Substrate feeders live in their food source and eat their way out. This is the strategy of many insect **larva**e. Fluid feeders suck nutrients from a host. Aphids or mosquitoes are examples. Bulk feeders eat large pieces of food.

Eating is called **ingestion** and it is the first step in food processing. **Digestion** is the second step and this involves the breaking down of food into **molecule**s. Digestion breaks down **chemical bond**s in a process called enzymatic **hydrolysis**. Absorption is the third step. Here the appropriate compounds are taken in by the animal's cells. Elimination is the final stage in which unused and waste material are passed out of the **digestive system**.

Food has three primary functions for animals. First, it acts as fuel for various reactions. This energy runs all the **chem**ical reactions and provides energy for locomotion. Food also provides the organic raw materials for the body to make new molecules. This is how animals get their source of carbon and nitrogen. And finally food provides the essential nutrients which animals can not make for themselves. Essential nutrients include some **amino acid**s. There are eight amino acids that must be ingested because the human body can not produce them themselves. Certain fatty acids are also essential. These are unsaturated fatty acids, an example of which is linoleic acid. **Vitamins** are another required nutrient. There are 13 vitamins required in a typical human diet. They act as coenzymes and are required in small amounts. Minerals are inorganic materials that are essential to proper nutrition. For example, they are involved in such things as bone and tooth formation, iron **metabolism** and muscle activity. Minerals that are required include such things as calcium, phosphorus, and sodium. There are over 17 different minerals that are required in a human diet.

O

OCHOA, SEVERO (1905-1993)
Spanish American biochemist

Severo Ochoa is best known for being the first to synthesize **ribonucleic acid (RNA)** outside the **cell.** He has also discovered several important metabolic processes. For his work with RNA, he received half of the 1959 Nobel Prize in physiology or medicine.

Ochoa was born in Luarca, Spain, where his father was a lawyer, and graduated from the University of Malaga in 1921. He received a medical degree in 1928 from the University of Madrid. After further studies in experimental biology, in 1940 he joined the Medical School faculty of Washington University in St. Louis. In 1942, he moved to New York University's College of Medicine, becoming chairman of the biochemistry department in 1954. He became an American citizen in 1956.

Ochoa's synthesis in 1955 of RNA was pure serendipity—an unexpected byproduct of his study of the way cells use glucose that is stored as ATP (adenosine triphosphate). Ochoa and a French associate, Marianne Grunberg-Manago, had purified an **enzyme** (now called polynucleotide phosphorylase) from the **bacteria** *Azotobacter vinelandii.* They were trying to study its reactions with ATP and other base-sugar combinations (called nucleosides) with one or three phosphate groups attached. No reaction occurred. However, when they added the enzyme and some magnesium to a nucleoside with two phosphate groups (diphosphate), over half of the nucleoside disappeared and some phosphorus was freed.

Ochoa traced the nucleoside to a new molecule that ultraviolet chromatography identified as a nucleotide. He then repeated the reaction with other nucleoside-diphosphates, in each case finding a nucleotide. Further analysis showed that the sugar was ribose, meaning that the reaction produced ribonucleic acid (RNA). Since the reaction was also reversible, Ochoa concluded that adding and removing phosphorus groups is a major mechanism in the synthesis and breakdown of nucleotide chains. Ochoa and other scientists used this method to decipher the genetic code. Later studies by others showed that RNA polymerase, not Ochoa's enzyme, is the main RNA synthesizing enzyme.

Ochoa is also known for his work on how the body uses carbon dioxide, and he helped identify a key compound in the metabolism of carbon dioxide. He also identified Krebs cycle reactions leading to energy storage in phosphate bonds.

ODUM, EUGENE PLEASANTS (1913-)
American ecologist and ornithologist

Ecologist and ornithologist Eugene Pleasants Odum is renowned for his views concerning the interrelationship between man and environment, having researched the **ecology** of birds, wetland ecology, landscape ecology, and vertebrate populations, as well as the general principles of ecology. The author of numerous works, including the widely used textbooks *Fundamentals of Ecology* and *Ecology,* Odum was the recipient of the 1987 Crafoord Prize from the Royal Swedish Academy of Science for his investigations into ecological issues (the Crafoord Prize is regularly awarded to scientists working in disciplines not addressed by the Nobel Prize). This honor was bestowed jointly upon Eugene P. Odum and his brother, Howard T. Odum, also an ecologist.

A native of Lake Sunapee, New Hampshire, Odum was born September 17, 1913. He received his undergraduate education at the University of North Carolina at Chapel Hill in 1934, then obtained his doctorate degree in ecology and **ornithology** from the University of Illinois in 1939. During his studies, Odum was assistant zoologist at the University of Georgia from 1934 to 1936. Upon completion of his doctoral studies, he was named resident biologist at the Edmund Niles Huyck Preserve in New York, a position he held from 1938 to 1940, and again during the summer of 1941.

Odum became an instructor at the University of Georgia in 1940 and was named assistant professor in 1942, then associate professor in 1945, receiving full professorship at the university in 1954. Odum also held two adjunct positions during his career: in the summers of 1942 and 1945, he carried out research at the Mountain Lake Biological Station at the University of Virginia, and from 1957 to 1961 he was an instructor of marine ecology during the summer training program of the Marine Biological Laboratory at Woods Hole, Massachusetts. Named Callaway Professor of Ecology at the University of Georgia in 1977, Odum also served as director of the university's Institute of Ecology from 1960 until 1984, when he became director emeritus.

Authors Primary Ecology Textbook

Odum wrote the influential textbook *Fundamentals of Ecology* in 1953, when ecological science was generally regarded as a subtopic within the field of biology. In the book, Odum describes the delicate balance of life among **plant**s, **herbivore**s, and **carnivore**s, and their interaction with **microorganism**s. These complex relationships play a crucial role in the recycling of **nutrients** and the continuation of each species.

By 1989, when Odum authored *Ecology and Our Endangered Life-Support Systems,* ecology had not only emerged as its own field of study, but was evolving into an integrative discipline which encompasses human life and the environment in which we live. In this volume, which was written for general readers as well as students of science, Odum declares that we have reached a ''turning point in history... when we cannot continue to postpone the environmental and human costs of development without incurring widespread damage to our global life-support systems.'' Reiterating basic ecological principles, Odum then demonstrates that each life form, rather than engaging in a merely competitive struggle, is connected with the others as part of a unified, dynamic process. Species of plants, herbivores, and carnivores maintain a beneficial interdependence upon one another. Odum also explains how human economic activities have disturbed both local and global ecosystems.

Discussing ecological and human systems, Odum notes that human communities are similar to organic communities, since each pass through phases from pioneering—when resources abound—to maturity, when resources are less plentiful and their consumption must be moderated. In *Ecology and Our Endangered Life-Support Systems* Odum describes biological organization within a hierarchical framework, applying the concept to such examples as endangered species, prescribed burning, **population** biology, and the **Gaia hypothesis,** which views the Earth as an organism able to regulate its own **biosphere.** Also included in the volume are sketches of prominent scientists in the field of ecology.

Odum has received numerous honors for his work in ecological science. In addition to the Crafoord Prize, he was awarded the prestigious Prix de l'Institute de la Vie by the French government in 1975. Odum was honored in 1956 with the Mercer Award of the Ecological Society of America and was presented with the Tyler Ecological Award by United States president Jimmy Carter in 1977. A member of the National Academy of Science, the American Academy of Arts and Sciences, and the American Society of Limnology and Oceanography, Odum also served as president of the Ecological Society of America from 1964 to 1965.

OMNIVORE

An omnivore (from Latin *omnis* = all, *vorare* = to devour) is an **animal** that obtains its **nutrition** by consuming both **producers** (**autotroph**s) and other **consumers** (**heterotroph**s). In other words, omnivores eat plant **matter** as well as meat from animals. Most humans are omnivores because we eat **fruits**, seeds and vegetables, which are from **plants**, in addition to meat from animals. Other examples of omnivores are some types of bears, which eat fruits, nuts, and leaves of plants, as well as fish and occasionally other animals. In addition, many **scavengers**, **organism**s that eat food that they did not kill, are omnivores. For example, crabs, which often scavenge for their food, eat both plant and animal matter.

Because omnivores are not able to make their own food, they are considered heterotrophs. They must obtain all **nutrients** and **energy** from the food they consume.

Omnivores hold an important position in food webs of many **ecosystem**s. Unlike producers, which are always at the base of the food web, or **herbivore**s, which only feed on the producers, omnivores can be found at several different levels of a food web. They can be considered primary or first-order consumers because they eat plants. However, they can also be considered secondary and sometimes even third-order consumers because they also eat animals. For example, when a bear eats fruits, nuts, and leaves of a plant, it is a primary consumer. However, this same bear would be considered a secondary or third-order consumer when it eats a fish. Because it is feeding on more than one level of the food web, this bear is considered an omnivore.

ONCOGENE

Oncogenes are segments of **genetic material** (**DNA**) that are able to induce **cancer** in **animal**s. They were first discovered in **retrovirus**es (**virus**es containing the **enzyme, reverse transcriptase,** and **RNA** rather than DNA) that were found to cause cancer in many animals. Studies of humans led to the discovery of related **gene**s called proto-oncogenes, which do not normally cause cancer, but have similar DNA sequences. Proto-oncogenes have since been found to play an important role in the growth, **differentiation** and proliferation of normal **cell**s. They may be converted into oncogenes by several mechanisms, the simplest of which is a **point mutation** that alters a single **nucleotide** base pair. They may also be converted to cancer-causing agents by chromosomal rearrangements that remove them from controlling elements present in unaltered **chromosome**s. The altered oncogenes can no longer function as control agents do, and the result is uncontrolled cell growth resulting in a cancerous tumor. About 60 oncogenes have been discovered in humans. They have been linked to malignancies of the breast, lung, colon, and **pancreas**.

Grizzly bears (*Ursus horribilis*), like humans, eat both plant and animal matter. *(U.S. Fish and Wildlife Service. Reproduced by permission.)*

The nomenclature in this field is somewhat confusing. Oncogenes were discovered first, and their name was derived from the Greek ONKOS, meaning bulk, or mass, because of their ability to cause tumor growth. Proto-oncogenes were named for their potential to become agents of malignant growth, and as a result, their important role as regulators of normal cellular activity is not properly emphasized.

ONE GENE-ONE POLYPEPTIDE HYPOTHESIS

In 1941, **George Beadle** and **Edward Lawrie Tatum** proposed the one gene-one enzyme theory. The four main tenets of this theory (as modified by Tatum in 1959) were:

- All biochemical processes in all living organisms are under genetic control.
- All biochemical reactions in an organism are resolvable into separate steps.
- Each step or reaction is under the control of a single gene.
- Mutation of a single gene results in the loss of function of the appropriate enzyme. In other words, each gene controls the reproduction, function, and specificity of a particular enzyme.

The theory was based on results originally obtained from *Neurospora crassa*, a fungus that was grown in a medium containing only the bare minimum of nutrients necessary (the fungus being capable of manufacturing the rest). After inducing mutations in the mold using radiation, some of the progeny were unable to grow on the medium. By testing with different supplements, it was found that the mutants had lost the ability to manufacture a single amino acid. By breeding the lab specimens with wild specimens, it was found that the mutation was transmitted in a simple Mendelian fashion. It was assumed that the ability to synthesize the appropriate amino acid was caused by the loss of a single **enzyme**. The work was supported by similar evidence found in humans, **plants**, and *Drosophila* (genus of fruit fly).

The hypothesis was further modified in 1962 by Vernon Ingram, and from it, the one gene-one polypeptide hypothesis was born. The modification arose from research conducted on sickle cell anemia and sickle cell trait. In 1949, it was proposed that sickling was caused by a single gene mutation, which was heterozygous in sickle **cell** trait individuals and homozygous in individuals with full sickle cell anemia. Simultaneously, it was also noted that the hemoglobin from normal individuals and that from sickle cell anemic individuals migrated differently on an electrophoresis plate, illustrating that there was a physical difference in the hemoglobin types and supporting the single **gene** mutation. A normal hemoglobin molecule is made of four different polypeptide chains—two identical alpha chains and two identical beta chains. All of the chains are approximately the same length, but they can be distinguished by

Lars Onsager, receiving the Nobel Prize in chemistry from King Gustav Adolf of Sweden, 1968. *(Archive Photos, Inc. Reproduced by permission.)*

their chemical and electrophoretic properties. Each of these chains contains approximately 140 amino acids, and Ingram analyzed them using a modified form of Frederick Sanger's protein analysis. This technique gave a fingerprint of the different hemoglobin types. The fingerprint showed that the differences between the two types of hemoglobin could be found in one peptide section of eight amino acids. When this section was isolated and analyzed, the only difference was in one amino acid (glutamic acid in normal and valine in sickle cell hemoglobin). The difference between these amino acids was one base in the triplet codon. Further analysis showed that amino acid changes in one chain were independent of changes in the other chain, suggesting that the genes determining the alpha and beta chains were located at different loci. The alpha and beta chains show independent assortment.

From this, it can be seen that hemoglobin is composed of two independent gene products, each of which is a separate polypeptide. The gene is a section of **DNA** that determines the amino acid sequence of a polypeptide. One gene codes for one polypeptide and several polypeptides may be required for a functional **protein** or enzyme.

ONSAGER, LARS (1903-1976)

American chemist

Lars Onsager was born in Oslo (then known as Christiania), Norway, in 1903. His parents were Erling Onsager and Ingrid Kirkeby Onsager.

Onsager excelled in chemistry from an early age, and his prodigious talent in the field was exhibited while still a student. One of the topics that caught his attention concerned the chemistry of solutions. In 1884, Svante Arrhenius had proposed a theory of ionic dissociation that explained a number of observations about the conductivity of solutions and, eventually, a number of other solution phenomena. Over the next half century, chemists worked on refining and extending the Arrhenius theory.

The Dutch chemist Peter Debye and the German chemist Erich Hückel, had proposed a revision of the Arrhenius theory that explained some problems not yet resolved—primarily, whether ionic compounds are or are not completely dissociated ("ionized") in solution. After much experimentation, Arrhenius had observed that dissociation was not complete in all instances.

Debye and Hückel realized that ionic compounds, by their very nature, already existed in the ionic state *before* they ever enter a solution. They explained the apparent incomplete level of dissociation on the basis of the interactions among ions of opposite charges and water molecules in a solution. The Debye-Hückel mathematical formulation almost perfectly explained all the anomalies that remained in the Arrhenius theory.

Almost perfectly, but not quite, as Onsager soon observed. The value of the molar conductivity predicted by the Debye-Hückel theory was significantly different from that obtained from experiments. By 1925, Onsager had discovered the reason for this discrepancy. Debye and Hückel had assumed that most—but not all—of the ions in a solution move about randomly in "Brownian" movement. Onsager simply extended that principle to *all* of the ions in the solution. With this correction, he was able to write a new mathematical expression that improved upon the Debye-Hückel formulation. Onsager had the opportunity in 1925 to present his views to Debye in person. Debye was sufficiently impressed with the young Norwegian to offer him a research post in Zurich.

In 1928, Onsager emigrated to the United States where he received appointments at Johns Hopkins, Brown, and Yale universities. In each case, Onsager proved to be a brilliant theoretician but a poor instructor. In his research at Brown University, Onsager attempted to generalize his earlier research on the motion of ions in solution when exposed to an electrical field. In order to do so, he went back to some fundamental laws of thermodynamics, including Hermann Helmholtz's "principle of least dissipation." He was eventually able to derive a very general mathematical expression about the behavior of substances in solution, an expression now known as the Law of Reciprocal Relations.

Onsager first published the law in 1929, but continued to work on it for a number of years. In 1931, he announced a more general form of the law that applied to other nonequilibrium situations in which differences in electrical or magnetic force, temperature, pressure, or some other factor exists. The Onsager formulation was so elegant and so general that some scientists now refer to it as the Fourth Law of Thermodynamics.

The Law of Reciprocal Relations was eventually recognized as an enormous advance in theoretical chemistry, earning Onsager the Nobel Prize in 1968. However, its initial announcement provoked almost no response from his colleagues. It is not that they disputed his findings, Onsager said many years later, but just that they totally ignored them. Indeed, Onsager's research had almost no impact on chemists until after World War II had ended, more than a decade after the research was originally published.

During the late 1930s, Onsager worked on another of Debye's ideas, the dipole theory of dielectrics. That theory had, in general, been very successful, but could not explain the special case of liquids with high dielectric constants. By 1936, Onsager had developed a new model of dipoles that could be used to modify Debye's theory and provide accurate predictions for all cases. Onsager was apparently deeply hurt when Debye rejected his paper explaining this model for publication in the *Physikalische Zeitschrift,* which Debye edited. It would be more than a decade before the great Dutch chemist, then an American citizen, could accept Onsager's modifications of his ideas.

In the 1940s, Onsager turned his attention to the very complex issue of phase transitions in solids. He wanted to find out if the mathematical techniques of statistical mechanics could be used to derive the thermodynamic properties of such events. Although some initial progress had been made in this area, resulting in a theory known as the Ising model, Onsager produced a spectacular breakthrough on the problem. He introduced a ''trick or two'' (to use his words) that had not yet occurred to (and were probably unknown to) his colleagues—the use of elegant mathematical techniques of elliptical functions and quaternion algebra. His solution to this problem was widely acclaimed.

Though his status as a non-U.S. citizen enabled him to devote his time and effort to his own research during World War II, Onsager was forbidden from contributing his significant talents to the top-secret Manhattan Project, the United State's research toward creating atomic weapons. Onsager and his wife finally did become citizens as the war drew to a close in 1945.

The postwar years saw no diminution of Onsager's energy. He continued his research on low-temperature physics and devised a theoretical explanation for the superfluidity of helium II (liquid helium). The idea, originally proposed in 1949, was arrived at independently two years later by Princeton University's Richard Feynman. Onsager also worked out original theories for the statistical properties of liquid crystals and for the electrical properties of ice. In 1951 he was given a Fulbright scholarship to work at the Cavendish Laboratory in Cambridge; there, he perfected his theory of diamagnetism in metals.

During his last years at Yale, Onsager continued to receive numerous accolades for his newly appreciated discoveries. In addition to his Nobel Prize, Onsager garnered the American Academy of Arts and Sciences' Rumford Medal in 1953 and the Lorentz Medal in 1958. Upon reaching retirement age in 1972, Onsager was offered the title of emeritus professor, but without an office. Disappointed by this apparent slight, Onsager decided instead to accept an appointment as Distinguished University Professor at the University of Miami's Center for Theoretical Studies. At Miami, Onsager found two new subjects to interest him, biophysics and radiation chemistry. In neither field did he have an opportunity to make any significant contributions, however, as he died in 1976.

ONTOGENY AND PHYLOGENY

Ontogeny refers to the development of an **organism**. With humans, ontogeny begins with the fertilized **egg** and continues through embryonic and fetal development, **birth**, maturation, and ultimately senescence. Clearly, many developmental phenomena in humans are found in other **animals** and seem to be related to events in the evolutionary history of the group. **Phylogeny** relates to the development of a group. Phylogeny is a history of a group of organisms from the beginning of life to the present time.

The terms ontogeny and phylogeny are grouped together frequently because of the fundamental biogenetic law espoused by the German scientist Ernst Heinrich Philip August Haeckel (1834-1919). Haeckel's law, also known as the theory of **recapitulation**, states that ontogeny is the short and rapid recapitulation of phylogeny. Haeckel, as had others before him, noted the similarity of embryonic forms within a group. Certainly a mouse, elephant, and human embryo appear remarkably similar in contrast to the vast differences in the adults of those forms. The similarity is thought to be due to common descent from a more primitive form. Some embryological similarities are ancient in origin. Consider the aortic arches of vertebrates. Aortic arches connect a ventral **aorta** to the dorsal aorta in the pharyngeal region of an early embryo. **Blood** is pumped from the beating **heart** via the ventral aorta to six **arteries** which arch around the pharynx. Blood collects in the dorsal aorta and is distributed throughout the body and head of the embryo. During early embryonic development there are six aortic arches in all vertebrates. In fish and amphibian **larva**e, the posterior arches sprout a capillary bed and ultimately, the arches and their capillaries become the respiratory gills of the organism. Reptilian, avian, and mammalian embryos, being mostly terrestrial, have absolutely no need for functional gills. Nevertheless, as stated previously, they develop six aortic arches. That includes humans. The first and second arches of humans are mostly lost, portions of the third pair form the common carotid arteries, the left fourth contributes to the arch of the aorta, the right fourth forms part of the right subclavian artery, the fifth pair are lost, and the sixth contribute to the pulmonary arteries. One cannot but be impressed that humans, standing at the peak of organic evolution, have a vascular system with an early development greatly similar with the early development of lowly fishes and amphibians. Most would argue that this suggests descent from some common ancestor. Other human organ systems can be shown to have similarities in their development with corresponding organs in lower vertebrate forms.

Is this similarity in developmental pattern recapitulation? The answer to that rhetorical question is a firm negative. At no time does a human recapitulate the ancestral forms to which the human may be related. Humans never have the **morphology** of a mature fish nor do they have the **anatomy** of a mature amphibian. And, of course, they never recapitulate a mature mouse or mature elephant form. Thus, ontogeny does NOT recapitulate phylogeny. Clearly, the history of **human evolution** is not repeated in detail and thus, Haeckel was wrong. However, the very existence of the six embryonic aortic arches in human embryos witnesses the close relationship of humans with all other vertebrate forms.

Phylogeny is a history. Not much historical information can be deduced from the study of embryos. Indeed, it may be stated that vertebrates are related and have similar early embryos. But, how are they related? **Comparative anatomy** which involves seeking homologous structures in contemporary related forms provides much better information on relationships and so too does the paleontological record.

OOCYTE

An oocyte is a developing ovum which undergoes two meiotic (maturation) divisions to give rise to the mature female **gamete**, the **egg**. A fetus destined to becoming a female has primordial germ **cell**s which migrate to the ovary. The primordial germ cells when in the ovary are known as oogonia. They proliferate in the ovary. In humans, oogonia vary in abundance with the largest number present at **birth** and a declining prevalence thereafter. Oogonia contain the same **chromosome** number, **diploid**, as body (somatic) cells. Oogonia give rise to primary oocytes which begin the first meiotic division. Primary oocytes in humans are found in the ovarian cortex and soon become surrounded by other ovarian cells to form primary non-fluid filled follicles. The primary oocyte grows in the follicle from a cell about 0.0008 in (0.02 mm) to a cell about 0.0055 in (0.14 mm) diameter while the follicle matures to its characteristic fluid filled structure. The great growth of the oocyte is associated with the fact that in **mammals** in general, and humans in particular, the fertilized ovum is enclosed in a capsule (the zone pellucida) and receives no nourishment from the mother until after transit of the uterine (Fallopian) tube, where **fertilization** occurs, and entrance into the uterus proper where implantation occurs. From fertilization through implantation in humans is about seven days and the **zygote** grows from one cell to a blastocyst containing the inner cell mass during that time. The primary follicles do not form mature follicles until the onset of puberty. At that time, certain oogonia contained within follicles may begin to grow. At about the time of ovulation, the primary oocyte finishes its first meiotic division and gives rise to a secondary oocyte and a first polar body. The secondary oocyte is lost if fertilization does not occur. However, if the secondary oocyte encounters a **sperm** and fertilization occurs, the second meiotic division proceeds resulting in a mature egg and the extrusion of the second polar body. The mature egg pronucleus is **haploid** and the sperm pronucleus is haploid and when they fuse, a zygote **nucleus** is formed which has the characteristic diploid chromosome number of the **species**. Oocytes disappear within a few years after menopause.

OOGENESIS

Oogenesis is the process by which **eggs** develop in the ovaries. This process begins during development of the **embryo**. Three months into embryonic development, primitive egg **cell**s called oogonia fill the ovaries of the developing female. Initially, these cells replicate through the process known as **mitosis**. Mitosis allows two identical daughter cells to be produced from a single original cell. All of these oogonia have the normal number of **chromosome**s (46).

After three months, however, the oogonia within the embryo begin to enter a new phase of development. Because the ultimate use of an egg is to unite with a **sperm** to create a new **zygote**, and because any such fertilizing sperm will carry 23 chromosomes of its own, the ultimate goal of the oogonia is to differentiate into an egg which itself carries 23 chromosomes. The union of the sperm and egg, then, will create an **organism** with the normal complement of 46 chromosomes.

To accomplish this "downsizing" of the chromosomes, oogonia utilize a process called **meiosis**. The first steps in the process of meiosis are begun during embryonic and fetal development, but the process is halted prior to **birth**. At its greatest, the number of egg cells reaches around 7 million; however, even before birth, the vast majority of these egg cells will regress and disappear. When a baby girl is born, her ovaries will contain about 400,000 primary **oocyte**s. The final steps of meiosis will not be completed until the girl enters puberty. After puberty, a single primary ooctye will enter further stages of meiosis, although the final stages of meiosis are only completed by any primary oocyte which actually undergoes **fertilization** by a sperm.

OPARIN, ALEKSANDR IVANOVICH (1894-1980)
Russian biochemist

Aleksandr Ivanovich Oparin was a prominent biochemist in the former Soviet Union whose achievements were recognized throughout the international scientific community. He is best known for his theory that life on earth originated from inorganic **matter**. Although a belief that life formed through spontaneous generation was prevalent up to the nineteenth century, that theory was disputed by the development of the **microscope** and the experiments of French scientist **Louis Pasteur**. Oparin's materialistic approach to the subject was responsible for a renewed interest in how life on earth originated. His book *The Origin of Life* outlined his basic theory, which was that life originated as a result of evolution acting on molecules created in the primordial atmosphere through energy discharges. In addition to his work on the origin of life, he played a major role in the development of technical botanical biochemistry in the Soviet Union.

Oparin was born near Moscow on March 2, 1894. He was the youngest child of Ivan Dmitrievich Oparin and Aleksandra Aleksandrovna. He had a sister, Aleksandra, and a brother, Dmitrii. His secondary education was marked by his achievements in science. He studied plant physiology at Moscow State University, graduating in 1917. He was a graduate student and teaching assistant there from 1921 to 1925. He also studied at other institutes of higher learning in Germany, Austria, Italy, and France, but it is thought that he never earned a graduate degree (he was awarded a doctorate in biological sciences in 1934 by the U.S.S.R. Academy of Sciences).

Aleksei N. Bakh, Oparin's mentor during his years of graduate study, was to have great influence on Oparin's later role in the development of Soviet biochemistry. Bakh was well known internationally for his research in medical and industrial chemistry, and played an important role in the organization of the chemical industry in Russia. After the Russian Revolution in 1917, Bakh helped develop the chemical section of the National Economic Planning Council (VSNKh) and founded its Central Chemical Laboratory. Oparin studied plant chemistry with Bakh in 1918, and from 1919 through 1925, he worked under Bakh at the VSNKh and the Central Chemical Laboratory. Bakh and Oparin cofounded the Institute of Biochemistry at the Academy of Sciences of the Soviet Union in Moscow in 1935. Oparin was appointed deputy director of the institute and held that position until 1946. After Bakh's death that same year, Oparin assumed the director's position, which he held until his death.

The practical aspects of Oparin's work during his association with Bakh in the early thirties involved biochemical research for increasing production in the food industry, work that was of extreme importance to the Soviet economy. Through his study of enzymatic activity in plants, he found that it was necessary for **molecule**s and **enzyme**s to combine in order to create starches, sugars, and other **carbohydrate**s and proteins. He was able to show that this biocatalysis was the basis for producing many food products in nature. He held a post from 1927 through 1934 as assistant director and head of the laboratory at the Central Institute of the Sugar Industry in Moscow, where he conducted research on tea, sugar, flour, and grains. During this same period, he also taught technical biochemistry at the D. I. Mendeleev Institute of Chemical Technology. As professor at the Moscow Technical Institute of Food Production from 1937 to 1949, he continued his research of plant processes and began the study of **nutrition** and **vitamins.**

Oparin's biochemical research on plant enzymes and their role in plant **metabolism,** so important for its practical application, would also be important for what was to be the focus of his career, the question of how life first appeared on earth. His first paper on this subject was presented to a meeting of the Moscow Botanical Society in 1922. This paper, which was never published, was revised and published in 1924 by the *Moscow Worker*. In it, Oparin discussed the problem of spontaneous generation, arguing that any differences between living and nonliving material could be attributed to physicochemical laws. This work went largely unnoticed, and Oparin did not seriously consider the topic again until the mid-thirties. In 1936, he published *The Origin of Life,* which modified and enlarged his earlier ideas. His ideas at this time were influenced not only by contemporary international thinking on astronomy, geochemistry, organic chemistry, and plant enzymology, but also by the dialectic philosophy espoused by Friedrich Engels, and the work of H. G. Bungenburg de Jong on colloidal coacervation. Translated into English in a 1938 edi-

Alexander Ivanovich Oparin. *(Science Photo Library, Photo Researchers, Inc. Reproduced by permission.)*

tion, *The Origin of Life* was also revised and updated in 1941 and 1957. Although the later versions amended the original, the concept that life arose through a natural evolution of matter remained central, and he often described this concept metaphorically by comparing life to a constant flow of liquid in which elements within are constantly changed and renewed.

The Origin of Life Theory

Oparin's theory that the origin of life had a biochemical basis was based on his suppositions concerning the condition of the **atmosphere** surrounding the primeval earth and how those conditions interacted with primitive organisms. It was his idea that the primeval atmosphere (consisting of ammonia, hydrogen, methane and water) in conjunction with **energy** (probably in the form of sunlight, volcanic eruptions, and lightning) gave this primitive matter its metabolic ability to grow and increase. He speculated that the first organisms had appeared in ancient seas between 4.7 and 3.2 billion years ago. These living organisms would have evolved from a nonliving coagulate, or gel-like, solution. Oparin argued that a separation process called coacervation occurred within the gel, causing nonliving matter at the multimolecular level to be chemically transformed into living matter. He further theorized that this chemical transformation was dependent upon protoenzymatic catalysts and promoters contained in the coacervates. From there, a process of natural selection began, which resulted in the formation of increasingly complex organisms and, eventually, primitive systems of life. Although others, such as de Jong and T. H. Huxley, would postulate that life arose from a kind of ''sea jelly,'' Oparin's theory that nonliving material was a catalyst for the formation of living organisms is considered by many to be his special contribution to the issue.

His suppositions on life's origins were not merely theoretical. In laboratory experiments, he showed how molecules might combine to produce the needed protein structure for transformation. Experiments of other scientists, such as **Stanley Lloyd Miller**, Harold Urey, and **Cyril Ponnamperuma,** confirmed his initial experiments on the chemical structure necessary to produce life. Ponnamperuna took the work a step further when he altered Oparin's original experiments and was able to easily produce nucleotides, dinucleotides, and adenosine triphosphate, which also contribute to the formation of life. Building on Ponnamperuna's research, Oparin was able to produce droplets of gel that he called protobionts. He believed these protobionts were living organisms because of their ability to metabolize and reproduce. Although later research of scientists in both the Soviet Union and the West would develop independently of Oparin's biochemical experiments, he must be given credit for putting the question of the origin of life into the realm of modern science. It has been said that his work in this area opened the door, and scientists in the West walked through.

Biochemistry in the Service of Dialectical Materialism

Oparin was a man of his time, and his thinking was greatly influenced by **Charles Darwin**'s theory of natural selection and the ideological climate of dialectical materialism which pervaded Soviet society during the 1930s. Although Oparin was never a Communist Party member, both his writings and his research methods reflect a bias toward dialectical materialism. However, it has been suggested that his denigration of the science of genetics and his support of Trofim Denisovich Lysenko and the Marxist-Leninist ideology which permeated and controlled Soviet **genetics** at that time may have resulted from political pressure and a desire to protect his career, as much as philosophical and scientific belief. Whatever the reasons, Oparin used his influence and prestige as chief administrator of the U.S.S.R. Academy of Sciences from 1948 through 1955 to implement policies that advanced Lysenko's views at the expense of the advancement of Soviet genetics. The influence of Lysenko and the Marxist-Leninist view of biology waned after Stalin's death in 1953. In 1956, as the result of a petition by 300 scientists calling for his resignation, Oparin was removed from his top position in the academy's biology division. He was replaced by Vladimir A. Engelhardt, a leading Soviet advocate of molecular biology. The 1950s saw an international explosion in the growth of molecular biology, but Oparin was severely critical of its principles. Although he considered the discoveries made by **James Watson** and **Francis Crick** concerning **DNA** to be important, he was skeptical of the idea of a genetic code, calling it ''mechanistic

reductionism.'' He did, however, support DNA research within his own Institute of Biochemistry during this time, and was a coauthor of papers discussing DNA and **RNA** in coacervate droplets.

Although Oparin's influence in Soviet science weakened in the early sixties, his international reputation, based on the origin of life theory, remained strong. This, coupled with his political reliability, led his government to send him abroad as a Soviet representative. Traveling by scientists in the Soviet Union was severely restricted in the 1950s, but Oparin was sent on official Soviet business not only to countries in the Eastern bloc and Asia, but to Europe and the United States as well. He also represented his country at international scientific and political conferences, such as the World Peace Council and the World Federation of Scientists.

His work brought him numerous honors. His awards from the Soviet Union include the A. N. Bakh Prize in 1950, the Elie Metchnikoff Gold Prize in 1960, the Lenin Prize in 1974, and the Lomonosov Gold Medal in 1979. The International Society for the Study of the Origin of Life elected him as its first president in 1970. He also was elected a member of scientific societies in Finland, Bulgaria, Czechoslovakia, East Germany, Cuba, Spain, and Italy.

Beginning in 1965 and continuing through 1980, the Soviet Union placed new emphasis on the science of genetics and molecular biology. However, Oparin's Institute of Biochemistry remained a stronghold of old-style biochemistry, and it eventually was bypassed by more progressive research institutions. Oparin died of heart disease in Moscow on April 21, 1980.

OPERON

An operon is a single unit of physically adjacent **gene**s that function together under the control of a single operator gene. The genes within an operon code for **enzyme**s or **protein**s that are functionally related and are usually members of a single enzyme system. The operon is under the control of a single gene that is responsible for switching the entire operon ''on'' or ''off.'' A repressor **molecule** that is capable of binding to the operator gene and switching it, and consequently the whole operon, off, controls the operator gene. A gene that is not part of the operon produces the repressor molecule. The repressor molecule is itself produced by a regulator gene. The repressor molecule is inactivated by a metabolite or signal substance (effector). In other words, the effector causes the operon to become active.

The lac operon in the bacterium *E. coli* was one of the first discovered and still remains one of the most studied and well known. The **deoxyribonucleic acid (DNA)** segment containing the lac operon is some 6,000 base pairs long. This length includes the operator gene and three structural genes (lac Z, lac Y, and lac A). The three structural genes and the operator are transcribed into a single piece of messenger ribonucleic acid (mRNA), which can then be translated. **Transcription** will not take place if a repressor protein is bound to the operator. The repressor protein is encoded by lac I, which is a gene located to the left of the lac promoter. The lac promoter is located immediately to the left of the lac operator gene and is outside the lac operon. The enzymes produced by this operon are responsible for the hydrolysis (a reaction that adds a water molecule to a reactant and splits the reactant into two molecules) of lactose into glucose and galactose. Once glucose and galactose have been produced, a side reaction occurs forming a compound called allolactose. Allolactose is the chemical responsible for switching on the lac operon by binding to the repressor and inactivating it.

Operons are generally encountered in lower organisms such as **bacteria.** They are quite commonly encountered for certain systems, suggesting that there is a strong evolutionary pressure for the genes to remain together as a unit. They have not yet been found in higher organisms, such as multicellular life forms.

A mutation in the operator gene which renders it nonfunctional would also render the whole operon inactive. As a direct result of inactivation, the coded pathway would no longer operate within the cell. Even though the genes are still separate individual units, they cannot function by themselves, without the control of the operator gene.

OPPORTUNISTIC ORGANISM

Opportunistic organisms commonly refer to **animals** and **plants** that tolerate variable environmental conditions and food sources. Some opportunistic **species** can thrive on almost any available nutrient source: omnivorous rats, bears, and raccoons are all opportunistic feeders. Many opportunists flourish under varied environmental conditions: the common house sparrow (*Passer domesticus*) can survive both in the warm, humid climate of Florida and in the cold, dry conditions of a Midwestern winter. Aquatic opportunists, often aggressive fish species, fast-spreading plankton, and water plants, frequently tolerate fluctuations in water salinity as well as temperature.

A secondary use of the term ''opportunistic'' signifies species that can quickly take advantage of favorable conditions when they arise. Such species can postpone reproduction, or even remain dormant, until appropriate temperatures, moisture availability, or food sources make growth and reproduction possible. Some springtime-breeding lizards in Australian **deserts,** for example, can spend months or years in a juvenile form, but when temperatures are right and a rare rainfall makes food available, no matter what time of year, they quickly mature and produce young while water is still available. More familiar opportunists are **virus**es and **bacteria** that reside in the human body. Often such **organisms** will remain undetected with a healthy host for a long time. But when the host's immune system becomes weak, resident viruses and bacteria seize an opportunity to grow and spread. Thus people suffering from malnutrition, exhaustion, or a prolonged illness are especially vulnerable to common opportunistic diseases such as the common cold or pneumonia.

Adaptable and prolific reproductive strategies usually characterize opportunistic organisms. While some plants can

Purple loosestrife *(Lythrum salicaria)* is an aggressive wetland plant species first introduced into the United States from Europe. As a highly adaptive and tolerant plant, loosestrife is often able to outcompete native plant species in many environments. *(Photograph by Robert J. Huffman, Field Mark Publications. Reproduced by permission.)*

reproduced only when pollinated by a specific, rare insect and many animals can breed only in certain conditions and at a precise time of year, opportunistic species often reproduce at any time of year or under almost any conditions. House mice (*Mus musculus*) are extremely opportunistic breeders: they can produce sizeable litters at any time of year. Opportunistic feeding aids their ability to breed year round; these mice can nourish their young with almost any available vegetable matter, fresh or dry.

The common dandelion (*Taraxacum officinale*) is also an opportunistic breeder. Producing thousands of seeds per plant from early spring through late fall, the dandelion can reproduce despite competition from fast-growing grass, under heavy applications of chemical herbicides, and even with the violent weekly disturbance of a lawn mower. Once mature, dandelion seeds disperse rapidly and effectively, riding on the wind or on the fur of passing rodents. The common housefly (*Musca domestica*) is also an opportunistic feeder and reproducer—it can both feed and lay eggs on almost any organic material as long as it is fairly warm and moist.

Because of their adaptability, opportunistic organisms commonly tolerate severe environmental disturbances. Fire, floods, drought, and pollution disturb or even eliminate plants and animals that require stable conditions and have specialized nutrient sources. Fireweed (*Epilobium angustifolium*), an opportunist that readily takes advantage of bare ground and open sunlight, spreads quickly after land is cleared by fire or by human disturbance. Because they tolerate, or even thrive, in disturbed environments, many opportunists flourish around human settlements, actively expanding their ranges as human activity disrupts the **habitat** of more sensitive animals and plants. Opportunists are especially visible where chemical pollutants contaminate habitat. In such conditions overall species diversity usually declines, but the population of certain opportunistic species may increase as competition from more sensitive or specialized species is eliminated. Because they are tolerant, prolific, and hardy, many opportunistic organisms, including the house fly, the house mouse, and the dandelion, are considered pests. Where they occur naturally and have natural limits to their spread, however, opportunists play important environmental roles. By quickly colonizing bare ground, fireweed and opportunistic grasses help prevent erosion. Cottonwood trees (*Populus spp.*), highly opportunistic propagators, are among the few trees able to spread into arid regions, providing shade and nesting places along stream channels in deserts and dry plains. Some opportunists that are highly tolerant of pollution are now considered indicators of otherwise undetected chemical spills. In such hard-to-observe environments as the sea floor, sudden population explosions among certain bottom-dwelling marine mollusks, plankton, and other invertebrates have been used to identify petrochemical spills around drilling platforms and shipping lanes.

ORDER

The term order is a taxonomical category used to classify **organism**s within one or more closely related families. Order names of plant families generally use the suffix -ales (ex. Rosales) and order names of **animal** families end with an -a (ex. Carnivora). Generally, the order of an organism is not considered during scientific discussion, but rather the more specific terms of **family**, **genus**, and **species** are used.

See also Class; Classification; Phylum

ORGAN

An **organ** is a multicellular, multitissue part of an **animal** or plant, which forms a discrete structural and functional unit. (**Tissue**s are formed of many **cell**s that are similar in structure and function, and are bound together as a unit by intercellular material.)

Large, advanced animals have many kinds of organs. Some examples include: **arteries** and **veins**, bones, the **brain**, eyes, the **heart**, intestines, kidneys, the liver, ovaries and testes, the skin, and the stomach. Some of these are organized into functional complexes known as organ systems, such as the **digestive system** that is responsible for processing and absorbing **nutrients** from food, and in typical **mammals** is composed of the stomach, large and small intestine, and organs that synthesize and secrete digestive **enzyme**s, such as the **pancreas**. Another example is the **skeletal system**, which is composed of numerous bones.

Plants also have many kinds of organs, such as leaves, stems, roots, and the various parts of **flower**s. An example of an organ system in a higher plant is a flower, which (depending on the **species**) may be composed of stamens (which are composed of anthers and a filament), pistils (containing stigma, style, and ovary), petals, bracts, and receptacle.

ORGANELLE

Cellular organelles are the **membrane**-bound or macromolecular structures that make up the internal architecture of the cell or provide compartments within which many metabolic processes take place. Among the most prominent membranous organelles of eukaryotic cells (those with membrane-bound nuclei) are **mitochondria** and **chloroplast**s (the primary **energy** sources for **animal**s and **plant**s respectively), **lysosome**s (the main digestive compartments of cells), smooth and rough **endoplasmic reticulum** (the site of synthesis of complex **lipids** and membrane or export **protein**s, respectively) and the **Golgi apparatus** (the major site for assembly, processing, sorting, and packaging of macromolecular products that will be shipped to other organelles or secreted from the cell). Among the most prominent macromolecular organelles are **ribosome**s (responsible for **protein synthesis**) and elements of the cellular **cytoskeleton** such as microtubules and microfilaments. Some cell biologists might also consider eukaryotic **chromosome**s to be organelles since they are large enough be seen during **mitosis** or **meiosis** in a **light microscope**. Prokaryotic cells (**bacteria**, some **fungi** and their kin) generally lack any of these organelles except ribosomes.

Organelles are important in establishing cell structure and provide a wide variety of individual spaces and surfaces within or on which different **chemical compound**s can be separated, organized, or stored. They also keep potentially incompatible reactions apart and allow higher plant and animal cells to simultaneously carry out a wide range of highly specialized metabolic operations.

ORGANIC COMPOUND

An organic compound is a compound containing carbon in combination with one or more elements. Organic compounds are generally characterized by chains of connected carbon **atom**s. **Hydrocarbon**s contain only carbon and hydrogen, while many other organic compounds contain carbon, hydrogen, and oxygen. Other major elements in naturally occurring organic compounds are **nitrogen**, phosphorus, and sulfur. Exceptions to the **classification** of carbon-containing compounds as organic compounds include **carbon dioxide**, **carbon monoxide**, carbon disulfide, carbon tetrachloride, hydrogen cyanide, carbonates, and carbides, which have traditionally been regarded as inorganic compounds.

It was commonly believed that compounds of carbon could only be produced by the *vital force* found in living **organism**s. However, in 1828, Friedrich Wöhler found, by accident, that the application of heat to ammonium cyanate, an inorganic compound, caused it to change to urea, a compound thought to be formed only by living organisms. Today many organic compounds are products of synthetic chemistry, with no similar products existing in **nature**.

Organic compounds are essential for the functioning of human life. Modern civilization is almost totally dependent on hydrocarbons that occur in the earth's crust as natural gas and petroleum, which provide the major sources of **energy** and also serve as raw materials for the manufacture of plastics and many other materials. Food substances consist of a variety of organic compounds that serve as a source of energy for living organisms. Examples of organic compounds used for food include **carbohydrate**s, **fats and oils**, and **protein**s and **amino acid**s. Detergents and soaps are organic compounds that are used as cleansing materials. Natural and synthetic organic **pesticides** are used extensively in **agriculture** to control undesirable forms of life. Both natural and synthetically produced organic compounds also serve as effective medicines for the treatment of disease.

Properties of organic compounds that differ from those of inorganic compounds include: (1) organic compounds are usually combustible; (2) organic compounds, in general, have lower melting and boiling points; (3) organic compounds are usually less soluble in **water**; (4) several organic compounds may exist for a given formula (known as **isomers**); (5) reactions of organic compounds are usually molecular in nature rather than ionic and so are often quite slow; (6) molecular weights of organic compounds may be very high (e.g., well over 1,000); and (7) most organic compounds can serve as a food source for **bacteria**.

Common types of organic compounds include aliphatic hydrocarbons (including alkanes, alkenes, and alkynes), aromatic hydrocarbons (e.g., benzene), **alcohol**s, esters, ethers, amines and amides (which contain nitrogen), and carboxylic acids.

ORGANISM

An organism is an individual, living entity. Organisms range in size and complexity from tiny **microorganism**s, to large, multicellular **plants** and **animals**. Biologists classify organisms into five **kingdom**s on the basis of their **cellular** and subcellular organization, **metabolism**, reproduction, and behavior. These kingdoms are described below, in order of their earliest appearance in the **fossil** record of life.

Monera are prokaryotic microorganisms, meaning their **genetic material** is not contained within a bounded **organelle** called a **nucleus**. This group includes the simplest organisms, including **virus**es, which consist of little more than a **protein** shell containing **nucleic acids**. Viruses are incapable of reproduction unless they parasitize the metabolism of an unrelated host cell. Blue-green **bacteria** and true bacteria are two other groups of monerans.

Protista are a diverse group of microorganisms, including the simplest eukaryotic organisms, which have an organized nucleus, one or more **flagellae**, and usually **mitochondria** and plastids. **Protozoa**ns are the most representative group, but some flagell ated **fungi** and **algae** are also considered protists.

Fungi are a diverse group of non-flagellated, unicellular or multicellular organisms. They range in complexity from single-celled **yeast**s, through multicellular but microscopic fungi growing as a thread-like mycelium, to relatively complex fungi that develop large mushrooms as reproductive structures.

Plantae, or green plants, are photosynthetic organisms that absorb solar radiation using chlorophyll and other pigments, and utilize the **energy** to fix simple mineral **nutrients** into energy-rich biochemicals. Organisms in this group range from unicellular algae, through multicellular but non-vascular algae, liverworts, and mosses, to vascular plants such as ferns, conifers, and flowering plants.

Animalia, or multicellular animals, are **heterotrophic** organisms, meaning they must ingest their food in the form of **biomass** of plants or other animals. Animals are capable of movement, often in response to sensory stimuli. They range in size and complexity from small sponges and arthropods to large vertebrates weighing ton.

To varying degrees, all organisms are related, and they share certain elements of their physiology and other functions. Many distinctive organisms have a relatively ancient evolutionary lineage that extends far back into the geological past, while others have evolved more recently. Some organisms are enormously more complex in biological organization than others. However, the modern interpretation of life suggests that none of Earth's organisms are "higher" or more "primitive" than any others, and none have greater intrinsic value. Evolution is not interpreted as a deterministic progression of kinds of organisms that represent a logical, directed **succession** from simple types such as viruses and bacteria, to more complex ones such as birds and **mammals** (including humans). Earth's enormous **biodiversity** of living organisms utilizes many body and metabolic plans of varying complexity, but all **species** represent successful **adaptation**s to the planet's habitable environments.

ORGANOGENESIS

Organogensis refers to that period of time during development when the organs are being formed. After an **egg** has been fertilized, and has been implanted in the uterus, the developing form is known as the **embryo**. Organogenesis takes place during this embryonic phase. In fact, most organogenesis has begun as early as week 5 in humans (remember that a normal human pregnancy lasts an average of 40 weeks). Therefore, damage to any of the organ systems of the body which may ultimately result in some type of **birth** defect usually strikes during this time frame.

By week five, the buds of **tissue** which will become the limbs are in place. The structures which will become the skeleton, **nervous system**, and **circulatory system** of the face, neck, and jaws are in place. A five week old embryo has the early developmental structures of the esophagus, stomach, intestine, liver, and **pancreas**. The **heart** is already functioning, and continues to develop and change over this period of time. The **respiratory system** begins developing, as do **blood** vessels, blood cells, nervous and endocrine organs. Clearly, the most crucial organs of the human form are developing during organogenesis. Essentially, the earlier the injury to these developing buds of tissue, the more severe the ultimate defect. This is because these tiny buds of tissue hold all the primitive cells which should differentiate into the myriad number of cells necessary to create all of the varied organs of the human body.

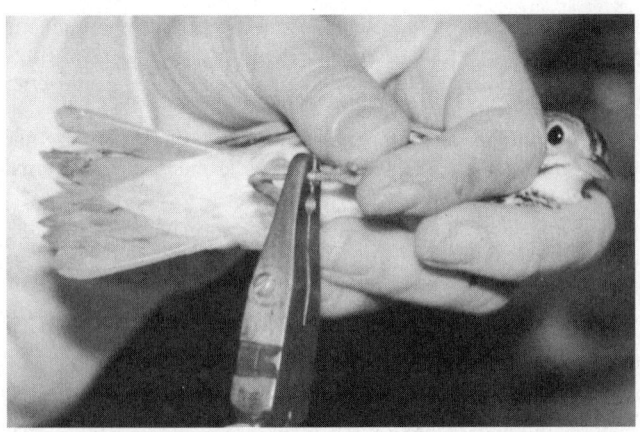

This ovenbird's legs are being banded, or ringed. *(Photograph by Robert J. Huffman, Field Mark Publications. Reproduced by permission.)*

It is an irony that, during this crucial period of development, when **toxins** from the outside world can have such devastating effects on the ultimate development of the embryo, many women are not even yet aware that they are pregnant, and are therefore not in the mindframe of protecting the developing embryo from exposure to such harmful substances as cigarette smoke, **alcohol**, certain drugs or medications, or extremes of heat (as could be experienced in a very hot Jacuzzi).

One of the most infamous agents responsible for widespread deformities during the period of organogenesis is a drug called Thalidomide. Thalidomide was administered to women (particularly in Europe in the 1950s) because it was thought to combat the nausea present in early pregnancy. Over time, however, it became evident that babies born of thalidomide-using mothers had very high rates of serious limb deformities. In particular, the long bones of the limbs were either absent or seriously deformed. Furthermore, many of these children had associated defects of the heart and intestine. Thalidomide was ultimately determined to be at fault, causing the most severe defects when given between weeks four and six of pregnancy: the period of organogenesis. Thalidomide was subsequently withdrawn from the market.

ORNITHOLOGY

Ornithology is the branch of **zoology** that deals with the study of birds. Birds are any **organism**s in the **class** Aves. They are warm-blooded (or homoiothermic) vertebrates that have feathers covering their body; forelimbs modified into wings; stouter hindlimbs used for walking, swimming, or perching; scaly legs and feet; jaws reduced to a toothless beak; and a four-chambered **heart**. Birds lay hard-shelled **egg**s from which their young hatch. Major subject areas in ornithology include: **anatomy**, physiology, behavior, **ecology**, **evolution**, and **classification** and **systematics**.

Birds evolved from a group of reptiles known as the dinosaurs (**order** Dinosauria), which first appeared during the late Triassic period (which ended about 210 million years ago). The earliest bird known in the **fossil** record is *Archaeopteryx*, from the mid-Jurassic period about 160 million years ago. These extremely early birds had many typical avian characteristics, such as feathers and a horny beak, and were very likely warm-blooded, but they also had reptilian features, such as teeth. In fact, modern birds and dinosaurs still have many characteristics in common, and some biologists and paleontologists believe that birds should be viewed, and classified, as "living dinosaurs."

There are 27 orders of birds, divided into about 166 families, and containing about 9,000 living **species**. Some prominent families from the Americas include the Gaviidae (loons), Podicipedidae (grebes), Pelecanidae (pelicans), Phalacrocoracidae (cormorants), Anatidae (swans, geese, ducks, mergansers), Cathartidae (vultures), Accipitridae (hawks, eagles), Falconidae (falcons), Tetraonidae (grouse, ptarmigan), Phasianidae (quail, pheasants), Ardeidae (herons, bitterns), Rallidae (rails, coots), Charadriidae (plovers, turnstones), Scolopacidae (sandpipers), Laridae (gulls, terns), Columbidae (pigeons, doves), Strigidae (owls), Trochilidae (hummingbirds), Psittacidae (parrots, macaws), Picidae (woodpeckers), Tyrannidae (flycatchers), Hirundinidae (swallows), Corvidae (jays, crows, raven), Paridae (chickadees, titmice), Troglodytidae (wrens), Turdidae (thrushes, bluebirds), Sturnidae (starlings), Vireonidae (vireos), Parulidae (wood warblers), Icteridae (blackbirds, orioles), and Fringillidae (grosbeaks, finches, sparrows).

Avian anatomy changes remarkably during development, from the relatively simple structures of the fertilized egg, to the much more complex adult form. Anatomy is also extremely variable among species. Size alone ranges from flightless ostriches weighing up to 330 lb (150 kg), to tiny hummingbirds weighing only 0.08 oz (2.25 g). Perhaps the most distinguishing anatomical characteristic of birds is their feathers, which provide insulation against the loss of body heat, and a broad, yet light, wing and tail surface for flight. The forelimbs of birds are highly modified as wings, especially through the extension of their "fingers" into an airfoil surface for active and/or gliding flight (a few species, such as the ostrich and emu, have secondarily lost the power of flight). Other important anatomical **adaptation**s for flying include the large breast muscles that are used to power flight, the large keeled sternum (or breastbone) to which the flight muscles attach, and the hollow bones and extensive air-sacs of most species of birds. Some birds have extraordinarily light bodies for their size (the magnificent frigatebird [*Fregata magnificens*] has a wingspan of 7 ft [2 m], but its skeleton weighs only 4 oz [113 g], less than the weight of its feathers).

Birds lay eggs with a hard shell that contain a fertilized **embryo**. In most species, the eggs are laid into a nest constructed by one or both of the parents, and the eggs are incubated by the parents. The newly hatched young of most bird species are born in a relatively early stage of development, and are unfeathered, ungainly, and virtually helpless. These almost incompetent young must be fed and otherwise tended by their parents for some time, until they finish development and learn to fly and forage for themselves. The young of some other species are more developed when born, and may be capable of immediately leaving the nest to live a semi-independent, or even fully independent life.

A distinguishing element of avian physiology is homoiothermy, or metabolic activity that maintains the body temperature within a narrow, warm range optimized for muscular functions and **enzyme** efficiency. (**Mammals** and some other vertebrates are also warm-blooded.) Homoiothermy is a crucial physiological trait that allows almost all bird species to have an extremely active lifestyle. It also allows some species to live year-round in cold environments.

One of the most notable elements of avian behavior is their use of song to proclaim a breeding **territory**, and other distinct sounds to organize their social system, keep flocks together, warn other individuals of predators, and for other kinds of communication. Some other classes of **animal**s are also rather vocal, particularly many mammals, but not to the same degree that most birds are. Some birds are also quite intelligent, second in this regard only to mammals. (Actually, corvids such as the raven (*Corvus corax*) are more intelligent than most species of mammals.) Another notable element of avian behavior is the habit of many species to undertake long-distance **migration**s between their breeding and wintering **habitat**s. The most extensive migrations are made by the arctic tern (*Sterna paradisaea*, some of whom breed at the most northerly limits of land on the northern tips of Greenland and Ellesmere Island, and then migrate to spend their winter foraging in Antarctic **water**s.

OSBORNE, THOMAS BURR (1859-1929)
American biochemist

Osborne, the son of a banker, was born in New Haven, Connecticut, and did both his undergraduate and graduate work at Yale University. After getting his Ph.D. in 1885, he joined the recently-established Connecticut Agricultural Experiment Station as an analytical chemist. Four years later, at the suggestion of the station's director, Osborne began an investigation into the proteins of **plant** seeds—an investigation which eventually became his lifelong work.

The young chemist began by studying oat kernels, in time managing to isolate an alcohol-soluble **protein** and a globulin from them. Intrigued, he turned to other seeds and, over the next three years, isolated the proteins of at least 32 different plant species, including nuts, legumes, and cereal grains. Subjecting them to an intensive chemical analysis, he found, to his surprise, that the proteins of different species were distinctly different from each other. The differences were especially marked, he noted, in the **amino acid** content of the various proteins. Although findings like these contradicted the well-known (and widely accepted) doctrine of Justus von Liebig—that only four kinds of protein existed in nature, albumin, casein, fibrin and gelatin, and that they were all pretty much alike—Osborne became increasingly convinced he was on the right track.

In 1909, Osborne invited another biochemist, **Lafayette B. Mendel** (1872-1935), then working at his alma mater, Yale,

to join him in his ongoing investigations, now directed toward probing into nutritional properties of plant proteins. Mendel accepted and the two biochemists proceeded to work together for almost twenty more years. They co-wrote roughly a hundred papers and made a number of important discoveries. For example, they found that two amino acids in particular, lysine and tryptophan, were essential for the normal growth of animals. Furthermore, laboratory rats were unable to manufacture these substances within their bodies and thus had to rely on dietary lysine and tryptophan to survive.

The outstanding accomplishment of the Osborne and Mendel collaborations was made in 1913 with the discovery of the substance that later proved to be vitamin A. In that year, the two researchers determined that butter contains a fat-soluble essential nutrient (a **nutrient** they also found in cod-liver oil). Unfortunately, they published their results three weeks after **Elmer McCollum** had announced his discovery of the same substance. McCollum, therefore, received most of the credit for the discovery.

OSMOSIS

Osmosis is a process by which a solvent (the liquid that dissolves another substance) in solution passes through a barrier. The solvent may pass through the barrier, but the solute (the substance dissolved in the solvent) either does not go through it, or passes through much more slowly than the solvent. The solvent will pass through the barrier until the concentration of solvent is the same on both sides of the barrier. The barrier is a **membrane** that is either permeable, allowing solvent and solute **molecule**s to pass through, or semipermeable, allowing only solvent molecules to pass through. The pressure of the **water** passing through the membrane is called osmotic pressure.

The process was first investigated by a French physicist, Abbé Jean Antoine Nollette (1700-1770), in 1748. Nollette covered a glass tube containing sugar water with a piece of paper. He placed the tube, paper end down, into the water. The level of liquid in the tube rose. The pure water passed through the paper faster than the sugar water could. More experiments on osmosis followed. René Dutrochet (1776-1847) investigated the phenomena in the 1820s and 1830s. In the 1840s and 1850s Thomas Graham and **Justus von Liebig** researched osmosis but could not develop a suitable theory to explain it. Graham did distinguish between those substances that passed through parchment, which he called crystalloids, and those that did not, which he called colloids. Graham's additional research led to the process of dialysis, used today in artificial kidney machines.

The next major advance in the field came in 1877 when Wilhelm Pfeffer (1845-1920), a German botanist, studied osmotic pressure. Again the test subject was sugar water. The sugar solution was placed in a porous clay vessel, which in turn was placed in a container filled with pure water. Using a manometer Pfeffer measured the osmotic pressure and discovered was inversely proportional to the volume of a solution and directly proportional to absolute **temperature** or PV = kT, where *P* is pressure, *V* is volume, and *T* is absolute temperature. The constant *k* was later used in defining the universal gas constant by Jacobus Henricus Van't Hoff and was used in other gas laws. It was determined that the osmotic pressure a solute displays is the same pressure it would exert as a gas at the same volume and temperature.

Hugo van Mohl (1805-1872), a German botanist, continued on Pfeffer's path and was the first to describe **cell division**. He also provided the first lucid explanation for osmosis. Pfeffer's clay pot was a semipermeable membrane, as are the membranes surrounding most **animal** and vegetable **cell**s. Studies in osmosis led to studies in cell physiology and solution purification.

Reverse osmosis is the process of applying a pressure greater than osmotic pressure on a solution. This reverses the process. In the end, a water-sugar water system would consist of pure water on one side of the barrier and a concentrated solution of sugar water on the other. This method is sometimes used for desalination, the process of removing salt from salt water to make it potable. The method is also used by hikers to remove harmful **microorganism**s from stream and lake water. Reverse osmosis is sometimes referred to as ultra-or hyper-filtration. The process is used to purify many liquids from milk to polio vaccines. Recently osmosis has been used to dehydrate fruit. **Fruits** slices are blanched; a sugar solution is pumped over the fruit, which is kept at 140°F (60°C) and the fruit slices lose most of their component water. They are then rinsed and dried briefly. The total process is faster than conventional dehydration, which takes about seven hours.

OUTBREEDING

Outbreeding is a term used to describe the union of genetically unrelated individuals within the same **species**. Outbreeding promotes genetic variability and vitality within a breeding **population** by eliminating homozygous individuals. Harmful recessive **allele**s have a better chance of being repressed by the dominant alleles in heterozygous individuals, thus reducing the number of negative **mutation**s.

Outbreeding promotes the genetic diversity and, hence, the overall **fitness** of a species. By producing new and varied combinations, outbreeding is a starting point of **natural selection** and the foundation of **evolution**. Outbreeding is an important tool in the continued survival of a species.

See also Genetics; Inbreeding

OVIPAROUS

The word oviparous refers to **animal**s that lay **egg**s which hatch externally. In other words the female retains the fertilized eggs in her body until they hatch, so that "live" young are born.

Depending on the **species**, the eggs of oviparous animals may be fertilized inside the body of the female, or exter-

nally. External **fertilization** is the more "primitive" condition (meaning it was the first to occur during the evolutionary history of animals). It involves the passage of the **sperm** to the ova through an ambient medium, which is almost always the **water** in which the unfertilized eggs are laid. Frogs, for example, achieve external fertilization of their eggs during amplexus (the male deposits sperm over the eggs as they are laid by the female. In many animals, external fertilization is a much-less controlled process, for example, in the case of corals, sea urchins and numerous other kinds of marine **invertebrates**, in which the sexes shed huge numbers of **gamete**s to the water at about the same time, so that the ova and sperm meet somewhat by chance. Many species of fish also have external fertilization. For instance, the sexes of salmonid fishes such as trout and salmon mix their unfertilized eggs (spawn) and milt (sperm) over a gravely riverbed. The eggs become externally fertilized, fall into spaces within the gravel, and develop and hatch into young fish fry. All animals that fertilize their eggs externally are oviparous.

In species that have internal fertilization of their eggs, the males must somehow pass their sperm into the female. Male salamanders, for example, deposit a sperm packet (or spermatophore) onto the bottom of their breeding pond, and then induce an egg-bearing female to walk over it. She picks up the spermatophore with the somewhat prehensile lips of her cloaca, and retains it inside her body where the eggs become fertilized. The internally fertilized eggs are later laid in the water and develop externally, representing oviparity.

Many species of fish, all species of reptiles and birds, and even a few primitive **mammals** (the platypus and echidnas) achieve internal fertilization of their eggs through copulation. If the eggs are then laid to develop externally, the process represents oviparity.

See also Viviparous

OVOVIVIPAROUS

The word ovoviparous refers to **animal**s that produce **egg**s, but retain them inside of the body of the female until hatching occurs, so that "live" offspring are born. Ovoviviparity is a much-less common reproductive strategy than oviparity (this is the external development of fertilized eggs have been laid by the female). Although infrequent, ovoviviparity is displayed by a rather wide diversity of animals, including **species** of insects, fish, lizards, and snakes.

Ovoviviparity in insects is rather simple, in that oxygen or nourishment are not provided to their developing eggs (the females are just a relatively safe brooding chamber. In contrast, species of ovoviviparous fish, lizards, and snakes may provide some oxygen and **nutrition** to progeny developing within the oviduct of the female (most nutrition, however, is provided by the yolk of the eggs). Moreover, the eggshell is essentially reduced to a **membrane** in these species, facilitating the exchange of oxygen and nutrition. Because the developing egg and **larva** receive some nutrition, and the eggshell is essentially absent, some zoologists interpret ovoviviparity in fish, lizards, and snakes to represent **viviparity**, or true live **birth**. This is a relatively narrow interpretation, however, and is not adopted here.

One familiar example of ovoviviparity is the guppy (*Lebistes reticulatus*), a small, freshwater fish indigenous to the West Indies and northern South America and commonly kept as a pet in aquaria. Ova of the guppy are internally fertilized by the male inserting **sperm** using a modified anal fin. The eggs are then retained in the oviduct of the female, where they hatch and develop, so that live, independent young are born.

Another case is the garter snake (*Thamnophis sirtalis*), a common and widespread reptile of North America. Their eggs are fertilized internally by copulation, incubated within the oviduct of the female. The young snakes are born live, enclosed in an amniotic sac, from which they quickly escape and slither into an independent life.

One of the most unusual cases of ovoviviparity involves the gastric-brooding frog (*Rheobatrachus silus*), an extremely rare (and possibly extinct) species of Australia. **Fertilization** has never been observed by scientists, but it is thought to occur externally, after which the female swallows the eggs and retains them in her stomach. The eggs hatch and develop over about a 37-day period, and are ''born'' as small, independent froglets emerging through the female's mouth. The young are almost identical in **morphology** to the adults, except for size. The female does not eat while she is brooding eggs, and the production of stomach acids and **enzyme**s are suppressed so as to not digest her progeny. The extraordinary case of the gastric-brooding frog is only known case of an externally fertilized, ovoviviparous species.

OWEN, RICHARD (1804-1892)
English biologist

Sir Richard Owen was a comparative anatomist, paleontologist, and zoologist. He was the Hunterian Professor of **Comparative Anatomy** and Physiology at the Royal College of Surgeons, London, from 1836 to 1856. He then became the superintendent of the Natural History Section of the British Museum in London in 1849, and was Superintendent of the entire museum from 1856 to 1883. He is best known as an influential paleontologist during an extremely exciting time in the nineteenth century, when the **fossils** of extinct dinosaurs were first discovered and their significance in chronicling Earth's biological history started to be understood. In fact, Owen coined the word ''dinosaur,'' and was largely responsible for kindling the dinosaur mania that began in the mid-nineteenth century, and continues today. His first, great popularizing event was the erection of a series of life-sized models of dinosaurs and other extinct creatures at the Crystal Palace in London in 1854, which created an absolute sensation among the Victorian population. Remarkably, a formal dinner partly was held within the body of one of the giant dinosaurs, a model of *Iguanodon*, as it was nearing completion (Owen sat at the ''head'' of the table, and in the head of the dinosaur!)

Along with his extensive work on extinct species of vertebrates, Owen also conducted some important studies of liv-

Rainbow trout eggs hatching. *(JLM Visuals. Reproduced by permission.)*

ing **animal**s. One of his works involved the confirmation of the earlier observations of James Paget, that the deadly **parasite** *Trichina spiralis* was the cause of trichinosis in humans, and was transmitted by eating inadequately cooked pork.

Owen was a strong opponent of the theory of **Charles Darwin** concerning **natural selection** as a critical force of **evolution**. To the end of his days Owen refused to accept the theories of evolution, and remained a determined creationist. Owen and other anti-Darwinians have badly lost this scientific and philosophical battle, as the ideas of Darwin and his followers have come to reign virtually supreme among modern biologists. Nevertheless, Richard Owen was perhaps the greatest vertebrate paleontologist of his time, and one of the most influential ever.

OXIDATION-REDUCTION (REDOX)

Oxidation-reduction (redox) reactions are chemical reactions in which there is a transfer of one or more electrons between **atoms**. These reactions often produce **energy** and are the primary pathway in which **molecule**s such as sugars, fats, and **protein**s are broken down in the body. One of the early scientists who studied redox reactions was the German chemist, Walther Nerst (1864-1941).

In a redox reaction, electrons are transferred from one reactant to another. Oxidation is the process by which one substance loses electrons. Reduction is the process by which a substance gains electrons. The term reduction refers to the fact that the amount of positive charge on the substance is reduced by the additional electrons. A simple example of a redox reaction is the one which creates sodium chloride (2NaCl) from elemental sodium (Na) and chlorine. In this reaction, the sodium is oxidized and chlorine is reduced. Since the sodium is the substance that is reducing chlorine, it is called the reducing agent. Chlorine, which is the substance that is oxidizing sodium, is known as the oxidizing agent. Since this type of electron transfer always requires a reducing and oxidizing substance, the process is called oxidation-reduction.

In the reaction of sodium with chlorine, there is a complete transfer of electrons between the two substances. However, this does not occur in all redox reactions. Some redox reactions change the amount of electron sharing that goes on in covalent bonds. An example of this is the combustion of methane in which it reacts with oxygen to form **carbon dioxide** and **water**. At the start of this reaction, electrons are shared equally between carbon and hydrogen in covalent bonds. When methane reacts with oxygen, a **covalent bond** is formed between the oxygen and carbon. Since oxygen is more electronegative than carbon, the electrons are shifted away from the carbon. This shifting of electrons releases potential energy that can be utilized by the **cell**.

The amount of energy that is released during a redox reaction is dependant on the relative electronegativities of the reactants. The electronegativity of an element is determined by its oxidation state or oxidation number. This value is determined by a set of rules which describe how electrons are positioned around an atom or molecule. Typically, it reflects the number of electrons an atom will accept or donate. For example, oxygen has a oxidation number of -2 because it has a tendency to accept two electrons.

Redox reactions are important in living systems because they allow energy in food molecules to be released and transferred to make **adenosine triphosphate (ATP)**. This process takes place during cellular respiration. In this reaction, glucose (or other food molecules) is combined with oxygen, or oxidized, to produce carbon dioxide, water, and energy. The energy is derived from the potential energy that is released as electrons are transferred to oxygen. The oxidation of glucose is done in a step-wise fashion by a series of enzymes which make up the **Krebs cycle** and the electron transport chain. One of the key enzymes in this process is nicotinamide adenine dinucleotide (NAD). The energy is released slowly because the complete release of all its stored energy would not be efficient.

OZONE

Ozone (O_3) is a highly reactive **molecule** composed of three linked oxygen **atoms**. It forms a colorless gas with a biting, acrid odor. Ozone is a strong oxidizing agent, meaning that it reacts with and damages vegetation, building materials (such as paint, rubber, and plastics), and sensitive **tissues** (such as eyes and lungs). Because ozone is a good disinfectant, it can be used to purify **water** or sterilize surgical instruments. Ozone is created by electric discharges such as lightning or arcing of electrical contacts as well as by high intensity, short wavelength light.

Ozone is a natural component of the **atmosphere**. It is both formed and destroyed by ultraviolet radiation in a photochemical reaction cycle in which reactive oxygen molecules are alternately added to and removed from ordinary diatomic oxygen molecules. Usually these reactions occur mainly in the stratosphere where the resulting ozone provides an important shield that protects life on the surface of the earth from dangerous levels of incoming ultraviolet radiation. Without this shield, most living things could only survive beneath the soil surface. Unfortunately, chlorine-containing chemicals such as the chlorofluorocarbons used as refrigerants, cleaning solvents, and blowing agents are diffusing into the stratosphere and catalyzing destruction of this protective ozone layer.

Ozone also is formed in ambient air (that surrounding us at ground level) as a result of photochemical reactions involving urban air **pollutants**. It is a major component of urban smog and is a threat to human health, wild and domestic **animals** and **plants**, and the built environment. In the presence of unburned **hydrocarbon**s and other volatile organic chemicals, nitric oxide (created from ordinary **nitrogen** molecules in the air during high-**temperature** combustion) is converted into nitrogen dioxide. Ultraviolet radiation in sunlight splits nitrogen dioxide into a highly reactive atomic oxygen atom and another molecule of nitric oxide. The oxygen atom binds to diatomic molecular oxygen to create ozone and the cycle starts over again.

Smog containing high levels of ozone is typical of urban areas with a dry, sunny climate, abundant motor vehicle traffic, and topographic or atmospheric features that keep air stagnant. Los Angeles, Denver, and Mexico City are well-known examples of these conditions. Nitrogen oxides and unburned hydrocarbons from internal combustion engines, dry cleaners, paints, and a variety of other urban sources are trapped in air by nearby mountains. Bare ground cools quickly at night in dry **desert** conditions creating an atmospheric **inversion** in which air near the ground is cooler than that at upper levels. Contaminants are concentrated in a dense layer about 1,000 ft (305 m) above the surface. When the sun comes up the next morning, photochemical reactions in this layer produce high ozone concentrations. As the atmosphere heats, mixing occurs and ground-level air quickly reaches unhealthy ozone levels. Since this series of atmospheric reactions was first recognized in the 1950s, attempts have been made to reduce urban **air pollution**. While many cities still have undesirable levels of photochemical oxidants, substantial progress has been made in controlling these air pollutants. The Environmental Protection Agency (EPA) reports that in spite of growing human **populations** and greatly increased traffic, urban areas in the United States have experienced about a 20% reduction in ambient ozone levels over the past 20 years.

See also Ozone depletion

OZONE LAYER DEPLETION

A highly reactive **molecule**, **ozone** (O_3) is composed of three linked oxygen **atoms**. Ozone is created by electric discharges such as lightning and as well as by high intensity light. It forms a colorless gas with a biting, acrid odor that is a major component of urban smog. A strong oxidizer, ozone in ambient air (that normally around us at ground level) burns eyes, degrades paint and rubber, kills vegetation, and has a number of other undesirable effects. Consequently, **air pollution** controls restrict actions that increase ozone levels in urban air. In the

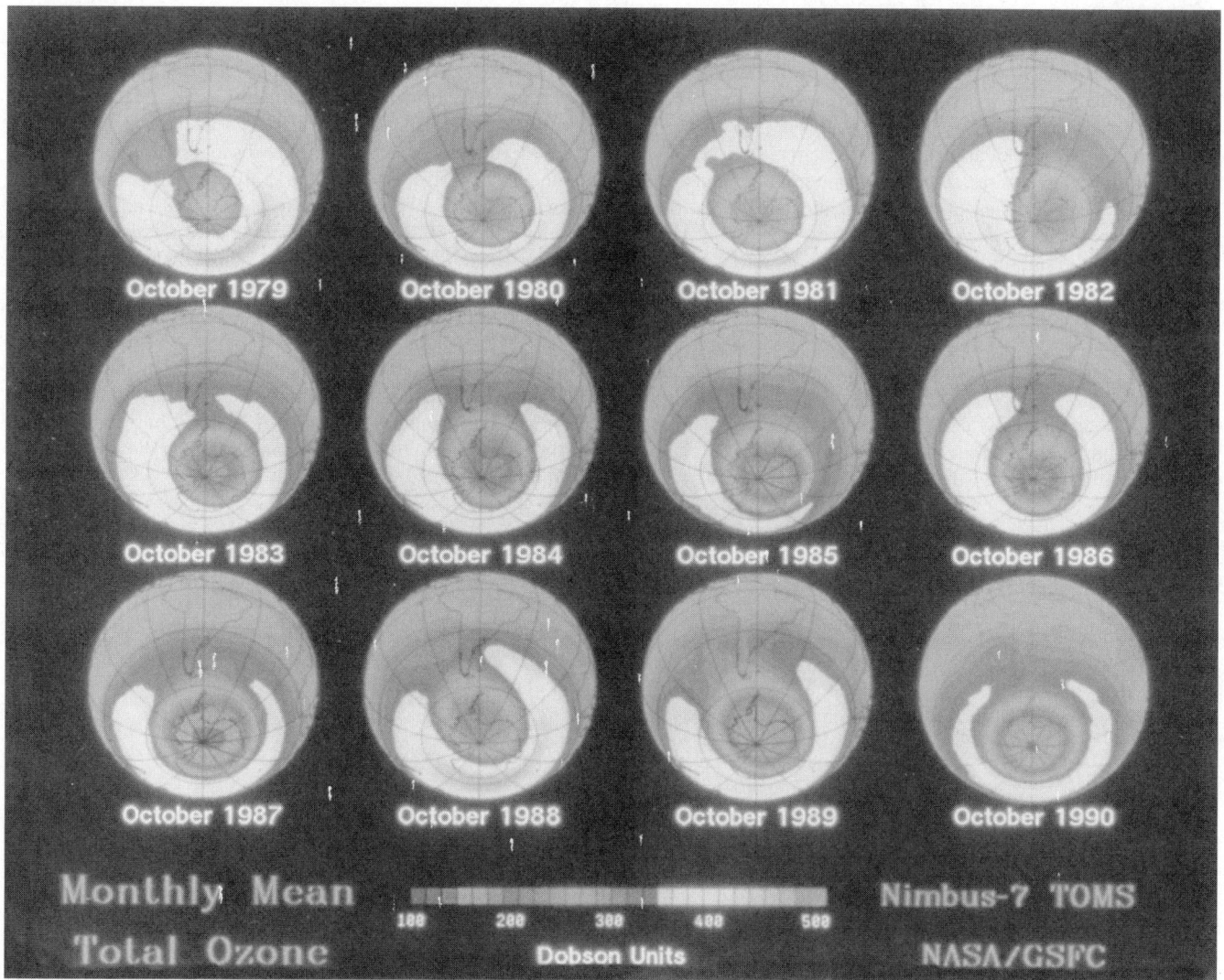

A series of NASA photos depicting ozone depletion over the South Pole from October 1979 to October 1990. *(Greenpeace. Reproduced by permission.)*

upper **atmosphere**, however, the situation is very different. Because it absorbs ultraviolet (UV) radiation very effectively, ozone in the stratosphere [that part of the atmosphere between about 9-30 mi (15-50 km) above Earth's surface] provides an irreplaceable protective shield. Without this layer, living **organisms** on Earth's surface would be subjected to life-threatening radiation burns and UV-induced genetic damage.

In 1985, a British Antarctic atmospheric research team announced a startling and disturbing discovery: stratospheric ozone levels over the South Pole were observed to drop precipitously during September and October every year as the sun reappeared at the end of the long polar winter. At its peak, almost all ozone disappeared at an altitude between 8.7-12.43 mi (14-20 km) over an area larger than the continental United States. This represents about 10% of all stratospheric ozone worldwide. Interestingly, this ozone depletion had been measured by sensors aboard satellites every year since 1960, but

it had been ignored by atmospheric scientists because they considered it impossible for so great a change to occur in such a remote and relatively pristine environment as the Antarctic.

Most scientists now believe that ozone depletion is caused by chlorinated **organic compound**s known as chloro-fluorocarbons (CFCs) released into the air primarily in the richer, industrialized countries far from the South Pole. Widely known by the trade name Freon, chlorofluorocarbons were once widely used as refrigerants, degreasing solvents, and spray-can propellants. A related family of compounds called halons were used mainly as fire extinguishers in places like libraries and museums where **water** sprinklers would damage sensitive materials. Inexpensive to manufacture, non-toxic, stable, and relatively inert, these chlorinated compounds were originally thought to be miracles of modern science. Their chemical stability turns out to be both a benefit and a scourge, however. It allows CFCs to persist for decades, even centuries in the atmosphere. They drift up into the stratosphere where conditions in the polar atmosphere result in an unusual set of **chemical reaction**s. Strong circumpolar winds that isolate Antarctica during most of the winter sweep up and concentrate contaminants. Exceptionally cold **temperature**s -121 to -130° F (-85 to -90° C) in Antarctica allow ice crystals to form at very high altitudes. These ice crystals bind contaminants and provide catalytic surfaces on which chemical reactions take place that would be slow or nonexistent elsewhere in the atmosphere. When the sun returns in the polar spring, ultraviolet radiation releases chlorine atoms that attack and destroy ozone molecules. Since the chlorine atoms are not themselves consumed in these reactions, they continue to deplete stratospheric ozone for years until they finally precipitate or are washed out of the air.

Although originally thought to be restricted only to Antarctica, it is now known that smaller amounts of stratospheric ozone depletion occurs in the Northern Hemisphere as well.

This is a serious concern because more people potentially will be exposed to excess UV radiation in the north than in the south. A 1% loss of ozone results in a 2% increase in UV reaching the earth's surface and is estimated to result in about a million excess human skin **cancer**s per year if it occurs in densely populated regions. The discovery of stratospheric ozone losses resulted in remarkably quick international response. At a 1989 conference in Helsinki, 81 nations agreed to phase out CFC production by the end of the century. There is some evidence that the CFC ban is already having an effect. The buildup of CFCs in the atmosphere is declining more rapidly than expected. If the phase out continues as promised and if developing nations join the rest of the world in replacing chlorinated compounds with alternatives, stratospheric ozone levels are expected to return to normal levels in about 70 years. This is an encouraging example of how nations can work together to solve environmental problems when the risks and solutions are clear.

P

PABA • See Para-aminobenzoic acid (PABA)

PALADE, GEORGE (1912-)
Romanian American cell biologist

George Emil Palade was born in 1912, in northeastern Romania. One of three children, Palade came from a professional family and earned his medical degree from the University of Bucharest in 1940. In 1941, he married Irina Malaxa; they eventually had two children together.

In 1945, after being discharged from the army, Palade obtained a research position at New York University. While there he met the eminent **cell** biologist **Albert Claude**, who had pioneered both the use of the electron **microscope** in cell study and techniques of cell fractionation (the separation of the constituent parts of cells by centrifugal action). The older scientist invited Palade to join the staff at the Rockefeller Institute (now Rockefeller University), and in 1946 Palade accepted a two-year fellowship as visiting investigator. Political instability in Romania caused Palade to stay in the United States permanently. He became a U.S. citizen in 1952 and a full professor of cytology at Rockefeller in 1958.

At the Rockefeller Institute, Palade and his collaborators reported groundbreaking descriptions of the fine appearance of the cell and of its biochemical function. Concentrating on the **cytoplasm**—the living material in the cell outside the **nucleus**—Palade was first attracted to larger **organelle**s (bodies of definite structure and function in the cytoplasm) which Claude had earlier called "secretory granules." Palade showed that these tiny sausage-shaped structures, **mitochondria**, are the site where energy for the cell is generated. **Animal** cells typically contain a thousand such mitochondria, each creating **adenosine triphosphate** (**ATP**), a high-energy phosphate **molecule**—through enzymic (**enzyme**-catalyzed) oxidation or breakdown of fat and sugar. The ATP is then released into the cytoplasm where it powers **energy**-requiring mechanisms such as nerve impulse conduction, muscle contraction, or **protein synthesis**.

Using the high-power electron microscope (a device that utilizes electrons instead of **light** to form images of minute objects), Palade next revealed a delicate tracery, subsequently termed the **endoplasmic reticulum**. The endoplasmic reticulum is a series of double-layered **membranes** present throughout all cells except mature erythrocytes, or red **blood** cells. Its function is the formation and transport of fats and **proteins**. By far Palade's most significant work was with so-called microsomes, small bodies in the cytoplasm that Claude had earlier identified and shown to have a relatively high **ribonucleic acid** (**RNA**) content. RNA is the genetic messenger in protein synthesis. Palade observed these microsomes both as free bodies within the cytoplasm, and attached to the endoplasmic reticulum. In 1956, using a high-speed **centrifuge**, Palade and his colleague **Philip Siekevitz** were able to isolate microsomes and observe them under the electron microscope. They discovered that these microsomes were made of equal parts of RNA and protein.

Palade assumed that these RNA-rich microsomes were in fact the factories producing protein to sustain not only the cell but the entire **organism**. The microsome was renamed the **ribosome**, and Palade and his team went to work to investigate the pathway of protein synthesis in the cell. Palade and Siekevitz began a series of experiments on ribosomes of the liver and pancreas, employing autoradiographic tracing, a sophisticated process similar to x-ray photography in which a picture is produced by radiation. Investigating in particular exocrine cells (those that secrete externally) of the guinea pig pancreas, the team was able, by 1960, to show that ribosomes do in fact synthesize proteins that are then transported through the endoplasmic reticulum. Further research elucidated the function of the larger ribosomes attached to the endoplasmic reticulum, establishing them as the site where **amino acids** assemble into **polypeptide**s (chains of amino acids).

Having completed his work on protein synthesis, Palade turned his attention to cellular transport—the means by which substances move through cell membranes. Working with Marilyn G. Farquhar, Palade demonstrated by electron mi-

Fossilized *Alethopteris*, a seed fern from the Pennsylvanian period. *(JLM Visuals. Reproduced by permission.)*

crography (images formed using an electron microscope) that molecules and ions were engorged by sacs or vesicles that move to the surface from within the cell. These vesicles actually merge with the outer membrane for a time, and then swallow up and bring the substances inside the cell. This vesicular model was in distinct contrast to the then current pore model whereby it was thought that molecules simply entered the cell through pores in the membrane.

Palade's later work at Yale University has been an attempt to establish links between defects in cellular protein production and various illnesses. In 1974 Palade shared the Nobel Prize in Physiology or Medicine with his former mentor, Albert Claude, and with **Christian R. de Duvé**, for their descriptions of the detailed microscopic structure and functions of the cell. In 1990, Palade left Yale to become the dean for scientific affairs at the University of California, San Diego.

PALEOBOTANY

Paleobotany is the study of **plants** that lived in prehistoric times. It largely involves the study of **fossils**, that is, impressions of plant parts that have been preserved in sedimentary rock or coal. The most ancient plant fossils are of microscopic **algae** that lived more than one billion years ago, during Precambrian times. Paleobotanists also study much younger plant fossils, such as pollen in recent lake sediment.

One of the goals of paleobotany is to discover the earliest occurrences of different kinds of plants in the geological record. This knowledge of sequential occurrence of taxa is then used to develop an understanding of the evolutionary relationships among groups of plants. Other paleobotanists are interested in determining what fossil plants were like, and the kinds of **animal**s that utilized them as food and **habitat**. This information can also be used to infer the characteristics of the ancient environment, including the type of climatic conditions

in which the plants grew. Sometimes paleobotanical knowledge can be used for quite practical purposes, such as the development of fossil-plant indicators that can be used to help locate underground deposits of coal or petroleum.

The most common samples studied by paleobotanists are microscopic **spore**s, pollen, and bits of **tissue**s. They also may identify larger, "macroscopic" remains, such as foliage and even substantial parts of fossil **tree** trunks. These various plant parts can usually be classified to their major group, such as an **order** or **family**. Plant fossils of **species** that still exist may be classified to the actual **genus** or even the species.

In addition to classifying the remains of their fossil samples, paleobotanists must know the context of fossils in terms of geological time. The relative age of fossil deposits can be inferred from geological stratigraphy (that is, the layers of rock or mud from which they were retrieved). An even more accurate process to measure age calculates the degree of **radioactive decay** of isotopes of certain elements.This carbon-dating method is based on the measurement of the amount of carbon-14 (compared with that of carbon-12) in an isotope, and is commonly used to date organic material up to 40-50 thousand years old. In some cases, recent fossils retrieved from deep lakes can be dated quite accurately, by calculating their position in annual layers of sediment (the highest, most recent layer corresponds to the present time).

Palynologists are paleobotanists who examine samples of lake sediment or bog peat of known age for fossil pollen. They carefully enumerate the pollen and classify it to the degree possible, which is often to species or genus. Palynologists use information on fossil pollen to make inferences about the types of local forest and other plant communities that may have occurred near where the fossils were found. Such interpretations must be made carefully, because species are not represented in the pollen record in the same proportions that they occurred in the local vegetation.The pollen of wind-pollinated species is much more abundant in lake sediment than that of insect-pollinated species. Reconstructions of local paleovegetation also allows inferences to be made about local climatic conditions at the time. For example, a typical palynological study might discover the following about the local environment around a lake in Wisconsin: about 15 thousand years ago the site supported plant species that are now typical of northern **tundra**; 10 thousand years ago the vegetation was a boreal forest of spruces and fir; and during the most recent centuries the pollen assemblage was dominated by oaks, maples, basswood, chestnut, and other relatively southern trees. These kinds of observations, coupled with knowledge of the present, climatically influenced distributions of these tree species, allows palynologists to make reasonable inferences about the historical plant communities and climates that occurred since the glaciers retreated from the region just over 15 thousand years ago.

PALEONTOLOGY

Paleontology is the study of plant and **animal** life in the geologic past. Paleontologists use **fossil**s to study past life forms.

They research what early **organism**s looked like, as well as their environments, their relationships to other life forms, and their correlations with modern **species**. Paleontology was crucial in building our current understanding of Earth's history. Prior to the **birth** of paleontology in the early nineteenth century, geologists had no way of determining the relative ages of rocks, but by comparing fossils that represented short-lived life forms, geologists were able to date rock strata more accurately, and put together a chronology of the earth. Paleontology also provided evidence for the theory of **evolution**.

The French biologist **Georges Cuvier** is considered the father of paleontology. Cuvier worked in the Museum of Natural History in Paris around the turn of the eighteenth century, and there he developed a keen interest in **comparative anatomy**. He was able to reconstruct partial skeletons, using his skill and intuition to posit the shape of missing bones. He perfected the **classification** system of **Carl Linnaeus**, grouping animals together according to their internal structure instead of their outer form. After completing his work with modern animals, Cuvier turned his attention to fossils. These skeletons of unknown animals frozen in stone still seemed to Cuvier to fit into his classification system. In 1796, he posited that a certain fossil skeleton was an extinct species of elephant, and later he related a giant fossil sloth to the smaller sloth species still living in South America. He identified and named the pterodactyl, though he apparently misidentified the first known discovery of dinosaur teeth as belonging to an extinct rhinoceros.

Unable to believe that ancient life forms could have evolved into modern ones, Cuvier argued instead that several prehistoric catastrophes, including the flood described in *Genesis*, had destroyed all life on Earth, and it had been created anew in differing forms. Cuvier sparked worldwide interest in fossils that eventually led to a non-Biblical interpretation of the formation of the earth and its creatures. Another Frenchman, Éduard Lartet (1801-1871) found a fossil mammoth tooth inscribed with a drawing in a cave in France, giving clear evidence that humans too had existed in the era of these strange extinct animals. Lartet's discovery made Cuvier's theory untenable, and opened the door for **Charles Darwin**.

Further studies unearthed fossils found to be over 600 million years old. These were the oldest fossils known until 1965, when an American paleontologist, Elso Sterrenberg Barghoorn (1913-1984) found bits of carbonized material in ancient rocks that seemed to be fossil **bacteria**. Barghoorn's microfossils were found to be perhaps 3,500 million years old, dating back to a billion years after the formation of the earth.

Prior to modern dating methods, paleontologists used relative dating techniques to determine the approximate age of a fossil. Stratigraphy is the study of layers of rocks or the objects embedded within those layers. It is based on the assumption (which nearly always holds true) that deeper layers were deposited earlier, and thus are older, than more shallow layers. The sequential layers of rock represent sequential intervals of time. Seriation is the ordering of objects according to their age. Artifact styles such as pottery types are seriated by analyzing their abundances through time. Seriation assumes that all differences in artifact styles are the result of different periods of

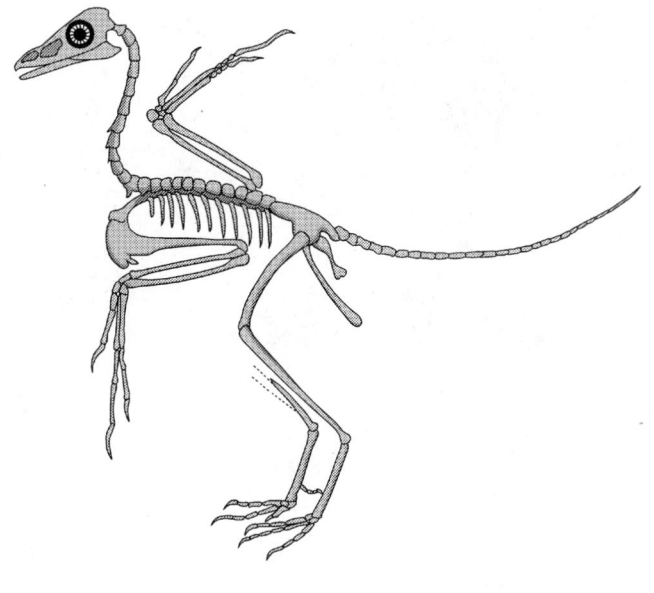

The pterosaur *Ramphorhynchus gemming*. (Illustration by Hans & Cassady, Inc.)

time, and are not due to the immigration of new **culture**s into the area of study. Faunal dating uses animal bones to determine the age of sedimentary layers or objects such as cultural artifacts embedded within those layers. Faunal dating works best if the animals belonged to species which evolved quickly, expanded rapidly over a large area, or suffered a mass **extinction**. Another relative dating technique that utilizes that fact that **plants** annually spread pollen over a given area and, in certain areas such as lake beds, that pollen is more likely to be preserved is pollen dating. Using pollen dating or **palynology**, scientists can develop a pollen chronology by noting which species of pollen were deposited earlier in time and fossilized within rock layers.

Modern paleontology uses the most advanced chemistry and microbiology, including **deoxyribonucleic acid** (**DNA**) research, to date samples. Known as absolute dating, such techniques include **amino acid** racimization, cation-ratio dating, thermoluminescence dating, **tree**-ring dating, **radioactive decay** dating, and uranium series dating. Popular during the 1970s, amino acid racimization is based on the principle that amino acids (except glycine) exist in two mirror image forms called stereoisomers. Living organisms synthesize and incorporate only the L-form into **protein**s. This means that the ratio of the D-form to the L-form is zero (D/L=0). When these organisms die, the L-amino acids are slowly converted into D-amino acids in a process call racimization.. The reversible re-

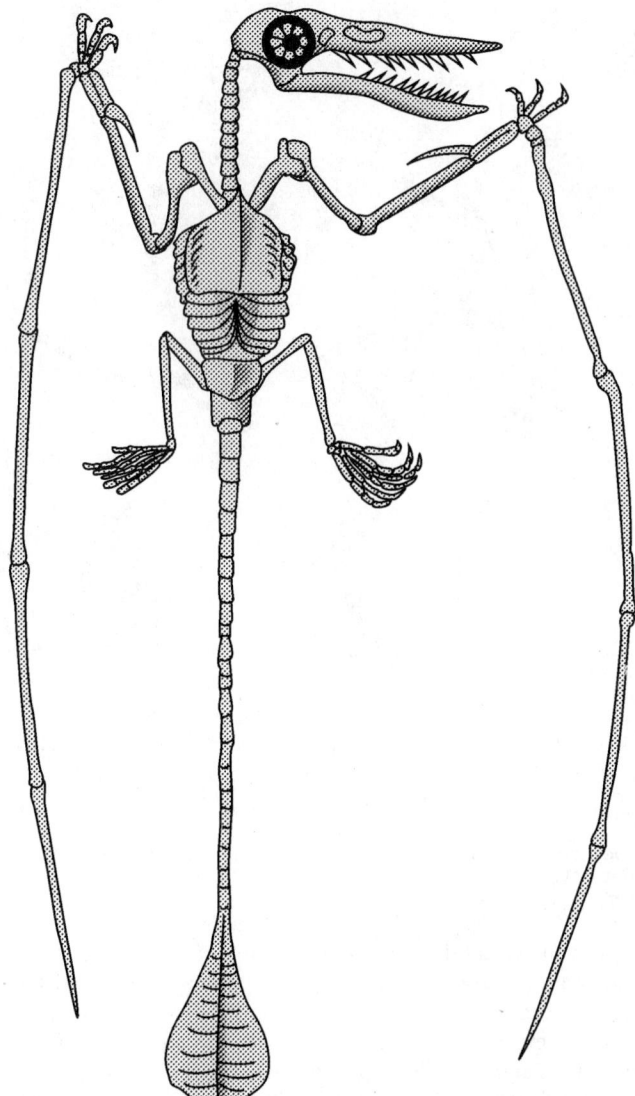

The first bird, *Archaeopteryx*. (Illustration by Hans & Cassady, Inc.)

the amount of emitted light, scientists calculate how much time has passed since the pottery was fired.

Also known as dendrochronology, tree-ring dating is based on the fact that trees produce one growth ring per year. The tree-ring patterns are all the same for a given species and geographical area. With the patterns from trees of different ages, a master pattern can be created and used to date buildings and archaeological sites. Tree rings are also used to date climate changes because each ring's thickness indicates the type of climate present for that particular year.

Radioactive decay is a process in which a radioactive form of an element is converted to a nonradioactive product at a constant rate. Radioactive decay dating techniques include potassium-argon dating and radiocarbon dating. When volcanic rocks are heated at very high temperatures, they release argon gas. As they cool, argon-40 (^{40}Ar) begins to accumulate. Argon-40 forms in the rocks from the radioactive decay of potassium-40 (^{40}K). The amount of ^{40}Ar formed is proportional to the decay rate of ^{40}K, which is 1.3 billion years. Since it takes a very long time to create a measurable amount of ^{40}Ar, this technique is only appropriate for rocks greater than three million years old. Radiocarbon dating is used to date charcoal, wood, and other biological materials. The range of conventional dating is 30,000-40,000 years, but with sensitive equipment this range can be extended to 70,000 years. Radiocarbon (^{14}C) spontaneously decays into **nitrogen**-14 (^{14}N). Both plants and animals contain carbon, and while alive, they sustain the same ratio of $^{14}C/^{12}C$ that is present in the **atmosphere**. When an organism dies, however, the ratio begins to change as the ^{14}C decays into ^{14}N. The rate at which the decay occurs is called half-life and refers to the time required for ^{14}C to decay into ^{14}N. The half-life of ^{14}C is 5,730 years, and by comparing the ratio of ^{14}C to ^{14}N within the organism's remains, the amount of time since the organism's **death** can be determined.

Uranium series dating utilizes the fact that radioactive uranium and thorium isotopes decay into a series of unstable, radioactive ''daughter'' isotopes until a stable lead isotope is formed. Parent isotopes have half-lives of several thousand million years, while daughter isotopes have half-lives ranging from a few hundred thousand years to only a few years. Daughter deficiency or daughter excess methods can be used to date a sample. In daughter deficiency situations, the parent radioisotope is initially deposited by itself. As the parent decays, it reaches the point where it contains the same amount as the daughter, and they are in equilibrium. The age of the deposit can be calculated by measuring how much the daughter has formed, providing that neither isotope has entered or exited the deposit after its formation. In the case of a daughter excess, a larger amount of the daughter is initially deposited than the parent. Over time, the excess daughter disappears as it is converted back into the parent, and by measuring the extent to which this has occurred, scientists can date the sample. Another form of uranium series dating is fission tracking dating, which considers the tracks made in volcanic minerals and glass from the fission of uranium-238 (^{238}U) **atoms**. The rate at which fission occurs is proportional to the decay rate of ^{238}U and its measurement is determined by the half-life of the element. The half-life of ^{238}U is 4.47×10^9 years.

action eventually creates equal amounts of L- and D-forms (D/L=1.0), which can help date a sample. Since amino acids react at different rates, accurate dating is often difficult.

Cation-ratio dating is a technique that examines how long rock surfaces have been exposed. Since cations move throughout the environment at different rates and the ratio of different cations to each other changes over time, cation ratio dating relies on the principle that the cation ratio ($K^+ + Ca^{2+}$)/Ti^{4+} decreases with increasing age of the sample.

Thermoluminescence dating is used to determine the age of pottery. When electrons from the **minerals** within pottery clay are exposed to radioactive substances that are present in the clay or burial medium, they are bumped out of their normal positions. When the pottery is then heated at a very high **temperature** (over 932°F or 500°C), the electrons fall back to their normal state, emitting **light**. The longer the exposure to radioactive material, the more light is emitted. By measuring

Presently, DNA has been examined by scientists in order to date fossil specimens. For example, paleontologists in the 1990s examined the rare carbon 12 and carbon 13 isotopes buried in ancient ocean sediment to trace early climactic activity. Other scientists have been using animal DNA as a "molecular clock" that can date when common species began to diverge. In many cases, the fossil record and the DNA record are at odds, sparking competitive theories between microbiologists and traditionally fossil-bound paleontologists. Much is still unknown about Earth's history. Thus paleontology, with ever more sophisticated tools at its disposal, is still a vital and intriguing field of study.

See also Evolutionary theory

PALYNOLOGY

Palynology is the study of **fossil** pollen (and sometimes plant **spore**s) extracted from lake sediment, peat bog, or other matrices. The most common goal of palynological research is to reconstruct the probable character of historical plant communities, inferred from the abundance of **species** in dated portions of the fossil pollen record. Pollen analysis is an extremely useful tool for understanding the character of ancient vegetation and its response to changes in environmental conditions, particularly in climate. Pollen analysis also has an economically important modern industrial use in the exploration for resources of **fossil fuels**. Palynology is also used to help reconstruct the probable **habitat**s and foods of ancient humans and of wild **animal**s.

Pollen consists of microscopic grains containing the male **gametophyte** of coniferous (cone-bearing) and angiosperm **plants**. Pollen of most species of plants undergoes a long-distance dispersal from the parent plant, so that **fertilization** can occur among individuals (instead of self-fertilization). A plant spore is a kind of reproductive grain capable of developing as a new individual, either directly or after fusion with another germinated spore, such as the kind produced by ferns, horsetails, and club-mosses. Spores with simpler functions are produced by mosses, liverworts, **algae**, **fungi**, and other less complex **organism**s.

The pollen of many plants can be classified by **genus**, and sometimes by species, on the basis of such characteristics as size, shape, and surface texture. In contrast, most spores can only be classified by higher taxonomic levels, such as **family** or **order**. Both pollen and spores are well preserved in lake sediment, peat bog, and many archaeological sites. Fossil pollen has even been identified from the bodies of extinct animals, such as mammoths discovered frozen in arctic permafrost (permanently frozen subsoil).

Plant species in the pollen record of lake sediment and peat are not represented in the same relative abundance they are in the nearby vegetation. Wind-pollinated plant species are most abundant, because these plants release huge amounts of pollen into the environment. For example, many species of pines, which are wind pollinated, are so prolific that during their flowering season a yellow froth of pollen may occur along the edges of lakes and ponds. Insect-pollinated plant species are more rare. The great differences in pollen production among plant species must be taken into account when interpreting the likely character of local vegetation on the basis of the fossil-pollen record.

Palynologists need to understand the historical context of their samples. The common method used to determine the age of samples of mud and peat is radiocarbon dating. This technique is based on the fact that after an organism dies, it no longer absorbs carbon-14 from the **atmosphere**. Because carbon-14 is a rare, radioactive isotope of carbon (that is, it "decays" into simpler isotopes or elements), its amount in dead **biomass** decreases progressively with time. This change can be used to estimate the age of organic material by calculating the ratio of carbon-14 to stable, non-radioactive carbon-12. Radiocarbon dating is effective for samples aged between 150 and 40-50 thousand years. Younger samples may be dated on the basis of their content of lead-210, and older samples using elemental isotopes with longer half-lives.

A typical palynological study might involve the collection of one or more cores of sediment or peat from a site. The layers occurring at various depths would be dated, and samples of the pollen grains contained in the layers would be extracted, classified, and enumerated. From the dated fossil pollen of various species or genera, the palynologist would develop inferences about the nature of the **forests** and other plant communities that may have occurred in the local environment at the time.

In the northern hemisphere, many palynological studies have been made of changes in vegetation occurring since the continental-scale glaciers melted back, beginning about 12-15 thousand years ago. A commonly observed pattern from the pollen record is that the oldest samples, representing recently deglaciated times, indicate plant species that are now typical of northern **tundra**, while somewhat younger samples suggest a boreal forest of spruces, fir, and birch. The pollen assemblage of younger, more recent samples is generally dominated by temperate **tree**s such as oaks, maples, basswood, chestnut, hickory, and other species that now have a relatively southern distribution. There may also, however, be indications in the pollen record of occasional climatic reversals, such as periods of cooling that interrupt longer, warm intervals. The most recent of these cool periods was the "Little Ice Age" that occurred between about 1550 and 1850. However, palynology has also detected more severe climatic deteriorations in the past, such as the "Younger Dryas" event that began about 11,000 years ago, causing a re-development of glaciers in many areas and temporarily reversing post-glacial vegetation development.

PANCREAS

The pancreas is a digestive **organ** which develops as an offshoot of the intestinal tract. It is unique in that it has both endocrine and exocrine functions. As an exocrine organ, it sends some of its chemical products through ducts (tubes) to their

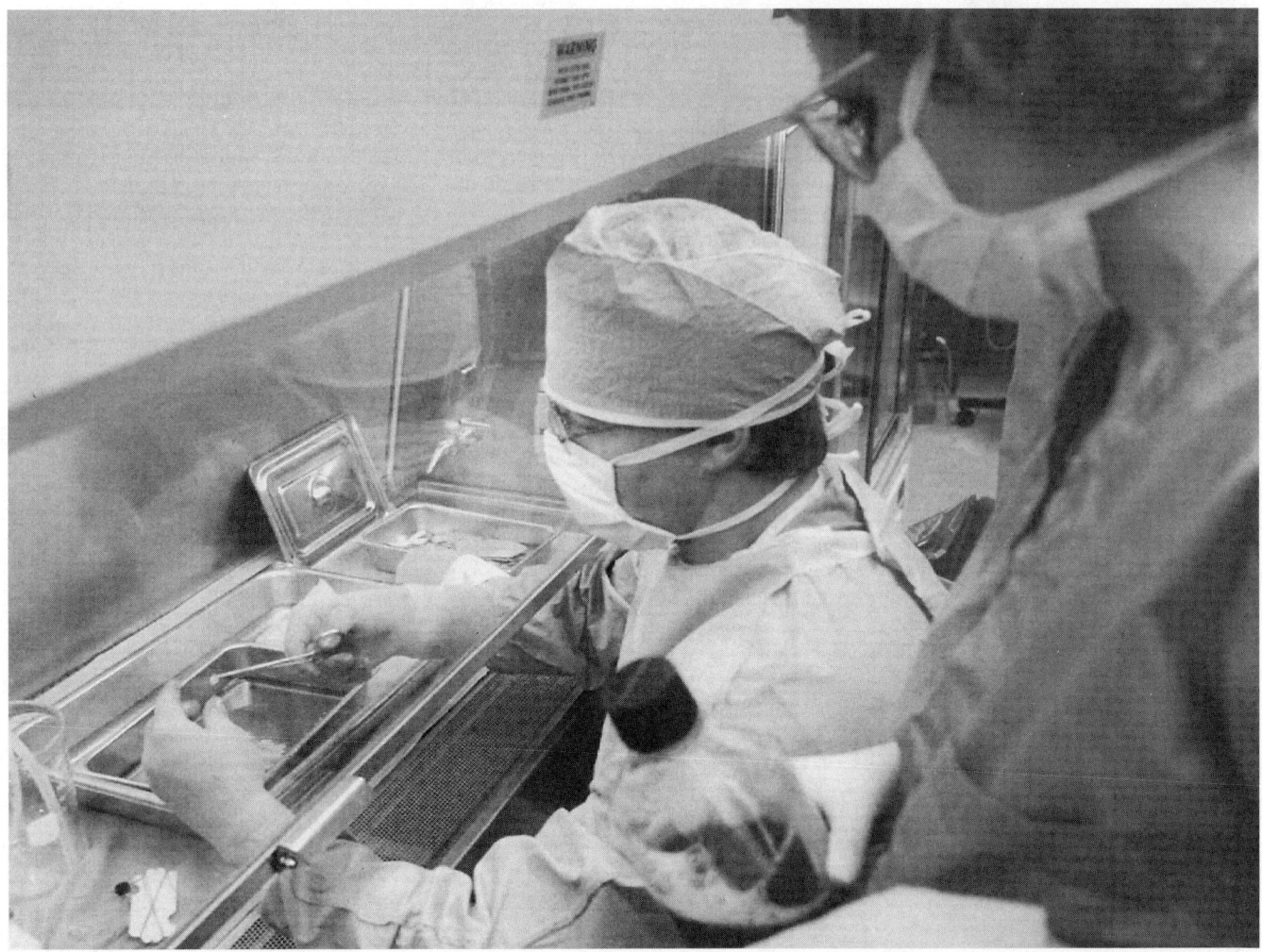

A surgeon harvests the islets of Langerhans from a donor pancreas. *(Photograph by Daniel Portnoy. AP/Wide World Photo. Reproduced by permission.)*

destination. As an endocrine organ, it sends some of its chemical products through the bloodstream to act on **tissue**s and organs at some distance from the pancreas itself.

The exocrine products of the pancreas enter the duodenum (the first part of the small intestine) through the pancreatic duct. The pancreatic juices are a combination of various **enzyme**s which help digest foods which have arrived at the duodenum from the stomach. Pancreatic enzymes break down **carbohydrate**s, **protein**s, and **fats**. The pancreas also produces bicarbonate, an alkaline substance which helps to neutralize the food which has become quite acidic in the environment created by the stomach.

As an **endocrine gland**, the pancreas plays a vital role in the processing of sugars (glucose). The pancreas contains specialized **cell**s, called the islets of Langerhans, which produce two important **hormone**s: **insulin** and glucagon. Both of these hormones serve to maintain the appropriate levels of glucose circulating within the bloodstream.

Diabetes is a common and serious condition. Some forms of diabetes occur when the islets of Langerhans are de-

creased in number or destroyed, and the pancreas can no longer produce enough insulin to adequately regulate glucose levels. Another form of diabetes occurs when the cells of the body become unresponsive to the presence of circulating insulin. Either way, the end result is that glucose levels in the bloodstream can reach extremely high levels, resulting in damage to a variety of organ systems throughout the body.

PANTOTHENIC ACID

Derived from the Greek word *pántothen*, meaning "from all quarters," pantothenic acid ($C_9H_{17}NO_5$), generally referred to as vitamin B_5, is an important factor in metabolic processes, particularly in the **Krebs cycle**. Historically, it has also been known as chick antidermatitis factor, Bios IIa, and the anti-gray-hair factor.

First discovered at the turn of the twentieth century and described as Bios, pantothenic acid was known to be essential for **yeast** growth. R.J. Williams and coworkers isolated crystalline Bios from yeast and named it pantothenic acid in 1933. A few years later, in 1938, Williams isolated the same substance from liver. By 1939, pantothenic acid was determined to be chemically identical to the antidermatitis factor found in chicks. Beta-alanine played an important role in pantothenic acid production, and in 1940, a method to synthesize and crystallize pantothenic acid was established. In 1950, acetyl **coenzyme** A, a derivative of pantothenic acid which reacts with enters at the start of the Krebs cycle and reacts with oxaloacetic acid to form citric acid, was discovered. Its structure and relationship to pantothenic acid was explained the following year.

Vitamin B_5 can be naturally derived from a variety of plant and **animal** sources. Yeasts, animal glands and **organ**s, ground peanuts, herring and cod, royal jelly from bees, and wheat bran and germ are typically rich in pantothenic acid. Moderate amounts are present in avocados, lima beans, broccoli, carrot, cauliflower, cheese, clam, lentils, mushroom, oats, rice, and soybeans. Small amounts can be found in **fruits** (such as apples, bananas, grapes, lemons, oranges, pineapples, plums, peaches, and pears); vegetables (including kidney beans, cabbage, lettuce, onions, peppers, potatoes, tomatoes, turnips, and watercress); and in some seafood (such as shrimp and lobster.) Pantothenic acid can also be produced synthetically via a **chemical reaction** involving the condensation of d-pantolactone with salt of p-alanine. To increase stability and make it more useable in a wide variety of products, B_5 is often sold in other forms including calcium pantothenate, dexpanthenol, and panthenol (which is also known as pantothenyl **alcohol** or provitamin B_5). Precursors in the **biosynthesis** of pantothenic acid include a-ketoiso-valeric acid (pantoic acid), uracil (p-alanine), and aspartic acid. Intermediates in the synthesis include ketopantoic acid, pantoic acid, and P-alanine.

Research has shown that pantothenic acid has several interesting properties. It promotes **amino acid** uptake; in combination with zinc, it can prevent hair from turning gray in rats; it can promote resistance to stress of cold immersion; it may be tied to tumor inhibition; it is necessary in the hatching process of **egg**s; it is useful in treating vertigo, postoperative shock, certain types of poisoning, and in accelerating wound healing; it is also useful in treating Addison's disease, liver cirrhosis, and diabetes; it is key to the **metabolism** of **carbohydrate**s, **fat**s, and **protein**s; and it is also involved in making fatty acids, **cholesterol**, steroid **hormone**s, and nerve regulators. Perhaps its most important role is in **cell**ular metabolism as a constituent of coenzyme A, which is a factor in the Krebs cycle.

Para-aminobenzoic acid (PABA)

One of the most important functions of para-aminobenzoic acid (PABA) is to serve as a nutrient for the intestinal tract's numerous **microorganism**s. When necessary—because of deficiencies in the diet, for instance—PABA also increases the intestinal synthesis of other B **vitamins**, particularly of folic acid, of which it is a structural unit.

But PABA has another important distinction. In 1940, the year **Richard Kuhn** isolated PABA, Kuhn and other researchers soon learned that the new compound appeared to reverse the action of sulfanilamide, one of the early sulfa drugs. When the vitamin and the antibacterial were compared, the two proved to have strikingly similar structures. Surprised but intrigued, Kuhn (and others) speculated that the two compounds might somehow be in **competition**—a competition that, in this case, PABA was clearly winning.

To help them find answers, researchers began to synthesize compounds that were almost chemically identical to other vitamins. They quickly learned that some of these new synthetics appeared to act as antivitamins: when fed to **animal**s, the synthetics caused a vitamin deficiency state that would, in many cases, be reversed by feeding the original vitamin. More importantly, the discovery of antivitamins made other beneficial advances possible. The synthesis of antivitamin K, for instance, led to the manufacture of dicumarol, an agent that helps prevent the formation of dangerous **blood** clots. The synthesis of anti-PABA compounds eventually produced sulfa drugs that could effectively treat **bacterial** infections in most cases, because they had the ability to successfully compete with PABA for positions in a **coenzyme** necessary for bacterial reproduction.

Parabiosis

Twins born with their bodies fused or joined together, known as Siamese twins, are a form of naturally occurring parabiosis. The area of fusion may be slight or extensive in these rare **birth**s and sometimes the twins may be successfully separated by extraordinarily skillful surgery. The term parabiosis refers to any two **organism**s that are cojoined and the term is frequently used to describe experimental parabiosis.

Sometimes it is desirable to produce parabiotic **animal**s for research purposes. Parabiosis as an experimental procedure dates back to the end of the last century. Amphibians have often been the animals of choice for such experimentation. Peter Volpe (1980) used the procedure extensively in an elegant series of studies concerning transplantation immunity. Volpe chose the common leopard frog, *Rana pipiens*, as an experimental animal because the well known reproductive biology of the **species** permits ease in obtaining many **embryo**s which develop outside of their mother's body. In Volpe's experiments, a portion of **ectoderm** is removed from the gill forming areas of two embryos. This surgery is performed under a dissecting **microscope**. The operated embryos are snugly held together in an operating dish and healing occurs in three to five hours. The choice of the gill area as the site of fusion is to insure that **blood** will intermingle in the parabiotic twins. Ordinarily, the parabiotic embryos remain attached through **metamorphosis** or beyond. Unrelated frogs usually reject transplanted **tissue** just as humans would do with transplanted tissue not protected with a immune suppressing drug. Unrelated frogs, joined in parabiosis, do not reject each others' grafted tissue. This shows that immune **tolerance** can be induced during embryonic development by means of shared blood resulting from parabiosis.

The technique has been used in studying, in addition to the development of immune tolerance, the migration of primordial germ cells, the migration of malignant cancer cells, the characterization of the stability of the differentiated stage, and many other problems. As long as there are experimenters with a steady hand and a creative mind, parabiosis will be one procedure that may serve to produce new understanding.

PARACELSUS (1493-1541)
German-Swiss physician and alchemist

The flamboyant physician, chemist, teacher and vagabond who called himself Paracelsus was born in Switzerland with the name Philippus Aureolus Theophrastus Bombastus Von Hohenheim. His mother died when he was very young, and he moved to Villach in southern Austria with his impoverished physician, chemist father. As a bright young student he dreamed that he might some day discover how to transmute lead into gold, a long sought goal of the alchemists. He left home at the age of 14, and wandered through Europe visiting university after university. By the time he was 20, he is said to have attended the universities of Basel, Tübingen, Vienna, Wittenberg, Leipziq, Heidelberg and Cologne. He was highly critical of all of them. His opinions angered academics of his day, which is not hard to understand when it is noted that at one point he wrote that he wondered how "the high colleges managed to produce so many high asses."

In spite of his low regard for the universities of his day, he is said to have earned a baccalaureate in medicine from the University of Vienna in 1510 at the age of 17. He claimed to have earned a doctoral degree from the University of Ferrara in 1516, although this can not be verified because university records for that year are missing. He was pleased to find that authorities at that university were critical of the accepted medical teaching of the Greek scholar, Galen, and medieval Arab teachers such as Avicenna. He felt free to criticize the popular view that stars and planets had a controlling influence on the human body.

After taking his doctoral degree, he traveled extensively quickly wearing out his welcome at many of the universities of his day. His travels took him to England, Scotland, Ireland, Russia, Lithuania, Hungary, Italy, Egypt, Arabia, the Holy Land, and Constantinople. Everywhere he went he sought out practitioners of practical alchemy, in an effort to discover the most effective medical treatments for various disorders. He wrote "the universities do not teach all things, so a doctor must seek out old wives, gypsies, sorcerers, wandering tribes, old robbers, and such outlaws and take lessons from them." He began calling himself para-Celsus apparently because he considered himself superior to Celsus, a noted first century Roman physician. Celsus medical treatise *De medicina* was considered a classic, though it had been overlooked by his contemporaries in first century Rome. It was discovered by Pope Nicholas V and was one of the first medical references published (1478) after the invention of the printing press. Thus Paracelsus was quite brash to place himself above this legendary figure.

When he returned home from his wandering in 1524 he found that his fame as a physician preceded him. Reports of his miraculous cures earned him a position as lecturer in medicine at the University of Basel. Students from all over Europe enrolled in his classes. When he advertised his classes with a public posting, and invited everyone, students and non-students alike, the authorities were enraged. Shortly thereafter, in the midst of cheering students, he burned the books of Avicenna and Galen in front of the university. Paracelsus stressed the importance of natural healing and railed against many of the popular methods of healing wounds, which he claimed prevented natural cures. He taught that nature would heal a wound by itself if infection were prevented

Ultimately, Paracelsus's outrageous behavior alienated nearly all of the authorities of his day, and he was forced to flee the University under threat to his life. In poverty, he wandered from place to place staying with friends, revising old manuscripts and writing new ones. In 1536 his efforts reached culmination with the publication of *Der Grossen Wundartzney*, a medical treatise that quickly restored and enhanced his reputation. His triumph was brief, however, as he died in 1541 under mysterious circumstances. It has been widely speculated that a jealous or offended colleague may have murdered him although no crime was ever proven.

Despite his difficult nature, Parcelsus's contributions to medicine were significant. He established a role for chemistry in medicine, introducing the use of chemicals rather than herbs for the treatment of disorders. For example, he wrote a definitive clinical description of syphilis and advocated treatment with measured doses of mercury compounds, a recommendation that preceded by many years a procedure that became standard medical practice in the early years of the twentieth century.

PARADIGM

In science and philosophy in the second half of the twentieth century, the word "paradigm" has become strongly associated with the 1962 book, *The Structure of Scientific Revolutions*, by Thomas Kuhn, in which he described a paradigm shift as a major change in the way science is viewed and practiced. But paradigm has been used for a long time in a variety of ways in biology and other sciences and has been applied at a number of levels of biological organization. Claims for the creation of paradigms occur too frequently in the literature of biology for all of them to be considered as shifts in the way science is done or as revolutions in a scientific world view. Instead, frequent claims are made in the normal on-going business of science for a paradigmatic analysis of this concept or a paradigmatic application of that idea.

Examples of recent specific claims for paradigms in biology include: *Drosophila* as a paradigm for genetic sex determination; self-incompatability (the cellular recognition system that limits inbreeding) as a paradigm for the study of cell-to-cell communications in plants; identification of the smallest cell genome to provide a useful paradigm for thinking about the origin of cells; the use of social tutoring in white-crowned sparrows as a paradigm of song development in some **species** of birds; the lactose permease of *Escherichia coli* as a paradigm for **membrane** transport **proteins**; the relationship between sea otters and kelp **forests** in Alaska as a paradigm for **community** ecology.

These claims are not intended to reconstruct a broad pattern for thinking about the world scientifically, nor are they claims for a paradigmatic shift in science. Instead, they are claims for normal biological science, for particular methods in particular experiments that the researchers believe will prove to be useful in similar work. Most biologists, perhaps most scientists, then, commonly use the word paradigm to describe methods they see as exemplars, patterns worthy of imitation, or models to be adopted by other researchers doing similar work.

Kuhn used the term paradigm in much the same way as in the examples above, stressing that paradigms provide models from which spring particular coherent traditions of scientific research. The study of such specialized paradigms is what prepares science students for membership in their particular scientific fields. But he also stressed paradigm shifts, a process in which the accumulating knowledge in a field slowly but clearly indicates weaknesses in a dominant or central paradigm of that discipline, resulting in its being discarded and replaced by a new dominant paradigm.

Such Kuhnian claims have been made for paradigm shifts in biology. The most obvious example is that most biologists now agree that **Charles Darwin**'s account of evolutionary change precipitated a paradigm shift in the way scientists view **nature** and how it works. A theory of autopoiesis has recently emerged as a paradigm meant to address the questions of what a living system is and what **cognition** is. **Holism** is considered by some scientists, especially in **ecology**, as an emerging paradigm complementing (though not replacing) the established reductionist world-view of many biologists. A recent issue of the journal *Nature* claimed that **molecular biology**, in which **DNA** sequences can now be worked out nearly completely, should be considered the latest such paradigm shift. If none of these fit exactly the sequence described by Kuhn, their emergence has certainly shaped the way research is done in the multiple disciplines of biology, and the way biologists view their science.

PARASITE

A parasite is a plant or **animal** that lives and feeds on another plant or animal. The existence of parasites was recognized as far back as 1500 B.C. The Ebes papyrus of that date mentions hookworm infections among the Egyptian royalty. Greek physicians also knew of parasites. **Hippocrates**, for example, wrote about intestinal disorders associated with flatworms. The association of parasites with so many kinds of infection led many physicians of the Middle Ages to conclude that parasites were actually created spontaneously as a result of disease. Thus, as scientists began to develop a better understanding of the role

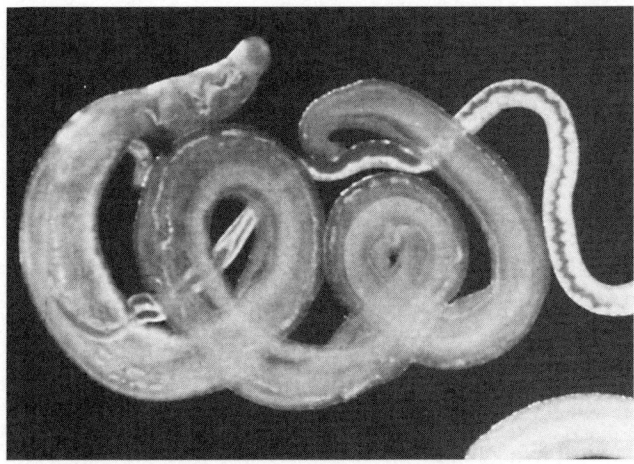

Male and female *Schistosoma masoni,* parasitic blood flukes that cause schistosomiasis. *(Photo Researchers, Inc. Reproduced by permission.)*

of parasites in causing disease in the eighteenth century, these discoveries also contributed to the downfall of the theory of **spontaneous generation**.

Probably the earliest argument for parasites as the cause of a disease appeared in a 1687 paper by the Italian biologist Giovan Cosimo Bonomo (1666-1696). Using a simple **microscope**, Bonomo found "minute living creatures" associated with the infection known as scabies. He described in detail how the animal reproduces, how it infects the skin, and how it is transmitted from person to person. Quite remarkably, Bonomo's research remained almost the only work on parasitic diseases for a century and a half.

Then after 1830, improved microscopes made possible a better understanding of such diseases, and the science of parasitology developed rapidly. The American biologist Joseph Leidy (1823-1891) discovered the presence of parasitic worms in hogs and cats in 1879 and recognized a parasitic amoeba at about the same time. The French chemist **Charles Louis Alphonse Laveran** isolated the parasites that cause malaria and traced their **life cycle** in and out of their hosts. The French zoologist Félix Dujardin (1801-1860) conducted research on flatworms that is generally regarded as the forerunner of modern parasitology.

Credit as founder of this field of science is often given to the German zoologist Karl Georg Friedrich Rudolf Leuckart (1822-1898). Leuckart carried out extensive research on the role of insects, tapeworms, flukes, and other **organisms** in human disease. He published his discoveries in a monumental two-volume work between 1862 and 1876. Other researchers focused on the etiology of specific parasitic diseases. The German zoologist Fritz Schaudinn (1871-1906), for example, developed new staining techniques that revealed to him the role of parasites in **syphilis**, dysentery, and a variety of tropical diseases. In one of the most remarkable of such discoveries, the Scottish physician Sir Patrick Manson (1884-1922) showed in the early 1900s that elephantitis is caused by parasitic worms.

Since the turn of the century, dozens of parasitic infections caused by **protozoa**ns, flatworms, and **invertebrates** have been studied. Causative agents, methods of transmission, and cures have been found for the vast majority of these diseases. New parasitic infections keep turning up, however. For example, in 1996, Stanford pathologist Luis Fajardo and his colleagues reported on a lethal infection that they think may be due to a new and particularly nasty tapeworm.

PARASYMPATHETIC NERVOUS SYSTEM ·

See Nervous system

PARTHENOGENESIS

In most living things, **fertilization** occurs when the **gamete**s of two compatible creatures are brought together. In **mammals**, these gametes are the (female) **egg** and the (male) **sperm**. When they meet, the egg and sperm will fuse into one body, allowing genetic information from the two parents to combine. As the offspring grows, it displays genetic traits similar to each of its parents, as well as traits uniquely its own. In this way, genetic diversity is preserved. However, there exists in **nature** a phenomenon known as *parthenogenesis* ("virgin birth"), in which an egg **cell** can develop into an offspring without being activated by a sperm. In such cases, the offspring, as well as its siblings, would be genetically identical to the mother. Parthenogenetic offspring can be male or female, and **haploid** (chromosomally normal) or **diploid** (chromosomally doubled), depending upon the **species**.

Natural parthenogenesis was discovered by the Swiss naturalist **Charles Bonnet** in the early 1740s, who noted the phenomenon in spindle-tree aphids. Still a very young man, he announced his discovery and quickly went on to other research. It was not until much later in life that Bonnet, his eyesight failing, turned his attentions toward the philosophical ramifications of his discovery. Since a sperm was not necessary for reproduction, he surmised, then the egg must be the key to the preservation of the species. He went on to hypothesize that within every egg there existed a daughter creature, perfectly formed, that contained within it another egg containing another daughter, *ad infinitum*.

While this theory explained the phenomenon of parthenogenesis, it seemed to indicate that every species was genetically fixed, without the possibility for evolutionary change. This was highly disputed by archaeologists, who had found **fossil** remains of creatures unlike anything seen in our time. Bonnet explained that these creatures had perished in worldwide catastrophes; these catastrophes happened periodically, wiping out all life on the planet. After each catastrophe, life on earth would "step up" a notch on the evolutionary ladder, remaining at that level until the next catastrophe. Bonnet continued by asserting that further catastrophes would advance apes into men and men into angels. This **evolutionary theory** became known as **catastrophism**.

In addition to its philosophical impact, the discovery of parthenogenesis created a stir among biologists who feared that, since eggs could reproduce alone, the male gender might be superfluous. As scientists began to learn more about **genetics** and **heredity**, however, it became clear that the parthenogenetic offspring were genetically inferior to those produced through fertilization.

Research on parthenogenesis continued for many years, and in the early years of the twentieth century, two American scientists, working separately, discovered that parthenogenesis could be triggered artificially. Working with near-transparent sea urchins, Jacques Loeb and **Ernest Just** each used a variety of methods to activate an egg cell without the presence of a sperm. They found that **temperature** changes, seawater solutions, diluted acids, and even thin needles could be used to stimulate parthenogenetic reproduction.

By observing the offspring of this artificial parthenogenesis, scientists were able to study the importance of the egg and sperm in the transfer of genetic information. For example, it is now known that the sperm plays little part in the formation of the **embryo**—its only essential task is to contribute **chromosome**s to the offspring.

PARTURITION

Parturition is a term that refers to labor and **birth**. **Mammals** in general and humans in particular give birth to living young. Accordingly, the procedure of bringing forth an infant from the uterus (womb) has been with human society as long as there have been humans.

Because birth involves great changes in the biology of the newborn and a significant change in the uterus, there is potential hazard associated with the process. The large **placenta** with a diameter of 6-8 in (15-20 cm) is physically separated from the richly vascular uterus. The separation process causes a potential hazard for maternal hemorrhage. The cervical opening of the uterus must dilate to about 4 in (10 cm) to permit passage of the baby's head to the exterior. The birth canal is subject to tears and other damage. In a very brief time, the **respiration** of the newborn must change from oxygenated **blood** provided to the fetus by the umbilical **veins** of the placental circulation to oxygenation by blood passing through the baby's lungs.

Ordinarily in the United States, parturition occurs in the safety of a hospital under the supervision of a trained physician or midwife. However, this is a relatively recent development because, up until the early part of this century, childbirth generally occurred at home. Women were attended by midwives who had acquired informal knowledge to assist in childbirth. While midwifery was recognized in ancient Greece and Rome and is referred to in the Old Testament (Exodus 1.15-22), midwives of the time lacked proper training and there was considerable mortality to both mother and infant. However, more recently, nurse midwifery programs have been established in universities in the United States and elsewhere. Admission to midwifery programs typically requires being a registered nurse and having a college degree. The graduate program in midwifery is two very intense years of training. It is not surprising that infant and maternal mortality associated with parturition has plummeted.

PASTEUR, LOUIS (1822-1895)

French chemist and microbiologist

Louis Pasteur was one of the most extraordinary scientists in history, leaving a legacy of scientific contributions which include an understanding of how **microorganism**s carry on the biochemical process of **fermentation**, the establishment of the causal relationship between microorganisms and disease, and the concept of destroying microorganisms to halt the transmission of communicable disease. These achievements led him to be called the founder of microbiology.

After his early education Pasteur went to Paris, studied at the Sorbonne, then began teaching chemistry while still a student. After being appointed chemistry professor at a new university in Lille, France, Pasteur began work on **yeast** cells and showed how they produce **alcohol** and **carbon dioxide** from sugar during the process of fermentation. Fermentation is a form of cellular **respiration** carried on by yeast **cells**, a way of getting **energy** for cells when there is no oxygen present. He found that fermentation would take place only when living yeast cells were present.

Establishing himself as a serious, hard-working chemist, Pasteur was called upon to tackle some of the problems plaguing the French beverage industry at the time. Of special concern was the spoiling of wine and beer, which caused great economic loss and tarnished France's reputation for fine vintage wines. Vintners wanted to know the cause of l'amer, a condition that was destroying the best burgundies. Pasteur looked at wine under the **microscope** and noticed that when aged properly the liquid contained little spherical yeast cells. But when the wine turned sour, there was a proliferation of bacterial cells which were producing **lactic acid**. Pasteur suggested that heating the wine gently at about 120°F would kill the **bacteria** that produced lactic acid and let the wine age properly. Pasteur's book *Etudes sur le Vin*, published in 1866 was a testament to two of his great passions—the scientific method and his love of wine. It caused another French Revolution—one in wine-making, as Pasteur suggested that greater cleanliness was need to eliminate bacteria and that this could be done with heat. Some wine-makers were aghast at the thought but doing so solved the industry's problem.

The idea of heating to kill microorganisms was applied to other perishable fluids like milk and the idea of pasteurization was born. Several decades later in the United States the pasteurization of milk was championed by American bacteriologist **Alice Catherine Evans** who linked bacteria in milk with the disease **brucellosis**, a type of fever found in different variations in many countries.

In his work with yeast, Pasteur also found that air should be kept from fermenting wine, but was necessary for the production of vinegar. In the presence of oxygen, yeasts and bacteria break down alcohol into **acetic acid**—vinegar. Pasteur also informed the vinegar industry that vinegar production could be increased by adding more microorganisms to the fermenting mixture. Pasteur carried on many experiments with yeast. He showed that fermentation can take place without oxygen (*anaerobic* conditions), but that the process still involved

Louis Pasteur. *(The Library of Congress. Reproduced by permission.)*

living things such as yeast. He did several experiments to show (as **Lazzaro Spallanzani** had a century earlier) that living things do not arise spontaneously but rather come from other living things. To disprove the idea of spontaneous generation, Pasteur boiled meat extract and left it exposed to air in a flask with a long S-shaped neck. There was no decay observed because microorganisms from the air did not reach the extract. On the way to performing his experiment Pasteur had also invented what has come to be known as sterile technique, boiling or heating of instruments and food to prevent the proliferation of microorganisms.

In 1862 Pasteur was called upon to help solve a crisis in another ailing French industry. The silkworms that produced silk fabric were dying of an unknown disease. So armed with his microscope, Pasteur went to the south of France in 1865. He found the tiny **parasite**s that were killing the silkworms and affecting their food, mulberry leaves. His solution seemed drastic at the time. He suggested destroying all the unhealthy worms and starting with new **cultures**. The solution worked and French silk scarves were back in the marketplace.

Pasteur then turned his attention to human and **animal** diseases. He had believed for some time that microscopic **organism**s cause disease and that these tiny microorganisms could travel from person to person spreading the disease. Other scientists had expressed this thought before, but Pasteur had more experience using the microscope and identifying different kinds of microorganisms such as bacteria and **fungi**.

In 1868, Pasteur suffered a stroke and much of his work thereafter was carried out by his wife Marie Laurent Pasteur. After seeing what military hospitals were like during the Franco-Prussian War, Pasteur impressed upon physicians that they should boil and sterilize their instruments. This was still not common practice in the nineteenth century.

Pasteur developed techniques for culturing and examining several disease-causing bacteria. He identified *Staphylococcus pyogenes* bacteria in boils and *Streptococcus pyogenes* in puerperal fever. He also cultured the bacteria that cause **cholera**. Once when injecting healthy chickens with cholera bacteria, he expected the chickens to get sick. Unknown to Pasteur, the bacteria were old and no longer virulent. The chickens failed to get the disease, but instead they received immunity against cholera. Thus Pasteur discovered that weakened microbes make a good vaccine by imparting immunity without actually producing the disease.

Pasteur then began work on a vaccine for **anthrax**, a disease that killed many animals and infected people who contracted it from their sheep and thus was known as "woolsorters' disease." Anthrax causes sudden chills, high fever, pain, and can affect the brain. Pasteur experimented with weakening or attenuating the bacteria that cause anthrax, and in 1881 produced a vaccine that successfully prevented the deadly disease.

Pasteur's last great scientific achievement was developing a successful treatment for rabies, a deadly disease contracted from bites of an infected, rabid dog. Rabies, or hydrophobia, first causes terrible pain in the throat that prevents swallowing, then brings on spasms, fever, and finally death. Pasteur knew that rabies took weeks or even months to become active. He hypothesized that if people were given an injection after being bitten, it could prevent the disease from manifesting. After methodically producing a rabies vaccine from the spinal fluid of infected rabbits, Pasteur sought to test it. In 1885 nine-year-old Joseph Meister, who had been mauled and bitten by a rabid dog, was brought to Pasteur, and after a series of shots of the new rabies vaccine, the boy did not develop any of the deadly symptoms of rabies. Pasteur's triumphant success was a great relief to many worldwide.

To treat cases of rabies, the Pasteur Institute was established in 1888 with monetary donations coming from all over the world. It later became one of the most prestigious biological research institutions in the world. When Pasteur died in 1895 he was well-recognized for his outstanding achievements in science.

PAULING, LINUS (1901-1994)
American chemist

One of the twentieth century's greatest chemists, Linus Pauling was born in Portland, Oregon, on February 28, 1901. He became interested in chemistry early in life as a result of his work with a home chemistry set, earning a bachelor's degree in chemistry from Oregon State College (now University) in 1922 and a doctorate in physical chemistry from California In-

Linus Pauling. *(The Library of Congress. Reproduced by permission.)*

stitute of Technology in 1925. During post-doctoral studies in Europe, Pauling worked with some of the most famous chemists and physicists of the time, **Erwin Schrödinger** in Zurich, Niels Bohr in Copenhagen, William Henry Bragg in London, and Arnold Sommerfeld in Münich.

Some of Pauling's earliest work dealt with the structure of **molecule**s. Molecular models of the 1920s generally assumed that chemical bonding occurred between stable electrons. Gilbert Newton Lewis, for example, drew valence electrons in an atom at the corners of a cube. Bonds formed when two cubic **atoms** got close enough to allow overlap of corners. Pauling recognized that a more accurate way to think of electrons was as particles moving through space in a wave pattern.

Louis Victor de Broglie had formulated the mathematics needed to describe such "electron waves" in 1923. Pauling showed how the wave patterns for electrons in two atoms might overlap to produce a new hybrid pattern that was less energetic and, therefore, more stable than either of the original wave patterns. This discovery provided the first rational explanation of chemical bond formation. A second consequence of Pauling's work was the theory of resonance. In some cases, the bonds in a molecule do not display the properties of a pure single bond or a pure double bond, but instead act as some type of intermediary structure. Pauling demonstrated that this behavior could be explained by the assumption that the electrons making up the bond resonate back and forth between two extreme positions. Yet a third implication of Pauling's analysis was the discovery that most chemical bonds are not, as had been supposed, purely ionic or purely covalent. Instead of thinking in terms of the complete loss or gain of electrons or the equal sharing of electron pairs, Pauling suggested that most bonds have a mixed character best described as polar covalent. That is, the pair of electrons shared between two atoms is more strongly attracted by one of the atoms than by the other. Pure ionic and pure **covalent bond**s, then, become only the extreme cases of polar bonds when one atom ''wins'' both electrons completely or when the two atoms share electrons equally. Pauling outlined his analysis of bonding in his influential book, *The Nature of the Chemical Bond*, in 1939.

In the 1930s, Pauling had become increasingly interested in the molecular structure of complex biological substances and studied the properties of the **hemoglobin** molecule and its role in the genetic disorder known as **sickle-cell anemia**. In the early 1950s, he and chemist Robert Corey demonstrated the helical structure of **protein**. After World War II, Pauling became concerned about the global implications of nuclear weapons testing. In 1958, he published those concerns in a book entitled *No More War!* and brought to the United Nations a petition bearing the names of more than eleven thousand scientists, asking for an end to nuclear weapons testing. Pauling was awarded the Nobel Prize in Chemistry in 1954 for his work on the **chemical bond** and in 1962 he won the Nobel Prize for Peace for his work on nuclear weapons testing. He taught at California Institute of Technology from 1927 until his retirement in 1958.

Later in his career Pauling became well known for his controversial claim that ascorbic acid— vitamin C—helps combat colds and disease. He outlined his findings in the 1970 book *Vitamin C and the Common Cold*, focusing in a later book on vitamin C and **cancer**. He has served as a professor emeritus at Stanford University beginning in 1974. Before his death in 1994 Pauling was a member of the Center for the Study of Democratic Institutions, a faculty member of Stanford University, and director fo research at the Linus Pauling Institute of Science and Medicine in Palo Alto, California.

PAVLOV, IVAN PETROVITCH (1849-1936)
Russian physiologist

Born on September 14, 1849, in Ryazan, Russia, Pavlov was the son of a village priest. He planned to follow family tradition by becoming a priest himself. While at a theological seminary, however, Pavlov read **Charles Darwin**'s *Origin of the Species* and found he really wanted a career in science instead. Soon afterward, in 1870, Pavlov transferred from the seminary to St. Petersburg University. There his professors included two renowned Russian chemists, Dmitri Mendeleev and Alexander Butlerov. Pavlov studied both chemistry and physiology. He obtained a medical degree from St. Petersburg Military Medical Academy in 1879, and a Ph.D. in 1883.

For the next few years, Pavlov studied cardiovascular and gastrointestinal physiology in Germany, then returned to the Medical Academy, where he was appointed Professor of Physiology and also conducted most of his research investigations. Pavlov's first major studies centered around the physiology of **digestion**. He was particularly interested in working out the nervous mechanism that controlled the secretion of the digestive tract's various glands.

In 1889, Pavlov designed one of his most important animal experiments: after severing a dog's gullet, he pulled the upper end out through an opening in the animal's neck. From then on, while the dog could be fed, his food would drop out through the open gullet rather than reach his stomach. Nevertheless, as Pavlov pointed out, after each feeding, the animal's gastric juices would flow, suggesting that nerves in his mouth must have been stimulated. These nerves must have then sent a message to the **brain** which, by way of other nerves, must have then stimulated the stomach's digestive glands, causing them secrete the juices. Pavlov performed a number of other experiments that not only helped demonstrate how digestion worked in a living animal, but also helped establish the importance of the autonomic **nervous system** in controlling the digestive process. For his work, Pavlov received the 1904 Nobel Prize in physiology or medicine.

Ironically, Pavlov then went on to design the series of animal experiments for which he is most famous: the ''salivating dog'' studies. In these studies, Pavlov confined a laboratory dog in a room that was kept soundproof in order to eliminate distracting noises. The dog was held in place by a loose harness, was fed by an automatic apparatus that was operated from outside the room, and had a small measuring tube attached to his cheek to collect the flow of saliva from his parotid gland. The dog's saliva was measured under varying situations and, before long, Pavlov was able to report that the dog's salivation began, as expected, as soon as he saw his food (a natural and unconditioned **reflex**). However, if a neutral sound, such as a bell, always accompanied the offering of his food, the dog began to salivate as soon as he heard the bell—even if the food *did not* immediately appear. Pavlov termed this second reaction a conditioned reflex—a reaction that was not really instinctive but had been learned through a sequence of associations.

Pavlov's continuing investigation of the conditioned reflex—although it took place in a laboratory and was conducted on animals—clearly had implications for human learning behavior as well. Psychiatrists and psychologists around the world began incorporating the concept into a number of different doctrines, particularly those relating to behavioral psychology. Pavlov continued his own studies, even after the Communist Revolution and, although he himself was an outspoken anti-Communist, he remained one of Russia's most highly treasured scientists until his death in Leningrad on February 27, 1936.

PEARSON, KARL (1857-1936)
English statistician

Karl Pearson is considered the founder of the science of statistics. He believed that a true understanding of **human evolution** and **heredity** required mathematical methods for analysis of the data. In developing ways to analyze and represent scientific observations, he laid the groundwork for the development of the field of statistics in the twentieth century and its use in medicine, engineering, anthropology, and psychology.

Pearson was born in London, England, on March 27, 1857, to William Pearson, a lawyer, and Fanny Smith. At the age of nine, Karl attended the University College School, but was forced to withdraw at sixteen because of poor health. After a year of private tutoring, he went to Cambridge, where the distinguished King's College mathematician E. J. Routh met with him each day at 7 a.m. to study papers on advanced topics in applied mathematics. In 1875, he was awarded a scholarship to King's College, where he studied mathematics, philosophy, religion, and literature. At that time, students at King's College were required to attend divinity lectures. Pearson announced that he would not attend the lectures and threatened to leave the college; the requirement was dropped. Attendance at chapel services was also required, but Pearson sought and was granted an exception to the requirement. He later attended chapel services, explaining that it was not the services themselves, but the compulsory attendance to which he objected. He graduated with honors in mathematics in 1879.

After graduation, Pearson traveled in Germany and became interested in German history, religion and folklore. A fellowship from King's College gave him financial independence for several years. He studied law in London, but returned to Germany several times during the 1880s. He lectured and published articles on Martin Luther, Baruch Spinoza, and the Reformation in Germany, and wrote essays and poetry on philosophy, art, science, and religion. Becoming interested in socialism, he lectured on Karl Marx on Sundays in the Soho district clubs of London, and wrote hymns for the Socialist Song Book. Pearson was given the name Carl at birth, but he began spelling it with a ''K,'' possibly out of respect for Karl Marx.

During this period, Pearson maintained his interest in mathematics. He edited a book on elasticity as it applies to physical theories and taught mathematics, filling in for professors at Cambridge. In 1884, at age twenty-seven, Pearson became the Goldsmid Professor of Applied Mathematics and Mechanics at University College in London. In addition to his lectures in mathematics, he taught engineering students, and showed them how to solve mathematical problems using graphs.

In 1885, Pearson became interested in the role of women in society. He gave lectures on what was then called ''the woman question,'' advocating the scientific study of questions such as whether males and females inherit equal intellectual capacity, and whether, in the future, the ''best'' women would choose not to bear children, leaving it to ''coarser and less intellectual'' women. He joined a small club which met to discuss questions of morality and sex. There he met Maria Sharpe, whom he married in 1890. They had three children, Egon, Sigrid, and Helga. Maria died in 1928, and Pearson married Margaret V. Child, a colleague at University College, the following year.

Develops Statistical Methods to Study Heredity

Pearson was greatly influenced by **Francis Galton** and his 1889 work on heredity, *Natural Inheritance.* Pearson saw that there often may be a connection, or correlation, between two events or situations, but in only some of these cases is the correlation due not to chance but to some significant factor. By making use of the broader concept of correlation, Pearson believed that mathematicians could discover new knowledge in biology and heredity, and also in psychology, anthropology, medicine, and sociology.

An enthusiastic young professor of **zoology**, W. F. R. Weldon, came to University College in 1891, further influencing Pearson's direction. Weldon was interested in **Charles Darwin**'s theory of **natural selection** and, seeing the need for more sophisticated statistical methods in his research, asked Pearson for help. The two became lunch partners. From their association came many years of productive research devoted to the development and application of statistical methods for the study of problems of heredity and evolution. Pearson's goal during this period was not the development of statistical theory for its own sake. The result of his efforts, however, was the development of the new science of statistics.

Remaining at the University College, Pearson became the Gresham College Professor of Geometry in 1891. His lectures for two courses there became the basis for a book, *The Grammar of Science,* in which he presented his view of the nature, function, and methods of science. He dealt with the investigation and representation of statistical problems by means of graphs and diagrams, and illustrated the concepts with examples from nature and the social sciences. In later lectures, he discussed probability and chance, using games such as coin tossing, roulette, and lotteries as examples. He described frequency distributions such as the normal distribution (sometimes called the bell curve because its graph resembles the shape of a bell), skewed distributions (for which the graphed design is not symmetrical), and compound distributions (which might result from a mixture of the two). Such distributions represent the occurrence of variables such as traits, events, behaviors, or other incidents in a given **population**, or in a sample (subgroup) of a population. They can be graphed to illustrate where each subject falls within the continuum of the variable in question.

Pearson introduced the concept of the ''standard deviation'' as a measure of the variance within a population or sample. The standard deviation statistic refers to the average distance from the mean score for any score within the data set, and therefore suggests the average amount of variance to be found within the group for that variable. Pearson also formulated a method, known as the chi-square statistic, of measuring the likelihood that an observed relation is in fact due to chance, and used this method to determine the significance of the statistical difference between groups. He also developed the theory of correlation and the concept of regression analysis, used to predict the research results. His correlation coefficient, also known as the Pearson r, is a measure of the strength of the relationship between variables and is his best-known contribution to the field of statistics.

Between 1893 and 1901 Pearson published thirty-five papers in the *Proceedings* and the *Philosophical Transactions* of the Royal Society, developing new statistical methods to deal with data from a wide range of sources. This work formed the basis for much of the later development of the field of statistics. He was elected to the Royal Society in 1896, was awarded the Darwin Medal in 1898, and, in 1903, was elected an Honorary Fellow of King's College and received the Huxley Medal of the Royal Anthropological Institute.

Establishes Journal and Compiles Statistical Tables

In 1901, Pearson helped found the journal *Biometrika* for the publication of papers in statistical theory and practice. He edited the journal until his death. His research often required extensive mathematical calculation, which was carried out under his direction by students and staff mathematicians in his biometric laboratory. Since high-speed electronic computers had not yet been invented, performing the calculations by hand was tedious and time-consuming. The laboratory staff produced tables of calculations which Pearson made available to other statisticians through *Biometrika,* and later as separate volumes. Access to these tables made it possible for others to carry out statistical research without the support of a large staff, and, again, proved to be a valuable contribution to the early development of the field of statistics.

Pearson became the Galton Professor of Eugenics in 1911, and headed a new department of applied statistics as well as the biometric laboratory and a eugenics laboratory, established to study the genetic factors affecting the physical and mental improvement or impairment of future generations. During World War I, Pearson's staff served Britain's interest by preparing charts showing employment and shipping statistics, investigating stresses in airplane propellers, and calculating gun trajectories. From 1911 to 1930, Pearson produced a four-volume biography of Francis Galton. In 1925, he founded the journal *Annals of Eugenics,* which he edited until 1933. In 1932, Pearson was the first foreigner to be awarded the Rudolf Virchow Medal by the Anthropological Society of Berlin. He retired in 1933 at age seventy-seven, and received an honorary degree from the University of London in 1934. Pearson died on April 27, 1936, in Coldharbour, Surrey.

Pearson produced more than three hundred published works in his lifetime. His research focused on statistical methods in the study of heredity and evolution but dealt with a range of topics, including albinism in people and animals, alcoholism, mental deficiency, tuberculosis, mental illness, and anatomical comparisons in humans and other primates, as well as astronomy, meteorology, stresses in dam construction, inherited traits in poppies, and variance in sparrows' eggs. Pearson was described by G. U. Yule as a poet, essayist, historian, philosopher, and statistician, whose interests seemed limited only by the chance encounters of life. Colleagues remarked on his boundless energy and enthusiasm. Although some saw him as domineering and slow to admit errors, others praised him as an inspiring lecturer and noted his care in acknowledging the contributions of the members of his lab group. For Pearson, scientists were heroes. The walls of his laboratory contained quotations from Plato, Pascal, Huxley and others, including

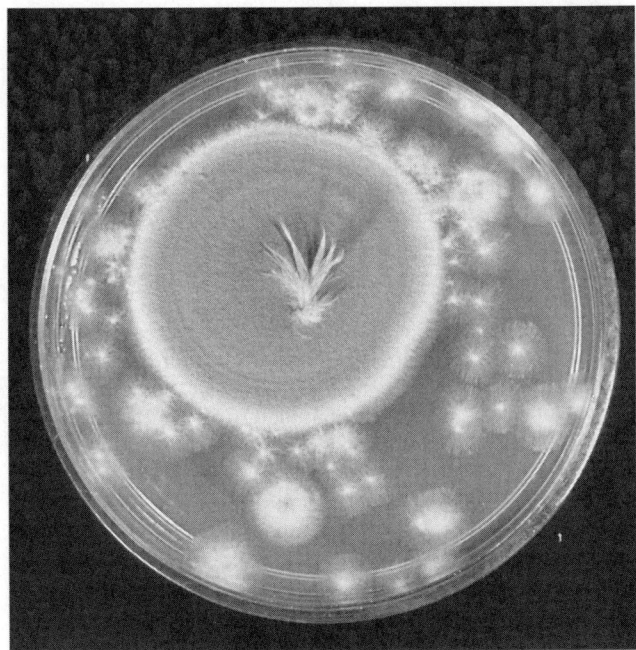

A penicillin culture. *(Photograph by P. Barber, Custom Medical Stock Photo. Reproduced by permission.)*

these words from Roger Bacon: ''He who knows not Mathematics cannot know any other Science, and what is more cannot discover his own Ignorance or find its proper Remedies.''

PENICILLIN

The discovery of penicillin may be one of the greatest accidents in medical history. Penicillin, found in common molds, is a potent antibacterial medication. Its discovery opened the door to a variety of new ''miracle drugs'' that have saved the lives of millions. In the early twentieth century, scientists had been looking for some kind of antibacterial agent to treat infections that could attack nearly every **organ** in the body. The only drugs that had been used to treat **bacterial** infections were quinine, **arsenic** and sulfa drugs, but these were highly toxic. Scottish bacteriologist **Alexander Fleming** discovered penicillin by accident in 1928. He was conducting research using several petri dishes of bacteria **culture**s, and one of the cultures was accidently left uncovered for several days. When Fleming found that the dish had become contaminated with a mold, he was about to discard the culture until he noticed that the mold was dissolving all the bacteria near it. Recognizing the importance of what was happening in the petri dish, Fleming kept a sample of the mold growing in a test tube and examined it with a **microscope**. He tested it against several types of bacteria and found that something in the mold inhibited the growth of the bacteria. The mold was from the **genus** *Penicillium*, so Fleming named the presumed antibacterial component penicillin. However, Fleming was unsuccessful in his attempts to isolate the bactericidal material and eventually abandoned his efforts. In 1935 at Oxford University, pathologist **Howard Florey** and biochemist **Ernst Chain** had been researching antibacterial substances when they stumbled across an article by Fleming about his work with penicillin. After obtaining a culture of Fleming's original mold, Florey and Chain were able to extract and purify the penicillin. Florey began testing the substance on **animal**s and found that it was nontoxic as well as an effective antibiotic. Furthermore, it did not harm living **cell**s or interfere with the activity of white **blood** cells. Chain and Forey began clinical trials in humans in 1941, and their patients by and large recovered completely from bacterial infections. One of the first patients to be treated with penicillin was an English policeman with a bacterial infection that had spread from his mouth to his face, shoulder, and lungs. For a time his condition improved, but the Oxford facility could not produce enough penicillin to overcome the infection and the policeman eventually died. In subsequent human trials the results were so successful that penicillin was rushed into production so it could be used to treat infection suffered by soldiers during World War II. England could not take on new manufacturing ventures due to its involvement in the war, so Florey traveled to the United States and convinced the government to sponsor research on the mass production of penicillin. Florey's effort bore fruit when American microbiologist Andrew J. Moyer, working with the U.S. Department of **Agriculture**'s Northern Regional Research Laboratory in Peoria, Illinois, soon developed an efficient method of mass-producing penicillin using **fermentation** and a cornstarch medium. His basic techniques are still used to produce many **antibiotics**. By 1943, several American pharmaceutical companies were producing the drug for use in treating war injuries. The availability of penicillin during the war prevented thousands of **death**s from gas gangrene and other infections. Once the penicillin production had begun, the race was on to discover its molecular structure so that it could be synthetically produced. In the mid-1940s, English researcher Dorothy Crowfoot Hodgkin began a single-crystal study of penicillin using x ray crystallography. This technique obtains data from repeated x rays to calculate the positions of the **atoms** in the basic unit of the crystal. Hodgkin was able to use an early IBM card-punch computer to complete the many mathematical computations required to carry out the experiment. Once Hodgkin determined the chemical structure of penicillin, the door was opened to other scientists to develop methods to synthesize it. Robert Burns Woodward, an organic chemist at Harvard University, completed the first penicillin synthesis in the 1950s. Still considered the most powerful of the antibiotics, penicillin is used to treat diseases such as **syphilis**, meningitis, and pneumonia. In addition, by eliminating the threat of bacterial infection during medical procedures, it has spurred the development of surgical operations, organ transplants, and open **heart** surgery, as well as improved the treatment of burns. The discovery and development of penicillin is rightly regarded as one of the greatest achievements in medical history, and many of the scientists who worked on it have been highly honored. Fleming, Florey, and Chain shared the 1945 Nobel Prize in Medicine for the development of penicillin. For their work with penicillin as well as other research, Hodgkin and Woodward also received the Nobel Prize in 1964 and 1965, respectively.

PEPTIDE

A peptide is an organic **molecule** consisting of two or more **amino acids** linked through an amide linkage called a peptide bond. Peptides are formed when the amino group ($-NH_2$) of one amino acid, links with the carboxyl group (-COOH) of an adjacent amino acid. An amide linkage (-CO-NH-) results. If two amino acids are linked, the compound is called a dipeptide, if three, a tripeptide, and so on. Longer chains containing a few amino acids are often called oligopeptides or **polypeptides** if they contain as many as 50. Still longer chains are called **proteins**. Peptides may have many different properties depending on the nature of the amino acids they contain, and the order in which they are linked. Some peptides function as **hormones**—a few as antibiotics and others as important participants in **metabolism**.

For example, an octapeptide with a hormone-like function, called angiotensin, is formed in the **blood** and targets the **circulatory system**, causing an increase **blood pressure**. Oxytocin, useful in obstetrics because it promotes uterine contraction, is an unusual form because it contains nine amino acids arranged in a ring structure. Two peptides, made by **bacteria**, with antibiotic activity, include gramicidin, which inhibits oxidative phosphorylation (the synthesis of **ATP** that obtains **energy** from electron transport occurring as a result of aerobic **respiration**), and bacitracin, used for treatment of bacterial infections of the eye.

Several peptides have neuroactive (stimulating neural **tissues**) properties and function similarly to other neurotransmitters but with some important differences. Classic transmitters like acetylcholine act on synaptic receptors (a cellular entity that is the mechanism for genetic cross over) for only milliseconds, while neuropeptides may act for seconds, hours, or even days. Neuropeptides are released in lower concentrations than other transmitters, and they have higher potency. Examples of neuropeptides include neurotensin, somatostatin, cholecystokinin, and the opioid peptides. Opioid peptides, including endorphins, enkephalins and dynorphins, have been widely studied because opiate drugs like morphine bind (latch on) to the same receptor sites and mimic their pain-killing action.

Peptides are formed routinely in the digestive tract of **animal**s when protein is broken down. As peptide bonds in the protein are hydrolyzed, the complex molecule is first converted to a number of smaller peptides. The peptides are subsequently converted to individual amino acids. Two enzymes involved in this digestive process are pepsin, released in the acid environment of the stomach, and trypsin, secreted by the **pancreas** and active in the duodenum (part of the small intestine) and upper jejunum (part of the small intestine just beyond the duodenum).

PERIPHERAL NERVOUS SYSTEM • See
Nervous system

PERT, CANDACE B. (1946-)

American neuroscientist and biochemist

Candace B. Pert is a leading researcher in the field of chemical receptors, places in the body where **molecules** of a drug or natural chemical fit like a key into a lock, thus stimulating or inhibiting various physiological or emotional effects. As a graduate student, Pert codiscovered the **brain**'s opiate receptors, areas that fit painkilling substances such as morphine. Her work led to the discovery of endorphins, the naturally occurring substances manufactured in the brain that relieve pain and produce sensations of pleasure.

Candace Dorinda Bebe Pert was born in New York City on June 26, 1946, to Mildred and Robert Pert. She went to General Douglas MacArthur High School in Levittown, New York. She attended Hofstra University but dropped out in 1966. That year she married Agu Pert and the couple moved to Philadelphia so that her husband could get a doctorate at Bryn Mawr College. In 1966, Candace Pert gave birth to the first of the couple's three children.

In 1967, to help support the family, Pert took a job as a cocktail waitress. On one occasion she chatted with a customer who turned out be an assistant dean at Bryn Mawr. The dean encouraged Pert to finish her B.A. at Bryn Mawr, and helped her through the admissions process. In 1970, Pert got her B.A. in biology and that year entered the doctoral pharmacology program at Johns Hopkins University in Baltimore.

Her first research assignment, working under Dr. Solomon Snyder, was to explore the mechanisms that regulate the production of **acetylcholine**, the body's most important **neurotransmitter**. Neurotransmitters are chemicals that stimulate or inhibit other **neurons** throughout the body, which in turn regulate the heart and other organs. Then in the summer of 1972, again working with Dr. Snyder, she embarked on her next project, the search for an opiate receptor. Opiate receptors were believed to exist, but finding them was another matter. Although techniques for locating receptors of hormones had been put into practice, many scientists thought it would be difficult, if not impossible, to transfer the technique to an opiate receptor.

Makes Surprising Discovery of Opiate Receptors

Receptors evolve from a chain of **amino-acid** molecules; these molecules are shaped by electrical forces into a three-dimensional shape with an electrically active indentation which recognizes correspondingly shaped molecules. These indentations are the points at which a receptor binds with a chemical substance or neurotransmitter. Using technology borrowed from identifying insulin receptors, Pert used radioactive drugs to identify receptor molecules that bonded with morphine and other opiate drugs in animal brain cells. The first report on her finding was published in *Science* in March 1973. Pert went on to investigate whether opiate receptors developed before **birth**. She used pregnant rats to evaluate the brains of the fetuses and found that during fetal development opiate receptors were present.

Pert and her colleagues mulled over why opiate receptors existed. It was certainly not that animals had evolved opiate receptors to interact with poppy plants, the natural source of opium. The scientist speculated that there might be an unknown neurotransmitter, naturally produced in the body, that fulfilled a similar function. Other experiments had already shown that stimulating the brainstem of rats caused pain relief, and that the best pain relief was obtained when a specific part of the brain was stimulated. After initial investigations proved inconclusive Pert turned to other areas of research. Eventually two Scottish scientists, John Hughes and Hans Kosterlitz, found the transmitters, which they called endorphins.

The discovery of endorphins led to the discovery of other types of receptors and corresponding chemicals in the brain. Uncovering the intricate system of chemicals changed the scientific conception of the brain as an organ that signals the rest of the body using just a few chemicals. Now it is understood that the nervous system uses many substances to signal pain, pleasure and emotions as well as sensory data. Many had mistakenly hoped that the discoveries would immediately result in a cure for drug addictions or a non-addicting pain killer for cancer patients, especially since the media had sensationalized these possibilities. Although these hopes proved overoptimistic, in 1978 Snyder, Hughes and Kosterlitz received the prestigious Lasker Award for their discoveries; Pert did not. The fact that the biochemist, who had received her Ph.D. in 1974, had not been recognized for her part in the discovery caused a controversy that even erupted on the editorial pages of the prominent journal *Science*.

Pert refused to become involved in any controversy, however, and continued on at Hopkins as a National Institutes of Health fellow from 1974 to 1975, as a staff fellow from 1975 to 1977, a senior staff fellow from 1977 to 1978, and then as research pharmacologist from 1978 to 1982. In 1982, she became chief of the section on brain chemistry at the National Institutes of Mental Health (NIMH). There, the neuroscientist turned her attention to Valium receptors in the brain and the receptors where the street drug PCP, or "angel dust," takes hold. In 1986, Pert led the NIMH team that discovered peptide T. Peptides are substances that are synthesized from amino acids and are intermediate in molecular weight and chemical properties between amino acids and proteins, and have been linked to the manifestation of emotions.

Pert left NIMH in 1987 and worked for laboratories in the private sector. She also started her own company, Peptide Design, to encourage research on peptides. The company was in existence from 1987 to 1990. Since then, Pert has become an adjunct professor in the department of physiology at Georgetown University. Among her other areas of research have been investigations into the immune system and the nature of the **human immunodeficiency virus** (**HIV**) that causes **AIDS**. Pert won the Arthur S. Fleming Award in 1979. She is a member of the American Society of Pharmacologists and Experimental Therapeutics; the American Society of Biological Chemists; the Society of Neuroscientists; and the International Narcotics Research Conference.

Since her first discovery of an opiate receptor, Pert has located endorphin receptors throughout the body, even in the pituitary gland. She suspects that the location of receptors in sites where there is no clear connection with conscious pain serves the function of signalling the central nervous system when there is a problem with an organ. She believes, as she told an *Omni* interviewer, that scientists will eventually be able to chart the various receptors of the brain and the reactions they produce. ''There's no doubt in my mind that one day— and I don't think that day is all that far away—we'll be able to make a color-coded map of the brain. A color-coded wiring diagram, with blue for one neurochemical, red for another, and so on—that's the neuroscientist's ambition.''

PESTICIDES

A pesticide is a chemical that is used to kill insects, weeds, and other **organism**s to protect humans, crops, and livestock. A broad-spectrum pesticide that kills all living organisms is called a biocide. Fumigants, such as ethylene dibromide or dibromochloropropane, used to protect stored grain or sterilize soil fall into this category. Generally, however, we prefer narrower spectrum agents that attack a specific type of pest: herbicides kill **plants**; insecticides kill insects; fungicides kill fungi; acaricides kill mites, ticks, and spiders; nematicides kill nematodes (microscopic roundworms); rodenticides kill rodents; and avicides kill birds. Pesticides can also be grouped by their method of application (fumigation, for example, is dispersal as a gaseous vapor) or by their mode of action (an ovicide kills the **egg**s of pests).

There are thousands of kinds of natural pesticides. Plants have been engaged for millions of years in chemical warfare with predators, most of which are insects. They have evolved a wide variety of complex protective mechanisms, many of which are toxic chemicals. Humans have probably known for a very long time that natural products such as nicotine from tobacco, turpentine from pines, pyrethrum from chrysanthemum **species**, and quinine from cinchona bark can provide protection from pests and **parasite**s. Our diet contains a large number of such chemicals but ordinarily we have mechanisms to detoxify or excrete them so that they are not a problem.

The modern era of chemical pest control began in 1934 with the discovery of the insecticidal properties of **DDT** (dichloro-diphenyl-trichloroethane) by Swiss chemist Paul Müller. It became extremely important during World War II in areas where tropical diseases and parasites posed greater threats to soldiers than did enemy bullets. DDT seemed like a wonderful discovery. It is cheap, stable, easily applied, and highly toxic to insects while being relatively nontoxic to **mammals**. It seemed like the magic bullet that would provide ''better living through chemistry.'' In 1948, Müller received a Nobel prize for his discovery. It was quickly discovered, however, that this magic bullet was not always benevolent. Within a short time, many beneficial organisms were exterminated by DDT, while the pests it was created to control had developed resistance and had rebounded to higher levels than ever. Furthermore, persistent chlorinated **hydrocarbon**s such as this tend to be taken up by living organisms and concentrated

through food chains until they reach toxic levels in the top **carnivore**s such as birds of prey or game fish. Species such as peregrine falcons, brown pelicans, osprey, and bald eagles disappeared from much of their range in the Eastern United States before DDT and similar persistent pesticides were banned. **Rachel Carson**'s *Silent Spring*, possibly the most influential book in all of American environmental history, presents the argument against excessive, widespread pesticide use.

In spite of continuing worries about the dangers of pesticides, we still depend heavily on them. The Environmental Protection Agency reports (EPA) that around 500,000 metric tons of pesticides are used in the United States every year. We rely on them for disease control, agricultural production, preservation of buildings and materials, elimination of biting and troublesome organisms, forest protection, and a host of other purposes. Herbicides account for about 59%, insecticides 22%, fungicides 11%, and all other types together about 8% of our total use. Pesticide advocates claim that without modern pesticides, we would lose as much as half of our harvest to pests and that the world would suffer widespread and calamitous famines. Pesticide opponents argue that we could use cultural practices and natural pest predators or repellents to accomplish many of these same goals more safely and more cheaply than we now do with toxic synthetic chemicals.

Several movements aimed at reducing pesticide use have gained adherents in the United States and around the world. Integrated pest management (IPM) is a flexible, ecologically-based, pest-control strategy that uses a combination of techniques applied at specific times and aimed at specific crops and pests. It does not shun pesticides entirely but uses them judiciously when, where, and only in the minimum amount needed. It also employs biological controls (natural predators, resistant crop species) and practices such as mechanical cultivation to reduce pest **population**s. Many consumers choose to buy organic foods grown without synthetic pesticides or fertilizers as a way of reducing their own personal exposure and to encourage growers to adopt environmentally sound production methods.

In addition to buying organic products, there are a number of things that individuals can do to reduce their exposure to dangerous pesticides. Plant ground cover that competes successfully with weeds. Install or repair screens on doors and windows to keep out insects. Wash house and garden plants with soapy **water** to get rid of pests. Plant pest-repelling species such as marigolds, garlic, basil, or peppermint around your sensitive garden crops. Put out a cup of stale beer to control slugs. Learn to accept slightly blemished **fruits** and vegetables. Trim away bad parts or give up a part of your crop rather than saturate your environment with toxic chemicals.

pH

pH refers to the amount of acid or base in an aqueous solution. It comes from a French word meaning ''hydrogen power'', or essentially the strength of the hydrogen ion. pH is defined as the negative log (in base 10) of the hydrogen ion concentra-

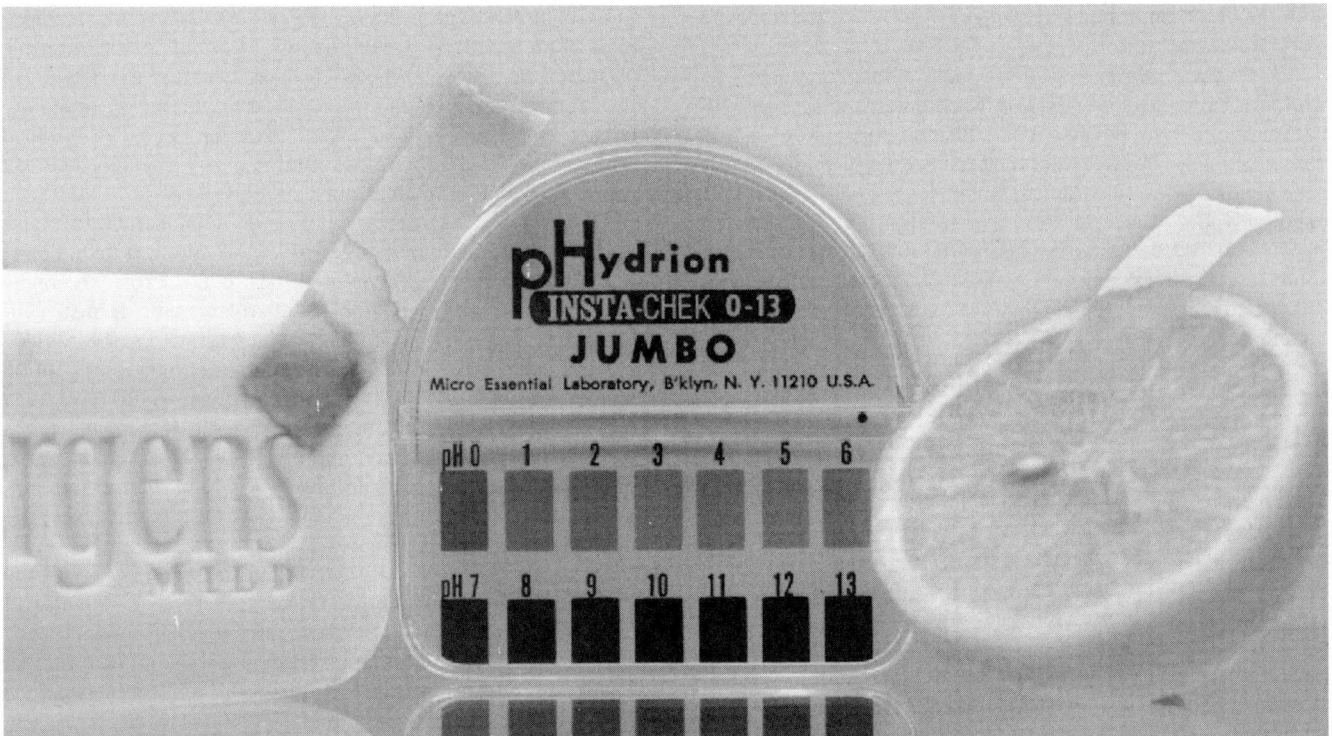

The special papers on this lemon and soap are treated with acid base indicators, which change color in response to changes in pH. The colors can then be checked against a chart. The bar of soap is alkaline, and the lemon is acidic. *(Phototake NYC. Reproduced by permission.)*

tion. A neutral solution has a pH of 7.0, acidic solutions are those with a pH less than 7, and basic solutions are those greater than 7. Since this is a logarithmic scale, the difference between each increment is a factor of 10. Thus, a solution of pH 5.0 is 10 times more acidic than a solution of pH 6.0, and 100 times more acidic than the solution of pH 7.0. Battery acid has a pH of nearly 1.0, normal stomach acidity ranges from 1.0 to 3.0, lemon juice is around 2.3, vinegar and soft drinks about 3.0, fresh tomatoes about 4.0, black coffee about 5.0, pure **water** 7.0, **blood** 7.3-7.5, sea water 7.8-8.3, baking soda and bleach 9.0, household **ammonia** and detergents 10.5-12.0, hair remover 12.5, and oven cleaner 13.5.

The pH of a solution can be measured using an electronic pH meter with electrodes, and by various indicators which change color depending upon the pH of the solution. For example, phenophthalein is colorless in an acidic solution but turns pink around pH 8.3.

Normal rain water has a pH of approximately 5.6 due to the reaction of **carbon dioxide** gas and water vapor in the **atmosphere**, producing weak carbonic acid. **Acid rain** is rain water with a pH less than this, and areas receiving high amounts of nitrous oxide and sulfur dioxide emissions receive more acidic rain water due to the production of the stronger nitric and sulfuric acids in the atmosphere. Acid deposition (both wet and dry) can cause considerable damage to unbuffered aquatic and terrestrial environments. It can also erode marble statues.

PHAGOCYTOSIS

Phagocytosis (from the Greek word *phagein*, meaning to eat, and *kytos*, meaning vessel) is the process in which cells engulf small solid particles, including, on occasion, whole cells. Because the cells are taking in solid particles, phagocytosis is sometimes called "cell eating."

Phagocytosis is one form of endocytosis, the process by which materials are engulfed into a cell. The other forms of endocytosis are pinocytosis and receptor-mediated endocytosis. In all three, there is an infolding of the cell membrane that pinches around substances, forming a vacuole or vesicle, and the materials are transported into the cytoplasm of the cell. Since all forms of endocytosis require energy to move the substances into the **cell**, some scientists consider them forms of active transport.

Some unicellular **organism**s such as amoeba and paramecia feed by phagocytosis. Their pseudopods or other cellular structures move around solid food particles that are engulfed into a **vacuole** or vesicle in the cytoplasm of the organism. The vacuole or vesicle then fuses with a **lysosome**, which is an organelle containing digestive enzymes. These digestive enzymes break down the food particles, and the useable nutrients such as sugars, **amino acids**, and nucleotides pass across the vacuole or vesicle membrane for use in the cytoplasm of the cell. In many less complex organisms, the remaining indigestible waste products are expelled from the cell by the process of exocytosis, when the vacuole or vesicle fuses with the cell membrane. Exocytosis is, in effect, the opposite of endocytosis.

Phagocytosis can occur in multicellular organisms as well. For example, some flatworms, cnidaria, and sponges obtain nutrients via phagocytosis. Phagocytosis even occurs within humans. Some types of specialized white blood cells called phagocytes (macrophages and leukocytes) in humans engulf invading **bacteria** and other foreign particles and destroy them. While phagocytosis is important for nutrition in some less complex organisms, these phagocytic cells are an important part of defense (immune) systems in some more complex organisms.

PHENOL

Phenols are **organic compound**s containing a six-carbon aromatic ring structure in which a hydroxyl group (-OH) is attached to one of the carbons in the ring. The term is also as the name of the simplest member of the group, monohydroxybenzene (C_6H_5OH), also called carbolic acid. Phenols are like **alcohol**s, and are sometimes called aromatic alcohols, but they have higher boiling points and are more soluble in **water** than simple alcohols of similar molecular weight. At room **temperature** they are either colorless liquids or white solids, with an odor that ranges from mildly agreeable to sharp and pungent. Like alcohols, phenols react with acids to form esters.

Phenol, or carbolic acid, has been widely used as an antiseptic to control **bacterial** growth and contamination, although other products have now largely replaced it. Its widespread use for this purpose led to the development of a concept called the phenol coefficient, used to rate the disinfectant quality of various antiseptics and germicides based on their effectiveness relative to phenol. It is also an important industrial chemical and is used as a starting material to make aspirin. It is an initial ingredient in making plastics such as Bakelite.

Phenol can be made from the distillation of crude petroleum or coal tar, but the **hydrolysis** of chlorobenzene or the oxidation of isopropylbenzene, both of which are first made from benzene, more commonly produces it. More complex naturally occurring phenols are found in essential oils derived from plant seeds and leaves.

PHENOTYPE

The word phenotype refers to the observable attributes of individual **organism**s, including their **morphology**, physiology, behaviour, and other traits. The phenotype of an organism is influenced by its specific genetic complement (that is, its **genotype**), plus any environmental influences on the expression of its genetic potential.

All organisms have unique genetic information, which is embodied in the particular **nucleotide** sequences of their **DNA** (**deoxyribonucleic acid**; this is the genetic biochemical of almost organisms, except for **virus**es and some **bacteria**). This

individual genotype is fixed and constant (except for very low rates of **mutation**). However, there is a certain degree of developmental flexibility in the phenotype, which is the actual expression of the genetic information in terms of **anatomy**, behavior, and biochemistry. This flexibility can occur because the expression of genetic potential is partly affected by environmental conditions and other circumstances.

Consider, for example, an individual geranium plant, with a fixed complement of genetic information. If that geranium is grown under well-**water**ed, fertile, non-crowded conditions, it will develop into a relatively tall, robust, and vigorously flowering specimen. However, if that same individual had been grown under drier, less-fertile, more competitive conditions, its productivity and growth form would have been quite stunted. Such varying growth patterns of the same genotype are referred to as phenotypic plasticity. Some traits of organisms, however, are fixed genetically, and their expression is not affected by environmental conditions. For instance, the flower colour of individual geraniums (which can be white, red, or pink) is genetically fixed, and does not vary with the conditions of cultivation. Moreover, the ability of **species** to exhibit phenotypically plastic responses to environmental variations is itself, to a substantial degree, genetically determined. Therefore, phenotypic plasticity reflects both genetic capability and varying expression of that capability, depending on circumstances.

PHENYLKETONURIA

Phenylketonuria (PKU) is an inherited disorder in which an **enzyme** (usually phenylalanine hydroxylase) crucial to the appropriate processing of the **amino acid**, phenylalanine, is totally absent or drastically deficient. The result is that phenylalanine cannot be broken down, and it accumulates in large quantities throughout the body. Normally, phenylalanine is converted to tyrosine. Because tyrosine is involved in the production of melanin (pigment), people with PKU usually have lighter skin and hair than other family members. Without treatment, phenylalanine accumulation in the **brain** causes severe mental retardation. Treatment is usually started during babyhood; delaying such treatment results in a significantly lowered IQ by age one. Treatment involves a diet low in phenylalanine (look for warnings aimed at people with PKU on cans of diet drinks containing the artificial sweetener aspartame, which is made from phenylalanine). PKU strikes about one out of every 20,000 newborns. Because it is so important to start treatment immediately, many states require that all infants be tested for the disease within the first week of life.

PHEROMONES

Pheromones are a specific type of chemical signal in which substances are produced by one **organism** that conveys a message to another organism usually of the same **species**. These intraspecific signals are of a specific composition, produced by special exocrine glands, and released at specific times to inhibit or stimulate specific biological functions in other organisms. Pheromones are released outside of the organism's body to coordinate behavioral and physiological activities among individuals of the population. This is in contrast to hormones which are internally-released chemicals that regulate body metabolism. Pheromonal communication is widespread among animals and microorganisms, but has been most extensively studied in insects and vertebrates.

Pheromones that directly influence behavior by stimulating an immediate change in behavior are termed releasers. In contrast, primer pheromones stimulate a long-term physiological effect that later influences a behavioral response. For example, a female moth may produce an airborne chemical mixture that causes conspecific male moths to fly toward the female. This is an example of a releaser pheromone. In contrast, the pheromones released by some social insect queens, that inhibit ovarian development of her female progeny, are called primers.

Pheromones may be categorized according to the type of behavior they coordinate. Stimulants elicit a particular behavioral response such as locomotion, **egg**-laying, or feeding, while deterrents inhibit one of these activities. Pheromones that cause an animal to make oriented movements toward the source of the pheromone are termed attractants, in contrast to repellents that stimulate movements away from the source. Pheromones that cause a decrease in locomotion or turning behaviors are termed arrestants.

Pheromones may also be grouped according to their function for the species. Sexual or courtship pheromones mediate attraction of mates and their courtship behaviors. Moth pheromones are good examples of the power and specificity of such sexual attractants. Males of some saturniid moths are able to locate females emitting pheromones as far as 7 mi (11.27 km) away. Aggregation pheromones function to attract both sexes to a particular area for feeding, egg-laying, or mating. For example, some beetles attacking trees will produce pheromones to attract both sexes to the site. Ants lay down chemical trails to a food source; such trail pheromones are common among social insects. Social animals may also release pheromones for defensive purposes to alert the group to defend themselves from an approaching intruder. Honey bee workers release such an alarm pheromone from their sting apparatus during their stinging attacks. Territorial or marking pheromones function to communicate the location of an individual's territory. Pheromones then are important means of regulating many behaviors among individuals in a population.

Investigations into chemical communication systems are revealing that mixtures of different chemicals released in different concentrations elicit different behavioral responses. Even a slight change to one chemical of a pheromonal blend may drastically alter the response of the receiving organism. Such studies indicate the great complexity of pheromonal communication systems. Also, most organisms integrate pheromones with other methods of communication, resulting in an even more complex communication system.

PHOSPHATE GROUP

A phosphate group is an ion that contains one phosphorous atom bonded to four oxygen **atoms**. It is typically derived from a phosphoric acid and is anionic, having a negative three charge. When a phosphate group is attached to an **organic compound** it is bonded to that **molecule** by an oxygen molecule. Organic molecules containing phosphate groups are found throughout the body and perform a variety of functions.

One area that phosphate groups are found extensively is in **nucleic acids**, including **DNA**, and **RNA**. These molecules consist of **nucleotide**s which contain a five carbon sugar, an organic base and a phosphate group. Nucleotides are formed in two steps; first, the sugar and the organic base react, then this unit reacts with phosphoric acid. Nucleic acids are formed by the **polymer**ization of nucleotide **monomer**s. Each of these monomers are connected through their phosphate groups. In this way, the phosphate group acts to stabilize the structure of the backbone of the nucleic acid.

Another important function of phosphate groups in living **organism**s is providing a source of **energy** for **cell**s to do work. Cells do three main types of work, mechanical, chemical and transport. Mechanical work includes such things as controlling the flow of **cytoplasm** within the cell, moving **chromosome**s during reproduction, or contracting during muscle movements. Chemical work involves the metabolic reactions that would not occur spontaneously. Examples are **protein synthesis** and glycogenesis. Transport work involves the pumping of substances in and out of a cell. In most cases, a phosphate-containing compound called **adenosine triphosphate (ATP)** provides the power for all types of cellular work.

ATP is an organic molecule that is similar to the nucleotides found in nucleic acids. It has a **nitrogen** containing base and a five carbon sugar. However, unlike a nucleotide, it contains three phosphate groups instead of one. These phosphates are connected to each other making this part of the ATP molecule unstable. When it reacts with **water**, a phosphate is released along with a certain amount of energy. This is the basic reaction used by cells to power **chemical reaction**s.

An example of how ATP works is found in protein synthesis. **Protein**s are composed of smaller units called **amino acid**s which do not react spontaneously. During protein synthesis, **enzyme**s in the cell cause a phosphate group to be transferred from ATP to an amino acid. This process, known as **phosphorylation**, energizes the amino acid and allows it to react with other amino acids to form a protein. A similar phosphorylation reaction occurs in nearly all other cases where a cell performs work.

While phosphate groups can energize reactions, cells also use them to store energy. This is done by converting energy from catabolic reactions such as those found during **cellular respiration** and **photosynthesis** to ATP. In this reaction, adenosine diphosphate (ADP) is reacted with a phosphate group to form ATP and water.

PHOSPHOLIPIDS

Phospholipids are complex **lipids** made up of fatty acids, alcohols, and **phosphate**. They are extremely important components of living **cells**, with both structural and metabolic roles. They are the chief constituents of most biological **membranes**. At one end of a phospholipid **molecule** is a phosphate group linked to an **alcohol**. This is a polar part of the molecule–it has an electric charge and is water-soluble (hydrophilic). At the other end of the molecule are fatty acids, which are non-polar, hydrophobic, fat soluble, and water insoluble. Because of the dual nature of the phospholipid molecules, with a water-soluble group attached to a water-insoluble group in the same molecule, they are called amphipathic or polar lipids. The amphipathic nature of phospholipids make them ideal components of biological membranes, where they form a lipid bilayer with the polar region of each layer facing out to interact with water, and the non-polar fatty acid ''tail'' portions pointing inward toward each other in the center of the membrane. The lipid bilayer structure of cell membranes makes them nearly impermeable to polar molecules such as ions, but proteins embedded in the membrane are able to carry many substances through that could not otherwise pass.

Phosphoglycerides, considered by some as synonymous for phospholipids, are structurally related to 3-phosphoglyceraldehyde (PGA), an intermediate in the catabolic metabolism of glucose, and the first product of carbon fixation by the 3-carbon pathway. Phosphoglycerides differ from phospholipids because they contain an alcohol rather than an aldehyde group on the 1-carbon. Fatty acids are attached by an ester linkage to one or both of the free hydroxyl (-OH) groups of the glyceride on carbons 1 and 2. Except in phosphatidic acid, the simplest of all phosphoglycerides, the phosphate attached to the 3-carbon of the glyceride is also linked to another alcohol. The nature of this alcohol varies considerably. The most abundant phosphoglycerides in **animals** and higher **plants** are ethanolamine phosphoglyceride (phosphatidyl ethanolamine) and choline phosphoglyceride (phosphatidyl choline) in which the alcohols are ethanolamine and choline, respectively. These two phosphoglycerides, often called cephalin and lecithin, are major lipid components of most animal membranes. Phosphatidyl serine and phosphatidyl inositol, in which the phosphate is linked to the **amino acid** serine or an alcohol with a ring structure, respectively, are two minor components of cells.

The formation of phospholipids begins with the synthesis of glycerol-3-phosphate by one of two routes. One method is the reduction of dihydroxyacetone phosphate, an intermediate in the glycolysis pathway for the metabolic breakdown of glucose. Another involves the direct phosphorylation of glycerol by adenosine triphosphate (ATP) and the enzyme glycerokinase. In **eukaryotes**, fatty acids are then added to the remaining carbons with the aid of **coenzyme** A.

See also Membrane structure

PHOSPHORUS CYCLE

Phosphorus is an essential **nutrient** needed in large amounts by both **plants** and **animals**, and is thus known as a macronutrient. The dry weight of most **organism**s is comprised of approximately 0.3% phosphorus. It is an essential element in the makeup of bones, teeth, deoxyribonucleic acid (**DNA**), ribonucleic acid (**RNA**), adenosine triphosphate (**ATP**), and even cell membranes. The phosphorus cycle is sedimentary in origin. Unlike the nitrogen and **carbon cycle**s, where the main reservoir is in the atmosphere, the main reservoir for phosphorus is in sediments. There is no gaseous phase for phosphorus, so rainwater contains minimal amounts of this nutrient. The main source of phosphorus in the soil is apatite—a calcium-phosphate mineral. Other sources include guano deposits, which are long-term accumulations of phosphate-rich manure produced by colonies of bats, or migrating fish-eating birds, such as pelicans and cormorants. Considerable environmental damage has resulted from mining these deposits on islands. Weathering of terrestrial deposits slowly releases phosphorus, some of which washes into nearby rivers, lakes, or the ocean. Plants take up phosphorus in the form of phosphate and manufacture organic **macromolecules** like ATP. Herbivores that graze on plants incorporate these organic chemicals into their tissues, and so on through the other links up the food chain. Phosphate is released to the soil or water by animal excretion. This form of recycling is a significant source of phosphorus in freshwater ecosystems, where it is often low in concentration. When organisms die, bacterial and fungal decomposers mineralize the organic phosphorus to inorganic forms, returning phosphorus to the soil, and the cycle continues.

Phosphorus is a **limiting nutrient** for plant growth in many soils and most freshwater ecosystems. Excess phosphorus that reaches lakes often leads to **eutrophication**, sometimes evidenced by dense blue-green algae (**Cyanobacteria**) blooms. Humans accelerate this process, known as cultural eutrophication, by adding excess fertilizers to residential lawns and farm crops, deforestation, and sewage effluent. Most cities have secondary sewage treatment plants, but these remove only a maximum of 30% of the phosphorus. In some rural areas, wastes from chicken and hog farms washes into nearby rivers and eventually into lakes downstream

PHOTOLYSIS

Photolysis is a term used to describe the mechanism of photochemical reactions initiated by the absorption of visible, infrared, or ultraviolet radiation. When a **molecule** absorbs a photon of radiant **energy**, it is energized to a higher level than is possible by ordinary heating. As a consequence, the molecules may be split into smaller entities in a manner different from normal thermal reactions. This initial splitting is the first stage in photolysis (splitting or loosening by **light**). The subsequent reactions of these energized components convert radiant energy into chemical energy. The immediate response to an absorbed photon is called the primary photochemical process. Subsequent changes are also a part of the photochemical reaction.

The best-known biological example of photolysis occurs in **photosynthesis**, in which two distinct photosystems, photosystem I and photosystem II, function in a coordinated manner to produce oxygen and fix **carbon dioxide**.

In photosystem I, light energizes an electron associated with a chlorophyll molecule and is captured by an electron acceptor that is thereby reduced. This initiates a series of reactions that eventually lead to the reduction of NADP+ (nicotine adenine dinucleotide phosphate) to reduced NADPH. NADPH, required for the synthesis of fatty acids, is used in a series of reactions that fix carbon dioxide and reduce it to sugar. An electron passed along by photosystem II replaces the electron lost by the **chlorophyll** molecule.

Light absorbed by a chlorophyll molecule in photosystem II excites an electron in the pigment to a higher energy level. The high-energy electron is captured by a system of molecules that function as electron carriers. The electron carrier molecules can alternately accept an electron (and become reduced), and donate an electron to the next carrier (to become oxidized). Some of the energy that the electron gradually loses in this process is captured and stored in the form of **adenosine triphosphate (ATP)**. The electron lost by the chlorophyll molecule is replaced from **water** in a series of reactions that lead eventually to the release of oxygen gas. This process is sometimes called the photolysis of water, although water molecules do not directly absorb the light energy.

The nature of the two photosystems has been extensively studied. It has been possible to separate the two systems by treating lamellar (thin-**tissue**d) fragments of disrupted **chloroplast**s with detergents. Further treatment with charged detergents, and separation of components with electrophoretic techniques has led to the identification of components of the two systems. Each photosystem contains a light-harvesting complex of pigment molecules and a core complex containing a reaction center where the photolytic reactions actually take place.

See also Photophosphorylation

PHOTOPERIODISM

Many **plants** and **animal**s have the ability to measure changes in the duration of day and night and respond in ways appropriate to climatic cycles accompanying these changes. It was first thought that **organism**s with this capacity were measuring photoperiod (day length) and thus the phenomenon was called photoperiodism. The name persists, although it is now known that some organisms are sensitive to the duration of the dark or night period as well.

In animals, photoperiodism influences many seasonal activities such as **migration**, reproduction, and changes in plumage. Changing periods of daylight affect the animal's **pituitary gland** and cause it to release **hormone**s that affect reproduction. By artificially manipulating day length, the mating season of a **species** can be changed. Thus, species that normally breed in the fall, such as goats and sheep, can be induced to breed at other times by exposing them to long periods of **light** followed by short periods.

Several aspects of plant development are influenced by the photoperiod, but the most widely studied is flowering. Plants are classified as long-day plants if they are induced to flower during day lengths longer than 14 hours, and short-day plants if the inducing light period is less than 10 hours. Day-neutral plants flower regardless of the photoperiod. The critical factor appears to be the length of the dark period rather than the length of the day. Thus, a short day plant can be prevented from flowering if it is exposed to light, for even a very brief period, during the night. A plant pigment called phytochrome has been shown to play a critical role in this reversal of expected behavior. Phytochrome exists in two forms. One form called P_r absorbs red light (660nm) and in the process is converted to form P_{fr} that absorbs far-red light (730nm). When P_r is illuminated, the short-day plant will not flower, but if the resulting P_{fr} is immediately illuminated and converted back to P_r, the plant will flower normally. Photoperiod is also an important factor in inducing dormancy in woody plants. The short days of autumn are processed by the leaves, which transmit a signal to the growing bud causing it to change into a cold, hardy, dormant form.

Photoperiodic responses, although reasonably predictable, are influenced by environmental circumstances in both plants and animals. **Temperature, nutrition**, and other factors can modify the manner in which an organism will respond to a particular light regime.

Photoperiodism has many practical applications. It can be used to control mating, **egg** production, and growth in chickens. Artificial lighting can be used to alter day length to control flowering and increase production in greenhouse crops. Knowledge of a plant's photoperiodic **nature** can be used to select optimal planting dates to increase yields.

PHOTOPHOSPHORYLATION

Photophosphorylation is the **light** powered production of **energy**-rich **adenosine triphosphate (ATP)** during **photosynthesis**. ATP is formed by **phosphorylation** of adenosine diphosphate (ADP) in a reaction that stores energy in the bond linking ADP with an added phosphoryl group. Light provides the needed energy and in the process is converted into chemical energy. The chemical energy stored in ATP is used to synthesize sugars and other biochemical products of photosynthesis. In green **plants** the process takes place in a plant **organelle** called the **chloroplast**. The primary light-absorbing pigment is called chlorophyll, and is the substance responsible for the plant's green color. Based on the type of electron flow involved there are two types of photophosphorylation, cyclic and non-cyclic. In cyclic photophosphorylation the electron associated with a chlorophyll **molecule** is excited by a light photon but than returns to the original chlorophyll molecule from which it was emitted, whereas in the non-cyclic process electrons are transferred from an emitting to an accepting chlorophyll molecule. In either case, intermediate electron carriers mediate the flow and capture the energy lost by the electron in returning to an unexcited state. The intricate energy conversion process re-

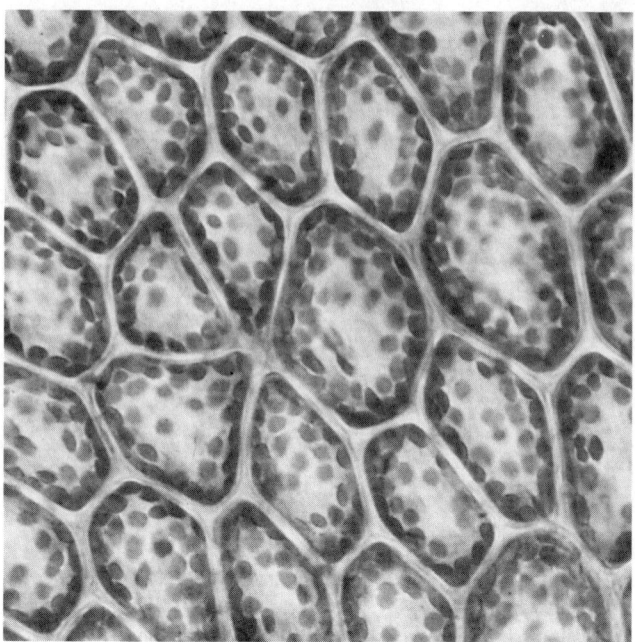

German botanist Julius von Sachs discovered chloroplasts, pictured above, in 1865. *(Photo Researchers, Inc. Reproduced by permission.)*

quires a precise structural organization of thylakoid **membrane**s in the interior of the chloroplast to successfully couple electron flow to the ATP producing process. Any disruption of the elaborate thylakoid structure leads to an uncoupling of electron flow from ATP production, and photosynthesis is no longer possible.

See also Carbon dioxide fixation; Chlorophyll function and structure

PHOTOSYNTHESIS

The process by which **plants** convert **carbon dioxide** and **water** into food via sunlight is called photosynthesis, which comes from the Latin words for ''formation in **light**.'' All our food comes from this process, either directly or indirectly. We eat green plants and their grains and **fruits** as well as the **animals** that feed on such plants.

It took scientists hundreds of years to understand what happens during photosynthesis. In this process, sunlight stimulates the **chlorophyll** found in structures in green plants called **chloroplast**s. In the chloroplasts, the sunlight reacts with carbon dioxide that the plant breathes in through microscopic holes in its leaves called stomata and with water that it absorbs through its roots. During a series of light and dark reactions, the water **molecule**s are broken down into hydrogen and oxygen, and the hydrogen combines with carbon dioxide to produce glucose, a simple sugar that is used as a building block for starch and other complex **carbohydrate**s. The excess oxygen is later released into the **atmosphere** during the processes of **respiration** and transpiration, and, in this manner, photosyn-

thesis is responsible for the renewal of the Earth's oxygen supply.

Research on photosynthesis has been closely linked to knowledge of the growth cycles and physical structure of plants. In the 1640s, the work of both **Johannes (Jan) Baptista van Helmont** and English clergyman and physiologist **Stephen Hales** indicated that plants require air and water to grow. In the 1700s, chemists began to identify the individual gases involved in the processes of combustion, respiration, and photosynthesis. **Joseph Priestley** demonstrated that green plants can replenish stale, or oxygen-poor, air so that it is capable of supporting combustion and respiration.

Dutch doctor and plant physiologist Jan Ingenhousz, inspired by Priestley's research, later learned that only the green parts of plants can revitalize stale air—that is, take in carbon dioxide and release oxygen—and that they do so only in the presence of sunlight. This was the first indication of light's role in the photosynthetic process. Ingenhousz also discovered that only the light of the Sun—and not the heat it generates—is necessary for photosynthesis.

In the nineteenth century, research on photosynthesis centered on the chemical processes in which carbon is ''fixed'' in carbohydrates. In the late 1800s, German botanist Julius von Sachs (1832-1897) suggested that starch is a product of carbon dioxide. He also argued in 1865 that, in the presence of light, chlorophyll catalyzes photosynthetic reactions, and he discovered the chlorophyll-containing chloroplasts. In the 1880s, German physiologist Theodor Wilhelm Engelmann (1843-1909) showed that the light reactions, which capture solar **energy** and convert it into chemical energy, occur within the chloroplasts and respond only to the red and blue hues of natural light.

It was not until the twentieth century that scientists began to understand the complex biochemistry of photosynthesis. Richard Willstätter recognized that there were two major types of chlorophyll in land plants: blue-green, or ''a'' type, and yellow-green, or ''b'' type. **Martin David Kamen**, a Canadian-born American biochemist, used the isotope oxygen-18 to trace the chemical's role in the process. He confirmed that the oxygen created during photosynthesis comes only from the water molecules. German biochemist **Otto Warburg** found that, under suitable conditions, the efficiency of the photosynthetic process can approach 100%, meaning that nearly all of the Sun's energy is converted to chemical energy.

In 1940, the discovery of carbon-14, a radioactive isotope of carbon isolated by Kamen, allowed for more detailed studies of photosynthesis. Using carbon-14, **Melvin Calvin** was able to trace carbon's path through the entire photosynthetic process. During the 1950s and 1960s, he confirmed that the light reactions involving chlorophyll instantly capture the Sun's energy. Then he studied the subsequent dark reactions, so-called because they can take place without sunlight, finding that carbohydrate molecules begin to form at this stage of the process. Working with green algal **cell**s, Calvin interrupted the photosynthetic process at different stages and plunged the cells into an **alcohol** solution. Then, using the laboratory technique called paper **chromatography**, he analyzed the cells and the

Plants respond to the direction and amount of light they receive. The seedling at the right was grown in normal, all-around light. The one in the center received no light. The plant at the left grew toward the light that it received on only one side. *(Photo Researchers, Inc. Reproduced by permission.)*

chemicals that had been produced, identifying at least ten intermediate products that had been created within a few seconds. This series of reactions is now called the **Calvin cycle**.

In 1998, scientists at Arizona State University announced that they had created an artificial photosynthetic energy system. The cell-like machine used light to power the synthesis of **adenosine triphosphate (ATP)**, a carrier of chemical energy in all **organism**s. The new technology could eventually lead to biological computers and new drugs.

PHOTOTROPISM

The tendency of **plants** to alter their growth in response to the direction of **light** shining on them is called phototropism. Generally the above ground shoot portion of plants tends to grow toward the light, exhibiting positive phototropism. This response increases the likelihood that the shoot will reach areas where it can maximize its ability to absorb light and carry out **photosynthesis**. As a result, the plant can grow more rapidly, and compete more successfully with its neighbors. Under some conditions a plant or a part of the plant such as a root may exhibit negative phototropism by growing away from the light. Phototropism is mediated by a yellow flavin pigment and responds to blue light.

A plant **hormone** called auxin stimulates the differential growth pattern that results in phototropism. Auxins are produced by **cell**s in the growing tip or apical meristem of the plant and diffuse down through the lower tissues. When light strikes the stem from one side it causes the auxin to concentrate on the opposite side, away from the light. Auxin stimu-

lates these cells to elongate. As a result, the tip bends toward the light. The bending reaction occurs in the zone of elongation, just below the rapidly dividing cells of the apical meristem

Some mobile lower forms of life in **algae**, **protozoa** and **bacteria** may respond to light by actually moving toward or away from it. This phenomenon is called phototaxis.

See also Plant Anatomy; Plant hormones

PHYLOGENETIC

The term phylogenetic (or phyletic) is used in reference to relationships among groups of **organism**s based on their shared evolutionary history (that is, on their phylogeny). One of the major goals of biological science (and of **paleontology**, the study of extinct organisms) is to understand the phylogenetic relations among living and extinct **species**, as well among higher taxa (this term refers to any taxonomic group, such as species, **genus**, **family**, **class**, or **phylum**). Phylogenetic knowledge is crucial to understanding the evolutionary history of life on Earth, including that of our own species, *Homo sapiens*.

Phylogenetic relationships among taxa are examined using various kinds of data, including information on the **comparative anatomy** of **fossil** species (and of higher taxa), and the **anatomy**, biochemistry, behavior, and **ecology** of living organisms. Complex sets of data on these sorts of variables are analyzed using certain mathematical procedures (such as cluster analysis and ordination) to determine groups of taxa based on their similarity and dissimilarity. In such analyses, a basic assumption is that taxa that are similar in terms of character variables (i.e., in their anatomy, biochemistry, behavior, and/or ecology) are likely to be more closely related than taxa that are dissimilar. (However, it must be borne in mind that **convergent evolution** can result in non-related taxa having similar characters. In addition, homologous characters can evolve if closely related taxa are subject to different regimes of **natural selection** (i.e., a particular attribute may differ greatly between related taxa, as is the case of the wings of birds and the forelimbs of their reptilian ancestors.)

The phylogeny of some groups of organisms has been worked out rather well by paleontologists. For example, an early progenitor of the lineage of modern horses was *Eohippus*, a relatively small, four-toed **animal** that lived in the Eocene epoch about 55 million years ago. The somewhat larger, three-toed *Mesohippus* lived during the Oligocene epoch (about 35 million years ago), apparently having evolved from the *Eohippus* lineage. Next to occur in this phylogenetic series was the larger, three-toed *Merychippus* of the Miocene epoch (25 million years ago), followed by the bigger, one-toed *Pliohippus* of the Pliocene (5 million years ago). Finally, this phylogenetic lineage led to the evolution of the even larger, one-toed, modern horses, of which there are several species, including the horse (*Equus caballus*), donkey (*E. asinus*), and zebras (including *E. zebra*).

PHYLUM

Phylum is a **taxon** or group within the hierarchical system of **taxonomy** located between **Kingdom** and **Class**. Since this is such a broad inclusion, members of the same phylum are quite diverse and generally have simple similarities. It is also presumed that they have a common evolutionary ancestry. Examples of phyla in the Kingdom Animalia include Porifera (sponges), Cnidaria (jellyfish and sea anemones), Platyhelminthes (flatworms), Nematoda (roundworms), Arthropoda (insects, crabs, etc.), Mollusca (snails, clams, octopi, etc.), Echinodermata (sea stars, sea urchins, etc.), and Chordata (fish, amphibians, reptiles, birds, and **mammals**). Within the Kingdom Plantae, the term Division is generally substituted for Phylum. Examples of Divisions include Coniferophyta (cone-bearing **plants** such as pine **tree**s) and Magniliophyta (flowering plants).

PHYSICS AND BIOLOGY

One of the most promising new directions in biology in the 1990s was the incorporation of physics into the field. Biologists began to borrow some of the mathematics of physics in order to explain biological phenomena. And similarly, physicists studied living biological systems as models for some particularly complex physical processes. The interface between biology and physics has come to be called biophysics.

Biologists have long used laboratory techniques that originated with physics. A prime example is the use of carbon

14 (C_{14}), the long-lived radioactive isotope that is found in all living things. Its value to the study of plant and **animal** systems and to **paleontology** is immeasurable. (C_{14}) was discovered in 1940 by a chemist working at one of the world's premier physics laboratories, the University of California's Radiation Laboratory. Other methods first devised or used by physicists and now commonly applied to biology include electron microscopy, **magnetic resonance imaging**, scanning, tunneling, microscopy, positron emission tomography, and other advanced testing and scanning processes and devices.

Perhaps the most influential work on physics and biology was Erwin Schrödinger's 1943 book, *What Is Life?* Schrödinger, a renowned physicist, applied his knowledge of quantum mechanics to the still-developing field of **genetics**, and suggested for the first time that **mutation** takes place in a specific **molecule**. His book directly inspired **James D. Watson** to search for the secret of **gene**s, and many other prominent scientists were influenced by Schrödinger's ideas.

However, many experiments in biophysics that seemed promising in the 1940s and 1950s were ultimately disappointing. The computers in use at the time were simply not powerful enough to handle the data, and because of this, interest in physics declined, and the idea of joining physics to biology became dormant for the next 30 or 40 years. And as each field became more specialized, the possibility of one scientist knowing enough about physics and biology to use both became more remote.

The 1990s, however, yielded some key experiments that showed that the latent field was again gathering momentum. In one case, a biologist, Neil Mendelson, discovered a unique mutant strain of **bacteria** that grew into twisted, ropy fibers. Although biologists were unimpressed by Mendelson's strange bacteria, called *Bacillus subtilis*, Mendelson worked with it for almost 20 years. In 1992, the mathematician Michael Tabor, inspired by a film of the growth of *Bacillus subtilis* filament, studied it for several years and used it to build a mathematical model of elasticity. Elasticity theory has wide-ranging uses. It models the twisting of **DNA** molecules as well as the coiling of magnetic field lines from a star, the cause of solar flares. So, this bacteria that was just a curiosity to biologists became of key importance to physicists, who hoped to use the math engendered by the *Bacillus subtilis* studies to explain, and perhaps predict solar flares.

Conversely, physicists trained to look for universal laws of motion or attraction have had a dramatic impact on our understanding of minute biological phenomena. One such instance is the study of special molecules, called "lock-and-key" molecules, on **cell membrane**s. Lock-and-key molecules allow certain cells to adhere, and this is pivotal to both normal cell development and to the growth of **cancer** cells and other diseases. Physicists viewed the lock-and-key molecules as a model of a particular kind of adhesion, and developed a mathematical description of how they work. This gave biologists a more general understanding of cell growth. Biophysicists in the 1990s also applied physics to the study of ion channels in cell membranes. These are tiny channels that open and close unpredictably in response to electric fields coming across the membrane. Biologists had known that ion channels were key to the extremely sensitive motion-sensing **organ**s of some animals such as sharks and crayfish. Biophysicists were able to use computer simulations and studies of ion channel **proteins** to develop a quantitative description of the way ion channels work at the molecular level. Using physics, researchers in the 1990s were able to carry their investigations to a molecular, mathematical level.

Studies like these, reported in scientific journals in the early 1990s, stirred more awareness. By the end of the 1990s, biophysics had generated great interest in both the fields of physics and biology. One area that generated many new studies was behavioral biology. Behavioral biology refers to the study of how cells, **organelle**s and **organism**s receive and act upon outside stimuli. This could mean mechanical, chemical, optical, thermal, electrical or magnetic signals. Physicists were already familiar with the complex physical laws governing these signals, and developed new hypotheses by studying the living sensory reception systems of such **species** as bats and electric fish.

The biggest barrier to the growth of biophysics at the end of the1990s was that biologists did not have the advanced mathematics background required of physicists. So, the impetus to connect the two fields seemed to come first from the physicist's side. As was the case of the Bacillus subtilis, biologists might have models of systems physicists are seeking to describe, and as each discipline becomes more exposed to the other in the future, more fruitful collaborations may occur.

See also Carbon dating; Electron microscope; Scanning electron microscope

PHYSIOLOGICAL ACOUSTICS · See Acoustics, physiological

PHYSIOLOGY, COMPARATIVE

While **anatomy** is the study of the structures of an organism, physiology is the science dealing with the study of the *function* of an organism's component structures. However, it often is not enough to know *what* an organ, tissue, or other structure does. Physiologists want to know *how* something functions. For example, physiological questions might ask: What is the function of human lung tissue? How can a seal survive under water without breathing for over ten minutes? How do camels survive so long without water? How do insects see ultraviolet light? Physiology examines functional aspects at many levels of organization, from molecules, to cells, to tissues, to organs, to organ systems, to an entire organism. It is the branch of biology that investigates the operations and vital processes of living organisms that enable life to exist.

Comparative physiology, then, is the comparison of physiological adaptations among organisms to diverse and changing environments. Comparative physiology, like **comparative anatomy**, attempts to uncover evolutionary relationships between organisms or groups of organisms. Comparative physiology seeks to explain the evolution of biological functions by likening physiological characteristics between and among organisms (usually animals). This branch of biology constructs **phylogenetic** relationships (or, more loosely, evolutionary connections) between and among groups of organisms. Comparative physiology, in conjunction with other comparative disciplines, enables us to trace the evolution of organisms and their unique structures and to view ourselves in a broader light. By comparing the physiology among living things, scientists can gain insights into how groups of organisms have solved the adaptive problems in their natural environments over time.

Comparative physiology compares basic physiological processes like **cellular respiration** and gas exchange, **thermoregulation**, circulation, water and ion balance, nerve impulse transmission, and muscle contraction. Because it focuses on function, comparative physiology can also be referred to as *functional anatomy*. The form of an organ, or other biological structure, is tied to its function much in the same way a tool is linked to its purpose. For example, the function of an **enzyme** (a protein molecule that speeds up a chemical reaction) depends heavily upon its three-dimensional shape. If the 3-D conformation of the enzyme molecule is altered (by heat or acid), the function of the enzyme will also be altered. If the shape of an enzyme is changed considerably, its biological activity will be lost.

A major theme dominating the topic of comparative physiology is the concept of **homeostasis**. The term is derived from two Greek words (*homeo*, meaning "same," and *stasis*, meaning "standing still") and literally means "staying the same." Homeostasis thus refers to the ability of animals to maintain an internal environment that compensates for changes occurring in the external environment. Only the surface cells of the human body, for example, and the lining of the gastrointestinal and respiratory tracts come into direct contact with the outside surroundings (like the atmosphere). The vast majority of cells of the body are enclosed by neighboring cells and the extracellular fluid (fluid found outside of cells) that bathes them. So the body in essence exists in an internal environment that is protected from the wider range of conditions that are found in the external surroundings. Therefore, to maintain homeostasis, the body must have a system for monitoring and adjusting its internal environment when the external environment changes. Comparative physiologists observe physiological similarities and differences in adaptations between organisms in solving identical problems concerning homeostasis.

Some of the problems that **animal**s face in maintaining physiological homeostasis involve basic life processes. **Energy** acquisition from food (digestion) and its expenditure, the maintenance of body temperature and metabolic rate, the use of oxygen or the ability to live in its absence, and the way body size affects **metabolism** and heat loss are examples of problems that require homeostatic systems. Comparative physiologists might, for example, compare the efficiency of the relative oxygen capturing abilities of mammalian **hemoglobin** (in red blood **cell**s) and insect hemolymph. Both groups of animals must maintain homeostasis and regulate the amount of oxygen reaching their tissues, yet each group solves the problem differently.

Comparative physiology makes specific measurements to obtain biologically relevant information from which to make comparisons. The kinds of processes that physiologists measure from anatomical structures to gain insight into their function include: rates (how fast something occurs), changes in rates, gradients (increasing or decreasing concentrations of substances), pressures, rate of flow (of a fluid such as air or **blood**), **diffusion** (the act of a substance moving from an area of high concentration to one of low concentration), tension (material stress caused by a pull), elasticity, electrical current, and voltage. For example, a comparative physiologist might measure the rate of diffusion of sugar **molecule**s across intestinal cell **membrane**s, or the pressure exerted on the walls of blood vessels that are close to the heart. In each case, the comparative physiologist is trying to gain information that will help explain how a particular structure functions and how it compares with similar structures in other organisms in solving the same homeostatic problem. The conclusions derived, then, tell us all about our evolutionary history.

PHYTOPLANKTON

Phytoplankton are microscopic, photosynthetic **organism**s that float in the **water** of the oceans and bodies of freshwater (the word phytoplankton is derived from the Greek for "drifting **plants**"). The most abundant organisms occurring within the phytoplankton are **algae** and blue-green **bacteria**, but this group also includes certain kinds of protists (especially **protozoa**ns) that contain symbiotic algae or bacteria.

Phytoplankton are responsible for virtually all of the primary production occurring in the oceans. Marine phytoplankton range in size from extremely small blue-green bacteria, to larger (but still microscopic) unicellular and colonial algae. Oceanic phytoplankton are grazed by tiny **animals** known as **zooplankton** (most of which are crustaceans). These are eaten in turn by larger zooplankton and small fish, which are fed upon by larger fish and baleen whales. Large predators such as bluefin tuna, sharks, squid, and toothed whales are at the top of the marine food web. Marine phytoplankton are much more productive near the shores of continents, and particularly in zones where there are persistent upwellings of deeper water. These areas have a much better nutrient supply, and this stimulates a much greater **productivity** of phytoplankton than occurs in the open ocean. In turn, these relatively fertile regions support a higher productivity of animals. This is why the world's most important marine fisheries are supported by the continental shelves (such as the Grand Banks and other shallow waters of northeastern North America, near-shore waters of western North and South America, and the Gulf of Mexico) and regions with persistent upwellings (such as those off the coast of Peru and elsewhere off western South America, and extensive regions of the Antarctic Ocean).

Some inland waterbodies occur in inherently fertile watersheds, and are naturally eutrophic, meaning they have a high productivity and **biomass** of phytoplankton (in shallow waters, larger aquatic plants may also be highly productive). So-called "cultural **eutrophication**" is a kind of **pollution** caused by nutrient inputs associated with human activities, such as the dumping of sewage waste and the runoff of fertilizer from agricultural land. Both fresh and marine waters can become eutrophic through increases in their nutrient supply, although the problem is more usually severe in freshwaters. The most conspicuous symptom of eutrophication is a large increase in the biomass of phytoplankton, which in extreme cases is known as an algal "bloom."

PIAGET, JEAN (1896-1980)
Swiss psychologist

Jean Piaget is considered the first person to study and identify how children learn. His studies in the early 1900s led to three well-known books, and an entirely new way to look at education and how teachers can best encourage **learning**.

Born in Neuchâtel, Switzerland on August 9, 1896, Piaget was initially interested in **zoology**. He studied this and philosophy at the University of Neuchâtel, earning his doctorate in 1918. His research focused on epistemology, or the theory of knowledge, and his studies included work with psychologists Carl Jung and Eugen Bleuler. In 1919 he began two years at the Sorbonne in Paris, where he created reading tests for children. By the end of his allotted time he had published his findings, which outlined the childrens' reasoning processes during learning.

Piaget combined these findings with those acquired in a subsequent study of 100 elementary-aged Swiss children to develop a theory of cognitive and moral development. He advanced this theory in his book, *The Moral Judgment of the Child*, in 1930. Piaget studied the childrens' attitudes towards the rules of the game marbles, and the moral judgments they expressed after listening to stories that included various types of dilemmas or conflicts. When the children played marbles, Piaget found two sequential moralities: one of constraint, in which the rules are considered obligatory and unbreakable, and thus obeyed without questions, or heteronomous thinking; and one of cooperation. In the latter, called autonomous thinking, rules are not seen as absolute because the players realize that the rules can change as a result of cooperation and consensus.

Piaget also identified a series of stages within these two sequential moralities, at which each successive stage the child takes on a broader perspective and is able to consider a greater number of variables of increasing complexity. Each stage builds on the previous ones, but takes on a new form as well. In the first stage called sensorimotor, at approximately two years of age, the child invents private rituals during playtime and games of make-believe. The rules are changed to fit this ritual or fantasy. At age five, the child progresses to the next two stages, which represent the true heteronomous level of reasoning. First, the child regards rules as permanent, external forces handed down from adults, which cannot be altered to suit personal needs. In the second substage, the child begins to establish a sense of mutual respect for peers, and thus increases the sense of moral obligation to accommodate the rules of others besides adults.

By the age of eight or nine, the child has moved into the fourth stage in which adult constraints are replaced by social cooperation with peers as the primary motive to obey rules. Finally, beginning around 11 years old, the child starts to acquire more extended reasoning abilities and can construct new rules to address all possible situations. At this final stage, the child has learned to consider and sort through the complexity of social, political and ethical issues, rather than focusing only on individual persons and interpersonal relationships.

Piaget became director of the Institut Jean-Jacques Rousseau in Geneva in 1921, where he completed his second round of studies on children. This was followed from 1926 to 1929 by a professorship of philosophy at the University of Neuchâtel, and then as professor of psychology at the University of Geneva. He continued in this post until his death on September 17, 1980.

PILTDOWN HOAX

On December 18, 1912, Charles Dawson (1865–1916) announced to the Geological Society in London that he had discovered skull fragments and a partial jaw in a gravel formation in Piltdown Common, Fletching, near Lewes, Sussex, England. The skull fragments were accompanied by bones of relatively recent hippopotamus, deer, beaver, and horse, as well as ancient bones of extinct mastodon and rhinoceros. Collected over a period of years, the skull fragments had an unusually thick **brain** case but were otherwise considered to be human. The jaw remnant was clearly primitive. This spectacular announcement was considered evidence, found in Britain, that supported the Darwinian **evolutionary theory** and provided a true representational link to modern man. Named in honor of its discoverer, Dawn man (*Eoanthropus dawsoni*), would eventually be known as Piltdown man, the most deceptive scientific hoax of the twentieth century that would take 40 years to disprove.

Initially, there was skepticism and scientists proposed that the jaw and cranium fragments were from two creatures, rather than one. However, in 1915, a second Piltdown man was discovered two miles from the original site. The second set of fossil remains seemed to indicate that the possibility of a human cranium and an ape jaw coming together purely by chance was unlikely. Clearly, both jaw and cranium fragments were from one type of human ancestor that provided evidence of an intermediary stage between ape and human, however when compared to other authentic prehuman **fossil**s, it was unclear where piltdown man fit in the evolutionary development of man.

Even with the lack of continuity between Piltdown man and other prehuman fossil remains, the authenticity of Piltdown man was not disproved until 1953, when dating techniques unequivocally proved it a fraud. Piltdown man was

Gregory Goodwin Pincus. *(The Library of Congress. Reproduced by permission.)*

merely a hoax made up of an ancient human skull and a contemporary orangutan jaw. The dark color of the fragments that was representative of fossil find in the area was artificial. The teeth in the orangutan jaw had been mechanically ground down to resemble humanlike wear, rather than that of apes. In 1912, accurate dating techniques were unavailable and the fervor to provide evidence to support the cherished belief that humans had first developed a big brain, and then later developed other human characteristics was great.

PINCUS, GREGORY GOODWIN (1903-1967)
American biologist

Gregory Goodwin Pincus' research in endocrinology resulted in pathbreaking work on **hormone**s and **animal** physiology. However, he is best known for developing the oral contraceptive pill. As his friend and colleague Hudson Hoagland remarked in *Perspectives in Biology and Medicine:* ''[Pincus'] highly important development of a pill... to control human fertility in a world rushing on to pathological overpopulation is an example of practical humanism at its very best.'' In addition, Pincus also participated in the founding of the Worcester Foundation for Experimental Biology and the annual Laurentian Hormone Conference.

Pincus was born in Woodbine, New Jersey, on April 9, 1903, the eldest son of Joseph and Elizabeth Lipman Pincus.

His father, a graduate of Storrs Agricultural College in Connecticut, was a teacher and the editor of a farm journal. His mother's family came from Latvia and settled in New Jersey. Pincus' uncle on his mother's side, Jacob Goodale Lipman, was dean of the New Jersey State College of Agriculture at Rutgers University, director of the New Jersey State Agricultural Experiment Station, and the founding editor of *Soil Science* magazine.

After attending a public grade school in New York City, Pincus became an honor student at Morris High School where he was president of the debating and literary societies. As an undergraduate at Cornell University, he founded and edited the *Cornell Literary Review*. After receiving his B.S. degree in 1924, he was accepted into graduate school at Harvard. He concentrated on **genetics** under W. E. Castle but also did work on physiology with animal physiologist W. J. Crozier. Pincus credited the two scientists with influencing him to eventually study reproductive physiology. He received both his Master of Science and Doctor of Science degrees in 1927 at the age of twenty-four. Pincus married Elizabeth Notkin on December 2, 1924, the same year he completed his undergraduate degree. They had three children—Alexis, John, and Laura Jane.

Pursues Research in Reproductive Biology

In 1927 Pincus won a three-year fellowship from the National Research Council. During this time, he travelled to Cambridge University in England where he worked with F. H. A. Marshall and John Hammond, who were pioneers in reproductive biology. He also studied at the Kaiser Wilhelm Institute with the geneticist Richard Goldschmidt. He returned to Harvard in 1930, first as an instructor in biology and then as assistant professor.

Much of the research Pincus did during the early part of his career concentrated on the inheritance of physiological traits. Later research focused on reproductive physiology, particularly sex **hormone**s and gonadotrophic hormones (those which stimulate the reproductive glands). Other research interests included geotropism, the inheritance of diabetes, relationships between hormones and **stress**, and endocrine function in patients with mental disorders. He also contributed to the development of the first successful extensive partial pancreatectomy in rats.

The development of the oral contraceptive pill began in the early 1930s with Pincus' work on ovarian hormones. He published many studies of living ova (**eggs**) and their **fertilization**. While still at Harvard he perfected some of the earliest methods of transplanting animal eggs from one female to another who would carry them to term. He also developed techniques to produce multiple ovulation in laboratory animals. As a consequence of this work, he learned that some phases of development of an animal's ovum were regulated by particular ovarian hormones. Next, he analyzed the effects of ovarian hormones on the function of the uterus, the travel of the egg, and the maintenance of the blastocyst (the first embryonic stage) and later the embryo itself. By 1939 he had published the results of his research on breeding rabbits without males by artificially activating the eggs in the females. This manipu-

lation was called "Pincogenesis," and it was widely reported in the press, but it was not able to be widely replicated by other researchers.

After returning from a year at Cambridge University in 1938, Pincus became a visiting professor of experimental **zoology** at Clark University in Worcester, Massachusetts, where he stayed until 1945. It was at Clark that Pincus began to work with Hoagland, though they had known each other as graduate students. Together they began to research the relationship between stress and hormones for the United States Navy and Air Force. Specifically, they examined the relationship between steroid excretion, adrenal cortex function, and the stress of flying. While at Clark University, Pincus was named a Guggenheim fellow and elected to the American Academy of Arts and Sciences.

Participates in the Founding of Scientific Organizations

In the spring of 1943, the first conference on hormones sponsored by the American Association for the Advancement of Science was held near Baltimore. Since the conference was held at a private club, African American scientist Percy Julian was excluded. Pincus protested to the management, and Julian was eventually allowed to join the conference. Although not an organizer the first year, Pincus was involved in reshaping the conference the following year, along with biochemist Samuel Gurin and physiological chemist Robert W. Bates. They held the conference in the Laurentian mountains of Quebec, Canada, and from then on the conference was known as the Laurentian Conference, and Pincus was its permanent chairperson. In addition to his administrative duties, he edited the twenty-three volumes of *Recent Progress in Hormone Research,* a compendium of papers presented at the annual conferences.

With Hoagland, Pincus also co-founded the Worcester Foundation for Experimental Biology (WFEB) in 1944. Hoagland served as executive director of the WFEB; Pincus served as director of laboratories for twelve years and then as research director. The WFEB served as a research center on steroid hormones and provided training for young biochemists in the methods of steroid biochemistry. From 1946 to 1950 Pincus was on the faculty of Tufts Medical School in Medford, Massachusetts, and then from 1950 until his death he was research professor in biology at Boston University Graduate School. Many of his doctoral students at these universities completed research at the WFEB.

Uses Hormone Research to Develop Oral Contraceptive

Pincus had been conducting research on sterility and hormones since the 1930s, but it was not until the 1950s that he applied his theoretical knowledge to the idea of creating a solution to the problem of overpopulation. In 1951 he was exposed to the work of Margaret Sanger, who had described the inadequacy of existing birth control methods and the looming problem of overpopulation, particularly in underdeveloped areas. By 1953, Pincus was working with Min-Chueh Chang at the WFEB, studying the effects of **steroids** on the fertility of laboratory animals.

Science had made it possible to produce steroid hormones in bulk, and Chang discovered a group of compounds called progestins which worked as ovulation inhibitors. Pincus took these findings to the G. D. Searle Company, where he had been a consultant, and shifted his emphasis to human beings instead of laboratory animals. Pincus also brought human reproduction specialists John Rock and Celso Garcia into the project. They conducted clinical tests of the contraceptive pill in Brookline, Massachusetts, to confirm the laboratory data. Pincus then travelled to Haiti and Puerto Rico, where he oversaw large-scale clinical field trials.

Oscar Hechter, who met Pincus in 1944 while at the WFEB, wrote in *Perspectives in Biology and Medicine* that "Gregory Pincus belongs to history because he was a man of action who showed the world that the population crisis is not an 'impossible' problem. He and his associates demonstrated that there is *a* way to control birth rates on a large scale, suitable alike for developed and underdeveloped societies. The antifertility steroids which came to be known as the 'Pill' were shown to be effective, simple, contraceptive agents, relatively safe, and eminently practical to employ on a large scale." Pincus spent much of the last fifteen years of his life travelling to explain the results of research. This is reflected in his membership in biological and endocrinological societies in Portugal, France, Great Britain, Chile, Haiti, and Mexico. His work on oral contraceptives was also recognized by awards such as the Albert D. Lasker Award in Planned Parenthood in 1960 and the Cameron Prize in Practical Therapeutics from the University of Edinburgh in 1966. He was elected to the National Academy of Sciences in 1965.

Pincus died before the issue of *Perspectives in Biology and Medicine* commemorating his sixty-fifth birthday was published. Although ill for the last three years of his life, he had continued to work and travel. He died in Boston on August 22, 1967, of myeloid metaplasia, a bone-marrow disease which some speculate was caused by his work with organic solvents.

PINEAL BODY

The pineal body (corpus pineale), or the pineal gland, is a small, pinecone-shaped **endocrine gland** located at the posterior portion of the third ventricle of the **brain**. The pineal gland, which was known to physicians in Antiquity, has been a mystifying subject to scientists. René Descartes, who, while maintaining that the soul and the body, including the brain, are two essentially different substances, identified the pineal body as the point of contact which enables the immaterial soul to affect the material body. At one time, the pineal gland was thought to be a vestigial third eye, due to the fact that in some lower **animal**s, such as fishes and amphibians, the pineal gland is comprised of **cell**s that are similar to the specialized, neuroepithelial cells of the retina that respond to **light**. Fossilized skulls of ostraderms, a type of jawless fish, reveal holes for three eyes, with the third eye hole located on top of the skull. Modern fishes and some amphibians and reptiles still have a remnant of that third eye which actually serves as a light receptor.

The pineal gland contains several **neurotransmitters**: the amines, melatonin, norepinephrine, serotonin, histamine, and the **peptide**s, thyrotropin-releasing **hormone**, somatostatin, luteinizing hormone-releasing hormone, and vasotocin. In addition, the pineal gland produces unique **enzyme**s that convert serotonin into melatonin. The primary **secretion** of the pineal gland is the hormone, melatonin, which is the only pineal neurotransmitter that is secreted directly into the **blood**stream. It has been found that the amount of light directly affects the formation of melatonin. Light entering the eyes stimulates retinal **neuron**s which transmits impulses to the **hypothalamus** gland and then the superior cervical **ganglion**, which in turn, stimulates the pineal gland to cease melatonin production. Darkness promotes the synthesis of melatonin and consequently, the highest level of melatonin found in the blood occurs at night.

The role of the pineal gland in human physiology is still not completely known, although it is accepted that our internal clocks, or circadian rhythms are partially regulated by this gland. Because melatonin is secreted in large amounts during darkness, it is postulated that melatonin induces sleepiness, thus regulating our sleep cycles. Research experiments have shown that even when pineal cells are grown under artificial conditions in the dark, for example, in a **culture** medium, melatonin is secreted on a 24-hour cycle. Experiments involving animals have shown that when the pineal glands is removed from a bird, the cyclic behavior, commonly manifested by the individual's **species**, stops. Furthermore, melatonin also has a suppressive affect on some, particularly gonadal, hormones. For example, when female rats are exposed to darkness for an extraordinarily long period, their estrus cycle slows down.

See also Biological rhythms

PINOCYTOSIS

Pinocytosis (from Greek *pinein*, "to drink" and *kytos*, "vessel") is the process in which **cell**s engulf liquids. These liquids may or may not contain dissolved materials needed by the cell. Because the cells are taking in liquid, pinocytosis is sometimes called "cell drinking."

Pinocytosis is one form of **endocytosis**, the process by which materials are engulfed into a cell. The other forms of endocytosis are **phagocytosis** and receptor-mediated endocytosis. In all three, there is an infolding of the cell **membrane**, which then pinches around substances, forming a vacuole or vesicle, and materials are transported into the **cytoplasm** of the cell. If the vacuole or vesicle is small it is called micropinocytosis. If the vacuole or vesicle is larger it is called macropinocytosis. The vacuole or vesicle often contains **water**, salts, and other substances from the exterior of the cell, in addition to the material that caused pinocytosis to occur. Once inside the cell the **vacuole**s or vesicles either break apart into smaller bundles or join to form larger ones. At this point, the materials that were engulfed into the cell can be used or processed. Since all forms of endocytosis require **energy** some scientists consider them forms of **active transport**.

Because all extracellular dissolved solutes will be taken in, cells are not able to select the materials that are transported. Thus, pinocytosis is unspecific in the materials that are transported.

Pinocytosis occurs in many types of cells of multicellular **organism**s. For example, as a human **egg** cell matures in the ovary, it is surrounded by other cells. These cells pass **nutrients** to the egg cell, which engulfs them using pinocytosis. Pinocytosis has also been observed in white **blood** cells (macrophages and leukocytes), kidney cells, epithelial cells of the intestine, and plant root cells.

PIONEER SPECIES

Pioneer species are the first **species** to enter an area after a disturbance. They are the colonizers. The pioneer species are the first stages of ecological **succession**, which is a series of changes over time in the types of **organisms** in an **ecosystem**. Common pioneer species include **bacteria**, **fungi** and lichens.

Pioneer species usually have the ability to live in harsh environments where other species cannot survive. These organisms are able to quickly colonize recently disturbed areas through rapid reproduction. They are well-adapted to dispersing their young to new locations. Pioneer species often slightly change the environments that they colonize. In some instances pioneer species such as lichens and **plants** break apart rock and add organic **matter** to soil. This frequently makes it easier for new organisms to enter the environment, survive and outcompete the pioneer species. Thus, organisms in the next stage of ecological succession eventually take over the **habitat** from the pioneer species.

An example of ecological succession, starting with a pioneer species, occurs when a volcano erupts and completely covers an area with lava. It would wipe out all plants and **animals** living there. After the lava cooled and hardened, it would be very difficult for plants to survive, because of the lack of soil. The first organisms to arrive and survive, the pioneer species, would be adapted to live in this environment. For example, lichens, a symbiotic relationship between a fungus and an alga, might be one of the first species to survive here. Lichens would be able to attach to the newly hardened lava because they have evolved to survive in areas with little to no soil. Over time, the lichens would slowly break apart the lava rock and create a small amount of soil. This would allow other organisms, such as a moss, to come into the habitat and outcompete the lichens. Thus, pioneer species do not last forever in one location. They are replaced as ecological succession continues.

PITUITARY GLAND

The pituitary gland, or hypophysis cerebri, is a smallendocrine gland, measuring approximately 0.5 in (1.3 cm), and is situated at the base of the **brain**, cradled in a fossa of the sphenoid bone. While the pituitary gland controls several other **endocrine glands**, it is controlled by the **hypothalamus**, to which it is

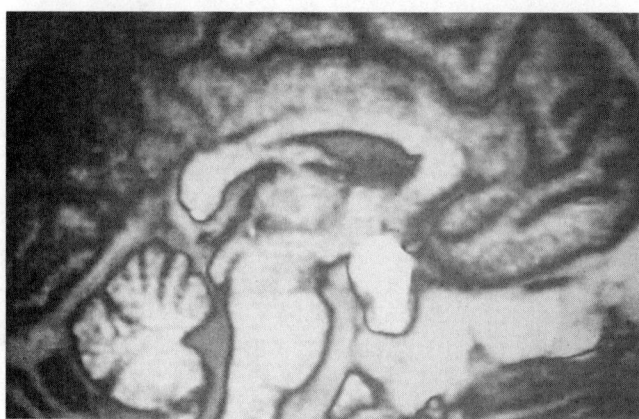

A magnetic resonance image (MRI) scan of brain revealing a pituitary gland tumor. *(The Stock Market. Reproduced by permission.)*

physically connected. There are three anatomical regions of the pituitary, the anterior and posterior lobes, and an intermediate zone whose function is not well understood.The pituitary develops in the **embryo** from the neural tube and the primitive digestive tube, forming the posterior and anterior lobes, respectively. Both lobes secrete **hormone**s responsible for maintaining **homeostasis**.

The posterior lobe, or neurohypophysis, directly receives neurosecretions from the hypothalamus by way of **axon**s that originate in the nerve **cell**s of the hypothalamus, so the posterior lobe actually does not synthesize hormones, but rather releases neurosecretions, or neurohormones, made in the hypothalamus. The two primary hormones released by the posterior lobe are vasopressin or antidiuretic hormone (ADH), which helps control **blood pressure**, and oxytocin, a hormone responsible for the release of breast milk and the contraction of smooth muscles, particularly the uterus.

The anterior pituitary, or adenohypophysis, is the larger of the two lobes, and makes up approximately three quarters of the entire pituitary gland. Its relationship to the hypothalamus is circulatory. The anterior lobe receives neurosecretions via capillaries from the hypothalamus. These neurosecretions regulate the production and the release of seven hormones which affect specific target **organ**s in the body. Other hormones secreted by the anterior lobe include, thyroid stimulating hormone (thyrotropin) which stimulates thethyroid gland, adrenocorticotropic hormone (ACTH), which acts on the **adrenal glands** to produce various steroid hormones, as well as two gonadotropic hormones, called FSH and LH, which stimulate **egg** and **sperm** production. Another hormone, prolactin, is thought to have appeared early in the **evolution** of vertebrates, and is responsible for stimulating milk production—it is not, however, to be confused with oxytocin, which stimulates the release of milk out of the mammary ducts. Melanostimulating hormone (MSH) influences the dispersion of pigment granules in melanocytes, determining the lightness or darkness of skin. Growth hormone, prolactin, and MSH are controlled by releasing hormones and release-inhibiting hormones produced by the hypothalamus. The other hormones produced by the anterior lobe are controlled through a negative feedback mechanism.

PLACENTA

The placenta is an **organ** that sustains and nourishes the fetus and is of both fetal and maternal origin. The placenta produce **hormone**s and thus has an endocrine function. The placenta has an immunological role in the suppression of fetal rejection. **Blood** travels from the fetus to the placenta and returns to the fetus via the umbilical cord which attaches near the center of the placenta. The placenta is expelled from the uterus following **birth** of the baby and is thus known as the afterbirth. The fully developed placenta is discoidal in shape with a diameter of 6-8 in (15-20 cm) and a thickness of about 1 in (2-3 cm).

The fetus has lungs but they have no respiratory function. The placenta serves in lieu of lungs. Fetal blood moves to the placenta via two umbilical **arteries**. Fetal blood loses **carbon dioxide** in the placenta and oxygen derived from the maternal blood diffuses across placental **membrane**s to oxygenate fetal blood. The oxygen restored blood returns to the fetal body via a single umbilical vein. During its nine month duration in the uterus, the fetus increases in mass from a tiny blastocyst to a newborn that may weigh an average of 7.5 lb (3.4 kg). All of this growth is sustained by **nutrients**, derived from the mother, which diffuse across the placental membranes. Living beings produce metabolic wastes and the fetus is no exception. Metabolic wastes in the blood of the fetus pass to the maternal blood via the placenta and are excreted by her kidneys. Perhaps the best known hormone produced by the placenta is human **chorion**ic gonadotropin (HCG) because of its use in the diagnosis of pregnancy. The production of progesterone is stimulated by HCG. While ovaries are the usual source of progesterone, HCG of the placenta stimulates other placental **cell**s to secrete progesterone also. Human chorionic somatomammotropin, also known as placental lactogen, aids in the preparation of the maternal mammary glands for milk production after the baby's birth.

Identical twins do not reject grafted **tissue** obtained from each other. Ordinarily, all other humans tend to reject tissue that is not identical with self. The rejection is a function of major histocompatibility **antigen**s. This is a problem that must be controlled with immunosuppressive drugs for the surgical transplantation of donor organs. How then is it that mothers do not reject the fetus that must live in the uterus for nine months? Mothers ordinarily reject transplanted organs of their children (after they are born) because of the expression of paternal antigens. It has been suggested that the placenta secretes **protein**s that block the production of antibodies. Further, the placenta stimulates production of **lymphocytes** concerned with blocking the immune response of the uterus. Increased knowledge of placental physiology as it relates to immunobiology may someday lead to enhanced organ transplantation.

As stated, the placenta is of dual origin. It is derived of fetal trophoblast cells and maternal decidua. **Fertilization** of the human **egg** occurs in the Fallopian (uterine) tubes. Implantation into the **endometrium** of the uterus does not occur until the sixth day. This gives the **zygote** time to cleave into a small cluster of blastomere cells. The cluster of cells continues to develop until the sixth day, at which time there is an inner cell mass that will become the **embryo** proper, and the surrounding trophoblast. The trophoblast cells become the embryonic component of the placenta. For nourishment (and other functions) to occur, the trophoblast must invade the endometrial lining of the uterus. The trophoblast becomes differentiated into a cellular layer, the cytotrophoblast, and a layer of trophoblast cells without cell membranes known as the syncytiotrophoblast. The syncytiotrophoblast cells are aggressively invasive and quickly digest their way past the epithelial lining of the uterus until they access and digest the walls of maternal blood vessels. Maternal blood then bathes the syncytiotrophoblast. The trophoblast cells in contact with maternal blood become invaded by branches of umbilical arteries which give rise to capillaries. It is at the capillary level that physiological exchange occurs. The refreshed blood returns to the embryo via the umbilical vein. It should be noted that fetal blood remains within the capillaries. The capillaries are in connective tissue covered with trophoblast cells. There is ordinarily no exchange of blood between the fetus and the mother because of the integrity of the umbilical capillaries and the covering of connective tissue. These tissues form an effective barrier against blood cell exchange.

PLANT

A plant is an **organism** in the **kingdom** Plantae. According to the five-kingdom **classification** system used by most biologists, plants have the following characteristics: they are multicellular during part of their life; they are eukaryotic, in that their **cell**s have nuclei; they reproduce sexually; they have **chloroplast**s with chlorophyll-a, chlorophyll-b, and carotenoids as photosynthetic pigments; they have cell walls with **cellulose**, a complex **carbohydrate**; they have **life cycle**s with an alternation of a **sporophyte** phase and a **gametophyte** phase; they develop **organ**s which become specialized for **photosynthesis**, reproduction, or mineral uptake; and most live on land during their life cycle. Biologists have identified about 500,000 **species** of plants, although there are many undiscovered species yet to be classified.

There are nine phyla within the kingdom Plantae. Bryophyta is a **phylum** with three classes, the largest of which is the mosses. Lycopodophyta contains about 1,000 species, most of which are abundant in the tropics. Sphenophyta has a single **genus**, *Equisetum*, with about 10 species. They are most commonly referred to as horsetails. Filicinophyta contains 11,000 species, which are commonly known as ferns. Cycadophyta includes 200 species, which grow in tropical and subtropical regions of the world. Ginkgophyta contains only one species, *Ginkgo biloba*, a **gymnosperm** that bears its seeds in green, fruit-like structures. Coniferophyta has approximately 600 species, including evergreen **tree**s such as pines, spruces, and firs. Gnetophyta is a phylum with three genera, *Gnetum*, *Ephedra*, and *Welwitxchia*. These three genera differ significantly from one another in their vegetative and reproductive structures, although all are semi-**desert** plants. Angiospermophyta is the largest plant phylum, with at least

300,000 species. All of its species reproduce by making **flowers**, which develop into **fruits** with seeds upon **fertilization**. Nearly all of the plant foods cultivated by humans and many drugs and other products come from angiosperms.

Due to the scarcity of **fossil** plants it is not easy to reconstruct their evolutionary history. By looking at the structure and form of extant plants general trends can be construed. In the most primitive plants, such as *Spirogyra*, all the cells are alike and all are capable of reproducing sexually. As plants became more complex, specialization of cells occurred. Two types of cells are now present in plants, the gonadic cells that are responsible for reproduction and somatic cells that carry out all of the vegetative functions of the organism. This can be seen even in such primitive groups of plants as the **algae**. This specialization of cells was one of the most major advances in plant evolution.

The life cycle of all plants consists of an alternation of generations, in which a **haploid** gametophyte (**tissue** in which each cell has one copy of each **chromosome**) alternates with a **diploid** sporophyte (tissue in which each cell has two copies of each chromosome). A major trend in plant evolution has been the increasing dominance of the sporophyte. Chlorophyta (green algae), the ancestors of land plants, have a dominate gametophyte and greatly reduced sporophyte. Bryophyta, the most primitive land plants, have a more elaborate sporophyte than Chlorophyta, although their gametophyte is still dominant. Free-sporing vascular plants (Filicinophyta, Lycopodophyta, and Sphenophyta) have a somewhat more dominant sporophyte phase than gametophyte phase. However, seed plants, the most advanced of the land plants, have a greatly reduced gametophyte, and a dominant sporophyte.

The seed plants are the dominant and most studied group of plants. Their leaves and other aerial portions are all covered with a cuticle, a waxy layer that inhibits **water** loss. Their leaves have stomata, microscopic pores which open in response to certain environmental cues for uptake of **carbon dioxide** and release of oxygen during photosynthesis. Leaves have veins, which connect them to the stem through a vascular system that is used for transport of water and **nutrients** throughout the plant.

As a plant grows, it undergoes developmental changes, known as morphogenesis, which include the formation of specialized tissues and organs. Most plants continually produce new sets of organs, such as leaves, flowers, and fruits, as they grow. The meristematic tissues of plants have the capacity for **cell division** and development of new and complex tissues and organs, even in older plants. Most of the developmental changes of plants are mediated by hormonal and other chemical changes, which selectively alter the levels of expression of specific genes.

A plant begins its life as a seed, a quiescent stage in which the metabolic rate is greatly reduced. Various environmental cues such as **light**, **temperature** changes, or nutrient availability, signal a seed to germinate. During early **germination**, the seedling depends on nutrients stored within the seed itself for growth. As it grows, the seedling begins to synthesize chlorophyll and turn green. Most plants become green when exposed to sunlight, because chlorophyll synthesis is light induced. As plants grow larger, new organs develop according to certain environmental cues and genetic programs of the individual.

Plants are what is known as primary **producers**. They are the basis of most food chains and food webs. Plants take simple, inorganic compounds such as water and carbon dioxide and convert them into simple sugars. This reaction is powered using **energy** from sun light and it takes place in the chloroplasts of the plant. **Animal**s then eat plants and their energy is incorporated into the animal. As a waste product plants produce oxygen which is used by all animals. As well as providing all food that humans eat, directly or indirectly, plants and plant products have many uses in modern society.

PLANT ANATOMY

Plant anatomy is a branch of plant **morphology** which studies the internal structure of **plants**. The plant body comprises three types of vegetative organs: the root, the stem, and the **leaf**; and a complex reproductive organ, or assemblage of organs called the **flower**. Precise separation of the plant into organs is often impossible. Seed plants, in particular, have a highly evolved body with many structural and functional **adaptation**s. Thus, it is difficult to clearly determine which is root, stem, or leaf, as the parts appear to blend into one another. The flower is even more difficult to distinguish and may be interpreted as being a modified shoot system with parts derived from leaves and stem.

There are several types of plant **tissue** and a number of **cell** types. One tissue, the epidermis, is composed of epidermal cells, and forms a continuous layer on the surface of young plants, enclosing all other cells within it. A significant feature of the tightly packed epidermal cells is that they tend to develop a protective layer, or cuticle, on their outer surface that reduces **water** loss from the plant. Parenchyma tissue, composed of relatively undifferentiated parenchyma cells, often occupies much of the plant interior. Parenchyma cells typically remain alive as the tissue develops, and may serve various functions, such as storing starch grains for use as **energy** by the growing plant. A specialized form of parenchyma, called chlorenchyma, is found in the mesophyll (the area between the upper and lower epidermal layers) of leaves. Chlorenchyma contains **organelle**s called **chloroplast**s where **light** is captured and **photosynthesis** occurs.

Two types of conductive, or vascular, tissue are found in plants. Xylem is the conducting tissue that carries water from the soil to all parts of the plant. Two types of cells, tracheids and vessel members, transport the water. Phloem, the second vascular tissue, conducts food materials throughout the plant in sieve cells and sieve-tube members. Both tissues also contain fibers that provide mechanical strength to the plant.

Two forms of apical meristem (cells at the tip of the root or stem) are responsible for a plant's primary growth. The main apex, found at the tip of the shoot, produces cells that differentiate into the leaves and stem, with a succession of nodes and internodes. Auxillary buds are formed at the nodes that give rise to branches and flowers. The main root apex produces cells that differentiate into the primary root. Small dicotyledonous annuals and most monocotyledons are able to survive and reproduce by primary growth alone. Most dicotyledons and **gymnosperm**s, on the other hand, develop a second stage of growth supported by the activity of a meristem called the vascular cambium. The vascular cambium adds conducting and supporting tissue and thereby causes an increase in the diameter of the stem or root. The vascular cambium is found near the outer surface of a stem, or branch. This possibility is lessened by the development of another layer called the cork cambium or phellogen. The phellogen forms a protective layer, called bark, around the circumference of the stem that shields the vascular cambium from harm.

See also Chloroplast; Leaf; Root system

PLANT HORMONES

Plant hormones are important naturally occurring chemicals that influence plant growth. They are often called plant growth regulators to distinguish them from **animal** hormones that were discovered first and differ somewhat in their mode of action. Like their animal counterparts, plant hormones are effective in very small amounts, and tend to be synthesized at one site and transported elsewhere before they become functional. The first plant hormone was isolated by the Dutch plant physiologist **Frits Went** in 1926. Because of its ability to stimulate growth, he named it auxin from a Latin word meaning ''to increase.'' Sensitive **bioassay**s enabled researchers to garner enough of the substance to chemically identify it as indoleacetic acid (IAA). IAA has been shown to play a role in a number of important features of plant growth including apical dominance, **phototropism**, gravitropism, and **differentiation** of vascular **tissue**. Several synthetic auxins have become important commercial products. The best known, dichlorophenoxyacetic acid (2,4-D), is widely used as a herbicide. Horticulturists use another, Naphthalenacetic acid (NAA), to induce the formation of roots on plant cuttings. An important group of plant hormones, called cytokinins, stimulate **cell division** and delay senescence in **aging** tissues. They are chemically related to the purine, adenine, found in **nucleic acids**. The gibberellins are a chemically complex **family** of plant hormones that influence shoot elongation as well as other aspects of growth. Perhaps the most unusual plant hormone is the gas ethylene, which stimulates fruit ripening. The hormone abscisic acid induces dormancy in perennial **plants**, and causes seeds to remain dormant. A hypothetical hormone called florigen has been proposed as a factor promoting flowering. It has never been isolated, however, and may not exist. The flowering response may instead result from the interaction of several of the known hormones.

PLANT PATHOLOGY

Plant pathology (or phytopathology) is the study of diseases, injuries, and other factors affecting the welfare of **plants**. It is an applied discipline, with the ultimate goal of achieving plant health, as indicated by a condition in which physiological functions (including growth) are carried out to the limits of the genetic potential. Plant pathologists seek to understand and control the ecological factors affecting the production of economically useful plants (plants that can be used profitably) and their products. These include the availability of appropriate regimes of environmental factors (such as **light, water, nutrients,** and **temperature**); the effects of **parasite**s and predators (or **herbivore**s); and infectious agents (or pathogens).

Plant pathology incorporates elements of many subject areas, including **botany** (the study of plants, including **anatomy**, physiology, **genetics**, and biochemistry); bacteriology (**bacteria**); mycology (**fungi**); nematology (nematodes); virology (**virus**es); chemistry; **ecology** (the influence of environmental factors on **organism**s and **ecosystem**s); forestry (cultivation of industrial **tree**s); horticulture (cultivation of ornamental plants); meteorology; physics; and soil science.

To facilitate study, plant diseases are divided into categories. Infectious diseases include those caused by transmissible biological agents, such as bacteria, fungi, or viruses. Non-infectious diseases (or physiological disorders) are caused by nutrient deficiency, mineral excess (or toxicity), a lack or excess of soil moisture, too high or too low a soil temperature, a lack or excess of light, a lack of oxygen, extreme soil acidity or alkalinity (pH disorders), and **air pollution** by sulfur dioxide, **ozone**, or other chemicals. Non-infectious biological agents that damage plants by acting as parasites or herbivores include arthropods, nematodes, and parasitic higher plants (e.g., mistletoes). In general, the most common factors severely affecting economic plants are pathogens, arthropod pests, and unfavorable weather.

The initial step in identifying the cause of a plant disease is determining whether it is being caused by a pathogen or a non-infectious environmental factor. In many cases, diseased plants show obvious symptoms, such as unusual pattern or color of foliage, abnormal growth of **tissue**s, or loss of **flower**s or foliage, which may indicate a particular causal agent. For example, an excess of the air pollutant ozone causes a banded chlorosis (bands of whitish and normal green coloration) on the foliage of pine trees. In other cases, however, the symptoms may not be clearly diagnostic, and may be as vague as decreased **productivity** (compared with what is ''normal'' for the area). Such cases require careful work in the field and laboratory to discover the cause of the disorder, and to determine appropriate methods for the management of plants and/or their environment.

Plant pathology involves the definitive identification of the cause of infectious diseases using the following, four-step process (these are sometimes referred to as **Koch**'s postulates): (1) the pathogen must be found in all cases of the disease (e.g., in a particular afflicted field or other study area); (2) the pathogen must be isolated from affected plants, grown in pure **culture**, and its characteristics described; (3) the pathogen from pure culture must be inoculated into healthy plants of the **species** of interest, and must cause the same disease symptoms as originally seen in **nature**; and (4) the pathogen must again be isolated from the affected, experimentally inoculated plants, grown in pure culture, and have the same characteristics as observed in step (2).

Plant pathology is an extremely important field of science. By aiding in the identification and effective treatment of plant diseases, phytopathology helps to provide food, medicine, materials, and other essential crops for the growing human **population**.

PLANT REPRODUCTION

Plant reproduction may be sexual, in which two parents produce a genetically different individual; or asexual, involving the propagation of **plants** that are genetically identical to the parent.

Sexual reproduction in plants involves the fusion of two **haploid gamete**s (or microspores; each has one set of **chromosome**s, signified as 1n, as a result of **meiosis**, or reduction division). The ''male'' gametes are found in pollen, which is produced by the anthers of plant **flower**s. The ''female'' gametes are contained within ovules, which are in the ovary of the pistil of plant flowers. **Pollination** is the process whereby pollen is transferred to the stigma of the pistil. This can occur by pollen dispersed into the **atmosphere**, or transferred by a pollinating insect, such as a bee or moth. The pollen germinates on the stigma and grows pollen tubes which penetrate through the style and into the ovary where **fertilization** of the ovule occurs, forming a **diploid** (or 2n) **zygote**, which develops into a seed. The ripe seed is then dispersed from the parent plant. If it encounters suitable environmental conditions the seed germinates to develop into a ''new'' individual that is closely related to its parents, but genetically different from both of them..

Asexual propagation can occur in various ways. Many plants produce genetically identical copies of themselves, through a mechanism referred to as ''**asexual reproduction.**'' However, botanists more properly refer to this mechanism as ''sexual propagation'' or ''vegetative propagation'' because many believe that only sexual reproduction should be referred to as true ''reproduction,'' since this is the only kind of propagation that results in the production of new, genetically unique individuals.

Another type of asexual propagation occurs when plants develop underground stems (or rhizomes) which grow outward, or new shoots which grow upward to form new shoots that are genetically identical to the ''parent.'' One such example is the trembling aspen (*Populus tremuloides*), which sometimes develops entire stands of ''trees'' growing out of the ground as seemingly individual stems, but are actually genetically identical and interconnected below-ground. Another example is the strawberry (*Fragaria virginiana*), although this **species** has its vegetative ''runners'' above-ground.

Other plants develop bulbils on their stems, which can detach, fall to the ground, and sprout to develop new plants that are genetically identical to the original one. One familiar species that does this is the tiger lily (*Lilium tigrinum*). Other

plants can propagate from twigs or stem pieces that fall from the parent, then lodge into a suitable site and develop into a new plant. The crack willow (*Salix fragilis*) can spread itself along **water**courses in this manner (as well as by disseminating seeds). Other non-sexual means of propagation include the production of underground bulbs, corms, and tubers that split into parts, each of which is capable of developing into a new plant. Plants that reproduce in this manner include irises and daffodils.

PLANT TRANSPIRATION • See Hales, Stephen

PLASMA MEMBRANE

The plasma membrane is a very thin, continuous sheet of **phospholipids** and **proteins** that surrounds all living **cell**s and separates them from their external environment. This membranous barrier maintains conditions within the cell and creates an internal chemical environment essential for life. It is one of the few structures common to every living cell. Invisible except in an electron **microscope**, a plasma membrane is typically only 4-5 nm thick. In spite of its thinness, however, the plasma membrane is surprisingly strong, flexible, and stretchable. It has a high electrical insulating capacity, and a remarkable ability to self-assemble to repair breaks and damage.

Structurally, almost all plasma membranes are composed of a double layer of phospholipid **molecule**s arranged with their hydrophobic (**water**-hating) lipid tails oriented toward the **membrane** interior and their hydrophilic (water-loving) heads making up the opposing membrane surfaces. A wide variety of globular proteins are embedded in or on this biomolecular lipid **leaf**let structure, some of them reaching completely across the membrane while others are implanted in or attached to either the extra-cellular or intra-cellular surface. These proteins not only help create the structure of the membrane but also carry out **enzyme** reactions or provide channels for transport of selected materials into or out of the cell. Integral membrane proteins can serve as recognition and binding sites for the cell to its neighbors or substrates. They also act as receptors for extracellular molecules that serve as signals, **nutrients**, or serve other important functions.

See also Endocytosis; Membrane structure

POINT MUTATION

A point mutation is a change in a single base in the coding part of the **deoxyribonucleic acid** (**DNA**). Point mutations are changes in one base pair of the DNA and they can vary in their effects. They can range from being unobservable to being fatal to the **organism**.

Due to the inbuilt redundancy of the code of DNA (whereby several different triplet codons can code for the same **amino acid**) some point mutations are neutral, that is they have no observable effect, the same amino acid is still produced in the same location. Other point mutations can substitute the amino acid produced in the **polypeptide** chain. This can produce a non-functioning **protein**, which can then affect the whole organism even leading to **death**. In some cases the protein produced under the action of the point mutation can actually perform better than the original form. Where this beneficial effect occurs it may lead to a greater ability to survive in the given environment, thus leading to a greater number of offspring potentially being produced with the concomitant chance of a wider spread over the whole **population** for this new form of the **gene**. Some point mutations change the section of DNA affected from sense to nonsense. When nonsense DNA is found it is a signal for the **cell** to stop making the polypeptide chain. This is the production of a stop codon.

POLLEN ANALYSIS · See Palynology

POLLINATION

Pollination is the process of transferring pollen grains from their production site in pollen sacs on male seed plant structures to a receptive female site on the same or a different plant. Specifically, the pollen grains, which contain male, or **sperm**, **cell**s, move from the anthers (the pollen-producing part of floral stamens, the male reproductive structure) of one **flower** to the stigma (the glandular female receptive portion) which is located in the pistil (female reproductive organ) of another flower. When a pollen grain lands on the female part of the flower, this male sex cell joins with the female sex cells in the flower in **fertilization** to form a seed from which a new plant can grow. The anthers and stigma can be on the same flower (self-pollination) or on different flowers (cross-pollination), but must be of the same **species**. Self-pollination is the simpler and more certain of the two fertilization processes, but a species producing such uniform offspring runs the risk of having its entire **population** wiped out by a single evolutionary event. Cross-pollinated species produce more diverse offspring that are better able to adapt to a changing environment, thus lessening the chances of **extinction**. Cross-pollinating **plants** also tend to produce more and better-quality seeds.

All higher plants, including the **gymnosperm**s, or naked-seeded plants, and the angiosperms, the enclosed-seeded or flowering plants, use pollination. Pollination can be accomplished by abiotic (or non-living) means, such as by wind and **water**, or by **animal**s, such as bees or bats. Many pollen grains are small (less than 0.002 in or 0.05 mm) and wind can carry them to other members of the species. To ensure propagation, many plants that are dependent on wind pollination grow in dense stands and produce millions of pollen grains. Wind-pollinated plants generally have small, inconspicuous flowers that dangle in the wind (e.g., willow catkins). Grasses have wispy plume-like flowers that catch grains floating in the air. Wind pollination may have evolved in plants that grow in

A honeybee becomes coated in pollen while gathering nectar and transports the pollen as it goes from flower to flower. *(The Stock Market. Reproduced by permission.)*

cooler, drier environments, where insect-pollinating species are more limited. Some water plants such as the hornwort have their pollen, which must be relatively water-resistant, transferred by water currents.

When animals transport pollen grains among plants, the process is referred to as biotic pollination and it requires a relationship between the pollinator and the flower to be pollinated. Such a relationship is usually established by some kind of direct attractant, such as nectar, sweet-tasting pollen, odor, or visual attraction (such as brightly colored flowers). There may also be an indirect attraction, such as when insects of prey visit flowers to catch other visiting pollinators.

Insects, including bees, beetles, flies, wasps, ants, butterflies, and moths, are common biotic pollinators. As an insect crawls in and out of flowers in its search for nectar or other food sources, it receives a dusting of pollen grains from the anther, the male part. When the insect visits another flower, the pollen rubs off on the stigma, the female part. If the pollen is left on the same species of flower, a long tube grows from each pollen grain down the stalk (style) of the stigma and into the ovule at the base, which contains the female **egg** cells. The male cells from the pollen grains pass along the tubes to the female cells and fertilize them. Plants with trumpet-shaped flowers, such as petunias, have nectar at the bottom, so only insects with long tube-like tongues can act as their pollinators.

The best-known and best-adapted biotic pollinator is the bee. This insect has a large demand for food both for itself and for its carefully looked-after brood. It normally gets all of its **nutrients** from flower blossoms. The bee has an ability to remember plant forms, which helps it find flowers.

Birds and bats may also act as pollinators. For example, hummingbirds feeding from the hibiscus flower carry pollen on their beak and heads. Bats hover in front of flowers that open at night, licking nectar and covering their faces with pollen.

Successful pollination is important for food production as well as maintenance of biological plant diversity. Pollination of plants is necessary for seed set, fruit yields, and reproduction of most food crops.

POLLUTANTS

The term **pollution** is derived from the Latin *pollutus*, which means to be made foul, unclean, or dirty. Pollutants, then, are factors that corrupt, degrade, or make something less valuable or desirable. In environmental terms, we consider pollutants to be chemical, physical, or social factors that have an undesirable effect on a particular environment. Air pollutants, for instance, include dust, smoke, haze, foul odors, noise, and volatile (airborne) chemicals such as sulfur dioxide, **carbon monoxide**, lead, **nitrogen** oxides, acids, ambient **ozone**, and a large number of toxic **organic compound**s. Among the major **water** pollutants are human and **animal** wastes; infectious agents; oxygen-demanding organic chemicals; plant **nutrients** such as phosphates and nitrogen; heavy metals such as lead and mercury, acids, salts, sediment; and excess heat. Other kinds of pollutants might include visual pollution (ugly billboards), exotic biological **species**, cultural pollution (fashions, practices, or trends that corrupt an existing culture), or even linguistic pollution (new words from a foreign language that affect an existing language).

What might be considered pollution by one might be regarded as a welcome change by another. A **toxic waste** for one **organism** might be a highly desirable resource for another species. Oftentimes, the definition of pollution is limited to anthropogenic (human-caused) environmental changes. This can be problematic, however, because most of the materials we consider major air and water pollutants, such as dust, sediment, carbon monoxide, organic acids, and infectious agents have both human and natural sources.

POLLUTION

The term pollution is derived from the Latin *pollutus*, which means to be made foul, unclean, or dirty. Anything that corrupts, degrades or makes something less valuable or desirable can be considered pollution. There is, however, a good deal of ambiguity and contention about what constitutes a pollutant. Many reserve the term for harmful physical changes in our environment caused by human actions. Others argue that any unpleasant or unwanted environmental changes whether natural or human-caused constitute pollution. This broad definition could include smoke from lightning-ignited forest fires, ash and toxic fumes from volcanoes, or bad-tasting **algae** growing naturally in a lake. Some people include social issues in their definition of pollution, such as noise from a freeway, visual blight from intrusive billboards, or cultural pollution when the worst aspects of modern society invade a traditional culture. As you can see, these definitions depend on the observer's perspective. What is considered unwanted change by one person might seem like a welcome progress to someone else. A chemical that is toxic to one **organism** can be an key nutrient for another.

The seven types of **air pollution** considered the greatest threat to human health in the United States, and the first regulated by the 1970 United States Clean Air Act, include sulfur dioxide, particulates (dust, smoke, etc.), **carbon monoxide**, volatile **organic compound**s, **nitrogen** oxides, **ozone**, and lead. In 1990, another 189 volatile **chemical compound**s from more than 250 sources were added to the list of regulated air **pollutants** in the United States. Air contaminants are divided into two broad categories: primary pollutants are those released directly into the air. Some examples include dust, smoke, and a variety of toxic chemicals such as lead, mercury, vinyl chloride, and carbon monoxide. In contrast, secondary pollutants are created or modified into a deleterious form after being released into the air.

A variety of chemical or photochemical reactions (catalyzed by **light**) produce a toxic mix of secondary pollutants in urban air. A prime example is the formation of ozone in urban smog. A complex series of **chemical reaction**s involving volatile organic compounds, nitrogen oxides, sunlight, and molecular oxygen create highly reactive ozone **molecule**s containing three oxygen **atoms**. Stratospheric ozone in the upper **atmosphere** provides an important shield against harmful ultraviolet radiation in sunlight. Stratospheric ozone depletion—destruction by chlorofluorocarbons (CFCs) and other anthropogenic (human-generated) chemicals—is of great concern because it exposes living organisms to dangerous ultraviolet radiation. Ozone in ambient air (that surrounding us), on the other hand, is highly damaging to both living organisms and building materials. Recent regulations that have reduced releases of smog-forming ozone in ambient air have significantly improved air quality in many American cities.

Among the most important types of water pollution are sediment, infectious agents, **toxins**, oxygen-demanding wastes, plant **nutrients**, and thermal changes. Sediment (dirt, soil, insoluble solids) and trash make up the largest volume and most visible type of **water** pollution in most rivers and lakes. Worldwide, erosion from croplands, **forests**, grazing lands, and construction sites is estimated to add some 75 billion tons of sediment each year to rivers and lakes. This sediment smothers gravel beds in which fish lay their **egg**s. It fills lakes and reservoirs, obstructs shipping channels, clogs hydroelectric turbines, and makes drinking **water** purification more costly. Piles of plastic waste, oil slicks, tar blobs, and other flotsam and jetsam of modern society now defile even the most remote ocean beaches.

Pollution control regulations usually distinguish between point and nonpoint sources. Factory smoke stacks, sewage outfalls, leaking underground mines, and burning dumps, for example, are point sources that release contaminants from individual, easily identifiable sources that are relatively easy to monitor and regulate. In contrast, nonpoint pollution sources

Pollution over the New Jersey turnpike. *(The Library of Congress. Reproduced by permission.)*

are scattered or diffuse, having no specific location where they originate or discharge into our air or water. Some nonpoint sources include automobile exhaust, runoff from farm fields, urban streets, lawns, and construction sites. Whereas point sources often are fairly uniform and predictable, nonpoint runoff often is highly irregular. The first heavy rainfall after a dry period may flush high concentrations of oil, gasoline, rubber, and trash off city streets, for instance. The irregular timing of these events, as well as their multiple sources, variable location, and lack of specific ownership make them much more difficult to monitor, regulate, and treat than point sources.

In recent years, the United States and most of the more developed countries have made encouraging progress in air and water pollution control. While urban air and water quality anywhere in the world rarely matches that of pristine wilderness areas, pollution levels in most of the more prosperous regions of North America, Western Europe, Japan, Australia, and New Zealand have generally been dramatically reduced. In the United States, for example, the number of days on which urban air is considered hazardous in the largest cities has decreased 93% over the past 20 years. Of the 97 metropolitan areas that failed to meet clean air standards in the 1980s, nearly half had reached compliance by the early 1990s. Perhaps the most striking success in controlling air pollution is airborne lead. Banning of leaded gasoline in the United States in 1970 resulted in a 98% decrease in atmospheric concentrations of

this toxic metal. Similarly, particulate materials have decreased in urban air nearly 80% since the passage of the U.S. Clean Air Act, while sulfur dioxides, carbon monoxide, and ozone are down by nearly one-third.

Unfortunately, the situation often is not so encouraging in other countries. The major metropolitan areas of developing countries often have appalling levels of air pollution, which rapid **population growth**, unregulated industrialization, lack of enforcement, and corrupt national and local politics only make worse. Mexico City, for example, is notorious for bad air. Pollution levels exceed World Health Organization (WHO) standards 350 days per year. More than half of all children in the city have lead levels in their **blood** sufficient to lower intelligence and retard development. The 130,000 industries and 2.5 million motor vehicles spew out more than 5,500 metric tons of air pollutants every day, which are trapped by mountains ringing the city.

While we have not yet met our national goal in the United States of making all surface waters ''fishable and swimmable,'' investments in sewage treatment, regulation of **toxic waste** disposal and factory effluents and other forms of pollution control have resulted in significant water quality increases in many areas. Nearly 90% of all the river miles and lake acres that are assessed for water quality in the United States fully or partly support their designed uses. Lake Erie, for instance, which was widely described in the 1970s as being ''dead,'' now has much cleaner water and more healthy fish **population**s than would ever have been thought possible 25 years ago. Unfortunately, surface waters in some developing countries have not experienced similar progress in pollution control. In most developing countries, only a tiny fraction of human wastes are treated before being dumped into rivers, lakes, or the ocean. In consequence, water pollution levels often are appalling. In India, for example, two-thirds of all surface waters are considered dangerous to human health. Hopefully, as development occurs, these countries will be able to take advantage of pollution control equipment and knowledge already available in already developed countries.

See also Carson, Rachel

POLYMER

Polymers are high molecular weight materials that are made up of smaller, repeating units called **monomer**s. They typically consist of a primary, long chain backbone **molecule** with attached side groups. Depending on the nature of the backbone, the polymer can be either linear or branched. Polymers, also called **macromolecule**s, may naturally occurring or synthetically produced.

The first polymers used by humans were naturally occurring. The ancient Greeks used many of these polymers even though they did not understand polymer chemistry. In fact, the term polymer is derived from the Greek words *poly* (many) and *meros* (parts). Early civilizations used polymers for many reasons including making clothing, shelter, and food. Aztec Indians used natural rubber to make waterproof clothing. Polymers were also used for making leather.

In nature, a wide variety of polymers can be found. In fact, most **organism**s are made up of polymers. The molecules that are the basis for life on earth are primarily polymers. Perhaps most important are the life-coding molecules **DNA** and **RNA**. They are made up of repeating monomer units called **nucleotide**s. They were first isolated from **cell** nuclei in 1896. Since they were found to be acidic, they became known as **nucleic acid**s. It was later discovered that nucleic acids provide the code for **protein**s, another type of important natural polymer.

Proteins are polymers that perform numerous functions in living organisms. They are catalysts that control most biological reactions essential for life. They also provide structure making up things such as hair, skin and nails. Proteins are made up of monomer units known as **amino acid**s. There are primarily twenty different types of amino acids.

Another type of natural polymers are polycarbohydrates. These molecules are used by living organisms as structural molecules and food storage. Polycarbohydrates include materials such as **cellulose**, starch and glycogen. Some naturally occurring polymers are highly specialized. For example, spider silk is a polymer, as is slug slime.

Numerous polymers have been produced synthetically. The first synthetic polymers were created to mimic the characteristics of naturally occurring polymers. Later, these synthetic copies were altered to emphasize desired properties. Types of synthetic polymers include plastics, elastomers, fibers, coatings, and adhesives.

POLYMORPHISM

Polymorphism refers to the presence of two or more distinct forms which exist together within a single breeding **population** of a **species**. The forms are discontinuous, meaning that the population lacks individuals that are intermediate. Polymorphisms are found in many plants and animals and often exist at a frequency too high to be maintained solely by **mutation**. Polymorphisms are known to exist either as obvious physical variations easily detected by examination of the body of the **organism**, **enzymes**, or **proteins** of those organisms, and as chromosomal or **deoxyribonucleic acid** (**DNA**) variants.

The term discontinuous is essential to an understanding of polymorphism. Most traits of organisms exist in a continuous gradation between extremes. Consider pigment and height in human populations. Skin pigment varies from a deep brown to various light hues. Similarly, height in humans varies from short to extremely tall. The existence of innumerable intermediate forms in both height and pigmentation qualifies those traits as being continuously variant. Other human traits have no intermediates. Consider the human ABO **blood type** polymorphism. Humans may be classified as blood type A, B, AB, or O. Intermediates of those blood types do not exist. Blood type is determined by **heredity** and this is true of other types of polymorphism. Environmental factors such as diet and ultraviolet radiation (uv) may affect height and pigment respectively, however environment does not change the genetic determinate for ABO blood types. Thus, they are not affected by what a human may or may not be exposed to.

Pigment pattern polymorphisms exist in populations of the northern leopard frog, *Rana pipiens*. The leopard frog ordinarily has spots on its back and limbs. There is a remarkable variation in the number and size and shape of the spots; it is appropriate to state that spotting is a continuously variable trait of the leopard frog. However, a spotless variant, known as the burnsi morph, exists in Minnesota and contiguous states. The spotless frog is not an albino. It has pigment **cell**s, but the cells do not aggregate into the well recognized spots of this common frog. Genetic studies reveal that the burnsi morph differs from ordinary spotted frogs by the possession of a single dominant Mendelian **allele**. The burnsi morph of the leopard frog seems to persist at a relatively stable frequency in many Minnesota populations, creating a polymorphic population. It is difficult to understand the biological significance of the balanced stability where it occurs, however it seems that the burnsi **gene** might convey an advantage when relatively rare but not when it becomes more common.

The kandiyohi morph is another example of pigment pattern polymorphism in some populations of *R. pipiens* in Minnesota. Kandiyohi frogs are identified by small mottled patches of pigment between the usual dark spots that characterize leopard frogs. The kandiyohi morph differs from ordinary spotted frogs by possession of a single dominant Mendelian allele. Intermediates with pigment patterns between ordinary spotted frogs and the kandiyohi frog do not exist. They are sharply distinct, and thus, kandiyohi is another example of polymorphism in leopard frogs.

POLYPEPTIDE

A polypeptide is a chain of **amino acid**s joined together by amide **linkage**s called **peptide** bonds. They are intermediate in length between oligopeptides containing several amino acids and **protein**s with more than fifty. The peptide bond (-CO-NH-) forms between the amino group (-NH$_2$) of one amino acid and the carboxyl group (-COOH) of the adjoining amino acid. As with other peptides and proteins, the properties of polypeptides depend on the kinds of amino acids they contain, and the order in which they are linked.

Certain polypeptides have important roles as **hormone**s. One example, ACTH (adrenocorticotropic hormone) is formed in the **pituitary gland** and regulates the activity of the adrenal cortex. Gastrin, secreted into the **blood**stream of **mammals** by the wall of the pyloric end of the stomach, stimulates **secretion** of gastric juices when food enters the stomach. Secretin, a polypeptide containing 27 amino acids, is found in the lining of the upper intestine, and encourages the flow of pancreatic juices into the duodenum when hydrochloric acid from the stomach arrives there. The pancreatic emissions serve to neutralize the stomach acid thus avoiding potential damage to the intestinal lining.

Some polypeptides have antibiotic properties. One type called polymyxin, synthesized by **species** of the soil bacterium *Bacillus*, is active against common gram-negative **bacteria** such as *Escherichia coli* and *Pseudomonas aeruginosa*. It functions by disrupting the **cell membrane**s of sensitive bacteria.

POLYPHYLOGENY

Polyphylogeny is a term used in the science of **systematics**. It describes a group of **species** that are classified in the same **taxon**, but do not share a single ancestor. Systematics is the study of biological diversity looked at through an evolutionary perspective. It was an outgrowth of **taxonomy** and has slowly become more sophisticated since the time **Charles Darwin** posed the **evolutionary theory**. Scientists who pursue this subject attempt to group **organism**s based on probable evolutionary relationships. They create genealogy diagrams, called phylogenic trees, that graphically show these relationships. A polyphylogenic group would contain species that seem related, but have different evolutionary ancestors. These species are grouped using a variety of characteristics such as **fossil** records, **comparative anatomy** and **DNA** and **protein** analysis. While most phylogenic trees contain polyphylogenic groups, the ideal in systematics is for each group to have a common ancestor or be monophylogenic. Unfortunately, a large amount of conflicting data is often collected. For example, two unrelated organisms may be grouped together because they independently evolved similar traits. Incidents of this type of **convergent evolution** create errors in phylogenic trees. However, as more data is collected about various species, scientists hope to achieve the goal of a unified phylogenic tree.

See also Ontogeny and Phylogeny

POLYSACCHARIDES

Polysaccharides are important biological **molecule**s which belong to a more general class of compounds called **carbohydrate**s. Carbohydrates are organic materials that contain carbon, hydrogen, and oxygen. They are the most abundant component of **plants** and are typically produced by **photosynthesis**. Polysaccharides have the largest molecular structure of any other type of carbohydrate. They are **macromolecule**s, or condensation **polymer**s, made up of thousands of monosaccharide units typically linked together by an oxygen **atom**s. The term polysaccharide refers to polymers that are composed of ten or more monosaccharide units. Polymers that consist of three to nine **monosaccharide**s are called oligosaccharides.

The **chemical reaction**s that produce polysaccharides from monosaccharides are reversible. When a polysaccharide is reacted with **water** and an acid, monosaccharides result. These are sugar molecules that are classified by their functional group. Depending on the location of their carbonyl group, the sugar can either be an aldose or a ketose. Monosaccharide sugars are also grouped according to the number of carbon atoms they contain. The smallest is triose which is a three carbon sugar. Pentose sugars have five carbon atoms. They include ribose and deoxyribose which are key components of **nucleic acids**. One of the most important types of monosaccharides are hexoses which contain six carbon atoms. These include sugars such as galactose, fructose and glucose.

Glucose is probably the most abundant **organic compound** found in **nature**. It is a monosaccharide that is present in honey and many fruit juices. In the body it is found in greatest concentration in the **blood**. In both plants and **animal**s, it provides the major source of **energy**. Several important polysaccharides are built from repeating glucose units. These include storage polysaccharides like starch and glycogen, and structural polysaccharides such as glycogen and **cellulose**.

Starch is a mixture of two types of polysaccharides. One is a straight chain polymer made up of glucose **monomer**s. It is called amylose and comprises 20% of the weight of the starch. The other is a branched polymer known as amylopectin. It is also made up of glucose monomers, and it makes up the other 80%. Starch is water soluble which makes it an important food source. In fact, starch is the most abundant polysaccharide that humans eat. A large quantity is found in cereals, potatoes, and vegetables. It is produced by plants and is used to store the energy collected during photosynthesis. When starch is digested, it is ultimately reduced to glucose.

Glycogen is also known as animal starch because it is the molecule in which animals store excess glucose. It is a branched polymer made up of glucose units and has a similar structure to the amylopectin in starch. However, it is more highly branched and exists in much greater molecular weights. Glycogen is found mainly in the liver and skeletal muscle. A small amount is found in almost all **tissue**s. In the body, excess glucose is transformed into glycogen by a process called glycogenesis. When the body needs it, the glycogen is rapidly reduced to glucose and used for energy.

Cellulose is another type of glucose-based polysaccharide. It is by far the most abundant polysaccharide in nature. It is found in plant **cell** walls and is also the main structural component of stems, leaves and bark. Unlike starch and glycogen, the glucose molecules in cellulose are connected through a beta **linkage**. This molecular configuration makes this polysaccharide rigid, fibrous, tough and insoluble in water. For this reason, it can not be digested by most **organism**s. Two notable exceptions are cows and termites who have **microorganism**s in their guts that can breakdown cellulose. This allows them to use cellulose as part of their diet. Cellulose is used for a variety of industrial applications. A modified version of it is used to make permanent press fabric. Rayon, cellophane, explosives, and celluloid (film) are also made with cellulose. In many consumer products, cellulose is used as a thickener.

A variety of other polysaccharides have important biological roles. Chitin is a structural polysaccharide that makes up the exoskeleton of insects and other arthropods. It is composed of an amino sugar which is a **nitrogen** containing glucose. Pectin is a cementing substance found in plants. And hyaluronic acid is a polysaccharide which helps cement connective tissue in the body.

PONNAMPERUMA, CYRIL ANDREW
(1923-1994)
Sri Lankan American biochemist

Ponnamperuma was born in Gale, Sri Lanka, an island near India formerly called Ceylon. He was educated at the University of Madras in India and received his B.A. in 1948. He then went on to the University of London where he earned a B.S. in 1959. Following this, he attended the University of California at Berkeley to pursue a doctorate in chemistry, which he earned in 1962. Four years later he became an American citizen.

Ponnamperuma worked as a research assistant at Lawrence Radiation Lab at the University of California at Berkeley. In 1963 he went on to work as a scientist at Ames Research Center, a facility funded by the National Aeronautics and Space Administration (NASA). Three years later, he took a position at the University of Maryland working as professor of chemistry and serving as Director of the Lab of Chemical Evolution.

Ponnamperuma's research at the University of Maryland helped him become one of the world's most noted scientific researchers investigating the origins of life. He furthered important theories set forth by renowned scientists **Aleksandr Oparin**, **Stanley Miller** and Harold Clayton Urey. Ponnamperu-

ma focused on producing compounds related to the **nucleic acid**s, and demonstrated that **nucleotide**s and dinucleotides can be formed by random processes alone. In another achievement, he showed the formation of ATP, a compound critical to the use of **energy** within a **cell**.

Ponnamperuma held various positions throughout his career, including that of principal investigator for the NASA Apollo Program and director to UNESCO's Institute of Early Education. He was also a foreign member of the Indian National Science Academy and was awarded an honorary degree from the University of Sri Lanka. In 1989 he established the Third World Foundation of North America, an organization devoted to scientific and governmental collaboration between undeveloped and developed countries.

POPULATION

A **population** is a geographically distinct group of individuals of the same **species** that is co-occurring in time and space, and can potentially interbreed with each other. Populations are an important element of evolutionary biology, and in the **ecology** of species and communities.

Evolution is sometimes defined as a change in the genetic information of populations over time. Individual **organism**s can be more or less successful in their reproduction, which means that they vary in "**fitness**." Because their fitness varies, individuals are the targets of **natural selection**. Individual organisms eventually die, and can not evolve. Populations and the species of which they are components are the units of evolution.

Populations of all species change in size over time in response to environmental factors that affect four population-related (or demographic) variables: **birth** rate (BR), immigration rate (IR), **death** rate (DR), and emigration rate (ER). The change in population size (P) during a unit of time (for example, one year) is described by the following equation: P = BR - DR + IR - ER. This demographic relationship is true of all populations, including humans. Population ecologists have developed mathematical models of population dynamics that account for the important influences of such factors as the intrinsic rate of population increase, the carrying capacity of **habitat**s, the effects of predators and disease, and the effects of unpredictable events of disturbance.

In some cases, isolated (or closed) populations do not add individuals through immigration, and do not lose individuals to emigration. Under such conditions, P is calculated as BR - DR, which is sometimes known as the natural rate of population change. P is expressed as a percentage change by dividing its value by the initial population size. For example, a population of 1000 individuals that increases by 100 in one year has a 10% per year growth rate. If the percentage change in any population is constant, there will be an accelerating rate of population increase or decrease, known as exponential change.

Imagine a circumstance in which a fertile pair of individuals discovers a "new" habitat, which is suitable for use but has not been previously utilized by their species. Under such conditions the founder population can increase over time. Initially, resources are abundant and do not constrain **population growth**, so the percentage rate of population increase is constant, being limited only by how quickly progeny can be produced and become fertile, and only countered by the death rate of individuals. This maximum rate of population growth, which is limited only by the biology of the species in the given environment and not by **competition** for resources, is referred to as the intrinsic rate of population increase. Any population growing for a number of generations at the intrinsic rate of population increase (or indeed at any fixed percentage rate) will quickly explode in abundance.

Eventually, however, the carrying capacity of the habitat is approached, and space and resources become limiting. (Carrying capacity is the abundance of the species that can be sustained without degrading the resources.) Once this happens, growth and reproduction become constrained by the availability of resources, and individuals in the population compete with each other. Intense competition results in physiological **stress**, which generally results in decreases in birth rates and increases in death rates. The rate of population increase may then decrease to zero. In other words, the birth rate equals the death rate, which is referred to as zero population growth or ZPG. If ZPG is maintained, the population size eventually levels off, perhaps to the carrying capacity of the habitat. However, if the earlier population growth results in an abundance exceeding what the habitat can support, the over-population causes environmental degradation by over-exploitation, resulting in a decrease in carrying capacity. The population would then decrease because of a rapid increase in the rate of mortality, or perhaps by emigration in search of new habitats. These factors could result in a rapid and uncontrolled "crash" in the size of the population, usually to a level below the carrying capacity, creating a circumstance for renewed population growth. In small habitats, however, the population crash could be massive enough to render a local population extinct.

POPULATION GENETICS

Population genetics is the study of the variations found within a population. Rather than examining the **genes** present within an individual, a study on population genetics looks at the gene frequencies within populations and also the selective factors that control their occurrence within natural populations. A powerful tool within this area of research is that of mathematical modeling. Mathematical models are used to predict what effects such things as selection, population size, **mutation**, and **migration** have upon the occurrence and frequency of both linked and unlinked genes.

Population genetics is an important study. Long term evolutionary changes are produced through the action of **natural selection** on populations, not individuals. A genetically isolated unit of population is a deme. A deme is self perpetuating by interbreeding within itself. All of the genes present in a deme make up the **gene pool**, and the future of the deme is dependent upon the gene pool. The **Hardy Weinberg equilibrium**

tells us that, in the absence of mutation, selection, and migration, the frequency of **allele**s in a population remains constant. This is a state of genetic equilibrium, and in such a state, there is no evolutionary change. Once there is some sort of change to the factors effecting the deme, evolution and natural selection are observable. It is these changes and the factors that cause them that are studied within population genetics.

Studies in population genetics are directed towards problems which can shed light on evolution. Some of the most important aspects of these are the relative importance of natural selection as compared to chance, the **rate of evolutionary change**, and the long-term effects of natural selection. All populations from **nature** to laboratory and even **fossil** record are used to answer these questions. Because population genetics studies real world populations, predictions are not always validated. The heterogeneity of the studied environment provides many factors that are too complex or not understood well enough to be represented with total accuracy. This does not detract from the worth of population genetics, but merely adds a margin of error to the study.

POPULATION GROWTH AND CONTROL (HUMAN)

The numbers of humans on Earth have increased enormously during the past several millennia, but especially during the past two centuries. By the end of the twentieth century, the global **population** of humans was 6.0 billion. That figure is twice the population of 1960, a mere 30 years earlier. Moreover, the human population is growing at about 1.5% annually, equivalent to an additional 89 million people per year. The United Nations Population Fund estimates that there will likely be about nine billion people alive in the year 2050.

In addition, the numbers of **animal**s that live in a domestic **mutualism** with humans have also risen. These companion **species** must be supported by the **biosphere** along with their human patrons, and can be considered an important component of the environmental impact of the human enterprise. The large domestic animals include about 1.7 billion sheep and goats, 1.3 billion cows, and 0.3 billion horses, camels, and **water** buffalo. Humans are also accompanied by a huge population of smaller animals, including 10-11 billion chickens and other fowl.

The biological history of *Homo sapiens* extends more than one million years. For almost all of that history, a relatively small population was engaged in a subsistence lifestyle, involving the hunting of wild animals and the gathering of edible **plants**. The global population during those times was about a million people. However, the discoveries of crude tools, weapons, and hunting and gathering techniques allowed prehistoric humans to become increasingly more effective in exploiting their environment, which allowed increases in population to occur. About ten thousand years ago, people discovered primitive **agriculture** through the domestication of a few plant and animal species, and ways of cultivating them to achieve greater yields of food. These early agricultural technologies and their associated socio-cultural systems allowed an increase in the carrying capacity of the environment for humans and their domesticated species. This resulted in steady population growth because primitive agricultural systems could support more people than a hunting and gathering lifestyle.

Further increases in Earth's carrying capacity for the human population were achieved through additional technological discoveries that improved capabilities for controlling and exploiting the environment. These included the discovery of the properties of metals and their alloys, which allowed the manufacturing of superior tools and weapons, and the inventions of the wheel and ships, which permitted the transportation of large amounts of goods. At the same time, further increases in agricultural yields were achieved by advances in the domestication and genetic modification of useful plants and animals, and the discovery of improved methods of cultivation. Due to innovations, the growth of the human population grew from about 300 million people in the year 1 A.D. to 500 million in 1650 A.D.

Around that time, the rate of population growth increased significantly, and continues into the present. The relatively recent and rapid growth of the human population occurred for several reasons. The discovery of better technologies for sanitation and medicine has been especially important, because of the resulting decreases in **death** rates. This allowed populations to increase rapidly, because of continuing high **birth** rates. There have also been great advances in technologies for the extraction of resources, manufacturing of goods, agricultural production, transportation, and communications, all of which have increased the carrying capacity of the environment for people. Consequently, the number of humans increased to 1 billion in 1850, 2 billion in 1930, 4 billion in 1975, 5 billion in 1987, and 6 billion in 1999. This rapid increase in the population has been labeled the ''population explosion.'' While there are clear signs that the rate of population increase is slowing, estimates show the number of humans on the planet to be nine billion in 2050.

Because the populations of humans and large domestic animals have become so big, some predict severe environmental damage caused by **pollution** and overly intense use of **natural resources**. If this were to happen, the carrying capacity for the human population would decrease, and famines could occur. A controversial movement in the latter years of the twentieth century for ''zero population growth'' advocates the widespread use of birth control, in order to maintain the birth rate at equal numbers to the death rate.

PORTER, KEITH ROBERTS (1912-1997)
Canadian-American cell biologist

Keith R. Porter was a true pioneer in the field of **cell** biology. During his five decades as a research professor at some of the leading academic institutions in the United States, Porter played a seminal role as the discipline came into existence and reached maturity. Many of the most important discoveries in this new and dynamic area of study were made by Porter himself or by students, colleagues, and collaborators working in his laboratories. A native of Nova Scotia, Porter graduated in 1934 from Acadia University in Halifax. In 1938, he received a Ph.D. from Harvard University and then moved to the Rockefeller Institute in New York City to begin his work on cells in tissue culture. By the early 1940s, the Rockefeller Institute had become the crucible of the new fields of cell fractionation and fine structure. Porter, as head of the laboratory of cytology was at the epicenter of this explosion of information and discovery. Together with colleagues such as **George Palade**, **Albert Claude**, Christian de Duve, G. C. Hogeboom, W. C. Schneider, and **Philip Siekevitz**, Porter nurtured the discipline and trained many of the students who became its leaders. One of Porter's first scientific inventions was the roller flask for culturing mammalian **cells**. Perhaps more important was that he discovered how to spread cultured cells thinly enough to be observed with a transmission electron **microscope**, a then-revolutionary instrument that uses high **energy** electrons rather than **light** for image formation. His photograph of a the interior structure of an intact cell became one of the most famous images in cell biology and triggered a scientific revolution in our understanding of cellular ultrastructure. To study solid tissues in the electron microscope, Porter and his assistant Joe Blum invented the first reliable microtome capable of cutting ultra-thin sections.

In the 1950s, Porter proposed a new journal for the rapidly expanding field of cell structure and function. Originally called the *Journal of Biophysical and Biochemical Cytology*, the publication later was renamed the *Journal of Cell Biology*. Porter served as its first editor and set the tone for what became the preeminent publication in the field. In 1961, Porter chaired the founding committee for the American Society for Cell Biology and became the organization's first president. At about the same time, he left Rockefeller to become chair of the Biology Department at Harvard where he continued his leadership in the growing field of cell studies. Among some of the most important discoveries made by Porter and his colleagues include the first ultrastructural description of a **virus** particle, the structure and function of rough and smooth **endoplasmic reticulum**, microtubules and microfilaments of the cell cytoskeleton, mechanisms of synthesis and assembly of collagen, the role of clathrin-coated vesicles in **endocytosis**, the nine plus two structure of ciliary axonemes, and the concept of compartmentalization of eukaryotic cells by internal membranes.

In 1968, Porter accepted the challenge of setting up a new department of molecular, cellular and developmental biology at the University of Colorado in Boulder. A highlight of this new position was a novel research tool, an enormous, million-volt transmission electron microscope that provided stunning images of whole cells and thick sections of **tissues**. His insights into the complex structure of the cell interior led him to propose a superorganization of the cytoskeleton and cytoplasmic membranes into what he called the ''microtrabecular lattice,'' an all-pervasive and all-controlling system that controls most cellular activities. During his years at Colorado, Porter also pioneered the use of the scanning electron microscope for the study of cell surfaces and tissue organization. He was fascinated with the elaborate ruffles, blebs, and ''bulbous excrescences'' visible on cell surfaces, and he speculated that they might tell us much about what is going on inside the cell.

An outstanding teacher and research mentor, Porter was renown for his dry wit and his penchant for skewering pomposity and pretension. He loved working with students. Porter retired from the University of Colorado in 1983, but his research career was not yet over. He spent four years as Distinguished Professor of Biological Sciences at the University of Maryland, and in 1987, at age 75, Porter moved once again to become the Distinguished Research Professor of Biology at the University of Pennsylvania. His indefatigable energy and curiosity inspired several generations of students and shaped the entire field of cell biology. During his 50-year career, Porter received numerous awards and prizes. Sadly, he did not share the Nobel prize awarded in 1974 to George Palade, Albert Claude, and Christian de Duve, an oversight that many of Porter's contemporaries felt was highly unjust since he worked closely with all three. Porter's legacy lives on, however, in his many important discoveries and in the numerous students whom he trained and inspired during his long career.

PORTER, RODNEY (1917-1985)
English biochemist

Rodney Robert Porter was born October 8, 1917, in Newton-le-Willows, near Liverpool in Lancashire, England. His mother was Isobel Reese Porter and his father, Joseph L. Porter, was a railroad clerk. He attended Liverpool University, where he earned a B.S. in biochemistry in 1939. During World War II he served in the Royal Artillery, the Royal Engineers, and the Royal Army Service Corps, and participated in the invasions of Algeria, Sicily, and Italy. After his discharge in 1946, he resumed his biochemistry studies at Cambridge University under the direction of Frederick Sanger.

Porter's doctoral research at Cambridge was influenced by Nobel laureate **Karl Landsteiner**'s book, *The Specificity of Serological Reactions,* which described the nature of antibodies and techniques for preparing some of them. Antibodies, at the time, were thought to be **protein**s that belonged to a class of **blood**-serum proteins called gamma globulins. From **Frederick Sanger**, who had succeeded in determining the chemical structure of **insulin** (a protein that metabolizes carbohydrates), Porter learned the techniques of protein chemistry. Sanger had also demonstrated tenacity in studying problems in protein chemistry involving **amino acid** sequencing that most believed impossible to solve, and he was a model for the persistence Porter would show in his later work on antibodies.

Fortunately, Porter chose rabbits to experiment on for his research. Although this was not known at the time, the antibody system is not as complex in this animal as it is in some. The most important antibody, or immunoglobulin, in the blood is called IgG, which contains more than 1,300 amino acids. The problem of discovering the active site of the antibody—the part that combines with the antigen—could be solved only by working with smaller pieces of the **molecule**. Porter discov-

ered that an **enzyme** from papaya juice, called papain, could break up IgG into fragments that still contained the active sites but were small enough to work with. He received his Ph..D. for this work in 1948.

Porter remained at Cambridge for another year, then in 1949 he moved to the National Institute for Medical Research at Mill Hill, London. There, he improved methods for purifying protein mixtures and used some of these methods to show that there are variations in IgG molecules. He obtained a purer form of papaya enzyme than had been available at Cambridge and repeated his earlier experiments. This time the IgG molecules broke into thirds, and one of these thirds was obtained in a crystalline form which Porter called fragment crystallizable (Fc).

Obtaining the Fc crystal was a breakthrough; Porter now was able to show that this part of the antibody was the same in all IgG molecules, since a mixture of the different molecules would not have formed a crystal. He also discovered that the active site of the molecule (the part that binds the antigen) was in the other two-thirds of the antibody. These he called fragment antigen-binding (or FAB) pieces. After Porter's research was published in 1959, another research group, led by **Gerald M. Edelman** at Rockefeller University in New York, split the IgG in another way—by separating amino acid chains rather than breaking the proteins at right angles between the amino acids as Porter's papain had done.

In 1960 Porter was appointed professor of immunology at St. Mary's Hospital Medical School in London. There he repeated Edelman's experiments under different conditions. After two years, having combined his own results with those of Edelman, he proposed the first satisfactory structure of the IgG molecule. The model, which predicted that the FAB fragment consisted of two different amino acid chains, provided the basis for far-ranging biochemical research. Porter's continuing work contributed numerous studies of the structures of individual IgG molecules. In 1967 Porter was appointed Whitley Professor of Biochemistry and chairman of the biochemistry department at Oxford University. In his new position, Porter continued his work on the immune response, but his interest shifted from the structure of antibodies to their role as receptors on the surface of cells. To further this research, he developed ways of tagging and tracing receptors. He also became an authority on the structure and **genetics** of a group of blood proteins called the complement, which binds the Fc region of the immunoglobulin and is involved in many important immunological reactions.

Porter was killed in an automobile accident a few weeks before he was to retire from the Whitley Chair of Biochemistry. He had been planning to continue as director of the Medical Research Council's Immunochemistry Unit for another four years; he had also intended to continue his laboratory work, attempting to crystallize one of the proteins of the complement system.

PORTMANN, ADOLF (1897-1982)
Swiss zoologist

Adolf Portmann was a Swiss zoologist best known for his controversial stance on evolution and his studies of **animal** consciousness. His highly philosophical writings and lectures attracted students from many disciplines in the sciences and from the humanities as well

Portmann was born in Basel on May 27, 1897. As a student, he attended universities all over Europe. He went from the University of Basel to colleges in Paris, Munich, Berlin, and Geneva. After his studies he returned to Basel, and became a professor of **zoology** at the university in 1926. In 1931 he was promoted to director of the university's Zoological Laboratory, a position he held until 1968.

Portmann's interest as a zoologist settled on areas other scientists had neglected, if not scorned. According to Portmann, the conventional scientific view described an animal as the sum of its survival strategies. Thus an animal's vital functions were defined by most zoologists as finding food, locating sexual partners, and avoiding enemies. Contrary to this accepted notion, Portmann saw animals as principally relating to others and to the world. This gave them a multiplicity of vital functions. He was interested in how animals appear to other animals, of the same and different species. He was also certain that animals have an inward life of complex mental processes not sheerly related to survival. To Portmann, the prevailing view of the animal, as derived from **Charles Darwin**, could not explain the richness of animals self-expression and the complexities of animals moods and abilities. His studies of birds concluded for instance that some bird songs serve the function of passing the time for the animal. This showed that birds have some kind of inner, mental life in their awareness of time, and that their actions express it.

The idea that animals play and take pleasure was central to Portmann, and put him at odds with conventional science. He also believed that mutation and natural selection alone could not solely account for species modification. This, too, put Portmann firmly on the margin of his discipline. In spite of his renegade views, however, which include the idea that biologists have every right to explore the spiritual dimension of life, Portmann's scientific research is highly regarded. In recognition of his accomplishments as a scientist, Portmann received many Swiss and international honors.

Portmann's best-known work in English is *The Animal As Social Being*, first published in 1956. He also published a book on animal camouflage and performed unusual studies on warblers. He released the songbirds inside a planetarium to study how they orient themselves using the stars when migrating.

Portmann's work is perhaps most notable for breaking down walls between biology and other disciplines. His radical views attracted students of psychology, philosophy, anthropology and even theology. His broader perspective on what animals are attempted to reconcile philosophical conceptions of life and soul with the beliefs of modern science. Though he took an unconventional stance on **evolution**, Portmann did not

Hailstones are often composed of concentric layers of clear and opaque ice. This is thought to be the result of the stone traveling up and down within the cloud during its formation. *(Photo Researchers, Inc. Reproduced by permission.)*

want to negate the accomplishments of twentieth century zoology. His aim was to give a broader understanding of living things.

PRECIPITATION

Precipitation refers to the deposition of atmospheric moisture from the atmosphere to the surface of **ecosystems**. Precipitation can occur as liquid rain, solid snow, or hail. In addition, vapor-phase **water** in the atmosphere can directly condense or freeze onto surfaces as dew or frost. Precipitation is a central element of Earth's hydrologic (or water) budget. Overall, the amount of water evaporated into the atmosphere from all oceanic and terrestrial surfaces is balanced by the amount returned to those surfaces by precipitation.

On the global scale, most evaporation to the atmosphere, about 86% of the total, occurs from the surface of the oceans. Much of it, about 77% of the total, precipitates back to the marine environment. The difference is transported by moving air-masses over the continents, resulting in a net import of evaporated water from Earth's oceans to the surfaces of the continental land-masses. The annual volume of precipitation can be especially large in hilly or mountainous terrain, where the prevailing winds carry warm, moisture-laden air inland from the oceans. The mountains act as a barrier that forces the moisture-laden air-masses to rise in altitude. This results in cooling of the air, or adiabatic cooling, and a decreased capacity to store moisture, which then precipitates as rain or snow. This phenomenon is known as orographic precipitation. Some places on the humid west coast of North America can have extremely high precipitation volumes, exceeding 33 ft per year (10 m/yr), because of the influence of orographic precipitation.

A female cheetah and her two cubs feasting on their kill. *(Photograph by Stan Osolinski, The Stock Market. Reproduced by permission.)*

In regions with a strongly seasonal climate, the amount and type of precipitation can vary greatly during the year. Some regions are characterized by pronounced wet and dry seasons, as occurs in much of coastal California, with its Mediterranean climate and winter rains. Other regions experience seasonally cold temperatures during much of the year. This is common in northern parts of North America, where the precipitation occurs as snowfall during the wintertime, and accumulates on the ground as a persistent snowpack. In addition, rare, severe events of precipitation may occur, during which huge amounts of rain can fall in a short period of time. The severest events are associated with hurricanes, typoons, and tropical storms, during which more than 1 foot per day of precipitation can fall (30 cm/day), causing severe flooding that threatens lives and causes great economic and ecological damage.

The amounts and types of precipitation have a great influence on the ecosystems that can develop in any region. In general, landscapes receiving less than about 10–12 in per year (25-30 cm/ yr) of precipitation can only support desert. Areas receiving about 12-28 in per year (25-70 cm/yr) support grassland, while those receiving more than about 16-30 in per year (40-75 cm/yr) can support forests of various kinds. These ranges overlap due to the influences of other climatic factors on the distribution of ecosystems, including mean annual temperature and the rate of evaporation of water back to the atmosphere, as well as disturbances such as wildfire.

The chemistry of precipitation can be significantly influenced by pollutants emitted to the atmosphere by industrial activities, and is considered an important environmental problem over large areas. Acid rain, or more accurately, acidified precipitation, is now common in extensive regions of North America, Europe, and Asia. Acid rain is caused by the presence of high concentrations of sulfate and nitrate in precipitation, originating from emissions of gaseous sulfur dioxide and oxides of nitrogen from automobiles, fossil-fueled electricity generation, and industrial processes. Acid rain is believed to be responsible for the acidification of tens of thousands of freshwater lakes. Acid rain causes much ecological damage, particularly the loss of fish populations, and may also be damaging forests over large areas.

PREDATION

Predation refers to the act of a predator killing its prey. All predators are **heterotroph**s, meaning they must consume the **biomass** of other **organism**s to fuel their own growth and reproduction. Predation may be important in regulating the size of **populations** of prey **species**, and in culling weak or diseased **animals**. The most common use of the term "predation" is to describe the act by which carnivorous animals catch, kill, and eat other animals. Sometimes, however, the term is also used in reference to herbivory, in which animals seek out and consume a "prey" of plant biomass, sometimes killing the plant in the process. In many cases only specific plant **tissues** or **organs** are consumed by the **herbivore**, and ecologists may refer to such feeding as, for example, seed predation or **leaf** predation.

There is a great diversity of carnivorous animals, ranging in size from tiny arthropods such as soil mites that eat other mites and springtails, to large mammalian **carnivore**s such as leopards and polar bears. With lions, wolves, orcas, and some other large species, predatory activity often occurs as a cohesive, group effort.

Most predation is undertaken by animals. However, a few **plants** and **fungi** are also predators. For example, pitcher plants and sundews are adapted to attracting, trapping, and digesting small arthropods as sources of **nutrients**. This unusual habit represents predation by these so-called carnivorous plants. A few types of fungi are also predatory, trapping tiny nematodes using anatomical devices such as sticky knobs or branches, or constrictive rings that close when their prey tries to move through. Once a nematode is caught, fungal hyphae surround and penetrate their victim, and absorb its nutrients.

PRIESTLEY, JOSEPH (1733-1804)
English scientist and clergyman

Priestley was born into a religious Calvinist family, and was encouraged by his parents to enter the ministry, not in the Church of England, but in the Dissenting church. In 1752 he enrolled in the new Dissenting Academy at Daventry, Northamptonshire, an educational institution founded by religious nonconformists. The academy offered a high-quality education, and encouraged his unorthodox approach to the curriculum, which he enriched by independently studying history, philosophy and science. In 1755 he became assistant minister to a Presbyterian congregation in Suffolk, but his heretical views gradually alienated the congregation. After three years he transferred to another more sympathetic congregation where he was allowed to open a highly successful day school with 36 students. Science experiments were an enticing feature of the curriculum. In 1761 his success as an educator led to his appointment as a tutor at Warrington Academy in Lancashire where he developed new courses emphasizing history, science, and the arts, in an effort to prepare students for the practical challenges of real life. This approach to learning was a sharp contrast to the classical university curriculum of his day. Through his efforts, Warrington Academy became the preeminent institution of its kind in England.

As Priestley's interest in science continued to grow, he sought to further his education by spending one month of each year in London. There he met leading scientists of his day including Benjamin Franklin who encouraged him in his experiments with electricity that led to his publication *The History and Present State of Electricity*. Soon Priestley's interests turned to the study of gases or "airs" as they were called in his day. Only three were known at the time: air, fixed air (**carbon dioxide**) and hydrogen. He soon claimed to have discovered ten more including nitrous air (nitric oxide), diminished nitrous air (nitrous oxide), and marine acid air (hydrogen chloride).

Priestley made a major contribution to biological science in 1771 when he accidentally discovered that **plants** could restore air "that had been injured by the burning of candles." On August 17, he "put a sprig of mint into air in which a wax candle had burned out and found that, on the 27th of the same month, another candle could be burned in this same air." He later found that a mouse could survive in the "restored" air, while it could not in the "injured" air. He concluded that green plants give off "dephlogisticated air," based on the prevailing theory of his day that held that air became unable to support combustion when it was saturated with a hypothetical substance called phlogiston. The French chemist Antoine-Laurent Lavoisier later found that the altered air was actually a gas that he called oxygen. Priestley's findings led to the discovery of **photosynthesis** when Dutch physician **Jan Ingenhousz** confirmed and extended his research. In 1779, Ingenhousz discovered that light must shine on the green tissues of the plant for the altered air (oxygen) to be produced. Ingenhousz also discovered that all living parts of the plant, green and non-green alike, could "damage" the air (respire) in darkness.

In 1774, Priestley was in the employ of the earl of Shelburne as a tutor to his two sons. With ample leisure time, he continued his experimentation. He heated red mercuric oxide, obtained a colorless gas, and found that a candle would burn in it "with a remarkably vigorous flame." He assumed that he had produced concentrated form of "dephlogisticated air". As he grew old, Priestley refused to accept Lavoisier's explanation that his observations were due to the production of oxygen. He became one of the last defenders of the phlogiston theory.

PRIGOGINE, ILYA (1917-)
Russian Belgian chemist

Ilya Prigogine was born in Moscow, Russia, in 1917. His father, Roman, was a chemical engineer and his mother, Julia had studied music at the conservatory in Moscow. Although Prigognine eventually elected to study chemistry, his highly philosophical approach to science reflected his numerous interests which included history, art, music, and law. Under the direction of Théophile De Donder, Prigogine received his Ph.D. from the Free University of Brussels in 1941 with the thesis, "The Thermodynamic Study of Irreversible Phenomena." De Donder was the founder of the Brussels school of thermodynamics (the branch of physics that deals with the behavior of heat and related phenomena).

Prigogine was particularly interested in the nature of time as it related to chemistry, and was therefore attracted to a study of the second law of thermodynamics that states that any spontaneous change in a closed system (one where neither matter nor **energy** flows into or out of the system) occurs in the direction that increases **entropy**, the measure of unavailable energy in a system or the measure of its disorder. This law indicates that as time passes in a closed system, disorder always increases, leading Sir Arthur Stanley Eddington to refer to the second law as supplying "the arrow of time." The move toward entropy described in the second law is irreversible, which contrasts with all other physical laws in which processes are reversible in time. This contrast begged the question of how the reversible, random workings of molecular and atomic motions could lead to processes that have a preferred direction in time. The second law also suggests that the universe is moving toward eventual decay, a point when all energy and matter will reach a uniform state of equilibrium known as heat death.

Intrigued by these issues, Prigogine moved his focus away from the ideal "closed" system described in the second law and instead studied open systems that exchange matter and energy with an outside environment. Prigogine's first success in dealing with irreversible processes and open systems not at equilibrium came in 1945. In his doctoral research, he showed that for systems not too far from equilibrium, changes take place so as to achieve a steady state in which the production of entropy is at a minimum. This is true near equilibrium where the flux (or flow) of energy or matter through the system is directly proportional to the force creating that flux; that is, the flux and the force are linearly related. But such a steady state, once established, is stable and continues unchanged; it cannot evolve into a new state.

In 1947 Prigogine succeeded De Donder to become full professor at the Free University of Brussels. He subsequently showed that far from equilibrium, where fluxes and forces are no longer linearly related, a system can become unstable and evolve new, organized structures spontaneously. Prigogine called these organizations dissipative structures and developed the mathematical means of describing them. Prigogine theorized that such structures can be maintained as long as the energy and material fluxes are kept up. The process by which a new order evolves is labeled self-organization. In a nonlinear system there exist points—Prigogine referred to these as moments of choice or bifurcation points—at which the system is unstable, and small fluctuations can grow to a macroscopic or large size, creating a new structure. Randomness enters at the bifurcation points, so that predictions with respect to outcomes can only be expressed as probabilities.

Prigogine's findings, while important, remained largely theoretical into the 1960s. Attempting to confirm his ideas, Prigogine worked with G. Nicolis and René Lefever to devise a simple mathematical model now called the Brusselator to better test his theories. Then in 1965 the Belousov-Zhabotinskii reaction, discovered in 1951 in the Soviet Union, became widely known abroad. One version of the reaction, in which the dissipative structures can be seen and do not have to be revealed by elaborate measurements, is a solution of malonic acid and bromate ion in sulfuric acid and ferrous phenanthroline (ferroin). Depending on the temperature and concentrations of the various species, the color of the solution may change back and forth from red to blue, or a pattern of red and blue may be formed that is either stationary or moves through the solution in a regular manner. These patterns gave striking visual proof of the existence of Prigogine's dissipative structures.

Prigogine's work is of great interest to many fields. In the fields of biology and biochemistry, it was suggested by Alan Mathison Turing in 1952 that instabilities in chemical re-

Ilya Prigogine. *(AP/Wide World Photos. Reproduced by permission.)*

action systems could explain the patterns of stripes on a zebra or spots on a leopard. On a still larger scale, the thermodynamics of irreversible systems may explain how evolution, a process that gives rise to ever more specialized forms, is compatible with a physical picture of the world in which systems inevitably move from an ordered to a disordered state.

Prigogine and others have also applied the principles of irreversible thermodynamics to such disparate systems as the development of traffic patterns on a highway in response to driving conditions and the buildup of giant termite mounds in which a large number of independent termites behave in an orderly, seemingly purposeful, and intelligent fashion. On a larger scale, Prigogine's research allows a somewhat different and brighter view of the universe's ultimate fate. As explained in *Omni*, the theory of dissipative structures "offers a guardedly optimistic alternative to the pessimistic view of mankind's future—that winding down of nature toward a kind of heat death."

Prigogine was named director of the Instituts Internationaux de Physique et de Chimie (the Solvay Institute) in 1959, a post in which he continued after his retirement from

Chimpanzees are the only primates whose genetic material closely matches that of humans. *(World Wildlife Fund. Reproduced by permission.)*

the Free University in 1985. In 1977 he became the first Belgian awarded the Nobel Prize. He and his wife Marina Prokopowicz, an engineer whom he married in 1961, live in Brussels.

PRIMATES

Primates are an **order** of **mammals**. Most primates are characterized by well-developed binocular vision, a flattened, forward-oriented face, prehensile digits, opposable thumbs (sometimes the first and second digits on the feet are also opposable), five functional digits on the feet, nails on the tips of the digits (instead of claws), a clavicle (or collarbone), a shoulder joint allowing free movement of the arm in all directions, a tail (except for apes), usually only two mammae (or teats), relatively large development of the cerebral hemispheres of the **brain**, usually only one offspring born at a time, and having a strong social organization. Most **species** of primates are highly arboreal (that is, they live in the forest canopy), but some live mostly on the ground. Primates first evolved early in the Cenozoic Era, about 60 million years ago. The ancestral stock of the primates is thought have been small, carnivorous **animals** similar to modern **tree** shrews (**family** Tupaiidae).

There are about 12 families and 60 genera of living primates (the numbers vary depending on the particular zoologi-

cal study being consulted). Most species of primates inhabit tropical and sub-tropical regions, and most occur in forested **habitat**s. Primates are divided into two sub-orders, the Prosimii (or prosimians) and the Anthropoidea (monkeys and apes). The families and examples of component species are given below.

Prosimii (This sub-order of primates has a relatively ancient evolutionary lineage, and includes several families of lemurs, lorises, and tarsiers, all of which have fox-like snouts, long tails, and inhabit **forests**).

Cheirogaleidae (This is a family of five species of dwarf or mouse lemurs, which only occur on the island of Madagascar, off Africa in the Indian Ocean). An example is the hairy-eared dwarf lemur (*Allocebus trichotis*).

Lemuridae (This is the largest family of lemurs, consisting of about 10 species, which only occur on Madagascar and the nearby Comoro Islands). Examples are the black lemur (*Eulemur macaco*) and the ring-tailed lemur (*Lemur catta*).

Megaladapidae (This is a family of two species of sportive lemurs, which also only occur on Madagascar). An example is the grey-backed sportive lemur (*Lepilemur dorsalis*).

Indridae (This is another family of prosimians of Madagascar, including four species known as wooly lemurs). An example is the indri (*Indri indri*).

Daubentoniidae (This family has only one species, the aye-aye (*Daubentonia madagascariensis*) which live in the forests of Madagascar).

Lorisidae (This prosimian family of 12 species occurs in forests of South Asia, Southeast Asia, and Africa). Examples are the slender loris (*Loris tartigradus*) of India and the potto (*Perodicticus potto*) of tropical Africa.

Tarsiidae (This family includes three species of small prosimians that inhabit forests of islands of Southeast Asia). One example is the Philippine tarsier (*Tarsius syrichta*).

Anthropoidea (This sub-order includes the Old World monkeys, the New World monkeys, the marmosets, and the apes). The various monkeys are relatively small, arboreal, and have tails, while the apes are larger, relatively intelligent, and lack a tail. Most species of anthropoid primates are arboreal and inhabit forests, but some do not.

Callitrichidae (This family includes about 33 species of small marmoset monkeys of tropical forests of South America and Panama). Examples include the golden-headed lion tamarin (*Leontopithecus chrysomelas*) and the pygmy marmoset (*Cebuella pygmaea*), both occurring in Brazil.

Cebidae (This family includes about 37 species of New World monkeys, distinguished by their prehensile (or grasping) tail, and nostrils separated by a relatively wide partition). Examples are the dusky titi monkey (*Cellicebus cupreus*) of northern South America and the Central American squirrel monkey (*Saimiri oerstedii*) of Costa Rica and Panama.

Cercopithecidae (This family includes about 60 species of Old World monkeys of Africa and Asia, characterized by non-prehensile tails, closely placed nostrils, and (usually) bare skin on the buttocks). Examples include the black colobus (*Colobus satanas*) of central Africa, the rhesus macaque (*Macaca mulatta*) of South Asia, the mandrill (*Mandrillus sphinx*) of West Africa, and the proboscis monkey (*Nasalis larvatus*) of Borneo.

Hylobatidae (This is a family of six species of gibbon apes, which are tail-less, highly arboreal and agile, and have loud, complex vocalizations (known as ''songs'')). Examples are the black gibbon (*Hylobates concolor*) of Southeast Asia and the siamang (*Symphalangus syndactylus*) of Malaysia and Sumatra.

Hominidae (This family includes five species of great apes, which are relatively large and robust, lack a tail, and are the most intelligent and socially complex species of primates). This group includes the gorilla (*Gorilla gorilla*) of Central Africa, the pygmy chimpanzee (*Pan paniscus*) of Congo, the chimpanzee (*Pan troglodytes*) of Central Africa, the orangutan (*Pongo pygmaeus*) of Borneo and Sumatra, and humans (*Homo sapiens*) who have worldwide distribution. All hominidae, with the exception of humans, only inhabit tropical forests.

Humans evolved about one million years ago. They are now by far the most widespread and abundant species of primate, living on all of the continents, including Antarctica. Humans are also the most intelligent species of primate, and probably of any species. Humans have undergone extremely complex cultural evolution, characterized by adaptive, pro-gressive discoveries of social systems and technologies that are allowing this species to use the products of **ecosystem**s in an increasingly efficient and extensive manner. Habitat changes associated with human activities, coupled with the harvesting of many species and ecosystems as resources, are now threatening the survival of numerous other species and natural ecosystems. This includes almost all other species of primates, whose **population**s have declined to the degree that the World Conservation Union (IUCN) considers them threatened by **extinction**.

PRION

Scrapie is a disease that attacks the **brain**s of sheep, producing strange behaviors such as the **animal**'s compulsion to scrape off its wool, which gives the disease its name. Scrapie has been known to scientists for more than 250 years, although virtually nothing about its **nature** or transmission was discovered until very recently. In 1986, a scrapie-like disorder was first noted in a herd of cows in Kent, England. Within five years, the disease had spread to cows throughout the country. Cows inflicted with the disease became very sensitive to sound and **touch**, walked unsteadily, and trembled and twitched. Scientists were able to determine that scrapie—one in a **family** of diseases termed spongiform encephalopathies—had been transmitted from sheep to cows. Studies of brains from infected animals showed some common features, especially the presence of rod-shaped particles that clumped together to form larger arrays.

What was the infectious agent in this disease? One possible answer had been suggested in late 1981 by neurologist **Stanley Prusiner** of the University of California at San Francisco. Prusiner had found that the scrapie agent was a very small particle, not much larger than a nucleic acid **molecule**. This finding suggested the existence of a **virus**, a simple pathogen consisting of **DNA** or **RNA** encased in a **protein** coat. However, none of the **enzyme**s that normally disable **nucleic acids** had any effect on the scrapie agent, while those enzymes that degrade protein did have an effect. Eventually, Prusiner concluded that he was dealing with an entirely new type of infectious particle, a protein with the ability to cause a contagious disease. He called this particle a prion, for ''infectious protein.''

Prusiner's hypothesis seemed unlikely to his colleagues at first. There was no precedent for the existence of an infectious agent that lacked a mechanism for its own **replication** (a nucleic acid). Prusiner found one possible answer to that objection in 1988, when he conducted a study of patients afflicted with Gerstmann-Strässler-Scheinker syndrome, a rare spongiform encephalopathy that affects humans. He found in their brain **tissue** a **gene** that codes for the production of an abnormal protein similar to the scrapie-like prion. Perhaps, he hypothesized, the defective gene was responsible for the manufacture of the disease-causing protein.

The validity of the prion hypothesis is still being tested and debated, particularly the question of whether or not the abnormal protein is infectious. However, more and more scientists have begun to accept that hypothesis. Today, the prion has been implicated in a half-dozen diseases: scrapie, transmissible mink encephalopathy, chronic wasting disease of mule deer and elk, kuru, Creutzfeldt-Jakob disease, and Gerstmann-Strässler-Scheinker syndrome. All of these diseases develop very slowly, over a period of many months or years. This pattern originally led scientists to suspect that the disease-causing agent was a slow-acting virus. The last three of the prion diseases listed above affect humans, although all are rare.

Kuru was once common among the Fore tribe in New Guinea. The disease was apparently transmitted through the ritual eating of the brain of dead relatives. The discovery of the mechanism by which kuru is transmitted was made by **D. Carleton Gajdusek**, who earned a share of the 1976 Nobel Prize for physiology or medicine for his work. Some scientists suspect that the prion may also be implicated in Alzheimer's disease and Parkinson's disease.

In the mid-1990s, a new variant of Creutzfeldt-Jakob disease attracted worldwide attention. From 1995 through 1997, this new form of the disease killed at least 19 people in Britain, all relatively young. What sets this type of Creutzfeldt-Jakob disease apart from the older form is that it is thought to be transmissible. Many scientists believe that its victims contracted the disease by eating beef from cows infected with bovine spongiform encephalopathy (BSE), or mad cow disease. In 1985, the first signs of a BSE epidemic began showing up in British cows that had eaten feed containing offals from scrapie-infected sheep. Due to the long incubation period of BSE, the epidemic didn't peak until 1992, and by 1997, about 37,000 animals a year were still being affected.

Although much more research on prions remains to be done, Prusiner saw his work receive the ultimate stamp of scientific approval in 1997, when he was awarded the Nobel Prize in physiology or medicine ''for his discovery of prions—a new biological principle of infection.'' Prusiner, born on May 28, 1942, received his M.D. degree from the University of Pennsylvania in 1968. In that same year, he began an internship at the University of California at San Francisco, where he has spent the bulk of his career.

PRODUCERS

Producers, sometimes called primary producers, are **organism**s that make their own food. Because these organisms change relatively simple inorganic **nutrients** into more complex, **energy**-rich, organic forms, they are **autotroph**s. Thus, they do not need any environmental source of organic material. Producers include green **plants**, **algae** and some **bacteria**.

Producers can use either sunlight or chemicals from inorganic **chemical reaction**s as their source of energy. The process of using sunlight to produce food is called **photosynthesis**. The vast majority of producers, including plants, algae and some bacteria, produce their food using photosynthesis. Producers that use chemicals from inorganic chemical reactions as their source of energy are using the process known as chemosynthesis. Chemosynthesis generally oc-

curs where there is no sunlight for photosynthesis, and bacteria are the only organisms that carry out this process. For example, in the deep ocean at hydrothermal vents, chemosynthetic bacteria are the base of the food web. All other organisms in this **ecosystem** rely on the chemosynthetic bacteria for food.

Producers hold an extremely important position in food webs in every ecosystem on Earth. In fact, without producers, an ecosystem would not be self-sustaining; it would not be able to exist on its own without the introduction of materials from another ecosystem. Producers are at the base of all food webs. They produce the organic nutrients upon which all other organisms in the ecosystem depend. Without these producers, there would be no way of capturing energy from the sun or chemical reactions, and therefore no new energy or organic material would enter the food web. When producers are consumed by **herbivore**s or **omnivore**s, organic material from the producers, as well as the stored energy they contain, is passed on to all other levels of the food web.

PRODUCTIVITY

Productivity is an ecological term referring to the fixation of solar **energy** by **plants** and **algae**, and the subsequent utilisation of that fixed energy (or **biomass**) by plant-eating **herbivore**s, **animal**-eating **carnivore**s, and detritivores that feed upon dead biomass. This complex of energy fixation and utilisation in an **ecosystem** is called a food web.

Ecologists refer to the productivity of green plants (and algae) as primary productivity, meaning it comprises the base of food webs. Gross primary productivity is the total amount of energy fixed by plants. Net primary productivity is smaller because it is adjusted for energy losses associated with plant **respiration**. If the net primary productivity of green plants in an ecosystem is positive, then the biomass of vegetation is increasing over time.

The productivity of herbivorous animals (which feed on plant biomass) is known as secondary productivity (both gross and net). That of carnivores (which feed on other animals) is known as tertiary productivity (again, both gross and net). The biomass of dead plants and animals in the ecosystem is consumed by decomposer **organism**s, in a detrital food web. (In some cases, environmental conditions may not allow **decomposition** to occur efficiently, usually because **water**logged soil restricts the supply of oxygen. In such ecosystems, dead biomass accumulates as peat or another kind of non-living organic **matter**).

Ecological productivity within food webs always has a pyramid-shaped structure. This means that the productivity of plants is much larger than that of herbivores, which in turn is much greater than that of their predators. Plants typically account for more than 90% of the total productivity of food webs, herbivores most of the rest, and carnivores less than 1%.

Because of differences in the availability of solar radiation, water, and **nutrients**, the world's ecosystems differ greatly in the amount of productivity that they sustain. The least productive ecosystems are **desert**, **tundra**, and the deep ocean.

These typically have a rate of energy fixation of less than 0.5 × 10³ kilocalories per square meter per year (thousands of kcal/m²/yr; a **calorie** is the amount of energy needed to raise the **temperature** of one gram of water by one degree C under standard conditions; there are 1000 calories in a kcal). Grassland, montane and boreal forest, water of the continental shelves, and rough **agriculture** typically have a productivity of 0.5-3.0 × 10³ kcal/m²/yr. Moist forest, moist prairie, shallow lakes, and typical agricultural systems have a productivity of 3-10 × 10³ kcal/m²/yr. The most productive ecosystems are fertile estuaries and marshes, coral reefs, terrestrial vegetation on moist alluvial deposits, and intensive agriculture, which can have a productivity of 10-25 × 10³ kcal/m²/yr.

PROKARYOTE

Prokaryotes are **cell**s or **organism**s that lack a nuclear **membrane** and membrane bound **organelle**s. Prokaryotic cells are characteristic of a metabolically diverse assortment of bacterial organisms grouped together in the **kingdom Monera**. Two major groups of bacteria, the **Eubacteria** and Archaebacteria, have been identified and because they differ in important ways, they are sometimes divided into separate phylogenetic kingdoms. Prokaryotic cells differ from eukaryotic cells in important ways. Prokaryotic cells are usually much smaller than eukaryotic cells, and are able to grow and divide rapidly. **Eukaryote**s, with a true membrane-bound **nucleus** and discrete membrane-bound organelles, are able to compartmentalize functional activity into structural components such as **mitochondria**, **chloroplast**s, **lysosome**s, **endoplasmic reticulum**, and Golgi complex. Prokaryote cells may carry out all of the activities assigned to organelles without the benefit of these specialized structures. Prokaryotes and eukaryotes differ to some extent in their biochemical composition, particularly with respect to their lipid composition. They also differ in many aspects of their **metabolism**. Some of the most significant of these differences involve the **nature** of **gene**s and the manner in which genetic information is expressed. The genes of prokaryotes are simpler and do not contain the large amounts of material incidental to **protein synthesis** found in eukaryotes. Prokaryotes have been present on Earth for much of its history, and in one form or another have occupied nearly all potential **habitat**s. Although some prokaryotes cause disease in **plant**s and **animal**s, the overwhelming majority do not. Many are necessary for timely recycling of dead organic **matter** that without them would tend to accumulate in the environment.

PROSTAGLANDINS

Prostaglandins are a type of chemical which can be found in almost all **tissue**s of **mammal**s. Prostaglandins exist in cyclic form, and are composed of fatty acids. They exert their functions both locally (in the basic vicinity of their production), as well as traveling through the **blood**stream to act on distant target **organ**s (in an endocrine manner). There are many types of prostaglandins, including those labeled PGA1, PGE1, PGE2, and PGI1.

Prostaglandins work on a number of different target tissues, including the uterus, ovaries, and fallopian tubes. Prostaglandins are known to be involved in the initiation and maintenance of uterine contractions during labor, as well as in the production of the uncomfortable cramps occurring during menstrual periods. Prostaglandins are also involved in regulating the size of the blood vessels of the kidney. Prostaglandins affect the actions of the blood **cell**s responsible for clotting (platelets). Prostaglandins are involved in the production of fever. Prostaglandins are also involved in the process of **inflammation**, and anti-inflammatory medications such as aspirin and ibuprofen work by interfering with the production of prostaglandins.

PROTEIN

Proteins are vital components of all forms of life, and are important to living **cell**s both structurally and functionally. They are made up of **amino acid**s linked together in a chain by a sequence of **peptide** bonds. A peptide bond links the amino group (-NH₂) of one amino acid with the carboxyl group (-COOH) of the next amino acid in the chain. Most simple proteins contain from 100 to 300 amino acid units and have molecular weights ranging from 12,000 to 36,000. Elaborate proteins with more than one **polypeptide** chain may have molecular weights of several hundred thousand.

The central role of proteins in living **organism**s has been recognized for nearly 200 years. They make up about 90% of the dry weight of **blood**, 80% of the dry weight of muscle, and lesser amounts of other **animal tissue**s and **organ**s. They are also important constituents of plant and bacterial cells. Although they are important quantitatively, because of their relative abundance, their functional importance is even more crucial. **Enzyme**s, the all-important cellular catalysts often present in minute amounts, are entirely or mostly protein.

Twenty different amino acids are commonly found in proteins. They all have the general formula RCH(NH₂)COOH, where C is carbon, H is hydrogen, N is **nitrogen**, and O is oxygen. The R represents a group called a side chain, and is different for each type of amino acid. Individual proteins contain their own characteristic component of amino acids arranged in a unique sequence. The properties of a protein are determined by the nature of its amino acid side chains, and the sequence in which the amino acids are arranged. The sequence of amino acids is called the primary structure of the protein, and by the interactions of side chains with each other and with the solvent (usually **water**), the shape and function of the protein is determined. Twelve of the amino acids can be made in cells; the eight remaining, which are called essential amino acids, are obtained through nutrition.

Proteins often exhibit a secondary, tertiary, and even quaternary structure. They develop a secondary structure due to the angles formed by the peptide bonds linking the amino acids. The bond angles are determined by weak links called hydrogen bonds that form between a nitrogen-bound hydrogen atom of one amino acid and an oxygen atom of another. Very often, the result is a helical secondary structure like a string spirally arranged around and imaginary cylinder. Proteins develop a tertiary structure by a bending and folding of the primary chain back upon itself. Common globular proteins, with a more or less spherical shape, are formed in this way. The side chains of amino acids are usually responsible for the patterns leading to tertiary structure. When side chains are large they interfere with the usual helical secondary structure of the protein causing bends and kinks. When they carry opposite electrical charges they attract one another, and when they have like charges they repel each other. Side chains containing carboxyl or hydroxyl groups (both hydrophilic) tend to form on the outside of the protein **molecule** in contact with an aqueous environment, whereas hydrophobic groups tend to avoid water by folding into the interior of the molecule. Some especially complex proteins are formed from more than one polypeptide chain, or subunit. The subunits are held together by the same kinds of forces involved in tertiary structure. **Hemoglobin** is a well known example of a protein with quaternary structure.

Although it has long been known that the properties of proteins are determined by their complement of amino acids and sequence in which the amino acids are arranged, the mechanism that determines which amino acids are incorporated into a given protein, and in what order, was discovered only recently. Not surprisingly, both the nature and sequence are determined by the genetic nature of the organism. Portions of the cellular **DNA** are able to code for a needed protein by the sequence of **nucleotide**s found in the applicable part of the DNA. The DNA code for the protein is transcribed into a complementary **RNA** nucleotide sequence. The segments of RNA serve as templates for protein synthesis with the aid of cellular components called **ribosome**s. Each group of three RNA nucleotides designates a specific amino acid. The sequence of amino acids is determined by the sequence of nucleotide triplets in the RNA molecule.

Proteins are classified as either simple or conjugated depending on whether they contain nonprotein components. Simple proteins are made up entirely of amino acids, whereas conjugated proteins contain a prosthetic group in addition to the polypeptide chain. The prosthetic group may be as simple as an ion, or as complex as a nucleotide. Proteins may also be classified according to their role. Two main types are structural proteins and functionally active proteins. Examples of the former are proteins found in skin, tendons, ligaments, hair, and nails. Functionally active proteins include enzymes, immunoglobulins (or antibodies), and protein **hormone**s.

See also DNA transcription; RNA function.

PROTEIN SYNTHESIS

Protein synthesis is the process by which **cell**s convert **amino acid**s into long chain **polymer**s called proteins. Proteins are **molecule**s which have a variety of functions in cells such as providing structure, storing **energy**, providing movement, transporting other substances, catalyzing biological reactions, and protecting against disease. They make up more than 50% of a cell's dry weight. Protein synthesis is programmed by **DNA**. During this process DNA is converted to **RNA** which is then translated into a protein by the **ribosome**s.

The theories that laid the foundation for our modern understanding of protein synthesis began in 1909 with **Archibald Garrod**. He was the first to suggest that genes were chemically expressed through **enzyme**s that catalyze specific **chemical reaction**s in the cell. He even theorized that an inherited disease reflected a person's inability to make a particular enzyme. Unfortunately, his ideas about **inheritance** were ahead of their time, and it took several decades before they were supported by further research.

This research came in the 1930s when **George Beadle** and **Edward Tatum** established the relationship between genes and enzymes. They discovered that in certain bread mold **species**, there were mutants which required extra **nutrients** to grow on a plate. These different mutants were thought to be deficient of certain metabolic enzymes. Through their research, Beadle and Tatum established the one gene-one enzyme hypothesis which states that the function of a gene is to dictate the production of a specific enzyme. This idea was later refined when it was found that genes also dictate the production of proteins.

Genes are the **genetic material** that dictate the production of proteins and enzymes. The genes are located in the **nucleus** of the cell and are composed of DNA. The DNA is composed of **nucleotide**s which are molecules made up of a sugar component, a **phosphate group** and a cyclic base. There are four different nucleotide bases in which all the information for making proteins is stored. A gene is typically hundreds or thousands of nucleotides long. Unlike DNA, proteins are made up of amino acids instead of nucleotides. To get from DNA to protein requires two steps, **transcription** and **translation**.

The first step in protein synthesis involves the transcription of DNA into messenger RNA (mRNA). While DNA and RNA are similar, there are subtle differences. For example the sugar component of DNA has one less hydroxyl group than ribose, the sugar component of RNA. Also, RNA uses the nucleotide uracil instead of thymine.

During the process of transcription an enzyme known as RNA polymerase separates an area of the double strands of DNA. It first binds to a region of DNA known as a promoter region. It then begins to transcribe, or copy, the DNA into RNA at an initiation site. Only one strand is used as a template for making the mRNA. This occurs by the **base pairing** of the nucleotides. The polymerase than moves along the DNA and elongates the RNA molecule. As the polymerase travels along the DNA it both unwinds, catalyzes copying and rewinds the DNA. When the polymerase comes upon a certain sequence of DNA, called the termination sequence, it stops transcription and releases the RNA. The resulting RNA contains an exact copy of the gene. This RNA is further processed by **RNA splicing** to remove introns. The result is mRNA which is transported out of the nucleus into the **cytoplasm** where it can be translated into a protein.

Translation is the actual production of the protein or **polypeptide**. It is the process by which the mRNA nucleotide code is translated into amino acids. It turns out that different combinations of three nucleotides code for the 20 amino acids. For example, the base triplet GCA is translated into the amino acid alanine. Thus, the **genetic code** is in the form of triplet codons. Since it takes three nucleotides to code for one amino acid, each gene must have three times as many nucleotides as amino acids that make up the protein. For example, a strand of mRNA that has 300 nucleotides would code for a protein that is 100 amino acids long.

The process of translation involves cellular **organelles** called ribosomes. The ribosome are made up two subunits which are composed of proteins and ribosomal RNA. During translation the small subunit of the ribosome attaches to the mRNA at the initiation site. This is a sequence of three nucleotides which is the same in every gene. The large subunit then converges on the structure carrying with it another type of RNA called transfer RNA tRNA). The tRNA has a reading portion which recognizes three nucleotides. It interacts with an enzyme called aminoacyl-tRNA synthetase which attaches specific amino acids to the tRNA.

After initiation, the amino acids are combined to create a growing polypeptide chain. The ribosome travels along the mRNA in blocks of three nucleotides, or codons. The tRNA reads the codon and new amino acids are added one by one. When the ribosome gets to a certain codon, called the stop codon, it ceases translation. The polypeptide chain is then modified by other enzymes in the cytoplasm. These posttranslational modifications can include the attachment of sugars or **lipids**, the removal of certain amonio acids, or the joining of one polypeptide chain with another to make a multiple unit protein.

Cracking the genetic code was done during the 1960s. The first experiment was done by **Marshall Nirenberg**. He created an artificial RNA molecule which contained only uracill. He put this in a test tube with ribosomes and the other materials needed to synthesize proteins. The result was a protein that contained only phenylalanine. This suggested that a codon of UUU specified the amino acid phenyalanine. After this experiment, a series of similar ones were run until the entire genetic code was determined. Now when scientists get the nucleotide sequence of a gene, they know exactly the amino acid sequence of the protein it codes for. The genetic code is nearly universal. It is shared by **organism**s as diverse as **bacteria** to humans. This is important because it has allowed the incorporation of human genes in bacteria to produce human proteins.

See also Transcription; Translation

PROTISTA

Protista refers to one of the **kingdom**s of **organisms** commonly called protists. They were first described by the German biologist **Ernst Haeckel** in the 1860s. Until this time, biologists recognized two kingdoms, the Plantae and Animalia. Haeckel observed a group of microscopic organisms with both plant and animal characteristics, such as **flagellated cells** with **chloroplasts**. At this time, the Kingdom Protista was proposed. Today, five kingdoms are recognized: Plantae, Animalia, Protista, **Fungi**, and **Monera** (**bacteria**). The protists are sometimes known as the Protoctista (meaning "very first" or "to know"). This kingdom includes approximately 27 phyla and more than 200,000 **species**.

The Kingdom Protista is a fairly heterogeneous group of unicellular **eukaryote**s. These organisms share several characteristics in common. First of all, they are unicellular, which means one cell functions as a whole organism where all life functions are carried out. Most protists are motile using pseudopodia ("false feet" extensions of the **protoplasm**), flagella (whip-like **organelle**s), or **cilia** (short, hair-like structures). Most parasitic species are non-motile. **Asexual reproduction** is most common through **mitosis**, but some species have **sexual reproduction** as exemplified by **conjugation** in ciliates. Most species are free living, but many form symbiotic relationships with other organisms including parasitism, **commensalism**, and **mutualism**.

Protists live in a wide variety of **habitat**s, including soil, freshwater, marine, and as ecto- and endoparasites. Taxonomically, protists are differentiated by the structure of their cell, types of organelles, and form of reproduction or **life cycle**. There are three general groups of protists. The first are the plant-like protists, sometimes known as the Protophyta. These include the diatoms or golden **algae** (**Phylum** Chrysophyta), dinoflagellates (Phylum Pyrrophyta), euglenoids (Phylum Euglenophyta), and cryptomonads (Phylum Cryptophyta). The animal-like protists are included in the Subkingdom **Protozoa**. These include the flagellated protozoans like the **genus** *Trypanosoma*Trypanosoma, a **parasite** that causes African Sleeping Sickness and Chaga's disease (Phylum Zoomastigina); amoeboid forms like *Amoeba*, radiolarians, and foraminiferans (Phylum Sarcodina); Ciliates like *Paramecium* (Phylum Ciliophora); and parasitic sporozoans like *Plasmodium*, a parasite that causes malaria (Phylum Apicomplexa). The fungus-like protista include slime molds, **water** molds, and mildews. These are sometimes included within this kingdom, but they are usually included in the Kingdom Fungi.

PROTOPLASM • See Cell

PROTOZOA

Protozoa are a very varied group of single-celled **organism**s, with more than 50,000 different types represented. The vast majority are microscopic, many measuring less than 1/200 mm, but some, such as the freshwater *Spirostomun*, may reach 0.17 in (3 mm) in length, large enough to enable it to be seen with the naked eye. Scientists have even discovered some **fossil** specimens which measured 0.78 in (20 mm) in diameter. Whatever the size, however, protozoans are well-known for their diversity and the fact that they have evolved under so many different conditions. One of the basic requirements of all

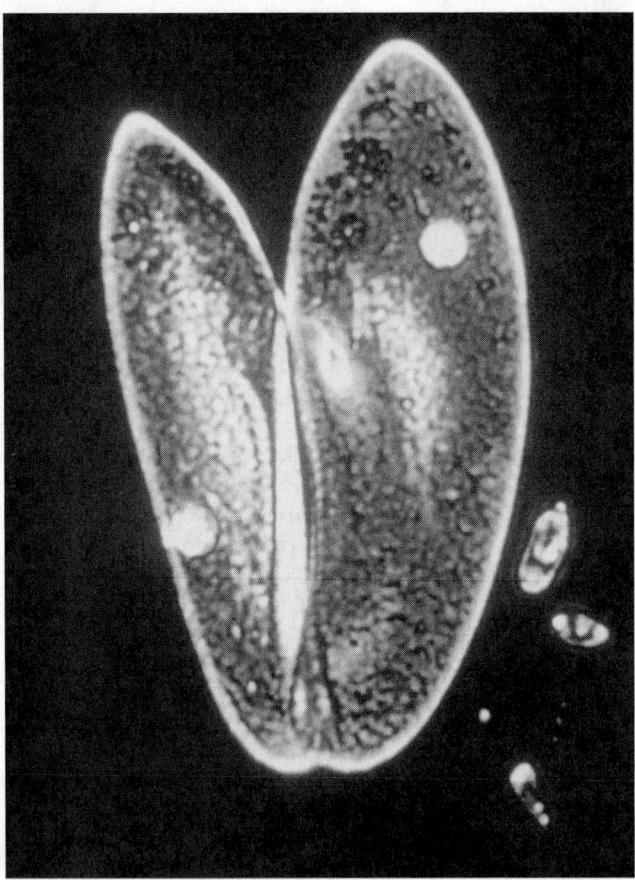

Two paramecia in sexual reproduction. Although most protozoans reproduce asexually, these paramecia are trading genetic information and each will then produce four daughter cells. *(Photo Researchers, Inc. Reproduced by permission.)*

Food vacuoles develop whenever food is ingested and shrink as digestion progresses. If too much water enters the cell, these vacuoles swell up, move towards the edge of the cell wall and release the water through a tiny pore in the membrane.

Some protozoans contain the green pigment chlorophyll more commonly associated with higher **plants**, and are able to manufacture their own foodstuffs in a similar manner to plants. Others feed by engulfing small particles of plant or animal **matter**. To assist with capturing prey items many protozoans have developed an ability to move around. Some, such as Euglena and Trypanosoma are equipped with a single whiplike flagella which, when quickly moved back and forth, pushes the body through the surrounding water body. Other protozoans such as Paramecium have developed large numbers of tiny cilia around the membrane; the rhythmic beat of these hairlike structures propel the cell along and also carry food, such as bacteria, towards the gullet. Still others are capable of changing the shape of their cell wall. The Amoeba, for example, is capable of detecting chemicals given off by potential food particles such as diatoms, algae, bacteria or other protozoa. As the cell wall has no definite shape, the cytoplasm can extrude to form pseudopodia (Greek: *pseudes*, false; *pous*, foot) in various sizes and at any point of the cell surface. As the Amoeba approaches its prey, two pseudopodia extend out from the main cell and encircle and engulf the food, which is then slowly digested.

Various forms of reproduction have evolved in this group, one of the simplest involves a splitting of the cell in a process known as binary fission. In species like amoeba, this process takes place over a period of about one hour: the nucleus divides and the two sections drift apart to opposite ends of the cell. The cytoplasm also then begins to divide and the cell changes shape to a dumb-bell appearance. Eventually the cell splits giving rise to two identical "daughter" cells which then resume moving and feeding. They, in turn, can divide further in this process known as **asexual reproduction**, where only one individual is involved.

Some **species**, which may reproduce asexually, may occasionally reproduce through sexual means, which involves the joining together, or fusion, of the nuclei from two different cells. In the case of paramecium, each individual has two nuclei: a larger macronucleus which is responsible for growth, and a much smaller micronucleus that controls reproduction. When paramecium reproduces by sexual means, two individuals join together in the region of the oral groove—a shallow groove in the cell membrane that opens to the outside. When this has taken place, the macronuclei of each begins to disintegrate, while the micronucleus divides in four. Three of these then degenerate and the remaining nucleus divides once again to produce two micronuclei that are genetically identical. The two cells then exchange one of these nuclei which, on reaching the other individual's micronucleus, fuses to form what is known as a "zygote nucleus." Shortly afterwards, the two cells separate but within each cell a number of other cellular and cytoplasmic divisions will continue to take place, eventually resulting in the production of four daughter cells from each individual.

Protozoans have evolved to live under a great range of environmental conditions. When these conditions are unfavor-

protozoans is the presence of **water**, but within this limitation they may live in the sea, in rivers, lakes or even stagnant ponds of freshwater, in the soil and even in some decaying matters. Many are solitary organisms, but some are colonial; some are free-living, others are sessile; and some species are even parasites of plants and animals—from other protozoans to humans. Many of them form complex, exquisite shapes and their beauty is often greatly overlooked on account of their diminutive size.

The **cell** body is often bounded by a thin pliable **membrane**, although some sessile forms may have a toughened outer layer formed of cellulose, or even distinct shells formed from a mixture of materials. All the processes of life take place within this cell wall. The inside of the membrane is filled with a fluidlike material called **cytoplasm**, in which a number of tiny organs float. The most important of these is the **nucleus**, which is essential for growth and reproduction. Also present are one or more contractile **vacuoles**, which resemble air bubbles, whose job it is to maintain the correct water balance of the cytoplasm and also to assist with food assimilation. Protozoans living in salt water do not require contractile vacuoles as the concentration of salts in the cytoplasm is similar to that of seawater and there is therefore no net loss or gain of fluids.

able, such as when food is scarce, most species are able to enter an inactive phase, where cells become non-motile and secrete a surrounding cyst that prevents desiccation and protects the cell from extreme temperatures. The cysts may also serve as a useful means of dispersal, with cells being borne on the wind or on the feet of animals. Once the cyst reaches a more favorable situation, the outer wall breaks down and the cell resumes normal activity.

Many species are of considerable interest to scientists, not least because of the medical problems that many cause. The tiny Plasmodium protozoan, the cause of malaria in humans, is responsible for hundreds of millions of cases of illness each year, with many deaths occurring in poor countries. This parasite is transferred from a malarial patient to a healthy person by the bite of female mosquitoes of the genus *Anopheles*. As the mosquito feeds on a victims' blood the parasite passed from its salivary glands into the open wound. From there, they make their way to the liver where they multiply and later enter directly into red blood cells. Here they multiply even further, eventually causing the blood cell to burst and release from 6-36 infectious bodies into the **blood** plasma. A mosquito feeding on such a patients blood may absorb some of these organisms, allowing the parasite to complete its life cycle and begin the process all over again. The shock of the release of so many parasites into the human blood stream results in a series of chills and fevers—typical symptoms of malaria. Acute cases of malaria may continue for some days or even weeks, and may subside if the body is able to develop an immunity to the disease. Relapses, however, are common and malaria is still a major cause of death in the tropics. Although certain drugs have been developed to protect people from Plasmodium many forms of malaria have now developed, some of which are even immune to the strongest medicines.

While malaria is one of the best known diseases known to be caused by protozoans, a wide range of other equally devastating ailments are also caused by protozoan infections. Amoebic dysentery, for example, is caused by Entamoeba histolytica; African sleeping sickness, which is spread by the bite of the tse-tse fly, is caused by the flagellate protozoan Trypanosoma; a related species *T. cruzi* causes Chagas' disease in South and Central America; Eimeria causes coccidiosis in rabbits and poultry; and Babesia, spread by ticks, causes red water fever in cattle.

Not all protozoans are parasites however, although this is by far a more specialized life style than that adopted by free-living forms. Several protozoans form a unique, nondestructive, relationship with other species, such as the those found in the intestine of wood-eating termites. Living in the termites' intestines the protozoans are provided with free board and lodgings as they ingest the wood fibers for their own nutrition. In the process of doing so, they also release proteins which can be absorbed by the termite's **digestive system**, which is otherwise unable to break down the tough cellulose walls of the wood fibers. Through this mutualistic relationship, the termites benefit from a nutritional source that they could otherwise not digest, while the protozoans receive a safe home and steady supply of food.

With such a vast range of species in this **phylum**, it is not surprising that little is still known about the vast majority

of species. Many protozoans serve as an essential food source for a wide range of other animals and are therefore essential for the ecological food webs of higher organisms. Many are also, of course, important for medical purposes, while others are now being used in a range of businesses that include purification of filter and sewage beds. No doubt as further research is undertaken on these minute organisms we shall learn how more of these species might be of assistance, perhaps even in combating some of the major diseases that affect civilization, including those caused by other protozoans.

PRUSINER, STANLEY B. (1942-)
American neurologist

Prusiner won the 1997 Nobel Prize in Physiology or Medicine for his ground-breaking, yet controversial, work on a type of **protein** particle, a **prion**, that he hypothesized was responsible for a number of fatal neurodegenerative disorders, including Creutzfeldt-Jakob disease (CJD) and mad cow disease.

Prusiner was born on May 28, 1942, in Des Moines, Iowa. He earned his B.A., *cum laude*, at the University of Pennsylvania, in 1964, and then went on to receive an M.D. in 1968 from the same university. Prusiner then served his internship and residency at the prestigious University of California at San Francisco (UCSF) from 1968 to 1969, where he later served a residency in neurology. Prusiner soon became interested in neurodegenerative diseases. This interest was in part developed after one of his patients died of CJD, a disease of the cerebral cortex which leads to dementia and eventual **death**.

As a result of this experience, Prusiner learned that an entire category of diseases was yet to be elucidated. At the time, researchers thought many neurodegenerative diseases were caused by so-called slow **virus**es, which would take years and sometimes decades to incubate in the host. As early as 1967, a British team working at Hammersmith Hospital had proposed the existence of an infective agent that lacked nucleic acid in the sheep disease known as scrapie (so called because infected sheep tend to scrape the wool off their bodies). Their hypothesis grew out of the fact that when the genetic substance was destroyed in known infected material, extracts from the infected material were still able to spread the disease. This led to the conclusion that perhaps the infective agents in such animal diseases as scrapie and kuru (a disease of the cannibalistic Fore people of New Guinea which had been traced to the ritual eating of the brains of departed relatives) were non-viral.

Prusiner combined a zeal for research with a disarming political sense to win grants for the study of such diseases, including CJD, scrapie, kuru, fatal familial insomnia, and bovine spongiform encephalopathy (BSE), or mad cow disease. Over three decades, he managed to gain funding totaling 56 million dollars from the National Institutes of Health (NIH).

Expanding on work by British researchers as well as the NIH's Rocky Mountain Laboratories, which had shown similarities between kuru and scrapie, Prusiner set up a research team at UCSF employing ultimately a quarter million mice infected with diseased **brain** matter in an attempt to isolate the infective agent in neurodegenerative diseases. Such research was laborious and time-consuming, for the incubation period in mice took upwards of 200 days. Early breakthroughs occurred when Prusiner switched from mice to hamsters, as the onset of illness in those animals would occur twice as fast. By 1981, Prusiner was able to conclude that a protein was the causative agent in these brain diseases, and he dubbed this agent the prion. Such proteins were resistant to any modification of nucleic acids. When he and his team added **enzyme**s that destroy nucleic acids in **gene**s, they discovered that there was no reduction in the infective power of prions.

By 1992, Prusiner's research more clearly demonstrated the nature of the interaction between prion proteins. Prusiner showed that when the gene encoding for the prion protein in mice was destroyed, such mice (called prion knock-out mice) proved resistant to the diseases when injected with preparations of disease-causing prion protein. Later, when the prion gene was re-activated and the same mice were injected with diseased matter, they again became susceptible to infection. Though the role of prion proteins is unclear, what has become clear is the close causative effect of prions in a variety of neural diseases.

The diseases which Prusiner studied are life-threatening to only a tiny fraction of the world's population—the annual death toll of CJD in the United States annually, for example, is about 225, less than the number of traffic fatalities in two days. However, as the Nobel committee noted, Prusiner's research could lead to new therapies for a larger array of neurological disorders such as Alzheimer's and Parkinson's disease, which may or may not be caused by protein-based infective agents. In the case of prion-based diseases, Prusiner has suggested gene therapy to curtail the production of prion proteins and thus eliminate the spread of such diseases, as was done with his prion knock-out mice.

PTERIDOPHYTA

Pteridophyta is a division of **plants** within the **classification** system that are all vascular in **nature** and reproduce by the means of **spore**s. The most common examples are ferns, horsetails, and club mosses.

Pteridophytes achieved prominence in the carboniferous period, along with the dinosaurs. At that time, many giant forms of pteridophytes superficially similar to extant **species** existed. The largest were some 60 ft (20 m) tall, whereas the largest living examples alive today are the giant horsetails *Equisetum giganteum* which grow up to 15 ft (5 m). The pteridophytes are responsible for many of the coal and oil deposits around the world, and as living **organism**s, they are popular as decorative plants.

Extant pteridophytes are widely distributed around the world, although their greatest range of diversity can be found in the tropics. They are mostly vascular and terrestrial, with no secondary thickening. They show alternation of generation with the **gametophyte** form being dominant. The spores are usually produced on the underside of laves or cones, and there is still a great reliance on **water** for the **fertilization** process when it occurs.

PUNCTUATED EQUILIBRIUM

Punctuated equilibrium is an **evolutionary theory** put forward by Niles Eldredge and Stephen Gould in 1972 that runs counter to **Charles Darwin**'s evolutionary theory of **gradualism**. Darwin's evolutionary theory maintains that **adaptation** in **species** are the result of a continuous process of gradual change. Within the tenants of gradualism, an adaptation would take many millions of years and all intermediate forms would exist for a long period, leaving a large likelihood for intermediate **fossil** types to exist. Surprisingly, intermediate forms are rare within the fossil record, and this lack of intermediate fossil evidence lead Eldredge and Gould to formulate the theory of punctuated equilibrium.

Contrary to Darwin's gradualism, punctuated equilibrium suggests that new species may have arisen rapidly over only a few thousand years and then remained unchanged for many millions of years before the next period of adaptation. Rather than occurring over the majority of the **population**, punctuated equilibrium postulates that change occurred in only a small part of the population. With change occurring rapidly within a small portion of the population, intermediate species would not likely be represented in the fossil record. Adaptation would be swift and profound to have such a major effect.

The sort of macro **mutation**s that Gould and Eldredge envisioned include mutations within regulator genes which simultaneously affect a whole **operon** and, thus, drastically affect the development of the **organism**. The incipient species would have to then supplant the original species and **natural selection** would eliminate intermediate types.

Darwin himself had recognized that such events may occur and contrary to some peoples beliefs punctuated equilibrium can be quite easily explained by and accepted into the system of natural selection.

PUNNETT, R.C. (1875-1967)
English morphologist and geneticist

Noted morphologist and geneticist R. C. Punnett was instrumental in introducing the field of **genetics** to lay audiences, especially to commercial breeders of livestock. His contributions significantly advanced knowledge of the genetics of fowl, ducks, rabbits, sweet pea plants, and humans; his research served as the foundation for poultry genetics for decades. Punnett was among the pioneering investigators who helped revolutionize scientific thought in the field of genetics after the rediscovery of **Gregor Mendel**'s work with genetics and heredity.

Reginald Crundall Punnett, the eldest of three children, was born on June 20, 1875, at Tonbridge in Kent, England, to George Punnett, the head of a Tonbridge building firm, and

Emily Crundall. He suffered from appendicitis as a child. During the treatment, which consisted of applying leeches to the lower stomach, and the daily bedrest required afterwards, he spent his time reading Jardine's *Naturalist's Library* and discovered a strong liking for natural history. He recovered and later was accepted to Cambridge University, where he developed an interest in human **anatomy**, human physiology, and **zoology**, and decided to pursue a career in zoology, not medicine as he originally had intended. As part of his zoological studies, Punnett observed sharks at the Zoological Station in Naples, Italy, for six months before graduating with first class honors from Cambridge's Caius College in 1898. He received his master's degree in 1902.

Studies Zoology of Worms

In 1899, Punnett was offered the position of demonstrator and part-time lecturer in the Natural History Department of St. Andrews University, where he stayed for three years. In 1901, he was elected a fellow of Caius College. During this time, his appendix had been troubling him sporadically and he decided to have it removed. As the scientific thought at the time was that worms caused appendicitis, he dissected the organ after the surgery, but found no worms. After he recovered from the operation, Punnett became unhappy with his teaching accommodations at St. Andrews and began a search for a new job. In 1902, he returned to Cambridge as a demonstrator in morphology in the Department of Zoology and remained in this position until 1904. While a demonstrator, Punnett had plenty of time to perform research and publish a number of papers on nemertines, a type of worm.

Discovers Several Mendelian Genetic Principles

Punnett then turned from the study of nemertines to genetics, working with **William Bateson**, an investigator also researching Mendelian principles, during a six-year period in which the pair produced noteworthy and lasting advances in Mendelian genetics. Gregor Mendel, an Augustinian monk, had used pea plants to demonstrate how genetic traits are inherited. Using the sweet-pea plant or fowl for their studies, Punnett and Bateson determined several basic classical Mendelian genetic principles, including the Mendelian explanations of sex-determination, sex-linkage, complementary factors and factor interaction, and the first example of autosomal linkage. It was early in his relationship with Bateson that Punnett was awarded the Balfour Studentship in the Department of Zoology, and he resigned as demonstrator. He held the Balfour Studentship position until 1908. Also, during this year he was awarded the Thurston medal of Gonville and Caius College. Punnett wrote *Mendelism,* the first published textbook on the subject of genetics, in 1905. As a reflection on his research, he was appointed superintendent of the Museum of Zoology in 1909. A year later, Punnett became professor of biology at the University of Cambridge. Punnett and Bateson started the *Journal of Genetics* in 1911 and edited it jointly until Bateson died in 1926. Punnett continued to edit the journal, credited with drawing numerous new students to the field of genetics, for twenty more years.

In 1912, the University of Cambridge changed the name of the chair of biology to the chair of zoology, and offered the position to Punnett, making him the first Arthur Balfour Professor of Genetics, a position that was the first of its kind in Great Britain. He held this prestigious position until his retirement at the age of sixty-five. Also in 1912, Punnett was elected a fellow of the Royal Society, and in 1922 he was awarded its Darwin Medal. Punnett's interests in genetics led him to a founding membership in the British Genetical Society; he served as one of the group's secretaries from 1919 to 1930, when he then became president.

During World War I, Punnett bought his expertise in poultry breeding to a position with the Food Production Department of the Board of Agriculture. He suggested that hens' plumage color could be used to determine the sex of the birds much earlier than previously was possible. This enabled the breeders to destroy most of the unwanted males, which were not used for food, and save precious resources for raising females for consumption. After the war ended, Punnett produced the first breed of poultry in which a trait is demonstrated uniquely in one sex or the other. This first auto-sexing breed, the Cambar, was followed a decade later by a second breed, the Legbar. Punnett's work laid the foundation for poultry breeding research for several more decades.

Punnett was married to Eveline Maude Froude, widow of Sidney Nutcombe-Quicke, at age forty-one; they had no children. In his leisure time, he enjoyed playing bridge, participated in many sports including cricket and tennis, and collected Japanese color prints, Chinese porcelain and old and rare biological and medical texts. Upon his death, his collection of Japanese prints was acquired by the Bristol Corporation for the city art gallery. Punnett died suddenly during a game of bridge in Bilbrook, Somerset, England, on January 3, 1967.

Punnett square

A Punnett square is a checkerboard representation of predictable **genotype**s from a second generation offspring of an experimental breeding. Named for geneticist **Reginald C. Punnett**, whom originally used the method to compute the results of a cross using only one gene. Punnett squares are now occasionally used for considering two genes and their **allele**s, however they quickly become unwieldy when used to calculate offspring of more than two alleles.

With a Punnett square, one gene is considered and the **genotype and phenotype** are predicted. Along the top of the checkerboard are the alleles found in the **gamete**s of one parent (usually the male by convention) and the alleles found in the other parent are written down the left-hand side. The products of the possible matings are then placed in the four boxes in the middle of the checkerboard. The relative numbers of the different progenies can then be calculated. For example a man heterozygous for ''A'' can produce gametes of ''A'' and ''a.'' If the man mates with a homozygous recessive female, who can only produce gametes of type ''a,'' their resulting offspring genotypes can be calculated using a Punnett square. The genotypes for the resultant offspring would either be ''Aa'' or ''aa.'' If ''A'' was dominant to ''a,'' then phenotypically half of the offspring would be "A" and half would be "a". If two heterozygotes mated, the phenotypical offspring would be produced in the ratio of three dominant to one recessive. In the latter case, the appropriate Punnett square would genotypically illustrate one quarter "AA," one half "Aa," and one quarter "aa."

A Punnett square can also be used to calculate the relative frequencies of a mating when two different genes are observed. The same rules are used, but there are four different gametes produced along the top and side, giving 16 different possibilities for the offspring. If both parents are "AaBb," then four different gametes are possible—"AB," "Ab," "aB," and "ab." Genotypically, the Punnett square shows us that in the 16 offspring one would be dominant homozygous, one would be recessive homozygous, and the remainder would be heterozygous for at least one of the gene pairs. If "A" coded for brown eyes, "a" for blue eyes, "B" coded for black hair, and "b" for blond hair, then the results would be in a 9:3:3:1 ratio (nine brown eyed and black haired, three brown eyed and blond haired, three blue eyed and black haired, and one blue eyed and blond haired).

A Punnett square considering two genes assumes the **law of independent assortment** is followed. First formulated by **Gregor Johann Mendel**, independent assortment is the random distribution of alleles to gametes during **meiosis**. If one particular allele goes to one gamete, this has no influence on the likelihood of any other allele going to the same gamete. All various allele combinations have an equal chance of representation in a gamete.

PURKINJE (PURKYNĚ), JAN EVANGELISTA (1787-1869)
Czech histologist and physiologist

As a boy growing up in Bohemia (now part of the Czech Republic), Jan Purkinje showed great promise. His father, an estate manager who encouraged his son's interests, died when Purkinje was six years old. At the age of ten, Purkinje, an only child, was admitted to a Piarist monastery near the Austrian border. (Such monasteries had been established in 1597 to educate the poor.) Purkinje became a choirboy and outstanding student at the monastery, quietly studying for the priesthood. Just before he was to be ordained a priest, however, Purkinje decided to take up the study of philosophy at Prague University. While there, he became interested in medicine. His research and tutoring during this time strengthened his physics background.

In 1818, Purkinje presented his graduate thesis, which described a visual phenomena now known as the *Purkinje effect*. He stated that, as the intensity of **light** decreases, different colored objects that appeared to be the same brightness in highly intense light appear to be unequally bright. In other words, as light intensity decreases, blue objects might appear to be brighter than red objects—even though they appeared to be the same brightness in the more intense light.

After graduating in 1819, Purkinje developed wide-ranging interests in the areas of experimental pharmacology and psychology, phonetics, **histology**, **embryology**, and physical anthropology. In 1823, the same year he took the position of Professor of Physiology and Pathology at the University of Breslau (now Wrocław, Poland), Purkinje published another paper that recognized fingerprints as a way to identify individuals. He noted that fingerprints seemed to follow nine general patterns, and he mentioned the ridge formations of the human palm.

In 1832, after obtaining a modern **microscope**, Purkinje began a new period of research. He found different ways to examine tissues under the microscope—fixing, sectioning, staining. With his techniques, he saw structures that other observers hadn't noticed. For instance, in 1837, Purkinje discovered large pear-shaped nerve **cell**s in the outer layer of the **brain** that had several branches. These are now called *Purkinje cells*. He was the first to describe these cells as formations in the central **nervous system** of vertebrates and pointed out that they play an important role in nervous activity. Two years later, as Purkinje was investigating the function of muscular organs, he discovered *Purkinje fibers*—special muscle fibers in the ventricles of the heart. Later it would be shown that Purkinje fibers have an important function: they conduct contraction to all parts of the heart.

That same year, Purkinje (no doubt influenced by his theological background) described the contents of animal embryos, using the theological term **protoplasm** in its scientific sense. To him, the term meant "first formed," but eventually it took on a more general meaning: the living material inside a cell. Purkinje also conducted comparative studies of animal and plant tissue, observing that "granules"—now termed cells—were present in both. These observations laid the groundwork for **Matthias Schleiden** and **Theodor Schwann**, who formulated their **cell theory** in 1839.

Purkinje went on to open an independent physiological institute in Breslau—the first of its kind. Although such an institute was very rare until the mid-nineteenth century, it soon became a regular part of medical schools. After 1850, Purkinje returned to Prague. He devoted the remainder of his life to the cause of Czech nationalism and making science more accessible to his countrymen.

PYRUVIC ACID

Pyruvic acid is a 3-carbon alpha-keto acid that is an intermediate in several important metabolic pathways. It is the simplest of all alpha-keto acids, containing a carbonyl group (a carbon atom doubly bonded to an oxygen atom) on the middle carbon. Pyruvic acid (in its salt form pyruvate) is formed in the normal **metabolism** of glucose as a product of the **glycolysis** metabolic pathway. In a series of **enzyme** catalyzed reactions, the six carbon sugar is converted to two **molecule**s of pyruvate that are further metabolized, with the loss of **carbon dioxide**, to two molecules of acetyl **coenzyme** A. Acetyl coenzyme A is fed into the **aerobic Krebs cycle** where it is further converted to carbon dioxide and **water**. In **anaerobic** conditions, pyruvate may accept hydrogen **atoms** that would otherwise be transferred to oxygen and thereby be converted to **lactic acid**. Lactic acid serves as an endpoint for glucose metabolism and may accumulate in **tissue**s, sometimes to toxic levels. Pyruvate can be converted to the **amino acid** alanine by replacing the car- bonyl group with an amino group (NH_2). In the **Hatch-Slack photosynthetic pathway**, pyruvate plays an important role as a precursor to phosphoenolpyruvate, the molecule that serves as an acceptor of carbon dioxide.

R

RADIOACTIVE DATING

In the nineteenth century, prominent scientists such as **Charles Lyell**, **Charles Darwin**, Sir William Thomson (Lord Kelvin), and Thomas Huxley, were in continual debate about the age of the earth. The discovery of the radioactive properties of uranium in 1896 by Henri Becquerel subsequently revolutionized the way scientists measured the age of artifacts and supported the theory that the earth was considerably older than what some scientists believed.

There are several methods of determining the actual or relative age of the earth's crust: examination of **fossil** remains of **plants** and **animals**, relating the magnetic field of ancient days to the current magnetic field of the earth, and examination of artifacts from past civilizations. However, one of the most widely used and accepted method is radioactive dating. All radioactive dating is based on the fact that a radioactive substance, through its characteristic disintegration, eventually transmutes into a stable nuclide. When the rate of decay of a radioactive substance is known, the age of a specimen can be determined from the relative proportions of the remaining radioactive material and the product of its decay.

In 1907, the American chemist Bertram Boltwood demonstrated that he could determine the age of a rock containing uranium-238 and thereby proved to the scientific community that radioactive dating was a reliable method. Uranium-238, whose half-life is 4.5 billion years, transmutes into lead-206, a stable end-product. Boltwood explained that by studying a rock containing uranium-238, one can determine the age of the rock by measuring the remaining amount of uranium-238 and the relative amount of lead-206. The more lead the rock contains, the older it is.

The long half-life of uranium-238 makes it possible to date only the oldest rocks. This method is not reliable for measuring the age of rocks less than 10 million years old because so little of the uranium will have decayed within that period of time. This method is also very limited because uranium is not found in every old rock. It is rarely found in sedimentary or metamorphic rocks, and is not found in all igneous rocks. Another method for dating the rocks of the earth's crust is the rubidium-87/strontium-87 method. Although the half-life of rubidium-87 is even longer than uranium-238 (49 billion years or 10 times the age of the earth), it is useful because it can be found in almost all igneous rocks. Perhaps the best method for dating rocks is the potassium-40/argon-40 method. Potassium is a very common mineral and is found in sedimentary, metamorphic, and igneous rock. Also, the half-life of potassium-40 is only 1.3 billion years, so it can be used to date rocks as young as 50,000 years old.

In 1947, a radioactive dating method for determining the age of organic materials, was developed by **Willard Frank Libby**, who received the Nobel Prize in chemistry in 1960 for his radiocarbon research. All living plants and animals contain carbon, and while most of the total carbon is carbon-12, a very small amount of the total carbon is radioactive carbon-14. Libby found that the amount of carbon-14 remains constant in a living plant or animal and is in equilibrium with the environment, however once the organism dies, the carbon-14 within it diminishes according to its rate of decay. This is because living organisms utilize carbon from the environment for metabolism. Libby, and his team of researchers, measured the amount of carbon-14 in a piece of acacia wood from an Egyptian tomb dating 2700-2600 BC. Based on the half-life of carbon-14 (5,568 years), Libby predicted that the concentration of carbon-14 would be about 50% of that found in a living **tree**. His prediction was correct.

Radioactive dating is also used to study the effects of **pollution** on an environment. Scientists are able to study recent climactic events by measuring the amount of a specific radioactive nuclide that is known to have attached itself to certain particles that have been incorporated into the earth's surface. For example, during the 1960s, when many above-ground tests of nuclear weapons occurred, the earth was littered by cesium-137 (half-life of 30.17 years) particle fallout from the nuclear weapons. By collecting samples of sediment, scientists are able to obtain various types of kinetic information based on the concentration of cesium-137 found in the samples. Lead-210, a naturally occurring radionuclide with a half-life of 21.4 years, is also used to obtain kinetic information about the earth. Radium-226, a grandparent of lead-210, decays to radon-222, the radioactive gas that can be found in some basements. Because it is a gas, radon-222 exists in the **atmosphere**. Radon-222 decays to polonium-218, which attaches to particles in the atmosphere and is consequently rained out—falling into and traveling through streams, rivers, and lakes.

Radioactive dating has proved to be an invaluable tool and has been used in many scientific fields, including geology, archeology, paleoclimatology, atmospheric science, oceanography, **hydrology**, and biomedicine. This method of dating has also been used to study artifacts that have received a great deal of public attention, such as the Shroud of Turin, the Dead Sea Scrolls, Egyptian tombs, and Stonehenge. Since the discovery of radioactive dating, there have been several improvements in the equipment used to measure radioactive residuals in samples. For example, with the invention of accelerator mass spectometry, scientists have been able to date samples very accurately.

See also Radioactive decay

RADIOACTIVE DECAY

The **nucleus** of each atom has a specific number of protons and neutrons and is either stable or unstable, depending on the relative number of each. The most stable **atoms** are those that have an equal number of protons and neutrons. Atoms that are unstable are radioactive. An atom that is radioactive can also be called a radionuclide. Of the known nuclides (approximately 2,000), only 264 are stable, and of the known radionuclides (approximately 1,700), only 70 occur in **nature**. The rest are man-made. Unstable atoms undergo a process called radioactive decay to reach a more stable state.

While a radionuclide is going through the process of decay, **energy** is released from the atom in one of three modes: alpha, beta, or gamma radiation. These modes may take several steps, involving only the nucleus or the entire atom. Each radionuclide has one or more characteristic modes of decay. The particular mode of decay determines the type of energy, or radiation, released from the atom, and consists of either subatomic particles, photons, or both.

Radionuclides are unstable to varying degrees. The more unstable a radionuclide is the faster it decays. The quantity of a radioactive substance is expressed as disintegrations per second, in units of Curies (Ci) named for Marie Curie, or if *Système International* is used, Becquerels (Bq) named for Henri Becquerel. The rate at which a radionuclide decays depends upon its half-life, the expected time required for half of the nuclei to decay to a stable state. The half-life is typically not affected by **temperature**, pressure, or gravitational, magnetic, or electrical fields.

When radioactivity was first discovered, it was thought that all the energy given off by the radionuclide was basically the same, with differences only in penetrating power. However-

er, research conducted by Becquerel and Pierre Curie proved that there were three distinct modes of radioactive decay, which differed not only in their ability to penetrate, but also in their velocity, as well as their susceptibility to magnetic fields.

Alpha and beta radioemissions are actually particulate **matter** that is thrown out from the nucleus. An alpha particle is two protons and two neutrons, or in other words, it is a helium atom without the electrons. After an alpha particle is emitted, the atomic mass decreases by four, and the number of protons and neutrons decrease by two. Alpha decay occurs in radionuclides with an atomic number greater than 83 and a mass number greater than 209. Alpha particles interact with negatively charged electrons in the environment, which consequently use up the energy in the particle, slowing it down and greatly diminishing its penetrating power. Even a sheet of paper can stop an alpha particle. The direction of an alpha particle is only slightly affected by a magnetic field because the particle has a balanced change. When a radionuclide decays by alpha radiation, it does not just disappear. Instead, the radionuclide transmutes into another radionuclide or nuclide. For example, uranium-238 transmutes into several other radionuclides, including radium-226 and radon-222, before ending up as lead-206, a stable nuclide.

Beta radiation, which also involves particulate emissions, can be either be negatively charged or positively charged. Beta particles are actually created in the nucleus by either a proton changing into a neutron (positron emission) or a neutron changing into a proton (negatron emission). A beta particle has a higher velocity than an alpha particle, and its path is markedly deflected by a magnetic field. When a negatron is emitted from an atom, the atomic mass of the atom is unchanged, the number of protons increases by one, and the number of neutrons decreases by one. The mass remains unchanged when a positron is emitted, the number of neutrons increases by one, and the number of protons decreases by one.

An atom usually becomes excited from either of the above-mentioned decay processes and sheds excess energy in the form of a gamma ray photon. With gamma emissions, the atomic mass, number of protons (atomic number), or the number of neutrons, remains unchanged. The velocity of a gamma ray is almost that of **light** and is not affected by magnetic fields.

RADIOACTIVE WASTE

Radioactive waste is what remains after the use or production of radioactive materials. Modern society uses a variety of products and services that depend on radioactive materials. Uses include the production of household smoke alarms, gauges and monitors used in industry to insure quality control, medicinal and pharmaceutical research, and to generate electricity.

Radioactivity consists of highly energetic subatomic particles emitted by atoms from their cores (nuclei). The handling of radioactive waste is a major challenge for modern society, because exposure to high levels of radiation can kill or cause cancer. However, scientists are unsure whether exposure to low levels of radiation has harmful effects. Everyone is exposed to low doses of radiation from the sun and naturally occurring elements in the earth. Levels to which we are exposed every day are called background levels.

Radioactivity decays (decreases) over time. The rate of decay is measured in units called half-lives. If a substance has a radioactive half-life of one hundred years, it will lose 50% of its radioactivity by the end of that time. Some radioactive substances (or isotopes) have half-lives of less than a second, but others have half-lives of hundreds, thousands, or even billions of years. Wastes from those with long half-lives create serious long-term storage problems.

Radioactive waste is categorized according to its half-life and level of radioactivity. Federal regulations are based on two broad categories: low-level radioactive waste (LLRW) and high-level radioactive waste (HLRW). The regulations include more detailed sub-categories and classification criteria. Government definitions of low-level waste are further divided into Classes ''A,'' ''B,'' and ''C.'' Class A LLRW generally includes wastes that have a half-lives of one hundred years or less. Some low-level radioactive waste can be kept in approved confinement areas while waiting for its radioactivity to decay. In most cases, fifty-five gallon drums are used. For example, hospitals are allowed to store waste until it decays to background radiation levels. Then it is placed with non-radioactive medical waste for disposal. Other LLRW cannot be handled in such a manner because of the long-half lives of the kinds of isotopes present in the waste and higher levels of radioactivity. Therefore, some wastes that are categorized as low-level must be sent to another site that is licensed to store them. Low-level radioactive waste includes items that have been contaminated as a result of exposure to radiation such as clothing, rags and mops used for cleaning, medical tubes and injection needles, equipment and tools, and animal tissues used in research. In 1997, about 315,000 cubic feet of commercial low-level radioactive waste were produced in the United States. Sixty-two percent came from nuclear reactors, 26% from industry, 9% from government (other than nuclear weapon facilities), 2.1% from educational institutions, and 0.4% from medical facilities.

High-level radioactive waste are materials that remain at dangerously high levels of radioactivity for hundreds or even thousands of years. High-level waste consists mainly of spent fuel rods from nuclear reactors and waste from the construction and destruction of nuclear weapons.

Radioactive waste is regulated by two federal agencies. The Department of Energy (DOE) regulates radioactive waste from nuclear weapons production and certain kinds of research. The federal Nuclear Regulatory Commission (NRC) regulates radioactive materials from non-military uses of nuclear material, including production of electricity. Some states, called Agreement States, have agreements with the NRC that authorize them to oversee regulation of radioactive materials within their own borders. Handling and storage of low-level radioactive waste requires licenses from the NRC or an Agreement State.

Containment of radioactive wastes is a major challenge. All radioactive waste goes to licensed and regulated facilities, but different facilities are used for differing levels and classifications of waste. As of 1998, there were only two facilities in the United States that accepted a range of low-level radioactive wastes. Sites in four other states that accepted such wastes in the past are closed. Most proposals to establish new LLRW containment facilities have been opposed by the public. Similarly, there is strong public opposition to siting of high-level waste facilities. In 1982, the U.S. Congress passed the Nuclear Waste Policy Act, calling for selection of a site and the construction of a high-level radioactive waste repository by 1998. In 1987, Congress directed the federal government to study one site, Yucca Mountain, Nevada. As of 1998, billions of dollars have been spent on studies of the site and preliminary construction. However, Nevada residents, environmental groups, and others continue to oppose the project. The site has not been officially approved by the NRC, and the estimated date for its opening has been moved to 2010. As of 1998, there were fewer than a dozen approved facilities in the United States that had HLRW on site. In addition, as of 1998 there are 110 nuclear reactors (not all operating) at seventy-three sites in thirty-four states. Each has a temporary storage site for radioactive wastes, each of which is full or nearly full.

Production and handling of radioactive wastes raise many questions that must be addressed. The decision to discontinue or limit certain activities that produce nuclear waste (such as the use of nuclear reactors to produce electricity) must be weighed against the fact that nuclear power provides an energy alternative that reduces the carbon emissions that lead to global warming. Furthermore, sites must be established for containment and long-term storage of radioactive wastes, with shipment of radioactive wastes from temporary storage sites to long-term facilities being done in a manner that maximizes protection for the public.

RAINFOREST

Rainforests receive more than 200 days of rain per year, or as much as 240 in (6 m) of **water**. They are characterized by thousands and even millions of **plants** and **tree**s growing thickly over the land, which soak up rainwater from the soil and return it to the air through transpiration. At least half of that water then falls back down onto the forest as rain again. Most rainforests are located in the central region of the earth, near the equator, where **temperature**s range from 70-90°F (21-32°C) or warmer. At least 50% of the northern half of South America is covered in rainforest, as is most of Central America, the middle portion of Africa and southern Asia, and even small parts of northern Australia. Together, rainforests cover approximately seven percent of the earth's land surface.

Some rainforests are believed to have existed for 100 million years. At least half of all of the world's **species** of plants and **animal**s make rainforests their **habitat**. For example, the largest rainforest, the Amazon in South America, has at least 1,600 species of birds alone, and approximately one mil-

Mountain rainforest in Kenya. *(Photograph by Robert J. Huffman, Field Mark Publications. Reproduced by permission.)*

lion types of insects. The Amazon occupies approximately 2.5 million square miles, or two-thirds the size of the United States, and makes up half of the world's remaining rainforest. Many of the plants found in the rainforest have been found to have medicinal value, accounting for more than one-fourth of all medicines. The diversity of life in a rainforest makes it a truly unique habitat compared with the rest of the world, where only five percent of this diversity is likely to exist.

These complex **ecosystem**s are vital to the world's climate, because of the large amount of rainfall transpired by the plants within them. While some of the rain falls back into the rainforest, the rest travels with warm currents of air to cooler sections of the earth to the north and south. The plants also absorb **carbon dioxide** (CO_2) from the **atmosphere**, and return vast amounts of oxygen (O_2) for other species to breathe.

A rainforest can be divided into horizontal layers according to the plants and animals living in them. The top layer includes the tallest trees that scatter themselves throughout the forest, called the emergents. They can grow to be 300 ft (914 m) tall and extend themselves above the next layer of forest growth, called the canopy. This most productive area of the forest includes trees that stand 60-150 ft (18-45 m) high, with branches and leaves that form an umbrella over the underlying forest. These trees grow so tightly together that rainfall reaches the ground only by running down the tree trunks or the stems of other plants. The canopy is alive with **flower** and animal life as well, such as iguanas, tree frogs, monkeys and bats, and countless birds and insects. The next section, called the understory, includes smaller trees, ferns, vines and palms, and smal-

ler bushes. Because the upper canopy traps much of the heat and moisture into the forest, the understory is extremely hot and humid with few flowering plants due to the lack of sunlight. The bottom layer, the forest floor, is filled with shade-loving mosses, herbs and **fungi**, and dead plants and animals. These decompose quickly, providing **nutrients** through the soil to the larger plants above. Thousands of insects and larger animals that feed on them, such as the anteater, live on the forest floor.

There are three primary types of rainforests: tropical, mangrove, and temperate. Tropical **forests** include those found closest to the equator, and are the wettest regions of the world. Some have wet and dry seasons, and others are found at the top of tropical mountains where clouds surround the mountain peaks most of the time. Mangrove rainforests grow along the oceans and soak up ocean and rainwater. Temperate rainforests exist in cooler climates along the western coast of North and South America, and in southern Australia and New Zealand. They include many of the ''old-growth'' forests such as the sequoia trees in northern California. Some of the sequoias are more than 1,000 years old.

While the rainforests have included people in small tribes for thousands of years, encroaching **population**s on these regions of the world have placed many rainforests in danger. Approximately 50 million acres (20 million hectares) of rainforests are disappearing each year, or 90 acres (36 hectares) every minute, as trees are harvested, and the land is cleared for farming, animal grazing, or for roads. In most cases, the trees are cut down and burned, a process called ''slash and burn,'' which produces tons of smoke and CO_2 gases. As fewer trees exist to absorb this CO_2, these warm gases increase and contribute significantly to the greenhouse or warming effect on the earth's atmosphere. Decreased vegetation also causes increased flooding in these areas, and subsequent erosion of vital topsoil. Habitat disappears for various animal and plant species. It is impossible to accurately predict how many species become extinct each year due to the rainforest's destruction. Efforts to stop this destruction include creating reserves or protected regions where limited harvesting of trees or **fruits**, plants and nuts is allowed.

See also Biodiversity; Climax communities; Ecosystem; Endangered species; Extinction; Greenhouse effect; Hydrologic cycle; Precipitation

RAMÓN Y CAJAL, SANTIAGO (1852-1934)
Spanish neurohistologist

The anatomical research of the Spanish neurohistologist Santiago Ramón y Cajal is central to the modern understanding of the **nervous system**. By adopting and improving the nervous-**tissue** staining process developed by the Italian scientist **Camillo Golgi**, Ramón y Cajal established that individual nerve cells, or neurons, are the basic structural unit of the nervous system. He also made important discoveries relating to the transmission of nerve impulses and the cellular structures of the brain. For his work in **histology**, the branch of anatomy concerned with minute tissue structures and processes, Ramón y Cajal shared with Golgi the 1906 Nobel Prize for physiology or medicine.

Ramón y Cajal was born on May 1, 1852, in the remote country village of Petilla de Aragon, Spain. He was the son of Justo Ramón y Casasús, a poor and self-educated barber-surgeon, and Antonia Cajal. The family subsequently moved to the university city of Zaragoza, where against considerable odds Ramón y Cajal's father earned a medical degree and became a professor of anatomy. As a young man, Ramón y Cajal was rebellious and independent-minded. He preferred drawing to studying, and although this passion for drawing would ultimately serve him well, it was vigorously opposed by his iron-willed father, who had determined that his son should become a doctor. As a disciplinary measure, his father apprenticed him to a barber and later to a shoemaker. During these apprenticeships, Ramón y Cajal also studied anatomy with his father—investigations which partially relied on bone specimens taken from a local churchyard.

When he was sixteen years old, Ramón y Cajal began medical studies at the University of Zaragoza, earning a degree in medicine in 1873. He then joined the army medical service and served as an infantry surgeon in Cuba for one year. He contracted malaria, however, which led to his discharge, and he returned to Spain. In 1879, still convalescent, he passed his examinations at Zaragoza and Madrid for his doctorate in medicine.

Ramón y Cajal was almost exclusively interested in anatomical research, and he embarked on an academic career. Beginning in 1879, Ramón y Cajal turned himself into a skilled histologist, initially working with an old, abandoned microscope he had found at the University of Zaragoza. He studied various anatomical tissues and began to publish articles on **cell** biology—complete with beautifully rendered ink drawings. His work was not immediately recognized in other countries, but the increasing prestige of his posts attests to his success in Spain. From 1879 to 1883, he directed the anatomical museum at the University of Zaragoza. In 1883, he assumed a professorship of descriptive anatomy at the University of Valencia, and in 1887 he became professor of histology at the University of Barcelona. In 1892, Ramón y Cajal assumed the chair of histology and pathologic anatomy at the University of Madrid, a post he retained until 1922.

Research Provides Evidence for Neuron Theory

Ramón y Cajal eventually turned to the most complex tissues, those of the nervous system. His research method now drew on Camillo Golgi's method of staining tissue samples to reveal their minute components. Under Golgi's method, a potassium dichromate-silver nitrate solution stained the nerve cells and fibers black, while the neuroglia, or supporting tissues, remained much lighter. By refining this staining technique and applying it to embryonic tissue samples, Ramón y Cajal was able to isolate the neuron as the basic component

Santiago Ramón y Cajal. *(The Library of Congress. Reproduced by permission.)*

of the nervous system; he also differentiated the neuron from the ordinary cells of the body. His work supported the neuron theory, which held that the nervous system consists of a network of discrete nerve fibers that end in terminal "buttons," which never actually touch the surrounding nerve cells. Up until that time, the majority of scientists were "reticularists," who held that the nervous system formed a continuous and interconnected system. Golgi was among these, and the rivalry between the two scientists was intense. Ramón y Cajal published fierce and relentless attacks both on this theory and on the scientists who held it.

Based on his studies, Ramón y Cajal became convinced that the conduction of nerve impulses occurs in one direction only—a postulate since formalized as the law of dynamic polarization. He also conducted important research on the tissues of the inner ear and the eye, as well as the tissues of the grey matter of the brain, establishing a cellular basis for the localization of different functions within the brain. This research has formed the physiological basis for the understanding of human psychology, intelligence, and memory.

Ramón y Cajal was a prolific writer and he published many articles, textbooks, and research monographs. In 1896, he established a journal of microbiology and published his *Manual de Anatomia Pathologica General* ("Manual of General Pathologic Anatomy"). His major neurohistological work, *Textura del Systema Nervioso del Hombre y de los Vertebrados* ("Texture of the Nervous System of Man and Vertebrates"), was published from 1899 to 1904. These publications were generally printed in Spanish, often at his own expense, and they were largely ignored by the international scientific community.

His struggle for due recognition of the importance of his work came to an end in 1906, when he shared the Nobel Prize in physiology or medicine with his rival Golgi for their work on the structure of the nervous system. In an apparent effort to emphasize what the two scientists had in common, rather than their area of disagreement, they were described by the prize committee as "the principal representatives and standard-bearers of the modern science of neurology." But the tension between them over the reticular doctrine was still evident on the awards platform.

Later Research and Writing

In the same year he received the prize, Ramón y Cajal turned to the problem of the degeneration of tissue in the nervous system and the regeneration of nerve fibers that had been severed. The result of these studies, the two-volume *Estudios Sobre la Degeneracion y Regeneracion del Sistema Nervioso* ("Studies on the Degeneration and Regeneration of the Nervous System"), was published in 1913 and 1914. In 1913, Ramón y Cajal also developed a gold-based method of staining neuroglia; he was able to use this to classify cell types in these tissues. This research provided the basis for the medical treatment of tumors and pathological tissues in the nervous system. A tireless researcher, Ramón y Cajal also studied the eyes and vision processes of insects.

Ramón y Cajal, a patriot, was always sensitive to the international and scientific reputation of Spain and the Spanish language—issues that had a significant impact on the dissemination of his research. It was thus fitting that in 1920 King Alfonso XIII commissioned the construction of the Instituto Cajal, which secured Madrid's position as an international histological research center. Ramón y Cajal worked at this institute named in his honor from 1922 until his death. In addition to sharing the Nobel Prize, Ramón y Cajal received numerous awards and honors, including the Fauvelle Prize of the Society of Biology in Paris in 1896; the Rubio Prize in 1897; the Moscow Prize in 1900; the Martinez y Molina Prize in 1902; the Helmholtz Gold Medal of the Royal Academy of Berlin in 1905; and the Echegaray Medial in 1922. He also received honorary degrees from various foreign universities and held memberships in scientific societies worldwide. The Spanish government bestowed an impressive series of posthumous honors on him, including the republication of his works.

Ramón y Cajal married Silveria Fananas Garcia in 1880. They had three daughters and three sons. In addition to drawing, his hobbies included chess and photography, which he pursued as single-mindedly as his research. In a merging of his work and recreational interests, Ramón y Cajal developed his own photographic process for the reproduction of his delicate histological drawings.

Between 1901 and 1917, Ramón y Cajal published the installments of his autobiographical *Recuerdos de mi Vida* ("Recollections of My Life"). His other published works include the anecdotal *Charlas de Cafe* ("Conversations at the Cafe") and *El Mundo Visto a los Ochenta Años* ("The World as Seen at Eighty"). Ramón y Cajal died in Madrid on October 18, 1934.

RAY, JOHN (1627-1705)
English naturalist

A predecessor of **Carl Linnaeus**, John Ray was the first naturalist to use the idea of **species** to distinguish different **organisms** from each other. Focusing primarily on the **classification** of **plants** and basing his system on the work of **Aristotle**, Ray divided plants into to groups: the moncotyledons and the dicotyledons. Both are still recognized today. In 1693, Ray published the final volume of *Histora Plantarum*, a complete classification of plants and one of the first natural systems of classification that was based on physical characteristics rather than origin and perceived use.

John Ray was born in Black Notley, Essex, England, to Roger Ray, a blacksmith, and Elizabeth Ray, an amateur herbalist and medical practitioner. He attended Trinity College at Cambridge from 1644-1651, receiving both a bachelor and masters degree. After graduation, he continued at Trinity as an appointed fellow of the college. He taught a number of courses, including Greek, mathematics, and humanities. Ray left his post at Trinity during the Reformation, when he refused to sign an oath required by the Act of Uniformity in 1662. It was at this time that his contribution to **taxonomy** flourished.

Without employment, Ray relied on the patronage of former students. One such patron was Francis Willughby, a wealthy contemporary from Cambridge. With the support of Willughby, Ray was able to expand his classification of plants from a part-time endeavor restricted to the indigenous species of Chambridgeshire to the whole of the British Isles and beyond. Willughby accompanied Ray on his many expeditions, and his interest in animals complimented Ray's own interests in plants. Ray's collaboration with Willughby ended in 1672, with the death of Willughby. That same year, Ray married Margaret Oakeley. They settled in Black Notley, where Ray continued his scientific endeavors.

As part of his work, Ray was able to convincingly show that **fossil**s represented extinct species. At the time, the link between fossils and extinct species was not an accepted model, however, Ray's evidence provided the basis for the formation of a more thorough system of **paleontology**. Such a view was unusual for a naturalist at this time, particularly considering Ray's strong religious beliefs.

John Ray never lost his love and wonder for the work of God and had no problem reconciling his views of the world with his views of religion. As well as publishing extensively on natural history, Ray also published many theological works, including *The Wisdom of God*, and he was only stopped from taking priestly orders by the English civil war and Reformation. According to Ray, the study of nature was a way to reveal the omnipotence of God and to be a naturalist was a way to work within divinity.

RÉAUMUR, RENÉ ANTOINE FERCHAULT DE (1683-1757)

French scientist

René Antoine Ferchault de Réaumur was a French scientist who made outstanding contributions to several branches of science. He is perhaps best recognized for his work which established how the process of **digestion** works in **animals**. Réaumur as born in 1683 in La Rochelle, France. Although there is little record of his early years or his education, it is known that at the age of 20 he moved to Paris. In 1708 he became a member of the French Academy of Sciences.

Early in his career Réaumur was put in charge of a government sponsored project known as *Description des arts et méitiers,* which required that he collect information on all of France's arts, industries and professions. This experience gave Réaumur a diverse background in science and technology which he put to good use. For example, his chemical knowledge helped him recognize the importance of carbon in the composition of steel. Using this fact, in 1720 he developed an improved method for manufacturing iron and steel which became known as the cupola furnace.

Réaumur applied his diverse expertise to a variety of fields. He compiled his knowledge of insects in a six volume series of memoirs known collectively as *Mé moires pour servir à l'histoire des insectes* which translates as Memoirs for Following Insect Study. This work, published between 1734 and 1742, is generally considered to be the first comprehensive publication describing insects. During the same period, he developed a new form of opaque porcelain, known as Réaumur's porcelain. He even devised the Réaumur **temperature** scale using a thermometer he had invented. This thermometer used a mixture of alcohol and water to register the freezing point of water as zero and boiling point as 80 degrees. He used 80 as the number of points on his scale because he believed it was ''a number convenient to divide into parts.'' The Réaumur scale was once widely used in Europe, particularly in France, but it has now been replaced by the Celsius scale.

In the world of biology, Réaumur is recognized for his studies of digestion in birds and animals. His research showed that digestion in higher animals was a chemical process and not strictly done by mechanical agitation. Réaumur showed that the stomach acted on food chemically by feeding meat-filled metal cylinders to a hawk and then examining the cylinders after the bird regurgitated them. He was eventually able to isolate these digestive chemicals by forcing the hawk to swallow a sponge and then squeezing the retrieved stomach liquid out of the regurgitated sponge. Réaumur continued his experiments until his death in 1757.

RECOMBINANT DNA · See DNA
(deoxyribonucleic acid)

RECOMBINATION

Recombination is a process by which **genetic material** is shuffled during reproduction. It is important from an evolutionary standpoint because it allows the mixing of different traits. This trait mixing has been crucial for organisms that have had to adapt to a changing environment. Recombination has also allowed for the separation of favorable and unfavorable genetic **mutations**. In this way, harmful mutations are minimized in the **gene pool**.

Recombination involves a physical exchange of **nucleotides** between duplicate strands of **deoxyribonucleic acid** (**DNA**). There are three known types of recombination—homologous recombination, specific recombination and transposition. Each type occurs under different circumstances.

In **eukaryotes**, homologous recombination typically occurs during **meiosis**, which is a special form of **cell division**. It takes place during the first phase of meiosis, and involves a physical exchange of parts between **chromosomes**. In most eukaryotic **cells**, genetic material is organized as chromosomes in the **nucleus**. During meiosis, every chromosome is duplicated, or replicated. Each duplicate condenses and forms two identical structures known as chromatids, which are joined at a point called the centromere. Homologous chromosomes pair up with each other and form bivalent structures. It is while the chromosomes are aligned that recombination occurs.

To begin the exchange of genetic pieces between chromosomes, a nick is made on the chromosomal DNA of corresponding strands. This breakage lets each strand move at the free ends. The broken strands then cross over, or exchange, with each other. The recombinant region is extended until a whole gene is transferred. At this point, further recombination can occur or it can be stopped. Both processes require the creation of another break in the DNA strand and subsequent sealing of the nicks.

Site specific recombination is the type of **crossing over** that usually occurs in **prokaryotes**. It is the mechanism by which phage, or viral, **genomes** are incorporated into bacterial chromosomes. When phage DNA infects a bacteria, it exists in two states: lysogenic or lytic. During the lytic stage, the phage DNA exists apart from the bacterial DNA. In the lysogenic stage, the two DNAs are combined. The integration of the bacterial and viral DNAs occurs at a specific location, called an attachment site, on the bacterial DNA. The attachment site has a homologous nucleotide sequence with the phage genome. When conditions are right, the two pieces of genetic material align and enzymes cause them to merge.

Transposition is a third type of recombination. It involves transposable elements called transposons–short segments of mobile DNA. They are found in both prokaryotes and eukaryotes. The process of transposition begins when an **enzyme** cuts DNA at a target site. This leaves a section that has unpaired nucleotides. Another enzyme called transposase inserts the transposon at this site and fills the gap. Through the process of transposition certain genes are made to move throughout the cell's genome. Transposition is the mechanism by which immunocytes manufacture the millions of antibodies required to protect vertebrates from antigens.

REDI, FRANCESCO (1626-1697)

Italian physician

Francesco Redi was born in Arezzo, Tuscany. Credited with the birth of modern experimentation he applied his enquiring, deductive mind and astute powers of observation to designing controlled experiments, the first of their type ever recorded. One magnificently contrived series of investigations led to the disproof of the centuries-old belief in **spontaneous generation**. Another, in what he called "unmasking of untruth," he discovered how vipers produce venom and inject it into their prey, and determined the venom's clotting effect on the the victim's blood. His ingenuity in designing these experiments has been compared to that of **Louis Pasteur** two hundred years later. His command of the written word also gained him fame in the literary world, primarily with his long poem published in 1685, *Bacco in Toscana,* (Bacchus in Tuscany).

Redi was born to Cecelia de'Ghinci and Gregorio Redi, a family of nobility. His father was physician to the Medician court of the Grand Duke Ferdinand II and his son, Cosimo III. As a lad, Redi lived with and was educated by Jesuit priests, whose teachings strictly adhered to the philosophy of **Aristotle**. He studied medicine at the University of Pisa where he earned his doctor's degree. He, too, became physician at the court after he was called to attend the Grand Duke who fell from his horse. Redi, who never married, became well respected and loved at the court. He was made superintendent of the pharmacy, the Duke sought his advice in diplomatic affairs, he mediated between the Duke and his somewhat rebellious son, and he shared his knowledge with scholars who sought his company. One renowned physician wrote of Redi, "Everyone burns incense to the Idol."

Although Redi lived in an era still deeply entrenched in Aristotelian philosophy, he was influenced by Galileo's theories and the works of Bruno and Kepler. Redi had also read works by Giuseppe Aromatari of Assisi who refuted the spontaneous generation theory, suspecting plants propagated from seeds and animals from eggs; and **William Harvey** who proposed that insects, worms, and frogs grew from seeds or eggs too small to seen. It was well known that maggots suddenly appeared in rotten meat, and Aristotelian philosophy implied this was due to spontaneous generation. Redi decided to find out. He put many types of meat into individual flasks, observed the maggots as they consumed the rotting meat, and discovered they went through a metamorphosis, ultimately becoming flies. He recalled that, before the maggots appeared, flies swarmed the rotten meat; he suspected they may somehow produce the maggots. Redi continued his experiment, once again placing meat into many glass flasks. This time he created a control group. He left half the flasks open and covered the other half with gauze, allowing air in and keeping flies out. Sure enough, maggots only developed in the open flasks. He correctly deduced the flies deposited eggs in the rotting meat. However, he was unable to deduce the same concept for insects which did not breed in rotting matter, such as gall flies—which bred in hollow twigs, and intestinal worms—which developed inside the body. These he still believed were spontaneously generated. It remained for one of his students, who continued his experiments, to debunk that belief. Even so, the entire concept was not readily accepted until 1859 when an ingenious experiment by Louis Pasteur finally closed the book on the subject.

Redi also debunked the wild folk tales and misconceptions of vipers. His controlled experiments showed that a viper's bile is not poisonous; that neither their venom nor teeth are poisonous if swallowed; that venom is fatal if rubbed into an open wound or injected under the skin with a broom straw; that venom is a yellow fluid produced by glands (and not a wild spirit) in the viper's head; that it is injected only through two teeth; and it will kill even if the snake has been dead a long time. He dispelled the myth that vipers drank wine and shattered wine flasks; that human saliva poisoned vipers; that venom's strength was determined by the viper's diet; and many other false ideas. Once again, it was not until a 1781 publication by Felice Fontana that Redi's conclusions were generally accepted.

Redi's health began failing in later years. He reportedly told a friend it was "...useless to be afraid of death as he had never observed that death could be kept away through fear." He died suddenly in his sleep in Pisa on March 1, 1697.

REDUCTIONISM

Harvard biologist, and two-time Pulitzer-Prize winner, **Edward O. Wilson**, in his 1998 book *Consilience*, stated that "the cutting edge of science is reductionism, the breaking apart of **nature** into its natural constituents." The journal *Nature*, in 1997, defined reductionism as the search "to explain the wide variety of natural phenomena by the behavior of limited numbers of simpler constituents subject to rigorous and simple laws."

A reductionist approach has been perhaps most successful in **molecular biology**, which, one author has argued "constitutes a research program that attempts to explain and understand biological systems completely in terms of the physical interactions of their parts." Reductionism in molecular biology has led to many advances in explaining the molecular basis, most spectacularly, of human disease, but also to sometimes startling advances in such fields as immunology and developmental biology.

Still, considerable debate has raged over the merits of reductionism. E.O. Wilson's ascendant field of **sociobiology** has been attacked for its genetic determinism, especially for its repeated assumption that the residual of phenomena unexplained by reductionism "shrinks as knowledge increases." The most controversial implication is that eventually all human behavior can be reduced to explanations at the level of the gene. Contrarily, most biologists would argue that it is not possible to reduce explanation of biological phenomena to the presently ultimate simplicity of subatomic particles.

The controversy has been intensified recently by the debate centered on the current **Human Genome Project** (HGP). Most biologists seem to agree that comprehensive mapping to locate genes precisely on **chromosome**s is a worthwhile project, so the controversy swirls around the purposes, ultimate value, and expense of total sequencing of the entire human **genome**. Some of the advocates of the HGP expect it to lead eventually to a better grasp of the interactions among genes and ultimately to "calculation" of the entire behavior of the **organism**. Skeptics argue that this is blind reductionism, that such expectations exhibit a reductionist naivete about biochemistry and organismal biology. As the authors of a recent article in *Perspectives in Biology and Medicine* suggest: "The HGP is the ultimate product of an extreme reductionist vision of biology that has held that to understand better one need only to go smaller."

The debate came into focus sharply in the Novartis Foundation symposium on the limits of reductionism in biology in 1997. Some consensus emerged: (1) on the belief that it is probably not very useful to try to reduce explanations of biological phenomena to the level of particles in physics (even if it could be done); (2) on the importance of whole organisms and of their current and past interactions with their environments; and (3) that adequate explanation and understanding depend on working at the appropriate level, not just the lowest level achievable.

Reductionism is now, and has been for some time, the dominant **paradigm** in biology and in all of the empirical sciences. Some scholars advocate systems theory, hierarchy theory, and **holism**, or some combination of these, as viable alternatives. These have become fundamental to some disciplines, e.g., **ecology**, but have not reached the level of acceptance among scientists that the reductionist approach has; neither have these alternatives been anywhere near as successful. Research in biology, and in all the other sciences, will no doubt continue to depend primarily on a reductionist approach, but will also increasingly incorporate alternative perspectives in a way complementary to reductionism.

Walter Reed. *(AP/Wide World Photos. Reproduced by permission.)*

REED, WALTER (1851-1902)
American physician and bacteriologist

Walter Reed, an Army surgeon and medical researcher, helped discover that mosquitoes transmitted yellow fever, an infectious, sometimes fatal, disease. During his career, Reed also made contributions toward the control of malaria and typhoid. Although some questioned his practice of using humans as test subjects for his yellow fever work, his findings saved thousands of lives. In honor of his efforts to control epidemics, the Army General Hospital in Washington, D.C., was named after Reed.

The youngest of five children, Reed was born on September 13, 1851 in Belroi, Virginia, to Lemuel Sutton Reed, a Methodist minister, and Pharaba White. His father's ministry took the family to different parishes every few years and, as a result, Reed's early education was somewhat sporadic. In 1865, however, Reed began two years of study under William R. Abbot. He entered the University of Virginia at age fifteen and, a year later, took a medical course. Reed received a medical degree in 1869.

Reed subsequently traveled to New York to pursue additional medical studies at Bellevue Hospital. He earned a second medical degree in 1870, but it was not official until 1872, when he turned twenty-one. In the meantime, Reed secured the position of assistant physician at New York Infants' Hospital, undertook residency at Kings County Hospital of Brooklyn and at Brooklyn City Hospital, and acted as district physician for the New York Department of Public Charities. For a year beginning in June of 1873, he served as sanitary inspector for the Brooklyn Board of Health.

In June of 1874, Reed received a commission as assistant surgeon, first lieutenant, with the U.S. Army Medical Corps and moved to Arizona. Before he left, he married Emilie Lawrence, a woman he had met while visiting his father in Murfreesboro, North Carolina. For the next eleven years Reed worked variously at bases in Arizona, Nebraska, Minnesota, and Alabama. During this time, Reed and his wife had two children, Lawrence and Blossom.

In the 1890s, Reed wished to pursue his interest in pathology. Because army bases did not offer appropriate facilities, he applied for a leave of absence to conduct advanced work in the field. His request was not granted; instead he was transferred to Baltimore to act as attending surgeon. There he took a brief clinical course at Johns Hopkins Hospital and met William Henry Welch, a pathologist who opened the first pathology laboratory in the United States. Under Welch's tutelage, Reed delved into pathology, performing autopsies, conducting experiments, and refining medical techniques. Reed specifically worked on the bacteriology of erysipelas (an acute fibroid disease accompanied by severe skin inflammation) and diphtheria. This work halted when Reed was sent to an army outpost at Fort Snelling, Minnesota, where he was promoted to major and made a full surgeon.

However, when George Sternberg became the nation's surgeon general, Reed returned to Washington as curator of the Army Medical Museum and also taught bacteriology and clinical microscopy at the Army Medical College. At this time, Reed began to make an impact in his field. When a malaria epidemic broke out at Fort Myer, Virginia in 1896, Reed proved that contaminated drinking water—as commonly believed—was not the cause. He noted that many areas of Washington, including the infected section, drew water from the Potomac. Reed also realized that malaria was striking the base's enlisted men, not officers. He traced this to the fact that the enlisted men often traveled to the city via a swamp trail. Reed postulated that "bad air" caused the disease (although it was later determined that mosquitoes spread malaria).

When the Spanish-American War erupted in 1898, Reed volunteered to serve in Cuba. To take advantage of his qualifications, he was instead appointed to chair a board investigating typhoid outbreaks in army camps. Hundreds of new cases—many of which proved fatal—were reported each day. In fact, the epidemic that killed more than fifty times as many soldiers as did combat. The bacillus, or rod-shaped bacterium, was believed to be transmitted by contaminated water, but the typhoid board found that it was passed by flies and contact with infected feces. The board further discovered that the infectious organisms were harbored by carriers—people who showed no signs of the disease. The typhoid board's two-volume report on its investigation is considered a model for epidemiologists.

Confronts Yellow Fever Epidemic

In 1900 Reed was selected to head an army board trying to discover the cause of yellow fever. This disease had spread among army troops in Cuba. In addition, annual outbreaks occurred along the East Coast and in the southern United States, killing thousands of people. Referred to colloquially as yellow jack, the disease was most prevalent in urban areas and was characterized by jaundice, hemorrhaging, fever, bloodshot eyes, hiccups, and dark-colored vomit. Yellow fever regularly hit the same cities during warm weather. By late autumn it was gone.

Alabama physician Josiah Nott postulated that mosquitoes caused the disease, but his evidence was scanty. In 1881 Carlos Finlay, a Cuban physician and epidemiologist who worked with the U.S. yellow fever commission in Havana, suggested that yellow fever was transmitted by *Culex fasciatus* (a mosquito now classified as *Aëdes aegypti*), but he was not taken seriously. Despite these suggestions, Italian physician Giuseppe Sanarelli maintained that *Bacillus icteroides* was the cause. Reed and American army physician James Carroll were assigned to investigate Sanarelli's claim, and they disproved

it. A rash of yellow fever subsequently broke out in Havana, killing thousands of soldiers. Reed traveled to Cuba to head a board including Carroll, Jesse W. Lazear and Aristides Agramonte—all physicians with the army medical corps. The board decided to test its theory that mosquitoes transmitted yellow fever.

Finlay secured mosquitoes and mosquito eggs to allow the group to raise the insects. Because animals were not affected by the disease, the board decided to use human test subjects. Participants in the study gave their consent and were paid $100, plus an additional $100 if they contracted the disease. Reed designed and conducted experiments that proved the *Aëdes aegypti* mosquito was a carrier, and not an originator, of the disease. The yellow fever board concluded that the female *Aëdes aegypti* mosquito could only become a carrier of yellow fever if it bit a victim during the first three days of the disease. The mosquito was unable to transmit the disease for two weeks, but could remain infectious for up to two months in a warm climate. The board also discovered that having had the disease provided immunity against further attacks.

During the course of the board's experiments, Lazear was accidentally bitten by an infected mosquito and died twelve days later. He left notes, however, to assist Reed and the others in their experiments. The board induced twenty-two other cases of yellow fever—none of which proved fatal. (Carroll became ill with the first experimental case but recovered.) Although a vaccination against the disease was not developed until the 1920s, yellow fever was virtually eradicated by 1902 in Cuba through mosquito control.

The board's accomplishment not only saved lives, but also paved the way for U.S. ventures in tropical regions of the world. (For instance, the U.S. government insisted that a way to control yellow fever was necessary before construction began on the Panama Canal.) Reed earned special recognition for heading the investigation. Harvard University awarded him an honorary masters degree for his work with the yellow fever board. Reed died November 23, 1902, in Washington following surgery for a ruptured appendix.

REFLEX

A reflex is an involuntary response to a specific **stimulus** (a stimulus is any act or agent which causes a reaction in a particular **tissue**). A reflex is considered a "stereotypical" response, in that it is a mechanical response brought about by the stimulation of a particular **neuron**, resulting in a particular action which is not varied through the voluntary control of the individual. An example of a reflex occurs when the correct location in the knee is tapped with a physician's hammer, resulting in the contraction of the thigh muscle. This causes the lower leg to kick out. Although one normally considers such movement to be completely at the will of the individual to initiate or prevent, this reflex occurs without the involvement of the individual's **brain**. It is always surprising to see one's body perform such a seemingly voluntary act, with no actual effort or desire on the part of the individual.

The reflex involved in this knee jerk (called the patellar reflex) occurs when a neuron is stimulated by a quick tap on the tendon of the knee. The sudden stretching of this tendon is transmitted quickly to the **spinal cord**. This neuronal impulse stimulates another neuron within the spinal cord which returns to the leg, causing the thigh muscle contraction and the straightening (kicking motion) of the lower leg. The neuronal impulses stay within neurons traveling to and from the spinal cord, never traveling to the level of the brain. Because this reflex is totally spinal, no voluntary control is possible over it.

Other reflexes do include the brain, yet they still remain at a level outside of voluntary control. For example, the pupillary reflex is a response of the muscles of the pupil to **light**. The presence of light causes the pupil to constrict (grow smaller), in order to protect the structures at the back of the eye from damage due to exposure to too much light. These pupillary reflexes should be brisk and equal between the two eyes. The pupillary reflex is tested when an individual is unconscious, to check on whether brain injury has occurred.

Anther reflex is called the withdrawal reflex. The withdrawal reflex occurs when someone touches something extremely hot. Before the individual can even consciously acknowledge the pain of the burn, the arm and hand will reflexively withdraw from the hot item. The neuron which senses the dangerously hot **temperature** passes this information on through the spinal cord to neurons which then command the muscles of the arm and hand to contract, quickly removing the hand from danger. The pathways which allow the brain to sense the pain of the burn are much slower, and the knowledge of burning pain comes after the hand has already pulled back from the burning item.

REGULATION GENE

Regulation genes are a type of gene that provides the instructions for creating **proteins** which help control the expression of other structural genes. They are a key part of the overall system of **gene expression**.

Gene expression is the process by which **cells** translate their **genetic code** into proteins. These proteins have a variety of functions such as providing structure, storing **energy**, providing movement, transporting other substances, catalyzing and controlling biological reactions, and protecting against disease. Depending on the type of cell and the environment, at any given time only a handful of genes are expressed in the cell. The other genes are suppressed by regulation genes.

Regulation genes were first discovered in 1961 by **Francois Jacob** and Jacques Monod in Paris. These scientists found that some of the genes in a **bacterial genome** were turned on or off depending on the environmental conditions. Their studies led to the development of the **operon** model. According to this model, regulation genes are present on the bacterial **chromosome** which control the expression of other structural genes. For example, certain bacteria have a set of genes which can produce the **amino acid** tryptophan through a variety of biosynthetic processes. When tryptophan is present in the cell's environment, it binds with the protein product of a regulation gene. This protein-tryptophan complex then binds to the

DNA of the genes that synthesize tryptophan and turns them off. When tryptophan is depleted from the environment, the **regulation** protein then ceases to bind to the DNA, thereby turning the tryptophan-synthesizing genes back on. This type of control system allows the cell to respond rapidly to a changing environment.

The proteins that are produced from regulation genes control the **transcription** of other genes by binding to particular sites on the DNA. This binding action can affect the target gene in either a positive or negative way by turning it on or off. The tryptophan operon is an example of negative control. Regulation genes are typically located upstream of the structural genes that they regulate. While the binding site of most regulation proteins vary, depending on the protein type, most bind upstream of the target genes.

While operon systems are common in **prokaryote** genomes, they are not generally found in **eukaryote**s. Instead, the regulation genes in a eukaryote may control genes in much more subtle ways. Evidence shows that regulation proteins can affect gene expression even when they bind to DNA that is thousands of base pairs away from the gene. Regulation proteins may also control gene expression in other ways such as binding to mRNA, reducing the speed at which mRNA is degraded, and self regulating. While a large amount of information has been gathered on the subject of eukaryotic gene expression, much is still unknown. Scientists hope that one day this information could lead to practical cures for **cancer**.

See also Structural gene

REICHSTEIN, TADEUS (1897-1996)
Polish Swiss organic chemist

It is now known that the **hormone**s of the **adrenal gland** are essential to controlling many challenges to the human body, from maintaining a proper balance between water and salt to responding to stress. Tadeus Reichstein is one of those responsible for this knowledge; **Edward Kendall** and **Philip Hench** also played an important role in these efforts, and the three men shared the 1950 Nobel Prize in physiology or medicine. Reichstein's work has had effects throughout medicine—in the treatments of Addison's disease and rheumatoid arthritis, for example, and in the understanding of the fundamental biochemical processes of steroid hormone metabolism.

The eldest son of engineer Gustava Reichstein and his wife, Isidor, Reichstein was born on July 20, 1897, near Warsaw in Poland. After moving first to Kiev in the Ukraine and then to Berlin, the family settled in Zürich and became Swiss citizens. Tadeus attended the Eidgenössiche Technische Hochshule and graduated in 1920 with a chemical engineering degree. He worked briefly in a factory, then returned to the Eidgenössiche Technische Hochschule where he earned his doctorate in organic chemistry in 1922.

For several years thereafter Reichstein continued to work with his doctoral advisor, Hermann Staudinger, who would later win the 1953 Nobel Prize in chemistry. Reich-

Tadeus Reichstein. *(Corbis Corporation (NY). Reproduced by permission.)*

stein's early work focused on identifying and isolating the chemical species in coffee that give it its flavor and aroma. This interest in plant products was to remain with Reichstein throughout his career. He had an early success when he discovered how to synthesize the newly discovered compound ascorbic acid (vitamin C). He published this method in 1933, and later that year Reichstein developed a second method of synthesis which is still widely used in the commercial production of this dietary supplement.

Isolates and Identifies Adrenal Cortical Hormones

In 1934, Reichstein began work on what he originally believed to be a single hormone produced by the cortex or outer layers of the adrenal glands. He soon realized, however, that the adrenals were producing a milieu of active substances. His work began with 1,000 kilograms (more than a ton) of adrenal glands that had been surgically removed from cattle. His first stage of purification resulted in one kilogram (about 2.2 pounds) of biologically active extract. He established that the extract was biologically active by injecting it into animals whose adrenal cortices had been removed; if the compound was active it replaced what was missing as a result of the oper-

ation and allowed the animal to survive. The next stage of purification reduced the kilogram of extract to 25 grams (less than one ounce), only about one-third of which proved to be the critical hormone mixture. Instead of one hormone, this sample contained no fewer than twenty-nine distinct chemical species.

Reichstein isolated the twenty-nine species and then individually examined them. He identified the first four which were found to be biologically active, and later synthesized one of them. It was also Reichstein who demonstrated that these compounds were all steroids. Steroids are a group of chemicals which share a particular structure of four linked carbon-based rings; other important compounds having steroid structure include the sex hormones, cholesterol, and vitamin D.

Synthesizes Steroid Hormones

Reichstein built on his earlier work with plant extracts to synthesize the steroid hormones. He and his colleagues developed several different methods to this end, though a process that used an animal waste product (ox bile) proved to be the most economical. One of the most important syntheses that Reichstein accomplished was that of aldosterone, which controls both water balance and sodium-potassium balance in the body. Aldosterone has been widely used in medical practice. Reichstein's work was also critical to the eventual syntheses of desoxycorticosterone, which for many years was the preferred treatment for Addison's disease, and cortisone, which is used for treating rheumatoid arthritis. It was principally for this latter accomplishment that Reichstein shared the 1950 Nobel Prize in chemistry.

Reichstein moved to the University of Basel in 1938 where he was appointed director of the Pharmaceutical Institute; in 1946 he became head of the organic chemistry division. Here he turned his attention to plant glycosides, a group of compounds with wide-ranging biological effects. They are the basis for a number of widely used drugs, and one of these, digitalis, has proven useful in controlling the heart rate. Reichstein was able to identify both the plants and the parts of the plants that contained glycosides, and his contributions were critical for initiating many botanical studies. He was one of the first researchers to realize the value of the tropical rain forests to the pharmaceutical industry. His work has also been pivotal in the field of chemical taxonomy, where the identities of plants are determined through their chemical composition—a method which has a higher degree of certainty than identification through visible characteristics. This technique has had broad applications in the development of both natural insecticides and drugs.

Reichstein was presented with an honorary doctorate from the Sorbonne in 1947. He received the Marcel Benoist Award in 1947, the Cameron Award in 1951, and a medal from the Royal Society of London in 1968. He is a foreign member of both the Royal Society and the National Academy of Sciences.

Reichstein married Henriette Louise Quarles van Ufford in 1927, while still at the Eidgenössische Technische Hochshule. They had one daughter. He retired from his academic posts in 1967, but continued to work in the laboratory until 1987. He died on August 1, 1996, in Basel, Switzerland, at the age of 99.

REPLICATION

Replication is the process by which **nucleic acids** such as **deoxyribonucleic acid** (**DNA**) or **ribonucleic acid** (**RNA**) are copied. The current method of replication was first proposed by James Watson and Francis Crick in 1953. They suggested that DNA was replicated by a semiconservative method in which each strand of DNA served as a template for new strands. In the late 1950s, this theory was confirmed through experiments performed by Matthew Meselson and Franklin Stahl. In their experiment, they grew *E. coli* **bacteria** in a medium, containing a heavy isotope of **nitrogen**. The bacteria naturally incorporated the heavy nitrogen into their DNA. The bacterial **cultures** were then placed in a lighter nitrogen, which enabled Meselson and Stahl to follow the production of new DNA and confirmed the semiconservative replication model.

DNA replication begins with a double strand of DNA. This structure, or parent **molecule**, is made up of two complementary strands of DNA that have paired **nucleotides**. Each nucleotide is paired with a specific partner, adenine (A) with thymine (T) and guanine (G) with cytosine (C). Replication begins at specific sites known as the origin of replication. These sites have a nucleotide sequence that is recognized by certain **proteins**. When replication is initiated, a short segment of DNA is untwisted by various **enzymes** called helicases at the origin. Next, the two DNA strands are separated, creating a replication bubble. The replication bubble is stabilized by another set of proteins that bind to DNA. This leaves segments of DNA that have a certain length of unpaired nucleotides. These unpaired nucleotides provide the template needed for the replication of the DNA strands. At each end of the replication bubble, there is a section called the replication fork where the new strands of DNA emerge as they are elongated.

During the replication process, key enzymes known as DNA **polymer**ases, bring new nucleotides onto each strand of DNA following the **base pairing** rules. DNA polymerases have the limitation of only being able to replicate DNA in one direction, namely 5-3 ft (1.5-0.9 m). However, a double strand of DNA is antiparallel, meaning the nucleotides on one strand are oriented in the opposite way as nucleotides on the other strand. One strand, called the leading strand, can be replicated easily by the polymerase. The other strand, called the lagging strand, employs short strands of replication called Okazaki fragments. These fragments are connected into a single DNA strand by another enzyme called DNA ligase.

Another limitation of DNA polymerase is that it requires a primer to initiate DNA replication. These primers are short stretches of RNA that are joined to the DNA by a primase enzyme. In **eukaryote**s, the primer is about 10 nucleotides long. With the primer attached to the DNA, the polymerase begins replication from that point. After replication, one DNA molecule has been made into two. Each of these DNA strands are made up of an old strand and a new one. The rate at which replication occurs can be as fast as 500 nucleotides per second in bacteria. In eukaryotes, the rate is more like 50 nucleotides per second.

DNA replication occurs during the S phase of a **cell**'s **life cycle**. At this time, duplicates of the **chromosome**s in the cell are visible. **Mitosis** separates the duplicate chromosomes and DNA is distributed between the two daughter cells. In this way, replication allows the transfer of **genetic material** from one cell to the next.

REPRODUCTION, ASEXUAL

Asexual reproduction is a method of reproduction that does not involve **meiosis** or the union of **gamete**s. As a result, the segregation and **recombination** of **genetic material** that are a normal part of **sexual reproduction** do not occur, and offspring are genetically identical or nearly identical with the parent **organism**. In normal sexual reproduction, offspring are genetically different from either of the parents, and as a result new traits may be introduced into the **population**. Some members of this diverse population may be able to adapt to changing conditions in ways that the parents could not, and the **species** as a whole is advantaged. In asexual reproduction these advantages are lost, but under some conditions the opportunity for rapid increase in numbers of individuals may compensate for this disadvantage.

Asexual reproduction is found in many types of organisms including **bacteria**, **algae**, **fungi**, **plants**, and some simple **animal**s. Although bacteria are capable of sexual union, they commonly reproduce asexually by **cell division**. Many fungi and algae form asexual reproductive structures, or **spore**s, that divide, grow, and develop into complete new organisms. Simple fragmentation of some simple forms of life may also give rise to new individuals. Higher plants have developed many mechanisms for asexual reproduction. Some plants may reproduce asexually when a mature plant is split into two or more parts and the fragments subsequently develop into new individuals. Plants like Strawberries (**Genus** *Fragaria*) and quack grass (*Agropyron repens*) form runners, or rhizomes, that spread out from the parent to form plantlets at their terminus. The plantlets become independent plants when they develop roots, and the connecting runners disintegrate. Other plants develop specialized structures that give rise to new organisms when released from the parent. One example is found in members of the genus *Gladiolus*, which form bulbs attached to the parent bulbs that are capable of producing new individuals. Members of the genus *Kalanchoe*, often called maternity plants, form specialized aerial plantlets attached to their leaves that eventually fall away to form new, genetically identical, individuals.

Asexual reproduction has been widely used in horticulture since ancient times. Many popular ornamental plants are reproduced almost exclusively with asexual techniques. Superior varieties of fruit and nut **tree**s are commonly reproduced this way. The practice can be assisted with the use of synthetic **plant hormones** like naphthaleneacetic acid and indolebutyric acid. When applied to the base of stem or **leaf** cuttings these chemicals stimulate the formation of adventitious roots, allowing the cuttings to develop into normal plants. Plants also may be asexually reproduced from undifferentiated plant calluses grown in **tissue culture**, or even from individual **cell**s grown

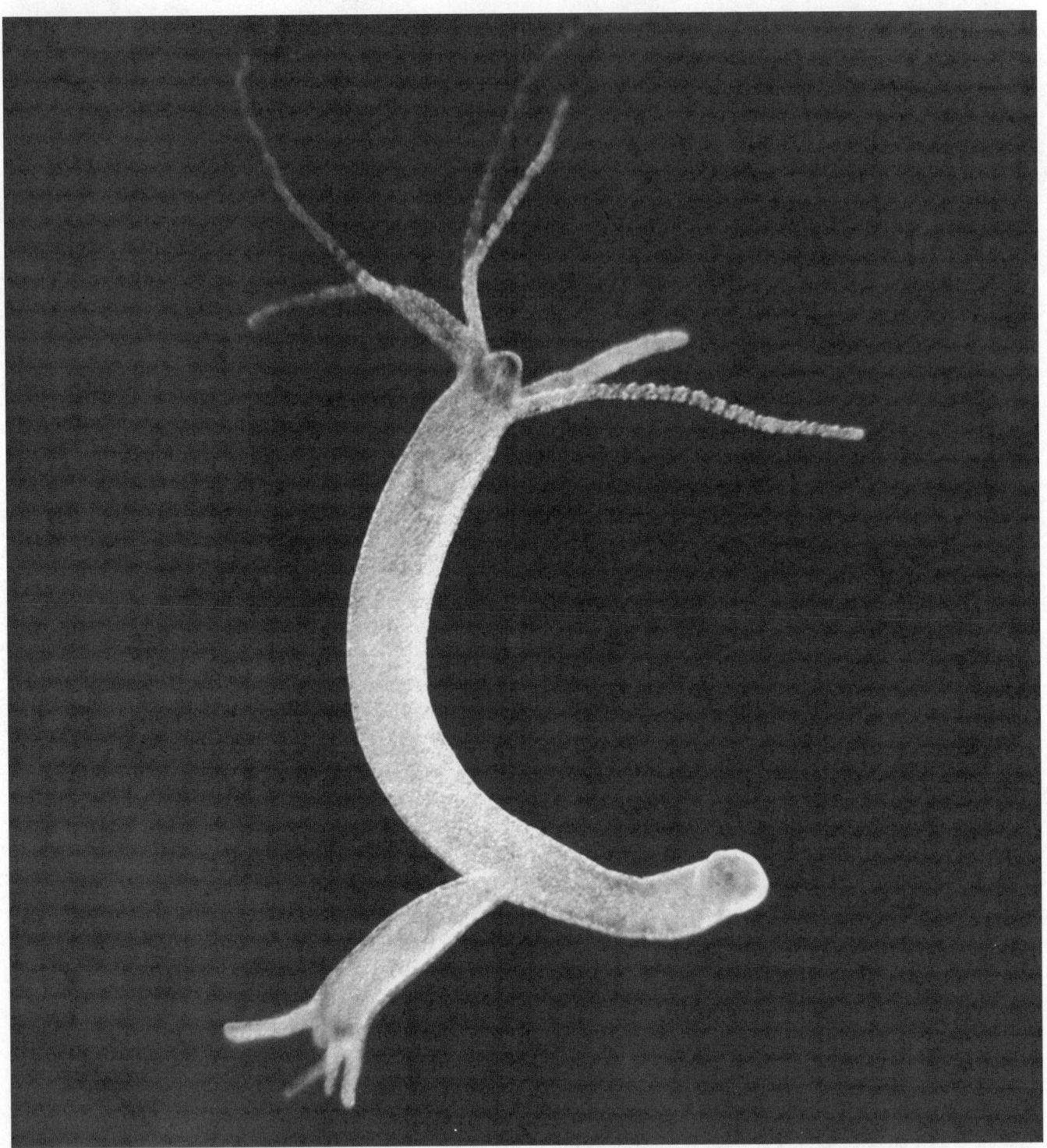

A freshwater hydra in process of budding, or reproducing. *(Photograph by James Bell. Photo Researchers, Inc. Reproduced by permission.)*

in **culture**. By suitably treating the cultured material with plant growth substances (**hormone**s), **differentiation** into complete individual plants, genetically identical with the original, can result.

Body cells of adult animals, including humans, are often asexually reproduced, or cloned, in research laboratories. Muscle cells, for example, can be removed from a donor animal and grown on suitable culture medium where they may go on growing and dividing, to produce an abundance of genetically identical descendants. Most of the cells of adult animals (and plants) contain the genetic information needed to produce a complete organism, but as cells mature into **tissue**s and **organ**s they tend to express only the information needed for their particular cell type. This has proven to be a particular challenge to scientists interested in the reproduction of complete organisms from adult animal cells. Nevertheless, it has been possible. By the 1950's scientists were able to clone frogs, and Scottish researchers reported successful **cloning** of sheep in 1996.

Unlike asexual reproduction in plants involving only somatic (non-sexually reproductive) cells, animal cloning has typically made use of enucleated **egg** cells to which the nuclear contents of somatic cells are added. The reconstituted egg cell is placed in suitable environment where it develops into an organism genetically identical with the organism from which the **nucleus** was taken. The process is difficult and complex. Despite these obstacles, the possibilities for large economic gain have led to much research in this field. Animal breeders, for instance, would welcome the opportunity to clone high quality livestock. As research advances, the possibilities for asexual **human reproduction** seem less remote. This has presented society with a very difficult ethical dilemma. Who will make decisions about the availability of this costly procedure? Will the advantages of genetic variation intrinsic to sexual reproduction be lost, making the human race more vulnerable to future challenges? Human society may have to face these and many similar questions in the near future.

See also Mitosis

REPRODUCTION, HUMAN

The human **reproductive system** is a set of **organ**s which work together to produce offspring. The male and female reproductive systems work together to best facilitate survival of the developing **embryo**.

The male **gamete**s are known as **sperm** or spermatozoa. The male reproductive system produces, stores, and releases these gametes. Males begin to produce sperm during puberty and continue to do so throughout most of their lives. Sperm have a structure which is directly related to their function. A large head contains the **DNA**, a narrow middle contains many **mitochondria** which produce the **energy** needed to move the flagellum, which propels the sperm through the female toward the **egg**.

Sperm are produced in the testes, two oval shaped organs contained within the scrotum. The scrotum is a pouch of skin formed from the lower part of the abdominal wall. This external storage of sperm provides a lower **temperature** which is optimum for their survival. The testes are composed of seminiferous tubules. These are tightly coiled tubes where germ **cell**s divide by **meiosis** and form sperm. The sperm are released into an elongated sac called the epididymis where they mature and are stored.

The sperm eventually leave the epididymis and travel through a duct called the vas deferens and into the urethra, a tube within the penis. In the urethra the sperm mix with **secretion**s from the seminal vesicles, the prostrate gland, and the Cowper's gland to make semen. These secretions nourish and protect the sperm; they contain a high amount of fructose which provides energy for movement, and they neutralize acidity in the urethra from any residues left from urine. During sexual intercourse, the semen is forcefully expelled by strong muscular contractions of the sperm ducts when the male reaches orgasm, a process called ejaculation.

The penis is the external portion of the male reproductive system. It contains the external portions of the urethra and is the organ by which sperm is introduced into the female. It is composed of spongy **tissue** which becomes turgid and erect when filled with **blood**, which occurs during sexual arousal.

The female gametes are known as eggs which are produced and stored in the ovaries. Each egg (also called an ovum) is 75,000 times larger than a sperm cell. Females are born with over 400,000 immature eggs and do not produce any more during their lifetime. In most females, only about 400 of these eggs actually mature. Other important structures of the female reproductive system include the Fallopian tubes, uterus, cervix, vagina, and external genitalia.

Each female has two Fallopian tubes, one next to each ovary. These tubes carry the eggs from the ovary to the uterus, and each can be the site of **fertilization**. The uterus is a muscular structure which houses the developing fetus if fertilization of the egg occurs. The cervix is the lower entrance to the uterus, and the vagina is a tube leading from the cervix to the outside of the body. It is the canal that accepts the penis during intercourse and through which the fetus passes during chldbirth. The female external genitalia are the inner and outer labia and the clitoris.

The female reproductive system undergoes a series of monthly changes called the **menstrual cycle**. The term "menstrual" comes from the Latin word *mensis* which means "month." Every 28 days an egg matures and is positioned to meet with a sperm cell in the Fallopian tube. If it is not fertilized, the egg is discharged through the vagina. The female menstrual cycle consists of four phases; the follicular phase, ovulation, the luteal phase, and menstruation. During the follicular phase, the cells surrounding the egg create a follicle which stimulates the maturation of the egg. Ovulation occurs when the follicle ruptures and releases the mature egg into the Fallopian tube. The luteal phase occurs when the ruptured follicle on the ovary develops into a new structure called the **corpus luteum**. Finally, menstruation occurs if the egg is not fertilized and is discharged through the cervix and vagina along with uterine tissues and blood. If fertilization of the egg occurs, the egg attaches to the uterine lining and menstruation does not occur.

Fertilization occurs when a sperm cell combines with an ovum in a Fallopian tube. Hundreds of millions of sperm cells are released at once during an ejaculation. The sperm move, propelled by the flagellum, through the uterus and into the Fallopian tube. Many sperm attach to the egg but only one actually enters it. Once a sperm penetrates the egg a **membrane** forms around the egg which prevents any other sperm from entering. The sperm **nucleus** then fuses with the egg nucleus. Because both the sperm and the egg are **haploid** cells containing half the number of **chromosome**s as somatic cells, when they fuse they create a **diploid zygote** with a total number of 46 chromosomes.

Immediately following fertilization, cleavage occurs. The zygote goes through many mitotic **cell division**s while still in the Fallopian tube. These divisions result in a ball of cells called the **morula**. The cells continue to divide and form a sphere called the blastocyst. Fluid is released into the center of the sphere and the outer layer of cells release an **enzyme** which breaks down the epithelial tissue of the uterus. This allows the blastocyst to embed itself into the thick uterine lining, which is called implantation. Once the embryo is implanted into the uterus, development continues for nine months until the fetus is born through the vagina.

See also Embryology; Oogenesis; Spermatogenesis

REPRODUCTION, SEXUAL

Sexual reproduction is the creation of new individuals resulting from the joining of the nuclei of two separate sex **cell**s. The separate sex cells are **gamete**s; the **sperm** and **egg**. **Fertilization** is the union of the two gametes, and the fertilized cell that is formed is called the **zygote**.

These gametes must have specific characteristics in order for sexual reproduction to work properly. Specifically, they must have only half the total number of **chromosome**s that normal cells of that **species** contain. Cells that have half the normal number of chromosomes are called monoploid or **haploid** cells. Thus, when the nuclei of the sperm and egg join together, instead of having twice the normal number of chromosomes, the resulting fertilized egg cell will have exactly the right number. A cell with the normal number of chromosomes is called a **diploid** cell. In order to form a monoploid sperm or egg cell, specific reproductive cells undergo a special type of **cell division** called **meiosis**. The sperm or egg cells that are formed as a result of meiosis are ready to be used as part of sexual reproduction. Meiosis forms the sex cells of all **organism**s that undergo sexual reproduction.

Sexual reproduction is very different from **asexual reproduction**, in which there is only one parent. In asexual reproduction, there are no gametes, and hence no fusion of the sperm and egg. Because there is only one parent, the offspring produced by asexual reproduction are genetically identical to the parent. This gives sexual reproduction a great advantage over asexual reproduction. Sexual reproduction produces variation among the offspring; the offspring are not exactly like the parents. If the environment changes, some individuals may be better able to survive than others, because of this variation. Those that survive will reproduce and pass these traits on to their offspring. In addition, these variations may give the species opportunities to enter and survive in new environments. Thus, sexual reproduction often gives a species more opportunities to survive because individuals are more varied.

While many organisms reproduce using asexual reproduction, some also have the ability to reproduce sexually. One of the simplest types of sexual reproduction is called **conjugation**. In this process, a connection between the **cytoplasm** of two unicellular organisms is formed and nuclear material is exchanged. While these simple organisms cannot really be considered males or females because they do not have testes or ovaries, respectively, there are different mating types. There are chromosomal and biochemical differences between the two mating types. Conjugation can occur in **bacteria** and some protists, such as **algae** and paramecia.

In more complex organisms that undergo sexual reproduction, individuals are usually either male or female; the sexes are usually separate. Males and females each have different types of gonads or sex **organ**s in which the gametes develop. The males produce sperm in gonads called testes and the females produce eggs (also called ova) in gonads called the ovaries. The sperm are generally smaller than the egg and must swim through a liquid to get to the egg in order for fertilization to occur. In most cases, when sexual reproduction occurs, the sperm and eggs originate from separate parents; one male and the other female. However, there are some organisms that can produce both sperm and eggs. These individuals that have both testes and ovaries are called hermaphrodites. However, it is relatively rare that the sperm from a hermaphroditic individual can fertilize its own eggs.

There are different strategies for sexual reproduction. Depending on where an organism lives, fertilization can occur inside the female (internal fertilization) or outside (external fertilization). Both strategies have advantages and disadvantages.

Aquatic organisms, those that live in **water**, more commonly use external fertilization. The sperm and eggs are released into the water where they meet and fertilization occurs. This is an advantage because the water is a liquid through which the sperm can swim to the egg. In addition, the female does not usually invest time and energy in care of her young. However, there is a very good chance that sperm and egg might not meet, they may succumb to **predation** by other organisms, or the environmental conditions might not be sufficient for survival of the gametes. In order to deal with these problems, organisms that use external fertilization release very large numbers of sperm and eggs into the water, and they usually release them at approximately the same time. For example, oysters release millions of sperm and eggs into the water at around the same time. Fertilization occurs externally, and the female provides no care for the developing offspring.

Terrestrial organisms, those that live on land, usually use internal fertilization. The sperm and egg meet inside the body of the female. With internal fertilization, both egg and sperm are protected, and there is a much greater chance that the gametes will meet. Thus, although there are still large numbers of sperm, fewer eggs are needed. In addition, the female's body usually provides the proper environmental conditions for survival of the gametes, unlike the environment outside of the body. Like external fertilization, timing of release of the gametes must be correct. There is often only a short period of time that the egg can be fertilized by the sperm, and mating must take place within this time period. In humans, for example, the egg is only able to be fertilized for approximately 24 hours during each 28 day cycle. Internal fertilization requires some specific **adaptation**s of the organisms that use it. For example, the female must provide a moist environment through which the sperm can swim to the egg, and there must be a specialized sex organ that places the sperm in relatively close proximity to the egg. After internal fertilization occurs, the zygote is either released from the body within some protective covering (i.e., a shell) or it remains within the body where development occurs.

Regardless of whether fertilization is internal or external, once it occurs, the unicellular zygote goes through a series of cell divisions. At a certain point, the cells begin to specialize and differentiate into the individual body parts of that organism. Eventually, as development continues, a fully formed individual results. Thus, the end product of sexual reproduction is a new individual.

REPRODUCTIVE SYSTEM

The reproductive system is a group of plant or **animal organ**s which are necessary for the production of offspring. The major structures which are involved in reproduction are known as the gonads. The basic units of sexual reproduction are the male and female germ **cell**s.

Germ cells are those cells which do not differentiate when an **embryo** is developing. Cells which develop into **tissue**s and organs are called somatic cells. In **invertebrates**, the germ cells congregate in the body cavity or the **circulatory system**, while in vertebrates, they move to tissues which eventually become the gonads, or reproductive organs. When an **organism** reaches sexual maturity, the germ cells undergo numerous **cell division**s and eventually develop, through a process called **meiosis**, into the mature reproductive cells, the **gamete**s. Gametes contain half the number of **chromosome**s of the somatic cells.

Male and female gametes are produced and stored in the gonads. Gonads play important regulatory roles as part of the **endocrine system**, but their primary function is the production of gametes. The male gametes are called **sperm**atozoa or sperm and are produced in the male gonads, the testes. The female gametes are called **egg**s or ova and are produced in the female gonads, the ovaries. In many invertebrates, an individual has both testes and ovaries and is called a hermaphrodite. In most vertebrates, an individual has either testes or ovaries, but not both. The gonads increase in size at sexual maturity because of the great number of germ cells produced. Some animals which experience a breeding season will undergo a seasonal increase in the size of their gonads for the same reason. The male and female gonads are both under the control of the **pituitary gland**.

The testes are composed of coiled, convoluted tubules called the seminiferous tubles, where the germ cells mature into spermatozoa. The mature spermatozoa are stored in the epididymis, a thick-walled, coiled canal, and then eventually discharged through a number of ducts called the efferent ducts.

The testes may remain within the body cavity during the entire lifetime of the animal, as in vertebrates below marsupials and in elephants, sea cows, and whales. The testes may also move into an external pocket (the scrotum) only during the breeding season, as in rodents, bats, and members of the camel **family**. In marsupials, and in most higher **mammals**, the testes are always enclosed in a scrotum. The descent of the testes into the scrotum keeps the germ cells at lower **temperature**s for optimum development.

Female germ cells originate as single cells in the embryonic tissue which will later develop into an ovary. At sexual maturity, the cells surrounding each ovum develop into follicle cells which eventually burst, releasing the ovum. At this point, the ovum is ready to be fertilized. The tissue which used to hold the follicle is then replaced by cells known as the **corpus luteum**, which secrete **hormone**s that prepare the uterus to receive a fertilized ovum.

Before leaving the body, the reproductive cells must travel from the gonads to an external body opening. In many invertebrates, and a few aquatic vertebrates, the reproductive cells are discharged directly from the gonads through pores in the body wall. In higher vertebrates the reproductive cells travel through ducts into urinary or cloacal **excretory system**s. In male vertebrates, the spermatozoa travel through a duct called the ejaculatory duct into the urethra, through which they are discharged. In female mammals the ovum falls into the oviduct or fallopian tube. In lower animals the oviducts lead to an opening called the cloaca. In marsupials and **placental** mammals, the oviducts lead to the uterus, where the fertilized ovum develops, and then the vagina, which is the external body opening.

In aquatic animals that discharge ova and spermatozoa into the **water**, the spermatazoa reach the ovum by following chemical signals. In terrestrial animals, the eggs are fertilized internally using reproductive organs known as genitals or genitalia. The male genital organ is the penis, which directs spermatozoa into the female cloaca or vagina. In all female marsupials and placental mammals, the female genital is the vagina.

Accessory glands in both the male and female play important roles in reproduction. In males, the prostrate gland produces a fluid in which the spermatozoa live, making semen. Lubricating glands in the female, including the cervical and Bartholin's glands, produce mucus to facilitate copulation. Female placental mammals also have uterine glands which prepare the uterus to receive a fertilized ovum. Accessory glands in both males and females produce **pheromones**, chemical scents which indicate reproductive readiness to members of the opposite sex. Reproductive glands can also function to nourish young, such as the placenta of placental mammals and the mammary glands of mammals. In female animals which lay eggs, glands produce albumin which nourishes the egg.

The evolutionary trend in vertebrate reproductive systems is toward internal **fertilization** and development. This pattern of reproduction best ensures survival of the young. Most fish and amphibians release eggs and sperm directly into the water, where fertilization takes place. In reptiles, birds, and mammals, fertilization takes place inside the female which increases the likelihood that the egg will be fertilized. Three patterns of reproduction based on whether development occurs internally or externally, and how the embryos are nourished, are recognized. **Oviparous** organisms have embryos which develop outside of the body protected by a shell or jelly-like layers, and are nourished by a yolk until they hatch. **Ovoviparous** organisms also develop in eggs which are nourished only by a yolk, but they remain inside the female until they hatch. Viviparous organisms have embryos which are nourished in the female through the placenta until they are born. The evolutionary trend is from oviparity (for example, many fish, amphibians, reptiles, and birds) to ovoviparity (for example, some higher fish, amphibians, and reptiles) to **viviparity** (for example, most mammals).

Reproductive systems in **plants** have male and female counterparts much like those of animals. Fertilization can occur internally or externally, and development of plant embryos can follow several different **life cycle**s.

See also Plant reproduction; Reproduction, asexual; Reproduction, human; Reproduction, sexual.

REPTILES • See Herpetology

RESOURCES, NATURAL

Natural resources, unlike man-made resources, exist independently of human labor. Natural resources can be viewed as an endowment or a gift to humankind. These resources are, however, not unlimited and must be used with care. Some natural resources are called ''fund resources'' because they can be exhausted through use, like the burning of **fossil fuels**. Other fund resources such as metals can be dissipated or wasted if they are discarded instead of being reused or recycled. Some natural resources can be used up like fund resources, but they can renew themselves if they're not completely destroyed. Examples of the latter would include the soil, **forests**, and fisheries.

Because of **population growth** and a rising standard of living, the demand for natural resources is steadily increasing. For example, the rising demand for **minerals**, if continued, will deplete the known and expected reserves within the coming decades.

The world's industrialized nations are consuming nonrenewable resources at an accelerating pace, with the United States ranking first on a per capita basis. With only 5% of the global **population**, Americans consumes 30% of the world's resources. Because of their tremendous demand for goods, Americans have also created more waste than is generated by any other country. The environment in the United States has been degraded with an ever-increasing volume and variety of contaminants. In particular, a complex of synthetic chemicals with a vast potential for harmful effects on human health has been created. The long-term effects of a low dosage of many of these chemicals in our environment will not be known for decades. The three most important causes for global environmental problems today are population growth, excessive resource consumption, and high levels of **pollution**. All of these threaten the natural resource base.

RESPIRATION

Respiration is the metabolic process by which living **organism**s produce needed **energy** by the controlled oxidation of **nutrient**s, such as **carbohydrate**s.

In the **enzyme**-controlled process, sequenced reactions slowly release the energy (essentially, the ability or capacity of an organism to perform work, an activity or function) in the nutrients. In most cases, oxygen, either molecular or that bound in **organic compound**s, is utilized, and **carbon dioxide** (CO_2), **water** (H_2O) and energy are produced. In the case of the sugar, glucose, the following formula typifies the reaction:

$$C_6H_{12}O_6 + 6O_2 \ 6CO_2 = 6H_2O + \text{Energy-dissipated Heat}$$

The process is often discussed as two separate processes:

- External respiration, such as breathing and the pumping water over gills, is a simultaneous mass-and-heat-transfer process whereby oxygen is moved into the organism, and carbon dioxide is respired. This involves the gas transfer of molecular oxygen into a **cell** and CO_2 by-product out of a cell by **diffusion** across **membrane**s.
- Internal (cellular or **tissue**) respiration is the enzyme-controlled **chemical reaction** (oxidation reduction) such as the Krebs and citric acid cycle, in which 3.74 **calorie**s per gram of glucose are liberated.

The eucharoyic organisms (which include all **animal**s, **plants**, **fungi** and **protista**) or cells are composed of many structures, each of which has evolved a specific purpose. They are composed of cells that have membrane-bounded nuclei, **organelle**s and **DNA** arranged in **chromosome**s. The **mitochondria** are organelles that are the oxygen-using part of the cell that possesses its own DNA, and replicates independently of the surrounding cell. These mitochondria (chondriosomes) contain enzymes systems (oxidases and dehydrogenases) that control the release of energy in the form of **adenosine triphosphate (ATP)**.

The Respiratory Quotient (R.Q.) is the ratio of carbon dioxide produced to a given volume of oxygen used up during respiration. The R.Q. varies depending on the characteristics of the nutrient being oxidized, and indicates whether the process is **aerobic** (requiring atmospheric or molecular oxygen) or **anaerobic** (obtains oxygen from organic compounds such as nitrates or sulfates in an anoxic environment).

Among other activities, life requires a continuous supply of certain chemicals and energy. The primary source for the energy is sunlight, and the chemicals are recycled from those found in the earth's crust. The program to manage the life process is contained in information and instructions learned over time and maintained in the genetic structures of each living organism. The chemicals must be taken in by the living organisms, and processed for use in maintaining life. These chemicals are bound up in different forms ("food" and "water") of **matter** such as **minerals** and organic compounds held together by weak and strong bonds that must be broken in order to release the chemicals necessary for life.

Autotrophic organisms, such as green plants, **algae** and **cyanobacteria**, can photosynthesize (photoauthrophy) the chemicals required for growth and energy from sunlight, soil chemicals, water and the carbon dioxide in the **atmosphere**. Together, photosynthetic organisms remove 250 billion tons of carbon annually as they produce the organization activity we call life. Certain sulfide, methane and **ammonia**-oxidizing **bacteria** use inorganic compounds (H_2S, CH_4 and NH_3) as an energy source.

Heterotrophic organisms in the animal **kingdom**, fungi and protista (cilliates, mastigotas, slime nets, etc.) must ingest (eat) organic plant and animal tissue composed of carbon, hydrogen and oxygen (together, carbohydrates) and other chemicals in order to form the chemical building blocks of life and to provide the energy to do so. In the process of living, heat is released and the by-products of **metabolism** are excreted. The preliminary breakdown of complex carbohydrates, **proteins** and **fats** into simple forms that can be used by cell enzymes is called "digestion." This is a progression of physical and chemical processes that render larger matter into successively smaller particles. The **circulatory system** transports these exceedingly small nutrient bundles to each cell of the organism to maintain its life processes The energy required to digest the food is provided by the chemicals in the food and from atmospheric oxygen.

Breaking the bonds of the food is called "metabolism." Enzymes (protein catalysts), such as those produced by cells, are the chemical machinery that do the work of digestion and metabolism, and control all of the chemical processes in living organisms. Living organisms digest and metabolize the food with specific enzymes tailored to transform food **molecules** into energy and to extract the chemicals they require for metabolism, growth and cellular maintenance.

To accomplish this activity, cells that make up living organisms require significant and continuous amounts of oxygen. Chemicals and specifically carbon in food are oxidized to carbon dioxide, water and energy. This is respiration.

RESPIRATION, CELLULAR

Cellular respiration is the process by which a living **cell** produces **adenosine triphosphate (ATP)**, **carbon dioxide**, and **water** from oxygen and organic fuel. It is a catabolic pathway that involves the release of stored **energy** from the break down of complex **molecules** to more simple ones. No single **chemical reaction** covers the entire process of cellular respiration. Instead it is the cumulative function of **glycolysis**, the **Krebs cycle** and electron transport. In **eukaryotes**, the **mitochondria** is the primary **organelle** that contains the **enzymes** that drive cellular respiration.

Nearly all eukaryotic cells contain some mitochondria. While there may be as few as one mitochondria in a cell, often there are hundreds or thousands. The number typically depends on the metabolic activity of the cell. The mitochondria is enclosed in a two **membrane** envelope in which a variety of **proteins** are embedded. Inside these membranes is the mitochondrial matrix which contains some of the enzymes that function in cellular respiration. Other enzymes including the one that makes ATP are attached to the inner membrane. This configuration provides an efficient way for cellular respiration to occur.

Although the mitochondria contains most of the enzymes related to cellular respiration, the process actually begins in the cytosol. This reaction, known as glycolysis, involves the breakdown of glucose into two molecules of a three carbon sugar called pyruvate. During this process, two molecules of ATP are consumed while four molecules of ATP are produced, resulting in a net gain of two ATP molecules. While this energy is beneficial to the cell, it pales in comparison to the amount produced by the later stages of cellular respiration.

After glycolysis, the pyruvate is transported across the mitochondrial membranes into the matrix. Here, it goes through a series of reactions called the Krebs cycle (also known as the citric acid cycle). First it is converted to acetyl CoA. It is then slowly oxidized into carbon dioxide and water. In the process, energy is transferred to storage molecules including three NADH and one $FADH_2$. Two molecules of ATP are also formed during this stage.

The final step in cellular respiration is the electron transport reactions. These reactions complete the oxidation of glucose and generate the greatest amount of energy. During this stage, each of the storage molecules transfers electrons to a series of **coenzymes** which then drive the production of ATP molecules. The actual production of ATP is the result of an enzyme called ATP synthase. This enzyme produces ATP from ADP by a process called oxidative **phosphorylation**. This phase of cellular respiration results in about 34 molecules of ATP.

In addition to glucose, many other compounds are used by the cell as a source of fuel. These include proteins, **carbohydrates** and **fats**. All of these complex molecules can be broken down to simpler ones which can then enter glycolysis or the Kreb's cycle at various points. For example, starch is hydrolyzed in the digestive tract producing a molecule that can be broken down by glycolysis. Similarly, glycogen can be hydrolyzed. Proteins are used as fuel, but only after they are reduced to their constituent **amino acids** and their amino groups are removed. Fats are the highest energy containing molecules. They are reduced to either glycerol or acetyl CoA before entering the cellular respiration reactions.

RESPIRATORY SYSTEM

It is amazing to think that almost all life forms require some type of respiratory system which allows them to take in oxygen and give off **carbon dioxide**. Going back through **evolution**, one can see an enormous variety of apparatuses which

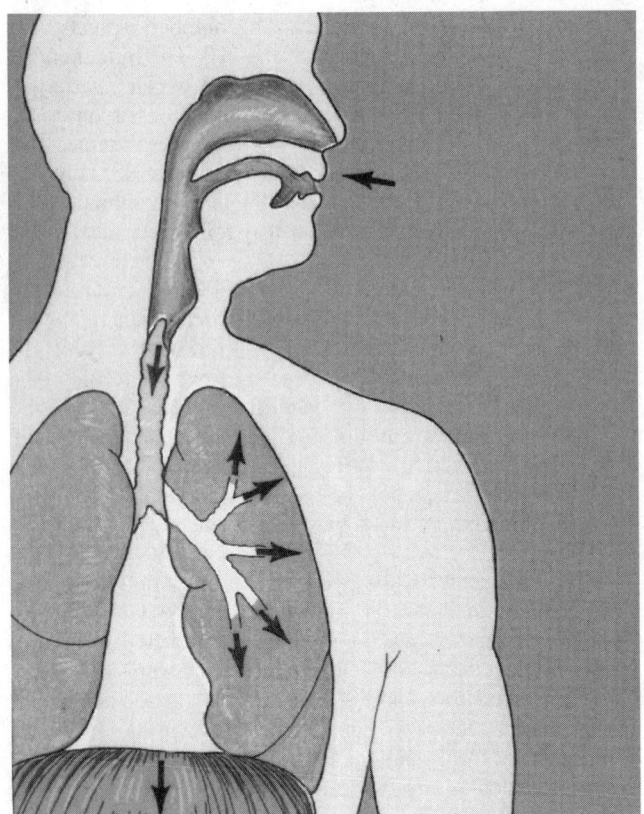

The human lung, which serves a supportive role in the respiration process. *(Illustration by June Hill Pedigo, Custom Medical Stock Photo. Reproduced by permission.)*

allow **organism**s to carry out this kind of gas exchange: starfish have pores distributed throughout their skeletons which serve as respiratory **organ**s; the gills of fish and other sea creatures are organs which expose thin-walled capillaries (tiny **blood** vessels) to the **water**y environment for gas exchange; earthworms and salamanders use their moist skin surfaces to exchange gases; amphibians use primitive lungs as well as their skin surfaces for gas exchange; birds have a highly complex system of air sacs throughout their bodies which increase their buoyancy in the air, as well as a complicated system whereby the actual beating of their wings while flying is partially responsible for air movement in and out of their lungs and air sacs. Although the mechanics of the types of respiratory systems found throughout the **animal kingdom** vary, the ultimate goal remains the same: the intake and distribution of oxygen, and the release of carbon dioxide.

The human respiratory system begins at the nose and mouth, where air is inhaled and and exhaled. The air tube extending from the nose is called the nasopharynx; the tube carrying air breathed in through the mouth is called the oropharynx. The nasopharynx and the oropharynx merge into the larynx. Because the oropharynx also carries swallowed substances, including food, water, and salivary **secretion**s which must pass into the esophagus and then the stomach, the larynx is protected by a trap door called the epiglottis. The epi-

glottis prevents substances which have been swallowed, as well as substances which have been regurgitated (thrown up(from heading down into the larynx and toward the lungs.

A useful method of picturing the respiratory system is to imagine an upside-down tree. The larynx flows into the trachea, which is the tree trunk and thus the broadest part of the respiratory tree. The trachea divides into two tree limbs, the right and left bronchi, each of which branches off into multiple smaller bronchi, which course through the **tissue** of the lung. Each bronchus divides into tubes of smaller and smaller diameter, finally ending in the terminal bronchioles. The air sacs of the lung, in which oxygen-carbon dioxide exchange actually takes place, are clustered at the ends of the bronchioles like the leaves of a tree, and are called alveoli.

Tiny thin-walled blood vessels called capillaries create vast networks surrounding the alveoli. The oxygen in inspired air can cross the very thin walls into the capillaries. The newly-delivered oxygen then travels through the capillaries and into the pulmonary vein, which carries it to the left side of the **heart** for delivery through the arterial system to every organ and tissue of the body. At the same time, **veins** have brought blood carrying the waste product carbon dioxide to the alveolar capillaries. The carbon dioxide crosses into the alveoli, and is expelled in subsequent exhalations.

The tissue of the lung which serves only a supportive role for the bronchi, bronchioles, and alveoli, is called the lung parenchyma.

Because the respiratory system brings in substances from the outside environment, there must be many safeguards to fight against any infecting agents which may be brought into the body along with the oxygen. The normal, healthy human lung is sterile, meaning that there are no normally resident **bacteria** or **virus**es (unlike the upper respiratory system and parts of the gastrointestinal system, where bacteria dwell even in a healthy state).

The first line of defense includes the hair in the nostrils, which serves as a filter for larger particles. The epiglottis is a trap door of sorts, designed to prevent food and other swallowed substances from entering the larynx and then the trachea. The acts of sneezing and coughing, both provoked by the presence of irritants within the respiratory system, help to clear such irritants from the respiratory tract.

Mucus, produced throughout the respiratory system, also serves to trap dust and infectious organisms. Tiny hair-like projections (**cilia**) from **cell**s lining the respiratory tract beat constantly, moving debris trapped by mucus upwards and out of the respiratory tract. This mechanism of protection is referred to as the mucociliary escalator.

Cells lining the respiratory tract produce several types of immune substances which protect against various organisms. Other cells (called macrophages) along the respiratory tract actually ingest and kill invading organisms.

See also Respiration

RESTORATION ECOLOGY

Although humans have probably tried to repair ecological damage and improve their local environment for as long as we have recognized the effects of careless use of our **natural resources**, restoration ecology is a relatively new field that attempts to combine practical knowledge of land management with the scientific insights of academic disciplines such as **ecology**, wildlife management, landscape architecture and horticulture. A pioneer in this field was the noted author and wildlife ecologist, Aldo Leopold, whose *Sand County Almanac* presents his ethical code for land stewardship together with practical experiments in restoring worn out, sandy farmland in central Wisconsin. The farm was far from a pristine wilderness, nor were the Leopold family merely spectators on it. They regarded themselves as active participants and citizens of the land community, seeking to restore the land to ecological health and beauty. ''**Conservation**,'' Leopold wrote, ''is the positive exercise of skill and insight, not merely a negative exercise of abstinence or caution.''

Modern restoration ecologists seek to extend Leopold's principles of practical land husbandry with new understandings of ecological processes to repair or reconstruct **ecosystem**s damaged by humans or natural forces. Recent laws requiring no net loss of **wetlands** and reclamation of **toxic waste** or surface mining sites have provided both incentive and finances to support this rapidly growing discipline. Restoration generally means to bring something back to a former condition. Ecological restoration involves active manipulation of **nature** to re-create **species** composition and ecosystem processes as close as possible to the state that existed before human disturbance. Rehabilitation refers to attempts to rebuild elements of structure or function in an ecological system without necessarily achieving complete recovery of its original condition. Often rehabilitation means to bring an area back to a useful state for human purposes rather than to a truly pristine state.

Remediation usually means the cleaning of polluted areas by physical or biological methods. Incineration, for instance, often is a good method of detoxifying oil-soaked soils. Living **organisms** are highly effective cleaning agents for many contaminants. Many plant species absorb toxic elements such as selenium or mercury from polluted **water** or soil. **Microorganisms** (**bacteria** and **fungi**) can be found in nature or engineered in the laboratory to destroy many dangerous chemicals. Reclamation is used to describe chemical or physical treatments of severely degraded areas such as open-pit mines. The Surface Mining Control and Reclamation Act, for example, requires mine operators to restore land surfaces to original contours and to replant vegetation. Many cities have old industrial sites known as brownfields with abandoned buildings and badly contaminated soils that prevent redevelopment. Reclaiming these sites is an important step in revitalizing inner city neighborhoods.

Re-creation attempts to construct a new biological community on a site so severely disturbed that there is virtually nothing left to restore. The new community may be modeled

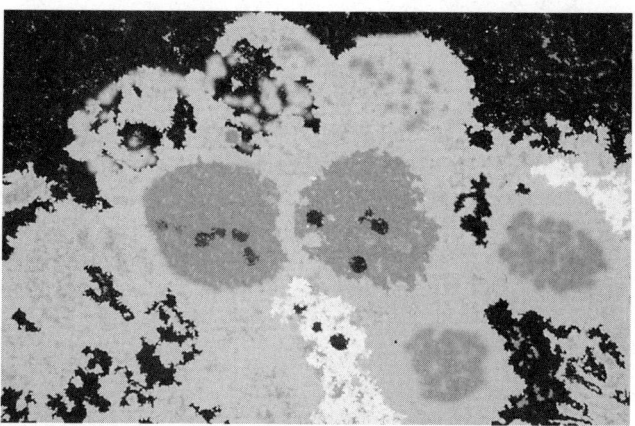

A transmission electron micrograph (TEM) scan of a retrovirus inside the mammary gland of a mouse. *(Photograph by Olin M. Pitts, Custom Medical Stock Photo. Reproduced by permission.)*

on what we think was there before human disturbance or it may be something that never existed on that site but that we think suits current conditions. Often developers are required to mitigate adverse effects of their projects by recreating a new ecosystem to replace one they have destroyed. A wetland destroyed by highway construction, for example, may have to be replaced by a newly constructed one in a new location. Environmentalists often oppose these policies. An artificially constructed ecosystem rarely has the complexity or diversity of a natural community. The purpose of restoration ecology, they claim, should be to repair previous damage, not to legitimize further destruction.

RETROVIRUS

A retrovirus is defined as any of a group of **virus**es that, unlike most other viruses and all cellular **organisms**, carry their genetic blueprint in the form of **ribonucleic acid (RNA)**. Retroviruses are responsible for certain **cancers** and virus infections of **animals** and cause at least one type of human cancer. They have also been identified as the cause of acquired immune deficiency syndrome (**AIDS**) in humans, and they have been linked to one form of human hepatitis.

Retroviruses are so named because, by means of a special **enzyme** called **reverse transcriptase**, they use RNA to synthesize **deoxyribonucleic acid (DNA)**. This establishes a reversal of the usual cellular process of **transcription** of DNA into RNA. The action of reverse transcriptase makes it possible for **genetic material** from a retrovirus to become permanently incorporated into the DNA of an infected cell.

The retrovirus HIV, the causal agent of AIDS, invades the target cell and infiltrates its DNA. The viral DNA is transcribed, or encoded, along with the cell's DNA into RNA, which then directs the production of **proteins**. The host cell doesn't recognize the viral DNA as invasive and the infection is permanent. The host continues to replicate the viral DNA and the disease spreads without impediment.

RETZIUS, ANDERS ADOLF (1796-1860)
Swedish anatomist

Andre Adolf Retzius was Swedish anthropologist known for his contributions to **comparative anatomy**, **histology** and anthropology. Retzius was born in Stockholm, Sweden on October 13, 1796, the son of Anders Jahan Retzius a professor of natural history at the University of Lund. The younger Retzius shared many of his father's interests and he eventually enrolled at the same school where his father taught. There he met an anatomy professor, Arvid Henrik Florman, who helped develop his interest in **zoology**. Retzius spent 1816 at Copenhagen with anatomist Ludwig Levin Jacobson, the physicist Han Christian Oersted, and zoologist J.H. Reinhard. In 1819, at age 23, he received a degree in medicine from the University of Lund; the same year his thesis *Observationes in Anatomiam Chondropterygium Praecipue Squali et Rajae Generum*, which dealt with the anatomy of cartilaginous fish, was published.

By 1823 Retzius was made professor of veterinary science at the Stockholm Veterinary Institute; a year later he was appointed professor of anatomy at the Karolinska Institute in Stockholm. He was appointed to the post of inspector of the Institute in 1830 and made many strong contributions to its development. Between the mid-1820s and late 1830s, Retzius studied a wide range of **species**, including primitive vertebrates such as slime eels. He made many important finding in comparative anatomy and embryology for these and related species. He is also noted for his work examining the anatomy of birds, dogs, and horses. Some of his early work involved comparisons of the structure and function of bird and reptile lungs.

Later in his career, when his eyesight began to fail him, Retzius was no longer able to use a **microscope** to study anatomical structures. Instead, he focused his efforts on the study of anthropology. By 1940 he had resigned his position at the Veterinary Institution to concentrate on his studies at the Karolinska Institute and in 1842 he embarked on a project to analyze racial differences by measuring skull sizes. He created a Cranial Index by multiplying the ratio of the width to the length of the skull by a factor of 100 and he classified skulls of various ethnic origin according to this index. He labeled skulls with a Cranial Index less than 80 as *dolichocephalic* (long headed) and those with values greater than 80 as *brachycephalic* (wide headed). He then attempted to sort different races according to this index. For example, he claimed that Europeans consisted of two subcategories: the Nordics who were dolichocephalic and the Mediterraneans who were brachycephalic. While Retzius' work in this field is not considered to be a valid analysis of racial types, it did set the stage for more valuable future studies.

Retzius continued to concentrated most of his work on the Karolinska Institute until his death in Stockholm on April 18, 1860.

REVERSE TRANSCRIPTION

Reverse **transcription** is an atypical method of synthesizing **DNA** from a template of **RNA**. Typically, transcription is a unidirectional process in which the **nucleotide** sequence of one strand of DNA is copied thereby creating a single strand of RNA with a nearly identical sequence. This RNA strand is then utilized by the **cell** for a variety of reasons related to the synthesis of **proteins** and **enzymes**.

In reverse transcription, the process is reversed. RNA is used as a template to make DNA. This process is performed by certain **retroviruses** whose **genetic code** is made up of single-stranded RNA **molecules**. It also requires a special enzyme known as a reverse transcriptase enzyme. When these **viruses** infect a cell, they inject it with their RNA. Instead of being utilized in **protein synthesis**, this RNA goes through the process of reverse transcription, and is converted into a single-stranded DNA molecule. This single-stranded DNA is further converted into a double-stranded DNA which then becomes integrated into the cell's **genome**. When these foreign genes are expressed, the cell's normal functions are altered and it becomes a manufacturing site for more viruses.

The existence of reverse transcription establishes the general rule that information stored in nucleic acid sequences as either RNA or DNA can be converted between either type. However, reverse transcription does not generally occur in the normal operations of a cell.

RH FACTOR

Rh factor describes **blood** type compatibility issues involving Rh **blood groups**. These play a critical role in transfusions and in obstetrics.

Rh blood groups were first discovered in 1940 by **Karl Landsteiner** and Alexander Wiener in their experiments with Rhesus monkeys, hence the name. Forty years earlier, Landsteiner had identified **blood types** A, B, AB and O, a discovery that made blood transfusions safe.

Landsteiner and Wiener noted that unexplained accidents in transfusions were attributed to this Rh blood group. The Rh blood group is also a correlary in hemolytic disease of the newborn, which is characterized by the breaking apart of red blood **cells**.

People with the Rh (rhesus) **protein** on the surface of their red blood cells are called Rh-positive (about 85 % of the **population**). Those who lack the protein are Rh-negative (about 15 %). The **immune system** of an Rh-negative person destroys Rh-positive blood, as it does any foreign **antigen**.

Specifically, Rh factor is a non-glycosylated, hydrophobic red blood cell surface protein with a size of 32 kDa and a structure similar to transporter glycoproteins. The Rh **gene** locus in humans consists of two structural genes, RHD and RHCE, and lies on **chromosome** 1.

When blood from an Rh-positive donor is used for transfusion into an Rh- negative recipient, the latter will develop specific antibodies which may produce a hemolytic reaction if

Rh-positive cells are again introduced in a transfusion. When an Rh-negative patient receives Rh-positive blood, the recipient produces anti-Rh antibodies to destroy the foreign cells by making them agglutinate, or clump together. This reaction in the body can cause serious injury or **death** to the patient.

Rh factor is also a critical to the blood compatibility of parents.If an Rh-negative woman gives **birth** to an Rh-positive child fathered by an Rh- positive man, there is a risk that she will become sensitized to the Rh factor in her baby's blood and begin to produce anti-Rh antibodies. Her first baby will not usually be affected, but in subsequent pregnancies, the mother may send enough damaging antibodies into the child's bloodstream to threaten its life.

Usually trophoblast cell layers in the **placenta** separate fetal cells from the mother's blood, however, some cells may still escape into the mother's circulation during late pregnancy or childbirth.

Anti-Rh antibodies, or IgG white blood cells, are not naturally present in the body, and the mother's initial **antibody** response is too late and too dilute to affect the fetus of the first pregnancy. In the second pregnancy, however, the mother's IgG cells are already present and have the ability to penetrate the trophoblast and to agglutinate the red blood cells of the fetus.

This condition is erythroblastosis fetalis, or hemolytic disease of the newborn (HDN), and consists of severe anemia, **brain** damage or even death. About one in 20 births conceived of Rh-positive mothers and Rh-negative fathers results in death. Before Rh factor was identified, it was a common cause of stillbirth pregnancies.

Parents with different ABO blood types have no Rh incompatibility problems because cells that enter the mother from the fetus are destroyed by antibodies, called IgM white blood cells, which already exist in the mother's blood.

IgM antibodies rarely cause agglutination in the fetus because they are unable to cross the trophoblast, and they are able to destroy the foreign red blood cells before they can trigger an IgG immune response.

To resolve general Rh-incompatibility, mothers receive anti-Rh antibodies after the first birth to destroy any Rh-positive fetal red blood cells that have entered the mother's circulation. This process of sensitization prevents the mother's immune system from producing IgG antibodies against the Rh factor.

RIBONUCLEASE

Ribonuclease is a type of **enzyme** which catalyzes the breakdown of bonds in **ribonucleic acid** (RNA). It plays an important role in the regulation of **protein** production in **cells**. There are a wide variety of ribonucleases found throughout **nature**. The most thoroughly studied of these is bovine pancreatic ribonuclease. It was the first enzyme whose two dimensional and three dimensional structures were determined.

Bovine pancreatic ribonuclease has been extensively studied because it is a relatively small enzyme, abundant, and

heat stable. It was first discovered in 1920 by W. Jones. He demonstrated that an extract from a bovine **pancreas** could hydrolyze **yeast** nucleic acid. In 1940, this ribonuclease was crystallized and isolated from the extract. It became the first enzyme for which structural information was obtained. Its primary structure was determined in 1962 and its three dimensional crystal structure was resolved by x-ray crystallography in 1967.

Ribonucleases help degrade ribonucleic acid (RNA). RNA is the material in a cell that codes for different proteins. When ribonuclease is present in the **cytoplasm**, RNA is degraded. This is important because it allows the cell to control the amount of protein produced by any RNA and to reuse the **nucleotide**s to create new RNA. While this process is designed to degrade unnecessary RNA, ribonucleases will degrade any type of RNA so a large amount is not present in the cell at any given time.

Bovine ribonuclease contains four disulfide bonds. It is the classic example of a protein whose activity is dependant on it higher order structure. When it is denatured, it loses its catalytic activity. This means that it no longer degrades RNA. When the denaturing process is reversed, it regains its activity. It is known as a ribonucleoprotein because it consists of both an RNA **molecule** and a protein. **Sidney Altman** discovered that the RNA component alone possesses enzymatic activity. It can interact to degrade transfer RNA. The protein portion helps maintain structure and is important for the proper function of the ribonuclease.

There are numerous types of ribonucleases distributed throughout nature, found in both **plant** and **animal species**. There are three different types of ribonucleases which are generally called ribozymes. A ribozyme is any RNA that has catalytic activity.

The different ribonucleases work in similar ways. Pancreatic ribonuclease catalyzes the **hydrolysis** of phosphodiester bonds between RNA chains. This is thought to occur in two steps. First, the enzyme breaks a bond between a phosphorus and oxygen on an RNA molecule. Next, **water** reacts with the free nucleoside permanently removing it from the RNA. This causes a breakdown of the RNA at specific sites, namely at pyrimidine bases. Other ribonucleases hydrolyze RNA at other sites. For example **bacterial** ribonucleases interact with purines. Typically, ribonucleases react with bases on hairpin loop structures in the RNA. Since ribonucleases breakdown RNA at different places, they have become important tools in the analysis of the sequence and structure of RNA.

RIBONUCLEIC ACID • See RNA (ribonucleic acid)

RIBOSOME

Ribosomes are **organelles** that play a key role in the manufacture of proteins. Found throughout the **cell**, ribosomes are composed of ribosomal ribonucleic acid (rRNA) and proteins. They are the sites of **protein synthesis**.

Although Robert Hooke first used a light microscope to look at cells in 1665, it was only in the last few decades that

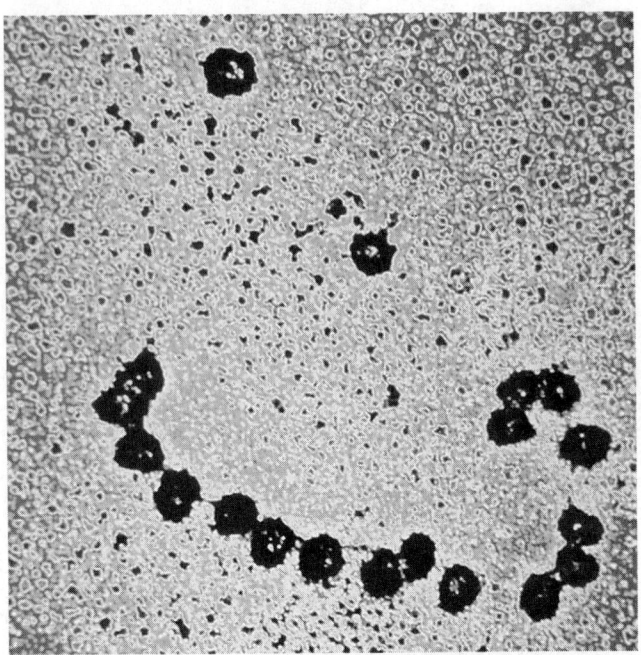

Ribosomes translating a messenger RNA strand to produce proteins. (Photo Researchers, Inc. Reproduced by permission.)

the cell's organelles were found. This is primarily because light **microscope**s do not have the magnifying power required to see these tiny structures. Using an electron microscope, scientists have been able to see most of the cells substructures, including the ribosomes.

Ribosomes are composed of a variety of proteins and rRNA. They are organized in two functional subunits that are constructed in the cell's nucleolus. One is a small subunit that has a squashed shape, while another is a large subunit that is more spherical. The large subunit is about twice as big as the small unit. The subunits usually exist separately, but join when they are attached to a messenger RNA (mRNA). This initiates protein synthesis.

Production of a protein begins with initiation. In this step, the ribosomal small subunit binds to the mRNA along with the first transfer RNA (tRNA). The next step is elongation, where the ribosome moves along the mRNA and strings together the amino acids one by one. Finally, the ribosome encounters a stop sequence and the two subunits release the mRNA, the polypeptide chain, and the tRNA.

Protein synthesis occurs at specific sites within the ribosome. The P site of a ribosome contains the growing protein chain. The A site holds the tRNA that has the next amino acid. The two sites are held close together and a **chemical reaction** occurs. When the stop signal is present on the mRNA, protein synthesis halts. The polypeptide chain is released and the ribosome subunits are returned to the pool of ribosome units in the cytoplasm.

Ribosomes are found in two locations in the cell. Free ribosomes are dispersed throughout the **cytoplasm**. Bound ribosomes are attached to a membranous structure called the **endoplasmic reticulum**. Most cell proteins are made by the free

ribosomes. Bound ribosomes are instrumental in producing proteins that function within or across the cell membrane. Depending on the cell type, there can be as many as a few million ribosomes in a single cell.

Since most cells contain a large number of ribosomes, rRNA is the most abundant type of RNA. rRNA plays an active role in ribosome function. It interacts with both the mRNA and **tRNA** and helps maintain the necessary structure. Transfer RNA is the molecule that interacts with the mRNA during protein synthesis and is able to read a three **amino acid** sequence. On the opposite end of the tRNAs, amino acids are bonded on a growing **polypeptide** chain. Generally, it takes about a minute for a single ribosome to make an average sized protein. However, several ribosomes can work on a single mRNA at the same time. This allows the cell to make many copies of a single protein rapidly. Sometimes these multiple ribosomes, or polysomes, can become so large that they can be seen with a light microscope.

The ribosomes in **eukaryotes** and **prokaryotes** are slightly different. Eukaryotic ribosomes are generally larger and are made up of more proteins. Since many diseases are caused by prokaryotes, these slight differences have important medical implications. Drugs have been developed that can inhibit the function of a prokaryotic ribosome but leave the eukaryotic ribosome unaffected. One example is the **antibiotic** tetracycline.

RICHARDS, JR., DICKINSON WOODRUFF (1895-1973)
American physician

In refining the technique of cardiac catheterization, Dickinson Woodruff Richards made significant contributions to the study of cardiopulmonary function in human patients. In collaboration with his colleague **André F. Cournand**, Richards elaborated upon earlier research by German physician **Werner Forssmann**, leading ultimately to the discovery of how pulmonary efficiency could be measured. For their work, Richards, Cournand, and Forssmann shared the 1956 Nobel Prize.

Richards was born October 30, 1895 in Orange, New Jersey to Sally (Lambert) and Dickinson Woodruff Richards. The legacy of the medical profession was established by his maternal forebears, the Lamberts. Richards' grandfather practiced general medicine in New York City, as did three of Richards' uncles; all either received their training or were otherwise affiliated with Bellevue Hospital or Columbia University's College of Physicians and Surgeons, where Richards himself would eventually study.

Richards received his A.B. from Yale University in 1917, and three months later enlisted in the United States Army, serving in France with the American Expeditionary Force. Upon his return to the United States, he entered the College of Physicians and Surgeons at Columbia; there he completed his M.A. in physiology in 1922 and his M.D. in 1923. Richards immediately received his license to practice medicine and interned for two years at Presbyterian Hospital in New York, spending two additional years there as a resident physician once his internship had expired. In 1927, Columbia University granted him a research fellowship to train at London's National Institute for Medical Research. From 1927 to 1928 he studied experimental physiology, working closely with Dr. Henry Hallett Dale, who Richards would later refer to as one of his greatest influences. He then returned to Columbia University's Presbyterian Hospital to study pulmonary and circulatory physiology. In 1930 he became engaged to Constance Riley, a Wellesley College graduate who worked as a technician in his research lab at Presbyterian Hospital. They married in September 1931.

Teams with Cournand to Refine Catheterization Technique

Richards' collaboration with André Cournand began in 1931 at Bellevue Hospital. Basing their research on Richards' concept "that lungs, **heart**, and circulation should be thought of as one single apparatus for the transfer of respiratory gases between outside atmosphere and working **tissue**s," these two physicians began a long and fruitful partnership. Their initial research involved the study of the physiological performance of the lungs and, in particular, a disorder known as chronic pulmonary insufficiency. Characterized by a malfunction in the heart's tricuspid and pulmonic valves, this defect causes **blood** to flow backward into the heart. Richards concluded, as had others before him, that it was necessary to be able to measure the amount of air in the lungs during different stages of pulmonary activity. Thus, he and Cournand unearthed studies done in 1929 by the German physician Werner Forssmann, wherein Forssmann had attempted to measure gases in the blood as it passed from the heart to the lungs.

Forssmann's technique was proven viable when he successfully inserted a narrow rubber catheter through a vein in his own arm and into the right atrium of his heart. This method gave access to blood as it entered the heart—blood that could then be examined in specific stages of pulmonary and cardiac activity and evaluated in terms of rate of flow, pressure relations, and gas contents. Catheterization would allow physicians to measure oxygen and carbon monoxide in blood returning from the right atrium, allowing for accurate measurement of blood flow through the lungs. Richards and Cournand sought to advance Forssmann's technique and to develop a safe procedure by first experimenting on animals. They began their research in 1936, and by 1941 they had successfully catheterized the right atrium of the human heart.

The measurements made possible through cardiac catheterization led Richards to other important assessments about functions of the heart and **circulatory system**. In 1941 he developed methods to measure the volume of blood pumped out of either ventricle (lower chamber) of the heart, and to measure blood pressure in the right atrium, the right ventricle, and the pulmonary artery, as well as total blood volume. More recent research has employed catheterization to diagnose abnormal exchange between the right and left sides of the heart, such as is present in some congenital cardiac defects. It has also contributed to the development of more sophisticated techniques such as angiocardiography (the X-ray examination of the heart after injection of dyes), which is used to determine whether normal circulation has resumed following a surgical procedure.

Richards and his colleagues also relied on their revolutionary research technique to study the effects of traumatic shock in heart failure and to identify congenital heart lesions. During World War II Richards and his colleagues were asked by the government to study the circulatory forces involved in shock, with Richards serving as chair of the National Research Council's subcommittee. The goal was to measure the effects of hemorrhage and trauma on the heart and cardiac circulation, and to evaluate various procedures for treatment. The most important result of this project was the discovery that whole blood, rather than just blood plasma, should be used in the treatment of shock to the cardiac system.

Richards was passionate about health issues in the social arena as well as in the laboratory: in 1957 he testified before the Joint Legislative Committee on Narcotics Study to suggest the construction of hospital clinics to legally distribute narcotics to recovering addicts; he lobbied for the building of a new hospital to replace the aging Bellevue; he spoke often about the need for constant reform within medical academia; and he supported the crusade to improve health care benefits for the elderly.

In 1945 Richards became the head of Columbia University's First Medical Division at Bellevue Hospital, and at the same time was promoted to the full-time Lambert Professorship of Medicine at the College of Physicians and Surgeons. He served as associate editor of *Medicine, Circulation, The Journal of the American Heart Association,* and of the *American Review of Tuberculosis.* His articles have appeared in many publications, including *Physiological Review, Journal of Clinical Investigation,* and *Journal of Chronic Diseases.*

Richards was elected to the National Academy of Sciences in 1958, and retired from practice in 1961, though he continued to lecture and publish frequent articles for several years. He died at his home in Lakeville, Connecticut on February 23, 1973, after suffering a heart attack.

Research Discoveries Rewarded with Nobel Prize and Other Honors

For their refinement of the catheterization procedure and the discoveries that followed, Richards, Cournand and Forssmann were awarded the Nobel Prize in physiology or medicine in 1956. In addition, Richards also received many individual honors and awards, including the John Phillips Memorial Award of the American College of Physicians (1960) and the Kober Medal of the Association of American Physicians (1970). He was made a chevalier of the Legion of Honor of France (1963), and was a fellow of the American College of Physicians, the American Medical Association, and the American Clinical and Climatological Association. He was offered numerous honorary degrees but accepted only two—Yale University, his alma mater, and Columbia University, where he did most of his work.

RICHET, CHARLES ROBERT (1850-1935)
French physiologist

Charles Richet was born on August 25, 1850, in Paris, the son of a professor of clinical surgery. As a student, Richet found himself drawn to both the humanities and the sciences. Swayed perhaps by family influence, he entered medical school at the University of Paris, although he continued to write poetry and drama on the side. While at medical school, Richet became involved in studies of hypnotism, of the gastric juices involved in **digestion**, and of the phenomenon of pain. Deciding on a career in physiology rather than surgery, Richet, after receiving his medical degree in 1877, earned his doctor of science degree in 1878. Soon afterward, he was named a professor of the Faculty of Medicine at the University of Paris and immediately began research on muscle contraction. By 1883, he had turned to studies on how warm-blooded animals maintain their constant body **temperature**, demonstrating that the larger the animal, the less heat it produces per unit of weight.

In 1880, Richet began investigating microbiology, the field in which he would make his great contribution. He observed **Louis Pasteur** demonstrate the inoculation of chickens against the fatal fowl cholera with a weakened strain of the cholera **bacteria**. Richet was struck by the idea that microbes might cause disease by producing a toxin, and that immune animals might carry a substance in their **blood** that counteracts the toxin. Richet reasoned that, if blood from immune animals were injected into nonresistant animals, the transfused toxin-resistant substance might confer immunity on the blood recipient. His applications of this theory to produce an immune serum for tuberculosis, however, failed.

Nevertheless, Richet continued his studies of toxicity, and in 1900, at the request of Prince Albert of Monaco, he began investigating the toxicity of sea anemone poison. He discovered "an extraordinary fact" completely opposite to what he had expected: when dogs that had been previously injected with toxin were reinjected with small doses of that toxin, the animals quickly died. The initial dose, instead of conferring immunity, produced fatal hypersensitivity. Richet called this reaction anaphylaxis, and in subsequent investigations, he and others found that it could occur as the result of exposure to a number of substances. Richet summarized anaphylaxis research in a 1911 monograph. For his work on anaphylaxis, he earned the 1913 Nobel Prize.

During World War I, Richet investigated blood plasma transfusion. After the war, he continued research in a wide range of areas. During the 1890s, he took part in the design and construction of one of the early airplanes. He was deeply interested in psychic phenomena, and he was also a dedicated pacifist who wrote several histories showing the malevolent effects of war. In later life, Richet also continued writing poems, plays, and novels. He died in Paris on December 4, 1935.

RICKETTS, HOWARD TAYLOR (1871-1910)
American pathologist

Howard Taylor Ricketts was an American pathologist who discovered the class of disease-carrying **microorganism**s now known as *Rickettsia*. These microorganisms have characteristics of both **viruses** and **bacteria**, and are responsible for lethal human diseases such as typhus and Rocky Mountain spotted fever. Though he died young, he made an important contribution to the study of vaccines and immunity.

Ricketts was born on a farm in Findlay, Ohio, in 1871. He attended Northwestern University and then graduated from the University of Nebraska. After receiving his bachelor's degree, he went back to Northwestern for medical school. He eventually became a pathologist at Chicago's Rush Medical College. He then went to Europe and studied microbiology in Berlin and at the Pasteur Institute. In 1902 he accepted a position as associate professor of pathology at the University of Chicago. There he published research on infection and immunity.

While Ricketts was on vacation in Montana in 1906, he began hearing of Rocky Mountain spotted fever, a deadly disease restricted to a small area of the West. In some outbreaks it killed more than 80% of those infected, and its cause was unknown. Ricketts worked intensely over the next three years to find the infectious agent. He eventually located the microorganism in the blood and eggs of a species of tick. Ricketts faced opposition from Western real estate agents and from the U.S. Department of Agriculture, who feared a drop in land prices if the tick was shown to be responsible. Nevertheless he began to formulate a plan to control the bug.

In 1910 Ricketts traveled to Mexico to investigate an outbreak of typhus. Ricketts found that typhus was similar to Rocky Mountain spotted fever, and was in fact caused by bacterial parasites transmitted by body lice. However, in his close work with typhus patients, Ricketts himself became infected and died.

His work was carried on by others after his death. In 1916 Henrique da Rocha-Lima named the class of microorganisms discovered by Ricketts *Rickettsia rickettsii*, and in the next several years another scientist made a complete study of them.

Modern medicine is indebted to Ricketts for his pioneering work with *Rickettsia*. *Rickettsia* is very small and difficult to see and stain using the methods available in Ricketts time, so it is a testament to his insight and grasp of theoretical microbiology that he was able to accomplish what he did in only four years of research. Ricketts also showed that laboratory animals could be used for inoculation experiments. Before his death he proved that monkeys could be infected with typhus, and that they had immunity to the disease after they recovered. His early work became the basis for many subsequent advances in vaccine and immunity research.

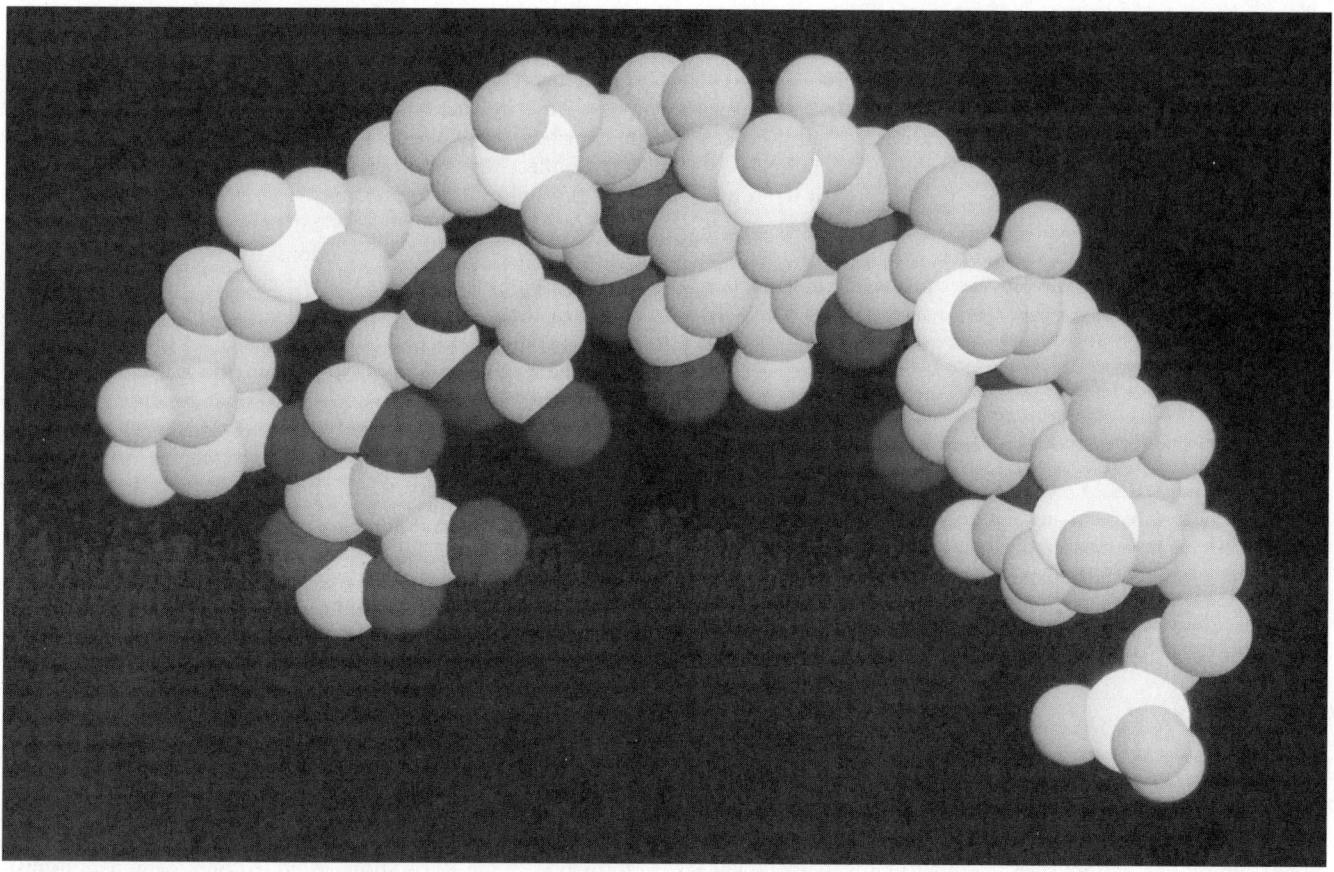

A computer-generated model of ribonucleic acid (RNA). *(Photo Researchers, Inc. Reproduced by permission.)*

RNA FUNCTION

RNA is a nucleic acid that has a variety of functions in a **cell**. Depending on the type of RNA it can function as a carrier for genetic information, a catalyst for biochemical reactions, an adapter **molecule** in **protein synthesis**, and a structural molecule in cell **organelles**.

Since the discovery of **DNA** in the early 1950s, scientists have studied the function and structure of **nucleic acids**. The various types and functions of RNA have been investigated by numerous researchers. Perhaps most famous is **Severo Ochoa** who received a Nobel prize for his work on RNA in 1959.

There are five major types of RNA that are found in the cells of **eukaryotes**. These include heterogeneous nuclear RNA, messenger RNA (mRNA), transfer RNA tRNA), ribosomal RNA (rRNA), and small nuclear RNA. Each has a different role in various cellular processes. In addition to these forms, RNA is a key component of certain **viruses**.

One of the primary functions of RNA is to facilitate the **translation** of DNA into a useful **protein**. This process begins in the **nucleus** of the cell with a series of enzymatic reactions that copy DNA, producing heterogeneous nuclear RNA. Since hnRNA is a direct copy of DNA, it contains **exons** and **introns** which are coding and noncoding regions of **nucleotides**. It undergoes **RNA splicing** which removes the introns and converts

it to mRNA. mRNA is transported out of the nucleus into the **cytoplasm** of the cell. In this way, it functions as a carrier for information from the cells DNA to the protein synthesizing organelles; the **ribosomes**.

The ribosomes interact with the mRNA and construct a protein based on the nucleotide sequence. Part of this process involves another type of RNA that is located in the ribosome called transfer RNA. tRNA is an adapter molecule which functions as a bridge between the codons on the mRNA and the **amino acids** that are used to construct the protein. During protein synthesis, one end of the tRNA interacts with three nucleotides on the mRNA. The other end of the tRNA carries an amino acid. This amino acid is then transferred and chemically bonded to a series of amino acids to produce a protein. Another type of RNA that is involved in protein synthesis is ribosomal RNA. rRNA has two functions in a ribosome. First, it provides the structure and shape for the catalytic areas of the ribosome. Second, it helps speed up, or catalyze, the action of the tRNA.

While DNA and RNA are very similar in their composition, RNA has a much more versatile role. This is because it can not only carrier genetic information, it can also catalyze reactions. RNA molecules that catalyze reactions are known as ribozymes. As previously mentioned, rRNA functions as a ribozyme during protein synthesis. Another form of RNA that acts as a ribozyme is small nuclear RNA (snRNA). During the process of RNA splicing snRNA catalyzes reactions in the spliceosome, a group of biomolecules that splice hnRNA. snRNA also plays a structural role in this process.

In addition to these beneficial cellular roles, RNA can also have a negative impact. Certain viruses contain RNA as their primary **genetic material**. These viruses are made up of a protein coat and a strand of RNA. They inject their RNA into a cell where it either gets translated by cellular **enzyme**s or gets incorporated into the cell's **genome**. When conditions are right, the virus genes are expressed causing the cell to malfunction. Other types of viruses known as viroids are composed entirely of circular RNA.

RNA (RIBONUCLEIC ACID)

Ribonucleic acid (RNA), generally abbreviated RNA, is an organic chemical substance in living cells that plays several essential roles in the transfer of genetic information from one generation to the next. The hereditary information itself is contained in a similar organic substance known as **deoxyribonucleic acid** (DNA). RNA is what enables this genetic information to be copied from the parent's DNA and inherited by the offspring.

Both RNA and DNA are nucleic acids, so called because they are found in cell nuclei. (RNA is also found in other parts of the cell.) Nucleic acids are the storehouse and delivery system of our genetic traits. The actual biological processes that they prescribe are carried out mostly by our proteins—our **enzymes**, **hormones** and muscles. In other words, nucleic acids are the instruction manual for life's protein-built operating equipment. Each nucleotide monomer molecule consists of a sugar part, a phosphate part, and an amine part. The main difference between RNA and DNA is that in RNA the sugar is ribose ($C_5H_{10}O_5$), while in DNA it is deoxyribose ($C_5H_{10}O_4$). The prefix "deoxy-" tells us that one oxygen atom is "missing" from the ribose.

Like other nucleic acids, RNA is built up from the nucleotides in much the same way that proteins are built up from **amino acids**; they even coil up into long spirals, as some protein molecules do. In a long nucleic acid polymer spiral, the backbone consists of alternating sugar and phosphate parts, with the amine parts sticking out like branches from the backbone. As to the amine parts (which are also referred to as **bases**), there are only four that are important in RNA: adenine, cytosine, guanine, and uracil. Scientists symbolize them as A, C, G, and U. DNA contains thymine instead of uracil: T instead of U.

In DNA, all the information about inherited characteristics exists in the form of genes—arrangements of the four amines in a specific order on a DNA molecule—just as the words in a sentence must be arranged in a specific order if they are to convey real information instead of nonsense. These sequences of amine "words" constitute a set of instructions for exactly which proteins must be manufactured in order to create a specific trait—either brown eyes or green eyes in a human, or a muscle cell for a lizard's tail or a brain cell for an elephant. RNA is what translates these instructions into action.

One kind of RNA "writes down" or transcribes the DNA's amine sequence onto its own molecule, like writing crib notes on the back of your hand. Then, a messenger RNA takes these instructions out of the nucleus and delivers them to the ribosomes—the cells' protein factories. Finally, transfer RNA collects the necessary amino acids and transfers them to the ribosomes for assembly into proteins. All of these processes are made possible by specific enzymes—chemicals that speed up vital chemical reactions in living things, making them go millions of times faster than they would otherwise.

In the first step of the **gene**-transmitting process, the DNA's double helix unwinds to produce two separated strands with their amines sticking out from the backbones. These strands of DNA then serve as an exposed pattern for the production of matching strands of RNA. That is, each protruding amine on the DNA strand picks up a partner amine to bond to according to its highly selective preference: cytosine and guanine (C and G) will always bond together, while adenine and thymine (A and T) will always bond together. In this way, a strand of RNA is built up (with U's instead of T's) that is exactly complementary to the amine sequence on the DNA: it has G's where the C's were and vice versa, and it has A's where the T's were and U's where the A's were. Similarly, if you hold a book up to a mirror, the image in the mirror will be complementary to the actual writing on the page. All of the information is still there, but it has been transcribed, or re-written, into a complementary or matching form. This first step of the process is therefore called transcription.

In the next step of the gene-transmitting process, the information in the RNA strand is edited or streamlined, to produce a strand of messenger RNA (mRNA) that is capable of escaping from the nucleus and carrying the essential genetic information to the ribosomes, which are out in the cell's cytoplasm.

In the cytoplasm are several kinds of smaller RNA molecules called transfer RNA (tRNA), which are swimming around in the pool of amino acids and other chemicals that surround the ribosomes. In the pool, each of these tRNA molecules carries around—is attached to—one particular kind of amino acid molecule, waiting to fill an order from the mRNA. The tRNA molecules read the instructions on the mRNA, and wherever the mRNA needs a particular amino acid, the corresponding tRNA molecule drags its attached amino acid into the protein factory. Thus the desired proteins are built up from the proper amino acids.

RNA SPLICING

RNA Splicing is a biological reaction in which **intron**s are removed from a transcribed **RNA** to create mRNA. This process occurs in conjunction with the **transcription** of **DNA** to mRNA.

The theory that RNA splicing occurs was only suggested recently. Before 1977, scientists were not aware that **eukaryote gene**s were dramatically different than **prokaryote**s. However, it was known that eukaryotes had significantly more DNA than prokaryotes. This difference, called the C-value paradox, led to the discovery that eukaryotes had interrupted genes. These are genes containing **exons and introns; nucleotide** sequences that are both coding and non-coding. Evidence for RNA splicing was obtained when nuclear RNA was compared to mRNA. It was found that nuclear RNA was much longer than mRNA suggesting mRNA was further processed before being transported to the **cytoplasm.**

RNA splicing is one step in the overall process in which the **genetic code** is transcribed into mRNA and then translated into **proteins**. It is known to occur in the **nucleus** of the **cell** where **DNA transcription** takes place. During the process of transcription, each nucleotide from the specific gene is translated into RNA. This results in an RNA **molecule** that contains sequence of both the exons and the introns. Since the introns do not code for proteins, they are removed by RNA splicing.

There are several types of known splicing systems. One system involves a spliceosome which is an array of proteins that function together. The human spliceosome has been found to contain 44 different components. Another type of system involves excision of introns by the RNA itself. Still another type involves the removal of introns by tRNA.

The spliceosome system has been one of the most thoroughly studied splicing system. It is responsible for removing an intron that is located between two exons. The process begins with a set of **enzyme**s that recognize and bind to the splice sites. These sites are located at the exon-intron boundaries and are made up of four nucleotides, two at each end of the intron. The bond between the left exon and intron is cut by enzymes in the spliceosome. This is a transesterification **chemical reaction**. Next, the intron is folded over on itself to the right, forming a molecular lariat structure. Another set of enzymes cuts the right end of the intron and finally joins the two exons.

The existence of a processing system for changing nuclear RNA to mRNA was surprising to scientists at first. It seems strange that a cell would waste the **energy** required to maintain an amount of **genetic material** that does not directly aid in the production of proteins. To explain this phenomena, researchers have attempted to find a function for introns. Current theories suggested that introns play a regulatory role in **gene expression**. Also, they may help cells produce multiple proteins from a single gene.

ROBBINS, FREDERICK (1916-)
American microbiologist

Frederick Chapman Robbins was born in 1916, in Auburn, Alabama. He was the eldest of three boys born to Dr. William Jacob Robbins and Christine F. (Chapman) Robbins. His father was a noted plant physiologist and was director of the New York Botanical Garden. As a medical student at Harvard University, Robbins roomed with **Thomas Weller** and studied virology under **John F. Enders**, the men with whom he would later share the Nobel Prize. Service in World War II interrupted his residency at Children's Hospital in Boston, but it provided the opportunity to study viruses and bacterial diseases. In 1948, Robbins married Alice Havemeyer Northrop, who had been Weller's assistant in the Enders laboratory.

That year, Robbins went to work in Enders's lab. Concentrating on pediatrics, he and Weller attempted to grow poliomyelitis in embryonic and intestinal **tissue**. Prior to this time, polio had only been shown to grow in neural and **brain** tissue of men or monkeys. Vaccinations from this type of growth were potentially deadly because of something present in this tissue which could not be refined out, so there was no vaccine for polio. Growth of viruses in tissue culture, or in vitro, had historically been difficult because of the threat of bacterial invasion into the cell cultures. By the 1950s, however, **antibiotics** had been developed and introduced into the laboratory, such as **penicillin** and streptomycin, which enabled scientists to begin to grow tissue cultures of viruses without the threat of a bacterial invasion.

Robbins and Weller, in their polio experiments, were taking advantage of the new antibiotics. The human intestine cultures grew, which proved for the first time that polio could grow outside neural tissue. This made the feasibility of a polio vaccine far greater, both because it provided a non-deadly vaccine source and because the supply could be grown more cheaply in vitro than in a live animal. This work was a major breakthrough for scientific research and led to the awarding of the Nobel Prize to Enders, Weller, and Robbins in 1954. Their development provided the technology needed to produce a vaccination for polio, which was done in 1953 by virologist **Jonas Salk**.

Robbins's career then took a turn from laboratory work to the health policy arena. He served as president of the Society for Pediatric Research in 1961 and 1962 and in 1965 became dean of the school of medicine at Case Western Reserve. Robbins also began an intense involvement in national committees on a wide range of topics including human experimentation, Third World health policies, and public food and safety policy. His contribution to science in terms of laboratory research has been memorialized by the receipt of the Nobel Prize. For all his research work, however, it is possible his greater legacy will be in the area of health policy.

ROBERTS, RICHARD J. (1943-)
English biochemist

Richard John Roberts was born in 1943, in Derby, England. His father, John Roberts, was a motor mechanic, while his mother, Edna (Allsop) Roberts, took care of the family and served as Richard's first tutor. After graduating with honors in 1965, Roberts remained at Sheffield University to study for his doctoral degree under David Ollis, his undergraduate professor of organic chemistry. After becoming interested in molecular biology, he moved to Harvard University in 1969, where he spent the next four years deciphering the sequence of nucleotides in a form of **ribonucleic acid** known as tRNA. Using a new method devised by English biochemist **Frederick Sanger** at Cambridge, he was able to sequence the RNA molecule, while teaching other scientists Sanger's technique. His creative work with tRNA led an invitation by genetic pioneer and Nobel laureate, **James Watson**, to join his laboratory in Cold Spring Harbor, Long Island, New York.

In 1972, Roberts moved to Long Island to research ways to sequence **DNA**. American microbiologists **Daniel Nathans** and **Hamilton Smith** had shown that a restriction enzyme, Endonuclease R, could split DNA into specific segments. Roberts thought that such small segments could be used for DNA sequencing and began looking for other new restriction enzymes to expand the repertoire. (Enzymes are complex proteins that catalyze specific biochemical reactions.) In 1977, he developed a series of biological experiments to "map" the location of various genes in adenovirus and found that one end of a messenger ribonucleic acid (mRNA) did not react as expected. With the use of an electron microscope, Roberts and his colleagues observed that genes could be present in several, well-separated DNA segments.

In 1986, Roberts married his second wife, Jean. He moved back to Massachusetts in 1992 to join New England Biolabs, a small, private company involved in making research reagents, particularly restriction enzymes. In 1993, Roberts was awarded the Nobel Prize for his discovery of "split genes." The Nobel Committee stated that, "The discovery of split genes has been of fundamental importance for today's basic research in biology, as well as for more medically oriented research concerning the development of cancer and other diseases."

RODBELL, MARTIN (1925-1998)
American biochemist

Rodbell was born on December 1, 1925 in Baltimore, Maryland. He attended a special Baltimore high school that accepted boys from all over the city and prepared them to enter college as sophomores. He entered Johns Hopkins University in 1943, pursuing his interest in chemistry. Not long after entering the university, Rodbell became bored with classes and felt (being Jewish) compelled to combat Hitler's armies. He spent the balance of World War II serving in the Navy, primarily in the South Pacific.

Rodbell returned to Johns Hopkins and received a B.A. in 1949. That same year he met his future wife, Barbara Lederman, a ballet dancer from Holland who had lost her family in the Auschwitz concentration camp. They married a year later, and Rodbell credits his wife for immersing him the world of the arts. Rodbell and his new wife traveled to Seattle, where Rodbell began his graduate studies in biochemistry at the University of Seattle. He studied the chemistry of **lipids** (the fatty substances in cells), and his thesis was on the biosynthesis of lecithin (fats found in cell membranes) in the rat liver. Unfortunately, his thesis was disproved by another scientist working on the same subject. This experience taught him not to assume that biological chemicals are pure, something that would help him later in his Nobel Prize-winning work.

Rodbell finished his Ph.D. in 1954 and then went to the University of Illinois for his post-doctoral fellowship. His research involved the biosynthesis of chloramphenicol, an **antibiotic**. When his fellowship advisor, Herbert Carter, asked him where he wanted to teach, Rodbell had to answer nowhere.

After having taught a lecture course to freshman, only a few of whom passed his exams, Rodbell decided that teaching was not his calling. He accepted a position at the National Heart Institute in Bethesda, Maryland, and continued his research into fats, identifying important **proteins** that pertained to diseases concerning lipoproteins.

In the 1960s he returned to his original interest in **cell** biology and was awarded a fellowship to work at the University of Brussels, where he learned new lab techniques. He returned to the United States and accepted a postion at the NIH Institute of Arthritis and Metabolic Diseases in the Nutrition and Endocrinology lab. There he developed a simple procedure that would separate and purify fat cells. He was also able to remove the fat from a cell, conserving most of the structure of the cell. He named these cells "ghosts."

In several groundbreaking experiments, Rodbell and his colleagues at the NIH showed that cell communication involves three different working devices: (1) a chemical signal; (2) a "second messenger" like a hormone; and (3) a transducer, something that converts energy from one form to another. Rodbell's major contribution was in discovering that there was a transducer function. He and his colleagues also speculated that guanine nucleotides, components of deoxyribonucleic acid (**DNA**) and ribonucleic acid (**RNA**), were somehow involved in cell communication, something that would later be confirmed by **Alfred Gilman**, the biochemist with whom he would share the Nobel Prize. Gilman searched for the chemicals involved with guanine nucleotides and discovered the G-proteins.

G-proteins are instrumental in the fundamental workings of a cell. They allow us to see and smell by changing light and odors to chemical messages that travel to the brain. Understanding how G-proteins malfunction could lead to a better understanding of serious diseases like cholera or cancer. Scientists have already linked improperly working G-proteins to diseases like alcoholism and diabetes. Pharmaceutical companies are developing drugs that would focus on G-proteins.

Rodbell served as director of the National Institute of Environmental Health Sciences in Chapel Hill, North Carolina, from 1985 until his retirement in 1994. Ironically, only a few months before receiving the Nobel Award, Rodbell opted for early retirement, because there were no funds to support the research he wanted to do. Upon receiving the Nobel Prize, Rodbell was vocal in his criticism of the government because of its unwillingness to provide adequate support for fundamental research. He criticized them for favoring projects that yield obviously tangible and potentially profitable results, like drug treatments. Rodbell's other awards include the NIH Distinguished Service Award in 1973 and the Gairdner Award in 1984.

ROOT SYSTEM

The root system, an **organ** of higher **plants** that is usually found underground, has several functions, including anchoring of a plant in the soil and absorption and transportation of **nutrients** and **water** into the plant. Roots may also store food and function in **asexual reproduction**.

The first root that forms is known as the radicle. It elongates during **germination** of the seed and forms the primary root. In many plants, root systems comprise a branching mass of similar-sized roots leading off into the soil in all directions. These fibrous roots (i.e., the roots of grasses) provide a large surface area for absorption of nutrients and water. Fibrous roots hold soil in place and prevent erosion. Some plants, however, (i.e., beets and carrots), have a primary food storage root, which is called a taproot because it is much larger than the secondary roots and penetrates deeper into the soil. Taproots may be fleshy, like that of a carrot, or woody, like a **tree** root. Some plants with a taproot cannot be transplanted, because if the taproot is broken, the plant may die. Adventitious roots (roots that occur in an unusual position), may arise from nodes, or junctions, on the stem, such as at the base of a corn stem. Adventitious roots may also form high on a stem. These roots, called aerial stems or prop roots, form above ground level and go down into the soil. Adventitious roots serve to support the stem and are found on the banyon and the mangrove. Aerial roots (i.e., roots of epiphytic orchids and mistletoe) are not in contact with the ground at all but absorb water from sources available above ground (i.e., from the air or from other plants). Underground plant parts such as tubers (i.e., potatoes), bulbs (i.e., onions), and corms (i.e., gladioli) are not roots but modified stems that store food.

A root comprises three types of **tissue**: the epidermis, or surface layer; the cortex, or root wall where water and food are stored; and the vascular core, located at the center of the root, which carries food and water into the stem. Root hairs are modified long, fine, tube-like epidermis **cells** that act as the absorptive surface of the root and also anchor the root to soil particles. Once absorbed, water and nutrients pass through the cortex to the vascular core where they are transported upward by xylem and phloem, which are complex tissues in the vascular system of higher plants. In a stem, xylem and phloem are grouped together in vascular bundles; but in a root, there is a central core of xylem with radial bands that extend outward toward the cortex. In between the bands are strands of phloem. In underground roots, the xylem core is solid; aerial roots can have a central zone of pith (spongy tissue, used to store nutrients).

Roots grow downward into the soil under the influence of gravity and the presence of water. However, if more water is available at the surface, downward growth will be inhibited. The tip of the root, which is protected from hard soil grains by a protective cap, penetrates into the soil through **cell division** and by the elongation of cells. In addition to the primary growth in length, a secondary growth occurs, in which xylem is added to the inside of the root, and phloem to the outside. This phloem helps form bark, which covers adult roots.

Roots help plants adapt to yearly cold seasons. In perennial plants, the whole plant survives the winter season (i.e., conifers), though in deciduous perennials, the foliage is shed and the plants use food reserves accumulated during other seasons that are stored in the stems and roots (i.e., flowering trees in the temperate zone). However, in herbaceous perennials (i.e., asparagus and dandelions), only the roots and a small

Martin Rodbell. *(Photograph by Martin Rodbell, AP/Wide World Photos, Inc. Reproduced by permission.)*

piece of underground stem survive the winter, using reserve food stored in the underground parts to survive and to start growth in the spring. In biennial plants (i.e., the carrot), leaves die off during the first winter after they have manufactured extensive food reserves that are stored in bulky roots. In the spring, a new plant develops from the roots and portions of stems that survived the winter. The second year, the plant flowers and forms seeds. By the second winter, the whole plant, including the roots, dies, while only the seeds survive to initiate a new two-year cycle. In annual plants, the plant flowers and produces seeds every year. The whole plant dies in the fall, and its seeds give rise to a new generation the following spring.

Roots of many plants are edible and contain food materials, especially starch. Important root crops include beets, turnips, sweet potatoes, carrots, parsnips, and cassava.

ROSS, RONALD (1857-1932)
Indian English physician and parasitologist

Ronald Ross is best known for his discovery of the method by which malaria is transmitted, research for which he was awarded the 1902 Nobel Prize in physiology or medicine. However,

Ross's true passion was the arts, and he became a doctor only because of his father's insistence. Ross's interest in bacteriology led him to study the causes of malaria, a disease that was widespread in India where he lived. His determination that the affliction was transmitted through a **parasite** common to mosquitos led to more advanced treatments for the condition and more effective means of preventing it. In addition to his Nobel Prize and other honorary awards, Ross was knighted in 1911. Ross was born in Almora, Nepal, on May 13, 1857. He was the first of ten children to be born to General Sir Campbell Claye Grant Ross, a British officer stationed in India, and the former Matilde Charlotte Elderton. General Ross was described by Paul DeKruif in his book *Microbe Hunters* as "a ferocious looking border-fighting English general with belligerent side-whiskers, who was fond of battles but preferred to paint landscapes."

First Passion Is for the Arts

In 1865 at the age of eight, Ross was sent to England for his schooling. When he returned to his family in India, he declared to his father that he wanted to pursue a career in the arts. General Ross's view was that the arts were a legitimate vocation but not a sensible career for a young man. Instead, he insisted that his son plan for a medical career in the Indian Medical Service. Ross returned to England in 1874 and began his medical education at St. Bartholomew's Hospital in London. He did poorly in his classes because he spent most of his time writing novels and reading. His father became so upset with his grades that he threatened to withdraw his son's financial support. In response, Ross took a job as a ship's doctor on Anchor Line ships plying the London-New York City route. DeKruif reports that Ross spent much of his time aboard ship "observing the emotions and frailties of human nature," which gave him more material for his novels and poems.

In 1879 Ross completed his course at St. Bartholomew's and was awarded his medical degree. He returned to India and held a series of posts in Madras, Bangalore, Burma, and the Andaman Islands. He soon became more interested in research than in the day-to-day responsibilities of medical practice and spent long hours working out new algebraic formulas.

Attacks the Problem of Malaria

An important turning point in Ross's life came with his first leave of absence in 1888. He returned to England and became interested in research on tropical diseases, many of which he had seen during his years in India. Ross took a course in bacteriology offered by E. Emanuel Klein and earned a diploma in public health. During this furlough he also met Rosa Bessie Bloxam, whom he married on April 25, 1889, just prior to returning to India. The Rosses later had four children: Charles Claye, Dorothy, Sylvia, and Ronald.

With his new found knowledge of bacteriology, Ross turned his attention to what was then the most serious health problem in India: malaria. In 1880 the French physician Alphonse Laveran had discovered that malaria is caused by a one-celled organism called *Plasmodium.* Two decades of research had produced further data on the organism's character-

istics, its means of reproduction, and its correlation with disease symptoms, but no one had determined how the disease was transmitted from one person to another.

Ross's original research led him to question Laveran's discovery, but for five years he made little progress in his studies. Then, on a second leave of absence in England during 1894, he met Patrick Manson, an English physician particularly interested in malaria. During Ross's year in England, he studied with Manson and became convinced that Laveran's theory was correct and that the causative agent for malaria was transmitted by mosquitoes.

When Ross returned to India in March of 1895, he was prepared to take up an aggressive research program on the mosquito-transmission theory. However, he was frustrated by working conditions in India—especially the lack of support from his superiors and the primitive equipment available to him—but with Manson's constant letters of support and encouragement, he eventually succeeded.

The key discovery came on August 20, 1897, when Ross first observed in the stomach of an *Anopheles* mosquito Anopheles a cyst with black granules of the type described by Laveran. Ross worked out the life cycle of the disease-causing agent, including its reproduction within human blood, its transmission to a mosquito during the feeding process, its incubation within the mosquito, and then its transmission to a second human during a second feeding (a "bite") by the mosquito.

Ross's work, however, was complicated by several factors. For example, in the midst of his research he was transferred to Rajputana, a region in which human malaria did not exist. He spent his time there instead working on the transmission of another form of the disease that affects birds. In addition, Ross was continually distracted by his passion for writing, and he produced a number of poems when he could no longer work on his battle against malaria.

Adding to Ross's frustration was the news he received late in 1898 that an Italian research team led by Battista Grassi had published reports on malaria closely paralleling his own work. Although little doubt exists about the originality of the Italian studies, Ross called Grassi's team "cheats and pirates." The dispute was later described by DeKruif as similar to a spat between "two quarrelsome small boys."

To some extent, the dispute was resolved in 1902 when the Nobel Prize committee awarded Ross the year's prize in physiology or medicine. By that time, Ross had retired from the Indian Medical Service and returned to England as lecturer at the new School of Tropical Medicine in Liverpool. There he worked for the eradication of the conditions (such as poor sanitation) that were responsible for the spread of malaria. In 1917, after eighteen years at Liverpool, Ross was appointed physician of tropical diseases at King's College Hospital in London. In 1926 he became director of a new facility founded in his name, the Ross Institute and Hospital for Tropical Diseases near London. He remained in this post until his death on September 16, 1932. Among the honors granted to Ross were the 1895 Parke Gold Medal, the 1901 Cameron Prize, and the 1909 Royal Medal of the Royal Society. He was knighted in 1911.

ROTHSCHILD, MIRIAM (1908-)
English naturalist

Miriam Rothschild's best-known work has been in the fields of **entomology** and parasitology, and she is considered the world's foremost authority on fleas. Although she has made numerous scientific contributions in such fields as marine biology, chemistry, horticulture, and **zoology**, her scientific background is unorthodox. Rothschild, though widely respected for her work and extensive knowledge of fleas, was never formally educated in these fields. In fact, her scientific endeavors are wholly a result of a natural curiosity about the physical world and the encouraging atmosphere of learning she grew up in.

Miriam Louisa Rothschild was born into the famed Rothschild banking family on August 5, 1908, at Ashton Wold, her parents' estate near Peterborough, England. The oldest of four children of Nathaniel Charles and Rozsika von Wertheimstein Rothschild, her own grandfather was the first Baron Rothschild. Although Nathaniel Charles Rothschild, her father, was a banker by profession, he was a zoologist by avocation; he founded the Society for the Promotion of Nature Preserves, and he studied moths, butterflies, and fleas for years. Rozsika Rothschild, her mother, was Hungarian by birth, and in addition to being astute in business, a champion in women's lawn tennis.

As a child, Rothschild spent six months of every year with her grandparents and uncle Walter at their estate outside London. Although all the Rothschilds expressed an interest in nature, it was Walter Rothschild who most sparked Miriam's interest in science. Walter Rothschild was a prolific collector of natural specimens, and his collection included more than two million butterflies, 300,000 bird skins, 200,000 bird eggs, and numerous other animals. And so, even as a young child of four, Rothschild began her own collection of ladybugs and caterpillars.

Rothschild had no formal education while growing up; her father believed formal education stifled creativity and natural curiosity. She read avidly and was tutored by her governess. When her father committed suicide after several years of chronic illness and depression, she lost interest in natural history, but her enthusiasm eventually returned, and at seventeen years of age, Rothschild enrolled herself in several evening classes at a local polytechnic institute.

A naturalist at the British Natural History Museum recommended her to the University of London in the late 1920s, and Rothschild became a researcher at the University's Biological Station located in Naples, Italy, where she studied marine life. She continued her studies when she went to the Marine Biological Station in Plymouth in 1932. It was at this time that she became interested in the study of **parasites** after finding out that some of the mollusks were infested with flatworms. She worked tirelessly, studying parasites, hosts, and other related marine animals, and collected numerous specimens and cultures. In 1939, however, the Germans bombed the research station during the Second World War, destroying Rothschild's laboratory completely. Rothschild now returned to Ashton Wold, which had been converted to a military hospital and air field during the war. At this time she was actively involved in the resistance movement, and she worked with mathematician Alan Turing on the top-secret British *Enigma* project, trying to crack the German code. She and her family also opened their home to European refugees.

Rothschild continued her scientific pursuits even while helping relocate many refugees after the war. Like her father, she had become interested in fleas. She studied many specimens and worked to catalog her father's collection—her findings were eventually amalgamated into six volumes and took twenty years to compile. She showed through her extensive research how fleas reproduce, how and why they choose their hosts, and the mechanics of how fleas can leap enormous distances. She also showed through research with Nobel laureate Tadeus Reichstein how the monarch caterpillar's diet of milkweed plants protects it (the glycosides in the milkweed are distasteful and possibly harmful to birds and other animals, who bypass monarchs for safer, tastier fare).

Rothschild was married to George Lane, a British soldier who had emigrated from Hungary, in 1943. The couple had four children and adopted two more. They divorced in 1957. In addition to science, Rothschild's other interests include travel, reading, and philanthropy. She has written and contributed to numerous articles about nature, and continues her research at Ashton Wold. Her 1983 book, *Dear Lord Rothschild*, honors her family and in particular her uncle Walter, who eventually became the second Baron Rothschild. Her interest in science has a mechanical side as well; she claims to be the first person to put seat belts in an automobile, in 1940.

ROUS, PEYTON (1879-1970)
American physician and pathologist

Francis Peyton Rous was born in 1879, in Baltimore, Maryland, to Charles Rous, a grain exporter, and Frances Wood, the daughter of a Texas judge. He pursued his biological interests at Johns Hopkins University, receiving a B.A. in 1900 and an M.D. in 1905. After a medical internship at Johns Hopkins, however, he decided to concentrate on research and the natural history of disease. In 1909, Simon Flexner, director of the newly-founded Rockefeller Institute in New York City, asked Rous to take over **cancer** research in his laboratory.

A few months later, a poultry breeder brought a Plymouth Rock chicken with a large breast tumor to the Institute and Rous, after conducting numerous experiments, determined that the tumor was a spindle-cell sarcoma. When he transferred a cell-free filtrate from the tumor into healthy chickens of the same flock, they developed identical tumors. Moreover, after injecting a filtrate from the new tumors into other chickens, a malignancy exactly like the original formed. Further studies revealed that this filterable agent was a **virus**, although Rous carefully avoided this word. Now called the Rous sarcoma virus (RSV) and classed as an **RNA retrovirus**, it remains a prototype of animal tumor viruses and a favorite laboratory model for studying the role of **gene**s in cancer.

Rous's discovery was received with considerable disbelief, both in the United States and in the rest of the world. His viral theory of cancer challenged all assumptions, going back to Hippocrates, that cancer was not infectious but rather a spontaneous, uncontrolled growth of cells and many scientists dismissed his finding as a disease peculiar to chickens. Discouraged by his failed attempts to cultivate viruses from mammal cancers, Rous abandoned work on the sarcoma in 1915. Nearly two decades passed before he returned to cancer research. During that time, Rous conducted breakthrough research on urgent medical problems such as emergency **blood** transfusions and culture-gathering techniques.

In 1933, a colleague's report stimulated Rous to renew his work on cancer. Richard Shope discovered a virus that caused warts on the skin of wild rabbits. Within a year, Rous established that this papilloma had characteristics of a true tumor. His work on mammalian cancer kept his viral theory of cancer alive. However, another twenty years passed before scientists identified viruses that cause human cancers and learned that viruses act by invading genes of normal cells. These findings finally advanced Rous's 1910 discovery to a dominant place in cancer research.

Meanwhile, Rous and his colleagues spent three decades studying the Shope papilloma to understand the role of viruses in causing cancer in mammals. Careful observations, over long periods of time, of the changing shapes, colors, and sizes of cells revealed that normal cells become malignant in progressive steps. Cell changes in tumors were observed as always evolving in a single direction toward malignancy.

The researchers demonstrated how viruses collaborate with carcinogens such as tar, radiation, or chemicals to elicit and enhance tumors. In a report co-authored by W. F. Friedewald, Rous proposed a two-stage mechanism of carcinogenesis, or the causilng of cancer, called initiation and promotion. He further explained that a virus can be induced by carcinogens or it can hasten the growth and transform benign tumors into cancerous ones. For tumors having no apparent trace of virus, Rous cautiously postulated that these "spontaneous" growths might contain a virus that persists in a "masked" or latent state, causing no harm until its cellular environment is disturbed. Rous eventually ceased his research on this project due to the technical complexities involved with pursuing the interaction of viral and environmental factors. He then analyzed different types of cells and their nature in an attempt to understand why tumors go from bad to worse.

In 1915, Rous married Marion de Kay, daughter of a scholarly commentator on the arts, and they had three daughters. Rous was appointed a full member of the Rockefeller Institute in 1920 and member emeritus in 1945. Though officially retired, he remained active at his lab bench until the age of ninety, adding sixty papers to the nearly three hundred he published. In 1966 he was awarded the Nobel Prize in Medicine. He died of abdominal cancer in 1970, in New York City.

Roux, Wilhelm (1850-1924)
German embryologist

Wilhelm Roux, the father of experimental **embryology**, was born in Jena, Germany in 1850 and he died at Halle, Germany

in 1924. His teachers were some of the finest scientists of the time, namely Ernst Haeckel at Jana, Rudolf Virchow at Berlin, and Friedrich von Recklinghausen (1833-1910) at Strasbourg. After receiving his university education, Roux taught at the University of Breslau 1879 until 1889, he became a professor at Innsbruck, Austria in 1889, and a professor and director of the anatomical institute at Halle, a position he held from 1889 until he retired in 1921.

Roux was a tireless worker who studied **embryos** seeking a causal connection between function and form. Rather than simply looking at embryos as his predecessors did, he often manipulated embryos to seek understanding of developmental patterns. Ultimately, he sought to describe development in physical and chemical terms. He focused on frog eggs and embryos and investigated many phenomena. An example: he studied the orientation of the plane of the first cleavage, i.e., he sought to ascertain where the cleavage furrow appears as the zygote begins its first **cell division**. He noted that when a sperm enters an egg, it leaves behind a track known as the "copulation path." Roux noted that the track was often the site of the first cleavage furrow. This appears to be true much of the time but there are exceptions. Roux was the first to describe the gray crescent in the **eggs** of *Rana*. This area of intermediate pigmentation (i.e., not as darkly pigmented as the animal hemisphere nor as light as the vegetal hemisphere, hence "gray") occurs after fertilization and is opposite the site of sperm entry. The gray crescent is the site of blastopore formation in gastrulation. Prior to fertilization, it may be said that the frog egg is radially symmetric; after fertilization and especially obvious with gray crescent formation, the egg is bilaterally symmetrical. The establishment of a left and right side of an embryo is crucial for future development and Roux demonstrated his penetrating insight by seeking understanding of this change in an egg.

However, it was the half embryos for which Roux is best remembered. Roux had marked eggs by pricking with a microneedle prior to his most famous experiment. Superficial egg pricking permits a tiny exovate (a small bleb of cytoplasm at the site of the prick) to form and this was a mode of marking a cell and following it in its development. In his most famous experiment, Roux plunged a hot needle into one of the blastomeres (a blastomere is an early embryonic cell) of a **zygote** at the two cell stage. The hot needle killed the operated cell. The question to be resolved was what developmental pattern would the remaining living cell follow. Roux reported that a half embryo was formed. He reasoned that genetic determinants were parcelled out with cell division and each of the two cells had received only the determinants to form a half embryo. Roux had studied the precision of nuclear division in mitosis. He reasoned that this precise mitotic mechanism functioned to divide the nucleus into qualitatively unequal halves. Mitosis thus segregated genetic determinants to daughter cells which were then unequal. His half embryo experiment supported this notion. Roux was incorrect. Hans Driesch separated sea urchin eggs at early cleavage stages. He got two whole but smaller larvae developing from each blastomere. Roux had not separated his frog blastomeres. It was later believed that the living frog blastomere developed only a half embryo because of the inhibiting effect of the dead blastomere next to it. Driesch attempted to separate frog eggs at the two cell stage but was unable to do so. However, several workers were successful in isolating frog blastomeres with results similar to those of Driesch with the sea urchin—these results are also obtained with mouse embryos at the two cell stage.

While it is true that Roux's most famous experiment led to an incorrect conclusion, it is nevertheless extraordinarily notable. Earlier embryologists looked but did not experiment. A notable example of this was William Harvey who described the development of the chick embryo in detail. However, Roux went to the embryo and sought answers to developmental questions by experimental means. He began the modern era of analytical experimental embryology. He is remembered also because he founded the first scientific journal devoted to experimental embryology, *Archiv fur Entwicklungsmechanik der Organism* which was renamed *Roux Archiv fur Entwicklungsmechanik* after his death, and is published at this time as *Development, Genes and Evolution*.

RUŽIČKA, LEOPOLD (1887-1976)
Croatian Swiss chemist

Leopold Ružička worked in what he referred to as the "borderland" between bio-organic chemistry and biochemistry. His studies of odorous natural products led to his discovery of carbon rings with many more carbon atoms than had been originally thought possible. His research also contributed important information on how living things biosynthesize some steroids and **sex hormones**. For this work he shared the 1939 Nobel Prize in chemistry.

Leopold Stephen Ružička was born on September 13, 1887, to Stjepan and Amalija (Sever) Ružička. Ružička attended elementary and high school in Osijek, Croatia (later part of Yugoslavia), where he received a classical education (Latin and Greek), and was initially determined to enter the Catholic priesthood. As a teenager, he changed his interests to chemistry, and upon graduation began to look for graduate schools in Germany and Switzerland. He eventually settled on the *Technische Hochschule* in Karlsruhe, Germany.

Ružička obtained his doctorate in only four years, under the direction of Hermann Staudinger at Karlsruhe. He then assisted Staudinger in research on the natural products in the *Chrysanthemum* **species**; these chemicals, called pyrethrins, were of particular interest as insecticides. In September of 1912, they both moved to ETH, where Ružička had originally considered studying, when Staudinger replaced Richard Willstätter as professor of organic and inorganic chemistry.

In 1916, Ružička started his own research program, supported financially by a Geneva perfume company. (His position at ETH carried no salary until 1925, two years after he was named a professor.) The University of Utrecht, in the Netherlands, offered him a job as an organic chemistry professor in

1926. After three years there, he went back to Zurich to take on the job of directing ETH. During much of his career he was also supported financially by the Rockefeller Foundation.

Ružička studied various organic compounds early in his career, but in 1921 his most fruitful work began—on the structure and synthesis of several natural compounds important to the fragrance industry. (His collaborations with the Swiss pharmaceutical and perfume industries was to continue throughout his working life.) Before Ružička's discoveries, chemists thought that ring structures containing more than eight carbons would be unstable, because no one had been able to synthesize large rings. Ružička's research on muscone (obtained from the male musk deer) and civetone (from both male and female civet cats), however, indicated rings with as many as seventeen carbons—a huge number. He was able to synthesize some of these very large rings with new procedures developed by his research group.

Another line of research dealt with isoprene. Biochemists are interested in how living things biosynthesize large molecules; they had known for some time that isoprene is one of nature's favorite building blocks. Ružička found many more large biochemicals that were constructed from isoprene units, and he formulated a rule of thumb called the "isoprene rule" for predicting biosynthesis based on this starting material. Ružička also synthesized testosterone and androsterone, the male sex hormones.

Ružička conducted research in an era when instrumentation was primitive by contemporary standards. The elucidation of molecular structure, therefore, depended entirely upon the observation of chemical reactions and the purification of reaction products. In this process an unknown compound would be exposed to various well-characterized reagents; if it reacted to give certain products, the chemist knew that the original molecule contained particular arrangements of atoms. Once these arrangements had been identified, the chemist would attempt to synthesize the original compound, and then compare the original and the synthetic. If they matched, that was taken as good evidence that the perceived structure was at least partly correct. This time-consuming, "wet" chemistry often gave ambiguous results, and polite arguments frequently occurred in scientific literature as the chemistry community debated the structure of a complicated new molecule. Often old rules had to give way when new discoveries were made

Ružička married Anna Housmann in 1912; they were divorced in 1950. In 1951, he married Gertrud Acklin. He was an avid gardener and collector of paintings, so much so that he once said his chemistry had suffered as a result of his hobbies. He established an important collection of Dutch and Flemish Masters of the seventeenth century, as well as an art library on that general period, which he later gave to the Zurich art museum. During World War II, he worked to secure the escape of several Jewish scientists from the Nazis, and founded the Swiss-Yugoslav Relief Society. He was instrumental in providing refuge in Switzerland to the future Nobel Laureate Vladimir Prelog, who succeeded Ružička as the director of ETH when the latter retired in 1957. Ružička died on September 26, 1976.

S

SABIN, ALBERT (1906-1993)
Russian American virologist

Albert Sabin, a noted virologist, developed an oral vaccine for polio that led to the once-dreaded disease's virtual extinction in the Western Hemisphere. Sabin's long and distinguished research career included many major contributions to virology, including work that led to the development of attenuated live-virus vaccines. During World War II, he developed effective vaccines against dengue fever and Japanese B encephalitis. The development of a live polio vaccine, however, was Sabin's crowning achievement.

Although Sabin's polio vaccine was not the first, it eventually proved to be the most effective and became the predominant mode of protection against polio throughout the Western world. In South America, "Sabin Sundays" were held twice a year to eradicate the disease. The race to produce the first effective vaccine against polio was marked by intense and often acrimonious competition between scientists and their supporters; in addition to the primary goal of saving children, fame and fortune were at stake. Sabin, however, allowed his vaccine to be used free of charge by any reputable organizations as long as they met his strict standards in developing the appropriate strains.

Albert Bruce Sabin was born in Bialystok, Russia (now Poland), on August 26, 1906. His parents, Jacob and Tillie Sabin, immigrated to the United States in 1921 to escape the extreme poverty suffered under the czarist regime. They settled in Paterson, New Jersey, and Sabin's father became involved in the silk and textile business. After Albert Sabin graduated from Paterson High School in 1923, one of his uncles offered to finance his college education if Sabin would agree to study dentistry. But during his dental education, Sabin read the *Microbe Hunters* by Paul deKruif and was drawn to the science of virology, as well as to the romantic and heroic vision of conquering epidemic diseases.

After two years in the New York University (NYU) dental school, Sabin switched to medicine and promptly lost his uncle's financial support. He paid for school by working at odd jobs—primarily as a lab technician—and through scholarships. He received his B.S. degree in 1928 and enrolled in NYU's College of Medicine. In medical school Sabin showed early promise as a researcher by developing a rapid and accurate system for typing (identifying) pneumococci, or the pneumonia viruses. After receiving his M.D. degree in 1931, he went on to complete his residency at Bellevue Hospital in New York City, where he gained training in pathology, surgery, and internal medicine. In 1932, during his internship, Sabin isolated the B virus from a colleague who had died after being bitten by a monkey. Within two years, Sabin showed that the B virus's natural habitat is the monkey and that it is related to the human herpes simplex **virus**. In 1934 Sabin completed his internship and then conducted research at the Lister Institute of Preventive Medicine in London.

Begins Polio Research

In 1935 Sabin returned to the United States and accepted a fellowship at the Rockefeller Institute for Medical Research. There, he resumed in earnest his research of poliomyelitis (or polio), a paralytic disease that had reached epidemic proportions in the United States at the time of Sabin's graduation from medical school. By the early 1950s, polio afflicted 13,500 out of every 100 million Americans. In 1950 alone, more than 33,000 people contracted polio. The majority of them were children.

Ironically, polio was once an endemic disease (or one usually confined to a community, group, or region) propagated by poor sanitation. As a result, most children who lived in households without indoor plumbing were exposed early to the virus; the vast majority of them did not develop symptoms and eventually became immune to later exposures. But after the public health movement at the turn of the century began to improve sanitation and more and more families had indoor toilets, children were not exposed at an early age to the virus and thus did not develop a natural immunity. As a result, polio be-

Albert Sabin. *(The Library of Congress. Reproduced by permission.)*

came an epidemic disease and spread quickly through communities to other children without immunity, regardless of race, creed, or social status. What made the disease so terrifying was that it caused partial or full paralysis by lodging in the brain stem and **spinal cord** and attacking the central **nervous system**. Often victims of polio would lose complete control of their muscles and had to be kept in a respirator, or what became known as an iron lung, to help them breathe.

In 1936, Sabin and Peter K. Olitsky used a test tube to grow some polio virus in the central nervous tissue of human embryos. Not a practical approach for developing the huge amounts of virus needed to produce a vaccine, this research nonetheless opened new avenues of investigation for other scientists. However, their discovery did reinforce the mistaken assumption that polio only affected nerve **cells**.

Although primarily interested in polio, Sabin was "never able to be a one-virus virologist," as he told Donald Robinson in an interview for Robinson's book *The Miracle Finders*. Sabin also studied how the immune system battled viruses and conducted basic research on how viruses affect the central nervous system. Other interests included investigations of toxoplasmosis, a usually benign viral disease that sometimes caused death or severe brain and eye damage in prenatal infections. These studies resulted in the development of rapid and sensitive serologic diagnostic tests for the virus.

During World War II Sabin served in the United States Army Medical Corps. He was stationed in the Pacific theater where he began his investigations into insect-borne encephalitis, sandfly fever, and dengue. He successfully developed a vaccine for dengue fever and conducted an intensive vaccination program on Okinawa using a vaccine he had developed at Children's Hospital of Cincinnati that protected more than 65,000 military personnel against Japanese encephalitis. Sabin eventually identified a number of antigenic (or immune response-promoting) types of sandfly fever and dengue viruses that led to the development of several attenuated (avirulent) live-virus vaccines.

After the war, Sabin returned to the University of Cincinnati College of Medicine, where he had previously accepted an appointment in 1937. With his new appointments as professor of research pediatrics and fellow of the Children's Hospital Research Foundation, Sabin plunged back into polio research. He and his colleagues began performing autopsies on everyone who had died from polio within a four-hundred-mile radius of Cincinnati, Ohio. At the same time, Sabin performed autopsies on monkeys. From these observations he found that the polio virus was present in humans in both the intestinal tract and the central nervous system. Sabin disproved the widely held belief that polio entered humans through the nose to the respiratory tract, showing that it first invaded the digestive tract before attacking nerve tissue. Sabin was also among the investigators who identified the three different strains of polio.

Sabin's discovery of polio in the digestive tract indicated that perhaps the polio virus could be grown in a test tube in non-nervous tissue as opposed to costly and difficult-to-work-with nerve tissue. In 1949 John Franklin Enders, Frederick Chapman Robbins, and Thomas Huckle Sweller grew the first polio virus in human and monkey nonnervous tissue cultures, a feat that would earn them a Nobel Prize. With the newfound ability to produce enough virus to conduct large scale research efforts, the race to develop an effective vaccine heated up.

Competes with Salk to Develop Vaccine

At the same time that Sabin began his work to develop a polio vaccine, a young scientist at the University of Pittsburgh—Jonas Salk—entered the race. Both men were enormously ambitious and committed to their own theory about which type of vaccine would work best against polio. While Salk committed his efforts to a killed polio virus, Sabin openly expressed his doubts about the safety of such a vaccine as well as its effectiveness in providing lasting protection. Sabin was convinced that an attenuated live-virus vaccine would provide the safe, long-term protection needed. Such a vaccine is made of living virus which is diluted, or weakened, so that it spurs the immune system to fight off the disease without actually causing the disease itself.

In 1953 Salk seemed to have won the battle when he announced the development of a dead virus vaccine made from cultured polio virus inactivated, or killed, with formaldehyde. While many clamored for immediate mass field trials, Sabin, Enders, and others cautioned against mass inoculation until further efficacy and safety studies were conducted. But Salk had won the entire moral and financial support of the National Foundation for Infantile Paralysis, and in 1954 a massive field trial of the vaccine was held. In 1955, to worldwide fanfare, the vaccine was pronounced effective and safe.

Church and town hall bells rang throughout the country, hailing the new vaccine and Salk. However, on April 26, just fourteen days after the announcement, five children in California contracted polio after taking the Salk vaccine. More cases began to occur, with eleven out of 204 people stricken eventually dying. The United States Public Health Service (PHS) ordered a halt to the vaccinations, and a virulent live virus was found to be in certain batches of the manufactured vaccine. After the installation of better safeguards in manufacturing, the Salk vaccine was again given to the public and greatly reduced the incidence of polio in the United States. But Sabin and Enders had been right about the dangers associated with a dead-virus vaccine; and Sabin continued to work toward a vaccine that he believed would be safe, long lasting, and orally administered without the need for injection like Salk's vaccine.

By orally administering the vaccine, Sabin wanted it to multiply in the intestinal tract. Sabin used Enders's technique to obtain the virus and tested individual virus particles on the central nervous system of monkeys to see whether the virus did any damage. According to various estimates, Sabin's meticulous experiments were performed on anywhere from nine to fifteen thousand monkeys and hundreds of chimpanzees. Eventually he diluted three mutant strains of polio that seemed to stimulate antibody production in chimpanzees. Sabin immediately tested the three strains on himself and his family, as well as research associates and volunteer prisoners from Chillicothe Penitentiary in Ohio.

Results of these tests showed that the viruses produced an immunity to polio with no harmful side effects. But by now the public and much of the scientific community were committed to the Salk vaccine. Two scientists working for Lederle Laboratories had also developed a live-virus vaccine. However, the Lederle vaccine was tested in Northern Ireland in 1956 and proved dangerous, as it sometimes reverted to a virulent state.

Although Sabin could not get backing for a large-scale clinical trial in the United States, he remained undaunted. He was able to convince the Health Ministry in the Soviet Union to try his vaccine in massive trials. At the time, the Soviets were mired in a polio epidemic that was claiming eighteen to twenty thousand victims a year. By this time Sabin was receiving the political backing of the World Health Organization in Geneva, Switzerland, which had previously been using Salk's vaccine to control the outbreak of polio around the world; they now believed that Sabin's approach would one day eradicate the disease. Sabin began giving his vaccine to Russian children in 1957, inoculating millions over the next several years. Not to be outdone by Salk's public relations expertise, Sabin began to travel extensively, promoting his vaccine through newspaper articles, issued statements, and scientific meetings. In 1960 the U.S. Public Health Service, finally convinced of Sabin's approach, approved his vaccine for manufacture in the United

States. Still, the PHS would not order its use and the Salk vaccine remained the vaccine of choice until a pediatrician in Phoenix, Arizona, Richard Johns, organized a Sabin vaccine drive. The vaccine was supplied for free, and many physicians provided their services without a fee on a chosen Sunday. The success of this effort spread, and Sabin's vaccine soon became "the vaccine" to ward off polio.

The battle between Sabin and Salk persisted well into the 1970s, with Salk writing an op-ed piece for the *New York Times* in 1973 denouncing Sabin's vaccine as unsafe and urging people to use his vaccine once more. But, for the most part, Salk was ignored, and by 1993, health organizations began to report that polio was close to extinction in the Western Hemisphere.

Sabin's drive and commitment (some called it stubbornness) to his work served him well during the scientific turmoil and infighting of the 1950s that surrounded the development of a polio vaccine. Described socially as mild-mannered, quiet, and unassuming, Sabin was known by his colleagues to be egotistical and possessive about his own work. Sabin often insisted that his vaccine was totally safe, despite the evidence that in very rare cases it could cause paralytic poliomyelitis. He continued his virology research focusing on the role of viruses in cancer.

Sabin's personal life largely remained behind the scenes. He married his first wife, Sylvia Tregillus, in 1935, and they had two daughters. After Sylvia Sabin died in 1966, Sabin married Jane Warner; the two later divorced. In 1972, he married Heloisa Dunshee De Abranches, a newspaperwoman.

Sabin continued to work vigorously and tirelessly into his seventies, traveling to Brazil in 1980 to help with a new outbreak of polio. He antagonized Brazilian officials, however, by accusing the government bureaucracy of falsifying data concerning the serious threat that polio still presented in that country. He officially retired from the National Institutes of Health in 1986. Despite his retirement, Sabin continued to be outspoken, saying in 1992 that he doubted whether a vaccine against the human immunodeficiency virus, or HIV, was feasible. Sabin died from congestive heart failure at the Georgetown University Medical Center on March 3, 1993. In an obituary in the *Lancet,* Sabin was noted as the "architect" behind the eradication of polio from North and South America. Salk issued a statement praising Sabin's work to vanquish polio.

SABIN, FLORENCE RENA (1871-1953)
American anatomist

Florence Rena Sabin's studies of the central **nervous system** of newborn infants, the origin of the **lymphatic system**, and the immune system's responses to infections—especially by the bacterium that causes tuberculosis—carved an important niche for her in the annals of science. In addition to her research at Johns Hopkins School of Medicine and Rockefeller University, she taught new generations of scientists and thus extended her intellectual reach far beyond her own life. In addition,

Florence Rena Sabin. *(The Library of Congress. Reproduced by permission.)*

Sabin's later work as a public health administrator left a permanent imprint upon the communities in which she served. Some of the firsts achieved by Sabin include becoming the first woman faculty member at Johns Hopkins School of Medicine, as well as its first female full professor, and the first woman to be elected president of the American Association of Anatomists.

Sabin was born on November 9, 1871, in Central City, Colorado, to George Kimball Sabin, a mining engineer and son of a country doctor, and Serena Miner, a teacher. Her early life, like that of many in that era, was spare: the house where she lived with her parents and older sister Mary had no plumbing, no gas and no electricity. When Sabin was four, the family moved to Denver; three years later her mother died.

After attending Wolfe Hall boarding school for a year, the Sabin daughters moved with their father to Lake Forest, Illinois, where they lived with their father's brother, Albert Sabin. There the girls attended a private school for two years and spent their summer vacations at their grandfather Sabin's farm near Saxtons River, Vermont.

Sabin graduated from Vermont Academy boarding school in Saxtons River and joined her older sister at Smith College in Massachusetts, where they lived in a private house near the school. As a college student, Sabin was particularly interested in mathematics and science, and earned a bachelor of science in 1893. During her college years she tutored other

students in mathematics, thus beginning her long career in teaching.

A course in zoology during her junior year at Smith ignited a passion for biology, which she made her specialty. Determined to demonstrate that, despite widespread opinion to the contrary, an educated woman was as competent as an educated man, Sabin proceeded to chose medicine as her career. This decision may have been influenced by events occurring in Baltimore at the time.

The opening of Johns Hopkins Medical School in Baltimore was delayed for lack of funds until a group of prominent local women raised enough money to support the institution. In return for their efforts, they insisted that women be admitted to the school—a radical idea at a time when women who wanted to be physicians generally had to attend women's medical colleges.

Begins Medical Career at Johns Hopkins

In 1893 the Johns Hopkins School of Medicine welcomed its first class of medical students; but Sabin, lacking tuition for four years of medical school, moved to Denver to teach mathematics at Wolfe Hall, her old school. Two years later she became an assistant in the biology department at Smith College, and in the summer of 1896 she worked in the Marine Biological Laboratories at Woods Hole. In October of 1896 she was finally able to begin her first year at Johns Hopkins.

While at Johns Hopkins, Sabin began a long professional relationship with Dr. Franklin P. Mall, the school's professor of anatomy. During the four years she was a student there and the fifteen years she was on his staff, Mall exerted an enormous influence over her intellectual growth and development into a prominent scientist and teacher. Years after Mall's death, Sabin paid tribute to her mentor by writing his biography, *Franklin Paine Mall: The Story of a Mind.*

Sabin thrived under Mall's tutelage, and while still a student she constructed models of the medulla and mid-brain from serial microscopic sections of a newborn baby's nervous system. For many years, several medical schools used reproductions of these models to instruct their students. A year after her graduation from medical school in 1900, Sabin published her first book based on this work, *An Atlas of the Medulla and Midbrain,* which became one of her major contributions to medical literature, according to many of her colleagues.

After medical school, Sabin was accepted as an intern at Johns Hopkins Hospital, a rare occurrence for a woman at that time. Nevertheless, she concluded during her internship that she preferred research and teaching to practicing medicine. However, her teaching ambitions were nearly foiled by the lack of available staff positions for women at Johns Hopkins. Fortunately, with the help of Mall and the women of Baltimore who had raised money to open the school, a fellowship was created in the department of anatomy for her. Thus began a long fruitful period of work in a new field of research, the embryologic development of the human lymphatic system.

Sabin began her studies of the lymphatic system to settle controversy over how it developed. Some researchers believed the vessels that made up the lymphatics formed independently from the vessels of the **circulatory system**, specifically the **veins**. However, a minority of scientists believed that the lymphatic vessels arose from the veins themselves, budding outward as continuous channels. The studies that supported this latter view were done on pig embryos that were already so large (about 3.5 in [9 cm] in length) that many researchers—Sabin included—pointed out that the embryos were already old enough to be considered an adult form, thus the results were inconclusive.

Embryo Research Yields Important Findings

The young Johns Hopkins researcher set out to settle the lymphatic argument by studying pig embryos as small as 23mm in length. Combining the painstaking techniques of injecting the microscopic vessels with dye or ink and reconstructing the three-dimensional system from two-dimensional cross sections, Sabin demonstrated that lymphatics did in fact arise from veins by sprouts of endothelium (the layer of cells lining the vessels). Furthermore, these sprouts connected with each other as they grew outward, so the lymphatic system eventually developed entirely from existing vessels. In addition, she demonstrated that the peripheral ends (those ends furthest away from the center of the body) of the lymphatic vessels were closed and, contrary to the prevailing opinion, were neither open to tissue spaces nor derived from them. Even after her results were confirmed by others they remained controversial. Nevertheless, Sabin firmly defended her work in her book *The Origin and Development of the Lymphatic System.*

Sabin's first papers on the lymphatics won the 1903 prize of the Naples Table Association, an organization that maintained a research position for women at the Zoological Station in Naples, Italy. The prize was awarded to women who produced the best scientific thesis based on independent laboratory research.

Back at Hopkins from her year abroad, she continued her work in anatomy and became an associate professor of anatomy in 1905. Her work on lymphatics led her to studies of the development of blood vessels and blood cells. In 1917 she was appointed professor of histology, the first woman to be awarded full professorship at the medical school. During this period of her life, she enjoyed frequent trips to Europe to conduct research in major German university laboratories.

After returning to the United States from one of her trips abroad, she developed methods of staining living cells, enabling her to differentiate between various cells that had previously been indistinguishable. She also used the newly devised ''hanging drop'' technique to observe living cells in liquid preparations under the microscope. With these techniques she studied the development of **blood** vessels and blood cells in developing organisms—once she stayed up all night to watch the ''birth'' of the bloodstream in a developing chick embryo. Her diligent observation enabled her to witness the formation of blood vessels as well as the formation of stem cells from which all other red and white blood cells arose. During these observations, she also witnessed the heart make its first beat.

Sabin's technical expertise in the laboratory permitted her to distinguish between various blood **cell** types. She was particularly interested in white blood cells called monocytes, which attacked infectious bacteria, such as *Mycobacterium tuberculosis,* the organism that causes tuberculosis. Although this organism was discovered by the German microbiologist **Robert Koch** during the previous century, the disease was still a dreaded health menace in the early twentieth century. The National Tuberculosis Association acknowledged the importance of Sabin's research of the body's immune response to the tuberculosis organism by awarding her a grant to support her work in 1924.

In that same year, she was elected president of the American Association of Anatomists, and the following year Sabin became the first woman elected to membership in the National Academy of Sciences. These honors followed her 1921 speech to American women scientists at Carnegie Hall during a reception for Nobel Prize-winning physicist Marie Curie, an event that signified Sabin's recognized importance in the world of science.

Although her research garnered many honors, Sabin continued to relish her role as a professor at Johns Hopkins. The classes she taught in the department of anatomy enabled her to influence many first-year students—a significant number of whom participated in her research over the years. She also encouraged close teacher-student relationships and frequently hosted gatherings at her home for them.

One of her most cherished causes was the advancement of equal rights for women in education, employment, and society in general. Sabin considered herself equal to her male colleagues and frequently voiced her support for educational opportunities for women in the speeches she made upon receiving awards and honorary degrees. Her civic-mindedness extended to the political arena where she was an active suffragist and contributor to the Maryland *Suffrage News* in the 1920s.

Immune System Research Continues at Rockefeller Institute

Sabin's career at Johns Hopkins drew to a close in 1925, eight years after the death of her close friend and mentor Franklin Mall. She had been passed over for the position of professor of anatomy and head of the department, which was given to one of her former students. Thus, she stepped down from her position as professor of histology and left Baltimore.

In her next position, Sabin continued her study of the role of monocytes in the body's defense against the tubercle bacterium that causes tuberculosis. In the fall of 1925, Sabin assumed a position as full member of the scientific staff at the Rockefeller Institute for Medical Research (now Rockefeller University) in New York City at the invitation of the institute's director, Simon Flexner. At Rockefeller Sabin continued to study the role of monocytes and other white blood cells in the body's immune response to infections. She became a member of the Research Committee of the National Tuberculosis Association and aspired to popularize tuberculosis research throughout Rockefeller, various pharmaceutical companies, and other universities and research institutes. The discoveries that she and her colleagues made concerning the ways in which the **immune system** responded to tuberculosis led her to her final research project: the study of antibody formation.

During her years in New York, Sabin participated in the cultural life of the city, devoting her leisure time to the theater, the symphony, and chamber music concerts she sometimes presented in her home. She enjoyed reading nonfiction and philosophy, in which she found intellectual stimulation that complemented her enthusiasm for research. Indeed, one of her co-workers was quoted in *Biographical Memoirs* as saying that Sabin possessed a "great joy and pleasure which she derived from her work... like a contagion among those around her so that all were stimulated in much the same manner that she was.... She was nearly always the first one at the laboratory, and greeted every one with a *joie de vivre* which started the day pleasantly for all of us."

Meanwhile, she continued to accrue honors. She received fourteen honorary doctorates of science from various universities, as well as a doctor of laws. *Good Housekeeping* magazine announced in 1931 that Sabin had been selected in their nationwide poll as one of the twelve most eminent women in the country. In 1935 she received the M. Carey Thomas prize in science, an award of $5,000 presented at the fiftieth anniversary of Bryn Mawr College. Among her many other awards was the Trudeau Medal of the National Tuberculosis Association (1945), the Lasker Award of the American Public Health Association (1951), and the dedication of the Florence R. Sabin Building for Research in Cellular Biology, at the University of Colorado Medical Center.

Plays Prominent Role in Denver's Public Health

In 1938 Sabin retired from Rockefeller and moved to Denver to live with her older sister, Mary, a retired high school mathematics teacher. She returned to New York at least once a year to fulfill her duties as a member of both the advisory board of the John Simon Guggenheim Memorial Foundation and the advisory committee of United China Relief.

Sabin quickly became active in public health issues in Denver and was appointed to the board of directors of the Children's Hospital in 1942 where she later served as vice president. During this time she became aware of the lack of proper enforcement of Colorado's primitive public health laws and began advocating for improved conditions. Governor John Vivian appointed her to his Post-War Planning Committee in 1945, and she assumed the chair of a subcommittee on public health called the Sabin Committee. In this capacity she fought for improved public health laws and construction of more health care facilities.

Two years later she was appointed manager of the Denver Department of Health and Welfare, donating her salary of $4,000 to the University of Colorado Medical School for Research. She became chair of Denver's newly formed Board of Health and Hospitals in 1951 and served for two years in that position. Her unflagging enthusiasm for public health issues bore significant fruit. A *Rocky Mountain News* reporter stated that "Dr. Sabin... was the force and spirit behind the Tri-County chest X-ray campaign" that contributed to cutting the death rate from tuberculosis by 50 percent in Denver in just two years.

But Sabin's enormous reserve of energy flagged under the strain of caring for her ailing sister. While recovering from her own illness, Sabin sat down to watch a World Series game on October 3, 1953, in which her favorite team, the Brooklyn Dodgers, were playing. She died of a heart attack before the game was over.

The state of Colorado gave Sabin a final posthumous honor by installing a bronze statue of her in the National Statuary Hall in the Capitol in Washington, D.C., where each state is permitted to honor two of its most revered citizens. Upon her death, as quoted in *Biographical Memoirs,* the Denver *Post* called her the "First Lady of American Science." Sabin's philosophy of life and work might be best summed up by words attributed to Leonardo da Vinci, with which she chose to represent herself on bookplates: "Thou, O God, dost sell unto us all good things at the price of labour."

SAGAN, CARL (1934-1996)
American astronomer and exobiologist

One of the first scientists to take an active interest in the possibility that life exists elsewhere in the universe, and an astronomer was both a best-selling author and a popular television figure, Carl Sagan was one of the best-known scientists in the world. He made important contributions to studies of Venus and Mars, and he was extensively involved in planning NASA's Mariner missions. Regular appearances on the *Tonight Show* with Johnny Carson began a television career which culminated in the series Sagan hosted on public television called *Cosmos,* seen in sixty countries by over 400,000,000 people. He is also one of the authors of a paper that predicted drastic global cooling after a nuclear war; the concept of "nuclear winter" affected not only the scientific community but also national and international policy, as well as public opinion about nuclear weapons and the arms race. Although some scientists considered Sagan too speculative and insufficiently committed to detailed scientific inquiry, many recognized his talent for explaining science, and acknowledged the importance of the publicity generates by Sagan's enthusiasm.

Sagan was born in Brooklyn, New York, on November 9, 1934, the son of Samuel Sagan, a Russian emigrant and a cutter in a clothing factory, and Rachel Gruber Sagan. He became fascinated with the stars as a young child, and was an avid reader of science fiction, particularly the novels by Edgar Rice Burroughs about the exploration of Mars. By the age of five he was sure he wanted to be an astronomer, but, as he told Henry S. F. Cooper, Jr., of the *New Yorker,* he sadly assumed it was not a paying job; he expected he would have to work at "some job I was temperamentally unsuited for, like door-to-door salesman." When he found out a few years later that astronomers actually got paid, he was ecstatic: "That was a splendid day," he told Cooper.

Sagan's degrees, all of which he earned at the University of Chicago, include an A.B. in 1954, a B.S. in 1955, an M.S. in physics in 1956, and a doctorate in astronomy and astrophysics in 1960. As a graduate student, Sagan was deeply interested in the possibility of life on other planets, a discipline known as exobiology. Although this interest then considered beyond the realm of responsible scientific investigation, he received important early support from scientists such as Nobel laureates Hermann Joseph Muller and Joshua Lederberg. He also worked with Harold C. Urey, who had won the 1934 Nobel Prize in chemistry and had been Stanley Lloyd Miller's thesis adviser when he conducted his famous experiment on the origin of life. Sagan wrote his doctoral dissertation, ''Physical Studies of the Planets,'' under Gerard Peter Kuiper, one of the few astronomers who was a planetologist at that time. It was during his graduate student days that Sagan met Lynn Margulis, a biologist, who became his wife on 16 June 1957. She and Sagan had two sons; they divorced in 1963.

From graduate school, Sagan moved to the University of California at Berkeley, where he was the Miller residential fellow in astronomy from 1960 to 1962. He then accepted a position at Harvard as an assistant professor from 1962 to 1968. On 6 April he married the painter Linda Salzman; Sagan's second marriage, which ended in a divorce, produced a son. From Harvard, Sagan went to Cornell University, where he was first an associate professor of astronomy at the Center for Radiophysics and Space Research. He was then promoted to professor and associate director at the center, serving in that capacity until 1977 when he became the David Duncan Professor of Astronomy and Space Science.

Suggestions About Mars and Venus Confirmed by Spacecrafts

Sagan's first important contributions to the understanding of Mars and Venus began as insights while he was still a graduate student. Color variations had long been observed on the planet Mars, and some believed these variations indicated the seasonal changes of some form of Martian plant life. Sagan, working at times with James Pollack, postulated that the changing colors were instead caused by Martian dust, shifting through the action of wind storms; this interpretation was confirmed by *Mariner 9* in the early 1970s. Sagan also suggested that the surface of Venus was incredibly hot, since the Venusian atmosphere of carbon dioxide and **water** vapor held in the sun's heat, thus creating a version of the ''greenhouse effect.'' This theory was also confirmed by an exploring spacecraft, the Soviet probe *Venera IV*, which transmitted data about the atmosphere of Venus back to Earth in 1967. Sagan also performed experiments based on the work of Stanley Lloyd Miller, studying the production of organic molecules in an artificial atmosphere meant to simulate that of a primitive Earth or contemporary Jupiter. This work eventually earned him a patent for a technique that used gaseous mixtures to produce **amino acid**s.

Sagan first became involved with spaceflight in 1959, when Lederberg suggested he join a committee on the Space Science Board of the National Academy of Sciences. He became increasingly involved with NASA (National Aeronautics and Space Administration) during the 1960s and participated in many of their most important robotic missions. He developed experiments for the Mariner Venus mission and worked as a designer on the *Mariner 9* and the Viking missions to

Carl Sagan. *(Corbis Images. Reproduced by permission.)*

Mars, as well as on the *Pioneer 10*, the *Pioneer 11*, and the *Voyager* spacecrafts. Both the Pioneer and the Voyager spacecrafts have left our solar system carrying plaques which Sagan designed with Frank Drake as messages to any extraterrestrials that find them; they have pictures of two humans, a man and a woman, as well as various astronomical information. The nude man and woman were drawn by Sagan's second wife, Linda Salzman, and they provoked many letters to Sagan denouncing him for sending ''smut'' into space. During this project Sagan met the writer Ann Druyan, the project's creative director, who eventually became his wife. Sagan and Druyan had two children.

Sagan continued his involvement in space exploration in the 1980s and 1990s. The expertise he developed in biology and genetics while working with Muller, Lederberg, Urey, and others, is unusual for an astronomer, and he extensively researched the possibility that Jupiter's moon, Titan, which has an atmosphere, might also have some form of life. Sagan was also involved in less direct searches for life beyond Earth. He was one of the prime movers behind NASA's establishment of a radio astronomy search program that Sagan calls CETI, for Communication with Extra-Terrestrial Intelligence.

A colleague of Sagan's working on the Viking mission explained to Cooper of the *New Yorker* that this desire to find extraterrestrial life is the focus of all of Sagan's various scientific works. ''Sagan desperately wants to find life someplace, anyplace—on Mars, on Titan, in the solar system or outside it. I don't know why, but if you read his papers or listen to his speeches, even though they are on a wide variety of seemingly unrelated topics, there is always the question 'Is this or that phenomenon related to life?' People say, 'What a varied career he has had,' but everything he has done has had this one underlying purpose.'' When Cooper asked Sagan why this was so, the scientist had a ready answer: ''I think it's because human beings love to be alive, and we have an emotional resonance with something else alive, rather than with a molybdenum atom.''

During the early 1970s Sagan began to make a number of brief appearances on television talk shows and news programs; Johnny Carson invited him on the *Tonight Show* for the first time in 1972, and Sagan soon was almost a regular there, returning to discuss science two or three times a year. However, it was *Cosmos,* which Public Television began broadcasting in 1980, that made him into a media sensation. Sagan narrated the series, which he wrote with Ann Druyan and Steven Soter, and they used special effects to illustrate a wide range of astronomical phenomena such as black holes. In addition to being extremely popular, the series was widely praised both for its showmanship and its content, although some reviewers had reservations about Sagan's speculations as well as his tendency to claim as fact what most scientists considered only hypotheses.

Warns About the Possibility of Nuclear Winter

Sagan was actively involved in politics; as a graduate student, he was arrested in Wisconsin for soliciting funds for the Democratic Party, and he was also involved in protests against the Vietnam War. In December, 1983, he published, with Richard Turco, Brian Toon, Thomas Ackerman, and James Pollack, an article discussing the possible consequences of nuclear war. They proposed that even a limited number of nuclear explosions could drastically change the world's climate by starting thousands of intense fires that would throw hundreds of thousands of tons of smoke and ash into the atmosphere, lowering the average temperature ten to twenty degrees and bringing on what they called a ''nuclear winter.'' The authors happened upon this insight accidentally a few years earlier, while they were observing how dust storms on the planet Mars cooled the Martian surface and heated up the atmosphere. Their warning provoked a storm of controversy at first; their article was then followed by a number of studies on the effects of war and other human interventions on the world's climate. Sagan and his colleagues stressed that their predictions were only preliminary and based on certain assumptions about nuclear weapons and large-scale fires, and that their computations had been done on complex computer models of the imperfectly understood atmospheric system. However, despite numerous attempts to minimize the concept of a nuclear winter, the possibility that even a limited nuclear war might well lead to catastrophic environmental changes was confirmed by later research.

The idea of nuclear winter not only led to the reconsideration of the implications of nuclear war by many countries, institutions, and individuals, but it also produced great advances in research on Earth's atmosphere. In 1991, when the oil fields in Kuwait were burning after the Persian Gulf War, Sagan and others made a similar prediction about the effect the smoke from these fires would have on the climate. Based on the nuclear winter hypothesis and the recorded effects of certain volcanic eruptions, these predictions turned out to be inaccurate, although the smoke from the oil fires represented about 1% of the volume of smoke that would be created by a full-scale nuclear war.

In 1994, Sagan with myelodysplasia, a serious bone-marrow disease. Despite his illness, Sagan kept working on his numerous projects. His last book, *The Demon-Haunted World: Science as a Candle in the Dark,* was published in 1995. At the time of his death, Sagan was co-producing a film version of his novel *Contact.* His partner in this project was his wife, Ann Druyan, who had co-authored *Comet.* Released in 1997, the film received popular and critical acclaim as a testimony to Sagan's enthusiasm for the search for extraterrestrial life. Sagan, who lived in Ithaca, NY, died at the Fred Hutchinson Cancer Research Center in Seattle.

Carl Sagan won a Pulitzer Prize in 1978 for his book on evolution called *The Dragons of Eden.* He also won the A. Calvert Smith Prize (1964), NASA's Apollo Achievement Award (1969), NASA's Exceptional Scientific Achievement Medal (1972), NASA's Medal for Distinguished Public Service (twice), the International Astronaut Prize (1973), the John W. Campbell Memorial Award (1974), the Joseph Priestly Award (1975), the Newcomb Cleveland Prize (1977), the Rittenhouse Medal (1980), the Ralph Coats Roe Medal from the American Society of Mechanical Engineers (1981), the Tsiolkovsky Medal of the Soviet Cosmonautics Federation (1987), the Kennan Peace Award from SANE/Freeze (1988), the Oersted Medal of the American Association of Physics Teachers (1990), the UCLA Medal (1991), and the Mazursky Award from the American Astronomical Association (1991). Sagan is a fellow of the American Association for the Advancement of Science, the American Academy of Arts and Sciences, the American Institute for Aeronautics and Astronautics, and the American Geophysical Union. Sagan was also the chairman of the Division for Planetary Sciences of the American Astronomical Society (from 1975 to 1976) and for twelve years was editor-in-chief of *Icarus,* a journal of planetary studies.

SAKMANN, BERT (1942-)
German physician and cell physiologist

Bert Sakmann, along with physicist **Erwin Neher**, was awarded the 1991 Nobel Prize in physiology or medicine for inventing the patch clamp technique. The technique made it possible to realize a goal that had eluded scientists since the 1950s: to be able to examine individual ion channels—pore-forming **proteins** found in the outer **membrane**s of virtually all **cell**s that serve as conduits for electrical signals. Introduced in 1976, the patch clamp technique opened new paths in the study of membrane physiology. Since then, researchers throughout the world have adapted and refined patch clamping, contributing significantly to research on problems in medicine and neuroscience. The Nobel Committee credited Sakmann and Neher with having revolutionized modern biology.

Sakmann was born in Stuttgart, Germany, on June 12, 1942. His later education involved much time around the laboratory. From 1969 to 1970, he was a research assistant in the department of neurophysiology at the Max Planck Institute for Psychiatry in Munich. Between 1971 and 1973, Sakmann studied biophysics with Nobel Laureate Bernard Katz at University College in London as a British Council scholar. In 1974 he received his medical degree from the University ofGöttingen. From that year until 1979 he was a research associate inthe department of neurobiology at the Max Planck Institute for Biophysical Chemistry in Göttingen.

In the 1950s and 1960s, the existence of ion channels that allow for the transmission of electrical charges from one cell to another was inferred from research since no one had been able to actually locate the sites of these channels. **Cell** physiologists were being drawn to thequestion of how electrically charged ions control such biological functions as the transmission of **nerve impulses**, the contraction of muscles, vision, and the process of conception. Sakmann's early interest in ion channels was stimulated by two papers published in 1969 and 1970 that gave strong evidence for the existence of ion channels. As stronger evidence began to accumulate for their existence, it became clear to Sakmann and Neher, who were sharing laboratory space at the Max Planck Institute, that they would have to develop a fine instrument to be able to locate the actual sites of the ion channels on the cell **membrane**.

Bedeviling efforts of researchers to that point was the electrical "noise" generated by the cell's membrane, which made it impossible to detect signals coming from individual channels. Sakmann and Neher set about to reduce the noise by shutting out most of the membrane. They applied a glass micropipette one micron wide and fitted with a recording electrode to a cell membrane and were able to measure the flow of current through a single channel. "It worked the first time," Sakmann recalled in *Science* magazine. The biophysical community was exultant.

Over the next few years, Sakmann and Neher refined their patch clamp technique. The refinements made it possible to measure even very small currents, and established the patch clamp as a tremendously versatile tool in the field of cell biology. Patch clamping has been instrumental in studies of cystic fibrosis, **hormone** regulation, and **insulin** production in diabetes. The technique has also made possible the development of new drugs in the treatment of heart disease, epilepsy, and disorders affecting the nervous and muscle systems. In 1991 Sakmann and Neher won the 1991 Nobel Prize in physiology or medicine for their work on ion channels.

Sakmann has continued to work with other research teams, altering the genes for identified ion channels in order to trace the molecules in the channel responsible for opening and closing the ion pore. Even though Sakmann expressed surprise at receiving the Nobel Prize, given all the other important work going on in cell physiology, the opinion of many of his colleagues was that the award was long overdue. Sakmann is married to Christianne, an ophthalmologist; they have three children.

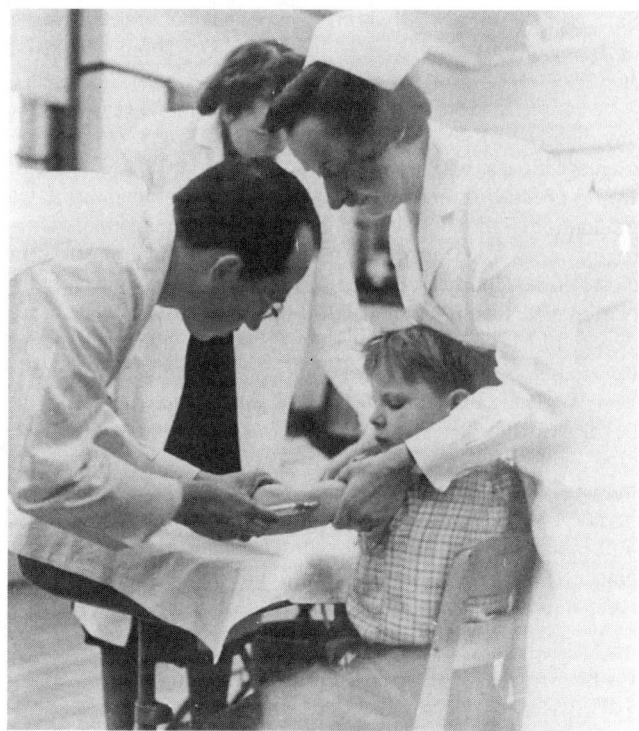

Jonas Salk inoculating a six year old child at school during the 1954 polio vaccine field trials. *(The Library of Congress. Reproduced by permission.)*

SALK, JONAS (1914-1995)
American microbiologist

Jonas Salk was one of the United States's best known microbiologists, chiefly celebrated for his discovery of the polio vaccine. His greatest contribution to immunology was the insight that a "killed **virus**" is capable of serving as an antigen, prompting the body's **immune system** to produce antibodies that will attack invading organisms. This realization enabled Salk to develop a polio vaccine composed of killed polio viruses, producing the necessary antibodies to help the body to ward off the disease without itself inducing polio.

The eldest son of Orthodox Jewish-Polish immigrants, Jonas Edward Salk was born in East Harlem, New York, on 28 October 1914. His father, Daniel B. Salk, was a garment worker, who designed lace collars and cuffs and enjoyed sketching in his spare time. He and his wife, Dora Press, encouraged their son's academic talents, sending him to Townsend Harris High School for the gifted. There, young Salk was both highly motivated and high achieving, graduating at the age of fifteen and proceeding to enroll in the legal faculty of the City College of New York. Ever curious, however, he attended some science courses and quickly decided to switch fields. Salk graduated with a bachelor's degree in science in 1933, at the age of nineteen, and went on to New York University's School of Medicine. Initially he managed on money his parents had borrowed for him; after the first year, however, scholarships and fellowships paid his way. In his senior year,

Salk met the man with whom he would collaborate on some of the most important work of his career, Dr. Thomas Francis, Jr.

On 7 June 1939, Salk was awarded his M.D. The next day, he married Donna Lindsay, a Phi Beta Kappa psychology major who was employed as a social worker. The marriage would produce three sons: Peter, Darrell, and Jonathan. After graduation, Salk continued working with Francis, and concurrently began a two-year internship at Mount Sinai Hospital in New York. Upon completing his internship, Salk accepted a National Research Council fellowship and moved to the University of Michigan to join Dr. Francis, who had been heading up Michigan's department of epidemiology since the previous year. Working on behalf of the U.S. Army, the team strove to develop a flu vaccine. Their goal was a "killed-virus" vaccine—able to kill the live flu viruses in the body, while simultaneously producing antibodies that could fight off future invaders of the same type, thus producing immunity. By 1943, Salk and Francis had developed a formalin-killed-virus vaccine, effective against both type A and B influenza viruses, and were in a position to begin clinical trials.

Gets Backing of National Foundation For Infantile Paralysis

In 1946, Salk was appointed assistant professor of epidemiology at Michigan. Around this time he extended his research to cover not only viruses and the body's reaction to them but also their epidemic effects in populations. The following year he accepted an invitation to move to the University of Pittsburgh School of Medicine's Virus Research Laboratory as an associate research professor of bacteriology. When Salk arrived at the Pittsburgh laboratory, what he encountered was not encouraging. The laboratory had no experience with the kind of basic research he was accustomed to, and it took considerable effort on his part to bring the lab up to par. However, Salk was not shy about seeking financial support for the laboratory from outside benefactors, and soon his laboratory represented the cutting edge of viral research.

In addition to building a respectable laboratory, Salk also devoted a considerable amount of his energies to writing scientific papers on a number of topics, including the polio virus. Some of these came to the attention of Daniel Basil O'Connor, the director of the National Foundation for Infantile Paralysis—an organization that had long been involved with the treatment and rehabilitation of polio victims. O'Connor eyed Salk as a possible recruit for the polio vaccine research his organization sponsored. When the two finally met, O'Connor was much taken by Salk—so much so, in fact, that he put almost all of the National Foundation's money behind Salk's vaccine research efforts.

Poliomyelitis, traceable back to ancient Egypt, causes permanent paralysis in those it strikes, or chronic shortness of breath often leading to **death**. Children, in particular, are especially vulnerable to the polio virus. The University of Pittsburgh was one of four universities engaged in trying to sort and classify the more than one hundred known varieties of polio virus. By 1951, Salk was able to assert with certainty that

all polio viruses fell into one of three types, each having various strains; some of these were highly infectious, others barely so. Once he had established this, Salk was in a position to start work on developing a vaccine.

Salk's first challenge was to obtain enough of the virus to be able to develop a vaccine in doses large enough to have an impact; this was particularly difficult since viruses, unlike culture-grown bacteria, need living cells to grow. The breakthrough came when the team of John F. Enders, Thomas Weller, and Frederick Robbins found that the polio virus could be grown in embryonic tissue—a discovery that earned them a Nobel Prize in 1954.

Salk subsequently grew samples of all three varieties of polio virus in cultures of monkey kidney tissue, then killed the virus with formaldehyde. Salk believed that it was essential to use a killed polio virus (rather than a live virus) in the vaccine, as the live-virus vaccine would have a much higher chance of accidentally inducing polio in inoculated children. He therefore exposed the viruses to formaldehyde for nearly 13 days. Though after only three days he could detect no virulence in the sample, Salk wanted to establish a wide safety margin; after an additional ten days of exposure to the formaldehyde, he reckoned that there was only a one-in-a-trillion chance of there being a live virus particle in a single dose of his vaccine. Salk tested it on monkeys with positive results before proceeding to human clinical trials.

Despite Salk's confidence, many of his colleagues were skeptical, believing that a killed-virus vaccine could not possibly be effective. His dubious standing was further compounded by the fact that he was relatively new to polio vaccine research; some of his chief competitors in the race to develop the vaccine—most notably **Albert Sabin**, the chief proponent for a live-virus vaccine—had been at it for years and were somewhat irked by the presence of this upstart with his unorthodox ideas.

As the field narrowed, the division between the killed-virus and the live-virus camps widened, and what had once been a polite difference of opinion became a serious ideological conflict. Salk and his chief backer, the National Foundation for Infantile Paralysis, were fairly lonely in their corner. But Salk failed to let his position in the scientific wilderness dissuade him and he continued, undeterred, with his research. To test his vaccine's strength, in early 1952 Salk administered a type I vaccine to children who had already been infected with the polio virus. Afterwards, he measured their antibody levels. His results clearly indicated that the vaccine produced large amounts of antibodies. Buoyed by this success, the clinical trial was then extended to include children who had never had polio.

In May 1952, Salk initiated preparations for a massive field trial in which over four hundred thousand children would be vaccinated. The largest medical experiment that had ever been carried out in the United States, the test finally got underway in April, 1954, under the direction of Dr. Francis and sponsored by the National Foundation for Infantile Paralysis. More than one million children between the ages of six and nine took part in the trial, each receiving a button that proclaimed them a "Polio Pioneer." A third of the children were given doses of the vaccine consisting of three injections—one for each of the types of polio virus—plus a booster shot. A control group of the same number of children was given a placebo, and a third group was given nothing.

At the beginning of 1953, while the trial was still at an early stage, Salk's encouraging results were made public in the *Journal of the American Medical Association.* Predictably, media and public interest were intense. Anxious to avoid sensationalized versions of his work, Salk agreed to comment on the results thus far during a scheduled radio and press appearance. However, this appearance did not mesh with accepted scientific protocol for making such announcements, and some of his fellow scientists accused him of being little more than a publicity hound. Salk, who claimed that he had been motivated only by the highest principles, was deeply hurt.

Despite the doomsayers, on 12 April 1955, the vaccine was officially pronounced effective, potent, and safe in almost 90 % of cases. The meeting at which the announcement was made was attended by five hundred of the world's top scientists and doctors, 150 journalists, and sixteen television and movie crews.

Instant Celebrity

The success of the trial catapulted Salk to instant stardom. He was inundated with offers from Hollywood and with pleas from top manufacturers for him to endorse their products. He received a citation from President Eisenhower and addressed the nation from the White House Rose Garden. He was awarded a congressional medal for great achievement in the field of medicine and was nominated for a Nobel Prize but, contrary to popular expectation, did not receive it. He was also turned down for membership in the National Academy of Sciences, most likely a reflection of the discomfort the scientific community still felt about the level of publicity he attracted and of continued disagreement with peers over his methods.

Wishing to escape from the glare of the limelight, Salk turned down the countless offers and tried to retreat into his laboratory. Unfortunately, a tragic mishap served to keep the attention of the world's media focused on him. Just two week after the announcement of the vaccine's discovery, eleven of the children who had received it developed polio; more cases soon followed. Altogether, about 200 children developed paralytic polio, eleven fatally. For a while, it appeared that the vaccination campaign would be railroaded. However, it was soon discovered that all of the rogue vaccines had originated from the same source, Cutter Laboratories in California. On May 7, the vaccination campaign was called to a halt by the Surgeon General. Following a thorough investigation, it was found that Cutter had used faulty batches of virus culture which were resistant to the formaldehyde. After furious debate and the adoption of standards that would prevent such a reoccurrence, the inoculation resumed. By the end of 1955, seven million children had received their shots, and over the course of the next two years more than 200 million doses of Salk's polio vaccine were administered, without a single instance of vaccine-induced paralysis. By the summer of 1961 there had been a 96% reduction in the number of cases of polio in the United States, compared to the five-year period prior to the vaccination campaign.

After the initial inoculation period ended in 1958, Salk's killed-virus vaccine was replaced by a live-virus vaccine developed by Sabin; use of this new vaccine was advantageous because it could be administered orally rather than intravenously, and because it required fewer "booster" inoculations. To this day, though, Salk remains known as the man who defeated polio.

Founds Institute for Biological Studies

In 1954, Salk took up a new position as professor of preventative medicine at Pittsburgh, and in 1957 he became professor of experimental medicine. The following year he began work on a vaccine to immunize against all viral diseases of the central nervous system. As part of this research, Salk performed studies of normal and malignant cells, studies that had some bearing on the problems encountered in **cancer** research. In 1960, he founded the Salk Institute for Biological Studies in La Jolla, California; heavily funded by the National Foundation for Infantile Paralysis (by then known as the March of Dimes), the institute attracted some of the brightest scientists in the world, all drawn by Salk's promise of full-time, uninterrupted biological research.

When his new institute finally opened in 1963, Salk became its director and devoted himself to the study of multiple sclerosis and cancer. He remained a driven man, thinking nothing of working sixteen to eighteen hours a day, six days a week. In 1968, his marriage ended in divorce, and he made the headlines again in 1970 when he remarried, this time to Françoise Gilot, Pablo Picasso's first wife and mother of two of the artist's children. During the 1970s Salk turned to writing, producing books about the philosophy of science and its social role. In 1977, he received the Presidential Medal of Freedom.

Despite the sense of expectancy that he seemed to encourage, Jonas Salk took his successes and failures in stride. In the early 1990s, many people looked to him as the one would might finally develop a vaccine against the HIV virus. But Salk, though continuing to strive toward scientific breakthroughs, seems content simply to work at his chosen craft. "I don't want to go from one crest to another," he once said, as quoted by Sarah K. Bolton in *Famous Men of Science.* "To a scientist, fame is neither an end nor even a means to an end. Do you recall what Emerson said?—' The reward of a thing well done is the opportunity to do more'. "

Salk died on 23 June 1995, at a San Diego area hospital. His death, at the age of 80, was caused by heart failure.

SAMUELSSON, BENGT (1934-)
Swedish biochemist

Bengt Samuelsson shared the 1982 Nobel Prize for physiology or medicine with his compatriot **Sune K. Bergström** and British biochemist **John R. Vane** "for their discoveries concerning **prostaglandins** and related biologically active substances."

Because prostaglandins are involved in a diverserange of biochemical functions and processes, the research of Bergström, Samuelsson, and Vane opened up a new arena of medical research and pharmaceutical applications.

Bengt Ingemar Samuelsson was born on May 21, 1934, in Halmstad, Sweden, to Anders and Kristina Nilsson Samuelsson. Samuelsson entered medical school at the University of Lund, where he came under the mentorship of Sune K. Bergström. Called "the father of prostaglandin chemistry," Bergström was on the university faculty as professor of physiological chemistry. In 1958, Samuelsson followed Bergström to the prestigious Karolinska Institute in Stockholm, which is associated with the Nobel Prize awards. There, Samuelsson received his doctorate in medical science in 1960 and his medical degree in 1961, and he was subsequently appointed as an assistant professor of medical chemistry. In 1961, he served as a research fellow at Harvard University, and then in 1962 he rejoined Bergström at the Karolinska Institute, where he remained until 1966.

At the Karolinska Institute, Samuelsson worked with a group of researchers who were trying to characterize the structures of prostaglandins. Prostaglandins are **hormone**-like substances found throughout the body, which were so named in the 1930s on the erroneous assumption that they originated in the prostate. They play an important role in the circulatory system, and they help protect the body against sickness, infection, pain, and stress. Expanding on their earlier research, Bergström, Samuelsson, and other researchers discovered the role that arachidonic acid, an unsaturated fatty acid found in meats and vegetable oils, plays in the formation of prostaglandins. By developing synthetic methods of producing prostaglandins in the laboratory, this group made prostaglandins accessible for scientific research world wide. It was Samuelsson who discovered the process through which arachidonic acid is converted into compounds he named endoperoxides, which are in turn converted into prostaglandins.

Prostaglandins have many veterinary and livestock breeding applications, and Samuelsson joined the faculty of the Royal Veterinary College in Stockholm in 1967. He returned to the Karolinska Institute as professor of medicine and physiological chemistry in 1972. Samuelsson served as the chair of the department of physiological chemistry from 1973 to 1983, and as dean of the medical faculty from 1978 to 1983, combining administrative duties with a rigorous research schedule. During 1976 and 1977, Samuelsson also served as a visiting professor at Harvard University and the Massachusetts Institute of Technology.

During these years, Samuelsson continued his investigation of prostaglandins and related compounds. In 1973, he discovered the prostaglandins which are involved in the clotting of the **blood**; he called these thromboxanes. Samuelsson subsequently discovered the compounds he called leukotrienes, which are found in white blood cells (or leukocytes). Leukotrienes are involved in asthma and in anaphylaxis, the shock or hypersensitivity that follows exposure to certain foreign substances, such as the toxins in an insect sting. In the wake of such research, prostaglandins have been used to treat fertility problems, circulatory problems, asthma, arthritis, menstrual cramps, and ulcers. Prostaglandins have also been used medically to induce abortions.

The importance of Samuelsson's research has been recognized by numerous awards and honors in addition to the Nobel Prize. Such acknowledgments include the A. Jahres Award in medicine from Oslo University in 1970; the Albert Lasker Medical Research Award in 1977; the Ciba-Geigy Drew Award for biomedical research in 1980; the Gairdner Foundation Award in 1981; and the Abraham White Distinguished Scientist Award in 1991. Samuelsson has published widely on the biochemistry of prostaglandins, thromboxanes, and leukotrienes.

SARGANT, ETHEL · See Cell division

SANGER, FREDERICK (1918-)
English biochemist

Frederick Sanger's important work in biochemistry has been recognized by two Nobel Prizes for chemistry. In 1958 he received the award for determining the arrangement of the **amino acids** that make up **insulin**, becoming the first person to thus identify a **protein molecule**. In 1980 Sanger shared the award with two other scientists, being cited for his work in determining the sequences of **nucleic acids** in **deoxyribonucleic acid (DNA)** molecules. This research has had important implications for genetic research, and taken in conjunction with Sanger's earlier work on the structure of insulin, represent considerable contributions to combatting a number of diseases.

Frederick Sanger was born in Rendcombe, Gloucestershire, England, on August 3, 1918. His father, also named Frederick, was a medical doctor, and his mother, Cicely Crewsdon Sanger, was the daughter of a prosperous cotton manufacturer. Young Frederick attended the Bryanston School in Blandford, Dorset, from 1932 to 1936 and was then accepted at St. John's College, Cambridge. By his own admission, Sanger was not a particularly apt student. Later in life he wrote in *Annual Review of Biochemistry* that "I was not academically brilliant. I never won scholarships and would probably not have been able to attend Cambridge University if my parents had not been fairly rich."

Upon arriving at Cambridge and laying out his schedule of courses, Sanger found that he needed one more half-course in science. In looking through the choices available, Sanger came across a subject of which he had never heard—biochemistry—but that sounded appealing to him. "The idea that biology could be explained in terms of chemistry," he later wrote in *Annual Review of Biochemistry,* "seemed an exciting one." He followed the introductory course with an advanced one and eventually earned a first–class degree in the subject.

Sanger rapidly discovered his strengths and weaknesses in science. Although he was not particularly interested in or skilled at theoretical analysis, he was a superb experimentalist.

He found that, as he later observed in *Annual Review of Biochemistry,* he could "hold my own even with the most academically outstanding" in the laboratory. This observation was to be confirmed in the ingenious experiments that he was to complete in the next four decades of his career.

Graduate Studies Lead to Protein Research

After receiving his bachelor's degree from St. John's in 1939, Sanger decided to continue his work in biochemistry. Though World War II had just begun, Sanger avoided service in the English army because his strong Quaker pacifist beliefs qualified him as a conscientious objector. Instead, he began looking for a biochemistry laboratory where he could serve as an apprentice and begin work on his Ph.D. The first position he found was in the laboratory of **protein** specialist, N. W. Pirie. Pirie assigned Sanger a project involving the extraction of edible protein from grass. That project did not last long as Pirie left Cambridge, and Sanger was reassigned to Albert Neuberger. Neuberger changed Sanger's assignment to the study of lysine, an amino acid. By 1943, Sanger had completed his research and was awarded his Ph.D. for his study on the **metabolism** of lysine.

After receiving his degree, Sanger decided to stay on at Cambridge, where he was offered an opportunity to work in the laboratory of A. C. Chibnall, the new Professor of Biochemistry. Chibnall's special field of interest was the analysis of amino acids in protein, a subject in which Sanger also became involved. The structure of proteins had been a topic of considerable dispute among chemists for many years. On the one hand, some chemists were convinced that proteins consisted of some complex, amorphous material that could never be determined chemically. Conversely, other chemists believed that, while protein molecules might be complex, they did have a structure that could eventually be unraveled and understood.

Probably the most influential theory of protein structure at the time of Sanger's research was that of the German chemist Emil Fischer. In 1902, Fischer had suggested that proteins consist of long chains of amino acids, joined to each other head to tail. Since it was known that each amino acid has two reactive groups, an amino group and a carboxyl group, it made sense that amino acids might join to each other in a continuous chain. The task facing researchers like Sanger was to first determine what amino acids were present in any particular protein, and then to learn in what sequence those amino acids were arranged. The first of these steps was fairly simple and straight-forward, achievable by conventional chemical means. The second was not.

The protein on which Sanger did his research was insulin. The reason for this choice was that insulin—used in the treatment of diabetes—was one of the most readily available of all proteins, and one that could be obtained in very high purity. Sanger's choice of insulin for study was a fortuitous one. As proteins go, insulin has a relatively simple structure. Had he, by chance, started with a more complex protein, his research would almost certainly have stretched far beyond the ten years it required.

Frederick Sanger. *(The Library of Congress. Reproduced by permission.)*

New Techniques Yield Structure of Insulin

In 1945, Sanger made an important technological breakthrough that made possible his later sequencing work on amino acids. He discovered that the compound dinitrophenol (DNP) will bond tightly to one end of an amino acid and that this bond is stronger than the one formed by two amino acids bonding with one another. This fact made it possible for Sanger to use DNP to take apart the insulin molecule one amino acid at a time. Each amino acid could then be identified by the newly discovered process of paper **chromatography**. This was a slow process, requiring Sanger to examine the stains left by the amino acids after they were strained through paper filters, but the technique resulted in the eventual identification of all amino acid groups in the insulin molecule.

Sanger's next objective was to determine the sequence of the amino acids present in insulin, but this work was made more difficult by the fact that the insulin molecule actually consists of two separate chains of amino acids joined to each other at two points by sulfur-sulfur bonds. In addition, a third sulfur-sulfur bond occurs within the shorter of the two strands.

Despite these difficulties, Sanger, in 1955, announced the results of his work: he had determined the total structure of insulin molecule, the first protein to be analyzed in this way. Sanger's work in this area was considered important because it involved proteins—"the most important substances in the human body," as Sanger described them in a *New York Times* report on his work. Proteins are integral elements in both the viruses and toxins that cause diseases and in the antibodies that prevent them. Sanger's research, in laying the groundwork for future work on proteins, greatly increased scientists' ability to combat diseases. For his important work on proteins, Sanger was awarded the Nobel Prize in chemistry in 1958.

Writing in *The Annual Review of Biochemistry,* Sanger referred to the decade after completion of the insulin work as the "lean years' when there were no major successes." Part of this time was taken up with various research projects aimed at learning more about protein structure. In one series of experiments, for example, he explored the use of radioactive isotopes for sequencing. The work was not particularly productive, however, and Sanger soon undertook a new position and a new area of research.

In 1962 he joined the newly established Medical Research Council (MRC) Laboratory of Molecular Biology at Cambridge, a center for research that included such scientists as Max Perutz, **Francis Crick**, and Sydney Brenner. This move marked an important turning point in Sanger's career. The presence of his new colleagues—and Crick, in particular—sparked Sanger's interest in the subject of nucleic acids. Prior to joining the MRC lab, Sanger had had little interest in this subject, but he now became convinced of their importance. His work soon concentrated on the ways in which his protein-sequencing experiences might be used to determine the sequencing of nucleic acids.

The latter task was to be far more difficult than the former, however. While proteins may consist of as few as 50 amino acids, nucleic acids contain hundreds or thousands of basic units, called **nucleotide**s. The first successful sequencing of a nucleic acid, a transfer **RNA** molecule known as alanine, was announced by Robert William Holley in 1965. Sanger had followed Holley's work and decided to try a somewhat different approach. In his method, Sanger broke apart a nucleic acid molecule in smaller parts, sequenced each part, and then determined the way in which the parts were attached to each other. In 1967, Sanger and his colleagues reported on the structure of an RNA molecule known as 5S using this technique.

Work on DNA Earns Second Nobel Prize

When Sanger went on to the even more challenging structures of DNA molecules, he invented yet another new sequencing technique. In this method, a single-stranded DNA molecule is allowed to replicate itself but stopped at various stages of replication. Depending on the chemical used to stop replication, the researcher can then determine the nucleotide present at the end of the molecule. Repeated applications of this process allowed Sanger to reconstruct the sequence of nucleotides present in a DNA molecule.

Successful application of the technique made it possible for Sanger and his colleagues to report on a 12 nucleotide sequence of DNA from **bacteriophage** λ in 1968. Ten years later, a similar approach was used to sequence a 5,386 nucleotide sequence of another form of bacteriophage. In recognition of his sequencing work on nucleic acids, Sanger was awarded his second Nobel Prize in chemistry in 1980, shares of which also went to **Walter Gilbert** and **Paul Berg**. Their work has been lauded for its application to the research of congenital defects and hereditary diseases and has proved vitally important in producing the artificial genes that go into the manufacture of insulin and **interferon**, two substances used to treat diseases.

In 1983, at the age of 65, Sanger retired from research. He was beginning to be concerned, he said in the *Annual Review of Biochemistry,* about "occupying space that could have been available to a younger person." He soon found that he very much enjoyed retirement, which allowed him to do many things for which he had never had time before. Among these were gardening and sailing. He also had more time to spend with his wife, Margaret Joan Howe, whom he had married in 1940, and his three children, Robin, Peter Frederick, and Sally Joan.

During his career, Sanger received many honors in addition to his two Nobel Prizes. In 1954, he was elected to the Royal Society and in 1963 he was made a Commander of the Order of the British Empire. He has been given the Corday-Morgan Medal and Prize of the British Chemical Society, the Alfred Benzons Prize, the Copley Medal of the Royal Society, and the Albert Lasker Basic Medical Research Award, among other honors.

SATURATED AND UNSATURATED FATS •
See Fats and oils

SAUSSURE, NICOLAS THÉODORE DE (1767-1845)
Swiss botanist

Nicolas Théodore de Saussure was a Swiss botanist and a pioneer in the field of phytochemistry, the study of plant chemistry. Nicolas, born in 1767, was the son of Horace Bénédict de Saussure, a Swiss geologist and meteorologist best known for his studies of the Alps. When Nicolas was in his early twenties he accompanied his father on several expeditions, including a 1787 climb of Mount Blanc to record weather observations and a 1789 climb of Mount Rosa to collect data on the weight of air. His findings from this expedition corroborated the observations of the French physicist and plant physiologist Edmé Mariotte (c. 1620-1684). The finding of **René Réaumur** helped Mariotte derive his own version of Boyle's law regarding gas behavior.

In his late twenties, Saussure lived in Geneva and began his work on plant physiology. One of his most important studies involved the effect of specific minerals on plant nutrition. The academic community recognized the value of his work and promised him of a chair of plant physiology at the

Vultures feeding on giraffe. *(JLM Visuals. Reproduced by permission.)*

Geneva Academy. Unfortunately, his plans were disrupted when the French revolution broke out, and instead of staying in Geneva he elected to move to England. When Saussure eventually returned to Geneva in 1802, the position was no longer available and instead he was named an Honorary Professor of Mineralogy and Geology. Although he held this title until 1835. he never actually taught at the Academy. Instead he continued with his studies of plant physiology.

Saussure performed a series of experiments to learn more about the role of chemistry in **plant** physiology. His key experiments included analysis to determine how much carbonic acid was present in plant **tissue**s and how much phosphorus there was in the seeds. He also studied specific biochemical processes in plants, like the way they convert starch into sugars, the way that flowers reach maturation, and the way plants germinate. One of his most important, yet most basic, findings was that plants need **water** as part of the process of **photosynthesis**. Much of his work in this area was published as *Recherches Chimiques sur la Végétation* in 1804. Before his death in 1845 he also contributed to studies on of **ecology** and soil science.

SCAVENGER

A scavenger is an **organism** that consumes dead and decaying plant and **animal matter**. Scavengers do not hunt and kill their food, but rather eat animals and **plants** that died by other causes. One of the most recognizable examples of a scavenger is a vulture. Earthworms are also scavengers, moving through the dirt eating dead plant and animal material found in the soil.

Some scavengers only consume dead plant matter, while others eat only dead animals. However, many consume both dead plant and animal matter and therefore are considered omnivorous. Other organisms will only scavenge for their food some of the time. For example, a crab will eat an already dead organism if it comes upon it. Yet other times the same crab acts as a predator, actively hunting for a live clam or mussel to eat. Crows are also only scavengers when dead meat is available. They will eat dead animals on the side of the road that have been killed by cars, but they will also eat farmers' live crops.

Because scavengers are not able to make their own food, they are considered **heterotroph**s. They must obtain all of their **nutrients** and **energy** from the food they consume.

Scavengers play a very important role in food webs. They are an important link between the **producers** and other consumers and the decomposers. Scavengers are an important first step in the recycling of nutrients back into a form that plants can use.

SCHALLY, ANDREW V. (1926-)

Polish American biochemist

Andrew V. Schally helped conduct pioneering research concerning **hormone**s, identifying three **brain** hormones and greatly advancing scientists' understanding of the function and interaction of the brain with the rest of the body. His findings have proved useful in the treatment of diabetes and peptic ulcers, and in the diagnosis and treatment of hormone-deficiency diseases. Schally shared the 1977 Nobel Prize with French-born American endocrinologist **Roger Guillemin** and **Rosalyn Yalow** (an American scientist whose work in the discovery and development of radioimmunoassay, the use of radioactive substances to find and measure minute substances—especially hormones—in **blood** and **tissue**, helped Schally and Guillemin isolate and analyze **peptide** hormones).

Andrew Victor Schally was born on November 30, 1926, in Wilno, Poland (now Vilnius, Lithuania), to Casimir Peter Schally and Maria Lacka Schally. His father served in the military on the side of the Allies during World War II, and Schally grew up during Nazi occupation of his homeland. The family later left Poland and immigrated to Scotland, where Schally entered the Bridge Allen School in Scotland. He studied chemistry at the University of London and obtained his first research position at London's highly regarded National Institute for Medical Research. Leaving London for Montreal, Canada, in 1952, Schally entered McGill University, where he studied endocrinology and conducted research on the **adrenal** and **pituitary gland**s. He obtained his doctorate in biochemistry from McGill in 1957. Also in 1957, Schally became an assistant professor of physiology at Baylor University School of Medicine in Houston, Texas. There he was able to pursue his interest in the hormones produced by the **hypothalamus.**

Expands on Geoffrey Harris's Discoveries

Scientists had long thought that the hypothalamus, a part of the brain located just above the pituitary gland, regulated the **endocrine system**, which includes the pituitary, **thyroid** and adrenal glands, the **pancreas**, and the ovaries and testicles. They were, however, unsure of the way in which hypothalamic hormonal regulation occurred. In the 1930s British anatomist Geoffrey W. Harris theorized that hypothalamic regulation occurred by means of hormones, chemical substances secreted by glands and transported by the blood. Harris was able to support his hypothesis by conducting experiments that demonstrated altered pituitary function when the blood vessels between the hypothalamus and the pituitary were cut. Harris and others were unable to isolate or identify the hormones from the hypothalamus.

Schally devoted his work to identifying these hormones. He and Roger Guillemin, who also worked at Baylor University's School of Medicine, were engaged in research to unmask the chemical structure of corticotropin-releasing hormone (CRH). Their efforts, however, were unsuccessful—the structure was not determined until 1981. The two then focused their work, independently, on other hormones of the hypothalamus. Schally left Baylor in 1962, when he became director of the

Endocrine and Polypeptide Laboratory at the Veterans Administration (VA) Hospital in New Orleans, Louisiana. Also that year, Schally became a U.S. citizen and took on the post of assistant professor of medicine at Tulane University Medical School.

Schally's first breakthrough came in 1966 when he and his research group isolated TRH, or thyrotropin-releasing hormone. In 1969 Schally and his VA team demonstrated that TRH is a peptide containing three **amino acid**s. It was Guillemin, though, who first determined TRH's chemical structure. The success of this research made it possible to decipher the function of a second hormone, called luteinizing-hormone releasing factor (LHRH). Identified in 1971, LHRH is a decapeptide and controls reproductive functions in both males and females. The chemical makeup of the growth-releasing hormone (GRH) was also discovered by Schally's team in 1971. Schally was able to show that GRH, a peptide consisting of ten amino acids, causes the release of gonadotropins from the pituitary gland. These gonadotropins, in turn, cause male and female sex hormones to be released from the testicles and ovaries. In conjunction with this, Schally was able to identify a factor that inhibits the release of GRH in 1976. Guillemin, however, had determined its structure earlier and named it somatostatin. Subsequent studies by Schally showed that somatostatin serves multiple roles, some of which relate to insulin production and growth disorders. This led to speculation that the hormone could be useful for treating diabetes and acromegaly, a growth-disorder disease.

The hormone research done by Schally and his colleagues was tedious and expensive. Thousands of sheep and pig hypothalami were required to extract the smallest amount of hormone. These organs were solicited from many area slaughterhouses and required immediate dissection to prevent the hormones from degrading. Their accomplishment of isolating the first milligram of pure thyrotropin-releasing hormone, Guillemin stated, cost many times more than the NASA space mission that brought a kilogram of moon rock back to earth.

Schally's intense years of hard work and accomplishment were capped by the Nobel Prize, but he has also received many other awards and honors. In 1974 he was given the Charles Mickle Award of the University of Toronto, and the Gairdner Foundation International Award. He received the Borden Award in the Medical Sciences of the Association of American Medical Colleges in 1975 and, that same year, the Lasker Award and the Laude Award. He has held memberships in the National Academy of Sciences, the American Society of Biological Chemists, the American Physiology Society, the American Association for the Advancement of Science, and the Endocrine Society. In the years prior to receiving the Nobel Prize, Schally and his colleagues published more than 850 papers. Married to Brazilian endocrinologist, Ana Maria de Medeiros-Comaru, Schally often lectures in Latin America and Spain. He and his first wife, Margaret Rachel White, have two children.

SCHICK, BELA (1877-1967)
American pediatrician

Bela Schick, the Hungarian-born American pediatrician, is best known for his child raising theories and his test for diphtheria. The *Schick* test was used to determine if a person was immune to the diphtheria toxin; if immunity was not evident, an antitoxin was given which prevented the disease. Although Schick did not isolate and identify the diphtheria bacterium, his test for diphtheria is considered one of the most important contributions to society in the twentieth century. The test and immunization were developed in 1913. There were several side effects associated with the inoculation but by 1923 a new antitoxin with fewer side effects was developed, which enabled physicians to start the practice of inoculating babies in their first year of life.

Schick was born on July 16, 1877 in Boglar, Austro-Hungarian Empire (now Hungary) to Jacob Schick and Johanna Pichler Schick. He attended the Staats Gymnasium in Graz, and received his medical degree at Karl Franz University. He served with the medical corps in the Austro-Hungarian Army, and afterward opened his own medical practice in Vienna. Schick devoted the rest of his career to teaching, research, and practicing pediatrics. In 1905, Schick published an important paper on the phenomenon of allergy, which was then called serum sickness. He was one of the first to describe allergy and ways to treat allergies. He worked at the University of Vienna as a pediatrician from 1902 to 1923, when he emigrated to the United States.

In 1923, when Schick moved to the United States, he found employment at Mt. Sinai Hospital in New York City as pediatrician-in-chief, a post that he held until 1943, when he retired. Also in 1923, Schick became an American citizen. He married Catherine C. Fries two years later. Schick worked at various other hospitals and institutions, and also enjoyed his popular private practice as well. He kept an impressive collection of toys and dolls in his office, as well as a piano. Schick was known to get down on the floor with the children and play with them, never bringing out his stethoscope until he felt the child was relaxed and calm. He was often quoted as saying, to be a good pediatrician, it helps to be a little childish yourself.

Bela and Catherine Schick never had any children, but he had definite beliefs regarding child-rearing. In 1932, Schick wrote his book, *Child Care Today*, in which he criticized corporal punishment, advocating non-violent discipline. He recognized that early trauma can have a lasting effect in a child's life. In retrospect, many of Schick's theories were ahead of their time. The Schicks were avid travellers, and during one cruise to South American, Bela became ill with pleurisy. He returned to New York, to Mt, Sinai Hospital, and died on December 6, 1967. He is remembered as the leading pediatrician during his time.

SCHICK TEST • See Diphtheria

Hungarian-born American physician Bela Schick receives an award for his development of the diphtheria test. *(Corbis-Bettmann. Reproduced by permission.)*

SCHIZOPHRENIA

Schizophrenia is a mental illness that interferes with normal thought processes, causing delusions, hallucinations, mental disorganization, and physical symptoms. Schizophrenia is perhaps the most severe common mental disorder. It affects men and women equally, and it is believed to afflict about 1% of the **population** of the United States. Its symptoms usually appear in late adolescence or young adulthood, though some young children are also afflicted. Though in rare cases a person may be diagnosed with schizophrenia and then recover with few after-effects, it is most usually a long-term illness with no definite cure. However, the symptoms can be treated successfully with drugs.

Schizophrenia was first described in 1896 by the German psychiatrist Emil Kraepelin. He called it dementia praecox, and divided it into three subtypes: paranoid, hebephrenic, and catatonic. In 1911 a Swiss psychiatrist, Eugen Bleuler, renamed the disease schizophrenia. Kraepelin's three categories are still retained in modern clinical practice. Paranoid schizophrenics typically suffer from delusions of persecution; the hebephrenic type has more disorganized thinking, is less able to communicate, and may have inappropriate emotional responses, such as laughing at a funeral; catatonic schizophrenics suffer from uncontrollable bodily movements. Any one patient may have symptoms from all three categories over the course of the illness.

Up until the 1950s, schizophrenics were considered untreatable, and many lived out their lives locked in mental hospitals. The first drug to effectively treat schizophrenia was reserpine. Reserpine came from an Indian plant that had been used in folk medicine to treat insomnia and insanity. The Swiss pharmaceutical company Ciba isolated the plant's active ingredient, and reserpine was first administered to schizophrenics in 1954. Another drug, chlorpromazine, better known by its brand-name, Thorazine, was developed initially as an anaesthetic. But schizophrenics given Thorazine in the 1950s showed marked improvement. Both these drugs were found to work by blocking dopamine, a neurotransmitter that overreacts in the **brain**s of schizophrenics. With the advent of drug treatment, schizophrenics were able to be released from hospitals, and a lifetime of confinement became a thing of the past. But the first generation of anti-schizophrenia drugs had serious side-effects. More than 10% of Thorazine users developed a syndrome called tardive dyskinesia in response to the drug, consisting of involuntary repetitive mouth and tongue movements and sometimes whole body twitching. A new generation of drugs to treat schizophrenia was first used in the United States in 1990. These drugs, including risperidone, olanzipine and clozapine, have far fewer side-effects than the earlier drugs. Many patients on these medications are able to lead normal lives.

Little is known about what causes schizophrenia, though advanced diagnostic tools gave researchers new insights into the disease in the late 1990s. Scientists at Rutgers University announced in 1997 that they had found a certain gene that appeared to be responsible for schizophrenic's psychotic symptoms. This gave more evidence to the long-held theory that schizophrenia was a genetic disorder. And the perfection of **magnetic resonance imaging (MRI)** gave scientists crucial glimpses into the brains of schizophrenics. Studies in the 1990s showed that schizophrenics had unusually large, fluid-filled spaces known as ventricles in their brains, and these tended to expand as the disease progressed. This evidence suggested that the various symptoms of schizophrenia were due to one underlying brain condition, and that brain changes were ongoing rather than set early in life.

SCHLEIDEN, MATTHIAS JACOB (1804-1881)

German botanist

Matthias Schleiden is credited, along with **Theodor Schwann**, with articulating the **cell theory**. Born in Hamburg, he began his career as a lawyer. He met with no great success in law, and, becoming increasingly depressed, attempted suicide. After recovering from the failed attempt, he returned to school to study medicine, specializing in **botany**.

Schleiden served as a professor first at Jena and later at Dorpat, then resigned and moved frequently from town to

town until he died in 1881. Possibly as a result of his previous career, Schleiden was impulsive, sharp and scornful of his opposition. He rejected the botanist as a glorified scientific librarian, opting for a focus on the **anatomy** and physiology of **plants**. "Most people of the world, even the most enlightened," he said, "are still in the habit of regarding the botanist as a dealer in barbarous Latin names, as a man who gathers flowers, names them, dries them, and wraps them in paper, and all of whose wisdom consists in determining and classifying this hay which he has collected with such great pains."

Schleiden's chief contribution to the cell theory was elaborated in an 1838 essay on the origins of the cell. First he concluded that plants structure was based on cells and that these cells were created in a common fashion. Schleiden argued that the cell developed from the growth of the **nucleus**, which he called the "cytoblast". He believed—and Schwann accepted his position—that the nucleus was spontaneously generated out of the **cytoplasm** or other unformed organic substances. Once the cell was fully formed, Schleiden believed, the nucleus dissolved.

That theory of cell formation was refuted by Robert Remak in 1852, who insisted—as we now understand—that cells are created by the division of other cells.

Despite its flaws, Schleiden's paper was extremely important to the world of biology. Firstly, his conclusion that plants consist entirely of cells or cell products focused attention on the cell as the basic unit of living organisms. Secondly, what Schleiden lacked in rigorous scientific foundation he made up for in forceful argument and ardent conclusions which, while many were later found to be wrong, laid the foundation for Schwann's broader, more comprehensive work on the cell theory. Together their work produced one of the most critical biological developments of their time. Schleiden and Schwann wedded empirical microscopical observations with the more speculative conclusions of natural philosophers to create a unifying theory on the structural similarity of plants and animals.

Matthias Schleiden also published a textbook on botany in 1842, outlining some of his own theories on natural science and criticizing other botanists of the age. Much of the book repeated general theories of the time, including his own work on cells, but it did attempt in its methodology to initiate comparitive investigations into plant evolution.

After his initial foray into cell theory, however, Schleiden did not pursue it to any great degree.

SCHRÖDINGER, ERWIN (1887-1961)

Austrian physicist

Erwin Schrödinger was one of the foremost physicists of our time, responsible for the theory of atomic motion known as wave mechanics or quantum mechanics. He is important to biology because of a slim volume he published in 1944 called *What Is Life?* In this book, he set out to apply the laws of quantum mechanics to the **molecule**s of living **cells**. From his rigorous analysis, he deduced that genetic **mutation**s must take place because of a change in a molecule. He also first described the **chromosome** fiber as a message written in code. Schrödinger's little book laid the groundwork for future breakthroughs in genetics. He was the first physicist to apply his field to biology, and thus was also the precursor of much recent work in biophysics.

Schrödinger was the only child of a well-to-do Viennese family. His father had studied chemistry and retained a lifelong interest in **botany**, though his profession was running a linoleum business. Schrödinger was a brilliant student from his earliest years. He entered the University of Vienna in 1906, and graduated with a degree in physics. He taught at various German-language universities, eventually arriving at the University of Zurich in 1922. In Zurich Schrödinger applied himself to working out his theory of wave mechanics. Albert Einstein and others had inferred that electrons have wave properties as well as particle properties. Schrödinger came up with a mathematical description of the wave properties within an atom. In 1933 he won the Nobel Prize in physics (shared with Paul Dirac) for his work on wave mechanics.

Schrödinger taught at the University of Berlin, and then at Oxford for a time, but returned to Vienna in 1936, apparently homesick for his native land. Hitler's army invaded Austria just two years later. Schrödinger was soon forced to flee for alleged anti-German activities, and he ended up in Dublin at the newly established Institute for Advanced Studies. It was there that Schrödinger gave a series of public lectures in 1943 that became his book *What Is Life?* He took up questions of biology, going back to interests he had picked up from his father and abandoned since his undergraduate days.

Schrödinger begins his book by explaining why **atoms** are so small, or why living **organisms** are so large by comparison. A small number of atoms behaves erratically, always subject to random heat motion. Only when a large number of atoms occur is randomness overcome, and physicists can predict what might happen to the group of atoms in a given situation. Living organisms, which thrive on orderliness, must therefore be made up of large amounts of atoms, in order to overcome the unpredictability of individual atomic behavior. Schrödinger next discussed **genetic theory** as it was known up to that time. He deduced that mutations—abrupt changes in an organism's offspring—must take place at a very small level, involving only a few million atoms. He introduced the idea that the **chromosome** fiber was actually a series of very small repeating units, like the dots and dashes of Morse code. *What Is Life?* begins with mathematics, statistics and genetics, and ends in mysticism. Schrödinger was ultimately unable to explain what life is any more completely than anyone else. But he showed that there might be unknown physical laws that applied to living things. This virtually opened the field of molecular biology. Both **Francis Crick** and **James D. Watson**, discoverers of the **DNA** code, were influenced by Schrödinger's book. Watson, in fact, claimed that reading *What Is Life?* as an undergraduate was what made him decide to study genetics and find out the secret of the **gene**.

Schrödinger returned to Vienna after the war. His later works were philosophical in nature, devoted to questions of consciousness and life after death. He retired from teaching in 1958, and he died in 1961.

See also Genetics; Physics and Biology

SCHWANN, THEODOR ABROSE HUBERT (1810-1882)
German biologist

Born near Düsseldorf, Germany, Schwann attended the University of Bonn, then did post-graduate work under **Johannes Müller** at the University of Würzburg and the University of Berlin. He later took a position as the Chair of Anatomy at the University of Louvain, Belgium, moving later to the University of Liège. He never married. Schwann was described as a gentle man who led a simple life and avoided controversy. Schwann's foremost contribution to biology was the generalized **cell theory** he published in 1839, when he was 29, which showed that animals, like **plants**, are made entirely of cells. Although Matthias Schleiden had published his findings on plant **cell** structure earlier, Schwann's work was more comprehensive. Having discussed Schleiden's work over dinner together in October 1838, Schwann said it sounded similar to what he had been seeing in animal cells. After dinner the two went to Schwann's lab and discovered that the cell structure in a dorsal cord was almost the same as that of plants. Schwann concluded that a cell structure was common to all living organic matter, a conclusion that united animal and vegetable biology as one science.

Schleiden and Schwann's cell theory was incomplete in many ways. It was also incorrect. It was incomplete in that it only focused on the cell structure, but didn't explore in great detail the cell contents, nor the chemical processes—dubbed ''**metabolism**'' by Schwann—carried out inside the cells. It was incorrect primarily in terms of cell development. Schwann wholeheartedly accepted Schleiden's assertion that cells were created by a process of crystalization, starting with a **nucleus** (which Schleiden called a ''cytoblast'') which gradually grew a protective wall around itself. Once cell formation was complete, they believed, the nucleus would dissolve, leaving a fluid interior. That theory, of course, was later disproved and replaced with the understanding we have today of **cell division**. After this work Schwann did not contribute much to the further development of **cytology**. Schwann's other primary contribution to the world of biology was his work on fermentation. As early as 1680, Leeuwenhoek had seen ''crystals'' forming in the process of **fermentation**. Schwann proved that those crystals were in fact living organisms. Most chemists at the time were convinced that alcoholic fermentation was a result of chemical instability. Through a series of controlled experiments, Schwann found that alcoholic fermentation seemed to be dependent on the presence of living organisms (yeasts) in the surrounding air. Other chemists were not quick to accept his conclusions until 1857, when the French chemist **Louis Pasteur** confirmed and expanded on Schwann's earlier work.

Sorted sediment in a gravel pit south of West Bend, Wisconsin. *(JLM Visuals. Reproduced by permission.)*

SEDIMENTATION

Sedimentation occurs when the velocity of flowing **water** is reduced, causing its load of suspended, insoluble materials to drop to the bottom (the faster water is flowing, the greater the amount of suspended material it can carry, and the larger the particle sizes).

In most cases, the insoluble materials suspended in a watercourse (such as a stream or river) are derived from the erosion of soil in the watershed (this is the area of land from which water drains into a waterbody). Erosion may be caused by natural processes, such as slumping of the earthen banks of unstable mountain slopes, particularly during times of the year when water flows are high. Rivers draining glaciers also have naturally high loads of suspended materials. However, in many other cases, erosion is caused by disturbances associated with some human activity, such as road-building across watercourses, deforestation, or improper methods of cultivation in **agriculture**.

In some situations, enormous amounts of sedimented material may accumulate. Deltas, for example, represent the accumulated load of riverine sediment. Some deltaic formations are extremely large, such as those of the Mississippi, Amazon, and Nile Rivers. Deltas are mostly composed of relatively fine-sized particles, such as clay, silt, and sand (depending on the velocity of flow at the site of deposition). Large sedimentary deposits may also accumulate in lakes that receive the flow of rivers; the extensive clay and silt plains of parts of the prairies and Great Lakes basin of North America were laid down in post-glacial lakebeds, which were much more extensive when the continental glaciers were melting back some 10-15,000 years ago.

Sedimentation can be an important environmental problem, particularly where human activities have greatly increased the intensity of this process. Excessive sedimentation can result in rivers having a much shallower basin than occurred historically, causing extensive flooding during high-flow periods during the springtime snow-melt, or after heavy rainfall events. The flooding can damage agricultural land, buildings, and roads, while endangering human lives. Sedimentation can also cause severe ecological damage, for example, by covering up the gravel spawning beds of fish (such as salmon and trout).

SEELEY, THOMAS DYER (1952-)
American biologist

The work of Thomas Dyer Seeley, professor of biology at Cornell University, focuses on understanding the evolutionary transition from **organism** to group as "...the highest level of functionally organized entity."

There are four transitional stages in **evolution**—replicator to **prokaryote**; prokaryote to eukaryote; **eukaryote** to multicellular organism; and multicellular organism to group. In order to understand what makes the latter transition so successful—what creates the strong cooperation between members of a group and how each member of that group interacts to create such a highly successful community—Seeley and his students investigate the behavior and habits of honey bee colonies. Seeley chose honey bee colonies for their incredible ability to adapt and survive in their environment, the highly cooperative interaction of each colony member, and the ease with which the colony and individuals within the colony can be observed.

One of the two questions Seely's research addresses is why there is so little conflict between the bees for reproduction rights: in other words, why only the queen bee produces offspring; why there is no competition between workers—all of which are female—to produce their own offspring; and how this lack of competition helps the queen produce strong, healthy offspring and creates cooperation and harmony between colony members.

The second question Seeley attempts to answer is how the colony is organized for effective food and water gathering, choosing a nesting site, building the nest, and protecting the nest and its inhabitants from predators. Seeley covers these subjects in great detail in his book *The Wisdom of the Hive.*

Seeley was born in Bellefonte, Pennsylvania on June 17, 1952. He earned his B.A. degree in Chemistry from Dartmouth College, graduating *summa cum laude,* then went on to earn his Ph.D. in biology from Harvard in 1978 where he remained until 1980 as a junior fellow. He moved to Yale University's Department of Biology in 1980 as assistant professor, leaving that institution as associate professor in 1986 to become assistant professor at Cornell University in their Section of Neurobiology and Behavior. In 1993-94 he was awarded a fellowship at the Institute for Advanced Study at Wissenschaftskolleg zu Berlin.

Among Seeley's outstanding list of honors and awards is the John S. Guggenheim Memorial Foundation Fellowship

(1992-93); J.I. Hambleton Award (Outstanding Researcher in Bee Biology), Eastern Apicultural Society (1994); and the Gold Medal Award for the best scientific book on bees published in 1996-97, from Apimondia, The International Federation of Beekeeping Organizations.

Seeley is on two editorial boards; is reviewer of books, manuscripts, and grants for dozens of institutions including Oxford University Press, Prentice Hall, Scientific American Press, Proceedings of the National Academy of Sciences, U.S. National Science Foundation, and National Geographic Society; is a prolific author; and has been awarded more than a dozen research grants totalling many thousands of dollars since 1979.

Seeley is married to Robin Hadlock. They have two children.

SEGREGATION

Originally formulated in **Gregor Johann Mendel**'s first **law of segregation**, segregation occurs during **meiosis** when a pair of **allele**s is split to form a single **gamete** (sex cell). Within meiosis, homologous **chromosome** pairs separate from each other, and consequently, a gamete receives only one of each type of chromosome. Occurring on the chromosomes, allele pairs split as the chromosomes split, naturally transmitting only one allele of a pair to a gamete.

When coupled with the observation of homologous chromosomes separating during meiosis, Mendel's law of segregation first indicated that **gene**s were located on chromosomes.

SEIBERT, FLORENCE B. (1897-1991)

American biochemist

A biochemist who received her Ph.D. from Yale University in 1923, Florence B. Seibert is best known for her research in the biochemistry of tuberculosis. She developed the **protein** substance used for the tuberculosis skin test. The substance was adopted as the standard in 1941 by the United States and a year later by the World Health Organization. In addition, in the early 1920s, Seibert discovered that the sudden fevers that sometimes occurred during intravenous injections were caused by bacteria in the distilled **water** that was used to make the protein solutions. She invented a distillation apparatus that prevented contamination. This research had great practical significance later when intravenous **blood** transfusions became widely used in surgery. Seibert authored or coauthored more than a hundred scientific papers. Her later research involved the study of **bacteria** associated with certain **cancer**s. Her many honors include five honorary degrees, induction into the National Women's Hall of Fame in Seneca Falls, New York (1990), the Garvan Gold Medal of the American Chemical Society (1942), and the John Elliot Memorial Award of the American Association of Blood Banks (1962).

Florence Barbara Seibert was born on October 6, 1897, in Easton, Pennsylvania, the second of three children. She was

Florence B. Seibert. *(The Library of Congress. Reproduced by permission.)*

the daughter of George Peter Seibert, a rug manufacturer and merchant, and Barbara (Memmert) Seibert. At the age of three she contracted polio. Despite her resultant handicaps, she completed high school, with the help of her highly supportive parents, and entered Goucher College in Baltimore, where she studied chemistry and zoology. She graduated in 1918, then worked under the direction of one of her chemistry teachers, Jessie E. Minor, at the Chemistry Laboratory of the Hammersley Paper Mill in Garfield, New Jersey. She and her professor, having responded to the call for women to fill positions vacated by men fighting in World War I, coauthored scientific papers on the chemistry of cellulose and wood pulps.

Although Seibert initially wanted to pursue a career in medicine, she was advised against it as it was "too rigorous" in view of her physical disabilities. She decided on biochemistry instead and began graduate studies at Yale University under Lafayette B. Mendel, one of the discoverers of Vitamin A. Her Ph.D. research involved an inquiry into the causes of "**protein** fevers"—fevers that developed in patients after they had been injected with protein solutions that contained distilled water. Seibert's assignment was to discover which proteins caused the fevers and why. What she discovered, however, was that the distilled water itself was contaminated—with bacteria. Consequently, Seibert invented a distilling apparatus that prevented the bacterial contamination.

Seibert earned her Ph.D. in 1923, then moved to Chicago to work as a post-graduate fellow under H. Gideon Wells at the University of Chicago. She continued her research on pyrogenic (fever causing) distilled water, and her work in this area acquired practical significance when intravenous blood transfusions became a standard part of many surgical procedures.

After her fellowship ended, she was employed part-time at the Otho S. A. Sprague Memorial Institute in Chicago, where Wells was the director. At the same time, she worked with Esmond R. Long, whom she had met through Wells's seminars at the University of Chicago. Supported by a grant from the National Tuberculosis Association, Long and Seibert would eventually spend thirty-one years collaborating on tuberculosis research. Another of Seibert's long-time associates was her younger sister, Mabel Seibert, who moved to Chicago to be with her in 1927. For the rest of their lives, with the exception of a year in Sweden, the sisters resided together, with Mabel providing assistance both in the research institutes (where she found employment as secretary and later research assistant) and at home. In 1932, when Long moved to the Henry Phipps Institute—a tuberculosis clinic and research facility associated with the University of Pennsylvania in Philadelphia—Seibert (and her sister) transferred as well. There, Seibert rose from assistant professor (1932–1937), to associate professor (1937–1955) to full professor of biochemistry (1955–1959). In 1959 she retired with emeritus status. Between 1937 and 1938 she was a Guggenheim fellow in the laboratory of **Theodor Svedberg** at the University of Uppsala in Sweden. In 1926 Svedberg had received the Nobel prize for his protein research.

Works on Unknown Aspects of Tuberculosis

Seibert's tuberculosis research involved questions that had emerged from the late-nineteenth-century work of German bacteriologist **Robert Koch**. In 1882 Koch had discovered that the tubercle bacillus was the primary cause of tuberculosis. He also discovered that if the liquid on which the bacilli grew was injected under the skin, a small bite-like reaction would occur in people who had been infected with the disease. (Calling the liquid "old tuberculin," Koch produced it by cooking a culture and draining off the dead bacilli.) Although he had believed the active substance in the liquid was protein, it had not been proven.

Using precipitation and other methods of separation and testing, Seibert discovered that the active ingredient of the liquid was indeed protein. The next task was to isolate it, so that it could be used in pure form as a diagnostic tool for tuberculosis. Because proteins are highly complex organic **molecule**s that are difficult to purify, this was a daunting task. Seibert finally succeeded by means of crystallization. The tiny amounts of crystal that she obtained, however, made them impractical for use in widespread skin tests. Thus, she changed the direction of her research and began working on larger amounts of active, but less pure protein. Her methods included precipitation through ultrafiltration (a method of filtering molecules). The result, after further purification procedures, was a dry powder called TPT (Tuberculin Protein Trichloracetic acid precipitated). This was the first substance that was able to be produced in sufficient quantities for widespread use as a tuberculosis skin test. For her work, Seibert received the 1938 Trudeau Medal from the National Tuberculosis Association.

At the Henry Phipps Institute in Philadelphia, Seibert continued her study of tuberculin protein molecules and their use in the diagnosis of tuberculosis. Seibert began working on the "old tuberculin" that had been created by Koch and used by doctors for skin testing. As Seibert described it in her autobiography *Pebbles on the Hill of a Scientist,* old tuberculin "was really like a soup made by cooking up the live tubercle bacilli and extracting the protein substance from their bodies while they were being killed." Further purification of the substance led to the creation of PPD (Purified Protein Derivative). Soon large quantities of this substance were being made for tuberculosis testing. Seibert continued to study ways of further purifying and understanding the nature of the protein. Her study in Sweden with Svedberg aided this research. There she learned new techniques for the separation and identification of proteins in solution.

Upon her return from Sweden, Seibert brought the new techniques with her. She began work on the creation of a large batch of PPD to serve as the basis for a standard dosage. The creation of such a standard was critical for measuring the degree of sensitivity of individuals to the skin test. Degree of sensitivity constituted significant diagnostic information if it was based upon individual reaction, rather than upon differences in the testing substance itself. A large amount of substance was necessary to develop a standard that ideally would be used world-wide, so that the tuberculosis test would be comparable wherever it was given. Developing new methods of purification as she proceeded, Seibert and her colleagues created 107 grams of material, known as PPD-S (the S signifying "standard"). A portion was used in 1941 as the government standard for purified tuberculins. Eventually it was used as the standard all over the world.

In 1958 the Phipps Institute was moved to a new building at the University of Pennsylvania. In her memoirs, Seibert wrote that she did not believe that the conditions necessary for her continued work would be available. Consequently, she and Mabel, her long-time assistant and companion, retired to St. Petersburg, Florida. Florence Seibert continued her research, however, using for a time a small laboratory in the nearby Mound Park Hospital and another in her own home. In her retirement years she devoted herself to the study of bacteria that were associated with certain types of cancers. Her declining health in her last two years was attributed to complications from childhood polio. She died in St. Petersburg on August 23, 1991.

In 1968 Seibert published her memoirs, which reveal her many friendships, especially among others engaged in scientific research. She particularly enjoyed international travel as well as driving her car, which was especially equipped to compensate for her handicaps. She loved music and played the violin (privately, she was careful to note).

SELECTIVE PRESSURE

Selective pressure is any phenomena which alters the behavior and **fitness** of living **organism**s within a given environment. It is the driving force of **evolution** and **natural selection**, and it can be divided into two types of pressure: biotic or abiotic.

Biotic pressure affecting an organism are living organisms within the same **ecosystem** that interact with the affected organism. Abiotic pressure is created by non-living factors within the organism's environment, such as **light**, wind, and soil. All of these factors interact with the organism to provide opposition to its continued survival. Some characteristics are better suited to survive with a particular set of conditions, and these conditions are the selective pressures that drive the course of evolution for the organism. Some selective pressure is neutral and an example of **stabilizing selection**. Other pressures that influence the development of an organism are examples of **directional selection** and **disruptive selection**.

See also Population

SELFISH DNA

Also known as junk **deoxyribonucleic acid** (**DNA**), selfish DNA are areas of DNA that have no apparent function, but are passed on from generation to generation. In some cases, the sequences of selfish DNA are repeated (repetitive DNA) and use the "host" **organism** as a means for survival (survival machine). This phenomenon is documented in the Selfish DNA theory, which states that the eukaryotic organisms carrying the replicating DNA are nothing more than survival machines that allow the DNA to survive and reproduce. This type of DNA is generally repetitive in its composition and it is typical of such regions as spacer DNA and satellite DNA.

Recently, the possibility that selfish DNA includes functional genes that are coded with viable characteristics has been suggested, and the selfish DNA theory has been expanded to the selfish DNA/**gene theory**. Championed by the British author Richard Dawkins in such books as *The Selfish Gene* and *The Blind Watchmaker*, the expanded theory proposes that the "host" organism is not merely a survival machine, but an organism that is affected by the selfish DNA/genes. The selfish DNA is functional, and therefore it can influence its own and its host's survival.

SEMMELWEIS, IGNAZ PHILLIP (1818-1865)

Hungarian physician

By the early nineteenth century, the field of obstetrics had emerged from the arena of general surgery as an independent area of specialization. Among the contributors to the new field, the Hungarian physician, Ignaz Semmelweis, is generally given the place of honor, because he fought valiantly, to his own detriment, to convince his colleagues that the cause of the high mortality rate among women giving birth was puerperal fever (also known as childbed fever), and because of his insistence on sterilizing the hands of those who assisted with the delivery of babies. The fact that puerperal fever is caused by infection was known to British physicians prior to Semmelweis's work: for example, the Scottish physician Alexander Gordon (1752-1799) is known to have provided empirical proof, refuting numerous popular, but utterly unscientific, theories about the disease. However, most physicians were revolted by the idea that a doctor's hands can carry any contagion, consequently opposing Gordon's conclusions. In America, Oliver Wendell Holmes (1809-1894) wrote an article alerting physicians to the contagiousness of puerperal fever, encountering derision in the medical community. But Semmelweis, whose battle against puerperal fever marked the beginning of an enlightened era in the history of obstetrics, suffered such vicious persecution that his mental health deteriorated. He died in a mental institution.

Semmelweis was born in Budapest, Hungary, in 1818. He studied at the University of Pest and then moved to Vienna, initially to study law, but subsequently switching to medicine. He received his medical degree in 1844, also earning a masters degree in midwifery.

Semmelweis found employment as an assistant to Johann Klein, at the First Obstetrical Clinic at the Vienna General Hospital. At that time, under Klein's direction, the mortality rate for women giving childbirth was approximately 30%. According to the current medical dogma, childbed fever was caused by numerous factors, including diet, the odor of certain flowers, even atmospheric and cosmic influences

Semmelweis observed that one division of the hospital staffed by medical students had three times as many deaths than another division of the hospital that was staffed by midwives. Realizing that the medical students would often assist in the delivery of babies after performing examinations on cadavers, he insisted that they first wash their hands in chlorinated lime. Consequently, the mortality rate fell to just over 1%. He also observed the autopsy of Jakob Kolletschka, Karl von Rokitansky's assistant, who died as a result of a wound which was inflicted during the dissection of a cadaver. Semmelweis noticed that the infected lesions on Kolletschka's body highly resembled those found on women with puerperal fever. By 1847, the year Kolletschka died, Semmelweis had confirmed his findings and adopted a method which called for the sterilization of hands, instruments, and dressings in the delivery room. In addition he demanded that the delivery rooms be cleaned with calcium chloride.

Semmelweis's views on maintaining asepsis were strongly contradicted by the leading obstetricians in Vienna, including Klein. He was eventually forced to resign from his position at the Vienna Krankenhaus. In 1851, Semmelweis returned to Budapest, where he was in charge of the maternity unit at St. Rochus Hospital until 1857. After moving to Budapest, Semmelweis married Marie Weidenhofer, with whom he had five children; only three survived infancy.

In 1861, Semmelweis finally presented his ideas in a book, *The Etiology, Concept, and Prophylaxis of Childbed*

Fever. Despite his convincing documentation, the work was savagely attacked. Even the great **Rudolf Virchow** dismissed Semmelweis's work, ostensibly, as Castiglione remarked, "on purely theoretical grounds." Some critics even condemned the book as obsolete. Depressed and enraged, Semmelweis retaliated with pamphlets and letters, reviling his critics and opponents, and branding those who refused to adhere to his methods as murderers. As time passed, Semmelweis's bitterness and depression deepened, and his wife sent him to a mental asylum, where he died in 1865, ironically, from an infected wound. Coincidentally, that year, **Joseph Lister**, who recognized Semmelweis's work, performed his first operation using his antiseptic method.

See also Antisepsis

SENEBIER, JEAN (1742-1809)
Swiss botanist

Jean Senebier was a Swiss botanist who is credited with being the first scientist to demonstrate the principle of **photosynthesis**. Senebier, the son of merchant Jean-Antoine Senebier, was born in Geneva, Switzerland, on May 6, 1742. Despite a strong interest in natural history, Senebier bowed to his family's wishes and became a minister. At age 23 he published a well-received thesis on polygamy and in 1765 he was ordained as a pastor of the Protestant church in Geneva. Still, young Senebier did not abandon his interest in the sciences and he spent a year in Paris where he became acquainted with a number of scientists and artists. One of his new friends, **Charles Bonnet**, was particularly influential in developing Senebier's interest in plant physiology and encouraged him to perform several experiments. Bonnet's influence also led Senebier to write a paper on the art of observing in response to a question raised by the Netherlands Society of Sciences in 1768. In 1769 Senebier became pastor of a church in Chancy, a small town near Geneva, but in 1773 he resigned this position to follow his interest in the sciences. He took a position as librarian for the Republic of Geneva.

In 1777 Senebier translated the first volume of work by **Lazarro Spallanzani**, the Italian physiologist who is one of the founders of experimental biology. Over the next several years he translated most of the other works by Spallanzani. In 1779, Senebier published his own work, *Action de la lumière sur la végétation*, which described his theories on the process of photosynthesis. This work established his credentials in the scientific community, and Senebier became the focus of a group of young like-minded scientists that included Pierre Huber, A.P. de Candolle, Jean-Antoine Colladon, and **Nicholas Saussure**.

Senebier was the first to discover that **plants** absorb carbonic acid gas and release oxygen. He described how plants exchange gasses in his 1788 paper *Expériences sur l'action de la lumière solaire dans la végétation*. Senebier also worked closely with Françoise Huber to study bees, and together they published a paper on the subject in 1801: *Influence de l'air dans la germination*. In 1802 he published *Essai sur l'art d'observer et de faire des expériences* which expanded his ideas on the art of observing and established much of the scientific method. Senebier died in Geneva on July 22, 1809.

SENSE ORGAN

A sense organ is any collection of **cells** which serve to receive sensory information from the environment either external or internal to a particular **organism**. That sensory information is then sent to the organism's **brain**, where it is processed. Some types of sensory information is consciously perceived by the organism; other types of sensory information may help the organism regulate various bodily functions, but may not actually be consciously perceived.

Sense organs require some type of receptor. The receptor is a collection of nerve cells which receive the information. Other **neuron**s carry that information to the appropriate location for processing. Although we usually think of humans as enjoying five basic senses (**hearing**, vision, **smelling**, tasting, and **touch**), in fact more specific categories of receptors exist.

Mechanoreceptors are those receptors which gather information about touch, pressure, gravity, stretch, or motion. These receptors help the body maintain its position in relation to gravity (for example, most **animals** prefer a head up, feet down position), as well helping the various parts of the body stay in position relative to other parts of the body. Mechanoreceptors include tactile receptors, which can exist on the skin, or at the base of a hair or bristle. When the hair or bristle is bumped, the tactile receptor at its base sends information about this sensation to the brain. Different types of tactile receptors exist to give information about such attributes as pressure, pain, **temperature**.

Some **invertebrates** have **organs** called statocysts which help them maintain their bodies in an appropriate position with regard to gravity. Fish have organs called lateral line organs which helps them sense obstacles in the water around them, which could represent either danger or the presence of prey.

Organs of proprioception are sense organs which provide an organism with continuous information with regard to the position of its muscles and joints. This allows an individual to locate his/her nose with the tip of a finger, even with eyes closed.

The sense of balance is maintained through the interaction of a number of different sense organs, including those of vision and proprioception. Organs within the inner ear also send vital information to the brain to allow an individual to maintain a sense of equilibrium, and therefore to allow that individual to stay balanced when walking or even when sitting.

Auditory receptors collect information from sound wave vibrations, and turn it into electrical impulses which go to the brain to provide information on sound in the environment. Some animals use their sense of sound to locate prey or enemies in their environment. Bats, for example, send high-frequency sound waves out into their environment. These sound wave bounce back off of anything which is in the bat's vicinity. The bat can use this information to very accurately track prey and avoid enemies.

Chemoreceptors are receptors which are stimulated by the presence of chemical substances in the environment. These include receptors which provide information for the senses of olfaction (smelling) and gustation (tasting). **Taste** receptors have only a four categories of taste which they are able to discern: sweet, sour, salty, and bitter. Of course, the mouth is also full of receptors which provide information on the mechanical attributes of the item being tasted, including texture, temperature, etc. The olfactory receptors can sense as many as 50 different types of smells, and can do so based on unbelievably tiny amounts of the material being smelled. A manufactured product which imitates the smell of a violet, for example, can be sensed by an individual at a concentration of only one unit per 30 billion units of air.

Thermoreceptors are those receptors which respond to the presence of heat. Thermoreceptors, for example, help mosquitoes find their next **blood** meal, and help snakes (who have notoriously poor vision) locate its prey and position itself to strike effectively. **Mammals** have thermoreceptors which sense temperature in the external environment, as well as specialized thermoreceptors to sense and regulate their own internal body temperature.

Photoreceptors are receptors which respond to **light** (the units of which are referred to as photons). Photoreceptors can be quite primitive, allowing some animals to sense light, but not distinguish actual objects. Other photoreceptors (as in the human eye) are quite sophisticated, providing information about light, objects, textures, colors, and depth **perception**. Insects have compound eyes, which are covered with multiple facets. These eyes do not sense form particulary well, but they are extremely acute at perceiving flicker (much more than are human eyes). This ability to perceive flicker allows flies, for example to sense the tiniest, beginning motion of a hand reaching for a fly swatter.

SENSORY NEURON

A sensory neuron is a **neuron** which collects or responds to sensory information in the periphery of an **organism**, and sends the information (itself or through associated neurons) toward the central **nervous system** (**spinal cord** and **brain**). Motor neurons (which are responsible for carrying impulses to direct movement) carry **nerve impulses** from the **brain** (the central nervous system) to the muscles (the peripheral nervous system). These neurons are considered to be efferent. Conversely, sensory neurons are considered to be afferent. That is, information in the form of a nervous impulse flows from the peripheral nervous system (eyes, ears, skin, nose, mouth) to the central nervous system.

Sensory neurons may themselves serve as the sensory receptor, or they may be closely associated with the actual receptor.

Sensory neurons are unique in that they are extremely specific. That is, they pick up on a very narrow type of sensation, and are essentially oblivious to any other type of sensation. Therefore, a sensory neuron may be particularly suited to

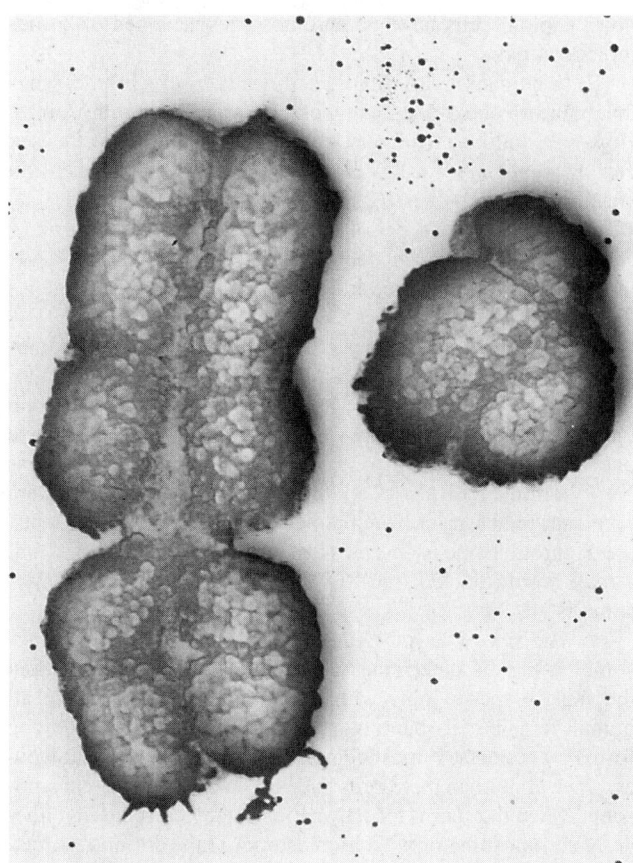

The human sex chromosomes X and Y. *(Photo Researchers, Inc. Reproduced by permission.)*

transmit information about **sight**. More specifically, that neuron may only respond to sight which involves specific wavelengths of **light**; that is, specific colors.

Sensory neurons have particular thresholds which must be met in order for impulses to be transmitted. Sensory information which does not meet the minimum threshold for a neuron's sensitivity will not be transmitted. In order to account for variations in intensity of sensation, sensory neurons may recruit other neurons in the area to transmit impulses simultaneously.

SEX CELLS • See Gamete

SEX CHROMOSOME

Sex chromosomes are the **chromosome**s within a **cell** that determine the sex of an **organism**. In all organism cells, with the exception of the **sperm** and **egg**s and the **gametophyte** generation in **plants**, chromosomes are arranged in pairs called homologous pairs. Each chromosome in the homologous pair comes from a different parent. A cell that has both chromosomes of each homologous pair is a **diploid** cell. In the human

•

body, diploid cells have 46 chromosomes arranged in 23 homologous pairs.

In humans, each diploid cell has two sex chromosomes that make up one of the 23 homologous pairs of chromosomes. However, unlike all of the other homologous pairs, the sex chromosomes are not always the same size and therefore do not always have genes for all of the same traits.

There are two types of sex chromosomes, the X chromosome and the Y chromosome. The X chromosome is larger than the Y chromosome, and therefore has spaces for genes that are not present on the Y chromosome.

In humans and many other **species**, the sex chromosomes of females are identical. Females have two X chromosomes (symbolized as XX). The chromosomes that make up the pair of sex chromosomes in males are different. Males have one X chromosome and one Y chromosome (symbolized as XY). In birds, moths, and butterflies the sex chromosomes are reversed; males are XX and females are XY. In some insects, the Y chromosome is absent. Thus, a female would be XX, but a male would be XO (the "O" indicates that the Y chromosome is absent).

The sex of an individual (based on what sex chromosomes it has) is determined at the time of **fertilization**, when the male's sperm joins with the female's egg. Because all human female cells have two X chromosomes, when the egg forms by **meiosis**, it must contain an X chromosome. However, since all human male cells have both an X and a Y chromosome, when the sperm form by meiosis, they can contain either an X chromosome or a Y chromosome. Thus, during fertilization, the egg always contributes an X chromosome and the sperm can contribute either an X or a Y chromosome. If the sperm has an X chromosome, the resulting offspring will have two X chromosomes and will be a female (XX). If, on the other hand, the sperm contributes a Y chromosome, the **zygote** that forms as a result of fertilization will have one X and one Y chromosome and will be a male (XY). In humans and many other species, it is the male that is responsible for determination of sex of the offspring, depending on which sex chromosome he contributes in his sperm.

SEX DETERMINATION

The sex of an individual is usually determined by the **sex chromosome**s that the **organism** has. Each **diploid cell** in the organism has one pair of sex chromosomes that differs from the other homologous pairs of **chromosome**s. For example, in humans, one of the 23 pairs of homologous chromosomes makes up the sex chromosomes. However, unlike all of the other homologous pairs, the sex chromosomes may not be the same as each other in size. There are two types of sex chromosomes, the X chromosome and the Y chromosome. The X chromosome is larger than the Y chromosome, and therefore has spaces for **gene**s that are not present on the Y chromosome.

In humans and many other **species**, females have two X chromosomes (symbolized as XX). However, the chromosomes that make up the pair of sex chromosomes in males are

different. Males have one X chromosome and one Y chromosome (symbolized as XY). In some species, such as birds, moths and butterflies the sex chromosomes are reversed; males are XX and females are XY. In some insects the Y chromosome is absent altogether. Thus, a female would be XX, but a male would be XO (the "O" indicates that the Y chromosome is absent).

The sex of an individual is based on its sex chromosomes. This is determined at the time of **fertilization**, when the male's **sperm** joins with the female's **egg**. For example, because all human female cells have two X chromosomes, when the egg forms by **meiosis**, it must contain an X chromosome. However, since all human male cells have both an X and a Y chromosome, when the sperm form by meiosis, they can contain either an X chromosome or a Y chromosome. Thus, during fertilization, the egg always contributes an X chromosome and the sperm can contribute either an X or a Y chromosome. If the sperm has an X chromosome, the resulting offspring will have two X chromosomes and will be a female (XX). If, on the other hand, the sperm contributes a Y chromosome, the **zygote** that forms as a result of fertilization will have one X and one Y chromosome and will be a male (XY). In humans and many other species, it is the sex chromosome contributed by the male that is responsible for determining the sex of the offspring. Since the sperm have a 50% chance of receiving an X chromosome and a 50% chance of receiving a Y chromosome, there are equal numbers of X and Y sperm. As a result, there is an almost equal chance of having a male or female offspring.

In some organisms, sex chromosomes are not wholly responsible for sex determination. Environmental conditions can also play a role. For example, in marine turtles and some other reptiles, **temperature** is an important factor. In some marine turtles, when eggs are incubated at higher temperatures, females will be produced. Under lower incubation temperatures, males will be produced. In some coral reef fish, such as wrasses, individuals can actually change sex. These fish live in small groups of females, with one dominant male. When the male dies, one of the females becomes a male. Her color and behavior change, and she begins to produce sperm. In addition, many types of organisms are hermaphroditic, possessing both male and female sex **organ**s. Thus, while sex is usually determined by the sex chromosomes an organism possesses, there are exceptions.

SEX HORMONES

Sex hormones are steroidal **hormone**s that determine both male and female sexual characteristics. The complex interaction of sex hormones is essential in **sexual reproduction**. Sex hormones consist of gonadotropin-releasing hormone (GnRH), follicle-stimulating hormone (FSH), luteinizing hormone (LH), **estrogen**, progesterone, prolactin, **prostaglandins**, androgens, and testosterone. The cyclic release of sex hormones begins in the **hypothalamus** situated near the **pituitary gland**, a pea-sized gland located at the base of the forebrain behind the bridge of the nose.

GnRH is pumped into **blood** vessels according to an extremely regulated pulse-pattern of the hypothalamus. During a woman's sexual cycle, the amount of GnRH released varies. Increased amounts of GnRH triggers the release of FSH and LH (known as gonadotropins) by the pituitary gland. In the male, FSH stimulates the formation of **sperm** in the testes; in the female, FSH stimulates a group of specialized **cells** which form a follicle surrounding each one of the millions of immature **egg**s (ovum) contained inside the ovaries.

When FSH stimulates these cells, the follicle develops, matures, and produces the female hormone estrogen (estradiol). Secondary sources of estrogen are fat cells that convert male hormones (androgens). During puberty, estrogen brings female sex **organ**s (breasts, vagina, uterus, and fallopian tubes) to maturity and stimulates bone growth. After women reach sexual maturity, estrogen produces mucus during ovulation, regulates the release of FSH and LH, lubricates vaginal **tissue**, stimulates **endometrium** growth and lactation, helps maintain bone density, regulates **cholesterol** levels, and effects cognitive and psychological functions. The male body also produces small amounts of estrogen, helping maintain normal bone growth and, according to research reported in *Nature* in 1997, plays an important role in sperm production and development.

Progesterone, manufactured by the ovary after an egg has been released, aids in breast development, encourages growth of the endometrium during the premenstrual stage of a woman's sexual cycle and early in pregnancy, helps regulate the release of FSH and LH, creates thick mucus within the cervix, is necessary in the continuation of pregnancy, reduces uterine contractions, and brings about the onset of premenstrual syndrome (PMS).

Also produced by the ovaries are the male hormones called *androgens*. Some androgens are converted into estrogen, others are not. Abnormally high androgen levels in women can cause acne and the growth of facial hair. In males, androgens—the most important being *testosterone*—are produced by the testes and **adrenal glands**, and they trigger the "change of voice," growth of facial hair, and development of muscle bulk during puberty. The importance of testosterone begins during the fetal stage, not only stimulating the growth of male genitals but affecting **brain**, kidney, liver, and muscle **tissue** development. During puberty, Leydig cells in the testes are stimulated by GnRH to increase testosterone production, resulting in masculine traits and sperm production. Testosterone is the primary influence behind the male's sexual desire and performance.

While lactation requires the interaction of many hormones, researchers believe the only role of *prolactin* is to facilitate lactation. Prolactin levels during breast feeding are high and drop upon cessation of nursing. Sometimes, however, levels are abnormally high in non-nursing women, often causing a milky discharge from the nipples.

Prostaglandins are a group of female hormones with differing functions, some of which include **regulation** of body **temperature**, ovulation, **blood pressure**, and uterine contractions. When the endometrium releases prostaglandins, menstrual cramps—often accompanied by nausea, vomiting, and diarrhea—may result.

In men, LH increase testosterone production; in women, it plays three major roles: it stimulates follicles to produce es-

trogen; a burst of LH in the middle of the **menstrual cycle** triggers ovulation; and it feeds the ruptured follicle (known as the **corpus luteum**) from which the ovum was released. The corpus luteum, in turn, produces estrogen and progesterone which "feed back" to the hypothalamus. Upon detecting these increased levels, the hypothalamus inhibits the release of GnRH which, in turn, signals the pituitary gland to inhibit the release of FSH and LH until progesterone and estrogen levels drop. This, once again, signals the release of GnRH by the hypothalamus, bringing about the onset of menstruation and the next sexual cycle.

During pregnancy, five hormones are secreted by the **placenta**—estrogen, progesterone, hCG and hCS (which stimulate breast milk production), and relaxin, allowing ligaments to relax during delivery.

In women, hormone production declines around midlife bringing about menopause with physical and emotional changes that can range from mild to severe. In men, **aging** reduces testosterone production by the Leydig cells that also triggers physical and emotional changes.

SEX-LINKED TRAITS

Sex-linked traits are characteristics that appear much more commonly in males then in females. The **gene**s for sex-linked traits are carried on the **sex chromosome**s.

Within an **organism**'s body **cells**, **chromosome**s are arranged in pairs called homologous pairs. The two chromosomes in each homologous pair are the same size and have corresponding genes for the same traits at the same locations. Each body cell has two sex chromosomes that make up one of the homologous pairs of chromosomes. In most **species**, there are two types of sex chromosomes, the X chromosome and the Y chromosome. Unlike other pairs of homologous chromosomes, the X chromosome is larger than the Y chromosome, and therefore the sex chromosomes do not always have corresponding genes for all of the same traits; the Y chromosome is lacking some genes that are present on the X chromosome.

In humans and many other species, females have two X chromosomes as their sex chromosomes (symbolized as XX), while males have one X chromosome and one Y chromosome (symbolized as XY). Since the X chromosome has spaces for genes that are not on the Y chromosome, males carry slightly less genetic information. A gene located on the X chromosome may not be located on the Y chromosome. Thus, for some traits, males will only carry one **allele** (form of a gene), while females carry two. The traits for which these genes code are the sex-linked traits. Because they only have one allele, males are neither homozygous (having two copies of the same allele) nor heterozygous (having two different alleles) for sex-linked traits.

Since females carry two alleles, they can mask recessive traits if they are heterozygous for that trait. The only way that a female will express the recessive trait is if she is homozygous for it, and this is a relatively rare occurrence. Males only have one allele for a sex-linked trait, and therefore have no opportunity to mask a recessive trait. Whatever trait they inherit on their X chromosome will be expressed. As a result, sex-linked traits are much more common in males then in females.

T.H. Morgan, at Columbia University, discovered sex-linked traits in the early 1900s, working with eye color in the **fruit fly**, *Drosophila melanogaster*. In his experiments, he observed that while most fruit flies had red eyes, occasionally, a white-eyed fly would appear. He noticed that most of the white-eyed flies were males. After much breeding of fruit flies, Morgan concluded that the eye color gene was carried only on the X chromosome; it was a sex-linked trait. Since the white-eyed gene was recessive, heterozygous females would have red eyes. In order for a female to have white eyes, she would have to be homozygous for this recessive trait. However, if a male received an X chromosome that had the white-eyed trait, he would express it; there would be no other allele to mask it.

While expression of sex-linked traits is more common in males, females can be carriers of sex-linked traits and males cannot. A heterozygous female genotypically (typical genetic makeup of an organism) possesses the sex-linked trait even though she is not expressing it. Since she can pass it on to her offspring, she is considered a carrier of that trait. Males cannot be carriers since they only have one allele for sex-linked traits. Males will pass on the one allele they have to their female offspring (since daughters inherit the X chromosome), but they will not pass any allele for that trait to their male offspring (since sons inherit the Y chromosome, which is missing that allele).

There are several sex-linked traits found in humans. For example, the genes for colorblindness are found on the X chromosome. Males who have defective genes on their one X chromosome are colorblind and have trouble distinguishing between red and green. If a male has uneffected genes on his one X chromosome, he will have normal vision. In females, the genes for normal vision are dominant over the genes for colorblindness. Females who are homozygous dominant or heterozygous for this trait have normal vision. Only females that are homozygous recessive for this trait, a relatively rare occurrence, are colorblind. However, heterozygous females are carriers, and may be able to pass this trait on to their daughters and sons. **Hemophilia**, a disease in which the **blood** does not clot properly, and muscular dystrophy, in which the muscles waste away, are also sex-linked traits that affect human males more commonly than females.

SEXUAL CYCLE

The term sexual cycle refers to the reproductive cycle. In human females, the sexual cycle, which begins at puberty and ends with menopause, is based on the cyclic development and release of the ovum (**egg**) by the ovaries. Each cycle lasts approximately 28 days and has two phases—the follicular phase, which begins with the first day of the **menstrual cycle** (*mensis* is Latin for month) and continues until ovulation; and the luteal phase, which follows ovulation. Sexual—or reproductive—

cycles are regulated by **sex hormones**, the release of which is controlled by the **hypothalamus** and **pituitary gland**. Although **stress** and illness can sometimes disrupt the cycle, the only normal interruption of a woman's sexual cycle is pregnancy.

During the follicular phase, luteinizing **hormones** (LH) and follicle stimulating hormones (FSH) released by the pituitary gland encourage follicles—an envelope of special **cell**s encasing each individual **oocyte** housed in the ovaries—to secrete estradiol, the primary ovarian **estrogen**. The follicles grow quickly at this time, producing increasing amounts of estrogen, and the endometrial cells (the internal lining of the uterus) multiply, causing the lining to become thicker. One follicle will grow more quickly than the rest, maturing in approximately 14 days. As this follicle—called the Graafian follicle—grows, the maturing oocyte it protects gets pressed against one side of the inner follicle wall. The wall bulges, becomes extremely thin, and eventually breaks, releasing the egg. This episode is called ovulation. The egg travels down the oviduct, where it may or may not become fertilized by male **sperm**, toward the uterus.

Now begins the luteal phase. Cells in the empty follicle, which is now filled with clotted **blood**, undergo a major change in both shape and color, forming luteal cells. The follicle now becomes the **corpus luteum** (yellow body). Reaching the peak of its functionality within four days but continuing to grow for another four or five days, the corpus luteum stores **cholesterol** and secretes estrogen and progesterone, the latter hormone in particular being essential in preparing the lining of the uterus for implantation of the egg should it be fertilized. If the egg is not fertilized and therefore does not implant into the uterine lining, the corpus luteum stops producing hormones, shrivels and decomposes. The decreased levels of estrogen and progesterone brings about menstruation and the beginning the next sexual cycle. During the menstrual stage, blood flow to the **endometrium** decreases, starving the blood vessels of oxygen. The degenerating endometrial **tissue** is flushed from the uterus with blood from the dying blood vessels, leaving only a thin layer of cells from which new endometrial tissue will grow during the following sexual cycle. If the egg is fertilized and implants, however, the corpus luteum continues to produce the progesterone and estrogen necessary to maintain pregnancy. Not until the pregnancy ends does the corpus luteum cease to function. The end of the pregnancy then signals the beginning of the next sexual cycle. Under normal circumstances, these cycles continue until menopause, a time when the limited supply of follicles—present from **birth**—is exhausted. Sexual cycles become increasingly irregular toward the onset of menopause until they cease altogether.

Only in humans, other **primates** such as apes and monkeys, and most rodents, does the sexual cycle continue year-round. Most other **animal**s cycle naturally only once a year, although some, including dogs, cycle twice a year. These cycles, often referred to as being ''in heat,'' are technically known as oestrus (or estrus). It is only during these cycles that females become sexually receptive and allow the males to mate with them.

SHARP, PHILLIP A. (1944-)
American biologist

Phillip A. Sharp has conducted research into the structure of **deoxyribonucleic acid** (**DNA** —the chemical blueprint that synthesizes **protein**s) which has altered previous views on the mechanism of genetic change. For his work in this area, Sharp was presented with the 1977 Nobel Prize in medicine along with **Richard J. Roberts**. Sharp was born in Falmouth, Kentucky, on June 6, 1944 to Katherin Colvin and Joseph Walter Sharp. He attended Union College in Barbourville, Kentucky, where he received a B.A. degree in chemistry and mathematics in 1966. Sharp earned his Ph.D. degree from the University of Illinois in 1969.

Sharp and Richard J. Roberts discovered in 1977 that, in some higher organisms, **gene**s may be comprised of more than one segment, separated by material which apparently plays no part in the creation of the proteins. Previously, most scientists believed that genes were continuous sections of DNA and that the string of coding information that makes up each gene was a single, linear unit. Sharp and Roberts, however, distinguished between the *exons,* the sequences that contain the vital information needed to create the protein, and the *introns,* incoherent biochemical information that interrupts the protein-manufacturing instructions. Each gene is apparently composed of fifteen to twenty exons, in between which introns may be located. During protein synthesis, exons are copied and spliced together, creating complete sequences, while the introns are ignored.

This discovery had not been made earlier largely because scientists had conducted most of their genetic research on prokaryotic organisms, such as **bacteria**, which do not have their genetic material located in clearly defined nuclei. Studies of bacteria had indicated that gene activity resulted in the transcription of double-stranded DNA into single-stranded messenger ribonucleic acid (mRNA); this is translated to the corresponding protein by ribosomes. Prokaryotic organisms have no introns, however, and therefore could not supply evidence for the existence, or the significance, of noncoding regions of DNA. Roberts and Sharp carried their research out on adenoviruses, the **virus** responsible for the common cold in humans. Although these are also prokaryotic organisms, Roberts and Sharp were able to take advantage of the fact that viruses reproduce themselves using the mechanisms of eukaryotic **cell**s. Since their genome has some similarities to the genetic material in human cells, their protein synthesis was therefore relevant to the study of the cells of higher organisms.

In their experiments, Sharp's team created hybrid **molecule**s in which they could observe mRNA strands binding to their complementary DNA strands. Electron micrographs allowed the scientists to identify which parts of the viral genomes had produced the mature mRNA molecules. What they discovered was that substantial sections of DNA were ignored in producing the final mRNA. This unexpected result gave evidence of a greater complexity of mRNA synthesis in eukaryotic organisms than in prokaryotic ones. Further research indicated that the mRNAs of eukaryotic organisms are synthesized as large mRNA precursor molecules; the introns are spliced out by means of enzyme activity to produce the mature mRNA that manufactures proteins. They found that a single gene could produce a variety of proteins—some defective—as a result of different splicing patterns.

It is now believed that many hereditary diseases are caused by imperfect splicing of the genetic material, leading to the creation of faulty proteins. This may occur if the copying and splicing of the exons is not carried out accurately. One such disease is beta-thalassemia, a form of anemia prevalent in some Mediterranean areas that is caused by a faulty protein responsible for the formation of **hemoglobin**. Because of the insight Sharp's and Roberts's research has produced into the mechanisms of cell reproduction, it has important ramifications for research on malignant tumors and the viruses responsible for their development. It has also led to an investigation of methods for stopping the replication of the **human immunodeficiency** virus type 1 (**HIV**–1), with potential benefits in the search for a treatment for **AIDS**.

Sharp and Roberts's work has also led to new theories on the nature of evolutionary change; rather than being the cumulative effect of genetic **mutation** over time, it is now believed that it may be the result of the shuffling of large segments of DNA into new combinations to produce new proteins.

In 1990, before his earlier work had led to his Nobel Prize, Sharp was offered, and accepted, the presidency of the Massachusetts Institute of Technology. A short time later, he decided not to accept the position in order to devote his time exclusively to research. He has remained active in the field of academic administration, however, and has lobbied for research funding. He has also been active in industry; he was one of the founders of Biogen, a corporation started in Switzerland and now operating in Cambridge, Massachusetts, that has employed techniques developed in genetic engineering to produce the drug interferon.

SHARPEY-SCHÄEFER, EDWARD ALBERT (1850-1935)
British physiologist

Sir Edward Albert Sharpey-Schäefer was a British physiologist and endocrinologist who is credited with discovering the effects of the **hormone adrenaline**. He was also the first to predict the existence of the hormone **insulin** although it wasn't actually discovered until 1922.

He was born Edward Albert Schäefer in London on June 2, 1850, to a local merchant. When he was twenty-one, Schäefer began attending University College in London and graduated with a degree in medicine three years later, in 1874. His professor was named William Sharpey and the two shared a great mutual respect for each other. In fact, Schäefer later named one of his sons after his mentor. When the elder professor retired in 1874, Schäefer became assistant professor, eventually taking the position as Jodrell professor in 1883. In 1876 he became one of the founding members of the Physiological

Society which he later wrote a history of. In 1885 he published *Essentials of Histology* which became an important medical text book.

One of the areas on which Schäefer focused his research was neurophysiology. Specifically he studied the theory of **brain** localization and he conducted numerous experiments on nerve **tissue**. He was also highly interested in endocrinology, and it is in this area that he is most recognized. In 1894 while working with George Oliver, Schäefer discovered that an **adrenal gland** extract could cause **blood pressure** to rise when it was injected into an animal. They linked this change to a vasoconstrictive effect, and they also noticed that the smooth muscles in the animal's bronchi relaxed. They surmised that these effects were due to a hormone that was produced in the medulla of the adrenal gland, which they named adrenaline. This hormone was finally isolated in 1901 by Japanese-American chemist Jokichi Takamine. Schäefer also correctly predicted that the islets of Langerhans in the **pancreas** produce a hormone that aids in carbohydrate metabolism. He named this proposed compound ''insulin'' even though scientists did not succeed in isolating it until 1922.

Schäefer left University College in 1899 and took a position as Professor of physiology at Edinburgh. It was there, in 1903, that he developed the classic method of artificial respiration, which requires the individuals receiving treatment to lie on their back. This method was soon adopted by the Royal Life Saving Society. Schäefer was also a strong supporter of some rather controversial causes including equal roles for women in medicine and the practice of vivisection as a learning tool.

He was elected president of the British Association in 1912 and received a knighthood in 1913. When both his sons were killed in World War I, Shäefer changed his name to Sharpey-Schäefer as a gesture of respect to his old mentor and his sons. He held his professorship at Edinburgh until his retirement in 1933. He died in North Berwick, Scotland on March 29, 1935.

SHELDRAKE, RUPERT (1942-)
English biochemist

Rupert Sheldrake, a British biochemist, is best known for his controversial hypothesis of ''formative causation,'' or the idea that nature itself has memory. According to Sheldrake's theory, every system in the universe—**molecules**, **cells**, crystals, **organism**s, societies—reacts in similar or established patterns in response to invisible fields of influence. This is known as ''morphic resonance.'' Sheldrake purports that the invisible field, known as a ''morphic field,'' is where established patterns collect to influence a like activity that may be taking place contemporaneously. An example that he often uses to convey this idea more readily is that of crystallization; in his book *The Rebirth of Nature*, Sheldrake explained morphic resonance thus: ''The development of crystals is shaped by morphogenetic fields with an inherent memory of previous crystals of the same kind. From this point of view, substances such as

penicillin crystallize the way they do not because they are governed by timeless mathematical laws but because they have crystallized that way before; they are following habits established through repetition.'' Sheldrake further claims that morphic resonance transcends time and space.

Born Alfred Rupert Sheldrake on June 28, 1942, in Newark Notts, England, Sheldrake received his Ph.D. in biochemistry from Cambridge University. He was a research fellow of the Royal Society and a fellow of and director of studies in cell biology and biochemistry at Clare College at Cambridge. He studied philosophy at Harvard from 1963 to 1964 as a Frank Knox Fellow in the special studies program. Beginning in 1974, Sheldrake conducted research on tropical plants at the International Research Institute in India, as well as in Malaysia. He is married to Jill Purce, has two sons, and lives in London. Sheldrake's father was an herbalist and pharmacist. Sheldrake credits his strong interest in plants and animals to both of his parents, who encouraged him in his studies.

Sheldrake developed the necessary emotional detachment required of one pursuing scientific study during his early years at Cambridge. He came to believe—as he was taught—that nature was, in fact, a lifeless mechanistic system without purpose. But a tension persisted between his scientific studies and his personal experiences. He felt the two bore little relationship to each other and were often irreconcilable. He later came to see this conflict as rooted in the mechanistic view of nature—nature as lifeless as opposed to nature as alive and evolving. Sheldrake's hypothesis of formative causation, with its morphic fields creating morphic resonance, subscribes to the latter view.

Sheldrake's books, *A New Science of Life* (1981), *The Presence of the Past* (1988), and *The Rebirth of Nature* (1991) address his theory of formative causation—one not openly embraced by the scientific community at large. Sheldrake's theory that nature has memory and is, therefore, alive challenges the basic foundations of modern science. According to Sheldrake, the conventional scientific approach has been unable to answer the questions relative to morphogenesis—how things come into being or take form—because of their mechanistic outlook.

Sheldrake's hypothesis has elicited much criticism from his contemporaries. Joseph Hannibal, writing for *Library Journal* on *The Rebirth of Nature,* stated, ''This new work is even more unorthodox—some might say outrageous—as Sheldrake attempts to combine scientific, religious, and even mystical views.'' Critic Patrick H. Samway wrote in *America,* ''Sheldrake's methodology parallels in many ways that of the Jesuit paleontologist Teilhard de Chardin, who formulated his view that the world in its entirety is developing toward the Omega Point.''

Some of Sheldrake's contemporaries who do not subscribe to the theory of formative causation do, nonetheless, believe that science must be open to new possibilities. ''Science is not threatened by the imaginative ideas of the Sheldrakes of the world,'' wrote fellow scientist James Lovelock in *Nature,* ''but those who would censor them.'' Lovelock went on to say, ''Sheldrake is a threat, but only to the established positions of those who teach and practice an authoritarian science. A healthy scientific community would accept or reject formative causation as the evidence appeared.'' Critic Theodore Roszak allowed in *New Science,* ''If for no better reason than to exercise their wits against a first class polemic, his critics should value this work. Finding answers to his questions will fortify their ideology.'' And though terms such as ''unrealistic,'' ''fanciful,'' and ''off-the-wall'' have been used to describe Sheldrake's hypothesis of formative causation, his theory has indeed received significant attention from the scientific community.

SHERRINGTON, CHARLES SCOTT
(1857-1952)
English neurophysiologist

Charles Scott Sherrington helped to found the discipline of neurophysiology by his research on how **nerve impulses** are transmitted between the central nervous system and muscles. Sherrington focused much of his career on understanding the structure and the function of the **nervous system**. Drawing on the research of Spanish neuroanatomist Santiago Ramón y Cajal, Sherrington proposed viewing nervous activity as part of an integrated and complex system. For his work on how the central nervous system elicits motor activity from muscles, Sherrington shared the 1932 Nobel Prize in physiology or medicine with **Edgar Douglas Adrian**.

Born November 27, 1857, in London, England, Sherrington was the son of James Norton and Anne (Brookes) Sherrington. James Sherrington died while his son was still very young, and later Sherrington's mother married Caleb Rose, Jr., a physician in Ipswich, England. Rose was broadly and classically educated, and his home served as a gathering place for artists, writers, and scholars. Exposure to these diverse arts influenced Sherrington and was reflected in his own broad interests in the humanities and the sciences. After attending Ipswich Grammar School, Sherrington began medical training in 1875 at St. Thomas's Hospital in London. In 1879 he enrolled in Caius College at Cambridge University. Two years later, Sherrington began work in the laboratory of Michael Foster, England's foremost physiologist. In Foster's laboratory, Sherrington also met John Newport Langley, Newell Martin, Walter Gaskell, and Sheridan Lea, individuals who would become important physiologists in their own right.

Proposes an Integrated Nervous System

After earning a bachelor's degree in medicine in 1884, Sherrington left Cambridge to pursue graduate studies in German laboratories. He remained abroad for three years, receiving training and conducting research in physiology, histology, and pathology, and working in the laboratories of **Rudolf Virchow**, **Robert Koch**, and Friedrich Goltz, with whom he studied the central nervous system. Upon returning to England, Sherrington assumed a post teaching systematic physiology to medical students at his training site, St. Thomas's Hospital in

London. He left this position in 1891 to become professor and superintendent of the Brown Institute for Advanced Physiological and Pathological Research. A year later, Sherrington married Ethel Mary Wright; their only child, Charles E. R. Sherrington, was born in 1897.

Sherrington accepted the physiology chair at the University of Liverpool in 1895. Seeking to understand the structures and the mechanisms that operated the nervous system, Sherrington began to draw on the work of Ramón y Cajal. Prior to the latter scientist's work in the late 1880s, neurophysiologists believed that nerve fibers formed a continuous network or system through the body. This proposition was known as the reticular theory. Ramón y Cajal refuted the reticular theory by using a silver-based dye developed by the Italian anatomist **Camillo Golgi**. Golgi's preparation stained individual nerve cells a black color and demonstrated to neuroanatomists that nerve cells were discrete entities and not part of a nexus as was previously thought. The new theory that saw nerve cells as independent units was called the **neuron** theory, or popularly, the neuron doctrine. Although nerve cells were discrete units, neurons in a series could form pathways through which information can be transmitted. Nerves—consisting of a bundle of fibers—relay sensations (like touch and smell) and instructions on motor activity (like moving an arm or a leg) by electrical impulses. Sherrington became interested in understanding how nerves formed integrative pathways between the central nervous system and muscles. He considered some simple reflexive behavior, such as the knee-jerk, and attempted to explain the neurophysiology of the phenomena. Finding that he had an insufficient knowledge of neural anatomy to conduct the research, Sherrington stoically devoted the next decade to mapping the pathways between the central nervous system and muscle groups and to identifying the sensory nerves that innervated muscle **tissue**.

Sherrington's commitment to understanding the neural pathways proved to have an important impact. He came to realize that a particular reflexive behavior was not controlled by a single pathway or an isolated response to a single stimulus. Rather, a simple reflex was the product of a complex process that involved the inhibition and excitation of many nerve **cells** in many different pathways. Sherrington concluded that the central nervous system was an integrated whole that coordinated multiple pathways to produce any single action. His contributions on this point were not only theoretical but experimental. He introduced seminal research strategies for studying questions of the central nervous system. For example, the spinal animal, an animal with a transected spinal cord, and the decerebrate rigid animal, an animal partially paralyzed by the excision of the cerebral cortex, were introduced as important approaches to exploring the activity of the nervous system. Sherrington's analysis of the hind limb scratch of a dog helped to elucidate neuronal action.

Continues Explorations of the Central Nervous System

Sherrington's study of the scratch reflex in dogs elucidated other important principles of how the central nervous system is organized. He concluded that reflexes can have "reciprocal innervation" so that inhibitory and excitatory reflexes are coordinated simultaneously. Sherrington also concluded that there are two levels on which actions are controlled—higher level control by the **brain** and lower level control by the muscle nerves. His most important idea perhaps reflected in the integrative scheme is that there is a break between one nerve cell and another, between brain and muscles, between inhibitory and excitatory processes. To describe this break, Sherrington coined the term "**synapse**." The idea of a synapse became important for two reasons. First, it acknowledged that nerve cells were not organized in the reticular fashion as it was previously argued. Second, understanding how synapses were transcended became the next challenge for twentieth-century neurophysiologists. Sherrington lucidly offered these ideas about the nervous system in his seminal work, *The Integrative Action of the Nervous System,* published in 1906.

In 1913 Sherrington left the University of Liverpool after eighteen years of service to assume the Waynflete Professorship of Physiology at Oxford University. The post offered Sherrington the opportunity to continue his research on the central nervous system, but the entry of Great Britain into World War I in August, 1914, meant Sherrington had to postpone his studies for some time. He joined the war effort, serving as chair of the Industrial Fatigue Board. Not satisfied with merely reading about the conditions of war-time industrial workers, in 1915 Sherrington worked incognito in a shell factory to experience first-hand the hardships and long shifts faced by workers. Although he managed to complete a textbook of physiology during the war period, Sherrington did not return to his normal research work until the mid–1920s. He successfully recruited a number of promising assistants, including E. G. T. Liddell and **John Carew Eccles**. Eccles would go on to win the 1963 Nobel Prize in physiology or medicine for research that had its roots in his stint in Sherrington's Oxford laboratory. Eccles, Liddell, and Sherrington's other students grew in reputation as the "Sherrington school," and their assistance allowed Sherrington to complete a minimum of an experiment a week.

Sherrington's research at Oxford after the 1920s differed from the work that he had been doing prior to World War I. Rather than studying the nervous system as a whole, Sherrington focused his attention on specific mechanisms in the central nervous system. He developed with Eccles the idea of a "motor unit" —a nerve cell that coordinates many muscle fibers. He also concluded that neuronal excitation and inhibition were separate and distinct processes; one was not merely the absence of the other.

Leads an Active Life During Retirement

Although Sherrington retired in 1936, four years after being named a Nobel Prize-winner, he maintained an active life after his formal retirement. He cultivated many of the interests that he had as child in the eclectic home of his stepfather, including poetry, history, and philosophy. In 1925 Sherrington wrote and published a book of poems titled *The Assaying of Brabantius.* His deep interests in philosophy and

history were reflected in two post-retirement publications, 1941's *Man on His Nature* and 1946's *The Endeavor of Jean Fernel.* In addition to being a popular and sought-after speaker, Sherrington was a trustee of the British Museum in London and served as governor of the Ipswich School from which he had graduated.

In addition to the Nobel Prize, Sherrington garnered virtually every honor that could be given to a British scientist. At the time of his death in 1952, he held memberships in more than forty scholarly societies and had been given honorary degrees from twenty-two universities. Most notably, Sherrington was a past president of the Royal Society of London (1920–1925), and recipient of the Knight Grand Cross of the British Empire in 1922 and the Order of Merit in 1924. He died on March 4, 1952, from heart failure.

SIBLEY, CHARLES GALD (? - ?)
AHLQUIST, JON (1917-)
American biologists

Charles Sibley, a Yale University professor of biology, and his colleague Jon Ahlquist developed a process for comparing **deoxyribonucleic acid** (**DNA**) between **species**. Not only did the procedure reveal the genetic similarity of two species, but it acted as a timeclock of evolutionary change. Applying the technique to primate evolutionary relationships, Sibley and Ahlquist showed that human and chimpanzee DNA are 99% identical and, therefore, that chimpanzees are the closest living relative to humans. They also proposed a relatively recent split for humans, chimpanzees and gorillas, possibly as recent as five million years ago.

Until the latter half of the twentieth century, anthropologists had relied on the fossil record to support their theories. The relative scarcity of fossils, the fragmentary nature of the **fossil** record and the limitations of using physical features to reconstruct human prehistory were obstacles faced by biochemists who began experimenting with the comparison of genetic material among living species.

Expanding on the work of earlier scientists who had experimented with using **blood** chemistry to determine genetic relatedness, Sibley and Ahlquist looked at the entire genetic makeup of a species. The double helix structure of DNA is held together by **chemical bond**s created through the attraction of its chemical subunits called **nucleotide**s. Sibley and Ahlquist separated the DNA strands by heating them and mixing the DNA of two species. The attraction between the two and the closeness of the match, was measured by again separating the strands. The heat required to break the chemical bonds measured how well they matched—a one percent reduction in temperature indicated a 99% match of nucleotides. The evolutionary clock was calibrated by comparing the DNA in species where a good fossil record of ancestral origins existed to known living descendants whose DNA could be analyzed. A **mutation** rate of one per cent every five million years was observed.

Because the analysis of DNA conflicted with the current interpretation of the fossil record, the anthropological community initially rejected Sibley's and Ahlquist's results. However, a 1979 fossil discovery which proved Ramapithecines ancestral to orangutans rather than humans reconciled the conflicting views.

SICKLE CELL ANEMIA

Sickle **cell** anemia is an inherited **blood** disorder striking almost exclusively individuals who are of African descent. Sickle cell anemia is a recessive trait, meaning that an individual only has the disease if he or she receives a **gene** for it from each parent. Individuals who have only one faulty gene are considered to have the sickle cell trait, but not the disease. Interestingly enough, sickle cell trait has been studied, and appears to provide some protection against other diseases, such as malaria (a disease which is rampant within the countries of Africa, where the sickle cell gene is believed to have originated).

The oxygen-carrying component of the red blood cell is called **hemoglobin**. In sickle cell anemia, a single incorrect **amino acid** is substituted at one point within the structure of the hemoglobin **molecule**. This results in red blood cells which have a characteristic sickle (or half-moon) shape, instead of the usual indented disc shape. These misshapen red blood cells are prone to damage, and are destroyed more quickly than are normal red blood cells, resulting in a low red blood cell count (anemia). The sickled cells also tend to move sluggishly through small vessels, sometimes blocking those vessels for a period of time. When a vessel is blocked, the **tissue**s or **organ**s usually served by that vessel are deprived of oxygen, resulting in damage and pain. Episodes of such pain are referred to as "crises." These crises can be precipitated by infection and by low-oxygen states.

SIGHT

Sight—one of a human being's five senses—begins when rays of **light** hit an object and reflect back into the eye through the cornea—the clear **membrane** covering the eyeball, through the pupil—the opening in the center of the eye, and through the lens situated behind the pupil. The lens focuses the rays onto the retina at the back of the eye where the light first penetrates a layer of **ganglion cell**s and a layer of bipolar **cell**s (the long **axon**s of which form the optic nerve to the **brain**), before stimulating rods and cones. These electromagnetic receptors contain photoreceptor **molecule**s which, when stimulated, produce **action potential**s which **synapse** to the bipolar cells which synapse to the ganglion cells. The signal travels along the optic nerve to the lateral geniculate **nucleus** and on to the primary visual cortex at the back of the brain. Some signals are carried to a "higher" cortex where shape, color, or motion are decoded.

Light—electromagnetic radiation—consists of differing wavelengths, or frequencies, measured in nanometers (nm). Humans detect light between 300-700 nm. Rods, pigmented

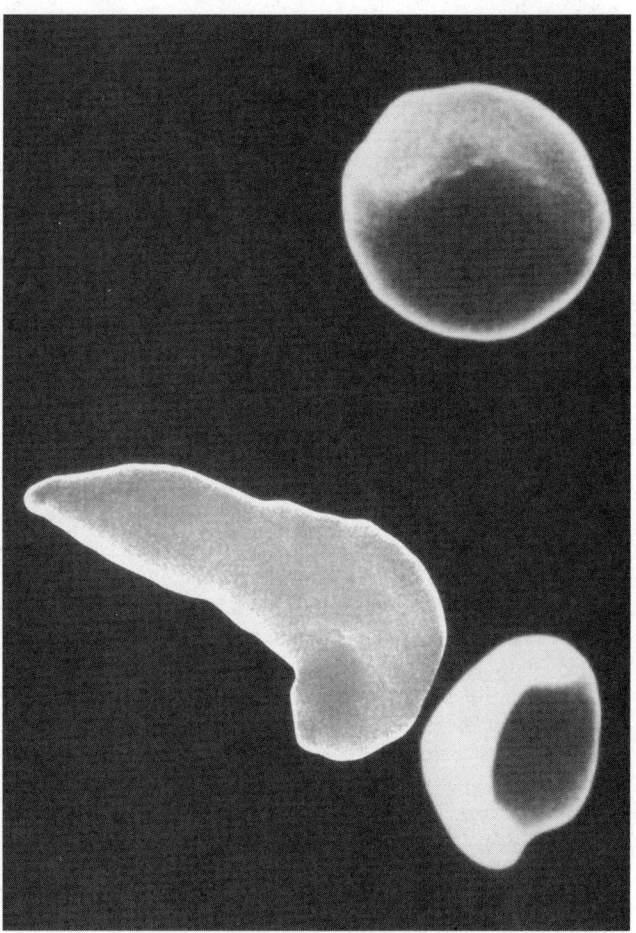

A scanning electron microscopy (SEM) scan of red blood cells taken from a person with sickle cell anemia. The blood cell at the top is normal; the diseased, sickle-shaped cells appear at the bottom. *(Photograph by Dr. Gopal Murti, Photo Researchers, Inc. Reproduced by permission.)*

with "visual purple" and stimulated by very low-level frequencies, do not function in bright light, nor do they differentiate color. They do, however, allow us to see by moonlight, but only in shades of grey. Rods are heavily concentrated around the outer edge of the retina, decreasing toward the center—the fovea—where they are nonexistent. Therefore, to see a dim star, we must look slightly away from it. Cones—outnumbered by rods twenty to one—do not respond to low frequency wavelengths, are concentrated toward the center of the retina, and densely packed in the fovea. Cones allow us to perceive color and see with high acuity. Three types of cones contain a different pigment, each of which only absorbs light waves of specific frequencies. Some absorb wavelengths between 445-450 nm (decoded as blue light); others wavelengths between 535-540 nm (green light); and others only long wavelengths (red light). Wavelengths of approximately 565 nm (yellow light) are perceived by both green and red cones, therefore yellow appears as a very bright color. Shades and hues are determined by the aggregate stimulation of these color receptors. Because pigment is destroyed as it processes light **energy** to chemical energy, it must regenerate. Vitamin A is essential to this regeneration process.

SKELETAL SYSTEM

The skeletal system is the framework of hard, articulated structures that provide physical support, attachment for muscles, and protection for the bodies of **animals**. The skeletal system of invertebrate animals, particularly those in the **phylum** Arthropoda, is on the outside of the body, and is referred to as an exoskeleton. That of vertebrate animals in the phylum Vertebrata is on the inside of the body, and is referred to as an endoskeleton.

The external skeleton of such arthropods as crustaceans, insects, and spiders consists of a large number of rigid plates and jointed appendages made of chitin (this is a complex polysaccharide **carbohydrate**, consisting of numerous linked monosaccharide units). The exoskeleton provides protection, and rigid structures to which muscles can attach, allowing the body parts to be moved in a directed fashion. The exoskeleton of arthropods is divided up into functional regions, such as the abdomen, thorax, and head of insects and crustaceans, and the abdomen and cephalothorax (fused head and thorax) of spiders. Each of these units typically has numerous jointed appendages, such as legs, antennae, or mouth parts. The numbers of structural units (such as appendages) and the size, shape, and color of parts of the exoskeleton all have functional significance in terms of the biology and **ecology** of arthropod **species**. These characteristics are commonly used as the principal characters to identify species of arthropods, and to study their evolutionary relationships.

The internal skeleton of vertebrate animals is composed of bone, a rigid, mineral-rich material laid down through the **metabolism** of the animal as part of its developmental biology. Bone is composed mostly of the mineral calcium phosphate, and to a much lesser degree, calcium carbonate. The structural elements of bone are: marrow, *periosteum*, and spongy bone. During the embryonic development of vertebrate animals, the skeleton is initially laid down as cartilage (a more flexible, elastic **tissue**), which later ossifies, or turns into bone. Even when fully mineralized, bones is not an ''inert'' tissue. Bones have a **blood** supply, are metabolically active, are capable of being remodeled throughout the life of an **organism**, and can be repaired if injured. Other materials are also part of the adult skeletal system, notably cartilage, which is generally present in the joints between bones. In addition, some vertebrate animals have elements of a so-called ''dermal'' skeleton, which provides protection to such animals as sturgeon, crocodiles, and turtles. Major elements of the skeletal system of vertebrate animals include the following (using a mammalian model):

- the vertebral column, composed of individual bones known as vertebrae, and providing axial (or lengthwise) support for the animal, while retaining a degree of lateral flexibility
- the ribs, which provide support and protection for the front of the thoracic cavity, and are important in the mechanics of breathing

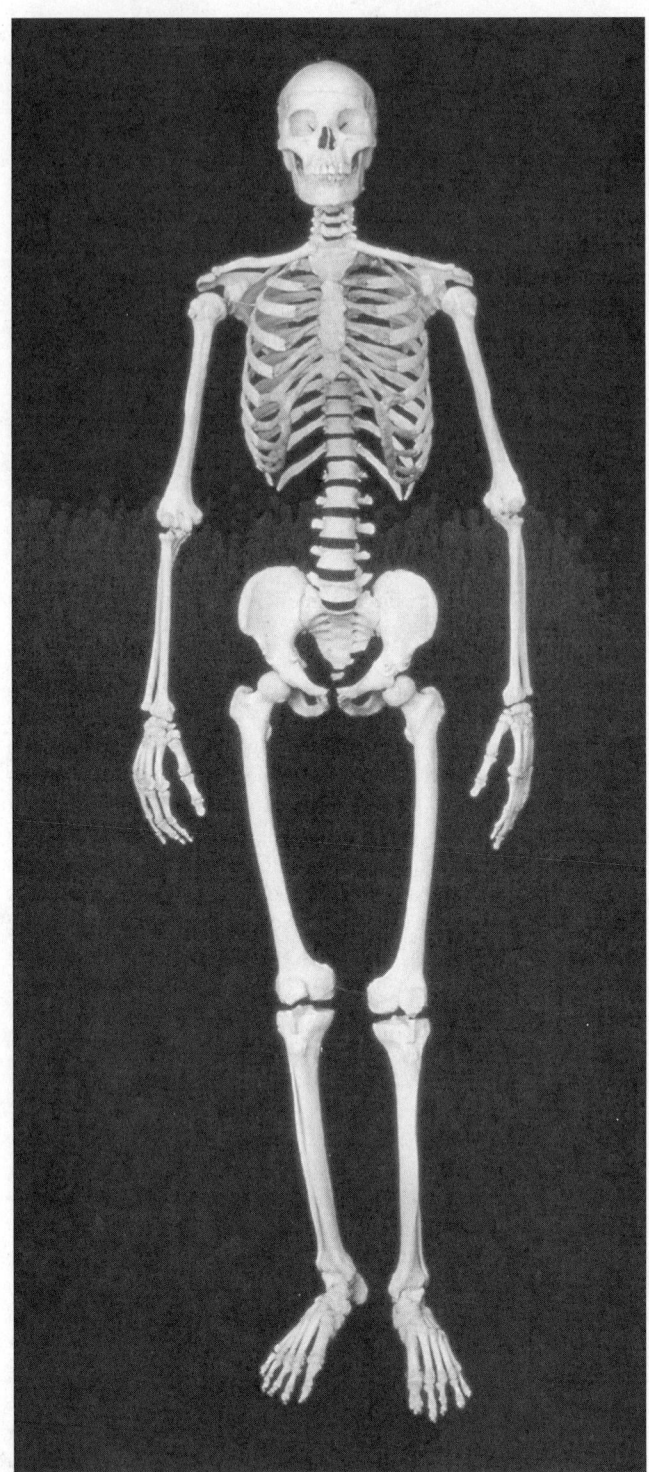

A frontal view of the human skeleton. *(Photo Researchers, Inc. Reproduced by permission.)*

- the skull, which includes the brain-case, enclosing and protecting the brain, the upper jaw, which is fused to the brain-case, and the separate lower jaw, opposite the upper jaw

- the appendicular skeleton, consisting of the bones of the forelimbs (arms and hands), the scapular (or shoulder) apparatus, to which these attach, the hindlimbs (legs and feet), and the pelvic bones, to which the hindlimbs attach
- the ligaments, consisting mainly of collagen, which keep the bones securely attached at the joints

The number, size, and shape of skeletal elements vary greatly within species during their development, and sometimes between sexes. They vary even more between species, and are commonly used as important characters when identifying species of vertebrates and studying their evolutionary relationships (this is particularly true of **paleontology**, or the study of **fossil** animals). Small species of fishes and salamanders have the tiniest bones, while large whales have the most massive ones. The human skeletal system contains 206 separate bones, the largest of which is the femur (upper leg bone), while the smallest are three bones of the inner ear known as the malleus, incus, and stapes (hammer, anvil, and stirrup; these are named after their presumed resemblance to these much-larger tools!).

SKINNER, BURRHUS FREDERIC (1904-1990)

American psychologist

B.F. Skinner was an American psychologist and an influential advocate of behaviorism, which views human **behavior** in terms of physiological responses to the environment and regards the controlled, scientific study of response as the most direct means of studying human behavior.

Skinner was attracted to psychology through the work of the Russian physiologist **Ivan Pavlov** on conditioned response, articles on behaviorism by Bertrand Russell, and the ideas of John B. Watson, the founder of behaviorism. After receiving his Ph.D. from Harvard University in 1931, he remained there as a researcher until 1936, when he joined the faculty of the University of Minnesota, Minneapolis, where he wrote *The Behavior of Organisms* (1938).

As professor of psychology at Indiana University, Bloomington (1945-48), Skinner gained public attention through his invention of the Air-Crib, a large, soundproof, germ-free, air-conditioned box designed to serve as a mechanical baby tender, supposed to provide an optimal environment for child growth during the first two years of life. In 1948 he published one of his most controversial works, *Walden Two*, a novel of life in a utopian community modeled on his own principles of social engineering.

As a professor of psychology at Harvard University from 1948 to 1974, Skinner's influence among his peers and students grew. A new generation of psychologists and psychological theory was born. Using various kinds of experimental equipment that he devised, he trained laboratory animals to perform complex and sometimes quite exceptional actions. A striking example was his pigeons that learned to play table tennis. One of his best-known inventions, the Skinner box, has

Burrhus Frederic Skinner. *(The Library of Congress. Reproduced by permission.)*

been adopted in pharmaceutical research for observing how drugs may modify animal behavior.

The Skinner Box was actually an adaptation of an earlier device invented by Edward L. Thorndike. Thorndike's invention, the "puzzle box," was an apparatus from which the animal could escape and obtain food only by pressing a panel, opening a catch, or pulling on a loop of string. Thorndike measured the speed with which the subject animal gained its release from the box on successive trials. He observed that on early trials the animal would behave aimlessly or even frantically, stumbling on the correct response purely by chance. However, with repeated trials, the animal eventually would execute the correct response proficiently within a few seconds of being placed in the box.

Skinner's refinement was to deliver food to the animal while still inside the box via an automatic delivery device and thus the probability rate at which the animal performed the designated response could be recorded over long periods of time without having to handle the animal. He also adopted some of Pavlov's terminology, referring to his procedure as instrumental, or operant, conditioning, and to the food reward as a reinforcer of conditioning. Skinner named the decline in response when the reward was no longer available as extinction. In Skinner's original experiments, a laboratory rat had to press a small lever protruding from one wall of the box in order to obtain a pellet of food. Subsequently, the "Skinner box" was adapted for use with pigeons, who were required to peck at a small, illuminated disk on one wall of the box in order to obtain some grain. With pigeons, Skinner developed the ideas of ''operant conditioning and ''shaping behavior. Operant conditioning is the rewarding of a partial behavior or a random act that approaches the desired behavior. If the goal is to have a pigeon turn in a circle to the left, a reward is given for any small movement to the left. When the pigeon catches on to that, the reward is given for larger movements to the left, and so on, until the pigeon has turned a complete circle before getting the reward. Skinner compared this learning with the way children learn to talk. They are rewarded for making a sound that is similar to a word until in fact they can say the word. Skinner believed other, more complicated tasks could be broken down in this way and taught. Indeed, these experiences in the step-by-step training of research animals led Skinner to formulate the principles of programmed learning.

Programmed learning in its most basic form deduces a subject into its component parts and arranges the parts in a sequential learning order. At each step in learning, the student is required to make a response and is told immediately whether or not the response is correct. The program is structured so that correct answers are likely to be extremely frequent (sometimes as much as 95% of the time). The theory is that this continual reward of correct responses encourages the student and serves as a motivator to continue. Reinforcement is the key element in Skinner's theories about learning. A reinforcer is anything that strengthens the desired response. It could be verbal praise, a good grade or a feeling of accomplishment or satisfaction.

The teaching machine itself is merely the device that provides the questions (stimulus) that require an answer (response) before the learner is allowed to continue. All teaching machines depend on a program, that is, a series of questions presented that provide a student with a certain amount of challenge as well as a chance to learn.

Unlike other psychologists before him (most notably Sigmund Freud) Skinner was not at all interested in understanding the human psyche. He sought only to determine how behavior is caused by external forces. He believed everything we do and are is shaped by our individual experiences of punishment and reward. He believed that the ''mind'' (as opposed to the brain) and other such subjective phenomena were simply matters of language and didn't really exist.

In addition to his widely read *Science and Human Behavior* (1953), Skinner wrote a number of other books, including *Verbal Behavior* (1957), *The Analysis of Behavior* (with J.G. Holland, 1961), and *Technology of Teaching* (1968). Another work that generated considerable controversy, *Beyond Freedom and Dignity* (1971), argued that concepts of freedom and dignity may lead to self-destruction and promoted the idea of a study of behavior comparable to that of the physical and biological sciences. Skinner published an autobiography in three parts: *Particulars of My Life* (1976), *The Shaping of a Behaviorist* (1979), and *A Matter of Consequences* (1983). The year before his death, *Recent Issues in the Analysis of Behavior* (1989) was published.

SKOU, JEN C. (1918-   )
Dutch biochemist

Jens Christian Skou was born October 8, 1918, in Lemvig, Denmark. He received his M.D. degree (cand.med.) from the University of Copenhagen in 1944. Ten years later, he received his Doctor of Medical Sciences degree (dr.med.) from Aarhus University.

After receiving his M.D., Skou went for clinical training at the Hospital at Hjørring and Orthopaedic Clinic at Aarhus, Denmark. He remained there until 1947, when he became an assistant professor in the University of Aarhus's Institute of Physiology. In 1954 the same year he received his Doctor of Medical Sciences degree, he became associate professor at the institute. In 1963 Skou became a full professor and was named chairman of the Institute of Physiology. From 1978 to 1988, he was professor of biophysics at the University of Aarhus.

Skou has devoted his career to both education and research. He has published more than 90 papers on his research, which has investigated the actions of local anesthetics and what mechanisms made them work, as well as the work that earned him the 1997 Nobel Prize, the transport of sodium and potassium ions through the **cell** membranes.

A cell's health depends on maintaining a balance between its inner chemistry and that of the cell's surroundings. This balance is controlled by the presence of the cell **membrane**, the wall between the cell's inner workings and its environment.

For more than 70 years, scientists have known that one of the delicate balances that are maintained involves ions (electrically charged particles) of the elements sodium (Na) and potassium (K). A cell maintains its inner concentration of sodium ions (Na+) at a level lower than that of its surroundings. Similarly, it maintains its inner concentration of potassium ions (K+) at a level higher than its surroundings.

This balance is not static, however. In the 1950s, English researchers **Alan Hodgkin** and Richard Keynes found that sodium ions rush into a nerve cell when it is stimulated. After the stimulation, the cell restores its original sodium/potassium levels by transporting the extra sodium out through its membrane. Scientists suspected that this transport involved the compound **adenosine triphosphate (ATP)**. Discovered in 1929, ATP was shown to carry **energy** in the cell. Scientists noticed that, when ATP's presence was inhibited, cells did not rid themselves of the extra sodium that they absorbed during stimulation.

In the 1950s, Skou began his investigations into the workings of ATP. For his experimental material he chose finely bound nerve membranes from crabs. He wanted to find out if there was an **enzyme** in the nerve membranes that degraded ATP, and that could be involved with the transport of ions through the membrane.

He did find such an ATP-degrading enzyme, which needed ions of magnesium. In his experiments, Skou found that he could stimulate the enzyme by adding sodium ions—but there was a limit to the stimulation he could achieve. Adding small amounts of potassium ions, however, stimulated the enzyme even more. In fact, Skou noted that the enzyme—called ATPase—reached its maximum point of stimulation when he added quantities of sodium and potassium ions that were the same as those normally found in nerve cells. This evidence made Skou hypothesize that the enzyme worked with an ion "pump" in the cell membrane.

Skou published his first paper on ATPase in 1957. In years of further experimentation, Skou learned more about this remarkable enzyme. He learned that different places on the enzyme attracted and bound ions of sodium and potassium.

When ATP breaks down and releases its energy, it become adenosine diphosphate (ADP) and releases a phosphate compound. Skou's work discovered that this freed phosphate bound to the ATPase as well, a process known as **phosphorylation**. The presence or absences of this phosphate changed the enzyme's interaction with sodium and potassium ions, Skou discovered. When the ATPase lacked a phosphate group, it became dependent on potassium. Similarly, when it has a phosphate, it became dependent on sodium.

This latter discovery was key to learning just how ATPase moved sodium out of the cell. ATPase **molecule**s are set into the cell membrane, and they consist of two parts, one which stabilizes the enzyme and the other which carries out activity.

Part of the enzyme pokes inside the cell. There, one ATP molecule and three sodium ions can bind to it at a time. A phosphorus group is taken from the ATP to bind to the enzyme, and the remaining ADP is released. The enzyme then changes shape, carrying the attached sodium ions with it to the outside of the cell membrane. There, they are released into the cell's surroundings, as is the attached phosphorus. In place of the three sodium ions, two potassium ions attach themselves to the enzyme, which again changes shape and carries the K+ into the cell's interior.

This activity uses up about one-third of the ATP that the body produces each day, which can range from about half of a resting person's body weight to almost one ton in a person who is doing strenuous activity.

Thanks to this molecular pump, the cell is able to maintain its balance of potassium ions on the inside and sodium ions on the outside, maintaining the electrical charges that allow cells to pass along or to react to stimulation from nerve cells.

This enzyme is important for other reason as well. For example, the pump's action on the balance of sodium and potassium makes it possible for the cell to take in nutrients and to expel waste products. If the molecular pump were to stop—as it can when a lack of nourishment or oxygen shuts down ATP formation—the cell would swell up, and it would be unable to pass along nerve impulse. If this were to happen in the brain, unconsciousness would rapidly follow.

Since Skou discovered ATPase, scientists have found other molecular pumps hard at work in the cell. They include H+, K+-ATPase, which produces stomach acid, and Ca2+-ATPase, which helps control the contraction of muscle cells.

SMELL

Smell—technically called olfaction—is one of a human being's five senses. Described as "sensitivity to substances in a gaseous phase," smell is closely related to **taste**, both of which are perceived and regulated by the process of chemoreception. When the sense of smell is defective, the ability to taste is greatly reduced. Unless there is a genetic deficiency, the sense of smell is present at **birth**. Humans smell with their noses; however, some **animal**s and insects smell with their tongues, feet, or antennae: a snake picks up chemical odor **molecule**s from the air by flicking its tongue; a butterfly senses sweetness with its feet and detects **pheromones** of the opposite sex with its antennae.

Smell is a highly complex processes which, in humans, begins when odor molecules are drawn up the nostrils. These molecules stimulate the approximately 50 million chemoreceptors concentrated in the olfactory **epithelium**—a tiny 1.2-1.9 in (3-5 cm) region in the mucus **membrane** located at the very top of the nasal cavity. Stimulation of these **neuron**s creates **action potential**s which travel along the **axon** of the neurons which are fibers of olfactory nerves. These fibers pass through minute openings in the criboform plate to **synapse** with second-order **cell**s, the axons of which travel to the olfactory bulbs located on either side of the ethmoid bone. Here, a "taste identity card" is created. The coded messages are transmitted along the olfactory tracts to the olfactory cortex located in the frontal lobes of the **brain** where they are decoded. Olfactory neurons allow us to distinguish a huge spectrum of fragrances and odors—perhaps as many as 2000. Researchers believe that special **gene**s may code specific receptor **protein**s creating more than 1000 different receptors, each individually coded to respond to a specific type of odor **molecule**. Olfactory **neuron**s are unique among neurons in that new ones are continuously being generated to replace those which die.

Smell plays a large role in forming life experiences and influencing our moods; odors associated with a pleasant experience instantly bring back fond memories while those associated with an unpleasant experience trigger negative **emotions**. Research also indicates that **memory** and **learning** can be enhanced by certain fragrances, giving credence to aroma therapy—the use of specific aromas to energize, relax, and provoke different moods. Interestingly, however, perception of an odor or fragrance decreases over time with continuous exposure. This phenomenon is called "adaption."

SMITH, HAMILTON O. (1931-)
American molecular biologist

Hamilton Othanel Smith shared the 1978 Nobel Prize in physiology or medicine with fellow biologists **Werner Arber** and **Daniel Nathans** for the set of linked discoveries that started off the boom in biotechnology. Because of these discoveries, researchers can more easily elucidate the structure and coding of **deoxyribonucleic acid** (**DNA**) molecules (the basic genetic map of an organism), and they hope to correct many genetic illnesses in the future. His research also made it possible to design new organisms, a controversial but potentially beneficial technology. Smith purified and explained the activity of the first restriction enzyme, which became the principal tool used by genetic engineers to selectively cut up DNA. (Arber had linked restriction and modification to DNA, and predicted the existence of restriction enzymes. Nathans, under Smith's encouragement at Johns Hopkins, developed techniques that enabled their practical use.)

Smith was born on August 23, 1931, in New York, New York, to Bunnie (Othanel) Smith and Tommie Harkey Smith. Smith graduated from University High School in three years, enrolling at a local university in 1948.

Smith came to the study of genetics by way of medicine. Initially a mathematics major at the University of Illinois, he transferred to the University of California at Berkeley in 1950 to study biology and graduated with a bachelor's degree in 1952. He obtained a medical degree from the Johns Hopkins School of Medicine in 1956. During the years 1956 to 1962, he held various posts, including an internship at Washington University in St. Louis, Missouri, a two-year Navy stint in San Diego, California, and a residency at Henry Ford Hospital in Detroit, Michigan. He gradually taught himself genetics and molecular biology in his spare time. In 1962 he began a research career at the University of Michigan on a postdoctoral fellowship from the National Institutes of Health, before finally returning to Johns Hopkins in 1965 as a research associate in the microbiology department. He was named a full professor of microbiology in 1973, and professor of molecular biology and genetics in 1981. In 1975, Smith was awarded a Guggenheim Fellowship for a year of study at the University of Zurich in Switzerland.

After his return to the United States, Smith purified the first Type II restriction endonuclease, which he obtained from the bacterium *Hemophilus influenzae,* and identified the nucleotide sequence which the enzyme would cut. He gave a supply of the enzyme to Daniel Nathans, who used it in his own work. The three men eventually won the 1978 Nobel Prize. The presenter of the prize noted that Smith proved Arber's hypothesis about restriction enzymes, pointing the way for future research.

Smith's exacting specificity of Class II restriction enzymes makes them useful because biotechnologists can now cut DNA apart selectively. Then they can add and subtract specific nucleotides, and reproducibly weld (recombine) the links back together in a new order. This new piece of DNA now codes for a different protein. The current and potential uses of these procedures are enormous. Biotechnologists can genetically engineer bacteria that produce a particular chemical; human insulin for the treatment of diabetes is now made by such recombinant bacteria. Other bacteria have been designed to clean up oil slicks. One of the tasks that biotechnologists would like to accomplish is the eradication of genetic illness by correcting the mistaken DNA codes that cause it.

SMITH, MICHAEL (1932-)

English Canadian biochemist

Michael Smith began his professional research career in salmon physiology and endocrinology, but returned to the chemical synthesis that had been his first interest, including the chemical synthesis of deoxyribonucleic acid (DNA). Smith experimented with isolating genes and invented site-directed mutagenesis, a technique for deliberately altering gene sequences. Smith's work was hailed as having tremendous implications for genetic studies and the understanding of how individual genes function, and already has been applied in the study of disease-producing viruses. In 1993 Smith shared the Nobel Prize in Chemistry independently with Kary Mullis. The Royal Swedish Academy of Sciences credited Smith and Mullis with having revolutionized basic research and saluted the possibilities offered by their research toward the cure of hereditary diseases.

Smith was born in Blackpool, England, on April 26, 1932. His parents were Rowland Smith, a market gardener, and Mary Agnes Armstead Smith, a bookkeeper who also helped with the market gardening. Smith was admitted to Arnold School, the local private secondary school, with a scholarship he earned based on his examination results (this examination was taken, at the time, by all English children when they finished their primary education). Without this scholarship, Smith would have had little opportunity for advanced education, as his parents did not have the money to pay for it. While at Arnold School, Smith became involved in scouting, which eventually led to a life-long interest in camping and other outdoor activities.

After graduating from Arnold School in 1950, Smith enrolled at the University of Manchester in order to study chemistry, realizing a natural inclination toward the ''hard'' sciences. He moved rapidly through school, receiving a B.Sc. in 1953, and a Ph.D. in chemistry in 1956, both sponsored by scholarship. Smith's desire following completion of his Ph.D. was to earn a fellowship on the West Coast of the United States. This did not work out, but he was accepted into biochemist Har Gobind Khorana's laboratory in Vancouver, Canada. Smith's original plan in migrating to Canada was to work for a year, then return to England and work for a chemical company. However, his experience working with Khorana, who would win the Nobel Prize in 1968 for his contributions to genetics, changed his plans. Smith decided university research was the path he wanted to take and that British Columbia, with its natural beauty, would be his home. Smith is now a Canadian citizen.

Smith stayed with the Khorana group and moved with it in 1960 to the Institute for Enzyme Research at the University of Wisconsin. (Smith had recently married Helen Christie. The couple later separated, but they had three children, Tom, Ian, and Wendy.) Until then, Smith's work in Canada had been in several different areas of chemical synthesis. In 1961 he decided it was time for a change and decided to re-locate to the West Coast. Smith accepted a position as head of the chemistry section of the Vancouver Laboratory of the Fisheries Research Board of Canada. His work there was mainly in salmon physiology and endocrinology, but he also continued to work in chemical synthesis.

Move to Academia Culminates in Nobel Prize

In 1966 Smith entered the academic field, taking an appointment as associate professor of biochemistry and molecular biology at the University of British Columbia (UBC), and bringing with him an interest in chemically synthesized DNA (the **molecule** of heredity). Also beginning in 1966 Smith held a concurrent position as medical research associate of the Medical Research Council of Canada. He was made full professor in 1970, and has continued his teaching duties ever since. In 1986 he was asked to establish a biotechnology laboratory on the campus of UBC, which he has headed since that time.

Smith has taken three sabbaticals from his duties at the University of British Columbia, spending three months in 1971 at Rockefeller University in New York, one year during 1975 and 1976 at the Medical Research Council laboratory in Cambridge, and eight months in 1982 at Yale University. The middle excursion was spent in English biochemist **Frederick Sanger**'s laboratory learning about DNA sequence determination, essential to Smith's later research.

Smith was first able to isolate genes using chemical synthesis in 1974. Slowly he developed what became known as site-directed **mutagenesis**, a technique that allows gene sequences to be altered deliberately. More specifically, it involves separating one strand of a piece of DNA and producing a mirror image of it. This mirror image can then be used as a probe into a gene. It can also be used with chemical **enzyme**s—proteins that act as catalysts in biochemical reactions—that are able to cut and splice DNA in living cells. Jeffrey Fox, editor of *Bioscience,* called this process the "intellectual bombshell that triggered protein engineering," as quoted in the Toronto *Globe and Mail.* Smith's findings were published in 1978 in *Journal of Biological Chemistry.* This paper lays the foundation of the research Smith has done since. The paper concludes, "This new method of mutagenesis has considerable potential in genetic studies. Thus, it will be possible to change and define the role of regions of DNA sequence whose function is as yet incompletely understood."

Smith, in demonstrating that biological systems are chemical, has allowed scientists to tinker systematically with genes, altering properties one at a time to see what effect each alteration may have on the gene's functioning. Genes are the building blocks for countless proteins that make up skin, muscles, bone, and hormones. Changes in the expression of these proteins reveal to the scientist how his or her tinkering has altered the gene function. This process has been used specifically to study disease-producing viruses, such as those that cause **cancer**. The eventual goal is to uncover the functioning of the genes, so drugs to combat the viruses can be developed.

After being several times a nominee, Smith was awarded the Nobel Prize in Chemistry in 1993 jointly with Kary Mullis from California. Their work was not collaborative, though both dealt with biotechnology. Announcing the award, the Royal Swedish Academy of Sciences credited Smith for having "revolutionized basic research and entirely changed researchers' way of performing their experiments," as quoted in the Toronto *Globe and Mail.* The academy further said

Smith's work holds great promise for the future with the "possibilities of gene therapy, curing hereditary diseases by specifically correcting mutated code words in the genetic material."

The award money from the Nobel Prize amounted to close to $500,000 Cdn for Smith. With it he established an endowment fund, half of which will be earmarked to aid research on molecular genetics of the central **nervous system**, specifically in relation to **schizophrenia** research. The other half is to be divided between general science awareness projects and the Society for Canadian Women in Science and Technology in an effort to induce more women to pursue careers in science. He also convinced both the provincial and federal governments to contribute to his funds.

In addition to his receipt of the Nobel Prize, Smith has garnered numerous other honors in the course of his career, including the Gairdner Foundation International Award in 1986, and the Genetics Society of Canada's Award of Excellence in 1988. He has assumed several administrative responsibilities, including becoming acting director of the Biomedical Research Center, a privately funded research institute, in 1991, and is a member of the Canadian Biochemical Society, the Genetics Society of America, and the American Association for the Advancement of Science. He is a fellow of the Chemical Society of London, the Royal Society of Canada, and the Royal Society of London, and has served on several medical committees, such as the advisory committee on research for the National Cancer Institute of Canada. He is a popular speaker, and has delivered over 150 addresses throughout the world during the course of his career. His scientific research articles number more than two hundred.

SNELL, GEORGE DAVIS (1903-1996)
American immunogeneticist

One of three children, Snell was born on December 19, 1903, in Bradford, Massachusetts. By 1922, he had enrolled at Dartmouth pursuing studies in biology.

He obtained a B.S. degree in that subject in 1926 and enrolled at Harvard that same year to study **genetics** under the renowned biologist William Castle, who was among the first American scientists to delve into the biological laws of **inheritance** regarding **mammals**. Snell received a Ph.D. in 1930 after completing his dissertation on linkage (the means by which two or more **gene**s on a **chromosome** are interrelated). That same year he became an instructor of **zoology** at Brown University, only to leave in 1931 to work at the University of Texas at Austin following receipt of a National Research Council Fellowship.

Snell's decision to accept the fellowship turned out to be a momentous one, as he began work for the famed geneticist **Hermann Joseph Muller**, whose research with fruit flies led to the discovery that x rays could produce **mutation**s in genes. At the university, Snell experimented with mice, showing that x rays could produce mutations in rodents as well. Although Snell left the University of Texas in 1933 to serve as assistant professor at the University of Washington, he ventured to the

Jackson Laboratory in Bar Harbor, Maine, in 1935 to return to research work. The laboratory, specializing in mammalian genetics, was well-known for its work in spite of its small size.

After continuing his work with x rays and mice, Snell decided to embark on a new study. Snell's project was concerned with the notion of transplants. Earlier scientific research had indicated that certain genes are responsible for whether a body would accept or reject a transplant. The precise genes responsible had not then been identified, however.

Snell began his experiments by performing transplants between mice with certain physical characteristics. He quickly discovered those mice with certain identical characteristics—in particular a twisted tail—tended to accept each other's skin grafts. In 1948 Peter Gorer came to Jackson Laboratory from London, England. Gorer, who had also conducted experiments on mice, developed an antiserum. He had discovered the existence of a certain antigen (foreign protein) in the **blood** of mice which induced an immune reaction when injected into other mice. Gorer had called this type of substance "Antigen II."

In collaboration, Snell and Gorer proved that Antigen II was present in mice with twisted tails, indicating that the genetic code for Gorer's antigen and the code found by Snell to be vital for **tissue** acceptance were identical. They called their discovery of this factor "H–2," for "Histocompatibility Two" (a term invented by Snell to describe whether a transplant would be accepted or rejected).

Later research revealed that instead of only a single gene being responsible for this factor, a number of closely related genes controlled histocompatibility. As a result, this was subsequently designated as the Major Histocompatibility Complex (MHC). The discovery of the MHC, and subsequent research by other scientists in the 1950s which proved it also existed in humans, made widespread **organ** transplantation possible. Donors and recipients could be matched (as had been done with blood types) to see if they were compatible.

Eventually Snell was able to produce what he called "congenic mice"—animals that are genetically identical except for one particular genetic characteristic. Unfortunately, the first strains of these mice were destroyed in a 1947 forest fire which burned down the laboratory. However, Snell's tenacity and dedication enabled him to rebound from this setback. Within three years he had created three strains of mice which differed genetically only in their ability to accept tissue grafts. The development of congenic strains of mice opened up a new field for experimental research, with Jackson Laboratory eventually being able to supply annually tens of thousands of these mice to other laboratories.

In 1952 Snell became staff scientific director and, in 1957, staff scientist at Jackson Laboratories. In those capacities he continued his research, particularly on the role that MHC plays in relation to **cancer**. Experiments he conducted with congenic mice found that on some occasions the mice rejected tumors which had been transplanted from their genetic twins. This "hybrid resistance" indicated that some tumors provoke an immune response, causing the body to produce antibodies to fight the tumor. This discovery could eventually be of great importance in developing weapons to fight cancer.

The success of Snell's work culminated in his winning the 1980 Nobel Prize in medicine or physiology for his work on histocompatibility. He shared this with two other immunogeneticists, **Jean Dausset** and **Baruj Benacerraf**. After being told of the Nobel committee's decision, Snell said there should have been a fourth recipient—his colleague Peter Gorer who died in 1962 and was thus ineligible to receive the prize.

SOCIAL DARWINISM

Social Darwinism is the theory that persons, groups, and races are subject to the same laws of **natural selection** as **Charles Darwin** had perceived in **plants** and **animals** in **nature**. According to the theory, which was popular in the late nineteenth and early twentieth centuries, the weak were diminished and their cultures delimited, while the strong grew in power and in cultural influence over the weak. Social Darwinists held that the life of humans in society was a struggle for existence ruled by "**survival of the fittest**," a phrase proposed by the British philosopher and scientist Herbert Spencer.

The theory of **evolution** by natural selection was proposed by Charles Darwin and **Alfred Russel Wallace** in 1858. They argued that **species** with useful **adaptations** to the environment are more likely to survive and produce progeny than are those with less useful adaptations, thereby increasing the frequency with which useful adaptations occur over the generations. The limited resources available in an environment promotes **competition** in which **organisms** of the same or different species struggle to survive. In the competition for food, space, and mates that occurs, the less well-adapted individuals must die or fail to reproduce, and those who are better adapted do survive and reproduce. In the absence of competition between organisms, natural selection may be due to purely environmental factors, such as inclement weather or seasonal variations.

The social Darwinists, notably Spencer and Walter Bagehot in England and William Graham Sumner in the United States, believed that the process of natural selection acting on variations in the **population** would result in the survival of the best competitors and in continuing improvement in the population. Societies, like individuals, were viewed as organisms that evolve in this manner.

The theory was used to support laissez-faire capitalism and political conservatism. Class stratification was justified on the basis of "natural" inequalities among individuals, for the control of property was said to be a correlate of superior and inherent moral attributes such as industriousness, temperance, and frugality. Attempts to reform society through state intervention or other means would, therefore, interfere with natural processes; unrestricted competition and defense of the status quo were in accord with biological selection. The poor were the "unfit" and should not be aided; in the struggle for existence, wealth was a sign of success. At the societal level, social Darwinism was used as a philosophical rationalization for imperialist, colonialist, and racist policies, sustaining belief in Anglo-Saxon or Aryan cultural and biological superiority.

Social Darwinism declined during the twentieth century as an expanded knowledge of biological, social, and cultural phenomena undermined, rather than supported, its basic tenets.

SOCIOBIOLOGY

Sociobiology, also called behavioral **ecology**, is the study of the **evolution** of social behavior in all **organisms**, including human beings. The highly complex behaviors of individual **animals** become even more intricate when interactions among groups of animals are considered. Animal behavior within groups is known as *social* behavior. Sociobiology asks about the evolutionary advantages contributed by social behavior and describes a *biological* basis for such behavior. It is theory that uses biology and **genetics** to explain why people (and animals) behave the way they do.

Sociobiology is a relatively new science. In the 1970s, **Edward O. Wilson**, now a distinguished professor of biology at Harvard University, pioneered the subject. In his groundbreaking and controversial book, *Sociobiology: The New Synthesis*, Dr. Wilson introduced for the first time the idea that behavior is likely the product of an interaction between an individual's genetic makeup and the environment (or culture in the case of human beings). Wilson's new ideas rekindled the debate of ''**Nature** vs. Nurture,'' wherein nature refers to **gene**s and nurture refers to environment.

Sociobiology is often subdivided into three categories: narrow, broad, and pop sociobiology. Narrow sociobiology studies the function of specific behaviors, primarily in non-human animals. Broad sociobiology examines the biological basis and evolution of general social behavior. Pop sociobiology is concerned specifically with the evolution of human social behavior.

Sociobiologists focus on reproductive behaviors because reproduction is the mechanism by which genes are passed on to future generations. It is believed that behavior, physically grounded in an individual's **genome** (or genes), can be acted upon by **natural selection**. Natural selection exerts its influence based upon the fitness of an organism. Individuals that are *fit* are better suited (genetically) to their environment and therefore reproduce more successfully. An organism that is fit has more offspring than an individual that is unfit. Also, **fitness** requires that the resulting offspring must survive long enough to themselves reproduce. Because sociobiologists believe that social behavior is genetically based, they also believe that behavior is heritable and can therefore contribute to (or detract from) an individual's fitness. Examples of the kinds of reproductive interactions in which sociobiologists are interested include **courtship**, mating systems like monogamy (staying with one mate), polygamy (maintaining more than one female mate), and polyandry (maintaining more than one male mate), and the ability to attract a mate (called *sexual selection*.)

Sociobiology also examines behavior that indirectly contributes to reproduction. An example is the theory of optimal foraging which explains how animals use the least amount of **energy** to get the maximum amount of food. Another example is altruistic behavior (**altruism** means selfless). Dominance hierarchies, territoriality, ritualistic (or symbolic) behavior, communication (transmitting information to others through displays), and **instinct** versus **learning** are also topics interpreted by sociobiology.

Sociobiology applied to human behavior involves the idea that the human **brain** evolved to encourage social behaviors that increase reproductive fitness. For example, the capacity for learning in human beings is a powerful characteristic. It allows people to teach their relatives (or others) important life skills that are passed-down from generation to generation. However, the ability to learn is also a variable trait. That is, not every person learns as quickly or as well as every other person. A sociobiologist would explain that individuals who learn faster and more easily have increased fitness. Another example is smiling. The act of smiling in response to pleasurable experiences is a universal social behavior among people. Smiling is observed in every culture. Furthermore, smiling is an example of an instinct that is modified by experience. Therefore, because the behavior is instinctual, it has a genetic and inheritable basis. Because it is altered by experience, the behavior is socially relevant. Sociobiologists might speculate, then, that since smiling is a visual cue to other individuals that you are pleased, people who tend to smile more easily are more likely to attract a suitable mate, and are therefore more fit.

The discipline of sociobiology is an important set of ideas because nearly every animal **species** spends at least part of its **life cycle** in close association with other animals. However, it is also riddled with debate, principally because it attempts to not only explain the behavior of animals but also of human beings. More dangerously, it tries to describe ''human nature.'' The idea that human behavior is subject to genetic control has been used in the past to justify racism, sexism, and class injustices. In this respect, sociobiology is similar to **Social Darwinism**. For this reason, sociobiology remains a controversial discipline. Further criticisms include the observation that sociobiology contains an inappropriate amount of anthropomorphism (giving human characteristics to animals) and it excessively generalizes from individuals to whole groups of organisms. Despite criticism, however, sociobiology is an enlightening new aspect of biology which, taken in context, can bridge the gap between life science and the humanities.

See also Evolutionary theory

SOLUTION, HYPERTONIC

A solution is hypertonic if it tends to gain **water** from a reference solution (or colloidal suspension) separated from it by a semipermeable **membrane**. This usually results from of a higher concentration of dissolved or dispersed substances in the hypertonic solution. It is assumed that the membrane is freely permeable to the solute (usually water) but impermeable, or nearly so, to substances in solution or suspension in the two liquids. When a living **cell** is immersed in a hypertonic solution, the cell will lose water to the solution. It may collapse if it is an **animal** cell, or plasmolyze by pulling away from the cell wall if it is a plant cell. However, as water flows out of the cell it dilutes the hypertonic solution, and causes the concentration of cellular **protoplasm** to increase. The ultimate result depends largely on the difference in solute concentration between the cell and the bathing liquid. When differences are small, equilibrium may be reached that does not damage the cell, but when they are large, harm is more apt to occur.

See also Osmosis; Plasmolysis; Solution, hypotonic; Solution, isotonic.

SOLUTION, HYPOTONIC

A solution is hypotonic if it tends to lose **water** to a reference solution (or colloidal suspension) separated from it by a semipermeable **membrane**. This usually results from of a lower concentration of dissolved or dispersed substances (and a higher effective concentration of water) in the hypotonic solution. The membrane is assumed to be freely permeable to the solute (water) but impermeable, or nearly so, to substances in solution or suspension in the two fluids. If a living **cell** is immersed in a hypotonic solution, water will move into the cell from the solution, causing the cell to expand. **Animal** cells may eventually burst if they are unable to contain all of the incoming water. Plant cells, with a rigid cell wall, will exert **turgor pressure** on the wall that normally will be balanced by a counteracting wall pressure, and the cell is said to be fully turgid. **Microorganism**s such as amebas and paramecia, which live in pond water that is hypertonic to their intracellular fluid, have evolved contractile **vacuoles** that are able to collect excess water from the **protoplasm** and pump it to the outside. Without this mechanism, the cells would quickly burst from the pressure developed by the entering water. This pumping mechanism requires **energy**, and the entering water will destroy cells that are poisoned, or for some other reason lose their ability to actively metabolize.

See also Osmosis; Solution, hypertonic; Solution, isotonic

SOLUTION, ISOTONIC

A solution is isotonic with another solution if it neither to gains **water** from, nor loses water to, the second solution when separated from it by a semipermeable **membrane**. This is the case when the solution has the same osmotic concentration as the second solution, For that reason such solutions are sometimes called isosmotic. The membrane is assumed to be freely permeable to the solute (usually water) but impermeable, or nearly so, to substances in solution or suspension in the two fluids. Growth medium used to **culture animal cell**s must be isotonic with the **protoplasm** of the cells to avoid damage. A solution of 0.9% sodium chloride, sometimes called "physiological saline," is isotonic to human cells.

See also Osmosis; Solution, hypertonic; Solution, hypotonic

SOUTHERN BLOTTING • See Genetic mapping

SPALLANZANI, LAZZARO (1729-1799)
Italian biologist and physiologist

Spallanzani was born on January 12, 1729, in Scandiano, Italy. He attended the University of Bologna and began his studies in law. However, his cousin, Laura Bassi, a professor of physics and mathematics, introduced him to a broad range of scientific studies. Spallanzani altered his educational course and, in 1754, he earned a Ph.D. in philosophy. He joined the priesthood to support himself while he studied natural phenomena, hoping to determine explanations for such events as a stone skipping on **water**, the regeneration of decapitated snail heads, and the electric discharge of torpedo fish. Over the course of his career, Spallanzani would examine the pits of spitting volcanoes, the world of reproduction, the waters of eels, the dark depths of the bat's home, and the intricacies of the vascular system.

Yet Spallanzani's greatest contribution was in the area of **spontaneous generation** of microorganisms. The theory of spontaneous generation held that living creatures could develop from lifeless matter, especially from decaying matter. For instance, **Aristotle** believed that **animal** life generated spontaneously from mud, dung, or decaying timber. Other scientists believed alligators arose from Nile River mud, worms came from Thames River mud, and mites came from cheese.

Francesco Redi, (1626-1697) an Italian physician and naturalist, conducted experiments in the seventeenth century that first dispelled the myths of spontaneous generation. Using the theory that decaying products only served as a nesting site for maggots to lay eggs, Redi showed that, in hot weather, maggots would appear on exposed meat or dead animals. If the fresh meat was placed in a jar covered with a fine gauze, no maggots appeared.

Spallanzani, meanwhile, set out in 1765 to prove that microorganisms existed because they were already present in some form in the solution, the container, or the air. He took solutions which he knew would "breed" organisms and boiled them for up to an hour. The flasks were hermetically sealed to keep out contaminated air. Nothing grew.

But proponents of the spontaneous generation theory dismissed Spallanzani's experiments, saying only that the boiling process had destroyed elements vital to the propagation of the organisms. It was not until **Louis Pasteur**'s experiments on bacteria a century later that Spallanzani was proved right. Spallanzani's work regarding spontaneous generation eventually led to means of food preservation through heat sterilization and canning.

Spallanzani also turned his attention to the **circulatory system**. Viewing the system of **blood** vessels within a hen's egg in 1771, he was able to determine that an arteriovenous network existed in a warm-blooded animal. With further study of the circulatory system, in which Spallanzani studied the changes that occur upon impending **death** as well as the effects of wounds on various parts of the system, he eventually developed a theory of **blood pressure**. He determined that the arterial pulse was not due simply to displacement of the cardiac muscle, but to an intentional and forceful push of blood against the vascular walls.

One of his next inquiries involved the **fertilization** of **egg**s. He began with the mating practices of frogs and toads. By 1785, when he was working with dogs, he induced the first case of artificial insemination. Spallanzani's curiosity surrounding natural phenomena took him on an expedition to the volcanoes of Vesuvius, Stromboli, Vulcano, and Etna. During his travels, he climbed to within five feet of red-hot lava in order to measure its flow. He suffered burned feet as he descended into the bowels of Vulcano. He was rendered unconscious by the gases at Etna. Spallanzani's volcanic studies earned him status as a pioneer in the volcanology.

One of Spallanzani's final investigations took him into the dark world of bats. He was fascinated by their ability to maneuver without light. Even blinded, the bats could travel and eat without interruption or hesitation. Spallanzani went through the senses one by one, trying to discover which one governed the habits of the bat. Through the process of elimination, he found that plugging up the bats' ears rendered them directionless. While Spallanzani accepted the theory of echolocation, this theory wasn't explained until 1941, when Donald R. Griffin described the bat's sensitivity to sound waves. Spallanzani died on February 11, 1799, in Pavia, Italy.

See also Blood circulation

SPECIATION

Speciation is the formation of a new **species** by dividing one species into two. Two species are considered to be separate if the individuals cannot reproduce to produce viable offspring (those that are fertile and cannot produce offspring of their own). Thus, if two individuals are reproductively isolated, they are considered different species.

Speciation often begins when members of one species are geographically isolated from each other. **Geographical isolation** can occur because of physical barriers such as a mountain, river, forest, **desert**, or island preventing movement of individuals and genetic mixing of members of the species. For example, a **population** of fish in a large lake might be split in two during a great drought. When **water** evaporated and two separate smaller lakes formed, some individuals would be isolated from the rest of the population. The two populations of fish would be geographically isolated from each other.

Two geographically isolated populations of the same species can evolve separately over time so that they are each able to survive in the specific conditions under which they live. For example, the hypothetical fish populations described above might each live in areas with different sources of food or different predators. One population could diverge from the other to take advantage of the different conditions. While one population might remain the same as it always was, the other population could evolve **adaptations** to exploit the different sources of food or to survive in the different environmental conditions of its lake.

If enough time goes by, and the populations diverged to a great enough extent, the two species might no longer be able to reproduce and have viable (fertile) offspring; they would be reproductively isolated. Reproductive isolation can occur because of changes in **anatomy**, behavior (different mating times or seasons, different **courtship** behaviors), or physiology (the **gamete**s are incompatible).

This peach-throated monitor (*Varanus jouiensis*) is a recently-discovered species. *(Photograph by Robert J. Huffman, Field Mark Publications. Reproduced by permission.)*

If in the hypothetical situation discussed above there were great rains after many thousands of years, the two populations could be once again living in one large lake. Because enough changes have occurred, even though they might no longer be geographically isolated, they would be reproductively isolated. Since they are reproductively isolated speciation has occurred. Individuals can no longer mate, and the two populations of the same species have evolved into two separate species.

In some cases, speciation occurs even though members of a population were not geographically isolated. Genetic changes such as polyploidy (more than two complete sets of **chromosome**s per **cell**) in **plants** can on rare occasions result in speciation if two individuals with the same genetic changes are able to mate and produce viable offspring.

SPECIES

The term species is used to describe a group of closely related, physically similar **organism**s that can interbreed.

In taxonomical **classification** of **plants** and **animals**, species is the most specific category, following in order **kingdom**, **phylum** (animals) or division (plants), **class**, **order**, **family**, and **genus**. Each identified species name is assigned a two part name, part of the binomial system popularized by **Carl Linnaeus**. The first part of the name is the genus of the species. The second part of the name is the specific epithet. The two parts together give a unique name for every type of living organism. The name is generally given in a Latinized form and there are rigid sets of rules to be followed in the description of a new species and the granting of a name. These rules are given in great detail in, for plants, *The International Code of Botanical Nomenclature* and for animals in *The Zoological Code*. Other descriptive rules exist for **bacteria** and other **microorganism**s.

The taxonomic species is the smallest unit of the classification system normally used. The common names associated with many organisms generally denote a particular species.

Many individual groups show an amount of regional variation, which sets them apart form other members of the same species. These different groups are still capable of breeding with each other but they are given the rank of subspecies. The dividing lines between species or between species and subspecies are sometimes unclear.

In order to distinguish species, biologists search for recognized diagnostic characteristics. However, species differ from one another not only by conspicuous features, but also by **habitat** and genetic makeup. Members of the same species share a common **gene pool**, as well as biochemical, morphological, and behavioral characteristics.

A species **population** has a range, the habitat within which it lives and a dimension in time. Back in time, the species merges with other species, and in the future, it may branch into several species. If it's future is questionnable, the species is considered endangered. **Extinction** is a result of an imbalance between a species and its range.

See also Allopatry and sympatry; Ecosystem

SPEMANN, HANS (1869-1941)
German embryologist

The son of a well known book publisher, Spemann was born on June 27, 1869, in Stuttgart, Germany. He was the eldest of four children in a family which was socially and culturally active, and lived in a large home that was well stocked with books (which helped shape the young Spemann's intellect). Upon entering the Eberhard Ludwig Gymnasium, Spemann first wished to study the classics. He later turned to **embryology**—the branch of biology that focuses on embryos and their development.

He entered the University of Heidelberg in 1891 to study medicine, however, his strict interest in medicine lasted only until he met German biologist and psychologist Gustav Wolff at the University of Heidelberg. Only a few years older than Spemann, Wolff had begun experiments on the embryological developments of newts and had shown how, if the lens of an embryological newt's eye is removed, it regenerates. Spemann remained interested and intrigued by both Wolff's finding and also in the newt, on which he based much of his future work. But more than the regeneration phenomenon, Spemann was interested in how the eye develops from the start. He devoted his scientific career to the study of how embryological **cells** become specialized and differentiated in the process of forming a complete **organism**.

Spemann left Heidelberg in the mid–1890s to continue his studies at the University of Munich; he then transferred to the University of Würzburg's Zoological Institute to study under the well-known embryologist **Theodor Boveri**. Spemann quickly became Boveri's prize student, and completed his doctorate in **botany**, **zoology**, and physics in 1895. Spemann stayed at Würzburg until 1908, when he accepted a post as professor at the University of Rostock. During World War I, he served as director of the Kaiser Wilhelm Institute of Biology (now the Max Planck Institute) in Berlin-Dahlem, and following the war, in 1919, he took a professorship at the University of Freiburg.

By the time Spemann began research at the Zoological Institute in Würzburg, he had already developed a keen facility and reputation for conducting well-designed experiments that centered on highly focused questions. His early research followed Wolff's closely. The eye of a newt is formed when an outgrowth of the **brain**, called the optic cup, reaches the surface layer of embryonic **tissue** (the ectoderm). The cells of the ectoderm then form into an eye. In removing the tissue over where the eye would form and replacing it with tissue from an entirely different region, Spemann found that the embryo still formed a normal eye, leading him to believe that the optic cup exerted an influence on the cells of the **ectoderm**, inducing them to form into an eye. To complete this experiment, as well as others, Spemann had to develop a precise experimental technique for operating on objects often less than 0.08 in (2 mm) in diameter. In doing so, he is credited with founding the techniques of modern microsurgery, which is considered one of his greatest contributions in biology. Some of his methods and instruments are still used by embryologists and neurobiologists today.

In another series of experiments—conducted in the 1920s—Spemann was able to conclude that at a certain stage of development, the future roles of the different parts of the embryo have not been fixed, which supported his experiments with the newt's eye. In an experiment conducted on older eggs, however, Spemann found that the future role of some parts of the embryo had been decided, meaning that somewhere in between, a process he called "determination" must have taken place to fix the "developmental fate" of the cells.

One of Spemann's greatest contributions to embryology—and the one for which he won the 1935 Nobel Prize in physiology or medicine—was his discovery of what he called the "organizer" effect. In experimenting with transplanting tissue, Spemann found that when an area containing an organizer is transplanted into an undifferentiated host embryo, this transplanted area can induce the host embryo to develop in a certain way, or into an entirely new embryo. Spemann called these transplanted cells organizers, and they include the precursors to the central **nervous system**. In vertebrates, they are the first cells in a long series of differentiations of which the end product is a fully formed fetus.

Spemann remained at the University of Freiburg until his retirement in the mid–1930s. When not busy with his scientific endeavors, he cultivated his love of the liberal arts. He died at his home near Freiburg on September 12, 1941.

SPERM

Sperm are **haploid** sex **cells** of the male. They are also known as spermatozoa. Unlike **eggs** which are large, non-motile, and generally few in number, sperm are tiny, motile, and produced in huge numbers. While the human sperm length (0.002 in [600 mm]) is relatively great due to its long tail, the volume of an entire sperm, tail and all, is only 1/85,000 of the egg.

Reproduction in humans occurs when the sperm are deposited in the vagina of a female near the cervical opening of the uterus. The haploid sperm move into the uterus and up the Fallopian (uterine) tube where an egg may be encountered. With **fertilization**, the egg finishes its second meiotic division to become haploid. The haploid sperm and mature egg together form a **diploid zygote** which is the beginning of a new individual.

The male gonad (testis singular, testes plural; testicle is derived from the diminutive of testis and perhaps is best used to describe the gonads of a little boy) produces the **hormone** testosterone and sex cells. Early in **embryo**nic development, primordial germ cells, which are diploid, migrate to the embryonic gonad. The primordial germ cells give rise to the diploid stem cells of the testis, known as spermatogonia. Each of the many spermatogonia, after the first meiotic division, form two primary spermatocytes which in turn, after the second meiotic division, form four haploid spermatids. In the process of forming mature sperm the spermatids lose much of their **cytoplasm** and develop a long, propulsive tail.

Motility of the sperm is due to the long tail which is a modified flagellum. **Cilia** and **flagella**, from **protozoa** through humans, all have a basically similar structure which has been intensively investigated since first described in early electron **microscope** studies. Microtubules that run the length of the sperm tail are arranged in a ring of nine pairs surrounding a pair in the center. Ciliary dynein is associated with each of the nine microtubule pairs. It is the interaction of the dynein with the microtubules which causes flagellar bending and thus propulsion.

As stated, sperm occur in large numbers. It is estimated that a quarter of a billion sperm are released in a single ejaculate in a healthy male. The exact number is not known and there is some evidence that numbers of sperm in the ejaculate of young men may be decreasing. The diminution in number of sperm, if it is real, may be due to toxic substances in the environment and/or tight clothing chosen by young men. **Spermatogenesis** requires lower than body **temperature**. Tightly fitting underwear and tight jeans hold the testes close to the body and may interfere with spermatogenesis because of body heat.

SPERMATOGENESIS

Spermatogenesis refers to the process by which mature **sperm** (spermatozoa) **cells** are produced. When a baby boy is born, his testicles contain cords of primitive sex cells (spermatogonia), but no mature spermatozoa. The sex cells are frozen in development until the boy reaches puberty. At puberty, as the **anatomy** of the testicles changes, the spermatogonia begin to undergo a process called **meiosis**.

Meiosis is a system where by sex cells replicate, creating cells with only half the normal number of **chromosomes** (23). This half number is all that's required in a sex cell, because its ultimate purpose is to unite with another sex cell during **fertilization**. Together, then, one sperm and one **egg** will create a single cell (**zygote**) with the normal complement of 46 chromosomes.

Roger W. Sperry. *(The Library of Congress. Reproduced by permission.)*

During spermatogenesis, a single spermatogonia will ultimately give rise to four separate spermatids. These spermatids are simple cells, and do not yet possess all of the characteristics of the mature sperm cell. These characteristics (including the **differentiation** of the cell to possess both a head and a whip-like tail which allows motility) are achieved through a separate process called spermiogenesis. This process of spermiogenesis, leading from spermatid to spermatozoa, takes about 61 days.

SPERRY, ROGER W. (1913-1994)
American psychobiologist

Roger Sperry was born on 20 August 1913, in Hartford, Connecticut. When Sperry was 11 years old, his father died and his mother returned to school and got a job as an assistant to a high school principal. Sperry attended local public schools through high school and then went to Oberlin College in Ohio on a scholarship. Although he majored in English, Sperry was especially interested in his undergraduate psychology courses with R. H. Stetson, an expert on the physiology of speech. Sperry earned his B.A. in English in 1935 and then worked as a graduate assistant to Stetson for two years. In 1937 he received an M.A. in psychology.

Thoroughly committed to research in the field of psychobiology by that time, Sperry went to the University of Chicago to conduct research on the organization of the central nervous system under the renowned biologist Paul Weiss. Before Weiss's research, scientists believed that the connections of the **nervous system** had to be very exact to work properly. Weiss disproved this theory by surgically crossing a subject's nerve connections. After the surgery was performed, the subject's behavior did not change. From this, Weiss concluded that the connections of the central nervous system were not predetermined, so that a nerve need not connect to any particular location to function correctly.

Sperry tested Weiss's research by surgically crossing the nerves that controlled the hind leg muscles of a rat. Under Weiss's theory, each nerve should eventually "learn" to control the leg muscle to which it was now connected. This did not happen. When the left hind foot was stimulated, the right foot responded instead. Sperry's experiments disproved Weiss's research and became the basis of his doctoral dissertation, "Functional results of crossing nerves and transposing muscles in the fore and hind limbs of the rat." He received a Ph.D. in **zoology** from the University of Chicago in 1941.

Sperry did other related experiments that confirmed his findings and further contradicted Weiss's theory that "function precedes form" (that is, the **brain** and nervous system learn, through experience, to function properly). From these and other experiments, Sperry deduced that genetic mechanisms determine some basic behavioral patterns. According to his theory, nerves have highly specific functions based on genetically predetermined differences in the concentration of chemicals inside the nerve cells.

In 1941, Sperry moved to the laboratory of the renowned psychologist Karl S. Lashley. A year later, Lashley became director of the Yerkes Laboratories of Primate Biology in Orange Park, Florida. Sperry joined him there on a Harvard biology research fellowship. While there, he disproved some Gestalt psychology theories about brain mechanisms, as well as some theories of Lashley's.

After World War II, in 1946, Sperry accepted a position of assistant professor at the University of Chicago in the school's anatomy department. By 1954, he transferred to the California Institute of Technology (Caltech). At Caltech, Sperry conducted research on split-brain functions that he had first investigated when he worked at the Yerkes Laboratory. It had long been known that the **cerebrum** of the brain consists of two hemispheres. In most people the left hemisphere controls the right side of the body and vice versa. The two halves are connected by a bundle of millions of nerve fibers called the corpus callosum, or the great cerebral commissure.

Neurosurgeons had discovered that this connection could be cut into with little or no noticeable change in the patient's mental abilities. After experiments on animals proved the procedure to be harmless, surgeons began cutting completely through the commissure of epileptic patients in an attempt to prevent the spread of epileptic seizures from one hemisphere to the other. The procedure was generally successful, and beginning in the late 1930s, cutting through the fore-brain commissure became an accepted treatment method for severe epilepsy. Observations of the split-brain patients indicated no loss of communication between the two hemispheres of the brain.

From these observations, scientists assumed that the corpus callosum had no function other than as a prop to prevent the two hemispheres from sagging. Scientists also believed that the left hemisphere was dominant and performed higher cognitive functions such as speech. This theory developed from observations of patients whose left cerebral hemisphere had been injured; these patients suffered impairment of various cognitive functions, including speech. Since these functions were not transferred over to the uninjured right hemisphere, scientists assumed that the right hemisphere was less developed.

Sperry's work shattered these views. He and his colleagues at Caltech discovered that the corpus callosum is more than a physical prop; it provides a means of communication between the two halves of the brain and integrates the knowledge acquired by each of them. They also learned that in many ways, the right hemisphere is superior to the left. Although the left half of the brain is superior in analytic, logical thought, the right half excels in intuitive processing of information. The right hemisphere also specializes in non-verbal functions, such as understanding music, interpreting visual patterns (such as recognizing faces), and sorting sizes and shapes.

Sperry started published technical papers on his split-brain findings in the late 1960s. The importance of his research was recognized relatively quickly, and in 1979 he was awarded the prestigious Albert Lasker Basic Medical Research Award, which included a $15,000 grant. The award was given in recognition of the potential medical benefits of Sperry's research, including possible treatments for mental or psychosomatic illnesses.

In 1981, Sperry was honored with the Nobel Prize in physiology or medicine. He shared it with two other scientists, **Torsten N. Wiesel** and **David H. Hubel**, for research on the central nervous system and the brain. In describing Sperry's work, the Nobel Prize selection committee praised the researcher for demonstrating the difference between the two hemispheres of the brain and for outlining some of the specialized functions of the right brain.

Spina Bifida

Spina bifida is the common name for a range of **birth** defects caused by problems with the early development of the spine. The term comes from the Latin "spina," meaning "thorn," or "spine," and "bifida," meaning "split into two parts." The main defect of spina bifida involves an abnormal opening in the bony column through which the **spinal cord** passes, called the **vertebral** column. In spina bifida, there is an abnormal opening somewhere along the vertebral column, which leaves the spinal cord unprotected, and vulnerable to either mechanical injury or invasion by infection.

Spina bifida occurs in one out of every 700 births to whites in North America, but in only one in every 3,000 births

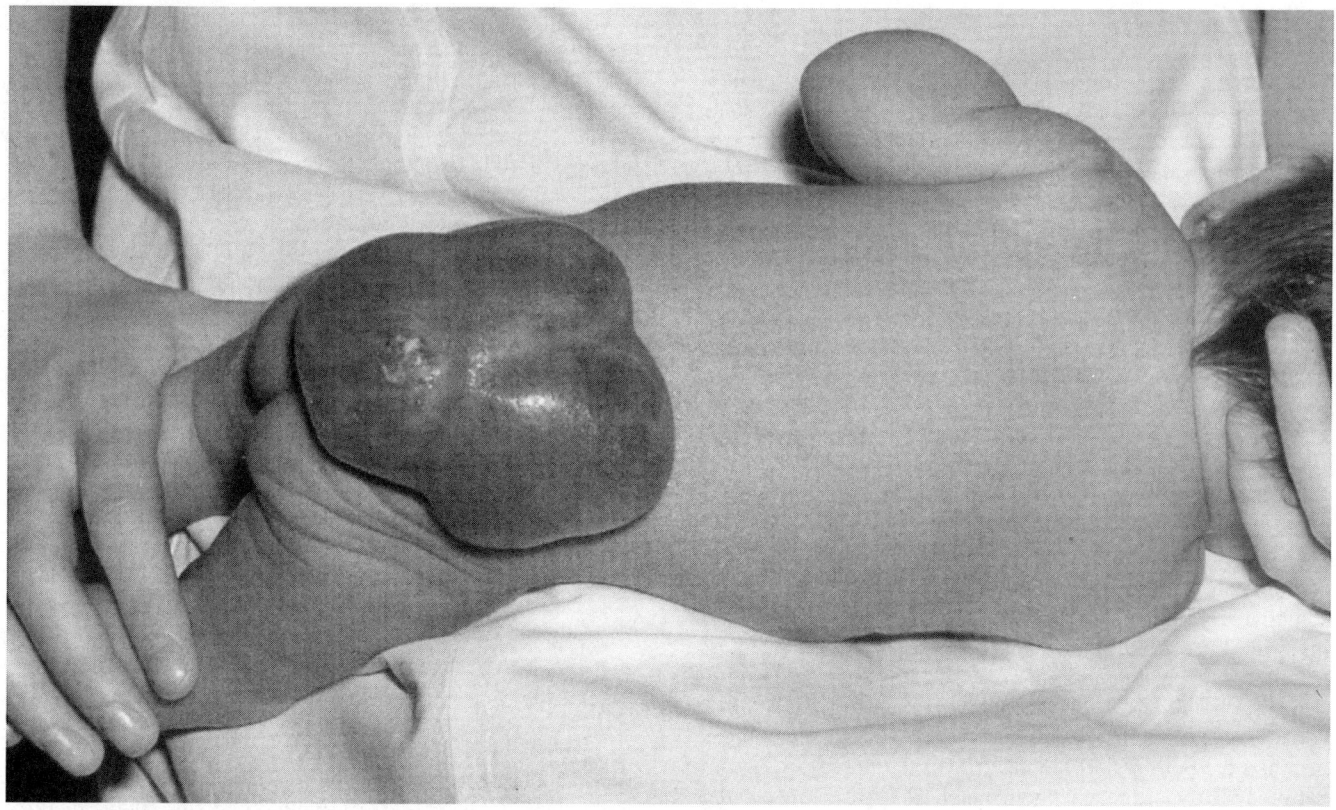

An infant with spina bifida. *(Photograph by Biophoto Associates, Photo Researchers, Inc. Reproduced by permission.)*

to African-Americans. In some areas of Great Britain, the occurrence of spina bifida is as high as one in every 100 births, leading to the hypothesis that some environmental factors may be at work.

The classic defect of spina bifida cystica is an opening in the spine, obvious at birth, out of which protrudes a fluid-filled sac. This sac may include either just the meninges, those **membrane**s which cover the spinal cord (a meningocele) or may include both the meninges and some part of the actual spinal cord (myelomeningocele). Often, the spinal cord itself has not developed properly. In spina bifida occulta, there may be some opening in the vertebrae, but no protruding sac. The entire defect may be covered with skin. At the other end of the spectrum of severity is rachischisis, in which the entire length of the spine may be open.

The problems caused by spina bifida depend on a number of factors, including where along the spine the defect occurs, what other associated defects are present, and what degree of disorganization of the spinal cord exists. The most severe types of spina bifida (raschischisis) often result in **death**, either by virtue of the greatly increased risk of infection (meningitis) due to the exposed meninges, or due to severe loss of function.

Because different levels of the spinal cord are responsible for different functions, the location and the size of the defect in spina bifida will affect what kind of disabilities an individual will experience. Most patients with clinically identi-

fiable spina bifida have some degree of weakness in the legs. This can be so severe as to be complete paralysis, depending on the spinal cord condition. The higher up in the spine the defect occurs, the more severe the disabilities.

People with spina bifida frequently face severe problems with both bladder and bowel function, because complete emptying of both bladder and bowels requires an intact spinal cord. Difficulty in completely emptying the bladder can result in severe, repeated infections, ultimately causing kidney damage, which can be life-threatening.

Other types of problems may accompany spina bifida, including changes in the architecture and arrangement of **brain** structures, resulting in the accumulation of fluid on the brain. Many people with spina bifida have other orthopedic complications, including clubfeet and hip dislocations, and abnormal curves and bends in the spine, resulting in a hunchbacked or twisted appearance. Intelligence in children with spina bifida varies widely, and certainly depends on the severity of the spinal defect and the presence of other associated defects. Children with brain abnormalities, as well as those who have had the misfortune of contracting meningitis, are most likely to have intellectual deficits. Children with spina bifida have also been noted to have a greatly increased risk of allergic sensitivity to latex. This sensitivity may cause only minor skin rashes, or may result in a major life-threatening reaction which interferes with breathing (**anaphylaxis**).

SPINAL CORD

Together, the spinal cord and the **brain** comprise the central **nervous system**. The spinal cord, a continuation of the brain stem located at the base of the brain, is protected by the bony **vertebral** column called the spine. Running approximately two-thirds of the way down the spine and ending at the small of the back, the spinal cord, like the brain, is made up of gray and white matter. The gray matter, the innermost portion of the spinal cord, is shaped like the letter "H" and formed primarily of **cell** bodies. Cell bodies in the posterior, (backward-facing) legs of the H—called the dorsal horns—are **sensory neuron**s which transmit information from the skin such as pain, pressure, **touch**, and stretch. **Neuron**s in the anterior (forward-facing) legs—called ventral horns—are motor neurons which signal muscle movement.

The white matter surrounds the gray matter and is made up of nerve tracts running to and from the brain: the corticospinal tracts carry motor (muscle) messages from the brain, dorsal columns carry sensory messages from the skin to the brain, and spinothalamic tracts carry pain and **temperature** messages to the brain. The white matter is surrounded by three layers of meninges—special **membrane**s that also envelope the brain—which form a fluid-filled protective environment for the spinal cord called the vertebral canal. The outer layer of the meninges is called the dura mater, the inner layer the pia mater, and in between the two is the arachnoid mater. The space between the pia mater and arachnoid mater—the subarachnoid space—is filled with cerebrospinal fluid which acts as a shock absorber for the spinal cord. Also, the major **blood** supply to the both the brain and spinal cord runs through this space. The space outside the dura mater, the epidural space, contains loose **tissue**. It is into the sacral area of the epidural space that anesthetics are sometimes injected to ease the pain of childbirth or for surgery performed below the waist. Branching off at intervals down the length of the spinal cord are 31 clusters of nerve roots from each of which a pair of nerve tracts emanate, one of each pair branching to the left and the other branching to the right side of the body. These nerves, which exit the vertebral canal through sleeve-like openings and the spinal column through spaces between vertebrae, form the peripheral nervous system (PNS).

The spinal column is thus the signal carrier between the brain and the PNS which has two pathways to the brain—the motor pathway and the sensory pathway. The motor pathways consist of a chain of two neurons called the upper (UMN) and lower (LMN) motor neurons. Cell bodies of the UMN are located in the gray matter of the brain and their long **axon**s run down the spinal cord. There, they **synapse** (transfer messages) to the LMNs, the long axons of which run in bundles from the spinal cord to muscles all over the body. Sensory pathways are somewhat more complex, involving a chain of at least three neurons from receptor to the brain. Instead of their **cell** bodies being located in the spinal cord, these lower sensory neurons connect to the nerve roots of the spinal column in clusters known as dorsal root **ganglion**s. Also, instead of axons, they transmit their signals via long dendrites which connect directly

to the cell body. These signals synapse to short axons which in turn synapse with intermediate sensory neurons. Some of these neurons have long axons extending up the spinal cord to the upper sensory neurons in the brain; however, some have very short axons which immediately "switch" the sensory message directly to outgoing motor neurons in the spinal cord. This is called a **reflex** arc because it bypasses the axons which carry messages to the brain, sending the sensory message directly to the muscles stimulating a reflex, an automatic, involuntary responses such as the withdrawal reflex which happens spontaneously when we touch a hot object, or the knee jerk response which makes the leg kick involuntarily when the kneecap is gently tapped. Because these incoming sensory messages trigger a response without involving the brain, reflex arcs are called "brainless responses." A reflex arc includes five parts—receptor, sensory neuron, interneuron, motor neuron, and effector.

Spinal cord injures; **virus**es such as poliomyelitis or meningitis; or cerebral vascular incidents such as strokes, hemorrhage, or tumors; can all cause motor deficits, uncontrollable reflex responses, paralysis, wasting of affected muscles, or **death**.

SPONTANEOUS GENERATION

Spontaneous generation, which is also called biopoesis or abiogenesis, is the process of living **organism**s arising *de novo* from non-living material. Until the nineteenth century, it was believed that spontaneous generation was the process by which many living creatures arose. Examples of spontaneous generation included mites arising from piles of non-living dust and **fungi** arising from dead wood, instead of merely taking advantage of the fact the dead wood is there and moving in to colonize it. Ancient Egyptians believed snakes arose from mud and the Greeks believed rats came from garbage. These misguided theories were logically based on physical observation of the relationship between the organism and its **habitat**.

By the time of the Renaissance, scientists did not believe that larger **animal**s arose from spontaneous generation, but they still clung to the belief that **bacteria** arose spontaneously. It was not until 1862, when **Louis Pasteur** offered proof to the contrary, that spontaneous generation was discarded. Pasteur showed that **microorganism**s were not spontaneously generated from non-living **matter**, but arose from preexisting microbes. By killing these preexisting microbes and not allowing new ones to grow, Pasteur illustrated that no microorganisms arose, even though the conditions prevailing had previously given rise to them. This is part of the **germ theory** of disease.

SPORE

Spores are tiny structures used by **organism**s to disperse into new environments, to survive temporarily unfavorable conditions, or to undergo **sexual reproduction**.

In **zoology**, spores are asexual, resting bodies, which depending on the **species** can be one-**cell**ed or multi-cellular.

Many **protozoa**ns have a life-cycle stage involving the development of a spore (or cyst) that is capable of surviving drought or other extreme environmental conditions. Most parasitic species, for example, must survive extremely inhospitable conditions during transmission from host to host, which often occurs through the ambient environment. Free-living protozoans also commonly utilize a spore stage to survive periods of severe environmental stress, as might occur when a pond becomes dry in late summer, or freezes during winter.

Many kinds of **bacteria**, **fungi**, actinomycetes, **yeast**s, and **algae** also develop spores, allowing them to survive periods of unfavorable conditions. Most species of fungus also develop spores as part of their generative process. Fungi have specialized **organ**s known as sporangia, which produce tiny, **diploid** spores which can germinate and develop into a mature organism. Such spores are extremely **light** and may be carried for great distances by wind or **water**, allowing long-distance dispersal to occur. Liverworts and mosses also produce tiny, diploid spores to disperse asexual progeny into new **habitat**s.

In **botany**, spores are reproductive cells that are capable of developing into a new individual plant, either directly or after fusion with another spore. Plant spores known as gonidia are developed by **mitosis** (i.e., the process of division and separation of **chromosome**s that occurs in a dividing cell). Mitosis produces two diploid daughter cells, each having the same chromosomal content as their parent cell. These kinds of spores are capable of growing into a mature organism, without undergoing fusion with another spore. The diploid spores of club-mosses and ferns, which are vascular **plants**, are bisexual structures used to propagate and disperse the plants.

Plant spores known as meiospores develop through the process of **meiosis** (i.e., the reduction division of a diploid cell, resulting in the formation of two **haploid** spores, each of which contains one of the two sets of chromosomes of the parent cell). Vascular plants produce two types of haploid spores. The megaspores are usually larger, and are regarded as the "female" spore in sexual reproduction of plants, because the female **gametophyte** (or ovule or "egg") develops from these types of spores. Microspores are smaller, and they develop into the "male" gametophyte (or pollen). Fusion of the male and female gametophytes leads to the development of plant seeds, which are diploid structures that culminate sexual reproduction in higher plants.

Spores are tiny structures used by organisms to disperse into new environments, to survive temporarily unfavorable conditions, or to undergo sexual reproduction. *(Photograph by Hugh Spencer, Photo Researchers, Inc. Reproduced by permission.)*

phase (2n) of the life cycle alternates with the gametophyte-producing phase (1n), in what is known as the "alternation of generations."

In many **algae**, the greater part of the life cycle consists of the gametophyte generation, with the sporophyte being restricted to a single-**cell**ed zygote. In contrast, bryophytes (mosses and liverworts) have two multi-celled generations of similar size and time of existence. In the vascular plants, the sporophyte phase is by far the dominant one, with the gametophyte restricted to an almost negligible part of the life-cycle.

SPOROPHYTE

A sporophyte is the **diploid** (i.e., it is 2n, having two sets of **chromosome**s), **spore**-producing generation in the **life cycle** of **plants**. A sporophyte is formed by the fusion of two **haploid gamete**s (or **gametophyte**s, these have one set of chromosomes, i.e., are 1n).

Sporophytes produce gametophytes through the process of **meiosis**, or reduction division of the chromosomal material. Two haploid gametophytes (one from each "parent") fuse (in a process known as **fertilization**) to produce a **zygote** (or **embryo**), representing a sporophyte. Therefore, the sporophyte

STABILIZING SELECTION

Stabilizing selection, also called normalizing or centripetal selection, is a form of **natural selection** that focuses on the removal of **genes** that produce any form of deviation from the mean of the **population**. It is a method of maintaining a particular set of characteristics within a population over successive generations, while reducing the amount of extreme variations.

See also Directional selection; disruptive selection

STANLEY, WENDELL MEREDITH (1904-1971)
American biochemist

Wendell Meredith Stanley was a biochemist who was the first to isolate, purify, and characterize the crystalline form of a **virus**. During World War II, he led a team of scientists in developing a vaccine for viral **influenza**. His efforts have paved the way for understanding the molecular basis of **heredity** and formed the foundation for the new scientific field of molecular biology. For his work in crystallizing the tobacco mosaic virus, Stanley shared the 1946 Nobel Prize in chemistry with **John Howard Northrop** and **James B. Sumner**.

Stanley was born in the small community of Ridgeville, Indiana, on August 16, 1904. His parents, James and Claire Plessinger Stanley, were publishers of a local newspaper. As a boy, Stanley helped the business by collecting news, setting type, and delivering papers. After graduating from high school he enrolled in Earlham College, a liberal arts school in Richmond, Indiana, where he majored in chemistry and mathematics. He played football as an undergraduate, and in his senior year he became team captain and was chosen to play end on the Indiana All-State team. In June of 1926 Stanley graduated with a bachelor of science degree. His ambition was to become a football coach, but the course of his life was changed forever when an Earlham chemistry professor invited him on a trip to Illinois State University. Here, he was introduced to Roger Adams, an organic chemist, who inspired him to seek a career in chemical research. Stanley applied and was accepted as a graduate assistant in the fall of 1926.

In graduate school, Stanley worked under Adams, and his first project involved finding the stereochemical characteristics of biphenyl, a **molecule** containing carbon and hydrogen atoms. His second assignment was more practical; Adams was interested in finding chemicals to treat leprosy, and Stanley set out to prepare and purify compounds that would destroy the disease-causing pathogen. Stanley received his master's degree in 1927 and two years later was awarded his Ph.D. In the summer of 1930, he was awarded a National Research Council Fellowship to do postdoctoral studies with Heinrich Wieland at the University of Munich in Germany. Under Wieland's tutelage, Stanley extended his knowledge of experimental biochemistry by characterizing the properties of some yeast compounds.

Stanley returned to the United States in 1931 to accept the post of research assistant at the Rockefeller Institute in New York City. Stanley was assigned to work with W. J. V. Osterhout, who was studying how living **cell**s absorb potassium ions from seawater. Stanley was asked to find a suitable chemical model that would simulate how a marine plant called *Valonia* functions. He discovered a way of using a water-insoluble solution sandwiched between two layers of **water** to model the way the plant exchanged ions with its environment. The work on *Valonia* served to extend Stanley's knowledge of biophysical systems, and it introduced him to current problems in biological chemistry.

Begins Chemical Studies on Tobacco Mosaic Virus

In 1932, Stanley moved to the Rockefeller Institute's Division of Plant Pathology in Princeton, New Jersey. He was primarily interested in studying viruses. Viruses were known to cause diseases in **plants** and **animals**, but little was known about how they functioned. His assignment was to characterize viruses and determine their composition and structure.

He began work on a virus that had long been associated with the field of virology. In 1892, D. Ivanovsky, a Russian scientist, had studied tobacco mosaic disease, in which infected tobacco plants develop a characteristic mosaic pattern of dark and light spots. He found that the tobacco plant juice retained its ability to cause infection even after it was passed through a filter. Six years later M. Beijerinck, a Dutch scientist, realized the significance of Ivanovsky's discovery: the filtration technique used by Ivanovsky would have filtered out all known **bacteria**, and the fact that the filtered juice remained infectious must have meant that something smaller than a bacterium and invisible to the ordinary light microscope was responsible for the disease. Beijerinck concluded that tobacco mosaic disease was caused by a previously undiscovered type of infective agent, a virus.

Stanley was aware of recent techniques used to precipitate the tobacco mosaic virus (TMV) with common chemicals. These results led him to believe that the virus might be a protein susceptible to the reagents used in **protein** chemistry. He set out to isolate, purify, and concentrate the tobacco mosaic virus. He planted Turkish tobacco plants, and when the plants were about six inches tall, he rubbed the leaves with a swab of linen dipped in TMV solution. After a few days the heavily infected plants were chopped and frozen. Later, he ground and mashed the frozen plants to obtain a thick, dark liquid. He then subjected the TMV liquid to various **enzyme**s and found that some would inactivate the virus and concluded that TMV must be a protein or something similar. After exposing the liquid to more than 100 different chemicals, Stanley determined that the virus was inactivated by the same chemicals that typically inactivated proteins, and this suggested to him, as well as others, that TMV was protein-like in nature.

Stanley then turned his attention to obtaining a pure sample of the virus. He decanted, filtered, precipitated, and evaporated the tobacco juice many times. With each chemical operation, the juice became more clear and the solution more infectious. The end result of two-and-one-half years of work was a clear concentrated solution of TMV which began to form into crystals when stirred. Stanley filtered and collected the tiny, white crystals and discovered that they retained their ability to produce the characteristic lesions of tobacco mosaic disease.

After successfully crystallizing TMV, Stanley's work turned toward characterizing its properties. In 1936, two English scientists at Cambridge University confirmed Stanley's work by isolating TMV crystals. They discovered that the virus consisted of ninety-four percent protein and six percent nucleic acid, and they concluded that TMV was a nucleoprotein. Stanley was skeptical at first. Later studies, however, showed that the virus became inactivated upon removal of the nucleic acid, and this work convinced him that TMV was indeed a nucleoprotein. In addition to chemical evidence, the first electron microscope pictures of TMV were produced by researchers in Germany. The pictures showed the crystals to have a distinct rod-like shape. For his work in crystallizing the tobacco mosaic virus, Stanley shared the 1946 Nobel prize in chemistry with John Howard Northrop and James Sumner.

Develops Influenza Vaccine

During World War II, Stanley was asked to participate in efforts to prevent viral diseases, and he joined the Office of Scientific Research and Development in Washington D.C. Here, he worked on the problem of finding a vaccine effective against viral influenza. Such a substance would change the virus so that the body's **immune system** could build up defenses without causing the disease. Using fertilized hen eggs as a source, he proceeded to grow, isolate, and purify the virus. After many attempts, he discovered that formaldehyde, the chemical used as a biological preservative, would inactivate the virus but still induce the body to produce antibodies. The first flu vaccine was tested and found to be remarkably effective against viral influenza. For his work in developing large-scale methods of preparing vaccines, he was awarded the Presidential Certificate of Merit in 1948.

In 1948, Stanley moved to the University of California in Berkeley, where he became director of a new virology laboratory and chair of the department of biochemistry. In five years Stanley assembled an impressive team of scientists and technicians who reopened the study of plant viruses and began an intensive effort to characterize large, biologically important molecules. In 1955 **Heinz Fraenkel-Conrat**, a protein chemist, and R. C. Williams, an electron microscopist, took TMV apart and reassembled the viral RNA, thus proving that RNA was the infectious component. In addition, their work indicated that the protein component of TMV served only as a protective cover. Other workers in the virus laboratory succeeded in isolating and crystallizing the virus responsible for polio, and in 1960 Stanley led a group that determined the complete **amino acid** sequence of TMV protein. In the early 1960s, Stanley became interested in a possible link between viruses and **cancer**.

Stanley was an advocate of academic freedom. In the 1950s, when his university was embroiled in the politics of McCarthyism, members of the faculty were asked to sign oaths of loyalty to the United States. Although Stanley signed the oath of loyalty, he publicly defended those who chose not to, and his actions led to court decisions which eventually invalidated the requirement.

Stanley received many awards, including the Alder Prize from Harvard University in 1938, the Nichols Medal of the American Chemical Society in 1946, and the Scientific Achievement Award of the American Medical Association in 1966. He held honorary doctorates from many colleges and universities. He was a prolific author of more than 150 publications and he co-edited a three volume compendium entitled *The Viruses*. By lecturing, writing, and appearing on television he helped bring important scientific issues before the public.

He served on many boards and commissions, including the National Institute of Health, the World Health Organization, and the National Cancer Institute.

Stanley married Marian Staples Jay on June 25, 1929. They had met at the University of Illinois, where they both were graduate students in chemistry. They coauthored a scientific paper together with Adams, which was published the same year they were married. The Stanleys had three daughters and one son. On June 15, 1971, while attending a conference on biochemistry in Spain, Stanley died from a heart attack.

STARLING, ERNEST H. (1866-1927)
English physiologist

An experimentalist who discovered a number of significant fundamental facts about the cardiovascular system, Ernest H. Starling is known for his discovery, along with long-time collaborator **William Maddock Bayliss**, of secretin. In 1902, Starling and Bayliss found that the release of digestive juices by the pancreas is caused by a chemical they named "secretin." Starling later suggested the name "**hormone**" for any chemical, such as secretin, that is released in one part of the body and causes an effect in another part of the body. In the course of his career, Starling also conducted studies into the **circulatory system**, as well as into the secretion of lymph and other body fluids. For his work, he was awarded the Medal of the Royal Society, the Baly Medal from the Royal College of Physicians, and several honorary degrees from such institutions as Trinity College, Dublin, the University of Strasburg, and the University of Sheffield.

Ernest Henry Starling was born in London on April 17, 1866. His father was Matthew Henry Starling, a clerk for the British government who served in Bombay and returned to England only once every three years. Rearing of the Starling children was the responsibility, therefore, of Matthew's wife, Ellen Mathilda Watkins. In *Dictionary of Scientific Biography*, essayist Carleton B. Chapman described the family as one "of limited financial means and fundamentalist religious beliefs."

Starling began his schooling at the age of six when he was enrolled at the Islington School. He then attended King's College School from 1880 to 1882 and, in the latter year, enrolled at Guy's Hospital Medical College in London. He interrupted his schooling at Guy's briefly in the summer of 1885 when he traveled to Heidelberg to study with German physiologist Wilhelm Kühne, who was known for his research into nerves and muscles. Kühne apparently had a significant impact on Stanley's growing view of the role of basic physiology in the understanding and treatment of medical disorders.

Even before receiving his medical degree in 1889, Starling achieved distinction by being appointed demonstrator in physiology at Guy's. At the time, the hospital had simple and inadequate research facilities, a condition that changed after Starling was promoted to head of the physiology department a few years later. By the time he left Guy's in 1899, Starling had overseen the construction and equipping of a new physiology building that ultimately earned a reputation as one of the best research facilities for physiology in London.

Starling was married in 1891 to Florence Amelia Wooldridge Sieveking, daughter of an eminent London physician and widow of another physiologist, Leonard Charles Wooldridge. The Starlings had four children, three daughters and a son; Florence assisted immensely with her husband's work.

The deplorable working conditions at Guy's prompted Starling to look elsewhere for research facilities even as he was working to improve those conditions at Guy's. Thus, in 1890 he was given a part-time appointment in the physiology laboratories of Edward Albert Schäfer at University College. There he met and began working with fellow physiologist William Maddock Bayliss, an association that was to last throughout Starling's life and, incidentally, resulted in the marriage of Starling's sister to Bayliss.

Starling's first important work resulted from an 1892 visit to the Breslau laboratories of German physiologist Rudolf Heidenhain, an authority on the study of lymph. During his stay in Breslau, Starling repeated many of Heidenhain's experiments (and conducted some of his own) and came to radically different conclusions about their meaning. He was able to demonstrate that a combination of hydrostatic **blood pressure** and osmotic forces could account for all of the observations made by Heidenhain and himself about the way lymph is formed and transported in the body. In *The Dictionary of National Biography, 1922–1930*, contributor J. Barcroft asserted that Starling's findings "so completely superseded previous work in this field as to put Starling, in his early thirties, into the first rank of experimental physiologists."

Bayliss Collaboration Results in Discovery of Secretin

Probably the discovery for which Starling is most famous occurred shortly after his return to London and his election to the Jodrell chair of physiology at University College in 1899. He and Bayliss began a study of the **secretion** of digestive juices by the pancreas. They eventually found that the process takes place under the control of a substance secreted by the small intestine, a substance they named "secretin." Starling suggested that the name "hormone" be given to any chemical, such as secretin, that transmits a message from one part of the body to another part. Although hormones had actually been known before the discovery of secretin in 1902, it was Starling who first clearly defined the concept and elucidated the role that such substances have in the body.

The year Starling and Bayliss discovered secretin also marked the beginning of the former scientist's research on the heart. In order to carry out this research, Starling used a "heart-lung preparation" consisting of a heart that has been isolated in an anesthetized **animal** from all other organs except the lungs. Starling focused his research on various factors—such as temperature and blood pressure—that affect the beating of the heart. As a result of his studies, he discovered a number of facts about the heart, including one that has become known as Starling's Law of the heart: the energy of contraction is a function of the length of the muscle fibers in the heart.

Starling's heart research was interrupted by the outbreak of World War I. After his enlistment, he first served as a medical officer at Herbert Hospital, then became a researcher on defensive mechanisms against poison gas. His contributions to the war effort were apparently somewhat limited, however, because of his "outspoken impatience with the obtuseness, where scientific matters were concerned, of his military superiors," as Chapman explained in *Dictionary of Scientific Biography*.

The intensity of Starling's research diminished after the war, but he continued to exert influence in the field of physiology. In 1922 he was appointed to the newly created post of Foulerton Research Professor at the Royal Society. He also became very much interested in the state of education in Great Britain, particularly with regard to the role of the natural sciences in a liberal education. His health began to deteriorate after the war, and he died on May 2, 1927, aboard a cruise ship outside Kingston, Jamaica, where he was buried.

STEIN, WILLIAM HOWARD (1911-1980)
American biochemist

William Howard Stein, in partnership with **Stanford Moore**, was a pioneer in the field of **protein** chemistry. Although other scientists had previously established that proteins could play such roles as that of **enzyme**s, antibodies, **hormone**s, and oxygen carriers, almost nothing was known of their chemical makeup. Stein and Moore, during some forty years of collaboration, were not only able to provide information about the inner workings of protein **molecule**s, but also invented the mechanical means by which that information could be extracted. Their discovery of how protein **amino acid**s function was accomplished through a study of ribonuclease (RNase), a pancreatic enzyme that assists in the digestion of food by catalyzing the breakdown of **nucleic acids**. But their work could not have been accomplished without the development of a technology to assist them in collecting and separating the amino acids contained in ribonuclease. Their invention of the fraction collector and an automated system for analyzing amino acids was of great importance in furthering protein research, and these devices have become standard laboratory equipment.

Stein and Moore began their collective work in the late 1930s under Max Bergmann at the Rockefeller Institute (now Rockefeller University). After Bergmann's death in 1944, the pair developed the protein chemistry program at the Institute and began their research into enzyme analysis. Except for a brief period during World War II when Moore served with the Office of Scientific Research and Development in Washington D.C., and the two years when Stein taught at the University of Chicago and Harvard University, the partnership continued uninterrupted until Stein's death in 1980. Their joint inventions and co-authorship of most of their scientific papers were said to make it impossible to separate their individual accomplishments. Their combined efforts were acknowledged in 1972 with the Nobel Prize in chemistry. According to Moore, writing about Stein in the *Journal of Biological Chemistry* in 1980, they received the award "for contributions to the knowledge of the chemical structure and catalytic function of bovine pancreatic ribonuclease." **Christian Anfinsen** shared the Nobel Prize with Stein and Moore for related research.

The son of community-minded parents, Stein was born in New York City on June 25, 1911. He was the second of three children. His father, Fred M. Stein, was involved in business and retired at an early age to lend his services to various health care associations in the community. The scientist's mother, Beatrice Borg Stein, worked to improve recreational and educational conditions for underprivileged children. From an early age, Stein was encouraged by his parents to develop an interest in science. He received a progressive education from grade school on, attending the Lincoln School of the Teacher's College of Columbia University, transferring at sixteen years of age to Phillips Exeter Academy for his college preparatory studies. He graduated from Harvard University in 1933, then took a year of graduate study in organic chemistry there. Finding that his real interest was biochemistry, he completed his graduate studies at the College of Physicians and Surgeons of Columbia University, receiving his Ph.D. in 1938. His dissertation concerned the amino acid composition of elastin, a protein found in the walls of veins and arteries. This work marks the beginning of his long search to understand the chemical function of proteins.

Improved Methodology Solves Amino Acid Puzzle

The successful research being done at the Rockefeller Institute under the direction of Max Bergmann caught Stein's attention. He pursued post-graduate studies there in 1938, spending his time improving analytical techniques for purifying amino acids. Moore joined Bergmann's group in 1939. There, he and Stein began work in developing the methodology for analyzing the amino acids glycine and leucine. Their work was interrupted when the United States entered World War II. Then, Bergmann's laboratory was given over to the study of the physiological effects of mustard gases, in the hope of finding a counteractant.

The group's efforts to find accurate tools and methods for the study of amino acid structure increased in importance when they assumed the responsibility of establishing the Institute's first program in protein chemistry. Looking for ways to improve the separation process of amino acids, they turned to partition **chromatography**, a filtering technique developed during the war by the English biochemists A. J. P. Martin and Richard Synge. Building on this technology, as well as that of English biochemist **Frederick Sanger**'s column chromatography and the ion-exchange technique of Werner Hirs, Stein and Moore went on to invent the automatic fraction collector and develop the automated system by which amino acids could be quickly analyzed. This automated system replaced the tedious two-week sequence that was previously required to differentiate and separate each amino acid.

From then on, the isolation and study of amino acid structure was advanced through these new analytical tools. Ribonuclease was the first enzyme for which the biochemical function was determined. The discovery that the amino acid sequence was a three-dimensional, chain-like structure that folds and bends to cause a catalytic reaction was a beginning for understanding the complex nature of enzyme catalysis.

Stein and Moore were certain that this understanding would result in crucial medical advances. By 1972, the year Stein and Moore shared the Nobel Prize, other enzymes had been analyzed using their methods.

Editorial Work Complements Research

Because he was extremely eager to see that research done in laboratories all over the country be disseminated as widely and as quickly as possible, Stein devoted many years in various editorial positions to the *Journal of Biological Chemistry*. Under his leadership, the journal became a leading biochemistry publication. He had joined the editorial board in 1962 and became editor in 1968. He only held the latter post for one year, however. While attending an international meeting in Denmark, he contracted Guillain-Barré Syndrome, a rare disease often causing temporary paralysis. In grave danger of dying, he managed to recover somewhat. The illness left him a quadriplegic, confined to a wheelchair for the rest of his life. Although he remained involved with the work of his colleagues both in the laboratory and at the *Journal,* he was unable to participate actively.

In addition to the Nobel Prize, Stein shared with Moore the 1964 Award in Chromatography and Electrophoresis and the 1972 Theodore Richard Williams Medal of the American Chemical Society. He served as chairperson of the U.S. National Committee for Biochemistry from 1968 to 1969, as trustee of Montefiore Hospital, and as board member of the Hebrew University medical school. He married Phoebe L. Hockstader on June 22, 1936. They had three sons: William Howard, Jr., David, and Robert. Stein died in Manhattan on February 2, 1980.

STEROIDS

Steroids are a type of lipid characterized by a seventeen carbon atom fused ring structure. This structure has four interconnected rings and a variety of functional groups attached depending on the class of steroid. Steroids are found throughout the human body and produce significant effects even in small amounts. There are four different classes of steroids including **cholesterol**, bile acids, adrenocorticoid **hormone**s and **sex hormones**. While each of these types of steroids have similar structures, they have vast differences in their effects.

Cholesterol is one of the most abundant and important type of steroid and is found in nearly every **organism** known. It is located in all **cell**s and in both **brain** and nervous **tissue** where it helps form myelin tissue. It has the typical steroid ring structure, but also contains a hydroxyl group which creates some unique functional characteristics. One implication of this hydroxyl group is that it gives the cholesterol **molecule** a degree of compatibility with an aqueous environment. Since cholesterol is generally lipid soluble, the added **water** compatibility has made it a critical material for stabilization of the cell **membrane**. Cholesterol is also an important precursor to a variety of other materials such as bile acids, steroid hormones and vitamin D.

The body's supply of cholesterol is obtained through the diet or through biosynthetic pathways. Although cholesterol is essential for life, excessive amounts can lead to health problems such as **heart** disease, gall stones, high **blood pressure**, and hypertension. These diseases are caused by a buildup of cholesterol deposits on the inside walls of the **arteries**. This condition, known as atherosclerosis or hardening of the arteries, is especially significant in arteries that supply **blood** to the heart. When these arteries are blocked the heart can become so damaged that **death** results. To correct this problem a procedure known as angioplasty is typically performed. Studies have shown that the amount of ingested saturated **fats** is related to the amount of cholesterol in the blood. This fact has led to the recommendation that dietary saturated fats be limited.

Bile acids are another type of steroid found in the body. They are produced from cholesterol, synthesized in the liver and stored in the gall bladder. The most abundant type of bile acid in humans is cholic acid. It is a material that helps digest fats in the intestine by emulsifying them. Bile acids also aid in the **digestion** of cholesterol in food and play a role in minimizing levels of body cholesterol.

Adrenocorticoid hormones are steroids which are produced by the **adrenal gland**. They are 21 carbon steroids that are involved in various biological processes such as the regulation of water and the **metabolism** of **protein**s and **carbohydrate**s. Steroids included in this class include cortisol, corticosterone, and aldosterone. Cortisol, also known as hydrocortisone, slows protein production which allows the liver to increase blood glucose levels. It also provides resistance to infections, trauma and hemorrhage.

The final class of naturally occurring steroids is the sex hormones. These materials are produced by special glands in the **endocrine system**, and are carried throughout the body in the bloodstream. They have a variety of effects. Testosterone is a male hormone that influences the development of male sex characteristics such as the growth of reproductive **organ**s, muscle structure, deep voice, and hair growth. Female hormones include progesterone and **estrogen**. These hormones control the events that make up the **menstrual cycle**. Estrogen produces feminine secondary sex characteristics such as body shape and axillary hair growth, and also leads to proper bone formation. Progesterone inhibits ovulation. This has led to the development of synthetic progesterone type molecules that function as birth control pills.

In addition to these naturally occurring steroids, a variety of synthetic steroids have been produced. These molecules are many times more powerful than natural hormones and are designed to have more specific effects. They are used to treat conditions such as **arthritis**, allergies, and rheumatic fever. Anabolic steroids, which are derivatives of male hormones, have been used by some athletes to increase muscle mass and weight. Due to health concerns and issues of unfairness, these materials have been banned by most sports. The use and misuse of synthetic steroids can lead to various undesirable side effects like weight gain, ulcers and hyperglycemia.

See also Adrenocortitropic hormone (ACTH)

STEVENS, NETTIE MARIA (1861-1912)
American biologist and cytogeneticist

Nettie Maria Stevens was a biologist and cytogeneticist and one of the first American women to be recognized for her contributions to scientific research. "She...produced new data and new theories," wrote Marilyn Bailey Ogilvie in *Women in Science*, "yet beyond these accomplishments passed along her expertise to a new generation.... illustrat[ing] the importance of the women's colleges in the education of women scientists." Although Stevens started her research career when she was in her thirties, she successfully expanded the fields of embryology and cytogenetics (the branch of biology which focuses on the study of **heredity**), particularly in the study of **histology** (a branch of **anatomy** dealing with plant and **animal** tissues) and of regenerative processes in invertebrates such as hydras and flatworms. She is best known for her role in **genetics**—her research contributed greatly to the understanding of **chromosome**s and heredity. She theorized that the sex of an organism was determined by the inheritance of a specific chromosome—X or Y—and performed experiments to confirm this hypothesis.

Stevens, the third of four children and the first daughter, was born in Cavendish, Vermont, on July 7, 1861, to Ephraim Stevens, a carpenter of English descent, and Julia Adams Stevens. Historians know little about her family or her early life, except that she was educated in the public schools in Westford, Massachusetts, and displayed exceptional scholastic abilities. Upon graduation, Stevens taught Latin, English, mathematics, physiology and zoology at the high school in Lebanon, New Hampshire. As a teacher she had a great zeal for learning that she tried to impart both to her students and her colleagues. Between 1881 and 1883, Stevens attended the Normal School at Westfield, Massachusetts, consistently achieving the highest scores in her class from the time she started until she graduated. She worked as a school teacher, and then as a librarian for a number of years after she graduated; however, there are gaps in her history that are unaccounted for between this time and when she enrolled at Stanford University in 1896.

Furthers Education at Stanford and Bryn Mawr

In 1896, Stevens was attracted by the reputation of Stanford University for providing innovative opportunities for individuals aspiring to pursue their own scholastic interests. At the age of thirty-five she enrolled, studying physiology under professor Oliver Peebles Jenkins. She spent summers studying at the Hopkins Seaside Laboratory, Pacific Grove, California, and pursuing her love of learning and of biology. During this time, Stevens decided to switch careers to focus on research, instead of teaching. While at Hopkins she performed research on the life cycle of *Boveria*, a protozoan **parasite** of sea cucumbers. Her findings were published in 1901 in the *Proceedings of the California Academy of Sciences*. After obtaining her masters degree—a highly unusual accomplishment for a woman in that era—Stevens returned to the East to study at Bryn Mawr College, Pennsylvania, as a graduate biology student in 1900. She was such an exceptional student that she was awarded a fellowship enabling her to study at the Zoological Station in Naples, Italy, and then at the Zoological Institute of the University of Würzburg, Germany. Back at Bryn Mawr, she obtained her doctorate in 1903. At this time, she was made a research fellow in biology at Bryn Mawr and then was promoted to a reader in experimental morphology in 1904. From 1903 until 1905, her research was funded by a grant from the Carnegie Institution. In 1905, she was promoted again to associate in experimental morphology, a position she held until her death in 1912.

Contributes to the Understanding of Chromosomal Determination of Sex

While Stevens' early research focused on **morphology** and **taxonomy** and then later expanded to **cytology**, her most important research was with chromosomes and their relation to heredity. Because of the pioneering studies performed by the renowned monk **Gregor Mendel** (showing how pea plant genetic traits are inherited), scientists of the time knew a lot about how chromosomes acted during **cell division** and maturation of germ cells. However, no inherited trait had been traced from the parents' chromosomes to those of the offspring. In addition, no scientific studies had yet linked one chromosome with a specific characteristic. Stevens, and the well-known biologist **Edmund Beecher Wilson**, who worked independently on this type of research, were the first to demonstrate that the sex of an organism was determined by a particular chromosome; moreover, they proved that gender is inherited in accordance with Mendel's laws of **genetics**. Together, their research confirmed, and therefore established, a chromosomal basis for heredity. Working with the meal worm, *Tenebrio molitor,* Stevens determined that the male produced two kinds of sperm—one with a large X chromosome, and the other with a small Y chromosome. Unfertilized eggs, however, were all alike and had only X chromosomes. Stevens theorized that sex, in some organisms, may result from chromosomal inheritance. She suggested that eggs fertilized by sperm carrying X chromosomes produced females, and those by sperm carrying the Y chromosome resulted in males. She performed further research to prove this phenomenon, expanding her studies to other species. Although this theory was not accepted by all scientists at the time, it was profoundly important in the evolution of the field of genetics and to an understanding of determination of gender.

Stevens was a prolific author, publishing some thirty-eight papers in eleven years. For her paper, "A Study of the Germ Cells of *Aphis rosae* and *Aphis oenotherae,* "Stevens was awarded the Ellen Richards Research Prize in 1905, given to promote scientific research by women. Stevens died of breast cancer on May 4, 1912, before she could occupy the research professorship created for her by the Bryn Mawr trustees. Much later, **Thomas Hunt Morgan**, a 1933 Nobel Prize recipient for his work in genetics, recognized the importance of Stevens' ground-breaking experiments, as quoted by Ogilvie in the *Proceedings of the American Philosophical Society,* "Stevens had a share in a discovery of importance and her name will be remembered for this, when the minutiae of detailed investigations that she carried out have become incorporated in the general body of the subject."

STIMULUS

A stimulus is an environmental change that triggers a response or reaction. **Organism**s use **sensory neuron**s to interpret these changes. Humans interpret their entire external environment through **energy** in the form of **light**, sound, pressure, chemicals, or heat. This process is called *exteroception*, in which energy external to the body triggers responses in excitatory **cell**s in the central **nervous system**. Different stimuli activate different receptors which trigger cells to produce **action potential**s. Action potentials send electrochemical signals from one sensory neuron to another via the spinal chord to the **brain** where the signal is decoded. Meanwhile, the peripheral nervous system senses and controls the body's internal environment such as **hormone** levels, **blood** sugar levels, **blood pressure**, and **temperature**. Internal stimulus and response is called *proprioception* and *interoception*. In all three, some type of stimulus is necessary to bring cells from resting **membrane potential** to an action potential. In exteroception, light stimulates rods and cones in the back of the eye to produce chemical responses which, when decoded by the visual cortex, produce visual images and thus we have **sight**. Chemical (odor) **molecule**s inhaled through the nostrils stimulate olfactory **neuron**s (chemoreceptors) whose action potentials carry information to the olfactory cortex to be decoded, giving us our sense of **smell**. The stimulus for neuroepithelial cells in **taste** buds on the tongue are chemical molecules which, when mixed with saliva, initiate action potentials which **synapse** to the medulla, the thalamus, and the postcentral gyrus, allowing us to taste. Tactile stimuli activates sensory neurons which carry messages to the somatosensory cortex, giving us the sense of **touch**. Sound waves stimulate hair cells inside the ear which convert the stimulus into electrochemical impulses which are transmitted to the brain via the **hearing** or auditory nerve, and we hear. Proprioception and interoception are also controlled by stimuli; however, unlike exteroceptive senses, we are unaware of the internal stimuli which trigger responses.

STREPTOCOCCUS • See Bacteria

STRESS

Stress can be defined as a force exerted when one body or body part presses on, pulls on, pushes against, or tends to compress or twist another body or body part. For example, geostress is the causal factor in all major landforms on the earth. The intensity of this mutual force is expressed in pounds per square inch. Stress can also refer to the deformation caused in a body by such a force.

Stress, in psychology and biology, is any strain or interference that disturbs the functioning of an **organism**. The human being responds to physical and psychological stress with a combination of psychic and physiological defenses. If the stress is too powerful, or the defenses inadequate, a psychosomatic disease, ailment or other mental disorder may result.

Stress is an unavoidable effect of living and is an especially complex phenomenon in modern technological society. There is little doubt that an individual's success or failure in controlling potentially stressful situations can have a profound effect on his ability to function. The ability to "cope" with stress has figured prominently in psychosomatic research. Researchers have reported a statistical link between coronary **heart** disease and individuals exhibiting stressful behavioral patterns designated "Type A." These patterns are reflected in a lifestyle characterized by impatience and a sense of time urgency, hard-driving competitiveness, and preoccupation with vocational and related deadlines.

Various strategies have been successful in treating psychological and physiological stress. Moderate stress may be relieved by exercise and any type of meditation (e.g., yoga or other Asian meditative forms). Severe stress may require psychotherapy to uncover and work through the underlying causes. A form of behavior therapy known as biofeedback enables the patient to become more aware of internal processes and thereby gain some control over bodily reactions to stress. Sometimes, a change of environment or living situation may produce therapeutic results.

STRUCTURAL GENE • See Operon

STURTEVANT, ALFRED HENRY (1891-1970)

American geneticist

Alfred Henry Sturtevant is the American geneticist recognized for introducing the principle of **gene mapping**. He was the first to suggest that the closer together genes are located on a **chromosome** the more likely they are to be inherited together.

Sturtevant was born in Jacksonville, Illinois, in 1891 and was raised on his parent's ranch. Even as a boy Sturtevant showed a strong interest in **genetics** and he prepared pedigrees of his father's horses. Alfred was encouraged by his brother, Edgar, to pursue his interests by studying genetics. In 1910 Sturtevant began working with professor **Thomas Hunt Morgan** at Columbia University.

At the time Morgan was already performing extensive experiments on *Drosophila* (the common **fruit fly**) and his results were consistent with Gregor Mendel's theories that hereditary factors are independent of one another. However, Morgan had also observed that some genes seemed to be linked together; in other words, they were not independent. For example, he had noticed that white-eyed fruit flies were almost always male. Morgan and his team developed a theory to explain this cross-over of linked traits. They proposed that **gene**s that remain together when passing from one generation to the next must be located on the same chromosome.

Sturtevant worked with Morgan in the "fly room," which was the nickname given to the lab where the fruit flies were kept. Here Sturtevant made his breakthrough discovery while researching the hereditary patterns of the flies. He took Morgan's theories one step further and proposed that genes which were close to one another on the chromosome could sometimes get passed along together because of their proximity. To support this theory, Sturtevent used his proficiency in solving mathematical puzzles to analyze how often certain genes were inherited together. From his analysis, he introduced the concept of genetic cross over. Applying his new theory, Sturtevant went home one evening in 1911 and drew the first chromosome map which detailed the gene positions on the four *Drosophila* chromosomes.

Sturtevant's work supported Morgan theories and together they laid the foundation for modern theories of genetic mapping. After receiving his B.A. in 1912 and his Ph.D. in 1914, Sturtevant (together with Morgan, Hermann Muller and C.B. Bridges) published *The Mechanism of Mendelian Inheritance in 1915*. This work established the basis for the chromosomal theory of heredity.

After leaving Columbia, Sturtevant spent his career at Caltech. He served as a professor of genetics from 1928 until 1947 and as professor of biology from 1947 until 1962. In 1965 he published *A History of Genetics*. It was his last work before his death in 1970.

SUBPHYLUM

A **subphylum** is a **classification** group used in **taxonomy**. The modern taxonomic system is based on ideas posed by **Carl Linnaeus** during the eighteenth century. He was a botanist who developed a method for classifying **plants** in an orderly manner. In his system, he gave every **organism** a two part name. He also grouped organisms into a hierarchy of increasingly general categories. His system has since been modified by discoveries in the areas of **genetics**, **evolution** and **morphology**. In the current system of taxonomy a subphylum is the third most general category after **kingdom** and **phylum**. Examples of subphylums include vertebrates and **invertebrates**.

SUCCESSION

Succession is the transformation or creation of a biological community as new **species** are introduced and modify the environment. A community does not remain in a fixed state throughout time. Even a small change can upset any balance that does exist, and this can then alter the whole structure of the community. Under normal circumstances, a community will start off with a simple structure and gradually become more complex with time. Primary succession is the progressive colonization of a previously barren area. Secondary succession is the recolonization of an area that has been disturbed by fire, hurricanes, field clearing, **tree** felling, or some other process that removes most plant and **animal** species. Intermediate successional communities are known as seral stages or seres.

One example of primary succession is a newly emerged volcanic island. Initially, the area is barren. The first **organ-**

isms to inhabit the area are lichens—one of the only living organisms capable of surviving on bare rock. While participating in their natural **life cycle**s, lichens break the volcanic rock down, allowing dust particles and **humus** to collect in crevices. This provides a **habitat** for mosses, ferns, and eventually grasses. With more available humus, the soil becomes more enriched and larger **plants** move into the area. With more plants come animal species. Eventually, the vegetation is dominated by one or two tree species. At this time the overall diversity in the system decreases, creating a climax community. The climax community is more or less stable and there is no new influx of species into the community. If there is a change in the environment it may alter the climax community, returning the whole area back to an earlier point in the successional series.

Succession alters with time, but it is possible to observe the stages of succession as changes occur within an environment. A classical example is the transition from an open sandy beach to sand dunes, and finally to the climax woodland. Within woodland, change is observed on a smaller scale when a tree falls. The gap in the forest canopy created by the fallen tree creates opportunity for less dominant woodland species that thrive on the extra sunlight. This type of succession is called gap succession.

SUGAR • See Carbohydrate

SULFUR CYCLE

Sulfur is an abundant element in Earth's crust, occurring mostly in **minerals**, but also as chemicals in **water**, the **atmosphere**, and **organism**s. The storage of sulfur in the various compartments of Earth and its **biosphere**, and the many transfers occurring among them, is referred to as the **sulfur cycle**.

Sulfur (S) occurs in the environment in many chemical forms, including organic and mineral compounds, either of which can be chemically transformed by both biological and inorganic processes. Sulfur dioxide (SO_2) is a gas that is toxic to **plants** at concentrations smaller than one part per million (ppm) in the atmosphere, and to **animal**s at larger concentrations. Natural sources of emissions of sulfur dioxide include volcanic eruptions and forest fires. Large emissions are also associated with human activities, especially the burning of coal and oil to generate electricity and the processing of sulfide metal ores. Gaseous hydrogen sulfide (H_2S), with a strong smell of rotten **egg**s, is also emitted to the atmosphere, usually in situations where organic sulfur compounds are being decomposed in the absence of oxygen. In the atmosphere, both sulfur dioxide and hydrogen sulfide become oxidized to sulfate (SO_4^{-2}), a negatively charged ion (or anion) that attracts positively charged cations, such as ammonium (NH_4^+), calcium (Ca^{+2}), and hydrogen ion (H^+). The resulting fine particulates (i.e., $CaSO_4$), can serve as condensation nuclei for the formation of ice crystals, which may settle from the atmosphere as rain or snow, delivering the sulfate to terrestrial and aquatic **ecosystem**s. If the sulfate is mostly balanced by hydrogen ion (as

H_2SO_4), the **precipitation** will be acidic, and high rates of input may damage some freshwater ecosystems. This environmental problem is sometimes called "**acid rain**."

Enormous amounts of sulfur occur in association with metals, as chemically reduced minerals (or sulfides). The most common of these are iron sulfides (such as FeS_2), but all heavy metals can occur as sulfide minerals. When metal sulfides are exposed to an oxygen-rich environment, certain **bacteria** known as *Thiobacillus thiooxidans* begin to oxidize the sulfide, generating sulfate as a product, and using **energy** liberated by the reaction to sustain their growth and reproduction (this **autotroph**ic process is called chemosynthesis). In situations where large amounts of sulfides are being oxidized in this way, an enormous amount of acidity is associated with the sulfate product. This environmental problem is sometimes called "acid-mine drainage."

Sulfur is an important nutrient for organisms, as it is a key constituent of certain **amino acid**s, **protein**s, and other biochemicals. Plants satisfy their **nutrition**al need for sulfur by assimilating simple mineral compounds from the environment. This occurs mostly as sulfate dissolved in water taken up by roots, or as gaseous sulfur dioxide absorbed by foliage from the atmosphere. Animals obtain sulfur by eating the **biomass** of plants or other animals. Sulfur is rarely a deficient nutrient for animals, but in some kinds of intensively managed **agriculture** its availability can be a **limiting factor** for plant **productivity**, and application of a sulfate-containing fertilizer may be beneficial.

SUMMATION

When an excitatory **neuron** is stimulated, the **cell** body (presynaptic neuron) creates an **action potential** which travels down the cell's **axon** to the **synapse**, releasing **neurotransmitters**. These chemical messengers stimulate the adjacent cell—the postsynaptic neuron—to rise above resting potential (or threshold), which is approximately -70 millivolts (mV). An action potential is thus stimulated in that neuron. Motor and sensory messages are carried throughout the **nervous system** in this manner.

In the central nervous system (CNS), however, one presynaptic action potential alone is seldom sufficient to stimulate the postsynaptic neuron above threshold. In the CNS there are two types of presynaptic potentials: excitatory (EPSP) which are positive but subthreshold and degrade (become weaker) as they travel; and inhibitory (IPSP), which are negatively charged and lower the cell **membrane**'s resting potential even further. *Summation* is the cumulative effect of many subthreshold and negative charges. If the aggregate is greater than the postsynaptic cell's **membrane potential**, the cell fires and its output signal travels down its axon to the next cell. There are two types of summation—temporal and spacial. Temporal summation is the convergence of many rapid-firing signals from a single synapse onto the postsynaptic axon hillock—the part of the axon closest to the cell body. When EPSP signals are sent in rapid succession, they cumulatively stimulate the postsynaptic cell. Spacial summation is the simultaneous convergence of one signal from several different synapses onto the axon hillock of a single postsynaptic neuron. In both types of summation, EPSPs and IPSPs can converge at the same time on the same postsynaptic axon hillock. In such instances, the cumulative strength of the EPSPs must cancel out the negative IPSPs and be strong enough to excite the cell above -55mV in order to trigger an all-or-none response.

SURVIVAL OF THE FITTEST

The term "**survival of the fittest**" was first used by the Victorian philosopher Herbert Spencer as a metaphor to help explain **natural selection**, the central element of **Charles Darwin**'s revolutionary theory of evolutionary change, first published in 1859 in his famous book, *The Origin of Species by Means of Natural Selection.*

In this extremely influential and important book, Darwin reasoned that all **species** are capable of producing an enormously larger number of offspring than actually survive. He believed that the survival of progeny was not a random process. (In fact, he described it as a "struggle for existence.") Rather, Darwin suggested that those progeny which were better adapted to coping with the opportunities and risks presented by environmental circumstances would have a better chance of surviving, and of passing on their favorable traits to subsequent generations. These better-adapted individuals, which contribute disproportionately to the genetic complement of subsequent generations of their **population**, are said to have greater reproductive "**fitness**." Hence the use, and popularization, of the phrase: "survival of the fittest." (Darwin also used another, more awkward expression to explain the same thing: the "preservation of favored races in the struggle for life." In fact, this is the subtitle that he used for *The Origin of Species by Means of Natural Selection.*)

Darwin's theory of evolution by natural selection is one of the most important concepts and organizing principles of modern biology. The differential survival of individuals that are more-fit, for reasons that are genetically heritable, is believed to be one of the most important mechanisms of evolution. And because of its clarity, the phrase "survival of the fittest" is still widely used to explain natural selection to people interested in understanding the evolution of life on Earth.

SUTHERLAND, EARL (1915-1974)
American biochemist

Earl Wilbur Sutherland, Jr., the fifth of six children in his family, was born on November 19, 1915, in Burlingame. His father, Earl Wilbur Sutherland, a Wisconsin native, had attended Grinnell College for two years and farmed in New Mexico and Oklahoma before settling in Burlingame to run a dry-goods business, where Earl Wilbur, Jr., and his siblings worked. Sutherland's mother, Edith M. Hartshorn, came from Missouri. She had been educated at a "ladies college," and had re-

ceived some nursing training. In 1933 Sutherland entered Washburn College in Topeka, Kansas. Supporting his studies by working as an orderly in a hospital, Sutherland graduated with a B.S. in 1937. Sutherland then entered Washington University Medical School in St. Louis, Missouri. There he enrolled in a pharmacology class taught by **Carl Ferdinand Cori**. Impressed by Sutherland's abilities, Cori offered him a job as a student assistant. This was Sutherland's first experience with research. The research on the sugar glucose that Sutherland undertook in Cori's laboratory started him on a line of inquiry that led to his later groundbreaking studies.

Sutherland received his M.D. in 1942, after which he worked for one year as an intern at Barnes Hospital while continuing to do research in Cori's laboratory. Sutherland was called into service during World War II as a battalion surgeon under General George S. Patton.

In 1945, Sutherland returned to Washington University in St. Louis where he decided to commit himself to a career in research. By 1953, Sutherland had advanced to the rank of associate professor at Washington University. During these years he came into contact with many leading figures in biochemistry, including **Arthur Kornberg**, **Edwin G. Krebs**, T. Z. Posternak, and others now recognized as among the founders of modern molecular biology. But Sutherland preferred, for the most part, to do his research independently. While at Washington University, Sutherland began a project to understand how an **enzyme** known as phosphorylase breaks down glycogen, a form of the sugar stored in the liver. He also studied the roles of the **hormone** adrenaline, also known as epinephrine, and glucagon, secreted by the pancreas, in stimulating the release of energy-producing glucose from glycogen.

Sutherland left Washington for Western Reserve (now Case Western) University in Cleveland in 1953. It was during the ten years he spent in Cleveland that Sutherland clarified an important mechanism by which hormones produce their effects. Scientists had previously thought that hormones acted on whole organs. Sutherland, however, showed that hormones stimulate individual **cell**s in a process that takes place in two steps. First, a hormone attaches to specific receptors on the outside of the cell **membrane**. Sutherland called the hormone a "first messenger." The binding of the hormone to the membrane triggers release of a molecule known as cyclic AMP within the cell. Cyclic AMP then goes on to play many roles in the cell's **metabolism**, and Sutherland referred to the **molecule** as the "second messenger" in the mechanism of hormone action. In particular, Sutherland studied the effects of the hormone adrenaline, also called epinephrine, on liver cells. When adrenaline binds to liver cells, cyclic AMP is released and directs the conversion of sugar from a stored form into a form the cell can use.

Sutherland made two more important discoveries while at Western Reserve. He found that other hormones also spur the release of cyclic AMP when they bind to cells, in particular, the adrenocorticotropic hormone and the thyroid-stimulating hormone. This implied that cyclic AMP was a sort of universal intermediary in this process, and it explained why different hormones might induce similar effects. In addition, cyclic AMP was found to play an important role in the metabolism of one-celled organisms, such as the amoeba and the bacterium *Escherichia coli,* which do not have hormones. That cyclic AMP is found in both simple and complex organisms implies that it is a very basic and important biological molecule and that it arose early in evolution and has been conserved throughout millennia.

In 1963 Sutherland moved to Vanderbilt University in Nashville, Tennessee, where he was able to devote more of his time to research. At Vanderbilt, Sutherland continued his work on cyclic AMP. He and other researchers continued to discover physiological processes in different **tissue**s and various animal species that are influenced by cyclic AMP, for example in **brain** cells and **cancer** cells. In the meantime, his pioneering studies had opened up a new field of research. By 1971, as many as two thousand scientists were studying cyclic AMP.

For most of his career Sutherland was well-known mainly to his scientific colleagues. In the early 1970s, however, a rush of awards gained him more widespread public recognition. Most notably, in 1971 he was awarded the Nobel Prize for ''his long study of hormones, the chemical substances that regulate virtually every body function.'' In 1973 Sutherland moved to the University of Miami. Shortly thereafter, he suffered a massive esophageal hemorrhage, and he died on March 9, 1974, after surgery for internal bleeding, at the age of 58.

SUTTON, WALTER STANBOROUGH (1877-1916)

American geneticist, cytologist, and biologist

Walter Stanborough Sutton was a surgeon and a biologist who advanced the findings and confirmed the genetic theories of **Gregor Mendel**. A physician in private practice for the last half of his life, Sutton discovered the role of **chromosomes** in **meiosis** (sex cell division) and their relationship to Mendel's laws of **heredity**. From his research with grasshoppers collected at his parent's farm in the summer of 1899, he went on to make a major contribution to the understanding of the workings of chromosomes in **sexual reproduction**.

The fifth of six sons, Sutton was born in Utica, New York, on April 5, 1877, to William Bell Sutton and Agnes Black Sutton. At age ten, Sutton moved with his family to Russell County, Kansas, where he attended public schools. He studied engineering at the University of Kansas, Lawrence, beginning in 1896. Following his younger brother's death from typhoid in 1897, however, he made a pivotal change in the course of his education that would eventually lead him to the study of medicine and to his discoveries in **genetics**.

Sutton earned a bachelor's degree from the University of Kansas School of Arts in 1900. He was elected to Phi Beta Kappa and Sigma Xi, the scientific fraternity. While an undergraduate, he met Clarence Erwin McClung, a zoology instructor. Their four-year association would greatly influence Sutton's later work. McClung persuaded Sutton to study **histology**, which led the young student to other areas of inquiry, including cytological examinations of the lubber (grass)hopper (*Brachystola magna*).

Publishes Significant Papers in Graduate School

Sutton's careful camera lucida drawings of the stages of **spermatogenesis** in the lubber hopper were described in his first paper, ''The Spermatogonial Divisions in Brachystola Magna,'' which appeared in the *Kansas University Quarterly* in 1900 and served as his master's thesis the following year. Following this work, McClung and other faculty members encouraged Sutton to pursue his doctoral studies under biologist **Edmund Beecher Wilson** at Columbia University. In 1901, Sutton began work there. He continued his cytological studies on chromosomal division in germ cells, and in 1902 detailed his research in ''On the Morphology of the Chromosome Group in Brachystola Magna.'' Earlier that year, based on his readings of work done by British biologist **William Bateson** on the relationship between **meiosis** in germ cells and body characteristics, Sutton made the connection between **cytology** and **heredity**, and thus opened the field of cytogenetics. Sutton's hypotheses and generalizations were later published in what has become a landmark work, ''The Chromosomes in Heredity'' in the *Biological Bulletin,* 1903.

In these papers, Sutton explained that through his observation of meiosis he found that all **chromosome**s exist in pairs very similar to each other; that each gamete, or sex cell, contributes one chromosome of each pair, or reduces to one-half its genetic material, in the creation of a new offspring cell during meiosis; that each fertilized **egg** contains the sum of chromosomes of both parent cells; that these pairs control heredity; and that each particular chromosome's pair is based on independent assortment, that is, the maternal and paternal chromosomes separate independently of each other. The result, Sutton found, was that an individual in a **species** may posses any number of random combinations of different pairs of maternal and paternal chromosomes. Sutton also hypothesized that each chromosome carries in it groups of genes, each of which represents a biological characteristic—a thought that contradicted the then prevalent theory that ascribed one inherited trait to each chromosome.

At the time Sutton's paper was published, Austrian scientist **Theodor Boveri** claimed he had reached the same conclusions. As a result, the biological generalizations of the association of paternal and maternal chromosomes in pairs and their subsequent separation, which makes up the physical basis of the Mendelian law of heredity, is called the Sutton-Boveri Hypothesis. This 1903 discovery, however, was to be Sutton's last in cytology; due to unknown reasons, he never completed his course of study at Columbia.

Returning to Kansas after his laboratory research, Sutton worked as a foreman in the Chautauqua County oil fields until 1905. While there he used his abilities to solve technical problems. He developed the first technique for starting large gas engines with high pressure gas. In 1907 he patented a device to raise an oil pump mechanism from a well when worn valve components required replacement. Sutton also began a design to use electric motors to run drilling devices, but did not complete it.

Pursues Further Study in the Life Sciences

Still fascinated with the intricacies of the life sciences, and at the request of his father, Sutton returned to the Columbia University College of Physicians and Surgeons. He earned his medical degree in 1907 and for the next two years served as surgical house officer at Roosevelt Hospital in New York City. During that time, Sutton designed and built a device to deliver rectal anesthesia to patients unable to inhale ether. In 1909 he moved back to Kansas to practice privately in Kansas City, and to teach in the department of surgery at the University of Kansas. Six years later, he took a leave of absence from the University to work at the American Ambulance Hospital in Juilly, France, during World War I. His experiences of working on injured soldiers led to a book chapter on wound surgery. In addition, Sutton developed a method of finding and removing foreign bodies in soft tissue involving the use of fluoroscopy and a simple device made from a hooked piece of wire. After the war, Sutton returned to his private practice and his teaching duties in Kansas. He continued his work there until his death from a ruptured appendix on November 10, 1916. He never married. Sutton's manual dexterity—evident in his surgical skills and handling of cells under a microscope—was nurtured by his love and talent for drawing and the mechanical abilities he practiced while working on farm machinery and oil wells. His mechanical repairs and inventions, home-built camera, and laboratory and surgical practice all bore obvious examples of his creativity, skill, and inventiveness.

SVEDBERG, THEODOR (1884-1971)
Swedish chemist

Theodor Svedberg, helped to turn the arcane field of colloid chemistry into a vigorous and productive field of study. In so doing, he developed the ultracentrifuge, one of the most basic and useful tools in the modern biomedical laboratory, and an achievement for which he won the 1926 Nobel Prize in chemistry. Svedberg's work was not only innovative but cross-disciplinary, having valuable applications in a variety of fields, beginning with colloid chemistry. Colloids, of which milk fat and smoke are examples, are substances dispersed (as opposed to being dissolved) in a medium; colloids cannot be observed directly under the microscope, nor do they settle out under the force of gravity. Svedberg's development of the ultracentrifuge to study solutions was of enormous importance to biologists, who believed that gaining an understanding of colloids would help them to create models of biological systems.

Theodor Svedberg—called "The Svedberg" by his colleagues—was born on August 30, 1884, in Fleräng, Sweden, a small town near Gävle on the eastern coast. The only child of Elias Svedberg, a civil engineer employed at the local ironworks, and Augusta Alstermark Svedberg, the young Theodor often accompanied his father on long trips through the countryside, and performed simple experiments in a small laboratory at the ironworks under Elias's guidance.

Theodor attended the Karolinska School in Örebro and showed a special aptitude for the natural sciences. Botany in particular piqued his interest, but he chose chemistry because of his interest in biological processes. His education prog-

ressed rapidly; he entered the University of Uppsala in January 1904, and received his B.S. in September of 1905 and his doctorate in 1907. He wrote his dissertation on colloids.

Until Svedberg's thesis describing his new method for producing colloidal solutions of metals, chemists made these mixtures by passing an electric arc between metal electrodes submerged in a liquid. Svedberg used an alternating current with an induction coil whose spark gap was submerged in a liquid to produce relatively pure colloidal mixtures of metals. The level of purity of these colloids, and the fact that the results were reproducible, permitted researchers to perform quantitative analyses during physicochemical studies. Svedberg's work propelled him quickly in the educational hierarchy at Uppsala, beginning with a lectureship in physical chemistry from 1907 to 1912. In 1912 he was awarded Sweden's first academic chair of physical chemistry, created by the University of Uppsala specifically for Svedberg and retained by him for thirty-six years.

Svedberg continued his work with colloids, using an ultramicroscope (a microscope that uses refracted light for visualizing specimens too small to be seen with direct **light**) to study the Brownian movement of particles. Brownian motion, the continuous random movement of minute particles suspended in liquid medium caused by collision of the particles with **molecule**s of the medium, was named for the British botanist **Robert Brown**, who observed the phenomenon among pollen grains in **water**. Brownian movement was of great interest to a number of other researchers, including two future Nobel Prize winners, Albert Einstein and Jean Perrin. Perrin's work had provided verification of the theoretical work of Einstein and Marian Smoluchowski, and established definitively the existence of molecules. Perrin determined the size of large colloidal particles by measuring their rate of settling, a time-consuming process.

Using the ultramicroscope, Svedberg showed that the behavior of colloidal solutions obeys classical laws of physics and chemistry. But his method failed to distinguish the smallest particle sizes or determine the distribution of colloidal particles—the constant collisions of particles with water molecules kept the particles from settling out. In 1923 Svedberg and his colleague Herman Rinde began determining particle size distribution by measuring sediment accumulation in colloidal systems suspended on a balance. Although the technique itself was not new, Svedberg and Rinde increased its resolution by controlling air currents and other factors that disturbed the balance scale. While further refining this technique, Svedberg was also contemplating other approaches, especially **electrophoresis** (separation of particles in an electric field based on size and charge), and centrifugation.

Centrifugation—spinning solutions around a fixed circumference at high speed—mimics the force of gravity. Centrifuges were already being used to separate milk from cream and red blood cells from plasma. But fat globules and red cells are relatively large and heavy, and thus relatively easy to force out of solution. In order to force the much tinier and lighter colloidal particles out of solution, Svedberg needed a stronger centrifuge than was currently available.

In 1923 Svedberg accepted the offer of an eight-month guest professorship at the University of Wisconsin, where he taught and continued his research into centrifugation, electro-phoresis, and diffusion of colloidal solutions. Working with J. Burton Nichols, Svedberg constructed the first ultracentrifuge, which could spin at up to thirty thousand revolutions per minute, generating gravitational forces thousands of times greater than earth's. This early ultracentrifuge was elaborate, equipped with both a camera and illumination for photographing samples during centrifugation. Using this device, Svedberg and Nichols determined particle size distributions and radii for gold, clay, barium sulfate, and arsenious sulfide.

Following his sabbatical in Wisconsin, Svedberg and his students continually increased the speed of successively higher-speed ultracentrifuges, pushing the limit from 100,000 g (gravitational force equivalent to that of earth's) at forty-five thousand revolutions per minute during the 1920s to 750,000 g by 1935. He used the machine to study **proteins**, which although huge molecules, retain their colloidal properties when in solution. Among the proteins whose weight and structure he studied using the ultracentrifuge were **hemoglobin**, pepsin, **insulin**, catalase, and albumin. The technique caught on, and Svedberg's invention became an invaluable tool used by most protein chemists. He extended his ultracentrifuge studies of large carbohydrate molecules, combining his interest in biomolecules with his interest in **botany** by undertaking a pioneering study of the complex sugars of the Lillifloreae family, which includes lilies and irises. His work contributed to the understanding of carbohydrate structure and provided a useful tool for later studies of **evolution** by biologists.

In 1926 Sweden, like much of Europe, was still recovering from the devastating effects of World War I, and research seemed destined to languish in an era of reduced government support and hopelessly outmoded facilities. Svedberg's Nobel Prize, however, gave Swedish science a boost, and led directly to the establishment in 1930 of Svedberg's proposed Institute of Physical Chemistry at Uppsala. Announced the same year that Perrin received the Nobel Prize for physics for his work with colloids, Svedberg's award greatly enhanced the recognition by science and society of the importance of the field of colloid chemistry to biological and physical processes. Svedberg became director of the new Institute for Physical Chemistry in 1931, allowing him to continue his research for the remainder of his career. During World War II, however, he was forced to switch his laboratory's research efforts to the development of polychloroprene (synthetic rubber), as well as other synthetic polymers. Despite this distraction from his main work, he was still able to devise ways to incorporate the use of the electron microscope and X-ray diffraction to study the properties of cellulose biomolecules. And he developed the so-called osmotic balance, which weighed colloid particles by separating particles through a permeable membrane.

On reaching mandatory retirement age in 1949, the Swedish government honored Svedberg with a promotion to emeritus professor and made a special exception to the retirement rule by appointing him lifelong director of the Gustav Werner Institute for Nuclear Chemistry; there he studied radiochemotherapy and the effects of radiation on macromolecules. Physical chemists also honored Svedberg by naming the so-called centrifugation coefficient unit after him: the svedberg unit, s, is equal to 1×10^{-13} seconds and represents the speed at which a particle settles out of solution divided by the force generated by the centrifuge. The coefficient depends on the density and shape of the particles, with specific values of s corresponding to specific masses measured in daltons, a unit that expresses relative atomic masses.

Svedberg was married four times, first to Andrea Andreen (1909), then to Jan Frodi Dahlquist (1916), Ingrid Blomquist Tauson (1938), and Margit Hallen Norback (1948); he had six daughters and six sons. He held memberships in the Royal Society, the American National Academy of Sciences, the Academy of Sciences of the USSR, among many other organizations. Svedberg received honorary doctorates from the universities of Delaware, Groningen, Oxford, Paris, Uppsala, Wisconsin, and Harvard. In addition, he was active in the Swedish Research Council for Technology and the Swedish Atomic Research Council. He died in Örebro, Sweden, on February 25, 1971.

SWAMMERDAM, JAN (1637-1680)
Dutch naturalist

Jan Swammerdam was a Dutch naturalist known for making several important biological observations, particularly ones related to the categorization of insects.

Swammerdam, the son of an apothecary, was born in February 12, 1637. His father maintained a museum of curiosities as a hobby and it is likely this pastime helped shape young Jan's interest in natural history. Although he studied medicine and received a degree from Leiden, Swammerdam never became a physician. Instead he decided to pursue his interest in the natural sciences.

These interests led Swammerdam to study th life cycles and anatomies of many species of insects, particularly honeybees, mayflies, and dragonflies. He contributed to the understanding of how mayflies and dragonflies begin to grow rudimentary wings while still in the aquatic nymph stage. On a broader scale he categorized insects into four major groups, three of which are still in use today. He also demonstrated that insects bodies have structure and internal organs, a notion which was contrary to the beliefs of his time.

In addition to insects, Swammerdam also studied **vertebrates** and he contributed to the understanding of the way their muscular and respiratory systems operate. For example he showed that lungs of newly born mammals undergo important changes just after birth. He also demonstrated that muscles removed from a frog could be stimulated to contract. An expert microscopist, in 1658 Swammerdam documented small oval particles in frog's **blood** which were probably the first recorded observation of red blood **cells**. Through his work on humans and mammals, Swammerdam discovered valves in the **lymphatic** system, which today are called Swammerdam valves. He was also interested in the fertilization of **eggs**, embryo growth, and other aspects of **sexual reproduction**. In addition, he was know for developing a dye injection technique for use in cadaver dissection.

A cape buffalo with an oxpecker on its back in Kenya. The relationship between the oxpecker and the buffalo is a type of symbiosis called mutualism; the oxpecker feeds from the supply of ticks on the buffalo, which in turn benefits from tick removal. *(JLM Visuals. Reproduced by permission.)*

Unfortunately, Swammerdam's father was not pleased with his son's work. The elder Swammerdam had always hoped his son would become a priest, and when he eventually saw this was not going to happen he withdrew his financial support. Although the younger Swammerdam continued his work he became chronically ill, both mentally and physically. In 1673 he came under the influence of a religious zealot, Antoinette Bourignon, and got caught up in a cult-like religious controversy. He died a few years later on February 15, 1680.

Almost 60 years after his death, in 1737, Swammerdam's works were translated into Latin and published by as the *Biblia Naturae* (or *Bible of Nature*). With this publication the world began to recognize the importance of Swammerdam's biological observations.

SYMBIOSIS

Symbiosis is a close association and interaction between **organisms** of two or more **species**. Symbiotic relationships that have occurred over long periods of time may have resulted in evolutionary changes in the organisms involved in the relation-

ship. In fact, some symbiotic relationships are considered necessary for the survival of the participating organisms.

Depending on the nature of the symbiotic relationship, there are three types of symbiosis:**mutualism, commensalism,** and parasitism. Mutualism is a symbiotic relationship that is beneficial to all participants and is often symbolized as (+, +). One example of mutualistic symbiosis is that of the **nitrogen-fixing bacteria,** *Rhizobia,* living in nodules on the roots of legumous **plants** such as clover, alfalfa, peas, and beans. These bacteria are able to take atmospheric nitrogen, a form that plants cannot use, and transform it into a form that is usable by plants. The relationship is mutualistic because both organisms benefit; the plants gain the important nutrient nitrogen in a usable form, and the bacteria get a source of **energy,** usually glucose, from the plants.

Commensalism is a type of symbiosis in which one participant benefits from the relationship, while the other remains neutral (it is neither helped nor harmed). This relationship is often symbolized as (+, 0). An example of commensalism is the innkeeper worm, which burrows in marine sediments, and the **animals** that live within its tube. The organisms living within the burrow benefit because they receive protection and food as currents created by the worm pass through the burrow. The innkeeper worm does not appear to be affected by the organisms.

Parasitism is a symbiotic relationship resulting in the harm of the host organism. This type of symbiosis is often symbolized as (+, -). In this type of symbiosis, the organism that benefits is called the **parasite**, while the organism in or on which the parasite lives is called the host. Tapeworms that live within the digestive **organ**s of **mammals** are an example of a parasite, while the mammal in which it lives is the host. This relationship is parasitic because the tapeworm benefits from obtaining **nutrition** from the host, while the host animal is harmed due to a decrease in **nutrients** and possibly even **tissue** damage.

SYMPATHETIC NERVOUS SYSTEM • See Nervous system

SYNAPSE

A synapse is the junction between one **neuron** and another in the **nervous system**. It is via the synapses that neurons communicate. The synapse is a microscopic gap—as minuscule as 100nm—between the terminal endings of the first **cell** (the pre-synaptic neuron), and the dendrites of the adjoining cell (the post-synaptic neuron). This gap, called the synaptic cleft, contains extracellular fluids consisting of ions and degrading **enzyme**s that are ''taken up'' as necessary by the pre- and post-synaptic neurons in order for them to transmit information effectively. The pre-synaptic terminals store different kinds of **neurotransmitters** in tiny capsules called synaptic vesicles; the post-synaptic dendrites are equipped with a multitude of receptors, each one designed to receive (or up-take) specific ions, **proteins**, or neurotransmitters. When an **action potential** in the pre-synaptic neuron travels down its **axon** and reaches the pre-synaptic terminals, a complicated electrochemical process causes the synaptic vesicles to burst open, releasing neurotransmitters into the synaptic cleft. Here they quickly accumulate and begin to bind to their appropriate receptor on the post-synaptic dendrites. These chemicals excite the post-synaptic neuron, stimulating an action potential in that cell, which causes it, in turn, to release neurotransmitters into the synaptic cleft between it and its adjacent neuron. Thus, the message is carried from neuron to neuron via the synapses to the target— for example, to muscles of the eye lid telling it to blink.

SYNERGISTIC EFFECTS

Synergistic effects are toxic effects caused by a combination of chemicals that produce total toxic effects that are greater than the effects of the sum of the effects of the individual chemicals. In effect, synergism is like one plus one equalling three or more.

The study of synergistic effects is important to government regulators and scientists who are working to establish

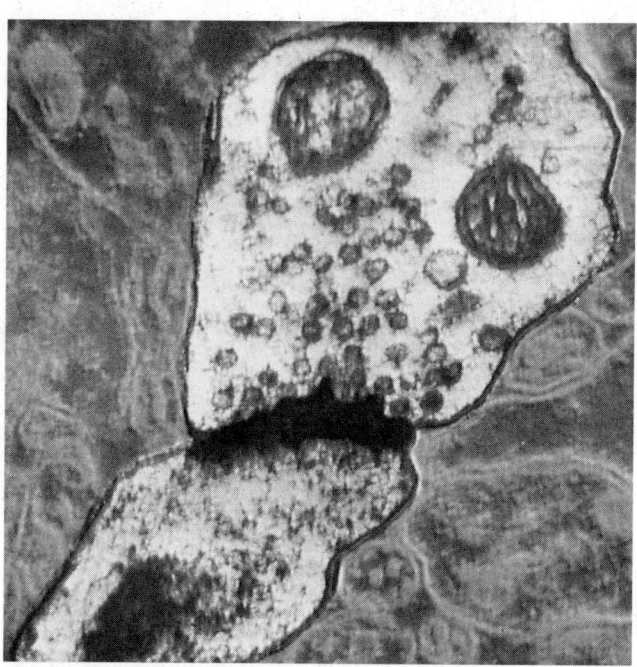

A synapse between two neurons in human cerebral cortex. *(Photo Researchers, Inc. Reproduced by permission.)*

standards that set legally-permissible levels for air, **water**, and ground **pollution**. For example, current air and water quality standards set by the United States Environmental Protection Agency (EPA) are based on studies of single metals and chemicals and how each affects plant, **animal**, and human life. **Pollutants** from a single factory could meet the EPA's air or water quality standards and still harm plant, animal, or human life because of synergist interactions among the chemicals in the substances being discharged. Environmentalists believe that air and water discharge regulations, which are based on studies of the effects of single chemicals, should be made more stringent because of the potential for synergistic effects.

One example of a synergistic effect is the effect of smoking on workers who are also exposed to asbestos in the workplace. The amounts of asbestos to which workers are exposed are limited pursuant to regulations made and enforced by the federal Occupational Safety and Health Administration (OSHA). Those standards allow for some exposure; they do not impose a complete ban on asbestos particles in the workplace. Yet, even when a factory's asbestos levels are within legally permissible levels, as a result of inhaling asbestos particles, workers may develop asbestosis or **cancer** of the lungs. (Asbestosis is a disease of the lining of the lungs.) Asbestos workers who smoke cigarettes may multiply their chances of developing lung cancer by about twelve times as compared to asbestos workers who do not smoke. Thus, while an asbestos worker who does not smoke is about five times more likely to develop lung cancer as a non-smoker who is not exposed to asbestos, a smoker who works with asbestos is about sixty times as likely to develop lung cancer than a non smoker who is not exposed to asbestos.

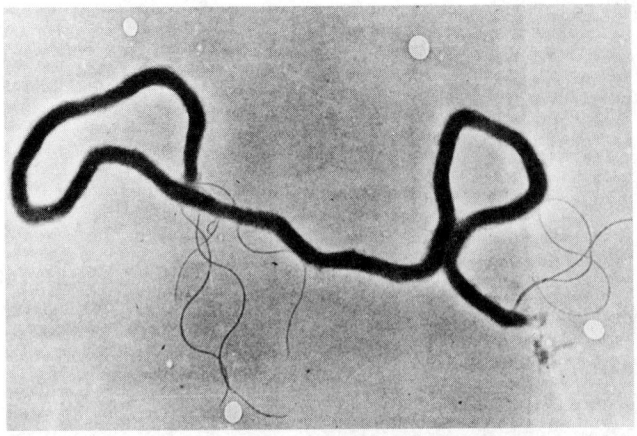

The spirochete *Treponema pallidum*, responsible for syphillis in humans. *(AP/Wide World Photos. Reproduced by permission.)*

Unfortunately, the study of a chemical's effects on the environment and humans is time-consuming and costly. Studies of synergistic effects are even more difficult and expensive. Therefore, such studies are only in beginning stages for a limited number of combinations of chemicals and have not been started for most combinations of chemicals.

SYPHILIS

Syphilis is a sexually transmitted disease caused by spiral-shaped **bacteria** called *treponema pallidum*. Syphilis is acquired during sexual contact with an individual already infected with the bacteria. The course of the disease caused by *t. pallidum* has four distinct phases: primary syphilis, secondary syphilis, latent syphilis, and advanced or tertiary syphilis.

Symptoms of primary syphilis crop up about three weeks after an individual acquires the causative bacteria. A small hard bump, followed by an ulcer (chancre) appears at the point of entry of the bacteria, usually on the penis in men or on the vulva, vagina, or cervix in women. This chancre lasts for about a month, during which time the individual is extremely contagious.

After the disappearance of the chancre, several months may pass, followed by a flu-like illness and rash. The individual during this phase is also extremely contagious, because of multiple lesions on within the mouth and on the genitalia.

Latent syphilis refers to the bacteria lying dormant within an individual's body, with no recognizable symptoms whatsoever. This phase can decades. During this phase, the body may overcome the bacteria, or the bacteria may take hold in multiple locations of the body, only to resurface as tertiary syphilis in the future.

Tertiary syphilis can occur in many different locations, but the worst results include insanity due to lesions on the **brain**, blindness, and severe destructive scarring lesions (called gumma) located in the bones, liver, spleen, **heart**, and skin. About 30% of all individuals with latent syphilis develop this serious, often fatal form of syphilis.

One of the most disturbing features of syphilis is its ability to cause disease in the babies of an infected mother. Symptoms of syphilis may be manifested in the newborn, or may remain latent, appearing later in life in the full-blown tertiary form.

All forms of syphilis are treatable with **penicillin antibiotics**, although tertiary syphilis may have already caused irreparable damage.

SYSTEMATICS

Systematics (or biosystematics) is the **classification** of living **organism**s into natural, ordered in a hierarchical (or layered) fashion that emphasizes their phylogenetic (or evolutionary) relationships. Systematics is related to **taxonomy**, although the latter is more directly concerned with the theory and practice of naming and describing **species** and other **taxon**omic units (or taxa).

Modern systematists believe that all organisms can be divided into five major group, known as **kingdom**s (these represent the highest level in the systematic organization of life. The five kingdoms are:

- **Monera**ns, including **bacteria** and cyanobacteria (or blue-green bacteria), which are the simplest **organism**s, being single-**cell**ed and lacking a **membrane**-bounded **organelle** called a **nucleus** (i.e., they are **prokaryote**s; all other kingdoms have a nucleus, and are **eukaryote**s).
- **Protista**, encompassing a wide diversity of simple, eukaryotic organisms, including unicellular and multicellular species of protozoans, foraminifera, slime moulds, single-celled **algae**, and multicellular algae.
- **Fungi**, including yeasts, which are single-celled **microorganism**s, and fungi, which are multi-celled and filamentous in their growth form.
- Plantae, encompassing multicellular, photosynthetic organisms that manufacture their own food by using the **energy** of sunlight to synthesize organic **molecule**s from inorganic ones (plants also differ from algae in having cell walls rich in cellulose, a mixture of photosynthetic pigments that includes chlorophylls *a* and *b* and carotenoids, and using starch as their principal means of storing energy).
- Animalia, which includes multicellular organisms that are mobile during at least some stage of their life history, and are **heterotroph**s that must ingest their food, ultimately consuming the photosynthetic products of plants or algae.

Below the level of kingdom, the hierarchical system used by systematists (and taxonomists) differs somewhat between **zoology** and **botany**. The major taxonomic elements (or taxa), listed in ascending order, are:

- Botany (twelve ranks): Kingdom, Division, **Class**, **Order**, **Family**, Tribe, **Genus**, Section, Series, Species, Variety, Form (some of the lower taxa are used infrequently, usually for genera containing very large numbers of species, or for economically important plants).
- Zoology (seven ranks): Kingdom, **Phylum**, Class, Order, Family, Genus, Species (additional groups may be rec-

ognized using the prefixes "super" and "sub," as in subspecies).

- The classification system can be illustrated using the following example, for the monarch butterfly of North America: Kingdom: Animalia; Phylum: Arthropoda; Class: Insecta; Order: Lepidoptera; Family: Danaidae; Genus: *Danaus*; Species: *Danaus plexippus*; Subspecies: *Danaus plexippus plexippus.*

Modern systematists examine as many biological characters as possible when trying to classify groups of organisms and determine their evolutionary relationships. Most commonly, structural (or anatomical) characters are used, because data on these are usually more readily available than for other elements. However, behavioral, biochemical, genetic, and ecological information may also be used in modern phylogenetic studies. To analyze their often large sets of data, systematists use various mathematical and statistical procedures. One frequently used procedure is cluster analysis, which divides taxa into groupings based on the similarities of their shared qualities. Species that cluster together in such an analysis are presumed to be more closely related and might (for example) be placed in the same **genus**.

The scientific goals of systematics are to develop catalogues of the world's species, to classify them into natural groups based on relatedness, and to understand their evolutionary relationships. Ultimately, this is founded in a desire to understand the patterns and processes of the natural world, and the pathways that species and other taxa have taken during the progression of life on Earth.

SZENT-GYÖRGYI, ALBERT (1893-1986)
American biochemist, molecular biologist, and physiologist

Albert Szent-Györgyi was a controversial, charismatic, and intuitive scientist whose career took many paths in the course of his life: physiologist, pharmacologist, bacteriologist, biochemist, molecular biologist. In 1937 he was awarded the Nobel Prize in physiology or medicine for his work in isolating vitamin C and his advances in the study of intercellular respiration; in 1954 he received the Albert and Mary Lasker Award from the American Heart Association for his contribution to the understanding of heart disease through his research in muscle physiology. In later years, Szent-Györgyi moved into the electron sphere, where he studied matter smaller than **molecule**s, seeking the substances that would define the basic building blocks of life. In his late seventies, he founded the National Foundation for Cancer Research.

Albert Szent-Györgyi von Nagyrapolt was born in Budapest, Austro-Hungarian Empire (now Hungary), on September 16, 1893, to Miklos and Josephine Szent-Györgyi von Nagyrapolt. He was the second of three sons. His father, whose family claimed a title and was said to have traced their ancestry back to the seventeenth century, was a prosperous businessman who owned a two-thousand-acre farm located outside Budapest. His mother came from a long line of notable Hungarian scientists.

Albert Szent-Györgyi. *(The Library of Congress. Reproduced by permission.)*

As a student, Szent-Györgyi did not begin to develop his potential until his last two years in high school, when he decided to become a medical researcher. In 1911 he entered Budapest Medical School. His education was interrupted by World War I, when he was drafted into the Hungarian Army. He was decorated for bravery; but in 1916, disillusioned with the country's leadership and the progress of the war, he deliberately wounded himself in his upper arm. He was released from the army and sent back home, where he resumed his medical studies. He received his medical degree in 1917 and that same year married Cornelia (Nelly) Demeny. Their daughter, Cornelia (Little Nelly), was born in 1918.

Hungary's Political Upheaval Forces Emigration

The political situation in Hungary after the Austrian defeat caused many families to lose all they had. Szent-Györgyi's family was no exception. With Budapest under Communist rule, Szent-Györgyi decided to leave and accepted a research position at Pozony, Hungary, one hundred miles away. It was there, at the Pharmacological Institute of the Hungarian Elizabeth University, that Szent-Györgyi gained experience as a pharmacologist. In 1919 war broke out between Hungary and the Republic of Czechoslovakia. The Czechs seized Pozony, renaming it Bratislava. In order to continue his scientific training, Szent-Györgyi joined the millions of intellectuals who left Hungary during this time.

In 1921 he accepted a position at the Pharmaco-Therapeutic Institute of the University of Leiden in The Netherlands. This began a period of intense productivity for Szent-Györgyi: by the time he was twenty-nine years old, he had written nineteen research papers, and his research spanned the disciplines of physiology, pharmacology, bacteriology, and biochemistry. Szent-Györgyi is quoted by Ralph W. Moss, author of his biography, *Free Radical: Albert Szent-Györgyi and the Battle over Vitamin C*, as saying: "My problem was: was the hypothetical Creator an anatomist, physiologist, chemist or mathematician? My conclusion was that he had to be all of these, and so if I wanted to follow his trail, I had to have a grasp on all sides of nature." The scientist added that he "had a rather individual method. I did not try to acquire a theoretical knowledge before starting to work. I went straight to the laboratory, cooked up some senseless theory, and started to disprove it."

It was while in The Netherlands, as assistant to the professor of physiology at Groningen, that he presented the first of a series of papers on **cellular respiration** (the process by which organic molecules in the **cell** are converted to carbon dioxide and **water**, releasing **energy**), a question whose answer was considered central to biochemistry. Competing theories put forth on this question (one citing the priority of oxygen's role in the process; the other championing hydrogen as having the primary role) had caused biochemists to take one side or the other. Szent-Györgyi's contribution was that both theories were correct: active oxygen oxidized active hydrogen. Szent-Györgyi's research into cellular respiration laid the groundwork for the entire concept of the respiratory cycle. The paper discussing his theory is considered to be a milestone in biochemistry. Here, also, was the beginning of the work for which he was eventually given the Nobel Prize.

Cambridge and Frederick Gowland Hopkins

While still at Groningen, Szent-Györgyi began studying the role of the **adrenal glands** (responsible for secreting **adrenaline** and other important hormones), hoping to isolate a reducing agent (electron donor) and explain its role in the onset of Addison's disease. This work was to occupy him for almost a decade, produce unexpected results, and bring him worldwide attention as a scientist. He was sure he had made a breakthrough when silver nitrate added to a preparation of minced adrenal glands turned black. That indicated a reducing agent was present, and he set out to explain its function in oxidative **metabolism**. He thought the reducing agent might be a **hormone** equivalent to adrenalin. Frustrated because scientists in Groningen seemed unconvinced of the importance of his discovery, he wrote to **Henry Hallett Dale**, a prominent British physiologist. As a result of their correspondence, Szent-Györgyi was invited to England for three months to continue his work.

Unfortunately, his testing proved a failure—the color change of the silver nitrate turned out to be a reaction of ad-

renaline with the iron in the mincer in which he ground the adrenal glands. Szent-Györgyi returned to The Netherlands, where he continued his work on cellular respiration in **plants**, writing a paper on respiration in the potato. But increasing friction with the head of the laboratory caused him to resign his position. Unable to support his wife and daughter, he sent them home to Budapest. In August 1926 he attended a congress of the International Physiological Society in Stockholm, Sweden. It was there that his luck turned. The chairman of the event was Sir **Frederick Gowland Hopkins**, considered to be the greatest living biochemist of his day. Much to Szent-Györgyi's surprise, Hopkins referred to Szent-Györgyi's paper on potato respiration in his address to the congress. After the address, Szent-Györgyi introduced himself to Hopkins, who invited him to Cambridge, where he was to remain until he returned to Hungary in 1932, eventually becoming president of the University of Szeged.

With the assurance of a fellowship from the Rockefeller Foundation (the foundation was to be a source of much financial support throughout his career), Szent-Györgyi sent for his family, rented a house, and set to work. Hopkins became his mentor—and the man Szent-Györgyi regarded as having the most influence on him as a scientist. While at Cambridge, he was awarded a Ph.D. for the isolation of hexuronic acid, the name given to the substance he had isolated from adrenal glands. One of the puzzling things about this substance was its similarity to one also found in citrus fruits and cabbage. Szent-Györgyi set out to analyze the substance, but the main obstacle to doing this was obtaining a sufficient supply of fresh adrenal glands. He finally was able to isolate a small quantity of a similar substance from orange juice and cabbage, learning that it was a carbohydrate and a sugar acid.

In 1929 Szent-Györgyi made his first visit to the United States. It was at this time that he visited the scientific community at Woods Hole, Massachusetts. He then went on to the Mayo Clinic in Rochester, Minnesota, where he had been invited to use the research facilities to continue his work isolating the adrenal substance. He managed to purify an ounce of the substance, and sent ten grams of it back to England for analysis. Nothing came of this, however, as the amount sent was too small. After almost ten years, the research appeared to be at a dead end. Szent-Györgyi took what remained of the purified crystals and returned to Cambridge.

Vitamin C and the Nobel Prize

In 1928 Szent-Györgyi had been offered a top academic post at the University of Szeged in Hungary. He accepted, but did not take up his duties there until 1931 because of delays in completing the Szeged laboratory. At Szeged, in addition to his duties as teacher, Szent-Györgyi continued his research, still trying to solve the puzzle of the adrenal substance, hexuronic acid.

It had been known since the sixteenth century that certain foods, especially citrus fruits, prevented scurvy, a disease characterized by swollen gums and loosened teeth. Although scurvy could be prevented by including citrus fruit in the diet, isolation of the antiscurvy element from citrus eluded researchers. It was not until after World War I that drug companies began a concentrated search for the antiscorbutic element (now called vitamin C). Scientists in Europe and the United States began competing to be the first to isolate this element. Vitamin C was not unfamiliar to Szent-Györgyi, and he had written of its possible connection with hexuronic acid. Now he was able to positively identify hexuronic acid as vitamin C and not an adrenal hormone, as he had previously thought. He suggested the compound be called ascorbic acid, and continued his study of its function in the body, using vitamin C-rich Hungarian paprika as the source material.

Although Charles Glen King had also isolated Vitamin C and made the connection between it and hexuronic acid—and announced his findings just two weeks before Szent-Györgyi made his report, in 1937 Szent-Györgyi was given the Nobel Prize. His acceptance speech, "Oxidation, Energy Transfer, and Vitamins," gave details of the extraordinary circumstances under which his discoveries were made.

Work during Nazi Occupation

In 1941 Szent-Györgyi and his wife were divorced. He married Marta Borbiro Miskolczy that same year. Bitterly opposed to Nazi rule in Hungary, he became an active member of the Hungarian underground. It was during the war years that he made some of his most important discoveries. His work during this time still concentrated on cellular respiration. His research in this area proved to be the basis for one of the fundamental breakthroughs in biology: the citric acid cycle. This cycle explains how almost all cells extract energy from food. It was during the war years that he also studied the chemical mechanisms of muscle contraction. His discoveries about how muscles move and function were fundamental to twentieth-century physiology, and made him a pioneer in molecular biology.

By 1944 Szent-Györgyi's outspoken opposition to Hitler's regime had put his life in danger. He and his wife went into hiding for the remainder of the war, surfacing in Budapest when the Russians liberated Hungary from the Nazis in 1945. Disillusioned with Soviet rule, he emigrated to the United States in 1947, and became an American citizen in 1954.

Woods Hole, Cancer Research, and the Vietnam War

Szent-Györgyi and his wife settled in Woods Hole, Massachusetts. Research facilities were provided for him at the Marine Biological Laboratories. He struggled to find backing to continue his work. With the help of five wealthy businessmen, he set up the Szent-Györgyi Foundation (later called the

Institute for Muscle Research), whose purpose was to raise money for muscle research and bring a group of Hungarian scientists to America to assist him. This endeavor met with partial success, but its full potential was not realized because of concerns about the legitimacy of the financial backing (and suspicion about the political loyalties of the Hungarians). As a result, Szent-Györgyi had a research team, but was unable to support them. To remedy this, he took a position in 1948 with the National Institutes of Health (NIH). He left there in 1950 for a short assignment at Princeton University's Institute for Advanced Studies. Then grants began to come in for his muscle research. Major funding came from Armour and Company (the Chicago meatpacking company), the American Heart Association, the Association for the Aid of Crippled Children, the Muscular Dystrophy Association, and NIH.

During these years, Szent-Györgyi and his team of researchers continued to make strides in the analysis of muscle protein. He also published three books: *Chemistry of Muscular Contraction, The Nature of Life,* and *Chemical Physiology of Contraction in Body and Heart,* and 120 scientific papers. These writings brought him to the attention of the American scientific community and had great influence on scientists worldwide.

Szent-Györgyi's wife Marta died of cancer in 1963. He married twice after her death. During the sixties and seventies, his opposition to the Vietnam War made him a hero to those connected to the peace movement. He wrote two books during this period that characterized his personal philosophy: *Science, Ethics, and Politics* and *The Crazy Ape* (which included his poem series, "Psalmus Humanus and Six Prayers"). He spoke out against the war on numerous occasions, both in public lectures and through letters to newspapers and periodicals.

Szent-Györgyi was almost eighty years old when he founded the National Foundation for Cancer Research. Funding from the NFCR supported his research until the end of his life. For more than forty years, his research had been concerned with the development of a basic theory about the nature of life. Szent-Györgyi called this new field of endeavor "submolecular biology." It was not just a cure for cancer that he was looking for, but a new way of looking at biology. He was convinced that his study of the structure of life at the level of electrons would not only make possible a cure for cancer but would also provide the knowledge to ensure the human body's optimum health.

Ralph Moss, the author of *Free Radical,* asked Szent-Györgyi for his philosophy of life shortly before the scientist's death of kidney failure on October 22, 1986. He scrawled on a piece of paper: "Think boldly. Don't be afraid of making mistakes. Don't miss small details, keep your eyes open and be modest in everything except your aims."

T

T CELLS

When a vertebrate encounters substances that are capable of causing it harm, a protective system known as the "**immune system**" comes into play. This system is a network of many different **organ**s that work together to recognize foreign substances and destroy them. The immune system can respond to the presence of a disease causing agent (pathogen) in two ways. Immune **cell**s called the B cells can produce soluble **proteins** (antibodies) that can accurately target and kill the pathogen. This branch of immunity is called "**humoral immunity**". In **cell-mediated immunity**, immune cells known as the T cells produce special chemicals that can specifically isolate the pathogen and destroy it.

The T cells and the B cells together are called the **lymphocytes**. The precursors of both types of cells are produced in the bone marrow. While the B cells mature in the bone marrow, the precursor to the T cells leaves the bone marrow and matures in the thymus. Hence the name, "T cells" for thymus-derived cells.

The role of the T cells in the immune response is to specifically recognize the pathogens that enter the body, and, to destroy them. They do this either by directly killing the cells that have been invaded by the pathogen, or by releasing soluble chemicals called "cytokines" which can stimulate other killer cells that are specifically capable of destroying the pathogen.

During the process of maturation in the thymus, the T cells are taught to discriminate between "self" (an individual's own body cells) and "non-self" (foreign cells or pathogens). The immature T cells, while developing and differentiating in the thymus, are exposed to the different thymic cells. Only those T cells that are "self-tolerant", that is to say, they will not interact with the **molecule**s normally expressed on the different body cells are allowed to leave the thymus. Cells that react with the body's own proteins are eliminated by a process known as "clonal deletion". The process of clonal deletion ensures that the mature T cells which circulate in the **blood** will not interact with, or, destroy an individual's own **tissue**s and organs. The mature T cells can be divided into two subsets, the T-4 cells (that have the accessory molecule CD4) or the T-8 (that have CD8 as the accessory molecule).

There are millions of T cells in the body. Each T cell has a unique protein structure on its surface known as the "T cell receptor" (TCR), which, is made before the cells ever encounter an **antigen**. The TCR can recognize and bind only to a molecule that has a complementary structure. It is kind of like a "lock-and key" arrangement. Each TCR has a unique binding site that can attach to a specific portion of the antigen called the "epitope". As stated before, the binding depends on the complementarity of the surface of the receptor and the surface of the epitope. If the binding surfaces are complementary, and the T cells can effectively bind to the antigen, then, it can set into motion the immunological cascade which eventually results in the destruction of the pathogen.

The first step in the destruction of the pathogen is the activation of the T cells. Once the T lymphocytes are activated, they are stimulated to multiply. Special cytokines called interleukins that are produced by the T-4 lymphocytes mediate this proliferation. It results in the production of thousands of identical cells, all of which are specific for the original antigen. This process of "clonal proliferation" ensures that enough cells are produced to mount a successful immune response. The large clone of identical lymphocytes then differentiates into different cells that can destroy the original antigen.

The T-8 lymphocytes differentiate into cytotoxic T-lymphocytes (CTLs) that can destroy the body cells that have the original antigenic epitope on its surface, e.g., **bacteria**l infected cells, viral infected cells, and tumor cells. Some of the T lymphocytes become "memory cells". These cells are capable of remembering the original antigen. If the individual is exposed to the same bacteria or **virus** again, these memory cells will initiate a rapid and strong immune response against it. This is the reason why the body develops a permanent immunity after an infectious disease.

Certain other cells known as the T-8 suppressor cells play a role in turning off the immune response once the antigen has been removed. This is one of the ways by which the immune response is regulated.

TAIGA

Taiga is a Russian word used to describe the subarctic evergreen forest of Siberia and other similar forest regions around the world. Characterized by long, cold winters and short, wet summers, this **biome** (broad biological **community**) is generally **species**-poor and low in **productivity**. **Tree**s only a few inches in diameter may be many decades old. Damage caused by forestry, mining, road-building, and other activities can take many years to heal. Conifers (cone-bearing evergreens such as pine, spruce, and fir) and a few deciduous species such as aspen, willow, birch, and alder are the main tree species in this community. The forest floor often is covered by mosses and other species adapted to short growing seasons, wet soil, and acidic conditions. Slow **decomposition** rates and lack of soil **organism**s lead to the formation of dense layers of peat in the extensive **wetlands** spread through the taiga. At its northern edge, the taiga forms a ragged, irregular border with the treeless arctic **tundra**. The sparsely scattered trees in this transition zone often are little more than a few feet tall and can be bent and tilted in weird patterns (sometimes called a drunken forest) where underlying permafrost prevents tree roots from gaining a firm hold in the ground. Among the predominant **animal** species of the taiga are woodland caribou, moose, bears, beaver, lynx, martin, and wolves. Swarms of biting insects during the summer can make the taiga a miserable place for both humans and other warm-blooded animals. Although it doesn't contain the highest amount of **biomass**, the taiga covers a larger area than any other forest type in the world.

TASTE

Taste—technically called gustation—is one of a human's five senses. The senses of **smell** and taste are extremely complex systems facilitated by chemoreception—special **neuron**s sensitive to chemical substances. Humans are born with the ability to taste, although taste preferences change with maturity, some are acquired, and some must be developed: babies and children seldom enjoy strong flavors like onions or mustard. Although we can taste a myriad of flavors, taste buds sense only four separate sensations: bitter, salty, sour, and sweet. Specific combinations of these, along with texture, **temperature**, and—in particular—smell, allow us to differentiate and detect the difference between cottage cheese and chocolate. Inhaling or exhaling as we eat or drink is highly important to our sense of taste, as some flavors—chocolate and coffee, for example—are primarily ''tasted'' by their smell. The sense of taste can be adversely affected by such things as zinc deficiency, smoking, a head cold or sinus problems, deficient sense of smell, or damage to certain parts of the **brain**.

Taste begins when food or drink mix with saliva in the mouth and stimulate taste buds on the surface of the tongue, soft palate, pharynx, larynx, and epiglottis. It was once thought that taste buds sensitive to sweetness were located near the tip of the tongue, those sensitive to salt along the front sides, those to sour further back along the sides, and bitter was detected by taste buds at the very back of the tongue. Researchers now discount this theory, and evidence suggests that even individual taste buds and receptor **cell**s are not specifically sensitive to an individual taste. Electrophysiological recordings show, instead, neuronal response to combinations of the four basic taste sensations, and that these combinations are decoded in the brain, not in the mouth. Humans have anywhere from 2,000-5,000 taste buds, each containing 50-150 neuroepithelial cells. Chemical stimulation initiates an **action potential** in these cells which interact through a complex system of electrical and synaptic responses with fibers within the central **nervous system** where they **synapse** to the medulla, the thalamus, and the postcentral gyrus. Neurons associated with smell and taste are the only nervous system cells to be replenished when old ones die or become damaged.

TATUM, EDWARD LAWRIE (1909-1975)
American biochemist

Edward Lawrie Tatum's experiments with simple **organism**s demonstrated that **cell** processes can be studied as chemical reactions and that such reactions are governed by **gene**s. With **George Beadle**, he offered conclusive proof in 1941 that each biochemical reaction in the cell is controlled via a catalyzing **enzyme** by a specific gene. The ''**one gene-one enzyme**'' **theory** changed the face of biology and gave it a new chemical expression. For the first time, the nature of life seemed within the grasp of science's quantitative methods. Tatum, collaborating with **Joshua Lederberg**, demonstrated in 1947 that **bacteria** reproduce sexually, thus introducing a new experimental organism into the study of molecular **genetics**. Spurred by Tatum's discoveries, other scientists worked to understand the precise chemical nature of the unit of **heredity** called the gene. This study culminated in 1953 with the description by **James Watson** and **Francis Crick** of the structure of **DNA** (deoxyribonucleic acid). Tatum's use of microorganisms and laboratory **mutation**s for the study of biochemical genetics led directly to the biotechnology revolution of the 1980s. Tatum and Beadle shared the 1958 Nobel Prize in physiology or medicine with Joshua Lederberg for ushering in the new era of modern biology.

Tatum was born on December 14, 1909, in Boulder, Colorado, to Arthur Lawrie Tatum and Mabel Webb Tatum. He was the first of three children; a younger brother and sister would follow. Both of Edward's parents excelled academically. His father held two degrees, an M.D. and a Ph.D. in pharmacology. Edward's mother was one of the first women to graduate from the University of Colorado. Presumably an interest in science and medicine ran in the Tatum family: Edward would become a research scientist, his brother a physician, and his sister a nurse. As a boy, Edward played the French horn and trumpet; his interest in music lasted his whole life. He also enjoyed swimming and ice-skating.

In 1925, when Tatum was fifteen years old, his father accepted a position as a pharmacology professor at the University of Wisconsin. Tatum studied at the University of Chicago Experimental School and for two years at the University of Chicago before transferring and completing his undergraduate work at the University of Wisconsin. He almost became a geologist before deciding in his senior year to major in chemistry.

Tatum earned his A.B. degree in chemistry from the University of Wisconsin in 1931. In 1932 he earned his master's degree in microbiology. Two years later, in 1934, he received a Ph.D. in biochemistry for a dissertation on the cellular biochemistry and nutritional needs of a bacterium. Understanding the biochemistry of microorganisms such as bacteria, **yeast**, and molds would persist at the heart of Tatum's career.

After receiving his doctorate, Tatum remained at the University of Wisconsin for one year as a research assistant in biochemistry. He married the same year he completed his Ph.D. In Livingston, Wisconsin, Tatum wed June Alton, the daughter of a lumber dealer, on July 28, 1934. They eventually had two daughters, Margaret Carol and Barbara Ann.

From 1936 to 1937, Tatum studied bacteriological chemistry at the University of Utrecht in the Netherlands while on a General Education Board fellowship for postgraduate study. In Utrecht he worked in the laboratory of F. Kogl, who had identified the vitamin biotin. In Kogl's lab Tatum investigated the nutritional needs of bacteria and **fungi**. While Tatum was in Holland, he was contacted by geneticist George Beadle. Beadle, seven years older than Tatum, had done genetic studies with the fruit fly *Drosophila melanogaster* while in the laboratory of **Thomas Hunt Morgan** at the California Institute of Technology. Beadle, newly arrived at Stanford University, was now looking for a biochemist who could collaborate with him as he continued his work in genetics. He hoped to identify the enzymes responsible for the inherited eye pigments of *Drosophila*.

Upon his return to the United States in the fall of 1937, Tatum was appointed a research associate at Stanford University in the department of biological sciences. There he embarked on the *Drosophila* project with Beadle for four years. The two men successfully determined that kynurenine was the enzyme responsible for the fly's eye color and that it was controlled by one of the eye-pigment genes. This and other observations led them to postulate several theories about the relationship between genes and biochemical reactions. Yet they realized that *Drosophila* was not an ideal experimental organism on which to continue their work.

Chooses Bread Mold Medium for Gene Experiments

Tatum and Beadle began searching for a suitable organism. After some discussion and a review of the literature, they settled on a pink mold that commonly grows on bread known as *Neurospora crassa*. The advantages to working with *Neurospora* were many: it reproduced very quickly, its nutritional needs and biochemical pathways were already well known, and it had the useful capability of being able to reproduce both

Edward Lawrie Tatum. *(UPI/Corbis-Bettmann. Reproduced by permission.)*

sexually and asexually. This last characteristic made it possible to grow **culture**s that were genetically identical and also to grow cultures that were the result of a cross between two different parent strains. With *Neurospora*, Tatum and Beadle were ready to demonstrate the effect of genes on cellular biochemistry.

The two scientists began their *Neurospora* experiments in March 1941. At that time, scientists spoke of "genes" as the units of heredity without fully understanding what a gene might look like or how it might act. Although they realized that genes were located on the chromosomes, they didn't know what the chemical nature of such a substance might be. An understanding of DNA was still twelve years in the future. Nevertheless, geneticists in the 1940s had accepted **Gregor Mendel**'s work with inheritance patterns in pea **plants**. Mendel's theory, rediscovered by three independent investigators in 1900, states that an inherited characteristic is determined by the combination of two hereditary units (genes), one each contributed by the parental cells. A dominant gene is expressed even when it is carried by only one of a pair of **chromosome**s, while a recessive gene must be carried by both chromosomes to be expressed. With *Drosophila*, Tatum and Beadle had taken genetic mutants—flies that inherited a variant form of eye color—and tried to work out the biochemical steps that led to the abnormal eye color. Their goal was to identify the variant enzyme, presumably governed by a single gene, that controlled the variant eye color. This proved technically very difficult, and as luck would have it, another lab announced the discovery of kynurenine's role before theirs did. With the *Neurospora* experiments, they set out to prove their one gene-one enzyme theory another way.

The two investigators began with biochemical processes they understood well: the nutritional needs of *neurospora*. By exposing cultures of *Neurospora* to x rays, they would cause genetic damage to some bread mold genes. If their theory was right, and genes did indeed control biochemical reactions, the genetically damaged strains of mold would show changes in their ability to produce **nutrients**. If supplied with some basic salts and sugars, normal *Neurospora* can make all the **amino acid**s and **vitamins** it needs to live except for one (biotin).

This is exactly what happened. In the course of their research, the men created, with x ray bombardment, a number of mutated strains that each lacked the ability to produce a particular amino acid or vitamin. The first strain they identified, after 299 attempts to determine its mutation, lacked the ability to make vitamin B_6. By crossing this strain with a normal strain, the offspring inherited the defect as a recessive gene according to the inheritance patterns described by Mendel. This proved that the mutation was a genetic defect, capable of being passed to successive generations and causing the same nutritional mutation in those offspring. The x ray bombardment had altered the gene governing the enzyme needed to promote the production of vitamin B_6.

This simple experiment heralded the dawn of a new age in biology, one in which molecular genetics would soon dominate. Nearly forty years later, on Tatum's death, Joshua Lederberg told the *New York Times* that this experiment "gave impetus and morale" to scientists who strived to understand how genes directed the processes of life. For the first time, biologists believed that it might be possible to understand and quantify the living cell's processes.

Tatum and Beadle were not the first, as it turned out, to postulate the one gene-one enzyme theory. By 1942 the work of English physician **Archibald Garrod**, long ignored, had been rediscovered. In his study of people suffering from a particular inherited enzyme deficiency, Garrod had noticed the disease seemed to be inherited as a Mendelian recessive. This suggested a link between one gene and one enzyme. Yet Tatum and Beadle were the first to offer extensive experimental evidence for the theory. Their use of laboratory methods, like x rays, to create genetic mutations also introduced a powerful tool for future experiments in biochemical genetics.

Research Leads to Mass-Production of Penicillin

During World War II, the methods Tatum and Beadle had developed in their work with pink bread mold were used to produce large amounts of **penicillin**, another mold. Their basic research, unwittingly, thus had a very important practical effect as well. In 1944 Tatum served as a civilian staff member of the U.S. Office of Scientific Research and Development at Stanford. Industry, too, used the methods the men developed to measure vitamins and amino acids in foods and **tissue**s.

In 1945, at the end of the war, Tatum accepted an appointment at Yale University as an associate professor of **botany** with the promise of establishing a program of biochemical microbiology within that department. Apparently the move was due to Stanford's lack of encouragement of Tatum, who failed to fit into the tidy category of biochemist or biologist or geneticist but instead mastered all three fields. In 1946 Tatum did indeed create a new program at Yale and became a professor of microbiology. In work begun at Stanford and continued at Yale, he demonstrated that the one gene-one enzyme theory applied to yeast and bacteria as well as molds.

Discovers Sexual Reproduction of Bacteria

In a second extremely fruitful collaboration, Tatum began working with Joshua Lederberg in March 1946. Lederberg, a Columbia University medical student fifteen years younger than Tatum, was at Yale during a break in the medical school curriculum. Tatum and Lederberg began studying the bacterium *Escherichia coli*. At that time, it was believed that *E. coli* reproduced asexually. The two scientists proved otherwise. When cultures of two different mutant bacteria were mixed, a third strain, one showing characteristics taken from each parent, resulted. This discovery of biparental inheritance in bacteria, which Tatum called genetic recombination, provided geneticists with a new experimental organism. Again, Tatum's methods had altered the practices of experimental biology. Lederberg never returned to medical school, earning instead a Ph.D. from Yale.

In 1948 Tatum returned to Stanford as professor of biology. A new administration at Stanford and its department of biology had invited him to return in a position suited to his expertise and ability. While in this second residence at Stanford, Tatum helped establish the department of biochemistry. In 1956 he became a professor of biochemistry and head of the department. Increasingly, Tatum's talents were devoted to promoting science at an administrative level. He was instrumental in relocating the Stanford Medical School from San Francisco to the university campus in Palo Alto. In that year Tatum also was divorced from his wife June. On December 16, 1956, he married Viola Kantor in New York City. Kantor was the daughter of a dentist in Brooklyn. Owing in part to these complications in his personal affairs, Tatum left the West Coast and took a position at the Rockefeller Institute for Medical Research (now Rockefeller University) in January 1957. There he continued to work through institutional channels to support young scientists, and served on various national committees. Unlike some other administrators, he emphasized nurturing individual investigators rather than specific kinds of projects. His own research continued in efforts to understand the genetics of neurospora and the nucleic acid metabolism of mammalian cells in culture.

Contributions to Biology Recognized with Nobel Prize

In 1958, together with Beadle and Lederberg, Tatum received the Nobel Prize in physiology or medicine. The Nobel Committee awarded the prize to the three investigators for their work demonstrating that genes regulate the chemical processes of the cell. Tatum and Beadle shared one-half the prize and Lederberg received the other half for work done separately from Tatum. Lederberg later paid tribute to Tatum for his role in Lederberg's decision to study the effects of x-ray-induced mutation. In his Nobel lecture, Tatum predicted that "with real understanding of the roles of heredity and environment, together with the consequent improvement in man's physical capacities and greater freedom from physical disease, will come an improvement in his approach to, and understanding of, sociological and economic problems."

Tatum had a marked interest in social issues, including population control. In 1965 and 1966 Tatum organized other Nobel laureates in science to make public endorsements of family planning and birth control. These included statements to Pope Paul VI, whose encyclical against birth control for Catholics was issued at this time.

Tatum's second wife, Viola, died on April 21, 1974. Tatum married Elsie Bergland later in 1974 and she survived his death the following year, on November 5, 1975. Tatum died at his home on East Sixty-third Street in New York City after an extended illness. In a memoir written for the *Annual Review of Genetics,* Lederberg recalled that Tatum's last years were "marred by ill health, substantially self-inflicted by a notorious smoking habit." Lederberg noted, too, that Tatum's "mental outlook" was scarred by the painful death of his second wife.

In addition to the Nobel Prize, Tatum received the Remsen Award of the American Chemical Society in 1953 for his work in biparental inheritance and sexual reproduction in bacteria. In 1952 he was elected to the National Academy of Sciences. He was a founding member of the *Annual Review of Genetics* and joined the editorial board of *Science* in 1957. Tatum's collected papers occupy twenty-five feet of space in the Rockefeller University Archives and span the years from 1930 to 1975.

TAXON

A taxon is a taxonomic group of any rank. When a new **organism** is discovered, its taxonomical placement may at first be unclear. Until the new organism is classified, it is normally referred to as merely a new taxon. As a twentieth century derivative of the word **taxonomy**, taxon can also refer to a group of organisms sharing similar characteristics when the hierarchical **classification** is unimportant to a scientific discussion.

TAXONOMY

Taxonomy is the study of the **classification** of **organism**s, according to their differences and similarities. The oldest form of taxonomy is what is now called classical taxonomy and it is concerned primarily with the description, naming, and classification of organisms based on their morphological characteristics. An adaptation of classical taxonomy is now taking into account molecular and biochemical (chemosystematics) data that is now available. Numerical taxonomy uses mathematical principles. This is done to try to remove the subjective **nature** of taxonomy that can lend more weight to certain characters than others, just because a structure is obvious and apparent it may not have much importance taxonomically or evolutionarily speaking. An example of this approach is cladistics. This looks at how organisms have evolved from a common ancestor and how particular structures have altered with time. Cladistics attempts to recreate the **phylogeny** of the organisms under study. Cytotaxonomy looks purely at the **chromosome**s of the organisms and traces the similarities in the chromosome numbers and **morphology** of the chromosomes. Even those who do not use cladistics have had their work affected by it. The start of cladistic analysis forced taxonomists to rethink concepts of characters and character states, weighting, and **homology**. Another approach is experimental taxonomy, which analyzes patterns of variation to try to show how they evolved. Experimental taxonomy also illustrates the genetic relationship between the variations and the environments that formed them.

One area of taxonomic research that many people feel needs to receive a much higher profile than it currently does is usually referred to as "completing the inventory". This refers to the fact that even today not all **species** are known or recorded by science. Many species, particularly in the tropics, have yet to be discovered. Many species are only known from one specimen and the variability within the species is largely unknown. These are in reality areas of high priority. Some of these species may already be on the brink of **extinction** and they may be lost to the world forever.

Within taxonomy the ultimate aim is to provide an evolutionary based classification system. Some groupings of organisms are referred to as a taxonomic species. This is merely the first look at a new group. A taxonomic species is a grouping of organisms based purely on physical characteristics and it does not necessarily represent any relationship. Within taxonomy **convergent evolution** is a problem it produces superficial similarities between organisms. These similarities can lead people to assume a relationship where none exists.

The taxonomic system that we employ today dates from the mid-eighteenth century when the Swedish naturalist **Carl Linnaeus** formalized the process. He enshrined the two part species name in the science of taxonomy. This system is known as the binomial system because each organism is identified by a two part specific name, usually in a Latin. Prior to the binomial system, scientific names were usually long scientific descriptions that were difficult to use.

Taxonomy is an important study. Without it, it would be harder to recognize natural relationships between organisms. Taxonomy is a database system, allowing us to access information about the species under study. Taxonomy and the associated classification is not a fixed and rigid system. Once new information is obtained it can be fed into the system allowing for constant refining. Taxonomy is a constant investigation to see what relationships exist between living organisms.

TEMIN, HOWARD (1934-1994)
American virologist

Howard Temin is an American virologist who revolutionized **molecular biology** in 1965 when he found that genetic information in the form of **ribonucleic acid** (**RNA**) can be copied into **deoxyribonucleic acid** (**DNA**). This process, called **reverse transcriptase**, contradicted accepted beliefs of molecular biology at that time, which stipulated that DNA always passed on genetic information through RNA. Temin's research also contributed to a better understanding of the role **virus**es play in the onset of **cancer**. For this, he was featured on the cover of *Newsweek* in 1971, which hailed his discovery as the most important advancement in cancer research in sixty years. In addition, Temin shared the 1975 Nobel Prize in physiology or medicine for his work on the Rous sarcoma virus. His discovery of the reverse transcriptase process contributed greatly to the eventual identification of the **human immunodeficiency virus** (**HIV**). Temin's later research focused on **genetic engineering** techniques. A vehement antismoker, he took every opportunity to warn against the dangers of tobacco, even in his acceptance speech for the Nobel Prize.

Howard Martin Temin was born in Philadelphia on December 10, 1934, to Henry Temin, a lawyer, and Annette (Lehman) Temin. The second of three sons, Temin showed an early aptitude for science and first set foot in a laboratory when he was only fourteen years old. As a student at Central High School in Philadelphia, he was drawn to biological research and attended special student summer sessions at the Jackson Laboratory in Bar Harbor, Maine. After graduation from high school, Temin enrolled at Swathmore College in Pennsylvania where he majored and minored in biology in the school's honors program. He published his first scientific paper at the age of eighteen and was described in his college yearbook as "one of the future giants in experimental biology."

After graduating from Swathmore in 1955, Temin spent the summer at the Jackson Laboratory and enrolled for the fall term at the California Institute of Technology in Pasadena. For the first year and a half, he majored in experimental **embryology** but then changed his major to **animal** virology. He studied under **Renato Dulbecco**, a renowned biologist in his own right, who worked on perfecting techniques for studying virus growth in **tissue** and developed the first plaque assay (a chemical test to determine the composition of a substance) for an animal virus. Temin received his Ph.D. in biology in 1959 and worked for another year in Dulbecco's laboratory. In 1960 he joined the McArdle Laboratory for Cancer Research at the University of Wisconsin—Madison, where he spent the remainder of his career as the Harold P. Rusch Professor of Cancer Research and the Steenbock Professor of Biological Sciences.

Studies in Viral Research Stir Controversy

Temin began studying the Rous sarcoma virus (RSV) while still a graduate student in California. First identified in the early twentieth century by Peyton Rous, RSV is found in some species of hens and was one of the first viruses known to cause tumors. In 1958 Temin and Harry Rubin, a postdoctoral fellow, developed the first reproducible assay *in vitro* (outside of an organism) for the quantitative measuring of virus growth. Accepting an appointment as assistant professor of oncology at Wisconsin in 1960, Temin continued his research with RSV. Using the assay method he and Rubin developed, Temin focused on delineating the differences between normal and tumor cells. In 1965 he announced his theory that some viruses cause cancer through a startling method of information transfer.

Scientists at the time thought that genetic information could only be passed from DNA to RNA. DNA is a long **molecule** comprised of two chains of nucleic units containing the sugar deoxyribose. RNA is a molecule composed of a chain of nucleic units containing the sugar ribose. For years, many of Temin's colleagues rejected his theory that some viruses actually reverse this mode of transmitting genetic information, and they cited a lack of direct evidence to support it. Temin, however, was convinced that RNA sometimes played the role of DNA and passed on the **genetic codes** that made a normal **cell** a tumor cell.

It took Temin several years, however, to prove his theory. Despite making further inroads in gathering evidence implicating DNA synthesis in RSV infection, many of his colleagues remained skeptical. Finally, in 1970, Temin, working with Satoshi Mitzutani, discovered a viral **enzyme** able to copy RNA into DNA. Dubbed "a reverse transcriptase virus," this enzyme passed on hereditary information by seizing control of the cell and making a reverse transcript of the host DNA; in other words, the enzyme synthesized a DNA virus that contained all the genetic information of the RNA virus. This discovery was made simultaneously by biologist **David Baltimore** at his laboratory at the Salk Institute in La Jolla, California.

The work of Temin and Baltimore led to a number of impressive developments in molecular biology and recombinant DNA experimentation over the next twenty years, including characterizing **retrovirus**es, a family of viruses that cause tumors in vertebrates by adding a specific gene for cancer cells. In 1975 Temin shared the Nobel Prize in physiology or medicine with his former mentor, Renato Dulbecco, and David Baltimore. These three scientists' research illustrated how separate avenues of scientific research could converge to produce significant advances in biology and medicine. Eventually, interdisciplinary research was to become a mainstay of modern science.

In 1987 Temin reflected on his discovery of viruses' roles in causing cancer. "I measure [my discovery's importance] by comparing what I taught in the experimental oncology course 25 years ago," said Temin in a University of Wisconsin press release, pointing out that the topic of viral carcinogenesis (the viral link to cancer) was rarely the focus of any lectures at that time. "Now, in the course we're teaching, between a third and half of the lectures are related directly or indirectly to viral carcinogenesis."

Research Leads to Understanding of AIDS

Temin's continuing work into the role viruses play in carcinogenesis had an important impact on **acquired immunodeficiency syndrome (AIDS)** research. Temin's discovery of reverse transcriptase provided scientists with the means to find and identify the AIDS virus. His interest in genetic engineering and the causes of cancer eventually led him to another exciting discovery. He found a way to measure the mutation rate in retroviruses (viruses that engage in reverse transcriptase), which led to insights on the variation of cancer genes and viruses, such as AIDS. Determining the speed at which genes and viruses change provided vital information for devising attempts to vaccinate or treat viral diseases. His discovery of reverse transcriptase also led to the development of standard tools used by biologists to prepare radioactive DNA probes to study the genetic makeup of viral and malignant cells. Another genetic engineering technique that arose from this research was the ability to make DNA copies of messenger RNA, which could be isolated and purified for later study.

Temin was also interested in such areas as **gene therapy**, which uses gene splicing techniques to "genetically improve" the host **organism**. As he began to apply genetic engineering techniques to his research, he recognized legitimate concerns about producing pathogens (microorganisms that carry disease) that could escape into the environment. He also served on a committee that drew up federal guidelines in human **gene** therapy trials.

Temin's research convinced him that science was making progress in the fight against cancer. Temin said in a 1984 United Press International release: "We know the names of some of the genes which are apparently involved in cancer. If past history is a guide, this understanding will lead to improvement in diagnosis, therapy, and perhaps prevention."

Throughout his career, Temin continued to teach general virology courses for graduates and undergraduates. He also worked with students in his laboratory. "I get satisfaction from a number of things—from discovering new phenomena, from understanding old phenomena, from designing clever experiments—and from seeing students and postdoctoral fellows develop into independent and outstanding scientists," he stated in a University of Wisconsin press release.

A scientist and family man who shunned the spotlight after winning the Nobel Prize (which he kept in the bottom drawer of a file cabinet), Temin was committed to quietly searching for clues into the mysteries of cancer-causing viruses. Temin married Rayla Greenberg, also a geneticist, in 1962, and the couple had two daughters, Miriam and Sarah. A familiar site on the Wisconsin-Madison campus, Temin bicycled to work every day on his mountain bike. Although he preferred not to attract attention so he could better concentrate on his work, Temin did not hesitate to speak out about his beliefs. For example, Temin said in an *On Wisconsin* article, "I enjoy teaching and believe I have gained a lot from doing it. As a

researcher, I'm able to present to students the newest work in certain areas. I see that as a benefit." Because of this dedication to academics, he became upset when researchers started to leave the University of Wisconsin-Madison in 1984 because of a state employee wage-freeze, even though his own salary was ensured through private and foundation support. Temin wrote the governor letters criticizing his lack of support for education and faculty researchers. Eventually, he reluctantly agreed to help the governor in developing salary proposals.

Temin also spoke out against cigarette smoking. During the award ceremonies for the Nobel Prize, he told the audience that he was "outraged" that people continued to smoke even though cigarettes were proven to contain **carcinogens**. He instructed that 80% of all cancers were preventable because they resulted from environmental factors, such as smoking. "It was the most important general statement I could make about human cancer," he said later in a *People* magazine interview. "And I realized the Nobel Prize would give me an opportunity to speak out that a person does not ordinarily have." Temin went on to testify before the Wisconsin legislature and congress in support of antismoking bills. His research efforts in AIDS led him to urge the federal government to increase funding for further research into the AIDS epidemic. Despite living a lifestyle designed to minimize the risk of cancer, Temin, who never smoked, developed lung cancer in 1992. His illness was a rare form of cancer called adenocarcinoma of the lung, which is not usually associated with cigarette smoking. He died of this disease on February 9, 1994. In addition to the Nobel Prize, Temin received many other awards for his research, including the prestigious Albert Lasker Award in Basic Medical Research in 1974 and the National Medal of Science in 1992.

TEMPERATURE

Temperature is a thermodynamic term (i.e., referring to the supply, use, or conversion of **energy**) referring to the average thermal energy in a system. (Thermal energy, or heat, is a type of kinetic energy, which is due to the rate of vibration of atoms and **molecule**s.) The distribution of temperature within a system determines the direction of any heat flow, which occurs from regions of higher to lower temperature (i.e., from higher to lower thermal density).

Temperature is expressed in terms of units of degrees (°), designated as a value on one of several scales. On the Fahrenheit (F) scale, the freezing point of pure **water** (at sea level) has a temperature value of 32°F, while the boiling point occurs at of 212°F. The Centigrade (or Celsius) scale is used by most of the world's scientists, and is the recommended unit of temperature in the *International System of Units* (or SI). On the Centigrade scale, water freezes at 0°C and it boils at 100°C. The Kelvin scale is less-well known, being used mostly by physicists. The Kelvin scale sets 0K at "absolute zero;" this is an extremely cold temperature at which no atomic or molecular vibration occurs (this is the coldest temperature that is physically possible).

The various scales for measuring temperature are all inter-convertible, as follows:

- degrees Fahrenheit = 9/5 (°C) + 32
- degrees Centigrade (or Celsius) = 5/9 (°F) - 32
- degrees Kelvin = (°Celsius) + 273.15
- 1°F = 0.555°C = 0.555 K

TEN PERCENT LAW

The so-called "ten percent law," also often called the "law of tens" refers to the idea that at each transfer of **energy** through a trophic structure, only a small percent of the energy remains available for use by the **organism** in the next level up in the system. The majority of the energy is lost to the surroundings in forms such as waste heat. Trophic structure refers to what is often called the **food chain** or food web, wherein green **plants** (the producer level) capture energy from the sun through the process of **photosynthesis** and transform it into energy available to other living organisms (the consumer levels).

The ten percent law suggests or implies that exactly 90% of the energy is lost in the transfer at each **trophic level**, and that only 10% is passed on as useable biological energy. That implied preciseness is misleading, however, and is one of the reasons that many scientists discount the concept, even label it a myth. Instead, it should be considered a rule of thumb, a teaching device, or perhaps best, a mnemonic device that serves as a reminder of the striking inefficiencies of natural systems.

The ten percent law does stand for a real fact of **nature**: most of the energy available at one level in an **ecosystem** is lost in the transfer to the next level. A very small percentage of the **light** energy that reaches the **leaf** surface of plants is actually assimilated, or turned by photosynthesis into **organic compounds**. Most of the light energy absorbed by plants is converted directly to heat and lost. Many plant parts are inedible by **herbivores** (the next trophic level) and is lost for energy transfer. The plants use much of the energy absorbed to live, as do the herbivores that eat them.

Energy lost at each level results in a pyramid of **biomass**: less energy available at the level of **carnivores**, for example, means less biomass at that level than at the level of herbivores. The growth and numbers of organisms in an ecosystem are closely (if not always directly) related to, and limited by, the amount of energy available, and that decreases markedly at each level of the food/energy system.

TERATOGEN

A teratogen is a substance that can cause physical malformations if a fetus or **embryo** is exposed to that substance while in the mother's womb. Scientists group teratogens with mutagens and **carcinogens** because all three cause **mutations** (changes) in the fetus or embryo. The three terms share the root "gen" which means birth or origin. While a teratogen causes physical changes, a carcinogen causes **cancer**, and mutagen causes genetic changes. In other words, it changes the genes within a **cell** of a fetus.

Prescribed in the early 1960s as a treatment for morning sickness, the drug thalidomide severely effected the infants born to mothers who used it. *(AP/Wide World Photo. Reproduced by permission.)*

When a fetus is exposed to a teratogen, results vary: (1) a low dose or exposure during a non-critical time of development of the fetus may result in no effect; (2) severe malformations may cause the fetus to die and the pregnancy will result in a miscarriage; or (3) the baby may be born with birth defects.

Birth defects are a significant concern to our society because they occur in 3-5% of newborns. Birth defects may include growth retardation (slow growth), mental retardation, structural defects (such as a missing or malformed limb), or functional damage. An example of functional damage is the effect of a drug called diethystilbestrl (DES). Female babies who were exposed to DES while in their mother's womb have suffered damage to their **reproductive system**s. As adults, the daughters who were exposed to DES while in their mother's womb have had trouble in becoming pregnant and have suffered unusually high rates of miscarriage.

Substances that can cause birth defects include viral diseases, physical agents, and chemicals. For example, rubella (German measles) is caused by a **virus**. If a mother becomes infected during pregnancy, the virus causes birth defects. Radiation is physical agent that causes birth defects. Exposure to radiation can occur naturally in the environment or through sources controlled by humans such as x rays. A few chemicals are known to be teratogenic in humans. They include anti-cancer drugs, steroid **hormone**s, and an infamous drug called thalidomide. Thalidomide was used in Europe in the 1960s to relieve nausea during pregnancy. However, it resulted in horrible defects in the mothers' babies. Thousands of babies were born with malformed or missing arms, legs, or both. The thalidomide disaster brought worldwide attention to the fact that the mother's womb does not protect the fetus from the effects of drugs. Today, many chemicals including many prescription drugs are suspected to be teratogens even if definite proof is unavailable. Therefore, pregnant mothers are usually advised to avoid all unnecessary drugs during pregnancy, especially during the first three months.

TERATOLOGY

Teratology is the study of abnormalities, malformations, monstrosities, and serious deviations from normal growth and development in **organism**s. The word teratogenesis is derived from the Greek *gennan*, meaning to produce, and *terata*, meaning monster. In humans, teratology is the study of chemicals, drugs, medications, **alcohol**, disease, or other environmental agents in relation to fetal abnormalities and **birth** defects. To be classified as a teratogen, an agent must cause either low birth weight, intrauterine growth retardation, or small size for gestational age; stillbirth or miscarriage; structural abnormalities; or functional defects (mental or developmental retardation). To cause birth defects, a teratogen must reach the developing fetus. Teratogens can also cause premature birth and fetal/newborn addiction to pharmacological agents. Thalidomide, tetracycline, valproic acid, cocaine, cigarette smoke, alcohol, lead, mercury, **syphilis**, herpes, and rubella are just a few examples of a teratogen.

Birth defects and abnormalities have baffled and frightened people throughout history: rock carvings and drawings from ancient civilizations around the world, including Australia and the Africas, depict individuals with deformities and malformations; writings from ancient Babylon describe them, as do clay tablets believed to date back to 2000 B.C. found in the library of an Assyrian king around 700 B.C.; and deformed human skeletal remains have been unearthed during archeological digs.

Before the advent of the modern study of birth defects in the 18th century, congenital abnormalities were viewed with superstition or as supernatural—perhaps punishment from the gods or the work of the devil. The first person to articulate the theory of ''developmental arrest'' was **William Harvey** in 1651 when he observed that a hair lip in a newborn infant closely resembled the normal condition of a fetus at a certain stage of development. **Albrecht von Haller** (1768) and **Kaspar Friedrich Wolff** (1759) followed a similar line of thinking. While not understanding the process, Ambroise Pare (1649) and **John Hunter** (1775) recognized familial and inherited traits. **Gregor Johann Mendel** began describing ''inheritance laws,'' now known as **genetics**, in 1865; Carl Correns, **Hugo de Vries**, and Erich von Tschermak rediscovered this theory in 1900.

In spite of today's advanced medical and scientific knowledge, 3-5% of all infants are born with birth defects. Approximately 5% of those appear to be due to teratogens, a small percentage of which are preventable through pre-pregnancy counselling and the assessment of risk factors in certain **populations**. In many countries, educational and informational organizations are working toward reducing the number teratogen-related birth defects, while research continues investigating the phenomenon. For example, recent evidence indicates a woman can reduce her chances of bearing a child with **spina bifida**, one of the most common types of birth defects, by 50% simply by consuming at least 400 micrograms of folic acid (a B vitamin) daily for least one month prior to conception.

TERRITORY

Biologists have accumulated abundant evidence to strongly indicate that many **species** of **animal**s exhibit territorial behavior, that is they demarcate and defend in some way a particular space or **territory**, especially against conspecifics (members of their own species). The range of species that exhibit such behavior is diverse, including numerous bird species, lizards and salamanders, wolves, deer, a variety of insect species, even fish. Most of the early studies of territorial behavior in animals were birds, work dating back to the mid-eighteen hundreds.

Territoriality is defined most simply as the attempt by an individual **organism** or group of organisms to control a specified area, i.e. territory is ''any defended area.'' The area or territory, once controlled, is often bounded by some kind of marker. And, control of territory usually means defense of territory, which may be enforced by the marker, and which may or may not be aggressive. Control of territory means benefits (access to limited resources, reproductive success); defense of territory incurs costs (e.g., stress, time for patrols, etc.).

Habitat selection is closely related to territorial behavior. Territories are usually small segments of larger habitats, and are chosen after assessing and moving in to an appropriate habitat. The spatial arrangement and relations of individuals in a given **population** are context dependent, and closely related to the dynamics of the population. Spatial relations are coupled to the age, size and sex of individuals, and change and fluctuate according to the appearance of new individuals, the departure or **death** of others, and the interactions with neighboring individuals, of the same and other species.

Territoriality is found in a wide range of organisms, probably including humans, though the degree to which the human species exhibits territorial behavior remains uncertain and controversial. Social scientists have studied a range of related concepts, including personal space, social space, crowding, privacy, boundary conflicts, and property. Biologists have also contributed to understanding of territorial behavior through the use of related concepts such as spacing and density, home range or domain, micro-**ecology**, refuging, body zones, and flight distance.

Walrus fighting over territory—a hauling-out spot—Bristol Bay, Alaska. *(JLM Visuals. Reproduced by permission.)*

THEILER, MAX (1899-1972)

South African American virologist

Max Theiler (pronounced Tyler) was one of the leading figures in the development of the yellow-fever vaccine. His early research proved that yellow-fever **virus** could be transmitted to mice. He later extended this research to show that mice which were given serum from humans or **animal**s that had been previously infected with yellow fever developed an immunity to this disease. From this research, he developed two different vaccines in the 1930s, which were used to control this incurable tropical disease. For his work on the yellow-fever vaccine, Theiler was awarded the Nobel Prize in medicine or physiology in 1951.

Theiler was born on a farm near Pretoria, South Africa, on January 30, 1899, the youngest of four children of Emma (Jegge) and Sir Arnold Theiler, both of whom had emigrated from Switzerland. His father, director of South Africa's veterinary services, pushed him toward a career in medicine. In part to satisfy his father, he enrolled in a two-year premedical program at the University of Cape Town in 1916. In 1919, soon after the conclusion of World War I, he sailed for England, where he pursued further medical training at St. Thomas's Hospital Medical School and the London School of Hygiene and Tropical Medicine, two branches of the University of London. Despite this rigorous training, Theiler never received the M.D. degree because the University of London refused to rec-

ognize his two years of training at the University of Cape Town.

Theiler was not enthralled with medicine and had no intention of becoming a general practitioner. He was frustrated by the ineffectiveness of most medical procedures and the lack of cures for serious illnesses. After finishing his medical training in 1922, the 23-year-old Theiler obtained a position as an assistant in the Department of Tropical Medicine at Harvard Medical School. His early research, highly influenced by the example and writings of American bacteriologist Hans Zinsser, focused on amoebic dysentery and rat-bite fever. From there, he developed an interest in the yellow-fever virus.

Yellow-Fever Work Generates Two Life-Saving Vaccines

Yellow fever is a tropical viral disease that causes severe fever, slow pulse, bleeding in the stomach, jaundice, and the notorious symptom, black vomit. The disease is fatal in 10-15% of cases, the cause of death being complete shutdown of the liver or kidneys. Most people recover completely, after a painful, extended illness, with complete immunity to reinfection. The first known outbreak of yellow fever devastated Mexico in 1648. The last major breakout in the continental U.S. claimed 435 lives in New Orleans in 1905. Despite the medical advances of the 20th century, this tropical disease remains incurable. As early as the 18th century, mosquitoes were thought to have some relation to yellow fever. Cuban physician **Carlos Finlay** speculated that mosquitoes were the carriers of this disease in 1881, but his writings were largely ignored by the medical community. Roughly 20 years later, members of America's Yellow Fever Commission, led by **Walter Reed**, the famous U.S. Army surgeon, concluded that mosquitoes were the medium that spread the disease. In 1901, Reed's group, using humans as research subjects, discovered that yellow fever was caused by a **blood**-borne virus. Encouraged by these findings, the Rockefeller Foundation launched a worldwide program in 1916 designed to control and eventually eradicate yellow fever.

By the 1920s, yellow-fever research shifted away from an all-out war on mosquitoes to attempts to find a vaccine to prevent the spread of the disease. In 1928, researchers discovered that the Rhesus monkey, unlike most other monkeys, could contract yellow fever and could be used for experimentation. Theiler's first big breakthrough was his discovery that mice could be used experimentally in place of the monkey and that they had several practical research advantages. When yellow-fever virus was injected into their brains, the mice didn't develop human symptoms. Instead, "when you give a mouse yellow fever, he gets not jaundice but encephalitis, not a fatal bellyache but a fatal headache," Theiler stated, according to Greer Williams author of *Virus Hunters.*

One unintended research discovery kept Theiler out of his lab and in bed for nearly a week. In the course of his experiments, he accidentally contracted yellow fever from one of his mice, which caused a slight fever and weakness. Theiler was much luckier than some other yellow-fever researchers. Many had succumbed to the disease in the course of their investigations. However, this small bout of yellow fever simply gave Theiler an immunity to the disease. In effect, he was the first recipient of a yellow-fever vaccine.

In 1930, Theiler reported his findings on the effectiveness of using mice for yellow fever research in the respected journal *Science.* The initial response was overwhelmingly negative; the Harvard faculty, including Theiler's immediate supervisor, seemed particularly unimpressed. Undaunted, Theiler continued his work, moving from Harvard University, where he was considered an upstart, to the Rockefeller Foundation in New York City. Eventually, yellow-fever researchers began to see the logic behind Theiler's use of the mouse and followed his lead. His continued experiments made the mouse the research animal of choice. By passing the yellow-fever virus from mouse to mouse, he was able to shorten the incubation time and increase the virulence of the disease, which enabled research data to be generated more quickly and cheaply. He was now certain that an attenuated live vaccine, one weak enough to cause no harm yet strong enough to generate immunity, could be developed.

In 1931, Theiler developed the mouse-protection test, which involved mixing yellow-fever virus with human blood and injecting the mixture into a mouse. If the mouse survived, then the blood had obviously neutralized the virus, proving that the blood donor was immune to yellow fever (and had most likely developed an immunity by previously contracting the disease). This test was used to conduct the first worldwide survey of the distribution of yellow fever.

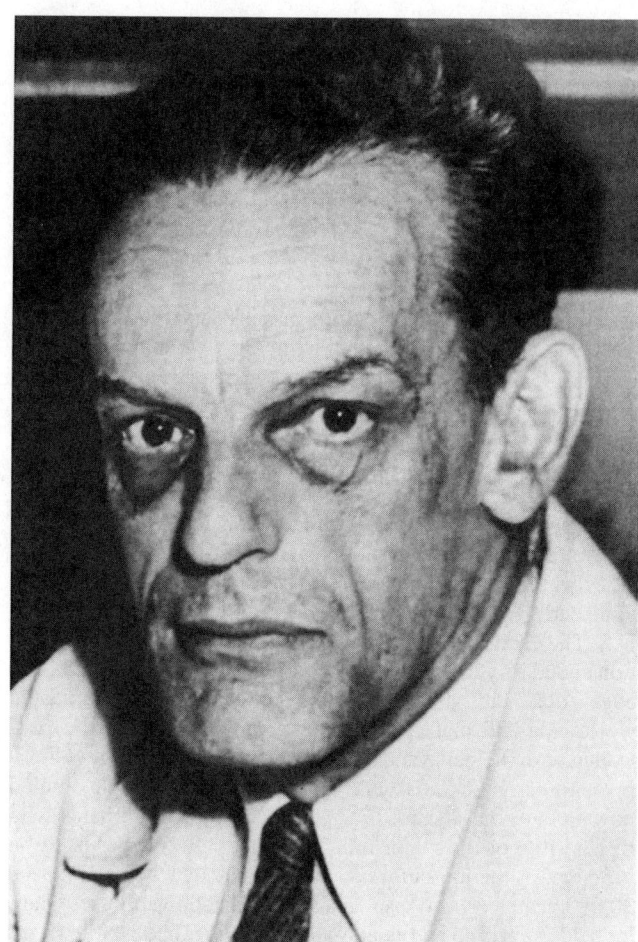

Max Theiler. *(The Library of Congress. Reproduced by permission.)*

A colleague at the Rockefeller Foundation, Dr. Wilbur A. Sawyer, used Theiler's mouse strain, a combination of yellow fever virus and immune serum, to develop a human vaccine. Sawyer is often wrongly credited with inventing the first human yellow-fever vaccine. He simply transferred Theiler's work from the mouse to humans. Ten workers in the Rockefeller labs were inoculated with the mouse strain, with no apparent side effects. The mouse-virus strain was subsequently used by the French government to immunize French colonials in West Africa, a hot spot for yellow fever. This so-called "scratch" vaccine was a combination of infected mouse brain tissue and cowpox virus and could be quickly administered by scratching the vaccine into the skin. It was used throughout Africa for nearly 25 years and led to the near total eradication of yellow fever in the major African cities.

Virus Work Leads to Nobel Prize

While he was somewhat pleased with the new vaccine, Theiler considered the mouse strain inappropriate for human use. In some cases, the vaccine led to encephalitis in a few recipients and caused less severe side effects, such as headache or nausea, in many others. Theiler believed that a "killed" vaccine, which used a dead virus, wouldn't produce an immune effect, so he and his colleagues set out to find a milder live strain. He began working with the Asibi yellow-fever strain, a form of the virus so powerful that it killed monkeys instantly when injected under the skin. The Asibi strain thrived in a number of media, including chicken embryos. Theiler kept this virus alive for years in **tissue** cultures, passing it from embryo to embryo, and only occasionally testing the potency of the virus in a living animal. He continued making subcultures of the virus until he reached strain number 176. Then, he tested the strain on two monkeys. Both animals survived and seemed to have acquired a sufficient immunity to yellow fever. In March 1937, after testing this new vaccine on himself and others, Theiler announced that he had developed a new, safer, attenuated vaccine, which he called 17D strain. This new strain was much easier to produce, cheaper, and caused very mild side effects.

From 1940 to 1947, with the financial assistance of the Rockefeller Foundation, more than 28 million 17D-strain vaccines were produced, at a cost of approximately two cents per unit, and given away to people in tropical countries and the U.S. The vaccine was so effective that the Rockefeller Foundation ended its yellow-fever program in 1949, safe in the knowledge that the disease had been effectively eradicated worldwide and that any subsequent outbreaks could be controlled with the new vaccine. Unfortunately, almost all yellow-fever research ended around this time and few people studied how to cure the disease. For people in tropical climates who live outside of the major urban centers, yellow fever is still a problem. A major outbreak in Ethiopia in 1960–62 caused 30,000 deaths. The World Health Organization still uses Theiler's 17D vaccine and is attempting to inoculate people in remote areas.

The success of the vaccine brought Theiler recognition both in the U.S. and abroad and even from his former employer, Harvard University. Over the next ten years, he received the Chalmer's Medal of the Royal Society of Tropical Medicine and Hygiene (1939), the Lasker Award of the American Public Health Association, and the Flattery Medal of Harvard University (1945).

In 1951, Theiler received the Nobel Prize in medicine or physiology "for his discoveries concerning yellow fever and how to combat it." According to Williams (in *Virus Hunters*), when Theiler was asked what he would do with the $32,000 Nobel award, he remarked, "Buy a case of scotch and watch the ole Dodgers."

After developing the yellow-fever vaccine, Theiler turned his attention to other viruses, including some unusual and rare diseases, such as Bwamba fever and Rift Valley fever. His other, less exotic research focused on polio and led to his discovery of a polio-like infection in mice known as encephalomyelitis or Theiler's disease. In 1964, he retired from the Rockefeller Foundation, having achieved the rank of associate director for medical and natural sciences and director of the Virus Laboratories. In that same year, he accepted a position as professor of epidemiology and microbiology at Yale University in New Haven, Connecticut. He retired from Yale in 1967.

Theiler married Lillian Graham in 1938. They had one daughter. His nonscientific interests included reading (mostly history and philosophy but absolutely no fiction) and watching baseball games, especially those involving his beloved Brooklyn Dodgers. Although he immigrated to the U.S. in 1923 and remained in America for the rest of his life, he never applied for U.S. citizenship. Theiler died on August 11, 1972, at the age of 73.

THEOPHRASTUS (372 B.C.-287 B.C.)
Greek philosopher

Theophrastus was a Greek Peripatetic philosopher and a disciple of **Aristotle**. He studied under Aristotle in Athens, and succeeded him as head of the Lyceum, the academy founded by Aristotle, in 323 B.C. Under his leadership, enrollment in the Lyceum reached its highest point. Although he wrote extensively on many subjects, few of his works remain. He wrote major treatises dealing with **plants**, and smaller treatises on fire, winds, signs of weather, scents, and sensations have also been attributed to him.

Because Theophrastus wrote extensively on the nature of plants, he is sometimes called the father of **botany**. It has been estimated that he wrote about 200 botanical treatises, but only a few Latin translations of the original Greek manuscripts survive. In *De causis plantarum* and *De historia plantarum* he presented his understanding of the **morphology, classification** and natural history of plants. His treatises were the principle source for botanical information for centuries. He identified and grouped over 500 plants, and developed a nomenclature for describing external plant parts, which he called **organ**s, and internal parts he referred to as **tissue**s. He described sexual reproduction in flowering plants, as well as seed **germination** and development.

Theophrastus wrote detailed directions for pollinating the date palm by hand. His description was based on a special ceremony of Arabs and Assyrians in which a participant climbed a male date tree, removed an inflorescence, and took it to the female tree where it was rubbed on the female flowers. The practice improved the yield of dates. Since pollen and its function were not known, he could not accurately explain why this procedure was effective.

His knowledge of plants was remarkable. He was familiar with monocots and dicots and was aware of differences in their seeds, stems and leaves. He was aware of the annual rings of both xylem and phloem. He distinguished between roots and rhizomes. His insights into the nature of hypogynous, perigynous and epigynous **flower**s and the possible nature of sepals and petals as modified leaves preceded modern science by two millennia. He was aware of the existence of flowers lacking petals. He described differences in seed structure between **gymnosperm**s and angiosperms and is credited with originating the term gymnosperm which in the Greek means "naked seed." His description of how the Greeks used crops of broad bean legume (*Vicia faba*) to enrich the soil predated by 2,000 years the modern practice of alternately planting non-leguminous crops with leguminous crops to maintain soil **nitrogen** fertility.

THEORELL, AXEL HUGO TEODOR
(1903-1982)
Swedish biochemist

Axel Hugo Theorell was born in Linköping, Sweden, on July 6, 1903. He received his bachelor of medicine degree (1924) and his doctor of medicine (1930) from the Karolinska Institute in Stockholm. He also studied at the Pasteur Institute in Paris. When a crippling attack of poliomyelitis made a career as a physician impractical, he decided instead to pursue research and teaching. His academic work while at Stockholm was an inquiry into the chemistry of plasma **lipids** (fatty acids) and their effect on red **blood** cells. A technique he developed at this time to separate the plasma **proteins** albumin and globulin was later to prove useful in his work on isolating **enzymes** (globular proteins) and **coenzymes**, which help to activate specific enzymes.

As professor of chemistry at Uppsala University from 1930–1936, Theorell expanded his research on plasma lipids to concentrate on myoglobin, a muscle protein whose oxygen-carrying capacities he compared to that of **hemoglobin** in the blood. By isolating (purifying) myoglobin, he was able to show its absorption and storage capacities, and to measure, using centrifugal force, its molecular weight. This determination of its physical properties showed that myoglobin was a separate protein from hemoglobin.

In 1933 Theorell received a grant from the Rockefeller Foundation that enabled him to further his study of enzymes with **Otto Warburg** at the Kaiser Wilhelm Institute (now the Max Planck Institute) in Berlin. Warburg had attempted without success to isolate the yellow enzyme. Using his own methods, Theorell accomplished the isolation. He further separated the yellow enzyme into two parts: the catalytic coenzyme and the pure protein apoenzyme. He also found that the main ingredient of the yellow enzyme is the plasma protein albumin. An important corollary to the research was Theorell's discovery of the chemical chain reaction necessary for cellular oxidation or **respiration**. These contributions brought a test-tube creation of life closer to reality, and advanced the study of the chemical differences between normal and cancerous **cells**.

Returning to Stockholm, Theorell became head of the biochemistry department at the Karolinska Institute, part of a Nobel Institute established for the purpose of providing Theorell with further research opportunities. Under his direction, the department acquired a reputation for excellence that attracted biochemists from all over the world. It was here that Theorell continued his research on cytochrome c, succeeding in his attempts to purify it by 1939. He furthered this study that same year in the United States with his colleague, **Linus Pauling**, who discovered the alpha spiral (protein **molecules** arranged in a twisted-atom chain).

After World War II, a collaboration with Britton Chance of the University of Pennsylvania elucidated steps in the oxidation (breakdown) of **alcohol** and gave the process a name—the Theorell-Chance mechanism. Theorell's study of the enzymes that catalyze the oxidation, alcohol dehydrogenases, provided a new method for determining the level of alcohol

in the bloodstream—a technique that came to be used by Sweden and West Germany to test the sobriety of their citizens. From a different perspective, Theorell's alcohol enzyme research pinpointed several **bacterial** strains, knowledge of which was thought to be useful in the treatment of tuberculosis. Theorell was awarded the 1955 Nobel Prize in physiology or medicine for ''his discoveries concerning the nature and mode of action of oxidation enzymes.'' Theorell retired from the Nobel Institute in 1970. Afflicted with a stroke in 1974, his health deteriorated over the following years. He died on August 15, 1982, while vacationing on an island off the coast of Sweden.

THERMODYNAMICS, LAWS OF

Thermodynamics is the study of **energy** exchange relationships involving heat and mechanical work. Broken down into its constituent fractions, the term thermodynamics means *heat motion*. It is the branch of physics dealing with the reversible conversion of heat energy into other forms, particularly mechanical energy. In order to discuss the behavior of heat energy in physical systems, the distinction between heat and **temperature** must be made.

Heat and temperature are fundamentally different quantities. The concept of heat can be understood on the basis of microscopic mechanical energy. Heat is the kinetic energy (energy of movement) and potential energy (energy in an inactive form) of the individual **molecules** of a material. More exactly, heat is the sum of all of the kinetic and potential energies of the particles in a quantity of **matter**. Temperature, then, is a measure of the *average* kinetic energy of the particles in a sample of matter. The greater the kinetic energy of the particles in a sample of matter, the higher the temperature and the hotter it feels. Heat, therefore, is energy while temperature is a measurable effect of the presence of heat.

As the term thermodynamics implies, heat energy is mobile. Prior to the eighteenth century, heat transfer was erroneously thought to be the flow of a weightless, invisible fluid called *caloric*. However, during the eighteenth and nineteenth centuries a relationship of heat to mechanical energy and work gradually emerged which discounted the theory of caloric. Count Rumford (1753-1814) studied the heat that formed while drilling cannon barrels. He noted that dull drills produced temperature increases. Later, Sir James Joule (1818-1889) showed that **water** is heated by stirring vigorously with a paddle wheel. He discovered a correlation between the temperature rise of the water and the work required to turn the wheel. Heat was thus identified as a form of energy, capable of doing work, and mechanical work results in the formation of heat. It was not until the middle of the nineteenth century that a theory of heat was set forth into scientific law. In 1842, **Hermann von Helmholtz** published a paper that described the first of three laws regarding heat.

The First Law of Thermodynamics is an extension of the law of conservation of energy that specifically includes heat. It states that heat energy can be transferred from one system to another, but can neither be created nor destroyed. It implies that energy can only change form. Put simply, when heat is transformed into any other form of energy, or when other forms of energy are transformed into heat, the total amount of energy in the system is constant.

The first law of thermodynamics has broad implications and is applicable to an infinite number of practical problems. For example, steam engines transform chemical potential energy (in the form of coal)into heat through combustion. The heat energy is transformed into mechanical energy that moves the pistons of the engine. Thus, heat energy is used to perform work by the engine (such as pulling a train.) The first law is also applicable in living systems. Muscle **tissue** utilizes energy from **metabolism** to do work. Chemical energy from food is converted to mechanical energy and heat in the form of muscle contraction.

The Second Law of Thermodynamics states that heat energy exhibits a tendency toward equilibrium. Heat always flows from a system containing more heat to a system containing less heat. It states that heat transfer is unidirectional unless work is added to the system. When a hot object contacts a cool object, heat will always move from the hot object to the cool one until thermal equilibrium is reached. Equilibrium is attained when both objects contain equal amounts of heat energy and there is no longer net movement of heat energy.

The tendency toward equilibrium is also an irreversible tendency toward increased **entropy**. Entropy is defined as the measure of disorder of a system. Uniformly distributed heat energy exhibits greater disorder (and thus greater entropy) than heat energy that is distributed unevenly. The Second Law states that the universe is continually moving toward increased entropy, or disorder. That is, given enough time, all matter and energy will be evenly distributed throughout the universe.

The Third Law of Thermodynamics describes the hypothetical state resulting from a total lack of heat energy in a system. *Absolute zero* is the temperature attained in the complete absence of all heat, and therefore, all thermal motion. The third law also predicts that at absolute zero, all kinetic energy, and thus motion, is zero. At absolute zero, heat is absent and molecular movement is nonexistent. The magnitude of absolute zero is defined as 0K, or -273°C. In theory, absolute zero can never be attained. However, recent advances have allowed scientists to reach temperatures much closer to absolute zero than previously thought possible.

While the **Laws of Thermodynamics** are the subject of physical science, they have important consequences in biological systems as well. For instance, life is possible because living **organism**s oppose entropy. Violation of the second law requires energy. Therefore, all living beings require energy to maintain their existence, whether in the form of sunlight as with **autotroph**s or digested plant or **animal** matter as with **heterotroph**s. Also, metabolic rate is a measure of the heat production of an animal. Metabolism includes all of the physical changes and **chemical reaction**s that take place in an organism. According to the laws of thermodynamics, all of the reactions and changes will ultimately release heat, defining its metabolic rate. Metabolic rate is thus related to the chemical, electrical,

and mechanical work of an organism and its cellular components. A final illustration of the relevance of the laws of thermodynamics in living systems involves temperature **regulation** or thermoregulation. Most vertebrate animals maintain a constant internal temperature. The temperature of an animal depends on the amount of heat generated from metabolic processes and heat exchange with the environment, involving the first and second laws of thermodynamics respectively.

THOMAS, E. DONNALL (1920-)
American physician

E. Donnall Thomas has pioneered techniques for transplanting bone marrow, an operation that has been utilized to treat patients with **cancer**s of the **blood**, such as leukemia. For proving that such transplants could save the lives of dying patients, Thomas was awarded the Nobel Prize in physiology or medicine in 1990.

E. Donnall Thomas was born on March 15, 1920, in the small town of Mart, Texas. After graduating from a high school class of approximately fifteen students, Thomas entered the University of Texas at Austin in 1937. He received a B.A. in 1941 and continued on for a master's degree, which was awarded in 1943.

After completing his master's degree, Thomas started medical school at the University of Texas Medical Branch in Galveston. After six months, however, he transferred to Harvard Medical School, where he received his M.D. in 1946. He became an intern and then a resident at Peter Bent Brigham Hospital in Boston and began to specialize in blood diseases. Thomas interrupted his formal medical training to serve as a physician in the United States Army (1948–1950). He then returned to the Boston area and did research on leukemia treatments for a year as a postdoctoral fellow at the Massachusetts Institute of Technology. In 1953 he worked as an instructor at Harvard Medical School.

Thomas moved to New York in 1955 to take the position of physician-in-chief at the Mary Imogene Bassett Hospital in Cooperstown. The next year he became, in addition, an associate clinical professor of medicine at the College of Physicians and Surgeons at Columbia University. During the next eight years Thomas had the opportunity to develop and research his ideas about bone marrow transplants, and he applied these concepts to treating cancers of the blood.

Leukemia is a type of cancer in which certain blood **cell**s, known generally as white blood cells, are produced in abnormally large numbers by the bone marrow. In other kinds of cancer, the diseased cells pile up into a tumor, which can often be treated by simply cutting out the lump. Leukemic blood cells, however, circulate throughout the body, making them much more difficult to eliminate. Furthermore, the white blood cells that become abnormal in leukemia are an important part of the body's **immune system**. Even if they could be destroyed by a means such as radiation, without them the patient would be vulnerable to infections.

In the 1950s, researchers showed that inbred laboratory mice could be irradiated, thus destroying the production of white blood cells by their bone marrow, and then saved from infection by a transplant of bone marrow taken from healthy mice. Inspired by these experiments, Thomas began similar studies on dogs, but he faced two important obstacles. First, the recipient **animal**'s immune system had to be prevented from attacking and destroying the transplanted bone marrow—such immune rejection has long been a problem for bone marrow as well as **organ** transplant surgery. And second, if the bone marrow transplant was successful and the donated marrow began to produce white blood cells, these cells were likely to attack the recipient's other **tissues**, perceiving them as foreign. Both of these problems had been avoided in the earlier studies with inbred mice because the mice were genetically identical, and hence, have identical immune systems. People are not so similar genetically, with the exception of identical twins. All attempts to graft bone marrow between a donor and recipient who were not identical twins failed. In 1956, Thomas performed the first bone marrow transplant to a leukemia patient from an identical twin. Although the patient's immune system did not reject the transplant, the cancer recurred.

Many researchers gave up working on organ transplants because the problems of immune rejection seemed insurmountable, but Thomas persisted. In 1963 he moved to Seattle to become a professor at the University of Washington Medical School. There he put together a team of expert researchers and began experimenting with new drugs that could suppress the recipient's immune system and thus prevent rejection of the new tissue. In the meantime, new methods were being developed by other researchers to identify people whose immune systems were similar, in order to match organ donors and recipients. The new methods of tissue typing were based on **molecule**s known as histocompatibility antigens. Thomas's team performed the first bone marrow transplant to a leukemia patient from a matched donor in March 1969. During the 1970s they developed and perfected a comprehensive procedure for treating leukemia patients: first the patients receive radiation, both to kill cancer cells and to weaken the immune system so that it does not reject the transplant; then their bone marrow is replaced with marrow from a compatible donor. The patients also are given drugs that continue to suppress their immune systems. Many patients had been cured of leukemia using this technique by the late 1970s. Since then Thomas and his colleagues have improved their success rate from about 12–50%. In addition to leukemia and other cancers of the blood, bone marrow transplants are used to treat certain inherited blood disorders and to aid people whose bone marrow has been destroyed by accidental exposure to radiation.

It was in 1990 that Thomas was awarded the Nobel Prize in physiology or medicine in 1990, a commendation he shared with Joseph E. Murray, another American physician who has done important work in the area of transplants. The Nobel Prize came as a surprise. Thomas told reporters that the award is more often given to scientists who do basic research than to those that develop clinical treatments. As reported in *Time* magazine, both men were cited by the Nobel committee for discoveries ''crucial for those tens of thousands of severely ill patients who either can be cured or given a decent life when other treatment methods are without success.''

THOMPSON, D'ARCY WENTWORTH (1860-1948)
Scottish zoologist

Sir D'Arcy Wentworth Thompson combined extensive knowledge of natural history with insight into mathematics to develop a new approach to **evolution** and the growth of living things. His 1917 work, *On Growth and Form,* represented a significant departure from the **zoology** of his day and has since contributed to **embryology, taxonomy, paleontology,** and **ecology,** as well as influencing artists, engineers, architects, and poets. Thompson was also trained in the classics from a young age, and he applied his knowledge of ancient Greek culture, thought, and natural history to his *A Glossary of Greek Birds* and *A Glossary of Greek Fishes.*

Thompson was born in Edinburgh, Scotland, on May 2, 1860. His father, also named D'Arcy Wentworth Thompson, was a classical master at the Edinburgh Academy, and then a professor of Greek at Queens College, Galway. The elder Thompson wrote books expressing liberal ideas, and delivered the Lowell Institute Lectures in Boston in the late 1860s, in which he espoused the cause of women's rights. Thompson's mother, Fanny Gamgee, who died when he was born, came from a family that was active in medicine and science. Young D'Arcy thus received a scientific background from his maternal grandfather as he was growing up, and a classical education from his father. As a result, he could read, speak, and write Greek and Latin fluently.

Thompson attended Edinburgh Academy, and studied medicine at the University of Edinburgh. He showed a bent for natural history, and at the age of 19, published papers in science journals on hydroid taxonomy—or **classification** of invertebrate **animal**s including corals, sea anemones, and jellyfishes—and on a Pleistocene **fossil** seal. He left Edinburgh for Trinity College, Cambridge, where he supplemented his finances by tutoring in Greek. While there, he translated **Hermann Müller**'s German work, *Die Befruchtung der Blumen durch Insekten,* as *The Fertilisation of Flowers;* it was published with a preface by the naturalist **Charles Darwin**, about which Thompson later said, as quoted in *Science* magazine, ''[It] is of peculiar interest as one of the very last of his writings.''

Embarks on a Career as a Zoologist

In 1884, at the age of 24, Thompson was appointed professor of biology at University College in Dundee, where he established a teaching museum of zoology. When the college was united with the University of St. Andrews in 1897, he became the chair of natural history. He would hold that position until his death at the age of eighty-eight. Beginning in 1885, Thompson wrote scientific papers on a wide variety of zoological subjects, including the **morphology** of vertebrate limbs, classification of the chameleon, the **nervous system** and **blood** cells of cyclostomes (jawless fishes including hagfishes and lampreys), the newly discovered ear of the sunfish, and a fossil mammal thought to be related to whales, which he showed to be more similar to seals. He also continued to work on such varied subjects as the bones of the parrot's skull, a rare cuttle-fish, the arrangement of feathers on the giant hummingbird, and a systematic survey of the sea spiders.

In l896, when a dispute arose between Great Britain and the United States over the fur-seal fisheries in the Bering Sea, Thompson was sent to Alaska to investigate the situation. After expeditions to the Pribilof Islands, he represented Britain at the International Conference in Washington the following year. In recognition of the success of his undertaking, he was awarded the title of Companion of the Order of the Bath in 1898. In that year, he was appointed scientific adviser to the Fishery Board for Scotland, a position he held until the Board was dissolved in 1939. Thompson issued a number of scientific reports in which he made biological, statistical, and hydrographical contributions. Starting in 1902, he began to serve as the British representative to the newly founded International Council for the Study of the Sea and regularly attended meetings in Copenhagen and elsewhere in Europe until 1947. During this time, he was chair of the Statistical Committee and editor of the *Bulletin Statistique,* writing many papers on oceanography and fishery statistics.

In 1895 Thompson drew upon his predilection for the classics and published his *A Glossary of Greek Birds.* Here, he revealed a learned understanding of ancient Greek literature, as well as medieval and modern **ornithology**. For many years afterward, he worked on a companion book, *A Glossary of Greek Fishes,* which was finally published in 1947, and which referred not only to fish mentioned in classic Greek literature, but also to other **species** listed under the heading of fish by the ancients, such as crabs, cuttlefish, and oysters. In both of these books, Thompson identified the bird or fish not only from a scientific point of view, but also classically, in terms of its interest to the poets and its relation to religion, folklore, and art. In 1910, he had issued an annotated translation of **Aristotle**'s *Historia Animalium;* he delivered the Herbert Spencer Lecture, "On Aristotle as a Biologist," three years later. He continued to express his dual interest in these subjects in his presidential address to the Classical Association, entitled *Science and the Classics,* which was published in *Nature.* In this lecture he drew connections between the two disciplines: "Science and the classics—both alike continually enlarge our curiosity, and multiply our inlets to happiness."

Blazes a New Trail in Biology

In 1908, Thompson published a paper in *Nature,* "On the Shapes of Eggs and the Causes Which Determine Them," which indicated a new direction in his explanation of morphology using mathematical interpretations. He continued with this concept in 1911 in his presidential address to section D of the British Association, entitled "Magnalia Naturae; or the Greater Problems of Biology." He had now departed from standard zoology, and its occupation with comparative morphology and evolution, and was blazing a new trail in which mathematics and physics were the tools for interpreting biological phenomena. Thompson's influential book *On Growth and Form* appeared in 1917, presenting his unorthodox new principles and explaining them with numerous illustrative examples from an-

cient and modern texts. The book deals with the development of form and structure in living things and how physical forces influence them in their lifetime. He demonstrated some of his ideas by showing that various natural phenomena, such as the repeated six-sided shape of cells in a bee honeycomb, the spirals in the arrangement of seeds in a sunflower, the curve in snailshells, and even the flight of a moth attracted to light, follow mathematical principles. Thompson postulated that these and other geometrical patterns evolved as ideal adaptations in the development of the **organisms**. By means of graphs of grid coordinates based on logarithmic projections, he compared the changes in growth and shape of various organisms during their development. In one instance, with respect to the structure of bone, he showed that the trabeculae, or lattice-work, of calcium deposition, is aligned in the most efficient placement to cope with the stresses placed on the bone. He proved this point by comparing the metal cross structures of a hoisting crane with the internal structure of a femur.

Thompson married Maureen Drury in 1901, and they had three daughters. As his career progressed, he was received many honors. He was elected to the Royal Society in 1916, was its vice-president from 1931 to 1933, and received the Darwin Medal in 1946. In 1928 he became president of the Classical Association of England and Wales, and from 1934 to 1939, he was president of the Royal Society of Edinburgh. Thompson was knighted in 1937, and a year later, the Linnean Society presented him with the Linnean gold medal. He delivered the Lowell Lectures in Boston in 1936, seventy-nine years after his father had had this honor. In 1946, he flew to India as a member of the Royal Society delegation to the Indian Science Congress at Delhi, but contracted pneumonia soon afterward and never recovered. Thompson died on June 21, 1948. He was remembered by his colleague, Robert Chambers, in *Science* magazine, as a "towering figure with massive sculptured head and [long] flowing beard," who had a ready sense of humor and a penchant for eloquent oratory.

THYROID GLAND

The thyroid gland is an **endocrine gland**, found in almost all vertebrate **animals**, that synthesizes, stores, and secretes two **hormones** that affect **metabolism** and growth. The hormones are thyroxine, also called tetraiodothyronine (T_4), and triiodothyronine (T_3). These hormones are composed of the **amino acid** tyrosine, and contain four and three iodine **atoms**, respectively. The thyroid gland constitutes only about 0.5% of the total human body weight, but holds about 25% of the total iodine in the body, which is obtained from dietary food and **water**. Iodine circulates in the **blood** as inorganic iodide, but is concentrated in the thyroid up to 500 times the level in the blood and is essential for synthesis of the hormones.

The thyroid gland is located below the larynx and consists of two lobes connected by an isthmus (band of **tissue**). It is brownish-red and normally weighs about 1 oz (28 g). The gland is composed of hollow sacs, called follicles, that are filled with colloid, a gelatin material, that contains thyroglobulin, the storage form of the hormones.

Secretion of thyroid hormones is controlled by the thyroid-stimulating hormone (TSH), which is secreted by the **pituitary gland**. TSH is regulated by a thyroid-stimulating hormone releasing factor (TRF), which is released by the **hypothalamus**. Increases in the level of thyroid hormones in the blood signals the pituitary to stop releasing TSH, which keeps the level of thyroid hormones in the **circulatory system** within a constant range.

Laboratory tests of the levels of thyroxine, T_3, and TSH are used to test the activity of the thyroid gland. Excessive production of thyroid hormones (i.e., hyperthyroidism) results in elevated metabolism and activity. Abnormalities of the eyes, including bulging eyes, may be present as well as weight loss, increased **heart** rate, higher **blood pressure**, more frequent bowel movements, muscle weakness, and trembling hands. In Graves' disease, hyperthyroidism is a result of an **autoimmune disease**. Nodules called toxic adenomas can also develop in the thyroid gland and secrete excess thyroid hormones. Treatment is accomplished by suppressing production of thyroid hormones through surgery, by using an antithyroid drug such as propylthiouracil or methimazole, or through a dose of radioactive iodine, which concentrates in the thyroid gland and destroys thyroid tissue.

Deficiency of thyroid hormones, or hypothyroidism, is characterized by lethargy, slower mental processes, reduced heart rate, increased sensitivity to cold, and tingling and numbness in the hands. Untreated hypothyroidism over a long period of time can result in a myxedema coma, a rare but potentially fatal condition. Hypothyroidism can be the result of thyroid destruction from the production of an **antibody** against thyroid tissues (i.e., Hashimoto's disease, which is an autoimmune disease). Hypothyroidism can also result from treatments for hyperthyroidism, in which the thyroid gland has been removed or destroyed, disorders of the pituitary gland, or from a deficiency of iodine in the diet, which causes a condition called goiter. Iodine is now added to table salt to prevent the development of goiters. Treatment of hypothyroidism is accomplished through a lifelong regimen of hormone replacement therapy. Synthetic forms of the thyroid hormone, such as levothyroxine, are usually used.

Congenital hypothyroidism, or cretinism, is an inherited deficiency in thyroid function that occurs in one of about 6000 **births**. Early treatment can prevent mental retardation and dwarfism that can occur with this condition.

Cancer of the thyroid gland is rare, but can occur, especially in people who have had radiation treatment to the head or neck earlier in life (e.g., for acne or ringworm). Treatment involves removing the cancerous tissue or the whole thyroid gland in a procedure called thyroidectomy.

TINBERGEN, NIKOLAAS (1907-1988)
Dutch English zoologist and ethologist

Nikolaas Tinbergen was born April 15, 1907, in The Hague, Netherlands. His older brother Jan studied physics but later turned to economics, winning the first Nobel Prize awarded in

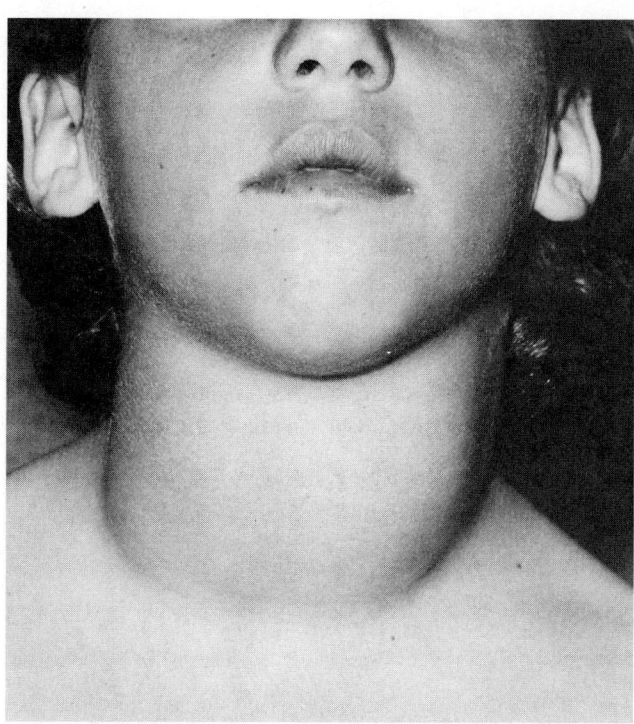

A symptom of hyperthyroidism is the enlargement of the thyroid gland. *(Photograph by Lester V. Bergman, Corbis Images. Reproduced by permission.)*

that subject in 1969. The Tinbergens lived near the seashore, where Tinbergen often went to collect shells, camp, and watch **animal**s, many of which he would later formally research.

After high school, Tinbergen worked at the Vogelwarte Rossitten bird observatory and later began studying biology at the State University of Leiden, Netherlands. For his dissertation, Tinbergen studied bee-killer wasps and was able to experimentally demonstrate that the wasps use landmarks to orientate themselves. Tinbergen first established the traditional routes of the wasps near their burrows, then altered the landscape to see how the wasps' behavior would be affected. Tinbergen was awarded his Ph.D. in 1932.

Shortly after his 1932 wedding to Elisabeth Rutten, the Tinbergens embarked on an expedition to Greenland, where Tinbergen studied the role of **evolution** in the behavior of snow buntings, phalaropes, and Eskimo sled dogs. When he returned to the Netherlands in 1933, he became an instructor at the State University, where he organized an undergraduate course on animal behavior. Tinbergen's work had been recognized in the field of biology but it was not until after he met **Konrad Lorenz**—the acknowledged father of **ethology**—that his work began to form a directed body of research. Tinbergen took his family to Lorenz's home in Austria for a summer so the two men could work together. Although they published only one paper together, their collaboration lasted a number of years.

During 1936, Tinbergen and Lorenz began constructing a theoretical framework for the study of ethology, which was then a fledgling field. They hypothesized that **instinct**, as opposed to simply being a response to environmental factors,

Nikolaas Tinbergen. *(Mary Evans Picture Library. Reproduced by permission.)*

arises from an animal's impulses. This idea is expressed by the concept of a fixed-action pattern, a repeated, distinct set of movements or behaviors, which Tinbergen and Lorenz believed all animals have. A fixed-action pattern is triggered by something in the animal's environment. In some species of gull, for instance, hungry chicks will peck at a decoy with a red spot on its bill, a characteristic of the gull. Tinbergen showed that in some animals learned behavior is critical for survival. Tinbergen and Lorenz also demonstrated that animal behavior can be the result of contradictory impulses and that a conflict between drives may produce a reaction that is strangely unsuited to the stimuli. Unfortunately, Tinbergen and Lorenz's work was disrupted by World War II.

Tinbergen spent much of the war in a hostage camp because he had protested the State University of Leiden's decision to remove three Jewish faculty members from the staff. After the war ended, he became a professor of experimental biology at the University. In 1949, Tinbergen traveled to Oxford University in England to lecture. He stayed at Oxford, establishing the journal *Behavior* with W. H. Thorpe and working in the University's animal behavior division. His 1951 book *The Study of Instinct* is credited with bringing the study of ethology to many English readers. The book summarized some of the newest insights into the ways signaling be-

havior is created over the course of evolution. In 1955, Tinbergen became an English citizen, and in 1966 he was appointed a professor and fellow of Oxford's Wolfson College. When the work of Tinbergen, Lorenz, and **Karl von Frisch**, who had demonstrated that honeybees communicate by dancing, received the Nobel Prize in 1973, it was the first time the Nobel Committee recognized work in **sociobiology** or ethology.

The ability of an **organism** to adapt to its environment is another element of Tinbergen's work. After he retired from Oxford in 1974, he and his wife attempted to explain autistic behavior in children to adaptability. The Tinbergens' assertion that autism may be caused by the behavior of a child's parents caused some consternation in the medical community. Tinbergen believed that much of the opposition to his work was caused by the unflattering view of human behavior it presented. Tinbergen died December 21, 1988, after suffering a stroke at his home in Oxford, England.

TISELIUS, ARNE (1902-1971)
Swedish chemist

Arne Tiselius was awarded the 1948 Nobel Prize in chemistry for his research in **electrophoresis** (the movement of **molecule**s based on their electric charge and their size) and for his investigations into adsorption, the inclination of certain molecules to cling to particular substances. Although the phenomenon of electrophoresis had been identified decades earlier, it did not become a useful technique for analyzing chemical compounds until Tiselius developed methods which delivered accurate results.

Arne Wilhelm Kaurin Tiselius was born in Stockholm, Sweden, on August 10, 1902, to Hans Abraham J. Tiselius, who was employed by an insurance company, and Rosa Kaurin Tiselius, the daughter of a Norwegian clergyman. Upon the death of Tiselius's father in 1906, Rosa relocated the family to Göteborg, Sweden, where Hans's family lived. Entering the gymnasium at Göteborg, Tiselius came under the tutelage of a chemistry and biology teacher who actively supported his student's interest in science. In 1921 Tiselius matriculated to the University of Uppsala—where his father had earned his degree in mathematics—and studied under the renowned physical chemist Theodor Svedberg. Earning his master's degree in chemistry, physics, and mathematics in 1924, Tiselius continued to work as Svedberg's research assistant in physical chemistry. Although Svedberg was interested in the electrophoretic properties of **protein**s, he turned the study of this over to his new assistant, and three years later Tiselius published his first paper jointly with Svedberg on the subject.

Tiselius would remain at Uppsala until his retirement in 1968, rising from researcher to full professor. His 1930 doctoral dissertation, which earned him a post as docent in the chemistry department, long stood as a standard in the field of electrophoresis. Sweden's first professorship in biochemistry was established for Tiselius at Uppsala in 1938. Besides his work in biochemistry, Tiselius had a strong interest in **botany** and **ornithology** and made frequent excursions into the Swedish countryside on photographic expeditions. On November 26, 1930, the year of his doctoral thesis, he married Ingrid Margareta Dalén, with whom he would have one son, Per, and a daughter, Eva.

Explores the Possibilities of Chromatography

Following his dissertation, Tiselius concentrated his attention in areas outside of chemistry. He expanded his research to include biochemical studies—not a typical element of the chemistry curriculum in those days—and became aware of the potential for exploiting the extremely specific electrical "signature" of proteins, as well as other substances. He became concerned, however, with the impurities in the substances under study, even those that had been carefully centrifuged, and turned to **chromatography** as a possible answer. In chromatographic analysis, **light** of a specific frequency is passed through a substance, and by using tables assembled over the course of many experiments, the "chromatic signature" of the particular sample can be detected. Tiselius applied this technique by looking into the properties of light **diffusion** through zeolite, a translucent mineral. While studying under Hugh S. Taylor from 1934 to 1935 at Princeton University's Frick Chemical Laboratory, Tiselius conceived of an accurate method to quantify the diffusion of **water** molecules through crystals of zeolite.

While at Princeton, Tiselius came to realize that a wealth of potential discoveries in the biochemical sciences awaited only the development of a method accurate enough to help separate and identify compounds. Returning to his original line of research, he completed a prototype of a new electrophoretic apparatus.

When Tiselius returned to Uppsala, he continued making improvements on his electrophoretic instrumentation. In one innovation, he filled a U-shaped tube with chemical solvents, added a solution containing the sample to be analyzed, then applied a charge to one end. As the elements migrated, they reached the solvents at different lengths along the tube. Tiselius constructed the tube so that test samples could be taken at various points along the path of migration and be analyzed to determine which of the original species had made it to that point. It was by using this technique that Tiselius was able to demonstrate that **blood** plasma contained a complex mix of different elements.

Tracking the movement of boundaries optically by a technique invented by August Toepler—the *Schlieren* method—Tiselius resolved the plasma into four distinct elements that showed up as separate bands in the tube. He was the first to isolate three of the blood proteins known as globulins, which he named *alpha, beta,* and *gamma.* These are important in many of the body's functions; the immunoglobulins, for example, are a critical factor in infection control. In the fourth band, located between those of *beta* and *gamma,* Tiselius discovered antibodies.

The method was a radical improvement but still dissatisfied Tiselius. At the time he was more interested in the breakdown products of **polypeptide**s than in blood compounds.

Arne Tiselius. *(The Library of Congress. Reproduced by permission.)*

Peptides represent some of the most important proteins in the body and for a clear understanding of their function, it is essential to know their types. However, when the long chain of a polypeptide is broken down, the individual peptides are so similar in nature that even Tiselius's improved electrophoretic technique could not distinguish between them. Faced with this problem, he turned to adsorption methods of analysis, using the then-common column method. In this procedure, a mixture which contains a substance with a specific affinity for absorbing one peptide or another is flushed through a column (a tube or cylinder). The peptides which had been in the original mix can then be determined by analyzing the eluate (the wash which passed through the column).

In 1943 Tiselius introduced a critical improvement in the process. Research to that point had been carried out using a "frontal analysis method," which revealed the concentration of the components in a mixture but was unable to separate them for further study. Elution could accomplish separation, but had a major setback, "tailing," which is the corruption of one part of a solution by molecules from the other. Tiselius demonstrated that a simple modification to the old technique could reduce tailing, and this new method became known as "displacement analysis."

Advises Government on Scientific Matters

Other important work came out of Tiselius's laboratory throughout the 1940s, such as research on paper electrophoresis and zone electrophoresis. However, increasing demands from other sources took over his time and, in the summer of 1944, Tiselius became an advisor to the Swedish government. His responsibilities included sitting on a committee established to help improve conditions for advancing scientific research, with a focus on basic research. This was the beginning of a long and distinguished relationship with the Swedish Parliament, an association that ended only when Tiselius suffered a **heart** attack following an important meeting in Stockholm. He died the next morning on October 29, 1971.

Up to his last day, Tiselius followed an active schedule. Having accepted the four-year chairmanship of the Swedish Natural Science Research Council in 1946, he was instrumental in the creation of the Science Advisory Council to the Swedish government. Tiselius was elected vice president of the Nobel Foundation with membership on the Nobel Committee for Chemistry in 1947, one year before he was awarded his own Nobel Prize. That same year, at the International Congress of Chemistry held in London, he was elected vice president in charge of the section for biological chemistry of the International Union of Pure and Applied Chemistry—a body which he led as president four years later.

Among other honors Tiselius received were the Bergstedt Prize of the Royal Swedish Scientific Society in 1926, the Franklin Medal of the Franklin Institute in 1956, and the Paul Karrer Medal in Chemistry from the University of Zurich in 1961. He was also presented with numerous honorary degrees from universities, including those of Stockholm, Paris, Glasgow, Madrid, California at Berkeley, Prague, Cambridge, and Oxford. Tiselius was always interested in fields beyond his own and was concerned with the environmental, social, and ethical implications of science and technology. As president of the Nobel Foundation in 1960, he established the Nobel Symposium, perceiving the foundation as the perfect vehicle for raising awareness of the need to promote science as a solution to mankind's problems. This organization gathered a mix of Nobel laureates to discuss the implications of their work during symposia in each of the five prize fields.

TISSUE

A tissue is a group of **cell**s that are integrated and have a common structure and function. The term tissue comes from the Latin word meaning "weave." This is reflective of the fact that tissues are often held together by an extracellular matrix that coats and weaves the cells together. In vertebrates, there are four main categories of tissue including: connective tissue, epithelial tissue, nervous tissue, and muscle tissue.

Connective tissues function mainly to bind and support other tissues and **organ**s. They are composed of a relatively small number of cells scattered throughout an extracellular matrix. Typically, this matrix is composed of some type of **protein** fiber embedded in a gelatinous substance. The main vertebrate connective tissues are: loose connective tissue, adipose tissue, fibrous connective tissue, cartilage, **blood** and bone. Loose connective tissue is the most abundant type and

it holds organs in place. It is made up of collagen and elastin fibers, fibroblast cells which secrete protein, and macrophage cells which protect and repair the tissue. Adipose tissue is a special type of loose connective tissue that stores fat.

The other connective tissues play important roles in the body. Fibrous connective tissue, which is more dense, holds muscles, bones and joints together. Cartilage is found between joints. It is a rubbery tissue that is embedded with collagen fibers and composed of chondrocyte cells which secrete chondroitin sulfate. Bone is a type of mineralized connective tissue. It is made up of osteoblast cells which release calcium phosphate and collagen. While blood appears different from other connective tissues it is categorized as such because it is made up of cells that are connected through an extracellular matrix. In this case the cells include red and white blood cells. The extracellular matrix is plasma which is a combination of **water**, salts and various proteins.

Epithelial tissue is composed of cells that are more tightly packed than connective tissue. It is found throughout the body lining the inner organs and covering the outside. The cells are typically connected to each other. This tight packing allows the epithelial tissues to act as a barrier to protect against injury, invading **microorganism**s, and regulate fluid loss. Skin and the mucous **membrane** are examples of epithelial tissues.

Nervous tissue is made up of nerve cells called **neuron**s. They are connected to each other by structures called dendrites and **axon**s. These structures allow the cells to transmit signals **nerve impulses**) throughout the body. Muscle tissue is composed of long contractible cells. Within the cells are a large number of microfilaments made up of proteins actin and myosin. Examples of muscle tissue include skeletal muscle, cardiac muscle, and smooth muscle.

See also Heart; Integumentary system; Muscular system; Skeletal system

TISSUE CULTURE AND STORAGE

Tissue culture is defined as the propagation of plant and **animal** cells through the placement of small amounts of **tissue** in an artificial environment. Given the appropriate conditions, most kinds of plant and animal **cell**s will live, even multiply and express differentiated properties in a tissue-**culture** dish.

Plant tissue culture was first used on a large scale by the orchid industry in the 1950s. Later, it became clear that any plant would respond to tissue culture as long as the right formula and the right processes were developed for its culture. In the case of plant tissue culture, a piece of plant, (which can be anything from a piece of stem, root, **leaf**, bud, or a single cell) is placed in a test tube. In a sterile environment (free from **microorganism**s) and in a balanced nutrient medium that bit of plant (explant) will form plantlets. These can multiply indefinitely if given proper care and later, can be taken out and planted normally. The medium for plant growth refers to a mixture of certain **chemical compound**s. By using tissue culture, one can control the environment and optimize it such that all the **plants** from the tissue culture are identical for the particular quality being sought, whether it be resistance to plant diseases or the production of a plant chemical.

Animal cell culture began in 1907, when a group of scientists decided to test the doctrine that states that each nerve fiber is the outgrowth of a single nerve cell and not a product of several cells. In order to test this, they placed small pieces of **spinal cord** on clotted tissue fluid in a warm moist chamber and observed it at regular intervals. After a day or so, individual nerve cells could be seen extending long thin processes into the clot. Thus the doctrine was validated and the foundation for cell-culture was laid.

Initially, small tissue fragments or explants were used for animal cell culture experiments. Today, however, cultures are more commonly made from suspensions of cells that have been dissociated from the tissues. These cells are propagated on a solid surface where they can grow and divide. Generally, a plastic culture dish is used to provide the mechanical support, and a nutrient medium is provided. Cells vary in their requirements and many will not grow unless their specific requirements are met. Liquid nutrient media containing well-defined mixtures of salts, **amino acid**s, **vitamins**, etc. are used for the propagation of cells. Most media also contain some proportion of a poorly defined biological material such as bovine serum, horse serum, or a crude extract from chick **embryo**s. Since the specific growth and nutrient requirements cannot be determined using such media, various chemically defined media were developed. In these media, each component is a known **molecule**. In addition to these known molecules, chemically defined media also contain one or more of the various growth factors that most cells need to proliferate, such as "nerve growth factor" for nerve cells etc. The cells can be watched under the **microscope** and analyzed biochemically if necessary. The effects of adding or removing specific molecules such as **hormone**s or growth factors can be investigated.

Cultures that are prepared directly from the tissues of an **organism** are called primary cultures. These primary cultures can be removed from the dish and seeded onto other dishes. This process known as "subculturing" can be used to form a large number of secondary cultures. They have to be repeatedly subcultured in this manner for weeks or even months. Cells cultured in this fashion often display the differentiated properties specific for their cell type, for e.g. fibroblasts will secrete collagen; nerve cells will extend **axon**s and make **synapse**s with other nerve cells; epithelial cells will form extensive sheets and so on.

Most vertebrate cells die after a finite number of divisions in culture, for e.g. human skin cells will typically last only several months in culture, dividing about 50-100 times. The limited life span is related to the limited life span of the animal from which they are derived. Occasionally, mutant cells (variants) arise that can be propagated indefinitely. These are known as "immortal cells". They are propagated as "cell lines" and grow best when attached to solid surfaces. They normally cease growing after they have formed a confluent layer over the surface of the culture dish.

Cell lines that are prepared from **cancer** cells differ from normal cell lines in that they grow without attaching to any cell

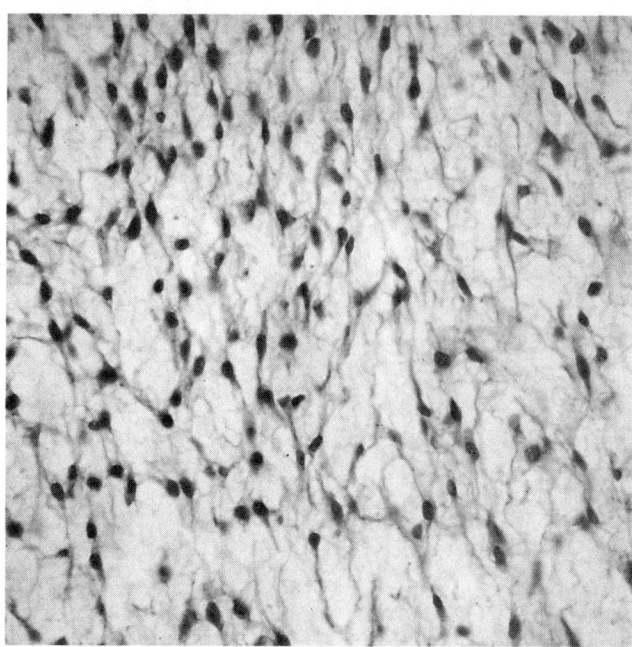

Human fetal tissue as seen through a microscope. *(Custom Medical Stock Photo, Inc. Reproduced by permission.)*

surface and they proliferate to a very high density. Such cell lines are called "transformed cell lines." Normal cell lines can be "transformed" by inducing them with certain chemicals or a tumor-inducing **virus**. Both transformed and untransformed cell lines are extremely useful in cell research. They can be stored at -158°F (-70°C) for an indefinite period and are still viable when thawed.

In tissue culture, it is even possible to fuse one cell with another to form a combined cell with two separate nuclei called a "heterocaryon". To do this, the cells are treated with certain chemicals or viruses which alters the **plasma membrane** of the cells and induces them to fuse. Such cells are useful for studying the interactions between the components of two different cells and they also provide a convenient method for assigning genes to human **chromosome**s.

TODD, ALEXANDER (1907-1997)
English chemist

Alexander Todd was awarded the 1957 Nobel Prize in chemistry for his work on the chemistry of **nucleotides**. He was also influential in synthesizing **vitamins** for commercial application. In addition, he invesitgated active ingredients in cannabis and hashish and helped develop efficient means of producing chemical weapons.

Alexander Robertus Todd was born in Glasgow, Scotland, on October 2, 1907, to Alexander and Jane Lowrie Todd. The family, consisting of Todd, his parents, his older sister, and his younger brother, was not well-to-do. Todd's autobiography, *A Time to Remember,* recalls how through hard work

Alexander Todd. *(The Library of Congress. Reproduced by permission.)*

his parents rose to the lower middle class despite having no more than an elementary education, and how determined they were that their children should have an education at any cost.

Education and Early Career

In 1918, Todd gained admission to the Allan Glen's School in Glasgow, a science high school; his interest in chemistry, which first arose when he was given a chemistry set at the age of eight or nine, developed rapidly. On graduation, six years later, he at once entered the University of Glasgow instead of taking a recommended additional year at Allan Glen's. His father refused to sign an application for scholastic aid, saying it would be accepting charity; because of superior academic performance during the first year, though, Todd received a scholarship for the rest of course. In his final year at university, Todd did a thesis on the reaction of phosphorus pentachloride with ethyl tartrate and its diacetyl derivative under the direction of T. E. Patterson, resulting in his first publication.

After receiving his B.S. degree in chemistry with first-class honors in 1928, Todd was awarded a Carnegie research scholarship and stayed on for another year working for Patterson on optical rotatory dispersion. Deciding that this line of research was neither to his taste nor likely to be fruitful, he went to Germany to do graduate work at the University of Frankfurt am Main under Walther Borsche, studying natural products.

Todd says that he preferred Jöns Berzelius's definition of organic chemistry as the chemistry of substances found in living **organism**s to **Leopold Gmelin**'s definition of it as the chemistry of carbon compounds.

At Frankfurt he studied the chemistry of apocholic acid, one of the bile acids (compounds produced in the liver and having a structure related to that of **cholesterol** and the **steroids**). In 1931, he returned to England with his doctorate. He applied for and received an 1851 Exhibition Senior Studentship which allowed him to enter Oxford University to work under Robert Robinson, who would receive the Nobel Prize in chemistry in 1947. In order to ease some administrative difficulties, Todd enrolled in the doctoral program, which had only a research requirement; he received his Ph.D. from Oxford in 1934. His research at Oxford dealt first with the synthesis of several anthocyanins, the coloring matter of **flowers**, and then with a study of the red pigments from some molds.

After leaving Oxford, Todd went to the University of Edinburgh on a Medical Research Council grant to study the structure of vitamin B $_1$ (thiamine, or the anti-beriberi vitamin). The appointment came about when George Barger, professor of medical chemistry at Edinburgh, sought Robinson's advice about working with B$_1$. At that time, only a few milligrams of the substance were available, and Robinson suggested Todd because of his interest in natural products and his knowledge of microchemical techniques acquired in Germany. Although Todd and his team were beaten in the race to synthesize B$_1$ by competing German and American groups, their synthesis was more elegant and better suited for industrial application. It was at Edinburgh that Todd met and became engaged to Alison Dale—daughter of Nobel Prize laureate **Henry Hallett Dale**—who was doing postgraduate research in the pharmacology department; they were married in January of 1937, shortly after Todd had moved to the Lister Institute where he was reader (or lecturer) in biochemistry. For the first time in his career, Todd was salaried and not dependent on grants or scholarships. In 1939 the Todds' son, Alexander, was born. Their first daughter, Helen, was born in 1941, and the second, Hilary, in 1945.

The Maturing of a Scientist

Toward the end of his stay at Edinburgh, Todd began to investigate the chemistry of vitamin E (a group of related compounds called tocopherols), which is an antioxidant—that is, it inhibits loss of electrons. He continued this line of research at the Lister Institute and also started an investigation of the active ingredients of the *Cannabis sativa* plant (marijuana) that showed that cannabinol, the major product isolated from the plant resin, was pharmacologically inactive.

In March of 1938, Todd and his wife made a long visit to the United States to investigate the offer of a position at California Institute of Technology. On returning to England with the idea that he would move to California, Todd was offered a professorship at Manchester which he accepted, becoming Sir Samuel Hall Professor of Chemistry and director of the chemical laboratories of the University of Manchester. At Manchester, Todd was able to continue his research with little interruption. During his first year there, he finished the work on vitamin E with the total synthesis of alpha-tocopherol and its analogs. Attempts to isolate and identify the active ingredients in cannabis resin failed because the separation procedures available at the time were inadequate; however, Todd's synthesis of cannabinol involved an intermediate, tetrahydrocannabinol (THC), that had an effect much like that of hashish on rabbits and suggested to him that the effects of hashish were due to one of the isomeric tetrahydrocannabinols. This view was later proven correct, but by others, because the outbreak of World War II forced Todd to abandon this line of research for work more directly related to the war.

As a member, and then chair, of the Chemical Committee, which was responsible for developing and producing chemical warfare agents, Todd developed an efficient method of producing diphenylamine chloroarsine (a sneeze gas), and designed a pilot plant for producing **nitrogen** mustards (blistering agents). He also had a group working on **penicillin** research and another trying to isolate and identify the ''hatching factor'' of the potato eelworm, a parasite that attacks potatoes.

Late in 1943 Todd was offered the chair in biochemistry at Cambridge University, which he refused. Shortly thereafter he was offered the chair in organic chemistry, which he accepted, choosing to affiliate with Christ's College. From 1963 to 1978, he served as master of the college. As professor of organic chemistry at Cambridge, Todd reorganized and revitalized the department and oversaw the modernization of the laboratories (they were still lighted by gas in 1944) and, eventually, the construction of a new laboratory building.

Wins Nobel Prize for Work on Nucleotides

Before the war, his interest in vitamins and their mode of action had led Todd to start work on nucleosides and nucleotides. Nucleosides are compounds made up of a sugar (ribose or deoxyribose) linked to one of four heterocyclic (that is, containing rings with more than one kind of atom) nitrogen compounds derived either from purine (adenine and guanine) or pyrimidine (uracil and cytosine). When a phosphate group is attached to the sugar portion of the **molecule**, a nucleoside becomes a nucleotide. The **nucleic acids** (**DNA** and **RNA**), found in cell nuclei as constituents of the **chromosome**s, are chains of nucleotides. While still at Manchester, Todd had worked out techniques for synthesizing nucleosides and then attaching the phosphate group to them (a process called phosphorylating) to form nucleotides; later, at Cambridge, he worked out the structures of the nucleotides obtained by the degradation of nucleic acid and synthesized them. This information was a necessary prerequisite to **James Watson** and **Francis Crick**'s formulation of the **double-helix** structure of DNA two years later.

Todd had found the nucleoside adenosine in some **coenzyme**s, relatively small molecules that combine with a **protein** to form an **enzyme**, which can act as a catalyst for a particular biochemical process. He knew from his work with the B vitamins that B_1 (thiamine), B_2 (riboflavin) and B_3 (niacin) were essential components of coenzymes involved in **respiration** and oxygen utilization. By 1949 he had succeeded in synthesizing adenosine—a triumph in itself—and had gone on to

synthesize adenosine di- and triphosphate (ADP and **ATP**). These compounds are nucleotides responsible for **energy** production and energy storage in muscles and in **plant**s. In 1952, he established the structure of flavin adenine dinucleotide (FAD), a coenzyme involved in breaking down **carbohydrate**s so that they can be oxidized, releasing energy for an organism to use. For his pioneering work on nucleotides and nucleotide enzymes, Todd was awarded the 1957 Nobel Prize in chemistry.

Todd collaborated with Dorothy Crowfoot Hodgkin in determining the structure of vitamin B_{12}, the antipernicious anemia factor, which is necessary for the formation of red **blood** cells. Todd's chemical studies of the degradation products of B_{12} were crucial to Hodgkin's X-ray determination of the structure in 1955.

Another major field of research at Cambridge was the chemistry of the pigments in aphids. While at Oxford and working on the coloring matter from some **fungi**, Todd observed that although the pigments from fungi and from higher plants were all anthraquinone derivatives, the pattern of substitution around the anthraquinone ring differed in the two cases. Pigment from two different insects seemed to be of the fungal pattern and Todd wondered if these were derived from the insect or from symbiotic fungi they contained. At Cambridge he isolated several pigments from different kinds of aphids and found that they were complex quinones unrelated to anthraquinone. It was found, however, that they are probably the products of symbiotic fungi in the aphid.

A Senior Scientist and Government Advisor

In 1952 Todd became chairman of the advisory council on scientific policy to the British government, a post he held until 1964. He was knighted in 1954 by Queen Elizabeth for distinguished service to the government. Named Baron Todd of Trumpington in 1962, he was made a member of the Order of Merit in 1977. In 1955 he became a foreign associate of the United States' National Academy of Sciences. He traveled extensively and been a visiting professor at the University of Sydney (Australia), the California Institute of Technology, the Massachusetts Institute of Technology, the University of Chicago, and Notre Dame University.

A Fellow of the Royal Society since 1942, Todd served as its president from 1975 to 1980. He increased the role of the society in advising the government on the scientific aspects of policy and strengthened its international relations. Extracts from his five anniversary addresses to the society dealing with these concerns are given as appendices to his autobiography. In the forward to his autobiography, Todd reports that in preparing biographical sketches of a number of members of the Royal Society he was struck by the lack of information available about their lives and careers and that this, in part, led him to write *A Time to Remember*. Todd died on 10 January 1997, in his home city of Cambridge, England. He was 89.

TOMOGRAPHY • See Computerized axial tomography

TONEGAWA, SUSUMU (1939-)

Japanese molecular biologist

In 1987, Susumu Tonegawa became the first Japanese recipient of the Nobel Prize for Physiology or Medicine for his study of the **immune system** and his subsequent discovery of the causes of antibody diversity—the ability of an antibody to resist infection from millions of different **virus**es and **bacteria**. Tonegawa provided direct evidence that a **gene**'s ability to encode antibody **protein**s is produced from separate, chain-like segments of **DNA** (**deoxyribonucleic acid**) molecules which mutate to code for different antibodies. Since 1981, Tonegawa has worked at the Massachusetts Institute of Technology (MIT), and was honored as Howard Hughes Medical Institute Investigator in 1988.

Tonegawa was born in Nagoya, Japan, on September 5, 1939, the second of four children born to Tsutomu Tonegawa and the former Miyoko Masuko. Tonegawa's father was an engineer whose work required him to move frequently from town to town across the country. As a result, Tonegawa and his older brother were sent to Tokyo to live with an uncle. In Tokyo, the boys attended the prestigious Hibiya High School, where Tonegawa eventually developed an interest in chemistry. After graduation, he entered the University of Kyoto in 1959 to pursue a degree in chemistry. He earned his degree in 1963 and began graduate studies in a then-emerging branch of biology—**molecular biology**—which is the study of **molecule**s that perform biological operations.

While he was still a student at Kyoto, Tonegawa learned about the field of molecular biology and decided that it was an area in which he wanted to specialize. In 1953, **James Watson** and **Francis Crick** had discovered the mechanism by which genetic information is stored in molecules. That discovery provided an exciting new way to understand biological phenomena in terms of atomic and molecular structure. Research in this promising new field developed very rapidly. In 1963, Tonegawa applied to the University of California at San Diego and began his graduate study in the Department of Biology under Masaki Hayashi. Tonegawa's research in genetic transcription in bacteriophages resulted in three scientific papers, published between 1966 and 1970 with Hayashi, and a Ph.D. in biology by 1968.

For his postdoctoral work, Tonegawa chose to stay in San Diego, working first with Hayashi from September, 1968, to April, 1969, and then moving to the Salk Institute in nearby La Jolla from May, 1969, to December, 1970. At the Salk Institute, Tonegawa studied genetic transcription in simian virus 40 (SV40), an important virus in genetic engineering, with Renato Dulbecco, who would go on to win a Nobel Prize in 1975.

Joins Institute of Immunology

In the fall of 1970, Tonegawa was confronted with a dilemma. His United States immigration visa was due to expire at the end of the year, and he had to decide where he was going to continue his studies. At that time, Tonegawa received a letter from Dulbecco notifying him of an opening for a molecular biologist at the Institute of Immunology in Basel, Switzerland.

Tonegawa had no formal training in immunology, but applied for the position, and was accepted. By February, 1971, Tonegawa found himself "surrounded by immunologists," as he was to point out in his Nobel Prize lecture many years later as cited in *Bioscience Reports*. It was a challenging position, and he soon became deeply involved in the research for which he was to win the Nobel Prize.

Biologists had long known that an individual vertebrate has the ability to generate millions of different antibodies before it ever encounters an antigen that stimulates a specific defense antibody. Biologists speculated on the mechanism by which the **organism**'s immune system adapts with two theories. According to the first, named the "germ line" theory, all the genes needed to make an antibody are part of the **genetic code**. The problem was that it seemed impossible for a single gene to carry that much information. A second theory, "somatic **mutation**," suggested that the antibody genes mutate readily, rearranging themselves in a variety of ways to code for different antibodies. According to this hypothesis, a relatively small number of genes would be able to generate a very large number of variants.

Solves the Antibody Diversity Puzzle

After half a decade of research, Tonegawa was able to report the first firm evidence on the antibody diversity debate. With Nobumichi Hozumi, a colleague, Tonegawa was able to prove that the somatic mutation theory was correct. The biologists demonstrated that the parts of a DNA molecule can rearrange themselves in many different ways—just as the fifty-two cards in a deck can be shuffled and rearranged—in response to an attack by a hostile organism. The antibody which is thus selectively produced then attacks the invader. As explained in a comment in *Nobel Prize Winners Supplement*, "DNA **recombination** and mutation could generate perhaps 10 billion different kinds of antibodies, more than enough to solve the diversity problem."

Among Tonegawa's contributions to molecular biology is the discovery, with Hozumi, that the DNA segments which undergo rearrangement are separated by seemingly inactive (noncoding) strands of DNA, now known as introns. Additionally, Tonegawa's research into the immune system resulted in the breakthrough discovery of a gene control element "enhancer" in the intron. Jean L. Marx, writing in *Science*, observed that Tonegawa's work has far-reaching significance: "One unexpected consequence of the antibody gene research was new information about the possible causes of **cancer**, especially the **blood** cancers known as lymphomas and leukemias."

Tonegawa stayed at Basel until 1981 when he was offered a position as professor of biology in the Center for Cancer Research and Department of Biology at MIT. In 1988, he was made Howard Hughes Medical Institute Investigator. In addition to the Nobel Prize in 1987, Tonegawa has received numerous honors and awards, including the Genetics Grand Prize of the Japanese Genetics Promotions Foundation in 1981, the V. D. Mattia Award of the Roche Institute of Molecular Biology in 1983, the Robert Koch Prize in 1986, the Albert and Mary Lasker Award for Basic Research in 1987, the Rabbi Shai Shacknai Memorial Prize in Immunology and Cancer Research in 1989, and Brazil's Order of the Southern Cross in 1991. In 1992, Tonegawa and his colleagues at MIT identified, for the first time, a specific gene which has an effect on the ability to learn. The *New York Times* reported that this research is "the first step toward discovering the entire repertoire of genes that affect **brain** function." Tonegawa was married to the former Mayumi Yoshinari on September 28, 1985. The couple has three children.

TOUCH

Touch is one of the five senses. Called the somatosensory—or body-sensing—system, tactile signals stimulate excitatory **neurons** which carry their messages through the central **nervous system** (CNS) to the somatosensory cortex in the **brain**. The sensory cortex is like an extremely intricate map of every **sensory neuron** in the entire body; fingers, legs, feet, **organs**— each one having a cluster of neurons in a very specific area of the cortex. There is a left and right cortex, each servicing the opposite side of the body. Damage to a specific part of the cortex will inhibit sensation from and movement in it's relative body part.

Touch is determined by sensory receptors designed to respond to specific stimuli. Exteroceptors, located primarily in the skin, include free nerve endings, which respond to touch, pressure, itching, **temperature**, and pain. Merkel's discs, found in extremely sensitive areas such as the fingertips, respond to touch and continuous steady pressure. Tactile hairs, located in all hairy parts of the body, activate receptors if a hair shaft moves. Ruffini's end organs, found in the skin, subcutaneous **tissue**, and joint capsules, detect touch, heavy pressure, and position of joints. Meissner's corpuscles, found primarily in the fingertips and lips, are especially sensitive to moving objects or to the fingertips moving over stationery objects (such as reading braille). Krause's bulb senses heat and cold. Interoceptors detect internal stimuli from **blood** vessels, body temperature, and stretching of tissue. Proprioceptors respond to muscle, ligament, tendon, joint, and connective tissue sensations; cutaneous mechanoreceptors respond to touch, pressure, vibrations, and stretch; termoreceptors detect temperature; and nociceptors detect pain. (Researchers believe pressure receptors work similarly to the eye by sending "pictures" to the brain. For example, a finger running over a braille letter creates indentations in the skin. Hundreds of neurons simultaneously send their own tiny piece of information to the brain, together forming a complete picture on the sensory cortex—much like a jig-saw puzzle. Vibratory receptors, however, are received more like vibrations from the auditory system.)

When sensory imput is received by the brain, the brain can determine how to signal the body to respond. For example, mechanoreceptors help the brain determine how hard to signal the hand to grip a tennis racket during a tennis game, or a newborn kitten. Touch also allows us to coordinate bodily movements—we unconsciously sense where our legs, arms, head, eyes, muscles, and internal organs are without having to pay conscious attention to them.

TOXINS

A toxin is a substance that is harmful to living **organisms**. Toxic effects can be the result of exposure to toxins. Since World War II, the use of chemicals, especially synthetic ones, has increased tremendously in our society. Consequently, people have become concerned about the toxic effects of those chemicals, and the relatively new area of science known as toxicology was developed. Toxicologists study the harmful effects of chemicals on **plants, animals**, and humans.

It is a common misconception that natural chemicals are safe while synthetic (human-made) chemicals are not. The origins of chemicals do not determine their toxicity. Toxins include naturally occurring substances as well as hundreds of thousands of human-made substances. For example, heavy metals such as cadmium, lead, mercury, and nickel are naturally occurring. We encounter them in our everyday lives yet they are dangerous toxins. We eat foods that contain natural toxins. Aflotoxins occur naturally in peanut butter and cause severe reactions in some people. Potatoes, peppers and tomatoes contain a toxin called **alkaloid** solanine, which some people must avoid. Human-made toxic chemicals in our environment include **pesticides** such as **DDT** (dichlorodiphenyltrichloroethane) and PCBs (polychlorinated biphenyls) found in industrial wastes. Many plastics made from petroleum are toxins. The toxicity of a chemical depends on the dose and an individual's susceptibility. An individual's susceptibility varies depending on his or her genetic background, age, weight, gender, overall health, and previous exposure to the toxin.

The effects of toxins are classified according to three time-based categories: acute, subacute, and chronic. Acute effects are those that show up immediately. If the effects appear over weeks or months they are called subacute effects. Damage to **organs** such as lungs is often acute or subacute. Effects that appear gradually and last over extended periods (even a lifetime and possibly, but not always, leading to **death**) are called chronic effects. Examples of chronic effects include carcinogenesis (**cancer**), **teratogen**esis (**birth** defects), and **mutagenesis** (changes in **genetic material**).

Since the early 1970s, the United States government and all fifty states have passed hundreds of statutes designed to control toxins in our environment. Most environmental protection laws in the United States are enforced by the federal Environmental Protection Agency (EPA) and its state-level counterparts. Toxins found inside the workplace are regulated by the federal Occupational Safety and Health Administration (OSHA) and its state level counterparts. Administrative agencies have made thousands of regulations that help implement the statutes passed by legislators.

Before administrators can make regulations banning a chemical, limiting its use, or controlling its storage or disposal, they need information on which to base their decisions. Toxicologists provide such information by performing tests and compiling toxicity assessments based on the results of those tests. Most testing is done on animals. It is done using massive single doses to determine what is a lethal dose (a dose that kills) or using lower doses for long periods of time to determine effects other than death. Another type of testing is through epidemiology. In an epidemiological study, a group of human beings in the general **population** is studied over a long period in an attempt to discover links between exposure to a chemical and a specific effect such as cancer. Asbestos provides a good example of how a toxin is regulated. Asbestos is a naturally occurring substance that was used extensively as insulation up until the 1970s. As a result of animal tests and epidemiological studies, most scientists agree that inhalation of asbestos fibers causes asbestosis (a disease in the lining of the lungs) as well as lung cancer. Therefore, OSHA developed regulations limiting the number of asbestos fibers that can be in air breathed by workers. In addition, the federal EPA regulated the use of asbestos in buildings.

There are hundreds of thousands of synthetic chemicals in our environment, yet it is estimated that only about twenty percent have been thoroughly tested for toxic effects, and about a third have not been tested at all. Furthermore, even when toxicity testing is conducted, the results represent scientific estimates and do not provide clear answers. Whenever scientists use animal testing, they must make assumptions as to how humans would react to the same chemical. Epidemiological studies are limited in that scientists can only observe what happens to humans, as it is considered unethical to deliberately expose human beings to measured doses of specific chemicals in order to study the effects of those chemicals. Therefore, during the twenty-first century, federal regulators and the scientists who work with them face massive tasks as they try to identify toxins and make regulations designed to protect us from their harmful effects.

TRANSCRIPTION

The genetic information that is passed on from parent to offspring is carried by the **DNA** of a **cell**. The genes on the DNA code for specific **protein**s that determines our appearance, different facets of our personality, our health etc. In order for the genes to produce the proteins, it must first be transcribed from DNA to **RNA** in a process known as transcription. Thus, transcription is defined as the transfer of genetic information from the DNA to the RNA.

The process of transcription occurs in the **nucleus** of the cell. There are three different phases involved: initiation, elongation, and termination.

To initiate the process of information transfer, one of the strands of the double stranded DNA serves as a template for the synthesis of a single strand of RNA that is complementary to the DNA strand. The **enzyme** RNA polymerase binds to a particular region of the DNA that is termed as the "promoter." The promoter is a particular unidirectional sequence that appears at the beginning of the genes, and tells the enzyme where to start the synthesis and which strand to synthesize. Once the enzyme is bound to the promoter, it unwinds the DNA and starts to make a strand of RNA with a base sequence complementary to the DNA template that is downstream of the RNA polymerase binding site. The strand from which it copies is

known as the template or the antisense strand, while the other strand to which it is identical is called the sense or the coding strand.

After initiation, is the process of elongation. The substrates for RNA polymerase are nucleoside triphosphates. The RNA polymerase matches a base on the DNA to an RNA **nucleotide** (by complementary base pair binding) and then adds that nucleotide to the elongating RNA strand. As a new ribonucleotide triphosphate forms a bond with the 3'- hydroxyl end of the growing strand, a pyrophosphate is given off. The **energy** that is needed for synthesizing RNA is derived from splitting up of the triphosphate into a monophosphate and releasing the other two inorganic phosphates.

The next phase is called "termination." Termination occurs when the RNA polymerase reaches a signal on the DNA template strand that tells it to stop. Once this termination signal is recognized by the RNA polymerase, it releases the DNA and transcription ceases. The newly synthesized RNA strand now undergoes "post-transcriptional processing."

Eukaryotic genes are not continuous. A typical gene consists of both coding sequences (**exon**s) and non-coding sequences (introns). The primary transcript that is formed at the end of the transcription is actually known as hnRNA and is an exact copy of the gene with both introns and exons. A process called "**RNA splicing**" occurs and the introns are removed. The remaining exons are joined together to form the final mRNA product which codes for a single protein. This post-transcriptional RNA processing take place in the nucleus. Besides splicing, the hnRNA strand also has to be capped and poly-adenylated before being transported to the **cytoplasm** for **translation** into proteins. The 5' capping of the hnRNA occurs soon after the beginning of transcription. A methylated G nucleotide that is believed to play an important role in the initiation of **protein synthesis** is added to the 5' end. For the addition of a poly A tail, one hundred to two hundred residues of adenylic acid are added by an enzyme known as poly-A polymerase. This tail helps to guard the RNA transcript against degradation and enables the transcript to exit from the nucleus to the cytoplasm where it can be translated into proteins.

TRANSLATION

Translation is the process where the **nucleotide** sequence of the **DNA** is translated to a sequence of **amino acid**s. The sequence of amino acids constitute the **proteins**.

A **molecule** known as the **ribosome** is the site of the **protein synthesis**. The ribosome is protein bound to a second **species** of **RNA** known as ribosomal RNA (rRNA). Several ribosomes may attach to a single mRNA molecule, so that many **polypeptide** chains are synthesized from the same mRNA. The ribosome binds to a very specific region of the mRNA called the promoter region. The promoter is upstream of the sequence that will be translated into protein.

The nucleotide sequence on the mRNA is translated into the amino acid sequence of a protein by adaptor molecules composed of a third type of RNA known as transfer RNAs (tRNAs). There are many different species of tRNAs, with each species binding a particular type of amino acid. In protein synthesis, the nucleotide sequence on the mRNA does not specify an amino acid directly, rather, it specifies a particular species of tRNA. Complementary tRNAs match up on the strand of mRNA every three bases and add an amino acid onto the lengthening protein chain. The three base sequence on the mRNA are known as "codons", while the complementary sequence on the tRNA are the "anti-codons".

The ribosomal RNA has two subunits, a large subunit and a small subunit. When the small subunit encounters the mRNA, the process of translation to protein begins. There are two sites in the large subunit, an "A" site, and a "P" site. The start signal for translation is the codon ATG that codes for methionine. A tRNA charged with methionine binds to the translation start signal. After the first tRNA bearing the amino acid appears in the "A" site, the ribosome shifts so that the tRNA is now in the "P" site. A new tRNA molecule corresponding to the codon of the mRNA enters the "A" site. A **peptide** bond is formed between the amino acid brought in by the second tRNA and the amino acid carried by the first tRNA. The first tRNA is now released and the ribosome again shifts. The second tRNA bearing two amino acids is now in the "P" site, and a third tRNA can now bind to the "A" site. The process of the tRNA binding to the mRNA aligns the amino acids in a specific order. This long chain of amino acids constitutes a protein. Therefore, the sequence of nucleotides on the mRNA molecule directs the order of the amino acids in a given protein. The process of adding amino acids to the growing chain occurs along the length of the mRNA until the ribosome comes to a sequence of bases that is known as a "stop codon". When that happens, no tRNA binds to the empty "A" site. This is the signal for the ribosome to release the polypeptide chain and the mRNA.

After being released from the tRNA, some proteins may undergo post-translational modifications. They may be cleaved by a proteolytic (protein cutting) **enzyme** at a specific site. Alternatively, they may have some of their amino acids biochemically modified. After such modifications, the polypeptide forms into its native shape and starts acting as a functional protein in the **cell**.

There are four different nucleotides, A, U, G and T. If they are taken three at a time (to specify a codon, and thus, indirectly specify an amino acid), 64 codons could be specified. However, there are only 20 different amino acids. Therefore, several triplets code for the same amino acid; for example UAU and UAC both code for the amino acid tyrosine. In addition, some codons do not code for amino acids, but code for polypeptide chain initiation and termination. The **genetic code** is non-overlapping; i.e. the nucleotide in one codon is never part of the adjacent codon. The code also seems to be universal in all living **organism**s.

TRANSLOCATION

In **plants**, the term translocation refers to the long-distance transport of **water**, **minerals**, or food. It is most often used to refer to the transport of food material from one part of the plant to another, or from source to sink. Thus, sugars produced in a source like photosynthesizing **cell**s in a **leaf** may be moved to sinks like developing **fruits** and seeds, or growing apical meristems. The sugars may also be moved to storage **organs** like tubers, corms, or bulbs. Storage organs may subsequently become sources when food is moved to plant meristems for use by growing and dividing cells. Food materials move primarily in conductive **tissue** called phloem by a mechanism that is not completely understood. The most widely accepted explanation for phloem transport is the mass flow, or pressure flow hypothesis. According to this theory, food material at a source is moved by **active transport** into phloem sieve tube elements, creating a **hypertonic solution** that attracts water from adjacent tissues and sets up a flow through the phloem driven by a buildup of hydraulic pressure. When foods reach the sink and are removed from the phloem, the transporting water flows out of the phloem and returns to the sink via the water-conducting xylem transport system.

In **genetics**, translocation refers to type of interchange of **chromosome** pieces following breakage, in which segments are transferred between nonhomologous chromosomes. When this exchange occurs without a net loss or gain of **genetic material**, it is called a balanced, or reciprocal, translocation, and there is no phenotypic change in the individual. When the exchange results in a deletion or duplication of chromosomal material, in **gamete**s or somatic cells, severe phenotypic changes may result.

See also Plant anatomy; Solution, hypertonic

TREE

A tree is a large, woody perennial plant with one main trunk and many branches. Trees are the most visible component of **forests**, a crucial component of the world's **ecology**.

Closely related **plants** may grow into large trees, or into much smaller plants. For example, the western yew (*Taxus brevifolia*) grows into a tree taller than 60 ft. (20 m) in the Pacific rainforests of western North America, while the closely related Canada yew (*T. canadensis*) of the northeastern forests is a shrub only 3-4 ft. tall (1 m). Many kinds of plants are capable of reaching tree size. This includes plants that do not develop true woody (or xylem) **tissue**s, such as tree-ferns and palms, as well as many **species** of coniferous and angiosperm trees.

Foresters and ecologists define trees strictly on the basis of their size. In some forests, for example, a tree might be defined as being taller than 10 ft. (3 m) and having a diameter at breast height (DBH) greater than 4 in. (10 cm). Any **ecosystem** that is dominated by tree-sized plants is known as a forest.

Trees began their evolution more than 400 million years ago. The oldest known **organism**s are certain individuals of bristlecone pine (*Pinus aristata*) growing in the Rocky Mountains, which can exceed 4,500 years of age. Some trees can be extraordinary in their age or size. The tallest organisms are

Trees, such as this white oak, are the dominant plants in the world's forests, providing critical habitats for the other species which live there. *(Photograph by Robert J. Huffman, Field Mark Publications. Reproduced by permission.)*

redwood trees (*Sequoia sempervirens*) of coastal California, one of which is more than 360 ft. tall (110 m). The widest tree is the "General Sherman Bigtree," a giant sequoia (*Sequoia gigantea*) growing in central California which has a circumference of 115 ft. (35 m). Trees are also the largest plant organisms (some stands of trembling aspen (*Populus tremuloides*) have developed from root-sprouts, so that "individual" trees, growing over large areas actually represent the same genetic organism, which can cover as many as 100 acres (40 ha) and contain tens of thousands of mature stems).

Many smaller plants, **animals**, and **microorganisms** rely on trees and/or forests for their necessary **habitat**. Trees are also responsible for immense amounts of **photosynthesis**, and are the dominant **energy** base of the food webs of all forests. Trees also store huge quantities of carbon in their **biomass**, and in this way reduce the amount of **carbon dioxide** present in the **atmosphere** (this is important in moderating the intensity of Earth's **greenhouse effect**). Intact forests also cleanse the air, **water**, and soil of other kinds of **pollutants**.

Trees also provide humans with important goods and services. These may be as simple as shade and landscape use, or as immense as the 10 million acres (25 million ha/yr) of forest that are harvested and cleared globally each year for timber, fuel, or to develop new agricultural land. In fact, ecologists have estimated that humans are now using about 28% of the global production of forested biomass. In many regions and on a global basis, humans are using trees faster than they are able to grow. Deforestation is now a common problem and overall, Earth's forest cover is now only about one-half of what it was several thousand years ago.

TRNA • See Ribonucleic acid (RNA)

TROPHIC LEVEL

One important concept in the study of **ecosystem**s is their trophic structure, which identifies the organization of the ecosystem by grouping its **organism**s according to where they get their **energy**. Each grouping is called a trophic level, and that level refers to their specific place within the ecosystem's trophic structure. The trophic structure, in turn, is the structure of all the energy flows between these different **population**s of organisms. A **community**'s trophic structure is also sometimes refered to as its food web or food pyramid, a series of interlinked food chains.

The bottom end of the structure is always a primary producer: a green plant or **autotroph** which captures solar energy and transforms it into **biomass**. That energy travels up the food chain to the second trophic level, **herbivore**s or primary **consumer**s, which feed upon organisms at the first level. Higher trophic levels are occupied by small **carnivore**s—who feed on the herbivores—on up to larger carnivores. At each consumer level of the chain are decomposers who recycle energy back to the primary **producer**s. There are some organisms, such as human beings, that feed at more than one trophic level.

At each step up the food chain, almost 90% of the chemical energy being transferred is lost as heat energy. As a result, in most ecosystems there is generally more biomass at lower trophic levels than at subsequent levels.

The study of trophic structure is not only useful for outlining the flow of energy in an ecosystem. It also allows us to predict the flow of **chemical element**s, which is increasingly important as we try to track the path and accumulation of **pollutants** in an ecosystem.

TROPHIC STRUCTURE

The trophic structure (from Greek *trophos*, meaning feeder) of an **ecosystem** is the arrangement of **organism**s based on their feeding relationships. All organisms interact with each other through the food web, based on what they eat and what eats them. Thus the trophic structure controls the passage of **energy** and **nutrients** from one organism to another in an ecosystem.

The trophic structure of an ecosystem can be broken up into various trophic, or feeding levels. At the base of the trophic structure are the **producers**, which are sometimes called primary producers. These **autotroph**ic organisms produce their own food using the processes of **photosynthesis** or chemosynthesis. The most common types of producers are green **plants** on land and **algae** in **water**. All other organisms in the trophic structure ultimately depend upon these producers for their energy and organic material. In a field ecosystem, the grasses would be an example of the producers.

Consumers are **heterotroph**ic organisms that cannot make their own food; they rely on eating other organisms to obtain their energy and other **nutrition**al requirements. **Herbi-**

vores, **animal**s that eat the producers, are the first level of consumers. They are often called primary or first-order consumers. A field mouse that eats grasses would be an example of one herbivore in a field ecosystem. **Carnivore**s are organisms that eat meat. There are two types of carnivores, predators, which actively hunt and kill the animals they eat, and **scavenger**s, which eat organisms that they did not kill. Both types of carnivores might eat herbivores or they might eat other carnivores. Those that eat herbivores, such as a snake eating our field mouse, are the secondary or second-order consumers. The tertiary or third-order consumers eat the secondary consumers. For example, a hawk that consumes the snake would be a tertiary consumer. In some ecosystems there are fourth and fifth level consumers, however trophic structure rarely gets to higher levels than this. In many cases, humans are at the top of the trophic structure.

Some organisms, such as most scavengers, can be on more than one **trophic level**, depending on what they eat. For example, a vulture in the example of the field used above, might eat the snake after it dies, making it a tertiary consumer. However, the same vulture could also eat the mouse, in which case it would be a secondary consumer. **Omnivore**s, organisms that eat both plants and animals, are another example of organisms that are found on more than one trophic level.

Decomposers are also part of the trophic structure. They are responsible for recycling nutrients from dead organisms and organisms' wastes back into a form that is usable by plants and other autotrophs. Decomposers are usually **bacteria** and **fungi**, and they play a dual role in an ecosystem's trophic structure: they both recycle nutrients and are a source of food for other organisms.

In any ecosystem, the amount of stored energy differs within the trophic structure. Stored energy of lower trophic levels, particularly the producers, is much greater than higher levels. This is because energy is "lost" between successive trophic levels. This energy is used up in a variety of ways, including by an organism's daily activities and metabolic processes (e.g., moving around, caring for young, **respiration**), and as excess body heat. Only a small portion of this energy (approximately 10%) is passed on to the next trophic level. As a result, there are usually far fewer top level consumers in an ecosystem than lower level consumers, and fewer lower level consumers than producers. The trophic structure of an ecosystem can be very complex, depending on the number of organisms in the ecosystem and their feeding relationships.

TROPISM

Most **plants**, and some **fungi**, **microorganism**s, and lower forms of **animal** life, respond to stimuli that act with greater intensity from one direction than another by orienting themselves with respect to the direction in which the stimuli are acting. These responses, called tropisms, may involve one of several kinds of stimuli and are named accordingly. Thus **phototropism** refers to a response to a directional source of **light**. Gravitropism, also called geotropism, is a response to gravity.

Chemotropism is a response to a source of chemical stimulation, and thigmotropism is a response to mechanical stimulation. Other kinds of tropism have been described including galvanotropism, or electrotropism (response to electric current), hydrotropism (response to **water**), and traumatotropism (response to a wound lesion). Tropic movements are usually directed toward or away from the source of the **stimulus** and are called positive or negative orthotropic responses, accordingly. Diatropic responses are at a right angle, and plagiotropic responses at some oblique angle to the direction of the stimulus.

Tropic responses typically result from shifts in an **organism**'s direction of growth, resulting in a change in its orientation with respect to the stimulus. Phototropism in plant shoots has been especially well studied, and has been found to result from changes in concentration of growth **hormone** on the side of the plant toward the light and the side away from the light. Auxin, a plant hormone that stimulates **cell** elongation, moves to cells on the side of the plant away from the light, causing those cells to expand more than cells closest to the light. This causes the plant to bend toward the light, a phenomenon well known to anyone who has attempted to grow plants in front of a windowpane. Plants in this circumstance have to be rotated regularly to maintain symmetrical growth. Gravitropism may also be explained by a redistribution of growth hormone within the root and stem, but the mechanism is less well understood. Plant shoots tend to grow upward, away from the force of gravity, exhibiting negative gravitropism, while roots grow downward, showing positive gravitropism. Tropisms in animals are found in sessile (i.e. sedentary) forms such as hydra that can direct growth-curvature movements in response to environmental stimuli.

A related phenomenon called taxis is found in animals and microorganisms that are able to move in response to a stimulus. Thus, motile **bacteria** that swim toward a directional source of light are said to exhibit positive phototaxis, and those that swim away show negative phototaxis. Microorganisms may exhibit a phenomenon called aerotaxis or movement in response to an oxygen gradient. Motile anaerobic bacteria may move away from an increased supply of oxygen showing negative aerotaxis, whereas aerobes may move toward the supply exhibiting positive aerotaxis. Microaerophyllic bacteria, intermediate in their affinity for oxygen, may accumulate between the anaerobes and aerobes in a region low in oxygen.

See also Aerobic/anaerobic

TRYPTOPHAN • See Amino acid

TSWETT, MIKHAIL (1872-1919)
Russian chemist and botanist

Although recognized only belatedly, Mikhail Tswett (sometimes spelled Tsvet) was the first to lay out in detail the methods of the separation technique called **chromatography**. Tswett himself regarded chromatography only as a tool in his chemical and biological studies; his purpose was to separate and identify the many different pigments in leaves and other plant parts, and he considered it merely an improvement on existing techniques such acid-extraction, base-extraction, and fractional crystallization. Since he first described this process, many kinds of chromatography have been developed, and no laboratory is considered complete without a number of chromatographic instruments.

Mikhail Semyonovich Tswett was born May 14, 1872, in Asti, in the northwest part of Italy about seventy miles from the Swiss border. His parents were Semyon Nikolaevich and Maria de Dorozza Tswett. His father was a Russian civil servant and his mother, who was very young, died soon after his birth. His father returned to Russia after her death, and left his son with a nurse in Lausanne. Tswett was educated in Lausanne and Geneva, becoming multilingual in the process. He received his secondary education at the Collège Gaillard in Lausanne and the Collège de St. Antoine in Geneva; he entered the University of Geneva in 1891, studying chemistry, **botany**, and physics. His baccalaureate in both physical and natural sciences was awarded in 1892. He began plant research during his undergraduate years, earning the Davy Prize while a doctoral student with a paper on plant physiology that was subsequently published. In 1896 he defended his thesis, ''Études de physiologie cellulaire,'' and received his doctoral degree.

Thereafter he moved to Russia, and in 1897 he began working at the laboratory of **plant anatomy** and physiology at the Academy of Sciences and the St. Petersburg Biological Laboratory. His academic horizon was limited by the fact that foreign degrees were not recognized in tsarist Russia, and he set to work earning another master's degree in botany at Kazan University. He finished in 1901, with a thesis in Russian whose title is translated ''The Physicochemical Structure of the Chlorophyll Grain.'' In 1902 Tswett became an assistant in the laboratory of plant anatomy and physiology at the University of Warsaw, which was under Russian control at that time, where he became a full professor in 1903. In 1907 he took on the additional task of teaching botany and microbiology at the Warsaw Veterinary Institute; a year later he was also teaching at the Warsaw Technical University. He resigned his teaching post at the University of Warsaw but took a second doctorate there in 1910 with a dissertation on plant and **animal** chromophils. This apparently led to his only book, published in the same year, whose title is translated as ''The Chromophils in the Animal and Vegetable Kingdoms.'' The book itself has never been translated. By 1914 Tswett's brief, brilliant research career was essentially at an end. The German invasion of Poland in 1915 forced the Technical University to move to Moscow, and then to Nizhni Novgorod in 1916. Tswett's time was largely consumed with organizing the work of the botanical laboratories after each of these moves. In 1917 he accepted a position at the University at Yuryev in Estonia, but that too was overrun by the German army a year later. The university moved to Voronezh in 1918, but Tswett's health, never robust, failed quickly, and he died of a heart ailment at age forty-seven, on June 26, 1919.

Tswett's strength as a scientist lay in how well he understood both chemistry and botany. He had always been interested in the internal molecular structures of **plants**, often inquiring what their purpose might be, and the work he did on chlorophyll was one of his most important research efforts. He had long doubted the contention, which was widely accepted at the time, that chlorophyll was a compound that actually existed in plants. He decided this belief was the result of a misunderstanding; he hypothesized that chemists had been confused either because chlorophyll was combined nearly inseparably with other **molecules** within the **leaf** or because a compound recovered by a particular separation technique might in fact be an artifact of the technique. He was able to demonstrate all of these misunderstandings in the work of others, both by his deployment of the chemical separation methods of the time (fractional solution and **precipitation**, **diffusion**, differential solution) and by the adsorption methods he developed, culminating in chromatography.

Develops Chromatography Process

''Adsorbent'' means holding molecules on the surface of the material, not in the body, and chromatography is a process which employs substances which have this property. It is a separation technique in which a very finely powdered adsorbent material is held in a vertical tube or ''column.'' The mixture to be separated is placed on the top of the column, dissolved in as small an amount of solvent as possible, so that it forms a narrow band of adsorbed mixture; then more solvent is allowed to flow through the column, top to bottom. The molecules in the mixture are more or less strongly held by the adsorbent; those weakly held are washed down the column most rapidly, and those strongly held move less rapidly. After a suitable development time, the components of the mixture separate into a series of bands spaced along the column. The plug of wet adsorbent is blown out of the column onto a plate, where the bands can be cut apart and the components recovered separately. As the mixtures separated in these early experiments were colored, and the bands absorbed light in the visible spectrum, Tswett named the process *chromatography* (''color-writing''), and the developed separation he called a *chromatogram.* Even though most mixtures are not colored, this terminology is retained; the components must be detected by some means other than the eye. Many sophisticated varieties of chromatography are in use today: paper, thin-layer, gas-liquid, and ion exchange, to name but a few. Still, Tswett's column method has not been totally displaced.

Tswett used this technique to demonstrate that chlorophyll indeed does not exist in the plant as a free molecule but is complexed with albumin. He named this complex ''chloroglobin,'' by analogy with the heme complex of the **blood, hemoglobin.** There was, however, widespread skepticism of his research methods, and this finding was sharply criticized. Tswett next analyzed the plant pigments themselves, which were understood at the time to be only two: green chlorophyll and yellow xanthophyll. Using not chromatography but the standard chemical methods of the time, he demonstrated that there are two chlorophylls: xanthophyll and carotene. This finding was hotly disputed, partly because chlorophyll passed the test of a single pure compound: it could be crystallized. Tswett was able to show that the "crystallizable chlorophyll" formed by lengthy extraction with hot ethanol was in fact another compound; it is known today as an ethyl ester formed by transesterification of one of chlorophyll's ester groups.

During the course of his pigment work Tswett had found that when he ground the plant leaves with powdered calcium carbonate to neutralize acids, all but carotene were adsorbed on the solid carbonate. He used this as a method to separate carotene. It is not clear that this led to his devising column chromatography, but once he had developed this technique he found that in addition to two chlorophylls there were four xanthophylls and, of course, carotene. These findings came to be accepted later, but mainly through the work of the German chemist **Richard Willstätter**.

The technique of column chromatography was not widely used in Tswett's lifetime, being regarded by his most vocal opponent, L. Marchlewski, as no more than a "filtration experiment." It was only later in the century that his work was re-evaluated and his status as one of the originators, though probably not the sole inventor, of chromatography, was confirmed. This is his legacy today, although some would consider the plant pigment work to be at least as important.

TUNDRA

Tundra is the land type of the Arctic and subarctic regions. With long cold winters, short cool summers, and low **precipitation**, the soils of the tundra are thin or absent, and the vegetation is sparse. The tundra is therefore the **ecosystem** most susceptible to environmental damage. Because of the small number of plant and **animal species** and the fragility of the food chains, damage to any element of the **habitat** may have an immediate and devastating chain reaction through the system. The permafrost is easily damaged by heavy equipment and by oil spills. The Inuit, who fish, hunt, and trap for a living, are directly affected by abuses of the **ecology**.

Tundra vegetation is quite varied, considering the climatic conditions. Although the rock **desert**s are almost devoid of vegetation, relatively fast-growing mosses often surround large rocks. In rock crevices such **plants** as the purple saxifrages survive, and the rock surfaces themselves may support lichens, some of the orange and vermilion species adding color to the landscape. Lichen tundra is found in the drier and better-drained parts. Mosses are common, and some species may dominate the landscape to such an extent that it appears snow-covered. The heath and alpine tundra support dwarf, often berry-bearing shrubs, while the ground between usually is covered with a thick carpet of lichens and mosses

The distinctive animals of the tundra are the seal and the polar bear, the latter feeding on seal; the musk-ox, caribou, and lemming, which feed on the tundra vegetation; and their predators, the Arctic wolf and white fox. Few birds make the tundra their year-round habitat, the great snowy owl and the ptarmigan being exceptions. Numerous birds that normally live in mild climates, however, often fly to the tundra for nesting. Two large birds that do this are the snow goose and the Canada goose.

Rocky tundra on Bear Island in the Barents Sea. *(Photo Researchers, Inc. Reproduced by permission.)*

TURNER'S SYNDROME

Turner's Syndrome (also referred to as gonadal dysgenesis) is a relatively common genetic disorder of females, which affects many body systems. Approximately one in every 2,000-5,000 female babies has Turner's Syndrome. About 98-99% of pregnancies with Turner's Syndrome abort spontaneously, usually during the first trimester of pregnancy. Approximately 10% of fetuses from pregnancies that have spontaneously aborted have Turner's Syndrome.

Described by Dr. Henry Turner in 1938, this disorder in due to a deficiency in the amount of **genetic material** on the X-**chromosome**, one of the two **sex chromosome**s. Diagnosis of Turner's Syndrome is made with a chromosome analysis. Turner's Syndrome is not related to advanced maternal age and is a sporadic event, with the risk of recurrence not increased for subsequent pregnancies.

Turner's Syndrome is associated with short stature and failure to mature sexually. Other problems may include **learning** difficulties, skeletal abnormalities (e.g., webbed neck, low posterior hair line), lymphedema (swelling of a part of the body due to an obstruction or deficiency of the lymphatic drainage system), **heart** and kidney abnormalities, **infertility**, obesity, formation of keloids (thick scars), and **thyroid gland** dysfunction (hypothyroidism). The type and amount of missing genetic material influence which specific **organ** abnormalities will be present, as well as the person's potential for growth. Since the syndrome is not always accompanied by distinctive features, Turner's Syndrome is often not diagnosed during infancy, but may be suspected during childhood because of the short stature of a child. During teenage years, Turner's Syndrome may be discovered due to delayed puberty and menarche, while in adult woman, anovulation and infertility may indicate Turner's Syndrome.

Short stature is almost always present in females with Turner's Syndrome. The causes are intrauterine growth retar-

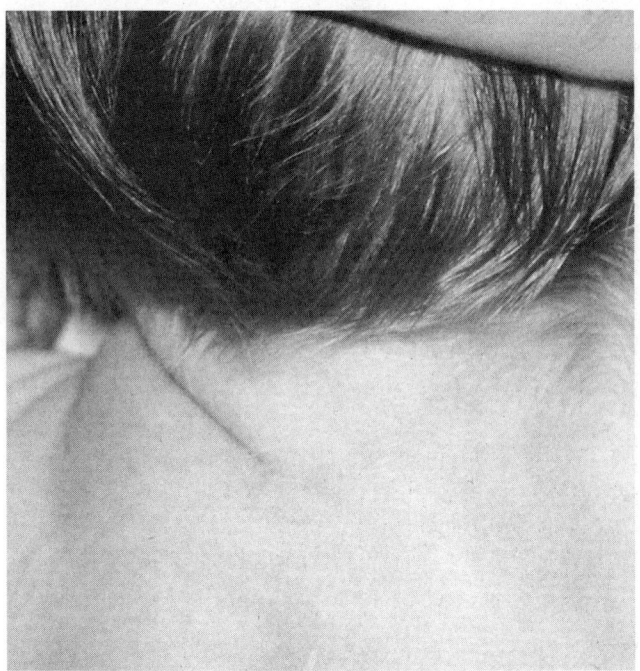

A low hairline at the back of the neck is one of several characteristics of Turner's syndrome. *(Custom Medical Stock Photo. Reproduced by permission.)*

dation, a gradual decline in growth rate during childhood, and the absence of a pubertal growth spurt. Females with Turner's syndrome have abnormal body proportions characterized by markedly shortened lower extremities. The ultimate height range is between 55-58 in (1.4-1.5m), though familial height may play a role in determining the ultimate height that will be reached. Recent studies have indicated that much of the growth deficit in females with Turner's Syndrome can be restored by injections of human growth **hormone**.

Normal pubertal development and spontaneous menstrual periods do not occur in the majority of females with Turner's Syndrome. Though 10% of females with Turner's Syndrome will go through puberty spontaneously, most will require the use of female hormone therapy for development of secondary sexual characteristics and menstruation. The time of initiation of therapy varies with each female but usually begins when she expresses concern about the onset of puberty, usually around 15 years of age. Delaying the use of **estrogen** therapy is recommended to maximize the height the female will achieve. Various estrogenic and progestational agents and schedules have been used as hormone therapy. Although infertility cannot be altered, pregnancy may be possible through *in vitro* **fertilization**.

Renal abnormalities occur in a third to a quarter of females with Turner's syndrome. The most common abnormality is a horse shoe kidney. Cardiac abnormalities are also common, with the coarctation of the **aorta** being the most common. There is an increased incidence of thyroid dysfunction, **diabetes mellitus**, and **carbohydrate** intolerance. Prevalence of me ntal retardation appears to be no greater than in the general

population, but Turner's Syndrome females may exhibit learning disabilities, especially with regard to spatial **perception**, visual-motor coordination, and mathematics. As a result, the nonverbal IQ in Turner's Syndrome tends to be lower than the verbal IQ. Females with Turner's Syndrome may also be socially immature for their age and may need support in developing independence and social relationships.

TWORT, FREDERICK (1877-1950)
English bacteriologist

As a pioneering bacteriologist, Frederick Twort was responsible for several important advances in his field. He discovered what would be known as **bacteriophage**s, **bacteria**-attacking **virus**es. This discovery led to the advent of **molecular biology**. Twort was the first scientist to grow the **organism** that caused Jöhne's disease, a deadly cattle infection, and his efforts contributed to its elimination. Twort also discovered a nutritional element later identified as vitamin K.

Frederick William Twort was born in Camberley, Surrey, England, on October 22, 1877. His father, William Henry Twort, was a doctor. Frederick was the oldest of ten siblings. He studied medicine in London at St. Thomas's Hospital Medical School. He became qualified and licensed in 1900, though he never actually practiced clinically. Soon after graduation Twort began his work as an assistant to Louis Jenner in London's St. Thomas's Hospital, working in their clinical laboratory. In 1902, Twort found work with William Bullock as an **anatomy** instructor in London Hospital. It was here that his first work in bacteriology began. He spent several years familiarizing himself with the bacteriology of hospitals and soon began his own experimentation.

Twort married Dorothy Nony Banister, who helped him with his work, and with her had a son and three daughters. His son, Antony, also became a doctor as well as his father's biographer.

Devotes Life to Research

Twort's own research work became of primary importance to him in 1907. In that year, he published one of his earliest significant papers on bacteria. In it, he outlined how bacteria adapted and mutated. Two years later, in 1909, he published on bacterial growth and related growth agents. In what became a common occurrence, Twort's results were basically ignored at the time, and found to be important only decades later.

In that same year Twort was named superintendent of the Brown Institution at the University of London, an animal hospital. While working here, Twort was able to devote all his time to research. His work was limited, however, by funding and support problems, which plagued him throughout his career. Still, Twort pushed ahead with his theories. His work was considered remarkable and original from the beginning. He believed that all pathogenic, or disease-causing, bacteria developed from organisms that lived freely, while most of his contemporary bacteriologists believed pathogens originated in the body.

Scrutinizes Jöhne's Disease

Twort's first important achievement was his in-depth study, with G.L.Y. Ingram, of Jöhne's Disease, the results of which were published in the early 1900s. Twort did the earliest **culture**s of the organism that caused the disease. He believed that there was a connection between tuberculosis and Jöhne's Disease, so he derived what he called his "essential substance" from dead tubercle bacilli. These bacilli, when incorporated in a culture medium, proved ideal for growing Jöhne's bacillus. Twort's study of Jöhne's disease directly led to the development of the Jöhnin test. His discovery also eventually proved important to biochemistry, specifically in the study of bacteria and their nutritional needs.

Discovers Bacteriophages

In 1915, Twort discovered what came to be known as bacteriophages. Twort's discovery was something of an accident. He spent several years using artificial media to grow viruses. Twort noticed that the bacteria infecting his plates kept becoming transparent. This was the earliest recorded proof of bacteriophages, though Twort called his discovery "transmissible lytic agent."

Twort published his results, but he was not certain about what he discovered. He made several guesses in his articles, but did not commit to any specific one, a hallmark of his career that lessened his findings in the eyes of his peers. Twort's experiments in this area were also overshadowed by World War I; he served in the Royal Army Medical Corps from 1915-18.

In 1917, Canadian bacteriologist Felix d'Hérelle made the same discovery, independent of Twort. D'Hérelle gave his findings their now common name, bacteriophages, which translates as bacteria eater. After d'Hérelle announced his findings, there was some controversy over who made the discovery first and when, in part because of Twort's published uncertainties. The results eventually carried both their names, and became known as the Twort-d'Hérelle phenomenon. Both scientists shared a life-long obsession with their discovery, and both wanted to use it to fight diseases plaguing humans.

Before antibiotics were developed, scientists were searching for ways to fight disease. Twort and d'Hérelle thought bacteriophages might be an answer, but the viruses did not work when used on human patients. The importance of the discovery of bacteriophages did not emerge until after Twort and d'Hérelle had died. In recent years, the idea has again come to light, as bacteria continues to develop resistance to antibiotics. In 1984, it was learned that illnesses in livestock and human illnesses such as meningitis can be curbed with bacteriophages.

Based on his many accomplishments, Twort was accorded some honors. Among other distinctions, he was appointed professor of bacteriology at the University of London in 1919 and in 1929, he was elected a fellow of the Royal Society. Twort's peers found him a difficult and remote man, which perhaps limited the acceptance of him and his ideas. Still, his rather unique ability to work independently and at a high level of technical aptitude contributed to his capacities as a scientific explorer.

Abrupt End to Research

As Twort's research progressed, he became obsessed with proving, in more specific terms, that bacteria evolved from viruses, and that these viruses had evolved from more primary cellular forms. Though he spent years on this idea, he did not publish anything of consequence. Twort's research was permanently interrupted in 1944 when the Brown Institution was destroyed by enemy fire during World War II. His laboratories and specimens were completely decimated. Twort spent his last years suffering greatly from this loss.

Twort died in the city of his birth, Camberley, on March 20, 1950. Posthumously, he was remembered for his scientific accomplishments, as well as his uncompromising belief that scientific funding should not be controlled by the government. His obituary in the leading journal *The Lancet* states: "An outstanding representative of the independent research-worker, he had made valuable contributions to bacteriology, but he felt himself increasingly uneasy in the closer relationship of research and the State."

Leaves a Complicated Legacy

There was more to Frederick Twort than his scientific accomplishments and related attitudes. In a review of his biography, *In Focus, Out of Step: A Biography of Frederick William Twort* written by son Antony Twort, Bernard Dixon describes the elder Twort as "a multifaceted man, who at various times made violins and began to design a more efficient internal combustion engine, who entered the Daily Mail contest for the biggest and best sweet pea in England, threw vegetables and meat each day into a large cooking pot of stew kept continually on the hob, and later developed considerable skills as an amateur radio constructor." This giant of science was a very complicated man working, as the title of his biography so succinctly states, in his own focus and quite out of step with his times.

V

VACUOLES

The term vacuole in biology refers to a **membrane** bound space, filled with liquid, often located in the center of a **cell**. Vacuoles are especially prominent in plant cells where they may occupy 95% or more of the cell volume. The vacuole was long considered to be merely a site in which **plants** disposed of waste material. This view began to change in the 1960s when researchers discovered that vacuoles often serve as **protein** storage centers. It was found that during **embryo** formation and seed development, proteins accumulate in the vacuoles of some cells in the cotyledons, or seed leaves, of dicotyledons (flowering plants with two cotyledons per embryo). When the seed germinates, **enzyme**s called proteases are synthesized in the **cytoplasm** and moved to the vacuole where they digest the stored protein, releasing **amino acid**s needed by the growing plant. It has also been discovered that some vacuoles store **carbohydrate**s for later use as an **energy** source for the plant. In more mature plant cells, the content of the vacuole, called cell sap, contain a **water** solution of salts and sugars at sufficiently high concentrations to attract the entry of water into the vacuole, via osmosis. This creates **turgor pressure** that presses the cell membrane against the cell wall, creating the force needed to expand the volume of growing cells. Turgor pressure also contributes to the rigidity of mature plant **tissue**.

Some **protozoa**ns contain an **organelle** called a contractile vacuole that enlarges with the influx of water and then discharges its contents to the exterior of the **organism** when it reaches a certain volume. This cycle of filling and discharging may occur as often as every 30 seconds. The contractile vacuole functions in osmotic regulation by preventing excess water from entering the organism and swelling it beyond a viable size.

VANE, JOHN R. (1927-)
English pharmacologist

John Robert Vane was born March 29, 1927, in Tardebigge, Worcester. His parents' Christmas gift of a chemistry set sparked Vane's interest in science when he was twelve, and his home became the site of numerous experiments. However, upon entering the University of Birmingham in 1944, he found that the work given him was not as challenging as he anticipated. At the advice of a professor, he decided to go to Oxford University to study pharmacology after receiving his B.S. in chemistry from Birmingham in 1946. Vane became a fellow on Oxford's Therapeutic Research Council for the next two years. He obtained a B.S. in pharmacology from Oxford in 1949, and earned his doctorate in 1953. After leaving Oxford, Vane came to America to teach pharmacology at Yale University. He returned to England in 1955 as a senior lecturer in pharmacology at the Royal College of Surgeons, at its Institute of Basic Medical Sciences.

Vane became interested in **prostaglandins** in the late 1950s. Discovered in the 1930s, they were originally thought to be secreted by the prostate gland, which is how they got their name. Prostaglandins are natural compounds, developed from fatty acids, and they control many bodily functions. Different prostaglandins regulate blood pressure and coagulation, allergic reactions to substances, the rate of **metabolism**, glandular secretions, and contractions in the uterus.

For many years after the discovery of prostaglandins, scientists were unaware of how they were produced and how they functioned. In the early 1960s Vane expanded upon the procedure known as biological assay (bioassay), by which the strength of a substance is measured by comparing its effects on an organism with those of a standard preparation. Vane developed the dynamic bioassay, which allows scientists to measure more than one substance in **blood** or body fluids. This method enabled Vane and his colleagues at the Royal College to prove that prostaglandins are produced by many tissues and organs in the body. Further research led the scientists to discover that, unlike hormones, certain prostaglandins are effective only in the areas where they were formed.

In 1966 Vane advanced to professor of experimental pharmacology at the Institute for Basic Medical Sciences and continued his studies. An experiment he conducted in 1969 resulted in the discovery of the methods by which aspirin alleviates pain and reduces inflammation. Using the lung tissue of guinea pigs, Vane found that aspirin inhibited the production of a certain prostaglandin that causes inflammation. He published the results in a June, 1971, issue of *Nature New Biology,* a science magazine.

In 1973 Vane resigned his post at the Institute to enter the business world as director of research and development at the Wellcome Foundation, a pharmaceutical company. Following up on research by the Swedish chemist Bengt Samuelsson (who found that a type of prostaglandin was responsible for allowing blood to clot), Vane discovered the existence of a prostaglandin with the opposite quality, which inhibits clot formation. With the assistance of the Upjohn Chemical Corporation, Vane isolated the secretion, which he named prostacyclin. This discovery proved to be of great assistance in dissolving clots blocking the blood supply in stroke and heart attack victims and is also useful for keeping blood from clotting during surgery. Scientists have discovered even more uses for prostaglandins, including the treatment of ulcers, alleviating pain from menstruation and gallstones, and stimulating contractions for childbirth.

Vane, along with Samuelsson and Swedish chemist Sune Bergström, was given the Albert Lasker Basic Medical Research Award in 1977 for his work on prostaglandins. Five years later, in 1982, the Nobel Committee gave the trio the Nobel Prize for medicine or physiology. After receiving the award, Vane predicted that future research on prostaglandins would create major breakthroughs in the areas of medicine. "In the next 20 years we should see a substantial attack on the disease process," *Time* quoted him as saying. "We will be able to find new drugs that have effects on cardiovascular disease, on asthma, on heart attack," and even health problems associated with old age, the magazine reported.

During the 1980s Vane embarked on a crusade for greater research on new drugs to fight both new diseases (such as acquired immunodeficiency syndrome, known as **AIDS**) and drug-resistant strains of old diseases, such as malaria. In articles for scientific and medical journals, he stressed the need for greater international cooperation in the search for a cure or vaccine for AIDS and advocated the creation of an Institute for Tropical Diseases to research new drugs to battle disease in the tropics.

VARMUS, HAROLD E. (1939-)
American microbiologist and virologist

Harold Eliot Varmus was born in Oceanside, New York, on December 18, 1939. He attended Amherst College, graduating with a B.A. degree in 1961 (twenty-three years later, Amherst

would award him an honorary doctorate). Varmus went on to perform graduate work at Harvard University, receiving an M.A. degree in 1962, then he studied medicine at Columbia University, receiving an M.D. in 1966.

Varmus practiced medicine as an intern and resident at the Presbyterian Hospital of New York City between 1966 and 1968. He then worked as a clinical associate at the National Institutes of Health in Bethesda, Maryland, from 1968 to 1970. Moving to California, Varmus served as a lecturer in the department of microbiology at the University of California in San Francisco, becoming an associate professor in 1974—the same year that he was named associate editor of *Cell and Virology*—then, in 1979, he was promoted to full professor of microbiology, biochemistry and biophysics. During the 1980s, Varmus began to accumulate a number of prestigious honors for his research, including the 1982 California Academic Scientist of the Year award and the 1983 Passano Foundation award; he was also the co-recipient of the Lasker Foundation award. In 1984, Varmus received both the Armand Hammer Cancer prize and the General Motors Alfred Sloan award, and the American Cancer Society made him an honorary professor of molecular virology. These honors were followed by the Shubitz Cancer prize and, in 1989, the Nobel Prize in physiology or medicine.

Varmus and J. Michael Bishop, his colleague from the University of California at San Francisco, were awarded the Nobel Prize in in honor of their 1976 discovery which showed that normal cells contain genes that can cause **cancer**. Varmus and Bishop, working with Dominique Stehelin and Peter Vogt, helped to prove the theory that cancer has a genetic component, demonstrating that oncogenes are actually normal genes that are altered in some way, perhaps due to carcinogen-induced mutations. Their research focused on Rous sarcoma, a virus which can produce tumors in chickens by attaching to a normal chicken **gene** as it duplicates within a **cell**. Since then, research has identified a number of additional "proto-oncogenes" which, when circumstances dictate, abandon their normal role of overseeing **cell division** and growth and turn potentially cancerous. Varmus's and Bishop's oncogene studies had a tremendous impact on the efforts to understand the genetic basis of cancer. The results of their work quickly found practical applications, especially in cancer diagnosis and prognosis.

Varmus was nominated by U.S. President Bill Clinton to the directorship of the National Institutes of Health and was confirmed in November 1993. The director of the NIH plays a vital part in setting the course for biomedical research in the United States. Varmus's nomination was strongly supported by biomedical scientists, but there was some opposition from AIDS activists. They—as well as others who were concerned with the health of women and members of minority groups—were concerned that Varmus would be more interested in basic biomedical research than in applied studies and feared that the medical research related to their specific concerns might be neglected. Varmus has argued that basic research in science, especially investigations of the fundamental properties of cells, genes, and **tissue**s, could eventually lead to cures for many diseases, such as **AIDS** and **cancer**. As director, Varmus is also interested in revitalizing the intramural research program at NIH. He believes that science education in the United States needs to be improved and that students should be exposed to a science curriculum sooner, in smaller classes, by better-informed teachers.

VEINS

William Harvey's classic work on the **circulatory system** in the seventeenth century first defined the function of veins. Through dissections and experimentation, Harvey was able to illustrate that veins are responsible for bringing **blood** to the **heart**.

Throughout the **animal kingdom**, veins are the blood vessels which carry oxygen-depleted (deoxygenated) blood from the **tissue**s and **organ**s of the body back to the heart. The only vein in the body which carries oxygen-rich (oxygenated) blood is the pulmonary vein, which carries blood from the lungs to the heart.

Veins are relatively thin-walled vessels, which can expand to hold the necessary quantity of blood. Many veins contain valves which prevent the back-flow of blood, insuring that blood flows solely towards the heart. Breathing and muscle contractions also help to push blood (often against gravity) back toward the heart.

Veins pick up blood from networks of tiny vessels called capillaries. Capillaries receive oxygenated blood from the **arteries**, and allow oxygen and other **nutrients** to pass through their single-celled walls to nourish the organs and tissues of the body. Capillary networks branch and re-branch, ultimately forming venules, which are very small veins. Venules become veins, which empty into the body's largest veins, the **vena cava**e

Diseases which can affect the veins include the development of varicose veins (overly-stretched, twisted, bulging veins which can occur in the legs or in the rectum [where they are called hemorrhoids]); blood clots which cut off the flow of blood through the veins (called venous thrombosis); swelling and **inflammation** of the veins which can decrease their ability to return blood to the heart (called vasculitis); and abnormal, excessively fragile new networks of vessels, often occurring due to cirrhosis of the liver (called varices).

VENAE CAVAE

The vena cavae are two very large-diameter **veins**. As with the other veins throughout the body (with the exception of the pulmonary vein), the venae cavae both carry oxygen-depleted (deoxygenated) **blood** from the **tissue**s of the body to the right atrium of the **heart**.

In humans, because we stand upright, one vena cava is located above the other. The upper vena cava is referred to as the superior vena cava; the lower vena cava is referred to as the inferior vena cava. In other **mammals**, however, these same vessels are referred to as the anterior and posterior venae cavae.

The superior vena cava receives blood from the upper part of the body, including the head, neck, arms, and chest. The inferior vena cava receives blood from the legs, pelvis, and all of the structures and **organ**s within the abdomen. Because the venae cavae carry such a large volume of blood, the diameter of each tube is the largest of any blood vessel within the body. For example, the average internal diameter of other veins is about 0.39 in (1.0 cm); the diameter of the venae cavae is about 2.39 in (6 cm). This is larger, even, than the radius of the body's largest artery, the **aorta**, which measures about 0.98 in (2.50 cm) in radius. The walls of the venae cavae are comparatively thin, and quite flexible.

VERTEBRA

A vertebra (plural, vertebrae) is one of a series of specialized bones that make up the spinal column (or backbone) of **animals** in the **phylum** Vertebrata. Like other bones, vertebrae are formed initially of cartilage during early stages of embryonic development, and then ossify into true bone (that is, the tough, elastic cartilage is converted into a rigid material composed mostly of calcium phosphate). The spinal column is the main axial (or length-wise) girder (or supporting structure) of the vertebrate body.

A typical vertebra contains a massive, roughly spool-shaped (or cylindrical), central structure, known as the centrum. It also has a structure on the top surface, known as the neural arch, that forms a tube through which the **spinal cord** passes. Many vertebra also have structures of various length that stick out from the sides, known as processes, and from the top, known as a neural spine, to which muscles and ribs can attach. The numerous vertebrae of the spinal column articulate (or interlock) together to give the entire structure length-wise rigidity and support, yet lateral flexibility. The vertebrae are separated and cushioned from each other by soft, flexible structures known as discs.

The sizes and shapes of vertebrae vary along the spinal column of any particular animal. The ones higher in the back tend to have larger processes than those further below. Vertebral size, shape, and number also vary greatly among **species**. Tiny fish have the smallest vertebrae, while whales have the most massive ones. Snakes and salamanders have the largest numbers (up to several hundred) of vertebra. In certain groups of **organism**s some regions of the spinal column have masses of fused vertebrae that form specialized structures, such as the urostyle of frogs and the synsacrum of birds.

VESALIUS, ANDREAS (1514-1564)
Flemish physician

Andreas Vesalius, a Flemish physician, was one of the first scientists to dissect human cadavers, and he is considered to be the founder of modern **anatomy**. He was born in Brussels on December 13, 1514 to a family of physicians originally from Wesel, Germany. His father was a pharmacist to Holy Roman

Andreas Vesalius. *(The Library of Congress. Reproduced by permission.)*

Emperor Charles V, and as a boy Vesalius expressed his interest in science by dissecting dead birds and mice. As he grew older he was educated at the University of Louvain and later studied medicine in Paris. Vesalius served as a military surgeon for a while and then moved to Padua where he earned his medical degree in 1537. He was then appointed as a lecturer in surgery and anatomy. Vesalius was known for giving entertaining lectures where he actually performed dissections himself rather than delegating them to an assistant.

Vesalius is best known for advancing the study of anatomy. In the early 1500s physicians were taught anatomy was in the tradition of Galen, the Greek physician who based all his research on the study of the Barbary ape. During this period in history, artists began demanding more accurate representations of the human body. Vesalius sought to more accurately portray the inner workings of the human body and he became the first scientist to seriously question Galen's teaching in over 1,300 years. For example, contrary to Galen, Vesalius believed that men did not have one fewer rib than women. He also pioneered the notion that the **brain** and **nervous system** were the center of the mind and emotion, not the heart as was postulated **Aristotle**.

Vesalius used talented artists to portray the anatomical features he discovered in his research and he compiled these drawing into a text book. The preparation of this manuscript took almost three years, from 1539 to 1542. It was eventually published as *De humani corporis fabrica* in 1543, while Vesalius was still teaching at Padua. This work introduced a new standard for anatomical textbooks. In addition to the high quality illustrations of bones and the nervous system, the text expressed Vesalius' belief that the scientific method should be used in the study of anatomy. However, because its ideas ran counter to the conventional wisdom of the time, *De humani corporis fabrica* came under criticism by the establishment. Because of this controversy, Vesalius gave up the study of anatomy and resigned his position at Padua. He then became court physician to Emperor Charles and later to his son Phillip II of Spain. While returning from a pilgrimage to Jerusalem, he died in a shipwreck off the coast of Greece on October 15, 1564.

VIGNEAUD, VINCENT DU (1901-1978)
American biochemist

Vincent du Vigneaud, an American biochemist, received the 1955 Nobel Prize for Chemistry for his breakthrough achievement of synthesizing oxytocin—a **hormone** released by the posterior **pituitary gland** used to induce labor and lactation in pregnant women, and for his work with sulfur. Throughout his career, du Vigneaud was recognized for isolating and synthesizing penicillin and the hormone vasopressin, which is used to suppress urine flow, identifying the chemical composition of **insulin**, discovering the structure of vitamin H, otherwise known as **biotin**, and his pioneering work with methyl groups.

Du Vigneaud was born in Chicago, Illinois, on May 18, 1901, to Alfred, an inventor and designer of machines, and Mary Theresa (O'Leary) du Vigneaud. Early in his high school education in Chicago's public school system, du Vigneaud demonstrated an aptitude for chemistry and physiology. He constructed a laboratory in his parents' basement, where he carried out his first experiments. Du Vigneaud enrolled in the University of Illinois as an organic chemistry major and graduated in 1923. He stayed on to earn a masters degree in 1924, studying under C. S. Marvel. Also in 1924, du Vigneaud married Zella Zon Ford. Both of their children went on to become doctors. Their son, Vincent du Vigneaud, Jr., became an obstetrician and gynecologist, and their daughter, Marilyn Renee Brown, became a pediatric gastroenterologist.

From 1924 to 1925, du Vigneaud was an assistant biochemist at the University of Pennsylvania's Graduate School of Medicine and also worked in Philadelphia General Hospital's clinical chemistry laboratory. He then moved to the University of Rochester in New York to study for his Ph.D. under John R. Murlin at the School of Medicine. For his doctoral research, he undertook an examination of the chemical makeup of insulin, the protein hormone and sulfur compound that is secreted by the islets of Langerhans located in the **pancreas**. Du Vigneaud's investigations, which were inspired by a lecture given by renowned biochemist W. C. Rose at the University of Illinois, sparked a lifelong interest in the range of sulfur compounds, but especially the sulfur-containing **amino acid**s methionine, homocystine, cystine, cysteine, and cystathionine.

After receiving his doctorate in 1927, du Vigneaud became a National Research Council fellow. He worked first at Johns Hopkins Medical School's Department of Pharmacology under John J. Abel, where he continued his research into the structure of insulin. His suspicion that insulin was a derivative of the amino acid cystine was justified when he succeeded in isolating cystine from insulin crystals. He was thereby able to prove that insulin consists only of amino acids and an ammonia by-product.

Du Vigneaud left the United States for Germany in 1928 on a brief overseas tour. He first stopped at the Kaiser Wilhelm Institute in Dresden, where he worked under Max Bergman, an expert on the chemistry of amino acids and **peptide**s (chains of amino acids). Du Vigneaud later turned down an assistantship position with Bergman to proceed to the University of Edinburgh's Medical School, where he worked with biologist George Barger. He also spent time at the University College Hospital Medical School at the University of London, where he worked with Charles Harrington.

Upon returning to the United States, du Vigneaud joined the University of Illinois's physiological chemistry staff under his mentor, W. C. Rose. In 1932, he left his alma mater to take up a position as head of the Department of Biochemistry at the George Washington University School of Medicine in Washington, DC. One of his innovations there was to add a course in biochemistry to the medical school curriculum. His own research lead him to investigate his hypothesis that insulin's blood sugar-lowering effects were related to disulfide bonds of cystine.

In 1936 he and his staff succeeded in artificially creating glutathione, a tripeptide containing the amino acids cysteine, glycine, and glutamic acid, that is widely occurring in plant and animal tissues and which plays a vital part in biological oxidation-reduction processes and the activation of some enzymes. He also continued to pursue his research into insulin. By the following year, he was in a position to prove that the amino acid cystine comprises insulin's entire complement of sulfur, and that insulin can be deactivated by the reduction of its bonds of insulin by cystine or glutathione. Also in the late 1930s, du Vigneaud's work with methionine revealed how the body shifts a methyl group (CH_3) from one compound to another.

In 1938 du Vigneaud was appointed head of Cornell University Medical College's biochemistry department. Within two years, he had succeeded in isolating biotin (vitamin H). He spent the next few years carefully studying the substance and by 1942, had figured out its structure. He next turned to the human posterior pituitary gland, especially the study of the hormones oxytocin and vasopressin that it produces. Oxytocin is known to stimulate the contraction of uterine muscles and the secretion of milk in women during labor. Vasopressin, also known as the antidiuretic hormone, is a polypeptide hormone responsible for causing increased blood pressure and de-

creased urine flow. Du Vigneaud and his colleagues managed to isolate a highly purified form of these hormones from the pituitary gland and set about discovering their chemical nature.

To his surprise, du Vigneaud discovered that oxytocin is made up of only eight amino acids. Most proteins are comprised of several hundred amino acids. It took du Vigneaud another ten years to determine their sequence in an oxytocin molecule. Once he had cracked this puzzle, he was finally able to synthesize oxytocin. The importance of du Vigneaud's achievement lay not only in its making available an unlimited supply of the protein, but also in the light it shed on the relationship between molecular structure and biological function. The synthetic protein was tested on pregnant women at the Lying-in Hospital of the New York Hospital-Cornell Medical Center, where it was found to be as effective in inducing labor and milk flow as pure oxytocin. In 1946 the journal *Science* announced another du Vigneaud breakthrough: his synthesis of penicillin. Although du Vigneaud carried out the decisive experiments at Cornell University, it was one of the greatest international efforts of its kind, said *Science,* the culmination of five years of concerted effort by thirty-eight teams of scientists in the U.S. and Britain.

Du Vigneaud's illustrious scientific career was widely recognized. In 1955, he was awarded the Nobel Prize for Chemistry for "his work on biochemically important sulfur compounds and especially for the first synthesis of a polypeptide hormone." Du Vigneaud's other awards include the Nichols Medal of the American Chemical Society in 1945, the Association of Medical Colleges' Borden Award in the Medical Sciences in 1947, the Public Health Association's Lasker Award in 1948, the Osborne and Mendal Award of the American Institute of Nutrition in 1953, Columbia University's Charles Frederick Chandler Medal, and the Willard Gibbs Medal of the American Chemical Society. In addition, he was a member of the American Philosophical Society, the National Academy of Sciences, and the New York Academy of the Arts and Sciences. Du Vigneaud's leisure interests included bridge and horse riding.

From 1967 to 1975, du Vigneaud served as Cornell University's professor of chemistry. In 1975 he advanced to the level of emeritus professor of biochemistry at Cornell. Du Vigneaud died at St. Agnes Hospital in White Plains, New York, on December 11, 1978.

VIRCHOW, RUDOLF LUDWIG KARL (1821-1902)
German pathologist

Many historians of science consider Rudolf Virchow the greatest of all pathologists. Virchow was best known as the leading pioneer in the field of cellular pathology. In medieval physiology it was thought that prominent bodily fluids, including **blood**, phlegm, choler, and bile determined the character and health of a person, and during the nineteenth century many scientists still relied on these humoral theories to explain diseases. Virchow's **cell** theory, the essence of which is brilliantly encapsulated in the axiom *Omnis cellula e cellula*, stating that every cell arises from a previously existing cell, virtually dismantled the humoral theories that had dominated the healing arts for more than twenty centuries, and laid the groundwork for a more rational and systematic mode of study.

Virchow was born on October 13, 1821, in Schivelbein, Pomerania, Prussia. He began his medical education in 1839 at the Friedrich Wilhelm Institute of the University of Berlin, where he studied with the famous natural scientist, **Johannes Müller**. He received his medical degree in 1843, was appointed prosector at the Berlin hospital, Charité in 1847, and taught pathology at the University of Berlin. In 1845, he published his classic paper about one of two of the earliest cases of leukemia and in 1847, he founded his journal, *Archives for Pathological Anatomy and Physiology, and for Clinical Medicine*, which became known simply as *Virchow's Archive*. The journal also included non-medical articles on subjects, such as oriental languages, archeology, and anthropology.

In 1848, the Prussian government sent Virchow to Upper Silesia to investigate an outbreak of typhus fever. Virchow blamed the extent of the outbreak on poor social conditions, a fact which made the government furious. 1848 was also a year of major political upheaval which resulted in social revolution, and after Virchow returned to Berlin from Upper Silesia, he joined in the struggle. After the revolution Virchow published a liberal newspaper which focused on medical reform. Virchow was labeled a radical socialist and consequently lost his post at Charité. Virchow then moved to Würtrzburg, where he was appointed chair of pathological anatomy. During Virchow's years at the University of Würtzburg, the number of medical students drastically increased and included several students who eventually became famous scientists, such as Elie Metchnikoff and Robert Koch. In 1850, Virchow married Rose Meyer, with whom he had three daughters and three sons.

In 1856, Virchow returned to Berlin, where he resided and worked for the rest of his life. A chair of pathological anatomy was established just for him and a new pathological institute was built to suit his needs. While in Berlin, Virchow became known as the Pope of pathology. His cell theory, expounded over a course of twenty lectures and published as *Cellular Pathology as Based upon Physiological and Pathological Histology*, completely modified scientific thought. For example, Virchow clearly demonstrated that in cell reproduction the nucleus and cytoplasm divided. Although Virchow seemed unwilling to recognize the limits of his cell theory, affirming, incorrectly, that all pathology is essentially cellular in nature, his work in pathology is nevertheless viewed as fundamental for the development of modern medicine and biology. Furthermore, he did pioneering work in several fields of research, including the study of degenerative diseases and vascular conditions such as thrombosis, embolism, and inflammation.

Virchow remained a political activist and in 1859 was elected to the Berlin City Council, focusing primarily on public health. He exercised his political power by designing new sewer systems; he supervised the design of new hospitals,

opened a nursing school, spoke out about the necessity of school hygiene and the importance of meat inspection. In 1861, he was elected to the Prussian Diet and founded the Progressive Party. During wartime, he shifted his attention to building military hospitals, and during the Franco-Prussian War, Virchow led the first hospital train to the front. An exceptional organizer, Virchow excelled in devising methods for dealing with a wide variety of practical and theoretical issues.

Concurrently with his practical and theoretical work in medicine, Virchow, an ardent adversary of the conservative Chancellor Otto von Bismarck, served as member of the German Reichstag from 1880 to 1893. In addition, Virchow, who until the end of his life most energetically pursued his remarkably rich, multifaceted, and eclectic career, contributed to various fields of knowledge outside his primary area of competence. He is generally regarded as the founder of physical anthropology.

VIRUS

A virus is an infectious agent made almost entirely of **protein** and **nucleic acids**. Viruses range in size from 0.02 to 0.25 micrometer (approximately 0.000000007 in) and can only be seen using an **electron microscope**. They are non-living assemblages of protein molecules and either **ribonucleic acid (RNA)** or **deoxyribonucleic acid (DNA)** that can replicate themselves only within the **cells** of their hosts. Viruses are known to cause many infectious diseases in humans, **animals**, and **plants**. Examples include diseases such as **chicken pox**, measles, encephalitis, hepatitis, herpes, **influenza**, viral meningitis, mumps, rabies, and even the common cold.

Viruses were discovered at the end of the nineteenth century when scientists began to understand the role that microorganisms played in the spread of many diseases. In 1892, the Russian botanist **Dmitry Ivanovsky**, discovered that the virus causing mosaic disease in tobacco plants could not be separated by filtration. This observation suggested that the organism was much smaller than **bacteria**. In 1898, Friedrich Loeffler and Paul Frosch discovered that the agents causing foot-and-mouth disease in cattle were also to small to filter. In addition, they showed that when small amounts of the filtered agent were injected into healthy cattle, its concentration increased over time. On this basis, they deduced that the infectious agent could reproduce itself inside the infected animal. These agents became known as filterable viruses and later simply as viruses.

Over the next 30 years, a number of viruses were discovered in animals, plants, insects, and bacteria. In the 1930s, scientists developed a number of new techniques for the study of viruses, including methods to grow viruses in mice and in chicken embryos which allowed the study of viral infections under controlled laboratory conditions. In addition, scientists learned how to use ultra high-speed **centrifuge**s to isolate the viral agents. Through chemical analysis, they were then able to determine that viruses consist primarily of nucleic acid and protein. In the 1940s, scientists used the newly developed electron microscope to study the size and shape of virus particles.

In 1952, A.D. Hershey and M. Chase found that when a virus infects a bacterium, only the viral nucleic acid enters the cell. By the late 1950s, techniques for growing animal cells in **tissue culture** were developed, making the study of viral growth possible. This research lead to the first synthesis of a virus in 1967. Today, hundreds of different viruses have been identified and they are believed to be one of the most abundant forms of life. Scientists now know that most viruses consist of small particles of nucleic acid, known as virons, which are surrounded by a protein coat, or capsid. Viruses are not living in the normal sense; they are not cells that reproduce by binary fission. Unlike bacteria, they reproduce by a synthesis process that first involves the duplication of their nucleic acids followed by assembly of the produced pieces. This replication process can occur only when the virus is inside a host cell.

Although many viruses are apparently harmless to their hosts, certain viruses are responsible for serious diseases of humans, other animals, and plants. Such diseases as chicken pox, the common cold, encephalitis, German measles, hepatitis, genital and oral herpes, influenza, measles, viral meningitis, mumps, poliomyelitis, rabies, shingles, smallpox, and yellow fever are all caused by viruses. Some viral infections can be prevented through the use of vaccines. A vaccine consists of either a dead or extremely weakened strain of the virus that can cause the body to form antibodies. These antibodies provide protection against subsequent infections. However, this protection is only against the specific immunizing virus and viruses closely related to it. Examples of viral diseases for which vaccines have been developed include small pox, polio, hepatitis, influenza, and rabies. However, vaccination is not always effective in fighting viral diseases. For viruses that change frequently through mutation, such as influenza virus and HIV, one preparation of vaccine may not be successful against all infections with that virus.

Viruses may also be very important in aquatic food webs. In coastal waters, there may be as many as 1×10^8 viruses in one milliliter. While these viruses are generally harmless to humans, they may be responsible for the death of one fourth of the bacterial population in coastal waters per day. Although little is known about viruses in aquatic systems, they may also be responsible for high mortality rates in some phytoplankton populations.

It is likely that as future biochemical research reveals more information about viruses, additional ways to combat them will be found. Drugs that will limit the growth of the virus, but not harm the growth of the host cells, have been difficult to develop because the virus is so closely connected to the host. However, scientists have had some success with drugs that inhibit viral growth processes that are not supplied by the host. For example, AZT, which is used to combat **HIV**, interferes with the replication of HIV nucleic-acids. Acyclovir, which is used against herpes simplex, inhibits that virus's replication **enzyme**.

See also AIDS; AIDS therapies and vaccines

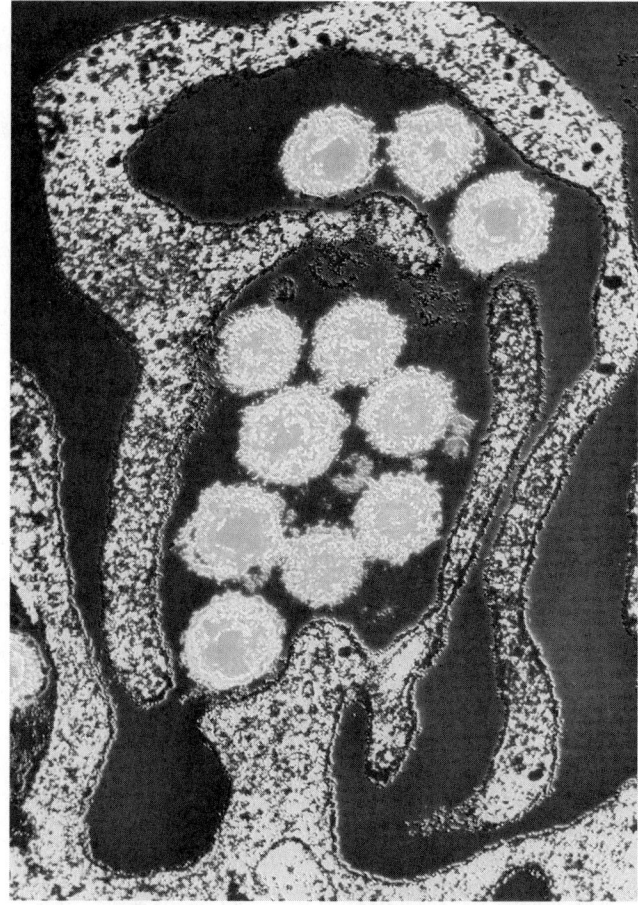

A tranmission electron micrograph (TEM) of human herpes virus type 6 (HHV6) infecting a human cell. This cell was isolated from an AIDS patient suffering a secondary infection with the virus. *(Photograph by A.B. Dowsett, Photo Researchers, Inc. Reproduced by permission.)*

VITALISM

Vitalism is a school of thought which postulates that life cannot be fully explained in physical material terms. According to vitalists, life, which in the material world is manifested as a physical process, emerges as a result of an immaterial impulse. **Aristotle**, who is regarded as the founder of scientific vitalism, believed that the soul, as a modality of life-**energy**, kept the **organism** alive. According to Aristotle, the soul affects the organism without being connected to it in a physical sense.

Historians of science often identify René Descartes as the great intellectual force that facilitated the switch from Aristotelian metaphysics to the more sober mechanistic-materialistic **paradigm** of modern mainstream science. A fervent Catholic, Descartes altered Aristotle's terminology, retaining, however, the fundamental idea that an organism, being a physical thing, receives direction from a spiritual entity.

Although the powerful mechanistic paradigm created by Isaac Newton (1642-1727) dominated the physical sciences, many natural scientists revolted against what they saw as a lifeless, cold, and rigid conception of the universe. Often branded as purely speculative thinkers, the greatest representatives of vitalism in biology were nevertheless brilliant researchers and practical scientists. For example, Newton's younger contemporary Georg Ernst Stahl (1660-1734), built a comprehensive medical theory and practice on vitalistic foundations. One of the greatest scientists of the eighteenth century was Marie-François-Xavier Bichat (1771-1802), the founder of **histology**, was a vitalist. Furthermore it was **Karl Ernst von Baer**, the eminent representative of nineteenth-century vitalism, who in 1827 made history with his discovery of the mammalian ovum.

Twentieth-century scientists and historians of science have often dismissed vitalism as basically obsolete, even unscientific, perhaps because it could not be proved. This insistence on empirical proof shows a profound misunderstanding of the essence of vitalism. Vitalism is an intellectual orientation, and not a mere hypothesis in need of material proof. During the first half of the twentieth century, vitalism's greatest exponents were Henri Bergson (1859-1941), who developed the concept of the *élan vital*, and **Hans Driesch** (1867-1941). While Bergson, who was primarily a philosopher, relied on secondary sources in biology, Driesch was a practicing biologist, who, in an experiment with sea urchins, showed that if a half of the **egg**, following the first division after **fertilization**, is destroyed, the remaining half will produce a complete, albeit smaller, **embryo**. In Driesch's view, this kind of regeneration clearly demonstrates that life follows a logic that is not determined by physical circumstances. Finally, the theory of morphogenetic fields, developed by **Rupert Sheldrake**, affirms the profoundly vitalistic idea that **nature** develops in harmony with invisible, immaterial, but powerful forces. According to Sheldrake, morphogenetic fields, like life itself, may not be detectable in a traditional sense, but biologists cannot afford to ignore them.

See also Portmann, Adolf

VITAMINS

Vitamins are **organic compound**s found in foods that are necessary in human and **animal** diets to sustain life and health. There are thirteen vitamins that have been identified as necessary for human health, and several other vitamin-like substances also contribute to good **nutrition**. The vitamins are named by letters—vitamin A, vitamin C, D, E, K, and eight B vitamins. Because the B vitamins were originally thought to be one vitamin, they were first differentiated by numeral subscripts. Currently B_6 and B_{12} retain the numeric names, and the rest of the B group vitamins are thiamine, riboflavin, niacin, **pantothenic acid**, **biotin**, and folic acid.

The discovery of vitamins was a patchwork affair, taking many scientists in distant communities years to piece together the story of the vitamins' role in nutrition. Many of the discoveries of vitamins came from doctors working on the cure of diseases such as rickets, scurvy and pellagra, which are now known to be caused by vitamin deficiency. Scurvy was a dis-

ease that particularly affected sailors on long voyages. A Scottish doctor, James Lind (1716-1794), found in 1747 that he could cure afflicted sailors by feeding them citrus **fruits**. Lind attempted to convince the British navy to change the standard sailor's diet to include citrus, but the navy took no action until forced to by a mutiny in 1795, a year after Lind's **death**. And more than a hundred years passed before it was understood that the vitamin C in citrus was responsible for curing scurvy, a vitamin-deficiency disease. The most exact theory of nutrition in the nineteenth century came from the English physician William Prout, who laid out the three essentials of the human diet in 1827 as "the oily, the saccharin, and the albuminous." The modern-day terms for these substances are **fats and oils**, **carbohydrate**s, and **protein**s. Until the early years of the 20th century, Prout's work was taken as the last word in nutrition. An English biochemist, Frederick Hopkins (1861-1947) and the Dutch physician **Christiaan Eijkman** (1858-1930) were instrumental in showing that something besides **fats**, carbohydrate and protein needed to be part of human and animal diets.

Eijkman worked in the Dutch East Indies beginning in 1886, studying the disease **beriberi**. Pasteur's **germ theory** of disease had just gained acceptance, and Eijkman and his team tried to discover the infectious agent in beriberi. Eijkman found that he could give the disease to chickens by feeding them polished rice, and cure them by giving them brown rice. He correctly surmised that beriberi was caused by a dietary deficiency of some element of rice husk. Hopkins, working in the first years of the twentieth century, found that laboratory mice could be made ill if deprived of the **amino acid** tryptophan. He articulated the theory that there were necessary trace substances in foods, and suggested that rickets and scurvy might be caused by lack of these essential ingredients. These essential substances were named "vitamines" in 1912 by a Polish-American biochemist **Casimir Funk** (1884-1967). Funk thought that all the various essential substances were amines, but when this turned out not to be true, the term was modified to simply vitamins. Eijkman and Hopkins shared a Nobel prize in 1929 for their early apprehension of the vitamin concept.

Research on vitamins proceeded quickly in the first decades of the twentieth century. Austrian-American physician Joseph Goldberger (1874-1929) identified pellagra as a vitamin deficiency disease, and the American biochemist Conrad Elvehjem (1901-1962), working at the University of Wisconsin, discovered in 1937 that niacin was the missing vitamin in this case. Another University of Wisconsin biochemist, Elmer McCollum (1879-1967), discovered that one vitamin was present in some fats, whereas Eijkman's anti-beriberi substance was **water**-soluble. This meant that they had to be quite chemically different substances. McCollum isolated the first of the named vitamins, vitamin A and vitamin B, and contributed to the discovery of vitamins D and E as well. Studies with animals on vitamin-supplemented diets later showed that something was missing from vitamin B, and the further B vitamins were discovered at various laboratories, with the last one isolated in 1948.

Lind's work had shown that some substance in citrus fruits prevented scurvy, and this was known as the antiscorbu-

ESSENTIAL VITAMINS

Vitamin	What It Does For The Body
Vitamin A (Beta Carotene)	Promotes growth and repair of body tissues; reduces susceptibility to infections; aids in bone and tooth formation; maintains smooth skin
Vitamin B_1 (Thiamin)	Promotes growth and muscle tone; aids in the proper functioning of the muscles, heart, and nervous system; assists in digestion of carbohydrates
Vitamin B_2 (Riboflavin)	Maintains good vision and healthy skin, hair, and nails; assists in formation of antibodies and red blood cells; aids in carbohydrate, fat, and protein metabolism
Vitamin B_3 (Niacinamide)	Reduces cholesterol levels in the blood; maintains healthy skin, tongue, and digestive system; improves blood circulation; increases energy
Vitamin B_5	Fortifies white blood cells; helps the body's resistance to stress; builds cells
Vitamin B_6 (Pyridoxine)	Aids in the synthesis and breakdown of amino acids and the metabolism of fats and carbohydrates; supports the central nervous system; maintains healthy skin
Vitamin B_{12} (Cobalamin)	Promotes growth in children; prevents anemia by regenerating red blood cells; aids in the metabolism of carbohydrates, fats, and proteins; maintains healthy nervous system
Biotin	Aids in the metabolism of proteins and fats; promotes healthy skin
Choline	Helps the liver eliminate toxins
Folic Acid (Folate, Folacin)	Promotes the growth and reproduction of body cells; aids in the formation of red blood cells and bone marrow
Vitamin C (Ascorbic Acid)	One of the major antioxidants; essential for healthy teeth, gums, and bones; helps to heal wounds, fractures, and scar tissue; builds resistance to infections; assists in the prevention and treatment of the common cold; prevents scurvy
Vitamin D	Improves the absorption of calcium and phosphorous (essential in the formation of healthy bones and teeth); maintains nervous system
Vitamin E	A major antioxidant; supplies oxygen to blood; provides nourishment to cells; prevents blood clots; slows cellular aging
Vitamin K (Menadione)	Prevents internal bleeding; reduces heavy menstrual flow

(Table by Standley Publishing.)

tic factor. But the factor was not isolated until 1928. The chemist Albert Szent-Györgi, working in Frederick Hopkins's laboratory, isolated what he at first called hexuronic acid in that year, and was able to prepare large quantities of it from Hungarian red peppers. Studies of Szent-Györgi's acid, which he re-named ascorbic (anti-scurvy) acid led several laboratories to come up with syntheses of it. This was vitamin C. By the late 1930s, cheap synthetic vitamin C was widely available.

Vitamin D's discovery is related to research into rickets, a bone-softening disease that affects humans and animals. The folk cure for it was exposure to sunlight. In the 1920s **Harry Steenbock**, another University of Wisconsin chemist, discovered that an anti-rickets factor could be added to foods by sub-

mitting them to ultraviolet radiation. Steenbock's process was patented for the enrichment of milk, and English and German researchers finally elucidated the structure of the substance, vitamin D, in the 1930s.

Vitamin E was isolated from wheat germ oil in 1936. Vitamin K, necessary for the coagulation of **blood**, was first discovered by a Danish scientist in 1929. Scientists at four different laboratories in the U.S. and in Zurich finally isolated vitamin K ten years later. Nutritionists in the 1930s and 1940s also identified other substances essential for human and animal health, such as sodium, potassium, magnesium, iron and zinc. Research into the known vitamins was essentially complete by 1956, when the complicated structure of vitamin B_{12} was finally determined. Vitamin research since that time has focused on the ways vitamins work in the body, and their role in fighting specific diseases. Vitamin research is still one of the most vibrant fields in medicine, as new studies suggest many complex and little-understood functions of vitamins in regulating human health.

VIVIPAROUS

Viviparous reproduction involves the fertilized **egg** (or **zygote**) being retained within the uterus of the female, where it is nourished by **nutrients** and gases (especially oxygen) passing from the mother across a specialized, **blood**-rich **tissue** known as the **placenta**, which is connected to the **embryo** by a cord-like umbilicus. Waste chemicals pass the other way across the placenta, from the developing embryo to the parent, who excretes them with her own metabolic wastes. The embryo grows and develops under these conditions of close, maternal nourishment. It is eventually born as a "live" young that can exist externally, although for some time it is still dependent upon its mother for nourishment and care. The time between **fertilization** of the egg and **birth** of the young is known as the gestation period.

Viviparous **animal**s are different from **oviparous** ones, which lay shelled eggs that incubate outside the mother's body. The developing embryos receive all of their nourishment from **protein**, fat, and **carbohydrate** stored in the yolk and albumin of the egg. Viviparous animals also differ from ovoviviparous **species**, in which the fertilized eggs are retained inside of the body of the female, where they develop (utilizing **nutrition** of the yolk and albumin) internally until fully independent, living young are born.

Almost all species of **mammals** are viviparous in their reproduction. The only exceptions are those in the **order** Edentata, which includes several species of echidnas and the platypus; these occur only in Australia and New Guinea. Edentates are oviparous mammals, laying shelled eggs that are incubated by the female parent. Among the viviparous mammals, the group known as marsupials gives birth to young that are in a relatively early stage of development. Examples of species within this group are kangaroos, wallabies, koalas, and opossums. In essence, marsupials give birth to embryos, which are then nurtured externally in a specialized pouch on the mother's belly (known as the marsupium). The marsupial young suckle and develop in the pouch until they are ready to live a more fully independent existence.

Other species are known as placental mammals. These retain the developing embryos within the body of the female parent until a much later stage of development is reached. However, after birth the young of placental mammals must be fed for some time by the mother. This is done by suckling on one of her teats to feed on a fat- and protein-rich material known as "milk." Familiar examples of placental mammals are cows, horses, dogs, and humans.

A few species of so-called "viviparous" **plants** retain their seeds on the parent, where they germinate and are then dispersed into the environment. One such example is the red mangrove (*Rhizophora mangle*), a **tree** which grows in flat, muddy places along tropical shores. The red mangrove retains its ripe seeds on its branches, where they germinate and grow a root-like radicle. The radicle may reach a length of 4-6 in (10-15 cm) before the germinated seed falls from the parent, to hopefully stick vertically into the mud and establish as a head-started seedling.

VON BEHRING, EMIL • See Behring, Emil von

VON HALLER, ALBRECHT • See Haller, (Victor) Albrecht von

VON HELMHOLTZ, HERMANN • See Helmholtz, Hermann von

VRIES, HUGO MARIE DE (1848-1935)
Dutch botanist

Hugo Marie de Vries was a Dutch botanist and geneticist who studied mutations in plant physiology and furthered the work of **Gregor Mendel** on the laws of heredity. The foremost botanist of his time, his theories provided the impetus for **Charles Darwin**'s work on **evolution**.

De Vries was born on February 16, 1848 in Haarlem, the Netherlands, the son of a Dutch Prime Minister. He studied medicine at Heidelberg and Leiden, graduating in 1870. De Vries taught at the newly established University of Amsterdam from 1871 until 1875 and became the first lecturer in plant physiology in 1877, following a brief period working with the Prussian Ministry of Agriculture in Würzburg from 1875 until 1877. He was made an assistant professor of botany in 1878 and a full professor in 1881 and taught at the University of Amsterdam until his retirement in 1918. Awarded eleven honorary degrees, he actively contributed to scientific research until he died on May 21, 1935 near Amsterdam.

De Vries first developed an interest in genetics in 1886 when he discovered variations in some specimens of plants. He devoted himself to cultivation of various types of evening primrose (*Oenothera lamarckiana*) and research into the origin and history of the European strain of the **species**. De Vries experimented with plant breeding which led him to rediscover the laws of **heredity** first set forth by in 1866 by Austrian monk Gregor Mendel. He further developed Mendelian genetics when he found that new varieties of evening primrose occurred spontaneously and recurred in subsequent generations. De Vries termed these new varieties variations. He later coined the term **mutation**s and divided the mutations into progressive mutants which contribute useful characteristics to the evolution of the species and retrogressive mutants which contribute useless or harmful characteristics. He also postulated that new species originated from successful mutations. The term mutation is used today to refer to any change in genetic material, whether it be to **gene**s or **chromosome**s.

In 1889, De Vries published *Intracellular Pangenesis* which theorized that hereditary traits in plants are carried by ''pangenes.'' His major work *Die Mutationstheorie* (translated into English as *The Mutation Theory*) which summarized his theories on heredity and mutations was published in 1901-1903. The latter was instrumental in establishing Charles Darwin's theory of evolution.

  De Vries's other contributions to botany include studies of the physiology of plant cells. He demonstrated **plasmolysis**, the shrinkage of **protoplasm** in a **cell** caused by loss of water through osmosis. He developed methods for studying turgor properties of plant cells. He also showed that the plasma membrane surrounding a plant cell is semi-permeable and permits the passage of small molecules.

W

WAGNER-JAUREGG, JULIUS (1857-1940)
Austrian physician

Julius Wagner-Jauregg was an Austrian psychiatrist whose experimental work in the first part of the twentieth century led to a new appreciation of the beneficial effects of bodily stress in the treatment of mental illness. In 1927 he became the first psychiatrist to win the Nobel Prize for his discovery that **syphilis**, a chronic, usually venereal disease caused by spirochete **bacteria**, could be cured by clinically induced malaria, which is characterized by symptoms of fever and chills.

Wagner-Jauregg was born Julius Wagner on March 7, 1857, in the village of Wels, Austria. He was the oldest son of Ludovika Ranzoni and Adolf Johann Wagner, a government official. The family name became "Wagner von Jauregg" when Adolf Johann was raised to the nobility, but following the collapse of the Austro-Hungarian empire in 1918, the "von" was dropped. After the early death of his mother, Julius Wagner-Jauregg was raised at home. In his youth he successfully fought off typhoid and tuberculosis to graduate from Vienna's prestigious Schottengymnasium.

While attending medical school at the University of Vienna, Wagner-Jauregg received thorough training in experimental biology and met the father of psychoanalysis, Sigmund Freud, who was studying at the Institute of General and Experimental Pathology. Despite Wagner-Jauregg's lack of interest in psychoanalysis, the two remained lifelong friends. In 1880 Wagner-Jauregg was awarded a medical degree for his thesis on the heart under conditions of acceleration.

Originally, Wagner-Jauregg hoped to practice general medicine, but when Vienna's two teaching hospitals turned him down, he reluctantly accepted a position as an assistant in the university's psychiatric clinic. Although he had little training in mental illness, he quickly became a qualified instructor in psychiatry and neurology. Wagner-Jauregg was a clinician, skilled in detailed observation and careful case analysis. Using the latest techniques of **animal** experimentation, he spent his life working to advance the biological understanding of mental illness. His first research entailed the investigation of how certain chemicals stimulate breathing after strangulation.

In 1889 Wagner-Jauregg was appointed professor of psychiatry at the University of Graz and for the next four years studied the effect of the **thyroid gland** on behavior. An ardent vivisectionist, he discovered that when the thyroid was removed from a cat, the animal's behavior became convulsive and violent. Cretinism in humans, Wagner-Jauregg put forth in an early paper, was due to a malfunction of the thyroid. During his years in Graz, he travelled frequently in central and southeastern Austria studying peasants with goiter and found that small amounts of iodine reduced their hugely swollen necks. He urged the sale of iodized salt in alpine regions, a measure the Austrian government undertook belatedly in 1923.

In 1893 Wagner-Jauregg was made a full professor at the University of Vienna and appointed director of the Hospital for Nervous and Mental Diseases and the State Mental Asylum. As a member of the Austrian Board of Health, he helped draft important legislation protecting the rights of the mentally ill and regulating the certification of the insane. At his urging, psychiatry became a compulsory subject in the undergraduate curriculum.

Discovers That Fever Can Cure the Mentally Ill

While still only a medical assistant, Wagner-Jauregg had studied the beneficial effect of high fever on psychotic patients. For a monograph that he published in 1888, he surveyed instances where epidemics of typhoid, malaria, smallpox, and scarlet fever had swept through mental asylums. In 30 cases reaching back to antiquity, he described how bouts of high fever had brought dramatic relief in cases of melancholy, mania, and paresis. At the end of his monograph, Wagner-Jauregg suggested that malaria might be used experimentally to induce a "fever cure" in psychotic patients, although at the time he lacked the authority to undertake so radical a treatment.

The monograph received little notice when it was published. In it, Wagner-Jauregg had formulated two bold hypotheses: first, that some psychoses were organic in nature, and second, that one disease might be employed to eradicate another disease. In Graz, he had produced fever with injections of tuberculin, a **protein** used to treat tuberculosis, until it was learned that tuberculin was unsafe. In Vienna, he injected paralytic patents with typhus vaccine and *Staphylococci* but was disappointed by the results. Most of the cures proved to be temporary, and patients soon relapsed.

It was not until World War I that conditions were ripe for a radical trial. By then a series of important discoveries had confirmed the link between paresis and syphilis. In 1905 researchers had identified the syphilis bacillus, *Spirochaete pallida*. A year later, the Wasserman test for syphilis revealed that paresis was a progressive disease of the brain caused by untreated syphilis. In Wagner-Jauregg's time, paresis accounted for fifteen percent of the patients confined to mental hospitals. The disease was thought to be incurable and invariably ended in insanity, paralysis, and **death** within three to four years.

In the final years of World War I, Wagner-Jauregg was treating victims of shell shock when he encountered a soldier suffering from malaria. On June 14, 1917, Wagner-Jauregg used **blood** drawn from the malarial soldier to infect nine patients suffering from paresis. Quinine, the medicine used to treat malaria, was withheld until each patient had endured seven to eleven attacks of fever. The results were astonishing. Six patients experienced a dramatic remission of symptoms, and three were able to return to normal life. In 1919 Wagner-Jauregg began full-scale clinical trials.

At first, Wagner-Jauregg's reports were greeted with considerable skepticism by the medical community. Some physicians considered it unethical to deliberately induce a disease as serious as malaria. Others feared the outbreak of malaria epidemics in major metropolitan centers. But trials elsewhere produced similar results. Employing only a mild strain of malaria easily cured by quinine, mortality remained low while complete recovery was experienced by thirty to forty percent of all patients. Patients who had only recently contracted syphilis could be cured completely when the "malaria cure" was used in conjunction with injections of Salvarsan and Neosalvarsan, two drugs used to treat early syphilis. In 1927 Wagner-Jauregg became the first psychiatrist to be awarded the Nobel Prize in physiology or medicine.

Safer methods of inducing fever were tried—preparations of colloidal sulfur, hot-water baths, and "fever cabinets"—but none had the high rates of success typical of malaria. Until the discovery of **penicillin** during World War II, malaria remained the preferred treatment for advanced syphilis. Medical opinion differed on just how the fever cure worked since it seemed unlikely that the fever killed all of the spirochete bacteria, which cause syphilis. Instead, it was believed that the stress produced by the malaria attack in some way strengthened the body's defenses against the syphilitic infection. Stress treatments such as electroshock continue to play a role in the treatment of psychiatric disorders.

In 1928, one year after receiving the Nobel Prize, Wagner-Jauregg retired at the age of seventy-one. In his youth he had been an avid mountaineer, and he was an accomplished chess player. During his long career he published some eighty papers and received several distinguished honors. In 1935 the University of Edinburgh awarded Wagner-Jauregg the Cameron Prize, and in 1937 he received the Gold Medal of the American Committee for Research on Syphilis. Julius Wagner-Jauregg died on September 27, 1940, in Vienna at age eighty-four, shortly before the discovery of penicillin made his fever cure obsolete. He was survived by his wife, Anna Koch, a daughter, Julia, and a son, Theodor, who became a distinguished professor of chemistry at the University of Vienna.

WAKSMAN, SELMAN (1888-1973)
Russian American microbiologist

Selman Waksman revolutionized medicine, thanks to his discoveries of life-saving antibacterial compounds. His investigations have also spawned further studies for other disease-curing drugs. Waksman isolated streptomycin, the first chemical agent that was effective against tuberculosis. Prior to his discovery, tuberculosis was a lifelong debilitating disease, and was fatal in some forms. Streptomycin effected a cure, and for this discovery, Waksman received the 1952 Nobel Prize in physiology or medicine. In pioneering the field of antibiotic research, Waksman had an inestimable impact on human health and well-being, creating both a new field of medicine and a new industry.

The only son of a Jewish furniture textile weaver, Selman Abraham Waksman was born in the tiny Russian village of Novaya Priluka on July 22, 1888. Life was hard in late-nineteenth-century Russia. Waksman's only sister died from **diphtheria** when he was nine. There were particular tribulations for members of a persecuted ethnic minority. As a teen during the Russian revolution, Waksman helped organize an armed Jewish youth defense group to counteract oppression. He also set up a school for underprivileged children and formed a group to care for the sick. These activities prefaced his later role as a standard-bearer for social responsibility.

Several factors led to Waksman's immigration to the United States. He had received his diploma from the *Gymnasium* in Odessa and was poised to attend university, but he doubtless recognized the very limited options he held as a Jew in Russia. At the same time, in 1910, his mother died, and cousins who had immigrated to New Jersey urged him to follow their lead. Waksman did so, and his move to a farm there, where he learned the basics of scientific farming from his cousin, likely had a pivotal influence on Waksman's later choice of field of study.

Begins Research on Soil Microbes

In 1911 Waksman enrolled in nearby Rutgers College (later University) of Agriculture, following the advice of fellow Russian immigrant Jacob Lipman, who led the college's bacteriology department. He worked with Lipman, developing a fascination with the **bacteria** of soil, and graduated with a B.Sc. in 1915. The next year he earned his M.S. degree.

Around this time he also became a naturalized United States citizen and changed the spelling of his first name from Zolman to Selman. Waksman married Bertha Deborah Mitnik, a childhood sweetheart and the sister of one of his childhood friends, in 1916. Deborah Mitnik had come to the United States in 1913, and in 1919 she bore their only child, Byron Halsted Waksman, who eventually went on to a distinguished career at Yale University as a pathology professor.

Waksman's intellect and industry enabled him to earn his Ph.D. in less than two years at the University of California, Berkeley. His 1918 dissertation focused on proteolytic **enzymes** (special **protein**s that break down proteins) in fungi. Throughout his schooling, Waksman supported himself through various scholarships and jobs. Among the latter were ranch work, caretaker and night watchman, and tutor of English and science.

Waksman's former advisor invited him to join Rutgers as a lecturer in soil bacteriology in 1918. He was to stay at Rutgers for his entire professional career. When Waksman took up the post, however, he found his pay too low to support his family. Thus, in his early years at Rutgers he also worked at the nearby Takamine Laboratory, where he produced enzymes and ran toxicity tests.

In the 1920s Waksman began to gain recognition in scientific circles. Others sought out his keen mind, and his prolific output earned him a well-deserved reputation. He wrote two major books during this decade. *Enzymes: Properties, Distribution, Methods, and Applications,* coauthored with Wilburt C. Davison, was published in 1926, and in 1927 his thousand-page *Principles of Soil Microbiology* appeared. This latter volume became a classic among soil bacteriologists. His laboratory produced more than just books. One of Waksman's students during this period was René Dubos, who would later discover the antibiotic gramicidin, the first chemotherapeutic agent effective against gram-positive bacteria (bacteria that hold dye in a stain test named for Danish bacteriologist **Hans Gram**). Waksman became an associate professor at Rutgers in the mid–1920s and advanced to the rank of full professor in 1930.

During the 1930s Waksman systematically investigated the complex web of microbial life in soil, humus, and peat. He was recognized as a leader in the field of soil microbiology, and his work stimulated an ever-growing group of graduate students and postdoctoral assistants. He continued to publish widely, and he also established many professional relationships with industrial firms that utilized products of microbes. These companies that produced enzymes, pharmaceuticals, vitamins, and other products were later to prove valuable in Waksman's researches, mass producing and distributing the products he developed. Among his other accomplishments during this period was the founding of the division of Marine Bacteriology at Woods Hole Oceanographic Institution in 1931. For the next decade he spent summers there and eventually became a trustee, a post he filled until his death.

Research Finds Practical Applications in Wartime

In 1939 Waksman was appointed chair of the U.S. War Committee on Bacteriology. He derived practical applications

Selman Waksman. *(The Library of Congress. Reproduced by permission.)*

from his earlier studies on soil microorganisms, developing antifungal agents to protect soldiers and their equipment. He also worked with the Navy on the problem of bacteria that attacked ship hulls. Early that same year Dubos announced his finding of two antibacterial substances, tyrocidine and gramicidin, derived from a soil bacterium (*Bacillus brevis*). The latter compound, effective against gram-positive bacteria, proved too toxic for human use but did find widespread employment against various bacterial infections in veterinary medicine. The discovery of gramicidin also evidently inspired Waksman to dedicate himself to focus on the medicinal uses of antibacterial soil microbes. It was in this period that he began rigorously investigating the antibiotic properties of a wide range of soil **fungi**.

Waksman set up a team of about fifty graduate students and assistants to undertake a systematic study of thousands of different soil fungi and other microorganisms. The rediscovery at this time of the power of **penicillin** against gram-positive bacteria likely provided further incentive to Waksman to find an antibiotic effective against gram-negative bacteria, which include the kind that causes tuberculosis.

In 1940 Waksman became head of Rutgers' department of microbiology. In that year too, with the help of Boyd Woodruff, he isolated the antibiotic actinomycin. Named for the actinomycetes (rod- or filament-shaped bacteria) from which it was isolated, this compound also proved too toxic for human use, but its discovery led to the subsequent finding of variant forms (actinomycin A, B, C, and D), several of which were found to have potent anti-**cancer** effects. Over the next decade Waksman isolated ten distinct **antibiotics**. It is Waksman who first applied the term antibiotic, which literally means against life, to such drugs.

Breakthrough with Isolation of Streptomycin

Among these discoveries, Waksman's finding of streptomycin had the largest and most immediate impact. Not only did streptomycin appear nontoxic to humans, but it was highly effective against gram-negative bacteria. (Prior to this time the antibiotics available for human use had been active only against the gram-positive strains.) The importance of streptomycin was soon realized. Clinical trials showed it to be effective against a wide range of diseases, most notably tuberculosis.

At the time of streptomycin's discovery, tuberculosis was the most resistant and irreversible of all the major infectious diseases. It could only be treated with a regime of rest and nutritious diet. The tuberculosis bacillus consigned its victims to a lifetime of invalidism and, when it invaded organs other than the lungs, often killed. Sanatoriums around the country were filled with persons suffering the ravages of tuberculosis, and little could be done for them.

Streptomycin changed all of that. From the time of its first clinical trials in 1944, it proved to be remarkably effective against tuberculosis, literally snatching sufferers back from the jaws of death. By 1950 streptomycin was used against seventy different germs that were not treatable with penicillin. Among the diseases treated by streptomycin were bacterial meningitis (an inflammation of membranes enveloping the **brain** and **spinal cord**), endocarditis (an inflammation of the lining of the heart and its valves), pulmonary and urinary tract infections, leprosy, typhoid fever, bacillary dysentery, **cholera**, and **bubonic plague**.

Waksman arranged to have streptomycin produced by a number of pharmaceutical companies, since demand for it soon skyrocketed beyond the capacity of any single company. Manufacture of the drug became a $50-million-per-year industry. Thanks to Waksman and streptomycin, Rutgers received millions of dollars of income from the royalties. Waksman donated much of his own share to the establishment of an Institute of Microbiology there. He summarized his early researches on the drug in *Streptomycin: Nature and Practical Applications* (1949). Streptomycin ultimately proved to have some human toxicity and was supplanted by other antibiotics, but its discovery changed the course of modern medicine. Not only did it directly save countless lives, but its development stimulated scientists around the globe to search the microbial world for other antibiotics and medicines.

Research Yields Other Antibiotics

In 1949 Waksman isolated neomycin, which proved effective against bacteria that had become resistant to streptomycin. Neomycin also found a broad niche as a topical antibiotic. Other antibiotics soon came forth from his Institute of Microbiology. These included streptocin, framicidin, erlichin, candidin, and others. Waksman himself discovered eighteen antibiotics during the course of his career.

Waksman served as director of the Institute for Microbiology until his retirement in 1958. Even after that time, he con-

tinued to supervise research there. He also lectured widely and continued to write at the frenetic pace established early in his career. He eventually published more than twenty-five books, among them the autobiography *My Life with the Microbes,* and hundreds of articles. He was author of popular pamphlets on the use of thermophilic (heat-loving) microorganisms in composting and on the enzymes involved in jelly-making. He wrote biographies of several noted microbiologists, including his own mentor, Jacob Lipman. These works are in addition to his numerous publications in the research literature.

On August 16, 1973, Waksman died suddenly in Hyannis, Massachusetts, of a cerebral hemorrhage. He was buried near the institute to which he had contributed so much over the years. Waksman's honors over his professional career were many and varied. A complete listing of his awards would fill many pages. Besides receiving the Nobel Prize in 1952, he was recognized by the French Legion of Honor, won the Lasker award for basic medical science, was elected a fellow of the American Association for the Advancement of Science, and received commendations from academies and scholarly societies in Brazil, Britain, Denmark, Italy, Japan, the Netherlands, Spain, and other countries. It can safely be said that Selman Waksman changed the face of modern medicine around the world.

WALD, GEORGE (1906-1997)
American biochemist

George Wald first won a place in the spotlight as the recipient of a Nobel Prize for his discovery of the way in which hidden biochemical processes in the retinal pigments of the eye turn **light** energy into sight. Among Wald's important experiments were the effects of vitamin A on sight and the roles played by rod and cone **cell**s in black and white and color vision. Outside the laboratory, his splendid lectures at Harvard to packed audiences of students generated great intellectual excitement. It was as a political activist during the turbulent 1960s, however, that Wald garnered further public recognition. Wald's personal belief in the unity of nature and the kinship among all living things was evidenced by the substantial roles he played in the scientific world as well as the political and cultural arena of the 1960s.

Wald's father, Isaac Wald, a tailor and later a foreman in a clothing factory, immigrated from Austrian Poland, while his mother, Ernestine Rosenmann Wald, immigrated from Bavaria. Most of Wald's youth was spent in Brooklyn, New York, where his parents moved after his birth on the Lower East Side of Manhattan on November 18, 1906. He attended high school at Brooklyn Tech, where he intended to study to become an electrical engineer. College changed his mind, however, as he explained for the *New York Times Magazine* in 1969, "I learned I could talk, and I thought I'd become a lawyer. But the law was man-made; I soon discovered I wanted something more real."

Wald's bachelor of science degree in zoology, which he received from New York University in 1927, was his ticket into the reality of biological research. He began his research career at Columbia University, where he was awarded a master's degree in 1928, working under Selig Hecht, one of the founders of the field of biophysics and an authority on the physiology of vision. Hecht exerted an enormous influence on Wald, both as an educator and a humanist. The elder scientist's belief in the social obligation of science, coupled with the conviction that science should be explained so the general public could understand it, made a great impression on the young Wald. Following Hecht's sudden death in 1947 at the age of 55, Wald wrote a memorial as a tribute to his colleague.

In 1932 Wald earned his doctorate at Columbia, after which he was awarded a National Research Council Fellowship in Biology. The two-year fellowship helped to support his research career, which first took him to the laboratory of Otto Warburg in Berlin. It was there, in 1932, that he discovered that vitamin A is one of the major constituents of retinal pigments, the light sensitive chemicals that set off the cascade of biological events that turns light into sight.

Warburg sent the young Wald to Switzerland, where he studied vitamins with chemist Paul Karrer at the University of Zurich. From there Wald went to Otto Meyerhof's laboratory of cell metabolism at the Kaiser Wilhelm Institute in Heidelberg, Germany, finishing his fellowship in the department of physiology at the University of Chicago in 1934. His fellowship completed, Wald went to Harvard University, first as a tutor in biochemistry and subsequently as an instructor, faculty instructor, and associate professor, finally becoming a full professor in 1948. In 1968, he became Higgins Professor of Biology, a post he retained until he became an emeritus professor in 1977.

Wald did most of his work in eye physiology at Harvard, where he discovered in the late 1930s that the light-sensitive chemical in the rods—those cells in the retina responsible for night vision—is a single pigment called rhodopsin (visual purple), a substance derived from opsin, a **protein**, and retinene, a chemically modified form of vitamin A. In the ensuing years, Wald discovered that the vitamin A in rhodopsin is "bent" relative to its natural state, and light causes it to "straighten out," dislodging it from opsin. This simple reaction initiates all the subsequent activity that eventually generates the sense of vision.

Wald's research moved from rods to cones, the retinal cells responsible for color vision, discovering with his co-worker Paul K. Brown, that the pigments sensitive to red and yellow-green are two different forms of vitamin A that co-exist in the same cone, while the blue-sensitive pigments are located in separate cones. They also showed that color blindness is caused by the absence of one of these pigments.

For much of his early professional life, Wald concentrated his energy on work, both research and teaching. His assistant, Brown, stayed with him for over 20 years and became a full-fledged collaborator. A former student, Ruth Hubbard, became his second wife in 1958, and they had two children, Elijah and Deborah. (His previous marriage to Frances Kingsley in 1931 ended in divorce; he has two sons by that marriage, Michael and David.) Wald, his wife, and Brown together became an extremely productive research team.

Research Efforts Receive Recognition with Nobel Prize

By the late 1950s Wald began to be showered with honors, and during his career he received numerous honorary degrees and awards. After Wald was awarded (with Haldan K. Hartline of the United States and Ragnar Granit of Sweden) the Nobel Prize in physiology or medicine in 1967 for his work with vision, John E. Dowling wrote in *Science* that Wald and his team formed "the nucleus of a laboratory that has been extraordinarily fruitful as the world's foremost center of visual-pigment biochemistry."

As Wald's reputation flourished, his fame as an inspiring professor grew as well. He lectured to packed classrooms, inspiring an intense curiosity in his students. The energetic professor was portrayed in a 1966 *Time* article that summarized the enthusiasm he brought to teaching his natural science course: "With crystal clarity and obvious joy at a neat explanation, Wald carries his students from protons in the fall to living organisms in the spring, [and] ends most lectures with some philosophical peroration on the wonder of it all." That same year, the *New York Post* said of his lectures, "His beginnings are slow, sometimes witty.... The talk gathers momentum and suddenly an idea *pings* into the atmosphere—fresh, crisp, thought-provoking."

Six days after he received the Nobel Prize, Wald wielded the status of his new prestige in support of a widely popular resolution before the city council of Cambridge, Massachusetts—placing a referendum on the Vietnam War on the city's ballot of November 7, 1967. Echoing the sentiments of his mentor Hecht, he asserted that scientists should be involved in public issues.

The Cambridge appearance introduced him to the sometimes stormy arena of public politics, a forum from which he has never retired. The escalating war in Vietnam aroused Wald to speak out against America's military policy. In 1965, during the escalation of that war, Wald's impromptu denunciation of the Vietnam war stunned an audience at New York University, where he was receiving an honorary degree. Shortly afterward, he threw his support and prestige behind the presidential campaign of Eugene McCarthy. His offer to speak publicly on behalf of McCarthy was ignored, however, and he became a disillusioned supporter, remaining on the fringe of political activism.

Political Activism Punctuated with "The Speech"

Then on March 4, 1969, he gave an address at the Massachusetts Institute of Technology (MIT) that, "upended his life and pitched him abruptly into the political world," according to the *New York Times Magazine*. Wald gave "The Speech," as the talk came to be known in his family, before an audience of radical students at MIT The students had helped to organize a scientists' day-long "strike" to protest the influence of the military on their work, a topic of much heated debate at the time.

Although much of the MIT audience was already bored and restless by the time Wald began, even many of those students who were about to leave the room stopped to listen as the Nobel laureate began to deliver his oration, entitled, "A Generation in Search of a Future." "I think this whole generation of students is beset with a profound sense of uneasiness, and I don't think they have quite defined its source," Wald asserted as quoted in the *New York Times Magazine*. "I think I understand the reasons for their uneasiness even better than they do. What is more, I *share* their uneasiness."

Wald's discourse evoked applause from the audience as he offered his opinion that student unease arose from a variety of troublesome matters. He pointed to the Vietnam War, the military establishment, and finally, the threat of nuclear warfare. "We must get rid of those atomic weapons," he declared. "We cannot live with them." Speaking to the students as fellow scientists, he sympathized with the their unease at the influence of the military establishment on the work of scientists, intoning, "Our business is with life, not death...."

The speech was reprinted and distributed around the country by the media. Through these reprints, Wald told readers that some of their elected leaders were "insane," and he referred to the American "war crimes" enacted in Vietnam. In the furor that followed, Wald was castigated by critics, many of whom were fellow academics, and celebrated by sympathizers. A letter writer from Piney Flats, Tennessee, was quoted in the *New York Times Magazine* as saying, "So good to know there are still some intellects around who can talk downright horsesense." Wald summed up his role as scientist-political activist in that same article by saying, "I'm a scientist, and my concerns are eternal. But even eternal things are acted out in the present." He described his role as gadfly as putting certain controversial positions into words in order to make it, "easier for others to inch toward it."

His role as a Vietnam war gadfly expanded into activism in other arenas of foreign affairs. He served for a time as president of international tribunals on El Salvador, the Philippines, Afghanistan, Zaire, and Guatemala. In 1984, he joined four other Nobel Prize laureates who went with the "peace ship" sent by the Norwegian government to Nicaragua during that country's turmoil.

In addition to his interests in science and politics, Wald's passion included collecting Rembrandt etchings and primitive art, especially pre-Columbian pottery. This complex mixture of science, art, and political philosophy was reflected in his musings about religion and nature in the *New York Times Magazine*: "There's nothing supernatural in my mind. Nature is my religion, and it's enough for me. I stack it up against any man's. For its awesomeness, and for the sense of the sanctity of man that it provides."

In addition to the Nobel Prize, Wald received numerous awards and honors, including the Albert Lasker Award of the American Public Health Association in 1953, the Proctor Award in 1955 from the Association for Research in Ophthalmology, the Rumford Premium of the American Academy of Arts and Sciences in 1959, the 1969 Max Berg Award, and the Joseph Priestley Award the following year. In addition, he was elected to the National Academy of Science in 1950 and the American Philosophical Society in 1958. He is also a member

of the Optical Society of America, which awarded him the Ives Medal in 1966. In the mid–1960s Wald spent a year as a Guggenheim fellow at England's Cambridge University, where he was elected an Overseas fellow of Churchill College for 1963–64. Wald also held honorary degrees from the University of Berne, Yale University, Wesleyan University, New York University, and McGill University.

Wald died on 12 April 1997, at his home in Cambridge, Massachusetts, at the age of 90.

WALKER, JOHN E. (1941-)
English biochemist

Walker was born on January 7, 1941, in Halifax, England. In 1960, he enrolled at St. Catherine's College, Oxford, and was awarded his B.A. in chemistry by St. Catherine's in 1964. Walker then spent four years as a research student at the Sir William Dunn School of Pathology at Oxford before earning his M.A. and D. Phil. degrees from Oxford in 1969.

Upon completion of his doctoral studies, Walker spent two years as a postdoctoral fellow at the School of Pharmacy at the University of Wisconsin. He then spent three additional years as a NATO Fellow at CNRS, in Gif-sur-Yvette, France, and as an EMBO Fellow at the Institut Pasteur in Paris. In 1974, Walker accepted an appointment as a member of the scientific staff at the Cambridge Laboratory of Molecular Biology of the Medical Research Council.

Adenosine triphosphate (ATP) is one of the most important molecules in the cells of living organisms. It has been described as an **"energy**-carrying" **molecule** because it provides the energy needed to drive many essential biochemical reactions.

The energy carried by an ATP molecule is stored in its phosphate bonds. A molecule of ATP is produced through a series of steps in which a **phosphate group** is first attached to a molecule of adenosine monophosphate (AMP) to form adenosine diphosphate (ADP). ADP then adds a second phosphate group to form adenosine triphosphate (ATP). As each phosphate group is added to the growing molecule, it brings with it energy stored in the form of the chemical bonds by which the phosphate is attached to the core molecule.

ATP acts as an energy-provider because it tends to break down to form first ADP plus a phosphate group, and then AMP and a second phosphate group. Each time one of these steps occurs, energy is released. That energy is transferred to some other set of chemical reactants in a **cell**, making it possible for those reactants to form new compounds.

Scientists have long been very interested in learning precisely how ATP is formed and how it carries out its biochemical functions. Discovered in 1929 by the German chemist Karl Lohmann, ATP was first synthesized two decades later by the Scottish chemist **Alexander Todd**. The role of ATP in providing energy to cell reactions was first elucidated by the German-American biochemist **Fritz Lipmann** in the period 1939-1941.

One important line of ATP research has focused on the mechanism by which the molecule is formed. In 1960, the

American biochemist Efraim Racker found a substance in the **mitochondria** of cells that appeared to be responsible for the synthesis of ATP. They called the enzyme F_0F_1 ATPase, although it is now better known as ATP synthase. The original name for the **enzyme** comes from the fact that it consists of two parts, an F_0 domain that is attached to a cell **membrane**, and an F_1 domain that protrudes from the membrane.

During the 1950s, the American biochemist Paul D. Boyer developed a theory to explain how ATP synthase is able to produce ATP. Essentially, he argued that a flow of hydrogen ions within the cell membrane causes the F_0 domain of the enzyme to rotate in much the same way that the wind causes the blades on a windmill to turn. Boyer hypothesized that the turning of the F_0 membrane sequentially exposed structurally different regions of the F_1 domain in such as way as to make possible the ADP + phosphate reaction to occur. Walker's research completely identified the **amino acid** sequence of ATP synthase and **protein**s attached to it and determined much of the three-dimensional structure of the molecule. It was for this research that he was awarded the 1997 Nobel Prize in chemistry. Walker's research dovetailed with similar work carried out by the second 1997 Nobel Laureate, **Paul Boyer**, who devised a theory that explained the process by which the ATP synthase enzyme operates.

WALLACE, ALFRED RUSSEL (1823-1913)

English naturalist

Alfred Russel Wallace was one of the greatest naturalists of all time, second in that regard perhaps only to **Charles Darwin**. Wallace's early training was as a surveyor and architect, but after 1845 he devoted his long and productive career to the study of natural history. He undertook several famous expeditions to the tropics, where he collected many **species** of **plants** and **animals** previously unknown to science. His most memorable expeditions were to the Amazon and Negro Rivers of South America (1848-1852), and to the East Indies (1854-1862), visiting peninsular Malaysia, Borneo, Java, Sumatra, various other islands of Indonesia, New Guinea, and Australia.

While in Southeast Asia, Wallace developed an outline of a theory of evolution by natural selection, which he shared with Charles Darwin in a letter in 1858. Wallace's idea astonished Darwin, who had been working on a similar theory for more than a decade, based on his own extensive observations during the famous voyage of the *HMS Beagle* (1831-1835), and on years of analyzing specimens afterwards. In fact, Darwin had been preparing a voluminous rationalization of his theory of **natural selection** for several decades. Initially, an enormously disappointed Darwin believed that Wallace's letter to him held priority over his own unpublished writings about natural selection. However, after consultation with **Charles Lyell** and **Joseph Hooker**, two famous biologists of the time, Darwin decided that his theory and Wallace's would be announced at the same time, at a meeting of the Linnaean Society in London in 1858. There, an unpublished manuscript that

Darwin had written thirteen years earlier, and Wallace's letter to Darwin, were read to the assembly. The result was a storm of scientific and social controversy, because the theory of evolution by natural selection was in direct opposition to the leading ideas of the time, which held that the Universe and everything it contained, including all life on Earth, had been created by God only a few thousand years previously. In 1859, Darwin published his famous book, *On the Origin of Species by Natural Selection*, which is now regarded as perhaps the most famous and influential work in biology. Today, the theory of **evolution** by natural selection is mostly credited to Charles Darwin, because of his painstaking documentation and analysis of supporting evidence. Wallace, however, should still be acknowledged as the co-discoverer of the theory.

Wallace is also famous for another theory that he developed about the biogeography of the East Indies. He observed that there is a distinct division of the flora and fauna among these many islands of this region. Some islands have a dominant influence of species of Australian affinity, while others are of Oriental (or Asian) origin. The "boundary" between these great biogeographical realms, whose natural intermixing had been prevented for millions of years by deep, submarine trenches, is today known as "Wallace's Line." It separates the **ecosystems** and species of Borneo, Sumatra, Java, Bali, the Philippines, and many smaller islands, from those of Sulawesi, Logbook, Timor, New Guinea, New Britain, and Australia.

Alfred Russel Wallace was a prolific writer. Among his ten books were: *Travels on the Amazon and Rio Negro* (1853), *The Malay Archipelago* (1869), *Contributions to the Theory of Natural Selection* (1870), *The Geographic Distribution of Animals* (1876), *Darwinism* (1889), *Man's Place in the Universe* (1903), and *The World of Life* (1910).

WARBURG, OTTO (1883-1970)

German biochemist

Otto Warburg is considered one of the world's foremost biochemists. His achievements include discovering the mechanism of **cell** oxidation and identifying the iron-**enzyme** complex, which catalyzes this process. He also made great strides in developing new experimental techniques, such as a method for studying the **respiration** of intact cells using a device he invented. His work was recognized with a Nobel Prize for medicine and physiology in 1931.

Otto Heinrich Warburg was born on October 8, 1883, in Freiburg, Germany, to Emil Gabriel Warburg and Elizabeth Gaertner. Warburg was one of four children and the only boy. His father was a physicist of note and held the prestigious Chair in Physics at University of Berlin. The Warburg household often hosted prominent guests from the German scientific community, such as physicists Albert Einstein, Max Planck, **Emil Fischer**—the leading organic chemist of the late-nineteenth century, and Walther Nernst —the period's leading physical chemist.

Warburg studied chemistry at the University of Freiburg beginning in 1901. After two years, he left for the University

Otto Warburg. *(The Library of Congress. Reproduced by permission.)*

just begun his work at the institute when World War I started. He volunteered for the army and joined the Prussian Horse Guards, a cavalry unit that fought on the Russian front. Warburg survived the war and returned to the Kaiser Wilhelm Institute for Biology in Berlin in 1918. Now 35 years old, he would devote the rest of his life to biological research, concentrating on studies of **energy** transfer in cells (cancerous or otherwise) and **photosynthesis**.

One of Warburg's significant contributions to biology was the development of a manometer for monitoring cell respiration. He adapted a device originally designed to measure gases dissolved in **blood** so it would make measurements of the rate of oxygen production in living cells. In related work, Warburg devised a technique for preparing thin slices of intact, living **tissue** and keeping the samples alive in a nutrient medium. As the tissue slices consumed oxygen for respiration, Warburg's manometer monitored the changes.

During Warburg's youth, he had become familiar with Einstein's work on photochemical reactions as well as the experimental work done by his own father, Emil Warburg, to verify parts of Einstein's theory. With this background, Warburg was especially interested in the method by which **plants** converted **light** energy to chemical energy. Warburg used his manometric techniques for the studies of photosynthesis he conducted on **algae**. His measurements showed that photosynthetic plants used light energy at a highly efficient sixty-five percent. Some of Warburg's other theories about photosynthesis were not upheld by later research, but he was nevertheless considered a pioneer for the many experimental methods he developed in this field. In the late 1920s, Warburg began to develop techniques that used light to measure reaction rates and detect the presence of chemical compounds in cells. His ''spectrophotometric'' techniques formed the basis for some of the first commercial spectrophotometers built in the 1940s.

Discovers Details of Cell Respiration

His work on cell respiration was another example of his interest in how living things generated and used energy. Prior to World War I, Warburg discovered that small amounts of cyanide can inhibit cell oxidation. Since cyanide forms stable complexes with heavy metals such as iron, he inferred from his experiment that one or more catalysts important to oxidation must contain a heavy metal. He conducted other experiments with carbon monoxide, showing that this compound inhibits respiration in a fashion similar to cyanide. Next he found that light of specific frequencies could counteract the inhibitory effects of carbon monoxide, at the same time demonstrating that the ''oxygen transferring enzyme,'' as Warburg called it, was different from other enzymes containing iron. He went on to discover the mechanism by which iron was involved in the cell's use of oxygen. It was Warburg's work in characterizing the cellular catalysts and their role in respiration that earned him a Nobel Prize in 1931.

Nobel Foundation records indicate that Warburg was considered for Nobel Prizes on two additional occasions: in 1927 for his work on **metabolism** of cancer cells, then in 1944 for his identification of the role of flavins and nicotinamide in

of Berlin to study under Emil Fischer, and in 1906 received a doctorate in chemistry. His interest turned to medicine, particularly to **cancer**, so he continued his studies at the University of Heidelberg where he earned an M.D. degree in 1911. He remained at Heidelberg, conducting research for several more years and also making several research trips to the Naples Zoological Station.

Warburg's career goal was to make great scientific discoveries, particularly in the field of cancer research, according to the biography written by **Hans Adolf Krebs**, one of Warburg's students and winner of the 1953 Nobel Prize in medicine and physiology. Although he did not take up problems specifically related to cancer until the 1920s, his early projects provided a foundation for future cancer studies. For example, his first major research project, published in 1908, examined oxygen consumption during growth. In a study using sea urchin **egg**s, Warburg showed that after **fertilization**, oxygen consumption in the specimens increased 600%. This finding helped clarify earlier work that had been inconclusive on associating growth with increased consumption of oxygen and energy. A number of years later, Warburg did some similar tests of oxygen consumption by cancer cells.

World War I Interrupts Research

Warburg was elected in 1913 to the Kaiser Wilhelm Gesellschaft, a prestigious scientific institute whose members had the freedom to pursue whatever studies they wished. He had

biological oxidation. Warburg did not receive the 1944 award, however, because a decree from Hitler forbade German citizens from accepting Nobel Prizes. Two of Warburg's students also won Nobel Prizes in medicine and physiology: Hans Krebs (1953) and **Axel Theorell** (1955).

In 1931 Warburg established the Kaiser Wilhelm Institute for Cell Physiology with funding from the Rockefeller Foundation in the United States. During the 1930s, Warburg spent much of his time studying dehydrogenases, enzymes that remove hydrogen from substrates. He also identified some of the cofactors, such as nicotinamide derived from vitamin B_3 (niacin), that play a role in a number of cell biochemical reactions.

Warburg conducted research at the Kaiser Wilhelm Institute for Cell Physiology until 1943 when the Second World War interrupted his investigations. Air attacks targeted at Berlin forced him to move his laboratory about 30 miles away to an estate in the countryside. For the next two years, he and his staff continued their work outside the city and out of the reach of the war. Then in 1945, Russian soldiers advancing to Berlin occupied the estate and confiscated Warburg's equipment. Although the Russian commander admitted that the soldiers acted in error, Warburg never recovered his equipment. Without a laboratory, he spent the next several years writing, publishing two books that provided an overview of much of his research. He also traveled to the United States during 1948 and 1949 to visit fellow scientists.

Survives Nazi Germany

Even though Warburg was of Jewish ancestry, he was able to remain in Germany and pursue his studies unhampered by the Nazis. One explanation is that Warburg's mother was not Jewish and high German officials "reviewed" Warburg's ancestry, declaring him only one-quarter Jewish. As such he was forbidden from holding a university post, but allowed to continue his research. There is speculation that the Nazis believed Warburg might find a cure for cancer and so did not disturb his laboratory. Scientists in other countries were unhappy that Warburg was willing to remain in Nazi Germany. His biographer Hans Krebs noted, however, that Warburg was not afraid to criticize the Nazis. At one point during the war when Warburg was planning to travel to Zurich for a scientific meeting, the Nazis told him to cancel the trip and to not say why. "With some measure of courage," wrote Krebs, "he sent a telegram [to a conference participant from England]: 'Instructed to cancel participation without giving reasons.'" Although the message was not made public officially, the text was leaked and spread through the scientific community. Krebs believed Warburg did not leave Germany because he did not want to have to rebuild the research team he had assembled. The scientist feared that starting over would destroy his research potential, Krebs speculated.

In 1950 Warburg moved into a remodeled building in Berlin which had been occupied by U.S. armed forces following World War II. This new site was given the name of Warburg's previous scientific home—the Kaiser Wilhelm Institute for Cell Physiology—and three years later renamed the Max Planck Institute for Cell Physiology. Warburg continued to conduct research and write there, publishing 178 scientific papers from 1950 until his death in 1970.

For all of his interest in cancer, Warburg's studies did not reveal any deep insights into the disease. When he wrote about the "primary" causes of cancer later in his life, Warburg's proposals failed to address the mechanisms by which cancer cells undergo unchecked growth. Instead, he focused on metabolism, suggesting that in cancer cells "**fermentation**" replaces normal oxygen respiration. Warburg's cancer studies led him to fear that exposure to food additives increased one's chances of contracting the disease. In 1966 he delivered a lecture in which he stated that cancer prevention and treatment should focus on the administration of respiratory enzymes and cofactors, such as iron and the B **vitamins**. The recommendation elicited much controversy in Germany and elsewhere in the Western world.

Warburg's devotion to science led him to forego marriage, since he thought it was incompatible with his work. According to Karlfried Gawehn, Warburg's colleague from 1950-1964, "For him [Warburg] there were no reasonable grounds, apart from death, for not working." Warburg's productivity and stature as a researcher earned him an exemption from the Institute's mandatory retirement rules, allowing him to continue working until very near to the end of his life. He died at the Berlin home he shared with Jakob Heiss on August 1, 1970.

WASTE DISPOSAL

The massive amount of waste generated every day is a hallmark of affluent, modern society. According to the Environmental Protection Agency (EPA), municipal solid waste—a combination of household and commercial refuse—amounts to about 180 million metric tons per year in the United States. That equals almost two-thirds of a ton of garbage for each individual every year. It represents nearly twice as much waste per capita as Europe or Japan, and five to ten times as much as most developing countries. The largest single category is paper and cardboard, which make up roughly 40% of the municipal waste stream. Another major category that constitutes about one quarter of United States garbage is organic materials including food wastes, plastics, and yard and garden wastes. Metal cans and glass bottles represent about 15% of total trash, and the remainder consists of miscellaneous refuse, including building materials, clothing, furniture, electronics, and paint.

Much of what's in the United States waste stream would be a valuable resource if it were not mixed with other garbage. Paper could be recycled and organic materials like food and yard waste could be composted. Plastics, glass, and metal containers could be melted and remanufactured into useful products. Building supplies and fabrics could be used over again for other purposes. One reason that other countries throw away so much less than the United States is that they can't afford to simply dump valuable commodities. They carefully sort, clean, and recycle many articles that are casually thrown away in the United States.

In most parts of the United States, the vast majority of all municipal waste is buried in sanitary landfills. Although

A household hazardous waste disposal day sponsored by the city of Livonia, Michigan. Residents are asked to bring toxic materials, such as paint, petroleum products, insecticides, and antifreeze, to a central location where they are combined and placed in barrels for disposal. *(Photograph by Robert J. Huffman, Field Mark Publications. Reproduced by permission.)*

these facilities are an advance over the older, open dumps in which garbage fires smoldered incessantly and rats and other vermin thrived, landfills can leak toxic contaminants into underlying groundwater aquifers. They also release methane, which is both explosive and a potent greenhouse gas. In newer landfills, rather than simply covering the garbage with a thin layer of soil, it is encased in impervious clay and plastic liners with drain pipes to catch any fluid effluent and vent pipes to draw off any methane or other volatile gases.

Faced with growing piles of trash and finding it more and more difficult to site new landfills, many communities are turning to waste incineration. Called **energy** recovery or waste-to-energy facilities, these incinerators burn garbage to produce steam that can be sold for space heating or used to generate electricity. Incinerators have met with considerable public resistance, however, because they often release **toxins** such as dioxins and heavy metals in their gaseous effluents and they concentrate other toxins in ash that must be handled and stored as a hazardous waste.

Some communities have had success with programs to encourage citizens to produce less waste or to separate and recycle specific components such as paper, glass, and metal. A combination of financial incentives and convenient alternatives can often reduce the waste stream by 50% or more. Some people who are especially conscientious get by without discarding anything at all. By avoiding excess packaging and re-

cycling, composting, or reusing everything they buy, they produce zero garbage.

See also Biodegradable; Greenhouse effect

WASTE, TOXIC

Toxins are poisonous materials that interfere with vital metabolic processes to sicken or kill living **organisms**. Toxins can be either general poisons that kill many types of **cells** and organisms, or they can be extremely specific in their target and mode of action. Some are extremely reactive and can be lethal even in very dilute concentrations. Ricin, for instance, is a **protein** found in castor beans, and is one of the most toxic **organic compounds** known. Three hundred picograms (trillionths of a gram) injected intravenously is enough to kill an average mouse. That means that a few teaspoonsful of this substance, if divided and delivered in individual doses, could potentially kill all the mice in the world. Put another way, an amount of supertoxins that is invisible to the naked eye, if delivered in the right way, could be lethal.

An important principle of toxicology (the study of poisons) is that "the dose makes the poison." This dictum, first pronounced by the German physician Paracelsus in the sixteenth century, means that almost everything is dangerous at some level. Even compounds like table salt and **water** that are essential parts of our diet in reasonable amounts could make you very sick or even kill you if ingested in excess. Contrarily, even the most toxic compounds generally have a threshold level below which they are effectively harmless. Toxicity depends on the amount, time, mode of delivery of the toxin, as well as age and physiological state of the target organism. Among the most dangerous toxins are **carcinogens** (cause **cancer**), mutagens (genetic damage), **teratogen**s (**birth** defects), and neurotoxins (nerve damage). Not all toxins are organic compounds. Many metals, such as lead, mercury, cadmium, and chromium, are highly poisonous as are elements such as **arsenic** and selenium, and **minerals** such as asbestos. Many people also assume that all human-made chemicals are poisonous, while all natural materials must be benign and wholesome. This is far from the truth. Many synthetic, industrial chemicals are relatively innocuous while perfectly natural materials, like some of those mentioned above, are extremely dangerous.

Toxic wastes, as their name implies, are unwanted materials known to be fatal to humans or laboratory **animal**s at low doses or that are carcinogenic, mutagenic, teratogenic, or neurotoxic to humans or other life forms. Radioactive materials are considered especially dangerous, and their use and disposal is tightly regulated. Modern societies produce, use, and discard a vast array of toxic chemical substances. According to the Environmental Protection Agency (EPA), the United States generates about 265 million metric tons of officially classified hazardous and toxic wastes each year. This amounts to about one ton per year for every individual in the United States. Fortunately, most of this material is stored, recycled, converted to non-hazardous forms, or otherwise disposed of safely. Shockingly, however, at least 40 million metric tons (22 billion pounds) of toxic and hazardous chemicals are released each year in the United States into the air, water, or land by unsound or illegal disposal methods. This represents an immediate health hazard to many people who live close to these disposal sites, and it may well represent a long-term health and ecological hazard to all of us. Scientists are discovering that persistent chemicals such as **pesticides**, dioxins, polychlorinated biphenyls (PCBs), and mercury can be carried over long distances and accumulate to levels that appear to be causing worrisome health effects in wildlife and human **population**s thousands of miles and dozens of years from their original source.

The preferred hierarchy of waste management is to reduce, reuse, recycle, detoxify, and—only as a last resort—store safely. Reducing waste amounts means not making it in the first place. Often we can find alternative products or industrial processes that avoid creating a particular waste. Reuse means using a material for some other purpose or process. What is one person's unwanted waste can be a valuable resource for someone else. Recycling and detoxification involve chemical, biological, or physical treatments to change toxins into harmless forms that could be used in beneficial ways. Storage of toxic wastes requires specialized facilities in which materials are isolated from the environment by secure metal containers, impermeable plastic liners, compacted clay cushions, and other coverings that prevent materials from ever escaping. Permanent, secure **waste disposal** sites are both very expensive to construct as well as difficult to site and maintain.

See also Bioaccumulation; Colburn, Theo

WATER

Water is a chemical compound composed of a single oxygen atom bonded to two hydrogen atoms (H_2O) which are separated by an angle of 105 degrees. Because of this asymmetrical arrangement, water **molecules** have a tendency to orient themselves in an electric field, with the positively charged hydrogen toward the negative pole and the negatively charged oxygen toward the positive pole. This tendency results in water having a large dielectric constant, which is responsible for making water an excellent solvent. Water is therefore referred to as the "universal solvent." Since mineral salts and organic materials can dissolve in water, it is the ideal medium for transporting life-sustaining **minerals** and **nutrients** into and through **animal** and plant bodies. Brackish and ocean waters may contain large quantities of sodium chloride as well as many other soluble compounds leached from the crust of the earth. For example, the concentration of mineral salts in ocean water is about 35,000 parts per million. Water is considered to be potable is it contains less than 500 parts per million of salts. Water can be reused indefinitely as a solvent because it undergoes almost no modification in the process.

Hydrogen bonding, which joins water molecule to water molecule, is responsible for other properties that make water a unique substance. These properties include its large heat capacity, which causes water to act as a moderator of **temperature** fluctuations; its high surface tension (due to cohesion among water molecules); and its adherence to other substances, such as the walls of a vessel (due to adhesion between water molecules and the molecules of a second substance). The high surface tension makes it possible for surface-gliding insects and broad, flat objects to be supported on the surface of water. Adhesion of water molecules to soil particles is the primary mechanism by which water moves through unsaturated soils.

Hydrogen bonding is also responsible for ice being less dense than water. If ice did not float, all bodies of water would freeze from the bottom up, becoming solid masses of ice and destroying all life in them. In addition, from season to season, frozen water bodies would remain frozen, resulting in large changes in climate and weather, such as decreased precipitation due to reduced evaporation. Ice floats because as the temperature of water is lowered to -24.8°F (4°C), the tendency of water to contract as its molecular motion decreases is overcome by the strength of hydrogen bonding between molecules. At 4°C, water molecules start to structure themselves directionally along the lines of the hydrogen bonds, at angles of 105 degrees. As the temperature drops toward -32°F (0°C), spaces develop between the lines until the open, crystalline form characteristic of ice develops. Its openness produces a density slightly less than that of liquid water, and ice floats on the surface, with approximately nine-tenths submerged.

Water is the only common substance that occurs naturally on earth in three different physical states. The solid state, ice, is characterized by a rigid crystalline structure occurring at or below -32°F (0°C) and occupying a definite volume (found as glaciers and ice caps, as snow, hail, and frost, and as clouds formed of ice crystals). The liquid state exists over a definite temperature range -32-148°F (0°-100°C), but is not rigid nor does it have a particular shape. In other words, it has a definite volume but assumes the shape of its container. Liquid water covers three-fourths of the earth's surface in the form of swamps, lakes, rivers, and oceans as well as found as rain clouds, dew, and ground water. The gaseous state forms at temperatures above 148°F (100°C), and neither occupies a definite volume nor is rigid. In other words, it takes on the exact shape and volume of its container. It occurs naturally as fog, steam, and clouds. One phase does not suddenly replace its predecessor as the temperature changes, but for a time at the melting or boiling point, two phases will coexist. As water changes from the gaseous form to the liquid form, it gives off heat at about 540 calories per gram, and as it changes from the liquid form to the solid form, it gives off about 80 calories per gram. The turbulence of thunderstorms is in large part due to the release of large amounts of energy into the atmosphere as water condenses into water droplets or into crystals of ice (i.e., hail). Pressure affects the transition temperature between phases. For example, at pressures below atmospheric, water boils at temperatures under 100°C, so food will take longer to cook at higher elevations.

Water can pass directly from the solid phase to the gaseous phase without going through the liquid phase. This process occurs at low temperatures and greatly reduced pressures through a process called sublimation. Dehydrated foods are produced by sublimation, in which foods are quick-frozen and then placed in evacuation chambers. Dehydration by sublimation requires less energy than other methods, reduces physical deterioration that accompanies prolonged or excessive heating, and decreases the loss of volatile aromatic compounds responsible for flavor.

Liquid water is critical to sustain life. Without water to drink, a human will die in less than a week. Water in the form of perspiration is important in maintaining thermal stability in the body by dispersing large quantities of heat from the surface of the skin into the atmosphere. As the principal constituent of **blood**, water is the medium by which red blood **cell**s transport oxygen (O_2) through the body, and by which carbon dioxide (CO_2) and other wastes are removed from the body. As a solvent, water in blood carries sugars and **protein**s, mineral salts, metabolites such as urea, **hormone**s, and compounds that cause blood to clot.

The principal use of water in agriculture is for irrigation of crops, while lesser amounts are used for watering of livestock and cleaning of produce. Most industrial and manufacturing processes use large quantities of water to provide energy, remove unwanted heat, serve as a solvent, wash away impurities, function as a transport mechanism, and serve as a raw material. Water is used to keep individuals, homes, and communities clean, thus improving public health. Water is used to flush, wash, and dilute wastes from both households and industries. Because water is an excellent solvent and a convenient repository for wastes, it may contain many impurities that, if present in sufficient concentrations, constitute pollution and may require reclamation and treatment.

Water is a major geologic agent of change for modifying the earth's surface through erosion by water and ice. Water is also an important recreational medium, supporting fishing, swimming, and boating, and is a major factor in the tourism industry.

WATER CYCLE • See Hydrologic cycle

WATER POLLUTION

Any physical, biological, or chemical change in **water** quality that adversely affects living **organism**s or makes water unsuitable for desired uses can be considered **pollution**. Often, however, a change that adversely affects one organism may be advantageous to another. **Nutrients** that stimulate growth of **bacteria** and other oxygen-consuming decomposers in a river or lake, for example, are good for the bacteria but can be lethal to game fish **population**s. Similarly, warming of waters by industrial discharges may be deadly for some **species** but may create optimal conditions for others. Whether the quality of the water has suffered depends on your perspective. There are natural sources of water contamination, such as **arsenic** springs, oil seeps, and **sedimentation** from **desert** erosion, but most environmental scientists restrict their focus on water pollution to factors caused by human actions and that detract from conditions and uses that humans consider desirable.

Water pollution control regulations usually distinguish between point and nonpoint pollution sources. Factories, power plants, sewage treatment facilities, underground mines and oil wells, for example, are classified as point sources because they release pollution from specific locations, such as drain pipes, ditches, or sewer outfalls. These individual, easily identifiable sources are relatively easy to monitor and regulate. Their unwanted contents can be diverted and treated before discharge. In contrast, nonpoint pollution sources are scattered or diffuse, having no specific location where they originate or discharge into water bodies. Some nonpoint sources include runoff from farm fields, feedlots, lawns, gardens, golf courses, construction sites, logging areas, roads, streets, and parking lots. Whereas point sources often are fairly uniform and predictable, nonpoint runoff often is highly irregular. The first heavy rainfall after a dry period, for example, may flush high concentrations of oil, gasoline, rubber, and trash off city streets, while subsequent runoff may have much lower levels of these contaminants. The irregular timing of these events, as well as their multiple sources, scattered location, and lack of specific ownership make them much more difficult to monitor, regulate, and treat than point sources.

Among the most important categories of water **pollutants** are sediment, infectious agents, **toxins**, oxygen demanding wastes, plant nutrients, and thermal changes. Sediment (dirt, soil, insoluble solids) and trash make up the largest volume and most visible type of water pollution in most rivers and lakes. Rivers have always carried silt, sand, and gravel down to the oceans but human-caused erosion now probably rivals the effects of geologic forces. Worldwide, erosion from croplands, **forests**, grazing lands, and construction sites is estimated to add some 75 billion tons of sediment each year to rivers and lakes. This sediment smothers gravel beds in which fish lay their **egg**s. It fills lakes and reservoirs, obstructs shipping channels, clogs hydroelectric turbines, and makes drinking water purification more costly. The most serious water pollutant in terms of human health worldwide is pathogenic (disease-causing) organisms. Among the most deadly waterborne diseases are **cholera**, dysentery, polio, infectious hepatitis, and schistosomiasis. Together, these diseases probably cause at least two billion new cases of disease each year and kill somewhere between six and eight million people. The largest source of infectious agents in water is untreated or insufficiently treated human and **animal** waste. The United Nations estimates that about half the world's population has inadequate sanitation and that at least one billion people lack access to clean drinking water.

Toxins are poisonous chemicals that interfere with basic **cellular metabolism** (the **enzyme** reactions that make life possible). Among some important toxins found in water are metals (lead, mercury, cadmium, nickel), inorganic elements (selenium, arsenic), acids, salts, and organic chemicals such as **pesticides**, solvents, and industrial wastes. Some of these materials are so toxic that exposure to extremely low levels (perhaps even parts per billion) can be dangerous. Others, while not usually found in toxic concentrations in most water bodies, can be taken up by living organisms, altered into more toxic forms, stored, and concentrated to dangerous levels through food chains. For example, fish in lakes and rivers in many parts of the United States have accumulated mercury (released mainly by power plants, **waste disposal**, and industrial processes) to levels that are considered a threat to human health for those who eat fish on a regular basis.

While we have not yet met our national goal in the United States of making all surface waters "fishable and swimmable," investments in sewage treatment, regulation of **toxic waste** disposal and factory effluents, and other forms of pollution control have resulted in significant water quality increases many areas. Nearly 90% of all the river miles and lake acres that are assessed for water quality in the United States fully or partly support their designed uses. Lake Erie, for instance, which was widely described in the 1970s as being "dead," now has much cleaner water and more healthy fish populations than would ever have been thought possible 25 years ago. Unfortunately, surface waters in developing countries have not experienced similar progress in pollution control. In most developing countries, only a tiny fraction of human wastes are treated before being dumped into rivers, lakes, or the ocean. In consequence, water pollution levels often are appalling. In India, for example, two-thirds of all surface waters are considered dangerous to human health.

See also Waste, toxic

Garbage washed up on the shore of the Hudson River. *(Phototake NYC. Reproduced by permission.)*

WATSON, JAMES D. (1928-)
American molecular biologist

James D. Watson won the 1962 Nobel Prize in physiology and medicine along with **Francis Crick** and **Maurice Wilkins** for discovering the structure of **DNA**, or **deoxyribonucleic acid**, which is the carrier of genetic information at the molecular level. Watson and Crick had worked as a team since meeting in the early 1950s, and their research ranks as a fundamental advance in molecular biology. More than thirty years later, Watson became the director of the **Human Genome Project**, an enterprise devoted to a difficult goal: the description of every human **gene**, the total of which may number up to one hundred thousand. This is a project that would not be possible without Watson's groundbreaking work on DNA.

James Dewey Watson was born in Chicago, Illinois, on April 6, 1928, to James Dewey and Jean (Mitchell) Watson. He was educated in the Chicago public schools, and during his adolescence became one of the original Quiz Kids on the radio show of the same name. Shortly after this experience in 1943, Watson entered the University of Chicago at the age of fifteen.

Watson graduated in 1946, but stayed on at Chicago for a bachelor's degree in **zoology**, which he attained in 1947.

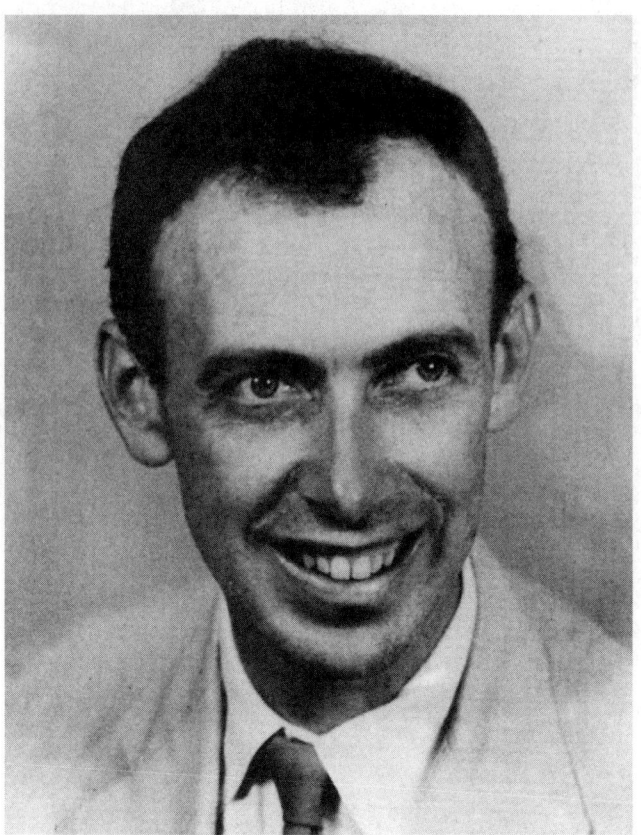

James D. Watson *(The Library of Congress. Reproduced by permission.)*

During his undergraduate years Watson studied neither genetics nor biochemistry—his primary interest was in the field of **ornithology**; in 1946 he spent a summer working on advanced ornithology at the University of Michigan's summer research station at Douglas Lake. During his undergraduate career at Chicago, Watson had been instructed by the well-known population geneticist Sewall Wright, but he did not become interested in the field of **genetics** until he read **Erwin Schrödinger**'s influential book *What is Life?;* it was then, Horace Judson reports in *The Eighth Day of Creation: Makers of the Revolution in Biology,* that Watson became interested in "finding out the secret of the gene."

Work with the "Phage Group"

Watson enrolled at Indiana University to perform graduate work in 1947. Indiana had several remarkable geneticists who could have been important to Watson's intellectual development, but he was drawn to the university by the presence of the Nobel laureate **Hermann Joseph Muller**, who had demonstrated twenty years earlier that X rays cause **mutation**. Nonetheless, Watson chose to work under the direction of the Italian biologist **Salvador Edward Luria**, and it was under Luria that he began his doctoral research in 1948.

Watson's thesis was on the effect of x rays on the rate of phagelysis (a phage, or **bacteriophage**, is a bacterial **virus**).

The biologist **Max Delbrück** and Luria—as well as a number of others who formed what was to be known as "the phage group"—had demonstrated that phages could exist in a number of mutant forms. A year earlier Luria and Delbrück had published one of the landmark papers in phage genetics, in which they established that one of the characteristics of phages is that they can exist in different genetic states so that the lysis (or bursting) of bacterial host cells can take place at different rates. Watson's Ph.D. degree was received in 1950, shortly after his twenty-second birthday.

Watson was next awarded a National Research Council fellowship grant to investigate the molecular structure of proteins in Copenhagen, Denmark. While Watson was studying enzyme structure in Europe, where techniques crucial to the study of macromolecules were being developed, he was also attending conferences and meeting colleagues.

From 1951-1953, Watson held a research fellowship under the support of the National Foundation for Infantile Paralysis at the Cavendish Laboratory in Cambridge, England. Those two years are described in detail in Watson's 1965 book, *The Double Helix: A Personal Account of the Discovery of the Structure of DNA.* (An autobiographical work, *The Double Helix* describes the events—both personal and professional—that led to the discovery of DNA.) Watson was to work at Cavendish under the direction of Max Perutz, who was engaged in the x-ray crystallography of **protein**s. However, he soon found himself engaged in discussions with Crick on the structure of DNA. Crick was twelve years older than Watson and, at the time, a graduate student studying protein structure.

Intermittently over the next two years, Watson and Crick theorized about DNA and worked on their model of DNA structure, eventually arriving at the correct structure by recognizing the importance of x-ray diffraction photographs produced by **Rosalind Franklin** at King's College, London. Both were certain that the answer lay in model-building, and Watson was particularly impressed by Nobel laureate **Linus Pauling**'s use of model-building in determining the alpha-helix structure of protein. Using data published by Austrian-born American biochemist **Edwin Chargaff** on the symmetry between the four constituent nucleotides (or "bases") of DNA molecules, they concluded that the building blocks had to be arranged in pairs. After a great deal of experimentation with their models, they found that the **double helix** structure corresponded to the empirical data produced by Wilkins, Franklin, and their colleagues. Watson and Crick published their theoretical paper in the journal *Nature* in 1953 (with Watson's name appearing first due to a coin toss), and their conclusions were supported by the experimental evidence simultaneously published by Wilkins, Franklin, and Raymond Goss. Wilkins shared the Nobel Prize with Watson and Crick in 1962.

Career Since the Discovery of DNA

After the completion of his research fellowship at Cambridge, Watson spent the summer of 1953 at Cold Spring Harbor, New York, where Delbrück had gathered an active group of investigators working in the new area of molecular biology. Watson then became a research fellow in biology at the California Institute of Technology, working with Delbrück and his colleagues on problems in phage genetics. In 1955, he joined the biology department at Harvard and remained on the faculty until 1976. While at Harvard, Watson wrote *The Molecular Biology of the Gene* (1965), the first widely used university textbook on **molecular biology**. This text has gone through seven editions, and now exists in two large volumes as a comprehensive treatise of the field. In 1968, Watson became director of Cold Spring Harbor, carrying out his duties there while maintaining his position at Harvard. He gave up his faculty appointment at the university in 1976, however, and assumed full-time leadership of Cold Spring Harbor. With John Tooze and David Kurtz, Watson wrote *The Molecular Biology of the Cell,* originally published in 1983 and now in its third edition.

In 1989, Watson was appointed the director of the Human Genome Project of the National Institutes of Health, but after less than two years he resigned in protest over policy differences in the operation of this massive project. He continues to speak out on various issues concerning scientific research and is a strong presence concerning federal policies in supporting research. In addition to sharing the Nobel Prize, Watson has received numerous honorary degrees from institutions, including one from the University of Chicago, which was awarded in 1961, when Watson was still in his early thirties. He was also awarded the Presidential Medal of Freedom in 1977 by President Jimmy Carter. In 1968, Watson married Elizabeth Lewis. They have two children, Rufus Robert and Duncan James.

Watson, as his book *The Double Helix* confirms, has never avoided controversy. His candor about his colleagues and his combativeness in public forums have been noted by critics. On the other hand, his scientific brilliance is attested to by Crick, Delbrück, Luria, and others. The importance of his role in the DNA discovery has been well supported by Gunther Stent—a member of the Delbrück phage group—in an essay that discounts many of Watson's critics through well-reasoned arguments.

Most of Watson's professional life has been spent as a professor, research administrator, and public policy spokesman for research. More than any other location in Watson's professional life, Cold Spring Harbor (where he is still director) has been the most congenial in developing his abilities as a scientific catalyst for others. His work there has primarily been to facilitate and encourage the research of other scientists.

WAXES

Waxes are pliable substances related to **fats**, but less greasy, and generally with a more firm consistency. They may be of plant, **animal**, mineral, or synthetic origin and generally contain high molecular weight fatty acids and **alcohol**s, or saturated **hydrocarbon**s. They tend to melt at moderate **temperature**s between 95°-212° F (35°-100°C). Waxes that form a hard film and can be polished to a high gloss are often used as polishes. Waxes of plant and animal origin are formed from fatty acids and alcohols or sterols (e.g., **cholesterol**). The fatty acids are

almost always saturated with no double bonded carbon to carbon links. They may vary in length from C_{12} to C_{38}. The alcohols are generally straight chain, and saturated with lengths from C_{12} to C_{36}.

Secreted by glands on the abdomen of the bee, beeswax is the most widely used animal wax. It is useful for its lubricating and waterproofing qualities, but not as a polish because of its soft consistency. Wool wax is obtained from raw wool during the scouring process. When pure, it is called lanolin, and is useful in pharmaceuticals and cosmetics because it is easily absorbed by human skin. Two waxes obtained from whales (**sperm** oil and spermaceti) are liquids at room temperature and useful as lubricants.

Several plant waxes have commercial importance. Carnauba wax, obtained from the surface of fronds of the fan palm, *Copernicia prunifera*, is used as a high gloss polish for automobiles and floors. The succulent stems of *Euphorbia antisyphilitica* yield candelilla wax, with properties much like carnauba. The plant grows in the Chihuahuan Desert along the United States-Mexico border, and has been so heavily harvested in some areas that it is in danger of **extinction**. These waxes are examples of plant cuticle waxes commonly found on the surface of stems and leaves. Although they are critically important to **plants** because they reduce water loss, few others have any commercial usefulness. The cuticle wax found on the outer surface of apples is interesting, however, because it can be polished to a high gloss making the fruit more enticing.

Petroleum is easily the most important source of commercial waxes. About 90% of commercial waxes are recovered by dewaxing lubrication-oil stocks. Paraffin, one form of petroleum wax, is widely used in candles, crayons, and industrial polishes. It is also used as an electrical insulator, and for waterproofing various materials. Microcrystalline wax, also derived from petroleum, is used to coat packaging paper. Petrolatum, yet another petroleum wax, is utilized in medicinal ointments and cosmetics. Synthetic waxes, blending petroleum waxes with ethylene glycol (the principle ingredient in automobile antifreeze), are used in various manufacturing processes.

WEISSMANN, AUGUST FRIEDRICH LEOPOLD (1834-1914)

German zoologist

August Weissmann was a German biologist who is best known for his contributions to **evolutionary theory**, specifically for his **germ**-plasm theory of **heredity**. Weissmann was born in Frankfurt, Germany in 1834. At the age of 18 he began his study of medicine at Göttingen University which he attended until 1856. After completing his schooling he practiced medicine until 1863, at which time he turned his attention to the study of biology and **zoology**. In1867 he became a professor of zoology at the University of Freiburg. He first served at the medical school associated with the university and later at a new Institute of Zoology. He remained in this capacity until 1912.

During this time Weissmann explored a variety of areas, most notably the **embryology** of insects and crustaceans. He is especially noted for his evaluations of the two-winged flies, the Diptera. He also conducted studies of Hydrozoa (a class of invertebrates which include jellyfish) that led to the development of his famous germ-plasm theory. In this theory Weissmann proposed that the **genetic code** for an **organism** is contained in the germ **cells** (the **egg** and **sperm**) which explains how the information can be transmitted unchanged from generation to generation. In accordance with this theory, Weissmann favored the **natural selection** ideas championed by **Charles Darwin**, who believed that specific inherited characteristics are passed from generation to generation. An opposing theory, supported by **Jean-Baptiste Lamarck** and others of the time, postulated that organisms acquired physical characteristics through exposure to their environment, and that they passed on these characteristics to their offspring. Weissmann was the first scientists to conduct experiments to disprove that acquired characteristics can be transmitted from parents to offspring. He did this using an experiment where he cut off the tails of mice for five consecutive generations. He then observed the offspring to determine if this "tailless"characteristic was passed on to subsequent generations. These studies confirmed that such environmental stimuli are not passed on in a hereditary fashion. In other studies Weissmann noted that the process of **meiosis** involved some form of reduction division because the genetic material did not double when the cell replicated.

Weissmann published his theories in a series of essays that were translated as *Essays upon Heredity and Kindred Biological Problems* which were published between 1889 and 1892. In 1902 he published *Vorträgeuber Descendenztheorie* which is still considered to be a valuable contribution to evolutionary theory. He was the first to provide a detailed explanation of the neuro-humoral organ that was later named after him: the Weissmann ring. He died at age 80 in 1914.

WELLER, THOMAS (1915-)

American pediatrician, parasitologist, and virologist

Thomas Weller was corecipient, with John F. Enders and Frederick Robbins, of the Nobel Prize in physiology or medicine in 1954. This award was given for the trio's successful growth of the poliomyelitis (polio) **virus** in a non-neural **tissue** culture. This development was significant in the fight against the crippling disease polio, and eventually led to the development, by **Jonas Salk** in 1953, of a successful vaccination against the virus. It also revolutionized viral work in the laboratory and aided the recognition of many new types of viruses. Weller also distinguished himself with his studies of human **parasite**s and the viruses that cause rubella and chicken pox.

Thomas Huckle Weller was born June 15, 1915, in Ann Arbor, Michigan. His parents were Elsie A. (Huckle) and Dr. Carl V. Weller. He received his B.S. in 1936 and M.S. in 1937, both from the University of Michigan, where his father was chair of the pathology department. He continued his studies at Harvard Medical School, where he met and roomed with his future Nobel corecipient Robbins. In 1938 Weller received a fellowship from the international health division of the Rockefeller Foundation, which allowed him to study public health in Tennessee and malaria in Florida, topics which first interested him during his undergraduate years.

Weller graduated from Harvard with magna cum laude honors in parasitology, receiving his M.D. in 1940. He also received a fellowship in tropical medicine and a teaching fellowship in bacteriology. He completed an internship in pathology and bacteriology (1941) at Children's Hospital in Boston. He then began a residency at Children's, with the intention of specializing in pediatrics, before enlisting in the U.S. Army during World War II.

Weller served in the Army Medical Corps from 1942-1945. He was initially given teaching assignments in tropical medicine, but he was soon made officer in charge of bacteriology and virology work in San Juan, Puerto Rico. His major research there related to pneumonia and the parasitic disease schistosomiasis, an infection that is centered in the intestine and damages tissue and the **circulatory system**. Before his military service ended, he moved to the Army Medical School in Washington D.C. Upon his discharge in 1945, Weller was married to Kathleen Fahey, with whom he had two sons and two daughters. Returning to Boston's Children's Hospital, he finished his residency and began a post-doctoral year working with Enders.

Helps Solve the Polio Puzzle

During 1948, Weller was working with the mumps virus, which Enders had been researching since the war. After one experiment, Weller had a few tubes of human embryonic tissue left over, so he and Enders decided to see what the virus poliomyelitis might do in them. A small amount of success prompted the duo, who had been joined in their research by Robbins, to try growing the virus in other biological mediums, including human foreskin and the intestinal cells of a mouse. The mouse intestine did not produce anything, but the trio finally had significant viral growth with human intestinal cells. This was the first time poliomyelitis had been grown in human or simian tissue other than nerve or **brain**. Using **antibiotics** to ward off unwanted bacterial invasion, the scientists were able to isolate the virus for study.

Once poliomyelitis was grown and isolated in tissue cultures it was possible to closely study the nature of the virus, which in turn made it possible for Salk to create a vaccine in 1953. Besides leading to an inhibitor against a debilitating disease, a major result of the trio's development was a decrease in the need for laboratory **animal**s. As Weller was quoted saying in the *Journal of Infectious Diseases,* "In the instance of poliomyelitis, one culture tube of human or monkey cells became the equivalent of one monkey." In times prior, viruses had to be injected into living animals to monitor their potency. Now, with tissue culture growth, cell changes were apparent under the **microscope**, showing the action of the virus and eliminating the need for the animals. The techniques for growing cells in tissue cultures developed by Weller and his associates were not only applicable to the poliomyelitis virus, however. They were soon copied by many other labs and scientists and quickly led to the identification, control, and study of several previously unrecognized virus types. For their work, and the improvements in scientific research it made possible, Weller, Enders, and Robbins shared the 1954 Nobel Prize in physiology or medicine.

Concurrent with his work with Enders and Robbins, Weller was named assistant director of the research division of infectious diseases at Children's Hospital in 1949. He held this position until 1954. At the same time, he began teaching at Harvard in tropical medicine and tropical public health, moving from instructor to associate professor. In 1953, Weller and Robbins shared the Mead Johnson Prize for their contributions to pediatric research. Then, in 1954, Weller was named Richard Pearson Strong Professor of Tropical Public Health and chair of the public health department at Harvard. As a consequence, he moved his research facilities to the Harvard Medical School. Later, he was appointed director of the Center for Prevention of Infectious Diseases at the Harvard School of Public Health.

Advances Knowledge of Parasites and Viruses

From the end of World War II until 1982 Weller also continued his research on two types of helminths, *Trichinella spiralis* and *Schistosoma mansoni*. Helminths are intestinal parasites, and these two cause, respectively, trichinosis, which can also severely affect the human musculature, and schistosomiasis. Weller was concerned with the parasites' basic biology and performed various diagnostic studies on them. His contributions to current understanding of these parasites are significant, advancing an understanding of the ailments they cause.

Weller spent a portion of the same period (1957-1973) establishing the basic available knowledge concerning cytomegalovirus (commonly known as CMV), which causes cell enlargement in various organs. Weller's most important finding in this area regarded congenital transmission of both CMV and rubella, a virus also known as German measles. A pregnant woman infected with either of these viruses may pass the infection on to her fetus. Weller showed that infected newborns excreted viral strains in their feces, providing another source for the spread of the diseases. His findings became significant when it was also learned that children born to infected mothers often risked birth defects.

In 1962 Weller, along with Franklin Neva, was able to grow and study German measles in tissue cultures. These two also went on to grow and isolate the chicken pox virus. Subsequently, Weller was the first to show the common origin of the varicella virus, which causes chicken pox, and the herpes zoster virus, which causes shingles. In 1971, Weller was the first to prove the airborne transmission of *Pneumocystis carinii*, a form of pneumonia that later appeared as a frequent side effect of the **human immunodeficiency virus** commonly known as HIV.

Weller was elected to the National Academy of Sciences in 1964. In addition, he served on advisory committees of the World Health Organization, the Pan American Health Organization, the Agency for Internation Development, and the Na-

tional Institute of Allergy and Infectious Disease. He continued his position at Harvard until 1985, when he became professor emeritus. While at Harvard, he helped establish the Public Health Department's international reputation. In 1988, Weller gave the first John F. Enders Memorial Lecture to the Infectious Disease Society of America. In addition to his Nobel Prize, Weller was the recipient of many awards and honorary degrees during his career.

WENT, FRITS (1903-1990)
Dutch American botanist

Lush, weed-free lawns owe their beauty in large part to American botanist Frits Went's discovery of the role of the plant growth **hormone** auxin. Went's 1927 discovery paved the way for the development of modern fertilizers and weed killers and the **genetic engineering** of **plants**. Went also developed a greenhouse that has enhanced botanical research by enabling scientists to control the plants' climate. Later in his career, the versatile scientist turned his attention toward environmental problems, including smog and the degradation of the Amazon rain forest.

A career in **botany** seemed virtually destined for Went, who was born in a garden and raised in a botany lab. He was born May 18, 1903, in Utrecht, Netherlands, the son of Catharina Jacomina (Tonckens) and Friedrich August Ferdinand Christian Went. The Wents lived in a 300-year-old house in the botany garden of the State University of Utrecht, where the senior Went worked as a professor and director of the garden and botany lab. Just across from the house was the newly rebuilt laboratory, which was considered one of the finest in the world.

The young Went was fascinated by the Venus fly traps, cacti, palms, and other exotic plants in the garden's greenhouse. Many hours of his boyhood years were spent there and in the laboratory. Went considered himself lucky to have been surrounded by such a variety of plants and some of the best minds in the field of botany. He credited his career choice to his boyhood at the university, although his father carefully avoided pushing him into science. Went earned his bachelor's, master's and doctoral degrees at the State University of Utrecht between 1920 and 1927.

Discovers Role of Plant Hormone

For the subject of his doctoral thesis, Went chose the plant growth hormone auxin. He was intrigued by **phototropism**, the tendency of plants to bend toward **light**. He knew they bent by growing faster on the dark side of the stem and slower on the side facing the light. He suspected the growth hormone auxin was responsible, but he was not sure how. Went conducted his research on oat seedlings. His most important finding was that auxin, which is produced at the tip of the stem, is unevenly distributed under unidirectional light. More auxin flows down the dark side, making it grow faster. His theory also explained why the phototropic curve in plants moves farther down the stem over time. Went made his discovery in

1927, at the same time as Russian botanist N. Cholodny. Since then, their theory of phototropism has been called the Cholodny-Went theory.

Knowledge of growth hormones in plants later gave rise to the field of agricultural chemistry. Many herbicides, fungicides, and fertilizers use auxins. These hormones are also used in genetic engineering to develop better plant **species**. Back in the 1920s and 1930s, however, Went's theory was rejected by many of his peers. The controversy disturbed the gentle, amiable scientist. Yet dispute seemed to follow him throughout his career. After earning his Ph.D. in 1927, Went accepted his first job as a botanist at the botanical garden in Bogor, Java. That same year he married Catharina Helena van de Koppel. They would have two children, Hans Went and Anneke (Went) Simmons.

Two years after arriving in Bogor, Went was appointed director of the Foreigners Laboratory there. He left in 1933 to become an assistant professor of plant physiology at California Institute of Technology in Pasadena. He was named a full professor in 1935, a position he held until he left the institute in 1958. While there, Went continued his research on auxins, which culminated in the 1937 publication of the book *Phytohormones,* which he wrote with K. V. Thimann.

The book's publication ended Went's research into the internal control of plant growth. By that time he had grown discouraged by the naysayers. "If a field becomes too controversial or too theoretical, I prefer to leave it, as I did the growth factor field in the early 1940s," he wrote in an introductory chapter for the *Annual Review of Plant Physiology.* "After Thimann and I had written *Phytohormones,* I felt that I degenerated to a policeman, overseeing the auxin field, checking doubtful statements or questionable results." Went shifted his studies from the internal factors affecting plant development to the external ones.

The laboratory for his new field was the first air-conditioned research greenhouse, which California Institute of Technology opened in 1939. By varying the **temperature**, Went learned that plants grew best when the daytime temperature was several degrees higher than at night. Since his discovery, commercial greenhouses have routinely varied the day and night temperatures for optimal growth. Went also learned that greenhouse plants cultivated under temperatures similar to those occurring in the wild would grow just like those in nature. The best way to research the effects of climate on plant growth, he reasoned, was to duplicate the natural climate. So he persuaded his friend H. Earhart from Michigan to finance a phytotron—a greenhouse that could duplicate the full range of naturally occurring temperature, lighting, wind, and humidity conditions—at the Institute. In June of 1949, the Institute opened the Earhart Plant Research Laboratory as the first phytotron. Soon, phytotrons became a fixture of the best botany departments at universities throughout the world.

In 1947, Went was elected to the prestigious National Academy of Sciences. He would be elected to the French Academy of Sciences in 1956, the Dutch Academy of Sciences in 1958, and the German Academy for Natural Sciences in 1977.

Theorizes Smog Comes From Plants

In the 1950s, Went again shifted his focus, this time toward the effects of plants on the environment. He began by analyzing the smog that hovers over Los Angeles. Until that time, it had been assumed that the smoky haze was sulfur dioxide. But Went rejected that assumption based on the reaction of plants to the smog. He organized a joint venture of the California Institute of Technology, the University of California, and the Los Angeles County Air Pollution Control District to identify the components of the haze. Went theorized that most of the smog in the **atmosphere** comes not from cars and factories, but from plants. During the process of **photosynthesis**, he claimed, the hydrocarbons in plants—known as terpene—decompose to produce a blue haze. This natural haze inspired the names of the Blue Ridge and the Smoky Mountains in Virginia and North Carolina. Went lacked the scientific data to prove his theory and had no desire to obtain it. He had grown away from the detailed analysis used in his early research on auxins. Yet his reluctance to research his hypothesis once again subjected his theory to widespread rejection.

"He was so far in front that mainstream scientists could shoot his theories full of holes," said Thomas Sharkey, Went's former colleague and chair of the department of botany at the University of Wisconsin in Madison. "The idea that most of **hydrocarbon** comes from plants turns out to be correct. Now the atmospheric scientists agree with him." Not all of Went's far-flung theories turned out to be correct. He theorized that the **greenhouse effect** leading to global warming was caused by the smoky vapors coming from vegetation, rather than **carbon dioxide**. In that case, Went was proven wrong. Carbon dioxide from **fossil fuels** contributes greatly to the greenhouse effect.

While spending his days with textbooks and microscopes, Went never lost his appreciation of the beauty of gardens. He was elected president of the California Arboretum Foundation and was the sponsor of the Los Angeles State and County Arboretum. In his *Annual Review* article, Went wrote that those positions "gave me the advantage of coming close to the living plant, to acquaint myself not only with its appearance and occurrence, but also with its workings. And it has prevented me from becoming a narrow specialist, spending my life on the response of a single plant or organ."

Went's second book, *The Experimental Control of Plant Growth,* was published in 1957. The following year, he left California to become a professor at Washington University in St. Louis and the director of the Missouri Botanical Gardens. Also in 1958, Went received the Stephen Hales Award from the American Society of Plant Physiologists for his outstanding career contributions. In 1959, he was awarded an honorary Ph.D. from McGill University.

Went used his knowledge of the phytotron in 1960 to build a large display greenhouse at the Botanical Gardens, where different climates could be replicated in different areas of the greenhouse. He ran the Botanical Gardens until 1963 and taught at the university until 1965. In 1964, his third book, *The Plants,* was published. Went returned to research in 1965 as a distinguished professor of botany at the University of Nevada's Desert Research Institute in Reno. That same year, the American Society of Plant Physiologists gave him the Charles Reid Barnes Life Membership Award.

Finds Key to Sustaining Amazon Rain Forest

In the late 1960s, Went made one of his most important discoveries on the floor of the Amazonian rain forest. He was travelling aboard the research vessel *Alpha Helix* and planned to conduct research on the Amazon basin. When Brazilian customs agents impounded his laboratory equipment, Went turned his attention toward the ground. There he found a garden of **fungi** among the dead leaves, branches, and other debris, making up a litter layer. Running throughout this layer was a network of tree roots. Went concluded that the fungi digest the litter and pass the extracted nutrients to the tree roots in a continuous nutrient cycle.

Converting this rich land to temperate-zone agriculture irreparably harms the rain forest, Went warned. He urged rain forest developers to avoid annual crops and instead plant Brazil nuts, oil palms, or cacao trees, which would perpetuate the rich forest. Although clear-cutting of the rain forest continues, by the late 1980s, Went's warning had become gospel for the growing movement to save the Amazonian rain forest.

In 1967, Went received the Hodgkins Award from the Smithsonian Institute for his contribution to the understanding of the environment. He retired from the Desert Research Institute as a research professor emeritus in 1975, but his devotion to studying the living world never ceased. Went spent his last years researching the effects of smog on weather. He concluded that most of the soot collects on the surface of cumulus clouds, then returns to the earth as dirty rain or snow. In 1989, months before his death, he received the Henry Shaw Medal from the Missouri Botanical Gardens.

The 86-year-old scientist died of a heart attack on May 2, 1990, during a visit from his retirement home in Portland, Oregon, to the Desert Institute in Reno. A manuscript he had written for a book about hydrocarbons and their relationship to thunderstorms was found in his suitcase. His children published the book posthumously to distribute to Went's friends and peers. It is titled *Black Carbon, Blue Sky.* Had he lived longer, Went most likely would have continued his research. In his 1974 article for the *Annual Review of Plant Physiology,* he wrote of his dream to delve into such uncharted fields as the sociology and physiology of ants and galls, or tumors, that insects develop on leaves. Went's son Hans, a retired zoology professor, observed in an interview that even in his leisure time, his father's sole interest was science. "He was always out looking at plants, analyzing, comparing," Hans said of his father. "Science was his life."

WETLANDS

Wetlands encompass an enormously diverse range of areas, with different water regimes, dominant plant **species**, and sediment or soil characteristics. However, all wetlands have **water** present for a significant period of time and **plants** adapted to their wet conditions. Wetlands encompass large areas along the shores of lakes and small prairie potholes found in the north central United States. Neither exclusively aquatic nor terrestrial, wetlands are in the zone between permanently wet and normally dry environments. While the crucial value of wetland **ecosystems** is widely recognized today, in the past they were considered swampland to be drained or filled in for agricultural and other uses. In the United States, some 117 million acres (47.4 million hectares) of wetlands, half the original acreage, have been lost to such conversion by the mid-1980s.

Modern society recognizes that wetlands play a major role in providing **habitat**s for numerous plants and **animals**, including waterfowl and other birds, fur-bearing **mammals**, reptiles, **fish**, shellfish. From wetlands wood and fibers are obtained. Wetlands help maintain water quality by retaining **nutrients**, metals, and other substances. They act as water absorbing sponges that control flooding and provide long term storage of surface water. In addition to these and other ongoing functions, wetlands are also recognized as having played an important role in human history. In the past, humans have depended on wetlands for food, clothing, and shelter, as shown by archaeological sites where the bones of waterbirds and the remains of fish spears indicate that people in Europe and elsewhere drew sustenance from wetlands.

Most of the wetlands in temperate parts of the world are freshwater marshes, but the broader group of marsh wetlands also includes tidal saltwater and tidal freshwater marshes. Marshes can be found in many places where frequent flooding is caused by streams or lakes, ground water, or water. Ninety percent of the wetlands in the lower 48 states are freshwater marshes. Tidal salt marshes with their salt tolerant plant species can be found along the Arctic and Atlantic seaboards of North America, the Gulf of Mexico, and the European coastline. Mangrove forests, which are mostly concentrated in the Indian Ocean and West Pacific region, are the subtropical equivalent of tidal salt marshes.

Reeds, grasses, sedges, and rushes grow in these highly productive ecosystems, depending on the hydrology. For instance, many marshes are dominated by what are called emergents, which have stems partly submerged and partly above the water. Where the water is deeper, submerged and floating water plants thrive. These ecosystems contain tidal creeks that provide a link between the marsh, the estuary, and the sea. Many marine organisms rely on these wetlands for spawning and feeding. Further inland can be found tidal freshwater marshes, which contain more diverse plants than in the salt marshes, including flowering plants.

Swamps, which differ significantly from marshes, are low-lying depressions that are poorly drained and are dominated by a single plant species. These ecosystems include North American bottomland hardwoods and cypress swamps, the swamp **forests** of the Amazon, and the swamps found along the shores of many large tropical lakes in Zambia, Uganda, and other African countries. Some South American and Southeast Asian swamp forests contain many species of **trees**, including some that are important sources of timber.

Floodplain wetlands are those flat areas on the borders of rivers that flood periodically, sometimes creating grassy re-

Sunset on a wetland near Lake Erie in Crane Creek State Park. During low water periods, mud flats are exposed providing nesting platforms for waterfowl like the Canadian goose shown above. *(Photograph by Robert J. Huffman, Field Mark Publications. Reproduced by permission.)*

gions that sustain large grazing populations. Often these wetlands spread into wide deltas, usually along the lower reaches of rivers but sometimes far inland. South America possesses some of the largest floodplains, including an area along the Paraguay River called the Gran Pantanal and along tributaries of the Orinoco River in Venezuela. Along the Lower Mekong River in Indochina and the Lower Mississippi in the United States are also found major floodplain areas.

Peatlands are another type of wetland. Water saturation, lack of oxygen, and other conditions can cause organic matter to decompose less slowly than it is produced, resulting in the accumulation of deposited organic matter. Peat deposits occur in marshes, swamps, floodplains, and other wetland types. When more than a foot (300 millimeters) of peat accumulates, it forms wetlands known as bogs and fens. Bogs can be either raised bogs, where the peat continues to accumulate and forms a dome, or blanket bogs, where the peat has spread beyond the borders of a lake or pond. Peatlands can be found on every continent. In North America, a unique peatland called a pocosin can be found in the flat, elevated regions between two rivers.

Marshes, swamps, and other wetlands receive protection under national and international laws. In the United States, for instance, the Federal Water Pollution Control Act amendments of 1972 authorized the U.S. Army Corps of Engineers and the Environmental Protection Agency to regulate pollution of the nation's waters, including wetlands. The federal jurisdiction has broadened since then to cover increasingly more acreage of wetlands. This has caused a political backlash by private companies who state that the wetlands regulations constitute a governmental "taking" of their property, and that they ought to be fairly compensated. Critics also accuse government agencies of defining wetlands so broadly that many non-wetlands were coming within the agencies' regulatory jurisdiction. Political battles became so intense during the 1990s that in 1993 Congress requested that the National Academy of Sciences create a committee to evaluate scientific wetlands issues, including how they are defined. In a 1995 report, *Wetlands Characteristics and Boundaries*, the academy addressed the various hydrology, vegetation, soil, and other factors that define wetlands. The report did not resolve the political controversy, leaving Congress to find a legislative compromise to address the concerns of all parties. Worldwide efforts to protect these precious areas are occurring through the 1971 Convention on Wetlands of International Importance especially as Waterfowl Habitat, or the "Ramsar Convention" as it is called after the place in Iran where it was adopted. More than 500 wetland sites have been placed on the Ramsar list, which designates wetlands of international importance whose ecological characteristics the treaty nations agree to maintain.

WEXLER, NANCY (1945-)
American neuropsychologist

Nancy Wexler's research on Huntington's disease has led to the development of a presymptomatic test for the condition as well as the identification of the **gene**s responsible for the disease. The symptoms of this fatal, genetically based disorder (for which Wexler herself is at risk) usually appear around middle age, and the disease leads to the degeneration of mental, psychological, and physical functioning. For her pivotal role in these achievements, Wexler was granted the Albert Lasker Public Service Award in 1993.

Nancy Sabin Wexler was born on July 19, 1945, to Milton Wexler, a Los Angeles psychoanalyst, and Leonore Sabin Wexler. She studied social relations and English at Radcliffe and graduated in 1967. Wexler subsequently traveled to Jamaica on a Fulbright scholarship and studied at the Hampstead Clinic Child Psychoanalytic Training Center in London.

In 1968 Wexler learned that her mother had developed the symptoms of Huntington's disease, a condition to which Wexler's maternal grandfather and three uncles had already succumbed. Efforts to fight the disease became a primary mission for Wexler and her family: Her father founded the Hereditary Disease Foundation in 1968, and Wexler herself, who was then entering the doctoral program in clinical psychology at the University of Michigan, eventually wrote her doctoral thesis on the ''Perceptual-motor, Cognitive, and Emotional Characteristics of Persons-at-Risk for Huntington's Disease,'' and received her Ph.D. in 1974.

After graduating from University of Michigan, Wexler taught psychology at the New School for Social Research in New York City and worked as a researcher on Huntington's disease for the National Institutes of Health (NIH). In 1976 she was appointed by Congress to head the NIH's Commission for the Control of Huntington's Disease and its Consequences. In 1985 she joined the College of Physicians and Surgeons at Columbia University.

In 1979 Wexler's research led her to Lake Maracaibo in Venezuela, where she studied a community which had a high incidence of Huntington's disease. Wexler kept medical records, took **blood** and skin samples, and charted the transmission of the disease within families. Wexler sent the samples she collected to geneticist James Gusella at Massachusetts General Hospital, who used the blood samples to conduct a study to locate the gene—the first such **genetic mapping** of a disease. Gusella eventually discovered a **deoxyribonucleic acid (DNA)** marker close to the Huntington's gene. Based on this study, Gusella introduced a test that was ninety-six percent accurate in detecting whether an individual bears the Huntington's gene. Because there was still no cure for the Huntington's disease, the test proved to be controversial, raising many

Nancy Wexler. *(The Library of Congress. Reproduced by permission.)*

issues involving patient rights, childbearing decisions, and discrimination by employers and insurance companies. In her interviews and writings Wexler has stressed the importance of keeping such genetic information confidential.

In 1993 the Huntington's gene was identified through research based on the Venezuelan blood samples and the work of the Huntington's Disease Collaborative Research Group. In October, 1993, Wexler received an Albert Lasker Public Service Award for her role in this effort. In addition, she has served as an advisor on social and medical ethics issues to the **Human Genome Project**—a massive international effort to map and identify the approximately 100,000 genes in the human body. Wexler also has assumed directorship of the Hereditary Disease Foundation founded by her father, to which she donated the honorarium that accompanied the Lasker Award.

WHIPPLE, GEORGE HOYT (1878-1976)
American pathologist

George Hoyt Whipple knew he would be a physician from the time he was in elementary school at the turn of the century. The son and grandson of doctors, Whipple followed the family tradition by choosing a career in medicine, researching the creation and breakdown of oxygen-carrying **hemoglobin** in the **blood**; this research resulted in not only a treatment for perni-

George Hoyt Whipple. *(The Library of Congress. Reproduced by permission.)*

cious anemia, but also in a share of the 1934 Nobel Prize. An industrious, hard-working Yankee from New Hampshire, Whipple authored more than 200 publications on anemia, pigment **metabolism**, liver injury and repair, and other related subjects. Yet in his last days, it was as an educator that he hoped to be remembered.

Whipple was born on August 28, 1878, in Ashland, New Hampshire, the son of Frances Anna Hoyt Whipple and Ashley Cooper Whipple, a general practitioner held in high esteem by his patients and colleagues. Whipple's father died of typhoid fever just two years after the birth of his son, and Whipple and his sister Ashley were brought up by their mother and grandmothers. His was an outdoor life in rural New Hampshire, and he took a love of hunting, fishing, and camping with him into adulthood. At the age of fourteen Whipple entered Phillips Academy in Andover, Massachusetts, enrolling at Yale College (now Yale University) as a premedical student four years later. At Yale, he was a star baseball player and was on the gymnastics and rowing teams, as well as an outstanding student. Though versed in the humanities in these years of public and private schools, he had always been attracted by science and mathematics. After graduating with high standing in

1900, Whipple spent a year teaching and coaching at Holbrook Military Academy in New York to earn money for medical studies, and in 1901 he entered Johns Hopkins University's School of Medicine.

Sets His Course for Research

During his years as a student at Johns Hopkins, Whipple earned his way with a paying instructorship. Initially Whipple had considered going into pediatrics, but upon receiving his M.D. in 1905 instead joined the Johns Hopkins staff as an assistant in pathology, working under the renowned pathologist William Henry Welch. It was as a 29-year-old assistant performing an autopsy on a missionary doctor that Whipple made his first notable medical contribution, describing a rare condition in the intestinal **tissue**s, which has since come to be called Whipple's disease. A year spent at a hospital in the Panama Canal Zone led to further notable advances in malaria and tuberculosis research.

When he returned to Johns Hopkins in 1908, Whipple turned his attention to studies in liver damage and the way in which liver **cell**s repair themselves. Studies with dogs led Whipple to realize the importance of bile, a substance manufactured in the liver by the breakdown of hemoglobin, a complex pigment in red corpuscles. In normal concentrations, bile helps to break down **fats** during digestion, but can produce jaundice when present in excessive amounts. Beginning his assistant professorship at Johns Hopkins in 1911, Whipple came to focus on the interrelationship of bile, hemoglobin, and the liver. In 1913, along with a talented medical student, Charles W. Hooper, Whipple was able to show that bile pigments could be produced outside of the liver, solely from the breakdown of hemoglobin in the blood. Using this experiment as a starting point, Whipple set a new course for his studies. Since bile pigments are formed from hemoglobin, Whipple reasoned that he should tackle the question of hemoglobin itself, beginning with how it is manufactured. It was a fateful decision.

In 1914 Whipple accepted a position as director of the Hooper Foundation for Medical Research at the University of California in San Francisco. In that same year he also married his long-time sweetheart, Katharine Ball Waring, and the couple moved to California. Though burdened with administrative duties, Whipple continued his researches into hemoglobin production. His assistant, Hooper, came with him to California and together with a new assistant, Frieda Robscheit-Robbins, they began experiments which would lead to a major breakthrough. By systematically bleeding laboratory dogs, Whipple and his team were able to induce a controlled anemic condition. They then tested various foods and their effects upon hemoglobin regeneration, finding that a diet of liver produced a pronounced increase in hemoglobin regeneration. While such short term effects were encouraging, they were still far from conclusive.

Research Proved Conclusive at Rochester

Though in 1920 Whipple was named dean of the University of California Medical School, he remained in California for just a year before accepting (somewhat reluctantly) a similar position at a new medical complex at the University of Rochester in New York—a facility heavily endowed by Kodak founder George Eastman and the Rockefeller Foundation. Courted enthusiastically by Eastman and university president Rush Rhees, Whipple moved home and laboratory to New York, bringing Robscheit-Robbins and the group of anemic dogs with him.

The next decade proved busy for Whipple: he directed the building and staffing of the University of Rochester School of Medicine and Dentistry, all the while directing further hemoglobin research. Perfecting their technique of bleeding the dogs, Whipple and Robscheit-Robbins induced long-term anemia and were able to prove conclusively that a liver diet was successful in counteracting its effects by increasing the production of hemoglobin. His results were published in 1925, and the pharmaceutical firm of Eli Lilly, with Whipple's cooperation, began producing a commercially available liver extract within a year. Whipple refused to patent his findings, and directed all royalties from the sales of the extract to fund additional research. Whipple's experiments paved the way for further studies by two Boston researchers, George Richards Minot and **William P. Murphy**, who used liver therapy to successfully treat pernicious anemia in 1926.

Whipple's work soon won international repute and in 1934 he received word that he, along with Minot and Murphy, was going to receive the Nobel Prize for Physiology or Medicine for their separate work in liver therapy. Whipple did not let fame slow him down. He continued his hemoglobin experiments, turning now to the study of iron in the body and utilizing the new technology of radioisotope elements to follow the distribution of iron in the body. He also made important contributions to the study of an anemic disorder peculiar to people of Mediterranean extraction, a disorder for which Whipple suggested the name *thalassemia*. Other studies involved the use of plasma or tissue **proteins** to rebuild hemoglobin in cases of anemia. A spin-off of this latter research was the development of intravenous feeding.

Despite the administrative and research duties that pressed upon him, Whipple did not forget his students, and took real pleasure in teaching. When in later years he was offered the position of Director of the Rockefeller Institute, he politely but adamantly declined, preferring his classes and his research. Whipple finally relinquished his chair as dean in 1953 at the age of 75, after a long and distinguished career that had seen the once-small university grow to more than 12,000 graduates in medicine and other related fields. He remained on the faculty of the University of Rochester teaching pathology until 1955. In 1963 he established a medical and dental library for the university valued at $750,000. In addition to the Nobel Prize, Whipple was also a trustee of the Rockefeller Foundation from 1927-43, a Kober Medal winner in 1939, and a recipient of the Kovalenko Medal of the National Academy of Sciences in 1962, among others.

Whipple's life was long and productive. He was an active outdoorsman well into his ninth decade. With his wife Katharine, he had two children: a son, Hoyt, who followed in the Whipple tradition of medicine, and a daughter, Barbara. He died in Rochester on February 1, 1976, in the hospital he had helped to build.

WHITTAKER, ROBERT HARDING (1920-1980)
American biologist

Robert H. Whittaker was an American ecologist, who was a professor at Cornell University. He studied terrestrial plant communities in the northeastern United States, particularly in parts of the Appalachian Mountains. He is best known for his contributions to the development of theories of the organization of ecological communities. In particular, Whittaker believed that species respond to variations of environmental factors on an individualistic basis. As such, the assembly of **species** within communities is somewhat arbitrary, depending on local environmental conditions, in combination with competitive and other ecological relationships occurring among the species present. As such, communities consist of biological and non-biological components, which interact closely, but do not have any inherently central, control system. This is a quite different idea from that of F.E. Clements, an earlier American ecologist, who suggested that communities represented a strongly integrated, organismic unit, which was even capable of **evolution**.

One of the powerful analytical tools used by Whittaker is known as gradient analysis, in which the distribution of species is plotted against variation of one or more environmental variables. For example, the abundance of a **tree** species (such as white pine, *Pinus strobus*) might be plotted against elevation in a mountainous region, or against moisture or nutrient content of the soil. Plots of this sort are known as response curves, and they reveal that species have idiosyncratic distributions, and that community composition varies continuously along gradients of environmental change, as occurs, for example, with changes in altitude up a mountain. Observations such as these were central to the rationalization of Whittaker's ideas of community organization.

In 1969, Whittaker proposed that the diversity of life should be divided among five **kingdom**s: **Monera**, **Protista**, **Plants**, **Animals**, and **Fungi**. This system of **classification** has become widely adopted by biologists around the world.

Robert Whittaker and his many students and other colleagues published numerous important papers and books in biology and ecology. Among his more prominent contributions were: *A Consideration of Climax Theory: the Climax as a Population and Pattern* (1953), *Vegetation of the Great Smoky Mountains* (1956), *Classification of Natural Communities* (1962), *Gradient Analysis of Vegetation* (1967), *Communities and Ecosystems* (1970), *Evolution in Natural Communities* (1972; with G.M. Woodwell), *Classification of Natural Communities* (1978), and *Ordination of Plant Communities* (1982).

WIELAND, HEINRICH (1877-1957)
German chemist

Heinrich Wieland was one of the greatest organic chemists of the century, admired for the breadth of his knowledge and his devotion to arduous, painstaking research. Wieland is known for his studies on the structures of important complex natural products, from toad poisons to butterfly pigments. He also made major contributions to biochemistry, especially in the study of the mechanism of biological oxidation. His most famous work, for which he was awarded the Nobel Prize in chemistry in 1927, was the determination of the molecular structure of the bile acids. This research combined superb experimental skill with precise deductive reasoning and remains a model of organic chemical investigation.

Heinrich Otto Wieland was born on June 4, 1877, in Pforzheim, Germany, to Theodor and Elise Blom Wieland. Theodor Wieland was a pharmaceutical chemist, and Heinrich studied the subject in school in Pforzheim. At that time, instead of studying at a single university to obtain a degree, a student enrolled at several universities, listening to the lectures of the best professors. Wieland spent 1896 at the University of Munich, 1897 at the University of Berlin, and 1898 at the Technische Hochschule at Stuttgart. In 1899 he returned to Munich to work toward his Ph.D. under the direction of Johannes Thiele, in the laboratory of Adolf von Baeyer. After he received his Ph.D. in 1901, Wieland remained at Munich to do research, eventually becoming a lecturer in 1904 and a senior lecturer in 1913. In 1917 he was appointed professor at the Technische Hochschule in Munich, but was granted leave to work for Fritz Haber's chemical warfare research organization at the Kaiser Wilhelm Institute in Berlin. At the end of World War I he returned to Munich, but left in 1921 to accept a professorship at the University of Freiburg. In 1925, Wieland returned to the University of Munich as professor and director of the Baeyer Laboratory, succeeding **Richard Willstätter**, who personally recommended Wieland for the position. By this time, Wieland was recognized as a world leader in organic chemistry, and he remained at Munich until his retirement in 1950.

Wieland's early research was concerned with the chemistry of organic nitrogen compounds. He explored the addition of dinitrogen trioxide and **nitrogen** dioxide to carbon-carbon double bonds. A large series of papers described the reactions of aromatic amines (a type of organic compound derived from ammonia), especially their oxidations. One line of experiments led to the discovery of nitrogen free radicals, unusually reactive short-lived species in which nitrogen is bonded to two atoms, instead of the usual three atoms. Wieland published almost one hundred papers on organic nitrogen chemistry, which in itself was a notable achievement.

Another series of experiments led to Wieland's 1912 theory of biological oxidation, a process by which biologic substances are changed by combining with oxygen or losing electrons. For years, the accepted theory involved some kind of change to molecular oxygen inside the cell in which the oxygen becomes "activated" and reacts with the oxidizable substance. Wieland proposed that the oxidizable substance itself becomes "activated" and loses hydrogen atoms in the oxidation process. Wieland published more than 50 papers from 1912-1943 on biological oxidation and was able to demonstrate that many reactions proceed through dehydrogenation and could proceed in the absence of oxygen. He was chal-

lenged, however, by the German physiologist Otto Warburg, who showed that respiratory enzymes which contain iron (sometimes copper) do activate oxygen, and both types of oxidation mechanism are found in nature. Warburg received the Nobel Prize in 1931 for his contribution to understanding oxidation, but Wieland's work has been recognized as equally significant by biochemists.

Determining Structure of Bile Acids Leads to Nobel Prize

In 1912, the year Wieland proposed his theory of biological oxidation, he published his first paper on the structure of the bile acids. This topic would occupy his interest for twenty years and earn him the Nobel Prize. Bile is a golden yellow liquid which is produced in the liver, stored in the gall bladder, and secreted in small amounts into the intestines. The sodium salts of bile acids, the principal constituent of bile, are essential to the digestion of **fats**. Although bile acids had been isolated early in the nineteenth century, their structural formulas were unknown when Wieland began his work. As the work progressed, it was shown by Adolf Windaus, a chemist at the University of Göttingen, that **cholesterol** and the bile acids share a common basic structure, allowing Windaus's research results on the structure of cholesterol (for which he won the Nobel Prize in 1928) to be used by Wieland, and vice versa. Later it was shown that the common basic structure, the **steroid** nucleus, is found in many naturally occurring sources, such as the sex **hormone**s, adrenal hormones (cortisone), digitalis (a plant cardiac poison, used medicinally as a stimulant), and toad poison. Steroid chemistry became essential to the development of many powerful medicines, as well as oral contraceptives. The pioneering work of Wieland and his students on the bile acids became a foundation of modern pharmaceutical chemical research.

The work on bile acids was an enormous challenge for organic chemistry in the first quarter of the century. First, a procedure for isolation and purification of the various acids, obtained from ox bile, was required. Then, each acid had to be characterized and chemically related to the others. The acids each contain 24 carbon atoms and differ in the number of hydroxyl (**alcohol**) groups. Wieland used the method of selective degradation to break the acids into simpler compounds, thus allowing him to identify the smaller **molecules**. Although his work was somewhat simplified because he could use the results of Windaus, Wieland admitted in his Nobel Lecture that "the task would appear to be a long and unspeakably wearisome trek through an arid desert of structure." In this lecture he outlined the course of his research, showing the failures as well as successes. Although the structures of the bile acids and cholesterol appeared to be solved when Wieland and Windaus received their Nobel prizes, in fact a conclusion which they had made based on analogous reactions was not correct, and the final, unequivocal structures were proposed by Wieland and others in 1932.

In addition to the bile acids, Wieland also investigated other natural products. He contributed to the determination of the structures of morphine, lobeline, and strychnine alkaloids, as well as butterfly wing pigments and mushroom and toad poisons. He had a wide range of interests, encompassing all areas of organic chemistry, and for twenty years he was editor of the major chemical journal *Justus Liebigs Annalen der Chemie*. His work was recognized throughout his career, and he was honored by scientific societies and universities in many countries. In 1955 he was named the first recipient of the German Chemical Society's Otto Hahn Prize for Physics and Chemistry.

Wieland remained at the University of Munich during World War II. He had little regard for the Nazi government in Germany and made no secret of it. He protected Jews in his laboratory and in 1944 testified on behalf of students who had been accused of treason.

Wieland married Josephine Bartmann in 1908. All three of their sons became scientists: Wolfgang, a pharmaceutical chemist; Theodor, a professor of chemistry; and Otto, a professor of medicine. Their daughter, Eva, married Feodor Lynen, a professor of biochemistry who won the Nobel Prize in physiology or medicine in 1964. In addition to his love of family and his work, Wieland also enjoyed painting and music. He died in Starnberg, Germany, on August 5, 1957, two months after his eightieth birthday.

WIESCHAUS, ERIC F. (1947-)
American biologist

Wieschaus was born in South Bend, Indiana, in 1947 but grew up in Alabama. He received his bachelor's degree in biology from the University of Notre Dame in 1969 and his doctorate from Yale in 1974. His doctoral dissertation involved using genetic methods to label the progeny (offspring) of single **cell**s in fly **embryo**s. He showed that even at the earliest cellular stages, cells were already determined to form specific regions of the body called segments.

Wieschaus began his Nobel-winning work in the latter part of the 1970s. The Alabama native spent three years with Christiane Nüsslein-Volhard in the European Molecular Biology Lab at the University of Heidelberg, Germany, tackling the question of why individual cells in a fertilized egg develop into various specific tissues. They elected to study *Drosophila*, or fruit flies, because of their extremely fast embryonic development. New generations of fruit flies can be bred in a week. In addition, fruit flies have only one set of **gene**s controlling development compared to the four sets humans possess. This means that testing each **fruit fly** gene individually takes one-fourth the time it would involve to test human genes.

To begin their experiment, Nüsslein-Volhard and Wieschaus damaged male fruit fly **deoxyribonucleic acid (DNA)** by applying ultraviolet light to the genes or by feeding the flies sugar water laced with chemicals. Then the team ''knocked out'' one gene from the fly, breeding generations of fruit flies without that particular piece of code. In this way, Nüsslein-Volhard and Wieschaus were able to isolate all the genes crucial to the early stages of embryonic development. When the flies were bred, the females produced dead embryos.

These lifeless embryos resulted from only 150 different **muta**tions of the 40,000 mutations applied. These 150 genes proved to be essential to the proper development of the fly embryo because, when damaged, the genes caused extraordinary deformities that killed the embryo. By viewing the fly embryos with a two-person **microscope**, Wieschaus and Nüsslein-Volhard were able to simultaneously view and classify a large quantity of malformations caused by gene mutations. Next, they identified 15 different genes, that, when mutated, eliminate specific body segments in the fly embryos. Wieschaus also established that systematic categorizing of genes that control the various stages of development could be accomplished.

Their first research results reported that the number of genes controlling early development was not only limited, but could also be classified into specific functional groups. They also identified genes that cause severe congenital defects in flies. After additional experimentation, the principles involved with the fruit fly genes were found to apply to higher animals and humans. This led to the realization that many similar genes control human development, and this finding could have a tremendous impact on the medical world. The applications of their research extend to in vitro **fertilization**, identifying congenital **birth** defects, and increased knowledge of substances that can endanger early stages of pregnancy.

It wasn't until 1995, however, that he won the Nobel Prize in Physiology or Medicine, along with Edward B. Lewis and Christiane Nüsslein-Volhard, for his work on identifying key genes that make a fertilized fruit fly egg develop into a segmented embryo. His research could help improve knowledge of how genes control embryonic development in higher **organism**s, including identifying genes that cause human birth defects.

WIESEL, TORSTEN (1924-)
Swedish American neurophysiologist

Torsten Nils Wiesel was born on June 3, 1924, in Uppsala, Sweden, the son of Anna-Lisa Bentzer Wiesel and Fritz S. Wiesel, the chief psychiatrist at the Beckomberga Mental Hospital in Stockholm. Wiesel entered medical school at the Karolinska Institute in Stockholm in 1941 and studied neurophysiology and psychiatry. In 1954, he received his medical degree, becoming an instructor at the institute as well as an assistant in the Department of Child Psychiatry at Karolinska Hospital. Wiesel then came to the United States in 1955 to do postdoctoral work at the Wilmer Institute of Johns Hopkins School of Medicine.

At Johns Hopkins, Wiesel worked under Stephen Kuffler, whose exhaustive work had proved that the vision of **mammals** is distinctly different from that of non-mammals. Wiesel became interested in the idea that the critical level of visual **perception** must take place in the **brain** of mammals. In 1958, Wiesel set off with **David Hubel** on the research that would result in a new theory of visual perception.

Wiesel and Hubel studied the striate or visual cortex which is located at the back of the brain. They discovered which **cell**s in the cortex responded to which pattern or level of light. They also conducted experiments to map the striate cortex by injecting the eyes of experimental **animal**s with radioactively labeled **amino acid**. These amino acids would be taken up by the cell bodies of the retina and transported to cells in the visual cortex. In some cases, the visual cortexes were dissected in order to see, by the use of autoradiographs or x-ray like photos, where the labeled amino acids actually ended up. Such experiments, begun in 1959, used both cats and macaque monkeys. That same year Kuffler was appointed a professor at the Harvard University Medical School, and Wiesel and Hubel joined him there. Wiesel was appointed assistant professor of physiology, and became a full professor in 1964.

The Wiesel-Hubel team soon began publishing the results of their experimental method, and it was clear that they had uncovered new complexities to the visual process. Within the visual cortex itself, Wiesel and Hubel made two important discoveries. First they showed that there is a hierarchy of types of cells in the cortex, ranking from simple to complex to hypercomplex, depending on the information each is able to process. They termed the process of putting the millions of building blocks of visual information back together into a picture "convergence." Their second major discovery was a further organization of the cortical cells into roughly vertical divisions of two types: orientation columns and ocular dominance columns. Within these columns are simple, complex, and hypercomplex cells working toward a progressive convergence of visualization. Until the time of Wiesel's and Hubel's work, it was assumed that all cells of the cerebral cortex were more or less uniform. Wiesel and Hubel showed that the visual cortex is constituted of a cell pattern, which appears to be designed specifically for vision. As a result of their discovery, current theory now posits that the rest of the cerebral cortex may follow this form-follows-function rule.

Wiesel and Hubel researched another experimental model in which they used kittens to study the effect of various visual impairments on development. They discovered that if one eye were deprived of certain or all visual stimuli at three to five weeks of age, the central functioning of that eye would always be suppressed from cortical processing. Kittens, and by extension mammals in general, though born with a complete visual cortex, must still "learn" to see. Even if an early impairment is later corrected, the repaired eye will still remain functionally impaired as far as the visual cortex is concerned. The realization that there is a critical stage for visual development revolutionized the field of pediatric ophthalmology, calling for the earliest possible intervention in cases of strabismus, or crossed eyes, and congenital cataracts.

By 1973 Wiesel succeeded Kuffler as chair of the Department of Neurobiology at Harvard, and was named the Robert Winthrop Professor of Neurobiology in 1974. In 1981, Wiesel and Hubel were awarded the Nobel Prize for Physiology or Medicine, sharing it with Sperry from Caltech. The Karolinska Institute in Stockholm, which administers the prize and where Wiesel began his professional career, praised Hubel and Wiesel for their discoveries concerning information processing in the visual system. Wiesel and Hubel continued their close working relationship until Wiesel left Harvard in 1984 to head the neurobiology lab at Rockefeller University where he continued his researches on vision. In 1992 he was named president of Rockefeller University.

Weisel's first marriage, to Teiri Stenhammer, ended in divorce after 14 years in 1970. Wiesel was married again in 1973, to Grace Yee. The couple had one child, Sara Elisabet. His second marriage also ended in divorce in 1981. Wiesel became a naturalized U.S. citizen in 1990.

WILKINS, MAURICE HUGH FREDERICK (1916-)
British physicist

Maurice Hugh Frederick Wilkins is a British born physicist known for helping to identify the structure of **deoxyribonucleic acid** (**DNA**). He was born on December 15, 1916, in Pongaroa, New Zealand, the son of a physician. Young Wilkins shared his father's interest in the sciences and he was taken to England at age 6 where he studied at the King Edward VI school in Birmingham and later at St. John's College in Cambridge. He graduated in 1938 with a degree in physics and two years later earned his doctorate from the University of Birmingham.

Almost immediately Wilkins went to work for the Ministry of Home Security and Aircraft Production where he used his expertise in physics to develop radar systems. Later, as part of a British team assigned to the Manhattan project, he worked for a time at the University of California to develop the atomic bomb. He soon became disillusioned with nuclear physics and, inspired by Schrödinger's book *What is Life*, Wilkins turned his interest instead to biophysics. He took a position with St. Andrews University in Scotland in 1945. The next year he settled in London working with the Medical Research Council's Biophysics Research Unit at King's College.

While working at King's Wilkins began to study the structure of the DNA molecule which had been identified earlier, in 1946. Using an x-ray diffraction method he exposed crystalline DNA **molecule**s to x-ray beams and then analyzed the pattern formed by the reflected radiation. Using this data, Wilkins determined that the DNA molecule has a double helical structure. One of his fellow researchers, **Rosalind Franklin**, determined that the phosphate groups of the molecule are located on the outside of the helix. Wilkins passed this information on to **James D.Watson** and **Francis Harry Compton Crick** who incorporated the **double helix** theory into their model of the DNA molecule. In 1962 Wilkins, Watson, and Crick shared the Nobel Prize for Physiology or Medicine for their discovery.

Wilkins became Assistant Director of the Biophysics Research Unit in 1950 and was promoted to Deputy Director in 1955. In 1970 he was appointed to Director, a post he held until 1972. Also in 1970, Wilkins was also appointed Professor of Biophysics and Head of Department at King's College.

See also Double helix

WILLIAMS, ANNA WESSELS (1863-1954)
American bacteriologist and public health pioneer

During a long professional career doing research at the New York City Department of Health Laboratory, Anna Williams pioneered laboratory techniques for the diagnosis of many diseases. Dr. Williams also wrote and lectured about her findings in an era when women in science were few and very rarely recognized for their contributions.

Anna Wessels Williams was born in 1863 in Hackensack, New Jersey and graduated from the Women's Medical College in New York City in 1891. She studied in Europe for a time, during which she visited Robert Koch, the discoverer of tuberculosis **bacteria**. When she returned to New York she began work for the Department of Health, where her first research assignment was in the diagnosis and treatment of diphtheria, a bacterial disease that was on the rise during the mid-1890s. Dr. Williams successfully isolated one strain of diphtheria bacteria, *Corynebacterium diphtheriae*, although her work went uncredited at the time. The strain became known as "Park 8," named after the director of the laboratory, Dr. W. H. Park, who was on vacation at the time of the discovery. During this period also, Williams worked with Dr. Alexander Lambert to develop a standardized test for typhoid fever.

In 1896 Dr. Williams briefly visited Paris where she learned valuable laboratory techniques for the diagnosis of rabies. Several years later, while examining microscope slides of smears taken from the brains of rabid animals, Dr. Williams discovered unusual "bodies" that appeared in all the cells. These characteristic features were named Negri bodies for an Italian scientist working at the same time in a laboratory in Italy. The two bacteriologists, Williams and Negri, communicated their findings, and Williams pioneered a technique that led to a faster diagnosis of rabies based on the appearance of the stained slides.

In 1904 New York City was experiencing an epidemic of pneumonia. In response, Williams and Park examined specimens from hundreds of pneumonia patients and observed that *pneumococcus bacteria* were present. Their findings occurred several years before discovery of the bacteria by **Oswald T. Avery** and others, who are usually credited with the diagnosis of the disease. Williams was also involved in tests for typhoid fever and helped to give laboratory confirmation of the diagnosis of "Typhoid Mary," a carrier of the disease who was subsequently apprehended and confined.

Williams worked closely with another pioneer in public health, Josephine Baker, who headed the Department of Children's Hygiene where great efforts were made to report and eradicate the children's diseases that often ran rampant in the city's slums. It was thought that *trachoma*, a chronic eye disease that can lead to a reduction in vision, was affecting thousands of New York City children. However, in her study from 1912 to 1913, Dr. Williams concluded that most cases had been misdiagnosed and that the children were suffering from the more common infection, conjunctivitis.

During the polio outbreak of 1916, Williams began research on that disease. 1918, however, marked the beginning of the "flu years" and much of her research time was devoted to finding the cause of influenza. In only two months of 1918 there were close to 11,000 reported deaths caused by influenza and almost 10,000 from pneumonia. Williams and others at their New York City laboratory set to work on the variety of bacterial agents that could cause flu. But their work was inconclusive and Williams made reference in her report of the difficulties in trying to look for a possible viral cause. Identification of viruses, which indeed cause the flu, remained elusive during this period of research.

In the 1920s, Williams did extensive studies on scarlet fever. The Dick test, which had just recently been developed by Gladys and **George Dick**, had been used to test thousands of school children for the disease. Williams surveyed hundreds of scarlet fever cases that had been positively diagnosed for the antitoxin that had been used.

In the 1930s Williams published a compilation of her work on bacterial diseases entitled *Streptococci in Relation to Man in Health and Disease*. She received honors for her pioneering work as a woman of science, but spoke of the difficulty and bias women faced in a field dominated by men. She retired in 1934 and was 91 when she died.

WILLSTÄTTER, RICHARD (1872-1942)
German chemist

A gifted experimentalist, Richard Willstätter's pioneering work on natural products, especially **chlorophylls** and anthocyanins (**plant** pigments), was honored with the 1915 Nobel Prize in chemistry. In 1924 Willstätter, who was Jewish, resigned from his position at the University of Munich in protest against the anti-Semitism of some of the faculty. This act of conscience seriously hampered his research activity. In 1939 the anti-Semitic policies of the Third Reich forced him to emigrate to Switzerland, where he spent the remaining few years of his life.

Education and Early Career

Richard Martin Willstätter was born in Karlsruhe, Germany, on August 13, 1872, the second of two sons of Max and Sophie Ulmann Willstätter. Willstätter's father was a textile merchant and his mother's family was in the textile business. Willstätter's education began in the classical Gymnasium in Karlsruhe. When he was eleven years old, his father moved to New York in search of better economic opportunities and to escape the circumscribed life in Karlsruhe; although this separation was meant to be short, it lasted seventeen years. Willstätter's mother took him and his brother to live near her family home in Nürnberg, a change to which Willstätter had difficulty adjusting, in part because of the more overt anti-Semitism he experienced there.

One effect of the move to a new school was that, although receiving good grades in his other subjects, he did poorly in Latin, the most important subject in the gymnasia of the time. A family council decided he should switch to the Realgymnasium and be educated for business instead of a profes-

Richard Willstätter. *(The Library of Congress. Reproduced by permission.)*

sion. Ironically, it was at this time, stimulated by some home experiments and good teachers, that he decided to become a chemist. In his autobiography, Willstätter observed that excellence in academic subjects caused one to be disliked, while athletic excellence resulted in popularity. He was also attracted to medicine and might have become a physician instead of a chemist, but because of the longer schooling required his mother would not permit him to change. An interest in biological processes remained with him, though, and is evident in the kinds of chemical problems he attacked. Much later, while teaching at Zurich, he still thought of studying physiology and internal medicine, but the death of his wife ended the idea.

In 1890 the eighteen-year-old Willstätter entered the University of Munich and also attended lectures at the Technische Hochschule. In 1893 he began his doctoral studies and was assigned to do his research under Alfred Einhorn on some aspects of the chemistry of cocaine. It was at this time that Adolf Baeyer, the leading organic chemist in Germany, began to take Willstätter under his wing. Although Willstätter never worked directly for Baeyer, he thought of himself as Baeyer's disciple. Willstätter completed his doctoral work in a year and stayed on doing independent research, becoming a privatdocent, or unsalaried lecturer, in 1896.

In his work with Einhorn, Willstätter had come to suspect that the structure assigned to cocaine by Einhorn and others was incorrect. When he started his independent research, Einhorn forbade him to work on the cocaine problem. Willstätter, with Baeyer's approval, decided to work instead on the

closely related chemical tropine, whose structure was suspected to be similar to that of cocaine; once the structure of tropine was known, the structure of cocaine could be easily derived. Willstätter showed that, indeed, the cocaine structure was not what it had been thought to be; for the remainder of his stay at Munich, Einhorn refused to speak to him. In 1902 Willstätter was appointed professor extraordinarius (roughly equivalent to associate professor), although Baeyer thought he should have accepted an industrial position. Baeyer, himself partly Jewish, also recommended that Willstätter be baptized, an act that would have removed the legal barriers he faced as a Jew. This Willstätter refused to consider. During Easter vacation in 1903 Willstätter met the Leser family from Heidelberg, and that summer he and Sophie Leser were married. Their son, Ludwig, was born in 1904 and their daughter, Margarete, in 1906.

Switzerland and Nobel Prize Work

In 1905 Willstätter accepted a call to the Eidgenössische Technische Hochscuhle in Zurich as professor of chemistry, beginning the most productive phase of his career. While at Munich he had begun an investigation into the chemical nature of chlorophyll, the green pigment in plants that converts light into energy through **photosynthesis**; at Zurich, he and his students made great strides in understanding this important material. They developed methods for isolating chlorophyll from plant materials without changing it or introducing impurities. Willstätter was then able to prove that the chlorophyll from different plants (he examined over two hundred different kinds) was substantially the same—a mixture of two slightly different compounds, blue-green chlorophyll a and yellow-green chlorophyll b, in a 3:1 ratio.

He also showed that magnesium, which had been found in chlorophyll by earlier workers, was not an accidental impurity but an essential component of these chlorophyll molecules, bonded in a way very similar to that in which iron is bonded in hemoglobin, the oxygen-carrying constituent of blood. The later work of others, especially **Hans Fischer**, in elucidating the detailed structures of the chlorophylls and hemoglobin would not have been possible without the pioneering work of Willstätter and his students. In 1913, Willstätter, in collaboration with his former student and good friend, Arthur Stoll, reviewed the work on chlorophyll in a book, *Untersuchungen über Chlorophyll*. In all, between 1913 and 1919, Willstätter published twenty-five papers in a series on chlorophyll. A preliminary step in the isolation of chlorophyll from plant materials yielded a yellow solution that on further study proved to contain carotenoid pigments. These had been described before, but Willstätter's work marked the beginning of our understanding of these materials that produce the color of tomatoes, carrots, and egg yolk.

In 1908, Willstätter suffered a devastating blow in the death of his wife after an operation for appendicitis had been delayed for thirty-six hours after the appendix had ruptured. He consoled himself with the care of his two children and with his work; in his autobiography he wrote that he took no vacations for the next ten years. During his stay at Zurich, Willstätter also did work on quinones and the mechanism of the oxidation of aniline to aniline black—a process of importance to the dye industry. He also completed a project begun eight years earlier, by synthesizing the chemical cyclooctatetraene and showing that it did not behave as an aromatic compound despite its structural similarities to benzene.

Berlin and the War

The Kaiser Wilhelm Institutes were founded in 1910 to afford outstanding scientists the chance to do research on problems of their own choosing, free of any teaching obligations. In 1911 Willstätter accepted the position of director of the Kaiser Wilhelm Institute of Chemistry and in 1912 moved into the new building at Berlin-Dahlem. The institute was situated next to the Institute for Physical Chemistry and Electrochemistry, headed by Fritz Haber, and a deep and lasting friendship developed between the two directors.

At Zurich, Willstätter had initiated a study of the pigments of various red and blue flowers, a class of compounds now known as anthocyanins. He began with dried cornflowers, or bachelor's button, because it was winter and they were commercially available. This choice, as it turned out, was not a good one; cornflowers only contained a percent or less of the pigment. In Berlin, Willstätter planted fields of double cornflowers, asters, chrysanthemums, pansies, and dahlias around the Institute and his residence. In these fresh flowers he found a much higher pigment content, up to 33% in blue-black pansies. Before World War I brought an end to this line of research, Willstätter published eighteen papers in an anthocyanin series between 1913 and 1916. He showed that the various shades of red and blue in these flowers as well as in cherries, cranberries, roses, plums, elderberries, and poppies all arose mainly from three closely related compounds, cyanidin, pelargonidin, and delphinidin chlorides, and were very dependent on the acidity or alkalinity of the flower. During the first year of the war, most of Willstätter's co-workers went into military service, and the flowers were taken to military hospitals instead of to the laboratory. Willstätter was bitterly disappointed by this interruption and could not bring himself to return to the problem after the war.

In 1915 Haber, who was in charge of Germany's chemical warfare work, asked Willstätter's assistance in developing the chemical absorption unit for a gas mask that would protect against chlorine and phosgene (a severe respiratory irritant). In five weeks, Willstätter came up with a canister containing activated charcoal and hexamethylenetetramine (also called urotropin). The use of charcoal was not new, but the use of hexamethylenetetramine was. When asked after the war how he had come to try so unusual a compound, he said that the idea had just popped into his head. For this work he received an Iron Cross, Second Class. He was also involved in an industrial research project with Friedrich Bergius on the hydrolysis of cellulose with hydrochloric acid to give dextrose, which could then be fermented to produce alcohol. The process, which was only perfected later, is now known as the Bergius-Willstätter process.

In the spring of 1915 Willstätter's ten-year-old son, Ludwig, died suddenly, apparently from diabetes. Willstätter wrote that his memory of the months following was blurred. Ironically, in November, while engaged in the work on gas masks, Willstätter learned that he had been awarded the 1915 Nobel Prize in chemistry in recognition of his work on chlorophylls and anthocyanins. Because of wartime conditions he did not travel to Stockholm to receive the prize until 1920, when a ceremony was held for a group of those who had been honored during the war. Willstätter made the trip in the company of fellow German awardees Max Planck, Fritz Haber, Max Laue, and Johannes Stark.

The Return to Munich and the Final Years

An offer of a full professorship to succeed Baeyer at Munich also came in 1915. This offer, recommended by Baeyer, was precipitated by an offer to succeed Otto Wallach, a pioneer in natural product chemistry, at Göttingen. Willstätter maintained that left to his own inclinations, he would have preferred Göttingen, because a medium-sized university would provide more contact with colleagues and greater interaction with different disciplines than was possible at large institutions. However, he accepted the appointment as professor and director of the state chemical laboratory in Munich and moved there in the spring of 1916.

He made two major demands before accepting the offer: that the old institute building be remodeled and a large addition to the chemical institute be built housing laboratories and a large lecture hall, and that a full professorship in physical chemistry be established. The first of these was contrary to the advice that the physical chemist Walther Nernst gave him before he left Berlin, "Don't ever build!" In fact, the construction, delayed by the war and post-armistice turmoil in Munich, was not completed until the spring of 1920.

At Munich, as before, Willstätter experienced the anti-Semitism that had troubled him during his earlier residence, and that finally brought about his resignation in 1924. The final straw was the refusal of the faculty to appoint the noted geochemist Victor Goldschmidt of Oslo, Norway, to succeed the mineralogist Paul von Groth, who had himself named Goldschmidt as the only one who could take his place. The sole reason for the refusal was that Goldschmidt was Jewish. When Willstätter's resignation became known, students and faculty joined in expressions of respect and confidence, urging him to reconsider. Nonetheless, he remained only for the time needed to see his students finish their research and to install Heinrich Wieland in his place. He received offers of positions at universities and in industry in Germany and abroad, but he declined all of them, finally leaving the university in September 1925 never to return.

Some of Willstätter's assistants continued work at the University, and in 1928 Wieland made room in what had been Willstätter's private laboratory for Willstätter's private assistant, Margarete Rohdewald, one of his former students. From 1929 until 1938 she collaborated with him in a series of eighteen papers on various aspects of enzyme research. It was an odd collaboration, conducted almost entirely over the telephone; Willstätter never saw her at work in the laboratory.

During the few years at Munich before his resignation, Willstätter began to concentrate his research on the study of enzymes. He had first encountered these biological catalysts in his early work on chlorophyll. Now he worked to develop methods for their separation and purification. His method for separation was to adsorb the materials on alumina or silica gel and then to wash them off using solutions of varying acidity, among other solvents. In this connection, Willstätter carried out a systematic study (comprised of nine papers) of hydrates and hydrogels during which he, with his assistants Heinrich Kraut and K. Lobinger, was able to show that aluminum hydroxide, silicic acid, ferric hydroxide, and stannic hydroxide do actually exist in solution and are not colloidal sols (dispersions of small solid particles in solution) of the corresponding oxides. Willstätter reported that this foray of an organic chemist into inorganic chemistry was not well received by inorganic chemists.

The enzyme studies were not as successful, in part because Willstätter thought that enzymes were relatively small molecules adsorbed on a protein or some other giant (polymer) molecule. The modern view, of course, is that enzymes are themselves proteins. Though Willstätter's chemical intuition failed him, there were positive results—for example, the enzymatic reduction of chloral and bromal resulted in the formation of trichloroethanol, a sedative (Voluntal), and tribromoethanol, an anesthetic (Avertin).

In 1938 the situation for Jews in Germany was becoming impossible. On a visit to Switzerland, Stoll tried to persuade Willstätter to stay, but he insisted on returning to Munich. There, after some trouble with the Gestapo, he was ordered to leave the country. After much red tape, which entailed the confiscation of much of his property, papers, and art collection, and an abortive attempt to leave unofficially, he entered Switzerland in March 1939 to stay for a while with Stoll and then to settle in the Villa Eremitaggio in Muralto. There he wrote his autobiography to pass the time. On August 3, 1942, Willstätter died of cardiac failure in his sleep. Among the honors received by Willstätter in addition to the Nobel prize were honorary membership in the American Chemical Society (1927), honorary fellowship in the Chemical Society (1927), the Willard Gibbs Medal for distinguished achievement in science from the Chicago Section of the American Chemical Society (1933), and election as foreign member of the Royal Society (1933). Willstätter's obituary by Sir Robert Robinson in *Obituary Notices of Fellows of the Royal Society,* has an eleven page bibliography, probably incomplete, listing over three hundred papers between 1893 and 1940.

WILMUT, IAN (1944-)

English embryologist

Ian Wilmut was born on July 7, 1944, in Hampton Lucey, England in Warwick. He attended the University of Nottingham, where he became fascinated with **embryology** after meeting G. Eric Lamming, a world-renowned expert in **reproduction**. The meeting became a turning point for Wilmut, who set out on a singular quest—to understand the **genetic engineering** of animals. He graduated from Nottingham in 1967, with a degree in agricultural science.

Wilmut continued his studies at Darwin College at Cambridge University in England. There he received his doctoral degree in 1973, awarded after he completed his thesis on the techniques for freezing boar semen. He immediately took a position at the Animal Breeding Research Station, an animal research institute supported by government and private funds. The research station eventually became the Roslin Institute. It is headquartered in Roslin, near Edinburgh, Scotland.

In 1973, after receiving his doctorate, Wilmut produced the first calf ("Frosty") born from a frozen **embryo** that had been implanted into a surrogate mother. The motivation for such an experiment was to harvest cows that provide the best meat and milk by implanting their embryos into other females. The average cow can birth five to ten calves during their lifespan. With the ability to transfer embryos, cattle breeders could increase the quality of their animal stock.

Wilmut continued his research during the 1980s, despite other scientists' growing discouragement in the possibility of **cloning**. In 1996, Wilmut overheard a story in an Irish bar, while attending a scientific meeting, that solidified his belief in cloning. The rumor that Wilmut heard was that Dr. Steen M. Willadsen of Grenada Genetics in Texas had cloned a lamb using a differentiated cell from an already developing embryo.

Like a fertilized egg that contains enough deoxyribonucleic acid (**DNA**) to build an entire organism, a differentiated cell carries a full complement of the genetic material for DNA, which forms a blueprint for an animal's characteristics. To clone an animal, an adult animal **cell** would have to be harvested, and the **nucleus** placed in an embryo cell, thereby replacing the nucleus of the embryo cell. Yet the problem was how to get the new nucleus to spawn growth in the embryo cell.

Keith Campbell, a biologist at the Roslin Institute, had an insight that proved to be crucial. He deduced that an **egg** probably will not use genetic material from a transplanted adult cell because the cycles of each cell are not synchronized. Cells go through specific cycles, growing and dividing and making an entirely new package of chromosomes each time. In order to synchronize the cells, Campbell slowed down adult **mammal** cells—in fact, nearly stopping them—so they would actually exist in synchrony with the embryos. Then each embryo could be joined with an adult cell, and in turn, they could join together and grow. To slow an already developing or adult cell down, Campbell forced it into a hibernating state by depriving it of nutrients. With this method, he and Wilmut were able to clone two sheep from developing embryo cells. They named the sheep Megan and Morag.

To clone an adult sheep, Wilmut and Campbell harvested udder, or mammary, cells from a six-year-old ewe. The cells were preserved in test tubes and starved by reducing their serum concentration for five days. Out of 277 attempts, Dolly's embryo was the only one to survive. They implanted Dolly's embryo when it was six days old into a surrogate mother, and on July 5, 1996, Dolly was born. She was named for the country singer Dolly Parton. They kept the lamb's birth secret until they had received a patent for the process that had created her. As a government employee working at Roslin,

Wilmut does not own the patent on his cloning procedure. The company which runs Roslin, called PPL P.L.C. Therapeutics, will benefit from the patent proceeds, and will produce new drugs through the procedure.

In late November 1997, Dolly was successfully mated with a four-year-old Welsh Mountain ram (David) and her lamb (Bonnie) was born on the morning of April 13, 1998. The birth of the lamb confirmed that while Dolly's origins were quite unique, she was able to breed normally and produce a healthy offspring.

Wilmut remains passionate about his work. His goal is to push the work of cloning animals forward, so that it can help solve some of the world's worst medical problems. He continues his cloning projects so he and other scientists can study genetic diseases for which there are presently no cures. He sees a day when genetic engineering and cloning can produce **proteins** like the clotting factor that hemophiliacs lack. Other diseases, resulting from the lack of a genetic material, might also be cured.

WILSON, EDMUND BEECHER (1856-1939)

American biologist

Edmund Beecher Wilson emphasized careful experimentation and analysis in biology at a time when the field was rife with theories based on little more than speculation. Indeed, Wilson's work was instrumental in transforming biology into a rigorous, scientific discipline. Although known for his meticulous approach to the study of the structure and function of the cell, he never lost sight of biology as a unified field that included **embryology**, **evolution**, and **genetics**. His influence in biology was felt through his position as a professor first at Bryn Mawr College and then at Columbia University, and through his highly influential textbook, *The Cell in Development and Inheritance.* His study of **chromosomes**, and especially his discovery of the sex chromosomes, helped lay the foundation for the study of genetics and evolution in the early-twentieth century. Many of the problems that Wilson tackled, including the details of cell development, remain unsolved today.

Edmund Wilson was born on October 19, 1856, in Geneva, Illinois. He was the second of four surviving children of Isaac Grant Wilson, a lawyer and eventually judge, and Caroline Louisa Clark, both of whom were originally from New England. When Edmund was two years old, his father was appointed a circuit court judge in Chicago. Rather than separate him from her childless sister and brother-in-law in Geneva, Edmund's mother left him to live with them while the rest of the family moved to Chicago. In this manner, he was "adopted" by Mr. and Mrs. Charles Patten and grew up counting himself very lucky to have two homes and four parents.

Shortly before he turned 16, Wilson taught school for one year from 1872 to 1873. As his older brother, Charles, had done the previous year, Wilson taught everything, including reading and arithmetic, to twenty-five pupils aged six to eighteen in a one-room schoolhouse. The following year he attend-

Ian Wilmut. *(AP/Wide World Photos, Inc. Reproduced by permission.)*

ed Antioch College (Yellow Springs, Ohio), following in the footsteps of an older cousin, Samuel Clarke. At Antioch, Wilson decided to devote himself to the study of biology, which, at that time, largely meant natural history.

In the fall of 1874, Wilson did not return to Antioch because he wished to prepare for studying at the Sheffield Scientific School of Yale University, which had been highly recommended to him by his cousin. To ready himself for Yale, Wilson moved to Chicago where he lived with his parents and took courses at the old University of Chicago from 1874-1875. He entered Yale in 1875 and received his bachelor's degree in 1878.

Lifelong Research Interest in Cell Development

Although Wilson's particular focus of research changed many times in his long career, his work was always concerned with gaining a better understanding of how the single fertilized egg gave rise to a complete individual, whether that individual be an earthworm, jellyfish, or human. This interest in the development of the organism led Wilson to study cell structure and function, heredity, and evolution.

During his years of graduate and postgraduate work, Wilson studied the embryology and morphology of earthworms, sea spiders, the colonial jellyfish (*Renilla*), and other invertebrates. After Yale, he again followed Sam Clarke's educational path, this time to Johns Hopkins University. A close friend, William T. Sedgwick, entered Johns Hopkins along with him. From 1878-1881, Wilson worked closely with William Keith Brooks, obtained his Ph.D. in 1881, and remained at Johns Hopkins for an additional year of postdoctoral work. In 1882 Wilson studied in Europe with the help of a loan from his older brother, Charles. He studied in Cambridge, and, with Thomas H. Huxley's recommendation, gave a paper on *Renilla* before the Royal Society in London. From England, he went to Leipzig, Germany, and then to the Zoological Station at Naples. Wilson worked for almost a year there and formed strong friendships with director Anton Dohrn and zoologist **Theodor Boveri**. (For Wilson, the embryos of marine **invertebrates** were more easily studied than those of terrestrial animals, and for almost 50 years, Wilson spent his summers working at the Marine Biological Laboratory in Woods Hole, Massachusetts.)

To visit Naples, Wilson had worked out an arrangement with Clarke, who was then teaching at Williams College (Massachusetts). The college would pay for a laboratory bench at Naples for two years as part of a professorship at Williams. Wilson would work at Naples the first year while Clarke taught at Williams, then the two would switch places. Wilson's stint at Williams College lasted between 1883 and 1884.

From Williams, Wilson moved to the Massachusetts Institute of Technology as an instructor from 1884 to 1885. There, he collaborated with his friend, William T. Sedgwick, in the creation of a textbook titled *General Biology* (1886). Wilson's next teaching appointment, unlike his previous two, offered him the time and opportunity to continue his research. M. Carey Thomas, the first dean of Bryn Mawr College (Bryn Mawr, Pennsylvania), invited Wilson to become the first professor of biology at the new women's college. he taught there between 1885 and 1891. While at Bryn Mawr, the scientist tackled the problem of cell differentiation—the way in which the fertilized egg gives rise to many kinds of specialized cells. To do this, he studied the cell-by-cell development of the earthworm and *Nereis,* a marine worm. This work, known as "cell lineage," established Wilson's reputation as a biologist of considerable skill. His 1890 and 1892 papers on *Nereis* demonstrated the value of cell lineage and inspired other scientists to pursue this fruitful avenue of research.

In 1891 Wilson accepted an appointment to become an adjunct professor of **zoology** in the new zoology department at Columbia University being organized by Henry Fairfield Osborn. He spent the rest of his career at Columbia, eventually becoming chair of the department, and retiring as DaCosta Professor in 1928. Before settling on campus, however, Wilson spent another fruitful year in Munich and Naples from 1891 to 1892. A series of lectures on the study of the cell that he gave during his first teaching year at Columbia formed the basis of his textbook *The Cell in Development and Inheritance,* published in 1896. Written before the fundamentals of heredity were understood, the book added a balanced, careful voice to the fierce debates over modes of inheritance and cell development that were occurring in biology at that time. The

book, which illuminated Wilson's penchant for observation and experimentation, was hugely influential and further cemented his already substantial reputation. The book was dedicated to Boveri, the Italian zoologist.

On September 27, 1904, Wilson married Anne Maynard Kidder. Kidder and her family lived in Washington, D.C., but spent their summers at their cottage in Woods Hole, and it was there that the two met. Their only child, Nancy, became a professional cellist. Wilson himself was an avid amateur musician, and his trips to Europe were warmly remembered as much for the music he heard as for the science he learned. A flutist as a young man, he began taking cello lessons while he was living in Baltimore. For the rest of his life, in Bryn Mawr and then New York, he always found himself a quartet of amateur musicians with which to play.

Helps Usher in Modern Era of Genetics

In 1900 the modern era of genetics was born. Three scientists, working independently from each other, stated that inherited characteristics were determined by the combination of two hereditary units, one from each parent. (Today, those two hereditary units are known as genes.) This theory had actually been published 36 years earlier by **Gregor Johann Mendel**, but had lain dormant until it was "revived" at the turn of the nineteenth century by **Hugo De Vries**, Karl Erich Correns, and Erich Tschermak von Seysenegg.

Wilson quickly saw the connection between the rediscovery of the laws of heredity and his own work with cells and cell structures. The laws of heredity stated that the fertilized egg received half of the blueprint for its own expression from each parent. Chromosomes, he theorized, were the cell structures responsible for transmitting the units of inheritance. By following instructions from the chromosomes, the fertilized egg gave rise to a complete individual.

In 1905 Wilson and Nettie Maria Stevens of Bryn Mawr College independently showed that the X and Y chromosomes carried by the sperm were responsible for determining gender: in many species, including humans, females had an XX pair of chromosomes while males had an XY pair. In eight papers published from 1905-1912 entitled "Studies on Chromosomes," Wilson brilliantly extended his study of the chromosomal theory of sex determination, and it is for this work with chromosomes that he is best remembered. He is also recognized for setting the stage for the zoology department's future excellence in genetics, as personified by **Thomas Hunt Morgan** and **Hermann Joseph Muller**.

In the last years of his career, Wilson continued his study of cell structures. Despite failing health, he also wrote the third edition of *The Cell in Development and Inheritance,* over 1,200 pages, which was published in 1925. In most respects, this was actually a completely new book that included the new discoveries in biology of the twentieth century. Wilson retired from Columbia University in 1928. He died in New York, on March 3, 1939, of bronchial pneumonia, and his ashes were buried in the churchyard of the Church of the Messiah in Woods Hole, Massachusetts.

WILSON, EDWARD O. (1929-)
American zoologist

World-renowned entomologist Edward O. Wilson is nick-named "Dr. Ant," but his achievements impact much of the field of biology. He is co-founder of the modern field of socio-biology, believed by some to be one of the great paradigms of science, which has touched off much controversy but also a great deal of research in animal and human social behavior. From his posts as Harvard Univeristy's Frank B. Baird, Jr. Pro-fessor of Science and Mellon Professor of Science, Wilson is the recipient of Sweden's Crafoord Prize (equal in stature to the Nobel Prize), a 1979 Pulitzer Prize for literature, and the 1977 National Medal of Science. He has influenced the field of animal **taxonomy** through his work in speciation theory, conducted research which led to the discovery of **phero-mones**—chemicals which cause behavior in **animals**—and has been a harbinger of the threat of mass extinction resulting from man's unchecked use of the environment.

Fateful Fishing Trip Determines Career

Edward Osborne Wilson was born on June 10, 1929, in Birmingham, Alabama. A descendant of farmers and shipown-ers in subtropical Alabama, Wilson had already decided to be-come a naturalist explorer by age seven. Fate intervened, however, when on a fishing trip he vigorously pulled his catch out of the water and its fin hit and damaged his right eye. He thus developed the habit of examining animals and objects close-up with his keen left eye, and when he subsequently read a National Geographic article entitled "Stalking Ants, Savage and Civilized" at age 10, the entomologist was born. Wilson later studied biology at the University of Alabama, obtaining a B.S. degree in this discipline in 1949 and an M.S. in 1950. In 1955, at age 26, he received his Ph.D. in biology from Har-vard. He gained full professorship in 1964, and became Frank B. Baird, Jr. Professor of Science in 1976.

The field of new systematics—the attempt to classify **species** based on the principles of **evolutionary theory**—occupied Wilson during the early years of his career. With his colleague William L. Brown, Wilson critiqued the utilization of the subspecies category, prompting revised procedures among taxonomists. In 1956, Wilson also co-developed the concept of "character displacement," which occurs when two similar species begin a process of genetic differentiation to avoid competition and cross-breeding.

During the mid- to late-1950s, Wilson traveled to Aus-tralia, the South Pacific islands, and Melanesia to further study and classify ants native to those regions. As a result of his field work in the Melanesian archipelagoes, he developed the con-cept of the **taxon** cycle, which has since been found among birds and other insects. Wilson described the taxon cycle of Melanesian ants as the process through which a species dis-perses to a new, harsher habitat and evolves into one or more new "daughter" species, which then adapt to the new habitat.

All the while, Wilson was developing the foundation for what would he would term "sociobiology" two decades later. In 1959, influenced by the rise of **molecular biology**, he proved

Edward O. Wilson. *(Courtesy of Edward O. Wilson. Reproduced by permission.)*

his hypothesis that social insects such as ants communicate through chemical releasers. Wilson crushed a venom gland ex-tracted from a fire ant and created a trail of the chemical near a colony of the same species. He had anticipated that a few ants would trace the chemical path. Instead, dozens of fire ants swarmed out of the colony to follow the trail, and were baffled at its end. "That night I couldn't sleep," Wilson notes. "I en-visioned accounting for the entire social repertory of ants with a small number of chemicals." Indeed, the chemicals came to be known as pheromones, and this discovery launched an "ex-plosion of research" on the behavior of social insects—research which continues still. Wilson wrote later that phero-mones were "not just a guidepost, but the entire message." These chemicals communicate complex instructions for fellow ants—everything from the location of food and how to obtain it to a call for help when in distress.

First to Identify Species Equilibrium Theory

In the early and middle 1960s, Wilson collaborated with Princeton University mathematician Robert H. Mac Arthur to develop the first quantitative theory of species equilibrium. Prior to their work in this area, it was believed that the regular-ity of species in a given area was maintained through incom-plete colonization. Wilson and his coauthor hypothesized that the number of species on a small island would remain constant, though the variety of species would undergo constant reshuf-fling.

Two factors affect the number of species in an ecosystem: extinction and immigration. In Wilson and MacArthur's island model, these factors are determined by the size and proximity of the islands—larger, less crowded islands typically have lower extinction rates, for example, and islands that are close together experience greater species immigration from one island to another. The "equilibrium hypothesis of island biogeography" describes the relationship between these factors in a mathematical model. The two determinants in the number of species are the rate of extinction of species (depicted by a positive sloping curve) and the rate of immigration of new species (indicated by a negative sloping curve). The actual number of species is found at the intersection of the two curves.

MacArthur and Wilson's hypothesis was borne out by a 1968 study by Wilson and biologist Daniel Simberloff, who examined the insect life on six islands off the Florida Keys. They first counted the number of insect species, then fumigated the islands and recounted eight months later. As Wilson had predicted, the number of species remained the same, while the composition of species was significantly different and did in fact evolve over time.

For this landmark work, Wilson received the Craoford Prize in 1990, awarded by the Royal Swedish Academy of Sciences. "This relatively simple idea transformed the study of species richness into a quantitative and experimental branch of biology," the academy noted. "Arguably, hardly a single important work in conservation biology is written today without the author making use of this theory as a launching ramp."

Founds Field of Sociobiology

But Wilson's greatest milestone probably was his 1975 book, *Sociobiology: The New Synthesis*. In it he defines sociobiology as "the systematic study of the biological basis of all social behavior." The term was in use prior to Wilson's landmark book, but he identified the interdisciplinary endeavor as one which was to change the way animal and human behavior is viewed and researched by the scientific community. Arthur Fisher, in *Society* magazine, declares, "Many biologists believe that sociobiology is indeed one of the great scientific paradigms, a powerful new tool for understanding some of the most baffling phenomena in the living world." Fisher compares the new framework to Darwin's theory of natural selection and Einstein's revolution of space/time theory. In fact, many of the tenets Wilson put forth in his book have gained widespread acceptance, and have aroused controversy over the ideological implications for human behavior.

The roots of Wilson's journey into this field lay in the beginning of his career. "In the forties and fifties," he says in *Society,* "we were in the midst of a very exciting development, called the new synthesis, which reinvigorated evolutionary biology by applying modern population genetics to what had previously been scattered and highly descriptive subjects.... It was a period of grand synthesis in which it seemed possible to understand some of the most intrinsically interesting phenomena."

Simultaneously, the field of molecular biology was gaining prominence, and Wilson observed that this threatened to relegate the softer study of animal behavior to a tiny corner of Harvard's biology department. Wilson began focusing on the significance of organisms as carriers of genetic information. Viewing the complex behavior of ants and other social insects in this framework prompted Wilson to describe behavior which served survival not of the individual, but of the population.

Thus Wilson was able to explain, in Darwinian terms, such characteristics as altruism, significance of kinship, communication, and specialization of labor—characteristics which had previously confounded scientists. Cooperation among individuals or between species was consistent with early evolutionary theory because it enabled individuals to survive and carry on the gene pool. But altruism (behavior in which one individual helps another at possible or certain cost to itself) and spiteful behavior (when an individual harms another and itself) were largely unexplained by biologists.

Explains Altruism in Ants

The answer lies in the broader view of population survival. In a colony of ants, sterile members will work for their family members who share similar genes and who will reproduce on their behalf. Wilson maintains that selflessness is a characteristic of most ant species. He describes their colonies as "superorganisms," in which the welfare of the colony—not the individual—is paramount. On the other end of the behavioral spectrum, a species of Malaysian ants will rupture glands of poison on their own bodies if invaded by enemies—killing themselves and their intruders, while signaling for help from members of their own colony. Other complex and intricate behavior is explained by Wilson's sociobiology. The European red amazon ant, for example, is an aggressive creature which actually invades the nests of more peaceful ant species, killing some individuals and capturing others for use as slaves in their own nests. The slave ants actually do "housework," digging chambers and feeding and nurturing the young Amazons.

Applies Sociobiology to Humans

It was the twenty-seventh chapter of *Sociobiology* which touched off a controversy that continues today. In "Man—From Sociobiology to Biology," Wilson argued for expanded research on the role of biology in human behavior. "There is a need for a discipline of anthropological genetics," he wrote. "By comparing man with other primate species, it might be possible to identify basic primate traits that lie beneath the surface and help to determine the configuration of man's higher social behavior."

Wilson noted that humans have always been characterized by "aggressive dominance systems, with males generally dominant over females." He also wrote that "a key early step in human social evolution was the use of women in barter." In a separate article, Wilson wrote: "In hunter-gatherer societies, men hunt and women stay at home. This strong bias... appears to have a genetic origin. Even with identical education and equal access to all professions, men are likely to continue to play a disproportionate role in political life, business, and science."

The anger with which Wilson's words were received led to noisy protests at a 1978 meeting of the American Association for the Advancement of Science. Intruders on the meeting first yelled a diatribe against him and then poured a pitcher of water over him. A letter of protest signed by, among others, two of Wilson's colleagues at Harvard, asserted that theories such as his in the past had led to the "sterilization laws and restrictive immigration laws by the United States between 1910 and 1930 and also for the eugenics policies which led to the establishment of gas chambers in Nazi Germany."

Wilson's worst detractors believed that the inevitable conclusion of his theories was "biological determinism." Harvard Professor Stephen Jay Gould (who had signed the letter of protest) sought a middle ground, since Wilson's theory didn't preclude the possibility that "peacefulness, equality, and kindness are just as biological as violence, sexism and general nastiness." But Gould maintained that there is no direct evidence in existence that specific human behaviors are genetically determined. In 1978, Wilson penned a follow-up to *Sociobiology,* the Pulitzer Prize–winning *On Human Nature.* In this volume he attempted to defend his hypotheses forwarded in chapter twenty-seven of *Sociobiology,* as well as to clear up certain areas that had become targets for controversy and prejudice. In particular, Fisher notes, Wilson "aimed for a fuller explanation of his views of the issues of free will, ethics, and development." Wilson's continuing research led to a collaboration with University of Toronto professor Charles Lumsden, with whom he penned 1981's *Genes, Mind and Culture* and 1983's *Promethean Fire*—the latter of which Wilson describes as his "last word on the subject" of human sociobiology.

Whatever the ramifications of Wilson's attempt to apply his entomological expertise to human behavior, his books and life's research represent great forward strides in the field of biology. His fascination with ants, begun in his childhood, culminated with the publication of 1990's *The Ants,* which he co-authored with German entomologist Bert Holldobler. Wilson, the world's leading authority on the creature with 8,800 species, believes they are essential to the world's ecosystems.

Wilson also is forthright in arguing for increased protection of the environment to minimize the mass species extinction now underway. He has warned that the current extinctions due to rainforest destruction will rival those which marked the end of the dinosaur age. Wilson has argued for surveyance of the earth's flora and fauna (the majority of which remain unclassified), the promotion of sustainable development, the wise use of the earth's plant and animal resources for food and medicine, and the restoration of terrains already damaged.

His many achievements include the Cleveland Prize (1967), the Mercer Award of the Ecological Society of America (1971), the Founders' Memorial Award from the Entomological Society of America (1972), the Leidy Medal (1978), the Carr Medal (1978), the L. O. Howard Award of the Entomological Society of America (1985), and the Tyler Prize for Environmental Achievement (1984). Wilson served on the World Wildlife Fund Board of Directors from 1984 to 1990. He is a member of the National Academy of Sciences, a fellow of the American Academy of Arts and Sciences, a fellow of the American Philosophical Society, the former president of the Society for the Study of Evolution, and an honorary member of the British Ecological Society.

Of his lifelong exploration of the insect world, Wilson says, "God is in the details." His contributions to species classification, biogeography, insect social organization, and his founding of sociobiology, have left a legacy the depth of which is yet to be measured.

See also Behaviorism

WOODWELL, GEORGE M. (1928-)
American ecologist

From the uproar in the 1960s over the insecticide dichlorodiphenyltrichloroethane (**DDT**), through the debate in the 1990s over global warming, American ecologist George M. Woodwell has been involved in nearly every environmental controversy of the late twentieth century. "I'm a citizen," Woodwell explained in a November, 1993, interview with Cynthia Washam, "and citizens have a role in steering the democracy." Woodwell has taken an active role in ecological issues throughout his career, holding such positions as founder of the Environmental Defense Fund, founding member of the Natural Resources Defense Council, president of the Ecological Society of America, founding trustee of the World Resources Institute, and chair of the World Wildlife Fund. Through his frequent articles and speeches, he has taken his plea to conserve the Earth's resources to politicians, fellow scientists, and laypeople.

While known to the general public as an activist, Woodwell also has earned the respect of his scientific colleagues. Walter Orr Roberts, director emeritus of the National Center for Atmospheric Research in Colorado, has stated that Woodwell "is characterized by solid scientific work that goes into tough issues with total objectivity," as cited by Denise Grady and Thomas Levenson in *Discover.* Woodwell has received four honorary doctoral degrees, won several scientific awards, and served on the editorial boards of three scientific journals. His research has focused on **ecosystems**, or communities of **plants** and **animals** and their environment, and he consistently delves into controversial issues that have significant implications for public policy, using the results of his research to support his activism.

Woodwell traces his interest in **ecology** to his childhood. Born October 23, 1928, in Cambridge, Massachusetts, he developed an appreciation of nature at his family's farm in Maine, where he spent his summers. There, along with his parents, Philip and Virginia (Sellers) Woodwell, both high school teachers, the young Woodwell cultivated potatoes, made maple syrup, and assisted neighbors on their farms. In 1946 Woodwell began his formal education in ecology at Dartmouth College, where he earned a bachelor's degree in **botany** in 1950. He joined the U.S. Navy as a lieutenant shortly after graduating and served for three years before returning to academia. While studying for his master's degree at Duke University, Woodwell met fellow graduate student Alice Katharine Rondthaler, who later dropped out of the program. In 1955 the couple married, and eventually they had four children: Caroline, John, Marjorie, and Jane. In 1956, Woodwell completed his master's degree in botany at Duke, and received his doctorate two years later. In the late 1950s he returned to New England as an assistant professor of botany at the University of Maine in Orono, and for several years he taught introductory ecology.

DDT Research Prompts Activist Role

During this time, Woodwell became involved in a project that changed the course of his career. He had been asked by the privately funded Conservation Foundation to study the effects of DDT on Maine forests. At first, he supported the use of the popular **pesticide**. He changed his mind a short time later, however, when he discovered that only about one half of the DDT sprayed on crops and forests actually settled on the soil—the rest was scattered by the wind. This drifting pesticide made its way into the food chain, as eagles, pelicans, ospreys, and other birds ate contaminated fish, then laid eggs with shells that were too fragile to survive. In 1966, Woodwell, along with some of his colleagues, filed the first of a series of lawsuits calling for a ban on the use of DDT. Their efforts eventually led to the Environmental Protection Agency's ban on the insecticide in 1972. The experience made Woodwell realize the value of taking environmental issues to court. To support further litigation, he founded the Environmental Defense Fund in 1967, which has become a thriving conservation law organization with more than two hundred thousand members. He also helped a group of Yale University law students establish the Natural Resources Defense Council.

By the early 1960s, Woodwell had left his post at the University of Maine to become an assistant scientist at Brookhaven National Laboratory in Upton, New York. At Brookhaven, he conducted an experiment that is recognized as a major contribution to ecological research. He planted radioactive cesium–137 in the center of a fourteen-acre oak and pine forest, and for the next eighteen years studied the radiation's effect on the ecosystem. The **forest** died in systematic stages: First the pine **trees** died, then the oaks, and then later the shrubs and grasses. Finally, only some mosses, **bacteria**, and lichens (plants made up of **algae** and fungi) were left. His experiment proved that the most sensitive species in an ecosystem died first and that only the most resistant ones survived. It also showed that, when under **stress**, a community could die in a much smaller amount of time than it had taken to develop.

Founding of Ecosystems Center Leads to Research into Deforestation

The ecologist eventually decided his opportunities for ecological research were limited at the physics laboratory, and in 1975 he left to establish the Ecosystems Center at the Marine Biological Laboratory in Woods Hole, Massachusetts. Woodwell's first major project there was studying sources of carbon dioxide in the **atmosphere**. Carbon dioxide is a greenhouse gas; in other words, it traps the sun's heat and contributes to dangerous global warming. Woodwell learned that atmospheric carbon dioxide is produced not only by industrial

and auto emissions, but also when forests are burned or plowed under: When trees are destroyed, he found, they release the carbon dioxide they normally absorb. Woodwell concluded that deforestation increases atmospheric carbon dioxide in two ways—first, by releasing carbon dioxide, and second, by destroying the forests' ability to absorb the gas. These findings prompted Woodwell to publicly condemn the destruction of the Earth's forests and call for drastic cuts in carbon dioxide emissions. "There's no chance that people will continue on an Earth that continues to warm," he told Washam. "It's important to see that the warming does not proceed. We would control it by reducing the use of fossil fuels by sixty percent and by stopping deforestation." Woodwell has taken his plea to the U.S. Congress several times, but has seen no action. The government's apparent reluctance to take strong steps to save the environment has been his greatest frustration. Still, he persists in fighting for conservation. In 1983, he served as chair of an international conference on the biological effects of nuclear war. There, he succeeded in gaining a consensus among more than one hundred scientists, who ultimately agreed that even a small-scale nuclear war would cause temperatures to drop below zero for months, a phenomenon called nuclear winter. In addition, in the early 1990s, Woodwell proposed the creation of an International Commission on the Conservation and Utilization of World Forests to stem the global destruction of forests. He had the support of several countries, including the United States. "I have no doubt the commission will exist," he told Washam. "The question is when."

Establishes Ecology Research Center

In 1985, Woodwell decided to leave the Ecosystems Center in order to develop his own ecology institute: the Woods Hole Research Center. With funding from federal grants and private foundations, the non-profit center focuses on the ecological impact of toxic waste, air pollution, deforestation, and other major environmental threats. Woodwell, as the director, uses the results of the center's studies to support his demand for an end to environmental destruction.

Woodwell's involvement with ecological issues has been recognized several times by his peers as well as by environmental organizations. In 1975, he garnered the Green World Award from the New York Botanical Garden, and in 1982 he received a Distinguished Service Award from the American Institute of Biological Sciences. He was elected to the National Academy of Sciences in 1990, won the Dartmouth College Class of 1950 Award in 1991, and received the Hutchinson Medal from Garden Clubs of America in 1993. He also has been awarded honorary doctorates from Williams College, Miami University, Carleton College, and Muhlenberg College.

While calling for action from industry and government, Woodwell takes care to monitor his own influence on the health of the planet. He walks to work every day and often travels around the village of Woods Hole on his bicycle. Several years ago, he and his son erected a twenty-eight-foot high structure containing more than twenty solar panels in order to heat the water in their home. Though Woodwell rarely takes a break from his work, he enjoys spending any free time on his boat, whose source of power is simply the wind.

WRIGHT, ALMROTH EDWARD (1861-1947)

English bacteriologist

Almroth Edward Wright made several significant contributions to science and is perhaps best known for introducing a vaccination against typhoid fever. Developed near the turn of the twentieth century, the vaccine was used on British soldiers during World War I and was responsible for saving many lives. The disease only claimed the lives of 1,191 British soldiers, instead of a projected 125,000 without the vaccination, according to estimates outlined in Leonard Colebrook's biography, *Almroth Wright: Provocative Doctor and Thinker.* Numerous honors were bestowed upon Wright for his scientific work, including a knighthood and election as a Fellow of the Royal Society of London, both of which were awarded in 1906.

Wright was born August 10, 1861, in Middleton Tyas, Yorkshire, England. He was the second son of Reverend Charles Henry Hamilton and Ebba Johanna Dorothea (Almroth) Wright. His father was an Old Testament scholar and a militant protestant. His mother was the daughter of a chemistry professor who was also governor of the Royal Mint in Stockholm. In his early years Wright was educated by tutors and lived in Germany and France where his father worked as a minister. Eventually, the family settled in Ireland, and Wright received his university education at Trinity College in Dublin, earning a degree in modern literature in 1882 and a degree in medicine in 1883. Winning a traveling scholarship to the University of Leipzig in Germany, Wright studied medicine there for a year.

Wright then returned to England, and was a bit unsure as to whether the future direction of his career led to literature or medicine; he soon decided to read law, and after two years took the civil service exam. Eventually Wright's interest in science took precedence over his other pursuits. After securing a fairly non-demanding position at the Admiralty in 1885, he also immediately began working evenings at the Brown Institution (University of London) as a science researcher on a volunteer basis. Wright was next offered a demonstratorship in the department of pathology at Cambridge in 1887, then soon after transferred to the department of physiology. Upon working in Germany for several months, Wright accepted a demonstratorship at the University of Sydney, in Australia, in 1889. That same year Wright married Jane Georgina Wilson, with whom he had two sons and a daughter.

In 1892 Wright was offered the chair of the pathology department at the Army Medical School in Netley, England. This was the first time Wright worked close to patients, and he claimed the atmosphere was productive since it never allowed the scientist to become too far removed from the ultimate goal of his work, which was to cure the sick. It was at this time Wright began his research on the phenomenon of blood coagulation, eventually linking clotting time to the presence of calcium in the **blood**. Laboratory instruments during this period were generally crude and home-made, so Wright—a pioneer in laboratory testing—made his own, de-

veloping and producing capillary tubes large enough to hold only a few drops of blood. These instruments could test blood without the necessity of drawing a great deal of it from a patient; all that would be required was a finger prick. Wright also recognized the importance of uniformity in laboratory testing, so he made sure each tube was identical.

Wright discovered that if blood was clotting too slowly, giving the patient a dose of calcium by mouth would speed up the process. Conversely, if clotting occurred too quickly, he found administering citric acid to the patient slowed it down. These same principles were also applied to a situation Wright was experiencing at home. Wright's young child seemed to experience distress when fed cow's milk; upon testing, Wright found cow's milk to have a greater concentration of calcium with harder, thicker clots than breast milk. Adding citrate of soda to the milk made the clots softer and thinner, rendering the milk easier for his child to digest, and thus decreasing digestive pains. Wright then tried feeding lemons to the family cow to see if it would change the concentration of calcium in the milk produced. The cow did not respond, but the housekeeper by this point had had enough and turned in her resignation.

Begins Work on Typhus

Near the turn of the century typhoid fever had a death rate of 10-30%. Although the disease had been partially eradicated with better sewage handling, Wright did not think this would eliminate the problem and believed these methods would break down during a war. Wright wanted to test the effects of injection with a heat-killed typhoid **culture**, to see if it would produce antibodies. He found it did, but there was what he termed a "negative phase"—a period of one to two days where **antibodies** seemed to decrease. Nonetheless, he believed his vaccine would be beneficial and set out trying to convince medical authorities of its merits. Wright convinced the War Office committee to set up an experimental situation, using military units over a three year period. Frequencies of inoculation and instances of typhoid records were measured, and in 1909 very positive results were published: Colebrook relates in *Almroth Wright* that deaths per 1,000 inoculated soldiers were 0.38, while for uninoculated were 3.93.

In 1906, prior to the publication of the typhoid inoculation results, Wright had been knighted and elected to Fellow of the Royal Society. After this success, he turned his lab over to serum production, so the vaccination would be available. Wright also wrote a long letter to the editor of the *New York Times* urging mandatory inoculation of troops. In 1914, only British troops entered World War I fully inoculated.

In the midst of his typhoid work, Wright had changed positions in 1902, from his professorship at the Army Medical School at Netley to pathologist and professor of pathology at St. Mary's Hospital in London. In 1911, Wright traveled to South Africa to help produce a pneumonia inoculation for the men who were working in the mines. The system Wright developed to inoculate the miners resembled the one he instituted earlier to fight typhus.

During World War I Wright served in France as head of a research lab which worked primarily on wound infections.

Wright developed at this time a method using a hypertonic salt solution to draw lymph into open wounds (lymph is a fluid derived from blood and which contains **lymphocytes**, a type of white blood cell which repels infection). Wright also developed a scientific basis for early wound closure, or suturing, which was not in practice up until that time. Several citations were presented to Wright after World War I, including a special medal of the Royal Society of Medicine in 1920 which credited him with providing the best medical work during the war.

Engages in Philosophical Debate

Wright's direct influence on scientific research seemed to taper off after World War I. His indirect influence was felt for many years, however, as several of his students went on to great fame, including Alexander Fleming, the scientist who discovered penicillin. For Wright this era was more a time for reflection and what Colebrook describes in *Almroth Wright* as the scientist's "search for truth".

Among the reasons contributing to Wright's declining influence may be his rather unpopular views. The treatise *The Unexpurgated Case Against Woman Suffrage,* for example, attempted to demonstrate the intellectual and psychological inferiority of women. Although the playwright George Bernard Shaw disagreed with this claim, he was, nevertheless, an admirer of Wright; the lead character in Shaw's play *The Doctor's Dilemma* is modeled after Wright, and the idea for the play came from the many discussions the writer shared with Wright as well as other members of the medical profession.

Wright published over 150 papers during his career. He advanced the truly scientific component of research to a great degree due to his insistence on the use of the scientific method, which involves several steps, including the formation of a hypothesis and the testing and confirmation of that hypothesis. Commonly accepted now, the scientific method was a revolutionary idea during Wright's time.

Wright continued his work at St. Mary's Hospital until 1946 and died shortly after in Buckinghamshire, England, on April 30, 1947. Wright was working—literally to the end—on a philosophical work *Alethetropic Logic* (in Wright's words, "a system of Logic which searches for the Truth"), which was published posthumously through the efforts of his grandson.

Y

YALOW, ROSALYN SUSSMAN (1921-)
American medical physicist

Rosalyn Sussman Yalow was co-developer of radioimmunoassay (RIA), a technique that uses radioactive isotopes to measure small amounts of biological substances. In widespread use, the RIA helps scientists and medical professionals measure the concentrations of **hormones**, **vitamins**, **viruses**, **enzymes**, and drugs, among other substances. Yalow's work concerning RIA earned her a share of the Nobel Prize in physiology or medicine in the late 1970s. At that time, she was only the second woman to receive the Nobel in medicine. During her career, Yalow also received acclaim for being the first woman to attain a number of other scientific achievements.

Yalow was born on July 19, 1921, in The Bronx, New York, to Simon Sussman and Clara Zipper Sussman. Her father, owner of a small business, had been born on the Lower East Side of New York City to Russian immigrant parents. At the age of four, Yalow's mother had journeyed to the United States from Germany. Although neither parent had attended high school, they instilled a great enthusiasm for and respect of education in their daughter. Yalow also credits her father with helping her find the confidence to succeed in school, teaching her that girls could do just as much as boys. Yalow learned to read before she entered kindergarten, although her family did not own any books. Instead, Yalow and her older brother, Alexander, made frequent visits to the public library.

During her youth, Yalow became interested in mathematics. At Walton High School in The Bronx, her interest turned to science, especially chemistry. After graduation, Yalow attended Hunter College, a women's school in New York that eventually became part of the City University of New York. She credits two physics professors, Dr. Herbert Otis and Dr. Duane Roller, for igniting her penchant for physics. This occurred in the latter part of the 1930s, a time when many new discoveries were made in nuclear physics. It was this field that Yalow ultimately chose for her major. In 1939

she was further inspired after hearing American physicist Enrico Fermi lecture about the discovery of nuclear fission, which had earned him the Nobel Prize the previous year.

Overcomes Sex Bias

As Yalow prepared for her graduation from Hunter College, she found that some practical considerations intruded on her passion for physics. At the time, most of American society expected young women to become secretaries or teachers. In fact, Yalow's parents urged her to pursue a career as an elementary school teacher. Yalow herself also thought it unrealistic to expect any of the top graduate schools in the country to accept her into a doctoral program or offer her the financial support that men received. "However, my physics professors encouraged me and I persisted," she explained in *Les Prix Nobel 1977*.

Yalow made plans to enter graduate school via other means. One of her earlier college physics professors, who had left Hunter to join the faculty at the Massachusetts Institute of Technology, arranged for Yalow to work as secretary to Dr. Rudolf Schoenheimer, a biochemist at Columbia University in New York. According to the plan, this position would give Yalow an opportunity to take some graduate courses in physics, and eventually provide a way for her to enter a graduate a school and pursue a degree. But Yalow never needed her plan. The month after graduating from Hunter College in January, 1941, she was offered a teaching assistantship in the physics department of the University of Illinois at Champaign-Urbana.

Gaining acceptance to the physics graduate program in the College of Engineering at the University of Illinois was one of many hurdles that Yalow had to cross as a woman in the field of science. For example, when she entered the University in September, 1941, she was the only woman in the College of Engineering's faculty, which included four hundred professors and teaching assistants. She was the first woman in more than two decades to attend the engineering college.

Rosalyn Sussman Yalow. *(The Library of Congress. Reproduced by permission.)*

Yalow realized that she had been given a space at the prestigious graduate school because of the shortage of male candidates, who were being drafted into the armed services in increasing numbers as America prepared to enter World War II.

Yalow's strong work orientation aided her greatly in her first year in graduate school. In addition to her regular course load and teaching duties, she took some extra undergraduate courses to increase her knowledge. Despite a hectic schedule, Yalow earned A's in her classes, except for an A- in an optics laboratory course. While in graduate school she also met Aaron Yalow, a fellow student and the man she would eventually marry. The pair met the first day of school and wed about two years later on June 6, 1943. Yalow received her master's degree in 1942 and her doctorate in 1945. She was the second woman to obtain a Ph.D. in physics at the University.

After graduation the Yalows moved to New York City, where they worked and eventually raised two children, Benjamin and Elanna. Yalow's first job after graduate school was as an assistant electrical engineer at Federal Telecommunications Laboratory, a private research lab. Once again, she found herself the sole woman as there were no other female engineers at the lab. In 1946 she began teaching physics at Hunter College. She remained a physics lecturer from 1946-1950, al-

though by 1947 she began her long association with the Veterans Administration by becoming a consultant to Bronx VA Hospital. The VA wanted to establish some research programs to explore medical uses of radioactive substances. By 1950, Yalow had equipped a radioisotope laboratory at the Bronx VA Hospital and decided to leave teaching to devote her attention to full-time research.

That same year Yalow met Solomon A. Berson, a physician who had just finished his residency in internal medicine at the hospital. The two would work together until Berson's death in 1972. According to Yalow, the collaboration was a complementary one. In Olga Opfell's *Lady Laureates* Yalow is quoted as saying, "[Berson] wanted to be a physicist, and I wanted to be a medical doctor." While her partner had accumulated clinical expertise, Yalow maintained strengths in physics, math, and chemistry. Working together, Yalow and Berson discovered new ways to use radioactive isotopes in the measurement of **blood** volume, the study of iodine **metabolism**, and the diagnosis of thyroid diseases. Within a few years, the pair began to investigate adult-onset diabetes using radioisotopes. This project eventually led them to develop the groundbreaking radioimmunoassay technique.

Diabetes Mystery Leads to a Discovery

In the 1950s some scientists hypothesized that in adult-onset diabetes, **insulin** production remained normal, but a liver enzyme rapidly destroyed the peptide hormone, thereby preventing normal glucose metabolism. This contrasted with the situation in juvenile diabetes, where insulin production by the pancreas was too low to allow proper metabolism of glucose. Yalow and Berson wanted to test the hypothesis about adult-onset diabetes. They used insulin "labeled" with iodine–131. (That is, they attached, by a chemical reaction, the radioactive isotope of iodine to otherwise normal insulin **molecule**s.) Yalow and Berson injected labeled insulin into diabetic and non-diabetic individuals and measured the rate at which the insulin disappeared.

To their surprise and in contradiction to the liver enzyme hypothesis, they found that the amount of radioactively labeled insulin in the blood of diabetics was higher than that found in the control subjects who had never received insulin injections before. As Yalow and Berson looked into this finding further, they deduced that diabetics were forming antibodies to the **animal** insulin used to control their disease. These antibodies were binding to radiolabeled insulin, preventing it from entering cells where it was used in sugar metabolism. Individuals who had never taken insulin before did not have these antibodies and so the radiolabeled insulin was consumed more quickly.

Yalow and Berson's proposal that animal insulin could spur antibody formation was not readily accepted by immunologists in the mid–1950s. At the time, most immunologists did not believe that antibodies would form to molecules as small as the insulin **peptide**. Also, the amount of insulin antibodies was too low to be detected by conventional immunological techniques. So Yalow and Berson set out to verify these minute levels of insulin antibodies using radiolabeled insulin as their marker. Their original report about insulin antibodies, however, was rejected initially by two journals. Finally, a compromise version was published that omitted ''insulin antibody'' from the paper's title and included some additional data indicating that an antibody was involved.

The need to detect insulin antibodies at low concentrations led to the development of the radioimmunoassay. The principle behind RIA is that a radiolabeled antigen, such as insulin, will compete with unlabeled antigen for the available binding sites on its specific antibody. As a standard, various mixtures of known amounts of labeled and unlabeled antigen are mixed with antibody. The amounts of radiation detected in each sample correspond to the amount of unlabeled antigen taking up antibody binding sites. In the unknown sample, a known amount of radiolabeled antigen is added and the amount of radioactivity is measured again. The radiation level in the unknown sample is compared to the standard samples; the amount of unlabeled antigen in the unknown sample will be the same as the amount of unlabeled antigen found in the standard sample that yields the same amount of radioactivity. RIA has turned out to be so useful because it can quickly and precisely detect very low concentrations of hormones and other substances in blood or other biological fluids. The principle can also be applied to binding interactions other than that between antigen and antibody, such as between a binding protein or **tissue** receptor site and an enzyme. In Yalow's Nobel lecture, recorded in *Les Prix Nobel 1977,* she listed more than one hundred biological substances—hormones, drugs, vitamins, enzymes, viruses, non-hormonal **protein**s, and more— that were being measured using RIA.

In 1968 she became a research professor at the Mt. Sinai School of Medicine, and in 1970, she was made chief of the Nuclear Medicine Service at the VA hospital. Yalow also began to receive a number of prestigious awards in recognition of her role in the development of RIA. In 1976, she was awarded the Albert Lasker Prize for Basic Medical Research. She was the first woman to be honored this laurel—an award that often leads to a Nobel Prize. In Yalow's case, this was true, for the very next year, she shared the Nobel Prize in physiology or medicine with **Andrew V. Schally** and **Roger Guillemin** for their work on radioimmunoassay. Schally and Guillemin were recognized for their use of RIA to make important discoveries about **brain** hormones.

Berson had died in 1972, and so did not share in these awards. Ecstatic to receive such prizes, Yalow was also saddened that her longtime partner had been excluded. According to an essay in *The Lady Laureates,* she remarked that the ''tragedy'' of winning the Nobel Prize ''is that Dr. Berson did not live to share it.'' Earlier Yalow had paid tribute to her collaborator by asking the VA to name the laboratory, in which the two had worked, the Solomon A. Berson Research Laboratory. She made the request, as quoted in *Les Prix Nobel 1977,* ''so that his name will continue to be on my papers as long as I publish and so that his contributions to our Service will be memorialized.''

Yalow has received many other awards, honorary degrees, and lectureships, including the Georg Charles de

Henesy Nuclear Medicine Pioneer Award in 1986 and the Scientific Achievement Award of the American Medical Society. In 1978, she hosted a five-part dramatic series on the life of French physical chemist Marie Curie, aired by the Public Broadcasting Service (PBS). In 1980 she became a distinguished professor at the Albert Einstein College of Medicine at Yeshiva University, leaving to become the Solomon A. Berson Distinguished Professor at Large at Mt. Sinai in 1986. She also chaired the Department of Clinical Science at Montefiore Hospital and Medical Center in the early- to mid–1980s.

By all accounts, Yalow was an industrious researcher, rarely taking time off. For example, some reports claim that she only took a few days off of work following the birth of her two children. In *The Lady Laureates,* Opfell reported that when the VA Hospital put on a party in honor of Yalow's selection for the Lasker Prize, Yalow herself ''brought roast turkeys from home and stood in the middle of a meeting peeling potatoes and making potato salad while fellows reported to her.''

The fact that Yalow was a trailblazer for women scientists was not lost on her, however. At a lecture before the Association of American Medical Colleges, as quoted in *Lady Laureates,* Yalow opined: ''We cannot expect that in the foreseeable future women will achieve status in academic medicine in proportion to their numbers. But if we are to start working towards that goal we must believe in ourselves or no one else will believe in us; we must match our aspirations with the guts and determination to succeed; and for those of us who have had the good fortune to move upward, we must feel a personal responsibility to serve as role models and advisors to ease the path for those who come afterwards.''

YEAST

Yeast is a unicellular fungus. There are many different **species** of yeast, but the majority of them are acomycetes. They all reproduce by a process of budding and they all have the ability to ferment sugars. Yeasts are a rich source of **protein** and **vitamins**, particularly of the B complex. There are two main industries that revolve around the use of yeast and their metabolic products. The first of these industries is baking, where **carbon dioxide** from the breakdown of sugar bubbles through the dough, making the bread rise. The second industry is the **alcohol** industry where the sugar is turned to alcohol by the action of the yeast.

One of the most common species encountered both in bread and alcohol production is *Saccharomyces cerevisiae.* Historically, this species has also been used in genetic research; it contains 17 **linkage** groups, which have some 150 known mutants associated with them. **Mitochondrial** mutants have also been extensively studied. Many other species of yeast exist and some companies have their own species that they use for the production of alcohol. (Different yeasts supposedly impart radically different tastes to alcohol.)

Yeasts are not a formal taxonomic group, but a growth form shown by a widely unrelated group of **organisms**. Some

Yeast (*Anthrocobia muelleri*). (*Phototake NYC. Reproduced by permission.*)

filamentous forms of **fungi** will show a yeast form under certain conditions. There are between five and six hundred species of yeast.

Z

ZAMECNIK, PAUL CHARLES (1912-)
American physician and genetic researcher

Paul Zamecnik was one of the first scientists to explore the field of **genetics**. Beginning his work shortly after the structure of **deoxyribonucleic acid (DNA)** was discovered in 1953, Zamecnik was the first to describe how the micromachinery of each **cell** turns the **genetic code** of DNA into the **proteins** that build cells and make them work. His groundbreaking research, spanning more than 60 years, opened up vast new areas of inquiry in genetics and pharmaceuticals, especially the hunt for "gene therapy" and drugs that work on the genetic level.

Son of a Musical Family

Zamecnik was born in Cleveland, Ohio, on November 22, 1912. His grandfather was a bandleader and composer who had emigrated from Prague, in what is now the Czech Republic, to Cleveland, where there was a significant **population** of what were then known as Bohemians. Zamecnik (pronounced ZAM-es-nick) remembers sitting on his father's shoulders as his grandfather's marching band paraded past on Cleveland's Euclid Avenue.

Zamecnik's father was a banker in Cleveland, and Paul decided he wanted to be a doctor. On the advice of a high school teacher whom he admired, he went to Dartmouth College in Hanover, New Hampshire. "I was a city boy, why not try the country? I liked it there," he told contributor Karl Leif Bates in an October 1997 telephone interview. Dartmouth enabled students to complete their premedical curriculum in three years and then go on for two years of medical school there, which Zamecnik did. He then transferred to Harvard University in 1933 to finish his last two years of medical school, earning his M.D. in 1936.

Upon earning this degree, Zamecnik took a departure from the normal course of study, which was to change his career. He took a research-oriented residency at C.P. Huntington Memorial Hospital, a 24-bed **cancer** facility owned by Har-

vard. "I thought I would become a surgeon, but I was rather dazzled by the prospect of working in a laboratory and seeing some medical patients," he told Bates. "This idea of pursuing research was rather interesting." But cancer research, at the time, was thought to be a dead-end career.

Just to be sure he wasn't missing anything in medicine, or Cleveland, Zamecnik took a general medicine internship at Western Reserve University Hospital in Cleveland. But from then on, his career has been strongly focused on the lab. Following the internship, he took a fellowship post at the Carlsberg Laboratories in Copenhagen, a biochemistry lab that was "a mecca for Americans at that time." One of his professors at Carlsberg was heard telling a colleague "that young man is throwing his medical education down the sink," Zamecnik remembers. He had to leave Copenhagen in 1940, as the Nazis began to take over Europe. Americans were neutral in the war at the time, but were having trouble getting their fellowship checks from home cashed.

He returned to the States to work at the Rockefeller Institute under biochemist Max Bergmann, who was trying to figure out how proteins were made in the cell and what their structures and functions were. Zamecnik thought it was a crucial area to study to advance medicine, because surgeons could operate on a disorder after the fact, "but had to leave the rest up to nature." With only one antibiotic on the market at the time, the biochemical approach to medicine was hardly considered.

Bergmann's idea was that the same **enzymes** involved in breaking proteins apart were also somehow responsible for putting proteins together. He had found a set of enzymes inside cells that were complementary to known break-down enzymes of the **digestive system**. Bergmann thought that the proteins were so specific that they might synthesize proteins under the proper conditions. Interestingly, Bergmann had initially rejected Zamecnik's query about working in his lab because he was a medical doctor, not an organic chemist. But after his experience in Copenhagen, "I had a little stardust on my shoulders from Carlsberg, so he decided to take me." When the war reached the United States in late 1941 and early 1942, though, plans changed. Zamecnik returned to Harvard to work in the Huntington Labs on war-related research, like separating and typing **blood**.

Still Puzzled by Source of Proteins

As the war ended, Zamecnik was still puzzling over **protein synthesis**. "Nothing was known about protein synthesis, there was no map to go by. It took up one paragraph in a biochemical textbook," he said. Why, he wondered, does a tumor cell continue to multiply wildly, when other cells seem to know when to quit? "So, I thought, 'there's something wrong with the **regulation**, but we've got to get inside the cell to figure out what it is.'"

Two key things had happened during the war. First, a scientist at Massachusetts General Hospital, **Fritz Lipmann** proposed that proteins are made by a different set of enzymes, and second, the government's Oak Ridge Labs had developed Carbon-14, a very useful radioactive probe. "So we decided to find out whether Bergmann was right or whether Lippman was right." Using a marker made by attaching C_{14} to an **amino acid**, Zamecnik and colleagues at Harvard found that aminos were the building blocks of proteins, but that assembly would require quite a bit of **energy**. He then identified where the energy for that assembly came from, **adenosine triphosphate, or ATP.**

"We felt quite confident that Lippman was right, and that there were a different set of enzymes involved." From 1948 to 1952 he worked on finding a cell-free system which produced protein in the presence of an energy donor. This work "created a whole new area of biochemistry. One pearl after another dropped out, as the days went on. It was very exciting."

A Glimpse into the Cell

In 1953, **James Watson** and **Francis Crick** first proposed the **double-helix** structure of deoxyribonucleic acid (DNA). Now scientists had a better idea about how DNA could convey information and duplicate itself reliably, but still no one knew how that information was turned into action inside the cell. In late 1955, and early '56, Zamecnik proposed the existence of "transfer **RNA**," a ribonucleic acid like DNA that could form a complement to the genetic strand and carry its information from the **nucleus** of the cell out into the **cytoplasm**, where proteins are made. Then he announced the discovery of **ribosomes**, small globular bodies in the cell's cytoplasm that appear to "read" the stretch of transfer RNA and bring in the amino acid building blocks it specifies to assemble the protein. "It seemed to be a spool on which the reaction took place," Zamecnik said. Then two competing labs were able to put the finishing touches on Zamecnik's idea, decoding the language of transfer RNA and identifying yet another code, messenger RNA, which worked between DNA and transfer RNA.

For the next 20 years Zamecnik turned his attention to research which created a whole new field of inquiry for the pharmaceutical industry, fighting diseases of protein synthesis by jamming the cell's DNA signals. The mechanism for this

was something now called "antisense DNA," a complementary short strand of DNA that can be used to bind to a piece of messenger RNA and stop it from working. Zamecnik targeted a **virus**, Rous Sarcoma Virus, which caused cancer in chickens. As he struggled to sequence the virus's genes and then make the complementary DNA sequences which would prevent it from copying itself, the tools of biotechnology took great strides forward.

By 1978, Zamecnik finally had his breakthrough against Rous Sarcoma, a strand of man-made DNA that blocked the virus' ability to copy itself. Other scientists had figured out how to sequence genes, chop them into manageable pieces, and make new sequences to order. Using these new tools, Zamecnik was finally able to make his antisense approach work. "Our results indicated that the small pieces could get in and affect the **metabolism** of cells. I was astonished that the (man-made sequences) did get into the cell, and blocked the **replication** of the Rouse's Sarcoma cells." He has been called "the father of antisense," but Zamecnik said he fought the use of the term for years, since it sounds so much like "nonsense."

Drug companies have been striving, since his 1978 breakthrough, to devise man-made antisense strands that will effectively block the genetic signals that cause protein-related diseases. While most drugs in use today treat the disorder after an errant protein has been manufactured by the cell, the antisense approach is believed to have great promise in the treatment of hepatitis, cancers, coronary artery disease and many other disorders, before they occur in the cell.

Zamecnik was a professor of oncologic medicine and director of the Huntington Lab at Harvard from 1956-1979. He was also a physician at Massachussets General Hospital for the same period. Upon retiring from Harvard, he joined the Worcester Foundation for Experimental Biology in Shrewsbury, Massachusetts, as principal scientist. In 1989, he founded a company, Hybridon, to pursue antisense drugs. He left the Worcester Foundation in July 1997 when it was acquired by the University of Massachusetts and went to work at Hybridon full time, keeping long hours in the lab well into his 80s. "I'm too old to retire now," he told Bates a month before his 85th birthday. "I've muffed it. I'm not very good at gardening nor at hammering nails and I don't consider this work. It's interesting. Like watching a horse race."

When he received the first ever-awarded Albert Lasker Award for Special Achievement in Medical Science in 1996, Zamecnik was cited for his "brilliant and original science that revolutionized biochemistry and created an entirely new field of scientific inquiry." He was also awarded a National Medal of Science in 1991, The National Cancer Society National Award in 1968, and several honorary degrees.

Zamecnik has been married to Mary Zamecnik since 1936, and she still assists him in the lab. They have three children.

ZINDER, NORTON (1928-)
American molecular geneticist

Norton Zinder is a molecular geneticist and John D. Rockefeller Jr. Professor of molecular **genetics** at Rockefeller University in New York City. He also serves as the university's dean of graduate and postgraduate studies. Zinder is known primarily for his research during the late 1940s and early 1950s, when he discovered a new mechanism of genetic transfer called bacterial transduction. This process refers to the transfer of **genetic material** between **bacteria** through bacterial **virus**es. The discovery has shed new light on the location and behavior of bacterial **gene**s. Zinder is the recipient of numerous awards and honors, including the 1962 Eli Lilly Award in Microbiology and Immunology from the American Society of Microbiology.

Norton David Zinder, the older of two boys, was born on November 7, 1928, in New York City to Harry Zinder, a manufacturer, and Jean (Gottesman) Zinder, a homemaker. He attended New York City public schools, graduating from the prestigious Bronx High School of Science, and went on to attend Columbia University, where he received his B.A. in biology in 1947. The following year, at the recommendation of Francis Ryan, a professor of **zoology** at Columbia and in whose laboratory he had worked, Zinder commenced his graduate career at the University of Wisconsin. There, he studied under American geneticist **Joshua Lederberg**, who had already discovered genetic **conjugation** (or "mating") a few years earlier and who would win a Nobel Prize in 1958 for his viral and bacterial research. Zinder focused his research on microbial genetics (the study of the genetics of microorganisms), at a time when the field was relatively new and when many basic phenomena were as yet undiscovered.

In 1946, Lederberg had researched mating in *Escherichia coli*— a bacterium that is found in the intestinal tract of **animal**s and which can cause bacterial dysentery. Zinder wished to continue Lederberg's investigations, and he chose to study the closely related **genus** of *Salmonella*— bacteria that cause illnesses such as typhoid fever or food poisoning in humans and other warm-blooded animals. For his work, Zinder needed to obtain large numbers of mutant strains, which were, at the time, acquired by randomly testing the survivors among bacteria that had been treated with mutagens, or agents that increase both the chance and extent of **mutation**. Zinder, however, wanted to experiment with a different method of acquisition: He knew that mutant bacteria will not grow in a nutritionally deficient medium and that antibiotic **penicillin** will kill only growing bacteria. So, he was able to collect bacteria into an environment that was nutritionally inadequate, then kill any normal bacteria by administering penicillin.

Discovers Genetic Transduction

Zinder obtained large numbers of mutant bacteria using this method, and he began his experiments to investigate conjugation in *Salmonella;* however, instead of observing conjugation, he stumbled upon a different method of genetic transfer in bacteria: genetic transduction. As Zinder continued his research, he determined that genetic material is transferred from one bacterial cell to another by means of a phage, or a virus that invades the bacterial cell, assumes control over the cell's genetic material, reproduces, then eventually destroys the cell. Zinder's discovery of this genetic transfer has led to further studies into the mapping and behavior of genes found in bacte-

Norton Zinder. *(Science Photo Library, Photo Researchers, Inc. Reproduced by permission.)*

ria. For example, Milislav Demerec and other researchers at New York's Cold Spring Harbor Laboratory later found that the bacterial genes that regulate biosynthetic steps are grouped in what have become known as "**operon**s," a term coined in 1960 to describe closely linked genes that function as an integrated whole.

In subsequent investigations, Zinder and his team also discovered the F2 phage, very small in size and the only virus known to contain **RNA (ribonucleic acid)** as its genetic substance. The researchers ascertained that the RNA generated by the virus contains codes for specific **amino acids**—the building blocks of protein **molecules**—as well as signals to control the termination and initiation of **protein** chains.

Zinder received his M.S. in genetics in 1949 from the University of Wisconsin and married Marilyn Estreicher in December of that same year; the couple eventually had two sons, Stephen and Michael. In 1952 Zinder completed his Ph.D. in medical microbiology, then accepted the post of assistant professor at Rockefeller University (then Rockefeller Institute for Medical Research). By 1964 he had become a full professor of genetics, and approximately ten years later he was named John D. Rockefeller Jr. Professor of Molecular Genetics; in 1993 he was appointed dean of graduate and postgraduate studies. The primary focus of Zinder's research has been in the molecular genetics of phages.

In addition to his positions at Rockefeller, Zinder also has been associated with other institutions. In the mid–1970s he began lengthy affiliations with the science departments of Harvard University, Yale University, and Princeton University, and, beginning in the same period, he also worked in the viral **cancer** program at the National Cancer Institute. In 1988 he assumed the position of chair of the program advisory committee for the National Institutes of Health (NIH) **Human Genome Project**, and remained in that capacity for three years. He has served in editorial capacities for scientific journals, such as *Virology* and *Intervirology,* and has published numerous articles in professional journals.

Throughout his career Zinder has received several honors, including the United States Steel Award in Molecular Biology from the National Academy of Sciences in 1966, the Medal of Excellence from Columbia University in 1969, and an honorary doctorate of science from the University of Wisconsin in 1990. He was named a fellow of the American Academy of Arts and Sciences, and is associated with such organizations as the National Academy of Sciences, the American Society of Microbiology, Genetics Society of America, the American Society of Virology, and HUGO (Human Genome Organization).

ZINKERNAGEL, ROLF M. (1944-)
Swiss immunologist and virologist

Zinkernagel was born on January 6, 1944, in Basel, Switzerland. In 1962, he attended the University of Basel, deciding to study medicine rather than chemistry—his other great interest—because the former profession offered the possibility of clinical or private practice as well as research. He passed his final boards in 1968 and in 1970, the university accepted his M.D. dissertation.

In 1969 Zinkernagel's work in the surgery department of a hospital in Basel failed to spark his interest. He began looking around for other possible career paths. From 1970-1973 he worked as a postdoctoral fellow at the University of Lausanne, Switzerland, in a laboratory studying the process by which the **immune system** kills **virus**-infected **cells**. Zinkernagel's project, trying to monitor the destruction of bacterial cells preloaded with radioactive chromium-51, was frustrating because the method never worked properly on the **bacteria**—but it gave him experience with a number of experimental techniques that were to prove crucial for his Nobel-winning research.

In 1972 Robert Blanden of the John Curtin School of Medical Research, Canberra, Australia, came to the Swiss university to teach a World Health Organization course on immunology. Intrigued by the course and encouraged by senior researchers at Lausanne, Zinkernagel applied for a fellowship with Blanden at the Curtin school. Thanks to a two-year Swiss Foundation for Biomedical Fellowships grant, Zinkernagel and his young family moved to Australia in 1973. While at the Curtin school, Zinkernagel earned a Ph.D. in immunology, finishing his dissertation in 1975.

A fortuitous accident led Zinkernagel to team up with another young postdoctoral fellow at the Curtin school, **Peter Doherty**. While the Blanden laboratory was cramped for space, Doherty had room in his assigned lab. Thanks in part to their shared love of operatic music—and Zinkernagel's penchant for singing it aloud while working—Zinkernagel began to work with Doherty on how white **blood** cells called killer T cells identify virus-infected host cells to attack. "He was tolerable, but loud," according to Doherty.

At the time, immunologists were very interested in a group of **genes** collectively called the major histocompatibility complex, or MHC. These genes, clustered together in the **DNA** sequence, encode a series of **proteins** called the MHC antigens, which determine whether a transplanted **organ** will be accepted or rejected by a recipient. If the MHC genes of the donor and the recipient match, the organ survives; if they do not, the organ is attacked by the recipient's immune system and dies.

A number of researchers had guessed that the rejection of MHC-mismatched organs was essentially the same process as the killing of virus-infected cells by killer **T cells**. Zinkernagel and Doherty demonstrated that this was true, and that the MHC antigens were necessary for killer T cells to tell friend from foe. But when they investigated further, they found something very unexpected; most immunologists had expected that when virus-infected cells and killer cells were poorly MHC matched, the immune cells' killing response would be strongest, much as in badly matched transplants. But the opposite was true. In order to get proper T-cell killing of the virus-infected cells, Zinkernagel and Doherty discovered, the cells' MHC regions had to match.

The two had discovered that T cells—indeed, the immune response in general—can only recognize viral proteins when they are displayed in the context of properly matched MHC antigens. The immune system, which had evolved to recognize "self" from "other" did not react most strongly to "other," but to a third state, "altered self." This discovery finally put transplant rejection into biological context. The body does not purposely reject mismatched organs because they are different, it rejects them because it mistakenly identifies the mismatched MHC antigens as "self" antigens that have been altered by interaction with viral proteins. The finding also opened the way to better methods for heading off transplant rejection, for creating vaccines, and for further unraveling the workings of immunity; vulnerability to certain infections; and **autoimmune disease**, where the body mistakenly attacks its own **tissue**s.

Zinkernagel's and Doherty's work together took place in a fairly short amount of time between 1973 and 1974. By 1976, both were moving on, with Zinkernagel going to the Scripps Clinic Research Institute in La Jolla, California, as an associate—a rank roughly equal to an assistant professor at a university. There he studied whether or not the thymus gland—long known to play a role in the "maturation" of infection-fighting white blood cells—used MHC antigens to select which white blood cells would mature and which would die before being released to the bloodstream. The work once again proved seminal, providing the first evidence that the thymus only allows killer cells that react against slightly altered self MHC antigens to survive. This helped explain how and why killer T cells recognize altered-self antigens most strongly. The thymus prevents autoimmune disease by killing off killer cells that would otherwise attack healthy tissues and prevents a too-weak immune response by destroying those that would fail to attack any but the most profoundly changed self antigens.

Zinkernagel became a member—the equivalent of a full professor—at Scripps in 1979. But later that year he returned to Switzerland, to take an associate professorship at the University of Zurich, followed by a full professorship in 1988. During that period, his work with Doherty began to receive growing international recognition, with an Ehrlich Prize in Germany in 1983 and a Gairdner Foundation International Award in Canada in 1986. In 1992 Zinkernagel was named head of the Institute of Experimental Immunology in Zurich and also received the Christoforo Colombo Award in Italy, to be followed by an Albert Lasker Medical Research Award—often a prelude to a Nobel—in 1995.

In 1996 Zinkernagel joined the ranks of the Nobel laureates because of his relatively early work with Doherty defining the system by which the immune system identifies friend and foe. His work since then has built upon this discovery, revealing how the thymus gland selects only white blood cells that react properly to virus-infected cells and investigating the complex interplay by which viruses and their hosts co-evolve.

ZINSSER, HANS (1878-1940)
American bacteriologist

Hans Zinsser was one of the leading bacteriologists and immunologists in the United States during the first half of the twentieth century. His work in advancing the understanding of typhus fever as well as a number of fundamental features of immunology remains central to this day. Zinsser was born on November 17, 1878, in New York City, and grew up in a household where German was the primary language that was spoken. Both of his parents had emigrated from Germany: His father, August Zinsser, was a wealthy manufacturing chemist originally from the Rhineland, and his mother, Marie Theresa (Schmidt), was from the Black Forest region, an area long dominated by French tradition. For this reason, the young Zinsser soon became fluent in a second language: French.

The youngest of August and Marie's four sons, Zinsser did not start formal schooling until age ten. At that time he was sent to a private school in New York City operated by Julius Sachs, and only then did Zinsser begin using English as his first language. The school emphasized the liberal arts, an area of learning especially valued by Zinsser, who had the fortune of spending some portion of every one of his first twenty years of life visiting the art galleries and concert halls of Europe. In 1895, at age seventeen, Zinsser entered nearby Columbia University, where he was intent on studying literature and pursuing a writing career. Studying under comparative literature specialist George Edward Woodberry, the already broadly educated Zinsser showed great promise in the writing of poetry. In an article in *Memoirs of the National Academy of Sciences,*

Zinsser's principal biographer, Simeon Burt Wolbach, noted that the "world of things and thoughts" had occupied Zinsser until that time. Although his intellectual life was about to change, Zinsser remained an accomplished poet and essayist as well as lucid writer of scientific prose for the remainder of his more than sixty years. On his deathbed, after battling leukemia for the final two years of his life, he wrote his last sonnet. The poem, which was published posthumously in his collection *Spring, Summer and Autumn,* ends with the lines: "Then, ageless, in your heart I'll come to rest / Serene and proud as when you loved me best."

Enters the Field of Bacteriology

It was only after his tutelage under biologists **Edmund Beecher Wilson** and Bashford Dean during his junior year at Columbia that Zinsser realized that the life sciences would be his career. He went on to Columbia's College of Physicians and Surgeons in 1899, deciding to devote his career to the application of his interest in biology to real human problems. Earning both an M.A. and an M.D. in 1903, he interned at Roosevelt Hospital, then began to practice medicine. He left that vocation after a short while, however, when Columbia offered him a post as instructor in bacteriology. In the meantime, he had married Ruby Handforth Kunz in 1905.

Zinsser taught bacteriology for a short time at Columbia and teamed with Philip Hanson Hiss, Jr., with whom in 1910 he coauthored *A Textbook of Bacteriology,* which has become a standard microbiology text. Simultaneously, he served as assistant pathologist at New York's St. Luke's Hospital. The same year *A Textbook of Bacteriology* was released, Zinsser moved his wife and first child, Gretel, to Palo Alto, California, to accept a position as associate professor of bacteriology and immunology at Stanford University. There, he set up a bacteriology laboratory with the most minimal of equipment in some space borrowed from the **anatomy** department. In 1913, Zinsser returned to Columbia University, where he concentrated his research in the field of immunology.

Encounters an Outbreak of Typhus

As a professor of bacteriology and immunology at Columbia, Zinsser experienced a decade that was both exciting and dismaying. In 1915, in the midst of World War I, Zinsser served first as a member of the Red Cross Typhus Commission and later as an officer in the U.S. Army Medical Corps. Arriving in Serbia in 1915, Zinsser had his first field contact with an epidemic of typhus—a disease that is caused by the family of **bacteria** known as rickettsia, and is characterized by stupors, delirium, high fevers, severe headaches, and dark rashes. Approximately one hundred and fifty thousand cases of typhus existed at the Belgrade front, with a fatality rate of about 60-70%. During their experiences in the Eastern Front, the scientists in the commission began to gain a rudimentary understanding of the bacteriology and pathology of the disease. For his contributions during the war, Zinsser was awarded the U.S. Distinguished Service Medal, the French Legion of Honor, and the Order of Sava, a major Serbian citation.

In 1918, Zinsser left the U.S. Army Medical Corps as a lieutenant colonel, and continued his professorial duties at

Columbia, where he specialized in immunology. In particular, Zinsser focused on discovering a way to immunize patients against the chronic and contagious disease **syphilis**. Though he did not succeed in his quest to discover a successful method of immunization, he did contribute to the existing knowledge of spirochete, a type of bacteria that causes syphilis and relapsing fevers. In addition, Zinsser continued to study typhus, since he had became an expert on military sanitation, especially with regards to typhus, during his service in the war. He wrote articles and books on the subject in the course of his career, and during his lifetime took a number of trips to distant lands to study epidemic typhus or **cholera**—a diarrheal disease caused by bacteria. Among his expeditions were excursions to the Soviet Union in 1923, to Mexico in 1931, and to China in 1938, where he lectured at the Peiping, Beijing, Medical College. His Columbia years came to an end in 1923 when, at the age of forty-five, he was offered a teaching position at Harvard University Medical School. Within two years he was named the Charles Wilder Professor of Bacteriology. Zinsser remained in Boston for the remainder of his life. The Zinsser family, along with their second and last child, Hans Handforth Zinsser (who later graduated from Columbia's College of Physicians and Surgeons), lived in a house in the city and traveled often to their country farm in Dover, Massachusetts. The farm became a retreat and entertainment site for Zinsser's colleagues and his medical students.

By 1930, Zinsser had decided to concentrate his studies on typhus fever research, and began a lengthy friendship with **Charles J. H. Nicolle**, the Nobel Prize–winning French physician and bacteriologist who discovered that typhus is transmitted by body lice. During the 1930s, Zinsser was able, either alone or with a variety of co-workers, to aid in the understanding of the cause of the several forms of typhus, including Brill's disease, named for American physician Nathan Edwin Brill, who investigated the malady. Zinsser was able to prove that the disease is caused by the microorganism *Rickettsia prowazekii* as opposed to *Rickettsia mooseri,* as was commonly believed, and hypothesized that Brill's disease is a form of recrudescent (or renewing) typhus. His theory was confirmed by later studies, and the disease has since been renamed Brill-Zinsser's disease. In addition, Zinsser worked on a vaccine against typhus and assisted in conceiving of a way to prepare the vaccine commercially, thus making the treatment available to large numbers of people. These endeavors have guaranteed him a place in the history of bacteriology and medicine.

In addition to his significant contributions to bacteriology, Zinsser also made advancements in the field of immunology. He discovered that it is not possible to create a grand conceptual unification for an understanding of the phenomenon of allergic reaction. Near the beginning of the twentieth century, Austrian pediatrician Clemens von Pirquet and Hungarian pediatrician **Bela Schick**, then leading figures in immunology, sought to explain allergic reactions as if all were antibody-mediated. They also believed that an allergic reaction was a typical step in the recovery process. Zinsser showed that certain forms of bodily responses to infection, including those involving the body's reaction to tuberculin (a substance that was later, and to this day, used in the test for tuberculosis infection), are fundamentally different from other types of allergic responses.

In several books and papers, Zinsser detailed his scientific studies in the fields of bacteriology and immunology; among the most well known of these volumes is his 1935 work *Rats, Lice and History.* An examination of the history of typhus, the book intermixes philosophy and wit along with scientific information. The book became a best-seller and was praised by several literary critics. Zinsser delved into his private life with his 1940 autobiography, *As I Remember Him: The Biography of R. S.* Made up of some of the author's thoughts regarding living with leukemia—the disease that eventually caused his death— *As I Remember Him* was a popular book whose somewhat odd subtitle was derived from Zinsser's use of a pseudonym for his literary writings. In these writings he often referred to himself as R. S. There is disagreement as to what R. S. stood for: Some say it meant "Romantic Self"; others believe that it was derived from a German author, Rudolf Schmidt, who in 1908 had written on pain and its significance in medicine.

During his lifetime, Zinsser won numerous awards and was actively involved in many scientific societies, such as serving as president of the American Association of Immunologists in 1919 and of the Society of American Bacteriologists in 1926. His major honors include the receipt of honorary doctorates from Columbia University in 1929, Western Reserve University in 1931, Lehigh University in 1933, Yale University in 1939, and Harvard University in 1939. Among his other accolades are his elections to the Harvey Society and Sigma Xi. His published articles number more than 270.

Zinsser possessed a life-long devotion to personal fitness—enjoying activities ranging from horseback riding to hounds (at which he was expert) to shooting. "Throughout his life," Wolbach reported in *Memoirs of the National Academy of Sciences,* "he carried the aura of youth." Zinsser died of leukemia in his native New York City on September 4, 1940.

ZOOLOGY

Zoology, which is one of two main branches of biology, is the science of living things. **Botany**, the other main branch of biology, involves the study of **plants**. Zoology includes the study of any kind of **animal**, ranging from a 180 ton blue whale to one-**cell**ed creatures such as **bacteria**. Animals include humans and other **mammals**, birds, insects, fish, reptiles, **invertebrates**, and even microscopic **organism**s. Presently, more than a million different **species** of animals have been identified, and new species continue to be identified every year.

The science of zoology began in the days of Greek scientist **Aristotle** (384-322 B.C.), who developed a system to classify animals. Eighteenth century botanist **Carl Linneaus** developed a hierarchical system for classifying plants and animals that is still used today. In his system, the basic unit of **classification** is called the species.

The science of classification was the primary focus of zoology for many centuries, but in the eighteenth and nineteenth centuries, scientific exploration expanded around the world, and the knowledge generated led to the development of many fields of study within zoology. For example, British scientist **Charles Darwin** traveled and studied in South America and Australia in the early 1830s. Based on his observations, he developed a theory of **evolution**. It is based on the idea that all plants and animals have developed gradually, originating from a few common ancestors. In the 1860s, Austrian monk **Gregor Mendel** discovered basic laws of **heredity** through his study of garden plants, and he developed the concept of **gene**s. In the mid-twentieth century, Austrian naturalist **Konrad Lorenz** helped establish the important new field called **ethology**, which is the study of animal behavior.

Because of the explosion of knowledge about animals in the nineteenth and twentieth centuries, zoology is now subdivided into many major fields. For example, **taxonomy** focuses on the classification of animals. There are at least three major divisions of science based on taxonomy: invertebrate zoology involves the study of multicellular animals without backbones; vertebrate zoology is the study of animals with backbones; and **paleontology** is the study of prehistoric organisms through their **fossil**s. Many branches of zoology are based on the study of a particular kind of animal. For example, **entomology** is the study of insects, **mammalogy** is the study of mammals, and **ichthyology** is the study of fish.

Several fields of zoology involve the study of structures or systems of animals. Those fields include **embryology** (the study of the development of individual animals), **anatomy** (the study of structure of the animal's body, and physiology (study of living processes within animals).

There are also major fields of zoology involving the study of animal behavior. They include natural history (the life and behavior of animals in **nature**), evolution (the origins and development of animal life) and **ecology** (the relationship of animals to their environment).

The study of zoology is helpful to human beings in many ways, and zoologists work in hundreds of careers in our society. For example, some teach and do research at colleges, universities, and medical schools. Because human beings and many animals have similar body parts and functions, the study of animals is helpful in the development of human medicine. Studies of bacteria help scientists discover ways to protect human beings from being infected by disease and to cure diseases.

Zoologists also work for private industry. For example, they may work for a chemical company to test the effects of a pesticide or fertilizer on animals or humans. Studies of domesticated animals such as cattle and chickens, are used to find ways to produce healthier animals and more animals for our food supply. Ichtyologists (those who study fish) work in the field of aquaculture, looking for ways to raise fish for food.

Other zoologists work as zoo managers, as game wardens, and for governmental agencies such as a state department of **natural resources** or the U.S. Environmental Protection Agency (EPA). Ecologists work to protect natural resources and minimize harm to plant and animal **population**s. The study of wild animals helps us learn to coexist with them

and to preserve **endangered species**. In the late twentieth century, the search for previously undiscovered species in the **rain forest**s of South America and elsewhere has become of great importance to environmentalists, who want to protect our planet, and to scientists in the field of human medicine, who are searching for cures for **cancer** and other diseases.

Zoologists also work in creative fields and the news media. They are writers, photographers, or illustrators.

ZOOPLANKTON

Zooplankton (from Greek *zoe* [life] *plankton* [wanderer]) are aquatic **animal**s that cannot completely control their own movements against waves or currents of **water**. They are part of the group of **organism**s called plankton, which are drifters. Thus, zooplankton are the animal drifters. However, there are many varieties of zooplankton that have a limited ability to move. For example, many types of copepods, one of the dominant marine zooplankton, carry out vertical **migration**s each day. They move to the surface waters at night to feed, and return to deeper waters in the morning to avoid being eaten. Despite these limited movements, these organisms are considered zooplankton because they are still not strong enough to swim any great distance against the currents. While the great majority of zooplankton are microscopic, many can be seen with the naked eye. For example, jellyfish, which are clearly visible without a **microscope**, are zooplankton, because they are primarily at the whim of the ocean currents.

Zooplankton can be divided into two groups based on how much of their life they spend as part of the plankton. Holoplankton are organisms that are drifters for their entire life. Salps and chaetognaths are examples of holoplankton. In the ocean, the dominant type of holoplankton are members of a group of crustaceans known as copepods. These tiny, shrimp-like organisms spend their entire life drifting with the currents. Meroplankton are drifters for only part of their lives, usually during **larva**l or juvenile stages. For example, crabs, barnacles, mussels, sea stars and most finfish each have larval stages in which they cannot control their movements against the currents. For this part of their lives, they are considered meroplanktonic. When these organisms complete this planktonic stage, they either settle and live a life on the bottom, or they become strong enough to swim against the currents. They are no longer considered zooplankton at this point.

Zooplankton play a very important role in marine and aquatic food webs. Most zooplankton are primary **consumer**s, eating **phytoplankton**, the plant drifters. Some types of zooplankton are secondary consumers, feeding upon other zooplankton. Because many marine and aquatic organisms spend only part of their lives in the zooplankton, they feed on a wide variety of prey items depending on their life stage. As a result, food webs in marine and aquatic **ecosystem**s are often very complicated. In addition to consuming food of their own, zooplankton are a major source of food for various life stages of **species** in higher **trophic level**s, including many species that are important for commercial and recreational fisheries. Without the zooplankton, these fisheries would not be sustained.

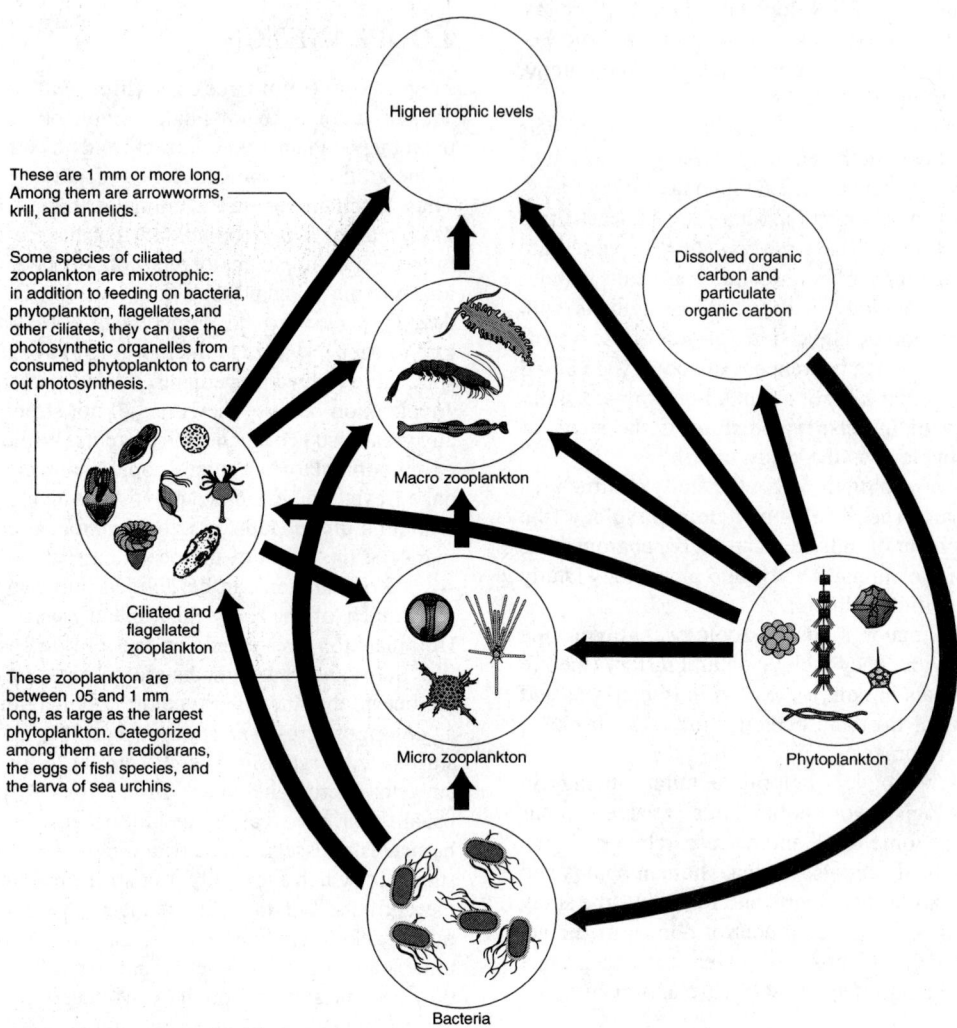

These are 1 mm or more long. Among them are arrowworms, krill, and annelids.

Some species of ciliated zooplankton are mixotrophic: in addition to feeding on bacteria, phytoplankton, flagellates, and other ciliates, they can use the photosynthetic organelles from consumed phytoplankton to carry out photosynthesis.

These zooplankton are between .05 and 1 mm long, as large as the largest phytoplankton. Categorized among them are radiolarans, the eggs of fish species, and the larva of sea urchins.

Higher trophic levels

Dissolved organic carbon and particulate organic carbon

Macro zooplankton

Ciliated and flagellated zooplankton

Micro zooplankton

Phytoplankton

Bacteria

The planktonic ecosystem. The arrows show the movement of biomass through the food chain. *(Illustration by Hans & Cassady, Inc.)*

ZYGOTE

In **animal**s, a zygote is a fertilized **egg**, formed by the fusion of a male **gamete** (or **sperm**) and a female gamete (or egg). Male and female gametes (collectively, these are referred to as sex **cell**s) are the unicellular products of **meiosis**, a kind of reduction cellular division that occurs in specialized **organ**s of sexually reproducing animals (the ovary of females, and the testes of males). Meiosis results in the formation of cells having only one of the two complementary (or homologous) sets of **chromosome**s possessed by animals (that is, gametes are **haploid** cells). However, the zygote formed from their union has two sets of chromosomes (i.e., it is **diploid**). Because one of the sets of chromosomes has been obtained each parent, the genetic information (or **genome**) of the offspring represents a unique combination of **DNA** (**deoxyribonucleic acid**, the genetic biochemical of animals). Zygotes are capable of developing into adult animals.

In **plants**, the product of meiosis is not gametes. Rather, this reduction division produces multicellular, haploid **organism**s, which then go on to produce haploid sex cells (or true gametes). The male plant gamete is known as pollen, and the female as an ovule. Fertilized ovules are diploid, and represent a unique genome, having obtained one of its two sets of chromosomes from each of its parents. Fertilized ovules develop into seeds, which are capable of germinating and growing in to an adult plant.

Sources Consulted

Alberts, B., et al. *Molecular Biology of the Cell*, 2nd ed. New York: Garland Publishing, Inc., 1989.

Alerstam, T.F. *Bird Migration*. U.K.: Cambridge University Press Cambridge, 1990.

Alexander, S.P. "Oasis Under the Ice." *International Wildlife* 18 (November-December 1988): 32-7.

Allaby, M. *MacMillan Dictionary of the Environment*. Hong Kong: MacMillan Press, 1991.

Allen, G.W. *William James: A Biography*. New York: The Viking Press, 1967.

Amberson, W.R., and D.C. Smith. *Outline of Physiology*, 2nd ed. New York: Appleton-Century-Crofts, Inc., 1948.

American Association for the Advancement of Science. *Liebig and After Liebig: A Century of Progress in Agricultural Chemistry*. Lancaster, PA: The Science Press Printing Co., 1942.

American Men & Women of Science 1998-99, 20th ed. New Providence, N.J.: R.R. Bowker.

Arcieri, D.T. "The Undesirable Alien—The House Sparrow." *The Conservationist* 46 (1992): 24-25.

Atkins, M.D. *Introduction to Insect Behavior*. New York: MacMillan Publishing Co., 1980.

Atlas, R.M., and R. Bartha. *Microbial Ecology*. Menlo Park, CA: Benjamin/Cummings, 1987.

Audubon, J.J. *Audubon's Western Journal: 1849-1850*. Irvine: Reprint Services, 1992.

Barbour, M.G. and W.D. Billings. *North American Terrestrial Vegetation*. New York: Cambridge University Press, 1988.

Barinaga, M. "New Imaging Methods Provide a Better View into the Brain." *Science*, June 27, 1997, p. 1,974.

Barry, R.J. *Atmosphere, Weather and Climate*. London: Methuen, 1976.

Bean, C.A. *Methods of Childbirth*. 2nd ed. Garden City, NY: Doubleday, 1990.

Beeson, P., W. McDermott, and J.B. Wyngarden (editors) *Cecil Textbook of Medicine*. 15th ed. Philadelphia: W. B. Saunders, 1979.

Begon, M., J.L. Harper, and C.R. Townsend. *Ecology. Individuals, Populations, and Communities*. Boston, MA: Blackwell Scientific Publications, 1998.

Begon, M., and M. Mortimer. *Population Ecology*. Sunderland, MA: Sinauer, 1986.

Berkow, R., et al. (editors). *The Merck Manual*. 16th ed. Rahway, N.J.: Merck & Co., Inc., 1992.

Berthold, P. *Bird Migration: A General Survey*. Oxford, U.K.: Oxford University Press, 1993.

Betteridge, K.J. (editor). *Embryo Transfer in Farm Animals*. Ottawa: Canada Department of Agriculture, 1977.

Biographical Dictionary of Scientists: Biologists. New York: Peter Bedrick Books, 1983.

Blakeslee, S. "Placebos Prove So Powerful Even Experts are Surprised." *New York Time*, October 13, 1998, p. D1.

Bonta, M.M. "Rachel Carson, Pioneering Ecologist." *Women in the Field: America's Pioneering Naturalists*. College Station, TX: Texas A&M University Press, 1991.

Boshoff, C., and R.A. Weiss. "Aetiology of Kaposi's sarcoma: Current Understanding and Implications for Therapy." *Molecular Medicine Today*, no. 3 (1997): 488-494.

Botkin, D.B., and E.A. Keller. *Environmental Science. Earth as a Living Planet*. New York: J. Wiley & Sons, 1995.

Bower, B. "Genetic Hint of Psychosis." *Science News*, no. 6, v. 153 (February 7, 1998): 91.

——. "Kids with Schizophrenia Yield Brain Clues." *Science News*, no. 17, v. 152 (October 25, 1997): 261.

Boyd, R. *General Microbiology*. St. Louis, MO: Mosby Year Book, 1988.

Bradley, R.A. *Husband-Coached Childbirth*. New York: Harper and Row, 1981.

Bradshaw, S. D., H. S. Giron, and F. J. Bradshaw. "Patterns of Breeding in Two Species of Agamid Lizards in the Arid Subtropical Pilbara Region of Western Australia." *General and Comparative Endocrinology* 82 (1991): 407-24.

Brady, N.C., and R.R. Weil. *The Nature and Property of Soils*, 12th ed. Upper Saddle River, N.J.: Prentice Hall, 1999.

Braverman, L.E. *Diseases of the Thyroid*. Totowa, N.J.: Humana Press, 1997.

Brock, T.D. *Biology of Microorganisms*. 3d ed. Englewood Cliffs, NJ: Prentice Hall, 1979.

Brock, W. *Justus Von Liebig: The Chemical Gatekeeper*. United Kingdom: Cambridge University Press, Cambridge, 1997.

Brooks, P. *The House of Life: Rachel Carson at Work*. Boston: Houghton Mifflin, 1972.

Browder, L.W., C.A. Erickson, and W.R. Jeffery. *Developmental Biology*. 3rd ed. Philadelphia: Saunders College Publishing, 1991.

Brown, P.R., and E. Grushka. (eds.) *Advances in Chromatography, vol. 39*. Monticello, N.Y.: Marcel Dekker, 1998.

Buchmann, S.L., and G.P. Nabhan. *Forgotten Pollinators*. Washington, D.C.: Island Press, 1996.

Callus, D.A. (editor) *Robert Grosseteste: Scholar and Bishop*. Oxford, U.K.: Oxford at the Clarendon Press, 1955.

Campbell, N.A. *Biology,* 4th ed. New York: The Benjamin/Cummings Publishing Company, Inc., 1996.

Carson, R. *Silent Spring*. Boston: Houghton Mifflin, 1962.

Cassirer, E. *The Problem of Knowledge: Philosophy, Science, and History since Hegel*. New Haven, CT: Yale University Press, 1950.

Castiglioni, A. *A History of Medicine*. New York: Alfred A. Knopf, 1941.

Causey, G. *The Cell of Schwann*. London: E&S Livingstone Ltd, 1960.

Christopher, G.W., et al. "Biological Warfare: a Historical Perspective." *JAMA*, vol. 278, no. 5 (August 6, 1997): 412.

Churchill, F.B. "Wilhelm Roux." *Dictionary of Scientific Biography*, vol XI. New York: Charles Scribner's Sons, 1975.

Clark, R.B. *Marine Pollution*. 3rd ed. Oxford, U.K.: Clarendon Press, 1992.

Coleman, W. *Biology in the Nineteenth Century* . New York: John Wiley & Sons, 1971.

Cook-Deegan, R. *The Gene Wars: Science, Politics, and the Human Genome*. New York: W.W. Norton & Company, 1994.

Cooper, N.G (editor). *The Human Genome Project: Deciphering the Blueprint for Heredity*. Mill Valley, CA: University Science Books, 1994.

Copleston, F. *A History of Philosophy*. Garden City, N.Y.: Image Books, 1985.

Craig, J.R. *Resources of the Earth*. Englewood Cliffs, N.J.: Prentice-Hall, 1988.

Culotta, E. "A Boost for Adaptive' Mutation." *Science* 265, no. 5170 (July 15, 1994): 318-19.

Curtis, H., and N.S. Barnes. *Biology*, 5th ed. New York: Worth Publishers, Inc., 1989.

Dampier, Sir W.C. *A History of Science and its Relations with Philosophy and Religion*. New York: The Macmillan Company, 1955.

Danzig, R., and P.B. Berkowsky. "Why Should we be Concerned about Biological Warfare?" *JAMA*, vol. 278, no. 5 (August 6, 1997): 431.

Darnell, J., H. Lodish, and D. Baltimore. *Molecular Cell Biology*. New York: Scientific American Books, Inc., 1986.

Darwin, C.R. *The Origin of Species*. London: John Murray, 1859.

Davis, R.E., et al. *Modern Chemistry*. New York: Holt, Rinehart, and Winston/Harcourt Brace & Company, 1999.

"Deciding to Be Born." *Discover* 13 (10 May 1992).

Dickinson, G., and K. Murphy. *Ecosystems*. New York: Routledge, 1998.

Downs, R.B. "Upsetting the Balance of Nature: Rachel Carson's *Silent Spring*." In *Books That Changed America*. New York: Macmillan, 1970.

Duellman, W.E. and L. Trueb. *Biology of Amphibians*. Baltimore: Johns Hopkins University Press, 1994.

Dunbar, R.I.M. *Primate Social Systems*. Ithaca, N.Y.: Cornell University Press, 1988.

Eberhard, G. *Bird Migration: Physiology and Ecophysiology*. Berlin: Springer-Verlag, 1990.

Eckert, R. *Animal Physiology: Mechanisms and Adaptations*. New York: W.H. Freeman, 1988.

Ehrlich, P.R. and A.H. Ehrlich. *The Population Explosion*. New York: Simon & Schuster, 1990.

Erdtman, G. *Handbook of Palynology*. Copenhagen: Scandinavian University Books, 1969.

Erwin, D.A. "The End-Permian Mass Extinction." *Annual Reviews of Ecology and Systematics*, (1990) 21: 69-91.

"Evolution Evolving." *Scientific American* (September 1997): 15-6.

Faegri, K., and J. Iversen. *Textbook of Pollen Analysis*. New York: Hafner Press, 1975.

Faegri, K., and L. van der Pijl. *The Principles of Pollination Ecology*. 3rd revised ed. Oxford, England: Pergamon Press, 1979.

Fischman, J. "Putting a New Spin on the Birth of Human Birth." *Science* 264 (20 May 1994): 1082.

Fishbein, S. (editor). *Our Continent*. Washington, D.C.: The National Geographic Society, 1976.

Fleagle, J.G. *Primate Adaptation and Evolution*. San Diego, CA: Academic Press, 1988.

Foster, H.D. *Health, Disease and the Environment*. Boca Raton, FL: CRC Press, 1992.

Frauenfelder, H., and H.C. Berg. "Physics and Biology." *Physics Today*, no. 2 (February 1994): 20-21.

Freedman, B. *Environmental Ecology*. San Diego, CA: Academic Press, 1995.

——. *Environmental Science*. Toronto: Prentice Hall, 1998.

Gardner, E. *Fundamentals of Neurology*. 6th ed. Philadelphia: W. B. Saunders, 1975.

Gardner, E.J. *History of Biology*. Minneapolis: Burgess Publishing Company, 1965.

Gates, D.M. *Energy and Ecology*. New York: Sinauer, 1985.

Gianelli, D.M. "Enhancement Gene Therapy Raises a New Ethical Dilemma." *American Medical News*, vol. 40, no. 37 (October 6, 1997): 3.

Gilbert, S.F. *Developmental Biology*. 5th ed. Sunderland, MA: Sinauer Associates, Inc., 1997.

Gillispie, C.C. (editor). *Dictionary of Scientific Biography*. New York: Scribners, 1970.

Glanz, J. "Physicists Advance into Biology." *Science*, vol. 272, no. 5262 (May 3, 1996): 646.

Goldemberg, J. "Energy, Technology, Development." *Ambio*, 21 (1992): 14-17.

Gopnik, A. "Audubon's Passion." *The New Yorker* 67 (25 February 1991): 96-104.

Gorgas, W.C. *Sanitation in Panama*. New York: D. Appleton and Company, 1915.

Graham, F., Jr. "Rachel Carson." *EPA Journal* 4 (November-December 1978): 5-7+.

Gray's Anatomy, 38th ed. New York: Churchill Livingston, 1995.

Gregory, R.L. (editor). *The Oxford Companion to the Mind*. Oxford: Oxford University Press, 1987.

Guttmann, B.S. *Biology*. Boston: WCB/McGraw Hill, 1999.

Hadley, M.E. *Endocrinology*. 4th ed. Upper Saddle River, N.J.: Prentice Hall, 1996.

Hartman, W., and R. Miller. *The History of the Earth*. New York: Workman Publishing, 1991.

Harvey, W. *De Motu Cordis (Movement of the heart and blood in animals: An anatomical essay)*. Translated by Kenneth J. Franklin. Oxford, U.K.: Blackwell Scientific Publications, 1957.

Hayes, T.M., and W.S. Wilson. *Humic Substances in Soils, Peats, and Waters: Health and Environmental Aspects*. Cambridge, U.K.: Royal Society of Chemistry, 1997.

Herbert, W. "Fearsom Madness: Schizophrenia Remains Frustratingly Hard to Control." *U.S. News & World Report*, no. 6, v. 125 (August 10, 1998): 53-4.

Huggins, C., and C.V. Hodges. "Studies on Prostate Cancer: The effect of castration, of estrogen and of androgen injection on serum phosphatase in metastatic carcinoma of the prostate." *Cancer Research* 1, (1941): 293-297.

Hynes, H.P. *The Recurring Silent Spring*. New York: Pergamon Press, 1989.

Jablonski, D. "Extinctions: A Paleontological Perspective." *Science*, 253 (1991): 754-757.

Jacobsen, S. "Lost World." *Discover* 13 (1992): 22.

Jeffers, J.N.R. "Ecological Concepts: Their Relevance to Human Nutrition" in Blaxter, K. (editor). *Food Chains and Human Nutrition*. London: Applied Science Publishers, 1980.

Joklik, W.K., et al (editors). *Zinsser Microbiology*. 19th ed. Norwalk, CT: Appleton & Lange, 1988.

Junqueira, L.C., J. Carneiro, and R.O. Kelly. *Basic Histology*. 8th ed. Stamford, CT: Appleton and Lange, 1995.

Kahn P., and A. Gibbons. "DNA from an extinct human." *Science* 277 (1997):176-8.

Kalat, J.W. *Biological Psychology*. 2nd ed. Belmont, CA: Wadsworth, 1984.

Kamen, Martin D. *Radiant Science, Dark Politics: A Memoir of the Nuclear Age*. Berkeley, CA: University of California Press, 1985.

Kamrin, M.A. *Toxicology: A Primer on Toxicology Principles and Applications*. Lewis Publishers, 1988.

Karmel, M. *Thank you, Dr. Lamaze*. New York: Harper and Row, 1993.

Kelley, W.N., et al (editors). *Textbook of Internal Medicine*. Philadephia: Lippincott-Raven Publishers.

Kestenbaum, D. "Gentle Force of Entropy Bridges Disciplines," *Science*, vol. 279, no. 5358 (March 20, 1998): 1,849.

Kimmel, C.A., and J. Buelke-Sam (editors).*Developmental Toxicology*. New York: Raven Press Ltd., 1994.

Knobil, E., and J.D. Neill. (editors). *The Physiology of Reproduction*, 2nd ed. New York: Raven Press, 1994.

Koch, E.D., and C.R. Peterson. *Amphibians and Reptiles of Yellowstone and Grand Teton National Parks*. Salt Lake City: University of Utah Press, 1995.

Kolata, G. *Clone: The Road to Dolly and the Path Ahead*. New York: Morrow, 1998.

Korte, D., R. Scaer, and L. Baker. *A Good Birth, A Safe Birth*. Boston: Harvard Common Press, 1992.

Krings M., A. Stone, R.W. Schmitz, H. Krainitzki, M. Stoneking, and S. Paabo. "Neandertal DNA sequences and the origin of modern humans." *Cell* 90 (1997):19-30.

Lamberg, L. "New Medications Aid Cognition in Schizophrenia." *JAMA* no. 11, v. 280 (September 16, 1998): 953.

Lee, T.F. *The Human Genome Project: Cracking the Genetic Code of Life*. New York: Plenum Press, 1991.

Lerman, M. *Marine Biology: Environment, Diversity, and Ecology*. Menlo Park, CA: The Benjamin/Cummings Publishing Company, Inc., 1986.

Lickey, M.E., and B. Gordon. *Medicine and Mental Illness*. New York: W.H. Freeman & Company, 1991.

Lindala, E. "Renewed Debate Surfaces around Human Genome Project." *Alternatives*, vol. 20, no. 4 (Sept.-Oct. 1994): 12.

Locy, W. *Biology and its Makers*. New York: Holt, Rinehart & Winston, 1960.

Long, L. *Geology*. New York: McGraw-Hill, 1974.

Lyons, A.S., and R.J. Petrucelli, II. *Medicine: An Illustrated History*. New York: Abradale Press, 1987.

MacMahon, J. *Deserts*. New York: Alfred A. Knopf, 1985.

Magner, L.N. *A History of Medicine*. New York: Marcel Dekker, Inc., 1992.

Marco, G. J., R.M. Hollingworth, and W. Durham. *Silent Spring Revisited*. Washington, D.C.: American Chemical Society, 1987.

Marrone, S.P. *William of Auvergne and Robert Grosseteste*. Princeton, N.J.: Princeton University Press.

Martin, R.D. *Primate Origins and Evolution: A Phylogenetic Reconstruction*. Princeton, N.J.: Princeton University Press, 1990.

Masterson, W.L., and C.N. Hurley. *Chemistry: Principles & Reactions*. Philadelphia: Saunders College Publishing, 1989.

Mayr, E. *Principles of Systematic Zoology*. New York: McGraw-Hill, 1991.

——. *This is Biology: the Science of the Living World*. Cambridge: Harvard University Press, 1997.

McKinnell, R.G. *Cloning, Nuclear Transplantation in Amphibia*. Minneapolis: University of Minnesota Press, 1978.

——. *Cloning of Frogs, Mice, and Other Animals*. Minneapolis: University of Minnesota Press, 1985.

McKinnell, R.G., R.E. Parchment, A.O. Perantoni, and G.B. Pierce. *The Biological Basis of Cancer*. New York: Cambridge University Press, 1998.

Mead, C. *Bird Migration*. New York: Facts on File, 1983.

Meadows, D. H., et al. *The Limits to Growth*. New York: Universe Books, 1972.

Meeuse, R.J.D., K. Faegri, and J. Iversen. *Textbook of Pollen Analysis*. New York: Hafner Press, 1989.

Mefien, S.V. *Fundamentals of Paleobotany*. New York: Chapman and Hall, 1987.

Mitford, J. *The American Way of Birth*. New York: Dutton, 1992.

Modern Biology. Orlando, FL: Holt Rinehart and Winston, Inc., 1991.

Monroe, J.S., and R. Wicander. *Physical Geology: Exploring the Earth*. St. Paul, MN: West Publishing Company, 1995.

Moreno, J. M., and W.C. Oechel. "Fire Intensity Effects on Germination of Shrubs and Herbs in Southern California Chaparral." *Ecology* 72 (1991): 1993-2004.

The New Book of Popular Science, vol. 3. Danbury, CT: Grolier, Inc., 1996.

"New Views of the Origins of Mammals." *Science* vol. 281 (August 7, 1998): 774-75.

Newman, H.H. "Evidences from Paleontology." In: *Evolution, Genetics, and Eugenics*. Chicago: University of Chicago Press, 1932.

Nordenskiöld, E. *The History of Biology: A Survey*. New York: Tudor, 1935.

Norman, A.W. and G. Litwack. *Hormones*. 2nd ed. San Diego: Academic Press, 1997.

Norstog, K., and A.J. Meyerriecks. *Biology*. Columbus. OH: Charles E. Merrill Publishing Co., 1983.

Nowak, R.M. *Walker's Mammals of the World*. Baltimore: Johns Hopkins University Press, 1991.

Odum, E.P. *Ecology and Our Endangered Life Support Systems.* New York: Sinauer, 1989.

Ottoboni, M.A. *The Dose Makes the Poison: A Plain-Language Guide to Toxicology.* Vincente Books, 1984.

Patton, T. *Anatomy & Physiology.* St. Louis: Mosby-Year Book, Inc., 1996.

Peckham, M. *The Origin of Species by Charles Darwin: A Variorum Text.* Philadelphia: University of Pennsylvania Press, 1959.

Pepper, I.L., C.P. Gerba, and M.L. Brusseau. *Pollution Science.* San Diego: Academic Press, 1996.

Porter, R. *The Greatest Benefit to Mankind: A Medical History of Humanity.* New York: Norton, 1997.

Post, W.M., T. Peng, W.R. Emanual, A.W. King, V.H. Dale, and D.L. DeAngelis. "The Global Carbon Cycle." *American Scientist,* 78 (1990): 310-326.

Postethwait, J.H., and J.L. Hopson. *The Nature of Life.* New York: Random House, Inc., 1989.

Potera, C. "Physics, Biology Meet in Self-assembling Bacterial Fibers." *Science,* vol. 276, no. 5318 (June 6, 1997): 1,499.

Priest, J. *Energy.* New York: Addison-Wesley, 1991.

Rand, G.M., and Petrocelli, S. R. *Fundamentals of Aquatic Toxicology: Methods and Applications.* New York: Hemisphere Publishing Corporation, 1985.

Raven, P.B., and G.B. Johnson. *Biology.* Toronto, ON: Mosby Year Book, 1992.

Read, G.D. *Childbirth Without Fear.* New York: Harper and Row, 1984.

Rennie, J. "Grading the Gene Tests." *Scientific American,* June 1994, pp. 89-97.

Reynolds, S.H. *The Vertebrate Skeleton.* Cambridge, U.K.: Cambridge University Press, 1982.

Robards, K., E. Patsalides, P. Haddad, and P. Jackson. *Principles and Practices of Modern Chromatography.* New York: Academic Press, 1994.

Robbins, L.L. (editor). *Golden's Diagnostic Radiology.* Baltimore, MD: The Williams & Wilkins Company.

Roitt, I.M., J. Brostoff, and D.K. Male. *Immunology.* London: Gower Medical Publishing, 1989.

Romer, A.S. *The Vertebrate Body.* Philadelphia: Saunders College Pub., 1986.

Rosenthal, M.S., and R. Volpe. *The Thyroid SourceBook: Everything You Need to Know.* Boston: Lowell House, 1996.

Rubin, J. "The Politics of Nuclear Waste." ABC News and Starwave Corporation website: http://www.ABC-NEWS.com.

———. "Will Nuke Waste Come Near You?" ABC News and Starwave Corporation website: http://www.ABC-NEWS.com.

Running Press Staff (editors). *Audubon Journal.* Philadelphia: Running Press, 1993.

Sager, R.J., et al. *Modern Earth Science.* Orlando, FL: Holt, Rinehart and Winston, 1998.

Schlesinger, W.H. *Biogeochemistry: An Analysis of Global Change.* San Diego: Academic Press, 1992.

Schrödinger, E. *What is Life? The Physical Aspect of the Living Cell.* Cambridge, U.K.: Cambridge University Press, 1967.

Scott, W.B. "Critique of the Recapitulation Theory." In: *Evolution, Genetics and Eugenics* (H.H. Newman, editor). Chicago: University of Chicago Press, pp. 114-123, 1932.

Sears, F.W., et al. *University Physics.* 7th ed. Reading, MA: Addison-Wesley Publishing Co., 1987.

Shorey, H.H. *Animal Communication by Pheromones.* New York: Academic Press, 1976.

Silverton, J.W. *Plant Population Biology.* London: Longman, 1987.

Simmons, I. G. *Earth, Air, and Water: Resources and Environment in the Late 20th Century.* London: Edward Arnold, 1991.

Singer, C.J., and E.A. Underwood. *A Short History of Medicine.* New York: Oxford University Press, 1962.

Snustad, D.P., M.J. Simmons, and J.B. Jenkins. *Principles of Genetics.* New York: John Wiley & Sons, Inc., 1997.

Solomons, G.W., and T.W. Solomons. *Organic Chemistry.* New York: John Wiley & Sons, 1995.

Spaulding, N.E., and S.N. Namowitz. *Earth Science.* Lexington, MA: D.C. Heath, 1994.

Speaker, S.L., and M.S. Lindee. *A Guide to the Human Genome Project: Technologies, People, and Institutions.* Philadelphia: Chemical Heritage Foundation, 1993.

Spence, A.P., and E.B. Mason. *Human Anatomy and Physiology.* St. Paul, MN: West Publishing Co., 1992.

Spencer, F. *The Piltdown Papers 1908- 1955.* London: Oxford University Press, 1990.

Stewart, W.N. *Paleobotany and the Evolution of Plants.* Cambridge, U.K.: Cambridge University Press, 1983.

Stevens, W.K. "A Theory: Too Cold, Too Hot, Then Just Right for Animals." *New York Times,* (September 1, 1998): F4.

The Story of Pollination. New York: The Ronald Press Company, 1961.

Strickberger, M.W. *Genetics.* New York: Macmillan Co.

Swanson, C.P. and P.L. Webster. *The Cell*. New Jersey: Prentice Hall, 1977.

Thompson, D.W. *On Growth and Form*. 2nd ed. New York: The Macmillan Compary, 1942.

Thurman, H.V. *Introductory Oceanography*. 5th ed. Columbus, OH: Merrill Publishing Co., 1988.

Tobin, A.J., and J. Dusheck. *Asking About Life*. New York: Saunders College Publishing, 1998.

Tortora, G.J. *Microbiology: An Introduction*. 4th ed. Redwood City, CA: The Benjamin/Cummings Publishing Company, Inc., 1992.

Tortora, G. J., and S.R. Grabowski. *Principles of Anatomy and Physiology*. 7th ed. New York: HarperCollins, 1993.

Tymoczko, D. "The Nitrous Oxide Philosopher." *The Atlantic Monthly* (May 1996): 93-101.

Uschmann, G. "Haeckel." *Dictionary of Scientific Biography*. VI:6-11. New York: Charles Scribner Sons, 1975.

Vander Meer, R.K. (editor) *Pheromone Communication in Social Insects*. Boulder, CO: Westview Press, 1998.

Veggeberg, S. *Medication of the Mind*. New York: Henry Holt and Company, 1996.

Villee, C.A., et al. *Biology*. Philadelphia: Saunders College Publishing, 1989.

Volpe, E.P. "Polymorphism in Anuran Populations." W. F. Blair (ed.) *Vertebrate Speciation*. Austin: University of Texas Press, 1961.

———. *The Amphibian Embryo in Transplantation Immunity*. Basal, Switzerland: S. Karger, 1980.

Wallace, R.A., et al. *Biology: The Science of Life*. 4th ed. New York: HarperCollins, 1996.

Walsh, J.E. *Unraveling Piltdown*. New York: Random House, 1996.

Weichert, C.K. *Anatomy of the Chordates*. New York: McGraw-Hill Book Company, Inc., 1951.

Weiner, J.S. *The Piltdown Forgery*. London: Oxford University Press, 1955.

Weiss, P. *Principles of Development*. New York: Henry Holt and Company, 1939.

Wellington, E.N, *Genetic Interaction among Microorganisms and the Natural Environment*. Pergamon Press, 1992.

Werkmeister, W.H. "Hans Driesch." *The Encyclopedia of Philosphy*, P. Edwards (editor), vol 2, pp. 418-420. New York: Crowell Collier and Macmillan, 1967.

Willier, B.H. and J.M. Oppenheimer. *Foundations of Experimental Embryology*. Englewood Cliffs, N.J.: Prentice-Hall, Inc., 1964.

Winch, D. *Malthus*. Oxford: Oxford University Press, 1987.

World Resources, 1990-91: A Report. New York: Oxford University Press, 1990.

Wood, L.D. *Your Thyroid*. New York: Ballantine Books, Inc., 1996.

Woodland, D.W. *Contemporary Plant Systematics*. Englewood Heights, N.J.: Prentice Hall, 1991.

Wyman, R.C. (editor). *Global Climate Change and Life on Earth*. New York: Routledge, Chapman & Hall, 1991.

Young, R., and R.M. Rowell. *Cellulose: Structure, Modification, and Hydrolysis*. New York: John Wiley & Sons, 1986.

c. 50,000 B.C.

Homo sapiens sapiens emerges as a conscious observer of nature.

c. 10,000 B.C.

Neolithic Revolution: transition from a hunting and gathering economy to agriculture.

c. 3500 B.C.

Sumerians describe date harvest.

c. 2800 B.C.

Manuscript attributed to the Chinese emperor Shen Nung discusses medicinal plants, including the soybean.

c. 2500 B.C.

Agriculture flourishes in northwest India.

c. 1800 B.C.

Male flower of the date plant mentioned in a Babylonian trade document.

c. 1000 B.C.

Iron used in agriculture and warfare.

c. 600 B.C.

Thales (c. 625-c. 547), the founder of the Ionian school of Greek philosophy, identifies water as the fundamental element of nature. Other Ionian philosophers construct different theories about the origin of nature.

c. 550 B.C.

Pythagoras (c. 582-500), philosopher and founder of a mystical brotherhood, defines the human soul as immortal and capable of reincarnation.

c. 500 B.C.

Alcmaeon, Pythagorean philosopher and naturalist, pursues anatomical research, concluding that humans are fundamentally different from animals.

c. 450 B.C.

Empedocles (c. 492-c. 432), Greek philosopher, identifies four fundamental elements underlying all of nature: earth, air, fire, and water.

c. 400 B.C.

Democritus (c. 460-c. 432), who believed that atoms are the building-blocks of nature, identifies the human brain as the organ of thought.

Hippocrates (c. 460-c. 362), Greek physician, founds a school of medicine on the Aegean island of Cos. Inspired by Empedocles's theory of four elements, the Hippocratic medical tradition eventually developed the theory of the four humors in the human body.

c. 350 B.C.

Building on earlier taxonomies, developed by Democritus, Aristotle (384-322) attempts to classify animals. Widely considered the founder of biology, particularly zoology, Aristotle was a keen observer of nature. Nevertheless, he believed that the brain was a blood-cooling organ.

Theophrastus (c. 372-c. 287), Aristotle's disciple and the founder of botany, identifies plants and animals as fundamentally different.

c. 300 B.C.

Herophilus (c. 325-c. 255), Alexandrian physician, studies the circulatory system; his investigation of the nervous system prompts him to identify the brain as the organ of thought.

c. 275 B.C.

Herophilus's younger colleague Eristratus (c. 310-250) declares that veins and arteries are connected.

c. 50 B.C.

Roman poet Lucretius (99 B.C.-44 B.C.), in his poem *On the Nature of Things*, proposes a materialistic, atomistic theory of nature, asserting that the soul does not survive death.

c. 70 Roman author and naturalist Pliny the Elder (23-79) writes his influential *Natural History*, a vast compilation combining observations of nature, scientific facts, and mythology. Naturalists will use his work as a reference book for centuries.

200 Galen (b. 130), the great medical authority dies. Combining extensive anatomical research with traditional religious and philosophical ideas, Galen created a philosophy of medicine that, despite many misconception about human anatomy, dominated the natural sciences until the sixteenth and seventeenth centuries. He believed in the divine origin of life.

529 Byzantine Emperor Justinian closes the Academy (founded by Plato) in Athens and forbids pagan scientists and philosophers to teach. This causes an exodus of scientists to Persia.

c. 850 Arab scholar Yaqub ibn-Ishaq al-Kindi (c. 800-870) develops a theory of vision.

c. 1150 Hildegard of Bingen (1098-1179), German mystic and visionary writes *The Book of Simple Medicine*, a treatise on the medicinal qualities of plants and minerals.

c. 1250 Arab scientist al-Qurashi ibn an-Hafis (d. 1288) discovers pulmonary circulation centuries before it is known in the West.

Albertus Magnus (c. 1200-1280), German philosopher, theologian, and naturalist, writes books on various aspects of nature.

Roger Bacon (1214-1292), English philosopher and scientist, declares that nature should be studied empirically.

1505 Leonardo da Vinci (1452-1519), Italian artist and scientist, makes the first wax cast of the brain ventricles (of an ox).

c. 1525 Paracelsus (1493-1541), Swiss physician and alchemist, uses mineral substances as medicines. Denying Galen's authority, he teaches that life is a chemical process.

1543 Andreas Vesalius (1514-1564), Flemish anatomist, publishes his epoch-making *The Fabric of the Human Body*. Regarded by many as the founder of modern anatomy, Vesalius corrects many of Galen's

misconceptions regarding the human body. Coincidentally, this is the publication year of another work that revolutionized science, Copernicus's *On the Revolutions of the Celestial Orbs*.

1553 Publication of *On the Restoration of Christianity* by the Spanish theologian and naturalist Michael Servetus (1511-1553). In this work, Servetus described pulmonary circulation. Persecuted by the Catholic Inquisition, Servetus was arrested and burned at the stake in Protestant Geneva.

1583 Andreas Cesalpino (1524-1603), Italian physician also known as Cesalpinus, writes *On Plants*. It is widely considered the first truly scientific textbook on botany because it defines the criteria for botanical taxonomy.

1599 Ulisse Aldrovandi (1522-1612) publishes a three-volume work on birds.

1628 William Harvey (1578-1657), English physician, publishes his *Anatomical Treatise on the Movement of the Heart and Blood*. This scientific classic traces the course of blood through the heart, arteries, and veins, present the first accurate description of blood circulation.

1651 William Harvey publishes an embryological work, *On the Generation of Animals*, which states that all living "things come from an egg." He also affirms that oviparous and viviparous generation are analogous to each other. Harvey's book contains the first mention of the term "epigenesis," which refers to the idea that generation starts with the egg.

1664 The idea of reflex action, formulated by René Descartes (1596-1650), French philosopher and mathematician, is made public. It is included in a French edition of his posthumously published work on animal physiology. Descartes applied his mechanistic philosophy to the analysis of animal behavior and first used the concept of reflex to denote any involuntary response the body makes when exposed to a stimulus.

1665 First drawing of a cell is made by Robert Hooke (1635-1703), English physicist. While observing a sliver of cork under a microscope, Hooke notices that it is composed of a tiny rectangular holes, which he names "cells," because each looks like an empty room. Although the cells he observes are not living, the name is retained.

1667 Following his discovery of ovaries in a shark, Nicolaus Steno (1638-1686), Danish anatomist and geologist introduces the term "ovary," stating that the female testes contain ova.

1668 Regnier de Graaf (1641-1673), Dutch anatomist, writes his treatise on the human sex organs, confirming Harvey's analogy between oviparous and viviparous reproduction.

Francesco Redi (1626-1697), Italian physician, publishes the results of his experiments to disproving the theory of spontaneous generation. He demonstrates that maggots are not born spontaneously, but come form eggs laid by flies.

1669   Jan Swammerdam (1637-1680), Dutch naturalist, begins his pioneering work on the metamorphosis of insects and the anatomy of the mayfly.

Swammerdam also makes the first published reference to "preformation theory," which negates the idea of epigenesis, maintaining that the organism is already formed in its germinal stage.

1677   Antoni van Leeuwenhoek (1632-1723), Dutch biologist and microscopist, discovers spermatozoa and describes them in a letter he publishes in *Philosophical Transactions* two years later.

1680   Posthumous publication of *On Motion in Animals* by Giovanni Alfonso Borelli (1608-1679), Italian mathematician and physicist. Borelli studied the human body from the standpoint of Descartes' mechanistic philosophy, describing physiology as a branch of physics.

1681   First use of the term "comparative anatomy" is made by Nehemiah Grew (1641-1712), English botanist and physician.

1682   John Ray (1628-1705), English naturalist, publishes his book *Methodus plantarum* in which he divides flowering plants into monocotyledons and dicotyledons.

1683   Antoni van Leeuwenhoek (1632-1723), Dutch biologist and microscopist, discovers different types of infusoria (minute organisms found in decomposing matter and stagnant water). He also discovers several protozoa.

1686   John Ray (1637-1705), English naturalist, publishes the first volume of his three-volume *General History of Plants*. This work introduces the idea of species groups of individual plants with similar seeds into botany and lays the groundwork for Linnaeus's systematic classification.

1694   Rudolph Jakob Camerarius (1665-1721), German botanist, first demonstrates, by experiment, that plants reproduce sexually.

1700   Joseph Pitton de Tournefort (1656-1708), French botanist, develops the binomial method of classification, preparing the way for Linnaeus's classification system.

1727   Stephen Hales (1677-1761), English botanist and

chemist, studies plant nutrition and measures water absorbed by plant roots and released by leaves; says that something in the air (carbon dioxide) is converted into food and that light was a necessary element of this process.

1735   Carl Linnaeus (1707-1778), Swedish botanist, publishes his *Systema Naturae, or The Three Kingdoms of Nature Systematically Proposed in Classes, Orders, Genera, and Species*, a methodical and hierarchical classification of all living beings. He develops the binomial nomenclature that for the classification of plants and animals. In this system, each type of living being is first given a generic (denoting the group to which it belongs), and then a particular, individual name.

1737   *The Bible of Nature*, by Dutch naturalist Jan Swammerdam (1637-1680) is published a 100 years after his birth. With Hermann Boerhaave's (1681-1738) rediscovery and publication of this massive work, Swammerdam is recognized as the founder of modern entomology.

1740   Abraham Trembley (1720-1793), Swiss naturalist, observes the phenomenon of regeneration when he cuts a hydra into several pieces which then turn into complete individuals.

1743   American Philosophical Society is founded by American statesman and inventor Benjamin Franklin (1706-1790). It becomes the first permanent scientific organization in the American colonies.

1745   Charles Bonnet (1720-1793), Swiss naturalist, publishes his *Insectology*, in which he details his discovery of parthenogenesis in aphids (aphids can reproduce by developing an unfertilized gamete).

1746   Pierre-Louis Moreau de Maupertuis (1698-1759), French mathematician, publishes his *Venus Physique* and foresees the chromosomal basis of heredity. He reject theories inspired by the idea of preformation, such as ovism and spermism on the grounds that children often resemble both parents.

1754   Pierre-Louis Moreau de Maupertuis (1698-1759), French mathematician, suggests that species transform over time.

1757   Albrecht von Haller (1708-1777), Swiss physiologist, publishes the first volume of his eight-volume *Elements of Physiology of the Human Body* (1757-1766) and distinguishes irritability or inherent muscular force from nerve, or nervous, force. This encyclopedic work heralds the rise of modern physiology.

1759   Kaspar Friedrich Wolff (1733-1794), German physiologist, publishes *Theory of Generation* which finally refutes the preformation thesis. He formulates a theory, in accordance with the idea of epigenesis, arguing that each germ or embryo is an entirely new creation. This book essentially marks the beginning of modern embryology.

1760   In his book on plant sexuality, Carl Linnaeus (1707-1778), Swedish botanist, discusses hybrids as originators of new species.

1762   Marcus Anton von Plenciz, Sr. (1705-1786), Austrian physician, expresses the idea that all infectious diseases are caused by living organisms, and that there is a particular organism for each disease.

1765   Lazzaro Spallanzani (1765-1777), Italian biologist, publishes his *Microscopical Observations*, offering minute details of experiments as refutation of the theory of the spontaneous generation of infusoria. He also succeeds in artificially inseminating a toad in 1780.

First drawing of cell division is made by Abraham Trembley (1710-1784), Swiss naturalist, who witnesses multiplication of protozoans by cell division. He also observes cell division in algae.

1771   Luigi Galvani (1737-1798), Italian anatomist, discovers the electric nature of nervous impulses.

1772   Joseph Priestley (1733-1804), English theologian and chemist, discovers that plants give off oxygen.

1774   Antoine-Laurent Lavoisier (1743-1794), French chemist, discovers that oxygen is consumed during respiration.

1779   Jan Ingenhousz (1739-1799), Dutch physician and plant physiologist publishes his *Experiments upon Vegetables* and shows that light is necessary for the production of oxygen, and that carbon dioxide is taken in by plants in the daytime and given off at night.

Domenico Cotugno (1736-1822), Italian anatomist, declares that cerebrospinal fluid, and not "animal spirit," as previously believed, fills the brain's cavities and ventricles.

Johann F. Blumenbach (1832-1840), German naturalist publishes his influential *Handbook of Natural History*.

1780   Antoine-Laurent Lavoisier (1743-1794), French chemist, and Pierre-Simon Laplace (1749-1827), French astronomer and mathematician, collaborate to demonstrate that respiration is a form of combustion. Breathing, like combustion, liberates heat, carbon dioxide and water.

George Adams (1750-1795), English engineer, devises the first microtome. This mechanical instrument cuts thin slice for examination under a microscope, thus replacing the imprecise procedure of cutting by s hand-held razor.

1785   Humphry Marshall (1722-1801), American botanist, publishes his *Arbustum Americanum: The American Grove*, the first American book devoted expressly to botany. It is published in Philadelphia.

1786   Adam Kuhn (1741-1817), native of Germantown, Pennsylvania, and a student of Linnaeus, becomes the first professor of botany and *materia medica* in America at the College of Philadelphia.

1789   Antoine-Laurent de Jussieu (1748-1836), French botanist, publishes his *Plant Genera,* a widely accepted book which incorporates the Linnean system of binomial nomenclature and is regarded as the foundation of the natural system of botanical classification. Following Ray, Jussieu classifies plants on the basis of cotyledons, dividing all plants into acotyledons, monocotyledons, and dicotyledons.

1793   Christian Konrad Sprengel (1750-1816), German botanist, clearly describes the role of insect in plant fertilization in his *Nature's Revealed Secret in the Structure and Pollination of Flowers.*

1794   First paper on systematic zoology published in America is written by American naturalist William Dandridge Peck (1763-1822). It is titled "Descriptions of Four Remarkable Fishes Taken Near the Pisquataqua in New Hampshire."

1796   Erasmus Darwin (1731-1823), English physician, publishes his *Zoonomia*. In this work, Charles Darwin's grandfather argues that evolutionary changes are brought about by the environment's direct influence on the organism.

1797   Georges-Léopold-Chrétien-Frédé Dagobert Cuvier (1769-1832), French anatomist, founds comparative zoology. With the publication of his firstbook, *Basic Outline for a Natural History of Animals*, Cuvier introduces his basic ideas about the way an animal's function and habits determine its form. The possibility of a reverse process is excluded.

1800   Marie-François-Xavier Bichat (1771-1802), French physician, publishes his first major work, *Treatise on Tissues*, effectively founding histology as a scientific discipline. Bichat distinguishes 21 kinds of tissue and relates particular diseases to particular tissues.

1802   Jean-Baptiste-Pierre-Antoine de Monet de Lamarck (1744-1829), French naturalist, and German naturalist Gottfried Reinhold Treviranus (1776-1837), propose the term "biology" to denote a general science of living beings.

1803   The first comprehensive survey of American flora, *Flora Boreali-Americana*, by the French botanist André Michaux (1746-1802) is published posthumously in Paris. The books include numerous plates drawn by the French artist Pierre-Joseph Redoute.

1805 William Dandridge Peck (1763-1822), American naturalist, is elected the first professor of natural history at Harvard University.

1809 Jean-Baptiste-Pierre-Antoine de Monet de Lamarck (1744-1829), French naturalist, introduces the term "invertebrate" in his *Zoological Philosophy*, which contains the first scientific theory of evolution. He also classifies organisms by function rather than by structure and is the first to use the genealogical tree.

1810 Franz Joseph Gall (1758-1828), German physician, first lays the basis for modern neurology with his dissections of the brain and his correct suggestions about nerve organization.

1811 Julien-Jean-César Legallois (1770-1814), French physiologist, locates the first physiological center in the brain.

Charles Bell (1774-1842), Scottish surgeon, differentiates between sensory and motor roots of spinal nerves in his *New Idea of Anatomy of the Brain*. He declares that each nerve carries either a motor or a sensory stimulus, and not both simultaneously.

1812 Georges-Léopold-Chrétien-Frédéric-Dagobert Cuvier (1769-1832), French anatomist, founds vertebrate paleontology with his *Investigations of the Fossil Bones of Quadrupeds*.

1813 William Charles Wells (1757-1817), American-English scientist, and his colleagues Prichard and Lawrence present theories of natural selection and reject Lamarckism (which asserts that acquired characteristics can be inherited).

1817 Georges-Léopold-Chrétien-Frédéric-Dagobert Cuvier (1769-1832), French anatomist, publishes his major work, *The Animal Kingdom*, which expands and improves Linnaeus's classification system by grouping related classes into broader groups called phyla. He is also the first to extend this system of classification to fossils.

1820 First United States *Pharmacopoeia* is published.

1821 Jean-Louis Prévost (1790-1850), Swiss physician, jointly publishes a paper with French chemist Jean-Baptiste-André Dumas (1800-1884), which demonstrates for the first time that spermatozoa originate in tissues of the male sex glands. In 1824, they also give the first detailed account of the segmentation of a frog's egg.

1822 François Magendie (1783-1855), French physiologist, publishes his paper "Functions of the Roots of the Spinal Nerves," which lays the foundation for the Bell-Magendie Law.

1824 René-Joachim-Henri Dutrochet (1776-1847), French physiologist, discovers that tissue is composed of living cells.

Marie-Jean-Pierre-Flourens (1794-1867), French physiologist, first proves that the brain's respiratory center lies low in the brainstem. A pioneer of the idea of nervous coordination, he is also the first the recognize the role of the inner ear in maintaining body equilibrium and coordination.

1825 Jan Evangelista Purkinje (1787-1869), Czech physiologist, first describes the "germinal vesicle," or nucleus, in a hen's egg.

1827 Karl Ernst von Baer (1792-1876), German-Estonian embryologist, publishes his *On the Mammalian Egg*, documenting his discovery, in 1826, of the mammalian egg. In his book *On the Developmental History of Animals* (2 volumes, 1828-1837), von Baer he demonstrates that mammal, and consequently human, development, is not fundamentally different from that of other animals. His books become the first text of comparative embryology.

1828 Friedrich Wohler (1800-1882), German chemist, first synthesizes urea, laying the foundations (along with Berzelius, Liebig, and Bunsen) of organic chemistry. His work also paves the way to the foundation of biochemistry. Wohler's preparation of an organic compound dispels the notion that organic compounds can only be produced by using living organisms.

Robert Brown (1773-1858), Scottish botanist, observes a small body within the cells of plant tissue and names it a "nucleus." He also discovers what becomes known as "Brownian movement" in plants. This phenomenon is explained a generation later by the kinetic theory of gases.

Luigi Rolando (1773-1831), Italian anatomist, achieves the first electrical stimulation of the brain.

First American book dealing exclusively with flowers, Robin Green's *Treatise on the Cultivation of Flowers*, is published in Boston.

1830 Boston Society of Natural History is founded. With Thomas Nuttall (1786-1859) as its first president, the Society incorporates the C. J. Sprague and Thomas Taylor lichens and the Francis Boott collections.

1831 Charles Robert Darwin (1809-1882), English naturalist, begins his historic voyage on the H.M.S. *Beagle* (1831-1836).

Robert Brown (1773-1858), Scottish naturalist, recognizes the cell nucleus as a regular feature of all plant cells.

1832 Anselme Payen (1795-1871), French physiologist, first isolates diastase, which he separates from barley. This is the substance that had the property of hastening the conversion of starch into sugar, and is an example of the organic catalysts within living tissue (which eventually come to be called enzymes).

1833 Johannes Peter Müller (1801-1858), German physiologist, first proposes his law of specific nerve energies. According to this law, every sensory nerve gives rise to one form of sensation, even it is excited by stimuli outside a normal range.

1836 Felix Dujardin (1801-1860), French zoologist, describes the "living jell" of the cytoplasm, calling it "sarcode."

Theodor Schwann (1810-1882), German physiologist, allows only heated air to enter a media flask and finds that growth occurs, thus invalidating the theory of spontaneous generation. He also demonstrates that alcoholic fermentation depends on yeast. The same conclusion is reached independently by Charles Caignard de la Tour (1777-1859).

Asa Gray (1810-1888), American botanist, publishes his first book, *Elements of Botany*, in New York.

1837 René-Joachim Dutrochet (1776-1847), French physiologist, publishes his research on plant physiology, which include pioneering work on osmosis. The first scientist to systematically investigates the process of osmosis (which he names), Dutrochet recognizes the importance of this phenomenon, as well as the fact that chlorophyll is necessary for photosynthesis.

Karl Theodor von Siebold (1804-1884), German zoologist, and Michael Sars (1805-1869), Norwegian priest and marine zoologist, describe the division of invertebrate eggs.

Robert Remak (1815-1865), German physician, accurately notes the relationship of nerve cells to nerve fibers, and first names the neurolemma (the myelin sheath around many nerve fibers. He later offers names that are still used for the three germinal layers in the developing embryo: ectoderm (outer skin), mesoderm (middle skin), and endoderm (inner skin).

1838 Matthias Jakob Schleiden (1804-1881), German botanist, applies, in an article, the emerging cell theory to plants, which he describes as a community of cells. He develops the ideas of Robert Brown, influencing Schwann to expand the cell theory even further.

Jan Evangelista Purkinje (1787-1869), Czech physiologist, first describes a large group of distinct cells in the cerebellum. These cells are now known as "Purkinje cells."

1839 Theodore Schwann (1820-1882), German physiolo-

gist, extends the theory of cells to include animals. In his book *Microscopical Researches into the Accordance in the Structure and Growth of Animals and Plants*, he states all that all living things are made up of cells, each of which contains certain essential components. He also coins the term "metabolism" to describe the overall chemical changes that take place in living tissue.

Jan Evangelista Purkinje (1787-1869), Czech physiologist, uses the term "protoplasm" to define the substance of the cell.

1840 Rudolf Albert von Kölliker (1817-1905), Swiss anatomist and physiologist proves (while still as student) that spermatozoa and eggs, or ova, are derived from tissue cells. He also applies the cell theory to embryology and histology.

Karl Bogislaus Reichert (1811-1883), German anatomist and embryologist, introduces the cell theory into embryology, showing that egg segments develop into cells, and that organs develop from cells.

Justus von Liebig (1803-1873), German chemist, shows that plants synthesize organic compounds from carbon dioxide in the air, but take their nitrogenous compounds from the soil. He also states that ammonia (nitrogen) is needed for plant growth.

John James Audubon (1785-1851), French-American ornithologist, publishes the first volume of his *Birds of America* (1840-1844). This illustrated work is considered among the most beautiful natural histories ever done. Although Audubon usually painted from dead specimens, he is regarded as one of the first American conservationists.

Friedrich Gustav Jacob Henle (1809-1885), German pathologist and anatomist, publishes the first histology textbook, *General Anatomy*. This book includes the first statement of the germ theory of communicable diseases.

1842 Charles Robert Darwin (1809-1882), English naturalist, writes the first abstract of his theory of evolution.

Theodor Ludwig Wilhelm Bischoff (1807-1882), German embryologist, publishes the first textbook of comparative embryology, *Developmental History of Mammals and Man*.

1843 Martin Berry (1802-1855), English physician and physiologist, observes the union of sperm and ovum of a rabbit.

1844 Robert Chambers (1802-1871), Scottish publisher, anonymously publishes his *Vestiges of the Natural History of Creation*. This best-selling book offers a sweeping view of evolution. Although quite unreli-able as a scientific work, the book paves the way for Darwin theory by familiarizing the public with evo-lutionary concepts.

1845   Karl Theodor Ernst von Siebold (1804-1885), German zoologist, recognizes that protozoa are sin-gle-celled organisms and is the first to define them as organisms.

1846   Giovanni Battista Amici (1786-1868), Italian physi-cist, definitely establishes sexuality in flowering plants.

1848   Karl Theodor Ernst von Siebold (1804-1885), German zoologist, publishes his *Textbook of Invertebrate Comparative Anatomy,* one of the first major textbooks on invertebrate anatomy.

1849   Rudolf Wagner (1805-1864), German anatomist and physiologist, and Karl Georg Friedrich Rudolf Leuckart (1822-1989), German zoologist, state that spermatozoa are a definite and essential part of the semen, and that the liquid merely keeps them in sus-pension. They also demolish the old hypothesis that spermatozoa are parasites, affirming that spermato-zoa are essential for fertilization.

Karl Friedrich Gärtner (1772-1850), German physi-cian, publishes the first comprehensive treatment of hybridization. It will be closely studied by Darwin and will influence Mendel.

1850   Wilhelm Friedrich Hofmeister (1824-1877), German botanist, erases the distinction between flowering and nonflowering plant, demonstrating the alterna-tion of sexual and nonsexual generations. This has been called "the greatest discovery that has ever been made in the realm of plant morphology and taxono-my."

1851   Hugo von Mohl (1805-1872), German botanist, pub-lishes his *Basic Outline of the Anatomy and Physiology of the Plant Cell*, in which he first pro-poses that new cells are created by cell division.

1854   George Newport (1803-1854), English naturalist and anatomist, performs the first experiment on animal embryos. He notes that the point of sperm entry determines the planes of the egg's segmentation.

Rudolf Ludwig Carl Virchow (1821-1902), German pathologist, first names the neuroglia, or supportive "glue cells" in the brain.

1855   Alfred Russell Wallace (1823-1913), English natu-ralist, publishes his paper "On the Law Which has Regulated the Introduction of New Species." Although Wallace has not yet conceived of the notion of natural selection, this work shows him in the process of anticipating Darwin.

Barolomeo Panizza (1785-1867), Italian anatomist, first proves that parts of the cerebral cortex are essential for vision.

1856   Nathanael Pringsheim (1821-1894), Silesian biolo-gist, first sees the sperm enter the female egg (of a freshwater algae plant).

1857   Louis Pasteur (1822-1895), French chemist, demon-strates that lactic acid fermentation is due to a living organism. Also, during the period between this year and 1880, he invalidates the idea of spontaneous generation by showing that organisms move through the air; discovers a cure for silk worm disease; uses attenuated viruses for resistance to fowl cholera; and develops immunization for anthrax and inoculation for rabies.

1858   Rudolf Ludwig Carl Virchow (1821-1902), German pathologist, founds cellular pathology with the pub-lication of his seminal paper "Cellular Pathology" in Berlin. This work includes his historic assertion *Omnis cellula e cellula* ("Every cell comes from a cell"). He also states that all disease is disease of the cells.

1859   Charles Robert Darwin (1809-1882), English natu-ralist, publishes his landmark book *On the Origin of Species by Means of Natural Selection*. The classic of science establishes the natural mechanism of the selection of favorable (inherited) traits or variation as the cornerstone of his theory of evolution.

1860   Max Johann Sigismund Schultze (1825-1874), first clearly describes the nature of the protoplasm, show-ing that is fundamentally the same for all life forms.

1861   Carl Gegenbaur (1826-1903), German biologist, proves that Schwann is correct and that all vertebrate eggs are single cells.

Louis Pasteur (1822-1895), French chemist, finally demolishes the theory of spontaneous generation. His ingenious experiment which allows air but no dust particles or spores into a flask containing a piece of meat shows that the meat does not spoil or decay.

Pierre-Paul Broca (1824-1880), French surgeon and anthropologist, first identifies a location in the brain's left hemisphere which (in most people) is associated with speech. It is later called "Broca's area."

Rudolf Albert von Kölliker (1817-1905), Swiss anatomist and physiologist, publishes his *Developmental History of Man and Higher Animals*, which is the first treatise on comparative embryolo-gy.

1862   Julius von Sachs (1832-1897), German botanist, states that sunlight is necessary for photosynthesis.

he also proves that starch is a product of photosynthesis and later (1865) finds that chlorophyll is contained in the chloroplasts.

1863 Thomas Henry Huxley (1825-1895), English biologist, extends Darwin's theory of evolution to include humans in his book *Evidence As to Man's Place in Nature*. He becomes the foremost defender and popularizes of Darwinism in England.

1864 Louis Pasteur (1822-1895), French chemist, invents a process of slow heating that kills bacteria and other microorganisms. Called pasteurization, it is first used to keep wine and beer from turning sour.

1865 Franz Schweiger-Seidel (1834-1871), German physiologist, first proves that a spermatozoon has a nucleus and cytoplasm.

Jules-Bernard Luys (1826-1897), French physician, describes a nucleus in the hypothalamus, forming a part of the descending pathway from the corpus striatum. It becomes known as the "nucleus of Luys."

1866 Johann Gregor Mendel (1822-1884), Austrian botanist and monk, discovers the laws of heredity and writes the first of a series of papers on heredity (1866-1869), which formulate the laws of hybridization. His work is disregarded until 1900, when de Vries rediscovers it. Unbeknownst to both Darwin and Mendel, Mendelian laws provide the scientific framework for the concepts of gradual evolution and continuous variation.

Ernst Heinrich Haeckel (1834-1919), German naturalist, makes him famous statement "ontogeny recapitulates phylogeny," which essentially means that particular stages of an individual organism's embryonic development correspond to stages of the species' evolutionary development. His book, *A General Morphology of Organisms,* in which he expounds his eclectic theory of life, is published in Berlin.

1868 Thomas Henry Huxley (1825-1895), English biologist, introduces the term "protoplasm" to the general public in a lecture, entitled "The Physical Basis of Life," which he delivers in Edinburgh.

Othniel Charles Marsh (1831-1899), American paleontologist, lays the foundation of American paleontology.

1869 Johann Friedrich Miescher (1844-1895), Swiss biochemist, discovers nucleic acid (by isolating the nucleus in which deoxyribonucleic acid, or DNA, will later be found).

Paul Langerhans (1847-1888), German physician, discovers irregular islands of cells in the pancreas which produce insulin. They become known as the "Isles of Langerhans."

1870 Gustav Theodor Fritsch (1838-1927), German anatomist and anthropologist, and Eduard Hitzig (1838-1907), German physiologist and neurologist, discover that electric shocks to one cerebral hemisphere of a dog's brain produces movement on the other side of the animal's body. This is the first clear demonstration of the existence of cerebral hemispheric lateralization.

1871 Charles Robert Darwin (1809-1882), English naturalist, publishes his *The Descent of Man, and Selection in Relation to Sex.* This work expands his theory of evolution to include humans.

1872 Ferdinand Julius Cohn (1850-1898), German botanist, publishes the first of four papers titled "Research on Bacteria," which mark the beginning of bacteriology as a distinct field. He systematically divides bacteria into genera and species.

1873 Walther Flemming (1843-1905), German anatomist, discovers chromosomes, observes mitosis, and provides the modern interpretation of nuclear division.

Franz Anton Schneider (1831-1890), German zoologist, first describes cell division in detail. His drawings include both the nucleus and chromosomal strands.

Camillo Golgi (1843-1926), Italian histologist, devises a way to satin tissue samples with inorganic dye (silver salts). He applies this new method to nerve tissues and is able to see details not visible before.

1874 Carl Wernicke (1848-1905), German neurologist, discovers the area of the brain associated with word comprehension. It is later called "Wernicke's area."

1875 Eduard Adolf Strasburger (1844-1912), German botanist, publishes his *Cell-Formation and Cell-Division*, in which he details mitosis in plant cell. This exhaustive work establishes cytology as a distinct branch of histology.

Theodor Wilhelm Engelmann (1843-1909), German physiologist, proves experimentally that the heartbeat is myogenic, which mean means that it originates in the heart muscle itself, and not from an external impulse.

1876 Edouard Gürard Balbiani (1825-1899), French biologist, observes the formation of chromosomes.

Wilhelm August Oscar Hertwig (1849-1922), German embryologist and anatomist, observes the fertilization of a sea urchin egg and establishes that genetic material is contributed to the offspring from both parents. He also shows that fertilization is due to a fusion of sperm with the egg nuclei.

Robert Koch (1843-11910), German bacteriologist, cultivates the anthrax bacteria outside the body. He then studies its life cycle and learns how to defeat it. Within the next six years, Koch isolates the *tubercle bacillus* and discovers the cause of cholera.

Robert Koch (1843-1910), German bacteriologist), describes his new techniques of fixing, staining, and photographing bacteria.

Alfred Russel Wallace (1823-1913), English naturalist, publishes his major work, *The Geographical Distribution of Animals*, which establishes zoogeography as a modern scientific discipline.

1878   Wilhelm Friedrich Kuhne (1837-1900), German physiologist, introduces the term "enzyme."

Charles-Emanuel Sedillot (1804-1883), French surgeon, introduces the term "microbe." It is accepted by Pasteur and becomes a universal term for a pathogenic bacterium.

1879   Hermann Fol (1845-1892), Swiss physician and zoologist, is the first to observe a spermatozoon penetrate an egg; he demonstrates that only one spermatozoon is needed for fertilization.

1880   David Ferrier (1843-1928), Scottish scientist, maps the region of the brain called the motor cortex and discovers the sensory strip.

The journal *Science* is first published by the American Association for the Advancement of Science.

1881   Wilhelm Roux (1850-1924), German zoologist, elaborates on existing mechanistic accounts of embryonic development, affirming that there is a "functional adaptation" in embryology. His efforts to determine how organs and tissues are assigned their structural form and function at fertilization make him the founder of experimental embryology.

1882   Walther Flemming (1843-1905), German anatomist, publishes his *Cell Substance, Nucleus, and Cell Division*, in which he details his observation of longitudinal division or splitting of chromosomes in animal cells. He is the first to observe this phenomenon.

Robert Koch (1843-1910), German bacteriologist, discovers the tubercle bacillus and enunciates "Koch's postulates," which define the classic method of preserving, documenting, and studying bacteria.

1883   Wilhelm Roux (1850-1924), German zoologist, postulates that chromosomes in the nucleus are bearers of hereditary determiners.

August F. Weismann (1834-1914), German biologist, begins work that culminates in his germ-plasm theory (1883-1889). This notion anticipates the knowledge eventually attained by DNA research, for the germ-plasm theory postulates that all living things contain a particular hereditary substance.

1884   Elie Metchnikoff (1845-1916), Russian-French physiologist, discovers the antibacterial activity of some white cells and formulates the theory of phagocytosis

1885   Edouard van Beneden (1846-1910), Belgian cytologist, proves that chromosomes persist between cell divisions. The first scientist to do a chromosome count, he discovers that each species has a fixed number of chromosomes. He also discovers that in the formation of sex cells, the division of chromosomes during one of the cell divisions was not preceded by a doubling. This is explained by the fact that each egg and sperm cell has only half the usual number of chromosomes.

Louis Pasteur (1822-1895), French chemist, inoculates a boy, Joseph Meister, against rabies. He had been bitten by a mad dog, and the treatment saves his life. This is the first time Pasteur uses an attenuated herb on a human being.

1888   Heinrich Wilhelm Gottfried Waldyer-Hartz (1836-1921), German anatomist, first introduces the term "chromosome."

Theodor Heinrich Boveri (1862-1915), German zoologist, discovers and names "centrosome" the mitotic spindle that appears during cell division. He is also the first to note that chromosomes split as part of the mechanism of reproduction.

Woods Hole Marine Biological Station is founded in Massachusetts. It eventually becomes headquarters for the Woods Hole Oceanographic Institution (1930) and the Marine Biological Laboratory.

1889   Richard Altmann (1852-1900), German histologist, isolates and names nucleic acid.

1891   Charles-Edouard Brown-Sequard (1817-1894), French physician, formulates the idea of internal secretions (hormones).

Hermann Henking (1858-1942), German zoologist, describes sex chromosomes and autosomes.

1892   August F. Weismann (1834-1914), German biologist, publishes his book *The Germ-Plasma: A Theory of Heredity*, which contributes to the founding of the science of genetics. Weismann's germ-plasm theory also lead him predict the discovery of meiosis (the splitting of paired chromosomes).

1895   John William Harshberger (1869-1929), American botanist, first introduces the term "ethobotany."

1896   Edmund Beecher Wilson (1856-1939), American

zoologist, publishes his major work, *The Cell in Development and Heredity*, in which he connects chromosomes and sex determination. He also discovers that chromosomes affect and determine other inherited characteristics as well.

1897   John Jacob Abel (1857-1938), American physiologist and chemist, isolates epinephrine (adrenalin). This is the first hormone to be isolated.

1898   Carl Benda (1857-1932), German physician discovers and names the mitochondria, previously seen by Richard Altmann (1852-1900), German physician.

Martin Wilhelm Beijerinck (1851-1931), Dutch botanist, discovers and names the causative agent of tobacco mosaic disease. This agent, which he describes as a new type of microscopically-visible organism, eventually comes to be known as a virus.

Charles Reid Barnes (1858-1910), American botanist, proposes the term "photosynthesis."

1899   Jacques Loeb (1859-1924), German-American physiologist, induces parthenogenesis in the sea urchin. He demonstrates what is essentially artificial insemination when he induces an unfertilized egg to develop to maturity by means of specific environmental changes.

1900   Hugo Marie de Vries (1848-1935), Dutch botanist, independently discovers the law of inheritance. Before publishing, he discovers Mendel's forgotten papers in which the laws of heredity are formulated. De Vries subsequently announces Mendel's earlier discovery and offers his own work only as confirmation. He presents the idea of mutations in *Mutation Theory*.

1901   Jokichi Takamine (1854-1922), Japanese-American chemist, and T. B. Aldrich first isolate epinephrine from the adrenal gland. Later known by the trade name Adrenalin, it is eventually identified as a neurotransmitter. This is also the first time a pure hormone has been isolated.

Santiago Ramon y Cajal (1852-1911), Spanish histologist, first discovers the nature of the connection between nerves, showing that the nervous system consists of a maze of individual cells. He demonstrates that neurons do not touch but that the signal somehow crosses a gap (now called a "synapse.")

1902   Ernest H. Starling (1866-1927) and William H. Bayliss (1860-1924), both English physiologists, discover and isolate the first hormone ("secretin," found in the duodenum).

1903   Walter Stanborough Sutton (1876-1916), American geneticist, writes a short paper in which he presents the chromosome theory of inheritance. This important theory, which states that the hereditary factors are located in the chromosomes, is independently developed by German biologist Theodor Boveri (1862-1915).

Archibald Edward Garrod (1857-1936), British physician, suggests that errors in genes cause hereditary disorders. His 1909 book *The Inborn Errors of Metabolism* is the first study in biochemical genetics demonstrating that errors on genes lead to hereditary disorders.

Emil G. Racoviça (1868-1947), Rumanian biologist, publishes his first finding dealing with animal populations dwelling in caves, His work founds modern biospeleology.

1905   Nettie Maria Stevens (1856-1939), American geneticist, discovers the connection between chromosomes and sex determination. She determines that there are two basic types of chromosome: X and Y.

Fritz Schaudinn (1871-1906), German zoologist, and Erich Hoffmann (1868-1959), German dermatologist, discover the *Spirochaeta pallida* of syphilis. This is a major step toward controlling the disease.

1907   Ivan Petrovich Pavlov (1849-1910), investigates the conditioned reflex (1904-1907). A great stimulus for behaviorist psychology, his work is the high point of physiologically-oriented psychology.

William Bateson (1861-1926), English biologist, first uses the term "genetics" to indicate the science of heredity.

1908   Godfrey Harold Hardy (1877-1947), English mathematicians, and German physician Wilhelm Weinberg (1862-1937) publish similar papers (six months apart) on a mathematical system that describes the stability of gene frequencies in succeeding generations of s population. Their resulting "Hardy-Weinberg Law" links the Mendelian hypothesis with actual population studies."

Margaret A. Lewis (1881-1970), American cytologist, first cultures mammalian cells *in vitro*.

1909   Wilhelm Ludwig Johannsen (1857-1927), Danish botanist, first proposes the terms "gene" (carrier of heredity), "genotype" (an organism's genetic constitution), and "phenotype" (the actual organism that is their product).

Phoebus Aaron Theodore Levene (1869-1940), Russian-American chemist, discovers the chemical difference between DNA (deoxyribonucleic acid) and RNA (ribonucleic acid).

Thomas Hunt Morgan (1866-1945), American geneticist, selects the fruit fly Drosophila to investigate genetics by observing mutations. He confirms the chromosome theory of heredity and notes the significance of the fact that certain genes tend to be transmitted together. He postulates the mechanism of "crossing over," which Alfred Henry Sturtevant (1891-1970), American geneticist, demonstrates in 1913.

Jean de Mayer, French physiologist, first suggests the name "insulin" for the hormone of the islet cells.

Korbinian Bordmann (1868-1918), German neurologist, publishes a "map" of the cerebral cortex, assigning numbers to particular regions.

1910 Paul Ehrlich (1854-1915), German bacteriologist, discovers a chemical treatment for syphilis. He introduces "salvarsan," or arsphenamine, and founds modern chemotherapy.

Harvey Cushing (1869-1939), American surgeon, and his team present the first experimental evidence of the link between the anterior pituitary and the reproductive organs.

1912 Casimir Funk (1884-1967), Polish-American biochemist, coins the term "vitamine." Since the dietary substances he discovers are in the amine group he calls all of them "life-amines" (using the Latin word for life, "vita").

First chair in Biochemistry in the United Kingdom goes to Frederick Gowland Hopkins (1861-1957).

1913 Elmer Verner McCollum (1879-1967), American biochemist, produces the first chromosome map, showing five sex-linked genes.

1914 Edward Calvin Kendall (1886-1972), American biochemist, extracts thyroxin from the thyroid gland (in crystalline form) and founds hormonology.

Frederick William Twort (1877-1950), English bacteriologist, and Felix H. D'Herelle (1873-1949), Canadian-Russian physician, discover, independently of each other, bacteriophages, viruses which destroy bacteria.

1917 D'Arcy Wentworth Thompson (1860-1948), Scottish zoologist, publishes his book *On Growth and Form*, which views the evolution of one species into another as a series of transformations involving the entire organism, rather than a succession of minors changes in the body parts.

Calvin Blackman Bridges (1889-1938), American geneticist, discovers the first chromosome deficiency.

1920 Frederick Grant Banting (1891-1941), Canadian physician, Charles Best (1899-1878), Scottish-American physiologist, and James B. Collip (1892-1965), Canadian biochemist, discover insulin. They

develop a method a method of extracting insulin from the human pancreas. The insulin is then injected into the blood of diabetics to lower their blood sugar.

1921 Otto Loewi (1873-1961), German-American physiologist, discovers that acetylcholine functions as a neurotransmitter. It is the first such brain chemical to be so identified.

1922 Frederick Banting (1891-1941), Canadian physician, and Charles Herbert Best (1899-1978), Canadian physiologist, makes the first clinical adaptation of insulin for the treatment of diabetes.

Herbert McLean Evans (1882-1971), American physician, and colleagues discover vitamin E.

Elmer Verner McCollum (1879-1967), American biochemist, discovers vitamin D.

1925 Johannes Hans Berger (1873-1941), German neurologist, records the first human electroencephalogram (EEG).

1927 Hermann Joseph Muller (1890-1967), American biologist, induces the first artificial mutations in the *Drosophila* fruit fly by exposing it to x rays. His work shows that mutations result from some type of chemical change. It also alerts him to the danger of excessive x rays.

1928 Alexander Fleming (1881-1955), Scottish bacteriologist, discovers penicillin. He observes that the mold *Penicillium notatum* inhibits the growth of some bacteria. This is the first antibacterial, and it opens a new era of wonder drugs.

Wilder Graves Penfield (1891-1976), Canadian neurosurgeon, first uses microelectrodes to map areas in the human cerebral cortex.

1929 Willard Myron Allen, American physician, and George Washington Corner, American anatomist, discover progesterpine. They demonstrate that it is necessary for the maintenance of pregnancy.

1931 Joseph Needham, English biochemist, publishes his landmark work *Chemical Embryology*, which shows the relation of biochemistry and embryology, thereby founding chemical embryology and modern molecular biology.

1932 Hans Adolf Krebs (1900-1981), German-British biochemist, first describes and names the citric acid cycle.

1935 Wendall Meredith Stanley (1904-1971), American biochemist, discovers that viruses are partly protein-based. By purifying and crystallizing viruses, he enables scientists to identify the precis molecular structure and propagation modes of several viruses.

1937   Richard Benedict Goldschmidt (1878-1958), German-American geneticist, theorizes that the gene is a chemical entity rather than a discrete physical structure. His theories cause a major reevaluation of the concepts governing the science of genetics.

        James W. Papez (1883-1958), American anatomist, suggest the name "limbic system" for the old mammalian part of the human brain that produces our emotions.

1940   George Wells Beadle (1903-1989), American geneticist, and Edward Lawrie Tatum (1909-1975), American biochemist, establish the formula "one gene-one enzyme." This discovery that each gene determines the production of only one enzyme lays the foundation for future DNA discoveries.

1944   Oswald Theodore Avery (1877-1947), Canadian biologist, Macklin McCarthy, Canadian physician, and Colin Munro Macleod (1909-1972), Canadian physician and physician and microbiologist, discover the "blueprint" function of DNA (that DNA carries genetic information).

1946   Joshua Lederberg, American geneticist, and Edward Lawrie Tatum (1909-1975), American biochemist and geneticist, first demonstrate genetic recombination of *E. coli* bacteria. This intermingling of genetic material of bacteria greatly expands the scope of genetic research.

1949   The role of mitochondria is finally revealed. These slender filaments within the cell, which participate in protein synthesis and lipid metabolism, are the cell's source of energy.

        Walter R. Hess (1881-1973), Swiss physiologist, receives the Nobel Prize for his experiments involving probes of deep-brain functions. Using microelectrodes to stimulate or destroy specific areas of the brain in experimental animals, he discovers the role played by particular brain areas in determining and coordinating the functions of internal organs.

1950   Erwin Chargraff, American-Austrian biochemist, establish a set of rules concerning the basic chemistry of DNA. His findings eventually lead to greater understanding of the chemical structure of nucleic acids.

1952   Alan L. Hodgkin and Andrew F. Huxley, both English physiologists, first work out the mechanism of nerve-impulse transmission, showing that a "sodium pump" system works to carry impulses.

        Maurice Hugh Frederick Wilkins, New Zealand biophysicist, and Rosalind Elsie Franklin (1920-1958), British physicist, take x-ray diffraction pictures of DNA. This information is of enormous help to Watson and Crick.

1953   Francis Harry Compton Crick, English biochemist, and James Dewey Watson, American biochemist, work out the double-helix, or double spiral, DNA model. They rely heavily on earlier work (which they fail to acknowledge) done by Rosalind Elsie Franklin (1920-1958). Their model explains how DNA transmits hereditary traits in living organisms.

1954   Frederick Sanger, English biochemist, determines the sequence of amino acids in insulin. This extraordinary achievement is a true breakthrough in protein chemistry, leading to knowledge of the exact structure of even more complex compounds.

1956   J. Hin Tijo and Albert Lavan show that the number and Albert Lavan show that the number of chromosomes in a human cell is 46, and not 48, as believed since the early 1920s.

1957   Alick Isaacs (1921-1967), Scottish virologist, and Jean Lindenmann, Swiss microbiologist, discover interferon. They find that the body naturally produces a protein that acts as the first line of defense against invading viruses.

1958   Francis Harry Compton Crick, English biochemist, predicts the discovery of transfer DNA.

1959   Jerome Lejeune, French pediatrician, and colleagues M. Gautier and R. A. Turpin, first identify a disease caused by a chromosome aberration. They discover that patients with Down's syndrome have 47 chromosomes instead of the normal 46, The extra is a third copy of chromosome 21.

        Arthur Kornberg, American biochemist, and Severo Ochoa, Spanish-American biochemist, share the Nobel Prize for their discovery of enzymes that produce artificial DNA and RNA.

1961   Marshall Warren Nirenberg, American biochemist, synthesizes a protein molecule by using artificial DNA.

        Marshall Warren Nirenberg, American biochemist, and Severo Ochoa, Spanish-American biochemist, work out the beginning of the RNA messenger code. RNA contains instructions for carrying out cellular processes, and it works with DNA, which contains the organism's genetic information.

1963   John Carew Eccles (1903-1997), Australian neurophysiologist, shares a Nobel Prize for his work on the mechanisms of nerve-impulse transmission. He also suggests that the mind is separate from the brain. The mind, he affirms, acts upon the brain by effecting subtle changes in the chemical signals that flow among brain cells.

1967   Charles T. Caskey, Richard E. Marshall, and Marshall Warren Nirenberg, American biochemist suggest that there is an universal genetic code shared by all life forms. Their statement is based on the fact that amino acids, bacteria, and guinea pigs have identical forms of messenger RNA.

Charles Yanofsky, American biologist, shows that the codon (genetic code) sequence in a gene determines the sequence of amino acids in a protein.

1968 Werner Arber, Swiss geneticist and biochemist, discovers that bacteria defend themselves against viruses by producing DNA-cutting enzymes. This insight later proves valuable to future work in genetic engineering.

Mark Steven Ptashne, American molecular biologist, and biologist Walter Gilbert independently identify repressor genes.

1969 Stanford Moore (1913-1982), and William H. Stein (1911-1980) both American biochemists, independently achieve the synthesis of an enzyme (ribonuclease from a bovine pancreas).

Jonathan R. Beckwith, American molecular biologist, and colleagues isolate a single gene.

1970 Har Gobind Khorana, Indian-American chemist, and colleagues announce the first complete synthesis of a gene.

Howard Martin Temin, American virologist, and David Baltimore, American biochemist, discover reverse transcriptase in viruses. This is an enzyme which causes the transcription of RNA into DNA.

1973 Genetic engineering is performed as Herbert Wayne Boyer, American microbiologist, and Stanley H. Cohen. American biochemist, cut DNA molecules using restriction enzymes, join the molecules together, and reproduce them into the bacterium *Escherichia coli.*

First report is made claiming a circadian variation in blood melatonin levels (pineal hormone) in humans. These variations affect mood and may cause the type of depression association with seasonal affective disorder (SAD).

1975 John R. Hughes, Scottish physiologist and others discover enkephalin. This first known opioid peptide, popularly called "brain morphine," occurs naturally in the brain, indicating that the brain's chemical block the transmission of pain signals.

Cesar Milstein, English nuclear biologist, discovers monoclonal antibodies.

1976 Har Gobinda Khorana, Indian-American chemist, and colleagues build a functional synthetic gene which has a complete system of regulatory mechanisms.

1977 Philip Allen Sharp and Richard John Roberts, both American molecular biologists, discover, independently of each other, that DNA in higher organisms contains a considerable amount of meaningless information (called an "intron") which is discarded during the process of making protein.

1980 The United States Supreme Court rules that a living organism developed by General Electric (a microbe used to clean up an oil spill) can be patented.

Martin Cline and colleagues successfully transfer a gene from one mouse to another. The gene remains functional.

1981 Karl Illmensee clones baby mice.

Researchers in China clone a fish.

First transfer of genes from one species of animal to another is completed.

1982 Gene from one mammal (rat growth hormone gene) functions for the first time in another mammal (mouse). The mouse grows to twice its normal size.

1983 First artificial chromosome is created by Andrew W. Murray and Jack William Szostak, Canadian biochemist.

1984 Steen A. Willadsen successfully clones sheep.

Allan Charles Wilson, American biochemist, and Russel Higuchi first clone genes from an extinct species as they remove a gene from the preserved skin of a quagga, a type of zebra.

1985 Alec Jeffreys first develops a method of "fingerprinting" with DNA. He shows that there are certain core sequences of DNA that are unique to each individual.

1986 First gene known to inhibit growth is produced by an American team led by molecular biologist Robert A. Weinberg. The is able to suppress the cancer retinoblastoma.

1987 David C. Page and colleagues discover the gene responsible for maleness in mammals. It is a single gene on the Y chromosome that causes the development of testes instead of ovaries.

1989 Seven cloned calves are born from one embryo.

1990 The Human Genome Project is officially launched.

1991 The gender of a mouse is changed at the embryo stage.

1992 Harry F. Noller, American biologist, demonstrates that RNA plays a greater role in protein synthesis than had been assumed.

1993 Scientists identify p53, a tumor suppressor gene, as

•

the crucial factor preventing uncontrolled cell growth. In addition, scientists find that p53 performs a variety of functions ensuring cell health.

Biologists find bacteria living at a depth of 1.67 mi (2.7 km). Subsurface bacteria can survive in what is known as "extreme environments," subsisting, for example, on sulphur only.

1994 Scientists create prions and discover prion-like proteins in yeasts.

Biologists discover that both vertebrates and invertebrates share certain developmental genes.

Researchers identify a metastasis-suppressor gene and determine that Tamoxifen, an anti-cancer drug, blocks a malignant tumor's blood supply.

Geneticists determine that DNA repair enzymes perform several vital functions, including preserving genetic information and protecting the cell from cancer.

1995 University of Copenhagen scientists Peter Funch and Reinhardt Møberg Kristensen decide that the minuscule invertebrate living in the mouths of Norwegian lobsters does not fit in any known phylum. They name the animal *Symbion pandora* and place it in the newly created phylum, Cycliophora.

Scientists report success in the genetic treatment of ADA-deficiency disease. When the body lacks the ADA enzymes, the T cells, a major element of the immune mechanism, are poisoned, which results in severely impaired immunity.

1996 American geneticist Gerard Schellenberg and colleagues discover the gene that causes Werner's syndrome, a condition which leads to premature aging. Scientists hope that this discovery will shed light on a variety of diseases associated with aging.

William R. Bishai, researcher at the Johns Hopkins School of Hygiene and Public Health, and co-workers find that SigF, a gene in the tuberculosis bacterium, enables the bacterium to enter a dormant stage.

Researchers C. Cheng and L. Olson discover that the spinal cord can be regenerated in adult rats. Experimenting on rats with a severed spinal cord, Cheng and Olson used peripheral nerves to connect white matter and gray matter.

New knowledge about BRCA1 and BRCA2, the genes responsible for some types of breast cancer, may lead to new treatments.

Scientists confirm findings that individuals with two mutant copies of the CC-CLR-5 gene are generally resistant to HIV infection.

Researchers find that abuse and violence can alter a

child's brain chemistry, placing him or her at risk for various problems, including drug abuse, cognitive disabilities, and mental illness, later in life.

Scientists discover a link between autoptosis (cellular suicide, a natural process whereby the body eliminates useless cells) gone awry and several neurodegenerative conditions, including Alzheimer's disease.

Chris Paszty, researcher at the Lawrence Berkeley National Laboratory, and co-workers, use genetic engineering to create mice with sickle-cell anemia, a serious human blood disorder. This is seen as a breakthrough in the search for treatments.

1997 Ian Wilmut, Scottish embryologist, and co-workers, successfully clone a lamb, Dolly, from a cell in a pregnant ewe's mammary gland.

Scientists discover p123, a protein component of telomerase, an enzyme which prevents chromosome shortening during cell division. Further research shows that if the protein's gene is mutated, chromosomes start shrinking. University of Colorado researcher Joachim Lingner stresses the importance of this discovery for cancer research.

Microbiologist William Jacobs and immunologist Barry Bloom create a biological entity which combines the characteristics of a bacterial virus and a plasmid (a DNA structure that functions and replicates independently of the chromosomes). This entity is capable of triggering mutations in *Mycobacterium tubercolosis*, thus enabling scientists to learn how to combat this pathogen.

While performing a cloning experiment, Christof Niehrs, researcher at the German Center for Cancer Research, identifies a protein responsible for the creation of the head in a frog embryo.

Researchers identify a gene which plays a crucial role in the left-right configuration during organ development.

Thomas Nagel, professor of philosophy at New York University, backed by several prominent biologists, presents arguments against reductionism, the view that fundamental elements explain complex biological phenomena.

Focusing on genetic mutations on humans and mice, researchers learn about the molecular signals that lead undeveloped neurons from inside the brain to their final position in the cerebral cortex.

Mickey Selzer, neurologist at the University of Pennsylvania, and co-workers, finds that in lampreys, which have a remarkable ability to regenerate a severed spinal cord, neurofilament messenger

RNA effects the regeneration process by literally pushing the growing axons and moving them forward.

1998 Scientists find that an adult human's brain can replace cells. This discovery heralds possible breakthroughs in neurology.

Immunologist Ellen Heber-Katz, researcher at the Wistar Institute in Philadelphia, reports than a strain of laboratory mice can regenerate tissue in their ears, closing holes which scientists had made for identification purposes. This discovery reopens the discussion on possible regeneration in humans.

British and American scientists complete the genetic map of the roundworm *(Caenorabditis elegans)*, a soil-dwelling worm measuring 0.04 in (1 mm). The genetic map, showing the 97 million genetic letters, in correct sequence, derived from the worm's 19,900 genes, is the first completed genome of an animal. Biologists declare that this accomplishment will vastly improve the understanding of human genetics, shedding light on a variety of diseases, include stroke, cancer, diabetes, and Alzheimer's disease.

Scientists in Korea claim to have cloned human cells.

1999 Danish researchers find what they believe is evidence of the oldest life on Earth—fossilized plankton from 3.7 billion years ago.

GENERAL INDEX

A

Abegg, Richard, 147
Abel, John Jacob, 386, 485
Abiogenesis. *See* Origins of life
Abiotic environment, **1,** 82, 90, 380–81, 539
Abiotic pressure, 697
Abnormalities. *See* Birth defects
ABO blood-group system, 12, **99–101**
    in blood transfusions, 236
    discovery of, 408, 453–54
    as polymorphism, 616
    Rh factor and, 660–61
Aborigines, 101–2
Abscisic acid, 611
Absolute dating, 579–80
Absolute zero, in thermodynamics, 756
Abundance, 83
Accessory minerals, 514
Accomodation theory of optics, 347
Acetanilide, 54
Acetate, 2, 96
Acetic acid, **1–2,** 3, 95, 153
Acetyl-coenzyme A, 96, 483–84, 583
Acetylcholine, **2–3**
    antihistamines, 37
    discovery of, 478, 544
    function of, 197, 243–44, 435–36
    neurotransmitters, 546
Acetylcholinesterase, 2
ACh. *See* Acetylcholine
Achondroplasia, 20
Acid-base equilibrium, 3
Acid-mine drainage
    sulfur cycle and, 731
Acid rain, **4,** 17–18, 595, 623
    from fossil fuels, 305
    nitrogen oxides in, 552
    sulfur cycle and, 731

Acidity
    bioremediation of, 88
    from coal mines, 305
    in pH balance, 595
    of surface water, 349
Acids, **3–4,** 403. *See also* specific acids, e.g., Amino acids
Acinar cells, 64
Acomycetes, 829
Acoustics
    in echolocation, 244
    physiological, **4–6,** 110
Acquired immunodeficiency syndrome. *See* AIDS
Acromegaly, 20, 690
ACTH. *See* Corticotropin
Actin, 194, 762
Actinomycetes
    spores and, 723
Actinomycin, 149, 791
Action potentials, **6**
    mathematical models of, 398
    membrane potential and, 500
    nerve impulses and, 317–18, 379, 398, 435, 542
    neurons and, 544
    in summation, 731–32
Active immunity, 411
Active transport, **6,** 769
    cell membrane and, 499, 500
    in endocytosis, 263
    in pinocytosis, 607
Actively acquired tolerance, 497
Aculeata. *See* Bees
Acyclovir, 149, 151, 254–55, 783
ADA deficiency, 323
Adams, Roger, 724
Adaptation (Biology), **7–8,** 760
    in abiotic environments, 1
    to biomes, 86
    to climate, 490
    evolutionary, 273, 279, 521, 538

population and, 620
   in rain forests, 807
   water in, 800
Ahlquist, Jon, **706–7**
Ahlquist, Raymond P., 94
AIDS, **13–17,** 395–96
   drugs for, 149, 255, 377–78
   gene splicing and, 703
   in hemophilia, 363
   immune system and, 377–78, 409
   latency period of, 396
   as retrovirus, 659
   reverse transcriptase and, 749
   vaccine, 16, 396
Air masses, 214
Air pollution, **17–18**
   acid rain from, 4
   amensalism of, 25
   biomagnification of, 86
   from fossil fuels, 305
   Greenhouse effect from, 343
   ozone and, 574
   plant diseases from, 611
   sources of, 46
Air pumps, 384
al-Qasim, Abu, 363
Alanine, 324, 583, 637
Alberts, Alfred W., 336
Albinism, 317
Albumin, 655, 755, 786
   chlorophyll and, 772
Alchemy, 3, **18,** 43, 361, 584
Alcoholism, 75, 365
Alcohols, **18**
   coenzymes and, 164
   cyclic AMP and, 190
   in dehydration synthesis, 210
   from dextrose, 817
   fermentation and, 287, 353–54, 587
   oxidation of, 755
   production of, 10, 11
   as teratogens, 751
   yeasts in producing, 829
Aldobionic acid, 52
Aldosterone, 281, 650, 728
Aldrin, 204, 207
Alemacon, 98
Algae, **19–20.** *See also* Marine algae
   asexual reproduction in, 651
   as autotrophs, 770
   blue-green (*See* Cyanobacteria)
   brown (*See* Phaeophceae)
   carbon cycle and, 125
   DDT and, 206
   in eutrophication, 275
   fire (*See* Dinoflagellates)
   golden, 19
   green, 19, 121, 610
   marine, 313
   mutualism and, 535
   nitrogen-fixing, 552

nutritional needs of, 556–57
   productivity of, 628–29
   as protista, 738
   red (*See* Rhodophyceae)
   spores and, 723
   sporophytes and, 723
   unicellular, 189
Algor mortis, 207
Alicyclic hydrocarbons, 400–1
Aliphatic hydrocarbons, 400–1
Alkaloid solanine, 767
Alkaloids, **20**
   bases of, 3
   for cancer, 149
   ergot, 352
   glycoside, 449
   synthesis of, 435
Alkaptonuria, 317
Alkylating agents, 149
Allantois, 154
Alleles, 635–36, 695, 701–2
   dominant *vs.* recessive genes, 232–33
   in eugenics, 274
   frequency of, 320
   in gene mapping, 325
   heterozygous, 371
   in independent assortment, 413
   in isolated populations, 175
   mutated, 157
   polymorphism in, 616
   variance in, 327
Allelic frequency. *See* Gene frequency
Allelomorph, **20**
Allergy, 37
   antibodies and, 836
   latex, 721
   steroids for, 728
Allison, Anthony C., 101
Allogamy. *See* Heterogamy
Allometry, **20–21**
Allopatry, **21–22**
Allopurinol, 254, 376
Alloway, James, 52
Alpha-carotene, 449
Alpha helix proteins, 212
Alpha interferons, 420
Alpha particles, 371
Alternation of generations, **22,** 316, 473, 723
Altman, Sidney, **22–24,** 661
Altmann, Richard, 222
Altruism (Biology), **24,** 296
   in insects, 822
   sociobiology and, 715
Aluminum hydroxide, 818
Alvarez, Luis, 138, 282
Alvarez, Walter, 282
Alveoli, **24–25,** 658
Alzheimer's disease, 12
   prions in, 628, 634
Amanita muscaria, 352
Amanita pantherina, 352

B

pituitary hormones in, 608
respiration and, 373
theories of, 716
Blood proteins, 101–2
Blood transfusions
AIDS from, 14
antibodies in, 101
blood groups and, 99, 100–101
blood type and, 660–61
development of, 133, 235–36, 453
reactions to, 102–3, 202
screening, 102
Blood types. *See* ABO blood-group system
Blood vessels. *See also* Arteries; Capillaries; Veins
cauterization of, 187
contraction and dilation of, 76
liver, 365
in respiration, 24
sickle cell anemia and, 707
structure and function of, 355–56
suturing, 132–33
Blum, Joe, 620
Blumberg, Baruch Samuel, **101–2**
Boas, Franz, 274
Boas, M.A., 91
Body language, 110–11
Body temperature, 664
of birds, 571
in hibernation, 373–74
integumentary system and, 419
of mammals, 490
metabolism and, 506
neurotransmitters and, 546
in osmosis, 572
photoperiodism and, 599
regulation of, 280
spermatogenesis and, 719
Bogs, 578, 581, 808
Bohr, Niels, 47, 371
Boiling point, 569
Boise, Charles, 459
Boltwood, Bertram, 639
Bombykol, 118
Bonaventura, Joseph, 97
Bonding. *See* Chemical bonds
Bone, 397, 708–9, 762
Bone marrow, 361
lymphocytes and, 743
rejection of, 757
Bone marrow transplantation, 756–57
Bonnet, Charles, **102**, 586, 698
Bonomo, Giovan Cosimo, 585
Bordet, Jules, **102–3**, 408
Boreal forests. *See* Taigas
Borel, Jean-François, 190–91
Borelli, Giovanni, 218
Borgognoni, Theodoric, 37
Borlaug, Norman E., 489
Bosch, Carl, 550
Botany, **103–4**
as branch of biology, 836

genetics and, 786–87
spores in, 723
systematics and, 738
Theophrastus and, 754
Boussingault, Jean, 550, 552
Boveri, Theodor, 72, **104**, 156, 279, 367, 718, 733, 820
Bovet, Daniele, 37, **104–6**, 231
Bovine spongiform encephalopathy, 628, 633–34
Boyd, William C., 100
Boyer, Herbert Wayne, **106**, 321
Boyer, Paul D., **106**, 795
Boyle, Robert, **106–8**
on acidity, 3
on atomic theory, 48, 253, 492
on fermentation, 287
Boyle's law, 107, 384, 688
Brachial plexus, 542
Braconnot, Henri, 25
Bradshaw, Ralph, 541
Brain, **108–10**. *See also* Central nervous system; Cerebellum; Cerebral cortex
area 17, 389
development, 258, 389
evolution, 239
function, 369–70, 488
hormones in, 345, 690
hypothalamus and, 404–5
vs. mind, 542, 780
prefrontal lobes, 249
receptors, 593–94
self-modification, 248
sound wave interpretation, 5
spinal nerves and, 486
structure and function of, 270–71, 388–89
visual cortex, 388–89
Brain death, 208
Brain diseases, 269, 314
Brain mapping, 389
Brain stem, 108, 110, 216
Brain surgery, 187, 249
Brandt, Georg, 43
Braun tube, 318
BRCA gene, 123–24
Breast cancer, 123–24, 149
Breast feeding, 14
Breeding
bacteria, 464
cryobiology for, 186
genetic engineering of, 324
inbreeding, 82, 175, **413**
linebreeding, 413
migration and, 510–11
in opportunistic organisms, 567
outbreeding, 276, 320, 413, **572**
plant (*See* Plant breeding)
territories, 571
Brenner, Sydney, 688
Bretonneau, Pierre-Fidele, 220
Bridges, Calvin B., 527, 730
Briggs, Robert, 237
Brill, Nathan Edwin, 836

C

Cancer metastasis, 270, 583
Candelilla wax, 803
Candidin, 792
Candolle, A.P. de, 698
Cane sugar, 523
Canis lupus. See Wolves
Cannabinol, 764–65
Cannabis, 763–65. *See also* Marijuana
Cannizarro, Stanislao, 146
Cannon, Walter, 382
Cannuba wax, 803
Canopy, in rainforests, 642
Cantell, Kari, 419–20
Capillaries, 158
    arteries and, 779
    in blood pressure, 100
    discovery of, 488
    in muscle contractions, 448
    in respiration, 24, 658
    umbilical, 609
    veins and, 779
Capitalism, Social Darwinism and, 714
Capsids, 783
Capsular polysaccharides, 53
Carbohydrates, **124–25**
    in biomass, 86
    biotin for, **91**
    calories from, 121
    cellulose as, 143
    energy from, 466–67
    fermentation for, 287
    glycogen and, 76
    in immune reactions, 52
    lactic acid and, 451
    metabolism of
        adrenal gland hormones and, 261
        carbon fixation and, 127
        glucose and, 52, 124, 125, 485
        insulin and, 176, 240–41, 418
        Kreb's cycle in, 446
        oxygen in, 10
        phosphates in, 476
        pituitary hormones in, 388
        sugar nucleotides in, 466–67
    photosynthesis and, 121, 127, 600
    side-chain theory of, 500
Carbolic, 38, 477, 596
Carbon, 253, 568–69
    in acetic acid, 1–2
    in radioactive dating, 639
    rings, 673
    in trees, 770
Carbon cycle, **125–26,** 127–28, 152
Carbon dating. *See* Radiocarbon dating
Carbon dioxide, **126–27**
    absorption, 642
    in acid rain, 4
    atmospheric, 45–46, 343
    from cellular respiration, 657
    in Cretaceous period, 183
    discovery of, 361

    excretion of, 280
    in fermentation, 116, 287, 587
    in Gaia hypothesis, 313
    in greenhouse effect, 806, 824
    in photosynthesis, 121, 600, 624
    in respiration, 24, 656–58
    trees and, 770
    water transport of, 800
Carbon fixation, **127–28,** 356, 444
Carbon 14, 433–34, 471, 580, 600
Carbon monoxide, **128,** 796
    in air pollution, 17, 614, 615
    atmospheric, 45
    discovery of, 361
    from fossil fuels, 305
Carbon 13, 488
Carbonate minerals, 514
Carbonation, 126
Carbonic acid, 128
Carcinogens, 122, 124, **128–30,** 534
    mutation from, 750
    smoking, 748–49
    as toxins, 799
Cardiac catheterization. *See* Heart catheterization
Cardiovascular system, 97–98, **158**
    arteries in, 38, 43–44, 779
    capillaries in, 779
    catheterization (*See* Heart catheterization)
    discovery of, 97, 271, 355–56
    heart in, 358–59
    lymph and, 725–26
    lymphatic system and, 679
    as nutrient carrier, 657
    study of, 716
    veins in, 779
Caribou, 509–10
Carnassial teeth, 130
Carnivores, **130–31**
    balance between, 560
    bioaccumulation in, 594
    in biotic communities, 83
    in food chains, 300, 370, 628
    vs. herbivores, 173, 188
    nutritional needs of, 557
    predation and, 623
    as trophic level, 750, 770, 771
Carnivorous plants, 173
Carnot, Nicolas-Léonard Sadi, 266
Caro, Heinrich, 550
Carotene, 448–49, 772–73
Carotenoid pigments, 816
Carotenoids, 434, 435, 448–49
Carotid arteries, 373
Carr, Archie Fairly, **131–32**
Carrel, Alexis, **132–33**
Carrel-Dakin method, 133
Carroll, James, 648
Carson, Rachel, **134–36,** 206, 594
Cartilage, 708, 762
Cartilaginous fishes, 407
Carver, George Washington, **136–37**

Conditioning (Psychology), 69–70, **170–71**
　　classical, 462–63
　　operant (*See* Operant conditioning)
　　reflex actions, 589
Cones (Retina)
　　in sight, 707
　　in vision, 792–93
Confocal light microscopy, 509
Coniferophyta, 609
Conifers
　　in biotic communities, 83, 86–87
　　in evolution, 301
　　as gymnosperms, 347
　　in taiga, 744
Conjugation, **171**, 654
Conjunctivitis, 284, 815
Connective tissue, 762
Consciousness, in animals, 622
Conservation, **171–73**
　　of ecosystems, 246
　　of energy, 755–56
　　Gaia hypothesis and, 313
　　Jacques Cousteau and, 182
　　Rachel Carson and, 134–35
　　restoration ecology and, 659
　　of species, 283
Conservatism (Political), Social Darwinism and, 714
Consumer Product Safety Commission, 128
Consumers, **173**
　　in ecosystems, 245
　　in food chains, 300
　　in food web, 750
　　heterotrophs as, 370
　　omnivores as, 560
　　primary *vs.* secondary, 83, 300
　　vs. production, 489
　　in trophic structure, 770–71
　　zooplankton as, 837
Contact dermatitis, 454
Contact lenses, 284
Continental drift, **173–74**, 183, 277
Continuous variation, **174–75**, 221
Contractile vacuoles, 632, 777
Convergent evolution, **276**, 278, 383, 602, 617
Coombs, Robert W., 378
Cooperation, 296, 694
Coordinate covalent bonds, 147
Copepods, 837
Coprolites, 303
Coral reefs, 343
Corey, Elias James, **175–76**
Corey, Robert, 589
Corey latone aldehyde, 175
Cori, Carl Ferdinand, **176**, 732
　　on glycogen, 124, 446
　　Krebs and, 445
　　Meyerhoff and, 508
Cori, Gerty Theresa Radnitz, **176**
　　on glycogen, 124, 446
　　Krebs and, 445
　　Meyerhoff and, 508

Cori ester, 176
Cormack, Allan M., 169–70, **176–77,** 387
Corms, 612, 669
Corn, 26, 495–96, 496. *See also* Grain crops
Cornea, 707
Cornforth, John, 154
Coronary artery bypass, 44
Corpus callosum, 110, 145, 720
Corpus luteum, **177–78,** 391, 653, 655, 701, 702
Correlation (Statistics), 188, 590
Correns, Karl Erich, 367, 751, 820
Corticol. *See* Hydrocortisone
Corticospinal tracts, 722
Corticosteroids. *See* Adrenal cortex hormones
Corticosterone. *See* Adrenal cortex hormones
Corticotropin
　　for arthritis, 364
　　cortisone production and, 178
　　discovery of, 386–87
　　isolation and synthesis of, 469–70
　　releasing hormone, **9**
Cortisol. *See* Hydrocortisone
Cortisone, **178–79,** 650. *See also* Hydrocortisone
　　for arthritis, 364
　　discovery of, 45
　　isolation and synthesis of, 153, 386, 436
Corynebacterium diphtheriae, 220–21
Cosmic rays, 471
Cosmology, 266
Coster, Dirk, 372
Cotton, 136, 163
Couper, Archibald Scott, 147
Cournand, André F., **179–80,** 302, 662–63
Courtois, Bernard, 37
Courtship (Animals), 273, 597, 715
Cousteau, Jacques, **180–83**
Cousteau, Philippe, 182
Covalency, **183**
　　coordinate bonds, 147
　　vs. ionic bonds, 422, 589
　　in molecules, 519
　　in oxidation reduction, 574
　　polar bonds, 183, 589
Covariance, 295
Cowbirds, 418
Cowper's gland, 653
Cowpox, 428–29
Cows, 618, 619. *See also* Cattle
CPX, 192
Craniology, 111
Craters (Meteorite), 282, 491
Creationism, 57, 138, 276, 278–79, 398
Creighton, Harriet, 495
Cretaceous period, **183–84,** 329
　　extinction during, 81, 282, 491
Cretinism, 759, 789
Creutzfeldt-Jakob disease, 314, 628, 633–34
Crick, Francis Harry Compton, **184,** 801–3
　　on adapter hypothesis, 334
　　Altman and, 23
　　Avery and, 51, 52

D

Devonian period, 329
Dew, 622
Dewar, James, 185
Dexpanthenol, 583
Dextrin, 124
Dextrose, 817
Diabetes mellitus, **214–15**
  adult-onset (*See* Non-insulin-dependent diabetes)
  cause of, 582
  epinephrine/norepinephrine in, 478
  insulin-dependent (*See* Insulin-dependent diabetes mellitus)
  insulin for, 63–64, 77–78, 418–19
  juvenile-onset (*See* Insulin-dependent diabetes)
  non-insulin-dependent (*See* Non-insulin-dependent diabetes mellitus)
  pantothenic acid and, 583
  somatostatin for, 690
Diagnostic tests, 12, 36, 53, 135
Dialysis (Chemical process), 354
Diamagnetism, 563
Diaphragm (Physiology), **215–16**, 318, 490
Diastase, 267
Diastolic pressure, 100, 359
Diatoms, 19
Diatropic response, 771
Diauxy, 521
Dichlorodiphenyltrichloroethane. *See* DDT
Dichlorophenoxyacetic acid, 611
Dichotomous key, 451
Dick, George, **216,** 228, 815
Dick, Gladys, 228, 815
Dick method, 216, 815
Dicotyledons, 300, 611, 644, 777
Dicumarol, 583
Didanosine, 15
Die Mutationstheorie, 787
Dieldrin, 207
Dielectrics, 563
Diencephalon, 370
Diet, 129, 304, 385, 556–57
Diet therapy, 45
Diethylstilbestrol, 391, 750
Differentiation, 141, **216–17,** 468
  in blastulas, 95
  centrioles in, 144
  in chimeras, 151
  corpus luteum and, 178
  in frogs, 255
  in mesoderms, 505
  proto-oncogenes in, 560
  of sperm, 719
Diffusion, **217**
  for artificial organs, 133
  in comparative physiology, 604
  facilitated, 499, 500
  pump, 134
  simple, 500
Digestion, **217–18,** 219, 557
  acids in, 361
  anaerobic, 88
  bacteria and, 58

bile acids and, 728
  of carbohydrates, 124
  in carnivores, 130
  in cellular respiration, 657
  of cellulose, 143
  chemical reactions in, 148, 645
  commensalism in, 168
  diaphragm in, 216
  enzymes in, 263, 484
  in herbivores, 130
  in heterotrophs, 657
  intracellular, 484, 508
  pancreatic enzymes in, 66
  peptides in, 592
  physiology of, 68, 76, 589
Digestive system, **218–19,** 507
Digitalis, **219–20,** 650, 812
Digitalis lanata, 220
Digitalis purpurea, 219–20
Digitoxin, 220
Digoxin, 220
Dihybrid crosses, 502
Dimethyl sulfide, 313, 333
Dinoflagellates, 19, 20, 85, 267
Dinosaurs, 184, 282, 491, 570, 573
Dinotherium, 188
Dinucleotides, 618
Dioptrics, 347
Dioxins, 130, 798, 799
Diphenhydramine, 37
Diphenylamine chloroarsine, 765
Diphtheria, **220–21**
  cause of, 815
  Schick test and, 691
  vaccine for, 691
Diphtheria antitoxin, 70, 118, 250, 411, 438
Diplohaplontic life cycle, 473
Diploid reproduction. *See* Sexual reproduction
Diploids, 20, **221,** 654, 699–700
  in human reproduction, 719
  in plant reproduction, 723
  in sexual reproduction, 654
  zygotes as, 839
Dipoles, 563
Dirac, Paul Adrien Maurice, 47, 693
Diradicals, 147
Directional selection, **221,** 222, 697
Discontinuous variation, **221**
Discordant ages, 331
Discs, vertebrae and, 779
Disease outbreaks. *See* Epidemics
Diseases. *See also* specific diseases, e.g., Cancer
  diagnosis of, 375–76
  genetic causes of, 394–95
  theories of, 331–32, 375
Displacement analysis, 761
Disruptive selection, **222,** 697
Dissection, 28, 270, 779–80
Dissipative structures, in thermodynamics, 625
Distillation, 18
Diuretics, 478

Engelmann, Theodor Wilhelm, 600
Engels, Friedrich, 565
Engines, steam, 756
Enkephalins, 546, 592
Enterobacteria, 553
Entoderm. *See* Endoderm
Entomology, **265–66**
    behavior in, 821–23
    in zoology, 837
Entomophagous predators, 266
Entropy, **266**
    in biological systems, 756
    in ecosystems, 246
    energy transformations and, 265
    in thermodynamics, 624–25, 756
Environment. *See* Ecology
Environmental Defense Fund, 823–24
Environmental degradation
    with DDT, 205
    from introduced species, 421
Environmental effects
    in cancer, 129
    of fossil fuels, 304–5
    of introduced species, 421
    in Lamarck's theory, 452
    as limiting factors, 473–74
    on opportunistic organisms, 567
    on organogenesis, 570
Environmental illness, 167
Environmental law, 4, 17, 614, 615
Environmental policy, 135
Environmental protection
    extinction without, 821, 823
    from toxins, 767
    of wetlands, 808–9
Environmental Protection Agency
    on air pollution, 18
    establishment of, 136
    on ozone, 574
    on pesticides, 594
    toxin regulation by, 767–68
Enzyme inhibitors, 79
Enzyme-linked immunosorbent assay, 14, 396
Enzymes, **30–31, 267–69.** *See also* Coenzymes; specific enzymes, e.g.,
    Amylases
    ATP synthase as, 795
    in biosynthesis, 90
    in carbohydrate metabolism, 240–41
    catalysis by, 755
    in cell membranes, 710–11
    in cellular respiration, 657
    chemical nature of, 267, 553
    classification of, 268
    coenzymes and, 755
    diseases from, 268
    DNA and, 224, 267
    in endergonic reactions, 281
    engineering of, **266–67**
    in gene splicing, 41, 106, 281–82, 321, 537–38
    genes and, 630, 744–47
    in genetic recombination, 645

    in genetic transcription, 319
    in heredity, 522
    induction of, 521–22
    latency, 241
    laws of thermodynamics and, 351
    mutagenesis of, 713
    for neurotransmission, 546
    one gene theory of, 66–68, 561–62
    oxidation and, 755, 795–97
    in protein production, 831–32
    proteins as, 629, 726–27
    proteolytic, 791
    in respiration, 656–57
    restriction, 712
    in reverse transcription (*See* Reverse transcriptase)
    in RNA splicing, 667
    separation of, 818
    synthesis of, 480
    in translation, 768–69
    types of
        anticholesterol, 336
        artificial, 268–69
        catalytic, 356, 403
        decomposing, 310
        digestive, 124, 217, 219, 263, 484
        fermentation, 116, 287
        lysosomal, 240–41
        metabolic, 317, 506, 521
        methylase, 41
        restriction, 61, 73, 321, 381
        thrombolytic, 99
        yellow, 755
Eocene epoch, 329
Eohippus, 602
Eosinophils, 482
Eounthropus dawsoni, 605
EPA. *See* Environmental Protection Agency
Ephedra, 609
Epidemics
    AIDS, 13–14
    as biological warfare, 83–84
    bubonic plague, 114–15
    cholera, 153
    diphtheria, 220
    influenza, 415–16
Epidemiology, toxins and, 768
Epidermal growth factors, 166
Epidermis, 457, 610
Epididymis, 653, 655
Epidural space, 722
Epigenesis, 102, 256, 258, **269,** 356
Epiglottis, 658
Epilepsy, **269–70,** 720
Epinephrine, 740
    in angina, 94
    as antihistamine, 37
    in diabetes, 478
    discovery of, 703–4
    in dominance relations, 232
    function of, 176, 196
    isolation and synthesis of, 386

●

liver cells and, 732
as neurotransmitters, 546
production of, 9, 593
Episodic growth, 116
Episomes, 425
Epithelial cells, 762, 763
Epithelium, **270,** 280
Epp, Otto, 211
Epstein-Barr virus, 123
Equilibrium, 144–45, 698
punctuated (*See* Punctuated equilibrium)
of species, 821–22
in thermodynamics, 625, 756
Equisetum, 609
Equus asinus. See Donkeys
Equus caballus. See Horses
Equus zebra, 602
Erasistratus, 28, 98, 218, **270–71,** 542
Ergosterol, 153
Ergot, 196, 352
Erlanger, Joseph, **271–72,** 317, 543
Erlichin, 792
Erosion, 301
sedimentation from, 694
by water, 800
in water pollution, 801
ERT, 273
Erwin, Terry, 80
Erythroblastosis fetalis, 99, 661
Erythrocruorin, 390
Erythrocytes. *See* Red blood cells
Erythromycin, 34, 220
Erythroxylon coca, 20
Escherichia coli
antibiotics for, 617
bacteriophages and, 60
commensalism in, 168
genetic engineering of, 73, 106, 321, 423–24, 464
genetic mapping of, 59, 224
insulin from, 418
reproduction of, 746
X174 and, 444
zygote induction of, 425
Eskimos, 448
Esophagus, 219
Essential amino acids, 26, 385, 502
Essential fatty acids, 286
Estradiol, 273, 505
Estriol, 118, 273
Estrogen replacement therapy, 273
Estrogens, **272–73,** 700–701, 728
in aging, 12
biological source of, 261, 702
in birth, 91–92
for cancer, 149
in dominance relations, 232
isolation and synthesis of, 118
Turner's syndrome and, 774
Estrone, 273
Estrus, 702
Ethanol, 18

Ethanolamine phosphoglyceride, 598
Ethics
of animal research, 339
Aristotle on, 43
of biodiversity loss, 81
of DNA patents, 394
of gene therapy, 323
of genetic engineering, 164, 274, 324, 465, 496
of recombinant DNA technology, 73–74, 226
Ethnic groups, 101
Ethology. *See* Animal behavior
Ethylene, 611
Ethylene glycol, 803
Eubacteria, **273–74**
Euchromatin, 155
Euclid, 345
Eugenics, **274,** 315, 528
Eukaryotes, 264
algae, 19
C-value paradox and, 666–67
cellular respiration in, 657
endoplasmic reticulum of, **274**
as evolutionary stage, 694
flagella of, 297
genes of, 326, 768
genetic recombination in, 645
messenger RNA and, 703
vs. prokaryotes, 629
protista as, 631
regulation genes in, 649
replication in, 651
respiration in, 656
ribosomes in, 662
RNA in, 665
selfish DNA in, 697
in systematics, 738
viruses and, 703
Euler, Ulf von, 53, 54, 74
Euler-Chelpin, Hans, 164, 547
Euphenics, **274**
European cuckoos, 418
Eustachian tubes, 358
Eustigmatophyceae, 20
Eutherians, 278
Eutrophication, 189, **275**
as limiting factor, 474
phosphorus in, 598
phytoplankton in, 604
species in, 189
Evans, Alice Catherine, 113, **275–76,** 587
Evans, Herbert, 386, 387, 469
Evaporation, 401
Even-toed ungulates, 489
Evergreen coniferous forests, 301
Evolution, **276.** *See also* Catastrophism; Natural selection; Survival of
the fittest
abiotic environment and, 1
adaptation in, 7–8, 273, 279, 521, 538
agriculture in, 12
allopatry and sympatry in, 21–22
altruism in, 24

F

FK-506, 291

Flagella, **297**

    of algae, 19

    vs. cilia, 157

    of eubacteria, 273

    in protists, 631

    in protozoa, 632

    in sperm, 719

Flatworms, 219, 280, 585

Flavin adenine dinucleotide, 765

Flavins, 796

Fleas, 115, 671

Fleming, Alexander, 32–34, 149, **297–98,** 299, 591–92, 826

Flemming, Walther, 141, 142, 156, **298,** 367

Flesh-eating bacteria. *See* Necrotizing fasciitis

Fletcher, Walter, 385

Flexner, Simon, 679

Fliess, Wilhelm, 89

Floodplains, 807–8

Floods

    precipitation and, 623

    from sedimentation, 694

    wetlands and, 807–8

Flora. *See* Plants

Florey, Howard Walter, 34, 145, 149, 298, **298–99,** 592

Florigen, 611

Florman, Arvid Henrik, 660

Flowering plants. *See* Angiosperms

Fluctuation test, 480

Fluid-electrolyte balance, 546

Fluorescein, 292

Fluoxetine, 214

Flying, 570, 606

Fog, 800

Fol, Hermann, 288

Foliage. *See* Leaves

Folic acid, 361, 784–85

Follicle stimulating hormones, 700–2

    corpus luteum and, 177–78

    from hypothalamus, 261

    ovulation and, 249

    pituitary gland and, 608

Fontana, Felice, 646

Food, dehydration of, 800

Food and Drug Administration

    on AIDS, 16

    on cyclosporin, 191

    on lovastin, 336

    on 6MP, 253

    on t-PA, 99

    on taxol, 149

Food chains, **300**

    aquatic, 783

    bioaccumulation in, 78

    carnivores in, 130, 300

    competition in, 370

    consumers in, 173

    DDT in, 206

    decomposers in, 628

    dichlorodiphenyltrichloroethane and, 824

    in grasslands, 341–42

    herbivores in, 300, 366, 370

    heterotrophs in, 370

    insects in, 265–66

    invertebrates in, 422

    nitrogen in, 550

    omnivores in, 560

    pesticides in, 594

    phytoplankton in, 604

    plants in, 300, 610

    producers in, 628

    trees in, 770

    trophic structure and, 750, 770–71

    viruses in, 783

    zooplankton in, 837–38

Food crops, 125. *See also* Grain crops

Food irradiation, 130

Food preparation, 13

Food preservation, 1

Food supply, 488–89

Forager bees, 309

Foramen magnum, 199

Forceps, 441

Forensic sciences, 225–26

Forest floor, 642

Forest products, 300

Forestier, Jacques, 45

Forests, **300–301.** *See also* Deforestation

    acid rain effects on, 4

    as biomes, 86

    boreal (*See* Taigas)

    climax community of, 162

    conifer, 83, 86–87, 301, 347

    conservation of, 172

    drunken, 744

    as ecosystems, 769–70

    ecotones between, 246

    introduced species in, 421

    national, 172

    old-growth, 162

    precipitation and, 623

    rain (*See* Rain forests)

    in swamps, 807

    taiga as, 744

    trees and, 769–70

    tropical (*See* Rain forests)

Forgetting, 500

Formative causation, 704–5

Forssmann, Werner, 179, **301–2,** 662–63

Fossey, Dian, 460

Fossil fuels, **304–6,** 400–401. *See also* Global warming

    acid rain from, 4

    air pollution from, 17

    in carbon cycle, 125

    in greenhouse effect, 806, 824

    natural resources depleted by, 656

Fossil humans

    Australopithecus, **49,** 383

    Australopithecus aethiopicus, 393

    Australopithecus afarensis, 393, 394

    Australopithecus africanus, 111, 198, 199–200, 392–94, 462

    Australopithecus anamensis, 393

G

Geotropism. *See* Gravitropism
Germ cells, 288, 583, 655
Germ-plasm theory of heredity, 803–4
Germ Theory of Disease, **331–32,** 408, 784
 antiseptics and, 38
 development of, 115, 306, 440
 spontaneous generation and, 722
German measles. *See* Rubella
Germination, **332–33**
Gerontology, 12
Gerstmann-Strässler-Scheinker syndrome, 627–28
Gestalt, **333**
Gestation period, 490, 786
Gibbs free energy, 281
Gilbert, Walter, **333–34,** 688
Gilman, Alfred Goodman, **334–35,** 668
Ginkgo, 175, 609
Ginkgo biloba. See Ginkgo
Giraffes, 452–53
Glaciers, 11
Gladstone, C.P., 32
Gland of Leydig, 469
Glands. *See* specific types of glands, e.g., Adrenal glands
Glass recording electrodes, 540
Glauber, Johann Rudolf, 3
Glaucoma, 284
Glial cells, 544, 545
Global cycles, 1, 183
Global warming. *See also* Greenhouse effect
 from carbon emissions, 126, 127
 from fossil fuels, 305–6
 Gaia hypothesis in, 313
 progress of, 204, 343–44
Globin, 361
Globulins, 755, 761
Glucagon, 190, 418, 423–24, 582
Glucocorticoids, 9
Glucosamine, 45
Glucose
 biosynthesis of, 567
 blood (*See* Blood glucose)
 in carbohydrates, 52, 124, 125
 in cellular respiration, 657
 insulin and, 418
 metabolism of, 95
 from proteins, 178
 pyruvic acid and, 636–37
 sensors, 419
 sources of, 523
 structure and function of, 356–57, 617
 synthesis of, 292–93
Glucose-1-phosphate, 176
Glutamic acid, 781
Glutathione, 385, 436, 781
Glycerol, 286, 598
Glycerol-3-phosphate, 598
Glycine, 25, 546, 781
Glycogen, 617–18
 in digestion, 76, 124
 discovery of, 466–67
 lactic acid and, 451, 508
 in metabolism, 506

Glycolysis
 in cellular respiration, 657
 pyruvic acid and, 636
Glycoproteins, 337, 386, 391, 395
Glycosides, 449, 650
Gmelin, Leopold, 218, 764
Gnetophyta, 609
Gnetum, 609
GnRH. *See* Gonadotropin-releasing hormone
Goats, 619
God, 40, 57
Godke, R.A., 163
Goiter, 441–42, 759, 789
Gold, 18
Goldberger, Joseph, 784
Golden algae, 19
Goldschmidt, Richard, 494
Goldschmidt, Victor, 817
Goldstein, Joseph L., 112, 335–36
Golgi, Camillo, **336–37**
 cell staining and, 142, 705
 on nerve transmission, 543
 on nervous system anatomy, 546
 Santiago and, 642–44
Golgi apparatus, 194, **337**
 discovery of, 142
 endoplasmic reticulum and, 264
 function of, 140
 lysosomes and, 484
Gonadal dysgenesis. *See* Turner's syndrome
Gonadotropin-releasing hormone, 405, 608, 700–701
Gonadotropins, 178, 690, 701
Gonads, 261, 497, 654, 655, 719
Gonidia, 723
Gonorrhea, 540
Goodall, Jane, **337–39,** 460
Gordon, Alexander, 38, 697
Gorer, Peter, 71, 714
Gorgas, William Crawford, **339–40**
Gosio, B., 32
Goss, Raymond, 802
Gould, Stephen, 634
Goulian, Mehran, 443
Gout, 44, 254
Government regulations of toxins, 767–68
Gowans, C.S., 409
Gowers, William Richard, 97
Graaf, Regnier de, 62, 386, 418
Graafian follicle, 178, 702
Gracile australopithecines, 383
Gradients
 analysis, 811
 concentration, 6, 499
 electrical, 6
Gradualism, 276, 278, 280, **340,** 634
Graft rejection, 118
Graham, Clarence H., 354
Graham, Thomas, 571
Grain crops, 13, 342, 420, 496
Gram, Hans Christian Joachim, 58, 142, **340–41,** 791
Gram calorie. *See* Calories

H

Half-life, 331, 471, 580, 640, 641
Halides, 514
Hall, Marshal, 543
Haller, Victor Albrecht von, 218, 351–52, 751
Hallucinations, 691–92
Hallucinogens, **352–53**
Halons, 575–76
Halpern, Bernard N., 37
Halsted, William, 38
Hampton, Caroline, 38
Hamsters, 363
Hand, foot, and mouth disease, 782
Hanson, Jean, 399
Haploid reproduction. *See* Sexual reproduction, haploid
Haplontic life cycle, 473
Haptens, 454
Harden, Arthur, 164, **353–54,** 506, 547
Hardness, 514
Hardy Weinberg equilibrium, 619
Harmine, 352
Harris, Geoffrey W., 346, 387, 690
Harris, Stuart, 419
Harrison, Ross G., 133
Hartline, Haldan Keffer, **354–55,** 793
Harvey, William, 97, 98, 100, 158, 488, 646, 672–73, 751, 779
Hashimoto's disease, 759
Hashish, 352, 763, 765
Hata, Sahachiro, 251
Hatch, M.D., 356
Hatch-Slack photosynthesis pathway, 152, **356,** 444–45, 637
Haworth, Walter, **356–57,** 435
Haworth formula, 356–57
Hayashi, Masaki, 766
Haycraft, J.B., 36
Hayem, Georges, 97
Hazardous wastes, **798–99**
Hazen, Elizabeth, **357–58**
HCG. *See* Human chorionic gonadotropin
HCI. *See* Hydrochloric acid
HCN. *See* Hydrocyanic acid
HCO3. *See* Bicarbonates
Hearing, **4–6,** 309, **358,** 729
Hearing aids, 6
Heart, 158, **358–59**
    aorta and, 38
    artificial, 133–34
    monitoring, 252
    muscles, 534
    Starling's law of, 726
    structure and function of, 355–56
    veins and, 779
Heart atrium, 358–59
Heart catheterization, 179, 301–2, 662–63
Heart diseases, 359
    antisense DNA for, 832
    arteriosclerosis, 44
    cholesterol and, 2, 728
    diagnosis of, 252, 271–72, 302
    LDLs in, 336
    stress and, 730
Heart-lung machines, 134

Heart valves, 359
Heart ventricles, 358–59
Heat
    in closed systems, 266
    excretion of, 280
    for microorganisms, 58
    from muscle contractions, 374
    mutations from, 527
    vs. temperature, 755
    as thermal energy, 749
    in thermodynamics, 624–25, 755–56
    thermoreceptors and, 699
    value of, 493
Heat-death theory, 266, 625
Heavy chains, 248
Heavy metals, 87–88
Heavy water, 372
Hecht, Selig, 793
Hedin, Sven, 25
Heidelberger, Charles, 149
Heidelberger, Michael, 409
Heidenhain, Rudolf, 726
Heisenberg, Werner Karl, 47
Helium, 45–46, 46–47, 563
Helmholtz, Hermann von, 5, 283, **359–60,** 493, 506, 755
Helminths, 805
Helmont, Johannes Baptista van, 126, **360–61,** 556, 600
Hematology, 250–51
Heme, 361
Hemin, 294
Hemoglobin, **361**
    in beta-thalassemia, 703
    bile and, 810–11
    carbon monoxide poisoning of, 128
    vs. chloroplasts, 152
    gene, 63
    liver and, 810
    mammalian, 603–4
    myoglobin and, 755
    oxygen and, 816
    plasma and, 811
    as protein, 630
    regeneration of, 809–11
    in respiration, 24
    in sickle cell anemia, 370, 589, 707
    structure and function of, 97, 386, 449
    synthesis of, 293
Hemolymph, 604
Hemolytic diseases, 660–61
Hemolytic streptococcus, 53
Hemophilia, **361–63,** 702
Hemophilus influenza, 411
Hemorrhage, 187, 198, 230, 231
Hemorrhoids, 779
Hemp, 352
Hench, Philip Showalter, 45, 178, **363–65,** 386, 436, 649
Henckel, J.F., 43
Henderson, Lawrence J., 179
Henle, Friedrich, 332
Heparin, 36, 78

I

Kraepelin, Emil, 691
Kranz anatomy, **444–45**
Krause's bulb, 767
Kraut, Heinrich, 818
Krebs, Edwin G., 445–46, 447
Krebs, Hans Adolf, **446**, 476, 506, 796–97
Kreb's cycle, 446, **446–47**, 741
    in cellular respiration, 657
    coenzymes in, 476
    in metabolism, 506
    oxidation reduction and, 574
    pantothenic acid in, 582–83
Kreel, Louis, 387
Krogh, August, **447–48**
Kuffler, Stephen, 389, 813
Kuhn, Richard, **448–49**, 583, 584
Kühne, Wilhelm, 218, 267, 725
Kulper, Gerald Peter, 681
Kunitz, Moses, 52
Kuru, 314, 628, 633–34
Küster, W., 294
Kymograph, 480
Kynurenine, 746

L

Labia, 653
Labor (Obstetrics)
    oxytocin and, 780–81
Lac operon, 566
Lactation, 786
    oxytocin and, 780–81
    prolactin and, 701
Lactic acid
    in carbohydrate metabolism, 446
    in fermentation, 287
    glycogen cycle and, 508
    Kreb's cycle and, 476–277
    in muscles, 385, 451
    pyruvic acid and, 637
Lactose, 124, 425, 426, 523, 567
Lagomorpha, 489
Laidlaw, Patrick Playfair, 37
Lake Erie, 275
Lake Victoria, 422
Lakes
    acidic, 4, 88
    DDT in, 206
    eutrophic (*See* Eutrophication)
    limiting factors of, 474
    pollen in, 204
    pollution of, 615–16
    sediment of, 578, 581
Lamarck, Chevalier. *See* Lamarck, Jean-Baptiste
Lamarck, Jean-Baptiste, **451–52**, 804
    on acquired traits, 201, 366, 452
    on gradualism, 278
    on human evolution, 392–93
Lamarckism, **452–53**
Lambert, Alexander, 815
Lamming, Eric, 818

Lancelots, 154
Lancisi, Giovanni Maria, 97
Land management, in restoration ecology, 658
Landfills, 798
Landforms, 173, 183
Landre-Beauvais, Augustin, 44
Landscape architecture, in restoration ecology, 658
Landsteiner, Karl, **453–55**, 660
    on antibodies and antigens, 35, 408
    on blood groups, 12, 99, 100
Langdon, Robert G., 96
Langerhans, Paul, 418, **455**
Langerhans cells, 454
Langesterol, 154
Langmuir, Irving, 147
Language
    cerebellum in, 145
    of chimpanzees, 338
    development of, 110–11, 165
Languesse, G.E., 455
Lanolin, 803
Laplace, Pierre, 126
Lartet, Eduard, 579
Larva, **455–56**, 506–7, 541
Larynx, 285, 658
Lashley, Karl S., 720
Lassone, Joseph Marie François de, 128
Latex allergy, 721
Latour, Charles Cagniard de, 287
Laue, Max, 210
Laval, Carl Gustaf Patrik de, 144
Laveran, Alphonse, **456**, 585, 670
Lavoisier, Antoine-Laurent, 624
    on acids, 3
    on chemistry, 253
    on combustion, 550
    on gases, 126, 361
    on metabolism, 124
    on respiration, 97
Law
    of Dominance (*See* Dominant genes)
    genetic engineering and, 394
    of Independent Assortment (*See* Independent assortment)
    of the Minimum, 323, 354, 472, **515**
    of Reciprocal Relations, 562
    of Segregation, 501, 502
    of tens (*See* Ten percent law)
    of Thermodynamics (*See* Thermodynamics, laws of)
    of Tolerance, 515
Lazear, Jesse W., 648
LDLs. *See* Low-density lipoproteins
Le Bel, Joseph, 147
Leaching, 397
Lead
    as teratogen, 751
    as toxin, 17, 18, 614, 799
    in water pollution, 801
Lead-210, 371–72
Leakey, Jonathan, 462
Leakey, Louis, 338, 394, **458–60**, 462
Leakey, Mary Douglas Nicol, 394, 459, **460–62**

Mosquitos
- DDT and, 205–6
- in malaria transmission, 339–40, 456, 670
- in yellow-fever transmission, 339–40, 647–48

Mosses, 609, 773
Motagnier, Luc, 377
Mother Earth. *See* Gaia hypothesis
Mothers, 256, 478–79
Moths, 437
Motion, 264–65
Motion sickness, 37
Motor system, 110
Motor units, 706
Motulsky, Arno G., 335
Mountains, 214
Movement, 65, 144–45, 562
Moyle, Jennifer, 516
MPF, 498
MRI. *See* Magnetic resonance imaging
mRNA. *See* Messenger RNA
Mucopeptides, 189
Mucous membranes, 762
Mucoviscidosis. *See* Cystic fibrosis
Mucus, 192
Muir, John, 172
Mulder, Gerardus, 25
Mules, 400
Muller, Franz, 513
Muller, Hermann Joseph, **526–28**
- on genes and chromosomes, 156, 279, 367, 525–26
- on mutations, 67
- Sagan and, 681
- Snell and, 713
- Sturtevant and, 730
- Thompson (D'Arcy Wentworth) and, 757
- Watson and, 802
- Wilson (Edmund Beecher) and, 820

Müller, Johannes Peter, **528,** 543, 693, 782
Muller, Otto Friedrich, 58
Müller, Paul, 204–5, **528–29,** 548, 594
Muller-Hill, Benno, 333
Mullerian mimicry, 513
Mullis, Kary, **529–31,** 712–13
Multiple sclerosis, 51, 536
Multivariate analysis, 295
Mumps, 411, 782–83
Murchison, Roderick, 330
Murphy, William P., **531–32,** 811
Murray, Joseph E., **532–33**
Muscimol, 352
Muscle contraction, 741
- ATP and, 8, 291
- capillaries and, 448
- electrical stimulation and, 10, 252, 315
- glycogen-lactic acid cycle in, 508
- heat from, 374
- Kreb's cycle in, 445
- nerves and, 352
- sliding filament theory of, 399–400

Muscle relaxants, 105
Muscle tissue, 762

Muscles
- contraction of (*See* Muscle contraction)
- heart, 534
- involuntary, 534
- lactic acid and, 385, 451
- nervous system and, 705–6
- neuron communication, 243–44
- skeletal, 534
- smooth, 534
- structure of, 399–400
- thermodynamics and, 756
- voluntary, 533–34

Muscone, 673
Muscular dystrophy, 702
Muscular system, **533–35.** *See also* Skeletal system
Mussels, 455
Mutagenesis, 526–28, **534,** 712–13
Mutagens, 750, 799
Mutation, **534–35,** 703
- of bacteria, 211–12, 425, 464
- beneficial, 535
- biotechnology and, 744
- cancer from, 122, 129, 130
- of DNA, 706, 813
- in *Drosophilia melanogaster,* 382, 383, 469, 525, 527
- in evolution, 278, 279, 443, 495, 525
- genetic recombination and, 645
- homeotic, **382–83**
- hypercholesteremia from, 335
- as molecular change, 692–93
- in natural selection (*See* Natural selection, mutations in)
- neutral, 320, 612–13
- oncogenes from, 778
- in One Gene, One Polypeptide theory, 561
- operons in, 567
- ozone layer and, 575
- in plants, 786–87
- point, 534, 560, **612–13**
- polymorphism from, 616
- population genetics and, 619
- punctuated equilibrium and, 634
- random, 480
- rate of, 350–51
- in retroviruses, 749
- of sex chromosomes, 65
- somatic, 766
- spontaneous, 369
- substitution, 382
- toxins and, 767
- from ultraviolet radiation, 575
- of viruses, 416, 783
- x rays and, 67, 495, 526–28, 713–14, 746–47

Mutualism (Biology), 90, 167, 520, **535–36,** 619, 736
Mycelium, 310
Mycobacterium tuberculosis, 440
Mycorrhizal fungi, 420
Mycosis, 358
Myelin, 727
Myelin sheath, **536**
- Schwann cells and, 545
- structure and function of, 55, 108, 542, 544

O

by cell nuclei, 554
DNA and, 323–24
engineering of, 267
Jacob-Monod hypothesis for, 427, 520–22
in plants, 152
ribosomes and, 661–62
transcription and, 768
triplet theory of, 274, 437
transcription of, 274
translation of, 630–31, 768–69
transport of, 6
types of
    alpha helix, 212
    beta-pleated, 212
    blood, 100–102
    catalytic, 616
    chaperone, 337
    dietary, 486
    G (*See* G-proteins)
    insect, 390
    kinases (*See* Kinases)
    photosynthesizing, 389–90
    purifying, 621
    random coil, 212
    repressor, 426
    reproductive, 23
    ribosomal, 553, 661–62
    seed, 571
    viral, 307, 724, 782–83
    in vacuoles, 777
Prothrombin, 36
Protists, 438, 569, **631,** 654, 738, 811
Proto-oncogenes, 123, 560–61
Protonephridia, 280
Protons, 48, 487, 640
Protophyta, 631
Protoplasm, 139, 787
Protozoans, **631–33**
    anaerobic, 10
    malaria from, 456
    mutualism in, 536
    spores in life cycle of, 723
    study of, 465–66
    vacuoles in, 777
Prout, William, 25, 218, 784
Prusiner, Stanley B., 627–28, **633–34**
Prymnesiophyceae, 20
Pseudomona aeruginosa, 617
Pseudomonas, 84
Pseudopodia
    in protists, 631
    in protozoans, 632
Psilocin, 352
Psilocybin, 352
Psychoimmunology, 409
Psychological adaptation, **7–8**
Psychology, 427–28, 709–10
Psychosomatic diseases, 729–30
Psychotherapy, 214, 730
Psylocybe mexicana, 352
Pteridophyta, **634**

Pterosaurs, 491
Ptyalin, 217
Puberty, Turner's syndrome and, 773–74
Public lands, 172
Puerperal fever, 38, 588, 697–98
Pulmonary artery, 359
Pulmonary vein, 658, 779
Pumps, 480
Punctuated equilibrium, 276, 278, 280, 340, **634**
Punishment, 170
Punnett, Reginald Crandall, 66, 279, **634–35,** 635
Punnett squares, 233, **635–36**
Pupil (Eye), 707
Purcell, Edward M., 486
Purdie, Thomas, 357
Purines, 254, 765
    in base pairs, 65, 146
    isolation of, 444, 554
    synthesis of, 292–93
Purkinje, Jan Evangelista, 140, 143, **636**
Purkinje cells, 636
Purkinje effect, 636
Purkinje fibers, 636
Pyrethrins, 673
Pyridoxine, 44, **91,** 784–85
Pyrilamine, 105
Pyrimethamine, 254, 376
Pyrimidine, 65, 146, 444, 765
Pyrroles, 293, 294
Pyruvate, 657
Pyruvic acid, **636–37**

Q

Quantum mechanics, 692–93
Quaternary period, 329
Quinine, 20, 486
Quinones, 817

R

R group, 25
Rabies, 588
    diagnosis of, 815
    vaccine, 408
    viral, 782–83
Rachischisis, 721
Racism, Social Darwinism and, 714
Racker, Efraim, 795
Radiata, 188
Radiation, **8,** 48
    adaptive, 277
    alpha, 640
    background levels of, 641
    beta, 640
    birth defects and, 750
    electromagnetic, 264, 473
    energy from, 640
    environmental impact of, 824
    gamma, 640
    infrared, 264, 598–99
    toxicity of, 641
    ultraviolet (*See* Ultraviolet radiation)

Sagan, Carl, **680–82**
Sakmann, Bert, 539, **682–83**
Salamanders, 541, 572
Saline waters, 349, 351
Salisbury, Jeffrey, 141
Saliva, 219, 377, 589
Salk, Jonas, **683–85**
  Enders and, 259, 260
  Landsteiner and, 454
  Robbins and, 667
  Sabin and, 676–77
  Weller and, 804
Salmon, 413, 510
Salmon, Daniel, 408
Salmons, Josephine, 199
Salps, 837
Salt metabolism, 178
Saltpeter, 293
Salvarsan, 148, 250, 251, 790
Samuelsson, Bengt, 74, **685–86**
Sanarelli, Giuseppe, 648
Sandfly fever, 676
Sandworms, 431
Sanger, Frederick, 667, **686–88,** 727
  Cesar Milstein and, 36
  on gene mapping, 325
  on genetic engineering, 30
  on insulin, 418, 524, 621
  on nucleotide sequence, 61
  on protein analysis, 561
Sanger, Margaret, 606
Sanitation
  for plague prevention, 116
  population and, 620
  for yellow fever prevention, 340
Sap, 351
Saquinavir, 16
Sarcomas, 15, 505
Sarin, 84
Sarrett, Lewis, 45
Sato, T., 283
Saturated fats, 287, 476
Saussure, Horace Bénédict de, 552
Saussure, Nicolas Théodore de, **688–89,** 698
Savannas, 342
Sawyer, Wilbur A., 753
Scarification, 333
Scarlet fever, 216, 229, 815
Scarpa, Antonia, 44
Scavengers, **689,** 771
  in biotic communities, 83
  in food chains, 300, 370
Schally, Andrew V., 345, 387, **690,** 829
Schaudinn, Fritz Richard, 540, 585
Scheele, Carl Wilhelm, 126, 286, 550
Schick, Bela, 221, **691,** 836
Schick test, 221, 691
Schistosomiasis, 801, 804–5
Schizophrenia, **691–92**
Schlanger, Jay, 284
Schleiden, Matthias Jacob, 140, 143, **692,** 693

Schlieren method, 761
Schmiedeberg, Oscar, 220
Schneider, Richard, 302
Schoenheimer, Rudolf, 95, 521–22
Scholfield, F.W., 36
Schönlein, Johann, 363
Schooley, Robert, 378
Schrödinger, Erwin, 47, **692–93,** 802
Schwann, Theodor Abrose Hubert, 140, 143, 267, 332, 506, 692, **693**
Schwann cells, 55, 536, 544, 545
Schwartz, Robert, 253, 533
Science Advisory Committee, 135
Scientific method, 826
  anatomy and, 780
  origins of, 40, 57, 107, 345, 365–66
  paradigms of, 584–85
  in psychology, 427–28
Scientific revolution, 57–58, 584–85
Scleroderma, 51
Scott, David, 36
Scout bees, 309
Scrapie, 314, 627–28, 633–34
Scriptures, 57
Scrotum, 655
Scurvy, 740–41, 784–85
Sea urchins, 237, 250, 431, 507
Seaborg, Glenn T., 472
Seafloor. *See* Ocean bottom
Seafood poisoning, 358
Seasons, 509–10, 599
Seaweed. *See* Marine algae
Second Law of Thermodynamics, 130, 246, 264, 266
Second messengers, 190
Secretin, 66, 386, 725–26
Secretory phase, 505
Sedgwick, Adam, 330
Sedgwick, William T., 820
Sediment, 204, 578, 581, 614
Sedimentation, 349, **694**
Seeds, 299–300, 310, 839
  germination of, 332–33
  naked, 347
  in plant evolution, 610
  pollination of, 613
  proteins in, 571
  psychoactive, 352
Seeley, Thomas Dyer, **694–95**
Segregation (Genetics), 501, 502, **695**
Seibert, Florence B., **695–96**
Seizures, 269
Selective pressures, 276, 328, 538, **697**
Selenium
  as toxin, 799
  in water pollution, 801
Selfish DNA, **697**
Selfish gene hypothesis, 24
Semen, 653, 655
  analysis of, 414
Semidesert, 214
Seminal reasons, 57
Seminal vesicles, 653

Skeletal system, **708–9.** *See also* Appendicular skeleton; Bone; Endoskeletons; Exoskeletons; Muscular system

Skin, 762
> electrical current across, 316
> integumentary system and, 419
> in respiration, 447

Skin cancer, 130, 290, 576

Skin grafts, 118, 532–33

Skinner, Burrhus Frederic, 69–70, 170, 543, **709–10**

Skinner Box, 709–10

Skou, Jen C., **710–11**

Skull
> fossils, 199
> kanjera, 458
> measurement of, 111
> size of, 660
> in skeletal system, 708

Slack, C.R., 356

Sleep, 388, 546

Sliding filament theory, 399–400

Slit lamp, 347

Sludge, 88

Small intestine, 76, 217–18, 582

Small nuclear RNA, 665–66

Smallpox, 290

Smallpox vaccine, 407, 416, 428, 783

Smell, **711**
> stimuli and, 729
> taste and, 744

Smiling, sociobiology and, 715

Smith, Elliot, 199

Smith, Grieg, 33

Smith, Hamilton O., 41, 537, 668, **711–12**

Smith, Hugh, 321

Smith, Michael, **712–13**

Smith, Sydney, 220

Smith, William, 330

Smog, 574. *See also* Air pollution
> amensalism of, 25
> plants and, 805–7
> weather and, 807

Smoking
> arteriosclerosis from, 44
> asbestos and, 737
> cancer and, 748–49
> cancer from, 129, 270
> synergistic effects of, 737
> as teratogenic, 751

Smoluchowski, Marian, 734

Snake venom, 381

Snakes, 572–73

Snell, George Davis, 71, 202, 409, **713–14**

Snow, 622

Snyder, Solomon, 593

SO2. *See* Sulfur dioxide

Social behavior
> aggression in, 479
> in children, 605
> of chimpanzees, 338
> pheromones and, 597

Social class, Social Darwinism and, 714

Social Darwinism, **714**
> insect behavior and, 822
> sociobiology and, 715

Social responsibility, of science, 793–94

Society for the Study of Social Biology, 274

Sociobiology, 273, 646–47, **715,** 760, 821–23

Soddy, Frederick, 371

Sodium
> in action potentials, 6
> cell membranes and, 710–11
> in cystic fibrosis, 192
> membrane potential and, 500
> in nerve impulses, 379, 398, 435, 542
> in nutrition, 556, 785

Sodium citrate, 36

Sodium nitrate, 550

Sodium phosphate, 36

Sodium-potassium pumps, 6

Sodium-24, 372

Soft tissue, 488

Soil chemistry, 34

Soil conservation, 13

Soil Conservation Service, 172

Soil erosion. *See* Erosion

Soil fertility, 136
> denitrification and, 213
> humus and, 396–97
> nitrogen in, 550, 552
> phosphorus in, 598

Soil pollution, 77–78

Soils, 608
> crop rotation and, 754
> microbiology of, 790–92
> in taiga, 744

Solanaceae, 20

Solar energy
> in ecosystems, 89, 246
> First Law of Thermodynamics and, 493
> human use of, 283
> photosynthesis of, 1
> in productivity, 628–29

Solid state devices, 43

Solids, 563

Solutions (Chemistry). *See* specific types of solutions, e.g., Hypotonic solutions

Solvents, water in, 799

Soma, 55

Somatic cells, 653–55

Somatic death, 207

Somatosensory system, 767

Somatostatin, 345, 346, 592, 690
> synthesis of, 423–24, 607

Somatotropin, 386, 388
> isolation of, 469–70
> synthesis of, 423–24

Songbirds, 417, 510–11, 571

Sorbitol, 523

Sound. *See* Acoustics

Sound waves, **4–6,** 698

Sounds of Korotkoff, 272

South African clawed toad, 368

T

Vauquelin, Louis-Nicolas, 25
Veins, 98, 158, **779**
    vs. arteries, 367
    hepatoportal, 365
    lymphatic system and, 679
    pulmonary (*See* Pulmonary vein)
    in respiratory system, 658
    valves in, 356
    varicose (*See* Varicose veins)
    venae cavae, 359, 779
Veldt, 342
Venae cavae, 359, **779**
Venereal diseases, 540. *See also* specific diseases, e.g., Syphilis
Venoms, 381, 646
Venous thrombosis, 779
Venter, Craig, 325
Ventral horns, 722
Ventral root, 542
Venules, 779
Venus, 680–81
Vernadsky, Vladimir, 89
Vertebrae, 708, **779**
Vertebral canal, 722
Vertebral column, 720–21, 722
Vertebrates, 779
    brain of, 108–9, 145
    circulatory system of, 158
    classification of, 188
    cloning of, 368
    digestion in, 219
    embryonic development in, 255, 258, 505
    excretion in, 280–81
    germ cells in, 655
    muscles of, 735
    musculoskeletal systems in, 708–9
    organizer effect in, 718
    respiration in, 735
    T cells and, 743
    temperature regulation of, 756
    tissues in, 762
Vesalius, Andreas, 28, 98, 285–86, 386, **779–80**
Vesicular model, 577–78
Vestibular apparatus, 64
Vestigial organs, 29, 117
Vibratory receptors, 767
Vibrio cholerae, 440
Viceroy butterfly, 513
Vierordt, Karl, 97
Vinblastine, 81, 149
Vincristine, 81, 149
Vinegar, 1, 3
Vipers, 646
Viral diseases. *See also* specific viral diseases
    bacteriophagic, 59–61
    cyclosporin for, 191
    drugs for, 149
    interferons and, 419–20
    transmission of, 415–16
Viral hepatitis, 363, 411, 659, 782–83
Viral meningitis, 782–83
Viral vaccines. *See* Vaccines

Virchow, Rudolf, 44, 143, 239, 672, 698, 781–82
Viroids, 666
Virons, 783
Viruses, **782–83**
    adenoviruses, 415
    attentuated live, 683–85
    avian, 416
    B, 675
    bacteriophagic (*See* Bacteriophages)
    for biological warfare, 84
    breeding, 464
    cancer and, 671–72, 725, 748–49
    cancer causing, 122, 123, 129, 481
    in cell theory, 143
    chemical nature of, 553
    crystallization of, 724–25
    cultures of, 667
    cytomegalovirus, 284, 805
    discovery of, 415
    DNA in, 240, 703, 832
    Epstein-Barr, 123
    as eukaryotes, 703
    fish, 407
    genetic engineering of, 748–49
    genetic recombination of, 212
    genome of, 703
    germ theory of, 331–32
    hepatitis, 123
    human immunodeficiency (*See* HIV)
    identification of, 815
    immune response to, 743
    interference phenomena in, 419
    killed, 676–77, 683–85
    lentivirus, 395
    mengoviruses, 62
    messenger RNA in, 703
    mutagenesis of, 712–13
    mutation of, 783
    picornaviruses, 62
    pig, 416
    plague assay for, 239–40
    pneumococcus, 52–53, 344
    polio, 259–60
    prions and, 627–28, 633–34
    as prokaryotes, 703
    Rauscher murine leukemia, 62
    replication of, 61, 62, 537
    retroviruses (*See* Retroviruses)
    RNA, 62, 307, 419, 666
    Rous sarcoma (*See* Rous sarcoma virus)
    simian, 537
    structure and function of, 307, 439, 620
    synthesis of, 480
    T cells and, 834–35
    tissue culture of, 804–5
    tobacco mosaic (*See* Tobacco mosaic virus)
    vaccines for (*See* Vaccines)
Vision, 283, 354, 360. *See also* Eye; Sight
Vision disorders. *See also* Eye disorders
    from computers, 284
Visual cortex, 388–89, 813–14

Visual-spatial determination, 165
Vital signs, 40
Vitalism, 189, 237, **783–84**
Vitamin A, 784–85
    discovery of, 496, 502, 571
    isolation and synthesis of, 434–35, 448
    in pigment regeneration, 708
    sight and, 792–93
Vitamin B, 75, 502, 547, 583
Vitamin B1. *See* Thiamine
Vitamin B2. *See* Riboflavin
Vitamin B3. *See* Niacin
Vitamin B5, 583
Vitamin B6. *See* Pyridoxine
Vitamin B12, 361, 532, 765, 784–85
Vitamin C, 356–57, 523, 589, 784–85
    isolation of, 741
    synthesis of, 650
Vitamin D, 153, 419, 496, 784–85
Vitamin E, 435, 764–65, 784–85
Vitamin H. *See* Biotin
Vitamin K, 197, 230–31, 435, 774–75, 784–85
Vitamins, **784–86**
    coenzymes and, 164
    deficiency of, 583
    in dietary lipids, 476
    dietary requirements for, 310–11, 556, 557
    discovery of, 385
    fat-soluble, 287, 496
    mutation and, 746
    structure and function of, 434–35
    synthesis of, 448–49, 763–65
    utilization of, 434–35
    yeasts as sources of, 829
Viviparity, 332, 656, **786**
Viviparous birth, 91
Vivisection, 789
Voelkel, Stephen, 163
Vogt, M., 196
Voit, Karl von, 124, 506
Volcanoes
    continental drift from, 174
    mass extinction from, 491
    origins of life and, 565
    pioneer species and, 608
    submarine, 181
Volpe, Peter, 583
Volta, Alessandro, 315
von Frisch, Karl. *See* Frisch, Karl von
von Helmholtz, Hermann. *See* Helmholtz, Hermann von
von Hevesy, Georg. *See* Hevesy, Georg von
von Hohenheim, Philippus Aurcolu Theophrastus Bombastus. *See* Paracelsus
von Liebig, Justus. *See* Liebig, Justus von
Vries, Hugo Marie de, 751, **786–87,** 820
    on Mendelian laws, 141, 279, 367, 534
    on mutations, 443, 525

W

Waardenberg, P.J., 234

Wagner-Jauregg, Julius, **789–90**
Waksman, Selman Abraham, 34, 149, **790–92**
Wald, George, **792–94**
Waldeyer-Hartz, Wilhelm von, 141, 156, 298, 337, 543
Walker, John E., **794–95**
Wallace, Alfred Russel, 279, 489, 714, **795**
Wallach, Otto, 817
Warblers, 511
Warburg, Emil, 795–96
Warburg, Otto, 193, 547, 600, 755, 793, **795–97,** 812
Warfarin, 36
Warm-blooded. *See* Homiothermy
Wasps, 759
Wasserman test, 135, 540
Wassermann, August von, 103, 540
Waste disposal, 80, 280, **797–98**, 799, 801
Waste management, 799
Waste treatment, 801
Water, **799–800**
    in abiotic environment, 1
    atmospheric, 45–46
    in biomes, 86–87
    chemical reactions in, 401
    distillation of, 351
    excretion of, 280
    Greek theory of, 360–61
    heavy, 372
    in humans, 372
    hydrotropism and, 771
    in hypertonic solutions, 715
    in hypotonic solutions, 716
    in isotonic solutions, 716
    in leaves, 457
    light refraction from, 345
    metabolism of, 178
    in photosynthesis, 121, 600
    protozoa and, 632
    saline (*See* Saline waters)
    solubility in, 569
    surface (*See* Surface water)
    in wetlands, 807–8
Water buffalo, 619
Water conservation, 172
Water cycle. *See* Hydrologic cycle
Water levels, 349, 401
Water molds, 310
Water pollution, 614, **800–801**
    bioaccumulation of, 77–78, 86
    cholera from, 153
    key indicators of, 266
    in lakes, 615–16
Water pollution control, 87, 166–67, 173
Water quality, 4
Water supply, 83, 153
Water vapor, 622
Watson, James D., 667, **801–3**
    Altman and, 23
    Avery and, 51, 52
    on base sequence, 333
    Chargaff's rules and, 146
    Crick and, 184

Delbrück and, 212
on DNA structure, 222–24, 233–34, 321
Franklin (Rosalind) and, 307–8
nucleic acids and, 554
on replication, 651
Schrödinger and, 693
Tatum and, 744
Todd and, 765
Tonegawa and, 766
Wilkins and, 814
Zamecnik and, 832
Watson, John B., 69–70, 709
Wave mechanics. *See* Quantum mechanics
Wave motion, 147, 385
Waxes, **803**
Weapons
    chemical, 763, 765, 817
    nuclear, 528, 589
Weather, 1, 314, 574, 807
Weather monitoring, 314
Wegener, Alfred Lothar, 173
Weigert, Carl, 142, 341
Weiss, Paul, 720
Weissman, Ernst, 288
Weissmann, August Friedrich Leopold, 288, 366, **803–4**
Weissmann, Charles, 420
Weissmann ring, 804
Welch, William Henry, 647–48
Welcher, Hermann, 97
Weldon, W.F.R., 590
Weller, Thomas, 150, 259–60, 667, **804–5**
Wells, H. Gideon, 695–96
Welwitxchia, 609
Went, Frits, 611, **805–7**
Werner, Abraham Gottlob, 138, 330
Werner, Alfred, 147
Wernicke's area, 111
Western Blot test, 396
Western grebes, 206, 207
Wetlands, 744, **807–9**
    carbon cycle and, 126
    conservation of, 173
    introduced species in, 421
Wexler, Nancy, **809**
Whales, 803
What Is Life?, 692–93
Whipple, George Hoyt, 97, **809–11**
Whipple's disease, 810
White, Charles, 38
White, David, 191
White blood cells, 229–30. *See also* IgM cells; T lymphocytes
    in immune system, 756
    in leukemia, 756–57
    leukotrienes (*See* Leukotrienes)
    monocytes (*See* Monocytes)
    as tissue, 762
Whittaker, Robert Harding, 161, 162, 438, **811**
WHO. *See* World Health Organization
Whooping cough vaccine. *See* Pertussis vaccine
Widal, Fernand, 11
Widal test, 12

Wieland, Heinrich Otto, 153, 724, **811–13,** 817
Wiener, Alexander S., 99, 660
Wiener, Norbert, 189
Wieschaus, Eric F., 555, **813**
Wiesel, Torsten N., 389, 720, **813–14**
Wilcox, K.W., 41
Wilderness Society, 173
Wildfires, 146
Wildlife management, 658
Wilkins, Maurice Hugh Frederick, 223, 234, 554, **814**
    Crick and, 184
    Franklin (Rosalind) and, 307
    Watson and, 801–2
Wilks, Samuel, 380
Willadsen, Steen M., 162–63, 818
Willard, Huntington, 156
Williams, Anna Wessels, 221, **815**
Williams, R.C., 725
Williams, Robert, 311
Williams, Roger John, 583
Willis, Thomas, 542
Willstätter, Richard, 152, 268, 600, 773, **815–18**
Willughby, Francis, 644
Wilmut, Ian, 163, **818–19**
Wilson, Alexander, 48
Wilson, Edmund Beecher, 104, 729, 733, **819–20,** 835
Wilson, Edward O., 273, 494, 646, 715, **821–23**
Wind, 576, 613
Windaus, Adolf, 812
Wine, 58, 287, 587
Withering, William, 219–20
Woese, Carl, 59
Wöhler, Friedrich, 423, 506, 568
Wolff, Gustav, 718
Wolff, Kaspar Friedrich, 751
Wollaston, William Hyde, 25
Wollman, Elie, 425–26
Wolves, 259, 510
Wood, 143
Woods Hole Research Center, 824
Woodward, Robert Burns, 152, 154, 178, 592
Woodwell, George M., **823–24**
Woody plants, 116, 214
Wool wax, 803
Woolly mammoths, 303
World Conservation Union, 82
World Health Organization
    on air pollution, 18, 615
    on cancer rates, 130
World Resources Institute, 82
World Wildlife Fund, 173
Worms, 280. *See also* Flatworms
Wounds, 825–26
Wright, Almroth Edward, 297–98, 408, **825–26**
Wright, Sewell, 228, 802

X

X chromosome, 319
    in cystic fibrosis, 191
    disorders of, 322

ADULT AND PEDIATRIC UROLOGY

Volume 1

Adult and Pediatric Urology

Jay Y. Gillenwater, M.D.

John E. Cole Professor and Chairman of Urology
University of Virginia School of Medicine
Charlottesville, Virginia

John T. Grayhack, M.D.

Herman L. Kretschmer Professor and Chairman of Urology
Northwestern University Medical School
Chief of Urology
Northwestern Memorial Hospital
Chicago, Illinois

Stuart S. Howards, M.D.

Professor of Urology and Physiology
University of Virginia
Chief, Division of Pediatric Urology
University of Virginia Hospital
Charlottesville, Virginia

John W. Duckett, M.D.

Professor of Urology in Surgery
University of Pennsylvania School of Medicine
Director, Division of Pediatric Urology
Children's Hospital of Philadelphia
Philadelphia, Pennsylvania

Volume 1

YEAR BOOK MEDICAL PUBLISHERS, INC.
CHICAGO • LONDON • BOCA RATON

1 2 3 4 5 6 7 8 9 0 K C 91 90 89 88 87

Library of Congress Cataloging-in-Publication Data

Adult and pediatric urology.

   Includes bibliographies and index.
   1. Genito-urinary organs—Diseases. 2. Urology.
3. Pediatric urology. I. Gillenwater, Jay Y. (Jay
Young), 1933–   . [DNLM: 1. Urologic Diseases.
2. Urologic Diseases—in infancy & childhood.
WJ 100 A2437]
RC871.A25   1987   616.6   86-29008
ISBN 0-8151-3476-2

Sponsoring Editor: Daniel J. Doody
Assistant Manager, Copyediting Services: Deborah Thorp
Production Project Manager: Max Perez
Proofroom Supervisor: Shirley E. Taylor

Contributors

DEMETRIUS H. BAGLEY, M.D., F.A.C.S.
Associate Professor of Urology and Radiology, Jefferson Medical College of Thomas Jefferson University, Thomas Jefferson University Hospital, Philadelphia, Pennsylvania

DAVID M. BARRETT, M.D.
Professor of Urology, Mayo Medical School and Mayo Foundation, Rochester, Minnesota

MARK F. BELLINER, M.D.
Visiting Associate Professor of Surgery (Urology), University of Pittsburgh School of Medicine, Pittsburgh, Pennsylvania

GEORGE S. BENSON, M.D.
Professor of Surgery (Urology), University of Texas Medical School at Houston, Hermann Hospital, Houston, Texas

MICHEL A. BOILEAU, M.D.
Associate Professor, Department of Urology, University of Washington School of Medicine, Seattle, Washington

JEFFREY P. BUCH, M.D.
Department of Urology, Baylor College of Medicine, Houston, Texas

ANTHONY A. CALDAMONE, M.D., M.M.S.
Associate Professor of Surgery (Urology), Brown University; Head, Division of Pediatric Urology, Rhode Island Hospital, Providence, Rhode Island

ROBERT M. CAREY, M.D.
Professor of Medicine and Dean, University of Virginia School of Medicine, Charlottesville, Virginia

WILLIAM J. CATALONA, M.D.
Professor and Chief, Division of Urologic Surgery, Washington University School of Medicine; Urologic Surgeon-in-Chief, Barnes Hospital, St. Louis, Missouri

CHRISTIAN G. CHAUSSY, M.D.
Professor of Surgery/Urology, UCLA School of Medicine; Director of UCLA Stone Center, UCLA Medical Center, Los Angeles, California

BERNARD M. CHURCHILL, M.D., F.R.C.S.(C.)
Associate Professor of Surgery, University of Toronto; Chief, Division of Urology, Hospital for Sick Children, Toronto, Ontario, Canada

ARNOLD H. COLODNY, M.D.
Associate Professor of Surgery, Harvard Medical School; Senior Surgeon, Associate Director, Division of Urology, Children's Hospital of Boston, Boston, Massachusetts

BO L. R. A. COOLSAET, M.D.
Professor of Urology, State University of Utrecht, Utrecht, The Netherlands

JOSEPH N. CORRIERE, Jr., M.D.
Professor and Director, Division of Urology, Department of Surgery, University of Texas Medical School at Houston, Houston, Texas

BARBARA Y. CROFT, Ph.D.
Assistant Professor of Radiology, University of Virginia, Charlottesville, Virginia

JEAN B. deKERNION, M.D.
Professor of Surgery/Urology, UCLA School of Medicine; Chief, Division of Urology, UCLA Medical Center, Los Angeles, California

WILLIAM C. DeWOLF, M.D.
Associate Professor of Surgery, Harvard Medical School; Assistant Surgeon in Urology, Beth Israel Hospital, Boston, Massachusetts

JOHN P. DONOHUE, M.D.
Professor and Chairman, Department of Urology, Indiana University Medical Center, Indianapolis, Indiana

JOHN W. DUCKETT, M.D.
Professor of Urology in Surgery, University of Pennsylvania School of Medicine; Director, Division of Pediatric Urology, Children's Hospital of Philadelphia, Philadelphia, Pennsylvania

RICHARD F. EDLICH, M.D., Ph.D.
Distinguished Professor of Plastic Surgery and Biomedical Engineering, University of Virginia School of Medicine, Charlottesville, Virginia

JACK S. ELDER, M.D.
Associate Professor of Surgery (Urology), Case Western Reserve University School of Medicine; Director of Pediatric Urology, Rainbow Babies and Children's Hospital, Cleveland, Ohio

LOULIE M. FISHER, M.D.
Senior Resident, Department of Radiology, University of Virginia Hospital, Charlottesville, Virginia

JACKSON E. FOWLER, Jr., M.D.
Clarence C. Saelhof Professor of Urology, University of Illinois College of Medicine; Chief, Division of Urology, University of Illinois Hospital, Chicago, Illinois

GERHARD J. FUCHS, M.D.
Visiting Professor of Surgery/Urology, UCLA School of Medicine; Co-Director, UCLA Stone Center, UCLA Medical Center, Los Angeles, California

JAY Y. GILLENWATER, M.D.
John E. Cole Professor and Chairman of Urology, University of Virginia School of Medicine, Charlottesville, Virginia

JOHN T. GRAYHACK, M.D.
Herman L. Kretschmer Professor and Chairman of Urology, Northwestern University Medical School; Chief of Urology, Northwestern Memorial Hospital, Chicago, Illinois

FARUK HADZISELIMOVIC, M.D.
Privat Dozent, University of Basle; Oberarzt, Children's Hospital, Basle, Switzerland

LAWRENCE J. HAYDEN, M.D., B.Sc., M.B., B.S.
Campbelltown, New South Wales, Australia

J. HAROLD HELDERMAN, M.D.
Professor of Internal Medicine, Medical Director, Renal Transplantation, University of Texas Health Science Center at Dallas, Southwestern Medical School; Attending Physician, Parkland Memorial Hospital, Dallas, Texas

TERRY W. HENSLE, M.D.
Associate Professor of Clinical Urology, College of Physicians and Surgeons of Columbia University; Director, Pediatric Urology, Babies Hospital/Columbia Presbyterian Medical Center, New York, New York

STUART S. HOWARDS, M.D.
Professor of Urology and Physiology, University of Virginia; Chief, Division of Pediatric Urology, University of Virginia Hospital, Charlottesville, Virginia

ROBERT HUBEN, M.D.
Department of Urology, State University of New York at Buffalo, Buffalo, New York

JOHN C. HULBERT, M.D., F.R.C.S.
Assistant Professor, University of Minnesota Hospitals; Clinical Staff, University of Minnesota, Fairview-St. Mary's Hospital, Minneapolis, Minnesota

ALAN D. JENKINS, M.D.
Assistant Professor of Urology, University of Virginia School of Medicine, Charlottesville, Virginia

JUDITH M. JOYCE, M.D.
Chief Resident, Radiology, University of Virginia, Charlottesville, Virginia

DONALD KAYE, M.D.
Professor and Chairman, Department of Medicine, The Medical College of Pennsylvania; Chief of Medicine, Hospital of the Medical College of Pennsylvania, Philadelphia, Pennsylvania

PANAYOTIS P. KELALIS, M.D., F.A.C.S., F.A.A.P.
Anson L. Clark Professor of Pediatric Urology and Chairman, Department of Urology, Mayo Clinic and Mayo Foundation, Rochester, Minnesota

CHARLES D. KELLUM, M.D.
Assistant Professor of Radiology, University of Virginia School of Medicine; Director of Uroradiology, University of Virginia Medical Center, Charlottesville, Virginia

STEPHEN A. KOFF, M.D.
Associate Professor of Surgery, Chief, Section of Pediatric Urology, Ohio State University College of Medicine; Section Chief, Division of Urology, Columbus Children's Hospital, Columbus, Ohio

STANLEY J. KOGAN, M.D.
Adjunct Professor of Urology, New York Medical College; Associate Clinical Professor of Pediatrics, Albert Einstein College of Medicine; Co-Director, Section of Pediatric Urology, Westchester County Medical Center, Valhalla, New York; Attending Pediatric Urologist, Albert Einstein College Hospital and the Montefiore Medical Center, Bronx, New York

JAMES M. KOZLOWSKI, M.D.
Department of Urology, Northwestern University Medical School, Chicago, Illinois

STEPHEN A. KRAMER, M.D.
Associate Professor of Urology and Pediatric Urology, Mayo Medical School; Consultant, St. Mary's Hospital, Rochester, Minnesota

JOHN N. KRIEGER, M.D.
Associate Professor, Department of Urology, University of Washington School of Medicine; Attending, University Hospital, Seattle, Washington

KENNETH A. KROPP, M.D.
Professor of Surgery and Pediatrics, Chief, Division of Urology, Department of Surgery, Medical College of Ohio, Toledo, Ohio

PAUL H. LANGE, M.D.
Professor and Vice-Chairman, Department of Urologic Surgery, University of Minnesota; Chief, Urology Section, V.A. Medical Center, Minneapolis, Minnesota

ROBERT M. LEVIN, Ph.D.
Division of Urology, University of Pennsylvania School of Medicine, Philadelphia, Pennsylvania

MARGUERITE C. LIPPERT, M.D.
Assistant Professor of Urology, University of Virginia Medical School, Charlottesville, Virginia

LARRY I. LIPSHULTZ, M.D.
Professor of Urology, Baylor College of Medicine, Houston, Texas

FRAY F. MARSHALL, M.D.
Professor of Urology, The Johns Hopkins University School of Medicine, Baltimore, Maryland

LESTER W. MARTIN, M.D.
Division of Pediatric Surgery and Urology, Children's Hospital Medical Center, Cincinnati, Ohio

JACK W. McANINCH, M.D.
Professor of Urology, University of California at San Francisco; Chief of Urology, San Francisco General Hospital, San Francisco, California

DAVID L. McCULLOUGH, M.D.
Professor of Surgery/Urology, Chairman of Section of Urology, Bowman Gray School of Medicine of Wake Forest University; Chief of Urology, Bowman Gray/North Carolina Baptist Medical Center, Winston-Salem, North Carolina

W. SCOTT McDOUGAL, M.D.
Professor and Chairman, Department of Urology, Vanderbilt University Medical School; Attending Urologist, Vanderbilt University Hospital, Nashville, Tennessee

GORDON A. McLORIE, B.Sc., M.D., F.R.C.S.(C.), F.A.A.P.
Assistant Professor, Department of Surgery, Division of Urology, University of Toronto; Staff Urologist, Division of Urology, Hospital for Sick Children, Toronto, Ontario, Canada

ELIAHU MUKAMEL, M.D.
Senior Lecturer, Tel Aviv University; Associate Professor of Surgery/Urology, Beilinson Medical Center, Petah Tiqva, Israel

HARVEY L. NEIMAN, M.D.
Clinical Professor of Radiology, University of Pittsburgh; Chairman, Department of Radiology, The Western Pennsylvania Hospital, Pittsburgh, Pennsylvania

H. NORMAN NOE, M.D.
Professor of Urology, Chief of Pediatric Urology, University of Tennessee, Memphis Health Science Center; Chief, LeBonheur Children's Medical Center, Memphis, Tennessee

THOMAS E. PALMER, M.D.
Bowman Gray School of Medicine of Wake Forest University, Winston-Salem, North Carolina

PAUL C. PETERS, M.D.
Professor and Chairman of Urology. The University of Texas Health Science Center at Dallas; Chief of Urology, Parkland Memorial Hospital, Dallas, Texas

RONALD RABINOWITZ, M.D., F.A.A.P., F.A.C.S.
Associate Professor of Urology and Associate Professor of Pediatrics, University of Rochester School of Medicine and Dentistry; Chief of Urology and Attending Pediatric Urologist, Rochester General Hospital, Rochester, New York

JOHN F. REDMAN, M.D.
Professor and Chairman, Department of Urology, Professor, Department of Pediatrics, University of Arkansas College of Medicine; Chief, Urology Service, University Hospital, Little Rock, Arkansas

MARTIN I. RESNICK, M.D.
Professor and Chairman, Division of Urology, Case Western Reserve University School of Medicine; Director of Urology, University Hospitals of Cleveland, Cleveland, Ohio

GEORGE T. RODEHEAVER, Ph.D.
Associate Professor, University of Virginia School of Medicine, Charlottesville, Virginia

RANDALL G. ROWLAND, M.D., Ph.D.
Associate Professor, Department of Urology, Indiana University School of Medicine, Indianapolis, Indiana

JOHN C. ROWLINGSON, M.D.
Associate Professor of Anesthesiology, Director, Pain Management Center, University of Virginia School of Medicine, Charlottesville, Virginia

ARTHUR I. SAGALOWSKY, M.D.
Associate Professor, University of Texas Southwestern Medical School; Surgical Director, Renal Transplant, Parkland Memorial Hospital, Dallas, Texas

ANTHONY J. SCHAEFFER, M.D.
Professor, Department of Urology, Northwestern University Medical School; Attending, Northwestern Memorial Hospital, Chicago, Illinois

CURTIS A. SHELDON, M.D.
Division of Pediatric Surgery and Urology, Children's Hospital Medical Center, Cincinnati, Ohio

ANUP SINGH, M.D.
Urology Fellow, The James Buchanan Brady Urological Institute, The Johns Hopkins Hospital, Baltimore, Maryland

ELLEN BLAIR SMITH, M.D.
Assistant Professor, University of Virginia School of Medicine, Department of Obstetrics and Gynecology, Charlottesville, Virginia

JOSEPH A. SMITH, Jr., M.D.
Associate Professor of Surgery, Division of Urology, University of Utah Health Sciences Center, Salt Lake City, Utah

BRENT W. SNOW, M.D.
Assistant Professor of Surgery, Assistant Professor of Pediatrics, University of Utah; Pediatric Urologist, University of Utah Medical Center, Salt Lake City, Utah

HOWARD McC. SNYDER III, M.D.
Associate Professor of Urology in Surgery, University of Pennsylvania School of Medicine; Associate Director, Division of Pediatric Urology, Children's Hospital of Philadelphia, Pennsylvania

JACK D. SOBEL, M.D.
Professor of Medicine, Chief, Infectious Diseases Division, Wayne State University School of Medicine; Chief, Infectious Diseases Section, Hutzel Hospital, Detroit, Michigan

R. ERNEST SOSA, M.D.
Department of Urology, New York Hospital, New York, New York

J. PATRICK SPIRNAK, M.D.
Assistant Professor of Urology, Case Western Reserve University; Director, Division of Urology, Cleveland Metropolitan General Hospital, Cleveland, Ohio

GERALD SUFRIN, M.D.
Professor and Chairman, Department of Urology, State University of New York at Buffalo; Director of Urology, Buffalo General Hospital, Buffalo, New York

CHARLES D. TEATES, M.D., F.A.C.R.
Professor of Radiology, University of Virginia School of Medicine; Director of Medical Imaging, Director of Nuclear Medicine, University of Virginia Hospital, Charlottesville, Virginia

CHARLES J. TEGTMEYER, M.D.
Professor of Radiology, Associate Professor of Anatomy, University of Virginia; Director of Angiography, Interventional Radiology and Special Procedures, Co-Director, Urologic Radiology, University of Virginia Medical Center, Charlottesville, Virginia

JOHN G. THACKER, Ph.D.
Associate Professor of Mechanical Engineering, University of Virginia, Charlottesville, Virginia

DAVID T. UEHLING, M.D.
Professor of Surgery/Urology, University of Wisconsin Medical School, Madison, Wisconsin

E. DARRACOTT VAUGHAN, Jr., M.D.
James J. Colt Professor of Urology, Cornell University Medical Center; Attending Urologist-in-Chief, The New York Hospital, New York, New York

ROBERT L. VOGELZANG, M.D.
Assistant Professor of Radiology, Northwestern University Medical School; Chief, Angiography and Interventional Radiology, Co-Director, Body CT, Northwestern Memorial Hospital, Chicago, Illinois

R. DIXON WALKER, M.D.
Professor of Surgery and Pediatrics, University of Florida College of Medicine; Chief of Pediatric Urology, Shands Hospital, Gainesville, Florida

ALAN J. WEIN, M.D.
Professor and Chairman, Division of Urology, University of Pennsylvania School of Medicine; Chief of Urology, Hospital of the University of Pennsylvania, Philadelphia, Pennsylvania

ROBERT M. WEISS, M.D.
Professor, Section of Urology, Yale University School of Medicine; Attending, Yale-New Haven Hospital, New Haven, Connecticut

SIR DAVID INNES WILLIAMS, M.D., F.R.C.S.
Pro-Vice-Chancellor, University of London; Consulting Surgeon, Hospital for Sick Children, London, England

RICHARD D. WILLIAMS, M.D., F.A.C.S.
Professor and Chairman, Department of Urology, University of Iowa Hospitals and Clinics, Iowa City, Iowa

B. DALE WILSON, M.D.
Clinical Professor of Dermatology, State University of New York at Buffalo School of Medicine; Clinician II, Roswell Park Memorial Institute, Buffalo, New York

HENRY A. WISE II, M.D.
Professor of Urology, Ohio State University, Columbus, Ohio

ARTHUR W. WYKER, Jr., M.D.
Professor of Urology, University of Virginia Hospital, Charlottesville, Virginia

Preface

Urology is one of the most interesting and diverse specialties in medicine. The amount of new information in urology is staggering and the half-life of current knowledge is short. *Adult and Pediatric Urology* is designed to give a complete description of urologic physiology, anatomy, pathology, disease processes, diagnosis, and therapy. Particular attention has been paid to the mechanisms of various diseases. In some areas, such as infection, we have purposefully invited differing points of view to reflect the actual state of knowledge. Chapter subjects were chosen to reflect the entire field and the chapters were written to serve as a reference text for urologists in training and in practice. Chapters on specialized subjects such as wound healing, pain management, gynecology, lasers, new imaging technologies, and new technologies for the treatment of calculus disease were included.

All textbooks contain dated information. We have addressed this problem in two ways. First, most of these chapters were written less than a year prior to publication. Second, we will be updating *Adult and Pediatric Urology* annually in the YEAR BOOK OF UROLOGY. The YEAR BOOK OF UROLOGY has been reorganized to function as a supplement to this textbook, in addition to being an annual survey of the best in the year's urological literature.

Any text is no better than its editors, authors, and publishers. Working with Jack Grayhack, Stuart Howards, and John Duckett has been both pleasant and intellectually stimulating. Each of these editors has worked closely with his authors to ensure the best coverage possible for a wide array of subjects. The authors enthusiastically responded to the challenge and produced scholarly, critical, and informative chapters. Daniel J. Doody and the Year Book Medical Publishers staff have also been invaluable partners in this enterprise. I know of no other group who could transform so much manuscript into a completed textbook in such a compressed time frame. The result is an up-to-date, thorough, and eminently readable textbook of general urology.

JAY Y. GILLENWATER, M.D.

Contents

Volume 1

Contents

Volume 2

PART I

Adult Urology

Chapter 1

Anatomy of the Genitourinary System

JOHN F. REDMAN, M.D.

The urologist is foremost a genitourinary surgeon. A working knowledge of anatomy is the basis for surgical competence. Over the millennia the anatomy of the human body has not changed, only the descriptions. They continue to change. Illustrations and descriptions obtained from cadaveric study frequently do not coincide with the living, pulsating, even bleeding tissue seen through the developing incision. For the surgeon, the study of anatomy is usually accomplished in piecemeal fashion. Only by utilizing each surgical procedure as an opportunity to learn by attention to anatomical detail will he continue to increase his understanding of this complex area.

THE ANATOMY OF THE BODY WALL

The abdominal wall is for the most part composed of muscle and its aponeuroses. Because of the degree of overlapping, it is best to assign groups arbitrarily to the muscles for the purpose of discussion. The groupings discussed include the musculature of the anterolateral wall, the posterior abdominal wall, the thoracic abdominal wall, the pelvic floor, and the inguinal canal.

ANTEROLATERAL ABDOMINAL WALL

The anterolateral abdominal wall is formed of large, flat muscle sheets, the external oblique, the internal oblique, the transversus abdominis, and the rectus abdominis. Two small muscles of the anterior abdominal wall are the pyramidalis and cremasteric.

The external oblique is the most superficial of the anterolateral group (Fig 1-1). It arises as seven or eight fleshy slips from the outer surface of the lower ribs (5th through 12th) (Fig 1-2,A). The slips interdigitate with the fleshy origins of the serratus anterior and the latissimus dorsi muscles. The fibers of the muscle, which inserts into the anterior half of the iliac crest, extend obliquely downward. A broad, flat aponeurosis extends medially to the midline, forming the anterior layer of

the rectus sheath. In the midline it fuses with the aponeurosis of the opposite side, forming the linea alba. The inferior free border termed the "inguinal ligament" stretches from the anterosuperior iliac spine to the pubic tubercle, being attached laterally to the iliopsoas fascia and medially to the pectineus fascia. The innervation of the external oblique is by the ventral rami of the lower six thoracic spinal nerves.

The internal oblique muscle lies between the external oblique and the transversus abdominis. It arises from the iliopsoas fascia, from the anterior two-thirds of the intermediate line of the iliac crest, and from the posterior lamina of the lumbodorsal fascia (Figs 1-2,B and 1-3). The fibers that run obliquely upward insert into cartilage of the lower 4th to 6th ribs. The broad aponeurosis of the internal oblique muscle divides cranial to the arcuate line (semilunar line of Douglas). The anterior lamina passes ventral to the rectus abdominis musculature to fuse with the aponeurosis of the external oblique to form the anterior rectus sheath. The posterior laminal passes dorsal to the rectus abdominis to fuse with the aponeurosis of the transversus abdominis to form the posterior rectus sheath. Caudal to the arcuate line all layers of the aponeurosis pass ventral to the rectus abdominis muscle (Fig 1-4). The innervation is by the ventral rami of the lower sixth thoracic and first lumbar spinal nerves.

The transversus abdominis muscle lies under the internal oblique. It arises from the iliopsoas fascia, the anterior two-thirds of the inner lip of the iliac crest, the anterior lamina of the lumbodorsal fascia, and the inner surface of the lower sixth costal cartilages, where it interdigitates with the fleshy slips of the diaphragm (Fig 1-2C, and 1-5). The broad aponeurosis fuses with the posterior lamina of the aponeurosis of the internal oblique to form the posterior rectus sheath, passing dorsal to the rectus abdominis muscle. However, caudal to the arcuate line the aponeurosis passes ventral to the rectus to join in the formation of the anterior rectus

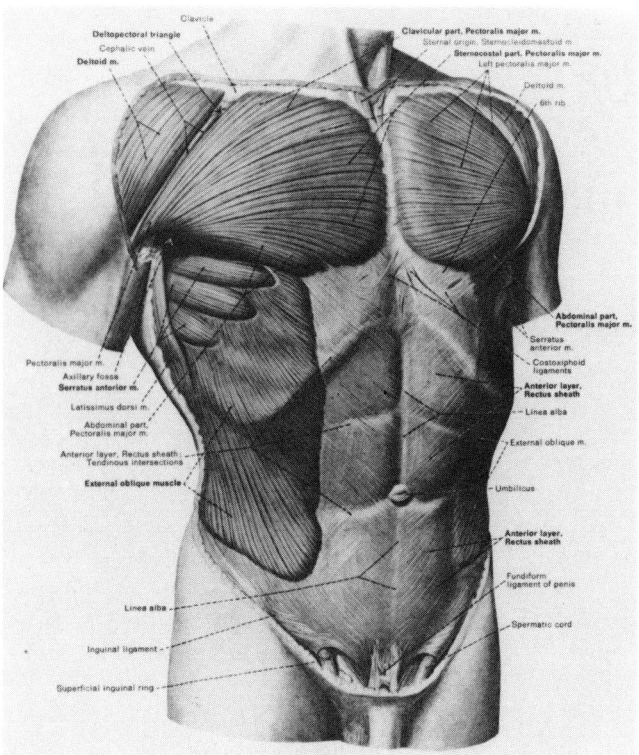

FIG 1–1.
Superficial musculature of the anterior abdominal and thoracic wall. (From Clemente D: *Anatomy—A Regional Atlas of the Human Body,* ed 2. (Baltimore, Urban & Schwarzenberg, 1981. Used by permission.)

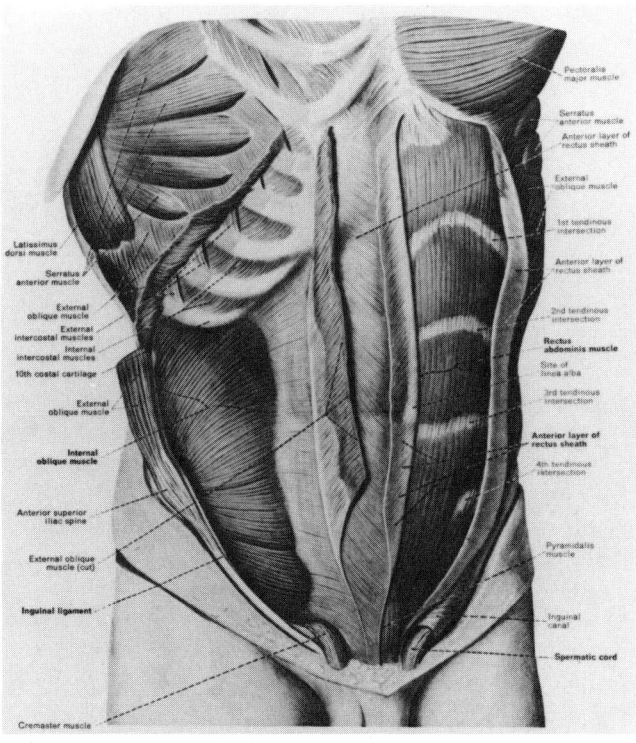

FIG 1–3.
Internal oblique muscle showing relationship to other abdominal wall musculature. (From Clemente D: *Anatomy—a Regional Atlas of the Human Body,* ed 2. (Baltimore, Urban & Schwarzenberg, 1981. Used by permission.)

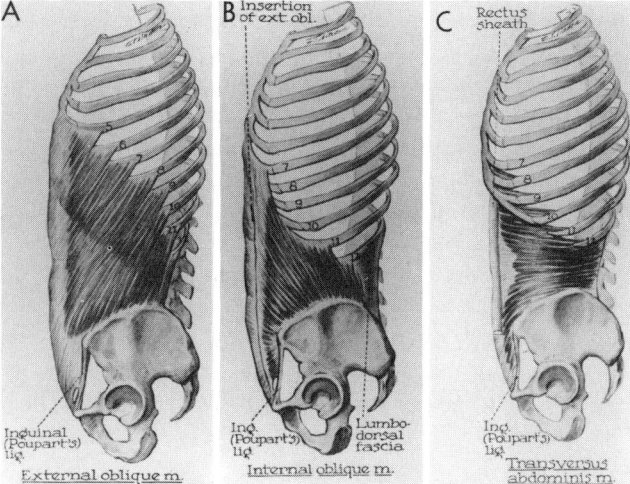

FIG 1–2.
A, the external oblique muscle. **B,** the internal oblique muscle. **C,** the transverse abdominis muscle. (From Thorek P: *Anatomy in Surgery,* ed 2. Philadelphia, JB Lippincott Co, 1962, pp 360–364. Used by permission.)

sheath. Only its dorsal investing fascia, the transversalis fascia, covers the dorsal surface of the rectus muscles. The innervation of the transversus abdominis muscle is by the ventral rami of the lower sixth thoracic and first lumbar spinal nerve.

The rectus abdominis muscles that are invested by the rectus sheath are long, broad muscular straps that extend from the thorax to the pubis on either side of the linea alba (see Fig 1–5). They arise from tendinous attachments on the pubis and invest by slips on the cartilage of the 5th, 6th, and 7th ribs and the xiphoid process. The area covered by the insertion is three times as wide as the origin on the pubis. The muscle fibers themselves are interrupted by three fibrous bands termed tendinous intersections. Innervation is by the ventral rami of the lower sixth or seventh thoracic spinal nerves.

The pyramidalis muscles are triangular muscles situated just ventral to the rectus abdominis muscles and covered by its sheath. They arise from the pubis and attach to the linea alba between the pubis and the umbilicus, one muscle frequently being larger than the other (see Fig 1–5). Innervation is by the subcostal nerve, the ventral ramus of the 12th thoracic nerve.

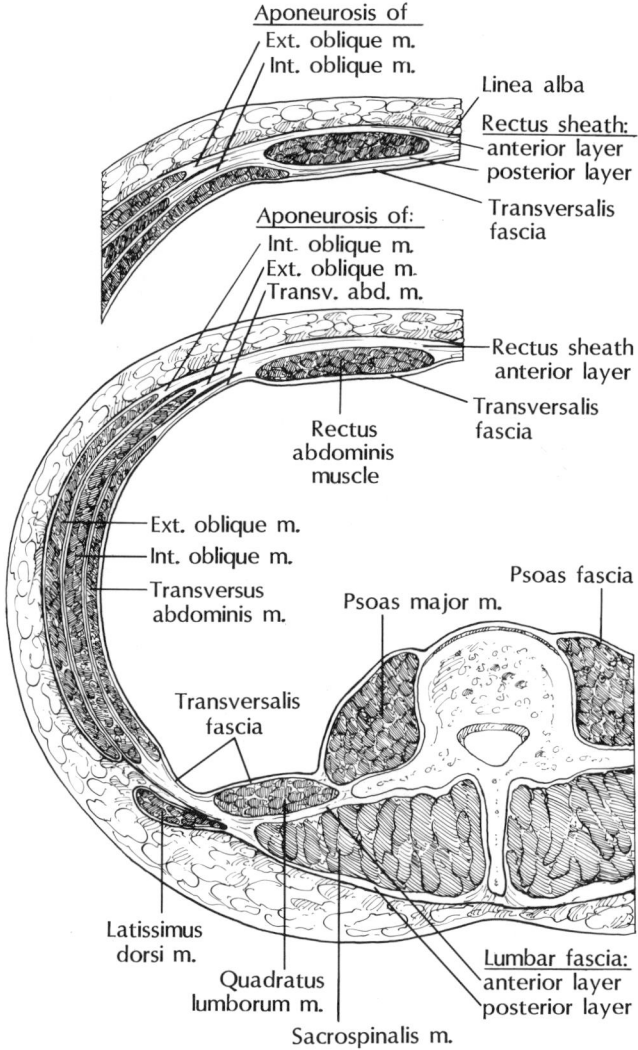

FIG 1–4.
Cross-section in the lumbar region showing lamina of the lumbar (lumbodorsal) fascia and the musculature and fusion of the anterior abdominal below the arcuate line. *Inset* shows composition of rectus sheath above the arcuate line. (Original drawing by Christine Young.)

The cremasteric muscle derives from the lower margin of the internal oblique and transversus abdominis muscles, which stretch from the inguinal ligament to the pubic tubercle (see Fig 1–5). Innervation is by the genital branch of the genitofemoral nerve.

POSTERIOR ABDOMINAL WALL

For discussion, the posterior abdominal wall musculature may be divided arbitrarily into three groups: A superficial group, an intermediate group, and a deep group. The superficial group is composed of the external oblique and the latissimus dorsi. The intermediate group consists of the serratus posterior inferior, the in-

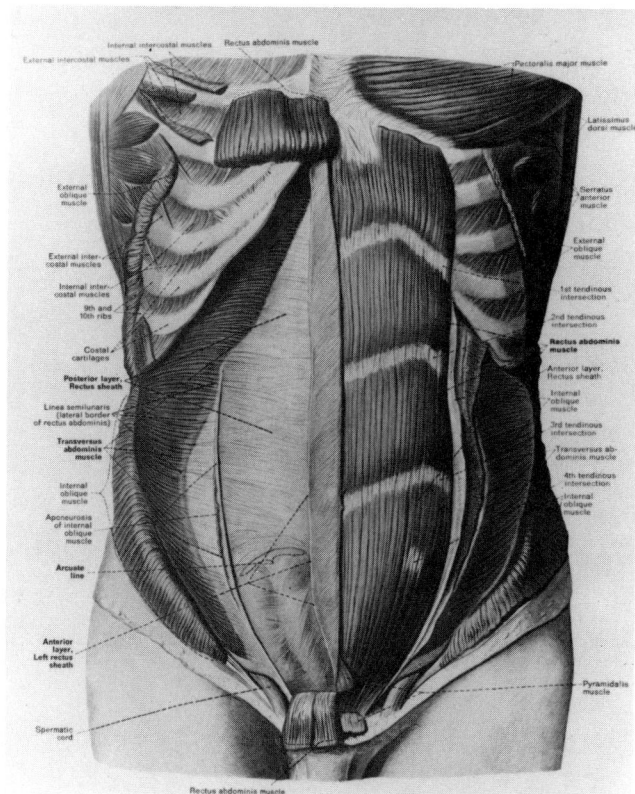

FIG 1–5.
Deep layer of abdominal musculature: transversus abdominis muscle. (From Clemente D: *Anatomy—a Regional Atlas of the Human Body,* ed 2. Baltimore, Urban & Schwarzenberg, 1981. Used by permission.)

ternal oblique and the sacrospinalis, while the deep group consists of the transversus abdominis, quadratus lumborum, psoas, iliacus, and diaphragm.

In regard to the superficial group, the external oblique has been described with the anterolateral abdominal wall musculature. The latissimus dorsi is a large, flat, triangular-shaped muscle that arises via the posterior layer of the lumbodorsal fascia from the lower sixth thoracic vertebra, the lumbar spine, and the upper sacral spine. It also has origin from the outer posterior lip of the iliac crest. The insertion is by a 1-in. broad tendon to the intertubercular sulcus of the humerus (Figs 1–4 and 1–6).

In the intermediate group the internal oblique has been described with the anterolateral group. The serratus posterior inferior arises from the spines of the T11–L2–3 vertebra. The fibers run obliquely upward to insert by digitations on the inferior outer border of the last four ribs just lateral to their angles. The sacrospinalis or erector spinae muscle is a large complex muscle that lies in the groove on the side of the vertebral column. It is covered by the posterior layer of the

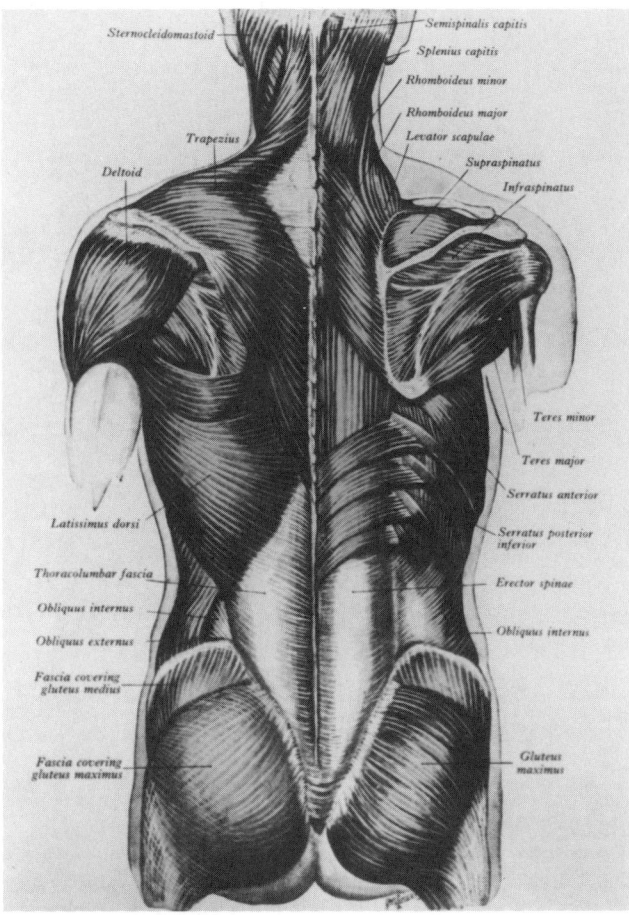

FIG 1–6.
Left, superficial musculature of the posterior abdominal wall. *Right,* with the removal of the latissimus dorsi and the external oblique are seen the intermediate group. (From Gray H: *Gray's Anatomy,* 36th British ed. Philadelphia, WB Saunders Co, 1980, p 565. Used by permission.)

lumbodorsal fascia and splits into three groups in the lumbar region. From medial to lateral they are the spinalis, the longissimus, and the iliocostalis. The portion of the iliocostalis embracing the ribs by inserting on the upper borders of the angles of the last six ribs is termed the "iliocostalis thoracis."

Of the deep group of posterior abdominal wall musculature, the transversus abdominis has been described with the anterolateral group (Fig 1–7). The quadratus lumborum arises from the posterior aspect of the iliac crest and, pursuing an upward and medially oblique course, inserts on the tips of the transverse processes of the upper four lumbar vertebra and the medial half of the lower border of the 12th rib. The psoas major may be considered a lumbar extension of the iliacus muscle. Its origin is from the T12–L5 vertebra, from which it passes laterally to pass under the inguinal ligament to insert ultimately on the lesser trochanter of the femur. The iliacus is a triangular-shaped muscle that originates

from the superior two thirds of the concavity of the iliac fossa, the inner lip of the iliac crest, and the upper part of the lateral aspect of the sacrum. It inserts in the tendon of the psoas and the femur. The diaphragm is a complex, domed, fibromuscular sheet that separates the thoracic and abdominal cavities. Basically, the diaphragm arises from the periphery of the caudal aspect of the thorax and inserts into the fibrous center, the central tendon of the diaphragm. The origins are from three parts. The sternal part arises from the posterior aspect of the xiphoid. The costal part arises from the internal surfaces of the lower six ribs and their cartilage. The lumbar part arises from the medial and lateral lumbocostal arcs or arcuate ligaments, the lateral covering the quadratus lumborum, the medial covering the psoas major. Two crura also provide origin for the diaphragm, a right and left arising from the anterolateral surfaces of the upper two and three lumbar vertebrae. They converge in the midline to form the median lumbocostal arc over the aorta (Gray, 1980b). The diaphragm has three apertures or hiatuses: the aortic, the esophageal, and the vena caval. Just craniolateral to the lumbocostal arc, the musculature of the diaphragm is frequently thin, the pleura being separated only by fascia.

THORACIC WALL

The thoracic wall of interest to the urologist is the caudal aspect. Muscles encountered include the external oblique, latissimus dorsi, serratus anterior, serratus posterior inferior, sacrospinalis, the intercostals and levators costarum (Fig 1–8). The external oblique, latissimus dorsi, serratus posterior inferior, and the sacrospinalis have been previously considered. The serratus anterior arises from muscular digitations on the upper 8–10 ribs and the intercostal fascia. The lower four digitations mesh with those of the external oblique. The serratus anterior inserts on the anteromedial aspect of the scapula. The intercostal muscles of the caudal aspect of the thorax include primarily the external intercostal muscles and the internal intercostal muscles. An innermost intercostal is quite variable. The external intercostal is a thoracic equivalent of the external oblique with fibers that pass obliquely downward. At the level of the cartilage the muscle is replaced by an aponeurosis termed the "external intercostal membrane." Likewise, the internal intercostal muscle is a thoracic equivalent of the internal oblique muscle with fibers that pass obliquely upward. At the angle of the ribs the muscle fibers are replaced by an aponeurosis termed the "internal intercostal membrane." If an innermost intercostal muscle is present corresponding to the transversus abdominis, it is ventral to the internal intercostal muscle. The intercostal artery, nerve, and vein course along the subcostal groove on the outer in-

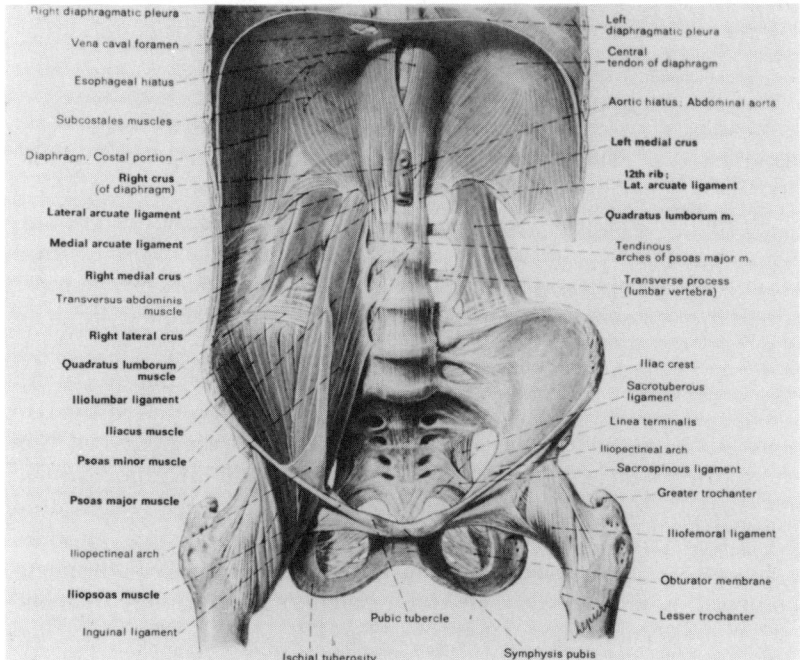

FIG 1–7.
The diaphragm and posterior abdominal wall musculature. (From Clemente D: *Anatomy—A Regional Atlas of the Human Body,* ed 2. Baltimore, Urban & Schwarzenberg, 1981. Used by permission.)

ferior surface of each rib ventral to the internal intercostal muscle. The levators costarum arise from the ends of the transverse processes and pass obliquely downward and laterally and insert on the upper edge of the rib immediately below the vertebra of their origin.

PELVIC FLOOR

The musculature of the pelvic floor or pelvic diaphragm consists of the levator ani, the coccygeus, the obturator internis, and the piriformis.

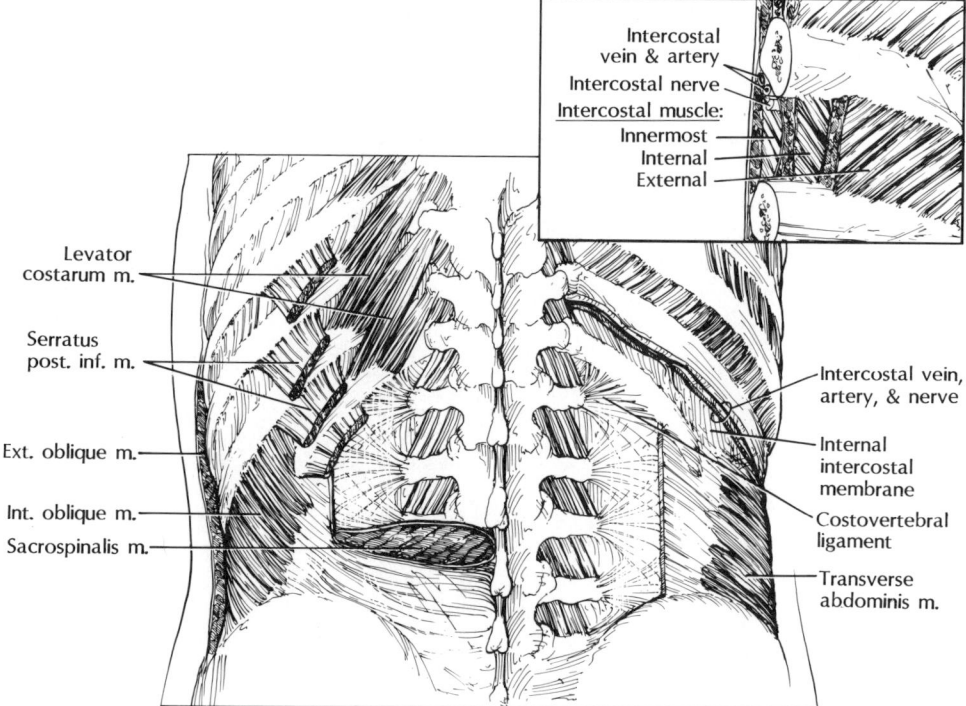

FIG 1–8.
Overview of thoracic musculature deep to the sacrospinalis, latissimus dorsi and external oblique. (Original drawing by Christine Young.)

The levator ani is a funnel-shaped muscle that can be subdivided into two primary portions—the pubococcygeus and iliococcygeus. The pubococcygeus is placed more medially whereas the iliococcygeus is placed laterally. Both portions of the levator ani arise from the tendinous arc of the pelvis. The pubococcygeus arises partially from the inner aspect of the pubis, surrounds the visceral tubes exiting the pelvis, and attaches to the sacrum and coccyx. The iliococcygeus inserts also into the sacrum and coccyx. The coccygeus is situated in the same tissue plane as the levator ani. It is triangularly shaped and originates from the ischial spine and inserts into the sacrum and coccyx.

The obturator internis muscle is partially situated in the true pelvis and partially related to the hip joint. Its origin in the pelvis is from the pubic rami, the ramus of the ischium, and the medial two thirds of the outer surface of the obturator membrane. The muscle ultimately inserts on the greater trochanter. As viewed from within the pelvis, the muscle is largely obscured by the covering levator ani.

The piriformis arises from the sacrum and inserts in the greater trochanter of the femur. In so doing, it forms part of the pelvic diaphragm. Lying in the same plane is the coccygeus. By virtue of its position, it is more of a muscle of the wall of the pelvis than of the floor (Fig 1–9). The piriformis is significant in that it is covered by the sacral plexus as it proceeds on its course beneath the coccygeus.

INGUINAL CANAL

The contents of the inguinal canal are covered by the aponeurosis of the external oblique, the internal oblique, the transversus abdominis, and the cremasteric muscle and fascia. The three flat muscles arise from the iliac crest and continue out onto the iliopsoas fascia. The muscles continue their trilaminar arrangement until they approach the vascular lacunae, where the internal oblique and transversus abdominis meld. Over the great vessels there are two arches: a superficial and a deep. The superficial arch is the free edge of the aponeurosis of the external oblique, the inguinal ligament. The deep arch is the free edge of the fused internal oblique and transversus abdominis muscles. The medial aspect of the arches attaches to the pectineus fascia. The arches in the midportion are attached to the femoral sheath. The lateral aspects are attached to the iliopsoas fascia (Doyle, 1971). The floor of the inguinal canal is formed of transversalis fascia, the inner investing fascia of the transversus abdominis. The lateral aspect of the canal is penetrated by the spermatic cord or round ligament. The fasciae of the abdominal

FIG 1–9.
Pelvic floor as seen from above. (From Kelly HA: *Operative Gynecology.* New York, Appleton-Century-Crofts, 1906, p 82. Used by permission.)

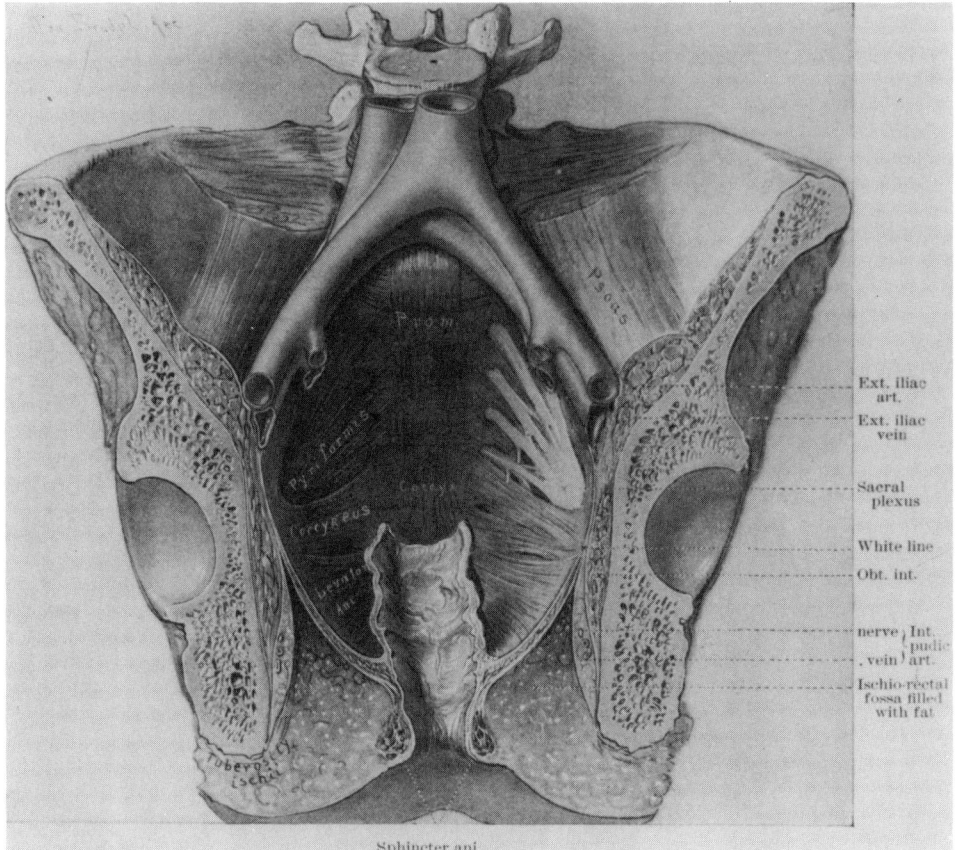

wall musculature then become the coverings of the cord. In the male the continuation of the transversalis fascia is termed the "internal spermatic fascia"; the continuation of the internal oblique, the "cremasteric fascia and muscle" (Griffith, 1984). The continuation of the external oblique fascia is noted from where the cord emerges from beneath the external oblique aponeurosis, the external ring, and is termed the "external spermatic fascia." It should be noted that the cremasteric muscle and fascia do not attach to the inguinal ligament but rather to the fused transversalis and iliac fascia, also termed the "iliopubic tract" (Fig 1–10).

The inguinal floor in addition is penetrated by the cremasteric or external spermatic vessels, which are accompanied by the genital branch of the genitofemoral nerve. The cremasteric vessels arise from the inferior epigastric vessels, which lie immediately beneath the transversalis fascia of the inguinal floor just medial to the internal ring.

PERINEUM

For discussion of the anatomy, the perineum is frequently divided into triangles that join by their bases,

the urogenital triangle and the anal triangle. The points of the triangles are the pubic symphysis, the ischial tuberosity and the tip of the coccyx. A line through the anterior ischial tuberosities forms the shared base of the triangles. The perineal body is the central fixation point of the perineum.

The urogenital triangle is the anterior triangle of the perineum. The deep layer of superficial fascia of the urogenital triangle is continuous with Scarpa's fascia of the anterior abdominal wall and is termed Colles' fascia. A fibrous septum joins Colles' fascia to the underlying perineal musculature. The potential space limited by Colles' fascia and the urogenital diaphragm is termed the "superficial perineal pouch" as opposed to the deep perineal pouch, which is the potential space limited by the superior and inferior fascia of the urogenital diaphragm.

The musculature of the urogenital triangle is also divided into superficial and deep. There are three superficial pairs of muscles: the median bulbocavernosus (bulbospongiosus), the right and left ischiocavernosus, and the superficial transverse perinei. In the male the superficial transverse perinei is a weak muscle, which

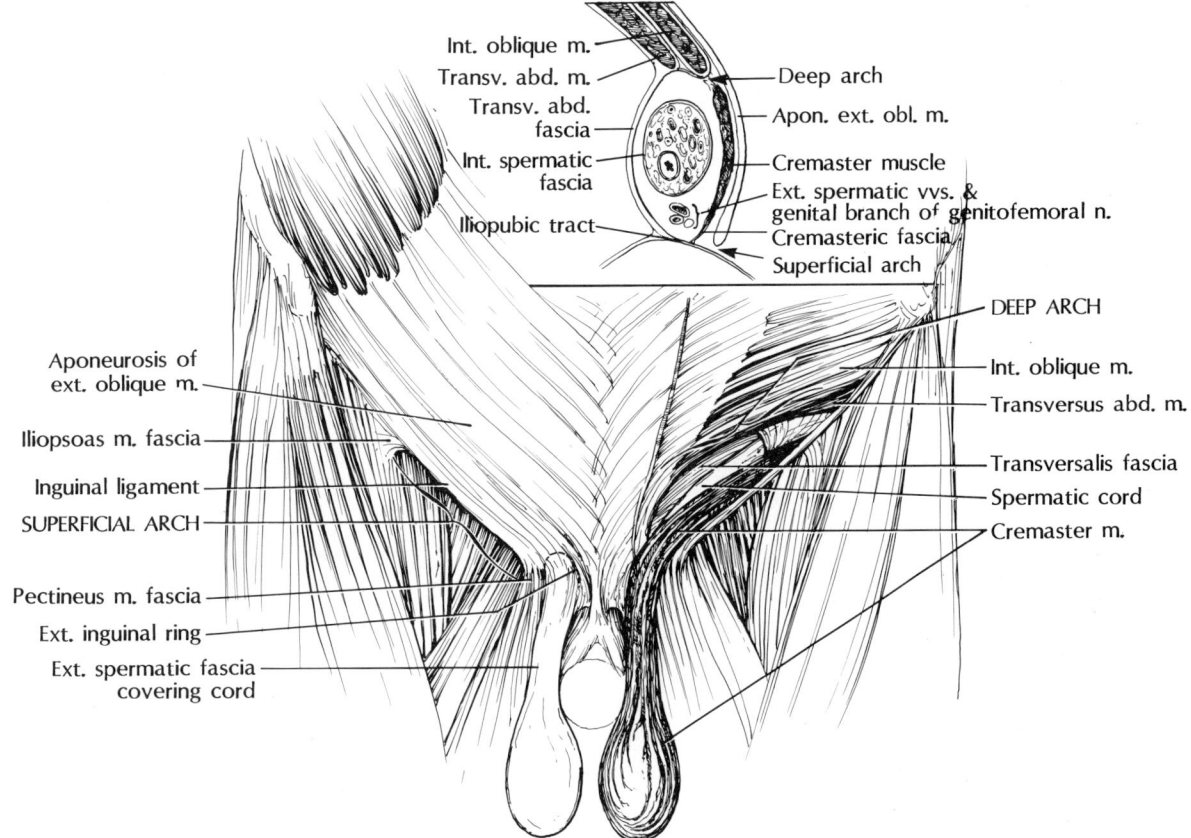

FIG 1–10.
Anatomy of the inguinal area. Superficial anatomy *(right);* deep anatomy *(left).* Inset, cremasteric through inguinal canal. (Original drawing by Christine Young.)

is sometimes absent and extends from the ischial tuberosity to the perineal body. The bulbocavernosus muscle arises from the perineal body and from the median raphe, which separates the two muscles on the inferior surface of the bulb of the corpus spongiosa. It inserts on the fascia of the dorsum of the corpus spongiosa and the ventrum of the corpus cavernosa. The ischiocavernosus muscle covers the crura of the penis. It is attached by fleshy and tendinous fibers to the ischial tuberosity and the ramus of the ischium (Fig 1–11).

In the female the superficial transverse perinei muscle is much the same as in the male. The bulbospongiosus muscle surrounds the vaginal introitus and covers the lateral part of the vestibular bulbs. It attaches to the corpora of the clitoris and the perineal body. The ischiocavernosus muscle covers the crura of the clitoris.

In the male the deep perineal musculature is essentially one sheet which is divided into two parts: the deep transverse perineal muscle and the urethral sphincter muscle. The origin of the deep transverse perinei is the ramus of the ischium. It joins the urethral and anal sphincters and the perineal body. Within the musculature of the deep transverse perinei lie the bulbourethral glands, which are two pea-shaped bodies that are located posterolateral to the membranous urethra. The ducts open into the bulb of the urethra and

are approximately 2.5 cm long. The urethral sphincter muscle surrounds the membranous urethra. In the female the deep transverse perinei is penetrated by the vagina but is otherwise like that of the male. The urethral sphincter muscle surrounds the urethra. The innervation of all of the urogenital musculature is by the perineal branch of the pudendal nerve.

In the anal triangle the superficial fascia, which is fat laden, fills the wedge-shaped space between the levator ani and the obturator internis. The space is termed the "ischiorectal fossa." In the lateral wall of the ischiorectal fossa covered by a sheath of fascia, termed the "pudendal canal" or "Alcock's canal," are located the internal pudendal vessels and their accompanying nerves.

ANATOMY OF THE RETROPERITONEAL CONNECTIVE TISSUE

The retroperitoneal connective tissue is basically the connective tissue that lies between the osteomuscular wall and the peritoneum (Redman, 1983). Tobin and associates (1946) subdivided the retroperitoneal connective tissue into three strata: outer, intermediate, and inner. The outer stratum is the inner investing fascia of the muscle, which is termed collectively the "transversalis fascia." The term transversalis fascia may be used

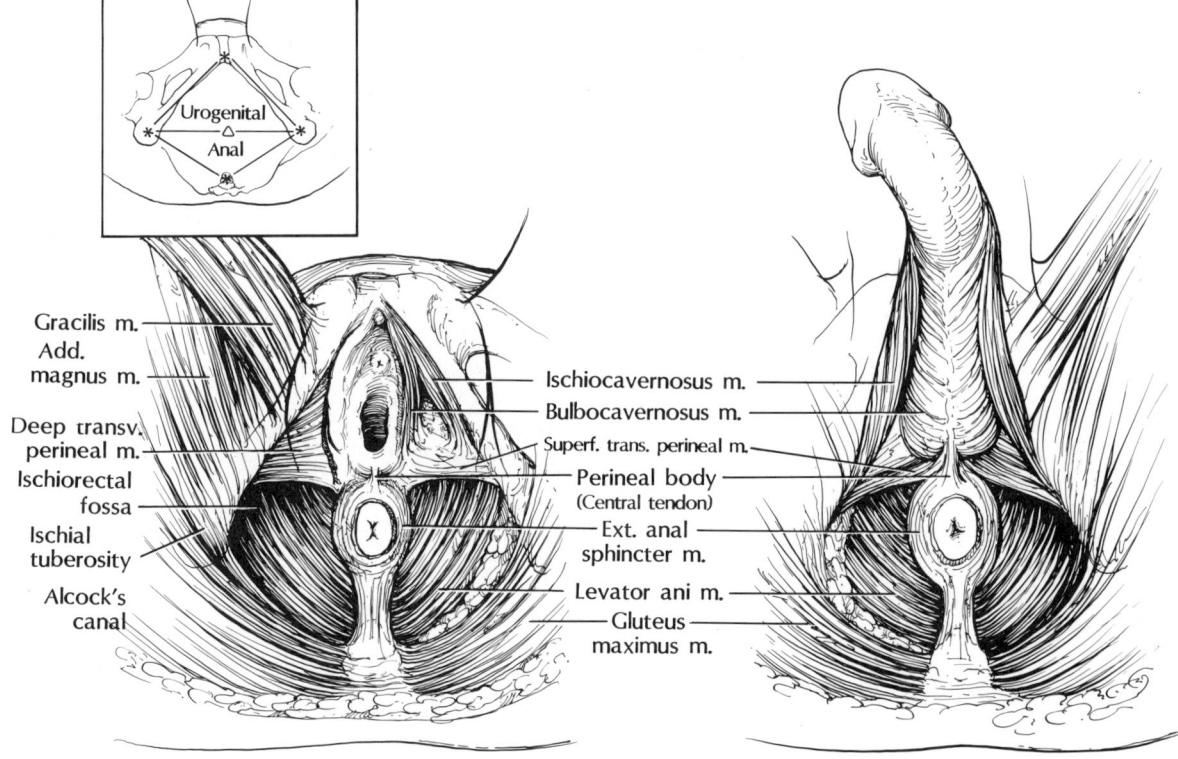

FIG 1–11.
The superficial muscles of the male and female perineum. (Original drawing by Christine Young.)

interchangeably with the investing fascia of the muscles composing the abdominal and pelvic walls—the iliacus, the levator ani, the obturator internis, the quadratus, the psoas, the diaphragm, the rectus abdominis, and the transversus abdominis.

The inner stratum of the retroperitoneal connective tissue is the connective tissue immediately adjacent to the peritoneum. This layer of peritoneum persists, even when the serosal surface itself disappears. When the ascending and descending colon come to rest against the parietal peritoneum, the serosal surface of the peritoneum of the dorsal serosal surface of the bowel and the mesentery disappears, leaving only the inner stratum of retroperitoneal connective tissue. The fusion fascia leaves a potential cleavage plane between the inner and intermediate stratum (Hayes, 1950). The intermediate stratum of the retroperitoneal connective tissue is composed of all of the retroperitoneal connective tissue between the transversalis fascia (outer stratum) and the connective tissue of the peritoneum (inner stratum). The intermediate stratum may vary in amount, depending on the obesity of the individual, and in character, depending on the structures it encompasses. Specialized thicknesses of the stratum around the kidney are termed renal (Gerota's) fascia (Tobin, 1944). It is apparent that the renal fascia forms a rather loose common investment of the kidney and the adrenal. The anterior and posterior layers of the renal fascia are fused superiorly and laterally, and contrary to a commonly held view, they are also united medially and inferiorly (Mitchell, 1950). The weakest spot in the renal fascia is in the area of the ureter. In the region of the kidneys, the intermediate stratum, particularly in obese individuals, may be in the form of two lamina with local thickenings of the ventral lamina forming the renal fascia with its contained perirenal fat. The dorsal lamina forms the pararenal fat and the areolar tissue between the ventral lamina and the transversalis fascia. In thinner individuals the intermediate stratum caudal to the region of the kidney will not be laminated. The ureter and the gonadal vessels are also contained within the intermediate stratum. The intermediate stratum is almost nonexistent between the peritoneum and the transversalis fascia investing the diaphragm (Figs 1–12 and 1–13). On the ventral abdominal wall immediately above and lateral to the umbilicus, the fascial strata are so intimately blended because of the paucity of intervening fatty and areolar tissue that the peritoneum appears to be attached to the transversalis fascia. As stated by Tobin and associates (1946): "Such an arrangement does not permit surgical manipulation of the various strata and has no doubt led to the current misunderstanding of structure and continuity of these strata into the dorsal part of the abdomen and into the pelvis—

since many of the original descriptions were made primarily by individuals interested in surgery of the ventral abdominal wall" (Fig 1–14).

In the pelvis the viscera, including the bladder, prostate, vagina, and rectum, are encased within the intermediate stratum of retroperitoneal connective tissue (Fig 1–15).

The inner stratum is found only over the dome of the bladder and in areas in which a serosal surface once existed, that is, around the rectum between the rectum, bladder, and prostate (Denonvilliers' fascia) and between the rectum and vagina. Denonvilliers' fascia or the rectogenital septum has been for years a point of anatomical controversy. Uhlenhuth and associates (1948) from their studies concluded that the rectogenital septum is of peritoneal origin and the result of fusion of the dorsal with the ventral wall of the rectogenital pouch. Tobin and Benjamin (1945) from their studies concluded that the fascia around the rectal musculature should be designated the posterior layer of Denonvil-

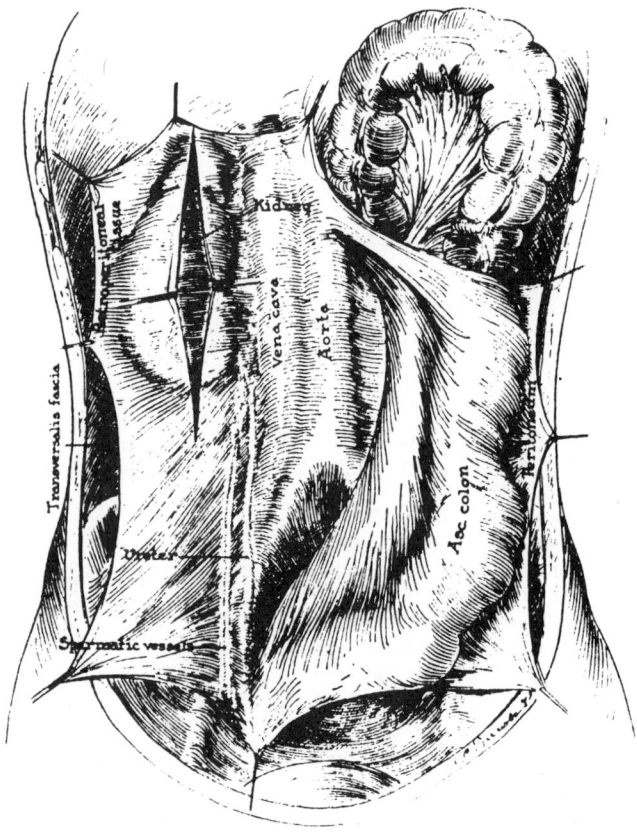

FIG 1–12.
Kidney, ureter, and spermatic vessels contained within intermediate stratum of retroperitoneal connective tissue, which is contiguous over great vessels to the contralateral side. (From Tobin C: The renal fascia and its relation to the transversalis fascia. *Anat Rec* 1944; 89:306. Used by permission.)

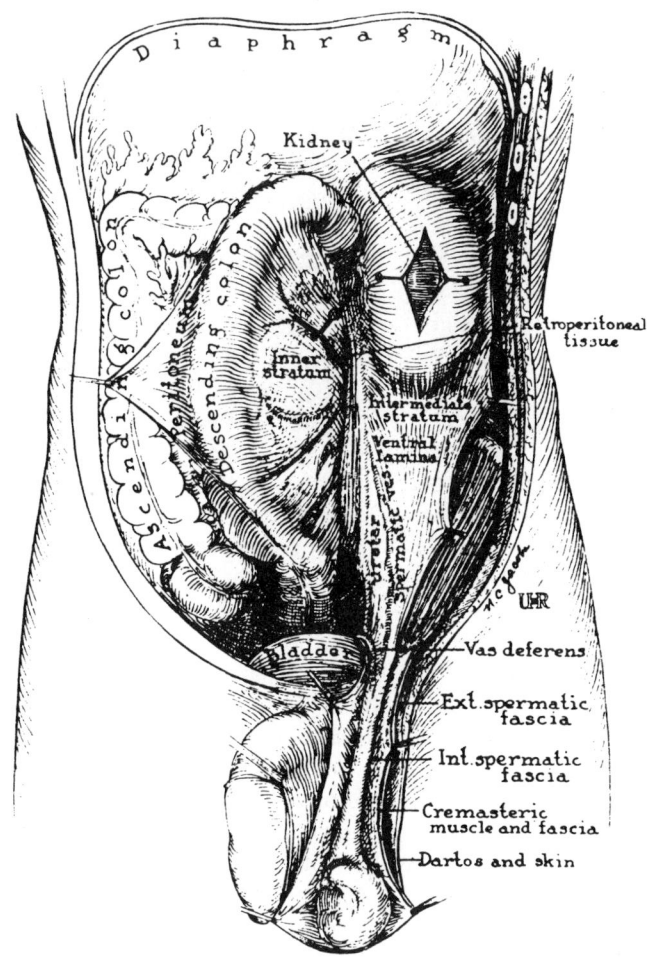

FIG 1–13.
Intermediate stratum extends into depths of scrotum. Note also inner stratum covering descending colon and mesentery. (From Tobin CE, Benjamin JA, Wells JC: Continuity of the fasciae lining the abdomen, pelvis and spermatic cord. *Surg Gynecol Obstet* 1946; 83:586. Used by permission.)

FIG 1–14.
Cross-sectional view of posterolateral abdominal wall and retroperitoneal connective tissue showing potential cleavage planes. (Original drawing by Christine Young, redrawn from Redman JF: An anatomic approach to the kidneys and retroperitoneum in Crawford ED, Borden TA: *Genitourinary Cancer Surgery,* Philadelphia, Lea & Febiger, 1982, p 4. Used by permission.)

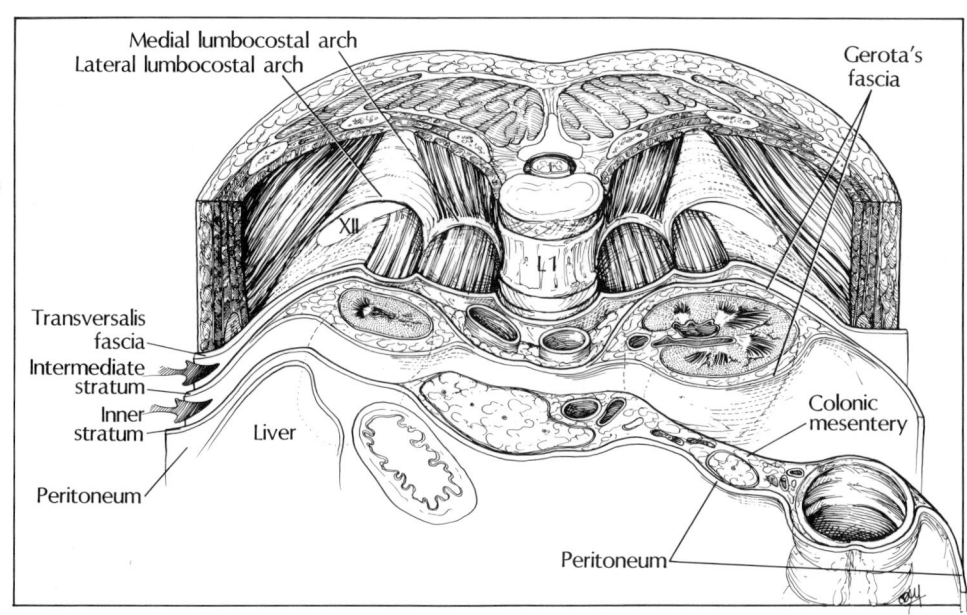

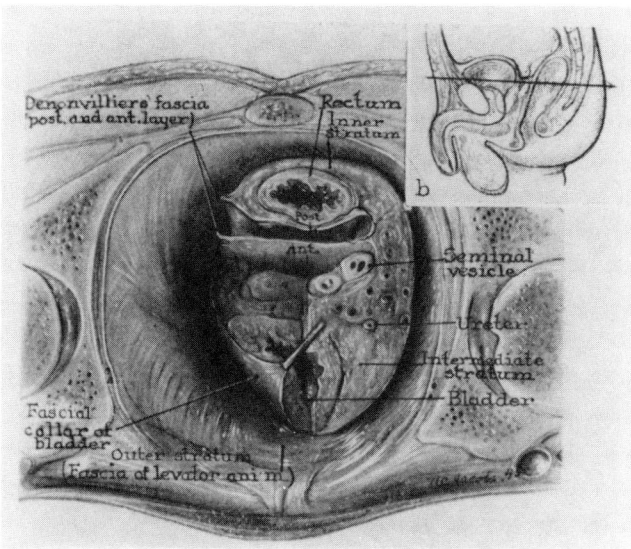

FIG 1–15.
Transverse section through male pelvis showing relationship of pelvic viscera to retroperitoneal connective tissue. (From Tobin CE, Benjamin JA, Wells JC: Continuity of the fasciae lining the abdomen, pelvis and spermatic cord. *Surg Gynecol Obstet* 1946; 83:589. Used by permission.)

liers' fascia and that the fibrous membrane that is derived from the pelvic cul-de-sac of peritoneum be designated the anterior layer of Denonvilliers' fascia. Jewett and associates (1972) in their studies could find no structure that corresponds to the textbook description of Denonvilliers' fascia, only a space that existed between the rectal wall and the prostate filled with areolar tissue containing vessels, nerves, ganglia, and a variable amount of fat.

Where the pelvic viscera passes through the musculature of the pelvic floor, the outer stratum (transversalis fascia) of the retroperitoneal connective tissue surrounds and invests these structures to form a fascial collar, which is also termed the "endopelvic fascia."

The strata of retroperitoneal connective tissue is found also in the layers of the spermatic cord. The fat and fibrous tissue that invests the spermatic vessels and the vas is contiguous with the intermediate stratum found in the pelvis and the upper abdomen. The outer stratum persists as the internal spermatic fascia surrounding the vessels, vas, and patent processus vaginalis, if present.

INCISIONAL ANATOMY

Incisional anatomy is applied anatomy of the body wall. The general classes of incisions include posterior lumbodorsal, flank, thoracoabdominal, anterior abdominal wall (upper), anterior abdominal wall (lower), inguinal, and perineal.

FLANK INCISIONS

The flank or lumbar incision traditionally used has been the lateral subcostal incision. Variations include incisions through the bed of the 12th or 11th rib, through the 11th intercostal space, or the dorsolumbar extrapleural flap incision.

In the lateral subcostal incision the initial muscle exposed is the external oblique and the latissimus dorsi musculature (Fig 1–16). Incision of the latissimus dorsi exposes the serratus posterior inferior muscle in the lateral extremity of the incision and the posterior layer of the lumbodorsal fascia. The subcostal nerve is prominent emerging from beneath the tip of the 12th rib and crossing the field caudally and obliquely. The internal oblique muscle is exposed with incision of the external oblique. With incision of the posterior layer of the lumbodorsal fascia, the quadratus lumborum and the erector spinae (sacrospinalis) muscles are exposed. After incision of the internal oblique, the transversus abdominis and its posterior aponeurosis are exposed. Incision of the transversus and its aponeurosis and the underlying transversalis fascia exposes the intermediate stratum of retroperitoneal connective tissue.

In the 12th or 11th rib resection incisions and the 11th rib intercostal incision, basically the same anatomy is involved, with a few exceptions. With incision of the latissimus dorsi muscle over the 12th or 11th rib, the serratus posterior inferior muscle is noted in the lateral extremity of the incision, and the rib is exposed to its tip. The intercostal nerves do not cross the incision but parallel it. With incision of the posterior periosteum of the rib, following its resection in the 12th rib incision, usually only the underlying diaphragm is exposed except in the lateralmost aspect of the incision. With the 11th rib incision the pleural reflection is generally noted lying between the posterior periosteum and the diaphragm (Fig 1–17). With the intercostal incision the external intercostal muscle and membrane are uncovered with incision of the latissimus dorsi. With incision of the external intercostal, the internal intercostal is exposed. Beneath its investing fascia are found the pleural reflection and the diaphragm. The neurovascular bundle is observed at the undersurface of the 11th rib.

The dorsolumbar flap incision uncovers the same anatomical layers as the 12th rib incision, except that the cranial limb of the incision exposes further the serratus posterior inferior muscle, which, when incised, exposes the tendinous attachment of the sacrospinalis on the lower borders of the ribs at the angles of the 11th and 10th ribs and their intervening intercostal musculature.

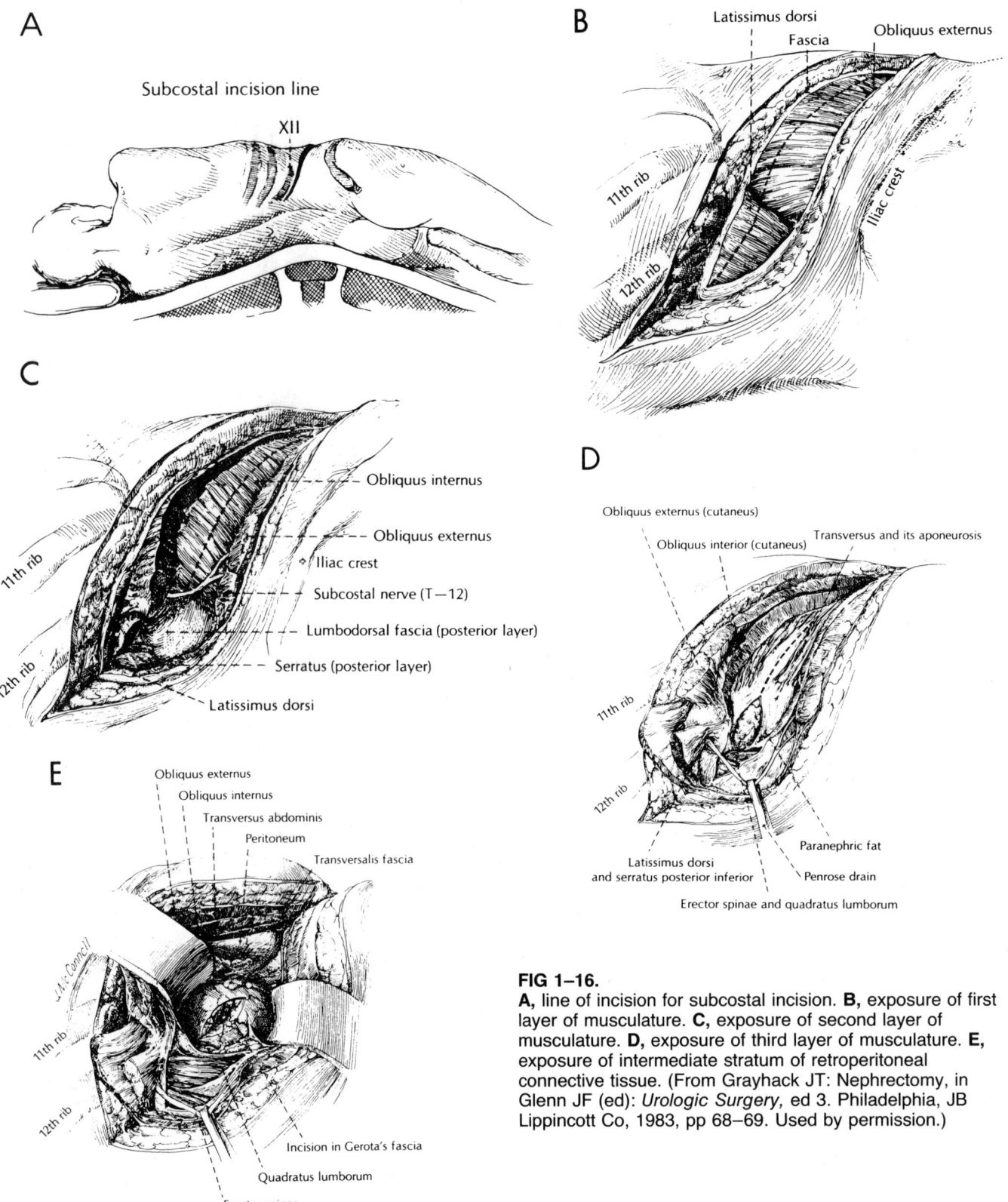

A Subcostal incision line

XII

B Latissimus dorsi
Fascia
Obliquus externus
11th rib
12th rib
Iliac crest

C
11th rib
12th rib
Obliquus internus
Obliquus externus
Iliac crest
Subcostal nerve (T—12)
Lumbodorsal fascia (posterior layer)
Serratus (posterior layer)
Latissimus dorsi

D
Obliquus externus (cutaneus)
Obliquus interior (cutaneus)
Transversus and its aponeurosis
11th rib
12th rib
Latissimus dorsi
and serratus posterior inferior
Erector spinae and quadratus lumborum
Paranephric fat
Penrose drain

E
Obliquus externus
Obliquus internus
Transversus abdominis
Peritoneum
Transversalis fascia
11th rib
12th rib
Incision in Gerota's fascia
Quadratus lumborum
Erector spinae

FIG 1–16.
A, line of incision for subcostal incision. **B,** exposure of first layer of musculature. **C,** exposure of second layer of musculature. **D,** exposure of third layer of musculature. **E,** exposure of intermediate stratum of retroperitoneal connective tissue. (From Grayhack JT: Nephrectomy, in Glenn JF (ed): *Urologic Surgery,* ed 3. Philadelphia, JB Lippincott Co, 1983, pp 68–69. Used by permission.)

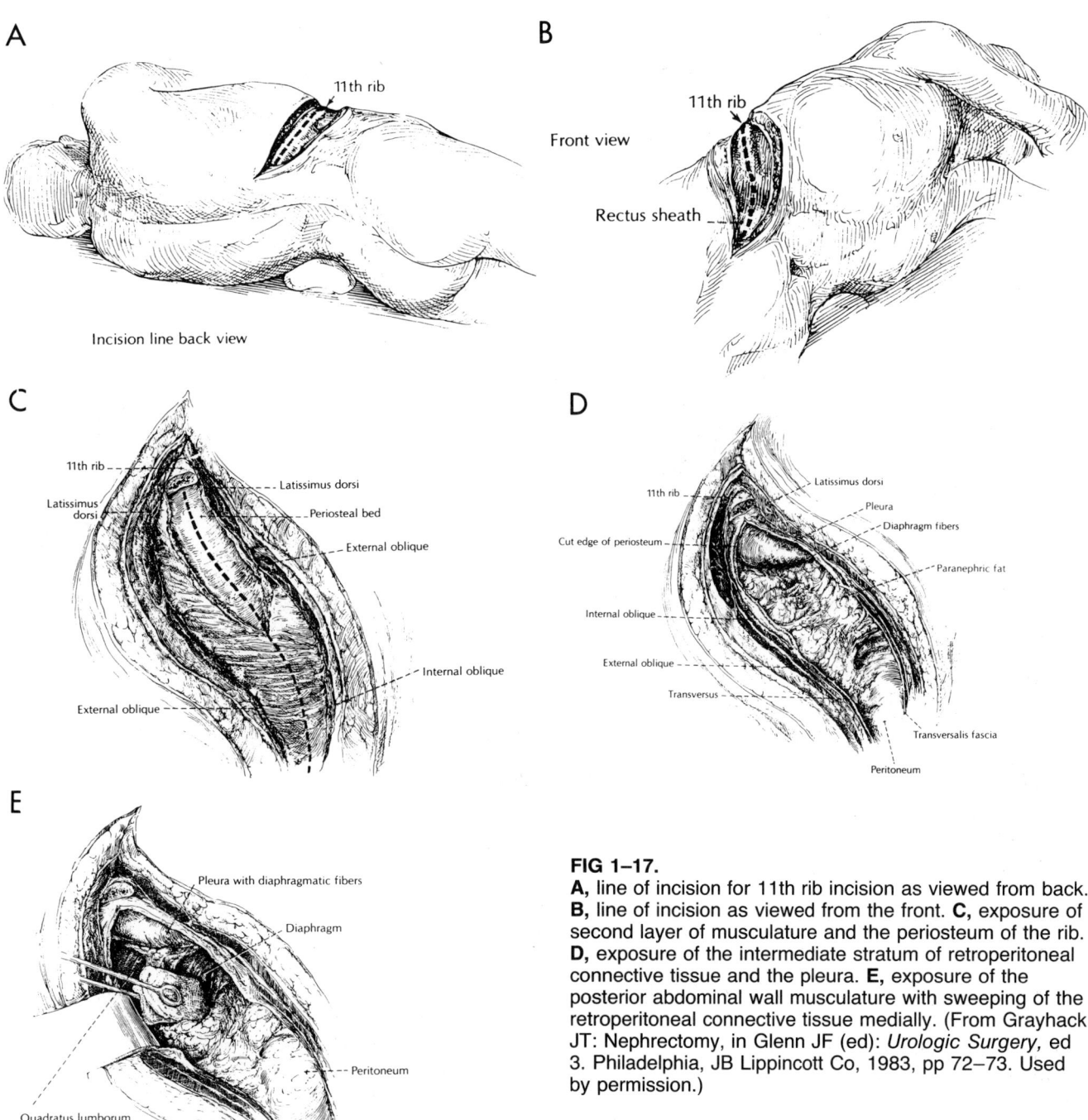

FIG 1–17.
A, line of incision for 11th rib incision as viewed from back. **B,** line of incision as viewed from the front. **C,** exposure of second layer of musculature and the periosteum of the rib. **D,** exposure of the intermediate stratum of retroperitoneal connective tissue and the pleura. **E,** exposure of the posterior abdominal wall musculature with sweeping of the retroperitoneal connective tissue medially. (From Grayhack JT: Nephrectomy, in Glenn JF (ed): *Urologic Surgery,* ed 3. Philadelphia, JB Lippincott Co, 1983, pp 72–73. Used by permission.)

With resection of a portion of each rib just medial to their angles, the flap is raised, exposing the underlying pleura (Fig 1–18).

THORACOABDOMINAL INCISIONS

The thoracoabdominal incision is a transpleural transthoracic incision which usually includes a 10th or 11th rib resection. The incision may be carried medially and transversely or extended in a more caudal fashion. The incision is much like an 11th rib resection flank incision, with the exception that the pleura is purposely incised, as is the underlying diaphragm, to gain access to the peritoneal cavity. The incision is carried to the rectus sheath, which is incised along with the rectus abdominis muscle either to the linea alba or across into

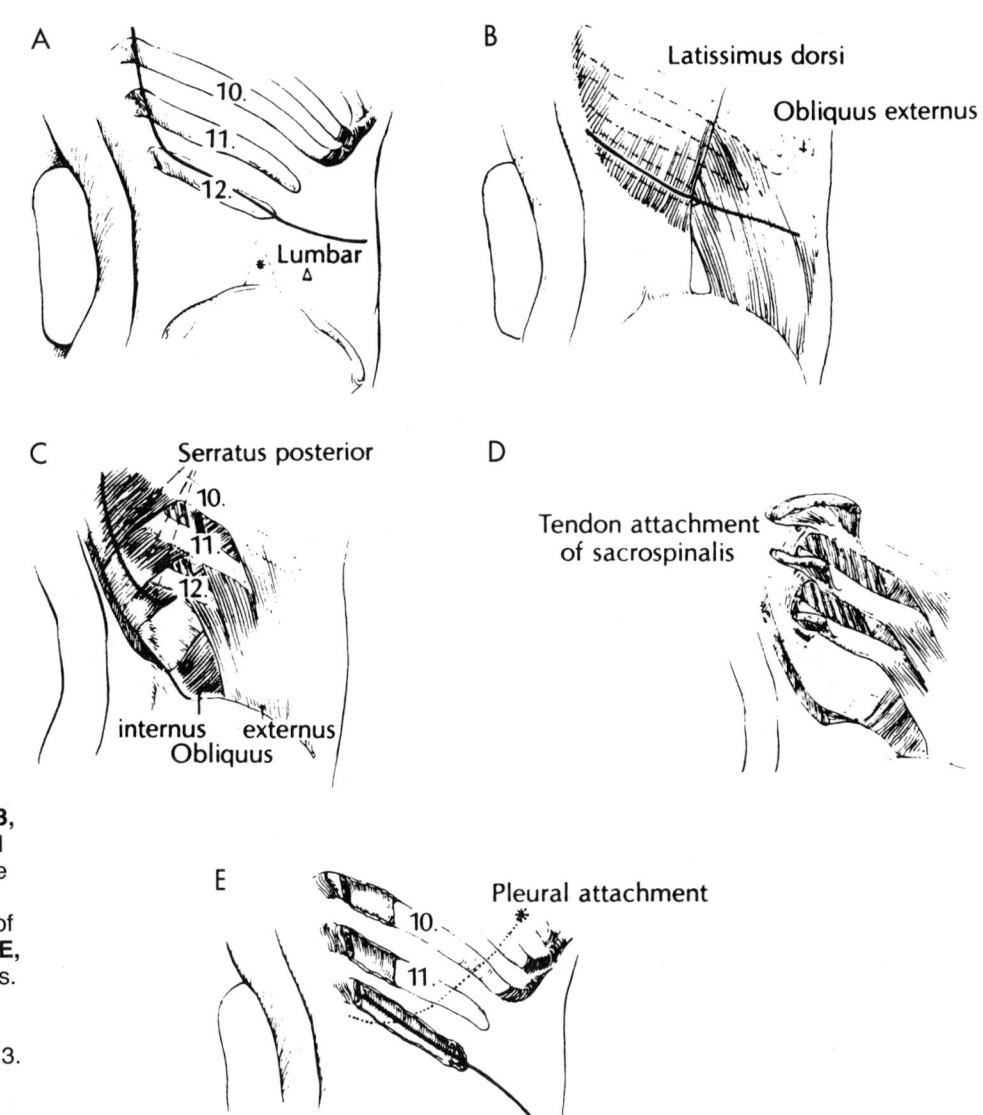

FIG 1–18.
A, line of incision for dorsolumbar flap incision. **B,** superficial musculature and line of incision. **C,** exposure of second layer of musculature. **D,** exposure of third layer of musculature. **E,** relationship of pleura to ribs. (From Grayhack JT: Nephrectomy, in Glenn JF (ed): *Urologic Surgery,* ed 3. Philadelphia, JB Lippincott Co, 1983, p 76. Used by permission.)

the contralateral rectus sheath and the underlying rectus (Fig 1–19).

ANTERIOR ABDOMINAL WALL (UPPER) INCISIONS

The anterior abdominal wall may be incised transversely in a paramedian fashion or in the midline. If confined to one side, the incision may be retroperitoneal.

The transverse upper abdominal incision is as described with the abdominal portion of the thoracoabdominal incision. If the peritoneum is not be entered, the freeing of the intermediate stratum of retroperitoneal connective tissue from the overlying transversalis fascia is carried out, as is done in the extraperitoneal flank incision.

In the paramedian incision the line of incision is made a few centimeters lateral to the midline. The anterior rectus sheath is incised, exposing the rectus abdominis muscle, which is retracted laterally, exposing the posterior rectus sheath. On incision of the posterior rectus sheath the peritoneum is exposed.

The midline incision requires only the incision of the linea alba to gain access to the peritoneal cavity.

ANTERIOR ABDOMINAL WALL (LOWER) INCISIONS

Lower anterior abdominal wall incisions are basically of three types: lateral abdominal wall, transverse, and midline.

The lateral abdominal wall incisions are usually oblique, curvilinear incisions which parallel the iliac crest and the inguinal ligament. Anatomically they are simple in that the first level of musculature exposed is

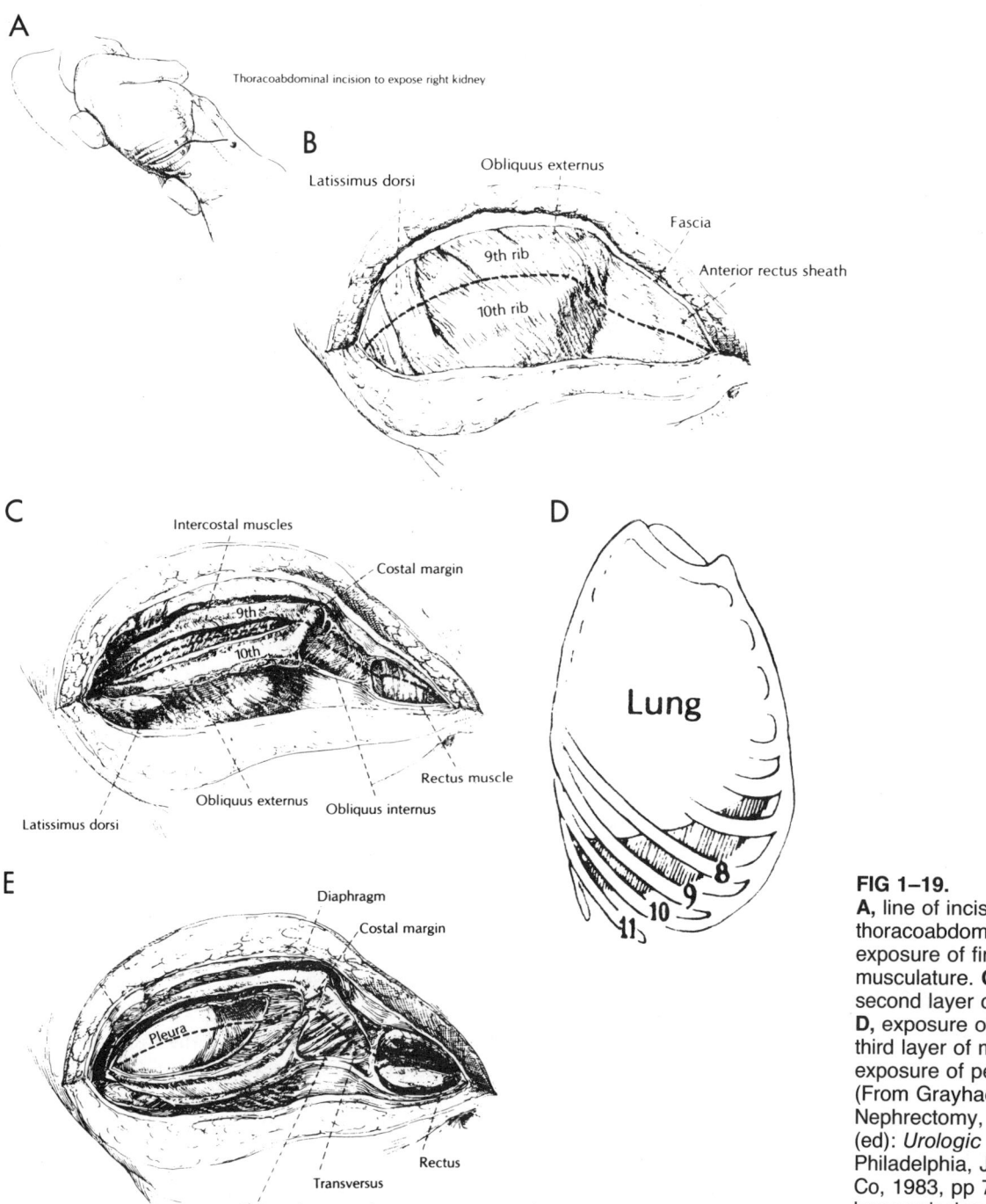

A

Thoracoabdominal incision to expose right kidney

B

Obliquus externus

Latissimus dorsi

Fascia

Anterior rectus sheath

9th rib

10th rib

C

Intercostal muscles

Costal margin

9th

10th

Rectus muscle

Latissimus dorsi

Obliquus externus Obliquus internus

D

Lung

8

9

10

11

E

Diaphragm

Costal margin

Pleura

Rectus

Transversus

Obliquus internus and externus

FIG 1–19.
A, line of incision for thoracoabdominal incision. **B,** exposure of first layer of musculature. **C,** exposure of second layer of musculature. **D,** exposure of pleura and third layer of musculature. **E,** exposure of peritoneum. (From Grayhack JT: Nephrectomy, in Glenn JF (ed): *Urologic Surgery,* ed 3. Philadelphia, JB Lippincott Co, 1983, pp 78–79. Used by permission.)

the external oblique and its aponeurosis, which may be sharply incised or separated in the direction of its fibers to expose the underlying internal oblique and its less extensive medial aponeurosis. The internal oblique may be either incised or separated to expose the transversus abdominis and its aponeurosis. An incision through its investing fascia, the transversalis fascia, exposes the intermediate stratum of retroperitoneal connective tissue.

The transverse lower abdominal incision (Pfannen-stiel's incision) and the midline lower abdominal incision differ anatomically essentially only in the direction of the incision of the skin, subcutaneous tissue and the anterior rectus sheath. The transverse incision is generally made below the arcuate line (semilunar line of Douglas). With incision of the anterior rectus sheath and reflection of the fascia cranially and laterally, the rectus abdominis muscle is exposed. In the most caudal aspect of the wound the pyramidalis is exposed. With separation of the rectus and the pyramidalis in the midline the transversalis fascia is exposed. An incision above the level of the arcuate line will expose the posterior rectus sheath (Fig 1–20). In older individuals the transversalis fascia and the underlying intermediate stratum of retroperitoneal connective tissue and frequently the peritoneum are melded together for several centimeters on either side of the midline. Therefore, an incision near the lateral border of the rectus muscle into the transversalis fascia will expose the intermediate stratum of retroperitoneal connective tissue as a separate entity. Just beneath the symphysis the melding is generally not evident. The transversalis may be incised in the midline along with the underlying peritoneum to gain access to the peritoneal cavity or the retroperitoneal connective tissue may be swept cranially off of the bladder, carrying the peritoneum with it.

INGUINAL INCISIONS

Inguinal incisions may be made transversely in the infra-abdominal crease or fold or may be made obliquely, paralleling the inguinal ligament. In the inguinal region the subcutaneous tissue is distinguished by the fascia-like character of the superficial connective tissue; the deeper whitish portion exposed is Scarpa's fascia. Incision of Scarpa's fascia exposes the loose fat-laden tissue, which when incised exposes the external oblique aponeurosis and its extension, covering the cord past the external ring, the external spermatic fascia. In the subcutaneous tissue are encountered two sets of vessels coursing from caudal to cranial and paralleling each other roughly. The medially placed vessels encountered are the superficial inferior epigastric artery and vein. The laterally placed vessels are the superficial circumflex iliacs.

With incision of the external oblique aponeurosis in the direction of its fibers, is exposed the internal oblique muscle and covering the cord structures the cremasteric muscle and fascia. Paralleling the inguinal canal are two distinctive nerves. The most cranial is the iliohypogastric. The most caudal is the ilioinguinal. The cremasteric muscle and fascia may be incised over the cord structures or along its intersection with the iliopubic tract. The cremasteric thus incised exposes the cord covered by internal spermatic fascia emerging from the internal ring. Beneath the cord is exposed the transversalis fascia of the inguinal floor. Perforating the floor, also through the internal ring or just caudal, are the external spermatic or cremasteric vessels and the accompanying genital branch of the genitofemoral nerve. Beneath the internal spermatic fascia lies the cord vasculature, the vas, and the patent processus vaginalis, if present, contained within the intermediate stratum of retroperitoneal connective tissue. Incision of

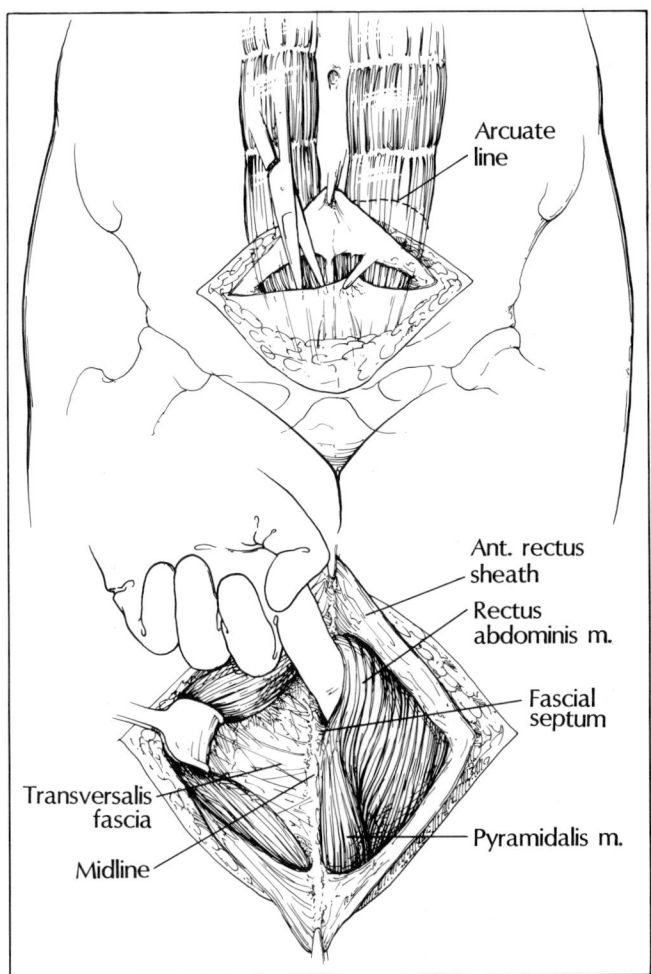

FIG 1–20.
The Pfannenstiel incision. The anterior rectus sheath has been incised transversely *(upper)*. (Original drawing by Christine Young.)

the transversalis fascia also exposes the intermediate stratum and at the medial aspect of the internal ring the inferior epigastric vessels. Sweeping of the intermediate stratum cranially exposes the external iliacs and femoral nerve and the pectineal line (Cooper's ligament).

PERINEAL INCISIONS

A curvilinear perineal incision exposes initially the subcutaneous connective tissue, Colles' fascia. Dissection on either side of the midline enters the wedge-shaped ischiorectal fossa between the obturator internis muscle and the levator ani muscle. The musculature exposed will depend on the cranial extent of the curvilinear incision and may include the levator ani, external anal sphincter, superficial transverse perineal, and bulbocavernosus. With incision of the attachments of the external anal sphincter from the perineal body, the rec-

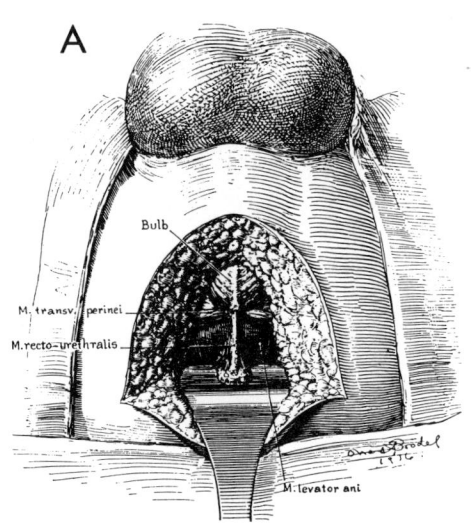

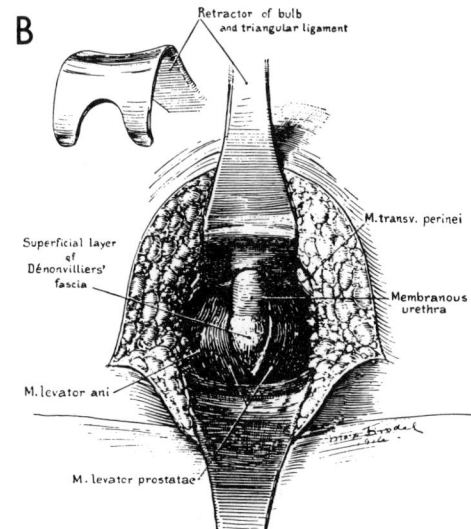

FIG 1–21.
A, exposure of rectourethralis muscle; **B,** exposure of Denonvilliers' fascia and membranous urethra following incision of rectourethralis muscle. (From Young H, Davis DM: *Young's Practice of Urology,* Philadelphia, WB Saunders Co, 1926, vol 2, p 519. Used by permission.)

tourethralis muscle is exposed. With its incision the space between the levator ani muscles is exposed with the underlying retroperitoneal connective tissue (Denonvilliers' fascia). Incision of the retroperitoneal tissue exposes the prostate and the rectum (Fig 1–21).

POSTERIOR INCISIONS

Posterior incisions parallel the sacrospinalis (erector spinae) musculature, either exactly or obliquely. Following incision of the skin and subcutaneous tissue, the first layer of musculature encountered is the latissimus dorsi and its aponeurosis (lumbodorsal fascia). With incision of the latissimus dorsi and its aponeurosis are exposed, in the cranial portion of the wound, the serratus posterior inferior muscle and the quadratus lumborum. Following incision of the serratus posterior inferior, the quadratus is retracted medially to expose the transversalis fascia, which, when excised, exposes the intermediate stratum of retroperitoneal connective tissue (Fig 1–22).

CROSS-SECTIONAL ANATOMY

Although the study of cross-sectional anatomy has been employed for years as a further way of understanding structure and relationships, the increasing use of imaging modalities which show the body in cross-section has made knowledge of this aspect of anatomy mandatory. A great many texts employing illustrations of cross-sectional anatomy and photographs of actual cross-sections of the body are now available for study in most medical libraries. For the urologist the cross-sections of most interest are those from T12 to the pubic ramus (Figs 1–23 through 1–33).

ORGANOLOGY

ADRENAL GLAND

Gross Anatomy

The adrenal or suprarenal glands are paired organs that differ from each other. In common, they are relatively flattened and have a lightly cobblestoned texture. The cadmium yellow coloration is distinctive and contrasts readily against the surrounding fat or pancreatic tissue. The right adrenal is triangular or pyramidal. The left adrenal is more flattened and crescentic or palm-leafed. Adrenal glands weigh 4–8 gm each, the left being slightly larger than the right (Cerny, 1977). Each gland has a concave aspect that represents the site of association with the anterocraniomedial aspect of the kidney. A prominent adrenal vein is conspicuous as it emerges from each gland. On cut section or on fracture, each gland is noted to have a completely encased medulla of brown coloration, which contrasts with the yellow of the cortex.

Dobbie and Symington (1966), following extensive study of the adrenal utilizing three-dimensional reconstruction models, concluded that many of the descriptive terms applied to the adrenal glands were uninformative and suggested new terms. As they described, the anterior surface of both glands is flat and has a shallow groove, the anterior groove, from which emerges the adrenal vein. On the posterior surface of both glands there is a ridgelike elevation or crest, which increases in prominence as it approaches the lateral tip of the gland. The crest is flanked by two winged portions—the ala; the linear central drainage region found in the ala between the two opposing layers of cortex where no medulla is present is referred to as the alar

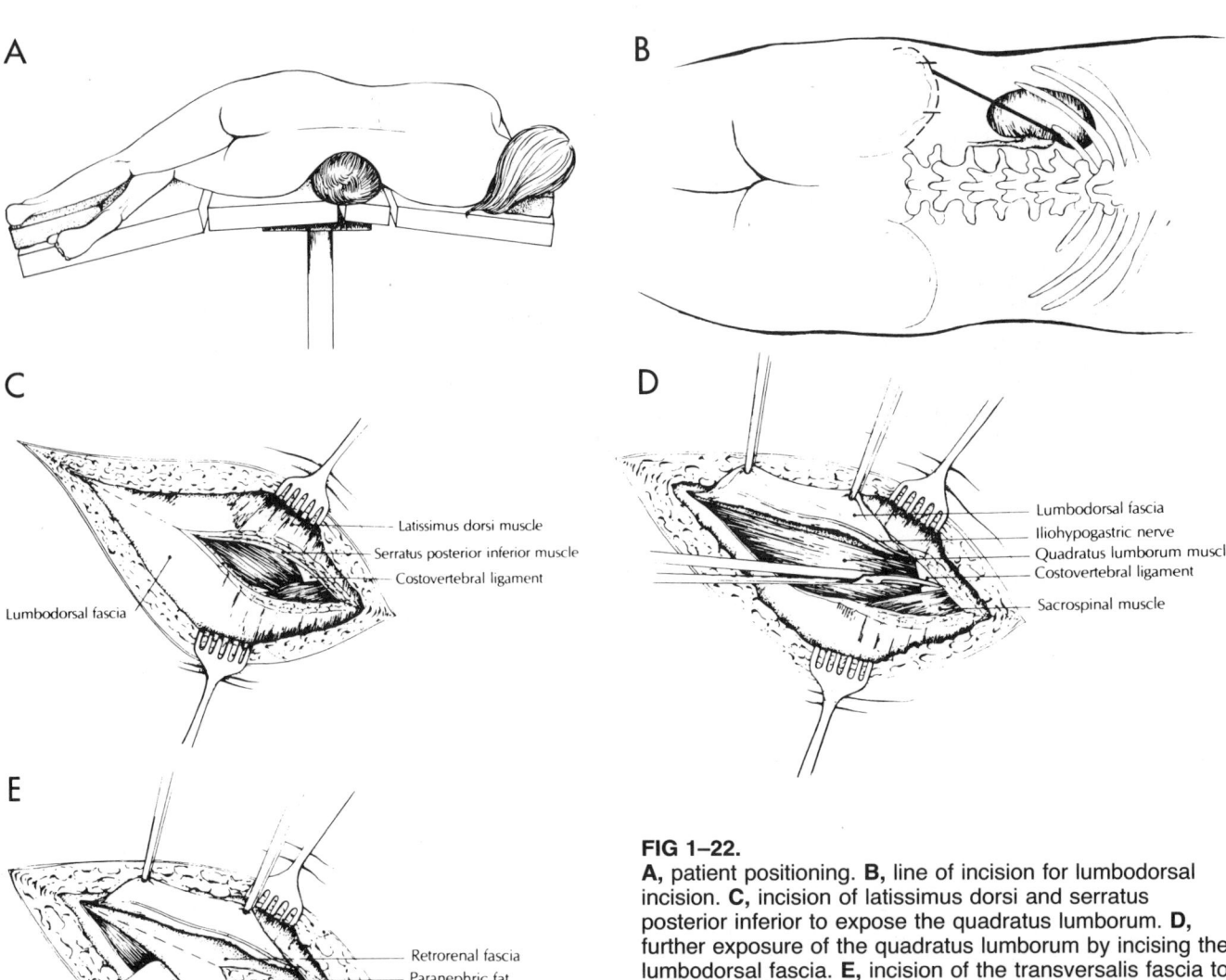

FIG 1–22.
A, patient positioning. **B,** line of incision for lumbodorsal incision. **C,** incision of latissimus dorsi and serratus posterior inferior to expose the quadratus lumborum. **D,** further exposure of the quadratus lumborum by incising the lumbodorsal fascia. **E,** incision of the transversalis fascia to expose the intermediate stratum of retroperitoneal connective tissue. (From Lutzeyer W: Lumbodorsal surgery, in Glenn JF (ed): *Urologic Surgery* ed 3. Philadelphia, JB Lippincott Co, 1983, pp 304–306. Used by permission.)

raphe. They noted that because of the definite distribution of medullary tissue within the gland, the adrenal had a tripartite structure and could be divided into three segments—head, body, and tail. The head anatomically is the most medial, the tail being in the most lateral position.

Relationships

Each adrenal gland is contained within the intermediate stratum of retroperitoneal connective tissue within the specialization of the connective tissue termed Gerota's fascia. Along its border are noted numerous arteries, which are generally small calibered.

On the right side the anteromedial aspect of the adrenal lies under the posterolateral wall of the inferior vena cava. The anterocranial portion of the right adrenal is bordered by the bare right lobe of the liver. The right adrenal is bordered posteriorly by the diaphragm. Inferiorly it is in association with the anterocranial aspect of the right kidney. On occasion the curvature of the duodenum or the renal vasculature will cover a portion of the anterior surface of the right adrenal (Fig 1–34).

On the left side the gland is distinguished by the large adrenal vein, which is joined by the left phrenic vein. The left adrenal is crossed on its anteroinferior border by the body of the pancreas and the splenic artery and vein. The cranioanterior surface is covered by

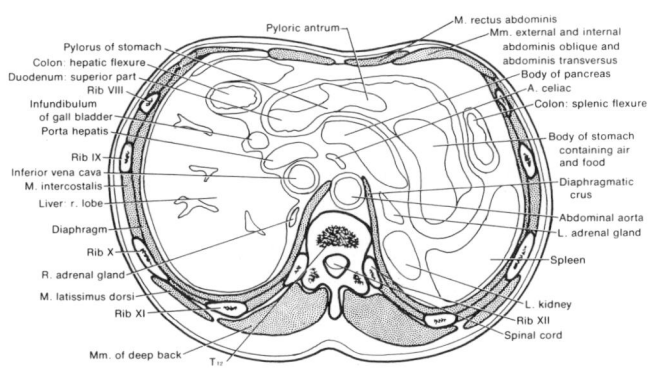

FIG 1–23.
Cross-section at T12 level. (From Ledley RS, Huang HK, Mazziotta JC: *Cross-sectional Anatomy—an Atlas for Computerized Tomography.* Baltimore, Williams & Wilkins Co, 1977, p 146. Used by permission.)

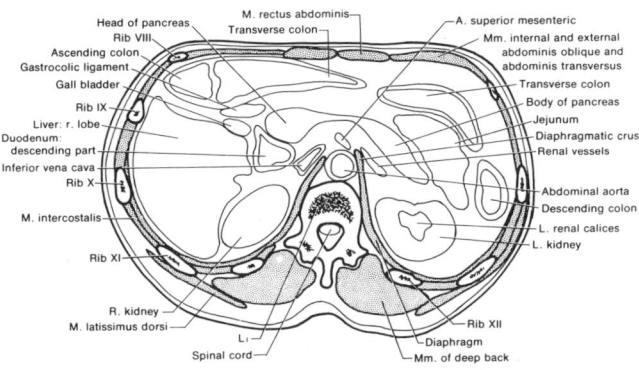

FIG 1–24.
Cross-section at L1 level. (From Ledley RS, Huang HK, Mazziotta JC: *Cross-sectional Anatomy—an Atlas for Computerized Tomography.* Baltimore, Williams & Wilkins Co, 1977, p 150. Used by permission.)

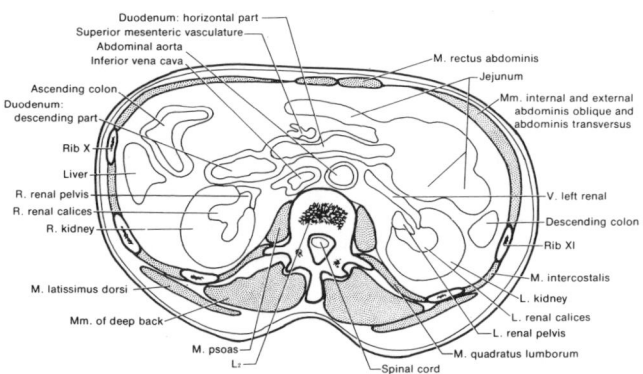

FIG 1–25.
Cross-section at L2 level. (From Ledley RS, Huang HK, Mazziotta JC: *Cross-sectional Anatomy—an Atlas for Computerized Tomography.* Baltimore, Williams & Wilkins Co, 1977, p 156. Used by permission.)

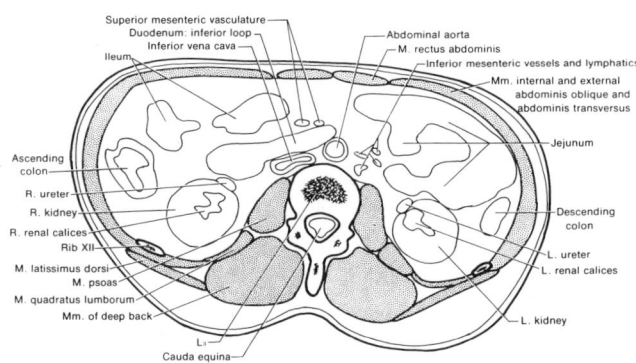

FIG 1–26.
Cross-section from L3 level. (From Ledley RS, Huang HK, Mazziotta JC: *Cross-sectional Anatomy—an Atlas for Computerized Tomography.* Baltimore, Williams & Wilkins Co, 1977, p 162. Used by permission.)

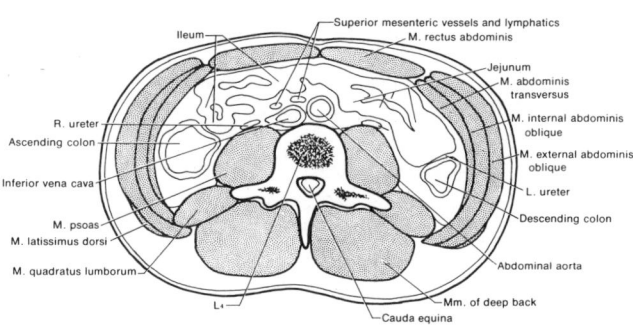

FIG 1–27.
Cross section from L4 level. (From Ledley RS, Huang HK, Mazziotta JC: *Cross-sectional Anatomy—an Atlas for Computerized Tomography.* Baltimore, Williams & Wilkins Co, 1977, p 168. Used by permission.)

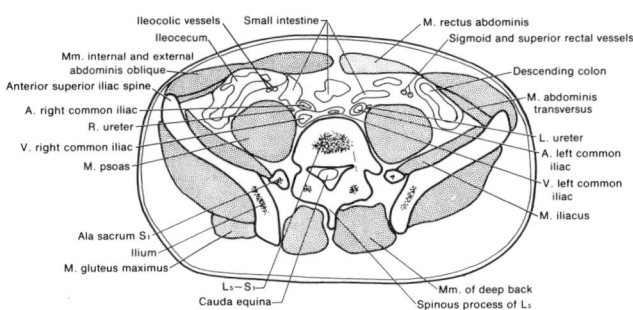

FIG 1–28.
Cross-section from L5 and S1 level. (From Ledley RS, Huang HK, Mazziotta JC: *Cross-sectional Anatomy—an Atlas for Computerized Tomography.* Baltimore, Williams & Wilkins Co, 1977, p 184. Used by permission.)

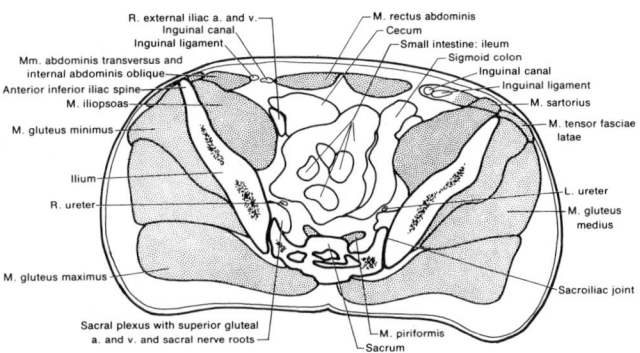

FIG 1–29.
Cross-section from midsacral level. (From Ledley RS, Huang HK, Mazziotta JC: *Cross-sectional Anatomy—an Atlas for Computerized Tomography.* Baltimore, Williams & Wilkins Co, 1977, p 196. Used by permission.)

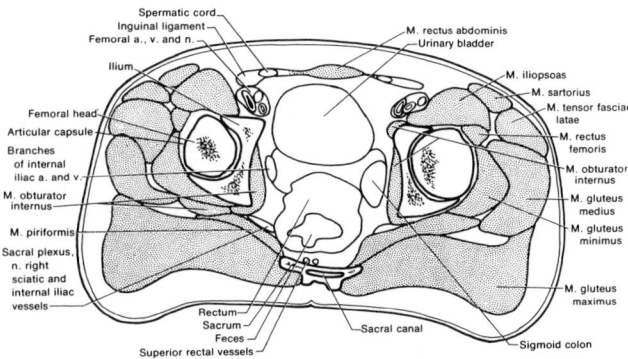

FIG 1–30.
Male pelvis. Cross-section through superior border of hip joint. (From Ledley RS, Huang HK, Mazziotta JC: *Cross-sectional Anatomy—an Atlas for Computerized Tomography.* Baltimore, Williams & Wilkins Co, 1977, p 206. Used by permission.)

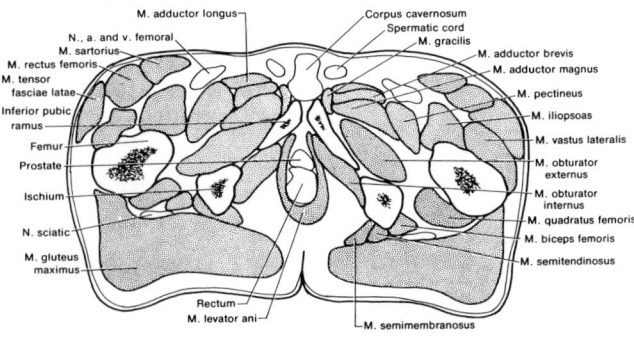

FIG 1–31.
Cross-section through male pelvis at the inferior pubic ramus. (From Ledley RS, Huang HK, Mazziotta JC: *Cross-sectional Anatomy—an Atlas for Computerized Tomography.* Baltimore, Williams & Wilkins Co, 1977, p 218. Used by permission.)

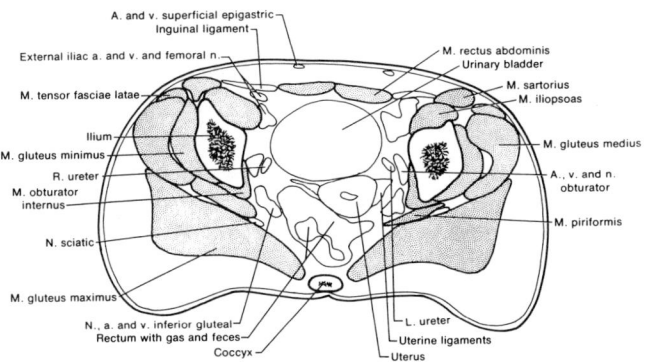

FIG 1–32.
Cross-section of female pelvis at the superior border of hip joint. (From Ledley RS, Huang HK, Mazziotta JC: *Cross-sectional Anatomy—an Atlas for Computerized Tomography.* Baltimore, Williams & Wilkins Co, 1977, p 242. Used by permission.)

the peritoneum of the omental bursa which separates it from the cardiac portion of the stomach. The posterior aspect of the left adrenal is in contact with the diaphragm and its crus, while the posterolateral aspect is associated with the craniomedial aspect of the left kidney.

Vasculature

The blood supply to the adrenal arises from three main groups of arteries: a superior group from the inferior phrenic artery and its branches, a middle group from the aorta, and an inferior group from the renal artery. The numerous branches arising from each group of arteries divide and redivide as they approach the adrenals so that as many as 50 small branches may invest the gland (Dobbie and Symington, 1966; Anson and Daseler, 1961).

The small branches of the adrenal arteries which

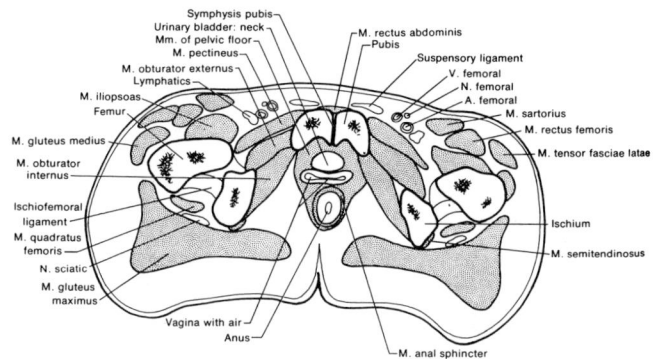

FIG 1–33.
Cross-section of female pelvis at the superior portion of the symphysis pubis. (From Ledley RS, Huang HK, Mazziotta JC: *Cross-sectional Anatomy—an Atlas for Computerized Tomography.* Baltimore, Williams & Wilkins Co, 1977, p 250. Used by permission.)

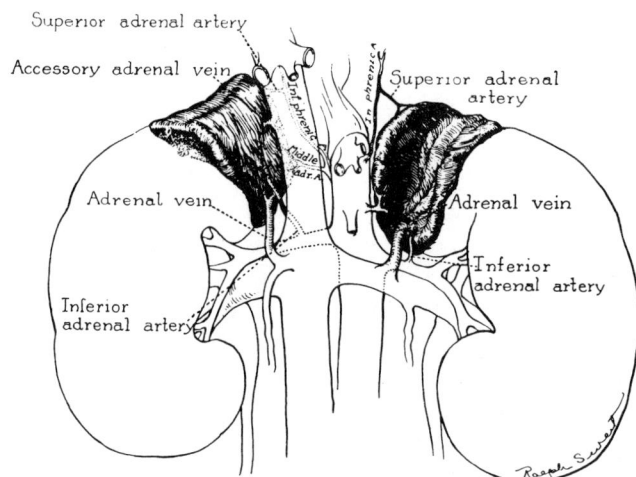

FIG 1–34.
Relation of adrenal glands to kidneys and great vessels. (From Hinman F, *Principles and Practice of Urology.* Philadelphia, WB Saunders Co, 1935, p 250. Used by permission.)

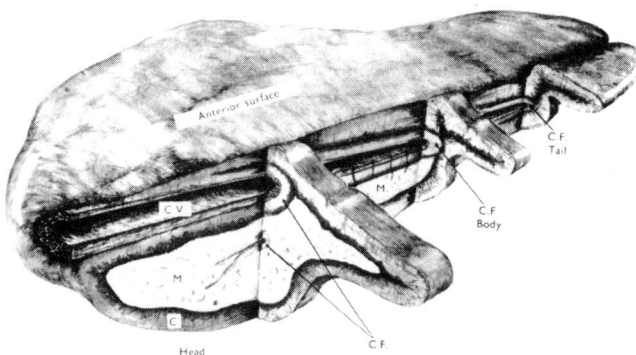

FIG 1–35.
A diagrammatic representation of a left adrenal from which segments have been removed to show the course of the central vein with its enveloping cortical cuff. *C.,* cortex; *C.F.,* cortical cuff; *C.V.,* central vein; *M.,* medulla. (From Dobbie JW, Symington T: The human adrenal gland with special reference to the vasculature. *J Endocrinol* 1966; 34:490 (facing), plate 4. Used by permission.)

ramify over the surface of the adrenal diverge to form a thin subcapsular plexus from which straight capillaries pass down through the zona fasciculata. In the zona reticularis they join to form a rich plexus, which ends abruptly at the corticomedullary junction to form a corticomedullary vascular dam. This rich vascular plexus (plexus reticularis) is drained by relatively few channels that pass into the medulla to join the venous sinuses (Dobbie and Symington, 1966).

As stated by Dobbie and Symington (1966) the adrenal venous system is remarkable for the complex structure of its muscular veins. The central vein carries with it into the gland an invaginated cuff of cortical tissue, which envelops it throughout its entire length and merges imperceptibly with the cortex in the tail (Fig 1–35). Cuffs of cortical cells encompass all main branches of the central vein, and only in the head region do small muscular venous radicals lie free in the medullary tissue unencompassed by a cortical sleeve. The central vein and its various branches drain all of the adrenal (Fig 1–36). In contrast to the great profusion of adrenal arteries, both adrenals have only one main vein. On the right side it is quite short and opens directly into the inferior vena cava on the posterolateral aspect. The left vein is much longer and receives on its course to the renal vein the inferior phrenic vein and branches from the capsule.

Innervation

Relative to its size the adrenal gland has among organs one of the richest nerves supplies. The medulla only is innervated and the origin of innervation is pri-

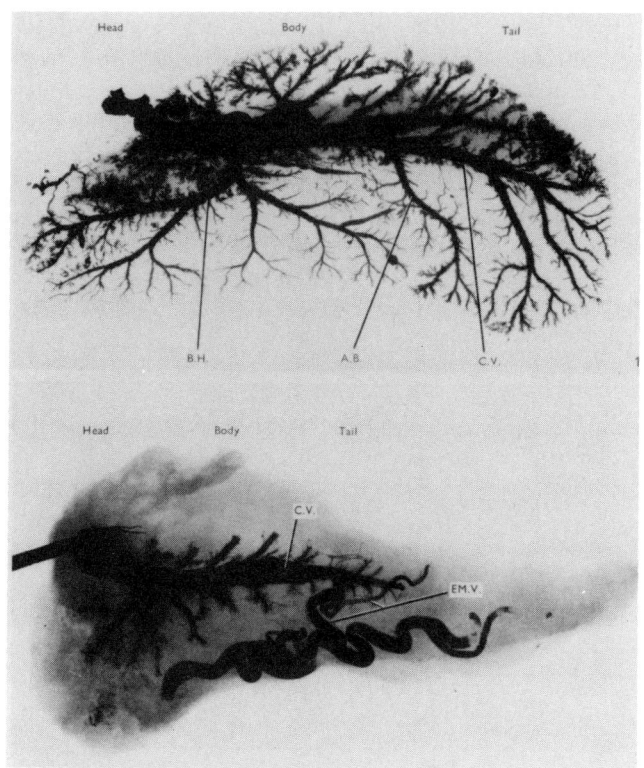

FIG 1–36.
A venogram of a left adrenal showing the distribution of the central vein and its branches. (From Dobbie JW, Symington T: The human adrenal gland with special reference to the vasculature. *J Endocrinol* 1966; 34:490 (facing), plate 5. Used by permission.)

marily of sympathetic derivation (Cerny, 1977). Sympathetic preganglionic fibers arise from the T10–L1 spinal cord segments carried by the greater splanchnic nerve and the celiac ganglion and terminate in synapses in continuity with large chromaffin cells within the renal medulla.

Lymphatics

The lymphatic drainage of the adrenal gland usually follows the course of the adrenal vein. The ultimate drainage is to the lateral aortic lymph nodes.

Microscopic Anatomy

The adrenal gland on cross-section is composed of vastly differing portions, as if there were an organ inside another, the inner organ being the medulla and the outer organ the cortex. The cortex is surrounded by a thick, collagenous capsule from which trabeculae pass to varying depths into the cortex.

Histologically, the cortex is seen to be composed of three zones of cells termed, from cortex to medulla, the zona glomerulosa, the zona fasciculata, and the zona reticularis. The zona glomerulosa consists of small, polyhedral cells arranged in rounded groups or curved columns (Gray, 1980d). Just internal to the zona glomerulosa is the zona fasciculata, which is broader and consists of large, polyhedral cells, that are arranged in straight columns two cells thick with fenestrated venous sinusoids coursing parallel and between the columns. The innermost layer of the cortex is termed the "zona reticularis," which consists of branching and anastomosing columns of rounded cells.

The adrenal medulla is composed of groups and columns of chromaffin cells or pheochromocytes with wide venous sinusoids permeating between them. Small groups of nerve cells or single nerve cells are scattered throughout the medulla.

KIDNEY

Gross Anatomy

The kidneys are paired, reddish brown, bean-shaped organs with a convex outer border and a concave inner border. Each kidney is approximately 10–11 cm long, 5–6 cm wide, and 2.5–3 cm thick. In man the renal weight is about 150 gm and in women 135 gm. Along the medial borders of the kidneys the contour is broken by a deep depression, the renal hilum, which opens into the renal sinus, which is surrounded by renal parenchyma. Through the renal hilum enters the renal vasculature and exits the renal pelvis.

On longitudinal sectioning of the kidney grossly it is evident that the renal parenchyma forms a crescent around the renal collecting structures, the interposing spaces being filled with fat and vasculature (Fig 1–37).

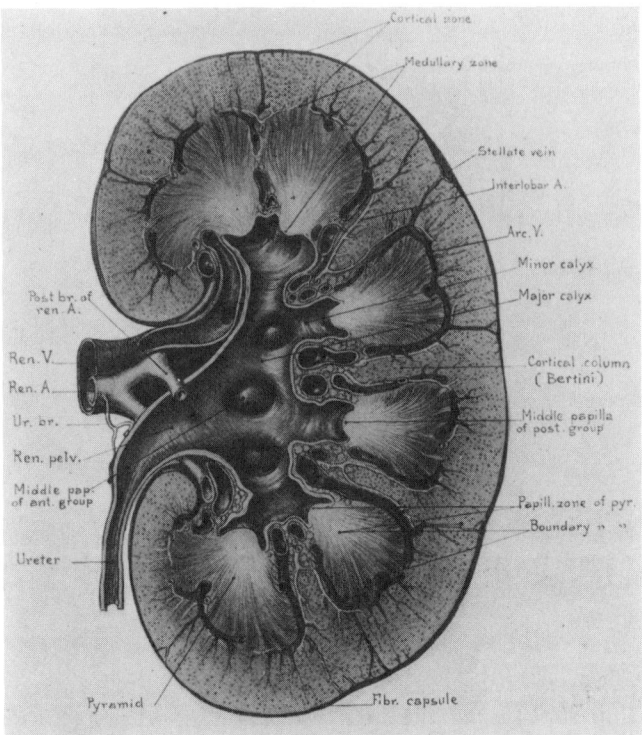

FIG 1–37.
Left kidney: longitudinal section showing major structural detail. Fat and vasculature have been removed. (From Kelly HA, Burnam CF: *Diseases of the Kidneys, Ureters and Bladder.* New York, Appleton-Century-Crofts, 1922, vol 1, p 151. Used by permission.)

The renal parenchyma is covered with a thin, whitish capsule which strips off easily except where perforated by capsular vessels. The substance of the kidney is noted to be composed of an inner medulla and an inner cortex. Approximately 1 cm beneath the capsule are noted on longitudinal section pyramidal-shaped, pale, stained masses which are termed the "renal pyramids" or "medulla." Between the pyramids, the granular-appearing cortical substance extends to the renal sinus as columns of Bertin. The tip of each pyramid is rounded and fits into a minor calyx. A tip is termed a renal papilla. Each pyramid, capped with cortical substance, is termed a lobe of the kidney, which range in number from four to 19 (Inker, 1981). Where lobes of the kidney join and the renal columns interface is termed the cortical arches or cortical lobules.

Relationships

The kidney is embedded in the intermediate stratum of retroperitoneal connective tissue (Tobin et al., 1946). The connective tissue may be in the form of two lamina, particularly in obese individuals. Local thickenings of the ventral lamina form the renal fascial layer of Gerota and the perirenal fat. The dorsal lamina forms the pararenal fat and the areolar tissue between the ventral lamina and the transversalis fascia. The amount of fat covering the ventral surface of the kidney is rather scant and patchy, whereas the amount of fat dorsally is substantial.

Because of the surrounding fibrofatty and areolar tissue, the kidneys are not in direct contact with any other structure. However, their position in the retroperitoneum is bordered dorsally by the osteomuscular wall and ventrally by the abdominal viscera. The relationship of the parietes and the viscera of the kidneys is best illustrated diagrammatically (Figs 1–38, 1–39).

Vasculature

The blood supply to the kidney generally arises as a single, large artery from the lateral aspect of the aorta just caudal to the midline origin of the superior mesenteric artery. By virtue of the position of the aorta, the right renal artery is longer. The origin of the left renal artery is frequently higher. The right renal artery passes behind the vena cava and the right renal vein. The left artery passes dorsal to the left renal vein. Accessory renal arteries are common. Graves stated (1955) that the term accessory is incorrect and instead the vessels should be termed normal segmental arteries of precocious origin. Common origins of the vessels, which are usually to the upper and lower pole, are the aorta, the adrenal arteries, or gonadal arteries (Anson and Daseler, 1961). Graves (1955), on the basis of plastic injection studies of the renal vasculature, concluded that the distribution of the arteries within the kidney forms a pattern that is constant and also that there is no

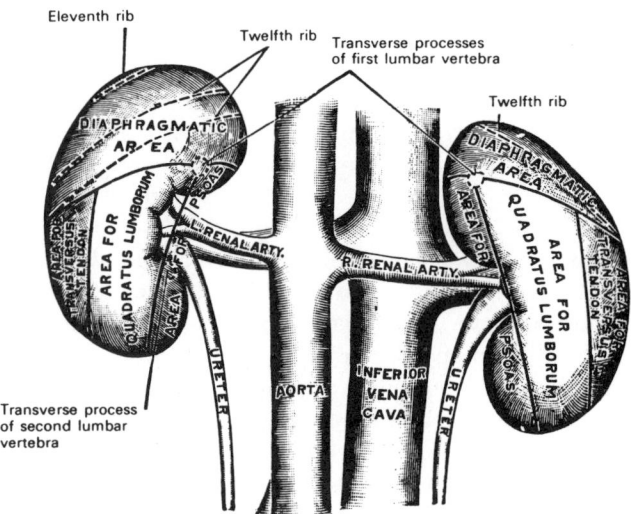

FIG 1–39.
The dorsal surface of the kidney showing the areas of relation to the parietes. (From Gray H: *Anatomy of the Human Body,* 30th American ed. Philadelphia, Lea & Febiger, 1985, p 1529. Used by permission.)

collateral supply between these segments. In 75%–80% of all kidneys, branching of the main renal artery occurs in the distal third of the vessel or in the renal sinus (Elkins and Resnick, 1982). The main renal artery in so branching forms an anterior division and a posterior division. The anterior division gives rise to the upper, middle and lower segmental arteries and the apical segmental artery usually arises from it. The posterior division supplies only the posterior segment but may give origin to the apical segment. Graves (1955) described five renal segments based on the blood supply: an apical segment, an upper segment, a middle segment, a lower segment, and a posterior segment. The artery to the apical segment usually arises from the anterior division or from the artery to the upper segment, but it may arise from the posterior division or even the aorta or the adrenal. The upper segment artery arises from the anterior division and gives rise to an upper and a horizontal branch. The artery to the middle segment arises from the anterior division. The artery to the lower segment usually arises from the anterior division and passes anterior to the pelvis or ureter, following which it divides into an anterior and a posterior branch. The anterior branch divides into smaller branches that supply the anterior surface of the lower pole. The posterior branch passes under the neck of the inferior calyx to supply the posterior aspect of the lower pole. The artery to the posterior segment is a continuation of the posterior division. It courses over the back of the renal pelvis and the superior calyx. The artery then courses caudally and gives off three groups of branches: an up-

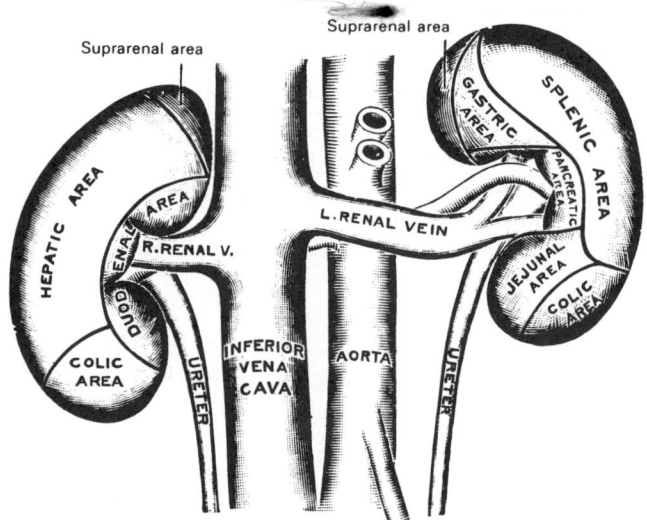

FIG 1–38.
The ventral surface of the kidney showing the areas of contact with the overlying viscera. (From Gray H: *Anatomy of the Human Body* 30th American ed. Philadelphia, Lea & Febiger, 1985, p 1529. Used by permission.)

per group, which supplies the posterior part of the upper calyx; a middle group, which is small but supplies the posterior aspect of the middle calyces; and a terminal group, which supplies the upper portion of the posterior aspect of the lower pole (Graves, 1982).

Not all authorities agree on the constancy of the renal vasculature. Fine and Keene (1966) stated: "We have found such variability in the volume of tissue supplied by particular branches of the renal artery that we were unable to define any segment of constant size common to most kidneys. In any kidney well-defined segments exist, but in any particular kidney the extent of the segments cannot be predicted from an inspection of the outer surface of the kidney or of its arteries in their extrarenal course" (Fig 1–40). The branches of the segmental arteries give rise to the lobar arteries, usually one for each renal pyramid. Prior to entering the kidney parenchyma, they divide again into two or three interlobular arteries to enter the cortex on either side of the pyramids (Fig 1–41). At the juncture of the cortex with the medulla, the interlobular arteries divide to form arcuate arteries, which they give off at right angles. Although the arcuate arteries pursue a course between the cortex and the medulla, they do not communicate. Branching from the arcuate arteries toward the cortex are interlobar arteries. The interlobar arteries are the origin of the afferent glomerular arterioles, which arise as side branches. Other interlobar arteries pass to the capsule where they anastomose with the capsular plexus of arteries, which may be derived from renal, adrenal, or gonadal arteries. From the majority of the glomeruli the efferent glomerular arterioles divide to form the peritubular capillary plexus, which

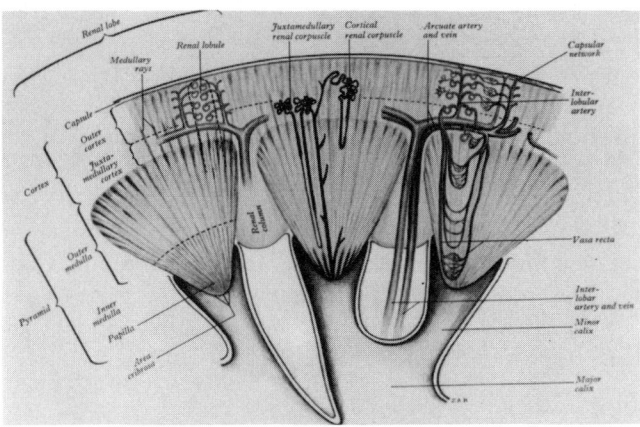

FIG 1–41.
Diagrammatic illustration of the major structures in the renal cortex and medulla *(left),* the position of cortical and juxtamedullary nephrons *(middle),* and the major blood vessels *(right).* (From Gray H: *Gray's Anatomy,* 36th British ed. Philadelphia, WB Saunders Co, 1980, p 1392. Used by permission.)

runs between and around the proximal and distal convoluted tubules (Gray, 1980d).

The main blood supply to the medulla is derived from the efferent arterioles of the juxtamedullary glomeruli. The vessels descend into the medulla by looping around the arcuate vessels, giving small branches to the loose capillary plexus in the outer stripe of the outer medulla. Near the arcuates the efferent arteriole breaks up abruptly into a large number of descending vasa recta (12 to 25). The vasa recta ultimately divide into capillaries (Moffatt, 1981). From the inner medulla the capillaries form ascending vasa recta joining with ascending vasa recta draining the outer medulla and then drain into the arcuate or interlobular veins. The cortex is drained by fine radicals which form from the peritubular capillaries and drain to the interlobar veins. The intrarenal veins accompany the intrarenal arteries and are similarly termed.

Although the renal arterial supply is segmental, the renal venous system is a system of longitudinal and horizontal arcades that maintain a free circulation throughout the kidney. Graves (1982) proposed an explanation based on the embryologic interrelationships between the medial subcardinal vein and the lateral posterior cardinal vein on either side of the developing mesonephros.

The renal veins eventually join the inferior vena cava on its lateral aspect. The left renal vein is three times as long as the right and much more frequently the recipient of accessory and anomalous venous drainages, which facilitates the development of collateral venous drainage if the renal vein is occluded.

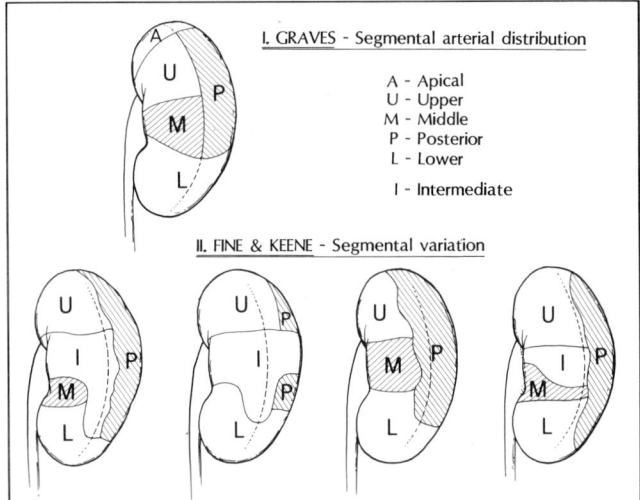

I. GRAVES - Segmental arterial distribution

A - Apical
U - Upper
M - Middle
P - Posterior
L - Lower

I - Intermediate

II. FINE & KEENE - Segmental variation

FIG 1–40.
Segmental renal arterial distribution as defined by *(I)* Graves and *(II)* Fine and Keene. (Original drawing by Christine Young.)

Innervation

The innervation of the kidney is via the renal plexus, which arises from the aortorenal ganglion, the middle and inferior splanchnic nerves, the intermesenteric nerves, and the lumbar sympathetic chain (Mitchell, 1935). The plexus is continued into the kidney accompanying the renal vasculature. Primarily the renal innervation is vasomotor in function.

Lymphatics

Lymphatic vessels accompany the arterial tree from the interlobular arteries to the renal artery. From the renal pedicle the lymphatic channels, usually four or five trunks, drain to lymph nodes along the inferior vena cava and to the lateral aortic nodes. The renal capsule and perirenal tissue have a rich lymphatic vascularization that communicates with the renal lymphatic vessels (Rouiller, 1969).

Microscopic Anatomy

Histologically the renal parenchyma is composed of uriniferous tubules contained within the supporting connective tissue through which is carried the vasculature, lymphatics, and innervation. The term "uriniferous" tubules includes a nephron and a collecting tubule. The term "nephron" includes the renal corpuscle and the renal tubule (Fig 1–42).

The renal corpuscle or malpighian corpuscle is composed of two portions, the glomerulus and the glomerular capsule (Fig 1–43). The renal corpuscle incorrectly is referred to as the glomerulus. The glomeruli are noted throughout the renal cortex with the exception of a narrow segment just under the renal capsule, the cortex corticis. The glomerulus itself is made up of a vascular capillary tuft that invaginates the blind proximal end of the renal tubule (Tighe, 1982). It seems apparent that the capillary loops anastomose. The capillary network arises from the afferent arteriole, which enters the capsule opposite the proximal end of the collecting tubule. In the same location the capillary terminates in the efferent arterioles. The site of entrance and exit of the arterioles is termed the "vascular pole" or "hilus of the capsule." The capillaries of the glomerular tuft are lined by endothelial cells that are of a finely fenestrated type. The epithelium of the visceral layer of the glomerular capsule surrounds the endothelial cells. The endothelial cells and the epithelial cells each have a basal lamina that fuses. The fused basal lamina are termed collectively the "glomerular basement membrane." The glomerular basement membrane is made up of three layers: the lamina rara externa, the lamina densa, and the lamina rara interna. Not all authors agree with the endothelial contribution to the glomer-

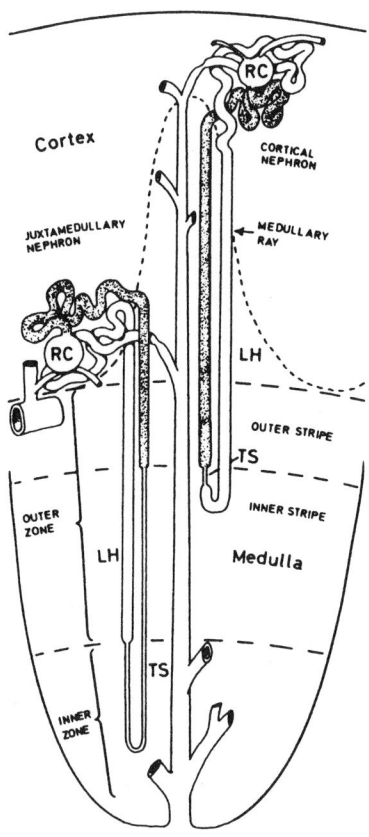

FIG 1–42.
The nephron with long loop of Henle (LH) and with short loop of Henle (TS). (From Krstic RV: *Illustrated Encyclopedia of Human Histology.* New York, Springer-Verlag, 1984, p 283. Used by permission.)

ular basement membrane. Since there are no pores in the membrane, it has been suggested that the lamina densa is a thixotropic gel that permits the passage of large molecules (Tanagho and Smith, 1966). The mesangium is a stalk that stems from the hilus of the glomerulus and gives support to the capillaries (Rouiller, 1969) (Fig 1–44).

The glomerular epithelial cells rest on the basement membrane by foot processes and so are called "podocytes." The parietal layer of the glomerular capsule is formed of epithelial cells of the parietal type. The cavity between the visceral and parietal epithelium is termed the "urinary space," which is continuous with the proximal convoluted tubule.

The renal tubule is composed of the following segments in order: the urinary space of the renal capsule, the proximal convoluted tubule, the straight portion of the proximal tubule, the thin or descending loop of Henle, the thick or ascending loop of Henle, the straight portion of the distal tubule, the distal convoluted tubule, the junctional tubule, and the collecting duct.

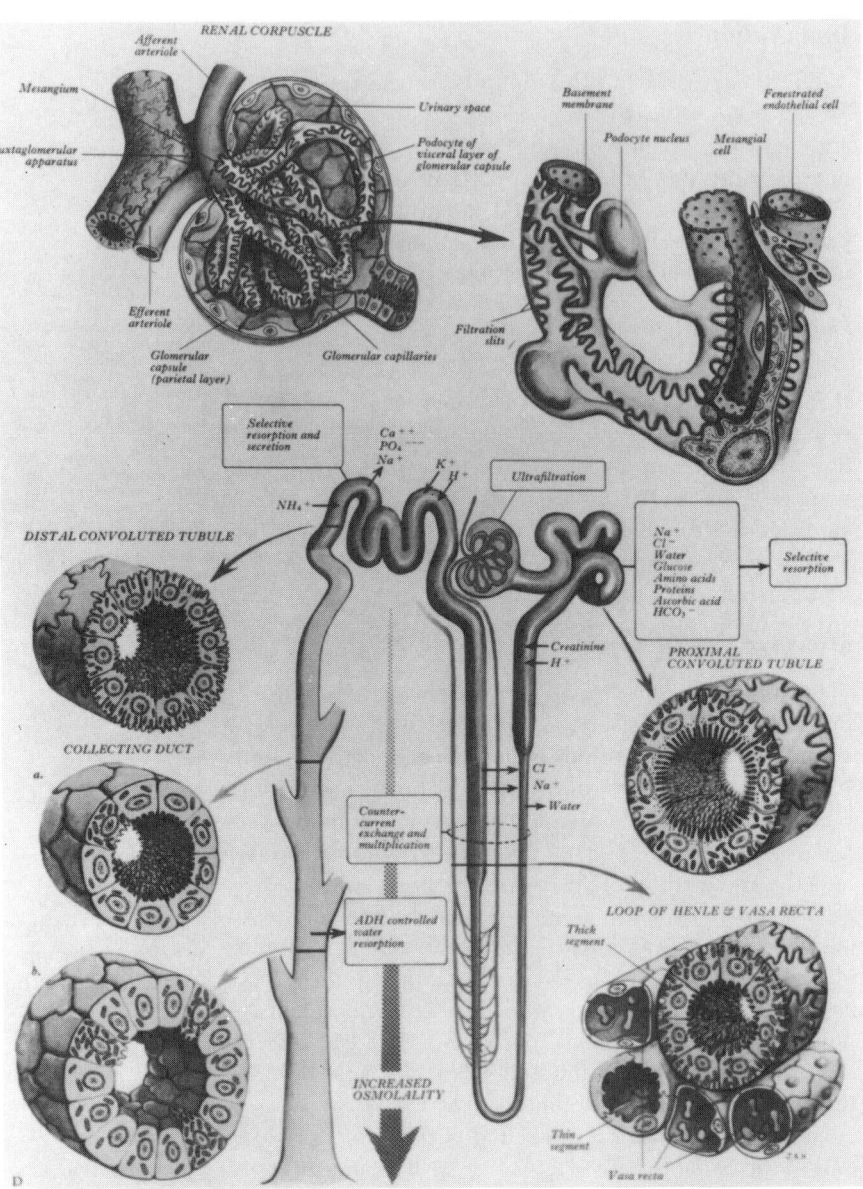

FIG 1–43.
Detail of renal corpuscle and visceral layer of the glomerular capsule. (From Gray H: *Gray's Anatomy,* 36th British ed. Philadelphia, WB Saunders Co, 1980, p 1393. Used by permission.)

The proximal convoluted tubule is lined by cuboidal epithelial cells with long microvilli on their luminal surfaces termed the "brush border." The straight portion of the convoluted tubule is similar in appearance but with shorter microvilli. In the thin limb of the loop of Henle, the lumen is lined by low cuboidal cells with only occasional short microvilla (Tighe, 1982). The thick loop of Henle is lined with cuboidal cells with short microvilli.

The distal tubule is subdivided into three parts: the pars recti, the pars convoluti, and the macula densa. The pars recti is lined by cuboidal cells that have only a few short microvilli. The distal convoluted tubule is lined by cells that are taller than in the straight portion. They are dome-shaped and devoid of microvilli. The macula densa forms the first part of the distal convoluted tubule and is a constituent of the group of specialized structures collectively termed the "juxtaglomerular apparatus."

The collecting ducts are composed of simple cuboidal columnar epithelium, which becomes taller on examination from the cortex to the medulla. Two types of cells are noted in the luminal wall, light cells and dark cells, of which the light cells predominate. The dark cells have more microvilli and more mitochondria. The collecting ducts eventually drain to the ducts of Bellini opening in the papillary tip, termed the "area cribrosa."

The juxtaglomerular apparatus describes a collection of structures at the hilum of the glomerulus, which includes afferent arterioles supplying the vascular tuft,

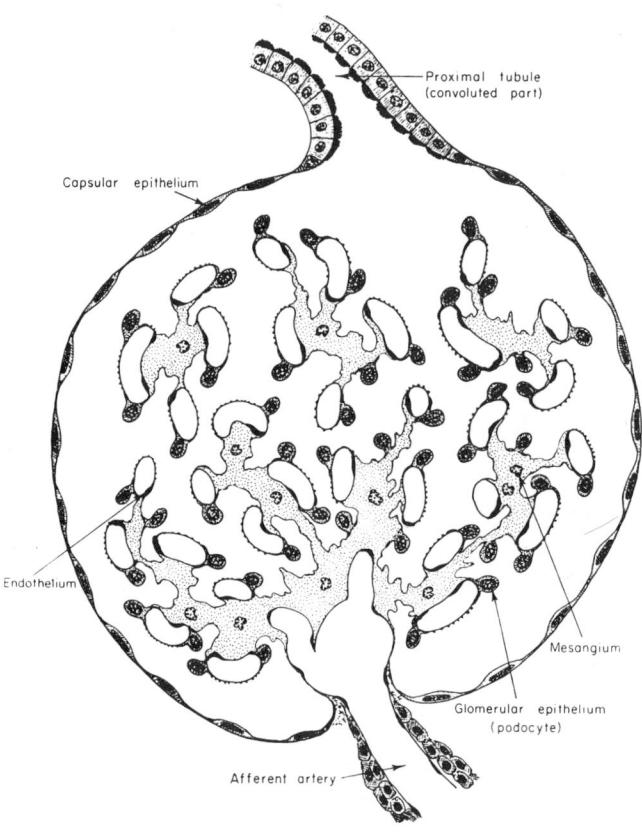

FIG 1–44.
Diagram of a section through a normal renal corpuscle. (From Rouiller C: *The Kidney.* New York, Academic Press Inc, 1969, vol 1, p 77. Used by permission.)

the efferent arteriole leaving the tuft, and macula densa of the distal convoluted tubule and the cushion of cells at the vascular pole termed "lacis cells," which are continuous with the mesangium (Fig 1–45).

UPPER URINARY TRACT (CALYCES, RENAL PELVIS, AND URETER)

Gross Anatomy

The calyces, pelvis, and ureter form the collecting structure bridge between the renal parenchyma and the bladder. The calyces, numbering from eight to 18, may be loosely grouped into those draining the upper and lower poles and two parallel rows of anterior and posterior calyces draining the midportion of the kidney. The upper and lower pole calyces are frequently compound; that is, they accommodate several papillae (Fig 1–46). The papillae themselves invaginate the calyces, the communication occurring through the ducts of Bellini, the perforations being known as the area cribrosa. On the periphery the calyces are sealed by the renal capsule, which, in addition to covering the kidney, also encloses the renal sinus, joining finally the adventitia of the calyces. The calyx per se consists of the funnel portion encompassing the papillae and a neck of varying length. The coalescence of several of these calyces into a larger chamber is called a major calyx or an infundibulum, which generally numbers two or three. The major calyces fuse to form the renal pelvis. An aberration of nomenclature has produced a confusion regarding the major and minor calyces. The calyx, which properly means "cup," is easily seen to relate to the portion of the collecting structures that encompasses the papillae. However, the neck of the calyx is also included in that

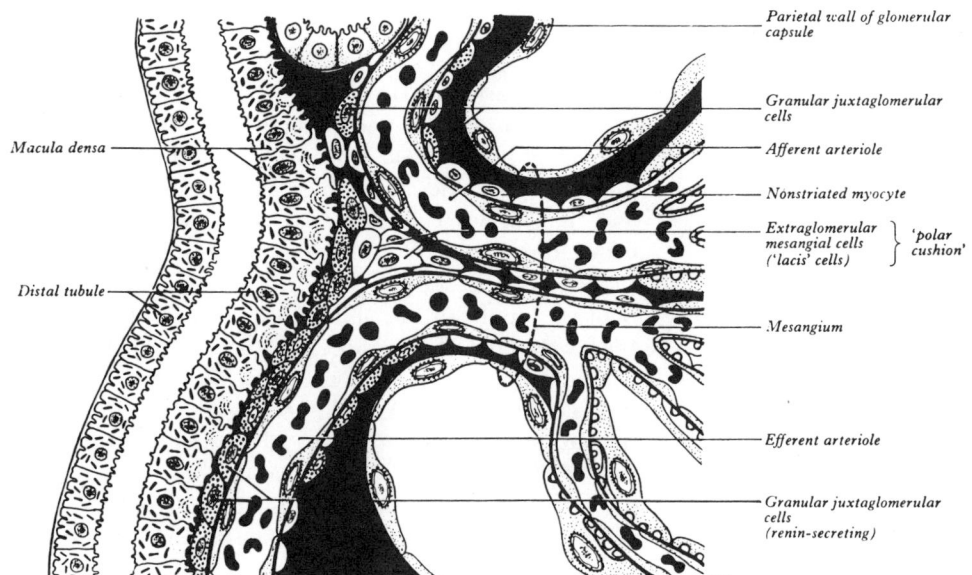

FIG 1–45.
Diagrammatic representation of constituents of the juxtaglomerular apparatus including the macula densa *(left),* granular juxtaglomerular cells *(middle),* and the vascular pole of the glomerular capsule *(right).* (From Gray H: *Gray's Anatomy,* 36th British ed. Philadelphia, WB Saunders Co, 1980, p 1395. Used by permission.)

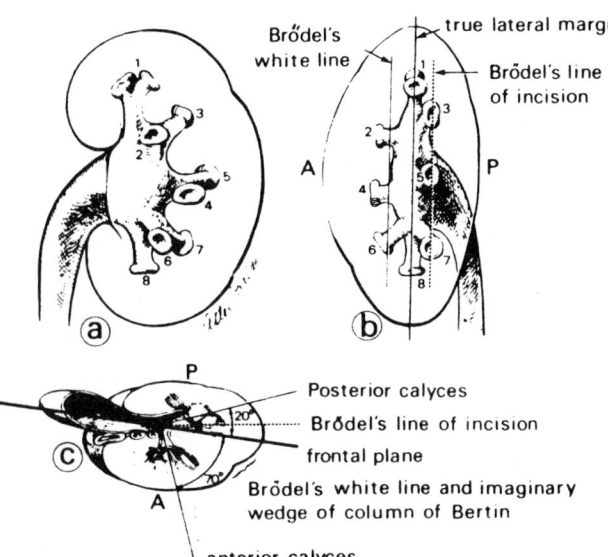

FIG 1–46.
Left kidney showing portion of pelvis and calyces. *A,* anterior view; *B,* lateral view. *C,* transverse section. *1* and *8* represent polar calyces; *2, 4,* and *6* are anterior calyces; *3, 5,* and *7* are posterior calyces. (From Kaye KW, Goldberg E: Applied anatomy of the kidney and ureter. *Urol Clin North Am* 1982; 9:9. Used by permission.)

term. The coalescence of these necks is called an infundibulum or major calyx, obviously not a cup at all; nonetheless, the term stands. The number and distribution of the calyces and infundibula may be quite variable. Brodel divided renal pelves into two types, a true pelvis and a divided pelvis (Kaye and Goldberg, 1982). The divided pelvis represents a duplication of the collecting structures. The pelvis may also vary considerably in size and shape from rather capacious structures to those in which the infundibula meld intrarenally into the ureter with scarcely any pelvic dilatation at all.

The ureters are paired muscular tubes that course through the intermediate stratum of retroperitoneal connective tissue. They are dynamic structures with relative freedom of peristaltic movement. Their course is serpentine from the renal pelvis to the bladder and is characterized by two artifactual waists, one at the ureteropelvic junction and the other at the crossing of the iliac vessels. Midway between the two waists, the ureter assumes a spindle-like configuration. The ureteral lumen from the ureteropelvic junction to the bladder is functionally of a uniform diameter. From the renal pelvis to the great vessels the ureter is referred to as the abdominal portion. From the great vessels to the bladder the term "pelvic ureter" is applied. The ureterovesical junction may be divided into three sections: the terminal portion of the ureter, termed the "juxtavesical ureter"; the intramural portion; and that lying under the mucosa of the bladder, the submucosal portion.

A common variant noted is duplication of the renal pelvis or the ureter. The duplicated ureters may either join or proceed separately to the bladder to enter by separate orifices, the upper ureter usually being drained by the lowermost orifice.

Relationships

The minor calyces and the renal pelvis are encased in fat separating them physically from the renal capsule of the renal sinus through which courses the vasculature of the kidney. The ureteropelvic junction on the right is covered ventrally by the curvature of the duodenum, whereas on the left side the duodenojejunal junction and the body of the pancreas cover the ureteropelvic junction on its ventral aspect. The abdominal portion of the ureter pursues a very slightly oblique course medially, completely encased in the intermediate stratum of retroperitoneal connective tissue, sandwiched between the psoas muscle dorsally and the posterior peritoneum ventrally on the right side and the mesentery of the descending colon on the left side. Along its course it is ventrally crossed obliquely by the gonadal vessels, the vessels coursing from medial to lateral. The pelvic portion of the ureter reaches its most medial portion just after crossing the iliac vessels, whereupon it courses laterally into the hollow of the pelvis before turning again medially to join the bladder. Midway in its course the pelvic ureter is covered ventrally by the lateral umbilical ligament and the leash of blood vessels proceeding to the lateral pelvic wall from the bladder. In the male the ureter is also crossed ventrally by the vas deferens.

Vasculature

The ureteric arteries are illustrated in Figure 1–47. In order, from the renal pelvis to the bladder, the sources of the ureteral arteries include the renal artery, internal spermatic, aorta, hypogastric, superior vesical, and inferior vesical arteries. Other blood supply may come from capsular, suprarenal, or uterine vessels. The richest arterial supply is to the pelvic ureter and the poorest supply to the abdominal portion; that is, from the lower pole of the kidney to the brim of the pelvis, where the aorta and the common iliac give off only a few segmentally arranged lumbar rami. Ureteric vessels tend to anastomose. A frequently noted anastomosis is that of the renal and hypogastric ureteric vessels. Plexuses of vessels are noted within the adventitia of the ureteral wall. The venous drainage is variable but basically follows the arterial supply.

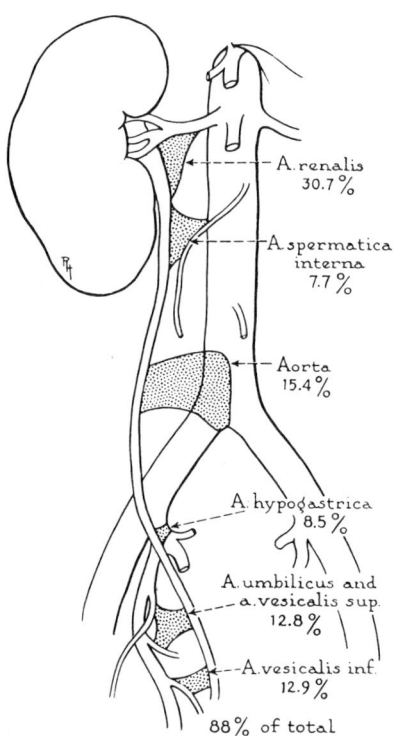

FIG 1–47.
Schematic of ureteric arteries. The sources of supply are indicated, with percentage of vessels from each of the important contributing vessels representing 88% of the vessels. (From Anson BJ, McCormack LJ: The accessory pulmonary lobe of the azygos vein. *Q Bull Northwestern University medical School* 1950; 24:292. Used by permission.)

Innervation

The renal calyces, pelvis and ureter are innervated by the autonomic nervous system, receiving afferent and efferent contributions from the sympathetic and parasympathetic pathways. The parasympathetic contribution is from the second, third, and fourth sacral segments of the spinal cord. The lower ureter is supplied by the pelvic splanchnic nerves via the pelvic plexus. The upper ureter and collecting structures are supplied by the hypogastric nerve. The sympathetic innervation from the T12–L2 segments of the spinal cord is by the thoracic and lumbar splanchnic nerves and then by the celiac, superior hypogastric, and renal plexuses. Unmyelinated nerve fibers have been observed in all layers of the ureteral wall. No autonomic ganglion cells have been observed anywhere in the ureter.

Lymphatics

Lymphatic capillaries in the wall of the ureter coalesce and pass diagonally outward through the muscular walls of the ureter. They then course proximally and distally for greater or lesser distances before exiting to pass to regional lymph nodes. There is no continuity of lymph channels from the bladder to the kidneys by the ureter. The regional lymph node groups that drain the ureter are the lateral abdominal chains, the common iliac, the external iliac, and the hypogastric group of nodes.

Microscopic Anatomy

Microscopically the ureteric wall is composed of three layers—an adventitia, a muscular layer, and a mucous membrane. The adventitia consists of an intertwined mass of collagen fibers containing blood vessels and nerve bundles. The muscle layer is composed of smooth muscle cells running in all directions. The mucous membrane consists of transitional cell epithelium separated from the muscle by fibrous lamina propria or submucosa (Notley, 1971). Although traditionally it has been recorded that the muscular coat of the ureter is characterized by three layers, it is apparent that this differentiation is difficult to find. In the renal pelvis, muscle fibers are noted to run obliquely separated by connective tissue. The ureteropelvic junction per se is defined histologically. It has been described that the upper abdominal ureter has a relatively thin muscular wall. Muscle bundles of different orientation lie side by side with the musculature consisting of braided bundles of muscle fibers arranged in interlacing spirals. In the pelvic ureter the muscular helices exhibit some degree of muscle layering (Hanna et al., 1976) (Fig 1–48).

The microscopic anatomy of the ureter and the ureterovesical junction has been controversial. For discussion, the musculature of the terminal ureter may be divided into a juxtavesical segment, an intramural segment, and a submucosal part. In the juxtavesical segment the muscle fibers tend to be layered with an inner longitudinal coat and an outer coat with well-developed oblique and circular fibers. The intramural portion is composed of longitudinal fibers separated from the adventitia by a fibromuscular, tissue-lined space termed Waldeyer's sheath. The submucosal ureter lies superficially in the bladder base and is covered only by mucous membrane. The longitudinal muscle fibers of the roof and the floor of the ureter pass medially and inferiorly to form the superficial ureteric trigone muscle (Figs 1–49 and 1–50). The trigone of the bladder then consists of three layers of muscle, a superficial or submucosal layer continuous with the ureteric muscle, a middle layer continuous with Waldeyer's sheath, and the deep trigonal layer, which is the bladder muscle wall itself.

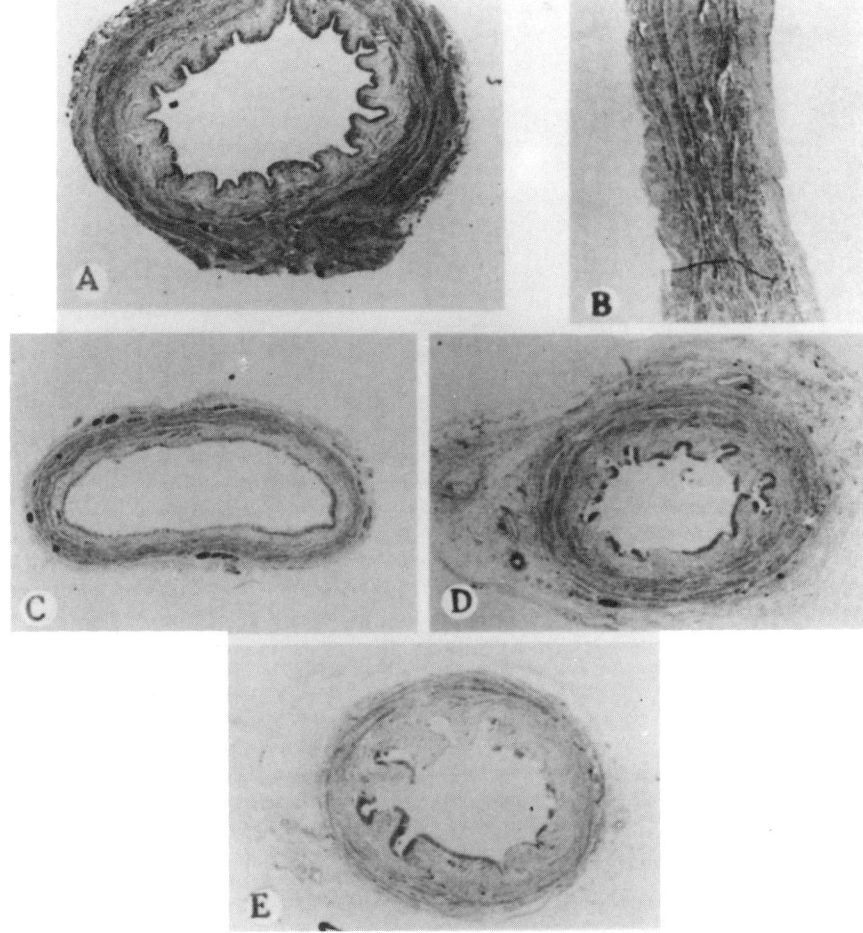

FIG 1–48.
Ureteral structure and muscle orientation. **A,** renal pelvis, transverse section; **B,** renal pelvis, longitudinal section; **C,** upper abdominal ureter; **D,** abdominopelvic ureter; **E,** lower pelvic (juxtavesical) ureter. Note absence of muscle layering, except in lower pelvic ureter in which muscle bundles are arranged into outer circular and inner longitudinal. (From Hanna K, Jeffs RD, Sturgess JM, et al: Ureteral structure and ultrastructure: Part I. The normal human ureter. *J Urol* 1976; 116:719. Used by permission.)

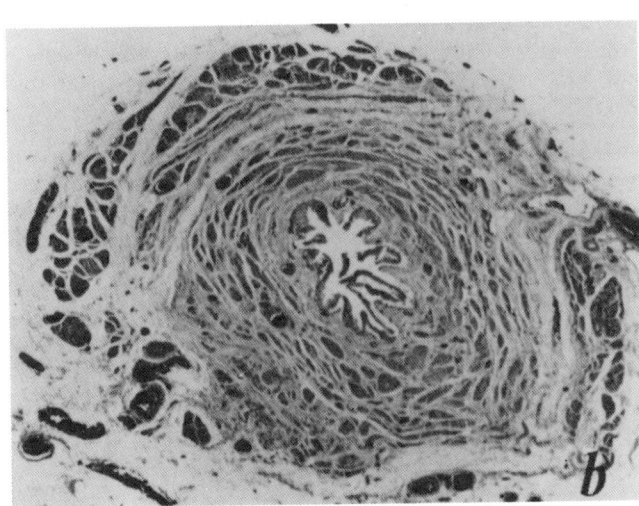

FIG 1–49.
Cross-section of the juxtavesical ureter, predominantly longitudinal muscle with well-developed oblique and circular fibers. Waldeyer's sheath is seen surrounding the ureter. (From Tanagho EA, Pugh RCB: The anatomy and function of the ureterovesical junction. *Br J Urol* 1963; 35:154. Used by permission.)

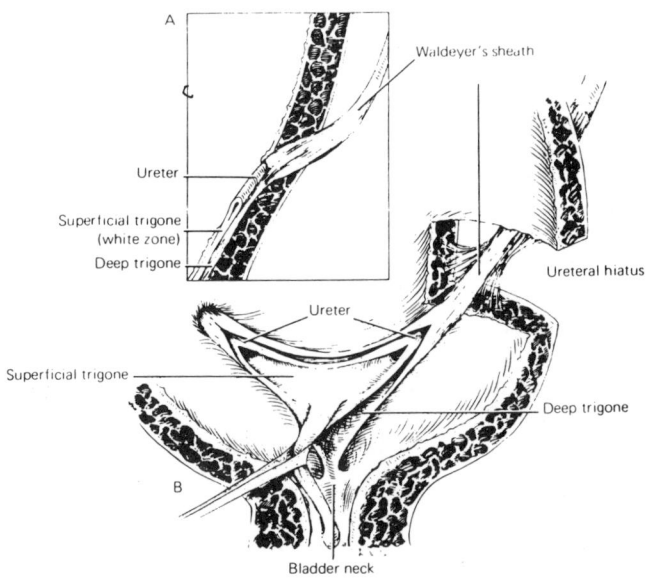

FIG 1–50.
Normal ureterovesico-trigonal complex. (From Tanagho EA: Anatomy of the lower urinary tract, in Walsh PC, Gittes RF, Perlmutter AD, et al: *Campbell's Urology,* ed 5. Philadelphia, WB Saunders, 1986, p 53. Used by permission.)

BLADDER

Gross Anatomy

The bladder in a full or distended state is a hollow, spherical, muscular organ that displays almost equal internal and external surfaces. In the empty or contracted state the bladder is usually described as having the form of a tetrahedron. The four surfaces are a superior surface, a posterior surface (fundus or base), and two inferolateral surfaces. The angles of the tetrahedron are formed by the ureterovesical junction, the urachovesical junction, and the urethrovesical junction. The juncture of the urachus and the bladder is termed the apex and the junction with the urethra the neck. The internal surface of the bladder is characterized, depending on the state of the distention of the bladder, by mesh-like folds formed by mucosal folds in the contracted state and mucosal-covered musculature in the distended state. In the base of the bladder is noted a triangular-shaped, smooth area termed the "trigone." The angles of the trigone are formed by three orifices, the paired ureteral orifices and the internal urethral orifice. In adult males the elevation of the posterior lip of the vesical neck produced by underlying prostatic tissue has been termed the "uvula of the bladder." The superior boundary of the trigone lying between the two ureteric orifices has been termed the interureteric crest or ridge or Mercier's bar.

The bladder is contained in a layer of retroperitoneal connective tissue that is most evident on the inferolateral aspects. The bulk of the bladder is formed of muscle, collectively termed the "detrusor muscle." The bladder musculature for discussion may be divided into that of the bladder wall and the trigone.

It should be stated that the anatomy of the musculature of the bladder, the vesical neck, and the ureterovesical junction has been extensively studied and is indeed complex. The serious student should consult the works of Hutch (1972), Tanagho and Pugh (1963), Uhlenhuth and Hunter (1953), and Woodburne (1960).

The bladder wall musculature is often described as having three layers, or coats. This layering only occurs at the bladder neck. The remainder of the bladder musculature is composed of fibers that run in many directions. Nonetheless, the terminology of the bladder wall musculature is given in layers—an inner longitudinal layer, a middle circular layer, and an outer longitudinal layer (Hutch, 1972). The inner longitudinal layer is located immediately beneath the vesical mucosa. In the bladder wall the fibers are widely separated and course multidirectionally. However, near the bladder neck, they assume a longitudinal pattern that is contiguous through the vesical neck and into the urethra as its inner longitudinal layer.

The middle circular layer is found throughout the bladder wall but is least developed in the dorsal midline. This layer terminates at the vesical neck and is not contiguous as a part of the urethral musculature. Near the vesical neck, the layer is thicker and much more prominent than at any other place in the bladder. In effect the middle circular layer forms concentric rings around the neck of the bladder and incorporates into the base of the bladder. The circular layer pierces the deep trigone and completes its arc through the deep trigone, forming a fusion between the detrusor and the trigone.

The outer longitudinal layer is most prominent along the anterior and posterior walls of the bladder and is quite thin on the lateral aspects of the bladder. As these fibers progress toward the vesical neck, they seemingly form groups, generally an anterior and a posterior group. The anterior group is a wide band of muscle that runs from the vesical neck to the urachus. The posterior group also runs from the urachus to the bladder neck and is stronger than the anterior group. Near the bladder neck it forms into a posteromedial group and a lateroposterior group. The posteromedial group of musculature inserts into the apex of the deep trigone. The lateroposterior group loops around the apex of the deep trigone to form the very superior part of the anterior wall of the urethra. This loop has been termed the "detrusor loop," or Heiss's loop. The anterior group and the detrusor loop attach by a fibrous band termed the "transverse precervical arc," which surrounds the anterior one third of the bladder neck (Fig 1–51).

The trigone is formed of two muscle layers superimposed on the detrusor muscle (Tanagho and Pugh, 1963). The superficial trigone is formed as a direct continuation of fibers in the roof and the floor of the intravesical ureter. The deep trigone is formed by the continuation of Waldeyer's sheath, a fibromuscular structure that completely encircles the distal 3–4 cm of the juxtavesical ureter and follows the ureter through the ureteral canal. The sheath (now the deep trigone) continues under the superficial trigone. The trigone and the dense ventral condensation, the middle circular layer of the detrusor, surround the bladder outlet. It should be stated that Woodburne believed that the concept of Waldeyer's sheath is one that Waldeyer never originated, that he was pointing out a space between the ureter and the detrusor fascicles that approach it and that if a sheath is to be described, it is composed only of the fascicles of bladder muscle (Woodburne, 1960).

Relationships

The bladder is encased in intermediate stratum of extraperitoneal connective tissue that carries through it

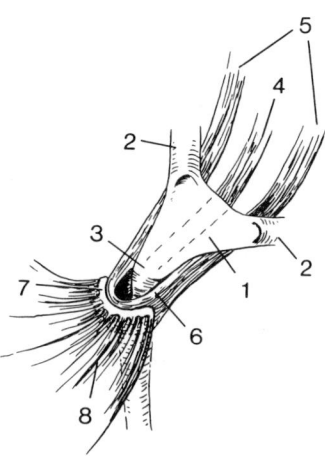

FIG 1–51.
Line drawing illustrating the relationship of the various muscles of the outer longitudinal layer to the deep trigone. *1,* deep trigone. *2,* Waldeyer's sheath. *3,* apex of the deep trigone. *4,* the medial posterior outer longitudinal muscle which runs under the posterior surface of the deep trigone to insert into the apex of the deep trigone. *5,* the right and left lateral posterior outer longitudinal muscle which lies under each Waldeyer's sheath and passes around the bladder neck to form the anterior and lateral walls of the urethra. Just below the base plate is the detrusor loop. *6,* insertion of the detrusor loop into the transverse precervical arch. *7,* transverse precervical arch. *8,* anterior outer longitudinal muscle, which also inserts into the transverse precervical arch. (From Hutch JA: *Anatomy and Physiology of the Bladder, Trigone and Urethra.* New York, Appleton-Century-Crofts, 1972, p 86. Used by permission.)

the rich plexus of vesical veins and also arteries. The superior and posterior surface of the bladder is covered with the peritoneum, which is closely adherent, with minimal extraperitoneal connective tissue. This is the only serosal layer of the bladder. Extending cranially from the apex of the bladder to the umbilicus is the urachus, the allantoic remnant composed of fibromuscular tissue. On the inferolateral aspects the bladder is in proximity to the obturator internis and levator ani muscles, although separated by the extraperitoneal connective tissue. The neck of the bladder in the male is intrinsic to the prostate. The vesical neck is attached to the overlying pubic symphysis by the puboprostatic ligaments. The anterior surface of the bladder lies under the pubis and the investing fascia of the rectus abdominis muscles. The potential prevesical space is termed the space of Retzius. Various terms have been applied to the specializations of the extraperitoneal connective tissue enclosing the bladder and associated structures. One term used is "umbilical fascia". Caudally this specialization ensheaths the bladder, seminal vesicles, and prostate (Hayes, 1950).

Posteriorly, in the male, the base of the bladder is in apposition to the rectum. Near the neck the ampulla of

the vas deferens and the seminal vesicles lie between the rectum and the bladder. All are separated by retroperitoneal connective tissue. In the female, posteriorly, the base of the bladder is in apposition to the vagina.

In speaking of the anatomical relationships of the bladder, the term "ligaments of the bladder" is frequently encountered. Condensations of the intermediate stratum of retroperitoneal connective tissue extend from the tendinous arc of the pelvic side wall to the bladder, carrying the vasculature to the bladder. This condensation has been termed the "lateral true ligament" of the bladder ventrally and the "posterior ligament" dorsally. The bladder is anchored from the neck of the bladder to the pubis in the male by the puboprostatic ligaments and in the female by the pubovesical ligaments which are, again, dense condensations of retroperitoneal connective tissue. The apex of the bladder is fixed to the umbilicus by the urachus, which is termed the "median umbilical ligament." Extending from the external iliac artery over the anterior surface of the bladder toward the umbilicus on each side are the obliterated umbilical arteries. Contained within extraperitoneal connective tissue and covered with peritoneum on their dorsal aspect, they are termed the "lateral umbilical ligaments."

Vasculature

Generally it is stated in anatomy texts that the bladder receives its arterial supply primarily from the superior and inferior vesical arteries. It is most probable that the number of vessels supplying the bladder is greater (Shehata, 1976) (Fig 1–52). A complete set of vesical arteries is found only in newborn infants. Obliteration of the umbilical artery and the occlusion of some of its branches lead to the reduced number of vessels seen in adults. Most of the vesical vessels derive from the umbilical artery trunk, which arises from the internal iliac artery, with the exception of the inferior vesical artery, which may arise directly from the internal iliac. A complete set of vessel branches from distal to proximal on the vascular tree would consist of a urachal, two superior vesicals, a middle vesical, one or two additional superior vesicals, a vesiculodifferential, and an inferior vesical. The middle vesical artery, which is not mentioned in some studies, is noted on the anterior surface of the bladder coursing to the neck, where it supplies the middle and lower part of the bladder. The vesiculodifferential artery in the male forms the main vascular supply to the trigone. In the female the trigone is vascularized by the uterine artery. The vaginal artery supplies the area which in men is supplied by the inferior vesical artery, that is, the inferolateral aspect of the bladder neck. Additional vasculature to the bladder

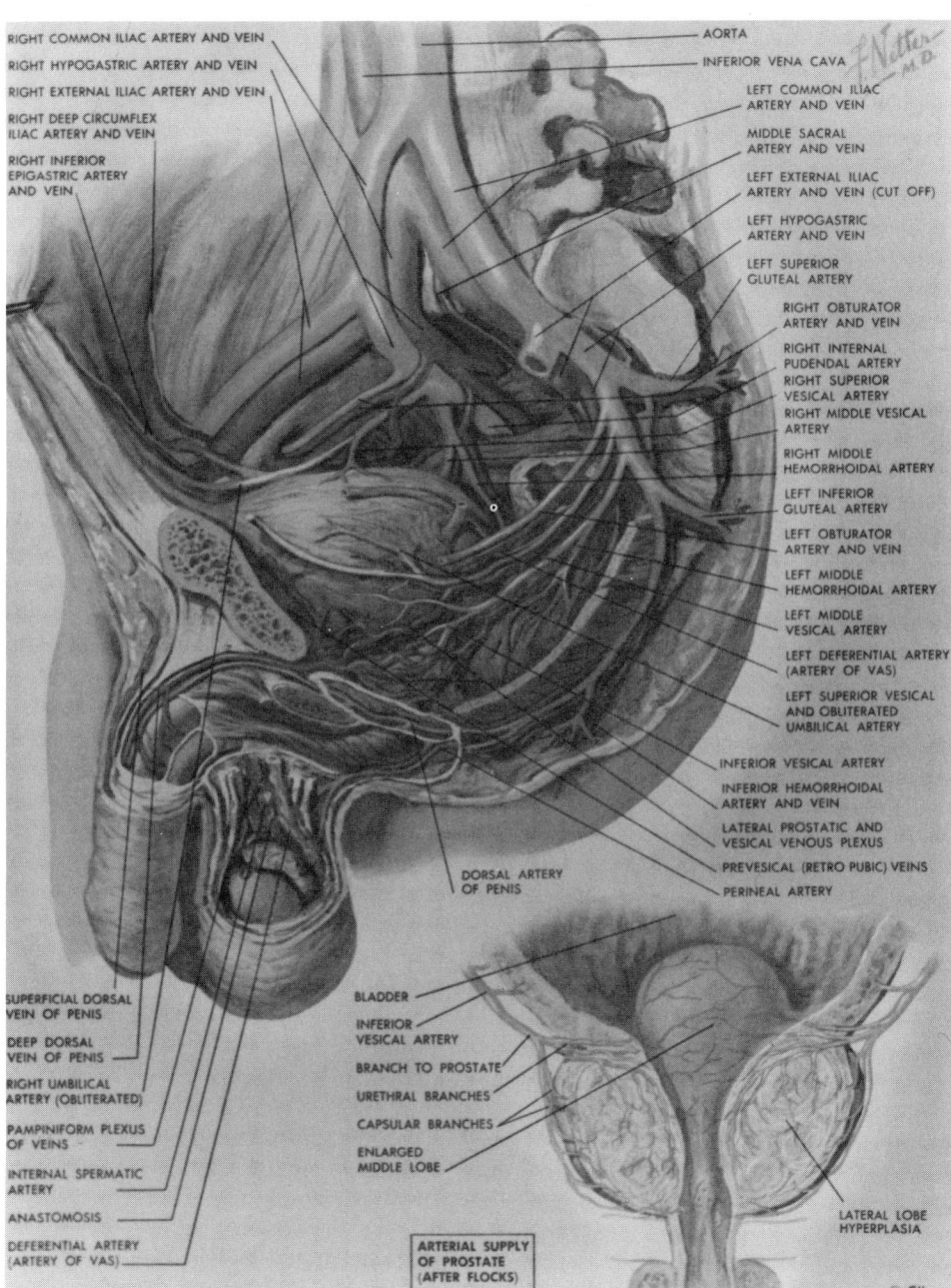

RIGHT COMMON ILIAC ARTERY AND VEIN
RIGHT HYPOGASTRIC ARTERY AND VEIN
RIGHT EXTERNAL ILIAC ARTERY AND VEIN
RIGHT DEEP CIRCUMFLEX
ILIAC ARTERY AND VEIN
RIGHT INFERIOR
EPIGASTRIC ARTERY
AND VEIN

AORTA
INFERIOR VENA CAVA
LEFT COMMON ILIAC
ARTERY AND VEIN
MIDDLE SACRAL
ARTERY AND VEIN
LEFT EXTERNAL ILIAC
ARTERY AND VEIN (CUT OFF)
LEFT HYPOGASTRIC
ARTERY AND VEIN
LEFT SUPERIOR
GLUTEAL ARTERY
RIGHT OBTURATOR
ARTERY AND VEIN
RIGHT INTERNAL
PUDENDAL ARTERY
RIGHT SUPERIOR
VESICAL ARTERY
RIGHT MIDDLE VESICAL
ARTERY
RIGHT MIDDLE
HEMORRHOIDAL ARTERY
LEFT INFERIOR
GLUTEAL ARTERY
LEFT OBTURATOR
ARTERY AND VEIN
LEFT MIDDLE
HEMORRHOIDAL ARTERY
LEFT MIDDLE
VESICAL ARTERY
LEFT DEFERENTIAL ARTERY
(ARTERY OF VAS)
LEFT SUPERIOR VESICAL
AND OBLITERATED
UMBILICAL ARTERY
INFERIOR VESICAL ARTERY
INFERIOR HEMORRHOIDAL
ARTERY AND VEIN
LATERAL PROSTATIC AND
VESICAL VENOUS PLEXUS
PREVESICAL (RETRO PUBIC) VEINS
PERINEAL ARTERY

SUPERFICIAL DORSAL
VEIN OF PENIS
DEEP DORSAL
VEIN OF PENIS
RIGHT UMBILICAL
ARTERY (OBLITERATED)
PAMPINIFORM PLEXUS
OF VEINS
INTERNAL SPERMATIC
ARTERY
ANASTOMOSIS
DEFERENTIAL ARTERY
(ARTERY OF VAS)

DORSAL ARTERY
OF PENIS

BLADDER
INFERIOR
VESICAL ARTERY
BRANCH TO PROSTATE
URETHRAL BRANCHES
CAPSULAR BRANCHES
ENLARGED
MIDDLE LOBE

ARTERIAL SUPPLY
OF PROSTATE
(AFTER FLOCKS)

LATERAL LOBE
HYPERPLASIA

© Ciba

FIG 1–52.
Overview of blood supply of pelvis. *Inset* demonstrates the blood supply of the prostate. (From Netter FH: *The Ciba Collection of Medical Illustrations.* Indianapolis, Curtis Publishing Company, 1965, vol 2, plate 6, p 14. Used by permission of the Ciba Pharmaceutical Company.)

may be supplied by the obturator and inferior gluteal arteries.

The venous drainage of the bladder is via extensive vesical and prostatic plexuses, which form two to five veins (usually three) that drain to the internal iliac vein. In some instances this extravesical plexus will be noted, but the veins will drain directly from the bladder wall to the internal iliac veins. In the female only vesical plexuses are found. All of the vasculature of the bladder is carried in the intermediate stratum of retroperitoneal connective tissue, which has been termed the lateral and posterior ligaments of the bladder.

Innervation

The innervation of the bladder should be considered in terms of the bladder itself and the bladder outlet. The bladder outlet consists of smooth muscle elements of the bladder base, bladder neck, and the prostatic urethra. The bladder receives innervation from the pelvic plexus, which is located as a rectangular network on the lateral aspect of the rectum. The innervation accompanies the vessels and utilizes the vasculature as a framework. The portion of the pelvic plexus that specifically supplies the bladder is sometimes termed the "vesical plexus." The plexus receives input from the S2–S4 spinal cord segments via the pelvic splanchnic nerve and additional input from the T10–L2 cord segments via the presacral nerve. The primary supply to the detrusor is by parasympathetic nerves, which are uniformly and diffusely distributed throughout the detrusor. The detrusor has an exceedingly sparse sympathetic nerve supply (Gosling, 1979).

The bladder neck smooth muscle, which is histologically and histochemically different from that of the detrusor, is supplied by a rich plexus of sympathetic nerve terminals. In contrast in the female the bladder neck and urethral muscle are supplied by numerous parasympathetic nerves that are identical to those innervating the detrusor. In the female the bladder muscle and urethra receive a poor supply of sympathetic innervation. In both sexes the pudendal nerve carries somatic fibers from the S2–S4 spinal cord segments, which supply the striated muscle of the pelvic floor. However, the external extrinsic urethral sphincter, which is striated, is supplied by the pelvic splanchnic nerve (Gosling, 1979).

Lymphatics

The lymphatic drainage of the bladder has been intensely studied by Parker (1936). The lymphatics of the bladder are generally divided into the lymph collectors of the anterior and posterior bladder wall (Fig 1–53). The lymphatic channels arise from an intramural network which forms lymphatic anastomoses between the right and left halves of the bladder. The anterior bladder wall drainage may be separated into anterosuperior drainage and anterior middle drainage. The anterosuperior drainage is along the lateral umbilical ligaments to the middle chain of the external iliac group between the external iliac vein and artery and to the internal

chain of the external iliac group situated around the obturator nerves. The middle bladder wall lymphatic drainage follows the inferior border of the lateral umbilical ligaments and terminates in the middle chain of the external iliac nodal group.

The posterior wall of the bladder has four primary lymphatic drainage channels from the superior, middle, and lower portion of the bladder exclusive of the base, and the base itself. The superior, middle, and lower portions of the bladder wall ultimately drain to the middle and internal group of the external iliac chain. The base of the bladder drains to nodes situated near the bifurcation of the iliacs, although the plexus of the trigone forms a continuous network with the lymphatics of the main body of the bladder.

Injection studies have shown the regional nodes of the bladder to include also nodes located along the hypogastric, common iliac, and lateral sacral chains.

Microscopic Anatomy

On microscopic section the bladder wall is composed of three layers; an outer adventitial layer of connective tissue, a smooth muscle coat (the detrusor muscle), and an inner layer of mucosa. The adventitia represents the intermediate stratum of retroperitoneal connective tissue, which is fibrofatty and demonstrates numerous venous channels, arteries, nerve bundles, and lymphatics. The posterior surface of the bladder is covered with peritoneum that would appear on section of that area as a simple layer of squamous mesothelial cells. The muscle layer of the bladder section is composed of relatively large-diameter interlacing bundles of smooth muscle cells arranged in a complex meshwork. It appears, because of the predominance of longitudinal fibers in the inner and outer layers, that three distinct layers—outer longitudinal, middle circular and inner longitudinal—

FIG 1–53.
The lymphatic drainage of the male and female bladder. (From Hinman F: *Principles and Practice of Urology.* Philadelphia, WB Saunders Co, 1935, p 189. Used by permission.)

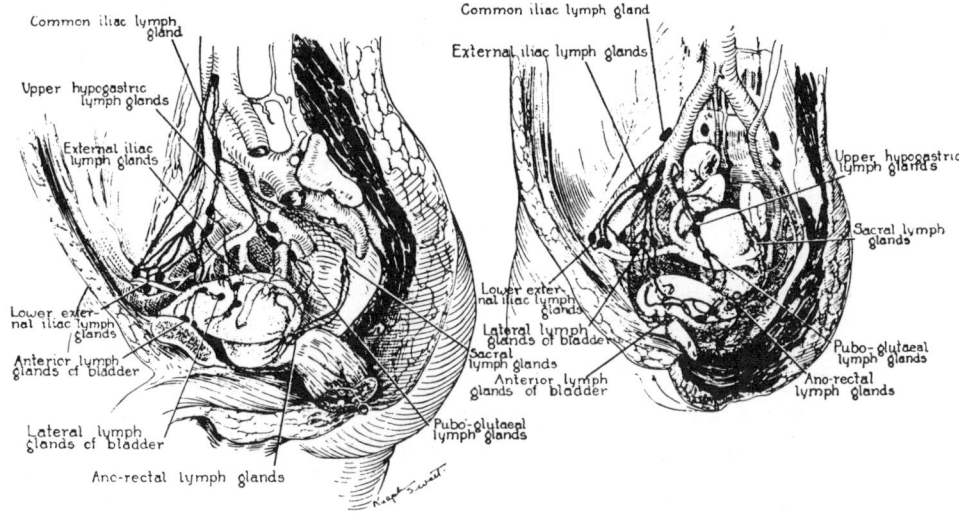

exist as discrete sheets of muscle, which is not the case. The muscle of the superficial trigone differs from the muscle of the detrusor and the deep trigone in that it is composed of relatively small-diameter muscle bundles (Gosling et al., 1982).

The mucosa consists of two layers, the epithelium itself and its supporting connective tissue, the lamina propria. The epithelium is transitional cell and is composed of five to eight layers of stratified cells. The basal cells are slender, pyramidal, or columnar, and by progressive differentiation they become flattened. The lamina propria, which is loose in its deeper layers, allows the mucosa in an empty state to be thrown into folds. The epithelium over the trigone may be only two to three cell layers thick. A commonly noted histologic appearance is that of proliferation of morphologically normal basal urothelial cells that project into the underlying connective tissue of the lamina propria, which are termed Brunn's nests. These are particularly common in the trigone and in some sections will appear as isolated nests of cells. Fluid within these nests will at times give a cystlike appearance. Particularly in females the epithelium of the trigone will be that of a nonkeratinizing squamous metaplasia.

PROSTATE

Gross Anatomy

The prostate in the adult male is roughly the size and shape of a chestnut. Its approximate pyramidal shape is described as having a base and an apex. To the touch the prostate is firm and only slightly pliable. The base is firmly adherent to the base of the bladder, whereas the apex is firmly adherent to the urogenital diaphragm. The prostate itself is well encircled with the intermediate stratum of extraperitoneal connective tissue. With dissection of this connective tissue cover, the thin prostatic capsule is noted. The substance of the prostate is traversed by three lumenal structures, the posterior urethra and the ejaculatory ducts. An incision through the anterior aspect of the prostate in the midline will enter the posterior urethra. On the floor of the posterior urethra from the bladder to the membranous urethra extends a ridge, the urethral crest. The urethral crest is wider at the vesical neck and tapers as it approaches the membranous urethra. In its midportion it is elevated into a small hillock termed the "verumontanum" or "colliculus seminalis." In the midportion of the veru is noted a shallow crypt, the utricle or vagina masculini. On either side of the veru and somewhat posterior are two smaller openings, the orifices of the ejaculatory ducts. The depression on either side of the urethral crest is termed the "prostatic sinus." Numerous small openings are noted in these sinuses, which

represent the openings of the urethral glands and the prostatic ductules.

Although not evident on holding the prostate, it is generally believed that the structure is composed of distinct lobes. The exact nature and number of the prostatic lobes have been the source of controversy and disparate opinion. The most frequently held opinion is that of Lowsley, who described six lobes: anterior, posterior, median, subcervical, and left and right lateral lobes (Lowsley, 1930). The discreteness of the lobes noted on fetal examination is for the most part lost by adulthood. It should be noted that a frequent misinterpretation of this lobular terminology has been applied to the lobes that result from benign intraurethral hyperplasia. The presentation of two opinions regarding the lobular nature of the prostate will give further insight into the state of controversy that exists. One opinion utilizes a terminology like that of Lowsley (Hutch and Rambar, 1970). The prostatic lobes are divided into two groups, intraurethral and extraurethral lobes. The intraurethral lobes include the anterior lobes, the two lateral lobes, and the subcervical lobe. It is stated that they are all situated in the wall of the urethra between the verumontanum and the vesical neck and lie between the inner longitudinal and outer smooth muscle wall of the urethra. In the normal state they are small and occupy little space. It is further stated that the extraurethral lobes, which include a posterior lobe and a median lobe, are outside the prostatic urethra. The posterior lobe is said to wrap around the posterior three fourths of the urethral wall. The median lobe lies between the ejaculatory ducts and the vesical neck or the posterior wall of the urethra.

The second position is that Lowsley's fetal lobes have no identity or significance in the adult (McNeal, 1972). McNeal proposed that there are in essence two glands within the same capsule: a central zone and a peripheral zone. The central zone is defined as a wedge-shaped area of glandular tissue with its apex at the verumontanum and its base exiting superiorly behind the bladder and bladder neck, forming much of the prostate. The peripheral zone then covers over most of the central zone except its very base. The apex of the peripheral zone extends to partly surround the urethra below the verumontanum (Fig 1–54).

Relationships

The base of the prostate is intimately related to the bladder. Indeed, the deep trigone extends into the midportion of the posterior urethra thus overlying the median lobe or central zone of the prostate. Cranially and posteriorly the bases of the seminal vesicles rest against the base of the prostate.

The apex of the prostate is melded to the urogenital

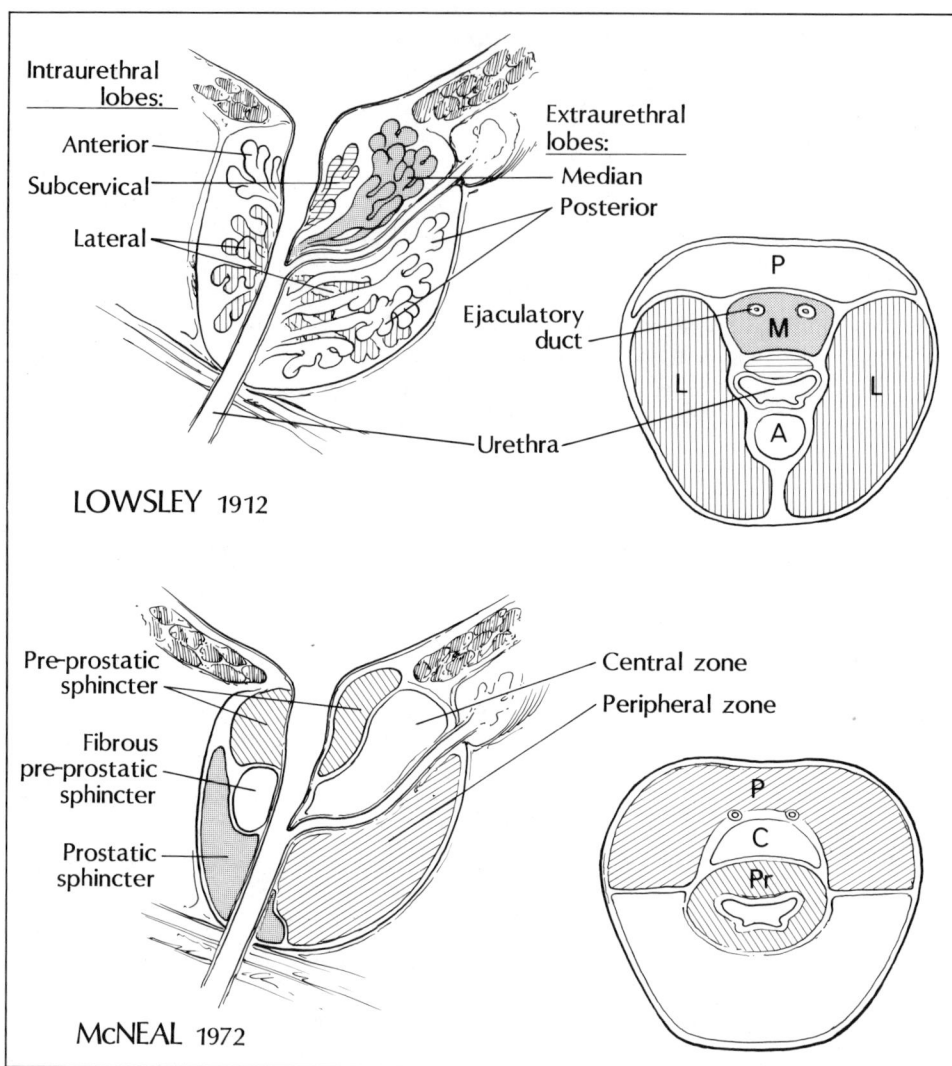

Intraurethral lobes:
Anterior
Subcervical
Lateral

Extraurethral lobes:
Median
Posterior
Ejaculatory duct
Urethra

P
M
L   L
A

LOWSLEY 1912

Pre-prostatic sphincter
Fibrous pre-prostatic sphincter
Prostatic sphincter

Central zone
Peripheral zone

P
C
Pr

McNEAL 1972

FIG 1–54.
Differing concepts of prostate lobes as described by Lowsley and McNeal. (Original drawing by Christine Young.)

diaphragm, the prostatic capsule and closely adherent extraperitoneal connective tissue being attached to the investing fascia of the deep transverse perineal muscles. The puboprostatic ligaments are dense condensations of extraperitoneal connective tissue that attach the prostate anteriorly at the prostatovesical junction to the overlying pubic symphysis just on each side of the cartilaginous symphysis. They contain no vessels and are composed primarily of collagenous fibers with variable amounts of smooth muscle deriving from the vesical neck (Albers et al., 1973). Coursing beneath the arch of the pubic symphysis through the arcuate pubic ligament, the dorsal vein of the penis emerges and then trifurcates, its branches being intimately contained within the intermediate stratum of retroperitoneal connective tissue covering the prostate on its anterior and lateral aspects (Redman, 1983) (Fig 1–55).

On its lateral aspect the prostate, covered by extraperitoneal connective tissue, is bordered by the levator

ani muscles. The portion of the muscle encroaching on the prostate has been termed the "levators prostatae." The inner investing fascia of the levator ani (endopelvic fascia) forms a fascial collar and is attached to the extraperitoneal connective tissue surrounding the prostate.

The relationships of the posterior aspect of the prostate have been the subject of considerable dissection and dissertation. The object of interest in this area is the existence or nature of a stratum termed "Denonvilliers' fascia" (Uhlenhuth et al., 1948; Tobin and Benjamin, 1945). It is generally convincingly stated that a plane of cleavage exists between the former leaves of the rectogenital septum. The studies and illustrations of Jewett and associates (1972), however, emphatically state that they "fail to demonstrate any structure that fits the testbook description of 'Denonvilliers' fascia.'" They further observed that the prostate is surrounded by a laminated capsule of dense connective tissue, the inner aspect of which is distinct and merges with the

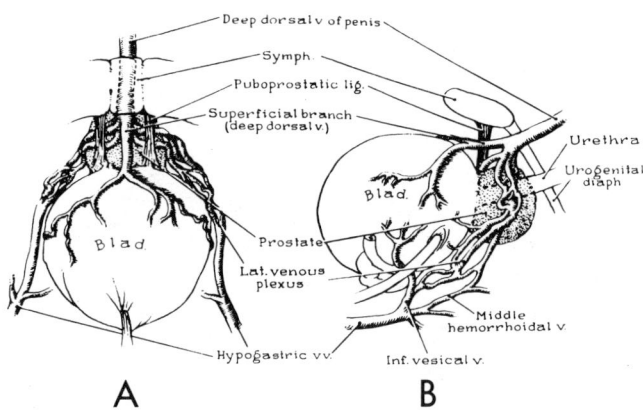

FIG 1–55.
A, view of trifurcation of dorsal veins of penis with patient in supine position. Relationship of venous branches to pubo-prostatic ligaments is depicted. **B,** lateral view showing anatomical relationships at trifurcation. (From Reiner WG, Walsh PC: An anatomical approach to the surgical management of the dorsal vein and Santorini's plexus during radical retropubic surgery. *J Urol* 1979; 121:199. Used by permission.)

stroma of the prostate. The capsule that they described is not a sharply defined structure like the skin of an apple. Therefore, they noted that in exposing the posterior surface of the prostate by stripping off the so-called posterior surface of Denonvilliers' fascia some layers of the laminated capsule can be easily peeled back. Their observation of practical significance to the urologist is that the lower anterior rectal wall is firmly attached to the apical portion of the prostate and proximal membranous urethra by the rectourethralis muscle, which spreads out somewhat onto the prostatic apex. A space exists between the anterior rectal wall and the prostate that is filled with areolar tissue containing vessels, nerves and ganglia, and a variable amount of fat. A filmy, delicate layer of connective tissue continuous with that outside the perivesical fat extends caudally to cover the posterior surface of the seminal vesicles and to fit snugly against the posterior prostatic capsule.

Vasculature

The arterial supply to the prostate may be divided for discussion into the arterial supply to the prostate and the vascular arrangements within the prostate. In regard to the arterial supply to the prostate, Clegg (1955) stressed that the prostate is supplied by a well-defined trunk, the prostatovesical artery, but that the trunk was of variable origin. The most common origin of the trunk was gluteopudendal. The course of the prostatovesical artery is fairly constant. It passes obliquely downward, forward, and medially on the anteroinferior surface of the bladder toward the prostate

gland. At a varying distance from the prostate gland it divides into two terminal branches—the inferior vesical and prostatic arteries. Clegg demonstrated that the subdivision was variable also in that there might occur a prostatic and inferior vesical artery, a large prostatic artery, and a small inferior vesical artery, or no inferior vesical arteries. The prostatic artery, however, is a constant branch of the trunk. It reaches the prostate on its anterolateral surface and passes down the lateral border, giving off five branches to the surface of the organ. A further finding was that the superior rectal artery supplied the prostate in 30% of the cases studied.

It should be emphasized that a distinct pedicle of vasculature to the prostate does not exist, only that the vessels are contained within the extraperitoneal connective tissue and may be intraoperatively bundled together as they pass to and from the posterolateral aspects.

The arterial distribution within the prostate as described by Flocks (1937) has been the most frequently utilized by urologists in the descriptions of operative techniques. Flocks described two groups of arteries within the prostate—an external capsular group and an internal urethral group (see Fig 1–52). Clegg, utilizing a more pervasive method of injecting the intraprostatic vasculature, described three zones of vessels—anterior (capsular) vessels, an intermediate zone, and a urethral plexus. Of practical significance in both investigators' descriptions is that most of the urethral vessels entered the prostatovesical junction perpendicular to the urethral surface to reach the region of the internal urethral orifice at 7 to 11 o'clock and 1 to 5 o'clock on its circumference. With distal progression, the vessels proceeded into the prostatic tissue parallel to the urethral surface. It should be noted that there are few anastomoses in the gland between the vessels of the two sides.

The venous drainage of the prostate is mainly to lateral capsular vessels, to irregular channels in the anteroinferior part of the prostate, and to the veins of the vas deferens. The capsular veins drain to the prostatic venous plexus sometimes termed Santorini's plexus. The ultimate drainage is to the vesical and internal iliac veins. The plexus is not visualized as such but is contained within the extraperitoneal connective tissue surrounding the prostate. Generally mentioned in a discussion of the prostatic venous plexus is the valveless venous communication between the prostatic plexus and the extradural venous plexus as a factor in metastases from the prostate. This extradural venous plexus is sometimes termed Batson's plexus (Batson, 1940).

Innervation

The innervation of the prostate derives from the lower part of the pelvic plexus (inferior hypogastric), which is a rectangular plate spreading over the lateral aspect of the rectum (Fig 1–56). The nerve fibers accompany blood vessels and are contained within the extraperitoneal connective tissue (Fig 1–57). Innervation is both parasympathetic, reaching the pelvic plexus through the pelvic splanchnic nerve, and sympathetic, reaching the pelvic plexus via the hypogastric nerve (Lue et al., 1984). The preganglionic efferent sympathetic fibers originate in the T10–L2 spinal cord segments. The preganglionic parasympathetic fibers arise in the S2–S4 spinal cord segments. The intrinsic innervation of the prostate is a dual autonomic innervation. Both adrenergic and cholinergic nerves are mainly concentrated in the smooth muscle layer around the prostatic acini and ducts (Vaalasti and Hervonen, 1980).

Lymphatics

The lymphatic drainage of the prostate is usually combined for consideration with that of the posterior membranous urethra. In the prostate, lymphatic chan-

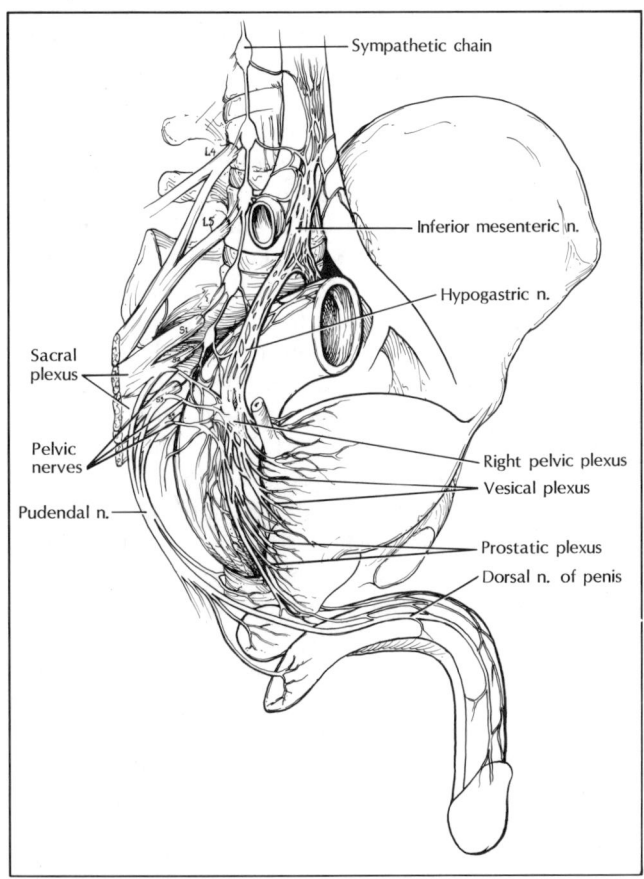

FIG 1–56.
The anatomical relationships of the pelvic innervation. (Original drawing by Christine Young, redrawn from Toldt C: *An Atlas of Human Anatomy.* New York, MacMillan, 1948, vol 1, p 890. Used by permission.)

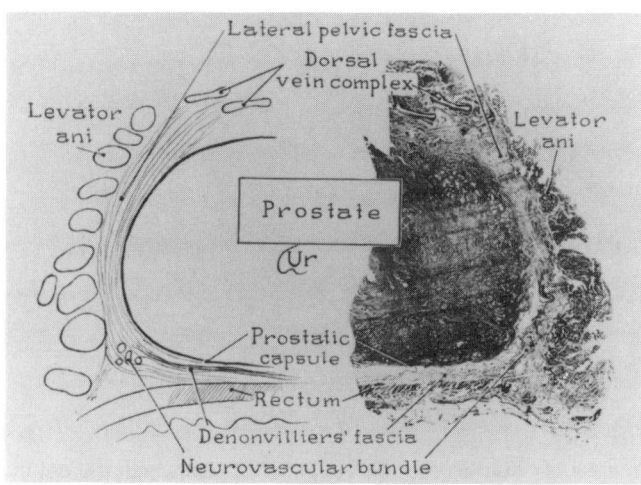

FIG 1–57.
Cross-section of adult prostate demonstrating position of neurovascular bundle, which carries innervation to corpora cavernosa. (From Walsh PC, Lepor H, Eggleston JC: Radical prostatectomy with preservation of sexual function: Anatomical and pathological considerations. *Prostate* 1983; 4:477. Used by permission.)

nels surround each prostatic acinus, which coalesce to form larger intraprostatic channels. These trunks then progress to the level of the prostatic capsule where, uniting, they form the periprostatic plexus. From this plexus major lymphatics arise and follow vascular channels. Several major groups of drainage are usually recognized. Lymphatics from the superolateral aspect of the prostate accompany the prostatic artery and drain via the inferior vesical channels to the internal iliac nodes. There is further posterior drainage to the presacral nodes and from the posterior surface of the prostate to the internal iliacs and occasionally the external iliac nodes (Raghaviah, 1979).

Microscopic Anatomy

The prostate on transverse section microscopically is noted to be surrounded by the intermediate stratum of extraperitoneal connective tissue containing lymphatics, nerves, arteries, and a relatively great number of veins of varying dimensions. Surrounding the substance of the prostate is a laminated capsule of dense connective tissue, the inner aspect of which is indistinct and merges with the stroma of the prostate. The bulk of the prostatic substance is noted to be composed of fibromuscular stroma and glandular tissue. A lobular pattern is not noted, but instead two concentric zones of glandular tissue surrounding the urethra in a semicircular fashion (Fig 1–58). The outermost glands are long and branching with ducts that empty into the prostatic sinuses and into the lateral aspect of the posterior urethra. Within the inner zone the so-called mucosal

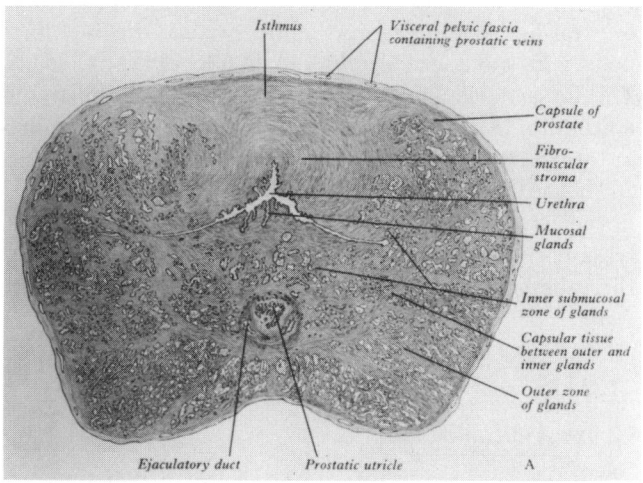

FIG 1–58.
Transverse section of prostate at level of the prostatic utricle and ejaculatory ducts. (From Gray H: *Gray's Anatomy*, 36th British ed. Philadelphia, WB Saunders Co, 1980, p 1421. Used by permission.)

glands are located and drained by ducts into the prostatic sinuses and through the colliculus seminalis. The areas of glandular tissue are separated by an intervening stroma which suggests a capsule. Histologically the glands of the prostate are tubuloalveolar, with small to relatively large glandular spaces lined by tall to low columnar epithelium, which is usually one cell layer thick. In many layers the epithelium demonstrates small villous projections of papillar inbuddings.

URETHRA (MALE)

Gross Anatomy

The male urethra from the urogenital diaphragm to the tip of the glans penis is regionally described as having a membranous part, the portion which traverses the urogenital diaphragm just prior to entering the corpora spongiosa, and a spongiose part. Collectively the membranous and spongiose urethra may be termed the anterior urethra as opposed to the posterior or prostatic urethra. For approximately 2 cm on the anterior or ventral surface the membranous urethra is covered only by fibers of the urethral sphincter. The spongiose part of the urethra may be further regionally divided into that traversing the root of the penis, that is, to the level of the convergence of the corpora cavernosa known as the bulbous urethra, and the portion traversing the pendulous portion of the penis, the pendulous urethra. There are two relative dilatations of the urethra. Within the glans penis the urethral dilatation is termed the "fossa navicularis." Within the bulb of the corpora spongiosa the urethral dilatation is termed the "intra-

bulbar fossa." Three relative narrowings of the urethra are noted, the membranous urethra itself, the external urethral meatus, and the juncture of the corpus spongiosum with the glans penis. At this juncture, plications of the mucosa are noted and are known as the valve of the fossa navicularis.

Within the lumen of the membranous urethra on the floor is noted a ridge that is contiguous with the verumontanum in the posterior urethra proximally and continuous into the bulbous urethra where it bifurcates. This ridge is termed the crista urethralis or urethral crest. Approximately 2–3 cm distal to the membranous urethra the paired orifices of the ducts of the bulbourethral glands (Cowper's glands) are noted on the floor. In the distal one half of the pendulous urethra are noted numerous small recesses termed "urethral lacunae." Within the midportion of the anterior aspect of the fossa navicularis occurs a large lacuna of varying size, the lacuna magna. Additional numerous small orifices are located on the floor of the spongiose urethra, which represent the openings of the mucosal and submucosa urethral glands (glands of Littre) (Fig 1–59).

Relationships

With the exception of the membranous urethra the remainder of the anterior urethra is contained within the corpus spongiosum lying in the ventral groove of the corpora cavernosa of the penis. Just posterior and lateral to the membranous urethra lying within the urogenital diaphragm are the bulbourethral glands (Cowper's glands).

Vasculature

The arterial supply of the urethra is primarily through the paired bulbourethral arteries, which arise as the first of three penile branches of the internal pudendal artery. The venous drainage is via emissary veins, which drain to the circumflex branches of the deep dorsal vein of the penis.

Innervation

Innervation to the urethral mucosa is via the urethrobulbar nerve, a branch of the nerve to the bulbocavernosus, which is a deep branch of the perineal nerve derived from the pudendal nerve. The pudendal nerve derives its fibers from the second, third, and fourth spinal nerves.

Lymphatics

The lymphatic drainage of the anterior urethra is to the superficial and deep inguinal lymph nodes and ultimately to the external iliac nodes. The membranous urethral drainage is generally to the hypogastric, obtu-

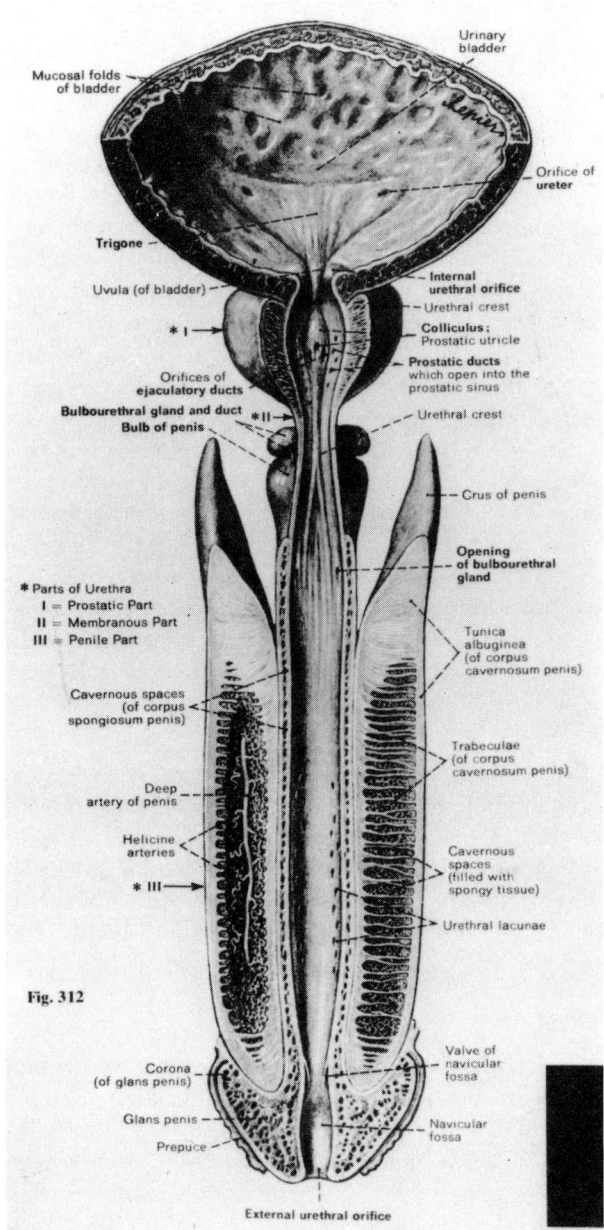

FIG 1–59.
The male urethra and its associated orifices. (From Clemente CD: *Anatomy—A Regional Atlas of the Human Body,* ed 2. Baltimore, Urban & Schwarzenberg, 1981. Used by permission.)

rator, and external iliac nodes but may also drain via the lymphatics of the anterior urethra.

Microscopic Anatomy

On microscopic cross-section the urethra is seen to be composed of a mucosa and a submucosa. The mucosa of the urethra is contiguous with the epithelium of the glans and the mucosa of the posterior urethra at the juncture with the membranous urethra. The cellular type of the mucosal surface is primarily pseudostratified or stratified columnar epithelium within the membranous and spongiose urethra to the level of the glans. The fossa navicularis is lined with stratified squamous epithelium, the epithelium extending into the openings of the respective orifices and lacunae.

The submucosa is composed of a thick stroma of a fibroelastic tissue rather firmly attached to the underlying corpus spongiosa. Minimal attachment is found of the membranous urethra, which is layered with smooth muscle.

TESTIS

Gross Anatomy

The testes are paired ovoid structures that measure approximately 4 cm long, 3.5 cm wide, and 3 cm deep. Their approximate volume is 30 ml. They are positioned so that the long axis is ventral. The anterolateral two thirds of the organ is free of any attachment. The posterolateral aspect is covered by the epididymis, connective tissue, and vasculature. The cranioposterior portion of the testes is called the mediastinum, the site of exit of the seminal conduits.

The capsule of the testis is termed the tunica albuginea. Immediately beneath the tunica albuginea is an inner vascular layer termed the tunica vasculosa. The tunica albuginea is thickest in the posterior aspect of the testis, where it forms the mediastinum of the testis (Nistal and Paniagua, 1984). Arising from the tunica albuginea are numerous trabeculae or septa, which divide the substance of the testis into approximately 250 lobules. The septa provide a framework for the vasculature and innervation of the testis.

The bulk of the testicular substance is made up of seminiferous tubules. The tubules themselves are U-shaped, with approximately two to four of the convoluted tubules composing a testicular lobule. Each tubular loop is located near the tunica albuginea, with its two ends joining together within the lobules as the straight duct or ductus recti near the mediastinum of the testis. The initial portion of the ductus recti is dilated to form an ampulla, whereas the more distal portion is thinner. The ductuli recti then drain into a system of anastomosing channels that course through the mediastinum to form the rete testis (Fig 1–60). The channels of the rete testis converge forming 10–15 efferent ductules or efferent cones, which eventually converge to form a single duct, the epididymal duct. In regard to the length of each outstretched seminiferous tubule, it has been stated that each tubule may extend for 1 m and that the total length of tubules in the human testis approaches a normal mean of 255 m with an SSD of 69 m (Kormano and Sioranta, 1971).

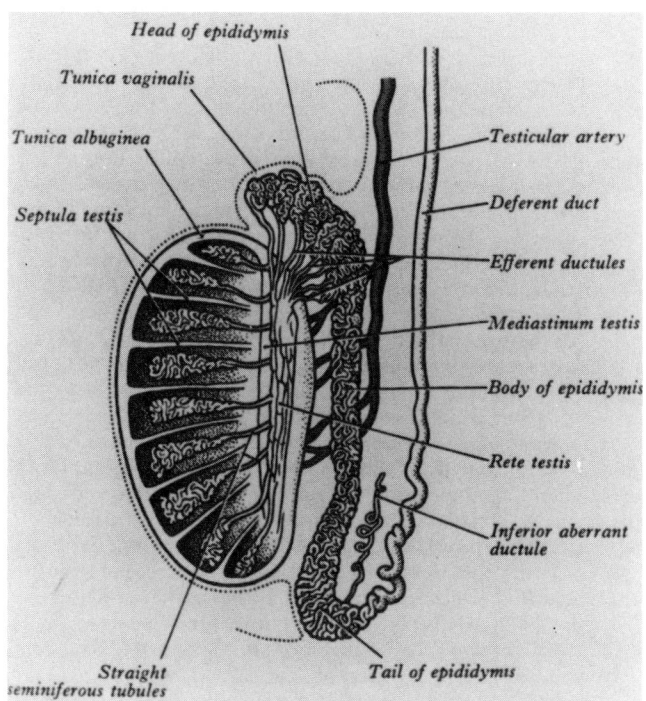

FIG 1–60.
Schematic illustration of seminiferous tubules and intratesticular collecting duct. (From Gray H: *Gray's Anatomy,* 36th British ed. Philadelphia, WB Saunders Co, 1980, p 1411. Used by permission.)

Relationships

The testis is housed in the scrotum, well insulated by the layers of scrotum derived from the investments of the abdominal wall. The most immediate investment is derived from the peritoneum and is termed the tunica vaginalis. The portion applied to the anterior and lateral aspects and the epididymis is termed the "visceral layer," while the outer wall of the serosal-lined space is termed the "parietal layer." In progression, the investments of the serosal sac surrounding the anterior three quarters of the testis are the internal spermatic fascia, cremasteric muscle and fascia, external spermatic fascia, dartos tunic, and skin. On the posterior aspect the testis is in apposition to the epididymis and the retroperitoneal connective tissue which accompanies the vasculature of the testis and epididymis.

Vasculature

The primary arterial supply of the testis begins as paired arteries, testicular arteries arising from the ventral aspect of the aorta approximately 2–3 cm caudal to the renal arteries. They progress caudally and laterally through the intermediate stratum of retroperitoneal connective tissue through the internal ring, where they accompany the spermatic cord to the testis. Either in the spermatic cord or on the surface of the testis the testicular artery divides into several branches termed the "main branches" of the testicular artery (Kormana and Suoranta, 1971). The single or divided testicular artery reaches the posterior border of the testis and passes into the thickness of the tunica albuginea obliquely. No main arterial branches pass through the mediastinum to the parenchyma of the testis (Harrison and Barclay, 1948). The main branches then divide into branches directed toward the rete testis, which are termed "centripetal arteries." Many of the major branches of the centripetal arteries run in an opposite direction and are therefore centrifugal arteries (Kormana and Suoranta, 1971). Both the centripetal and centrifugal arteries further divide, terminating in interlobular arterioles, which are located between the seminiferous tubules. The capillaries that form the capillary network mainly inside the columns of interstitial tissue are termed "intertubular capillaries," whereas the rope-like capillaries running near the tunica propria and communicating with the neighboring collection of intertubular capillaries are termed the "peritubular capillaries."

It should be noted that although the primary blood supply derives from the testicular arteries, there exist numerous anastomotic communications between the cremasteric, vasal, epididymal, and testicular arteries in man (MacMillan, 1954). The tunica albuginea has a capillary network within its layers unconnected with the testicular microvasculature. It is instead supplied from branches of the testicular arterial vessels from the region of the rete testis and branches of the epididymal artery.

The venous drainage of the testis begins as dense capillaries within the testicular substance. The intratesticular postcapillary venules join to form collecting tubules, which join to form veins directed toward the rete testis (centripetal) or toward the surface (centrifugal). The centrifugal group of veins drains only to the peripheral part of the testis and connects with the larger venous channels on the surface of the testis, which then run toward the rete on the deep surface of the tunica albuginea (Kormana and Suoranta, 1971). The centripetal veins drain the greater part of the testis running directly to the rete testis, where they form larger channels located around the poorly vascularized rete. Outside the testis the two groups of veins form the pampiniform plexus. The veins of the pampiniform plexus coalesce ultimately to form a single vein, usually within the extraperitoneal connective tissue between the internal ring and its termination. The right testicular vein drains into the vena cava on its lateral aspect 4–5 cm caudal to the renal vein. The left testicular vein drains into the caudal aspect of the left renal vein. A valve is varyingly present at the termination of the testicular veins but is least common on the left side.

Innervation

Innervation of the testis is primarily from sympathetic postganglionic and visceral afferent fibers. The innervation generally follows the vasculature. Mitchell (1935) recognized three different sources of innervation to the testis and the epididymis. The innervation includes a superior spermatic group derived from the inner mesenteric nerves and from the renal plexus, an intermediate or middle spermatic group from the superior hypogastric plexus or from the upper end of the hypogastric nerve, and an inferior spermatic group from the inferior hypogastric plexus.

Once the nerves reach the testis they surround it, branching out at the level of the tunica albuginea innervating the interlobular septa. They reach the interstitium along the blood vessels (Nistal and Paniagua, 1984). The tunica albuginea possesses sensory innervation in the form of encapsulated endings similar to Meissner and Pacini's corpuscles.

Lymphatics

The lymphatics of the testis begin as lymphatic capillaries which are seen only in the interlobular septa, but not around the seminiferous tubules (Nistal and Paniagua, 1984). Prominent lymphatic ducts in the spermatic cord accompany the vasculature and drain to para-aortic nodes. Jamieson and Dobson (1910) presented evidence of lymphatic drainage to nodes located between the renal vessels and the aortic bifurcation. Lymphatic drainage based on the appearance of metastatic lesions from testis cancer provides further evidence of the drainage of the testicular lymphatics. Donahue and associates (1982) found that for primary right-sided testis tumors, the most common site of metastasis (93%) was to nodes of the inner aortocaval zone just below the renal vein, whereas for left-sided lesions the most common nodes involved were the preaortic (88%) and the left para-aortic group (86%). There is apparent abundant crossover of lymphatics and subsequent superhilar node drainage.

Microscopic Anatomy

The tunica albuginea is generally described as a dense connective tissue consisting mainly of fibroblasts and collagen. On electron microscopy it is seen to be formed of primarily long-branching smooth muscle cells. Beneath the tunica albuginea is a heavily vascularized layer of loose connective tissue, the tunica vasculosa.

Microscopic examination of the testicular substance shows it to be composed of seminiferous tubules and interstitial tissue (Fig 1–61).

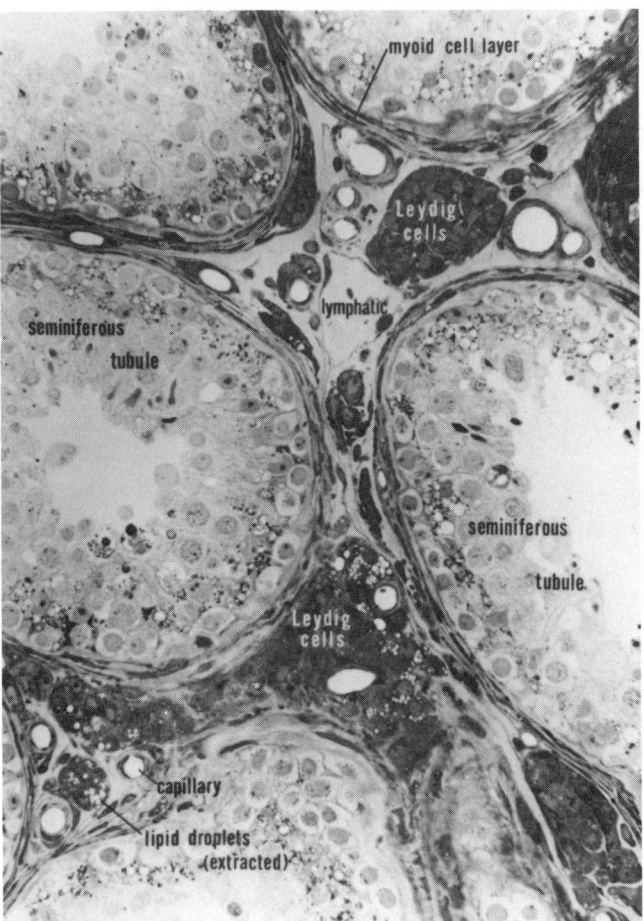

FIG 1–61.
Light photomicrograph of human testis demonstrating relationships of seminiferous tubules and interstitium. (From Christensen AK: Leydig cells, in Hamilton DW, Greep RO: *Handbook of Physiology, Section 7, Endocrinology, Male Reproductive system.* Bethesda, American Physiological Society, 1975, vol 5, p 60. Used by permission.)

The interstitium is a loose connective tissue which contains abundant lymphatics, blood vessels, and nerves. The most prominent component of the interstitium is the Leydig cell, the intratesticular source of testosterone. It is found either singly or in groups and is generally associated with capillaries.

The seminiferous tubules have an outer layer composed of a thin basement membrane surrounded by scattered myofibroblasts or myoid cells that separate the tubules from the interstitium (Nistal and Paniagua, 1984). The seminiferous tubules themselves contain two types of cells, nonproliferating Sertoli's cells and proliferating germinal cells.

Sertoli's cells or supporting cells are roughly columnar and extend from the basement membrane to the lumen of the seminiferous tubules. From the trunk extend lamellae or thin sheets between the associated germ cells and partially or often completely surround them (Fawcett, 1975) (Fig 1–62). It is through the lattice-like network of cellular extensions that the germ cells proceed through the stages of spermatogenesis.

The germ cells are proliferative. Resting on the basement membrane are type A spermatogonia, which progress from those with a pale nucleus (Ap) to those with a dark nucleus (Ad). The dark spermatogonia give rise to type B spermatogonia, which have only partial contact with the basement membrane. Type B spermatogonia give rise to primary spermatocytes by mitotic division, the so-called preleptotene stage. By meiotic division the primary spermatocytes progress through the following stages: leptotene, zygotene, pachytene, and diakinesis. The resulting two cells are termed secondary spermatocytes. Further meiotic division occurs, producing two spermatids that develop without further division into spermatozoa.

The cell types of the germinal line can be identified in random sections of any seminiferous tubule. The arrangement of the germinal cells is organized in six cellular associations. The six associations occur in such a way that in a single cross-section of several seminiferous tubules a combination of up to four of these associations may be seen (Nistal and Paniagua, 1984). The pattern of associations is apparently irregular.

EPIDIDYMIS, VAS DEFERENS, AND SEMINAL VESICLES

Gross Anatomy

The seminal conduit is a continuum of the efferent ducts through the epididymis, the vas deferens, and the ejaculatory ducts. The epididymis has a crescentic shape and is generally regionalized into the head (caput epididymis), the body (corpus epididymis), and the tail (cauda epididymis). The bulbous head of the epididymis is also termed the "globus major," whereas the tail is termed the "globus minor." Several hydatids, the appendices epididymis, are frequently noted on the anterior margin of the epididymal head. The bulk of the epididymis is composed of the tightly convoluted epididymal duct which may approach 6–8 m long. By virtue of the loose areolar tissue around the coils of the epididymal duct, the epididymis itself appears lobular (Glover, 1982).

The tail of the epididymis turns cranially to become the convoluted portion of the vas or ductus deferens, a thin, firm, muscular, whitish tube (Fig 1–63). From the tail of the epididymis the vas continues through the scrotum, the inguinal canal, and the internal abdominal ring to converge on its mate near the posterior aspect of the bladder. At its terminus the vas deferens is dilated and is termed the ampulla of the vas. The terminal portion of the vas is contiguous with the ejaculatory duct. A thin-walled, lobulated diverticulum of the ampulla is noted rising from the lateral aspect, termed the seminal vesicle, which measures 5–10 cm long. Three types of seminal vesicles based on degree of tortuosity and presence of side branches have been recognized (Aboul-Azin, 1979).

Relationships

The epididymis is closely applied to the posterior aspect of the testis and is firmly attached to the head and the tail. It is covered by the visceral layer of the tunica albuginea, except on its posterior aspect, where it is covered by the fibrofatty connective tissue contiguous with the spermatic cord.

The vas deferens occupies a posterolateral position in the spermatic cord and accompanies the vasculature of the testis covered by the fascial investments derived from the abdominal wall as it courses through the scrotum and the inguinal canal. At the internal abdominal ring the vas deferens courses medially, contained

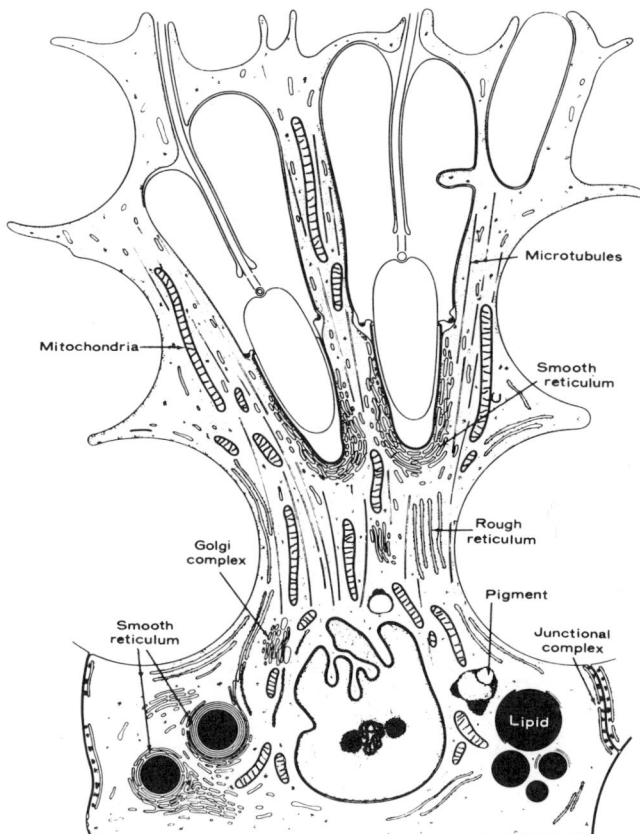

FIG 1–62.
Drawing of a typical Sertoli cell showing its shape and relationship to germ cells as well as the function and distribution of its principal organelles and inclusions. (From Fawcett DW: Ultrastructure and function of the Sertoli cell, in Hamilton DW, Greep RO: *Handbook of Physiology,* section 7, *Endocrinology, Male Reproductive System.* Bethesda, American Physiological Society, 1975, vol 5, p 23. Used by permission.)

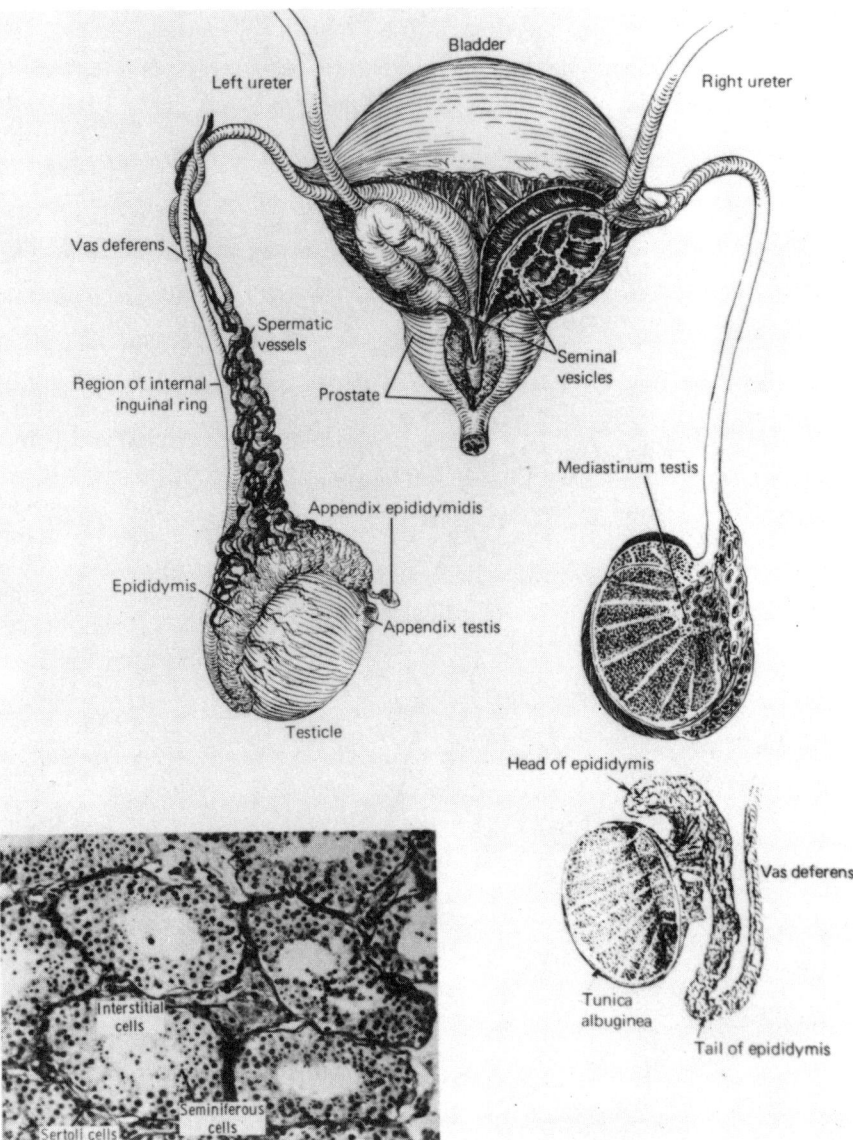

FIG 1–63.
Posterior view of anatomical relationship of testis, epididymis, vas deferens, and seminal vesicles. (From Tanagho EA: Anatomy of the genitourinary tract, in Smith DR (ed): *General Urology,* ed 11. Los Altos, Calif, Lange Medical Publications, 1984, p 8. Used by permission.)

within the intermediate stratum of retroperitoneal connective tissue to the posterior aspect of the bladder. As the vasa converge toward the midline, they pass ventral to the ureter and the posterior wall of the bladder. At the level of the interureteric ridge, the vasa lie quite close together. The ampullae of the vas and the seminal vesicles are situated between the bladder and the anterior rectal wall, being separated from these structures by retroperitoneal connective tissue (Jewett et al., 1972).

The vasa continue through the prostate substance as the ejaculatory ducts, which terminate on the lateral aspects of the verumontanum lateral to the utricle.

Vasculature

The arterial blood supply to the epididymis is via epididymal branches of the testicular artery, which vary in position and number. The richest blood supply is to the head of the epididymis. The epididymal blood supply is characterized by tortuosity of the vessels and the large number of anastomosing vessels (MacMillan, 1954). Particularly in the tail of the epididymis are noted communications between the epididymal, vasal, cremasteric, and testicular arteries.

The vas deferens is supplied by the vesiculodifferential artery, which generally arises from the umbilical artery usually in the angle between the umbilical vessel and the terminal part of the anterior division of the internal iliac artery (Braithwaite, 1952). Likewise, the ampullae of the vas and the seminal vesicles are supplied by the vesiculodifferential artery. Communication with this artery and the inferior vesical artery is sometimes seen.

The venous drainage of the epididymis is through a

scattered superficial plexus which stretches over the epididymis (MacMillan, 1954). The veins of the head and portion of the body drain to the ventrolateral aspect of the pampiniform plexus. The body and tail are drained by channels that form veins termed "marginal veins" and "epididymal veins," which ultimately join the pampiniform plexus. There are anastomoses which occur between the epididymal vein and the vasal, testicular, and cremasteric veins.

The venous drainage of the vas and the seminal vesicle are via vesiculodifferential veins and the inferior vesical plexus veins.

Innervation

The innervation of the epididymis is via the intermediate spermatic nerves arising from the superior portion of the hypogastric and the inferior spermatic nerves arising from the vesical plexus (Mitchell, 1935). The innervation primarily accompanies the vasculature with little innervation of the walls themselves noted, except in the tail of the epididymis. Innervation of the smooth muscle in the seminal vesicle and vas deferens is via adrenergic fibers from the hypogastric nerve (Mawhinney, 1983).

Lymphatics

The lymphatic drainage of the epididymis is much like that of the testis, with the lymphatics accompanying the testicular artery ultimately draining to nodes which are para-aortic and preaortic. The lymphatic drainage of the vas deferens is primarily to the external iliac nodes.

The lymphatic drainage of the seminal vesicles and ampullae of the vas is to both the external and internal groups of nodes.

Microscopic Anatomy

The epididymis from the efferent ductules to the vas deferens differs in histologic appearance as the examination moves from proximal to distal (Fig 1–64). The cells of the efferent ducts rest on a basement membrane, which is covered with circularly arranged myocytes and fibroblasts. The epithelium appears scalloped because of a combination of cuboidal and columnar cells (Nistal and Paniagua, 1984). A general statement is that the microvilli or stereocilia of the cells are shorter as the epididymal duct progresses distally and the muscular coat becomes thick. The epididymal duct acquires a longitudinal muscle coat to cover the circularly arranged myocytes as the vas itself is approached.

The epididymal duct is primarily lined with columnar epithelium with abundant microvilli termed "principal cells." Smaller basal cells are also noted in the basement membrane. Stratification and pseudostratification of the epithelium are characteristic (Glover, 1982).

The vas deferens in cross-section demonstrates a thick, muscular wall that consists of three layers, an inner and an outer longitudinal layer with a middle layer of circular muscle fibers. The epithelium is pseudostratified with stereocilia and a layer of basal cells and tends to be thrown into folds.

The ampulla of the vas appears histologically similar to the remainder of the vas with the exception of more abundant elastic fibers (Nistal and Paniagua, 1984).

The seminal vesicle is marked by enfoldings of the epithelium that range from columnar to cuboidal. Its thin muscular coat is arranged in two layers—an inner circular and an outer longitudinal.

PENIS

Gross Anatomy

The penis may be divided into two parts for discussion—a pendulous portion (the portion that is visible) and a perineal portion. The pendulous portion is termed the "corpus," or body, of the penis, while the perineal portion is termed the "radix" or root of the penis. Three masses of erectile tissue form the bulk of the penis, the paired cavernosa and the midline ventrally placed corpus spongiosum. When viewed in their entirety ventrally, the corpora cavernosa resemble an elongated pyramid with the base in the perineum. The proximal two thirds of the corpora are divergent and are attached laterally to the rami of the pubis and the medial border of the ischium. This divergent portion of each corpora is termed the "crus." At the level of the pubic symphysis the corpora cavernosa join each other, remaining uniform in size until near the terminal end, where they narrow. In the body of the penis a cross-section demonstrates a shallow groove in the dorsal midline and a deeper groove in the ventral midline. The midline dorsal groove carries the neurovascular bundle of the penis, but most prominently the deep dorsal vein of the penis. The midline ventral groove supports the corpora spongiosa which in cross-section contains the urethra. The walls of the corpora cavernosa themselves, termed the "tunica albuginea," are thick, grayish white, and have a wood-grained appearance. The cavernous bodies themselves are filled with vascular spongy tissue. The tunica albuginea is characterized by deep and superficial fibers. The superficial fibers are longitudinal. The deep fibers are circular and are arranged around each corpus, forming a midline septum between them. In the root of the penis this is more complete, while in the body of the penis it assumes a comblike appearance termed the "pectiniform septum."

The corpus spongiosum in the root of the penis is bulbous and is termed the bulb of the penis. From the convergence of the corpora cavernosa the corpus spongiosum continues distally in a ventral groove of the cor-

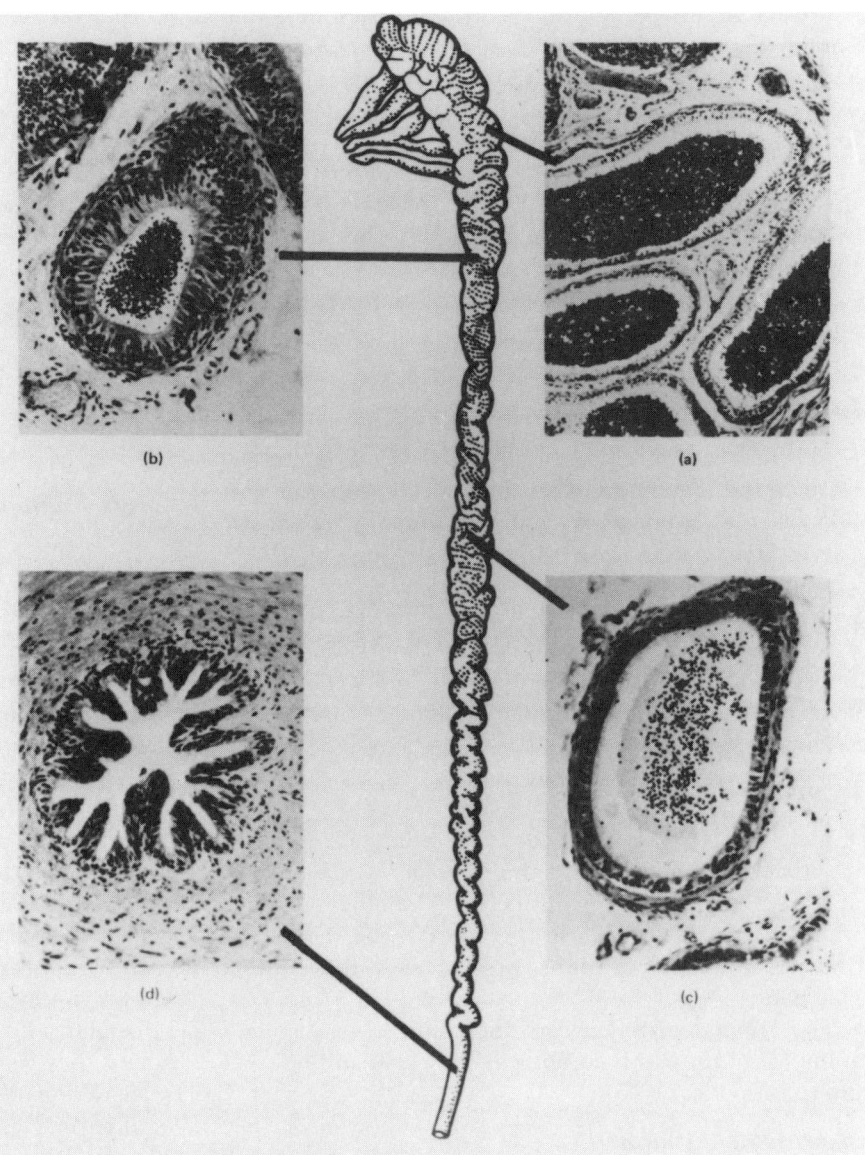

FIG 1–64.
The epididymis and the histologic appearance of the duct at different levels. *(a)* Entry of efferent ductules. *(b)* Middle segment (body). *(c)* Terminal segment (tail). *(d)* Ductus deferens (vas). (From Glover TD: The epididymis, in Chisholm GD, Williams DI (eds): *Scientific Foundations of Urology,* ed 2. Chicago, Year Book Medical Publishers, 1982, p 552. Used by permission.)

pora cavernosa to terminate as the glans of the penis, the term meaning "acorn." Another term applied to the glans is "balanus." The glans penis covers the distal one third of the corpora cavernosa. The dorsum and lateral aspects of the glans project over the body of the penis termed the "corona" of the glans. The dorsolateral juncture of the glans with the body of the penis is termed the "retroglandular sulcus," or the neck of the penis.

The investments of the corporal bodies are more easily understood if the portions of the penis are considered separately. The body of the penis is covered by skin that is characteristically thin and underlain with little fat. Unless surgically removed, the skin folds on itself at the level of the neck of the penis to form a covering for the glans termed the "prepuce" or foreskin. On the ventrum of the glans, the skin is attached

5 mm proximal to the urethral meatus along the ventrum of the glans to create a tether termed the "frenulum." Beneath the skin the body of the penis is covered with the dartos layer, which is continuous with the superficial layer (Scarpa's fascia) of the anterior abdominal wall fascia, and the dartos tunic of the scrotum and Colles' fascia of the perineum (Fig 1–65). The superficial or cutaneous dorsal veins of the penis are contained within this layer. Deep to this layer is Buck's fascia (deep fascia of the penis), which contains the neurovascular bundle of the penis dorsally. Buck's fascia covers the corpora cavernosa of the penis circumferentially and surrounds the corpora spongiosa.

The root of the penis is covered by muscle. The corpora spongiosa is covered by the bulbocavernosus muscle from the urogenital diaphragm to the convergence

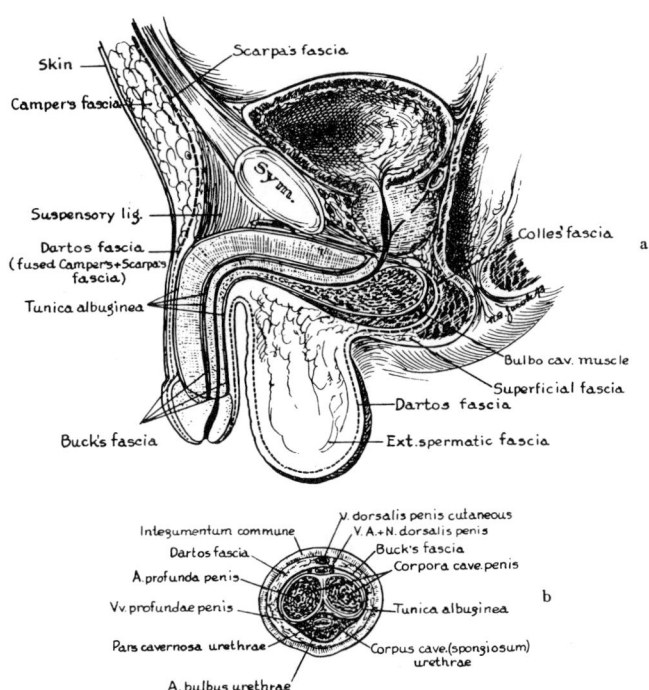

FIG 1–65.
Longitudinal **(a)** and cross-sectional **(b)** illustration of penile investments. (From Tobin CE, et al: Anatomical study and clinical consideration of the fasciae limiting urinary extravasation from the penile urethra. *Surg Gynecol Obstet* 1944; 79:199. Used by permission.)

of the corpora cavernosa. The corpora cavernosa are covered from their origin on the ischial rami to their convergence by the ischiocavernosus muscles, which are distinguished by their medial aponeurosis. Buck's fasia per se is derived from the investing fascia of the bulbocavernosus muscle.

Relationships

The root of the penis merges with the abdominal wall, the dartos tunic of the penis melding with the dartos layer of the scrotum and the Scarpa's fascia of the abdominal wall. The hollow of the pubic arch from which the penis emerges is covered on either side by the aponeurosis of the external oblique musculature, which melds with the external spermatic fascia of the cord and the fascia lata covering the pubic rami. The root of the penis is attached to the anterior abdominal wall by an extensive thickening of Scarpa's fascia in the midline termed the fundiform ligament. Inferior to the fundiform ligament the root of the penis is attached to the symphysis pubis by a fibrous extension of Buck's fascia termed the suspensory ligament. In clinical parlance the fundiform and suspensory ligaments are often together termed the suspensory ligament. At the level

of the urogenital diaphragm, Buck's fascia, the inner investing fascia of the bulbocavernosus muscles, melds with the fascia of the deep transverse perinei muscle.

The bulbocavernosus muscle is crossed on its medial aspect by the perineal artery, vein, and nerve in a longitudinal fashion as it courses from the pudendal artery, vein, and nerve in Alcock's canal.

Vasculature

The arterial supply of the penis is via the terminations of the internal pudendal artery. Three or four pairs of arteries supply the penis, the bulbourethral arteries (or bulbar and urethral), the profunda or cavernous arteries, and the dorsal arteries. The bulbourethral arteries are the first branch of the internal pudendals to reach the penis. The point of entrance is just distal to the urogenital diaphragm. The artery traverses the corpora spongiosa and ultimately provides some of the arterial supply to the glans. The second branch of the internal pudendal artery is a deep or cavernous artery that pursues a central location in each corpora cavernosa. The third branch is a dorsal artery of the penis, which gives off circumflex branches that perforate the tunica albuginea, particularly near the distal end of the corpora, and also supplies a portion of the glans penis. These arteries give off branches that form the vascular complex of the corpora cavernosa. They are described as irregular with intermittently large, distended and then narrower portions along their courses. They communicate among themselves and form the vascular lagoons of the cavernous plexus. In addition, some of the branches anastomose with branches of the artery of the opposite side (Alvarez-Morijo, 1967). Others have described these branches as having a spiraled appearance, thus the term "helicine" arteries (Fig 1–66).

A term that has frequently appeared in the literature regarding the penile vasculature is "polsters," or cushions, occurring in the lumen of penile arteries and veins. It is presumed that the term should have been translated as "bolster," which means cushion. It is now generally assumed that these cushions represent early evidence of vascular degeneration (Benson et al., 1981). The venous drainage of the penis is complex and intercommunicating. The primary vessels include a superficial dorsal vein coursing deep to the dartos tunic; a deep dorsal vein located beneath Buck's fascia and the tunica albuginea; and a short, deep or cavernous vein or veins that emerge from the proximal ends of each corpora cavernosa. The superficial dorsal vein drains primarily to the saphenous and pudendal veins. The deep dorsal vein that arises from a retrocoronal plexus is joined by circumflex veins (four to 14 in number) that drain the cavernous sinus via emissary veins that perforate the tunica albuginea both dorsally and ventrally

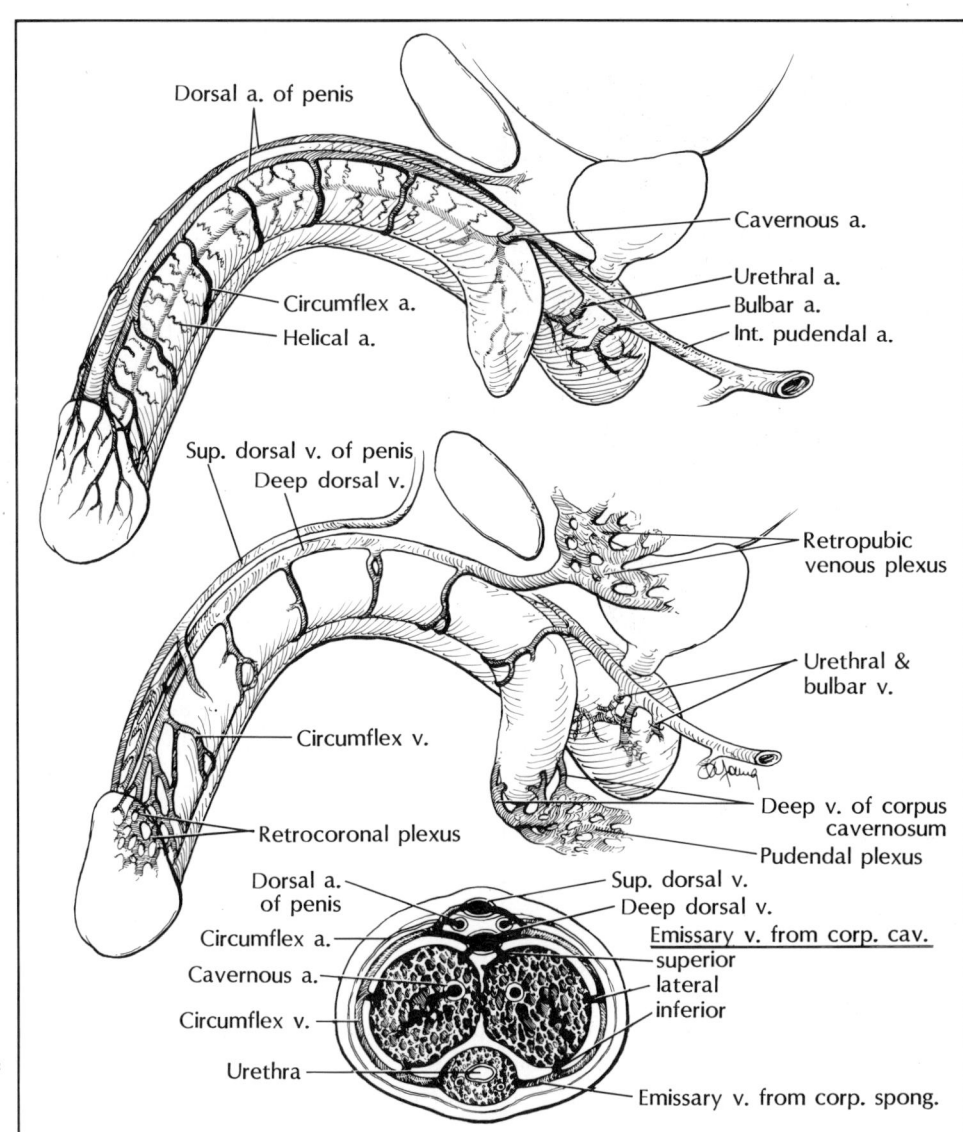

FIG 1–66.
Illustration of vasculature of penis. (Original drawing by Christine Young.)

to join the circumflex veins. In like fashion the corpora spongiosa is drained in part by emissary veins via the circumflex veins. The deep dorsal and superficial dorsal veins anastomose proximal to the coronal sinus and just distal to the pubic symphysis. The primary venous drainage of the corpora cavernosa is via the deep or cavernous veins. The primary veinous drainage of the corpora spongiosa is through the urethral and bulbar veins. The deep dorsal penile vein passes under the pubic arch to trifurcate. The lateral venous plexuses thus formed ultimately drain to the internal iliac veins. The cavernous, bulbar, and urethral veins drain to the pudendal plexus in the pelvis (Newman and Tchertkoff, 1980).

Innervation

The somatosensory afferent innervation of the penis is through the dorsal penile nerve, the perineal nerves,

the pudendal nerves, and branches of the ilioinguinal nerves. The autonomic motor innervation of the penis is from the pelvic plexus formed by visceral efferent parasympathetic preganglionic fibers from the two, three, and four sacral nerve roots and by sympathetic postganglionic fibers from the thoracolumbar ganglia (T11–L2). The parasympathetic nerves reach the pelvic plexus through the pelvic splanchnic nerve (nervi erigentes). The sympathetics reach the plexus primarily through the hypogastric nerve. The pelvic plexus forms a fenestrated rectangular plate which is found retroperitoneally by the rectum and is perforated by the branches of the inferior vesical arteries and veins (Walsh and Donker, 1982). The visceral branches of pelvic plexus that innervate the corpora cavernosa are located dorsolaterally in the connective tissue between the prostate and rectum and reach the corpora cavernosa by continuing their dorsolateral course immediately adjacent to the membranous urethra, where they traverse the urogenital diaphragm and pass behind the dorsal penile artery and nerve before entering the corpora cavernosa (Walsh et al., 1983) (Fig 1–67).

Lymphatics

The lymphatic drainage of the penile skin and prepuce accompanies the superficial vasculature of the penis and drains to the superficial inguinal lymph nodes. The lymphatic drainage of the glans and corpora spongiosa accompanies the deep dorsal vein to the root of the penis, then to the deep inguinal lymph nodes, and subsequently to the external iliac chain. Debate continues as to the existence of direct lymphatic drainage from deep penile structures to pelvic lymph nodes.

Microscopic Anatomy

The areas of most interest microscopically are the spongy tissue of the corpora cavernosa and the corpus spongiosa. The tunica albuginea per se is composed of thick collagenous fibers. In the flaccid state they have a wavy appearance, as opposed to a straight orientation on erection. Emanating from the tunica albuginea and

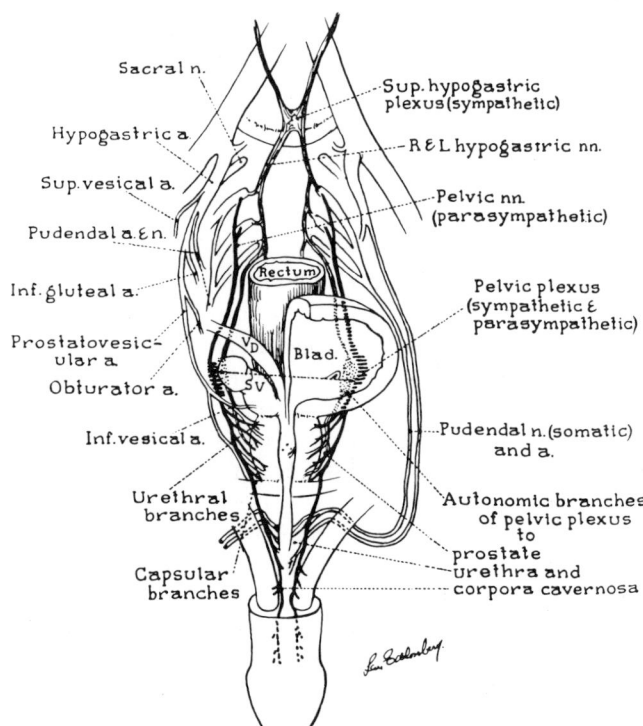

FIG 1–67.
Schematic diagram of the arterial blood supply of the prostate *(left)* and the relationship of these vessels to the pelvic plexus and its branches to the corpora cavernosa *(right)*. (From Walsh PC, Lepor H, Eggleston JC: Radical prostatectomy with preservation of sexual function: Anatomical and pathological considerations. *Prostate* 1983; 4:476. Used by permission.)

its midline septum is a fibrous skeleton or trabeculae, which surrounds the arteries, nerves, and smooth muscle fibers. The trabeculae divide the substance of the corpora into cavernous spaces that are lined with a continuous layer of flattened endothelial cells. The trabeculae are thicker at the periphery of the corpora, whereas the sinuses in the central portion are larger. The corpora spongiosa has a vaguely similar appearance. However, the trabeculae are finer and the sinuses more uniform. The tunica albuginea is considerably thinner and more elastic (Goldstein et al., 1982).

FEMALE GENITALIA

Gross Anatomy

The female genitalia may be divided for discussion into internal and external genitalia. The internal genitalia include the ovaries, the fallopian tubes, uterus, and vagina. The external genitalia include the vulva and clitoris. The ovaries, fallopian tubes, and uterus are intraperitoneal structures and as such their gross appearance is imparted by the serosal coverings (Fig 1–68).

The ovary following the start of ovulation has a grayish coloration and an ovoid, irregular contour that measures approximately 3 cm long, 1.5 cm wide, and 1 cm thick.

The fallopian or uterine tubes are approximately 10 cm long. Each tube joins the uterus medially at the superior angle of the cavity of the uterus. From medial to lateral the tube is described by regions: the isthmus, the ampulla, and the fimbriated end, which is open into the peritoneal cavity.

The uterus is roughly pear-shaped, smooth, and hollow. The domelike portion of the uterus is termed the "fundus." The caudal portion is the cervix, which protrudes into the anterior wall of the vagina. The centrally located opening of the cervix is termed the cervical os, while the canal from the vagina to the uterine cavity is the cervical canal. The tissue between the fundus and the cervix is termed the "corpus," or body, of the uterus.

The vagina, which is embedded in retroperitoneal connective tissue, is hollow and muscular. Its lumen is characterized by two median, longitudinal ridges, one anterior and the other on the posterior wall of the vagina, which are termed "vaginal columns." The recesses between the intravaginal portion of the cervix and the vaginals walls are termed the "fornices."

The clitoris consists of two paired, erectile bodies that diverge at the pubic arch to be applied to the inferior pubic rami.

Relationships

To understand the relationships of the internal female genitalia requires an understanding of the sur-

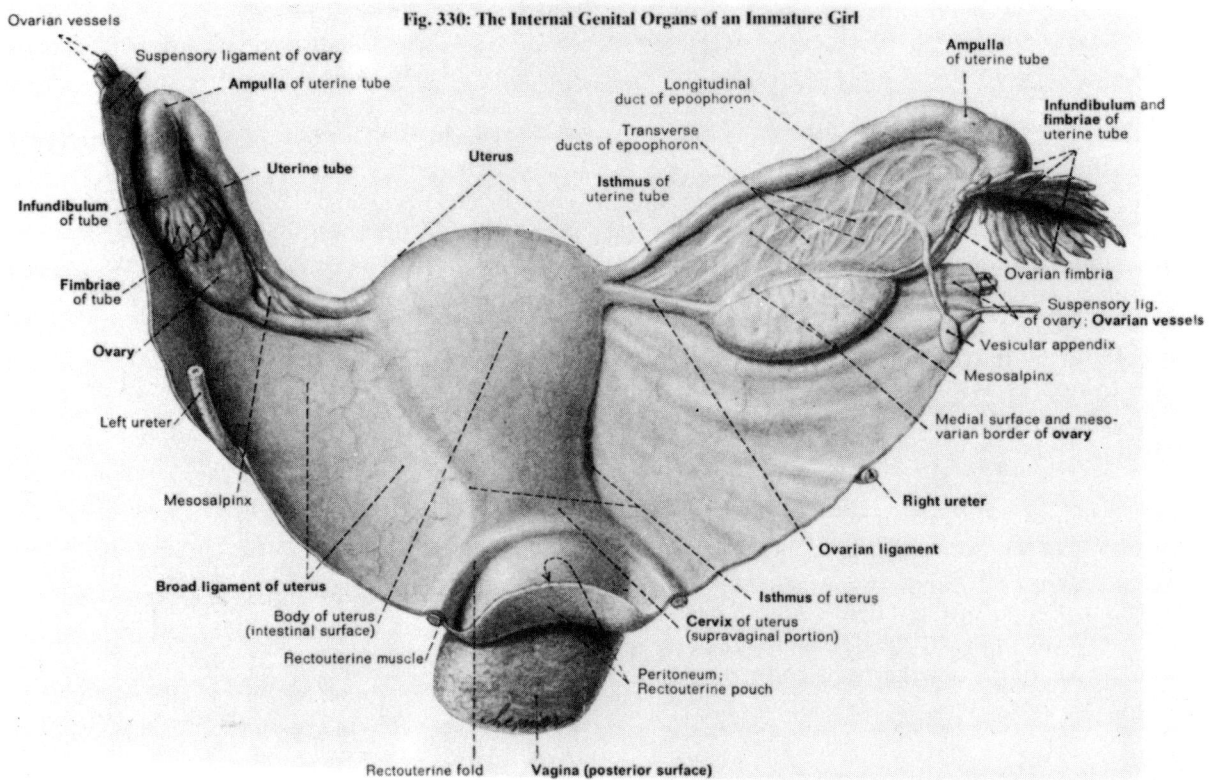

Fig. 330: The Internal Genital Organs of an Immature Girl

FIG 1–68.
Internal female genital organs. (From Clemente CD: *Anatomy—A Regional Atlas of the Human Body,* ed 2. Baltimore, Urban & Schwarzenberg, 1981. Used by permission.)

rounding retroperitoneal connective tissue and its overlying peritoneal coverings. From the craniolateral aspect of the ovary is noted the junction of the vascular pedicle overfolded with peritoneum. It is termed the suspensory ligament of the ovary or infundibulopelvic ligament. Sweeping on either side of the uterus to the pelvic side walls and encompassing the fallopian tubes, the ligaments of the ovaries and uterus, and the uterine vasculature is the so-called broad ligament, which represents extraperitoneal connective tissue covered by two layers of peritoneum. Between the fallopian tube and ovary, the intervening broad ligament is termed the "mesosalpinx," which contains the paroophoron and epoophoron. Between the ovary, ovarian ligament, and the uterus, the intervening broad ligament is termed the "mesometrium." Adjacent to the uterus the broad ligament is referred to as the parametrium. The base of the broad ligament is termed Mackenrodt's ligament or cardinal ligament or the transverse cervical ligament which primarily contains veins. There are three fibromuscular ligaments or cords that support the uterus on each side—the round ligament, the ovarian ligament, and the uterosacral ligament. The round ligament

passes from the superolateral angle of the uterus through the internal ring ultimately to reach the labium majus and is contained within the broad ligament. The ovarian ligament extends from the lower pole of the ovary to the lateral aspect of the uterus. The uterosacral ligament passes backward from the posterior aspect of the upper end of the cervix on each side of the rectum and terminates in the rectum. It is thus frequently said that there are six ligamentous supports of the female pelvic viscera: the broad, the round, the uterosacral, Mackenrodt's, the ovarian, and the infundibulopelvic ligaments (Fig 1–69).

Of utmost surgical significance is the relationship of the pelvic ureter to the uterus and its vasculature. The ureter in its terminal portion traverses the lowest portion of the broad ligament (Mackenrodt's ligament) surrounded by the uterine venous plexus and accompanied for a distance of 2.5 cm by the uterine artery. The ureter passes the supravaginal part of the cervix approximately 1.5 cm from its side to reach the bladder (Fig 1–70). The ureter is also crossed by the gonadal vein near the infundibulopelvic ligament.

The broad ligament separates the pelvis into two

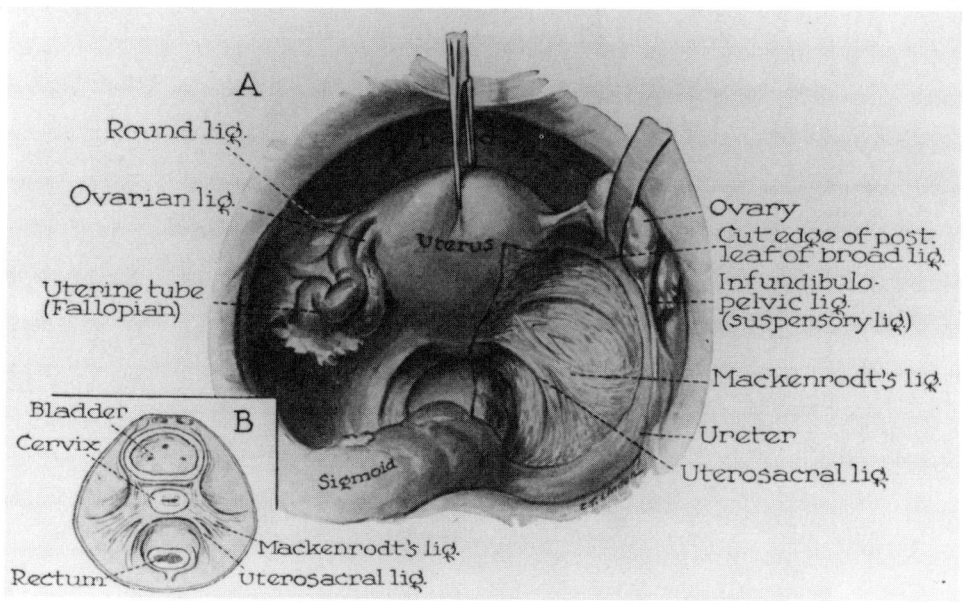

A

Round lig.

Ovarian lig.

Uterine tube (Fallopian)

Uterus

Ovary

Cut edge of post. leaf of broad lig.

Infundibulo-pelvic lig. (suspensory lig)

Mackenrodt's lig.

Bladder

Cervix

B

Sigmoid

Ureter

Uterosacral lig.

Mackenrodt's lig.

Rectum

Uterosacral lig.

FIG 1–69.
The six ligamentous supports of the female pelvic viscera. (From Thorek P: *Anatomy in Surgery,* ed 2. Philadelphia, JB Lippincott Co, 1962, p 556. Used by permission.)

halves, the anterior part containing the bladder and the posterior containing the rectum.

The vagina is intimately related anteriorly to the urinary bladder and the urethra, being separated by the extraperitoneal connective tissue. On its posterior aspect the vagina is intimately related to the rectum, being separated by the rectovaginal septum. In its terminal portion the vagina is separated from the rectum by the perineal body. As the vagina passes through the pelvic floor through the levator ani musculature, the

inner investing fascia of the levators sweep up and around the vagina to form the fascial collar of endopelvic fascia. As it passes through the pubic arch, it passes through the deep transverse perinei muscle. Just distal to the urogenital diaphragm, the vagina is in association on each side with the bulb of the vestibule and the bulbospongiosus muscle.

The clitoral crura are for most of their length on the pubic arch covered by the ischiocavernosus muscles.

Vasculature

The arterial supply to the ovary, the ovarian arteries, derive from the ventral aspect of the aorta a few centimeters distal to the origin of the renal arteries. The artery anastomoses freely with the uterine artery. Venous drainage is via the pampiniform plexus, which accompanies the artery. On the right side the pampiniform plexus coalesces to form a single gonadal vein, which joins the inferior vena cava on its anterolateral aspect near the renal vein. On the left side the gonadal vein joins the renal vein on its inferior aspect just opposite the adrenal vein.

The fallopian tube receives blood supply to its lateral one third from the ovarian artery and to the medial two thirds from the uterine artery.

The primary arterial supply to the uterus is from the uterine arteries that derive from the anterior trunk of the internal iliac artery. The uterine arteries run a tortuous course between the two layers of the broad ligament along the lateral side of the uterus and turn laterally at the junction of the uterus and fallopian tube to run toward the hilum of the ovary and terminate by joining the ovarian arteries (Farrer-Brown et al., 1970) (Fig 1–71).

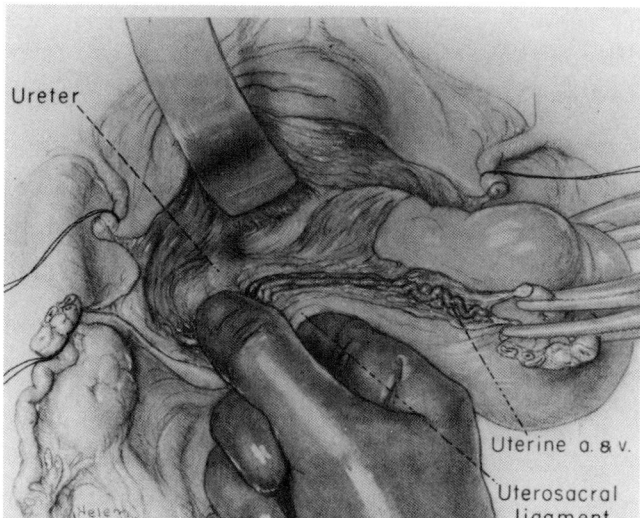

Ureter

Uterine a. & v.

Uterosacral ligament

FIG 1–70.
Intraoperative demonstration of identification of the ureter prior to clamping the uterine vessels at the level of the uterosacral ligaments. (From Burch JC, Lavely HT: *Transactions of the American Surgical Association.* Philadelphia, JB Lippincott Co, 1952, vol 70, p 388). Used by permission.)

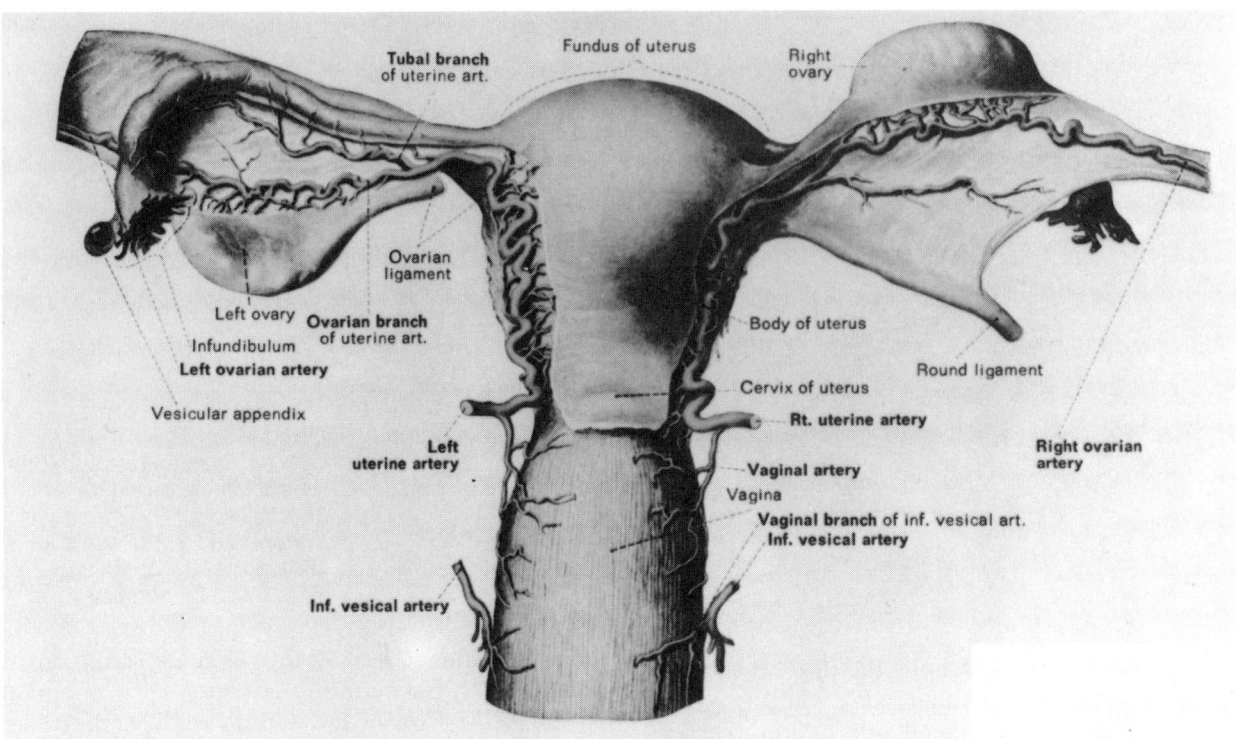

FIG 1–71.
Arterial supply to female pelvic genital organs. (From Clemente CD: *Anatomy—A Regional Atlas of the Human Body,* ed 2. Baltimore, Urban & Schwarzenberg, 1981. Used by permission.)

In the broad ligament each uterine artery supplies lateral branches that immediately enter the uterus and give off tortuous arteries and posterior arcuate divisions that run circumferentially in the myometrium approximately at the junction of its outer and middle thirds.

The uterine artery supplies the cervix and sends branches to the vagina, which anastomose with the vaginal artery to form two median longitudinal vessels, the azygous arteries of the vagina, one of which descends on the anterior aspect and the other on the posterior aspect of the vagina.

The venous drainage begins with a capsular venous system in the surface of the endometrium, which then runs a relatively straight course through the endometrium. The veins converge at the base of the endometrium to form a main collecting vein in the inner part of the myometrium (Farrer-Brown et al., 1970). The myometrial veins join with the arcuate veins, which drain to the uterine veins, forming a uterine plexus that extends along the lateral side of the uterus within the broad ligament, where it communicates with the veins of the ovarian and vaginal plexuses. The uterine plexus of veins ultimately coalesces to form uterine veins that drain into the internal iliac vein. The vagina, in addition to the uterine artery contribution, is also supplied by several vaginal arteries that derive from the anterior trunk of the internal iliac artery, corresponding to the inferior vesical arteries in the male. The venous drainage of the vagina is via the vaginal venous vein plexus, which coalesces to form vaginal veins that drain into the internal iliac vein.

The labia and vulva are supplied by branches of the perineal arteries termed "posterior labial branches" and arteries to the bulb, respectively. The deep artery of the clitoris deriving from the pudendal artery supplies the corpus cavernosum. The dorsal artery of the clitoris supplies the dorsum and the glans. The venous drainage of the clitoris is through the dorsal vein, which courses under the pubic arch to join the vesical plexus.

Innervation

The innervation of the ovary is via the ovarian plexus by efferent and afferent sympathetic fibers. Sympathetic fibers reach the ovary by the inferior hypogastric plexus.

The innervation of the fallopian tube reaches the tube along the ovarian and uterine arteries, which receive both sympathetic and parasympathetic supplies.

Uterine nerves arise from the inferior hypogastric plexus primarily from a part of the plexus that lies in the base of the broad ligament termed the "uterovaginal plexus." Innervation to the uterus or vagina then accompanies the respective arteries. Nerves passing to the cervix form a plexus sometimes containing a large ganglion, the uterine cervical ganglion. Efferent preganglionic sympathetic fibers supplying the uterus derive from the T12–L1 spinal cord segments. Preganglionic parasympathetic fibers arise in the S2–S4 spinal cord segments.

The vaginal nerves arise from the lower part of the inferior hypogastric plexus, uterosacral plexus, and pelvic splanchnic nerves and follow the vaginal arteries and their branches to supply the walls of the vagina, the vestibular bulbs, and the clitoris (Cerny, 1977).

Lymphatics

The lymphatic drainage of the ovary accompanies the ovarian artery to terminate in the lateral aortic and preaortic lymph nodes.

The fallopian tube and the fundus and upper body of the uterus drain by lymphatics that follow the ovarian vessels to terminate in the lateral and preaortic lymph nodes. The body of the uterus is also drained by lymphatics that terminate in the external iliac nodes. The lymphatics of the portion of the uterus near the junction with the round ligament follow lymphatics accompanying the round ligament to drain to superficial inguinal lymph nodes. The lower part of the body of the uterus drains by lymphatics that terminate in the external iliac nodes. The cervix is drained by three lymphatic pathways. Lymphatics travel through the parametrium to the external iliac nodes posterolaterally to the internal iliac nodes and posteriorly in the sacrogenital fold to the rectal and sacral nodes.

The vagina is drained regionally. The upper portion of the vagina is drained by lymphatics that accompany the uterine artery to terminate in the external iliac and internal iliac nodes. The lymphatics draining the middle third of the vagina accompany the vaginal artery and terminate in the internal iliac nodes, while the vaginal introitus, vulva, and skin of the perineum drain to superficial inguinal nodes. The clitoris and labia minora are drained to deep inguinal nodes. The basilar aspects of the clitoris are drained to internal iliac nodes (Gray, 1980a).

Microscopic Anatomy

The ovary on cross-section is noted to be composed of a stroma, consisting of spindle-shaped cells, which is richly vascularized. The stroma is covered by a thin tunica albuginea and a covering of cuboidal cells, the germinal epithelium. Just beneath the tunica albuginea are numerous small vesicles of primary follicles that constitute the cortex of the ovary. The central stroma is called the medulla. The stroma is filled with follicles in varying stages of maturity or involution (Gray, 1985).

The fallopian tube on cross-section is composed of three coats. The outer serosal coat is peritoneum. The middle muscular coat consists of an external longitudinal layer of smooth muscle and an internal circular layer contiguous with the musculature of the uterus. The inner mucous coat is composed of ciliated columnar epithelium, which forms longitudinal folds.

The uterus also has three coats on section—peritoneum, muscle, and mucosa. The fundus portion of the body of the uterus is covered with peritoneum. The middle layer of muscle or myometrium has in general an interlacing three layer configuration. The musculature of the cervix contains more elastic and collagen fibers than that of the body. The mucous membrane is termed the "endometrium." It consists of nonciliated columar epithelium with a subepithelial layer of highly vascularized connective tissue.

The vagina on section is noted to have an inner mucosa and an outer muscular coat consisting of two layers. The mucosa is composed of stratified squamous epithelium that rests on a subepithelial layer of loose connective tissue that contains a plexus of large veins. The muscle layer is composed of an external longitudinal layer contiguous with the superficial musculature of the uterus and an inner circular layer.

URETHRA (FEMALE)

Gross Anatomy

The female urethra measures 3–4 cm in length, and extends from the internal urethral orifice to the external urethral meatus. On incision anteriorly through its length, it is noted to be a muscular tube with a mucosal lining. A longitudinal mucosal fold runs the length of the posterior wall of the urethral lumen and is termed the "posterior urethral crest." Numerous openings representing the orifices of the urethral glands are noted as well as small recesses termed "lacunae." The ducts of paired periurethral mucosal glands drain near the lateral margins of the urethral meatus.

Relationships

From the vesical neck to the pelvic floor the urethra is surrounded by extraperitoneal connective tissue. Anteriorly it lies beneath the pubic symphysis and is firmly attached with the pubourethral ligaments, condensations of extraperitoneal connective tissue. The dorsal vein of the clitoris passes between the ligaments to join the pelvic venous plexus. Near the junction of the urethra and pelvic floor the investing fascia of the levator ani form a fascial collar around the urethra and the vagina (endopelvic fascia). Posteriorly the urethra is intimately related to the anterior surface of the vagina. The urethra passes through the urogenital diaphragm before terminating as the external urethral meatus.

Vasculature

The arterial supply to the female urethra is from branches of the vaginal artery. The venous drainage is via the pelvic venous plexus.

Innervation

The smooth muscle of the urethra in the female is primarily innervated by numerous parasympathetic nerves draining from the pelvic plexus. There is minimal sympathetic innervation. The striated muscles of the external intrinsic striated muscle fibers are branches of the pelvic splanchnic nerves (Gosling, 1979).

Lymphatics

The lymphatic drainage of the female urethra is primarily to the internal iliac nodes, with some drainage to the external iliac chains.

Microscopic Anatomy

On transverse section the female urethra is seen to be composed of an outer muscular coat and an inner mucosa. The mucosa is formed of a stratified squamous nonkeratinizing epithelium, except near the internal urethral orifice, where the epithelium is transitional. The squamous cells are approximately three cell layers thick. The basal cells are cuboidal or columnar, while the intermediate cells are polyhedral. The epithelium is supported by a lamina propria that contains an abundance of longitudinally and circularly oriented elastic fibers. The profusion of numerous thin-walled veins in the lamina propria in the past was incorrectly likened to erectile tissue.

The muscular coat consists of the external urethral sphincter, which is separate from the adjacent periurethral striated muscle of the pelvic floor. The muscle cells are striated and are arranged in a circular fashion around the urethra. In a spindle-like fashion the muscle is thickened in the midportion of the urethra, particularly anteriorly, and thinnest near the vesical neck and the distal end. It is also thinner on the posterior aspect adjacent to the vagina. The smooth muscle coat lies just under the lamina propria and consists of slender muscle bundles running obliquely and longitudinally. Adjacent to the midportion of the striated layer are some circular fibers.

ANATOMY OF SPECIAL INTEREST TO UROLOGISTS

ANTERIOR FEMORAL REGION

The anterior femoral region or inguinal femoral region is roughly triangular shaped. In muscular individuals the triangle appears as a flattened, depressed area. On the medial aspect the triangle is formed by the pectineus and abductor longis muscles. On the lateral aspect the sartorius muscle forms a side, with the inguinal ligament forming the base. The triangle thus formed is known as the femoral triangle or triangle of Scarpa (McVay, 1984; Gray, 1957) (Fig 1–72).

The deep fascia that covers the musculature of the thigh is marked by a large opening which is frequently oval near the pubic tubercle and the inguinal ligament termed the "fossa ovalis." The fossa itself is covered by loose fenestrated fascia, the cribriform fascia. Within the superficial fascia are noted the superficial lymph nodes surrounding the vessels and the saphenous vein which passes through the fossa ovalis to drain into the femoral vein. Lymph nodes surrounding the vessels to the deep fascia are termed the "deep inguinal nodes."

A space exists between the inguinal ligament and the iliopsoas and pectineus muscles. Through the space pass from lateral to medial, the femoral nerve, artery, and vein. A potential space termed the "annulus femoralis" exists medial to the femoral vein.

The primary branch of the femoral artery deep in the femoral triangle is the profunda femoris artery, which bifurcates from the femoral several centimeters caudal to the inguinal ligament. Likewise, the femoral vein

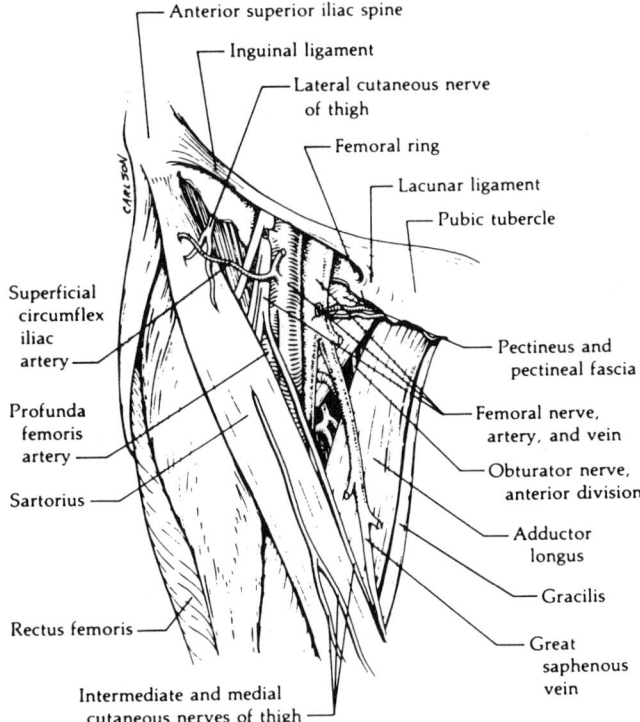

FIG 1–72.
Anatomy of the femoral triangle. (From Johnson DE, Ames FC: *Groin Dissection*. Chicago, Year Book Medical Publishers, 1985, p 5. Used by permission.)

bifurcates to receive drainage from the profunda femoris vein. Emerging from the fossa ovalis are several tributaries that travel together, the arteries from the femoral artery and the veins draining into the femoral vein. These vessels include the medially placed external pudendal, the cranially placed superficial epigastric, and the laterally placed superficial circumflex iliacs.

OMENTUM

The greater omentum is a double-leafed apron that extends from the greater curvature of the stomach to the transverse mesocolon. It is covered by peritoneum and is heavily vascularized and fat laden. The blood supply arises from the gastroepiploic arteries, the left gastroepiploic artery arising from the splenic artery and

the right from the gastroduodenal artery. Branches termed "gastric branches" supply the stomach and arise as short perpendicular vessels. The gastroepiploic vascular arcade becomes relatively small and variable in continuity towards its left extremity, which favors a pedicle based on the right gastroepiploic artery in most instances (Turner-Warwick, 1976). Ventrally the bulk of the omentum is supplied by parallel vessels that run perpendicularly and caudally from the gastroepiploic, which are termed "epiploic vessels." The epiploics may vary greatly in size and form a long, looping arcade. A potential cleavage plane exists between the omentum and anterior surface of the transverse colon and its mesocolon (Fig 1–73).

LIVER

The liver, being an intraperitoneal organ, is almost entirely covered by peritoneum. Those areas not covered are referred to as the bare areas of the liver, which are located in the cranial and posterior aspect. The

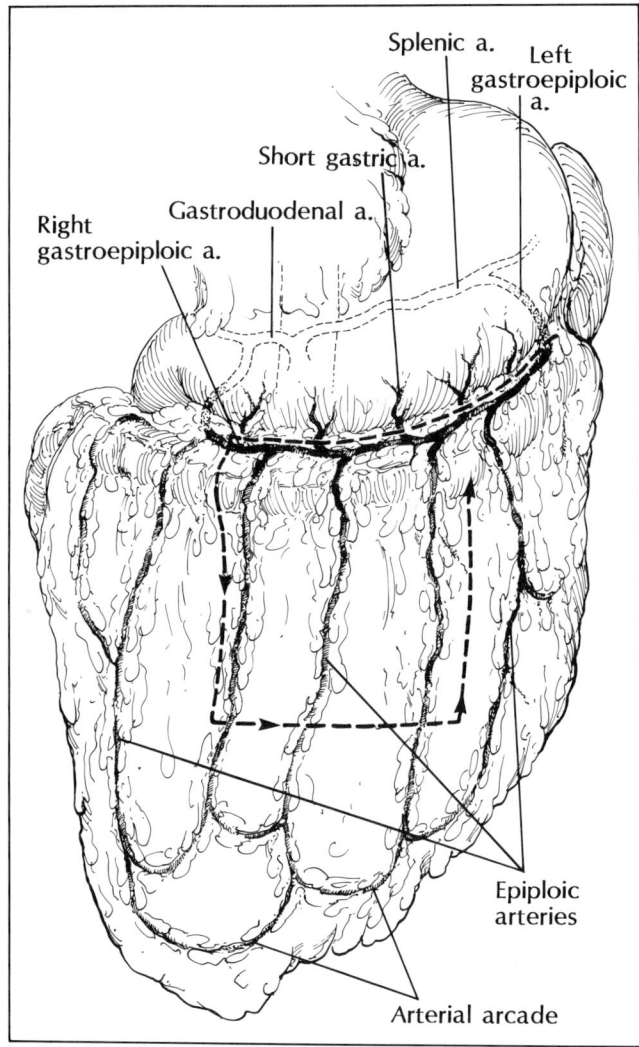

FIG 1–73.
Basic arterial anatomy of omentum. *Broken lines,* route by which omentum can be divided without loss of viability. (Original drawing by Christine Young.)

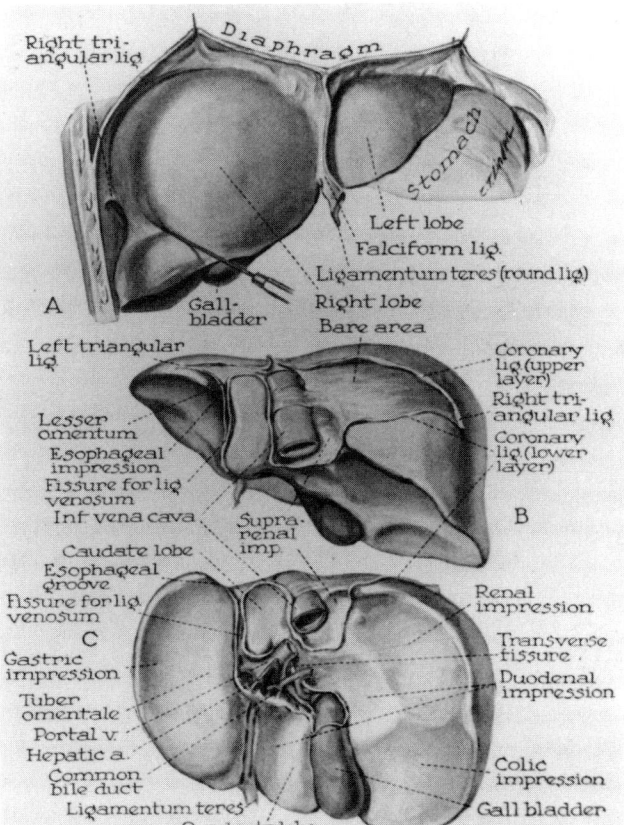

FIG 1–74.
The liver: *(A),* seen from in front—the right lobe is retracted medially to place the right triangular ligament on stretch; *(B),* the posterior aspect; *(C),* the inferior surface. (From Thorek P: *Anatomy in Surgery,* ed 2. Philadelphia, JB Lippincott Co, 1962, p 480. Used by permission.)

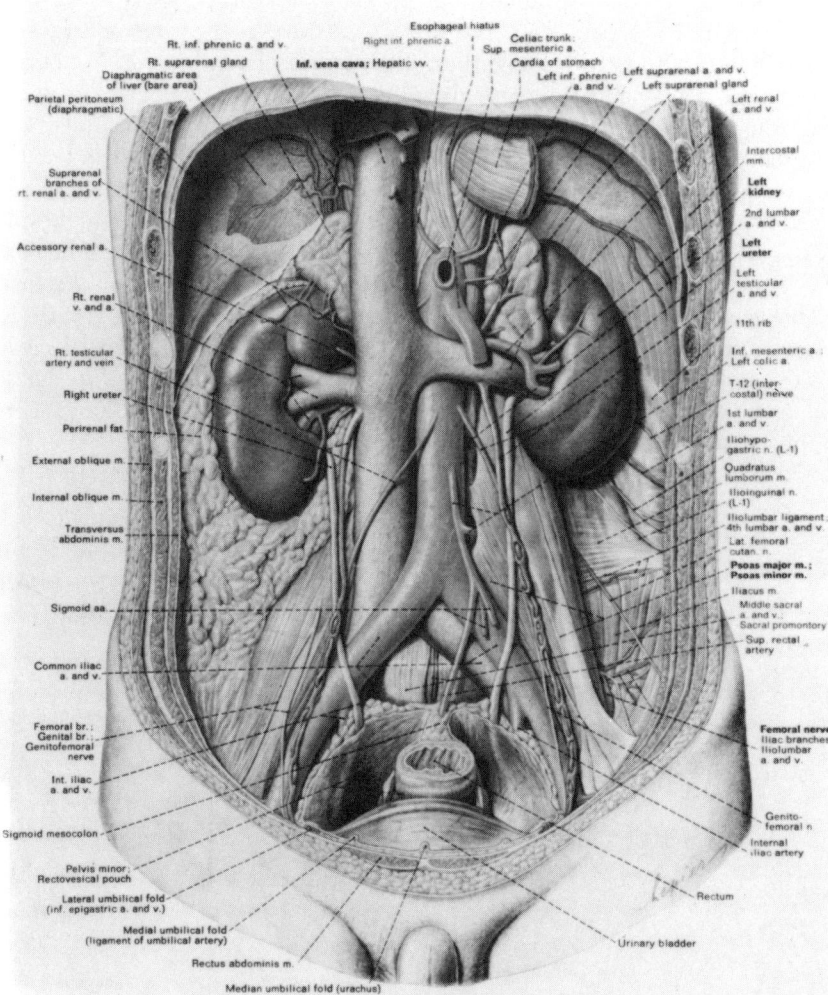

FIG 1–75.
Great vessels of the abdomen and pelvis. (From Clemente CD: *Anatomy—A Regional Atlas of the Human Body,* ed 2. Baltimore, Urban & Schwarzenberg, 1981. Used by permission.)

thickened reflections of the peritoneum onto the diaphragm are termed "ligaments" and include the right and left triangular ligaments and the upper and lower coronary ligaments (Fig 1–74). The falciform ligament containing the ligament of teres (the obliterated umbilical vein) divides the liver into lobes, the right being six times larger than the left.

The blood supply of the liver, the hepatic artery, enters the liver accompanied by the portal vein and the common bile duct, in the hepatoduodenal ligament ventral to the epiploic foramen (of Winslow), which is termed the "porta hepatis" or "portal triad." The hepatic venous drainage is via the posteriorly placed hepatic veins, which drain into the vena cava on its anterior aspect. The larger hepatic veins drain near the diaphragm. The vena cava runs in a deep groove on the posterior aspect of the liver and is variably covered by hepatic tissue or a fibrous band for a short distance termed the "ligament of the vena cava."

GREAT VESSELS

The distribution of the great vessels that lie within the intermediate stratum of retroperitoneal connective tissue is best illustrated pictorially (Fig 1–75). These include the aorta, the celiac trunk, the vena cava, the renal vessels, and the common and external iliac vessels.

COLONIC VASCULATURE

The colonic blood supply is best illustrated pictorially (Fig 1–76). It should be noted that although the colon is well supplied by a marginal artery, which is formed by anastomotic channels between the adjacent branches of colonic vessels, the junction between the left colonic artery and the left branch of the midcolic artery near the splenic flexure is occasionally minimal or absent, a critical consideration when ligation of the inferior mesenteric artery is planned.

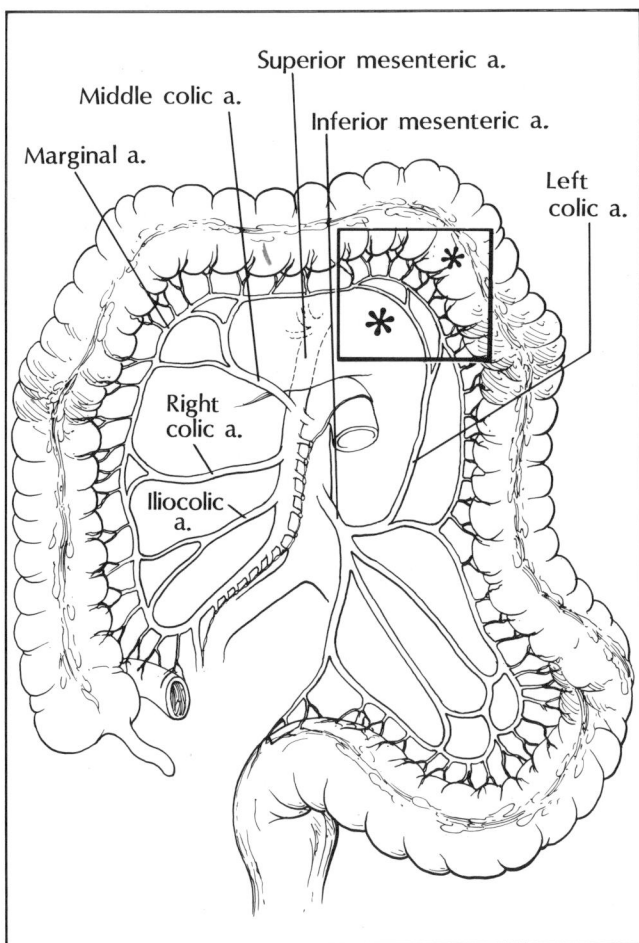

FIG 1–76.
Arterial supply of the colon. The ileocolic, right colic, and midcolic arteries arise from the superior mesenteric artery. The left colic, sigmoidal, and superior hemorrhoidal arteries arise from the inferior mesenteric artery. The anastomosis of midcolic and left colic branches at the splenic flexure is a point of union between the superior and inferior mesenteric arteries *(inset)*. (Original drawing by Christine Young.)

REFERENCES

1. Aboul-Azin TE: Anatomy of the human vesicles and ejaculatory ducts. *Arch Androl* 1979; 3:287.
2. Albers DD, Faulkner KK, Cheatham WN, et al: Surgical anatomy of the pubovesical (puboprostatic) ligaments. *J Urol* 1973; 109:388.
3. Alday ES, Goldsmith HS: Surgical technique for omental lengthening based on arterial anatomy. *Surg Gynecol Obstet* 1972; 135:103.
4. Alvarez-Morijo A: Terminal arteries of the penis. *Acta Anat* 1967; 67:387.
5. Anson BJ, Daseler EJ: Common variations in renal anatomy, affecting blood supply, form, and topography. *Surg Gynecol Obstet* 1961; 112:439.
6. Batson OV: The function of the vertebral veins and their role in the spread of metastases. *Ann Surg* 1940; 112:138.
7. Benson GS, McConnell JA, Schmidt WA: Penile polsters: Functional structures or atherosclerotic changes? *J Urol* 1981; 125:800.
8. Braithwaite JL: The arterial supply of the male urinary bladder. *Br J Urol* 1952; 24:64.
9. Cerny JC: Anatomy of the adrenal gland. *Urol Clin North Am* 1977; 4:169.
10. Clegg EJ: The arterial supply of the human prostate and seminal vesicles. *J Anat* 1955; 89:209.
11. Dobbie JW, Symington T: The human adrenal gland with special reference to the vasculature. *J Endocrinol* 1966; 34:479.
12. Donohue JP, Zachary JM, Mynard BR: Distribution of nodal metastases in nonseminomatous testis cancer. *J Urol* 1982; 128:315.
13. Doyle JF: The superficial inguinal arch, a re-assessment of what has been called the inguinal ligament. *J Anat* 1971; 108:297.
14. Elkins IB, Resnick MI: Intrarenal anatomy, in Resnick MI, Parker MD (eds): *Surgical Anatomy of the Kidney.* Mt Kisco, NY, Futura Publishing Co, 1982, pp 17–34.
15. Farrer-Brown G, Beilby JOW, Tarbit MH: The blood supply of the uterus: 1. Arterial vasculature. *J Obstet Gynecol* 1970; 77:673.
16. Farrer-Brown G, Beilby JOW, Tarbit MH: The blood supply of the uterus: 2. Venous pattern. *J Obstet Gynecol* 1970; 77:682.
17. Fawcett DW: Ultrastructure and function of Sertoli cells, in Greep RO, Astwood EB, Hamilton AW, et al (eds): *Handbook of Physiology.* Washington DC, American Physiological Society, 1975, section 7, vol V, pp 21–55.
18. Fine H, Keen EN: The arteries of the human kidney. *J Anat* 1966; 100:881.
19. Flocks RH: The arterial distribution within the prostate gland: Its role in transurethral prostatic resection. *J Urol* 1937; 37:524.
20. Glover TD: The epididymis, in Chisholm GD, Williams DI (eds): *Scientific Foundations of Urology,* ed 2. Chicago, Year Book Medical Publishers Inc, 1982, pp 544–555.
21. Goldstein AM, Meehan JP, Zakhary R, et al: New observations on microarchitecture of corpora cavernosa in man and possible relationship to mechanism of erection. *Urology* 1982; 20:259.
22. Gosling JA, Dixon JS, Humpherson JR: *Functional Anatomy of the Urinary Tract.* Baltimore, University Park Press, 1982.
23. Gosling JA: The structure of the bladder and urethra in relation to function. *Urol Clin North Am* 1979; 6(1):31.
24. Graves FT: The anatomy of the intrarenal arteries and its application to segmental resection of the kidney. *Br J Surg* 1955; 42:132.
25. Graves FT: The vascular tree, in Chisholm GD, Williams DI (eds): *Scientific Foundation of Urology,* ed 2. Chicago, Year Book Medical Publishers Inc, 1982, pp 1–9.

26. Gray DB, Bailey HA: A new technic for radical ilio-in-guinal lymph node dissection. *Ann Surg* 1957; 145:873.

27. Gray H: Angiology: The lymphatic drainage of the abdomen and pelvis, in Williams PL, Warwick R (eds): *Gray's Anatomy,* 36th British ed. Philadelphia, WB Saunders Co, 1980a, p 793.

28. Gray H: Myology, in Williams PL, Warwick R (eds): *Gray's Anatomy,* 36th British ed. Philadelphia, WB Saunders Co, 1980b, pp 506–619.

29. Gray H: Neurology, in Williams PL, Warwick R (eds): *Gray's Anatomy,* 36th British ed. Philadelphia, WB Saunders Co, 1980c, p 1135.

30. Gray H: Splanchnology, in Williams PL, Warwick R (eds): *Gray's Anatomy,* 36th British ed. Philadelphia, WB Saunders Co, 1980d, pp 1228–1464.

31. Gray H: The urogenital system (female genital organs), in Clemente CD (ed): *Gray's Anatomy,* 30th American ed. Philadelphia, Lea & Febiger, 1985, p 1556.

32. Griffith CA: The Marcy repair revisited. *Surg Clin North Am* 1984; 64(2):215.

33. Hanna MK, Jeffs RD, Sturgess JM, et al: Ureteral structure and ultrastructure: Part I. The normal human ureter. *J Urol* 1976; 116:718.

34. Harrison RG, Barclay AE: The distribution of the testicular artery (internal spermatic artery) to the human testis. *Br J Urol* 1948; 20:57.

35. Hayes MA: Abdominopelvic fascia. *Am J Anat* 1950; 87:119.

36. Hutch JA: *Anatomy and Physiology of the Bladder, Trigone and Urethra.* New York, Appleton-Century-Crofts, 1972.

37. Hutch JA, Rambo ON Jr: A study of the anatomy of the prostate, prostatic urethra and the urinary sphincter system. *J Urol* 1970; 104:443.

38. Inke G: Gross internal structure of the human kidney. *Prog Clin Biol Res* 1981; 598:71.

39. Jamieson JK, Dodson JF: The lymphatics of the testicle. *Lancet* 1910; 1:493.

40. Jewett HJ, Eggleston JC, Yawn DH: Radical prostatectomy in the management of carcinoma of the prostate: Probable course of some therapeutic failures. *J Urol* 1972; 107:1034.

41. Kaye KW, Goldberg ME: Applied anatomy of the kidney and ureter. *Urol Clin North Am* 1982; 9(1):3.

42. Kormano M, Suoranta H: Microvascular organization of the adult human testis. *Anat Rec* 1971; 170:31.

43. Lennox B, Ahmad KN: The total length of tubules in the human testis. *J Anat* 1970; 107:191.

44. Lowsley OS: Embryology, anatomy and surgery of the prostate gland: With report of operative results. *Am J Surg* 1930; 8:526.

45. Lue TF, Zeinch SJ, Schmidt RA, et al: Neuroanatomy of penile erection: Its relevance to iatrogenic impotence. *J Urol* 1984; 121:273.

46. MacMillan EW: The blood supply of the epididymis in man. *Br J Urol* 1954; 26:60.

47. Mawhinney MG: Male accessory sex organs and androgen action, in Lipshultz LI, Howards SS (eds): *Infertility of the Male.* New York, Churchill Livingstone, 1983, pp 135–163.

48. McNeal JE: The prostate and prostatic urethra: A morphological study. *J Urol* 1972; 107:1008.

49. McVay CB: Thigh, in McVay CB (ed): *Anson and McVay Surgical Anatomy,* ed. 6. Philadelphia, WB Saunders Co, 1984, pp 1176–1216.

50. Mitchell GAG: The innervation of the kidney, ureter, testicle, and epididymis. *J Anat* 1935; 70:10.

51. Mitchell GAG: Renal fascia. *Br J Surg* 1950; 37:257.

52. Moffatt DB: New ideas on the anatomy of the kidney. *J Clin Pathol* 1981; 34:1197.

53. Newman HF, Tchertkoff V: Penile vascular cushions and erection. *Invest Urol* 1980; 18:43.

54. Nistal M, Paniagua R: Adult testis, in Nistal M, Paniagua R (eds): *Testicular and Epididymal Pathology.* New York, Thieme-Stratton, Inc, 1984, pp 26–31.

55. Notley RG: The structural basis for normal and abnormal ureteric motility: The innervation and musculature of the human ureter. *Ann R Coll Surg Engl* 1971; 49:250.

56. Parker AE: The lymph collectors from the urinary bladder and their connections with the main posterior lymph channels of the abdomen. *Anat Rec* 1936; 65:443.

57. Raghaviah NV, Jordan WP Jr: Prostatic lymphography. *J Urol* 1979; 121:178.

58. Redman JF: Anatomy of the retroperitoneal connective tissue. *J Urol* 1983; 130:45.

59. Reiner WG, Walsh PC: An anatomical approach to the surgical management of the dorsal vein and Santorini's plexus during radical retropubic surgery. *J Urol* 1979; 121:198.

60. Rouiller C: General anatomy and histology of the kidney, in Rouiller C, Miller AF (eds): *The Kidney.* New York, Academic Press, 1969, vol 1, pp 61–156.

61. Shehata R: The arterial supply of the urinary bladder. *Acta Anat* 1976; 96:128.

62. Tanagho EA, Pugh RCB: The anatomy and function of the ureterovesical junction. *Br J Urol* 1963; 35:151.

63. Tanagho EA, Smith DR: The anatomy and function of the bladder neck. *Br J Urol* 1966; 38:54.

64. Tighe JR: Histology and ultrastructure, in Chisholm GD, Williams DI (eds): *Scientific Foundation of Urology,* ed 2. Chicago, Year Book Medical Publishers, Inc, 1982, pp 10–20.

65. Tobin CE: The renal fascia and its relation to the transversalis fascia. *Anat Rec* 1944; 89:295.

66. Tobin CE, Benjamin JA: Anatomical and surgical restudy of Denonvillier's fascia. *Surg Gynecol Obstet* 1945; 80:373.

67. Tobin CE, Benjamin JA, Wells JC: Continuity of the fasciae lining the abdomen, pelvis and spermatic cord. *Surg Gynecol Obstet* 1946; 83:575.

68. Turner-Warwick T: The use of the omental pedicle graft in urinary tract reconstruction. *J Urol* 1976; 116:341.

69. Uhlenhuth E, Hunter de WT: *Problems in the Anatomy of the Pelvis.* Philadelphia, JB Lippincott, 1953.

70. Uhlenhuth E, Wolfe WM, Smith EM, et al: The rectogenital septum. *Surg Gynecol Obstet* 1948; 86:148.

71. Vaalasti A, Hervonen A: Autonomic innervation of the human prostate. *Invest Urol* 1980; 17:293.

72. Walsh PC, Donker PJ: Impotence following radical prostatectomy: Insight into etiology and prevention. *J Urol* 1982; 128:492.

73. Walsh PC, Leper H, Eggleston JC: Radical prostatectomy with preservation of sexual function: Anatomical and pathological considerations. *Prostate* 1983; 4:473.

74. Woodburne RT: Structure and function of the urinary tract. *J Urol* 1960; 84:79.

Chapter 2
Standard Diagnostic Considerations

ARTHUR W. WYKER, Jr., M.D.

The majority of patients seen by a urologist are referred for evaluation of one or more urologic complaints. To arrive at the correct diagnosis, he relies primarily on a skillfully taken history supplemented by a careful physical examination and appropriate laboratory studies. When a careful and detailed history of the patient's chief complaints is taken, more often than not the probable diagnosis is indicated even before a physical examination is made or any laboratory test is performed.

Believing that a thorough understanding of urologic signs and symptoms is critical to proper evaluation, they are analyzed here in some detail. A special effort has been made to explain the mechanism of each sign or symptom and to discuss the differential diagnosis.

SIGNS AND SYMPTOMS

FREQUENCY

Frequency is usually defined as voiding at least every two hours, and it is the most common urologic symptom. When it occurs at night, it is particularly disturbing and often is the reason for medical consultation.

Etiology and Mechanism

There are two primary causes of urinary frequency—a small bladder capacity and high urine output (polyuria). An excellent screening test is the determination of *urine output per voiding*. It will be low when the bladder capacity is reduced and normal or high with polyuria.

A reduced bladder capacity may be secondary to inflammation of the bladder, changes in the bladder wall induced by infravesical obstruction, or extravesical lesions pressing on the bladder. Normal bladder mucosa is pain and pressure sensitive, and when inflamed, its threshold is markedly decreased so that it takes fewer stimuli to initiate the desire to void. Acute bacterial cystitis is by far the most common cause of bladder inflammation. Other causes include: cancer, stones, for-eign bodies, drugs (e.g., cyclophosphamide), nonbacterial cystitis (viral, fungal, parasitic, interstitial), and inflammatory processes in the adjacent bowel or vagina.

Infravesical obstruction causes a reduced bladder capacity in a variety of ways. In the early phase, the bladder muscle hypertrophies because of the increased effort necessary to empty the bladder. This hypertrophied muscle has a higher resting tonus, so that smaller than normal volumes of urine initiate the desire to void. These patients will often note a decrease in the size and force of their urinary stream. Later, the bladder muscle, initially able to compensate for obstruction by more forceful contraction, eventually becomes fatigued and is unable to empty the bladder. The bladder is decompensated and the resultant residual urine decreases the functional capacity of the bladder with more frequency. Benign prostatic hyperplasia is the most common cause of infravesical obstruction, accounting for over 90% of all cases. Other causes include carcinoma of the prostate, bladder neck obstruction, urethral stricture, posterior urethral valves (in boys), and neurogenic bladder.

Extravesical lesions pressing on the bladder may cause frequency by mechanically interfering with normal bladder expansion or by causing an irritable focus in the bladder wall. Common causes include pregnant uterus, fibroids, and ovarian masses.

When polyuria is the cause of urinary frequency, the urine specific gravity is usually less than 1.010, urine volume per voiding is more than 150 ml, and the daily urine output exceeds 2,000 ml. There are usually no other associated urinary symptoms.

The causes of polyuria are ingestion of excess fluid, voluntarily or on a psychogenic basis, diabetes mellitus, chronic renal failure, and diabetes insipidus.

In diabetes mellitus, the unreabsorbed glucose is an osmotic diuretic causing the polyuria. These patients often complain of excessive thirst, and dipstick tests of the urine reveal glucose.

In chronic renal failure, there is an increased flow of water in each nephron accompanied by an increased flow of solute. This osmotic diuresis washes out the medullary concentration gradient. The resultant impaired ability to concentrate the urine causes polyuria.

Diabetes insipidus means the passage of large volumes of dilute urine. Verney in his classic experiments (Verney, 1947) noted three fundamental causes: (1) inadequate production or release of antidiuretic hormone (ADH) by the pituitary; (2) inadequate water reabsorption by the kidneys despite normal amounts of ADH in the blood; (3) inhibition of ADH secretion from the posterior pituitary gland due to chronic excessive ingestion of fluid. Based on these studies, diabetes insipidus may be classified as pituitary, nephrogenic, or psychogenic. Clinically, the psychogenic form, often called "compulsive water drinking," is not usually thought of as a form of diabetes insipidus, since the ADH mechanism is intact.

Water balance and plasma osmolality are normally maintained by an intact thirst mechanism and by varying the rate of secretion of ADH. When fluid intake is adequate and the ADH mechanism is intact, plasma osmolality is around 285 mOsm/kg. In pituitary and nephrogenic diabetes insipidus, the mean plasma osmolality is significantly *higher than normal,* because the initial disturbance is polyuria and the excessive drinking is a normal response to the contraction and concentration of body fluids. In sharp contrast, the mean plasma osmolality in psychogenic polydipsia or compulsive water drinking is significantly *lower than normal,* since the initial disturbance is excessive drinking, and the polyuria is the normal response to the expansion and dilution of body fluids. These compulsive water drinkers have an intact ADH mechanism, so they respond appropriately to dehydration and injection of ADH with increases in urinary osmolality. Patients with the pituitary form of bladder insipidus with ADH deficiency cannot raise their urinary osmolality above the plasma osmolality when dehydrated, but they often double their urinary osmolality after the injection of ADH. In nephrogenic diabetes insipidus, there is no significant change in urinary osmolality with dehydration or after injections of ADH. The hereditary form is very rare, fortunately, for it probably is the most severe form of diabetes insipidus known. The acquired form with a nephrogenic concentrating defect is common. Examples are electrolyte disorders (hypokalemia and hypercalcemia), kidney disorders (pyelonephritis and obstructive nephropathy), drugs (lithium, phenacetin, and other analgesics), and sickle cell anemia and trait.

Emotional stress often causes urinary frequency by inducing a significant rise in intravesical pressure. Straub and associates performed cystometrograms while interviewing normal subjects (Straub et al. 1949). Stressful topics evoked a significant rise in the intravesical pressure, usually accompanied by a desire to void. Following reassurance and relaxation, the intravesical pressure returned to normal. Interestingly, discussion of emotion-laden material produced a bladder response only when it *disturbed the subject.* Some subjects seemed to enjoy venting their feelings, and they experienced no rise in their intravesical pressure.

PAIN ON URINATION

Pain on urination is usually due to inflammation of the lower urinary tract. Patients may localize their discomfort to the suprapubic area or to the end of the penis or urethra. Milder degrees of pain are often described as a burning sensation. The most common cause is bacterial infection. Other causes include nonbacterial cystitis, cancer, stone or foreign body in the bladder or urethra, and excess phosphates in the urine (phosphaturia).

It is important to determine when pain or burning is noted during urination. If it begins with the onset and stops abruptly at the end, the primary pathology is probably in the urethra. When the bladder is primarily involved, as in acute bacterial cystitis, patients experience some discomfort during urination, but often the most severe pain occurs after voiding has ceased.

URGENCY

Urgency of urination is a sudden, strong desire to void. Clinically, urgency is often classified as motor or sensory, the major difference being the presence or absence of detrusor contractions.

Motor urgency is associated with detrusor contractions, so urge incontinence is almost invariably present. When there is no evidence of neurologic abnormality, it has been called "unstable bladder." It should be noted that involuntary detrusor contractions may occur without any accompanying sensation of urgency.

Sensory urgency is caused by inflammatory lesions of the bladder and urethra and in most cases is not associated with detrusor contractions, so urge incontinence is uncommon.

In patients with urgency, look for evidence of irritative lesions of the lower urinary tract, infravesical obstruction, and neurologic disorders.

ENURESIS

Enuresis is the repeated, involuntary loss of urine during sleep. If present since birth, it is called primary enuresis; if it follows a significant "dry" interval, it is called secondary enuresis. Bedwetting spontaneously ceases with increasing age, being present in 15% of 5-year-olds, 5% of 10-year-olds, and 1% of 15-year-olds.

Etiology

Since enuresis is a symptom, not a disorder, it has more than one cause. Most authorities believe that primary enuresis is mainly due to maturational lag of the CNS with delayed development of inhibitory control of the bladder. This delayed maturation theory is supported by the following findings in enuretics: (1) the high incidence of spontaneous cures with time, (2) the presence of bladder capacity smaller than that in normal children, with resultant increased frequency of voiding, and (3) the documented hereditary aspect of enuresis.

To develop bladder control, children go through three stages. In the first stage, from birth to age 2, voiding is reflex in nature, with bladder filling evoking a bladder contraction with complete emptying of the bladder. In the second stage, age 2–3, the child becomes aware of the desire to void, but voiding is still precipitous and uncontrolled. It is during the third stage, age 3–4, when most children become able to postpone voiding. This ability is the key to normal bladder control.

Urinary tract obstruction and emotional disorders were once thought to be the chief causes of enuresis, but we now know that this is not true, for together they account for only about 5% of all cases of enuresis. Emotional disorders play their chief role in acquired or secondary enuresis.

Sleep Disorders

Since most enuretics sleep through their bedwetting episodes, it has been suggested that sleep disorders may play a role in enuresis. Earlier studies suggested that enuretics slept more deeply than nonenuretics and that their bedwetting episodes tended to occur during the deeper stages of sleep. The current studies show that deep sleep occurs with about equal frequency in enuretic and nonenuretics. Kales demonstrated that enuretic episodes occur randomly during the night, with the frequency in any given sleep stage being proportionate to the amount of time spent in that stage (Kales et al., 1977).

Bladder Function

Esperanca and Gerrard studied bladder function in normal children and enuretics and found that the fundamental disability of the enuretic was a decreased functional bladder capacity (Esperanca and Gerrard, 1969). In addition, they noted that enuretics had significant urinary frequency, voiding approximately twice as many times as normal children.

Physiologic Maximum Bladder Capacity in Children

Esperanca and Gerrard (1969) believed that a bladder capacity of over 300 ml was necessary for a child to go through the night without voiding. Notice that all of the normal children had a bladder capacity of 300 ml or more, whereas all the enuretic children except the 10-year-olds had a bladder capacity of less than 300 ml (Table 2–1). Although enuretics usually have a reduced functional bladder capacity, their bladder capacity during anesthesia is normal, indicating that there is no structural abnormality of the bladder.

Frequency of Voiding in 24 Hours

Normal children over 4 years of age usually void 5–6 times a day, whereas enuretics usually void 9–10 times a day (Table 2–2).

Cystometrograms are abnormal in about 75% of children with primary enuresis (Pompeius, 1971; Whiteside and Arnold, 1975). The most common findings were small bladder capacity and uninhibited detrusor contractions.

Evaluation

If bedwetting is the only symptom, urologic evaluation is not indicated. When bedwetting is accompanied by daytime frequency, urgency, and incontinence, urologic studies have usually demonstrated an unstable detrusor with involuntary detrusor contractions. Urologic evaluation is primarily reserved for children with a history of urinary infections or a weak urinary stream.

Structural abnormalities of the urinary tract that cause enuresis do so by obstructing the lower urinary tract. Bladder muscle hypertrophies because of the increased effort necessary to empty the bladder. This thickened muscle is more rigid and has a higher resting

TABLE 2–1.

Physiologic Maximum Bladder Capacity*

AGE, YR	NORMAL	ENURETICS
4	296	180
5	301	238
6	359	279
7	394	217
8	428	272
9	457	281
10	473	353

*From Esperanca M, Gerrard JW: Nocturnal enuresis: Studies in bladder function in normal children and enuretics. Can Med Assoc J 1969; 101:324. Used by permission.

TABLE 2–2.

Frequency of Voiding in 24 Hours*

AGE, YR	NORMAL	ENURETICS
4	5.3	11.9
5	5.7	11.0
6	6.4	10.0
7	5.5	8.4
8	5.3	9.7
10	4.6	10.7

*From Esperanca M, Gerrard JW: Nocturnal enuresis: Studies in bladder function in normal children and enuretics. Can Med Assoc J 1969; 101:324. Used by permission.

tonus than normal so that smaller than normal volumes of urine may trigger voiding.

If you are asked to evaluate a child with enuresis, your primary responsibility is to rule out infection and lower urinary tract obstruction. Infection is easily excluded by obtaining a clean voided urine for microscopic examination of the sediment and culture. Obstruction can be assessed by the two following *noninvasive* studies. The simplest and least costly study is a urinary flow rate obtained with a full-bladder voiding. Alternatively, ultrasound may be used to scan the kidneys for hydronephrosis, the bladder for wall thickness and postvoid residual urine. If the findings are inconclusive, a voiding cystourethrogram can be obtained without catheterization by using the contrast material present in the bladder at the end of the IV urogram. When one of these studies indicates obstruction, cystourethrograms and cystoendoscopy should be performed to pinpoint the site and degree of obstruction.

Management

If your evaluation reveals no evidence of urinary obstruction or infection, allay anxiety by presenting the following information to the parents and their child.

1. The child's urinary tract is normal.

2. Bedwetting is *involuntary,* so the child should not be punished or humiliated.

3. Be patient, for bedwetting spontaneously stops in time.

4. Effective treatment is available.

Two treatment regimens have had significant impact on enuresis: drug therapy and behavioral conditioning (Wagner et al., 1982). When nocturnal enuresis is accompanied by daytime frequency, urgency, and incontinence, involuntary detrusor contractions are usually present. They may be prevented by anticholinergic drugs. Imipramine is the most effective drug used to treat enuresis. Although its mechanism of action is not completely understood, it increases functional bladder volume and raises urethral pressure, both actions tending to eliminate enuresis.

Behavioral conditioning using the urine alarm is more effective than drug therapy, but it requires considerable patience, since successful treatment usually takes about four months. When a few drops of urine activate the lightweight alarm system attached to the child's pajamas, a buzzer goes off. The child awakens and completes voiding in the toilet. The alarm evokes two responses, awakening and inhibition of the voiding reflex. In time, conditioning results in the child's awakening before voiding, and enuresis stops.

PAIN

Mechanism

The chief cause of pain in the urinary tract is distention, with increased intraluminal pressure. The severity of the pain is not primarily related to the degree of distention but to the rapidity with which it develops. Sudden distention of the ureter, renal pelvis, or calyces causes severe pain, whereas gradual distention of the same structures causes little or no pain.

Intraluminal pressure seems to be the critical determinant of pain. When intrapelvic pressures have been measured in patients with an acutely obstructing ureteral calculus, the higher the pressure, the more severe the pain. With long-standing obstruction, distention may be marked, but the pressure in the renal pelvis is normal or only minimally elevated, and the patients experience little or no pain.

This concept that pain equals distention is extremely useful clinically. Urologists are often asked to evaluate patients with pain in the back, flank, or abdomen. When an IV urogram at the time of a pain episode is normal with no dilation of the ureter, pelvic, or calyces, you can state with confidence that pain is nonurologic in origin or is not present. Narcotic addicts mimicking renal colic are most commonly identified in this way.

Two other causes of pain are distention of the renal capsule and acute renal ischemia. Both produce steady, usually mild pain in the costovertebral angle (CVA) region.

Type and Location of Pain

Pain due to distention may be steady or intermittent. Intermittent pain, particularly when it is severe, is often called renal colic and is most commonly caused by an obstructing stone in the lower ureter. With each peristaltic wave, more urine is pumped into the obstructed portion of the ureter, with a resultant increase in the hydrostatic, intraluminal pressure and in the intensity of the pain.

Steady pain, uniformly present with distention of the renal capsule and acute renal ischemia, also occurs in up to 50% of patients with distention of the ureter, renal pelvis or calyces. This steady pain in the face of obstruction and distention appears to be due to the absence of ureteral pressure waves. When the ureter is acutely obstructed, there is an increase in the frequency of ureteral contractions and in the amplitude of the generated pressure waves. This response, however, is short-lived, for soon ureteral contractions cease and the intraluminal ureteral pressure remains chronically elevated and steady. The distribution of urinary tract pain has been carefully mapped out by distending the renal pelvis and various portions of the ureter with small balloon catheters (McLellan and Goodell, 1942). The location of renal and ureteral pain is shown in Figure 2–1.

Notice that urinary tract pain does not occur in the central portion of the abdomen, but is lateralized to the outer abdomen, flank, and CVA region. Ureteral pain followed a line along the lateral edge of the rectus muscle.

Primary pain sites   **Referred pain sites**

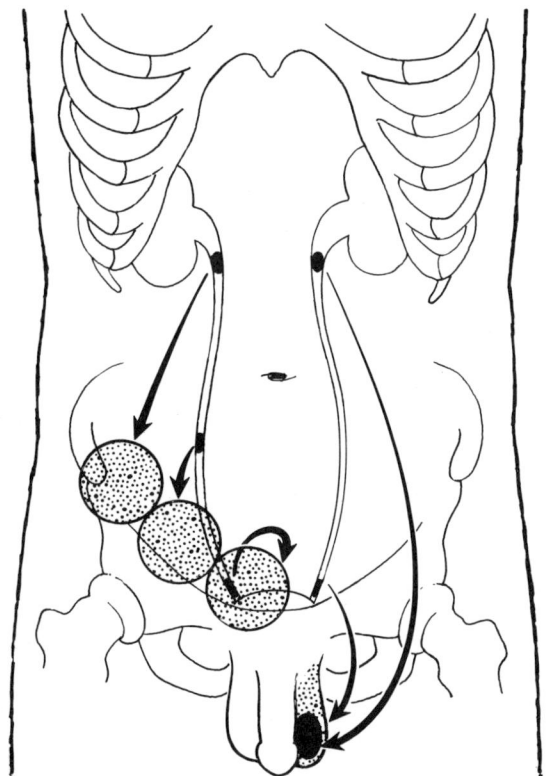

FIG 2–1.
Sites of urologic pain. (From Wyker A, Gillenwater JY: *Method of Urology*. Baltimore, Williams & Wilkins Co, 1975. Used by permission.)

SITE OF BALLOON DISTENTION	SITE OF PAIN
Renal pelvis	CVA
Upper ureter	Flank
Mid-ureter	Mid-inguinal canal
Lower ureter	Suprapubic area

Pain Pathways and Referred Pain

Kidney and ureter pain impulses are carried in vesical afferent fibers that accompany the sympathetic nerves and enter the spinal cord via the posterior spinal roots. Renal and ureteral pain can be abolished by sympathectomy of T7–L3 or by sectioning posterior spinal roots $TH_{11, 12} L_{1, 2}$.

Referred pain is pain projected to an area distant from the point of stimulation. It may occur in addition to or in the absence of true visceral pain. The exact mechanism of referred pain is unknown. In the pain sensory pathway, the visceral afferent fibers are in close proximity to the somatic neurons, and the ureteral or renal pain impulse may be diverted to the somatic neuron so that the brain interprets the pain as having come from the skin segment. They share a common segmental innervation so that the location of referred pain depends on the spinal cord segment involved. Spinal cord segments $T_{11, 12}$ receive sensory fibers from both the upper ureter and testis, so that distention of the upper ureter may cause referred pain to the ipsilateral testis. In like fashion, distention of the lower ureter may cause referred pain to the ipsilateral scrotum.

Pain in the Lower Urinary Tract

Bladder distention initially causes fullness, then pain in the suprapubic region and end of the penis associated with an intense desire to void (urgency). Normal bladder mucosa is sensitive to stimuli, which is greatly increased by inflammation. Stimulation of the trigone, ureteral orifices, or anterior urethra causes referred pain to the end of the penis. The posterior urethra is relatively unresponsive to stimuli.

URINARY INCONTINENCE

Urinary incontinence is the *involuntary* loss of urine. Blandy (1978) described it as follows: "There is almost no symptom more degrading or miserable than incontinence of urine: it inevitably brings, at any stage after early childhood, shame, stink, soreness, and ostracism. It is hard to visit and love him."

Mechanism

From the urodynamic point of view, urine is retained in the bladder for one and only one reason: intraurethral pressure is greater than bladder pressure. Liquids never flow from an area of low pressure to one of higher pressure, so urine remains in the bladder because bladder pressure is usually in the 10–20 cm H_2O range, whereas urethral pressure is usually in the 80–100 cm H_2O range. The difference between these two pressures—approximately 70–80 cm H_2O—acts as a protecting pressure gradient, or closure pressure. When this pressure gradient is lost and bladder pressure exceeds urethral pressure, urinary incontinence occurs. This may result from an abnormal increase in bladder pressure, an abnormal decrease in urethral pressure, or both.

Classification of Urinary Incontinence

The key to diagnosis of urinary incontinence is a careful, detailed history.

STRESS INCONTINENCE.—Most patients with this form of urinary incontinence have a characteristic history. They leak only after coughing, sneezing, jogging, or other activities that cause a sudden rise in intra-abdominal pressure. When leakage occurs, it does so in jet fashion, often in sizable quantities.

These patients are usually women who have lost some of their urethral support. With stress, the now mobile urethra herniates into the vagina, so less pressure is transmitted to the urethra than to the bladder. This deficient transmission of intra-abdominal pressure to the urethra is the primary cause of stress incontinence. Normally, stress transmits equal pressure to both the bladder and urethra, preserving the protective pressure gradient and preventing stress incontinence.

OVERFLOW INCONTINENCE.—This form resembles stress incontinence in that urine leakage is triggered by a rise in intravesical pressure, but the etiology and mechanism differ. In stress incontinence, residual urine is low, and the primary defect is loss of urethral support. Overflow incontinence is due to *chronic failure of bladder emptying* secondary to obstruction or, less commonly, inadequate detrusor contractions. Long-standing infravesical obstruction may cause bladder decompensation with resultant chronically high residual urine, often over 1,000 ml. The distended bladder is usually easily palpable and percussible in the lower abdomen. The intravesical pressure is consistently elevated so that slight increases in intra-abdominal pressure may raise the intravesical pressure enough to overcome the urethral resistance, with escape of urine. Clinically, this form of urinary incontinence is most commonly seen in men with obstruction from benign prostatic hyperplasia.

URGENCY INCONTINENCE.—In this form, the patient has a sudden, strong desire to void, but is unable to reach the bathroom in time. The key point here is that the *desire to void* precedes urinary leakage. The cause of urine leakage is an involuntary detrusor contraction, so leakage tends to be intermittent and often of significant quantity. The usual causes of urgency incontinence are inflammation in or adjacent to the bladder, long-standing infravesical obstruction, and varying forms of neurogenic bladder.

Clinically, this is where the cystometrogram is most useful, demonstrating involuntary detrusor contractions.

NONRESISTANCE, OR 'LEAKY FAUCET' INCONTINENCE.—This form of incontinence differs significantly from the others. These patients have virtually no pressure gradient between the bladder and the urethra, so that urine leakage occurs day and night and is *independent of stress or desire to void*. The urinary sphincters may be absent (epispadias), bypassed (urinary fistula or ectopic ureter in females), or damaged (postprostatectomy).

When an ectopic ureter in a female causes urinary incontinence, the history is virtually diagnostic. Despite a normal voiding pattern, they are continuously wet both day and night. Urinary incontinence occurs because the ectopic ureter opens distal to the urinary sphincters, most commonly in the outer urethra or in the vestibule adjacent to the urethral meatus. It may also open into the vagina, cervix, or uterine cavity. The ectopic ureter is usually associated with complete ureteral duplication, with the ectopic ureter going to the upper segment of the kidney. Urinary incontinence does not occur in males with ectopic ureters because all wolffian duct derivatives—posterior urethra, vas deferens, epididymis, and seminal vesicle—are proximal to the urinary sphincters. Because of its wolffian duct origin, the ectopic ureter must open somewhere along the course of the wolffian duct.

Postprostatectomy incontinence most commonly results from damage to that portion of the smooth muscle urinary sphincter that is distal to the verumontanum.

HEMATURIA

Hematuria is a dramatic indicator of pathology in the urinary tract, yet it is often ignored by patients and physicians alike. The passage of blood-stained urine may be the first sign of serious disease in the urinary tract, and a single episode of hematuria warrants a thorough urologic investigation.

The presence of blood is established by finding an abnormal number of red blood cells (RBCs) in the urinary sediment. A normal individual usually excretes about 30,000 RBCs/hr, but may excrete up to 100,000 RBCs/hr, or about 1,600 RBC/min. If the urine output is 50 ml/hr (1,200 ml/day), up to 2,000 RBCs/ml may be excreted (100,000/50), and this concentration of RBCs gives urine findings of occasional to 1 RBC per high-power field (HPF). Therefore, greater than 1 RBC/HPF may be considered microscopic hematuria.

There is one common clinical situation, hypotonicity, in which hematuria may be present, yet the urinary sediment shows no RBC. When the urine specific gravity is less than 1.008, RBCs rupture, with release of hemoglobin, which can be detected with a dipstick test for hemoglobin. These impregnated cellulose strips are designed to detect free hemoglobin, hemoglobin contained in RBC, and myoglobin.

Osmotic Rupture of RBCs by Hypotonic Urine

Urine containing 100,000 RBCs/ml with a urinary sediment finding of 30 RBCs/HPF was used as the standard.

URINE SPECIFIC GRAVITY	RBCs/HPF	DIPSTICK	% LYSIS
1.001	0	+ + + +	100
1.005	0	+ + + +	100
1.007	0	+ + + +	100
1.010	30	+	0
1.015	30	+	0
1.020	30	+	0
1.025	30	+	0
1.028	30	+	0

Notice that the dipstick test is positive in all cases of hematuria regardless of the urine specific gravity. Since osmotic lysis may occur whenever the urine specific gravity is 1.007 or less, hematuria cannot be ruled out unless you examine fresh urinary sediment for RBC *and* test the supernatant with dipstick.

Nonhematuric Red Urine (Pseudohematuria)

Certain urinary pigments may impart a pink-to-red color to the urine mimicking hematuria, but the urinary sediment shows no RBC and the dipstick test is negative.

Common causes of nonhematuric red urine are (1) anthocyanins in beets and berries (beeturia), (2) phenolphthalein (present in some laxatives), (3) vegetable dyes (used for food coloring), (4) heavy concentration of urates, (5) presence of phenazopyridine (Pyridium), and (6) *Serratia marcescens* infection in infants (red diaper syndrome).

Myoglobinuria is also a cause of nonhematuric red urine, but although the urinary sediment shows no RBCs, the dipstick test is positive.

Classification of Hematuria

It is helpful clinically to classify hematuria in two ways: first by quantity and second by the time of its appearance during voiding. Quantitatively, it is called gross hematuria if it is evident to the naked eye and microscopic if demonstrable under the microscope. Microscopic hematuria is more commonly nephrologic in origin, whereas gross hematuria is more commonly urologic in origin.

Blood noted chiefly at the beginning of urination is called *initial hematuria*, and it indicates pathology in the urethra. Blood noted chiefly at the end of urination is called *terminal hematuria*, and it indicates pathology near the bladder neck or in the prostatic urethra. Uniformly bloody urine, *total hematuria*, indicates pathology in the bladder, ureters, or kidneys. When blood is noted only between voidings or as stains on underclothing or pajamas and the voided urine is clear, the pathology is at the urethral meatus or in the anterior urethra.

Significance of Hematuria

Hematuria is a sign or symptom of a pathologic process in the urinary tract, and it should be investigated, not treated. The significance of hematuria varies with the age of the patient (Table 2–3).

Hematuria and Tumors of the Urinary Tract

Gross total painless hematuria is also the first manifestation of a urinary tract tumor, particularly a bladder tumor. The episodic nature of the bleeding and the absence of other symptoms may lull both patient and physician into a false sense of security, so that urologic in-

---

TABLE 2–3.

The Most Common Causes of Hematuria By Age and Sex

0–20 Years
  Acute glomerulonephritis
  Acute urinary tract infection
  Congenital urinary tract anomalies with obstruction
20–40 Years
  Acute urinary tract infection
  Stones
  Bladder tumor
40–60 Years (Males)
  Bladder tumor
  Stones
  Acute urinary tract infection
40–60 Years (Females)
  Acute urinary tract infection
  Stones
  Bladder tumor
60 Years (Males)
  Benign prostatic hyperplasia
  Bladder tumor
  Acute urinary tract infection
60 Years (Females)
  Bladder tumor
  Acute urinary tract infection

vestigation is often recommended only after multiple episodes of hematuria.

In one review of 1,000 cases presenting with gross hematuria (Lee and Davis, 1953), tumors were found in 21.5%, and two thirds of these were bladder tumors. In another evaluation of 500 cases of asymptomatic microscopic hematuria (Greene et al,. 1956), tumors were found in only 2.2%. In a more recent study from the same institution (Carson et al., 1979) evaluating 200 consecutive cases of asymptomatic microscopic hematuria, tumors were found in 12.5%. This current figure reflects the newer diagnostic modalities available today, particularly the Papanicolaou urine test.

Other Causes of Hematuria

In addition to the most common causes of hematuria—infection, stones and tumors—there are a number of less common causes including: (1) trauma, (2) sickle cell disease, (3) glomerulonephritis, (4) benign prostatic hyperplasia, (5) tuberculosis, (6) renal infarction, (7) renal vein thrombosis, (8) coagulation and platelet deficiencies, (9) exercise-related (jogging), (10) hypercalciuria, and (11) vasculitis.

The Diagnostic Approach to Hematuria

After a careful history and physical examination, the fundamental means of diagnosis are examination of the urine; cystoendoscopy, preferably at the time hematuria is present; and IV urography.

Urinalysis

Glomerular bleeding is indicated if RBC casts are found in the urinary sediment or if the RBCs are dysmorphic on phase contrast microscopy (Fairley and Birch, 1982; Fassett et al., 1982) (Fig 2–2). Proteinuria and an abnormal number of casts also suggest renal parenchymal disease.

Significant bacteriuria (over 10 rodlike bacteria/HPF in fresh urinary sediment) with or without pyuria indicates urinary tract infection.

Pyuria (leukocyturia), the indicator of the body's response to inflammation of the urinary tract, is less helpful diagnostically because it may be present with infection, stones, tumors, or glomerulonephritis.

Cystine crystals in the urine indicate cystinuria and the probable presence of a cystine stone. Other stone-forming crystals may be found in normal urine, so they are not usually diagnostic.

Cystoendoscopy

This procedure is extremely informative if performed while the patient is experiencing hematuria for it may be the only chance to discover the source of bleeding. Later x-ray studies may be inconclusive, particularly

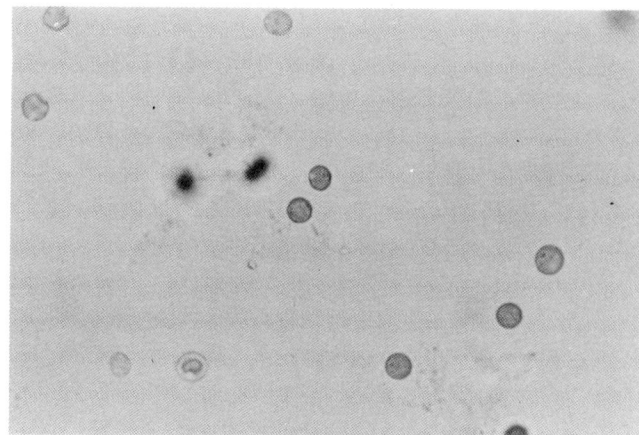

Normal Red Blood Cells

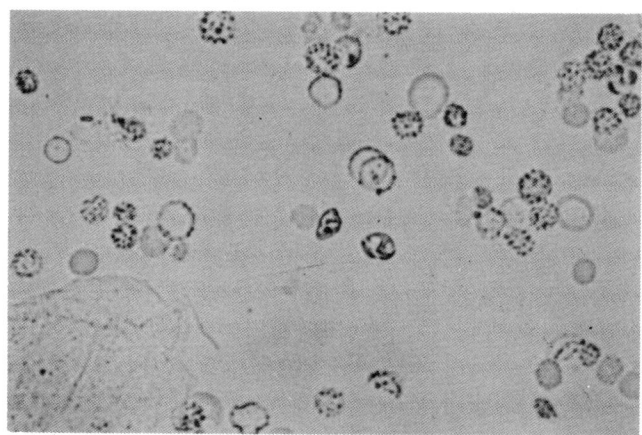

Dysmorphic Blood Cells

FIG 2–2.
Normal and dysmorphic red blood cells. (From Stamey TA, Kindrachuk RW: *Urinary Sediment and Urinalysis: A Practical Guide for the Health Science Professional.* Philadelphia, WB Saunders Co, 1985. Used by permission.)

when the bleeding is due to a small lesion in one kidney. Cystoendoscopy is not usually indicated in children with hematuria, in the presence of infection or after major trauma.

Intravenous Urogram

An intravenous urogram is indicated in virtually all cases not clearly due to acute glomerulonephritis or acute hemorrhagic cystitis. If bleeding has ceased, this study should precede cystoendoscopy.

After these fundamental studies have been completed, the diagnosis will be apparent in about 75% of cases. For the remaining 25% with unexplained hematuria, additional special studies may be indicated. Although hematuria cannot occur without a cause, in 5%–10% of cases no definite cause can be found.

PYURIA

Pyuria means increased WBC excretion in the urine (Little, 1962; Mabeck, 1969). This can be quantitated by determination of the WBC excretion rate or of the WBC concentration in the urine. Urologists usually count the number of WBCs in the centrifuged, unstained urinary sediment. If the urinary flow rate is in the normal range, there is a good correlation between the WBC excretion rate and the WBC concentration in the urine. However, when the urinary flow rate is high, an increased excretion rate of WBCs may not be reflected in the urinary sediment because the excessive urine output may keep the urinary WBC concentration in the normal range. A count of 400,000 WBCs/hr may be considered the upper limit of normal WBC excretion. If this excretion rate of 400,000 WBCs/hr is divided by the average urine output per hour of 50 ml, the upper limit of WBC concentration in the urine would be 400,000/50, or 8,000 WBCs/ml, giving urinary sediment findings of 4–5 WBCs/HPF. For clinical purposes, pyuria may be defined as more than 5 WBCs/HPF. It should be emphasized that this method of quantitating pyuria is imprecise. Factors that may alter the microscopic findings significantly include (1) presence or absence of contamination of the urine specimen, (2) urine flow rate at the time the urine specimen was obtained, and (3) the specifics of preparing and examining the urinary sediment.

Significance of Pyuria

Pyuria is the body's response to *inflammation* of the urinary tract. Though bacterial infection is the most common cause of inflammation and pyuria, other significant causes include tumors, stones, glomerulonephritis, foreign bodies, drugs (e.g., cyclophosphamide [Cytoxan]), and tuberculosis.

Pyuria and Urinary Infection

To many physicians pyuria (literally, pus in the urine) means urinary infection. This concept is treacherous for patients with tumors, stones, or tuberculosis, since they may be treated for long periods with antibiotics for a supposed infection. This clinical error in judgment could be significantly reduced if the term "leukocyturia" were substituted for "pyuria."

Pyuria is a poor indicator of infection, being absent in about 50% of documented urinary tract infections. Bacteria are responsible for the infection, so they are always present, whereas pyuria is a reaction to the presence of bacteria and is present only half the time.

The diagnosis of urinary infection may be made by finding 10 rodlike bacteria/HPF in fresh urinary sediment. This indicates significant bacteriuria, and urine cultures will usually show 100,000 colonies/ml of urine.

If fresh urinary sediment shows pyuria but no bacteriuria, keep in mind the possibility of tumor, stone, or tuberculosis.

PROTEINURIA

Proteinuria means the urinary excretion of more than 150 mg of protein/24 hr. Normal individuals excrete a very small amount of protein in the urine, about 80 mg/24 hr, two thirds being derived from the plasma and one third contributed by the tubules. The tubular portion is primarily Tamm-Horsfall mucoprotein, a globulin secreted by the thick ascending limb of the loop of Henle. It is the major protein constituent of urinary casts.

The glomerular capillaries behave *qualitatively* like other capillaries, being freely permeable to water and crystalloids but relatively impermeable to colloids such as proteins. Richards and associates in 1921, using a tiny micropipette, collected fluid from Bowman's capsule of the frog. When analyzed, this fluid was virtually identical to plasma but contained no protein. This glomerular filtrate in fact is not protein-free but does contain a very small amount of protein, about 10–20 mg/L. If the glomerular filtration rate (GFR) is normal (180 L/24 hr), 2–3 gm of protein will be filtered daily. Since less than 0.1 gm is normally excreted in the urine, more than 95% of the filtered protein is reabsorbed by the proximal tubules.

The glomerulus acts like a very fine filter with the chief determinants of passage through it being (1) the size, shape, and electrical charge of the molecule, and (2) the electrophysical properties of the filter (pore size, net negative charge).

Molecular size is a major factor in glomerular restriction of molecules. On the basis of "sieving" of different-sized molecules experimentally, the "pores" in the filter appear to have a radius of around 30 A. Small molecules with radii of 20 A or less passed through the filter easily; intermediate-sized molecules such as albumin with a radius of 37 A passed through in very small quantities; and large molecules with radii over 40 A did not pass through at all. The traces of albumin that pass through the filter account for over 99% of the protein in the glomerular filtrate. Most proteins in plasma are larger than albumin.

Electrical charge is also an important factor in determining whether a particular molecule passes through the filter. Because the glomerular wall is negatively charged, for any given size, negatively charged molecules are restricted more than neutral or positively charged molecules from moving through the wall. Most proteins are negatively charged and so are subject to this electrical hindrance.

Detection of Proteinuria

Because of its simplicity and convenience, the dipstick test is the most commonly employed screening test for urinary protein. It is appropriately sensitive, detecting protein concentrations of 10 mg/dl, so that normal supernatant urines usually give a negative test for protein. The daily protein excretion of normal individuals averages about 80 mg/day and is always less than 150 mg, so that urine concentrations of protein rarely exceed 10 mg/dl.

The dipstick has a pH-sensitive indicator dye that changes color with various concentrations of protein. Although more sensitive to negatively charged protein (albumin) than to positively charged ones (Bence-Jones), it does give a positive test when the concentration of Bence-Jones protein is greater than 50 mg/dl. Also, albumin is the most clinically significant protein, since it is the one primarily excreted in the urine when there is glomerular disease.

The sulfosalicylic acid test, which depends on acid precipitation of the protein, is sometimes preferred because it detects *all proteins* in the urine and is slightly more sensitive than the dipstick test, detecting protein at a concentration of 5 mg/dl.

When a qualitative test for proteinuria is performed, the specific gravity of the urine should be measured, because these tests determine the urine concentration of protein, not the amount excreted. Significant proteinuria may be missed if the urine is sufficiently dilute.

Both the dipstick and the sulfosalicylic acid test are semi-quantitative, so when they are positive, determine the 24-hr protein excretion in the urine.

Types of Proteinuria

There are four types of proteinuria: glomerular, tubular, overflow, and functional.

GLOMERULAR.—This is the most common form of proteinuria, found in all patients with significant glomerular damage. A defect in the glomerular filter permits increased filtration of normal plasma proteins, and since albumin has the highest concentration in the plasma, glomerular proteinuria is predominantly an albuminuria. In the other types of proteinuria, protein excretion rarely exceeds 2 gm/day, so whenever there is massive proteinuria with excretion of 4 gm/day, it is glomerular proteinuria. This heavy proteinuria is characteristic of the nephrotic syndrome.

TUBULAR.—Proteinuria secondary to tubular or interstitial disorders of the kidney occurs when the proximal tubules are unable to reabsorb the normally filtered proteins. In tubular proteinuria, one usually finds increased amounts of low molecular weight proteins,

smaller than albumin, in the urine, and the total protein excretion is usually 1–2 gm/day.

OVERFLOW.—Overflow proteinuria is due to the presence in plasma of abnormal quantities of low molecular weight proteins that are filtered across the normal glomerular capillary wall and saturate the proximal tubular reabsorptive mechanism. Examples include Bence-Jones proteinuria, myoglobinuria, and hemoglobinuria. This is the least common type of proteinuria.

FUNCTIONAL.—When proteinuria occurs in the absence of any clear-cut renal or systemic disease, it is called functional or physiologic proteinuria. The mechanisms responsible for this type of proteinuria are unknown but are probably hemodynamic. Functional proteinurias are characteristically both intermittent and mild, with protein excretion rarely exceeding 1 gm/day.

Clinical states that may cause this type of proteinuria include fever, exercise, emotional stress, and renal venous hypertension (congestive heart failure).

Orthostatic (postural) proteinuria is a special form of functional proteinuria seen in healthy young adults. These individuals excrete protein in their urine in the upright position but excrete little or none on recumbency. To make the diagnosis of orthostatic proteinuria, compare the protein excretion between collections made after ambulation and after recumbency. The level should not exceed 1.5 gm/L after ambulation and should be less than 0.1 gm/L after recumbency. Long-term follow-up studies of patients with orthostatic proteinuria demonstrate a benign course.

PNEUMATURIA

Pneumaturia is the passage of gas in the urine and almost always is due to a fistula between the intestinal tract and the urinary tract. Since gas is much lighter than water, it always rises to the top of the bladder and consequently is passed at the end of urination (terminal pneumaturia).

This fistula may occur at any level, from the stomach to the rectum of the intestinal tract and from the kidney to the urethra of the urinary tract, but the large majority involve the bladder and the sigmoid colon or terminal ileum. The most common causes are diverticulitis of the sigmoid colon, carcinoma of the colon, and Crohn's disease.

Urinary tract infections caused by gas-forming bacteria and prior introduction of air into the bladder via insertion of a catheter or cystoscopy are occasional causes of pneumaturia.

CHYLURIA (MILKY URINE)

Chyluria is the passage of lymph or chyle in the urine due to the presence of an intrarenal, lymphatic-urinary fistula. The cause of this fistula is obstruction of the lymphatics superior to the kidney, usually the thoracic duct. With obstruction, there is increased back pressure in the retroperitoneal and renal lymphatics, and eventually they rupture into a calyceal fornix.

Etiology

The most common cause of chyluria is filariasis due to *Wuchereria bancrofti*. The adult filarial worms invade the suprarenal lymphatics, causing obstruction and severe inflammation. Less common causes include posterior mediastinal and retroperitoneal tumors, tuberculosis, and trauma.

Clinical Picture

Patients with chyluria usually give a history of living in or visiting a tropical or semitropical country where filariasis is endemic. The passage of milky urine occurs intermittently, varying with the amount of fat ingested and sometimes with the patient's posture. Chylous urine contains fibrinogen, and fibrin clots may cause renal colic or urinary retention.

Diagnosis

The appearance of the urine is usually diagnostic. The urine has a milky color that varies in intensity according to the amount of fat present. Small amounts of fat produce a cloudy urine, difficult to distinguish from phosphaturia or gross pyuria. However, the cloudiness remains after adding acid and heating the urine, and the urinary sediment shows no WBCs.

The presence of fat in the urine may be established by the following simple test. The urine is mixed with an equal volume of petroleum ether and then shaken or centrifuged. The fat separates from the urine and can be seen as a ring between the petroleum ether above and the slightly turbid urine below. A small amount of this fat is removed and mixed on a glass slide with Sudan III or other fat dyes, allowing easy identification of the fat droplets under the microscope.

To establish the kidney involved, cystoendoscopy may be performed after the patient has ingested several glasses of milk while the urine is milky. Creamy, white efflux can be seen coming from the ureteral orifice on the side of the fistula. Pedal lymphangiography will demonstrate the exact location of the intrarenal lymphatic-urinary fistula, usually a calyceal fornix. Intravenous urograms and retrograde ureteropyelograms are usually normal, even when performed while the urine is milky white. Occasionally, the fistula may be demonstrated on retrograde pyelography.

Treatment

Most patients respond to a high-fluid, low-fat diet. Surgical excision of the renal lymphatic is reserved for those patients not responding to medical treatment.

PHYSICAL EXAMINATION

KIDNEY

Inspection

A kidney must be grossly enlarged or displaced to cause a perceptible bulge in the upper abdomen or flank. If a perinephric abscess is suspected, the patient should be examined in the knee-elbow position. The normal shallow depression below the lowermost rib may be obliterated by fullness and edema secondary to an underlying abscess.

Palpation

Because of their location in the uppermost portion of the abdominal cavity, normal kidneys are usually not palpable. An exception is the newborn in whom both kidneys may be palpable during the first 48 hours of life due to the hypotonicity of all muscles. In about 10% of adults, usually thin women, the lower pole of the right kidney can be felt.

Two maneuvers aid palpation, deep breathing and bimanual examination. The normal kidney is movable because it is fixed only by its vascular pedicle, and with deep inspiration the descending diaphragm pushes the kidney down towards the examining fingers. The kidney is most easily felt between the fingers of both hands as a firm, smooth mass slipping upward as expiration starts.

The posterior hand lifts up the soft tissue in the CVA, and the anterior hand presses deeply into the abdomen. A renal mass is characteristically ballottable, unlike the liver and spleen, which usually can only be palpated with the anterior hand.

Tenderness due to inflammation in or around the kidney is best detected by exerting firm pressure in the CVA region.

Auscultation

Particularly in patients with hypertension it is important to listen over the renal artery areas for the presence of a bruit. These murmurs are best heard anteriorly after complete exhalation.

Transillumination

This examination is occasionally helpful in newborns or small children with large, easily palpable abdominal masses. With the room as dark as possible, manipulate the mass against the abdominal wall with one hand, while firmly applying a high-intensity light (e.g., fiberoptic light cord) to the mass with the other hand. If the mass is cystic rather than solid, it will transilluminate, and there will be a reddish glow.

URETER

In males the ureter is not palpable by either abdominal or rectal examination, but in females the lower ureter can be felt on vaginal examination. One or two fingers are gently pushed upward and outward, and at the limit of your fingertips the ureter lies close to the bony pelvic wall and lateral to the ovary. From this point, the ureter can be followed to its junction with the bladder by carrying the fingers downward and inward. If the ureter is normal, it is usually not identifiable because it is soft and nontender. If a stone is present, both the stone and ureter are usually palpable. The ureter can be felt as a tender tubular mass proximal to the stone.

URINARY BLADDER

The normal empty or nearly empty bladder is neither palpable nor percussible because of its anatomical location in the pelvis. When it contains about 125 ml, it rises out of the pelvis into the lower abdomen projecting one fingerbreadth above the pubis. With further filling, it rises progressively toward the umbilicus.

Inspection

If the bladder contains over 500 ml, it may be identifiable as a bulge in the midlower abdomen. This swelling rising out of the pelvis is best appreciated by observing from the side, with the eyes more or less level with the lower abdomen.

Percussion

Whether or not a mass is visible, the lower abdomen should be percussed from the umbilicus to the pubis. If the patient has a distended bladder, the normally resonant note is replaced by dullness. Percussion over a distended bladder may also cause the patient to experience a desire to void because of the sudden induced rise in intravesical pressure.

Palpation

The distended bladder may be palpated as a firm, round, movable mass rising out of the pelvis into the lower abdomen. On bimanual examination, the mass can often be ballotted between the two hands. For assessing the presence and extent of bladder tumors, bi-

manual examination is performed using anesthesia with the bladder empty.

PROSTATE AND SEMINAL VESICLES

The prostate and seminal vesicles are palpated through the anterior rectal wall (Fig 2–3). Rectal examination may be performed with the patient in a variety of positions. Older patients are best examined in the lithotomy position, whereas younger men may be examined while bending over or in the knee-elbow position.

Two distinct lobes about the size of the distal segment of the thumb are separated by a shallow median furrow. The normal gland is smooth, slightly movable and nontender, and has a rubbery consistency.

Above the prostate, if your finger is long enough, you may be able to feel the soft, tubular seminal vesicles. Coming off the base of the prostate somewhat obliquely, they are most easily felt when they are distended or tender.

SCROTUM AND TESTES

After inspecting the scrotal skin and perineum, palpate the testes and epididymides between the thumb and finger. The comma-shaped epididymis is closely attached to the posterolateral side of the testis. You can get your fingers into the groove between the epididymis and testis everywhere except superiorly, where the two structures are anatomically joined. In many

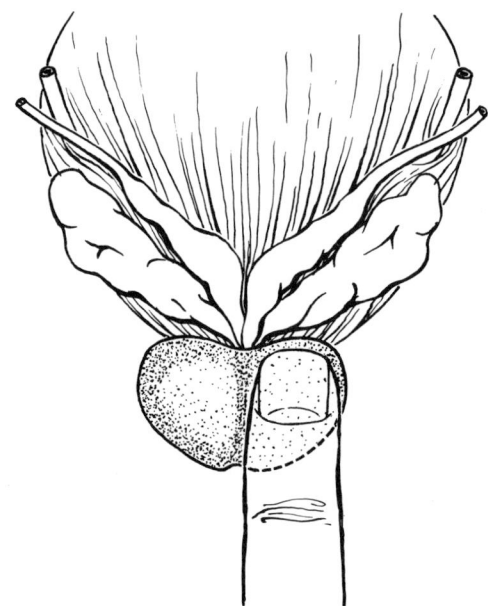

FIG 2–3.
Palpation of the prostate and seminal vesicles. (From Wyker A, Gillenwater JY: *Method of Urology.* Baltimore, Williams & Wilkins Co, 1975. Used by permission.)

men, a small ovoid lump, the rudimentary appendix testis, can be felt in or near the groove between the upper pole of the testis and the epididymis. The cord structures at the neck of the scrotum should be sifted through your fingers. The solid cordlike vas is easily identified and followed to its junction with the tail of the epididymis. The other soft, stringy structures in the cord cannot be defined. To rule out the presence of the gravity-dependent varicocele, the cord must also be examined with the patient standing. If a varicocele is present, the intrascrotal varicosities secondary to valvular incompetence of the internal spermatic vein become distended in the upright position and feel like a bag of worms.

PENIS

If the patient is uncircumcised, retract the foreskin to rule out phimosis with an obstructing, small aperture. To inspect the urethral meatus, pinch the glans between the thumb and finger placed at the 6 and 12 o'clock positions. If the urethral meatus is not in the normal location, it can be found by following the midline raphe to its end on the undersurface of the penis. The shaft of the penis is palpated, looking particularly for the firm, fibrous plaques of Peyronie's disease. Palpate the floor of the urethra from the corona to the bulb, looking for induration secondary to a stricture.

URINALYSIS

A standard urinalysis consists of: (1) determination of the physical characteristics of the urine, (2) dipstick chemical tests, and (3) microscopic examination of the urinary sediment. This key examination should not be delegated to uninterested laboratory personnel but should be performed by the physician. For reliable urinalyses, the urine must be collected properly and examined promptly.

COLLECTION OF THE URINE SPECIMENS

Timing

Most urinalyses of necessity are performed on a random specimen freshly voided by the patient. More information can be obtained if the first morning specimen is examined, because it is the best one for detecting formed elements in the urine and for determining if urinary infection is present. The formed elements—RBCs, WBCs, and casts—are preserved in this characteristically acid and concentrated urine, but they may be lysed and disappear in dilute or alkaline urine. Bacterial colony counts are usually highest in this specimen because the bacteria have had more time to multiply during overnight incubation in the bladder.

Methods

The external genitalia, perineum, and urethra of both males and females have bacteria on the surface, so if the urine specimen is to reflect the true bacteriologic status of the urinary tract, these contaminating bacteria must be excluded. This is most commonly accomplished by obtaining a clean-catch, midstream specimen. In selected cases urine may be obtained by urethral catheterization or by suprapubic aspiration of the bladder.

Clean-Catch

The reliability of a urine specimen obtained in this fashion varies directly with the adequacy of the cleansing procedure. If a female patient cleanses herself in a perfunctory manner, the urine specimen may contain bacteria from the vagina or external genitalia with a resultant reduction in reliability. However, if this same patient is placed in the lithotomy position, the perineum and periurethral areas thoroughly cleansed by a trained nurse, and with the labia held apart while a midstream specimen is collected, the reliability approaches 100%. In males, a midstream specimen collected after cleansing the glans penis is a very reliable one.

The midstream portion is collected to reduce contamination of the urine, since over 90% of all urethral bacteria are voided in the initial 10 dl.

URETHRAL CATHETERIZATION

This technique, used primarily in women, has a reliability of about 95%, but it has several drawbacks. Passage of a catheter may carry urethral bacteria back into the bladder, causing contamination of the urine specimen or, less commonly, a frank urinary infection (1%–2%). It also causes some discomfort in all patients and produces urethral edema in a few, with resultant voiding difficulty.

In infants of either sex, urethral catheterization may be performed with minimal risk if a well lubricated no. 5 feeding tube is passed carefully.

SUPRAPUBIC ASPIRATION OF THE BLADDER

The reliability of this technique approaches 100%, but it is not practical for routine use. It is used most commonly in premature infants, infants, and small children because urine specimens obtained with a strap-on plastic collection device are often unsatisfactory. Since the bladder is largely an abdominal organ at this age, aspiration can be accomplished easily and safely with minimal discomfort. In adults, the bladder must contain more than 125 dl before it rises above the pubis and is accessible for aspiration.

PROMPT EXAMINATION

For best results, urine should be examined within 30 minutes of collection, for, if it is allowed to stand, bacterial growth takes place rapidly, alkalinizing the urine with resultant destruction of RBCs, WBCs, and casts. If the urine cannot be examined promptly, it should be refrigerated, for this is the best method of preservation.

PHYSICAL CHARACTERISTICS OF URINE

Color

The color of normal urine is determined by the concentration of urochrome, an endogenously formed yellow-brown pigment excreted at a uniform rate. Since the amount of pigment excreted each hour is the same, the color of the urine varies directly with the urine output. With high urine flow rates, the urine is very pale, almost water-colored, whereas with low urine flow rates it is a deep yellow color.

The most common cause of an abnormal urine color is medication.

Clarity

Freshly voided urine is usually clear. Cloudiness of the urine is fairly common, however, and is usually due to excessive amounts of crystals—amorphous phosphate in alkaline urine or amorphous urates in acid urine. This crystalluria is usually not clinically significant. The crystals can be dissolved, rendering the urine clear by adding acid to dissolve the amorphous phosphates or by heating to dissolve the amorphous urates.

Heavy pyuria, usually secondary to a bacterial infection, is a less common cause of cloudy urine.

Specific Gravity

The specific gravity of a solution is the measure of its density and it is easily determined with a hydrometer. The specific gravity of water is 1.000, plasma 1.010, and urine 1.003–1.040. Normally functioning kidneys conserve and excrete water as needed, accounting for the wide range of 1.003–1.040, but poorly functioning kidneys lose this ability, so their urine specific gravity remains fixed at around 1.010, plasma level.

Clinically, determination of the specific gravity of a random urine specimen is most useful in assessing the significance of dipstick chemical test and urinary sediment findings. Substances such as WBCs, RBCs, and casts are generally excreted into the urine at a fairly constant rate, so the urine concentration of these substances and the related urinary sediment findings vary directly with urine volume.

When the urine specific gravity is less than 1.007 due to a high urinary flow rate, the urine concentration of protein, RBCs, WBCs, casts, and bacteria will be so reduced that the urinary sediment findings are almost meaningless. Also, in this hypotonic environment, RBCs are lysed, so patients with hematuria will show no RBCs in their urinary sediment.

CHEMICAL TESTS

Dipstick tests have greatly simplified urinalyses. These firm cellulose strips have test areas impregnated with chemicals that react with abnormal substances in the urine to produce color changes. Within one minute, using this single dipstick, you can get a semiquantitative estimation of nine parameters—pH, protein, glucose, ketones, bilirubin, blood (RBCs, hemoglobin, and myoglobin), urobilinogen, nitrite, and WBCs. If the leukocyte esterase test for WBCs is used, increased strip sensitivity may be obtained by examining the strip at five minutes (Shaw et al., 1985).

URINARY SEDIMENT

Dilute urine lyses RBCs and lowers the urinary concentration of *all* excreted substance, so check the urine specific gravity before examining the urinary sediment. When urine stands in a bottle, the formed elements gradually fall to the bottom, so before taking a sample, mix the urine thoroughly. If your sample is cloudy and alkaline, acidify it to dissolve amorphous phosphates; if it is cloudy and acid, heat it to dissolve urates. Centrifuge 15 dl for 3–5 minutes, pour off the supernatant, and examine the sediment under the microscope using reduced light. Staining of the urine may be helpful in selected cases (e.g., identifying gonococci).

BACTERIA

A special effort should be made to detect bacteria, for if more than 10/HPF are present, infection is usually present. If this urine is cultured, the colony count is usually 1×10^5 bacteria/ml or greater (Kunin, 1961).

WHITE BLOOD CELLS

Leukocytes are lysed in alkaline urine but not in hypotonic urine. The rate of pH-induced lysis increases with increasing alkalinity and with time but is clinically not significant until the pH reaches 8 or higher. When the urine pH is 7.1–7.9, fewer than 10% of WBCs are lysed (Gadeholt, 1968). Urea-splitting infection due to Proteus is the most common cause of a urine pH of 8 or higher. When this occurs, often no identifiable

WBCs can be seen in the urinary sediment, only sludge.

Acidification of nonacid urine samples accentuates the nuclei of the leukocytes and aids in their identification.

RED BLOOD CELLS

Red blood cells are relatively resistant to lysis in alkaline urine, even when the pH is over 8, but they are readily lysed in hypotonic urine, 100% disappearing when the urine specific gravity is 1.007 or lower.

It is clinically important to know whether the red blood cells present in the urinary sediment are glomerular or non-glomerular in origin because glomerular RBCs are diagnostic of glomerulonephritis. There are two findings that identify the glomeruli as the source of the red cells—RBC casts and dysmorphic RBCs. Since casts are formed only in the kidney, anything entrapped within a cast must have been present within the kidney. Red blood cell casts are more likely to be found in the first morning specimen, because casts are preserved in this characteristically concentrated acid urine but largely disappear in dilute, alkaline urine. Fairley (Fairley and Birch, 1982), using a phase microscope to study RBC morphology, reported that these dysmorphic RBCs were characteristic of glomerular bleeding. The marked RBC membrane distortions are thought to be caused by osmotic and physical changes during the passage of RBCs through the nephron. These findings have been confirmed by others, and Stamey in his excellent manual *Urinary Sediment and Urinalysis* showed many fine photographs contrasting dysmorphic RBCs of glomerular origin with RBCs of nonglomerular origin (see Fig 2–2). He also demonstrated that these dysmorphic RBCs could be identified under standard light microscopy as well as phase-contrast microscopy.

CASTS

The basic foundation of all renal casts is a special protein, Tamm-Horsfall globulin, secreted by the tubular epithelial cells. A small number of these basic hyaline casts are excreted normally, so their presence in the urinary sediment is not clinically significant. If RBCs, WBCs, or sloughed tubular epithelial cells are present in the renal tubular lumina, they may become incorporated within the cast. If these casts are detected in the urinary sediment, the presence of renal pathology is established. Casts containing sloughed epithelial cells may be granular or waxy.

Since Tamm-Horsfall protein is soluble at a pH of 7.1 or higher, *all* casts disappear in alkaline urine.

REFERENCES

1. Blandy J: *Operative Urology.* Oxford, Blackwell Scientific Publications, 1978.
2. Carson CC III, Segura JW, Greene LF: Clinical importance of microhematuria. *JAMA* 1979; 241:149.
3. Esperanca M, Gerrard JW: Nocturnal enuresis: Studies in bladder function in normal children and enuretics. *Can Med Assoc J* 1969; 101:324.
4. Fairley KF, Birch DF: Hematuria: A simple method of identifying glomerular bleeding. *Kidney Int* 1982; 21:105.
5. Fassett RG, Horgan BA, Mathew TH: Detection of glomerular bleeding by phase-contrast microscopy. *Lancet* 1982; 1:1432.
6. Gadeholt H: Persistence of blood cells in urine. *Acta Med Scand* 1968; 183:49.
7. Greene LF, O'Shaughnessy EJ Jr, Hendricks ED: Study of five hundred patients with asymptomatic microhematuria. *JAMA* 1956; 161:610.
8. Kales A, Kales JD, Jacobsen A, et al: Effects of imipramine on enuretic frequency and sleep stages. *Pediatrics* 1977; 60:431.
9. Kunin CM: The quantitative significance of bacteria visualized in the unstained urinary sediment. *N Engl J Med* 1961; 265:589.
10. Lee LW, Davis E Jr: Gross urinary hemorrhage: A symptom, not a disease. *JAMA* 1953; 153:782.
11. Little PJ: Urinary white-cell excretion. *Lancet* 1962; 1:1149.
12. Mabeck CE: Studies in urinary tract infections: IV. Urinary leucocyte excretion in bacteriuria. *Acta Med Scand* 1969; 186:193.
13. McLellan AM, Goodell H: Pain from the bladder, ureter and kidney pelvis. *Res Publ Assoc Nerv Ment Dis* 1942; 23:252.
14. Pompeius R: Cystometry in paediatric enuresis. *Scand J Urol Nephrol* 1971; 5:222.
15. Shaw ST Jr, Poon SY, Wong ET: Routine urinalysis, is the dipstick enough? *JAMA* 1985; 253:1596.
16. Stamey TA, Kindrachuk RW: *Urinary Sediment and Urinalysis, A Practical Guide for the Health Science Professional.* Philadelphia, WB Saunders Co, 1985.
17. Straub LR, Ripley HS, Wolf S: Disturbances of bladder function associated with emotional status. *JAMA* 1949; 141:1139.
18. Verney EB: The antidiuretic hormone and the factors which determine its release. *Proc R Soc Lond (Biol)* 1947; 135:25.
19. Wagner W, Johnson SB, Walker D, et al: A controlled comparison of two treatments for nocturnal enuresis. *J Pediatr* 1982; 101:302.
20. Whiteside CG, Arnold EP: Persistent primary enuresis: A urodynamic assessment. *Br Med J* 1975; 1:364.

Chapter 3

Imaging

Chapter 3A / Excretory Urography

CHARLES D. KELLUM, M.D.
LOULIE M. FISHER, M.D.
CHARLES J. TEGTMEYER, M.D.

Excretory urography remains the basic radiologic examination of the urinary tract and is the foundation for the evaluation of suspected urologic disease. Despite development of the newer diagnostic modalities such as isotope scanning, ultrasonography, CT, and magnetic resonance imaging (MRI), excretory urography has maintained a prominent role in uroradiology. Some indications have been altered and will continue to change with the newer imaging modalities, but the initial evaluation of suspected urinary tract structural abnormalities, hematuria, pyuria, and calculus disease is best performed with excretory urography. The examination is relatively inexpensive and simple to perform, with few contraindications. Excretory urography, when properly performed, can provide valuable information about the renal parenchyma, pelvicalyceal system, ureters, and urinary bladder. As with all radiologic procedures, careful monitoring with close attention to detail is important and will increase the diagnostic yield.

CONTRAST MEDIA
CHEMICAL PROPERTIES AND EXCRETION

The first compounds introduced for clinical use in excretory urography during the early 1930s were diiodinated compounds that were hampered by low radiopacity and moderate toxicity. More often than not, retrograde pyelography was necessary to evaluate the urinary tract properly. The ability to achieve adequate opacification of the pelvicalyceal system and the renal parenchyma resulted from the introduction of the triiodinated benzoic acid derivatives in the early 1950s. Since then, several salts of triiodinated benzoic acid

compounds have been marketed for excretory urography with slight structural differences but similar physiologic and radiographic properties (Table 3A–1).

The compounds are water-soluble salt solutions that are hypertonic with an osmolality 2.5–6 times that of plasma. The cations are either sodium or meglumine. The sodium salts are slightly more radiopaque than the meglumine salts because of their smaller molecules, and they contain relatively more iodine. They also appear to be better concentrated in the urine. The lower viscosity of the sodium salts makes it easier to inject through smaller-bore needles and catheters. On the other hand, meglumine salts are thought to be slightly less toxic than sodium salts (Fischer and Cornell, 1965). With the doses used in present-day excretory urography, the advantages and disadvantages are minimal. Good-quality urography can be achieved using either of the salts or compounds utilizing a combination of the salts.

Contrast material is almost entirely excreted by glomerular filtration with little or no tubular excretion. Approximately 0.5%–2% is excreted by the liver and bowel. Most of this is excreted by the biliary system, although the small intestine, stomach, and salivary glands can excrete small amounts. "Vicarious excretion" is a term used when extrarenal excretion is apparent on the radiographs (Becker, 1968). Contrast medium may be seen in the gallbladder and colon 24 hours after injection in patients with decreased renal function, in whom extrarenal routes of excretion take on greater importance. Occasionally, this phenomenon is also seen in patients with normal renal function after they have received large doses of contrast medium (Fig 3A–1).

TABLE 3A–1.
Commonly Used Contrast Media for Excretory Urography

TRADE NAME	GENERIC NAME	CONCENTRATION, %	SODIUM CONTENT, MEQ/ML	IODINE CONTENT, MG/ML
Renografin-60	Meglumine and sodium diatrizoate	60	.16	288
Reno-M-60	Meglumine diatrizoate	60	.04	282
Reno-M-Dip	Meglumine diatrizoate	30	—	141
Hypaque-50	Sodium diatrizoate	50	.8	300
Hypaque-M-60	Meglumine diatrizoate	60	—	282
Conray-60	Meglumine iothalamine	60	—	282

The amount of contrast excretion by the kidney and collecting system with resultant radiopacity is related to plasma concentration and glomerular filtration rate (GFR) (Cattell, 1970). Since the GFR is fixed in the individual, only the plasma concentration can be manipulated. Increasing the amount of contrast material will increase the radiopacity of the urinary tract to a certain extent. At doses greater than 2 ml/kg, the radiopacity will not continue to increase significantly, while the risk of toxicity is increased. The method of injection will also alter the quality of the urogram. The intensity

of the nephrogram is related to the plasma concentration. Studies have shown that after bolus injection a peak plasma concentration is achieved, which then declines rapidly. Using a drip infusion, the plasma concentration plateau is slowly reached and falls off over a longer period. However, the drip infusion technique never achieves a plasma level as high as the bolus injection and is of no benefit save convenience.

Following glomerular filtration, approximately 85% of the water resorption that concentrates the urine and increases radiopacity occurs in the proximal convoluted tubules. Additional resorption occurs in the distal convoluted tubules and collecting ducts depending on the state of hydration. However, because of this small contribution, the effectiveness of dehydration in increasing radiopacity is minimal, especially in light of the higher contrast doses used today. Thus, dehydration is unnecessary and is likely to increase the risk of contrast-induced nephrotoxicity.

PHYSIOLOGIC CONSIDERATIONS

Following IV injection of contrast material, there are several subjective effects, the intensity of which varies with the rate of injection and the individual. A sense of warmth and a "flushing" sensation are almost universal and may be misinterpreted as a reaction. A metallic taste and circumoral tingling are frequently observed. Sensations of pelvic and perineal warmth are not unusual and may be distressing to some patients.

On the other hand, several physiologic responses have been noted with intravascular injection of contrast media. There is usually a mild and transient decrease in the systemic blood pressure along with a transient decrease in the heart rate (Stanley and Pfister, 1976). Pulmonary arterial pressure has been found to be transiently increased by contrast medium (Peck et al., 1983). Renal blood flow is initially increased but is followed by a decrease in renal blood flow proportional to the dose administered (Forrest et al., 1981). A depressant effect on the myocardium has been shown with decreased myocardial contractility that is thought to be secondary to calcium binding and a depressant effect on

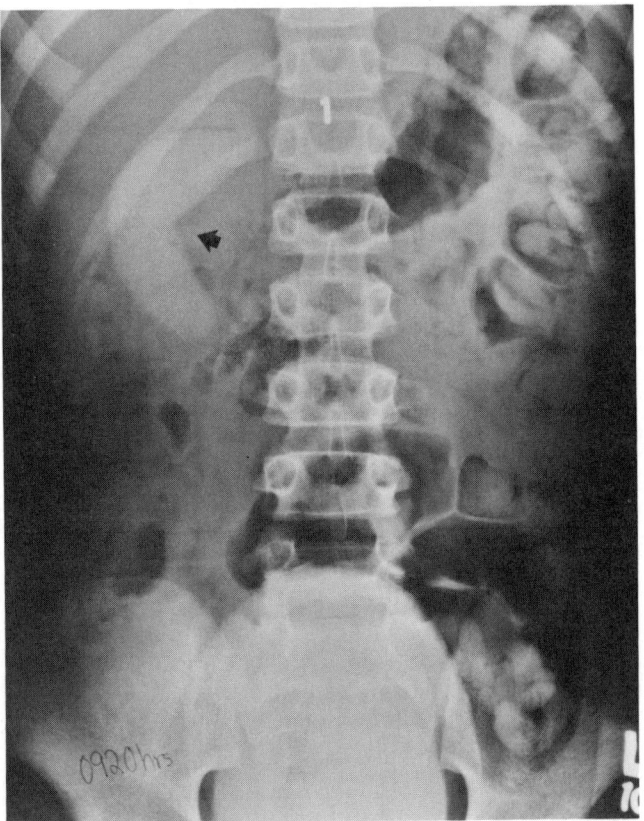

FIG 3A–1.
Vicarious excretion of contrast material with opacification of the gallbladder *(arrow)*. X-ray taken supine 24 hours after large dose of IV contrast material.

the SA and AV nodes (Popio et al., 1978; Katzberg et al., 1983). It has been suggested that subclinical bronchospasm occurs in most subjects (Rosenfield et al., 1977). These effects are all transient and seem inconsequential but may play a part in the untoward reactions seen in a small percentage of subjects.

CONTRAST MEDIUM REACTIONS

INCIDENCE AND CLASSIFICATION

Adverse reactions to intravascular contrast media are well known and are a source of concern, but fortunately the majority of cases are mild and insignificant. Several large series of patients have demonstrated an overall incidence of contrast reactions of all types ranging from 5%–8% (Shehadi and Toniolo, 1980; Shehadi, 1982; Witten et al., 1973). However, major respiratory or cardiovascular reactions are reported to occur in fewer than 1% of patients.

The reported frequency of fatal reactions ranges between 1 in 14,000 and 1 in 75,000. The former figure is unlikely to be accurate, and the true rate probably ranges between 1 in 40,000 and 1 in 75,000 (Hartman, et al., 1982; Ansell, 1970; Shehadi, 1975).

Contrast reactions are usually classified as mild, moderate, or severe. In most series, 95% or more of all reactions are classified as mild to moderate. A sense of heat and flushing, along with most cases of nausea and vomiting, should not be considered adverse reactions but common side effects of the contrast material. Mild reactions include mild urticaria and minimal respiratory symptoms. Moderate reactions include extensive urticaria, angioneurotic edema, and bronchospasm. Severe reactions include intense bronchospasm with laryngeal edema along with potentially lethal cardiovascular responses such as marked hypotension, pulmonary edema, and ventricular arrhythmias that can lead to cardiovascular collapse.

ETIOLOGY

Although adverse reactions to contrast media have been extensively studied over the years, their pathogenesis remains somewhat obscure. Most evidence points toward a multifactoral concept in which several poorly understood responses to contrast media may lead to an adverse reaction.

Certain reactions such as urticaria, erythema, and bronchospasm appear to be allergic. They resemble an allergic reaction and respond as an allergic reaction to antihistamines and epinephrine. The allergy concept, however, is controversial, with many researchers disputing any validity. Failure to find the classic IgE-type antibodies to contrast media in humans, and the fact that individuals may react without prior exposure to contrast media does not support the allergy model (Las-ser, 1968, 1974; Gillenwater, 1971). However, work by Brasch (1980) supports the theory that contrast material may act as haptens and induce specific antihapten antibodies. Anticontrast media antibodies of both the IgG and IgE classes have been produced in rabbits. Contrast reactions in individuals without prior exposure are explained on the basis of antibody cross-reactivity, where the contrast media is similar to other antigens to which the individual has had prior exposure, such as other halogenated benzene rings seen in food additives, pesticides, etc.

Some of the most severe reactions, however, do not resemble allergic reactions. Sudden major cardiovascular responses such as profound hypotension, pulmonary edema, ventricular arrhythmias, and myocardial infarction involve responses that are poorly understood. Hypertonicity and direct chemotoxicity of the contrast material have been suggested as etiologies (Fischer and Cornell, 1965; Pfister, 1975). Enhanced vagal tone as the result of stimulation of the vasomotor center of the medulla and accentuation of contrast medium-induced myocardial toxicity has been suggested as the etiology in the sudden deaths of fluid-depleted dogs following IV injection of contrast material (Katzberg et al., 1983).

Lalli (1974, 1980) has developed an interesting hypothesis based on the effect of contrast medium on the CNS after crossing the blood-brain barrier. He thinks that all reactions are ultimately neurogenic, centered on the hypothalamus and medulla. The combination of anxiety and contrast materials stimulating the hypothalamus can lead to a variety of neurogenic responses such as hypotension, ventricular arrhythmias, and pulmonary edema. He thinks that anxiety plays a large role in contrast reactions and that examinations performed in a quiet, nonthreatening manner will reduce the number of adverse reactions.

RELATIONSHIP TO PREVIOUS REACTIONS AND HYPERSENSITIVITY STATES

Studies have shown that individuals with known hypersensitivity states have contrast reactions on the order of twice that of the normal population, 10%–12% vs. 5%. Individuals who have had a previous contrast material reaction have a threefold increase; 15%–16% of these individuals have a recurrent contrast reaction (Witten et al., 1973; Shehadi, 1982). The nature of the recurrent reaction is variable, with minor reactions being repeated more often than life-threatening reactions.

VALUE OF PRETESTING AND PREMEDICATION

Intravenous injection of a test dose of contrast medium has not been proved useful as a screen for potential contrast reactions. Yocum and associates (1978) demonstrated some value in high-risk patients with pretesting, but it involved a sophisticated protocol using multiple serial dilutions of the contrast material. A standard 0.5–1-ml IV test injection is of little use and may indeed precipitate life-threatening reactions (Shehadi, 1982).

Likewise, premedicating a patient with a history of contrast material reaction is controversial. Many authorities believe that premedicating patients with antihistamines and corticosteroids is of little value. However, other investigators have shown that the incidence of recurrent reactions can be reduced by one-half or greater of the expected incidence with prophylactic antihistamines and corticosteroids (Kelly et al., 1978; Rocum et al., 1978; Greenberger et al., 1985). Until definite evidence is produced to the contrary, it seems prudent to continue to premedicate those with a history of significant contrast material reaction.

Several pretreatment protocols have been described using corticosteroids and antihistamines. Hydrocortisone, 100 mg, should be administered IV at least 12 hours before contrast material injection. Prednisone, 50 mg administered orally every six hours for three doses, ending one hour before contrast material injection, is an alternative. Diphenhydramine should be given, 50 mg IM one hour before the study or IV five minutes before the study.

TREATMENT OF CONTRAST MATERIAL REACTIONS

Contrast reactions usually occur immediately or within minutes following the injection. However, severe reactions can occur after 15 minutes or even later. All physicians, nurses, and technologists involved in urography should be well trained in the recognition of the early signs of a contrast reaction and in resuscitation. Prompt recognition of a reaction and immediate treatment can be lifesaving. In all excretory urograms, the IV needle or catheter used for delivering the contrast should be left in place for the duration of the examination in case a reaction develops and IV access is needed. Blood pressure and ECG monitoring equipment as well as oxygen should always be close at hand. In each room, there should be ready access to epinephrine, atropine, and diphenhydramine. In case of cardiovascular collapse, a "crash cart" containing a defibrillator, endotracheal set, and drugs useful during cardiac arrest should be nearby.

Mild reactions usually do not require treatment except reassurance and comforting. One must cast a jaundiced eye on all reactions, however, as they may progress to a more serious nature. If the urticaria is symptomatic, 25–50 mg of diphenhydramine given IM or IV is effective.

Moderate reactions, including significant urticaria, erythema, and mild bronchospasm should be treated with diphenhydramine and 0.3–0.5 mg of 1:1000 epinephrine subcutaneously. Severe bronchospasm and evidence of laryngeal edema should be treated immediately with IV epinephrine, because such a reaction may suddenly worsen, and subcutaneous epinephrine at this point would be useless because of poor absorption. Doses of 0.3–0.5 ml of 1:10,000 epinephrine should be given IV. Epinephrine is short-lived, and repeated doses may be necesary at 5–15-minute intervals.

Severe cardiovascular reactions that involve hypotension, loss of consciousness, and airway obstruction require immediate therapy. In the hypotensive patient, it is extremely important to check the pulse rate. Bradycardia in a hypotensive patient indicates a vasovagal type of response, and treatment with 1 mg of atropine IV is usually extremely effective. Two milligrams of atropine can be given over a 10–20 minute period. On the other hand, tachycardia associated with hypotension dictates the use of IV epinephrine. In total cardiovascular collapse, additional IV lines and establishment of an adequate airway are mandatory when embarking on cardiopulmonary resuscitation.

It should be again emphasized that IV epinephrine is the treatment of choice for anaphylactoid reactions and should be given immediately, at the first signs of the reaction, before anything else is done. Prompt therapy can reverse an anaphylactoid reaction within minutes, while epinephrine given after full development of the reaction may not reverse it for hours.

Corticosteroids are of no value during the immediate reaction but may be helpful if the reaction takes a prolonged course.

NEPHROTOXICITY

Acute renal insufficiency following the use of intravascular contrast medium is a well-known potential complication that is fortunately uncommon in excretory urography. Acute renal insufficiency may be manifested by a transient rise in the serum creatinine or by acute oliguric renal failure. The true incidence of nephrotoxicity in excretory urography is not clear but is probably lower than 5%. Some data suggest that while a small percentage of individuals will demonstrate a subclinical transient rise in the serum creatinine following excretory urography, oliguric renal failure is uncommon (Teruel et al., 1981).

An increased risk of developing acute renal failure after excretory urography is seen in patients with preexisting renal insufficiency, diabetes mellitus, and generalized atherosclerosis (Davidson, 1970, 1985). Diaz-Buxo et al. (1975) found acute renal failure in 2 of

1,000 diabetic, compared with 0 in 100,000 excretory urograms in nondiabetic patients. Dehydration potentiates the risk of renal insufficiency and should be avoided in patients receiving contrast material. On the other hand, hydration may reduce, although not totally eliminate, the risk of contrast-induced renal failure.

Acute renal insufficiency associated with contrast medium is usually self-limited. The serum creatinine usually peaks in 2–3 days and returns to the preexisting level within 1–3 weeks. The majority of patients will have complete recovery, with few requiring dialysis.

The theories on the pathogenesis of contrast-induced acute tubular necrosis include direct toxic effects of the contrast media on the proximal tubular cells and ischemia resulting from damage to the renal microcirculation. Nephrotoxicity has been attributed to the hypertonicity of the contrast media, resulting in a unique vacuolization of the cytoplasm of the proximal tubular epithelium termed "osmotic nephrosis" (Moreau et al., 1975). Others think that acute tubular necrosis may result from ischemia due to vasoconstriction or sludging of the RBCs in the renal microcirculation. Katzberg et al. (1983) suggested that the hypertonicity of the contrast media reduces the GFR by increasing the hydrostatic pressure in Bowman's capsule and the proximal tubules. A prolonged response in some patients could explain the pathogenesis of contrast-induced acute tubular necrosis. In certain cases, precipitation of urie acid within the proximal tubules may lead to acute tubular necrosis. Contrast material has a known uricosuric effect and may result in acute urate nephropathy in individuals with hyperuricemia.

A few cases of acute oliguric renal failure resulting in the death of patients with multiple myeloma have been reported following excretory urography. Because of these reports, some authorities considered multiple myeloma an absolute contraindication to excretory urography. However, most authorities today think that excretory urography in patients with multiple myeloma is probably no different from that in patients with other forms of renal impairment (Vix, 1966; Myers and Witten, 1971). When necessary, excretory urography may be performed with cautiion, and dehydration should be avoided.

Most authorities recommend that in the high-risk patient who requires a contrast examination dehydration and large contrast medium loads be avoided. Intravenous hydration performed overnight as well as following the examination may be helpful in avoiding renal problems. Nephrotoxicity is thought to be dose related, and contrast media volume should be kept to the minimum that is necessary to provide the diagnosis. Sequential diagnostic studies that employ intravascular contrast material should not be scheduled without allowing adequate intervals for observation and recovery.

NEWER CONTRAST AGENTS

Over the past few years, several new contrast media compounds have been developed and are now becoming available for clinical use. Their advantage is that they contain the same iodine concentration as the conventional contrast agents but are much less hypertonic. This feature will be of benefit in angiography, where the lower osmolality will decrease the pain on intra-arterial injections. The impact on excretory urography and other modalities using IV contrast material is less clear (Palma et al., 1982). The incidence of systemic effects and reactions to the new agents appears to be no greater than with conventional agents and may indeed be lower. Theoretically, the new agents should demonstrate less renal toxicity than the conventional agents because of the lower hypertonicity. Preliminary studies have shown no distinct advantage of the new agents over the conventional agents in this regard; however, larger series which include patients with underlying renal insufficiency are needed for complete evaluation (Gale et al., 1984). A distinct disadvantage is the prohibitively high cost of the new agents compared with the conventional agents at this time.

TECHNIQUE

OVERVIEW

Several parameters must be considered to properly perform excretory urography. An adequate number of well-trained personnel should be employed to obtain good-quality radiographs as well as to closely monitor the patient during the entire procedure. Emergency drug boxes and monitoring equipment should be available in every room where urography is performed. Only a physician should administer IV contrast material, and he should remain close by in case complications arise. Regularly scheduled x-ray machine maintenance with established quality control guidelines for the technologists is essential.

Before the examination, the physician should have a completed x-ray request outlining the current problem as well as pertinent past medical history (diabetes, nephrolithiasis, allergies, etc.). Laboratory values to assess renal function, BUN and creatinine, are important baselines, especially in debilitated or elderly individuals. Old films should be requested and reviewed. Routine bowel preparation is necessary to eliminate overlying bowel gas and stool, which can obscure renal detail. A typical regimen begins on the afternoon before the examination by giving the patient an unlimited clear liquid diet. One 10-oz. bottle of magnesium citrate is administered, followed two hours later by three Dulcolax tablets. Dehydration is no longer considered a requirement for excretory urography and may actually be detrimental in the pediatric or elderly patient.

CONTRAST MEDIA: DOSAGE AND ADMINISTRATION

Several contrast agents are commercially available (see Table 3A–1). Most authors recommend doses of 50% diatrizoate sodium or 60% meglumine diatrizoate ranging from 30 ml (10 gm iodine) to 100 ml (23 gm iodine), depending on body surface area (BSA) (Witten et al., 1977). An average dose of 0.5–1 ml of contrast material per pound of patient's weight, with a maximum dose of 100 ml, is a generally accepted rule of thumb. Intravenous contrast is rapidly administered through a 19- or 21-gauge "butterfly" scalp vein needle, which is taped and left in place until the procedure is finished. Thus, venous access is established throughout the examination in case of reaction. The contrast media should be administered as quickly as possible to demonstrate a good nephrogram phase, particularly in the

work-up of a renal mass. The antecubital fossa and dorsum of the hand are preferable sites for contrast injection, but central lines may be used as well. If venous access is unavailable, a foot vein may be cannulated. However, after the contrast material is given, the leg should be elevated, and 100 ml of saline solution flushed through the line to decrease the possibility of thrombophlebitis. Contrast media is very irritating, and care should be taken to avoid extravasation into the subcutaneous tissues, as skin slough can occur.

RADIOGRAPHS

Various radiographic views and maneuvers allow the physician the flexibility to "tailor" (Lalli, 1980) each examination to a patient's particular problem. The one mandatory film, however, is "the scout" (KUB, preliminary). This view is a supine abdominal roentgenogram (14 x 17 in.) taken before contrast injection, which includes the kidneys to the pubic symphysis (Figs 3A–2,A and B). Two exposures may be necessary in large patients to cover the entire anatomy. A careful review of the film for technique and underlying abnormalities is required. The size, shape, and position of both renal

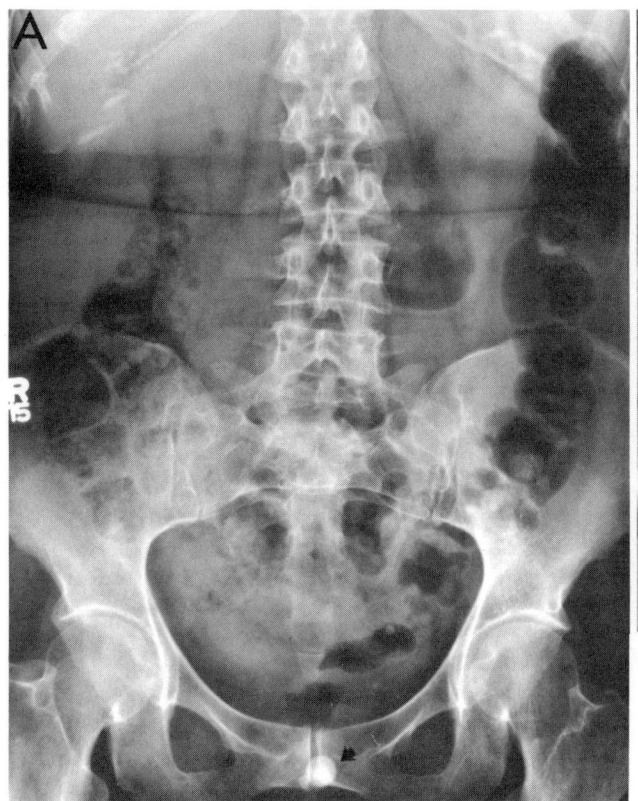
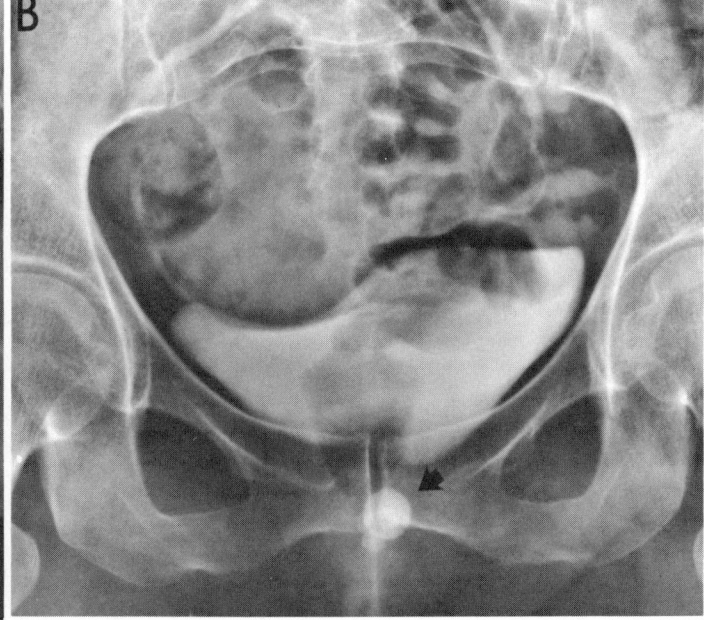

FIG 3A–2.
A, the preliminary radiograph should include area from kidneys to pubic symphysis. Calcification overlying pubic symphysis *(arrow)* is a calculus within a urethral diverticulum. **B,** coned-down bladder radiograph following contrast media injection demonstrating urethral diverticulum. The urethral calculus would have been missed if the radiograph before contrast injection did not include the pubic symphysis.

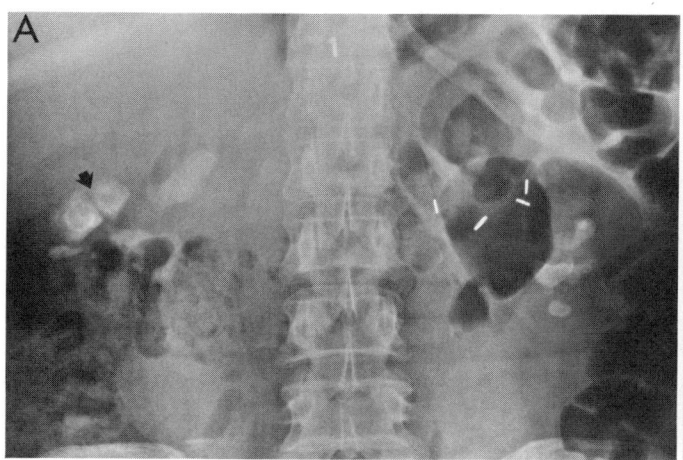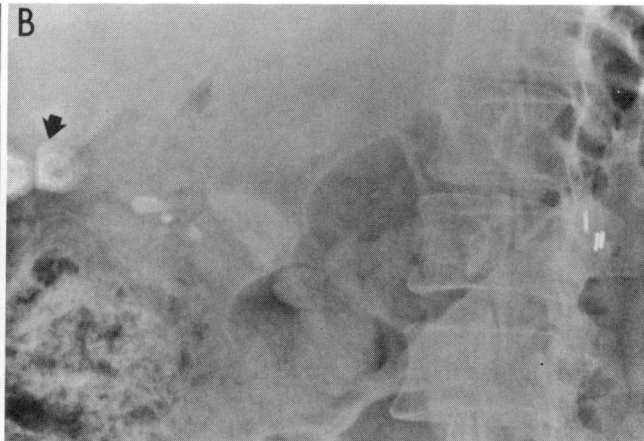

FIG 3A–3.
A, preliminary radiograph demonstrates calcific densities overlying both renal shadows. Two laminated calcific densities *(arrow)* overlying the right renal shadow are likely gallstones. **B,** radiograph in the right posterior oblique projection confirms the presence of gallstones *(arrow)* in addition to renal calculi.

outlines should be observed. Abnormal calcifications (renal and ureteral calculi, gallstones, aneurysm, fibroids, etc.) should be identified, and oblique views before contrast administration may be needed to prove that a calcification is indeed renal in origin (Figs 3A–3,A and B). Bony architecture should be scrutinized for pathology such as metastasis. The bowel gas pattern should be evaluated for possible obstruction and abnormal gas collections (Fig 3A–4). The lung bases are often included on the film and may reveal occult pulmonary disease. Radiopaque substances such as tablets, foreign bodies, and recently administered barium will be demonstrated and may even preclude the examination.

The nephrogram phase is most commonly evaluated by a coned-down one-minute film (11 × 14 in.) postinjection. The nephrogram phase demonstrates the densest opacification of the kidney parenchyma and should be obtained within 1 minute after injection (Fig 3A–5). Again, the kidneys are compared and evaluated for size, shape, and position. The nephrogram can be delayed (obstruction), misplaced (pelvic/thoracic/kidney), irregular (mass), or prolonged (hypotension). In lieu of a one-minute film, however, nephrotomography is often preferred, particularly in the workup of a possible renal mass. Three tomograms at 1.5–2-cm intervals in the supine position are taken at 30, 60, and 90 seconds after injection, providing excellent visualization of the renal contours (Fig 3A–6).

A 3–5-minute coned-down film (11 × 14 in.) is obtained to demonstrate the pyelogram phase with opacification of both collecting systems and proximal ureters (Fig 3A–7,A). Calyceal distortion, irregularity, or filling defects may signal underlying pathology. Tomography may also be helpful to better delineate the calyces. Usually, the nephrotomogram which best demonstrated

both kidneys at one minute is selected, and this cut is taken after the 3–5-minute view.

Relative obstruction of the ureters with an inflatable compression device placed across the lower abdomen after the nephrogram phase achieves optimal distention

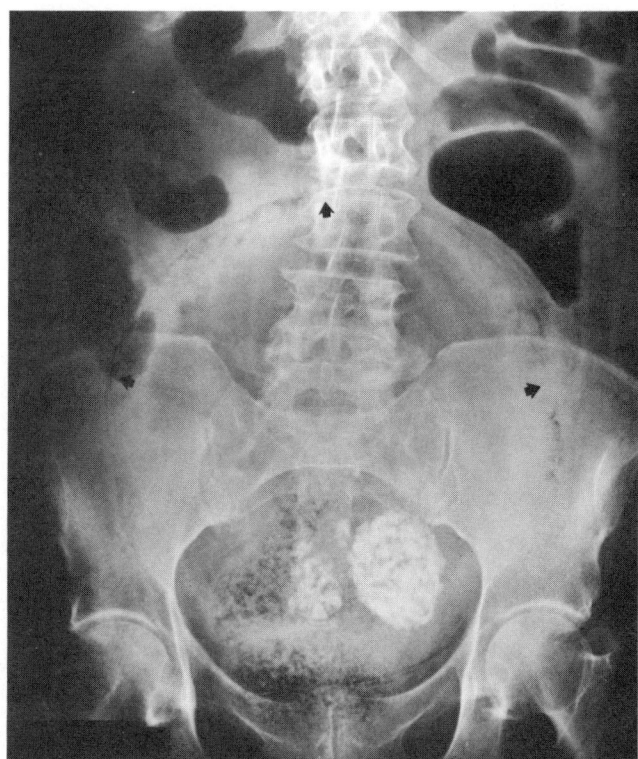

FIG 3A–4.
Preliminary radiograph demonstrates marked distention of the bladder with air within the wall *(arrows)* in a patient with emphysematous cystitis. Large calcified uterine fibroids are noted within the pelvis.

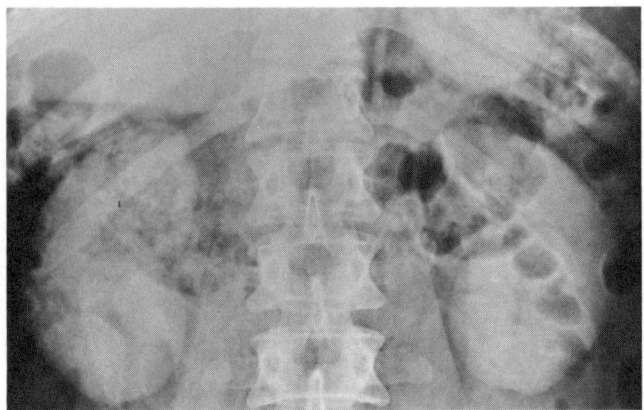

FIG 3A–5.
The nephrogram phase demonstrates parenchymal opacification of both kidneys. Kidney size and contour abnormalities are best evaluated during the nephrogram phase.

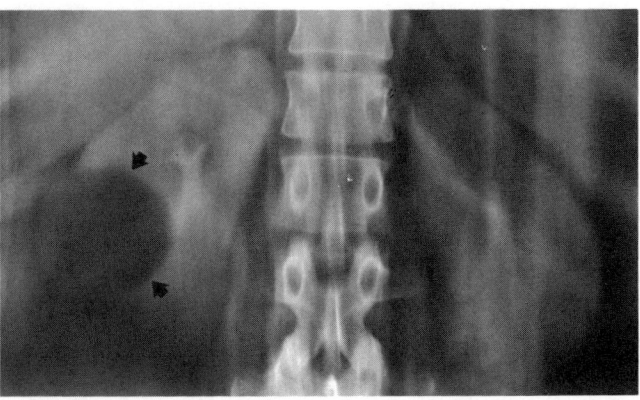

FIG 3A–6.
Nephrotomography demonstrates a large, well-circumscribed mass in the lower pole of the right kidney *(arrows)* consistent with a renal cyst. Nephrotomograms provide excellent detail of renal contour abnormalities.

of the collecting systems and proximal ureters (Fig 3A–7,B). Contraindications include suspected ureteral obstruction, abdominal aortic aneurysm, and recent abdominal or renal surgery. Another alternative, however, is to place the patient in the Trendelenburg position after the one-minute film to enable better filling of the collecting system. Both the compression device and Trendelenburg position are maintained until the large 10-minute film is taken (14 × 17 in.). This view visualizes the kidneys to the bladder. The patient is brought up into "reverse" Trendelenburg, or the compression device is released, and an immediate film is taken in hopes of catching the contrast media flowing down both ureters to the bladder. Usually, the distal ureters are well demonstrated, and no additional exposures are required. The large 10-minute film is often the sole film taken during emergency evaluation of

trauma patients to grossly assess renal integrity.

The bladder is further evaluated with a coned-down view (11 × 14 in.) for better anatomical detail. Postvoid films are useful to evaluate benign prostatic hypertrophy and other bladder abnormalities (Fig 3A–8,A and B). The lower ureteral segment and ureterovesical junction can be better delineated with oblique views either before or after voiding (Figs 3A–9,A and B, 3–10,A and B). If there is inadequate definition or a suspicious finding in the distal ureters or bladder, a prone film (11 × 14 in. or 14 × 17 in.) will distend the anterior portions of the collecting system as they become dependent, i.e., the middle and distal portions of the ureters. This is often helpful in localizing an obstruction (Fig 3A–11,A and B). Lesions involving the anterior wall of the bladder are better visualized as well.

Delayed views are important when hydronephrosis is

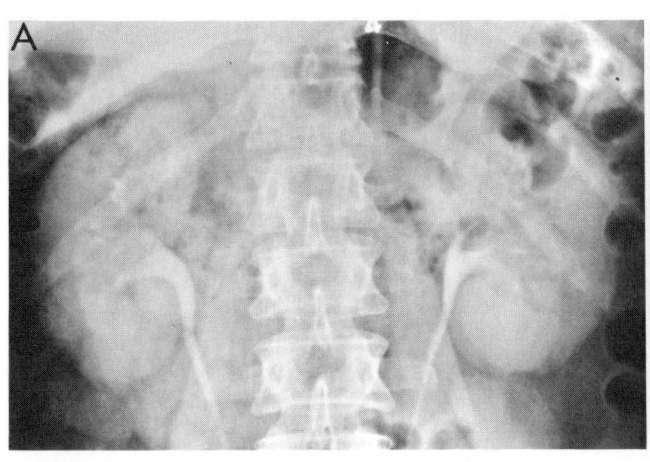
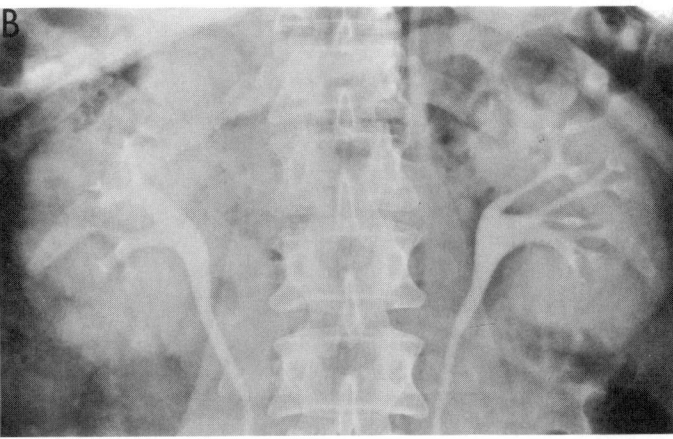

FIG 3A–7.
A, radiograph taken five minutes following contrast injection demonstrates the pyelogram phase with opacification of the collecting system and proximal ureters. **B,** radiograph fol-

lowing the use of a compression device in the same patient. Note distention of collecting systems with improved visualization of the calyces and infundibula.

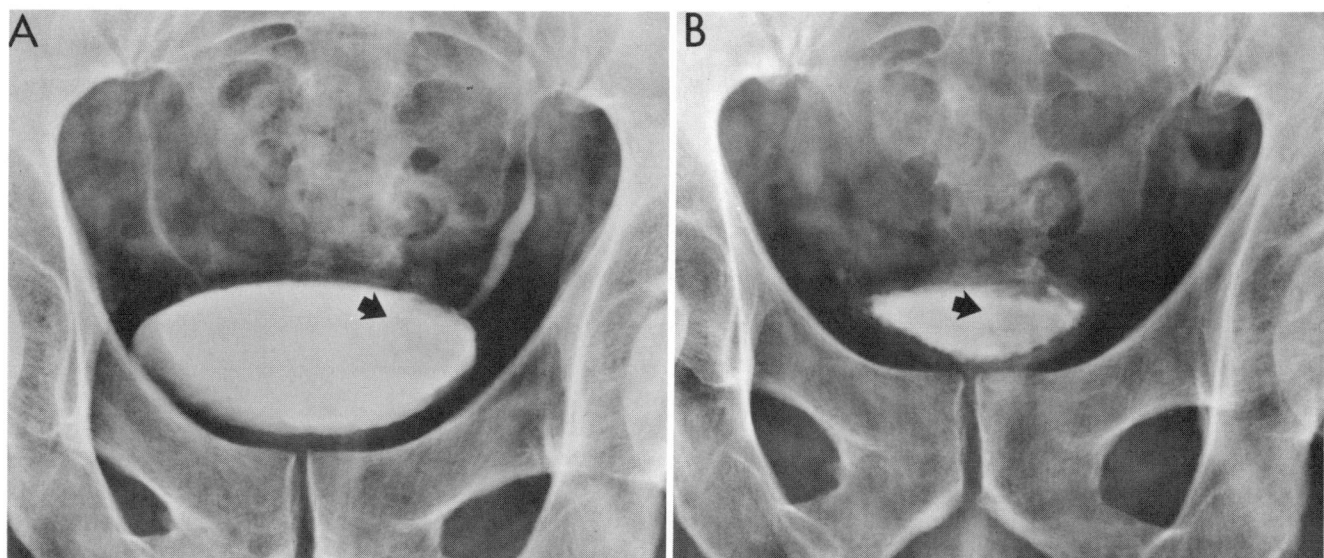

FIG 3A–8.
A, the filled bladder suggests a filling defect near the left bladder wall *(arrow).* **B,** the postvoid radiograph confirms the defect *(arrow)* proven to be bladder carcinoma.

present. Films taken at 12 and 24 hours postinjection may demonstrate the delayed pyelogram and level of obstruction.

Oblique views can be taken at any time during the examination for further anatomical clarification. Calcification overlying the kidney shadows can be assessed to determine whether they are renal or extrarenal in origin before contrast material administration. Oblique views are also helpful in projecting purely anterior or

posterior renal masses which may be missed on tomography or conventional films. Again, the ureterovesical junction is well seen on oblique views.

Lateral views (14 × 17 in.) are used on occasion to evaluate extrinsic mass effects on the kidneys. Calcified abdominal aortic aneurysms, spinal disease, and possible enterovesical fistulas can also be shown with this view.

A double-exposure inspiration/expiration film can be

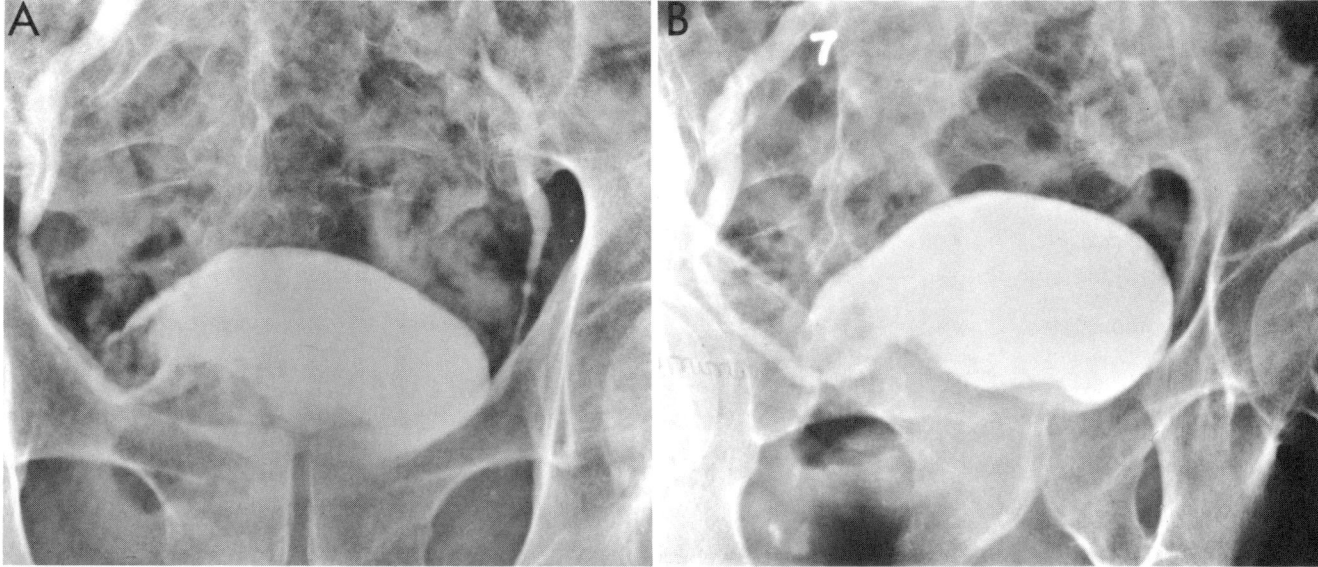

FIG 3A–9.
A, coned-down radiographs of the bladder suggest a mass involving the inferior aspect of the right wall of the bladder. **B,** radiograph taken in the left posterior oblique projection confirms irregular mass near the insertion of the right ureter consistent with carcinoma.

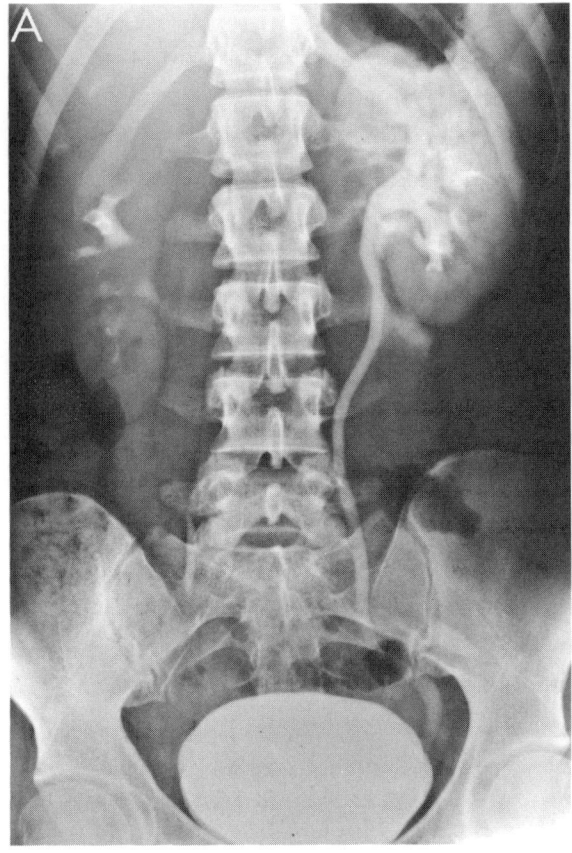

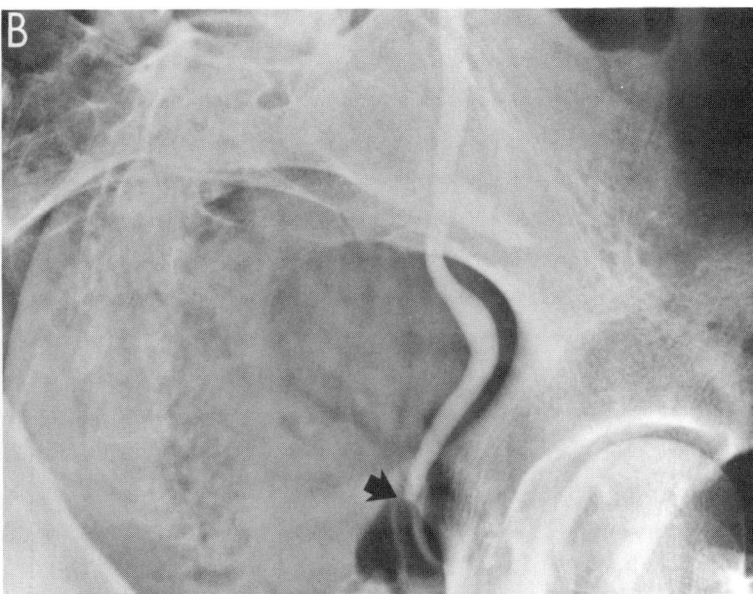

FIG 3A–10.
A, two-hour radiograph in a patient with left-sided renal colic. There is dilatation of the left ureter down to the level of the bladder. The point of obstruction is not demonstrated. Note the peripelvic extravasation of contrast resulting from forniceal rupture. **B,** postvoid radiograph in the left posterior oblique projection demonstrates the level of obstruction *(arrow)* due to a uric acid stone.

obtained to measure renal excursion if there is a suspicion that the kidney is fixed in position as in abscess or tumor. The normal kidney will be blurred on the film, and the abnormal kidney will stay in focus, i.e., not changing in position with respiratory movement.

Fluoroscopy often provides additional information about urinary tract physiology. In combination with spot filming, fluoroscopy can reveal renal dynamics and relationships not seen with standard static radiographs.

THE PEDIATRIC PATIENT

Excretory urography in children requires special attention to several technical factors. First, establishment

FIG 3A–11.
A, ten-minute radiograph in a patient with right renal colic. There is hydronephrosis and dilatation of the ureter with poor definition of the level of the obstruction. **B,** prone radiograph demonstrates excellent opacification of the ureter with identification of the level of the obstruction *(arrow).*

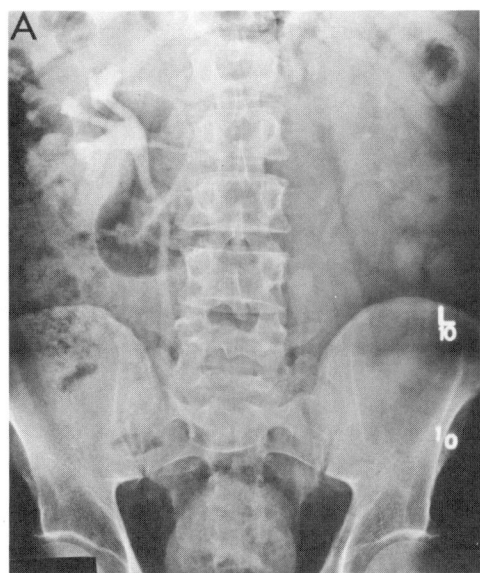

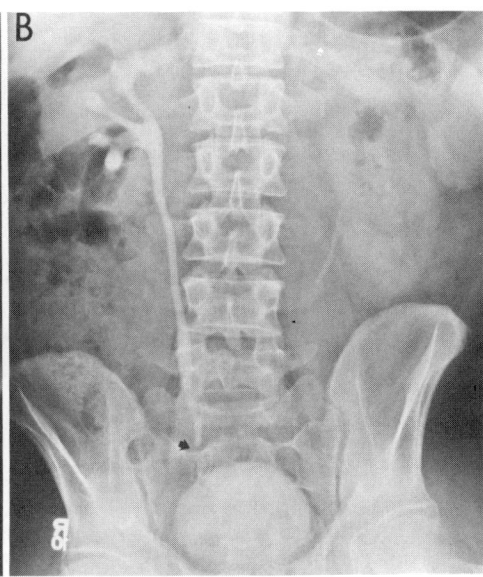

of rapport among the physician, child, and parents is crucial before obtaining radiographs. Explaining the procedure in a step-by-step fashion helps to alleviate some of the anxiety regarding the examination. In the older age group, the clinician should direct the explanation primarily to the child.

Anesthesia or sedation is rarely indicated in excretory urography. In infants and preschool children, a "cradle" device is placed directly on the x-ray table for immobilization of the patient. Heat lamps, pediatric drug boxes, and resuscitative equipment should be available in each x-ray room.

Bowel preparation is not indicated in children. An adequate state of hydration is essential, especially in infants and younger children, to avoid dehydration from the diuresis caused by the excess solute load of the contrast medium. Abstinence from food of about 5 hours is sensible in most children to prevent hypoglycemia as well as to keep the stomach empty in case of emesis after contrast administration (Chrispin et al., 1980).

The contrast dosage in children is basically the same as for adults. Between 0.5–1 ml/lb (1–3 ml/kg) of patient's weight, with a maximum of 100 ml, is a safe and effective amount of contrast medium to give a satisfactory urogram. Venous access in young children may be difficult to obtain. The veins on the dorsum of the hand or foot are usually the most accessible for cannulation with a 25-, 23-, or 21-gauge "butterfly" needle. Other sites include the antecubital fossa and scalp veins in infants. Pneumothorax is a complication of external jugular vein injections, and the femoral vein should be avoided because of the reported risk of septic arthritis of the hip (Chrispin et al., 1980). Rapid bolus injection technique is recommended, and the needle should remain in place until the study is terminated. Immediately after giving the bolus, a carbonated beverage or juice may be given to the patient. This allows distention of the gastric bubble for more optimal visualization of the kidneys as well as providing calories and fluid intake for the child.

The filming sequence in the pediatric age group is aimed at obtaining the most information from a minimum number of radiographic exposures. Collimation should be used whenever possible.

A preliminary film, which includes the kidneys to the pubic symphysis, should be carefully scrutinized for underlying skeletal anomalies such as sacral agenesis and spinal dysraphism. A search for abdominal masses, an abnormal bowel gas pattern, and any calcifications should also be conducted.

In most cases, the entire pediatric excretory urogram can be completed with two postcontrast films. The nephrogram phase is demonstrated with a 1-minute coned-down view of the kidneys. A large 10-minute film will usually demonstrate both collecting systems, ureters, and the bladder. A postvoid bladder film is frequently not obtained in children. However, additional views may be indicated for a particular clinical history or to clarify further abnormalities seen while performing the urogram.

INDICATIONS FOR EXCRETORY UROGRAPHY

The indications for excretory urography are extensive and varied. Almost all patients with major urologic signs or symptoms require excretory urography as do some patients with nonurologic disease. The development of the newer imaging modalities has somewhat altered the indications for excretory urography, but for the most part excretory urography continues to be the initial, and often the only, examination in the patient with urologic disease. Urinary tract structural anomalies, hematuria, infectious and calculus disease, and obstructive uropathy can be properly evaluated with excretory urography.

In the past excretory urography was used extensively for the evaluation of nonurologic conditions such as retroperitoneal masses and retroperitoneal lymphadenopathy. In such cases, CT and ultrasonography have proved to be more valuable and have replaced excretory urography as the initial examination. The excretory urogram continues to have a role in the preoperative evaluation of these patients as well as in patients with gynecologic neoplasms to provide accurate depiction of the ureters and any structural anomalies.

The initial evaluation of hematuria and a suspected renal mass is performed with excretory urography. Further characterization of a detected renal mass is usually required and best performed with ultrasonography, CT, and angiography. Since the urographic findings in a renal mass are related to contour abnormalities and calyceal displacement, tomography and oblique views are necessary to achieve optimal results. An exophytic mass arising from the anterior or posterior surface of the kidney may demonstrate only subtle changes and can be missed (Kass et al., 1983). Although the false negative rate of excretory urography is not known, the addition of CT should detect these small, exophytic lesions. The use of ultrasonography or CT as the initial examination for hematuria and suspected renal mass has some support. Hematuria, however, may be due to lesions involving the collecting system, ureters, and bladder, as well as a parenchymal mass, and these structures are best visualized initially with excretory urography.

The diagnosis of obstruction in patients with renal failure is best performed with ultrasonography. Chronic obstruction can almost always be accurately depicted with ultrasonography without the use of contrast material, with its attendant risk in a patient with renal insufficiency. Exceptions include large, obstructing staghorn calculi that may give a false negative result with ultrasonography and problems with distinguishing hydronephrosis from a normal but mainly extrarenal pelvis (Talner et al., 1981). There are problems, however, in the evaluation of acute obstruction with ultrasonography (Laing et al., 1985). Acute obstruction may be missed by ultrasonography because of minimal or no hydronephrosis in patients with a partially obstructing calculus or forniceal rupture leading to a decompressed but obstructed collecting system. Excretory urography remains the examination of choice in the evaluation of patients with acute flank pain in whom acute obstruction may be present.

Calculus disease of the urinary tract is best evaluated with excretory urography. Ultrasound has made some contributions with regard to radiolucent calculi, but the excretory urogram remains the basic examination. The development of extracorporeal shock wave lithotripsy (ESWL) has radically altered the management of patients with calculus disease and will increase the number of patients treated for urinary tract stones. The evaluation and management of the ESWL patient is largely dependent on plain film radiography and excretory urography. Accurate pre-ESWL evaluations as well as follow-up evaluations for function and obstruction are performed with excretory urography.

The role of excretory urography in the trauma patient has been questioned during the past few years. Computerized tomography has made a large impact on the evaluation of the abdomen and retroperitoneal structures following trauma. CT can provide excellent depiction of renal trauma that clearly surpasses excretory urography with regard to diagnostic information. Most urologists agree that all patients with posttraumatic hematuria should have the urinary tract screened for injury at some point. In cases where there is a high probability of a significant renal injury, CT or angiography should be considered as the first step. On the other hand, excretory urography is a simple and relatively inexpensive method of excluding a significant renal injury. Although a CT examination may detect minor injuries not seen on an excretory urogram, in most cases these will not be clinically significant. The evaluation of the trauma patient with microscopic hematuria and no other clinical signs of renal injury usually can be performed with excretory urography alone.

The indications for excretory urography in children have undergone changes in the past few years. Ultrasonography and CT have virtually replaced the excretory urogram in the evaluation of abdominal and renal masses, hydronephrosis, and abdominal pain and as a screen for associated renal anomalies in children with other congenital anomalies (Lebowitz and Ben-Ami, 1983). Ultrasonography is evolving into the initial screening procedure for pediatric urologic disease because of excellent detail in children and lack of radiation exposure.

Excretory urography is indicated in evaluating urinary tract infection in children in whom reflux has been demonstrated on voiding cystourethrography. The excretory urogram provides better definition of cortical scarring associated with vesicoureteral reflux (Leonidas et al., 1985). Excretory urography is also indicated in calculus disease, nonmedical hematuria, and in children who have recently undergone urologic surgery.

REFERENCES

1. Ansell G: Adverse reactions to contrast agents: The scope of the problem. *Invest Radiol* 1970; 5:374.
2. Becker JA: Vicarious excretion of urographic media. *Radiology* 1968; 90:243.
3. Brasch RC: Allergic reactions to contrast media: Accumulated evidence. *AJR* 1980; 134:797–801.
4. Cattell WR: Excretory pathways for contrast media. *Invest Radiol* 1970; 9:473.
5. Chrispin AR, Gordon I, Hall C, et al: *Diagnostic Imaging of the Kidney and Urinary Tract in Children.* Berlin, Springer-Verlag, 1980.
6. Davidson AJ: *Radiology of the Kidney.* Philadelphia, WB Saunders Co, 1985.
7. Davidson AJ, Becker J, Rothfield N, et al: An evaluation of the effects of high-dose urography on previously impaired renal and hepatic function in man. *Radiology* 1970; 97:249.
8. Diaz-Buxo JA, Wagoner RD, Hattery RR, et al: Acute renal failure after excretory urography in diabetic patients. *Ann Intern Med* 1975; 83:155–158.
9. Fischer HW, Cornell SH: The toxicity of the sodium and methylglucamine, salts of diatrizoate, iothalamate, and metrizoate: An experimental study of their circulatory effects following intracarotid injection. *Radiology* 1965; 85:1013.
10. Forrest JB, Howards SS, Gillenwater JY: Osmotic effects of intravenous contrast agents on renal function. *J Urol* 1981; 125:147.
11. Gale ME, Robbins AH, Hamburger RJ, et al: Renal toxicity of contrast agents: Iopamidol, iothalamate, and diatrizoate. *AJR* 1984; 142:333–335.
12. Gillenwater JY: Reactions associated with excretory urography—current concepts. *J Urol* 1971; 106:122.
13. Greenberger PA, Patterson R, Tapio CM: Prophylaxis against repeated radiocontrast media reactions in 857 cases: Adverse experience with Cimetidine and safety of β-adrenergic antagonists. *Arch Intern Med*, 1985; 145:2197–2200.
14. Hartman GW, Hattery RR, Witten DM, et al: Mortality during excretory urography: Mayo Clinic experience. *AJR* 1982; 139:919–922.
15. Kass DA, Hricak H, Davidson AJ: Renal malignancies

with normal excretory urograms. *AJR* 1983; 141:731–734.

16. Katzberg RW, Morris TW, Schulman G, et al: Reactions of intravenous contrast media: Part I. Severe and fatal cardiovascular reactions in a canine dehydration model. *Radiology* 1983; 147:327–330.

17. Katzberg RW, Morris TW, Schulman G, et al: Reactions to intravenous contrast media. Part II. Acute renal response in euvolemic and dehydrated dogs. *Radiology* 1983; 147:331–334.

18. Kelly JF, Patterson R, Lieberman P, et al: Radiographic contrast media studies in high-risk patients. *J Allergy Clin Immunol* 1978; 62:181–184.

19. Laing FC, Jeffrey RB, Wing VW: Ultrasound versus excretory urography in evaluating acute flank pain. *Radiology* 1985; 154:613–616.

20. Lalli AF: Urographic contrast media reactions and anxiety. *Radiology* 1974; 112:267.

21. Lalli AF: Contrast media reactions: Data analysis and hypothesis. *Radiology* 1980; 134:1–12.

22. Lalli AF: *Tailored Urologic Imaging*. Chicago, Year Book Medical Publishers, 1980.

23. Lasser EC: Basic mechanisms of contrast media reactions: Theoretical and experimental considerations. *Radiology* 1968; 91:63.

24. Lasser EC, Walters AJ, Lang JH: An experimental basis for histamine release in contrast material reactions. *Radiology* 1974; 110:49–59.

25. Lebowitz RL, Ben-Ami T: Trends in pediatric uroradiology. *Urol Radiol* 1983; 5:135–147.

26. Leonidas JC, McCauley RGK, Klauber GC, et al: Sonography as a substitute for excretory urography in children with urinary tract infection. *AJR* 1985; 144:815–819.

27. Moreau J, Droz D, Sabto J, et al: Osmotic nephrosis induced by water-soluble triiodinated contrast media in man. *Radiology* 1975; 115:329.

28. Myers GH, Witten DM: Acute renal failure after excretory urography in multiple myeloma. *AJR* 1971; 113:583.

29. Palma LD, Rossi M, Stacul F, et al: Iopamidol in urography. *Urol Radiol* 1982; 4:1–3.

30. Peck WW, Slutsky RA, Hackney DB: Effects of contrast media on pulmonary hemodynamics: Comparison of ionic and non-ionic agents. *Radiology* 1983; 149:371–374.

31. Pfister RC: Reactions to urographic contrast media. Syllabus for the Categorical Course in Genitourinary Radiology. Chicago, Radiological Society of North America, Nov 30–Dec 5, 1975. Syracuse, NY, Radiological Society of North America, 1975.

32. Popio KA, Ross AM, Oravec JM, et al: Identification and description of separate mechanisms for two components of Renografin cardiotoxicity. *Circulation* 1978; 58:520.

33. Rosenfield AT, Littner MR, Ulreich S, et al: Respiratory effects of excretory urography: A preliminary report. *Invest Radiol* 1977; 12:295.

34. Shehadi WH: Adverse reactions to intravascularly administered contrast media: A comprehensive study based on a prospective survey. *AJR* 1975; 124:145.

35. Shehadi WH, Toniolo G: Adverse reactions to contrast media. *Radiology* 1980; 137:299–302.

36. Shehadi WH: Contrast media adverse reactions: Occurrence, recurrence, and distribution patterns. *Radiology* 1982; 143:11–17.

37. Stanley RJ, Pfister RC: Bradycardia and hypotension following use of intravenous contrast media. *Radiology* 1976; 121:5.

38. Talner LB, Scheible W, Ellenbogen PH, et al: How accurate is ultrasonography in detecting hydronephrosis in azotemic patients? *Urol Radiol* 1981; 3:1–6.

39. Teruel JL, Marcen R, Onaindia JM, et al: Renal function impairment caused by intravenous urography: A prospective study. *Arch Intern Med* 1981; 141:1271–1274.

40. Vix VA: Intravenous pyelography in multiple myeloma. *Radiology* 1966; 87:896.

41. Witten DM, Hirsh FD, Hartman GW: Acute reactions to urographic contrast medium: Incidence, clinical characteristics and relationship to history of hypersensitivity states. *AJR* 1973; 119:832.

42. Witten DM, Myers GH, Utz DC: *Emmett's Clinical Urography*, ed 4. Philadelphia, WB Saunders Co, 1977.

43. Yocum MW, Heller AM, Abels RI: Efficacy of intravenous pretesting and antihistamine prophylaxis in radiocontrast media-sensitive patients. *J Allergy Clin Immunol* 1978; 62:309–313.

Chapter 3B / Renal Arteriography and Computed Tomography

ROBERT L. VOGELZANG, M.D.
HARVEY L. NEIMAN, M.D.

RENAL ARTERIOGRAPHY

The role of angiography in the evaluation of the urologic patient has changed significantly in the past several years. Initially, angiography served as a key examination in the differential diagnosis of renal masses, provided useful information in evaluating infection, hydronephrosis, renal trauma, or renal anomalies, and was valuable in the initial workup of hypertension. Only the last role remains significant. Since ultrasound and CT now serve as the primary triage techniques, angiography assumes the role of providing preoperative staging for renal tumors, providing interventional techniques such as embolotherapy and angioplasty, and being a diagnostic modality only when other, less invasive procedures are found wanting.

TECHNIQUES OF ANGIOGRAPHY

Angiography is a problem-oriented procedure that requires diagnostic acumen, technical skills, and an understanding of the clinical problem so that an efficient and accurate examination can be performed.

Before the procedure, it is essential that the angiographer discuss at length the procedure, risks, benefits, and possible alternatives with the patient. Preangiographic orders include no solid foods after midnight, although liquids are strongly encouraged. Administration of IV fluids is routine to prevent dehydration, particularly in patients with compromised renal function. Preoperatively, patients are sedated with meperidine hydrochloride and secobarbital or other equivalent drugs.

Percutaneous transfemoral catheterization is the preferred access route for arteriography using the Seldinger technique. New catheter materials have recently become available that permit use of smaller catheter diameters (4 and 5 F), which may decrease puncture site complications. Occasionally translumbar or axillary approaches must be used. Multiple catheter shapes have been designed for performance of aortography and selective renal arteriography. Generally, aortography is performed with a pigtail-shaped or straight catheter with multiple side holes. Selective arteriography can be carried out using a variety of preformed catheters (Nei-

man and Yao, 1985; Johnsrude and Jackson, 1979; Reuter and Redman, 1979; Gerlock and Mirfakhraee, 1985).

Angiography using any technique carries a risk of morbidity and mortality (Lang, 1963; Sigstedt and Lunderquist, 1978; Silverman and Wexler, 1976; Beinart et al., 1983; Cochran et al., 1983). Therefore, these procedures should be reserved for those individuals in whom the diagnostic or therapeutic result may significantly aid in the patient's management. In patients with atherosclerotic peripheral vascular disease, the high incidence of concomitant coronary artery and cerebral vascular disease increases the risk for this group. The mortality from angiographic procedures is about 0.05%. The incidence of serious complications from catheter techniques has varied from 0.5% to 2.3% in different series (Lang, 1963; Sigstedt and Lunderquist, 1978). The true incidence of complications is probably closer to the higher figure. Translumbar techniques have been reported to involve major complication rates of 0.5% (Szilogyi et al., 1977). Allergic reactions are not common with arterial injections, but in an individual with a documented history of contrast material allergy, we have followed the policy of performing arteriography when indicated, following pretreatment at 13, 7, and 1 hours prior to the examination with prednisone, 50 mg orally, and diphenhydramine hydrochloride (Benadryl), 50 mg IM.

Acute renal failure, however, is a serious complication of angiography (Cochran et al., 1983). The most important factor indicating a potential case of renal failure is diabetes mellitus. Dehydration is also a significant risk factor. The use of the newer ionic and nonionic contrast materials may not only decrease patient discomfort from angiographic procedures, but may also increase the safety of the examination. These agents have been shown to be less nephrotoxic and less injurious to the CNS and myocardium (Felder et al., 1977; Gordon et al., 1984; Widrich et al., 1983; Lund et al., 1984; Tornquist et al., 1984).

ANATOMY

The renal angiogram consists of three phases: arterial, nephrographic, and venous.

The arterial phase begins with injection of contrast material into the abdominal aorta or selectively into the renal artery. The main renal artery originates from the abdominal aorta at the level of L1–L2. The right main renal artery usually is slightly more cephalad than the left and both arise from the lateral aspect of the aorta. When single, the main renal artery in the adult varies between 5 and 10 mm in diameter. Contrast injections given at the rate of 6–8 ml/sec will equal the renal blood flow in the majority of healthy kidneys. Multiple renal arteries are seen frequently in these cases, the diameter being proportionally smaller. Occasionally, a small accessory renal artery is seen supplying a pole. The main renal artery usually divides into two branches that pass anteriorly and posteriorly to the renal pelvis. Usually five segmental arteries arise from these two branches. The entire upper and lower poles are each supplied by a single segmental artery; the other segmental arteries supply either the anterior or the posterior portion of the middle of the kidney. These segmental renal arteries in turn branch into multiple levels of interlobar arteries. Each lobe of the kidney is supplied by at least two interlobar arteries. Six to eight arcuate arteries arise and enter the renal tissue at the junction between the cortex and medulla (Hodson, 1978). The arcuate arteries in turn give off the interlobular arteries, which can be seen on magnification angiograms. Significant variation does exist in the branching pattern of the renal arteries (Boijjsen 1961).

During the arterial phase, adrenal, renal pelvic, and renal capsular arteries can also be noted. The usefulness of the perirenal and pelvic arterial supply as a diagnostic tool has been described (Boijjsen and Folin, 1961; Eliska, 1968; Yune and Klatte, 1976).

The nephrogram is composed of four phases: the cortical arteriogram, the glomerulogram, the cortical nephrogram, and the general nephrogram. The first three occur within approximately six seconds following the end of the contrast material injection. The general nephrogram lasts approximately 20 seconds. The venous phase occurs with opacification of the renal veins first appearing approximately four seconds after the onset of the injections (Boijjsen, 1983) (Fig 3B–1).

APPROACH TO THE RENAL MASS

Diagnostic ultrasound and CT have replaced angiography and percutaneous cyst aspiration as the examinations of choice for differentiation of the renal cyst from renal tumor (Mauro et al., 1982; Weyman et al., 1980). The individual with a rounded anechoic, sharply marginated renal mass with acoustic enhancement on ultrasound need not undergo further tests for this renal cyst. Those with equivocal ultrasound findings should have CT performed. Angiography is reserved as a preoperative staging procedure for those individuals with renal cell carcinoma or other suggested neoplasms. Even in this clinical situation, current-generation CT scanners probably preclude the need for an invasive procedure,

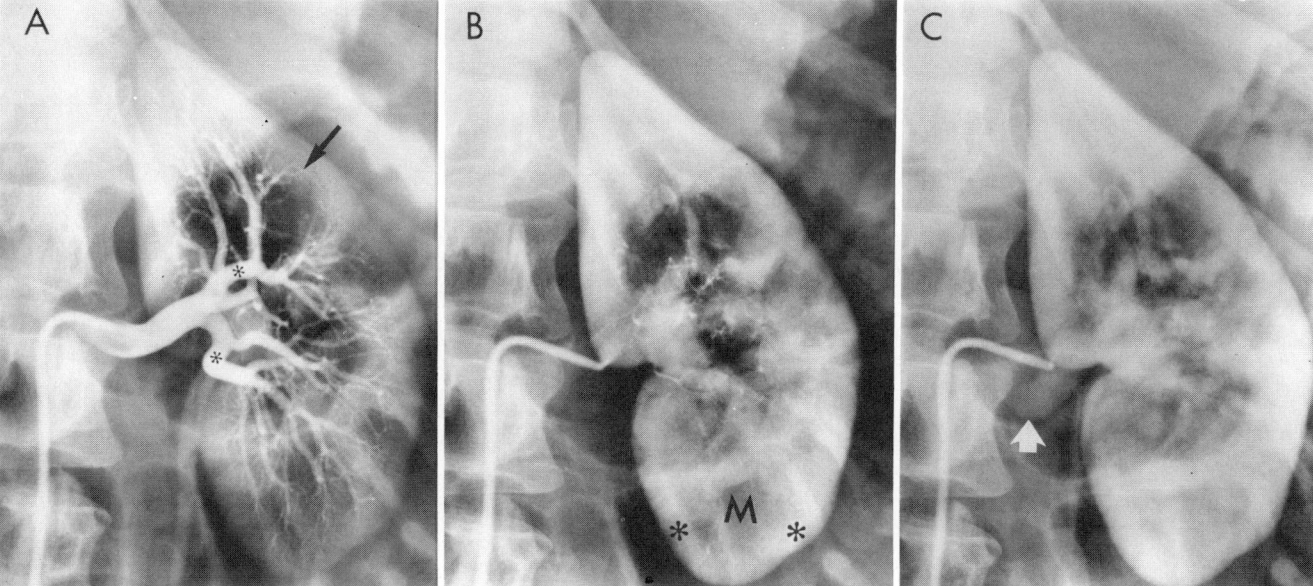

FIG 3B–1.
Normal renal angiogram. Midarterial phase **(A)** demonstrates normal intrarenal arterial anatomy including segmental arteries *(asterisks)* and multiple levels of interlobar vessels arising from these segmental vessels. Arcuate arteries *(arrows)* at terminus of interlobar vessels. Nephrogram **(B)** immediately after arterial phase demonstrates some granularity consistent with glomerular filling. There is also some differentiation between cortex *(asterisks)* and medulla *(M)*. Venous phase **(C)** demonstrates diminution in nephrographic density and main renal vein *(arrow)* filling. Corticomedullary differentiation is present.

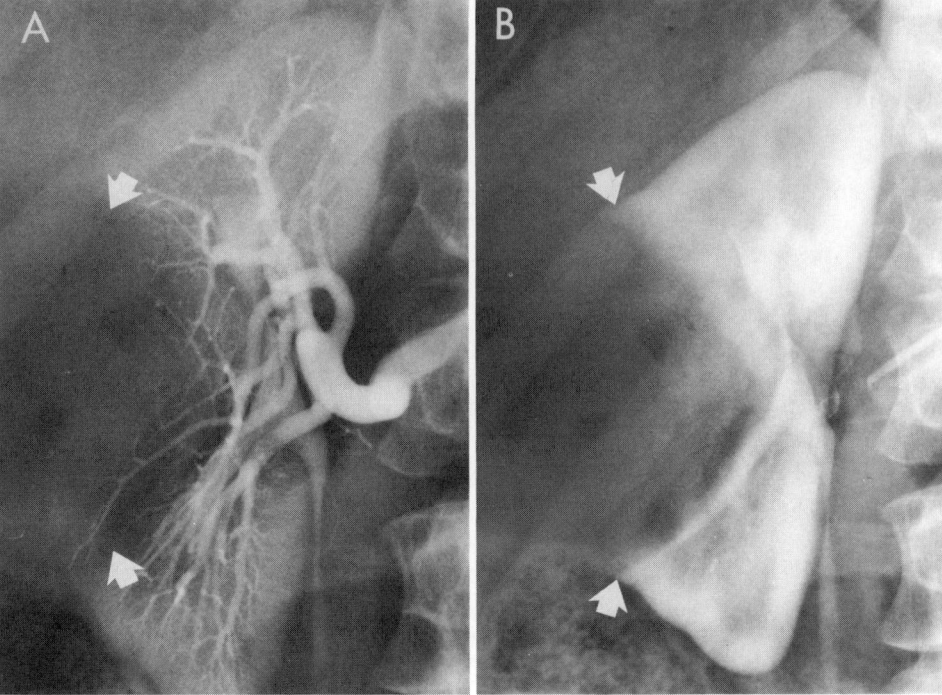

FIG 3B–2.
Renal cysts. Arterial phase **(A)** demonstrates avascular mass *(arrows)* with displacement of intrarenal vessels. Nephrogram **(B)** shows absence of contrast enhancement and a "beak" *(arrows)* of renal tissue displaced by the cyst.

since CT is better able to document extension into the perirenal soft tissues, lymph node involvement, contralateral renal metastatic foci, and distant metastases. CT and angiography are equally accurate in identifying venous invasion.

RENAL CYSTS

The angiographic criteria for simple renal cysts are well established. An avascular, sharply margined, radiolucent mass with a pencil-thin wall that is well defined utilizing multiple oblique views is a cyst. Frequently, the mass can be seen to form an acute angle at the junction of the cyst wall and renal cortex, demonstrating the so-called beak sign (Fig 3B–2). The diagnosis of a renal cyst is one of exclusion, and the possibility of an avascular renal tumor must be constantly considered and ruled out angiographically. As we previously indicated, the diagnosis of cyst should not be made angiographically, but when renal arteriography done for another purpose reveals these common masses, careful and stringent use of the diagnostic criteria outlined above is mandatory.

RENAL CELL CARCINOMA

The diagnostic accuracy of angiography for renal cell carcinoma is reported as 94%–97% (Eliska, 1968; Watson et al., 1968; Meany, 1969). At the time of diagnosis, the mass may vary from 1 to 20 cm or larger. Approximately 90%–95% of these lesions demonstrate markedly increased vascularity, although 5%–10% of lesions

are hypovascular to avascular. In addition to increased vascularity, innumerable abnormal vessels (neovascularity, tumor vascularity) are seen in which there is a random distribution of tortuous, irregular-sized channels which appear to start from no specific point and go

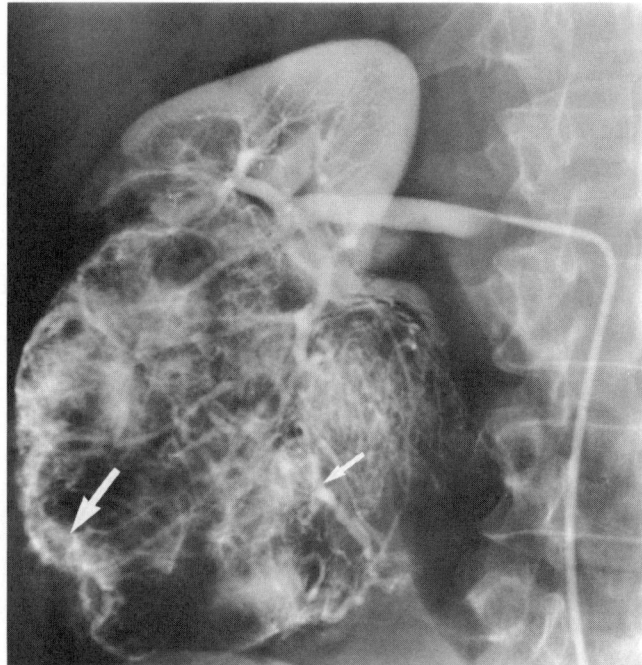

FIG 3B–3.
Bulky hypervascular left lower pole renal cell carcinoma demonstrating neovascularity *(arrows)* and dense contrast staining.

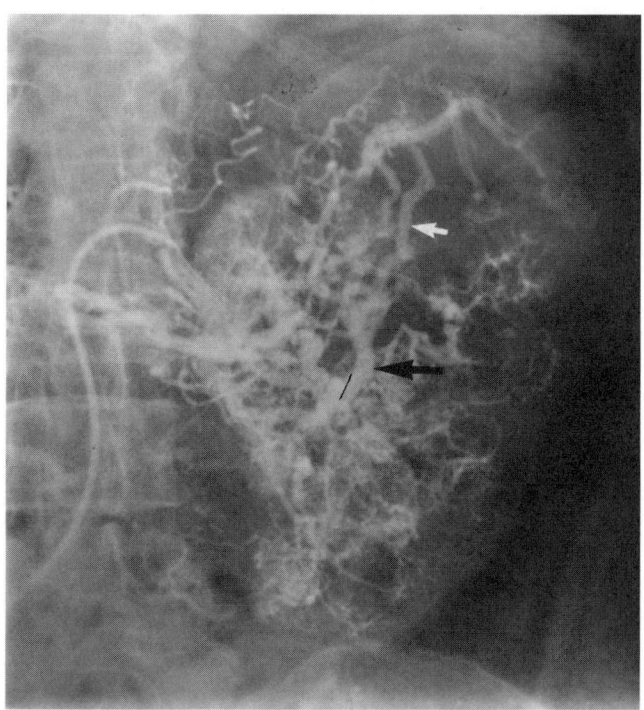

FIG 3B–4.
Virtual replacement of the left kidney by a hypervascular renal cell carcinoma. Diffuse neovascularity and very prominent, irregular draining veins *(arrows)*.

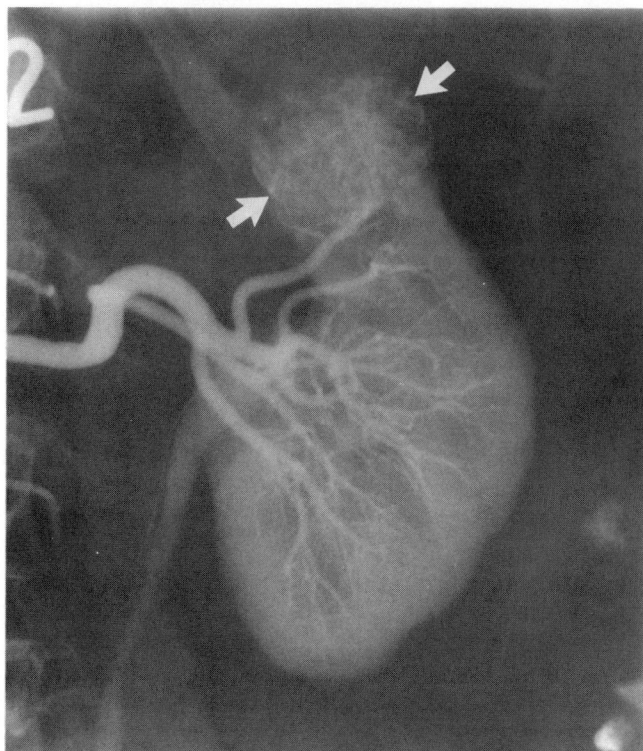

FIG 3B–5.
Small left upper pole renal cell carcinoma *(arrows)* with only modest contrast enhancement and little neovascularity.

nowhere (Figs 3B–3 to 3B–5). Frequently, there may be abnormal arteriovenous communications with early and prominent venous collateral vessels and drainage (Fig 3B–6). Pooling of contrast material may also be noted in so-called vascular lakes. The margins of the mass with the kidney and adjacent perirenal structures are often very poorly defined. As part of any arteriographic study for renal cell carcinoma, the opposite kidney should be examined. Celiac arteriography may be performed to evaluate for hepatic metastases. For renal venous evaluation a high-volume contrast material injection (25–35 ml) is performed once the diagnosis of renal cell carcinoma is made (Fig 3B–7). However, it should be stressed that direct catheterization of the ipsilateral renal vein and inferior vena cava should be carried out for definitive documentation of venous involvement if the arterial examination is suboptimal or inconclusive.

Avascular renal cell carcinomas can generally be

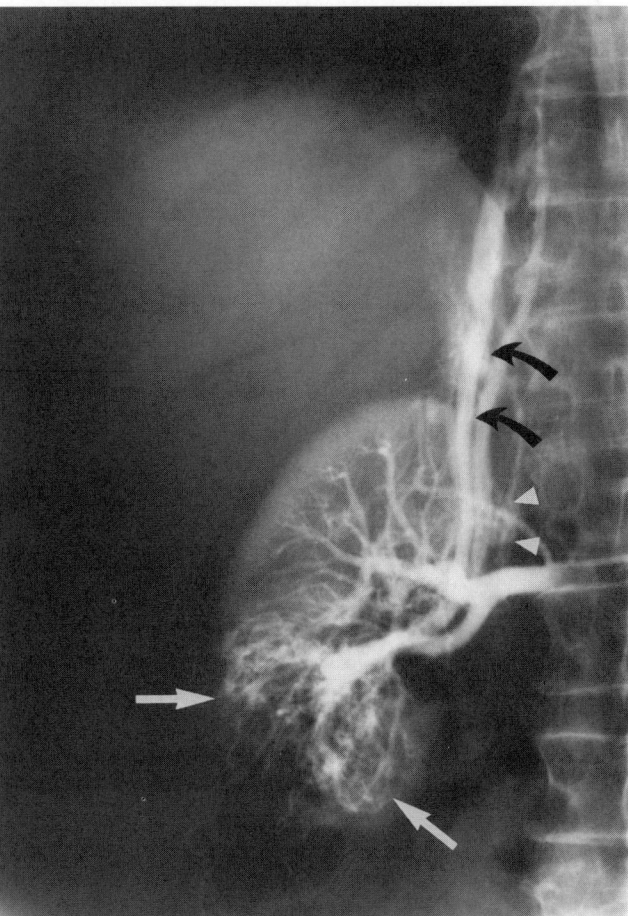

FIG 3B–6.
Rapid arteriovenous shunting through an extremely hypervascular right lower pole renal cell carcinoma *(white arrows)*. Visualization of inferior vena cava in midarterial phase. Strands *(curved arrows)* and tumor vascularity *(arrowheads)* within the main renal vein and inferior vena cava indicate venous invasion.

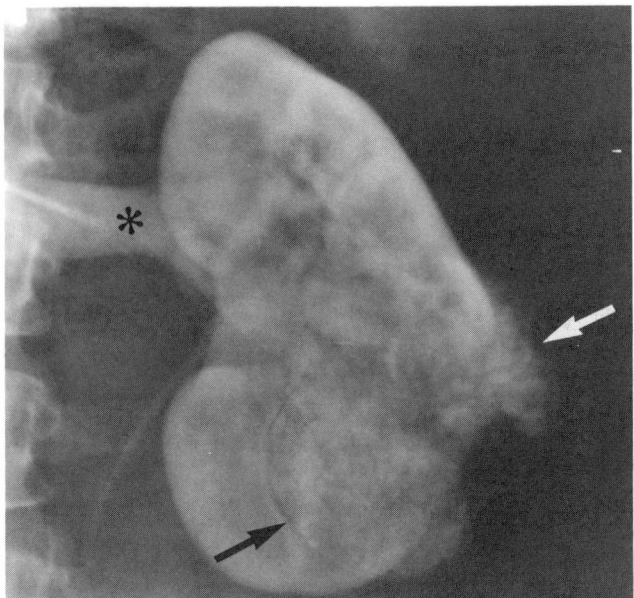

FIG 3B–7.
Left lower pole renal cell carcinoma *(arrows)* fairly well defined by a hypovascular rim. Good visualization of renal vein *(asterisk)* using high-dose arterial injection.

demonstrated to lack one or more of the features of a renal cyst. Geometric magnification filming is obligatory to exclude conclusively the possibility of abnormal vessels in the lesion (Fig 3B–8).

Potential metastatic involvement of renal cell carcinoma to bone is best evaluated by radionuclide studies. Pulmonary metastases are detected by chest radiography and/or CT.

The differential diagnosis of renal cell carcinoma includes angiomyolipoma, inflammatory lesions, and lymphomas. Angiomyolipoma may appear identical to renal cell carcinoma. Small aneurysms of the renal arteries may suggest the diagnosis of angiomyolipoma (Fig 3B–9). In a patient with tuberous sclerosis this is probably true; however, in other individuals the differential diagnosis cannot be conclusively made. Inflammatory lesions, particularly xanthogranulomatous pyelonephritis and occasionally renal abscess, may mimic a renal cell carcinoma, in which case the clinical history may be helpful. Xanthogranulomatous pyelonephritis may present with calcifications and other signs of pyelonephritis of the kidney. While abnormal vascularity may be seen in these lesions, arteriovenous shunting is generally not present. Lymphoma may also mimic renal cell carcinoma, although the clinical diagnosis is generally apparent before onset of renal involvement. In lymphoma, a palisade-like appearance has been described; however, this sign has not proved to be consistently useful in our experience (Seltzer et al., 1967).

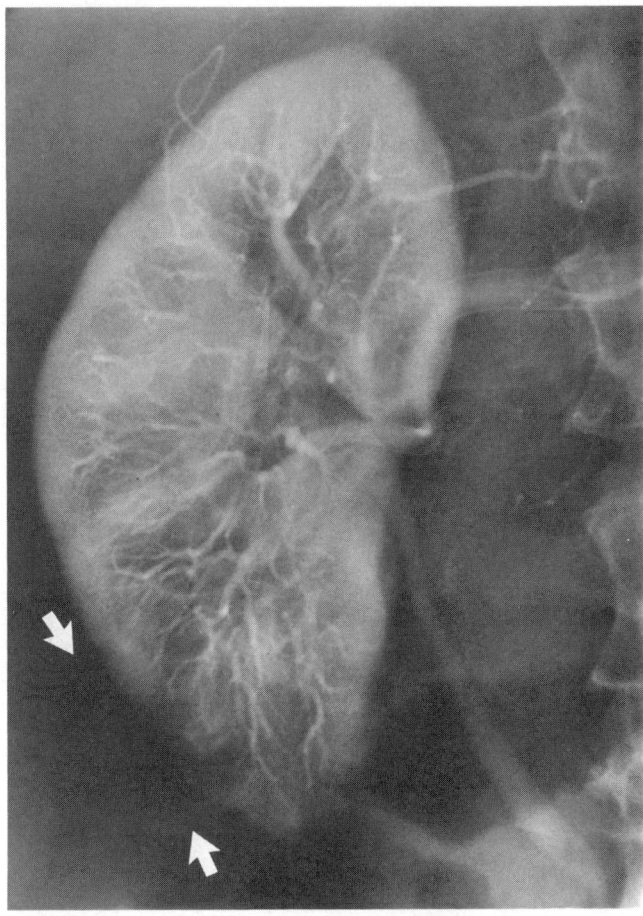

FIG 3B–8.
Small peripheral right lower pole hypovascular renal cell carcinoma *(arrows)*. Note disruption of nephrogram and very minimal contrast enhancement of peripheral portion of the tumor.

OTHER ASPECTS OF RENAL CELL CARCINOMA, INCLUDING EMBOLIZATION

The metastatic foci of renal cell carcinoma tend to have the same angiographic appearance as the primary lesion. In the liver these lesions can be quite hypervascular, sometimes with a central lucency (Hellekant and Nyman, 1979). Hemangiomas of the liver may have a similar appearance but are usually solitary, frequently in the lateral aspect of the right lobe of the liver, and may present as a signet ring-shaped structure. The contrast enhancement is from peripheral to central, with an extremely long and persistent stain (Niendorf et al., 1974) (Fig 3B–10). Hepatocellular carcinoma, liver cell adenomas, and focal nodular hyperplasia have a different appearance, although also quite hypervascular (Goldstein et al. 1974). The nephrogenic hepatic dysfunction syndrome seen rarely in patients with chronic renal disease, including neoplasms, may simu-

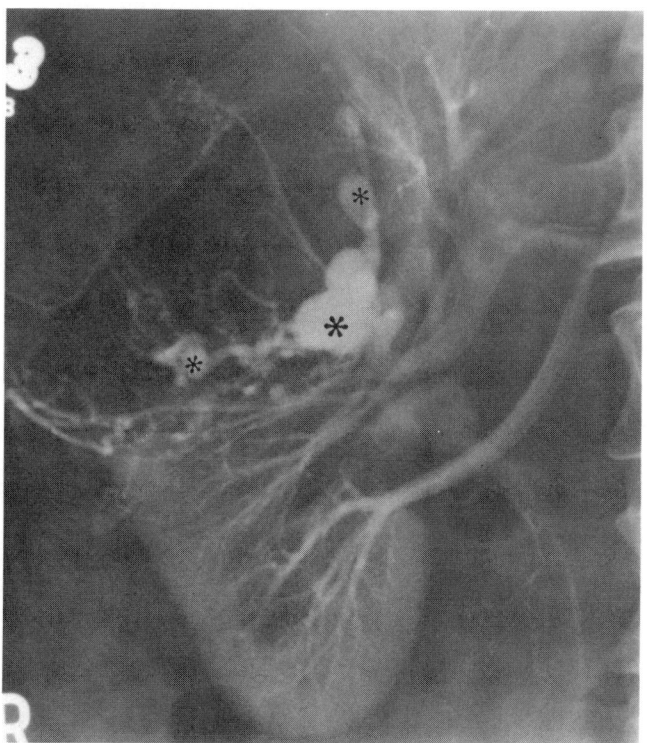

FIG 3B–9.
Angiomyolipoma of right upper pole. Arterial aneurysms *(asterisks)* are associated with this mass and are quite characteristic of this tumor.

late metastatic foci, but the appearance is that of a diffuse hypervascular liver (Mena et al., 1974).

The role of transcatheter embolization in the management of renal cell carcinoma is unclear. Advocates of the technique generally suggest that embolization be performed 24 hours before nephrectomy in individuals with large renal cell carcinomas to devascularize the neoplasm and reduce its size, facilitating surgical resection (Anderson et al., 1979; Bank and Kerber, 1979; Castaneda-Zuniga et al., 1980; Chuang et al., 1979; Eason et al., 1979; Freeny et al., 1979; Gomes et al., 1978; Kaufman, 1979; Mazer et al., 1981; White et al., 1979). Embolization has also been advocated as a palliative procedure in those individuals who have a contraindication to surgical removal of a renal cell carcinoma.

A number of agents have been utilized as embolic material. These include particulate matter such as surgical gelatin and polyvinyl alcohol, tissue adhesives, and larger occlusive agents such as stainless steel coils (Gianturco's coils) and detachable balloons. Liquid agents such as hot contrast and absolute alcohol have been used more recently.

Complications of renal infarction have been described (Bergreen and Woodside, 1976; Gang et al., 1977; Levin et al., 1980; Rankin, 1979; Struthers et al., 1980; Tisnado et al., 1979), including distal embolization of the embolic material into the lower extremities, contralateral kidney, and spinal arteries. Ethanol may float to the nondependent portion of the vascular system and embolize the mesenteric circulation in the supine patient. Malpositioning of a Gianturco coil, renal infection, and renal failure secondary to large volumes of contrast material used during the embolic placement have also been described.

RENAL ADENOMA

Renal adenomas are the most common benign renal tumors, many being found incidentally at autopsy. In the clinical setting of angiography, however, they are quite uncommon. Angiography of these lesions demonstrates increased vascularity in some cases, but the vascular pattern is less bizarre and more ordered than

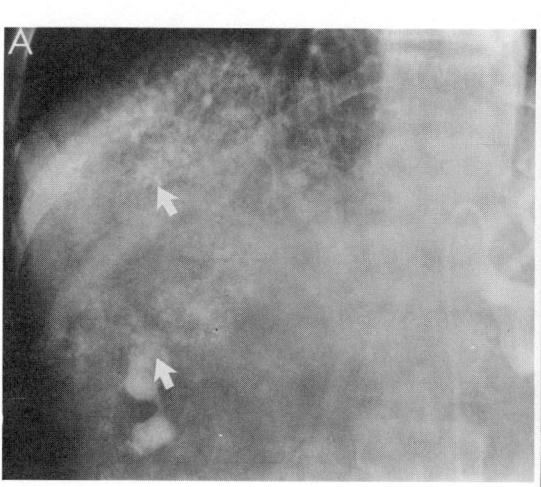

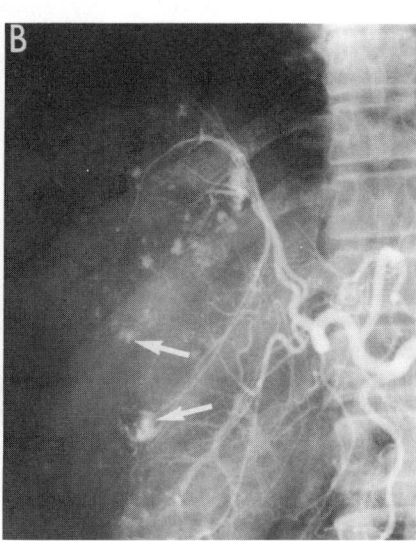

FIG 3B–10.
Multiple hepatic hemangiomas. Arterial **(A)** and venous **(B)** phases demonstrate densely enhancing lesions *(arrows),* which persist late into venous phase.

in carcinoma. Hypovascular lesions may also occur, with a faint blush being noted during the nephrographic phase. A third "pattern" has been noted in which the kidney is relatively normal (Holt et al., 1975).

Fibromas, lipomas, and neurogenic tumors such as neurofibroma or schwannoma have also been reported but without distinct appearances.

Metastases to the kidneys occur in 10% of all patients who die of carcinoma, most commonly of breast or lung origin. The renal involvement presents no distinct pattern, with the vascularity resembling that of the primary lesion (Abrams et al., 1950).

RENAL PSEUDOTUMORS

Frequently, a prominent column of Bertin presents as an area that can be confused with a renal mass. Calycine stretching may be noted, but on angiography there are no abnormal vessels, and the appearance of the renal parenchyma is that of the normal adjacent tissue. Normal fetal lobulation may also simulate a tumor and is seen most often in the middle segment along the lateral margin of the kidney. Renal sinus lipomatosis may produce significant stretching and deformity of the collecting system, suggesting a tumor, but with CT this is rarely a problem.

RENAL PELVIC TUMORS

These lesions are rarely seen on angiography, because they are most frequently diagnosed by excretory and retrograde urography. Angiography is occasionally helpful but frequently of less value than in renal cell carcinoma. The lesions are generally hypovascular, but occasionally fine tumor vessels may be defined on magnification angiography (Fig 3B–11). Additionally, encasement of segmental and interlobar arteries may be noted. There may be subtle distortion of the renal architecture, since these lesions frequently infiltrate the kidney (Rabinowitz et al., 1972).

Renal venography has been suggested as a means of establishing the diagnosis of transitional cell carcinoma of the pelvis; however, the great variability of the venous anatomy makes demonstration of venous encasement and compression difficult.

WILMS' TUMORS

Wilms' tumor is by far the most common renal mass in infants and children. Arteriography is an adjunct to IV urography, ultrasound, and CT. On arteriography, the lesion varies from small to enormous with a variable portion of the kidney involved. It may appear sharply marginated or diffusely infiltrating and is generally avascular to hypovascular, occasionally moderately hypervascular. There is frequently stretching and dis-

placement of vessels, indicative of a mass. Encasement and occlusion of arteries may also be present. At times, metastasis to the liver and retroperitoneal lymph nodes may be identified at the time of arteriography. Inferior venacavography and renal venography are useful in demonstrating venous involvement. Included in the differential diagnosis is infantile polycystic disease, which is usually bilateral and can be differentiated by ultrasound. Hydronephrosis should also be differentiated by ultrasound or CT; there is an absence of abnormal vascularity with the latter process. Neuroblastoma usually displaces the kidney with little, if any, intrarenal alteration, although the mass may infiltrate the kidney, making the differential diagnosis difficult. Calcification is common in neuroblastoma.

RENAL PARENCHYMAL DISEASE

Angiography is rarely utilized in patients with acute and chronic renal infection or other renal parenchymal diseases such as glomerulonephritis. In those instances where an imaging procedure is necessary, urography, ultrasound, and CT suffice. However, occasionally patients suspected of having renal cancer may prove to

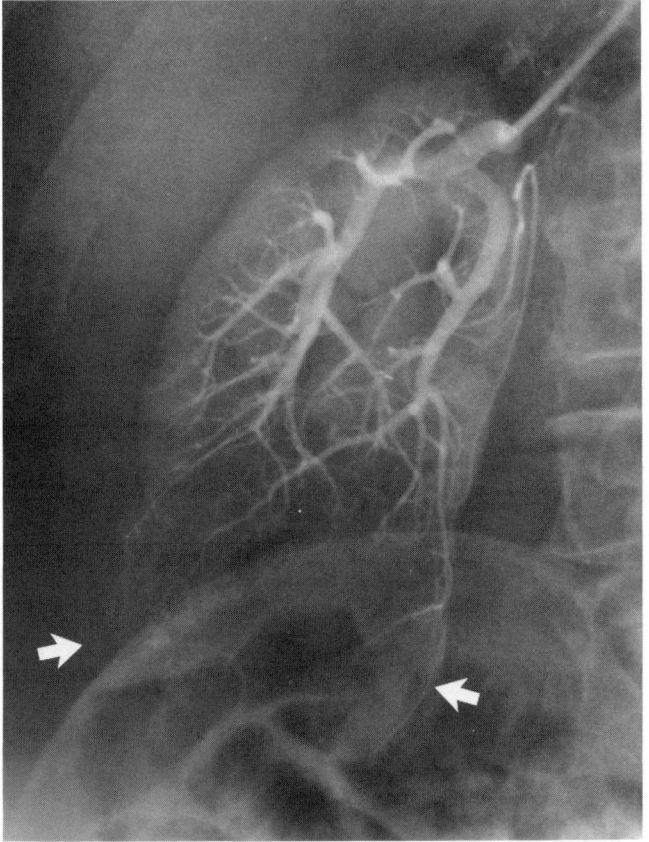

FIG 3B–11.
Right lower pole transitional cell carcinoma *(arrows)*. Note minimal fine neovascularity within tumor.

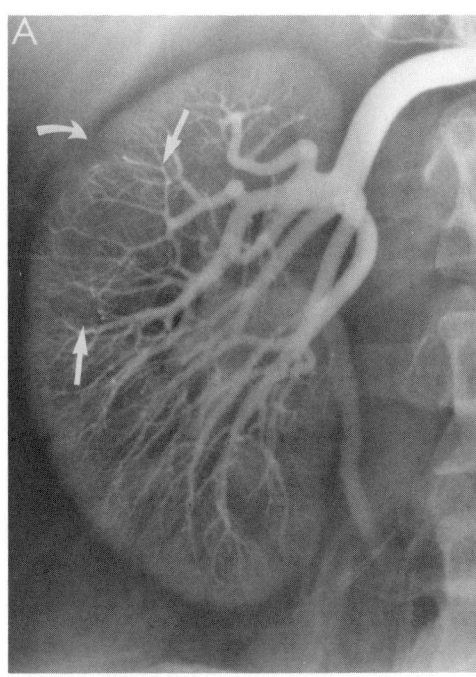
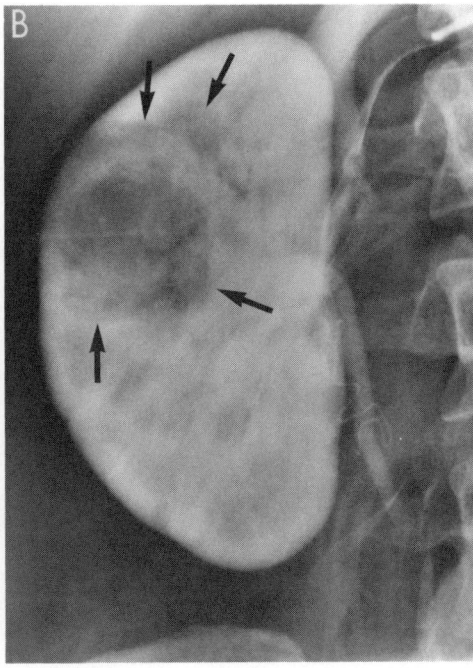

FIG 3B–12.
Renal abscess. Arterial **(A)** and nephrogram **(B)** phases. Arterial phase **(A)** demonstrates a mass *(arrows)* with displacement of intrarenal vessels and disruption of cortex *(curved arrow)*. Nephrogram **(B)** shows irregular mass *(arrows)* without sharply defined borders consistent with abscess.

have an inflammatory mass, in which instance the radiographic findings are important.

Renal Infection

Arteriography demonstrates a wide spectrum of findings in acute pyelonephritis depending on the severity of the disease and whether it is generalized or focal. Signs of parenchymal edema may be noted including stretching and displacement of intrarenal arteries and loss of definition of the corticomedullary junction. In patients with acute lobar nephronia, small radiolucencies representing microabscesses may be noted, producing a mottled appearance on the nephrographic phase. A striated or corrugated appearance of alternating lucent and radiodense stripes in the cortex may be seen (Bart et al. 1976; Wichs and Thornbury, 1979). When the disease process progresses to a true abscess (renal carbuncle), the findings may be more dramatic. The angiographic appearance is variable, depending on the extent of the abscess. Occasionally, the lesion is not seen, with only the edematous changes in the parenchyma appreciated (Himmelfarb et al., 1972). With development of a mass, however, there is stretching of vessels and displacement about the mass with a hypervascular zone surrounding the abscess and presumably representing compressed normal renal parenchyma (Fig 3B–12). The margin of the mass may be poorly defined with respect to adjacent kidney. Abnormal vascularity

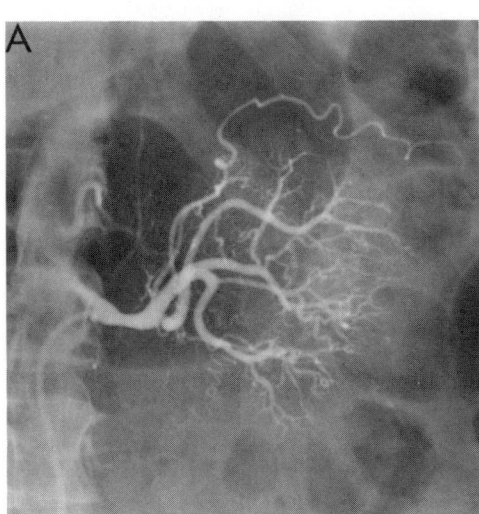
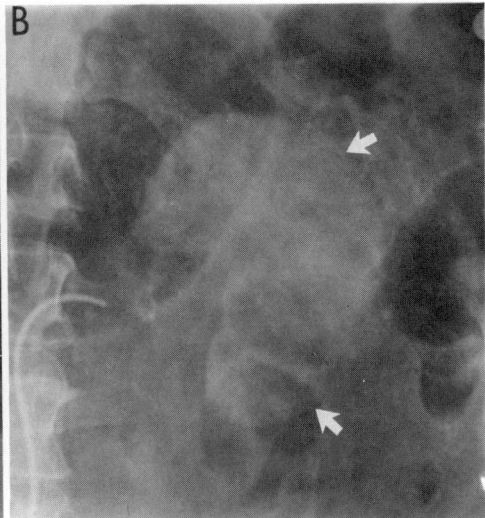

FIG 3B–13.
Xanthogranulomatous pyelonephritis. Arterial phase **(A)** demonstrates significant intrarenal vascular irregularity and diffusely abnormal vessels. Nephrogram phase **(B)** well defined *(arrows)* with diffuse perinephric neovascularity and contrast enhancement which somewhat obscures the nephrogram.

may be present, frequently demonstrating a fine reticulated pattern. The appearance of a hypovascular inflammatory mass with hypervascular rim and abnormal vascularity may simulate a hypovascular renal cell carcinoma.

The other type of inflammatory disease that may simulate carcinoma is xanthogranulomatous pyelonephritis. The diffuse variety, which may replace the entire kidney, is the more common form and usually is associated with an obstructing renal pelvic calculus (Beachley et al., 1974; Vinik et al., 1969). Hydronephrosis is seen in addition to stretching and marked displacement of intrarenal vessels (Vinik et al., 1969; Gingell et al., 1973). The nephrogram is usually unhomogeneous with scattered lucencies. The corticomedullary junction is usually ill-defined, and prominent capsular and ureteric vessels are noted. Abnormal vascularity is frequently seen (Fig 3B–13).

The focal or tumefactive variety frequently simulates the appearance of a hypervascular renal cell carcinoma (Malek and Elder, 1978). The central portion of the mass is hypovascular to avascular with significant hypervascularity on the periphery. The mass is usually ill defined with respect to the rest of the kidney, and there is no associated calculus.

Other Renal Parenchymal Disease

The angiographic findings of chronic renal parenchymal disease have been well described (Mena et al., 1973a, 1973b, 1974; Foster et al., 1965; Friedenbury et al., 1965; Ekelund et al., 1973; Chueng and Reuter, 1974). Although angiography is only occasionally performed, selective arterial catheterization with magnification filming is mandatory. The arteriogram is not only able to detail vascular anatomy, but also can suggest renal blood flow in a semiquantitative fashion. Arterio-

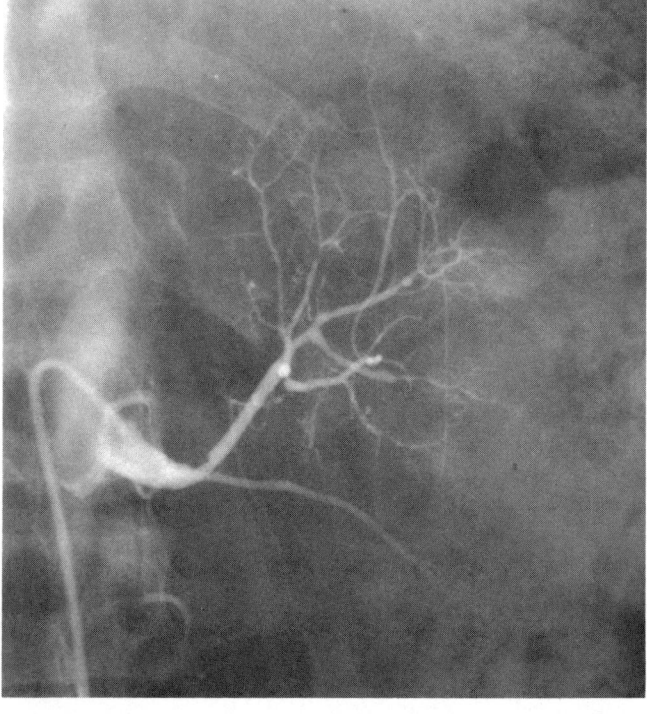

FIG 3B–14.
Arteriolar nephrosclerosis. Severe interlobar pruning and tortuosity with absence of interlobular arteries.

lar nephrosclerosis demonstrates severe pruning and tortuosity of interlobar arteries, cortical thinning, and microinfarcts (Fig 3B–14). The main renal artery may be normal in caliber. The nephrographic phase usually remains homogeneous, but sometimes coarse lucencies secondary to multiple cortical infarcts appear. Chronic glomerulonephritis demonstrates more moderate degrees of pruning of the intralobar arteries and slowed velocity of flow of the contrast material. Tortuosity and

FIG 3B–15.
Chronic glomerulonephritis. Arterial phase **(A)** and nephrogram **(B)**. Arterial phase demonstrates mild to moderate pruning and some crowding and tortuosity of interlobar arteries. Interlobular arteries are not visualized. Nephrogram **(B)** shows "lucent cortex" (C). Nephrographic defects in upper and lower poles represent small cysts.

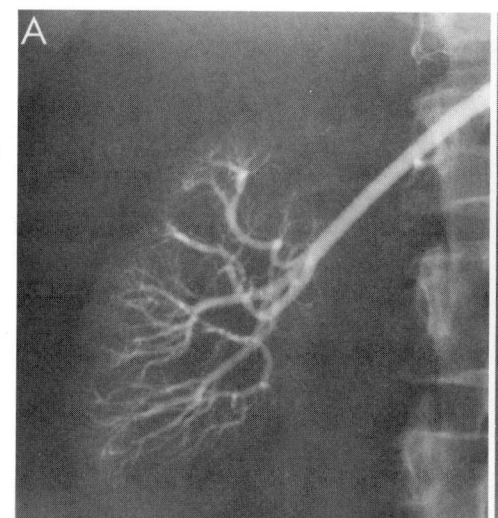
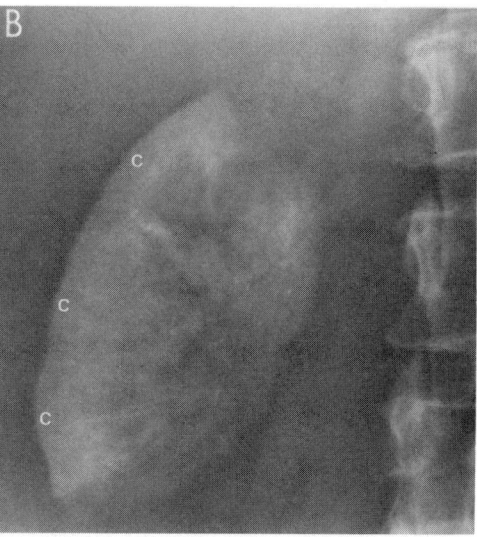

crowding of the interlobar arteries are more prominent than in arteriolar nephrosclerosis. The main renal artery is usually diminished in diameter. Interlobular arteries are usually not visualized, producing the so-called lucent cortex sign (Mena et al., 1973) (Fig 3B–15). Chronic pyelonephritis produces asymmetric involvement of both kidneys and uneven involvement within each kidney. Interlobar arteries are very tortuous, especially in areas of focal scars. In less involved areas, or those with regenerative cortical nodules, the appearance is that of increased radiodensity. The main renal artery is usually small, and renal blood flow is diminished as noted by decreased velocity of contrast flow.

HYDRONEPHROSIS

Angiography is rarely utilized as a primary imaging technique in patients with hydronephrosis. Occasionally it may be useful in treatment planning to determine the salvageability of a chronically obstructed kidney. The size of the main renal artery and the amount of functioning parenchyma left remaining may thus be evaluated.

In patients with acute obstruction, the findings are usually minimal. The interlobar branches may be stretched with slowed flow in the intrarenal arteries. With greater severity, there is significant stretching, displacement and pruning of vessels as they pass around the dilated pelvis and calices (Fig 3B–16). The differential diagnosis includes polycystic kidney in which the findings are almost always bilateral and there is significantly greater distortion of the arterial branch-

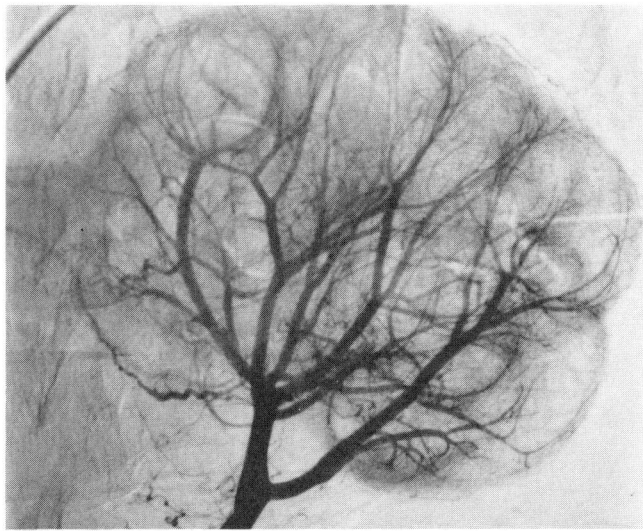

FIG 3B–16.
Hydronephrosis. Subtraction angiogram shows renal enlargement and smooth displacement of intrarenal vessels around a severely hydronephrotic collecting system.

ing pattern by the markedly disorganized parenchyma secondary to the innumerable cysts. Multiple simple cysts may also cause displacement, stretching, and pruning of intrarenal arteries; however, in the nephrographic phase, the diagnosis is usually evident.

HYPERTENSION

Angiography retains an essential role in the workup and management of patients with suspected renal vascular hypertension (Scott et al., 1983). However, the exact place of angiography is changing as newer imaging techniques emerge. The selection of patients also remains somewhat controversial as newer drug regimens have been developed concurrently with the development of percutaneous transluminal angioplasty.

Digital subtraction angiography (DSA) has replaced the intravenous urogram as the screening imaging procedure of choice (Goldstone et al., 1981; Gomes et al., 1983; Smith et al., 1982). Intravenous DSA is performed using a central catheter, usually a 5 F pigtail, placed in the superior vena cava. Injection rates are usually 15–25 ml/sec for a total of 30–40 ml. In two series, the sensitivity of IV DSA was 83% and 85.5%, respectively, the specificity 79%–85%, and the accuracy 80%–85% (Smith et al., 1982; Clark and Alexander, 1983). Renal vein specimens for renin determination can be obtained at the same time as the DSA from the right and left renal veins and lower inferior vena cava. An alternative use of digital subtraction techniques is the performance of intra-arterial examinations for screening purposes.

The decision to proceed to invasive arterial catheter studies is based on the analysis of the noninvasive studies and depends on whether the patient will be a candidate for angioplasty or surgery if an appropriate lesion is found. Certainly, arteriography should be performed on all operative candidates who have positive screening tests. Young adults and children particularly should be evaluated by arteriography. Patient selection is also influenced by the availability and success of angioplasty and/or surgery as well as results with current medical regimens such as the use of captopril.

ANGIOGRAPHIC FEATURES OF DISEASES CAUSING RENAL HYPERTENSION

Atherosclerosis

Atherosclerosis is the most common cause of renal artery stenosis. Bilateral involvement is the rule. The proximal third of the renal artery is the segment most frequently involved, but distal and segmental involvement may be seen. Frequently, there are changes of atherosclerosis in the abdominal aorta and other vessels. The process causes stenosis of varying degree, in-

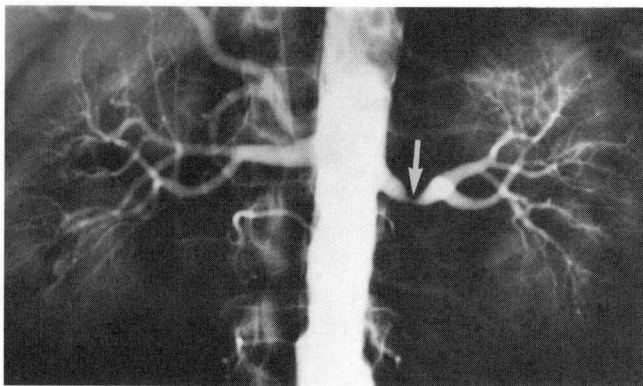

FIG 3B–17.
Atherosclerotic left renal artery stenosis *(arrow)* with some poststenotic dilatation.

cluding occlusion, that is generally irregular and asymmetric. Poststenotic dilatation may be observed (Fig 3B–17).

Atheromatous renal artery stenosis may be an "incidental" finding in patients who are normotensive. It was noted in an autopsy series that moderate or severe lesions were present in 49% of 256 normotensive and in 77% of 39 hypertensive patients (Halley et al., 1964).

Fibromuscular Dysplasia

Fibromuscular dysplasia is a general term for a group of idiopathic fibrous and fibromuscular lesions of the renal arteries which are generally classified into three types (Hunt et al., 1962): (1) intimal, (2) medial, and (3) adventitial or periarterial. The disease predominates in females by a ratio of 3 to 1. Usually the lesion involves the distal two-thirds of the main renal artery and proximal segmental renal arteries. Bilateral involvement is common, occurring in 43% in one series.

In the most frequent variety, medial fibroplasia with aneurysms, the arteriogram demonstrates the main renal artery to have multiple, very short stenotic segments with intervening normal or dilated portions, giving a beaded appearance (Fig 3B–18). Rarely, a diaphragm-like stenosis involving a very short segment may be seen. Smooth segmental narrowing may also occur somewhat similar to atherosclerosis. However, the peripheral involvement and the younger age, as well as lack of abdominal aortic involvement, usually allow a correct diagnosis. Fibromuscular dysplasia can also occur in the internal carotid artery, usually the cervical portion. An increased incidence of intracerebral aneurysms has been reported with this type.

Segmental Renal Artery Stenosis or Occlusion

These lesions may be secondary to atherosclerosis, fibromuscular dysplasia, arteritis, or embolism. Magni-

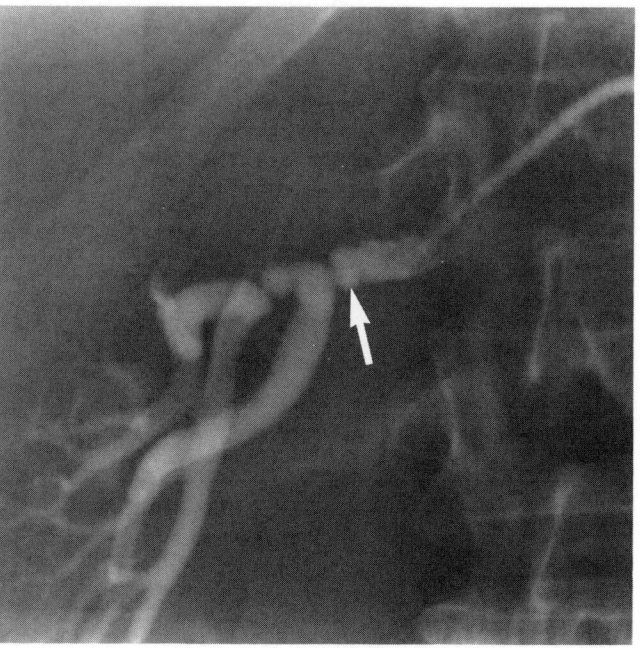

FIG 3B–18.
Fibromuscular disease. Short-segment stenoses with intervening dilated segments *(arrow)* give a "string-of-beads" appearance.

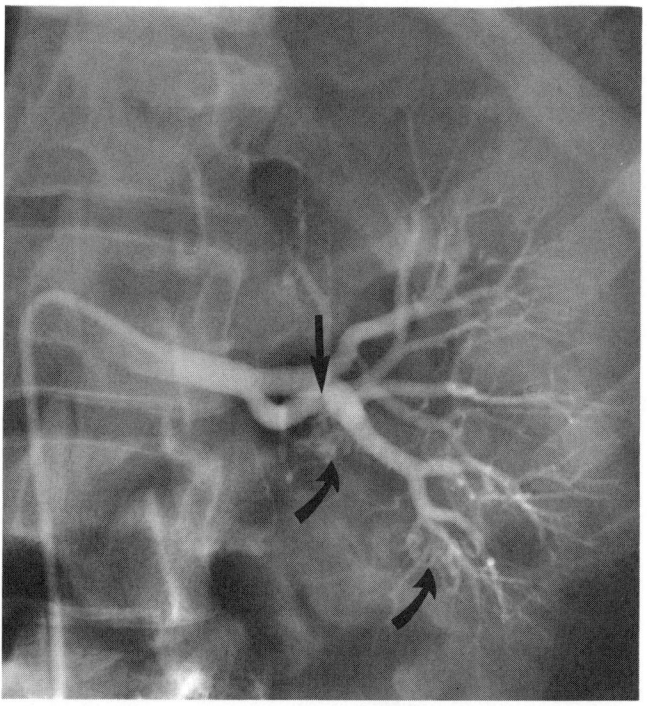

FIG 3B–19.
Segmental renal artery stenosis *(straight arrow)* is partially obscured by overlapping vessels. Multiple intrarenal collaterals *(curved arrows)* indicate hemodynamically significant lesion. These collaterals occasionally have been misdiagnosed as intrarenal vascular malformations.

fication angiography is essential for appropriate diagnosis. Because of the development of collateral vessels, these lesions have been misdiagnosed as arteriovenous malformations within the kidney (Fig 3B–19). In our experience, localized segmental renal artery disease is usually secondary to fibromuscular dysplasia.

Arteriovenous Fistulas

Arteriovenous fistulas may be either congenital or acquired. The congenital variety is extremely rare. The acquired form usually follows biopsy or trauma. There is significant increased velocity of flow of the blood through the fistula with early filling of an enlarged vein (Fig 3B–20). A segmental area of ischemia may be noted.

Takayasu's Arteritis

Segmental stenosis of the aorta or its branches is a characteristic finding of this disease. Arterial wall calcification, aneurysm, occlusion, and dissection are associated features (Chuang and Ernst, 1985). In addition to the stenosis or aneurysm in the abdominal aorta documented by arteriography, segmental or interlobar artery occlusion may occur in the kidney, causing ischemia and hypertension.

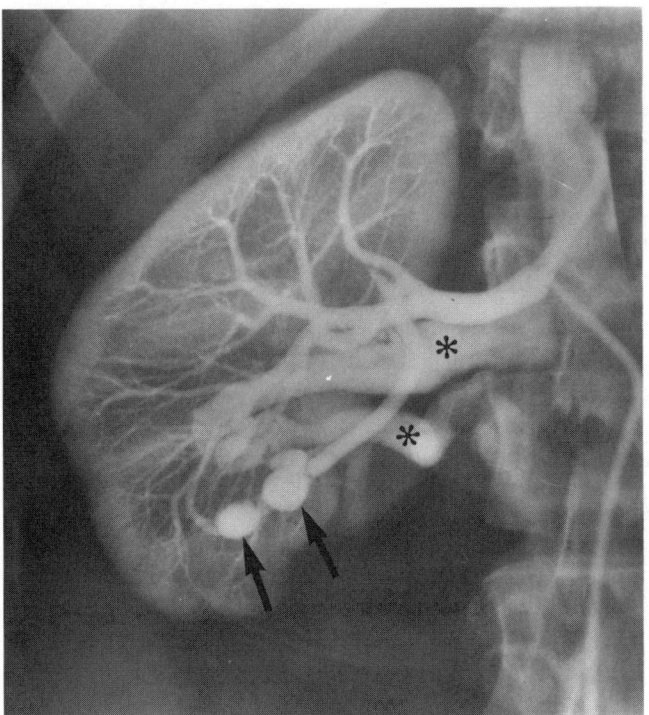

FIG 3B–20.
Renal arteriovenous fistula following renal biopsy. Brisk, dense early filling of renal vein (asterisks) during arterial injection. Other dilated intrarenal veins (arrows) are seen.

Neurofibromatosis

This is generally seen in the pediatric age group, and stenosis of the main renal artery may be documented.

PERCUTANEOUS TRANSLUMINAL ANGIOPLASTY

The results to date indicate that percutaneous transluminal angioplasty of the renal artery is a significant and effective means for improvement or permanent cure of the hypertension of many individuals with renal artery stenosis (Schwarten, 1980; Sos et al., 1983; Tegtmeyer et al., 1984).

In one series of 90 patients with 119 renal artery stenoses, over a mean follow-up of 22 months after angioplasty, the mean diastolic pressure decrease was 36.7 mm Hg. Thirty-one patients were considered cured, 59% improved, and only 10% were nonresponders (Tegtmeyer et al., 1984). In another series with an average of 16 months of follow-up, angioplasty was technically successful in 87% of cases with fibromuscular dysplasia, in 57% with unilateral atheromatous diseases, and only 10% with bilateral atheromatous stenoses. So few of the patients with bilateral atherosclerotic disease had bilaterally successful angioplasties that long-term results were statistically nonsignificant. More important, however, the results indicate that the latter group may not be candidates for angioplasty. However, cure or improvement was maintained in 93% of patients with fibromuscular dysplasia and 84% with unilateral atheromatous disease. In this series there were no deaths. Eight of the 89 patients required reconstructive vascular surgery after a complication of angioplasty, but in only one case was the ultimate outcome worse than if angioplasty had not been initially attempted (Sos et al., 1983). The results of this and other series indicate that angioplasty of the renal artery is an effective treatment for hypertension related to renal artery stenosis, secondary to unilateral or bilateral fibromuscular dysplasia and unilateral, nonostial atherosclerosis.

Percutaneous transfemoral renal artery embolectomy has also been attempted with some success (Maxwell and Mispireta, 1982). Streptokinase infusion for embolization has also been used.

COMPUTED TOMOGRAPHY

BASICS OF COMPUTED TOMOGRAPHY

The production of a CT image is a complex process. An x-ray tube has its beam narrowly confined or collimated so that x-rays are only emitted in a thin beam, which is directed at a patient lying directly beneath. This beam passes through the patient and is sensed on the opposite side by a series of x-ray detectors. That same slice is then repeatedly imaged from different angles, with all the information following the imaging of that slice composing a complete set of measurements of density or x-ray attenuation of the subject (Hounsfield, 1973; McCullough et al., 1974; Christensen et al., 1978; New and Scott, 1975; Hounsfield, 1980; Winter and King, 1983). The technology used to obtain these multiple angles of view has changed significantly over the years. Most of the refinements have increased the speed of rotation of the x-ray tube and detectors in order to stop motion. Detector number and resolution have also improved. So-called first-generation machines possessed a single x-ray tube and a single detector, which rotated in discrete increments about the patient. Second-, third-, and fourth-generation machines utilized x-ray beams of greater coverage, increased numbers of detectors, and significantly faster rotation to obtain the images (Fig 3B–21). Modern machines obtain a single slice in 2 seconds compared with up to 60 seconds on early units. Currently, most machines in use are of third- or fourth-generation technology. The ref-

erence to third or fourth generation implies only the time frame in which these machines were introduced. Both machines are in wide use and have advantages and disadvantages. Suffice it to say that both third- and fourth-generation machines are state of the art (Hounsfield, 1973; Winter and King, 1983).

IMAGE RECONSTRUCTION AND MANIPULATION

After the large amount of data is collected from a single slice, high-speed computers analyze the measurements and assign precise density values to individual points within the body. The two factors that allow the discrimination of small differences in density are the use of complex computer algorithms and the ability to view each slice multiple times. The main strength of CT lies within this contrast resolution, not in its spatial resolution, which in reality is inferior to that of conventional x-rays. Density values are standardized through the use of the Hounsfield scale, named after Sir Godfrey Hounsfield, the inventor of CT (Christensen et al., 1978; Winter and King, 1983; Cormack, 1973). The

FIG 3B–21.
A-D, diagrammatic representation of scan production in four generations of CT imaging. In third- **(C)** and fourth-generation **(D)** scans the linear traverse by x-ray tube is eliminated to increase scan speed. Also note number of detectors increases substantially in third and fourth-generation instruments.

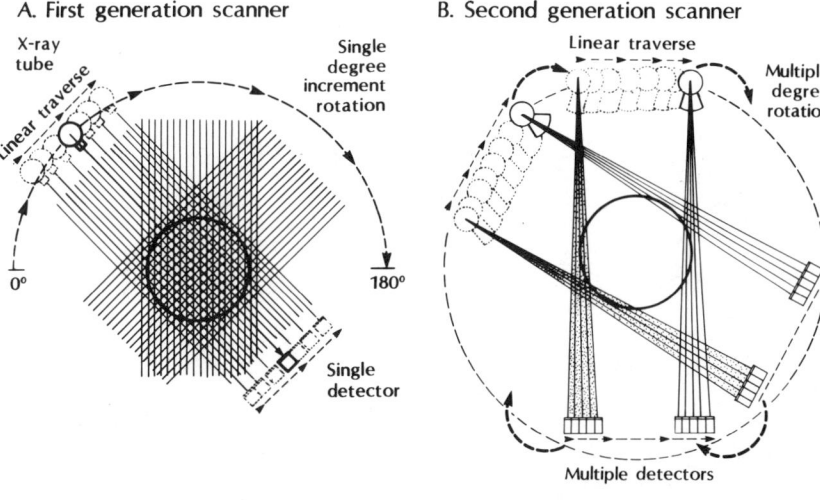

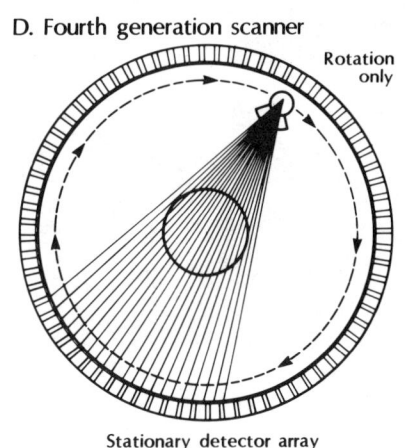

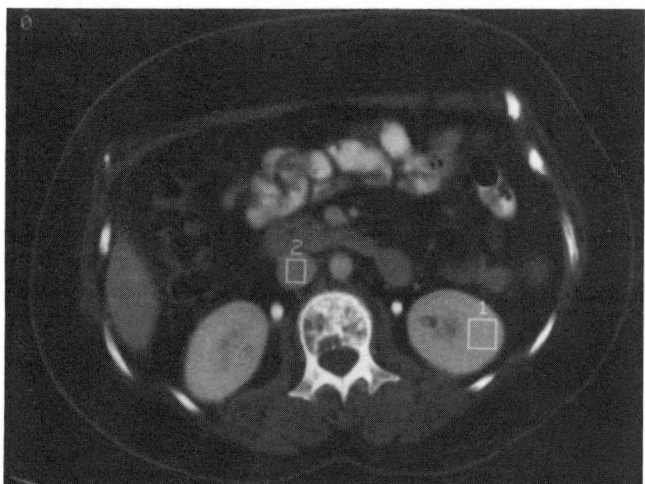

FIG 3B–22.
Determination of the attenuation coefficients of tissues. Cursor boxes numbered *1* and *2* are seen over regions of interest in the left kidney *(1)* and vena cava *(2)*. Tissue within the cursor box will be analyzed and an average attenuation value in Hounsfield units produced.

Hounsfield scale extends from $-1,000$ to $+1,000$ Hounsfield units (HU), with air fixed at $-1,000$ HU, water at 0 HU, and dense bone at $+1,000$ HU. Typical values for tissue seen in the body include fat at approximately -50 HU, soft tissue arund 40 HU, and clotted blood in the range of 70 HU. On modern CT scanners a measurement of the attenuation values of tissue is easily performed by placing a small cursor box over the area of interest and instructing the computer to give a reading of the average Hounsfield value of that region (Fig 3B–22). Obviously the cursor box should be placed in an area of homogeneous density. For example, a reading taken in an area that includes both fat and soft tissue will yield a falsely low average value.

The viewer may also make structures more or less visible by changing the brightness and the contrast of the image. The brightness control, usually called the *window level*, selects the mid-gray value of the picture. The contrast control or the *window width* selects the range of densities around this mid-gray value that will be shown from black to white in the image. For example, the brightness, or window level, for viewing most abdominal images will be 0 to $+40$ HU. The window width is generally $+400$ to $+600$. A window width of $+600$ with a window level of 0, for example, means that anything below -300 HU will appear black and anything above $+300$ HU will appear white. For viewing lung parenchyma the window level should be set at about -100 HU and the window width approximately 1,000. The window width and level controls are extremely important in viewing of CT images, and how

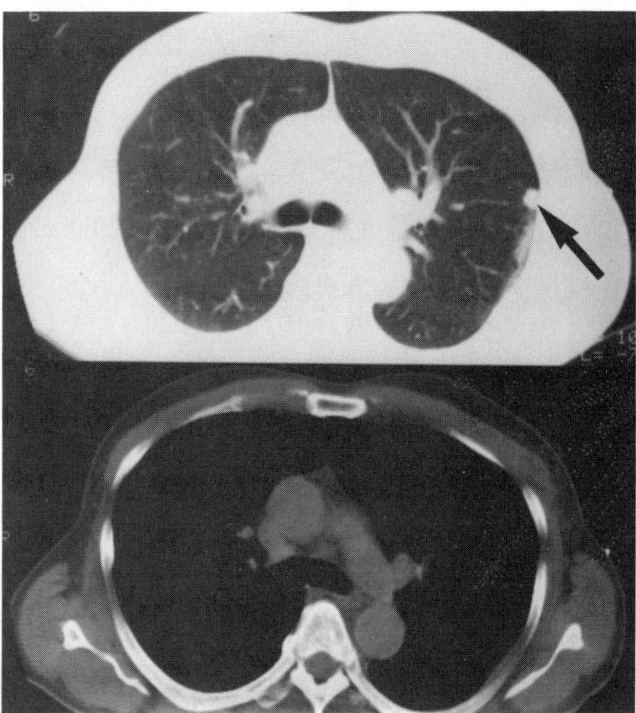

FIG 3B–23.
The effect of varying the window width and window level controls are illustrated on this section viewed at settings appropriate for pulmonary parenchyma *(top)* and mediastinal structures *(bottom)*. In the top image, window level is -700 HU and the window width is 1,000 HU. The lower image is photographed at a window level of $+40$ HU and a window width of $+5000$ HU. Pulmonary nodule in the left chest *(arrow)* is only visible on the settings appropriate for pulmonary parenchyma.

they are used will determine to a large degree how the image appears on the screen (Christensen et al., 1973; Hounsfield, 1980; Winter and King, 1983) (Fig 3B–23).

COMPUTED TOMOGRAPHY OF THE KIDNEYS

Computed tomography (CT) has created a revolution in the imaging and workup of disease processes since its introduction in the early 1970s. Initially, applications of CT were limited to the head, since long scan times of about one minute did not permit examination of the body. Prolonged breath-holding was impossible, and motion artifact severely degraded the images. In addition, spatial resolution was poor. With the introduction of a truly practical whole-body scanner, the EMI 5005, which had 18-second scan time, evaluation of the body became a reality. This instrument, however, had significant limitations in image quality compared with current scanners, since motion due to inadvertent respiratory movement or bowel peristalsis yielded prominent streak artifacts. The limited number of detectors also did not permit good spatial resolution, particularly in patients without a large amount of intra-abdominal fat. Despite these drawbacks, many investigators immediately saw the promise of CT, particularly since good visualization of the perirenal spaces was afforded and identification of retroperitoneal structures not previously amenable to investigation was possible. A landmark article by Sagel et al. (1977) accurately predicted the value of CT in characterization of renal masses and a variety of other conditions. Other papers rapidly followed, indicating that the process was indeed of great value. It was not, however, until the introduction of third- and fourth-generation scanners capable of fast scan times and better spatial resolution as a result of improved x-ray detector technology that CT of the kidneys was widely shown to have a major impact on the workup of renal and perirenal processes, including renal cell carcinoma. Weyman et al. (1980) first identified that CT of the kidneys in renal cell carcinoma could substitute for arteriography. That view has been widely

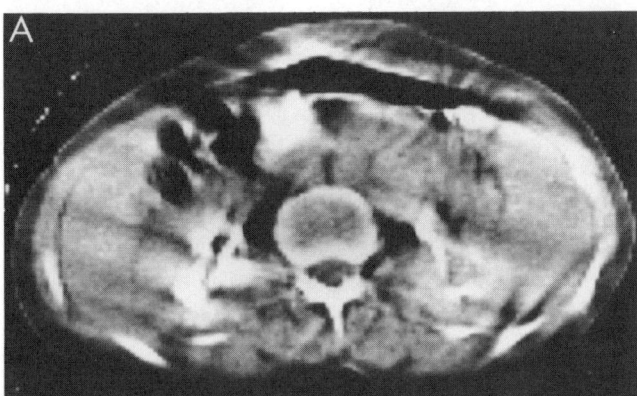

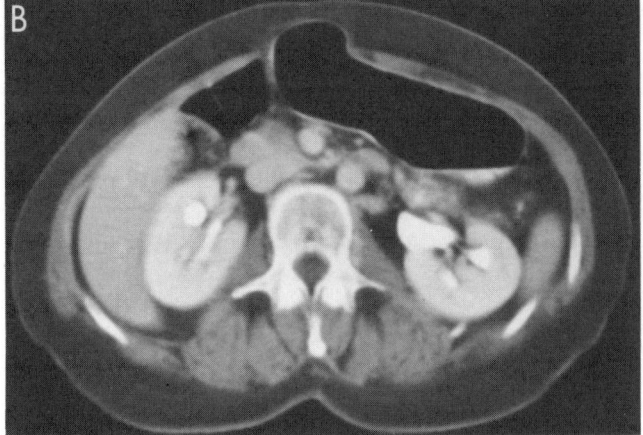

FIG 3B–24.
Improvement in spatial and contrast resolution is dramatic when scans on the same patient performed on an EMI 5005 18-second scanner **(A)** are compared with images obtained on a third-generation GE 9800 **(B)** with 2-second scan time and increased numbers of detectors. Newer scanners virtually eliminate streak artifact caused by respiratory motion, high-density objects, and bowel peristalsis.

upheld. Indeed, in most centers with high-quality CT, renal arteriography is infrequently done except for arterial or venous pathology. Further improvements in CT technology have brought us to current third- and fourth-generation instruments which are capable of two-second scan times with outstanding spatial and contrast resolution.

The introduction of these fast scanners has opened a broad new area of medical research and investigation. Scans of the abdomen yield highly diagnostic information in almost 100% of patients; even patients who minimally cooperate can receive a diagnostic examination. Depiction of abdominal anatomy is virtually free of streak artifacts from bowel gas and peristalsis. Metal clip artifacts have almost been eliminated. The change in the quality of the examination is startling when older and newer scans are compared (Fig 3B–24). The addition of "dynamic" scan programs capable of rapid acquisition of images in a short period has allowed the surveillance of blood flow within renal masses and within the normal kidney; attempts have been made to translate this information to physiologic information concerning disease processes (Fuld et al., 1984). Finally, reconstruction of the images obtained is now virtually instantaneous, making high-volume patient loads feasible and allowing screening of a large number of patients in a relatively short time. Most examinations now require 20–30 minutes. Nowhere have the benefits of modern imaging technology been as dramatically realized as in CT of the whole body and kidneys.

TECHNIQUE

Renal CT requires careful monitoring by a radiologist who is familiar with the clinical problem and has viewed results of any other urologic imaging procedures performed. In general, oral contrast medium is administered the evening before the study to opacify the colon. Sixteen to twenty ounces of diluted contrast medium (barium or water soluble contrast) given 30–60 minutes before the examination usually opacifies all bowel loops and avoids confusion of fluid-filled bowel with abdominal masses or lymph nodes. In those patients being studied specifically for high-density ureteral calculi, oral contrast material should be omitted.

Intravenous contrast material should be given in almost all cases; postcontrast studies allow better characterization and delineation of renal masses, permit evaluation of renal venous and vena caval patency in renal neoplasm, and help differentiate vascular structures from solid masses such as retroperitoneal lymphadenopathy. In addition, opacification of the collecting system is helpful in identifying urothelial masses and thickenings. Whether noncontrast-enhanced images should be routinely obtained is controversial. Engelstad et al. (1980) concluded that regular use of nonenhanced images was probably not indicated except in cases of renal calcification, hemorrhage, or contrast extravasation. On the other hand, certain high-density renal cysts or cystic neoplasms may be mistaken for simple cysts on postcontrast views (Dunnick and Korobkin, 1984). These abnormalities may be unsuspected, and their accurate detection mandates the routine inclusion of precontrast scans. Although the number of such abnormalities is small, and many institutions perform noncontrast scans only in specified cases, some authors argue for the routine inclusion of noncontrast scans. Certainly, patients with renal failure or documented contrast material allergy should be scanned without contrast material, since most abnormalities can be adequately seen.

Dynamic scanning (CT angiography) can be performed in specific cases when more information about the vascularity or contrast enhancement characteristics of processes, such as arteriovenous fistulas and renal cell carcinoma, is required (Tada et al., 1979). Lang et al. (1984, 1985) stated that the use of this technique yields greater accuracy in the evaluation of renal trauma and in the staging of renal cell carcinoma.

CT of the kidneys must be performed with contiguous or slightly overlapping slices, usually 10 mm thick; 5-mm slice thickness (collimation) may be helpful for increasing detail and avoiding the so-called partial volume average effect, in which inclusion of a portion of adjacent tissue or abnormality in the slice under consideration gives spurious information about a lesion or structure (Segal and Spitzer, 1979; LiPuma and Haaga, 1983). The entire kidney including adrenals must be scanned in cases of renal mass; inclusion of the ureters is routine if pathology is suspected or observed there. If distal ureteral obstruction is present, scanning is continued to the level of the abnormality; prone positioning helps fill the anteriorly placed ureters with contrast material to delineate more precisely the level and nature of the occlusion.

Our contrast enhancement regimen consists of administration of a rapid bolus of 50 ml of 60% contrast medium followed by a pressurized infusion of 300 ml of 30% contrast medium to yield intense nephrographic opacification as well as good venous opacification for the evaluation of patency of that structure. Repeated scans after 5 minutes allow study of the collecting system. Alternately, 25–50 ml of contrast material may be administered 10 minutes before initiation of the rapid infusion to opacify the collecting structures adequately.

ANATOMY

The relationship of the kidneys to the various retroperitoneal compartments has been well defined by several authors, including the now-classic descriptions of the three renal compartments by Meyers (Meyers, 1976; Love et al., 1981; Pariety et al., 1981; Vogelzang and Neiman, 1983). CT has allowed the routine assessment of these areas and disease processes that originate within them and/or involve them secondarily. Their identification is vital in the staging of neoplasms and infectious processes.

Anterior Pararenal

This area is bounded by the posterior parietal peritoneum anteriorly and posteriorly by the anterior renal fascia. It contains the pancreas, retroperitoneal duodenum, and ascending and descending colons.

Perirenal

The kidney, adrenal, and associated structures (vessels and the collecting system) are contained within the renal fascia, which is termed Gerota's fascia anteriorly and Zuckerkandl's posteriorly. The two fasciae fuse laterally to form the lateroconal fascia.

Posterior Pararenal

This space contains only fat and is bounded by posterior renal and transversalis fasciae.

The relationship of the kidneys to adjacent abdominal organs has been well described, but awareness of important relationships must be maintained, since disease processes in other organs may cause changes in renal position, produce hydronephrosis, or invade the kidney. The most significant are the relationships of the adrenal glands. The right adrenal gland lies superior to the upper pole and is best found immediately dorsal to the inferior vena cava. The left gland is more anteriorly and medially positioned. An important and constant anatomical landmark is the splenic vein just ventral to it (Karstaedt et al., 1978; Hattery et al., 1981). Other significant relationships to the right upper pole include the descending duodenum and the right lobe of the liver; the right lower pole lies medial to the ascending colon. The left upper pole is intimately associated with the spleen lying anterolaterally and the tail of the pancreas and splenic vessels anteriorly; the descending colon is placed laterally. The psoas muscles are located medially (Vogelzang and Neiman, 1983).

In cross section, the kidneys are oval or round with a centrally located collecting system "complex" consisting of vessels, fat, and collecting structures; abundant amounts of fat may give a low-density appearance, but generally the density of the collecting complex is similar to renal parenchyma on unenhanced scans. After contrast administration, collecting structures are densely opacified and calyces, infundibula, and renal pelvic structures can be identified. The renal hilum is an anteromedially directed break in the renal outline from which the renal vessels arise. Renal vascular relationships are important, relatively constant, and easily visualized. The veins lie ventral to the arteries on both sides. On the left the longer left renal vein runs between the aorta and superior mesenteric artery to the inferior vena cava, which is often oval or slitlike at this level. On the right the vein is shorter, more obliquely oriented, and enters the cava directly; its course is usually seen on several contiguous sections and as a result it may not always be adequately imaged (Fig 3B–25). In our experience, an equivocal CT requiring renal venography to rule out tumor thrombus usually occurs on the right. The renal arteries originate laterally from the aorta, the left running directly to the kidney and the right coursing behind the cava. In general, visualization of the renal veins requires less attention to techniques than seeing the smaller renal arteries. Overlapping sections may be required to demonstrate the origins of these vessels (Sagel et al., 1977; Dunnick and Korobkin, 1984; LiPuma and Haaga, 1983; Williamson et al.,

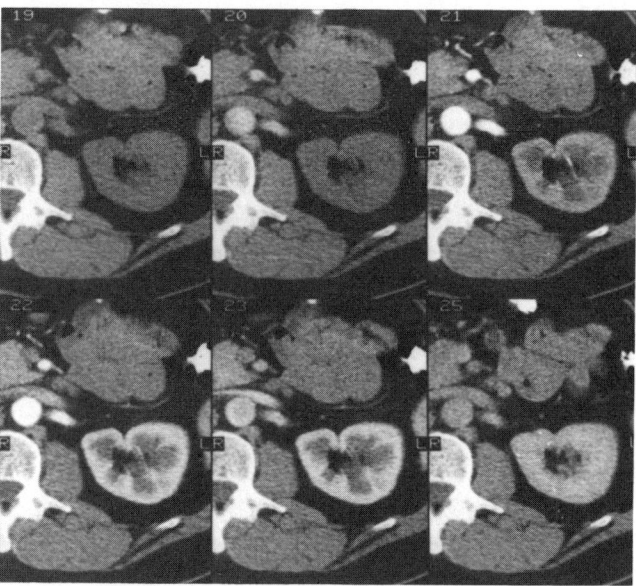

FIG 3B–26.
Phases of renal contrast enhancement. Before contrast enhancement *(left upper frame),* renal parenchyma is homogeneous. After bolus administration, there is excellent corticomedullary differentiation *(left lower frame).* Within 30 seconds *(right lower frame),* there is homogeneous uniform enhancement of renal parenchyma.

1978; Churchill et al., 1979; Elkin, 1980).

The renal parenchyma is homogeneous, with an attenuation coefficient of 30–60 HU on precontrast scans. Immediately after bolus administration of contrast material, there is good differentiation between cortex and medulla; later scans show even homogeneous contrast enhancement to 80–120 HU (Fig 3B–26).

INDICATIONS

Modern indications for CT have been consistently broadening as the efficacy of this remarkable tool becomes more and more widely appreciated. CT serves as a tertiary examination following a preliminary study with excretory urography and ultrasound, with the notable exception of abnormalities confined to the urothelium, in which case retrograde pyelography is probably the most helpful. The following areas are considered current indications for CT of the kidneys.

1. Detection or delineation of a renal mass.
   A. Solid mass on ultrasound requires staging (i.e., venous involvement or perinephric extension) or characterization (detection of fat within an angiomyolipoma).
   B. Indeterminant mass on ultrasound.
   C. Strong clinical suspicion of a mass despite normal IVP and ultrasound.
   D. Detection of local recurrence after nephrectomy for renal cell carcinoma.

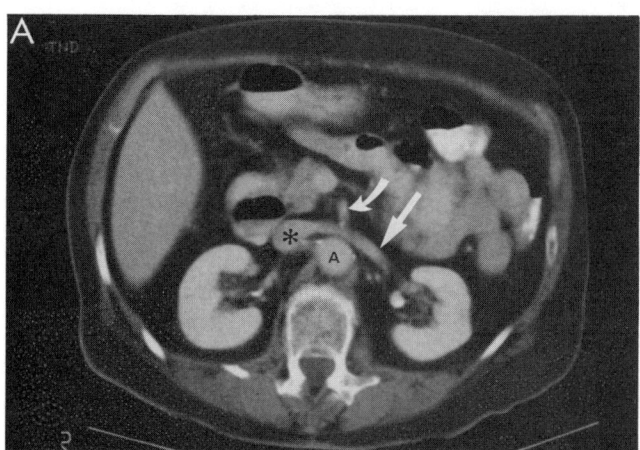

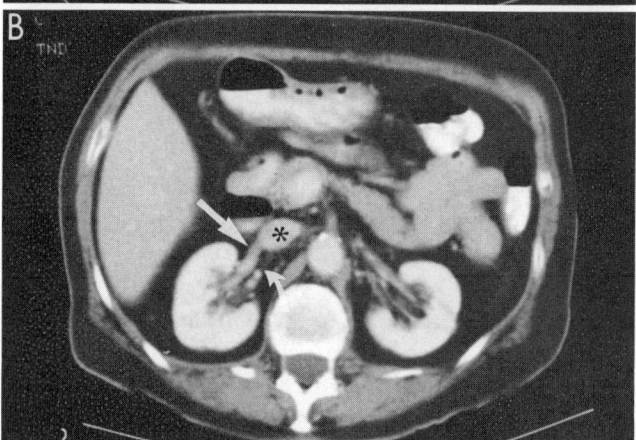

FIG 3B–25.
Normal arterial and venous relationships. **A,** left renal vein *(arrow)* passes between aorta *(A)* and superior mesenteric artery *(curved arrow)* into vena cava *(asterisk).* **B,** right renal vein P *(arrow)* is shorter and enters cava *(asterisk)* somewhat more obliquely. Note posterior relationship of right renal artery *(curved arrow).*

2. Characterization of renal inflammatory processes.
   A. Assessment of extent of perinephric or pararenal abscesses or fluid collections.
   B. Detection of focal inflammatory processes (abscess, focal pyelonephritis).
3. Determination of the level and etiology of hydronephrosis.
4. Characterization of noncalcified collecting system filling defects (differentiation of stone from tumor).
5. Imaging and staging of renal trauma.
   A. Urinary extravasation and urinoma detection.
   B. Assessment of renal viability.
6. Congenital lesions of the kidney.
   A. Masses.
   B. Alterations in rotation or fusion.
   C. Variants of size and number.
7. Evaluation of renal transplants.
8. Guidance of radiologic intervention.
   A. Needle biopsy and cyst puncture.
   B. Abscess and fluid drainage.

APPROACH TO THE RENAL MASS

The most common indication for any uroradiologic test is the presence or suspicion of a renal mass. Excretory urography remains the primary diagnostic modality, since it reveals a great deal of diagnostic information about the entire urinary tract, particularly the collecting system and bladder. In most cases, the urographic study is the only test required for appropriate detection of a renal mass.

The large majority of renal masses are simple cysts of no clinical significance. For this reason, if a mass is found or suspected on IVP, ultrasound should be the next test of choice. Ultrasound possesses an extremely high degree of accuracy in differentiating simple cysts from other cystic or noncystic abnormalities, as demonstrated by Pollack et al. (1982), who found that the false negative rate (misdiagnosing renal neoplasms as renal cysts) in a group of 163 cysts was 0%, and the overall accuracy of a diagnosis of cyst was 98%, with 2% being due to hematomas, hydronephrosis, or septated cysts. Their false positive rate was high, 48%, but since the consequences of failure to detect renal cell carcinoma are serious, that rate is probably acceptable. The authors also point out that the false negative rate among experienced observers is probably 20%–30% (Pollack et al., 1982). If the mass is a simple cyst, the workup ends. If, however, the mass is indeterminant or solid, or the scan is technically inadequate due to obesity for example, CT should be performed. CT determination that a mass is a simple cyst again terminates the workup. Solid masses must be presumed to be malignant except when CT detects fat within a mass as in the benign angiomyolipoma or definitively documents that the mass is inflammatory (Sherman et al., 1981; Lee et al. 1980). Most solid masses will require surgery, since most are renal cell carcinomas.

If a patient has a normal excretory urogram but there is strong suspicion of a mass, or hematuria without cause continues, CT or ultrasound should be done. We prefer ultrasound, but the decision which to use depends on the availability of the equipment and technical expertise at a given institution. Angiography should be reserved for attempts to exclude or include purely vascular abnormalities such as arteriovenous malformation or polyarteritis nodosa, although some authors have described characteristic secondary changes seen with CT within the kidneys in cases of polyarteritis nodosa (Hekali and Kivisaari, 1985).

RENAL MASSES

Renal Cysts

Because it is the most common renal mass and is usually of no clinical significance, the simple cyst must be differentiated from masses which are not cysts and therefore require biopsy and/or surgery. Ultrasound is very accurate in differentiation between these masses but also is unable to clearly identify a cyst as such in about 20%–40% of cases if the criteria for cyst are stringently applied (Pollack et al., 1982). Many of these cases of "indeterminant" masses will be secondary to ultrasound artifacts and to the difficulty in imaging specific areas such as the left upper pole. CT, on the other hand, has no such anatomically difficult areas and is not hampered by bowel gas artifacts. In addition, faster scan times (2 seconds) have markedly reduced the number of examinations limited by respiratory motion. CT can thus function to completely clarify most ultrasound indeterminant masses and to clearly separate these lesions into benign cysts and noncystic (i.e., presumably solid) masses. McClennan et al. (1979) showed that CT had 100% accuracy in 56 lesions shown to be cysts by percutaneous aspiration and recommended the abandonment of renal cyst puncture for typical renal cysts. We agree with this concept.

To achieve a high degree of accuracy in identifying simple cysts, all of the following criteria must be met: (1) uniform water density contents with coefficients of attenuation no greater than 15–20 HU; (2) rounded or oval, sharply marginated mass without a perceptible wall; (3) absence of contrast enhancement (Fig 3B–27) (Sagel et al., 1977; Dunnick and Korobkin, 1984; LiPuma and Haaga, 1983; Elkin, 1980; Pollack et al., 1982; McClennan et al., 1979; Magilner and Ostrum, 1978). Using these criteria, most peripheral cysts are not difficult to diagnose, but occasionally small cysts, as

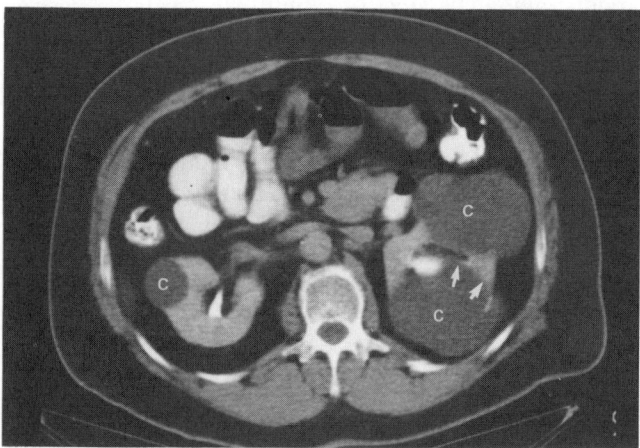

FIG 3B–27.
Bilateral renal cysts *(C)* are of water density and have an imperceptible wall. There is absence of contrast enhancement as well. Margins of cysts with central renal parenchyma on the left *(arrows)* are not sharp due to partial volume averaging. Other sections will usually demonstrate the margin to be well defined.

well as parapelvic ones, may present diagnostic problems when adjacent structures cause artificially thick walls or artifactually raised CT numbers, when partial volume averaging occurs (Sagal, 1979; LiPuma and Haaga, 1983; Hidalgo et al., 1982; Morag et al., 1983). Volume averaging is a phenomenon seen when an area of adjacent tissue is included on the same slice as the cyst or edge of a cyst and CT numbers are raised or the wall is thickened spuriously. Review of other slices may allow identification of the process, but most helpful is the use of thinner (3- or 5-mm) slices to eliminate inclu-

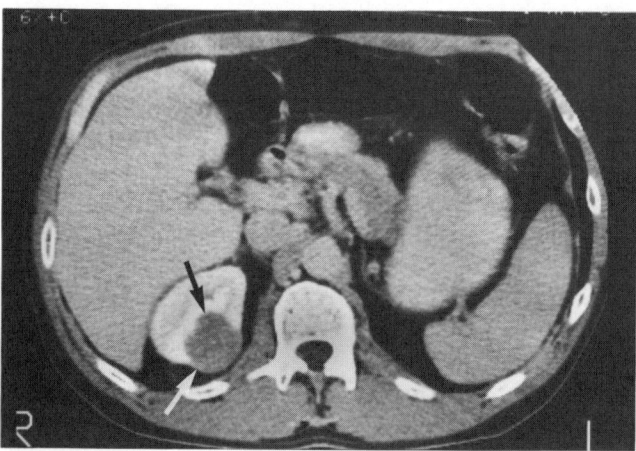

FIG 3B–28.
"Cyst-like" right renal cell carcinoma. Although the mass *(arrows)* is fairly well defined and homogeneous, the attenuation coefficient is similar to that of paraspinous musculature, and the mass definitely should be considered solid rather than cystic.

sion of adjacent renal parenchyma in the edge area or in small cysts. If these criteria are not met, the mass *must* be considered solid or indeterminant and further workup initiated, since necrotic or low-density solid masses can appear very much like a cyst (Fig 3B–28).

Balfe et al. (1982) examined the problem of indeterminant masses on CT and identified three types of non-cystic masses and their ultimate outcome. Of 26 lesions that appeared to be cysts but were poorly imaged or were not classic because of technical artifacts, all were shown to be benign. Twenty-four lesions that were "cyst-like" but possessed features such as wall thickening, irregular contour, increased density, or peripheral calcification were also studied. Fourteen (58%) of these were complicated cysts, and 10 (42%) were either solid masses or abscesses. The third group consisted of eight patients with apparently solid masses with complex features. Three were renal cell carcinoma, and four were xanthogranulomatous pyelonephritis. One case was undiagnosed due to loss of follow-up. From these data, it appears that if a mass is not obviously a cyst and there are no technical problems related to imaging, a high percentage will indeed be solid neoplasms requiring surgery or biopsy.

The introduction of CT scanners with fast scan times and large numbers of x-ray detectors has clarified the issue even further by almost completely eliminating technically inadequate examinations. At Northwestern University and Northwestern Memorial Hospital non-cystic masses are treated as solid and either surgically explored or biopsied except renal abscess, which may be percutaneously drained. We believe that state-of-the-art CT functions as a "radiologic" biopsy, since the CT determination of abnormal contents of the cyst ensures abnormality of some sort on aspiration or biopsy.

Recently, a specific type of lesion, the so-called hyperdense renal cyst, has been discussed in the radiology literature (Coleman et al., 1984; Dunnick et al., 1984; Zirinsky, et al., 1984; Sussman et al., 1984). These masses have the well-marginated characteristics of renal cysts but possess uniform, high-density contents on precontrast scans. Most have been determined to be hemorrhagic or filled with fluid that has a high protein content. The appearance has been suggested as being typical but the number reported is small; one group reported a high density of renal cell carcinoma with an appearance very similar to that of benign hyperdense cysts. Caution is warranted with these lesions. Careful follow-up or cyst puncture is currently recommended (Fig 3B–29).

The problem of calcification within a renal mass has been widely discussed. Until the advent of CT, virtually any calcification in a mass detectable by routine radiography was considered to put that mass at higher

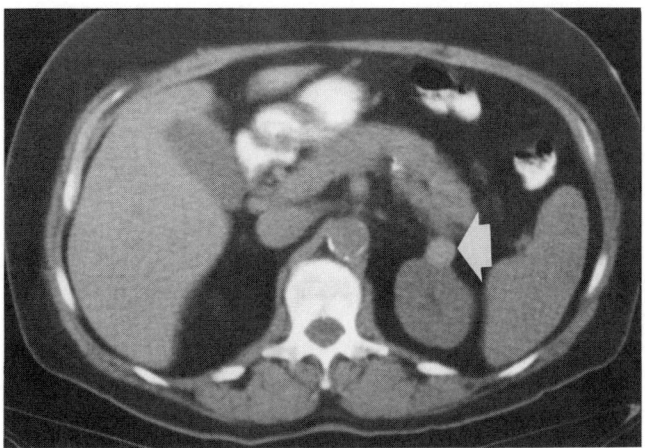

FIG 3B–29.
Tiny left-sided hyperdense renal cyst *(arrow)*. Unenhanced scan shows the cyst to be of high density. Follow-up over a two-year period showed no change in the size or configuration of this cyst. Increased density may be due to either hemorrhage or elevated protein content of cysts.

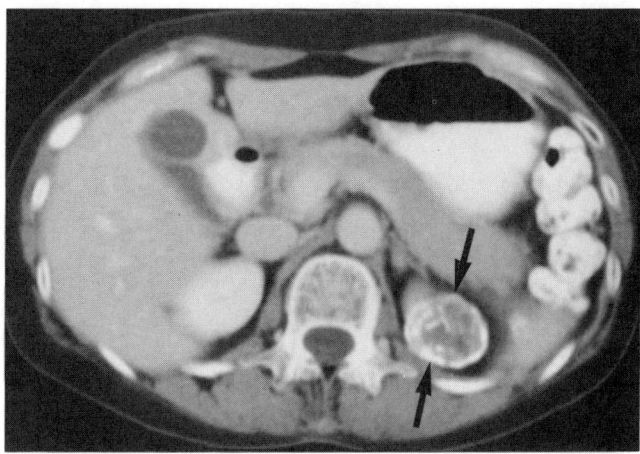

FIG 3B–31.
Calcified renal cell carcinoma. Left upper pole calcified mass *(arrows)* contains internal as well as peripheral calcifications. On resection the mass was shown to be renal cell carcinoma. Such masses are inevitably avascular on renal angiography.

risk for malignancy (Snickerman et al., 1979; Phillips et al., 1963; Kikkawa and Lasser, 1969; Daniels et al., 1972). This also included those masses with peripheral calcification. Weyman et al. (1982) presented convincing data that the location of calcification as precisely determined by CT is a good prognostic indicator of malignancy. They found that lesions with linear calcification confined to the wall of a uniform water density mass with no detectable soft tissue mass were benign cysts in 10 of 10 cases. Nine patients with calcified renal cell carcinomas had nonmural calcification or soft tissue masses demonstrable on CT (Weyman et al., 1982). From these data it seems likely that if a mass appears

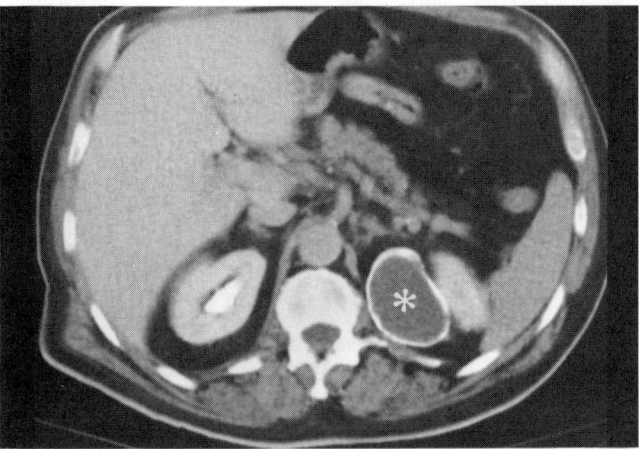

FIG 3B–30.
Peripherally calcified simple cyst. Left upper pole water density mass *(asterisk)* has a smoothly and entirely peripherally calcified rim. Aspiration of this mass revealed clear yellow fluid without evidence of malignant cells.

to be a simple cyst with a calcified wall, it is just that; any deviation from that finding, such as central calcification or soft tissue mass, requires further investigation (Figs 3B–30 and 3B–31). Similar findings concern septations within a cyst. Rosenberg et al. (1985) found that thin septations are of no importance, but if solid elements are noted, further evaluation is necessary. Finally, the issue of alleged coexisting tumor within a simple cyst requires clarification. Although Emmett et al. (1963) claimed an incidence of about 1%, the study did not reflect the accuracy of current imaging techniques. That study also more than likely reflects the fact that cysts occur in 50% of all patients (Ackerman, 1974), some of whom have renal cell carcinomas. We believe that if renal cell carcinoma does occur within a cyst wall, that mass will be abnormal on CT, and solid elements or increased attenuation of its contents would be seen, thus alerting the clinician to the presence of a solid mass.

Polycystic renal disease is usually detected by ultrasound but will show similar signs on CT. The kidneys are markedly enlarged and essentially replaced by myriad cysts (Levine and Grantham, 1981; Grantham, 1979; Lawson et al., 1978). CT and ultrasound may detect hepatic and pancreatic cysts as well. High-density or calcified cysts are common and may be seen in as many as 68% of patients (Fig 3B–32). In all likelihood the appearance is due to cyst hemorrhage followed by clot retraction and protein concentration. Hemorrhage may also be secondary to trivial trauma to the enlarged kidney or to spontaneous small vessel rupture. Hypertension or heparinization during dialysis were also cited as plausible explanations by Levine and Grantham (1985).

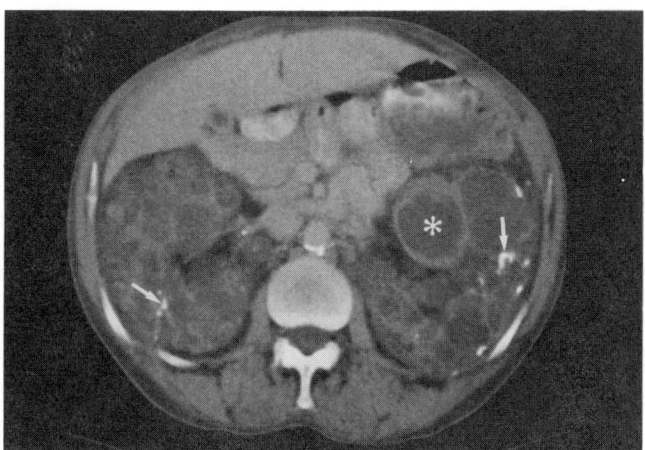

FIG 3B–32.
Polycystic renal disease. Bilaterally enlarged kidneys with multiple cysts. Several cysts are calcified or have calcified walls *(arrows)*. In addition, one of the cysts on the left (*) has a thick wall. Such abnormal cysts are common in polycystic renal disease and should be followed with CT or treated according to clinical symptoms and findings.

These authors evaluated seven of these high-density cysts and found no pathologic evidence of carcinoma. They believe that such cysts are common and should be followed with CT, unless significant findings such as irregular interfaces or heterogeneity are seen. Calcification is also often present and is not of the same concern as in solitary masses.

Renal Cell Carcinoma

Renal cell carcinoma typically appears as a solid mass of lower density than enhanced renal parenchyma. On unenhanced scans the tumor typically has attenuation values similar to renal parenchyma (30–60 HU). Necrosis and/or hemorrhage frequently are seen as areas of lower and higher attenuation, respectively. The overall appearance is one of tissue inhomogeneity. Speckled or irregular calcification may also be seen. After IV infusion of contrast material, the tumor enhances but to a lesser degree than the surrounding renal parenchyma, which concentrates the contrast agent, making these neoplasms much easier to see postinfusion (Sagal et al., 1977; Dunnick and Korobkin, 1984; LiPuma and Haaga, 1983; Vogelzang and Neiman, 1983). If dynamic scanning after bolus injection of contrast material is used the vascular nature of the renal cell carcinoma can be seen transiently (Fig 3B–33). However, *vascularity* and *contrast enhancement* depend on separate factors, such as extracellular accumulation of contrast, accounting for the disparity between the intensely vascular renal cell carcinoma on angiography vs. the relative lack of contrast enhancement during equilibrium phase scanning (Fig 3B–34) (Young and Rumbaugh, 1978; Kormano and Dean, 1976).

STAGING OF RENAL CELL CARCINOMA.—Before the advent of CT, angiography was advocated for accu-

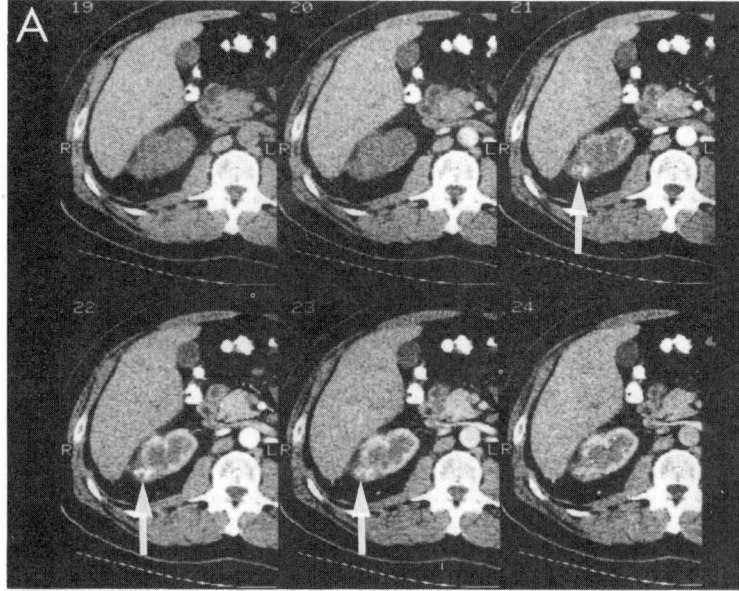

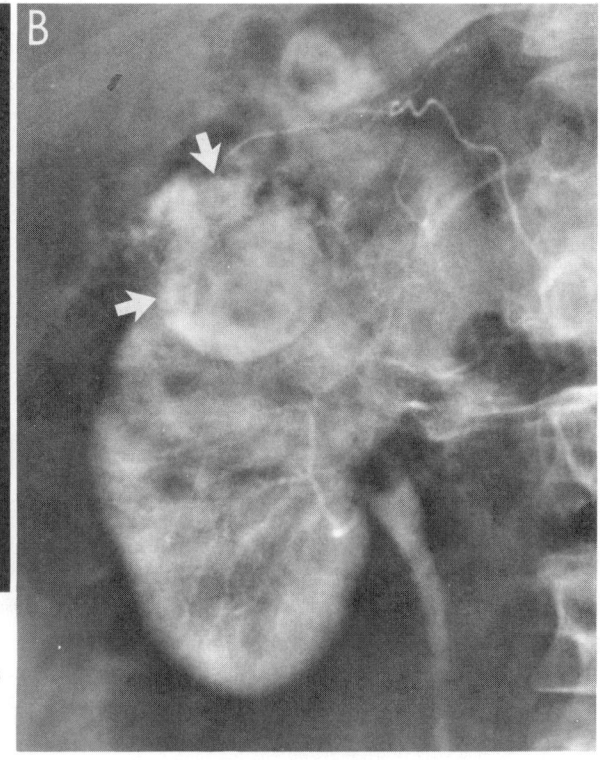

FIG 3B–33.
A, dynamic CT of a small right renal cell carcinoma. Small, right upper pole carcinoma *(arrows—frames 21, 22, and 23)* enhances transiently. **B,** angiogram in late arterial phase shows densely enhancing small renal cell carcinoma *(arrows).*

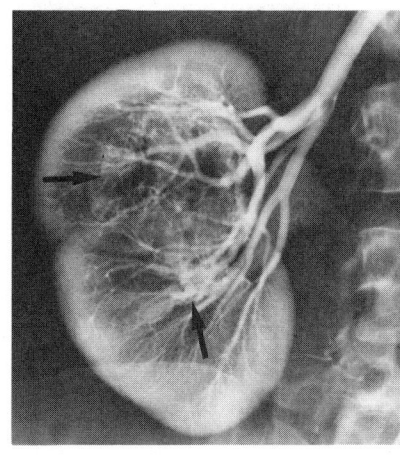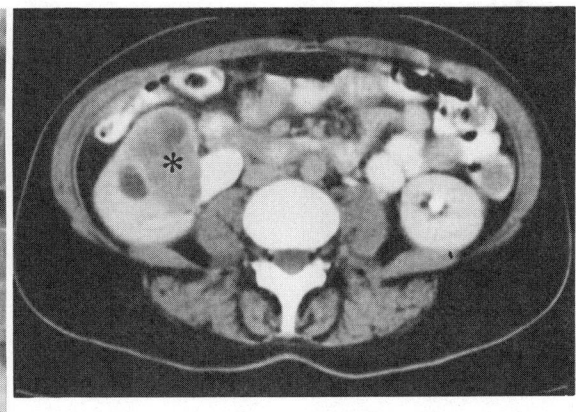

FIG 3B–34.
Vascular renal cell carcinoma (**A**, *arrows*) shows relatively decreased enhancement on contrast-enhanced CT (**B**, *asterisk*).

rate preoperative staging of renal cell carcinoma. Lang (1971, 1973) showed that renal arteriography demonstrated a high degree of accuracy in arteriographic staging of renal cell carcinoma; however, others have found that arteriography has far less accuracy, Bracken and Josson (1979) documenting 36% accuracy and Das et al. (1977) only 60% accuracy. Most angiographic errors related to the lack of ability to predict perirenal and nodal involvements, since "parasitic" blood supply from adjacent vessels may mimic tumor invasion (Buist, 1974). When body CT became widely available, it was immediately apparent to most workers that the direct depiction of the perirenal spaces and retroperitoneum was superior in assessing tumor extent to the relatively indirect signs of involvement such as contrast staining and vascularity (Sagel et al., 1977). Reports of the ability of CT to see inferior vena caval clot (Steele et al., 1978; Marko et al., 1978; Ferris et al., 1979) also provided evidence that CT would be able to outperform angiography on a regular basis. Ultrasound has been advocated as a tool in staging but is generally less accurate than CT in identifying retroperitoneal lymph nodes and generally cannot detect extension into perinephric fat or musculature (Levine et al., 1980).

It can now be stated unequivocally that contrast-enhanced CT is *the* definitive tool for identifying, localizing, and determining the extent of renal cell carcinoma. It possesses greater accuracy than any other modality (Weyman et al., 1980; Levine et al., 1979; Jaschke et al., 1982; Hata et al., 1983; Richie et al., 1983).

Weyman et al. (1980) retrospectively compared angiography and CT in staging of renal cell carcinoma in 62 patients. CT was more accurate than angiography in determining perinephric extension (83% vs. 64%). It was more sensitive in detecting nodal metastases (73% vs. 33%) and equally accurate in noting venous (renal vein and caval) involvement. Jaschke et al. (1982) prospectively evaluated 125 patients with proof of renal

cell carcinoma and found that perirenal extension was correctly predicted in 79%, lymph node involvement in 87%, renal vein and inferior vena caval involvement in 91% and 97%, respectively, and adjacent organ involvement in 96%. The relatively low degree of accuracy in perirenal extension was accounted for by 12 examples in which penetration of the capsule could be seen only microscopically. False positive diagnoses were made when peritumoral densities or strands were mistaken for tumor. The misdiagnosis of nodal metastases or lack thereof was usually secondary to false positive diagnoses of enlarged nodes that at surgery contained not tumor but inflammation (Jaschke et al., 1982). Lang recently advocated the use of dynamic CT or CT during selective renal intra-arterial injection of contrast material to increase the accuracy of CT for local extension to 100% (29 of 29) and local nodal involvement (6 of 8 patients) (Lang, 1984); the method has not been confirmed by others.

CT STAGING IN RENAL CELL CARCINOMA.—CT staging of renal cell carcinoma is generally discussed relative to Robson's classification (Sagel et al., 1977; Weyman et al., 1980; Dunnick and Korobkin, 1984; LiPuma and Haaga, 1983; Love et al., 1981; Magilner and Ostrum, 1978; Levine et al., 1979; Jaschke et al., 1982; Hata et al., 1983; Richie et al., 1983; Robson et al., 1969).

Stage 1: Changes in renal contour and a (usually) well circumscribed and contained solid mass can be seen. The perirenal fat is usually well preserved and shows no evidence of increased density or tumor extension. Gerota's fascia, if seen, is unthickened (Fig 3B–35). *Stage 2:* The perirenal fat is invaded by tumor mass or may exhibit streaky density within. If the mass is bulky, it is well defined by the sharp outer margin of Gerota's fascia, and there is good preservation of fat planes between the tumor and adjacent organs (Fig 3B–36). The

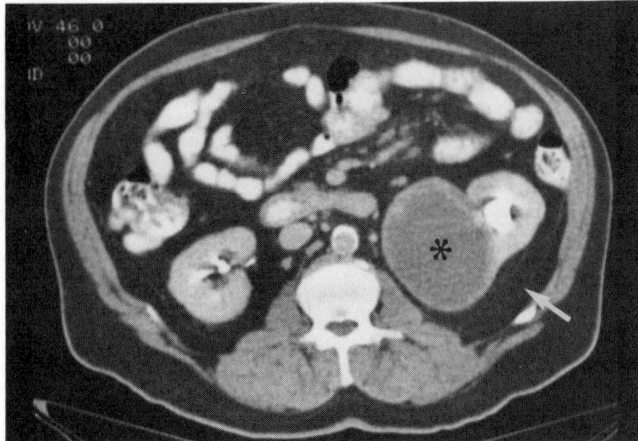

FIG 3B–35.
Stage 1 left renal cell carcinoma. Low-density mass *(asterisk)* is well defined and entirely intrarenal. Gerota's fascia *(arrow)* is well identified and the perirenal fat is undisturbed.

absence of fat plane visualization may not indicate true tumor invasion, however, since compression by a large mass may obliterate the plane and the lack of correlation between tumor size and the disease stage is well known (Dao et al., 1977). Other sources of error in CT differentiation of stages 1 and 2 include visualization of perirenal densities, which represent not tumor but collateral veins or lymphatics (Baert et al., 1982; Fererstein et al., 1984). *Stage 3:* Retroperitoneal adenopathy in the perihilar or periaortic chains can be seen (Fig 3B–37). Any node larger than 1 cm in diameter should be considered positive, although false positive diagnoses will result from enlarged nodes due to hyperplasia or reaction. Likewise, falsely negative reports will

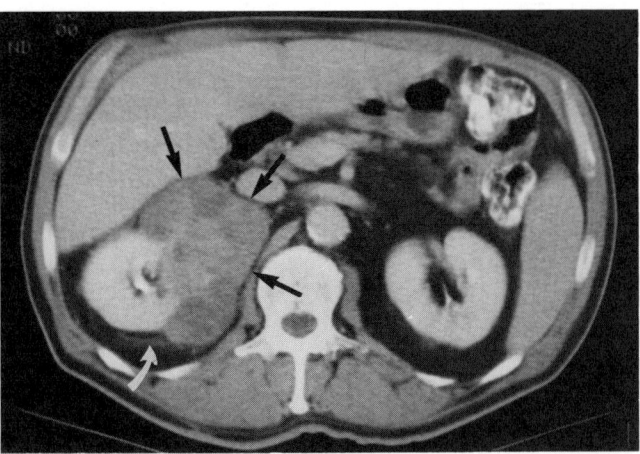

FIG 3B–36.
Stage 2 renal cell carcinoma. Bulky right renal mass shows good preservation of margin between it and adjacent organs *(arrows)*. Gerota's fascia and perirenal fat show some thickening and/or infiltration *(curved white arrow)*.

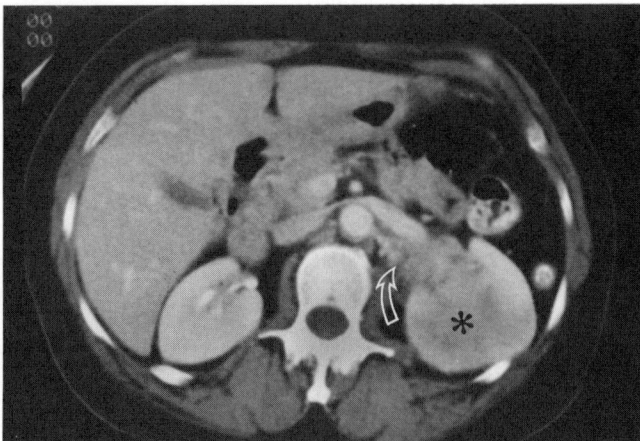

FIG 3B–37.
Left renal cell carcinoma *(asterisk)* with renal hilar adenopathy *(curved arrow)*.

be seen in those nodes that are only microscopically involved and not enlarged. CT is about 75% accurate in correctly assessing lymph node involvement (Weyman et al., 1980).

Venous involvement is displayed by CT with between 82% and 91% accuracy for the renal veins and about 95% accuracy for the inferior vena cava (Weyman et al., 1980; Hata et al., 1983). Venous thrombus is best seen with high doses of intravascular contrast and careful scanning at the renal vein level. The left renal vein is easily seen throughout its longer, more horizontal course, but the shorter, more obliquely oriented right renal vein can be difficult to visualize. We have found a rapid infusion during nondynamic scanning to be very reliable, but others advocate bolus injection and dynamic scan techniques (Lang, 1984). Tumor thrombus is recognized as a low-density filling defect within the opacified vein (Fig 3B–38) or as venous nonopacification with or without enlargement. It should be pointed out that rapid arteriovenous shunting through a tumor can enlarge a renal vein or inferior vena cava without tumor involvement. The diagnosis is most firmly made when a filling defect is seen, although the presence of collateral vessels or a venous enlargement may be very helpful (Jasche et al., 1982; Glazer et al., 1984).

The vena cava is usually easily seen and the size and extent of low-density thrombus demonstrated. Another less reliable sign of thrombus is the presence of vena caval enlargement (Weyman et al., 1980; Steele et al., 1978; Marks et al., 1978; Ferris et al., 1979). Occasionally thrombus may be seen as far cephalad as the right atrium. Although there are those who have advocated foot vein infusion (Marks et al., 1978), we and others believe layering and laminar flow of dense contrast media cause false positive diagnoses (Glazer et al., 1982).

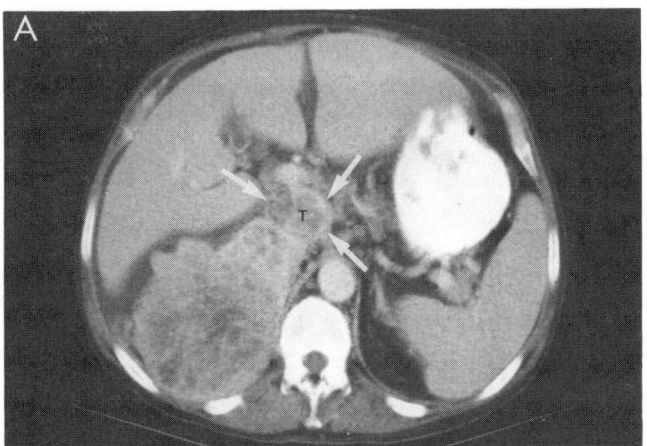

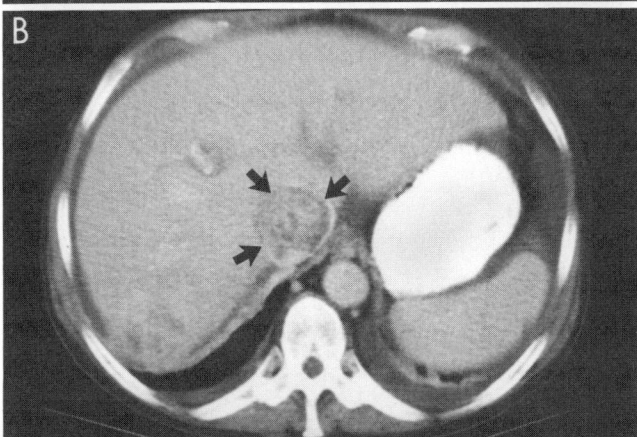

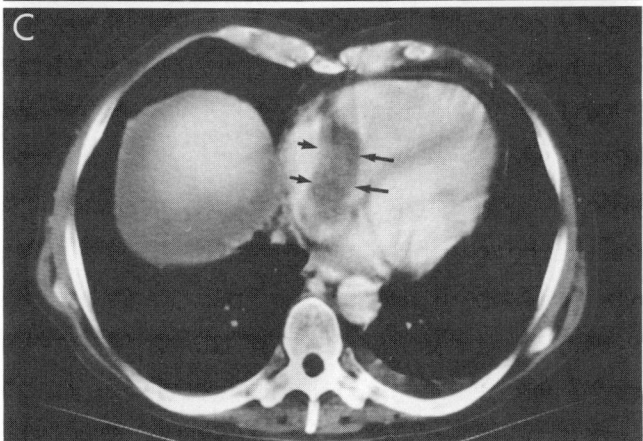

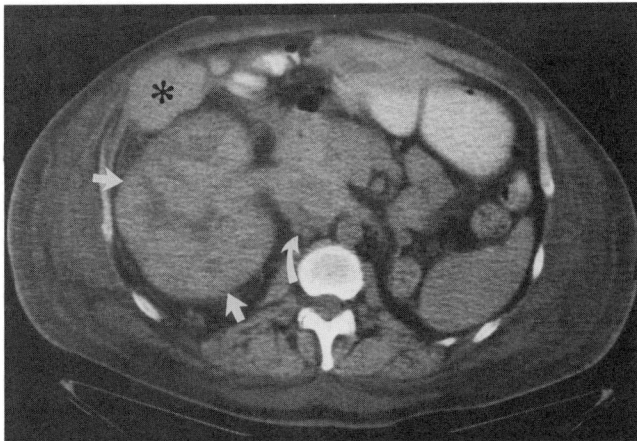

FIG 3B–39.
Right renal cell carcinoma *(arrows)* with periaortic adenopathy *(curved arrow)* and involvement of the body wall *(asterisk)*.

ample of confusion may occur if the vena cava is severely compressed or displaced by tumor and intraluminal clot cannot be excluded. Finally, it should be stressed that venography should be performed if any doubt exists as to the status of the venous structures on CT.

Stage 4: Involvement of surrounding organs and the body wall is seen when tumor actually invades these structures (Fig 3B–39). Care must be taken to identify tumor invasion and not simple adherence. Lang advocated dynamic CT or CT with selective arterial injection for the identification of this sometimes difficult to detect finding (1984). Metastatic disease to liver, brain, and lung is also best screened with CT. Bone scan possesses sensitivity superior to CT and plain films.

FIG 3B–38.
Renal venous and vena caval involvement with extension to right atrium. Massive right renal cell carcinoma replaces right kidney **(A).** Low-density thrombus *(T)* fills vena cava *(arrows)* and extends up enlarged intrahepatic vena cava **(B,** *arrows)* and into right atrium **(C,** *arrows).*

For this reason, arm vein infusion is preferred, but even a "pseudoclot" can be seen here in the early stages of rapid arm vein infusion and should be recognized when an ill-defined defect is seen at the level of the renal veins (Vogelzang et al., 1985). Another ex-

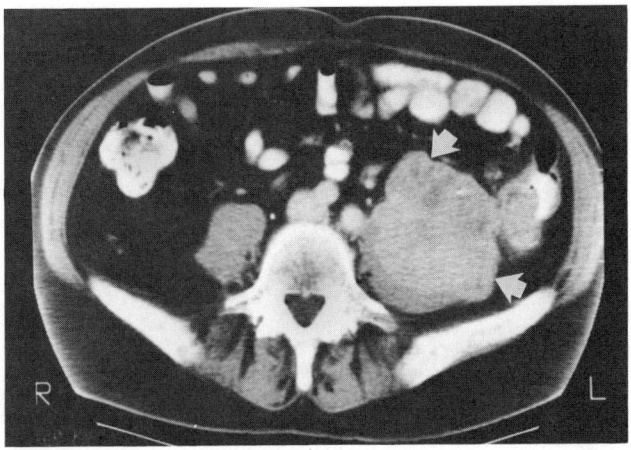

FIG 3B–40.
Recurrent renal cell carcinoma. Large mass *(arrows)* in the inferior aspect of the left renal fossa represents a bulky recurrence after nephrectomy.

OTHER USES.—Recurrence of renal cell carcinoma after nephrectomy is best examined with CT (Fig 3B–40) (Bernardino et al., 1979). In the empty renal fossa bowel loops may simulate masses; opacification by oral contrast will avoid this pitfall. Scans performed in the immediate postoperative periods may be confusing when irregularity or masses caused by hematoma, inflammation, or scar simulate residual or recurrent tumor. Follow-up scans or percutaneous biopsy under CT guidance is often helpful. Finally, the opposite kidney and adrenal gland may be sites of metastatic disease and should be scrutinized carefully. CT can also be very helpful in screening patients with the von Hippel-Lindau syndrome who are at risk for the development of bilateral renal cell carcinoma, as reported by Levine et al. (1979, 1983).

Transitional Cell Carcinoma

Transitional cell carcinoma of the renal pelvis and ureter is best imaged using excretory urography, retrograde pyelography, and urinary cytology. CT scan, however, plays a role in staging and identification (Baron et al., 1982; Pollack et al., 1981; Breatnach et al., 1984). When visualized, the tumor is usually an irregular, mucosally based defect in the contrast-filled collecting system (Fig 3B–41). These tumors are usually of the papillary type. Nonpapillary transitional cell carcinoma usually does not produce filling defects (Lang, 1984). Larger tumors may present as soft tissue masses in a central pelvic location with obliteration of the peripelvic fat or infiltration and invasion of the renal parenchyma. In such circumstances differentiation from renal cell carcinoma may be difficult. Transitional cell carcinoma, however, has a greater tendency to produce hydronephrosis and nonfunction and less tendency to change the renal outline and may be distinguished on

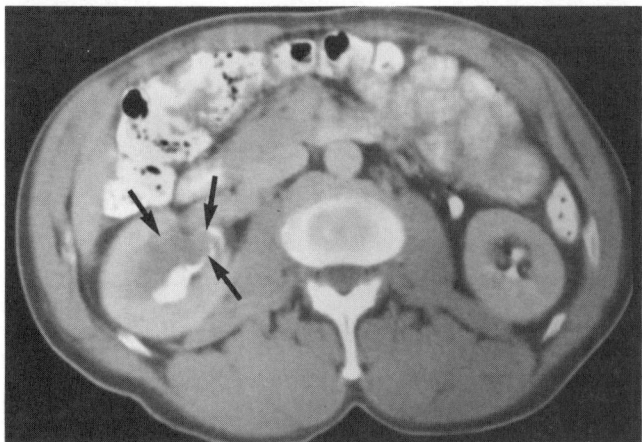

FIG 3B–41.
Transitional cell carcinoma *(arrows)* with pelvic filling defect and some intraparenchymal extension.

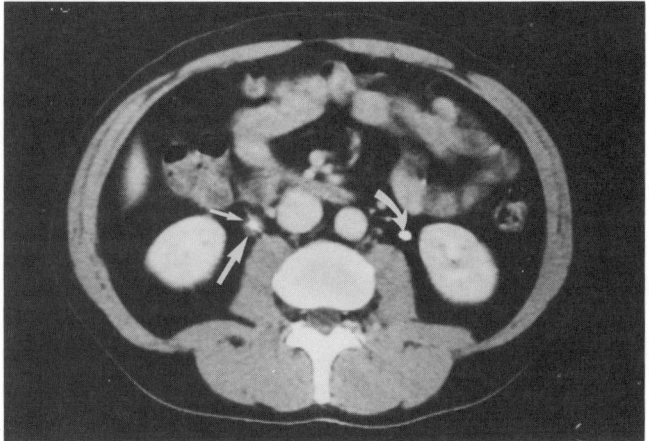

FIG 3B–42.
Proximal right transitional cell carcinoma demonstrating focal ureteral wall thickening *(arrows)*. Compare the appearance of the right ureter with normal left ureter *(curved arrow)*.

that basis in most cases. When transitional cell carcinoma is ureteral in origin, focal ureteral thickening or an intraureteric mass can be seen (Figs 3B–42 and 3B–43). Gross spread into the retroperitoneum and a periureteric mass may also be identified. The tumor may occasionally calcify. The main value of CT in the management of transitional cell carcinoma is in staging when bulky extraureteral or extrarenal spread or adenopathy may be visualized. Baron et al. (1982) found CT to be 83% accurate in staging in a group of 24 patients with transitional cell carcinoma and also noted the utility of CT when excretory urography showed nonfunction or poor function. Another significant use of CT is in the evaluation of nonopaque renal pelvic filling defects where CT can easily identify a mass as densely calcified or noncalcified or minimally calcified (Pollack et al., 1981; Parienty et al., 1982). Noncalcified masses must be assumed to be transitional cell carcinoma unless proved otherwise. In cases of ureteral obstruction or nonfunction, CT can detect the level of obstruction and identify that obstruction as stone disease or mass (Megibow et al., 1982).

Renal Lymphoma

Renal involvement by lymphoma is probably common but often clinically silent. The current incidence is not known, but a report from an era prior to modern chemotherapy showed an occurrence rate of one third (Richmond et al., 1962; Martinez-Madoado et al., 1966). The routine use of CT in patients with lymphomas of all types has significantly increased our awareness of the variety of forms which renal lymphoma takes. Lymphoma is almost always of lower attenuation

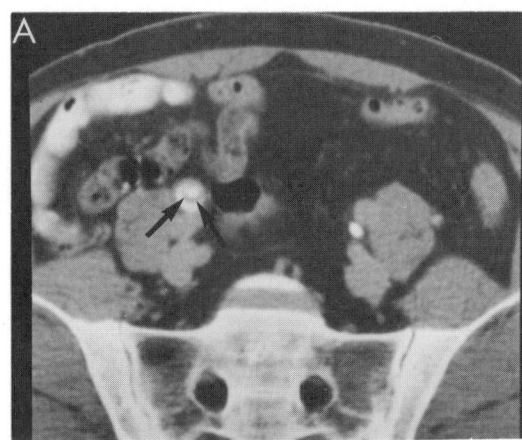

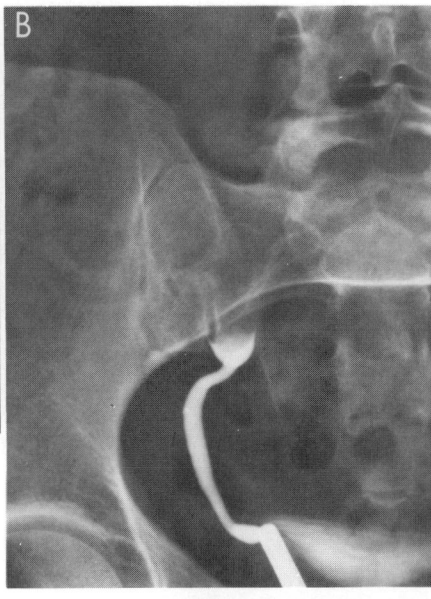

FIG 3B–43.
Intraureteric mass of transitional cell carcinoma. **A,** CT shows enlarged right ureter with filling defect *(arrows).* **B,** retrograde pyelogram shows an obstructing mass.

than the contrast enhanced renal parenchyma and may be seen in several forms: (1) parenchymal renal nodules, single or multiple and sometimes bilateral (Fig 3B–44); (2) diffuse renal infiltration and enlargement (Fig 3B–45); (3) engulfment by contiguous bulky retroperitoneal masses (Hartman et al., 1982; Chilcote and Borkowski, 1983; Jafri et al., 1982). When low attenuation nodules are seen, differentiation from other tumors including metastatic disease can be difficult. A diagnosis is often suggested by the concurrent clinical diagnosis of lymphoma. One common feature that aids in the diagnosis is the presence of retroperitoneal adenopathy, which was seen in all 31 patients with renal lymphoma whose CT appearance was reported in two recent publications (Chilcote and Borkowski, 1985; Jafri et al., 1982).

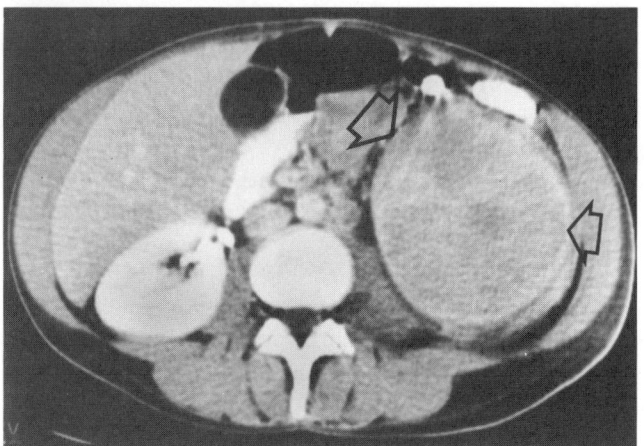

FIG 3B–45.
Massive enlargement and infiltration of the left kidney *(arrows)* in a patient with Hodgkin's disease.

It also seems that the appearance of renal lymphoma is almost entirely based on the stage and extent of disease at the time of imaging, with nodular disease probably progressing to diffuse infiltration. Although it has been stated that Hodgkin's disease has a lower incidence of renal involvement, Chilcote and Borkowski (1983) had equal numbers of lymphomas represented in their report on renal lymphoma and CT. It is probable that some differences between tumor types exist, but this feature has not been adequately studied in the modern era of CT and chemotherapy.

Angiomyolipoma

Angiomyolipoma or renal hamartoma is composed of variable amounts of smooth muscle, fat, and blood vessels and is most common in women (Shannon et al.,

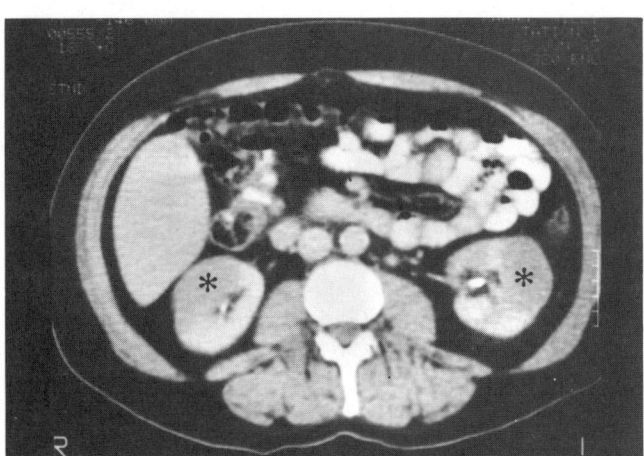

FIG 3B–44.
Bilateral low-density masses *(asterisks)* of lymphomatous involvement.

1981). On ultrasound the tumor is usually echogenic; it is hypervascular arteriographically. Both of these characteristics may give an erroneous preoperative diagnosis of renal cell carcinoma (Meany, 1969; Hartman et al., 1981). The presence of fat, however, has allowed the preoperative identification of these lesions with a high degree of certainty; lesions which previously required nephrectomy can now best be treated with local resection as a result of this accuracy (Hansen et al., 1978; Bosniak, 1981). The presence of fat is highly specific for this tumor. Bosniak (1981) pointed out that there are no reported cases of renal cell carcinoma containing fat. Lipomas will appear fatty but as benign lesions present no difficulty. He also noted that the extremely rare liposarcomas seen in and around the kidney usually contain enough high-density masses to allow differentiation, although well-circumscribed, low-grade tumors might cause differential diagnostic problems. In our opinion, avoidance of a nephrectomy for this benign, easily diagnosed tumor makes preoperative CT mandatory for all solid renal masses.

CT usually demonstrates a large, low-density, fatty renal tumor which often extends into the perinephric space but not beyond Gerota's fascia. The tumor usually has areas of intermixed tissue density that represent myomatous and angiomatous elements or hemorrhage (Fig 3B–46). Poorly marginated tumors usually contain significant hemorrhage within or around the lesion pathologically. Indeed, hemorrhage is often the presenting complaint of a patient with angiomyolipoma and may be massive and life-threatening. Large amounts of blood may also obscure many of the fatty elements. Atypical tumors composed entirely of muscle and vessel

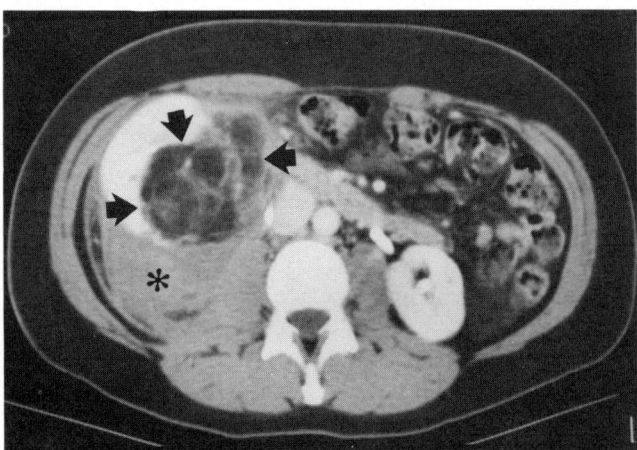

FIG 3B–46.
Right renal angiomyolipoma. Low-density fatty tumor *(arrows)* has areas of intermixed tissue density which represent myomatous or angiomatous elements. Perirenal hemorrhage *(asterisk)* is seen posteriorly as a result of spontaneous hemorrhage of the tumor.

do not permit an accurate diagnosis. Such lesions were seen in 3 of 17 cases analyzed recently by authors at the Armed Forces Institute of Pathology but are probably more atypical than this selected series indicates (Sherman et al., 1981). A group of French authors recently described the appearance of small, asymptomatic angiomyolipomas discovered on CT or ultrasound (Bret et al., 1985). This group of tumors should probably be observed using imaging techniques and not operated on. Multiple bilateral tumors are usually only seen in patients with tuberous sclerosis, but in about 5% of all patients with angiomyolipoma there will be bilateral lesions without coexistent tuberous sclerosis. Extension into the vena cava may rarely occur; CT has been used to detect fatty tumor in the vein (Brantley et al., 1985).

Renal Oncocytoma

Oncocytoma is a solid lesion that has little tendency to recur or metastasize. Radiologically it has been difficult to diagnose, with the diagnosis relying almost exclusively on the finding of a "spoke-wheel" configuration of tumor vessels on angiography (Ambos et al., 1978). Recently, however, Quinn et al. (1984) suggested that the diagnosis may have a greater chance of being made preoperatively if the following imaging criteria are met: (1) small, well-marginated mass with *homogeneous* appearance on CT or ultrasound and a central scar; and (2) "spoke-wheel" angiographic pattern with a homogeneous blush. Using these criteria retrospectively, the authors concluded that the correct diagnosis would have been made in 16 of their 18 cases (Quinn et al., 1984). Cohan et al. (1984) also found similar CT features in six tumors but thought that the diagnosis could not be made reliably on CT alone. Levine and Huntrakoon (1983) had similar views. A single case of calcified oncocytoma has also been reported, as has a case of bilateral tumors (Wasserman and Ewing, 1983; Hara et al., 1982). In our experience the tumor usually presents as an incidental finding, but thus far we have been unable to make the diagnosis on imaging criteria alone in the three tumors we have seen with CT.

OTHER RENAL MASSES.—Many unusual and rare kidney tumors have also been reported, but most appear solid and do not permit accurate diagnosis without surgery or percutaneous needle biopsy. These masses include various mesenchymal tumors such as fibromas, lipomas, sarcomas, and multilocular cystic nephromas. Lipomas are fatty and easily identified. Multilocular cystic nephroma is a congenital intrarenal multiseptated cystic mass with calcification (Parienty et al., 1981). Juxtaglomerular renin-secreting tumors have also been reported but are identified only as solid renal masses without distinguishing features (Dunnick et al., 1983).

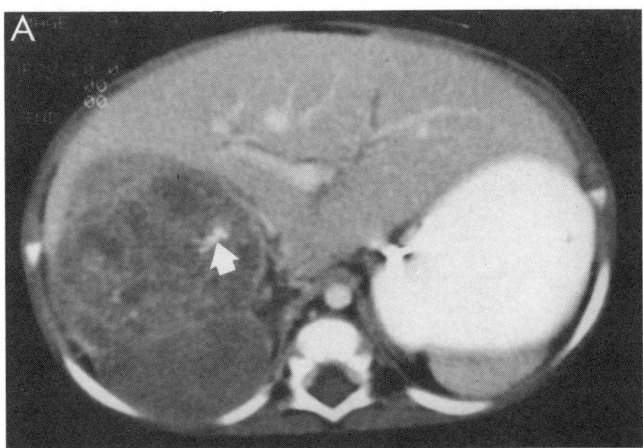

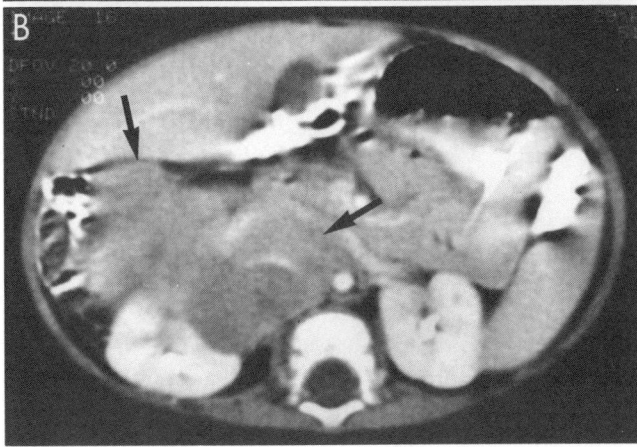

FIG 3B–47.
Neuroblastoma. **A,** large low-density tumor is seen in the suprarenal area. The mass contains calcification *(arrow)*. **B,** section at a lower level shows displacement of right kidney by the mass *(arrows)*. There is no obvious intrarenal extension indicating an extrarenal origin of the mass.

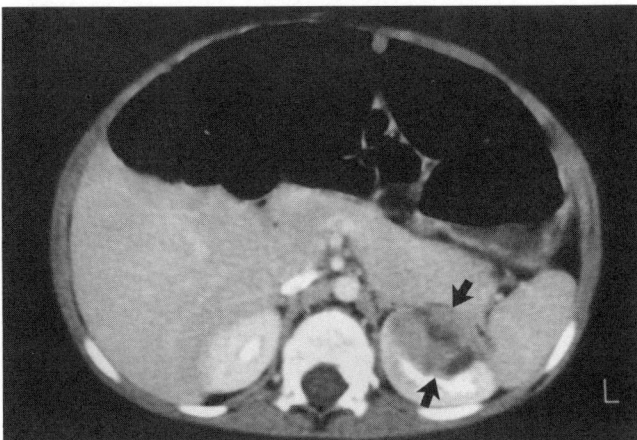

FIG 3B–48.
Small left-sided Wilms' tumor *(arrows)* seen as an entirely intrarenal mass. Low-density areas represent necrosis.

In children Wilms' tumor and neuroblastoma commonly involve the kidney or perirenal space. Neuroblastoma is generally extrarenal and displaces the kidney, although invasion can be seen. Calcification is often present (Fig 3B–47). Wilms' tumor, the chief differential lesion in the child with a renal or perirenal mass, is generally intrarenal and may contain necrosis (Fig 3B–48). CT is valuable in identifying contralateral disease and metastasis (Brasch et al., 1982).

Metastatic disease may also involve the kidney and have unusual appearances; a history of such malignancies as squamous cell carcinoma or breast cancer that are known to give such masses may be helpful in a patient with multiple renal abnormalities (Nishitani et al., 1984).

RENAL INFLAMMATORY DISEASE

Renal inflammation has been classified and categorized by many authors and includes a wide spectrum of acute and chronic processes, which include acute diffuse pyelonephritis, focal bacterial nephritis, pyelonephritis, renal and perirenal abscesses, xanthogranulomatous pyelonephritis, and the sequelae of many of these processes, i.e., scarring and loss of renal parenchyma (Kincaid-Smith, 1979; Heptinstal, 1974; Dunnill, 1976).

Acute pyelonephritis is usually demonstrated on CT as renal enlargement secondary to renal edema. The nephrogram may be decreased. Subtle changes in the parenchyma consist of focal wedge-shaped zones of diminished enhancement after contrast administration that also demonstrate "striations" probably due to slowing of tubular flow, while adjacent peritubular capillaries are filled with contrast (Dunnill, 1976; Gold et al., 983; Rauschkolb et al., 1982). *Focal bacterial nephritis* occurs when inflammation predominantly involves a single renal lobe or lobes without liquefaction or abscess formation (Morehouse et al., 1984), although Gold et al. (1983) suggested that "focal bacterial nephritis is only one phase of the spectrum of acute pyelonephritis and that the rest of the kidney is often involved with areas of inflammation." When seen, however, focal bacterial nephritis (sometimes referred to as lobar nephronia) has a distinctive CT appearance consisting of a masslike area of diminished contrast enhancement which may be poorly defined (Rosenfield et al., 1979; Lee et al., 1980) (Fig 3B–49). Aspiration of this area will inevitably yield no frank pus (Hekali et al., 1985). When localized inflammation persists, ultimately liquefaction and frank pus formation are seen, and an abscess will form. CT will demonstrate a well-delineated, low-attenuation mass which may have a thick, irregular wall (Fig 3B–50). Some renal abscesses may be loculated or contain septae and may extend into the perirenal or pararenal spaces. *Perinephric abscesses* are

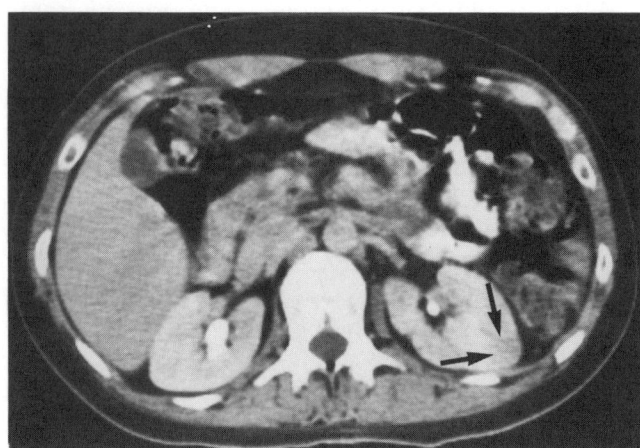

FIG 3B–49.
Focal bacterial nephritis. Low-density wedge-shaped area in the left kidney *(arrows)* showing diminished contrast enhancement.

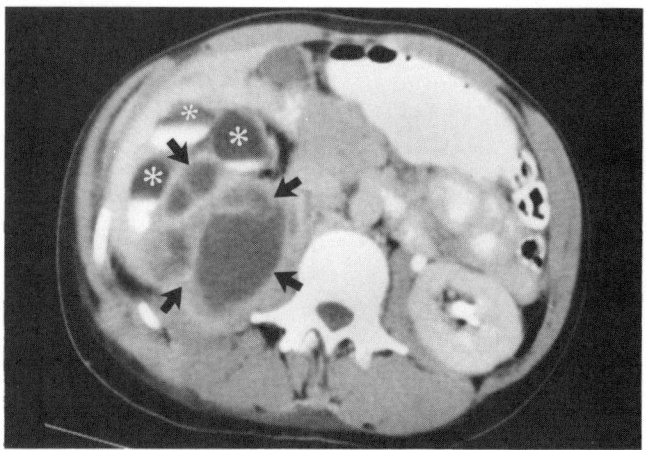

FIG 3B–51.
Perinephric abscess *(arrows)* demonstrating displacement of right kidney and hydronephrosis *(asterisks)*.

soft tissue or fluid masses which displace or deform the renal contour. Gas bubbles may be present, and Gerota's fascia is inevitably thickened (Gold et al., 1983; Rauschkolb et al., 1982; Morehouse et al., 1984; Rosenfield et al., 1979; Lee et al., 1980; Zaonta et al., 1985; Hoddick et al., 1983; Mendez et al., 1979) (Fig 3B–51). *Pyonephrosis* refers to the presence of infected hydronephrosis together with the suppurative destruction of the kidney, although many authors currently use the term to mean simple infected hydronephrosis (Rauschkolb et al., 1982; Morehouse et al., 1984; Rosenfield et al., 1979; Lee et al., 1980; Zaonta et al., 1985; Hoddick et al., 1983). The CT findings are those of chronic hydronephrosis. Stones may be present in the collecting system, which usually contains fluid of

density greater than urine. *Xanthogranulomatous pyelonephritis* is an uncommon form of renal inflammation seen in female patients with chronic infection and stone disease in which granulomatous inflammation destroys and replaces the renal parenchyma with sheets of lipid-laden macrophages. The CT appearance is somewhat variable, as a radiologic pathologic study by Goldman et al. (1984) indicated. Features commonly seen included a contracted fibrotic renal pelvis containing stones with replacement of the parenchyma by multiple, rounded, low-density cortical cavities that represented debris-filled dilated calyces or parenchymal cavities. A common feature of this disease is evidence of extrarenal extension to the perirenal or pararenal spaces or adjacent structures such as the psoas or diaphragm

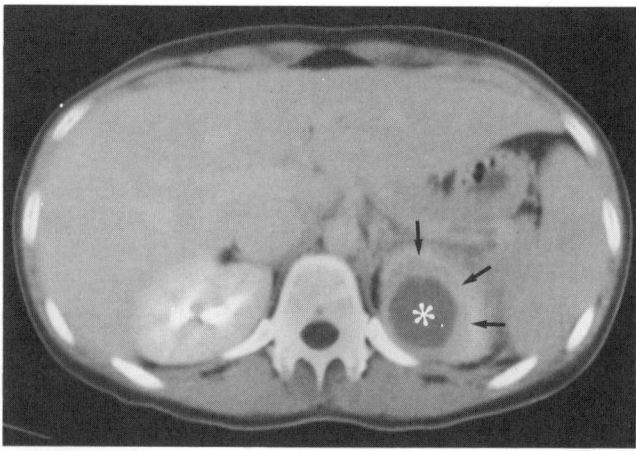

FIG 3B–50.
Renal abscess. Left upper pole abscess *(asterisk)* with surrounding rim of low-density edema *(arrows)*. *Escherichia coli* was cultured from the aspirate of this mass.

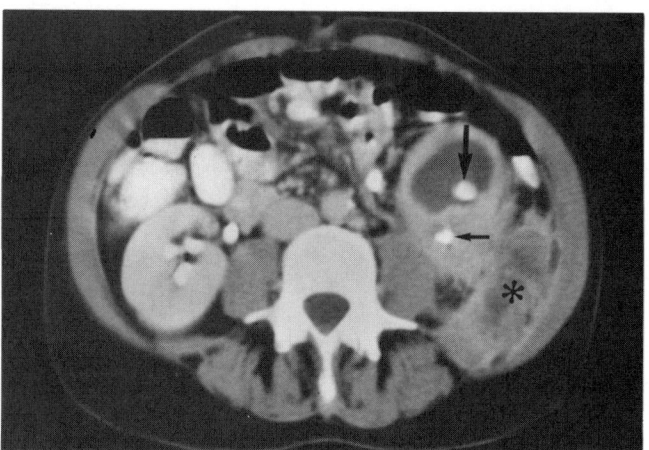

FIG 3B–52.
Xanthogranulomatous pyelonephritis. Calculi *(arrows)* within a hydronephrotic left kidney. Perinephric abscess *(asterisk)* seen in soft tissues of the left flank. Marked infiltration and stranding of surrounding soft tissues.

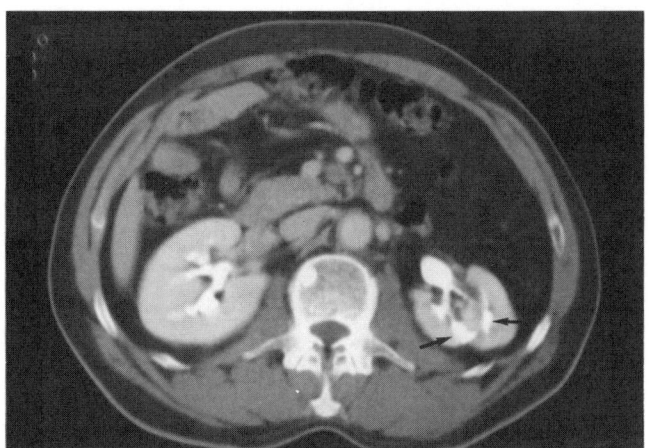

FIG 3B–53.
Chronic left pyelonephritis with scarring. Note renal parenchymal loss and some degree of calyceal clubbing *(arrow)*.

(Fig 3B–52). The kidney may also lack significant contrast enhancement or function. The focal form of xanthogranulomatous pyelonephritis can appear as a local low-density mass adjacent to a stone-containing calyx. *Chronic pyelonephritis* usually appears as focal or global renal atrophy and scarring, as would be expected from the urographic findings in this disease. Calyceal clubbing may be present as well (Fig 3B–53).

In *renal tuberculosis*, the findings are usually nonspecific but may be identified as thick-walled abscess cavities which may be calcified (Dunnick et al., 1984). Psoas abscesses or spinal disease may be a helpful ancillary sign in recognition.

HYDRONEPHROSIS

CT is neither the most sensitive nor the least expensive method to detect hydronephrosis, but inevitably

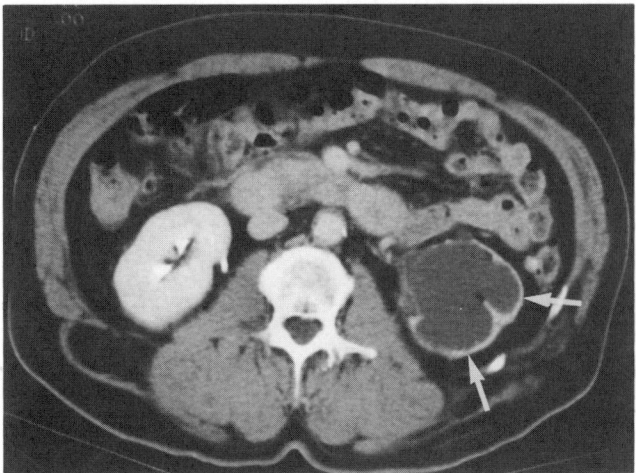

FIG 3B–55.
Severe hydronephrosis. Long-standing left ureteropelvic junction obstruction with severe renal parenchymal thinning. Renal parenchyma does exhibit some contrast enhancement *(arrows)*.

other renal and nonrenal conditions cause obstruction which may be recognized during CT examination. *Mild hydronephrosis* shows slight obliteration of renal fat and sinus structures by the dilated renal pelvis and calyces. *Moderate hydronephrosis* includes more prominent intrarenal collecting system dilatation. A helpful sign in early and moderate hydronephrosis may be the presence of a urine-contrast layer in the collecting system (Fig 3B–54). *Severe hydronephrosis* shows varying degrees of marked calyceal dilatation with renal parenchymal thinning surrounding the low-density central hydronephrotic mass (Fig 3B–55). Contrast excretion is usually delayed or absent. In *acute obstruction* peripelvic extravasation of contrast may be demonstrated (Fig 3B–56). The main advantage of CT in renal obstruction

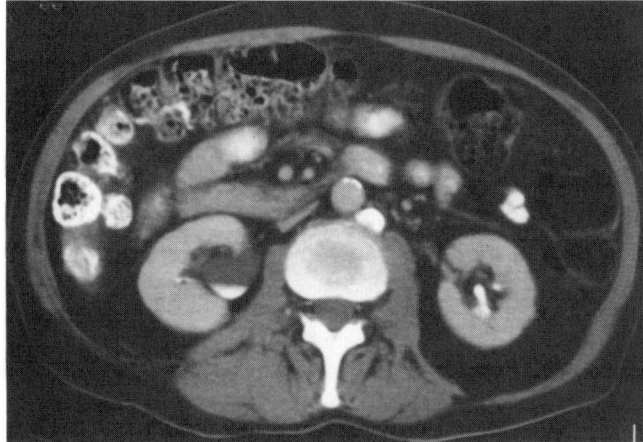

FIG 3B–54.
Moderate hydronephrosis. Urine-contrast layer in right renal pelvis.

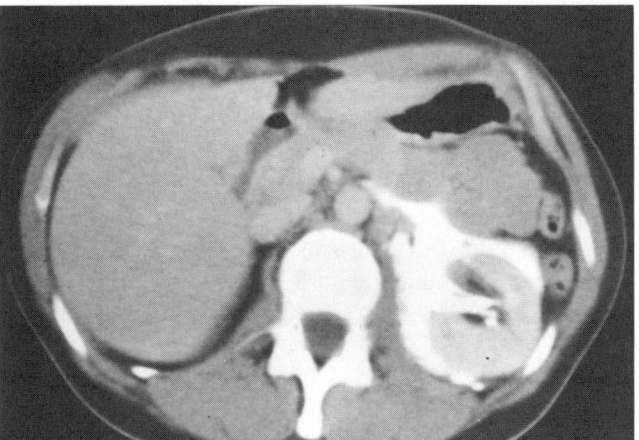

FIG 3B–56.
Acute left ureteral obstruction with perirenal extravasation of contrast.

is in identifying the nature and level of the process (Megibow et al., 1982; Goldman et al., 1984). In identifying the cause of obstruction, the dilated ureters usually can be readily traced to the point of occlusion. Delayed scans or prone position will allow better opacification of the ureters. Bosniak et al. (1982) studied 36 patients with ureteral obstruction of unknown cause on IVP and found CT helpful in 33 (92%). CT readily detects tumor or other mass, although fibrosis and strictures are not readily seen, except in retroperitoneal fibrosis, when a contrast enhancing mass is generally identified (Dalla-Palma et al., 1981). Calculus obstruction may be obscured by contrast administration and, when suspected, should be sought with noncontrast-enhanced scans.

Some sources of confusion with hydronephrosis do exist such as large parapelvic cysts on unenhanced scans. Contrast opacification of the kidney then prominently displays the unenhanced cysts surrounding the compressed spidery collecting structures. Likewise, an extrarenal pelvis may be a problem, but contrast enhancement shows maintenance of sharp delineation of intrarenal calyces. In any questionable case, an IVP or ultrasound usually solves the diagnostic dilemma.

URINARY CALCULI

CT is an extremely sensitive tool for the detection of calculus disease. Generally, plain films show the calculus or calculi to excellent advantage, but uric acid stones are usually radiolucent, and cystine stones have limited radiodensity. Differentiation of these calculi from tumor or clot is of major importance when a filling defect is present on excretory urography or when ureteral obstruction is present and the cause is not apparent (Pollack et al., 1981; Parienty et al., 1982; Megibow et al., 1982). Similarly, tiny opaque calculi may be too small to be visible or can be obscured by bowel gas. CT is exceptionally helpful in both of these circumstances, since without exception calculi are high-density objects. Uric acid and cystine stones usually have densities between 100 and 300 HU, with oxalate and struvite stones greater than 400 HU and as high as 1,200 HU (Resnick et al., 1984; Hillman et al., 1984; Newhouse et al., 1983). Uric acid and cystine stones can be accurately differentiated on the basis of their lower CT densities, but calcium-containing stones cannot be reliably distinguished (Hillman et al., 1984; Newhouse et al., 1983). When examining for calcified masses, IV contrast should obviously not be given.

RENAL TRAUMA AND VASCULAR INJURY

In the past five years, CT has become recognized as the most accurate and reliable imaging tool for the detection and staging of renal injuries mainly based on the San Francisco General Hospital's experiences as reported by Federle and co-workers (Federle et al., 1981, 1982; McAninch and Federle, 1982; Steinberg et al., 1984; Heswel and Smith, 1974; Cass, 1975; Morrow and Mendez, 1979; Mahoney and Persky, 1968; Kasmin et al., 1969; Cass, 1979; Mendez, 1977; Lang, 1975; Elkin et al., 1966; Erturk et al., 1985). Initial radiologic evaluation of suspected renal trauma has historically included IVP with tomography, which is fairly sensitive but not specific (Resnick et al., 1984). Cass (1975) found that a spectrum of abnormalities existed for even minor injuries with most of these abnormalities consisting of poor visualization or absence of function. A specific diagnosis is usually not possible although "accuracies" of 60%–85% have been reported by various authors (Morrow and Mendez, 1979; Mahoney and Persky, 1968; Kazmin et al., 1969; Cass, 1979; Mendez, 1977). Angiography was shown to have excellent accuracy for most traumatic renal injuries, and it was used as the definitive examination when an abnormality was seen on IVP (Lang, 1975; Elkin et al., 1966). Angiography, however, is invasive and organ specific; the presence of associated organ injuries in 20% of blunt and 80% of penetrating renal injuries made the use of CT extremely attractive as an imaging tool, particularly in light of the emphasis on conservative management of renal injuries (Federle et al., 1981).

Renal injuries are conveniently and typically divided into three major categories. Category 1 consists of contusion and lacerations of varying degrees not involving the collecting system. Category 2 includes lacerations or fractures which extend to the collecting system and usually include urinary extravasation. Category 3 is the catastrophic variety of injury that involves the renal pedicle and/or shatters the entire kidney. Federle et al. (1981) and McAninch and Federle (1982) found that CT clearly separated minor from major injuries and significantly influenced the decision to perform surgery and allowed a confident decision in cases in which conservative management was elected (Sherman et al., 1981; Federle et al., 1981). Erturk et al. (1985) showed similar findings in 22 patients and also noted a large reduction in renal angiography as a result (only 9% of their patients had arteriography).

CT Findings in Renal Trauma

CATEGORY 1.—In incomplete laceration or contusion, subcapsular and/or perirenal hemorrhage is seen consistently; subcapsular hematomas show good preservation of pararenal fat and flattening of underlying renal parenchyma, while perirenal hemorrhage infiltrates the pararenal fat (Erturk et al., 1985; Sandler and Toombs, 1981; Schaner et al., 1977). Gerota's fascia contains perirenal hemorrhage, which may have a

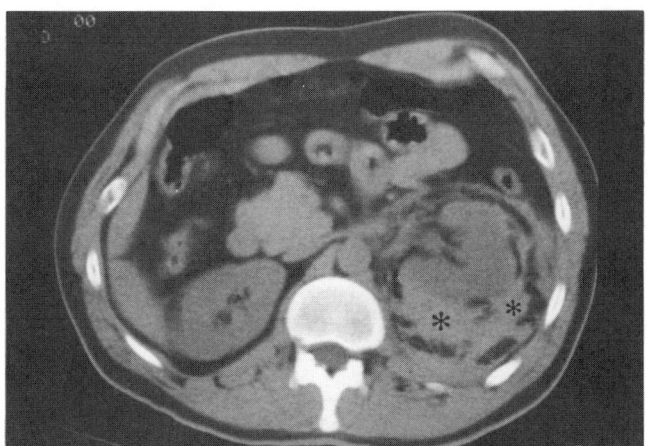

FIG 3B–57.
Incomplete left renal laceration with high-density perinephric blood *(asterisks)* mixed with low-density perirenal fat giving a "bubbly" appearance.

"bubbly" appearance due to fat interspersed with high-density fresh blood (Fig 3B–57). Intrarenal hematoma and minor lacerations show focal masses usually exhibiting lack of normal contrast enhancement. Contusions may have prolonged accumulation or staining of contrast within the interstitium of the renal parenchyma (Fig 3B–58 and 3B–59). A striated nephrogram has also been seen in these areas secondary to renal tubular stasis of contrast.

CATEGORY 2.—Deep and complete renal lacerations and fractures which usually extend into the collecting system are always accompanied by large perirenal collections of blood. The laceration can usually be seen except in cases in which the fracture plane runs parallel to the scan plane; the use of sagittal or coronal recon-

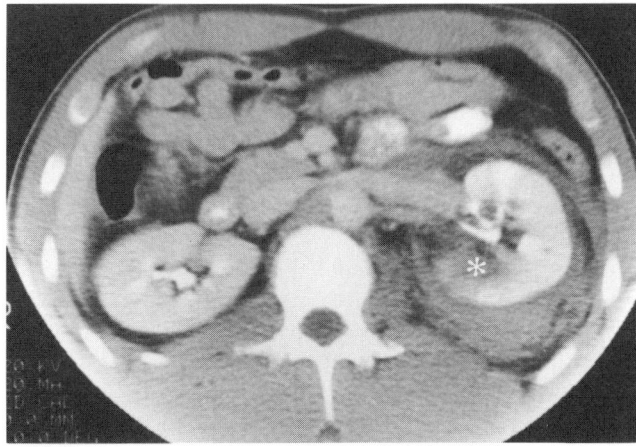

FIG 3B–58.
Left intrarenal hematoma *(asterisk)* with perinephric hemorrhage. Note lack of normal contrast enhancement of the intrarenal hematoma.

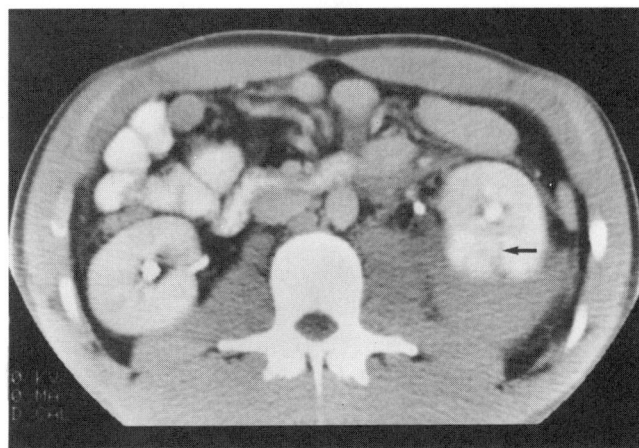

FIG 3B–59.
Minor left renal laceration *(arrow)* with perinephric hemorrhage. Increased contrast enhancement of posterior aspect of the left kidney in the region of laceration is a typical appearance due to tubular stasis.

struction is helpful in this circumstance (Fig 3B–60). Urinary extravasation is best identified by the presence of contrast-enhanced urine (Mitty, 1980). McAninch and Federle (1982) found CT to be much more sensitive for this important finding.

CATEGORY 3.—Complete shattering and disruption of the kidney is seen as multiple clefts with fragments

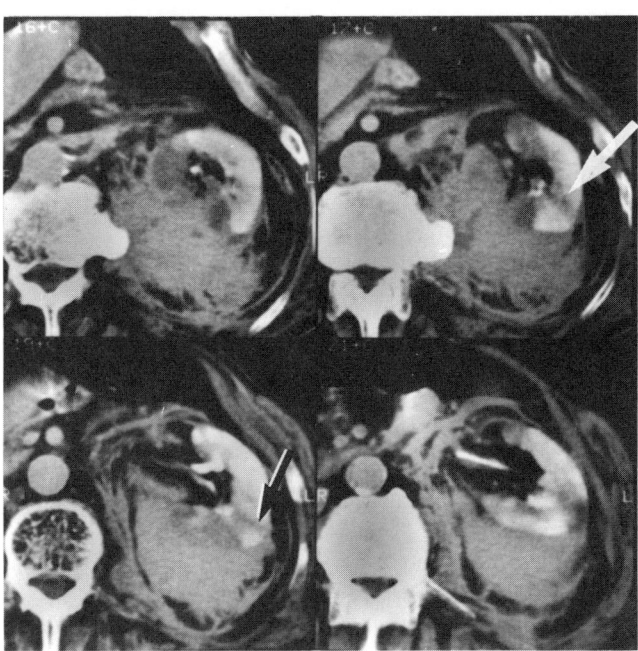

FIG 3B–60.
Category 2 renal injury. Renal lacerations *(arrows)* extend into the collecting system. Associated large perinephric hematoma. (Case courtesy of Robert Churchill, M.D., Loyola University Medical Center, Maywood, Ill.)

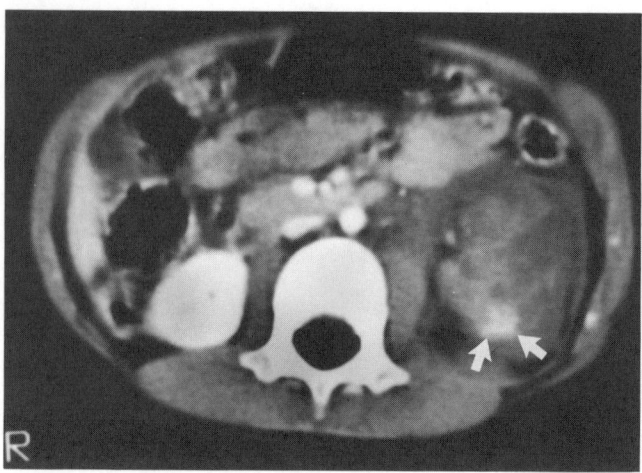

FIG 3B–61.
Shattered kidney. Large left perinephric hematoma with small fragment of enhancing renal parenchyma *(arrows)*. (Case courtesy of Robert Churchill, M.D., Loyola University Medical Center, Maywood, Ill.)

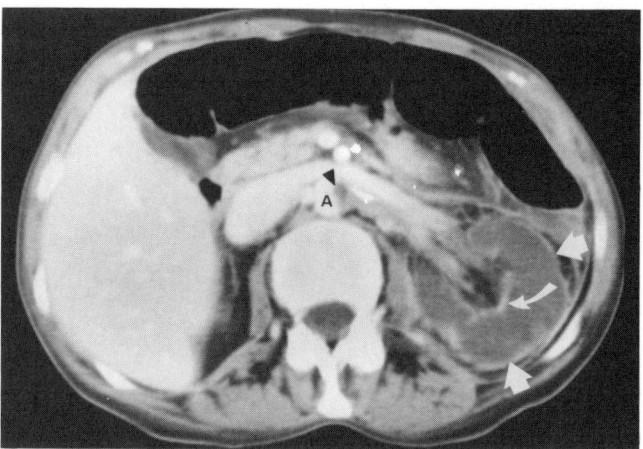

FIG 3B–62.
Global left renal infarction from aortic thrombosis demonstrating only a thin rim of cortical enhancement *(arrows)* and peripelvic enhancement *(curved arrow)* resulting from cortical and peripelvic collaterals. Note filling defect *(black arrowhead)* within contrast-enhanced aorta *(A)*.

and a large hematoma (Fig 3B–61). The renal pedicle is invariably disrupted as well.

Renal Pedicle Injury and Vascular Occlusions

The appearance of both segmental and global renal infarctions is quite characteristic on contrast-enhanced CT (Glazer et al., 1983; Wong et al., 1984; Reagan et al., 1984; Pazmino et al., 1983; Haynes et al., 1987; Haaga and Morrison, 1980; Hilton et al., 1984). Focal arterial occlusion is seen as a wedged-shaped zone of absent contrast enhancement corresponding to the areas supplied by the occluded segmental branch. Global infarction of the kidney also demonstrates absence of enhancement, with minimal perfusion of a rim of cortical tissue secondary to collaterals (Fig 3B–62). This "rim sign" is highly specific and is the CT equivalent of the sign described on excretory urography (Glazer et al., 1983). It was seen in 47% of patients by Wong et al. (1984). The sign may also be seen in some segmental infarcts. Based on these findings, it appears likely that CT can accurately identify most renal artery occlusions and pedicle injury. Caution should be exercised, since the degree of collateral formation may yield a false negative CT, as reported by Ishikawa et al. (1982). Lang et al. also believe that conventional enhanced CT may occasionally misdiagnose renal contusion and hematoma as renal infarction, as it did in six patients in a series of 130 patients with renal injury. They recommended dynamic CT as a more accurate means of differentiating infarct from hematoma. These authors pointed out that the lack of reliable CT criteria for arterial injuries is a serious shortcoming and argue for a more liberal use of arteriography (Lang et al., 1985).

CT OF RENAL TRANSPLANTS

CT is effective in defining complications associated with renal transplant surgery (Nakstad et al., 1982; Kittredge, 1980). These complications are mainly associated with fluid collections including urinoma, abscess, lymphocele, and hematoma. Urinomas appear early in the postoperative course and are uniform, low-attenuation fluid collections as are lymphoceles, from which they cannot be reliably differentiated unless IV administered contrast material leaks into the urinoma. Hematomas are of higher mixed attenuation and show consistent changes over time as clot lysis proceeds. Abscesses have an enhancing rim and may contain gas. In many of these problems positive identification of the nature of the fluid may require needle aspiration.

ANATOMICAL VARIANTS

Because of its superior display of intra-abdominal relationships, CT can clarify confusing renal congenital abnormality and anomalies of rotation, fusion, or migration. While not the test of first choice in these problems, CT can be helpful, and we recommend its use in situations such as horseshoe kidney or pelvic kidneys when relationships require clarification or tumor must be excluded. Congenital renal absence may be reliably diagnosed on CT and differentiated from other causes of nonfunction such as arterial occlusion, atrophy, or hydronephrosis.

CT OF THE PELVIS

CT plays a role in assessment of pelvic pathology and has found wide application in staging of genitourinary malignancy in both male and female patients. In addition, tumors which secondarily involve the pelvis are well demonstrated. The main limitation of CT has been its inability to detect small or microscopic metastases in normal-sized lymph nodes and minimal or microscopic tumor invasion of peripelvic fat and contiguous surfaces such as the interface between the bladder and rectum. CT is also limited by its inability to detect the normal ovary (Walsh, 1983). It is in this regard that sonography has been a major competitor in imaging of the pelvis. Sonography is able to obtain multiple planes of section and does not utilize radiation, making it ideal for women of childbearing age. Sonography also is extremely sensitive in detecting small pelvic fluid collections and can easily separate the uterus from the adnexae (Walsh et al., 1978; Fleischer et al., 1982; Stanley et al., 1978; Raskin, 1980; Sawyer et al., 1985; Lee and Balfe, 1983). Despite these limitations, CT will continue to be an extremely valuable tool in the detection of pelvic disease by virtue of its comprehensive display of osseous soft tissue and fluid structures and their relationships.

Technique

The technique for pelvic CT is similar to that utilized for abdominal CT. Contrast opacification of the bowel is of great significance, since unopacified loops of small bowel or colon may simulate masses. The oral administration of diluted barium the evening before the examination ensures that the colon is filled with contrast; oral administration of contrast material 1–2 hours before initiation of the examination ensures filling of the small bowel. IV contrast is routinely used to opacify the ureters and bladder. Other techniques may be helpful, including decubitus and prone scans for visualization of tumor fixed to the bladder or rectum.

Bladder Cancer

Staging of bladder cancer by CT is somewhat limited, since CT cannot distinguish the Jewett-Strong-Marshall stages O, A, B1, or B2. CT findings of bladder tumors include irregular wall thickening, polypoid mural masses, and masses filling the bladder lumen (Walsh, 1983; Lee and Balfe, 1983) (Fig 3B–63). Perivesical extension can be seen when that fat is invaded, smudged, or obliterated. Obliteration of the normal seminal vesicle angle may also be used, although we have not found this sign to be of value. When adenopathy is obvious and lymph node enlargement is bulky, CT can be helpful; however, the tendency of bladder carcinoma to cause microscopic metastases limits its role in this regard. In addition, we have found limited usefulness in nodal metastases, particularly to the internal iliac chain.

The overall accuracy of CT staging compared with surgical staging varies from 59% to 88% (Walsh, 1983; Koss et al., 1981; Hamlin et al., 1981; Seidelmann et

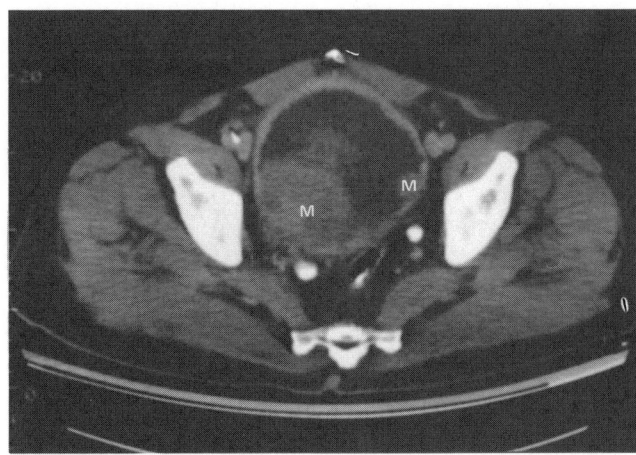

FIG 3B–63.
Transitional cell carcinoma of the bladder. Bulky masses *(M)* within bladder. Significant thickening of the bladder wall, in this patient secondary to extensive mural infiltration.

al., 1978; Yu et al., 1979; Jeffrey et al., 1981). The accuracy of lymph node assessment varies from 79% to 92%, with a true positive rate of 60%–75%, a false negative rate of 25%–40%, a true negative rate of 86%–100%, and a false positive rate of 0%–14% (Koss et al., 1981; Morgan et al., 1981). Overstaging can also be a problem in stage B tumors, when the loss of fat planes between the seminal vesicles and posterior bladder wall is obliterated, but no true invasion is seen on CT. CT's role is thus primarily limited to advanced tumors and to demonstrate gross lymph node metastases.

Prostate Cancer

The utility of CT in prostate cancer is similar to the problems associated with staging of bladder tumors. CT

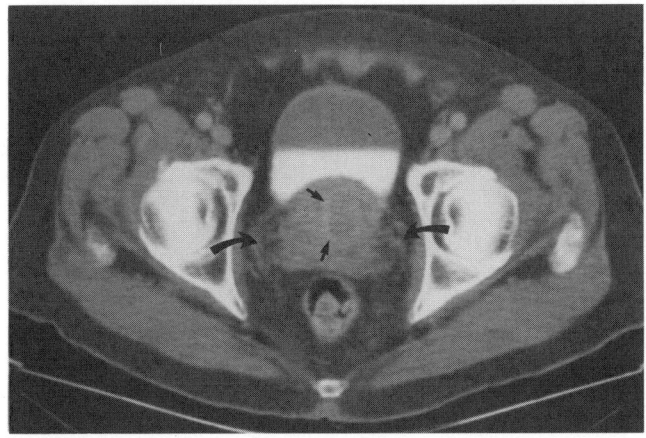

FIG 3B–64.
Prostate carcinoma. The prostate is enlarged, indenting bladder base. Inhomogeneity within the gland *(arrows)* and soft tissue stranding *(curved arrows)* extending laterally. The soft tissue infiltration probably represents lymphatic congestion, although in such cases tumor extension cannot accurately be excluded.

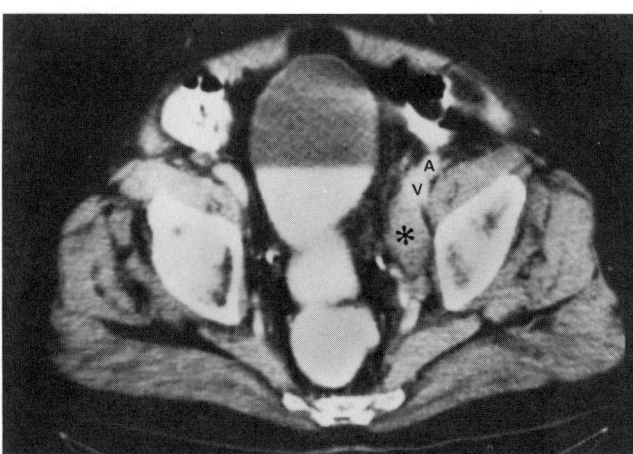

FIG 3B–65.
Iliac adenopathy in prostatic carcinoma. Left iliac adenopathy *(asterisk)* is demonstrated posterior to the iliac artery *(A)* and vein *(V)*. Note normal appearance on the right side.

cannot distinguish stage A or B and in general only demonstrates enlargement of the gland (Morgan et al., 1981) (Fig 3B–64). Differentiation of specific sites of tumor from normal gland is often difficult or impossible. When there is advanced extraprostatic disease, invasion of the bladder base or seminal vesicles may be seen. Lymphatic metastases are as difficult to visualize as they are in bladder cancer. The major limitations of the technique are again that CT cannot distinguish normal prostate tissue from prostatitis, prostatic adenoma, or carcinoma (Sukov et al., 1977; Van Engelshoven and Kreel, 1979; Price and Davidson, 1979). CT also has a tendency to overestimate the volume of the gland because it cannot completely separate the prostatic capsule from the gland or separate the prostate from the pelvic diaphragm inferiorly.

Staging errors are high, varying between 20% and 50%. They usually occur on the basis of overstaging, which in one study was as high as 29% (Morgan et al., 1981). The detection of lymph node metastases is also extremely variable, although bulky disease is well seen (Fig 3B–65).

Testis Cancer

In this disease CT plays its most important role in detecting metastases to retroperitoneal lymph nodes and lung. The overall accuracy of CT is between 76% and 87%. False negative results are seen in 3%–12% and false positive in 5%–9% (Walsh, 1983; Burney and Klatte, 1979; Williams et al., 1980). CT is more accurate for stages 2B and 3 than for stages 1 and 2A.

The CT findings in metastatic disease usually consist of lymph node enlargement, which is sometimes very bulky. CT can also be used for biopsy of suspicious nodal groups.

CT is also useful in observing regression of disease and in monitoring response to treatment including recurrences. Cystic changes may occur in lymph node masses as a result of radiation or chemotherapy and are considered a normal response to successful treatment. The major roles of CT, therefore, in testicular tumor are to demonstrate tumor volume and monitor treatment response, plan radiation therapy portals, define operability and surgical approach, localize sites of recurrence when tumor markers are positive, and to identify patients with bulky disease who should receive chemotherapy.

CT has been found to be very useful in the localization of undescended impalpable testis (Lee et al., 1980). It can easily detect an oval soft tissue mass in the inguinal area or lower pelvis within the inguinal canal or superficial inguinal pouch. An intra-abdominal location in the region of the external iliac vessels also may be visualized.

Gynecologic Masses

UTERINE TUMORS.—Endometrial carcinoma is usually demonstrated on CT as a mass of decreased density within the uterus or as a contrast enhancing lesion in the myometrium (Hamlin et al., 1981) (Fig 3B–66). Again, CT has significant limitations in detection of lymph node metastases. The accuracy of CT staging of endometrial malignancy is about 85% (Walsh and Goplerud, 1982). Other tumors that can be seen are lobulated uterine myomas, which may calcify.

CERVICAL CARCINOMA.—The primary use of CT in cervical cancer is in detection of pelvic side wall extension (stage 3) and the presence of nodal metastases or metastases elsewhere in the abdomen (Fig 3B–67). Differentiation between stage 1 and stage 2B can be diffi-

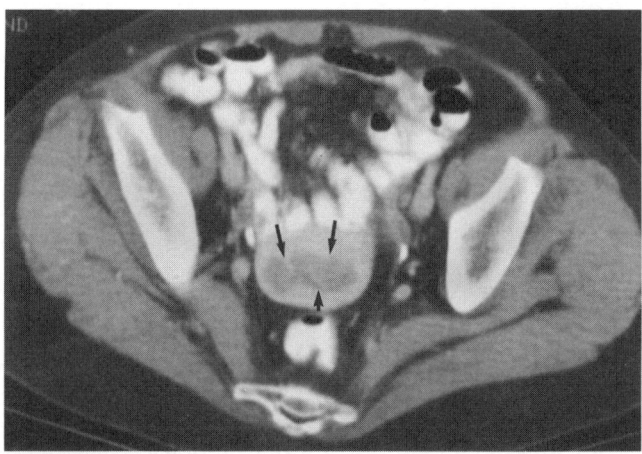

FIG 3B–66.
Endometrial carcinoma. Intrauterine mass *(arrows)* of decreased attenuation. The appearance is typical of endometrial carcinoma.

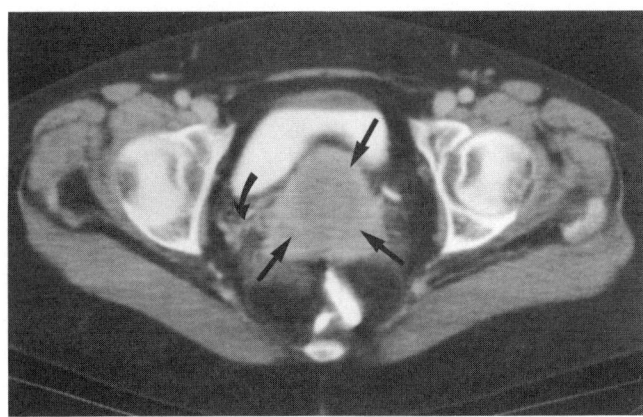

FIG 3B–67.
Cervical carcinoma. Large low-density mass *(arrows)* in the cervix. Parametrial extension *(curved arrow)* is seen on the right.

cult, since this diagnosis is often based on size alone. In stages 3B, 4A, and 4B, however, CT has a 92% accuracy (Hamlin et al., 1981). CT may also demonstrate incidental findings such as hydronephrosis or impingement on adjacent structures.

CONCLUSION

CT plays a role in pelvic disease, but that role has not been nearly as great nor as accurate as in the definition of renal pathology. These limitations are secondary to overstaging and understaging errors and an inability to detect small nodal metastases with accuracy. CT will continue to be a useful tool, although findings on CT necessarily require corroboration with surgical and clinical findings.

REFERENCES

1. Abrams HL, Spiro R, Goldstein N: Metastases in carcinoma: Analysis of 1,000 autopsied cases. *Cancer* 1950; 3:74.
2. Ackerman, LV, Rosal J: *Surgical Pathology*, ed 5. St Louis, CV Mosby Publishing Co, 1974, p 639.
3. Ambos MA, Bosniak MA, Valensi QJ, et al: Angiographic patterns in renal oncocytomas. *Radiology* 1978; 129:615–622.
4. Anderson JH, Wallace, S, Gianturco C, et al: "Mini" Gianturco stainless steel coils for transcatheter vascular occlusion. *Radiology* 1979; 132:301.
5. Androulakakis PA: Pyonephrosis: A critical review of 131 cases. *Br J Urol* 1982; 54:89–92.
6. Baert AL, Wilms G, Usewils R, et al: Dynamic CT of the urogenital tract. *Urol Radiol* 1982; 4:69–83.
7. Balfe DM, McClennan BL, Stanley RJ, et al: Evaluation of renal masses considered indeterminate on computed tomography. *Radiology* 1982; 142:421–428.
8. Bank WO, Kerber CW: Gelfoam embolization: A simplified technique. *AJR* 1979; 132:299.
9. Barbaric ZL, Davis RS, Frank IN, et al: Percutaneous nephropyelostomy in the management of acute pyohydronephrosis. *Radiology* 1976; 118:567–573.
10. Baron RL, McClennan BL, Lee JKT, et al: Computed tomography of transitional-cell carcinoma of the renal pelvis and ureter. *Radiology* 1982; 144:125–130.
11. Barth KH, Lightman NI, Ridolfi RL, et al: Acute pyelonephritis simulating poorly vascularized renal neoplasm: Non-specificity of angiographic criteria. *J Urol* 1976; 116:650.
12. Beachley MC, Ranniger K, Roth FJ: Xanthogranulomatous pyelonephritis. *AJR* 1974; 121:500.
13. Beinart C, Sos TA, Soddekni S: Arterial spasm during renal angioplasty. *Radiology* 1983; 149:97–99.
14. Bergreen W, Woodside J: Distal embolization complicating therapeutic renal infarction. *N Engl J Med* 1976; 294:1406.
15. Bernardino ME, deSantos LA, Johnson DE, et al: Computed tomography in the evaluation of post-nephrectomy patients. *Radiology* 1979; 130:183–187.
16. Boijsen E: Anatomic and physiologic considerations, in Abrams H (ed): *Abrams Angiography*. Boston, Little Brown and Co, 1983, pp 1107–1122.
17. Boijjsen E, Folin J: Angiography in the diagnosis of renal carcinoma. *Radiology* 1961; 1:173.
18. Bosniak MA: Angiomyolipoma (hamartoma) of the kidney: A preoperative diagnosis is possible in virtually every case. *Urol Radiol* 1981; 3:135–142.
19. Bosniak MA, Megibow AJ, Ambos MA, et al: Computed tomography of ureteral obstruction. *AJR* 1982; 138:1107–1113.
20. Bracksen B, Jonsson K: How accurate is angiographic staging of renal carcinoma? *Urology* 1979; 14:96–99.
21. Brantley RE, Mashni JW, Bethards BE, et al: Computerized tomographic demonstration of inferior vena caval tumor thrombus from renal angiomyolipoma. *J Urol* 1985; 133:836–837.
22. Brasch RC, Randel SB, Gould RG: Follow-up of Wilms tumor: Comparison of CT with other imaging procedures. *AJR* 1982; 137:1005–1009.
23. Breatnach ES, Stanley RJ, Lloyd K: Focal obstructive nephrogram: An unusual CT appearance of a transitional cell carcinoma. *J Comput Assist Tomogr* 1984; 8:1019–1022.
24. Bret PM, Bretagnolle M, Gillard D, et al: Small, asymptomatic angiomyolipomas of the kidney. *Radiology* 1985; 154:7–10.
25. Buist TAS: Parasitic arterial blood supply to intracapsular renal cell carcinoma. *AJR* 1974; 120:653–659.
26. Burney, BT, Klatte EC: Ultrasound and computed tomography of the abdomen in the staging and management of testicular carcinoma. *Radiology* 1979; 132:415–419.
27. Cass AS: Renal trauma in the multiple injured patient. *J Urol* 1975; 114:495–497.
28. Cass AS: Immediate radiological evaluation and early surgical management of genitourinary injuries from external trauma. *J Urol* 1979; 122:772.
29. Castaneda-Zuniga W, Zollikofer C, Barreto A, et al: A

new device for the safe delivery of stainless steel coils. *Radiology* 1980; 136:230.

30. Chilcote WA, Borkowski GP: Computed tomography in renal lymphoma. *J Comput Assist Tomogr* 1983; 7:439–443.

31. Christensen EE, Curry TS III, Dowdey JE: Computed tomography, in Christensen EE, Curry TS III, Dowdey JE (eds): *An Introduction to the Physics of Diagnostic Radiology.* Philadelphia, Lea & Febiger, 1978, p 329.

32. Chuang VP, Ernst CB: Angiography for renal hypertension, in Neiman HL, Yao JST (eds): *Angiography of Vascular Disease.* New York, Churchill Livingstone, 1985.

33. Chuang VP, Reuter SR: Angiographic features of Alport's syndrome: Hereditary nephritis. *AJR* 1974; 121:539–543.

34. Chuang VP, Wallace S, Swanson D, et al: Arterial occlusion in the management of pain from metastatic renal carcinoma. *Radiology* 1979; 133:611.

35. Churchill RJ, Reynes CJ, Love L, et al: CT imaging of the abdomen: Methodology and normal anatomy. *Radiol Clin North Am* 1979; 17:13–24.

36. Clark RA, Alexander ES: Digital subtraction angiography of the renal arteries: Prospective comparison with conventional arteriography. *Invest Radiol* 1983; 18:6–10.

37. Cochran ST, Wong WS, Roe DJ: Predicting angiography-induced acute renal function impairment: Clinical risk model. *AJR* 1983; 141:1027–1030.

38. Cohan RH, Dunnick R, Degesys GE, et al: Computed tomography of renal oncocytoma. *J Comput Assist Tomogr* 1984; 8:284–287.

39. Coleman BG, Arger PH, Mintz MC, et al: Hyperdense renal masses: A computed tomographic dilemma. *AJR* 1984; 143:291–294.

40. Cormack AM: Reconstruction of densities from their projections with applications in radiological physics. *Phys Med Biol* 1973; 18:195–207.

41. Dalla-Palma L, Rossetti SR, Pozzi-Mucelli RS: Computed tomography in the diagnosis of retroperitoneal fibrosis. *Urol Radiol* 1981; 3:77–83.

42. Daniels WW, Hartmen GW, Witten DM, et al: Calcified renal masses: A review of ten years' experience at the Mayo Clinic. *Radiology* 1972; 103:503–508.

43. Das G, Chisholm GD, Sherwood T: Can angiography stage renal carcinoma? *Br J Urol* 1977; 49:611–614.

44. Dunnick NR, Hartman DS, Ford KK, et al: The radiology of juxtaglomerular tumors. *Radiology* 1983; 147:321–326.

45. Dunnick NR, Korobkin M: Computed tomography of the kidney. *Radiol Clin North Am* 1984; 22(2):297–313.

46. Dunnick NR, Korobkin M, Clark WM: CT demonstration of hyperdense renal carcinoma. *J Comput Assist Tomogr* 1984; 8:1023–1024.

47. Dunnick NR, Korobkin M, Silverman PM, et al: Computed tomography of high density renal cysts. *J Comput Assist Tomogr* 1984; 8:458–461.

48. Dunnill MS: *Pathological Basis of Renal Disease.* London WB Saunders Co, 1976.

49. Eason AA, Cattolica EV, McGrath TW: Massive renal angiomyolipoma: preoperative infarction by balloon catheter. *J Urol* 1979; 121:360.

50. Ekelund L, Kaude J, Lindhold T: Angiography in glomerular disease of the kidney: A correlation with clinical findings and the stage of the disease. *AJR* 1973; 119:739–747.

51. Eliska O: The perforating arteries and their role in the collateral circulation of the kidneys. *Acta Anat* 1968; 70:184.

52. Elkin M: Computed tomography of the urinary tract, in Elkin M (ed): *Radiology of the Urinary System.* Boston, Little, Brown & Co, 1980, pp 1119–1139.

53. Elkin M, Meng CH, de Paredes RG: Roentgenologic evaluation of renal trauma with emphasis on renal angiography. *AJR* 1966; 98:1.

54. Emmett JL, Levine SR, Woolner LB: Co-existence of renal cyst and tumor: Incidence in 1007 cases. *Br J Urol* 1963; 35:403–410.

55. Engelstad BL, McClennan BL, Levitt RG, et al: The role of pre-contrast images in computed tomography of the kidney. *Radiology* 1980; 136:153–155.

56. Erturk E, Sheinfeld J, DiMarco PL, et al: Renal trauma: Evaluation by computerized tomography. *J Urol* 1985; 133:946–949.

57. Federle MP, Crass RA, Jeffrey RB, et al: Computed tomography in blunt abdominal trauma. *Arch Surg* 1982; 117:645–650.

58. Federle MP, Kaiser JA, McAninch JW, et al: The role of computed tomography in renal trauma. *Radiology* 1981; 141:455–460.

59. Felder E, Pitre D, Tirone P: Radiopaque contrast media: Preclinical studies with a new nonionic contrast agent—Iopamidol II. *Farmaco Ed Sci* 1977; 32:835–844.

60. Ferris RA, Kirschner LP, Mero JH, et al: Computed tomography in the evaluation of inferior vena caval obstruction. *Radiology* 1979; 130:710.

61. Fererstein IM, Zeman RK, Jaffe MH, et al: Perirenal cobwebs: The expanding CT differential diagnosis. *J Comput Assist Tomogr* 1984; 8:1128–1130.

62. Fleischer AC, Walsh JW, Jones HW III, et al: Sonographic evaluation of pelvic masses: Method of examination and role of sonography relative to other imaging modalities. *Radiol Clin North Am* 1982; 20:397–412.

63. Foster RS, Shuford WH, Weens HS: Selective renal arteriography in medical diseases of the kidney. *AJR* 1965; 95:349.

64. Freeny PC, Bush WH Jr, Kidd R: Transcatheter occlusive therapy of genitourinary abnormalities using isobutyl 2-cyanoacrylate (Bucrylate). *AJR* 1979; 133:647.

65. Friedenberg MJ, Eisen S, Kissane J: Renal angiography in pyelonephritis, glomerulonephritis, and arteriolar nephrosclerosis. *AJR* 1965; 95:349.

66. Fuld IL, Matalon TA, Vogelzang RL, et al: Dynamic CT in the evaluation of physiologic status of renal transplants. *AJR* 1984; 142:1157–1160.

67. Gang DL, Dole KB, Adelman LS: Spinal cord infarction following therapeutic renal artery embolization. *JAMA* 1977; 237:3841.

68. Gerlock AJ Jr, Mirfakhraee M: *Essentials of Diagnostic and Interventional Angiographic Techniques.* Philadelphia, WB Saunders Co, 1985.

69. Gingell JC, Roylance, J, Davies ER, et al: Xanthogran-

ulomatous pyelonephritis. *Br J Radiol* 1973; 46:99.

70. Glazer GM, Callen PW, Parker JJ: CT diagnosis of tumor thrombus in the inferior vena cava: Avoiding the false-positive diagnosis. *AJR* 137:1265–1267.

71. Glazer GM, Francis IR, Brady TM, et al: Computed tomography of renal infarction: Clinical and experimental observations. *AJR* 1983; 140:721–727.

72. Glazer GM, Francis IR, Gross BH, et al: Computed tomography of renal vein thrombosis. *J Comput Assist Tomogr* 1984; 8:288–293.

73. Gold LP, McClennan BL, Rottenberg RL: CT appearance of acute inflammatory disease of the renal interstitium. *AJR* 1983; 141:343–349.

74. Goldman SM, Hartman DS, Fishman EK, et al: CT of xanthogranulomatous pyelonephritis: Radiologic-pathologic correlation. *AJR* 1984; 141:963–969.

75. Goldstein HM, Neiman HL, Mena E, et al: Angiographic findings in benign liver cell tumors. *Radiology* 1974; 110:339–343.

76. Goldstone J, Osborne RW Jr, Hillman BJ, et al: Digital video subtraction angiography: Screening technique for renovascular hypertension. *Surgery* 1981; 90:932–939.

77. Gomes AS, Pais SO, Barbaric ZL: Digital subtraction angiography in the evaluation of hypertension. *AJR* 1983; 140:779–782.

78. Gomes AS, Rysavy JA, Spadaccini CA, et al: The use of the bristle brush for transcatheter embolization. *Radiology* 1978; 129:345.

79. Gordon IJ, Skoblar RS, Chicatelli PD, et al: A comparison of iohexol and Conray-60 in peripheral angiography. *AJR* 1984; 142:563–565.

80. Grantham JJ: Polycystic renal disease, in Early LE, Gottschalk CW (eds): *Strauss and Welt's Diseases of the Kidney*, ed 3. Boston, Little, Brown & Co, 1979, pp 1123–1146.

81. Grim CE, Luft FC, Yune HY, et al: Percutaneous transluminal dilatation in the treatment of renal vascular hypertension. *Ann Intern Med* 1981; 95:439–442.

82. Haaga JR, Morrison, SC: CT appearance of renal infarct. *J Comput Assist Tomogr* 1980; 4:246–247.

83. Hamlin DJ, Burgener FA, Beechman JB: CT of intramural endometrial carcinoma: Contrast enhancement is essential. *AJR* 1981a; 137:551–554.

84. Hamlin DJ, DiSaint Agnese PA, Kerp HM, et al: Updating computed tomography of bladder carcinoma in assessing response to immunotherapy and attenuated irradiation. *Urology* 1981b; 17:622–627.

85. Hansen GC, Hoffman RB, Sample WF, et al: Computed tomography diagnosis of renal angiomyolipoma. *Radiology* 1978; 128:789–791.

86. Hara M, Yoshida K, Tomita M, et al: A case of bilateral renal oncocytoma. *J Urol* 1982; 128:576–578.

87. Hartman DS, Davis CJ Jr, Goldman SM, et al: Renal lymphoma: Radiologic-pathologic correlation of 21 cases. *Radiology* 1982; 144:759–766.

88. Hartman DS, Goldman SM, Friedman AC, et al: Angiomyolipoma: Ultrasonic-pathologic correlation. *Radiology* 1981; 139:451–458.

89. Hata Y, Tada S, Kato Y, et al: Staging of renal cell carcinoma by computed tomography. *J Comput Assist Tomogr* 1983; 7:828–832.

90. Hattery RR, Sheedy PF, Stephens DH, et al: Computed tomography of the adrenal gland. *Semin Roentgenol* 1981; 16:290–300.

91. Haynes JW, Walsh JW, Brewer WH, et al: Traumatic renal artery occlusion: CT diagnosis with angiographic correlation. *J Comput Assist Tomogr* 1984; 8:731–733.

92. Hekali P, Kivisaari L, Standertskjold-Nordenstam CG, et al: Renal complications of polyarteritis nodosa: CT findings. *J Comput Assist Tomogr* 1985; 9:333–338.

93. Hellekant C, Nyman U: Routine celiac angiography in patients with renal cell carcinoma. *J Urol* 1979; 122:17.

94. Heptinstal RH: *Pathology of the Kidney*. Boston, Little, Brown & Co, 1974.

95. Heswel SJ, Smith EH: Renal trauma: A comprehensive review and radiologic assessment. *CRC Crit Rev Diagn Imaging* 1974; 5:251–293.

96. Hidalgo H, Dunnick NR, Rosenberg ER, et al: Parapelvic cysts: Appearance on CT and sonography. *AJR* 1982; 138:667–671.

97. Hillman BJ, Drach GW, Tracey P, et al: Computed tomographic analysis of renal calculi. *AJR* 1984; 142:549–552.

98. Hilton S, Bosniak MA, Raghavendra BN, et al: CT findings in acute renal infarction. *Urol Radiol* 1984; 6:158–163.

99. Himmelfarb EH, Rabinowitz JG, Kinkhabwala MN, et al: The roentgen features of renal carbuncle. *J Urol* 1972; 108:846.

100. Hoddick W, Jeffrey RB, Goldberg HI, et al: CT and sonography of severe renal and perirenal infections. *AJR* 1983; 140:517–520.

101. Holley KE, Hunt JC, Brown AK Jr, et al: Renal artery stenosis: A clinical-pathologic study in normotensive and hypertensive patients. *Am J Med* 1964; 37:14.

102. Hodson CJ: The logic of the blood supply to the kidney, in Margulis AR, Gooding GA (eds): *Diagnostic Radiology*. San Francisco, University of California Press, 1978.

103. Holt RG, Neiman HL, Korsower JM, et al: Angiographic features of benign renal adenoma. *Urology* 1975; 6:764.

104. Hounsfield GN: Computerized transverse axial scanning (tomography). *Br J Radiol* 1973; 46:1016.

105. Hounsfield GN: Computed medical imaging. *Med Phys* 1980; 7:283–290.

106. Hunt JC, Harrison EG Jr, Kincaid OW, et al: Idiopathic fibrous and fibromuscular stenoses of the renal arteries associated with hypertension. *Mayo Clin Proc* 1962; 36:707–712.

107. Ishikawa I, Matsuura H, Onouchi Z, et al: CT appearance of the kidney in traumatic renal artery occlusion. *J Comput Assist Tomogr* 1982; 6:1021–1024.

108. Jafri SZH, Bree RL, Amendola MA, et al: CT of renal and perirenal non-Hodgkin lymphoma. *AJR* 1982; 138:1101–1105.

109. Jaschke W, van Kaick G, Peter S, et al: Accuracy of computed tomography in staging of kidney tumors. *Acta Radiol Diagn* 1982; 23:593–598.

110. Jeffrey RB, Palubinskas AJ, Federle MP: CT evaluation

of invasive lesions of the bladder. *J Comput Assist Tomogr* 1981; 5:22–26.

111. Johnsrude IS, Jackson CD: *A Practical Approach to Angiography*. Boston, Little, Brown & Co, 1979.

112. Karstaedt N, Sagel SS, Stanley RJ, et al: Computed tomography of the adrenal gland. *Radiology* 1978; 129:723–730.

113. Kaufman SL: Simplified method of transcatheter embolization with polyvinyl alcohol foam (Ivalon). *AJR* 1979; 132:853.

114. Kazmin MH, Brosman SA, Cockett ATK: Diagnosis and early management of renal trauma: A study of 120 patients. *J Urol* 1969; 101:783.

115. Kikkawa K, Lasser EC: Ring-like or "rim-like" calcification of renal cell carcinoma. *AJR* 1969; 107:737–742.

116. Kincaid-Smith P: Pyelonephritis, chronic interstitial nephritis and obstructive uropathy, in Hamburger J, Crosnier J, Grunfeld JP (eds): *Nephrology*. New York, John Wiley & Sons, 1979, pp 553–582.

117. Kittredge RD: Computed tomography in the renal transplant patient: Problems in interpretation. *J Comput Tomogr* 1980; 4:118–120.

118. Kormano M, Dean PB: Extravascular contrast material: The major component of contrast enhancement. *Radiology* 1976; 121:379–382.

119. Koss JC, Arger PH, Coleman BG, et al: CT staging of bladder carcinoma. *AJR* 1981; 137:359–362.

120. Lang EK: A survey of the complications of percutaneous retrograde arteriography. *Radiology* 1963; 81:257–263.

121. Lang EK: Arteriographic assessment and staging of renal cell carcinoma. *Radiology* 1971; 101:17–27.

122. Lang EK: Arteriography in the diagnosis and staging of hypernephromas. *Cancer* 1973; 32:1043–1052.

123. Lang EK: Arteriography in the assessment of renal trauma: The impact of arteriographic diagnosis on preservation of renal function and parenchyma. *J Trauma* 1975; 15:553.

124. Lang EK: Angio-computed tomography and dynamic computed tomography in staging of renal cell carcinoma. *Radiology* 1984; 151:149–155.

125. Lang EK, Sullivan J, Frentz G: Renal trauma: Radiologic studies—comparison of urography, computed tomography, angiography, and radionuclide studies. *Radiology* 1985; 154:1–6.

126. Lawson TL, McClennan, BL, Shirkhoda A.: Adult polycystic kidney disease: Ultrasonographic and computed tomographic appearance. *J Clin Ultrasound* 1978; 6:279–302.

127. Lee JKT, Balfe DM: Pelvis, in Sagel SS, Stanley RJ (eds): *Computed Body Tomography*. New York, Raven Press, 1983, pp 393–414.

128. Lee JKT, McClennan BL, Melson BL, et al: Acute focal bacterial nephritis: Emphasis on gray scale sonography and computed tomography. *AJR* 1980; 135:87–92.

129. Lee JKT, McClennan BL, Stanley RJ, et al: Utility of computed tomography in the localization of the undescended testis. *Radiology* 1980; 135:121–125.

130. Levin DC, Beckmann CF, Hillman B: Experimental determination of flow patterns of gelfoam emboli: Safety implications. *AJR* 1980; 134:525.

131. Levine E, Grantham JJ: The role of computed tomography in the evaluation of adult polycystic kidney disease. *Am J Kidney Dis* 1981; 1:99–105.

132. Levine E, Grantham JJ: High-density renal cysts in autosomal dominant polycystic kidney disease demonstrated by CT. *Radiology* 1985; 154:477–482.

133. Levine E, Huntrakoon M: Computed tomography of renal oncocytoma. *AJR* 1983; 141:741–746.

134. Levine E, Lee KR, Weigel J: Preoperative determination of abdominal extent of renal cell carcinoma by computed tomography. *Radiology* 1979a; 132:395–398.

135. Levine E, Lee RK, Weigel JW, et al: Computed tomography in the diagnosis of renal carcinoma complicating Hippel-Lindau syndrome. *Radiology* 1979b; 130:703–706.

136. Levine E, Maklad NF, Rosenthal SJ, et al: Comparison of computed tomography and ultrasound in abdominal staging of renal cancer. *Urology* 1980; 16:317–322.

137. Levine E, Weigel JW, Collins DL: Diagnosis and management of asymptomatic renal cell carcinomas in von Hippel-Lindau syndrome. *Urology* 1983; 2:146–150.

138. LiPuma JP, Haaga JR: The kidney, in Haaga JR, Alfidi RJ (eds): *Computed Tomography of the Whole Body*. St Louis, CV Mosby Co, 1983, pp 706–752.

139. Love L, Meyers MA, Churchill RJ: CT of extraperitoneal spaces. *AJR* 1981; 136:781–789.

140. Lund G, Rysany J, Salomonwitz E, et al: Nephrotoxicity of contrast media assessed by occlusion arteriography. *Radiology* 1984; 152:615–618.

141. Magilner AD, Ostrum BJ: Computed tomography in the diagnosis of renal masses. *Radiology* 1978; 126:715–718.

142. Mahoney SA, Persky L: Intravenous drip nephrotomography as an adjunct in the evaluation of renal injury. *J Urol* 1968; 99:153.

143. Malek RS, Elder JS: Xanthogranulomatous pyelonephritis: A critical analysis of twenty-six cases of the literature. *J Urol* 1978; 119:589.

144. Marks WM, Korobkin M, Callen PW, et al: CT diagnosis of tumor thrombosis of the renal vein and inferior vena cava. *AJR* 1978; 131:843–846.

145. Martinez-Madoado M, Ramirez de Arellano GA: Renal involvement in malignant lymphomas: A survey of 49 cases. *J Urol* 1966; 95:485–488.

146. Mauro MA, Wadsworth DE, Stanley RJ, et al: Renal cell carcinoma: Angiography in the CT era. *AJR* 1982; 139:1135–1140.

147. Maxwell DD, Mispireta LA: Transfemoral renal artery embolectomy. *Radiology* 1982; 143:653–654.

148. Mazer MJ, Baltaxe HA, Wolf GL: Therapeutic embolization of the renal artery with Gianturco cells: Limitations and technical pitfalls. *Radiology* 1981; 138:37.

149. McAninch JW, Federle MP: Evaluation of renal injuries with computerized tomography. *J Urol* 1982; 128:456–460.

150. McClennan BL, Stanley RJ, Melson RG, et al: CT of the renal cyst: Is cyst aspiration necessary? *AJR* 1979; 133:671–675.

151. McCullough EC, Baker HL, Houser OW, et al: An evaluation of the quantitative and radiation features of a scanning x-ray transverse axial tomography: The EMI scanner. 1974; *Radiology* 111:709.

152. Meaney TF: Errors in angiographic diagnosis of renal masses. *Radiology* 1969; 93:361–366.

153. Megibow AJ, Mitnick JS, Bosniak MA: The contribution of computed tomography in the evaluation of the obstructed ureter. *Urol Radiol* 1982; 4:95–104.

154. Mena E, Bookstein JJ, Gikas PW: Angiographic diagnosis of renal parenchymal disease. *Radiology* 1973a; 108:523–532.

155. Mena E, Bookstein JJ, Holt JF, et al: Neurofibromatosis and renovascular hypertension in children. *AJR* 1973b; 118:39–45.

156. Mena E, Bookstein JJ, McDonald FD, et al: Angiographic findings in renal medullary cystic disease. *Radiology* 1974; 110:277–281.

157. Mena E, Bull FE, Bookstein JJ, et al: Angiography of the nephrogenic hepatic dysfunction syndrome. *Radiology* 1974; 111:65–68.

158. Mendez G Jr, Isikoff MB, Morillo G: The role of computed tomography in the diagnosis of renal and perirenal abscesses. *J Urol* 1979; 122:582–586.

159. Mendez R: Renal trauma. *J Urol* 1977; 118:698.

160. Meyers MA: *Dynamic Radiology of the Abdomen.* New York, Springer Verlag, 1976, pp 195–235.

161. Mitty HA: CT for diagnosis and management of urinary extravasation. *AJR* 1980; 134:497–501.

162. Morag B, Rubinstein ZJ, Hertz M, et al: Computed tomography in the diagnosis of renal parapelvic cysts. *J Comput Assist Tomogr* 1983; 7:833–836.

163. Morehouse HT, Weiner SN, Hoffman JC: Imaging in inflammatory disease of the kidney. *AJR* 1984; 143:135–141.

164. Morgan CL, Calkins RF, Cavalcanti EJ: Computed tomography in the evaluation, staging, and therapy of carcinoma of the bladder and prostate. *Radiology* 1981; 140:751–761.

165. Morrow JW, Mendez R: Renal trauma. *J Urol* 1979; 104:649.

166. Nakstad P, Kolmannskog F, Kolbenstvedt A, et al: Computed tomography in surgical complications following renal transplantation. *J Comput Assist Tomogr* 1982; 6:286–289.

167. Neiman HL, Yao JST: *Angiography of Vascular Disease.* New York, Churchill Livingstone, 1985.

168. New PFT, Scott WR: *Computed Tomography of the Brain.* Baltimore, Williams & Wilkins Co, 1975.

169. Newhouse JF, Prien EH, Amis ES Jr, et al: Computed tomographic analysis of urinary calculi. *AJR* 1983; 142:545–548.

170. Niendorf DC, Philipps E, Palmer JM: Angiographic pseudometastasis to the liver in renal cell carcinoma. *Urology* 1974; 4:764.

171. Nishitani H, Onitsuka H, Kawahira K, et al: Computed tomography of renal metastases. *J Comput Assist Tomogr* 1984; 8:727–730.

172. Parienty RA, Ducellier R, Pradel J, et al: Diagnostic value of CT numbers in pelvocalyceal filling defects. *Radiology* 1982; 145:743–747.

173. Parienty RA, Pradel J, Imbert MC, et al: Computed tomography of multilocular cystic nephroma. *Radiology* 1981; 140:135–139.

174. Parienty RA, Pradel J, Picard JD, et al: Visibility and thickening of the renal fascia on computed tomograms. *Radiology* 1981; 139:119–124.

175. Pazmino P, Pyatt R, Williams E, et al: Computed tomography in renal ischemia. *J Comput Assist Tomogr* 1983; 7:102–105.

176. Phillips TL, Chin FG, Palubinskas AJ: Calcification in renal masses: An eleven year survey. *Radiology* 1963; 80:786–794.

177. Pollack HM, Arger PH, Banner MP, et al: Computed tomography of renal pelvic filling defects. *Radiology* 1981; 138:645–651.

178. Pollack HM, Banner MP, Arger PE, et al: The accuracy of gray-scale renal ultrasonography in differentiating cystic neoplasms from benign cysts. *Radiology* 1982; 143:741–745.

179. Price JM, Davidson AJ: Computed tomography in the evaluation of the suspected carcinomatous prostate. *Urol Radiol* 1979; 1:39–42.

180. Quinn MJ, Hartman DS, Freidman AC, et al: Renal oncocytoma: New observations. *Radiology* 1984; 153:49–53.

181. Rabinowitz J, Kinkhabwala M, Himmelfarb E, et al: Renal pelvic carcinoma: An angiographic re-evaluation. *Radiology* 1972; 112:551.

182. Rankin RN: Gas formation after renal tumor embolization without abscess: A benign occurrence. *Radiology* 1979; 103:317.

183. Raskin MR: Combination of CT and ultrasound in the retroperitoneum and pelvis examination. *CRC Crit Rev Diagn Imaging* 1980; 13:173–228.

184. Rauschkolb EN, Sandler CM, Patel S, et al: Computed tomography of renal inflammatory disease. *J Comput Assist Tomogr* 1982; 6:502–506.

185. Raz S: A simple method of pyonephrosis drainage. *Lancet* 1971; 2:529–530.

186. Reagan K, Beckmann CF, Larsen CR, et al: Renal infarction: Computerized tomographic appearance with angiographic correlation. *J Urol* 1984; 132:331–334.

187. Resnick MI, Kursh ED, Cohen AM: Use of computerized tomography in the delineation of uric acid calculi. *J Urol* 1984; 131:9–10.

188. Reuter SR, Redman HC: *Gastrointestinal Angiography.* Philadelphia, WB Saunders Co, 1979.

189. Richie JP, Garnick MB, Seltzer S, et al: Computerized tomography scan for diagnosis and staging of renal cell carcinoma. *J Urol* 1983; 129:1114–1116.

190. Richmond R, Sherman RS, Diamond HD, et al: Renal lesions associated with malignant lymphomas. *Am J Med* 1962; 32:184–207.

191. Robson CJ, Churchill BM, Anderson W: The results of radical nephrectomy for renal cell carcinoma. *J Urol* 1969; 101:297–301.

192. Rosenberg ER, Korobkin M, Foster W, et al: The sig-

nificance of septations in a renal cyst. *AJR* 1985; 144:593–595.

193. Rosenfield AT, Glickman MG, Taylor KJW, et al: Acute focal bacterial nephritis (acute lobar nephronia). *Radiology* 1979; 132:555–561.

194. Sagel SS, Stanley RJ, Levitt RG, et al: Computed tomography of the kidney. *Radiology* 1977; 124:359–370.

195. Sandler CM, Toombs DB: Computed tomographic evaluation of blunt renal injuries. *Radiology* 1981; 141:461–466.

196. Sawyer RW, Vick CW, Walsh JW, et al: Computed tomography of benign ovarian masses. *J Comput Assist Tomogr* 1985; 9:784–789.

197. Schaner EG, Balow JE, Doppman JL: Computed tomography in the diagnosis of subcapsular and perirenal hematoma. *AJR* 1977; 129:83–88.

198. Scott JA, Rabe FE, Becker GJ, et al: Angiographic assessment of renal artery pathology: How reliable? *AJR* 1983; 141:1299–1302.

199. Schwarten DE: Transluminal angioplasty of renal artery stenosis: 70 experiences. *AJR* 1980; 135:969–974.

200. Segal AJ, Spitzer RM: Pseudo thick-walled renal cyst by CT. *AJR* 1979; 132:827–828.

201. Seidelmann FE, Cohen WN, Bryank PJ, et al: Accuracy of CT staging of bladder neoplasms using the gas-filled method: Report of 21 patients with surgical confirmation. *AJR* 1978; 130:735–739.

202. Seltzer RA, Wenlund DE: Renal lymphoma: Arteriographic studies. *AJR* 1967; 101:692.

203. Sherman JL, Hartman DS, Friedman AC, et al: Angiomyolipoma: Computed tomographic-pathologic correlation in 17 cases. *AJR* 1981; 137:1221–1226.

204. Sigstedt B, Lunderquist A: Complications of angiographic examinations. *AJR* 1978; 130:445–460.

205. Silverman JF, Wexler L: Complications of percutaneous transfemoral coronary arteriography. *Clin Radiol* 1976; 27:317–321.

206. Smith DW, Winfield AC, Price RR, et al: Evaluation of digital venous angiography for the diagnosis of renovascular hypertension. *Radiology* 1982; 144:51–54.

207. Sniderman KW, Kreiger JN, Seligson GR, et al: The radiologic and clinical aspects of calcified hypernephroma. *Radiology* 1979; 131:31–35.

208. Sos TA, Pickering TG, Sniderman K, et al: Percutaneous transluminal angioplasty in renovascular hypertension due to atheroma or fibromuscular dysplasia. *N Engl J Med* 1983; 309:274–297.

209. Stanley RJ, Sagel SS, Fair WR: Computed tomography of the genitourinary tract. *J Urol* 1978; 119:700–782.

210. Steele JR, Sones PJ, Heffner LT Jr: The detection of inferior vena caval thrombosis with computed tomography. *Radiology* 1978; 128:385–386.

211. Steinberg DL, Jeffrey RB, Federle MP, et al: The computerized tomography appearance of renal pedicle injury. *J Urol* 1984; 132:1163–1164.

212. Struthers NW, Samu P, Chalvardjian A: Renal artery aneurysm: A complication of Gianturco coil embolization of renal adenocarcinoma. *J Urol* 1980; 133:324.

213. Subramanyam BR, Raghavendra BN, Bosniak MA, et al:

214. Sukov RJ, Scardino PT, Sample WF, et al: Computed tomography and transabdominal ultrasound in the evaluation of the prostate. *J Comput Assist Tomogr* 1977; 1:281–289.

215. Sussman S, Cochran ST, Pagani JJ, et al: Hyperdense renal masses: A CT manifestation of hemorrhagic renal cysts. *Radiology* 1984; 150:207–211.

216. Szilogyi ED, Smith RF, Elliott JP, et al: Translumbar aortography. *Arch Surg* 1977; 112:339–408.

217. Tada S, Fukuda K, Aoyagi Y, et al: CT of abdominal malignancies: Dynamic approach. *AJR* 1979; 135:455–461.

218. Tegtmeyer CJ, Kofler TJ, Ayers CA: Renal angioplasty: Current status. *AJR* 1984; 142:17–21.

219. Tisnado J, Beachley MC, Cho SR, et al: Peripheral embolization of stainless steel coil. *AJR* 1979; 133:324.

220. Tornquist C, Haltos S: Renal angiography with isohexol and metrizoate. *Radiology* 1984; 150:331–333.

221. Van Engelshoven JMA, Kreel L: Computed tomography of the prostate. *J Comput Assist Tomogr* 1979; 3:45–51.

222. Vinik M, Freed TA, Smellie WAB, Weidner W: Xanthogranulomatous pyelonephritis: Angiographic considerations. *Radiology* 1969; 92:537.

223. Vogelzang RL, Gore RM, Neiman HL, et al: Inferior vena cava CT pseudothrombus produced by rapid arm-vein contrast infusion. *AJR* 1985; 144:843–846.

224. Vogelzang RL, Neiman HL: Computed tomography of the kidneys, in Greenberg M (ed): *Essentials of Body Computed Tomography.* Philadelphia, WB Saunders Co, 1983.

225. Walsh JW: Pelvis, in Greenberg M, Greenberg BM, Greenberg IM (eds): *Essentials of Body Computed Tomography.* Philadelphia, WB Saunders Co, 1983, p 345.

226. Walsh JW, Goplerud DR: Prospective comparison between clinical and CT staging in primary cervical carcinoma. *AJR* 1981; 137:997–1003.

227. Walsh JW, Goplerud DR: Computed tomography in primary, persistent, and recurrent endometrial malignancy. *AJR* 1982; 139:1149–1159.

228. Walsh JW, Rosenfield AT, Jaffee CC, et al: Prospective comparison of ultrasound and computed tomography in the evaluation of gynecologic pelvic massees. *AJR* 1978; 131:955–960.

229. Wasserman NR, Ewing SL: Calcified renal oncocytoma. *AJR* 1983; 141:747–749.

230. Watson RC, Fleming RJ, Evans JA: Arteriography in the diagnosis of renal carcinoma. *Radiology* 1968; 91:888.

231. Watt I, Roylance J: Pyonephrosis. *Clin Radiol* 1975; 27:513–519.

232. Weyman JW, McClennan BL, Lee JKT, et al: CT of calcified renal masses. *AJR* 1982; 138:1095–1099.

233. Weyman MA, McClennan BL, Stanley RJ, et al: Comparison of computed tomography and angiography in the evaluation of renal cell carcinoma. *Radiology* 1980; 137:417–424.

234. White RI Jr, Kaufman SL, Barth KH, et al: Therapeutic

Sonography of pyonephrosis: A prospective study. *AJR* 1983; 140:991–993.

embolization with detachable silicone balloons. *JAMA* 1979; 241:1257.

235. Wicks JD, Thornbury JR: Acute renal infections in adults. *Radiol Clin North Am* 1979; 17:245.

236. Widrich WC, Beckman CF, Robbins AH, et al: Iopamidol and meglumine diatrizoate: Comparison of effects on patient discomfort during aortofemoral arteriography. *Radiology* 1983; 148:61–64.

237. Williams RD, Feinberg SB, Knight LC, et al: Abdominal staging of testicular tumors using ultrasonography and computed tomography. *J Urol* 1980; 123:872–875.

238. Williamson B, Hattery RR, Stephens DH, et al: Computed tomography of the kidneys. *Semin Roentgenol* 1978; 13:249–255.

239. Winter J, King W III: Basic principles of computed tomography, in Greenberg M, Greenberg BM, Greenberg IM (eds): *Essentials of Body Computed Tomography.* Philadelphia, WB Saunders Co, 1983, p 1.

240. Wong WS, Moss AA, Federle MP, et al: Renal infarction: CT diagnosis and correlation between CT findings and etiologies. *Radiology* 1984; 150:201–205.

241. Yoder IC, Pfister RC, Lindfors KK, et al: Pyonephrosis: Imaging and intervention. *AJR* 1983; 141:735–740.

242. Young SW, Rumbaugh CL: Time-related contrast enhancement of blood, normal muscle, and B2 carcinoma in the rabbit as determined by CT scanning. *Invest Radiol* 1978; 13:334–336.

243. Yu WS, Sagerman RH, King GA, et al: The value of computed tomography in the management of bladder cancer. *Int J Radiat Oncol Biol Phys* 1979; 5:135–142.

244. Yune HY, Klatte EC: Collateral circulation to an ischemic kidney. *Radiology* 1976; 119:539.

245. Zaonta MR, Pahira JJ, Wolfman M, et al: Acute focal bacterial nephritis: A systematic approach to diagnosis and treatment. *J Urol* 1985; 133:752–757.

246. Zirinsky K, Auh YH, Rubenstein WA, et al: CT of the hyperdense renal cyst: Sonographic correlation. *AJR* 1984; 143:151–156.

Chapter 3C / Magnetic Resonance Imaging/Spectroscopy and Positron Emission Tomography

RICHARD D. WILLIAMS, M.D.

Within the past decade, profound advances in diagnostic body imaging have significantly altered the approach to the diagnosis of urologic diseases. Computed tomography (CT), one of the earlier of these innovations, has now become essential to many urologic evaluations. Magnetic resonance imaging (MRI) and positron emission tomography (PET) have been heralded as diagnostic techniques that are expected to attain comparable status to CT. Magnetic resonance imaging and CT have in common their ability to evaluate pathophysiology but differ widely in their basic physical principles and perhaps in their ultimate utility. Magnetic resonance imaging is becoming readily available in large- and medium-sized communities, whereas PET is likely to remain available only in centers near a cyclotron where the rapidly dissipating isotopes required for its use are produced. Magnetic resonance spectroscopy (MRS) promises to provide specific physiologic information about normal and diseased tissue, but it is not yet routinely available for clinical studies. In this discussion we will review the basic principles of each modality and describe current and projected uses.

MAGNETIC RESONANCE IMAGING

PRINCIPLES

Magnetic resonance imaging is based on complex physical principles only briefly reviewed here; detailed reports are available elsewhere in the literature (Pykett, 1982; Budinger and Lauterbur, 1984; Paushter et al., 1984; Williams and Hricak, 1984; Young, 1984; Johnston et al., 1985). Current clinical MRI relies on the natural abundance of protons (hydrogen ions; Fig 3C–1,A) in living tissues that, because of their uneven atomic number, develop a small magnetic field around themselves. These multiple minute magnets within the human body are normally in random thermal motion and thus cancel each other, so that no magnetic force can be measured (Fig 3C–1,B). When placed inside a MR imager (strong electromagnet), the vectors of the magnetic fields of the protons are aligned along the field of the external magnet in their lowest energy state (Fig 3C–1,C). Protons shift their alignment to a higher

energy state (Fig 3C–1,D) when they are perturbed by radiofrequency (RF) pulses of specific frequency. The effective frequency varies with the strength of the imager's magnet. When the RF pulse ceases, the protons return to the low-energy aligned state within the magnet (characterized by two time constants known as T_1 and T_2 relaxation times) and release the previously stored energy, producing a current or signal that can be detected by RF receiving coils (Fig 3C–1,E). Additional magnetic gradient coils placed within the aperture of the imager magnet allow preferential changes in the orientation of tissue protons in isolated planes and thus produce an MR signal encoded with spatial information. Computer transformation converts the MR signals into an image that reflects the anatomical distribution of tissue protons.

MRI is not limited to protons, since isotopes of other atomic species with uneven atomic numbers such as ^{13}C, ^{23}Na, and ^{31}P are potentially useful for imaging. Proton imaging is currently used for clinical studies because of (1) the relative abundance of hydrogen in tissue water and lipids and (2) the requirement of a much stronger magnet than that used for proton MRI to image other species satisfactorily.

MR imaging is most often accomplished using spin-echo (SE) techniques involving repetitive RF pulses that cause a 90° shift in proton orientation followed by 180° refocusing pulses. When this occurs, the signal echo is computer processed to provide an image. The time between sequential 90° RF pulses is designated as the repetition time (TR). The interval from 90° RF pulses to signal recording is the echo delay time (TE). Clinical MRI uses TR values of 0.5–3.0 seconds and TE from 28 to 140 msec. Usually two TR and two TE values are used for each MR study.

MRI can provide information concerning the density, relaxation characteristics (T_1 and T_2), and bulk motion (flow) of protons in the tissues examined. Signal intensity with TR and TE held constant is greater as proton density or T_2 increases or as T_1 decreases. Tissues or body fluid with a characteristic long T_1 (urine and cortical bone) will show a low-intensity signal (black rather than white) with short TR. Intensity increases as TR is

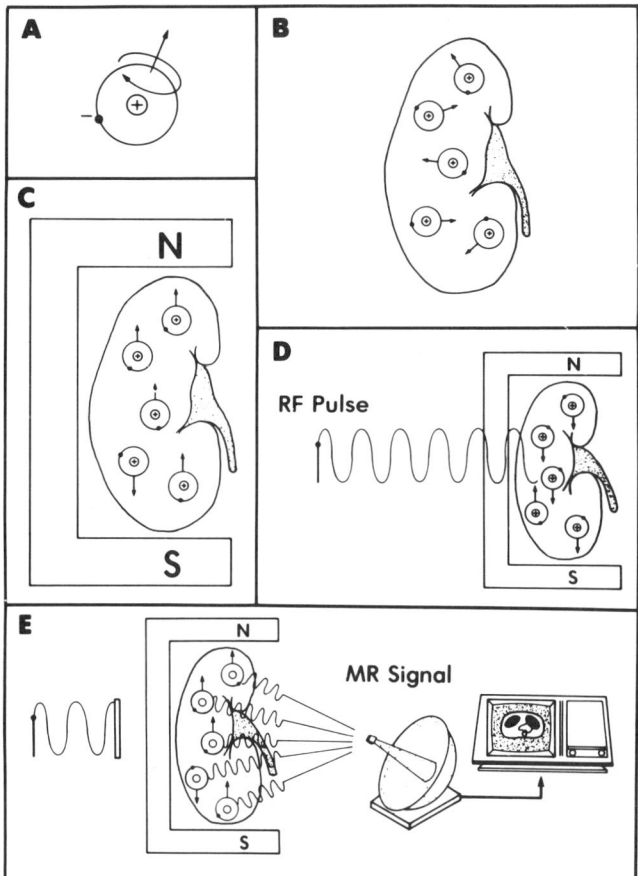

FIG 3C–1.
A, a proton is shown indicating angular momentum and the resultant magnetic force vector. **B,** in undisturbed tissue proton vectors are in random motion, and thus cancel each other's magnetic force. **C,** within a strong magnet the proton force vectors are aligned with the majority in their lowest-energy state. **D,** radiofrequency (RF) pulses shift the proton vectors to a higher-energy state within the magnet. **E,** when the radiofrequency pulse ceases, the proton vectors release the absorbed energy as they return to the low-energy state within the magnet. MR indicates magnetic resonance.

prolonged. A high-intensity signal is revealed in tissues with short T_1 (fat and bone marrow), regardless of short or long TR. As TE is increased, tissues of long T_2 may have equal or increased intensity. Tissues of short T_2 decrease in intensity with prolonged TE.

Magnetic field strengths for MRI are measured in terms of tesla (T). Diagnostic MRI currently uses magnets of 0.1–1.5 T (1 T = 10,000 gauss), which is 1,700–25,000 times the magnetic field of the earth. In general, the stronger the magnet, the greater the contrast resolution achieved by the imager.

MRI has been shown to have several distinct advantages over other diagnostic imaging methods (Table 3C–1). Contrast resolution, particularly with respect to soft tissues, is superior using MRI. Images in coronal,

TABLE 3C–1.

Advantages and Disadvantages of MRI vs. Conventional Radiologic Methods

Advantages
  Superior soft tissue contrast resolution
  Direct multiplanar images
  Lack of ionizing radiation
  Bone and small metal clip artifact minimal
  Blood flow measurable
Disadvantages
  Lack of detection of tissue calcium
  Image degradation with bowel and respiratory motion
  Imaging of pacemaker patients contraindicated
  Extended examination time
  Claustrophobia (1%–3%)

sagittal, and transaxial planes can be obtained directly, obviating the loss of resolution that occurs in the reconstruction of images as with CT Figs 3C–2 to 3C–4). MRI does not involve generation of ionizing radiation as do radiologic techniques and thus far has shown no

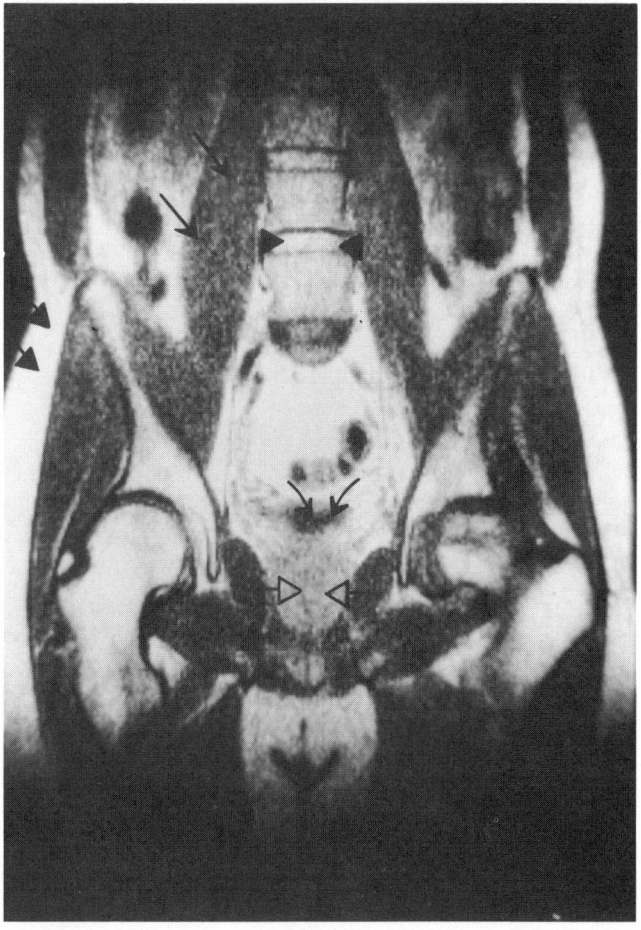

FIG 3C–2.
Coronal MRI of the pelvis depicting subcutaneous fat (short solid arrows), psoas muscle (long arrowheads), intervertebral disk (solid arrowheads), bladder (curved arrows), and prostate (open arrows).

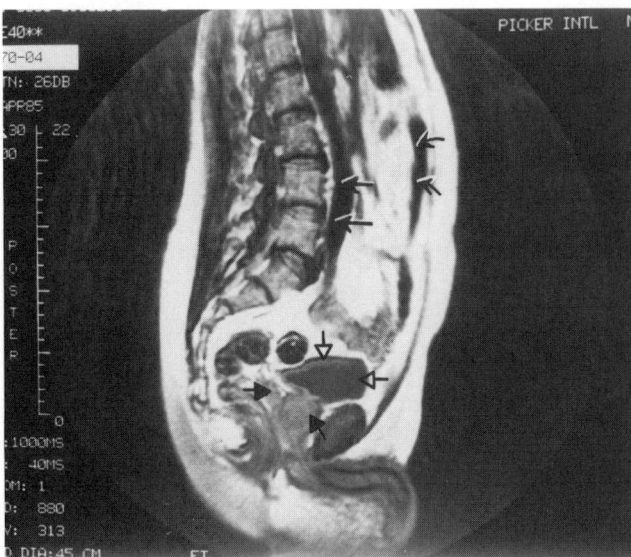

FIG 3C–3.
Sagittal MRI of pelvis depicting bladder *(open arrows)*, prostate and seminal vesicle *(closed arrows)*, vena cava *(straight arrows)*, and gas within small bowel *(curved arrows)*.

harmful effects. Artifacts produced by bone and surgical clips are minimal with MRI. Importantly, because flowing blood perturbed by RF waves does not emit the stored energy in the same frame as stationary tissues, blood flow can be estimated by rapidly repetitive studies. Sluggish blood flow, tumor thrombi, or intravascular blood thrombi can easily be detected and differentiated from one another.

Disadvantages of MRI include: the inability to detect tissue calcium either in tumors or urinary calculi; the possible inhibition of cardiac pacemakers by the mag-

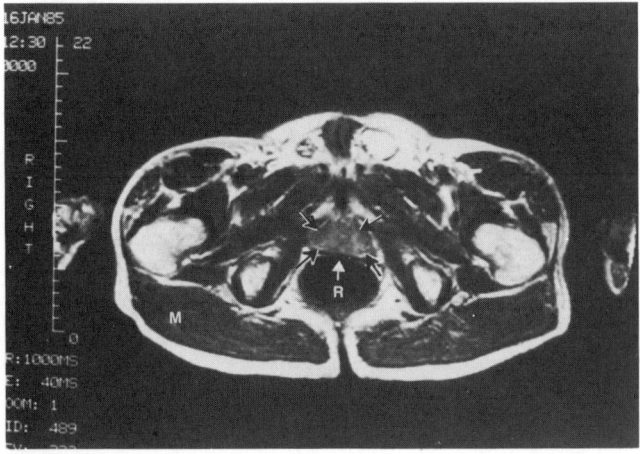

FIG 3C–4.
Transaxial MRI of pelvis depicting prostate *(straight arrows)*, rectum *(R)*, muscle *(M)*, and Denonvilliers' fascia *(white arrow)*.

TABLE 3C–2.
MRI Relative Gray Scale

High-intensity—white
Fat
Bone marrow
Nonflowing blood
CNS
Medium-intensity—gray
Viscera
Muscle
Low-intensity—black
Flowing blood
Cortical bone
Air

net, thus contraindicating MRI for pacemaker patients; the degrading of MR images caused by bowel, respiratory, or cardiac motion with resultant poor contrast and spatial sensitivity; slow imaging time (1–1½ hours) compared with CT; and, in some patients, claustrophobia that precludes MR examination. In this last instance, however, experience and patient education have decreased the incidence to below 3%.

MRI tissue contrast is variable, depending on many parameters. Using the SE technique, high-intensity signals appear white, while absent or low-intensity signals appear black (see Figs 3C–2 to 3C–4). Fat yields the highest-intensity signals followed by bone marrow, nonflowing blood, brain and spinal cord, viscera, striated muscle, ligaments and tendons, rapidly flowing blood, cortical bone, and air in decreasing order of intensity (Table 3C–2). These contrast intensities depend on imaging technique (pulse sequences), characteristics of the tissues examined, and the magnetic strength of the imager, and are thus less absolute than with CT.

CLINICAL MRI

When first introduced, MRI was expected to revolutionize diagnostic imaging and perhaps replace CT and ultrasound. Three years of rapidly accumulating experience derived from patients with a variety of pathologic states have tempered the early enthusiasm. There are, at present, only a few areas of the body and specific pathologic conditions in which MRI has been proved superior to contrast-enhanced CT. These loci and conditions primarily involve the CNS, cerebral edema, demyelinating diseases, and some CNS tumors. Further objective and comparative studies are needed to determine conclusively the usefulness of MRI in the evaluation of urologic organs and diseases. For the most part, reports in the literature have been anecdotal, although recently investigations comparing MR and CT images using pathologic data for confirmation have been published. The following sections summarize clinical data on MRI available in the literature.

Renal Anatomy and Pathology

Anatomical features of the retroperitoneal space including the kidneys and adrenals are clearly defined by MRI (Hricak and Newhouse, 1984; Leung et al., 1984; Lipuma, 1984.). In technically adequate examinations using appropriate SE methods, MRI scans are similar in quality to nonenhanced CT scans. The renal parenchyma is easily distinguished with a medium-intensity signal in sharp contrast to the high-intensity signal of the renal sinus and perinephric fat (Fig 3C–5). Renal parenchyma can also be separated into cortex and medulla, with the former having a higher-intensity signal than the latter, particularly using short TR values. Renal capsule is rarely imaged in the normal state. Owing to the lack of signal detection from flowing blood, renal vessels are imaged as tubular structures with low intensity. This finding represents a distinct advantage over CT, where renal vessels are not clearly defined without rapid contrast injection (dynamic) studies. Renal pelvis and calyceal structures and ureters are also seen as low-intensity structures, particularly if images are obtained while the patient is in a well-hydrated or diuretic state. Coronal images allow differentiation of vessels and collecting systems, as they can be followed from their origins in sequential scans. Similarly, the renal outline can best be clarified on coronal scans. Gerota's fascia is inconsistently seen as a thin, low-intensity structure surrounding the perirenal fat.

To date, the most extensively studied renal abnormality using MRI has been the parenchymal mass. Renal cysts are readily identified on MRI owing to their

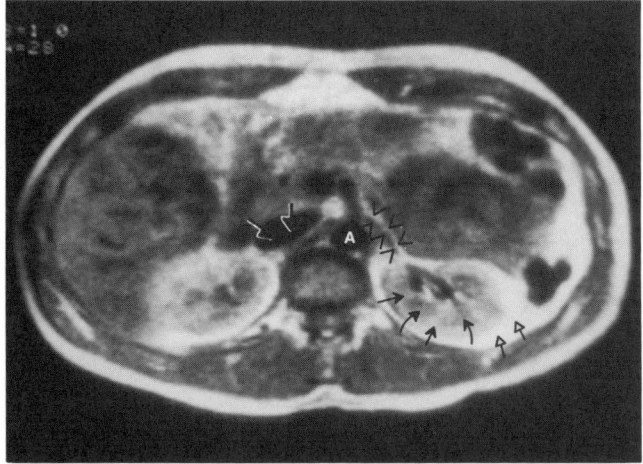

FIG 3C–5.
Transaxial MRI of kidney area. The renal cortex *(straight arrows)* is easily distinguished from renal medulla *(curved arrows)*. Perirenal fat *(open arrows)* is prominent. Aorta *(A)* and vena cava *(large closed arrows)* are clearly seen as dark areas indicative of rapid flow. Left renal vein *(arrowheads)* is well demarcated.

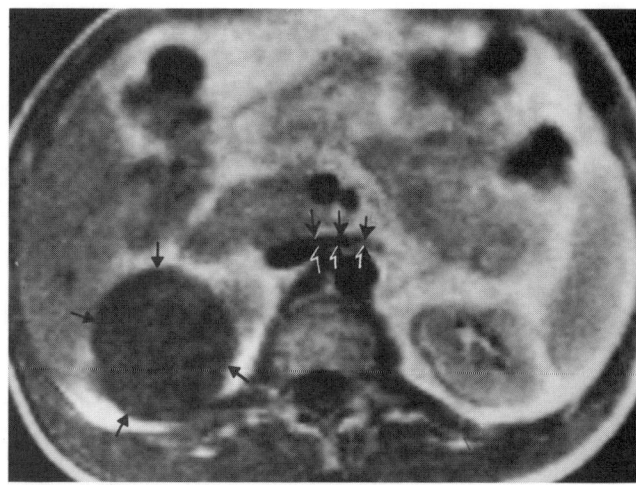

FIG 3C–6.
Transaxial MRI of left simple renal cyst *(closed arrows)* showing low signal intesity and smooth outline. Left renal vein *(contiguous arrows)* is seen entering the vena cava.

homogeneous, low image intensity (Kulkarni et al., 1984). Simple cysts have smooth outlines with well-defined margins (Fig 3C–6) (Choyke et al., 1984). Cysts containing blood have a high image intensity in all pulse sequences used, and thus MRI is clearly superior to either CT or ultrasound in defining hemorrhagic cysts (Hricak and Newhouse, 1984; Leung et al., 1984). Lipuma has determined a change from high to low intensity of hemorrhagic cysts over time, suggesting that MRI may allow periodic reexamination to supplant interventional diagnostic techniques (Lipuma, 1984). Magnetic resonance imaging of simple renal cysts at present, however, does not provide more information than either ultrasound or CT and thus is not recommended for their definition.

Solid renal masses are readily depicted by MRI and easily distinguished from normal renal parenchyma. Transaxial scans are best used to detect masses (Figs 3C–7, A and B), but coronal scans can be helpful in clearing renal outlines of suspected masses (Fig 3C–8). There have been few (two reported) benign renal tumors imaged, but both revealed a homogeneous appearance, with a signal intensity lower than that of renal parenchyma (Leung et al., 1984). Magnetic resonance imaging of angiomyolipomas exhibits areas of high-intensity signal presumably corresponding to areas of fat within the mass (Leung et al., 1984; Choyke et al., 1984). As yet, investigators have been unable to consistently identify distinguishing characteristics of primary malignant tumors by MRI (Choyke et al., 1984; Hricak and Williams, 1984; Kulkarni et al., 1984; Williams and Hricak, 1984; Demas et al., 1985; Hricak et al., 1985). In general, renal cancers reveal a medium-intensity heterogeneous signal that tends to be higher

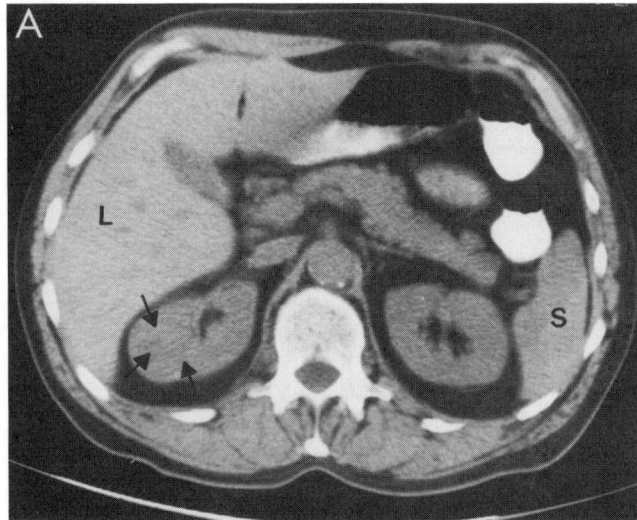

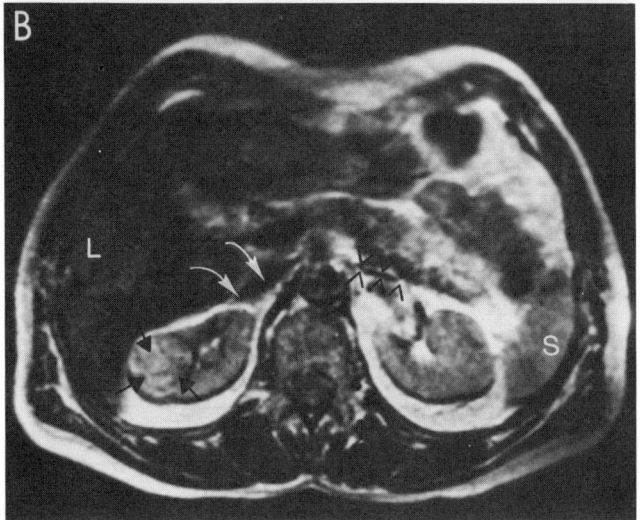

FIG 3C–7.
A, transaxial noncontrast-enhanced CT showing right renal mass *(arrows)*, liver *(L)*, and spleen *(S)*. **B,** transaxial MRI of same patient showing heterogeneous medium intensity right renal cancer *(straight arrows)*, liver *(L)*, and spleen *(S)*. The vena cava *(curved arrows)* and left renal vein *(open arrows)* are clearly seen.

than surrounding parenchyma (see Figs 3C–7,B, and 3C–8). Tumors with central necrosis (Fig 3C–9) tend to have lower signal intensity in the center than in the periphery of the tumor (Choyke et al., 1984). Tumors metastatic to kidney have shown very low-intensity homogeneous signals in the two patients reported (Choyke et al., 1984), but further documentation is required. A recently reported study of 31 patients with known renal masses revealed that CT and MRI were equally accurate in detecting solid renal masses (Hricak et al., 1985). A major disadvantage of MRI in this regard is that contrast enhancement, which is capable of defining characteristics typical of primary renal cancer

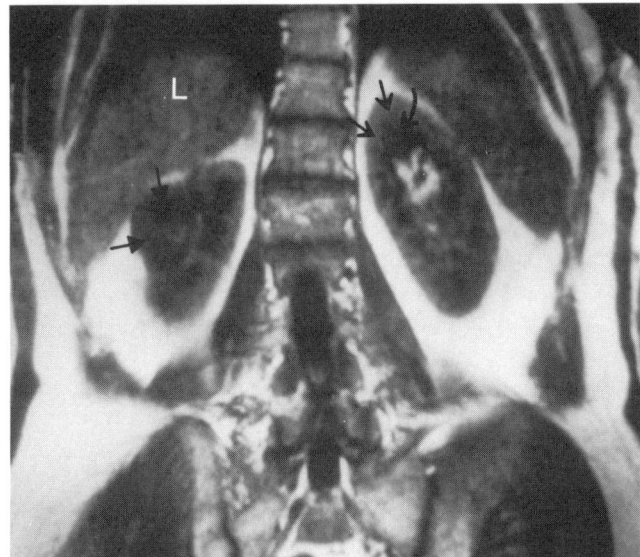

FIG 3C–8.
Coronal MRI of same patient as in Fig 3–7, A and B. The renal cancer *(closed arrows)* is clearly surrounded by a fat plane and is thus confined to the renal capsule. Renal cortex *(straight arrowheads)* and medulla *(curved arrow)* are easily differentiated. *L* indicates liver.

by CT, is not yet routinely available for MRI studies. Prototype paramagnetic contrast agents have recently been shown to enhance renal tumors in the few patients studied thus far (Leung et al., 1984). Definitive reports of MRI contrast enhancement are anticipated shortly.

While the diagnosis of renal cancer is not as yet improved by MRI, local staging appears to be. Various investigators have described the superior ability of MRI

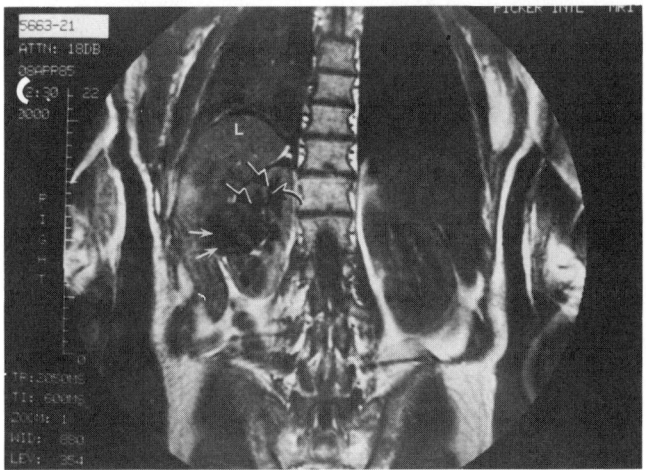

FIG 3C–9.
Coronal MRI of patient with right renal cancer. Tumor shows central necrosis *(small white arrows)*. Renal cortex *(straight arrows)* and medulla *(curved arrow)* are seen. *L* indicates liver.

compared with ultrasound to detect (1) local tumor extension, (2) retroperitoneal lymphadenopathy, and (3) vascular involvement (renal vein and vena cava) (Choyke et al., 1984; Kulkarni et al., 1984; Hricak et al., 1985). Hricak and others directly compared CT and MRI for staging renal cancer. Pathology confirmed 96% staging accuracy with MRI compared with 70% with CT (Hricak et al., 1985). In this study, MRI was particularly accurate in determining the presence of tumor thrombi in renal veins and vena cava, and imaging minimally enlarged lymph nodes (Fig 3C–10,A and B). CT tended to overstage patients with large upper pole tumors thought to be invading the liver on transaxial scans, whereas sagittal MRI was capable of delineating a clear tissue plane between the liver and the renal mass (Fig 3C–11,A and B). Further studies are needed to corroborate these results, but currently MRI appears superior to CT in staging renal cancer. In practical use

at present, however, MRI is not recommended unless CT is equivocal with respect to vena caval involvement or enlarged para-aortic lymph nodes. Refinement of pulse sequences and the availability of contrast enhancement may alter this view in the near future.

With respect to other renal pathologic conditions few MRI studies have been reported. Leung and associates (1984) observed a loss of corticomedullary differentiation, decreased overall parenchymal signal intensity, and small-volume kidneys with increased sinus lipomatosis in patients with chronic renal failure. Patients imaged with acute tubular necrosis have shown distinct corticomedullary differentiation with an increased medulla thickness of very low signal intensity (Lipuma, 1984). Transplanted kidneys undergoing acute or chronic rejection have shown a total loss of corticomedullary differentiation and overall low signal intensity on MR imaging (Geisinger et al., 1984; Leung et al., 1984; Lipuma, 1984). These latter findings are distinct from those of acute tubular necrosis and are expected to allow MRI to clinically differentiate these two entities. Definitive studies to include transplant kidney biopsy for pathologic confirmation will be required, however, for final conclusions to be established. Nonetheless, it does appear that the improved corticomedullary differentiation provided by MRI may represent a distinct advantage over CT examinations.

Adrenal Anatomy and Pathology

Adrenal glands image with homogeneous intermediate intensity generally lower than that of the adjacent liver when short TR values are used (Moon et al., 1983; Schultz et al., 1984). The adrenals are clearly outlined by the higher intensity periadrenal fat. Normal adrenal glands, however, are demonstrated as frequently with MRI as with CT examinations (Hricak and Williams, 1984; Schultz, et al., 1984).

The adrenal cortex is clearly distinguished from the medulla in cases of adrenal hyperplasia and is occasionally noted in patients with normal adrenals; the medulla appears less intense than the cortex (Schultz et al., 1984). In all studies to date contrast resolution was better imaged with MRI, but sharper definition of the adrenals was seen on CT.

Limited experience has been gained with adrenal pathology. Although adrenal masses (metastatic and primary tumors, myelolipoma, and hemorrhagic cysts) have been imaged (Fig 3C–12), MRI cannot distinguish benign from malignant lesions and does not appear to provide information superior to current-generation CT studies (Schultz et al., 1984; Johnston et al., 1985).

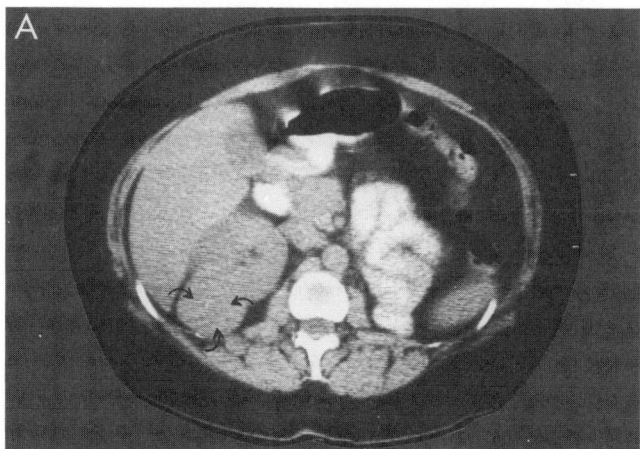

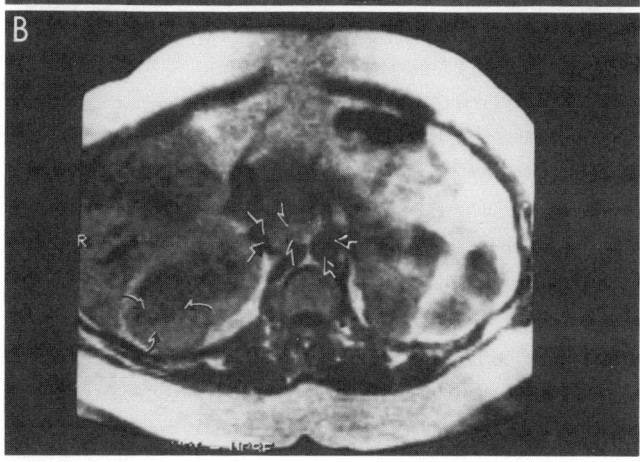

FIG 3C–10.
A, transaxial CT of patient with recurrent renal cancer in right kidney *(curved arrows)* and poor delineation of vena cava. **B,** transaxial MRI of same patient showing renal cancer *(curved arrows),* and vena cava compressed by an enlarged lymph node *(long closed arrows).* Normal aorta and superior mesenteric artery *(short open arrows)* are seen.

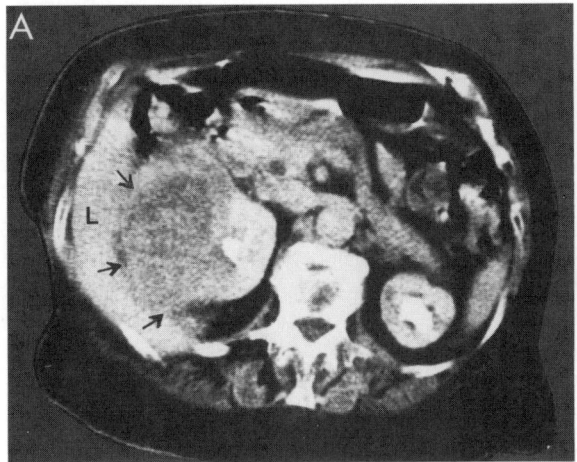
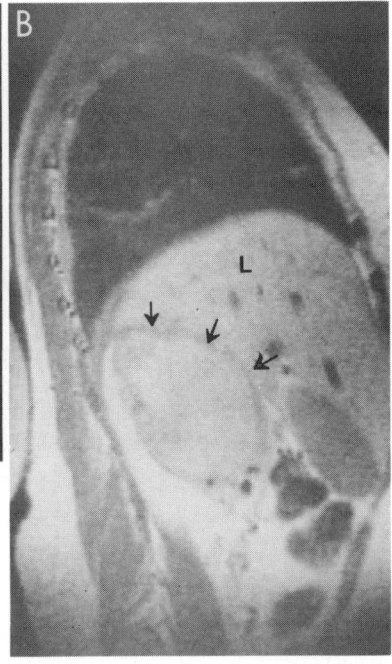

FIG 3C–11.
A, transaxial CT of patient with large right renal cancer. The tumor appears to be invading the liver *(L)* at the *arrows.* **B,** sagittal MRI of the same patient. The renal cancer is clearly separated *(arrows)* from the liver *(L).*

Retroperitoneal Anatomy and Pathology

MRI of the retroperitoneum is equivalent to CT in general, yet the availability of coronal imaging that allows precise delineation of the retroperitoneal vessels and surrounding lymph nodes could prove to be an advantage (Fig 3C–13,A and B). In our early experience MRI was capable of differentiating primary retroperitoneal tumors such as sarcomas from primary renal masses (Hricak and Williams, 1984; Williams and Hricak, 1984b). This was primarily due to a homogeneous, medium-intensity signal of the mass and clear tissue planes between the mass and other retroperitoneal structures such as kidney, adrenal, and pancreas. We

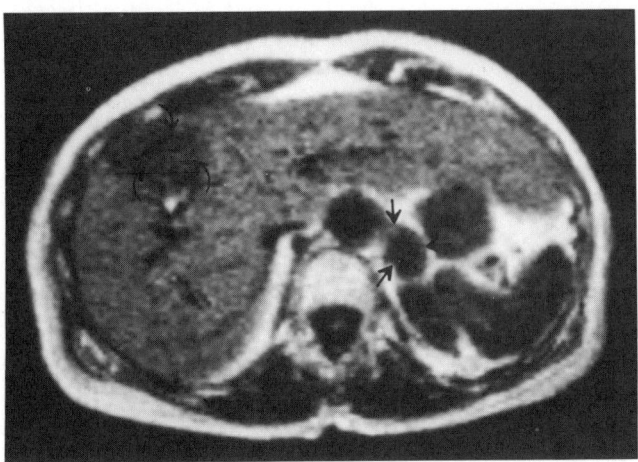

FIG 3C–12.
Transaxial MRI of patient with cancer metastatic to the left adrenal *(straight arrows)* and liver *(curved arrows).*

studied one patient with a primary leiomyosarcoma of the wall of the inferior vena cava who had an extensive caval tumor thrombus but an empty renal vein on the side of the mass. Magnetic resonance imaging clearly eliminated a primary renal cancer as the source of the caval thrombus (Fig 3C–14,A and B). Perivascular lymph nodes image as medium-intensity nodular structures but are not routinely visualized unless they are enlarged. Ellis and associates (1984) have compared MRI and CT in the retroperitoneal nodal staging of 25 patients with nonseminomatous testicular cancer, all of whom had retroperitoneal lymphadenectomy for confirmation. Magnetic resonance imaging and CT were equivalent, since 80% of patients were correctly staged by MRI and 84% by CT. While multiplanar imaging and superior demonstration of vascular anatomy are distinct advantages of MRI of the retroperitoneum, further technical advances and perhaps oral contrast agents will be required to prove MRI superior to CT.

Anatomy of the Pelvis

Magnetic resonance imaging of the pelvis is particularly promising, because bowel and respiratory motion and bone artifact in the pelvis are minimal. CT has been the modality of choice in evaluation of the pelvis, but its ability to differentiate pelvic organs and to detect and stage urologic tumors has been disappointing. MRI of the pelvis has further advantages over CT in that even minor amounts of pelvic fat allow distinct contrast between pelvic planes and organs, and pelvic vessels are clearly differentiated from pelvic muscles, fat, and lymph nodes.

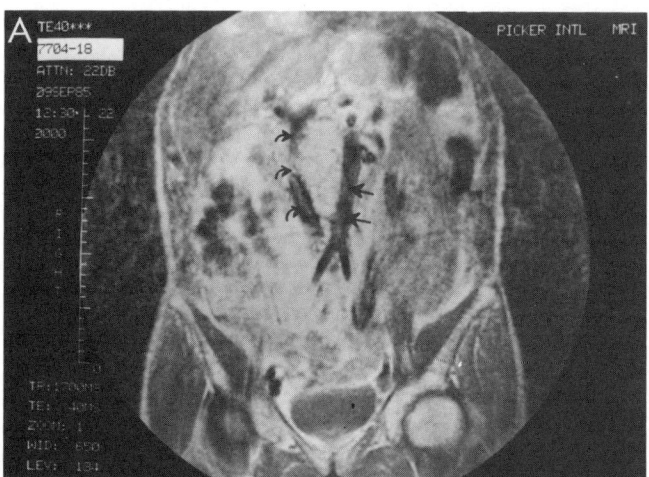

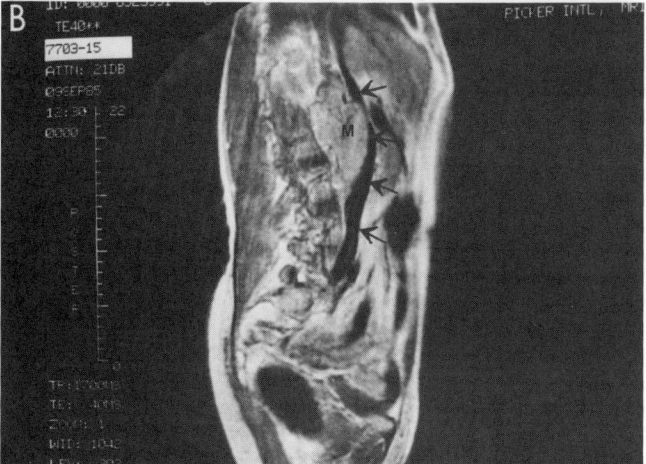

FIG 3C–13.
A, coronal MRI of patient with metastatic renal pelvis cancer to lymph nodes between vena cava *(curved arrows)* and aorta *(straight arrows).* **B,** sagittal MRI of same patient showing the lymph node mass *(M)* elevating the vena cava *(arrows)* anteriorly.

The muscles of the pelvis provide superb landmarks to assist MRI evaluation and differentiation of pelvic viscera. Skeletal muscle images with a low intensity clearly distinguishable from lower-intensity bone and higher-intensity pelvic fat. Pelvic blood vessels with normal flow are easily depicted because of their lack of signal in relation to the surrounding adipose tissue, lymph nodes, and muscles. Coronal planes are particularly striking in their ability to image the distribution of major vessels. Normal pelvic lymph nodes are located along the vessels and are best appreciated as elliptical structures with a higher intensity than skeletal muscle, particularly on coronal sections. The urinary bladder is best imaged while distended with urine because the signal intensity of urine is low due to its long T_1 and T_2; the intensity increases from dark to medium gray as TR is increased (Hricak et al., 1983; Fisher et al., 1985a;

Schmidt et al., 1985). The smooth muscle of the bladder wall is easily distinguished as a medium-intensity structure between the lower-intensity urine and higher-intensity pervesical fat. Alternatives in pulse sequences change the signal intensity of both urine and bladder wall, but the contrast difference is maintained.

In the female pelvis, the uterus on SE images reveals the myometrium on short TR to be medium intensity, which is higher than urine but only slightly higher than skeletal muscle (Hricak, 1984). Endometrium can be delineated from myometrium with long TR and TE. The cervix is easily distinguished, exhibiting inner and outer layers of high intensity with a low intensity middle layer. The vagina on MRI is best seen on coronal and sagittal scans with signal intensity similar to the uterus on short pulse sequence intervals and more eas-

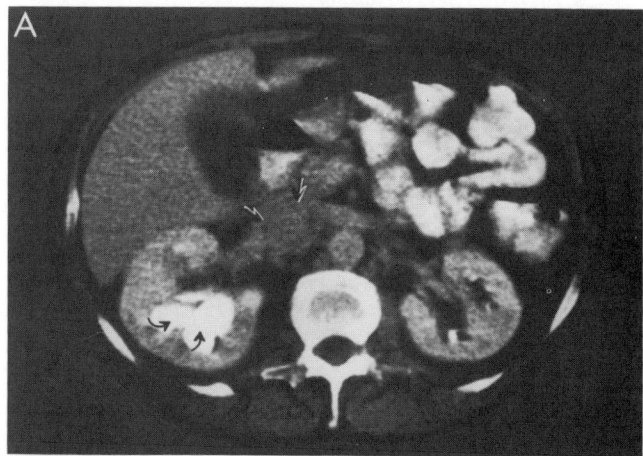

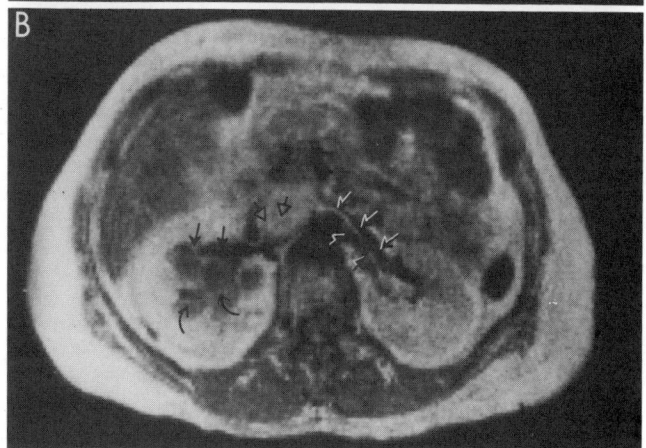

FIG 3C–14.
A, transaxial CT of patient with retroperitoneal leiomyosarcoma. The right renal calyces *(curved arrows)* are dilated and an enlarged vena cava *(straight arrows)* presumably contains tumor. **B,** transaxial MRI of same patient shows dilated calyces *(curved arrows),* open right renal vein *(straight arrows)* and tumor thrombus in vena cava *(open arrows).* The left renal vein and artery are easily seen *(opposing arrows).*

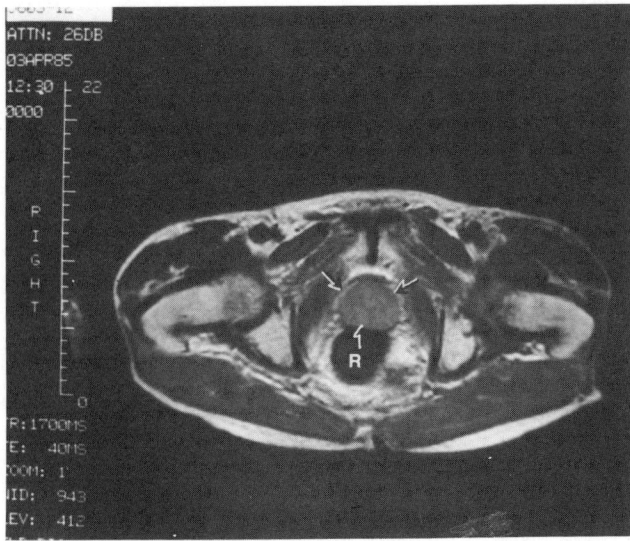

FIG 3C–15.
Transaxial MRI of normal prostate *(arrows)*. R indicates rectum.

ily differentiated on longer TR values. Ovaries are low-to medium-intensity composition on T_1-weighted images.

In the male pelvis, the normal prostate is imaged with a homogeneous medium intensity signal with short TR. Hricak and colleagues (1983) used MRI to differentiate anatomical zones of the prostate, but similar findings are not appreciated by others (Poon et al., 1985). Coronal and sagittal images are best for complete evaluation of the prostate (Figs 3C–15 to 3C–17). Denonvilliers' fascia is clearly seen as a low-intensity line

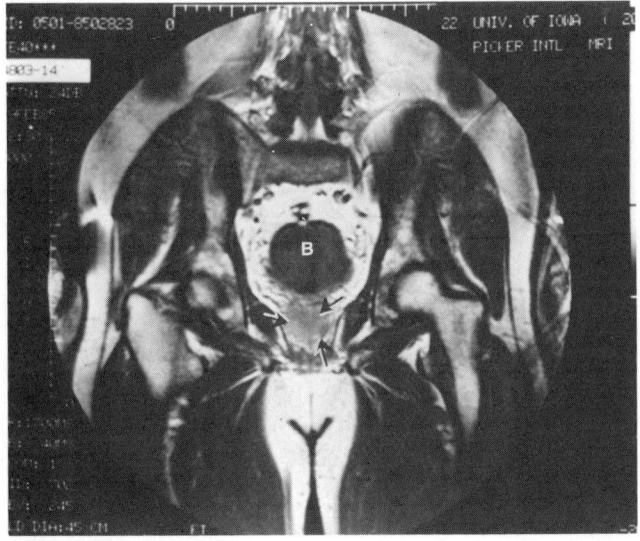

FIG 3C–16.
Coronal MRI of normal prostate *(arrows)*. B indicates bladder.

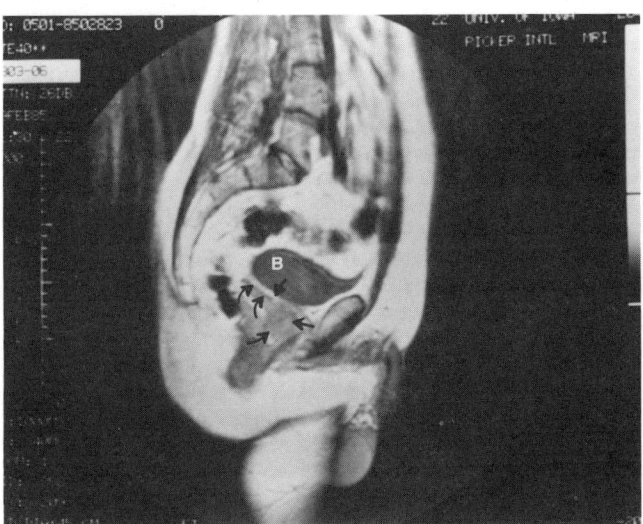

FIG 3C–17.
Sagittal MRI of normal prostate *(long arrows)*, seminal vesicle *(curved arrows)*. The normal bladder seminal vesicle angle *(short arrow)* is easily seen. B indicates bladder.

posterior to the prostate on sagittal images. The seminal vesicles are readily seen on transaxial and sagittal images, exhibiting a medium-signal intensity, with the surrounding fat providing superb contrast, particularly using short TR. The angles between prostate, seminal vesicles, and the bladder are exquisitely seen on sagittal images (see Fig 3C–17).

Bladder Pathology

The few bladder studies conducted to date have centered on malignancy as the primary focus of interest. Hypertrophy of the bladder wall secondary to bladder outlet obstruction reveals a uniform, medium-intensity signal that increases slightly with longer TR (Fisher et al., 1985b). Bladder neoplasms can easily be distinguished from the normal bladder wall because of their higher signal intensity (Fig 3C–18,A and B). The combination of axial and sagittal images may reveal the depth of penetration of bladder tumors, emphasizing the great potential of MRI in the local staging of bladder tumors (Fisher et al., 1985b) (Fig 3C–19). Perivesical fat infiltration can be seen as the lower signal tumor encroaches on the high-intensity fat. Local lymph nodes are best seen on coronal images, although only enlarged nodes are appreciated. For the most part, bladder neoplasms have only been anecdotally described. However, three recently reported studies (Resnick et al., 1984; Fisher et al., 1985b; Amendola et al., 1985) have correlated CT and MRI findings with pathologic data. In a preliminary study, Resnick and associates (1984) evaluated three patients and found that the diagnostic capabilities of MRI and CT were equal in determining

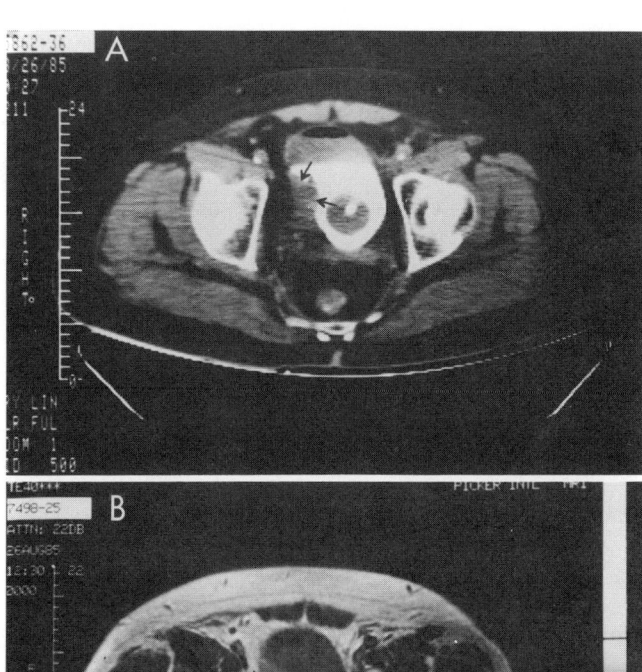

FIG 3C–18.
A, transaxial CT of patient with exophytic bladder cancer *(arrows).* The extent of the tumor in the bladder wall is presumed to be quite deep but is uncertain. There is a Foley catheter in the bladder. **B,** transaxial MRI of the same patient. The tumor *(arrows)* infiltrates deep into the perivesical fat.

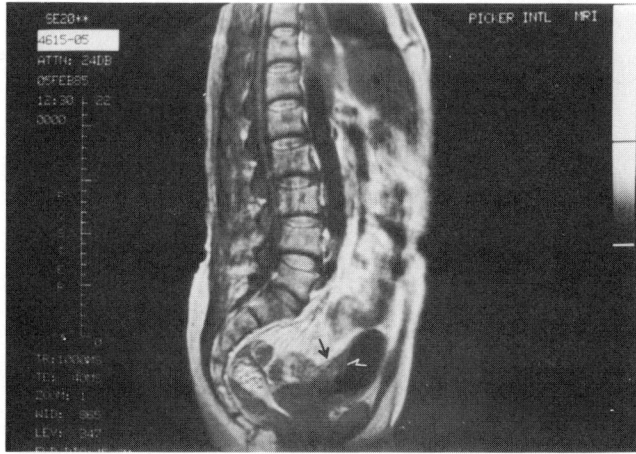

FIG 3C–19.
Sagittal MRI of patient with muscle invasive bladder cancer *(arrows).*

the presence of bladder cancer. However, perivesical fat extension was slightly more apparent in MRI. They did not find depth of penetration into the bladder wall to be accurate with either modality. Fischer and associates evaluated 14 patients with malignant bladder tumors (only 11 had concomitant CT) and found that tumors smaller than 1.5 cm were not imaged by MRI (Fisher et al., 1985). The only patient with carcinoma in situ did not show discernible findings on MRI. Eighty-five percent of the patients had correct staging by MRI compared with only 63% by CT. Amendola and others (1985) evaluated ten patients with transitional cell bladder cancer who had MRI and CT followed by radical cystectomy. Their results showed all tumors were visualized by both modalities, but lesions at the bladder base and dome were better seen by MRI. CT and MRI detection of bladder cancer was nearly equal, but pelvic staging was 90% accurate using MRI compared with 60% with CT. The difference between the accuracy of the two modalities was due to the superior MRI depiction of lymph node or prostate involvement by the tumor. The Amendola study revealed that MRI was not capable of determining depth of penetration within the bladder wall (differentiating B_1 from B_2 lesions). These initial studies indicate that MRI has no advantage over CT for initial diagnosis but is superior to CT for local pelvic staging of bladder cancer. Prospective studies of large numbers of patients comparing CT, MRI, and pathology are needed to corroborate these findings before determining whether MRI will emerge as the staging method of choice for bladder cancer.

Although adequate study of inflammatory bladder conditions has not been undertaken, it appears that inflammation is not clearly distinguishable from neoplasms, since both prolong T_1 and T_2 and may cause asymmetric abnormalities in the bladder wall (Fisher et al., 1985b).

Prostate Pathology

Benign prostatic hyperplasia (BPH) as imaged on MRI (Fig 3C–20) reveals an enlarged gland with a uniformly homogeneous, medium-intensity signal similar to that of normal prostate (Williams and Hricak, 1984a; Poon et al., 1985). The adenoma can be appreciated to elevate the bladder floor, and generally the surrounding prostatic capsule is seen as a lower-intensity band. Multiplanar images allow accurate volumetric measurements, clear depiction of intravesical extension, and the determination of prostatic length, all of which could be helpful in determining the operative approach to BPH in selected patients.

MRI of patients with acute prostatitis has also shown the prostate to image with a homogeneous, medium-

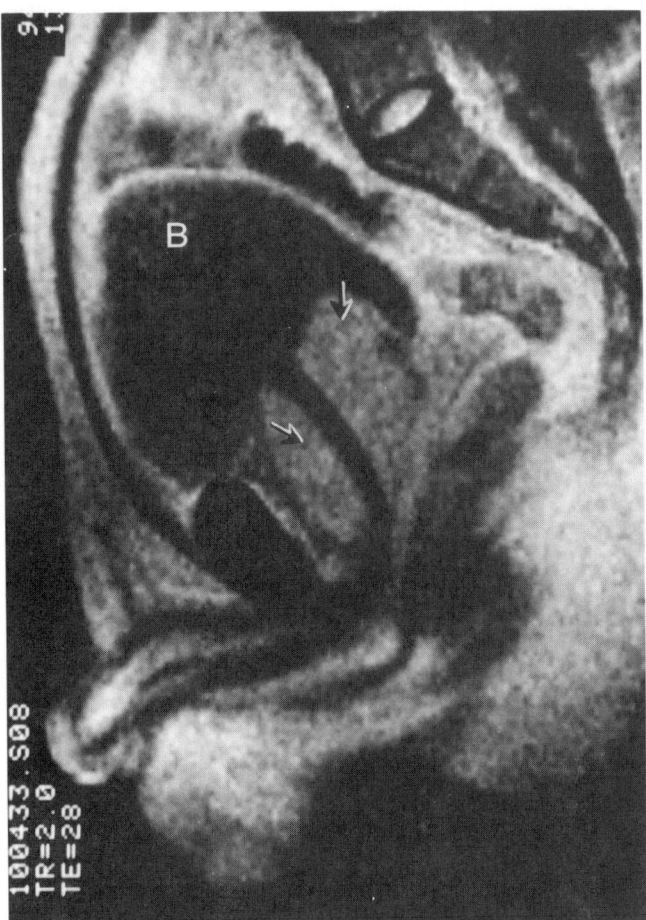

FIG 3C–20.
Sagittal MRI of patient with benign prostatic hyperplasia (arrows). There is a catheter in the urethra. B indicates bladder.

intensity signal. MRI of chronic prostatitis, however, reveals an inhomogeneous appearance of medium signal intensity with scattered areas of increased signal (Poon et al., 1985).

More than 50 patients with various stages of prostate cancer have been imaged by MRI and the results reported (Bryan et al., 1983; Buonocore et al., 1984; Hricak et al., 1983; Hricak, 1984; Nijhout et al., 1985; Poon et al., 1985). Unfortunately, few of these studies have included direct pathologic confirmation of the abnormal areas on MRI or correlation with ultrasound or CT scans. In our early studies of localized prostate cancer MRI showed a homogeneous, medium signal intensity laced with focal (usually multiple) areas of higher-intensity signal (Fig 3C–21) (Williams and Hricak, 1984a). Seminal vesicle involvement was defined by virtue of increased intensity signals in seminal vesicles and obliteration of the bladder-seminal vesicle angle on sagittal images (Fig 3C–22). Extension of tumor outside the prostate was determined by an obliteration of the

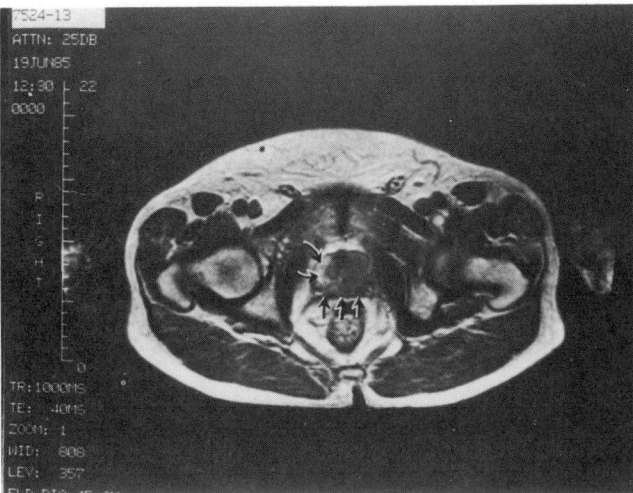

FIG 3C–21.
Transaxial MRI of patient with B₂ cancer of the prostate (curved arrows). Denonvilliers' fascia is intact (straight arrows).

capsular margin and high-intensity extension into the levator ani of the pelvis sidewalls. These changes are best appreciated by using more than one planar projection (Runge et al., 1984). Various reports have confirmed our findings, suggesting that intraprostatic cancer can be detected by MRI by virtue of the inhomogeneous image. MRI is not superior to ultrasound or CT in the diagnostic screening for prostate cancer, however, because it is unable clearly to differentiate the similar findings exhibited by chronic prostatitis and prostate cancer. Definitive studies of prostate

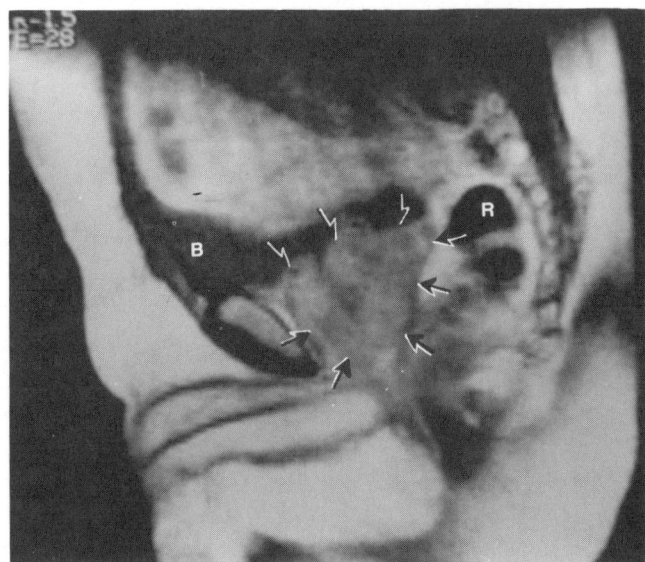

FIG 3C–22.
Sagittal MRI of patient with stage C cancer of the prostate (arrows). The seminal vesicle is involved with tumor and cannot be delineated. B indicates bladder; R, rectum.

cancer nodule mapping on MRI need to be correlated with precise anatomical location and biopsy to determine whether MRI has any role in prostate cancer screening. Further, because intraprostatic hemorrhage may exhibit high-intensity areas, studies by MRI before biopsy are indicated, although none has yet been reported.

Based on studies reported to date, MRI appears promising for staging prostate cancer. The use of multiplanar imaging allows determination of capsular penetration (coronal scans) and seminal vesicle involvement by loss of the bladder-seminal vesicle angle (sagittal scans) (Bryan et al., 1983; Buonocore et al., 1984; Demas et al., 1985). There have been no studies to our knowledge correlating MRI, CT, and pathologic data. It is speculated that local pelvic staging of prostatic cancer will be improved by MRI, but conclusions await definitive studies.

The spread of prostate carcinoma to bone can also be assessed by MRI. MRI is capable of showing lesions of higher intensity than cortical bone that can be delinated from nonmalignant lesions (Baker et al., 1985). Although radionuclide scanning will continue to be used as the primary modality for screening, MRI can be used in equivocal cases to detect or confirm metastases.

Penis and Scrotum

The anatomy of the penis is seen well on MRI (Hricak et al., 1983). The corpora cavernosa and spongiosum are of medium intensity with a short TR but reveal increased intensity on long TR (Fig 3C–23, A and B). The tunica albuginea surrounding the corpora is a low-intensity line owing to its collagen content. The bulbospongiosus muscle surrounding the corpus spongiosum at the base of the penis images with low intensity. The testicles image with a medium-intensity homogeneous signal (see Fig 3C–23,A), which increases as TR is elongated (Hricak and Williams, 1984). There has not been to our knowledge any systematic study of penis and scrotal pathology, so little information is available. Although anatomy of this region is well imaged by MRI, it is not expected that major improvements over ultrasound or CT examinations will be demonstrated.

PARAMAGNETIC CONTRAST AGENTS

Perhaps the most promising adjunct to MRI is the possibility of utilizing paramagnetic contrast agents to enhance spatial and contrast resolution. A variety of agents currently is under study, which basically fall into two categories, nitroxide-stable free radicals and metal ion chelates (Runge et al., 1984; Carr, 1985). The effect of these agents is primarily an increase in signal inten-

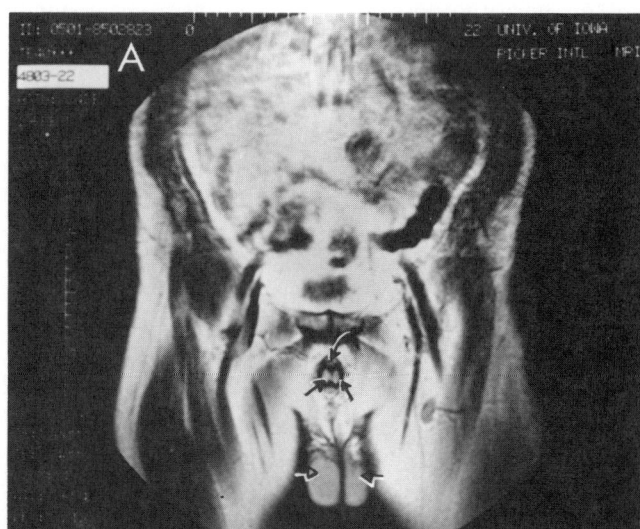

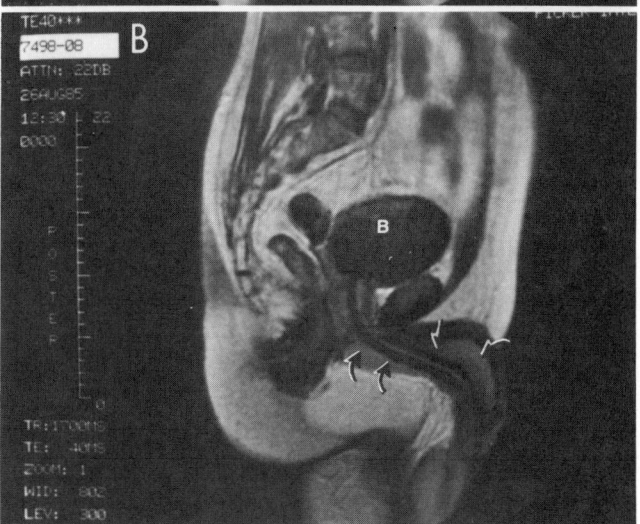

FIG 3C–23.
A, coronal MRI of normal pelvis showing corpus spongiosum *(curved arrow)*, corpora cavernosa *(straight arrows)*, and testicles *(open arrows)*. **B,** sagittal MRI of normal pelvis showing corpus cavernosum *(straight arrow)* and bulbocavernosus muscle *(curved arrows)*. There is a catheter in the bladder *(B)*.

sity by indirectly enhancing relaxation of protons surrounding the contrast agent. Studies have been done on animals using a variety of such agents. Nitroxide-stable free radicals such as TES and ethylene diamine tetra-acetic acid (EDTA) or gadolinium diethylene triamine penta-acetic acid (DTPA) have shown renal enhancement and excretory urogram-like studies in animals (Braasch et al., 1983, 1984; Ehman et al., 1985). We have previously shown that nitroxide-stable free radicals are capable of enhancing experimental human renal tumors in nude mice (Braasch et al., 1984). Recently gadolinium DTPA studies in human renal cancer have shown tumor enhancement (Leung et al., 1984). These studies raise the possibility that contrast enhancement may improve the ability to image urologic abnormalities more definitively in the future.

MRI FUTURE PERSPECTIVES

Magnetic resonance imaging is still in its developmental stages with respect to determining the optimum imaging device and defining the parameters and methods for clinical imaging. The potential of MRI in urology is particularly encouraging. Studies with first-generation MRI equipment and methods have produced images comparable in resolution to current-generation CT scans.

The usefulness of MRI is expected to increase with advances in imagers, including stronger magnets capable of improved contrast resolution, surface coils providing better images of specific anatomical regions, and faster scanning times. Technical advances in pulse sequencing for selected organs or disease states promise to provide substantial improvement of the art. The use of contrast agents and determination of blood and urine flow will likely raise MRI from an experimental to an essential technique. Whether MRI will be extended to imaging atoms other than hydrogen to provide further physiologic information is not yet known but is anticipated. Currently MRI is not sufficiently superior to other imaging modalities in the urinary tract to replace their use routinely, although staging of renal, bladder, and prostate tumors by MRI has improved. The early promise that MRI would be capable of preferentially distinguishing benign from malignant tissues has not been realized, but technical advances may change this view. As with CT, research into the uses of MRI is expected to advance rapidly. With increased experience, more sophisticated imaging equipment, and studies designed to compare MRI with other imaging modalities, the precise value of MRI in urologic patient care is expected to emerge in the very near future.

MAGNETIC RESONANCE SPECTROSCOPY

Magnetic resonance spectroscopy (MRS) is an analytic method that has been used in physics and chemistry to determine the conformation of organic molecules and to monitor in vitro metabolic reactions (Budinger and Lauterbur, 1984). Although MRS techniques actually predate the clinical use of MRI, the applications of MRS have been extended only recently to in vivo studies on humans. The major advantages of MRS include: accurate determination of the amounts and precise forms of carbohydrate, nitrogenous and high-energy phosphate compounds within tissues; measurement of intracellular pH; MRS is noninvasive and

nondestructive; and MRS studies can be sequentially and rapidly repeated (Johnston et al., 1985). The ability to monitor human intraorgan and intracellular metabolism in normal and pathologic studies in vivo has just become available; however, findings thus far suggest that MRS will become an extremely valuable technique for diagnosis and the evaluation of therapeutic response in a wide variety of human diseases.

The basic principles of MRS are similar to those described for MRI. The primary difference is that MRI uses encoded MR signals to provide a visual image directly related to the number and position of susceptible nuclei (primarily protons) within the tissue examined, whereas MRS uses the same information to provide the identity of specific chemical groups and their relative amounts within the tissue examined (Morgan and Hendee, 1984). If, for example, MRS of ^{31}P compounds in a skeletal muscle group is studied, each phosphorous form present will emit an MR signal (after RF pulsing) that is slightly different from the others, so that a spectrum (chemical shift spectra) of MR signals can be measured and recorded by the spectroscopic device. Even though the ^{31}P nuclei of each form are similar, small magnetic field modifications and shielding effects of the surrounding molecules (intrinsic to the metabolic state of the tissue examined) will cause the chemical shift spectra of ^{31}P to be detectable (Rosen and Brady, 1983). The ^{31}P chemical shift spectra shown in Figure 3C–24 identify various forms of ^{31}P in skeletal muscle as separate spectral peaks, which include sugar phosphate (SP), inorganic phosphate (Pi), phosphocreatine (PC), and the multiple forms of ATP (γ, α, and β). The rela-

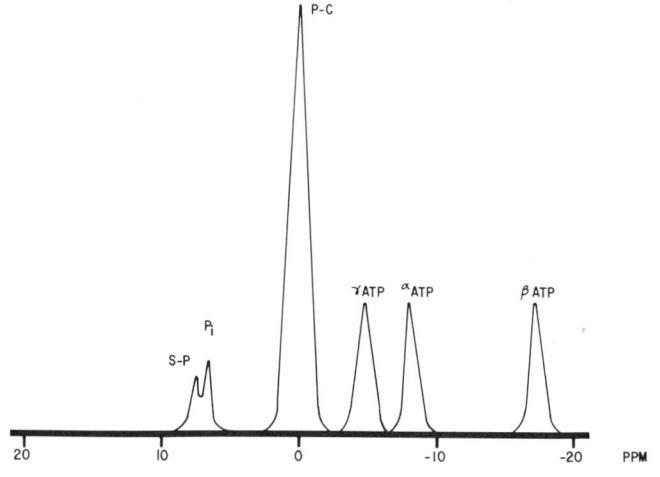

^{31}P chemical shift spectra from human muscle.

FIG 3C–24.
^{31}P MRS chemical shift spectra of normal human muscle. Sugar phosphate *(S-P)*, inorganic phosphate *(Pi)*, phosphocreatine *(P-C)*, adenosine triphosphate α, β, γ *ATP)*.

tive position of each form of ^{31}P is reasonably constant in multiple studies and thus allows their differentiation. Additionally, the area under each spectral peak is indicative of the concentration of each form of ^{31}P present (Weiner and Adam, 1985).

As with MRI, nuclei susceptible to MRS are those with an uneven atomic number—^{1}H, ^{13}C, ^{19}F, and ^{23}Na and ^{31}P. The relative abundance of each isotopic species in biologic tissues, however, dictates that only ^{1}H, ^{13}C, and ^{31}P are sufficient for clinical study. To determine the various forms of each nuclei in minute concentrations, a magnet strength greater than 1.5 T is required compared with the 0.15–0.6-T magnets used for current MRI (Budinger and Lauterbur, 1984; Nunnally, 1983). In addition, specialized RF sources (surface coils), acting as both RF generators and receivers, must be placed within the magnet and surround the subject preferentially to excite nuclei and detect MR signals from specific organs (Budinger and Lauterbur, 1984). Until recently, the strong magnets and surface coils necessary for MRS were available with only a small aperture or central bore, so that studies were restricted to appendages or small children. Devices with stronger magnets and a larger bore are currently in developmental stages, and the possibility of combined MRI and MRS studies is imminent. In the brief discussion that follows, the various clinical applications of MRS are described.

^{31}P MRS

Phosphorous MRS has been the most studied and is the most promising for clinical use. Phosphorus 31 MRS is of interest because: (1) phosphorous compounds are the major sources of free energy to drive cellular functions; (2) maintenance of adequate levels of high-energy phosphates indicates continuing cellular metabolic function; (3) high-energy phosphate levels can reflect metabolic alterations associated with genetic defects or neoplasia; and (4) intracellular pH can be estimated by measuring cellular Pi levels (Nunnally, 1983). Magnetic resonance spectroscopy research in the past was limited primarily to animals; the few human studies focused on skeletal and cardiac muscle exercise physiology, diagnosis of muscle disorders, and renal physiology. Experimental muscle ischemia has shown spectra with decreased phosphocreatine and elevated inorganic phosphate levels that return to normal with return of blood flow (Nunnally, 1983). Myocardial responses to exercise in patients with abnormal coronary blood flow may be monitored by MRS in the future. Altered cardiac muscle phosphates have been used to diagnose congenital myocardial disease (Ross et al., 1981).

The study of renal physiology by MRS is receiving

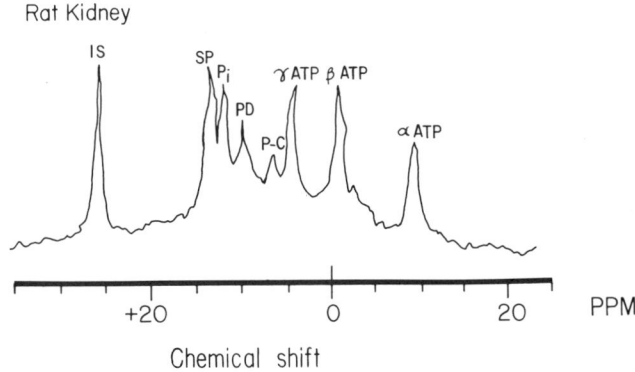

FIG 3C–25.
^{31}P MRS chemical shift spectra of normal rat kidney. Internal standard *(IS)*, sugar phosphate *(S-P)*, inorganic phosphate *(Pi)*, phosphodiester *(PD)*, phosphocreatine *(P-C)*, adenosine triphosphate *(α, β, γ ATP)*.

considerable attention. The renal chemical shift spectra of ^{31}P (Fig 3C–25) exhibit phosphodiesters (thought to represent phosphorylcholine and phosphoethanolamine) presumed to be confined to the renal medulla (Weiner and Adam, 1985). Presence of phosphodiesters will allow easy differentiation of kidney from overlying muscle and may permit the study of cortex and medulla separately. Several studies have documented that intracellular pH of the kidney can be monitored by MRS (Weiner and Adam, 1985). The spectal peak of inorganic phosphate has been shown to shift to the right with acidosis and to the left with alkalosis. The pH estimated by MRS is 7.34–7.39, correlating well with isolated tubule measurement: it is slightly higher than that measured in isolated perfused kidneys, and considerably higher than that observed in muscle (pH 7.1). The advantages of pH measurement by MRS are that it is a rapid, noninvasive method, and other determinations such as extracellular fluid volume are unnecessary. Unfortunately, calibration is difficult, since there is no practical way to reproduce the internal cellular environment for control studies. In addition, the pH measured is a mean of all renal cell types as well as that from plasma, interstitial fluid, tubular fluid and urine. Nonetheless, the evidence is convincing that intracellular pH is accurately reflected in the recorded Pi level and spectal peak shifts. Studies of the control of renal ammoniagenesis and hydrogen ion excretion should thus be facilitated by the use of MRS.

Acute renal ischemia has been shown to lower ATP and pH levels and raise Pi levels when monitored by MRS. Hypotension or noradrenalin infusion causes similar changes, which can be reversed by the return of normal blood pressure and flow. Prolonged ischemia and the development of acute renal failure results in continued alteration of Pi and ATP levels. Induction of acute renal failure by $HgCl_2$ causes no effect on the MR spectra, suggesting that neither acidosis nor ATP depletion are causative in this form of kidney failure.

Similar findings of low ATP and pH levels by MRS have been directly correlated with the development of acute tubular necrosis and subsequent viability of human kidneys in cold storage before renal transplantation (Chen et al., 1981). Kidneys with preserved ATP and low Pi levels were shown to have an improved level of renal function after transplantation. Animal studies correlating MRS of kidneys perfused for extended periods have corroborated these findings and determined a time-dependent decay in sugar phosphate-Pi and NAD-Pi ratios (Bretan et al., 1985). Because MRS can easily be accomplished in isolated cadaver kidneys during perfusion, ^{31}P MRS may allow preoperative prediction of the functional success of human cadaver renal transplantation and perhaps permit salvage of some kidneys thought to be nonviable by less optimal techniques. It is also probable that MRS can provide a useful method to study renal preservation techniques.

Damadian originally described MR signal differences between benign and malignant tissue (Damadian, 1971). MRI has not proved specific for human tumors, but it has been suggested that the intermediary metabolism of tumors differs from normal tissues due to anaerobic glycolysis, and thus an increase in lactate and local tumor acidosis may indicate neoplasia. Human tumor studies using MRS are very limited, with only two reports thus far. Ross and associates (1984) studied the ^{31}P spectra of isolated perfused human kidneys from 12 patients immediately following removal of a primary renal tumor (ten renal adenocarcinomas and two Wilms' tumors) and two normal human cadaver kidneys. In seven of the kidneys containing adenocarcinoma and one with a Wilms' tumor, a unique spectral peak not present in normal kidneys was found. They speculated the peak was due to Pi in the highly acidic environment found in metabolizing tumors. Furthermore, in one of five tumorous kidneys infused with chemotherapeutic agents, the abnormal Pi peak increased in intensity four- to five-fold. It was suggested that the Pi increase may indicate drug sensitivity of the tumor, and thus MRS may be useful to monitor therapeutic responses in selected tumors. Recently Moris et al. (1985) studied two children with neuroblastoma metastatic to liver by ^{31}P MRS and reported that both primary tumor and metastases had an elevated phosphomonoester (PME) peak when compared and normalized to the ATP β peak. They were unable to determine whether the PME peak was specific to the tumor, but because it is present in low levels in the normal liver, it is likely that only the quantity of PME is important. They determined that the PME peak was composed primarily of phosphorylethanolamine and phosphorylcholine, both known to be high in the neonatal brain, suggesting that increased phospholipid synthesis is the cause for the PME elevation. Further, spontaneous regression or favorable response to chemotherapy and/or radiation therapy correlated positively with normalization of the PME/ATP ratio. They concluded that ^{31}P MRS can be used to detect tumors and to monitor their response to therapy. These two studies are interesting, but much more research is required to determine whether MRS will be useful in oncology.

^1H, ^{13}C, and ^{14}N MRS

Proton MRS can be conducted at a magnetic field strength lower than other atoms because of its relative abundance in tissues. Proton MR spectra have been used to monitor lactate levels in ischemic and hypoxic animal brains with some success. Tissue lipids monitored by MRS (particularly in the liver in patients with fatty livers and in bone marrow) have diagnostic promise, but too few studies are available to be overly optimistic. Although no human data exist now, it is also predicted that ^1H MRS may be valuable for investigating renal ischemia and malignancy because of its ability to detect changes in lactate.

Carbon 13 has had very limited study to date but appears to be useful to study carbohydrates, particularly glycogen metabolism within the liver (Johnston et al., 1985). Studies of ^{14}N MRS are also few but suggest that nitrogen compounds such as urea, trimethylamines, amino acids, and NH^4 in the kidney and other tissues can be accurately measured (Balaban and Kuepper, 1983).

MRS FUTURE PERSPECTIVES

Magnetic resonance spectroscopy offers great promise to the clinician, particularly with respect to ^{31}P spectra. It is very likely that MRS will become a useful technique to study renal perfusion and metabolism and perhaps the response of renal tumors to therapy. Serious limitations to the routine use of MRS include the requirement of a magnet of sufficient field strength and aperture to study humans and the availability of surface coils or other methods to limit RF wave detection to the organ of interest. The most anticipated development is the combination of MRI and MRS to provide anatomical and biochemical information simultaneously. Considerable work and some success have already been achieved using MR spectra to produce an improvement in MRI spatial and contrast resolution. The future of MRS is bright; however, it will probably be several years before it becomes routinely available in the clinical setting.

POSITRON EMISSION TOMOGRAPHY

PET scanning is a relatively recent innovation in tissue and organ imaging that promises to provide information concerning regional tissue physiology. The basis of PET scanning relies on the exogenous administration of a positron-emitting radionuclide in the form of a biologic tracer. The radionuclides used commonly are short-lived isotopes of common biologic elements such as ^{11}C, ^{15}O, and ^{13}N. Fluorine 18 behaves similar to ^{1}H (which does not emit positrons) and is thus used as a substitute. These isotopic radionuclides decay rapidly while emitting positrons (positive electrons). Positrons move a few millimeters in soft tissue and then interact with a negative electron, causing the emission of two 511-keV gamma rays, which move 180° apart and can easily penetrate tissues. These paired rays are coincidentally detected externally by specially constructed gamma detectors (Phelps and Mazziatta, 1985; Brownwell et al., 1982).

Fundamental enthusiasm for PET scanning is related to the facts that: (1) the substrates that could be radiolabeled are limitless, because the basic elements are part of nearly all biomolecules and drugs; (2) the method will provide high spatial resolution and detection efficiency; (3) the regional concentration of radioisotopes can be determined in measurable units, because the PET system can correct for signal loss by tissue attenuation; and (4) minute amounts of tracer can be utilized to provide accurate measurements. Current resolution appears to be in the range of 6–8 mm, but 2-mm resolution scanners are expected within the near future. Disadvantages of this technique include: (1) the extremely short physical half-life of radioisotopes—^{15}O, 2 min; ^{11}C, 20 min; ^{13}N, 10 min; and ^{18}Fl, 110 min— necessitating being in close proximity to a center where the isotopes are produced (the cyclotron); and (2) the limitations of partial volume effects and collected data statistics.

A variety of compounds have been labeled with positron-emitting radionuclides, including: ^{68}Ga-labeled EDTA used for cerebral blood volume measurements; ^{18}F-labeled ethanol used for cerebral blood flow calculations; ^{11}C deoxyglucose used for glucose transport and metabolism studies; ^{11}C leucine used for protein synthesis studies; and a variety of ^{11}C and ^{18}F drug tracers used for receptor studies.

To date, the majority of laboratory studies have been focused on brain metabolism and physiologic function, although a few studies have determined that neoplasms (particularly those primary in the brain and breast and colon metastatic to liver) can be detected using ^{18}F deoxyglucose as a tracer (Beaney, 1984). Other investigators have utilized labeled amino acid uptake for imaging with some success. Labeling of monoclonal antibodies with positron emitters is being explored for PET scanning of specific tumors, but no data have yet been reported.

There are as yet no studies on human urologic tumors reported, although recent studies by Fair and Kadmon (1983) suggest that polyamines labeled with positron-emitting radionuclides such as ^{11}C putrescine may be extremely useful for prostate imaging. Their early work has shown that the polyamines, which are known to be in high concentration in the human prostate, can be selectively taken up by normal dog prostate and rat prostate tumors by manipulating the polyamine metabolic enzymes (Kadmon et al., 1985). Positron emission tomography scanning of the normal dog prostate and spontaneous dog tumors has been successful (Miller et al., 1978). Current research relating to development of appropriate radiolabeled putrescine analogs for human PET scanning is under way, but no human studies have been completed.

Positron emission tomographic scanning for human urologic imaging is in only the earliest developmental stages, but the potential of the PET method has broad implications for the study of urologic organ and tumor physiology.

REFERENCES

1. Amendola MA, Glazer GM, Grossman HB, et al: Staging of bladder carcinoma: MRI-CT surgical correlation. *AJR* 1986; 146:1179.
2. Baker HL Jr, Berquist TH, Kispert DB, et al: Magnetic resonance imaging in a routine clinical setting. *Mayo Clin Proc* 1985; 60:75–90.
3. Balaban RS, Kuepper MA: Nitrogen-14 nuclear magnetic resonance spectroscopy of mammalian tissues. *Am J Physiol* 1983; 245:c439–c444.
4. Beaney RP: Positron emission tomography in the study of human tumors. *Semin Nucl Med* 1984; 14:324.
5. Braasch RC, London DA, Wesbey GE, et al: Work in progress: Nuclear magnetic resonance study of a paramagnetic nitroxide contrast agent for enhancement of renal structures in experimental animals. *Radiology* 1983; 147:773–779.
6. Braasch RC, Weinemann HJ, Wesbey GE: Contrast-enhanced NMR imaging: Animal studies using gadolinium-DTPA complex. *AJR* 1984; 142:625–630.
7. Bretan PN, Vigneron DB, James TL, et al: Assessment of renal viability by ^{31}Phosphorus magnetic resonance spectroscopy. *J Urol* 1986; 135:866.
8. Brownwell GL, Budinger TF, Lauterbur PC, et al: Positron tomography and nuclear magnetic resonance imaging. *Science* 1982; 215:619–626.
9. Bryan PJ, Butler HE, Lipuma JP: NMR scanning of the pelvis: Initial experience with a 0.3T system. *AJR* 1983; 141:1111–1118.
10. Budinger TF, Lauterbur PC: Nuclear magnetic resonance

technology for medical studies. *Science* 1984; 226:288.

11. Buonocore E, Hesemann C, Pavlicek W, et al: Clinical and in vitro magnetic resonance imaging of prostatic carcinoma. *AJR* 1984; 143:1267–1272.

12. Carr DH: The use of proton relaxation enhancers in magnetic resonance imaging. *JMRI* 1985; 3:17–25.

13. Chen L, French ME, Gadian DG, et al: Studies of human kidneys prior to transplantation by phosphorus nuclear magnetic resonance, in Pegg, Halasz, Jacobsen (eds): *Organ Preservation*, Lancaster, Pa, Lancaster MTP Press, 1981, vol 3.

14. Choyke PC, Kressel HY, Pollack HM: Focal renal masses: Magnetic resonance imaging. *Radiology* 1984; 152:471–477.

15. Damadian R: Tumor detection by nuclear magnetic resonance. *Science* 1971; 171:1151–1153.

16. Demas BE, Hricak H, Williams RD: Magnetic resonance imaging in the evaluation of urologic malignancies. *Semin Urol* 1985; 3:27–33.

17. Ehman Rl, Wesbey GE, Moon KL, et al: Enhanced MRI of tumors utilizing a new nitroxyl spin label contrast agent. *JMRI* 1985; 3:89.

18. Ellis JH, Bies JR, Kopecky KK, et al: Comparison of NMR and CT imaging in the evaluation of metastatic retroperitoneal lymphadenopathy from testicular carcinoma. *J Comput Assist Tomogr* 1984; 8:709–719.

19. Fair WR, Kadmon D: Carcinoma of the prostate: Diagnosis and staging. *W J Urol* 1983; 1:3.

20. Fisher MR, Hricak H, Crooks LE: Urinary bladder MR imaging: Part I. Normal and benign conditions. *Radiology* 1985a; 157:467–470.

21. Fisher MR, Hricak H, Tanagho EA: Urinary bladder MR imaging: Part II. Neoplasm. *Radiology* 1985b; 157:471–477.

22. Geisinger MA, Risius B, Jordon ML, et al: Magnetic resonance imaging of renal transplants. *AJR* 1984; 143:1229–1234.

23. Hricak H: MR imaging of the retroperitoneum and pelvis. *Br Med Bull* 1984; 40:197–201.

24. Hricak H, Demas BE, Williams RD, et al: Magnetic resonance imaging in the diagnosis and staging of renal and perirenal neoplasms. *Radiology* 1985; 154:709–715.

25. Hricak H, Newhouse JH: MR imaging of the kidney. *Radiol Clin North Am* 1984; 22:287–296.

26. Hricak H, William RD: Magnetic resonance imaging and its application in urology. *Urology* 1984; 23:442–454.

27. Hricak H, Williams RD, Spring DB, et al: Anatomy and pathology of the male pelvis by magnetic resonance imaging. *AJR* 1983; 141:1101–1110.

28. Johnston DC, Lin P, Wismer GL, et al: Magnetic resonance imaging: Present and future applications. *Can Med Assoc J* 1985; 132:765–777.

29. Kadmon D, Mahle D, Heston WDW, et al: Effect of estrogen and androgen administration on alpha DFMO-enhanced putrescine uptake by the rat prostate. *Prostate* 1985; 6:343.

30. Kulkarni MV, Shaff MI, Sandler MP: Evaluation of renal masses by MR imaging. *J Comput Assist Tomogr* 1984; 8(5):861–865.

31. Leung AWL, Bydder GM, Steiner RE, et al: Magnetic resonance imaging of the kidneys. *AJR* 1984; 143:1215–1227.

32. Lipuma JP: Magnetic resonance imaging of the kidney. *Radiol Clin North Am* 1984; 22:925–941.

33. Miller TR, Sieget BA, Fair WR, et al: Imaging of canine tumors with ^{11}C-Methyl-Putrescine. *Radiology* 1978; 129:221.

34. Moon KL, Hricak H, Crooks LE, et al: NMR of the adrenal gland: A preliminary report. *Radiology* 1983; 147:155–160.

35. Moris JM, Evans AE, McLaughlin AC, et al: ^{31}P nuclear magnetic resonance spectroscopic investigation of human neuroblastoma in situ. *N Engl J Med* 1985; 312:1500–1505.

36. Morgan CJ, Hendee WR: Magnetic resonance spectroscopy, in *Introduction to Magnetic Resonance Imaging*. Denver, Multi-Media Publishing, 1984, p 159.

37. Nijhout MA, Falke THM, Jones B, et al: Magnetic resonance imaging of the bladder and prostate. *World J Urol* 1985; 3:66–72.

38. Nunnally RL: In vitro monitoring of metabolism with nuclear magnetic resonance spectroscopy. *Semin Nucl Med* 1983; 13:377–382.

39. Paushter DM, Modic MT, Borkowski GP: Magnetic resonance principles and application. *Med Clin North Am* 1984; 68:1393–1421.

40. Phelps ME, Mazziatta JC: Positron emission tomography: Human brain function and biochemistry. *Science* 1985; 228:799.

41. Poon PY, McCollum RW, Henkelman MM: Magnetic resonance imaging of the prostate. *Radiology* 1985; 154:143–150.

42. Pykett IL: NMR imaging in medicine. *Sci Am* 1982; 246:78.

43. Resnick MI, Kursh ED, Bryan PD: Nuclear magnetic resonance imaging and bladder cancer, in Kuss R, et al (eds): *Bladder Cancer, Part A: Pathology, Diagnosis and Surgery*. New York, Alan R Liss, Inc, 1984, pp 255–265.

44. Ross ED, Radda GK, Gadian DG, et al: Examination of a case of suspected McArdle's syndrome by ^{31}P nuclear magnetic resonance. *N Engl J Med* 1981; 304:1338–1343.

45. Ross B, Smith M, Marshall V, et al: Monitoring response to chemotherapy of intact human tumors by ^{31}P nuclear magnetic resonance. *Lancet* 1984; 1:641–646.

46. Rosen BR, Brady TJ: Principles of nuclear magnetic resonance for medical application. *Semin Nucl Med* 1983; 13:308–318.

47. Runge VM, Clanton JA, Herzer WA, et al: Intravascular contrast agents suitable for magnetic resonance imaging. *Radiology* 1984; 153:171–176.

48. Schmidt HC, Tschakaloff D, Hricak H, et al: MR image contrast and relaxation times of solid tumors in the chest, abdomen and pelvis. *J Comput Assist Tomogr* 1985; 9:738–748.

49. Schultz CL, Haaga JR, Fletcher BD, et al: Magnetic resonance imaging of the adrenal glands: A comparison with computed tomography. *AJR* 1984; 143:1235–1240.

50. Weiner MW, Adam WR: Magnetic resonance spectros-

copy for evaluation of renal function. *Semin Urol* 1985; 3:34–42.

51. Williams RD, Hricak H: Magnetic resonance imaging in urology. *J Urol* 1984a; 132:641–649.

52. Williams RD, Hricak H: Nuclear magnetic resonance imaging in urologic oncology, in Ratliff TL, Catalona WJ (eds): *Urologic Oncology.* Boston, Martin Nijhoff Publishers, 1984b, pp 317–354.

53. Young SW: *Nuclear Magnetic Resonance Imaging: Basic Principles.* New York, Raven Press, 1984.

Chapter 3D / Ultrasound

J. PATRICK SPIRNAK, M.D.
MARTIN I. RESNICK, M.D.

The development of ultrasound instrumentation effective in the evaluation and treatment of many medical conditions dates back to the 1940s, when Firestone first described a technique using ultrasonic waves to detect flaws in metal castings (Martin, 1984). During World War II, the principles of ultrasound were used to develop SONAR (*Sound Navigation And Ranging*), a tracking technique useful in identifying submerged enemy submarines. The first medical use of ultrasound was reported by an Austrian psychiatrist, Karl Dussik (1942), who attempted to locate brain tumors using two opposing ultrasound transducers. In the early 1950s Howry and Bliss (1952), using discarded naval equipment, studied and recorded the echo patterns obtained from a variety of soft tissue structures (Howry and Bliss, 1952). They initially used water immersion techniques, and it was not until the early 1960s that they began experimenting with hand-held scanners applied directly to the body surface. Since the early 1970s, with the development of real-time capability and gray-scale imaging, ultrasound has gained an important role in the clinical evaluation of many urologic abnormalities.

BASIC PRINCIPLES OF ULTRASOUND

Ultrasound consists of sound waves with frequencies beyond the audible range of the human ear (>20,000 cps). For medical purposes sound waves with frequencies of 2–8 million cps (2–8 MHz) are used. The ultrasound waves used in the examination are formed in the transducer following the excitation of a piezoelectric crystal with electrical energy (Fig 3D–1,A). They are transmitted in short bursts lasting 10 msec. The thickness of the crystal determines the frequency of the generated sound waves. The sound waves are focused using an acoustical lens, which produces a narrow beam of only a few millimeters in diameter with good lateral resolution.

The body is the molecular medium used to propagate the generated sound waves. In the presence of a homogeneous fluid medium, the sound waves are propagated in an uninterrupted fashion (anechoic). If the sound waves encounter a different density of tissue, a portion will be reflected back to the source (Fig 3D–1,B). The reflecting boundary is called an acoustic interface, and it exists because of the differing acoustic impedances between the two different mediums. Acoustic impedance is defined as the product of the medium's density and the speed of sound in the medium. The greater the difference in acoustic impedance between two mediums, the greater the amount of sound reflected back from the interface. Typically, less than 1% of the incident energy is reflected.

Whether or not the sound waves are reflected also depends on the relative size of the interface and the frequency of the generated wavelength. Sound waves of higher frequencies are generally reflected by smaller surface interfaces but have less depth of penetration, thereby giving better spatial resolution at the expense of decreased sound wave penetration. The use of lower-frequency sound waves sacrifices spatial resolution but makes it possible to image deeper structures. Appropriate transducer selection is critical for the success of the evaluation, and it is generally best to select the highest frequency that permits adequate tissue penetration and visualization of the desired structures.

During the course of the examination, the ultrasound transducer also functions as a receiver of the sound waves that have been reflected from the different tissue interfaces. The sound wave energy is converted in the transducer by the piezoelectric crystal to electrical energy, which is then processed and displayed as a dot on a cathode ray tube (oscilloscope). Amplitude or A-mode records reflected echoes as spikes arising from a horizontal baseline (Fig 3D–2). The main application of this mode is to measure the distances from the margins of a mass to the skin surface or to determine the internal architecture of a mass. The brightness modulated display, or B-mode, is a composite two-dimensional image obtained while the transducer is moved along a given arc; the resultant display is of a cross-sectional image of the examined area. With the recent development of scan converters, images are now displayed in varying shades of gray (gray scale). Gray-scale imaging offers a significant improvement in the quality of images compared with those obtained on early B-scanners.

The images available for interpretation in most diagnostic ultrasound departments are gray-scale ultrasono-

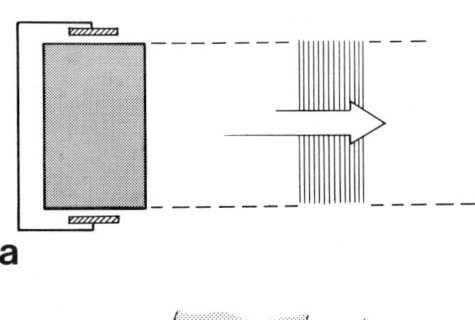

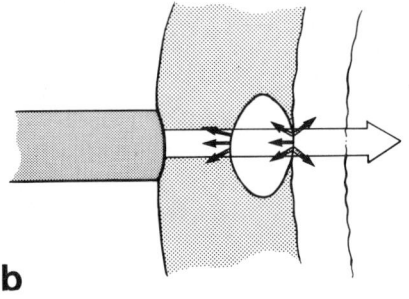

FIG 3D–1.
A, high-frequency sound waves are transmitted following excitation of a piezoelectric crystal. **B,** transmitted sound waves are reflected at tissue interfaces where there is a change in acoustic impedance. Reflected sound waves obey laws of optics—the angle of incidence is equal to the angle of reflection.

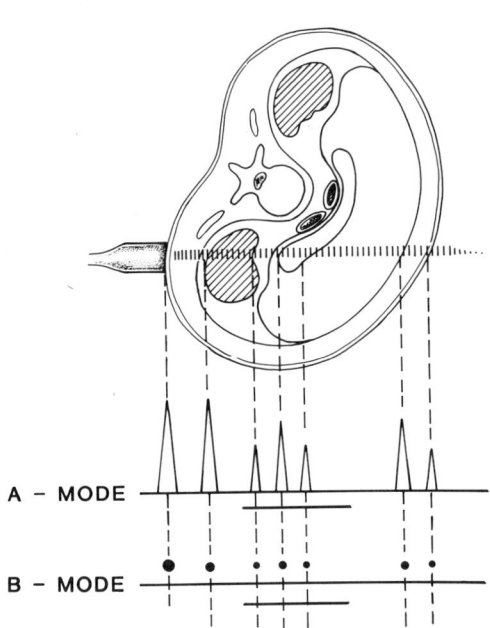

FIG 3D–2.
Amplitude mode (A-mode) delineates reflected echoes as spikes arising from a horizontal baseline. The echo amplitude is plotted against distance between transducer and reflecting tissue interface. Brightness modulated display (B-mode) is a composite two-dimensional image obtained while transducer is moved along a given arc. Echoes are shown as multiple dots or points on the oscilloscope screen, the brightness of which is dependent on the intensity of the echo.

grams and stop-action views obtained from a real-time examination. Real-time ultrasonography offers the advantage of dynamic imaging similar to that obtained during fluoroscopy. The obtained images are representations of a spectrum of echoes arising from the surface and substance of the given organ in two dimensions.

With current scanners, a complete cross section of the body is formed one line at a time. After one line is drawn, the transducer is moved slightly and pulsed again to draw another line. Typically, 1,000/sec are drawn, enabling a complete body cross section to be formed in 1–10 seconds. A permanent record of a scan can be recorded on either Polaroid or radiographic film. Echo display initially limited to shades of gray on a white background can now also be obtained in a white-on-black format depending on the preference of the examining physician.

RENAL ULTRASOUND

Patients can be scanned in the supine, decubitus, or prone positions, and images are obtained in the transverse and longitudinal planes (Fig 3D–3,A and B). In the past, the kidneys were usually examined with the patient in the prone position. However, the frequent

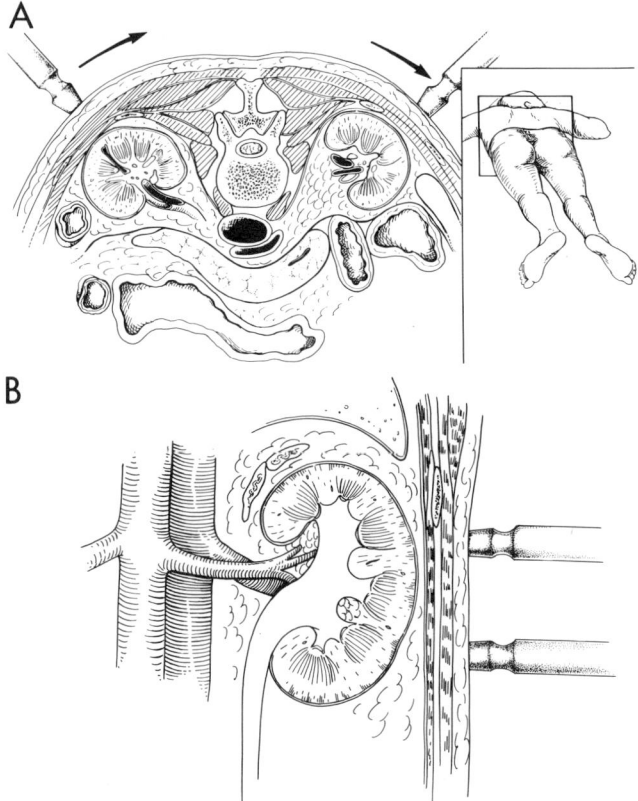

FIG 3D–3.
Image planes of a transverse renal scan **(A)** and a longitudinal renal scan **(B).**

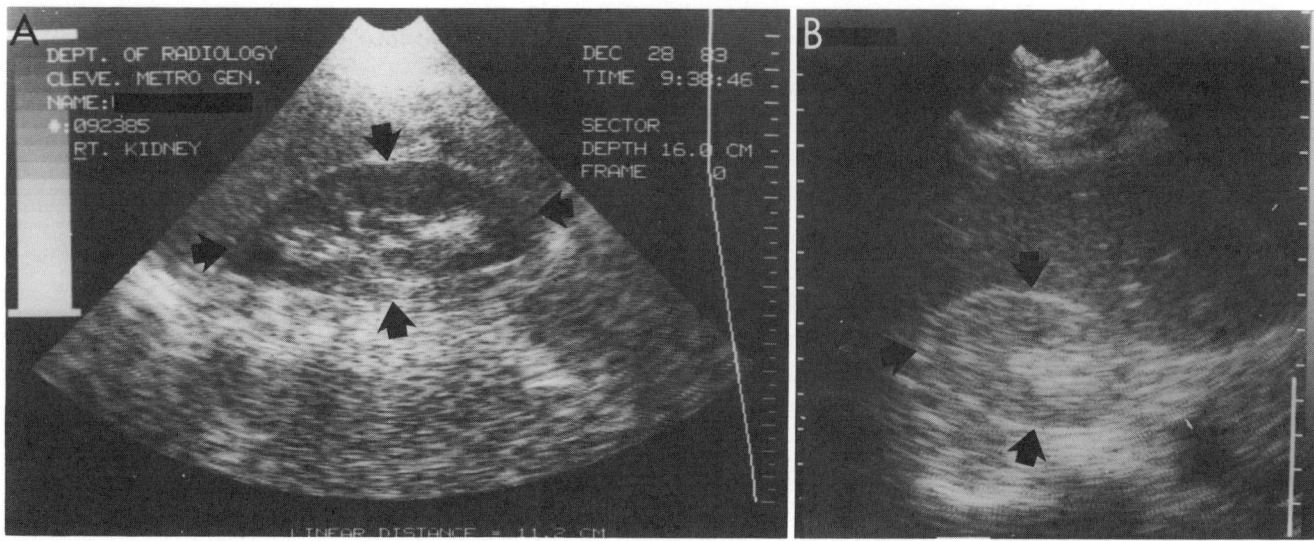

FIG 3D–4.
Longitudinal **(A)** and transverse **(B)** scans of normal kidney. Note sharp demarcation of renal capsule *(arrows)*; pelvic-calyceal system noted centrally as echogenic area.

location of the upper poles beneath the ribcage often led to incomplete or unsatisfactory studies. Currently the right kidney is best examined with the patient supine, utilizing the liver as an acoustic window. The best images of the left kidney are obtained with the patient in the right lateral decubitus position, using the spleen as an acoustic window (Albarelli and Lawson, 1978). When indicated, different views can also be obtained in the oblique plane by moving the transducer through different arcs relative to the longitudinal and transverse planes.

RENAL ANATOMY

The adult kidneys are paired retroperitoneal structures lying between the 12th thoracic and second lumbar vertebrae. Each kidney is approximately 10 cm long, 5 cm wide, and 2–5 cm thick. The right kidney is typically lower than the left (Bo and Krueger, 1984; Brandt et al., 1982). The upper pole of the right kidney is in contact with the adrenal superomedially, the duodenum anteriomedially, and the posterior aspect of the right lobe of the liver overlies its anterior surface. Overlying the lower pole of the right kidney is the hepatic flexure of the colon and loops of small bowel. The anterior surface of the left kidney is in contact with the spleen, left adrenal, stomach, pancreas, descending colon, and small bowel. Posteriorly the kidneys are in contact with the 12th rib, diaphragm, transversus abdominis muscle, quadratus lumborum muscle, and the psoas major muscle.

Each kidney is surrounded by a closely adherent, dense, fibrous capsule, which is surrounded by peri-nephric fat and enclosed in Gerota's fascia. On ultrasonic examination, the renal outline is determined by the acoustic interface formed by the renal capsule and surrounding fat. Transverse scans show the kidney to be round or bean-shaped, whereas longitudinal images reveal an elliptical configuration (Fig 3D–4,A and B). The renal parenchyma is relatively homogeneous, with the cortex appearing more echogenic than the medulla. Located in the center of the longitudinal renal scan is a dense echocomplex consisting primarily of peripelvic fat, renal vessels, lymphatics, and normal collecting system. The renal medullary pyramids appear somewhat less echogenic (more sonolucent) than the renal cortex and are frequently identified abutting on the dense sinus fat. The arcuate arteries separate the cortex from the renal medulla and when visualized allow for an accurate assessment of the renal cortical thickness. The renal vascular pedicle often can be identified using real-time scanning techniques. The ureters, when not dilated, are rarely visualized (Bodney et al., 1983).

INDICATIONS FOR RENAL ULTRASOUND

RENAL MASS

The question of a renal mass is usually raised by an abnormality found on excretory urography. In the adult, simple renal cysts are the most commonly diagnosed renal mass lesion. They are rarely seen in children. They occur with increasing frequency beyond the age of 30, and in an autopsy series about 50% of all patients older than 50 years were found to have at least

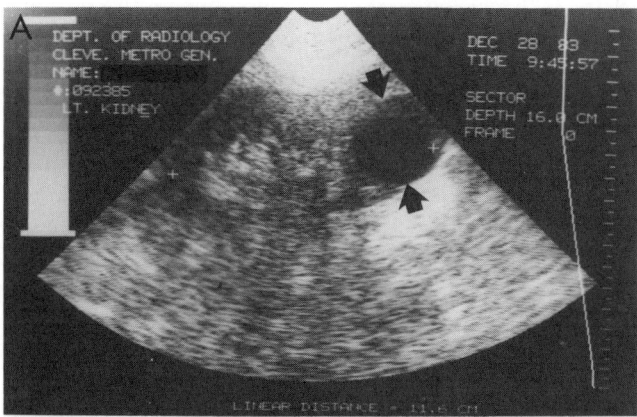

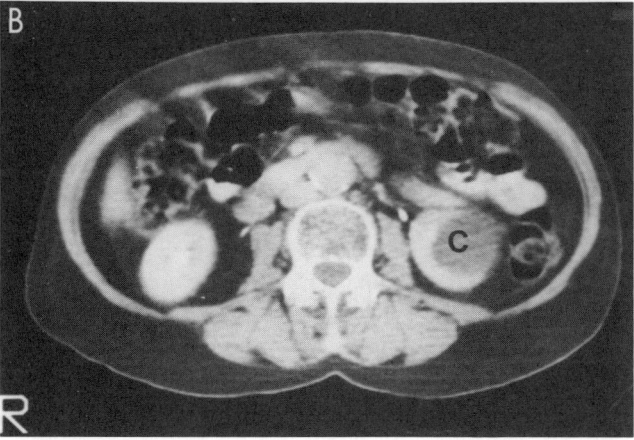

FIG 3D–5.
A, longitudinal renal scan showing characteristic findings of a simple renal cyst (arrows). **B,** cyst (C) demonstrated on CT scan.

one renal cyst (Kissane, 1974). The pathogenesis of a simple renal cyst is still unknown; however, experimental studies have shown that cysts will rapidly develop in the presence of tubular obstruction and ischemia (Hepler, 1930).

Ultrasound has been found to be extremely valuable in differentiating a simple renal cyst from other renal masses and is, therefore, the first study obtained in the evaluation of a suspected renal mass. On ultrasound a renal cyst is typically spherical or slightly ovoid, demonstrates no internal echoes, has a clearly identifiable thin wall separate from the surrounding parenchyma, and allows enhancement of ultrasound transmission beyond the cyst (Green and King, 1976) (Fig 3D–5,A and B). When these criteria are strictly adhered to, diagnostic accuracy rates approaching 100% have been reported (Pollack et al., 1982). If a renal mass meets all of the above criteria and the patient is asymptomatic, no further evaluation is required. If, however, the patient is symptomatic (pain and/or hematuria), cyst aspiration under ultrasound guidance is performed to confirm the diagnosis. Aspirated fluid is routinely sent for cytologic study, and lactate dehydrogenase (LDH), culture, and cholesterol determinations. While performing cyst puncture, contrast material and air may be injected into the cyst to define better the cyst's margins and contours of the wall.

The finding on ultrasound of a solid or indeterminate (complex) renal mass raises the possibility of the presence of a renal malignancy and dictates that additional evaluation be performed. Ultrasound characteristics of a solid renal mass include the presence of internal echoes, poor delineation of the posterior wall, and the lack of enhancement through transmission (Fig 3D–6,A and B). Charboneau and associates (1983) reviewed the

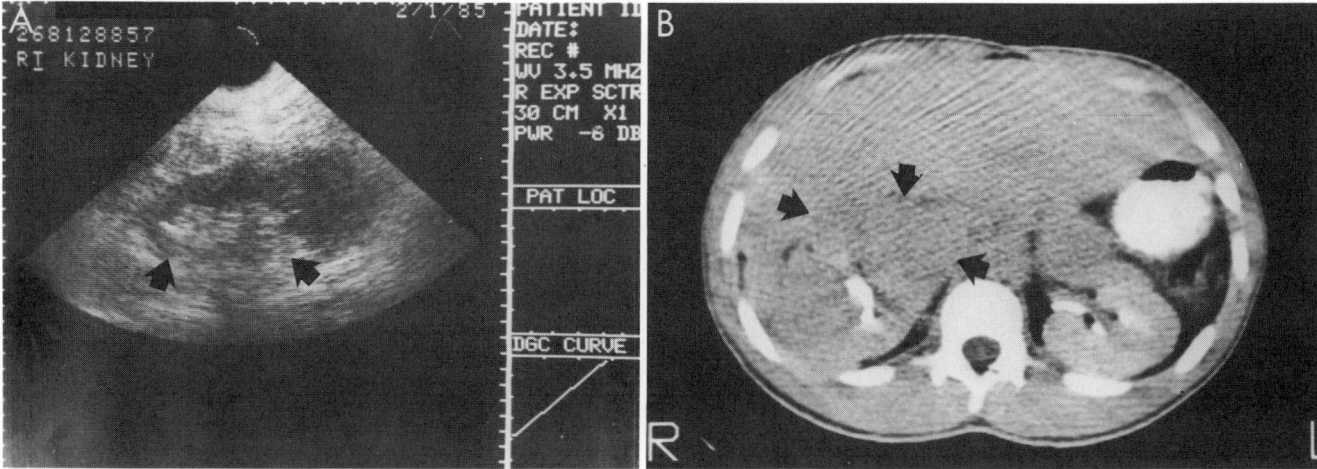

FIG 3D–6.
Longitudinal renal ultrasound **(A)** and CT **(B)** demonstrating findings of a hypernephroma (arrows).

ultrasound findings in 125 patients with renal masses not fulfilling the criteria of a simple renal cyst and identified 102 cases of renal carcinoma. In their experience 86 patients demonstrated internal echogenicity similar to that found in either liver or normal renal parenchyma. In ten patients the echogenicity of the mass was less than that of normal renal parenchyma. Four patients demonstrated echogenicity similar to that of renal sinus fat. Acoustic transmission was unchanged in 82, increased in 13, and decreased in 5 compared with normal parenchyma; those tumors with increased enhancement through transmission had pathologically confirmed cystic areas within the mass. Ninety-four of the tumors had ill-defined or undefined walls, whereas six had well-defined walls. Two tumors had internal calcifications, and both showed high-level echoes with acoustic shadowing. There were 9 angiomyolipomas, 3 multiloculated cysts, 3 renal abscesses, 2 transitional cell carcinomas, and 1 each of cystic nephroma, oncocytoma, malacoplakia, focal interstitial nephritis, and renal artery aneurysm.

The ultrasonic diagnosis of solid lesions is not as accurate as that of cystic lesions, and a false negative rate has been reported to be as high as 14% (Sherwood, 1975). The diagnosis of a solid lesion therefore requires further evaluation and possibly exploration.

Ultrasound can identify cystic masses as small as a few millimeters in diameter. Solid mass lesions 1–2 cm in diameter also are routinely identified. The ability to identify a renal mass lesion is dependent on the depth of the lesion beneath the skin and the body habitus of the patient.

Angiomyolipoma, a benign renal tumor, seems to be the only solid renal mass with typical ultrasonographic features that allow presumptive diagnosis and safe differentiation from other solid mass lesions (Lee et al.,

1978). If the lesion contains a high concentration of fat, it will be extremely echogenic, even more so than a solid renal mass. Hyperechogenicity coupled with the CT findings of fat lucency allows the diagnosis to be made with certainty (Pode et al., 1985) (Fig 3D–7).

In the presence of a complex cystic lesion, the most likely diagnosis is necrotic tumor, hematoma, or abscess. The clinical course coupled with the findings on angiography and/or CT allow an accurate, certain diagnosis. When the diagnosis remains obscure, aspirational biopsy and/or exploration may be required.

COLLECTING SYSTEM MASSES

Renal ultrasound has been found to be extremely helpful in evaluating a noncalcified mass of the collecting system detected on excretory urography. A noncalcified or poorly calcified renal calculus (uric acid) demonstrates classic ultrasound findings. Typically these calculi demonstrate increased echogenicity with acoustic shadowing (Pollack et al., 1978) (Fig 3D–8). Other common soft tissue masses of the collecting system include tumors primarily of transitional cell origin, blood clots, and sloughed renal papilla (Subramanyam et al., 1982) (Fig 3D–9). These abnormalities, although usually easily distinguishable from a nonopaque stone, frequently require additional urologic studies to arrive at the correct diagnosis.

RENAL FAILURE

Classically, renal failure may be due to prerenal, renal, or postrenal causes. The patient who presents with acute renal failure represents a diagnostic challenge and requires immediate evaluation not to miss a surgically correctable cause. Renal ultrasound is usually the first study obtained and has been found to be ex-

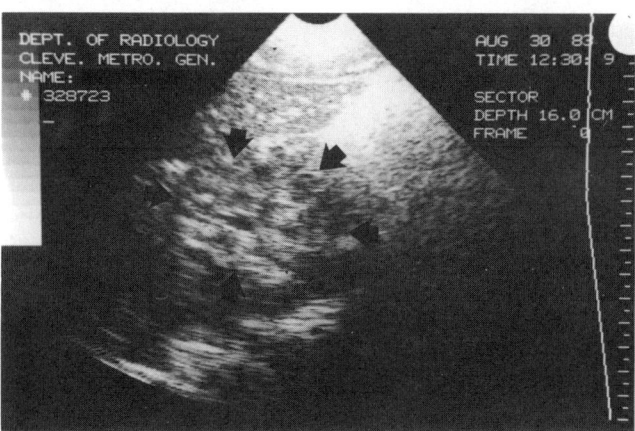

FIG 3D–7.
Transverse renal scan showing highly echogenic renal mass *(arrows)*. Exploration revealed angiomyolipoma.

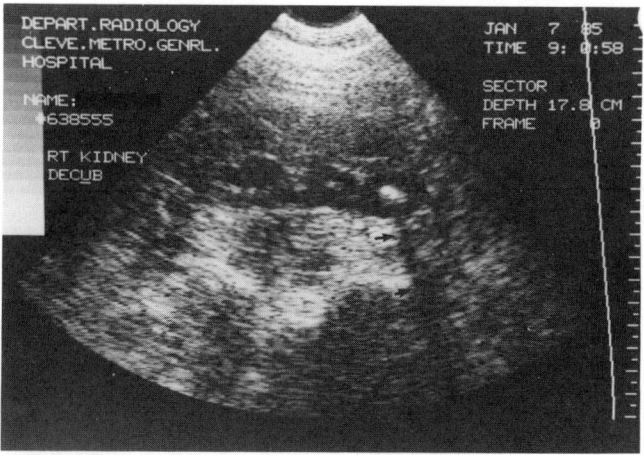

FIG 3D–8.
Longitudinal scan demonstrating stone in the lower pole calyx. Stone is highly echogenic. Note area of acoustic shadowing *(arrows)*.

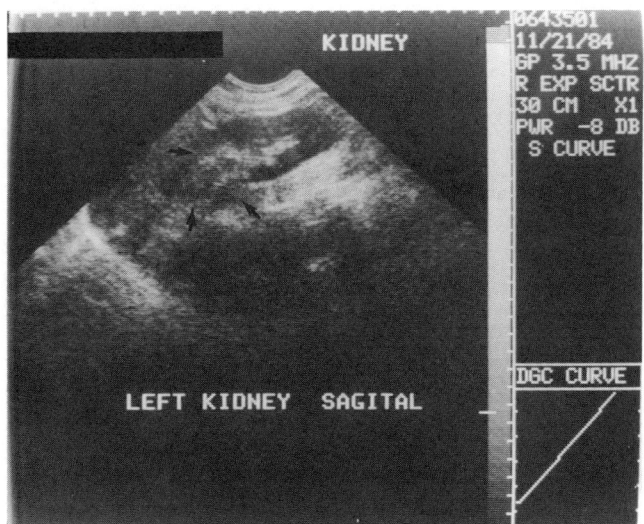

FIG 3D–9.
Longitudinal renal scan demonstrating upper pole mass *(arrows)*. Surgical exploration revealed transitional cell carcinoma.

tremely valuable in differentiating medical from surgically correctable causes of acute renal failure.

Although accounting for only about 5% of all cases of renal failure, ureteral obstruction is potentially reversible and is usually accompanied by hydronephrosis. As the collecting system becomes dilated, the distended calyces and infundibula become readily visible as hypoechoic areas, and the central echo complex becomes distorted (Fig 3D–10,A and B). With massive long-standing obstruction, the central echo complex frequently becomes unidentifiable, and it may become difficult to differentiate chronic obstruction with marked calyceal dilation from large cystic lesions (Russell and Resnick, 1979).

As a screening study to rule out obstruction as a cause of renal failure, ultrasound has a sensitivity approaching 100% (Ellenbogen et al., 1978; Talner et al., 1981). However, a normal study does not totally exclude obstruction and additional urologic evaluation may be required in selected cases.

Patients with renal failure resulting from parenchymal disease characteristically demonstrate small kidneys with a diffuse increase in echogenicity (Moccia et al., 1980).

RENAL TRANSPLANTATION

The transplanted kidney, owing to its superficial position in the pelvis, is particularly well suited for ultrasound evaluation, which has also proved to be useful in evaluating posttransplant diminishing urine output. The acute changes in renal volume associated with rejection can usually be differentiated from an obstructive cause

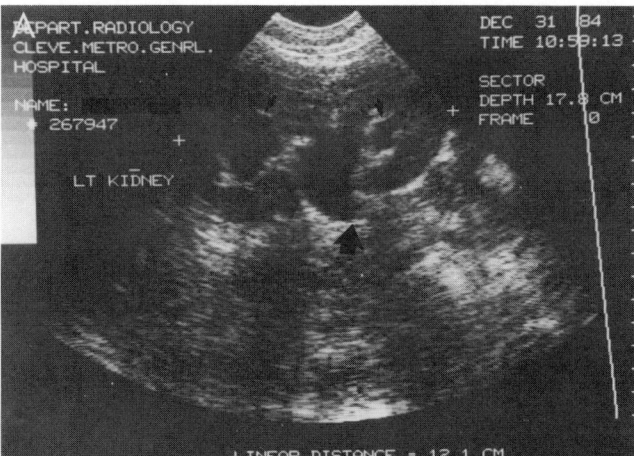

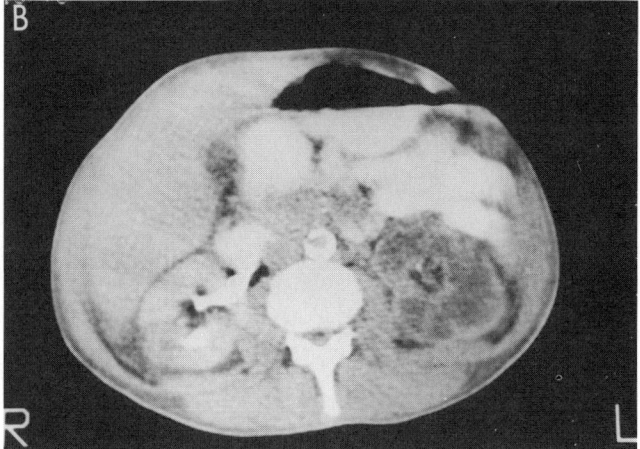

FIG 3D–10.
A, longitudinal scan demonstrating hydronephrosis secondary to ureteropelvic junction obstruction *(large arrow)*. Note dilated calyces *(small arrows)*. **B,** CT scan of same patient showing hydronephrotic kidney.

(Rosenfield and Taylor, 1976). It is also possible to detect the presence of perirenal hematomas, lymphoceles, and urinomas that may occur in the transplant patient.

PERIRENAL EVALUATION

Ultrasonography is useful in the evaluation of the perirenal space. Perirenal fluid collections may represent hematoma or urinoma, which usually are associated with an antecedent history of trauma or prior renal surgery (Fig 3D–11,A and B). Urinomas tend to be cystic, whereas hematomas may be solid, cystic, or a combination of both, depending on the stage of clot breakdown. Ultrasound offers an accurate, noninvasive, radiation-free means of following the presence and course of perinephric hematomas resulting from blunt trauma.

The presence of a perinephric abscess is usually suggested by a febrile course unresponsive to the usual antibiotic regimen. Sonographically the perinephric col-

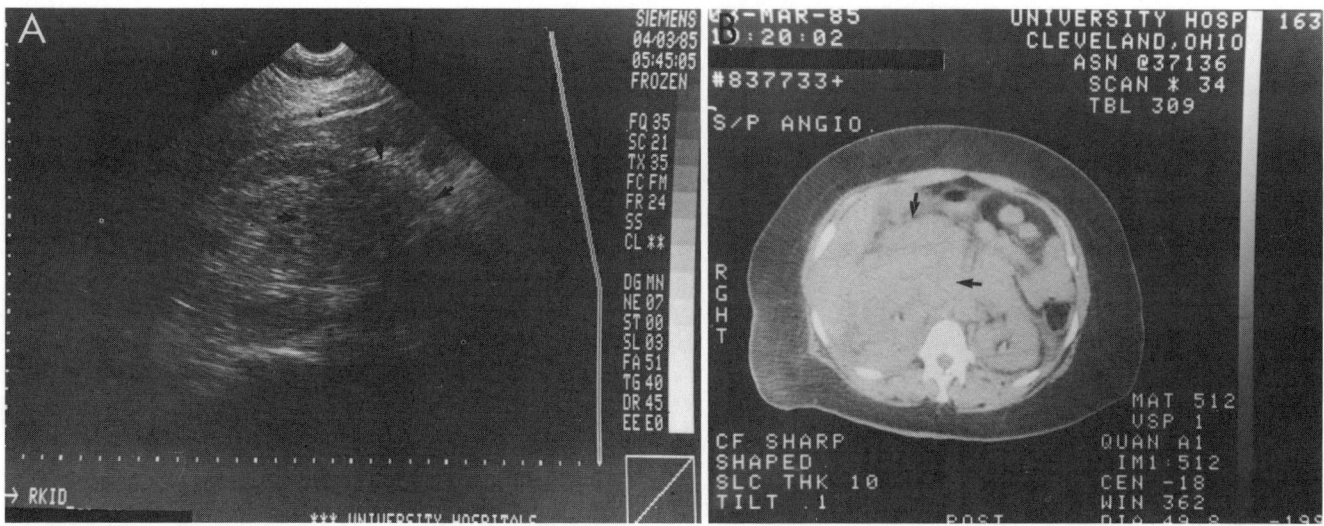

FIG 3D–11.
A, large perirenal hematoma following blunt abdominal trauma *(arrows).* **B,** CT of same patient.

lection appears as a complex mass with the internal architecture dependent on the amount and type of pus and the presence or absence of gas within the abscess (Kressel and Filly, 1978) (Fig 3D–12).

SCREENING PROCEDURES

Renal ultrasonography has a role in the evaluation of suspected adult polycystic renal disease. The sonographic appearance is that of enlarged kidneys with randomly distributed cysts (Fig 3D–13,A and B). Early cystic renal changes are ultrasonically visualized prior to their detection by excretory urography; therefore, screening of family members is useful in identifying affected individuals in their early years.

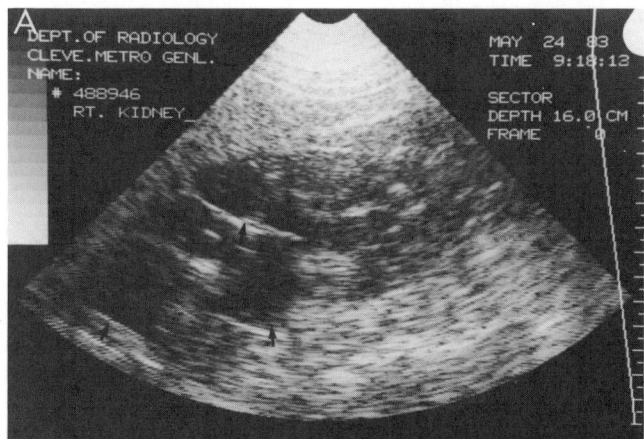

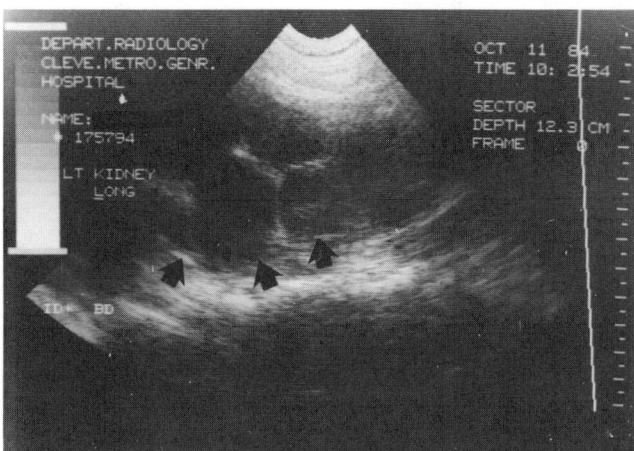

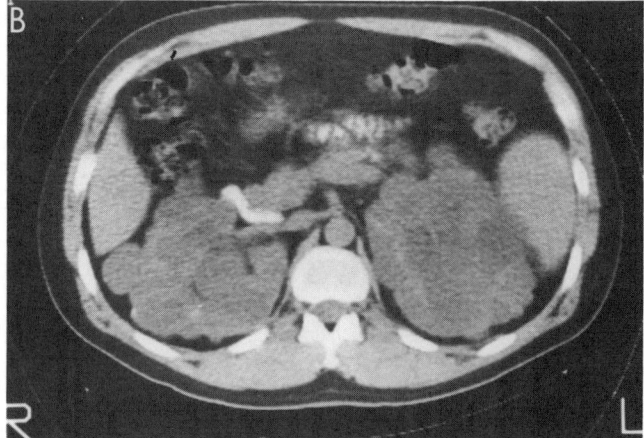

FIG 3D–12.
Transverse scan showing complex perirenal mass, proved to be multiloculated perinephric abscess. Note multiple loculations *(arrows).*

FIG 3D–13.
A, transverse scan showing multiple cysts *(arrows)* in an enlarged kidney. Changes typical of adult polycystic kidney disease. **B,** CT scan of same patient.

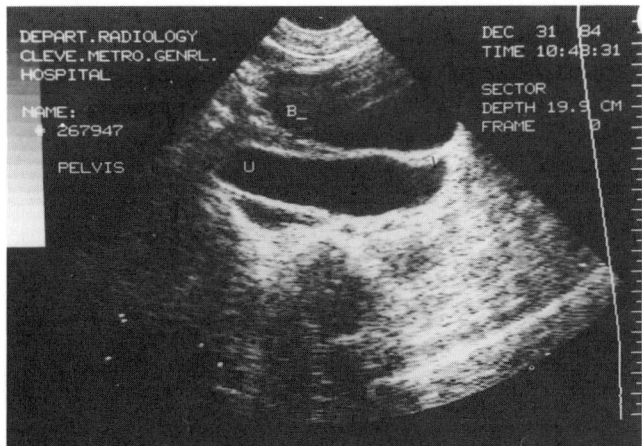

FIG 3D–14.
Longitudinal pelvic scan showing dilated megaureter. *B* indicates bladder; *U, ureter.*

FETAL RENAL ULTRASOUND

Obstetric ultrasonography is frequently performed during the course of pregnancy and is useful in the evaluation of fetal size, gestational age, fetal lie, and placental position. Recent advances in ultrasound technology have improved the visualization of fetal anatomy and led to increased identification of unsuspected congenital anomalies. Between 15 and 17 weeks of gestation, it is possible to identify the fetal kidneys by ultrasound about 50% of the time (Lawson et al., 1981). By 20 weeks of gestation, it is unusual not to be able to identify the kidneys accurately. It is, therefore, currently possible to diagnose accurately the presence of in utero hydronephrosis; however, in most instances it is still impossible to differentiate its cause as posterior urethral valves, primary megaureter (Fig 3D–14), ure-

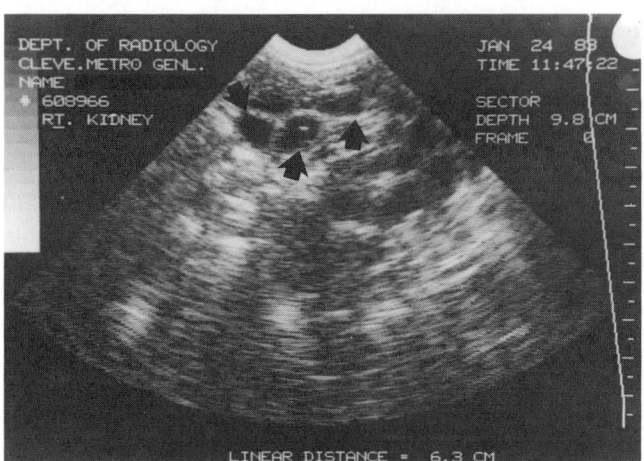

FIG 3D–15.
Ultrasonic changes of multicystic kidney. Note multiple cysts *(arrows).*

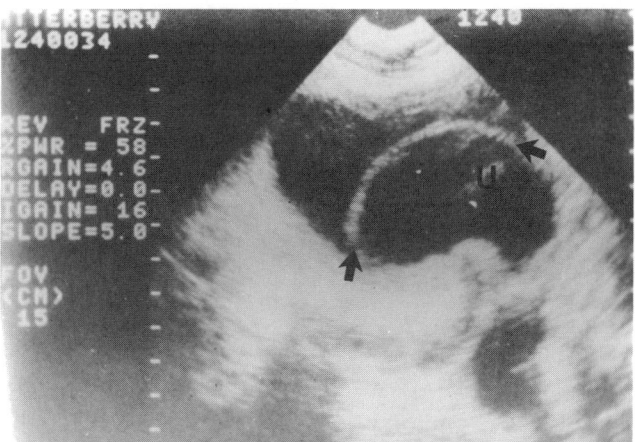

FIG 3D–16.
Longitudinal scan of bladder demonstrating presence of ureterocele *(U).* Note thin wall of ureterocele *(arrows).*

teropelvic junction obstruction, multicystic kidney (Fig 3D–15), duplication anomalies with or without ureteroceles (Fig 3D–16), or vesicoureteral reflux.

The diagnosis of fetal hydronephrosis best serves to alert the primary physician to the need for additional urologic evaluation in the neonatal period. When indicated, appropriate surgical intervention may be expeditiously performed (Spirnak et al., 1984).

RETROPERITONEAL SCANNING

The retroperitoneal space is a difficult area to examine using conventional radiographic methods but is well suited for sonographic studies. It is possible to identify nodal enlargement, primary retroperitoneal tumors, and aortic aneurysms with a considerable degree of accuracy (Fig 3D–17,A to C). Ultrasound may be helpful in determining the cause of ureteral deviation in patients with obstruction detected on excretory urography. In patients with retroperitoneal nodal metastases, ultrasound remains an excellent way to follow up the response to chemotherapeutic regimens without the risk of repeated radiation exposure.

Ultrasonic examination of the retroperitoneum has also been used to identify and locate accurately the undescended testicle. The study is helpful in identifying the inguinal testis but is less accurate in localizing the intra-abdominal testis.

ULTRASONOGRAPHY OF THE ADRENAL GLAND

Due to their small size and protected location beneath the ribcage, it is frequently difficult to study the normal adrenal glands sonographically (Becker and Schneider, 1977). Similarly, the evaluation of small (<3

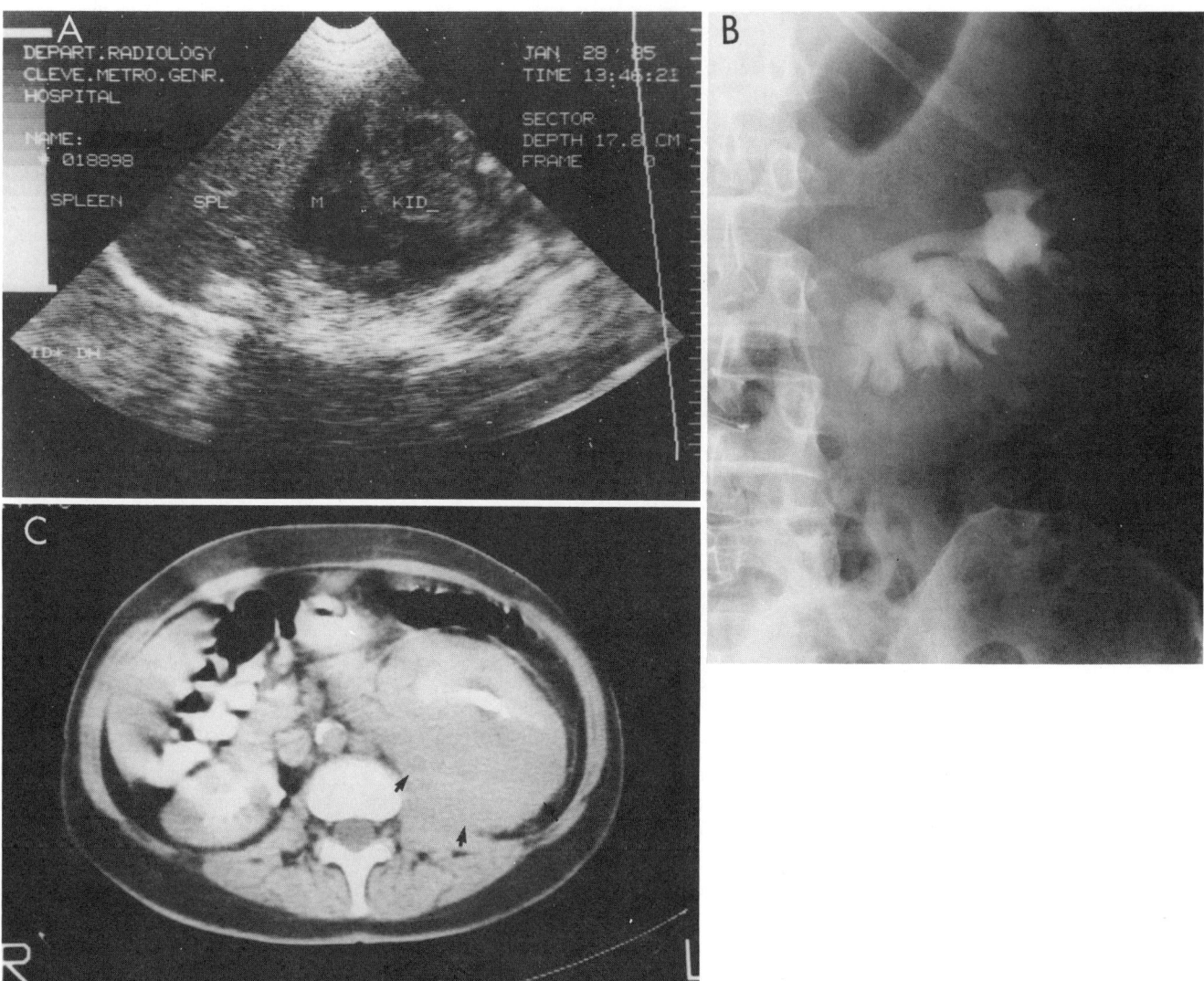

FIG 3D–17.
A, longitudinal scan of retroperitoneum demonstrating left perirenal mass. *M* indicates mass; *KID,* kidney; and *SPL,* spleen. Surgical exploration revealed lymphoma. Intravenous urogram **(B)** and CT **(C)** from same patient.

cm) adrenal masses is probably best performed using other radiographic techniques.

SCROTAL ULTRASONOGRAPHY

The superficial location and the maneuverability of the intrascrotal contents make this anatomical area well suited for sonographic evaluation. The primary use of scrotal ultrasound is in differentiating intratesticular lesions (tumors) from abnormalities arising from the paratesticular tissues. Using the currently available high-frequency transducers and gray-scale real-time techniques, a complete scrotal examination can be performed in minutes while providing excellent tissue resolution. The procedure is usually performed by first immobilizing the testicle with the free hand and then placing the transducer directly in contact with the scrotal skin. Mineral oil or acoustic gel is routinely used as a coupling agent to help reduce artifact. Recently water immersion techniques have been described (Friedrich et al., 1981). By interposing a water surface between the transducer and scrotal wall, improved delineation of the size and topography of the intrascrotal contents is now possible while avoiding the problem associated with direct contact scanning techniques, namely, reproducibility of scan planes and poor visualization of superficial areas.

NORMAL ULTRASONIC SCROTAL ANATOMY

Sonographically the scrotal wall appears as a hyperechoic strip 3–4 mm thick. Between the scrotal wall and the testicle is an anechoic area, usually no more

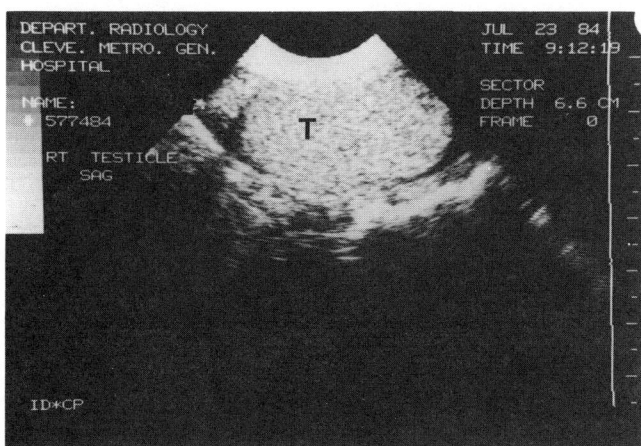

FIG 3D–18.
Scrotal ultrasound showing sonographic appearance of normal testicle (T).

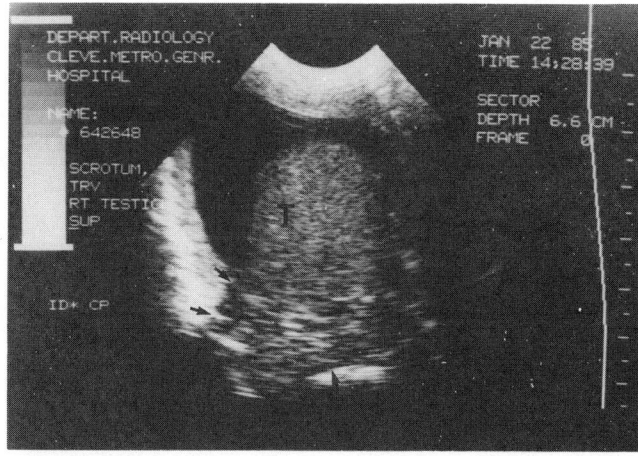

FIG 3D–19.
Scrotal ultrasound showing changes of acute epididymitis (arrows). Note reactive hydrocele surrounding normal testicle (T).

than a few millimeters thick, which represents a small amount of fluid normally present between the visceral and parietal layers of the tunica vaginalis. The normal testicle measures about 4 × 3 × 3 cm and is characterized by a homogeneous, fine granular echo pattern (Fig 3D–18). The tunica albuginea appears as a thin hyperechoic layer surrounding the testicle. Posteriorly the tunica albuginea reflects into the testicle to form the mediastinum testis, a structure which is usually seen only with the newer water-coupled scanners (Roberts et al., 1985). The epididymis runs along the posterolateral aspect of the testicle. The head or globus major is seen above the superior pole of the testicle and when normal appears to be either hyperechoic or of the same echogenicity as the testicle itself. The body and tail of the epididymis sonographically appear as a coarser echo pattern than the testicle itself. The vas deferens, when visualized, appears as a circular hyperechoic area on transverse scans.

Patients presenting with scrotal swelling and pain represent a diagnostic challenge and require prompt evaluation. Scrotal ultrasound has been found to be helpful in differentiating the presence of inflammatory lesions (orchitis and/or epididymitis) from spermatic cord torsion, traumatic testicular disruption, and testicular or paratesticular tumors (Pintauro et al., 1985). Sonographically epididymitis appears as a hypoechoic enlargement (inflammatory edema) usually involving the globus major. A reactive hydrocele may also be present (Fig 3D–19). In the presence of epididymo-orchitis the testicle adjacent to the involved epididymis may demonstrate hypoechoic areas. Individuals with presumed epididymo-orchitis who fail to respond to antibiotic therapy may demonstrate a testicular or scrotal abscess on ultrasound examination. When acute scrotal swelling occurs as a result of trauma, ultrasound has

been found helpful in differentiating testicular rupture from a normal testicle surrounded by hematoma (Fig 3D–20). Phillips and associates (1983) have reported a 98% accuracy rate using ultrasound to diagnose spermatic cord torsion. Acutely the ischemic testicle is enlarged, with decreased echogenicity compared with the normal testicle.

A hydrocele is a fluid collection that occurs between the visceral and parietal layers of the tunica vaginalis, making accurate testicular examination difficult to perform. Scrotal ultrasound has been found to be helpful in identifying hydrocele secondary to a malignant process. In the presence of a hydrocele the quality of the testicular sonogram may be improved owing to the fluid, which acts as a biologic water bath. Typically a simple hydrocele appears as an anechoic area surrounding the testes anterolaterally (Fig 3D–21). The pres-

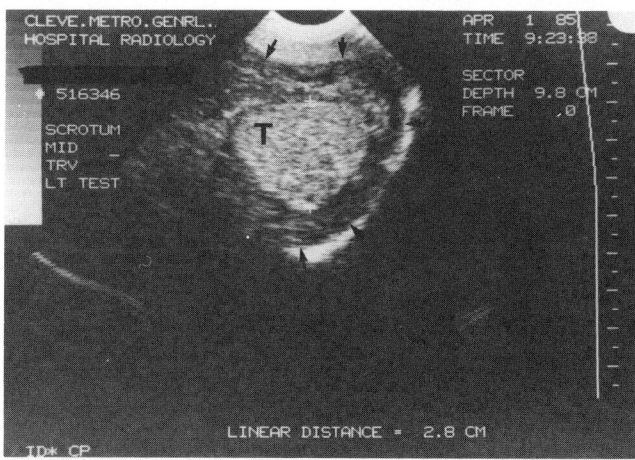

FIG 3D–20.
Scrotal ultrasound demonstrating large hematoma (arrows) surrounding normal testicle (T).

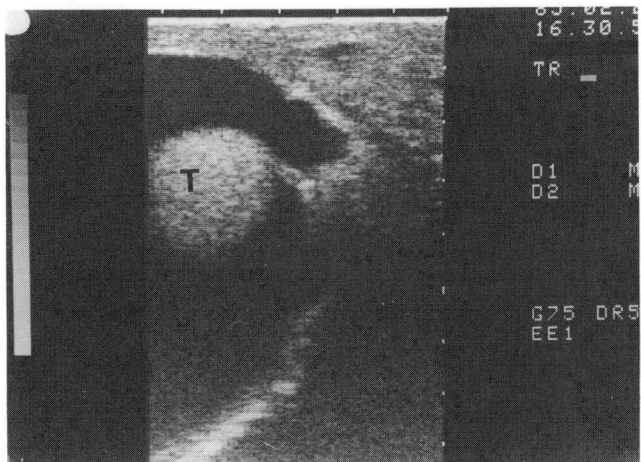

FIG 3D–21.
Scrotal ultrasound demonstrating characteristic findings of normal testicle (T) surrounded by large hydrocele.

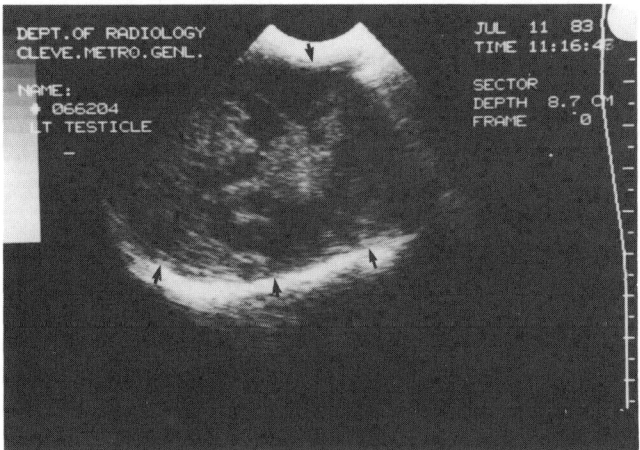

FIG 3D–22.
Scrotal ultrasound demonstrating large inhomogeneous testicular mass (arrows). Pathology revealed embryonal cell carcinoma.

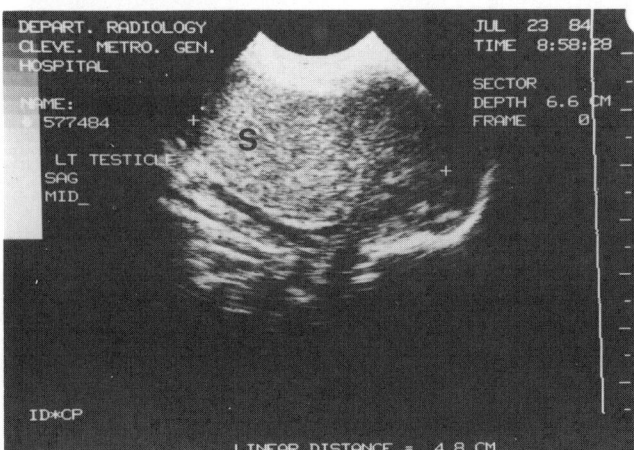

FIG 3D–23.
Homogeneous testicular mass (S) pathologically proved seminoma.

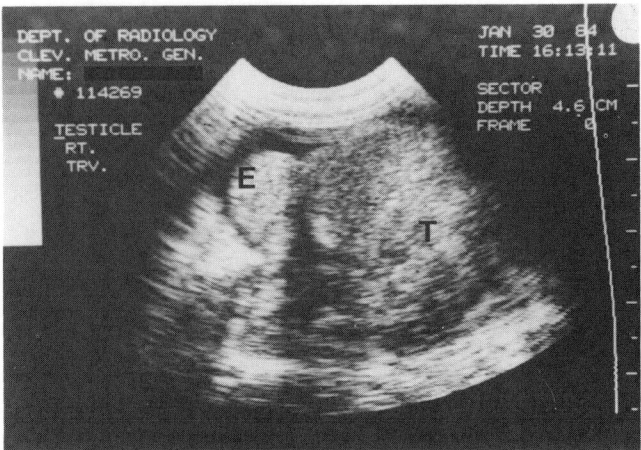

FIG 3D–24.
Scrotal ultrasound demonstrating changes typical of chronic epididymitis. Note enlarged hyperechoic epididymis (E) and normal testicle (T).

ence of a hypoechoic inhomogeneous mass in the testicle is suggestive of a malignancy and warrants surgical exploration (Fig 3D–22). Unlike embryonal cell carcinoma, seminoma sonographically has a more homogeneous pattern (Fig 3D–23). Individuals with repeated bouts of epididymal infection may also present with scrotal pain and swelling. Ultrasonically a diagnosis of chronic epididymitis is suggested by an enlarged hyperechoic epididymis (Fig 3D–24).

Recently scrotal ultrasound has been found helpful in identifying the subfertile male patient with a nonpalpable or subclinical varicocele (McClure et al., 1985). The clinical significance of this diagnosis is yet to be determined.

TRANSRECTAL ULTRASONOGRAPHY OF THE PROSTATE

Since its inception in the 1960s, abdominal and scrotal ultrasound has gained an important role in the evaluation of many urologic disorders. However, due to the relatively inaccessible pelvic location of the bladder and prostate, ultrasonic evaluation of these organs using conventional techniques has not been as widely accepted.

In the late 1950s Wild and Reid were the first to develop and test a transrectal ultrasonic probe (Holm and Gammelgaard, 1982). Although their initial experience with the transrectal approach resulted in visualizing only the rectal wall, the possible usefulness of this technique as a means to evaluate the lower urinary tract was clearly demonstrated. In 1964, Takahashi and Ouchi, using a transrectal probe equipped with a radial scanning device, obtained poor-quality tomographic pictures of the prostate which were of no clinical use. It was not until 1967 that Watanabe and associates obtained the first clinically useful transrectal ultrasonotomograms of the prostate. Since then, additional advances in instrumentation and technique have allowed for more reliable ultrasonic visualization of the pelvic organs. Today the use of transrectal ultrasonography is helpful in the diagnosis, staging, and treatment of carcinomas involving the prostate and bladder and in the clinical evaluation of voiding dysfunction.

Instruments currently in use consist of a transrectal transducer, radial scanner, and imaging screen. Most units consist of a two-part probe: a freely moving inner assembly and a stationary outer assembly. The outer unit has inflow and outflow parts necessary to inflate the polyethylene condom that covers the tip, and when inflated acts to protect the rectal mucosa while providing a coupling medium between the transducer and rectal wall. The inner assembly moves in a cephalad–caudad direction, rotating circumferentially to permit imaging of multiple serial sections of the bladder and prostate. The ultrasonic beam is emitted at right angles to the probe. Instruments may be fitted with either a 3.5- or 7.0-MHz transducer. The higher frequency transducer has proved useful in examining the prostate and seminal vesicles while the 3.5-MHz transducer is used for bladder evaluation.

The examination may be performed in the lithotomy, lateral decubitus, supine, or sitting positions. Prior to performing the study, a cleansing enema is administered. The rectal probe is inserted 3–9 cm above the anal verge, and the outer rubber condom is inflated with water to provide a tight fit against the rectal wall. Beginning at the level of the bladder, serial transverse sonograms are obtained at 0.5–1-cm intervals. Permanent records are made with a Polaroid or multiformat camera. The entire examination takes only 10–15 minutes, can be performed on an outpatient basis without anesthesia, and results in minimal patient discomfort.

Sekine and associates (1982) have developed a transrectal electronic linear scanner that allows one to obtain longitudinal sonograms of the lower urinary tract. Using real-time imaging, dynamic studies obtained while the patient is voiding may provide valuable diagnostic information on voiding dysfunction (Shapeero et al., 1983).

Further advances in instrument design have allowed the placement of a needle guide into the probe (Holm and Gammelgaard, 1982). Using this device it is now possible to improve the accuracy of prostatic needle biopsy by placing the needle under ultrasonic guidance directly into any suspicious or abnormally appearing areas. Recently, Holm and associates (1983) have further expanded the use of the needle guide and have developed a template that fits directly into the probe and allows the accurate placement of transperineal radioactive seeds into proved areas of carcinoma.

CLINICAL APPLICATIONS

Studies have shown high false positive and low false negative rates in detecting prostatic carcinoma (Fritzsche et al., 1981; Watanabe et al., 1980). Therefore, the routine use of transrectal prostatic ultrasonography as a screening procedure in asymptomatic elderly men is not now warranted. The low specificity results primarily from the inability of the study to differentiate accurately the abnormal echo pattern of carcinoma from that of other benign prostatic findings such as calculi, inflammation and postoperative fibrosis (Spirnak and Resnick, 1984a).

Although ultrasonic imaging of the prostate seems to be of little value as a screening procedure, it has proved to be clinically useful in the patient with a palpable nodule (Peeling and Griffiths, 1984). Using transrectal ultrasound guidance and the needle guide to obtain a biopsy of the questionable area should decrease the false negative rate associated with blind prostatic needle biopsy (Resnick, 1981) (Fig 3D–25).

Transrectal prostatic ultrasonography is also useful as a staging modality in patients with biopsy-proved adenocarcinoma (Braeckman and Denis, 1983; Fujino and Seardino, 1985). Tumor extension into the seminal vesicles, bladder neck and prostatic capsule not palpable on rectal examination may be accurately identified using ultrasound imaging (Resnick et al., 1980; Spirnak and Resnick, 1984b).

Ultrasonic prostatic imaging also appears to have a role in the treatment of localized prostatic carcinoma. Holm and associates (1983) have used a special punc-

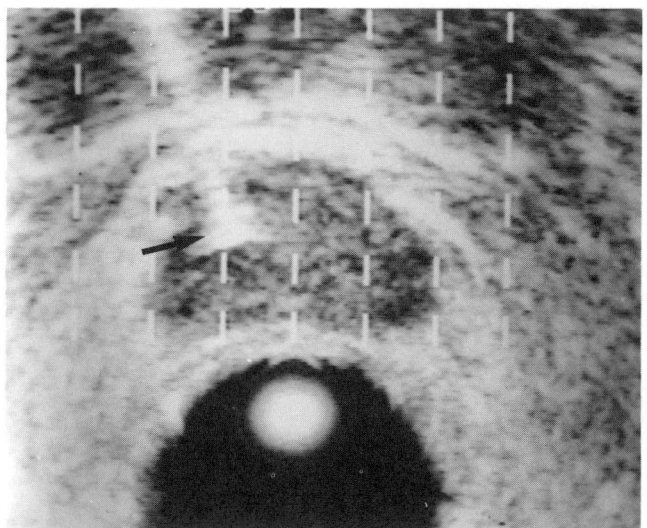

FIG 3D–25.
Transrectal sonogram demonstrating use of technique in facilitating needle biopsy of prostate. Biopsy grid helps localize abnormal areas of interest. Needle appears as highly echogenic area *(arrow).*

ture attachment fitted to the transrectal probe to help accurately place 125 iodine seeds into the area of localized carcinoma. Advantages over the more conventional methods of seed placement include: (1) sparing the patient the morbidity of an open surgical procedure, (2) more accurate dose planning based on serial ultrasonotomograms, and (3) more accurate placement allowing for higher-dose radiation to be delivered to the tumor with a simultaneous decrease in radiation-associated complications to the surrounding tissues. Further studies and additional clinical experience are required to determine whether this treatment modality will alter the prognosis in patients with localized disease.

Transrectal prostatic ultrasonography also offers a method to objectively monitor tumor response to hormonal or cytotoxic therapy and may be of use in assessing the clinical response to new chemotherapeutic agents. Patients who have undergone radical extirpative surgery may have recurrence at a sight inaccessible to the examining finger. The routine use of this imaging modality may identify local recurrence before it is clinically manifest (Resnick et al., 1977).

The use of transrectal ultrasonography also has clinical application in the patient with obstructive symptoms due to benign disease. Preoperatively it is possible to determine accurately the gland's weight and help decide whether to proceed with an open or endoscopic prostatectomy (Watanabe et al., 1974).

SONOGRAPHIC FINDINGS

The normal prostate gland has a triangular configuration, with the anterior-posterior diameter being smaller than the bilateral diameter. The parenchyma is

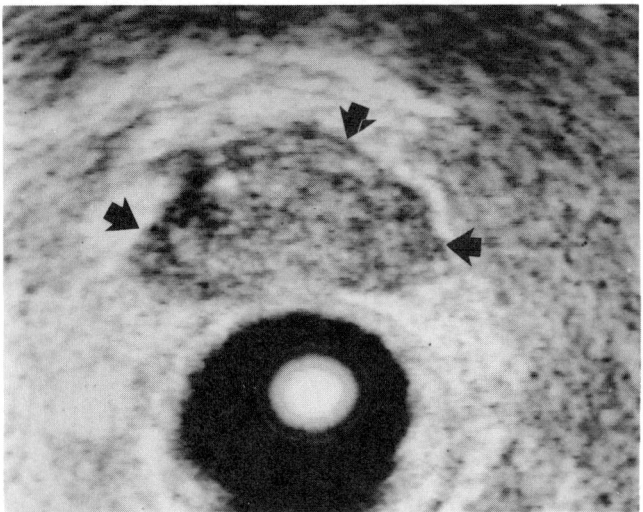

FIG 3D–26.
Transrectal ultrasound demonstrating typical findings of a normal prostate. Prostatic capsule is well demarcated (arrows).

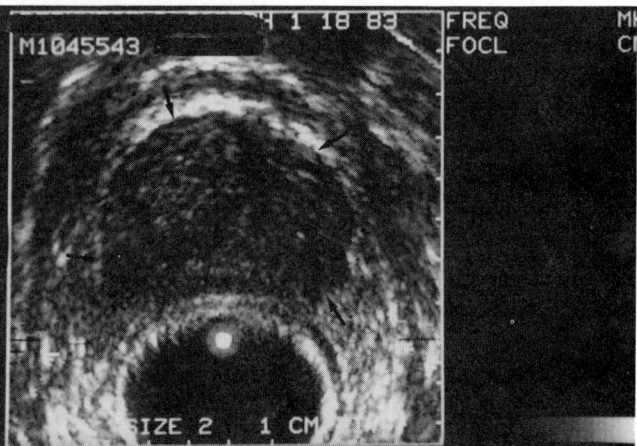

FIG 3D–27.
Ultrasonic findings typical of benign prostatic hyperplasia. Note increased AP diameter and well-demarcated capsule (arrows).

composed of multiple fine homogeneous echoes which represent the acoustic interfaces created by the periurethral glands, and these areas generally fade with increased instrument attenuation (Resnick, 1980). The prostatic capsule is usually well defined, highly echogenic, and circumferentially continuous (Fig 3D–26).

Benign prostatic hyperplasia appears as diffuse enlargement with the greatest increase in size occurring in the AP diameter. The prostatic capsule remains well defined but thicker than that found in the normal gland pattern. The hyperplastic portion of the gland is composed of multiple fine homogeneous echoes believed to represent small adenomas located in fibrous tissue between glandular elements and multiple microcysts created by dilated prostatic ducts containing secretions, corpora amalacea, and microcalculi (Fig 3D–27).

Ultrasonically the malignant prostate appears asymmetrically enlarged. The gland demonstrates areas of increased, decreased, or mixed echogenicity (Fig 3D–28). The interpretation of these changes has been an area of much controversy, but it appears that no one characteristic that has yet been identified is specific for carcinoma. The presence of early localized carcinoma demonstrated by localized dense areas without capsular distortion may be easily confused with inflammatory changes or prostatic calculi. Prostatic carcinoma involving the capsule will show deformed or irregular capsular echoes. With tumor extension into the seminal vesicles, their appearance will become distorted and may be obliterated (Resnick, 1981).

Prostatic calcifications occur with surprising frequency in the general population (Peeling and Griffiths, 1984). On ultrasound examination calculi may be confused with prostatic carcinoma. Ultrasonically, prostatic calculi appear as focally dense, highly echogenic areas

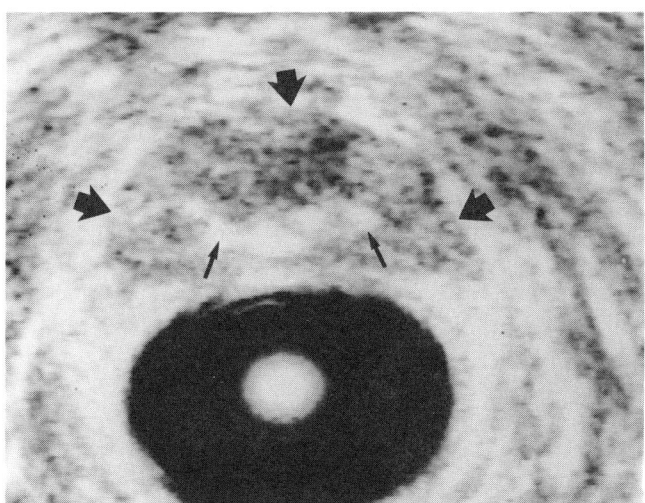

FIG 3D–28.
Ultrasonic changes typical of carcinoma. Note increased echogenicity *(small arrows)* and distorted prostatic capsule *(large arrows)*. Carcinoma can also appear as an area of low or mixed echogenicity.

that produce sonic shadows (Fig 3D–29). The capsule remains symmetric and intact.

The ultrasonic appearance of chronic prostatitis will frequently mimic the changes of carcinoma and may result in a false positive study. In chronic prostatitis the ultrasonic patern shows irregularly distributed heterogeneous densities that often extend laterally from the urethra and may distort or obscure the prostatic capsule (Fig 3D–30). Unlike carcinoma and calculi, sonic shadowing seldom occurs, and the anterior aspect of the gland is nearly always visualized.

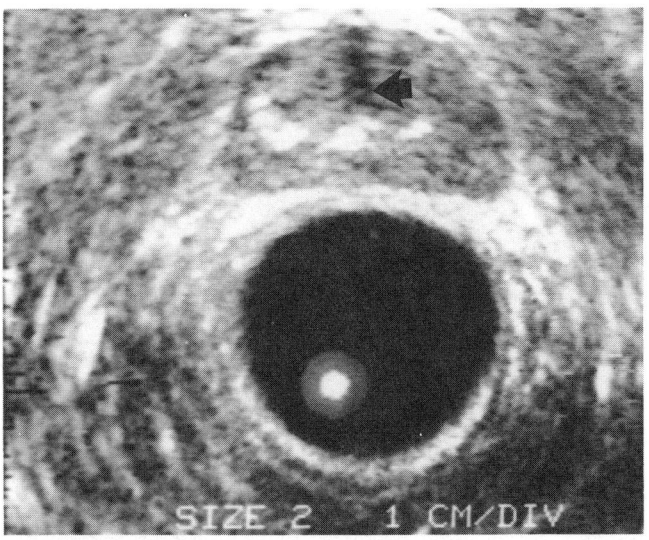

FIG 3D–29.
Transrectal prostatic sonogram demonstrating changes secondary to prostatic calculi. Note sonic shadowing *(arrow)*.

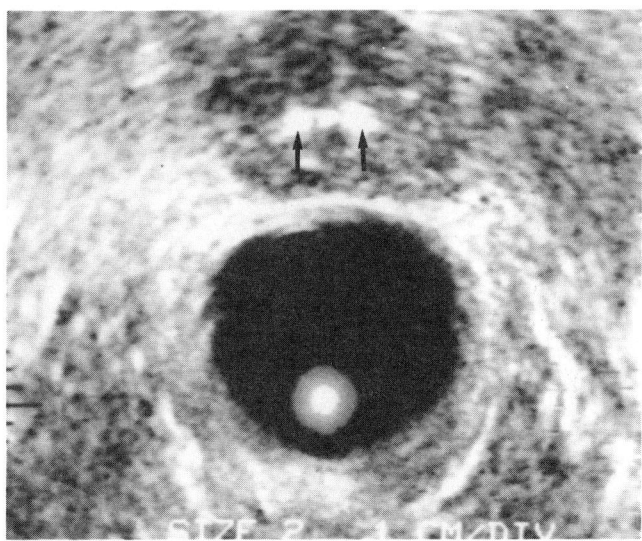

FIG 3D–30.
Transrectal prostatic sonogram demonstrating changes typical of prostatitis. Note increased periurethral echogenicity *(arrows)*.

Recently Shapeero and associates (1983) utilized a transrectal linear scanner to evaluate patients with voiding dysfunction. They studied the lower urinary tract in 32 men with known vesical dysfunction and compared the information to that obtained by conventional voiding cystourethrographic studies. In addition to avoiding the complications associated with repeated radiation exposure, they found the sonographic voiding studies to provide more clinically useful information than the conventional studies. The linear array scanner also provides a visual means whereby the effects of various pharmacologic agents on the urinary tract could be monitored.

ULTRASONOGRAPHY OF THE URINARY BLADDER

The urinary bladder is an extraperitoneal musculomembranous sac that functions primarily as a urinary storage reservoir. In children younger than 6 years the bladder is predominantly an intra-abdominal organ lying beneath the anterior abdominal wall. In adults it is a pelvic organ lying, when empty, beneath the symphysis pubis. When distended, the bladder assumes a globular shape and is easily studied using the transabdominal approach. Recently transrectal and transurethral approaches have been used to study and define subtle bladder abnormalities.

TRANSABDOMINAL SCANNING

Transabdominal ultrasound imaging of the bladder is usually performed with the patient in the supine position and the bladder distended. Although it is not nec-

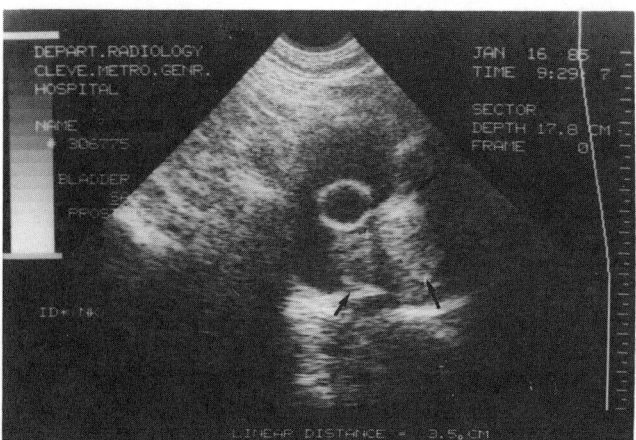

FIG 3D–31.
Longitudinal abdominal scan showing urinary bladder and enlarged prostate *(arrows)*. Note Foley catheter balloon at bladder neck.

essary to have a Foley catheter in place, its presence is helpful in identifying the bladder neck (Fig 3D–31). Usually a 3.5-MHz transducer is used to obtain transverse as well as longitudinal scans.

TRANSRECTAL SCANNING

Transrectal ultrasonic imaging of the bladder is performed using the same equipment and technique as previously described (Fig 3D–32).

TRANSURETHRAL SCANNING

Although both transabdominal and transrectal approaches have been found to be helpful in evaluating bladder abnormalities, subtle changes in the bladder wall—e.g., muscular invasion by tumors—are fre-

quently not discernible. Recent experience by Gammelgaard and Holm and others suggests that the bladder may be best studied using the transurethral approach (Gammelgaard and Holm, 1980; Nakamura and Nijima, 1981; Resnick and Kursh, 1985).

The scanner for transurethral ultrasonography consists of a motor that rotates a long rod, which is connected at the opposite end to an interchangeable transducer. The scanner fits within a standard resectoscope sheath and is interchangeable with the usual optic system. Two 5.5-MHz transducers are currently available. One emits the ultrasonic beam at 90° to the instrument, while the other emits the beams at 135° to the probe in retrograde fashion and is useful in studying the bladder neck region. Using these two transducers, complete visualization of the entire bladder is possible. During routine cystoscopy, when further evaluation of a detected bladder lesion is desired, the telescope is removed from the sheath and is replaced with the sterilized scanner. Dynamic scans utilizing the two transducers are obtained. The entire bladder wall can be rapidly scanned in a matter of minutes.

NORMAL BLADDER

Ultrasonically, the normal bladder appears as a globular structure, the shape of which varies depending on the patient's position and degree of distention. The bladder wall is hyperechoic and appears as a symmetric, smooth surface. When distended, the fluid-filled bladder is anechoic (Fig 3D–33).

CLINICAL APPLICATION

Bladder tumors are a common urologic problem, the treatment of which depends on an accurate assessment

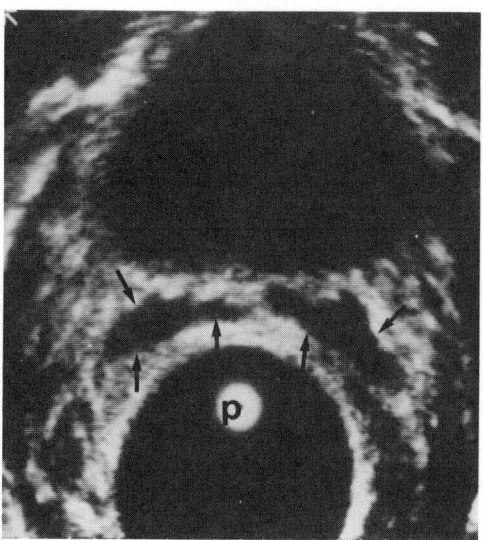

FIG 3D–32.
Transrectal scan of normal bladder and seminal vesicles *(arrows)*. Rectal probe *(p)* is in place.

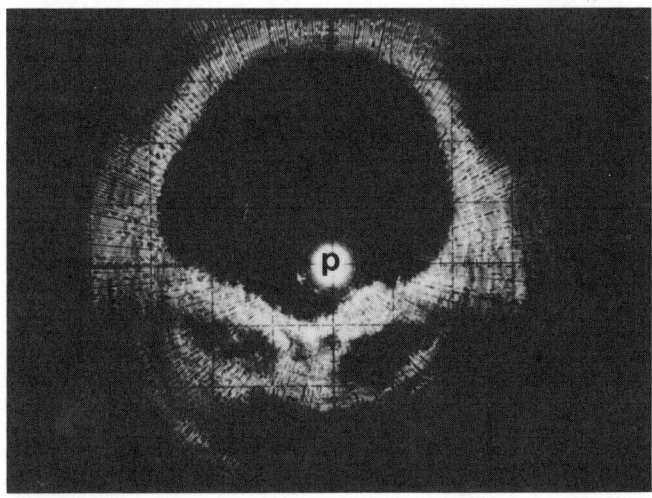

FIG 3D–33.
Transurethral scan of normal urinary bladder. *p* indicates transurethral probe. Note seminal vesicles posterior to bladder.

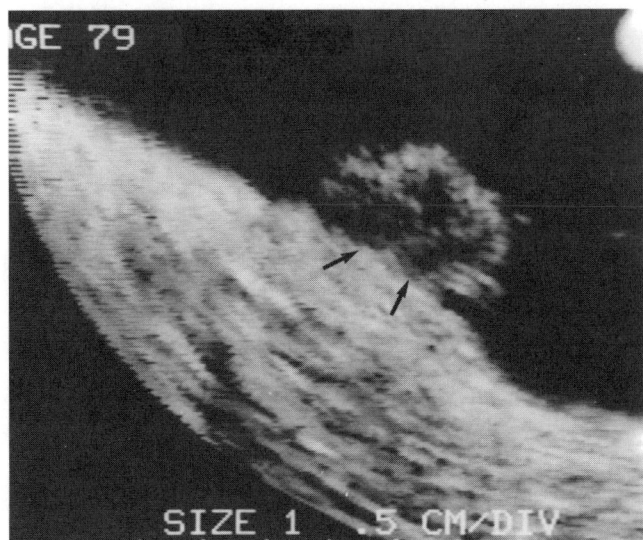

FIG 3D–34.
Transurethral bladder scan demonstrating large exophytic bladder tumor.

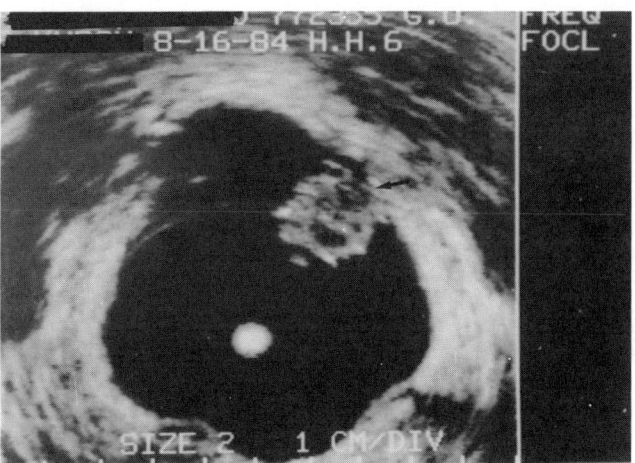

FIG 3D–35.
Transurethral bladder scan demonstrating superficial bladder tumor. Note lack of bladder wall invasion *(arrow)*.

of the grade and stage of the primary tumor. Using currently available staging modalities, errors may occur in nearly 50% of cases (Bodner et al., 1984). The use of ultrasound has been reported to decrease the staging error in this patient population (Resnick and Kursh, 1985).

Large exophytic tumors ultrasonically appear as echogenic masses projecting into the lumen of the echofree bladder (Fig 3D–34). These areas are fixed to the wall, and unlike blood clots or stones, do not move as the patient changes position. Ureteroceles, although fixed to the bladder wall, are easily recognized by their thin echogenic wall surrounding a relatively anechoic center. Bladder stones readily move about with changes in intravesical volume and patient position and like renal stones appear as dense hyperechoic areas associated with sonic shadowing.

Superficial bladder (stage O–A) tumors do not cause distortion or fixation of the bladder wall, and they demonstrate a well-defined base (Fig 3D–35). Multiple scans obtained during bladder filling will demonstrate free movement of the bladder wall.

Infiltrative tumors tend to be broad based, and the bladder wall may be fixed and distorted. Stage C tumors that have completely extended through the bladder wall will be visualized as an extravesical mass.

Recent reports indicate that transurethral scanning is more accurate in staging bladder tumors than either transabdominal or transrectal techniques (Gammelgaard and Holm, 1980; Holm and Northeved, 1974; Nakamura and Nijima, 1981). Our experience has been similar (Resnick and Kursh, 1985). Transurethral scanning offers the experienced clinician the ability to accurately

differentiate tumors with superficial muscle invasion (B) (Fig 3D–36) from those more deeply invasive lesions (B₂) (Fig 3D–37). The ability to accurately discriminate B_1 from B_2 lesions could prove helpful in treatment planning.

Ultrasonography of the urinary bladder has also been found to be helpful in estimating total bladder capacity and postvoid residual urine when urethral instrumentation is not desired or is contraindicated (Garrison et al., 1975; Orgaz et al., 1981). Using the formula 12.56 times the radius times the height ($12.56 \times r \times h$), Orgaz and associates (1981) were able to calculate the bladder capacity with an average error of only 12.9%.

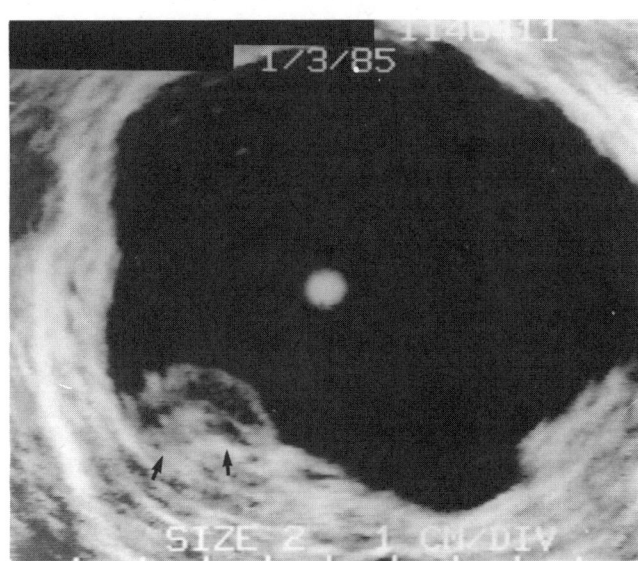

FIG 3D–36.
Transurethral bladder scan demonstrating minimally invasive tumor *(arrows)*.

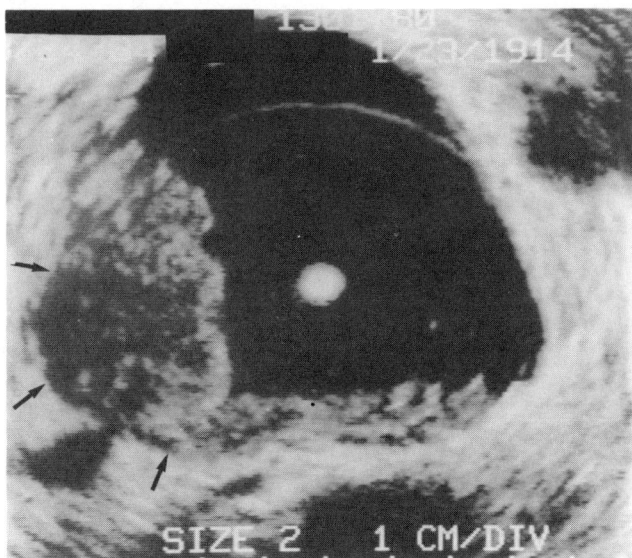

FIG 3D–37.
Transurethral bladder scan demonstrating tumor invasion beyond bladder wall *(arrows).*

INTRAOPERATIVE ULTRASONOGRAPHY

Schlegel and associates (1961) reported the first use of intraoperative ultrasound to aid in the localization of renal calculi. However, owing to difficulties in interpreting the intraoperative studies, its use fell into disfavor. It was not until 1977, when Cook and Lytton utilized real-time B-mode (intraoperative) ultrasound that the technique became an accepted and useful adjunct in renal stone surgery.

Intraoperative real-time scanning displays a cross section of renal tissue and, utilizing multiple images, allows accurate, three-dimensional stone localization. Most intraoperatively used portable ultrasound units utilize a frequency of 7–10 MHz and are modifications of the small ophthalmic ultrasound probe.

After adequately mobilizing the kidney, it is scanned in multiple planes until the stone is identified by its characteristic dense echo pattern and the presence of acoustic shadowing (Sigel et al., 1982) (Fig 3D–38, A to C). Fine-needle probes are then passed to the stone and a nephrostomy made directly over the stone. Using this technique, an experienced ultrasonographer can identify stones 2 mm or more in diameter (Fried, 1984). Lytton (1983) reviewed a total of 69 patients who underwent attempted intraoperative localization of renal calculi, and in 12 cases the stone was not identified. Clearly, intraoperative ultrasound can be a valuable adjunct in identifying the elusive stone fragment.

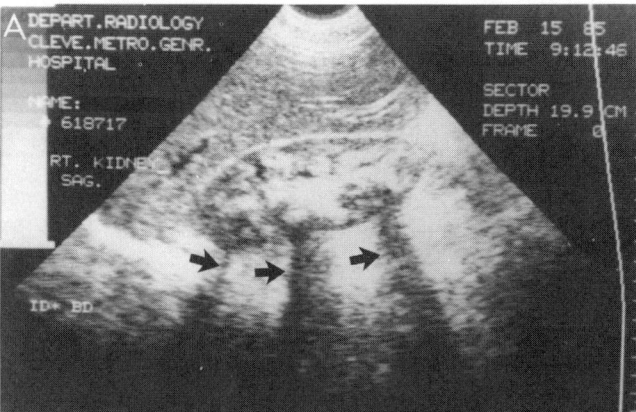
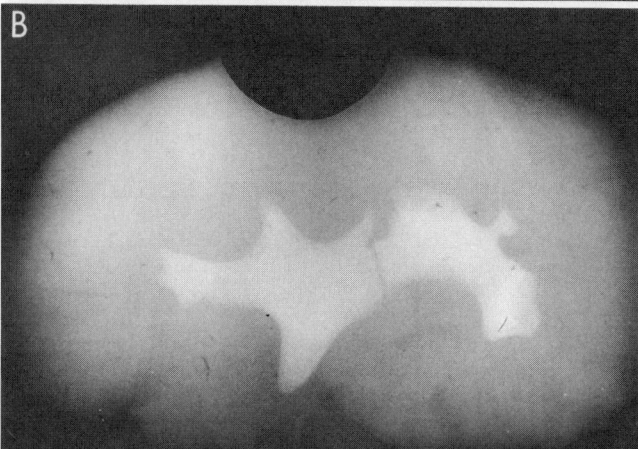
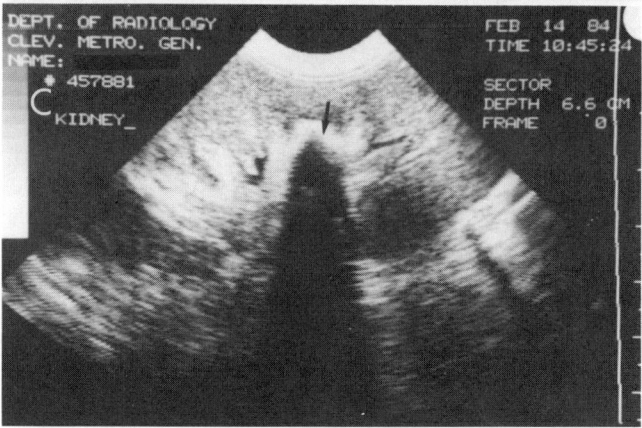

FIG 3D–38.
A, preoperative renal scan of large staghorn calculus. Note multiple areas of acoustic shadowing *(arrows).* **B,** intraoperative radiograph prior to nephrotomy. **C,** intraoperative ultrasound localizing small retained stone fragment *(arrow).*

REFERENCES

1. Albarelli JN, Lawson TL: Renal ultrasonography advantages of the decubitus position. *J Clin Ultrasound* 1978; 6:73.

2. Becker JA, Schneider M: Techniques and applications of sonography and computed tomography, in Witten DM, Myers GH, Utz DC (eds): *Emmett's Clinical Urography*, ed 4. Philadelphia, WB Saunders Co, 1977, pp 214–376.

3. Bo WJ, Krueger WA: *Cross-sectional Anatomy of the Male Urogenital System*, ed 2. Baltimore, Williams & Wilkins Co, 1984, chap 3, pp 29–65.

4. Bodner D, Bryan PJ, Lipuma JP, et al: Ultrasonography of the urinary bladder, in Resnick MI, Sanders RC (eds): *Ultrasound in Urology*, ed 2. Baltimore, Williams & Wilkins Co, 1984, pp 209–238.

5. Bodner D, Lipuma J, Resnick MI: Basic principles and utilization of ultrasound in urological malignancies, in Javadpour N (ed): *Principles and Management of Urologic Cancer*, ed 2. Baltimore, Williams & Wilkins Co, 1983, chap 10, pp 239–264.

6. Braeckman J, Denis L: The practice and pitfalls of ultrasonography in the lower urinary tract. *Eur Urol* 1983; 9:193.

7. Brandt TD, Neiman HL, Dragowski MJ, et al: Ultrasound assessment of normal renal dimensions. *J Ultrasound Med* 1982; 1:49.

8. Charboneau JW, Hattery RR, Ernst EC III, et al: Spectrum of sonographic findings in 125 renal masses other than benign renal cysts. *AJR* 1983; 140:87.

9. Cook JH III, Lytton B: Intraoperative localization of renal calculi during nephrolithotomy by ultrasound scanning. *J Urol* 1977; 117:543.

10. Dussik KT: Uber die Moglichkeit hoch frequente mechanische Schwingungen als diagnostisches Hilfsmittel zu verwenden. *Z Ges Neurol Psych* 1942; 174:153.

11. Ellenbogen PH, Scheible RW, Talner LB, et al: Sensitivity of gray scale ultrasound in detecting urinary tract obstruction. *AJR* 1978; 130:731.

12. Fried FA: Intraoperative localization of renal calculi with ultrasound, in Resnick MI, Sanders RC (eds): *Ultrasound in Urology*, ed 2. Baltimore, Williams & Wilkins, 1984, pp 373–385.

13. Friedrich M, Claussen CD, Felix R: Immersion ultrasonography of scrotal and testicular pathology. *Eur J Radiol* 1981; 1:60.

14. Fritzsche PJ, Axford PD, Ching VD, et al: Correlation of transrectal sonographic findings in patients with suspected and unsuspected prostatic disease. *J Urol* 1983; 130:272.

15. Fujino A, Scardino PT: Transrectal ultrasonography for prostatic cancer: Its value in staging and monitoring the response to radiotherapy and chemotherapy. *J Urol* 1985; 133:806.

16. Gammelgaard J, Holm HH: Transurethral and transrectal ultrasonic scanning in urology. *J Urol* 1980; 124:863.

17. Green WM, King DL: Diagnostic ultrasound of the urinary tract. *J Clin Ultrasound* 1976; 4:55.

18. Garrison NW, Parks C, Sherwood T: Ultrasound assessment of residual urine in children. *Br J Urol* 1975; 47:805.

19. Hepler AB: Solitary cysts of the kidney: A report of seven cases and observations on the pathogenesis of these cysts. *Surg Gynecol Obstet* 1930; 50:668.

20. Holm HH, Gammelgaard J: Ultrasonically guided precise needle placement in the prostate and seminal vesicles. *J Urol* 1982; 125:385.

21. Holm HH, Juul N, Pedersen JF, et al: Transperineal [125]iodine seed implantation in prostatic cancer guided by transrectal ultrasonography. *J Urol* 1983; 130:283.

22. Holm HH, Northeved A: A transurethral ultrasonic scanner. *J Urol* 1974; 111:238.

23. Howry DH, Bliss WR: Ultrasonic visualization on soft tissue structures of the body. *J Lab Clin Med* 1952; 40:579.

24. Kissans JM: Congenital malformations, in Heptinstall RH (ed): *Pathology of the Kidney*, ed 2. Boston, Little, Brown & Co, 1974, pp 69–119.

25. Kressel HY, Filly RA: Ultrasonographic appearance of gas-containing abscesses in the abdomen. *AJR* 1978; 130:71.

26. Lawson TL, Foley WD, Berland LL, et al: Ultrasonic evaluation of fetal kidneys. *Radiology* 1981; 138:153.

27. Lee TG, Henderson SC, Freeny PC, et al: Ultrasound findings of renal angiomyolipoma. *J Clin Ultrasound* 1978; 6:150.

28. Lytton B: Intraoperative ultrasound for nephrolithotomy. *J Urol* 1983; 130:213.

29. Martin JF: History of ultrasound, in Resnick MI, Sanders Rc (eds): *Ultrasound in Urology*, ed 2. Baltimore, Williams & Wilkins Co, 1984, Chapter 1.

30. McClure RD, Hoddick W, Abber JC, et al: Scrotal ultrasound before and after varicocelectomy. Houston, American Urological Association Office of Education, Abstract 198, 1985, p 163A.

31. Moccia WA, Kaude JV, Wright PG, et al: Evaluation of chronic renal failure by digital grayscale ultrasound. *Urol Radiol* 1980; 2:1.

32. Nakamura S, Niijima T: Transurethral real-time scanner. *J Urol* 1981; 125:781.

33. Orgaz RE, Gomez AZ, Ramirez CT, et al: Applications of bladder ultrasonography: I. Bladder content and residue. *J Urol* 1981; 125:174.

34. Peeling WB, Griffiths GJ: Imaging of the prostate by ultrasound. *J Urol* 1984; 132:217.

35. Phillips C, Abrams HJ, Kumari-Subaiya S: Scrotal ultrasonography. *Ultrasound Annu* 1983; 2:207.

36. Pintauro WL, Klein FA, Vick CW III, et al: The use of ultrasound for evaluating subacute unilateral scrotal swelling. *J Urol* 1985; 133:799.

37. Pode D, Meretik S, Shapiro A, et al: Diagnosis and management of renal angiomyolipoma. *Urology* 1985; 25:461.

38. Pollack HM, Arger, PH, Goldberg BB, et al: Ultrasonic detection of nonopaque renal calculi. *Radiology* 1978; 127:233.

39. Pollack HM, Banner MP, Arger PH, et al: The accuracy of gray-scale renal ultrasonography in differentiating cystic neoplasms from benign cysts. *Radiology* 1982; 143:741.

40. Resnick MI: Ultrasonic evaluation of the prostate and bladder. *Semin Ultrasound* 1980; 1:69.

41. Resnick MI: Non-invasive techniques in evaluating patients with carcinoma of the prostate. *Urology* 1981; 17(suppl):25.

42. Resnick MI: Ultrasonography in the detection and diagnosis of carcinoma of the prostate, in Ablin RJ (ed): *Prostatic Cancer.* New York, Marcel Dekker, 1981, p 101.

43. Resnick MI, Kursh ED: Transurethral ultrasonography in staging bladder cancer. American Urological Society Annual Meeting, Atlanta, Georgia, May 1985, abstract 561, p 254A.

44. Resnick MI, Willard JW, Boyce WH: Recent progress in ultrasonography of the bladder and prostate. *J Urol* 1977; 117:444.

45. Resnick MI, Willard JW, Boyce WH: Transrectal ultrasonography in the evaluation of patients with prostatic carcinoma. *J Urol* 1980; 124:482.

46. Roberts SD, Lipuma HP, Resnick MI: Ultrasonography of the scrotum and retroperitoneum, in Javadpour N (ed): *Testicular Cancer.* New York, Thieme-Stratton, 1986, pp 178–192.

47. Rosenfield AT, Taylor KJ: Obstructive uropathy in the transplanted kidney: Evaluation by gray scale ultrasonography. *J Urol* 1976; 116:101.

48. Russell JM, Resnick MI: Ultrasound in urology. *Urol Clin North Am* 1979; 6:445.

49. Schlegel JU, Diggdon P, Cuellar J: The use of ultrasound for localizing renal calculi. *J Urol* 1961; 86:367.

50. Sekine H, Oka K, Takehara Y: Transrectal longitudinal ultrasonotomography of the prostate by electronic linear scanning. *J Urol* 1982; 127:62.

51. Shapeero LG, Friedland GW, Perkash I: Transrectal sonographic voiding cystourethrography: Studies in neuromuscular bladder dysfunction. *Am J Radiol* 1983; 141:83.

52. Sherwood T: Renal masses and ultrasound. *Br Med J* 1975; 4:682.

53. Sigel B, Coelho JCU, Sharifi R, et al: Ultrasonic scanning during operation for renal calculi. *J Urol* 1982; 127:421.

54. Spirnak JP, Mahoney S, Resnick MI, et al: Incidental fetal hydronephrosis: Clinical implication. *Urology* 1984; 24:105.

55. Spirnak JP, Resnick MI: Clinical staging of prostatic cancer: New modalities. *Urol Clin North Am* 1984a; 11:221.

56. Spirnak JP, Resnick MI: Transrectal ultrasonography. *Urology* 1984; 23:461.

57. Subramanyam BR, Raghavendra BN, Madamba MR: Renal transitional cell carcinoma: Sonographic and pathologic correlation. *J Clin Ultrasound* 1982; 10:203.

58. Takahashi H, Ouchi T: The ultrasonic diagnosis in the field of urology on the diagnosis of prostatic disease, in *Proceedings of the 4th Meeting of the Japanese Society Ultrasound Medicine.* Osaka, 1964, p 35.

59. Talner LB, Scheible W, Ellenbogen PH, et al: How accurate is ultrasonography in detecting hydronephrosis in azotemic patients? *Urol Radiol* 1981; 3:1.

60. Watanabe H, Date S, Ohe H: A survey of 3000 examinations by transrectal ultrasonotomography, in *Proceedings of the 25th Annual Convention of the American Institute of Ultrasound in Medicine.* p 137, Sept 14–19, 1980.

61. Watanabe H, Igari D, Tanahashi Y, et al: Development and application of new equipment for transrectal ultrasonography. *J Clin Ultrasound* 1974; 2:91.

Chapter 3E / Nuclide Studies

Judith M. Joyce, M.D.
Barbara Y. Croft, Ph.D.
Charles D. Teates, M.D.

Radioactive tracers have been used for over 20 years to measure renal function and to image the urinary tract. Because of the kidneys' rich vascularity, unique function, and high metabolic rate, a number of radiopharmaceuticals have been used to study this organ system. The anatomical placement of the urinary tract is favorable for external counting and imaging, except that the bladder is anterior and the kidneys posterior. Nuclear medicine has made unique and valuable contributions to the study of the genitourinary tract, but it cannot compete with other modalities such as radiography and sonography for high-resolution anatomical imaging. On the other hand, none of the other imaging modalities has the ability of nuclear medicine for functional imaging or measurements.

This chapter outlines the pharmaceuticals used in genitourinary evaluation, radiation dose from nuclear procedures in relation to radiographic procedures, instrumentation, and clinical applications. Nuclear medicine procedures applicable to patients with genitourinary disease that are not specific to the genitourinary system will not be discussed in any detail. Included among the latter techniques are procedures for imaging infection (gallium citrate Ga 67 scans and indium In 111 oxine-labeled WBC scans), bone imaging (e.g., ^{99m}Tc methylene diphosphonate), lung scanning (^{99m}Tc macroaggregated albumin and xenon 127 gas or xenon 133 gas), and cardiovascular studies (for instance, thallium-201 images of myocardial perfusion or ^{99m}Tc-tagged RBCs for evaluation of cardiac function). These and other nuclear procedures are beyond the scope and the intent of this chapter.

RENAL RADIOPHARMACY

PHARMACEUTICALS

Chemical Structures

The chemical structures for the various compounds in use in nuclear medicine renal work vary from the simple to the complex. The compounds themselves vary in their pharmacology from the general to the specific. We shall first discuss the chemical structure, then the pharmacology.

Xenon 127 and xenon 133 are the simplest compounds in use; they are also nonspecific. Xenon is a monatomic gas, belonging to the noble, or "inert," gas group. It is sparingly soluble in water or isotonic saline and quite soluble in fat. Xenon 127 has the better half-life for storage, the better energy for imaging, and confers the smaller radiation dose, but xenon 133 has been the more available and cheaper isotope.

The radioactive nuclide most used in nuclear medicine is technetium 99m (^{99m}Tc), which has a six-hour half-life and a 140-keV gamma ray energy. It is ideal for examinations taking less than 1 day using the Anger camera, the most common nuclear medical imaging instrument today. The radionuclide is obtained in the pertechnetate form, TcO_4^-, from a "generator." A new generator is delivered to most laboratories weekly. There is a supply of sterile, pyrogen-free ^{99m}Tc in most nuclear medicine laboratories at all times. The pertechnetate form may itself be used in studying renal blood flow. Other technetium radiopharmaceuticals are compounded from purchased kits and available pertechnetate.

Most of the compounds in use have been chosen because of specific interaction with the kidneys. For example, EDTA (ethylenediamine tetra-acetic acid) is a chelating compound with the ability to bond a positive metal ion; the nitrogens provide electrons for the covalent bonds, as do the oxygens of the acetic acids. The compound in routine use in nuclear medicine is DTPA (diethylene triamine penta-acetic acid). It bonds its three nitrogens and five acidic oxygens to positively charged metal ions. EDTA and DTPA are shown schematically in Figure 3E-1. The method for naming these compounds has been to assume that biomedical people could not remember the chemical names and that the initials were not sonorous enough, so parts of the chemical names or initials have been turned into the simpler names edetate and pentetate.

Commercial kits are available for compounding ^{99m}Tc DTPA, ^{99m}Tc glucoheptonate (GH) (Fig 3E-2), and

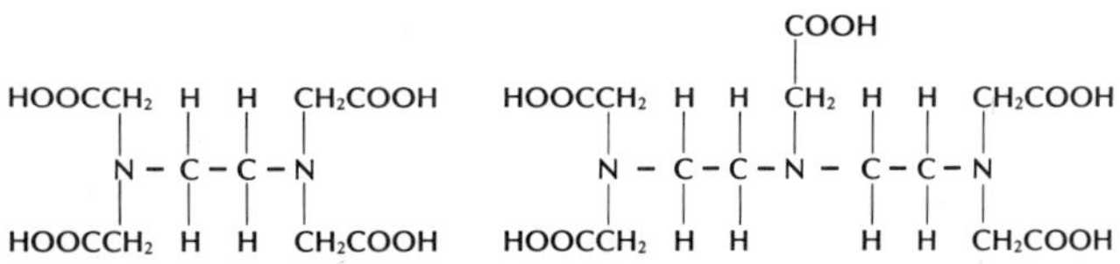

FIG 3E–1.
Chemical structures of ethylene diamine tetra-acetic acid (EDTA) and diethylene triamine penta-acetic acid (DTPA).

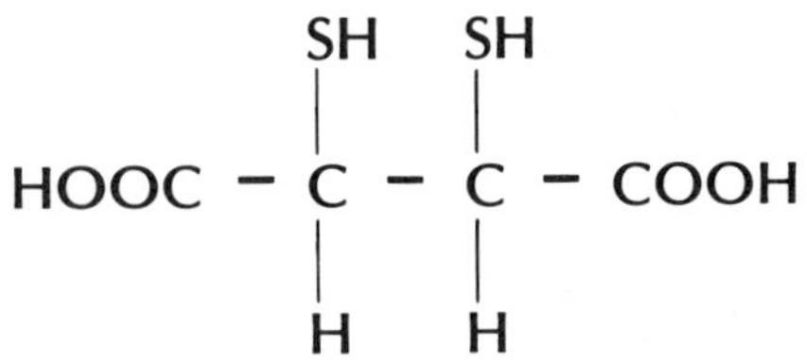

FIG 3E–3.
Chemical structure of dimercaptosuccinic acid (DMSA).

^{99m}Tc meso-2,3-dimercaptosuccinic acid (DMSA) (Fig 3E–3). All three kits contain stannous chloride as a reducing agent, so that Tc^{+4} will be the positive ion species chelated by the organic compound. The package inserts should be consulted for special instructions in using the radiopharmaceuticals.

The iodine isotopes, ^{123}I and ^{131}I, have been attached to the ortho position of hippuric acid to create compounds for renal tubular function studies. The structure is given in Figure 3E–4. Since ^{123}I has a 13.2-hour half-life, ^{123}I hippurate (Hippuran) must be made on site with recently purchased ^{123}I or be purchased for the studies for that day. The ^{131}I Hippuran, with an eight-day half-life, may be kept on hand at all times if demand for its use is sufficient.

BIOLOGIC BEHAVIOR

Xenon is more soluble in fat than in blood. If the patient is caused to rebreathe from a closed system containing radioactive xenon mixed with air or oxygen, the patient becomes more and more radioactive, with fat accumulating the major part of the activity. If the patient is then caused to breathe air alone, the washout of the xenon from the tissues can be observed with nuclear medical instruments. The rate of washout correlates with the blood flow to the organ or area. This technique has been used in many organs in the body.

The specific behavior of ^{99m}Tc DTPA will be discussed below in the discussion of GFR. A variable amount (10% or less) of injected ^{99m}Tc DTPA is protein-bound.

Technetium 99m glucoheptonate is injected IV. The material is found in the plasma, and blood clearance is rapid (Boyd, 1973; Arnold, 1975). The glucoheptonate clears by both tubular secretion and glomerular filtra-

tion. About 70% of the material is excreted in the urine of normal subjects. About 6% of the dose is retained in each kidney, largely in the cortex, permitting imaging of the cortex 3–4 hours after injection. A normal variation is clearance by the liver into the gallbladder and intestines.

Technetium 99m DMSA is slowly injected IV. It is distributed in the plasma, loosely bound to plasma proteins. The activity clears from the plasma with a half-time of 60 minutes and concentrates in the renal cortex. Approximately 16% of the activity is excreted within two hours, increasing to 25% by six hours. At two hours, 15% is concentrated in each kidney; this increases to 20% by six hours (Arnold, 1975). Imaging is best performed three hours after injection or later.

Iodinated hippurate has been extensively used in the examination of renal function; there is a voluminous literature on the use of this compound. The biologic behavior of iodinated hippurate is discussed below, in the section on effective renal plasma flow. Most of the hippurate is actively secreted by renal tubules. In a normal person, 70% of the compound is excreted in the urine in 30 minutes.

DOSIMETRY

Both nuclear studies and radiographic procedures result in patient exposure to ionizing radiation. The major difference between the two procedures is that radioactive pharmaceuticals produce patient exposure from internal sources, whereas that from a radiographic diagnostic procedure is due to external irradiation. Exposure from radiographic procedures can be accurately determined if the unit is accurately calibrated and the exposure factors are known. In nuclear procedures, the patient dosage is determined by the biologic

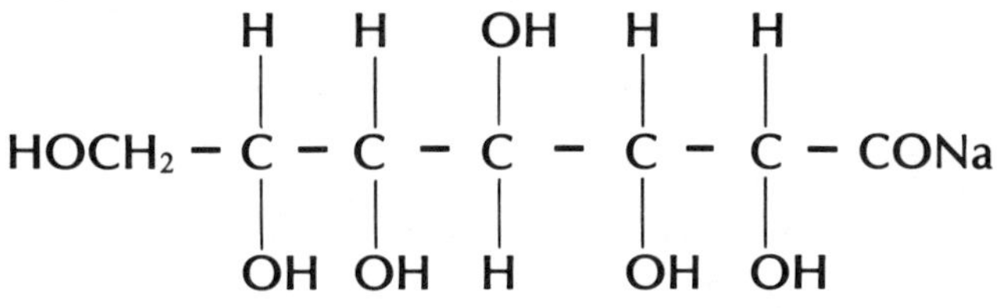

FIG 3E–2.
Chemical structure of glucoheptonic acid.

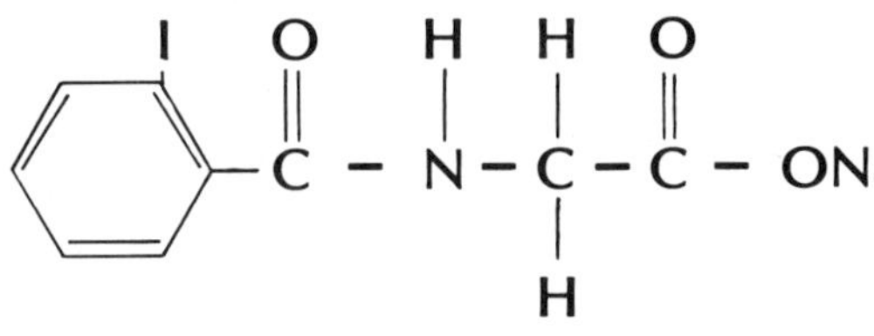

FIG 3E–4.
Chemical structure of sodium o-iodohippurate (OIH).

TABLE 3E–1.

Absorbed Dose in Rad for Nuclear Imaging

AGENT	USUAL DOSE, mCi	KIDNEY	BLADDER WALL	GONADS	WHOLE BODY
99mTc DTPA					
Adult[1]	15	1.35	1.73*	0.15*	0.09
10-year-old[6]	9.75	0.68	7.8*	—	0.29
99mTc GH					
Adult[2]	15	2.55	4.2*	0.2*	0.11
10-year-old[6]	9.75	1.95	7.8*	0.2*	0.07
99mTc DMSA					
Adult[3]	5	3.8	1.4*	0.1*	0.08
10-year-old[6]	3.25	2.3	0.98*	0.07*	0.07
131I Hippuran					
Adult[4]	0.2	0.02	0.06*	0.03*	0.02†
10-year-old[6]	0.13	0.01	0.04*	0.003*	0.008
99mTc cystography[5]	1	—	0.07	0.002	

*Bladder and gonad doses determined by frequency of voiding.
†Unblocked thyroid dose as high as 8.7 rad.
References:
  1. Package insert, Syncor International Corporation, 1980.
  2. Package insert, New England Nuclear, 1978.
  3. Esser et al., 1984.
  4. Package insert, E. R. Squibb & Sons, Inc., 1982.
  5. Dimitriou et al., 1984.
  6. Koenigsberg et al., 1978.

handling of the pharmaceutical as well as the physical behavior of the radionuclide and patient dosage. Alterations in patient physiology, as for example in renal failure, affect the absorbed dose of radiation markedly, if the radiopharmaceutical behavior is altered by the disease process. In recent years, patient exposure has been reduced in both areas by improvements in radiopharmaceuticals, nuclear instrumentation, radiographic and fluoroscopic equipment, and film/screen sensitivity. The patient exposures listed in Tables 3E–1 and 3E–2 are considered average exposures and may not be accurate for a given patient or a particular institution. Personnel exposures should be low with both types of procedures, if appropriate protective measures such as syringe shields and equipment shielding are utilized. The procedure potentially resulting in the highest exposure is fluoroscopy, since a typical fluoroscope exposes the patients and personnel to as much as 10 rad/min.

Certain terms must be defined to understand exposure to ionizing radiation. The patient dose of a radiopharmaceutical is measured by the number of atoms disintegrating per second. The term most frequently used in nuclear medicine is millicurie (mCi), defined as 3.7×10^7 disintegrations per second. A microcurie (μCi) is 3.7×10^4 disintegrations per second. Obviously, a radionuclide with a long half-life in the patient will usually expose the patient to more ionizing radiation than another radiopharmaceutical with a shorter effective half-life. The effective half-life is influenced both by the physical decay rate of the nuclide and the biologic turnover of the pharmaceutical.

Two terms are commonly used as measures of x-ray and patient exposure. The roentgen (R) is a measure of ionization of air by x-rays or gamma rays. There is no physical difference between an x-ray and a gamma ray. An x-ray originates from atomic electrons, and a gamma ray originates in the nucleus of an atom. One R results in 2.082×10^9 ion pairs in 1 cc of air at standard atmospheric pressure. The term "roentgen" is usually used to express the output of an x-ray machine. The amount of energy absorbed by a patient's tissue is expressed in rad. The rad is defined as 100 ergs absorbed per gram of tissue. In most tissues exposure to 1 R results in approximately 1 rad of absorbed energy. The two terms are used somewhat interchangeably, but it

TABLE 3E–2.

Adult Doses From Radiographic Procedures*

PROCEDURE	DOSE
Intravenous urograms	
(No. of films/exam)	(5.31)
Mean exposure/exam	3,133 R
Mean marrow dose/exam	0.103 rad
Mean male gonadal dose/exam	0.207 rad
Mean female gonadal dose/exam	0.588 rad
Mean dose to total body/exam	0.278 rad
Fluoroscopy	
Exposure/min	2–10 rad

*From Gorson RO, Lassen M, Rosenstein M: Patient dosimetry in diagnostic radiology, in Waggener RG, Kereiakes JG, Shalek RJ (eds): *CRC Handbook of Medical Physics*. Boca Raton Fla, CRC Press Inc, 1984, vol 2, pp 474, 487, 488. Used by permission.

should be remembered that the roentgen is a measure of exposure while the rad is a measure of energy absorbed by tissue.

Table 3E–1 lists typical absorbed doses in patients receiving the five most common radionuclide studies of the urinary tract. The bladder receives the highest exposure from most of these pharmaceuticals, but this dose can be reduced considerably by frequently emptying the bladder. Free iodine, present to some extent in ^{131}I hippurate, will result in exposure to the thyroid. The dose of the thyroid can be reduced by more than 90% by giving Lugol's solution, saturated solution of potassium iodide, or other blocking agents prior to the study.

The doses listed from radiographic procedures in Table 3E–2 are based on a survey performed by the Bureau of Radiologic Health between 1964 and 1970. The patient exposures can be reduced somewhat from the values listed by using better collimation of the x-ray beam, improved films and screens, and by reducing the number of films or fluoroscopic exposures.

INSTRUMENTATION

Two types of instruments are most often used to detect ionizing radiation in clinical applications. The oldest and simplest technique uses a gas detector. In this type of instrument, a charge is placed across electrodes in a chamber containing some type of gas. Gamma rays or x-rays cause the gas to ionize, thereby resulting in current flow across the electrodes. The amount of voltage between the electrodes determines the amount of current flow, the sensitivity of the detector, and the useful range of radiation that the instrument can accurately measure. The walls of the chambers can be altered to allow detection of very poorly penetrating radiation or highly penetrating radiation.

Two types of gas detectors are commonly used in nuclear medicine laboratories. A Geiger-Müller (GM) survey meter is usually employed for detecting contamination in the nuclear medicine laboratory. This gas detector operates with a relatively high voltage, making the detector sensitive for small amounts of radiation. A typical range of usefulness is between 0 and 50 millirads per hour (1 rad = 1,000 mrads). The ionization chamber is usually attached to a rate meter and electronic package by an electrical cord that allows survey of work areas. The probe is relatively nondirectional. The standard survey meter is intended for detection of x-rays and gamma rays and is not very suitable for low-energy β-ray detection.

The second type of gas detector usually found in the lab is a dose calibrator. At one time, most of the radiopharmaceuticals administered in nuclear medicine lab-

oratories had relatively long half-lives and were purchased as needed from a pharmaceutical supplier in precalibrated doses. Patient doses were withdrawn from a precalibrated vial, and activity was calculated based on the known activity per cubic centimeter. As discussed above, many of the radiopharmaceuticals carry the radioactive label ^{99m}Tc. ^{99m}Tc is eluted daily from a shielded ion exchange column containing molybdenum 99. The elution is unpredictable, and therefore the patient dose must be accurately measured in the laboratory. Highly accurate dose calibrators currently available allow the measurement of bulk quantities or individual patient doses of radiopharmaceuticals. The dose calibrators use a gas-filled ionization detector. Physically the detector is a well chamber that allows the vial or syringe to be inserted for high-efficiency counting. These instruments are capable of accurately measuring over a wide range of activity, typically from approximately 1 μCi to over 2,000 mCi. One millicurie equals 1,000 microcuries. The dose calibrators are designed to give a digital readout of activity and are usually accurate to within 5%. The dose calibrators must be checked daily against known standard amounts of activity. These chambers operate with a relatively low voltage to allow measurements of high radiation intensities, and are not accurate for measuring low-energy x-rays, low-energy gamma rays, or β rays.

The other major category of detectors of ionizing radiation uses solid crystals. In most laboratories, the detector is sodium iodide, doped with an impurity such as thallium. Sodium iodide is hygroscopic, and therefore the crystal is completely enclosed in a barrier impervious to water, such as aluminum or glass. Ionizing radiation is absorbed by the crystal, converting the energy to visible light. Thus, the sodium iodide crystal is called a scintillation crystal or scintillation detector. The light produced by each ionizing event is quite small and is detected by a nearby photomultiplier (PM) tube, which has a photocathode that converts light energy into a small electronic pulse. The electronic pulse is amplified over 1 million times within the PM tube, making it large enough to be amplified and processed in standard electronic circuits. Thus, an ionizing event in the crystal produces light that is converted into an electronic pulse. The magnitude of the pulse is related to the amount of energy contained in the ionizing event. The pulse can be analyzed to determine the energy of the x-ray or gamma ray that hit the crystal. Pulses can be integrated over a variable time to give counts per second or counts per minute. Ultimately, the count rate can be displayed on a meter, or a digital printout, or stored by a computer.

A single crystal detector is usually mounted in some type of lead or other heavy-metal shielding so that the

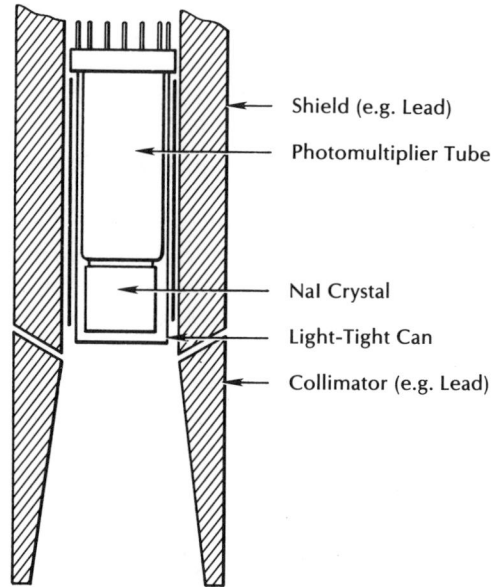

FIG 3E–5.
Typical probe detector system. The sodium iodide (NaI) crystal and photomultiplier (PM) tube are shielded by lead to restrict the field of view to the desired anatomical region. The crystal is enclosed in a polished can to increase the reflection of light to the PM tube and protect the crystal from degradation by moisture.

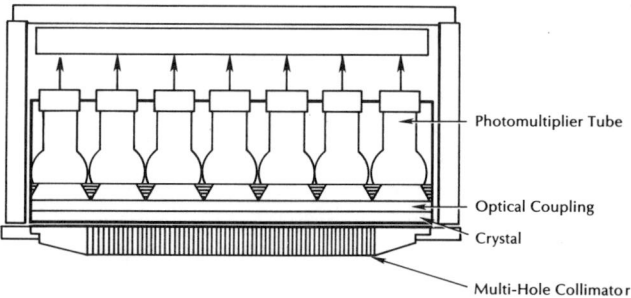

FIG 3E–6.
Diagram of a cross section through a 37 PM tube gamma camera. The PM tubes are arranged in an hexagonal array, with three rings of tubes surrounding a central tube. Gamma rays or x-rays pass through the collimator and interact with the crystal, producing light. The location of the scintillation and the total energy absorbed are determined from the signal output by the PM tubes. The amplified signal is analyzed for energy levels and sent to the cathode ray tube and/or computer for recording.

crystal is sensitive to x-rays or gamma rays from a localized region (Fig 3E–5). Typically, a single crystal-single PM tube system is used for measuring count rate from the thyroid for thyroid uptake determination or monitoring the count rate in the kidneys during a renogram. Under these circumstances, the probe and its collimator are positioned so that the crystal monitors the count rate in the organ and nearby tissues being surveyed. The successful use of this detector assumes that the organ can be accurately localized from anatomic landmarks.

A sodium iodide scintillation detector can be manufactured in the form of a "well counter." This is used for counting in vitro samples. In effect, the sample is inserted into a hole in the protected crystal so that high counting efficiency is achieved. As with the scintillation probe, the activity can be counted over time to determine sample activity.

The imaging device used in most departments is the Anger camera. In the camera there is a single crystal measuring 13–20 in. in diameter, ¼ to ⅜ in. thick. Behind to the crystal, there is a matrix of PM tubes, typically numbering anywhere from 37 to 91, that look at the intensity of light from each scintillation in the crystal (Fig 3E–6). The light striking the multiple PM tubes is analyzed so that the location of the scintillation is determined as well as the total energy from the x-ray or gamma ray. The front of the crystal is protected by a

collimator. The multiple holes in the collimator determine the origin of the photons that hit the crystal. Ordinarily, most Anger cameras are used with a parallel-hole collimator that sees a field of view 10–18 in. in diameter, depending on the size of the system. The thickness of lead septa between holes is selected for the appropriate energy of the photons being imaged, i.e., low-energy (up to 160 keV) and medium energy (160–360 keV). The hole diameter and collimator thickness are selected for desired resolution and counting efficiency.

The camera electronics include an energy discriminator (spectrometer) to be sure that only the appropriate energy is accepted. The crystal will detect stray cosmic rays, scattered x-rays, and so forth, as well as the desirable photons emitted by the nuclide being imaged. The image quality will suffer unless the unwanted radiation is excluded by energy discrimination.

The final image that is viewed depends on where the signal is sent once it leaves the PM tubes and analyzers. It may enter a cathode ray tube (CRT) that puts a series of dots on the screen. By using time-lapse photography of the CRT, an image is generated on Polaroid film or transparency film. The same signals may be sent to a computer for storage and later manipulation.

Computers have been used in nuclear medicine for a dozen years, to digitize and store the information from the scintillation probe or Anger camera for later manipulation. Over the years, the storage capacity and sophistication of programs have improved markedly. In utilizing a computer to store and manipulate data, it is essential that the information be stored in an adequate format. If images are integrated for one minute each, for instance, the data cannot be later analyzed at one-second intervals. Therefore, the appropriate prescription for the storage must be determined prior to initiating the study. Most commercial computers now have several software programs that allow histogram displays of activity vs. time. These curves can be stripped or analyzed to allow washin and washout rate determinations, fractionation of individual renal function, and analysis of blood clearance rates to determine GFR and effective renal plasma flow. In addition, computer manipulation is essential to enhance images from an Anger camera and perform specialized procedures such as emission computerized tomography (ECT). ECT creates tomographic images in much the same way as CT scanning by combining images from multiple positions as the camera rotates about the patient.

PROCEDURES

The following sections describe the range of procedures available in nuclear medicine to assist the renal diagnostician. For easy reference, the examinations, radiopharmaceuticals, and typical doses of radioactivity are outlined in Table 3E–3.

RENAL IMAGING

Renal images are typically performed with Anger cameras. The arrival of activity can be recorded in the form of renal vascular studies. This information can also be analyzed by computer to compare the washin rates for the two kidneys. Most of the technetium pharmaceuticals can be used for the vascular sequence, provided adequate amounts of activity (15–20 mCi) are injected, but the later static views show varying features depending on the pharmaceutical used. The vascular sequence is typically recorded on film and by the computer at one frame per two seconds, but the framing rate can be tailored to the patient's age and the injected dose. In general, faster sequences are desired for children. The limiting factor on the

TABLE 3E–3.
Radiopharmaceuticals and Activity Dosage for Renal Examinations

EXAMINATION	AGENT	DOSE OF ACTIVITY
Renal blood flow imaging	^{99m}Tc compound	15 mCi
Effective renal plasma flow	^{131}I hippurate	
Imaging		200 μCi
Blood sampling only		30 μCi
Glomerular filtration rate	^{99m}Tc-DTPA	
Imaging		15 mCi
Blood sampling only		300 μCi
Diuretic renography	^{99m}Tc-DTPA	15 mCi
Renal cortical imaging	^{99m}Tc-DMSA	5 mCi
	^{99m}Tc-GH	15 mCi
Cystography	^{99m}Tc-pertechnetate	1 mCi
Scrotal imaging	^{99m}Tc-pertechnetate	20 mCi

renal vascular sequences is generally the low information density due to a small number of counts per image, so higher imaging rates may actually cause deterioration of the images. Because of its ability to add frames together, the computer is often a more satisfactory method than film for recording and displaying this information.

The static images of the kidneys are collected immediately after the vascular sequence and up to several hours later. Static images may contain as many as 1 million counts and require up to several minutes' accumulation time. ^{99m}Tc DTPA is filtered and concentrated in the tubules, then is excreted through the collecting system. Activity in the calyces, renal pelvis, and ureters decreases after 5–10 minutes, and delayed views beyond 30 minutes have little value unless the patient has obstruction. However, ^{99m}Tc glucoheptonate and DMSA show progressive accumulation in the kidneys over several hours; delayed views will show better images of the renal cortex after the background and collecting system activity have decreased. Patient hydration will influence the washout rates from the kidneys just as with a renogram. This effect is ulitized in the "Lasix renogram," to be discussed under Hydronephrosis.

RENOGRAMS

Initially renograms were performed using multiple single-crystal probes positioned over the patient's back. The probes were positioned by external anatomical landmarks, but this often did not locate the kidneys in the field of view of the probes. Later workers positioned the probes by injecting a small amount of renal tracer (e.g., ^{203}Hg chlormerodrin) and finding the maximum count rate prior to injection of ^{131}I hippurate. In more recent years, most renograms have been performed with Anger cameras, even though the sensitivity of a camera is somewhat less than that of a probe. The versatility of data recorded from a camera more than offsets the disadvantage of the higher pharmaceutical dose required. A renogram may be performed using external probes with as little as 30 μCi of ^{131}I hippurate. Renograms performed on Anger cameras typically use 200 μCi of the same agent. With the camera, the accumulated information is usually stored by a computer for later analysis of individual kidney count rates and generation of renogram curves. The camera-generated data are acquired by the computer at intervals of 10–15 seconds for as long as 30 minutes to an hour.

The term "renogram" simply indicates that an activity-vs.-time graph is being generated from the kidney activity. Classically the renogram study was performed with ^{131}I hippurate, but other agents such as ^{99m}Tc

DTPA may be used. The shape of the renal curve is obviously affected by the pharmaceutical employed. The shape is also affected by patient preparation and positioning. For instance, the classic renogram is performed with the patient mildly dehydrated and seated in front of the Anger camera. Drainage from the upper collecting system is more consistent in the upright position, but unfortunately there is more tendency for the patient to move. Patients who cannot maintain this position comfortably and consistently may be imaged in the supine position. Follow-up patient studies should always be performed in the same position if possible.

Hydration state affects the timing and shape of the renogram curves. Phase I, lasting approximately 30 seconds, represents the arrival of the IV injected pharmaceutical in the blood pool of the kidney and adjacent tissues. There will be a phase I increase in count rate regardless of the status of the kidney, because there is blood pool activity in all tissues. The activity during this phase is poorly related to renal blood flow. Phase II typically lasts 4–6 minutes and terminates when activity in the kidney reaches a maximum. As the kidney extracts the radiopharmaceutical from blood, the count rate gradually rises. Because the blood levels fall rapidly during the first few minutes, activity in the urine is maximum initially and gradually falls with time. When this most active urine leaves the region of interest being analyzed, the count rate will start to fall. Thus, the rate of increase in count rate during phase II is directly related to the renal blood flow and renal function. The rate of increase of activity is not affected by hydration, but the duration of phase II is inversely related to the urine formation rate.

Phase III starts at the peak of renal activity and illustrates a gradual fall in count rate as the most concentrated activity is washed from the collecting system. The time that is required for the count rate to fall to half of the peak value is called the half-time for washout, and this time is affected by urine formation rate. Numerous other aspects of the shape and timing of the renogram curve have been analyzed, but none is specific for disease process. However, if hydration state and positioning are consistent, changes in the slope of phase II, the time to peak, and the half-time for washout do reflect changes in renal status.

MATHEMATICAL MODEL— TWO-COMPONENT MODEL

To be able to quantitate the results of functional studies, a model is developed to describe the organ function. A connection is made between the numbers that are available from the noninvasive examination and the properties that it is desired to measure. In the case of the kidneys, the quantities to be measured include renal plasma flow, GFR, and individual kidney function. Whether renal plasma flow or GFR is measured depends on the pharmaceutical employed in the measurement.

The estimation of renal blood flow and function can be computed by the Fick principle, which when applied to the kidney gives the formula

$$\text{Clearance} = \frac{Uv}{A - V}$$

where U is the concentration of the test substance in the urine, v is the volume of urine, A is the concentration of the substance in the renal artery, and V is the concentration of the substance in the renal vein. If the substance is completely removed by the kidney, then the renal vein concentration can be assumed to be zero; the arterial concentration may be assumed to be equal to the peripheral venous concentration. Thus the formula is simplified to

$$\text{Clearance} = \frac{Uv}{A}$$

To perform a clearance measurement using Fick's principle, blood is sampled, and a total urine collection is made during continuous IV infusion of the agent for three 20-minute periods. The assumption is made that the body reaches a steady state in which the input of the measured substance is the same as the output. One can further assume that measurements of the quantity of the substance in the infusion and the peripheral venous concentration are sufficient. If the separate function of each kidney is desired, catheters must be placed in each ureter.

It was discovered by Sapirstein and co-workers (1955) that a single injection of the test substance could be substituted for continuous infusion with no compromise in the total functional information.

In a renal function examination using a radiopharmaceutical, the numbers that are available are the relative concentration in the kidneys as attenuated by tissue and the amount of activity per unit volume in the blood during the examination. The relative concentration in the urine in the bladder during the examination may be compared with the absolute concentration in voided urine at the end of the examination, especially in transplant patients whose kidney and bladder are close enough together for imaging at the same time. Note that unless complex techniques are resorted to, only samples removed from the patient can be quantitated in an absolute way.

It has been observed that the radioactivity per unit volume of plasma of radiopharmaceuticals injected for renal examination decreases according to a biexponen-

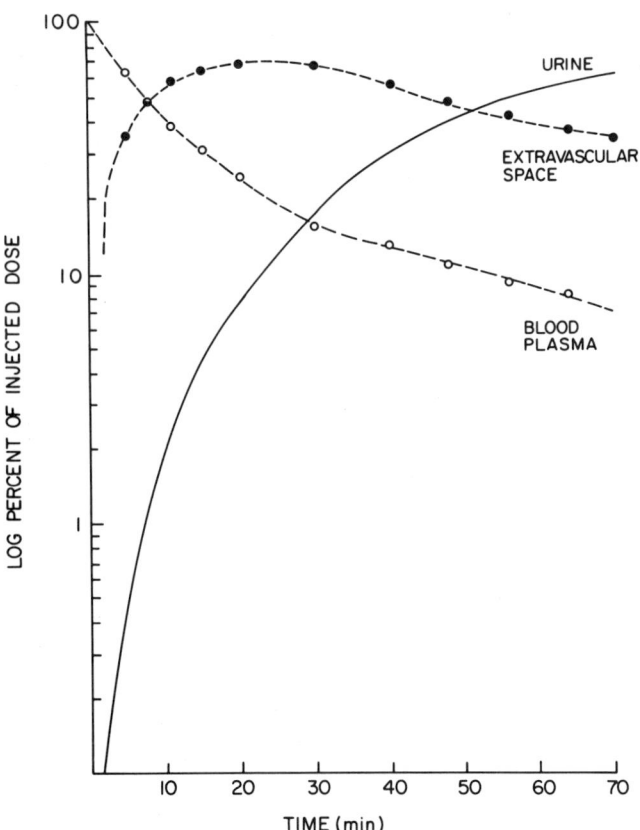

FIG 3E–7.
The semilog graph of percent of injected dose vs. time illustrates the amount of sodium o-iodohippurate in the blood, urine, and extravascular space.

tial curve as a function of time. This means that the curve of plasma activity as a function of time shown in Figure 3E–7 can be decomposed into two straight lines (on a semilogarithmic scale as shown in Figure 3E–8). Each of these lines is described by its intercept with the activity axis, which is its concentration at time equals zero, and its half-time, or the amount that disappears per unit time. Thus, one curve yields four numbers.

If, in turn, we consider a model of renal function (Fig 3E–9), it is possible to write differential equations describing the loss of material from one compartment and the gain of an equal amount by another, following the arrows of the model and using the language and symbolism of reaction-rate chemistry. The model pictured is called an open two-compartment mamillary model (Matthews, 1957). Such differential equations can be solved to yield mathematical functions that describe the concentration in each compartment as a function of time. For this model, the function that describes the concentration in the vascular compartment as a function of time is a biexponential curve. Thus, the kinetic variables in the model can be related to the numbers gen-

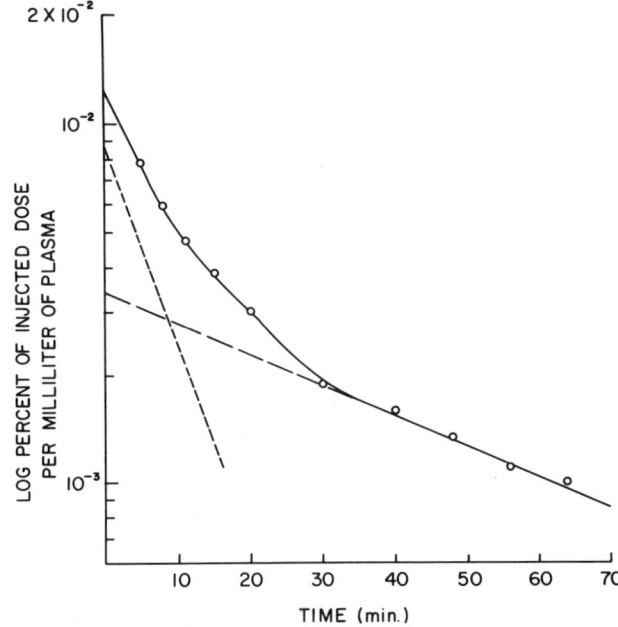

FIG 3E–8.
The semilog graph of percent of injected dose per milliliter of plasma vs. time shows a biexponential functionality. The two exponential parts are shown.

erated from patient plasma sampling. The renal clearance is a function of all four numbers, which come from both parts of the curve:

$$\text{Clearance} = \ln (2)/(A_1 \times t_1 + A_2 \times t_2)$$

where A_1 and A_2 are the fractions of the injected dose per milliliter for each of the two parts of the curve at time zero, and t_1 and t_2 are the half-times for the two parts of the curve.

The measurements of the plasma disappearance curve may be made in several different ways. One method is to take plasma samples over a time suitable for defining the two parts of the curve; 10 samples are sufficient to permit curve-fitting and the discrimination

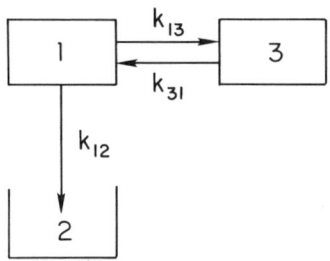

FIG 3E–9.
The open two-compartment mamillary model. Compartment 1 is the plasma, compartment 3 is the extravascular space, and compartment 2 is the urine. There is kinetic equilibrium between compartments 1 and 3, while there is no return from compartment 2. *k* indicates the rate constant for transfer from compartment i to compartment j.

of poor samples. Some laboratories have simplified this method to taking one blood sample at a particular time with the subsequent calculation of renal clearance by a relationship connecting the clearance to the single sample activity and several constants. A second method involves the measurement of a major blood pool in the body, such as the cranium or the heart, with a probe detector, combined with a blood sample that is compared with a standard to permit calibration of the plasma disappearance curve.

Clearly, using 8–10 blood samples to define the biexponential curve is the more accurate method, although simplification of the test procedure is necessary in some instances. In such a case, a method that accurately defines the longer half-life part of the curve should be sufficient to eliminate the difficulties of a single sample. In this case, the clearance equation becomes:

$$\text{Clearance} = \ln(2)/(A \times t)$$

where A is the fraction of the injected dose per milliliter of plasma at time zero, and t is the half-time of the line.

It should be pointed out that the simplified methods may not allow the accurate comparison of one patient to an absolute scale, but may permit the progress of that patient's disease to be observed with changes in the renal function value, since a particular patient's measurement is reproducible.

It seems that urine sampling might be as successful as blood sampling and even less invasive. The problem is that the urine curve depends on the transport of the urine through the collecting system to the bladder; transport may be inhibited so that the urinary activity does not accurately reflect renal function.

COMPUTER PROCESSING FOR FRACTIONATION

The fractionation of renal clearance into values for the individual kidneys is accomplished by reference to the images acquired by the Anger camera at 15-second intervals after injection of the tracer dose. Regions of interest are drawn around the kidneys; background-corrected activity-vs.-time curves are generated. From this point there are two different methods for completing the calculation.

In the slope method, the slopes of the curves for the two kidneys between 60 and 150 seconds after injection are calculated using least-square methods. The two slope values are added together and the fractional contribution of each kidney's slope to the total is calculated. This is the fractional contribution that each kidney makes to the clearance.

In the integrated count method, the counts for each kidney between 60 and 150 seconds after injection are found, after background correction; this is a relative measure of the activity multiplied by time in each kidney. The two integrated counts are added, and the fractional contribution of each kidney to the total is calculated. Once again, this is the fractional contribution that the kidney makes to the clearance.

Both of these methods give similar results for kidneys that function well. The results for poorly functioning kidneys are open to more uncertainty for several reasons: the selection of the background region becomes more critical; the difference between kidneys and background is small and shows great variability; and the statistical uncertainties of radioactive detection are more serious for lower counts, so all the calculations have greater uncertainty.

When percentages are used for comparisons, they must always add up to 100%. This means that if one kidney remains the same after some elapsed time but the other improves in clearance, the one that remains the same will appear to have lost function on a percentage basis. It is thus incumbent on the observer to look at both the percentages and the absolute clearance values.

GLOMERULAR FILTRATION RATE

The glomeruli produce an ultrafiltrate of the plasma by means of a physical process which is nonselective for substances of low molecular weight. The volume of this ultrafiltrate, expressed in ml/min, is defined as the GFR. The formation of the filtrate is regulated by hydrostatic pressure and diffusion secondary to a concentration gradient.

To measure GFR, the agent must (Smith, 1951):

1. Be nontoxic, physiologically inert, and chemically stable;
2. Be easily and accurately measured in blood and urine;
3. Be fully filterable through the glomerular membrane;
4. Not combine with plasma proteins;
5. Not be resorbed, synthesized, destroyed, or excreted by tubules;
6. Have a constant clearance with high or low urinary flow, and greater or lesser concentrations of the agent in the plasma;
7. Have a clearance equal to that of other tracers already proved adequate for GFR measurement, such as inulin;
8. Be eliminated exclusively by the kidneys.

The standard for comparison for GFR measurements is inulin clearance in a protocol that includes continuous IV infusion and three 20-minute complete urine collections and three blood samples drawn during the urine sampling periods. The protocol is not practical for clinical use.

The use of radioactively labeled compounds has simplified GFR measurement by simplifying the measurement of the agent. Of the radioactively labeled materials, ^{99m}Tc DTPA comes closest to fulfilling the criteria for GFR measurement. It can also be used in combination with an imaging examination to visualize renal and ureteral anatomy and to quantitate the fraction of the GFR attributable to each kidney (fractionation). Inulin, labeled with the beta-minus emitter carbon 14, can also be used with the standard continuous infusion method or in a plasma sampling mode, as can sodium iothalamate and other radiographic contrast agents, labeled with ^{125}I or ^{131}I. Chromium 51 DTPA has also been used. The ready availability of ^{99m}Tc DTPA makes its use simpler than any of the other materials mentioned here.

The measuring process has been described above. Ten plasma samples obtained during 1½–2 hours of blood sampling after a single IV injection should be sufficient to define the two parts of the biexponential curve. The single sample technique uses a function of the form

$$GFR = A(S-B)^{1/2} - C$$

where A, B, and C are constants, and S is the fraction of the injected dose per milliliter of plasma in the three-hour sample (Constable et al., 1980).

The radionuclide technique described by Gates (1983) is an alternative method of measuring the GFR. The calculation is based on the renal uptake on Anger camera images during the 2 to 3 minute interval following tracer (^{99m}Tc-DTPA) arrival in the kidneys. The GFR is computed by using the formula

$$GFR = A(L+R) - B$$

where L and R represent the depth-corrected percentages of renal uptake for the left and right kidneys and A and B are constants. The formula was derived from linear regression analysis comparing the renal uptake of ^{99m}Tc-DTPA with 24-hour creatinine clearance in 51 adult studies.

The advantage of this method is that it allows rapid determination of split renal function as well as total GFR without blood samples. The accuracy of this method, however, has shown variable results. Ginjaume (1986) compared four methods of measuring GFR and found that the effective volume technique using one blood sample taken at 2 hours was the best compromise between accuracy and convenience and that the Gates method presented practical problems due to uncertainty in background subtraction and kidney depth approximation.

The mean value of GFR in the normal adult is approximately 130 ml/min in the male and 120 ml/min in the female, with an uncertainty of 10%. The GFR of the newborns is between 20% and 40% of the adult value and increases progressively until, at the age of 1 year, it becomes equal to adult values in relation to the standard surface area.

EFFECTIVE RENAL PLASMA FLOW

If the renal tubules can be assumed to remove a substance totally from the blood during perfusion, a study of the concentration of that substance could yield values of renal blood flow. Most of the substances used in such a measurement are concentrated in the plasma, so the measurement is of renal plasma flow (RPF).

The ideal substance for estimating RPF should:

1. Be nontoxic, physiologically inert, and chemically stable;
2. Be easily and accurately measured in blood and urine;
3. Be fully secreted by renal tubules;
4. Be readily dissociated from any plasma protein complex in its transit through the kidney;
5. Not be resorbed, synthesized, or destroyed by tubules;
6. Have a saturable clearance, so that high concentrations have lesser clearance values;
7. Have nearly total renal extraction;
8. Be eliminated exclusively by the kidneys.

Both because only a portion of renal blood flow is presented to renal secretory tissue as opposed to the small fraction which normally perfuses the nonsecretory tissue (perirenal fat, pelvis, and capsule) and because no substance perfusing the kidney will be totally extracted, the calculated clearance will be less than the total renal plasma flow, so it is called effective renal plasma flow (ERPF). Extraction efficiency of the tubules may be decreased in disease, as may renal blood flow.

Para-aminohippurate (PAH) was found to meet the above criteria best. Sodium iodohippurate (OIH) was found to be similar in behavior to PAH; in addition it can be labeled with radioactive isotopes of iodine such as ^{131}I, ^{123}I, and ^{125}I, making detection of the material simpler than in the previously used chemical methods. ^{131}I and ^{123}I have the added advantage of ready imaging with the Anger camera.

Since ^{131}I OIH disappears from the blood according to a biexponential function, the open two-compartment mamillary system is appropriate for analysis. Again, the examination protocol may be combined with imaging, so that the images and fractional ERPF for each kidney are obtained along with the total ERPF value. The protocol may be based on serial blood sampling between 5 and 70 minutes after IV injection of ^{131}I OIH, to define the biexponential curve, or a less accurate single sampling method can be used. Tauxe and co-workers (1971) performed an elaborate analysis which suggested that it was feasible to use a single blood sample, at 44 minutes after injection, and a polynomial to calculate the ERPF:

$$ERPF = A + B/S + C/S^2$$

where A, B, and C are constants, and S is the fraction of the injected dose per milliliter of plasma.

Normal ERPF values lie above 600 ml/min, with a 10% uncertainty. Approximately 50% of the function should be attributable to each kidney. Extraction fraction is the comparison of GFR and ERPF measurements. The normal value is approximately 0.2. Recent publication documents ERPF values in normal patients and patients with unilateral nephrectomy (Tauxe, 1985).

NUCLEAR CYSTOGRAPHY

Although cystograms may be performed after the bladder has filled following an IV injection of a radiopharmaceutical, the study is more accurate in detecting ureteral reflux if the tracer is placed directly in the urinary bladder. Three-tenths to 1 mCi of ^{99m}Tc pertechnetate or DTPA is mixed with 250–500 ml of sterile saline in an IV bottle. The IV bottle should not be more than 3 ft above the bladder. After attaching the tubing to the catheter, the bladder is slowly filled, and the time and volume are recorded. At the same time, sequential images are recorded by the camera and computer. The patient is then instructed to void with the catheter in place; if voiding is impossible, the catheter is withdrawn.

The recording sequence varies somewhat among laboratories, but in general frames are recorded on film and in the computer at 15–30-second intervals. Posterior projections are used. Images or computer curves may reveal reflux up the ureters. The severity of reflux is usually gauged by the volume of infused fluid required to produce significant reflux. Increasing volumes instilled before reflux occurs implies improvement.

SCROTAL IMAGING

Nuclear imaging to evaluate scrotal pathology has been utilized since 1973 with excellent accuracy reported. Its primary role is the differentiation of testicular torsion from epididymitis in patients presenting with an "acute" scrotum. The standard radionuclide employed is ^{99m}Tc pertechnetate; however, other technetium compounds including DTPA, GH, and labeled RBCs could equally well evaluate scrotal perfusion.

In our procedure, 30 minutes before radionuclide dose, the patient is given an oral dose of potassium perchlorate to block thyroid uptake. Potassium iodide can also be used. Positioning includes taping the penis to the abdominal wall and supporting the testicle on a tape sling to rest the testicle on a slightly higher plane than the thigh. A large-field-of-view camera is positioned anteriorly, and no shielding is employed.

A bolus of 20 mCi of ^{99m}Tc pertechnetate is injected intravenously. A flow sequence of 3 seconds per frame at 70-mm image size is used, then static images with a LEAP collimator are obtained immediately and at 5 and 10 minutes for 500,000 counts per view. The analysis is described in the Clinical Applications section.

CLINICAL APPLICATIONS

ACUTE RENAL FAILURE

Acute renal failure may be due to anatomical or physiologic abnormalities. Anatomical causes of renal failure include occlusion of renal arteries and veins and obstruction of the urinary tract at any level. Physiologic causes of acute renal failure include blood volume depletion ("prerenal") and acute tubular necrosis.

The first step in evaluation of acute renal failure is sonography to evaluate the status of the collecting system (Bell et al., 1981). Contrast studies are not recommended because of their potential adverse effect on renal function. If the sonogram shows no evidence of dilatation of the collecting system, the next step is dynamic renal scintigraphy.

The flow portion of dynamic imaging is the essential part in the evaluation of renal arterial blood flow. Analysis is based on observing the symmetry and the intensity of kidney visualization. The peak of activity in the kidney should be no more than three seconds after the peak of activity in the aorta. The intensity of activity in a normal-sized kidney should equal or exceed the early activity in the spleen (Freeman, 1984).

Unilateral delay or decrease in kidney visualization on the flow study signifies a vascular abnormality. The possible vascular abnormalities include renal artery occlusion secondary to embolism, thrombosis, or dissection of an aortic aneurysm; renal artery laceration secondary to trauma; renal artery stenosis; and renal vein thrombosis. Unilateral delay can also be caused by severe unilateral ureteral obstruction, but is excluded by the sonographic findings.

Bilateral delay or decrease in renal perfusion is less specific. This pattern could be due to bilateral vascular compromise, severe prerenal circulatory failure, severe renal causes of failure including acute tubular necrosis, and severe bilateral ureteral obstruction (Fig 3E–10). Obstruction is again excluded by the sonographic findings.

The static images are analyzed for renal size and position along with symmetry, uniformity, and prompt-

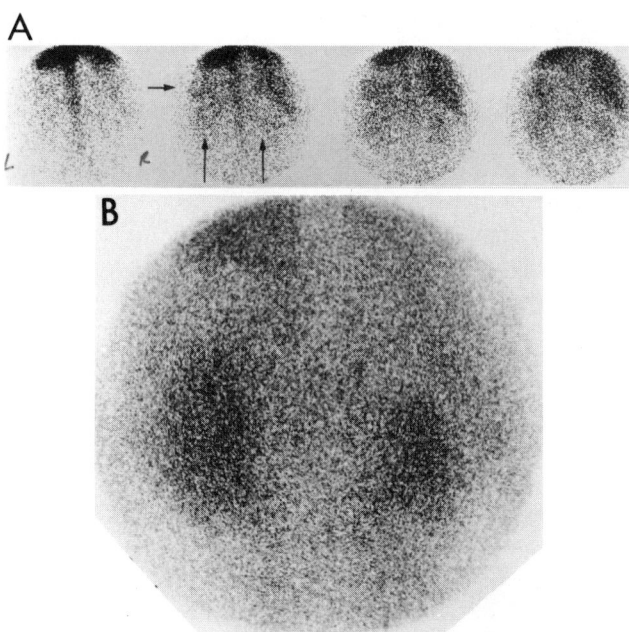

FIG 3E–10.
Woman, 64 years old, with acute tubular necrosis following partial resection of her small bowel. **A,** ^{99m}Tc-DTPA vascular sequence shows symmetrically poor renal blood flow. *Vertical arrows,* kidneys; *horizontal arrow,* spleen. **B,** static image at 20 minutes demonstrates moderate bilateral uptake. No bladder activity was demonstrated.

ness of uptake bilaterally. In normal kidneys, the immediate postdynamic images demonstrate symmetric, homogeneous activity in the renal cortices. Within five minutes, background activity lessens, and the collecting systems are well visualized.

Renal size and position can suggest the etiology as well as prognosis of the disease process. Bilaterally normal-sized kidneys suggest recent onset and potential reversibility. Small kidneys indicate chronic disease, congenital or acquired, and irreversibility. Large kidneys can be seen in polycystic disease, an infiltrating disease such as amyloidoisis, or renal vein thrombosis (Sherman and Byan, 1982).

Evaluation of symmetry and uniformity of uptake on the static views compared with the flow study can also suggest underlying etiologies. Asymmetry of uptake with a corresponding flow abnormality correlates with a vascular problem. Lack of uniformity of uptake is caused by localized parenchymal disease. Wedges or segments of decreased activity that correlate with vascular distributions imply vascular abnormalities. Mass lesions may also cause nonuniformity and are discussed in a later section.

Promptness of uptake is also an important indicator. Delay in uptake in the kidneys suggests poor renal function secondary to parenchymal compromise.

Causes include prerenal circulatory failure, renal disease such as acute tubular necrosis, or postrenal obstruction. When severe enough, this compromise can decrease the flow as described above.

In the severely oliguric or anuric patient, renal function may be so severely compromised that technetium compound scans are inadequate. In that case, ^{131}I hippurate is recommended because renal concentration can occur with as little as 3% of normal function (O'Reilly et al., 1979).

CHRONIC RENAL FAILURE

In evaluating the azotemic patient, excretory urography becomes inadequate when plasma creatinine level exceeds 5 mg/100 ml. High bolus doses of organic iodides may improve visualization but adversely affect already-compromised renal function. ^{131}I hippurate is the recommended agent for evaluation of chronic renal failure. Technetium compounds may be chosen when vascular problems are suspected, because flow studies can be performed. As in acute renal failure, the size, position, promptness, symmetry, and uniformity of renal uptake of the kidneys are important in guiding investigation and narrowing the diagnostic possibilities.

The scintigram can help in the preparation for biopsy, by indicating the lesser functioning kidney, which is preferentially sampled. The hippurate study can also be a predictor of the prognosis of renal function in episodes of renal failure. Good concentration of hippurate indicates the probability of eventual improvement of renal function. Poor concentration predicts permanent failure of life-sustaining renal function (Staab et al., 1973).

MASSES AND PSEUDOMASSES

In evaluating a renal mass, the purpose of noninvasive tests is to narrow the diagnostic possibilities and avoid intervention if the mass is benign. The current resolution of scintigraphy is not as good as that of radiography. Lesions as small as 1 cm have been detected on phantoms, but a larger lesion centrally placed in the kidney may not be detected by scintigram. On the other hand, peripheral lesions or those obscured by fat or bowel gas may be better visualized. Therefore, radiologic and radionuclide techniques are complementary in evaluation of renal masses.

Ultrasound is generally the recommended first procedure for initial characterization of a renal mass. If lobulation is noted, a normal variant or "pseudomass" of the kidney may be present, such as a hypertrophied column of Bertin, splenic impression, dromedary hump, or fetal lobulation. The next step is nuclear imaging with ^{99m}Tc glucoheptonate. ^{99m}Tc DMSA has also been used because of its concentration in the paren-

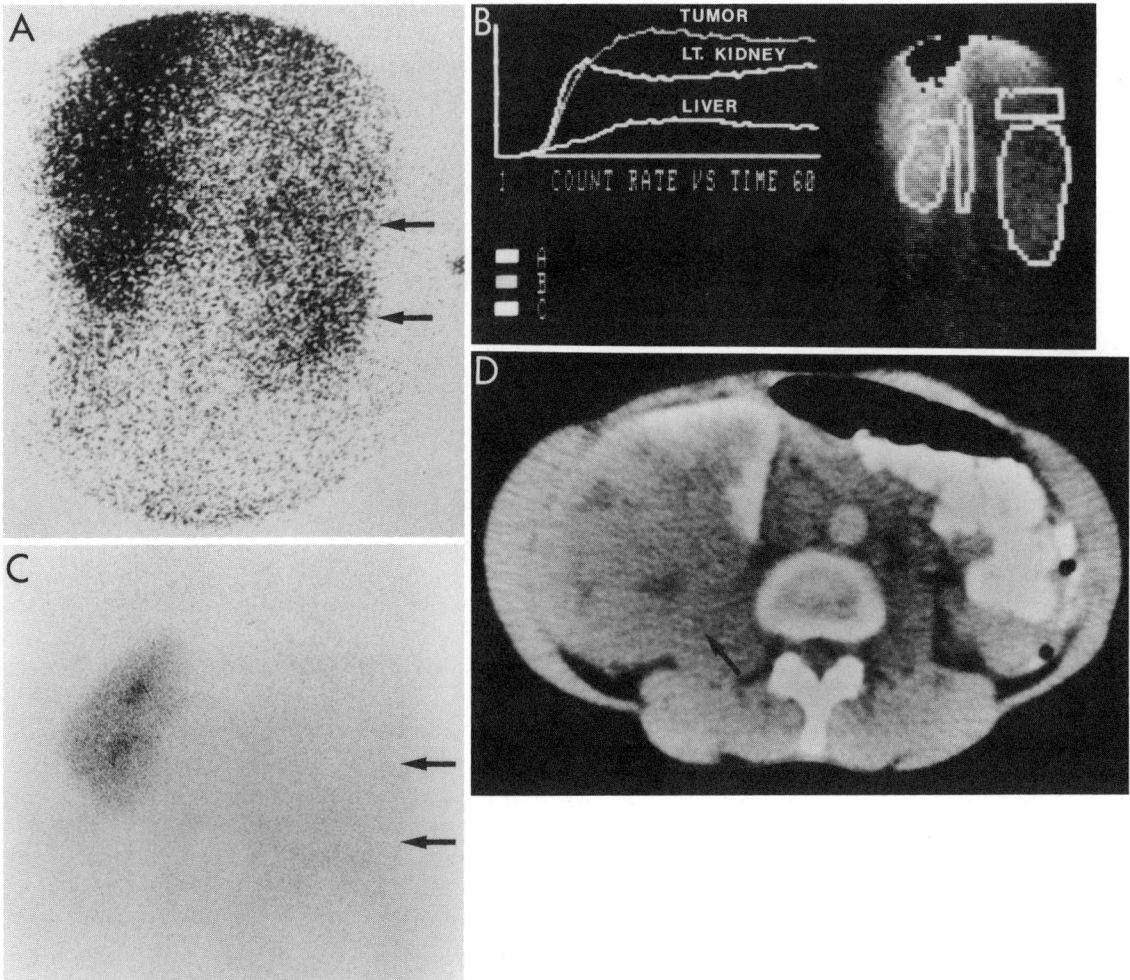

FIG 3E–11.
Woman, 38 years old, with right hypernephroma. ^{99m}Tc-glucoheptonate study showed initial delay of flow to right kidney (0–30 seconds). **A,** view at 40 seconds shows definite vascularity of tumor *(arrows).* **B,** histogram of blood flow to abdominal organs demonstrates initial delay in blood flow to tumor, then increased activity compared with left kidney. Normal delay in flow to liver is seen, corresponding to portal venous supply. **C,** static views at 5 minutes show minimal function of right kidney, particularly in the upper pole. **D,** CT shows large right renal mass *(arrow)* with crescentic area of functioning renal tissue present along anteromedial aspect. Bowel loops in the left abdomen are opacified with oral contrast agent.

chyma. If the lobulation is secondary to a pseudomass, the flow and static images will demonstrate normally functioning parenchyma. In some cases, the pseudomass may actually be more intense because of the increased thickness of the parenchyma (Older et al., 1980).

If lobulation is not seen sonographically and the mass is atypically cystic, complex or solid, CT or angiography is recommended. In determining whether a mass is vascular or nonvascular, radionuclide scans are 80%–85% accurate, a rate lower than the other two modalities.

Whether the mass is neoplasm, infarction, abscess, cyst, or localized pyelonephritis, the renal study will be abnormal due to replacement of normal parenchyma that concentrates radionuclide. A typical cyst or infarct will demonstrate no activity on blood flow or on early and delayed images. A typical renal carcinoma will demonstrate increased activity on blood flow and early images but a photon-deficient area on delayed images (Fig 3E–11). An abscess or localized pyelonephritis may be similar to a cyst or tumor, depending on the size and amount of hyperemia.

HYDRONEPHROSIS AND HYDROURETER

One of the most important applications of renal scintigraphy today is in the evaluation of urinary tract obstruction. Renal imaging can help in the diagnosis, determination of timing for surgical intervention, and evaluation of therapy. It is also valuable in assessing renal function, which cannot be evaluated reliably by IVP. In the setting of acute obstruction, any renal function implies salvageability, whereas in chronic obstruction poor function suggests permanent damage.

When the ultrasound examination shows urinary tract dilatation, the next question is whether obstruction is present. Conventional urography and radionuclide scanning are unreliable in differentiating obstructive from nonobstructive hydronephrosis.

Perfusion studies introduced by Whitaker (1973) obtaining pressure/flow relationships have provided a more functional approach. The procedure requires placing a catheter into the renal pelvis to obtain pressure readings. This is an invasive procedure that can involve a significant radiation dose when done fluoroscopically.

By using parenteral diuretics, radionuclide renography can be modified to obtain similar information by noninvasive methods. Published reports indicate a high degree of accuracy in distinguishing the dilated obstructed system from the nonobstructed (Thrall et al., 1981).

The success of diuretic renography is directly dependent on strictly following the procedure. First, the patient must be well hydrated either by p.o. or IV methods. The patient needs to void just before injection, or have a Foley catheter inserted for constant drainage. We prefer ^{99m}Tc DTPA because of its excellent visualization of the collecting system, and inject 15 mCi (or

the appropriate pediatric dose). Images are taken every 5 minutes until the *entire* collecting system is filled with radionuclide. (If this takes over 1 hour, the study decreases in reliability). The patient then voids and is injected with furosemide (Lasix), 1 mg/kg, up to 40 mg IV. Images are taken every 5 minutes for the next 30 minutes.

Computer analysis of the renal collecting system activity is performed, resulting in computer curves of the counts. Three curve patterns are possible. A definite decrease in counts with time after Lasix administration represents a nonobstructive pattern (Fig 3E–12), whereas a definite increase is obstructive (Fig 3E–13). A plateau signifies an indeterminate pattern which could be secondary to poor renal function and poor response to Lasix, and/or a very large atonic hydronephrotic sac.

VESICOURETERAL REFLUX

Reflux nephropathy is a recognized problem in the pediatric population; and is evaluated in the work-up of urinary tract infection. It is also recognized in some adults as a cause of hypertension, proteinuria, and renal failure. The traditional study is a voiding cystourethrogram, which can deliver a radiation dose of several hundred millirads to several rads. The retrograde radionuclide cystogram gives a fraction of this radiation dose depending on several factors, including the amount of material instilled and the time the solution remains in the bladder before voiding (Conway et al., 1972).

The traditional voiding cystourethrogram can catego-

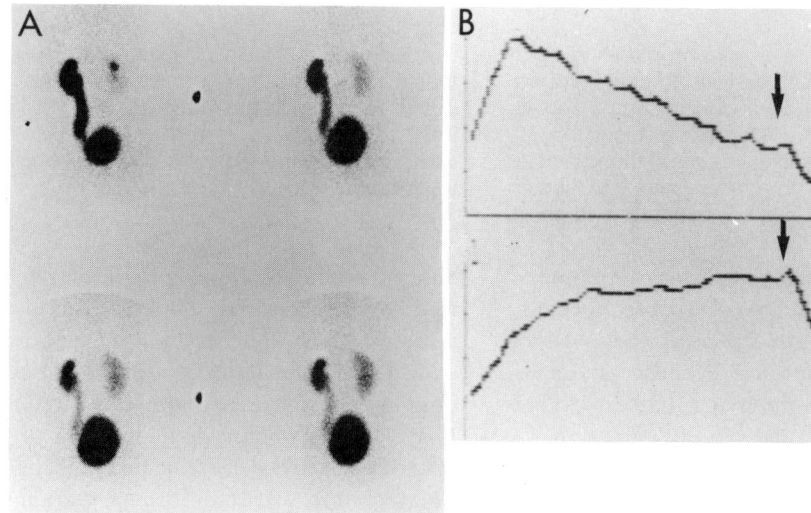

FIG 3E–12.
Boy, 2 years old, following left ureteral reimplantation. ^{99m}Tc DTPA study. **A,** posterior images demonstrate left hydronephrosis and hydroureter and a normal right kidney. The top left image was taken 30 minutes after radionuclide injection, at the time of Lasix administration. Subsequent images were made at 5-minute intervals showing radionuclide excretion. **B,** the right kidney *(top curve),* has excreted most of the radionuclide prior to Lasix administration. The *bottom curve* shows decreasing counts in the left kidney only after Lasix *(arrow)* signifying no evidence of obstruction.

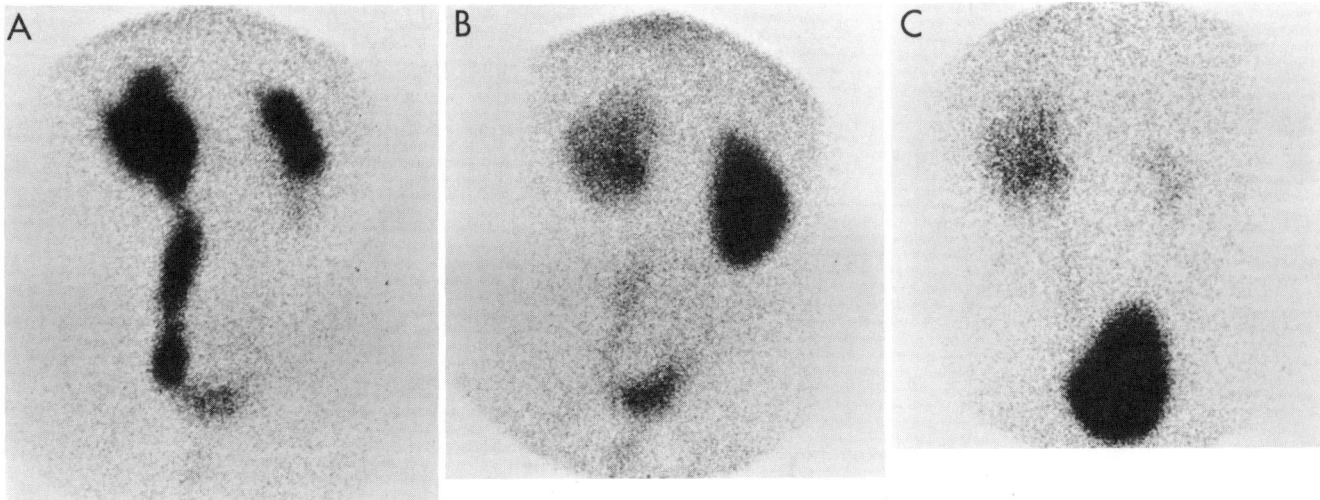

FIG 3E–13.
Boy, 5 years old, with dilated, nonobstructed collecting system on the left and a dilated, obstructed collecting system on the right. Lasix ⁹⁹ᵐTc DTPA study. **A,** posterior image taken at 30 minutes after radionuclide injection, at time of Lasix administration. **B,** 20 minutes after Lasix, the nonobstructed left system has emptied and the obstructed right side has accumulated more. **C,** postoperative image at 25 minutes after Lasix administration shows resolution of obstructed right system.

rize reflux into grades of severity by visualizing the morphology of the urinary tracts. The radionuclide grading is determined by the volume in the bladder when reflux occurs and the amount of reflux. The anatomy of the collecting system, as well as the base of the bladder and urethra, cannot be assessed (Fig 3E–14). However, the residual volume after voiding can be accurately determined.

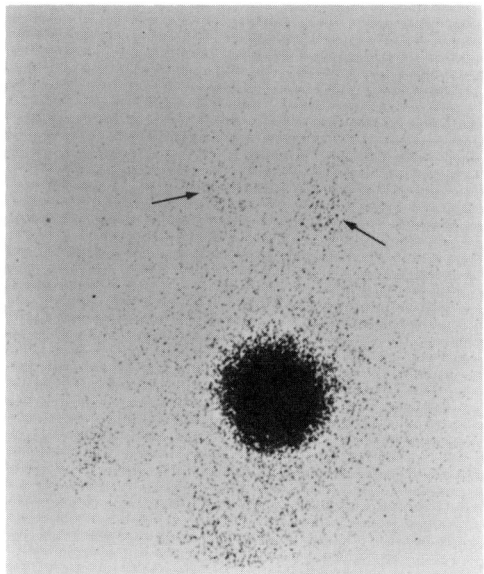

FIG 3E–14.
Retrograde ⁹⁹ᵐTc pertechnetate voiding cystourethrogram demonstrates activity in bilateral ureters, right greater than left, compatible with bilateral vesicoureteral reflux.

The quantity of reflux that is significant in a single study is not yet defined. Serial studies can be of value for comparison in order to determine if improvement has occurred. Accurate records of the volumes of solution instilled resulting in reflux are important, as is quantitation of reflux by visual and computer analysis. An increasing volume instilled before reflux implies improvement.

RENOVASCULAR ABNORMALITIES

Renal artery stenosis not only plays a role in hypertension but also is a potentially treatable cause of renal failure. In the majority of patients, renal damage has already occurred. Damage can cause abnormalities on the renal images that resemble other diseases, making the results nonspecific. However, this test can still be valuable in following up patients with documented renal artery stenosis.

As with renal artery embolism, dynamic radioisotope scanning with technetium compounds is preferred (Fig 3E–15). On the initial flow images, decreased perfusion of the affected kidney is noted along with reduced concentration followed by a prolonged parenchymal transit time. The sensitivity is about 85%. Serial studies can guide in timing for intervention and evaluation of open surgical or angioplastic treatment.

Certain centers prefer hippurate scanning to evaluate unilateral renovascular disease. Emphasis is on observing differences in the time interval between injection and peak activity (transit time) and on the symmetry of

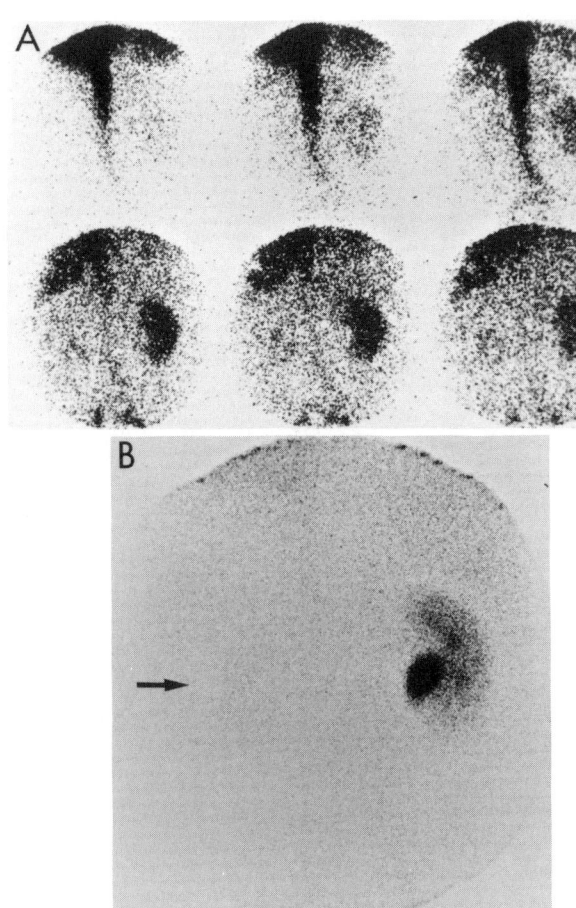

FIG 3E–15.
Male, 65 years old, with sudden onset of left flank pain and hematuria. **A,** ^{99m}Tc glucoheptonate flow study shows good flow to the right kidney and minimal flow to the left. **B,** delayed view demonstrates minimal left renal uptake *(arrow),* while the right kidney appears normal. The combination of significantly decreased flow to the entire left kidney and delayed uptake is compatible with left renal artery embolus.

the downslope of the third phase on the renogram curve. The affected kidney will show a prolonged transit time and a slower decline in the excretory phase. A difference of 20% or more in these parameters is generally considered the criterion for positive diagnosis. Accuracy from 87% to 96% has been claimed (McAfee et al., 1967). Analysis of ERPF and fractionation of renal function allow measurement of individual renal flow (Fig 3E–16).

Renal vein thrombosis is an uncommon entity that can be imaged with radionuclides. Technetium compounds are preferable because flow studies and detailed static views can be obtained. Typically, in renal vein thrombosis, the involved kidney is enlarged, with decreased flow and poor, delayed uptake (Fig 3E–17).

TRANSPLANT EVALUATION

Radionuclide imaging is a routine part of the assessment of renal function following renal transplant. Our technique involves injection of the patient with ^{131}I hippurate. Routine images are taken as previously described in the anterior position because of the kidney's position. Computer analysis produces the renogram curves.

We look at the total transplant counts, shape of the renogram curve, bladder:kidney count ratio at 30 minutes, and the static views. Total transplant counts are helpful to compare in serial studies; we have no absolute value for determination of function based on one individual scan.

The bladder-to-kidney ratio is determined by drawing regions of interest around the kidney and the bladder and obtaining total counts at a specific time (Hayes et al., 1972). The normal ratio at 30 minutes is 3:1 to 5:1. For this parameter to be useful, the Foley catheter must be clamped, which cannot usually be done until after the first week. Also, native kidneys may still have some function and falsely imply good transplant function.

In the early postoperative period (up to four weeks), serial radionuclide scanning and sonography are helpful in following renal function and determining the cause of oliguria or anuria. Total absence of flow and function, along with a photopenic area on the static images, can be caused by renal artery or vein thrombosis, hyperacute rejection, or severe urinary obstruction. Obstruction can usually be diagnosed by ultrasound; however, dilatation of the collecting system can occasionally occur after transplantation. Diuretic renography may be helpful in differentiating obstruction from nonobstruction if adequate function is present.

Diminished early radionuclide uptake with progressively increasing activity in later views and poor excretion can be seen with acute tubular necrosis, acute rejection, and urinary obstruction. This pattern implies preserved blood flow but decreased concentrating and excreting capability. A kidney with acute tubular necrosis (ATN) typically can extract hippurate from the blood but has difficulty transporting it into the tubular lumen ("tubular block"). Cadaveric transplants virtually always show an element of ATN in the early postoperative period. Postoperative ATN will resolve without therapy, after one day to several weeks.

As the kidney improves after ATN, the total excreted counts will increase and the bladder:kidney ratio will improve (Fig 3E–18). The shape of the curve may not change initially but contains more counts; after further improvement, the curve will start peaking in the first ten minutes. Any reversal of this sequence or failure of improvement implies rejection. After six months, a re-

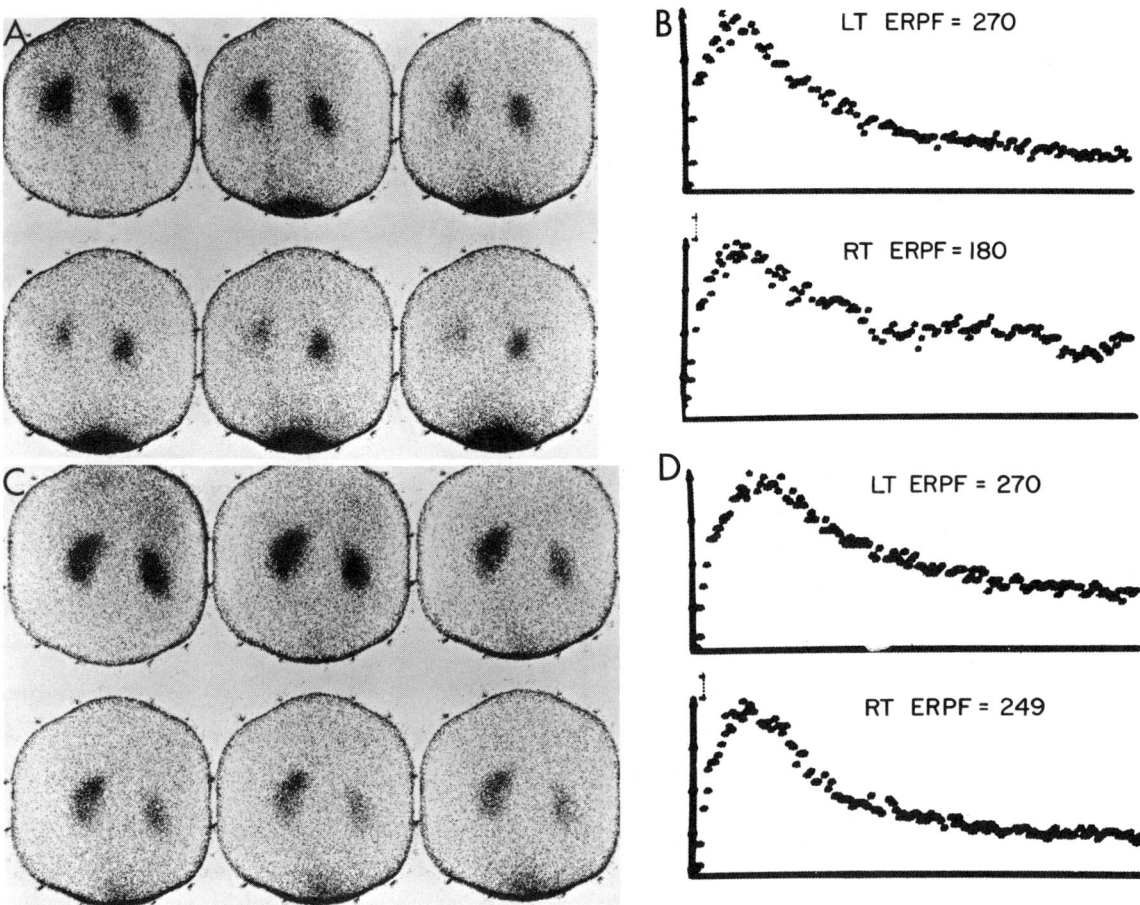

FIG 3E–16.
Angioplasty of renal artery. **A** and **B**, preangioplasty for right renal artery stenosis. [131]I hippurate scan shows delayed clearance of the right kidney on the static views and the renogram curve. The effective renal plasma flow (ERPF) was diminished on the right (180 ml/min) compared with the left. **C** and **D**, postangioplasty there was improvement in clearance of the right kidney and the effective renal plasma flow on the right increased to 249 ml/min.

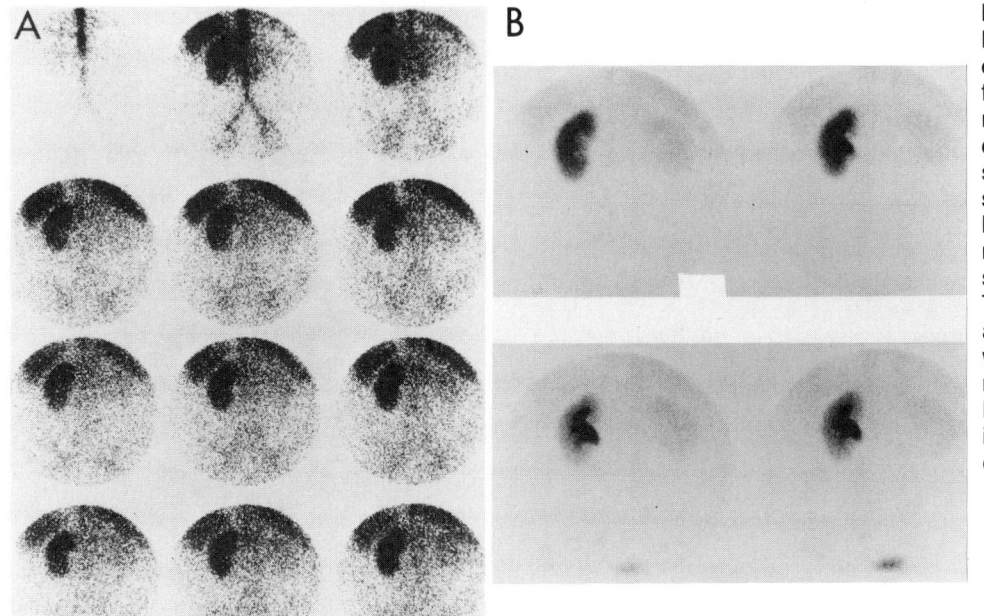

FIG 3E–17.
Man, 27 years old, who developed hematuria following retroperitoneal resection for testicular cancer. [99m]Tc glucoheptonate study. **A,** renal flow study shows poor flow to right kidney. **B,** static images at 5-minute intervals postinjection show poor right renal uptake. The combination of asymmetric poor flow along with poor function implies renal vein thrombosis. However, usually the involved kidney is also enlarged.

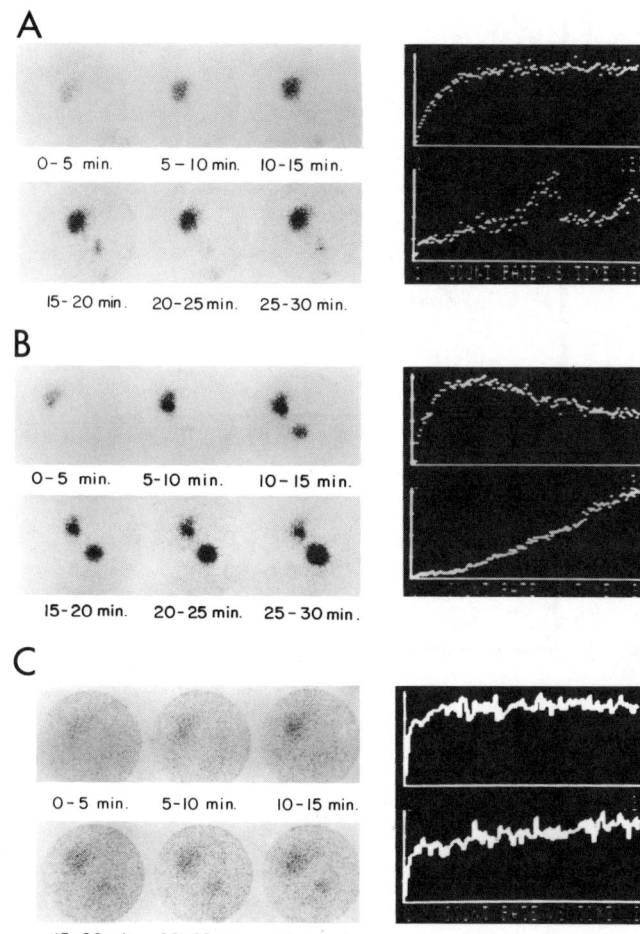

FIG 3E–18.
Man, 54 years old, with end-stage renal disease who received a cadaver transplant kidney in the right iliac fossa. A series of renograms were performed using 200 μCi of ^{131}I hippurate. **A,** baseline study 24 hours after transplantation. Sequential camera images *(left)* show progressive accumulation of activity in kidney with transport to urinary bladder. The renogram curves *(right)* reflect progressive kidney accumulation, the plateau occurring after approximately 10 minutes as the tracer emptied into the urinary bladder. The lower curve, from the urinary bladder, shows intermittent emptying because the Foley catheter was not clamped. **B,** the transplant function gradually improved as the acute tubular necrosis (ATN) cleared. On this study performed six weeks after transplantation, the renogram curve *(above)* shows a peak at approximately 8 minutes with a gradual fall in count rate during phase III. Tracer gradually accumulated in the bladder, with a bladder-to-kidney ratio of 2.8:1 at 30 minutes. **C,** repeat study at three months after transplantation. Serum creatinine levels were increased with marked deterioration in renal function. Note the very flat renogram curve *(above)* and minimal accumulation of activity in the urinary bladder. Not only has the bladder-to-kidney ratio deteriorated to 0.6:1, but the total excreted activity in the kidney and bladder fell to approximately 25% of prior values. The patient underwent therapy for rejection with a subsequent improvement in renal function.

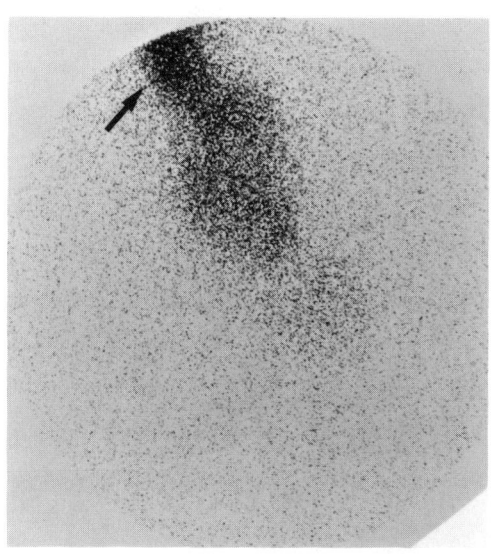

FIG 3E–19.
Man, 24 years old, three weeks after a cadaveric transplant. ^{131}I hippurate study shows extravasation of radionuclide at the superior margin of the transplant *(arrow)* representing a urinary anastomotic leak.

nogram is of little value unless serial studies are maintained; serial laboratory values (creatinine and BUN) are more pertinent and economical.

Leaks may develop at the ureterovesical anastomosis, cystotomy site, or from a renal biopsy site (Fig 3E–19). These can be well demonstrated by collection of the radionuclide outside of the urinary tract. Hematomas or lymphoceles (Fig 3E–20) appear as photon-deficient areas around the kidney or between the kidney and bladder.

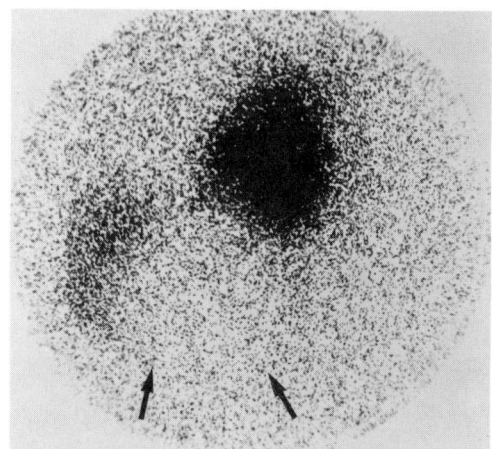

FIG 3E–20.
Man, 42 years old, who developed nephrotic syndrome following transplant. ^{131}I hippurate study shows large photopenic area *(arrows)* compressing the left lateral aspect of the bladder, which proved to be a lymphocele obstructing urinary excretion.

Some centers use technetium compounds to evaluate renal transplants, including evaluation of renal blood flow dynamically. In the normally perfused transplant, activity on the flow or dynamic images should:

1. Appear in the graft within six seconds of its appearance in the adjacent iliac artery;
2. Obtain maximum activity per unit area equal to or greater than that in the adjacent artery;
3. Clearly fall after the peak.

Renal blood flow is usually preserved in ATN but deteriorates in rejection. The evaluation of the static views is similar to that of the hippurate images. We have found the hippurate renogram to be more sensitive for detection of rejection in serial studies.

Other agents used include ^{99m}Tc sulfur colloid, gallium 67 citrate, and labeled platelets. Although such studies may be useful, they are not routine and are beyond the scope of this chapter.

CONGENITAL ANOMALIES

Congenital anomalies such as horseshoe kidney (Fig 3E–21), crossed fused ectopia, and ectopic kidneys are easily demonstrated by radionuclide scanning. ^{99m}Tc DTPA or glucoheptonate is the agent of choice.

A renal mass in the infant or child should first be evaluated by ultrasound. A solid mass should then be studied by CT. A cystic mass can be either a multicystic dysplastic kidney or hydronephrosis, which can sometimes be difficult to differentiate sonographically. A DTPA scan is the next recommended study. Nonvisual-

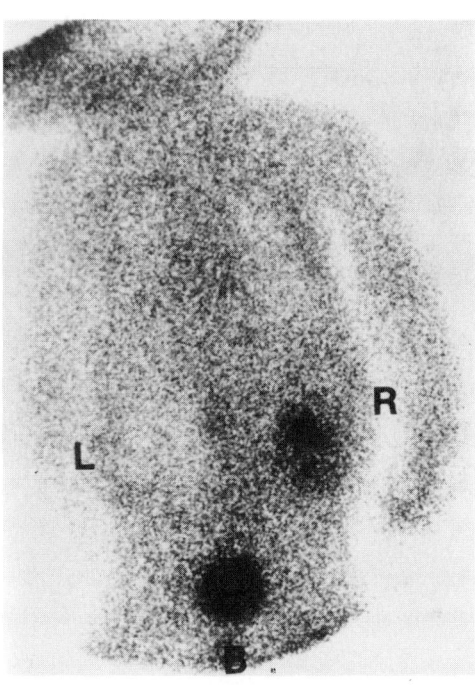

FIG 3E–22.
Boy, 3 days old, with cystic left kidney on ultrasound. ^{99m}Tc glucoheptonate study demonstrates a normal right kidney and a photon-deficient area in the region of the left kidney which remained photopenic, compatible with a multicystic kidney. Bladder activity (B) was evident on this delayed static image.

ization of the corresponding kidney on early and delayed views confirms dysplasia (Fig 3E–22). However, delayed films may show some activity secondary to a small amount of residual functioning tissue, which is usually irregularly located in a photopenic mass. A more uniform cortical rim sign with gradual filling of the central collecting system is typical of hydronephrosis. The time of appearance of these features during the scanning period depends on the severity of obstruction.

TRAUMA

The excretory urogram is the traditional mode of initial evaluation of the kidneys in the trauma setting. Many reports have shown the sensitivity of radionuclide image for detection of extent of damage and the evaluation of renal function. However, we advocate CT scanning for initial evaluation because other organs and the retroperitoneum can be evaluated simultaneously. Nuclear imaging can be valuable in following renal function after trauma.

The agent of choice is ^{99m}Tc glucoheptonate or DTPA, because blood flow to the kidneys can be rapidly evaluated. If there is no immediate blood flow to either kidney on the flow study or initial static images,

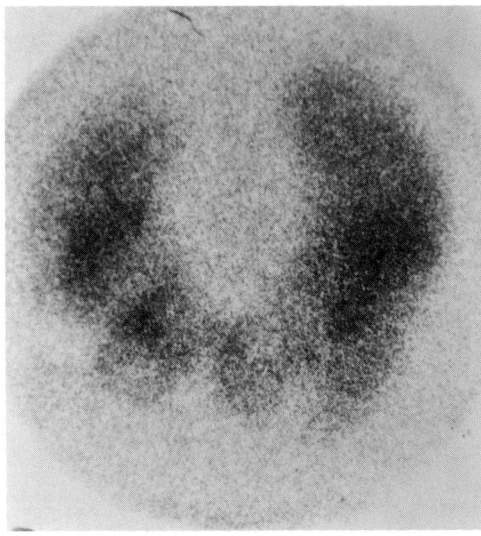

FIG 3E–21.
Boy, 8 years old, with recurrent urinary tract infections. ^{99m}Tc iron hydroxide DTPA study demonstrates a horseshoe kidney. This anterior view shows the bridging tissue better than a posterior view.

immediate angiography or surgical exploration may be necessary. Extravasation and/or obstruction can also be detected by scanning.

SCROTAL IMAGING

Scrotal imaging provides an accurate means of distinguishing between the most common causes of "acute" scrotum: epididymitis and testicular torsion. Less common etiologies include hydroceles and testicular tumors.

In classic acute epididymitis, hyperemia occurs in the head, body, and tail of the epididymis. The radionuclide angiogram shows markedly increased perfusion through the spermatic cord vessels. On the static views, increased tracer activity usually extends laterally, corresponding to the location of the epididymis (Fig 3E–23).

If the inflammatory process spreads to the testis, it is called epididymo-orchitis. The radionuclide activity then extends also medially to involve the testis. Exten-

sive scrotal swelling or rotation of the scrotum in positioning may produce medial activity in the absence of testicular involvement.

If the inflammation is confined to a small area of the epididymis, focal hyperemia occurs only in the infected part, and the radionuclide angiogram may often show "normal," barely perceptible activity. Later images may show a focal "spot" of increased activity.

Torsion of the spermatic cord has been described as displaying four radionuclide patterns (Holder et al., 1981). The first is seen if spontaneous detorsion occurs within four hours following torsion, which usually results in a normal scan. Occasionally, mild hyperemia is present throughout the entire hemiscrotum.

In early testicular torsion of less than seven hours' duration, the testicle is often viable. The radionuclide angiogram shows no increase of perfusion through the vessels or to the scrotum. The scrotal scans demon-

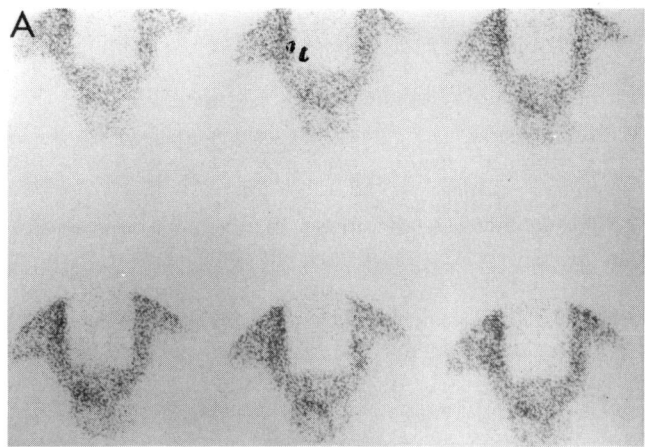

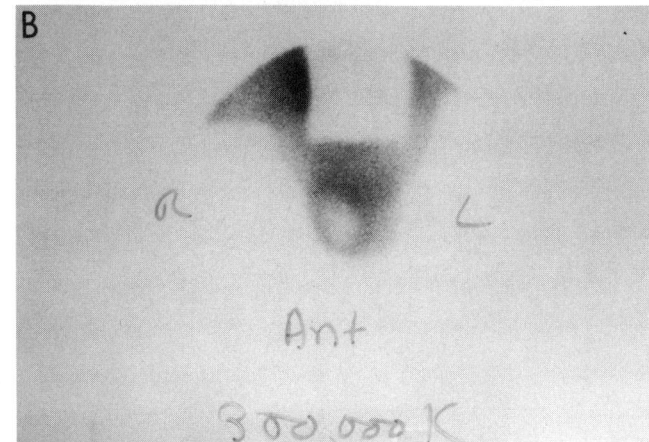

FIG 3E–23.
Boy, 11 years old, with left testicular swelling. **A,** ^{99m}Tc pertechnetate flow study shows increased blood flow to the left epididymis. **B,** increased activity along the lateral aspect of the left scrotum on the static views is present, compatible with epididymitis. The left testis *(arrow)* was not involved; therefore it appears relatively photopenic.

FIG 3E–24.
Missed testicular torsion. **A,** flow study shows no hypervascularity but a developing focal area of increased uptake along superior aspect of right testes on later views. **B,** delayed view shows photopenic area of right testes with surrounding increased uptake along all borders. Because the shield obscured the spermatic vessels, we no longer advocate its use.

strate a cold area in the location of the testicle. In this second pattern, minimal if any activity is seen in the dartos, making the cold area often difficult to see.

Midphase testicular torsion occurs approximately 7–24 hours following torsion, when the testicle may still be viable. A reactive edema and erythema are present, resulting in increased activity in the scrotal region. This results in a halo of activity around the cold testicle (Fig 3E–24), the third pattern.

The final pattern of "missed testicular torsion" occurs usually more than 24 hours following torsion. The testicle may still be viable if the twist is less than 360°. The radionuclide angiogram will often show dartos perfusion. An intense halo of activity with a cold center is present, and occasionally increased activity in the region of the spermatic cord. This symmetric halo should not be confused with the asymmetric curvilinear activity in epididymitis.

A hydrocele is a collection of fluid between the layers of tunica vaginalis. It may be primary or secondary to epididymitis, testicular torsion, tumor, trauma, or posthernia repair. On the radionuclide angiogram, the perfusion is normal or a reflection of the underlying cause. The scrotal scan demonstrates a lucency around the central nidus of testicular tissue.

Radionuclide scanning of testicular tumors is usually performed to exclude other causes of the pain and swelling. The radionuclide angiogram may show normal or moderately increased perfusion. If increased, the scrotal perfusion is diffuse rather than linear or halolike. A cool area may represent a focal area of central necrosis or relative tissue avascularity (Holder et al., 1981).

REFERENCES

1. Arnold RW, Subramanian G, McAfee JG, et al: Comparison of Tc-99m complexes for renal imaging. *J Nucl Med* 1975; 16:357.
2. Bell EG, McAfee JG, Makhuli ZN: Medical imaging of renal diseases: Suggested indications for different modalities. *Semin Nucl Med* 1981; 11:105.
3. Boyd RE, Robson J, Hunt FC, et al: Tc-99m gluconate complexes for renal scintigraphy. *Br J Radiol* 1973; 46:604.
4. Constable AR, Hussein MM, Albrecht MP, et al: Renal clearance determination from single plasma samples, in Hollenberg NK, Lange S (eds): *Radionuclides in Nephrology*. New York, Thieme-Stratton, Inc, 1980, pp 62–66.
5. Conway JJ, Lowell RK, Belman AB, et al: Detection of vesicoureteral reflux with radionuclide cystography. *AJR* 1972; 115:720.
6. Dimitriou P, Fretzayas A, Nicolaidou P, et al: Estimates of dose to the bladder during direct radionuclide cystography: Concise communication. *J Nucl Med* 1984; 25:792.
7. Esser PJ, McAfee JG, Subramanian G: Appendix: radio-

active tracers, in Freeman LM (ed): *Freeman and Johnson's Clinical Radionuclide Imaging*, ed 3. Orlando, Fl, Grune & Stratton, 1984, p 1519.
8. Freeman LM: The kidneys, in Freeman LM (ed): *Freeman and Johnson's Clinical Radionuclide Imaging*, ed 3. Orlando Fla, Grune & Stratton, 1984, pp 725–834.
8a. Gates GF: Split renal function testing using Tc-99m DTPA. *Clin Nucl Med* 1983; 8:400.
8b. Ginjaume M, Casey M, Barker M, et al: A comparison between four simple methods for measuring glomerular filtration rate using Technetium-99m DTPA. *Clin Nucl Med* 1986; 11:647.
9. Gorson RO, Lassen M, Rosenstein M: Patient dosimetry in diagnostic radiology, in Waggener RG, Kereiakes JG, Shalek RJ (eds): *CRC Handbook of Medical Physics*. Boca Raton, Fla, CRC Press, Inc, 1984, vol 2, pp 474, 487, 488.
10. Hayes M, Moore TC, Taplin GV: Radionuclide procedures in predicting early renal transplant rejection. *Radiology* 1972; 103:627.
11. Holder LE, Melloul M, Chen D: Current status of radionuclide scrotal imaging. *Semin Nucl Med* 1981; 11:232.
12. Koenigsberg M, Freeman LM, Blaufox MD: Radionuclide and ultrasound evaluation of renal morphology and function, in Edelman CM Jr (ed): *Pediatric Kidney Disease*. Boston, Little, Brown Co, 1978, p 238.
13. Matthews CME: The theory of tracer experiments with ^{131}I-tagged sodium o-iodohippurate. *Phys Med Biol* 1957; 2:36.
14. McAfee JG, Reba RC, Chodos RB: Radioisotopic methods in the diagnosis of renal vascular disease: A critical review. *Semin Roentgenol* 1967; 2:198.
15. Older RA, Korobkin M, Workman J, et al: Accuracy of radionuclide imaging in distinguishing renal masses from normal variants. *Radiology* 1980; 136:443.
16. O'Reilly PH, Shields RA, Testa HJ: Renovascular hypertension and renal failure, in O'Reilly PH, Shields RA, Testa HJ (eds): *Nuclear Medicine in Urology and Nephrology*. London, Butterworths, 1979, pp 81–85.
17. Rollo FD, Patton JA: Instrumentation and information portrayal, in Freeman LM (ed): *Freeman and Johnson's Clinical Radionuclide Imaging*. Orlando, Fla, Grune & Stratton, Inc, 1984, vol 1, pp 207–222.
18. Sapirstein LA, Vidt DG, Mandel MJ, et al: Volumes of distribution and clearance of intravenously injected creatinine in the dog. *Am J Physiol* 1955; 181:330.
19. Sherman RA, Byan KJ: Nuclear medicine in acute and chronic renal failure. *Semin Nucl Med* 1982; 12:265.
19a. Smith HW: *The Kidney Structure and Function in Health and Disease*. New York, Oxford University Press, 1951, p 47.
20. Staab EV, Hopkins J, Patton DD, et al: The use of radionuclide studies in the prediction of function in renal failure. *Radiology* 1973; 106:141.
21. Taplin GV, Nordyke RA, Tauxe WN: Kidney function and disease, in Blahd WH (ed): *Nuclear Medicine*, ed 2. New York, McGraw-Hill Book Co, 1971, pp 383–391.
21a. Tauxe WN: Prediction of residual renal function after unilateral nephrectomy, in Tauxe WN, Dubovsky EV (eds):

Nuclear Medicine in Clinical Urology and Nephrology. Norwalk, Conn, Appleton-Century-Crofts, 1985, pp 279–285.

22. Tauxe WN, Maher FT, Taylor WF: Effective renal plasma flow: Estimation from theoretical volumes of distribution of intravenously injected I-131 orthoiodohippurate. *Mayo Clin Proc* 1971; 46:524.

23. Thrall JH, Koff SA, Keyes JW Jr: Diuretic radionuclide renography and scintigraphy in the differential diagnosis of hydroureteronephrosis. *Semin Nucl Med* 1981; 11:89.

24. Whitaker RH: Methods of assessing obstruction in dilated ureters. *Br J Urol* 1973; 45:15.

Chapter 4

Biology of Wound Repair and Infection

RICHARD F. EDLICH, M.D., PH.D.
GEORGE T. RODEHEAVER, PH.D.
JOHN G. THACKER, PH.D.

The urologist's ultimate goal is to restore the physical integrity and function of the injured or diseased tissue. While this accomplishment could best be achieved by regeneration, wound restoration usually results from the synthesis of scar tissue. Unless controlled by the urologist, the fibrous tissue synthesis stage of healing can be detrimental to the host itself. By having a detailed knowledge of the biology of wound repair, the urologist can achieve wound closure with minimal deformity and dysfunction.

In the quest to reconstitute the impaired tissue, the urologist must appreciate the consequences of its devastation. A study of the mechanism of injury will provide a reliable indication of its ravages. Whether the tissue injury will be limited to the initial wounding depends on the outcome of the interaction between the contaminants and the surgical wound. In the event that the contaminants are very reactive, a relatively insignificant wound may become a catastrophe. This outcome can be averted by the implementation of well-devised surgical care based on the biology of wound infection.

The major focus of this chapter is on the biology of wound repair and infection, a subject which we have investigated in our research laboratory over the past 25 years (Edlich, 1985). The lessons learned from these studies are applicable to all surgical wounds. This chapter has been written specifically for students of urology who view themselves as artisans cultivating and practicing the "art of surgery." As with any master craftsperson, the urologist must understand the tools of the profession. The linkage between a urologist and surgical equipment is a closed kinematic chain in which the urologist's power is converted into finely coordinated motion. The ultimate goal of this interlinking is perfection of the surgical discipline.

Students of urology must not read this chapter looking for proof of prejudices, but rather in search for a way in which they can effect changes that might improve their results. We have tried to identify surgical principles which are fundamental truths that guide urologic practice. These truths must always stand the test of time or be replaced by new and improved concepts of wound care.

MECHANISM OF INJURY

In the language with which urologists label their world, they unconsciously separate the concept of traumatic injury from that of urology. This division of thought may provide the urologist with a false sense of confidence that the consequences of injury are indeed distinct from those of surgery, when, in fact, they are identical in most aspects.

The outcome of both injury and urology can be predicted by applying concepts of power, work, and force that were first appreciated in the 17th century. Urologists simply employ various principles of energy transfer, mechanics, thermodynamics, fluid dynamics, and electrical phenomena in a planned procedure to achieve an anticipated result. Although the traumatic injury is caused by the same energy sources, the incident is unexpected, unplanned, and uncontrolled. If the urologist does not appreciate and control the sources of energy employed, an operation becomes an assault similar in nature to a traumatic injury. The "complete" urologist must harness and control the sources of energy and force and appreciate their potentially destructive effects. The urologist must also be aware of the consequences of applying uncontrolled amounts of energy to tissues, as occurs in accidental injury, and treat the resultant wound appropriately. Clinically, one of the most important consequences of any wounding process is that the divided edges of the wound are more susceptible to infection than unwounded tissue (Edlich et al.,

1982). The magnitude of this enfeebled resistance to infection will vary with the mechanism of wounding.

MECHANICAL ENERGY

Of all energy sources employed, the urologist is most familiar with mechanical energy, which is the energy required to move an object. There are three mechanical forces which can lead to soft tissue injury: shear, tension, and compression. When performing surgery, the urologist applies a carefully controlled force of a planned magnitude to divide tissue. The shearing force is delivered to the tissue by scissors or scalpel. To divide tissue with scissors, the urologist applies a shearing force of equal magnitude in opposite directions, in two adjacent parallel planes separated by a small distance. A scalpel cuts tissue by applying a shearing force in only one direction. Because the volume of tissue contacted by these instruments is extremely small, very little total energy is required to produce tissue failure.

When cutting tissues, the urologist must use a scalpel that divides tissue with the least trauma. Consequently, the urologist judges the performance of the scalpel by the sharpness of the blade, which is dependent on the radius of curvature of its ultimate edge. The ultimate edge of the scalpel blade is generated by a series of grinding processes that expose it to grinding wheels of different coarseness (grit). The mechanics of the grinding process influence the configuration of the "ultimate edge" and the performance of the blade. In combination grinding, each side of the blade is ground separately. Each grinding step generates a curl of wire at the edge which is removed by buffing. A side effect of the buffing process is the rounding of the ultimate edge, diminishing blade sharpness. In rotary grinding, both sides of the blades are ground simultaneously, after which they are polished. The ultimate edge of the blade made by this process has a symmetric and triangular edge. This pointed configuration of the ultimate edge of the blade subjected to rotary grinding facilitates cutting of tissue. When the urologist applies equal force to the surgical blade, blades ground by the rotary grinding can cut to a greater depth than blades ground by the combination process.

Another criteria for the performance of a scalpel blade is its susceptibility to breakage. This parameter is particularly important when lateral loading is applied to the scalpel; e.g., when the urologist firmly presses the blade against a hard surface such as bone. As a result of this loading, the blade will bend until it fractures. The fractured segment frequently behaves as a missile becoming lodged in the tissues. The search for this embedded foreign body is usually fruitless, and it becomes a constant reminder to the urologist, and occa-sionally to the patient, of the operative procedure. The resistance of the scalpel to breakage can be enhanced substantially by tempering the blade after the edge has been created. Prototypes made by this process are nearly 50% more resistant to breakage than the standard scalpel blades.

The configuration of the cutting edge of a scalpel is designed to accomplish a specified surgical task using a prescribed technique. The cutting edge of the no. 10 blade is predominantly straight, except for its distal end. It is best to hold its knife handle like a violin bow, so that its long, straight cutting edge contacts the skin (Fig 4–1,A). One sweep of the blade results in a deep, straight incision. If the inexperienced urologist holds the no. 10 knife handle as a pencil, he will find it more difficult to control the movement of the blade and will cut jagged incisions (Fig 4–1,B).

The design of a no. 15 blade allows the urologist to cut tortuous or angulated incisions that often must follow irregular anatomical landmarks. Since its curved distal end is the major portion of its cutting edge, the urologist gains optimal control of this blade by holding its handle as a pencil (Fig 4–2,A). When its handle is held as a violin bow, the urologist will find it more difficult to cut tissue, since the dull, straight portion of the blade becomes its cutting edge (Fig 4–2,B).

Despite all the technical advances in scalpel design, the ultimate performance of the scalpel rests with the urologist's technical skill. The experienced urologist who is cognizant of the scalpel's performance and the anatomy of tissue can cut to the desired depth with one sweep of the blade. The resultant wound is very resistant to the development of infection, with 10^6 or more bacteria being necessary to elicit infection (Edlich et al., 1982). This level of host resistance is comparable to that encountered in nearly 80% of the traumatic soft tissue injuries—those due to a piece of metal or glass and recognized by their characteristic *linear* configuration (Fig 4–3). The urologist who does not appreciate the potential of his instrument and is unfamiliar with the structural configuration of the tissue will generally cut with multiple strokes of the knife. Such repeated passages of the scalpel through tissue damage local defenses and invite infection (Edlich et al., 1982).

While the decision to discontinue the use of a scalpel blade during an operation is usually related to the blade's dulling, some urologists discard a blade after it has made contact with what they believe to be a source of contamination. A ritual practice in surgery has been to discard the sharp blade following its use on skin, fearing that it introduces skin contaminants into the depths of the wound. Jacobs (1974) reported that after cutting skin, scalpels are almost always sterile and need not be discarded. The influence of the use of only one

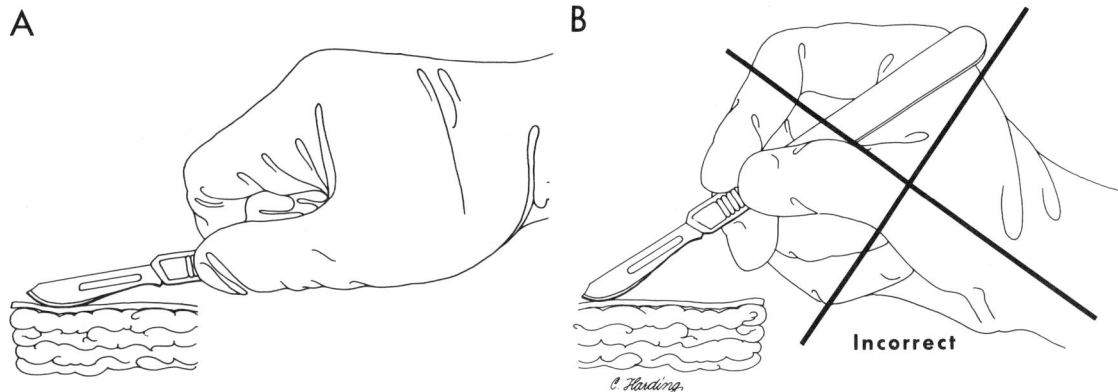

FIG 4–1.
Recommended technique for cutting skin with a no. 10 blade **(A)**. Incorrect surgical technique for cutting skin with a no. 10 blade **(B)**.

knife, instead of two knives, on the postoperative wound infection rate was not evaluated in this study. Hasselgren et al. (1984) initiated the first controlled clinical trial where the incidence of wound infections was recorded following the use of one and two knives for the skin incision. When the skin incision was made by only one knife, the incidence of infection in 277 patients was 3.6%. A comparable rate of infection (5.5%) was encountered in 309 patients whose skin incisions were made by two separate knives. This difference between the infection rates in these two groups of patients was not significantly different. Therefore, the custom of discarding a sharp scalpel following its use on the skin should no longer be a mandatory practice imposed on the urologist.

TENSION AND COMPRESSION

When a wound is caused by a collision of two bodies, the mechanisms of injury are predominantly compression and/or tension rather than shear. In each case, two forces of equal magnitude are applied in opposite directions to the tissues. Unlike shear forces, compression and tension forces act in the same plane. The mechanism of injury is mainly tension when a flat body collides against soft tissue not supported by underlying bone. For impact injuries due to a collision of a flat body against soft tissue overlying bone, tissue failure is due primarily to compressive forces.

The energy requirement for tissue failure as a result of these forces is considerably greater than for shear forces because the energy is distributed over a larger volume of tissue. The amount of energy absorbed by tissue during an impact can be calculated by the following equation:

$$T = \frac{MV^2}{2}$$

where T = kinetic energy (joule), M = mass of impacting object (kg), and V = relative velocity between the impacting object and tissue (m/sec). The extent of

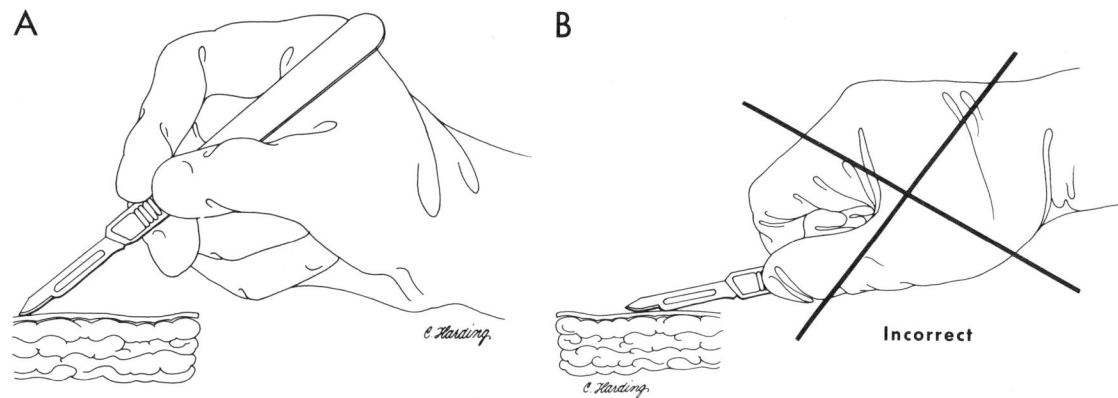

FIG 4–2.
Recommended techniques for cutting skin with a no. 15 blade **(A)**. Incorrect surgical techniques for cutting skin with a no. 15 blade **(B)**.

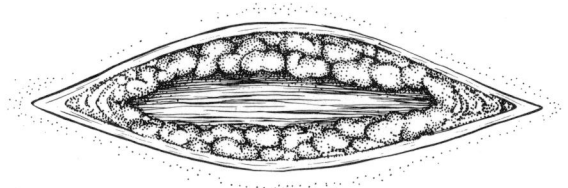

FIG 4–3.
Linear lacerations caused by shear forces.

compressive or tensile injuries will vary with the size (mass) of the object impacting the body surface and its velocity when it makes contact. Changes in the relative velocity (V) of the object that impacts have a greater influence on the level of kinetic energy (T) than do variations in the mass (M) of the object.

BLUNT TRAUMA

When the mechanism of injury is compression, tissue failure occurs at energy levels of 2.52 joule/cm^2, which is comparable to the energy encountered when a car going 8 km/hr strikes a tree (Cardany et al., 1976). The momentum of the collision causes the victim's head, weighing approximately 4 kg, to hit the dashboard with an area of impact approximately 8 cm^2. This absorbed level of energy disrupts the skin resulting in a characteristic *stellate* laceration (Fig 4–4). In addition, the wound edges are damaged and rendered relatively ischemic. Wounds caused by impact injuries are 100-fold more susceptible to infection than wounds caused by shear forces, and immediate antibiotic treatment is warranted to suppress the growth of the bacterial contaminants. Even then, antibiotic efficacy is substantially less than in contaminated wounds caused by shear forces.

While impact injuries with energy levels lower than 2.54 joule/cm^2 do *not* result in division of the skin, they can result in disruption of the vessels in the skin and underlying tissue. The vascular injury to the skin manifests itself clinically as ecchymosis. Disruption of ves-

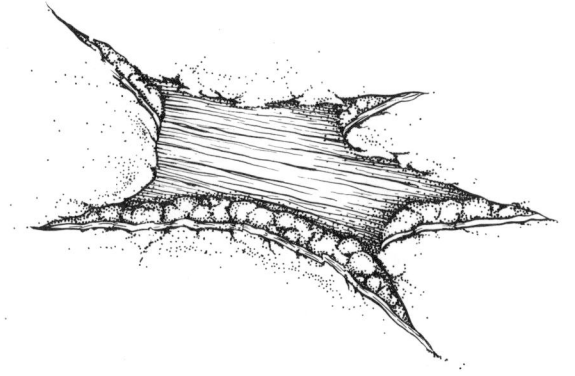

FIG 4–4.
Stellate laceration.

sels in the underlying tissue results in a hematoma. Some hematomas will resorb, but those that become encapsulated usually require surgical treatment. Left untreated, such hematomas may result in permanent subcutaneous deformity. When still in the currant jelly stage, a hematoma is best treated by incision and drainage. As further liquefaction occurs, aspiration with a large-bore needle (18-gauge or larger) may be possible.

MISSILE INJURY

A collision between a missile and the human body represents a considerably higher level of energy absorption per unit volume of tissue than that encountered in automobile accidents. The interaction of missile and man combines shear, tensile, and compressive forces. The severity of the injury is directly proportional to the amount of kinetic injury lost by the missile in the tissue. The efficiency with which kinetic energy is deposited is influenced by: (1) the velocity and orientation of the missile along its trajectory; (2) deformation of the missile within the tissue; and (3) the elasticity and density of tissue in the missile pathway.

Firearms include two basic types: rifled firearm and shotgun (Sturdivan et al., 1984). Rifled firearm, pistol and rifle, has spiral grooves in its barrel and discharges hot gases, soot, powder grains, and a bullet. Rifled bullets can be divided into two groups on the basis of their deposited energy in the tissue. Missiles that are classified as low-energy depositors include .22 rimfire bullets and all handgun bullets, except the .44 magnum pistol bullets, and expend less than 400 joule of kinetic energy into the wound. Low-energy deposit projectiles and simple stab wounds cause similar injuries, the tissue destruction being confined to the permanent wound tract. They usually do not require debridement and may be treated conservatively because the profile of the wound tract is narrow and unlikely to be contaminated significantly by debris drawn in from the outside. Lacerated vessels and vital organs are more serious sequelae of these missiles that will require immediate treatment.

In contrast, bullets classified as high-energy depositors expend 400 joule of kinetic energy or more into the wound and include all centerfire rifle bullets as well as the .44 magnum pistol bullets. Because they cause tissue destruction that extends beyond the permanent wound tract, extensive tissue debridement of the devitalized tissue in an operating theater is necessary. Since the magnitude of tissue destruction is extensive and difficult to ascertain soon after injury, the wound is left open to be reexamined daily in the operative theater to debride any residual devitalized tissue.

Shotguns differ from rifles and pistols in that they have a smooth barrel that discharges hot gases, wad, and either multiple projectiles or a single projectile (rifled slug) (DeMuth, 1971). Shot charges containing multiple projectiles spread out from the muzzle in a conelike pattern. The distance from the muzzle of the shotgun to the point of impact of the projectiles (impact range) is a key determinant of the magnitude of injury. At a short range, less than 7 m, the shot charge containing multiple projectiles results predominantly in a single hole inshoot wound (diamter ≤ 6 cm) that communicates with a deep underlying wound with massive tissue destruction. When the impact range exceeds 7 m, the multiple projectiles result in numerous discrete wounds that are not associated with underlying massive tissue destruction.

Shotguns also can discharge rifled slugs that are high depositors of energy (>400 joule) producing wounds in which the tissue destruction extends beyond the permanent wound tract, similar to that encountered from high-energy deposit rifled bullets. While the rifled bullets maintain their velocity over 45 m, the rifled slugs experience a marked decrease (approximately 25%) in velocity over that distance. This rather dramatic inverse relationship between the velocity of the slug and the deposited energy has obvious effects on the impact energy and the ultimate tissue injury.

ELECTRICAL ENERGY

Electricity is another source of energy that the urologist has learned to use constructively in surgery. The concepts of electricity are based on the accepted theoretical structure of the atom, which supposes the presence of electrons orbiting around a nucleus. Electricity is the flow of electrons from one atom to another. The electrons set in motion by the electric force (voltage) may collide with each other and generate heat. This process transforms electrical energy into thermal energy. The amount of heat developed by a conductor varies directly with its resistance. The magnitude of resistance to electron flow varies widely in different tissues. The high resistance of bone and low resistance of muscle to electron flow are cases in point. The control and localization of the heating effect of electrical current are the fundamental basis for electrosurgery.

Electrical power is usually generated with a continuous reversal of the direction of electrical pressure (voltage). The pressure in the conductor first pushes and then pulls electrons (AC). The frequency of the current in hertz (Hz) or cps is the time in which the complete cycle of positive and negative pressure occurs. The usual wall outlet in the United States provides a current with 120 reversals of the direction of flow occurring each second. Passage of 60 cycles of alternating current through a patient is extremely dangerous, since it can induce ventricular fibrillation as well as muscle contractions.

Involuntary spasmodic contractions of muscle in response to a low-frequency electrical stimulus subside as the frequency of the applied current exceeds 60 Hz. At frequencies between 0.5 and 1.0 megacycle, no muscle response is noted. In addition, high-frequency current can flow along paths that virtually block the 60-cycle current. The frequency of the current generated in electrosurgery is 250,000–2 million Hz. Research in electrosurgery has focused predominantly on electronic technology with only theoretical clinical application. In the future, comprehensive research investigations must be undertaken to relate these modern technologic advances to clinical performance.

The ability of the high-frequency current to damage tissue depends on its concentration or density. As the current density increases, its heating effect becomes more pronounced. The size of the active monopolar electrode is deliberately kept small so that concentrated heating will occur at its point of contact with tissue (Fig 4–5). Following contact, the current is dispersed to the return (ground) electrode, which encounters low current density and no tissue heating. The distribution of the current can be even more precisely controlled by making the active electrode bipolar rather than monopolar. The use of bipolar electrodes, usually in the form of forceps, delineates the tissue through which the current will pass. Bipolar electrodes contain two electrodes and contact the tissue at two points; current flows into the tissue through one arm of the forceps and back out through the other. The entire current is confined to the small area between the two ends of the forceps, and no return electrode is needed.

When undamped high-frequency currents are passed

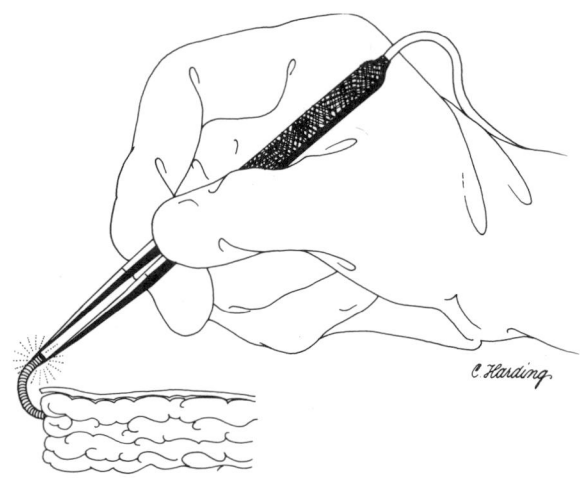

FIG 4–5.
Bipolar electrosurgical electrode.

through tissue, the active electrode acts as a bloodless knife (Fig 4–6). The cells at the edge of the resultant wound literally disintegrate. Away from the plane of cutting, one can see elongated tissue cells as well as histologic evidence of a mild thermal injury. Blood vessels at the wound edge are usually thrombosed, accounting for the hemostatic effect of the high-frequency current. This histologic evidence of tissue damage is also associated with an increased susceptibility of the wound to infection. The wound made by electrosurgery in experimental studies was approximately three times more susceptible to infection than wounds made with the stainless steel scalpel (Madden et al., 1970). In a prospective clinical study by Cruse and Foord (1973) the use of electrosurgery almost doubled the infection rates of surgical wounds. Although there is no prospective study comparing electrosurgery to the scalpel for cutting the fascia, Greenberg et al. (1979) believed that cutting the fascia with electrosurgery may have caused fascial neurosis in six of their eight most recent dehiscences. *The increased susceptibility of such wounds to infection and dehiscence mitigates against the use of electrosurgery for cutting skin, subcutaneous tissue, and fascia.*

In massive excisional surgery (e.g., large soft tissue tumors, debridement of third-degree burns), the threat of blood loss frequently outweighs the potential problems of subsequent infection. In burn wound excisions, Levine and associates (1975) reported that the operative blood loss during electrosurgery excision was approximately 50% less than that encountered during scalpel excision. The operative time was decreased by half by eliminating the additional time required to obtain hemostasis after the scalpel was used.

When the oscillations are damped, the current ac-

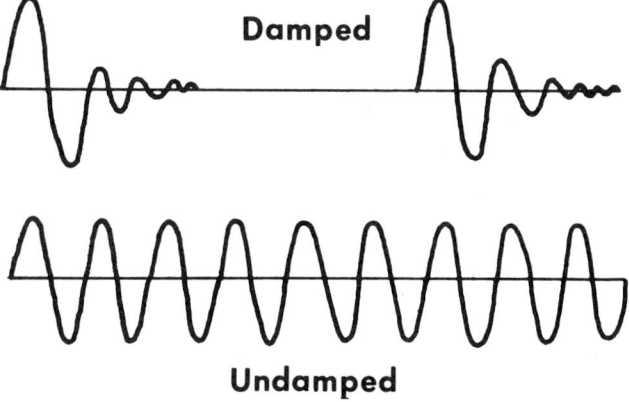

FIG 4–6.
Schematic representation of damped and undamped high-frequency electrical currents. Undamped high-frequency electrical current acts as a hemostatic knife, while damped current exhibits a hemostatic effect without cutting.

complishes hemostasis without cutting (see Fig 4–6). This type of current causes a rapid dehydration of living cells, and the affected tissue is fused into a structureless homogeneous mass with a hyalinized appearance. The bleeding vessels within the tissue become thrombosed, resulting in hemostasis. We prefer pinpoint electrosurgical coagulation of small bleeding vessels over suture ligation, but the power should be kept to the absolute minimum needed for vessel thrombosis.

The technique of electrocoagulation has considerable influence on the magnitude of injury. The use of *bipolar coagulation* is a more *precise method* of hemostasis that limits the tissue injury encountered with the more traditional monopolar coagulation. Ferguson (1971) noted that an equivalent current passed through a monopolar electrode caused approximately three times as much necrosis of the surrounding tissue as the use of bipolar coagulation.

Bleeding from cut ends of large vessels over 2 mm in diameter can rarely be controlled by electrocoagulation. In such cases, hemostasis can be achieved easily with a suture ligature of nonreactive materials. The intact vessel should be isolated over a short length and clamped with small hemostats applied contiguously (Fig 4–7,A). Subsequently, the vessel is divided between the ligatures. This technique is preferred over cutting the vessel first and then clamping the retracted vessel along with the contiguous bloodstained tissue (Fig 4–7,B). In the latter case Ferguson (1971) reported that the amount of strangulated tissue was about five times greater than with the vessel-isolating technique.

It is important to emphasize that hemostasis should be accomplished with the least amount of electrocoagulation and with a minimum of suture material. The best method to control oozing and minor bleeding is to gently exert continual pressure on the bleeding surface with sponges moistened with cool saline. Rubbing or abrading the skin must be avoided, since that dislodges thrombi and may cause further bleeding. Furthermore, urologists should never resort to hot (150° F) wet sponges for hemostasis. In experimental studies reported by McDowell (1959), this treatment resulted in a hyperthermic injury to tissue that potentiated the development of infection.

When the mechanism and source of electrical energy are not controlled during its application to the human body, the consequences can be disastrous. The development of an accidental acute electrical injury is a result of the reciprocal transformation of electricity into heat generated by the electrical current (Nichter et al., 1984). Accidental electrical injuries differ from electrosurgical wounds only in the relative frequency and voltage of the current delivered to the tissues.

From surgery's beginning and well into the 19th cen-

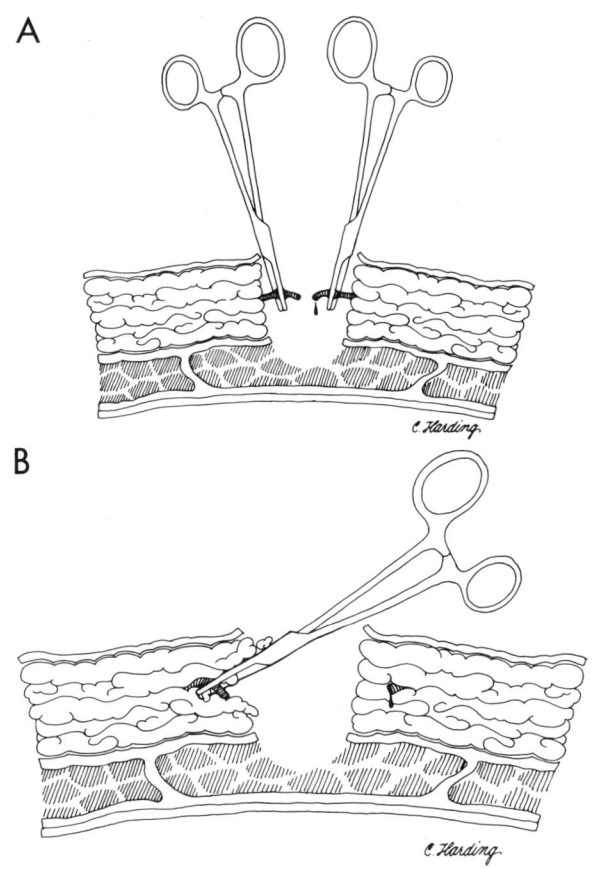

FIG 4–7.
Clamping of the vessel before its division **(A)** results in considerably less tissue injury than clamping the divided retracted bleeding vessel **(B)**.

tury, cautery heated over a bed of hot coals was regularly prepared to control hemorrhage as well as to serve as an antiputrefactive agent. Urologists can now cut tissue with a new heated scalpel whose temperature can be controlled within narrow limits. Levenson and colleagues (1982) reported that the heated scalpel allowed excision of third-degree burns in pigs and human subjects with much smaller blood loss than with the usual cold scalpel. Skin grafts applied immediately after excision had excellent rates of success, similar to those of grafts applied immediately after excision with the unheated scalpel. These investigators reported that the heated knife did not interfere with the wounds' ability to resist infection. In wounds in experimental animals made by either a heated or unheated scalpel and then contaminated with 100 million *Pseudomonas aeruginosa* or *Staphylococcus aureus* organisms, no wound infection developed in either group. The absence of infection in wounds made either by the heated or unheated scalpel is difficult to explain because the inoculum used was 100 times greater than the infective dose for these pathogens in soft tissues. Consequently, inoculation of

100 million organisms should have resulted in gross evidence of infection in all wounds. Moreover, Levenson and colleagues (1982) reported that the heated scalpel produced limited adverse effects on wound repair. Using a temperature of 180°C, the breaking strengths of wounds made by the hot scalpel did not differ significantly from those in wounds made by the cold scalpel 7, 14, 25, 28, and 42 days after wounding. The only statistically significant difference ($P < 0.05$) between the breaking strengths of incisions made by the unheated scalpel and the heated scalpel was noted at 21 days and was modestly in favor of the conventional scalpel. The innocuous effects of the heated scalpel could not be confirmed by our laboratory (Keenan et al., 1984). Using scalpels heated to a considerably lower temperature than that employed in the study of Levenson et al. (1982), we found that the heated knife damaged the wounds' resistance to infection and interfered with healing.

RADIANT ENERGY

As a result of recent scientific advances, urologists can now employ energy from light as a scalpel. This is one of the many forms of radiant or electromagnetic energy. Such energy consists of photons that are both waves and particles. Once it is absorbed by tissue, radiant energy is converted into heat that rapidly increases the temperature of a small volume of tissue. This precise thermal injury results in a relatively bloodless division of tissue. The concept of light as a source of energy is realized in lasers. Light waves emitted from lasers are coherent and are so nearly parallel that they can travel for miles in a straight line without spreading apart or converging. This coherent light provides tremendous pulses of power that do not diminish over great distances.

This energy source is now being used by some urologists to cut tissue. Lasers used in urology get their energy from rotation and vibration of electrons in the CO_2 molecule with a resultant emission of light having a wavelength of 10.6 μ. These infrared waves are then directed along an articulated arm and into a handpiece by means of mirrors located in precision rotary joints. A lens in the handpiece focuses the energy to a point less than 1 mm in diameter. This high concentration of energy at the focal point allows the beam to cut through skin and soft tissue. Despite this technologic advance, laser surgery is limited in usefulness by the cumbersome design of the surgical arm as well as an insufficient level of power. When maneuverability is required, laser surgery is difficult and time consuming.

The hemostatic effect of the laser scalpel makes it especially suitable for massive surgical excisions. In a clinical series of 26 patients subjected to burn wound excision, Levine et al. (1975) reported that the blood loss encountered by scalpel excisions was nearly 3.3 times greater than that following laser surgery. Electrosurgical excision had 1.67 times the blood loss of laser excision. The superior hemostatic effect of the laser over that of electrosurgery is associated with increased damage to the tissue defenses. Experimental wounds made by a laser were approximately tenfold more susceptible to infection than those made by electrosurgery (Madden et al., 1970). This infection-potentiating effect of the laser scalpel argues against its use for incisional surgery. Fortunately, the tissue damage resulting from either electrosurgery or the laser does not interfere with the "take" of either autografts or homografts on wound beds with low bacterial counts ($< 10^6$ bacteria/gm of tissue).

BIOLOGY OF WOUND REPAIR

The urologist can achieve wound closure with minimal deformity and dysfunction by following surgical principles that are based on the biology of wound repair. Repair is the response of living tissues to injury and is the keystone on which urology is founded. Many urologists have not grappled fully with the concepts of the biology of wound repair and do not appreciate that wound healing processes can now be effectively altered to reduce the pathophysiologic consequences of the disease or injury.

ANATOMY

To understand fully the pathophysiologic consequences of disruption of the integrity of skin, one must appreciate the organization and morphological features of uninjured skin. The largest organ of the body, skin, has three major tissue layers. The outermost layer, epidermis, is stratified epithelium whose thickness is relatively uniform in all areas of the body (75–100 μ), except in the palms and soles, where it is particularly thick (0.4–0.5 mm). Beneath the epidermis is the dermis, which is composed primarily of a dense fibroelastic connective-tissue stroma with collagen and elastic fibers and an extracellular gel called the ground substance. This amorphous gel is composed of acid mucopolysaccharide protein as well as salts, water, and glycoproteins. The functions of the ground substance are not known with certainty, but it is believed that it contributes to salt and water balance, serves as a support for other components of the dermis and subcutaneous tissue, and participates in collagen synthesis. The dermal layer encloses an extensive vascular and nerve network as well as special glands and appendages that communicate with the overlying epidermis. The dermis can be subdivided into two parts. Its most superficial portion, the papillary dermis, is molded against the epidermis and encloses superficial elements of the microcirculation of the skin. It consists of a relatively cellular, loose connective tissue with collagen and elastic fibers smaller in diameter and fewer in number than the underlying reticular dermis. Within the papillary dermis, there are dermal elevations that indent the inner surface of the epidermis. Between the dermal papillae, the downward projections of epidermis appear peglike and are referred to as rete pegs. In the reticular portion of the dermis, the collagen and elastic fibers are thicker and greater in number. There are fewer cells and less ground substance in the reticular dermis than in the papillary dermis.

The thickness of the dermis varies from 1 to 4 mm in different anatomical regions and is thickest in the back, followed by the thigh, abdomen, forehead, wrist, scalp, palm, and eyelid. Its thickness varies with the individual's age. In the very old, the dermis is often atrophic, whereas in the very young it is not fully developed. The third layer of the skin is the subcutaneous tissue, composed of areolar and fatty connective tissue. It shows great regional variations in thickness and adipose content. The subcutaneous layer insulates underlying tissue and protects it from external trauma.

The color of normal human skin is related to the number, type, size, and distribution of melanosomes. These specialized organelles are products of unicellular glands (melanocytes) that transfer them into the epithelial cells. The concentration of melanosomes varies considerably within an individual and between individuals. The pigmentation or coloring of a patient's skin is an important consideration in reconstructive surgery. Ideally, the color of the reconstructed tissue should be similar (color match) to that of contiguous tissue. When reconstructing the facial or cervical region, skin taken from the scalp, upper part of the chest, or upper part of the back provides the most superior color match (Edgerton and Hansen, 1960).

The muscles of facial expression are connected to the overlying skin by superficial connective-tissue strands. During the process of aging, natural wrinkle lines appear perpendicular to the direction of contraction of the underlying muscles of facial expression. For example, the wrinkle lines in the perioral region lie perpendicular to the direction of contraction of the orbicularis oris muscle. It has been recommended that the lines of either excision or incision coincide with these natural wrinkle lines, resulting in a more aesthetically pleasing scar (Cox, 1941; Kraissl, 1951).

Repair of skin wounds represents a highly dynamic, integrated series of cellular, physiologic, and biochemical events. Although these processes of repair of skin wounds occur concomitantly, examining all processes simultaneously can be confusing. For the purposes of discussion, several natural components of wound healing will be reviewed separately.

INFLAMMATION

It is almost axiomatic that injury is followed by inflammation that can be characterized by vascular and cellular responses designed to protect the body against alien substances and to dispose of dead and dying tissue preparatory to the repair process. In either a planned (surgery) or unplanned (trauma) assault on the body that results in disruption of skin, there is a physical interruption of the blood vessels, leading to immediate hemorrhage. The body's response to the injury regardless of the cause is basically the same, and the most significant element of this response is seen in the local vasculature. The immediate response of the small vessels is vasoconstriction, lasting 5–10 minutes, which limits the magnitude of blood loss.

Almost immediately after injury, leukocytes "stick" to the endothelium of the injured vessels. At the same time, RBCs adhere to each other and form rouleaux, which tend to plug the cut ends of capillaries. Since platelet thrombi do not form until later, this occlusion can be reversed. Coagulation is initiated by agglutinated platelets that adhere first to the subendothelial collagen of the cut ends of injured vessels. Within minutes, fibrinogen is converted to fibrin, which then forms a mesh at the vessel opening. The fibrin screen aggregates more platelets, and their developing pseudopods contract the fibrin mesh to form a firmer plug, sealing the ends of the blood vessels.

Immediately after injury, local vasoactive substances are released into the extracellular compartment of the injured tissue that act directly on the microvasculature. These intracellular materials cause active vasodilation and increased permeability of the local vasculature. The fluid that estravasates from the blood vessels has the same composition as plasma with its full complement of macromolecules, including fibrinogen. Once outside the vessel, the fluid exudate enters the extracellular space and wound. Much of the protein content of the exudate is reabsorbed slowly by lymphatics, with the exception of fibrinogen, which partly polymerizes to form fibrin.

Initially, histamine was considered the primary mediator of the inflammatory vascular response. Histamine, liberated from mast cells, polymorphonuclear cells, and platelets, produces local vasodilation and increases vascular permeability. However, it acts for short periods (less than 30 minutes) and local sources are depleted rapidly. Recently, the kinins, a series of biologically active peptides, and the prostaglandins have been implicated in this local inflammatory re-

sponse. The prostaglandins appear to be the final mediator of the vascular responses and may play a chemotactic role for fibroblasts and white cells as well (Weeks, 1972). After the leukocytes adhere to the endothelium these cells move through the vessel wall by a process of diapedesis and concentrate at the injury site. The fluids that escape from the vessels combined with migrating leukocytes constitute an inflammatory exudate.

The magnitude of this increase in vascular permeability can be correlated with the duration of exposure of the wound (Edlich et al., 1973). Closure of the wound within one hour after injury appears to reduce vascular permeability. If the wound is left open for three hours, there is a dramatic increase in vascular permeability that results in the development of a relatively thick inflammatory exudate on the wound surface. This exaggerated inflammatory response to prolonged exposure (longer than three hours) limits the therapeutic value of antibiotics (Fig 4–8). This fibrinous coagulum forms a protective cover around bacteria on the surface of the wound and prevents antibiotics administered systematically or topically from contacting them. Similarly, this developing inflammatory exudate accounts for the diminished effectiveness of delayed antibiotic treatment. Moreover, it is the reason for the success of preoperative antibiotics in the prevention of infection. In the absence of the inflammatory exudate, the antibiotic can gain access to the bacteria and prevent the subsequent development of wound infection.

Paradoxically, the fibrinous wound coagulum that limits the effectiveness of antibiotics is an important

RESISTANT TO ANTIBIOTIC TREATMENT

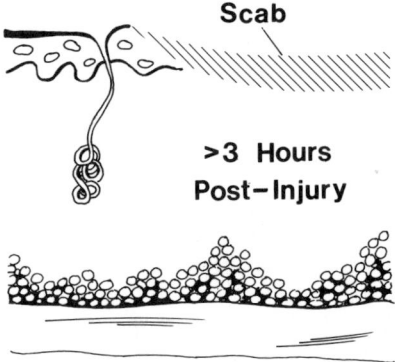

FIG 4–8.
Within three hours after injury, an inflammatory exudate develops on the surface of the open wound that becomes a protective barrier for bacteria against topically or systemically administered antibiotics. This developing coagulum (scab) accounts for the resistance of the open wound to antimicrobial prophylaxis.

factor in the open wound's defenses against infection (Edlich et al., 1969). This coagulum serves as a plug in the transected ends of lymphatics in the wound, thereby becoming an obstacle to the invasion of bacteria, and, in part, enhancing the resistance of an open wound to systemic sepsis. While the fibrinous coagulum prevents the lymphatic invasion of bacteria, the underlying wound gradually gains sufficient resistance to infection to permit an uncomplicated closure. The reparative process of the open wound, associated with the developing resistance to infection, is characterized by the appearance of capillary buds and young, fibrous tissue referred to as granulation tissue. On or after the fourth postoperative day, the margins of the open wound without devitalized tissue can be approximated with minimal risk of infection.

The cellular population of the wound varies considerably during repair. During the initial migration of leukocytes into the extravascular space, the ratio of polymorphonuclear cells to monocytes is very similar. Polymorphonuclear cells are short-lived, however, and they die releasing acid hydrolases into the tissue. As the polymorphonuclear cells disappear, monocytes predominate. The different survival rates of the white cell populations seem to account for the changes in cell population (Willoughby, 1970). The white cells are involved in a scavenging process that includes removal of bacteria, cell fragments, and debris. The precise role of each white cell type in the healing process has been elucidated in studies using specific anticellular antisera (Heppelston and Styles, 1967; Leibovitch and Ross, 1975). Monocytes must be present for wound healing to proceed normally, while the absence of polymorphonuclear cells and lymphocytes does not significantly alter the healing process.

Virtually any condition that contributes to a decrease in delivery of WBCs to an area of bacterial contamination will promote the development of infection. These conditions usually are caused by a diminution in blood flow, as may be found in vasocclusive states, in hypovolemic shock, and following the use of vasopressors. The infection-potentiating effects of tissue ischemia are documented with local anesthetic agents containing epinephrine (Stevenson et al., 1975). This drug is a potent vasoconstrictor that limits the clearance of local anesthetic agents from the tissue, thereby prolonging the duration of anesthesia. This beneficial effect of epinephrine is, however, associated with damage to the local wound defenses. The infection-potentiating effect of this powerful vasoconstrictor is proportional to its concentration and results from its vasoactivity. This damage to the tissue defenses argues against the use of epinephrine in heavily contaminated wounds.

Capillary budding begins within 36 hours after injury, and thereafter a large fibroblastic population becomes evident. The origin of fibroblasts is primarily from undifferentiated mesenchymal cells that begin to differentiate into migratory fibroblasts. These mesenchymal cells arise from loose areolar tissue located subepidermally, sleeving appendages, and, perhaps especially, cuffing all blood vessels in the papillary and reticular dermis and in the subcutaneous tissue. Fibroblasts are generally accepted as the source of ground substance, elastin, and collagen.

EPITHELIALIZATION

After an incision, the divided parts of the epithelium are closed by cellular migration and mitosis. Within hours after injury, the peri-incisional epithelium reacts by thickening and begins to migrate across the incisional gap as two long sheets of epithelium, one moving along each side of the entire length of the wound. The normal rate of mitosis increases from about 7% to 30% at the margin of the wound (Hell and Cruishank, 1963). The maximum activity is seen just a short distance (a few cell widths) back from the incision. The two thickening advancing surface epithelial sheets invert by growing down in the upper part of the incision. This downward growth of epithelial cells bridges the wound and unites with each other across the incision. Such epithelial union most commonly occurs at the level of the reticular dermis, creating an epithelium-lined furrow along the entire length of the incision.

The time in which this epithelial bridge is complete depends on the technique of closure. With slight eversion of the coapted skin edges, epithelial bridging of the skin edges occurs within 18–24 hours. End-to-end approximation of the skin edges results in an additional 12-hour delay in the formation of an epithelial bridge. If the wound edges are inverted, epithelial bridging is complete 72 hours after closure. Even after the incision is bridged by epithelium, some further downgrowth occurs in the dermal breach, forming epithelial spurs. Short secondary epithelial spurs invade laterally and downward from the sides of the simple inversion of epithelium.

This epithelial bridge forms a protective barrier against the invasion of bacteria into the wound. Prior to its formation, the wound is susceptible to infection by bacterial contamination on the surface of the wound. Experimental studies performed in our laboratory demonstrated that as they heal, sutured wounds are susceptible to infection following surface contamination during the first 48 hours after wound closure (Schauerhamer et al., 1971). Such surface contamination on the third postoperative day did not produce gross infection in the sutured wound. This susceptibility to infection during the early postoperative period confirms the apparent value of dressings to protect the sutured wound from surface contamination.

Cell division is seen consistently in the advancing tongue of migrating epithelium, indicating that migration and mitosis occur concurrently at the leading edge of epithelium. Cells at the surface of the tongue of advancing epithelium do not differentiate into keratinized and cornified cells, until either migration slows down or epithelial union of the advancing epithelium occurs. At some distance back from the incision, the cells continue their usual progression toward the surface of the epithelium.

Once the epithelial cells bridge the gap, they begin to acquire some of the structural features of the adjacent uninjured epithelium. They become columnar in shape, commence mitosis, and migrate upward toward the surface of the epithelium with the development of keratin in the uppermost epithelial cells. By 10–15 days, the invasive epithelium begins to resorb. Until the 20th day, the inversion of epithelium is distinguishable microscopically; thereafter it becomes less obvious as the epithelial surface becomes level.

The organization of scar epithelium has little resemblance to that of uninjured skin. Normally, epithelium is bound tightly to the underlying dermis through its undulating basement membrane and epidermal appendages. Unlike normal skin, the scar epithelium is relatively atrophic without rete pegs and epidermal appendages. Since the junction between scar epithelium and underlying dermis has limited surface contact, it is very prone to shear forces that result in epidermal dermal dysjunction.

Downward growth of epithelial cells occurs not only at the incision, but also at any interruption of the skin, such as suture tracts (Ordman and Gillman, 1966). In the presence of percutaneous sutures, epithelial cells migrate downward forming a perisutural cuff (Fig 4–9, A and B). Within this connective-tissue milieu, the epithelial cells undergo cellular differentiation, eventually ending in keratinization, which induces an intense inflammatory reaction and scar formation. If the percutaneous sutures are removed before the eighth postoperative day, the invasive spurs of epithelium regress, leaving no discernible deformity. After that time, the suture tract's reaction to the foreign body becomes so intense that permanent deformity results (i.e., needle puncture scars), which will be a constant and unnecessary reminder to the urologist and patient of the operative intervention.

Crikelair (1958) enumerated factors that predispose to the development of suture scars. The anatomical lo-

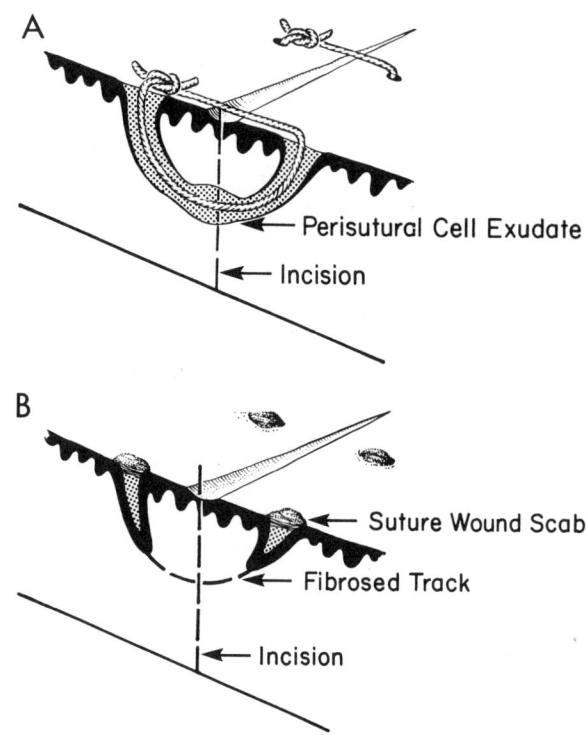

FIG 4–9.
(A), epithelial cells migrate downward, forming a perisutural cuff. **(B)**, if percutaneous sutures are not removed before the eighth day after surgery, the invasive spurs of epithelium results in needle puncture scars.

cation of the skin was an important factor. Skin of the eyelids, palms of the hand, and soles of the feet seldom showed suture puncture scars. By comparison, the skin of the back, chest, upper part of the arms, and lower extremities was more likely to develop suture puncture tracts. Patients with an inherent tendency to keloidal formation were prone to suture puncture scars. Of the variable factors under the control of the urologist, the length of time the skin suture remained in place was the most important determinant. The size of either the suture needle or suture was not significant. Surprisingly, the inherent skin tension or the tightness of the tied sutural loop did not predispose to the development of sutural scars.

DERMAL REPAIR

The fibroblasts within dermis, which have a full complement of metabolic pathways, synthesize collagen, elastin, and ground substance. Collagen is the principal component of the scar tissue in the dermis and accounts for the tensile strenth of the normal and repaired skin. A knowledge of the molecular structure and organization of collagen is essential for an understanding of the anatomical and physical properties of scars of the dermis.

Four distinct types of collagen have been identified. Type 1 collagen, found in the adult dermis, fascia, and bone, is the most common. There are four distinct levels of the molecular organization of dermal collagen (Peacock et al., 1976). The primary structure of collagen refers to the precise sequence in which amino acids are arranged. The amino acid composition of collagen is distinctly different from that of other proteins. Glycine accounts for approximately one third of the amino acid residues in the molecule. Another salient feature is the high concentration of the two amino acids, proline and hydroxyproline.

The collagen molecule consists of three peptide chains of approximately equal length, each containing about 1,000 amino acids. At least one of these chains has a different amino acid composition from the others. The tertiary structure of collagen refers to the unique spatial arrangement of three chains in the molecule; each of the collagen molecules exists as a triple helical coil. This unit molecule, composed of these three chains, is called tropocollagen. Its quarternary structure refers to the manner in which tropocollagen is aggregated into a stable biologic unit of mechanical strength. The tropocollagen molecules are oriented in the same direction, but they overlap by about one fourth of the length (quarter-stagger arrangement). It is this overlapping arrangement that accounts for the presence of bands, spaced approximately 680 A apart, that are detected by electron microscopy. Strong intermolecular and intramolecular covalent bonds join the individual tropocollagen molecules together, creating a huge polymer, the collagen fiber. The ultimate strength of scars depends on this extracellular bonding.

End-to-end attachment of the tropocollagen units produces a filament that is 14–200 A in diameter and of unlimited length (Bryant et al., 1968). These filaments are longitudinally organized into fibrils that measure approximately 2,000 A in diameter. A bundle of collagen fibrils forms a primitive collagen fiber whose diameter is approximately 20,000 A. Aggregation of the latter results in fibers that are 100,000 A in diameter and detectable with a light microscope.

In the uninjured skin, the collagen fibers are unbranched and extremely long structures (Thacker et al., 1975). Since there are no known anatomical linkages between the collagen fibers, it is assumed that the elastic fibers are so interlaced with them to restore the collagen network to its normal position after a mechanical load has been removed from the skin. In the relaxed skin dermis, the fibers appear randomly in convoluting intertwining coils. As a load is applied to the skin, the fibers become aligned until eventually most are lying parallel to the direction of maximum skin stretch. After the fibers have become uncoiled and oriented, the further elastic property of skin resembles that of collagen.

From the urologist's point of view, the rate of gain of strength of the wound is a key determinant of many decisions including when the sutures can be removed, the level of patient activity, as well as the selection of the incision. The answers to these questions are found in the results of bioengineering studies of the strength of wounds. Even though collagen fibers are evident on the third day postinjury, the skin wound has negligible tensile strength (Levenson et al., 1965). During the first eight days after closure, the wound is held together by blood vessels crossing the wound, epithelialization and a fibrinous coagulum. If the percutaneous sutures are removed at this time, the wound may be disrupted easily unless supported by dermal sutures and/or skin closure tapes. Over the next 13 days (8–21 days after injury), there is a rapid gain in strength of skin wounds. They continue to gain strength at a relatively rapid and constant rate for four months and at a slower rate for one year. The strength of repaired skin incisions never reaches that of uninjured skin. Adamsons and Kahan (1970) demonstrated that rabbit skin wounds closed with a continuous 4–0 silk suture regained only 40% of the strength of unwounded tissue 120 days after wounding. In the dog, Van Winkle et al. (1975) noted that skin wounds approximated by different percutaneous sutures developed 70% of their normal strength by 120 days. Consequently, the skin wound remains a relatively brittle structure that is capable of absorbing much less energy than normal skin.

The diminished tensile strength of wounded skin as compared to normal skin can be correlated with their histologic appearances (Forrester, 1980). The morphological features of scarred collagen differ distinctly from collagen in unwounded tissue, particularly collagen bundle size. Wound collagen bundles are narrower than normal collagen. Polarized light studies indicate that there is also a more generalized disorganization of the wound collagen. Normal collagen is birefrigent, while wound collagen is clearly a nonbirefrigent material, which indicates a relatively disorganized structure at the molecular or small fibril level. Physical irregularities in fiber shape and "weave" are more readily appreciated by scanning electron microscopic examination. Scanning electron micrographs of a normal collagen fiber show that it is made up of bundles of cross-banded fibrils, characteristically organized into an interlacing network. In the healing wound, the fibers lie in a haphazard pattern. As time goes by, the randomly dispersed collagen bundles coalesce to form irregular masses of collagen. Close examination of the wound shows no evidence of the collagen fibril structure. Unfortunately, scar collagen appears to be fixed irretrievably in this haphazard arrangement.

Through the years, imaginative biologists have suggested methods to accelerate healing. To date, this avenue of research has resulted in important findings on the repair of dehisced and resutured wounds. Incised wounds allowed to heal for short periods, then dehisced and immediately resutured, developed strength at a significantly faster rate than the primary wound (Botsford, 1941). Experiments in animals demonstrated that strength gained in secondary wound healing correlated with rate of collagen synthesis at the time of dehiscence rather than the collagen content of the wounds (Madden and Smith, 1970). Interestingly, excision of the wound edges and reapproximation of the debrided edges eliminated this acceleration of healing (Dunphy and Jackson, 1962). Such debridement of the wound edges of a dehisced wound is, therefore, clearly an error in surgical judgment. The benefits of secondary wound healing can be realized in patients requiring surgical intervention soon after the first procedure. Patients with wounds exhibiting gross malapposition of skin edges should be returned immediately to the operating theater to reapproximate these edges. The development of complications, such as ischemia of the wound edges or hematoma, also warrants reexploration of the wound. When performed between the first and fourth week after injury, the secondary wound exhibits a greater breaking strength than the first. This accelerated healing is associated with enhanced resistance of the wound to infection (Edlich et al., 1969).

SCAR REMODELING

During healing, scar remodeling occurs with progressive changes in its configuration. As the scar matures, it contracts in all dimensions. The collagen becomes more dense, the fibroblast population decreases, and many vascular channels are obliterated. As the collagen becomes more dense and the vascular supply reduced, the mature scar becomes flatter, thinner, and paler, making it less conspicuous and more esthetically pleasing. Since these changes usually evolve over 12 months, it is best to wait at least one year before making the decision as to whether surgical revision is really necessary. Consequently, experienced urologists consider the biologic processes of remodeling of scars to be their allies that usually improve the scar's appearance. Patients will unknowingly attribute these pleasing changes in the appearance of their scars to the initial surgery—a misconception that reinforces their confidence in their urologists.

Remodeling occurs by a process of resorption that occurs concomitantly with synthesis. In most scars, a balance between production of collagen by fibroblasts and

degradation by epidermal collagenase is reached. The rate of collagen synthesis and degradation is high for about six months and then diminishes greatly. When the rate of collagen synthesis remains greater than degradation for extended periods, hypertrophic scars or keloids develop. The mechanisms for controlling remodeling are still poorly understood. Clinical observations in man suggest that the biomechanical properties of the skin at the site of surgical incision or excision play an important role.

BIOMECHANICAL PROPERTIES

The ultimate appearance and function of a scar can be predicted by the static and dynamic skin tensions on the surrounding skin (Thacker et al., 1975). The static skin tensions are the forces that stretch the skin over the underlying bony framework when the body remains motionless. These inherent forces are dependent partly on the natural characteristics of dermal collagen fibers and partially on the pattern in which they are woven. Clinical evidence of these tensions is the retraction of the edges of the wounds, permitting visualization of the underlying tissue.

Static skin forces differ considerably in their magnitude and direction within the same person and between individuals. Large differences are noted between various anatomical sites. The skin in one region may be relatively taut; in others it is lax. In one human volunteer, the static skin tensions were fivefold greater in his extremities than in his abdominal skin. In some regions of the body there is a directional orientation of static skin tensions. This was first appreciated by Dupuytren, in 1834, when he examined a suicide victim who sustained three self-inflicted puncture wounds made by an awl. He noted that the wounds assumed an elliptical shape similar to the shape of skin wounds caused by a knife. He concluded that skin tensions in the long axis of the defect were substantially greater than in its short axis, distorting the wound accordingly.

In 1861, Langer (1978a) published a more comprehensive study on the biomechanical properties of human skin. His observations were made on the skin of cadavers, which were lying in the normal anatomical position, by inserting an awl 2.0 mm in diameter to a depth of 2.5 mm. As Dupuytren reported (1834), the circular defect was drawn into an ellipse. By drawing lines between the major axes of the ellipses, Langer identified the direction in which the tension predominated. These static lines of maximal skin tensions are known as "Langer's lines." In 1892, 31 years later, Kocher (1907) advocated that surgical incisions should follow Langer's lines. For nearly 100 years, surgeons referred to Langer's lines as the most appropriate guides for incisions that would heal with minimal scarring. It is now realized that the charts of Langer's lines appearing in textbooks and articles have little practical application, since they are erroneous in most cases and do not consider the highly important effect of the dynamic skin tensions on a healing scar.

The static skin tensions continually pull on the wound edges, resulting in the development of a visible scar. The width of this scar is proportional to the magnitude of skin tensions (Wray, 1983). Incisions made in skin subjected to strong skin tensions usually heal with wide, unattractive scars (Fig 4–10). In contrast, narrow, fine scars usually result from the repair of incisions made in skin with weak static skin tensions.

Dynamic skin tensions also have considerable impact on static skin tensions as well as on the magnitude and extent of scar formation. These changing tensions are caused by a combination of forces that are associated with either joint movement or mimetic muscle contraction. In the face, the dynamic skin tensions are perpendicular to natural skin wrinkles and parallel to the direction of contraction of the underlying mimetic muscles. The clinical significance of dynamic tensions is apparent in skin of changing dimensions where elasticity is needed for normal function. In general, a linear scar intersecting the wrinkle lines or lying parallel to the dynamic skin tensions can result in a serious contracture, because the scar does not stretch or recoil like uninjured skin.

PLANNING INCISIONS

In planning incisions, the urologist's primary responsibility is to identify a site through which the operative procedure can successfully and safely be completed. The plan must also allow adequate skin coverage of the surgical defect. If a variety of plans can achieve the same result, the urologist's selection should be based on the resultant cosmetic and physical deformity. Without sacrificing the goals of the urologic procedure, the incision should be placed so that the healing scar is the most esthetically pleasing. Hiding the scar within hair-bearing skin (i.e., scalp or pubic region), pigmented skin (i.e., areolae) or within hidden recesses of the body (i.e., axilla, nose, or oral cavity) is the best approach to minimizing deformity.

Unfortunately, incisions cannot always be placed in such hidden positions, and the choice of the site where the scar will be as inconspicuous as possible must be made. It is now generally recognized that the most esthetically pleasing scar occurs when the long axis of the scar is in the direction of maximal skin tension.

For incisions, their directional orientation should be parallel to the natural wrinkles or perpendicular to the dynamic skin tensions. Clinical observation of the wound after the incision is made provides a reliable

STATIC SKIN TENSIONS

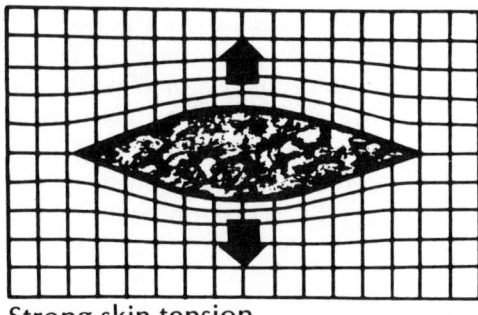

Strong skin tension

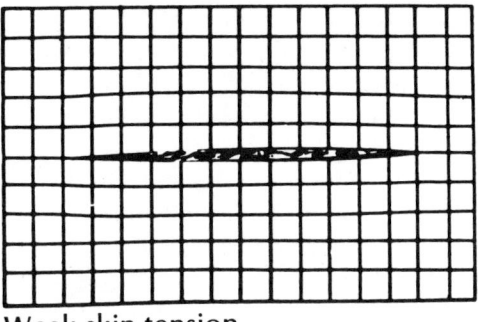

Weak skin tension

FIG 4–10.
The degree to which the wound edges retract can be correlated with the magnitude of static skin tensions. **Top,** wound exhibits marked retraction of its edges being subjected to strong static skin tensions. **Bottom,** minimal separation of the edges of the wound is consistent with the presence of weak static skin tensions.

method of predicting the appearance of the healing scar after closure. The degree to which the divided skin edges retract provides a rough estimate of the magnitude of static skin tensions. Wounds with marked retraction of the wound edges are subjected to strong static skin tensions and heal with wide scars. When there is minimal separation of the wound edges, wound repair will be accomplished with fine scars. When treating a patient with a gaping wound with marked retraction of its edges, it is best to warn the patient that the wound may heal with a wide scar, which may require revisional surgery 12 or more months after injury, if the patient so desires. This information should be shared with the patient as well as his family so that he can appropriately credit the esthetic result to the biology of wound repair.

Unfortunately, an accidental injury does not gain the benefit of this preplanning and may result in a laceration in which the long axis is parallel to the dynamic skin tensions and perpendicular to the natural wrinkle lines. This directional orientation of the wound predisposes to repair by a wide, unattractive scar. In such cases, it is imperative to warn the patient of the impending development, and referral to a plastic surgeon for follow-up examination and treatment is recommended. Failure to anticipate these developments may make the patient think the cause of the unattractive scar is proximal to the needle holder rather than distal. At a later date (beyond 12 months), the scar can be revised by either a W-plasty or Z-plasty so that the orientation of a portion of the wound becomes perpendicular to the dynamic skin tensions. The success of these scar revisions is also related to the increase in the length of the perimeter of the wound, accounting for lower skin tensions per unit length of wound.

This same inverse relationship between the length of wound perimeter and the magnitude of static skin tension accounts for the excellent esthetic results encountered following closure of uneven, jagged-edged lacerations. In such wounds, the perimeter of the wound is considerably longer than that in the linear incision. Consequently, the magnitude of static tensions per unit length of a jagged-edged wound is less than that for a linear laceration. Plastic surgeons have learned that meticulous reapproximation of the jagged edges of the wound yields a gratifying result with a narrow scar. Urologists, unaware of the biomechanical principles involved in wound repair, may elect to convert the jagged wound edges into a linear wound. This decision adds insult to the injury. This ill-conceived debridement eliminates the potential benefits of the long wound perimeter and leaves a lenticular-shaped defect that is considerably wider than the initial wound.

Reapproximation of the edges of the debrided wound will require greater closing forces than would have been needed prior to debridement, accounting for a development of a wide, unattractive scar. Faced with this physical deformity, the patient may seek the advice of a plastic surgeon. Twelve or more months after the injury, the plastic surgeon may elect to revise the scar by either a W-plasty or Z-plasty. The revised incision, the shape of which is reminiscent of the original laceration, will usually heal with a narrower scar than the initial one.

There are certain similarities between the healing of skin wounds and that of other organ systems, as well as individual differences peculiar to the organ involved. In general, the regain in prewounding strength is inversely proportional to the tensile strength of the unwounded tissue. The modern urologist must welcome objective data regarding the biology of repair of each organ system that will allow him to select the most appropriate closure technique. Influencing decisions are data on the normal strength of the unwounded tissue, the rate with which wounds gain strength, the strength of the closure technique, the rate at which the closure

technique loses strength in tissues, and the interactions between the closure technique and the tissue.

FASCIA

Early studies on the healing of musculoaponeurotic incisions utilized excised segments and were specifically designed to place the greatest stress on the wound itself rather than distributing tension throughout the entire abdominal wall. Most of these studies demonstrated that the excised segments of tissue were never as strong as that of unwounded tissue. In the study by Fast et al. (1947) of the gain in breaking strength of sutured paramedian incisions in rabbits, the wound with sutures present was 41.9% as strong as the undisturbed opposite side immediately after closure. This value dropped very slightly during the next three days and then rose sharply, approaching 80% at 15 days. At six weeks after surgery, the sutured abdominal incisions never regained their postoperative strength, remaining at about 80% of the value. Using a similar experimental model in the rabbit, Nelson and Dennis (1951) assessed the strength of healing paramedian incisions. The results from the groups of animals in which the suture materials were removed prior to testing demonstrated that the suture contributed a major amount of strength to the healing wound during the first 14 days postwounding, after which the suture had a negligible role in wound repair. In the six weeks of observation, the healing abdominal wall wounds never attained the full strength of the unwounded side, remaining at about 80% of that value.

In an investigation of the healing of lumbodorsal aponeurotic incisions in rabbits, Douglas noted that the strength of any of the wounds could not be detected until the sixth day postwounding. All wounds showed measurable strength by the eighth day, and thereafter a rapid increase until about the end of the second month, when the curve of healing began to flatten out. Subsequently, a slow increase in strength was detectable, which continued throughout the duration of the study (one year). At the end of two weeks, the wound approached 20% of that of unwounded tissue; at the end of the 1 month, 50%; at 2 months, 60%–80%; 1-year values, up to 90%. This investigator concluded that the aponeurosis seldom reaches the strength of the unwounded tissue and never before four months after the incision.

Adamsons and Kahan (1970) demonstrated that musculoaponeurotic wounds in rabbits gained most of their strength during the first 30 days postwounding, after which the increase was minimal. Despite the most rapid gain in strength during the first 30 days, the muscle wound regained only 57% of the strength of the uninjured contralateral muscle. The tensile strength measured over the same cross-sectional area was identical in wounds of skin and muscle during the first nine days of healing. The subsequent pattern of repair in these two tissues was, however, different. The skin wound gained tensile strength at a rapid rate, while the wounds of muscle demonstrated a slow increase.

In studies of incised fascial wounds in rabbits, Lichenstein and co-workers (1970) reported that the wound strength after suture removal markedly increased during the first two weeks postwounding, after which the rate of gain was gradual, increasing to 41% at two months. The wounds with nonabsorbable sutures intact were as strong the moment the operation was completed as they were two months later, 70% of the strength of the unwounded tissue.

This relative weakness of the musculoaponeurotic wounds in the rabbits may be species specific since it was not encountered in the studies by Adamsons et al. (1965) on repaired paramedian incisions in guinea pigs. Healing paramedian incisions in guinea pigs tested with nonabsorbable sutures in situ regained the strength of original uninjured tissues by the ninth day after wounding and exceeded it significantly by the 45th day. Healing paramedian incisions tested with sutures removed regained 80% of the strength of the original uninjured tissue by the ninth day after wounding and 114% by the 45th day.

Experimental models in which the peritoneal cavity of the intact animal is distended to measure the burst strength of the abdominal wound simulate the clinical situation more closely than the breaking or tensile strength studies. Using this abdominal wound burst model, Kon et al. (1984) reported that abdominal disruption often occurred at a site distinct from the wound at 1, 2, and 6 months after wounding, indicating that the wound was stronger than the unwounded abdominal tissue.

Most prospective studies of wound disruption report incidence rates of 1%–3% following major abdominal operations (Poole, 1985). The mortality after dehiscence is very high, ranging from 9.4% to 43.8%. Factors associated with an increased risk of wound dehiscence include male sex, advanced age, hypoproteinemia, malnutrition, obesity, malignant disease, treatment with steroids, and uremia (Stein and Wiersum, 1959). Experimental studies have confirmed that uremia has an adverse effect on wound repair (Nayman, 1966). Kursh et al. (1977) confirmed that uremia had a deleterious effect on wound repair by measuring the tensile strength of skin wounds and the amount of collagen formation in polyvinyl sponges implanted subcutaneously. The uremic rats had a significant reduction in wound tensile strength and collagen accumulation compared with that in the control animals. It was also noted that

the uremic animals had a significant decrease in caloric intake and final body weight, suggesting that nutritional factors were responsible for the demonstrated poor healing capacity associated with uremia. In another experimental study by Kursh et al. (1978), a group of nonuremic animals was parallel-fed (parafed) a diet and caloric intake identical to that of the uremic animals. The parafed rats were found to have a wound strength that closely approximated the values of the uremic rats and was significantly less than that of the controls. Accordingly, this investigation lends further evidence that the mechanism of reduced wound healing capacity associated with uremia is on the basis of reduced nutrition.

However, local and mechanical factors are probably more important in dehiscence than these systemic factors. Bringing a drain or stoma through the wound will obviously compromise the closure and contaminate the wound. Wound infection has been frequently implicated as a contributing factor to wound dehiscence and the development of incisional hernia. Smith and Enquist (1967) found that a standardized staphylococcal wound infection produced significantly weaker fascial wounds than the controls. Wound dehiscence is clearly associated with causes of increased intra-abdominal pressure to include abdominal complications (vomiting, ileus, or obstruction), pulmonary problems (atelectasis, bronchitis or pneumonia), or the nature of the operation (repair of diaphragmatic hernia) (Bitterman et al., 1967). Since the direction, length, and location of the abdominal incision are not important determinants of wound disruption, primary consideration should be given to the location of the incision that provides adequate exposure to perform the operation.

PERITONEUM

The most important principle of healing of peritoneum is that the entire surface is covered by mesothelial cells simultaneously, not gradually from the border as in epidermization of the skin wound (Ellis, 1971). Regardless of the size of the defect, a mesothelial surface indistinguishable from normal peritoneum will be reconstituted within a few days (3–4 days) in the rabbit, rat, dog, and human. The mesothelial defect is covered primarily by the free-floating peritoneal macrophages that contact the denuded surface. A centripetal spread of uninjured cells from the surrounding mesothelium as well as transformation of stem cells in the base of the wound contributes to a lesser degree to the repair process.

Peritoneal repair can be complicated by the development of intra-abdominal adhesions, which are the most common cause of acute intestinal obstruction. Adhesions that develop after abdominal operations represent a vascular response by surrounding structures to the stimulus of ischemic tissue or a foreign material within the peritoneal cavity. Foreign materials that elicit adhesions are gauze, swabs or cotton wool, glove lubricants (corn starch and talc), and drains. Under controlled conditions, in every species that has been studied, the fibrous band produced between the peritoneal scar and some other structure can take origin from a suture. Conolly and Stephens (1968) reported that in rat laparotomies closed without sutures, the peritoneum healed with a decreased incidence of adhesions to the wound. Hubbard et al. also demonstrated that the only discernible effect of peritoneal closure by sutures was to increase the formation of adhesions. The incidence of adhesion formation is not reduced by the use of less reactive sutures (Holtz, 1982). Some urologist justify sutural closure of peritoneum, suggesting that dehiscence is more likely if no sutures were used in the peritoneum. Karipineni et al. (1976), Ellis and Heddle (1977), Kapur et al. (1979), and McFadden and Peacock (1983) have shown the absence of sutural closure of the peritoneum does not reduce wound strength or predispose to wound dehiscence.

The therapeutic implications of these observations on the biology of peritoneal repair are being ignored by most urologists. Careful avoidance of all foreign materials is mandatory. No attempt should be made to reconstruct a peritoneal defect. Whenever possible, serosal defects should be left open rather than being pulled together under tension. The peritoneal portion of abdominal incisions should never be approximated by sutures.

URINARY TRACT

Studies on the urinary tract have focused primarily on total bladder regeneration and its epithelium. Most urologists have thought that the bladder is an exception to the principle that only tissues, rather than compound organs, regenerate. The idea that the bladder has regenerative capacity is probably due to the fact that a uroepithelial lined hollow scar is produced after partial extirpation of the bladder. Bohne et al. (1955), Johnson et al. (1962) and Baker et al. (1959) have found that excision of 75% of the bladder, followed by stenting, resulted in the development of a bladder whose capacity may be as great as 300 ml. When the residual bladder was closed over a 5-ml Foley catheter balloon, the size of the cavity increased to 30 ml in a few weeks. Its capacity was 300 ml within three months. Continence was noted, being due to the voluntary skeletal muscle contraction in the pelvic diaphragm. Ureteral reflux may occur after resecting a major part of the bladder, even with an intact ureter and trigone, because an increase in intracavitary pressure may disrupt the valvular action of the intramural portions of the ureter.

Following complete extirpation of the bladder, no uroepithelial lined bladder will form if urinary diversion is carried out. However, preservation of urinary flow with the aid of ureteral and urethral catheters complemented by a stent will result in a reservoir whose capacity may vary from 50 to 300 ml. In such cases, the ureteral orifices will invariably gape to permit reflux and the development of some degree of hydroureter and hydronephrosis.

Urologists realize that bladder epithelium regenerates rapidly after even 80%–90% is removed. McMinn and Johnson (1955), Sanders et al. (1958), and Baker et al. (1965) all have shown that mucosal regeneration in the bladder is rapid and virtually complete within 16 weeks following near-total uroepithelial denudation. In contrast, smooth muscle within the urinary conduit system appears to be associated with a negligible level of proliferation. There is no evidence that regeneration of a muscular coat occurs or that the biologic activity of the smooth muscle of the bladder might encourage such activity. This was confirmed by Ross et al. (1966), who could not detect smooth muscle regeneration using tritiated thymidine in a variety of extirpative procedures.

Experimental studies by Hastings et al. (1975) demonstrated that dog cystotomies have a high potential for repair. These wounds showed a rapid gain in wound strength, attaining 100% of the strength of unwounded tissue in 14–21 days. The rate of collagen synthesis in bladder wounds reached a peak at five days and returned to that of unwounded normal tissue by 70 days. In contrast, colon and stomach wounds had elevated rates of collagen synthesis, even at 120 days. As with colon and stomach wounds, absorbable sutures appeared to lower the strength of both the wound and unwounded tissue. In the stomach and colon, this effect was only observed during the first 21 days of healing. The effect of the absorbable sutures on the strength of the bladder was observed throughout the 120-day observation. Examination of the cystotomy wounds showed typical scar formation without evidence of smooth muscle regeneration in the wound. Consequently, it was concluded that wound healing of the bladder wall occurred by synthesis, deposition and remodeling of collagen to form a scar. Uroepithelium usually covered the wound as well as the surface of exposed sutures by the fifth day postwounding.

Braided synthetic absorbable sutures, polyglycolic acid, and polyglactin 910 sutures appear to be ideally suited for closure of the incised wounds of the urinary conduit. They maintain their tensile strength for approximately 21 days, during which time the healing tissue is rapidly regaining its strength. These sutures are completely resorbed by 80 days, reducing the likelihood of calculus formation on the suture surface. Because the rate of degradation of chromic gut sutures is unpredictable, we do not recommend their use in the urinary tract.

BIOLOGY OF INFECTION

The development of infection is a function of the interaction of several forces: nature and degree of local contamination, local tissue features, and the general resistance of the patient to infection as well as the therapeutic measures. The urologist with modern biologic training can distinguish easily between infection and contamination. The term "infected" refers to a wound that exhibits the classic signs of inflammation (calor, rubor, tumor, dolor) as a result of bacterial contamination. When these signs are present, even the most skilled urologist will be frustrated in attempting to convert the wound into a clean wound that can be closed primarily. In contrast, contamination refers only to the viable and nonviable foreign bodies within the wound which have, as yet, failed to elicit information.

Since Lister's time, bacterial contamination has been accepted as a prerequisite for the development of infection. Even though all surgical wounds are to some extent contaminated, it appears that the degree of bacterial contamination is related directly to risk of infection in clinical surgery. The relative level of bacterial contamination can be estimated roughly by the criteria incorporated into the definitions of five classes of surgical operations, which were originally described by the Ad Hoc Committee of the Committee on Trauma of the National Academy of Sciences (1964). Refined clean wounds, which were least likely to be contaminated from endogenous sources, exhibited the lowest infection rate (3.3%). These were elective, primarily closed, undrained, and uninfected operative wounds in which neither the GI tract, bronchi, nor genitourinary tract was entered. If the above operative procedures were drained, these wounds were judged to be refined clean. The increased hazard of exogenous contamination in this group of wounds accounted for the higher infection rate (7.4%), which was more than twofold greater than that of the clean wound.

Elective operative wounds in which the bronchus, GI tract or oropharyngeal cavity were entered were judged to be clean-contaminated. The incidence of infection in these wounds was 10.8%, over three times that in clean wounds. Wounds encountering acute, nonpurulent inflammation as well as traumatic wounds were classified as contaminated. The high infection rate in these wounds (16.3%) was attributed to a higher level of bacterial contaminants. The highest rate of infection (28.6%) was encountered in traumatic wounds or those involving abscesses or perforated viscera that were con-

sidered dirty. This classification of operative procedures is intended to yield a semiquantitative estimate of the bacterial contamination of the wound at the time of operation.

The concept that the number of infecting organisms influences the incidence of clinical infection is fundamental in laboratory investigations of surgical infection. A critical number of bacteria appears to be necessary to elicit infection in soft tissue wounds. In experimental animals, the infective dose of obligate and facultative aerobic bacteria in wounds is 10^6 bacteria or greater/gm of tissue (Edlich et al., 1982). When the aerobic bacterial counts are below this level, the wounds will heal consistently without infection. This remarkable resistance to infection has been identified in all soft tissues tested (Roettinger, 1973).

The type of aerobic bacteria contaminating a wound surface plays a lesser role in the development of infection than does the number of bacteria. Wound infections develop when the number of either gram-positive or gram-negative aerobic bacteria is 10^6 or greater/gm of tissue. The critical number of anaerobic organisms that will elicit soft tissue infections has not been documented. Unless quantitative microbiologic techniques are performed under strict anaerobic conditions (rarely achieved in hospital laboratories), the presence and number of these organisms cannot be appreciated. Specific strains of bacteria have been observed to elicit disease more frequently and are regarded as more virulent. However, this virulence may be based on ecological advantages possessed by the pathogens and may not reflect differences in the host-parasite relationship.

This important clinical relationship between bacterial counts and clinical wound infection has been a stimulus for the development of quantitative tissue bacteriologic techniques (Magoc et al., 1977). These measurements are initiated by excising a $2 \times 1 \times 0.5$-cm sample of tissue which weighs 0.5 gm (Fig 4–11). Since some anaerobic bacteria may not tolerate exposure to oxygen, the tissue specimen is placed immediately in an anaerobic transport vial. All fluids, including pus and exudates, are collected also in a sterile syringe that is introduced separately into the anaerobic transport vial. The specimens are taken promptly to the bacteriology laboratory for quantitative bacterial measurements and antibiotic sensitivity testing. The specimens are processed under anaerobic conditions, preferably in an anaerobic glove box. They are weighed in a measured amount of sterile prereduced salt solution that will also be used for the serial dilutions.

The tissue specimens may be macerated with a knife, although homogenization with a sterile rotor knife blade is simpler. During homogenization, the tube containing

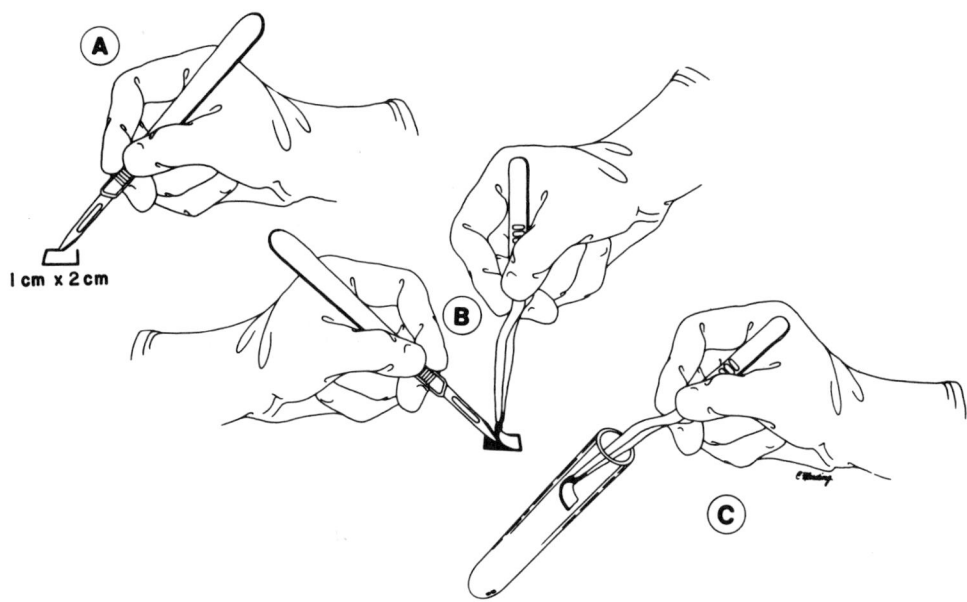

FIG 4–11.
The excisional biopsy is performed by excising a 2 × 1-cm sample of tissue **(A, B)**. Biopsy specimens for either aerobic bacterial quantitation or histologic examination are transported immediately to the laboratory in sterile glass tubes

(C). Special anaerobic containers are used for specimens that will be subjected to anaerobic bacterial isolation and culturing.

the bacterial suspension is immersed in an ice-water bath to keep the temperature low enough to prevent hyperthermic injury to the bacteria. The homogenate is then subjected to direct microscopic examination, qualitative and quantitative culturing procedures and immediate antibiotic sensitivity testing.

Direct microscopic examination of an aliquot of the bacterial suspensions provides an estimate of the total number and morphology of the viable and dead bacteria in the suspension (Fig 4–12) (Magee et al., 1977). The major advantage of this technique is the speed with which the results are available to the urologist, within

20 minutes after obtaining the specimen. These findings will suggest an appropriate antimicrobial therapy until the culture and antimicrobial susceptibility results are available, and also provide an appropriate means of quality control. For example, if morphotypes observed in the smear do not appear in the culture, the collection, culturing, or subculturing procedures may have been defective. Results of the microscopic examination should be reported immediately to the urologist.

In this measurement, a designated amount (0.01 ml) of the undiluted suspension, homogenate, and/or a serial dilution are spread uniformly over a delineated 1-cm^2 area of a glass slide, which is placed on a warmer to dry the smear. The smear is then subjected to the improved Gram stain technique developed in our laboratory (Carr-Scarborough Microbiologicals, Stone Mountain, Ga.) (Fig 4–13) (Magee et al., 1975). This technique provides a more reliable and accurate method of differentiating gram-negative from gram-positive organisms than the conventional Gram staining procedure. The major pitfall of the conventional technique is that gram-positive organisms are decolorized too easily by alcohol and judged to be gram-negative.

In the improved method, fixation of the bacteria is accomplished by the addition of methanol onto the surface of the warm slide (Magee et al., 1975). Bacteria fixed by methanol are more resistant to decolorization than are those fixed by drying or heat. When the methanol evaporates, the slide is then flooded with buffered

GRAM POSITIVE		GRAM NEGATIVE
COCCI ○	BACILLI ⬭	BACILLI ⬭
AEROBE*	AEROBE*	AEROBE* ANAEROBE
Streptococcus	Bacillus	E. coli Bacteroides
Staphylococcus		Enterobacter Fusobacterium
		Proteus
ANAEROBE	ANAEROBE·	Providencia
Peptococcus	Clostridium	Serratia
Peptostreptococcus	Actinomyces	Klebsiella
		Edwardsiella
		Citrobacter
		Pseudomonas
		Alcaligenes

* Facultative

FIG 4–12.
Morphology and Gram staining characteristics of bacteria.

GRAM STAIN TECHNIQUE

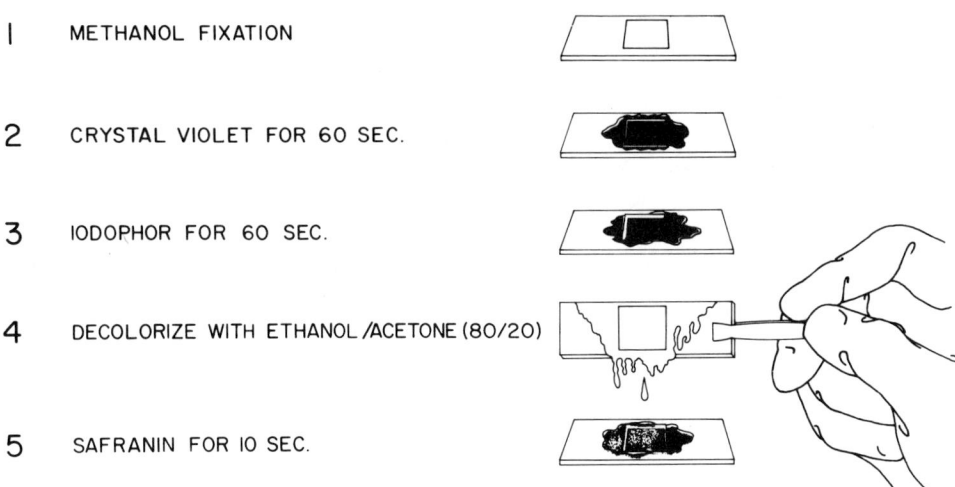

1   METHANOL FIXATION

2   CRYSTAL VIOLET FOR 60 SEC.

3   IODOPHOR FOR 60 SEC.

4   DECOLORIZE WITH ETHANOL/ACETONE (80/20)

5   SAFRANIN FOR 10 SEC.

FIG 4–13.
Improved Gram stain technique.

crystal violet, which is allowed to stand for 60 seconds. The primary stain is then poured off the slide with the iodophor mordant. The slide is then flooded with more mordant that remains on the slide for 1 minute. The aqueous mordant I_2KI employed in the conventional technique is unstable and rapidly loses its iodine content during storage. The degree to which iodine is lost is increased by elevating the room temperature and repeated exposure to the environment. As the concentration of iodine in the mordant solution is reduced, bacterial smears become susceptible to decolorization. The problems of loss of iodine from the mordant can be remedied by employing an iodophor as the mordant. This iodine complex is stable and has a long shelf life.

The slide is then decolorized uniformly with an acetone alcohol solution until the solvent flows colorlessly from the slide. Decolorization of the smear usually takes 5–10 seconds. Excess solvent is removed by rinsing the slide with water. The counterstain, safranin, is then added to the slide for 60 seconds before it is washed off with water. Each slide is allowed to dry and is examined under oil immersion using a $100\times$ objective. Ten separate fields of each smear are reviewed and the average number of bacteria per field are recorded (Fig 4–14). The average number of bacteria per field is multiplied by the number of fields in the 1 cm^2 area (4,000), giving the total number of bacteria in the 0.01-ml aliquot of the undiluted suspension. This number is multiplied by 50, if the original suspension volume is 5 ml, to give the total number of bacteria in the suspension. When the number of bacteria in the smear of the undiluted suspension is too numerous to count,

the number of bacteria per field in the first tenfold suspension dilution is checked. In these cases, the dilution factor of the homogenate is taken into account in the final calculations of the number of bacteria in the undiluted suspension. The shape, Gram staining characteristics, and number of bacteria per sample size are reported in the final results.

The rapid slide technique gives accurate and reliable measurement when 400 or more bacteria are in the 0.01 ml suspension delivered to the slide ($>2.5 \times 10^5$ organisms/gm of tissue) (Magee et al., 1977). When fewer than this number of bacteria are added to the slide, bacteria are not detected on microscopic examination. The development of this technique does not replace quantitative serial dilution and plating techniques. These latter techniques are always performed concomitantly with the rapid slide technique, since they allow speciation of the pathogen and antibiotic sensitivity testing. Aliquots of the specimen fluid tissue homogenate are incubated in air, 10% carbon dioxide (CO_2), and in the absence of oxygen to determine if the organisms are aerobic, microaerophilic, or obligate anaerobes. Incubation at 37°C for 24 hours usually will reveal the facultative organisms; however, continued incubation may be necessary for microaerophilic bacteria and obligate anaerobes. Antibiotic susceptibility of the cultured isolates then can be determined and species identification made. Quantitative bacteriology, consisting of rapid slide technique as well as serial dilution and plating, is used now routinely by surgeons in our medical center to predict the safety of wound closure—both primary (Duke et al., 1972; Marshall et al.,

RAPID BACTERIAL QUANTITATION

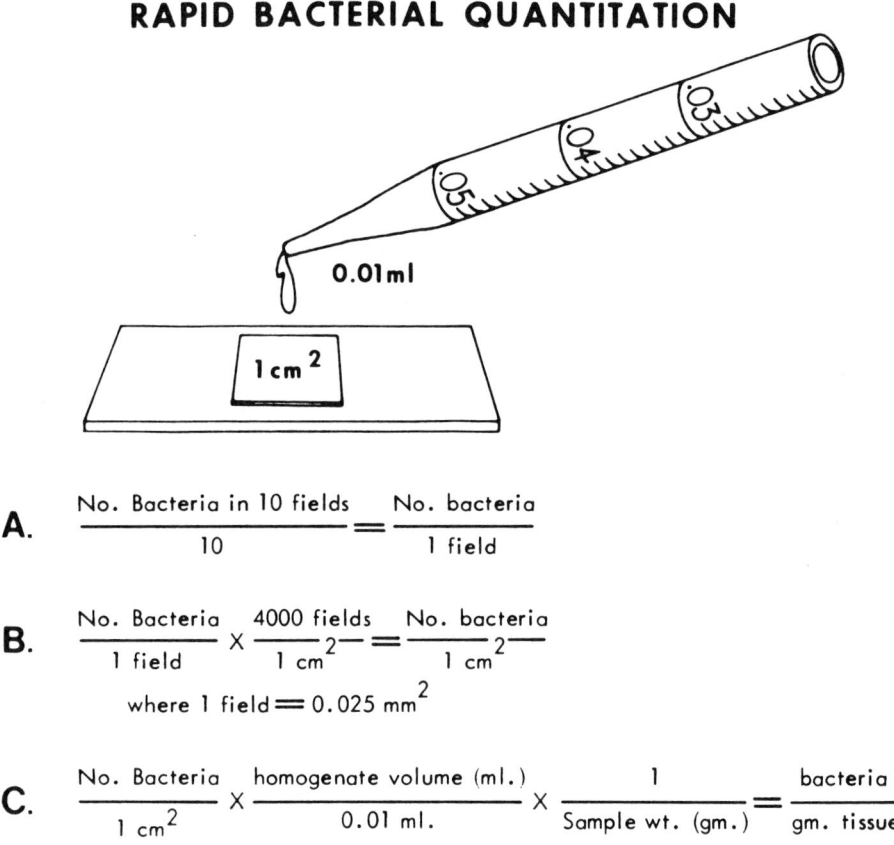

FIG 4–14.
Calculation of the results of the direct microscope count of bacteria.

1976) and delayed primary (Robson et al., 1968)—to determine graft bed receptivity and to diagnose the onset of burn wound sepsis (Edlich et al., 1977).

In our laboratory, antibiotic sensitivity testing also is performed under aerobic conditions directly on the bacterial suspension prepared from the wound specimen rather than on strains isolated from the tissue (Verklin et al., 1977). Performing the antibiotic sensitivity test on the tissue sample allows the urologist to receive the test results seven hours after receiving the specimen rather than 38–52 hours later, a delay encountered with the conventional technique. Use of this modification does not alter most of the standards recommended by the FDA, since there is no inclusive change in the medium, agar depth, or the antibiotic disc (Department of Health, Education and Welfare, 1970).

Aliquots of the bacterial suspensions prepared by the previous procedures are streaked in three directions onto the surface of Mueller-Hinton agar plates (5 × 150 mm) using sterile cotton swabs. After a three-minute delay, antibiotic disks are applied to the surface of the agar with an automatic dispenser and pressed onto the surface with sterile forceps. After incubation at 37°C for 7 and 18 hours, the zone of inhibition around each disc is measured with a ruler. The zone diameters, as recommended by the FDA, are used to interpret the susceptibility of the bacteria to the antibiotic.

In clinical and experimental studies, the changes in the test necessitated by using the clinical suspension did not alter significantly the interpretation of the antibiotic susceptibility (Verklin et al., 1977). Even when larger numbers of bacteria were present in the suspension (10^6–10^9), variation in the inoculum size did not change appreciably the results of the antibiotic susceptibility tests. Reducing the standard Kirby-Bauer antibiotic susceptibility test to seven hours also did not limit the accuracy of the test. As expected, the variable most difficult to standardize is the heterogeneous inoculum containing larger numbers (10^7) of different organisms. A zone of inhibition interpreted as sensitive with one organism was masked occasionally by the presence of the confluent growth of another organism whose zone of inhibition was considered resistant. We did not encounter the circumstances in which a number of sensitivie species gave reactions interpreted as resistant when tested in combination. However, even in these cases it is possible that the results of mixed culture sensitivities may provide the most valid information for treating mixed infections, since they most closely simulate the clinical situation. The merit of direct antibiotic sensitivity testing of clinical specimens must await further experimental and clinical studies in which the results of this proposed sensitivity test are shown to correlate with the clinical response to treatment.

An infective dose of bacteria may be derived either from an exogenous source (e.g., wounding instrument) or from the endogenous microflora of the patient. The role of commensal bacteria in the development of infection is a debated issue. Ecologic studies in humans are much needed. The neglect of this scientific subject has resulted in large gaps in our knowledge. Experimentation has focused more on the killing of inconsequential numbers of bacteria (<1,000 bacteria/gm of tissue) rather than documenting their role in infection. This obsession with the destruction of bacteria at any cost has disturbing ecologic penalties. The presence of resistant strains of bacteria as a consequence of the use of antimicrobial agents is a case in point.

In the healthy individual, bacteria are localized almost exclusively on the skin covering the surface of the body and on the mucous membranes (respiratory passages, alimentary tract, and genitourinary tract) that line all of those cavities and canals of the body that connect with the exterior (Rosebury, 1962). These membranes, composed of epithelium and underlying connective tissue, are effective barriers against bacterial invasion.

MICROFLORA OF SKIN

The skin, presenting a surface of about 2 m² in the adult, is an effective barrier to the penetration of most bacteria. Normal skin is also highly resistant to the growth of a wide variety of bacteria to which it is exposed. Bacteria applied to the skin in most anatomical areas disappear rapidly, often in minutes. This apparent "self-disinfecting" property is really a manifestation of desiccation. This same precipitous decline in bacterial count can be observed on inanimate objects, like glass. In contrast, skin hydration encourages bacterial growth (Marples, 1965). By covering the surface of the skin with an occlusive cover, a dramatic increase in bacteria is encountered. This proliferation may have important implications when wound epidermization is incomplete. *During this time, heavily contaminated skin under occlusive tape or drapes may be a potential source for infection.*

Through the interplay of the moisture content on the skin surface as well as of more obscure factors, certain bacterial species are successful in colonizing the skin surface, while others are excluded rapidly from permanent residence. The outcome of the different factors thought to be involved in the local resistance of the skin is a normal resident skin microflora that varies in concentration and type in different anatomic areas.

In general, the composition of the skin microflora allows the body to be subdivided into three anatomical areas (Table 4–1) (Kligman, 1965). Over most of the body surface, trunk, upper arms and legs, the density of the bacterial population is quite low, not millions per cm², but more often a few thousand or fewer. The moist areas of the body, such as the axillae, perineum, toe webs, and intertriginous areas, harbor millions of bacteria. The exposed anatomical areas of the body compose the third anatomical region and display a bacterial density numbering in the millions. No error is more frequent than the habit of generalizing about the microflora in this anatomical region. The dissimilarities in the density and diversity of organisms which inhibit the unique province of this region are truly profound. Normally, the organisms are quite sparse on the palms and dorsa of the hands, numbering in the hundreds per cm². The majority of organisms (10,000–100,000) on the hands reside beneath the distal end of the nail plate or adjacent to the proximal or lateral nail folds. These recesses frustrate our efforts to disinfect the hands of the surgical team, which are the leading vector of pathogenic bacteria in both the operating surgeon and the surgical nurse. The scalp and forehead skin also exhibit luxurious bacterial growth numbering in the millions.

The predominant organisms encountered on the human skin are staphylococci and diphtheroids. These aerobic organisms greatly outnumber strict anaerobes (10:1), except in the sebum-rich area where *Corynebacterium acnes* is present. *Staphylococcus epidermidis* (albus) accounts for the vast majority of staphylococci. *Staphylococcus aureus* is not a predominant member of the skin microflora, being encountered in only 5%–20% of the individuals. Gram-negative bacteria are

TABLE 4–1.

Composition and Concentration of Skin Microflora

	TOTAL BACTERIAL CONCENTRATIONS*	RATIO ANAEROBE-AEROBE
Skin		
Moister areas	10⁴–10⁶	1:10
Axillae, perineum		
Drier areas	10¹–10³	1:5–10
Trunk, upper arms and legs		
Exposed areas	10⁴–10⁶	1:5–10
Head, face, feet†		

*Per cm² of tissue.
†Anaerobes may outnumber aerobes in the skin of the cheeks, upper back, and presternum.

found uncommonly in most regions of the body, with the exception of the moister areas. This notable absence of gram-negative organisms in the drier regions can be changed dramatically by moisture (Marples, 1965). When these drier regions are covered by an occlusive dressing, the gram-negative population increases and may account for 10% of the total bacterial population. These gram-negative species, which appear under an occlusive cover, are the same organisms that have been implicated increasingly in nosocomial infections, such as infected venous cutdown sites.

In most anatomical regions, bacterial colonization is limited to the horny layer of skin, which is composed of a sloughing mass of dead cells, full of cracks that harbor bacteria. Beneath this horny layer, the stratum corneum, composed of tightly packed cells, provides an effective barrier against bacterial invasion. The horny layer of pilosebaceous appendages that line the infundibulum of hair follicles forms a receptacle for bacteria. However, these bacteria rarely descend any further than the entrance of the sebaceous ducts. Similarly, the depths of apocrine glands and sweat glands are devoid of bacteria. The organisms that reside less than 250 μ beneath the surface are within reach of topically applied antiseptic agents. *As a result of topical antisepsis, sterility or near-sterility can be achieved in most skin areas of the body* (Pecora et al., 1968).

Residence of bacteria in the depths of epidermal appendages has been identified in unique follicles found in areas which are susceptible to acne—the cheeks, presternum, and upper back. This follicle has been called a sebaceous follicle, since its piliary apparatus is a trivial rudimentary structure. Large numbers of gram-positive organisms, namely *Corynebacterium acnes*, occupy its lumen and are not susceptible to disinfection by topical treatment.

MICROFLORA OF MUCOUS MEMBRANES

Anaerobes are the predominant organism colonizing the surface of most mucous membranes (Table 4–2). The vast majority of investigations of the endogenous microflora of mucous membranes have provided qualitative rather than quantitative data. The relatively few quantitative studies required varied and complex media, extensive identification procedures, special environmental conditions, and fastidious technique. The diverse microflora of some regions, 100–300 bacterial strains, defies complete speciation, with some bacterial strains never being identified (see Table 4–2).

Respiratory System

Relatively few organisms are recovered from the membranes lining the respiratory passages. The principal habitat of bacteria is the nasal vestibule, the slightly

TABLE 4–2.

The Normal Microflora on Mucous Membranes

	TOTAL BACTERIAL CONCENTRATIONS	RATIO ANAEROBE:AEROBE
Respiratory system		
Nasal cavity	10^2	1:1
Paranasal sinuses	0	0
Trachea	0	0
Bronchi	0	0
Digestive system		
Saliva	10^6*	3–5:1
Tooth surface	10^{11}	1:1
Gingival crevice	10^{11}	100–1,000:1
Stomach	0–10^5	1:1
Proximal small bowel	10^2–10^4	1:1
Ileum	10^5–10^8	1:1
Colon	10^9–10^{11}	1,000–10,000:1
Urogenital system		
Male urethra	10^2–10^3*	Aerobes
Female urethra	10^2–10^3	Aerobes
Endocervix	10^8–10^9	5–10:1
Vagina	10^8–10^9	5–10:1

*Per gram of exudate.

expanded portion of the nasal cavity beneath the alae (Watson et al., 1962). The mucous membrane portion of the nose is often devoid of bacteria as shown by swab cultures of the nasal mucosa which are frequently sterile. The paranasal sinuses that communicate with the nasal passages are also usually sterile (Bjorkwall, 1950).

The trachea and bronchi in a healthy person are contaminated by few bacteria ($<100/cm^2$). The anatomical relationship of the nasopharynx to the lower airway and the contamination of air which is breathed preclude sterility. However, the self-cleaning action of the respiratory system maintains the tracheobronchial tree relatively free of bacteria. The wafting motion of the cilia in the mucous membrane of the respiratory system moves particulate matter unidirectionally into the nose and pharynx. A local antibody and, in the lower respiratory system, the alveolar macrophages are additional defense mechanisms.

Digestive Systems

Throughout the human digestive system, the types and numbers of bacteria vary considerably (see Table 4–2). The oral cavity serves as a microbial incubator which supports the growth of at least 200 species of facultative organisms and obligate anaerobes (Nolte, 1973). Concentrations of bacteria within the oral cavity vary widely in different anatomical sites. The pharynx, tongue, buccal mucosa, tooth, and gingival crevice have a distinct microflora. The major determinants of these unique microflora are attributed to the different environmental conditions (oxidation-reduction potential

[EH]) as well as to the propensity of bacteria to attach to different cell types.

The largest numbers of organisms are encountered in the gingival crevices and in plaque on the teeth (Nolte, 1973). The debris removed from the crevices and the plaque on the teeth are composed primarily of bacteria in the range of 10^{11}/gm net weight. The composite of microorganisms inhabiting the gingival pocket is different from that found in plaque. Plaque contains many facultative streptococci, neisseria, and lactobacilli whose numbers are approximately equal to that of the obligate anaerobic species. In the gingival crevices, the obligate anaerobes outnumber aerobes by a factor of 100–1,000:1. Gingival material is composed of large numbers of anaerobic streptococci, veillonellae, fusobacteria, spirochetes, and *Bacteroides melaninogenicus*. *Bacteroides fragilis* has rarely been identified in the oral cavity, being found predominantly in the lower part (colon) of the digestive system. The salivary microbial population represents those that have been dislodged from all oral surfaces consequent to the rinsing of these surfaces by saliva (10^6 bacteria/ml).

Several factors tend to alter the number and types of bacteria in the oral cavity; oral hygiene is one of the most important factors. Good oral hygiene may be deterred by a combination of regular and proper use of a toothbrush, periodic removal of calculus and dental plaque by the dentist, and such orthodontic measures as the individual's dentition requires. Under such conditions, the total number of microorganisms decreases and is composed predominantly of microorganisms tolerant of oxygen. Neglect of oral hygiene results in an increase of the total microbial flora with an anaerobic and putrefactive character. This increase is due possibly to an accumulation of food and debris in the gingival sulci and to an increased plaque formation.

The density of the oral microflora changes temporally during the day. One responsible factor is the flow of saliva, which is greater under the stimuli of the waking hours than during the sleeping hours. In contrast, aptyalism, or suppression of salivary secretion, results in an increase in the total microbial population, probably because of the undue accumulation of food and debris and the loss of the mediating factors inherent in the saliva. The complete loss of teeth will result in a decrease in the total number of oral bacteria compared with normal dentition or complete dentures. Anaerobic bacteria are, however, regularly present even in the edentulous person. The oral microflora also is influenced considerably by antibiotic therapy. These drugs usually suppress the normal flora with increased rates (50%) of colonization of coliforms. Normally, coliforms such as *Escherichia coli* are found in the oral cavity of only 3% of healthy persons.

Most of the oral microflora ingested are destroyed primarily by gastric acid and possibly other factors associated with the mucosa (Drager et al., 1968; Nichols and Smith, 1975; Nichols and Condon, 1975). Consequently, the stomach is sterile in most fasting individuals (Drager et al., 1969). In normal subjects, the upper small intestine is also virtually free of bacteria, except after a meal. Bile acids may also contribute to this low bacterial density, since they are toxic to a number of oropharyngeal bacteria. Following meals, bacteria number 10^4 or fewer per ml and usually colonize the upper bowel (duodenum, jejunum, or upper ileum). This sparse microflora consists mainly of gram-positive organisms that are acid-resistant, i.e., streptococci, aerobic lactobacilli, and fungi (Gorbach, 1971). This same microflora has been identified in the distal portion of the ileum in one third of normal individuals. However, the majority of healthy adults exhibit a striking change in the microflora in ths area with the appearance of gram-negative organisms, such as aerobic coliform and anaerobic bacteroides. The concentration of bacteria in the distal portion of the ileum in these people ranges from 10^5 to 10^8/ml.

The list of diseases associated with heavy contamination of the stomach and small bowel is increasing. In patients with primary gastric achlorhydria or induced hypochlorhydria secondary to operative therapy, bacterial colonization of the stomach and small bowel with approximately 10^7 organisms is recorded. (Gorbach, 1971). Intragastric colonization also is encountered routinely in patients with bleeding or obstructing ulcers (Ellis, 1971). Any interruption of the propulsive activity of the small bowel provides a fertile field for the growth of bacteria. This increased density of bacteria is associated with a shift toward a colonic-type flora with high concentrations of anaerobes. Examples of such conditions include diverticula, strictures, excessively long afferent loop or denervated segment of bowel, as well as small bowel obstruction (Gorbach, 1971).

Across the ileocecal valve, the concentration of bacteria and the numerical proportion of anaerobic bacteria to facultative species increase. The total microbial concentration of the contents of the intra-abdominal colon approaches 10^9 bacteria/gm net weight, which increases to 10^{11} bacteria/gm net weight in the passed stool, with 20%–30% of the fecal weight being a solid mass of bacteria. The colonic microflora of healthy individuals is relatively stable, so that only minimal changes are detected after periodic sampling, even after major alterations in diet (Gorbach, 1971). According to current estimates, the colon harbors more than 200–400 distinctive bacterial species. Obligate anaerobes outnumber facultative species such as coliforms by 1,000–10,000:1 (Nichols and Condon, 1975). The anaerobic organisms, bacteroides, clostridia, and lactobacilli, become the major constituents of the luxuriant microbial flora. The major types of coliforms, especially *E. coli*, streptococci, enterococci and *Bacillus* sp., compose the considerably smaller number of aerobes. In clinical intra-abdominal infections following colonic perforation, only a fraction of the normal colonic microflora (average of 5 microbes) is recovered from the site of infection. The factors that account for the survival of these organisms from the infected sites is not understood completely.

Mechanical cleansing of the colon removes gross stool and facilitates the surgical procedure. After only mechanical cleansing, however, the residual contents show the same concentration of both aerobic and anaerobic microorganisms as found in stool. Nichols and associates (1973) demonstrated that the addition of neomycin and erythromycin base to the mechanical cleansing produced suppression of fecal bacteria. This reduction in bacterial count was associated with a reduced wound infection rate in patients undergoing colonic surgery.

Urogenital System

The urogenital system in the normal individual is devoid of bacteria, except in the vagina and the urethra of both sexes. Diphtheroids, *Staphylococcus albus*, and streptococci (including enterococci) compose the bulk of the urethral flora of the normal adult male and are encountered in small numbers, usually 1,000 bacteria/ml of urine (Stamey, 1974). Gram-negative enteric bacteria are rare. Thousands of gram-negative bacteria can reside under the uncircumcised foreskin, even though it is easily retractable and visibly clean. Stamey (1974) indicated that these foreskin bacteria can be a major source of confusion in localizing the site of infection in the lower urinary tract.

The endogenous flora of the female genital tract is restricted to the vagina and proximal endocervical canal. The lining of the uterine cavity and fallopian tubes are usually sterile in normal individuals. When compared to the flora of the GI tract, the vaginal flora is relatively simple, with only 5–8 bacteria species in each cervical and vaginal specimen. The microflora of the female genital tract displays considerable instability. Periodic sampling of the microflora demonstrates wide variation in the bacterial species, with 25–35 different bacteria being recovered from multiple specimens. Using quantitative techniques, Bartlett et al. (1977) reported mean bacterial concentrations of $10^{8.1}$/gm for aerobic bacteria and $10^{9.1}$/gm for anaerobic bacteria. These data indicate that anaerobes outnumber aerobes by a factor of 5–10:1.

These high concentrations of bacteria persist in healthy women until menopause, at which time a sparse, varied flora and alkaline vaginal secretion becomes apparent. This luxuriant microflora of the female genital tract of women of reproductive age can tolerate the low pH of vaginal secretions, which is controlled hormonally. The predominant aerobic bacteria are *Hemophilus vaginalis*, *Staphylococcus epidermidis*, streptococci, coliforms, and lactobacilli (Doderlein's). These same aerobic organisms found in the vagina are observed in the urethra, but the colony counts are considerably lower ($10^2–10^3$). Lactobacilli are able to ferment glycogen with the production of lactic acid. This production of lactic acid contributes to the low pH of the vaginal secretions, but is not responsible primarily for it, since there is no correlation between the recovery of lactobacilli and the vaginal pH. The predominant obligate anaerobic bacteria are peptococci, peptostreptococci, anaerobic lactobacilli, eubacteria and *Bacteroides* sp. Interestingly, *B. fragilis*, which is seldom found in the normal vaginal flora, is involved frequently in infections of the upper female genital tract.

Preferably, the urologist must focus attention on anatomical regions of the normal healthy human body containing concentrations of organisms sufficient to elicit soft tissue infection (Edlich, 1982). When planning for surgical procedures in these sites, preoperative suppression of the microbiota is essential. Strict adherence to aseptic technique is mandatory to minimize the spread of endogenous contaminants. In the remaining regions, the microbiota of the healthy individual is usually sparse. However, damage to the host's defenses can change these bacteriologic deserts into a teeming jungle of pathogens.

RESISTANCE TO INFECTION

The quantitative relationship between the host's resistance to infection and the number of bacteria can be upset by systemic and local factors. Among the systemic factors, age appears to be one of the most important. In general, clinical reports reveal an increased infection rate with advancing age. Metabolic and nutritional conditions have also been incriminated as factors that significantly influence the host's resistance. Infection rates are reported to be higher in patients with diabetes mellitus, obesity, malnutrition, or those receiving steroid therapy. Shock, remote trauma, or distant infection have also been demonstrated to enhance wound infection rates (Miles and Niven; 1050; Conolly et al., 1970; Dineen, 1961). In addition, rare immunologic deficiencies, both primary and secondary, contribute to a diminished host's resistance (Alexander and Good, 1970).

Local factors can influence this interaction between the host and bacteria. Whether present by accident or intention, any foreign body in the wound damages the local tissue's defenses and invites infection. The magnitude of this damage appears to be related to the chemical reactivity of the foreign body. In traumatic injuries, soil and dirt are frequent contaminants. Although it has been widely recognized for centuries that severe bacterial infection often develops in wounds containing dirt and soil, there was little knowledge until recently of the role of the components of soil in this infection process.

Interdisciplinary research in this area has clarified the role of soil in the development of infection (Rodeheaver et al., 1974). Specific infection-potentiating fractions (IPF) have been identified in soil which include its organic components as well as inorganic clay fractions. For wounds contaminated by these fractions, only 100 bacteria/gm of tissue are necessary to elicit infection. Their ability to enhance the incidence of infection appears to be related to their damage to host defenses (Haury, 1977). In the presence of these IPF, leukocytes are not able to ingest and kill bacteria. This deleterious effect on WBC function is a result of a direct interaction between the highly charged soil particles and WBCs. Soil IPF also have considerable influence on nonspecific humoral factors. Exposure of fresh serum to these fractions eliminates its bactericidal activity. As expected, these particles, which are highly charged anions, also react chemically with amphoteric and basic antibiotics, limiting their activity in contaminated wounds (Roberts et al., 1979).

The concentration of these fractions in soil can be correlated with their location. Environmental conditions in swamps, bogs, and marshes encourage the production of soil with as much as 98% organic IPF. The major inorganic infection-potentiating particles are the clay fractions which reside in heaviest concentration in the subsoil rather than in topsoil. Consequently, traumatic soft tissue injuries occuring in swamps or excavations run a high risk of being contaminated by these fractions, which predispose the wound to serious infection.

A corollary to these observations is that some soil contaminants, such as sand grains, are relatively innocuous. This fraction, which has a large particle size and a low level of chemical reactivity, exerts considerably less damage to tissue defenses than do the infection-potentiating fractions. Surprisingly, the black dirt on the surface of highways also appears to have minimal chemical reactivity.

TECHNICAL FACTORS

When caring for a wound, the urologist must make judgments that frequently tip the balance in favor of either infection or healing per primum. A large number of clinical and experimental studies have provided evidence of the influence of various surgical decisions on the ultimate fate of the wound. Based on these findings, we make some specific recommendations regarding the management of the surgical wound.

LOCAL ANESTHESIA

Cleansing of bacteria, soil and other debris from traumatic injuries and debridement cannot be accomplished without anesthesia. The ideal local anesthetic agent should have rapid onset of local action and should not impair the wound's ability to resist infection. The effect of the local anesthetic agent on the viability of microorganisms is another important consideration. For infected wounds, an agent that displays antimicrobial activity may kill the pathogen and interfere with its identification.

Lidocaine hydrochloride does not exhibit antimicrobial activity or damage the local wound defenses. Its clinical usefulness can sometimes be enhanced by adding the vasoconstrictor epinephrine, which slows the clearance of lidocaine, thereby prolonging the duration of anesthesia. Epinephrine will damage the local wound defenses. The infection-potentiating effect of this powerful local vasoconstrictor is proportional to its concentration and results from its vasoactivity (Stevenson et al., 1975). As a result of its vasoconstrictive action, epinephrine may result in hypoxic conditions that limit white blood cell function. In vitro studies by Hohn et al. (1976) and Mandell (1974) demonstrated that hypoxia retards killing of *Staphylococcus aureus* by wound leukocytes. Consequently, this increased infection rate of epinephrine-treated wounds may result from impaired killing of bacterial contaminants by wound leukocytes in the ischemic wounds. This damage to tissue defenses argues against the use of epinephrine in potentially heavily contaminated wounds.

HAIR REMOVAL

The practice of hair removal may influence the likelihood of surgical infection. Since the days of Lister and Semmelweis, skin hair has been regarded as a source of contamination, and recent work supports this belief (Dineen and Drusin, 1973). Removal of hair also facilitates operative surgery by preventing hair from becoming entangled in sutures and the wound during closure. However, the practice of razor preparation of the operative site had been challenged in recent studies.

In a prospective clinical study, Seropian and Reynolds (1971) reported that the infection rate of surgical patients after razor preparation was 5.6% compared with 0.6% after a depilatory. These findings agree with those reported by Cruse and Foord (1973). In a five-year prospective study of 23,649 surgical wounds, the infection rate was 2.3% in the patients shaved. The incidence of infection fell to 0.9% in the patients who were not shaved or clipped.

In a recent study, Alexander et al. (1983) demonstrated that the technique and timing of hair removal has considerable influence on the incidence of postoperative infection. In their prospective study, Alexander et al. (1983) compared the wound infection rate in patients subjected to hair removal by either an electric clipper or razor. Patients were prospectively randomized to be either shaved or clipped the night before or the morning of the operation. Clipping the hair the morning of the operation was associated with significantly fewer infections than were encountered in patients whose hair was removed either by electric clippers the night before the operation or by shaving. These investigators concluded that preoperative shaving was deleterious and that the practice should be abandoned. Moreover, removal of hair with electric clippers immediately before surgery was recommended. The increased incidence of infection after razor shaves is probably related to the trauma inflicted by the razor. A safety razor consists of a blade held in a fixed geometry by the head (Fig 4–15). The exposure of the blade with respect to the razor head is the most important determinant of the blade's performance. The exposure of the surgical preparatory razor blade is so great that the infundibulum of the hair follicle is transected, so that the wounded hair follicles provide access and substrate for bacteria. In addition, the impermeable corneal layer is damaged, and the exudate provides a moist field for bacterial proliferation. Inoculation of shaved skin results in dermatitis. In contrast, skin shaved with a recessed blade is refractory to bacterial contamination.

Hair removal by depilatories has been reported to be an effective method that does not enhance the risk of infection (Seropian and Reynolds, 1971). This potential benefit of depilatories must be weighed against some shortcomings. A depilatory should not be used near a patient's eyes or genitalia, since it can cause serious irritation to these tissues. Inadvertent spillage of the depilatory into the wound must be avoided, since it can give rise to a prolonged inflammatory response and delayed healing (Almersjö et al., 1967). In rare cases (1.3%), the depilatory causes severe dermatitis that necessitates postponement of surgery (Seropian and Reynolds, 1971).

Recent advances in the design of surgical clippers have made them especially suitable for preoperative hair removal. The clipper blade assembly can be easily removed from its mounting assembly on the electric motor so that it can be cleaned and sterilized. Steriliza-

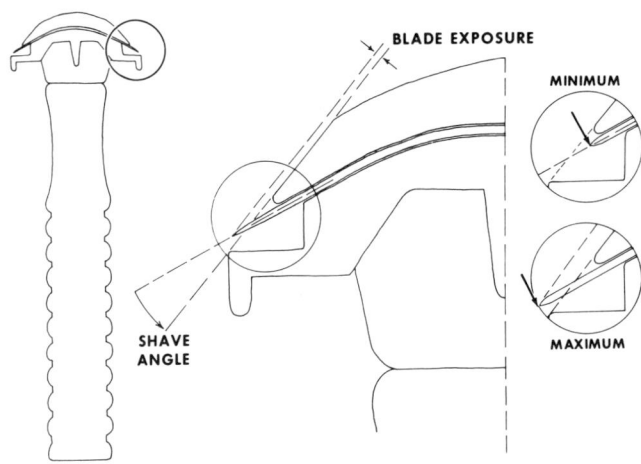

FIG 4–15.
Geometry for surgical preparation razors. If the blade exposure is minimal, the blade will cut the hairs above their infundibula. The infundibulum of the hair follicle is transected when the blade is exposed maximally beyond the razor head.

tion of the clipper blade assemblies for electric clippers eliminates its high level of bacterial contamination encountered after repeated use, a potential source for infection (Masterson et al., 1984). The fine teeth of its 0000 blade assembly cut hair close to the skin surface without nicking the skin. Since this atraumatic removal of hair by electric clippers is associated with a lower incidence of infection, we now employ only electric clippers to remove hair before surgery and have abandoned the use of preoperative shaving.

ANTISEPSIS

The term "antisepsis" refers to the use of antimicrobial chemicals on human tissue, while "disinfection" applies to the use of these agents on inanimate objects. In addition to an antiseptic agent, surgical scrub solutions contain a detergent or surface active agent, which facilitates removal of the surface contaminants.

The clinical efficacy of an antiseptic agent can be evaluated by several parameters including: (1) the effect of storage, (2) its spectrum of activity, (3) its duration of antimicrobial activity, (4) the degree of local and systemic damage to the host, and (5) the influence of the carrier on the performance of the agent. First, if it is rapidly inactivated during storage, the antiseptic agent must be freshly prepared before each use. Second, the antiseptic agent should exhibit antimicrobial activity against a broad spectrum of organisms. If active against gram-positive organisms alone, topical treatment of contaminated tissue may result in a potentially harmful shift of the normal flora. The widespread use of antimicrobials that act only against gram-positive organisms is associated with a tremendous increase in infections caused by gram-negative organisms in hospitals. Third, the antimicrobial activity should be fast-acting and substantive. A substantive effect is due to the retention of the agent by binding to a tissue (e.g., stratum corneum) after rinsing. The bound antiseptic agent limits the proliferation of the residual bacteria. Fourth, the degree to which the agent damages the host, both locally and systemically, must also be known. Finally, the influence of the vehicle or carrier (i.e., surfactant or detergent) on the performance of the antiseptic agent is of great importance to the urologist. The inactivation of cationic surface active agents by anionic surface agents is a case in point.

Not all antiseptic agents are used for the same purpose, nor should the requirements for their effectiveness be identical. Specific definitions for the antimicrobial product categories were established in a report by the advisory review panel on over-the-counter antimicrobial drug products for repeated daily use (FDA, 1968). This report provides detailed information regarding patient preoperative skin preparation, hand washing, and surgical skin wound cleansers.

SKIN WOUND CLEANSERS

Skin wound cleansers are designed to aid in the removal of bacterial and other contaminants from superficial wounds, and may or may not contain an antimicrobial agent. While cleansing the wound, the agent must not damage the wound or systemic defenses or deter healing.

Dilute solutions (1:750) of quaternary ammonium salts (quats) satisfy many of these requirements. These compounds are cationic surface-active agents, which are basically organically substituted ammonium compounds. Although they affect cell membrane potential, the consequences are clinically apparent only at concentrations higher than those recommended for clinical use.

Gram-positive microorganisms are generally more susceptible to quats than gram-negative bacteria. The gram-negative *Pseudomonas* sp. are usually resistant and even proliferate in the stored solution, accounting for occasional serious nosocomial outbreaks of gram-negative infection (Dixon et al., 1976).

In contrast, the commercially available surgical scrub solutions containing iodophors and hexachlorophene are not safe for use in surgical wounds. These solutions contain toxic anionic detergents that damage the tissue defenses and potentiate the development of infection (Custer et al., 1971). Until this observation was made, detergents and surfactants in surgical scrub solutions were considered by many to be innocuous ingredients.

Pluronic polyol F-68 (Calgon Corp., St. Louis), a nonionic surfactant, is an excellent substitute for toxic detergents. It belongs to a family of surfactants made of a series of block copolymers with a water-soluble polyoxyethylene group at both ends of a water-insoluble polyoxypropylene chain (Edlich et al., 1973). It has a molecular weight of 8,350 and contains 80% ethylene oxide. Long-term toxicity studies on animals and humans reveal it to be safe for IV use. Concentrated solutions containing as much as 40% Pluronic polyol F-68, when applied topically to the wound, do not damage resistance to infection. This surfactant also has negligible effect on bacterial viability.

Pluronic polyol F-68 has been approved by the FDA (1978) as a skin wound cleanser. Unlike all other commercial scrub solutions, this surfactant is so innocuous that it "does not bring tears to a baby's eye." Patients with painful partial-thickness burns or abrasions washed with it find it soothing. In our emergency medical service, this agent has completely replaced all other surgical scrub solutions for use in traumatic wounds.

Topical iodophors, even without addition of a surfactant, seem to be poor choices as skin wound cleansers. Application over a large surface area often results in a substantial elevation of serum free-iodide (Lavelle et al., 1975). Unexplained abnormalities have occurred in some patients, including renal failure, metabolic acidosis, and elevation of SGOT. It is conceivable that large iodide loads were, at least in part, responsible for these abnormalities. Iodophors and many other antiseptics depend on an oxidative reaction between the antiseptic and the organism. But iodine reacts as easily with mammalian as bacterial cells, and within seconds after instillation of a lethal dose of iodophor into the sterile peritoneal cavity, the titratable iodine is undetectable, having all reacted with the normal tissues (Ahrenholz and Simmons, 1979). In the heavily contaminated peritoneal cavity, bacterial counts are rapidly reduced by iodophors, but not all bacteria are killed (Bolton et al., 1979). The iodine can react with normal mesothelium and cause a chemical peritonitis, which is fatal in experimental animals even in the absence of bacteria. Repeated treatment of an open wound with this iodophor may delay healing.

Povidone-iodine (PVP-I) is cytocidal to human neutrophils at commercially available concentrations (1% titratable iodine). Even dilutions of 1:100 result in irreversible loss of a neutrophil's ability to respond to chemotactic stimuli. Topical antibiotics have a broad spectrum of activity without toxicity to mammalial tissue and continue to kill bacteria for prolonged periods. They appear to be safer agents for wound irrigation than most antiseptics.

Unaware of these potential toxic manifestations of PVP-I complexes, many surgeons have continued to apply these agents to contaminated wounds with surprisingly good results. Gilmore and Martin (1974) have made well-controlled studies of the efficacy of PVP-I powders in the treatment of contaminated wounds. In a series of 451 consecutive cases of appendectomy, topical PVP-I treatment reduced the wound infection rate and was found to be superior to an antiseptic spray. These beneficial effects of PVP-I in the treatment of contaminated wounds were later reported by Stokes and associates (1977) in a prospective randomized study of abdominal surgical patients. When topical prophylactic PVP-I was compared with topical cephaloridine in another series, however, antibiotic-treated wounds escaped infection far more often than those treated with the antiseptic agent (Pollock and Evans, 1975).

HAND WASHING

Hands are the leading vector of pathogenic bacteria to the wound. Hands can transmit an infective dose of bacteria during any patient care activity (shaving, marking, lifting) conducted before and after the operation or through punctured gloves during the procedure. Glove punctures are found after 5%–60% of all operations (Walter and Kundsin, 1969).

Normally, the organisms are quite sparse on the palms and dorsa of the hands, numbering in the hundreds per cm². The majority of organisms (10,000–100,000) on the hands reside beneath the distal end of the nail plate or adjacent to the proximal or lateral nail folds. These recesses frustrate efforts to disinfect the hands.

Before and after patient contact, hand washing is mandatory, even though the risk of acquiring an infective dose of bacteria from patients during routine physical examination seems to be negligible. Washing one's hands for 15 seconds with any of a variety of agents, including tap water, removes large numbers of bacteria and significantly reduces the bacterial count (Sprunt et al., 1973). The hands should be rinsed and then dried. Because rings and chipped nail polish make removal of organisms more difficult, operating room personnel should wear neither (A Report to the Medical Research Council, 1968).

Repeated use of an antiseptic agent before and after patient contact appears to be unwarranted and may occasionally lead to the development of dermatitis (Steere and Mallison, 1975). Ordinary bar soap appears to be an excellent alternative to antiseptic agents for routine hand washings, as long as the bars are small and kept on drainage racks between uses.

We favor antiseptic agents for special circumstances in which invasive procedures will be performed (Steere and Mallison, 1975). Hand washing with antiseptic agents should last at least 15 seconds, while surgical procedures require the greatest degree of hand antisepsis.

A variety of agents can be employed. Semmelweis in 1861 (Slaughter, 1950) decreased the incidence of puerperal fever by ordering the medical students to wash their hands in chlorinated lime. Lister degermed his hands in a 1:20 solution of carbolic acid. Until the 1950s, tincture of green soap followed by an alcohol rinse was used. Currently, three antiseptic agents have been advocated for hand washing before operation as well as preoperative patient skin preparation—hexachlorophene, iodophor, and chlorhexidene. Surgical scrubs of hands and forearms of adults, five times a day, with a 3% hexachlorophene preparation leaves blood levels of 0.5 mg/ml or higher in selected individuals after 10 days (US General Services Administration, 1974). Because these levels are potentially toxic, hexachlorophene should not be used routinely by operating room personnel, and should be dispensed only by prescription. Fortunately, both other solutions appear to be safe (Smylie et al., 1973; Lowbury and Lilly, 1973). Following hand washing with iodophor or chlorhexidene solutions, the bacterial counts of the skin are suppressed, and the build-up of microbial counts under the occlusive surgical glove is limited. Neither agent is superior.

The surgical hand wash should begin by cleaning the fingernails with a plastic stick. The hands should then be scrubbed with a sponge or brush for 5 minutes with a chlorhexidene or iodophor solution. The sponge causes less irritation to the skin (Bornside et al., 1968; Michaud et al., 1972).

Dermatitis from any cause, including washing, results in a high concentration of skin bacteria (Walter, 1965). Afflicted personnel who are members of the operating team should be limited to patient care not directly related to the operative procedure. During patient contact, personnel with dermatitis should wear gloves. The best treatment of this condition is to refrain from using an antiseptic agent or wearing gloves for longer than 5 minutes. Emollients added to hand creams should decrease or prevent dermatitis and dry skin, but there have been reports of contaminated hand creams associated with nosocomial infections (Steere and Mallison, 1975). Consistent use of emollients on the hands after hospital duty rather than during the work day might be efficacious in preventing dermatitis while reducing the risk of these infections.

The choice of the surgical glove for use in surgery is another important consideration. Commonly employed surgical gloves are coated with corn starch or talc powdered granules, which can cause granulomatous reactions in patients (Sheikh et al., 1984). Consequently, the FDA requires every glove manufacturer to place a warning label on the package to "remove powder by wiping the gloves thoroughly with a sterile wet sponge, sterile wet towel, or other effective methods." Unfortunately, these methods of powder removal are ineffective and result in larger clumping of the powder, which could predispose to a significant granulomatous reaction.

A new method of injection molding of gloves (Cooper Vision) has been developed that results in a very smooth surface, free of powder and other foreign bodies (Villasenor et al., 1984). The ease of donning and tactile sensitivity are comparable to that of standard powdered surgical gloves.

PATIENT PREOPERATIVE SKIN PREPARATION

Immediately before the operative procedure, the site should be thoroughly cleansed with a sterile sponge soaked in solution containing 20% Pluronic F-68 to remove superficial flora and debris, which could interfere with antisepsis. Then application of an antiseptic solution is necessary to kill or inhibit more adherent, deep, resident flora. Any of the agents used for antiseptic hand washing, except hexachlorophene, can be used to prepare the operative site, but tincture of chlorhexidene or iodophors offer advantages over the others (*Medical Letter*, 1976). Both have an excellent level of activity against bacteria. Tincture of chlorhexidene has a rapid onset of activity (due to alcohol) and persistent antibacterial activity. Iodophors also act rapidly and have persistent antibacterial activity if they are not wiped off. Tincture of iodine is also an excellent surgical prepping agent but can cause skin burns. Alcohols act rapidly, but have no persistent effect after they evaporate.

The FDA has continued to receive reports of patients suffering thermal burns during electrosurgery after prepping the skin with antimicrobial preparation containing 0.5% chlorhexidene gluconate and 70% isopropyl alcohol (FDA, 1985). Consequently, the skin must be completely dry without pooling of this antimicrobial agent around the patient's skin before electrosurgery is started. Pooling of an iodophor solution should also be avoided, since it can result in a chemical burn.

Hexachlorophene has been used as a preoperative skin prep, but it has poor activity on gram-negative organisms and requires repeated use for maximum effectiveness. Hexachlorophene can be absorbed from the skin and large amounts have toxic effects on the CNS, particularly in premature infants.

SURGICAL DEBRIDEMENT

Debridement is probably the most important single factor in the management of the contaminated wound (Jones and Shires, 1974). First, it removes tissue heavily contaminated by soil infection–potentiating fractions and bacteria, protecting the patient from invasive infection. Second, it removes permanently devitalized soft tissues that, if left in a wound, would damage its defenses and encourage the development of infection (Haury et al., 1978). The capacity of devitalized fat, muscle, and skin to enhance bacterial infection is comparable. The infection-potentiating effect of skin is further enhanced by exposing it to a dry thermal injury. This change in the capacity of skin to damage the wound's defenses may be related to the development of a burn toxin. This observation is consistent with the experimental findings of Allgöwer et al. (1973), who identified a toxin in skin that was subjected to dry heat. This toxin appears to be generated by a polymeric dehydration of a substance(s) that occurs naturally in the skin.

There are at least three mechanisms by which devitalized soft tissue enhances infection: (1) it acts as a culture medium promoting bacterial growth, (2) it inhibits leukocyte phagocytosis and subsequent bacterial kill, and (3) the anaerobic environment within devitalized tissue must also limit leukocyte function. (At low oxygen tension, the killing of certain bacteria by leukocytes is impaired [Hohn et al., 1976; Mandell, 1974].) While the need for debridement of devitalized tissue is undisputed, identification of the exact limits of devitalized tissue in wounds remains a challenging problem, especially in muscle. Traditionally, the viability of muscle is determined by its contractility, vascularity, color, and consistency. The best criterion is its capacity to contract after being stimulated (Scully et al., 1956), but all clinical indicators are more accurate when the wound is examined 4–5 days after the initial operation.

The viability of skin is considerably easier to judge than that of muscle. At 24 hours after injury, a sharp line of demarcation is often apparent between the devitalized and viable skin. In fresh skin wounds in which this demarcation is not precise, the distribution of an intravenously injected fluorescein dye within the tissues may prove helpful (Myers, 1967). Early staining of the injured skin by fluorescein is evidence of tissue viability. At times, active bleeding from the distal dermal margin may be present and indicates viability.

In some anatomical sites, such as the trunk, debridement is best accomplished by a more liberal excision of the skin and deep tissues. The soft tissue here is relatively free of tissues (e.g., nerves or tendons) that perform important physical functions, and cosmetic considerations are less important.

The adequacy of debridement may be monitored by forcibly packing the wound with gauze or by coloring the wound surface with a vital dye. Flooding the field with a mixture of methylene blue and peroxide will encourage deep penetration of the dye into distant cavities. Complete excision of the wound, back to a margin of normal tissue, is judged by dissecting in a plane that will not expose the gauze or the blue dye. Suturing the skin edges of the wound prior to excision may further minimize mechanical spread of the wound contaminants into uninjured tissue.

When a heavily contaminated wound contains specialized tissues, such as the ureter or kidney, complete excision is often not feasible. In such cases, high-pressure irrigation, followed by excision of all fragments of tissue that are not clearly viable, is indicated. In a compound wound of the abdominal cavity, selective debridement of nonviable fascia is tedious but essential as the site constitutes a favorable environment for bacterial growth.

One must not debride certain specialized tissues that perform important physical functions, regardless of their viability (Peacock and Van Winkle, 1970). Tissues such as dura, fascia, and tendon may survive as free grafts without living cells if immediately covered by healthy pedicle flaps. If these tissues can be rendered surgically clean, they should be left in the wound. Retraction of the wound edges during debridement is best accomplished by hooks. Kocher clamps or bulldog forceps should contact the wound only when the piece of tissue grasped are destined to be removed. Similarly, compressive retractors should be avoided; for prolonged exposure of the wound, its edges can be retracted by stay sutures placed in the dermis.

MECHANICAL CLEANSING

The urologist commonly employs mechanical forces to rid the wound of surface bacteria and other particulate matter retained by adhesive forces. Because these forces must exceed the adhesive forces of the contaminants, the two basic modes used are hydraulic forces and direct contact.

In irrigation, the hydraulic forces of the irrigating stream act on particulate matter in the wound. The total force component exerted on the particle by the moving stream is defined as drag. The total drag due to the fluid pressure and stress is expressed by the following equation:

$$\text{drag} = CA\,\rho\,\frac{V^2}{2}$$

where C is an experimentally derived drag coefficient dependent in part on the configuration of the particle, A is the projected area of the particle on a plane perpendicular to flow, ρ is the density of the fluid, and V is the relative velocity of the fluid with respect to the particle. It takes significantly smaller hydraulic pressures to rid the wound of large foreign bodies than it does to remove bacteria.

Irrigation will be most efficient if the velocity of the irrigating stream is raised by increasing the irrigation pressure and enlarging the internal diameter of the needle. The pressure exerted by fluid delivered through a 19-gauge needle by a 35 ml-syringe is 8 psi (Stevenson et al., 1976). Such high-pressure irrigation (> 8 psi) successfully cleanses the wound of small particulate matter, bacteria, and soil infection-potentiating fractions, thereby reducing the infection rate of experimentally contaminated wounds. In contrast, low-pressure syringe irrigation (as with a bulb syringe), even with large volumes of fluid, does not remove small particulate matter.

Despite the advantages of high-pressure irrigation, several objections have been raised against its routine use. One commonly expressed concern is that foreign bodies on the surface of the wound may be disseminated more deeply into the wound as a result of high-pressure irrigation, although experimental studies indicate that this fear is unfounded (Wheeler et al., 1976). Consequent to high-pressure irrigation, the bacteria remain at the surface of the wound, even though the irrigant solution may disseminate deeply into the tissues. The tissue penetration of a high-pressure irrigating stream is predominantly lateral, similar to a jet parenteral injection.

The concern that high-pressure irrigation can damage tissue defenses, however, appears to be justified. Pulsatile or syringe irrigation results in trauma to the tissues, which makes the wound more susceptible to experimental infection. High-pressure irrigation, therefore, should not be performed indiscriminately, but should be reserved for use in heavily contaminated wounds.

In the clinical setting, high-pressure irrigation is accomplished with an inexpensive disposable irrigation assembly consisting of a 19-gauge plastic needle attached to a 35-ml syringe (Fig 4–16). Sterile electrolyte solution (250 ml of 0.9% sodium chloride) is delivered through a one-way valve attached to the syringe barrel via standard IV plastic tubing. The tip of the needle, fastened to the syringe filled with saline, is placed perpendicular and as close as possible to the surface of the wound; then the urologist exerts maximal force to the syringe plunger delivering the irrigant to the wound.

Another force to cleanse a wound is direct mechanical contact. An example is scrubbing a dirty wound with a sponge. Although this technique removes bacteria from wounds, it does not decrease the incidence of infection. Also, trauma to the tissue inflicted by the sponge impairs the wound's ability to resist infection and allows the residual bacteria to elicit an inflammatory response (Rodeheaver et al., 1975). The magnitude of the damage to the local tissue resistance is correlated

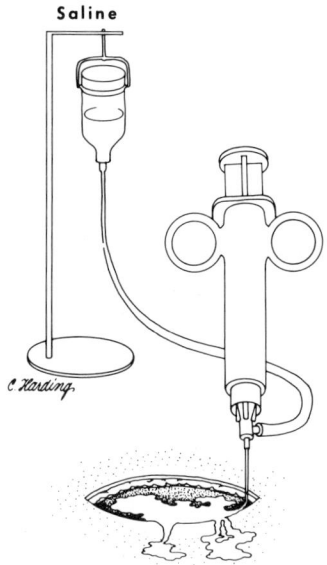

FIG 4–16.
High-pressure syringe irrigation assembly. Note that the needle is held as close as possible and perpendicular to the surface of the wound during wound irrigation.

with the porosity of the sponge—the less porous sponges are more abrasive and cause more damage to the wound. The addition of a nontoxic surfactant, such as Pluronic F-68, to a sponge minimizes the tissue damage it inflicts while maintaining the bacterial removal efficiency of mechanical cleansing. Consequently, the use of a surfactant-soaked sponge reduces the incidence of infection in contaminated wounds in experimental animals. Pluronic F-68 has been used to cleanse traumatic lacerations in over 100,000 patients without an adverse effect.

These fine-pore cell size sponges soaked in this surfactant may not, however, provide sufficient frictional forces to remove dirt and foreign bodies that are embedded in skin abrasions. The latter injury is due to shear forces that are applied almost parallel to the skin surface that result in disruption of the epidermis and papillary dermis. If the foreign body is not removed from the abrasion, wound repair will be accomplished by epithelial regeneration, with the embedded foreign body becoming a pigmented tattoo.

The most advantageous time to remove the embedded foreign debris in the abrasion is immediately. Removal of the embedded foreign particles will require either local or regional anesthesia, depending on the magnitude of injury. A natural fiber scrub brush soaked in Pluronic F-68 will remove the embedded debris from most wounds. Power-driven abraders with stainless steel cylinders coated with tungsten carbide grit can also be used to dislodge these foreign bodies. When

isolated foreign bodies are embedded deeply into the deep dermal tissue, a no. 11 scalpel blade can be used as a spud to tease or excise these particles individually. Furnas and Somers (1976) advocate prompt, aggressive, and meticulous removal of the particles visualized by an operating room microscope. They scrape particles from the depth of the wound with a no. 67 Beaver scalpel blade while stabilizing the skin edges with fine tooth eye forceps (Bonn forceps with 0.1-mm teeth). Any recalcitrant particles are excised with the no. 65 Beaver blade. Wound closure is accomplished with a single 8–0 monofilament synthetic nonabsorbable suture. The debrided wound is covered by fine mesh gauze (Type I) that is impregnated with bacitracin. This antibiotic is reapplied every six hours for approximately three to four days, at which time the gauze spontaneously separates from the underlying healing epidermis. If the foreign debris is not removed immediately, it becomes impossible to trace the paths and remove the individual particles from the depths of the dermis, leaving a permanent traumatic tattoo.

ANTIBIOTICS

The relative success of antibiotic therapy in the prevention of wound infection appears to be influenced by several factors. The timing of administration influences the success of such therapy. In laboratory and clinical studies, antibiotic therapy is significantly more effective when the drug is initiated preoperatively rather than intraoperatively or postoperatively (Edich et al., 1973). Delay in antibiotic treatment consistently diminishes its therapeutic merit. When there is an unavoidable delay in administering these drugs, the length of time during which the open wound is exposed plays an important role. As a result of this exposure, a sequence of events occurs that substantially limits the therapeutic value of antibiotics (Edlich et al., 1973).

When any wound is left open, its vessels exhibit a marked increase in vascular permeability. Fluids from the intravascular space extravasate and fill the wound crater. This exudate is rich in a wide variety of proteins, including fibrinogen. Once outside the vessels, much of the protein exudate is reabsorbed slowly by the lymphatics, with the exception of fibrinogen, which partly polymerizes to form fibrin. It is our belief that the resulting fibrinous coagulum surrounds the bacteria and protects them from contact with the antibiotic (Edlich et al., 1973). The cause of this exaggerated inflammatory response in the open wounds has not been defined. However, it may be related to environmental conditions. The temperature of the operating theater is usually considerably below the systemic body temperature, encouraging loss of heat from the wound. In addition, evaporation of fluid from the wound surface results in further heat loss and cooling of the tissues. A consequence of fluid loss from the wound is desiccation. Warming the operating theater or covering the wound with wet sponges should reduce these environmental effects. Paradoxically, the fibrinous wound coagulum which limits the effectiveness of antibiotics may be a crucial positive factor in the host's defense against infection. The coagulum may serve as a plug in the open mouths of lymphatics, preventing dissemination of bacteria. Occlusion of lymphatics by the coagulum then becomes an obstacle to the invasion of bacteria, and in part accounts for the resistance of an open wound to systemic sepsis.

The surface coagulum may be disrupted by a variety of procedures. Gentle scrubbing of the surface of the wound with a gauze sponge allows an antibiotic to gain intimate contact with the bacteria (Edlich et al., 1971). Consequently, the therapeutic effectiveness of systemic antibiotics is measurably enhanced by this treatment. Enzymatic digestion is a less traumatic and more selective means of disrupting the coagulum (Rodeheaver et al., 1974, 1975). The in vitro fibrinolytic capacity of certain enzymes provides an accurate measure of their ability to potentiate antimicrobial activity. Travase, a proteolytic enzyme produced by *B. subtilis* is the most effective enzymatic adjunct to antibiotic treatment. Hydrolysis of the protein coagulum can be accomplished within 30 minutes by the topical application of an appropriate solution of this proteolytic enzyme. Such a brief exposure does not damage the wound's defenses or its healing capacity.

The bacterial count of the wound can influence the outcome of antibiotic treatment. When the wound is contaminated by exceedingly large numbers of organisms (greater than 10^9), infection will develop despite systemic antibiotic treatment. This circumstance is encountered when the wound surface is contacted by either pus or feces.

The indications for systemic antibiotic treatment are dependent on the mechanisms of injury, the age of the wound, the total bacterial count, the presence of soil infection-potentiating fractions, and concomitant illnesses that predispose to infection and the operative procedure. An antibiotic should be administered to all patients with impact injuries (Cardany et al., 1976). In these wounds, the weakened local tissue defenses make them susceptible to a relatively small inoculum of bacteria (10^4/gm of tissue). It is fortuitous that systemically administered antibiotics can gain contact with pathogens in the ischemic tissue of impact injuries.

Another indication for antibiotic treatment is a traumatic wound in which treatment has been delayed for three or more hours. During this time, bacteria can proliferate to a level that will result in infection. Con-

currently, a thick, fibrinous exudate appears on the wound surface, which becomes a protective barrier against topically or systemically administered antibiotics.

Antibiotic treatment is also mandatory in wounds containing pus, feces, saliva, or vaginal secretions. While antibiotic treatment significantly reduces these levels of heavy contamination, residual viable bacteria are often sufficient to elicit infection after primary closure. Consequently, open wound management of these heavily contaminated wounds should supplement antibiotic treatment.

The presence of soil infection-potentiating fractions in wounds also has considerable influence on the efficacy of specific antibiotics (Roberts et al., 1979). The basic (e.g., gentamicin) and amphoteric (e.g., tetracycline) antibiotics are inactivated by these negatively charged fractions. The acidic antibiotics (e.g., cephalosporins and penicillin) do not bind with these fractions and exert their antibacterial effect in contaminated wounds.

More than 100 case reports provide reasonably good evidence that bacteremias originating from the genitourinary tract may cause endocarditis, especially when urologic or gynecologic operations are carried out in the presence of urinary or pelvic infections (Vosti, 1977). Those underlying conditions that display a relative high risk of developing endocarditis following a transient bacteremia include prosthetic valves, arteriovenous fistula, patent ductus arteriosus, Fallot's tetralogy, ventricular septal defect, coarctation of the aorta, aortic valve disease, mitral valve disease, Marfan's syndrome, and previous infective endocarditis. Antibiotic prophylaxis in such patients should include a combination of gentamicin and penicillin that is administered 30–60 minutes before the operative procedure (Table 4–3). Convincing cases of hematogenous infections of other prostheses are uncommon. Late hematogenous infections of arterial grafts (those appearing more than 60 days after surgery) are nearly unknown (Talkington and Thompson, 1982). Instead almost all reported infections have probably originated from contamination at surgery, from contiguous areas of purulence or from the development of a fistula between the graft and GI tract. Even in hip arthroplasties, the sources of infection have been distant sites of established suppuration, such as tonsillitis, cholecystitis, dental abscess, pyoderma, suppurative parotitis or cystitis, rather than procedure-induced transient bacteremia (D'Ambrosia et al., 1976). These findings emphasize the importance of prompt and thorough treatment of distant infections in patients with indwelling prosthetic devices.

Finally, antibiotics must be administered to patients with wounds in which the magnitude of tissue injury is

TABLE 4–3.
Antibiotic Prophylaxis of Infective Endocarditis in Patients*

Adults: Aqueous crystalline penicillin G, 2,000,000 units IM or IV, or ampicillin, 1.0 gm IV
       plus
      Gentamicin, 1.5 mg/kg IM or IV, or streptomycin, 1.0 gm IM. If gentamicin is used, give the same doses of gentamicin and penicillin (ampicillin) every 8 hr for 2 additional doses. If streptomycin is used, give the same dose of streptomycin and penicillin (or ampicillin) every 12 hr for 2 additional doses.
Children: Aqueous crystalline penicillin G, 30,000 units/kg IM or IV, or ampicillin, 50 mg/kg IM or IV
       plus
      Gentamicin, 2.0 mg/kg IM or IV, or streptomycin, 20 mg/kg IM
For patients allergic to penicillin
Adults: Vancomycin, 0.5–1.0 gm IV, infused slowly over 1 hr
       plus
      Streptomycin, 1.0 gm IM, given 30 min to 1 hr before the procedure. Both drugs may be repeated once, 12 hr later
Children: Vancomycin, 20 mg/kg IV, infused slowly over 1 hr
       plus
      Streptomycin, 20 mg/kg IM

*For brief procedures such as a change of urinary catheter or cystoscopy, one dose of these combination regimens will probably suffice. Conversely, in the case of unusually prolonged or repeated procedures or delayed healing, several additional doses may be given.

extensive and difficult to ascertain accurately soon after injury. In such cases, open management of the wound is the best treatment, with subsequent additional debridement as dictated by its appearance. It must be emphasized that antibiotic therapy is an adjunct to debridement, not a replacement.

The value of prophylactic antibiotics in elective urologic surgical procedures has not been adequately defined. In Chodak and Plaut's critical review (1979) of the literature on antibiotic prophylaxis in urologic operations, they reported that the design of most clinical studies are too flawed to yield useful data on the value of prophylaxis. Even well-designed studies were weakened by a variety of factors, the major ones being small or dissimilar study groups, scant attention to microbiologic techniques, and imprecise definition of infections and variations in catheter care. Support for the use of prophylactic antibiotics was lacking for open and transurethral prostatectomy, nephrectomy, and other transurethral procedures.

The immediate selection of the specific antimicrobial agent is based on consideration of the results of direct microscopic examination of a wound biopsy, the normal bacterial flora harbored in different parts of the body, and the pathogens usually encountered in various diseases or conditions. Later, the results of immediate antibiotic sensitivity testing can also influence the urolo-

gist's choice of antibiotic. In our laboratory, antibiotic sensitivity testing is performed under aerobic conditions directly on the bacterial suspension prepared from the wound biopsy specimen rather than on strains isolated from the tissue (Verklin et al., 1977).

HEMOSTATIC AGENTS

Topical hemostatic agents have found wide applications in a variety of bleeding sites. They can effectively control bleeding from large areas of cancellous bone (Cobden et al., 1976). These agents are often used to stop bleeding from vascular anastomoses before placing additional sutures (Abbott and Austen, 1974). Bleeding from splenic, hepatic and renal injuries can be also arrested by application of these materials (Hanisch and Guerriero, 1978; Morgenstern, 1977; Horsley, 1978).

A wide variety of topical hemostatic agents have been used clinically to reduce bleeding and include the following: Beeswax,* stabilized fibrin and collagen paste,† oxidized cellulose (OC),‡ oxidized regenerated cellulose (ORC),§ absorbable gelatin foam (AGF),‖ microfibrillary collagen (MC),¶ and bovine thrombin (BT).# OC is an absorbable oxidized cellulose that is available in either cotton type pledgets or gauzelike pads or strips. ORC is an absorbable knitted fabric prepared by the controlled oxidation of regenerated cellulose. AGF is a pliable, nonantigenic foam from specifically treated, purified gelatin solution and is capable of absorbing and holding within its meshes many times its weight of whole blood. MC is an absorbable agent prepared as a dry, fibrous, water-insoluble, partial hydrochloric acid salt of purified bovine corium collagen. In its manufacture, swelling of the native collagen fibrils is controlled by ethyl alcohol to permit noncovalent attachment of hydrochloric acid to amine groups on the collagen molecule and preservation of its essential morphology. BT is produced through a conversion reaction in which prothrombin is activated by tissue thromboplastin in the presence of calcium chloride. After standardization of potency by dilution, thrombin solution is sterilized by filtration, poured into vials, dried from the frozen state, and sealed under aseptic conditions.

Beeswax has been used almost exclusively to arrest bleeding from the cut edges of cancellous bone. It was first described by Horsley in 1882. His antiseptic bone wax consisted of 7 parts beeswax, 1 part almond oil, and

---

*Ethicon, Inc.
†Ethnor, Ethicon, Ltd.
‡Oxycel, Deseret Medical Inc.
§Surgicel, Johnson and Johnson Products, Inc.
‖Gelfoam, The Upjohn Co.
¶Avitene, American Critical Care.
#Thrombostat, Parke, Davis & Co.

salicylic acid 1%. The most commonly used bone wax currently available has a formula of refined beeswax 88% (w/w) and isopropyl palmitate 12% (w/w). The hemostatic action of bone wax is physical as the wax tamponades the bleeding edges of the cut bone. Its hemostatic benefit occurs at the expense of a foreign-body giant cell response, persistence of wax at the bony site for years, and a physical barrier to bony union (Brightmore, 1975). Thus, the wax appears to be a nonabsorbable foreign body impermeable to components of healing tissue (Howard and Kelley, 1969). Acting as a foreign body, bone wax significantly impairs the ability of cancellous bone to kill bacteria (Johnson and Fromm, 1981).

In 1977, Lawrie reported a new absorbable bone hemostatic sealant containing stabilized fibrin, soluble collagen, glycerol, and dextran that had the consistency of "putty." Its hemostatic effect appears to be by tamponade, but both the fibrin and collagen may contribute to local hemostasis. It was a more efficient hemostatic agent than bone wax and it did not interfere with bone healing (Harris and Cappernauld, 1980). Absorption was complete within three weeks. The effect of this stabilized fibrin and collagen paste on bacterial clearance has not been examined.

In experimental studies, Cobden et al. (1976) evaluated the efficacy of microcrystalline collagen, thrombin-soaked gelatin foam, and thrombin powder as topical hemostatic agents for bleeding cancellous bone. All three agents significantly reduced bleeding compared with the controls, the microcrystalline collagen being the most effective. At three months, there was no evidence that microcrystalline collagen and thrombin-gelatin interfered with bone healing. Their influence on infection in this clinical setting was not investigated.

Most experimental studies have demonstrated that the absorbable hemostatic agents damaged tissue defenses and invited the development of infection. Jenkins et al. (1946a, 1946b) pointed out that in the presence of contamination or infection AGF may act as a culture medium which may influence the development and propagation of infection. The experiments of Cipolla and Narat (1948) demonstrated that AGF and OC intensified the infection due to *Staphylococcus aureus* in the subcutaneous tissue of dogs. Moistening these absorbable agents with penicillin was not able to prevent infection. However, intramuscular administration of penicillin-in-oil inhibited the potentiation of infection caused by the absorbable hemostatic agents. Hinman and Babcock (1949) reported that OC and AGF enhanced the incidence of infection of contaminated nephrotomy incisions as compared to control wounds without this hemostatic agent subjected to a comparable amount of the suspension of feces.

Hill (1978) demonstrated that absorbable gelatin sponge powder and MC markedly increased the rate of experimental anaerobic infection when mixed with a normally subinfective dose of the bacterial inoculum that was injected into the peritoneal cavity. The degree to which these topical hemostatic agents enhanced infection did not differ significantly. Using a standard subcutaneous wound infection model, Scher and Coil (1982) demonstrated that both ORC and MC increased the frequency of infection as compared with the wound that did not contain any foreign body. This infection-potentiating property of these topical hemostats was more dramatic when MC was employed. It was possible to produce a dose-response curve for ORC by varying the quantity employed and by maintaining a constant bacterial inoculum. This data emphasizes the importance of using the minimal amount of ORC consistent with satisfactory hemostasis. No such dose response curve could be obtained for MC because significant infection occurred at even the smallest doses studied. Since ORC and MC were equally effective in controlling bleeding from raw surfaces (Hait et al., 1973), the advantage of ORC over MC in terms of infection makes ORC the superior hemostatic agent.

Laufman and Method (1948) showed deleterious effects of placing absorbable gelatin sponge and OC in contact with the anastomoses of the large intestine in dogs. They believed that the increased incidence of perforation and peritonitis with topical hemostatic agents was due to their ability to act as a culture medium for bacteria. In another experimental study, Uhrich (1949) demonstrated that OC did not prevent leakage from the anastomotic line when a single through-and-through suture end to end anastomosis was performed in dogs. In addition, the presence of OC appeared to promote the growth of bacteria in the infected peritoneal cavity. Chamberlain et al. (1951) demonstrated that absorbable gelatin sponge in apposition to intestinal anastomoses increased the risk of peritonitis and fatal perforation in dogs. Impregnation of this sponge with crystalline penicillin was effective in preventing perforation of the anastomosis and local abscess formation.

In contrast to the other investigations that document an infection-potentiating effect of topical hemostatic agents, Dineen (1976) reported that ORC had the ability to reduce the bacterial population *in vitro* and *in vivo*. Its mode of antibacterial action was explained, in part, by its low pH. *In vitro*, most of its activity was blocked by the addition of sodium hydroxide. Dineen (1966, 1977) and Kuchta and Dineen (1983) demonstrated that ORC prevented infection using several different experimental models. In standard infected wounds in guinea pigs, he found that ORC had in vivo activity as compared to control wounds as well as those with absorbable gelatin sponges (1976). Dineen (1977) reported that ORC also prevented sepsis following an IV bacterial challenge. In dogs with Teflon aortic patches, wrapping the area of the Teflon patch with ORC before the bacterial challenge reduced the bacterial contamination of the Teflon patches compared with animals without ORC. Similarly, ORC decreased the level of contamination in splenotomy sites as compared to absorbable gelatin sponge following an IV injection of a bacterial inoculum (Dineen, 1977). In another experimental model that simulated the clinical situation of infected subdiaphragmatic blood clots, animals treated with ORC showed longer survival times and no development of abscesses and/or peritonitis following an IV bacterial challenge compared with control animals or those treated with either absorbable gelatin sponge or MC who died within five days of frank peritonitis and/or abdominal abscess (Kuchta and Dineen, 1983).

WOUND DRAINAGE

Use of surgical drainage requires a delicate weighing of potential benefits and harmful effects. The obvious beneficial effect of drainage is its ability to evacuate or at least provide tracts for the elimination of potentially harmful collections of certain fluids such as pus, blood, bile, and gastric and pancreatic juices from wounds or body cavities. Pus is detrimental to the healing of wounds. Even sterile pus injected under the skin impairs healing of a distant wound (Carrel and Hartmann, 1916; Carrel, 1921). Smith and Enquist (1967) reported that subacute experimental staphylococcal wound infection in animals initiated profound gross, histologic, and biochemical changes in local and distant tissue. Pus within a wound or body cavity exerts many deleterious effects on the host and should be removed whenever a localized collection can be drained.

Sterile collections of blood per se are not major irritants to tissues, but hematomas enhance bacterial virulence (Lee et al., 1979; Krizek and Davis, 1965). Krizek and Davis (1965) found that when RBCs were injected into the same subcutaneous site as *E. coli*, a fatal infection often occurred, whereas the injection of bacteria alone was relatively innocuous. The precise mechanisms are unknown but crude hemoglobin preparations impair leukocyte function (Hau, 1977) and hemoglobin may provide iron essential for bacterial proliferation (Lee et al., 1979). Fibrin clot has a paradoxical role in tissue infection, preventing the systemic dissemination of wound bacteria on the one hand, and isolating it from phagocytic defense mechanisms on the other.

The location and contents of the harmful collections of fluids will dictate which drainage technique is the treatment of choice. Surgical intervention in abdominal infections is indicated in the following conditions: an infected organized hematoma, a parapancreatic phlegmon, thick fungal infections and a necrotic infected tumor metastasis (Pruett et al., 1984). Successful management of these infections includes incision into the abscess cavity, debridement of devitalized tissue, postoperative drainage and antimicrobial therapy appropriate for the pathogens in the abscess cavity. The efficacy of the postoperative drainage system is directly related to the efficacy of removing the collections of fluid in the abscess cavity. Nonvented suction drainage creates a vacuum that causes tissue encroachment on the holes of the drainage tube, resulting in impaired drainage and tissue damage. This problem was resolved by designing a drainage tube with an air vent that permitted access of air into the area of drainage (Edlich et al., 1985). A two-staged filter was attached to the vent lumen that removed particulate matter and bacteria from the air that passed through the filter. A clinical evaluation of this tube confirmed the superiority of the filtered sump tube as compared with closed suction drainage in the removal of fluids from wounds or cavities.

The performance of this drainage tube was improved further by coating the surfaces of the drain with a new hydrogel coating (Axiom Medical Inc., Paramount, Calif.) (Pearce et al., 1984). This hydrogel was formed by reacting polyvinylpyrrolidone with an isocyanate prepolymer. By absorbing water, the hydrogel creates a drain surface with a low coefficient of friction. The results of our experimental studies demonstrated that the drain coating reduced the adherence of blood clots to the drain and facilitated the removal of the drain from the wound. On the basis of these studies, the FDA has allowed this hydrogel-coated drain to be used in clinical patients.

Recent advances in interventional radiology have demonstrated the value of percutaneous aspiration (PCA) and drainage (PCD) in patients with abscesses. Catheter drainage is performed by a modified Seldinger technique, with the passage of the catheter being monitored by either ultrasound or CT (Van Sonnenberg et al., 1982). The indications for the technique can be diagnostic, palliative, and therapeutic. Cytologic and microbiologic examination of the aspirate can detect the presence of malignancy as well as the responsible pathogen. Palliation is defined as temporary control of the infection in which the clinical infection is relieved, but the infectious process is not cured until an elective operative procedure is performed. PCD is therapeutic in a well-localized, unilocular bacterial abscess with a success rate of 96% (Pruett et al., 1984). While PCD appears to be the preferred method of management of the latter abscess, it is uniformly unsuccessful in treating collections that contain semi-solid or solid particulate matter (e.g., organized hematoma, parapancreatic phlegmon, fungal infections, necrotic infected tumor metastases).

In instances where no definite collection of fluid exists, drainage must be considered to be prophylactic. In these circumstances, its potentially harmful effects become more important. Drains act as retrograde conduits through which skin contaminants gain entrance into the wound. Cerise et al. (1970) performed a splenectomy in rabbits and inoculated the skin around the drain tract with type 6 *Streptococcus*, taking care not to inoculate the drain. Twenty percent of the animals had positive intraperitoneal cultures at 24 hours, and 56% at 72 hours, compared to positive culture results in only 5% of the undrained animals. Nora et al. (1972) performed laparotomies in dogs and inserted Penrose drains separately into the splenic fossa and in Morison's pouch. Gross intra-abdominal infection was detected in nine of the ten dogs with drains in the splenic fossa. No infections occurred in the undrained dogs. In a clinical study, the same investigators detected skin contaminants on the intra-abdominal portion of drainage tubes in 17 of 50 patients. *S. epidermis*, a skin contaminant, was found in 14 of the 17 cases. Of the 17 infected patients, 12 had minimal egress of fluid from the drain tract, and neither the GI nor genitourinary tract had been opened at operation.

Raves et al. (1984) initiated an experimental study that demonstrated that closed suction drainage was associated with less bacterial migration into the splenic bed than Penrose drains. After inoculating the drain skin exit site with a strain of streptococcus, 90% of the cultures of splenic bed were positive in animals with Penrose drains. Only 20% of the animals with closed suction drains had streptococcus recovered from their splenic beds ($P < .001$).

The presence of drains impairs the resistance of tissues to infection. Drains placed within experimental wounds exposed to subinfective inoculations of bacteria enhanced the rate of infection (Magee et al., 1976). Both Silastic and Penrose drains dramatically increased the infection rate of soft tissue wounds in guinea pigs. The rate of infection when the drain was brought out through the wound was similar to the rate when the drain lay entirely within the wound, suggesting a deleterious effect of the drain.

Studies of the impairment of wound healing in the presence of drains have been undertaken with intestinal anastomoses. Berliner et al. (1964) performed proximal and distal intestinal anastomoses in dogs, draining one anastomosis in each dog. Three of the 20 undrained anastomoses leaked compared with 11 of 20 drained

anastomoses; of these 11, anastomotic disruption proved fatal in four cases. Manz et al. (1970) confirmed the damaging effects of drains on colonic anastomoses in dogs. Of 20 dogs with Penrose drainage at their anastomoses, nine died of anastomotic disruption and peritonitis, and the remainder had extensive adhesions, as well as varying degrees of stricture formation. All dogs with drainage had evidence of bacterial contamination at the site of the anastomosis; the control animals had only filmy adhesions and no stricture formation. In another experimental study, Crowson and Wilson (1973) reported that low canine anastomosis healed best when bacterial contamination was minimal, the anastomosis was not drained, and the anastomosis was intraperitoneal. A drain made intraperitoneal sepsis worse and was harmful to the peritoneal cavity. Silastic drains were less harmful than either latex or vinyl drains.

It has been postulated that an adjacent drain may block the desquamated mesothelial cells from contacting the anastomosis, thereby interfering with its healing. The drain may also act as a retrograde conduit for bacteria that then contaminate the anastomotic site, thereby predisposing to infection and leakage.

FASCIAL CLOSURE

Important considerations in abdominal fascial closure are the type of suture, the tying technique and the configuration of the suture loops. Selection of a suture material is based on its biologic interaction with the wound as well as its mechanical performance in vivo and in vitro. Measurements of the in vivo degradation of sutures separate them into two general classes (Edlich et al., 1973). Sutures that undergo rapid degradation in tissues, losing their tensile strength within 60 days, are considered absorbable sutures. Those that maintain their tensile strength for longer than 60 days are nonabsorbable sutures. This terminology is somewhat misleading, because even some nonabsorbable sutures (e.g., silk, cotton, and nylon) lose some tensile strength during this 60-day interval. Postlethwait (1970) measured the tensile strength of implanted nonabsorbable sutures during a period of two years. Silk lost approximately one half of its tensile strength in one year and had no strength at the end of two years. Cotton lost 50% of its strength in six months, but still had 30%–40% of its strength at the end of two years. Nylon lost approximately 25% of its original strength throughout the two-year observation period.

The nonabsorbable sutures can be classified according to their origin. Nonabsorbable sutures made from natural fibers include silk, cotton, and linen. Metallic sutures are derived from stainless steel. Modern chemistry has developed a variety of synthetic fibers from polyamides (nylon), polyesters (Dacron), and polyolefins (polyethylene, polypropylene) to polybutester. The latter suture is a block copolymer that contains poly (butylene) terephthalate (84%) and poly (tetramethylene ether) glycol terephthalate (16%).

The nonabsorbable sutures may also be characterized by their physical configuration. Sutures constructed from one filament are called a monofilament suture; sutures containing multiple fibers are called multifilament. Only nylon and stainless steel sutures are available as a monofilament or multifilament suture. Multifilament stainless steel and cotton sutures are formed by winding one filament around another, forming a twisted suture. Long continuous strands of stainless steel are twisted together to form different-gauge sutures. Noncontinuous natural fibers of cotton are combined into yarns, which are twisted into plies. The plies in the cotton suture are twisted together to form various-gauge sutures. The other multifilament sutures are formed by intertwining three or more filaments. Several very fine silk fibers are twisted together to form yarns, which are then braided. The number of silk fibers used regulates the suture gauge. With braids made from synthetic filaments, the large gauge suture can be made by either increasing the number of filaments or enhancing the size of the filaments.

The absorbable sutures are made from either collagen or synthetic polymers. The collagen sutures are derived from the submucosa of sheep or bovine small intestine (gut). This collagenous tissue is treated in an aldehyde solution, which crosslinks and strengthens the suture and makes it more resistant to enzymatic degradation. Suture materials treated in this way are called plain gut. If the suture is additionally treated in chromium trioxide, it becomes chromic gut, which is more highly crosslinked than plain gut and more resistant to absorption. The plain gut and chromic gut sutures are composed of several plies that have been twisted slightly, machine ground, and polished, yielding a relatively smooth surface and diameter that is monofilament-like in appearance. Salthouse et al. (1969) demonstrated that the mechanism by which gut resorbs is due to sequential attacks by lysosomal enzymes. In most locations, this degradation is started by acid phosphatase, with leucine aminopeptidase playing a more important role later in the absorption period. Collagenase is also thought to contribute to the enzymatic degradation of these collagen sutures.

A search for a synthetic substitute for collagen sutures began in the 1960s. Soon, procedures were perfected for the synthesis of high molecular weight polyglycolic acid and the next homologue in this series of alpha polyesters, polylactic acid, which led to the development of the polyglycolic acid and polyglactin 910 sutures (Rodeheaver et al., 1983). The copolymers of polyglactin 910 are prepared by polymerizing nine parts of glycolide with one part of lactide, while the polyglycolic acid sutures are produced from the homopolymer. Because of the inherent rigidity of these copolymers, monofilament sutures produced from polyglactin 910 or polyglycolic acid sutures are too stiff for surgical use. These polymers can be used as a monofilament suture only in the very finest size. Consequently, these high molecular weight polymers are extruded into thin filaments and braided. The polyglycolic acid and polyglactin 910 sutures degrade in an aqueous environment through hydrolysis of the ester linkage.

The surfaces of these synthetic sutures have been coated to decrease their coefficient characteristics. (Ray et al., 1981). The coating on the polyglycolic acid suture is an absorbable surface lubricant, poloxamer 188. The polyglactin 910 suture has been coated with an absorbable mixture of calcium stearate and a copolymer of lactic acid (65%) and glycolic acid (65%).

A new monofilament absorbable suture, polydioxanone, has recently been developed by polymerizing the monomer in the presence of a suitable catalyst (Ray et al., 1981). This polymer is processed into small granules and melt extruded through appropriate dies into monofilaments of any desired use.

A distinction must be made between the rate of absorption and the rate of tensile strength loss of that material. The terms are not interchangeable. Although the rate of absorption is of some importance with regard to late suture complications such as sinus tracts and granulomas, the rate of tensile strength loss is of much greater importance to the urologist considering the primary function of the suture—maintaining tissue approximation during healing.

Herrmann (1973) evaluated the tensile strength and knot security of knotted absorbable suture loops implanted in the subcutaneous tissue of rats and rabbits. Polyglycolic acid sutures possessed tensile strength and knot security almost equal to polyester sutures following implantation. The polyglycolic acid sutures retained their strength superiority over comparable sizes of chromic gut during the first 20 days after implantation. Chromic gut and polyglycolic acid sutures had the same period of strength retention, lasting about 25 days. Plain gut had the lowest knot break strength of the absorbable sutures prior to implantation and had no detectable knot break strength at 14 days. Although knot security of the gut sutures was excellent in a dry state, the three throw square knots became insecure when exposed to body fluids.

Ray et al. (1981) reported that the strength of the polydioxanone suture prior to implantation was comparable to that of either the polyglycolic acid suture or polyglactin 910 sutures. At 4, 6, and 8 weeks, the monofilament polydioxanone retained an average strength of 58%, 41%, and 14% of its original strength, respectively following in vivo residence in the rat subcutis. The prolonged retention of breaking strength was accompanied by a slower rate of absorption than encountered with either polyglycolic acid or polyglactin 910 sutures.

The initial breaking strength of an absorbable suture is considerably influenced by the conditions of implant site that differ in different species. Postlethwait (1975) reported that the rate of strength loss of two synthetic absorbable sutures in canine stomach approximated their rate of loss in muscle, whereas the breaking strength of gut was markedly decreased in the stomach. Perey and Watier (1975) reported that the breaking strength of chromic gut fell rapidly within one day of exposure to human gastric or duodenal juices. Beyond 24 hours, gut had lost a large part of its initial strength, while the strength of polyglycolic acid remained relatively unchanged for at least five days. Using baboons, Everett (1920) measured the tensile strength of no. 00 chromic gut sutures in the sheath of rectus abdominis muscle, colon, and ileum. At five days postimplantation, there was virtually no change in strength of the gut sutures in the sheath of the rectus abdominis muscle, whereas the suture in the ileum had no strength at three days, and the colon suture exhibited only 15%–20% of its original strength at five days.

Results of laboratory and clinical studies have shown that synthetic absorbable sutures maintain their strength and integrity when infection is present (Laufman and Rubel, 1977). This is in contrast to the behavior of gut, which disintegrates and loses strength in infected tissues. Results of two investigations showed that bacterial growth is inhibited in the areas of absorbing polyglycolic acid sutures (Edlich et al., 1973; Lilly et al., 1973). Even though polyglycolic acid and its degradation products are not bactericidal, it was theorized that the pH at the polyglycolic acid suture sites was such that it retarded the growth of bacteria.

Higgins et al. (1969), Haxton (1965), Kirk (1972), Goligher et al. (1975) and Leaper et al. (1976) found that fascial closure with chromic gut was associated with a significantly higher incidence of wound dehiscence than was encountered with closure with nonabsorbable sutures. Standeven (1955) and Alexander and Prudden (1966) noted that wound disruptions were usually caused by suture absorption or breakage. This dissolution of the chromic gut suture reduces its tensile strength, which accounts for the increased incidence of wound dehiscence. On the basis of this experience, the use of chromic catgut for fascial closure should be abandoned.

Since the advent of synthetic absorbable sutures,

polyglycolic acid and polyglactin 910, several randomized prospective studies have demonstrated that they are equal to nonabsorbable sutures, including wire, in ensuring wound integrity. The recently introduced monofilament absorbable suture, polydioxanone, may be another suitable alternative to nonabsorbable suture because it retains 50% of its tensile strength five weeks after implantation (Ray et al. 1981). In a long-term experimental study, this suture has been shown to provide a degree of protection against wound bursting that is comparable to that of polypropylene or Teflon-coated dacron sutures (Kon et al. 1984).

With nonabsorbable and synthetic absorbable sutures, wound disruptions are caused by the fascia tearing at the site of the suture. Most experimental studies demonstrated that placing the suture farther from the cut edges of the fascia reduced the risk of wound disruption. Sanders et al. (1977) reported that placing sutures 5 mm from the cut edges of the fascia resulted in a higher wound bursting strength than 1 to 2 mm from the fascial edges. Leaper et al. (1977) recorded the suture holding strength of abdominal wall structures in cadavers and noted the holding strength of sutures placed 1 cm from the fascial edge was 7.16 kg compared with 3.93 kg for sutures placed 0.5 cm from the wound edge. In a similar study of midline incisions, Tera and Aberg (1976) demonstrated that the holding strength of sutures placed lateral to the transition zone between the linea alba and the rectus sheath (22.9 kg) was more than twofold greater than the tissue holding power of sutures passed through the linea alba (10.8 kg). Most urologists who use the continuous running suture for midline incision closure advocate placing the suture 1.5 cm from the divided edge of the fascia, a distance that would be beyond this transition zone.

In 1938, Whipple and Elliot (1938) indicated that tying sutures too tightly caused strangulation of the tissue with ischemic necrosis and was the most common error in abdominal wound closure. This suggestion was confirmed by studies of Haxton (1965), Nelson and Dennis (1951) and Sanders et al (1977). These investigators reported that tight tying of interrupted sutures resulted in a lower wound strength than sutures tied when the wound edges were approximated.

The selection of a wound closure technique must also take into account the dynamic changes in wound length during distention (Jenkins, 1976). Measurements of the abdominal girth and xiphoid pubic distance before and after closure demonstrate that abdominal distention may lengthen the wound by 30%. When the stitch interval is 1 cm, it will become 1.33 cm when the wound is lengthened by 30% during abdominal distention. The continuous suture can accommodate to this increase in length of the incision by having an adequate reserve of suture length in the wound. Consequently, the continuous suture distributes its tension throughout the wound, limiting the forces on the tissues encircled by the suture. With interrupted closure, the suture cannot easily accommodate these changes in incisional length, and the tension remains isolated to each suture loop. Using an intact animal model, Poole et al. (1984) demonstrated that the continuous suture technique was associated with greater wound bursting pressures than the simple interrupted suture or figure-of-8 mattress suture. There are several clinical reports in which continuous sutures have been used with excellent results. In a prospective randomized trial of both vertical and oblique incisions, Richards et al. (1983) found no difference in the wound dehiscence rate or incidence of incisional hernia between continuous and interrupted (Smead-Jones) suturing. In addition, Stone et al. (1983) showed that continuous suturing resulted in a comparable incidence of dehiscence as interrupted sutures and had the average saving of 26 minutes in anesthesia time.

The value of retention sutures as compared to other wound closure techniques has not been answered. Haxton (1965) was unable to demonstrate any benefit from the use of retention sutures in either paramedian or midline wounds when assessing wound bursting strength in animals. In a randomized prospective clinical study, Hubbard and Rever (1972) reported that three of 203 patients whose midline laparotomy incisions were reinforced with no. 28 stainless steel retention sutures had dehiscence, whereas none of the 209 patients whose midline laparotomy incisions were closed without retention sutures had dehiscence.

ABDOMINAL WALL DEFECTS

Abdominal defects resulting from invasive infection, trauma hernia, or tumor are best reconstructed with myocutaneous flaps. If there is insufficient autogenous tissue for adequate wound closure, synthetic meshes are being employed to resurface the residual defect. Controversy exists concerning the advantages of different porous materials for reconstruction of the abdominal wall. Of the synthetic materials, nonabsorbable knitted polypropylene (P)* (filament diameter 150 µ, pore size 620 × 620 µ) is the most widely used for reconstruction of abdominal wall defects. Another nonabsorbable mesh, polytetrafluoroethylene† (PTFE) (filament diameter 140 µ, pore size 500 × 500 µ), used first as an expanded cardiovascular patch, is now being evaluated for abdominal wall reconstruction. An absorbable synthetic mesh, polyglactin 910‡ (filament di-

*Marlex, R. Bard, Inc.
†Gore-Tex, W.L. Gore.
‡Vicryl, Ethicon Inc.

ameter 140 µ, pore size 400 × 400 µ) is also available.

Lamb et al. (1983) assessed these synthetic meshes for reconstruction of abdominal wall defects in rabbits by measuring the bursting strength, fibrous tissue incorporation and inflammatory reaction. The bursting strength of abdominal wound defects closed with PTFE and PP was very similar to that of the control defects closed with a vascularized myofascial flap during the 12-week period of study. At three weeks, polyglactin 910 mesh had a bursting strength comparable to that of the control flaps, but at 12 weeks was significantly weaker. Fibrous tissue incorporation was better in PTFE mesh than the PP mesh. Adequate fibrous tissue incorporation into the polyglactin 910 mesh did not occur before hydrolysis, making it an unsatisfactory synthetic material for permanent abdominal wall reconstruction.

Fistula formation and intense adhesion formation when in contact with bowel are disadvantages of any prosthesis. In an experimental animal model, Jenkins et al. (1983) found that adhesion formation was moderate to maximal during the eight-week period of evaluation of PTFE and PP. Early adhesion formation with polyglactin 910 was minimal to moderate and decreased as the prosthesis was absorbed. The investigators concluded that absorbable polyglactin 910 provided the best long-term protection against adhesions. However, it should be emphasized that their control animals without synthetic meshes had no adhesion formation.

These latter experimental models do not simulate the clinical situation in which the abdominal defect has bacterial contamination. Using an experimental model in which the synthetic mesh was contaminated by *Staphylococcus aureus*, Brown et al. (1986) reported that PTFE was superior to PP for abdominal wall reconstruction. Fewer organisms adhered to PTFE than to PP regardless of whether antibiotics were administered. In all instances, PTFE produced fewer adhesions and was more easily removed than PP.

DEAD SPACE

The importance of dead space in the potentiation of wound infection has long been recognized and can be demonstrated experimentally (Ferguson, 1968; de Holl et al., 1974). Although dead space is bad, suture closure of dead space is worse (Ferguson, 1968; de Holl et al., 1974). The mechanism by which dead space potentiates the infectivity of a subinfective bacterial inoculum is not clear. One possibility is that even a bloodless exudate potentiates wound infection. Wood et al. (1951) noted that phagocytosis of bacteria suspended in a fluid medium was difficult unless the bacteria have been opsonized. In the same way, Ahrenholz (1979) demonstrated that sterile saline aggravated experimental peritonitis unless the bacteria were preopsonized. Alexander et al. (1976) have reported that wound exudates were depleted of opsonins. Suspension of contaminating bacteria in opsonin-depleted exudate may well interfere with antibacterial phagocytosis. The presence of the suture material appears to potentiate the development of infection. The harmful effects of suture closure of dead space in experimental animals also occur in surgical incisions not involving muscle. Obliteration of the potential space between the cut edges of adipose tissue by sutures potentiated the incidence of infection.

These studies, combined with clinical observations, form the basis for the recommendation that suture closure of dead space should be avoided in contaminated wounds. Although dead spaces in wounds should definitely be avoided, the collapse of such space can be achieved by physiologic methods such as relaxing incisions, rotation of distal flaps, and splinting dressings that may provide gentle surface pressure. The closure of dead space by sutures produces localized areas of wound ischemia and necrosis, and the presence of additional suture material adds further danger of wound complication. Whether suction drainage will achieve the effect of obliterating dead space without realizing the known effect of drains to potentiate infection is unknown (Alexander et al., 1976).

SKIN CLOSURE

The technique of skin closure selected depends on the type of wound. There are essentially two types of skin wounds. One type is characterized by a loss of tissue, while the other type has no evidence of tissue loss. In the latter, primary closure can be accomplished simply by reapproximating the divided skin edges by a variety of methods. For wounds with associated tissue loss, grafts or flaps are often required to close the defect.

A skin graft is a segment of epidermis which has been removed from its own blood supply and donor site attachment. It is transplanted to another area of the body which has less than 10^6 bacteria/gm of tissue (Robson and Krizek, 1973; Bacchetta et al., 1975) and sufficient blood supply to vascularize the skin graft. Cortical bone denuded of its periosteum, cartilage without perichondrium, tendon without paratenon, and nerve without its perineurium do not have adequate blood supply to maintain the viability of skin grafts. The only exception to this rule is when skin grafts survive on a small (<0.5 cm) avascular area (bridging phenomenon) (Gingrass et al., 1975). In such cases, collateral circulation connects the vessels in the skin graft overlying the avascular area to the surrounding vascularized peripheral skin graft on granulation tissue.

Flaps are usually required for covering recipient sites with poor vascularity; for padding bony prominences; for reconstructing full-thickness defects of eyelids, lips, ears, nose, and cheeks; and when it is necessary to reoperate through the wound at a later date to repair underlying structures. A flap is a composite of tissue attached by a vascular pedicle through which it receives its arterial blood supply and discharges venous effluent. The various types of flaps include skin, omentum, muscle, and musculocutaneous tissue. A flap with arterialized tissue possessing a longitudinal vascular supply can have its blood supply detached and then reattached to a new anatomical area (free flap). The selection of a flap is based on the reliability of flap survival, its resultant function, the appearance at the site of coverage, and the donor site availability.

When the area requiring coverage is heavily contaminated, it is of paramount importance that the flap chosen should have the capability to contain and eliminate local infection; otherwise the flap will succumb due to the underlying infection. In experimental studies, Chang and Mathes (1982) demonstrated that musculocutaneous flaps demonstrated a greater resistance to bacterial inoculation than random pattern skin flaps on both its cutaneous and muscular surfaces. Differences in the pattern of oxygen delivery to random pattern skin flaps and musculocutaneous flaps may in part explain the greater reliability of musculocutaneous flaps when transposed in the presence of infection. Gottrup et al. (1984) reported that tissue oxygen tensions were significantly higher in the musculocutaneous flaps than in random pattern skin flaps up to six days after operation. They concluded that oxygen tension may govern the resistance of the tissue to infection. Efforts to enhance oxygen tension and resistance of tissue to infection by using musculocutaneous tissues are having direct clinical applications in the management of complicated infected wounds. May et al. (1982) reported that radical debridement, IV antibiotics, and microvascular transfer of vascularized tissues for immediate closure resulted in resolution of the infection in 18 patients who had infected bone exposure wounds of the distal extremity.

TIMING OF SKIN CLOSURE

The timing of the closure is also critical. A decision must be made as to whether the closure should be immediate or delayed. Immediate closure should be reserved for (1) wounds resulting from elective procedures classified as being either refined-clean, clean, or clean-contaminated, and (2) traumatic wounds that have not been contacted by feces, saliva, purulent exudate, or soil infection-potentiating fractions. Immediate approximation of the skin edges of this group of wounds should be accompanied by an extremely low infection rate (less than 5%, regardless of the closure technique employed).

Open wound management with delayed primary closure is recommended for wounds that exhibit a high risk of infection following primary closure. The fundamental bases for delayed primary closure were the experience of military surgeons (Heaton et al., 1966), who learned repeatedly over the centuries that immediate closure of battle wounds frequently resulted in infection. These wounds were best left open until delayed primary closure could be undertaken. All wounds resulting from high-energy deposit missile injuries, regardless of their appearance, are candidates for delayed primary closure. In these cases, the wound should be explored to remove foreign bodies, to rule out the presence of damage to specialized structures (e.g., vessels, nerve) and to relieve increased compartmental pressure that may follow edema or slow bleeding into a fascia-enclosed muscle compartment. The removal of devitalized tissue is advisable, but in practice is difficult as its definition is unclear. Wounds contacted by either feces, saliva, purulent exudates, or soil infection are also candidates for open wound management. Since a delay in wound care that lasts longer than six hours is associated with an increased risk for infection, open wound management is also recommended.

The rationale for delayed primary closure is that the healing open wound will gain resistance to infection and permit an uncomplicated closure. The reparative process of open wounds associated with this developing resistance to infection in the open wound undergoing primary closure is associated with accelerated skin healing. In addition, Johnson et al. (1982) showed that secondary closure of the subcutaneous tissue and skin results in stronger fascial strength than that encountered following primary closure.

In a group of patients with contaminated battle wounds in Vietnam, Surgeon General Heaton reported an astoundingly low 2.5% incidence of infection (Heaton et al., 1966). Using experimental models, we have confirmed the superiority of delayed closure in the treatment of contaminated wounds (Edlich et al., 1969). The optimal time for closure of the contaminated wound is on or after the fourth postwounding day.

While the type of open wound management must be individualized for each wound, aseptic technique is mandatory. For an infected wound filled with purulent discharge, the major objective is to remove the inflammatory exudate that will interfere with wound repair. Packing the wound with sterile coarse mesh gauze (Type 8) is a reliable method of absorbing the purulent exudate from the crevices of the wound. Periodic dressing changes are usually necessary every 4–6 hours until a dry granulating wound bed becomes evident. The presence of residual necrotic tissue, foreign bodies, or soil infection-potentiating fractions demands additional meticulous debridements in an operating theater as dic-

tated by the appearance of the wound. This reinspection of the wound with debridement must be continued until the wound is free of devitalized tissue and foreign bodies. Management of wounds heavily contaminated by bacteria is accomplished by packing the wound with sterile, dry, fine meshed gauze (Type 1), which is then covered by a sterile dressing. This wound should not be disturbed for the first four postoperative days unless the patient develops an unexplained fever. Unnecessary inspection during this period increases the risk of contamination and subsequent infection. On or after the fourth day, the wound margins can be approximated with minimal risk of infection. The selection of the technique for delayed primary closure will be based on the same considerations as used in primary closure. If percutaneous sutures are selected for wound closure, they should be passed through the wound edges at the end of the initial operation and left untied until the time of delayed primary closure. This step spares the patient an additional local or general anesthetic agent, which is required for sutural closure. The occasional wound that is destined to develop infection after delayed closure can be identified by using quantitative microbiology. When the bacterial count of the tissue is lower than 10^5 organisms/gm of tissue, delayed closure can be accomplished without infection (Robson et al., 1968). With the proved merit of open wound management, it is inconceivable that some urologists resort to primary closure of contaminated, dirty wounds and risk life-threatening consequences.

TECHNIQUE OF SKIN WOUND CLOSURE

There are several different methods to provide an accurate and secure approximation of the skin edges—tissue adhesives, staples, tapes, or sutures. Ideally, the choice should be based on the biologic interaction of the materials employed, the tissue configuration, and the biomechanical properties of the wound. The tissue should be held in apposition until the tensile strength of the wound is sufficient to withstand stress. A common theme of the few reportable investigations is that all biomaterials placed within the tissue damage the host defenses and invite infection.

Tissue Adhesives

Cyanoacrylate tissue adhesives have been advocated for repair of organs, or as hemostatic agents in emergency or mass combat casualty situations (Marable and Wagner, 1962; Mathes and Terry, 1963; Morgenstern et al., 1966). Mathes and Terry (1963) demonstrated that methyl-2-cyanoacrylate adhesive resuled in successful hemostasis in 14 of 16 longitudinal nephrotomies in dogs. The adhesive failed to achieve adequate hemostasis in two nephrotomies and capsule sutures were required to stop bleeding. Microscopic examination of the nephrotomies treated with the adhesive revealed extensive scar formation. One kidney contained several calculi, measuring 2–4 mm in diameter, that probably resulted from the adhesive entering the renal calyx. Fein et al. (1970) compared the efficacy of suture repair of a kidney in experimental animals to that effected by n-butyl cyanoacrylate. The principal advantage of cyanoacrylate closure was the speed with which hemostasis and repair was achieved. The disadvantages included failure of hemostasis caused by excessive bleeding, increased adhesions at the surgical site and increased incidence of calculi. It was concluded that the tissue adhesive will be a useful adjunct in the management of renal injuries once a more suitable monomer is found.

Furka et al. (1976) described an experimental procedure for closing longitudinal nephrotomy incisions that employed a tissue adhesive and eliminated the need for sutures in the renal parenchyma. The tissue adhesive, histoacryl-N-blau, permitted approximation of the two halves of the organ without sutures. The incision was then covered by regenerated cellulose, which was impregnated with a minimal amount of tissue adhesive. During the incision and closure, the hilum was temporarily constricted with non-crushing clamps. This technique produced rapid and reliable hemostasis that was followed by complete absorption of the hemostatic agents.

The use of a tissue adhesive, however, is not indicated for skin closure (Edlich et al., 1971). The polymer acts as a barrier between the growing edges of the wound, which delays healing and increases susceptibility to infection.

Clips and Staples

The urologist's burgeoning interest in clips and staples for wound closure has provided the impetus for an increasing number of scientific studies on the influence of staples and clips on the biology of wound repair and infection. In 1971, Stephens et al. (1971) compared the breaking strength of wounds closed by nickel silver Michel clips to that of wounds approximated by continuous sutures of 4–0 monofilament stainless steel wire using rats as the experimental animal. Seven days after closure, the breaking strength of the wounds closed by Michel clips (602 ± 29.8 gm) was significantly greater than that of wounds approximated by wire (488.8 ± 27.7 gm) (P <.01). Early removal of wire sutures on the third postoperative day was associated with a greater breaking strength (674.0 ± 27.0 gm) than that encountered in wounds in which the sutures were left in the wound for seven days (506.7 ± 23.0 gm) (P <.001). In similar studies, Myers et al. (1969) reported that early

removal of sutures was associated with an increase in breaking strength of the skin wounds.

The results of the investigation by Stephens et al. (1971) contrast dramatically with those reported by Harrison et al. (1975), demonstrating a deleterious effect on wound healing caused by skin clips in piglet wounds examined at the 5th, 7th, and 14th postoperative days. Wounds approximated by clips had a significantly lower tensile strength at all stages than those approximated by interrupted 2–0 monofilament nylon sutures. Comparison of the strength of 14-day wounds in which clips had been removed after only seven days showed them to be significantly stronger than similar wounds in which the clips had been retained for the entire 14-day period. Retention of the nylon sutures during the second week of healing produced no significant change in strength as compared to similar wounds in which the sutures were removed early. Harrison et al. (1975) noted progressive ulceration at the site of contact of clip teeth to the skin that may account for its deleterious effects on wound healing.

More recent studies by Jewell et al. (1983) compared the healing of sutured wounds to that of stapled wounds in the rat. The skin incisions in one experimental group were closed with regular-sized Proximate (Ethicon, Inc., Somerville, N.J.) staples implanted at 5-mm intervals, while skin incisions in the remaining group were approximated by interrupted 4–0 monofilament nylon sutures placed at 5-mm intervals. Staples and sutures were removed from the wounds on the seventh postoperative day. They reported no significant difference between the mean breaking strength of wounds with either staples or sutures at 10, 42, and 80 days after closure. However, there was a significant difference in the breaking strength at 21 days, with the sutured wounds being slightly stronger than the stapled wounds.

Chavpil and colleagues (1985) examined the influence of staple configuration on the magnitude of tissue injury and the biology of wound repair. They tested the hypothesis that arcuate shaped staples exerted different tissue reactions than rectangular staples using finite element and morphometric analyses of the staple tracts. The staples examined in this study were the arcuate shaped wide Accustaple (Deknatel), and the rectangular shaped wide Proximate and Premium (U.S. Surgical Co.) staples. Stress distribution of the rectangular and arcuate staples was made in a Swanson Analysis system with no finite memory, using linear statistics Ausys Program. Finite element analysis indicated a more uniform distribution of stresses in the nonlinear elastic environment around the arcuate staple and accumulation of stresses (crowding of the tissue elements) at the rectangular corner of a staple. The results of the finite element analysis were in agreement with the morphometric analysis of the tissue reactions to these samples. The magnitude and distribution of the inflammatory reaction along the staple tract was characteristic for each staple. The Accustaple resulted in a smooth, round staple tract. A limited inflammatory reaction developed on the outside of this staple, mainly in its upper and lower portions. The rectangular staples inflicted an irregular shaped tract with evidence of tears. Inflammatory cells accumulated mostly along the outer side of the Proximate staple, being more abundant at the upper layer of skin and at the rectangular bend of the staple. The Premium staple caused a pronounced infiltration of inflammatory cells all along the staple, but mainly at the outer side of the staple. Chavpil suggested that the limited tissue reaction to the Accustaple and Proximate staples may reflect the depth of penetration of the staple legs, which are shorter than that of the Premium staple.

It was surprising that Chavpil reported that the magnitude of tissue injury inflicted by the staples did not correlate with the magnitude of the breaking strength of skin incisions in the neck, abdomen, and groin in pigs closed separately by the three different staples. Wounds closed by the Premium exhibited a significantly greater breaking strength than that of wounds approximated by either the Accustaple or Proximate staples, irrespective of the anatomical location of the incision.

The limited tissue damage inflicted by the arcuate staple as compared to that of the rectangular staple did not appear to minimize the pain of staple penetration or removal. In a clinical study by Silloway et al. (1985), rectangular and arcuate staples were implanted into the medial aspect of the forearms of eight healthy volunteers. After staple implantation and extraction, the patients were asked to grade the severity of the pain using a numerical scale. The pain associated with staple implantation was significantly greater than that caused by staple removal. However, the configuration of the staple did not influence the magnitude of pain associated with either staple implantation or removal. While the subjects tolerated implantation of four staples, they were reluctant to have implantation of additional staples without infiltration anesthesia.

There is uniform agreement that wounds closed by staples exhibit a superior resistance to infection than wounds contaminated by the least reactive suture. In an experimental study, Johnson et al. (1981) compared the resistance to infection of contaminated wounds approximated by tapes, Premium staples, or sutures. Wounds closed by tape exhibited the greatest degree of resistance to infection, followed by the stapled wounds and then the wounds approximated by sutures. The su-

periority of tape closure was evident at all levels of contamination except 5×10^7, at which all wounds were destined to develop infection regardless of the closure technique. In the presence of lower bacterial inocula, wounds approximated by staples exhibited a lower rate of infection than the least reactive nonabsorbable suture, monofilament nylon. The infection rate in these wounds correlated with the wound bacterial counts. Sutured wounds exhibited the highest bacterial counts, followed by stapled wounds and then taped wounds. When the infection rate of wounds reached 100%, the bacterial counts of the wounds did not differ significantly.

The superior resistance of stapled wounds to infection compared with sutured wounds was confirmed by the experimental study of Stillman et al. (1984). In contaminated wounds in mice, stapled wounds displayed a lower incidence of infection than wounds approximated by either percutaneous sutures (4–0 silk, 4–0 monofilament nylon, 4–0 polyglycolic acid suture) or subcuticular sutures (4–0 polyglycolic acid).

In a recent study, we examined the influence of the staple configuration as well as its depth of implantation on the resistance of the wound to infection in experimental animals. The results of this investigation demonstrated that the configuration of the staple as well as its depth of implantation did not significantly alter the tissue's resistance to infection. Even though we were not able to demonstrate any deleterious effects of deeply implanted staples on the tissue's defenses, it is important to remember that staple cross members that are flush with the skin can result in permanent transverse scars ("cross-hatching").

A variety of disposable skin stapling devices are now commercially available for use in surgery. The selection of a skin stapler by a urologist will be determined primarily by its mechanical performance. Ideally, the device should be designed so that it does not obstruct the urologist's view of the wound edge. Moreover, the stapler should have a prepositioning mechanism that permits the urologist to hold the staple securely during its formation. The configuration of the stapler should also allow the position of its cartridge to be adjusted manually to facilitate placement of the staple. In addition, the stapler should have an ejector spring that automatically releases the staple. Finally, the handling characteristics of the stapler should be such that the urologist can easily implant a large number of staples without becoming fatigued.

Sutures

Sutures remain the most common method of approximating the divided edges of tissue. All sutures damage the local tissue defenses to infection, and several mechanisms are implicated:

1. The trauma of inserting a needle is sufficient to cause an inflammatory response.
2. The urologist's suturing technique is crucially important. Sutures tied too tightly impair tissue defenses and invite infection (Edlich et al., 1968).
3. Sutures that penetrate the intact skin provide an avenue for wound contamination via the perisutural cuff.
4. The presence of the suture material itself increases the tissue's susceptibility to infection. The magnitude of this local injury to defenses is related to the quantity of suture within the wound (e.g., diameter, length) and to its chemical composition.

The infection-potentiating effects of suture materials are listed in Table 4–4. Polyglycolic and polyglactin 910 sutures elicit the least inflammatory response of the absorbable sutures, followed by plain gut and then chromic gut. Of the nonabsorbable sutures, nylon and polypropylene are the least reactive. The effect of polybutester and polydioxanone on tissue defenses has not been investigated.

The relatively high infection rates encountered with either monofilament or multifilament stainless steel sutures may be the result of their chemical or physical configuration. Stainless steel is not generally as inert as pure polymers and undergoes degradation in vivo. In addition, metallic sutures are so stiff that movement induces tissue damage and impairs the wound's ability to resist infection.

Sutures made of natural fibers potentiate infection more than any other nonabsorbable sutures, which correlates with the tissue's reaction to these sutures in clean wounds (Edlich et al., 1973). It would appear from these experimental studies that the use of silk and cotton should be avoided in wounds having known gross bacterial contamination.

Surprisingly, our studies indicate that the physical configuration of the suture has a relatively insignificant role in the development of infection. Although the incidence of infection in contaminated tissue containing monofilament sutures was lower than in those containing multifilament sutures, these differences were not statistically significant.

Using a model similar to that reported by our laboratory, Sharp et al. (1982) reported that synthetic sutures potentiated less infection than the nonsynthetic sutures. The synthetic monofilament sutures were associated with less infection than that encountered with the multifilament synthetic sutures exposed to the same bacterial inocula. The superiority of the synthetic absorbable sutures over that of the naturally occurring gut sutures was also evident using this model.

In general, the techniques of sutural closure of skin can be divided into two types: percutaneous sutures and dermal (subcuticular) sutures. The selection of the technique for closure is influenced by the wound's configuration and biochemical properties as well as other special circumstances.

Percutaneous sutures of either monofilament nylon or polypropylene are excellent for the immediate closure of clean, refined-clean, and clean-contaminated skin wounds, because these suture materials exert the least damage to the wound's defenses (Edlich et al., 1973). The polybutester suture has unique performance characteristics that may be advantageous for wound closure. This new monofilament suture exhibits distinct differences in elongation compared with other sutures. With the polybutester suture, low forces yield significantly greater elongation than that of other sutures. In addition, its elasticity is superior to that of other sutures, allowing the suture to return to its original length once the load is removed. In a clinical setting in which the tied suture loops are extended by the edema of the wound and yet are expected to return to their original length once the edema disappears, the performance of the polybutester suture would be expected to be superior to that of other sutures. Sutures with less extensibility under low forces, like nylon, polypropylene, polyester, or silk, will frequently lacerate or necrose the encircled tissue, thereby increasing its susceptibility to infection. Percutaneous sutures must be removed before the eighth postwounding day to prevent the development of needle puncture scars. Immediately thereafter, the wound edges should be reinforced with tape skin closures to prevent wound dehiscence. During approximation of the wound, the skin edges should not be grasped or crushed.

Dermal sutures can be used alone or as an adjunct to the percutaneous sutures in wounds subjected to strong skin tensions, to serve as an added precaution against disruption of the wound. Because the magnitude of the suture's damage to the local tissue defenses is related

TABLE 4–4.

Infection-Potentiating Effect on Surgical Sutures*

ABSORBABLE	NONABSORBABLE
Polyglycolic = polyglactin =	Nylon = polypropylene
Plain gut	Dacron (coated = noncoated)
Chromic gut	Metal
	Silk = cotton

*The least reactive suture materials are listed first.

to the quantity of the suture within the wound (e.g., diameter, length), we use the narrowest-diameter suture (5–0 or 6–0) whose strength is sufficient to resist disruption of the skin wound. Dermal sutures do not improve the cosmetic appearance of the scar (Winn et al., 1977).

In special circumstances, percutaneous sutures should be totally avoided in favor of a continuous dermal suture: (1) in infants frightened at the prospects of suture removal, (2) when follow-up appointments are difficult to keep, (3) when wounds are covered by casts, and (4) in patients prone to the development of keloids. When dermal closure alone is used, it is advisable to immediately apply tape skin closures to the wound edges to provide a more accurate approximation of the epidermis.

The most severe limitation of dermal skin closure is that it potentiates the susceptibility to infection in clean-contaminated and contaminated wounds more than percutaneous sutures (Foster et al., 1977). This increased rate appears to be related to the large quantity of suture material required for a continuous dermal skin closure. After infection develops, the collecting purulent exudate spreads preferentially between the divided edges of fat rather than penetrating the tightly sutured skin edges. By the time the infection becomes clinically apparent, it has involved the entire extent of the wound. This circumstance is distinct from the localized collections of purulent discharge encountered in infected taped closed wounds. In the latter instances, the purulent discharge first exits between its wound edges before spreading between the divided layers of adipose tissue.

Tape Skin Closure

Taped wounds demonstrate far superior resistance to infection when compared to sutured wounds (Edlich et al., 1974). The ease with which wounds can be closed by tape varies according to the anatomical and biomechanical properties of the wound site. Linear wounds in skin subjected to minimal static and dynamic tensions are easily approximated by tape. The relatively lax skin of the face and abdomen makes it amenable to wound closure by tapes. Contrary to the usual expectation, approximation and eversion of skin edges with tape is easily accomplished in obese patients. The taut skin of the extremities, subjected to frequent dynamic joint movements, requires that tape skin closures be supplemented with dermal sutures. The copious secretions from the skin of the axilla, palms, and soles also discourage tape adherence.

There are numerous advantages to tape closure at linear wounds subjected to weak skin tensions: (1) the cosmetic results are excellent, (2) the discomfort of su-

ture removal is avoided, (3) suture puncture scars are eliminated, and (4) no local anesthetic agent is required for suturing. This closure technique is being used on approximately 10% of the patients with lacerations seen in our emergency medical service.

An ideal surgical wound closure tape should have the following performance characteristics. It should be strong enough, even when wet, to withstand the forces disrupting the wound. Its adhesive should be sufficiently aggressive to adhere securely to the skin and maintain approximation of the wound edges. While a secure bond to skin is necessary for wound security, it also appears to be beneficial for the tape to stretch slightly under constant stress. This ability to elongate under moderate loads reduces the shear forces on the underlying edematous skin, thereby preventing blister formation. Finally, the tape construction should provide an environment on the skin surface that is antithecal to bacterial growth. Bacterial proliferation under a tape may be a source of infection during the time in which wound epidermization is incomplete (Schauerhamer et al., 1971). On the basis of the results of in vivo and in vitro investigations, the nonwoven microporous tape (Curistrip, Kendall Co., Boston) is recommended for skin closure (Rodeheaver et al., 1985).

Even this tape will not adhere to excessively wet skin. An adhesive adjunct such as compound benzoin tincture, if applied to the skin prior to tape application, reduces the chance for dislodgment. Inadvertent spillage of this adjunct will impair the wound's ability to resist infection (Panek et al., 1972). To minimize the chance of contamination, the adhesive adjunct is applied with applicator sticks in a thin film at the wound edge.

POSTOPERATIVE WOUND CARE

Postoperative wound care should optimize healing and should be tailored to the type of wound. Unfortunately, a 10-minute discussion of technical aspects of postoperative care should probably include nine minutes of silence, because the literature dealing with this aspect of wound management deals more with testimonials than with scientific fact.

WOUND DRESSINGS

The manner in which a dressing functions is determined by its physical and chemical composition (Scales, 1963). There are eight types of absorbent cotton gauze, each type defined by the number of warp and woof threads per square inch (Table 4–5). The degree of dressing adherence to a wound is directly related to the size of dressing interstices. The larger the interstice size, the greater the chance that the dressing will be

TABLE 4–5.
Types of Gauze Dressings

	THREADS/IN.^{2}	
TYPE	WARP	WOOF
I	41–47	33–39
II	30–34	26–30
III	26–30	22–26
IV	22–26	18–22
V	20–24	16–20
VI	18–22	14–18
VII	18–22	10–14
VIII	12–16	8–12

penetrated by the granulation tissue. If debridement is the objective, the urologist should use the dressing with greater interstice size, at least larger than type I. Absorption of wound exudates is another important function of a dressing. The beneficial effects of absorbency are: (1) the bacteria contained within the absorbed fluid are removed, (2) the exudate itself is removed, depleting the wound of bacterial nutrients, and (3) tissue maceration is prevented. Absorption is due to the capillary forces of attraction exerted by the capillary spaces between the dressing fibers. The magnitude of the capillary forces depends on the surface tension of the aqueous solution, as well as the chemical bonds that form between the molecules of the exudate and dressing. The rate and amount of fluid absorbed are easily measurable parameters of the absorbency of a dressing. The sinking test described in the British Pharmaceutical Code is a standardized measurement that records the rate of absorbency of dressings. High absorbency is incompatible with nonadhesion, because the serous exudate forms a powerful and coherent glue as it dries. Removal of the absorbent dressing disrupts the fibrinous scab and any granulation tissue that has become entrapped in the dressing. Absorbent dressings are therefore useful for the debridement of open wounds.

In primarily closed wounds, the surgical dressing acts as a barrier against exogenous bacteria. Soaking dressings with serum permits passage of bacteria through the dressing (Colebrook and Hood, 1948; Lowbury and Hood, 1952). Saturation of a dressing with fluid, which wets both inner and outer surfaces of the dressing, is called fluid strike-through. As long as its outer surface remains dry, however, a dressing will remain an effective barrier to bacterial contamination.

The length of time that dry dressings should cover the closed wound is based on the knowledge of the period during which the wound is susceptible to bacterial penetration. Warren (1963) indicated that the wound edges sealed rapidly with a coagulum, thereby elimi-

nating the need for dressings on primarily closed wounds. Other surgeons recommended that the dressing remain undisturbed, as long as it was dry, until the sutures or staples were ready for removal.

Sutured wounds, as they heal, become increasingly resistant to the development of infection following surface contamination (Schauerhamer et al., 1971). Swabbing the surface of the wound with either S. auerus or E. coli during the first 48 hours after closure caused localized gross infections. Contamination after the third postoperative day did not produce gross infection in the sutured wound. Thus, barrier dressings are useful to protect the fresh incision from surface contamination in the first few days. Thereafter, removal of the dressing permits daily inspection and palpation of the wound.

Wounds closed with tape had a greater capacity to resist infection than did sutured wounds. Even immediate contamination seldom caused infection in any taped wounds. This resistance to infection of taped wounds after surface contamination reduces the need for protective dressings during the postoperative period in wounds free of sutures. In a real sense, the skin suture has the objectionable features of a small drain.

Another important purpose of some dressings is to exert pressure on the underlying tissues. A pressure dressing minimizes the accumulation of the intercellular fluid and limits the dead space. The application of pressure dressings is easiest on convex surfaces (i.e., skull, extremity). Maximal pressure should be applied to the wound site, as well as distal to it. Proximal to the wound, the pressure applied is decreased to minimize any chance of compromising the venous or lymphatic return.

A pressure dressing, by the very nature of its bulk, will immobilize what it covers. Immobilization of the site of injury is of great value—lymphatic flow is reduced, thereby minimizing the spread of the wound microflora. Furthermore, immobilized tissue demonstrates the best resistance to the growth of bacteria. Whenever possible, the site of injury should be elevated above the patient's heart to limit the accumulation of fluid in the wound interstitial spaces.

Most importantly, dressings must provide a physiologic environment that is conducive to epithelial migration from the wound edges across the surface of the fresh wound. When an area of epidermis is lost, water vapor begins at once to evaporate from the exposed dermal tissue. The exudate on the surface dries and becomes the outer layer of the scab, which does not prevent water from evaporating from the dermis underneath. The surface of the dermis itself progressively dries (within 18 hours). This dry scab and dried dermis resist migration of epidermal cells, which must seek the underlying fibrous tissue of the upper reticular layer of dermis where enough moisture remains to support cellular viability (Winter, 1962).

When the wound is covered by a dressing that prevents or delays evaporation of water from the wound surface, the scab and underlying dermis remain moist. Epidermal cells can easily migrate through the moist scab over the surface of the dermis. Under such dressings, epithelialization is more rapid, and no dry dermis is sacrificed.

The totally occlusive dressing would seem to be ideal for coverage of primarily closed wounds, and has been usefully employed in the treatment of donor sites, mesh grafts, and dermabraded skin (Rovee et al., 1972). Unfortunately, excessive exudate may make it difficult to keep the fully occlusive dressing in place, and the moist exudate, which provides an ideal medium for epidermal repair, is also a suitable culture medium for the multiplication of microorganisms. Consequently, Scales (1963) suggested that an ideal wound dressing would be a compromise between occlusion and nonocclusion.

Many dressings are commercially available for wound coverage. Unfortunately, their performance has not been documented by scientific studies, so that selection must be based on testimonials. In our clinical service, primarily closed wounds (with the exception of those located on the face) are covered by nonwoven microporous polypropylene dressings, which are attached to surrounding skin by wide strips of microporous tape with no reinforcing fibers.

In facial lacerations the development of blood clots between the edges of the sutured wounds is of more concern than the potential dangers of surface contamination. These clots will be replaced by a healing scar that can be easily avoided by swabbing the wound with half-strength hydrogen peroxide every six hours until the wound edge is free of blood. The sutures will lose their color and can be easily removed before the eighth day after closure.

In abraded skin this method of suture line care is ineffective. Despite washing the wound with hydrogen peroxide, a scab develops that makes suture removal tedious and often painful. In such cases, we swab the wound and its adjacent edges with a water-soluble ointment, such as polyethylene glycol, which dissolves the wound exudates, thereby encouraging their exodus from the wound. These sutures also must be removed before the eighth postoperative day because needle puncture scars can develop. The wound edges then should be supported by sterile, microporous tape skin closures until their adhesive bond weakens.

Ship and Weiss (1985) indicated that elimination of exposure to sun in the early postoperative or postinjury period after abrasion of the skin to prevent hyperpigmentation of the skin is a sine qua non of patient management. In the event of inadvertent exposure to sunlight, the skin should be protected with a sun-blocking agent with a sun protection factor of 15.

A whole new host of synthetic dressing materials is now available. Op-Site is an adhesive drape that is based upon either Hydron (polyhydroxyethylmethacrylate) or an elastomeric polyurethane with an adhesive backing for attachment to the adjacent dry skin (Seymour et al.). Because of its hydrophilic nature, it is permeable to water vapor, making it suitable for use on abrasions and donor sites. Op-Site is impermeable to bacteria and thereby prevents exogenous contamination. Unfortunately, the water vapor permeability rate is low, and fluid build-up beneath the dressing can eventually cause wound maceration and detachment of the dressing (Lamke et al., 1977; James and Watson, 1975). A more permeable film with similar mechanical characteristics would be of interest to the urologist.

REFERENCES

1. Abbott WM, Austen WG: Microcrystalline collagen as a topical hemostatic agent for vascular surgery. *Surgery* 1974; 75:925–933.
2. Adamsons RJ, Kahan SA: The rate of healing of incised wounds of different tissues in rabbits. *Surg Gynecol Obstet* 1970; 130:837–846.
3. Adams RJ, Musco F, Enquist IF: The relative importance of sutures to the strength of healing wounds under normal and abnormal conditions. *Surg Gynecol Obstet* 1963; 117:396–401.
4. Ad Hoc Committee of the Committee on Trauma, Division of Medical Sciences, National Academy of Sciences—National Research Council Report: Postoperative wound infections: The influence of ultraviolet irradiation of the operating room and various other factors. *Ann Surg* 1964; 160(suppl):1–192.
5. Ahrenholz DH: Effect of intraperitoneal fluid on mortality of *Escherichia coli* peritonitis. *Surg Forum* 1979; 30:483–484.
6. Ahrenholz DH, Simmons RL: Povidone-iodine in peritonitis: I. Adverse effects of local instillation in experimental *E. coli* peritonitis. *J Surg Res* 1979; 26:458–463.
7. Alexander HC, Prudden JF: The causes of abdominal wound disruption. *Surg Gynecol Obstet* 1966; 122:1223–1229.
8. Alexander JW, Fischer JE, Boyajian M, et al: The influence of hair-removal methods on wound infections. *Arch Surg* 1983; 118:347–351.
9. Alexander JW, Good RA: *Immunobiology for Surgeons.* Philadelphia, WB Saunders Co, 1970.
10. Alexander JW, Korelitz J, Alexander NS: Prevention of wound infections: A case for closed suction drainage to remove wound fluids deficient in opsonic proteins. *Am J Surg* 1976; 132:59–63.
11. Allgöwer M, Cueni LB, Stadtler K, et al: Burn toxin in mouse skin. *J Trauma* 1973; 13:95–111.

12. Almersjö O, Hulten L, Rydberg B, et al: Wound healing after depilation with a keratolytic cream: A tensiometric and histological study in the rat. *Acta Chir Scand* 1967; 133:355–362.

13. Bacchetta CA, Magee W, Rodeheaver GT, et al: Biology of infection of split thickness skin grafts. *Am J Surg* 1975; 130:63–67.

14. Baker R, Maxted WC, Dipasquale N: Regeneration of transitional epithelium of the human bladder after total surgical excision for recurrent multiple bladder cancer: Apparent tumor inhibition. *J Urol* 1965; 93:593–597.

15. Baker R, Tehan T, Kelly T: Regeneration of the urinary bladder after subtotal resection for carcinoma. *Ann Surg* 1959; 25:348–352.

16. Bartlett JG, Onderdonk AB, Drude E, et al: Quantitative bacteriology of the vaginal flora. *J Infect Dis* 1977; 136:271–277.

17. Berliner SD, Burson LC, Lear PE: Use and abuse of intraperitoneal drains in colon surgery. *Arch Surg* 1964; 89:686–690.

18. Bitterman W, Gemer M, Lutwak EM: Wound dehiscence: Increased intra-abdominal pressure after repair of diaphragmatic hernia. *Arch Surg* 1967; 94:178–180.

19. Bjorkwall T: Bacteriological examinations in maxillary sinusitis. *Acta Otolaryngol [Suppl]* (Stockh) 1950; 83:1–58.

20. Bohne AW, Osborn RW, Hettle PJ: Regeneration of the urinary bladder in the dog, following total cystectomy. *Surg Gynecol Obstet* 1955; 100:259–264.

21. Bolton JS, Bornside GH, Cohn I Jr: Intraperitoneal povidone-iodine in experimental canine and murine peritonitis. *Am J Surg* 1979; 137:780–785.

22. Bornside GH, Crowder VH Jr, Cohn I Jr: A bacteriological evaluation of surgical scrubbing with disposable iodophor-soap impregnated polyurethane scrub sponges. *Surgery* 1968; 64:743–751.

23. Botsford TW: The tensile strength of sutured skin wounds during healing. *Surg Gynecol Obstet* 1941; 72:690–697.

24. Brightmore TGJ: Haemostasis and healing following median sternotomy. *Br J Surg* 1975; 62:152.

25. Brown GL, Richardson JD, Malangoni MA, et al: Comparison of prosthetic materials for abdominal wall reconstruction in the presence of contamination and infection. *Ann Surg* 1985; 201:705–711.

26. Bryant WM, Greenwell JE, Weeks PM: Alterations in collagen organization during dilatation of the cervix uteri. *Surg Gynecol Obstet* 1968; 126:27–39.

27. Cardany CR, Rodeheaver GT, Thacker JG, et al: The crush injury: A high risk wound. *J Am Coll Emerg Physicians* 1976; 5:965–970.

28. Carrel A: Cicatrization of wounds: XII. Factors initiating regeneration. *J Exp Med* 1921; 34:425–434.

29. Carrel A, Hartmann A: Cicatrization of wounds: I. The relation between the size of a wound and the rate of its cicatrization. *J Exp Med* 1916; 24:429–470.

30. Cerise EJ, Pierce WA, Diamond DL: Abdominal drains: Their role as a source of infection following splenectomy. *Ann Surg* 1970; 171:764–768.

31. Chamberlain BE, Delmonico JE Jr, Gregg RO: Effect of gelfoam on the integrity of intestinal anastomosis. *Am J Surg* 1951; 82:462–465.

32. Chang N, Mathes SJ: Comparison of the effect of bacterial inoculation in the musculocutaneous and random flaps. *Plast Reconstr Surg* 1982; 70:1–9.

33. Chlorhexidine and other antiseptics. *Med Lett* 1976; 18(21):85–86.

34. Chodak GW, Plaut ME: Systemic antibiotics for prophylaxis in urologic surgery: A review. *J Urol* 1979; 121:695–699.

35. Chvapil M: Personal communication, 1985.

36. Cipolla AF, Narat JK: Effect of absorbable sponges on infection: Experimental study. *Surgery* 1948; 24:828–831.

37. Cobden RH, Thrasher EL, Harris WH: Topical hemostatic agents to reduce bleeding from cancellous bone: A comparison of microcrystalline collagen, thrombin, and thrombin-soaked gelatin foam. *J Bone Joint Surg [Am]* 1976; 58:70–73.

38. Colebrook L, Hood AM: Infection through soaked dressings. *Lancet* 1948; 2:682–683.

39. Conolly WB, Hunt TK, Sonne M, et al: Influence of distant trauma on local wound infection. *Surg Gynecol Obstet* 1969; 128:713–718.

40. Conolly WB, Stephens FO: Factors influencing the incidence of intraperitoneal adhesions: An experimental study. *Surgery* 1968; 63:976–979.

41. Cox HT: The cleavage lines of the skin. *Br J Surg* 1941; 29:234–240.

42. Crikelair GF: Skin suture marks. *Am J Surg* 1958; 96:631–638.

43. Crowson WN, Wilson CS: An experimental study of the effects of drains on colon anastomoses. *Am Surg* 1973; 39:597–601.

44. Cruse PJE, Foord R: A five-year postoperative study of 23,649 surgical wounds. *Arch Surg* 1973; 107:206–209.

45. Custer J, Edlich RF, Prusak M, et al: Studies in the management of the contaminated wound: V. An assessment of the effectiveness of pHisoHex and Betadine surgical scrub solutions. *Am J Surg* 1971; 121:572–575.

46. D'Ambrosia RD, Shoji H, Heater R: Secondarily infected total joint replacements by hematogenous spread. *J Bone Joint Surg [Am]* 1976; 58:450–453.

47. de Holl D, Rodeheaver G, Edgerton MT, et al: Potentiation of infection by suture closure of dead space. *Am J Surg* 1974; 127:716–720.

48. DeMuth WE Jr: The mechanism of shotgun wounds. *J Trauma* 1971; 11:219–229.

49. Department of Health, Education, and Welfare, Food and Drug Administration. OTC topical antimicrobial products. *Fed Reg* 1978; 43:1210.

50. Department of Health, Education, and Welfare, Food and Drug Administration, Proposed Rule Making. *Fed Reg* 1970; 36:6899.

51. Dineen P: Antibacterial activity of oxidized regenerated cellulose. *Surg Gynecol Obstet* 1976; 142:481–486.

52. Dineen P: The effect of oxidized regenerated cellulose

on experimental infected splenotomies. *J Surg Res* 1977; 23:114–116.

53. Dineen P: The effect of oxidized regenerated cellulose on experimental intravascular infection. *Surgery* 1977; 82:576–579.

54. Dineen P: A critical study of 100 consecutive wound infections. *Surg Gynecol Obstet* 1961; 113:91–96.

55. Dineen P, Drusin L: Epidemics of postoperative wound infections associated with hair carriers. *Lancet* 1973; 2:1157–1159.

56. Dixon RE, Kaslow RA, Mackel DC, et al: Aqueous quaternary ammonium antiseptics and disinfectants: Use and misuse. *JAMA* 1976; 236:2415–2417.

57. Douglas DM: The healing of aponeurotic incisions. *Br J Surg* 1952; 40:79–84.

58. Draser BS, Shiner M, McLeod GM: Studies on the intestinal flora: I. The bacterial flora of the gastrointestinal tract in healthy and achlorhydric persons. *Gastroenterology* 1969; 56:71–79.

59. Duke WF, Robson MC, Krizek TJ: Civilian wounds, their bacterial flora and rate of infection. *Surg Forum* 1972; 23:518–520.

60. Dunphy JE, Jackson DS: Practical applications of experimental studies in the care of the primarily closed wound. *Am J Surg* 1962; 104:273–281.

61. Dupuytren JF: *Traite theorique et practique des blessure par armes de guerre.* Paris, JB Baillere, 1834, vol 1.

62. Edgerton MT, Hansen FC: Matching facial color with split thickness skin grafts from adjacent areas. *Plast Reconstr Surg* 1960; 25:455–464.

63. Edlich RF: The biology of wound repair and infection: A personal odyssey. *Ann Emerg Med* 1985; 14:1018–1025.

64. Edlich RF, Haines PC, Pearce RSC, et al: Evaluation of a new, improved surgical drainage system. *Am J Surg* 1985; 149:295–298.

65. Edlich RF, Madden JE, Prusak M, et al: Studies in the management of contaminated wounds: VI. The therapeutic value of gentle scrubbing in prolonging the limited period of effectiveness of antibiotics in contaminated wounds. *Am J Surg* 1971; 121:668–672.

66. Edlich RF, Panek PH, Rodeheaver GT, et al: Physical and chemical configuration of sutures in the development of surgical infection. *Ann Surg* 1973; 117:679–687.

67. Edlich RF, Rodeheaver G, Kuphal J, et al: Technique of closure: Contaminated wounds. *J Am Coll Emerg Physicians* 1974; 3:375–381.

68. Edlich RF, Rodeheaver GT, Spengler M, et al: Practical bacteriologic monitoring of the burn victim. *Clin Plast Surg* 1977; 4:561–569.

69. Edlich RF, Rodeheaver GT, Thacker JG: Technical factors in the prevention of infections, in Simmons RL, Howard RJ (eds): *Surgical Infectious Diseases.* New York, Appleton-Century-Crofts, 1982, pp 449–472.

70. Edlich RF, Rogers W, Kasper G, et al: Studies in the management of the contaminated wound: I. Optimal time for closure of contaminated open wounds: II. Comparison of the resistance to infection of open and closed wounds during healing. *Am J Surg* 1969; 117:323–329.

71. Edlich RF, Schmolka IR, Prusak MP, et al: The molecular basis for toxicity of surfactants in surgical wounds: I. EP:PO block polymers. *J Surg Res* 1973; 14:277–284.

72. Edlich RF, Smith QT, Edgerton MT: Resistance of the surgical wound to antimicrobial prophylaxis and its mechanisms of development. *Am J Surg* 1973; 126:583–591.

73. Edlich RF, Thul J, Prusak M, et al: Studies in the management of the contaminated wound: VIII. Assessment of tissue adhesives for repair of contaminated tissue. *Am J Surg* 1971; 122:394–397.

74. Edlich RF, Tsung MS, Rogers W, et al: Studies in the management of the contaminated wound: I. Technique of closure of such wounds together with a note on a reproducible model. *J Surg Res* 1968; 8:585–592.

75. Ellis H: The cause and prevention of postoperative intraperitoneal adhesions. *Surg Gynecol Obstet* 1971; 133:497–511.

76. Ellis HJ, Heddle R: Does the peritoneum need to be closed at laparotomy? *Br J Surg* 1977; 64:733–736.

77. Everett WG: Suture materials in general surgery. *Prog Surg* 1970; 8:14–37.

78. Fast J, Nelson C, Dennis C: Rate of gain in strength in sutured abdominal wall wounds. *Surg Gynecol Obstet* 1947; 84:685–688.

79. Fein RL, Matsumoto T, Soloway HB: Renal injury: Suture versus n-butyl-cyanoacrylate tissue adhesive spray repair. *Invest Urol* 1970; 8:12–20.

80. Ferguson DJ: Clinical application of experimental relations between technique and wound infection. *Surgery* 1968; 63:377–381.

81. Ferguson DJ: Advances in the management of surgical wounds. *Surg Clin North Am* 1971; 51:49–59.

82. Food and Drug Administration: Burns with Hibitane tincture. *FDA Drug Bull* 1985; 15(1):9.

83. Forrester JC: Collagen morphology in normal and wound tissue, in Hunt JK (ed): *Wound Healing and Wound Infection: Theory and Surgical Practice.* New York, Appleton-Century-Crofts, 1980, pp 118–134.

84. Foster GE, Hardy EG, Hardcastle JD: Subcuticular suturing after appendicectomy. *Lancet* 1977; 1:1128–1132.

85. Furka I, Bornemissza GY, Miko I: Use of absorbable material for closure of experimental longitudinal nephrotomy. *Intern Urol Nephrol* 1976; 8:107–112.

86. Furnas DW, Somers G: Microsurgery in the prevention of traumatic tattoos. *Plast Reconstr Surg* 1976; 58:631–633.

87. Gilmore PJA, Martin TDM: Aetiology and prevention of wound infection in appendicectomy. *Br J Surg* 1974; 61:281–287.

88. Gingrass P, Grabb WC, Gingrass RP: Skin graft survival on avascular defects. *Plast Reconstr Surg* 1975; 55:65–70.

89. Goligher JC, Irvin TT, Johnston D, et al: A controlled clinical trial of three methods of closure of laparotomy wounds. *Br J Surg* 1975; 62:823–829.

90. Gorbach SL: Intestinal microflora. *Gastroenterology* 1971; 60:1110–1129.

91. Gottrup F, Firmin R, Hunt TK, et al: The dynamic properties of tissue oxygen in healing flaps. *Surgery* 1984; 95:527–536.

92. Greenburg AG, Saik RP, Peskin GW: Wound dehiscence: Pathophysiology and prevention. *Arch Surg* 1979; 114:143–146.

93. Hait MR, Robb CA, Baxter CR, et al: Comparative evaluation of Avitene microcrystalline collagen in experimental animal wounds. *Am J Surg* 1973; 125:284–287.

94. Hanisch MG, Guerriero WG: The use of microfibrillar collagen hemostat for control of renal bleeding. *J Urol* 1978; 119:312–315.

95. Harris P, Cappernauld I: Clinical experience in neurosurgery with Absele: A new absorbable haemostatic bone sealant. *Surg Neurol* 1980; 13:231–235.

96. Harrison I, Williams DF, Cuschieri A: The effect of metal clips on the tensile properties of healing skin wounds. *Br J Surg* 1975; 62:945–949.

97. Hasselgren P-O, Hagberg E, Malmer H, et al: One instead of two knives for surgical incision: Does it increase the risk of postoperative wound infection? *Arch Surg* 1984; 119:917–920.

98. Hastings JC, Van Winkle W Jr, Barker E, et al: The effect of suture materials on healing wounds of the bladder. *Surg Gynecol Obstet* 1975; 140:933–937.

99. Hau T, Nelson RD, Fiegel VD, et al: Mechanisms of the adjuvant action of hemoglobin in experimental peritonitis: 2. Influence of hemoglobin on human leukocyte chemotaxis in vitro. *J Surg Res* 1977; 22:174–180.

100. Haury BB, Rodeheaver GT, Pettry D, et al: Inhibition of nonspecific defenses by soil infection-potentiating factors. *Surg Gynecol Obstet* 1977; 144:19–24.

101. Haury B, Rodeheaver G, Vensko J, et al: Debridement: An essential component of traumatic wound care. *Am J Surg* 1978; 135:238–242.

102. Haxton H: The influence of suture materials and methods on the healing of abdominal wounds. *Br J Surg* 1965; 52:372–375.

103. Heaton LD, Hughes CW, Rosegay H, et al: Military surgical practices of the United States Army in Viet Nam. *Curr Probl Surg* 1966; 3:19.

104. Hell EA, Cruikshank CND: The effect of injury on the uptake of 3H-thymidine by guinea pig epidermis. *Exp Cell Res* 1963; 31:128–139.

105. Heppelston AG, Styles JA: Activity of a macrophage factor in collagen formation by Silica. *Nature* 1967; 214:521–522.

106. Herrmann JB: Changes in tensile strength and knot security of surgical sutures in vivo. *Arch Surg* 1973; 106:707–710.

107. Hill GB: Enhancement of experimental anaerobic infections by blood, hemoglobin, and hemostatic agents. *Infect Immun* 1978; 19:443–449.

108. Higgins GA Jr, Antkowiak JG, Esterkyn SH: A clinical and laboratory study of abdominal wound closure and dehiscence. *Arch Surg* 1969; 98:421–427.

109. Hinman F Jr, Babcock KO: Local reaction to oxidized cellulose and gelatin hemostatic agents in experimentally contaminated renal wounds. *Surgery* 1949; 26:633–640.

110. Hohn DC, MacKay RD, Halliday B, et al: Effect of O_2 tension on microbicidal function of leukocytes in wounds and in vitro. *Surg Forum* 1976; 27:18–20.

111. Holtz G: Adhesion induction by suture of varying tissue reactivity and caliber. *Int J Fertil* 1982; 27:134–135.

112. Horsley V: Historical note: Horsley and bone wax. *Surg Neurol* 1978; 9:366.

113. Howard TC, Kelley RR: The effect of bone wax on the healing of experimental rat tibial lesions. *Clin Orthop* 1969; 63:226–232.

114. Hubbard TB Jr, Khan MZ, Carag VR Jr, et al: The pathology of peritoneal repair: Its relation to the formation of adhesions. *Ann Surg* 1967; 165:908–916.

115. Hubbard TB Jr, Rever WB Jr: Retention sutures in the closure of abdominal incisions. *Am J Surg* 1972; 124:378–380.

116. Jacobs HB: Skin knife-deep knife: The ritual and practice of skin incisions. *Ann Surg* 1974; 179:102–104.

117. James JH, Watson ACH: The use of Opsite, a vapour permeable dressing, on skin graft donor sites. *Br J Plast Surg* 1975; 28:107–110.

118. Jenkins HP, Janda R, Clarke J: Clinical and experimental observations on the use of gelatin sponge or foam. *Surgery* 1946a; 20:124–132.

119. Jenkins HP, Senz EH, Owen HW, et al: Present status of gelatin sponge for the control of hemorrhage: With experimental data on its use for wounds of the great vessels and the heart. *JAMA* 1946b; 132:614–619.

120. Jenkins SD, Klamer TW, Parteka JJ, et al: A comparison of prosthetic materials used to repair abdominal wall defects. *Surgery* 1983; 94:392–397.

121. Jenkins TPN: The burst abdominal wound: A mechanical approach. *Br J Surg* 1976; 63:873–876.

122. Jewell ML, Sato R, Rahija R: A comparison of wound healing in wounds closed with staples versus skin sutures. *Contemp Surg* 1983; 22(1):29–32.

123. Johnson, A, Rodeheaver GT, Durand LS, et al: Automatic disposable stapling devices for wound closure. *Ann Emerg Med* 1981; 10:631–635.

124. Johnson AJ, Kinsey DL, Rehm RA: Observations on bladder regeneration. *J Urol* 1962; 88:494–502.

125. Johnson BW, Scott PG, Brunton JL, et al: Primary and secondary healing in infected wounds: An experimental study. *Arch Surg* 1982; 117:1189–1193.

126. Johnson P, Fromm D: Effects of bone wax on bacterial clearance. *Surgery* 1981; 89:206–209.

127. Jones RC, Shires GT: Principles in the management of wounds, in Schwartz SI (ed): *Principles in Surgery*. New York, McGraw-Hill, 1974, p 204.

128. Kapur BML, Daneswar A, Chopra P: Evaluation of peritoneal closure at laparotomy. *Am J Surg* 1979; 137:650–652.

129. Karipineni RC, Wilk PJ, Danese CA: The role of the

peritoneum in the healing of abdominal incisions. *Surg Gynecol Obstet* 1976; 142:729–730.

130. Keenan KM, Rodeheaver GT, Kenney JG, et al: Surgical cautery revisited. *Am J Surg* 1984; 147:818–821.

131. Kirk RM: Effect of method of opening and closing the abdomen on incidence of wound bursting. *Lancet* 1972; 2:352–353.

132. Kligman AM: The bacteriology of normal skin, in Maibach HI, Hildick-Smith G (eds): *Skin Bacteria and Their Role in Infection.* New York, McGraw-Hill Book Co, 1965, pp 13–21.

133. Kocher T: *Circurgishe Operationslehr.* Jena, Gustav Fisher, Germany, 1907.

134. Kon ND, Meredith JW, Poole GV Jr, et al: Abdominal wound closure: A comparison of polydioxanone, polypropylene, and Teflon-coated braided Dacron sutures. *Am Surg* 1984; 50:549–551.

135. Kraissl CJ: Selection of appropriate lines for elective surgical incisions. *Plast Reconstr Surg* 1951; 8:1–8.

136. Krizek TJ, Davis JH: The role of the red cell in subcutaneous infection. *J Trauma* 1965; 5:85–94.

137. Kuchta N, Dineen P: Effect of absorbable hemostats on intraabdominal sepsis. *Infect Surg* 1983; 2(6):441–444.

138. Kursh ED, Klein L, Persky L, et al: A comparison of the healing processes in uremic and parallel-fed rats. *Invest Urol* 1978; 15:328–330.

139. Kursh ED, Klein L, Schmitt J, et al: The effect of uremia on wound tensile strength and collagen formation. *J Surg Res* 1977; 23:37–42.

140. Lamb JP, Vitale T, Kaminski DL: Comparative evaluation of synthetic meshes for abdominal wall replacement. *Surgery* 1983; 93:643–648.

141. Lamke LO, Nilsson GE, Reithner HL: The evaporative water loss from burns and the water vapor permeability of grafts and artificial membranes used in the treatment of burns. *Burns* 1977; 3:159–165.

142. Langer K: On the anatomy and physiology of the skin: I. The cleavability of the cutis. *Br J Plast Surg* 1978a; 31:3–8.

143. Langer K: On the anatomy and physiology of the skin: II. Skin tension. *Br J Plast Surg* 1978b; 31:93–106.

144. Laufman H, Method H: Effect of absorbable foreign substance on bowel anastomosis. *Surg Gynecol Obstet* 1948; 86:669–673.

145. Laufman H, Rubel T: Synthetic absorbable sutures. *Surg Gynecol Obstet* 1977; 145:597–608.

146. Lavelle KJ, Doedens DJ, Kleit SA, et al: Iodine absorption in burn patients treated topically with povidone-iodine. *Clin Pharmacol Ther* 1975; 17:355–362.

147. Lawrie P: Unpublished work. Ethicon, Ltd, Research Unit, Edinburgh.

148. Leaper DJ, Pollock AV, Evans M: Abdominal wound closure: A trial of nylon, polyglycolic acid and steel sutures, *Br J Surg* 1977; 64:603–606.

149. Leaper DJ, Rosenberg IL, Evans M, et al: The influence of suture material on abdominal wound healing assessed by controlled clinical trials. *Eur Surg Res* 1976; 8(suppl 1):75–76.

150. Lee JT Jr, Ahrenholz DH, Nelson RD, et al: Mecha-

nisms of the adjuvant effect of hemoglobin in experimental peritonitis: V. The significance of the coordinated iron component. *Surgery* 1979; 86:41–48.

151. Leibovich SJ, Ross R: The role of the macrophage in wound repair: A study with hydrocortisone and antimacrophage serum. *Am J Surg* 1975; 78:71–92.

152. Levenson SM, Geever EF, Crowley LV, et al: The healing of rat skin wounds. *Ann Surg* 1965; 161:293–308.

153. Levenson SM, Gruber DK, Gruber C, et al: A hemostatic scalpel for burn debridement. *Arch Surg* 1982; 117:213–220.

154. Levine NS, Peterson HD, Hugh D, et al: Laser, scalpel, electrosurgical and tangential excisions of third degree burns: A preliminary report. *Plast Reconstr Surg* 1975; 56:286–296.

155. Lichenstein IL, Herzifoff S, Shore JM, et al: The dynamics of wound healing. *Surg Gynecol Obstet* 1970; 130:685–690.

156. Lilly GE, Obson DB, Hutchinson RA, et al: Clinical and bacteriologic aspects of polyglycolic acid sutures. *J Oral Surg* 1973; 31:103–105.

157. Lowbury EJL, Hood AM: A disinfectant barrier in dressings applied to burns. *Lancet* 1952; 1:899–901.

158. Lowbury EJL, Lilly HA: Use of 4% chlorhexidine detergent solution (Hibiscrub) and other methods of skin disinfection. *Br Med J* 1973; 1:510–515.

159. McDowell AJ: Wound infections resulting from the use of hot sponges. *Plast Reconstr Surg* 1959; 23:168–174.

160. McFadden PM, Peacock EE Jr: Preperitoneal abdominal wound repair: Incidence of dehiscence. *Am J Surg* 1983; 145:213–214.

161. McMinn RMH, Johnson FR: The repair of artificial ulcers in the urinary bladder of the cat. *Br J Surg* 1955; 43:99–103.

162. Madden JE, Edlich RF, Custer JR, et al: Studies in the management of the contaminated wound: IV. Resistance to infection of surgical wounds made by knife, electrosurgery, and laser. *Am J Surg* 1970; 119:222–224.

163. Madden JW, Smith HC: The rate of collagen synthesis and deposition in dehisced and resutured wounds. *Surg Gynecol Obstet* 1970; 130:487–493.

164. Magee C, Haury B, Rodeheaver G, et al: A rapid technic for quantitating wound bacterial count. *Am J Surg* 1977; 133:760–762.

165. Magee CM, Rodeheaver G, Edgerton MT, et al: A more reliable gram staining technic for diagnosis of surgical infections. *Am J Surg* 1975; 130:341–346.

166. Magee C, Rodeheaver GT, Golden GT, et al: Potentiation of wound infection by surgical drains. *Am J Surg* 1976; 131:547–549.

167. Mandell GL: Bactericidal activity of aerobic and anerobic polymorphonuclear neutrophils. *Infect Immun* 1974; 9:337–341.

168. Manz CW, LaTendresse C, Sako Y: The detrimental effects of drains on colonic anastomoses: An experimental study. *Dis Colon Rectum* 1970; 13:17–25.

169. Marable SA, Wagner DE: The use of rapidly polymerizing adhesives in massive liver resection. *Surg Forum* 1962; 13:264–266.

170. Marples RR: The effect of hydration on the bacterial flora of the skin, in Maibach HI, Hildick-Smith G (eds): *Skin Bacteria and Their Role in Infection*, New York, McGraw-Hill Book Co, 1965, pp 33.

171. Marshall KA, Edgerton MT, Rodeheaver GT, et al: Quantitative microbiology: Its application to hand injuries. *Am J Surg* 1976; 131:730–733.

172. Masterson TS, Rodeheaver GT, Morgan RF, et al: Bacteriologic evaluation of electric clipper for surgical hair removal. *Am J Surg* 1984; 148:301–302.

173. Mathes GL, Terry JW Jr: Non-suture closure of nephrotomy. *J Urol* 1963; 89:122–125.

174. May JW Jr, Gallico GG, Lukash FN: Microvascular transfer of free tissue for closure of bone wounds of distal lower extremity. *N Engl J Med* 1982; 306:253–257.

175. Michaud RN, McGrath MB, Goss WA: Improved experimental model for measuring skin degerming activity on the human hand. *Antimicrob Agents Chemother* 1972; 2:8–15.

176. Miles AA, Niven JSF: The enhancement of infection during shock produced by bacterial toxins and other agents. *Br J Exp Pathol* 1950; 31:73–95.

177. Morgenstern L: Microcrystalline collagen used in experimental splenic injury: A new surface hemostatic agent. *Arch Surg* 1974; 109:44–47.

178. Morgenstern L, Kahn FH, Weinstein IM: Subtotal splenectomy in myelofibrosis. *Surgery* 1966; 60:336–339.

179. Morgenstern L, Michel SL, Austin E: Control of hepatic bleeding with microfibrillar collagen. *Arch Surg* 1977; 112:941–943.

180. Myers MB: Prediction and prevention of skin sloughs in radical cancer surgery. *Pacific Med Surg* 1967; 73:315–318.

181. Myers MB, Cherry G, Heinburger S: Augmentation of wound tensile strength by early removal of sutures. *Am J Surg* 1969; 117:338–341.

182. Nayman J: Effect of renal failure on wound healing in dogs. Response to hemodialysis following uremia induced by uranium nitrate. *Ann Surg* 1966; 164:227–235.

183. Nelson CA, Dennis C: Wound healing: Technical factors in the gain of strength in sutured abdominal wounds in rabbits. *Surg Gynecol Obstet* 1951; 93:461–467.

184. Nichols RL, Smith JW: Intragastric microbial colonization in common disease states of the stomach and duodenum. *Ann Surg* 1975; 182:557–561.

185. Nichols RL, Broido P, Condon RE, et al: Effect of preoperative neomycin-erythromycin intestinal preparation on the incidence of infectious complications following colon surgery. *Ann Surg* 1973; 178:453–459.

186. Nichols RL, Condon RE: Role of the endogenous gastrointestinal flora in postoperative wound sepsis. *Surg Annu* 1975.

187. Nichter LS, Bryant CA, Kenney JG, et al: Injuries due to commercial electric current. *J Burn Care Rehab* 1984; 5:124–137.

188. Noe JM, Kalish S: The problem of adherence in dressed wounds. *Surg Gynecol Obstet* 1978; 147:185–188.

189. Nolte WA: Oral ecology, in Nolte WA (ed): *Oral Microbiology*, ed 2. St Louis, CV Mosby Co, 1973, pp 3–44.

190. Nora PF, Vanecko RM, Bransfield JJ: Prophylactic abdominal drains. *Arch Surg* 1972; 105:173–175.

191. Ordman LJ, Gillman T: Studies in the healing of cutaneous wounds: III. A critical comparison in the pig of the healing of surgical incisions closed with sutures or adhesive tape based on tensile strength and clinical and histological criteria. *Arch Surg* 1966; 93:911–928.

192. Panek PH, Prusak MP, Bolt D, et al: Potentiation of wound infection by adhesive adjuncts. *Am Surg* 1972; 38:343–345.

193. Peacock EE Jr, Van Winkle W Jr: Repair of skin wounds, in *Surgery and Biology of Wound Repair*. Philadelphia, WB Saunders Co, 1970, p 71.

194. Peacock EE Jr, Van Winkle W Jr: *Surgery and Biology of Wound Repair*, ed 2. Philadelphia, WB Saunders, 1976, pp 76–89.

195. Pearce RSC, West LR, Rodeheaver GT, et al: Evaluation of a new hydrogel coating for drainage tubes. *Am J Surg* 1984; 148:687–691.

196. Pecora DV, Landis RE, Martin E: Location of cutaneous microorganisms. *Surgery* 1968; 64:1114–1118.

197. Peinert RD, Courtiss EH: Excision from a distance: A technique for removal of benign subcutaneous lesions. *Plast Reconstr Surg* 1983; 72:94–96.

198. Perey B, Watier A: Effect of human tissues on the breaking strength of catgut and polyglycolic acid sutures. *Chir Gastroenterol* 1975; 9:87–91.

199. Pollock AV, Evans M: Povidone-iodine for the control of surgical wound infection: A controlled clinical trial against topical cephaloridine. *Br J Surg* 1975; 62:292–294.

200. Poole GV Jr: Mechanical factors in abdominal wound closure: The prevention of fascial dehiscence. *Surgery* 1985; 97:631–639.

201. Poole GV Jr, Meredith JW, Kon ND, et al: Suture technique and wound-bursting strength. *Am Surg* 1984; 50:569–572.

202. Postlethwait RW: Long term comparative study of nonabsorbable sutures. *Ann Surg* 1970; 171:892–897.

203. Postlethwait RW: Rate of breaking strength of absorbable sutures in the stomach. *Surgery* 1975; 78:531–533.

204. Pruett TL, Rotstein OD, Crass J, et al: Percutaneous aspiration and drainage for suspected abdominal infection. *Surgery* 1984; 96:731–735.

205. Raves JJ, Slifkin M, Diamond DL: A bacteriologic study comparing closed suction and simple conduit drainage. *Am J Surg* 1984; 148:618–620.

206. Ray JA, Doddi N, Regula D, et al: Polydioxanone (PDS), a novel monofilament synthetic absorbable suture. *Surg Gynecol Obstet* 1981; 153:497–507.

207. A Report to the Medical Research Council by the Subcommittee on Aseptic Methods in Operating Theatres of their Committee on Hospital Infection: Aseptic methods in the operating suite. *Lancet* 1968; 1:705–714.

208. Richards PC, Balch CM, Aldrete JS: Abdominal wound closure: A randomized prospective study of 571 patients comparing continuous vs. interrupted suture techniques. *Ann Surg* 1983; 197:238–243.

209. Roberts AH, Rye GD, Edgerton MT, et al: Activity of

antibiotics in contaminated wounds containing clay soil. *Am J Surg* 1979; 137:381–383.

210. Robson MC, Lea CE, Dalton JB, et al: Quantitative bacteriology and delayed wound closure. *Surg Forum* 1968; 19:501–502.

211. Rodeheaver GT, Edgerton MT, Elliott MB, et al: Proteolytic enzymes as adjuncts to antibiotic prophylaxis of surgical wounds. *Am J Surg* 1974; 127:564–572.

212. Rodeheaver GT, McLane M, West L, et al: Evaluation of surgical tapes for wound closure. *J Surg Res* 1985; 39:251–257.

213. Rodeheaver GT, Marsh D, Edgerton MT, et al: Proteolytic enzymes as adjuncts to antimicrobial prophylaxis of contaminated wounds. *Am J Surg* 1975; 129:537–544.

214. Rodeheaver G, Pettry D, Turnbull V, et al: Identification of the wound infection-potentiating factors in soil. *Am J Surg* 1974; 128:8–14.

215. Rodeheaver GT, Smith SL, Thacker JG, et al: Mechanical cleansing of contaminated wounds with a surfactant. *Am J Surg* 1975; 129:241–245.

216. Rodeheaver GT, Thacker JG, Edlich RF: Mechanical performance of polyglycolic acid and polyglactin 910 synthetic absorbable sutures. *Surg Gynecol Obstet* 1981; 153:835–841.

217. Rodeheaver GT, Thacker JG, Owen J, et al: Knotting and handling characteristics of coated synthetic absorbable sutures. *J Surg Res* 1983; 35:525–530.

218. Roettinger W, Edgerton MT, Kurtz LD, et al: Role of inoculation site as a determinant of infection in soft tissue wounds. *Am J Surg* 1973; 126:354–358.

219. Ross G Jr, Thompson IM, Bynum WR, et al: The role of smooth muscle regeneration in urinary tract repair. *J Urol* 1966; 95:541–546.

220. Rosebury T: Distribution and development of the microbiota of man, in Rosebury T (ed): *Microorganisms Indigenous to Man.* New York. McGraw-Hill Book Co, 1962, pp 310–350.

221. Rovee DT, Kurowsky CA, Labun J, et al: Effect of local wound environment on epidermal healing, in Maibach HI, Rovee DT (eds): *Epidermal Wound Healing.* Chicago, Year Book Medical Publishers, 1972, p 159.

222. Salthouse TN, Williams JA, Willigan DA: Relationship of cellular enzyme activity to catgut and collagen suture absorption. *Surg Gynecol Obstet* 1969; 129:691–696.

223. Sanders AR, Schein CJ, Orkin LA: Total mucosal denudation of the canine bladder: Experimental observations and clinical implications: Final report. *J Urol* 1958; 79:63–77.

224. Sanders RJ, DiClementi D, Ireland K: Principles of abdominal wound closure: I. Animal studies. *Arch Surg* 1977; 112:1184–1187.

225. Scales JT: Wound healing and the dressing. *Br J Ind Med* 1963; 20:82–94.

226. Schauerhamer RA, Edlich RF, Panek P, et al: Studies in the management of the contaminated wound: VII. Susceptibility of surgical wounds to postoperative bacterial contamination. *Am J Surg* 1971; 122:74–77.

227. Scher KS, Coil JA Jr: Effects of oxidized cellulose and microfibrillar collagen on infection. *Surgery* 1982; 91:301–304.

228. Scully RE, Artz CP, Sako Y: The criteria for determining the viability of muscle in war wounds, in *Battle Wounds.* Surgical Research Team, Army Medical Service Graduate School, Washington, DC, 1956, vol 3.

229. Seropian R, Reynolds BM: Wound infections after preoperative depilatory versus razor preparation. *Am J Surg* 1971; 121:251–254.

230. Seymour DE, DaCosta NM, Hodgson ME, et al: British patent #1, 280, 631.

231. Sharp WV, Belden TA, King PH, et al: Suture resistance to infection. *Surgery* 1982; 91:61–63.

232. Sheikh KMA, Duggal K, Relfson M, et al: An experimental histopathologic study of surgical glove powders. *Arch Surg* 1984; 119:215–219.

233. Ship AG, Weiss PR: Pigmentation after dermabrasion: An avoidable complication. *Plast Reconstr Surg* 1985; 75:528–532.

234. Silloway KA, Morgan RF, Kenney JG, et al: Arcuate staple: Its influence on pain of staple penetration and removal. *Am J Surg* 1985; 150:612–614.

235. Slaughter FG: *Immortal Magyar: Semmelweis Conqueror of Child-bed Fever.* New York, Schuman, 1950, p 3.

236. Smith M, Enquist IF: A quantitative study of impaired healing resulting from infection. *Surg Gynecol Obstet* 1967; 125:965–973.

237. Smylie HG, Logie JRC, Smith G: From PhisoHex to Hibiscrub. *Br Med J* 1973; 4:586–589.

238. Sprunt K, Redman W, Leidy G: Antibacterial effectiveness of routine hand washing. *Pediatrics* 1973; 52:264–271.

239. Stamey TA: Diagnosis of bacteriuria, in Stamey TA (ed): *Urinary Infections.* Baltimore, Williams & Wilkins Co, 1974, p 24.

240. Standeven A: Rupture of laparotomy wounds. *Lancet* 1955; 1:533–535.

241. Steere AC, Mallison GF: Handwashing practices for the prevention of nosocomial infections. *Ann Intern Med* 1975; 83:683–690.

242. Stein AA, Wiersum J: The role of renal dysfunction in abdominal wound dehiscence. *J Urol* 1959; 82:271–273.

243. Stephens FO, Hunt TK, Dunphy JE: Studies of traditional methods of care on the tensile strength of skin wounds in rats. *Am J Surg* 1971; 122:78–80.

244. Stevenson TR, Rodeheaver GT, Golden GT, et al: Damage to tissue defenses by vasoconstrictors. *J Am Coll Emerg Phys* 1975; 4:532–535.

245. Stevenson TR, Thacker JG, Rodeheaver GT, et al: Cleansing the traumatic wound by high pressure syringe irrigation. *J Am Coll Emerg Phys* 1976; 5(1):17–21.

246. Stillman RM, Marino CA, Seligman SJ: Skin staples in potentially contaminated wounds. *Arch Surg* 1984; 119:821–822.

247. Stokes EJ, Howard E, Peters JL, et al: Comparison of antibiotic and antiseptic prophylaxis of wound infection

in acute abdominal surgery. *World J Surg* 1977; 1:777–782.

248. Stone HH, Hoefling SJ, Strom PR, et al: Abdominal incisions: Transverse vs. vertical placement and continuous vs. interrupted closure. *South Med J* 1983; 76:1106–1108.

249. Sturdivan LM, Sacco WJ, Edgerton MT, Edlich RF: Firearm injuries: Medicolegal considerations, in Wolcott BW, Rund DA (eds): *Emergency Medicine Annual*, Norwalk, Conn, Appleton-Century-Crofts, 1984, p 125–168.

250. Talkington CM, Thompson JE: Prevention and management of infected prostheses. *Surg Clin North Am* 1982; 62:515–530.

251. Tera H, Aberg C: Tissue strength of structures involved in musculoaponeurotic layer sutures in laparotomy incisions. *Acta Chir Scand* 1976; 142:349–355.

252. Thacker JG, Iachetta FA, Allaire PE, et al: Biomechanical properties—their influence on planning surgical excisions, in Krizek TJ, Hoopes PE (eds): *Symposium on Basic Science in Plastic Surgery*, St Louis, CV Mosby Co, 1975, vol 15, pp 72–79.

253. Uhrich GI: Effect of oxidized cellulose in the protection of the suture line in intestinal anastomoses in dogs. *Arch Surg* 1949; 59:326–336.

254. US General Services Administration: O-T-C topical antimicrobial products and drug and cosmetic products. *Fed Reg* 1974; 39(179):33118.

255. vanSonnenberg E, Ferrucci JT Jr, Mueller PR, et al: Percutaneous drainage of abscesses and fluid collections: Technique, results, and applications. *Radiology* 1982; 142:1–10.

256. Van Winkle W Jr, Hastings JC, Barker E, et al: Effect of suture materials on healing skin wounds. *Surg Gynecol Obstet* 140:7–12, 1975.

257. Verklin R, Rodeheaver GT, Hudson R, et al: Rapid antibiotic disk sensitivities of burn eschar and infected wounds. *Surg Gynecol Obstet* 1977; 144:507–511.

258. Villasenor RA, Harris DF II, Barron GJ, et al: Powder-free surgical gloves. *Ophth Surg* 1984; 15:241–243.

259. Vosti KL: Special problems in prophylaxis of endocarditis following genitourinary tract and obstetrical and gynecological procedures. American Heart Association Monograph, no. 52, 1977, pp 75–79.

260. Walter CW: Disinfection of hands. *Am J Surg* 1965; 109:691–693.

261. Walter CW, Kundsin RB: The bacteriologic study of surgical gloves from 250 operations. *Surg Gynecol Obstet* 1969; 129:949–952.

262. Warren R: *Surgery*. Philadelphia, WB Saunders Co, 1963, p 43.

263. Watson ED, Hoffman NJ, Simmers RW, et al: Aerobic and anaerobic bacterial counts of nasal washings: Presence of organisms resembling *Corynebacterium acnes*. *J Bacteriol* 1962; 83:144–148.

264. Weeks JR: Prostaglandins. *Annu Rev Pharmacol* 1972; 12:317–336.

265. Wheeler CB, Rodeheaver GT, Thacker JG, et al: Side-effects of high pressure irrigation. *Surg Gynecol Obstet* 1976; 143:775–778.

266. Whipple AO, Elliott RHE Jr: The repair of abdominal incisions. *Ann Surg* 1938; 108:741–756.

267. Willoughby DA: Some views on the pathogenesis of inflammation, in Montagna W, Bentley JP, Dobson R (eds): *The Dermis: Advances in the Biology of the Skin*. New York, Appleton-Century-Crofts, 1970, vol X, pp 221–230.

268. Winn HR, Jane JA, Rodeheaver G, et al: Influence of subcuticular sutures on scar formation. *Am J Surg* 1977; 133:257–259.

269. Winter GD: Formation of the scab and the rate of epithelization of superficial wounds in the skin of the young domestic pig. *Nature* 1962; 193:293–294.

270. Wood WB Jr, Smith MR, Perry WD, et al: Studies on the cellular immunology of acute bacteriemia: I. Intravascular leucocytic reaction and surface phagocytosis. *J Exp Med* 1951; 94:521–533.

271. Wray CR: Force required for wound closure and scar appearance. *Plast Reconstr Surg* 1983; 72:380–382.

Chapter 5

Urinary Tract Infections

JACK SOBEL, M.D.
DONALD KAYE, M.D.

A normal urinary tract is one without bacteria, obstruction, inflammation, stones, tumors, vesicoureteral reflux, or urologic disorders. "Bacteriuria" is a commonly used term and literally means bacteria in the urine. The probability of the presence of infection in the urinary tract as opposed to bacterial contamination of a specimen can be ascertained by means of quantitating numbers of bacteria in voided urine, suprapubic aspiration, and in urine obtained via urethral catheter. "Significant bacteriuria" has a clinical connotation and is used to describe the number of bacteria in voided urine that exceed the number usually due to contamination from the anterior urethra. In the past two decades the standard for significance with regard to bacteriuria has been $\geq 10^5$ bacteria/ml, the implication being that in the presence of 10^5 bacteria/ml, infection must be seriously considered. Recently, it has become apparent that patients with symptomatic lower urinary tract infections may have true bacterial infection with between 10^2 and 10^4 bacteria/ml; however, the standard of $\geq 10^5$ still applies to asymptomatic patients. "Asymptomatic bacteriuria" refers to significant bacteriuria in patients without symptoms.

Urinary tract infections may involve only the lower urinary tract or may involve both the upper and lower urinary tracts simultaneously. The term "cystitis" has been used to describe the clinical syndrome caused by bacteria accompanied by dysuria, frequency, urgency, and occasionally suprapubic tenderness. These symptoms, however, may be related to lower urinary tract inflammation involving the urethra only. Thus, the term "cystourethritis" is preferable. It has become popular to refer to this clinical entity as the "frequency-dysuria syndrome," as it may be caused by both bacterial and nonbacterial etiologies (for example, gonorrhea or chlamydial urethritis). Furthermore, the presence of symptoms of lower tract infection without upper tract symptoms by no means excludes upper tract involvement, which is often simultaneously present.

"Acute pyelonephritis" describes the clinical syndrome characterized by flank pain and/or tenderness with or without fever, often associated with dysuria, urgency, and frequency. However, these symptoms can occur in the absence of infection (e.g., in renal infarction or renal calculus). A more rigorous definition of acute pyelonephritis is the above syndrome accompanied by significant bacteriuria and acute infection in the kidney. Acute pyelonephritis is associated with bacteriuria of 10^5 or more bacteria/ml in > 93% of the cases.

Urinary tract infection may occur de novo or may be recurrent. Recurrences may be either *relapses* or *reinfections*. Relapse of bacteriuria refers to recurrence of bacteriuria with the same infecting microorganism present before therapy was started, due to persistence of the organism in the urinary tract. Reinfection is a recurrence of bacteriuria with a different microorganism, a new infection. Occasionally reinfection may occur with the same microorganism, which may have persisted in the vagina or feces. This can be mistaken for a relapse.

The term "chronic urinary tract infection" has little meaning. True chronic infection should refer to the persistence of the same organism for months or years with relapses after treatment. The term "chronic pyelonephritis" is difficult to define and means different things to different authors. To some, chronic pyelonephritis refers to pathologic changes in the kidney due to infection only. In pathological terms, rigid criteria have been used to describe areas of normal kidneys with patchy interstitial nephritis associated with inflammation of the pelvis and calyces. Unfortunately, similar pathologic alterations are found in several other entities such as chronic urinary tract obstruction, analgesic nephropathy, hypokalemic nephropathy, vascular disease, and uric acid nephropathy. Pathologic descriptions frequently do not differentiate between the changes produced by infection versus those produced by these other entities.

Papillary necrosis from infection is an acute complication of pyelonephritis usually in the presence of diabetes mellitus, urinary tract obstruction, sickle cell disease, or analgesic abuse. Papillary necrosis can occur in the absence of infection in some of these conditions. The necrotic renal papillae may slough and cause unilateral or bilateral ureteral obstruction. *Intrarenal abscess* may result from bacteremia or may be a complication of severe pyelonephritis. *Perinephric abscess* occurs when microorganisms from either the renal parenchyma or blood are deposited in the soft tissues surrounding the kidneys.

PATHOGENESIS

ROUTES OF INFECTION

There are three possible routes by which bacteria can invade and spread within the urinary tract. These are the lymphatic, hematogenous, and ascending pathways. By far the overwhelming majority of urinary infections result from the ascending route and only infrequently arrive via the hematogenous or perhaps the lymphatic route.

HEMATOGENOUS ROUTE.—Infection of the kidney by the hematogenous route is uncommon; however, the kidney is occasionally secondarily infected in patients with *Staphylococcus aureus* bacteremia or *Candida fungemia*. Animal studies have shown that following *Escherichia coli* bacteremia, these organisms are rapidly eliminated from the urine unless the urinary tract is obstructed by ligation of the ureter (Guze and Beeson, 1956; Pratt et al., 1965). Thus, it appears that in humans infection of the kidney with gram-negative bacteria rarely occurs by the hematogenous route.

LYMPHATIC ROUTE.—Although there have been demonstrations of lymphatic connections between the bowel and ureter and kidney in animals as well as the observation that increased pressure in the bladder can cause lymphatic flow to be directed toward the kidney, evidence for a significant role of renal lymphatics in the pathogenesis of pyelonephritis is unimpressive.

ASCENDING ROUTE.—Considerable clinical evidence suggests that the ascent of bacteria within the urinary tract represents the most common pathway of infection in the urinary tract. The ascent is initiated within the urethra. In the male, the length of the urethra together with the antibacterial properties of prostatic secretions are effective barriers to invasion by this route. In contrast, in the female, the short length of the urethra together with the tendency to periurethral contamination with pathogenic bacteria may explain why females have a higher frequency than males of urinary tract infection (Cox and Hinman, 1961; Stamey, 1973a). Studies on females using suprapubic puncture techniques revealed the occasional presence of small numbers of microorganisms in the bladder urine of uninfected persons (Bran et al., 1972). Massage of the urethra in women (Bran et al., 1972) and presumably sexual intercourse (Buckley et al., 1978; Kelsey et al., 1979) can force bacteria into the bladder. Further evidence for the role of the ascending route is suggested by studies in catheterized patients. A straight catheterization of the bladder results in urinary tract infections in about 1% of ambulatory patients (Turck et al., 1962b). An infection develops within three or four days in virtually all patients with indwelling catheters with open drainage systems (Kass, 1956). Recently, Schaeffer and Chmiel (1983a) as well as Daifuku and Stamm (1984) in prospective studies of catheterized patients observed an excellent correlation between a rectal and periurethral culture and the organisms subsequently found to be the cause of urinary tract infection. Similarly, Stamey et al., (1971) showed in women prone to recurrent bouts of cystitis that the organisms that cause urinary tract infection colonize the vaginal introitus before the infection develops. The source of *E. coli* has been found to be the feces, in which it occurs in very large numbers. This relationship has been substantiated using a typing system for *E. coli* (Gruneberg et al., 1969; Stamey et al., 1971). Within the bladder, bacteria may multiply and then pass up the ureters, especially if cystoureteral reflux is present, to reach the renal pelvis and parenchyma.

Experimental confirmation of the ascending route of infection has been provided by several investigators. Although it is difficult to produce pyelonephritis by introduction of bacteria into the normal bladder, it may be achieved by means of a large inoculum of uropathogens (Freedman and Beeson, 1958; Hepinstal, 1964). Introducing a foreign body into the bladder facilitates experimental bladder and renal infection and reduces the bacterial inoculum required. The most conclusive experimental evidence for ascending infection was provided by unilateral ureteral ligation in rats and then infecting the bladder. In these animals, infection was confined to the kidney that was not obstructed (Vivaldi et al., 1965; Smellie et al., 1975a). In contrast, the obstructed kidney became highly susceptible to hematogenous infection.

Once bacteria reach the urinary tract, three factors determine whether infection ensues: (1) virulence of the microorganism, (2) the inoculum size, and (3) the adequacy of host defense mechanisms. These factors will also determine the anatomical level of urinary tract infection. In general, recurrent urinary infection mainly reflects a failure of normal host defense mechanisms rather than microbial characteristics.

URINARY PATHOGENS

The majority of infections are caused by facultative anaerobes usually originating from the flora of the bowel. Other pathogens such as Group B streptococci, *S. epidermidis*, and *Candida albicans* originate in the flora of the vagina or perineal skin in females.

Overall, *E. coli* is by far the most common urinary pathogen, accounting for 85% of community-acquired urinary tract infections (Maskell et al., 1983b; Bryan and Reynolds, 1984a). Far less commonly, other enteric gram-negative bacteria such as *Proteus* sp., *Klebsiella* sp., etc., as well as *S. saprophyticus*, are responsible for community-acquired infections. The distribution of urinary pathogens in hospitalized patients is different, with *E. coli* accounting for about 50% of infections, and *Klebsiella, Enterobacter, Citrobacter, Serratia, Pseudomonas aeruginosa, Providencia*, enterococcus, and *S. epidermidis* accounting for most of the rest (Kennedy et al., 1965). Fungal infections occur almost exclusively in hospitalized patients. Indwelling catheters, cross infection, instrumentation of the urinary tract, and selection of a resistant bowel and environmental flora by antimicrobial therapeutic agents contribute to the altered microbiology of nosocomial urinary tract infections.

Virtually every organism has been associated with urinary tract infection, particularly in the hospital setting, often in association with instrumentation and catheterization. Included in the growing list of suspected urinary pathogens are *Lactobacillus, Gardnerella vaginalis*, and *Mycoplasma* species including *Ureaplasma urealyticum* (Maskell et al., 1983b). Group B streptococcal infections are predominantly found in diabetics, and *S. epidermidis* is commonly found in catheterized patients. In contrast, *S. faecalis* frequently occurs in noninstrumented subjects.

It is apparent that some urinary pathogens are relatively avirulent and opportunistic and only capable of inducing urinary infection when natural host defense mechanisms are compromised. Other organisms appear capable of causing infection and invading the lower and/or upper urinary tract in the absence of obstruction, other structural abnormalities, or urinary tract catheterization. These true uropathogens appear capable of gaining access to the urinary tract and withstanding the normally highly efficient defense mechanisms. Certain species of bacteria (e.g., *E. coli*) are the dominantly found uropathogens. Even within these species, a few serogroups of *E. coli* (01, 02, 04, 06, 07, 075) cause a high proportion of urinary infections (Roberts et al., 1984). Some have reported that the frequency exceeds the anticipated frequency based on the distribution frequency of *E. coli* strains found in the stool (Vosti et al., 1964). Recently, additional studies by Brooks et al.

(1980) have reported that serogroups 02, 04, 06, 08, 018ab, and 072 were significantly more common in patients with urinary infections than among periurethral isolates in healthy individuals. Additional studies of pyelonephritogenic strains of *E. coli* reveal the presence of a limited number of virulence clones of *E. coli* possessing several virulence factors, such as P-fimbriation and hemolysin production, which are only expressed in certain O groups (Vaisanen-Rhen et al., 1984). This has led to the concept of uropathogenic *E. coli* whereby certain strains are selected from the fecal flora, not by chance, but by the presence of virulence factors that enhance colonization and facilitate invasion of the urinary tract. Other virulence factors aid in persistence of the organisms within the urinary tract and provide these organisms with the capacity to induce inflammation in the urethra, bladder, or renal parenchyma.

Recognized *E. coli* virulence factors include increased adherence to uroepithelial cells (Svanborg-Eden et al., 1981; Svanborg-Eden et al., 1983), resistance to serum bactericidal activity (Bjorksten and Kaijser, 1978), high quantity of K antigen (Roberts et al., 1984), and hemolysin production (Hughes et al., 1983). These virulence factors are less frequently found among serotypes of *E. coli* in the fecal flora that in turn are less likely to produce infection (Vosti et al., 1964).

The mean generation time of six isolates of *E. coli* in shake culture in urine was significantly shorter than that of 14 isolates of less commonly encountered pathogens, including *Proteus, Pseudomonas, Klebsiella, S. saprophyticus*, and *Streptococcus faecalis* (O'Grady et al., 1968; Rocha, 1972). In an experimental in vitro model of the lower urinary tract which used urine as a medium and simulated the flushing effect of the human bladder, *E. coli* outgrew other organisms in mixed cultures. Rapid generation may therefore be an additional factor determining the pathogenicity of *E. coli* in the urinary tract.

In the past, urinary isolates of coagulase negative staphylococci were considered contaminants even when found in titers greater than 10^5 bacteria/ml. Their pathogenic role has now been acknowledged, and in the United States approximately 10% of symptomatic lower urinary tract infections in young, sexually active adult females are recognized as being caused by *S. saprophyticus* (Latham et al., 1983). This is still somewhat lower than in Sweden and Europe, where approximately 27% of community-acquired urinary tract infections are caused by this organism (Hovelius and Mardh, 1977). *S. saprophyticus* is usually identified by novobiocin resistance. In contrast to the situation in young females, *S. saprophyticus* rarely causes infection in males or elderly persons, although infection of struvite stones has been reported (Fowler and Mariano, 1984). This is not surprising, since *S. saprophyticus* synthesizes urease, which converts urea to ammonia with a resultant increase in urinary pH enhancing struvite and carbonate apatite crystallization. After *E. coli, S. saprophyticus* is now the second most common cause of urinary tract infection in young sexually active females, resulting in both cystitis and pyelonephritis. A pathogenesis similar to that of *E. coli* is suggested by the association of rectal, vaginal, and urethral colonization with *S. saprophyticus*. Antibody coating of bacteria tests suggests that half the urinary tract infections caused by *S. saprophyticus* involve the upper urinary tract. Low bacterial count (10^2–10^4 CFU/ml) frequency-dysuria may also occur. *S. saprophyticus* may have a predilection for causing urinary tract infection by virtue of its avid adherence to uroepithelial cells. Recently, Hovelius and Mardh (1984) suggested that the attachment adhesin has a lactosamine structure. *S. saprophyticus* appears uniformly sensitive to most antimicrobials used to treat urinary tract infection. Recurrent infection due to *S. saprophyticus* occurs in approximately 10% of patients after appropriate therapy (Latham et al., 1983). A seasonal variation with a later summer/fall peak has been reported (Hovelius and Mardh, 1977).

BACTERIAL ADHERENCE IN THE PATHOGENESIS OF URINARY TRACT INFECTIONS

Adherence of bacteria to uroepithelial cells is a prerequisite for colonization and infection, particularly in a system of continuous urinary flow, including the powerful effect of micturition. Pathogens must bind to the epithelial surface to cause disease (Mulholland, 1979). Bacterial adherence or attachment is used to refer to the specific binding of bacteria to epithelial cell surfaces and not to passive mechanical trapping.

Numerous studies have emphasized the critical role of bacterial adherence in selecting uropathogens from the enormous numbers of bacteria in the fecal flora potentially capable of causing urinary infections (Svanborg-Eden et al., 1981; Svanborg-Eden et al., 1983), as well as in determining the level of infection in the urinary tract. Adherence is, however, only one of many virulence factors bacteria must possess to induce inflammation and damage within the urinary tract.

Although many in vitro studies indirectly support the concept that bacterial adherence is a prerequisite for colonization and infection, some authors have not accepted this basic premise and argue that all bacteria can potentially cause urinary tract infection and that tissue tropism of uropathogens is based on specific factors other than the capacity to adhere to urinary mucosal surfaces (Harber et al., 1984). Unfortunately, models

suitable for studying in vivo attachment are relatively rare. Recently, however, the experimental model of ascending pyelonephritis in mice and primates has been found to be useful in defining the role of bacterial adherence and bacterial adhesins (Roberts and Phillips, 1979; Hagberg et al., 1983; Svanborg-Eden et al., 1983).

Indirect evidence supporting the importance of bacterial adherence has been the observed relationship between in vitro adherence of *E. coli* and severity of urinary tract infection in vivo (Varian and Cooke, 1980). Svanborg-Eden et al. (1976) reported that *E. coli* from the urine of children with acute pyelonephritis attached better than *E. coli* from urine of children with acute cystitis, which in turn attached better than those from urine of children with asymptomatic bacteriuria or *E. coli* derived from the stools of healthy children. This concept equates anatomical level of infection with severity of infection and, unfortunately, neglects to some extent the fact that localization of infection depends on other bacterial virulence factors as well as host susceptibility and defense mechanisms. In other studies, however, no relationship was seen between infection site and intensity of bacterial adherence in adult females with complicated urinary tract infections (Fowler and Stamey, 1977). Nevertheless, Svanborg-Eden and associates confirmed their original observations in an additional study of adult females without underlying obstruction or defects in urine flow (Svanborg-Eden et al., 1981, 1983). The low attachment capacity of *E. coli* urinary isolates in patients with asymptomatic bacteriuria (even when chronic and originating from the kidney) is unexplained. Possibly after reaching the kidney, bacteria may alter to ensure survival. At this stage, mucosal adhesion may no longer be required.

Evidence supporting the role of adherence capacity in selecting for uropathogens is suggested by studies of various staphylococcal species. *S. saprophyticus*, a common cause of cystitis in sexually active females, adheres in greater numbers in vitro to uroepithelial cells than *S. aureus*, a rare cause of urinary infection. Furthermore, *S. epidermidis*, a common cause of catheter-associated urinary infection, possesses the unique capacity to adhere to foreign bodies in contrast to other staphylococcal species. While *P. mirabilis* adheres in large numbers to uroepithelial cells in vitro, Svanborg-Eden et al. (1981) noted that *P. mirabilis* strains attach only to squamous epithelial, not transitional epithelial, cells in vitro.

BACTERIAL ADHERENCE PROPERTIES.—Bacterial fimbriae are believed to be the most common bacterial surface adhesins or ligands responsible for attachment. Methods such as electron microscopy as well as erythrocyte agglutination have been used to characterize fimbriae, and several specific morphological and functional types have been identified. However, several bacterial species are known to adhere in the absence of fimbriae (Bruce et al., 1983). Adhesive capacity of the same bacterial strain with or without fimbriae has been shown to correlate with the presence of fimbriae. Purified fimbriae have been prepared and type specific antibodies produced.

Preincubation of target epithelial cells with fimbriae dramatically reduced bacterial attachment as has preincubation of bacteria with fimbrial type-specific antibodies (Svanborg-Eden et al., 1982a, 1983). Hemagglutination patterns of isolated fimbriae and whole fimbriated bacteria are identical.

Most of the studies on bacterial fimbriae were conducted on E. coli. Type I fimbriae cause agglutination of guinea pig erythrocytes. The agglutination is inhibited by the presence of mannose and is thus called mannose sensitive (MS) (Ofek et al., 1977). MS fimbriae are found on most E. coli isolates, and studies suggest that virtually all E. coli strains have at least the capacity to express these fimbriae in vitro (Duguid et al., 1955). Contradictory data has been forthcoming concerning the role of MS fimbriae in attachment to uroepithelial cells (Schaeffer et al., 1979; Svanborg-Eden et al., 1981). Svanborg-Eden et al. (1981) and Hagberg et al. (1981) found that E. coli strains expressing MS fimbriae attached poorly to uroepithelial cells but adhered well to buccal epithelial cells. In contrast, Ofek et al. (1981) and Schaeffer et al. (1984) found significant MS attachment to uroepithelial cells. The difference in these observations may relate to the different methodologies employed. It is clear, however, that type I (MS) fimbriae mediate attachment of E. coli to exfoliated buccal and vaginal epithelial cells (Schaeffer et al., 1981, 1984) and to uromucoid (Orskov et al., 1980).

A second fimbrial type has been described which causes mannose-resistant (MR) agglutination of human erythrocytes. These fimbriae have been termed P-fimbriae and can be expressed by only certain E. coli strains under certain environmental growth conditions. P-fimbriae do not attach to uromucoid, and bacteria possessing P-fimbriae adhere in large numbers to uroepithelial cells (Kallenius et al., 1981; Vaisanen et al., 1981; Svanborg-Eden et al., 1983). P-fimbriae are so called because they bind specifically to P-blood group antigens present on human erythrocytes and uroepithelial cells (Leffler and Svanborg-Eden, 1981; Lomberg et al., 1981; Vaisanen et al., 1981). Both MS and MR fimbriae may co-exist on the same bacterial strains. A third E. coli fimbrial type has been described by Schoolnick et al. (1984) termed X-fimbriae, which, while resembling P-fimbriae in terms of hemagglutination, appears

to attach to a totally different human epithelial cell receptor.

When the adhesive properties of urinary E. coli isolates were studied, some interesting observations emerged. The vast majority of pyelonephritic strains of E. coli adhered extremely well to exfoliated uroepithelial cells. Adherence of pyelonephritogenic strains was considerably higher than that observed with cystitis strains and the lowest level of adherence was found with fecal strains of E. coli (Svanborg-Eden et al., 1981; Svanborg-Eden et al., 1983; Svenson et al., 1984). Capacity to adhere correlated directly with the presence of P-fimbriae in that >90% of pyelonephritogenic strains expressed P-fimbriae in contrast to only 16% of fecal strains of E. coli (Kallenius et al., 1981). No correlation was found between severity of urinary infection and type 1 fimbriae. This does not negate any role for these ligands in colonizing the urinary tract. On the contrary, human and animal studies suggest that type 1 (MS) fimbriae facilitate colonization of the vaginal introitus and lower urinary tract including the bladder, whereas P- and X-fimbriae appear essential in colonization and infection of the upper urinary tract (Svanborg-Eden et al., 1983; Iwahi et al., 1983). Not all investigators support the role of fimbrial adherence as a critical virulence factor. In particular, limited studies by Harber et al. (1982, 1984) and Guze et al. (1983) failed in human and mouse studies to verify the importance of fimbrial adherence in the pathogenesis of urinary tract infections.

EPITHELIAL CELL RECEPTORS.—The inhibitory effect of D-mannose on E. coli type 1 fimbrial induced hemagglutination and attachment to buccal epithelial cells suggested that oligosaccharide residues containing D-mannose could serve as a receptor for E. coli expressing these ligands. Ofek et al. (1977) have found that D-mannose residues or receptors are widely found in most tissue systems and studies of the urinary tract have shown D-mannose receptors throughout the urinary tract. It is therefore unclear why E. coli possessing only type 1 (MS) fimbriae adhere relatively poorly to exfoliated human uroepithelial cells in vitro in contrast to buccal epithelial cells. Noteworthy, uromucoid or Tamm-Horsfall protein, a normal constituent of urine rich in mannose residues, is strongly adhered to by E. coli possessing these fimbriae (Orskov et al., 1980).

Considerable evidence has accumulated that a special class of glycosphingolipid, the globoseries glycolipids, act as receptors for uropathogenic E. coli possessing P-fimbriae (Leffler and Svanborg-Eden, 1981). Gal α 1–4 Gal, a fraction of the glycolipid, inhibits attachment of P-fimbriae E. coli to uroepithelial cells (Svanborg-Eden and Leffler, 1980; Kallenius et al., 1981) as well as inhibiting bacterial hemagglutination. Similarly, nonagglutinable erythrocytes are made agglutinable after passive coating with synthetic D-Gal α 1–4 Gal disaccharide, and animal studies have shown that a soluble receptor analogue globotetraose reduced experimental urinary tract infection (Svanborg-Eden et al., 1982b). Finally, P-fimbriated *E. coli* failed to adhere to uroepithelial cells of P̄ humans (lacking P antigens). The globoseries glycolipid receptors are antigens in the P blood group system. Persons of blood group P₁ phenotype have a higher density of receptor glycolipids in their erythrocyte membrane than individuals of the P₂ phenotype (Lomberg et al., 1981). It has been hypothesized that P₁ group phenotype may increase host susceptibility to upper urinary tract infection (Kallenius, 1980; Lomberg et al., 1983). In one study (Lomberg et al., 1984) the P₁ blood phenotype was overrepresented among children with recurrent pyelonephritis without reflux (97% compared with 75% in healthy children). In children with recurrent pyelonephritis and reflux, no increased prevalence of P₁ blood group phenotype was found (Lomberg et al., 1983). It is probable that in these patients, reflux that facilitates both the ascent of bacteria from the bladder to the kidney as well as the persistence of bacteria within the urinary tract is a more important risk factor than the P₁ blood group phenotype. About 70 people in the world are known to lack P antigens, i.e., be of phenotype P̄. Pyelonephritic P-fimbriated *E. coli* did not adhere to erythrocytes or epithelial cells from these subjects (Kallenius, 1980; Korhonen et al., 1982).

The epithelial receptor for *E. coli*, X-fimbriae, is as yet undefined but appears to be distinct from Gal α 1–4 Gal, since the analogues of the latter fail to inhibit attachment of organisms with X-fimbriae in vitro (Schoolnick et al., 1984). Similarly, little is known about receptors for other Enterobacteriaceae and nonfimbrial adhesins. Finally, it is conceivable that in addition to specific receptor binding, nonspecific attachment may occur based on hydrophobicity and surface charge characteristics as well as alterations in the surface structure of the urinary mucosa, e.g., fibrin or fibronectin deposition, etc. Whatever the primary mechanism of adherence, once bacteria are attached and able to multiply, additional bacterial surface alterations may occur, such as further development of the bacterial glycocalyx. With some bacterial species (e.g., *P. aeruginosa*) microcolonies of protected bacteria can become established on the mucosal surface (Marrie et al., 1979).

ANTIADHERENCE MECHANISMS IN THE URINARY TRACT.—There is scant knowledge available regarding antiadherence mechanisms in the urinary tract. Mechanisms may be specific, e.g., immunoglobulins, or completely nonspecific, interfering with colonization of all organisms. A number of possible mechanisms may prevent attachment and hence colonization. The normal bacterial flora of the vaginal introitus, periurethral region, and urethra may cause stearic hindrance and make receptors less available. Lactobacilli have been shown to interfere with *E. coli* adherence to uroepithelial cells (Chan et al., 1985).

Uromucoid or urinary slime (Tamm-Horsfall protein) rich in mannose residues avidly binds *E. coli* and may prevent attachment to uroepithelial cells (Orskov et al., 1980). Immunoglobulins IgG, IgA, and SIgA in the urine of patients with pyelonephritis have been shown to inhibit adherence of the responsible strain of *E. coli* to uroepithelial cells (Svanborg-Eden and Svennerholm, 1978; Svanborg-Eden et al., 1982a). Similarly, antibodies prepared in animals against purified P-fimbriae inhibit, in a dose-related manner, attachment of the strain from which the fimbriae were isolated as well as those strains with antigenically cross-reacting fimbriae (Svanborg-Eden et al., 1983).

Another important natural antiadherence mechanism has been identified in the bladders of several species of animals—dogs, rabbits, rats, and mice (Parson et al., 1975, 1977, 1978, 1979; Parsons and Schmidt, 1980; Sobel and Vardi, 1982; Vardi et al., 1983). Under experimental conditions, the bladder is resistant to colonization in spite of inoculation of large numbers of pathogenic bacteria. After brief treatment of the bladder with dilute hydrochloric acid, adherence and hence colonization increases dramatically for all bacterial strains tested (Parsons and Schmidt, 1980). These observations of Parsons and co-workers have been confirmed by several investigators (Sobel and Vardi, 1982; Vardi et al., 1983) and support the concept of a superficial acid-sensitive natural antiadherence mechanism. Within 24 hours of acid treatment, the antiadherence function returns to the intact bladder. Histologic studies in animals and humans have identified the presence of a thin layer of mucopolysaccharide coating the transitional epithelial cells of the bladder mucosa (Parsons et al., 1974, 1977, 1979). After acid treatment, histochemical staining techniques have revealed depletion of the mucopolysaccharide. This layer becomes visible once more after 24 hours and is thought to be regenerated locally by the transitional cells (Parsons et al., 1978). Chemical analysis of the mucopolysaccharide has revealed a glycosaminoglycan, which is hydrophilic and attracts an aqueous film of water or urine onto its surface (Parsons and Schmidt, 1980). How this mucopolysaccharide interferes with adherence is unclear apart from the need of the mobile organisms to penetrate the layer en route to attaching to specific cell receptors. In limited studies in canines, Jarvinen and Sandholm

(1980) described the natural presence of urinary oligosaccharides with the potential to cause detachment of epithelially bound *E. coli* as well as to prevent attachment of bacteria to uroepithelial cells by binding of the sugars to the microorganisms, i.e., oligosaccharides in urine presumably resemble sugar moieties on the epithelial cell surface. Finally, the mechanical effect of flushing during bladder emptying is probably essential in preventing adherence.

INCREASED SUSCEPTIBILITY TO BACTERIAL ADHERENCE.—The majority of children with severe and recurrent urinary infections have gross structural abnormalities in the urinary tract accompanied by obstruction, stasis, and frequently vesicoureteric reflux. Other children without gross defects as well as many young adult women with radiologically normal tracts may also suffer from recurrent infections. Most of these recurrent urinary tract infections occurring in the absence of structural changes involve the lower tract only. As yet, there is no satisfactory explanation for this increased susceptibility in an apparently normal urinary tract. Theories include minor degrees of urinary obstruction, bladder overdistention, defective immune and antiadherence mechanisms, and, finally, increased receptivity of uroepithelial cells to bacterial attachment.

An increased receptivity to bacterial attachment has been observed with vaginal (Fowler and Stamey, 1977), periurethral (Kallenius and Winberg, 1978), and uroepithelial cells (Svanborg-Eden et al., 1983) obtained from patients with recurrent urinary tract infection when the patient's own infecting strain was used. Parsons and Schmidt (1980), however, failed to confirm these observations in adult females. One postulated mechanism for increased epithelial cell receptivity is increased density of surface receptors. Several investigators (Cruz-Coke et al., 1965; Kinane et al., 1982) have shown that women of blood groups B and AB who are nonsecretors of blood group substances have a significantly higher risk of developing urinary tract infection when compared with women of other blood groups (or with those who are secretors).

As mentioned previously, Lomberg et al. (1983) suggested that individuals who possess the P_1 blood group run a higher risk of developing recurrent upper urinary tract infection (pyelonephritis) than P_2-positive individuals. In contrast, in a recent study of blood groups in premenopausal women prone to recurrent lower urinary tract infection (cystitis), patients were found to have a normal distribution in the ABO system, but unexpectedly 85% were found to be of the P_2 phenotype (P_1 negative) compared with the expected frequency of 21% in the general population (Mulholland et al., 1984). The mechanism for this phenomenon is obscure;

nevertheless, genetic differences at the cellular level may possibly influence bacterial adherence and make certain women more prone to urinary tract infection and also influence the anatomical level of infection.

The density or availability of receptors may depend on the maturity of the epithelial cells. Reid et al. (1983) reported cyclical changes in uroepithelial cell receptivity which correlated with in vivo estrogen levels. A similar observation has been reported in a comparison of elderly females with young healthy female controls, with the latter demonstrating significantly higher attachment avidity (Sobel and Muller, 1984). An influence of hormonal status of the donor on adherence has also been demonstrated for *P. mirabilis*, streptococci, and *E. coli*.

The recent observation of reduced SIgA urinary concentrations in patients with acute urinary tract infection may be relevant to reduced resistance to bacterial colonization in these patients (Riedasch et al., 1983). Antibodies in vaginal secretions may prevent bacterial colonization of the introitus. Several studies have failed to show reduced concentrations of immunoglobulins in vaginal secretions in women prone to recurrent urinary tract infection (Kurdydyk et al., 1980); however, Fowler and Stamey (1977) demonstrated reduced antibody coating of vaginal strains of *E. coli* in such patients. They concluded that reduced immunoglobulins in cervicovaginal secretions could result in increased susceptibility to adherence and hence explain the increased coliform introital colonization in at-risk patients. Another interesting concept, developed by Tullus et al. (1984), suggests that increased susceptibility to urinary tract infection in children results not from increased cell receptor density, but rather from acquiring a fecal flora that is dominated by P-fimbriated strains of *E. coli*. Screening both the fecal flora as well as the periurethral flora in susceptible children revealed dominantly P-fimbrated *E. coli* isolates, which were thought to increase the risk of subsequent invasion of the urinary tract (Tullus et al., 1984).

EXPERIMENTAL STUDIES IN ANIMALS.—Although several animal species have been useful for studying experimental urinary tract infection, several problems exist which to some extent do not allow extrapolation of animal data to human subjects. No animal model mimics the specificity and the early colonization phase of human urinary tract infection; however, using the ascending model of nonobstructed pyelonephritis in mice, considerable information has been accumulated (Hagberg et al., 1983). This model, together with use of genetically manipulated *E. coli* strains, in which expression of adhesins could be altered and controlled, has confirmed the in vivo significance of bacterial adherence in the pathogenesis of urinary tract infection. Following cloning experiments, mutants with both adhesins (P and type 1 fimbriae) remain in the bladder in higher numbers than do bacteria with either adhesin alone. In contrast, pyelonephritis is dependent mainly on possession of P-fimbriae (Hagberg et al., 1983). Similar results have been obtained using a monkey primate model (Roberts and Phillips, 1979). Nevertheless, the contribution of the adhesive properties to virulence in the mouse model is also dependent on other pathogenic factors in the organism (serotype, serum resistance, etc.). Animal experiments have not permitted conclusions to be made about the relative importance of different virulence factors.

Using animal models, the importance of in vivo bacterial adherence has also been evaluated by the administration of soluble receptor sugars or analogues which competitively inhibit adherence in vitro. Accordingly, Arson et al. (1979) were able to reduce in vivo attachment of E. coli expressing type 1 fimbriae by administration of D-mannose into the bladders of mice, and similar results were obtained by Michaels et al. (1983) in rats. Similarly, Svanborg-Eden et al. (1982b) by adding globotetraose to a bladder bacterial inoculum decreased the infectivity of a strain of E. coli which bound to this receptor. Monoclonal antibodies to fimbriae passively administered also protected against infection with homologous strains of E. coli (Svanborg-Eden et al., 1983).

The monkey model has been useful in that, like in humans, surface glycosphingolipids corresponding to the human P-blood group are present in kidney tissue and blood cells. Consequently, P-fimbriated E. coli adhere well to uroepithelial cells and in contrast to non-P-fimbriated E. coli cause acute and chronic pyelonephritis (Roberts and Phillips, 1979).

Finally, numerous animal experiments have verified the value of active immunization of the test animal with a variety of fimbrial and other antigens aimed at preventing experimental pyelonephritis with the homologous strain of bacteria (Mattsby-Baltzer et al., 1982; Silverblatt et al., 1982; Kaijser et al., 1983; Layton and Smithyman, 1983; Jensen et al., 1984). Vaccination with highly purified P-fimbriae reduced and modified renal damage in mice and primates.

Since the adherence of uropathogens is mediated by bacterial surface structures, antimicrobial agents (which at low concentrations may alter the bacterial surface) are likely to affect attachment. Subinhibitory concentrations of ampicillin, amoxicillin, streptomycin, tetracycline, and trimethoprim reduce both the mannose-binding capacity of E. coli as well as the ability of the organisms to adhere to uroepithelial cells (Stenqvist et al., 1982) by either decreasing synthesis of or less effi-

cient expression of fimbriae. Accordingly, some of the beneficial effect of long-term, low-dosage trimethoprim prophylaxis in women with recurrent urinary tract infection may be due to the effect of low urinary and vaginal concentrations of the antimicrobial agent on decreasing bacterial adherence and thus reducing in vivo colonization.

NATURAL DEFENSES OF THE URINARY TRACT

THE URINE.—Numerous characteristics of normal urine constitute important antimicrobial defense mechanisms both in vivo and in vitro. The chemical composition of urine is such that most bacteria can multiply because urine for the most part lacks complement, humoral, and cellular defenses against bacteria. Urine is usually hyperosmolar in relation to plasma and will inhibit leukocyte phagocytosis (Bryant et al., 1973). Nevertheless, urine from normal individuals may be inhibitory and capable of killing microorganisms responsible for urinary infection, especially when the inoculum is small (Kaye, 1968). The most inhibitory factors are the osmolality, urea concentration, organic acid concentration, and pH. Bacterial growth is inhibited by a very dilute urine, and a high osmolality when associated with a low pH is highly inhibitory.

The presence of glucose in the urine provides a better medium for multiplication of E. coli. This may be a factor in influencing the frequency and severity of infections in diabetics (Asscher et al., 1968). Urine obtained from women is more often at a suitable osmolality and pH for growth of E. coli than urine from men (Asscher et al., 1973). Urine of pregnant women exhibits a more suitable pH for growth of E. coli at all stages of gestation (Asscher et al., 1973). Evaluation of all the antibacterial components of urine suggests that the urea content is more crucial a determinant of antibacterial activity than osmolality, organic acid concentration, or urinary pH (Kaye, 1968).

Following the observation in men that urine obtained after prostatic massage is more inhibitory to bacterial growth in urine, it has been suggested that prostatic secretions contain antibacterial substances (Stamey et al., 1968a). This antibacterial activity was shown to be related to a heat-stable compound. Fair et al. (1973) suggested that one additional factor may be a zinc salt, and several investigators have reported reduced levels of zinc in the prostatic secretions of patients with chronic bacterial prostatitis. In general, anaerobic bacteria and other fastidious organisms that normally colonize the urethra will not multiply in urine and are rarely responsible for urinary infections (Asscher et al., 1968; Kaye, 1968). Current data based on measurement of oxygen tension in the urine suggest that obligate anaerobes would be unable to survive except under particular conditions in which oxygen tension might be abnormally low, e.g., renal scar tissue, bladder tumors, necrotic renal papillae, and possibly in the prostate. There are only occasional reports of anaerobic and capnophilic bacteria being recovered from catheterized patients and rarely from suprapubic bladder punctures (Finegold et al., 1965).

URETHRA AND PERIURETHRAL REGION.—The normal flora of the urethra, periurethral area, introitus, and vagina usually does not contain recognized uropathogens, but instead includes lactobacilli, coagulase-negative staphylococci, corynebacteria and streptococci (Pfau and Sacks, 1977; Marrie et al., 1978; Fair et al., 1981). These organisms collectively account for about 75% of the CFUs in the distal urethra of young women. Anaerobic bacteria account for the remaining 25% of the flora (Marrie et al., 1978).

The initial step in the pathogenesis of urinary infection is the arrival in the periurethral region of uropathogens, predominantly *E. coli* selected from the fecal reservoir. Little is known about the factors that predispose to urethral colonization with these organisms or the natural history thereof in otherwise healthy adult females. The proximity of the urethral meatus to the vulvar and perianal areas suggests that contamination is frequent. The nature of urethral defense mechanisms other than flow of urine is largely unknown. Bacterial multiplication in the normal urethra may be inhibited by the resident normal flora, which may also interfere with initial adherence of pathogenic bacteria (Chan et al., 1984). Recently, Chan et al. have shown that whole lactobacilli and cell wall fragments interfere in vitro with *E. coli* adherence to uroepithelial cells by stearic hindrance (Chan et al., 1985). Kunin et al. (1980) postulated that antimicrobial agents administered for any reason, by virtue of the high concentrations present in the urine, might secondarily alter the normal flora of the urethra and periurethral region, facilitating the acquisition of gram-negative bacteria.

After several years of intensive epidemiologic study, Stamey and co-workers concluded that in the absence of structural or anatomical defects in the urinary tract, women with recurrent cystitis have a biologic predisposition to urinary infection as a result of a more frequent and more prolonged colonization of the perineum, including the periurethral region, with coliform bacteria. Support for this hypothesis was based on longitudinal studies of the periurethral flora in women prone to recurrent infection (Stamey et al., 1971; Stamey, 1973a). Episodes of bacteriuria were found to be associated with, or were frequently preceded by, colonization of the vaginal introitus and periurethral area

with bacteria from the fecal flora. Moreover, between episodes of bacteriuria, women and children with recurrent urinary tract infections showed both a higher prevalence and a greater density of perineal colonization with urinary pathogens than did healthy controls (Stamey et al., 1971; Stamey, 1973a). Similarly, other studies have shown a higher prevalence of *E. coli* isolated from the introitus and urethra in asymptomatic abacteriuric females prone to recurrent bouts of cystitis and the urethral syndrome (Cox et al., 1968; Pfau et al., 1983). The critical host biologic factor that facilitates this abnormal colonization is thought to be the result of enhanced susceptibility of human eukaryotic cells (vaginal and uroepithelial) to bacterial adherence (Fowler and Stamey, 1977; Schaeffer et al., 1981).

Furthermore, reduced titers of cervical IgA bathing the introital cells were thought to contribute to the increased attachment (Stamey et al., 1978). Similarly, Kallenius and Winberg (1978) and Svanborg-Eden et al. (1982a) demonstrated increased adherence of uropathogenic *E. coli* to periurethral and uroepithelial cells, respectively, obtained from children with urinary tract infections. They postulated an increased density or expression of receptor sites to explain the enhanced adherence. Since Schaeffer et al. (1981) also found bacterial receptivity of buccal cells to be increased in infection-prone females, they suggested that genetic factors might influence the receptivity of vaginal, buccal, and uroepithelial cells. This concept was further supported by the observation that the HLA-A3 subtype was more prevalent in women with recurrent urinary tract infection than in women who had never had urinary infections (Schaeffer et al., 1983b). Stamey and Timothy (1975) suggested that low vaginal pH was another factor discouraging coliform, introital colonization and that women with recurrent cystitis tended to have a higher pH. Furthermore, serogroups of *E. coli* that were more likely to cause urinary tract infection were more resistant to low pH than serogroups that did not commonly cause infection (Stamey and Timothy, 1975).

In contrast to the above observations, Parsons et al. (1980) found no increased bacterial adherence to vaginal epithelial cells in women with recurrent urinary tract infections. Furthermore, Kurdydyk et al. (1980) found no difference in IgG and IgA levels in cervicovaginal washings between women prone to infection and those with no history of urinary tract infection.

Although few doubt that periurethral carriage of uropathogens is a prerequisite for the development of urinary tract infection, not all investigators agree that increased frequency of and more prolonged colonization of the periurethral region are the most important factors in the pathogenesis of recurrent urinary tract infection (Cattell et al., 1974; Elkins and Cox, 1974; Marsh et al., 1975; Kunin et al., 1980). These investigators claim that introital colonization with coliforms and other uropathogens is as common in women not prone to infection as in those who are infection prone. In this context, Kunin et al. (1980) stated that all women who do not have a structural or neurologic problem in the voiding mechanism are approximately at the same risk of having a first urinary tract infection. Once established, each infection sets the stage for the next episode, because infection itself may perpetuate abnormal colonization unless abnormal periurethral colonization is also eradicated by therapy. The longer the periods between infection, the less likelihood there is for recurrence (Kunin et al., 1980). This suggests that absence of infection and absence of treatment increases the resistance to colonization.

According to Kunin and associates, it is probably not the colonization of the periurethral area per se but rather the frequency with which uropathogens ascend into the urethra and bladder together with the avidity with which they adhere to bladder mucosal cells that determines the frequency of infection.

As with the urethra, little is known regarding the normal host perineal resistance mechanisms other than the concept of colonization resistance offered by the normal bacterial flora. It is unclear whether sexual intercourse and urethral trauma constitute risk factors for periurethral colonization with enteric uropathogens.

THE BLADDER.—For bladder infection to occur, uropathogens colonizing the urethra and periurethral tissue must ascend the urethra and enter the bladder. A defect in the bladder defense mechanism is not necessary for the production of infection, particularly in the presence of a large inoculum, but it is undoubtedly permissive in allowing infection to develop.

The normal bladder appears to be inherently resistant to infection (Mulholland, 1979). Several studies have shown that bacteria normally present in the urethra and periurethral region frequently make their way into the bladders of women. In particular, sexual intercourse is a frequent factor facilitating the ascent of bacteria into the bladder (Buckley et al., 1978; Kelsey et al., 1979; Nicolle et al., 1982). Inoculation of urethral organisms into the bladder after sexual intercourse usually produces a self-limited state of low-level bladder bacteriuria without establishment of infection. Within 24 hours of introduction, these organisms are usually completely eliminated from the bladder. Whether bacteria introduced into the bladder persist, multiply, and ultimately result in symptomatic infection or asymptomatic bacteriuria depends on the bacterial inoculum, bacterial virulence factors, and bladder defense mechanisms.

Human and animal studies have demonstrated that after large bacterial inocula (i.e., 10–100 million organisms) reach the bladder urine, they are usually promptly eliminated (Cox and Hinman, 1961). The mechanisms within the bladder lumen responsible for this action are complex and as yet incompletely understood. Possible factors are: (1) removal of bacteria by urination, (2) inability of urine to support growth of bacteria or ability actually to destroy them, (3) antiadherence mechanisms of urine or of the bladder mucosa (see section on adherence), (4) effectiveness of antibacterial properties of intact bladder mucosa, and (5) efficiency of phagocytosis. All cannot be of equal importance. For example, voiding is extremely important in removing the vast majority of an inoculum. However, it may be unable to totally eliminate all of the bacteria. A film of urine always remains behind, and ample time exists for multiplication of bacteria before the next voiding.

Micturition and the efficient emptying of the bladder are critically important defense mechanisms (Cox and Hinman, 1961). In vitro bladder models confirm the role of both dilution and displacement of infected urine in preventing the initiation of and maintenance of bladder infection (Cox and Hinman, 1961). Furthermore, it should be stressed that (except when there is gas in the urinary tract) there is always a direct connection between the bladder urine and urine in the renal pelvis. Studies have shown that even under normal conditions, particles may make their way up the ureter against the flow during normal ureteral peristalsis.

It has long been suggested that the bladder possesses intrinsic antibacterial mechanisms whereby bacterial multiplication is inhibited. Surprisingly little progress in this area has been made clinically or in experimental animals since the original studies of Vivaldi et al. (1965). Working with an isolated pouch of rabbit bladder mucosa, these investigators demonstrated that bacteria that had been applied to the surface of the pouch disappeared. It was postulated that mucosal cells produced antibacterial substances (Mulholland, 1979). The true nature of intrinsic antibacterial properties of the bladder, also suggested by Norden and Kass (1968) using a guinea pig bladder model, remains an enigma. In contrast, however, once the bacteria actually invade the bladder mucosa, they induce a mucosal inflammatory reaction characterized by the infiltration of polymorphonuclear leukocytes into the mucosa and ultimately into the urine. Cobbs and Kaye (1967) studying the rat bladder demonstrated a relationship between clearance of bacteria and the migration of leukocytes within the bladder mucosa. This defense mechanism is probably important in the self-limiting nature of most cases of bacterial cystitis. Leukocytes, however, probably play no role in prevention or the initial superficial phase of bladder infection.

THE KIDNEY.—Numerous experimental studies indicate the greater the number of organisms delivered to the kidney by any route, the greater the chance of producing infection (Guze et al., 1961; Cotran et al., 1963). The kidney itself is by no means uniformly susceptible to infection. Freedman and Beeson (1958) demonstrated that very few organisms are required to infect the medulla, whereas 10,000 times as many are needed to infect the cortex. The medulla and the renal papilla have been identified as the regions most susceptible to infection and the most frequent sites where infection is initiated. The greater susceptibility of the medulla may be due to its high osmolality, low pH, and low blood flow (factors adversely affecting leukocyte chemotaxis) and to the higher concentration of ammonia, which is thought to inactivate complement (C4) (Beeson and Rowley, 1959).

The hyperosmolality of the renal medulla also favors the conversion of bacteria to a cell-wall–deficient state (L forms), which is resistant to cell-wall active antimicrobials and may lead to persistence of infection in spite of therapy (Gutman et al., 1965). Thus, the medulla and papilla are the sites where the decisive struggle takes place determining whether infection in the kidney parenchyma ensues (Rocha et al., 1958).

Uropathogenic *E. coli* that gain access to the kidney by the ascending route share many virulence factors, including the enhanced ability to adhere to uroepithelial cells in numbers significantly higher than those of *E. coli* derived from patients with cystitis (Varian and Cooke, 1980; Svanborg-Eden et al., 1983). This increased adherence is due to the presence, not of common type 1 fimbriae, which bind to mannose-containing cell receptors, but of MR-fimbriae, which avidly bind to uroepithelial cells (Vaisanen et al., 1981). P-fimbriae in particular are usually present or expressed by pyelonephritic strains of *E. coli* (Kallenius et al., 1981; Svanborg-Eden et al., 1983; Svenson et al., 1984). These fimbriae attach to epithelial cell receptors containing globoseries glycolipid (Kallenius et al., 1981; Leffler and Svanborg-Eden, 1981; Korhonen et al., 1982), which are distributed throughout the urinary tract, particularly in the kidney (Leffler and Svanborg-Eden, 1981). Type 1 fimbriae appear to play little role in kidney colonization and infection. Further, they could constitute a disadvantage because defending polymorphonuclear leukocytes are equipped with mannose receptors for type 1 fimbriae that facilitate phagocytosis of the invading organisms in the absence of immune opsonization (Perry et al., 1983). This may explain why uropathogenic *E. coli* tend to lose or stop expressing type 1 fimbriae on arrival in the kidney (i.e., type 1 fimbriae no longer constitute a biologic advantage in the kidney and in fact increase susceptibility to removal). Similar observations of loss of fimbriae were made by Silverblatt (1974) in studying the fimbrial role of *P. mirabilis* invasion of the rat renal pelvis. This phenomenon of altered fimbrial expression and surface composition of microorganisms is termed phasic variation.

No natural barrier or defense mechanism against bacterial adherence has been identified in the kidney similar to the antiadherence mechanisms of the bladder. This is not surprising, because bacterial ascent into the kidney is far less frequent than into the bladder. The role of tubular secretion of Tamm-Horsfall protein in this context may be important but requires further study. The flow of urine is undoubtedly important in washing bacteria down the ureters. Subjects who lack the antigen of blood group P_1 phenotype appear less susceptible to urinary infection, especially pyelonephritis, perhaps because they have reduced Gal α 1–4 β-containing globoseries or glycosphingolipid receptor molecules for P fimbriae on their epithelial cells (Lomberg et al., 1983). Recently, a globoside-binding *E. coli* P-fimbrial vaccine was shown to prevent pyelonephritis in monkeys (Roberts et al., 1984).

During pyelonephritis, an acute inflammatory exudate consisting predominantly of polymorphonuclear leukocytes is present. Although the inflammatory response is aimed at limiting bacterial spread and persistence within the kidney, the infiltrating phagocytic cells may also contribute to the local tissue damage and resultant renal scarring. Bille and Glauser (1982), in experimental animals, reduced parenchymal kidney destruction by inducing neutropenia as well as by using agents that interfere with leukocyte chemotaxis. The clinical significance of these observations is doubtful, since any interference with the normal inflammatory response could be expected to facilitate bacterial spread in the kidney and contribute to bacteremia.

URINARY TRACT IMMUNE MECHANISMS

Recent studies suggest that the genitourinary tract forms part of the secretory immune system. In vitro studies have shown that antigen presentation by accessory cells, e.g., macrophages, is essential for immunoregulation and that Ia antigens are important in the regulation of the immune response. Ia-expressing cells, many of them of a similar morphology and analogous to the Langerhans cells of the skin, have been found in urinary tract tissue from different species (Hjelm, 1984). In rats, only a few cells of this type have been identified in the bladder and kidney, and little information concerning these accessory immune cells in the urinary tract of humans is available. As a result of infection, Ia antigens are expressed on renal tubular epithelial cells (Hjelm, 1984).

Most of the human and experimental animal studies have focused on the immune response to upper tract bacterial infections. Several investigators have documented that kidney infection is accompanied by both serum (systemic) and local kidney immunoglobulin synthesis with the appearance of type-specific antibodies in the urine. Antibodies in serum against the O antigen and to a lesser extent the K antigen of the infecting *E. coli* strain have been found. Recently, serum antibodies directed at bacterial ligands or adhesins such as type 1 and P-fimbriae have also been identified following acute pyelonephritis (Rene et al., 1982). IgM antibodies dominate the initial response to upper tract infection and are followed by IgG and IgA response. High titers of IgG antibodies against the lipid A component of gram-negative bacterial lipopolysaccharide tend to correlate well with increased severity of renal infection and progression of renal parenchymal destruction (Mattsby-Baltzer et al., 1982). In pyelonephritis, IgG and SIgA (secretory) immunoglobulins also appear in the urine and may become evident before antibody is detected in the serum. These latter antibodies are synthesized locally within the kidney. The presence of antibodies within the renal parenchyma enhances bacterial opsonization and ingestion of the invading microorganisms by local phagocytic cells. Biological amplification of the local inflammatory response is achieved by the rapid activation of the complement system by IgG and IgM. Thus, local antibody presence serves to reduce local tissue damange and to restrict bacterial spread. The other potential beneficial or protective effect of urinary antibody synthesis is prevention of reinfection by the same strain in the future. This effect was suggested by clinical observations and *E. coli* serotyping in longitudinal studies. Eighty percent of reinfections of the urinary tract were caused by new and different serogroups of *E. coli* or other species of bacteria (Stamey et al., 1971; Stamey, 1973a). Several experimental animal studies of hematogenous and ascending pyelonephritis have shown that these animals, after recovering from pyelonephritis, became more resistant to attempts at reinfection with the same strain of microorganism (Kaijser et al., 1983). Similarly, systemic immunization with bacterial components of the test organism may also result in renal resistance to recurrent urinary infection (O'Hanley et al., 1983). Apart from the role of antibodies interacting with phagocytic cells within the kidney, it has been suggested that immunoglobulins, particularly those found within the urine, may operate in other ways to prevent infection. In particular, attention has been directed at the protective function of SIgA. Svan-

borg-Eden and Svennerholm showed that commercial gamma globulin and IgG and SIgA derived from the urine of patients with acute pyelonephritis reduced in vitro adherence of the same strain of *E. coli* to uroepithelial cells (Svanborg-Eden and Svennerholm, 1978). Similarly, immunization with *E. coli* P-fimbriae resulted in immunoglobulin production in experimental animals which prevented ascending pyelonephritis by reducing the adhesive capacity of the invading autologous uropathogenic *E. coli* (Roberts and Phillips, 1979; O'Hanley et al., 1983). Silverblatt (1982), using the ascending model of pyelonephritis in rats, concluded that immunization of rats with purified *E. coli* type 1 fimbriae could provide substantial protection from pyelonephritis.

The role of urinary immunoglobulins in preventing bladder infections is less clear. Using monoclonal antibodies, IgA-producing lymphocytes were demonstrated in the submucosa of infected rat bladders (Hjelm, 1984). Infection of the lower urinary tract is usually associated with a reduced or nondetectable serological response, reflecting the superficial nature of the infection. Absence of urinary antibodies results in failure of antibody coating of the infecting bacteria in urine which forms the basis of the localization test currently used (Jones et al., 1974). In particular, antifimbrial antibodies are absent from the urine in lower tract infection despite the critical role of fimbriae in colonization of the bladder and urethra (Rene et al., 1982). Accordingly, if cystitis fails to stimulate an effective local antibody response as indicated by the lack of detectable antifimbrial antibodies in the urine, the bladder would be susceptible to recolonization and reinfection by the same strain of bacteria. This, however, is not the case as indicated by epidemiologic studies (Stamey et al., 1971). Furthermore, in spite of the low-level or undetectable urine humoral response, Uehling and Wolf (1969) showed that bladder immunization of rats with bacterial antigens decreased in vivo adherence of *E. coli* to bladder mucosa. Bladder immunization with killed bacteria may also protect against pyelonephritis in rats in the absence of serum antibodies (Uehling et al., 1978). The converse may also occur in that Aronson et al. (1979) showed reduced bladder involvement in mice reinfected several weeks after primary induced ascending pyelonephritis. Infection penetrating the bladder mucosa may result in both a urinary and systemic antibody response. These studies suggest a possible protective role for urinary immunoglobulins in preventing both experimental bladder and kidney infections.

The protective function of urinary immunoglobulins was further emphasized by the observation by Riedasch et al. that lower urinary levels of SIgA were associated with an increased risk of urinary tract infections in humans (Riedasch et al., 1983). Of interest is a recent report in which IgA proteases capable of cleaving IgA were isolated from several gram-negative bacterial species capable of causing urinary tract infection (Milazzo and Delisle, 1984).

Several studies have demonstrated the persistence of bacterial antigens in the kidney of experimental animals following pyelonephritis with accompanying chronic antibody stimulation (Hanson et al., 1981; Mattsby-Baltzer et al., 1982). In particular, E. coli cell wall O antigen has been shown to stimulate B lymphocytes (Hanson et al., 1981). Accordingly, it has been postulated that the kidney may then become the site of prolonged antigen-antibody interaction, and this immune reaction including local complement activation may result in chronic renal interstitial damage long after cessation of bacteriuria.

Several investigators have drawn attention to a persistent immune response directed at Tamm-Horsfall protein antigen in animal and human pyelonephritis (Work and Andriole, 1980; Mayrer et al., 1983). Although normally secreted by tubular cells of the ascending loop of Henle into the urine, Tamm-Horsfall protein may gain access to the renal interstitial space during bacterial pyelonephritis, possibly stimulating a potent and persistent humoral and cellular immune response, which may further damage the kidney. In particular, tubular obstruction and stasis as in nephrolithiasis and reflux nephropathy result in the highest humoral and T-cell immune response directed at Tamm-Horsfall protein antigens (Mayrer et al., 1983). The highest titers of serum IgG and IgA antibodies against Tamm-Horsfall protein were seen in patients with vesicoureteral reflux even in the absence of bacteriuria (Mayrer et al., 1983). No elevated serum titers of antibodies directed against Tamm-Horsfall protein were found in patients with cystitis. Cross-reactivity between Tamm-Horsfall protein and gram-negative bacteria was reported by Fasth et al. (1980), raising the possibility of an immune response induced by gram-negative bacteria but subsequently directed against renal tubular cells containing Tamm-Horsfall protein even after elimination of bacteria. As indicated below, evidence has accrued that a cellular immune mechanism directed against Tamm-Horsfall protein rather than a humoral response is the main mechanism of renal damage.

Renal autoantigens other than Tamm-Horsfall protein released or altered during the initial phase of bacterial infections may also be responsible for eliciting an immune response and subsequent renal interstitial damage. Studies by Holmgren and Smith (1975) have shown a cross-reaction between certain strains of E. coli and renal tissue. Antibodies produced in response to certain bacterial antigens were cytotoxic for fetal kidney cells in tissue culture. Subcutaneous injection of selected killed strains of E. coli which cross-react with renal tissue resulted in enlargement of the kidneys (Holmgren and Smith, 1975). Nevertheless, there is no evidence in humans that renal disease is more severe in patients infected with O serogroups, which cross-react with renal tissue. It is generally thought that renal damage resulting from immunopathologic mechanisms is small in relation to the direct effect of bacterial invasion and the effects of ureterovesical reflux. Even less information is available about the protective role, if any, of T lymphocytes and cellular-mediated immunity (Miller and North, 1974; Smith, 1980). During experimental pyelonephritis induced by either the ascending or IV route, both T and B lymphocytes accumulate in the kidney (Hjelm, 1984). Infected kidneys contain prominent T-cell infiltrates, whereas only a transient T-cell infiltration has been observed early in bladder infection in rats. Following ascending E. coli urinary tract infections in rats, an increased number of irregular Ia-expressing cells are detected in the bladder submucosa, in renal tubular cells, and in the intertubular space in the kidney (Hjelm, 1984). A slight increase of T-helper cells in the bladder submucosa as detected by immunohistochemical methods was noted during the initial phase of infection, but appeared to disappear fairly rapidly. Thus, evidence is accumulating of a local T-cell response in both bladder and kidney infections in animals although the extent and quality of the response differs markedly in each organ. T-cell accumulation tends to correlate with chronicity of inflammation (Miller, 1983). Similarly, T-cell activation distant from the kidney has been observed in pyelonephritis (Miller and North, 1974; Miller et al., 1978; Miller, 1983).

Evidence is still lacking of a protective role of cell-mediated immunity in urinary infection. Clinical observation of patients with profound depression of or even absent T-cell function has not suggested an increased frequency of urinary tract infections or an altered course of infection. However, experimental studies in thymectomized mice did show late increased scarring of kidneys following hematogenous enterococcal pyelonephritis (Pitchon, 1984). In rat studies, Miller et al. have shown depression of cell-mediated immunity due to suppressor T lymphocytes during the course of ascending pyelonephritis (Miller et al., 1978, 1983). Furthermore, Miller (1983) demonstrated that the administration of cyclophosphamide led to the eradication of bacterial infection in experimental pyelonephritis in rats by enhancing the resistance of lymphocytes to the activity of suppressor cells.

ABNORMALITIES INTERFERING WITH DEFENSE MECHANISMS IN THE URINARY TRACT

Obstruction to urine flow at all anatomical levels of the entire urinary tract from the urethral meatus to the renal tubules is the most important predisposing factor to urinary infection. Obstruction inhibits the normal flow of urine, and the resultant stasis compromises bladder and renal defense mechanisms.

Obstructed urinary flow emphasizes the primary importance of the "flushing" effect of urinary flow in preventing infection. Obstruction not only permits the multiplication of bacteria in urine but also increases the capacity of bacteria to multiply within the kidney parenchyma, allowing infection to be transmitted from one part of the kidney to another. Renal cortical scars have little effect on susceptibility to infection, whereas renal papillary scars, by causing intrarenal tubular obstruction, result in dramatic increased susceptibility (Rocha et al., 1958; Rocha, 1972). In the animal model of experimental hematogenous pyelonephritis, the kidney is relatively resistant to infection unless the ureter is ligated. Under these circumstances only the obstructed kidney becomes infected (Guze and Beeson, 1956). It is thought that increased retrograde pressure resulting from obstruction alters the pattern of intrarenal blood flow. This not only impairs the delivery of phagocytic inflammatory cells but also the ischemic papilla and medulla are rendered more susceptible to bacterial invasion, multiplication and spread.

Clinical observations support the significance of obstruction in the pathogenesis of urinary infections. An increased incidence of pyelonephritis is associated with obstruction in male infants, female children, pregnant women, and obstruction in both sexes in the elderly. Congenital anomalies which obstruct urinary flow include valves, bands, and stenosis, together with calculi, bladder neck obstruction, and extrinsic compression of ureters (tumors, retroperitoneal fibrosis). All are accompanied by a high frequency of pyelonephritis. Similarly, localized intrarenal obstruction to urinary flow caused by nephrocalcinosis, uric acid nephropathy, chronic potassium depletion, polycystic kidney, analgesic nephropathy, and sickle cell disease are associated with an increased frequency of pyelonephritis (Rocha, 1972).

Calculi may increase susceptibility to urinary tract infection by producing partial or complete urinary obstruction. However, not all calculi obstruct, and calculi (particularly those in the bladder) may serve as irritants, facilitating adherence and colonization by bacteria. In addition, bacteria survive deep within the stones and are extremely difficult to eradicate, accounting for the well-known clinical difficulty of curing urinary tract infection in the presence of stones. Relapse usually occurs after antimicrobial therapy is stopped because bacteria grow out from the protected deep interstices of the calculus.

Conversely, calculi may be the result of urinary infection (and then perpetuate it). In particular, *Proteus* and other urea-splitting organisms contribute to the formation of magnesium-ammonia phosphate (struvite) stones (Cotran et al., 1963).

There is general acceptance of the important association of vesicoureteral reflux with urinary tract infection, and recently the role of reflux in producing renal damage in the absence of infection has been emphasized (Smellie and Normand, 1968, 1975b; Bailey, 1973). Clinically, vesicoureteral reflux has been observed in patients with congenital anomalies of the urinary tract, particularly children with recurrent urinary infection, patients with bladder neck obstruction and some adults with long-standing urinary tract infections. These individuals with significant vesicoureteral reflux often have recurrent episodes of ascending pyelonephritis. In patients with unilateral reflux, pyelonephritis tends to develop on the side where reflux exists. Both experimental and clinical observations in children have shown that bladder infection may induce reflux during the active phase of infection and reflux may disappear once the infection is controlled. A vicious cycle is established whereby infection produces or aggravates reflux, which in turn perpetuates or maintains infection by producing residual urine with a failure of the urinary tract to empty completely after voiding (Smellie and Normand, 1975b). Substantial evidence has accumulated suggesting that reflux, especially in young children with growing kidneys, plays an important role not only in producing upper tract infection but also in scarring the kidney.

Follow-up studies in children with primary reflux have shown in all but the grosser forms that if infection is controlled, the reflux tends to diminish or disappear with time, possibly as a result of the ureteral and bladder wall muscle growing thicker with age. Accordingly, there is no need for routine surgery for reflux. Surgery should be reserved for the few cases with markedly dilated ureters who have progressive renal scarring. In 70%–80% of children with reflux no other anatomical abnormality, e.g., distal obstruction, is found.

Recently, Mayrer et al. (1983) suggested that in addition to the mechanical damage from pressure of reflux another possible mechanism may exist to explain the progressive renal destruction that accompanies severe reflux. The presence of Tamm-Horsfall protein, normally present in the renal tubular lumen, was detected in the renal interstitium in patients with severe reflux. Within the renal parenchyma, Tamm-Horsfall protein elicited both a humoral and cellular immune response, inducing an interstitial nephritis that further damaged the kidney.

Many mechanical and neurologic factors may lead to incomplete emptying of the bladder. These include obstruction of the bladder neck caused by prostatic hypertrophy, or obstruction from urethral valves and strictures or severe bladder prolapse, and neurogenic bladder from poliomyelitis, diabetic or other neuropathy, and spinal cord injuries. All of these conditions predispose to recurrent urinary tract infections, usually with multiple different microorganisms, complicated by the need for bladder instrumentation or catheterization. The resulting bladder overdistention may interfere with local defense mechanisms; however, the most important pathogenetic factor is the increased residual volume, which both predisposes to and maintains bladder infection by providing a continuous pool of media suitable for bacterial growth. Vesicoureteral reflux and the formation of bladder stones further complicate these cases and add to both the severity of infection and difficulty in its eradication.

Diabetes mellitus is said to be associated with an increase in the frequency of urinary tract infection. However, much of this evidence can be traced to uncontrolled or poorly controlled surveys. Recent evidence indicates that patients with diabetes who have no neurologic complications affecting bladder emptying and who have not undergone instrumentation are not at greater risk of developing urinary tract infection (Gocke, 1980). Following catheterization, however, or in the presence of autonomic neuropathy involving the bladder, ascending infection is more frequent and more severe. In particular, parenchymal damage is more extensive, with a higher incidence of severe local complications. Renal damage appears to correlate best with the severity of the underlying nephrosclerosis, which delays the protective inflammatory response and increases the possibility of papillary necrosis. Leukocyte dysfunction has been described in poorly controlled, insulin-deficient diabetes; leukocyte dysfunction also contributes to the severity of renal parenchymal destruction.

PATHOLOGIC CHARACTERISTICS

ACUTE PYELONEPHRITIS.

In severe acute pyelonephritis, the kidney is somewhat enlarged; and discrete, yellowish raised abscesses are apparent on the surface of the kidney. The patho-

gnomonic histologic feature is suppurative necrosis or abscess formation within the renal parenchyma (Susin and Becker, 1972).

CHRONIC PYELONEPHRITIS

The pathologic picture of chronic pyelonephritis can be described as follows (Susin and Becker, 1972). One or both kidneys contain gross scars, but even when involvement is bilateral, the kidneys are not equally damaged. This uneven scarring is useful in differentiating chronic pyelonephritis from diseases that cause symmetrical contracted kidneys, for example, chronic glomerulonephritis. There are inflammatory changes in the pelvic wall, with papillary atrophy and blunting. The parenchyma shows interstitial fibrosis with an inflammatory infiltrate of lymphocytes, plasma cells, and occasionally neutrophils. The tubules are dilated or contracted, with atrophy of the lining epithelium. Many of the dilated tubules contain colloid casts, which suggest the appearance of thyroid tissue (thyroidization of the kidney). There is also concentric fibrosis about the parietal layer of Bowman's capsule (termed "periglomerular fibrosis") and vascular changes similar to those of benign or malignant arteriolar sclerosis.

Several studies (Freedman, 1967, 1975; Murray and Goldberg, 1975) have found little correlation between these pathologic findings and evidence for past or present urinary tract infection. Clearly a better term for this pathologic entity would be "chronic interstitial nephritis" to encompass all the clinical states that can cause these changes. To incriminate infection as the sole cause of chronic interstitial nephritis, evidence is required of past or present urinary tract infection and the absence of any other condition that can cause the pathologic picture of chronic interstitial nephritis. These criteria are seldom met, and even if they are, it is frequently impossible to establish whether infection is complicating interstitial nephritis of some unrecognized etiology. For example, analgesic nephropathy was not recognized until the 1950s.

PAPILLARY NECROSIS CAUSED BY INFECTION

Frequently both kidneys are affected, and one or more pyramids may be involved (Susin and Becker, 1972). The pyramids are replaced by wedge-shaped areas of yellow necrotic tissue with the base located at the corticomedullary injunction. As the lesion progresses, a portion of the necrotic papilla may break off, producing a calyceal deformity that results in a recognizable radiologic filling defect. The sloughed portion may be voided and in some instances can be recovered from the urine. Microscopically, edema of the interstitium is initially seen. Eventually, the lesion resembles an infarct, with coagulation necrosis involving the entire pyramid. The collecting tubules are filled with bacteria and polymorphonuclear leukocytes.

EPIDEMIOLOGY

BACTERIURIA IN CHILDREN

Urinary tract infections not infrequently occur in neonates. The prevalence of bacteriuria in the neonate is about 1%, much more common in boys (Boineau and Lewy, 1975). Infection often results in bacteremia and neonatal sepsis. Many of these neonates have severe congenital structural abnormalities. Autopsy studies have also revealed a predominance of infant boys with pyelonephritis (Neumann and Pryles, 1962).

In the next age group following infancy, preschool children, urinary infection is more common in girls than in boys (Boineau and Lewy, 1975). When infection occurs in preschool boys, it is once more often associated with serious congenital abnormalities. With repeated studies, including a follow-up conducted over one year, the period prevalence of significant bacteriuria in this age group was reported to be 4.5% for girls and about 0.5% for boys (Randolph and Greenfield, 1964). Radiologic abnormalities including reflux are found in about 40% of bacteriuric preschool girls (Whitaker and Sherwood, 1984). Infection during this period may be symptomatic or asymptomatic. This period is considered the critical period of kidney growth. It is during this crucial growth phase that obstruction, reflux, and infection as independent factors may cause serious renal damage (Smellie et al., 1975a, 1975b).

Kunin et al. and others studied the epidemiology and natural history of urinary tract infection in schoolchildren (Kunin, 1970a, 1976; Gillenwater et al., 1979). The onset of infection at this later age was associated less frequently with underlying disease, and moreover was less likely to cause severe or permanent renal damage. Hence, routine screening for bacteriuria in schoolchildren has largely been abandoned. It was found that bacteriuria in girls in this population is often asymptomatic and frequently recurs. For example, the prevalence of bacteriuria among school girls was about 1.2%, and about 5% of the girls had significant bacteriuria at some time before leaving high school. About one third of these patients had some symptom referable to the urinary tract when the bacteriuria was first detected. It was shown that each year about 0.3%–0.4% of the female population (25% of those infected) were either cured spontaneously or with antimicrobial agents and were replaced by an equal number who developed bac-

teriuria. Bacteriuria was rare in school boys (prevalence 0.03%).

These studies also provided an opportunity to treat the patients and follow their clinical course. Patients were initially treated for ten days to two weeks. Girls with frequent infections were given longer courses of therapy (1–3 months). Caucasian girls tended to have more frequent reinfection than black girls. With each course of therapy, about 20% of white girls went into long-term remission. However, when many of these girls were married or became pregnant, bacteriuria recurred at a rate far above that expected for the general population. Over 50% developed bacteriuria within three months after marriage. Thus, the presence of bacteriuria in childhood defines a population at higher risk for the development of bacteriuria in adulthood.

BACTERIURIA IN ADULTS

Once adulthood is reached, the prevalence of bacteriuria increases in the female population. The prevalence of bacteriuria in young nonpregnant women is about 1%–3% (Freedman, 1965; Kass, 1960a, 1960b; Kass et al., 1965). Each year about 25% of bacteriuric women clear their bacteriuria and are replaced by an equal number who have become infected (often women who have had urinary infection previously). At least 10%–20% of the female population experience a symptomatic urinary tract infection at some time during their life (Kass et al., 1965; Sanford, 1975).

The prevalence of bacteriuria in adult men is low (0.1% or less) until the later years, when it rises. The increase in bacteriuria in older men is probably mainly related to prostatic disease and the resultant instrumentation. Men with bacteriuria frequently have anatomic abnormalities of the urinary tract.

BACTERIURIA IN THE ELDERLY

A marked increase in the prevalence of bacteriuria is observed in the elderly in both sexes (Bentzen and Vejlsgaard, 1980). At least 10% of men and 20% of women over 65 have bacteriuria. In contrast to young adults, in whom bacteriuria is 30 times more frequent in women than men, over the age of 65 the ratio changes dramatically, with a progressive decrease in the female-male ratio (Kaye, 1980; Romano and Kaye, 1981). Possible reasons for the high frequency of urinary infection in the elderly include obstructive uropathy from the prostate and loss of bactericidal activity of prostatic secretions in men, poor emptying of the bladder due to prolapse in women, soiling of the perineum from fecal incontinence in demented women, and neuromuscular diseases and increased instrumentation and bladder catheter usage in both sexes (Kaye, 1980; Romano and Kaye, 1981). The spectrum of microorganisms is unaltered in the noninstitutionalized elderly.

CLINICAL MANIFESTATIONS

SYMPTOMS

The classic clinical manifestations of upper urinary tract infection include fever (often with chills and rigors), flank pain, flank tenderness, and frequently lower tract symptoms. Nausea and vomiting are common. At times, the lower tract symptoms antedate the appearance of fever and upper tract symptoms by one or two days. The symptoms described, while classic, may vary greatly. In fact, pyelonephritis may show protean clinical manifestations, especially in children. Flank tenderness or discomfort is frequent in upper tract infection in adults and is more intense when urinary obstruction is present. Severe pain with radiation into the groin is rare in acute pyelonephritis per se and suggests the presence of a renal calculus. Pain originating in the kidney is occasionally felt in or near the epigastrium with widespread radiation, and renal infection may present with ileus. These manifestations often result in a difficult differential diagnosis and may suggest gallbladder disease or appendicitis.

Lower tract symptoms result from bacteria inducing irritation and inflammation of urethral and vesical mucosa, causing frequent and painful urination (dysuria) of small amounts of turbid urine. Cystitis is further suggested by a sense of urgency and suprapubic heaviness or pain. Occasionally, the urine is grossly bloody or shows a blood tinge at the end of micturition. In contrast to upper tract infection, fever tends to be absent in infection limited to the lower tract.

Urinary tract infection in children tends to produce different symptoms depending on the age of the child (Goven, 1974; Boineau and Lewy, 1975; Margileth et al., 1976). Symptoms in neonates and children less than 2 years of age are nonspecific, including failure to thrive, vomiting, and unexplained fever. When children over 2 years of age (and especially over 5 years) develop infection, they are more likely to display localizing symptoms such as frequency, dysuria, and abdominal or flank pain.

The vast majority of elderly patients with urinary tract infection are asymptomatic (Kaye, 1980; Romano and Kaye, 1981). Unfortunately, even when lower urinary tract symptoms are present, they are often not specific or diagnostic, since noninfected elderly patients often experience frequency, dysuria, hesitancy and incontinence. Nevertheless, typical symptoms may occur, and less frequently acute pyelonephritis develops, usually necessitating hospitalization. Gleckman et al. (1982) observed a much higher frequency of bacteremia (61%)

associated with pyelonephritis in the elderly than is seen in the young, and shock more frequently supervenes. The effect of asymptomatic bacteriuria on the general sense of well-being, urinary continence, and associated diseases such as control of diabetes is unknown.

Upper tract infection may result in only lower tract symptoms or no symptoms at all. Patients with urinary tract infection in the presence of an indwelling urinary catheter usually have no lower tract symptoms, but flank pain or fever may occur. Urinary tract infection is the most common source of bacteremia produced by gram-negative bacilli. Bacteremia may occur with no urinary symptoms, especially in the presence of an indwelling catheter.

ALTERATION IN RENAL FUNCTION

For the most part, renal function remains unaltered in patients with urinary tract infection. In experimentally produced pyelonephritis, the only consistent abnormality of renal function is the inability to concentrate the urine maximally (Kaye and Rocha, 1970). The mechanism of the concentrating defect is not clear but seems to be related in experimental animals to inflammation and possibly to increased production of prostaglandins (Levison and Levison, 1976; Levison et al., 1980). The concentrating defect occurs early in the course of experimental infection and is rapidly reversible with antibiotic therapy and following the administration of prostaglandin inhibitors (Kaye and Rocha, 1970; Levison et al., 1980).

Progressive destruction of the kidney rarely occurs in the absence of obstruction or intrarenal reflux. Infection may contribute to renal insufficiency when the latter conditions coexist. Bilateral papillary necrosis occasionally leads to rapidly progressive renal failure (Hellebusch, 1969). With these exceptions, urinary tract infection rarely if ever results in renal insufficiency.

DIAGNOSIS

PRESUMPTIVE DIAGNOSIS

The first step in the laboratory diagnosis of urinary tract infection is microscopic examination of the urine for leukocytes. A clean-catch midstream urine specimen is obtained, and a drop of unspun urine should be examined as well as the sediment from a specimen obtained by centrifugation for 5 minutes at 2,000 rpm. Measurement of leukocyte excretion rate is too time consuming and correlates with a random unspun specimen as measured in a counting chamber.

In a centrifuged specimen each leukocyte seen under high power represents about 5–10 cells/mm^3 of urine;

10–50 WBCs/mm^3 have been stated to be the upper limit of normal (Brumfitt, 1964). With this criterion, 5–10 leukocytes/HPF in the sediment from a clean-catch midstream urine specimen is the upper limit of normal, as they represent 50–100 cells/mm^3. This method has been criticized because of numerous variables, most important of which is the widely varying volumes used for resuspension after centrifugation (Stamm, 1983a). Stamm relied on quantitative leukocyte counts in a standard counting chamber using unspun urine. In asymptomatic abacteriuric patients, a count of >10 WBC/mm^3 was found in <1% of clean-catch specimens. In symptomatic bacteriuric patients (≥10^5 bacteria/ml), 96% of patients with lower urinary tract infections had >10 WBC/mm^3 in both sexes (Musher et al., 1976; Stamm, 1983a). In symptomatic abacteriuria (urethral syndrome), 70% of the patients had ≥10 WBC/mm^3, and the majority of patients with pyuria had demonstrable infection (either 10^2–10^5 bacteria/ml bladder urine or *Chlamydia*) (Stamm et al., 1980a). In nonpregnant women with asymptomatic bacteriuria, Musher et al. (1976) found that most women had >10 WBC/mm^3, suggesting that such women have true infection and not urinary tract colonization. In pregnant women with asymptomatic bacteriuria, 49% of subjects had ≥10 WBC/mm^3 (Williams et al., 1965). Similarly, in catheterized patients, bacteriuria is generally accompanied by pyuria (≥10 WBC/mm^3), representing not simply bladder colonization but true inflammation or infection (Musher et al., 1976). Accordingly, in all types of patients there exists a striking association between pyuria and bacteriuria. Pyuria is extremely useful in identifying infected patients with <10^5 bacteria/ml urine (Stamm, 1982). Whether quantitative leukocyte counts as measured by a counting chamber should replace conventional methods using spun urine is controversial (Stamm, 1983a; Stamm and Turck, 1983b; Brumfitt and Percival, 1964). It should be emphasized that some patients with and without pyuria may or may not have infection (Thysell, 1969; Brumfitt and Percival, 1964).

Microscopic or sometimes gross hematuria is occasionally seen in patients with urinary tract infection (i.e., hemorrhagic cystitis). However, RBCs may be indicative of other disorders such as calculi, tumor, vasculitis, glomerulonephritis, and renal tuberculosis. White cell casts in the presence of an acute infectious process are strong evidence for pyelonephritis, but their absence does not rule out upper tract infection. White cell casts can also be seen in renal disease in the absence of infection.

Proteinuria is a common although not universal finding in urinary tract infection. Most patients with urinary tract infection excrete less than 2 gm of protein in 24 hours; excretion of 3 gm or more suggests glomerular disease.

In the presumptive diagnosis of urinary tract infection, an extremely useful test is the microscopic examination of a specimen for bacteria. The ability to identify bacteria in the urine depends on whether the specimen has been centrifuged and on whether it has been stained with Gram or methylene blue stain (Table 5–1). Smaller numbers of bacteria can be detected microscopically in a stained than in an unstained specimen and smaller numbers can be detected in a centrifuged than in uncentrifuged urine. Using a Gram stain the presence of at least one bacterium per oil immersion field correlates with ≥10^5 bacteria/ml of urine if midstream, clean-catch uncentrifuged urine is examined. The absence of bacteria in several fields in a stained sedimented specimen indicates the probability of less than 10^4 bacteria/ml. For detection of ≥10^5 bacteria/ml, Gram stain of urine smears is rapid, accurate, and inexpensive, with 94% sensitivity, >70% specificity, and a negative predictive value of 99% (Pezzlo, 1983). Nevertheless, it is apparent that low-count bacterial infection of the lower urinary tract may be associated with a negative Gram stain.

Biochemical tests have been devised to detect bacteriuria for presumptive diagnosis. The Griess test (a diazotization reaction) detects the presence of nitrite in the urine that is formed when bacteria reduce the nitrate that is normally present. Chernow et al. (1984) reported high sensitivity of reagent strip leukocyte esterase for detecting pyuria, making it a valuable screen test for infection, with high sensitivity but moderate specificity (76%). The positive nitrate reaction is more specific but has lower sensitivity (Chernow et al., 1984). Recently, the nitrate/nitrite reaction has been combined with leukocyte esterase-sensitive dipsticks, aimed at detecting significant bacteriuria (and pyuria). Although these strips have been useful in office practice, they pick up numerous false positive results (positive predictive value, 40%) but few false negatives (Bartlett and Trieber, 1984; Smalley and Dittman, 1983). Their greatest use may be in screening urine samples submitted to the laboratory for routine culture, since about 80% of such urine specimens are culture-negative. Thus, an inexpensive, rapid, and accurate method is needed to identify infected urine requiring culture.

TABLE 5–1.

Correlation of Observing Bacteria on Direct Examination of Urine With Quantitative Cultures

	UNSTAINED	STAINED
Uncentrifuged	≥10^6 (400×)	≥10^5 (1,000×)
Centrifuged	≥10^5 (400×)	≥10^4 (1,000×)

DIAGNOSIS BY CULTURE

Urine in the bladder is normally sterile but usually becomes contaminated by organisms from the urethra or external surfaces during voiding. Therefore, the presence of bacteria in the urine is of little use in diagnosis of urinary tract infection. Quantitative studies by Kass permitted separation of those subjects with urinary tract infection and consistently high densities of bacteria in the urine from normal people who had insignificant bacteriuria, i.e., low densities of bacteria that were contaminants (Kass, 1956). By quantitating bacteria in midstream, clean voided urine, it is possible statistically to distinguish between bacterial infection of the urinary tract and contamination or urinary colonization. Criteria depend on the fact that density of bacteria in infected urine is usually several orders of magnitude higher than the density of bacteria in contaminated urine. Most quantitative definitions have concerned gram-negative rods and females.

For acute pyelonephritis and asymptomatic bacteriuria a criterion of $\geq 10^5$ bacteria/ml provides optimal separation of infection from contamination of voided urine (Kass, 1956). These patients with infection usually have $\geq 10^5$ bacteria/ml in urine in the bladder, and therefore voided urine contains $\geq 10^5$ bacteria/ml.

If there are more than 10^5 bacteria/ml in a clean-catch urine specimen from an asymptomatic woman, there is an 80% probability that this represents true bacteriuria. If two different specimens demonstrate at least 10^5 of the same bacterium per milliliter, the probability increases to 95%. Thus, two clean-catch specimens should be obtained in an asymptomatic woman to confirm the diagnosis. When the number of bacteria per milliliter is between 10^4 and 10^5 in an asymptomatic woman, a confirmatory second specimen will contain $\geq 10^5$ bacteria/ml in only 5% of instances. Thus, in asymptomatic women, 95% of the time, 10^4–10^5 bacteria/ml represents contamination with occasional lower tract infection manifested by $< 10^5$ bacteria/ml urine. In men, in whom contamination is less likely, 10^4 organisms/ml is more suggestive of infection (Stamey et al., 1965). In contrast to asymptomatic patients, in patients with symptoms of urinary tract infection, one titer of $\geq 10^5$ bacteria/ml carries a 95% probability of true bacteriuria.

In the frequency-dysuria syndrome (urethral syndrome) it is important to remember that about one fourth of women with symptomatic lower tract infection have $< 10^5$ bacteria/ml urine. In young women presenting with frequency and dysuria, a colony count of $> 10^2$ Enterobacteriaceae/ml is now accepted as a valid criterion for diagnosis (Stamm et al., 1980a).

Accordingly, interpretation of culture results de-pends on both the clinical setting (symptomatic or asymptomatic, upper or lower urinary tract symptoms) and the manner in which the specimen is obtained. Contamination occurs when organisms in the distal urethra, vagina, labia, or periurethral skin inadvertently enter the collecting vessel. Contamination should be suspected in the presence of squamous epithelial cells seen on urinalysis and when cultures grow more than one organism in high titer, especially microorganisms normally found in the distal urethra and on perineal skin. In asymptomatic patients, contamination is suggested by $< 10^5$ bacteria/ml of urine, whereas in symptomatic patients with frequency and dysuria, low counts may be due to contamination or true low-count bacterial infection. In these patients repeated culture and urinalysis for the detection of pyuria will help to separate these two entities. Nevertheless, in women with frequency-dysuria, 10^2–10^4 Enterobacteriaceae/ml of urine more often indicates true infection than contamination. The sensitivity is 95% and the specificity is 85% (Stamm, 1982). Unequivocal separation requires a suprapubic aspiration.

Several methods can be employed to quantitate bacteria present in urine. The serial dilution and pour plate method is the most accurate (Andriole, 1972a) but is cumbersome and not suitable for routine use in a busy clinical laboratory. Calibrated loops serve as a simple, inexpensive way to examine quantitatively the bacteriologic characteristics of urine specimens (Hoeprich, 1960). Platinum loops that deliver 0.01 ml and 0.001 ml are used to streak urine onto agar plates. After incubating at 37°C for 24 hours, the number of CFUs is counted, and the total number of organisms originally present in the specimen is estimated by multiplying the colony count by 10^2 or 10^3, respectively. A further refinement of the technique involves the use of differential agars such as desoxycholate or MacConkey's (for gram-negative bacilli) and phenylethyl alcohol or colistin-nalidixic acid agar (for gram-positive organisms) to allow isolation from mixed cultures and more rapid identification.

Other methods of quantitative culture include: (1) the flood plate method, which is similar to the calibrated loop method but involves pipeting a volume of urine onto a plate (Andriole, 1972a); (2) the filter paper method (Andriole, 1972a) in which a given volume of urine is absorbed in a piece of filter paper and then put on a plate; and (3) the dip inoculum method (Dunin, 1975a) in which an agar-coated glass slide is dipped into urine. The dip inoculum method and its variants have excellent correlation with pour-plate techniques and are available for office use at inexpensive prices (Margileth et al., 1976).

Acceptable methods for urine collection include: (1)

midstream clean-catch, (2) catheterization, and (3) suprapubic aspiration. The clean-catch method is preferred for routine collection of urine for culture. It avoids the risk of infection inherent in catheterization. The patients must be instructed in the proper technique of obtaining the urine; this is especially important in women. The woman should wash her hands, straddle the commode (facing the back of the commode), wash the vulva from front to back four times with four different sterile gauze pads soaked in green soap or another appropriate cleansing agent, and then rinse with two more sponges soaked in sterile distilled water. She should then spread the labia and void, discarding the first portion of urine and collecting the second. The urine should be processed immediately, or if refrigerated at 4°C, can be cultured within 24 hours. In men the prepuce should be retracted, and thereafter the technique is similar. In infants and small children sterile bags have been used for collection of urine, but contamination is common (Hardy et al., 1976).

In patients unable to cooperate, such as those with an altered sensorium, or those who are unable to void for neurologic or urologic reasons, catheterization may be necessary. When catheterization is performed, scrupulous aseptic technique should be observed.

The suprapubic aspiration method has been established as a safe technique in premature infants, neonates, children, adults, and even pregnant patients (Monzon et al., 1958; McFayden and Eykyn, 1968; Boineau and Lewy, 1975; Margileth et al., 1976). With this method the bladder must be full. The patient refrains from voiding until the bladder can be percussed above the pubis, and suprapubic pressure causes the urge to void. After preparation of the skin, the bladder is then punctured above the symphysis pubis with a 22-gauge needle on a syringe (local anesthesia is not required). Following the procedure, self-limited hematuria may be observed. Suprapubic aspiration may be indicated in special clinical situations such as in pediatric practice, when urine is difficult to obtain. Another situation is the rare adult in whom infection is suspected, results obtained from more routine procecures have been confusing or equivocal, and diagnosis is critical.

The quantitative criteria described above apply only to the Enterobacteriaceae. Gram-positive organisms, fungi, and bacteria with fastidious growth requirements may not reach titers of 10^5 /ml in patients with infection and may be in the 10^4–10^5/ml range (Andriole, 1972a). False positive cultures are caused by contamination or incubation of urine before processing. False negative cultures may be due to use of antimicrobial agents, soap from the preparation falling in the urine, total obstruction below the infection, infection with a fastidious organism, renal tuberculosis, and diuresis (Cattel et al.,

1968). Mixed infection occurs in about 5% of the cases.

Samples obtained by catheterization from noninfected patients are less likely to become contaminated enough to demonstrate 10^5 bacteria/ml. For example, one catheterized specimen in an asymptomatic patient that contains 10^5 or more organisms/ml has a 95% chance of indicating infection, and counts between 10^4 and 10^5/ml (which are uncommon) are significant at least 50% of the time. The contamination is presumably from the urethra. Bladder urine obtained by suprapubic aspiration is either sterile or contains significant growth even if bacterial numbers are below 10^5/ml. The practice of forcing fluids before the procedure tends to reduce titers (Goldberg et al., 1965). In fact, almost 50% of such specimens contain $<10^5$ organisms/ml. However, small numbers of bacteria may be found in aspirated urine from presumably noninfected persons (Monzon et al., 1958). This suggests that bladder urine may occasionally be contaminated from the urethra.

Asymptomatic colonization of the bladder and lower urinary tract is a concept not widely accepted. As opposed to contamination, colonization refers to multiplication of microorganisms in urine without invading mucosal tissue or causing tissue injury. Colonization would occur in the absence of signs and symptoms of infection, absence of pyuria, and absence of serological response to the infecting organism. Since infection is often asymptomatic, pyuria would have to be the mainstay of differentiating colonization and infection (Stamm, 1983a). The possibility of asymptomatic colonization is supported by data of Nicolle et al. (1982) and Kunin et al. (1980), who found that a proportion of patients with asymptomatic bacteriuria ($\geq 10^5$ bacteria/ml) actually had a transient self-limited state.

Recently, several automated methods have been introduced that are designed to detect bacteria in the urine within 30 minutes to nine hours. Detection of growth is based on changes in light transmission (photometry or bioluminescence). Most of these systems as screening methods have acceptable sensitivity and specificity for diagnosis of bacteriuria $\geq 10^3$ bacteria/ml urine with relatively few false positives (Pezzlo, 1983).

LOCALIZATION OF SITE OF INFECTION

Proper therapy and response to therapy of urinary tract infection depends on the site or level of infection. Clinical signs and symptoms are often unreliable and misleading. Several methods have been used to determine whether infection is confined to the urinary bladder or whether the upper tract is also involved. Needle biopsy specimens of the kidney have been cultured (Brun et al., 1965). However, this is both an unnecessarily invasive and unreliable approach because pyelonephritis is a focal disease and specimens obtained by needle biopsy may miss the area of infection.

The most reliable method of localization of infection involves obtaining urine directly from the ureter for quantitative cultures. In one study (Stamey et al., 1965) using this method, 95 women and 26 men with bacteriuria were evaluated. Approximately one half had infection limited to the bladder. Turck et al. (1968) using similar techniques demonstrated that in women with recurrent urinary tract infection, relapse was associated with upper tract involvement and reinfection with lower tract infection.

Fairley and colleagues (1967) devised a technique for assessing ureteral bacteriuria that involves Foley catheterization and bladder washout only. However, results are equivocal in about 10%–20% of patients (Fairley et al., 1971; Sanford, 1975). As in the ureteral catheterization studies, about 50% of the patients have renal infection regardless of lack of signs or symptoms (Fairley et al., 1971). Methods are also available for localization of bacteria in the prostate gland (Meares and Stamey, 1968) and are discussed later.

Several studies (Clark et al., 1969; Ronald et al., 1969) have reported the association of a defect in renal concentrating ability with upper tract infection. As might be expected, patients with unilateral ureteral bacteriuria have an ipsilateral defect in concentrating ability (Ronald et al., 1969). However, there are too many false positive and false negative responses to allow the use of concentrating ability for localization of urinary tract infection. Even the water-loading test with administration of furosemide has not found widespread application (Dontas et al., 1974).

The immune response has been used as a means of localizing the site of infection. The presence of high titers of serum antibody directed against the infecting organism has been correlated with the presence of upper tract infection in patients undergoing ureteral catheterization (Reeves and Brumfitt, 1968). Although there is a good association of high antibody titers and renal infection, there is a high incidence of false positive and false negative results (about 20% each).

The measurement of urinary leukocyte enzymes has been stated to be of some value in detecting urinary tract infections and the differentiation of upper versus lower tract involvement. Various enzymes such as lactic dehydrogenase isoenzymes IV and V, alkaline phosphatase, catalase, β-glycouronidase, transaminase, leucineaminopeptidase, and lysozyme have been evaluated (Turck, 1975). From these studies, it is apparent that determination of urinary enzyme activity adds little if anything to the diagnostic approach of localization of infection, because several inflammatory processes as well as upper or lower tract infection can result in increased enzyme activity (Turck, 1975; Andriole, 1982).

Gallium 67 has been used to localize the site of urinary tract infection. In one study (Kessler et al., 1974), selective renal uptake of this isotope was reported in all patients with acute pyelonephritis. Another study compared this procedure with ureteral catheterization and bladder washout techniques and reported an accuracy of greater than 85% (Hurwitz et al., 1976).

Thomas et al. (1974) first described a noninvasive, simple, and inexpensive direct immunofluorescence technique for detection of antibody-coated bacteria (ACB) in the urine sediment of bacteriuric patients to localize infection. The theoretical basis for the test is that the kidney itself produces specific antibodies against bacteria invading the renal parenchyma (unrelated to specific serum antibodies). In contrast, bladder infections, being relatively superficial, fail to elicit the same antibody response in the urine. Fluorescein-conjugated antihuman globulin is added to urine containing bacteria and the bacteria are examined for fluorescence. The sensitivity of the test for ACB has been established by several studies as 88% (range 72%–100%) with a specificity of 76% (range 50%–100%), correlating the presence of ACB with upper tract infection when confirmed by established direct methods (Jones et al., 1974; Hellerstein et al., 1978; Riedasch et al., 1978). According to Thomas and Forland (1982), the predictive value of associating a positive test for ACB with upper tract infection is about 83%, the predictive value of associating a negative test result for ACB with bladder bacteriuria is 83%.

Discrepancies have resulted from lack of standardization of criteria of what consitutes a positive test result. In the original description, a test was considered positive if ≥20% of bacteria/HPF were brilliantly fluorescent, whereas subsequent investigators have used values of 1%–25%. Other problems include interobserver disagreement of interpretation. False negative results have been obtained in: (1) acute pyelonephritis when duration of infection is too short to elicit and detect an immune response; this was the case in 20% of acute renal infections seen in pregnancy (Thomas and Forland, 1982); (2) extremes of age (Hellerstein et al., 1978); (3) low quantities of antibodies (Ratner et al., 1981); (4) mucoid *Pseudomonas* infection where extracellular polysaccharide blocks antibody adherence to the bacterial surface (Marrie et al., 1979); and (5) in patients with neurogenic bladder and kidney infection (Merrit and Keys, 1982). However, in immunosuppressed patients with kidney infection in whom one might expect false negatives, the kidney does not lose its ability to produce antibodies (Riedasch et al., 1978).

False positive ACB test results occur in approximately 20%–30% of patients with lower urinary tract infection. In men, the most important cause of false positive results is the presence of bacterial prostatitis (Jones et al., 1974; Thomas and Forland, 1982). In females, false positive results may be the consequence of contaminaation of urine samples by small numbers of ACB and yeast from the vaginal vestibule of patients without urinary tract infection (Montplaisir et al., 1976). Yeasts and pseudomonads may fluoresce even if uncoated (Jones, 1979). Deep bladder invasion, particularly when accompanied by hemorrhagic inflammation, bladder tumors, and bladder stones may be associated with false positive ACB tests (Forsum et al., 1976).

There has been a major discrepancy in infants between ACB localization and direct as well as other indirect localization techniques (Montplaisir et al., 1976; Hellerstein et al., 1978; Kwasnik et al., 1979). Difficulty obtaining clean voided specimens, particularly in small girls, could result in contamination with small numbers of ACB of perineal origin (Montplaisir et al., 1976). However, Hellerstein et al. (1978), using catheter-obtained specimens, still observed high false positive and false negative rates.

In patients with short-term bladder catheterization, a positive ACB test is of some value in detecting clinically inapparent pyelonephritis (Giamerellou et al., 1982); however, if the catheter remains in situ for more than a few days, the value of this test declines significantly with many false positives, probably due to chronic bladder invasion. Low sensitivity but particularly low specificity accompanies the use of long-term indwelling catheters (Merrit and Keys, 1982; Kuhlemeier et al., 1982; Hutler et al., 1984; Hooton et al., 1984). Furthermore, no relationship was seen between localization of infection (ACB testing) and response to therapy in patients with indwelling catheters with neurogenic bladders (Hooton et al., 1984). ACB testing has been found to be reasonably accurate during pregnancy (Harris et al., 1975; Thomas et al., 1975). Harris et al. found that 50% of asymptomatic bacteriuric women in pregnancy could be assigned to upper tract localization by ACB testing. In the same study, Harris and associates reported that a higher percentage of pregnant women with ACB-positive bacteriuria had elevated serum creatinine and reduced glomerular filtration, and intrauterine growth retardation was more frequently seen in infants born to mothers with positive ACB tests.

Several investigators have expressed serious doubts about the reliability of ACB testing in defining the site of infection (Mundt and Polk, 1979; Gleckman, 1979). Given the acknowledged limitations of this localization test, the major clinical usefulness of a positive ACB test

(if appropriately standardized) and in selected patients (young women) is to indicate to the clinician that the problem is more than a simple lower uirnary tract infection. It may influence both the selection and duration of antimicrobial therapy. In this population, a positive ACB test in midstream urine identifies patients who are at high risk of treatment failure, particularly single-dose therapy (Fang et al., 1978; Gargan et al., 1983). A negative ACB test in the absence of a clinical picture of acute pyelonephritis provides reasonable assurance that renal infection is absent. Currently, ACB testing is not widely available to physicians, nor is there good evidence to conclude that this assay has a major role in the routine management of patients with urinary tract infections. Its main use is as an epidemiologic tool and in the study of the pathogenesis and treatment of urinary tract infection.

Outcome of therapy can also be used in a crude but useful manner to separate those with upper and lower tract infection. Virtually all patients with infection restricted to the lower tract can be cured with a short course of antimicrobial therapy. However, the relapse rate with upper tract infection is appreciable, even with 7–10 days of therapy.

At present, only ureteral catheterization studies can reliably predict the site of infection in the urinary tract. However, this procedure is not without risk and cannot be justified for routine use. The Fairley bladder washout procedure is also quite reliable but gives equivocal results in about 10%–20% of the patients. It also involves catheterization. The determination of presence of ACB in the urine is practical, looks promising as a method of demonstrating upper tract infection, and is noninvasive. However, in the clinical management of patients, it is rarely important to localize infection definitively to either the upper tract or the lower tract.

FREQUENCY-DYSURIA SYNDROME

Numerous studies have indicated that approximately 50% of women with acute onset of frequency and dysuria lack "significant" bacteriuria, and approximately one quarter may have sterile urine (Stamm et al., 1980a; Stamm, 1982) (Table 5–2). The term "urethral syndrome" has been used to refer to this entity of acute onset of dysuria and frequency in young females whose voided urine is either sterile or contains $<10^5$ bacteria/ml (Gallagher et al., 1965). About 20% of adult females may experience an episode of acute dysuria each year (Waters et al., 1970). Dysuria in the absence of urgency and frequency has low diagnostic specificity and is one of the most common clinical problems seen by clinicians in developed nations (Komaroff, 1984), accounting for 3 million office visits per year in the United States, and is often due to vaginitis. The conventional clinical

TABLE 5–2.

Relative Frequency of Causes of Acute Onset of Frequency-Dysuria in Young Women

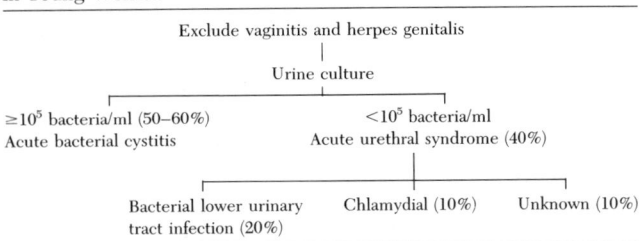

approach to acute onset of dysuria has been based upon the following misconceptions: (1) the vast majority of such women have bacterial cystitis; (2) the responsible microorganisms are almost always coliforms; (3) the single most important test is a urine culture; (4) finding of quantitative cultures of $\geq 10^5$ bacteria/ml constitutes the only proof of urinary tract infection; and (5) patients with positive cultures should receive 7–14 days of antimicrobial therapy. The majority of women with acute dysuria-frequency have infections of the bladder, urethra, or vagina.

Komaroff et al. (1978) reported that vaginitis due to the common vaginal pathogens was an extremely common cause of dysuria (up to 60% of cases), and accordingly patients should be carefully questioned regarding vaginal symptoms, particularly if the complaint of burning is described as "external" such as pain felt in the vaginal labia during micturition. Vaginitis is suggested by the presence of a vaginal discharge, pruritus, dyspareunia, and vaginal irritation. If these symptoms are present, a pelvic examination is mandatory (Komaroff et al., 1978; Komaroff, 1984; Berg et al., 1984). In vaginitis it is uncertain whether dysuria is due to simultaneous infection of the urethra or urethral meatus or related to vulvar and labial inflammation only.

In the absence of vaginal symptoms or when urinary frequency and urgency are also present, the cause of the symptoms lies with inflammation of the bladder (cystitis) or urethra (urethritis) (see Table 5–2). Urethritis in females, unlike males, almost never presents with a urethral discharge, but with dysuria and frequency. *Neisseria gonorrhoeae* has long been known to cause dysuria in women. This observation was based on the finding of positive cultures for *N. gonorrhoeae* from the cervix, urethra, urine, or rectum in a small percentage of women presenting with the urethral syndrome. Similarly, Paavonen et al. in 1978 first suggested that *Chlamydia* could cause the urethral syndrome. The pathogenetic role of *C. trachomatis* in the urethral syndrome was substantiated by Stamm et al. (1980a). *C. trachomatis* was found in 5% of asymptomatic female

controls, 3% of symptomatic females with bacterial cystitis, 6% of females with the urethral syndrome without pyuria, and in more than 60% of women with the urethral syndrome with pyuria but sterile bladder urine.

Herpes simplex genitalis may affect the urethra, usually in association with vulvar and cervical involvement. Dysuria has been described in 10% of women with initial genital herpes infection (Stamm, 1982). Stamm et al. (1980a) have shown that about one third of young women with the urethral syndrome ($<10^5$ bacteria/ml) have bacteria in bladder urine as demonstrated by suprapubic puncture. In fact, about a quarter of young women with symptomatic infection localized to the lower urinary tract have $<10^5$/ml bacteria in the urine. Significant bacteriuria in this context may be considered as any number $\geq 10^2$ bacteria/ml (Stamm et al., 1980a). Additional evidence supporting the causal relationship of low-level bacteriuria and symptomatic disease is suggested by response to therapy. Furthermore, longitudinal studies over days and weeks have shown that many of these patients will develop significant bacteriuria ($\geq 10^5$/ml).

The microorganisms responsible for low-bacterial-count cystourethritis are identical to those causing higher-titer infection ($\geq 10^5$/ml), coliforms and *S. saprophyticus*. Bacteria causing low-count cystitis also possess the identical virulence factors expressed by organisms causing classic cystitis. The distribution of causes of the urethral syndrome varies considerably according to the population studied. As yet there is still inconclusive evidence that *Ureaplasma urealyticum*, *Mycoplasma hominis*, viruses, lactobacilli, and other fastidious anaerobes have a causal role in the urethral syndrome. However, *Chlamydia*-negative patients with sterile urine often respond to empirical antibiotic treatment, suggesting that other microorganisms may be involved.

In spite of the progress made in the understanding of the pathogenesis of the urethral syndrome, a study by Berg et al. (1984) emphasizes the current limitations of our knowledge and diagnostic methods. Of 204 females presenting to a family practice office with different types of genitourinary symptoms, an etiologic diagnosis was only made in one third of patients using routine laboratory tests. (One third of the patients with dysuria could not distinguish between external and internal dysuria, and therefore this differentiation is frequently of little value to the physician.) A diagnosis could be further made in an additional one third of the symptomatic women if very selected, nonroutine laboratory procedures were used. Nevertheless, in the final one third no etiologic diagnosis was realized. This study emphasized that a large proportion of females presenting with "genitourinary symptoms" in whom a diagnosis could be made suffered from vaginitis—77%, which correlates well with the findings of Komaroff of 72% (1984). Noninfectious etiologies remain enigmatic but include hormonal, traumatic, allergic, chemical, and psychological mechanisms.

NATURAL HISTORY OF URINARY TRACT INFECTION

CHILDREN

In general, children with urinary tract infection without obstruction or vesicoureteral reflux have a very good prognosis (Smellie et al., 1975a; Smellie and Normand, 1975b). In contrast, in the presence of obstruction (e.g., urethral valves) or severe reflux, severe destruction of renal parenchyma can occur, resulting in chronic pyelonephritis and renal failure.

Reflux is found in 20%–50% of preschool children with asymptomatic or symptomatic bacteriuria (Boineau and Lewy, 1975; Smellie et al., 1975a; Whitaker and Sherwood, 1984). Reflux may be caused by bladder neck or urethral obstruction with increased pressure in the bladder, delayed development of the ureterovesical junction, a short intravesical ureter, and/or inflammation of the vesicoureteral junction. Reflux in the presence of infection is associated with the development of scarring detected by intravenous pyelography (IVP) (Smellie et al., 1975a).

Infants and young preschool children are at the highest risk for the development of progressive renal scarring (Bailey, 1973; Boineau and Lewy, 1975; MacGregor and Freeman, 1975). These children frequently have severe degrees of reflux (grades III and IV) with repeated infections, and some develop end-stage renal disease and hypertension. Obstruction (most common in infant boys with congenital anomalies) is likely to be associated with marked reflux (Bailey, 1973; Siegel et al., 1973; Smellie et al., 1975a; Smellie and Normand, 1975b; Cohen, 1976; Burbige et al., 1984). Bailey (1973) contends that preschool children of both sexes should undergo urologic investigation after the first urinary tract infection. The likelihood of finding a radiologic abnormality is greatest in children less than 2 years of age (up to 90%) and is less in children between the ages of 2 and 5 years (approximately 40%))McKerrow et al., 1984).

It should be emphasized that the contribution of reflux alone compared with reflux plus infection in the progression of renal scarring has not been clearly delineated. Reflux alone can apparently lead to renal damage and insufficiency (Bakshandeh et al., 1976; Salfatierra and Tangaho, 1977; Andriole, 1982). Studies in uninfected animals (Hodson et al., 1975) have demonstrated that reflux alone and in particular intrarenal reflux can produce "pyelonephritic" scars. It has also been shown that the immature kidneys of infants are more prone to intrarenal reflux (Rolleston et al., 1974). The term "reflux nephropathy," infected or uninfected, has been suggested to emphasize the primary role of reflux in scarring (Bailey, 1973). However, it is probable that reflux is more likely to lead to severe damage and scarring when infection is also present (Hodsen et al., 1975). It is also clear that infection tends to produce reflux or at least to make it more severe.

After the age of 5 years, children (predominantly girls) with bacteriuria frequently have renal scars, presumably acquired during the preschool years. Many of these children also have reflux. Reflux tends to decrease with elimination of bacteriuria. In addition, mild to moderate degrees of reflux are likely to disappear with the passage of time, probably related to maturation of the vesicoureteral junction (Edwards et al., 1977). Progression of scars already present or development of new ones is uncommon after the age of 5 (Blank, 1973; Edwards et al., 1977; Cardiff-Oxford Bacteriuria Study Group, 1978; Gillenwater et al., 1979). In fact, some investigators (Dodge et al., 1974) have questioned the need for detecting and treating bacteriuria in school-aged children. However, it is clear that progression does occur in some of these children, especially in the presence of severe reflux (Edwards et al., 1977).

ADULTS

Urinary tract infections are much more common in women than in men. Many of these patients previously had urinary tract infections as children and continue to have infections as adults (Gillenwater et al., 1979). Once a woman has had an infection, she is more likely to develop subsequent infections than a patient who has had no previous infections.

Longitudinal studies in young women with symptomatic recurrent urinary tract infections were performed by Kraft and Stamey (1977). The overall attack rate was 0.2 infections per month. Significant bacteriuria ($\geq 10^5$ bacteria/ml) was present in 70% of symptomatic episodes. Infections tended to occur in clusters, with an increased attack rate of 0.5% per month. These periods of more frequent infection were followed by a remission, or infection-free interval, that averaged about 13 months. However, most remissions were followed by further clusters of infection. Thus, in many women it is more correct to use the term "remission" rather than "cure" of urinary tract infection. It may be a simple matter to cure an individual episode, but reinfection is common.

It is clear that urinary tract infection in adults can lead to progressive renal damage in the presence of obstruction (Freedman et al., 1975). However, recurrent infection in adults in the absence of obstruction rarely, if ever, leads to renal failure.

Autopsy studies (Pawlowski et al., 1963; Freedman, 1967) have shown that it is difficult to implicate infection per se (i.e., in the absence of other renal abnormalities) as an important pathogenic factor in the production of severe renal disease in adults. One exception might be severe papillary necrosis secondary to infection. In fact, some authors (Schechter et al., 1971) have been unable to find any cases of uncomplicated pyelonephritis that progressed to end-stage renal disease among 173 patients admitted to dialysis programs. In prospective studies (Gower et al., 1968; Bullen and Kincaid-Smith, 1970; Zinner and Kass, 1971; Asscher et al., 1973; Freedman and Andriole, 1974; Gaches et al., 1976; Gillenwater et al., 1979), hundreds of patients have been followed up for years with persistent or recurrent infections without documented progression of renal disease from infection alone.

The role of infection in the progression of clinically or radiographically diagnosed interstitial renal disease has also been examined (Johnson and Smythe, 1969; Gower, 1973; Murray and Goldberg, 1975). In general, these studies indicate that infection is rarely, if ever, the major factor leading to further renal decompensation. However, infection may occasionally accelerate the progression of the primary underlying disease process (Gower, 1973). In summary, except for rare instances, there is no evidence to indicate that uncomplicated urinary tract infection alone produces renal failure in adults (Freedman, 1975).

THE ELDERLY

Althouth asymptomatic bacteriuria in the elderly has been considered benign and not cost-effective or risk-benefit-effective to treat, Dontas et al. in a recent study concluded that asymptomatic bacteriuria in the elderly was associated with reduced survival of 30%–50% (1981). The methodology in this study has been criticized, and as yet there is no evidence to suggest that treatment of asymptomatic bacteriuria in elderly patients reduces mortality or morbidity.

MANAGEMENT

GENERAL CONSIDERATIONS

Symptoms are not a reliable indication of infection (Kass, 1956; Gallagher et al., 1965; Stamm et al., 1980a). In the asymptomatic patient, the diagnosis of infection should be made by no fewer than two cultures of clean-voided, midstream urine in which the same microorganism is present in significant titers. If the pa-

tient is symptomatic, one specimen for culture will suffice, and therapy should be started. In the past, in the office strategy of management of lower tract symptoms, urine cultures were routinely obtained before initiating therapy. Antibiotics were empirically selected under these circumstances prior to the results of the cultures becoming available to the physician. Currently, there is considerable controversy about cost-efficacy and the necessity of routine pretherapy cultures (see "Frequency-Dysuria Syndrome"). Ideally, antimicrobial agents should be administered only when there is reasonable evidence of infection in the urinary tract.

A rational approach to treatment of urinary tract infection depends on an appreciation of the prognosis of untreated infection and the long-term results to be expected from treatment. The side effects, costs, and inconvenience of different therapeutic regimens must also be considered. As the prognosis of urinary tract infection in nonpregnant, adult women seems to be quite good, and since reinfection is common, therapy probably makes little contribution to the patient's general well-being or prognosis other than eradicating symptoms (Johnson and Smythe, 1969).

Whether asymptomatic bacteriuria in the elderly is causally related to mortality is controversial (Kaye, 1980). While urinary tract infection probably serves as a marker for debilitating disease, it remains to be determined whether infection contributes to mortality. This point is important, since asymptomatic bacteriuria is very common in the elderly, and many of these patients become reinfected or relapse after antimicrobial therapy. Furthermore, a higher frequency of side effects from antibiotics would be expected in an older age group. Considering the lack of evidence of benefit of therapy of urinary tract infection in the elderly (other than relief of symptoms) and the large number of patients involved, antibiotic therapy may lead to an unwarranted financial burden and risk of drug toxicity.

In contrast, asymptomatic bacteriuria in preschool children with vesicoureteral reflux (especially if congenital anomalies are present) can result in stunted growth of the kidney, renal scar formation, and, rarely, renal failure. Bacteriuria in pregnancy may also have serious implications. Accordingly, treatment of children and pregnant women is most likely to be beneficial, both with regard to symptomatic and asymptomatic infection. Furthermore, it is feasible to treat all these patients, since the prevalence of bacteriuria is relatively low in these groups.

It is usually necessary to treat all symptomatic patients regardless of age even when infection is likely to recur. Some patients have such frequent symptomatic episodes (either relapses or reinfections) that they are almost chronically incapacitated. In these patients it may be necessary to give prolonged therapy or prophylaxis to prevent recurrent episodes.

NONSPECIFIC THERAPY

HYDRATION.—Forcing fluids has been advocated in therapy of urinary tract infection. Hydration produces rapid dilution of bacteria and removal of infected urine by frequent bladder emptying, which may offset the logarithmic growth of gram-negative bacilli. Forcing fluids usually results in rapid reduction of bacterial titers. Hydration alone is inadequate. In most patients, bacterial counts return to the original level when hydration is stopped (O'Grady et al., 1968).

Medullary hypertonicity tends to inhibit leukocytic migration into the renal medulla, and the high concentration of ammonia tends to inactive complement (Beeson and Rowley, 1959; Rocha and Fekety, 1964). Abolition of medullary hypertonicity by diuresis would be expected to reverse these effects. In addition, a reduction in bacterial counts in the urine by hydration would enhance the effect of factors otherwise overwhelmed by large numbers of bacteria (e.g., bladder mucosal defenses or the effect of relatively low concentrations of antimicrobial drugs).

Hydration may also have some disadvantages. Increased fluid intake could theoretically result in increased vesicoureteral reflux and possibly cause acute urinary retention in the partially obstructed bladder. The larger urine output results in dilution of antibacterial substances normally present in the urine as well as lower urinary concentrations of antimicrobial agents. Water diuresis also decreases urinary acidification, which enhances the antibacterial activity of urine and certain antimicrobial agents.

As there is no evidence that hydration improves the results of appropriate antimicrobial therapy, and because continuous hydration is inconvenient, we are not in favor of this approach.

URINARY pH.—Spontaneous antibacterial activity of urine results mainly from high urea concentration and high osmolality and is pH-dependent, being greater at a lower pH (Kaye, 1968). The pH-dependent activity may be related to a high concentration of various weakly ionizable organic acids such as hippuric and β-hydroxybutyric acids (Kass and Zangwill, 1060c). The antibacterial activity of these organic acids is related to the concentration of the undissociated molecules that probably penetrate better than the ionized form into the bacterial cell. As these organic acids have a relatively low pKa (the pH at which 50% of the molecules are undissociated), the lower the urinary pH, the greater the concentration of undissociated molecules and the greater the antibacterial activity of the organic acid.

Hippuric acid is a common constituent of urine, being the glycine conjugate of dietary benzoic acid, and is bacteriostatic in proportion to the concentration of undissociate molecules (Levison and Kaye, 1972). The production of antibacterial activity in urine by ingestion of large volumes of cranberry juice (if the urinary pH level is kept low) results from the appearance in the urine of high concentrations of hippuric acid derived from precursors in the berry. The successful use of mandelic acid, another organic acid, is also dependent on maintenance of a low urinary pH level.

The urinary pH level affects the antibacterial activity of many chemotherapeutic agents used in the treatment of urinary tract infections. The activity of methenamine results from the release of formaldehyde as the urinary pH level is decreased to below 5.5. Clinically, methenamine is used in the form of its mandelic acid salt (methenamine mandelate) or its hippuric acid salt (methenamine hippurate). The antibacterial activity to these salts is related to the formation of the un-ionized oganic acid and formaldehyde, which is highly dependent on maintenance of a urinary pH of 5.5 or less. The effectiveness of nitrofurantoin (pKa 7.2) is also greater at a low urinary pH level. In contrast, the aminoglycoside antibiotics such as gentamicin, tobramycin, and amikacin are more effective in alkaline urine. Erythromycin and other macrolides, generally considered to be effective primarily against gram-positive bacteria, are known to have increased activity against gram-negative bacilli at an alkaline pH (e.g., 8.5) (Zinner et al., 1969).

Although different antimicrobial agents have maximum effectiveness at different pH levels, most agents exhibit adequate antibacterial activity at usual urinary pH levels. The major exceptions are mandelic and hippuric acids and methenamine. Maintenance of urine at the low pH level required for effective, antibacterial activity of organic acids and methenamine can be accomplished by administration of ascorbic acid or methionine. Acidification of the urine can result in precipitation of urate stones, and since oxalate is a metabolite of ascorbic acid, large doses of ascorbic acid can cause formation of oxalate stones (Levison and Kaye, 1972).

To acidify the urine, it is often necessary to modify the diet by restriction of agents that tend to alkalinize the urine, for example, milk, fruit juices (except cranberry juice), and sodium bicarbonate. Another major problem with acidification is that patients with renal insufficiency are unable to excrete an acid load and may become systemically acidotic when urinary acidification is attempted. It may be impossible to acidify urine infected with urea-splitting organisms such as *Proteus* sp. because of production of ammonia from urea (Musher et al., 1975).

We do, on occasion, recommend acidification for long-term antimicrobial therapy in patients with normal renal function, but only with concomitant use of organic acids or methenamine. However, urinary acidification is frequently difficult to achieve (Vainrub and Musher, 1977).

ANALGESICS.—Urinary analgesics such as phenazopyridine hydrochloride (Pyridium) have little place in the routine management of symptomatic infections. The dysuria of urinary tract infection usually responds rapidly to antibacterial therapy and requires no local analgesia. If flank pain or dysuria is severe, systemic analgesics can be used. Analgesics such as phenazopyridine hydrochloride may be useful in the management of certain patients with dysuria but without infection.

PRINCIPLES OF ANTIMICROBIAL THERAPY

Selection of appropriate antibiotics has become difficult because of the increasing number of agents available, each with its characteristic spectrum and toxic properties. Fortunately, in most cases, any of many available agents is quite satisfactory. Ideally, the agent chosen should have the narrowest possible spectrum of antibacterial activity and the least toxicity. There is no evidence to support any superiority of bactericidal drugs over bacteriostatic agents in urinary tract infection. However, there may be theoretical reasons for using bactericidal agents in the treatment of relapsing urinary tract infection.

SERUM, TISSUE, AND URINE CONCENTRATIONS OF ANTIMICROBIAL AGENTS.—A poor correlation exists between response of bacteriuria and blood levels of antimicrobial agents (McCabe and Jackson, 1965; Stamey et al., 1965, 1974). Many oral antimicrobial agents, in the dosages commonly used for urinary tract infection, do not achieve serum levels above the minimal inhibitory concentration for most urinary pathogens.

Disappearance of bacteriuria is closely correlated with the sensitivity of the microorganism to the concentration of the antimicrobial agent achieved in the urine (McCabe and Jackson, 1965; Stamey et al., 1965, 1974). Inhibitory urinary concentrations are achieved after oral administration of essentially all commonly used antimicrobial agents. While blood levels do not seem to be important in treatment of urinary tract infection, they may be critical in patients with bacteremia and may be important in the cure of patients with renal parenchymal infection who have relapse.

In patients with renal insufficiency, dosage modifica-

tions are necessary for agents that are excreted primarily by the kidneys and cannot be cleared by another mechanism. In renal failure, the kidney may not be able to concentrate an antimicrobial agent in the urine, and difficulty in eradicating bacteriuria may occur. This may be an important factor in failure of therapy of urinary tract infection with aminoglycosides.

In addition, high concentrations of magnesium and calcium as well as a low pH level can raise the minimal inhibitory concentrations of aminoglycosides for gram-negative bacilli to levels above those achievable in the urine of patients with renal failure (Minuth et al., 1976). In general the penicillins and cephalosporins attain adequate urine concentrations despite severely impaired renal function and are the agents of choice in renal insufficiency (Kunin and Finkelberg, 1970b).

RESPONSE TO THERAPY.—Since symptoms of lower urinary tract infection may abate spontaneously without chemotherapy (even though bacteriuria persists), results of therapy can only be determined by follow-up urine cultures. There are four patterns of response of bacteriuria to antimicrobial therapy: cure, persistence, relapse, and reinfection. Quantitative bacterial counts in urine should decrease within 48 hours after initiation of an antimicrobial agent provided that the microorganism is sensitive in vitro. If titers do not decrease within this time, it is unlikely that continued therapy will be successful.

Cure is defined as negative urine cultures on chemotherapy and during the follow-up period (usually 1–2 weeks). However, it must be understood that some of these patients will develop reinfection at a later time.

Persistence has been used in two ways to describe response to therapy: (1) persistence of significant bacteriuria (i.e., $\geq 10^5$/ml) after 48 hours of treatment and (2) persistence of the infecting organism in low numbers of urine after 48 hours. Significant bacteriuria usually persists only if urinary levels of the antimicrobial agent are below the concentration of the drug needed to inhibit the microorganism. This can occur when the infecting strain is resistant to the urinary levels usually attained (i.e., a resistant organism) or because levels are inordinately low (i.e., from not taking the agent, insufficient dosage, poor intestinal absorption, or poor renal excretion as in renal insufficiency). Persistence of the infecting microorganism in low titers in voided urine may mean persistence in the urinary tract or contamination from the urethra or vagina. Bladder puncture cultures would be needed to evaluate the significance of low titers of bacteria obtained on therapy, and we do not routinely recommend this procedure. Also worth noting is that bacteria may persist within the urinary tract during therapy without excretion of organisms in the urine. Sites of persistence within the urinary tract are the renal parenchyma, calculi, and the prostate. The simplest way of determining the significance of persistence of the organism in low titers in the urine is to obtain follow-up urine cultures after therapy has been stopped. Prompt relapse of significant bacteriuria usually follows persistence of the organism in the urinary tract.

Relapse usually occurs within 1–2 weeks after cessation of chemotherapy and is often associated with renal infection, with structural abnormalities of the urinary tract, or with chronic bacterial prostatitis. Relapse indicates that the infecting microorganism persisted in the urinary tract during therapy. However, an apparent relapse can be related to reinfection (new infection) with the same microorganism. In spite of eradication from the urinary tract, the original infecting organism may still be present in the intestine, vagina, or external urethra and then may cause a new infection. Markedly delayed relapses (more than one month after stopping therapy) are much more likely due to this phenomenon or to chronic bacterial prostatitis than to true relapse. Relapses occurring within 1–2 weeks are usually true relapses. One postulated but unsubstantiated mechanism of relapse following treatment with cell-wall active antibiotics (e.g., penicillins and cephalosporins) is persistence of osmotically fragile, cell-wall deficient forms in the hypertonic renal medulla during therapy, with reversion to normal bacteria after therapy is stopped (Gutman et al., 1965).

After initial sterilization of the urine, *reinfection* may occur during administration of chemotherapy (also called superinfection) or at any time thereafter. Reinfection is easy to identify when there is a change in bacterial species. However, there may be reinfection with a different serotype of the same species (usually *E. coli*) or even the same serotype.

ACUTE PYELONEPHRITIS

Patients who are severely ill with acute pyelonephritis should be hospitalized promptly (Fig 5–1). Although mild to moderate illness responds well to orally administered antibiotics, nausea and vomiting may make oral therapy impossible. If gram-negative bacillary bacteremia is suspected to be complicating pyelonephritis (high fever, shaking chills, hypotension), empirical parenteral antibacterial therapy should be initiated immediately and is directed at life-threatening bacteremia (Platt, 1983b). In all seriously ill patients, the spectrum of antibacterial activity of the initial chemotherapy should include all potential pathogens.

At the time of initial antibiotic selection, a Gram stain of the infected urine should have indicated the morphology of the infecting organism (e.g., gram-neg-

ative bacilli, gram-positive coccus), but the precise identity and antimicrobial susceptibility pattern are usually unknown. When streptococci are suspected, ampicillin is probably the agent of choice. When staphylococci are implicated on Gram stain, cephalosporins (such as cefazolin) are appropriate agents.

In the severely ill patient with gram-negative bacilli on Gram stain of the urine, appropriate initial therapy in adults is a combination of a parenteral cephalosporin (e.g., cephalothin 6–12 gm/day or cefazolin 3–6 gm/day) and a parenteral aminoglycoside (e.g., gentamicin 1.7 mg/kg body weight every 8 hours). Alternatively, a third-generation cephalosporin alone—such as cefotaxime, 6–12 gm/day, or ceftizoxime, 3–6 gm/day, or ceftriaxone, 2 gm/day—is reasonable as it permits avoidance of an aminoglycoside. Once the susceptibility pattern of the infecting organism is known, therapy can be altered to less expensive and less toxic single-agent therapy. After defervescence, oral therapy can be used and should be continued for a total of 14 days.

In the less severely ill patients with pyelonephritis, single-drug therapy can be used with an oral or parenteral cephalosporin, gentamicin alone, or oral trimethoprim-sulfamethoxazole, trimethoprim alone, ampicillin, amoxicillin, or a tetracycline. The doses are listed under "Lower Urinary Tract Infection." Therapy should be continued for 14 days.

The excellent results obtained with aminoglycosides and trimethoprim may relate to the high concentrations that these agents achieve in the renal medulla, the initial and main site of renal parenchymal infection. The new quinolones, which similarly concentrate well in all parts of renal parenchyma, may become useful in the treatment of pyelonephritis.

Epidemiologic considerations influence initial selection of antibiotics such as age of the patients (increased frequency of bacteremia in the elderly), hospital acquisition of the infection, recurrent infection, presence of obstruction or calculi, or previous antibiotic therapy (more resistant organisms).

Effective therapy results in a marked decrease in bacterial titers in the urine within 48 hours after onset of treatment. Urine cultures should ideally be obtained after three or four days of therapy. Microscopic examination of the urine for significant bacteriuria or inexpensive screening bacteriologic methods such as the dip inoculum method can be substituted. Antimicrobial agents are sometimes effective in vivo, even when disk sensitivity tests indicate drug resistance, because most antimicrobial agents are excreted in the urine in concentrations much higher than tested for by disk sensitivity testing. Therefore, even if disk tests indicate resistance, a chemotherapeutic agent may be continued if there has been bacteriologic response (Stamey et al.,

Management of Urinary Tract Infection

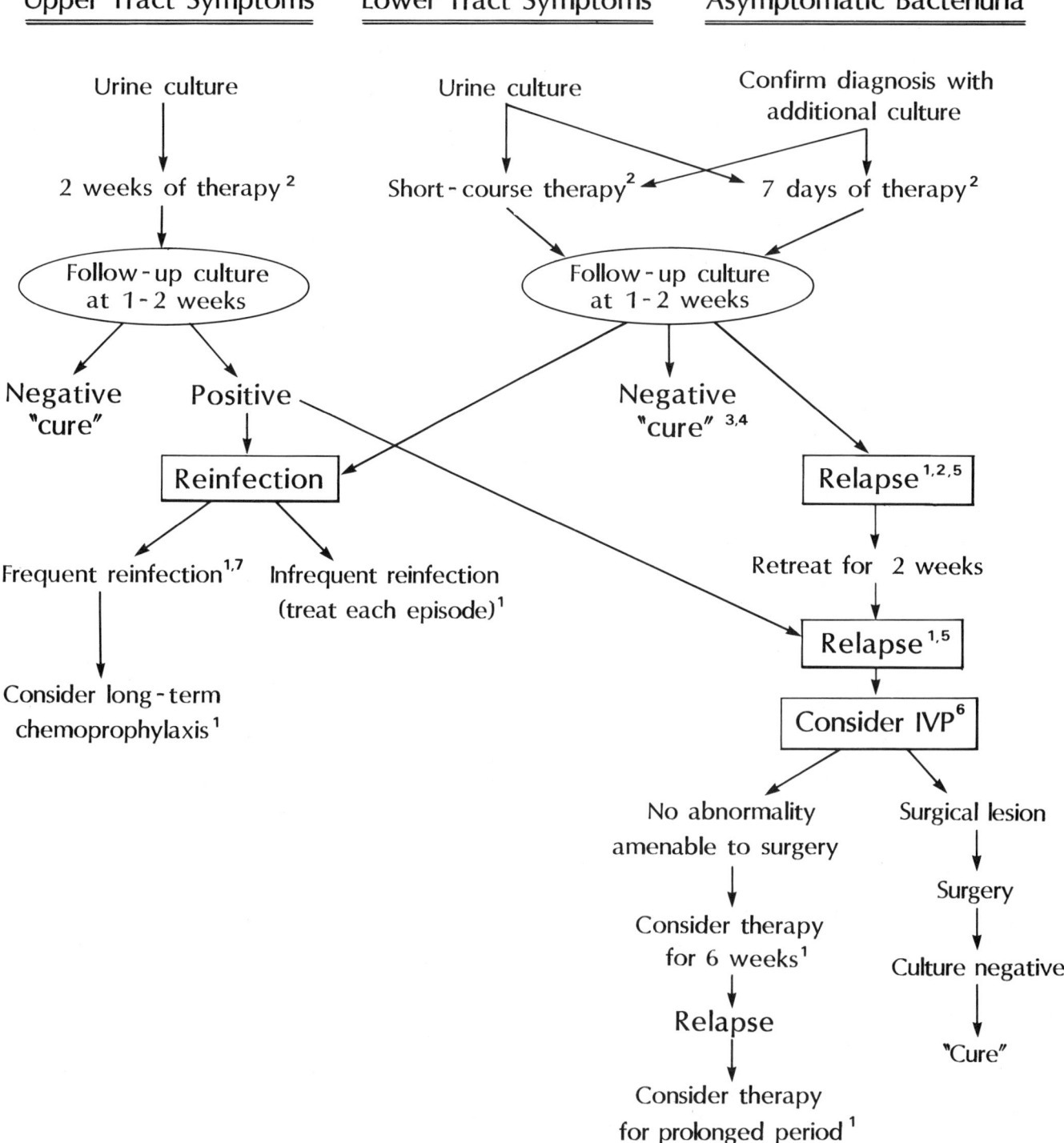

FIG 5–1.
Management of urinary tract infection. Footnote cross-references are as follows: [1]Consider no therapy in nonpregnant adults without uropathy or symptoms of urinary tract infection. [2]Consider urologic evaluation in children and men. [3]Or at least long-term infection-free interval. [4]Follow-up cultures monthly in pregnant women and at six weeks and six months in children. [5]Evaluate men for chronic bacterial prostatitis. [6]Delay two months postpartum in pregnant women. [7]Consider urologic evaluation after three or four reinfections in women.

1974). If bacteriologic response does not occur by 48 hours, there is no point in continuing the same regimen. Therapy is then changed to one of the alternate drugs on the basis of sensitivity tests (e.g., from the initial isolate). The finding of continuing positive blood cultures or persistent high fever and toxicity past the first three days indicates the need for investigation to exclude urinary obstruction or intrarenal or perinephric abscess formation. Investigation should include renal ultrasound, CT scan, and, according to the findings, perhaps an IVP examination. The availability of sensitive noninvasive studies has resulted in early diagnosis of intrarenal or perinephric abscess formation that may respond to antibiotic therapy alone.

LOWER URINARY TRACT INFECTION

CONVENTIONAL THERAPY.—Acute uncomplicated lower urinary tract infection is one of the most common forms of bacterial infection affecting adult women (see Fig 5–1). Approximately 6% of women come to medical attention each year because of symptomatic bacterial urinary tract infection (Sanford, 1975). In the past because of difficulties in differentiating clinically between upper and lower tract infection, all bacterial lower urinary tract infections were conventionally treated with 7–14 days of oral antimicrobial agents (Sanford, 1975; Fang et al., 1978; Harbord and Gruneberg, 1981). Although no one chemotherapeutic agent is unequivocally the drug of choice for conventional therapy, a short-acting oral sulfonamide such as sulfisoxazole (Gantrisin) has been preferred by many authorities. Long-acting sulfonamides that more frequently have been associated with hypersensitivity reactions should be avoided (Levison and Kaye, 1972). The dose of sulfisoxazole is 1 gm 4 times a day in adults and 150 mg/kg/24 hours daily in four divided doses (not to exceed 6 gm/24 hours) in children. The sulfonamides should not be used in newborn infants (especially prematures) or in pregnant women near term because of the possibility of producing kernicterus in the newborn. When taken with proper precautions, short-acting sulfonamides have few side effects and are inexpensive. Equally effective alternate choices of oral chemotherapeutic agents are ampicillin, amoxicillin, cephalexin, cephradine, nitrofurantoin, tetracycline, trimethoprim-sulfamethoxazole, and trimethoprim (Levison and Kaye, 1972; Charlton et al., 1976). The doses are (1) ampicillin and amoxicillin, 500 mg, 1 gm 4 times a day in patients weighing over 20 kg, and 100 mg/kg/day in 4 divided doses in patients under 20 kg; (2) cephalexin or cephradine, 250–500 mg 4 times a day in adults, and 25–50 mg/kg/day in 4 divided doses in children; (3) nitrofurantoin, 100 mg 4 times a day in adults, and 5–7 mg/kg/day in 4 divided doses in children; (4) tetracycline, 250–500 mg 4 times a day in adults, and 20 mg/kg/day in 4 divided doses in children (WARNING: tetracycline may cause dental staining in children); (5) trimethoprim-sulfamethoxazole, 2 single-strength tablets twice a day in adults, and 8 mg/kg/day trimethoprim and 40 mg/kg/day sulfamethoxazole in 2 divided doses in children; and (6) trimethoprim 100 mg twice a day in adults, and 8 mg/kg/day in children.

Conventional therapy has been associated with a 10%–30% rate of side effects, relatively poor compliance, and a relatively high failure rate of 10%–20% (Ronald et al., 1976; Sanford, 1975; Fang et al., 1978; Savard-Fenton et al., 1984). The latter observation is not surprising, since approximately 20% of women with acute cystitis have organisms originating from one or both kidneys.

SINGLE-DOSE THERAPY.—The use of single-dose therapy is based on the rationale that lower tract infection is a superficial mucosal infection only. Single-dose therapy of lower tract infection with amoxicillin (Bailey and Abbott, 1977; Fang et al., 1978; Rubin et al., 1980b; Harbord and Gruneberg, 1981; Savard-Fenton et al., 1984), trimethoprim-sulfamethoxazole (Bailey and Abbott, 1978; Harbord and Gruneberg, 1981; Counts et al., 1982; Tolkoff-Rubin et al., 1982), trimethoprim (Harbord and Gruneberg, 1981), kanamycin (Ronald et al., 1976), sulfonamide (Kallenius and Winberg, 1979; Ludwig et al., 1980), tetracycline (Rosenstock et al., 1983), and cefonicid (Pontzer et al., 1983) appears to give results comparable to those achieved with conventional therapy, a 75%–95% success rate. The regimens used were amoxicillin, 3 gm orally; trimethoprim-sulfamethoxazole, 2 double-strength tablets orally; trimethoprim, 200 mg orally; kanamycin, 0.5 gm IM, sulfadoxine, 2 gm orally; sulfamethoxypyridazine, 2 gm orally; sulfisoxazole, 2 gm orally; tetracycline, 2 gm orally, and cefonicid, 1 gm IM. Approximately 90%–95% of patients with ACB negative organisms in the urine are cured by single-dose therapy, whereas only 50%–65% of patients with ACB-positive organisms in the urine are cured by similar treatment (Ludwig et al., 1980; Rubin, 1980b; Counts et al., 1982; Tolkoff-Rubin et al., 1982; Savard-Fenton et al., 1984). However, in individual patients, the ACB test is a relatively poor predictor of response to therapy, and single-dose treatment can be administered safely in women with urinary tract infection with only lower tract symptoms without knowledge of the results of the ACB assay or other infection localizing tests.

Single-dose therapy with amoxicillin was also highly successful in eradicating infection in patients with symptomatic lower urinary tract infection due to low count bacteriuria (10^2–10^4 bacteria/ml). In contrast, fewer than 20% of symptomatic patients with pyuria but without bacteriuria were cured (Tolkoff-Rubin et al., 1984).

The advantages of single-dose therapy include lesser expense, ensured compliance, fewer side effects, and perhaps less intense selective pressure for emergence of resistant organisms in gut, urinary, or vaginal flora. Possible deleterious effects include a poorer outcome of infections that are actually in the upper tract, as a delay in appropriate therapy may result in more deeply seated infection and impair the response to subsequent, more prolonged therapy. Finally, it should not be assumed that every antibiotic administered as a single dose will be effective even with regard to susceptible organisms. Results depend on high, sustained urinary concentrations of the antimicrobial agent. For example, a 2-gm oral dose of cefaclor resulted in a 57% failure rate (Greenberg et al., 1981).

With regard to the above concerns, it should be pointed out that women who failed or relapsed on single-dose therapy developed symptoms or signs no worse than the initial presenting illness. No bacteremias, hospitalizations, renal failure or fatalities have been reported (Bailey and Abbott, 1977; Fang et al., 1978; Tolkoff-Rubin et al., 1982, 1984). Interestingly enough, single-dose amoxicillin therapy of infections caused by amoxicillin-resistant organisms has been remarkably effective in that about half of attacks caused by resistant organisms were cured by a 3-gm oral dose of amoxicillin. Presumably this is because high amoxicillin levels were delivered to bladder mucosa (Fang et al., 1978). In one study, approximately half the patients who relapsed after single-dose amoxicillin (amoxicillin-sensitive organisms) also failed to be cured by conventional therapy. All these patients were ultimately cured by a six-week course of therapy (Tolkoff-Rubin et al., 1984). Three studies failed to demonstrate that single-dose therapy was accompanied by a tendency to select resistant strains (Anderson et al., 1979; Kallenius and Winberg, 1979; Lincoln et al., 1979). In fact, Counts et al. (1982) found a less suppressive effect on vaginal bacterial flora when single-dose therapy was used.

A by-product of single-dose therapy is that failure to eradicate a urinary tract infection after a single dose of an agent may indicate in which patients further investigation should be considered. Cure with single-dose therapy appears comparable with a negative ACB test in localizing the site of infection to the lower tract (Stamm, 1980c).

Before using single-dose therapy, it would be best to identify those patients who are most likely to respond to single-dose therapy. Although the absence of antibody-coated bacteria in the urine seems to correlate well with responsiveness to single-dose therapy, this test is not widely available or cost-effective to perform. Certain patients should be excluded from single-dose therapy either because this approach is largely ineffective or data of efficacy are still not available. Included in this category are patients with signs of upper tract involvement, males (high incidence of deep tissue infection of prostate, kidney, or both), catheterized patients, diabetics, neonates, and patients with structural or functional abnormalities of the urinary tract. Any factor that increases the likelihood of upper tract infection would mitigate against using single-dose therapy. There are insufficient data regarding the use of single-dose therapy in pregnancy; however, its use appears reasonable. Follow-up cultures in this population are critical. In children over the age of 2 years, cure rates with single-dose therapy and conventional treatment were similar (Stahl et al., 1984), corroborating previous studies (Wientzen et al., 1979; Kallenius and Winberg, 1979; Shapiro and Wald, 1981).

In spite of all the enthusiasm for single-dose therapy, it should be pointed out that several authors have reported a higher recurrence rate (due to relapse) in patients who were subjected to a longer period of follow-up (Schultz et al., 1984; Leibovici et al., 1984; Hooton et al., 1985). Furthermore, patients with more prolonged symptoms (>7 days), by virtue of the increased chance of upper tract involvement, respond less well to single-dose therapy. In addition, patients presenting to an emergency room or sexually transmitted disease clinic with symptoms of lower urinary tract infection have a higher prevalence of having ACB in the urine (40%–60%) (Wong et al., 1984) and are best treated with conventional therapy, especially since long-term follow-up cannot be guaranteed.

Agents with a short half-life in serum are less likely to be effective in single doses because of a shorter duration in the urine. This may be the explanation for the high failure rate with cefaclor (Greenberg et al., 1981). However, when these agents are given in conventional doses for 24 hours, they should be equally effective as single-dose therapy with proved agents. It would probably be best to use the term "short-course therapy" to refer to both single-dose therapy and conventional dose therapy given for 24 hours.

FOLLOW-UP CULTURES

A follow-up urine culture should ideally be obtained 1–2 weeks after discontinuing therapy to detect relapses. Although Winickoff et al. (1981) suggested that routine posttreatment cultures were not cost-effective, most authorities (Ronald et al., 1976; Savard-Fenton et al., 1984) remain convinced of the importance of recognizing those patients whose bacteriuria is usually of renal origin. It is our belief that in the nonpregnant adult who remains asymptomatic after therapy, follow-up urine cultures are not necessary. In children and other patients at high risk of renal damage, subsequent cultures should be obtained periodically to detect reinfection.

OFFICE STRATEGY FOR FREQUENCY-DYSURIA SYNDROME IN WOMEN

The ideal theoretical goals of management include identification of the site of infection, identification of the specific etiologic agent, and finally ensuring that infection is limited to the lower genitourinary tract.

Clinicians should attempt to differentiate the various causes of frequency-dysuria on the basis of a good history, physical examination, and urinalysis. Firstly, clinical upper tract involvement should be excluded. Upper tract involvement is suggested by fever, chills, nausea, vomiting, and flank pain.

Dysuria without frequency is more often caused by vaginitis than urinary tract infection. A history of vaginal symptoms (discharge, pruritus, etc.) together with the absence of urgency and suprapubic pain will identify about two thirds of women with vaginitis presenting with dysuria. Dysuria with frequency makes urinary tract infection or urethritis more likely. Urethritis caused by *C. trachomatis* and *N. gonorrhoeae* may be accompanied by mucopurulent cervicitis and in some cases pelvic inflammatory disease. Patients with urethritis due to these organisms tend to have a history of a recent new sexual partner, milder dysuria, and more gradual onset of symptoms, which have often been present for more than one week before seeking attention. Herpes urethritis is generally accompanied by other external manifestations and is usually only seen with the initial herpetic infection; patients may have fever, malaise, inguinal lymphadenopathy, and meningismus. Bladder infections tend to begin more abruptly, with more severe dysuria, and suprapubic pain and tenderness may be present. Most patients with cystitis seek medical attention within 2–3 days of onset of symptoms and frequently have a history of previous urinary tract infections.

All patients with frequency-dysuria syndrome should have their temperatures measured and undergo an examination testing for suprapubic and costovertebral tenderness. In the presence of vaginal symptoms a pelvic examination is essential. A pelvic examination is also required if a urinalysis is negative. Thereafter, a clean voided midstream urine specimen should be obtained, and fresh unspun urine should be examined microscopically. The finding of 1 or more gram-negative rods per oil immersion field correlates with $\geq 10^5$ bacteria/ml; however, one third of women with bladder infections have low titer bacterial infection with $< 10^5$ bacteria/ml,

and therefore failure to visualize organisms does not exclude bacterial infection. The presence of microscopic hematuria strongly suggests bacterial cystitis.

The most important aspect of urinalysis is detection of pyuria. Ideally a hemocytometer chamber should be used to quantitate the leukocytes. A count of more than 8 leukocytes/mm^3 on a fresh unspun midstream urine specimen is abnormal (Stamm, 1982). Pyuria when present is still somewhat nonspecific in that it may be caused by bacterial cystitis (high or low bacterial count), chlamydial urethritis and far less frequently by gonococcal or herpetic urethritis and finally rarely by trichomonal vaginitis/urethritis. The necessity for routine pretreatment culture is controversial, especially when cost-efficacy is considered (Carlson and Mulley, 1985). Many physicians now believe that in typical cases of cystitis, particularly those confirmed by positive Gram stain, culture confirmation is unnecessary and expensive. They further advocate culture only when diagnosis is unclear or when the patient fails to respond to initial empirical antimicrobial therapy. According to Stamm et al. (1981b, 1982) one should always do a microscopic examination of the urine and, where possible, a Gram stain. Urine cultures should ideally be obtained when a diagnosis cannot be substantiated by initial positive Gram stain and where differentiation of cystitis from *Chlamydia* or *N. gonorrhoeae* infection may be difficult. Urine culture should also be obtained when infection is complicated by urologic abnormalities, multiple previous infections, previous infection with resistant strains, diabetes, pregnancy, and possible pyelonephritis. We believe that an initial culture is useful (but not essential), since the primary goal of therapy is minimizing morbidity, and reducing costs is only a secondary consideration. The information obtained by pretreatment cultures allows more rapid modification of initial empirical therapy in patients who remain symptomatic and planning an appropriate strategy in cases with recurrent urinary tract infection. Office cultures (e.g., dip slide) can be performed very inexpensively.

A practical strategy based on urinalysis and particularly the presence or absence of pyuria is outlined in Fig 5–2. If initial Gram stain shows bacteria, single-dose (or short-course) antibiotic therapy as previously outlined is recommended. Our preference is to give two double-strength trimethoprim-sulfamethoxazole tablets. If bacteria are not visualized, in the presence of pyuria, single-dose (or short course) antimicrobial therapy should be given to patient in whom low titer bacterial infection is suspected. If chlamydial or gonococcal infection is suspected, one week's oral therapy with tetracycline, 500 mg four times daily, is appropriate. If the clinician is unsure of the cause of the urethral syndrome in the presence of pyuria, one option is to use single-

Office Approach to Treatment of Dysuria and Frequency in Sexually Active Women[1-4]

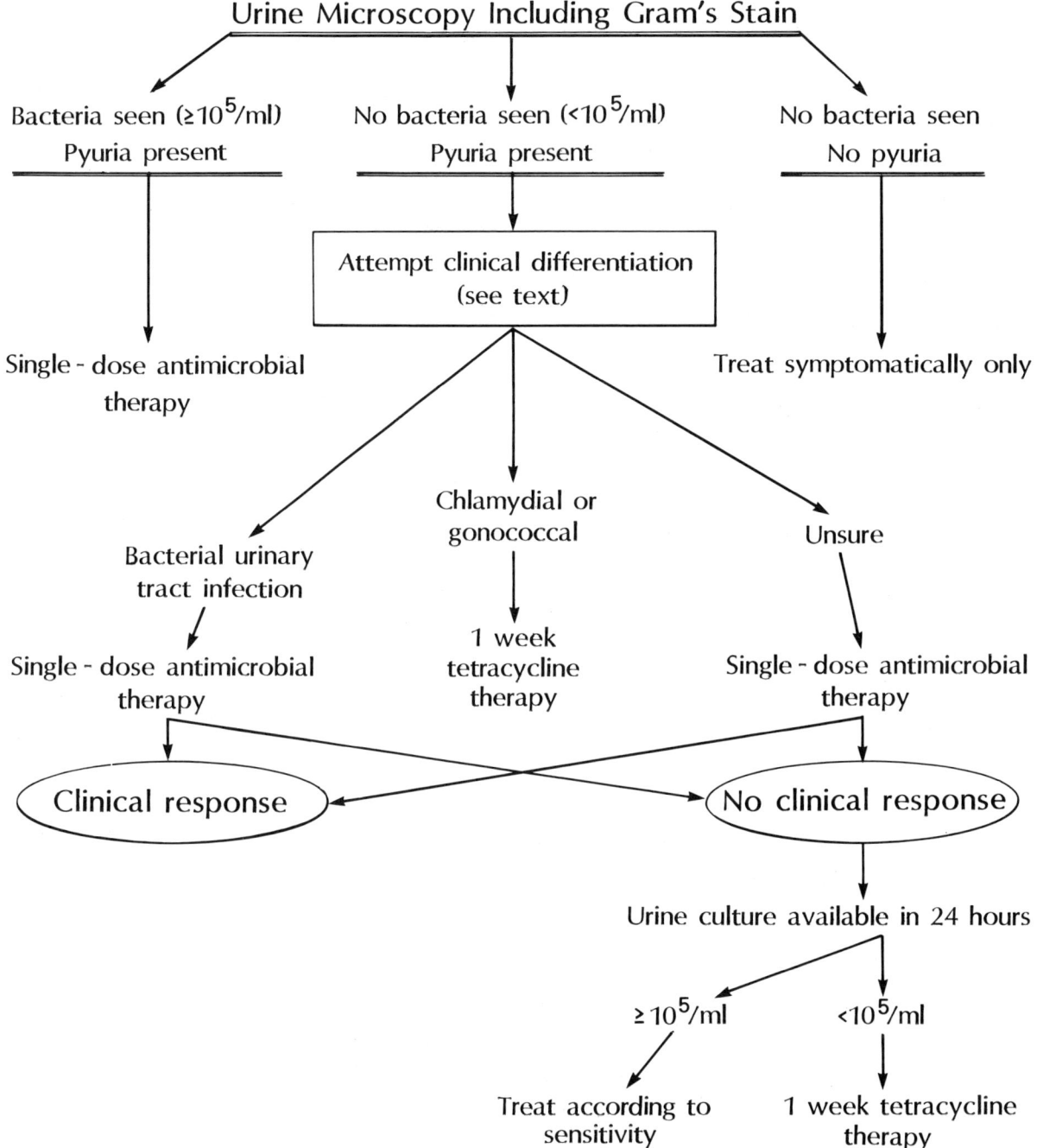

FIG 5–2.
Office approach to dysuria and frequency in sexually active women. Footnote cross-references are as follows: [1]Vaginitis must be excluded. [2]Only applies in the absence of upper tract signs and symptoms. [3]Outline is based on lack of routine availability of methods for detecting *Chlamydia* and the fact that results of bacterial cultures are usually unknown at the time of initiation of therapy. [4]Most microbiological laboratories do not quantitate bacterial growth $< 10^5$/ml and report as "no significant growth."

dose (or short-course) antimicrobial therapy or to treat with an agent which is reasonably active against all the possible pathogens (e.g., tetracycline).

In females without pyuria, there is little to justify the use of antibiotics. Pyridium should be given to relieve symptoms and the patient asked to return in 48 hours if symptoms persist. Usually symptoms resolve. If they do not, a repeated examination is in order, including evaluation for pyuria and culture. In spite of advocates who contend that posttherapy urine culture is unnecessary and not cost-effective, some authorities still think that routine posttherapy cultures are of value to confirm bacteriologic cure. In the patient who is clinically cured, posttherapy cultures are of doubtful use except in pregnancy and in children. In patients with urethritis due to *Chlamydia* or *N. gonorrhoeae*, proof of cure is required.

ASYMPTOMATIC BACTERIURIA

Most patients with asymptomatic bacteriuria are women and are in the older age groups (see Fig 5–2). Although cure may result following treatment, relapse and especially reinfection are common. The approach to asymptomatic bacteriuria depends on the age of the patient. In all children, therapy should be given as described for symptomatic infection. A trial of single-dose therapy is reasonable. In contrast, therapy for asymptomatic bacteriuria in the adult is by no means mandatory in the absence of obstruction except during pregnancy. Nonpregnant women can be treated providing that a nontoxic antimicrobial agent is used. If the infecting microorganism is resistant to all but toxic agents, treatment should not be instituted in the nonobstructed patient.

At present, most physicians believe that asymptomatic bacteriuria in the elderly is a benign disease and need not be treated, especially since with vigorous treatment many people will be exposed to drug toxicity.

When dealing with asymptomatic bacteriuria, there is no urgency in treating. Therapy should be delayed until two cultures have been obtained for confirmation of presence of bacteriuria. By that time, the identity and antimicrobial susceptibility pattern of the infecting organism will have been determined.

RECURRENT URINARY TRACT INFECTIONS
(see Table 5–3)

RELAPSING URINARY TRACT INFECTION.—If the patient relapses after therapy for symptomatic urinary tract infection or for asymptomatic bacteriuria, one should immediately recheck antimicrobial sensitivity and investigate patient compliance. If it turns out that the relapse has occurred in spite of taking appropriate

doses of appropriate chemotherapy then the most likely possibilities are: (1) renal parenchymal involvement; (2) structural abnormality of the urinary tract (e.g., calculi); or (3) chronic bacterial prostatitis (see section on prostatitis).

Relapses at all ages, especially in the absence of structural abnormalities, often may be related to renal parenchymal infections that require a longer duration of therapy for cure. Patients who relapse after a single dose or a seven-day course of therapy should be given a two-week course of treatment. In patients who relapse after two weeks of therapy, Turck et al. demonstrated that a six-week course of therapy resulted in a higher cure rate than a two-week course. Subsequently, similarly good results with 6–12 weeks of trimethoprim-sulfamethoxazole treatment have been obtained with fewer posttherapy relapses (Gleckman et al., 1980). Men with relapsing urinary tract infection should be evaluated for chronic bacterial prostatitis.

Structural abnormalities of the urinary tract predispose to relapse. Urinary tract infection in the presence of obstruction is likely to be associated with renal involvement, a tendency for renal functional impairment, and bacteremia. Obstructive lesions can be corrected surgically and should be sought in the evaluation of patients with relapsing infection. Ultimate success of chemotherapy is dependent on the removal of stones (Kass and Zangwill, 1960c).

Some patients continue to relapse despite surgical correction of urologic abnormalities. In others, surgical correction may not be indicated or feasible or no abnormalities may be found. In those patients who relapse after two weeks of antimicrobial therapy, a six-week course should be considered.

If relapses occur after a six-week course, courses lasting six months or more may be considered. Only carefully selected patients such as (1) repeatedly symptomatic adults, (2) children, and (3) adults who are at high risk of developing progressive renal damage (e.g., noncorrectable obstruction) should be considered for continuous therapy (termed "suppressive prophylaxis"). Asymptomatic nonpregnant adults without obstruction should not receive six-week and longer courses.

Some of the oral agents that can be used for longterm therapy are ampicillin, amoxicillin, sulfisoxazole, cephalexin, cephradine, trimethoprim-sulfamethoxazole, and trimethoprim in the usual doses already described; nitrofurantoin in full dosage for one week and then half the usual dose; nalidixic acid (1 gm four times daily in adults and 55 mg/kg/day in four divided doses in children); carbenicillin indanyl sodium (2 tablets four times daily in adults); and methenamine mandelate (1 gm four times daily in adults and 0.5 gm four times daily in children over 6 years, and 0.25 gm/30 lb body weight four times daily in children under 6 years), with methionine or ascorbic acid to acidify the urine.

Patients receiving methenamine mandelate (or hippurate) are instructed to avoid alkalinizing foods such as milk and all fruit juices other than cranberry juice. Bicarbonate of soda must also be avoided. In addition, they are given nitrazine paper to test that their urinary pH is maintained at 5.5 or below. Methionine or ascorbic acid is added as needed to reduce urinary pH. Methenamine mandelate is difficult to use because it requires major changes in diet; therefore, this drug is reserved for unusually refractory cases. Recently, new oral quinolone agents have been found to be highly effective in suppressive prophylaxis (Schaeffer et al., 1982).

Any antimicrobial agent being used for long-term therapy is continued only as long as significant bacteriuria is absent. If bacteriuria persists or relapses during chemotherapy (indicating that the infecting organism is now resistant to that agent), the agent is altered. The aim is to achieve continuous suppression of bacteriuria for the entire course of therapy. If relapse occurs after discontinuation of the antimicrobial agent, therapy can be reinstituted with the same or another drug. If deemed necessary, this agent is administered for an additional 6–12 months (if bacteriuria remains suppressed). All patients are followed up with urine cultures at least monthly during therapy.

Long-term therapy or even repeated courses of short-term therapy should be reserved for children, symptomatic patients of any age, and patients at high risk of developing progressive renal damage. A creatinine clearance determination and an IVP initially and yearly (or at least every two years) should be obtained on patients receiving long-term therapy to determine GFR and structural changes in the kidneys. Blood counts, urinalyses, and liver chemistries (when indicated) are also obtained periodically as tests for drug toxicity.

REINFECTION OF THE URINARY TRACT.—Patients with reinfection can generally be divided into two groups: (1) those who have relatively infrequent reinfections (fewer than three attacks a year) and (2) those who develop frequent reinfections (three or more each year). An extreme example of the latter group is patients who become reinfected during or shortly after each course of antimicrobial therapy. With infrequent reinfections, each episode can be approached with therapy as if it were a new episode of either symptomatic or asymptomatic infection. Short-course therapy may be particularly useful in women with lower tract symptoms. Patients with frequent reinfection fall into two further categories. Many are middle-aged or elderly

women in whom infection is limited to the lower urinary tract. Moreover, many of these episodes of reinfection are asymptomatic and may be self-limiting. Hence, they should not be treated, because the frequent use of antibiotics in this group is apt to result in toxic side effects and because progressive renal destruction is rare. If, however, the episodes are symptomatic or if the likelihood of renal damage is increased, these patients should be treated with a long-term prophylaxis. The second category consists mainly of young healthy sexually active women who develop symptomatic reinfection so frequently that they can become incapacitated. In fact 80% of recurrent urinary tract infections in young otherwise healthy adult females result from exogenous reinfection with a newly acquired strain (Kunin, 1970a, 1975b; Stamey et al., 1971).

In some women these symptomatic reinfections are associated with sexual activity. Voiding immediately after intercourse may help prevent infection. However, single-dose postcoital prophylaxis is a more effective method of decreasing episodes (Vosti, 1975; Pfau et al., 1983). No specific contraceptive method has been clearly linked to recurrent urinary tract infections; however, the diaphragm has recently been associated with increased frequency of cystitis (Stamm, 1984; Gillespie, 1984).

In other patients with frequent symptomatic reinfections, no precipitating event is apparent, and investigation usually fails to reveal any underlying structural, anatomical, or radiologic abnormality. These unfortunate women end up undergoing excretory urograms, voiding studies, and cystoscopy with extremely low yield, making investigation not only cost-ineffective but also unjustified (Engel et al., 1980; Fowler and Pulaski, 1981; Mogenson and Hanson, 1983). Accordingly, IVP is not indicated unless obstructive symptoms are present or there is a history of renal calculi or unexplained hematuria, past history of repeated pyelonephritic attacks, evidence of neurogenic bladder dysfunction, or poor response to chemoprophylaxis. In these patients, long-term chemoprophylaxis should be instituted. Three approaches are currently available for patients with frequently recurrent urinary tract infections unrelated to coitus. The most widely accepted methods are those of continuous daily low-dosage chemoprophylaxis or thrice-weekly antibiotic prophylaxis (Harding et al., 1979), and recently the value of intermittent self-administered treatment of established reinfection has been recognized (Wong et al., 1985). The latter approach of self-diagnosis and self-treatment with an agent such as trimethoprim-sulfamethoxazole is effective and economical. However, it should be applied only in selected, reliable subjects capable of accepting the responsibility of diagnosis and therapy and in whom the likelihood of symptomatic episodes' being due to cystitis is high. This experience with patient-initiated therapy at home requires confirmation by other authors (Wong et al., 1985).

In contrast, an enormous experience has been accumulated with long-term daily or intermittent chemoprophylaxis (Bailey et al., 1971; Ronald et al., 1975; Smellie et al., 1976; Stamey et al., 1977; Harding et al., 1979; Stamm et al., 1980b). Continuous prophylaxis has been shown not only to reduce dramatically the attack rate of recurrent infections but also to be cost-effective (Stamm et al., 1981b) in that the annual cost of continuous prophylaxis is lower than the cost of therapy of two episodes of cystitis treated with conventional physician-initiated therapy. These investigators demonstrated that low-dosage antimicrobials given either daily or thrice weekly reduced rates of infection in susceptible women from 2–4 to nearly 0 per year. Trimethoprim-sulfamethoxazole, nitrofurantoin, or trimethoprim alone are particularly useful for long-term prophylaxis (Ronald et al., 1975). Before prophylaxis is initiated, the patient should receive a standard course of therapy with an appropriate antimicrobial agent. Full antimicrobial dosage is not necessary for successful prophylaxis. One 50-mg tablet of nitrofurantoin or one half tablet of trimethoprim-sulfamethoxazole nightly and even three times a week is effective. Sulfisoxazole, nalidixic acid, methenamine mandelate, and other agents have also been used with good results (Levison and Kaye, 1971).

It is unclear exactly how long-term chemoprophylaxis prevents symptomatic recurrences. Some agents (e.g., trimethoprim-sulfamethoxazole) eradicate periurethral coliform and uropathogen carriage, whereas nitrofurantoin has no effect on the fecal and perineal flora. Both these agents probably act by providing antibacterial activity in the bladder urine and in subinhibitory concentrations may reduce bladder colonization by interference with bacterial adherence. Long-term antibiotic prophylaxis has been effective in older men and women as well as children (Freeman et al., 1975; Smellie et al., 1976). In addition to use in symptomatic patients, long-term chemoprophylaxis should be considered for asymptomatic patients who have reinfection frequently and are at risk of developing renal parenchymal damage with each reinfection (for example, young children with vesicoureteral reflux and children and adults with obstructive uropathy). In these groups, keeping the patient abacteriuric helps to protect the kidneys. Several studies in patients with frequent reinfections indicate that such prolonged chemotherapy reduces the frequency of reinfections.

One of the concerns regarding long-term prophylaxis has been the fear of the development of widespread antimicrobial resistance. For the most part, this has not been a problem (Stamm et al., 1980b). However concern has been expressed about using trimethoprim alone (Svensson et al., 1982), and several reports have appeared describing increasing frequency of rectal and urinary isolates resistant to trimethoprim (Pearson et al., 1979; Dornbusch and Toivanen, 1981; Maskell, 1983a). In contrast, emergence of resistance to trimethoprim-sulfamethoxazole has infrequently been reported.

Patients receiving long-term prophylaxis should be followed with monthly urine cultures or more often if interim symptomatic episodes develop. If bacteriuria persists or recurs during administration of the antimicrobial agent, therapy is altered, using response of bacteriuria as a parameter of adequacy of therapy. Long-term prophylaxis is effective only as long as the active drug is taken and does not influence the natural history of infection (Stamm et al., 1980b). Accordingly, after initial eradication of bacteriuria in patients with three or more urinary tract infections per year, prophylaxis may be considered for six months and then stopped. Usually following the cessation of prophylaxis, the attack rate returns to the previous state, and frequently new reinfections appear within the ensuing months. However, the history of such patients is further complicated by the fact that infections may occur in clusters, and the periodicity may decline after completing six months of prophylaxis as women move from being high- to low-risk subjects (Kraft and Stamey, 1977). If necessary, prophylaxis may be given for years.

URINARY TRACT INFECTION IN PREGNANCY

PHYSIOLOGIC ALTERATIONS IN THE URINARY TRACT

There is dilatation of the ureters and renal pelves during pregnancy with marked reduced ureteral peristalsis. These changes begin as early as the seventh week of gestation and progress to term (Andriole, 1975a). The bladder also decreases in tone, so that late in gestation it can contain twice its normal contents without causing discomfort and is associated with increased residual bladder urine. The changes in the ureters and renal pelves are more marked on the right side and are more likely to occur during the first pregnancy or when pregnancies occur in rapid succession. The urinary tract tends to revert to normal by the second month following delivery (Popky and Pollack, 1972; Andriole, 1975a).

Changes similar to those of pregnancy have been described in the urinary tracts of women taking oral contraceptives (Guyer and Delaney, 1972). Because of this observation, it has been suggested that the urinary tract alterations may be at least in part related to "hyperestrogenism" (Andriole, 1975a). Other possible explanations for the alterations are obstruction of the ureters by the gravid uterus and hypertrophy of muscle bundles at the lower end of the ureter (Andriole, 1975a). To investigate the effects of estrogens on these changes, Andriole and Cohn (1973) treated nonpregnant female and male rats with estrogens and obtained IVPs before and during treatment. Hydroureter and marked increased susceptibility to *E. coli* pyelonephritis were observed in both male and female animals after receiving estrogens.

EPIDEMIOLOGY

Asymptomatic bacteriuria occurs in approximately 5% of pregnant women (range 4%–7%) (Norden and Kass, 1968; Norden, 1972). Pregnant women of higher socioeconomic status have a lower frequency of bacteriuria than women of lower socioeconomic status (Turck et al., 1962a). The prevalence of bacteriuria also rises with parity and age (Kass, 1959). For example, in low-income populations, the prevalence of bacteriuria is about 2% in primiparas under age 21 as compared with 8%–10% in grandmultiparas over age 35 (Norden, 1972). Most women (about 75%) who develop bacteriuria during pregnancy have infection at the first prenatal visit. However, 1%–1.5% of pregnant women or about 25% of those with bacteriuria of pregnancy develop infection in the later trimesters (Norden and Kass, 1968; Norden, 1972). The development of *symptomatic* pyelonephritis late in pregnancy usually follows asymptomatic bacteriuria that was present earlier in parturition. The marked dilatation of the ureters and pelves during the later stages apparently facilitates the development of symptomatic pyelonephritis.

Approximately 20% of women with bacteriuria early in pregnancy develop acute symptomatic pyelonephritis later in pregnancy (Kass, 1959; Kincaid-Smith and Bullen, 1965). In contrast, fewer than 1% of women whose urine is uninfected early in pregnancy develop pyelonephritis. Accordingly, >75% of the cases of acute pyelonephritis may be prevented by eliminating asymptomatic bacteriuria in the early stages of pregnancy (Norden, 1972). Patients whose bacteriuria fails to respond to therapy are at highest risk of developing symptomatic infection (Condie et al., 1968). Lack of cure of bacteriuria is probably an indication of upper as opposed to lower tract infection.

An association between acute pyelonephritis or pregnancy and premature delivery was well-known in the preantibiotic era (Andriole, 1975a). The rate of prematurity can be as high as 20%–50%. Kass (1959) reported an association between asymptomatic bacteriuria and prematurity, and that the eradication of bacteriuria significantly reduced the rate of premature delivery. Since then, there have been conflicting studies both supporting and denying these observations (Condie et al., 1968; Norden, 1972; Andriole, 1975a). In general, it seems that premature delivery is increased in patients with asymptomatic bacteriuria. However, it does not necessarily follow that asymptomatic bacteriuria is a cause of prematurity. It is possible that certain patients are predisposed to both bacteriuria and to deliver premature infants. Some investigators have reported that elimination of bacteriuria decreases the frequency of prematurity (Condie et al., 1968; Norden, 1972; Andriole, 1975a). However, other studies have failed to show a decrease in prematurity or fetal wastage with elimination of asymptomatic bacteriuria (Condie et al., 1968; Norden, 1972; Andriole, 1975a). Neonates of patients refractory to multiple courses of therapy have been reported to have a significantly lower birth weight than infants of those who respond; this phenomenon may be related to the presence of upper tract infection in nonresponders (Condie et al., 1968; Gruneberg et al., 1969; Harris et al., 1975). Several studies have attempted to relate asymptomatic bacteriuria to the development of hypertension in pregnancy, but results have been unclear (Norden, 1972).

Even though the data relating asymptomatic bacteriuria of pregnancy to prematurity are not clear-cut, the relationship of asymptomatic bacteriuria to later development of acute pyelonephritis is indisputable. As acute pyelonephritis has possible serious consequences for both mother and fetus, screening for and treatment of bacteriuria of pregnancy seem justified. Quantitative urine cultures should be obtained in all pregnant patients at the initial prenatal visit, and those who have significant bacteriuria should be treated.

Postpartum studies of patients with bacteriuria of pregnancy demonstrate a high frequency of bacteriuria even with treatment during the pregnancy (Leigh, 1968). Postpartum IVP of these patients has shown that 10%–30% have radiologic changes of "chronic pyelonephritis" and other abnormalities (Leigh et al., 1968; Zinner and Kass, 1971). These abnormalities are most common in patients in whom renal bacteriuria has been demonstrated or in whom bacteriuria during pregnancy was difficult to eradicate with antimicrobial therapy (Leigh et al., 1968; Williams et al., 1968). However, pyelographic abnormalities should not be attributed to the infection that occurred during the pregnancy. In fact, these abnormalities probably antedated the pregnancy and in most cases are related to childhood infection. Treatment of bacteriuria of pregnancy has little effect on the long-term course of the patient. When patients who had bacteriuria of pregnancy were studied 10–14 years later, there were no differences between those who were treated and those who were not. About 25% of the women in each group had bacteriuria at time of follow-up (Zinner and Kass, 1971).

TREATMENT OF BACTERIURIA OF PREGNANCY

All pregnant women should be screened for bacteriuria during the first trimester and once again during the third trimester. First trimester screening and eradication of bacteriuria will markedly reduce the expected cases of acute pyelonephritis in pregnant women. Screening in the third trimester will prevent a few additional cases (Andriole, 1982). When asymptomatic bacteriuria is found, it should be confirmed by a second urine culture. Treatment with an appropriate antimicrobial agent is recommended for all pregnant patients found to have significant bacteriuria (Norden, 1972; Andriole, 1982). Therapy is aimed at maintaining sterile urine throughout gestation and thereby avoiding the complications associated with urinary tract infection during pregnancy. The administration of a relatively nontoxic drug for seven days (e.g., a sulfonamide, ampicillin, cephalexin, nitrofurantoin) eradicates bacteriuria in 70%–80% of patients (Andriole, 1972a, 1975a, 1982). Failure of treatment is most commonly seen in patients with renal infection or radiologic abnormalities of the urinary tract (Andriole, 1972b, 1975a, 1982). Sulfonamides should not be administered in the last few weeks of gestation because of danger of producing hyperbilirubinemia and kernicterus in the newborn. Tetracyclines should be avoided during pregnancy.

There are relatively few studies evaluating the efficacy of single-dose antimicrobial therapy for symptomatic bacteriuria during pregnancy. Harris et al. (1982) noted a cure rate of 71%–75% after 3 gm of ampicillin together with probenicid, 200 mg of nitrofurantoin, or 2 gm of sulfisoxazole. Although these results appear to be inferior to conventional therapy, single-dose therapy may be a reasonable first option in both symptomatic and asymptomatic infection in an attempt to decrease drug administration in pregnancy.

After successful treatment, follow-up cultures should be obtained monthly until delivery and recurrences treated. If bacteriuria recurs multiple times, continuous low-dosage antibiotic prophylaxis for the remainder of pregnancy has been advocated by some, although it is doubtful whether this approach is preferable to careful follow-up. Catheterization should be avoided at the time of delivery. If relapses or multiple reinfections occurred during pregnancy, radiologic evaluation should be considered two months or more postpartum.

PROSTATITIS

The route by which bacteria reach the prostate is largely unknown (Meares, 1975a), but possibilities include ascending infection via the urethra, hematogenous route, and lymphatic spread from the rectum. Urethral instrumentation including catetherization and prostatic surgery are known but infrequent causes of prostatitis, and most patients have no history of a precipitating event.

Stamey and colleagues (1981) have noted that male sex partners of women with vaginal colonization by gram-negative bacilli may develop transient urethral colonization with the same organisms. They postulated that sexual intercourse might play an important role in infection of the prostate. Prostatic fluid normally has substantial antibacterial properties (Fair et al., 1973; Stamey et al., 1968a). However, the prostatic secretions of some patients with chronic bacterial prostatitis have been shown to lack such activity (Fair et al., 1973).

The syndromes of acute and chronic bacterial prostatitis are different and distinct. Acute prostatitis does not usually result in chronic prostatitis, and chronic bacterial prostatitis is not usually antedated by acute prostatitis. Acute prostatitis is an acute infectious disease and is similar to an acute localized infection in any other organ, producing local heat, tenderness, and fever. In contrast, chronic bacterial prostatitis often produces few or no symptoms related to the prostate, which just serves as a nidus of low-grade infection. Some patients with chronic bacterial prostatitis have persistent symptoms such as perineal pressure, low back pain, or difficulty urinating. Symptoms of acute cystitis or pyelonephritis occur occasionally when bacteria, which repeatedly invade the bladder, overcome the defense mechanisms of the bladder.

Bacteria originating in the prostate may be coated with antibody and, therefore, are a cause of a false positive ACB test (Jones, 1979; Thomas and Forland, 1982).

ACUTE BACTERIAL PROSTATITIS

In the preantibiotic era most cases of acute bacterial prostatitis were caused by N. gonorrhoeae. Subsequently, gram-negative enteric urinary pathogens, E. coli, Proteus sp., Klebsiella sp., and Pseudomonas sp. became the most frequent pathogens (Meares, 1975a). The only important gram-positive organism causing acute prostatitis is S. faecalis, although in some studies S. aureus has been incriminated (Giamerellou et al., 1982).

Pathologically, acute bacterial prostatitis is characterized by inflammation of part or all of the gland with marked cellular infiltrate (predominantly polymorpho-nuclear leukocytes), diffuse edema, and hyperemia of the stroma. Microabscesses may occur and may be followed by large, clinically apparent collections of pus.

Acute bacterial prostatitis is characterized by high fever, chills, perineal and back pain, and symptoms of urinary tract infection such as frequency, urgency, and dysuria (Meares, 1975a). The patient may have urinary retention due to bladder outlet obstruction. The prostate gland is warm, swollen, and extremely tender on rectal examination. Expressed prostatic fluid contains many polymorphonuclear leukocytes and the infecting organism can frequently be seen on Gram stain. However, massage of the acutely infected prostate gland can precipitate bacteremia and should be discouraged. Since most patients also have bacteriuria, the infecting organism can usually be isolated by midstream urine culture. Clinical diagnosis of acute bacterial prostatitis is made on the basis of the clinical picture, tender prostate (not specific, however), urinalysis and urine culture. To distinguish clearly acute prostatitis from acute urethritis, the three glass test should be performed (see below). Many antibiotics diffuse well into the acutely inflamed prostate, and acute bacterial prostatitis responds well to appropriate antimicrobial therapy. Complications such as bacteremia, prostatic abscess, epididymitis, seminal vesiculitis, and pyelonephritis may occur.

If the urine culture is negative and pyuria is present, nonbacterial pathogens should be considered. Suspected microorganisms include *Chlamydia trachomatis* and *Ureaplasma urealyticum;* however, their role in causing prostatitis as opposed to urethritis remains speculative.

CHRONIC BACTERIAL PROSTATITIS

Chronic bacterial prostatitis is most commonly caused by *E. coli* (80%), but *Klebsiella-Enterobacter, P. mirabilis,* and enterococci are also common causes (Meares, 1973). *S. epidermidis, S. aureus,* and diphtheroids have been frequent isolates in some series (Drach, 1975). Fifteen to twenty percent of chronic infections are polymicrobial.

The histologic finding of chronic bacterial prostatitis is focal, nonacute inflammation. Similar findings may be noted in patients without evidence of bacterial infection and are therefore not diagnostic of bacterial prostatitis.

Many men with chronic infection of the prostate are totally asymptomatic. However, some have perineal discomfort, low back pain, or dysuria. Symptoms of acute urinary tract infection may appear periodically. In fact, chronic bacterial prostatitis is probably the most common cause of relapsing urinary tract infection in men. Fever, if present, tends to be low grade unless pyelonephritis occurs. The results of rectal examination and IVP are unremarkable unless the patient also has an enlarged prostate gland from benign prostatic hypertrophy or carcinoma.

Because of the focal nature of chronic bacterial prostatitis, needle biopsy of the prostate gland for culture of tissue is unreliable (Schmidt and Patterson, 1966). Demonstration of leukocytes in prostatic fluid is not specific for bacterial prostatitis. Meares and Stamey (1968) described a quantitative localization technique for making the bacteriologic diagnosis. Because bacteria present in the urethra can contaminate prostatic secretions obtained by prostatic massage, accurate diagnosis requires simultaneous quantitative cultures of: (1) urethral urine (VB_1), (2) midstream urine (VB_2), (3) prostatic secretions expressed by massage (EPS), and (4) the urine voided after massage (VB_3). An ejaculate is probably preferable to the EPS.

The specimens must be cultured immediately after collection, and methods of quantitating small numbers of bacteria must be used. The study should be done at a time the patient does not have significant bacteriuria. If bacteriuria is present, ampicillin, cephalexin, or nitrofurantoin should be given for 2–3 days to sterilize the urine; these agents will not affect bacterial counts in the prostate in chronic bacterial prostatitis. If chronic bacterial prostatitis is present, the number of bacteria in the EPS or ejaculate will exceed those in VB_1 or VB_2 urine by at least tenfold. If no EPS or ejaculate can be obtained, the bacterial counts in the VB_3 specimen should be at least tenfold higher than the VB_1 or VB_2 samples. The three glass test urine specimens can be analyzed in terms of the leukocytes present. A threefold increase in polymorphonuclear leukocytes is found in urine after massage (VB_3) when compared to urethral urine (VB_1). In addition to increased leukocytes, lipid-filled macrophages may be observed and are highly suggestive of chronic bacterial prostatitis.

Chronic bacterial prostatitis is very difficult to cure, since few antimicrobial agents penetrate well into the noninflamed prostate. Furthermore, the nidus of infection in some patients may be small prostatic calculi that presumably are difficult to sterilize. Chronic bacterial prostatitis is therefore likely to persist and cause relapsing urinary tract infection. Unlike classic urinary tract infection, relapses may occur after long periods without bacteriuria (e.g., months). Management may be difficult (see "Therapy").

THERAPY

A dog model has been used to measure diffusion of antimicrobial agents into the noninflamed prostate (Stamey et al., 1968b). In this system, antimicrobial agents are infused, giving high and constant plasma levels, and prostatic secretions are simultaneously collected. Although the basic macrolides such as erythromycin penetrated well into prostatic secretions, penicillins, cephalosporins, tetracyclines, nitrofurantoin, and vancomycin did not. The explanation given was that only lipid-soluble and basic compounds are capable of entering the acid milieu of the prostate gland. Trimethoprim has been shown to diffuse into prostatic fluid in high concentrations (Stamey, 1973b).

Acute bacterial prostatitis frequently responds dramatically to antibiotic therapy. The intense diffuse inflammatory reaction of acute prostatitis is thought to facilitate the diffusion of antimicrobial agents from plasma into the prostate (Meares, 1975a). Accordingly, in the management of acute bacterial prostatitis, antibiotics should be given in doses that achieve therapeutic concentrations in the blood. Therapy should be selected on the basis of urine culture and sensitivity studies. In acutely ill patients requiring hospitalization, parenteral therapy is advisable. Antimicrobial agents frequently used include trimethoprim-sulfamethoxazole, tetracyclines, amoxicillin, and ampicillin in combination with aminoglycosides. Oral therapy should be continued for at least 30 days. Measures such as hydration, analgesic therapy, bed rest, and stool softeners may be helpful. Urethral instrumentation should be avoided. If acute urinary retention occurs, suprapubic needle aspiration of the bladder is recommended for drainage of urine (Meares, 1975a). When prolonged bladder drainage is required, a suprapubic catheter is preferred.

Chronic bacterial prostatitis is very difficult to cure. The primary approach to therapy is an attempt at cure with antimicrobial therapy. Although occasional cures have been achieved with penicillins, cephalosporins, tetracyclines, or aminoglycosides, the focus of infection in the prostate has usually persisted, resulting in relapse after therapy was discontinued. Superior results have been reported with trimethoprim-sulfamethoxazole (1 double-strength tablet twice daily) administered orally for 12 weeks (Meares, 1975b). Oral carbenicillin indanyl sodium alone or in combination with rifampin has been shown to be effective in therapy of pseudomonal and enterococcal infections that fail on trimethoprim-sulfamethoxazole treatment. Cure rates have varied from one third to most of the patients treated with these agents for 1–3 months. The sulfonamide component of trimethoprim-sulfamethoxazole probably contributes little, and rifampin may be more suitable than sulfamethoxazole as a partner for trimethoprim. Rifampin has a broad antibacterial spectrum but has the major drawback of rapid development of resistance. Although trimethoprim prevents emergence of resistance to rifampin (Giamerellou, 1982), the role of rifampin in prostatitis remains to be determined. At present, the initial regimen of choice is trimethoprim-sulfamethoxazole. The carboxyquinolone agents currently under study appear to be extremely promising in the treatment of chronic bacterial prostatitis.

Transurethral prostatectomy is curative only if all the infected tissue is removed. About one third of the patients are cured by this procedure (Meares, 1975a). Whereas total prostatectomy is contraindicated because of the complications (sexual impotence and incontinence), Meares has expressed greater satisfaction with a radical transurethral prostatectomy aimed at resecting all infected tissue and calculi, particularly the posterior prostate near the external urethral sphincter. Such radical surgery should be considered only when all attempts at medical cure with antibiotics have failed. Other alternatives include the treatment of each acute exacerbation of urinary tract infection or the use of chronic suppressive therapy using antimicrobial agents. There is no evidence that zinc therapy or regular prostatic massage is helpful in management.

PROSTADYNIA

Prostadynia or prostatosis is a syndrome that Meares (1973) thinks is more common than bacterial prostatitis. It is commonly seen in young men. The symptoms are those of perineal pressure, pelvic pain with or without dysuria, urgency, and/or low back pain, symptoms that can also be caused by chronic bacterial prostatitis. However, bacterial pathogens cannot be demonstrated using quantitative cultures, and urinary tract infection does not occur. Other clinical features include painful erection and postejaculatory discomfort, with varying degrees of obstructive voiding. The etiology is somewhat obscure. Some authorities believe that prostadynia is actually prostatitis caused by as yet unidentified microorganisms or by low numbers of gram-positive or gram-negative bacteria (Drach, 1975; Weidner et al., 1980). Currently, most believe that symptoms are caused by spasm of the pelvic floor musculature (Segura et al., 1979). Because the etiology is unknown, therapy is extremely difficult and consists of reassurance, hot sitz baths and the possible use of α-blockers (phenoxybenzamine), anticholinergics and anti-inflammatory agents.

INTRARENAL ABSCESS AND PERINEPHRIC ABSCESS

INTRARENAL ABSCESS

Although intrarenal abscess may develop as a consequence of bacteremia (often caused by coagulase positive staphylococci), these lesions usually result from classic acute pyogenic pyelonephritis. The clinical picture is that of acute pyelonephritis with high fever and severe flank pain and tenderness but with no response or very slow response to appropriate antimicrobial therapy. Most patients with intrarenal abscess respond, although slowly, to antimicrobial therapy, but fever and severe flank pain may persist for days. Occasionally, drainage is necessary.

PERINEPHRIC ABSCESS

Perinephric abscess is an uncommon complication of urinary tract infection (Thorley et al., 1974). It usually occurs secondary to obstruction of an infected kidney or calyx or occasionally secondary to bacteremia. It may occur insidiously, and up to one third of cases may not be diagnosed until autopsy. The infecting bacteria are usually gram-negative enteric bacilli and occasionally gram-positive cocci when the infection is of hematogenous origin.

The patients have a syndrome suggestive of acute pyelonephritis with fever, abdominal and flank pain (usually unilateral), and often symptoms of lower tract infection. The patient often has been ill for two or more weeks. The diagnosis should be strongly considered in any patient with a febrile illness and unilateral flank pain who does not respond to therapy for acute pyelonephritis. A palpable mass may or may not be present. About one half of the patients have an abnormal plain x-ray film of the abdomen (e.g., abdominal mass, a calculus, a poorly defined renal shadow), and 85% have abnormal IVPs.

DIAGNOSIS AND TREATMENT

The introduction of renal ultrasonagraphy and in particular CT scans has added a new dimension of sensitivity and specificity, permitting earlier diagnosis of suppurative complications of acute pyelonephritis (Gerzof and Gale, 1982; Rauschkolb et al., 1982) (Figs 5–3 through 5–6). These studies also permit subsequent monitoring of the progress of renal infections (Pollack et al., 1979).

A renal abscess contains debris and therefore sonographically results in fine internal echoes with good sound transmission or a fluid-debris level. The wall of the abscess is generally thick and irregular. Occasionally, gas in an abscess produces increased echogenicity. Peripheral enhancement in the wall may be observed on CT scan (Kitchner and Rosenblatt, 1984). Soon after its introduction, sonography proved to be superior to urography in diagnosis and evaluation of renal abscess, pyonephrosis, and perinephric abscess, also replacing angiography, retrograde pyelography, and nephrotomography. In general, ultrasound detects only abscesses 2 cm or larger in diameter (Hoddick et al., 1983). In the acute phase, no wall enclosing the abscess is seen, but with time and chronicity, the wall becomes more prominent and is surrounded by edematous parenchyma.

CT scan of an abscess mass reveals a thick wall, attenuated values above water, and, in many cases, en-

FIG 5–3.
CT scan of the right kidney after injection of iodinated contrast material. Intrarenal abscess with possible gas formation *(arrow)*. (Courtesy of George Popky, M.D.)

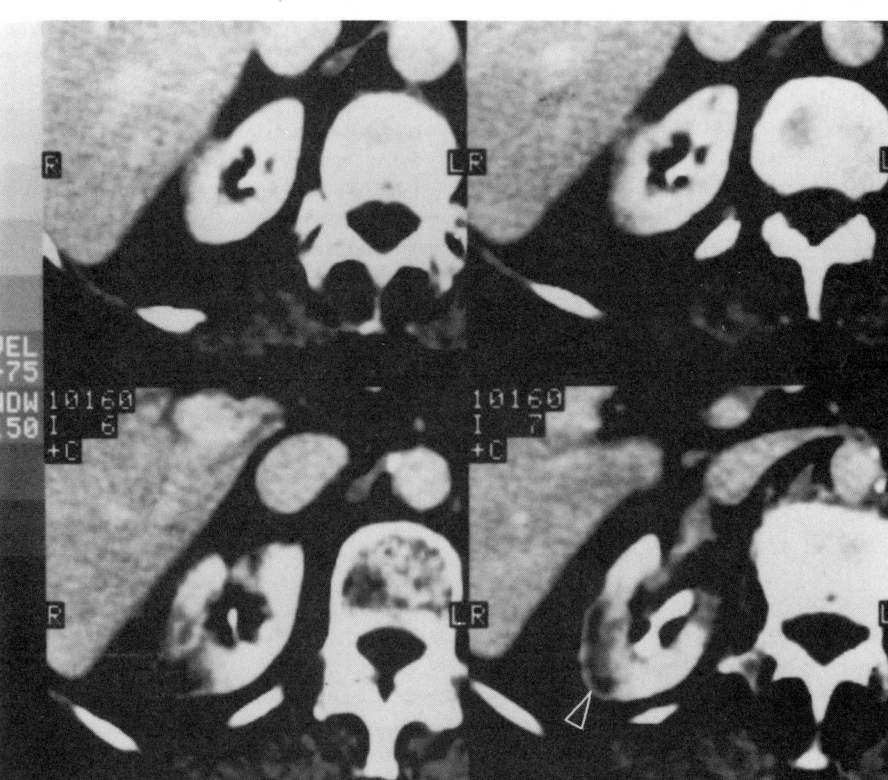

hancement on postcontrast studies. The picture may resemble that of a tumor and require needle aspiration for differentiation.

In patients with a clinical or radiographic suspicion of perinephric abscess, diagnostic needle aspiration can be safely performed by using ultrasound or CT scan guidance. When an abscess is confirmed, small catheters can be introduced percutaneously via the diagnostic aspiration route to provide immediate decompression as well as continuous and definitive drainage without need for surgery (Gerzof and Gale, 1982; Rauschkolb et al., 1982). Advantages to guided percutaneous drainage compared with open surgical drainage include earlier diagnosis and treatment, avoidance of general anesthesia and surgery, less expensive therapy, easier nursing care, and greater patient acceptance of closed drainage. Accordingly, it is now recommended that after starting antimicrobial therapy directed against the most likely pathogens, a trial of percutaneous drainage should be the initial mode of therapy for perinephric abscess. Surgical intervention should be undertaken only when percutaneous drainage fails or is contraindicated. While parenteral antimicrobial therapy directed against the infecting organism isolated from blood or urine should be initiated before drainage, if additional organisms are isolated at the time of surgery, treatment directed against these organisms must be added. Therapy must also be used for the underlying disease (for example, staphylococcal bacteremia or obstructive uropathy).

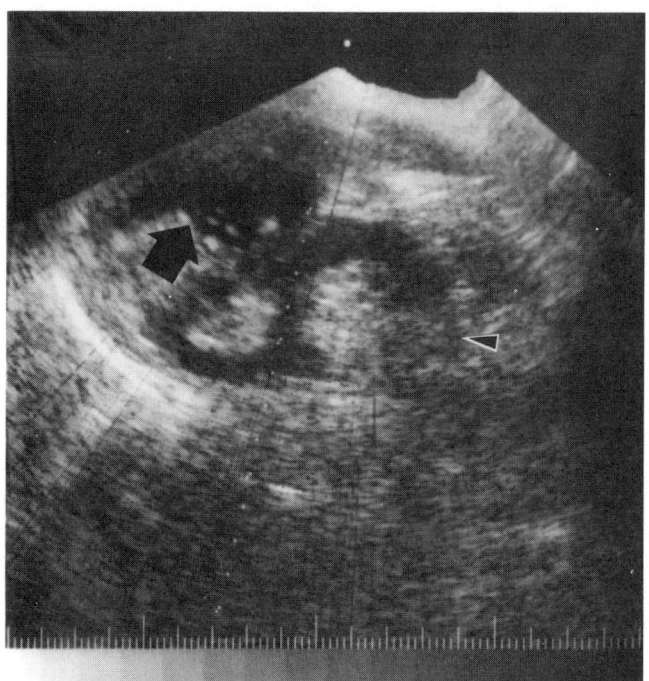

FIG 5–4.
Ultrasound of right kidney with a large perinephric abscess. *Small arrow,* edge of the kidney; *large arrow,* the abscess. (Courtesy of George Popky, M.D.)

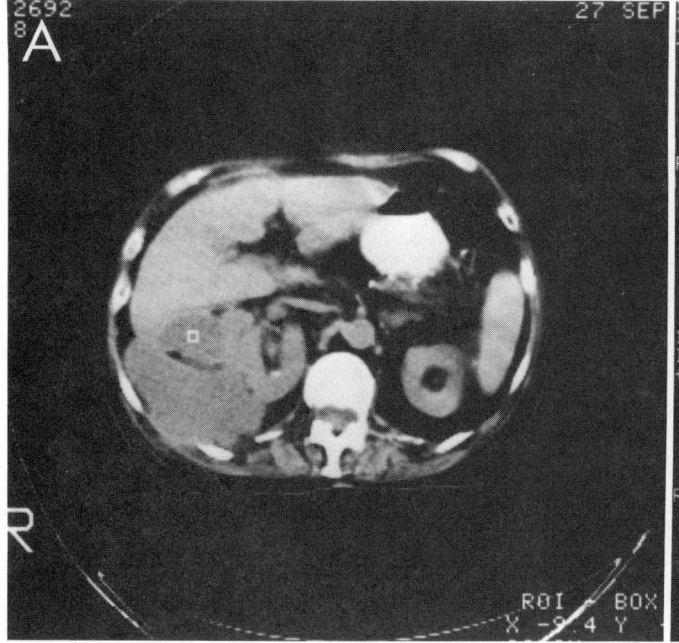
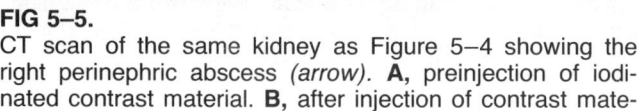
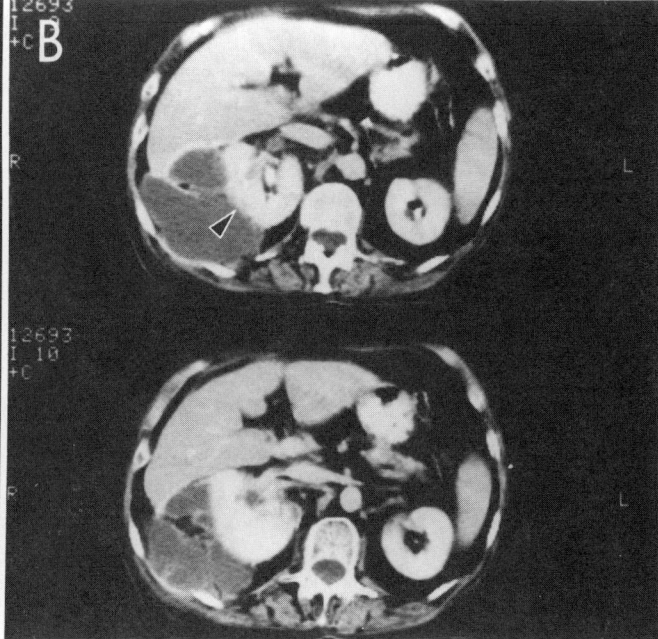

FIG 5–5.
CT scan of the same kidney as Figure 5–4 showing the right perinephric abscess *(arrow).* **A,** preinjection of iodinated contrast material. **B,** after injection of contrast mate- rial. The demarcations between the kidney and the wall of the abscess *(arrow)* are much better seen after injection of contrast. (Courtesy of George Popky, M.D.)

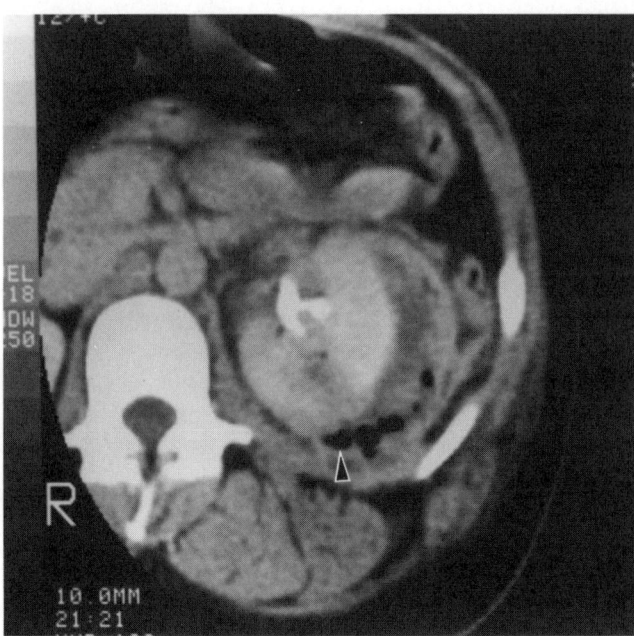

FIG 5–6.
CT scan shows a right perinephric abscess with gas formation *(arrow)*. (Courtesy of George Popky, M.D.)

Percutaneous drainage has been effective in drainage of renal abscesses and infected renal cysts, avoiding the previous approach of open surgical drainage or nephrectomy (Gerzof and Gale, 1982; Finn et al., 1982; Costello et al., 1983).

RADIOLOGIC EVALUATION

Radiologic procedures play an important role in the management of patients with urinary tract infection (Filly, 1983). The most important contribution provided by radiology is the detection of surgically correctable abnormalities of the urinary tract. Calculi, urinary tract obstruction, and many congenital anomalies can best be recognized by radiographic means. Therefore, urography should be considered in patients at greatest risk of having surgically correctable abnormalities. Persons included in this high-risk category are all children and patients whose infection has been complicated by bacteremia. The study should consist of an IVP with a postevacuation view of the bladder to evaluate for residual urine. Ultrasound examination of the bladder is also useful for determining the volume of residual urine and does not involve radiation exposure or injection of contrast material (Filly, 1983). It is not necessary to obtain radiologic studies on adult women experiencing their first urinary tract infection. After three or four reinfections, evaluation may be indicated. Women with bacteriuria of pregnancy in whom eradication of infection is difficult should be evaluated radiologically. These pa-

tients should not be studied until at least two months after delivery, by which time the physiologic alterations of the urinary tract that occur during pregnancy should be reversed (Popky and Pollack, 1972; Andriole, 1975a).

During acute bacterial pyelonephritis, several abnormalities may be seen on IVP. These include focal or diffuse renal enlargement, diminished nephrographic density, delayed calyceal appearance time, faint visualization of the collecting system, pyelocaliectasis, and acute lobar nephronia (McDonough et al. 1981). The latter refer to focal bacterial nephritis with focal swelling or mass-effect that occurs in the absence of frank local suppuration; i.e., solid inflammatory mass that often suggests the possibility of a renal abscess. However, ultrasound and CT examination usually distinguish this solid inflammatory process (not requiring drainage) from that of frank abscess (Fig 5–7). The IVP may be normal in as many as 30% of patients with acute unobstructed pyelonephritis.

Radiologic investigation is often indicated in patients with unresponsive pyelonephritis (particularly if bacteremic) to identify local complications such as renal and perinephric abscess formation. As an alternative to urography, noninvasive ultrasound examination in patients with acute pyelonephritis is useful in excluding urinary obstruction and residual urine in the bladder. A nuclear scan with furosemide to increase urine flow is useful in determining whether there is structural as opposed to functional ureteropelvic junction obstruction. CT scans are of particular value in the diagnosis of local suppurative complications (Gerzof and Gale, 1982).

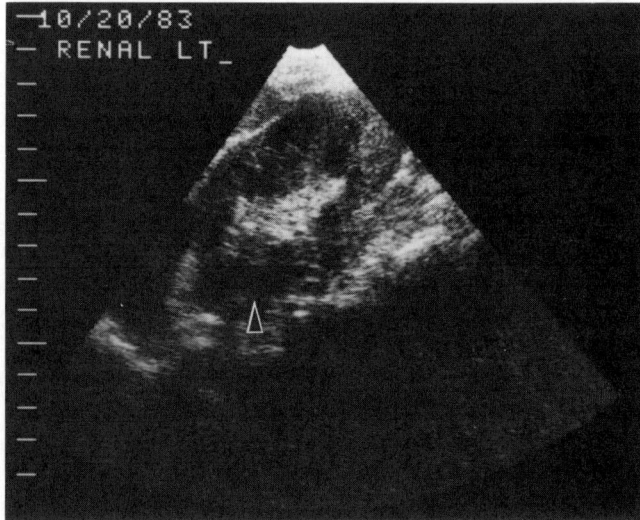

FIG 5–7.
Enlarged kidney with an intrarenal infiltrative relatively sonolucent process *(arrow)* in a patient with pyelonephritis. (Courtesy of George Popky, M.D.)

Although scintiphotography with gallium-67 has been reported to be useful in diagnosis of pyleonephritis and renal abscess, it is uncommonly required and may be positive in noninfectious entities (Schardijn, 1984). Scintigraphy with indium-111 oxine-labeled granulocytes (an experimental technique) appears promising (Filly, 1983).

In additional to delineating lesions amenable to surgical correction, urography frequently provides information previously unknown to the patient or physician. For example, unsuspected renal scarring may be seen, suggesting the presence of undiagnosed urinary tract infection in childhood. Occasionally, an unusual or unsuspected type of renal infection such as tuberculosis, papillary necrosis, or xanthogranulomatous pyelonephritis may be discovered (Grainger and Longstaff, 1982). The latter is a severe, chronic form of kidney inflammation in which areas of renal parenchyma are replaced by an inflammatory granulomatous reaction characterized by lipid-laden macrophages (foam cells) (Popky and Pollack, 1972). Renal calculi and obstruction are often associated with this lesion (Popky and Pollack, 1972). Two major radiologic patterns are seen—that of localized mass and that of diffuse nodularity. When a mass lesion is present, differentiation from pyogenic abscess, tuberculous abscess, or avascular carcinoma may not be possible.

Cystourethrography traditionally has been thought to be indicated in all children with urinary tract infection. However, the voiding cystourethrogram is unpleasant for the patient, requires catheterization, and is poorly tolerated (Fairley, 1976). An alternative method that is tolerated better is radionuclide evaluation. A radionuclide is inserted into the bladder, and scans are performed. The procedure involves less radiation exposure than cystourethrography. Since vesicoureteral reflux is particularly important in preschool children and is related to renal scarring, all preschool children with proved urinary tract infection probably should undergo a voiding cystourethrogram (Fairley, 1976). This procedure should be avoided in older children unless IVPs show evidence of renal scars. In the presence of scars, cystourethrography may be indicated. However, even with scars, if serial urographic evaluation demonstrates stability of upper tract lesions, the need for studying the lower tract is questionable. Fairley (1976) suggested that it may be possible to avoid cystourethrography by taking a late roentgenogram (4–6 hours) after an IVP. By this time the ureters should no longer contain contrast material, but the bladder will be filled. A voiding film taken at this time may demonstrate presence or absence of reflux. Fairley believes that if no reflux is demonstrated by this method, it is doubtful that cystourethrography will add much information in older chil-

dren and adults. When reflux is found, it should be graded as minimal (grade I) to severe (grade IV), so progression or improvement can be quantitated and decisions on surgery can be made (Smellie and Normand, 1975b).

CATHETER-ASSOCIATED URINARY TRACT INFECTION

The urinary tract is the most common site of nosocomial infections and most of these hospital-acquired infections occur in patients who have undergone urologic manipulation, particularly catheterization (Turck et al., 1962b). Approximately 10%–15% of hospitalized patients undergo urinary catheterization with an estimated 25% risk of developing a urinary tract infection, resulting in approximately 600,000 episodes of catheter-associated infection annually (Turck and Stamm, 1981). Moreover, it has been estimated that 1%–3% of these infections result in bacteremia (Turck and Stamm, 1981), with a case fatality rate of greater than 30%.

Urinary catheters are now recognized as the most frequent source of nosocomial bacteria (Kreger et al., 1980, 1983a, 1983b). Platt et al. (1982) determined that nosocomial urinary tract infections are associated with an approximately threefold increase in patient mortality during hospitalization by as yet undefined mechanisms, although the relatively few patients who die of these infections have extensive underlying disease (Gross et al., 1980). Bryan and Reynolds (1984b) recently analyzed 221 nosocomial bacteremic urinary tract infections that were accompanied by an overall mortality of 30.8%. The mortality attributed specifically to bacteremic urinary tract infection was 12.7%, in two thirds of whom there was serious underlying disease. If clinical shock supervened, the mortality was 49%. Another important aspect of catheter-associated infection is that microorganisms associated with such infections constitute a vast reservoir of antibiotic-resistant hospital pathogens, with frequent cross-infection to other catheterized patients. Not surprisingly, therefore, clusters or outbreaks of catheter-associated urinary tract infection have been described in hospitals due to specific strains of *S. marcescens* (Maki et al., 1973), *P. rettgeri* (Lindsey et al., 1976), and *Providencia stuartii* (Fierer and Ekstrom, 1981). Such nosocomial outbreaks are more likely in long-term care facilities and in crowded intensive care units. In addition to the mortality and morbidity of catheter-associated urinary infection, there are added costs of tests, antimicrobial agents, and excess hospitalization. The mean additional duration of hospitalization in surgical patients with postoperative urinary tract infections was 2.4 days (Kreger et al., 1983b).

Risk factors associated with catheter-related infection include female sex, advanced age, and increasing degree of underlying illness (Garibaldi et al., 1974). Other alterable factors associated with increased risk are (1) lengthy duration of catheterization, (2) improper nonsterile insertion technique, (3) failure to provide closed sterile drainage and (4) suboptimal type of drainage system (Turck and Stamm, 1981).

Bacteria may enter the bladder at the time of catheterization, especially if faulty technique is used. Therefore, two pathways have been postulated by which bacteria invade the bladder through indwelling catheters. The most important pathway with open-drainage systems is probably via the catheter lumen including the upward movement of air bubbles, by motility of the bacteria, or by capillary action (Weyrauch and Bassett, 1951; Gillespie et al., 1960; Andriole, 1972b). The second route is through the exudative sheath that surrounds the catheter in the urethra; this route is probably more important with properly managed closed-drainage systems (Garibaldi et al., 1980). Recent prospective studies confirm the significance of the latter route in that meatal colonization with gram-negative rods precedes the development of bacteriuria, and these uropathogens can be identified in urethral and perianal flora 2–4 days before the onset of bacteriuria (Daifuku and Stamm, 1984). Relatively little is known about the specific bacterial virulence factors that enhance both colonization of the periurethral region and ascent into the bladder. Once catheter-associated bacteriuria develops, as in patients with asymptomatic bacteriuria, it is difficult to distinguish between upper and lower tract infection. Kostiak et al. (1981) showed that within five weeks, 70% of bacteriuric catheterized patients had positive ACB tests. An indwelling catheter not only provides a portal of entry for bacteria but also seriously compromises the normal bladder defense mechanisms in that the flushing effect of micturition is lost. In addition, the foreign body acts as a source of irritation, causing no inconsiderable trauma to the bladder mucosa. The catheter acts as a nidus for bacterial adherence and growth, with encrustations forming around the balloon and within the catheter lumen. The catheter as a foreign body thus maintains its own bacterial ecosystem (Rubin et al., 1980a).

The "open system" of indwelling catheters consists of a catheter with two lumens—one for balloon inflation and one for urine that drains into an open receptacle. Fifty percent of the patients with sterile urine before catheterization develop significant bacteriuria within 24 hours with this system, and virtually all are infected at four days (Kass, 1956). Utilizing the closed-drainage system considerably reduces the risk of infection, especially with short-term catheterization. The risk of in-fection after a single catheterization is about 1% (Turck, 1962b), but it is higher in elderly or debilitated patients, in patients with urologic abnormalities and in pregnant women (Brumfitt et al., 1961; Turck, 1962b).

Although the majority of catheter-associated infections are caused by *E. coli*, any microorganism may induce infection, particularly hospital-acquired multiresistant bacteria. Studies by Warren et al. (1982) and Breitenbacher (1984) indicate that in patients with long-term indwelling bladder catheters, the bacterial flora is predominantly polymicrobial, with rapid and constant change in species and number negating the usefulness of surveillance cultures.

Although quantitative cultures for the diagnosis of catheter-associated urinary infection usually require $\geq 10^5$ bacteria/ml for confirmation, Stonk and Maki (1984) showed that low-level bacteriuria (10^2–10^4/ml) may also be relevant. Such patients invariably are found to have $\geq 10^5$ bacteria/ml within 24–48 hours. Thus, lower titers should not be disregarded in symptomatic patients.

PREVENTION

Most patients who require indwelling urethral catheters need them for only short periods (less than one week). The system of closed sterile drainage has been shown to reduce the risk of infection. In particular, in a randomized prospective study, catheters with preconnected sealed junctions were associated with significantly fewer infections and deaths (Platt et al., 1983a). The use of systemic antibiotic prophylaxis fails to prevent catheter-associated infection (other than temporarily) and leads to the emergence of resistant strains such as *Pseudomonas*, *Serratia*, and *Citrobacter* (Sanford, 1964; Breitenbacher, 1984). The application of antibiotics into lubricants and the impregnation of the catheter itself with antibiotics have met with little success in preventing infection (Butler and Kunin, 1968a; Andriole, 1972b; Garibaldi et al., 1980; Burke et al., 1983).

In contrast, the use of antibacterial bladder rinses with the open drainage system using a triple-lumen catheter has been shown to delay significantly the development of bacteriuria (Martin and Bookrajian, 1962; Martin et al., 1965; Andriole, 1972b). Several rinse solutions have been used, including 0.25% acetic acid, nitrofurazone, and neomycin plus polymyxin. All three are capable of substantially delaying the development of bacteriuria (to beyond ten days in most patients), but the neomycin-polymyxin rinse is probably the most effective. Most patients have sterile urine after catheterization for up to ten days using this system (Andriole, 1975b); however, irrigation may result in colonization with enterococci and yeasts (Thornton et al., 1966).

When a closed-drainage system is in place, the use of antimicrobial irrigants to eradicate organisms after bacteriuria was established was not found to be effective (Warren et al., 1978).

Sterile closed-drainage systems are also capable of preventing bacteriuria in most patients for up to ten days without the use of antibiotic rinses provided the system is kept closed (Andriole, 1975b). Indiscriminate opening and flushing of the sterile closed-catheter system are common causes of contamination and infection. Antibiotic rinses and ointments add little to the protective effect of a closed system (Butler and Kunin, 1968b; Gladstone and Robinson, 1968; Garibaldi et al., 1974; Andriole, 1975b; Warren et al., 1978). However, some studies indicate that there may be an advantage in using systemic antibiotics when the closed system is used for only a few days (Garibaldi et al., 1974).

Closed-drainage systems and three-way catheter systems with a neomycin-polymyxin rinse are comparable in preventing infection. However, our preference is for the closed-drainage system, because (1) it is less expensive, (2) it is easier to maintain, and (3) if infection occurs, there is a reasonable chance that it will be with an antibiotic-susceptible organism. However, if for some reason frequent irrigation of the catheter is needed, an antibiotic drip is preferable. When catheters are required for many months or permanently, no system prevents infection.

With regard to catheter design, there is no evidence that silicone vs. latex has any advantage (Kunin, 1984). Of great significance was the observation by Warren et al. (1978) that a major risk factor for the development of catheter-associated infection was the disconnection of the catheter drainage tube junction. The urine collecting bag may be an important source of infection. Bags should be used with devices to prevent reflux of urine from the drainage bag retrogradely into the drainage system. Similarly, the downward direction of urine drainage must be maintained. Although one study (Desautels et al., 1982) suggested that addition of 3% hydrogen peroxide to the drainage system might prevent acquisition of bacteriuria, a subsequent study (Thompson et al., 1984) failed to detect any benefit. Although condom catheters tend to have a lower risk of urinary tract infection, the condom may frequently serve as a reservoir for bacteria (Fierer and Ekstrom, 1981). Furthermore, infections are particularly common in uncooperative patients who tend to pull on the condom (Hirsh et al., 1979).

The use of intermittent catheterization, particularly in patients with neurogenic bladders, although reducing the frequency of urinary infection compared with indwelling catheters, has little application in most acute-care facilities (Kunin, 1984). Long-term use, however, has been associated with fewer episodes of infection or sepsis and decreased stone formation (Lapides et al., 1974; Kunin, 1984). Although some authors think that urinary antiseptics should be instilled following each catheterization procedure, most authorities are not in favor of long-term prophylaxis (Kunin, 1984).

Recommendations for urinary catheter care are listed in Table 5–3 (Stamm, 1975).

TREATMENT

Although systemic antimicrobial therapy is effective in eliminating catheter-associated bacteriuria, treatment is not recommended in the absence of symptoms. This is because reinfection rapidly occurs during or following cessation of treatment and usually with more resistant organisms (Butler and Kunin, 1968b). Even if the catheter is to remain in situ for a short time only, therapy is best deferred until after the catheter is removed. If fever, flank pain, or other symptoms of infection occur, therapy must be started even though the catheter remains in place. In patients undergoing prostatectomy who have postoperative catheter-associated bacteriuria, the bacteriuria frequently persists when caused by coliforms, but often clears spontaneously when caused by *S. epidermidis* (Gordon et al., 1983).

Candida sp. are frequently isolated from urine spec-

TABLE 5–3.
Prevention of Urinary Infection Associated With Urinary Catheters

1. Avoid use when not essential; for care (not for convenience) remove as soon as possible.
2. Educate personnel in correct techniques of catheter insertion and care.
3. Emphasize handwashing.
4. Insert catheters using stringent aseptic technique.
5. Secure catheter properly.
6. Maintain sterile closed-drainage. Avoid disconnection of the catheter and drainage tube.
7. Maintain unobstructed urine flow including "downhill" continuous drainage. The catheter should always be below the level of bladder, and the bag should be emptied regularly.
8. Obtain urine samples aseptically.
9. Use smallest suitable bore catheter.
10. If closed sterile drainage system is violated, replace immediately.
11. Replacement of indwelling catheter is unnecessary unless concretions are felt or system is obstructed.
12. Spatially separate infected and uninfected patients with indwelling catheters.
13. Routine bacteriologic monitoring not required; however, post-catheterization urine should be obtained for culture.
14. Continuous irrigation, meatal antibiotics, and use of local and systemic antimicrobial prophylaxis not recommended.
15. Patients with cardiac disease that predisposes to bacterial endocarditis should receive antibiotic prophylaxis at the time of catheter insertion and removal.
16. Consider (reconsider) introduction of intermittent urinary catheterization.

imens obtained from catheterized subjects, especially when antimicrobial therapy has been used. In asymptomatic patients, even when isolated in pure culture and in large numbers they rarely are of clinical significance and can be ignored (Thornton et al., 1966). Nevertheless, candiduria can represent invasive lower urinary tract infection in catheterized subjects or reflect renal candidiasis in patients with disseminated candidiasis. Under these circumstances, antifungal therapy becomes necessary. Irrigation of the bladder with amphotericin B is effective for treatment of lower tract fungal infection as well as in differentiating upper and lower tract sources of candiduria (Dudley and Barriere, 1981). If the candiduria persists after appropriate bladder irrigation with amphotericin B, renal candidiasis should be suspected, and systemic therapy may be indicated. Some patients even develop fungus balls that may obstruct a calyx, ureter, or the bladder. *Candida*-associated urinary tract infections may account for 10% of nosocomial urinary tract infections (Krieger et al., 1983b).

REFERENCES

1. Anderson JD, Aird MY, Johnson AM, et al: The use of a single 1g dose of amoxicillin for treatment of acute urinary tract infections. *J Antimicrob Chemother* 1979; 5:481.
2. Andriole VT: Diagnosis of urinary tract infection by culture, in Kaye D (ed): *Urinary tract Infection and Its Management*. St Louis, CV Mosby Co, 1972a, p 28.
3. Andriole VT: Care of the indwelling catheter, in Kaye D (ed): *Urinary Tract Infection and Its Management*. St Louis, CV Mosby Co, 1972b, p 56.
4. Andriole VT: Urinary tract infection in pregnancy. *Urol Clin North Am* 1975a; 2:285.
5. Andriole VT: Hospital acquired urinary tract infections and the indwelling catheter. *Urol Clin North Am* 1975b; 2:451.
6. Andriole VT: Advances in the treatment of urinary infections. *J Antimicrob Chemother* 1982; 9(suppl A):163.
7. Andriole VT, Cohen GL: The effect of diethylstilbestrol on the susceptibility of rats to hematogenous pyelonephritis. *J Clin Invest* 1973; 43:1136.
8. Aronson M, Medalia O, Schori L, et al: Prevention of colonization of the urinary tract of mice with Escherichia coli by blocking of bacterial adherence with methyl-α-D-mannopyranoside. *J Infect Dis* 1979; 139:329.
9. Asscher AW, Chick S, Radford N, et al: Natural history of asymptomatic bacteriuria in nonpregnant women, in Brumfitt W, Asscher AW (eds): *Urinary Tract Infection*. London, University Press, 1973, p 51.
10. Asscher AW, Sussman M, Weiser R: Bacterial growth in human urine, in O'Grady F, Brumfitt W (eds): *Urinary Tract Infection*. London, Oxford University Press, 1968, p 3.
11. Bailey RR: The relationship of vesicoureteric reflux to urinary tract infection and chronic pyelonephritis-reflux nephropathy. *Clin Nephrol* 1973; 1:132.
12. Bailey RR, Abbott GD: Treatment of urinary tract infection with a single dose of amoxicillin. *Nephron* 1977; 13:316.
13. Bailey RR, Abbott GD: Treatment of urinary tract infection with a single dose of trimethoprim-sulfamethoxazole. *Can Med Assoc J* 1978; 118:551.
14. Bailey RR, Roberts AP, Gower PL, et al: Prevention of urinary tract infection with low-dose nitrofurantoin. *Lancet* 1971; 2:1112.
15. Bakshandeh K, Lynne C, Carrion H: Vesicoureteral reflux and end stage renal disease. *J Urol* 1976; 116:427.
16. Bartlett RC, Treiber N: Clinical significance of mixed bacterial culture of urine. *Am J Clin Pathol* 1984; 82:319.
17. Beeson PB, Rowley D: The anticomplementary effect of kidney tissue: Its association with ammonia production. *J Exp Med* 1959; 110:685.
18. Bentzen A, Vejlsgaard R: Asymptomatic bacteriuria in elderly subjects. *Dan Med Bull* 1980; 27:101.
19. Berg AO, Heidrich FE, Fihn SD, et al: Establishing the cause of genitourinary symptoms in women in a family practice: Comparison of clinical examination and comprehensive microbiology. *JAMA* 1984; 251:620.
20. Bille J, Glauser MP: Protection against chronic pyelonephritis in rats by suppression of acute suppuration: Effect of colchicine and neutropenia. *J Infect Dis* 1982; 146:220.
21. Bjorksten B, Kaijser B: Interaction of human serum and neutrophils with Escherichia coli strains: Differences between strains isolated from urine or patients with pyelonephritis or asymptomatic bacteriuria. *Infect Immun* 1978; 22:308.
22. Blank E: Caliectasis and renal scars in children. *J Urol* 1973; 110:225.
23. Boineau FG, Lewy JE: Urinary tract infection in children: An overview. *Pediatr Ann* 1975; 64:515.
24. Brachman PS: Epidemiology of nosocomial infections, in Bennett JV, Brachman PS (eds): *Hospital Infections*. Boston, Little, Brown & Co, 1979, p 9.
25. Bran JL, Levison ME, Kaye D: Entrance of bacteria into the female urinary bladder. *N Engl J Med* 1972; 286:626.
26. Breitenbacher RB: Bacterial changes in the urine samples of patients with long-term indwelling catheters. *Arch Intern Med* 1984; 144:1585.
27. Brooks H, O'Grady F, McSherry A, et al: Uropathogenic properties of Escherichia coli in recurrent urinary tract infection. *J Med Microbiol* 1980; 13:57.
28. Bruce, A, Chan R, Pinkerton D, et al: Adherence of gram negative uropathogens to human uroepithelial cells. *J Urol* 1983; 130:293.
29. Brumfitt W, Davies BL, Rosser E: The urethral catheter as a cause of urinary tract infection in pregnancy and puerperium. *Lancet* 1961; 2:1059.
30. Brumfitt W, Percival A: Pathogenesis and laboratory diagnosis of nontuberculous urinary tract infection: A review. *J Clin Pathol* 1964; 17:482.

31. Brun C, Raschou F, Eriksen KR: Simultaneous bacteriologic studies or renal biopsies and urine, in Kass EH (ed): *Progress in Pyelonephritis*. Philadelphia, FA Davis Co, 1965, p 461.

32. Bryan CS, Reynolds KL: Community-acquired bacteremic urinary tract infection: Epidemiology and outcome. *J Urol* 1984a; 132:490.

33. Bryan CS, Reynolds KL: Hospital-acquired bacteremic urinary tract infection: Epidemiology and outcome. *J Urol* 1984b; 132:494.

34. Bryant RE, Sutcliffe MC, McGee FA: Human polymorphonuclear leukocyte function in urine. *Yale J Biol Med* 1973; 46:113.

35. Buckley RM, McGuckin M, MacGregor RR: Urine bacterial counts following sexual intercourse. *N Engl J Med* 1978; 298:321.

36. Bullen M, Kincaid-Smith P: Asymptomatic pregnancy bacteriuria: A follow-up study 4–7 years after delivery, in *Renal Infection and Renal Scarring*. Melbourne, Mercedes Publishing, 1970, p 33.

37. Burbige KA, Retik AB, Colodny AH, et al: Urinary tract infections in boys. *J Urol* 1984; 132:541.

38. Burke JP, Jacobson JA, Garibaldi RA, et al: Evaluation of daily meatal care with polyantibiotic ointment in prevention of urinary catheter-associated bacteriuria. *J Urol* 1983; 129:331.

39. Butler HK, Kunin CM: Evaluation of polymyxin catheter lubricant and impregnated catheters. *J Urol* 1968a; 100:560.

40. Butler HK, Kunin CM: Evaluation of specific antimicrobial therapy in patients while on closed catheter drainage. *J Urol* 1968b; 100:567.

41. Cardiff-Oxford Bacteriuria Study Group: Sequelae of covert bacteriuria in schoolgirls: A four year follow-up study. *Lancet* 1978; 1:889.

42. Carlson KJ, Mulley AG: Management of acute dysuria: A decision-analysis model of alternative strategies. *Ann Intern Med* 1985; 102:244.

43. Cattell WR, McSherry MA, Northeast A, et al: Periurethral enterobacterial carriage in pathogenesis of recurrent urinary infection. *Br Med J* 1974; 4:136.

44. Cattell WR, Sardeson JM, Sutcliffe MB, et al: Kinetics of urinary bacterial response to antibacterial agents, in O'Grady F, Brumfitt W (eds): *Urinary Tract Infection*. London, Oxford University Press, 1968, p 212.

45. Chan RCY, Bruce AW, Reid G: Adherence of cervical, vaginal, and distal urethral normal microbial flora to human uroepithelial cells and the inhibition of adherence of gram negative uropathogens by competitive exclusion. *J Urol* 1984; 131:596.

46. Chan RCY, Reid G, Irvin RT, et al: Competitive exclusion of uropathogens from human uroepithelial cells by Lactobacillus whole cells and cell wall fragments. *Infect Immun* 1985; 47:84.

47. Charlton CAC, Crowther A, Davies JG, et al: Three-day and ten-day chemotherapy for urinary tract infections in general practice. *Br Med J* 1976; 1:124.

48. Chernow B, Zaloga GP, Soldano S, et al: Measurement of urinary leukocyte esterase activity: A screening test for urinary tract infections. *Ann Emerg Med* 1984; 13:150.

49. Clark H, Ronald AR, Cutler RE, et al: The correlation between site of infection and maximal concentrating ability in bacteriuria. *J Infect Dis* 1969; 120:47.

50. Cobbs CG, Kaye D: Antibacterial mechanisms in the urinary bladder. *Yale J Biol Med* 1967; 40:93.

51. Cohen M: The first urinary tract infection in male children. *Am J Dis Child* 1976; 130:810.

52. Condie AP, Williams JD, Reeves DS, et al: Complications of bacteriuria in pregnancy, in O'Grady F, Brumfitt W (eds): *Urinary Tract Infection*. London, Oxford University Press, 1968, p 148.

53. Costello AJ, Blandy JP, Hately W: Percutaneous aspiration of renal cortical abscess. *Urology* 1983; 21:201.

54. Cotran TS, Vivaldi E, Zangwill DP, et al: Retrograde pyelonephritis in rats. *J Pathol* 1963; 43:1.

55. Counts GW, Stamm WE, McKevitt M, et al: Treatment of cystitis in women with a single dose of trimethoprim-sulfamethoxazole. *Rev Infect Dis* 1982; 4:484.

56. Cox CE, Hinman F Jr: Experiments with induced bacteriuria, vesical emptying and bacterial growth on the mechanism of bladder defense to infection. *J Urol* 1961; 86:739.

57. Cox CE, Lucy SS, Hinman F Jr: The urethra and its relationship to urinary tract infection: II. The urethral flora of the female with recurrent urinary infection. *J Urol* 1968; 99:632.

58. Cruz-Coke R, Parades L, Montengro A: Blood groups and urinary microorganisms. *J Med Genet* 1965; 2:185.

59. Daifuku R, Stamm WE: Association of rectal and periurethral colonization with urinary tract infection in patients with indwelling catheters. *JAMA* 1984; 252:2028.

60. Desautels RE, Chibaro EA, Lang RJ: Maintenance of sterility in urinary drainage bags. *Surg Gynecol Obstet* 1982; 154:838.

61. Dodge WF, West EF, Travis LB: Bacteriuria in school children. *Am J Dis Child* 1974; 127:364.

62. Dontas AS, Kasviki-Charvati P, Panayiotis CL, et al: Bacteriuria and survival in old age. *N Engl J Med* 1981; 304:939.

63. Dontas AS, Marketos SG, Papanayiotou PC, et al: Simplified water-loading test in bacteriuria. *Nephron* 1974; 12:121.

64. Dornbusch K, Toivanen P: Effect of trimethoprim or trimethoprim-sulfamethoxazole usage on the emergence of trimethoprim resistance in urinary tract pathogens. *Scand J Infect Dis* 1981; 13:203.

65. Drach GW: Prostatitis: Man's hidden infection. *Urol Clin North Am* 1975; 3:499.

66. Dudley MN, Barriere SL: Antimicrobial irrigations in the prevention and treatment of catheter-related urinary tract infection. *Am J Hosp Pharm* 1981; 38:59.

66a. Duguid JP, Smith IW, Dempster G, et al: Nonflagellar filamentous appendages ("fimbriae") and hemagglutinating activity in Bacterium coli. *J Pathol Bacteriol* 1955; 70:335.

67. Edwards D, Normand ICS, Prescod N, et al: Disap-

pearance of vesicoureteric reflux during long-term prophylaxis of urinary tract infection in children. *Br Med J* 1977; 2:285.

68. Elkins IB, Cox CE: Vaginal and urethral bacteriology of young women: I. Incidence of gram negative colonization. *J Urol* 1974; 111:88.

69. Engel G, Schaeffer AJ, Crayhack JT, et al: The role of excretory urography and cystoscopy in the evaluation and management of women with recurrent urinary tract infection. *J Urol* 1980; 123:90.

70. Fair WR, Couch J, Wehner N: The purification and assay of the prostatic antibacterial factor (PAF). *Biochem Med* 1973; 8:329.

71. Fair WR, Timothy MM, Millar MA, et al: Bacteriologic and hormonal observations of the urethra and vaginal vestibule in normal, premenopausal women. *J Urol* 1981; 104:426.

72. Fairley KG: The investigation and treatment of urinary tract infection. *Med J Aust* 1976; 2:305.

73. Fairley KG, Bond AG, Brown RB, et al: Simple test to determine the site of urinary tract infection. *Lancet* 1967; 2:427.

74. Fairley KG, Carson NE, Gutch RC, et al: Site of infection in acute urinary tract infection in general practice. *Lancet* 1971; 2:615.

75. Fang LST, Tolokoff-Rubin NE, Rubin RH: Efficacy of single-dose and conventional amoxicillin therapy in urinary tract infection lcoalized by the antibody-coated bacteria technique. *N Engl J Med* 1978; 298:413.

76. Fasth A, Ahlstedt S, Hanson LA, et al: Cross reaction between Tamm-Horsfall glycoprotein and Escherichia coli. *Int Arch Allergy Appl Immunol* 1980; 63:303.

77. Fierer J, Ekstrom M: An outbreak of Providencia stuartii urinary tract infections. *JAMA* 1981; 245:1553.

78. Filly R: Ultrasonography, in Friedland GW, Filly R, Goris ML, et al (eds): *Uroradiology: An Integrated Approach.* New York, Churchill Livingstone, 1983, p 311.

79. Finegold SM, Miller L, Merrill SL, et al: Significance of anaerobic and capnophilic bacteria isolated from the urinary tract, in Kass EH (ed): *Progress in Pyelonephritis.* Philadelphia, FA Davis Co, 1965, p 159.

80. Finn DJ, Palestrant AM, DeWolf WC: Successful percutaneous management of renal abscess. *J Urol* 1982; 127:425.

81. Forsum U, Jhelm E, Jonsell G: Antibody-coated bacteria in the urine of children with urinary tract infections. *Acta Paediatr Scand* 1976; 65:639.

82. Fowler JE, Mariano M: Longitudinal studies of prostatic fluid immunoglobulin in men with bacterial prostatitis. *J Urol* 1984; 131:363.

83. Fowler JE, Pulaski ET: Excretory urography, cystography, and cystoscopy in the evaluation of women with urinary tract infection. *N Engl J Med* 1981; 304:462.

84. Fowler JE Jr, Stamey TA: Studies of introital colonization in women with recurrent infections: VII. The role of bacterial adherence. *J Urol* 1977; 117:472.

85. Freedman L: Chronic pyelonephritis at autopsy. *Ann Intern Med* 1967; 66:697.

86. Freedman LR: Natural history of urinary tract infection in adults. *Kidney Int* 1975; 8:S96.

87. Freedman LR: The epidemiology of urinary tract infections in Hiroshima. *Yale J Biol Med* 1965; 37:262.

88. Freedman LR, Andriole VA: The long term follow-up of women with urinary tract infection. *Proc 5th Int Congr Nephrol (Mexico City)* 1974; 3:20.

89. Freedman LR, Beeson PB: Experimental pyelonephritis: IV. Observations on infections resulting from direct inoculation of bacteria in different zones of the kidney. *Yale J Biol Med* 1958; 30:406.

90. Freeman RB, Smith WM, Richardson JA, et al: Long term therapy for chronic bacteriuria in men; US Public Health Service Cooperative Study. *Ann Intern Med* 1975; 81:133.

91. Gaches CGC, Miller KW, Roberts BM, et al: The Bristol pyelonephritis registry: 10 years on. *Br J Urol* 1976; 47:721.

92. Gallagher DJ, Montgomerie JZ, North JD: Acute infections of the urinary tract and the urethral syndrome in general practice. *Br Med J* 1965; 5435:622.

93. Gargan RA, Brumfitt W, Hamilton-Miller JMT: Antibody-coated bacteria in urine: Criterion for a positive test and its value in defining a higher risk of treatment failure. *Lancet* 1983; 2:704.

94. Garibaldi RA, Burke JP, Britt MR, et al: Meatal colonization and catheter-associated bacteriuria. *N Engl J Med* 1980; 303:316.

95. Garibaldi RA, Burke JP, Dickman ML, et al: Factors predisposing to bacteriuria during indwelling urethral catheterization. *N Engl J Med* 1974; 291:215.

96. Gerzof SG, Gale ME: Computed tomography and ultrasonography for diagnosis and treatment of renal and retroperitoneal abscesses. *Urol Clin North Am* 1982; 9:185.

97. Giamerellou H, Kosmidis J, Leonidas M, et al: A study of the effectiveness of rifampin in chronic prostatitis caused mainly by Staphylococcus aureus. *J Urol* 1982; 128:321.

98. Gillenwater JY, Harrison RB, Kunin CM: Natural history of bacteriuria in schoolgirls: A long-term case-control study. *N Engl J Med* 1979; 30:396.

99. Gillespie L: The diaphragm: An accomplice in recurrent urinary tract infections. *Urology* 1984; 14:25.

100. Gillespie WA, Linton KB, Miller A, et al: The diagnosis, epidemiology and control of urinary infection in urology and gynecology. *J Clin Pathol* 1960; 13:187.

101. Gladstone JL, Robinson CG: Prevention of bacteriuria resulting from indwelling catheters. *J Urol* 1968; 99:458.

102. Gleckman R: A critical review of the antibody-coated bacteria test. *J Urol* 1979; 122:770.

103. Gleckman R, Blagg N, Hibert D, et al: Acute pyelonephritis in the elderly. *South Med J* 1982; 75:557.

104. Gleckman R, Crowley M, Natsios GA: Recurrent urinary tract infections in men: An assessment of contemporary treatment. *Am J Med Sci* 1980; 279:31.

105. Gocke TM: *Infection in the Abnormal Host.* New York, Yorke Medical, 1980.

106. Goldberg LM, Vosti KL, Rantz LA: Microflora of the

urinary tract examined by voided and aspirated urine culture, in Kass EH (ed): *Progress in Pyelonephritis.* Philadelphia, FA Davis Co, 1965, p 545.

107. Gordon DL, McDonald PJ, Bruno A, et al: Diagnostic criteria and natural history of catheter-associated urinary tract infection after prostatectomy. *Lancet* 1983; 2:1269.

108. Govan D: Investigation and management of urinary tract infection in children. *Urol Clin North Am* 1974; 1:397.

109. Gower PE: A long term study of renal function in patients with radiological pyelonephritis and other allied radiological lesions, in Brumfitt W, Asscher AW (eds): *Urinary Tract Infection.* London, Oxford University Press, 1973, p 74.

110. Gower PE, Haswell B, Sidaway ME, et al: Follow-up of 164 patients with bacteriuria of pregnancy. *Lancet* 1968; 1:990.

111. Grainger RG, Longstaff AJ: Xanthogranulomatous pyelonephritis: A reappraisal. *Lancet* 1982; 1:1398.

112. Greenberg RN, Sanders CV, Lewis AC: Single-dose therapy for urinary tract infection with cefaclor. *Am J Med* 1981; 71:841.

113. Gross PH, Neu HC, Asuropokee P, et al: Deaths from nosocomial infections: Experience in a university hospital and a community hospital. *Am J Med* 1980; 63:219.

114. Gruneberg RN, Leigh DA, Brumfitt W: Relation of bacteriuria to acute pyelonephritis, prematurity and fetal mortality. *Lancet* 1969; 2:1.

115. Gutman LT, Turck M, Peterdorf RG, et al: Significance of bacterial variants in urine of patients with chronic bacteriuria. *J Clin Invest* 1965; 44:1945.

116. Guyer PB, Delaney DJ: Overdistensibility of the female urinary tract. *Br J Radiol* 1972; 45:392.

117. Guze LB, Beeson PB: Experimental pyelonephritis: I. Effect of ureteral ligation on the course of bacterial infection in the kidney of the rat. *J Exp Med* 1956; 104:803.

118. Guze LB, Goldner BH, Kalmanson GM: Pyelonephritis. I. Observation on the course of chronic non-obstructed enterococcal infection in the kidney of the rat. *Yale J Biol Med* 1961; 33:372.

119. Guze LB, Silverblatt F, Montgomerie J, et al: Lack of significance of pili in experimental ascending Escherichia coli pyelonephritis. *Scand J Infect Dis* 1983; 15:57.

120. Hagberg L, Hull R, Hull S, et al: Contribution of adhesion to bacterial persistence in the mouse urinary tract. *Infect Immun* 1983; 40:265.

121. Hagberg L, Jodal U, Korhonene TK, et al: Adhesion, hemagglutination and virulence of Escherichia coli causing urinary tract infections. *Infect Immun* 1981; 31:564.

122. Hanson LA, Fasth A, Jodal U, et al: Biology and pathology of urinary tract infection. *J Clin Pathol* 1981; 34:695.

123. Harber M, Mackenzie R, Chick S, et al: Lack of adherence to epithelial cells by freshly isolated urinary pathogens. *Lancet* 1982; 1:586.

124. Harber M, Mackenzie R, Asscher A: Comments on the role of bacterial adherence in the pathogenesis of urinary tract infections. *Contrib Nephrol* 1984; 39:273.

125. Harbord RB, Gruneberg RN: Treatment of urinary tract infection with a single-dose of amoxicillin, cotrimoxazole or trimethoprim. *Br Med J* 1981; 303:409.

126. Harding GKM, Buckwold FJ, Marrie TJ, et al: Prophylaxis of recurrent urinary tract infection in female patients. *JAMA* 1979; 242:1975.

127. Hardy JD, Furnell PM, Brumfitt W: Comparison of sterile bag, clean catch, and suprapubic aspiration in the diagnosis of urinary tract infection in early childhood. *Br J Urol* 1976; 48:279.

128. Harris RE, Gilstrap LC, Pretty A: Single-dose antimicrobial therapy for asymptomatic bacteriuria during pregnancy. *Obstet Gynecol* 1982; 59:546.

129. Harris RE, Thomas VL, Shelokov A: Asymptomatic bacteriuria in pregnancy: Antibody-coated bacteria renal function and intrauterine growth retardation. *Am J Obstet Gynecol* 1975; 126:20.

130. Hellebusch AA: Renal papillary necrosis: A urological emergency. *JAMA* 1969; 210:1098.

131. Hellerstein S, Kennedy E, Nussbaum L, et al: Localization of the site of urinary tract infections by means of antibody-coated bacteria in the urinary sediment. *J Pediatr* 1978; 92:188.

132. Hepinstal RH: Experimental pyelonephritis: Ascending infection of the rat kidney by organisms residing in the urethra. *Br J Exp Pathol* 1964; 45:436.

133. Hirsh DD, Fainstain V, Musher DM: Do condom catheter collecting systems cause urinary tract infections? *JAMA* 1979; 242:340.

134. Hjelm EM: Local cellular immune response in ascending urinary tract infection: Occurrence of T-cells, immunoglobulin-producing cells, and Ia-expressing cells in rat urinary tract tissue. *Infect Immun* 1984; 44:627.

135. Hoddick W, Jeffrey RB, Goldberg HI, et al: CT and sonography of severe renal and perirenal infections. *AJR* 1983; 14:517.

136. Hodson J, Maling TMJ, McManamon PS, et al: Reflux nephropathy. *Kidney Int* 1975; 8:550.

137. Hoeprich P: Culture of the urine. *J Lab Clin Med* 1960; 56:899.

138. Holmgren J, Smith JW: Immunological aspects of urinary tract infections. *Prog Allergy* 1975; 18:289.

139. Hooton TM, O'Shaughnessy EJ, Clowers, D, et al: Localization of urinary tract infection in patients with spinal cord injury. *J Infect Dis* 1984; 150:85.

140. Hooton TM, Running K, Stamm WE: Single-dose therapy for cystitis in women: A comparison of trimethoprim-sulfamethoxazole, amoxicillin, and cyclacillin. *JAMA* 1985; 253:387.

141. Hovelius B, Mardh PA: On the diagnosis of coagulase-negative staphylococci with emphasis on Staphylococcus saprophyticus. *Acta Pathol Microbiol Scand* 1977; 85B:427.

142. Hovelius B, Mardh PA: Staphylococcus saprophyticus as a common cause of urinary tract infection. *Rev Infect Dis* 1984; 6:328.

143. Hughes C, Hacker J, Roberts A, et al: Hemolysin production as a virulence marker in symptomatic and

asymptomatic urinary tract infections caused by Escherichia coli. *Infect Immun* 1983; 39:546.

144. Hurwitz SR, Kessler WO, Alazrake NP, et al: Gallium-67 imaging to localize urinary infections. *Br J Radiol* 1976; 49:156.

145. Hutler HN, Borchardt KA, Mahood JA, et al: Localization of catheter-induced urinary tract infections: Interpretation of bladder washout and antibody-coated bacteria tests. *Nephron* 1984; 38:48.

146. Iwahi T, Abe Y, Nakao M, et al: Role of type 1 fimbriae in the pathogenesis of ascending urinary tract infection induced by Escherichia coli in mice. *Infect Immun* 1983; 39:1307.

147. Jarvinen A, Sandholm M: Urinary oligosaccharides inhibit adhesion of E. coli onto canine urinary tract epithelium. *Invest Urol* 1980; 17:443.

148. Jensen J, Uehling DT, Kim K, et al: Enhanced immune response in the urinary tract of the rat following vaginal immunization. *J Urol* 1984; 132:164.

149. Johnson CW, Smythe CM: Renal function in patients with chronic bacteriuria. *South Med J* 1969; 62:81.

150. Jones SR: The current status of urinary tract infection localization by the detection of antibody-coated bacteria in the urinary sediment, in Gilbert DA, Sanford JP (eds): *Infectious Diseases: Current Topics*. New York, Grune & Stratton, 1979, vol 1, p 97.

151. Jones SR, Smith JW, Sanford JP: Localization of urinary tract infection detection of antibody-coated bacteria in urine sediment. *N Engl J Med* 1974; 290:591.

152. Kaijser B, Larson P, Olling S, et al: Protection against acute pyelonephritis caused by Escherichia coli in rats, using isolated capsular antigen conjugated to bovine serum albumin. *Infect Immun* 1983; 39:142.

153. Kallenius G: The p^k antigen as receptor for the hemagglutination of pyelonephritic Escherichia coli. *FEMS Microbiol Lett* 1980; 7:297.

154. Kallenius G, Mollby R, Svensson SB, et al: Occurrence of P-fimbriated Escherichia coli in urinary tract infections. *Lancet* 1981; 2:1369.

155. Kallenius G, Winberg J: Bacterial adherence to periurethral epithelial cells in girls prone to urinary tract infection. *Lancet* 1978; 3:540.

156. Kallenius G, Winberg J: Urinary tract infections treated with a single dose of short-acting sulphonamide. *Br Med J* 1979; 2:1175.

157. Kass EH: Asymptomatic infections of the urinary tract. *Trans Assoc Am Physicians* 1956; 69:57.

158. Kass EH: Bacteriuria and pyelonephritis of pregnancy. *Trans Assoc Am Physicians* 1959; 72:257.

159. Kass EH: Bacteriuria and the pathogenesis of pyelonephritis. *Lab Invest* 1960a; 9:110.

160. Kass EH: The role of asymptomatic bacteriuria in the pathogenesis of pyelonephritis, in Quinn EL, Kass EH (eds): *Biology of Pyelonephritis*. Boston, Little, Brown & Co., 1960b, p 399.

161. Kass EH, Savage W, Santamarina BAG: The significance of bacteriuria in preventive medicine, in Kass EH (ed): *Progress in Pyelonephritis*. Philadelphia, FA Davis Co, 1965, p 3.

162. Kass EH, Zangwill DP: Principles in the long-term management of chronic infection of the urinary tract, in Quinn EL, Kass EH (eds): *Biology of Pyelonephritis*. Boston, Little, Brown & Co, 1960c, p 663.

163. Kaye D: Antibacterial activity of human urine. *J Clin Invest* 1968; 47:2374.

164. Kaye D: Urinary tract infection in the elderly. *Bull NY Acad Med* 1980; 56:209.

165. Kaye D, Rocha H: Urinary concentrating ability in early experimental pyelonephritis. *J Clin Invest* 1970; 49:1426.

166. Kelsey MC, Mead MG, Gruneberg RN, et al: Relationship between sexual intercourse and urinary tract infection in women attending a clinic for sexually transmitted diseases. *J Med Microbiol* 1979; 12:511.

167. Kennedy RP, Plorde JJ, Petersdorf RG: Studies on the epidemiology of Escherichia coli infections: IV. Evidence for a nosocomial flora. *J Clin Invest* 1965; 44:193.

168. Kessler WO, Gittes RF, Hurwitz SR, et al: Gallium-67 scans in the diagnosis of pyelonephritis. *West J Med* 1974; 121:91.

169. Kinane DF, Blackwell CC, Brettle RP, et al: ABO blood group, secretor state and susceptibility to recurrent urinary tract infection in women. *Br Med J* 1982; 285:7.

170. Kincaid-Smith P, Bullen M: Bacteriuria in pregnancy. *Lancet* 1965; 1:395.

171. Kitchner R, Rosenblatt R: Sonographically guided percutaneous renal interventional procedures. *JAMA* 1984; 251:3126.

172. Komaroff AL, Pass TM, McCue JD, et al: Management strategies for urinary and vaginal infections. *Arch Intern Med* 1978; 138:1069.

173. Komaroff AL: Acute dysuria in women. *N Engl J Med* 1984; 310:378.

174. Korhonen T, Vaisanen V, Saxen H, et al: P-antigen recognizing fimbriae from human uropathogenic Escherichia coli strains. *Infect Immun* 1982; 37:286.

175. Kostiak AI, Nyren P, Jokinen P, et al: Prospective study on the appearance of antibody-coated bacteria in patients with an indwelling urinary catheter. *Nephron* 1981; 30:279.

176. Kraft JK, Stamey TA: The natural history of symptomatic recurrent bacteriuria in women. *Medicine (Baltimore)* 1977; 56:55.

177. Kreger DE, Craven DE, Carling PC, et al: Gram negative bacteremia: III. Reassessment of etiology, epidemiology and ecology in 612 patients. *Am J Med* 1980; 68:332.

178. Krieger JN, Kaiser DL, Wenzel RP: Nosocomial urinary tract infections: Secular trends, treatment and economics in a university hospital. *J Urol* 1983a; 130:102.

179. Krieger JN, Kaiser DL, Wenzel RP: Urinary tract etiology of bloodstream infections in hospitalized patients. *J Infect Dis* 1983b; 148:57.

180. Kuhlemeier KV, Lloyd LK, Stover SL: Localization of upper and lower urinary tract infections in patients with neurogenic bladders. *Model Systems' SCI Digest* 1982; 4:29.

181. Kunin CM: New method of detecting urinary tract infections. *Urol Clin North Am* 1975a; 2:423.

182. Kunin CM: The natural history of recurrent bacteriuria in school girls. *N Engl J Med* 1970a; 282:1443.

183. Kunin CM: Urinary tract infections: Flow charts (algorithms) for detection and treatment. *JAMA* 1975b; 233:458.

184. Kunin CM: Urinary tract infections in children. *Hosp Pract* 1976; 11:91.

185. Kunin CM, Finkelberg Z: Oral cephalexin and ampicillin: Antimicrobial activity, recovery in urine, and persistence in blood of uremic patients. *Ann Intern Med* 1970b; 72:349.

186. Kunin CM, Polyak F, Postel E: Periurethral bacterial flora in women: Prolonged intermittent colonization with Escherichia coli. *JAMA* 1980; 243(2):134.

187. Kunin CM: Genitourinary infections in the patients at risk: Extrinsic factors. *Am J Med* 1984; 76:131.

188. Kurdydyk LM, Kelly K, Harding GKM, et al: Role of cervicovaginal antibody in the pathogenesis of recurrent urinary tract infection in women. *Infect Immun* 1980; 29:76.

189. Kwasnik I, Klauber G, Tilton RC: Clinical and laboratory evaluation of the antibody-coated bacteria test in children. *J Urol* 1979; 121:658.

190. Lapides J, Diokno AC, Lowe BS, et al: Follow-up on unsterile, intermittent self-catheterization. *J Urol* 1974; 111:184.

191. Latham RH, Running K, Stamm WE: Urinary tract infection in young adult women caused by Staphylococcus saprophyticus. *JAMA* 1983; 250:3063.

192. Layton GT, Smithyman AM: The effect of oral and combined parenteral/oral immunization against an experimental Escherichia coli urinary tract infection in mice. *Clin Exp Immunol* 1983; 54:305.

193. Leffler H, Svanborg-Eden C: Glycolipid receptors for uropathogenic Escherichia coli binding to human erythrocytes and uroepithelial cells. *Infect Immun* 1981; 34:930.

194. Leibovici L, Alpert G, Laon A, et al: Single-dose treatment of urinary tract infection in young women: Data indicating a high rate of recurrent infection during a short follow-up. *Isr J Med Sci* 1984; 20:257.

195. Leigh D, Gruneberg R, Brumfitt W: Long-term follow-up of bacteriuria in pregnancy. *Lancet* 1968; 1:503.

196. Levison ME, Kaye D: Management of urinary tract infection, in Kaye E (ed): *Urinary Tract Infection and Its Management.* St. Louis, CV Mosby Co, 1972, p 188.

197. Levison SP, Levison ME: The effect of indomethacin and sodium meclofenamate on the renal concentrating defect in experimental enterococcal pyelonephritis in rats. *J Clin Lab Med* 1976; 88:958.

198. Levison SP, Pitsakis PG, Levison ME: Free water reabsorption during saline diuresis in experimental enterococcal pyelonephritis in rats. *J Clin Lab Med* 1980; 99:474.

199. Lincoln K, Lidin-Janson G, Winberg J: Resistant urinary infections resulting from changes in resistance pattern of faecal flora induced by sulphonamide in hospital environment. *Br Med J* 1979; 3:305.

200. Lindsey JO, Martin WT, Sonnenwirth AC, et al: An outbreak of nosocomial Proteus rettgeri urinary tract infection. *Am J Epidemiol* 1976; 103:261.

201. Lomberg H, Jodal V, Svanborg-Eden C, et al: P₁ blood group and urinary tract infection. *Lancet* 1981; 1:551.

202. Lomberg H, Hanson LA, Jacobsson B, et al: Correlation of P₁ blood group, vesicoureteral reflux, and bacterial attachment in patients with recurrent pyelonephritis. *N Engl J Med* 1983; 308:1189.

203. Lomberg H, Hellstrom M, Jodal U, et al: Virulence-associated traits in Escherichia coli causing first and recurrent episodes of urinary tract infection in children with or without vesicoureteral reflux. *J Infect Dis* 1984; 150:561.

204. Ludwig P, Buckwold F, Harding G, et al: Single-dose therapy of acute cystitis in adult females: Prospective randomized comparison of four regimens, in Nelson JD, Grassi C (eds): *Current Chemotherapy and Infectious Diseases.* Washington, DC, 1980, ASM, p 1297.

205. MacGregor ME, Freeman P: Childhood urinary infection associated with vesicoureteric reflux. *Q J Med* 1975; 175:481.

206. Maki DG, Hennekens, CK, Phillips CW, et al: Nosocomial urinary tract infection with Serratia marcescens: An epidemiologic study. *J Infect Dis* 1973; 128:579.

207. Margileth AM, Pedreira FA, Hirschman GH, et al: Urinary tract bacterial infections: Symposium on Pediatric Nephrology. *Pediatr Clin North Am* 1976; 23:71.

208. Marrie TJ, Harding GKM, Ronald AR: Anaerobic and aerobic urethral flora in healthy females. *J Clin Microbiol* 1978; 8:67.

209. Marrie TJ, Harding GK, Ronald AR, et al: Influence of mucoid antibody-coating of Pseudomonas aeruginosa. *J Infect Dis* 1979; 139:357.

210. Marsh FP, Murray M, Panchamia P: The relationship between bacterial cultures of the vaginal introitus and urinary infection. *Br J Urol* 1975; 44:368.

211. Martin CM, Bookrajian EN: Bacteriuria prevention after indwelling urinary catheterization: A controlled study. *Arch Intern Med* 1962; 110:703.

212. Martin CM, Vaquer F, Meyers MS, et al: Prevention of gram negative rod bacteremia associated with indwelling urinary tract catheterization, in Sylvester JC (ed): *Antimicrobial Agents and Chemotherapy—1963.* Washington, DC, ASM, 1964, p 617.

213. Maskell R: Trimethoprim resistance in gram negative urinary pathogens. *Br Med J* 1983a; 286:1182.

214. Maskell R, Pead L, Sanderson RA: Fastidious bacteria and the urethral syndrome: A 2 year clinical and bacteriological study of 51 women. *Lancet* 1983b; 2:1277.

215. Mattsby-Baltzer I, Hanson LA, Kaijser B, et al: Experimental Escherichia coli ascending pyelonephritis in rats: Changes in bacterial properties and the immune response to surface antigens. *Infect Immun* 1982; 35:639.

216. Mayrer AR, Miniter P, Andriole VT: Immunopathogenesis of chronic pyelonephritis. *Am J Med* 1983; 75:59.

217. McCabe WR, Jackson GG: Treatment of pyelonephritis:

Bacterial drug and host factors in success or failure among 252 patients. *N Engl J Med* 1965; 272:1037.

218. McDonough WD, Sandler CM, Benson GS: Acute focal bacterial nephritis: Focal pyelonephritis that may simulate renal abscess. *J Urol* 1981; 126:670.

219. McFayden IR, Eykyn SS: Suprapubic aspiration of urine in pregnancy. *Lancet* 1968; 1:1112.

220. McKerrow W, Davidson-Lamb N, Jones PF: Urinary tract infection in children. *Br Med J* 1984; 289:299.

221. Meares EM: Bacterial prostatitis vs. "prostatosis": A clinical and bacteriologic study. *JAMA* 1973; 224:1372.

222. Meares EM: Long-term therapy of chronic bacterial prostatitis with trimethoprim-sulfamethoxazole. *Can Med Assoc J* 1975b; 112:225.

223. Meares EM: Prostatitis: A review. *Urol Clin North Am* 1975a; 2:3.

224. Meares EM, Stamey TA: Bacteriologic localization patterns in bacterial prostatitis and urethritis. *Invest Urol* 1968; 5:492.

225. Merritt JL, Keys TF: Limitations of the antibody-coated bacteria test in patients with neurogenic bladders. *JAMA* 1982; 247:1723.

226. Michaels EK, Chmiel JS, Plotkin BJ, et al: Effect of D-mannose and D-glucose on Escherichia coli bacteriuria in rats. *Urol Res* 1983; 11:97.

227. Milazzo FH, Delisle GJ: Immunoglobulin A proteases in gram negative bacteria isolated from human urinary tract infections. *Infect Immun* 1984; 43:11.

228. Miller TE, North JD: Host response in urinary tract infections. *Kidney Int* 1974; 5:179.

229. Miller TE, Scott L, Stewart E, et al: Modification by suppressor cells and serum factors of the cell-mediated immune response in experimental pyelonephritis. *J Clin Invest* 1978; 61:964.

230. Miller TE: Immunomodulatory interactions of suppressor cells, cell-mediated immunity and cyclophosphamide in experimental pyelonephritis. *J Infect Dis* 1983; 148:1096.

231. Minuth JN, Masher DM, Thorsteinsonn SB: Inhibition of the antibacterial activity of gentamicin by urine. *J Infect Dis* 1976; 133:14.

232. Mogenson P, Hanson LK: Do intravenous urography and cystoscopy provide important information in otherwise healthy women with recurrent urinary tract infection. *Br J Urol* 1983; 55:26.

233. Montplaisir S, Cote P, Martinelli B, et al: Localization du site de l'infection urinaire chez l'enfant par la recherche de bacteries recouvretes d'anticorps. *Can Med Assoc J* 1976; 115:1096.

234. Monzon OT, Ory EM, Dobson HL, et al: A comparison of bacterial counts of the urine obtained by needle aspiration of the bladder, catheterization and midstream-voided methods. *N Engl J Med* 1958; 259:764.

235. Mulholland SG: Lower urinary tract antibacterial defense mechanisms. *Invest Urol* 1979; 17:93.

236. Mulholland SG, Mooreville M, Parsons CL: Urinary tract infection and P blood group antigens. *Urology* 1984; 24:232.

237. Mundt KA, Polk BF: Identification of site of urinary tract infections by antibody-coated bacteria assay. *Lancet* 1979; 2:1172.

238. Murray T, Goldberg MD: Etiologies of chronic interstitial nephritis. *Ann Intern Med* 1975; 82:453.

239. Musher DM, Griffith DP, Yawn D, et al: Role of urease in pyelonephritis resulting from urinary tract infection with Proteus. *J Infect Dis* 1975; 131:177.

240. Musher DM, Thorsteinsson SB, Airola VM: Quantitative urinalysis: Diagnosing urinary tract infection in men. *JAMA* 1976; 236:2069.

241. Neumann CG, Pryles CV: Pyelonephritis in infants and children: Autopsy experiences at the Boston City Hospital 1933–1960. *Am J Dis Child* 1962; 104:215.

242. Nicolle LE, Harding GKM, Preiksaitis J, et al: The association of urinary tract infection with sexual intercourse. *J Infect Dis* 1982; 146:579.

243. Norden CW: Significance and management of bacteriuria of pregnancy, in Kaye D (ed): *Urinary Tract Infection and Its Management*. St. Louis, CV Mosby Co, 1972, p 171.

244. Norden CW, Kass EH: Bacteriuria of pregnancy: A critical appraisal. *Annu Rev Med* 1968; 19:431.

245. Ofek I, Mirelman D, Sharon N: Adherence of Escherichia coli to human mucosal cells mediated by mannose receptors. *Nature* 1977; 265:623.

246. Ofek I, Mosek A, Sharon N: Mannose-specific adherence of Escherichia coli freshly excreted in the urine of patients with urinary tract infections and of isolates subcultured from the infected urine. *Infect Immun* 1981; 34:708.

247. O'Grady F, Gauci CL, Watson BW, et al: In vitro models simulating conditions of bacterial growth in the urinary tract, in O'Grady F, Brumfitt W (eds): *Urinary Tract Infection*. London, Oxford University Press, 1968, p 80.

248. O'Hanley PD, Lark D, Falkow S, et al: A globoside binding E. coli pilus vaccine prevents pyelonephritis (abstract). *Clin Res* 1983; 31:372A.

249. Orskov I, Ferencz A, Orskov F: Tamm-Horsfall protein or uromucoid in the normal urinary slime that traps type 1 fimbriated Escherichia coli. *Lancet* 1980; 1:887.

250. Paavonen J, Kousa M, Saikku P, et al: Examination of men with nongonococcal urethritis and their sexual partners for Chlamydia trachomatis and Ureaplasma urealyticum. *Sex Transm Dis* 1978; 5:93.

251. Parsons CL, Greenspan C, Mulholland SG: The primary antibacterial defense mechanism of the bladder. *Invest Urol* 1975; 123:72.

252. Parsons CL, Greenspan C, Moore S, et al: Role of surface mucin in primary antibacterial defense of bladder. *Urology* 1977; 9:48.

253. Parsons CL, Mulholland SG, Anwar H: Antibacterial activity of bladder surface mucin duplicated by exogenous glycosaminoglycan (heparin). *Infect Immun* 1979; 24:552.

254. Parsons CL, Schrom SH, Hanno P, et al: Bladder surface mucin: Examination of possible mechanism for its antibacterial effect. *Invest Urol* 1978; 6:196.

255. Parsons CL, Schmidt JD: In vitro bacterial adherence to

vaginal cells of normal and cystitis prone women. *J Urol* 1980; 123:184.

256. Pawlowski JM, Bloxdorf JW, Kimmelstein P: Chronic pyelonephritis: A morphologic and bacteriologic study. *N Engl J Med* 1963; 268:965.

257. Pearson NJ, Towner KJ, McSherry AM, et al: Emergence of trimethoprim-resistant enterobacteria in patients receiving long-term cotrimoxazole for the control of intractable urinary tract infection. *Lancet* 1979; 1:1205.

258. Perry A, Ofek I, Silverblatt FJ: Enhancement of mannose-mediated stimulation of human granulocytes by type 1 fimbriae aggregated with antibodies on Escherichia coli surfaces. *Infect Immun* 1983; 39:1332.

259. Pezzlo MT: Automated methods for detection of bacteriuria. *Am J Med* 1983; Infectious Diseases Symposium Supplement A:71.

260. Pfau A, Sacks T: The bacteria flora of the vaginal vestibule, urethra and vagina in the normal premenopausal woman. *J Urol* 1977; 118:292.

261. Pfau A, Sacks T, Englestein D: Recurrent urinary tract infections in premenopausal women: Prophylaxis based upon an understanding of the pathogenesis. *J Urol* 1983; 129:1153.

262. Pitchon H, Glassock R, Kalmanson GM, et al: Experimental pyelonephritis: Effect of T-cell deficiency on the course of hematogenous enterococcal pyelonephritis in the mouse. *Am J Pathol* 1984; 115:25.

263. Platt R: Diagnosis and empiric therapy of urinary tract infection in the seriously ill patients. *Rev Infect Dis* 1983; 5:S65.

264. Platt R, Polk BF, Murdock B, et al: Mortality associated with nosocomial urinary tract infection. *N Engl J Med* 1982; 307:637.

265. Platt R, Polk BF, Murdock B, et al: Reduction of mortality associated with nosocomial urinary tract infection. *Lancet* 1983; 1:893.

266. Pollack HM, Banner MP, Arger PH, et al: Comparison of computed tomography and ultrasound in the diagnosis of renal masses, in Rosenfield AT (ed): *Genitourinary Ultrasonography*, vol 2 in *Clinics in Diagnostic Ultrasound*. New York, Churchill Livingstone, 1979, p 25.

267. Pontzer RE, Krieger RE, Boscia JA, et al: Single-dose cefonicid therapy for urinary tract infections. *Antimicrob Agents Chemother* 1983; 23:814.

268. Popky GL, Pollack HW: Radiologic evaluation of patients with urinary tract infection, in Kaye D (ed): *Urinary Tract Infection and Its Management*. St Louis, CV Mosby, 1972, p 84.

269. Pratt V, Hatala M, Venesova D, et al: Pathogenicity of various strains of Escherichia coli for the intact rabbit kidney and the effect of repeated passage on renal tissue, in Kass EH (ed): *Progress in Pyelonephritis*. Philadelphia, FA Davis, 1965, p 135.

270. Randolph MF, Greenfield M: The incidence of asymptomatic bacteriuria and pyuria in infancy. *J Pediatr* 1964; 65:57.

271. Ratner J, Thomas VA, Sanford BA: Bacteria specific an-

tibody in the urine of patients with acute pyelonephritis and cystitis. *J Infect Dis* 1981; 143:404.

272. Rauschkolb EN, Sandler CM, Patel S, et al: Computed tomography of renal inflammatory diseases. *J Comput Assist Tomogr* 1982; 6:502.

273. Reeves DS, Brumfitt W: Localization of urinary tract infection, in O'Grady F, Brumfitt W (eds): *Urinary Tract Infection*. London, Oxford University Press, 1968, p 53.

274. Reid T, Brooks HJL, Bacon DF: In vitro attachment of Escherichia coli to human uroepithelial cells: Variation in receptivity during the menstrual cycle and pregnancy. *J Infect Dis* 1983; 148:412.

275. Rene P, Dinolfo M, Silverblatt FJ: Serum and urogenital antibody response to Escherichia coli pili in cystitis. *Infect Immun* 1982; 37:749.

276. Riedasch G, Ritz E, Dreikorn K, et al: Antibody-coating of urinary bacteria in transplanted patients. *Nephron* 1978; 20:267.

277. Riedasch G, Heck P, Rautenberg E, et al: Does low urinary SIgA predispose to urinary tract infection? *Kidney Int* 1983; 23:759.

278. Roberts AP, Phillips R: Bacteria causing symptomatic urinary tract infection or bacteriuria. *j Clin Pathol* 1979; 32:492.

279. Roberts JA, Hardaway K, Kaack B, et al: Prevention of pyelonephritis by immunization with P-fimbriae. *J Urol* 1984; 131:602.

280. Rocha H: Pathogenesis and clinical manifestations of urinary tract infection, in Kaye D (ed): *Urinary Tract Infection and Its Management*. St Louis, CV Mosby Co. 1972, p 6.

281. Rocha H, Fekety FR: Acute inflammation in the renal cortex and medulla following thermal injury. *J Exp Med* 1964; 119:131.

282. Rocha H, Guze LB, Freedman LR, et al: Experimental pyelonephritis: III. The influence of localized injury in different parts of the kidney on susceptibility to bacillary infection. *Yale J Biol Med* 1958; 30:341.

283. Rolleston GI, Maling TMJ, Hodson CJ: Intrarenal reflux and the scarred kidney. *Arch Dis Child* 1974; 7:531.

284. Romano JM, Kaye D: UTI in the elderly: Common yet atypical. *Geriatrics* 1981; 36:113.

285. Ronald AR, Boutrous P, Mourtada H: Bacteriuria localization and response to single-dose therapy in women. *JAMA* 1976; 23:1854.

286. Ronald AR, Cutler RE, Turck M: Effect of bacteriuria on renal concentrating mechanisms. *Ann Intern Med* 1969; 70:723.

287. Ronald AR, Harding GKM, Mathias R, et al: Prophylaxis of recurring urinary tract infection in females: A comparison of nitrofurantoin with trimethoprim-sulfamethoxazole. *Can Med Assoc J* 1975; 112:135.

288. Rosenstock J, Smith LP, Gurney M, et al: Comparison of single-dose tetracycline HCl to conventional therapy of urinary tract infection. Abstract 522. Proceedings of the 23rd Interscience Conference on Antimicrobial Agents and Chemotherapy, Las Vegas, 1983, p 178.

289. Rubin M: Effect of catheter replacement on bacterial

counts in urine aspirated from indwelling catheters. *J Infect Dis* 1980a; 142:291.

290. Rubin RH, Fang LST, Jones SR, et al: Single-dose amoxicillin therapy for urinary tract infection. *JAMA* 1980b; 244:561.

291. Salfatierra O, Tangaho E: Reflux as a cause of end stage kidney disease: Report of 32 cases. *J Urol* 1977; 117:441.

292. Sanford JP: Hospital-acquired urinary tract infections. *Ann Intern Med* 1964; 60:90.

293. Sanford JP: Urinary tract symptoms and infection. *Annu Rev Med* 1975; 26:485.

294. Savard-Fenton M, Fenton BW, Reller LB, et al: Single-dose amoxicillin therapy with follow-up urine culture: Effective initial management for acute uncomplicated urinary tract infection. *Am J Med* 1984; 78:808.

295. Schaeffer AJ, Jones JM, Dunn JK: Association of in vitro Escherichia coli adherence to vaginal and buccal epithelial cells with susceptibility of women to recurrent urinary tract infection. *N Engl J Med* 1981; 304:1062.

296. Schaeffer AJ, Amundsen SK, Schmidt LN: Adherence of Escherichia coli to human urinary tract epithelial cells. *Infect Immun* 1979; 24:753.

297. Schaeffer AJ, Jones JM, Flynn SS: Prophylactic efficacy of cinoxacin in recurrent urinary tract infection: Biological effect on the vaginal and fecal flora. *J Urol* 1982; 127:1128.

298. Schaeffer AJ, Chmiel JS: Urethral meatal colonization in the pathogenesis of catheter-associated bacteriuria. *J Urol* 1983a; 130:1096.

299. Schaeffer AJ, Chmiel JS, Duncan JL, et al: Mannose-sensitive adherence of Escherichia coli to epithelial cells from women with recurrent urinary tract infections. *J Urol* 1984; 131:906.

300. Schaeffer AJ, Radvany RM, Chmiel JS: Human leukocyte antigens in women with recurrent urinary tract infections. *J Infect Dis* 1983b; 148:604.

301. Schardijn GHC, von Epps LWS, Pau W, et al: Comparison of reliability of tests to distinguish upper from lower urinary tract infections. *Br Med J* 1984; 289:284.

302. Schechter H, Leonard CD, Scribner BH: Chronic pyelonephritis as a cause of renal failure in dialysis candidates. *JAMA* 1971; 216:514.

303. Schmidt JD, Patterson MC: Needle biopsy study of chronic prostatitis. *J Urol* 1966; 96:519.

304. Schoolnick G, Labigne-Roussel AF, Lark D, et al: Cloning and expression of a fimbrial adhesin (AFA-I) responsible for P blood group-independent, mannose-resistant hemagglutination from a pyelonephritic Escherichia coli strain. *Infect Immun* 1984; 46:251.

305. Schultz HJ, McCaffrey LA, Keys TF, et al: Acute cystitis: A prospective study of laboratory tests and duration of therapy. *Mayo Clin Proc* 1984; 59:391.

306. Seguar JW, Opitz J, Green LF: Prostatosis, prostatitis, or pelvic floor tension myalgia? *J Urol* 1979; 122:168.

307. Shapiro ED, Wald EF: Single-dose amoxicillin treatment of urinary tract infection. *J Pediatr* 1981; 99:989.

308. Siegel SR, Sokoloff B, Siegel B: Asymptomatic and symptomatic urinary tract infection in infancy. *Am J Dis* 1973; 125:45.

309. Silverblatt FS, Weinstein R, Rene P: Protection against experimental pyelonephritis by antibodies to pili. *Scand J Infect Dis (Suppl)* 1982; 33:79.

310. Silverblatt FS: Host-parasite interaction in the rat renal pelvis: A possible role of pili in the pathogenesis of pyelonephritis. *J Exp Med* 1974; 140:1696.

311. Smalley DL, Dittman AW: Use of leukocyte esterase-nitrate activity as predictive assays of significant bacteriuria. *J Clin Microbiol* 1983; 18:1256.

312. Smellie JM, Normand ICS: Experience of follow-up of children with urinary tract infection, in O'Grady F, Brumfitt W (eds): *Urinary Tract Infection.* London, Oxford University Press, 1968, p 123.

313. Smellie JM, Edwards D, Hunter N, et al: Vesicoureteral reflux and renal scarring. *Kidney Int* 1975a; 8:565.

314. Smellie JM, Normand ICS: Bacteriuria, reflux, and renal scarring. *Arch Dis Child* 1975b; 50:581.

315. Smellie JM, Gruneberg RN, Leakey A, et al: Long-term low-dose cotrimoxazole in prophylaxis of childhood urinary tract infection: Clinical aspects. *Br Med J* 1976; 2:203.

316. Smith IW: Role of suppressor cells in experimental pyelonephritis. *J Infect Dis* 1980; 142:199.

317. Sobel JD, Vardi Y: Scanning electron microscopy study of Pseudomonas aeruginosa in vivo adherence to rat bladder epithelium. *J Urol* 1982; 128:414.

318. Sobel JD, Muller G: Pathogenesis of bacteriuria in elderly women—Role of E. coli adherence to vaginal epithelial cells. *Gerontology* 1984; 38:682.

319. Stahl GE, Topf P, Fleisher GR, et al: Single-dose treatment of uncomplicated urinary tract infections in children. *Ann Emerg Med* 1984; 13:705.

320. Stamey TA: The role of introital enterobacteria in recurrent urinary infections. *J Urol* 1973a; 109:467.

321. Stamey TA, Bushby SRM, Bragonse J: The concentration of trimethoprim in prostatic fluid: Nonionic diffusion or active transport. *J Infect Dis* 1973b; 128(suppl):686.

322. Stamey TA, Condy M, Mihara G: Prophylactic efficacy of nitrofurantoin macrocrystals and trimethoprim-sulfamethoxazole in urinary infection. *N Engl J Med* 1977; 296:780.

323. Stamey TA, Fair WR, Timothy MM, et al: Antibacterial nature of prostatic fluid. *Nature* 1968a; 218:444.

324. Stamey TA, Wehner N, Mihara G, et al: The immunologic basis of recurrent bacteriuria: Role of cervicovaginal antibody in enterobacterial colonization of the introital mucosa. *Medicine* 1978; 57:47.

325. Stamey TA: Prostatitis. *J R Soc Med* 1981; 74:22.

326. Stamey TA, Fair WR, Timothy MM, et al: Serum versus urinary antimicrobial concentrations in cases of urinary tract infections. *N Engl J Med* 1974; 291:1159.

327. Stamey TA, Govan DE, Palmer JM: The localization and treatment of urinary tract infections: The role of bactericidal urine levels as opposed to serum levels. *Medicine* 1965; 44:1.

328. Stamey TA, Meares EM, Winningham EF: Diffusion of antibiotics from plasma into prostatic fluid. *Nature* 1968b; 219:139.

329. Stamey TA, Timothy MM: Studies of introital colonization in women with recurrent urinary infections: I. The role of vaginal pH. *J Urol* 1975; 114:261.

330. Stamey TA, Timothy MM, Millar M, et al: Recurrent urinary infections in adult women: The role of introital enterobacteria. *Calif Med* 1971; 115:1.

331. Stamm WE: Guidelines for prevention of catheter-associated urinary tract infections. *Ann Intern Med* 1975; 82:386.

332. Stamm WE: Management of the acute urethral syndrome. *Drug Ther* 1982; 12:155.

333. Stamm WE: Measurement of pyuria and its relation to bacteriuria. *Am J Med* 1983a; 75:53.

334. Stamm WE: Prevention of urinary tract infections. *Am J Med* 1984; 76:148.

335. Stamm WE: Single-dose treatment of cystitis. *JAMA* 1980c; 244:591.

336. Stamm WE, Counts GW, Wagner KF, et al: Antimicrobial prophylaxis of recurrent urinary tract infection: Double-blind placebo control trial. *Ann Intern Med* 1980b; 92:770.

337. Stamm WE, McKevitt M, Counts GW, et al: Is antimicrobial prophylaxis of urinary tract infections cost-effective? *Ann Intern Med* 1981b; 94:251.

338. Stamm WE, Running K, McKevitt M, et al: Treatment of the acute urethral syndrome. *N Engl J Med* 1981a; 304:956.

339. Stamm WE, Turck M: Urinary tract infection. *Adv Intern Med* 1983b; 28:141.

340. Stamm WE, Wagner KF, Amsel R, et al: Causes of the acute urethral syndrome in women. *N Engl J Med* 1980a; 303:409.

341. Stenqvist K, Sandberg T, Ahlstedt S, et al: Effects of subinhibitory concentrations of antibiotics and antibodies on the adherence of *Escherichia coli* to human uroepithelial cells in vitro. *Scand J Infect Dis (Suppl)* 1982; 33:104.

342. Stonk RB, Maki DG: Bacteriuria in the catheterized patient: What quantitative level of bacteriuria is relevant? *N Engl J Med* 1984; 311:560.

343. Susin M, Becker EL: The pathology of pyelonephritis, in Kaye D (ed): *Urinary Tract Infection and Its Management.* St Louis, CV Mosby, 1972, p 65.

344. Svanborg-Eden C, Hanson LA, Jodal U, et al: Variable adherence to normal human urinary tract epithelial cells of Escherichia coli strains associated with various forms of urinary tract infection. *Lancet* 1976; 2:490.

345. Svanborg-Eden C, Leffler H: Glycosphingolipids of human urinary tract epithelial cells as possible receptors for adhering Escherichia coli bacteria. *Scand J Dis (Suppl)* 1980; 24:144.

346. Svanborg-Eden C, Hagberg L, Hanson LA, et al: Adhesion of Escherichia coli in urinary tract infection. *CIBA Found Symp* 1981; 80:161.

347. Svanborg-Eden C, Marild S, Korhonen TK: Adhesion inhibition by antibodies. *Scand J Infect Dis (Suppl)* 1982a; 33:72.

348. Svanborg-Eden C, Gotschlich EC, Korhonen TK, et al: Aspects of structure and function of pili of uropathogenic E. coli. *Prog Allergy* 1983; 33:189.

349. Svanborg-Eden C, Freter R, Hagberg L, et al: Inhibition of experimental ascending urinary tract infection by an epithelial cell-surface analogue. *Nature* 1982b; 298:560.

350. Svanborg-Eden C, Svennerholm AM: Secretory immunoglobulin A and G antibodies prevent adhesion of Escherichia coli to human urinary tract epithelial cells. *Infect Immun* 1978; 22:790.

351. Svanson S, Kallenius G, Korhonen TK, et al: Initiation of clinical pyelonephritis—the role of P fimbriated mediated bacterial adhesion. *Contrib Nephrol* 1984; 39:252.

352. Svensson R, Larsson P, Lincoln K, et al: Low dose trimethoprim prophylaxis in long-term control of chronic recurrent urinary tract infection. *Scand J Infect Dis* 1982; 14:139.

353. Thomas VL, Forland M: Antibody coated bacteria in urinary tract infections. *Kidney Int* 1982; 21:1.

354. Thomas VL, Harris RE, Gilstrap LC III, et al: Antibody-coated bacteria in the urine of obstetrical patients with acute pyelonephritis. *J Infect Dis* 1975; 131(suppl):557.

355. Thomas VL, Shelokov A, Forland M: Antibody-coated bacteria in the urine and the site of urinary tract infection. *N Engl J Med* 1974; 290:588.

356. Thompson RL, Haley CE, Searay MA, et al: Catheter-associated bacteriuria: Failure to reduce attack rates using periodic instillations of a disinfectant into urinary drainage system. *JAMA* 1984; 251:747.

357. Thorley JD, Jones SR, Sanford JP: Perinephric abscess. *Medicine (Baltimore)* 1974; 53:441.

358. Thornton GF, Lytton B, Andriole VT: Bacteriuria during indwelling catheter drainage. *JAMA* 1966; 195:179.

359. Thysell H: Evaluation of chemical and microscopical methods for mass detection of bacteriuria. *Acta Med Scand* 1969; 185:393.

360. Tolkoff-Rubin NE, Weber D, Fang LST, et al: Single-dose therapy with trimethoprim-sulfamethoxazole for urinary tract infection in women. *Rev Infect Dis* 1982; 4:444.

361. Tolkoff-Rubin NE, Wilson ME, Zuromskis P, et al: Single-dose amoxicillin therapy of acute uncomplicated urinary tract infection in women. *Antimicrob Agents Chemother* 1984; 25:626.

362. Tullus K, Horlin K, Svenson S, et al: Epidemic outbreaks of acute pyelonephritis caused by nosocomial spread of P fimbriated Escherichia coli in children. *J Infect Dis* 1984; 150:728.

363. Turck M, Goffe B, Petersdorf RG: Bacteriuria of pregnancy. *N Engl J Med* 1962a; 266:857.

364. Turck M: Localization of the site of recurrent urinary tract infection in women. *Urol Clin North Am* 1975; 2:433.

365. Turck M, Goffe B, Petersdorf RG: The urethral catheter and urinary tract infection. *J Urol* 1962b; 88:834.

366. Turck M, Ronald AR, Petersdorf RG: Relapse and reinfection in chronic bacteriuria: II. The correlation be-

tween site of infection and pattern of recurrence in chronic bacteriuria. *N Engl J Med* 1968; 278:422.

367. Turck M, Stamm WE: Nosocomial infection of the urinary tract. *Am J Med* 1981; 70:651.

368. Uehling D, Wolf I: Enhancement of the bladder defense mechanism by immunization. *Invest Urol* 1969; 6:520.

369. Uehling DT, Mitzutani K, Bolish E: Effect of immunization on bacterial adherence to uroepithelium. *Invest Urol* 1978; 16:145.

370. Vainrub B, Musher DM: Lack of effect of methenamine in suppression of, or prophylaxis against chronic urinary infection. *Antimicrob Agents Chemother* 1977; 12:625.

371. Vaisanen V, Tallgren L, Makela P, et al: Mannose-resistant haemagglutination and P antigen recognition are characteristic of Escherichia coli causing primary pyelonephritis. *Lancet* 1981; 2:1369.

372. Vaisanen-Rhen V, Elo J, Vaisanen E, et al: P-fimbriated clones among uropathogenic Escherichia coli strains. *Infect Immun* 1984; 43:149.

373. Vardi Y, Meshulam T, Obedeanu N, et al: In vivo adherence of Pseudomonas aeruginosa to rat bladder epithelium. *Proc Soc Exp Biol Med* 1983; 172:449.

374. Varian S, Cooke E: Adhesive properties of Escherichia coli from urinary tract infections. *J Med Microbiol* 1980; 13:111.

375. Vivaldi E, Munoz J, Cotran R, et al: Factors affecting the clearance of bacteria within the urinary tract, in Kass EH (ed): *Progress in Pyelonephritis*. Philadelphia, FA Davis Co, 1965.

376. Vosti KL, Goldberg LM, Monto AS, et al: Host-parasite interaction in patients with infections due to Escherichia coli from intestinal and extraintestinal sources. *J Clin Invest* 1964; 43:2377.

377. Vosti KL: Recurrent urinary tract infection: Prevention by prophylactic antibiotics after sexual intercourse. *JAMA* 1975; 231:934.

378. Warren JH, Platt R, Thomas RJ, et al: Antibiotic irrigation and catheter-associated urinary tract infection. *N Engl J Med* 1978; 299:570.

379. Warren JH, Terry JW, Hoopes JM, et al: A prospective microbiologic study of bacteriuria in patients with chronic indwelling urethral catheters. *J Infect Dis* 1982; 146:719.

380. Waters WE, Elwood PC, Asscher AW, et al: Clinical significance of dysuria in women. *Br Med J* 1970; 2:75.

381. Weidner W, Brunner H, Krause W: Quantitative culture of Ureaplasma urealyticum in patients with chronic prostatitis or prostatosis. *J Urol* 1980; 124:62.

382. Weyrauch HM, Bassett JB: Ascending infection in an artificial urinary tract: An experimental study. *Stanford Med Bull* 1951; 9:25.

383. Whitaker RH, Sherwood T: Another look at diagnostic pathways in children with urinary tract infection. *Br Med J* 1984; 288:839.

384. Wientzen RL, McCracken GH Jr, Petruska ML, et al: Localization and therapy of urinary tract infections of childhood. *Pediatrics* 1979; 63:467.

385. Williams JD, Leigh DA, Rosser E, et al: The organization and results of a screening program for the detection of bacteriuria of pregnancy. *J Obstet Gynaecol Br Emp* 1965; 72:327.

386. Williams JD, Reevers DS, Condie AD, et al: The treatment of bacteriuria in pregnancy, in O'Grady F, Brumfitt W (eds): *Urinary Tract Infection*. London, Oxford University Press, 1968, p 160.

387. Winickoff RN, Wilner SI, Gall G, et al: Urine culture after treatment of uncomplicated cystitis in women. *South Med J* 1981; 74:165.

388. Wong ES, McKevitt M, Running K, et al: Management of recurrent urinary tract infections with patient-administered single-dose therapy. *Ann Intern Med* 1985; 102:302.

389. Wong ES, Stamm WE, Fennell CL: Urinary tract infection among women attending a clinic for sexually transmitted diesases. *Sex Transm Dis* 1984; 11:18.

390. Work J, Andriole VT: Tamm-Horsfall protein antibody in patients with end-stage kidney disease. *Yale J Biol Med* 1980; 53:133.

391. Zinner SH, Sabath LD, Casey JI, et al: Erythromycin plus alkalinization in treatment of urinary infections. *Lancet* 1971; 1:1267.

392. Zinner S, Kass EH: Long term (10 to 14 years) follow-up of bacteriuria of pregnancy. *N Engl J Med* 1971; 285:820.

Chapter 6

Management of Pain in Urologic Patients

JOHN C. ROWLINGSON, M.D.

The phenomenon of pain is multifaceted. Pain is described differently by different people, although the stimulus may be the same, and it can have varying significance even to the same person under different circumstances. Pain is said to be the most common complaint of patients presenting to the health care professional for treatment. Therefore, few medical specialists will escape having patients with this ubiquitous experience. It is crucial that pain be effectively treated not only because of the pathophysiologic impact on the body and the adverse effects on morbidity and mortality but also because of the incalculable suffering and agony it can provoke.

Pain has most recently been defined as "an unpleasant sensory and emotional experience associated with actual or potential tissue damage or described in terms of such damage" (Mersky, 1979). This definition eliminates the archaic, shortsighted view of pain as a purely physical phenomenon and mandates that all pain now be considered as representing an intense interaction of numerous kinds of input. This concept must be extended to patients whether their pain is acute as with epididymitis or ureteral obstruction by stones, chronic as with intractable cystitis or perineal pain, or secondary to devastating illness such as cancer.

Acute pain is usually a biologically necessary physiologic response (Bonica, 1982). Its teleologic function is to alert the organism to ongoing or impending tissue damage and to provoke an escape reaction that subsequently reduces or eliminates the painful stimulus. Given by evolution the ability to vocalize and communicate, man additionally can use descriptions of pain for diagnostic purposes if the etiology is not obvious, i.e., external trauma, penile pain secondary to Peyronie's disease, or testicular pain that is associated with torsion. That the "pain" associated with a broken heart or the loss of a loved one can be as unpleasant and excruciating as pain secondary to physical damage is a simple reminder that all pain need not be directly associated with true pathologic changes.

When pain is not acute, but lingers despite treatment, it is called chronic pain (Bonica, 1985). Chronic pain may be associated with an ongoing, progressive disease such as arthritis or cancer, or it may seem to defy etiologic diagnosis. Yet, the complaints of pain and their consequences in the patient's life can be disruptive or devastating (Levine, 1984; Reuler et al., 1980). Arbitrarily, chronic pain is defined as that which lasts longer than six months or beyond the normal course of the disease or injury (Bonica, 1985).

Distinguishing between acute and chronic pain is crucial, because this will necessarily influence the appropriate evaluation scheme and the therapeutic options considered. Much experience in the management of acute episodes of pain is gained during the training of most health care professionals. However, serious difficulties arise when the principles of treatment of acute pain are applied to chronic pain problems. Treating all pain the same regardless of its etiology is not logical, nor will it reliably result in successful pain control. Other significant differences between acute and chronic pain are listed in Table 6–1.

The concepts about the mechanisms of pain are constantly evolving (Loeser, 1985; Melzack and Wall, 1965; Melzack, 1982). As the appreciation for neuroanatomy and neurophysiology grows, the understanding about valid definitions of pain, the mechanisms of specific pain syndromes, and their particular treatment will be forthcoming. The single biggest clinical problem yet encountered is that there is no all-conclusive diagnostic test to document how much pain a patient has. In accepting the contemporary concept that pain is at least a sensory and emotional experience the expectation that a blood test or a chemical assay can be developed to reflect accurately the patient's total pain may be unfulfilled. The lack of a scientific, mathematical method to quantitate pain denies physicians the assessment of the pain in the numerical terms with which most are comfortable. What is crudely measurable is the patient's behavior that results from the pain. However, pain is a

TABLE 6–1.

Characteristics of Acute and Chronic Pain

ACUTE PAIN	CHRONIC PAIN
Common experience in clinical medicine such that many physicians can become proficient at treating it	Less skill is acquired in treating chronic pain effectively because physicians use the same treatments as for acute pain and they do not work
Acute pain signals the body that damage is occurring or is about to occur	Has less signal function because many factors other than tissue damage influence the complaints of "pain"
A biologically useful symptom, since acute pain usually reflects tissue damage	Complaints have less specific association with ongoing tissue damage
Anxiety frequently associated with the complaints of acute pain	Anxiety is less likely—frustration and some form of depression are more likely
Diagnosis is usually straightforward	Diagnosis is more vague because of the many factors that influence the complaints of "pain"
Short-term treatment course implied (days)	Long-term treatment likely (weeks to months)
Drug abuse unlikely because of the short treatment period	Polypharmacy can be a major problem because of multiple symptoms and the protracted treatment course
Cure of pain expected	Cure of all the pain may be unrealistic

personal, subjective experience that will influence the patient's behavior (use of drugs, activity, etc.). The shortcomings of conclusions about the patient's level of pain documented by behavior are obvious—patients who are of stoic constitution will persevere in the face of tremendous pain, whereas those with a more histrionic personality can provoke excessive treatment for seemingly minor complaints.

Other problems with pain include the lack of experimental models for suitable research outside the clinical arena, too little crosstalk among personnel of many specialty backgrounds in which diverse research is done, inadequate teaching about pain management in health care professional curricula, and, unfortunately, insufficient application in everyday practice of what is known about pain control. These factors will contribute to the perpetuation of pain mismanagement until solutions are found. All health care professionals must acknowledge these issues in pain and work to resolve them.

THE EVALUATION OF THE PATIENT WITH PAIN

Because pain is some combination of physical and nonphysical components, the appropriate evaluation scheme takes into account more than just the history of the pain and its character. One must be thorough, because most complaints of pain have a vast differential diagnosis, yet efficient, so as not to delay treatment and extend the patient's suffering. Clearly, the evaluation

process must result in a diagnosis as to the etiology of the pain complaints. Only after reaching this vital plateau will suitable treatment be likely to follow. In the absence of an all-diagnostic test for pain, the physician must rely on the traditional tools of diagnosis—history, physical examination, and laboratory test results (Rowlingson and Stehling, 1982).

When pain complaints are perceived as acute, the patient is questioned about his general medical condition, the particulars of the pain (circumstances of onset, description, location, impact on activity level, associated symptoms, and factors that make the pain better or worse), the pertinent surgical history, and the treatment record to date. Insight into whether this episode is new or is not entirely different from past occurrences of pain is necessary. One must get to know the patient to assess whether the pain represents progression of a chronic disease or complications of past treatment.

If the pain complaints are more chronic, the historical investigation must be broadened to include not only the general medical, pain, and treatment histories mentioned above, but also the psychosocial, occupational, and economic consequences of the pain. Usually, in patients with chronic pain complaints, their "pain" is a blend of pathologic changes, frustration, desperation, work interruption or cessation, income compromise, and family stress or distress. The evaluation routine must elicit the presence or absence of these potent modifiers of the pain, lest the database on which the patient's management will be predicated be incomplete.

In the busy practice of urology a previsit questionnaire or other standardized format for information collection streamlines the history taking, allows pertinent information to be obtained on a consistent basis, and permits a time-efficient, focused interview. The time thus saved can be profitably invested in giving the patient an unpressured opportunity to discuss what is important to him, i.e., compromise of sexual function, painful urination, or explaining the rationale for the therapeutic plan based on the diagnosis.

Physical examination of the patient with pain allows for the association of the patient's complaints with anatomical possibilities and helps to identify pathologic abnormalities of the genitourinary and related structures. Serial examinations may detect progression of illness (as in patients with cancer) or regression, as should occur with treatment. Specific aspects of appropriate physical examination techniques are mentioned in other chapters.

The final major source of information will be the results of laboratory tests and diagnostic investigations. The patient's past records must be reviewed to minimize the duplication of such tests and prevent unnecessary discomfort and economic hardship. The indications for specific tests such as renal or prostatic biopsy, cystoscopy, IVP, etc., are well-known and discussed elsewhere.

As the patient's complaints of pain become chronic, the history becomes long, and findings on physical examination are few, the usefulness of the laboratory for diagnostic purposes may diminish. One must acknowledge that diagnostic tests for every conceivable cause of pain do not exist. However, negative results do *not* necessarily mean that there is no pain. A more realistic approach to chronic pain may be to order only those tests whose results will truly change individual management. As with the physical examination, some laboratory tests, when performed in a serial fashion, may be helpful in documenting the elimination of pathologic reasons for pain and/or improvement with treatment.

One must not forget that the patient is entitled to his opinion as to the cause of the complaint. Physicians must not be too quick to disregard the patient's expressed beliefs. For example, a middle-aged man with the onset of new perineal pain may become concerned that the pain is secondary to prostatic cancer or may be using pain as an excuse to avoid intercourse with an overweight or unattractive spouse. A tactful, compassionate physician will be able to sort out fact vs. fantasy, discuss the diagnostic possibilities based on the results of the workup, and explain the rationale for proposed treatment. No physician can shirk the obligation of listening to the pain patient's concept of his illness and then dealing with it in a sensitive, forthright manner.

The stated purpose of the evaluation scheme is that of establishing a diagnosis. By the modern definition of pain, this must take into account not only the physical causes of pain but also the nonphysical factors that can so strongly influence the patient's complaints. Once the diagnosis has been made, the treatment program that addresses the specific needs of the patient can be created. For example, in the patient who complains of low back pain postoperatively, an analgesic may not be required if questioning discloses that the patient has a past history of postural low back pain for which a simple change in position will suffice to treat the complaint. A patient with the recent onset of bladder pain who during careful interview reveals his father is dying of cancer of the prostate and has a similar pain will warrant a workup different from another patient with fewer possible etiologic causes. The evaluation scheme should determine the likely etiology of the pain and any extraneous modifying influences. Furthermore, one may gain insight into the unique significance the patient is attaching to the pain and more clearly understand what the pain means to him.

Because pain is more than pathologic changes in the tissues of the body, the lack of a physical cause does not necessarily imply the presence of a psychological one (Giddon and Rabinovitz, 1980). Every patient is entitled to have an emotional reaction to the physical stress and suffering that results from pain. Classic research reveals that observance of and reaction to these additional influences on the pain will significantly reduce its intensity and impact. For example, Egbert et al. have shown that postoperative pain control begins in the preoperative period (Egbert et al., 1963, 1964). Particular attention to providing the surgical patient with information about his pain, its cause, and what can and will be done about it lessens the physical and emotional consequences of the pain, allows more cooperation with postoperative stir-up regimens and can result in earlier discharge from the hospital. Cancer is obviously not a mental disease, but who can deny the influence of a patient's mood and spirit, indeed the very will to live, on the course of the disease? The frustration and despair that accompany most chronic pain are yet another common example of nonphysical consequences on physical pain that must be identified in the thorough, comprehensive evaluation.

GENERAL COMMENTS ABOUT PAIN TREATMENT

It is a matter of philosophy as to whether pain treatment is necessary because of the pathophysiologic consequences it can have or because of the anguish and distress it causes the human condition. Generally, the treatment of (acute) postoperative pain seems to exemplify the former and the management of chronic pain the latter. Patients with pain due to cancer encompass both schools of thought.

Before treatment is contemplated, a diagnosis has been made, thought has been given to identifying modifying factors in the pain, and the distinction between acute and chronic pain has been clarified. With this frame of reference, one is ready to explain to the patient the diagnosis and the rationale for the options of therapy. There must be an understanding that many of the options so successfully used in acute pain control make chronic pain worse. Thus, treating all pain in the same manner is not practical or effective. Even the goals of treatment are fundamentally different. Before specific pain problems are considered, general comments about the therapeutic options for pain are necessary.

Patients who have had experience with acute pain have the realistic expectation that their pain will be completely eliminated after a period of treatment. Indeed, this is a model of pain we are all familiar with from our upbringing. Past experience tells us that after a short period of attention, sympathy, and treatment, the symptoms of illness subside, and life returns to normal. However, when patients develop a condition with which chronic pain or intermittent relapsing pain is associated, the expectation of complete cure is frequently not met. Despair and frustration can become intermingled with the physical impact of the chronic pain and the primary disease. In these patients new understanding must be attained, because treatment is aimed first at decreasing the frequency and intensity of the pain but not necessarily totally eliminating it. If the pain can be decreased, adjustments can be made in the use of potent medication, considerations of surgery to relieve the pain rethought, and the utilization of therapies conceived in desperation minimized (Aronoff, 1982).

Furthermore, the ongoing treatment implied as being necessary to manage a chronic pain problem must accomplish more than the maximal decrease in pain possible. Since some of the patient's complaints may persist despite all treatment, some therapeutic energy must be reserved for helping the patient to cope with and understand his residual pain. This is accomplished largely by tactful, honest, informative discussions. The intent is to increase the patient's insight into the physical and nonphysical causative factors in the pain complaints, and, through encouragement and support, to increase the patient's functional status in spite of residual pain. It is helpful to use the following scheme when discussing treatment with the patients (Rowlingson and Stehling, 1982).

- Medications, surgery, nerve blocks, and stimulation techniques (i.e., TENS, or transcutaneous electrical nerve stimulation) are used to decrease pain.
- Generally, once the pain begins to diminish, it is expected that the patient will increase his physical activity. This could be manifested by cooperation with stir-up regimens in the postoperative period or in patients with chronic pain by their participating in rehabilitative physical exercise programs. If the patient admits to a decrease in the pain but does not manifest a more aggressive attitude about resumption of activity, further investigation into his reasons for the self-imposed restrictions is necessary.
- The pain may not improve at a consistent rate or disappear as fast as the patient feels it should. At all times during the treatment course the patient will benefit from encouragement and support. As progress is made with physical rehabilitation and the pain reduction plateaus, the patient with chronic pain complaints may be confronted with the reality that all of the pain is not going to disappear. It is at this time that formal psychological-supportive therapy techniques may be necessary to encourage the patient's continued compliance with the treatment program and provide alternate methods for coping with the residual pain.
- The ultimate goal of treatment is rehabilitation, as defined as a restoration of former capacity, condition of health, or useful activity (*Webster's New Collegiate Dictionary*, 1980). In patients with acute pain, such as those with fleeting episodes of renal colic or postoperative pain, the pain will only be a short-term deterrent, and a return of the patient to full activity is expected. In patients with chronic pain the complete rehabilitative process is much more complicated and the likelihood of success is not unexpectedly diminished.

The basic concept in pain management is that as the pain becomes more complicated, any one treatment modality taken alone is not likely to be "the" answer. Rather, the percentage decrease in pain provided by a single option can be supplemented by using other alternative treatments that also contribute a percentage decrease in the pain, and the total reduction is then likely to be of significance to the patient.

OPTIONS FOR PAIN MANAGEMENT

MEDICATIONS.—The innumerable drug products that are available, the ease of prescribing them, and the propensity of most patients to "take a pill" for their complaints make medications the most common form of treatment for pain. The fact that the prescribing physician does not need to know exactly why the patient is complaining of pain to use medications further enhances the likelihood of their administration. Unfortunately, many cases of acute pain are inadequately treated because of an overemphasis on narcotic drug dangers and side effects relative to their beneficial effects (Marks and Sachar, 1973; Angell, 1982). The unfounded fears of inducing addiction in patients limits the dosages and duration so that the patients suffer (Porter and Jick, 1980). Acute pain treatment can continue up to 21 days, after which the need for potent medication will pass or the lack of improvement will prompt a workup that clarifies the diagnosis and suggests more effective therapy.

Table 6–2 lists some of the common analgesic drugs and categorizes them by the severity of the pain to be treated (Amadio, 1984; Inturrisi, 1984; Stimmel, 1985). The physician must be careful to match the potency of the drug chosen for treatment with the degree of the patient's pain. The projected length of treatment will also influence the specific drug chosen, since long-term use of potent analgesics can have detrimental consequences. The correct dosage is that which, when repeated often enough, results in the desired analgesic effect yet avoids drug-related side effects. The drug therapy is started, its effect on the pain assessed, the dosage titrated to produce an acceptable level of analgesia, and the maximal dose limited by the onset of intolerable side effects.

Patients with chronic pain problems, on the other hand, are frequently passively overtreated (Ready et al., 1982; Turner et al., 1982). That is, they are provided habituating medications long past their therapeutic usefulness, perhaps because it is something that the physician thinks can be done for the patient that will not prolong the office visit time frame. As time passes and the pain continues, the prescription drugs (barbiturates, narcotics, sedative/hypnotics) cause familiar drug-induced side effects, not the least of which is a chemically induced depression. Medication use may be supplemented by the patient with alcohol. The subsequent significant impairment of cognitive function, manifested by short attention span and inability to concentrate, will interfere with the patient's ability to handle and react to environmental stressors.

When such drugs are used for months at a time, tolerance develops, pain persists, and the patient has only drug side effects to show for the therapy, it is logical to withdraw the potent medications and substitute drugs that are safer over a prolonged period (Khantzian and McKenna, 1979). This is done with two assumptions: (1) the more benign medications, even in combinations, are not expected to take away all of the pain; and (2) the concurrent use of therapeutic options (other than medications alone) in a coordinated program of treatment will likely result in a significant reduction of the patient's pain (Aronoff, 1982). Patients may occasionally require formal, in-hospital detoxification but more commonly need only to be advised to slowly decrease the use of the potent medications over time, with careful, frequent follow-up.

Alternative drugs may include acetaminophen, nonsteroidal anti-inflammatory drugs (NSAIDs) (Simon and Mills, 1980), tricyclic antidepressants (TCAs) (Hollister, 1978; Rosenblatt et al., 1984; Ward et al., 1979), phenothiazines (Kanner, 1983; Nathan, 1978), and anticonvulsants (Swerdlow, 1984) (Table 6–3). The prescription of any of these alternative medications aims to provide

TABLE 6–2.

Common Analgesic Drugs

DRUG	DOSAGE REGIMEN*	
Mild pain		
Aspirin	325–975	mg every 4 hr
Acetaminophen	325–975	mg every 4 hr
Ibuprofen	300–400	mg every 4–6 hr
Indomethacin	25	mg every 6 hr
Naproxen	250	mg every 12 hr
Mild-moderate pain		
Codeine	65–120	mg every 4–6 hr
Propoxyphene	65–130	mg every 4–6 hr
Pentazocine	50	mg every 4–6 hr
Moderate-severe pain		
Oxycodone	5	mg every 3–4 hr
Meperidine	50	mg every 4 hr
Hydromorphone	2–4	mg every 4–6 hr

*All drugs given orally.

TABLE 6–3.

Alternative Drugs for Pain Management

DRUG	DOSAGE REGIMEN*
Acetaminophen	650–975 mg every 4 hr
Aspirin	650–975 mg every 4 hr
NSAIDs[†]	Dose varies depending on the specific drug chosen
Amitriptyline	50–150 mg 1 hour before sleep
Doxepin	50–150 mg 1 hour before sleep
Trazodone (Desyrel)	100–300 mg 1 hour before sleep
Fluphenazine dihydrochloride (Prolixin)	1 mg 2–3 times daily
Chlorpromazine HCl (Thorazine)	25 mg 2–3 times daily
Hydroxyzine	25 mg 2–3 times daily
Haloperidol	1 mg 2–3 times daily
Phenytoin sodium (Dilantin)	100–200 mg 3 times daily
Carbamazepine (Tegretol)	250 mg 2–3 times daily
Valproic acid (Depakene)	250 mg 2–3 times daily

*All drugs given orally.
[†]Nonsteroidal anti-inflammatory drugs.

maximal analgesia with less risk of drug misuse or habituation than that associated with routinely given analgesic drugs.

The NSAIDs can have potent effects on the kidney and therefore on patients with renal disease. As specialists with a unique orientation to renal function, urologists are in a crucial position to suggest safe pharmacologic treatment in patients with chronic pain and concurrent medical conditions that could be aggravated by the effects of NSAIDs (Henrich, 1983; Stillman et al., 1984). As a class, these drugs inhibit cyclooxygenase such that arachidonic acid in the cell membrane is not converted to prostaglandins. The prostaglandins have potent effects on the kidney at the afferent and efferent arterioles, mesangium of the glomerulus, distal convoluted tubule and collecting tubules. Therefore, when prostaglandin generation is blocked by the use of NSAIDs, the following effects may be seen:

- With the vasodilatation of the prostaglandins missing, unopposed vasoconstriction may cause an increase in renal vascular resistance and a decrease in renal blood flow to the point of producing acute renal failure secondary to ischemia.
- Fluid retention secondary to sodium retention.
- Hyperkalemia (caution should intervene in patients concurrently taking potassium-sparing diuretics)
- Hypertension
- An idiosyncratic reaction manifested by acute interstitial disease or nephrotic syndrome

Thus, all physicians should be careful when prescribing NSAIDs in patients with renal disease, nephrosis, depletion of plasma volume (as in patients taking diuretics), or systemic lupus erythematosus (SLE). Because of the potential fluid retention and electrolyte abnormalities, patients with a history of congestive heart failure or cirrhosis might also be at risk with NSAID therapy.

SURGERY.—Surgery is as much an option in pain control as anything else, so long as it is used in the logical place in the grand scheme of treatment. Many surgical procedures are done for specific pathologic indications and with realistic expectations at both ends of the scalpel. Declaration of appropriate surgical procedures is dealt with in other chapters. Suffice it to say that surgical procedures can be of tremendous benefit in relieving some pain, e.g., that secondary to ureteral obstruction or to tumor expansion with compression of nerves. However, intractable pain may develop after surgery, e.g., incisional neuralgias (Loeser, 1985), ilioinguinal neuropathy (Hameroff et al., 1981), pelvic or penile pain secondary to adhesion formation, and scar tissue and neuromas acting as persistent pain-generators in an operative area as healing occurs.

Selective neurosurgical procedures aimed at interrupting neural transmission pathways are of limited benefit to most patients with pain problems except those with short life spans associated with advanced cancer. These patients are not likely to outlive the beneficial effects of such surgery, which after weeks to months usually give way to more severe pain than that originally treated. The modern-day appreciation for the multisynaptic and neurochemical nature of pain impulse transmission and processing helps us understand why simple transection of the nervous system cannot provide permanent pain relief (Loeser, 1985).

NERVE BLOCKS.—Locally given anesthetic drugs of varying potency and duration can be used for diagnostic, prognostic, and therapeutic nerve block procedures. The epidural anesthetic performed for surgery can be continued into the early postoperative period if a catheter technique was used. In this way the local anesthetics have the advantage of providing continuous analgesia (Pflug et al., 1974). However, this is achieved at the expense of some persistent sensory loss and orthostatic hypotension secondary to the persistent sympathetic blockade. Also, there is the risk of a toxic reaction to the local anesthetic. These drugs are short acting, result in only temporary pain relief, and must be given by intermittent injection or continuous infusion. Traditional narcotic drugs now can be administered in the epidural space (to be discussed later), and provide high-quality analgesia that is not associated with any other neurosensory disruption and that lasts longer (12–24 hours) than that produced by local anesthetics (Bromage et al., 1980; Cousins and Mather, 1984). These techniques will be used increasingly in the future, especially when the mechanisms of their significant side effects are clarified and reliable methods for control or elimination of such effects are established.

The advantages of nerve blocks include their nonaddictive nature, being repeatable; having lower risks and fewer side effects than surgery; and provision of prompt, albeit temporary, relief of pain. Some clinical examples of nerve block use follow. Continuous epidural anesthesia (average duration, 32.4 hours) was reported by Romagnoli and Batra (1973) to have been effective in treating the pain of impacted ureteric stones in 75% of 22 patients. Patient acceptance was very good, secondary to the rapid relief of pain and its quality. Intercostal nerve blocks can be performed before closure of flank wounds to decrease early postoperative pain (Noller et al., 1977) or are done as diagnostic and therapeutic procedures in the management of troublesome postoperative intercostal neuralgias. Sympathetic nerve blocks may delineate a visceral etiology of a patient's pain complaints by selective blockade of the sympathetic nervous system and may be useful in ultimate pain control of selected problems (Boas, 1978). Paravertebral blocks at the T12–L2 levels may help decrease groin and scrotal pain.

Nerve block procedures in chronic pain may be part of the diagnostic or therapeutic program. For example, differential spinal blocks are advocated by some as beneficial in establishing the etiology (sympathetic, sensory, central, psychogenic) of chronic scrotal, testicular, or penile pain (Ghia et al., 1979; Winnie and Collins, 1968). The prompt decrease in pain wrought by nerve blocks may bias the patient's attitude about getting better, relieve the need to use narcotic medications, and encourage the patient to participate actively in rehabilitation programs. In experienced hands the complications from nerve blocks should be few. This becomes of paramount importance when consideration is given to the injection of neurolytic agents in the management of pain due to malignancy or in selected cases of chronic, nonmalignant pain. Simon et al. (1982) revealed the beneficial effects of selected transsacral blocks with phenol in a study group of 15 patients. All of the patients had bladder pain, frequency, and urgency associated with a painful, spastic, contracted bladder. Eleven of the 15 patients had interstitial cystitis. Once the dominant sacral root was identified by a decrease in symptoms following a block with local anesthetic, phenol was subsequently used to provide significant symptomatic relief in 53% of the patients, with morbidity and mortality rates much lower than those associated with pharmacologic or surgical alternatives.

STIMULATION TECHNIQUES.—The Gate Control Theory presented by Melzack and Wall in 1965 declared that there were two basic kinds of sensory input to the CNS—that from large (nonpain) and that from small (pain-transmitting) fibers (Melzack and Wall, 1965; Melzack, 1982). Activation of large fibers was proposed to be capable of "drowning out" the pain information coming into the CNS by the small fibers. Thus, intentional hyperstimulation of the nervous system could decrease pain. We all have used this theory—when the thumb is hit with a hammer, the common reaction is to suck the wounded digit. This activates large fiber input that seems to diminish the intense pain. Although the first clinical application of this concept was through operative placement of electrode wires into the substance of the dorsal column of the spinal cord, a simpler, more practical form is embodied in TENS.

TENS has a significant number of advantages, listed in Table 6–4. If one could describe the pain management modality of choice, many of the desirable qualities would be answered by TENS (Long, 1983). McGuire et al. (1983) reported that TENS over the posterior tibial or common peroneal nerve was used beneficially to in-

hibit detrusor muscle activity and relieve the sense of urgency in 15 patients with a variety of neural lesions.

Acupuncture may be an ancient form of hyperstimulation of the nervous system. Its use is limited by variable licensure requirements for practitioners, the need for a special therapist, and its invasive nature (Ulett, 1981). Research in Western cultures has not documented the success such treatment is stated to have in China. Further research will no doubt more clearly define its precise role in pain management.

PHYSICAL THERAPY AND REHABILITATION.—Pain frequently limits physical activity. When pain is expected to be of short duration, actually a few days or weeks of rest may be therapeutic. After this, however, pain-induced immobility begins to compromise the patient's energy, spirit, and body functions, i.e., bowel movements, urination, appetite, and vitality. Thus, treatment must have as part of its purpose the objective of getting the patient mobile. Indeed, this is the intent of stir-up regimens used in the early postoperative period and the modern aggressive treatment of sports injuries.

Many patients with chronic pain have needlessly prolonged (weeks to months) the recommendation to rest-up-and-it-will-get-better from the time of the acute injury. Over time these patients develop postural limitations to movement, loss of range of motion of joints, muscle weakness, weight gain, and at least a perceived intolerance to physical activities. Even if the patient's pain cannot be modified greatly with treatment, the physical deterioration cannot be allowed to persist. The workup is crucial, since it *must rule out* that the patient's ongoing pain reflects ongoing tissue damage. This information, coupled with the fact that weeks to months of inactivity have not improved that patient's pain, mandates that increased physical activity be prescribed. This is best accomplished with recommendations for gradual, progressive increases in the patient's activity up to and including cooperation with formal physical therapy programs. Fordyce et al. (1981) dem-

TABLE 6–4.

Advantages of Transcutaneous Electrical Nerve Stimulation*

Noninvasive
Simple, easy to use, no special therapist
No systemic side effects (e.g., sedation, urinary retention, orthostatic hypotension, respiratory depression)
No psychological or physiologic dependence
Enables the *patient* to affect his own level of pain relief
Emphasizes the body's natural mechanisms for handling pain
High patient acceptance
Decreases the use of narcotics, time to ambulation, period of hospital confinement, and perhaps health care costs

*From Rowlingson JC: Current concepts of postoperative pain. *Infect Surg* 1984; 3:527–535. Used by permission.

onstrated the lack of correlation between the patients' complaints of chronic pain and their performance of prescribed exercises and physical activity.

As progress is made with treatment and the pain decreases to its maximal level, the patient must consider increased physical capabilities with regard to maximal rehabilitation. For some patients this will mean a return to complete and full activity, e.g., as after the treatment of kidney stones and ureteral colic with lithotripsy. In others, e.g., those with pain due to cancer or chronic pain, it may mean switching jobs, changing to part-time work, or retiring altogether. Unfortunately, many social support programs do not currently serve the needs of the injured worker, and persistence of pain may be the only means by which a patient will have any income. Catchlove et al. (1980) have shown the positive effects of realistic approaches to compensation and the return to work.

Prescription of activity schedules that intersperse fixed periods of rest between active intervals may be necessary in some patients with chronic pain to avoid their pushing themselves to the point of exhaustion and collapse with which frustration, reinforcement of negative attitudes, and decreased pain tolerance may be associated. The treating physician will do well to remember that success with treatment, as defined by relief of the patient's subjective complaints of pain, may be harder to come by than when success is measured by positive changes in drug use, renewed physical abilities, and restoration of productive activity.

PSYCHOLOGICAL INTERVENTIONS.—By current definition, pain is more than physical complaints and can exist in the absence of a medically identifiable cause (Gibbon and Rabinovitz, 1980). The limitations of contemporary medical diagnosis contribute to this being so. With the renewed interest in the phenomenon of pain, we have moved well away from the antiquated concept that a patient's pain is "all in his head." The term "psychological factors" really is used as a generic one that represents all of the nonphysical components, be they psychological, economic, social, or whatever. Pain, and particularly chronic pain, is not just psychological. What is obvious is that pain distorts normal behavior, and if the pain persists, frequently the behavior will become entrenched, unless formal treatment intervenes. Thus, the workup of the patient with acute or chronic pain must include an assessment of their pain-associated behavior.

Formal therapeutic intervention is not frequently required. However, when necessary, the patient *must* understand that the involvement of a psychologist or a psychiatrist is a positive, progressive step, not an implication that the pain is all imagined. This is yet another method by which insight can be gained, understanding fostered, and pain conquered. For example, the patient with pelvic floor tension myalgia who complains of perineal pain and symptoms not unlike prostatitis may benefit from biofeedback and relaxation training. These techniques are introduced in an attempt to eliminate the causative habitual contraction and spasm of the muscles of the pelvic floor (Segura et al., 1979).

ISSUES IN POSTOPERATIVE PAIN

As a physical experience, pain has a critical signal function—that of alerting an organism of tissue damage. Most acute postoperative pain is an exception to this concept because it serves no useful purpose except rarely when it forbodes a complication, e.g., ischemia or wound breakdown. Rather, postoperative pain provokes anxiety, suffering, and pathophysiologic consequences that jeopardize the patient in terms of morbidity and mortality and impede his recovery (Beecher, 1969; Craig, 1981; Pflug and Bonica, 1977). The pain input to the CNS from the incisional area incites vascular, metabolic, endocrine, muscular, and sympathetic nervous system activity. The overall condition of the patient is further influenced by his concept of the significance of the pain and his understanding of it. The pathophysiologic consequences, well detailed by Pflug and Bonica (1977), include such changes as hypertension, tachycardia, hypoxia, hypercarbia, nausea, vomiting, inhibition of urination, and fatigue.

The preoperative preparation of the patient can have tremendous impact on his postoperative status. This preparation theoretically begins with the surgeon's initial recommendation that surgery is necessary. It is continued by the ward nursing staff who can bias the patient's attitudes about sickness, health and the proposed surgery by verbal reinforcement, information sharing, and supportive behavior or lack thereof (Guerra and Aldrete, 1980). There is good evidence that an informative preoperative visit by the anesthesiologist is another potent intervention that adds to or detracts from the many factors that influence the patient's perception of postoperative pain (Egbert et al., 1963, 1964). All involved in the preoperative care of patients must be sensitive to the potential for what they say and do to influence the patient's pain (Rowlingson and Chalkley, 1984). This supports again the concept that nonphysical factors influence the perception of pain.

Clearly, medications are the mainstay of postoperative pain control, with the narcotic drugs being most commonly used. Morphine is frequently given, starting at a dose of 0.1 mg/kg every 3–4 hours IM. Equivalent doses of other narcotics can be substituted after one checks an equivalence chart so that an equipotent dose for the planned route of administration is given (Table 6–5). In spite of the perceived potency of narcotic drugs in comparison to the degree of pain a patient experiences, 80% of postoperative patients complain that the duration of analgesia is too short, and 5%–20% of patients simply do not get adequate relief (Tammisto, 1978). In part, these deplorable numbers may be due to the ordering of routine doses of narcotics at fixed intervals, regardless of the effect of the drug and dose on the patient. As Austin et al. have shown, meperidine, 100 mg IM every 4 hours, exceeded the measured minimal plasma concentration of meperidine necessary to provide analgesia only 33% of the time (Austin et al., 1980). The effect of the last dose given should determine the interval a₁ 1 milligram dosage of the next dose. Forgotten pharmacology is frequently an influential factor on the care patients receive in terms of pain control postoperatively (Weis et al., 1983).

Because absorption from intramuscular depots is erratic (Austin et al., 1980) and actual delivery of the drugs to the patient from the nursing staff is fraught with misconceptions and interruptions (Cohen, 1980; Mather and Mackie, 1983), alternate techniques for drug delivery are being developed. Patient-controlled analgesia (PCA) systems allow the patient to self-administer IV a preset dose of narcotic while providing a subsequent lockout period during which no further drug can be delivered (Bennett et al., 1982; Graves et al., 1983; Tamsen et al., 1982). This obviously bypasses the absorption kinetics of the chosen drug and perhaps decreases the amount of nursing time involved in actual drug administration.

In the future analgesic drug preparations compatible with sublingual/buccal or transdermal administration may be available. Clearly the use of adjunctive drugs such as phenergan, 25 mg IM every 4–6 hours (Keeri-Szanto, 1961), vistaril, 100 mg IM (Hupert et al., 1980), and even dextroamphetamine, 5–10 mg IM (Forrest et al., 1977) have an important role in postoperative pain management. Their concurrent use can decrease the total dose of narcotics necessary because of potentiation of the narcotic analgesia. Indirectly, then, these adjunctive drugs limit narcotic side effects. More effective oral medications may also become available (Okun, 1982) (see Table 6–2).

One need only to refer to Table 6–4 to be reminded of the advantages of TENS and to see why this modality might have great utility in the management of postoperative pain. Studies reveal that up to 70% of patients can benefit with TENS (Lim et al., 1983; Solomon et al., 1980; Tyler et al., 1982) (the 30% failure rate is not much different from the 20% figure for medications). This noninvasive therapy, which the patient can manipulate to meet his individual needs and has no systemic side effects, is a definite boon in the armamentarium for postoperative pain control.

The use of local anesthetic drugs and regional anesthetic techniques can provide at least a temporary reduction in postoperative pain. This allows a calm emergence from general anesthesia, early readiness for discharge from the recovery room or ambulatory surgery setting, and a positive attitude toward a rapid and uneventful recovery. One must always be aware of the recommended maximum doses of local anesthetic drugs, i.e., lidocaine, 7–8 mg/kg, or bupivacaine, 2–3 mg/kg (DiFazio, 1985), and the common symptoms of a toxic reaction related to high blood levels of the local anesthetic drugs, i.e., ringing in the ears, circumoral numbness, slurred speech, tremors, sedation, seizures, and cardiovascular collapse (Carron et al., 1985). It is interesting that the prodromal symptoms are more common with lidocaine toxicity than with bupivacaine.

A common example of these principles is the well-documented beneficial effect of continuous epidural analgesia on postoperative pulmonary function and the subsequently diminished need for potent analgesics (Craig, 1981; Pflug et al., 1974). A simple application of local anesthetic use is the technique of wound margin infiltration. Bupivacaine 0.25%–0.5%, without epinephrine (to avoid intense vasoconstriction that would impair wound healing) is injected along the length of the wound to minimize incisional pain for 4–8 hours. Penile block is another example of an effective use of regional anesthesia for postoperative pain control (Goulding, 1981; Lunn, 1979). The dorsal nerves of the penis are blocked with 3–8 ml of 0.25%–0.5% plain bupivacaine at the 10 o'clock and 2 o'clock positions at the base of the penis just after the nerves emerge from underneath the symphysis pubis (Carron et al., 1985). Blockade of the ilioinguinal and iliohypogastric nerves can decrease pain and muscle spasm (if concentrations of local anesthetic drug that block motor function are

TABLE 6–5.

Equianalgesic Doses of Narcotics*

DRUG	IM (MG)	PO (MG)	DURATION (HR)
Morphine	10	60	4–6
Meperidine	75	300	3–4
Methadone	10	20	5–6
Hydromorphine	1.5	7.5	4–6
Codeine	130	200	4–5
Oxycodone	—	30	4–6
Buprenorphine	0.4	—	4–6
Pentazocine	60	180	4–6

*Adapted from Foley 1985, Inturrisi CE: Role of opioid analgesics. *Am J Med* 1984;77:27–37.

used) after surgery in the inguinal and scrotal areas. These nerves are readily accessible as they course between the superficial muscle layers of the abdominal wall 2.5–3 cm medial and 2–3 cm caudal to the anterior superior iliac spine. Five to ten milliliters of 0.25%–0.5% bupivacaine is injected at this location as the needle is carefully moved in and out of the muscle layers. This block could be repeated every 8–12 hours if needed. (Carron et al., 1985). Genitofemoral nerve block may be more useful when scrotal surgery has been performed.

Intercostal nerve blocks can provide remarkable relief of pain after flank incisions. These blocks may be performed at the time of surgery from the "inside" before the wound is closed or as a percutaneous procedure. Noller et al. (1977) showed that patients so treated used fewer narcotics, ambulated sooner, resumed oral intake sooner, and could cough and deep breathe more effectively. Three to five milliliters of 0.5% bupivacaine with epinephrine is placed at the neurovascular bundle under each rib to be blocked (Carron et al., 1985). Katz et al. (1979) proposed the use of cold thermal nerve blocks (done with a cryoprobe) as an alternative beneficial procedure for postoperative pain control.

ISSUES IN PAIN RELATED TO CANCER

Fear abounds in the lay public over the diagnosis of cancer (Bonica, 1982; Foley, 1985). But it is not only the possibility of facing the end of one's life that is so threatening. Many patients with cancer are afraid of having to endure severe pain as they die. Available treatments for pain control are often inadequately applied, and thus the patient spends his final days and weeks suffering unnecessary discomfort (Twycross, 1980). Patients with pain due to cancer must have concern for their quality of life placed ahead of all other considerations.

The common causes of pain in patients with cancer are well known (Rowlingson and Chalkley, 1985). Pain directly related to the oncologic process (i.e., metastasis to bone, ureteral obstruction) commonly presents as acute pain, whereas that following treatment (i.e., after surgery or following chemotherapy or radiation therapy) is more likely to present as chronic pain (Bonica, 1982). Usually the causative diagnosis has been determined and reasons for the complaints of pain are not in question. Thus, less time is spent in the formal evaluation process compared with some other pain syndromes, and energy should be applied to creating an effective treatment program. The physical and emotional deterioration that so commonly characterizes patients with nonmalignant pain is exaggerated in those with cancer.

The diagnosis of cancer is so charged with emotion that these nonphysical factors cannot be ignored in treatment planning.

Considerations of treatment in patients with pain due to cancer must be influenced by the *patient's* choice and the quality of life that will result. One must weigh the risk/benefit ratio of treatment, the amount of additional suffering the patient will incur, the patient's probable life expectancy, and the severity and duration of the current pain. Preservation of life at all costs is not appropriate in patients with far-advanced cancer.

The scheme for treatment presented earlier provides a convenient framework in which to discuss therapeutic options for cancer pain control. Surely, definitive and palliative surgical procedures can be very useful, as discussed in other chapters. Neurosurgical procedures that interrupt pathways of neural transmission may find utility in patients with pain due to malignancy because the patient is not likely to live long enough to suffer from the frequent complications. Percutaneous anterolateral cordotomy is valuable when intractable pain is unilateral.

Narcotic medications become the mainstay of treatment in most patients with acute and chronic pain secondary to malignancy (Beaver, 1980; Walsh, 1984; Foley, 1985). As in the management of acute pain, concerns about iatrogenic addiction to medications are given too much publicity, and the patient suffers from an unnecessary restriction of medication (Marks and Sachar, 1974; Porter and Jick, 1980). On the other hand, just because the diagnosis of cancer has been made, it does not mean that narcotics are the first-line drugs. For instance, the NSAIDs have a potent prostaglandin inhibitory action and can, therefore, have an enormous positive effect on pain from metastatic disease to bone, which is the result of localized induction of prostaglandin synthesis and a subsequent enhancement of neural transmission of pain.

The analgesic effects of NSAIDs, acetaminophen, and the drugs for mild to moderate pain (see Table 6–2) can be potentiated by the concurrent use of phenothiazines, antidepressants, and butyrophenones (see Table 6–3). The use of effective combinations of drugs will decrease pain, minimize drug-related side effects, and delay the need for potent narcotic medications (Beaver, 1980; Kanner, 1983; Moertel, 1980; Pilowsky et al., 1982; Twycross, 1980). This may allow the cancer patient with pain to function more normally and without undue sedation, maintain a positive attitude in the face of his serious illness, and postpone the development of tolerance to narcotic drugs. Prescription of medications ideally will be made by a single physician so that illogical combinations of drugs, i.e., narcotic agonist-antagonists given concurrently with narcotics, will not disrupt the pharmacologic management of the patient. Fewer medications should be necessary as other nonmedication treatment options are also effectively used. Potent narcotic drugs should *not* be withheld from any patient with terminal illness when medical science has nothing humane left to offer the patient.

Nerve blocks are another technique that can eliminate neuropathic pain (Long, 1980; Lund, 1982). The shortcomings of continuous local anesthetic injections or infusions—the temporary relief of pain, development of tolerance (called tachyphylaxis), and the need for direct physician supervision if not administration—have been described. Generally they preclude the use of regional anesthesia techniques with local anesthetics except as diagnostic tools in patients with cancer pain. Thus, consideration is given to the use of neurolytic agents such as alcohol or phenol for injection, with the expectation that the analgesia thus obtained would be of at least 4–8 weeks' duration (Wood, 1978). For example, transsacral blocks with phenol can be very effective in relieving rectal or perineal pain secondary to prostatic cancer. Since neurolytic agents are not selective in their neural destruction, sensory modalities other than pain may be disturbed, with persistent numbness and/or paresis not satisfactory to the patient.

An exciting recent technique for pain control combines the traditional narcotic drugs with routine regional anesthetic techniques (Cousins and Mather, 1984; Findler et al., 1982). The placement of poorly lipid-soluble narcotics—e.g., morphine, 5–7.5 mg—into the epidural space provides a depot supply of drug that slowly diffuses into the subarachnoid fluid and bathes specific narcotic receptors in the substantia gelatinosa area of the dorsal horn of the spinal cord. This "topical" application of the drug provides analgesia of excellent quality but does not result in any change in motor or proprioception modalities or sympathetic dysfunction. Thus, patients provided analgesia with epidural narcotics have excellent pain relief yet remain ambulatory and generally free of sedation. The predictable and perhaps tolerable side effects of itching, nausea, vomiting, and urinary retention are generally not manifested in patients with some degree of tolerance to narcotic drugs (as would be typical of patients with pain due to cancer and who have had previous narcotic treatment). The risk of severe respiratory depression, which currently limits the widespread application of this technique to pain control, is also diminished in previously tolerant individuals, patients receiving epidural narcotics (as compared to subarachnoid administration), and those in whom narcotic agonist/antagonist drugs or naloxone are given simultaneously. The duration of the analgesia and the occurrence of the side effects are dose related. In the very near future, tech-

niques using more narcotic receptor-specific drugs and/or continuous infusion of highly soluble narcotic drugs, i.e., fentanyl (that have a shorter duration of action but less spread in the CSF and fewer side effects) will become more common. Patients with pain due to cancer can have catheters implanted in the epidural space to which injection ports or drug-delivery reservoirs can be attached. Thus, pain can be managed on an outpatient basis with the patient remaining at home in a comfortable, familiar, supportive environment.

TENS is yet another adjunctive modality that patients may find useful, especially since it is not invasive and is under their control. Restoration or preservation of physical function will be important to the motivated patient with pain due to cancer, as will the approval for him to participate in any and all appropriate activities. No health care professional can disregard the emotional needs of patients with terminal or preterminal illness. Pain may be the patient's way of expressing innumerable emotions such as fear, anxiety, or fatigue, and compassionate dialogue may be critically important in bolstering the patient's well-preserved mind even as his body is ravaged by malignancy (Twycross, 1980).

PAIN MANAGEMENT CENTERS AND CHRONIC PAIN

There has been a growing awareness that patients with chronic pain are unique and complex. Specialized facilities dedicated to their care have become increasingly popular (Aronoff, 1983; Hallett and Pilowsky, 1982; Lutz et al., 1983). It has been popular in the past 10–15 years for specialists with a common interest in helping patients afflicted with chronic pain to organize themselves into dedicated pain treatment clinics or centers. The referring specialist must have a rudimentary knowledge about the purpose and intent of such a facility so that realistic expectations will be encouraged in the referred patient.

Pain centers are divided by their inpatient and outpatient design. The former can be expected to be more expensive (given the common 2–6 weeks of hospitalization) and more suitable to those patients with severe drug, behavioral, or psychosocial problems associated with their pain. Outpatient centers, although less costly, place the burden for improvement on the patients by demanding that they develop self-help attitudes that are manifested by cooperation with the therapeutic recommendations. The staff of pain centers are interested in patients with chronic pain, have knowledge of the significant differences between acute and chronic pain, and can devote the time necessary to the *patient* to gain satisfaction in the telling of his complete story. As many facilities are associated with university medical centers, it is likely that more evaluation and/or treatment consultations can be expeditiously arranged than in the patient's home area. The purpose of the pain center is to evaluate thoroughly patients with pain, associate their physical complaints with anatomical possibilities, gain insight into psychosocial factors that interplay with the physical pain, and encourage the patients to more fully understand their pain. Further, the staff will assist in the creation of therapeutic programs that combine reputable treatment options in a logical and coordinated way.

A pain center does not challenge the authenticity of the patient's pain. Rather, the pain is accepted as described by the patient, and the tenor of treatment is that the patient must become actively involved in his own treatment program in spite of the presence of pain. When the patient understands that the ongoing pain does not indicate continued tissue damage, that just taking the pain away will not answer all of his problems with pain, and that the apparent diminution in the potency of medications and the increase in daily activity do not reflect denial of the severity of pain (but rather concessions to the dangers of unnecessarily prolonged medication use and inactivity), he can be encouraged to cooperate with the proposed treatment. The ultimate goal is to return referred patients to their primary physicians with a shared understanding of the diagnosis and rationale for the specific therapeutic program outlined. Because there is often more than simple tissue damage involved in the perpetuation of the pain, emphasis is placed on comprehensive but nonsurgical approaches to pain management.

CONCLUSION

Pain depletes the patient's energy and spirit and tries the physician's patience. No specialist can ignore addressing the comprehensive needs of a patient with a pain problem. To make a diagnosis from which appropriate treatment will follow, the evaluation scheme must be thorough and systematic, a degree of realism must intervene in the ordering of pertinent laboratory tests, and nonphysical factors related to the pain must be recognized. Many patients with urologic complaints will have acute pain problems that respond well to traditional treatment. Other patients will complain of chronic pain, and the urologist must be prepared to manage them or make an appropriate referral (Nocks, 1985).

One cannot treat all complaints of pain in the same way or make the naive assumption that the absence of an obvious physical cause indicates the predominance of a psychological one. The patient's "pain" must be interpreted in the context of his unique presentation. In treatment the goal is for early, positive intervention to minimize the influence that inadequately managed pain can have on the patient's physical and emotional status. One must effectively treat acute pain and base further treatment on the patient's response to the initial modalities chosen. As pain becomes chronic in spite of treatment and begins to lose its vital signal function, the physician must help the patient set realistic goals for pain treatment, embark on a program of management that is maximally effective yet avoids iatrogenic complications, and help the patient to live within the reasonable physical and social restrictions of his pain.

REFERENCES

1. Amadio P Jr: Peripherally acting analgesics. *Am J Med* 1984; 77:17–26.
2. Angell M: The quality of mercy. *N Engl J Med* 1982; 306:98–99.
3. Aronoff GM: The use of non-narcotic drugs and other alternatives for analgesia as part of a comprehensive pain management program. *J Med* 1982; 13:191–202.
4. Aronoff GM, Evans WO, Enders PL: A review of followup studies of multidisciplinary pain units. *Pain* 1983; 16:1–11.
5. Austin KL, Stapleton JV, Mather LE: Multiple intramuscular injections: A major source of variability in analgesic response to meperidine. *Pain* 1980; 8:47–62.
6. Beaver WT: Management of cancer pain with parenteral medication. *JAMA* 1980; 244:2653–2657.
7. Beecher HK: Anxiety and pain. *JAMA* 1969; 209:1080.
8. Bennett RL, Batenhorst RL, Graves D, et al: Patient-controlled analgesia: A new concept of postoperative pain relief. *Ann Surg* 1982; 195:700–705.
9. Boas RA: Sympathetic blocks in clinical practice, in Stanton-Hicks M (ed): *International Anesthesiology Clinics—Winter 1978*. Boston, Little, Brown & Co, 1978, pp 149–182.
10. Bonica JJ: Introduction (to Narcotic analgesics in the treatment of cancer and postoperative pain). *Acta Anaesthesiol Scand* 1982; 26 (suppl 74):5–10.
11. Bonica JJ: Management of cancer pain. *Acta Anaesthesiol Scand* 1982; 26 (suppl 74):75–82.
12. Bonica JJ: Introduction, in Aronoff GM (ed): *Evaluation and Treatment of Chronic Pain*. Baltimore-Munich, Urban & Schwarzenberg, 1985, pp xxxi–xliv.
13. Bromage PR, Camporesi E, Chestnut D: Epidural narcotics for postoperative analgesia. *Anesth Analg* 1980; 59:473–480.
14. Carron H, Korbon GA, Rowlingson JC: *Regional Anesthesia: Techniques and Clinical Applications*. New York, Grune & Stratton, 1985.
15. Catchlove R, Cohen K: Effects of a directive return to work approach in the treatment of workman's compensation patients with chronic pain. *Pain* 1982; 14:181–191.
16. Cohen FL: Postsurgical pain relief: Patient's status and nurses' medication choices. *Pain* 1980; 9:265–274.

17. Cousins MJ, Mather LE: Intrathecal and epidural administration of opioids. *Anesthesiology* 1984; 61:276–310.

18. Craig DB: Postoperative recovery of pulmonary function. *Anesth Analg* 1981; 60:46–52.

19. DiFazio CA: Pharmacology of local anesthetics and choice of agent, in Carron H, Korbon GA, Rowlingson JC (eds): *Regional Anesthesia. Techniques and Clinical Applications.* New York, Grune & Stratton, 1985, pp 1–7.

20. Egbert LD, Battit GE, Turndorf H, et al: The value of the preoperative visit by an anesthetist. *JAMA* 1963; 185:553–555.

21. Egbert LD, Battit GE, Welch CE, et al: Reduction of postoperative pain by encouragement and instruction of patients. *N Engl J Med* 1964; 270:825–827.

22. Findler G, Olshwang D, Hadani M: Continuous epidural morphine treatment for intractable pain in terminal cancer patients. *Pain* 1982; 14:311–315.

23. Foley KM: Adjuvant analgesic drugs in cancer pain management, in Aronoff GM (ed): *Evaluation and Treatment of Chronic Pain.* Baltimore-Munich, Urban & Schwarzenberg, 1985, pp 425–434.

24. Foley KM: The treatment of cancer pain. *N Engl J Med* 1985; 313:84–95.

25. Fordyce W, McMahon R, Rainwater G, et al: Pain complaints-exercise performance relationship in chronic pain. *Pain* 1981; 10:311–321.

26. Forrest WH Jr, Brown BW Jr, Brown CR, et al: Dextroamphetamine with morphine for the treatment of postoperative pain. *N Engl J Med* 1977; 296:712–715.

27. Ghia JN, Toomey TC, Mao W, et al: Towards an understanding of chronic pain mechanisms: The use of psychological tests and a refined differential spinal block. *Anesthesiology* 1979; 50:20–25.

28. Giddon DB, Rabinovitz SL: The psychological aspects of treatment of chronic pain patients. *Regional Anesthesia* 1980; 5:16–23.

29. Goulding RJ: Penile block for postoperative pain relief in penile surgery. *J Urol* 1981; 126:337.

30. Graves DA, Foster TS, Batenhorst RL, et al: Patient-controlled analgesia. *Ann Intern Med* 1983; 99:360–366.

31. Guerra F, Aldrete JA: *Emotional and Psychological Responses to Anesthesia and Surgery.* New York, Grune & Stratton, 1980.

32. Hallett EC, Pilowsky I: The response to treatment in a multidisciplinary pain clinic. *Pain* 1982; 12:365–374.

33. Hameroff SR, Carlson GL, Brown BR: Ilioinguinal pain syndrome. *Pain* 1981; 10:253–258.

34. Henrich WL: Nephrotoxicity of non-steroidal anti-inflammatory agents. *Am J Kidney Dis* 1983; 2:478–484.

35. Hollister LE: Medical intelligence-tricyclic antidepressants (part 1). *N Engl J Med* 1978; 299:1106–1109.

36. Hollister LE: Medical intelligence-tricyclic antidepressants (part 2). *N Engl J Med* 1978; 299:1168–1172.

37. Hupert C, Yacoub M, Turgeon LR: Effect of hydroxyzine on morphine analgesia for the treatment of postoperative pain. *Anesth Analg* 1980; 59:690–696.

38. Inturrisi CE: Role of opioid analgesics. *Am J Med* 1984; 77(3A):27–37.

39. Kanner R: Psychotropic drugs in the management of pain. *Curr Concepts Pain* 1983; 1:11–15.

40. Katz J, Nelson WL, Forest R, et al: Prolonged post-thoracotomy analgesia by cryoprobe nerve block. *Anesthesiology* 1979; 51:S233.

41. Keeri-Szanto M: The mode of action of promethazine in potentiating narcotic drugs. *Br J Anaesth* 1961; 33:422–431.

42. Khantzian EJ, McKenna GJ: Acute toxic and withdrawal reactions associated with drug use and abuse. *Ann Intern Med* 1979; 90:361–372.

43. Levine J: Pain and analgesia: The outlook for more rational treatment. *Ann Intern Med* 1984; 100:269–276.

44. Lim AT, Edis G, Kranz H, et al: Postoperative pain control: Contribution of psychological factors and transcutaneous electrical stimulation. *Pain* 1983; 17:179–188.

45. Loeser JD: Pain due to nerve injury. *Spine* 1985; 10:232–235.

46. Long DM: Stimulation of the peripheral nervous system for pain control. *Clin Neurosurg* 1983; 31:323–343.

47. Long DM: Relief of cancer pain by surgical and nerve blocking procedures. *JAMA* 1980; 244:2759–2761.

48. Lund PC: The role of analgesic blocking in the management of cancer pain: Current trends, a review article. *J Med* 1982; 13:161–182.

49. Lunn JN: Postoperative analgesia after circumcision. *Anaesthesia* 1979; 34:552–554.

50. Lutz RW, Silbret M, Olshan N: Treatment outcome and compliance with therapeutic regimes: Long-term follow-up of a multidisciplinary pain program. *Pain* 1983; 17:301–308.

51. Marks RM, Sachar EJ: Undertreatment of medical inpatients with narcotic analgesics. *Ann Intern Med* 1973; 78:173–181.

52. Mather L, Mackie J: The incidence of postoperative pain in children. *Pain* 1983; 15:271–282.

53. McGuire EJ, Shi-Chun Z, Horwinski ER, et al: Treatment of motor and sensory detrusor instability by electrical stimulation. *J Urol* 1983; 129:78–79.

54. Melzack R, Wall PD: Pain mechanisms: A new theory. *Science* 1965; 150:971–979.

55. Melzack R: Recent concepts of pain. *J Med* 1982; 13:147–160.

56. Melzack R, Wall PD: *The Challenge of Pain.* New York, Basic Books, 1982.

57. Mersky H: Pain terms: A list of definitions and notes on usage. *Pain* 1979; 6:249–252.

58. Moertel CG: Treatment of cancer pain with orally administered medications. *JAMA* 1980; 244:2448–2450.

59. Nathan PW: Chlorprothixene (Taractan) in post-herpetic neuralgia and other severe chronic pains. *Pain* 1978; 5:367–372.

60. Nocks BN: Pain in the male genitalia, in Aronoff GM (ed): *Evaluation and Treatment of Chronic Pain.* Baltimore-Munich, Urban & Schwarzenberg, 1985, pp 393–405.

61. Noller DW, Gillenwater JY, Howards SS, et al: Intercostal nerve block with flank incision. *J Urol* 1977; 117:759–761.

62. Okun R: Analgesic effects of oral nalbuphine and codeine in patients with postoperative pain. *Clin Pharmacol Ther* 1982; 32:517–524.

63. Pflug AE, Murphy AM, Butler SH, et al: The effects of postoperative peridural analgesia on pulmonary therapy and pulmonary complications. *Anesthesiology* 1974; 41:8–17.

64. Pflug AE, Bonica JJ: Physiopathology and control of postoperative pain. *Arch Surg* 1977; 112:773–781.

65. Pilowsky I, Hallett EC, Bassett DL, et al: A controlled study of amitriptyline in the treatment of cancer pain. *Pain* 1982; 14:169–179.

66. Porter J, Jick H: Addiction is rare in patients treated with narcotics. *N Engl J Med* 1980; 302:123.

67. Ready LB, Sarkis E, Turner JA: Self-reported versus actual use of medications in chronic pain patients. *Pain* 1982; 12:285–294.

68. Reuler JB, Girard DE, Nardone DA: The chronic pain syndrome: Misconceptions and management. *Ann Intern Med* 1980; 93:588–596.

69. Romagnoli A, Batra MS: Continuous epidural block in the treatment of impacted ureteric stones. *Can Med Assoc J* 1973; 109:968.

70. Rosenblatt RM, Reich J, Dehring D: Tricyclic antidepressants in treatment of depression and chronic pain: Analysis of the supporting evidence. *Anesth Analg* 1984; 63:1025–1032.

71. Rowlingson JC, Stehling L: Anesthesia update #10: The evaluation and treatment of the patient with chronic pain. *Orthopaed Rev* 1982; 11:79–85.

72. Rowlingson JC: Current concepts of postoperative pain. *Infect Surg* 1984; 3:527–535.

73. Rowlingson JC, Chalkley JD: Common pain syndromes—diagnosis and management. *Semin Anesth* 1985; 4:223–230.

74. Segura JW, Opitz JL, Greene LF: Prostatosis, prostatitis or pelvic floor tension myalgia? *J Urol* 1979; 122:168–169.

75. Simon LS, Mills JA: Non-steroidal anti-inflammatory drugs. *N Engl J Med* 1980; 302:1179–1185.

76. Simon LS, Mills JA: Non-steroidal anti-inflammatory drugs. *N Engl J Med* 1980; 302:1237–1243.

77. Simon D, Carron H, Rowlingson JC: Treatment of bladder pain with transsacral nerve block. *Anesth Analg* 1982; 61:46–48.

78. Solomon RA, Viernstein MC, Long DM: Reduction of postoperative pain and narcotic use by transcutaneous electrical nerve stimulation. *Surgery* 1980; 87:142–146.

79. Stillman MT, Napier J, Blackshear JL: Adverse effects of nonsteroidal anti-inflammatory drugs on the kidney. *Med Clin North Am* 1984; 68:371–385.

80. Stimmel B: Pain, analgesia, and addiction: An approach to the pharmacologic management of pain. *Clin J Pain* 1985; 1:14–22.

80a. Swerdlow M: Anticonvulsant drugs and chronic pain. *Clin Neuropharmacol* 1984; 7:51–82.

81. Tammisto T: Analgesics in postoperative pain relief. *Acta Anaesthesiol Scand* 1978; 70(suppl):47–50.

82. Tamsen A, Hartvig P, Fagerlund D, et al: Patient-controlled analgesic therapy: Clinical experience. *Acta Anaesthesiol Scand* 1982; 74:157–160.

83. Turner JA, Calsyn DA, Fordyce WE, et al: Drug utilization patterns in chronic pain patients. *Pain* 1982; 12:357–363.

84. Twycross RG: The relief of pain in far-advanced cancer. *Reg Anaesth* 1980; 5:2–11.

85. Tyler E, Caldwell C, Ghia JN: Transcutaneous electrical nerve stimulation: An alternative approach to the management of postoperative pain. *Anesth Analg* 1982; 61:449–456.

86. Ulett GA: Acupuncture treatments for pain relief. *JAMA* 1981; 245:768–769.

87. Walsh TD: Oral morphine in chronic cancer pain. *Pain* 1984; 18:1–11.

88. Ward NG, Bloom VL, Friedel RO: The effectiveness of tricyclic antidepressants in the treatment of co-existing pain and depression. *Pain* 1979; 7:331–342.

89. Weis OF, Sriwatanakul K, Alloza JL, et al: Attitudes of patients, house-staff and nurses toward post-operative analgesic care. *Anesth Analg* 1983; 62:70–74.

90. Winnie AP, Collins VJ: Differential neural blockade in pain syndromes of questionable etiology. *Med Clin North Am* 1968; 52:123–129.

91. Wood KM: The use of phenol as a neurolytic agent: A review. *Pain* 1978; 5:205–229.

Chapter 7

Urologic Laser Surgery

Joseph A. Smith, Jr., M.D.

Although the principles of laser energy were established with the theories of Bohr in 1898 and Einstein (1917), the first beam of visible laser light was generated as recently as 1960 (Maiman, 1960). Mulvaney and Beck (1968) investigated a ruby laser and a CO_2 laser in urology in 1968 and found them to be of minimal value in destroying kidney stones but projected possible use for tumor ablation. After that, little attention was paid to the potential surgical application of lasers, particularly in urology, until 1974, with the development of the neodymium:YAG (Nd:YAG) laser and a suitable fiber delivery system.

Clinical experience with lasers in urologic surgery is expanding rapidly. Certain areas have already been identified in which lasers appear to offer no practical or therapeutic advantages over standard treatment methods. In other settings, lasers appear to offer technical improvements over existing treatments or the capability of expanding therapeutic options. At any rate, it seems inevitable that lasers will play at least some role in the future of urologic surgery and that they may become an integral part of urologic clinical practice.

LASER PHYSICS

To apply laser energy safely and effectively as a surgical tool, a basic understanding of the physics and tissue effects of lasers is essential. The tissue destructive properties of a laser beam can be therapeutically beneficial if properly used but can also produce disastrous complications if misdirected or applied with inadequate knowledge or experience.

The word LASER is an acronymn for Light Amplification by the Stimulated Emission of Radiation. White light from an incandescent bulb is composed of multiple wavelengths that scatter in different directions. In contrast, light from a laser consists of a single wavelength (monochromatic) that travels in a unidirectional manner and can be deflected for projection onto tissue surfaces.

In theory, the beam is totally coherent, although there is a small angle of divergence from most surgical laser fibers.

Surgical lasers are powered by electricity which is used to ignite a flashlamp. Atoms in the active medium are stimulated from the ground state to an excited state. When they spontaneously return to the ground state, a photon is emitted. The photons are deflected from a totally reflecting mirror at one end of the laser cavity and exit as a parallel beam through a partially reflecting mirror at the other end (Fig 7–1). The beam may pass directly from the laser or be transmitted by a flexible quartz fiber. The active medium, i.e., the source from which the photons are emitted, determines the wavelength of a particular laser. It may be a gas (carbon dioxide, argon), a liquid (rhodamine-B), or a solid (neodymium).

In general, minimal modifications are necessary to prepare an operating room or cystoscopic area for laser use (Fig 7–2). 220-V electrical outlets, ideally with isolated transformers, are required for argon or Nd:YAG lasers although 110-V outlets will suffice for a CO_2 laser. The internal system of most lasers is cooled by water obtained through a hose connected to a running water source. Controlled access to the room should be provided to prevent inadvertent entry during laser use. Protective shades should cover all windows. As with any procedure using electrically powered equipment, the operating room floor should be kept relatively dry.

Deflection of the laser beam into the eye can produce a retinal injury with the Nd:YAG or argon laser and a corneal burn with the CO_2 laser. Therefore, eye protection is mandatory for all persons in the operating room during laser use. Glasses or goggles with green-tinted lenses are used for the Nd:YAG laser, amber lenses for the argon laser, and clear lenses for the CO_2 laser. During endoscopic laser surgery, a simple lens cap over the eyepiece of the telescope may suffice once the fiber is inserted.

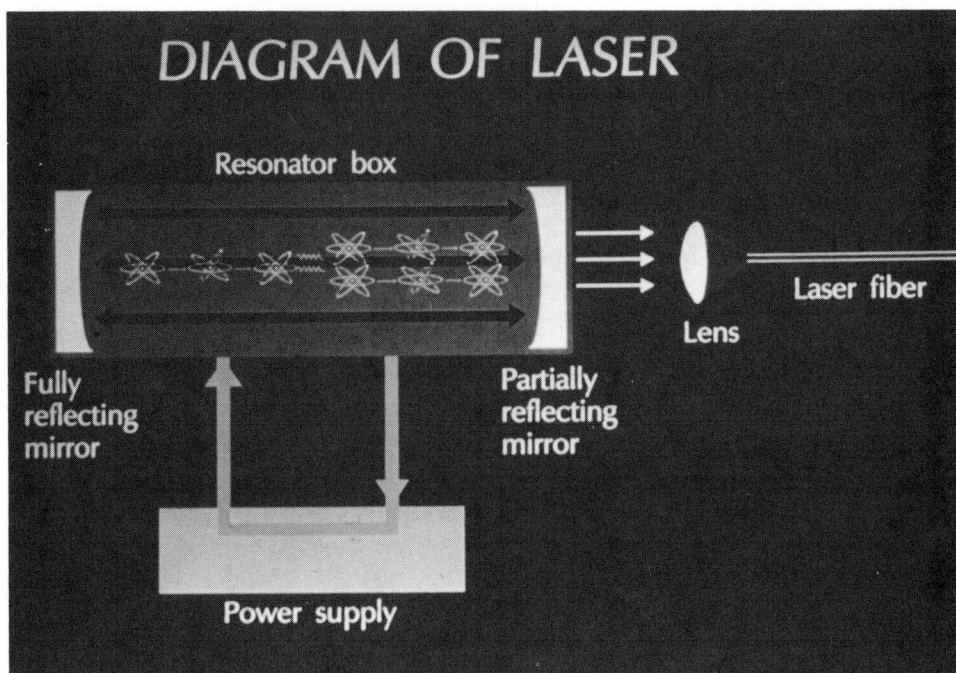

FIG 7–1.
Most surgical lasers use electricity as a power source. The active medium is contained within the resonator box. Light may exit directly from the laser or be focused onto a fiber. (Photo courtesy of Cooper LaserSonics, Sunnyvale, Calif.)

TISSUE EFFECTS OF LASER ENERGY

The tissue effects and the surgical potential of a laser result from transformation of light energy into heat. Cellular destruction generally is not evident when temperatures less than 60° C are maintained for only a few seconds. Above 60° C, protein denaturation ensues, although minimal volatilization and tissue vaporization occur below 100° C. Above 100° C, cellular water evaporates, and burning and tissue vaporization are observed.

Several important factors influence the extent of thermal destruction that occurs when a laser beam is projected onto tissue surfaces. The most obvious and the one most easily controlled is the wavelength of the laser (Fig 7–3). A CO_2 laser has a wavelength of 10,600

FIG 7–2.
Outpatient cystoscopy area adapted for laser treatment. 220-volt electrical power is available as well as water hoses for internal cooling of the laser.

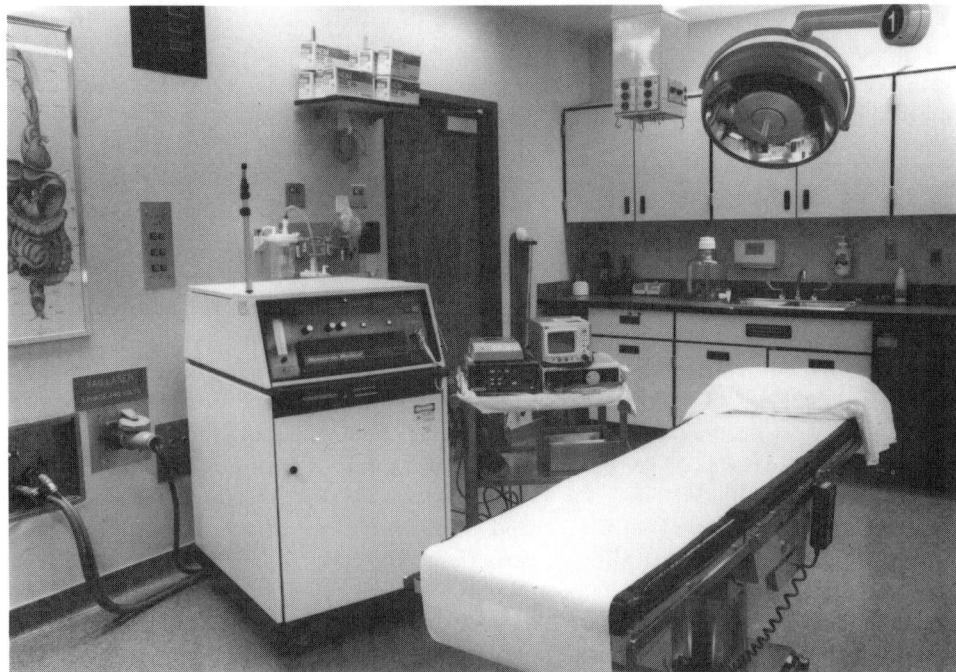

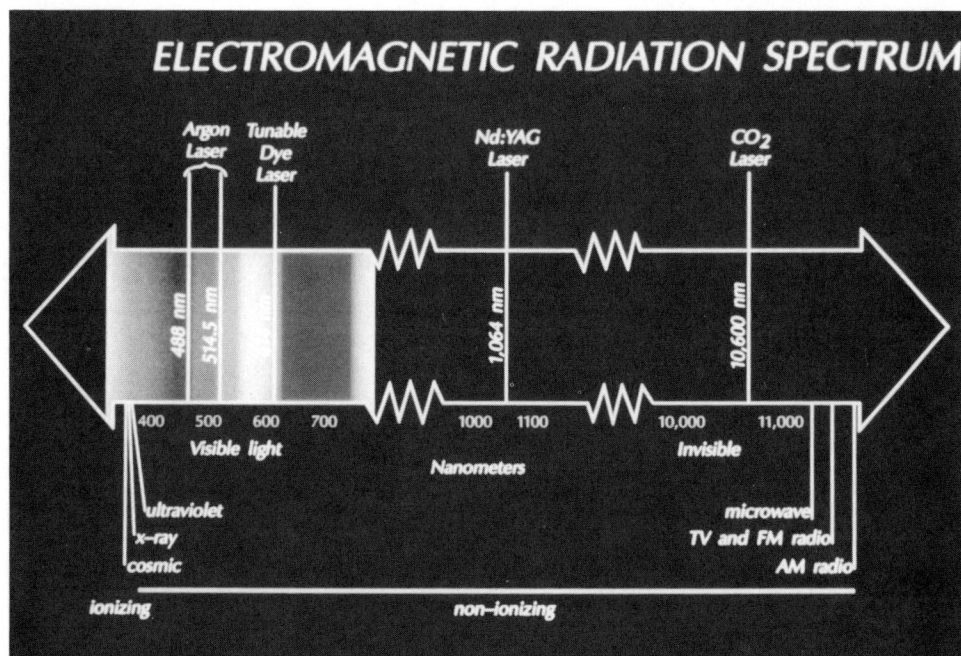

FIG 7–3.
Currently used surgical lasers emit light in the visible or infrared portion of the electromagnetic spectrum. (Photo courtesy of Cooper LaserSonics, Sunnyvale, Calif.)

nm in the near, infrared portion of the electromagnetic spectrum. Light at this wavelength is readily absorbed by water, the primary component of body tissue. Thus, the laser energy is absorbed rapidly, and the tissue penetration is minimal. Intense surface heating occurs with vaporization and smoke production, but the thermal injury usually extends less than 1 mm into the tissue. A fiber for conduction of CO_2 laser light is not commercially available, and CO_2 cystoscopes have not been perfected. Thus, endoscopic delivery of CO_2 laser energy is difficult, and use in urology generally has been restricted to lesions of the external genitalia (Stein and Kendall, 1984).

An argon laser produces a spectral emission between 488 and 514 nm and is visible as a blue-green color. A small flexible quartz fiber transmits the beam. Although an argon laser is minimally absorbed by water, light at this wavelength corresponds to the peak absorption of hemoglobin and melanin. This allows selective treatment of pigmented or erythematous lesions. Tissue penetration is around 1 mm.

The Nd:YAG laser has proved to be the most versatile for endoscopic urologic surgery and some open applications. It produces an invisible beam transmitted by a flexible quartz fiber that can easily be inserted through the working element of a cystoscope or ureteroscope. The point of impact is defined by an aiming beam from a helium neon laser (red) or xenon flash lamp (white). The Nd:YAG laser wavelength of 1,060 nm is poorly absorbed by both water and body pig-

ments. Thus, the thermal effect extends deeply into tissue. Surface heating is less than with a CO_2 laser and the primary tissue effects of a Nd:YAG laser are through coagulation rather than vaporization. Thus, the walls of a hollow organ such as the bladder may maintain their structural integrity even after a transmural Nd:YAG laser injury (Frank et al., 1982).

Besides wavelength, other significant factors influence the tissue effects. The amount of energy delivered to a given area of tissue, i.e., the energy density, is determined by the formula: power (joules) × duration (seconds)/surface area (cm^2). Certain factors in this equation are easily controlled; others are not. The power output is selected from the laser control panel. A Nd:YAG laser is capable of generating an output of up to 100 W, although powers greater than 50 W seldom are necessary for urologic applications. A precise pulse duration may be selected from the control panel or by a foot pedal. The denominator of the equation, however, the surface area of tissue treated, depends on several factors. Most laser fibers have a small angle of beam divergence (5°–8°), so the distance between the fiber tip and the target tissue, or any angulation of the laser beam is important. In addition, damage to the fiber tip may affect beam divergence (Fig 7–4). These variables are difficult to control with precision in a clinical endoscopic setting, so calculation of energy density becomes an approximation based on theoretical considerations and clinical experience (Smith and Dixon, 1985).

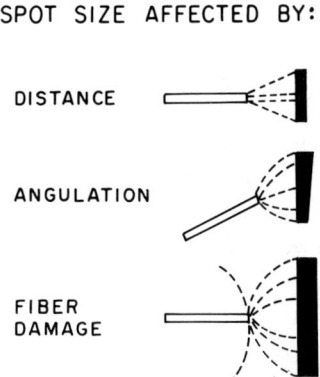

SPOT SIZE AFFECTED BY:

DISTANCE

ANGULATION

FIBER
DAMAGE

FIG 7–4.
Several factors affect the spot size of a laser beam. The larger the surface area being treated, the less the power density.

CLINICAL APPLICATIONS

The thermal effects of laser energy have been used for treatment of a variety of urologic lesions. Clinical experience is expanding, and results from large series are now available (Table 7–1). In certain situations, lasers do not appear to be as effective as standard therapy, or, at least, they offer no particular advantages. In other circumstances, lasers provide significant practical and therapeutic improvements over existing treatment methods. Finally, the unique physics and tissue effects of laser energy have created new treatment opportunities for selected problems in urologic surgery.

EXTERNAL GENITALIA

Direct application of CO_2 laser energy to lesions of the external genitalia is performed either through a coupling microscope or, more commonly, with a handpiece and a series of articulating arms (Fig 7–5). Fiber conducted argon or Nd:YAG light can be directed either by holding the fiber in the hand or by using a

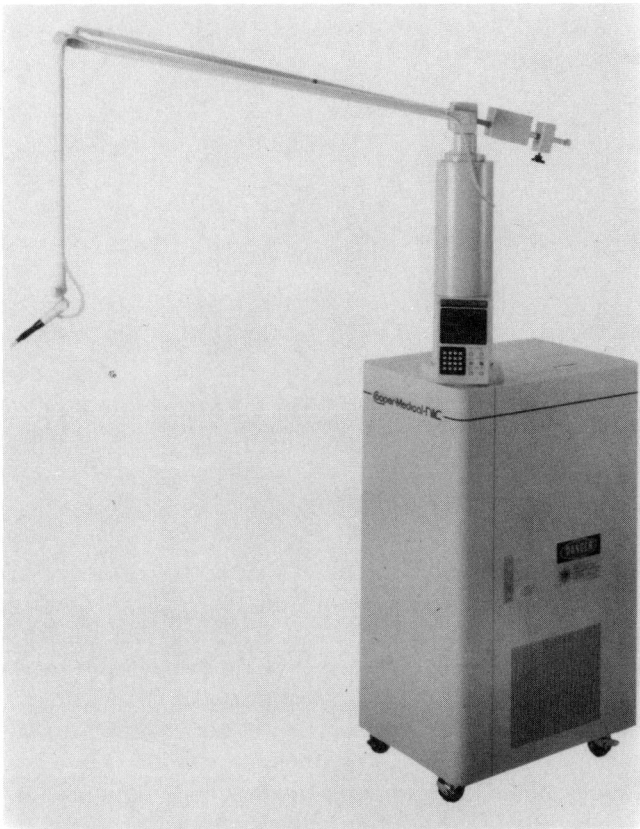

FIG 7–5.
Carbon dioxide laser. The light is conducted via a series of articulated arms to a handpiece.

handpiece. All three lasers have proved to be of use in treatment of a variety of lesions of the external genitalia.

Condyloma Acuminata

Lasers are well established as effective treatment for condyloma acuminata. Both CO_2 and Nd:YAG lasers have been used, although most clinial experience is

TABLE 7–1.

Areas in Which Lasers Have Been Investigated in Urologic Surgery

CARBON DIOXIDE	ARGON	Nd:YAG
Condyloma acuminata	Penile hemangioma	Condyloma acuminata
Erythroplasia of Queyrat	Urethral stricture	Carcinoma of penis
Balanitis obliterans xerotica	Urethral melanoma	Urethral caruncle
Vasovasostomy	Bladder hemangioma	Urethral stricture
Urethral stricture	Papillary bladder tumor	Bladder neck contracture
Partial nephrectomy	Carcinoma in situ of	Superficial bladder tumors
Urethral caruncle	bladder	Invasive bladder cancer
	Ureteral tumors	Interstitial cystitis
	Genital and urethral	Prostatectomy
	telangiectasia	Ureteral tumors
	Posterior urethral valves	Ureteropelvic junction obstruction
		Partial nephrectomy

with the former. With a CO_2 laser, the lesion literally is vaporized, and large amounts of smoke are produced. Application of laser energy to the skin is painful, and local anesthetics are injected subcutaneously before treatment. For large or more extensive lesions, a penile block may be preferable. The skin is cleansed with an iodine based or equivalent compound. The power output chosen is based to some extent on the size of the lesion, but the lowest output that will successfully vaporize the condyloma is desirable. Usually, 5–10 W of power will suffice. Rosemberg et al. (1981) described a technique of vertial, horizontal, and oblique application of the beam to ensure complete removal. As successive portions of the condyloma are treated, the carbonized surface is removed with a saline-soaked sponge. Bleeding does not occur until the deeper layers of the dermis are reached, and care should be taken to avoid a full-thickness skin injury. An antibacterial cream may be applied topically after treatment, but analgesics are usually unnecessary. Lesions of the urethral meatus can be treated satisfactorily with a CO_2 laser, but the lack of a fiber delivery system has precluded treatment of more proximal portions of the urethra. In most patients, excellent cosmetic results can be anticipated.

Lundquist and Lindstedt (1983) treated over 150 patients with condyloma resistant to podophyllin therapy, of whom 95% were cured, although some required a second treatment. Similar results have been reported by Bellina (1983). Rosemberg (1983) found a 88% cure rate in 61 patients undergoing only a single treatment.

The Nd:YAG laser has also proved to be effective in treatment of condyloma acuminata (Hofstetter and Frank, 1983). Unlike after CO_2 laser treatment, the lesions do not undergo vaporization. Rather, they are coagulated and may be removed with forceps or allowed to slough secondarily. Penetration depths may be more difficult to control than with the CO_2 laser, but treatment proceeds more rapidly when large lesions are encountered.

Carcinoma of the Penis

Although results of large clinical series are not available, laser treatment of selected carcinomas of the penis has been successful and may avoid the need for partial penectomy in some patients. A CO_2, argon, or Nd:YAg laser should be capable of eliminating carcinoma in situ (erythroplasia of Queyrat) with good cosmetic results (Rosemberg and Fuller, 1980). For an invasive carcinoma, a Nd:YAG laser seems to be the most appropriate, since its greater depth of penetration would be desirable. Rothenberger and co-workers (1981) have reported 19 patients with selected squamous cell tumors of the penis who underwent primary therapy with

a Nd:YAG laser. Cosmetic and functional results were excellent, and there were no instances of local recurrences.

Urethral Strictures

Experimental evidence suggests that a thermal injury from a Nd:YAG laser heals with more elastic fibers and less collagen deposition than a comparable electrocautery injury (Hofstetter and Frank, 1983). In addition, there may be a more even distribution of the thermal damage. These data have formed the basis for laser treatment of benign urethral strictures. There seems little doubt that laser treatment of urethral strictures can provide short-term symptomatic improvement, but delayed results have been less satisfactory.

Two basic techniques for Nd:YAG laser treatment of urethral strictures have been described. In both, a metal guide wire is inserted through the urethral lumen to maintain orientation. Circumferential laser energy may be delivered to the entire stricture in anticipation of secondary slough of this tissue and nonstrictured healing. Using this technique and no postoperative Foley catheter, Smith and Dixon (1984) reported a recurrence rate of 56% at 6 months and considered this method generally to offer no advantages over standard therapy (Fig 7–6). Shanberg and Tansey (1985) have described radial incisions through the stricture using direct contact between the fiber and the tissue. Results in five patients with bladder neck contractures were good. However, recurrences were noted in 58% of the patients with urethral strictures.

Argon laser energy has also been used in some series. Rothauge (1980) treated 41 patients, but long-term results were not reported. The coagulative effects on the tissue would be similar to those obtained with a

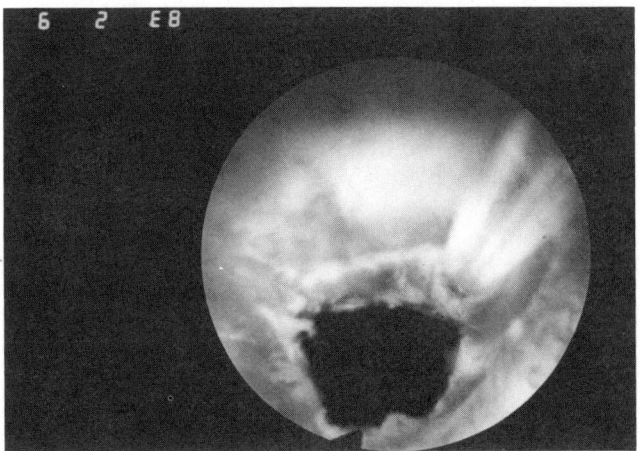

FIG 7–6.
Urethral strictures after Nd:YAG laser treatment. The laser fiber is seen *(left lower)*. The interface between the laser coagulated tissue and the normal urethra is visible.

Nd:YAG laser, but the limited power output of an argon laser appears to make the procedure quite tedious and technically unsatisfactory for strictures with dense scar tissue.

Based on physics and tissue effects, the CO_2 is the most appealing laser for treatment of urethral strictures. Theoretically, the scar could be vaporized with little effect on the underlying urethra. However, until a CO_2 laser fiber or cystoscope is perfected, treatment requires gaseous distention of the urethra, a setting in which fatal air embolus has been reported (Voure'h et al., 1982).

BLADDER CANCER

Clinical experience from centers in Europe, the United States, and Japan has established the safety and feasibility of laser treatment of bladder cancer. Less certain are whether laser treatment offers any substantial practical or therapeutic advantages over electrocautery resection in the treatment of superficial bladder tumors and what role lasers may have in the cost-effective management of recurrent bladder tumors. In addition, although there is theoretical justification for laser treatment of invasive bladder cancer, the efficacy of this approach is not established.

SUPERFICIAL BLADDER TUMORS

Either Nd:YAD or argon laser energy is easily delivered to bladder tumors via a small flexible fiber which can be inserted through a cystoscope and manipulated in a manner similar to that used with a ureteral catheter (Fig 7–7). The endoscopic expertise of urologists and the sophisticated instrumentation available have facilitated training of urologists in laser treatment of bladder cancer. Although laser treatment of superficial bladder tumors has been proved to be safe and effective, the obvious issue that arises is whether laser treatment is safer or more effective than an already established treatment technique, i.e., electrocautery resection.

Despite recent improvements in intravesical chemotherapy and immunotherapy, the recurrence rate of superficial bladder tumors remains disturbingly high. After electrocautery resection alone, recurrence rates of

some 50%–70% can be anticipated. Although there are multiple potential explanations for this high rate of recurrence, experimental evidence and clinical observations suggest that at least some recurrences are due to implantation of viable tumor cells that may become dislodged at the time of electrocautery resection (Soloway and Masters, 1979). Treatment of superficial bladder tumors with a Nd:YAG laser is performed in a noncontact fashion. Protein denaturation and thermal coagulation of the tumor cells occur in situ, and the lesions slough secondarily. Thus, it has been theorized that the recurrence rate of bladder cancer will be less after laser treatment than after electrocautery resection.

Randomized studies to support this theory are incomplete, and comparison of uncontrolled populations of patients with superficial bladder tumors is subject to error from differences in patient selection criteria, follow-up, and analysis of treatment results. Nevertheless, Hofstetter and Frank found a decreased recurrence rate in patients treated by laser compared with electrocautery (Hofstetter and Frank, 1983; Hofstetter et al., 1981). Malloy and colleagues (1981) found a recurrence rate of 18% at one year in patients undergoing Nd:YAG laser treatment. Eighteen of 50 patients (36%) at the University of Utah with superficial bladder tumors treated with a Nd:YAG laser have developed a recurrence in remote areas of the bladder within one year (Table 7–2). Only two patients had tumor persistence at the site of laser treatment (Smith and Middleton, 1985). Some of the differences between this and other series may be explained by patient selection. All of the Utah patients had recurrent tumors before therapy, and lasers were not used as treatment of an initial bladder tumor. At any rate, although these series demonstrate the effectiveness of a Nd:YAG laser in the treatment of superficial bladder tumors, superiority to electrocautery resection in terms of decrease in the rate of tumor recurrence has not been established.

Since tissue is not retrieved for histologic examination after laser therapy, the issue of tumor staging and grading must be addressed. Clearly, treatment goals, technique, and follow-up depend on the depth of tumor invasion in the bladder wall, and adequate staging is mandatory. Although a preliminary electrocautery resection of the tumor prior to laser therapy has been

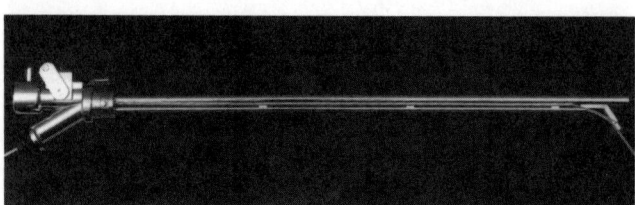

FIG 7–7.
Modified Albarràn element for endoscopic laser treatment. The Nd:YAG laser fiber is inserted through the working channel and stabilized by the deflector mechanism.

TABLE 7–2.

University of Utah Selection Criteria for Laser Treatment of Superficial Bladder Cancer

Prior documented history of superficial bladder tumors
Normal or low-grade malignant cells on voided cytology
Papillary low-grade-appearing tumor at cystoscopy
Tumor less than 3 cm in diameter

mentioned in some reports as a means of tumor staging, such an approach appears to obviate many of the potential practical advantages of lasers. Futher, it would negate any theoretical benefit laser treatment has in decreasing tumor recurrence due to implantation. Cold cup biopsies from the tumor base may be useful in confirming the histologic diagnosis and in determining grade but do not adequately define the stage. Finally, submission for pathologic examination of the coagulated tumor mass after laser treatment has also been suggested. Although some gross architectural detail is maintained by the tumor after Nd:YAG laser treatment, this method has been unsatisfactory in our hands and has been insufficient for determination of tumor stage.

On the other hand, at the University of Utah, we have established patient selection criteria for laser treatment which appear to address satisfactorily the issue of tumor staging. Patients are selected for laser therapy if they have a history of low-grade, low-stage bladder tumors, normal or low-grade malignant cells on voided cytology, and what appears visually to be a superficial papillary tumor at cystoscopy (Fig 7–8). In addition, tumors should be less than 3 cm in diameter. If these criteria are satisfied, histologic examination of the tumor appears to be unnecessary and the risk of significant understaging low. On the other hand, patients with high-grade malignant cells on cytology, a sessile appearing tumor, or those presenting with their initial lesion have been excluded from laser therapy and have undergone definitive electrocautery resection (Fig 7–9). Although this more conservative approach differs from that of others, it has allowed adequate tumor staging when necessary and effective laser therapy without compromise in other patients (Table 7–3). Importantly,

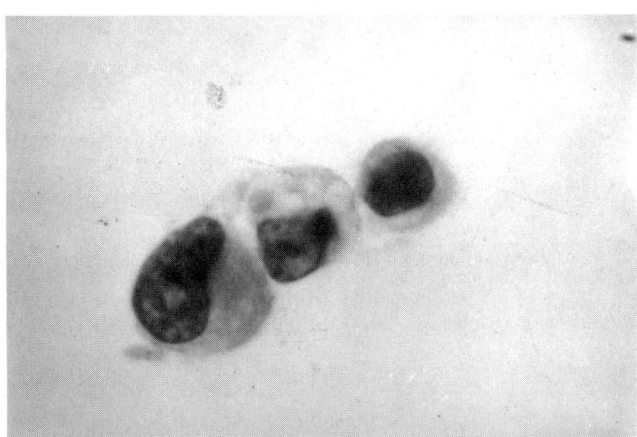

FIG 7–9.
High-grade, malignant transitional cells on voided cytology. Patients with such cytologic findings generally should undergo an electrocautery resection for definitive staging.

no patient selected by these criteria has experienced tumor recurrence at a higher stage.

Treatment Technique

Although definitive therapeutic advantages for laser treatment of superficial bladder tumors compared with electrocautery resection have not been established, several potential practical advantages appear to reduce patient morbidity and perhaps overall treatment cost. Small papillary bladder tumors can be cauterized successfully without anesthesia using a Bugbee electrode or electrical fulguration. However, electrical resection of bladder tumors is usually performed in an operating room using general or spinal anesthesia. The irregular thermal conduction of electrical energy within the bladder wall is poorly tolerated without anesthesia. A Nd:YAG laser produces a rapid, intense, and more defined thermal injury within the bladder wall. Although patients are able to perceive the laser energy and may describe it as a "burning sensation," anesthesia is rarely necessary.

Laser treatment may be performed in any suitably

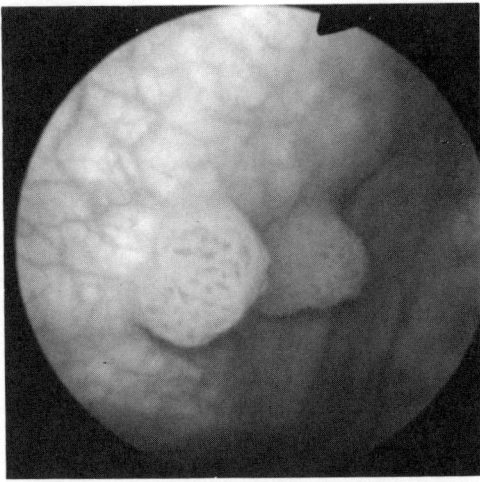

FIG 7–8.
Papillary, low-grade–appearing bladder tumor. In this setting, laser treatment frequently can be performed without biopsy.

TABLE 7–3.
Practical Advantages of Laser Treatment of Superficial Bladder Tumors Compared With Electrocautery Resection

Small cystoscopic sheath (17–21 F)
Treatment possible with flexible cystoscopes
No obturator nerve stimulation
Excellent hemostasis
Minimal risk of bladder perforation
Local anesthesia
Foley catheter unnecessary
Ambulatory, outpatient procedure
Postoperative bladder spasm minimal

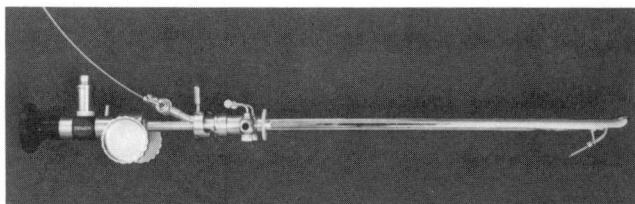

FIG 7–10.
Retrograde viewing telescope with exaggerated deflection mechanism. Such instruments or flexible cystoscopes are useful for laser treatment of tumors at the bladder neck.

equipped operating room or cystoscopic area. The laser fiber is inserted through a modified cystoscope, which provides stabilization of the fiber tip and a watertight seal. Treatment is possible through a flexible cystoscope or nephroscope, and a retrograde viewing lens may be desirable, especially for lesions at the bladder neck (Fig 7–10). Lidocaine jelly may be instilled into the urethra to lessen the discomfort of cystoscopy, but no other anesthetics are used.

After the tumor is visualized, the fiber tip is positioned approximately 3–5 mm from the tumor surface. A power output of 35–45 W is chosen, depending on the size of the tumor. Treatment is applied to a given area until a white discoloration of the tissue indicative of adequate thermal necrosis is evident (Fig 7–11). Generally, this requires 2–3 seconds for application. The pulse duration usually is controlled by the foot pedal rather than from the instrument control panel. Treatment adequacy is determined by visible changes in the tumor instead of an absolute figure for total energy. For larger tumors, the tip of the cystoscope may

be used to dislodge the superficial, thermally coagulated tissue and expose deeper portions of the lesion. Laser energy does not stimulate the obturator nerve and cause spasm of the bladder wall as may occur with electrical energy. Bleeding is usually minimal or nonexistent. A Foley catheter is not used postoperatively, and patients may be discharged to home directly from the cystoscopy area. Tumor slough should be complete within six weeks (Fig 7–12).

An argon laser may be used in a similar fashion to treat bladder tumors. The limited power output of an argon laser restricts its use to small tumors, usually those smaller than 1 cm. On the other hand, the superficial penetration imparts a high margin of safety (Smith and Dixon, 1984).

At the energy levels described, laser treatment of superficial bladder tumors has proved to be safe and generally free of complications. Because tissue destruction with a Nd:YAG laser occurs due to thermal coagulation rather than vaporization, the bladder wall maintains its structural integrity even after a full-thickness injury. Therefore, the risk of bladder perforation or urine extravasation is low. Of more concern is forward scattering of the laser energy with damage to the small bowel. Nevertheless, the incidence of this complication has been quite low. Hofstetter and Frank (1983) treated over 500 tumors and have reported only two instances of small bowel perforation, at least one in which excessive energy levels were used inadvertently. We have not observed this complication in any of the 112 laser treatments of superficial bladder tumors that we have performed. Significant bleeding either at the time of laser treatment or as a delayed phenomenon has not

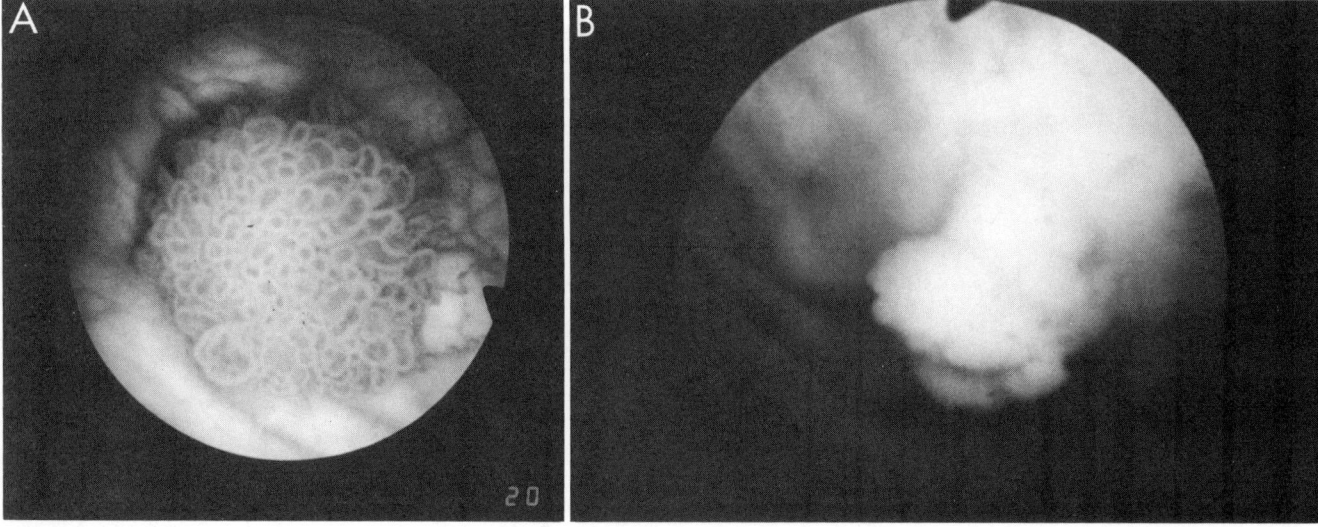

FIG 7–11.
A, papillary bladder tumor before treatment. **B,** after application of Nd:YAG laser energy, white discoloration indicative of adequate thermal necrosis is evident.

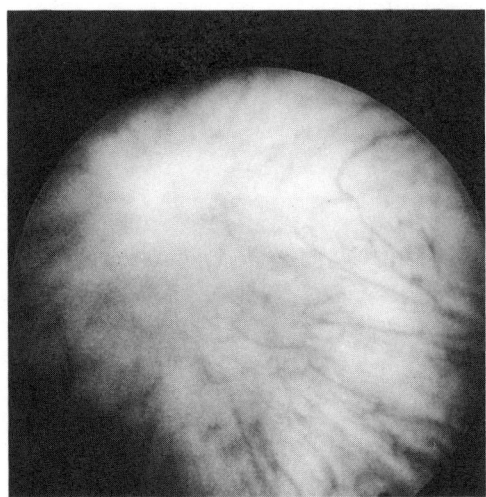

FIG 7–12.
Pale, white scar is visualized six weeks after laser treatment. Biopsy specimens show no residual tumor.

been seen in any patient. In general, postoperative discomfort from bladder spasm has been minimal and apparently less than that observed after electrocautery resection of superficial bladder tumors.

All of these factors appear to impart substantial practical benefits for laser treatment of superficial bladder tumors. On an individual patient basis, the technique can be cost-effective, as the need for anesthesia or hospitalization is eliminated (see Table 7–3). A Nd:YAG laser is an instrument used by medical and surgical specialties throughout the hospital. Thus, laser costs usually are diluted by larger patient volume. Whether or not a laser dedicated to use only in urology would be cost-effective depends on the frequency of its use.

INVASIVE BLADDER CANCER

Based on physics and tissue interactions, several properties of a Nd:YAG laser make it an attractive potential treatment for muscle-invading bladder cancer. In a controlled setting a Nd:YAG laser can produce a transmural coagulation of the bladder wall without perforation. Temperatures at the serosal surface of the bladder do not exceed 60°C, the point at which thermal protein denaturation occurs (Pensel et al., 1981). In addition, there is some experimental evidence to suggest that a Nd:YAG laser is capable of sealing lymphatic vessels within the bladder wall (Zimmermann et al., 1984).

An important consideration is that laboratory studies dealing with energy density are difficult to transpose to a clinical endoscopic setting. Multiple factors affect the penetration depth. Some of these, including selection of laser wavelength, power output, and pulse duration, can be controlled with precision in a clinical situation. On the other hand, one important factor that determines energy density, the spot size of the laser beam or the surface area being treated, is more difficult to control. Thus, reliable, predictable depths of thermal injury are difficult to achieve. Nevertheless, the potential for definitive endoscopic management of muscle-invading bladder tumors with a Nd:YAG laser has prompted limited clinical studies.

Patient Selection and Treatment Techniques

Because of uncertainties regarding the efficacy of this approach, laser treatment of invasive bladder cancer has been limited in most centers to patients who are not candidates for cystectomy due to age or health or to those who refuse radical surgery. Ideally, a preliminary transurethral electrocautery resection of the tumor should be performed. This not only establishes the histologic diagnosis of invasive bladder cancer but debulks the tumor prior to laser therapy. Several days should elapse between the time of electrocautery resection and laser treatment to allow all bleeding to cease and the overlying clot to lyse. Because of the higher energy densities and longer duration of the procedure, laser treatment of invasive bladder cancer is best performed using general or spinal anesthesia.

The laser fiber is positioned approximately 2–5 mm from the tissue surface. Higher energy densities are used than with treatment of superficial bladder tumors, since greater penetration depths are desired. In general, a power output of 45–50 W is maintained in a given area for three or four seconds (Table 7–4).

The tumor bed is generally a shaggy and irregular crater before treatment, and this precludes effective visual control of treatment adequacy. Therefore, a systematic application of laser energy to the entire surface of the tumor bed as well as the surrounding tissue should be performed (Fig 7–13). Although bleeding is usually negligible, a Foley catheter is used for 24 hours after treatment.

Healing and reepithelialization of the tumor bed may take up to two months. Unless persistent tumor is evident, the area heals as a white-colored scar. In some patients, even in the absence of tumor, a dystrophic

TABLE 7–4.

Energy Specifications for Laser Treatment of Bladder Cancer

	DISTANCE, MM	POWER, W	DURATION, SEC	VISIBLE CHANGES
Superficial tumors	2–5	35–40	2–3	White discoloration indicates adequate thermal necrosis
Invasive cancer	2–5	45–50	4	Systematic treatment—not dependent on visible changes

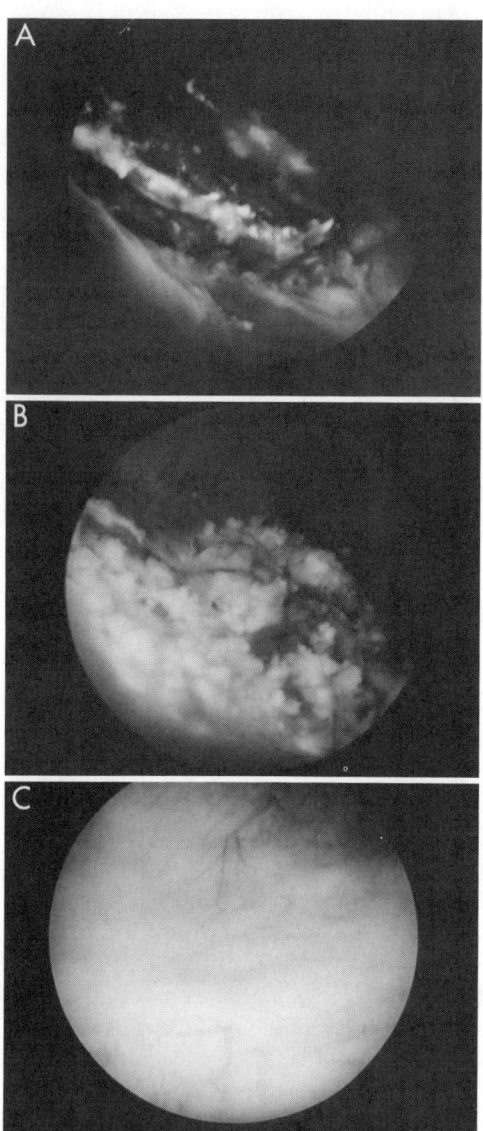

FIG 7–13.
A, tumor crater after electrocautery resection of a high-grade transitional cell carcinoma. **B,** after Nd:YAG laser treatment, irregular thermal injury is evident. **C,** three months later, healing is complete without apparent residual tumor.

calcification of the bladder wall has occurred although there were no related symptoms. Determination of treatment adequacy relies on periodic cystoscopic examinations and voided cytologies. A concern is that re-epithelialization of the tumor surface could occur despite the presence of viable tumor in deeper layers of the bladder. Although this potential does exist and detection of this event may be difficult, it has not been observed to date in any of our patients.

Only limited clinical series are available from which to evaluate results of laser treatment of invasive bladder cancer. Overall, reported results have been relatively encouraging in clinical stage B_1 (T_2) tumors. Five of six such patients whom we have treated remain apparently free of tumor from six months to three years after treatment (Smith, 1986). Libertino (1986) has performed laser therapy prior to cystectomy. There was no histologic evidence of tumor in the majority of the patients with clinical stage B_1 tumors.

As may be anticipated, results are less satisfactory with tumors that invade the deeper layers of the bladder wall. Only three of the six patients with clinical stage B_2 tumors whom we have treated remain tumor free, and 80% of patients with stage C lesions have experienced tumor recurrence. Laser treatment of bulky tumors up to stage T_4 can only be performed with palliative intent. Although objective evaluation of results has not been performed, most investigators have concluded that the Nd:YAG laser was at least moderately successful in decreasing irritative voiding symptoms in these patients and diminishing the frequency of bleeding episodes in patients with unresectable stage T_4 tumors.

URETER

Recent advances in rigid instrumentation have allowed assessment of virtually the entire ureter and upper collecting system. In addition, these instruments have a working channel that allows therapeutic intervention and will easily accept a Nd:YAG or argon laser fiber. Selection of lesions of the ureter and upper collecting systems that may be amenable to laser therapy is more difficult. There are ongoing studies investigating the use of laser energy for destruction of urinary calculi. Pulsed laser energy can mechanically disrupt urinary calculi but thermal effects may be excessive. Other laser modalities such as the eximer laser being investigated by Watson (1985) hold promise, but clinical utility has not yet been demonstrated. Laser incision of ureteral strictures and ureteropelvic junction obstructions has been suggested, but the coagulative effects of a Nd:YAG or argon laser appear to be unsuitable for this purpose, and the lack of a fiber delivery system for a CO_2 laser precludes its use through ureteroscopes.

Laser treatment of transitional cell tumors of the ureter and upper collecting system not only seems feasible but has been accomplished successfully in limited numbers of patients. The hemostatic capabilities of a Nd:YAG laser may be particularly advantageous in this setting, since minimal amounts of bleeding can obscure vision through a ureteroscope. Control of penetration depths in the thin-walled ureter may be difficult. A power output of 35 W applied for two seconds, an energy level that is safe within the bladder, produced a ureteral perforation in dogs with thermal injury extending into the periureteral tissue (Smith, 1983) (Fig 7–14). These laboratory results are not directly applicable to a clinical situation with tangential application of the laser beam through a ureteroscope. Nevertheless, they do mandate caution in laser treatment of the ureter.

Hofstetter et al. (1983) successfully treated six patients with tumors of the ureter using 35 W of power output. Ureteral perforation was not observed, although postoperative ureteral stents were used in all patients. We performed ureteroscopic laser treatment on four patients with transitional cell tumors of the intramural ureter where the backing of the bladder wall decreases the risk of perforation. Malloy (1985) applied Nd:YAG laser energy through an open pyelotomy to transitional cell tumors in solitary kidneys in two patients prior to partial nephrectomy.

Overall, the current indications for laser treatment of the ureter are limited. Visualization and application of laser energy to the surface of tumors can be somewhat difficult, and ureteral perforation is a possibility. However, the blood-free visual field and the potential for stricture-free healing of the ureter justify continued investigations.

KIDNEY

Scattered reports in the literature have suggested a role for lasers in the performance of partial nephrectomy. Barzilay and Fine (1981) performed partial nephrectomy in dogs using a CO_2 laser. They reported a reduced blood loss compared with a control group as well as a substantial decrease in the overall operating time. On the other hand, Hall (1982) found the CO_2 laser to be of little value for partial nephrectomy.

Overall, it seems valid to conclude that the use of lasers alone for partial nephrectomy has been disappointing. The rapid absorption of a CO_2 laser beam and the intense surface effect allow its use as a "surgical knife." A cutting action is observed, and hemostasis of small blood vessels can be achieved. However, the cutting action is slow and the hemostatic properties are insufficient for the larger, high-pressure blood vessels within the renal parenchyma. On the other hand, the primary effects of a Nd:YAG laser are through tissue coagulation. It is an ineffective cutting device, and lateral propagation of the thermal damage causes excessive injury before tissue separation occurs. Finally, when attempts have been made to attain hemostasis with a Nd:YAG laser after a guillotine-type amputation of the kidney, the vessels retract within the parenchyma, and excessive destruction of viable renal tissue may occur (Fig 7–15).

A Cooper Ultrasonic Surgical Aspirator (CUSA) is an instrument that converts electricity into ultrasonic vibration and mechanical energy. High collagen density structures such as blood vessels and nerves as well as the renal capsule are relatively resistant to the action of a CUSA. On the other hand, renal parenchyma is readily separated and aspirated. Blood vessels can be isolated and skeletonized. Prior reports of CUSA partial nephrectomy have relied upon surgical steel clips to secure the blood vessels prior to division (Chopp et al., 1983). Recently, the combined use of a Nd:YAG laser and CUSA to achieve hemostasis after partial nephrectomy has been reported (Melzer et al., 1985). A cold knife incision of the renal capsule is performed and the

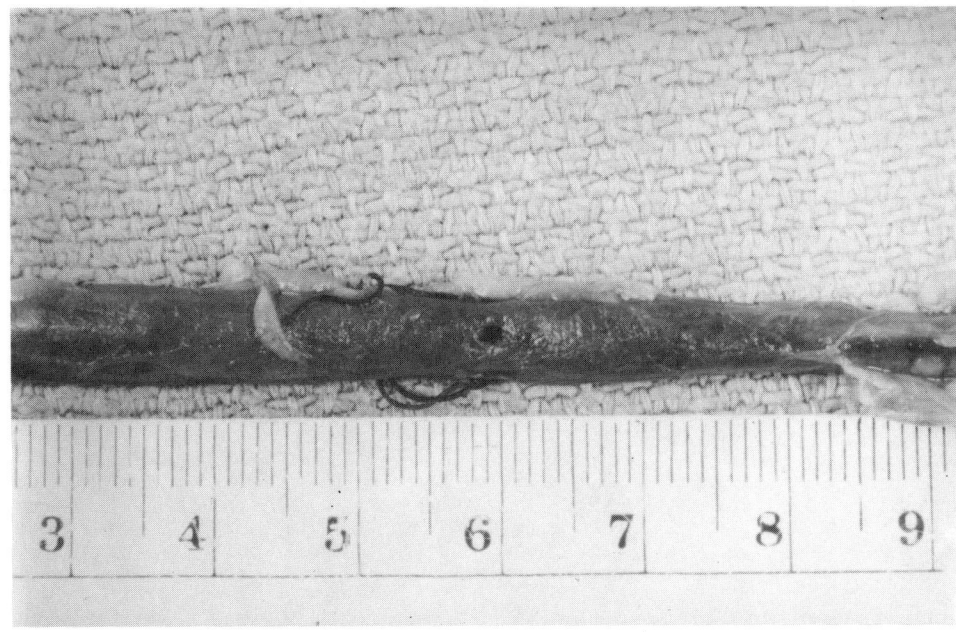

FIG 7–14.
Gross perforation of canine ureter from Nd:YAG laser energy, 35 W × 2 seconds.

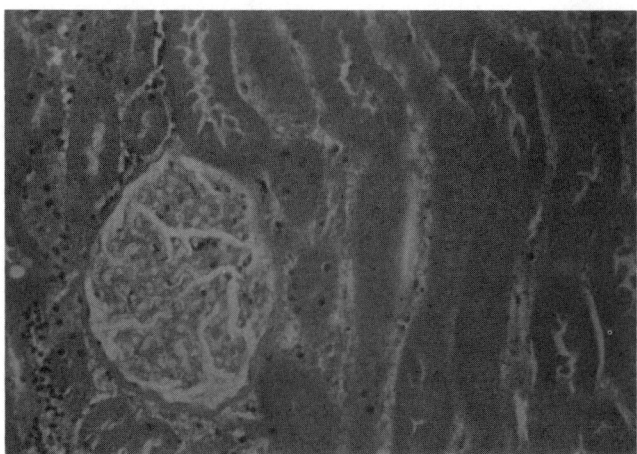

FIG 7–15.
Thermal injury to renal parenchyma after partial nephrectomy with Nd:YAG laser.

CUSA is used to separate the renal parenchyma and skeletonize blood vessels. The Nd:YAG laser is used to coagulate these vessels prior to their division. In this manner, effective coagulation can be achieved with minimal thermal damage to the surrounding renal parenchyma (Fig 7–16).

Although a Nd:YAG laser functions poorly as a cutting device, its coagulative abilities may be useful to increase the margin of resection after partial nephrectomy or enucleation of renal cell carcinomas. Percutaneous application of laser energy through nephroscopes with destruction of tumors within the renal pelvis or collecting system appears feasible and relatively easy to accomplish. However, concerns regarding spillage of tumor cells and implantation along the nephrostomy tract or in the perinephric space have limited clinical application of this technique.

HEMATOPORPHYRIN SENSITIZATION

Previous discussions in this chapter have dealt with the pure thermal effect of lasers. The thermal injury that occurs is rather nonspecific and dependent on beam direction. A potentially promising form of tumor treatment relies not on the thermal activity of lasers but on a chemical reaction in photosensitized cells.

Several potential photosensitizing agents have been developed. The one studied most extensively has been hematoporphyrin derivative (HpD). The mechanism of the photocytotoxic action of HpD is uncertain but may occur at both a cellular and a vascular level. Excitation of HpD-sensitized cells produces singlet oxygen, which is toxic to cellular components.

After IV administration, HpD concentrates in dysplastic or frankly malignant cells. The mechanism is not completely defined but probably is due to prolonged retention of the drug within the cell rather than selective uptake. At any rate, this phenomenon has been observed in several experimental tumor models as well as in clinical experience. This property has prompted investigation of HpD phototherapy as a means of both treatment and detection of multifocal carcinoma in situ of the bladder as well as other tumor systems.

Excitation of HpD can be produced by any light

FIG 7–16.
Skeletonization of intrarenal blood vessels during partial nephrectomy using an ultrasonic surgical aspirator.

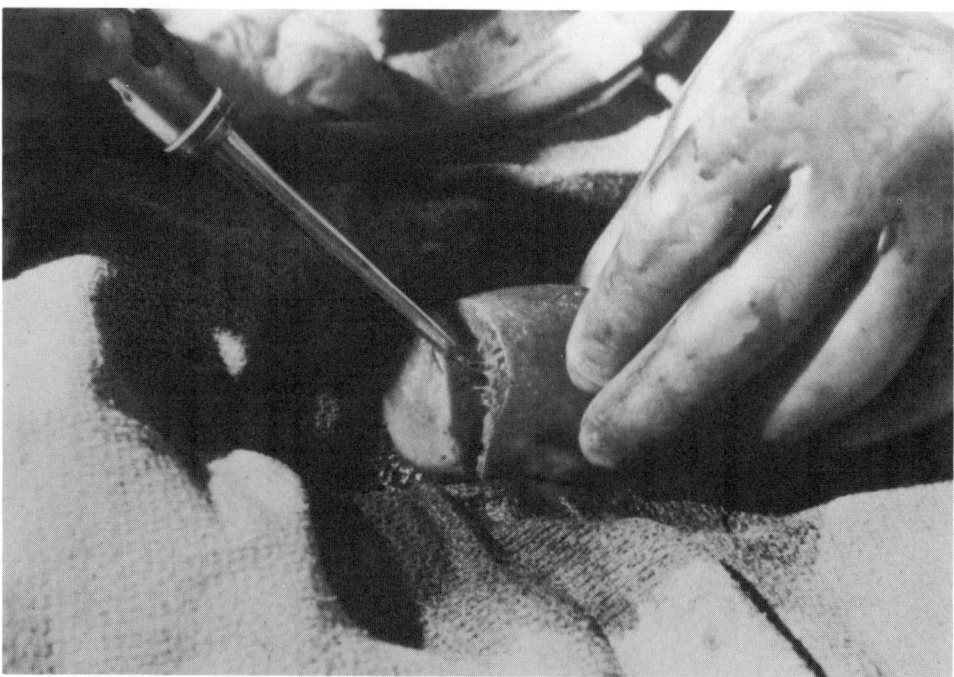

source that emits a spectrum corresponding to the absorption characteristics of HpD. The peak absorption of HpD occurs in the blue (approximately 400 nm) or blue/green (approximately 500 nm) portion of the electromagnetic spectrum. However, light at these wavelengths penetrates tissue only 1–2 mm, a depth that may be insufficient for clinical use. Although red light (630 nm) corresponds to only a minor absorption peak of HpD, it penetrates tissue approximately 1 cm and has been used most frequently with hematoporphyrin phototherapy. Usually, an "argon-dye" laser is used for production of the appropriate wavelength red light. The active medium is a rhodamine-B dye, and the machine is powered by an argon laser. The energy is transmitted via a flexible quartz fiber.

Although HpD is selectively localized in malignant and premalignant cells, uptake in other organs occurs with systemic administration, and the drug may also be retained in inflamed tissue. The primary morbidity of HpD phototherapy is skin photosensitization. After therapeutic doses of HpD are administered IV, HpD absorption by the skin produces a marked photosensitization. Exposure to direct sunlight within four weeks of drug administration can produce a severe skin burn. The drug has been poorly absorbed by the bladder epithelium after intravesical administration. The active component of HpD is uncertain, and ongoing trials may allow more selective treatment of the bladder without skin toxicity.

RESULTS

The fluorescence emitted by cells that have concentrated HpD can be used for both the diagnosis and detection of areas of dysplasia or carcinoma in situ. Benson and associates (1983) as well as Lin and co-workers (1984) described a device that consists of "chopped" light. A fluorescence detector transposes the fluorescence into an audible tone. An audible signal is emitted when the laser light is directed at dysplastic or malignant cells.

The optimal interval between HpD administration and laser excitation is undetermined, but most clinical studies have used 48–72 hours. Treatment can be area specific, especially when coupled with an audio detector. However, the potential for simultaneous whole-bladder therapy by means of diffusion of the light appears to be of more practical value. This can be accomplished by rounding the fiber tip into a bulb or through the use of an intravesical diffusion medium such as a lipid solution. Benson et al. (1985) treated nine patients with diffuse carcinoma in situ resistant to alternative therapy using a bulb-tip fiber. Three milligrams per kilogram of body weight of HpD was administered IV, and treatment with red light from an argon pumped

dye laser was performed at three hours and at 48 hours after drug administration. Energy delivery to the entire bladder approximated 50 joule/cm^2, and seven of nine patients had complete disappearance of their disease at three months (Benson, 1985).

Many questions remain unanswered regarding HpD phototherapy. The mechanism of action of the drug, the active component, the optimal timing between drug administration and laser excitation, and the appropriate energy levels all are uncertain. In addition, long-term clinical results and side effects are not known. Bladder contracture has been reported after hematoporphyrin phototherapy, but the incidence and relation of this complication to treatment is poorly defined. Nevertheless, the potential for selective sensitization of malignant cells using HpD and the ability to deliver simultaneous whole bladder therapy using an argon dye laser create a unique and potentially exciting treatment technique for patients with diffuse carcinoma in situ of the bladder.

SUMMARY

Lasers were first investigated in urologic surgery almost 17 years ago, but it has only been in the past several years that interest has increased. Not only is there a greater awareness of these instruments by urologists but also there have been quantum improvements in laser technology. Laser surgery is still in its infancy, and undoubtedly major changes will occur in the future. Further clinical investigation will allow identification not only of specific areas in which laser treatment may be indicated but also of conditions in which lasers appear to offer no benefit over standard therapy.

The unique physics and tissue effects of laser energy create the potential for new therapeutic options in urology or improvements on existing ones. As more urologic surgeons become familiar with the technique and the instruments become more widely available, clinical experience will increase. It is hoped that objective assessment and a critical evaluation of treatment results will allow further definition of the proper role of lasers in urologic surgery.

REFERENCES

1. Barzilay BI, Fine H: Experimental and clinical application of a CO_2 laser beam in renal parenchymal surgery. Proceedings of the 4th Congress International Society of Laser Surgery, Inter-Groups, 1981.
2. Bellina JH: Lasers in gynecology. *World J Surg* 1983; 7:692.
3. Benson RC Jr: Hematoporphyrin photosensitization and the argon-dye laser, in Smith JA (ed): *Lasers in Urologic Surgery.* Chicago, Year Book Medical Publishers, 1985, p 114.

4. Benson RC Jr, Farrow GM, Kensey JH, et al: Detection and localization of in situ carcinoma of the bladder with hematoporphyrin derivative. *Mayo Clin Proc* 1983; 57:548–555.

5. Chopp RT, Shah BB, Addonizio JC: Use of ultrasonic surgical aspirator in renal surgery. *Urology* 1983; 22:157.

6. Einstein A: Zur Quantentheorie der Strahlung. *Physiol Z* 1917; 18:121–128.

7. Frank F, Keiditsch E, Hofstetter A, et al: Various effects of the CO_2, the neodymium:YAG, and the argon laser irradiation on bladder tissue. *Lasers Surg Med* 1982; 2:89–96.

8. Hall RR: Report to the standing committee on urological instruments: Lasers in urology. *Br J Urol* 1982; 54:421.

9. Hofstetter A, Frank F, Keiditsch E, et al: Endoscopic neodymium:YAG laser application for destroying bladder tumors. *Eur Urol* 1981; 7:278.

10. Hofstetter A, Frank F: Laser use in urology, in Dixon JA (ed): *Surgical Applications of Lasers*. Chicago, Year Book Medical Publishers, 1983.

11. Hofstetter A, Bowering R, Keiditsch E, et al: Ders Uretertumoren mit dem Nd:YAG laser. *Fortschr Med* 1983; 14:619.

12. Libertino JA: Personal communication, 1986.

13. Lin CW, Bellnier DA, Prout GR Jr, et al: Cystoscopic fluorescence detector for photodetection of bladder carcinoma with hematoporphyrin derivative. *J Urol* 1984; 131:587–590.

14. Lundquist SA, Lindstedt EM: Laser treatment of condylomata acuminata (abstract). *Lasers Surg Med* 1983; 3:152.

15. Maiman TH: Stimulated optical radiation in ruby. *Nature* 1960; 187:493–494.

16. Malloy TA, Wein AJ, Shanberg A: Superficial transitional cell carcinoma of the bladder treated with a neodymium:YAG laser: A study of recurrence rate within the first year. *J Urol* 1984; 131:251.

17. Malloy TR: Laser treatment of the ureter, in Smith JA Jr (ed): *Lasers in Urologic Surgery*. Chicago, Year Book Medical Publishers, 1985, pp 85–90.

18. Melzer RB, Wood TW, Landau ST, et al: Combination of C.U.S.A. and Nd:YAG laser for canine partial nephrectomy. *J Urol* 1985; 134:620.

19. Mulvaney WP, Beck CW: The laser beam in urology. *J Urol* 1968; 99:112–115.

20. Pensel J, Hofstetter A, Frank F, et al: Temporal and spatial temperature profile of the bladder serosa in intravesical neodymium:YAG laser irradiation. *Eur Urol* 1981; 7:298.

21. Rosemberg SK, Fuller TA: Carbon dioxide rapid superpulsed laser treatment of erythroplasia of Queyrat. *Urology* 1980; 16:181.

22. Rosemberg SK, Fuller TA, Jacobs H: Continuous wave carbon dioxide laser treatment of urethral condylomata. *Urology* 1981; 17:149.

23. Rosemberg SK: The use of the CO_2 laser in urology (abstract). *Lasers Surg Med* 1983; 3:114.

24. Rothauge CF: Urethroscopic recanalization of urethral stenosis using an argon laser. *Urology* 1980; 16:158.

25. Rothenberger K, Hofstetter A, Pensel J, et al: Laserbehandlung maligner tumoren des penis. *Fortschr Med* 1982; 10:806.

26. Shanberg AM, Tansey LA: Laser treatment of urethral strictures, in Smith JA Jr (ed): *Lasers in Urologic Surgery*. Chicago, Year Book Medical Publishers, 1985, p 47.

27. Smith JA Jr: Neodymium:YAG laser photoradiation of canine ureters: An analysis of penetration depth and subsequent healing. *Surg Forum* 1983; 34:696.

28. Smith JA Jr, Dixon JA: Argon laser phototherapy of superficial transitional cell carcinoma of the bladder. *J Urol* 1984; 131:655.

29. Smith JA Jr, Dixon JA: Neodymium:YAG laser treatment of benign urethral strictures. *J Urol* 1984; 131:1080.

30. Smith JA Jr, Middleton RG: Bladder cancer, in Smith JA Jr (ed): *Lasers in Urologic Surgery*. Chicago, Year Book Medical Publishers, 1985, p 53.

31. Smith JA Jr, Dixon JA: Tissue effects of lasers in the genitourinary system, in Smith JA Jr (ed): *Lasers in Urologic Surgery*. Chicago, Year Book Medical Publishers, 1985.

32. Smith JA Jr: Laser treatment of invasive bladder cancer. *J Urol* 1986; 135:55–57.

33. Soloway MS, Masters S: Implantation of transitional tumor cells on the cauterized murine urothelial surface. *Proc Am Assoc Cancer Res* 1979; 20:256.

34. Stein BS, Kendall AR: Lasers in urology: II. Laser therapy. *Urology* 1984; 23:411.

35. Voure'h G, Bereti E, Trichet B, et al: Two unusual cases of gas embolism following urethral surgery with laser. *Intensive Care Med* 1982; 8:239.

36. Watson GM: Laser fragmentation of urinary calculi, in Smith JA Jr (ed): *Lasers in Urologic Surgery*. Chicago, Year Book Medical Publishers, 1985, p 136.

37. Zimmermann I, Stern J, Frank F, et al: Interception of lymphatic drainage by Nd:YAG laser irradiation in rat urinary bladder. *Lasers Surg Med* 1984; 4:167.

Chapter 8

Malignancy: Underlying Concepts of Etiology, Natural History, and Treatment

WILLIAM C. DEWOLF, M.D.

Cell growth and division are fundamental properties of life. For survival, an organism must create new cells at a rate as fast as that at which its cells die. In an adult human, for example, millions of cells must divide every hour simply to maintain the status quo. Although this process of cell division appears strikingly simple, in reality it is a complex process in which each individual cell is required to double its mass and duplicate all of its contents. When growth regulatory mechanisms are deranged for whatever reason, cell growth and division may either stop or proceed unrestricted, producing a cancer and eventual death of the host.

That cancers exist at all in nature is not surprising. For example, in unicellular organisms, such as bacteria and protozoa, there is a strong selective pressure for each individual cell to grow and divide as rapidly as possible. For this reason, the rate of cell division is generally limited only by the rate at which nutrients can be taken up from the medium and converted to cellular materials. The situation is quite different in multicellular organisms in which cell growth for various tissues is strictly regulated so that their numbers and function can be kept at a level that is optimal for the organism as a whole: it is the survival of the organism that is paramount, not the survival of the individual cells. As a result, the 10^{13} cells in the human body divide at very different rates. For example, some cells, such as mature neurons, skeletal muscle cells, and RBCs, do not divide at all. On the other hand, cells such as epithelial cells divide continuously and rapidly. Therefore, with the understanding that growth and division is such a common, fundamental process in the human body and by nature is subject to deregulation, it is easy to appreciate the likelihood of malignant transformation of some cells at some point in the lifetime of one human being. In fact, cancer is the second leading cause of death in the United States, exceeded only by

heart disease; in children, however, it is the leading cause of death in the 3–14-year-old age group.

Malignancy therefore requires a broad and realistic understanding of biology, ranging from molecular genetics and cell biology to general medicine and psychology. A correct and appropriate approach to the diagnosis and management of urologic cancer can provide a challenge at all levels of medical interest. The purpose of this chapter is to identify and review basic problems and concepts of cancer management so as to stimulate thought and controversy with respect to recommendations set forth in subsequent chapters dealing with various urologic malignancies.

Historically the study and management of urologic cancer has provided unique information to the oncologic community and has contributed to the formation of many guiding principles of cancer management. The earliest and most well-known example concerns the work of Percival Pott, an English surgeon, who in 1775 described a cancer of the scrotum frequently occurring in chimney sweeps. The significance of this finding, of course, is that it represents the first cause-and-effect relationship between a carcinogenic stimulus and the development of cancer in humans. The carcinogen was eventually found when Yamagiwa and Ichikawa, working from 1915–1918, experimentally produced cancers by painting rabbits with coal tar (DeVita, 1978).

Another major step forward in the management and understanding of malignancy came with the knowledge of hormone sensitivity by some cancers. This concept was first put to practical use by Charles Huggins, a urologist who received a Nobel prize for describing the sensitivity of prostate cancer to androgen stimulation. This became a true clinical entity when the work of Huggins and Hodges demonstrated the value of orchiectomy and administration of estrogens in the palliative treatment of advanced cancer of the prostate (Huggins and Hodges, 1941). In their classic paper, which took only 2 hours to write and required no revision, they reported on 25 patients with metastatic prostate cancer and the symptomatic and biochemical response to hormone therapy. With this success, an entirely new branch of oncology was born, which has made its presence known in many other branches of basic and clinical medicine.

The etiology and recognition of paraneoplastic syndromes is another major aspect of cancer management that is familiar to those in the urologic community because of experience with renal cell carcinoma. This tumor may present insidiously as one of many paraneoplastic syndromes; because of this, renal cell carcinoma has become of special interest to those in many areas of clinical medicine as well as basic research. For example, fever and anemia are frequent presenting factors, which, when present, are not necessarily related to tumor necrosis or bleeding; it is thought, for example, that lactoferrin produced by the tumor complexes with iron, making it unavailable for use by hemoglobin, resulting in anemia (Loughlin et al., 1985). Other, quite opposite manifestations may also occur, including polycythemia, which occurs as a result of excessive production of erythropoietin. Interestingly, the polycythemia may disappear with removal of the tumor and return with development of metastasis. The tumor may also produce parathormone, resulting in hypercalcemia and alkalosis with associated lassitude and psychosis. Finally, patients may also present with hepatic dysfunction without evidence of liver metastasis. The etiology is still unknown but is thought to represent another paraneoplastic syndrome of unknown mechanics.

The development of urologic oncology has provided other dramatic advances in modern cancer therapy that continue to serve as models for other clinicians and researchers alike. Perhaps the best demonstration involves the use of tumor markers, which is now refined to such a degree that its use in the treatment of testicular cancer is in some cases replacing surgery as a staging modality. The use of tumor markers as developed by the urologic community embodies a certain amount of fascination, because it represents the clinical fulfillment of a hypothesis of malignant transformation. For example, the concept that fetal cells and tumor cells are related has been a recurrent theme in cancer research for the past 100 years during which malignancy has been thought to be a disease of abnormal differentiation; in today's terminology, this translates to "disease of gene regulation." The basic idea is that malfunction of certain genetic mechanisms would allow derepression of genes in a manner that is prohibited in a normal

differentiated cell. In a sense, cancer would then be explained by the untimely or abnormal expression of "normal" genes usually expressed during embryonic and fetal development. Because cancer cells express so many of the properties of embryonic cells (invasiveness, metastasis, rapid cell growth, escape from host immune systems), there should be no surprise at the unique appearance of embryonic-fetal antigens in association with neoplastic tissue. The most extensively studied example is α-fetoprotein produced by hepatocellular carcinomas and embryonal carcinomas of the testis and ovaries. The development of techniques for the detection of minute amounts of α-fetoprotein, including enzyme immunoassay, radioimmunoassay, immunoperoxidase, and immunofluorescence, has led to the use of α-fetoprotein as a model diagnostic tool. Several other markers are in use and will be discussed in this and other chapters.

A final, perhaps futuristic example of advanced state-of-the-art cancer management cites the clinical and experimental use of immunotherapy. The best example, discussed in more depth in Chapter 31, utilizes intravesical bacille Calmette-Guérin vaccine (BCG) in the treatment of bladder cancer. The theoretical basis for this type of immunotherapy depends on the fact that certain substances, such as mixed bacterial toxins and fractions of the tubercle bacillus, have the ability nonspecifically to enhance host resistance to most viral, fungal, and bacterial agents. Although the exact mechanism is unknown, these agents seem to stimulate immune response to a wide variety of antigens including tumor antigens. Originally intradermal BCG therapy was shown to be successful in controlling melanoma on a local basis, with less success in controlling systemic disease. Subsequent trials, however, as an intravesical agent to control "localized" bladder cancer whether invasive or superficial have also been encouraging, to the point that some practitioners are using BCG routinely for treatment of superficial disease.

ETIOLOGIC FACTORS AND THE BIOLOGY OF CANCER

One of the defining features of cancer cells is that they respond abnormally to growth control mechanisms that regulate the division of normal cells. As a result, cancer cells continue to divide in an uncontrolled fashion until they kill the host. This absence of restraint has provided great incentive for studying the control of cell division. While the actual molecular mechanisms remain elusive, it is clear that cancer cells are less subject to most feedback mechanisms that control normal cell division both in living tissue and in culture.

GROWTH CONTROL, VIRUSES, AND ONCOGENES

Under normal circumstances, the rate of cell division in tissues is controlled by unknown mechanisms that allow cells to divide when new cells are needed. The classic example cites liver cells that are usually quiescent but are stimulated to divide when there is loss of hepatic parenchyma; when normal liver mass has been regenerated, the cells stop dividing. This phenomenon is observed even in tissue culture, where cells often grow only when there is room to grow. As soon as a confluent monolayer is formed in which neighboring cells touch, the cells stop dividing—a process called contact inhibition (Folkman and Moscona, 1978). The mechanism responsible for this all-important process is unknown. It has been shown, however, that cell shape and configuration directly affect protein synthesis, perhaps through alteration in the cytoskeleton, and that these changes may be responsible for shutdown of growth.

Besides the basic property of contact inhibition, at least two other phenomena should be mentioned. First, cells have a definite positional requirement for proper cell growth (French et al., 1976; Bryant et al., 1977). For example, numerous examples indicate the existence of a complex pattern of cell-cell interaction that control cell division. Second, growing cells have been shown to require minute amounts of specific growth factors, down to 10^{-10} mole/L (Sato, 1979). These growth factors can be protein or small molecules, such as peptides or steroids, which probably act both systemically and locally in cell-cell interaction.

Cancer cells, then, are cells that have lost their normal growth control. Although the exact molecular mechanisms are not yet known, it is clear that cancer cells are less subject to most of the feedback mechanisms that control normal cell division, both in tissues and in culture (Pierce and Lox, 1978). For example, when normal cells are put in culture, they will divide a limited but variable number of times and then die. In a sense, they have a programmed cell death (a trait that could be evolutionarily useful). In contrast, cancer cells, as a population, have the capacity to divide indefinitely without regard for contact inhibition or position effect. In addition, cancer cells require fewer growth factors than do normal cells for survival and, in fact, may produce their own growth factors.

Much of our understanding of normal growth and its regulation has come from the study of viruses and their interaction with normal tissues to produce cancer. It is now well established that many viruses cause tumors in a variety of vertebrates ranging from reptiles to monkeys. Although it seems reasonable that certain specific viruses should be found as the cause of specific cancers, in fact the search has proved difficult, the results suggesting that viruses are not a major cause of cancer in man. Nevertheless, the study of such tumor viruses has proved rewarding because it provides insight into normal and abnormal growth regulation in cells.

The first tumor virus to be properly identified was the Rous sarcoma virus, a virus of chickens whose genetic component is an RNA molecule. It was then discovered that its RNA genome is transcribed into DNA that subsequently becomes inserted into host chromosomal DNA. Many RNA tumor viruses (retroviruses) are known that are valuable to research because of their ability to transform a variety of normal cell types to their malignant counterparts (Weiss et al., 1982). For example, one of the mechanisms of virally induced malignant transformation involves the effect of *oncogenes*. Oncogenes are genetic segments capable of transforming normal cells into malignant cells and were originally found by studying the malignant component of retroviruses. About 20 oncogenes are known to exist, which have been identified by their ability to produce morphological transformation in tissue culture cells or by their ability to induce tumor formation in animals. Of special interest is that DNA sequences homologous to most oncogene sequences that are found to be part of the retroviruses (V-onc) are also present in normal, uninfected cells including those of man (Table 8–1) (Heldin and Westermark, 1984). These genes, termed cellular oncogenes (C-onc), or protooncogenes, appear to be the evolutionary progenitors of the viral oncogenes (V-onc), simply because their gene sequence can be acquired by retroviruses. To the casual reader it may seem curious that the "malignant element" (V-onc) of viruses should be present in similar form in normal cells (C-onc). It appears, however, that the function of protooncogenes (C-onc) are in general related to regulation of cell proliferation. Likewise, their gene products seem to be associated with constitutive expression of specific steps or products along the cell proliferation pathway such as production of growth factor itself, the membrane receptor that serves as a transducer of the extracellular signal for growth factors, or the intracellular signal system that ultimately leads to the initiation of DNA synthesis in cell division (Fig 8–1). If, then, malignant transformation is associated with deregulation of normal growth control (which is the composite product of several C-onc), it seems likely that malignancy may be associated with an abnormality or "mutation" of one or more C-onc in a particular cell lineage. This, in fact, has been found. For example, the activated C-ras gene found in a bladder carcinoma line dif-

TABLE 8–1.

Common Oncogenes and Their Normal Cellular Homologues

ONCOGENE	VIRAL ORIGIN	VIRAL GENE PRODUCT	NORMAL CELLULAR HOMOLOGUE	ACTIVITY	SUBCELLULAR LOCALIZATION
myc	Avian myelocytomatosis virus MC29	$p110^{gag\text{-}myc}$	$p58^{c\text{-}myc}$	Binds DNA	Nucleus
H-ras	Harvey murine sarcoma virus	$p21^{v\text{-}H\text{-}ras}$	$p21^{c\text{-}H\text{-}ras}$	Threonine kinase, binds GDP or GTP	Plasma membrane
K-ras	Kirsten murine sarcoma virus	$p21^{v\text{-}K\text{-}ras}$	$p21^{c\text{-}K\text{-}ras}$	—	—
sis	Simian sarcoma virus	$p28^{sis}$	PDGF B-chain	PDGF agonist	Cytoplasm
src	Rous sarcoma virus	$pp60^{v\text{-}src}$	$pp60^{c\text{-}src}$	Tyrosine kinase	Plasma membrane
erbB	Avian erythroblastosis	$gp65^{erbB}$	Truncated EGF receptor	—	Plasma membrane

fers from its counterpart in normal cells by a single base change in codon 12. This results in the substitution of a valine residue for the normal glycine residue. This seemingly minor change appears capable of initiating a cascade of events leading to deregulation of its role in growth regulation with subsequent malignant transformation. The information is important when considering future methods of diagnosis and management of patients with cancer. For example, with the development of clinically useful probes to specific nucleic acids or

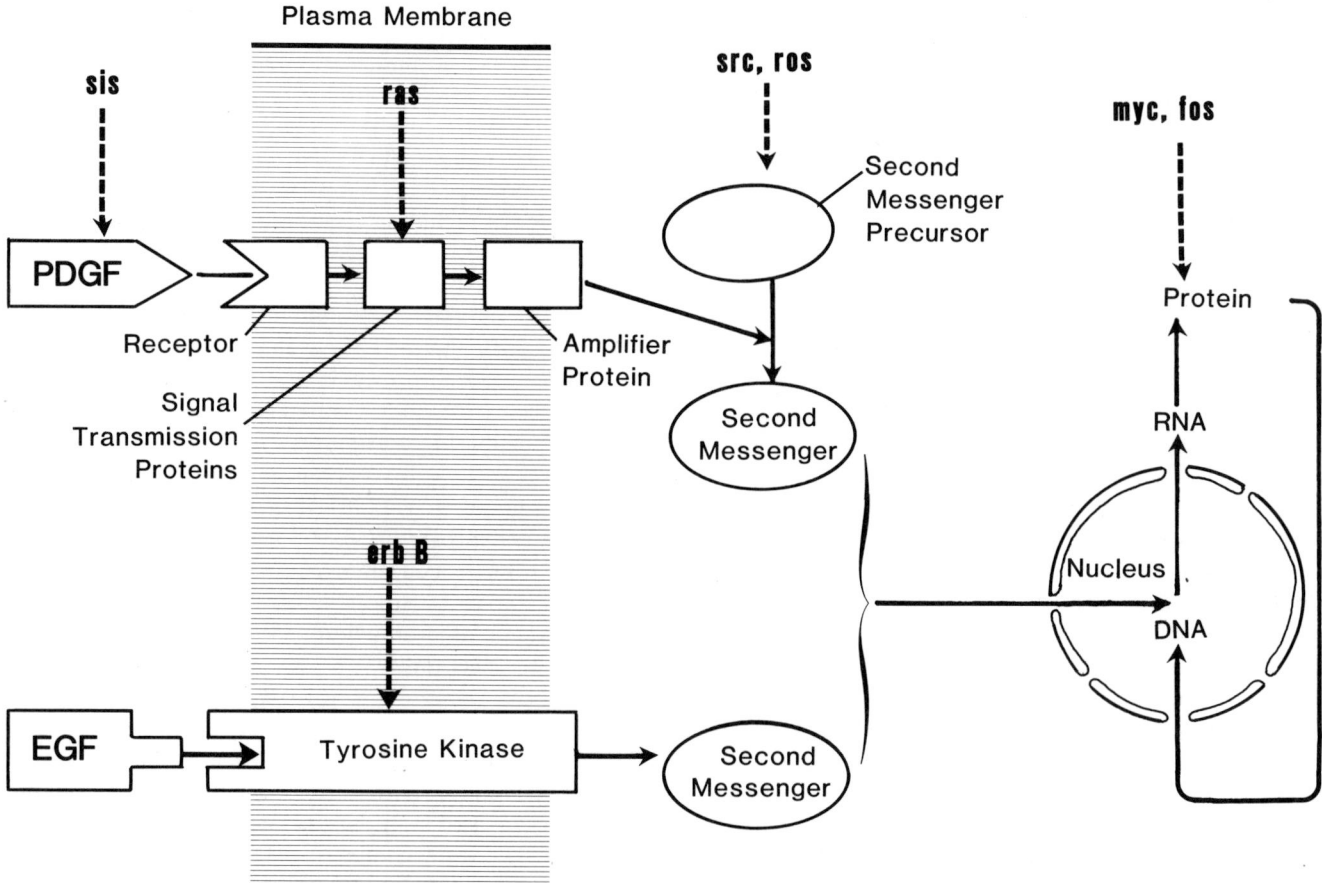

FIG 8–1.
Schematic diagram of possible locations where oncogenes may disrupt regulation of cell growth. External signals in the form of growth factors such as platelet derived growth factor *(PDGF)* or epidermal growth factor *(EGF)* act on plasma membrane receptors *(left)* and eventually act to initiate DNA transcription.

antibodies to oncogenes themselves, specific tumors could be diagnosed by laboratory means with DNA technology (Slamon et al., 1984; Thor et al., 1984).

TUMOR ANGIOGENESIS

Exponential growth of tumor cells requires the support of nutrient blood vessels. Quite logically it has been shown that the vascular framework supporting a growth tumor is formed under the guidance and direction of a diffusible tumor product. This process by which defined capillary in-growth is directed by tumor growth is called tumor angiogenesis (Folkman and Havdenschild, 1980; Chodak, 1984). The importance of this subject is recognized by the fact that without organized vascularization, solid tumors are incapable of growing beyond 1–2 mm in diameter (Folkman and Cotran, 1976).

Research directed at this process began with the hope of learning more about malignant transformation as well as understanding aspects of "anti-angiogenesis" for use in therapeutics. With models developed which utilize naturally transparent structures such as the cornea of the eye, several interesting facts about normal angiogenesis became known. For example, it is possible for capillary tubes to develop in a pure culture of endothelial cells that obviously do not contain blood and are otherwise empty. These and other experiments demonstrate that blood pressure and blood flow are not necessary for the formation of a capillary network. A second important observation is that, in general, endothelial cells form new capillaries only where there is a need for them, and their growth can be induced, for example, by local infections and damaged tissue. This subject has recently become of intense interest because of a tumor angiogenesis factor that has been isolated from some tumors. The clinical implication of this knowledge is particularly important because it is known that a fragment of heparin, which by itself is nonanticoagulant, when administered with nonimmunosuppressive angiotrophic steroids, produces a potent angiogenesis inhibition. With this combination, it is possible to cause tumor regression and prevent metastasis in some experimental tumor systems including bladder carcinoma (Folkman et al., 1983).

Tumor angiogenesis factor may also play a role in the diagnosis and identification of potentially malignant lesions. It has been shown, for example, that angiogenesis activity may be a distinguishing factor between benign and malignant bladder tissue from humans (Chodak et al., 1980). For example, when urine is collected from humans with a wide variety of diseases, samples from those patients with urothelial tumors contain significantly more migration-stimulating activity than samples from nontumor-bearing patients (Chodak

et al., 1981). When tumor angiogenesis factor may be more conveniently identified, this tool may become important in the diagnosis of early or recurrent malignant disease.

CHEMICAL CARCINOGENS

It is generally believed that the development of clinical cancer is a multiple-step process involving both initiating and promoting factors. An initiator is a carcinogen that can transform a normal cell into a malignant cell. A promoter is a cocarcinogen that provides selection advantages for growth of the transformed or malignant cell. Some factors are considered to be complete carcinogens by supplying both initiating and promoting influences for cancer development, and such factors may be of environmental or "host" (genetic) origin.

Carcinogenic agents were actually known from clinical observation and experience long before common use of laboratory investigation. The classic example is probably the observation of Percival Pott in 1775 (Pott, 1775), when he described scrotal carcinoma in men exposed to constant contact with soot. Since then, numerous carcinogenic agents and classes of compounds have been described. Of particular interest to urologists are the aromatic amines and azo dyes. The importance of such compounds was originally identified in 1895, when Rehn described an increased incidence of bladder cancer in workers of the aniline dye industry. It was subsequently shown that prolonged subcutaneous injection of 2-naphthylamine compounds in dogs led to the appearance of papillomas and carcinomas of the bladder (Hueper, 1942). This effect was subsequently shown to be due to the effect of the carcinogen transported in the urine and not the blood. Although this and other carcinogenic associations have been described, the mechanism of action is poorly understood, meaning that the critical cellular targets in the induction of neoplastic transformation by chemical carcinogens have never been clearly identified. Unfortunately, the result is that the outcome of this research has little impact on clinical practice of urology beyond the phenomenological aspect. In contrast, certain parasitic infections have been shown to be carcinogenic which does have clinical significance. For example, the association of cancer of the bladder and infestation with *Schistosoma haematobium* was known from clinical experience for almost 100 years. This finding can now be experimentally reproduced in subhuman primates following *Schistosoma haematobium* infection (Kuntz et al., 1972).

The presence of carcinogenic agents and their clinical impact on malignancy is also demonstrated by epidemiologic studies. An example is the above-mentioned higher than expected incidence of bladder cancer in Egypt, which is associated with infection with *Schisto-*

soma. Conversely, the abnormally low incidence of cancer of the prostate in Japan may imply an association between a physical agent and cancer of the prostate. These studies emphasize the relative contributions of genetics and environment to malignancy. It has been estimated, for example, that up to 80% of cancers in the United States may be due in part to environmental factors and are therefore potentially preventable. This figure was derived from differences in cancer incidence among communities, changes in incidence with migration, and changes over time (Costanza et al., 1982). Most of these factors relate to life-style.

As previously mentioned, the molecular mechanism behind observed carcinogenic activity is not known. It is generally agreed, however, that malignancy is induced by a change in cellular genetic material. This is known as the somatic mutation theory of cancer. A somatic mutation has been defined as an inheritable alteration in DNA sequence that could be brought about by point mutation, deletion, chromosomal rearrangement, or chromosomal loss (Siminovitch, 1976). A large body of evidence now supports a genetic mechanism of cancer, demonstrating that carcinogens interact with DNA and that the extent of interaction correlates with their carcinogenic potential. Supporting this concept are recent data suggesting that a single base change within a cellular transforming gene or insertion of a promoter sequence next to an oncogene may play a role in converting a normal cell to its malignant counterpart (Hayward et al., 1981).

With these DNA changes in mind, it is not surprising to learn that nature has taken great care to ensure the fidelity of DNA; in fact, DNA stability is probably the most fundamental property of survival in multicellular organisms. This is usually accomplished by (1) extremely accurate mechanisms for copying DNA and (2) a mechanism for repairing the many accidental lesions that occur spontaneously in DNA. Most spontaneous DNA changes, or mutations, that occur in proteins are deleterious and are eliminated by natural selection (Jukes, 1980). This genetic stability is so reliable that the genetic heritage of man, which began with the evolution of mammals (about 3×10^8 years ago), has resulted in evolutionary changes that are observed, for the most part, only in morphological appearance, not in the molecular structure of materials of which mammals are made.

The force behind this remarkable stability is DNA repair. There are several mechanisms for repair, but they all depend on the existence of two copies of the genetic information, one on each strand of the DNA double helix. For example, the altered strand of DNA is recognized and removed by one set of enzymes and then replaced in its original form by another enzyme (DNA polymerase), which copies the information stored in the "good" strand by complementary base pairing. Finally, an enzyme called DNA ligase seals a nick that remains in the DNA helix to complete the restoration of an intact DNA strand. Single-stranded DNA can also store genetic information; in fact, some very small viruses have single-stranded genomes. DNA repair processes, however, cannot operate on such DNA, and the mutation rates of these viruses are high.

IMMUNOBIOLOGY

General Principles and the Immune Surveillance Hypothesis

All vertebrates have an immune system. Any child born with a severely defective immune system will die under normal circumstances. The general purpose of the immune system is to destroy and eliminate invading organisms and any toxic molecules produced by them. During the execution of this function, the host must retain the ability to distinguish self from foreign molecules, which is a fundamental feature of the immune system. Failure to make distinction results in autoimmune reactions that can be fatal. Another function of the immune system, perhaps less well understood, serves to protect the individual against neoplasms—the so-called immune surveillance hypothesis. The original idea came from Burnet (1970), who postulated that the teleology of the immune system was to recognize the foreignness of tumor-specific antigens on neoplastic cells and then to produce an immune response capable of eliminating them.

In this context clinical cancer would represent a failure of the mechanisms for immunologic destruction. This concept is reinforced by the well-known fact that people with naturally occurring immunodeficiencies have an abnormally high incidence of tumors. The most common histologic type is non-Hodgkin's lymphoma, although there is an increase in all types of tumors. Current confusion regarding support for this concept stems from observations in nude mice, which are congenitally hairless (hence their name) and markedly T-cell deficient (because they have abnormally developed thymus glands). When kept under infection-free conditions, these mice do not have a higher incidence of spontaneous tumors than normal mice, nor are they more prone to develop tumors when they are treated with chemical carcinogens. It is probable then that T-cell dependent immunity normally plays little part in the control of most spontaneously or chemically induced cancer. On the other hand, the immune system does play an important part in defending vertebrates against the great majority of virally induced tumors, at least in experimental animals.

Another perhaps more realistic mechanism for an immune surveillance hypothesis involves the work of natural killer (NK) cells. Several years ago, as an outgrowth of studies investigating the generation of specific cell-mediated immunity to tumors, it was noted that the lymphocytes of individuals, including those unrelated or not exposed to cancer patients, were found to react to leukemic cells or against cell lines derived from tumors (Rosenberg et al., 1972; Oldham and Herberman, 1973). During this early period, most investigators attributed this "extra" reactivity to in vitro artifact; but despite persistent effort, it was not possible to eliminate this cytotoxic reactivity of normal individuals, and gradually investigators accepted its presence. The cells responsible for this natural cytotoxicity were found to represent a small population of nonadherent lymphocytes that were termed, logically, "natural killer" (NK) cells. Because of the general belief that NK cells may represent a first line of defense against the development of tumors, studies of these cells have expanded. Their importance is underscored by the fact that activated NK cells (activated by exposure to interleukin-2, lectins, or pooled alloantigens) recognize and lyse fresh autologous (in the human) or syngeneic (in the mouse) cancer cells but do not lyse normal cells. These cells also effectively decrease melanoma lung metastasis when transferred to mice used in adoptive transfer assays (Mazumber and Rosenberg, 1984). The exact cell lineage of this effector lymphocyte population as well as its mechanism of action in terms of target specificity is not known. When these issues are better understood and cloning techniques are refined, purified effector cells may become available as a therapeutic modality. A brief characterization of these cytotoxic cells is given in Table 8–2.

AGENTS THAT MODIFY THE IMMUNE RESPONSE TO CANCER.—In a traditional sense, cancer-related manipulations of the immune system have been either active or passive. Active immune manipulation utilized immunostimulants such as contact allergens, bacterial products or tumor cell vaccines to be given directly to patients, with resultant stimulation of their endogenous immunity. By contrast, passive immunomanipulation involves transfer of antibodies, lymphocytes, RNA, or transfer factors from immune donors to patients with cancer to provide exogenous immunity. Current urologic practice makes use of such agents in both clinical and experimental ways.

Interferon has recently received a great deal of interest with regard to treatment of urologic and other human tumors. Interferon was originally described in 1957 as a protein that induced resistance to infection by a wide variety of viruses (Isaacs and Lindenmann, 1957). It is now known that interferon exists in three forms. α- and β-interferons are produced by leukocytes and fibroblasts respectively, after stimulation by viruses or other nonviral inducers. γ-Interferon is produced by immunocompetent lymphocytes in response to sensitizing antigen or mitogen stimulation (Ratliff et al., 1984).

TABLE 8–2.

Major Cytolytic Systems

	CYTOLYTIC CATEGORY		
CHARACTERISTIC	NK*	AK†	CTL‡
Developmental kinetics	Fresh nonstimulated PBL	Requires 2–5 days of activation (in vitro)	Requires 5–6 days sensitization to specific cellular antigen
Stimulus	None	Alloantigen Mitogen IL-2	Viral antigen Alloantigen (transplantation antigen)
Specificity of cytotoxicity	Bone marrow; some cultured cells; leukemia; not PBL or lymphoblasts	Autologous or allogeneic fresh solid tumors; some cultured cells (some shared with NK); not PBL or lymphoblasts	PBL and lymphoblast; viral antigen, transplantation antigen
Serologic phenotype of effector	OKM.1+ OKT.3− HNK−1+	OKM.1− OKT.3+ HNK−1−	OKM.1− OKT.3+
Radiation sensitivity	—	Radioresistant	Radiosensitive
Memory cells	No	No	Yes
Kills autologous B lymphoid	No	Yes	No

*Natural killer.
†Activated killer.
‡Cytotoxic T lymphocyte.

Since its discovery, interferon has been principally to have important effects on the molecular biology of the cell. Studies on the mechanism by which these interferons inhibit virus multiplication have led to the discovery that interferons also inhibit cell growth (Gresser and Tovey, 1978), modify cell structure and differentiation (Pfeffer et al., 1980), and stimulate some immune functions and inhibit others (DeMaeyer, 1981). Which of these effects is important in antitumor activity remains to be established. It is likely, however, that a direct effect on gene regulation will prove to be the deciding mechanism. Studies have demonstrated the capability of γ-interferon actually to induce the transcription and expression of HLA antigens in K562 erythroleukemia cancer cells, which ordinarily do not express HLA molecules (Rosa and Fellous, 1984). Other similar studies showed that γ-interferon is able to induce new proteins on human fibroblasts (Weil et al., 1983) and to inhibit transformation of mouse L cells by exogenous cellular or viral genes (Dubois et al., 1983). Although these and other in vivo laboratory studies have been encouraging, clinical trials as a whole have been only equivocal in patients with transitional cell carcinoma and renal carcinoma (Ratliff et al., 1984). Whether interferon will be of clinical use is not known, but it has already provided valuable insight into basic mechanisms of cancer biology, especially with respect to gene activation and differentiation.

BCG is another example of an immunomodulating agent that currently represents the most successful form of immunotherapy in man, especially its use in treatment of bladder cancer. Despite its extensive use and success, the exact mechanism of action of BCG remains elusive. It is generally accepted that BCG acts singly by being a nonspecific immunostimulant, thereby activating those immune factors (whatever they are) responsible for antitumor activity. The foundation for this belief is that tumor antigens exist as does an immune response to them. Whether this response is humoral (mediated by antibodies), cellular (mediated by T cells or NK cells), or a combination of both is not understood. Nonetheless, it is only logical that the use of a pharmacologic means to bolster the response to those antigens should represent a reasonable therapy against cancer. Clinical experience with BCG and bladder cancer follows the same basic guidelines as those set forth by the primary laboratory animal study reported in 1974 (Zbar and Rapp, 1974). The original study reported on the effects of intralesional BCG therapy on intradermal inocula of hepatoma cells in guinea pigs. The authors concluded that for BCG to be effective: (1) there must be close contact between the tumor cells and BCG; (2) immunocompetence of the host must be intact, for example, to stimulation by cutaneous an-

tigens; (3) there must be limited tumor burden; and (4) there must be an adequate number of BCG organisms. Amazingly, current data suggest that these original observations apply to human tumors. Results of PPD skin testing prior to BCG therapy suggest that the best candidates for BCG immunotherapy may be patients without prior *Mycobacterium tuberculosis* exposure who have a positive PPD skin test after therapy (Kelley and Catalona, 1984). Another point is that contact between the tumor and BCG may be important, since systemic immunostimulation seems less effective for most tumors in general (Bast, 1982). Likewise, there are also questions regarding the strain of BCG to be used, its frequency, and dose (Kelley and Catalona, 1984). Specific details of the use of BCG will be covered in Chapter 31; important here is that the success of its use is one of the best indicators that immunomanipulation is a clinical reality and will probably improve with time.

Interferon and BCG represent two classes of drugs that have undergone clinical trials with at least some favorable results. Other agents are also available which, although not so far along in their developmental stages, are promising, especially when considering the mechanism of action that they utilize. For example, the use of differentiating agents is currently under way. One example is vitamin A (retinol) and some of its derivatives (retinoids), which represent a class of physiologic substances that can control the proliferation and differentiation of a wide variety of epithelial and mesenchymal cell types. Moreover, retinoid substances often exert a regulatory influence on the immune response (Malkovsky et al., 1983). Only little is known, however, about the immunologic mechanism underlying the observed retinoid-mediated augmentation of specific antitumor immunity. Unlike BCG, nonspecific immune activation is believed to be negligible (Dennert and Lotan, 1978), although there is an increase in NK and cytotoxic T-cell activity (Goldfarb and Herberman, 1981). The intriguing aspect of vitamin A in cancer therapy makes use of its propensity to promote differentiation of some experimental undifferentiated stem cells. This is particularly appealing to some investigators, because it makes use of the old concept that malignancy is a result of abnormal gene regulation and is like a stem cell population that ultimately enlarges with impairment in its ability to differentiate. In a simplistic sense, if cells could be induced to differentiate by retinoic acid or by other compounds, such as DMSO (Fishman et al., 1981) and hexamethyl bisacetamide (Tapiero et al., 1980), they would change their program of growth control to one of a more "ordered" style found in differentiated tissues. Although this idea is of historical interest, there are few present day data to support it.

Immune properties of the genitourinary system are of interest from another point of view. Unlike the rest of the body, male germ cells and those tissues that surround it make a tremendous effort to suppress or bypass the immune response. The basis for this generalization is not clearly understood. In all likelihood, the story begins with one of the greatest central questions in nature: how does the conceptus survive the immunologic attack of the mother and thus seem to violate the basic rules of immunogenetics? There is no doubt that the survival of the early fetus is not due to the absence of histocompatibility antigens or to depression of the maternal immune response. It is now hypothesized that certain hormones and perhaps selected plasma proteins act as immunosuppressive factors active at the microenvironmental level. The relationship of the male reproductive tract to the immune system is very similar; the existence of immunosuppressive factors in the male reproductive tract is already known. For example, seminal plasma contains extremely potent low molecular weight factors that inhibit NK cell function; likewise other reports show that whole seminal plasma or semen components suppress T cell activation, in vivo humoral responses, complement activity, and macrophage activity (Anderson and Harkins, 1985; Byrd et al., 1977). Undoubtedly, the basis for this immunosuppressive activity is related to the highly immunogenic capacity of germ cells and seminal components, which can result in an autoimmune phenomenon. One of the most fascinating aspects of this self-protective system (immunosuppressive factors surrounding a highly immunogenic system, i.e., sperm) is the blood-testis barrier, which serves to isolate male germ cells from the remainder of the immune system. The testis can be considered an immunologically privileged site shown experimentally to be equivalent to other well-known privileged sites such as the brain and anterior chamber of the eye (Whitmore and Gittes, 1979). The traditional explanation for this phenomenon pointed to an impairment in lymphatic drainage; however, the testis is exceptional in this sense because it has a rich lymphatic network (Whitmore and Gittes, 1975). Available data show that the testis maintains this immunoprivileged state by interfering with the effector limb, not the sensitizing limb, of the immune system (Whitmore and Gittes, 1979). Contributing to this effect is the ability of germ cells to stimulate or activate T suppressor lymphocytes, which in turn inhibit proliferation of other helper lymphocytes (Hurtenbach et al., 1980). Despite this finding, the entire picture of the role of the testis and germ cells with respect to the immune system and malignant transformation is still incompletely understood.

In addition to an immunologic barrier, spermatogenic cells are protected by an "anatomical barrier" in the form of tight junctions of the Sertoli cells that line the seminiferous tubules, thus mechanically preventing contact between immunocompetent cells and the developing spermatozoa in the tubular lumen. The tight junctions also prevent the passage of circulating antibodies into the seminiferous tubules.

Why have evolutionary forces maintained the male reproductive tract to be immunologically distinct from the remainder of the body? It is generally thought that tolerance to most self-constituents is the result of the elimination of self-reactive clones, either during fetal development or shortly after birth. Because spermiogenesis does not commence until long after this tolerance, several alternative mechanisms are necessary to ensure that the adult male does not become sensitized to differentiation and other potential autoantigens associated with his own sperm; hence, the "blood-testis" barrier. This is not the whole answer; however, eventual solutions to these and related questions should provide insight into the biology and natural history of male reproductive tract tumors and their treatment.

METASTASIS

Metastasis is a major cause of mortality in cancer. The tumor cells involved are not merely extensions of the original cancer but have their own specific characteristics, which as a whole differ from the primary tumor; in fact, the propensity to metastasize is found only in a minority of the cells in a malignant tumor (Poste and Fidler, 1980). To metastasize, tumor cells must first enter, then leave the circulatory system, evade host defenses, and proliferate in a secondary site. Not all cells within a tumor are capable of this. Current evidence suggests that fibronectin suppresses the invasive potential of cells, while laminin increases invasiveness (Terranova et al., 1985). Efforts are now being made to determine whether cells capable of metastasis possess specific regulatory genes that are responsible for this property. The elucidation of such "oncogenes of metastasis" would have great impact on understanding the natural history, diagnosis, and management of cancer.

Exemplifying this point are current results demonstrating a marked difference in the stability of the metastatic phenotype in cloned and uncloned populations. In polyclonal populations, the various clonal subpopulations might somehow interact to stabilize their relative proportions within the population (Poste et al., 1981). If this is true, the therapeutic elimination of a major fraction of the subpopulations, through either surgery or chemotherapy, may destroy the stabilizing equilibrium that exists between the subpopulations, resulting in enhanced instability of the remaining cells. The end result is the generation of new subpopulations with more malignant phenotypes. This process tends to have a snowball effect in that as a tumor progresses to an advanced state of malignancy, as shown by its ability to metastasize, it contains cells that are more mutable than cells that did not progress as far and are still non-metastatic (Cifone and Fidler, 1981). These concepts support the notion that evolution of tumors from a benign to a malignant state could be the consequence of genetic instability in addition to other specific oncogenic factors.

Another important aspect of the biology of metastasis concerns its relationship to differentiation. This is especially important to the clinician treating germ cell tumors. For example, it is not uncommon to find pure teratoma in metastasis following chemotherapy for non-seminomatous germ cell tumors (NSGCT). This is in contrast to the rare appearance of pure teratoma in the metasasis of untreated NSGCT. The question then arises as to the etiology and significance of this lesion. There are two possible mechanisms for the appearance of teratomatous elements in metastasis following chemotherapy: (1) selection of differentiated or differentiating cells, or (2) induction of differentiation of embryonal carcinoma cells (de novo induction of differentiation). Studies comparing the histology of primary NSGCT with the histology of their retroperitoneal lymph node metastasis in patients with or without chemotherapy strongly suggest that chemotherapy does not induce differentiation de novo in tumors that lack an inherent capacity for differentiation (Oosterhuis et al., 1983). Further studies investigating the effects of cis-diamine-dichloroplatinum (CDDP) on various in vitro and in vivo mouse teratocarcinoma cells substantiate these findings to indicate that histologic maturation of murine tumors most likely results from chemotherapeutic elimination of malignant tumor stem cells rather than induction of their differentiation in somatic tissues (Oosterhuis et al., 1983). The central question behind these studies searches out the true nature of these lesions, i.e., what is their malignant potential? Patients found to have mature teratoma in surgically removed metastatic tumor have a favorable prognosis regardless of postoperative chemotherapy (Einhorn et al., 1981). However, regrowth of teratomatous tumors has been reported (Peckham, 1981). This was evaluated experimentally, again with mouse teratocarcinoma models, where it was found that treated and presumably benign lesions reverted to contain foci of embryonal carcinoma (EC) (Oosterhuis and Damjanov, 1983). Although not completely understood, it is generally believed that some EC cells might survive treatment, whether surgical or chemotherapy, and thus serve as the source of recurrent malignancy. Surgical removal of all residual tumor after chemotherapy thus seems justifiable.

TUMOR MARKERS

Perhaps the area in medicine that best links basic science to clinical practice is that of tumor markers. The fascination derives not only from theory regarding why tumor markers should work, but also from speculation regarding the existence of new markers. Tumor markers are a heavily used modality in urologic practice, and more thorough descriptions are provided elsewhere. A few principles and generalizations deserve mention.

The best definition for a tumor marker is a broad one, i.e., any substance present in the body that registers either qualitatively or quantitatively the presence of malignant disease. This presumes that tumors possess distinctive antigens that set them apart from normal tissue. Indeed, early experimental work demonstrated the presence of such antigens when tumors were either transplanted or used as immunogens. Within inbred strains of mice, recipient mice developed resistance to the tumors, providing evidence for tumor-specific antigens. Additional studies have shown that many "tumor specific antigens," or biomarkers, are also produced at some stage during embryogenesis or fetal development and are thus nonimmunogenic (Coggin and Anderson, 1974). Although this circumstance may frustrate efforts at immunotherapy, it has allowed clinicians the opportunity to use the detected presence of such antigens as evidence for the presence of cancer; hence many current tumor markers are "oncofetal proteins."

The hypothesis that fetal cells and cancer cells are related is a recurrent theme in cancer research, at least 100 years old. This is not unexpected, because malignant cells and embryonic cells share several unique biologic properties. For example: (1) *invasiveness* is a prime characteristic of the trophoblast as well as tumor cells. One contributing factor is thought to be due to secretion of a protease, plasminogen activator, which has been isolated from both trophoblast and malignant cells (Pollack et al., 1974); (2) *metastasis* is another shared event, because it may be likened to specific cell migration during fetal embryonic development (Coggin and Anderson, 1974); (3) similar changes in malignant and embryonic cells membranes can occur as evidenced by changes in lectin-binding properties (Barger and Noonan, 1970; Moscona, 1971); (4) tumors have long been known to produce a factor which stimulates the host to provide a blood supply; this angiogenesis factor is also found in the placenta (Folkman, 1972); (5) there is extensive literature on isoenzyme homology between fetal and malignant tissue (Stein et al., 1978); (6) there is an increasing number of embryonic and fetal antigens found to be associated with neoplastic tissue. The hypothesis behind this expected association between cancer and development is that malignancy is a result of abnormal gene regulation manifested by diminished or disordered cell differentiation. This is usually thought to involve the participation of a "stem cell" population and differentiative events which normally occur during embryogenesis and early development rather than in the adult (Mintz and Fleishman, 1981). This view that malignant conversion parallels conversion of a pleuripotent stem cell population, with associated impairment in its ability to differentiate, has received considerable attention and has led Pierce and Cox to refer to some neoplasms as "caricatures of tissue renewal" (1978). Cancer, in other words, may be associated with the untimely or abnormal expression of "normal" genes. In a general way this observation is being confirmed by oncogene data as covered earlier in this chapter. The relationship of malignant and embryonic cells is most

clearly demonstrated by the mouse teratocarcinoma system. For example, teratocarcinoma cells and embryo cells, under appropriate experimental conditions, may be exchanged reversibly to assume a malignant or developmental phenotype. Embryo cells can be induced by culture conditions to become teratocarcinoma, and teratocarcinoma cells can be successfuly integrated into the developmental program of a *normal* embryo (Fig 8–2) (Narayan and DeWolf, 1984).

Tumor markers can also be molecules with abnormal tissue expression in contrast to abnormal developmental expression. In other words, malignant cells may express genetic information normally repressed prior to the onset of cancer but not fetus related, as exemplified by the production of hormones by nonendocrine tumors. The most common of these tumors is bronchogenic carcinoma, which secretes ACTH. This phenomenon is of

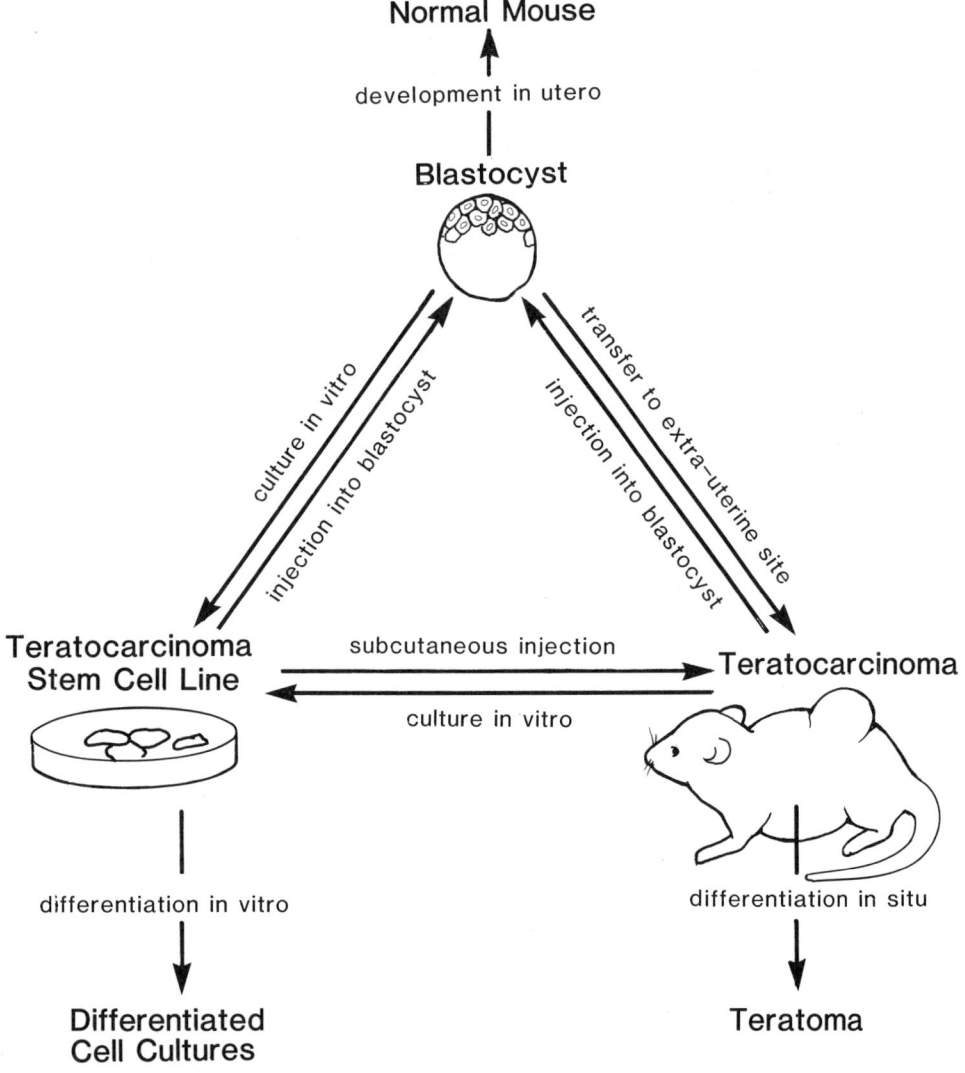

FIG 8–2.
Mouse teratocarcinoma model demonstrating the relationship between embryonic and malignant cells.

urologic interest because of the similar behavior of renal cell carcinoma. Again, this provides more evidence that the manifestations of cancer can be associated with expression of genes normally active only in other cell types.

Therefore, the questions and ideas generated by the concepts associated with tumor markers brings us to the central question in molecular biology: What is the basis for gene control and gene regulation, especially as it applies to differentiated cells? One specific experimental example pointing out this problem (which also may have implications regarding hormone therapy for cancer of the prostate) concerns an explanation for the different patterns of expression of steroid-inducible genes in different cell types, all of which contain appropriate steroid hormone receptors. For example, the chicken ovalbumin gene is induced by estrogen in oviduct cells, in which the vitellogenic gene is permanently inactive; yet in liver cells the ovalbumin gene is permanently inactive and the vitellogenic gene is estrogen inducible. The answer to this question—of how a gene can be regulated by steroid hormones in some cells and not in others, even though both categories of cells may contain perfectly functional steroid hormone receptors—may lead the way to improved hormone treatment of estrogen-resistant prostate cancer.

The actual markers that are clinically relevant to urologic practice will be discussed in corresponding chapters, but a few generalizations regarding their clinical application should be mentioned here.

As mentioned earlier in this section, the clinical definition of a tumor biomarker is any substance present in a body fluid that reflects either quantitatively or qualitatively the presence of malignant disease. Most clinically useful markers are synthesized by cancer cells, i.e., peptide hormones and oncofetal antigens. However, some tumor biomarkers such as hydroxyproline present products that are derived from the destruction of cells. As the clinical use of tumor markers is described, several considerations should be kept in mind. For example, can the marker be used for mass screening? Two classic problems are often raised when dealing with this issue; the first concerns the assumption that earlier detection from mass screening will result in improved prognosis. Although this seems *a priori*, there is a current suggestion that improved survival has not resulted from detection programs for cervical cancer, breast cancer, and other common malignancies (Miller et al., 1976; Capone, 1978; Sackett, 1975). The second concern is that almost all blood tests currently in use seem to be much less effective than anticipated when they are used as screening tools.

To understand this dilemma, one must first be familiar with the three important variables that determine the predictive value of a screening test: the sensitivity of the test, the specificity of the test, and the prevalence of the disease in the examined population. *Sensitivity* refers to the true positive rate, which is determined by dividing the experimental (tumor marker) positive number by the number of patients who actually have the tumor; an insensitive test will produce false negative results (Table 8–3). *Specificity*, on the other hand, concerns itself with the accuracy of the tumor marker in determining which patients are truly free of neoplasm; it is a measure of the true negative rate, that is, the experimental (tumor marker) negative number divided by the true negative number. Failures of specificity are termed false positive results. An ideal test would yield 100% sensitivity and 100% specificity (no false negatives and no false positives). However, laboratory standards and procedures tend to make sensitivity inversely linked with specificity. For example, selecting a level that will give a quantitative screening test a high sensitivity, in general, will diminish the degree of specificity, whereas selection of maximum specificity tends to decrease sensitivity. The third factor in the equation determining the predictive value of a screening test is *prevalence*, which is the number of patients per 100,000 population who have the disease.

These three factors—sensitivity, specificity, and prevalence—become important elements in the equation to equal the predictive value of a screening test. Let us consider a hypothetical example. A new tumor marker with a 95% specificity and a 95% sensitivity is applied to a population of 100,000. The prevalence rate of the disease in question is 5%. Under these conditions, of 5,000 people who have tumor, 4,750 are correctly detected, and an identical number of patients are detected as having tumor who really do not have tumor. The predictive value is then only 50%, despite the fact that the quality of the marker appeared excellent at the outset. This example points out some of the public health problems that must be considered before routine use of tumor markers can be considered.

NATURAL HISTORY

This section will look at some of the fundamental properties of cancer behavior that form the basis for the final section on management.

TABLE 8–3.

Quality of Screening Tests Using Tumor Markers

Sensitivity	$\dfrac{\text{Observed positive}}{\text{True positive}}$	$\times 100$
Specificity	$\dfrac{\text{Observed negative}}{\text{True negative}}$	$\times 100$

Consideration of the *extent* of the patient's tumor at any given time is of fundamental importance to researchers and clinicians alike. In the past numerous staging systems were operational, which led to confusion. This understandably led to the formation of international and national committees to solve the problem of establishing meaningful nosology. Since 1950 the International Union Against Cancer has devoted special attention to the problem of staging with a resultant unified system for all nations (Commission on Clinical Oncology, 1968). This has been called the TMN system, because it measures the tumor in terms of the primary tumor (T), presence or absence of node metastasis (N), and the presence or absence of distant metastasis (M). In general, increasing numbers after the TNM designation indicate increasing size and hence severity of disease, i.e., T_1 usually signifies disease limited to the mucosa, while T_2 indicates muscularis involvement. Likewise, the absence of nodal disease is indicated by N0, while the presence of nodal disease is indicated by N_1, and for more extensive nodal disease additional numbers may be used. Finally, distant metastases are indicated by adding a subscript 1 following M for metastasis or the subscript 0 for their absence. Thus, a lesion limited to the mucosa of the bladder would be designated as $T_1N_0M_0$; a lesion that involved the superficial muscularis and involved the regional nodes without distant metastasis would be termed $T_2N_1M_0$. Details of staging for each organ system are given in the appropriate chapters.

GROWTH RATE

There is no better plea for the development of new tumor markers than the fact that approximately two thirds of the growth of human neoplasms occurs before they are clinically detectable by size. If one assumes that a cancer begins from a single cell, it takes about 30 exponential divisions to produce a 1-cm nodule (1 billion cells); at 45 exponential divisions, the patient is apt to be dead from the bulk of the tumor alone. The growth rate of tumors can be expressed by the tumor doubling time, i.e., the time it takes a tumor to double its volume. This usually varies from 8 to 600 days, with most tumors falling in the range of 20–100 days. On the basis of growth rate dynamics, then, most human tumors have been present in the body for at least 1 year, and some for as long as 10–15 years, prior to their clinical detection. It is this long "latent period," i.e., from the time of inception to the time of detection, that should allow the clinician adequate time for detection, treatment and cure. Future methodology should therefore be designed to detect microscopic disease.

An introduction to the concepts of cancer growth is fundamental to understanding the various modalities of treatment, simply because the main rationale in noninvasive treatment of cancer is to achieve selective killing of tumor cells, which usually depends on differential growth characteristics of malignancy. To review, all cells undergoing DNA renewal go through specific growth phases known as the *cell cycle*. At the completion of mitosis (M), the cell spends a variable period in a "resting" phase (G1), during which the synthesis of RNA and protein continues normally. In late G1, an unknown signal initiates a burst of RNA synthesis, and shortly thereafter the period of DNA synthesis (S) begins and the cell is committed to undergo division. Prior to M phase, the cell enters G2 phase where, again, RNA and protein synthesis continues. In mitosis (M), the rates of protein and RNA synthesis diminish abruptly, while the genetic material is segregated into daughter cells. Occasionally cells are taken out of the cycle for variable periods and are said to be in G0 resting phase. Cells in G0 phase are generally (but not always) refractory to chemotherapy or radiotherapy. Accordingly, chemotherapeutic agents can be divided into three categories (Bruce et al., 1966). The first group, or nonphase-specific, are equally toxic for both proliferating and resting (G0) cells. Examples include nitrogen mustard, nitrosoureas, and gamma irradiation. The second class of drugs are phase-specific agents and kill cells only during a specific part of the cell cycle. They do not affect "resting" or G0 cells if the exposure time is short, and they generate dose-survival curves which reach plateau levels. Examples include vinblastine, cytarabine, and methotrexate. The third class of drugs are cell cycle specific. Cycling cells are killed in preference to resting cells. The toxic effect lasts throughout the cell cycle. Examples include fluorouracil, actinomycin D, and cyclophosphamide. With elemental knowledge of the cell cycle, the dosage scheduling can be logically deduced, i.e., drugs belonging to the first or third group (nonphase or cycle dependent) are usually given in short, intensive courses using the maximum tolerated dose. Phase-specific drugs (group 2), however, should be given repeatedly or as an infusion to expose all tumor cells as they arrive in the sensitive phase of the cycle.

THE MORPHOLOGICAL DIAGNOSIS OF CANCER

Despite our increasing understanding of cell biology and factors important in malignant transformation, the mainstay of cancer diagnosis and classification is microscopic examination of tissues or exfoliated cells. The identification of malignancy usually focuses on nuclear features that differ from normal cells. For example, malignant nuclei are large, hyperchromatic, and have bizarre configurations; the nuclear chromatin is frequently clumped and irregular; mitotic figures are easily found and are often abnormal. Also, the nucleus-to-cytoplasm ratio is increased. Benign neoplasms, as expected, are generally composed of cells with cytologic features similar to those of normal tissues and typically lack the malignant nuclear features. In the past, it was customary to place every neoplasm into one of two pigeonholes—benign or malignant. However, it has become evident that invasive carcinomas, especially of the testis and bladder, may be preceded or associated with intraepithelial morphological changes. Specifically, the normal-appearing epithelium in the involved site is replaced in varying degrees by either cytologically highly abnormal cells or malignant-appearing cells. The usual terms used to describe these preinvasive epithelial changes include atypia, dysplasia, and carcinoma-in-situ (CIS), with CIS representing the end point in the spectrum of epithelial atypia. CIS then represents a cancer confined to the area normally occupied by the cell of origin and behaves as a tumor that has not yet gained the capacity to invade local tissues. This disease is among the most insidious, because the tissue of origin may look perfectly normal while the cells themselves may gain metastatic potential with no warning. The main concerns center on the consideration of doing a radical operation for a tumor that has not yet demonstrated local invasiveness.

Although CIS has been described in association with several organs of the urinary system, most of our understanding concerning the natural history of this disease comes from our knowledge of CIS of the bladder. Clinical evidence thus far suggests that this disease is unique with respect to the usual noninvasive papillary tumor disease. For example, CIS tends to be associated with a "field change," often extensive and, importantly, involving normal-looking bladder mucosa. In general, this disease is very peculiar in that while it exhibits cellular characteristics of malignancy, it does not demonstrate associated growth or metastatic quality. According to several studies of the natural history of this disease, the majority of cases of untreated CIS progress to frank malignant disease with invasion. The trigger mechanism for this change, as well as specific markers to presage it, is currently not at hand. This disease may also occur in the testis, in which case it may be called an intratubular germ cell tumor (Waxman, 1976). CIS is often seen at the edge of invasive tumors (Jacobson et al., 1981), and although sometimes it is found accidentally in the absence of an obvious tumor, most often it is found in association with testicular biopsies in the study of infertility or in orchiectomy specimens in cases of cryptorchidism. These changes have been associated with the eventual appearance of invasive neoplasm in

two series of patients (Skakkebaek, 1978). They have also been described in the contralateral testis in patients with testicular tumors (Berthelsen et al., 1979). The existence of CIS, however, is enough to support the concept that metastasis and invasiveness are distinct components of malignancy with corresponding distinct genetic controls. CIS, of course, would represent "half a cancer." If this hypothesis is true, then it should be only a matter of time before these elements can be controlled and highly malignant tumors prevented from metastasizing.

DIFFERENTIATION

Differentiation is a process by which cells change their physical characteristics so that they are capable of a specialized function. We usually think of this process as occurring in one irreversible direction. Once a cell lineage has been "committed" to exist, as for example, an epithelial cell, it generally remains as such without changing to another cell type. This is true simply because it is the most economical and organized way to do things; during embryonic development, a complex multicellular organism is formed consisting of differentiated cells arranged in a precise pattern. Generally, the body plan is set up on a miniature scale and then simply grows. If cells changed during growth, the body plan would change. Therefore, differentiated cells are usually quite stable. A few minor exceptions to this rule exist and are currently found in nature. For example, cells of one epithelial type may change to another in response to a noxious stimulus. This is familiar to urologists who find foci of squamous epithelium in the prostate following focal infarction. This change from one mature cell type to another is commonly known as *metaplasia* and in fact can be found in relation to drug use, growth of embryonic rests, response to injury, and other induced states.

The scientific basis for metaplasia is poorly understood. Most differentiated cells still contain a complete genome, even though only a small, specialized part is being used. An explanation for how one "program" of genes is turned off and another turned on is still unknown. Experimentally, metasplasia can be induced by substances that interfere with DNA methylation (Razin and Riggs, 1980; Felsenfeld and McGhee, 1982), but this is only a superficial observation considering the magnitude of the question, one of the most fundamental questions in biology today—namely, exactly what dictates and controls gene activation and how does it relate to differentiation? Although the molecular mechanisms responsible for control and initiation of differentiation are poorly understood, it is known that a cell's state of differentiation can be affected by extracellular matrix as well as directly by surrounding cells. Usually, changes associated with metaplasia are not great and are interconversions between closely related cell phenotypes. These tissue modulations may depend on short-range interactions between neighboring cells similar to those that control cell character in the embryo. This effect is demonstrated by in vivo experiments demonstrating that the development of the *epidermis* can be governed by the *dermis* beneath it (Billingham and Silvers, 1967).

Metaplasia represents one kind of abnormal differentiation. More dramatic examples can occur in other unnatural situations, such as those associated with malignancy; differentiation associated with germ cell tumors invites the best examples. That such tumors differentiate is no surprise based on the origin of the tumor, but some puzzling clinical situations may be better understood with some background on differentiation. It is not uncommon to observe differentiated metastasis from a primary embryonal cell carcinoma of the tesis. There is overwhelming evidence that these "mature metastases" are due not to the induction of differentiation by chemotherapy, but to selection of benign elements that are less sensitive to chemotherapy. Similarly, "benign" teratoma appearing one or more years following chemotherapy must be assumed to have originated from reactivation of a malignant stem cell. Finally, better understanding of factors controlling differentiative events may lead to successful clinical manipulation of the differentiative state by pharmacologic agents—i.e., differentiation therapy.

PARANEOPLASTIC ENDOCRINOPATHIES

A fascinating manifestation of altered differentiation associated with malignant transformation are the paraneoplastic endocrinopathies. Under these circumstances, determined cells of one differentiative phenotype produce ectopic hormones (and other substances) in response to malignant transformation. Although the mechanism of action is unknown, the phenomenon provides presumptive evidence that malignant transformation may be associated with gene rearrangement and/or altered transcriptional regulation. This is true because all cells contain the same genetic information; the real differences between cells are determined by those unknown forces that regulate which of the genes will be activated. Examples of humoral substances elaborated by cancers include peptide hormones, peptide hormone precursors, prostaglandins, fetoproteins (e.g., carcinoembryonic antigen, α-fetoprotein), and enzymes (e.g., fetal isoenzymes of alkaline phosphatase; thymidine kinase). At least 21 hormones or their precursors have been described to be produced by cancers, and almost all are protein or peptide in nature. The average urologic practice is exposed to this phenomenon through management of renal cell carcinoma, which can be considered the prototype malignancy for ectopic hormone production. For example, renal cell carcinoma has been shown to produce parathyroid hormone, erythropoietin, gonadotropins, placental lactogen, enteroglucagon, insulin-like substances, prostaglandin A, and prolactin (Altaffer and Chenault, 1979). These highly unusual and varied differentiated forms of renal cell carcinoma may be related to a second unusual and unique feature of renal cell carcinoma, its notorious reputation for metastasis to unusual locations. For example, many urologists have had the experience of evaluating metastatic renal cell carcinoma to the thyroid presenting as a thyroid nodule, either as the initial presenting symptom or many years following removal of the primary tumor. Metastatic disease may also appear in other unusual sites such as the testis (Talerman and Kniestedt, 1974) as well as in the more customary sites such as the brain, bone, and lung. If one assumes that the property of metastasis is a programmed event that depends on numerous variables, including appropriate tumor cell surface properties, so that the tumor interacts with the normal tissue at the metastatic site, it is evident that renal cell carcinoma has the capacity of programming itself to one of several modes, depending on which genetic program is selected. The altered transcriptional changes in renal cell carcinoma, compared with normal renal cells, are not strictly random, because the associated endocrinopathies are limited compared to those possible. Ectopic ACTH secretion, which is a common endocrinopathy associated with other tumors, is not even found with renal cell carcinoma. In contrast, prostatic carcinoma, which is less commonly associated with endocrinopathies, has been associated with ectopic ACTH formation (Newmark et al., 1973). This provides suggestive evidence that these changes do not represent random gene "derepression" but rather are associated with specific DNA or DNA transcriptional changes.

SPONTANEOUS REGRESSION OF CANCER AND IMMUNE SURVEILLANCE

Equal to the mystery of how new genes are activated with malignant transformation is how malignant cells will suddenly disappear—commonly known as spontaneous regression. The urologic community has been confronted with an unusually high number of such cases, generating considerable interest. In trying to understand and evaluate this phenomenon, it must be pointed out that spontaneous regression may involve the same factors (although in reverse) as those responsible for sudden activation of tumors many years after dormancy, as found with cancer of the prostate or renal cell carcinoma metastasis. Simply put, some tumors have a metabolic "on-off" switch that may be spontaneously activated without warning.

The incidence of spontaneous regression is interesting because most reported cases involve patients in the younger age group (including the first year of life for neuroblastoma), despite the fact that most urologic tumors occur in the sixth to seventh decades of life. This suggests that tumors arising in younger people are under different constraints, with perhaps different regulatory mechanisms than in otherwise identical patients who are older. The four most common types of tumors involved with spontaneous regression are hypernephroma, neuroblastoma, melanoma, and choriocarcinoma, and although there have been rare case reports of spontaneous regression of bladder cancer (Smith and Herr, 1980) and testicular cancer (Mueh et al., 1980), there has been no case of spontaneous regression of prostate cancer. The regression observed is often not permanent, with reports describing reactivation of tumor as long as 20 years following inactivation (Fairlamb, 1981).

The mechanism of spontaneous regression is unknown, although at least one of three factors appears to be present in the majority of cases presented, i.e., infection, hormonally induced changes, or altered immune status (Stephenson et al., 1971). The evidence that implicates a hormonal influence in spontaneous regression of renal cell carcinoma is circumstantial and derived from epidemiologic studies, laboratory experiments, and trials with a variety of hormonal agents. For example, menopause and pregnancy seem to provide a protective effect against renal cell carcinoma (Garfield and Kennedy, 1972). Likewise, experimentally it has been shown that long-term administration of estrogen to the golden Syrian male hamster will result in renal cortical tumors. This effect can be prevented by the administration of androgens and will not occur at all if the male is first castrated.

Infection and fever have also been associated with spontaneous regression of tumor. Initial reports of this association described a frequent association between regression and infections involving pneumonia, smallpox, malaria, tuberculosis, and erysipelas (Rohdenburg, 1918). Subsequent studies have shown an increased five-year survival (>50%) in patients who suffered empyema after pulmonary resection for lung cancer (Takita, 1970). This concept has also been supported by experimental evidence. Rabbits with Brown Pearce carcinoma who simultaneously were subjected to mass infection with hemolytic streptococci had no evidence of metastatic disease in contrast to 50% of controls who

did suffer metastasis (Christensen, 1959). It is possible, thus, that spontaneous regression may be mediated through a stimulatory effect on the immune system by either the invading organism or its endotoxin. It is also possible that the organism may alter specific antigenicity of the tumor cells, allowing increased host responsiveness.

Although activation of the immune system has been a favorite hypothesis to explain spontaneous regression, our understanding of a responsible mechanism is as yet rudimentary. As mentioned earlier in the chapter, the notion that the immune system plays a major role with respect to spontaneous regression has its conceptual foundation in the positive relationship between the immune system and bacterial infections. Yet, despite the logical and theoretical relationship between cancer and the immune system, a functional clinical application of these findings is not available, which is leading to doubts as to whether the immune system has a relationship with cancer strong enough to allow it to be used as a predictable and therapeutic tool in the clinical approach to cancer management.

MANAGEMENT CONCERNS

SURGICAL THERAPY

More patients are cured of cancer by surgery than by any other therapeutic modality. The amount is growing even larger with improved diagnostic techniques and improved adjuvant protocols. It is therefore apparent that the role of the surgeon is changing with the new problems and questions. Reviewing some of the basic foundations in light of newer modalities is the purpose of this section.

Classic cancer surgery is based on the idea that cancer begins as a local disease and spreads in an orderly fashion from the primary site to adjacent tissue by lymphatics and through blood vessels. This led to the concept of the en bloc resection technique, which represented one of the foremost advances in surgical technique. In fact, surgical cure rates have changed little in the past 2–3 decades, because the limits of surgical resection per se have been reached. Most examples of improved survival have been associated with improved perioperative care as well as adjuvant therapy. The proper use of surgery with respect to cancer treatment and the development of surgery as an improved modality in cancer treatment takes into account several important management concerns that will be reviewed.

One of the primary concerns of surgical manipulation of malignant tissue is dissemination of cancer cells and cancer cell implantation. Classic examples of this problem are represented by reports of blood-borne metastasis or wound seeding following manipulation of the primary tumor and wound seeding following biopsy of a malignant mass. When considering the problem of blood-borne metastasis, it is important to realize that the ability to metastasize is not a given property for all tumor cells; metastatic cells have unique, yet-to-be-identified features that allow them to implant, which makes them different from the primary tumor. Therefore, a tumor embolism does not necessarily result in a tumor metastasis. There is, however, a correlation between the presence of tumor cells seen in blood during the operative procedure and prognosis. Likewise, manipulation of the tumor at any time in the surgical procedure can greatly increase the number of cancer cells recovered from the blood. Good surgical technique takes these factors into account; when possible, the tumor should be removed by the "no-touch" technique and there should be early ligation of the vascular pedicle if appropriate for the operation. Considerations have led to new innovations in surgical technique such as preoperative embolization in the treatment of renal cell carcinoma.

In addition to a decrease in blood-borne tumor cells, en bloc resection with no-touch technique decreases the likelihood of wound contamination with tumor cells. Special wound precautions usually need not be taken if the excised tumor is not exposed and is well contained in the resected specimen. However, if tumor itself is transected, the operative field must be isolated with drapes and the cut surface of the tumor cauterized. Finally, consideration may be given to wound irrigation in an effort to sterilize the operative site. However, agents such as nitrogen mustard, thiotepa, and sodium hypochlorite solution have been tried with no definite success. A circumstance of special interest to the urologist concerns the fate of tumor cells disseminated in the retroperitoneum from a bladder perforation during bladder tumor resection. Such cells are capable of tumor implantation, especially if the original tumor is found to be either high grade or invasive. Dissemination of tumor cells from a low grade noninvasive tumor is generally considered to present a minimal chance of wound implantation. A similar role applies for operations that combine transurethral prostatectomy with TUR bladder tumor; both procedures may be done simultaneously when the malignancy appears to be low grade.

Wound seeding is also a consideration when planning to biopsy a malignant tumor. Fortunately, clinical experience with urologic tumors has, for the most part, formulated operative policy regarding biopsy. These policies are not always consistent but seem to provide a workable approach to urologic malignancy. For example, needle biopsy of the prostate is considered routine,

while needle biopsy of a testicular mass is rarely if ever performed. This, of course, is based on the well-known foundings that transscrotal open biopsy of testicular tumor in many cases resulted in wound contamination (Markland et al., 1973). In contrast, biopsy of cancer of the prostate rarely leads to wound seeding and metastasis. Likewise, needle biopsy of renal tumors is rarely performed, because other diagnostic techniques, such as CT scan and arteriography, are sufficiently developed. Needle aspirations of malignant renal cysts seldom, if ever, result in wound seeding. The substance of most of these observations again goes back to the original statement that metastasis is not an inherent property of all malignant cells, and when it does occur it may be variable depending on the origin of the tumor.

Cytoreductive Surgery

The role of cytoreductive surgery, or "debulking," as it is commonly called, should be at least mentioned because it represents a unique aspect of the surgical management of malignancy. The foundations for its use rest on the clinical and experimental observations that smaller tumor volumes respond better to radiotherapy or chemotherapy (Donohue et al., 1980; Griffiths and Fuller, 1978). Its routine clinical use is therefore predicated on the existence of a reliable form of adjuvant therapy for the tumor in question. Therefore, it is not surprising that most of our understanding of cytoreductive surgery is drawn from testicular tumor data. Cytoreductive surgery has been used in two general clinical circumstances in the treatment of nonseminomatous germ cell tumor: (1) prior to any systemic chemotherapy, and (2) after 3–4 cycles of chemotherapy when obviously persistent disease is present—i.e., as demonstrated by elevated markers or persistent mass. Although this approach fits the motif of surgical management of disease, there is still considerable debate as to its true value in a pretreatment setting. Some studies have shown that cytoreductive therapy does not improve response to subsequent chemotherapy compared with treatment with chemotherapy alone (Javadpour et al., 1982). In contrast, cytoreductive surgery may be of value in treatment of residual disease persisting after several cycles of chemotherapy. In this setting, however, the debulking procedure may be acting as more of a diagnostic and staging tool rather than a therapeutic maneuver (Einhorn, 1980). These issues will be further discussed in the appropriate chapters.

THE IMMUNE SYSTEM

The discipline of immunology has produced some of the most significant technological advances in approach to disease in the past 20 years. Its impact has been important not only in helping to understand malignancy, but perhaps more important, to lay the groundwork for more advanced approaches to research and therapeutics as represented by molecular genetics. As scientific understanding moves from the golden era of the phenomenology of immunology to the molecular understanding of cell biology and genetics, it is apparent that an understanding of the immune system has touched on our clinical approach to malignancy in several ways.

Monoclonal Antibodies

Antibodies were discovered through their ability to neutralize bacterial toxins, and for almost 100 years have been used in the form of polyspecific heteroantisera for both experimental and therapeutic purposes. As science and medicine became more exacting, difficulties began to develop with this type of antibody "mixture," in that it was not specific enough for exact molecular analysis and that repeated injections of supposedly innocuous "foreign" antibodies often lead to harmful hypersensitivity reactions.

In 1976, many of these problems, particularly antibody heterogeneity, were overcome by a new technique described by Kohler and Milstein that revolutionized the use of antibodies as tools for medicine and science and resulted in a 1984 Nobel Prize for its authors (Kohler and Milstein, 1976). The technique involves cloning a single antibody secreting B-lymphocyte (which secretes only one kind of antibody), so that an unlimited amount of identical antibody molecules can be obtained. Briefly, the classic technique involves immunizing a mouse with the appropriate immunogen, removing the spleen which contains the sensitized and antibody-secreting B-lymphocytes, and fusing the spleen cells with myeloma (B-cell tumor) cancer cells. The fusion is done to "immortalize" the B cells, which otherwise have a limited life span. The *hybrid* cells are then cloned, and it is determined which hybrid is producing the exact correct antibody, i.e., different hybridomas can produce a slightly different antibody against the same immunizing molecule. Informative hybridomas are propagated as individual clones, each of which provides a permanent and stable source of a single monoclonal antibody that is uniform with regard to its antigen-binding site (Milstein, 1980).

This type of controlled antibody production with uniform, exact specificity has provided new inroads into clinical and experimental approaches for treating and understanding malignancy. An interesting example involves the aforementioned ongoing search for tumor specific antigens. This subject always stimulates lively conversation, because it is not exactly clear whether such an entity does or should exist. The development and use of monoclonal antibodies could help resolve

this issue, although great care must be taken to search for the antigen on normal tissues; the smaller the epitope recognized by the monoclonal antibody, the greater the chance that the molecular structure of the target will be present on more than one tissue in the body. For example, the leukemia antigen CLLA (common lymphatic leukemia antigen) is also found on a minority of cells in immature but normal bone marrow (Ritz et al., 1980).

Another related and interesting use of monoclonal antibodies makes use of its tumor localizing capabilities. The idea of localizing tumor deposits with isotopes is not new but has only recently become a practical proposition with the advent of subtraction scanning. Such manipulation is required because, even though the tumor contains a concentrated amount of label compared with surrounding tissues, the actual amount is very little compared with the total amount injected (Goldenberg et al., 1978). Under these conditions background counts become significant. To resolve this problem, the vast quantities of radiolabeled antibodies in the vascular and extravascular compartments are "subtracted" from the tumor scan. The image subtraction is determined by measuring emission from another isotope that localizes to these tissue spaces. Preliminary results with ^{131}I-tagged anti-AFP and anti-HCG monoclonal antibodies demonstrated promising results, with some results bettering nodal detection by CT scanning in cases of metastatic AFP- or HCG-positive testicular cancer (Bradwell et al., 1981; Begent et al., 1981). Likewise, successful imaging of a mouse teratocarcinoma was performed with ^{131}I-labeled monoclonal antibody to stage specific embryonic antigen-1 (SSEA-1) by gamma ray scintigraphy (Ballou et al., 1979).

Current limitations and problems include relatively poor resolution by the gamma counter; localization is possible only with tumor deposits greater than 2 cm in diameter. Also, with the use of other antibodies, cross reactivity with normal tissues decreases the security of the study.

Despite the theoretical advantages of in vivo imaging, a major problem exists in the consideration of the in vivo use of monoclonal antibodies. After a few trials it was determined that repeated injections of a supposedly innocuous "foreign" antibody (remember that the antibody is of mouse origin) can lead to harmful hypersensitivity reactions. Currently, three approaches have been studied. The first considers induction of specific unresponsiveness to rodent immunoglobulin before or during therapy. The second approach requires the use of human-derived monoclonal antibodies produced either from human hybridomas or from Epstein-Barr virus—immortalized human B-lymphocyte cell lines. However, despite considerable efforts, the success of this approach has been limited, and there is no generalized way of obtaining human monoclonal antibodies of the right specificity. The third approach, the most recent and perhaps most intriguing, makes use of the techniques of genetic engineering; it may be possible to obtain antibodies in which the antigen-binding site is defined by genetic sequences from a rodent monoclonal antibody of the right specificity, whereas the rest of the molecule is as "human" as possible. In a sense, chimera antibodies have been made containing immunoglobulin light chains and heavy chains in which the variable regions (or the antigen binding site) are of mouse origin and the constant regions (or the Fc portions) are human (Morrison, 1984). It is possible that the immune system will not recognize such chimeric molecules as foreign, but even if it does, it should be easier to induce unresponsiveness to rodent variable regions when they are part chimeric antibodies.

Another fascinating yet problematic outgrowth of the monoclonal era concerns the potential of such antibodies to deliver toxic drugs or tumoricidal substances to malignant cells. Some of the reports have been intriguing. For example, the simplistic approach involving the use of unmodified monoclonal antibodies has been shown to be extremely effective in depleting cells in vivo (in mice) and can be used for selective manipulation and depletion of different compartments of the immune system (Cobbold, 1984). Most monoclonal reagents, however, cause only transient reductions in their target cells in vivo, which led many workers to develop the antibodies as targeting agents for highly cytotoxic drugs. Most antibody-directed cytotoxic agents have been produced primarily by preparing drug or toxin conjugates with antibodies. Other approaches have included monoclonal antibody-targeted liposomes, shown successfully to bind and deliver encapsulated methotrexate gamma aspartate to L929 fibroblasts with a high degree of specificity (Heath et al., 1983). The development of monoclonal antibodies has also allowed resurrection of older ideas in cancer treatment. For example, more than 30 years ago it was suggested that boron conjugated to antitumor antibodies might provide a selective and effective method of cancer therapy because of the property of boron-10 nuclei to absorb thermal neutrons (Bale, 1952). This reaction releases an α particle and lithium 7 ion, which have limited penetration and thus a restricted cytotoxic range. Tissues not containing boron-10 have low neutron capture values for thermal neutrons, and thereby selective tissue cytotoxicity can be achieved in tissues containing boron-10 that received irradiation with thermal neutrons (Mizusawa et al., 1982). Studies have now shown that boron-10 can be coupled to antitumor monoclonal antibodies such as anti-CEA without destroying the immunoreactivity of

the antibody (Mizusawa et al., 1982). Further work is necessary to adjust the amount of boron-10 delivered to the tumor to be sufficient to achieve destruction by neutron capture.

It is clear that monoclonal antibodies offer promise in the therapy of malignant disease. It is equally clear that their true role remains to be established and that they will certainly be used in conjunction with other effective modalities of treatment rather than as a substitute for them. Before that role can be found, the limitations, as described above, must be resolved.

Immunosurveillance and Immunotherapy

Immunosurveillance can best be considered the "natural" form of immunotherapy and refers to a mechanism whereby a host produces an immune response against antigens expressed by tumors. Immunosurveillance as a mechanism for the control of cancer was first suggested by Green in 1954. This idea was further amplified by Burnet, who said that a major function of the immunologic mechanism in mammals is to recognize and eliminate foreign patterns of behavior arising in the body by somatic mutation or some other equivalent process (Burnet, 1970). This concept, of course, assumes the presence and existence of tumor-specific antigens.

Indirect evidence for the existence of a surveillance system is provided by the recognized increased incidence of malignancy in patient populations with a history of transplantation and immunosuppression (Sheil et al., 1985). For example, the incidence of cancer in 31 of 40 sites normally classified in the WHO International Classification of Diseases was higher in transplant patients than in age-matched controls. The greatest increased risks were for cancers in which viruses are thought to be implicated, including cancer of the cervix and Kaposi's sarcoma. Two cancers were demonstrated to have a decreased incidence, one of which is cancer of the prostate. This abnormal distribution of tumors suggests that the oncogenic mechanism in the transplant immunosuppressed population is different from that of the normal population. The findings may be partially explained by the suggestion that immunosuppressed patients are subject to viral infections, some of which are oncogenic; also, the antigen load of transplanted cells may contribute to the development of tumors. Of additional importance are reports of tumor regression with renal graft rejection following cessation of immunosuppressive therapy.

Evidence for effectiveness of surveillance in vivo is partial; in general, experimental data show that tumors with high immunogenicity are difficult to induce. Numerous attempts to artificially induce a state of surveillance in the name of immunotherapy have been undertaken since the turn of the century. Although occasional striking regressions have been reported, in most cases the results were not consistent. A rational basis for cancer immunotherapy was provided only during the past several years, as evidence was accumulated indicating the participation of immune response in human cancer. It was found that cancer patients develop two types of immune response to the antigens of their neoplasm: humoral antibodies and cell-mediated immune reactions. The relative importance of these two types of immune responses has been and still is incompletely understood and controversial.

Several approaches to immunotherapy have been tried that may be *specific* and designed to cope with a particular tumor or *nonspecific* when the overall immunologic reactivity of the host will be changed. Specific immunotherapy can be: (1) *passive* when antiserum specific to a particular tumor is used; (2) *adoptive*, where donors are sensitized and cytotoxic cells transferred; or (3) *active*, where tumor cells or part of them are given to stimulate the host to provide an exaggerated response against a particular antigen.

The greater effectiveness of active immunization over passive immunization for protection against infectious diseases provided a strong stimulus for studies of active immunotherapy against cancer. The rationale for this approach is based on animal studies that showed that a growing tumor does not induce a maximum immune response in the host. In this method of treatment, efforts are made to increase the patient's immunity either by altering the tumor specific antigen in such a way that it becomes more antigenic or by stimulating the patient's lymphoreticular system with immunologic adjuvants. Active specific immunotherapy then usually has involved preparation of tumor-specific vaccines that involve injection of tumor cell membranes or whole tumor cells inactivated by a variety of different methods to render the cells incapable of proliferation (such as freeze-thaw, heat, mitomycin-C treatment, or radiation). The efficacy of such vaccines in humans has been difficult to prove, even with attempts to increase the antigenity of tumor vaccines by coupling highly antigenic carrier proteins to the tumor cells or by nonspecific enhancement of the host immune system by nonspecific immunologic adjuvants such as BCG vaccine, *Corynebacterium parvum,* and Freund adjuvant (Droller, 1985). It has been considered that renal artery embolization (angioinfarction) before nephrectomy in patients with metastatic renal cell carcinoma has been considered a form of active specific immunotherapy. Initial reports demonstrated improvement and regression of pulmonary metastasis (Wallace et al., 1981); however, later studies suggested minimal objective response with no significant change in survival (Swanson

et al., 1983; Mebust et al., 1984). The same conceptual reasoning, as well as results, has been invoked relative to cryotherapy for prostate cancer. It was thought that controlled freezing of malignant prostatic tissue not only would improve mechanical voiding and other local disease, but also would stimulate immune response. The overall results, however, have not been satisfactory (Soanes et al., 1970).

Passive immunotherapy is another form of specific immunotherapy, and the best example has already been mentioned, i.e., the use of monoclonal antibodies. This technology is especially important, because earlier antitumor sera produced in a foreign species usually proved fairly toxic, because it contained antibodies against normal tissue antigens of the host. This approach has many great theoretical advantages, especially considering the development of monoclonal antibodies specific for antigens that are thought to be expressed specifically on renal cell and transitional cell tumors (Begun and Grossman, 1984; Fradet et al., 1984). Clinical use of such antibodies is still not available because of problems with host sensitivity to antibodies of different species (i.e., the origin of monoclonal antibodies is usually mouse) and inexact knowledge of antibody reactivity with respect to tissue specificity for clinical use. When these problems are answered, the use of monoclonal antibodies will be important in both diagnosis and treatment. Another form of passive specific immunotherapy involves adoptive transfer of immunized lymphocytes. As a form of cancer treatment, this is really of more laboratory interest than clinical importance with respect to urologic tumors; it has been used mainly for a few bladder cancer patients, with mixed results (Cockett et al., 1982). Currently, the best example of adoptive transfer involves immune modification of the patient's own lymphocytes with reinfusion. This technique has been used in patients with metastatic renal cell carcinoma where autochthonous lymphocytes are sensitized to the patient's own tumor and the suppressor cells removed. The resulting lymphocytes are then reinfused back into the patient. Results have indicated three of eight patients had objective responses (Carpinito et al., 1984). This approach is still in question; however, it is likely that adoptive immunotherapy with in vitro nonspecific stimulation of lymphocytes will have significant clinical and antitumor effect.

In addition to specific immunotherapy, nonspecific immunotherapy is also of use. The theoretical approach to nonspecific immunotherapy is based on the observation that nonspecific immunostimulants, such as mixed bacterial toxins and the tubercle bacillus, have the ability to nonspecifically enhance host resistance to most viral, fungal, and bacterial agents. Although the mechanism is unknown, these agents also appear to have the ability to stimulate immune response to tumor antigens in addition to other antigens. One of the first examples of this form of therapy was reported by Bradford Coley at the turn of the century. He reported a complete regression of an inoperable sarcoma of the neck following an attack of erysipelas. This observation led to the development of Coley's toxins, which was a mixture of killed bacterial toxins, which was then injected directly into the tumor lesions. In recent-day therapeutics, Coley's toxin has been replaced by BCG, whose first real consistent effect was observed in dealing with metastatic malignant melanoma. As many as 90% of the melanoma lesions would disappear following intradermal injection with BCG in patients who were immunologically competent as judged by dinitrochlorobenzene sensitization. It is apparent, therefore, that whatever effect BCG has, there is probably no direct antitumor effect because it usually does not cause tumor regression in tuberculin negative patients. The primary use of BCG with respect to urologic disease, of course, is for bladder cancer, where it has been found to be effective against superficial disease (Lamm et al., 1981). Its success in bladder cancer raises even more questions, because some reports suggest that oral treatment may be as effective as intravesical, while other reports indicate that its effectiveness is based on a simple inflammatory response in the bladder, with passive sloughing of the tumor. This perhaps is related to the aforementioned increased survival afforded by empyema to patients undergoing lung resection for cancer. The association among inflammation, malignant growth, and the immune system is still poorly understood. Despite these questions, this type of immunotherapy is becoming increasingly important in the treatment of bladder tumors and should provide a platform on which we can learn more about active nonspecific immunotherapy to apply it to other tumors.

REFERENCES

1. Altaffer LF III, Chenault DW Jr: Paraneoplastic endocrinopathies associated with renal tumors. *J Urol* 1979; 122:573–577.
2. Bale WF: *Proc Natl Cancer Conf* 1952; 2:967–976.
3. Ballou B, Levine G, Hakala T, et al.: Tumor location detected with radioactivity labeled monoclonal antibody and external scintigraphy. *Science* 1979; 206:844–847.
4. Barger M, Noonan KD: Restoration of normal growth by covering of agglutinin sites on tumor cell surface. *Nature* 1970; 228:512–515.
5. Bast R Jr: Tumor immunology, in *Cancer Manual. American Cancer Society.* Boston, American Cancer Society, 1982, pp 72–82.
6. Begent RH, Searle F, Stanway G, et al: Radioimmunolocalization of malignant teratoma using radiolabelled antibody directed against human chorionic gonadotrophin, in Anderson CK, Jones WG, Ward AM (eds): *Germ Cell Tumors.* New York, Alan R Liss, 1981, pp 264–265.
7. Begun FP, Grossman HB: Murine monoclonal antibodies reactive with normal kidney and renal carcinoma cells. *J Urol* 1984; 131:130A.
8. Berthelsen JG, Skakkeback NG, Morgensen P, et al: Incidence of carcinoma in situ of germ cells in contralateral testis of men with testicular tumors. *Br Med J* 1979; 2:363–364.
9. Billingham RE, Silvers WK: Studies on the conservation of epidermal specificities of skin and certain mucosas in adult mammals. *J Exp Med* 1967; 125:429–446.
10. Bradwell AR, Fairweather DS, Dykes PW: Isotope localization of germ cell tumors, in Anderson CK, Jones WG, Ward AM (eds): *Germ Cell Tumors.* New York, Alan R Liss, 1981, pp 260–263.
11. Bruce WR, Meeker BE, Valeriote FA: Comparison of the sensitivity of normal hematopoietic and transplanted lymphoma colony forming cells to chemotherapeutic agents administered in vivo. *JNCI* 1966; 37:233.
12. Bryant PJ, Bryant SV, French V: Biological regeneration and pattern formation. *Sci Am* 1977; 237(1):67–81.
13. Burnet FM: The concept of immunological surveillance. *Prog Exp Tumor Res* 1970; 13:1–28.
14. Byrd WJ, Jacobs DM, Amos MS: Synthetic polyamines added to cultures containing bovine sera reversibly inhibit in vitro parameters of immunity. *Nature* 1977; 267:621.
15. Capone PP: A lesson from the mammography issue (editorial). *Ann Intern Med* 1978; 88:703–704.
16. Carpinito GA, Krane RJ, Osband ME, et al: Effective treatment of metastatic carcinoma with in vitro immunized autologous lymphocytes and cimetidine. Presented at Annual Meeting, New England Section, American Urological Association, September 1984.
17. Chodak G: Tumor angiogenesis, in Catalona W, Ratliff T (eds): *Urologic Oncology.* Boston, Martinus Nijhoff, 1984, pp 1–13.
18. Chodak GW, Handenschild C, Gittes RF, et al: Angiogenic activity as a marker of neoplastic and preneoplastic lesions of the human bladder. *Ann Surg* 1980; 192:762.
19. Chodak GW, Scheiner CJ, Zetter BR: Urine from patients with transitional cell carcinoma stimulates migration of capillary endothelial cells. *N Engl J Med* 1981; 305:869–874.
20. Christensen EA: Infection and malignant tumors: I. Growth of Brown-Pearce carcinoma in rabbits treated with living or killed haemolytic streptococci. *Acta Pathol Microbiol Scand* 1959; 46:285.
21. Cifone M, Fidler I: Increasing metastatic potential is associated with increasing genetic instability of clones isolated from murine neoplasms. *Proc Natl Acad Sci USA* 1981; 78:6949–6952.
22. Cobbold SP, Jayasolla A, Nash A, et al: Therapy with monoclonal antibodies by elimination of T cell subsets in vivo. *Nature* 1984; 312:548–551.

23. Cockett ATK, diSant'Agnese PA, Hamlin DJ, et al: Porcine sensitized lymph node cells (immunotherapy) and attempted irradiation for infiltrative transitional cell carcinoma of the bladder. *Urology* 1982; 19:593–598.

24. Coggin JH Jr, Anderson NC: Cancer, differentiation and embryonic antigens: Some central problems. *Adv Cancer Res* 1974; 19:105.

25. Commission on Clinical Oncology of the Union International Contre le Cancer [International Union Against Cancer (IUCC)]: TNM Classification of malignant tumors. Geneva International Union Against Cancer, 1968.

26. Costanza M, Li F, Greene H, et al: Cancer prevention and detection, in *Cancer Manual, American Cancer Society*, Boston, American Cancer Society, 1982, p 2.

27. DeMaeyer E: in DeMaeyer E, Galasso G, Schellekens H (eds): *The Biology of the Interferon System*. Amsterdam, Elsevier, 1981, pp 203–209.

28. Dennert G, Lotan R: Effects of retinoic acid on the immune system: Stimulation of T killer cell induction. *Eur J Immunol* 1978; 8:23–29.

29. DeVita VT: The evolution of therapeutic research in cancer. *N Engl J Med* 1978; 298:907–910.

30. Donohue JP, Einhorn LE, Williams SD: Cytoreductive surgery for metastatic testis cancer: Considerations for timing and extent. *J Urol* 1980; 123:876–880.

31. Droller M: Immunotherapy in genitourinary neoplasia. *J Urol* 1985; 133:1–5.

32. Dubois MF, Vignal M, LeCunff M, et al: Interferon inhibits transformation of mouse cells by exogenous cellular or viral genes. *Nature* 1983; 303:433–435.

33. Einhorn LH: The role of surgery in disseminated testicular cancer (abstract). *Proc Am Soc Clin Oncol* 1980; 21:159.

34. Einhorn LH, Williams SD, Mandelbaum I, et al: Surgical resection in disseminated testicular cancer following chemotherapeutic cytoreduction. *Cancer* 1981; 48:904–908.

35. Fairlamb SJ: Spontaneous regression of metastasis of renal cancer. *Cancer* 1981; 47:2102–2106.

36. Felsenfeld G, McGhee J: Methylation and gene control. *Nature* 1982; 296:602–603.

37. Fishman M, Dragsten PR, Spector I: Immobilization of concanavalin A receptors during differentiation in neuroblastoma cells. *Nature* 1981; 290:781.

38. Folkman J: Antiangiogenesis: New concept for therapy for solid tumors. *Ann Surg* 1972; 175:409–446.

39. Folkman J, Cotran R: Relation of vascular proliferation to tumor growth. *Int Rev Exp Path* 1976; 16:207.

40. Folkman J, Haudenschild C: Angiogenesis in vitro. *Nature* 1980; 288:551–556.

41. Folkman J, Langer R, Linhardt RJ, et al: Angiogenesis inhibition and tumor regression caused by heparin or a heparin fragment in the presence of cortisone. *Science* 1983; 221:719–725.

42. Folkman J, Moscona A: The role of cell shape in growth control. *Nature* 1978; 273:345–349.

43. Fradet Y, Cordon-Cardo C, Whitmore WF Jr, et al: Bladder tumor heterogeneity defined by tissue typing with monoclonal antibodies. *J Urol* 1984; 131:278A.

44. French V, Bryant PJ, Bryant SV: Pattern regulation in epimorphic fields. *Science* 1976; 193:969–981.

45. Garfield DH, Kennedy BJ: Regression of metastatic renal cell carcinoma following nephrectomy. *Cancer* 1972; 30:190–196.

46. Goldenberg DM, Deland F, Kim E, et al: Use of radiolabelled antibodies to CEA for the detection and localization of diverse cancers by external photoscanning. *N Engl J Med* 1978; 298:1384–1388.

47. Goldfarb R, Herberman RB: Natural killer cell reactivity: Regulatory interactions among phorbol ester interferon, choleratoxin and retinoic acid. *J Immunol* 1981; 126:2129.

48. Green HN: An immunological concept of cancer: A preliminary report. *Br Med J* 1954; 2:1374–1380.

49. Gresser L, Tovey MG: *Biochem Biophys Acta* 1978; 516:281–287.

50. Griffiths CT, Fuller AT: Intensive surgical and chemotherapy management of ovarian cancer. *Surg Clin North Am* 1978; 58:131–142.

51. Harkins H, Anderson DJ: Seminal plasma factors released by sperm inhibit cytotoxic activity of natural killer cells. *Biol Rep* 1984; 30:1395.

52. Hayward WS, Neel BG, Astrin SM: Activation of cellular onc gene by promotor insertion of ALV induced lymphoid leukosis. *Nature* 1981; 290:475–480.

53. Heath T, Montgomery J, Piper J, et al: Antibody-targeted liposomes: Increase in specific toxicity of methotrexate-gamma-aspartate. *Proc Natl Acad Sci USA* 1983; 80:1377–1381.

54. Heldin CH, Westmark B: Growth factors: Mechanism of action and relation to oncogenes. *Cell* 1984; 37:9–20.

55. Hueper WC: *Occupational Tumors and Allied Diseases*. Springfield, Ill, Charles C Thomas Publisher, 1942.

56. Huggins C, Hodges CV: Studies of prostatic cancer: I. The effect of castration, of estrogen and of androgen injection on serum phosphatases in metastatic carcinoma of the prostate. *Cancer Res* 1941; 1:293–297.

57. Hurtenbach U, Morgenstern F, Bennett D: Induction of tolerance in vitro by autologous murine testicular cells. *J Exp Med* 1980; 151:827–838.

58. Isaac A, Lindenmann J: Virus interference: I. The interferons. *Proc Rev Soc* 1957; B147:258–267.

59. Jacobson GK, Hendriksen OB, Van der Masse H: Carcinoma in situ of testicular tissue adjacent to malignant germ cell tumors: A study of 105 cases. *Cancer* 1981; 47:2660–2662.

60. Javadpour N, Ozols RF, Anderson T, et al: A randomized study of cytoreductive surgery followed by chemotherapy versus chemotherapy alone in bulky stage III testicular cancer with poor prognostic features. *Cancer* 1982; 50:2004–2010.

61. Jukes TH: Silent nucleotide substitutions and the molecular evolutionary clock. *Science* 1980; 210:973–978.

62. Karnik S, Dekernion J: The diagnosis and management of urothelial atypia, in Resnick M (ed): *Current Trends*

in Urology. Baltimore, Williams & Wilkins, 1981, vol 1, p 123.

63. Kelley D, Catalona W: BCG therapy for superficial bladder cancer, in Ratliff T, Catalona W (eds): *Urologic Oncology*. Boston, Martinus Nijhoff, 1984, pp 169–184.

64. Kohler G, Milstein C: Continuous cultures of fused cells secreting antibody of the defined specificity. *Nature* 1975; 256:495–497.

65. Kuntz RE, Cheever AW, Myers BJ: Proliferative epithelial lesions of the urinary bladder in nonhuman primates infected with *Schistosoma haematobium. JNCI* 1972; 48:223–235.

66. Lamm DL, Thor DE, Winters WD, et al: BCG immunotherapy of bladder cancer inhibition of tumor recurrence and associated immune responses. *Cancer* 1981; 44:82.

67. Loughlin K, Gittes R, Partridge D, et al: The relationship between lactoferrin and the anemia of renal cell carcinoma. *J Urol* 1984; 131:131A.

68. Malkovsky M, Dore C, Hunt R, et al: Synergy of ricin A chain-containing immunotoxins and ricin B chain-containing immunotoxins in in vitro killing of neoplastic human B cells. *Proc Natl Acad Sci USA* 1983; 80:6322–6326.

69. Markland D, Kedia K, Fraley EE: Inadequate orchiectomy for patients with testicular tumors. *JAMA* 1973; 224:1025–1028.

70. Moscona AA: Embryonic and neoplastic cell surfaces: Availability of receptors for concanavalin A and wheat germ agglutinin. *Science* 1971; 171:905–907.

71. Mazumber A, Rosenberg S: Successful immunotherapy of natural killer resistant established pulmonary melanoma metastasis by the intravenous adoptive transfer of syngeneic lymphocytes activated in vitro by interleukin-2. *J Exp Med* 1984; 159:495–507.

72. Mebust WK, Weigel JW, Lee KR, et al: Renal cell carcinoma—angioinfarction. *J Urol* 1984; 131:231–235.

73. Miller AB, Lindsay J, Hill GB: Mortality from cancer of the uterus in Canada and its relationship to screening for cancer of the cervix. *Int J Cancer* 1976; 17:602–611.

74. Milstein C: Monoclonal antibodies. *Sci Am* 1980; 243:66–74.

75. Mintz B, Fleishman RA: Teratocarcinomas and other neoplasms as developmental defects in gene expression. *Adv Cancer Res* 1981; 34:211–278.

76. Mizusawa E, Dahlman HL, Bennett SJ, et al: Neutron capture theory of human cancer: In vitro results on the preparation of boron-labeled antibodies to carcinoembryonic antigen. *Proc Natl Acad Sci USA* 1982; 79:3011–3014.

77. Morrison SL: Chimeric human antibody molecules: Mouse antigen binding domains with human constant region domains. *Proc Natl Acad Sci USA* 1984; 81:6851–6855.

78. Mueh JR, Greco CM, Green MR: Spontaneous regression of metastatic testicular cancer in a patient with bilateral sequential testicular tumor. *Cancer* 1980; 45:2908–2912.

79. Narayan P, DeWolf WC: Differentiation in teratocarcinoma: Its implication in therapy, in Catalona WJ, Ratliff T (eds): *Urologic Oncology*. Boston, Martinus Nijhoff, 1984, pp 189–204.

80. Newmark SR, Dluhy RG, Bennett AH: Ectopic adrenocorticotropic syndrome with prostatic carcinoma. *Urology* 1973; 2:666–668.

81. Oldham RK, Herberman RB: Evaluation of cell mediated cytotoxic reactivity against tumor associated antigens with ^{125}I iododeoxyuridine labeled target cells. *J Immunol* 1973; 111:1862–1871.

82. Oosterhuis JW, Damjanov E: Treatment of primary embryo derived teratocarcinomas in mice with cis diamminedichloroplatinum. *Eur J Cancer Clin Oncol* 1983; 19:695–699.

83. Oosterhuis JW, Suurmeijer AJ, Sleijfer D, et al: Effects of multiple drug chemotherapy (cis-diamine-dichloroplatinum, bleomycin, actoid vinblastine) on the maturation of retroperitoneal lymph node metastasis of non seminomatous germ cell tumors of the testis: No evidence for de novo induction of differentiation. *Cancer* 1983; 51:408–416.

84. Peckham MJ: Non-seminomas: Current treatment results and future prospects, in Peckham MJ (ed): *The Management of Testicular Tumors*. London, Edward Arnold, 1981, pp 218–239.

85. Pfeffer LM, Wang E, Tamm I: Interferon effects on microfilament organization cellular fibronectin distribution and cell motility in human fibroblasts. *J Cell Biol* 1980; 85:9–17.

86. Pierce GB, Cox WF Jr: *Cell Differentiation and Neoplasia*. New York, Raven Press, 1978, pp 57–66.

87. Pierce GB, Shikes R, Fink LM: Cancer: A problem, in *Developmental Biology*. Englewood Cliffs, NJ, Prentice-Hall, 1978.

88. Pollack R, Risser R, Conlon S, et al: Plasminogen activator production accompanies loss of anchorage regulation in transformation of primary rat embryo cells by simian virus 40. *Proc Natl Acad Sci USA* 1974; 17:4792–4796.

89. Poste G, Doll J, Fidler IJ: Interactions among clonal subpopulations affect stability of the metatatic phenotype in polyclonal populations of B16 melanoma cells. *Proc Natl Acad Sci USA* 1981; 78:6226–6230.

90. Poste G, Fidler IJ: The pathogenesis of cancer metastasis. *Nature* 1980; 283:139–146.

91. Pott P: Chirurgical observations relative to the cataract, the polypus of the nose, the cancer of the scrotum, the different kinds of rupture, and the mortification of the toes and feet. London, L Hawes, Clarke, and Collins, 1775.

92. Ratliff T, Shapiro A, Catalona W: Interferon as an antitumor agent for urologic tumors, in Ratliff T, Catalona W (eds): *Urologic Oncology*. Boston, Martinus Nijhoff, 1984, pp 211–238.

93. Razin A, Riggs AD: DNA methylation and gene function. *Science* 1980; 210:604–610.

94. Ritz J, Pesando JM, Notis-McConarty J, et al.: A mono-

clonal antibody to acute lymphoblastic leukaemia antigen. *Nature* 1980; 283:583–585.

95. Rohdenburg GL: Fluctuations in the growth energy of malignant tumors in man with special reference to spontaneous regression. *J Cancer Res* 1918; 3:193–225.

96. Rosa F, Fellous M: The effect of gamma interferon on MHC antigens. *Immunol Today* 1984; 5:261–262.

97. Rosenberg EB, Herberman RB, Levine PH, et al: Lymphocyte cytotoxicity reactions to leukemia associated antigens in identical twins. *Int J Cancer* 1972; 9:648–658.

98. Sackett D: Periodic examination of patients at risk, in Schottenfeld D (ed): *Cancer Epidemiology and Prevention.* Springfield, Ill, Charles C Thomas Publisher, 1975.

99. Sato G: *Hormones and Cell Culture.* New York, Cold Spring Harbor Laboratory, 1979.

100. Sheil AG, Flavel S, Disney AP, et al: Cancer in dialysis and transplant patients. *Transplant Proc* 1985; 17:195–198.

101. Siminovitch L: On the nature of heritable variation in cultured somatic cells. *Cell* 1976; 7:1–11.

102. Skakkebaek NE: Carcinoma in situ of the tesis: Frequency and relationship to invasive germ cell tumors of infertile men. *Histopathology* 1978; 2:157–170.

103. Slamon DJ, Dekernion JB, Verma IM, et al: Expression of cellular oncogenes in human malignancies. *Science* 1984; 224:256–262.

104. Smith JA Jr, Herr HW: Spontaneous regression of pulmonary metastasis from transitional cell carcinoma. *Cancer* 1980; 46:1499–1502.

105. Soanes WA, Albin RJ, Gander MJ: Remission of metastatic lesions following cryosurgery in prostatic cancer: Immunologic considerations. *J Urol* 1970; 104:154–159.

106. Soloway AH: *Progress in Boron Chemistry.* New York, MacMillan, 1964, vol 1, pp 203–234.

107. Stein G, Stein J, Thompson J: Chromosomal proteins in transformed and neoplastic cells. *Cancer Res* 1978; 38:1181–1201.

108. Stephenson HE Jr, Delmez JA, Renden DI, et al: Host immunity and spontaneous regression of cancer evaluated by computerized data reduction study. *Surg Gynecol Obstet* 1971; 133:649–655.

109. Swanson DA, Johnson DE, VonEschenbach AC, et al: Angioinfarction plus nephrectomy for metastatic renal cell carcinoma—an update. *J Urol* 1983; 130:449–452.

110. Takita H: Effect of postoperative empyema on survival of patients with bronchogenic carcinoma. *J Thorac Cardiovasc Surg* 1970; 59:642–644.

111. Talerman A, Kniestedt WF: Testicular tumor as the first manifestation of renal carcinoma. *J Urol* 1974; 111:584–586.

112. Tapiero H, Fourcade A, Billard C: Membrane dynamics of Friend leukemia cells. II. Changes associated with cell differentiation. *Cell Differ* 1980; 9:211–218.

113. Terranova V, Williams JE, Liotta LA, et al: Modulation of the metastatic activity of melanoma cells by laminin and fibronectin. *Science* 1985; 226:982–985.

114. Thor A, Horan-Hand P, Wunderlich D, et al: Monoclonal antibodies define differential vas gene expression in malignant and benign colonic diseases. *Nature* 1984; 311:562–564.

115. Wallace S, Chuang VP, Swanson DA, et al: Embolization of renal carcinoma. *Radiology* 1981; 138:563–570.

116. Waxman M: Malignant germ tumor in situ in a cryptorchid testis. *Cancer* 1976; 38:1452–1456.

117. Weil J, Epstein C, Epstein LB: A unique set of polypeptides induced by alpha interferon in addition to those indicated in common with alpha and beta interferons. *Nature* 1983; 301:437–439.

118. Weiss R, Teich N, Varmus H, et al: *RNA Tumor Viruses.* New York, Cold Spring Harbor, 1982.

119. Whitmore WF III, Gittes RF: Afferent lymphatics: their importance to immunologically privileged sites. *Surg Forum* 1975; 26:338–340.

120. Whitmore WF III, Gittes R: Intratesticular grafts: The testis as an exceptional immunologically privileged site. *Trans Am Assoc Genitourinary Surg* 1979; 70:76–80.

121. Zbar B, Rapp HJ: Immunotherapy of guinea pig cancer with BCG. *Cancer* 1974; 34:1532–1540.

Chapter 9

Calculus Formation

ALAN D. JENKINS, M.D.

The surgical treatment of urolithiasis has been revolutionized by percutaneous lithotripsy, extracorporeal shock wave lithotripsy (ESWL), and ureteroscopic lithotripsy. These new techniques permit the rapid removal of upper urinary tract calculi without the discomfort and postoperative recuperative period associated with traditional open surgical procedures.

For a few patients, especially those with solitary or infrequent recurrent stone formation, intermittent ESWL treatment may be preferable to lifelong preventive measures. Most patients, however, do not understand the nature of their disease and want to know why they form kidney stones. Other patients may have an easily identifiable etiology, such as cystinuria or renal tubular acidosis, for their stone formation. Specific and effective medical treatment programs exist for these patients. Other patients will benefit from such simple measures as increased fluid intake and dietary moderation. Patients with idiopathic calcium urolithiasis and frequent recurrent stone formation will require a medical regimen that is designed to correct the physicochemical abnormalities commonly associated with all forms of calcium urolithiasis.

The purpose of this chapter is to impart an understanding of the basic pathophysiology of urinary tract stone formation and to enable the urologic practitioner to blend successfully the medical and surgical management of his urolithiasis patients.

BASIC PRINCIPLES

Biologic mineralization involves the precipitation of a poorly soluble salt, usually in association with an organic matrix. Supersaturation of the precipitating phase must be present before crystallization can occur. If at least local supersaturation is not present, then crystallization is thermodynamically impossible. Furthermore, most medical treatment programs depend on a reduction of supersaturation to prevent further stone formation. The solubility concept is one of the most important aspects of biologic mineralization (Robertson, 1982).

The physical chemistry of ions in an aqueous solution, such as urine, involves four basic concepts: ion activity, ion pairing, solubility, and relative supersaturation. The effective concentration of an ion, such as Ca^{++}, in solution is different from its actual concentration. This effective concentration, the *chemical activity*, depends on the ionic strength of the solution. Through electrical field effects, the other ions in the solution will affect the true chemical activity of a particular ion. The ionic strength is a measure of the magnitude of this electrical field and increases as the concentration of ions increases and their valence or charge increases. The activity of an ionic species decreases as the ionic strength increases. For a given total calcium concentration, the activity of Ca^{++} would be greater in distilled water, which contains only the Ca^{++} and its accompanying anions, than in urine, which contains many different ions.

The relationship between ion activity, $\{Na^+\}$, and ion concentration, $[Na^+]$, is given by

$$\{Na^+\} = [Na^+] \times a_1$$

where a_1 is the activity coefficient for Na^+. Activity coefficients are always less than 1 and approach unity as the concentration of the solution decreases. The ionic strength of a physiologic salt solution, such as urine, is approximately 0.15 M. The corresponding activity coefficients for the divalent ions Ca^{++} and $C_2O_4^{--}$ (oxalate) are approximately 0.3. The activity of each of these ions is less than one third of the concentration.

Specific ions of opposite charge, such as Ca^{++} and SO_4^{--}, can interact to form soluble *ion pairs* or complexes. These interactions effectively reduce the "ionized" concentrations of the ions involved. Many such interactions are possible in urine. Ionized calcium can be measured directly with an ion-selective electrode, but no satisfactory method has been developed to measure the free ion concentrations of those anions (especially oxalate) that participate in urinary tract mineralization.

Computer programs have been written to calculate the free ion concentrations of the major ionic species in urine, given the total concentrations of the ions, the pH, and the stability constants of the various ion pairs (Finlayson, 1977; Werness et al., 1985). (The stability constant is a measure of the strength of the association between the ions forming an ion pair.) These algorithms also permit the calculation of ionic strength, activity coefficients, and activity products. The *activity product* of a salt, such as NaCl, is given by the product of the activities of Na^+ and Cl^-: $\{Na^+\} \times \{Cl^-\}$. An estimate of supersaturation can be made if the activity product is compared with the thermodynamic solubility product.

A solid salt added to an aqueous solution will dissolve to an extent determined by the thermodynamic *solubility product* of the particular compound. If the solution is at equilibrium with the solid phase, the numerical value of the solubility product is equal to the product of the activities of the constituent ions of the salt. For NaCl,

$$K_{sp} = \{Na^+\} \times \{Cl^-\}$$

where $\{Na^+\}$ and $\{Cl^-\}$ are the activities of Na^+ and Cl^- in equilibrium with pure, solid NaCl. K_{sp} is a constant at a given temperature and pH.

Within the range of physiologic urine pH, the solubilities of two common stone salts, calcium phosphate and uric acid, are pH sensitive. For calcium phosphate,

$$Ca_5(PO_4)_3OH = 5Ca^{++} + 3PO_4^{---} + OH^-$$
$$K_{sp} = \{Ca^{++}\}^5 \times \{PO_4^{---}\}^3 \times \{OH^-\}$$
$$H_2PO_4 = HPO_4^{--} + H^+$$
$$HPO_4^{--} = PO_4^{---} + H^+$$

As the pH increases, $\{OH^-\}$ increases and hydrogen ion activity, $\{H^+\}$, decreases. More phosphate exists as PO_4^{---}, and the solubility of $Ca_5(PO_4)_3OH$ decreases. Clinically, calcium phosphate urolithiasis tends to occur in alkaline urine.

The pK_a at 38°C of the first dissociable proton in uric acid is 5.5 (Finlayson and Smith, 1974). In an aqueous solution of uric acid at pH 5.5 and 38°C, one half exists as dissolved uric acid (HU) and half exists as urate ion (U^-): $HU = H^+ + U^-$. This physicochemical property of uric acid is responsible for the sensitive pH dependence of uric acid solubility in urine. As urinary pH increases, more of the uric acid exists as urate ion, and urate is more soluble than uric acid. The solubility of uric acid as a function of pH is shown in Table 9–1. The dramatic increase of uric acid solubility with increasing pH is the cornerstone of the medical treatment of uric acid lithiasis. The ratio of the calculated activity

TABLE 9–1.

Solubility of Uric Acid
as a Function of pH

PH	SOLUBILITY, MG/L
5.0	60
6.0	200
7.0	1,600

product (AP) to the thermodynamic solubility product (K_{sp}) is the *relative supersaturation* (RSS or SS):

$$\text{Relative supersaturation (SS)} = \frac{AP}{K_{sp}}$$

If solid crystals of a salt are added to a solution of the salt, they will dissolve if SS <1 and grow if SS >1. If SS = 1, the crystals will neither grow nor dissolve. Supersaturation (relative supersaturation ratio >1) of the precipitating salt must be present for stones to form and grow.

The initial step in the actual formation of a crystal is *nucleation* or the birth of crystals from solution (Walton, 1979). If the supersaturated solution is very pure, nucleation may occur *homogeneously* at a critical level of supersaturation. Urine, however, contains many foreign surfaces, such as cell membranes, that can act as *heterogeneous* nuclei. Heterogeneous nucleation occurs at a lower level of supersaturation than does homogeneous nucleation and is the most common, if not the only, type of nucleation that occurs in biologic systems (Garside, 1982). The level of supersaturation at which nucleation occurs is often referred to as the *formation product*. The formation product is not as precisely defined as the solubility product and is most accurately described as a range of supersaturation that permits nucleation.

The range of supersaturation between the solubility product and the formation product is called the metastable zone (Fig 9–1). Spontaneous nucleation will not occur in this zone, but preformed crystals will grow until the relative supersaturation is reduced to one.

Ideal crystals are composed of identical units arranged in a repetitive pattern. These units may be atoms, molecules, ions, or groups of these particles. In real crystals, these units are not always identical, and the pattern is not strictly repetitive. Deviations from periodicity, called dislocations, commonly occur in crystals formed in biologic systems. All crystalline substances, however, have an approximate periodic structure, or lattice, that can be characterized with x-ray diffraction. If the lattice structure of one crystal is very similar to that of a different crystal, the second crystal may be able to nucleate and grow on the first crystal.

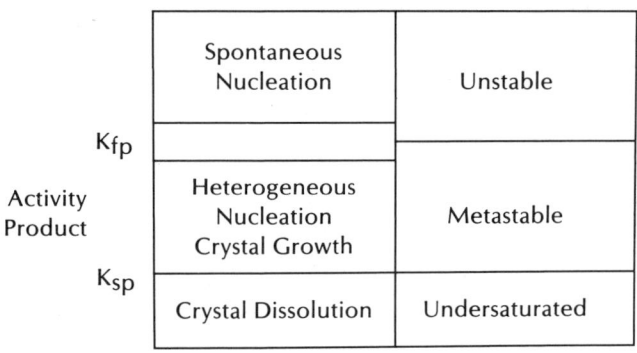

K_{fp}	Spontaneous Nucleation	Unstable
Activity Product	Heterogeneous Nucleation Crystal Growth	Metastable
K_{sp}	Crystal Dissolution	Undersaturated

FIG 9–1.
Schematic representation of states of saturation. K_{sp} indicates solubility product; K_{fp}, formation product (range).

This oriented overgrowth is called *epitaxy* (Lonsdale, 1968; Modlin, 1967). The concept of epitaxy was offered several years ago as a possible explanation for the growth of mixed urinary stones. The clinical association of disorders of uric acid metabolism (hyperuricosuria and hyperuricemia) with calcium oxalate urolithiasis motivated the intensive study of a possible epitaxial relationship between urate crystallization and calcium oxalate crystallization (Coe et al., 1975; Pak and Arnold, 1975; Koutsoukos et al., 1980). Other investigators have questioned the specific importance of epitaxial crystal growth in clinical urolithiasis (Meyer et al., 1976; Burns and Finlayson, 1980; Meyer, 1981). It is likely that mixed stone formation in the urinary tract occurs by heterogeneous nucleation and overgrowth, not by the highly specific mechanism of epitaxy.

Nuclei will grow to form larger crystals if the urine remains supersaturated for the precipitating phase. The "growth units" of the crystal are added to growth sites on the crystal surface (Nielsen and Christoffersen, 1982). Available evidence suggests that the growth sites of urinary crystals are screw dislocations (Fig 9–2). As the crystal grows, the step will wind itself into a spiral with the center fixed at the dislocation. Since the step does not disappear during growth, the crystal can grow continuously at a very low supersaturation. The crystal

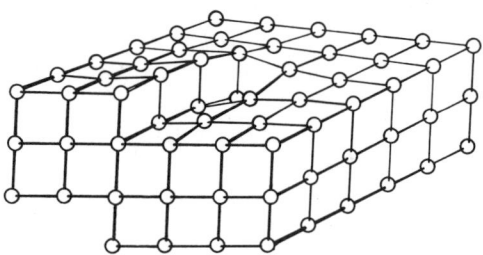

FIG 9–2.
Schematic representation of a crystal with a screw dislocation.

will grow as long as the bathing solution remains supersaturated for the precipitating phase.

Many collisions occur between the small crystals in an aqueous solution. Some of these crystals may "stick" together to form larger crystalline masses. This process is called *aggregation* or agglomeration. It is another mechanism by which crystals can increase in size to form a stone. Crystal growth produces a more dense particle than does aggregation (Finlayson, 1978). The density of actual uroliths is similar to that of pure stone crystals. This implies that urinary tract stones increase in size primarily through crystal growth, not aggregation.

CLINICAL STONE FORMATION

Why do some people form many kidney stones while others do not form a single stone over a 70-year lifetime? The answer requires the consideration of four factors: urinary supersaturation, crystal growth inhibition, particle retention, and matrix (Burns and Finlayson, 1983). Our understanding of the pathophysiology and effective treatment of urolithiasis is based on these four factors.

SUPERSATURATION

Urinary supersaturation for the precipitating stone salt is a necessary condition for stone formation. It is thermodynamically impossible for a stone to grow if the urine is not supersaturated for the particular salt. Continuous supersaturation does not have to be present. Periods of intermittent supersaturation will permit the growth of urinary tract stones. Urine may become supersaturated after meals, especially after those that contain large amounts of stone constituents, such as calcium or oxalate. This effect may be accentuated after the evening meal, because the lack of fluid consumption during sleep permits a decrease in overnight urine volume. Stone growth can occur only at night, even though daytime fluid intake is sufficient to prevent supersaturation.

INHIBITORS

Urinary supersaturation alone does not explain the presence of stone disease in some individuals and its absence in others. Urine is commonly supersaturated for stone salts, especially calcium oxalate, in healthy individuals (Robertson et al., 1968; Gill et al., 1974). Small urinary crystals are often passed by individuals who have never formed a stone (Robertson et al., 1971; Werness et al., 1981). One explanation for this apparent paradox is the presence of crystallization inhibitors in urine (Fleisch, 1978). Inhibitors of crystal growth and aggregation have been isolated from human urine, and inhibitors of nucleation may exist (Robertson et al., 1973; Ryall et al., 1981). Inhibitors of crystal growth have received the most attention, but the methods used to study these phenomena are controversial (Fleisch, 1985). Inhibitors are usually classified according to their ability to inhibit the growth of calcium phosphate or calcium oxalate. Known inhibitors of calcium phosphate crystal growth are pyrophosphate, citrate, and magnesium. Pyrophosphate and citrate also inhibit calcium oxalate crystal growth, but most of the calcium oxalate crystal growth inhibition in urine is provided by larger molecular weight polyanions: glycosaminoglycans, such as chondroitin sulfate, and RNA fragments (Meyer and Smith, 1975; Scurr and Robertson, 1985). Heparin, although not found in urine, is a potent inhibitor of *in vitro* calcium oxalate crystal growth. Most recently, acidic glycoproteins have been isolated from human urine and human kidney tissue culture medium (Ito and Coe, 1977; Nakagawa et al., 1978; Nakagawa et al., 1981). There is evidence that patients with calcium oxalate nephrolithiasis have intrinsically abnormal acidic glycoproteins (Nakagawa et al., 1985). These glycoproteins from healthy persons contain γ-carboxyglutamic acid and are strong inhibitors of calcium oxalate crystal growth at low concentrations (10^{-7} M). Glycoprotein crystal growth inhibitor from patients does not contain γ-carboxyglutamic acid and is a functionally poor inhibitor. This molecular disorder may be responsible for some component of nephrolithiasis.

Citrate has received much attention in the recent urologic literature. Citrate, through its ability to complex calcium and inhibit the growth of calcium salts, may play a role in the pathogenesis of calcium urolithiasis. Citrate is a tricarboxylic acid which, when totally ionized above pH 6.5 at 37°C, has a charge of minus three:

$$\begin{array}{c} COO^- \\ | \\ CH_2 \\ | \\ HO-C-COO^- \\ | \\ CH_2 \\ | \\ COO^- \end{array}$$

The following reaction occurs with calcium in urine at a physiologic pH:

$$Ca_3Cit_2 = Ca^{++} + 2CaCit^-$$
$$CaCit^- = Ca^{++} + Cit^{--}$$

As the concentration of citrate increases, the reaction is shifted toward the left with increased complexation of calcium. This effect will decrease calcium oxalate super-

saturation and the potential for crystallization. Citrate will weakly inhibit the growth of preformed calcium oxalate crystals, but this action is not as important as complexation.

Citrate is filtered at the glomerulus and reabsorbed primarily in the proximal tubule (Simpson, 1982). The ability of renal mitochondria to metabolize citrate via the tricarboxylic acid cycle is thought to control the renal clearance of citrate. Metabolic acidosis increases the entry of citrate into the matrix space of mitochondria and decreases the exit of citrate from mitochondria, thereby allowing mitochondrial oxidation of citrate. Cytoplasmic citrate levels fall, reabsorption of citrate from tubular fluid is enhanced, and less citrate appears in the final urine. Metabolic alkalosis has an opposite effect.

Recent studies have shown that the luminal membrane of the proximal tubules (brush border membrane, or BBM) is equipped with a Na^+-gradient-dependent transport system that is highly specific for intermediates of the tricarboxylic acid cycle, including citrate (Kippen et al., 1979). Metabolic acidosis caused by dietary acid loading increases the intrinsic capacity of proximal tubular BBM to transport citrate from the tubular lumen into the cell interior (Jenkins et al., 1985). A corresponding decrease in urinary citrate excretion was seen. A dramatic increase in urinary citrate excretion was seen with dietary alkali loading (with $NaHCO_3$), but BBM transport of citrate was unchanged. These studies demonstrate that urinary citrate excretion is exquisitely sensitive to manipulation of systemic acid-base balance through effects on renal cellular function.

Urine from patients with recurrent calcium oxalate nephrolithiasis tends to have greater calcium oxalate supersaturation and lower inhibitor levels than urine from healthy individuals (Robertson et al., 1971; Dent and Sutor, 1971). However, considerable overlap exists between these two groups. It is not possible to predict consistently who will and who will not be a recurrent calcium oxalate stone former based on supersaturation or inhibition alone. A combination of these two factors, the saturation-inhibition index, has accurately discriminated between healthy individuals and patients with recurrent stone formation (Robertson et al., 1976). The saturation-inhibition index is a mathematical combination of relative calcium oxalate supersaturation and inhibition of crystal growth and aggregation (as measured by the change in particle-size distribution in an in vitro crystal growth system). Patients with the greatest saturation-inhibition indexes had the highest recurrence rates. Since supersaturation and inhibition are difficult to measure, this method has not had widespread clinical application.

An extension of the saturation-inhibition index is the concept of urinary risk factors. This concept attempts to account for the multifactoral nature of calcium urolithiasis by considering six factors (Robertson et al., 1978, 1981; Pak et al., 1985): urine volume, urine pH, and urinary excretion of oxalate, uric acid, calcium, and alcian blue precipitable polyanions (a measure of acid mucopolysaccharides or large molecular weight inhibitors). Although no solitary abnormality may distinguish a stone former, clear discrimination can be made by the presence of several risk factors. A low urine volume was found to have the greatest risk, followed by a high urinary oxalate excretion. A high urine pH or uric acid excretion was associated with a higher probability of forming stones. Hypercalciuria was the least important risk factor.

Certain combinations of urinary supersaturation and inhibition will permit the nucleation and growth of small crystals in urine from healthy individuals. Freshly voided urine samples from most healthy persons will intermittently contain these small crystals (usually calcium oxalate dihydrate), but only 5%–10% of such persons will ever develop an actual kidney stone. These crystals probably form in the papillary collecting ducts and are flushed out with the urine before they grow large enough to become lodged in the lumen. Anatomical abnormalities or adherence to the epithelium may prevent these particles from leaving the kidney. This increased particle retention will permit the crystals to grow larger, further reducing the likelihood that they will be passed spontaneously (Burns et al., 1984). This scenario could easily predispose to the formation of kidney stones. An attractive hypothesis is that patients with stone disease may have an increased particle retention time, possibly due to an abnormal tendency for small crystals to adhere to the epithelial lining of the upper urinary tract.

Urinary crystal retention may occur by two theoretical mechanisms: free particle or fixed particle. The free particle mechanism assumes that nucleation and initial crystal growth occur in the tubular lumen. The crystalline particles grow with sufficient rapidity that they become trapped in the papillary collecting ducts, where they grow to form a macroscopic stone (Vermeulen and Lyon, 1968). The rate of calcium oxalate crystal growth in the distal tubule has been estimated from calcium oxalate supersaturation found in human urine. Calcium oxalate crystals cannot grow rapidly enough to occlude the lumen before they are washed out of the collecting ducts by the flow of urine (Finlayson et al., 1978). It seems that free particle calcium oxalate stone disease cannot occur in the renal tubule. However, some investigations have suggested that rapid aggregation of small crystals would permit free particle trapping to occur.

It is more likely that clinical stone formation occurs by a fixed particle mechanism. Several different means of particle fixation have been proposed. Rats with magnesium deficiency will develop nephrocalcinosis and stone formation (Oliver et al., 1966). Small stones were found attached to normal-appearing tubular epithelium near the bend of the loop of Henle. This intranephronic calculosis may have occurred by crystal nucleation on the luminal membrane or attachment of a passing crystalline particle. Carr (1969) suggested that crystals floating in the renal pelvis could be trapped in forniceal lymphatics and grow to macroscopic size. Papillary tip stones could originate from crystals that had become attached to the epithelial lining of the distal collecting ducts and had grown out of the lumen to form a papillary cup (Vermeulen et al., 1967). Randall's plaques (1937) are macroscopic subepithelial deposits of calcium crystals. Although the correlation between the incidence of stone disease and the incidence of Randall's plaques is not good, the epithelium over the plaque may erode, and a calyceal stone could develop from crystal growth on the plaque (Prien, 1975). Investigators have induced calcium oxalate nephrolithiasis in rats with intraperitoneal injections of oxalate (Khan et al., 1982). Calcium oxalate crystals were found in the tubular lumina, in the intercellular spaces between cells, and attached to the tubular epithelial basal lamina. Necrosis of tubular cells was responsible for exposure of the tubular basal lamina. Much evidence supports the importance of a fixed particle mechanism in the pathogenesis of stone disease.

In addition to crystalline constituents, calculi contain a variable amount of organic material called matrix. The matrix content of most urinary calculi is only 2.5% by weight (Boyce, 1968). Cystine stones contain approximately 10% matrix. The rare "matrix" calculus is a soft, radiolucent body that occurs in patients whose upper urinary tracts are infected with urea-splitting bacterial organisms (Pyrah, 1968). The matrix content of these calculi is variable, but averages 62%.

Macroscopic examination of whole renal calculi reveals concentric laminations and radial striations (Boyce, 1985). Organix matrix has such a close structural relationship with these gross physical characteristics that a definitive role in the architectural development of stones has been assumed. Scanning electron microscopic studies of fractured calcium oxalate calculi demonstrate fibrous material bridging adjacent crystals (Stacholy and Goldberg, 1985). These findings support the proposal that matrix acts as a ground substance and thereby controls crystallization within it.

Other investigators believe that the presence of matrix in urinary stones is serendipitous, because crystallization occurs in the presence of urinary macromolecules (Finlayson et al., 1961). Nonspecific physical adsorption of these organic compounds on growing crystals may account for at least some of the matrix found in calculi (Leal and Finlayson, 1977). Electron microscopic examination of calcium oxalate crystals incubated with γ-globulin or albumin has revealed an amorphous coat of material covering the crystals (Khan et al., 1983). This continuous coat is consistent with simple adsorption.

Few studies have attempted to isolate and precisely identify the chemical composition of matrix. The best known investigations found similarities between urinary mucoproteins and matrix material that was extracted from renal stones with ethylenediaminetetraacetic acid (EDTA). A mucoprotein material, matrix substance A, was identified in urine from patients with recurrent stone disease (Boyce et al., 1962). This organic compound comprised approximately 85% of the total organic matrix of kidney stones. One third of matrix substance A was carbohydrate and two thirds was protein. Aspartic and glutamic acids were the most common amino acids found in the protein component. The carbohydrate component contained galactose, mannose, methylpentose, glucosamine, and galactosamine (Boyce, 1970). More recent chemical studies have been restricted to dialyzed ultrafiltrates of matrix that have been solubilized by EDTA. In addition to aspartic and glutamic acids, γ-carboxyglutamic acid has been identified using alkaline instead of acid hydrolysis (Lian et al., 1977). Proteins containing this amino acid have a strong affinity for calcium ions. In spite of these extensive studies, the exact role of matrix in the pathogenesis of clinical stone formation has not been clarified (Malagodi and Moye, 1981).

EPIDEMIOLOGY

Urolithiasis in the United States and other technologically developed countries most commonly occurs as upper urinary tract stones, while bladder stones are more common in less-developed countries (Asper and Schmucki, 1984). Epidemiologic data suggest that climate, geology, and diet are important factors in the pathogenesis of urolithiasis. The best-known example of this influence is the apparent existence of "stone belts" or geographical areas that are associated with an especially high occurrence of renal and ureteral calculi (Higgins, 1954). One of the earliest studies that supported this concept was a questionnaire survey of hospitals (Boyce et al., 1956). During 1952, it was estimated that 0.95 persons per 1,000 population were admitted to a hospital with a diagnosis of urinary calculi. A rate of 1.93/1,000 population in South Carolina and 0.43 in Missouri provided evidence of geographical variability. Each of the southeastern states had a high rate. A sim-

ilar survey was repeated for the year 1974 (Sierakowski et al., 1978). This more recent study found that 1.64 persons per 1,000 population were admitted to a hospital with the diagnosis of urolithiasis, an increase of 75% over the 22-year period. High rates were again found in the southeastern states, especially the Carolinas (North Carolina: 3.0/1,000 population, and South Carolina: 2.7/1,000 population), but the differences were not statistically significant.

Studies of hospitalization rates may underestimate the number of patients with urolithiasis, because not all patients with renal or ureteral calculi are hospitalized (Johnson et al., 1979). A study of residents of Rochester, Minn., found that 51% of patients with stone disease were seen only as outpatients. These investigators precisely defined their epidemiologic terms and the population under study. *Incidence* was the *first* symptomatic and diagnosed episode in a person's life. The *incidence rate* was the ratio of the number of persons who experienced such initial episodes during a specified period to the size of the population at the midpoint of the period. *Prevalence* was the number of people who had had at least one symptomatic episode, while *recurrence* referred to episodes that followed the initial episode. Patients with asymptomatic stones, urinary tract infections, or struvite calculi were exluded from this study.

Six hundred seventy-two persons had their first episode of symptomatic urolithiasis while a resident of Rochester, Minn.: 468 (70%) men and 204 (30%) women. The first episode tended to occur between the ages of 30 and 60 years. The annual age-adjusted incidence rate for males was 1.1/1,000 population and for females was 0.36/1,000 population. The incidence rates for females were stable over the 25-year study period (1950–74), but the male rate per 1,000 population increased from 0.8 to 1.24. This increase was statistically significant and was most apparent in the 50–70-year-old group of men. Prevalence was estimated to increase to a peak of approximately 12% in males more than 70 years old. Prevalence in females was less than 5%. Recurrences tended to occur during the first year: 15.9% in males and 12.4% in females. Annual recurrence rates for subsequent years was 3.7% for males and 2.0% for females.

A more recent study of the membership of a large prepaid health plan found that 26.2/1,000 persons had at one time been told that they had a urinary tract stone (Hiatt et al., 1982). This would correspond to a prevalence of 2.6% in the study population. Upper urinary tract calculi were first diagnosed in 1.22/1,000 members per year at an outpatient facility, while 0.36/1,000 members were discharged from a hospital each year with the diagnosis of urolithiasis. Incidence was best estimated by the data from the outpatient facility, because all patients who were hospitalized were first seen as outpatients.

Many explanations have been offered for the increasing incidence of kidney stones in the world population. These include increased dietary protein intake, increased intake of refined sugar, and decreased dietary fiber. Even affluence appears to be associated with calcium urolithiasis. Recurrent stone formation is associated with a greater expenditure on food (Zechner and Scheiber, 1981). There is a positive correlation between monthly income and urinary excretion of calcium, uric acid, and inorganic phosphorus. Stone-forming patients may consume less dietary fiber than those who do not form stones (Robertson, 1985). One explanation is that fiber, possibly through its phytate content, binds calcium in the GI tract and prevents its absorption. The relationship between stone disease and sugar consumption is more controversial. Increased dietary sugar can increase urinary calcium excretion (Thom et al., 1978), but epidemiologic data show an inverse relationship between sugar consumption and hospitalization rates for stone disease (Robertson, 1985).

High consumption of animal protein seems to correlate best with affluence and stone disease (Robertson et al., 1981). Individuals on a high-protein diet excrete more urinary calcium, cAMP, and hydroxyproline (Fellstrom et al., 1984). The increased fixed acid load provided by a high-protein diet may cause mild resorption of bone and reduced renal tubular reabsorption of calcium. Gastrointestinal absorption of calcium is not affected. Calculated urinary supersaturation for calcium oxalate does not change, but urinary citrate exretion and pH fall. The reduced effectiveness of crystal growth inhibitors at the lower urinary pH would allow the growth of larger crystals.

The relationship between the composition of drinking water and urolithiasis has been examined in several studies. Dissolved calcium and magnesium are primarily responsible for the hardness of water. The incidence of urolithiasis tends to be higher in areas of the United States that have softer drinking water (Churchill et al., 1978; Sierakowski et al., 1979). The formation of insoluble calcium oxalate complexes in the intestinal lumen may not occur with a diet low in calcium. Free oxalate would more readily be absorbed and urinary oxalate and calcium oxalate supersaturation would increase. This hypothesis has been tested. Low urinary calcium excretion and high urinary oxalate excretion were not seen in individuals consuming water with a very low calcium content (Churchill et al., 1981). Lower urinary excretion of magnesium was found in female stone formers. Reduced urinary levels of this crystal growth inhibitor could permit the growth of larger crystals. No

difference was found between male stone formers and control subjects.

A more recent study examined two specific geographical regions: the Carolinas (North and South Carolina), which had soft water and a high stone incidence, and the Rockies (Colorado, Idaho, Montana, Nevada, Utah, and Wyoming), which had hard water and a low stone incidence (Shuster et al., 1982). No significant differences were found for the concentration of calcium, magnesium, or sodium in home tap water. An incidental finding was that individuals drinking private well water had a greater risk of stone formation than those drinking public water. The authors concluded that water hardness should be a minor concern with respect to stone formation. The role of phosphate and sulfate in drinking water has not been examined but may be important (Vahlensieck, 1985). Orthophosphates are used to treat patients with idiopathic calcium urolithiasis. Sulfate will form soluble complexes with calcium in urine, thereby reducing free calcium ion activity and calcium oxalate supersaturation.

Another study found a positive association between urinary stone disease and consumption of carbonated beverages (sugared cola) (Shuster et al., 1985). A negative association existed between coffee and beer consumption and stone disease in the Rockies. Primary intake of milk, water, or tea was not associated with urinary stone disease. The relationship between stone disease and soda consumption is being examined in greater detail.

The widespread consumption of large amounts of iced tea in the southeastern United States could be related to the high incidence of urolithiasis in that area of the country. Although dietary oxalate is responsible for only 10%–15% of total urinary oxalate (Hodgkinson, 1977), urinary oxalate excretion will increase after the ingestion of oxalate-rich foods such as spinach (Strenge et al., 1981). A case control study from Newfoundland examined tea consumption in stone formers but found no evidence to support the suggestion that tea drinking is a risk factor for calcium oxalate urolithiasis (Churchill et al., 1981). These investigators calculated that one cup of tea would add only 0.5 mg of oxalate to total urinary excretion.

ETIOLOGY

A classification of the etiologies is given in Table 9–2 (Smith, 1979a). The syndrome of idiopathic calcium urolithiasis syndrome accounts for 70%–80% of the stone disease in industrialized nations, while inherited enzyme disorders or renal tubular syndromes are found in fewer than 1% of stone-forming patients. Primary hyperparathyroidism is the most common hypercal-

TABLE 9–2.
Etiology of Urolithiasis*

Renal tubular syndromes
  Renal tubular acidosis
  Cystinuria
Hypercalcemic disorders
  Primary hyperparathyroidism
  Immobilization
  Milk-alkali syndrome
  Sarcoidosis
  Hypervitaminosis D
  Neoplastic diseases
  Cushing's syndrome
  Hyperthyroidism
Uric acid lithiasis
  Idiopathic
  Gout
  Low–urine-output states
  Myeloproliferative diseases
Enzyme disorders
  Primary hyperoxaluria
  Xanthinuria
  2,8-Dihydroxyadeninuria
  Enteric hyperoxaluria
Secondary urolithiasis
  Infection
  Obstruction
  Medullary sponge kidney
  Urinary diversion
  Drugs
Idiopathic calcium urolithiasis
  Hypercalciuria
  Normocalciuria

*Adapted from Smith LH: Urolithiasis, in Gottschalk CW, Earley LE (eds): *Strauss and Welt's Diseases of the Kidney,* ed 3. Boston, Little, Brown & Co, 1979, pp 893–931.

cemic state associated with urolithiasis and is responsible for stone formation in 5% of patients.

RENAL TUBULAR ACIDOSIS

Renal tubular acidosis (RTA) is a syndrome of disordered renal acidification that causes a hypokalemic hyperchloremic metabolic acidosis. The inability to excrete normal amounts of acid into the urine may be responsible for the entire syndrome, because the administration of sodium bicarbonate will correct the hyperchloremic acidosis and the excessive urinary losses of potassium, calcium, and phosphorus (Seldin and Wilson, 1978).

Two basic types of defective urinary acidification have been identified in these patients. Patients with type 2 RTA (proximal) have a defect in the reabsorption of filtered bicarbonate, a process that occurs in the proximal tubule (Morris, 1969). When the plasma bicarbonate is only moderately reduced, the urinary pH is inappropriately high, but with the development of a more severe systemic acidosis, the bicarbonaturia disappears, and the urinary pH decreases to a normal minimum. Patients with the most frequently studied disorder, type 1, or classic, RTA (distal) have a normal capacity to reabsorb filtered bicarbonate but cannot lower the urine pH below 6.0, regardless of the severity of the systemic acidosis (Stinebaugh et al., 1981).

Classic RTA may exist as a primary or secondary form. The primary form may be subdivided into infantile and adult types. Adult, or persistent, primary RTA occurs predominantly in females. Most of the cases are sporadic, but the disease may be inherited as an autosomal dominant trait (Buckalew et al., 1974). The reclamation of filtered bicarbonate is intact in the proximal tubules, but the distal tubule is unable to generate or maintain steep lumen-peritubular hydrogen ion gradients. Ammonia production is normal.

The electrolyte abnormalities are responsible for the symptoms. Chronic acidosis may contribute to the impaired growth of children with type 1 RTA (McSherry, 1978). Although urinary wasting of calcium and phosphorus may lead to osteomalacia, the hyperchloremic acidosis is so readily detected that patients are rarely left untreated long enough to develop this complication. Urinary potassium wasting may result in severe hypokalemia and a flaccid paralysis.

Nephrolithiasis occurs in 70% of patients with distal renal tubular acidosis (Van den Berg et al., 1983). Multiple calculi are usually present in both kidneys. Nephrocalcinosis is found in approximately three fourths of adults with type 1 RTA (Brennen et al., 1982). Stone formation is related to the hypercalciuria, relatively alkaline urine, and very low urinary citrate excretion. Reduced urinary excretion of pyrophosphate, sulfate, and inhibitors of hydroxyapatite crystal growth may also contribute to clinical stone formation.

The diagnosis of distal RTA is made when systemic acidosis (serum bicarbonate <20 mEq/L) is present and urine pH is greater than 5.5. An ammonium chloride (NH_4Cl) load is often used as a stress test to confirm the diagnosis. Liquid NH_4Cl (100 mg/kg—as a 5% solution) is given in the evening. The patients voids at 6 AM on the following day, drinks three 8-oz glasses of water, and voids again at 7:30 AM. The pH of this latter urine specimen is measured using a pH meter. The pH of this urine sample will be between 5.0 and 5.5 in a healthy individual. If the pH is greater than 5.5 and the serum bicarbonate is greater than 20 mEq/L, the NH_4Cl load is repeated. The urinary pH measurement is repeated in four hours.

The distal tubular acidification mechanism is intact in type 2 or proximal RTA, but the reabsorption of filtered bicarbonate is reduced in the proximal tubule (Morris, 1981). Proximal RTA is usually associated with an underlying disorder of proximal tubular function, such as the Fanconi syndrome, hereditary fructose intolerance, Wilson's disease, or multiple myeloma. A dietary deficiency of vitamin D may be associated with proximal RTA in children. The most common cause of proximal RTA in adults is intestinal malabsorption that leads to vitamin D deficiency, hypocalcemia, secondary hyperparathyroidism, and hypophosphatemia (Muldowney et al., 1970). Proximal RTA is not associated with stone formation.

Infants and children with a distal acidification defect and bicarbonate wasting were said to have had type 3 RTA. This term is no longer used, because the bicarbonate wasting is thought to reflect the same defect in renal acid excretion that causes type 1 RTA in adults (McSherry et al., 1972). The reduction in acid excretion is secondary to the renal bicarbonate wasting as a cause of the acidosis. One to three percent of filtered bicarbonate is excreted in adults with type 1 RTA, while infants and children with this disorder will excrete 6%–14% of filtered bicarbonate.

Type 4 RTA has been applied to an acidification defect that may accompany a reduction in the renal clearance of potassium. This disorder is thought to involve the cation-exchange segment of the distal nephron, where aldosterone stimulates hydrogen ion secretion. Type 4 RTA may be the most common form of RTA and is often associated with the syndrome of hyporeninemic hypoaldosteronism (Batlle et al., 1981). Hyperkalemic distal RTA has also been found in patients with obstructive uropathy (Batlle et al., 1981). Type 4 RTA is not associated with nephrolithiasis.

Some patients with recurrent calcium urolithiasis are not systemically acidotic but are unable to lower their urine pH after an ammonium chloride load (Buckalew et al., 1968). Both proximal and distal acidification defects have been identified in stone formers with the syndrome of incomplete RTA (Backman et al., 1980; Tessitore et al., 1985). Patients with these disorders tend to develop stone disease at an earlier age, have more frequent recurrences, and grow larger stones that often require surgical intervention. Hypocitraturia and hypercalciuria are usually present. The hypercalciuria is unexplained, because these patients do not have systemic acidosis. Serum electrolytes are normal, and a standardized acid loading study is required for the diagnosis of incomplete RTA.

Urolithiasis is a complication of long-term treatment with carbonic anhydrase inhibitors (Parfitt, 1969). Carbonic anhydrase, which catalyzes the hydration of CO_2, is present in both the proximal and distal nephron. Since carbonic anhydrase aids the proximal reclamation of filtered bicarbonate and the distal secretion of hydrogen ions, the administration of an inhibitor of the enzyme, such as acetazolamide, causes a proximal and dis-

tal RTA. Treatment with acetazolamide produces alterations in the ionic composition of urine, such as a low citrate concentration, that closely resemble those found in untreated distal RTA (Harrison and Harrison, 1955; Simpson, 1964). All of the changes can be reversed by stopping administration of the carbonic anhydrase inhibitor.

CYSTINURIA

Cystinuria is an inherited disorder of amino acid metabolism in which there is defective transport of cystine, ornithine, lysine, and arginine in the renal tubule and GI tract. This disorder would be a metabolic curiosity if it were not for the relative insolubility of cystine in urine and the subsequent formation of renal calculi (Dahlberg et al., 1977). Cystine is the least soluble of the naturally occurring amino acids (Crawhall and Watts, 1968). The defect is transmitted as an autosomal recessive trait. Homozygotes excrete large amounts of cystine, lysine, arginine, and ornithine in their urine. The mnemonic "COAL" or "COLA" can be used to remember these four amino acids.

Cystinosis, another recessively inherited metabolic disorder, may be confused with cystinuria. Cystinosis is characterized by the intracellular accumulation of excessive quantities of cystine (Schneider et al., 1978). Crystal deposition occurs in the cornea, conjunctiva, bone marrow, lymph nodes, leukocytes, and internal organs. All patients with nephropathic cystinosis have a generalized amino aciduria, but the daily excretion of cystine is only 5%–10% of that found in patients with cystinuria. Children with cystinosis also tend to produce a relatively alkaline urine. Consequently, cystine stone formation rarely occurs in patients with cystinosis.

Cystine is a disulfide composed of two cysteine molecules (Fig 9–3). The pK_a of cystine is 8.0. Approximately 300 mg of cystine is soluble in 1 L of urine at a pH of 7.0. The solubility of cystine more than doubles as the pH rises above 7.5 (Fig 9–4). This provides the scientific basis for the medical treatment of cystinuria.

Three types of homozygous cystinuria have been identified based on in vitro intestinal transport of amino acids, response to oral amino acid loading, and urinary

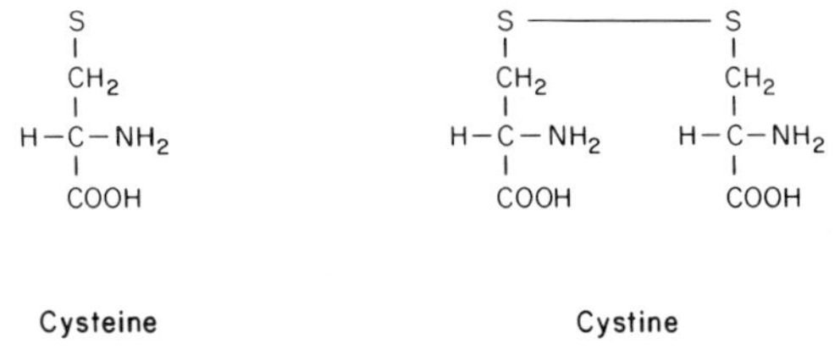

FIG 9–3.
Structures of cysteine and cystine (a disulfide).

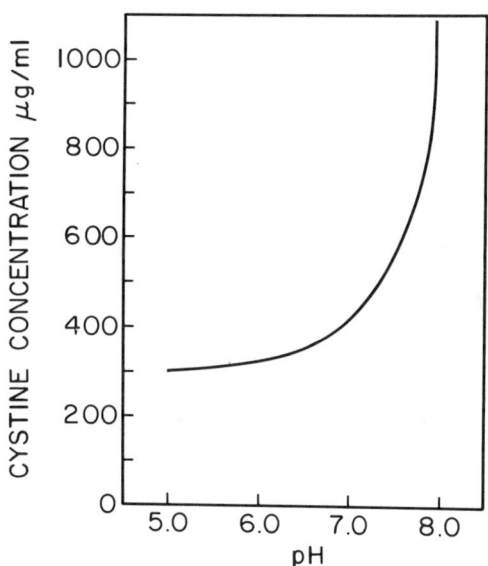

FIG 9–4.
The pH dependence of the solubility of cystine. (Adapted from Dent CE, Senior B: Studies on the treatment of cystinuria. *Br J Urol* 1955; 27:317–332.)

amino acid excretion in relatives with the heterozygous disorder (Thier and Segal, 1978). The intestinal transport defect for the four amino acids has been demonstrated in vivo by oral loading tests and intestinal perfusion studies and in vitro by transport studies of mucosal biopsies. The nature of the renal tubular lesion has not been clearly defined. The clearance of cystine in patients with cystinuria frequently exceeds the GFR. This implies the occurrence of secretion.

Cystinuria is manifested clinically by the formation of homogeneous radiodense calculi. Cystine stones often have a branched configuration. Multiple small satellite stones may accompany a large stone. The stones have the gross appearance of maple sugar and tend to be hard and tough. Hexagonal cystine crystals may be present in a voided urine sample, especially one that is concentrated and acidified. The diagnosis is confirmed by the analysis of a stone and the finding of increased urinary cystine on an amino acid analysis. Urine can be screened with the cyanide-nitroprusside test, which is positive if more than 75–125 mg of cystine per gram of creatinine is present (Dahlberg et al., 1977). This test will not differentiate homozygotes from heterozygotes. A new test kit based on the reaction of cystine with nickel ion and sodium hyposulfite has been marketed by Mission Pharmacal Company (George and Politzer, 1970). This convenient method can be used to screen patients who are interested in ESWL but whose stones have a radiographic appearance strongly suggestive of cystine.

Urinary excretion of cystine is less than 30 mg/day in healthy adults. Heterozygous adults excrete less than 400 mg of cystine per day and usually do not form stones. Daily urinary cystine excretion is usually greater than 400 mg in homozygous cystine stone formers. No overlap in cystine excretion has been demonstrated between well-confirmed homozygotes and heterozygotes (Crawhall et al., 1969).

HYPERCALCEMIC DISORDERS

Primary hyperparathyroidism is the most common disorder associated with hypercalcemia and urolithiasis. It is found in approximately 5% of patients with stone disease (Arnaud et al., 1973). The diagnosis of hyperparathyroidism is based on the presence of hypercalcemia with an inappropriately elevated parathyroid hormone (PTH) level.

PTH is synthesized in the chief cells of the parathyroid gland and is split into at least two major fragments after being secreted into the circulation (Kao, 1982). The N-terminal fragment is responsible for the biologic activity and has a short half-life. The C-terminal fragments have longer half-lives but no biologic activity. PTH increases bone resorption, increases renal reabsorption of calcium, decreases renal reabsorption of phosphate, and augments renal conversion of 25-hydroxyvitamin D to 1,25-dihydroxyvitamin D, thereby increasing intestinal absorption of calcium. All of the effects of PTH tend to increase the serum concentration of calcium. Serum calcium exerts negative feedback control over the secretion of PTH.

A single adenoma, usually chief cell, is responsible for the disease in 80% of patients (Granberg et al., 1985). Chief cell hyperplasia is found in most of the remaining 20%. The hypercalcemia and inappropriately high PTH levels found in patients with primary hyperparathyroidism are due to the relatively autonomous function of the adenomatous or hyperplastic tissue.

Patients with hypercalcemia and neoplastic diseases usually have very low PTH levels, while patients with chronic renal failure have high PTH levels but low serum calcium levels. The phosphaturia in patients with primary hyperparathyroidism may lead to hypophosphatemia, but most patients have normal serum phosphate concentrations. A liberal dietary phosphate intake may compensate for the phosphaturia.

The clinical presentation of primary hyperparathyroidism has changed over the past 20 years. Generalized osteitis fibrosa cystica is extremely rare, and the incidence of urolithiasis has fallen. A population study in Rochester, Minn., from 1965 through 1976 found that the average annual incidence of cases of primary hyperparathyroidism rose from 7.8 to 51/100,000 population (Heath et al., 1980). This dramatic increase in the apparent incidence occurred immediately after routine measurement of serum calcium was begun in 1974. The frequency of urolithiasis fell from 51% to 4%. The proportion of patients without symptoms or complications rose from 18% to 51%.

The definition of hypercalcemia depends on the normal range for serum calcium. A normal range of 9–11 mg/dl had been used in some laboratories, but up to one third of patients with surgically proved hyperparathyroidism will have a mean preoperative serum calcium of less than 11 mg/dl (Purnell et al., 1971). Authors of another study from the Mayo Clinic found that the normal range varied from 8.9 to 10.1 mg/dl in their laboratory (Keating et al., 1969).

The reason for stone formation in patients with hyperparathyroidism and urolithiasis is not known. Patients have urine that is supersaturated with respect to calcium stone salts whether or not stone disease is present (Pak et al., 1981). The magnitude of supersaturation is not greater for those patients with urolithiasis. A possible explanation is that patients with hyperparathyroid-induced stone formation may have been predisposed to form stones. The advent of hypercalcemia and hypercalciuria may have promoted crystallization. The hypercalciuria may have no ill effects in those who lack the predisposing factors.

Immobilization may be complicated by hypercalcemia, hypercalciuria, and stone formation. The hypercalcemia and hypercalciuria may be especially severe in adolescents with active bone growth. Approximately 10% of patients with traumatic spinal cord injuries will develop renal calculi (DeVivo et al., 1984). The risk of stone formation is greatest during the first three months after injury. Urinary calcium excretion exceeds normal levels at about the fourth week of immobilization and reaches maximal levels at 16 weeks. The hypercalciuria may persist for 12 months but resolves by 18 months. Resorption of bone appears to be the primary process. This suppresses the parathyroid–1,25 dihydroxyvitamin D axis, thereby minimizing the possibility that skeletal calcium loss will lead to hypercalcemia. Nevertheless, serum calcium levels are elevated or in the high-normal range. British investigators found that paraplegic patients excrete less citrate, orthophosphate, and pyrophosphate than do healthy individuals (Burr et al., 1985). Stone-forming paraplegics excreted more urate than non-stone-forming paraplegics.

Hypercalciuria and hyperphosphaturia occur during the weightlessness of space travel, but there has been no evidence of clinical stone formation (Lutwak et al., 1969). Long-term bed rest is used as a model to study the effects of weightlessness on metabolism. Such stud-

ies at simulated high altitudes disclosed that urinary losses of calcium were significantly smaller at a higher altitude. Exercise does not prevent the hypercalciuria (Issekutz et al., 1966).

Nephrocalcinosis and renal insufficiency may occur with the milk-alkali syndrome and vitamin D intoxication. Stone formation may also be seen in sarcoidosis, where there is increased intestinal calcium absorption, hypercalcemia, and hypercalciuria (Heneman et al., 1956). Hypercalcemic patients with sarcoidosis have elevated serum levels of 1,25-dihydroxyvitamin D. Healthy persons produce this active metabolite of vitamin D only in the renal tubule. There is evidence that patients with sarcoidosis convert 25-hydroxyvitamin D to the active compound in the granulomas (Lemann and Gray, 1984). Hypercalcemia also occurs in patients with other granulomatous diseases, such as tuberculosis, berylliosis, and coccidioidomycosis. Some patients with neoplastic diseases develop hypercalcemia, but stone formation is unusual. Patients with Cushing's syndrome or hyperthyroidism may have increased blood and urine calcium levels.

URIC ACID LITHIASIS

Uric acid is the end product of purine metabolism in man. Uric acid has two dissociable protons (Fig 9–5), the first with a pK_a of 5.5 and the second with a pK_a of 10.3 (Finlayson and Smith, 1974). The limited solubility of this weak acid accounts for its propensity to form renal calculi (Fig 9–6). Uric acid solubility is approximately 15 mg/dl at pH 5, but 200 mg/dl at pH 7.

Uric acid excretion depends on the biosynthesis of new purines and, to a lesser extent, on preformed dietary purines (Gutman and Yu, 1968; Fellstrom, 1981). A series of enzymatic reactions leads to the formation of the mononucleotide inosine monophosphate (IMP). Dietary nucleic acids can be catabolized to form two other mononucleotides: adenosine monophosphate (AMP) and guanine monophosphate (GMP). Cleavage of the phosphate group by nucleotidases forms the corresponding nucleosides: inosine, guanosine, and adenosine. The action of nucleoside phosphorylases forms the purine bases: hypoxanthine (from inosine), guanine, and adenine. Xanthine oxidase converts hypoxanthine to xanthine and uric acid (see Fig 9–5). The purine salvage enzymes, hypoxanthine-guanine-phosphoribosyltransferase (HGPRT) and adenine-phosphoribosyltransferase (APRT), can reconvert the purine bases to the mononucleotides. This is the purine salvage pathway.

An X-linked deficiency of HGPRT is responsible for two clinical syndromes (Wilson et al., 1983). Enzyme activity is virtually absent in the Lesch-Nyhan syn-

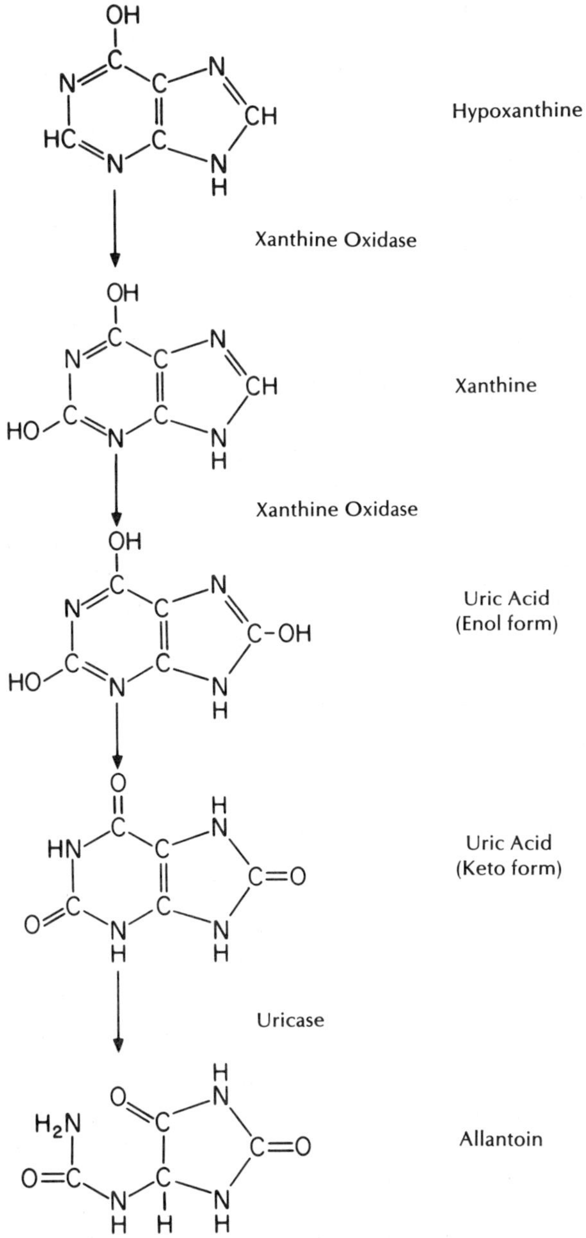

OH

Hypoxanthine

Xanthine Oxidase

OH

Xanthine

Xanthine Oxidase

OH

Uric Acid
(Enol form)

Uric Acid
(Keto form)

Uricase

Allantoin

FIG 9–5.
The metabolic pathway for conversion of hypoxanthine to xanthine and uric acid. Most ureotelic mammals, except man, have hepatic uricase that converts uric acid into the more soluble allantoin.

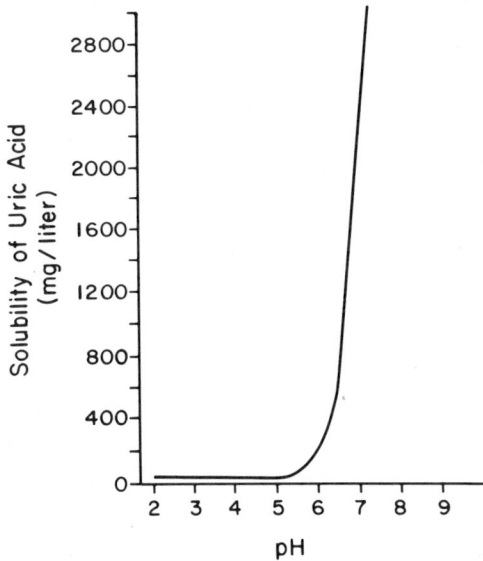

FIG 9–6.
The pH dependence of the solubility of uric acid at 38° C. (Adapted from Finlayson B, Smith A: Stability of first dissociable proton of uric acid. *J Chem Eng Data* 1974; 19:94–97.)

drome, a disease characterized by uric acid overproduction and a CNS disorder (mental retardation, spasticity, choreoathetosis, and self-mutilation). HGPRT activity is partially deficient in the second syndrome, which is characterized by uric acid overproduction and severe gout but no neurologic abnormality. Uric acid lithiasis may occur in both syndromes.

Biosynthesis of purines from amino acids also provides a way to eliminate waste nitrogen as uric acid

(Gutman and Yu, 1968). This is the major pathway of waste nitrogen disposal in birds and uricotelic reptiles. Loss of water and electrolytes is minimized by discharging a semisolid uric acid mass through a cloaca. In most ureotelic mammals, except man, urate that enters the glomerular filtrate is reabsorbed in the proximal tubule and recycled through the liver for conversion to water-soluble allantoin by hepatic uricase. Allantoin is freely excreted by the kidney. Uricase is not present in humans, and uric acid cannot be converted to the more soluble allantoin. Plasma and urine uric acid levels in man are of an order of magnitude greater than those in most mammals. The high concentrations of uric acid are precariously held in solution, a situation that predisposes man to gout and uric acid lithiasis.

The Dalmatian coach hound also is predisposed to uric acid urolithiasis (Yu et al., 1966). Normal quantities of uricase are present in the liver, but uric acid conversion to allantoin is slow and incomplete. This, together with defective tubular reabsorption of urates, leads to hyperuricosuria and stone formation.

Uric acid lithiasis accounts for 5%–10% of stones formed in the United States. Approximately one quarter of patients with primary gout will develop uric acid stones. Uric acid lithiasis is found in approximately 40% of stone-forming persons in Israel (Atsmon et al., 1963). Uric acid bladder calculi are a common problem in children of rural Southeast Asia.

Patients with uric acid lithiasis may excrete too much uric acid or excessively acid urine. Overproduction of uric acid, as occurs in some patients with gout, may be responsible for the hyperuricosuria. Purine and protein gluttony may also increase urinary acid excretion. Most patients with uric acid lithiasis do not have gout or any recognizable disorder of purine metabolism. Serum and urine uric acid levels are usually normal in patients with idiopathic uric acid urolithiasis. Many of these patients have a persistently low urine pH. Although some investigators have found an isolated defect in renal tubular ammonia secretion, the mechanism of the low urinary pH is still poorly understood. Urinary pH also tends to be low in gout (Gutman and Yu, 1965).

Gouty patients who are treated with uricosuric agents may be at risk for uric acid stone formation. This can be prevented with an increased fluid intake and urinary alkalinization. Patients with myeloproliferative disorders are also at risk for uric acid stone formation, especially with the initiation of chemotherapy or radiotherapy. The increased purine load may even lead to intratubular precipitation of uric acid and anuria.

Disorders that reduce urine volume will be associated with uric acid precipitation in an acid urine. Patients with ileostomies or chronic diarrhea can lose large amounts of fluid and bicarbonate. Maintenance of an adequately dilute urine can be difficult for these patients, because oral fluids and electrolytes are not well absorbed. The diarrhea or ileostomy output may in-

crease as fluid intake increases. Small bladder calculi composed of uric acid may be found in men with prostatism. These patients reduce their urinary frequency by reducing their fluid intake. Pure uric acid is radiolucent, but gradual incorporation of impurities (usually metals, such as calcium) will make larger stones (>2 cm diameter) faintly radiopaque. Identification of smaller stones can be accomplished with excretory urography, sonography, or CT. Large uric acid stones may have a branched configuration (Fig 9–7).

XANTHINURIA

Xanthine is less soluble than uric acid (see Fig 9–4). Its solubility increases with increasing urine pH, but the effect is not as great as that with uric acid. The pK_a of the first dissociable proton of xanthine is 7.7. Like uric acid, xanthine is radiolucent. Xanthine calculi have been reported to occur in xanthinuria, a rare, inherited deficiency of xanthine oxidase (Dent and Philpot, 1954). Serum and urine uric acid levels are very low, but urinary excretion of hypoxanthine and xanthine is elevated. Stone formation has occurred in approximately one third of patients.

Urinary excretion of hypoxanthine and xanthine is more commonly elevated in patients who are being treated with a xanthine oxidase inhibitor such as allopurinol (Fig 9–8). Xanthine stone formation during al-

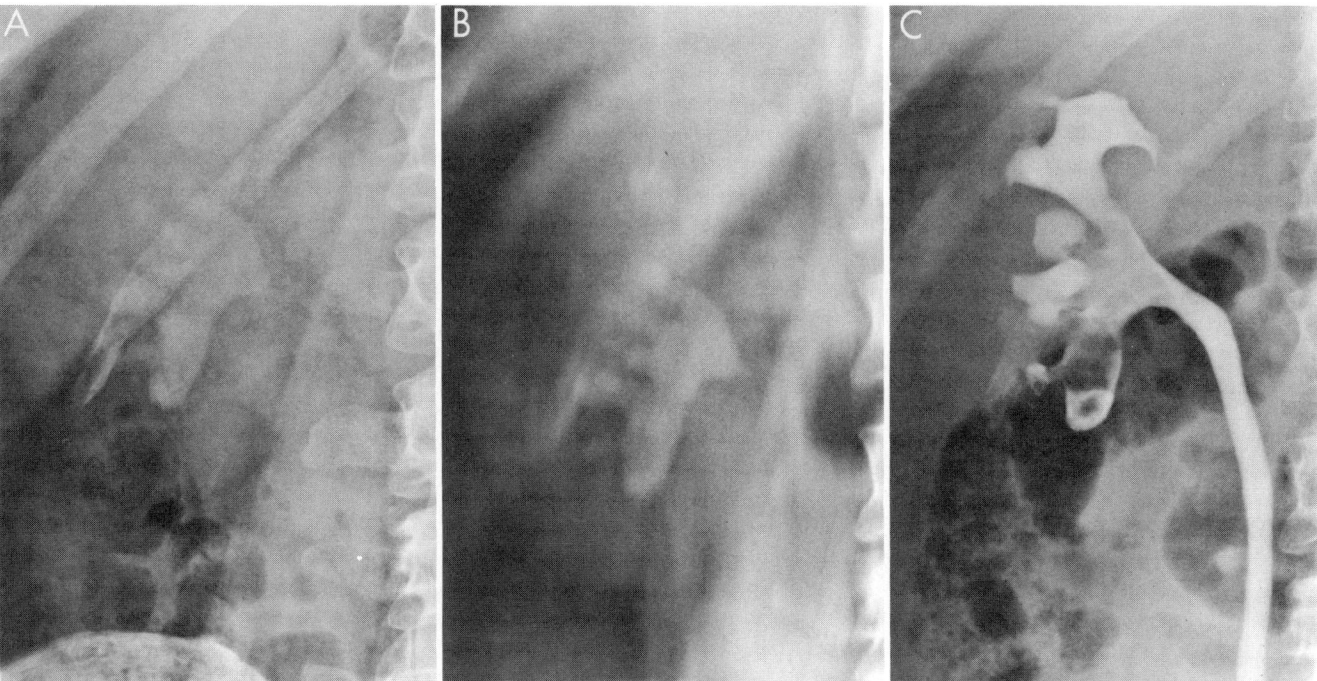

FIG 9–7.
A, plain abdominal radiograph that demonstrates a large branched right renal calculus. **B,** plain tomographic film. **C,** retrograde pyelogram that demonstrates a filling defect.

Analysis of the removed stone material revealed 98% uric acid.

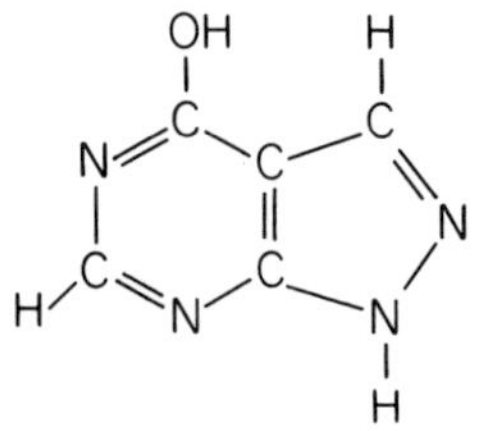

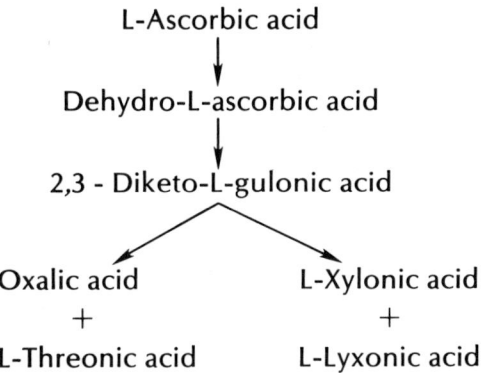

Allopurinol

FIG 9–8.
Structure of allopurinol, an analogue of hypoxanthine and xanthine.

lopurinol administration has been reported in patients with Lesch-Nyhan syndrome and in patients with myeloproliferative disorders who are receiving chemotherapy (Brock et al., 1983).

2,8-DIHYDROXYADENINURIA

This is an inherited disorder caused by a defect of the purine salvage enzyme APRT (Gault et al., 1985). Adenine is converted to 2,8-dihydroxyadenine (2,8-DHA), which has a very low solubility. Stone formation and renal failure have been described, usually in children. 2,8-DHA stones give falsely positive reactions for uric acid with standard wet chemical colorimetric procedures, and, like uric acid, are radiolucent. Enzymatic analysis with uricase will avoid mistaken identification as uric acid, while infrared spectroscopy will confirm the diagnosis. "Uric acid" stones in children must always be suspect and subjected to a sophisticated analysis.

PRIMARY HYPEROXALURIA

This is a rare, inherited disorder of glyoxylate metabolism. Clinical manifestations include recurrent calcium oxalate nephrolithiasis, nephrocalcinosis, and chronic renal failure. Extrarenal deposits of oxalate, or oxalosis, will develop in the presence of renal failure.

Although oxalate is absorbed from the GI tract, most urinary oxalate is derived from endogenous metabolism. Oxalate is produced in mammals as an end product of the oxidative metabolism of ascorbic acid (Fig 9–9) and by oxidation of glyoxylic acid (Fig 9–10). The major precursor of oxalate in humans is glyoxylate. Glyoxylate is formed primarily from glycine, glycolic acid, and α-keto-γ-hydroxyglutamic acid.

Type 1 primary hyperoxaluria, or glycolic aciduria, is due to a deficiency of the cytoplasmic enzyme α-ketoglutarate:glyoxylate carboligase (reaction 1 in Figure 9–10). This enzyme is also present in mitochondria, but its activity is unimpaired. Urinary excretion of oxalate and glycolate is elevated. This is the most common form of the disease. Type 2 primary hyperoxaluria, or L-glyceric aciduria, is due to a deficiency of the enzyme

L-Ascorbic acid
↓
Dehydro-L-ascorbic acid
↓
2,3 - Diketo-L-gulonic acid

Oxalic acid   L-Xylonic acid
+   +
L-Threonic acid   L-Lyxonic acid

FIG 9–9.
The oxidative metabolism of ascorbic acid leads to the production of oxalate.

D-glyceric dehydrogenase (reaction 2 in Figure 9–10) (Williams and Smith, 1968). Urinary excretion of oxalate and L-glyceric acid is elevated, but glycolate excretion is normal. The cause of the hyperoxaluria in these patients is not directly explained by this enzyme defect. Both types of defects lead to increased endogenous production of oxalate and hyperoxaluria.

Primary hyperoxaluria is a very rare disorder. An extensive review in 1964 reported on 63 typical and 47 atypical cases (Hockaday et al., 1964). The disease had become clinically manifest in more than half of the patients by 4 years of age. Almost half had died from renal failure by the age of 20.

The disease should be suspected in patients who are young, have a family history of urolithiasis, and have

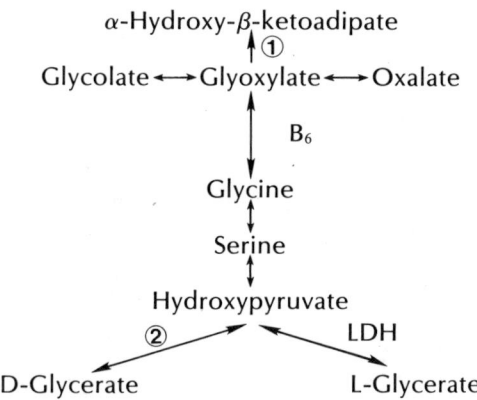

FIG 9–10.
Pathways of oxalate metabolism in man. The enzyme defect responsible for type 1 primary hyperoxaluria is at *1,* and that for type 2 is at *2*. Pyridoxine (vitamin B_6) is a cofactor for the conversion of glyoxylate to glycine. (Adapted from Williams HE, Smith LH Jr: Primary hyperoxaluria, in Stanburg JB, Wyngaarden JB, Fredrickson DS (eds): *The Metabolic Basis of Inherited Disease,* ed 4. New York, McGraw-Hill Book Co, 1978, pp 182–204.)

nephrocalcinosis or large, radiodense calculi on plain abdominal roentgenograms. Urinary oxalate excretion is almost always greater than 100 mg/24 hours, except in the presence of renal failure. Differentiation between types 1 and 2 is made with urinary glycolate measurements. Glycolate excretion is normal in type 2 but elevated in type 1.

ENTERIC HYPEROXALURIA

Dietary oxalate is a relatively minor source of urinary oxalate in healthy persons. Patients with small-bowel disorders, however, may develop hyperoxaluria and recurrent calcium oxalate stone disease. Gastrointestinal absorption of dietary oxalate is more avid in these patients. Enteric hyperoxaluria was first described in the early 1970s and was responsible, in part, for the abandonment of jejunoileal bypass surgery as a means to control obesity (Smith et al., 1972). Major sources of dietary oxalate are the green leafy vegetables: rhubarb, spinach, and kale. Tea, cocoa, chocolate, and pepper also have a high oxalate content. The average daily diet contains 100–900 mg of oxalate (Williams, 1976).

The acid medium in the stomach releases oxalate from foodstuffs. After combining with free dietary calcium in the alkaline medium of the small intestine, most of the oxalate exists as insoluble calcium oxalate. The small amount of free oxalate that does exist can be absorbed. Oxalate absorption can occur throughout the GI tract, but the colon seems to be the major absorptive area. Patients with ileostomies very rarely develop hyperoxaluria (Dobbins and Binder, 1977). The reaction between intestinal calcium and oxalate prevents the absorption of no more than 10% of dietary oxalate by healthy persons.

Patients with a variety of chronic GI disorders, such as small-bowel resection, inflammatory small-bowel disease, chronic pancreatitis, or a jejunoileal bypass, may malabsorb fat. The intraluminal concentration of fatty acids will increase, and calcium will be bound to form calcium-fatty acid soaps (Earnest et al., 1974). Less calcium will be available to bind oxalate, and more free oxalate will be available for absorption. There is also evidence that malabsorbed fatty acids or bile acids will increase colonic permeability to oxalate (Dobbins and Binder, 1976). Patients with enteric hyperoxaluria may absorb up to one third of their dietary oxalate.

Increased availability of free oxalate with colonic hyperabsorption is the major mechanism of enteric hyperoxaluria, but increased urinary oxalate excretion may persist even on a very low-oxalate diet (Hoffman et al., 1983). This implies the presence of oxalate precursors in the diet. The identity of these "oxalogenic" components is not known, but oxalate excretion decreased when protein-rich foods were removed from the diet.

Tissue or bacterial production of oxalate from dietary constituents may play a role in this phenomenon.

The incidence of nephrolithiasis in inflammatory bowel disease is 2%–3%, but ileal resection increases the risk to 10% (Dobbins, 1985). The primary risk factor is increased urinary excretion of oxalate. In comparison with the primary hyperoxalurias, excretion of L-glyceric acid and glycolic acid is normal.

Elevated urinary oxalate is not the only risk factor in these patients (Smith et al., 1979). The multiple risk factors are shown in Table 9–3. All will increase the propensity for calcium oxalate precipitation. Reduced urinary volume and pH may promote uric acid stone formation.

Another form of enteric hyperoxaluria is that possibly associated with ascorbic acid. The metabolic pathway for in vivo conversion of ascorbate to oxalate is shown in Figure 9–6. In vitro conversion of ascorbate to oxalate may occur in saline solutions or pooled urine samples, especially those that have not been acidified (Rundquist et al., 1981). This may produce a factitious hyperoxaluria in patients who are consuming large quantities of vitamin C. A further confounding factor is that some methods used to measure oxalate may not be able to distinguish ascorbate metabolites from oxalate. Patients should be advised not to take large amounts of vitamin C when they are collecting a 24-hour urine specimen for chemical analysis.

SECONDARY UROLITHIASIS

Stone formation may be associated with a group of unrelated conditions: urinary tract infection with bacterial organisms that produce urease, obstructive uropathy, structural anomalies such as medullary sponge kidney, urinary diversion, and pharmacologic agents. Although all of these disorders may be associated with stone formation, they may not be its primary cause. It is important to search for underlying disorders, such as

TABLE 9–3.

Enteric Hyperoxaluria Risk Factors for Urolithiasis*

MALABSORBED COMPONENT	URINE COMPOSITION
Fatty acids, bile acids	Increased oxalate
Water	Decreased volume
Electrolytes	Decreased ionic strength
Bicarbonate	Decreased pH, decreased citrate
Magnesium	Decreased magnesium
Protein	Decreased sulfate, decreased phosphate, decreased pyrophosphate

*Adapted from Smith LH, Werness PG, Wilson DM: Enteric hyperoxaluria: Associated abnormalities that promote formation of renal calculi, in Rose GA, Robertson WG, Watts RWE (eds): *Oxalate in Human Biochemistry and Clinical Pathology*. London, Wellcome Foundation, 1979, pp 224–230.

primary hyperparathyroidism, cystinuria, renal tubular acidosis, or idiopathic hypercalciuria. Prevention of further stone formation requires treatment of any underlying metabolic disorder and the immediate cause of stone formation.

Infected renal lithiasis refers to the pathologic occurrence of stones composed of magnesium ammonium phosphate ($MgNH_4PO_4 \cdot 6H_2O$, or struvite). Stones caused by infection are not pure struvite—careful crystallographic analysis of infected stone material from humans has revealed a mixture of struvite, carbonate apatite ($Ca_{10}[PO_4]_6 \cdot CO_3$), and hydroxyapatite ($Ca_{10}[PO_4]_6[OH]_2$) (Griffith, 1978).

Infection of the urinary tract with urease-producing bacterial organisms is a necessary prerequisite for the formation and growth of struvite stones (Griffith et al., 1976). The enzyme urease catalyzes the formation of ammonia and CO_2 from urea:

$$H_2N—C—NH_2 + H_2O \rightarrow 2NH_3 + CO_2$$
$$\underset{O}{\overset{\|}{}}$$

The formation of ammonia leads to an increase in urinary pH:

$$NH_3 + H_2O \rightarrow NH_4^+ + OH^- \quad (pK = 9.0)$$

When urine is physiologically alkaline, ammonia levels are low, but when urease is present, the urine is alkaline and ammonia levels are high.

The higher pH also leads to the further dissociation of phosphate:

$$H_2PO_4^- \rightarrow H^+ + HPO_4^{--} \quad (pK = 7.2)$$
$$HPO_4^{--} \rightarrow H^+ + PO_4^{---} \quad (pK = 12.4)$$

Under these conditions, urine is supersaturated for magnesium ammonium phosphate, and precipitation of this material occurs.

Because all of the reactions occur in aqueous solution, the CO_2 exists as carbonic acid ($CO_2 + H_2O = H_2CO_3$). In the presence of an alkaline pH, the carbonic acid dissociates to form bicarbonate ($H_2CO_3 = H^+ + HCO_3^-$; pK = 6.3), and the bicarbonate dissociates to form carbonate ($HCO_3^- = H^+ + CO_3^{--}$; pK = 10.2). The CO_3^{--} can precipitate with PO_4^{---} and Ca^{++} to form carbonate apatite.

Proteus species are most commonly associated with infection stones, but some species of *Klebsiella, Pseudomonas,* and *Staphylococcus* may produce urease (Griffith, 1978). Virtually all *Proteus* species produce urease, while *E. coli* rarely, if ever, produces urease. Even some anaerobic bacterial organisms will produce urease.

Infection stones account for 15%–20% of all urinary stones. Struvite stones frequently have a branched or staghorn configuration, but not all branched calculi are infection induced.

Cystine and uric acid stones also may have a branched configuration. Some patients with metabolic stone disease may have urinary tract infections with bacteria that do not produce urease. The stones do not contain struvite, and the infection is not responsible for the stone formation. The infection can be treated successfully in half of these patients with antibiotics alone.

Some patients with metabolic stone disease may develop recurrent urinary tract infections with urease-producing bacterial organisms. Struvite may then precipitate on a core of calcium oxalate or cystine. Some investigators have found an underlying metabolic disorder in over 50% of patients with struvite stone formation (Resnick, 1981; Segura et al., 1981). Prevention of further stone formation requires the identification and treatment of any such metabolic disorder. Other investigators found that recurrent stone formation was negligible in a group of patients who had operative removal of their stone material, specific antimicrobial therapy, and postoperative irrigation of the collecting system with hemiacidrin solution (Silverman and Stamey, 1983). This study implies that metabolic disorders are relatively unimportant in the pathogenesis of struvite urolithiasis.

Stone formation may be associated with obstructive uropathy. The most common example is uric acid bladder stone formation with bladder outlet obstruction, usually from prostatic hyperplasia. Upper urinary tract obstruction may delay the normal washout of crystal aggregates and gravel. These particles may continue to grow and form macroscopic stones. Persistent infection with urease-producing bacterial organisms may occur in the presence of urinary stasis and result in struvite stone formation. Most patients with obstruction, however, do not form kidney stones. It is likely that such patients have a more fundamental metabolic basis for their stone formation.

Medullary sponge kidney is characterized by dilated collecting tubules in one or more renal papillae. In the absence of complications, it is an asymptomatic and benign condition, but urinary tract infection and nephrolithiasis are frequent complications (Yendt, 1982). The diagnosis is made by excretory urography. Linear or spherical tubules in the renal papillae are filled with contrast medium. Some authors think that a "papillary blush" is found in patients with the mildest form of this disorder.

The role of medullary sponge kidney in the pathogenesis of nephrolithiasis has not been clarified, but 20% of patients with calcium urolithiasis may have this disorder (Yendt, 1982). Other investigators found medullary sponge kidney in fewer than 5% of their calcium stone-forming patients (O'Neill et al., 1981). Medullary sponge kidney is found more commonly in women than men. Patients with medullary sponge kidney and nephrolithiasis have the same spectrum of metabolic abnormalities as the overall population of calcium stone formers (O'Neill et al., 1981; Parks et al., 1982), although patients with concurrent medullary sponge kidney and hyperparathyroidism have been described (Gremillion et al., 1977). This suggests that renal hypercalciuria from disordered nephron function may lead to parathyroid hyperplasia or adenoma formation.

Early studies reported that the most common stone constituent in patients with medullary sponge kidney was calcium phosphate. More recent studies found that calcium oxalate was the most common stone salt (Harrison and Rose, 1979). A superimposed infection with urease-producing bacterial organisms will lead to an explosive progression of stone formation. Struvite will then be found in addition to the usual calcium salts.

Stone formation and recurrent infections may be associated with urinary tract diversion. Urease-producing bacterial infections may lead to struvite stone formation. The diversionary procedure may also produce metabolic abnormalities that encourage stone formation. Gastrointestinal bicarbonate loss and hyperchloremic acidosis are known side effects of ureterosigmoid anastomoses (McConnell et al., 1979). The systemic acidosis may lead to hypercalciuria and hypocitraturia. Osteomalacia has been reported to develop in patients with ureterosigmoidostomies and a metabolic acidosis (Siklos et al., 1980). Occasional patients with ileal conduits will develop a hyperchloremic acidosis (Koff, 1975; Dretler, 1973). Most stones form in the presence of *Proteus* infections and are composed of struvite. Since excess conduit length may contribute to the bicarbonate loss and chloride absorption, these metabolic derangements may occur more freqeuntly with the new continent diversionary procedures. The use of nicotinamide to block intestinal chloride transport may be a useful preventive measure (Koch and McDougal, 1985).

The metabolic effects of some drugs will lead to stone formation. Acetazolamide, a carbonic anhydrase inhibitor, produces changes in urine composition that are very similar to those found in distal renal tubular acidosis. Kidney stones have formed in patients who have received this drug for the treatment of glaucoma. Very rarely, xanthine stones have formed in patients treated with allopurinol.

Drugs or their metabolites may have limited solubility in urine. These compounds may be absorbed onto calculi already present in the urinary tract or may precipitate to form new stones. Approximately 50% of an ingested dose of allopurinol is excreted as oxypurinol, an oxidative metabolite. Oxypurinol, like xanthine, is much less soluble than allopurinol or hypoxanthine. The solubility of oxypurinol decreases with a decrease in pH. Radiolucent oxypurinol stone formation has been reported in a patient with regional enteritis who was receiving allopurinol for the prevention of recurrent uric acid lithiasis (Stote et al., 1980). Persistent oliguria and aciduria contributed to the precipitation of oxypurinol.

Triamterene and its metabolites have been identified in renal calculi. Some investigators have suggested that their precipitation may be a causative factor in the formation of calcium stones (White and Nancollas, 1982). Other investigators found that triamterene and its metabolites adsorb to stone matrix (Werness et al., 1982). Since triamterene-containing stones have an unusually high matrix content, triamterene and its metabolites may be a passive constituent of renal calculi. Clinical studies of patients receiving Dyazide (a combination of triamterene and hydrochlorothiazide) suggest that nephrolithiasis is not a clinically significant side effect (Jick et al., 1982).

Sulfonamides were one of the first classes of drugs to precipitate in the urinary tract, but numerous other drugs and their metabolites have been identified in renal calculi (Daudon and Reveillalud, 1985). Several factors can predispose to the urinary precipitation of a drug: a high renal excretion rate, a low solubility of a drug or its metabolites, a low urine volume, and prolonged treatment at a high dose. The appearance of synthetic compounds in renal calculi will become more common as new drugs are introduced and analytical methods of stone analysis become more sophisticated.

Urinary calculi may also form on foreign bodies in the urinary tract (Dalton et al., 1975). Foreign-body stones in the upper urinary tract have been associated with ureteral catheters, nephrostomy tubes, sutures, biliary calculi, shrapnel, and acupuncture needles. Renal papillary necrosis with calcification of the sloughed papillae may mimic nephrolithiasis (Fig 9–11).

IDIOPATHIC CALCIUM UROLITHIASIS

This diagnosis is one of exclusion and is applicable in 70%–80% of North American and Western European patients with urolithiasis. If pure uric acid lithiasis, cystinuria, renal tubular acidosis, primary and secondary hyperoxaluria, and the hypercalcemic states have been excluded in a patient with calcium oxalate or mixed calcium oxalate/calcium phosphate stone formation, this diagnosis may be applied. Unfortunately, a patient may have more than one disorder. Primary hyperparathyroidism and idiopathic calcium urolithiasis (ICU) or cystinuria and ICU may occur in the same patient. The diagnosis of ICU is usually applied to patients who have

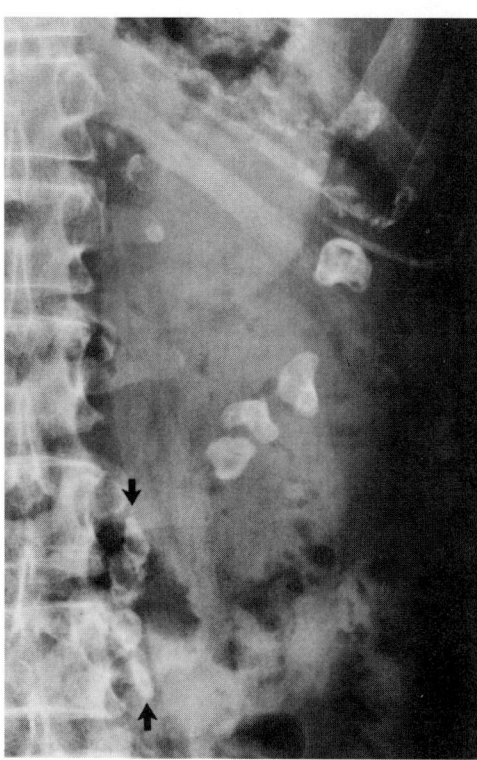

FIG 9–11.
Multiple calcified bodies in left kidney and ureter *(between arrows)*. The radiolucent centers could represent uric acid or necrotic renal papillae *(arrows)*. Histologic examination revealed necrotic calcified renal papillae.

calcium stone formation that is not the result of a specific, well-defined metabolic disorder.

The ICU syndrome includes several as yet ill-defined disorders. Careful metabolic studies of patients with ICU reveal a multiplicity of abnormalities: hypercalciuria, minimal hyperoxaluria, hyperuricosuria, crystal growth inhibitor deficiencies, hypocitraturia, and "incomplete" renal tubular acidosis. One or more of these abnormalities may be found in the same patient, and some of the abnormalities may be related to the others, such as hypocitraturia to "incomplete" RTA, or hyperuricosuria to deficient crystal growth inhibition. Several of these abnormalities may be indicative of a more generalized renal tubular dysfunction. Some patients with ICU may have no clearly recognizable disorder other than calcium stone formation.

From 50% to 70% of these patients will have hypercalciuria (Robertson and Morgan, 1972). Most healthy men receiving a 1-gm calcium diet excrete less than 275 mg of calcium over 24 hours. The corresponding figure for women is 250 mg/24 hours. A more convenient way to remember is that urinary calcium excretion should not exceed 4 mg/kg/24 hours. Although the hypercalciuria of ICU is often called idiopathic, the pathogenesis

of the hypercalciuria has, in part, been revealed by several investigators. Most patients with ICU and hypercalciuria are thought to have primary intestinal hyperabsorption of dietary calcium (Pak et al., 1974). Increased absorption of dietary calcium may slightly increase the serum calcium concentration and suppress parathyroid hormone secretion. The increased filtered load of calcium and decreased tubular reabsorption results in hypercalciuria. The urinary calcium loss compensates for the enhanced intestinal absorption, thereby maintaining serum calcium within a normal range. In the past, patients with absorptive hypercalciuria have been separated into three categories that depended on the serum phosphorus level and the level of dietary calcium at which hypercalciuria ensued.

Patients with absorptive hypercalciuria have been studied with in vivo intestinal perfusion (Brannan et al., 1979). Net calcium absorption in the jejunum was markedly increased, while ileal absorption was only mildly increased. The absorptive defect was specific for calcium, since magnesium absorption was normal. A possible criticism of these studies is that the perfusate was a simple electrolyte solution mixed with mannitol and polyethyleneglycol. Absorption from a more physiologic solid meal may have been different.

There is a positive correlation between urinary calcium excretion and sodium excretion (Rao et al., 1985). Although there is some question as to the magnitude of this correlation, a recent study found that a group of hypercalciuric subjects on a 200-mEq sodium diet converted to normocalciuria on an 80-mEq sodium diet (Muldowney et al., 1982). The sensitive relationship between urinary calcium and sodium and the normal day-to-day variation of sodium intake may confound the results of standard 24-hour urine collections. The induction of idiopathic hypercalciuria by a habitually high sodium intake is an intriguing, but as yet unstudied, possibility.

It is not known if absorptive hypercalciuria is due to a primary intestinal defect or to a more complex metabolic disorder. Several investigators have found elevated vitamin D metabolites in one third to one half of patients with absorptive hypercalciuria (Shen et al., 1977). Some patients with absorptive hypercalciuria have low serum phosphorus levels. The stimulation of 1,25-dihydroxyvitamin D synthesis in the kidney by the hypophosphatemia could lead to increased intestinal calcium absorption (Gray et al., 1977). The primary event would be a "renal leak" of phosphate. Other investigators, however, do not think that vitamin D and its metabolites play a critical role in the pathogenesis of absorptive hypercalciuria (Pak, 1979; Netelenbos et al., 1985).

A smaller group of patients with hypercalciuria may

have impaired renal tubular reabsorption of calcium (Pak, 1979; Coe et al., 1973). Fasting urinary calcium excretion will be elevated, the serum concentration of calcium will be reduced, and parathyroid function will be stimulated. The elevated parathyroid hormone level tends to restore serum calcium to normal levels by mobilizing bone calcium and enhancing intestinal absorption. Renal hypercalciuria and absorptive hypercalciuria are differentiated by measuring parathyroid hormone levels and fasting urinary calcium excretion. Parathyroid hormone levels should be elevated in renal hypercalciuria but normal in absorptive hypercalciuria. Fasting urinary calcium excretion should be high in renal hypercalciuria but normal in absorptive hypercalciuria.

Early studies found renal hypercalciuria in more than 50% of patients with hypercalciuria, but this proportion fell to 10% with later studies. This discrepancy may have been due to differences in the specific immunoassays used to measure parathyroid hormone. Furthermore, more intensive fasting tends to reduce the proportion of patients with renal hypercalciuria. Renal and absorptive hypercalciuria may represent extremes of a variable disorder rather than distinct clinical entities. A uniform elevation of intestinal calcium absorption and a variable defect of renal calcium reabsorption may better explain the data than separate absorptive and renal forms of hypercalciuria (Coe et al., 1982).

Many studies have examined calcium metabolism in nephrolithiasis, but relatively few have investigated the role of oxalate in ICU. The paucity of such studies is related to the difficult nature of oxalate measurement in urine. Several investigators have found a mild but definite increase in urinary oxalate excretion in a subset of patients with ICU (Robertson, 1980; Baggio et al., 1983). A mildly elevated urinary oxalate is a greater risk factor for the precipitation of calcium oxalate than is a mildly elevated urinary calcium, because normal urinary oxalate levels are tenfold lower than normal urinary calcium levels. A small increase in urinary oxalate will lead to the precipitation of a greater volume of crystals than will a comparable increase in urinary calcium.

Patients with mild hyperoxaluria do not appear to have a specific disturbance of glyoxylate metabolism. The hyperoxaluria is thought to be secondary to intestinal hyperabsorption of oxalate and calcium (Marangella et al., 1982). The mechanism is analogous to that for enteric hyperoxaluria, but intraluminal intestinal calcium is reduced by hyperabsorption, not by complexation with fatty acids. Less calcium is available to bind oxalate, thereby permitting more avid absorption of free oxalate. A similar situation may occur when patients with recurrent calcium lithiasis are instructed to limit their dietary intake of calcium-containing foods. Isolated dietary calcium restriction may increase the risk of stone formation by allowing oxalate excretion to increase without a commensurate decrease in calcium excretion (Bataille et al., 1985).

Intestinal hyperabsorption of oxalate may not be solely secondary to calcium hyperabsorption and reduced complexation. A more widespread defect in cellular transport of oxalate may exist. A recent series of studies examined the transport of oxalate across RBC membranes from healthy persons and patients with recurrent calcium oxalate nephrolithiasis (Baggio et al., 1984, 1986). The mean transmembrane oxalate flux rate in the stone-forming patients was triple that in the healthy controls. This meant that oxalate could cross the RBC membrane much faster in patients with ICU. A similar situation in the luminal membrane of the GI tract could provide another mechanism for oxalate hyperabsorption.

Some patients with ICU will have a concurrent disorder of uric acid metabolism, such as gout or hyperuricosuria. These patients tend to have a more severe form of stone disease, manifestated by more frequent stone formation and a greater need for surgical intervention (Coe, 1983). Specific epitaxial overgrowth of calcium oxalate on uric acid or monosodium urate has been suggested as a possible explanation, but it is doubtful that this occurs in the urinary tract. Heterogeneous nucleation could occur, but freshly voided urine specimens from patients with ICU rarely contain uric acid or urate crystals (Werness et al., 1981a). The binding of macromolecular inhibitors by another form of urate (colloidal) was proposed and has been supported by some studies (Ryall et al., 1985). The reduction of hyperuricosuria with allopurinol also increased calcium oxalate crystal growth inhibition (Pak et al., 1978), and hyperuricosuric calcium oxalate nephrolithiasis has successfully been treated with alkalinizing agents (Pak and Peterson, 1986).

A quantitative or qualitative disorder of crystal growth inhibition has been proposed as a possible reason for stone formation in patients with ICU. As discussed in the section on clinical stone formation, citrate is a complexor of calcium and a weak inhibitor of calcium oxalate crystal growth. Some investigators have found a subgroup of patients with ICU who have low urinary citrate levels (Menon and Mahle, 1983; Rudman et al., 1982). Hypocitraturia in patients with distal RTA may enhance stone growth. Although the most beneficial action of citrate is through complexation of calcium, its weak inhibitory effects would help prevent further stone formation. Pharmacologic citrate preparations, through their alkalinizing property, have successfully prevented further stone formation in patients with hypocitraturic calcium lithiasis (Pak et al., 1985). Not all investigators, however, have been able to demonstrate an association between urinary citrate levels and clinical stone formation (Hosking et al., 1985).

Persistently alkaline urine has been found in another subset of patients with ICU but no clearly identifiable acid-base disorder (Robertson et al., 1972). Careful analysis of stone material formed by patients with ICU reveals the presence of calcium phosphate, albeit in small amounts, in most of the stones. Since calcium phosphate will preferentially precipitate at an alkaline pH, small calcium phosphate crystals may provide heterogeneous nuclei for the overgrowth of calcium oxalate. This mechanism may become more important as alkalinizing agents are more commonly used to treat patients with ICU.

The syndrome of incomplete RTA was discussed in the section on renal tubular acidosis. Patients with this disorder are not systemically acidotic, but they have an impaired ability to acidify their urine after an ammonium chloride load. This implies the presence of a renal tubular defect in hydrogen ion secretion. The hypocitraturia found in some patients with ICU may be the result of a subtle intracellular acidosis.

Defective tubular reabsorption of calcium is thought to be the cause of renal hypercalciuria. All of these disorders may be different manifestations of a more widespread tubular dysfunction. Evidence for this was provided by a study of the effects of hydrochlorothiazide and acetazolamide on the urinary excretions of calcium, sodium, and magnesium (Sutton et al., 1980). These diuretic agents were chosen for their known actions at different sites in the nephron. Hydrochlorothiazide augmented sodium, calcium, and magnesium excretion to a greater extent in patients with ICU than in control subjects. With acetazolamide, the increase in sodium excretion was less in the patients than in the controls. The abnormal responses to both diuretics were most marked in patients with hypercalciuria during fasting. Such studies implicate a disorder of renal tubular transport as the primary abnormality in ICU. Gastrointestinal hyperabsorption of calcium would be a secondary phenomenon.

A more recent study found evidence of renal tubular dysfunction in a group of patients with different etiologies for their urolithiasis (Jaeger et al., 1985). These investigators concluded that the tubular dysfunction was a result of the stone formation, not the primary cause. The stone disease could be exacerbated, however, by the tubular disorder after the initial episode of stone formation. Much research remains to be done before the pathogenesis of ICU is completely understood.

EVALUATION

HISTORY

The clinical history may help to determine the cause of stone formation. The important elements in the history are outlined in Table 9–4. Inherited disorders, such as primary hyperoxaluria, cystinuria, and renal tubular acidosis, are clearly associated with urolithiasis, but even ICU tends to occur within families. Idiopathic stone formation and uric acid lithiasis are more common in males, while primary hyperparathyroidism and renal tubular acidosis are more common in females. Although ICU is the most common type of stone disease in children and adults, the classic inherited disorders associated with stone formation are more common in children than adults.

Past residence in a geographical area with a high incidence of stone disease, such as the southeastern United States, may have some epidemiologic importance. More pertinent would be temporary residence in a arid climate, such as the Middle East or southwestern United States. People who are not acclimated to these conditions may have a low urine output state as a result of large fluid losses from perspiration and respiration. Long-distance runners may have the same problem. There is evidence to suggest that stone formation may be more common in marathon runners than in the general population (Milvy et al., 1981). Urine volume tends to be low in patients who must perform manual labor in a hot environment such as a steel mill. Easy access to fluids and bathroom facilities may not be possible with some occupations. A low urine output state is a by-product of many occupations.

Excess dietary intake of calcium, oxalate, or protein and a low fluid intake may encourage the formation of stones. Stone formation is also associated with certain medications: vitamin D, absorbable alkali (calcium carbonate), and carbonic anhydrase inhibitors (acetazol-

TABLE 9–4.

Clinical History

Age at onset
Sex
Family history
Geographical residence
Occupation
Fluid intake
Diet
Medications
Previous stone passage
Interventional procedures
Previous stone composition
Urinary tract infections

amide). Patients need not consistently overindulge to form a stone. Only brief periods of dietary indiscretion or poor fluid intake may initiate crystallization. A similar consideration applies to medications, particularly those associated with a nutritional fad.

A patient's past experience with stone formation and passage may provide clues as to the etiology of the stone disease and the patient's motivation to pursue a lifelong treatment program. A history of multiple stone passages and rapid recurrent stone formation may indicate a fundamental metabolic etiology, such as primary hyperparathyroidism or renal tubular acidosis. The necessity for multiple interventional procedures, whether endoscopic or open surgical, implies the growth of larger stones. This may be indicative of a more malignant form of stone disease, such as infected lithiasis. Struvite lithiasis should also be suspected if there is a history of infection with urease-producing bacterial organisms. The composition of previously analyzed stones should be sought. If old stones have not been analyzed, they should be examined and sent for formal analysis. It is often possible to determine a stone's composition just by looking at it. Cystine and uric acid stones are good examples.

PHYSICAL EXAMINATION

Recognizable physical abnormalities may be present in a few disorders associated with urolithiasis (see Table 9–2): sarcoidosis, Cushing's syndrome, hyperthyroidism, and gout. Patients with neurogenic bladder dysfunction, whether from traumatic spinal cord injuries or dysraphism, may be at risk for recurrent urinary tract infections and infected stone formation. Calcium stones may form in a recently immobilized patient. Scars may provide a more accurate record of previous surgical procedures and may refresh a patient's memory. Extensive abdominal scarring will be seen in some patients with small-bowel disorders and should raise the possibility of enteric hyperoxaluria. Most patients, however, will not have physical findings related to their stone disease, and laboratory studies are required to uncover the etiology.

LABORATORY EVALUATION

The extent of the laboratory evaluation of a particular patient depends on the severity of the stone disease. A 40-year-old man who has passed several dozen calculi over the past ten years will need a more extensive metabolic evaluation than a 60-year-old woman who has a 1-cm asymptomatic renal pelvic stone but no previous history of stone formation or passage. Patients who have formed a single calcium stone have the same range of metabolic abnormalities as do patients with recurrent

stone formation (Pak, 1982; Strauss et al., 1982). One half to two thirds had hypercalciuria, approximately 5% had primary hyperparathyroidism, and 20%–30% had no identifiable metabolic disorder. The patients with single stones were older when they passed their stones, had a higher incidence of urinary tract infection, and were more likely to have had surgical or endoscopic intervention. These authors recommended that single-stone formers have the same evaluation and be treated no differently from other patients with stone disease.

Many patients who have formed a single stone do not want to undergo the inconvenience or expense of an extensive metabolic evaluation, particularly when it will not alter their treatment program. These patients may be willing to modify their diets and increase their intake of fluids, but it is unlikely that they will unfailingly take some kind of medication for the rest of their lives. A utilitarian evaluation of a single-stone former would include an assessment of calcium metabolism and renal function (usually obtained as a multichannel chemical analysis of serum), a urinalysis and culture, and a stone analysis. These studies should detect obvious hyperparathyroidism, infected stone formation, and most of the disorders for which specific therapy is especially beneficial (uric acid lithiasis and cystinuria). A more extensive evaluation is usually reserved for those patients who prove to have metabolically active stone disease.

A complete metabolic workup should include the studies listed in Table 9–5. An isolated elevated serum calcium level should be confirmed on two or three sep-

TABLE 9–5.

Laboratory Evaluation

Serum
Calcium
Phosphorus
Uric acid
Creatinine
Protein electrophoresis
Alkaline phosphatase
Urine
Urinalysis
Culture
Fasting pH
24-hr volume
Urine chemistry (24-hr volume)
Calcium
Phosphorus
Uric acid
Oxalate
Cystine
(Citrate)
(Sodium)
(Magnesium)
Stone analysis

arate occasions. The normal range will vary among laboratories, but one should be suspicious of a serum calcium greater than 10.1 mg/dl. Total serum calcium includes ultrafilterable calcium and protein-bound calcium. Ultrafilterable calcium, that which enters the glomerular filtrate, is composed of ionized calcium and a small amount of complexed calcium. Approximately half of total serum calcium is protein bound, primarily to albumin. The critical fraction that is controlled by the homeostatic mechanisms of the body is ionized calcium, but it is very difficult to measure reliably ionized calcium. Total serum calcium is usually measured and corrected for serum albumin.

Serum phosphorus varies with age, sex, renal function, and diet, but may be low in patients with primary hyperparathyroidism and some patients with ICU. Serum alkaline phosphatase activity may be elevated in patients with hyperparathyroidism. Elevated serum uric acid levels are found in some patients with uric acid lithiasis, but calcium stone formers also may have hyperuricemia. A serum bicarbonate measurement is usually included in a multichannel analysis and may be decreased in patients with renal tubular acidosis or with small-bowel disorders associated with malabsorption.

A urinalysis should be done promptly after collection. The pH is best measured in a morning urine specimen after an overnight fast. A pH less than 5.5 eliminates distal renal tubular acidosis. A short urinary acidification test can be performed if the pH is not less than 5.5. Uric acid stone formers may have a persistently low pH. A very high urine pH (> 8.0) will be found if the urine is infected with a urease-producing bacterial organism. The infection must be eliminated to accurately assess the pH. Ingestion of alkali or citrate preparations also will raise the pH.

Red blood cells and WBCs are usually seen with urolithiasis, and bacteria may be visible if an infection is present. A urine culture will confirm the presence of an infection, but some patients with struvite stone formation will have negative bladder urine and even renal pelvic urine cultures. Documentation of the infection and identification of the organism require culture of the stone (Fowler, 1984). Bipyramidal-shaped calcium oxalate dihydrate crystals can be found in healthy persons. Multiple small, platelike or dumbbell-shaped calcium oxalate monohydrate crystals are often found in patients with hyperoxaluria, usually the primary disorder. Calcium oxalate monohydrate crystals are birefringent and will appear as bright specks with polarized microscopy. The presence of hexagonal cystine crystals is virtually diagnostic of cystinuria. A 24-hour urine collection is the traditional mainstay of a metabolic stone evaluation. The most important, and often most neglected, measurement is the volume. Many patients with recurrent calcium stone formation will have a 24-hour urine volume that is approximately 1 L. The basic principles of the medical treatment of urolithiasis are dietary moderation and a consistently high urine volume. Patients can easily monitor their progress by measuring their 24-hour urine volume at home. This may be a minor nuisance, but it is safe and inexpensive.

Urinary excretion of calcium, phosphorus, oxalate, and uric acid is a function of dietary intake. Urinary phosphorus excretion is very sensitive to diet and is best used to monitor patient compliance with orthophosphate therapy. Increased urinary oxalate is found in patients with primary or secondary hyperoxaluria. Dietary oxalate content has its greatest impact in patients with enteric hyperoxaluria. Uric acid excretion may be elevated in uric acid or calcium oxalate stone-forming patients. Urinary calcium excretion has received much attention. Elaborate protocols have been devised to differentiate the renal and absorptive hypercalciurias and the different types of absorptive hypercalciuria (Pak et al., 1975). The goal has been to institute specific pharmacologic therapy, but these calcium tolerance tests have achieved limited clinical utility (Lein et al., 1983).

Urinary cystine excretion can be screened with the traditional cyanide-nitroprusside reaction or newer spot tests (George and Politzer, 1970). A positive reaction should be followed by a quantitative amino acid analysis. Urinary citrate excretion will be decreased in patients with distal renal tubular acidosis and intestinal malabsorption. Citrate is low in the presence of a urinary tract infection, because bacteria metabolize citrate. The clinical importance of citrate measurements in patients with ICU has not been clarified, but some investigators advocate the use of citrate preparations in those patients with hypocitruria. Urinary magnesium excretion is of some interest to investigators but is not clinically useful. Urinary sodium excretion may be elevated in some patients with hypercalciuria.

Computer programs have been written to calculate urinary saturations for the major stone-forming salts (Robertson, 1982; Finlayson, 1977) and have become available commercially (Pak et al., 1985). These programs are very useful in research, but their widespread clinical utility has not been demonstrated. Examination of the 24-hour urine volume and the total excretions of the major ions should provide enough information to make sound clinical decisions.

Stone material that has been passed or removed should be analyzed formally. Several commercial laboratories perform stone analyses for a very modest charge. Many different crystalline components have been identified in urinary calculi (Table 9–6), and several methods have been used to identify accurately

TABLE 9–6.
Crystalline Constituents of Human Urinary Calculi*

SUBSTANCE	MINERALOGIC NAME	FORMULA
Calcium oxalate monohydrate	Whewellite	$CaC_2O_4 \cdot H_2O$
Calcium oxalate dihydrate	Weddellite	$CaC_2O_4 \cdot 2H_2O(to\ 2.5H_2O)$
Magnesium hydrogen phosphate trihydrate	Newberyite	$MgHPO_4 \cdot 3H_2O$
Magnesium ammonium phosphate hexahydrate	Struvite	$MgNH_4PO_4 \cdot 6H_2O$
Hydroxyapatite	Hydroxyapatite	$Ca_{10}(PO_4)_6(OH)_2$
Carbonate-apatite	Carbonate-apatite	$Ca_{10}(PO_4)_{6-x}(OH)_{2-y}(CO_3)_{x+y}$
Calcium hydrogen phosphate dihydrate	Brushite	$CaHPO_4 \cdot 2H_2O$
Tricalcium phosphate	Whitlockite	$\beta\text{-}Ca_3(PO_4)_2$
Octacalcium phosphate		$Ca_4H(PO_4)_3 \cdot 2.5H_2O$
Uric acid		$C_5H_4N_4O_3$
Uric acid dihydrate		$C_5H_4N_4O_3 \cdot 2H_2O$
Ammonium acid urate		$C_5H_3N_4O_3NH_4$
Sodium acid urate monohydrate		$C_5H_3N_4O_3Na \cdot H_2O$
Cystine		$[-SCH_2CHNH_2COOH]_2$
Xanthine		$C_5H_4N_4O_2$
Calcium sulphate dihydrate	Gypsum	$CaSO_4 \cdot 2H_2O$

*Adapted from Sutor DJ, Scheidt S: Identification standards for human urinary calculus components, using crystallographic methods. *Br J Urol* 1968; 40:22–28.

these constituents (Sutor and Scheidt, 1968). The composition of some calculi, such as cystine, can be identified with simple macroscopic or microscopic analysis. Whole stones covered with calcium oxalate dihydrate crystals will have a burr or "hair-on-end" appearance. If the whole stone is available, an attempt should be made to separate the nucleus from the rest of the stone, because the composition of the nucleus may differ from that of the outer layers. Optical properties (crystal system, optical sign, refractive index, angle of extinction, and birefringence) can be measured using polarization microscopy (Prein, 1941). The limited number of crystalline components in urinary calculi can be identified from these optical properties. Chemical tests have been used for qualitative identification of stone components, but quantitation is inaccurate (Laskowski, 1965; Sutor and Wooley, 1971). Closely related compounds, such as uric acid and 2,8-DHA, may not be differentiated by standard chemical tests. X-ray powder diffraction has been the standard method for stone analysis, because it enables almost absolute identification of crystalline materials and mixtures of these materials (Morriss and Beeler, 1967). The equipment is expensive, and the procedure is time consuming.

Infrared spectroscopy is becoming the most widely used method of stone analysis (Beischer, 1955; Hesse and Bach, 1982). Finely powdered stone material is pressed into a transparent tablet with optically pure potassium bromide. The infrared spectrum is recorded and compared with a library of standard spectra. Infrared spectroscopy is rapid, relatively inexpensive, and permits the identification of noncrystalline components and artifacts. This method is especially useful for identifying drugs and their metabolites. Other methods of stone analysis have been proposed, but they appear to have limited applicability: thermogravimetric technique (Rose, 1982), scanning electron microscopy with energy dispersive x-ray analysis (Khan and Hackett, 1986), and CT (Hillman et al., 1984).

ACTIVITY

The propensity to form stones will vary from patient to patient and will vary in a particular patient over time. An arbitrary method for assessing the activity was developed over 20 years ago at the Mayo Clinic (Smith, 1983b). The concept of stone activity enables one to evaluate the need for long-term pharmacologic therapy and the response to such therapy. Two basic categories of activity are defined: surgical activity and metabolic activity. A patient with renal colic, obstruction, or infection associated with a stone has surgically active urolithiasis. Surgical intervention is often necessary in these patients.

Metabolic activity refers to the precipitation of stone material. Urolithiasis is metabolically active in patients who have had growth of old stones, formation of new stones, or the passage of gravel within the past year. These criteria must be documented radiographically. If a patient's stone disease meets none of these criteria, the urolithiasis is metabolically inactive. If previous radiographs are unavailable or of poor quality, the metabolic activity is said to be indeterminate. Such patients are instructed to increase their fluid intake and avoid dietary excesses. They are followed up until the metabolic activity can be determined.

The distinction between surgical and metabolic activity is very important, because a symptomatic stone may have formed several years before. No specific medical treatment may be needed after the immediate surgical problem is resolved. Metabolically active urolithiasis may exist without symptoms. No surgical intervention may be needed in this situation, but specific medical therapy may be required to prevent further stone formation.

One of the goals of urolithiasis research has been the development of a method to predict stone activity when a patient is first evaluated (Hallson and Rose, 1978). Several measures have been proposed, including the previously discussed saturation-inhibition index, but none has had greater utility than this simple, time-dependent method of assessing activity. Recurrent stone formation is common, but it cannot be predicted from standard laboratory evaluations in individual patients (Ljunghall and Danielson, 1984).

RADIOGRAPHIC EVALUATION

Patients who are passing a ureteral calculus should have an excretory urogram to document the diagnosis. A plain x-ray of the abdomen is taken before the patient is given any contrast material. It is frequently possible to identify a small calculus along the presumed course of the ureter. The presence of other calculi in the calyces or renal pelvis should be noted. Oblique films may be needed to confirm that a calculus is within the kidney. Prompt bilateral nephrograms should be seen after the injection of contrast, but a delay in the appearance of the pyelogram implies obstruction of the affected kidney. The collecting system of the affected kidney is often dilated, and delayed films may show contrast in the ureter down to the level of the obstructing stone. A small forniceal tear occasionally permits extravasation of contrast material into the retroperitoneal space. This has a striking appearance but is of little clinical significance. Films should be exposed until the point of obstruction is accurately defined or all of the contrast has been excreted.

The assessment of metabolic activity depends on the availability of high-quality radiographs. Tomographic views will provide more detail than standard radiographs, because the overshadowing effects of bowel gas and intestinal contents are lessened. The disadvantages of tomograms are that more radiation is delivered, and small stones may be missed if the spacing between cuts is too large. Further detail can be obtained with CT scans, but the expense and radiation dose limit their routine use.

The radiographic appearance of a stone depends on the composition of the stone, its thickness and orientation, surrounding tissues or bowel contents, and radiographic technique. Radiographic techniques have been standardized, and bowel cleansing is commonly used. A stone may rotate and give the illusion of a change in size. This rotation cannot be prevented, but its possible occurrence should be remembered when radiographs are examined. All stones, except uric acid, are clinically radiopaque. Even large uric acid stones will appear faintly radiopaque from incorporated impurities. Some stone materials, such as cystine, struvite, and uric acid, are not as radiopaque as iodinated contrast agents and will appear as filling defects on an excretory urogram. Relative radiodensities of the common stone salts have been measured and in decreasing order are apatite, whitlockite, brushite, whewellite, weddellite, cystine, struvite, and uric acid (Roth and Finlayson, 1973).

MEDICAL MANAGEMENT

With rare exceptions, the goal of medical treatment is to prevent the formation of new stones or the further growth of old stones. This prophylaxis must be effective and continuous. Patients must understand that prevention of further stone formation will probably require lifetime treatment.

The therapy of urolithiasis is based on two basic principles. First is a reduction of urinary supersaturation, and second is an increase in net inhibitory activity. The latter can be achieved by increasing the quantity of inhibitors, by increasing the potency of inhibitors, or by corresponding decreases in promoter activity.

The purpose of a high fluid intake is to lower urinary supersaturation. The dilution will reduce ionic strength, complexation, and the concentration of inhibitors, but these side effects are more than offset by the reduction of supersaturation. It is impossible for stone formation to occur in urine that is undersaturated for the particular stone salt. All patients with renal calculi should be counseled to increase their fluid intake. An 8-oz glass of fluid should be consumed hourly while awake, and 8–16 oz of fluid should be consumed if the patient is up at night. Approximately one half of the fluid should be water. The patient should produce at least 2,500 ml of urine per 24 hours. The fluid intake should be consistent. A liberal intake of fluids during the day but poor intake during the night will not uniformly lower supersaturation, especially after a heavy evening meal. Patients can use a container of known volume to monitor inexpensively their 24-hour urine output. A fixed numeric goal is very helpful, because most patients are poor estimators of their fluid intake and urine output.

A dietary history should be taken and dietary excesses eliminated (Pak et al., 1984). The traditional recommendation has been a low-calcium diet, but this increases urinary oxalate excretion. Urinary supersaturation can be lowered with a low-calcium and low-oxalate diet, but patients are less likely to adhere to a more complex program. The same criticism applies to a low-carbohydrate diet, a high-fiber diet, and a low-animal protein diet, although the last may be useful in the treatment of idiopathic uric acid lithiasis. Since dietary therapy requires long-term patient compliance, the encouragement of dietary moderation may be the best advice.

Fluid and dietary therapy should be used in all patients with urolithiasis and should be the only initial therapy in patients with ICU. One hundred eight patients with ICU of indeterminate metabolic activity were initially treated with fluid and dietary therapy at the Mayo Clinic (Hosking et al., 1983). No stone growth or new stone formation was seen in 58% of these patients during a mean follow-up period of over five years. Seventy percent of those patients with hypercalciuria alone proved to have metabolically inactive stone disease. The existence of metabolically active stone formation should be proven before a patient is committed to lifelong pharmacologic therapy (Smith, 1983a).

RENAL TUBULAR ACIDOSIS

The metabolic abnormalities of patients with distal renal tubular acidosis are corrected with replacement of sodium, potassium, and bicarbonate. Daily, 90–150 mEq of base is usually required and may be given as sodium bicarbonate or citrate (Bicitra, Polycitra, or Urocit-K). Total body potassium is low in untreated patients, and the serum potassium may fall as the acidosis is corrected. Potassium should be replaced while monitoring serum levels. Urinary calcium excretion will decrease, and urinary citrate will rise to a normal level with correction of the systemic acidosis. If stone formation does not cease with an adequate program of fluids and electrolyte replacement, neutral orthophosphate may be added to provide 1.5–2.0 gm of elemental phosphorus per day.

CYSTINURIA

The goal of therapy is to reduce the urinary supersaturation of cystine. Twenty-four-hour urine volume should be maintained at 3–4 L. The solubility of cystine in normal urine is approximately 300 mg/L but increases as the pH increases (see Fig 9–4). If the 24-hour urinary cystine excretion of a patient is known, the information in Figure 9–4 can be used to calculate the urine volume and pH required to solubilize adequately all of the cystine. The pH usually must be raised to 7.5–7.8. This degree of alkalinity should be maintained with a citrate preparation that provides 15–30 mEq of base four times daily. Citrate is preferable to sodium bicarbonate, because a uniform alkalinization will be achieved. It may be helpful to have patients monitor their urinary pH with pH paper.

A carbonic anhydrase inhibitor such as acetazolamide occasionally has been used to maintain urinary alkalinity throughout the night. These agents probably should be avoided, because they induce changes in urinary composition that favor the precipitation of calcium phosphate. If a cystine stone becomes covered with a layer of calcium phosphate, further attempts at dissolution will be fruitless.

Cystine is two cysteine molecules linked by a disulfide bond (see Fig 9–3). The drug D-penicillamine (Cu-

primine) will react with cystine to form penicillamine-cysteine, a mixed disulfide that is more soluble than cystine in urine. The dose of D-penicillamine is 250–500 mg four times daily. It is given 30 minutes before meals and at bedtime. D-penicillamine is a potentially toxic drug that should be used for attempted stone dissolution or in patients whose disease cannot be controlled with hydration and alkalinization. Adverse reactions include skin rashes, fever, arthralgias, and lymphadenopathy. A potential pyridoxine deficiency can be avoided with prophylactic pyridoxine (50 mg twice daily) therapy. Stone dissolution may require several months to one or two years. Urinary cystine excretion can be lowered by limiting dietary methionine. This therapy severely limits protein intake and is rarely, if ever, used.

HYPERCALCEMIC DISORDERS

The primary cause of the hypercalcemia will determine the therapy. Since primary hyperparathyroidism is the most common hypercalcemic disorder responsible for stone formation, surgical correction will usually prevent further stone formation. If surgical treatment of a stone is contemplated in a patient with primary hyperparathyroidism, a parathyroidectomy should be the first procedure. New stones may grow in the postoperative recovery period if stone removal is accomplished before cervical exploration. Patients should be followed up carefully after parathyroidectomy to ensure that the hypercalcemia has been corrected and stone formation has ceased. Persistent stone growth may be caused by another disorder. Concurrent ICU may be present in a patient with primary hyperparathyroidism.

An adequate intake of fluids should be encouraged in all immobilized patients. Oral orthophosphates will decrease urinary calcium excretion in immobilized patients and may be used if fluid therapy is unsuccessful (Goldsmith et al., 1969). Use of corticosteroids should reduce the hypercalciuria in patients with sarcoidosis.

URIC ACID LITHIASIS

The medical treatment of uric acid lithiasis is very satisfying, because uric acid stones can be dissolved. Fluid intake should be increased to achieve a 24-hour urine output of 3 L. The solubility of uric acid can be increased tenfold by raising the urine pH to 6.5 (see Fig 9–6). A sodium bicarbonate or citrate preparation can be used to accomplish this. It may be helpful if the patient monitors his urine pH with nitrazine paper. At a daily dose of 300 mg, allopurinol, a xanthine oxidase inhibitor (see Figs 9–5 and 9–8), will reduce the amount of uric acid excreted in the urine.

When all three elements of this program are used,

most uric acid stones can be dissolved within three months. The allopurinol can be stopped after the stones are dissolved. New stone formation usually can be prevented by maintaining an alkaline urine. Overexuberant alkalinization may permit the precipitation of a layer of calcium phosphate over the stone. Dissolution then becomes impossible. Use of carbonic anhydrase inhibitors should be avoided for the same reason.

XANTHINURIA

Patients who form xanthine calculi while taking allopurinol should discontinue use of this drug. Patients with xanthinuria secondary to an inherited deficiency of xanthine oxidase should maintain a high urine volume and restrict their dietary intake of purine-containing foods. Since the pK_a of the first dissociable proton of xanthine is 7.7, the solubility of xanthine cannot be increased significantly by physiologic alkalinization.

2,8-DIHYDROXYADENINURIA

Xanthine oxidase is responsible for the oxidation of adenine to 2,8-DHA (Gault et al., 1981). Treatment consists of fluids and allopurinol without urinary alkalinization. The solubility of 2,8-DHA is not affected by urinary pH. Dietary adenine may contribute to stone formation and should be reduced to a minimum. Lentils and other grains have a high adenine content (Clifford and Story, 1976).

PRIMARY HYPEROXALURIA

Large doses of pyridoxine (50 mg four times daily) will reduce oxalate excretion in 20%–50% of patients (Gibbs and Watts, 1970). Neutral orthophosphate (1.5–2.0 gm/24 hr of elemental phosphorus in four divided doses) may halt the growth of existing stones and prevent the formation of new stones (Smith et al., 1969). Renal transplantation has been attempted in patients with primary hyperoxaluria and renal failure, but oxalate mobilization from preexistent oxalosis may lead to deposition of oxalate in the transplanted kidney (David et al., 1983). Transplantation should be done soon after the onset of renal failure, because oxalate is not removed efficiently by hemodialysis.

ENTERIC HYPEROXALURIA

Dietary intake of oxalate and fat should be restricted. An additional advantage of a low-fat (50-gm) diet is that the bothersome steatorrhea will be reduced. Dietary calcium supplementation has been used to increase precipitation of calcium oxalate within the GI lumen (Stauffer et al., 1974), but this may increase urinary calcium. The stone disease usually can be controlled without resorting to calcium supplementation. Cholestyramine (12 gm daily in three or four divided doses) will bind acidic compounds, including oxalate, in the colonic lumen. Oxalate absorption will decrease, steatorrhea will decrease, and water absorption and urine volume may increase. Intestinal bicarbonate loss will reduce urinary pH and citrate excretion. Some patients may even have a mild metabolic acidosis. These abnormalities can be corrected with base replacement (preparations of bicarbonate or citrate). If metabolically active stone formation persists in patients with an ileal bypass, then the normal anatomy of the GI tract should be restored (Dickstein and Frame, 1973).

SECONDARY UROLITHIASIS

Surgical removal of a struvite stone is almost always necessary to preserve renal function and reduce long-term morbidity and potential mortality (Resnick, 1981). Urine cultures should be obtained and specific bactericidal therapy should be started 48 hours before surgery. Urine from the renal pelvis should be cultured at the time of surgery, all infected stone material should be removed, and the stone should be cultured. Antibiotic treatment should be continued for 10–14 days after surgery. Postoperative irrigation of the collecting system with hemiacidrin has been advocated to dissolve any minute retained stone fragments (Nemoy and Stamey, 1971).

The patients should be maintained on antibacterial prophylaxis for 3–12 months. Adjunctive urinary acidification with ammonium chloride (2 gm/day) has been used in conjunction with long-term antimicrobial therapy (Zinsser et al., 1968). Small retained fragments may even dissolve with this regimen. The proper antibacterial agents will suppress bacterial growth, and because of the decrease in urease production, urinary acidification can be achieved with the ammonium chloride.

Acetohydroxamic acid, a urease inhibitor, may prevent the further growth of struvite stones and rarely may lead to dissolution of the stone material (Williams et al., 1984). Up to 50% of patients may experience side effects from the drug: tremulousness, headache, or deep-vein thrombosis.

The Shorr regimen combines a low-phosphorus diet with aluminum hydroxide capsules to achieve selective dietary phosphorus depletion (Lavengood and Marshall, 1972). A few uncontrolled studies suggest that the Shorr regimen is an effective therapy for struvite urolithiasis, but the diet is difficult to follow. Low-phosphorus diets also increase urinary calcium excretion and promote crystalluria (Werness et al., 1981).

Most investigators report recurrent stone growth in 25%–30% of patients (Griffith, 1978). If recurrent metabolic stone formation is excluded, recurrent struvite stone formation occurs in 10%–15% of patients. The lowest reported recurrence rate was only 2% (Silverman and Stamey, 1983). The standard treatment was open surgical removal of the stone material, appropriate antimicrobial therapy, and selected postoperative irrigation with hemiacidrin. Alternatives to open surgical removal have been developed, including primary percutaneous dissolution with hemiacidrin irrigation (Dretler and Pfister, 1984), percutaneous ultrasonic nephrolithotripsy (Smith, 1984), and ESWL. The combination of percutaneous and extracorporeal lithotripsy is receiving the greatest attention (Kahnoski et al., 1986), but recurrence data are not available. One study examined the recurrence rates of bladder calculi following three different methods of treatment: vesicolithotomy, litholapaxy, and electrohydraulic lithotripsy (Short et al., 1984). Suprapubic vesicolithotomy was associated with fewer recurrences and had a much longer stone-free interval compared with litholapaxy or electrohydraulic lithotripsy. It remains to be shown that these newer techniques are better than or even comparable to standard surgical methods in the treatment of struvite urolithiasis.

When patients are free of infection, they should be evaluated for the presence of an underlying metabolic disorder. The prevention of further stone formation requires the identification and treatment of any such disorder.

Patients with obstructive uropathy or medullary sponge kidney should have a metabolic evaluation. A more basic disorder may be responsible for the stone formation. If drugs or their metabolites are found on a stone analysis, administration of the responsible drug should be discontinued or its dose reduced.

IDIOPATHIC CALCIUM UROLITHIASIS

Patients who consume adequate amounts of fluid and in whom dietary excesses have been eliminated may continue to have metabolically active stone formation. Several effective treatment programs are available.

Thiazide Diuretics

Thiazides are particularly effective when significant hypercalciuria is present. Hydrochlorothiazide, 50 mg twice daily, or trichlormethiazide, 2 mg twice daily, will reduce urinary calcium excretion, crystalluria, and urinary supersaturation for calcium oxalate and calcium phosphate (Yendt and Cohanim, 1985). The mechanism of action is thought to be stimulation of renal tubular calcium reabsorption by the extracellular volume contraction. A high-sodium diet can blunt or prevent the hypocalciuric effect of thiazides. Moderate sodium restriction may be needed to reduce urinary calcium excretion. Up to 90% of patients who are treated with thiazides will cease further stone formation.

Thiazides also may prevent stone formation in nor-

mocalciuric persons. This is a controversial finding but may be related to a reported reduction in oxalate excretion after long-term thiazide therapy (Cohanim and Yendt, 1980; Yendt and Cohanim, 1986). Side effects occur in up to 10% of patients: fatigue unrelated to the hypokalemia, hypomagnesemia, muscle weakness and cramping, decreased libido, impotence, and abnormalities in serum calcium, glucose, and uric acid.

Orthophosphate

Orthophosphate is the treatment of choice in patients with normocalciuria (Smith, 1983a). It also is effective in patients with hypercalciuria. Orthophosphate decreases urinary calcium excretion and increases inhibitor activity (Wilson et al., 1985). Increased calcium complexation and decreased free calcium ion activity are a result of increased urinary phosphate, citrate, and pH. Calcium oxalate supersaturation decreases, but the supersaturation of hydroxyapatite or brushite does not change. Excretion of the inhibitors pyrophosphate and citrate increases, and the more alkaline pH increases the potency of pyrophosphate as a crystal growth inhibitor. Orthophosphate also may reduce 1,25-dihydroxyvitamin D synthesis (Van dan Berg et al., 1980).

The drug is usually given as the neutral or mildly alkaline salt. The total daily dose should provide 1.5–2.0 gm of elemental phosphorus. Patients will lose previously formed stone material during the first 3–6 months of therapy. If radiographs document that stone mass is being lost and no new stone material is precipitating, patients should be reassured that the medication is preventing further stone growth and that the troublesome stones were formed before the drug therapy was started. Calcium stone formation will cease in 90% of patients (Thomas, 1978).

Diarrhea is the most common complication of orthophosphate therapy. The dose may be halved until the diarrhea subsides and then gradually increased to a therapeutic level. Orthophosphates should not be used in the presence of secondary urolithiasis due to infection or obstruction or in the presence of renal failure (GFR < 30 ml/min).

Cellulose Phosphate

The purpose of cellulose phosphate therapy, unlike orthophosphate therapy, is not to provide absorbable phosphate. Cellulose phosphate binds calcium in the intestinal lumen and decreases calcium absorption (Pak et al., 1974). Cellulose phosphate should be used only in patients with absorptive hypercalciuria, because it may cause a negative calcium balance in those patients with normal intestinal calcium absorption or renal hypercalciuria. The usual dose is 5 gm two or three times daily.

When given alone, cellulose phosphate may be ineffective or even increase calcium oxalate crystalluria (Backman et al., 1980). Urinary oxalate excretion will increase with cellulose phosphate therapy alone, probably because intestinal absorption increases. Less calcium is available to complex dietary oxalate in the intestinal lumen. Cellulose phosphate also will bind magnesium and decrease urinary magnesium excretion. Supplementation of cellulose phosphate therapy with magnesium (1–1.5 gm magnesium gluconate per day), a low-oxalate diet, and a high fluid intake will effectively halt stone formation in almost 80% of patients (Pak, 1981).

Magnesium

Magnesium oxide (Melnick et al., 1971) and magnesium hydroxide (Danielson, 1985) have been advocated for the treatment of ICU, but the effectiveness of this therapy has not been demonstrated systematically. (The same criticism may be applied to clinical trials of thiazides and orthophosphates (Churchill, 1985). Anticipated benefits of increased magnesium excretion are increased complexation of oxalate and phosphate and a slight increase in crystal growth inhibition. Some investigators have reported prevention of recurrent stone formation in 80% of treated patients. Approximately 1 gm of magnesium oxide or 250–750 mg of magnesium hydroxide have been given in two or three divided daily doses.

One group of investigators from the southeastern United States suggested that their patients might have had a magnesium deficiency because the soil in this area of the country is low in magnesium (Melnick et al., 1971). Urinary magnesium excretion was lower during the initial period of magnesium therapy and increased as the length of treatment increased. Replenishment of depleted magnesium stores may have been responsible for this phenomenon.

Allopurinol

The use of this drug in patients with hyperuricosuric calcium urolithiasis is controversial (Finlayson et al., 1985). The usual daily dose is 300 mg. Early reports were very encouraging (Coe and Kavalach, 1974), but later reports have been less enthusiastic (Miano et al., 1985). Two adverse effects of hyperuricosuria have been proposed: heterogeneous nucleation of calcium oxalate on urate crystals and binding of large molecular weight calcium oxalate crystal growth inhibitors to colloidal urate (Fellstrom, 1985). An allopurinol-induced reduction of urate excretion could preclude either mechanism. Allopurinol has even been reported to decrease urinary oxalate excretion (Scott et al., 1978). Allopurinol does not appear to have a direct effect on calcium oxalate precipitation. The most conservative approach may be to use allopurinol as a secondary drug in those patients who have abnormal uric acid metabolism.

Potassium Citrate

Potassium citrate has previously been available in a liquid preparation with a mixture of sodium citrate or citric acid. The recent introduction of a tablet containing only potassium citrate (Urocit-K, Mission Pharmacal, 5 mEq potassium citrate per tablet) may be more convenient and may avoid the disadvantages of those preparations that contain sodium.

This compound has been used to treat patients with ICU and hypocitraturia (Pak and Fuller, 1986). The alkalinizing effects of citrate will increase urinary pH and urinary citrate excretion. Complexation of calcium with citrate will increase, and ionized calcium and calcium oxalate supersaturation will decrease. Inhibition of calcium oxalate crystal growth will increase slightly (increased citrate excretion and greater potency of pyrophosphate at a more alkaline urinary pH). Further stone formation ceased in almost 90% of ICU patients with hypocitraturia. Citrate excretion rose to the normal range, while urinary pH was maintained at 6.5–7.0.

Several liquid citrate preparations of citrate contain sodium. The effects of sodium citrate and potassium citrate on urine composition have been compared (Sakhaee et al., 1983). Urinary pH and citrate excretion increased with both compounds, but calcium excretion decreased only with potassium citrate. The higher urinary pH increased brushite (calcium phosphate) saturation in both groups of patients, but a supersaturated state was obtained only in those patients treated with sodium citrate. The failure to reduce urinary calcium excretion may have contributed to this latter effect.

Long-term treatment with potassium citrate (20 mEq three times daily) reduced stone formation in 98% of patients, and stone formation ceased in 80% (Pak et al., 1985). Minor GI complaints were the most common side effects. No patients had melena or occult fecal blood. Although potassium citrate therapy of ICU is becoming more popular, not all investigators think that it is more beneficial than thiazides or even a placebo (Park and Spector, 1986).

Investigational Agents

Some patients with ICU and mild hyperoxaluria have been treated with pharmacologic doses of pyridoxine (200–400 mg/day) (Balcke et al., 1983). Oxalate excretion fell by approximately one third. The proposed mechanism is stimulation of pyridoxal-5-phosphate-dependent transaminases that are responsible for the conversion of glyoxalate to glycine (see Fig 9–10). Less glyoxalate would be available for conversion to oxalate. Pyridoxine also decreased urinary glycolate excretion and increased erythrocyte glutamic oxaloacetic transaminase activity. Individuals who consume very large quantities of pyridoxine (2–6 gm/day) may develop a sensory neuropathy (Schaumburg et al., 1983).

Pentosan polysulfate is a structural analogue of heparin but has very little anticoagulant activity (Ryde et al., 1981). Up to 4% of an oral dose is excreted in the urine. This agent was used initially to treat patients with interstitial cystitis (Parsons et al., 1983). Pentosan polysulfate also is a potent inhibitor of calcium oxalate crystal growth (Martin et al., 1984). A 50% reduction in the rate of calcium oxalate crystal growth was achieved with a concentration of 2.4×10^{-9}M. A daily dose of 300 mg should provide a urinary concentration of 10 mg/L and increase inhibition by one third.

More recently, sodium thiosulfate has been used to treat patients with ICU (Yatzidis, 1985). Increased urinary sulfate excretion will theoretically retard stone formation by reducing calcium ion activity and calcium oxalate supersaturation. These investigators found that urinary ionized calcium, as measured with an ion-selective electrode, decreased when urinary thiosulfate increased. Calcium thiosulfate solubility was three orders of magnitude greater than the solubility of calcium citrate. Reduction of calcium oxalate supersaturation may be achieved more efficiently by increasing urinary sulfate excretion than by increasing urinary citrate excretion. The role of each of these investigational agents in the treatment of ICU has yet to be defined.

BLADDER CALCULI

Bladder calculi are commonly found in children who live in lesser developed countries and in adult males with bladder outlet obstruction. Bladder stones are usually freely movable in the bladder but may be fixed to a bladder wall suture placed during a previous surgical procedure. The usual symptoms of bladder outlet obstruction may be present: urinary hesitancy, frequency, and nocturia. Patients may also have hematuria, dysuria, suprapubic pain that frequently radiates to the tip of the penis, and an interrupted urinary stream.

The diagnosis of a bladder stone is usually confirmed by radiographic examination or cystoscopy. Since many bladder stones are composed of radiolucent uric acid, the absence of a radiopaque shadow on a plain abdominal radiograph does not exclude the presence of a stone. An accurate diagnosis usually requires cystoscopy.

PROSTATIC CALCULI

The majority of prostatic calculi are found in men 50–65 years of age. Prostatic calculi are formed by the deposition of calcium salts on corpora amylacea. Corpora amylacea are laminated organic structures that are thought to form around desquamated epithelial cells in prostatic alveoli. Prostatic calculi may be associated with prostatic hyperplasia or prostatitis or may be asymptomatic. They may be seen on a plain roentgenogram of the pelvis and frequently can be palpated by rectal examination. Asymptomatic prostatic calculi require no treatment, but they often are removed during a transurethral prostatectomy for benign prostatic hyperplasia.

CHILDHOOD UROLITHIASIS

Bladder stone disease was very common in children in 18th century Europe (Rose, 1982). Boys younger than 10 years of age typically were affected. The prevalence of pediatric bladder stone disease became less common as Europe became industrialized. The occurrence of bladder stones in children was virtually unknown in Norway by 1831 (Salliner, 1959/1960). The disappearance of this disease is probably related to the concurrent improvement in nutrition. The disease still exists in less well-developed nations in Asia and the Middle East.

Bladder stone disease has been studied extensively in Thailand (Valyasevi and Dhanamitta, 1977; Dhanamitta et al., 1977). Boys younger than 10 years of age usually are affected. The stones are composed of a mixture of calcium oxalate and ammonium acid urate. Concurrent renal lithiasis typically is not present. A low intake of breast milk and early rice supplementation provide a diet low in protein and minerals. Urinary excretion of phosphate, sulfate, sodium, potassium, and magnesium is low. Urinary excretion of oxalate, calcium, uric acid, and ammonia is high. The high oxalate excretion is thought to be due to consumption of local vegetables and plant leaves that have a high oxalate content. The hot climate and low fluid intake also contribute to crystallization.

Two to three percent of patients with urinary calculi are children (Vahlensieck and Bastian, 1976). Bladder stone disease is now uncommon in Europe and the United States. Most children with urolithiasis present with renal or ureteral calculi. A majority of patients will have stones composed of calcium oxalate, calcium phosphate, or mixtures of these stone salts (Malek and Kelalis, 1975; Sinno et al., 1979). Most children with urolithiasis will have bacteriuria that is secondary to stone formation, but most of these organisms do not produce urease. One study found struvite stone formation in one third of their patients (Malek and Kelalis, 1975). All had a history of multiple urologic procedures, diversionary procedures, or indwelling drainage devices. *Proteus* was the most common bacterial organism. Other investigators have found a lower incidence of infection-induced stone formation.

The spectrum of etiologies is the same in children as it is in adults. One third to one half of patients will have ICU. One study reported finding primary hyperparathyroidism in 6% of their pediatric patients (Malek and Kelalis, 1975), but most authors think that hyperparathyroidism as a cause of urinary stone formation in children is extremely rare. Some of the discrepancy may be explained by the rarity of any kind of urolithiasis in children. Other metabolic causes of stone formation in children are distal renal tubular acidosis, cystinuria, and primary hyperoxaluria. Inflammatory bowel disease also is responsible for recurrent calcium oxalate stone formation in children (Clark et al., 1985). The pathophysiology is the same as that for enteric hyperoxaluria in adults.

Classic renal colic occurs in a minority of children. Gross or microscopic hematuria is the most frequent clinical manifestation, followed by flank or back pain, nausea and vomiting, or irritative bladder symptoms. Some children may simply pass a stone.

Episodic gross hematuria may precede overt urolithiasis in children with hypercalciuria (Stapleton et al., 1984). Hypercalciuria was present in approximately 40% of children with gross hematuria but without urinary infection or proteinuria. Many children had concurrent abdominal or suprapubic pain, dysuria, and urinary frequency. Three fourths of the children with hypercalciuria had a family history of urolithiasis. The hematuria resolved during anticalciuric therapy with hydrochlorothiazide or dietary calcium restriction. The long-term risk of clinical stone formation in these children is not known.

REFERENCES

1. Arnaud CD, Wilson DM, Smith LH: Primary hyperparathyroidism, renal lithiasis and the measurement of parathyroid hormone in serum by radioimmunoassay, in Cifuentes Delatte L, Rapado A, Hodgkinson A (eds): *Urinary Calculi—Recent Advances in Aetiology, Stone Structure and Treatment*. Basel, S Karger AG, 1973, pp 346–353.
2. Asper R, Schmucki O: Socio-economic aspects of urinary stone disease in Eurasia in the 19th and 20th century, in Ryall R, Brockis JG, Marshall V, et al (eds): *Urinary Stone*. New York, Churchill Livingstone, 1984, pp 18–23.
3. Atsmon A, deVries A, Frank M: *Uric Acid Lithiasis*. Amsterdam, Elsevier, 1963.
4. Backman U, Danielson BG, Johansson G, et al: Treatment of recurrent calcium stone formation with cellulose phosphate. *J Urol* 1980; 123:9–13.
5. Backman U, Danielson BG, Johanssen G, et al: Incidence and clinical importance of renal tubular defects in recurrent renal stone formers. *Nephron* 1980; 25:96–101.
6. Baggio B, Gambaro G, Marchini F, et al: An inheritable anomaly of red-cell oxalate transport in "primary" calcium nephrolithiasis correctable with diuretics. *N Engl J Med* 1986; 314:599–604.
7. Baggio B, Gambaro G, Favaro S, et al: Prevalence of hyperoxaluria in idiopathic calcium oxalate kidney stone disease. *Nephron* 1983; 35:11–14.
8. Baggio B, Gambaro G, Marchini, G, et al: Raised transmembrane oxalate flux in red blood cells in idiopathic calcium oxalate nephrolithiasis. *Lancet* 1984; 2:12–13.
9. Balcke P, Schmidt P, Zazgornik J, et al: Pyridoxine therapy in patients with renal calcium oxalate calculi. *Proc Eur Dial Transplant Assoc* 1983; 20:417–421.
10. Bataille P, Pruna A, Gregoire I, et al: Critical role of oxalate restriction in association with calcium restriction to decrease the probability of being a stone former: insufficient effect in idiopathic hypercalciuria. *Nephron* 1985; 39:321–324.
11. Batlle DC, Shey JT, Roseman MK et al: Clinical and pathophysiologic spectrum of acquired distal renal tubular acidosis. *Kidney Int* 1981; 20:389–396.
12. Batlle DC, Arruda JAL, Kurtzman NA: Hyperkalemic distal renal tubular acidosis associated with obstructive uropathy. *N Engl J Med* 1981; 394:373–380.
13. Beischer DE: Analysis of renal calculi by infrared spectroscopy. *J Urol* 1955; 73:653–659.
14. Boyce WH: Organic matrix of human urinary concretions. *Am J Med* 1985; 45:673–683.
15. Boyce WH, Garvey FK, Strauscutter HE: Incidence of urinary calculi among patients in general hospitals, 1948 to 1952. *JAMA* 1956; 161:1437–1442.
16. Boyce WH: Proteinuria in kidney calculous disease, in Manuel Y, Revillard JP, Betuel H (eds): *Proteins in Normal and Pathological Urine*. Baltimore, University Park Press, 1970, pp 235–243.
17. Boyce WH, King JS Jr, Fielden ML: Total nondialyzable solids (TNDS) in human urine: XIII. Immunological detection of a component peculiar to renal calculous matrix and to urine of calculous patients. *J Clin Invest* 1962; 41:1180–1189.
18. Brannan PG, Morawski S, Pak CYC, et al: Selective jejunal hyperabsorption of calcium in absorptive hypercalciuria. *Am J Med* 1979; 66:425–428.
19. Brennen RJ, Spring DB, Sebastian A, et al: Incidence of radiographically evident bone disease, nephrocalcinosis, and nephrolithiasis in various types of renal acidosis. *N Engl J Med* 1982; 307:217–221.
20. Brock WA, Golden J, Kaplan GW: Xanthine calculi in the Lesch-Nyhan syndrome. *J Urol* 1983; 130:157–159.
21. Buckalew VM Jr, Purvis ML, Shulman MG, et al: Hereditary renal tubular acidosis. *Medicine* 1974; 53:229–254.
22. Buckalew VM Jr, McCurdy DK, Ludwig GD, et al: Incomplete renal tubular acidosis. *Am J Med* 1968; 45:32–42.
23. Burns JR, Finlayson B: Why some people have stone disease and others do not, in Roth RA, Finlayson B (eds): *Stones: Clinical Management of Urolithiasis*. Baltimore, Williams & Wilkins, 1983, pp 3–7.

24. Burns JR, Finlayson B: The effect of seed crystals on calcium oxalate nucleation. *Invest Urol* 1980; 18:133–136.

25. Burns JR, Finlayson B, Gauthier J: Calcium oxalate retention in subjects with crystalluria, in Ryall R, Brockis JG, Marshall V, et al (eds): *Urinary Stone.* New York, Churchill Livingstone, 1984, pp 253–257.

26. Burr RG, Nuseibeh I, Abiaka CD: Biochemical studies in paraplegic renal stone patients: Renal excretion of citrate, inorganic pyrophosphate, silicate, and urate. *Br J Urol* 1985; 57:275–278.

27. Carr RJ: Aetiology of renal calculi: Micro-radiographic studies, in Hodgkinson A, Nordin BEC (eds): *Renal Stone Research Symposium.* London, J & A Churchill, 1969, pp 123–132.

28. Churchill DN, Morgan J, Gault MH: Tea drinking—a risk factor for urolithiasis? in Schwille PO, Smith LH, Robertson WG, et al (eds): *Urolithiasis and Related Clinical Research.* New York, Plenum Publishing Corp, 1981, pp 789–794.

29. Churchill DN: Appraisal of methodology in studies of either thiazide or orthophosphate therapy for recurrent calcium urolithiasis, in Schwille PO, Smith LH, Robertson WG, et al (eds): *Urolithiasis and Related Clinical Research.* Plenum Publishing Corp, New York, 1985, pp 479–481.

30. Churchill D, Bryant D, Fodor G, et al: Drinking water hardness and urolithiasis. *Ann Intern Med* 1978; 88:513–514.

31. Churchill DN, Black DP, Maloney CM, et al: Urine chemistry in renal stone formers in an area with soft drinking water, in Smith LH, Robertson WG, Finlayson B (eds): *Urolithiasis—Basic and Clinical Research.* Plenum Publishing Corp, New York, 1981, pp 349–352.

32. Clark JH, Fitzgerald JF, Bergstein JM: Nephrolithiasis in childhood inflammatory bowel disease. *J Pediatr Gastroenterol Nutr* 1985; 4:829–834.

33. Clifford AJ, Story DL: Levels of purines in foods and their metabolic effects in rats. *J Nutr* 1976; 106:435–442.

34. Coe F: Uric acid and calcium oxalate nephrolithiasis. *Kidney Int* 1983; 24:392–403.

35. Coe FL, Lawton RL, Goldstein RB, et al: Sodium urate accelerates precipitation of calcium oxalate in vitro. *Proc Soc Exp Biol Med* 1975; 149:926–939.

36. Coe FL, Canterbury JM, Firpo JJ, et al: Evidence for secondary hyperparathyroidism in idiopathic hypercalciuria. *J Clin Invest* 1973; 52:134–142.

37. Coe FL, Kavalach AG: Hypercalciuria and hyperuricosuria in patients with calcium nephrolithiasis. *N Engl J Med* 1974; 291:1344–1350.

38. Coe FL, Favus MJ, Crockett T, et al: Effects of low-calcium diet on urine calcium excretion, parathyroid function and serum $1,24(OH)_2D_3$ levels in patients with idiopathic hypercalciuria and in normal subjects. *Am J Med* 1982; 72:25–32.

39. Cohanim M, Yendt ER: Reduction of urine oxalate during therapy in patients with calcium urolithiasis. *Invest Urol* 1980; 18:170–173.

40. Crawhall JC, Watts RWE: Cystinuria. *Am J Med* 1968; 45:736–755.

41. Crawhall JC, Purkiss P, Watts RWE, et al: The excretion of amino acids by cystinuric patients and their relatives. *Ann Hum Genet* 1969; 33:149–169.

42. Dahlberg PJ, Van den Berg CJ, Kurtz SB, et al: Clinical features and management of cystinuria. *Mayo Clin Proc* 1977; 52:533–542.

43. Dalton DL, Hughes J, Glenn JF: Foreign bodies and urinary stones. *Urol* 1975; 6:1–5.

44. Danielson BG: Drugs against kidney stones: Effects of magnesium and alkali, in Schwille PO, Smith LH, Robertson WG, et al (eds): *Urolithiasis and Related Clinical Research.* Plenum Publishing Corp, New York, 1985, pp 525–532.

45. Daudon M, Reveillalud RJ: Drug nephrolithiasis: An unrecognized pathology, in Schwille PO, Smith LH, Robertson WG, et al (eds): *Urolithiasis and Related Clinical Research.* Plenum Publishing Corp, New York, 1985, pp 371–374.

46. David DS, Cheigh JS, Stenzel KH, et al: Successful renal transplantation in a patient with primary hyperoxaluria. *Transplant Proc* 1983; 15:2168–2171.

47. Dent CE, Senior B: Studies on the treatment of cystinuria. *Br J Urol* 1955; 27:317–332.

48. Dent DF, Sutor DJ: Presence or absence of inhibitor of calcium-oxalate crystal growth in urine of normals and of stone formers. *Lancet* 1971; 2:775–778.

49. Dent CE, Philpot GR: Xanthinuria: An inborn error (or deviation) of metabolism. *Lancet* 1954; 1:182–185.

50. DeVivo MJ, Fine PR, Cutler GR et al: The risk of renal calculi in spinal cord injury patients. *J Urol* 1984; 131:857–860.

51. Dhanamitta S, Valyasevi A, Susilavorn B: Research report on bladder stone disease, Thailand, in van Reem R (ed): *Idiopathic Urinary Bladder Stone Disease.* Washington, DC, Department of Health, Education, and Welfare, publication 77–1063, 1977, pp 151–171.

52. Dickstein SS, Frame B: Urinary tract calculi after intestinal shunt operations for the treatment of obesity. *Surg Gynecol Obstet* 1973; 136:257–260.

53. Dobbins JW, Binder HJ: Importance of the colon in enteric hyperoxaluria. *N Engl J Med* 1977; 296:298–301.

54. Dobbins JW: Nephrolithiasis and intestinal disease. *J Clin Gastroenterol* 1985; 7:21–24.

55. Dobbins JW, Binder HJ: Effect of bile salts and fatty acids on the colonic absorption of oxalate. *Gastroenterology* 1976; 70:1096–1100.

56. Dretler SP, Pfister RC: Primary dissolution therapy of struvite calculi. *J Urol* 1984; 131:861–863.

57. Dretler SP: The pathogenesis of urinary tract calculi occurring after ileal conduit diversion: I. Clinical study. II. Conduit study. III. Prevention. *J Urol* 1973; 109:204–209.

58. Earnest DL, Johnson G, Williams HE, et al: Hyperoxaluria in patients with ileal resection: An abnormality in dietary oxalate absorption. *Gastroenterology* 1974; 66:1114–1122.

59. Fellstrom B: Urate metabolism and renal calcium stone

disease. *Scand J Urol Nephrol [Suppl]* 1981; 62:1–59.

60. Fellstrom B: Allopurinol treatment in urolithiasis, in Schwille PO, Smith OH, Robertson WG, et al (eds): *Urolithiasis and Related Clinical Research.* Plenum Publishing Corp, New York, 1985, pp 505–512.

61. Fellstrom B, Danielson BG, Karlstrom B, et al: Effects of high intake of dietary animal protein on mineral metabolism and urinary supersaturation of calcium oxalate in renal stone formers. *Br J Urol* 1984; 56:263–269.

62. Finlayson B, Smith A: Stability of first dissociable proton of uric acid. *J Chem Eng Data* 1974; 19:94–97.

63. Finlayson B: Physicochemical aspects of urolithiasis. *Kidney Int* 1978; 13:344–360.

64. Finlayson B, Vermeulen CW, Stewart EJ: Stone matrix and mucoprotein from urine. *J Urol* 1961; 86:355–363.

65. Finalyson B, Reid F: The expectation of free and fixed particles in urinary stone disease. *Invest Urol* 1978; 15:442–448.

66. Finlayson B, Newman RC, Hunter PT: The role of urate and allopurinol in stone disease, in Schwille PO, Smith LH, Robertson WG, et al (eds): *Urolithiasis and Related Clinical Research.* Plenum Publishing Corp, New York, 1985, pp 499–503.

67. Finlayson B, Smith A: Stability of first dissociable proton of uric acid. *J Chem Eng Data* 1974; 19:94–97.

68. Finlayson B: Calcium stones: Some physical aspects, in David DS (ed): *Calcium Metabolism in Renal Failure and Nephrolithiasis.* New York, John Wiley & Sons, 1977, pp 337–382.

69. Fleisch H: Inhibitors and promoters of stone formation. *Kidney Int* 1978; 13:361–371.

70. Fleisch H: Round table discussion on the comparison of models for the study of inhibitory activity in urine, in Schwille PO, Smith LH, Robertson WG, et al (eds): *Urolithiasis and Related Clinical Research.* New York, Plenum Publishing Corp, 1985, pp 903–910.

71. Fowler JE: Bacteriology of branched renal calculi and accompanying urinary tract infection. *J Urol* 1984; 131:213–215.

72. Garside J: Nucleation, in Nancollas GH (ed): *Biological Mineralization and Demineralization.* New York, Springer-Verlag, 1982, pp 23–35.

73. Gault MH, Simmonds HA, Sneeden W, et al: Urolithiasis due to 2,8-dihydroxyadenine in an adult. *N Engl J Med* 1981; 305:1570–1572.

74. Gault MH, O'Toole T, Wilson JM, et al: Urolithiasis in a large kindred deficient in adenine phosphoribosyl transferase (APRT). in Schwille PO, Smith LH, Robertson WG, et al (eds): *Urolithiasis and Related Clinical Research.* New York, Plenum Publishing Corp, 1985, pp 9–12.

75. George RJ, Politzer WM: A new method for the detection of cystine and its mechanism. *Clin Chim Acta* 1970; 30:737–740.

76. Gibbs DA, Watts RWE: The action of pyridoxine in primary hyperoxaluria. *Clin Sci* 1970; 38:277–286.

77. Gill WB, Silvert MA, Roma MJ: Supersaturation levels and crystallization rates from urines of normal humans and stone-formers determined by a ^{14}C-oxalate technique. *Invest Urol* 1974; 12:203–209.

78. Goldsmith RS, Killian P, Ingbar SH, et al: Effect of phosphate supplementation during immobilization of normal men. *Metabolism* 1969; 18:349–368.

79. Granberg PO, Cedermark B, Farnebo LO, et al: Parathyroid tumors. *Curr Prob Cancer* 1985; 9:1–52.

80. Gray RW, Wilz DR, Caldas AE, et al: The importance of phosphate in regulating plasma 1,25-(OH)-vitamin D levels in humans: Studies in healthy subjects, in calcium-stone formers and in patients with primary hyperparathyroidism. *J Clin Endocrinol Metab* 1977; 45:299–306.

81. Gremillion DH, Kee JW, McIntosh DA: Hyperparathyroidism and medullary sponge kidney—a chance relationship? *JAMA* 1977; 237:799–800.

82. Griffith DP: Struvite stones. *Kidney Int* 1978; 13:372–382.

83. Griffith DP, Musher DM, Itin C: Urease: The primary cause of infection-induced urinary stones. *Invest Urol* 1976; 13:346–350.

84. Gutman AB, Yu T-F: Urinary ammonia excretion in primary gout. *J Clin Invest* 1965; 44:1474–1481.

85. Gutman AB, Yu T-F: Uric acid nephrolithiasis. *Am J Med* 1968; 45:756–779.

86. Hallson PC, Rose GA: A new urinary test for stone "activity." *Br J Urol* 1978; 50:442–448.

87. Harrison HE, Harrison HC: Inhibition of urine citrate excretion and the production of renal calcinosis in the rat by acetazolamide (Diamox) administration. *J Clin Invest* 1955; 34:1662–1670.

88. Harrison AR, Rose GA: Medullary sponge kidney. *Urol Res* 1979; 7:197–207.

89. Heath H III, Hodgson SF, Kennedy MA: Primary hyperparathyroidism: Incidence, morbidity, and potential impact in a community. *N Engl J Med* 1980; 302:189–193.

90. Heneman PH, Dempsey EF, Carroll EL, et al: The cause of hypercalciuria in sarcoid and its treatment with cortisone and sodium phytate. *J Clin Invest* 1956; 35:1229–1242.

91. Hesse A, Bach D: Stone analysis by infrared spectroscopy, in Rose GA (ed): *Urinary Stones—Clinical and Laboratory Aspects.* Baltimore, University Park Press, 1982, pp 87–105.

92. Hiatt RA, Dales LG, Friedman GD, et al: Frequency of urolithiasis in a prepaid medical care program. *Am J Epidemiol* 1982; 115:255–265.

93. Higgins CC: Urolithiasis, in Campbell M (ed): *Urology.* Philadelphia, WB Saunders Co, 1954, pp 767–842.

94. Hillman BJ, Drach GW, Tracey P, et al: Computed tomographic analysis of renal calculi. *Am J Radiol* 1984; 142:549–552.

95. Hockaday TDR, Clayton JE, Frederick EW, et al: Primary hyperoxaluria. *Medicine* 1964; 43:315–354.

96. Hodgkinson A: *Oxalic Acid in Biology and Medicine.* London, Academic Press, 1977.

97. Hoffman AF, Laker MF, Dharmsathaphorn K, et al:

Complex pathogenesis of hyperoxaluria after jejunoileal bypass surgery. *Gastroenterology* 1983; 84:293–300.

98. Hosking DH, Erickson SB, Van den Berg C, et al: The stone clinic effect in patients with idiopathic calcium urolithiasis. *J Urol* 1983; 130:1115–1118.

99. Hosking DH, Wilson JWL, Liedke RR, et al: Urinary citrate excretion in normals and patients with idiopathic calcium urolithiasis, in Schwille PO, Smith LH, Robertson WG, et al (eds): *Urolithiasis and Related Clinical Research.* New York, Plenum Publishing Corp, 1985, pp 367–370.

100. Issekutz B Jr, Blizzard JJ, Birkhead NC, et al: Effect of prolonged bed rest on urinary calcium output. *J Appl Physiol* 1966; 21:1013–1020.

101. Ito H, Coe FL: Acidic peptide and polyribonucleotide crystal growth inhibitors in human urine. *Am J Physiol* 1977; 233:F455–463.

102. Jaeger P, Portmann L, Burckhardt P: Tubular dysfunction in renal lithiasis: Cause or consequence? *Schweiz Med Wochenschr* 1985; 155:160–162.

103. Jenkins AD, Dousa TP, Smith LH: Transport of citrate across renal brush border membrane: effects of dietary acid and alkali loading. *Am J Physiol* 1985; 249:F590–595.

104. Jick H, Dinan BJ, Hunter JR: Triamterene and renal stones. *J Urol* 1982; 127:224–225.

105. Johnson CM, Wilson DM, O'Fallon WM, et al: Renal stone epidemiology: A 25-year study in Rochester, Minnesota. *Kidney Int* 1979; 16:624–631.

106. Kahnoski RJ, Lingeman JE, Coury TA, et al: Combined percutaneous extracorporeal shock wave lithotripsy for staghorn calculi: An alternative to anatrophic nephrolithotomy. *J Urol* 1986; 135:679–681.

107. Kao PC: Parathyroid hormone assay. *Mayo Clin Proc* 1982; 57:596–597.

108. Keating FR Jr, Jones JD, Elveback LR et al: The relation of age and sex to the distribution of values in healthy adults of serum calcium, inorganic phosphorus, magnesium, alkaline phosphatase, total proteins, albumin and blood urea. *J Lab Clin Med* 1969; 73:825–834.

109. Khan SR, Finlayson B, Hackett RL: Experimental calcium oxalate nephrolithiasis in the rat—role of the renal papilla. *Am J Pathol* 1982; 107:59–69.

110. Khan SR, Hackett RL: Identification of urinary stone and sediment crystals by scanning electron microscopy and x-ray microanalysis. *J Urol* 1986; 135:818–825.

111. Khan SR, Finlayson B, Hackett RL: Stone matrix as proteins absorbed on crystal surfaces: a microscopic study. *Scan Electron Microscope* 1983; 1:379–385.

112. Kippen I, Hirayama B, Klinenberg JR, et al: Transport of tricarboxylic acid cycle intermediates by membrane vesicles from renal brush border. *Proc Natl Acad Sci USA* 1979; 76:3397–3400.

113. Koch MO, McDougal WS: Nicotinic acid: Treatment for the hyperchloremic acidosis following urinary diversion through intestinal segments. *J Urol* 1985; 134:162–164.

114. Koff SA: Mechanism of electrolyte imbalance following urointestinal anastomosis. *Urology* 1975; 5:109–114.

115. Koutsoukos PD, Lan-Erwin CY, Nancollas GH: Epitaxial considerations in urinary stone formation: I. The urate-oxalate-phosphate system. *Invest Urol* 1980; 18:178–184.

116. Laskowski DE: Chemical microscopy of urinary calculi. *Anal Chem* 1965; 37:1399–1404.

117. Lavengood RW, Marshall VF: The prevention of urinary phosphatic calculi by the Shorr regimen. *J Urol* 1972; 108:368–371.

118. Leal JJ, Finlayson B: Adsorption of naturally occurring polymers onto calcium oxalate crystal surfaces. *Invest Urol* 1977; 14:278–283.

119. Lein JW, Keane PM: Limitations of the oral calcium loading test in the management of the recurrent calcareous renal stone former. *Am J Kidney Dis* 1983; 3:76–79.

120. Lemann J Jr, Gray RW: Calcitriol, calcium, and granulomatous disease. *N Engl J Med* 1984; 311:1115–1117.

121. Lian JB, Prien EL Jr, Glincher MJ, et al: The presence of protein-bound gamma-carboxyglutamic acid in calcium-containing renal calculi. *J Clin Invest* 1977; 59:1151–1157.

122. Ljunghall S, Danielson BG: A prospective study of renal stone recurrences. *Br J Urol* 1984; 56:122–124.

123. Lonsdale K: Epitaxy as a growth factor in urinary calculi and gallstones. *Nature* 1968; 217:56–58.

124. Lutwak L, Whedon GD, Lachance PA, et al: Mineral, electrolyte and nitrogen balance studies of the Gemini-VII fourteen day orbital space flight. *J Clin Endocrinol* 1969; 29:1140–1150.

125. Malagodi MH, Moye HA: Physical and chemical characteristics of renal stone matrix. *Urol Surv* 1981; 31:87–91.

126. Malek RS, Kelalis PP: Pediatric nephrolithiasis. *J Urol* 1975; 113:545–551.

127. Marangella M, Fruttero B, Bruno M, et al: Hyperoxaluria in idiopathic calcium stone disease: Further evidence of intestinal hyperabsorption of oxalate. *Clin Sci* 1982; 63:381–385.

128. Martin X, Werness PG, Bergert JH, et al: Pentosan polysulfate as an inhibitor of calcium oxalate crystal growth. *J Urol* 1984; 132:786–788.

129. McConnell JB, Murison J, Stewart WK: The role of the colon in the pathogenesis of hyperchloraemic acidosis in ureterosigmoid anastomosis. *Clin Sci* 1979; 57:305–312.

130. McSherry E, Sebastian A, Morris RC Jr: Renal tubular acidosis in infants: The several kinds, including bicarbonate-wasting, classic renal tubular acidosis. *J Clin Invest* 1972; 51:499–514.

131. McSherry E: Acidosis and growth in nonuremic renal disease. *Kidney Int* 1978; 14:349–354.

132. Melnick I, Landes RR, Hoffman AA, et al: Magnesium therapy for recurring calcium oxalate urinary calculi. *J Urol* 1971; 105:119–122.

133. Menon M, Mahle CJ: Urinary citrate excretion in patients with renal calculi. *J Urol* 1983; 129:1158–1160.

134. Meyer AS, Finlayson B, DuBois L: Direct observation

of urinary stone ultrastructure. *Br J Urol* 1971; 43:156–163.

135. Meyer JL: Nucleation kinetics in the calcium oxalate-sodium urate monohydrate system. *Invest Urol* 1981; 19:197–201.

136. Meyer JL, Bergert JH, Smith LH: The epitaxially induced crystal growth of calcium oxalate by crystalline uric acid. *Invest Urol* 1976; 14:115–119.

137. Meyer JL, Smith LH: Growth of calcium oxalate crystals: II. Inhibition by natural urinary crystal growth inhibitors. *Invest Urol* 1975; 13:36–39.

138. Miano L, Petta S, Galatioto GP, et al: A placebo controlled double-blind study of allopurinol in severe recurrent idiopathic renal lithiasis—preliminary results, in Schwille PO, Smith LH, Robertson WG, et al (eds): *Urolithiasis and Related Clinical Research.* New York, Plenum Publishing Corp, 1985, pp 521–524.

139. Milvy P, Colt E, Thornton J: A high incidence of urolithiasis in male marathon runners. *J Sports Med* 1981; 21:295–298.

140. Modlin M: The aetiology of renal stone: A new concept arising from studies on a stone-free population. *Ann R Coll Surg Engl* 1967; 40:155–178.

141. Morris RC Jr: Renal tubular acidosis. *N Engl J Med* 1981; 304:418–419.

142. Morris RC Jr: Renal tubular acidosis. *N Engl J Med* 1969; 281:1405–1413.

143. Morriss RH, Beeler MF: X-ray diffraction analysis of 464 urinary calculi. *Am J Clin Pathol* 1967; 48:413–417.

144. Muldowney FP, Donohoe JP, Freaney R, et al: Parathormone-induced renal bicarbonate wastage in intestinal malabsorption. *Irish J Med Sci* 1970; 3:221–231.

145. Muldowney FP, Freaney R, Moloney MF: Importance of dietary sodium in the hypercalciuria syndrome. *Kidney Int* 1982; 22:292–296.

146. Nakagawa Y, Abram V, Parks JH, et al: Urine glycoprotein crystal growth inhibitors: Evidence for a molecular abnormality in calcium oxalate nephrolithiasis. *Clin Invest* 1985; 76:1455–1462.

147. Nakagawa Y, Margolis HC, Yokoyama S, et al: Purification and characterization of a calcium oxalate monohydrate crystal growth inhibitor from human kidney tissue culture medium. *J Biol Chem* 1981; 256:3936–3944.

148. Nakagawa Y, Kaiser ET, Coe FL: Isolation and characterization of calcium oxalate crystal growth inhibitors from human urine. *Biochem Biophys Res Comm* 1978; 84:1038–1044.

149. Nemoy NJ, Stamey TA: Surgical, bacteriological, and biochemical management of "infection stones." *JAMA* 1971; 215:1470–1476.

150. Netelenbos JC, Jongen MJM, van der Vijgh WJF, et al: Vitamin D status in urinary calcium stone formation. *Arch Intern Med* 1985; 145:681–684.

151. Nielsen AE, Christoffersen J: The mechanisms of crystal growth and dissolution, in Nancollas GH (ed): *Biological Mineralization and Demineralization.* New York, Springer-Verlag, 1982, pp 36–66.

152. O'Neill M, Breslau NA, Pak CYC: Metabolic evaluation of nephrolithiasis in patients with medullary sponge kidney. *JAMA* 1981; 245:1233–1236.

153. Oliver J, MacDowell M, Whang R, et al: The renal lesions of electrolyte imbalance: IV. The intranephronic calculosis of experimental magnesium depletion. *J Exp Med* 1966; 124:263–277.

154. Pak CYC, Fuller C, Sakhaee K, et al: Long-term treatment of calcium nephrolithiasis with potassium citrate. *J Urol* 1985; 134:11–19.

155. Pak CYC, Ohata M, Lawrence EC, et al: The hypercalciurias—causes, parathyroid functions, and diagnostic criteria. *J Clin Invest* 1974; 54:387–400.

156. Pak CYC, Fuller C: Idiopathic hypocitraturic calcium-oxalate nephrolithiasis successfully treated with potassium citrate. *Ann Intern Med* 1986; 104:33–37.

157. Pak CYC, Delea CS, Bartter FC: Successful treatment of recurrent nephrolithiasis (calcium stones) with cellulose phosphate. *N Engl J Med* 1974a; 290:175–180.

158. Pak CYC, Nicar MJ, Peterson R, et al: A lack of unique pathophysiologic background for nephrolithiasis of primary hyperparathyroidism. *J Clin Endocrinol Metab* 1981; 53:536–542.

159. Pak CYC, Smith LH, Resnick MI, et al: Dietary management of idiopathic calcium urolithiasis. *J Urol* 1984; 131:850–852.

160. Pak CYC, Kaplan R, Bone H, et al: A simple test for the diagnosis of absorptive, resorptive and renal hypercalciurias. *N Engl J Med* 1975; 292:497–500.

161. Pak CYC: Should patients with single renal stone occurrence undergo diagnostic evaluation? *J Urol* 1982; 127:855–858.

162. Pak CYC, Fuller C, Sakhaee K, et al: Long-term treatment of calcium nephrolithiasis with potassium citrate. *J Urol* 1985; 134:11–19.

163. Pak CYC, Peterson R: Successful treatment of hyperuricosuric calcium oxalate nephrolithiasis with potassium citrate. *Arch Intern Med* 1986; 146:863–867.

164. Pak CYC, Barilla DE, Holt K, et al: Effect of oral purine load and allopurinol on the crystallization of calcium salts in urine of patients with hyperuricosuric calcium urolithiasis. *Am J Med* 1978; 65:593–599.

165. Pak CYC, Skurla C, Harvey J: Graphic display of urinary risk factors for renal stone formation. *J Urol* 1985; 134:867–870.

166. Pak CYC: Physiological basis for absorptive and renal hypercalciurias. *Am J Physiol* 1979; 237:F415–F423.

167. Pak CYC, Arnold LH: Heterogeneous nucleation of calcium oxalate by seeds of monosodium urate. *Proc Soc Exp Biol Med* 1975; 149:930–932.

168. Pak CYC: A cautious use of sodium cellulose phosphate in the management of calcium nephrolithiasis. *J Urol* 1981; 19:187–190.

169. Parfitt AM: Acetazolamide and sodium bicarbonate induced nephrocalcinosis and nephrolithiasis. *Arch Intern Med* 1969; 124:736–740.

170. Park GD, Spector R: Hypocitraturic calcium-oxalate nephrolithiasis (letter). *Ann Intern Med* 1986; 104:723–724.

171. Parks JH, Coe FL, Strauss AL: Calcium nephrolithiasis and medullary sponge kidney in women. *N Engl J Med* 1982; 306:1088–1091.

172. Parsons CL, Schmidt JD, Pollen JJ: Successful treatment of interstitial cystitis with sodium pentosan polysulfate. *J Urol* 1983; 130:51–53.

173. Prien EL: Use of polarized light in analyses of calculi and study of crystals in tissue. *J Urol* 1941; 45:765–769.

174. Prien EL: The riddle of Randall's plaques. *J Urol* 1975; 114:500–506.

175. Purnell DC, Smith LH, Scholtz DA et al: Primary hyperparathyroidism: A prospective clinical study. *Am J Med* 1971; 50:670–678.

176. Pyrah LN: *Renal Calculus*. New York, Springer-Verlag, 1979.

177. Randall A: The origin and growth of renal calculi. *Ann Surg* 1937; 105:1009–1027.

178. Rao PN, Faraghar EB, Buxton A, et al: Is salt restriction necessary to reduce the risk of stone formation?, in Schwille FO, Smith LH, Robertson WG, et al (eds): *Urolithiasis and Related Clinical Research*. New York, Plenum Publishing Corp, 1985, pp 429–432.

179. Resnick MI: Evaluation and management of infection stones. *Urol Clin North Am* 1981; 8:265–276.

180. Robertson WG, Peacock M, Nordin BEC: Calcium oxalate crystalluria and urine saturation in recurrent renal stone formers. *Clin Sci* 1971; 40:365–374.

181. Robertson WG, Peacock M, Heyburn PJ, et al: The risk of calcium stone formation in relation to affluence and dietary animal protein, in Brockis JG, Finlayson B (eds): *Urinary Calculus*. Littleton, Mass, PSG Publishing, 1981, pp 3–12.

182. Robertson WG, Peacock M, Heyburn PJ, et al: A risk factor model of stone formation: Application to the study of epidemiological factors in the genesis of calcium stones, in Smith LH, Robertson WG, Finlayson B (eds): *Urolithiasis, Clinical and Basic Research*. New York, Plenum Publishing Corp, 1981, pp 303–307.

183. Robertson WG, Peacock M, Heyburn PJ, et al: Risk factors in calcium stone disease of the urinary tract. *Br J Urol* 1978; 50:449–454.

184. Robertson WG, Peacock M, Marshall RW, et al: Saturation-inhibition index as a measure of the risk of calcium oxalate stone formation in the urinary tract. *N Engl J Med* 1976; 294:249–252.

185. Robertson WG, Peacock M, Nordin BEC: Activity products in stone-forming and non-stone forming urine. *Clin Sci* 1968; 34:579–594.

186. Robertson WG: The solubility concept, in Nancollas GN (ed): *Biological Mineralization and Demineralization*. New York, Springer-Verlag, Dahlem Konferenzen, pp 5–21.

187. Robertson WG, Peacock M, Nordin BEC: Measurement of activity products in urine from stone-formers and normal subjects, in *Urolithiasis: Physical Aspects*. Washington, DC, National Academy of Sciences, 1972, pp 79–95.

188. Robertson WG, Peacock M, Nordin BEC: Inhibitors of the growth and aggregation of calcium oxalate crystals in vitro. *Clin Chim Acta* 1973; 43:31–37.

189. Robertson WG, Peacock M: The cause of idiopathic calcium stone disease: Hypercalciuria or hyperoxaluria? *Nephron* 1980; 26:105–110.

190. Robertson WG Morgan DB: The distribution of urinary calcium excretions in normal persons and stone-formers. *Clin Chim Acta* 1972; 37:503–508.

191. Robertson WG: Dietary factors important in calcium stone formation, in Schwille PO, Smith LH, Robertson WG, et al (eds): *Urolithiasis and Related Clinical Research*. New York, Plenum Publishing Corp, 1985, pp 61–68.

192. Rose GA: Stone analysis by thermogravimetric technique, in Rose GA, (ed): *Urinary Stones—Clinical and Laboratory Aspects*. Baltimore, University Park Press, 1982, pp 77–85.

193. Rose GA: An overview of some problems in urolithiasis, in Rose GA, (ed): *Urinary Stones—Clinical and Laboratory Aspects*. Baltimore, University Park Press, 1982, pp 1–21.

194. Roth R, Finlayson B: Observations on the radiopacity of stone substances with special reference to cystine. *Invest Urol* 1973; 11:186–189.

195. Rudman D, Kutner MH, Redd SC II, et al: Hypocitraturia in calcium nephrolithiasis. *J Clin Endocrinol Metab* 1982; 55:1052–1057.

196. Rundquist RT, Smith LH, Werness PG: Factitious hyperoxaluria from ascorbic acid (abstract). Central Society for Clinical Research, Midwest Meeting 1981.

197. Ryall RL, Harnett RM, Marshall VR: The effect of urine, pyrophosphate, citrate, magnesium and glycosaminoglycans on the growth and aggregation of calcium oxalate crystals in vitro. *Clin Chim Acta* 1981; 12:349–356.

198. Ryall RL, Harnett RM, Marshall VR: The effect of uric acid on the inhibitory activity of glycosaminoglycans and urine, in Schwille PO, Smith LH, Robertson WG, et al (eds): *Urolithiasis and Related Clinical Research*. New York, Plenum Publishing Corp, 1985, pp 855–858.

199. Ryde M, Eriksson H, Tangen O: Studies on the different mechanisms by which heparin and polysulfated xylan (PZ 68) inhibit blood coagulation in man. *Thromb Res* 1981; 23:435–445.

200. Sakhaee K, Nicar M, Hill K, et al: Contrasting effects of potassium citrate and sodium citrate therapies on urinary chemistries and crystallization of stone-forming salts. *Kidney Int* 1983; 24:348–352.

201. Salliner, A: Some aspects of urolithiasis in Finland. *Acta Chim Scand* 1959/1960; 118:479–487.

202. Schaumburg H, Kaplan J, Windebank A, et al: Sensory neuropathy from pyridoxine abuse—a new megavitamin syndrome. *N Engl J Med* 1983; 309:445–448.

203. Schneider JA, Schulman JD, Seegmiller JE: Cystinosis and the Fanconi syndrome, in Stanbury JB, Wyngaarden JB, Fredrickson DS (eds): *The Metabolic Basis of Inherited Disease*, New York, McGraw-Hill Book Co, 1978, pp 1660–1682.

204. Scott R, Paterson PJ, Mathieson A, et al: The effect of allopurinol on urinary oxalate excretion in stone formers. *Br J Urol* 1978; 50:455–458.

205. Scurr DS, Robertson WG: Studies on the mode of action of polyanionic inhibitors of calcium oxalate crystallization in urine, in Schwille PO, Smith LH, Robertson WG, et al (eds): *Urolithiasis and Related Clinical Research*. New York, Plenum Publishing Corp, 1985, pp 835–838.

206. Segura JW, Erickson SB, Wilson DM, et al: Infected renal lithiasis: Results of long-term surgical and medical management, in Smith LH, Robertson WG, Finlayson B (eds): *Urolithiasis—Basic and Clinical Research*. New York, Plenum Publishing Corp, 1981, pp 195–198.

207. Seldin DW, Wilson JD: Renal tubular acidosis, in Stanbury JB, Wyngaarden JB, Fredrickson DS (eds): *The Metabolic Basis of Inherited Disease,* New York, McGraw-Hill Book Co, 1978, pp 1618–1633.

208. Shen FH, Baylink DJ, Nielson RL, et al: Increased serum 1,25-dihydroxyvitamin D in idiopathic hypercalciuria. *J Lab Clin Med* 1977; 90:955–962.

209. Short KL, Amin M, Harty JI, et al: Comparison of recurrence rates of calculi of the bladder in patients with indwelling catheters following vesicolithotomy, litholapaxy, and electrohydraulic lithotripsy. *Surg Gynecol Obstet* 1984; 159:247–248.

210. Shuster J, Finlayson B, Schaeffer RL, Sierakowski R, et al: Primary liquid intake and urinary stone disease. *J Chron Dis* 1985; 38:907–914.

211. Shuster J, Finlayson B, Schaeffer R, et al: Water hardness and urinary stone disease. *J Urol* 1982; 128:422–425.

212. Sierakowski R, Finlayson B, Landes RR, et al: The frequency of urolithiasis in hospital discharge diagnoses in the United States. *Invest Urol* 1978; 15:438–441.

213. Sierakowski R, Finlayson B, Landes R: Stone incidence as related to water hardness in different geographical regions of the United States. *Urol Res* 1979; 7:157–160.

214. Siklos P, Davie M, Jung RT, Chalmers TM: Osteomalacia in ureterosigmoidostomy: Healing by correction of the acidosis. *Br J Urol* 1980; 52:61–62.

215. Silverman DE, Stamey TA: Management of infection stones: The Stanford experience. *Medicine* 1983; 62:44–51.

216. Simpson DP: Citrate excretion: A window on renal metabolism. *Am J Physiol* 1982; 244:F2323–F2324.

217. Simpson DP: Effect of acetazolamide on citrate excretion in the dog. *Am J Physiol* 1964; 206:883–886.

218. Sinno K, Boyce WH, Resnick MI: Childhood urolithiasis. *J Urol* 1979; 121:662–664.

219. Smith LH: New treatment for struvite urinary stones. *N Engl J Med* 1984; 311:792–794.

220. Smith LH, Jones SD, Keating KR Jr: Primary hyperoxaluria, in *Proceedings of the Renal Stone Research Symposium.* London, J and A Churchill, 1969, pp 297–307.

221. Smith LH: Medical treatment of idiopathic calcium urolithiasis. *Kidney* 1983a; 16:9–15.

222. Smith LH: Stone activity, in Roth RA, Finlayson B (eds): *Stones: Clinical Management of Urolithiasis.* Baltimore, Williams & Wilkins Co, 1983b, pp 183–186.

223. Smith LH: Urolithiasis, in Gottschalk CW, Earley LE (eds): *Strauss and Welt's Diseases of the Kidney,* ed 3. Boston, Little, Brown & Co, 1979, pp 893–931.

224. Smith LH, Werness PG, Wilson DM: Enteric hyperoxaluria: Associated abnormalities that promote formation of renal calculi, in Rose GA, Robertson WG, Watts RWE (eds): *Oxalate in Human Biochemistry and Clinical Pathology.* London, Wellcome Foundation, 1979, pp 224–230.

225. Smith LH, Fromm H, Hoffman AF: Acquired hyperoxaluria, nephrolithiasis, and intestinal disease: Description of a syndrome. *N Engl J Med* 1972; 286:1371–1375.

226. Stacholy J, Goldberg EP: Microstructural matrix-crystal interactions in calcium oxalate monohydrate kidney stones. *Scan Electron Microscopy* 1985; 2:781–787.

227. Stapleton FM, Roy S III, Noe HN, et al: Hypercalciuria in children with hematuria. *N Engl J Med* 1984; 310:1345–1348.

228. Stauffer JQ, Stewart RJ, Bertrand G: Acquired hyperoxaluria: Relationship to dietary calcium content with severity of steatorrhea. *Gastroenterology* 1974; 66:783.

229. Stewart AF, Adler M, Byers CM, et al: Calcium homeostasis in immobilization: An example of resorptive hypercalciuria. *N Engl J Med* 1982, 306:1136–1140.

230. Stinebaugh BJ, Schloeder FX, Tan SC, et al: Pathogenesis of distal renal tubular acidosis. *Kidney Int* 1981; 19:1–7.

231. Stote RM, Smith LH, Budd JW, et al: Oxypurinol nephrolithiasis in regional enteritis secondary to allopurinol therapy. *Ann Intern Med* 1980; 92:384–385.

232. Strauss AL, Coe FL, Parks JH: Formation of a single calcium stone of renal origin—clinical and laboratory characteristics of patients. *Arch Intern Med* 1982; 142:504–507.

233. Strenge A, Hesse A, Bach D, et al: Excretion of oxalic acid following the ingestion of various amounts of oxalic acid-rich foods, in Smith LH, Robertson WG, Finlayson B (eds): *Urolithiasis—Basic and Clinical Research.* New York, Plenum Publishing Corp, 1981, pp 789–794.

234. Sutor DJ, Wooley SE: Urinary tract calculi—a comparison of chemical and crystallographic analyses. *Br J Urol* 1971; 43:149–153.

235. Sutor DJ, Scheidt S: Identification standards for human urinary calculus components, using crystallographic methods. *Br J Urol* 1968; 40:22–28.

236. Sutton RAL, Walker VR: Responses to hydrochlorothiazide and acetazolamide in patients with calcium stones. *N Engl J Med* 1980; 302:709–713.

237. Tessitore N, Ortalda V, Fabris A, et al: Renal acidification defects in patients with recurrent calcium nephrolithiasis. *Nephron* 1985; 41:325–332.

238. Thier SO, Segal S: Cystinuria, in Stanbury JB, Wyngaarden JB, Fredrickson DS (eds): *The Metabolic Basis of Inherited Disease,* ed 4. New York, McGraw-Hill Book Co, 1978, pp 1578–1492.

239. Thom JA, Morris JE, Bishop A, et al: The influence of refined carbohydrate on urinary calcium excretion. *Br J Urol* 1978; 50:459–464.

240. Thomas WC Jr: Use of phosphates in patients with calcareous renal calculi. *Kidney Int* 1978; 13:390–396.

241. Vahlensieck W: Influence of water quality on urolithiasis, in Schwille PO, Smith LH, Robertson WG, et al (eds): *Urolithiasis and Related Clinical Research.* New York, Plenum Publishing Corp, 1985, pp 97–103.

242. Vahlensieck W, Bastian HP: Clinical features and treatment of urinary calculi in childhood. *Eur Urol* 1976; 2:129–134.

243. Valyasevi A, Dhanamitta S: A general hypothesis concerning the etiological factors in bladder stone disease, in van Reem R (ed): *Idiopathic Urinary Bladder Stone Disease.* Washington DC, Department of Health, Education, and Welfare, 1977, pp 135–150.

244. Van den Berg CJ, Kumar R, Wilson DM, et al: Orthophosphate therapy decreases urinary calcium excretion and serum 1,25-dihydroxyvitamin D concentrations in idiopathic hypercalciuria. *J Clin Endocrinol Metab* 1980; 51:998–1001.

245. Van den Berg CJ, Harrington TM, Bunch TW, et al: Treatment of renal lithiasis associated with renal tubular acidosis. *Proc Eur Dial Transplant Assoc* 1983; 20:473–476.

246. Vermeulen CW, Lyon ES, Ellis JE, et al: The renal papilla and calculogenesis. *J Urol* 1967; 97:573–582.

247. Vermeulen CW, Lyon ES: Mechanisms of genesis and growth of calculi. *Am J Med* 1968; 45:684–692.

248. Walton AG: The formation and properties of precipitates. Huntington NY, Robert E Krieger, 1979,

249. Werness PG, Brown CM, Smith LH, et al: Equil 2: A basic computer program for the calculation of urinary saturation. *J Urol* 1985; 134:1242–1244.

250. Werness PG, Bergert JH, Smith LH: Crystalluria. *J Crystal Growth* 1981a; 53:166–181.

251. Werness PG, Knox FG, Smith LH: Low phosphate diet in rats: A model for calcium oxalate urolithiasis, in Smith LH, Robertson WG, Finlayson B (eds): *Urolithiasis—Clinical and Related Research.* New York, Plenum Publishing Corp, 1981, pp 731–734.

252. Werness PG, Bergert JH, Smith LH: Triamterene urolithiasis: Solubility, pk, effect on crystal formation, and matrix binding of triamterene and its metabolites. *J Lab Clin Med* 1982; 99:254–262.

253. White DJ, Nancollas GH: Triamterene and renal stone formation. *J Urol* 1982; 127:593–597.

254. Williams HE: Oxalic acid: Absorption, excretion, and metabolism, in Fleisch H, Robertson WG, Smith LH, et al (eds): *Urolithiasis Research,* New York, Plenum Publishing Corp, 1976, pp 181–188.

255. Williams HE, Smith LH Jr: L-Glyceric aciduria: A new genetic variant of primary hyperoxaluria. *N Engl J Med* 1968; 278:233–239.

256. Williams HE, Smith LH Jr: Primary hyperoxaluria, in Stanbury JB, Wyngaarden JB, Fredrickson DS (eds): *The Metabolic Basis of Inherited Disease,* New York, McGraw-Hill Book Co, 1978, pp 182–204.

257. Williams JJ, Rodman JS, Peterson CM: A randomized double-blind study of acetohydroxamic acid in struvite nephrolithiasis. *N Engl J Med* 1984; 311:760–764.

258. Wilson JM, Young AB, Kelley WN: Hypoxanthine-guanine phosphoribosyl transferase deficiency. *N Engl J Med* 1983; 309:900–910.

259. Wilson JWL, Werness PG, Smith LH: Effect of orthophosphate treatment on urine composition in idiopathic calcium urolithiasis, in Schwille PO, Smith LH, Robertson WG, et al (eds): *Urolithiasis and Related Clinical Research.* New York, Plenum Publishing Corp, 1985, pp 491–493.

260. Yatzidis H: Successful sodium thiosulphate treatment for recurrent calcium urolithiasis. *Clin Nephrol* 1985; 23:63–67.

261. Yendt ER, Cohanim M: Absorptive hyperoxaluria: A new clinical entity—successful treatment with hydrochlorothiazide. *Clin Invest Med* 1986; 9:44–50.

262. Yendt ER, Cohanim M: The prevention of calcium stones with thiazides, in Schwille PO, Smith LH, Robertson WG, et al (eds): *Urolithiasis and Related Clinical Research.* New York, Plenum Publishing Corp, 1985, pp 463–470.

263. Yendt ER: Medullary sponge kidney and nephrolithiasis. *N Engl J Med* 1982; 306:1106–1107.

264. Yu TF, Berger L, Gutman AB: Defective conversion of uric acid to allantoin in the Dalmatian dog. *Arthritis Rheum* 1966; 9:552–553.

265. Zechner O, Scheiber V: The role of affluence in recurrent stone formation, in Smith LH, Robertson WG, Finlayson B (eds): *Urolithiasis—Basic and Clinical Research.* New York, Plenum Publishing Corp, 1981, pp 309–313.

266. Zinsser HH, Seneca H, Light I, et al: Management of infected stones with acidifying agents. *NY State J Med* 1968; 68:3001–3010.

Chapter 10

Perioperative Care

W. Scott McDougal, M.D.

The risk of an operative procedure must be weighed against its benefit so the patient can be given a realistic view of the probable outcomes of both the nonoperative and the operative approaches. Although determining risk is simple in theory, it is extremely difficult in practice. Because of the wide variety of surgical procedures performed and the uniqueness of each patient, attempts to quantitate the risk factor have not met with great success. Many investigations have been unable to correlate accurately risk with any specific factor; however, a loose association has been shown with poor general health, advanced age, emergency operation and the site of the surgical procedure (Goldman et al., 1978).

In an attempt to quantitate the risk associated with an anesthetic, irrespective of procedure performed, the American Society of Anesthesiologists has proposed a classification of physical status: Class I: a normal healthy person; Class II: patients with mild to moderate systemic disease; Class III: patients with severe systemic diseases that are not incapacitating; Class IV: incapacitating systemic disease that is a constant threat to life; and Class V: a moribund patient not expected to survive more than 24 hours without an operation. This classification allows a general assessment but is not specific enough to accurately quantitate risk.

The most important factor in evaluating the risk of an operation is the functional status of the cardiovascular system. Pulmonary function also significantly affects the risk of anesthesia; perhaps the most common postoperative complication involves the respiratory tree. Hypoxemia secondary to respiratory dysfunction, when combined with myocardial disease, carries with it a markedly increased chance of myocardial infarction and death following the procedure.

The death rate from anesthesia, taking into account all operative procedures, has been estimated to be about 0.3%. Of patients who die as a result of the anesthetic and operative procedure, about one tenth will die in the period of induction, about one third during the operative procedure itself, and the remainder within the first 48 hours postoperative (Feigal and Blaisdell, 1979).

Preoperative assessment is undertaken both to determine the risk of the proposed procedures and to identify abnormalities that can be corrected to reduce morbidity and mortality. The assessment should include an evaluation of the blood count and blood volume; the integrity of the hemostatic mechanisms; an evaluation of the cardiac function; an assessment of pulmonary function; and a review of the metabolic status, which should include an assessment of the liver, the kidneys, the immune system, the nutritional status of the patient, the integrity of the adrenals, and an evaluation of any systemic diseases such as diabetes, obesity, and thyroid disease.

BLOOD COUNT

All preoperative patients require the determination of a packed cell volume or hematocrit. A value between 30 and 50 is acceptable. As the hematocrit falls below 30%, the viscosity of the blood is reduced, and flow characteristics through the small vessels improve. This, however, is offset by the decreased oxygen carrying capacity of the blood. Patients with chronic renal failure often have hematocrits of 20–24 and tolerate surgery well, provided there has not been a recent acute change in the hematocrit. The disadvantage is that there is little margin for error in terms of major blood loss. Patients with hematocrits above 55 have marked increased viscosity in their blood and are prone to thrombosis during periods of major fluid shifts. Therefore, except under extenuating circumstances, a hematocrit between 30% and 50% should be sought in the preoperative preparation of the patient.

BLOOD VOLUME

The status of the patient's blood volume must also be assessed. Hypovolemia may be a consequence of secondary hyperaldosteronism, Addison's disease, acute blood loss, vomiting, diarrhea, pancreatic fistula, ileostomy, intestinal obstruction, pheochromocytoma, and neuroblastoma. When these disorders are present, the blood volume must be restored through preoperative fluid administration and/or pharmacologic manipulation. α-Adrenergic blockade allows for volume expansion through relaxation of peripheral vasoconstriction with homeostatic volume restoration over a several-week period in patients with a pheochromocytoma and in hypertensive patients with a neuroblastoma. Fluid loss into the bowel, as occurs in bowel obstruction, diarrhea, or hemorrhagic blood loss, is not appropriately corrected by pharmacologic means, although drugs may be used as temporary blood pressure stabilizers until fluid volumes can be restored—generally by the IV mode when such losses are sufficient to cause cardiovascular instability.

The central venous pressure (CVP) is an indirect measure of the blood volume and competency of the heart to receive and propel blood. Provided there is no significant heart disease, it is an accurate measure of volume status. A normal CVP ranges between 4 and 8 cm H_2O with reference to the left atrium (4 cm below the angle of the sternum). A low CVP suggests hypovolemia, and an elevated CVP volume overload. A pulmonary artery catheter (Swan-Ganz) is necessary to determine accurately the volume status in patients with cardiac disease, those with chronic obstructive pulmonary disease, and in patients in whom the CVP does not correlate with the clinical status (a high CVP in the presence of hypoperfusion). Although the CVP is adequate, a pulmonary artery catheter is preferred in the patient with multisystem trauma, septic shock, decompensated cirrhosis, severe pancreatitis or peritonitis, and those receiving massive transfusions. In these circumstances, major fluid shifts can be more accurately titrated with a Swan-Ganz catheter. A normal pulmonary wedge pressure ranges between 12 and 15 cm of H_2O. This catheter also allows for the measurement of cardiac output and peripheral vascular resistance. Restoration of circulating volume and correction of metabolic disorders may require several weeks of proper fluid and pharmacologic manipulation.

A kidney in the diuretic state is less prone to injury; therefore, patients whose urinary tracts are operated upon should be volume replete prior to and during surgery. The usual practice of NPO past midnight requires an intraoperative catch-up of fluids. This can be obviated by beginning the IV fluid administration the night before surgery and giving the patient normal replacement fluids (see Fluid and Electrolytes).

HEMOSTASIS

The competency of the hemostatic mechanisms is assessed by history that addresses bleeding tendencies, bruisability, or a family history of bleeding disorders and serum studies. A platelet count, a prothrombin time (PT), and a partial thromboplastin time (PTT) serve as a good screen for major surgery. If there is any indication that there may be a qualitative platelet defect, as in uremic patients and those taking aspirin, a bleeding time is also obtained. If these values are normal, it is unlikely that a bleeding diathesis is present.

CARDIAC FUNCTION

Cardiac function is evaluated by history, physical examination, an ECG, and, in selected cases, a gated blood pool scan to provide ejection fraction. A previous myocardial infarction is of particular significance. As a group, patients with a history of a myocardial infarction have a tenfold increase in the probability of having a subsequent postoperative infarction. A further analysis reveals that the time elapsed since the infarction is of prime importance. One third of patients operated on within three months of their myocardial infarction will have another. Patients operated on six months or longer after their myocardial insult have their risk of postoperative infarction reduced to about 6%–8%. A postoperative infarction generally happens within the first seven postoperative days and carries a 50%–75% mortality (Tarhan et al., 1972). Other factors that appear to be particularly important in predicting myocardial dysfunction in the operative and perioperative period include: (1) an S_3 gallop and/or jugular venous distention, (2) a preoperative cardiac rhythm other than sinus, (3) more than five premature ventricular contractions per minute, (4) an intraperitoneal, intrathoracic, or aortic operation, (5) an age greater than 70 years, (6) aortic valvular heart disease, (7) the necessity for emergency operation, and (8) poor general medical condition as reflected by an arterial blood gas abnormality, decreased renal function, and/or evidence of hepatic disease (Goldman et al., 1977). Clearly, the most important predictors are the presence of congestive heart failure with jugular venous distention and/or an S_3 gallop and a history of a myocardial infarction within the preceding six months. These two findings carry with them a significant risk of death from the anesthetic.

Hypertension also increases the risk of an operative procedure, particularly if associated with coronary artery disease. Hypertensive patients have wider and more frequent blood pressure swings during anesthesia and are more likely to have associated cerebral vascular and coronary artery blood flow compromise. Patients should continue antihypertensive medicine if they are normotensive, or the dosages should be adjusted to achieve a normal blood pressure. In general, medications should not be discontinued in the preoperative period. The monoamine oxidase (MAO) inhibitors (such as guanethidine) may be an exception, since they can interfere with the anesthetic management.

Patients thought to have severe cardiac disease are best prepared preoperatively with a Swan-Ganz catheter placed in the pulmonary artery, ionotrophs or antiarrhythmics given as needed, and careful fluid resuscitation. An evaluation of both the cardiac and respiratory status can be performed by noting the patient's resting pulse, having him walk briskly a short distance, then rechecking the pulse. If the pulse rate does not rise to twice the resting level, the patient is probably not at significant risk.

RESPIRATORY FUNCTION

Respiratory status is evaluated by history and a determination of exercise tolerance. A smoking history of more than 20 pack-years is particularly significant. If there is any question about the respiratory status, simple spirometry is an excellent screen. If the FEV_1 is under 15 ml/kg and the vital capacity is less than 1.5 L or the maximum voluntary ventilation is under 50%, further pulmonary function tests and additional preoperative preparation are in order.

METABOLIC STATUS

The metabolic status of the patient is assessed with regard to the liver, renal, and immune function as well as the nutritional and adrenal status. If indicated by history, liver function is determined by liver enzymes, prothrombin time, and albumin. In patients with compromised hepatic function, preoperative preparation includes improving the nutritional status (see Nutrition) and in those with an impaired clotting mechanism the administration of fresh-frozen plasma and vitamin K. Cirrhotic patients with ascites are prepared preoperatively by improving nutritional status, limiting sodium intake to 1–2 gm/day, and reducing ascitic fluid by judicious use of spironolactone with or without other diuretics. Renal dysfunction is addressed in another chapter; however, renal function abnormalities generally do not significantly impair operation, provided fluid and electrolyte balance is corrected.

The immune status of the patient, if in question, may be examined by determining whether the patient is anergic to cutaneous skin tests. Unfortunately, little can be done to alter the immune status unless it is due to nutritional deficiency. A nutritional assessment also should be performed (see Nutrition). Medications of particular significance include antihypertensives and steroids. Adrenal insufficiency owing to intrinsic disease or suppression secondary to exogenous steroid administration demands the administration of preoperative, intraoperative, and postoperative steroids. The dose should be approximately ten times the resting non-stressed level and administered in three equally divided doses. Since the normal basal production of cortisol is about 35 mg/day, a replacement of an equivalent of approximately 300 mg of cortisone is indicated. One hundred milligrams of cortisol equivalent is administered several hours before the operation, 100 mg intraoperatively, and 100 mg infused 8 hours postoperatively. The dosage is gradually tapered to maintenance levels over several days, or if the adrenals function normally, it may be discontinued altogether. However, if postoperative complications occur, the ten-times basal level dosage should be continued until the complications have resolved.

SYSTEMIC DISEASE

The evaluation of systemic disease generally involves a determination of the carbohydrate (diabetes) and thyroid status. In diabetic patients, it is important to avoid hypoglycemic ketosis or hyperosmolarity. This is best accomplished if the blood glucose level is maintained between 125 and 250 mg/dl. If the patient is taking insulin, one half the usual dose is given on the morning of surgery and an IV dextrose-containing solution begun. During the operative procedure, blood glucose levels are determined and regular insulin administered as necessary. In the postoperative period, blood glucose levels must be determined every four hours, and regular insulin administered IV or subcutaneously for diabetics with major alterations in glucose metabolism. For patients who are particularly labile, an insulin infusion pump may be utilized in the postoperative period. The infusion of insulin is given in a concentration of 1 unit/ml of normal saline at a rate of approximately 2–3 units/hour. Continuous monitoring of blood glucose levels must be performed throughout the period of administration, particularly when metabolic aberrations, sepsis, and trauma coexist, since these can markedly alter glucose utilization and cause wide fluctuations in serum levels.

METABOLIC RESPONSE TO INJURY

Basal Metabolic Rate

Uninjured man at rest expends a definable amount of energy to perform physiologic work—the work required to maintain cardiac output, endocrine function, body temperature, liver function, respiration, renal function, etc. The amount of energy required to maintain these functions at rest is called basal metabolic rate and is expressed in calories per hour per square meter of body surface. Basal metabolism is the energy expended by the cells that constitute the active mass of the body; fat, extracellular fluid, and bone make no direct contribution to the metabolic rate. The energy used in physiologic work is derived from chemical energy, which in the course of the performance of the work is converted to heat and lost from the body. Therefore, the basal metabolic rate or energy expenditure may be determined by direct calorimetry in which the heat lost from the body is directly measured. Direct calorimetry is difficult to perform and often impractical in the critically ill. Therefore, indirect methods of estimating the basal metabolic rate are more commonly employed. The heat lost may be directly estimated from (1) oxygen consumption, CO_2 production, and nitrogen excretion; (2) measurement of energy intake and losses coupled with the measurement of changes in body composition; or (3) measurement of insensible water loss, assuming that this represents 25% of the total heat loss. The basal metabolic energy requirement depends on the age, sex, and lean body mass. A 1-month-old infant requires about 40 kcal/kg body weight/day. The requirement increases with age and body mass, so that boys between 15 and 18 years old consume 1,700 kcal/day and girls between 12 and 18 years old require 1,400 kcal/day. The energy requirement remains relatively stable during the active years of life. However, with progressive aging, it decreases to 1,300 kcal/day for men and 1,100 kcal/day for women (Passmore, 1969). The difference in metabolic rate between men and women probably relates to the fact that women have a larger proportion of fat per unit body weight. Indeed, if basal metabolic rate is expressed per unit fat-free body weight, it is remarkably similar for both males and females over an age span of 20–60 years (1.3 kcal/hr/kg).

Hypermetabolism

The response to injury is characterized by an increase in the basal metabolic rate, even though the patient remains at rest. The intensity of this hypermetabolic response depends on the severity of the injury, the nutritional status of the patient, and the presence or absence of infection. Patients in good nutritional balance undergoing an elective operation will have a change in metabolic rate of no more than 10% in the postoperative period, provided there are no complications. In contrast, patients with multiple fractures have an increase of their resting energy expenditure by 10%–25%, those with major infections by 20%–50%, and those with major thermal burns by 50%–125%. In the early posttraumatic period, satisfaction of these energy demands results in degradation of body protein (manifested by a negative nitrogen balance), depletion of energy reserves, and loss of body weight. The initial catabolic response in which body protein, fat, and carbohydrate are depleted is gradually reduced and reversed later in the recovery period. An anabolic phase ensues, provided adequate nutritional intake occurs, in which new protein is laid down and carbohydrate and fat reserves are repleted. The hypermetabolic response also gradually diminishes as the wounds heal and as infection is eradicated. In the early postinjury period, energy requirements are satisfied by glucose derived mainly from liver and muscle glycogen and by fatty acids released from adipose tissue. Glycogen reserves are rapidly depleted (usually within 48 hours). However, fat stores, which supply the bulk of the energy requirement during this period, continue to supply fatty acids for many days. Protein catabolism also occurs, even though it may not serve as a primary energy source. Amino acids are released primarily from skeletal muscle and are essential for maintenance of cellular metabolism. Moreover, the gluconeogenic amino acids serve as a source of new glucose and provide the basic glucose structure which is necessary for normal metabolism. Rapid depletion of glucose stores coupled with the inability of the two carbon fragments of the fatty acids to serve as a source for gluconeogenesis make protein the only available substrate from which new glucose can be synthesized. Protein catabolism can be reduced but not eliminated by providing exogenous glucose (nitrogen-sparing effect of glucose).

Catecholamines and Glucocorticoids

Posttraumatic catecholamine levels are elevated and promote hepatic glycogenolysis, inhibition of insulin release, stimulation of glucagon production, and stimulation of fat hydrolysis with the release of free fatty acids. This hormone is produced by the adrenal medulla almost exclusively and has been implicated as the mediator of the hypermetabolic response in injured patients (Wilmore et al., 1974).

The glucocorticoids are also characteristically elevated and remain so throughout the recovery period. The increased production is the direct result of an increased ACTH secretion by the anterior pituitary. It is presumed that the pituitary is stimulated to release ACTH by the action of the nerves from the periphery responding to the traumatic injury and perhaps by a direct effect of the elevated epinephrine levels. Patients who have sustained severe trauma that requires a prolonged period of convalescence, such as thermal burns, often have marked adrenal hyperplasia. Glucocorticoids promote gluconeogenesis and inhibit the action of insulin.

Insulin

In the immediate postinjury period, circulating levels of insulin and glucose are elevated. Insulin facilitates glucose transport across cell membranes and inhibits both the release of amino acids from muscle and the release of free fatty acids from adipose tissue. In the posttraumatic period hyperglycemia is common, possibly due to the development of insulin resistance. Indeed, glucose tolerance curves performed during this period simulate those observed in diabetes and have led many to refer to this state as the "diabetes of injury" (Kinney, 1977).

The mechanism of the hyperglycemia, however, may not be one of insulin resistance. More recent evidence indicates that glucose oxidation is unimpaired, and the hyperglycemia is due to an increase in gluconeogenesis rather than a reduction in peripheral utilization. Moreover, glucose flow studies have clearly demonstrated that glucose turnover rate is increased above normal (Long et al., 1971; Wilmore et al., 1976). Others suggest that for the concentration of glucose, the amount of circulating insulin is inadequate.

Glucagon

The increased production of catecholamines postinjury is known to stimulate the alpha cell of the pancreatic islets, resulting in increased circulating levels of glucagon, a hormone which promotes glycogenolysis and gluconeogenesis. The relationship between the concentration of insulin and glucagon (the insulin glucagon molar ratio) determines whether the major influence is toward glucose breakdown or glucose formation. In the early postinjury period, although insulin is elevated, glucagon is disproportionately increased, and the ratio of the two favors gluconeogenesis at the expense of protein formation. As healing occurs, the molar ratio reverses, favoring glycolysis and protein formation. Postinjury, the hypermetabolic response and hormonal balance direct the metabolism and utilization of carbohydrate, fat, and protein. They set the stage for a negative nitrogen balance and weight loss, both of which are related to the extent of injury. The protein that is catabolized to satisfy energy and substrate requirements is derived mainly from skeletal muscle. Alanine and glutamine released from the muscle are transported to the liver, where they are converted to glucose (gluconeogenesis). With protein breakdown, urinary excretion of nitrogen (predominantly as urea), potassium, phosphate, creatinine, magnesium, zinc, and sulfate are greatly increased, reflecting catabolism of protoplasmic mass.

NUTRITION

The metabolic response to the trauma of operation or injury is characterized by hypermetabolism, which, if left unchecked, results in increased tissue breakdown, loss of lean body mass, and depletion of essential intracellular constituents. The potential for malnutrition in both preoperative and postoperative surgical patients is great. Indeed, it has been suggested that nutritional depletion is present in 40%–60% of hospitalized surgical patients.

Malnutrition has been shown to increase the incidence of clean wound infections, to prolong postoperative ileus, to impair wound healing, to depress immunocompetence, to increase the patient's susceptibility to sepsis, to inhibit vital organ function, and to increase the risk of respiratory infections and respiratory insufficiency (Steiger, 1973; Law et al., 1973; McDougal et al., 1977b). Moreover, severe malnutrition results in loss or malfunction of various intestinal enzymes. Lactase seems to be particularly sensitive to variations in nutritional status. Thus, a patient who develops a lactase deficiency because of malnutrition will be incapable of absorbing supplements that contain milk or milk by-products. This, of course, greatly limits the type of supplements that can be administered orally.

If malnutrition is allowed to persist, there is a marked increase in morbidity and mortality. Loss of body protein appears to be the critical factor determining the point at which the nutritional depletion compromises the ability of the host to respond appropriately to the injury. Mortality and morbidity associated with weight loss and starvation are directly related to the loss of essential protein stores: death occurs with the loss of one fourth of body nitrogen or one third total body weight (Montemurro and Stevenson, 1960). Therefore, it is essential to limit and ultimately reverse the loss of body protein to promote early recovery and reduce the incidence of the life-threatening complications.

Nutritional Requirements

The daily caloric requirement for resting man is approximately 25 kcal/kg body weight. Hypometabolic, starved man may require somewhat less, but it is generally at the expense of limited organ function. Most surgical patients are hypermetabolic and will require an additional 5–60 kcal/kg body weight. Thus, when calculating the total caloric requirement in the average surgical patient, 30–35 kcal/kg body weight is employed. Daily protein requirements are normally 1–1.5 gm/kg/day, but under conditions of acute stress may be as high as 2.5–3.5 gm/kg body weight per day. While protein provides only 4 kcal/gm, it is the number of grams of protein that is important, with satisfaction of caloric requirements being provided by the addition of carbohydrate or fat. For each gram of protein administered, 25 kcal (in the form of carbohydrate or fat) should be provided. Protein consists of amino acids which are either essential or nonessential. The former cannot be manufactured by the body, whereas the latter can. In selected circumstances, it may be important to limit the total intake of protein or to alter the protein composition of the infusion. In patients with renal and hepatic failure, excessive amounts of protein may result in an excessively elevated BUN and/or serum ammonia and systemic manifestations of uremia in the former or hepatic encephalopathy in the latter. In these cases, 0.5–1.0 gm protein/kg body weight may be appropriate. Alteration of the composition of the protein infusion may also be helpful. Essential amino acid infusions without their nonessential counterparts may lessen the rise in serum urea in patients with acute renal failure, and branched-chain amino acids, which are metabolized by muscle and do not require the liver for metabolism, may be more appropriate in patients with hepatic disease.

Fats have a high caloric value and provide 9 kcal/gm. They are also classified as either nonessential or essential, depending on whether the body can manufacture them. Fats that cannot be manufactured by the body include linoleic, arachidonic, and linolenic. Only linolenic is absolutely essential, since a fatty acid deficiency will not develop if it alone is provided. Vitamins must be provided daily, particularly the water-soluble vitamins, as they are depleted rapidly. Sodium, potassium, calcium, magnesium, and phosphate must also be provided in sufficient quantities on a daily basis. Trace elements such as zinc, copper, iodine, manganese, and selenium must be provided over the long term.

A stable weight is often a good indication that basal needs are being met, provided that loss of lean body mass is not hidden by an increase in extracellular water. If the patient is incorporating exogenous protein into endogenous stores, he is said to be in a positive nitrogen balance. Nitrogen balance may be grossly calculated by assuming that 1 gm/day of nitrogen is lost in the feces if the patient is stooling, and about a quarter of a gram of nitrogen is lost through the skin. The remaining nitrogen loss occurs in the urine. Approximately 80% of the nitrogen lost in the urine is lost as urea nitrogen. Therefore, it becomes apparent that one may quickly calculate the nitrogen balance of any patient by measuring the 24-hour urine and its urea content. The amount of urea nitrogen excreted is multiplied by 1.25 to give the total nitrogen excreted in the urine. If one adds an additional 1–2 gm for skin and stool loss and subtracts this quantity from the protein nitrogen intake, one will obtain a positive or negative number. These collections are appropriate for patients in whom bowel is not interposed in the urinary tract. Patients with bowel interpositions alter urea excretion so that this measurement cannot be accurately performed. If the nitrogen balance is positive, the patient

is laying down or retaining protein (anabolism); and if negative, body protein is being broken down for energy requirements (catabolism).

Nutritional Status

Several modalities have been used to determine whether the patient is malnourished, and, if so, to quantitate the degree of nutritional deprivation. Among the indices most often used are weight loss, reactivity to skin test antigens, creatinine excretion index, mid-arm circumference measurement, tricep skinfold thickness, and the measurement of lymphocyte count, serum albumin, serum transferrin, retinal binding protein, and thyroxin binding prealbumin. A history of a recent weight loss is particularly important in determining current nutritional status. It must be remembered, however, that this may well underestimate loss of lean body mass, since when body protein is metabolized, there is an obligate increase in extracellular fluid. The weight loss, therefore, may not accurately reflect the loss of lean body mass. A patient who has lost fewer than 10 lb in the preceding 3 months is said to be mildly malnourished; between 10 and 20 lb, moderately malnourished; and more than 20 lb, severely malnourished. A patient's current weight can be compared with height/weight tables and a percentage deviation from normal obtained.

Reactivity to skin test antigens has been used frequently to determine a patient's immune status. Indirectly, the response to these antigens reflects the patient's nutritional status. If a patient is known to be allergic to one of the antigens and is anergic when tested or if anergic to agents that normally cause a response, severe malnutrition is present. Antigens commonly used are dermatophytin, mumps, PPD, streptokinase, and streptodornase. The importance of this index is illustrated by the fact that in several series, cancer patients did not respond effectively to chemotherapy or surgery if their skin test antigens were negative. If, however, the patients were nutritionally repleted, approximately one half converted to a positive status and then responded to either chemotherapy or surgery (Daly et al., 1980;. Copeland et al., 1976). The creatinine excretion index measures the lean body mass or muscle protein stores as does a measurement of mid-arm circumference. Triceps skinfold measurements indicate the status of fat stores. Lymphocyte count is a measure of visceral protein status and should normally be above 2,000/mm^3. If the lymphocyte count is between 1,200 and 2,000/mm^3, the patient is said to be mildly nutritionally depleted; between 800 and 1,200/mm^3, moderately nutritionally depleted; lower than 800/mm^3, severely nutritionally depleted.

Albumin levels are another measure of visceral protein status. Serum albumin has been demonstrated to accurately reflect the patient's nutritional status. Indeed, in one series the only parameter that correlated with an increased hospital stay in nutritionally depleted patients was their albumin status (Anderson and Wochos, 1982). Others have found it to be an accurate predictor of the success of a nutritional regimen as well as an indicator of the degree of nutritional repletion (Ching et al., 1980). If the albumin level is between 3 and 3.5 gm/dl, the patient is said to be mildly nutritionally depleted; if it is between 2.5 and 3.0 gm/dl, moderately nutritionally depleted; if it is less than 2.5 gm/dl, severely nutritionally depleted.

Serum transferrin levels have also been used as an indicator of visceral protein status. Transferrin determinations may not be readily available but can be calculated if one obtains a total iron binding capacity (TIBC). The formula for calculation is: serum transferrin $= (0.8 \times TIBC) - 43$. If the level is between 150 and 200 mg/dl, the patient is mildly nutritionally depleted; between 100 and 150 mg/dl, moderately nutritionally depleted; less than 100 mg/dl, severely nutritionally depleted. Retinal binding protein and thyroxin binding prealbumin are also measures of visceral protein status and are helpful in selected cases.

In practice it is often difficult for the busy clinician to seek out the various tables required for a complete nutritional assessment. Three modalities are conveniently utilized to assess the patient's status: weight loss, lymphocyte count, and albumin level. With these indices the patient can be placed into one of four categories: normal nutritional status, or mildly, moderately, or severely nutritionally depleted. If one is still unsure, skin test antigens may be employed. If the patient is anergic to a battery of skin test antigens, he is considered severely malnourished irrespective of the above indices (Table 10–1).

The amount of calories required per day for repletion must also be calculated so that appropriate amounts may be administered. The patient's weight $\times$ 25 gives the basal metabolic requirement of the patient. Depending on the severity of the insult, an additional 5–60 kcal/kg are added. In children the amount of kiloca-

lories metabolized varies according to weight. For the first 10 kg, 100 kcal/kg is metabolized; for the second 10 kg, 50 kcal/kg; and for each kilogram over 20, 20 kcal/kg (Table 10–2). The success of the regimen is determined by the return of the serum values to normal, the return of an allergic response to the skin test antigens if previously anergic, and weight gain. Moreover, periodic crude measurements of nitrogen balance as described above, if positive, will confirm that the amount and composition of the calorie load are appropriate. Having determined how much the patient requires, one must determine the route of administration: enteral, IV, or a combination.

Enteral Feedings

Enteral feedings are preferred if at all possible, since complications are decreased, and a more balanced, physiologic diet can be provided. Enteral diets are either nonelemental or elemental. Nonelemental enteral feedings consist of undigested and minimally digested protein hydrolysates, fat, and carbohydrates, whereas elemental diets consist of medium chain triglycerides, glucose, and amino acids. Enteral feedings can be provided either by a small feeding tube, a gastrostomy, or a feeding jejunostomy. Generally nonelemental enteral feedings are preferred, provided the gut is not diseased, because they have lower osmolality and are cheaper. The advantages of an elemental diet include its bulk-free and lactose-free composition (Fairfull-Smith et al., 1980). Therefore, patients with severely diseased bowel or patients who have been chronically starved are probably better served with an elemental diet—at least at the outset.

One should begin with the enteral feeding diluted to half strength at a rate of 50–75 ml/hour in the adult. If this is tolerated, the volume is increased to 2,500–3,000 ml/24 hours, and if tolerated, osmolality is increased to full osmotic content. For example, approximately 1 kcal/ml will be administered at 500–1,000 mOsm/kg. Complications include abdominal cramps, diarrhea, and diaphoresis. If any of these symptoms occur, the infusion is slowed, the osmolality is reduced, or both. Once the symptoms disappear, a gradual return to the

TABLE 10–1.

Nutritional Assessment: The Categorization of Patients Into Mild, Moderate, or Severely Nutritionally Depleted

MEASUREMENT	MILD	MODERATE	SEVERE
Weight loss, lb	<10	10–20	>20
Lymphocyte count, cells/mm^3	1,200–2,000	800–1,200	<800
Albumin, gm/dl	3.0–3.5	2.5–3.0	<2.5
Serum transferrin, mg/dl	150–200	100–150	<100
Skin test antigens	+	+	Anergy

TABLE 10–2.

Basic Caloric Requirements of Uninjured Humans at Rest

	KCAL/KG OF BODY WEIGHT
Children	
First 10 kg (0–10)	100
Second 10 kg (10–20)	50
Each additional kg >20	20
Adults	25

strength and rate desired is begun. If diarrhea continues to be poorly controlled, the administration of paregoric, 5 ml, in divided doses may be helpful.

Isosmotic Intravenous Nutrition

Carbohydrate, protein, and fat substrates may be infused individually or in combination in near-isosmotic concentrations. Because of their isosmotic character, not only may they be administered by peripheral vein, but also the rate of administration may be rapidly changed to satisfy a change in fluid requirements or even stopped so that medications, colloid, and blood may be administered. These properties make such solutions advantageous during periods of critical care when instability of the patient is not uncommon.

An understanding of the metabolic effects of each type of caloric source with respect to energy provision and its potential for endogenous protein preservation and maintenance of optimal organ function is essential if the proper substrate or combination of substrates is to be administered to acutely ill patients. The provision of glucose in doses up to 100 gm/day decreases protein loss as measured by the loss of urinary nitrogen. This protein-sparing effect is directly proportional to the quantity of calories administered. Infusion of larger quantities of glucose results in disproportionately lesser reductions in nitrogen losses. Supplying 700 protein-free calories to fasting normal man results in maximal reduction of protein losses. Increasing the nonprotein caloric intake is without further effect in the sparing of body protein. Indeed, positive nitrogen balance cannot be achieved even with high-dose glucose infusions. Starved, unstressed man given approximately 700 gm of glucose maintains a negative nitrogen balance of about 1.5 gm/m^2/day. However, if the same total caloric load is given, but part of the glucose calories are replaced by an equivalent amount of amino acid calories, the negative balance is eliminated (Wolfe et al., 1977). Supplying dietary protein with calories further improves nitrogen balance. Thus, on a fixed adequate protein intake, energy level is the deciding factor in nitrogen balance; and at a fixed adequate caloric intake, nitrogen intake is the determinant of nitrogen balance

(Calloway and Spector, 1954). Similarly, in critically ill traumatized patients and in those with superimposed bacteremia, at low-dose levels (or those which can be easily achieved employing isosmotic solutions), glucose has the same effect on nitrogen sparing as does an equivalent caloric load of amino acids (McDougal et al., 1977a). Thus, at low-dose levels, total caloric load, whether derived from protein or carbohydrate, determines the degree of nitrogen sparing, whereas when the caloric load is increased, amino acid intake becomes a more dominant determinant of nitrogen balance.

Fat emulsions may also be administered by peripheral vein and have the advantage of providing high caloric loads in relatively small volumes, since fat provides 9 kcal/gm, whereas glucose and protein provide a little less than 4 kcal/gm. The effect of fat emulsion on nitrogen sparing, however, is not equivalent to equal caloric amounts of glucose or amino acids. In normal unstressed man, equivalent caloric amounts of infused fat emulsion and glucose result in a lesser degree of nitrogen sparing for the former. When amino acids are added to equivalent caloric amounts of fat emulsion or glucose, the latter combination results in a less negative nitrogen balance. In severely injured man, some investigators failed to observe any reduction in nitrogen sparing with the infusion of large doses of soybean fat emulsions. On the other hand, more recent studies have demonstrated adequate utilization of fat calories in the immediate posttraumatic and postoperative period. In view of the degree of nitrogen sparing observed in normal and injured man, when used in combination with amino acids, fat emulsions seem helpful in sparing nitrogen, but glucose and amino acids are much more effective than fat, calorie for calorie.

Although there are no distinct differences in nitrogen balance in comparing low-dose isosmotic administration of glucose and amino acids, there are clear advantages to infusing a medium-dose combination of the two substrates. First, their effect is augmentative, and the impact on calories is not limited when protein is provided. Second, amino acids administered alone cause a constant rise in the BUN which is not observed when glucose is added. Third, altered liver and renal transport occur in critically ill patients who are given amino acids as their sole caloric source. The altered transport properties may be restored to normal by glucose addition. An appropriate combination that provides maximal nitrogen sparing per gram of nutrient administered while maintaining optimal hepatic, renal, and cardiac function consists of a liter solution in which one half is provided as 10% dextrose and the other half as a 7%–8% amino acid solution. Fat emulsions given in 500-ml amounts once or twice a day provide additional calories. Since these solutions may be administered by peripheral vein, and since alterations in infusion rate and even abrupt cessation of infusion can be accomplished without untoward effects so that blood, antibiotics, and other medications can be given, these infusates are an ideal means of preserving normal metabolic function of vital organs while limiting nitrogen loss in early post-traumatic, postoperative, and unstable critically ill patients. Unfortunately, it is generally not possible to achieve positive nitrogen balance with isosmotic solutions, since the amount of calories required would necessitate excessive fluid administration. If illness is protracted and oral alimentation not possible, positive nitrogen balance is achieved by administration of hyperosmotic solutions (hyperalimentation).

Hyperosmotic Intravenous Nutrition

Hyperosmotic IV solutions, which are capable of providing enough nitrogen and calories in an acceptable volume, are made up of equivalent amounts of a 50% dextrose and 7%–8% amino acid solution, providing approximately 1 kcal/ml of solution. Electrolytes including potassium, sodium, calcium, magnesium, and phosphate are added as required (Table 10–3). Trace elements (zinc, copper, manganese, chromium) and multivitamins are also added as required. In addition, essential fatty acids are provided by the administration of 500 ml of fat emulsion 2–3 times a week. Other serum components can be provided by administration of one to two units of fresh-frozen plasma weekly.

Because the hyperalimentation solution is hyperosmotic, it must be administered through a central venous line and its rate of administration rigidly controlled. If long-term or home hyperalimentation is to be administered, the solution should be given through a long-term indwelling catheter such as a Broviac or Hickman catheter. The catheters are placed in the superior vena cava, tunneled subcutaneously under the skin beneath the anterior chest wall, and brought out at about the level of the nipple—between it and the sternum. We have kept catheters functional as long as 21

TABLE 10–3.

Composition of Hyperalimentation Solutions

INGREDIENT	CONCENTRATION
Amino acid (7%–8%)	500 ml
Dextrose (50%)	500 ml
Potassium (KCl)	60–150 mEq
Sodium (NaCl)	60–180 mEq
Calcium (Ca gluconate)	5–15 mEq
Magnesium (MgSO$_4$)	8–24 mEq
Phosphate (K$_2$HPO$_4$)	15–20 mM
Trace elements	—
Multiple vitamins	—

months in adults and 14 months in children. In most cases, the hyperalimentation will be administered for limited periods, and a percutaneous central line is placed. The central venous line must be placed and maintained using assiduously sterile technique; otherwise infection will invariably follow. The skin over the puncture site is initially defatted with ether and then cleansed with an iodophor. The IV line is placed in the superior vena cava, either by a subclavian or internal jugular puncture. A sterile dressing is applied with an antibiotic ointment placed over the catheter entrance site. On alternate days the dressing is changed, and a new sterile dressing, iodophor preparation, and iodophor ointment are applied. A millipore filter is placed in line. The filter is changed and cultured every 24 hours. The IV tubing is changed with each bottle. The hyperalimentation solutions are made fresh daily and stored refrigerated. No medications, blood, or other fluid should be administered through the hyperalimentation line, nor should CVP measurements be made utilizing the catheter through which the hyperalimentation solution is being administered. Fluid should be administered intially at low rates (50 ml/hr) until tolerance has been achieved (blood glucose remains below 200 mg/dl), after which the rate of infusion may be gradually increased until the proper caloric load is achieved. Usually 3–4 L/day is given, providing 3,000–4,000 kcal/24 hr. Urine must be monitored for glucose every four hours; when present, either the infusion rate must be reduced or insulin administered. Blood glucose levels may be particularly difficult to control in the immediate postoperative and posttraumatic period. Insulin is given IV every four hours, adjusting it to blood glucose levels, usually suffices. Occasionally, blood glucose levels can be better controlled with a continuous insulin infusion. Two to three units of regular insulin are administered per hour in a concentration of one unit/ml of saline. If these simple measures do not control the glucose level or large amounts of insulin are required, the solution should be tapered or fat should be substituted for the glucose. It is apparent that high-dose insulin can lower serum glucose, but it is probable that under these conditions abnormal metabolism at the mitochondrial level occurs, obviating any beneficial effect of lowering the serum glucose. Similarly, after IV fat administration, the serum must be observed for clearance. Normally, immediately following the infusion of IV lipid, the serum will be lipemic. It should be totally clear approximately 8 hours after infusion. This may be checked by obtaining serum cholesterol and triglyceride levels or grossly at the bedside by drawing a blood sample, spinning it down, and observing the serum for lipemia. Plasma osmolality and sodium, potassium, chloride, CO$_2$, BUN, creatinine, calcium, phosphate, and glucose should also be monitored on a frequent periodic basis—initially daily. Magnesium levels should be monitored regularly, but somewhat less frequently.

When discontinuing the infusion, the rate should be gradually tapered over a 24–36-hour period. A prolonged infusion should never be abruptly stopped, for severe hypoglycemia may ensue. On the other hand, many patients with home hyperalimentation tolerate abrupt cessation of the infusion well. Indeed, these patients generally infuse their hyperalimentation solution over a 12-hour period while sleeping and immediately cease infusion when ambulatory. Critically ill and hospitalized patients are generally better served by tapering rather than an abrupt cessation. Complications are not infrequent with hyperalimentation and can be life-threatening. Placement of the central venous catheter has resulted in pneumothorax, hemothorax, hydrothorax, brachial plexus injury, venous thrombosis, and embolism. Because of the nature of the solution infused and the central venous location of the catheter, infectious complications have been reported and must be recognized and immediately corrected. Alterations in glucose metabolism may result in hyperglycemia, glucosuria with an osmotic diuresis, and in severe cases hyperosmolar nonketotic dehydration and coma. The mortality in the last situation can be as high as 50%.

Early recognition of hyperglycemia and its treatment by reducing the rate of infusion and/or administration of insulin correct these complications. A sudden change in glucose concentration in a patient who is receiving hyperalimentation should suggest sepsis. Patients with nonketotic hyperosmolar coma characteristically have blood glucose levels in excess of 500 mg/dl and must be treated aggressively, often with large doses of insulin and fluid. Ketoacids in diabetic patients given inadequate insulin for the additional carbohydrate load and postinfusion hypoglycemia resulting from rapid withdrawal of glucose in the face of persistently elevated endogenous insulin levels may also occur and are treated by increasing insulin and glucose infusions, respectively. Rarely, the complete metabolism of glucose for energy needs results in an elevated PCO_2 and carbon dioxide narcosis with respiratory insufficiency. Blood gas determinations confirm the diagnosis. Treatment is directed at reducing the amount of CO_2 produced for energy requirements by providing a greater share of the caloric load as fat. Alterations in amino acid metabolism may result in hyperchloremic metabolic acidosis, plasma and amino acid imbalances, hyperammonemia, and elevated BUN levels (Dudrick et al., 1972). Alterations in the content of the infusion or rate of infusion must be made. Hypophosphatemia and hypercalcemia and hypocalcemia are corrected by addi-

tion or removal from the infusate of the appropriate inorganic ion. Vitamin deficiencies, essential fatty acid deficiencies, trace mineral deficiencies, and abnormal plasma potassium, sodium, and magnesium levels all have been reported and are corrected by appropriate addition or removal of the particular substance. Trace minerals and essential fatty acid deficiencies are not encountered when fresh-frozen plasma and fat emulsions are administered as described above. Finally, alterations in liver enzymes, colostatic jaundice, and fatty infiltration of the liver complicate long-term administration. Indeed, the administration of more calories than required results in lipogenesis and fatty infiltration of the liver. This may be eliminated in patients receiving long-term hyperalimentation by administering it in a cyclic manner. Glucose is periodically withheld and only amino acids and fats are infused for an eight-hour period. This has been shown to reduce and, in fact, to clear fatty livers.

Once the amount of calories for maintenance of basal needs and restitution of adequate nutrition is determined, the route may be chosen. If the amount of calculated calories can be administered by oral or tube feedings, this is preferred. It appears that patients require fewer calories to maintain body functions and normal weight status if given orally than if given IV. If the gut cannot be used and the total caloric requirement must be replaced, IV hyperosmolar solutions must be employed. If the gut is temporarily unavailable and will be functional soon, isosmotic IV solutions may be used to limit catabolism until the gut is functional.

Intravenous Nutrition and Renal Failure

Surgical patients who have sustained acute renal failure have been successfully managed by providing them with hyperalimentation solutions. Providing adequate amounts of calories in these patients can be difficult, since fluid administration may be limited. This can be obviated by either frequent dialyses or continuous plasma filtration. Unfortunately, dialysis and plasma filtration result in 6–10 gm of amino acids lost per day. This amount must be added to the usual requirement to attain optimal balance. The mortality in surgical patients with acute renal failure is exceedingly high. According to some reports, the mortality may be reduced significantly by the provision of adequate calories and essential amino acids (Abel et al., 1973). Hyperalimentation in patients with acute renal failure, unfortunately, does not reduce the frequency of dialyses; it may have some effect on lessening the duration of renal failure. Essential amino acids have been found to be more efficacious than a combination of essential and nonessential regimens by some groups. It appears that those given essential amino acids have a less rapid rise in BUN and a reduced mortality compared with a similar group given essential and nonessential amino acids in their hyperalimentation solution (Freund et al., 1980).

Intravenous Nutrition and Hepatic Failure

Patients who have hepatic failure or hepatic encephalopathy are often nutritionally depleted. Straight-chain amino acids are metabolized by the liver, whereas branched-chain amino acids are metabolized by muscle. Amino acid profiles examined in patients with hepatic failure reveal that the concentrations of straight-chain amino acids are elevated, while the branched-chain amino acids are diminished. The assumption is that of the two major sites of amino acid metabolism, liver and muscle, the liver is incapable of metabolizing amino acids. Moreover, it has been proposed that if those amino acids that the liver metabolizes are withheld while those metabolized by muscle are given, the nutritional status of the patient might be improved. In one series, patients with cirrhosis given branched-chain amino acids had an 87% improvement in their condition, while 75% of those with hepatitis were improved after the administration of branched-chain amino acids (Freund et al., 1982). Patients with significant hepatic disease probably should have a limited protein intake, ranging between 0.5 and 1 gm/kg body weight. Simultaneous administration of oral neomycin and lactulose, by reducing gut flora and ammonia metabolism, may enhance protein tolerance.

FLUIDS AND ELECTROLYTES

Body Fluid Compartments

The treatment of many fluid and electrolyte disorders requires a knowledge of the body fluid compartments and the ability to calculate them in a given individual. Indeed, specific therapy of postobstructive diuresis, dehyration, hypovolemia, and water intoxication requires that knowledge for appropriate therapy. Total body water constitutes approximately 60% of the body weight in males and 50% in females. For a very lean person, 10% is added; for a very obese person, 10% subtracted, since muscle cells contain the greatest fraction of water. The total body water is divided into compartments: extracellular, intracellular, and transcellular (Table 10–4). The extracellular fluid compartment composes approximately 20% of the body weight and is divided into plasma (4.5%), interstitial fluid (16%), and lymph (2%). Since the blood volume is made up of solids as well as plasma, it can be calculated by multiplying the body weight by 7%. Intracellular fluid composes 30%–40% of the total body weight and is accessible only by freely diffusible molecules. Transcellular fluid constitutes 1%–

TABLE 10–4.
Body Fluid Compartments

COMPARTMENT	%
Total body water	50–60
Extracellular fluid	20–22
Blood plasma	4.5
Interstitial fluid	16
Lymph	2
Intracellular fluid	30–40
Transcellular fluid	1–3

3% of the total body weight and is composed of pleural, peritoneal, cerebrospinal, intraocular, salivary and digestive secretions. It is in equilibrium with the extracellular fluid, and in pathologic conditions may increase in amount, particularly during trauma. This is the so-called third space, fluid that is unavailable to the intravascular compartment in situations of injury.

The osmolality of the plasma is an indication of the endogenous substances contained in that compartment and is helpful clinically in situations of dehydration and hypervolemia. The osmolality is generally measured by freezing point depression but can be conveniently calculated by doubling the sum of the sodium and potassium concentrations and adding to that quantity the blood glucose level divided by 20 plus the blood urea level divided by three. These latter two substances are osmotically active and may contribute significantly to the total osmotic content:

$$\text{Serum Osm} = 2\text{Na} + 2\text{K} + \text{glucose}/20 + \text{urea}/3$$

Basic Fluid Requirements

The amount of fluid necessary to maintain homeostasis is equivalent to the urine output plus insensible loss plus abnormal losses minus the water produced by the metabolism of fat, carbohydrate, and protein (Table 10–5). Each of these entities must be calculated for the individual patient if optimal fluid balance is to be achieved.

TABLE 10–5.
Fluid Requirements*

	REQUIREMENT
Urine output (UO)	= 30–50 ml/hr (adult)
	1–2 ml/kg/hr (child)
Insensible loss	= 10–15 ml/kg/24 hr (adult)
	25–45 ml/100 kcal (child)
Abnormal loss	= Measured external or estimated third space loss
Water of metabolism	= 0.1 × 25 × body weight (adult)
	0.1 × kcal met (child)

*Basic fluid requirement = (UO + insensible loss + abnormal loss) − H_2O metabolism.

The amount of urine necessary to maintain proper balance is dictated by the physiologic limits of the kidney for solute and water excretion. In resting man, the products of normal metabolism produce a solute load which requires a minimum of 400–600 ml of urine for excretion. Traumatized and critically ill patients are hypermetabolic and may produce twice the normal solute load, necessitating a minimum of 800–1200 ml of urine excretion per day. On the other hand, excessive output may lead to a washout of the renal medullary osmotic gradient, resulting in impaired concentrating capabilities of the kidney. The fluid intake required to produce large urine outputs also may result in fluid retention and vascular overload. Therefore, there are limits between which urine output should be maintained. The adult kidney is most efficient in maintaining balance when fluid intake is sufficient to produce a urine output of 800–1200 ml/day, or 30–50 ml/hour. In children, urine output should be maintained between 1 and 2 ml/kg body weight/hour.

Insensible loss refers to the water lost from the respiratory tract and skin. The amount lost depends on the ambient temperature and humidity as well as the patient's BSA and body temperature. The normothermic adult in a comfortable environment loses 800–1,000 ml/day (10–15 ml/kg body weight/24 hours). Insensible losses in children are conveniently related to caloric consumption. A child will lose 25–45 ml of water for each 100 kcal metabolized. The amount of calories consumed per day may be calculated by multiplying the body weight by 100 for each of the first 10 kg, by 50 for each of the second 10 kg, and by 20 for each additional kg body weight in excess of 20 kg (see Table 10–2). Therefore a 10-kg child loses about 350 ml/day and a 20-kg child about 420 ml. Insensible losses increase by about 10% for each degree centigrade of temperature elevation above normal.

The water produced by metabolism is calculated from the caloric expenditure of the patient. The amount of water produced in milliliters is numerically equal to 10% of the total amount of kilocalories consumed. The resting adult metabolizes approximately 25 kcal/kg of body weight/24 hours, whereas the child's caloric expenditure is calculated on a graduated basis as described above. An 80-kg adult will consume 2,000 kcal (25 kcal/kg × 80 kg) and produce 200 ml of water, whereas a 12-kg child will consume 1,100 kcal (10 kg × 100 kcal/kg + 2 kg × 50 kcal/kg) and produce 110 ml of water. Since these volumes are small, the water of metabolism is often disregarded in total fluid calculations in patients with functioning kidneys.

Abnormal losses refer to fluids lost from the body by nasogastric suction, fistula drainage, vomiting and diarrhea, or from the vascular system by third-space seques-tration (retroperitoneal edema, operative trauma, ascites, bowel obstruction, etc.). The volumes of these losses are measured when external drainage occurs or estimated when sequestration is present and added to the total daily fluid requirements.

The total daily fluid requirement in an adult is calculated by adding the desired urine output, the insensible loss adjusted for temperature elevations, and measured or estimated abnormal losses. By monitoring the patient's urine output and weight, the appropriateness of the calculated fluid requirement may be determined. The patient who is not receiving total caloric replacement should lose approximately 1 lb/day unless, as in the immediate postoperative period, obligate third-space sequestration of fluid is occurring. If the urine output or weight status is inappropriate, the fluid administered is adjusted upward or downward accordingly. An example of total fluid replacement calculation follows: an 80-kg febrile (38°C) adult with a nasogastric tube draining 400 ml/day would require 1,200 ml urine output (50 ml/hr × 24 hr) plus 1,320 ml insensible loss (15 ml/kg of body weight/24 hr × 80 kg + 10% of this quantity for the one-degree temperature elevation) plus 400 ml abnormal loss (NG output), for a total of 2,920 ml/24 hr.

In children, total fluid replacement may be calculated from the calories metabolized. One milliliter of fluid is administered for each kilocalorie metabolized. Thus, a 25-kg child would require 1,000 ml for the first 10 kg, 500 ml for the second 10 kg, and 100 ml for the final 5 kg, for a total of 1,600 ml/day (see Table 10–2).

Basic Electrolyte Requirements

The average young adult eating a regular diet receives approximately 70–120 mEq of sodium, 60–80 mEq of potassium, 15–24 mEq of magnesium, and 80–140 mEq chloride per day. Although the kidney is extremely effective in conserving sodium, since it can reabsorb in excess of 99% of that filtered, total renal function is better preserved if enough sodium is administered so that maximum conservation of that filtered is unnecessary. Potassium, on the other hand, is not as efficiently conserved and therefore must be provided to avoid potassium depletion. Since magnesium is stored, patients who are in good nutritional balance prior to their illness do not require replacement, provided the period of IV therapy will be limited. Patients whose nutritional status is marginal or who have alcoholic cirrhosis will often require magnesium replacement at a rate of 5–20 mEq/day. Intravenous sodium, potassium, and chloride requirements may be satisfied by providing the adult with about 75 mEq of sodium chloride and 40 mEq of potassium chloride per day.

Baseline electrolyte requirements for children are

best calculated on the basis of caloric expenditure. Minimum 24-hour requirements are: 3 mEq of sodium/100 kcal, 2 mEq of chloride/100 kcal, and 2 mEq of potassium/100 kcal. Thus, the 30-kg child would require 1,700 kcal. The sodium requirement is $17 \times 3 = 51$ mEq, the chloride requirement is $17 \times 2 = 34$ mEq, and the potassium requirement is $17 \times 2 = 34$ mEq.

To these basic requirements, abnormal losses must be added. The fluid lost from the body may be analyzed for its electrolyte content to determine accurate replacement (Table 10–6). Fluid sequestered in a third space generally mimics the electrolyte content of plasma and therefore can be replaced accordingly.

Anuria

The fluid and electrolyte requirements for patients who are anephric or anuric are calculated from insensible and abnormal losses. Insensible loss: an adult should receive 10–15 ml/kg of body weight/24 hours, whereas a child should receive 25 ml of fluid/100 kcal expended. The caloric expenditure is estimated on the basis of weight as described above. One half of the fluid is administered as 10% dextrose in water and the other half as 5% dextrose in 0.2N saline. Potassium is generally not administered unless serum studies indicate the need for replacement. Abnormal losses are added to these basic requirements. These calculations are merely estimates of the patient's needs, and, therefore, fluid and electrolyte therapy must be continuously adjusted according to serum electrolyte analyses, patient weight, and, when appropriate, urine output.

Volume and Sodium Disturbances

DEHYDRATION.—When the minimal fluid requirements of the patient are not met, a water deficit accompanied by weight loss occurs. A 4% loss of body weight due to a water deficit requires emergent rehydration. A 6% loss of body weight resulting from lack of hydration results in a life-threatening condition often manifested by signs and symptoms of shock. The water lost may be relatively isotonic, in which case the dehydration is normonatremic, or hypotonic, in which case hypernatremia occurs.

TABLE 10–6.
Average Electrolyte Content of Gastrointestinal Losses (mEq/L)

	SODIUM	POTASSIUM	CHLORIDE
Gastric	60	10.0	90
Jejunum	105	5.0	100
Ileum	120	10.0	105
Cecum	80	20.0	50
Bile	175	5.0	100
Pancreas	170	4.5	75

Dehydration in urologic patients may be due to postobstructive diuresis, prolonged vomiting and diarrhea, and diabetes insipidus. Postobstructive diuresis may be either physiologic or pathologic. Physiologic diuresis occurs when volume overload precedes the relief of the obstruction or when serum urea is elevated. The kidney responds appropriately. when it is unobstructed by excreting the excess volume in the former circumstance or undergoing an osmotic diuresis due to the excess urea in the latter situation. The diuresis is self-limited, as it ceases when the kidney has returned the body to a homeostatic condition. Rarely, a pathologic diuresis is superimposed on the physiologic diuresis and occurs as a result of specific defects which cause a decreased proximal and distal tubule sodium reabsorption and lack of concentrating capabilities of the kidney due to a reduced medullary osmotic gradient (McDougal and Wright, 1972). Such patients do not respond to antidiuretic hormone or mineralocorticoid administration. Appropriate therapy consists of replacing insensible losses as outlined above in addition to measured losses. Since the sodium content in the urine usually ranges between 50 and 70 mEq/L, urine output is replaced with 0.5N saline. It is important not to overhydrate these patients, since such therapy will often perpetuate the pathologic diuresis and prevent reestablishment of the medullary osmotic gradient, prolonging the concentrating defect. As the concentrating ability of the kidney returns, fluid therapy is reduced accordingly. It is often helpful in patients who have severe diuresis to follow serum osmolalities as well as daily weights, adjusting therapy accordingly.

Gastric losses are repleted with 0.5N saline or normal saline to which potassium chloride is added. Diarrheic losses are replaced with lactated Ringer's solution. Diabetes insipidus may occur as a result of lack of antidiuretic hormone (ADH), collecting duct unresponsiveness to endogenous ADH, or lack of a medullary osmotic gradient. The patients are often slightly hypernatremic and mildly dehydrated. The urinary sodium content is generally low. Fluid administered should have a relatively modest sodium content until results of urinary electrolytes become available. The diagnosis should be sought expeditiously, since specific therapy will often correct the disorder.

VOLUME EXCESS.—Volume overload may be the result of either hypotonic or isotonic fluid excess. Hypotonic fluid excess results in the water intoxicaiton syndrome manifested by clouded sensorium, irrational behavior, changing neurologic signs, stupor, seizures, and coma. Water intoxication most commonly occurs as a result of hypotonic irrigant absorption through the prostatic bed in patients undergoing transurethral resections. The first signs of volume overload are usually a rising CVP and mental confusion. Blood pressure changes are not often noted initially. If the amount of water absorbed is minimal and CNS symptoms are not present, fluid restriction with judicious use of diuretics will suffice. In more severe cases where significant hyponatremia and CNS disorders occur, the treatment should include fluid restriction coupled with hypertonic sodium chloride (3%) administration. Diuretics are not effective in the presence of significant hyponatremia and should be used sparingly until the serum sodium is returning toward normal. Inappropriate secretion of ADH, often as a consequence of malignant tumors, is another cause of hypotonic fluid excess and also may result in hyponatremic fluid overload. These patients are often successfully managed by fluid restriction.

Isotonic fluid excesses are usually iatrogenic and are a consequence of excessive administration of sodium-containing fluids. The development of congestive heart failure, inappropriate weight gain, or peripheral edema depends on the severity of the overload. Such patients are treated with diuretics and fluid restriction.

Potassium Disorders

Hyperkalemia usually occurs in urologic patients as a result of acute renal insufficiency, addisonian crisis, trauma, shock, and diabetic acidosis. Life-threatening elevations in serum potassium often produce significant ECG alterations. Peaking of the T wave, lengthening of the P–R interval, prolongation of the QRS complex, and loss of the P wave occur with progressive hyperkalemia. As potassium continues to rise, the ECG may ultimately resemble a sine wave. Calcium gluconate protects the heart from the adverse effects of potassium and may be administered in severe cases of hyperkalemia. Treatment consists of IV administration of hypertonic sodium bicarbonate, which results in a shift of hydrogen ion out of the cell and concomitant movement of potassium into the cell, causing a temporary lowering of serum potassium. Glucose and insulin therapy (10 units of regular insulin plus 50 gm of glucose) is recommended only in extremely urgent situations. Potassium is temporarily bound during glucose transport, thus effectively removing it from the serum. Since these measures are only temporary, simultaneous institution of therapy to lower the serum potassium permanently is mandatory. This may be accomplished with ion exchange resins (Kayexalate given per os or per rectum), peritoneal dialysis, or hemodialysis.

Hypokalemia usually results from excessive upper GI losses, diuretic therapy, steroid administration, and hyperaldosteronism. Metabolic alkalosis is often associated. Therapy is directed at eliminating the disorder and replacing the loss. Administration of sodium-containing solutions in the face of hypokalemia may promote renal potassium loss, particularly in patients with primary hyperaldosteronism. When sodium losses occur concomitantly with potassium deficits, both must be replaced simultaneously. If rapid replacement therapy is required, the ECG should be continuously monitored. If ECG changes occur, the infusion must be slowed or stopped until the abnormalities resolve.

Hypercalcemia

Hypercalcemia in urologic practice is generally the result of metastatic tumor to bone, hydrochlorothiazide therapy, and hyperparathyroidism. The symptoms of hypercalcemia include anorexia, weakness, somnolence, polyuria, and coma. Initial therapy involves establishment of a sodium diuresis by administering IV saline and nonthiazide diuretics. If the diuresis must be prolonged, careful monitoring of serum potassium and magnesium concentrations is essential to prevent deficiencies.

Inorganic phosphate administration will rapidly lower serum calcium, but results in metastatic calcification of soft tissues. Ethylenediamine-tetra-acetate (EDTA), a chelating agent, also rapidly lowers serum calcium but has many associated complications. These two agents should not be used unless life-threatening hypercalcemia occurs. On rare occasions, emergency therapy is required in patients with hyperparathyroidism who have uncontrollable serum calcium levels. Emergency parathyroidectomy may be lifesaving when the hyperparathyroid crisis cannot be controlled medically.

Mithramycin and steroids also have been used to control hypercalcemia. With both agents, a fall in the serum calcium does not usually occur for 24–48 hours. Mithramycin, a cytotoxic agent used in the treatment of malignancies, has many serious side effects including bone marrow depression, renal failure, and hepatic toxicity. It should be used only in patients with neoplasms and then only when more conventional methods fail. The dose is 1–2.5 mg/day for three days. Steroids are somewhat less effective and are best employed in disorders in which vitamin D sensitivity is etiologic. Dialysis also may be employed, particularly in patients with associated renal failure.

A saline diuresis should be the first modality tried, reserving the other forms of therapy for specific indications. It is important that diagnostic studies be instituted early to define the cause of hypercalcemia so that definitive therapy may proceed without delay.

Hypermagnesemia

Hypermagnesemia interferes with neuromuscular transmission both peripherally and centrally. The signs and symptoms include deterioration of mental function, drowsiness, muscular paralysis, and, in severe cases, coma. Nausea, vomiting, peripheral vasodilation, and hypotension may also occur. The ECG shows prolongation of the Q–T interval. Persistent hypermagnesemia may cause soft tissue calcification and interfere with bone mineralization. With normal renal function, hypermagnesemia rarely occurs. Patients with decreased renal function who ingest magnesium-containing medications may be particularly prone to the disorder. Urologic patients in whom magnesium-containing solutions are used to irrigate the urinary system for dissolution of stones may also manifest hypermagnesemia. This is particularly true when the irrigant, such as Suby's solution G or Renacidin, is used at increased pressures in areas where vascular beds are exposed. When the symptoms are severe, emergency treatment using calcium gluconate may be necessary. However, in patients with normal renal function, hydration and furosemide administration generally will suffice. Rarely, patients with decreased renal function or those with severe neurologic symptoms require hemodialysis to return magnesium levels to normal.

Hypomagnesemia may be a complication of aminoglycoside therapy, liver disease, and nutritional deficiency. Symptoms include somnolence and weakness.

Anion Gap

The anion gap is defined as the difference between the sum of the major cations, sodium and potassium, minus the sum of the major anions, bicarbonate and chloride, and is normally about 16 mEq/L. The anion gap is of considerable importance in distinguishing among several types of acid-base disturbances. Most of the acidoses commonly found in urologic practice do not have an increased anion gap: hyperchloremic metabolic acidosis, renal tubular acidosis, uremic acidosis, and the acidosis accompanying diarrhea or excessive ileostomy drainage. An increased anion gap is most commonly associated with ketoacidosis, lactic acidosis, hyperosmolar hyperglycemic nonketotic coma, and occasionally uremic acidosis. Rarely, hyperchloremic metabolic acidosis due to enteric urinary absorption may present with an anion gap.

Acid-Base Disturbances

ACID-BASE BALANCE.—Under normal dietary conditions, man generates acid at 70–100 mEq/day, equal to 1 mEq/kg of body weight/24 hours. This acid load is buffered by both extracellular and intracellular buffers that include hemoglobin, protein, inorganic phosphate, organic phosphate, and the bicarbonic-carbonic acid buffer system. Organic phosphate represents the principal intracellular buffer, whereas the bicarbonate buffer system constitutes the major extracellular buffer. Under normal circumstances, these buffers can accommodate approximately 15 mEq/kg body weight of hy-

drogen ion without causing major shifts in systemic pH. Acid-base disturbances are classified into one of four basic types: (1) metabolic acidosis, (2) metabolic alkalosis, (3) respiratory acidosis, and (4) respiratory alkalosis. Irrespective of the primary etiology of the acid-base abnormality, the body compensates by establishing an acid-base disturbance. Thus, in patients with metabolic acidosis there is compensatory respiratory alkalosis, and in those with respiratory alkalosis there is compensatory metabolic acidosis, and so on. The compensatory mechanisms are generally incomplete, so that despite compensation, the pH generally remains on the side of the primary disturbance. Therefore, if the pH is below 7.38, the primary disorder is acidosis; if above 7.42, the primary disorder is alkalosis. The arterial PCO_2 distinguishes between respiratory and metabolic. A patient with a pH of 7.44 and a PCO_2 of 48 would have a primary alkalosis, which would be metabolic in view of the elevated PCO_2. The disturbance is a primary metabolic alkalosis with a secondary compensatory respiratory acidosis. The lungs are the primary mediator of the respiratory compensation and the kidneys the mediator of metabolic compensation. Clinically, the four types of acid-base disturbances are not classified as pure and are generally compensated as described. However, for illustrative purposes, each will be discussed individually.

METABOLIC ACIDOSIS.—Metabolic acidosis may be due to a decreased extracellular bicarbonate concentration or may be a consequence of an increased extracellular volume without a proportional increase in bicarbonate content. The latter explains the acidosis of dilutional hyponatremia and water intoxication. A shift in hydrogen ion from the cell to the extracellular fluid compartment and net loss of body bicarbonate account for the acidosis. Decreased extracellular bicarbonate concentration may be due to either a decrease in renal bicarbonate reclamation or to the consumption of serum bicarbonate due to an excessive acid load to the systemic circulation. In the former the acidosis develops slowly, whereas in the latter it is of acute onset. Decreased generation of bicarbonate by the kidney occurs in uremic acidosis, renal tubular acidosis, and in aldosterone deficiency. An increased production of metabolic acid may occur because of increased protein intake or an increased rate of tissue catabolism, increased production of endogenous organic acids such as lactic acid (shock) and ketoacids (diabetes), the administration of exogenous acids such as ammonium chloride, and finally extrarenal losses such as GI losses of bicarbonate in patients with diarrhea or an ileostomy. Excessive drug ingestion (salicylate intoxication) may also cause an acidosis. The effects of an acute acidosis include dilatation of arterioles, impairment of cardiac contractility, and systemic venous constriction, while chronic acidosis results in depletion of bone alkaline stores due to the buffering capacity of the bone. The bone generally buffers 30–40 mEq of acid per day in chronic acidotic states. The differential between a renal or extrarenal etiology can conveniently be made by checking the urinary pH. If the acidosis is due to exogenous or nonrenal mechanisms, the urinary pH will be acidic. On the other hand, if the mechanism is due to lack of bicarbonate generation and reclamation, the urinary pH will be persistently alkaline.

METABOLIC ALKALOSIS.—When metabolic alkalosis occurs, there is a net addition or increase in extracellular bicarbonate concentration. Since the kidney has a great capacity to excrete bicarbonate, two conditions must be met for metabolic alkalosis to occur. First, there must be a mechanism for increasing extracellular bicarbonate concentration, and second, there must be a mechanism to prevent the kidney from excreting the excess bicarbonate. This may occur as a result of increased loss of hydrochloric acid from the stomach following vomiting or nasogastric tube drainage; increased loss of hydrochloric acid in the stool in patients who have a defect in chloride reabsorption in the ileum and colon (congenital chloridorrhea); potassium depletion, which results in a shift of hydrogen ion into the cell; contraction of the extracellular volume without a proportional decrease in bicarbonate content; an increased renal production of bicarbonate due to the use of a diuretic; respiratory acidosis; primary hyperaldosteronism; and hypoparathyroidism. Excessive ingestion of alkali such as the milk alkali syndrome may also be etiologic.

RESPIRATORY ALKALOSIS.—Hyperventilation results in respiratory alkalosis and is always a result of stimulation of the respiratory center. This may be due to hypoxia, drugs, toxins, CNS disorders, psychogenic causes, and as a compensatory mechanism for metabolic acidosis. When primary, sepsis must always be suspected.

RESPIRATORY ACIDOSIS.—This results from hypoventilation and may be caused by depression of the respiratory center either due to CNS disease, drugs, defects in nerves and muscles of the respiratory center due to disease or injury, thoracic cage disorders, airway obstruction and chronic pulmonary disease.

FLUID AND ELECTROLYTE ABNORMALITIES ASSOCIATED WITH GENITOURINARY IRRIGANTS

Water Intoxication

Water intoxication in urologic practice is generally a result of excessive nonelectrolyte irrigant absorption during endoscopic or endourologic procedures. As the nonelectrolyte fluid is absorbed, volume expansion and dilutional hyponatremia occur. The clinical manifestations of this syndrome (TUR syndrome) were first described in 1946 following a transurethral resection ("TUR") of the prostate in which the patient was noted to become restless. There was dark red discoloration of the serum, progressive oliguria, azotemia, and pulmonary edema, followed by death. The development of the syndrome was associated with an 18.5% mortality (Creevy and Webb, 1947). As more experience has been gained with transurethral resections, this syndrome has become better defined. Following significant volume expansion with resultant hyponatremia, patients generally complain of nausea, become mentally confused or restless, have sensory disturbances, and if allowed to progress, blindness, convulsions, hypotension, coma, oliguria, and death supervene.

During a transurethral resection, some fluid is invariably absorbed; however, usually this is not sufficient to cause the clinical manifestations of water intoxication. The amount of fluid absorbed is directly related to: (1) the pressure of the irrigant, which is a function of the height the bag is placed above the prostatic fossa; (2) the intravesical pressure; (3) the intraprostatic pressure; and (4) the CVP. The duration of the resection, the size of the gland, the quality of resection and the type of resection equipment utilized also play a role. Blood loss, like volume absorption, has been correlated with the time of resection. As a rule, 20 ml/min of fluid is absorbed and 2 ml/min of blood is lost during resection. A continuous-flow resectoscope is more likely to produce significant water intoxication than the inflow-outflow method. Perhaps most important is the quality of resection, since frequent capsular penetration and entrance into venous sinuses predispose the patient to the development of water intoxication. The first sign that should arouse suspicion is an increase in CVP or left atrial pressure (Swan-Ganz). A change in blood pressure is generally not an early sign and in fact may not be noted, even when the syndrome is severe enough to cause blindness. The pathophysiology of the TUR syndrome has been reasonably well elaborated. Following nonelectrolyte irrigant absorption, serum sodium and chloride fall as volume overload occurs. Tachypnea, hypertension, and subsequently bradycardia due to the carotid reflex may occur. As electrolyte changes occur due to dilution, cardiac output falls, plasma volume is variable, and body weight increases. As the extracellular osmolality decreases, cerebral edema occurs, causing confusion, seizures, and blindness. Pulmonary edema also may occur with hypoxemia, cyanosis, and acidosis, and alteration of the clotting factors may occur with hemorrhage, hemolysis, anemia, and shock. Patients at increased risk for development of this syndrome are those who have cardiac disease and have been on a low-salt diet with diuretic supplementation, patients with hydronephrosis, salt-losing nephritis, urinary retention, and chronic illness and malnutrition.

Treatment of Water Intoxication Syndrome

Patients who are symptomatic and manifest severe neurologic abnormalities require rapid correction of their electrolyte status. This may be accomplished by the simultaneous administration of a potent diuretic such as furosemide with the restoration of serum sodium content by the infusion of hypertonic saline. The sodium deficit is calculated by subtracting the current serum sodium from the desired serum concentration and multiplying the value by the quantity $0.2 \times$ body weight in kilograms (extracellular fluid volume). This gives the mEq amount required to return the serum sodium to the desired value. One half of this amount is administered rapidly, following which a second serum sodium concentration is obtained and the therapy modified accordingly. For example, an 80-kg man whose serum sodium is 110 mEq/L would require 400 mEq to return serum sodium to 135 mEq/L. One half, or 200 mEq (about 400 ml of 3% NaCl), would be given rapidly, monitoring the neurologic, pulmonary, and cardiac status. The remainder would be given as needed, depending on a repeated serum measurement and clinical status. For patients who are less symptomatic and have a less severe hyponatremia, the administration of a potent diuretic and the infusion of normal saline may suffice.

Other Problems With Irrigants

If the area of instrumentation is infected, the irrigant may carry with it bacteria, resulting in bacteremia occasionally followed by sepsis. Other electrolyte abnormalities depend on the specific irrigant used. Currently, five types of isotonic fluids are in general use: sorbitol, 3.3%; glucose, 5.4%; glycine, 1.5%; mannitol, 3.0%; and urea, 1.8%. Water is also used; however, it is not isotonic.

WATER.—Because water is not isotonic, it can result in significant hemolysis of RBCs when it enters the systemic circulation. This hemolysis results in an increased level of serum potassium and, during a typical transurethral resection, elevated serum levels of free hemoglobin. It has been shown, however, that hemoglobin is not toxic in levels up to 600 mg, provided it is not combined with abnormal serum proteins. In the latter case it may in fact be nephrotoxic.

3.3% SORBITOL.—Sorbitol is metabolized either completely to carbon dioxide and water or to dextrose. It may also be excreted by the kidneys. The use of this solution may result in an elevated level of serum glucose, which may be particularly severe in noncompensated diabetics. The same propensity to water intoxication occurs with this fluid as with any other urologic nonelectrolyte irrigant fluid. Rarely, sorbitol may result in an osmotic diuresis with dehydration and a hyperosmolar state. This solution should be used with caution in diabetics.

GLUCOSE.—Glucose solutions are no longer generally used because they are sticky and not are comfortably handled in the urologic suite. However, if used, they have the same complications as sorbitol.

1.5% GLYCINE.—Ammonia intoxication has been reported following the use of glycine. Patients particularly prone to this complication are those with impaired liver function who presumably are unable to clear the ammonia generated from glycine metabolism. If the patient receives large doses of glycine, it may also cause salivation, nausea, and lightheadedness.

5% MANNITOL.—The main side effect of mannitol is an osmotic diuresis that can result in dehydration and hyperosmolality. Occasionally, this solution may cause systemic acidosis.

UREA.—Urea is no longer conventionally used because of its permeability both to the intracellular and to the extracellular space. Its use results in elevated serum urea concentrations and may also cause an osmotic diuresis.

Volume Deficits

With significant loss of body fluids, dehydration, hyperosmolality, and neurologic disturbances occur. If allowed to persist, hypotension and death supervene. With a 4% loss of body weight due to a water deficit, emergency rehydration is in order. These patients are generally symptomatic. A 6% loss of body weight due to a water deficit results in shock and often death. Common urologic disorders that lead to dehydration include post-obstructive diuresis, which results in a urine with a sodium content of 50–70 mEq/L and a potassium content of 10–40 mEq/L. These patients' urinary fluid losses may be accurately replaced with ½N saline. Prolonged vomiting or diarrhea may also result in significant dehydration. The former is replaced with ½N saline plus 40 mEq/L of potassium chloride/L if renal function is normal, and the latter replaced with Ringer's lactate. Finally, diabetes insipidus may result in significant dehydration. This syndrome is discussed in detail in the section on polyuria. During replacement therapy and in the follow-up period, daily body weight, serum sodium, and serum osmolality are measured to determine the efficacy of therapy.

Intestinal Conduits

The intestine's primary function is to selectively absorb electrolytes and nutrients. This function makes it less than an ideal structure as a urinary conduit or storage vehicle. Urine exposed to its surface will be altered as some constituents are reabsorbed and others are diluted due to intestinal secretion. The extent to which the fluid is altered depends on the surface area to which it is exposed, the time of exposure, and the composition of the urine.

Electrolyte transport occurs throughout all segments of the bowel. In the jejunum sodium absorption is coupled with glucose transport, whereas in the ileum and colon, sodium and chloride are actively absorbed. In these segments of bowel, there may be an exchange between sodium and hydrogen and chloride and bicarbonate. Thus with sodium absorption hydrogen is secreted, and with chloride absorption bicarbonate is secreted. The movement of these ions depends on cAMP, and therefore blocking this enzyme alters transport properties (Johnson, 1981). Metabolic disorders of intestinal urine transport are treated by choosing specific drugs that block this enzyme system. Water moves according to its concentration gradient; thus, hyperosmotic urine results in the movement of water into the conduit from the systemic circulation. Hypoosmotic urine, on the other hand, may result in a net movement of water into the extracellular space. The flux of water is particularly prominent in the jejunum, less so in the ileum, and least in the colon. Metabolic abnormalities brought about by intestinal alteration of urine are often compensated for by the kidney's ability to increase and alter its excretion rates of unwanted solutes. Severe metabolic disturbances often manifest themselves only when renal function is compromised—particularly when the serum creatinine concentration exceeds 2 mg/dl.

THE SYNDROME OF HYPERCHLOREMIC METABOLIC ACIDOSIS.—The ingestion of acetazolamide, or acid, diarrhea, and intestinal fistulas can cause hyperchloremic acidosis; however, the most common cause in urologic practice is intestinal interposition. This syndrome occurs in patients who have ileum or colon interposed in the urinary tract; is most common in patients in whom the urine remains in contact with the intestinal mucosa for extended periods, particularly those with ureterosigmoidostomies; and is most severe in the face of compromised renal function. The acidosis does not usually manifest an anion gap; however, recent evidence suggests that an anion gap may rarely occur in those with rectal bladders and other bowel substitutes in which the urine remains in contact with the intestinal mucosa for extended periods. This anion gap is most likely due to an increased absorption of phosphate, sulfate, and ammonium, resulting in decreased total body elimination of these anions.

Four hypotheses have been proposed to explain the acidosis: (1) renal tubule acidification defect, (2) intestinal absorption of ammonium, (3) intestinal bicarbonate secretion, and (4) active intestinal chloride transport. A renal tubule acidification defect seemed likely, since many of these patients had ascending pyelonephritis. Pyelonephritis affects the distal tubule, and it was proposed that this interfered with the kidney's ability to acidify (Lapides, 1951). It is clear that a distal tubule acidification defect will make the acidosis much worse; however, it is unlikely that it is the primary mechanism, since many patients with normal renal function also manifest the acidosis (Stamey, 1956). The second hypothesis proposed that the urea excreted in the urine is split to ammonium by intestinal bacteria, thereby increasing ammonium absorption, which accounts for the acidosis (Boyce and Vest, 1952; Rosenberg, 1953). Since ammonia derived from urea is not an acid (indeed, when hydrated with water, it is a base), this mechanism does not explain proton addition (hydrogen ion) to the systemic circulation. It may be, however, that ammonia serves as a proton acceptor from acid secreted in the urine and then accompanies, in a passive manner, actively reabsorbed anion. The third mechanism proposed is bicarbonate secretion by the intestine. Clearly, the ileum and large bowel are capable of significant bicarbonate losses, since patients with ileostomies and those with severe diarrhea can lose enough bicarbonate to become acidotic. Several experiments have also demonstrated bicarbonate loss in measurable quantities in patients with ureterosigmoidostomies (D'Agostino et al., 1953). The amount of bicarbonate lost to the body in most cases of intestinal interposition, however, seems insufficient to explain the severity of the acidosis. Indeed, experimental evidence in animals suggests that this plays only a very minor role in most cases (Koch and McDougal, 1985c). The final mechanism proposed, that active chloride transport was the primary etiology of the acidosis (Farris and Odel, 1950), currently seems to have the most evidence to support it. The hypothesis is that chloride is actively transported and then requires a cation to preserve electrical neutrality to diffuse across the membrane. The cation may either be hydrogen or ammonium, whichever is most available. Thus, in effect hydrochloric acid and/or ammonium chloride enter the systemic circulation in sufficient quantities to account for the acidosis. The evidence to support this hypothesis comes from experiments that showed that when ion fluxes are measured across the intestinal segment exposed to urine, net hydrogen uptake is directly proportional to chloride transport. Moreover, thorazine and nicotinic acid, two drugs known to block chloride transport through their effect on cAMP, correct this acidosis (Koch and McDougal, 1985a, 1985b).

Other electrolyte abnormalities associated with the hyperchloremic metabolic acidosis include hypokalemia, hyperkalemia, hypocalcemia, hypomagnesemia, and hypersulfatemia. Hypokalemia may be severe, since the chronic acidosis falsely elevates the serum potassium and therefore obscures the severity of the total body potassium deficiency. Upon correction of the acidosis, these patients often require considerable replacement of potassium to replenish body content. The intestinal loss of potassium can be reduced by the administration of spironolactone.

The mechanism of hypocalcemia and hypomagnesemia is not clear. It has been suggested that the chronic acidosis results in depletion of bicarbonate stores in the skeleton with release and loss of calcium initially through the urinary tract. The hypocalcemia is merely a reflection of total body depletion. Reports of rickets and growth retardation in some of these patients (Boyd, 1983) suggest that abnormal calcium metabolism may be clinically significant. Large losses of calcium in the urine with stimulation of calcium reabsorption by the tubule through the mechanism of parathormone would result in magnesium excretion by the binding of all available transport sites for calcium under maximal parathormone stimulation. Thus, excessive magnesium loss through the urinary tract results in hypomagnesemia. Initial experiments in our laboratory suggest that the increased intestinal absorption of sulfate results in that ion blocking renal reabsorption of calcium and magnesium. Whatever the mechanism, it is clear that hypocalcemia and hypomagnesemia occur most commonly when renal function is significantly impaired. In patients with normal renal function, serum concentrations of these two ions are rarely more than slightly depressed.

Urea and creatinine are also reabsorbed by the bowel and may result in elevated serum levels, which do not accurately reflect renal function. Since transport properties vary somewhat for various portions of the bowel, electrolyte abnormalities specific for each segment are described subsequently.

JEJUNUM.—The jejunum is seldom employed for intestinal diversion because of the severe fluid and electrolyte abnormalities that occur when it is used as a conduit or storage vehicle for urine. These segments lose large quantities of sodium chloride and absorb significant amounts of potassium and urea. With the extensive sodium and chloride loss into the lumen of the intestine, the osmotic gradient this creates necessitates the movement of water from the systemic circulation into the lumen of the intestine. This results in significant fluid losses to the body. Some hydrogen absorption and bicarbonate loss may also occur. These patients become hyponatremic, hypochloremic, hypovolemic, hyperkalemic, and azotemic. They may also have a mild acidosis and an associated serum hyperosmolality (Clark, 1974; Jolinbur and Morales, 1975). In response to these abnormalities, there is an increased secretion of aldosterone and renin, thus further reducing urine chloride concentration and thereby increasing the concentration gradient of chloride between serum and lumen, enhancing secretion by the jejunum.

ILEUM.—Ileum is perhaps the most commonly employed segment for urinary diversion. Chloride is actively absorbed and bicarbonate secreted by this segment. Although rarely a clinical problem, these patients may develop a hyperchloremic metabolic acidosis which may be associated with hypokalemia (Creevy, 1960). The metabolic abnormality is more common in patients with compromised renal function. Although water does move into the conduit if urine flowing through it is hypertonic, since the ileum does not secrete large amounts of osmotically active agents such as sodium and chloride, the amount of fluid lost is not as extensive as for the jejunum.

COLON.—The colon also actively transports chloride, and patients with colon segments are therefore prone to develop hyperchloremic metabolic acidosis. These patients are more likely to become hypokalemic, hypomagnesemic, and hypocalcemic than patients with ileal segments. They are less likely, however, to become hypovolemic, for fluid fluxes are less prominent in the colon than in ileal or jejunal segments. Indeed, patients with colon conduits can actually concentrate to about 400 mOsm/kg, as opposed to ileal segments in which the gradient generally does not exceed 350 mOsm/kg (McDougal and Koch, 1986).

TREATMENT.—The treatment of hyponatremic-hypochloremic, hyperkalemic, metabolic acidosis found in jejunal segment patients consists of oral sodium chloride, bicarbonate, and fluids. When severe, IV saline administration in large volumes and the establishment of a diuresis corrects both the hyponatremia and hyperkalemia as well as the hypovolemia.

Hyperchloremic metabolic acidosis is treated by oral chloride restriction, bicarbonate replacement, either in the form of sodium bicarbonate or polycitra, and drainage of the area of storage. In patients with a Kock pouch, rectal bladder, or ureterosigmoidostomy, this requires catheterization of the segment. Recent evidence indicates that thorazine and nicotinic acid may be effective. These drugs block active chloride transport in the intestine and so prevent the acidosis. This has been demonstrated both in experimental animals and in patients (Koch and McDougal, 1985a). The advantage to these drugs is that a sodium or potassium load is not given as when bicarbonate or polycitra is used. This may be particularly useful in patients who have congestive heart failure or those with severely compromised renal function in whom an excessive sodium or potassium load would be inappropriate. High doses and prolonged use of chlorpromazine may result in the development of extrapyramidal symptoms including tardive dyskinesia. Nicotinic acid's side effects are usually minimal but may consist of agitation and flushing. Because of the CNS effects of chlorpromazine, we prefer to use nicotinic acid in children and chlorpromazine in adults.

BOWEL PREPARATION

Urologic operations in which bowel is utilized as a conduit or storage vehicle for urine are invariably elective, and therefore proper preparation of the bowel may be planned preoperatively. The major risks involved in operating on the bowel include wound infection, anastomotic breakdown, and peritoneal infections. In unprepared bowel, the wound infection rate varies between 32% and 58%. This incidence may be reduced to 6%–9% if mechanical and antibiotic bowel preparation are employed (Clarke et al., 1977). Breakdown of the intestinal anastomosis is more common when performed on unprepared bowel. Antibiotics appear to have a protective effect on the anastomosis, particularly if the blood supply is compromised. In experimental animals, an unprepared bowel that is devascularized results in perforation and death, whereas an antibiotically prepared bowel that is devascularized heals. Finally, intraperitoneal abscesses are more common when unprepared bowel is operated on.

In mechanical preparation of the bowel the enteric content is reduced, and in antibiotic preparation the bacterial organisms are reduced. The mechanical bowel preparation reduces the amount of bacteria but not the number of bacteria per milliliter of feces. Spillage of enteric contents is less likely, since there is less of it to spill; however, with a spill, the inoculum is the same as if the spill occurred in unprepared bowel. The conventional method for preparing the bowel involves the patient's beginning a low-residue diet 3–5 days prior to surgery. Two days preoperatively, clear liquids are begun, and citrate of magnesia or phosphosoda is given orally. On the day before surgery, enemas are given until clear. This preparation may be exhausting to the patient, and in nutritionally unstable patients may exacerbate nutritional depletion. Indeed, significant weight loss and a negative nitrogen balance occur in all patients undergoing this type of preparation. To prevent the nutritional depletion, the administration of low-residue elemental diets has been suggested. These diets, however, are incapable of producing a clean colon and do not reduce the amount of bacteria. If given in sufficient amounts, they are effective in limiting nutritional depletion.

A second method of mechanical preparation involves whole bowel irrigation. A nasogastric tube is inserted through which saline warmed to 37° C is infused, 3–4 L/hr. This infusion is preceded by 10 mg of metoclopramide hydrochloride given IV to stimulate bowel activity. Forty to 60 minutes after beginning the irrigation, bowel motion begins. The irrigant is continued for one hour after clear fluid is passed per rectum. This usually takes 2–3 hours, consuming a total amount of irrigant of 9–12 L. The advantage of this regimen is that there is a better reduction of aerobic flora and a higher percentage of collapsed bowel (Christensen and Kronborg, 1981). In theory anastomotic leaks are less common when large and small bowel are collapsed for surgery. The disadvantages of whole bowel irrigation include a 3–4-lb weight gain, nausea, vomiting, and abdominal pain. Indeed, the catharsis is so active that the patient needs to remain on the commode during the irrigation process. Because of the significant fluid gain, this type of preparation is contraindicated in patients over 65 and those with renal disease, cirrhosis, heart failure, and bowel obstruction (Glass et al., 1981). This preparation, however, can be instituted on the day before surgery, thus lessening nutritional depletion.

A third type of mechanical preparation involves the oral administration of a mannitol or an electrolyte plus nonreabsorbed sugar solution (Hares et al., 1982). Four liters of 5% mannitol or 2 L of a 10% solution are taken orally over a four-hour period. The advantages of this method are that no nasogastric tube is required, and if the 5% solution is used, significant fluid abnormalities do not occur. However, with the 10% solution, the patient may become significantly dehydrated and require IV supplementation. Because mannitol is a sugar, it can serve as a nutrient to bacteria and theoretically predispose to infection (Table 10–7).

The mechanical preparation reduces the amount of enteric contents, not the number of bacteria in a single inoculum. Therefore, the mechanical bowel preparation is supplemented with an antibiotic bowel preparation. Adding an antibiotic reduces risk of wound infection and other infectious complications from 30% to 6% when antibiotics are combined with the mechanical preparation. Antibiotics also protect vulnerable and ischemic bowel. The disadvantage to giving antibiotics in preparation for bowel surgery is that it increases the incidence of tumor implantation on the anastomosis and

TABLE 10–7.

Mechanical Bowel Preparation

PREOPERATIVE DAY	CONVENTIONAL			SALINE BOWEL IRRIGATION		ORAL MANNITOL	
	DIET	CATHARTIC	ENEMA	DIET	IRRIGATION	DIET	ORAL MANNITOL
3	Low-residue plus supplements	250 ml citrate of magnesia p.o.		Regular plus supplements		Regular plus supplements	
2	Clear liquids	250 ml citrate of magnesia p.o.	SSE	Low-residue plus supplements		Low-residue plus supplements	
1	Clear liquids	250 ml citrate of magnesia p.o.	SSE until clear	Clear liquids	3–4 L/hr for 2–3 hr + 10 mg metoclopramide IV	Clear liquids	4 L 5% mannitol or 2 L 10% mannitol p.o.

might play a role in the development of pseudomembranous enterocolitis. The former is generally of little consequence to the urologist; however, the latter can be a significantly morbid and sometimes fatal complication. It seems unlikely, however, that antibiotics are responsible for the development of the latter syndrome, since there has been no change in its incidence in the preantibiotic and postantibiotic eras. It was originally thought to be caused by a staphylococcus but recently *Clostridium difficile* and the endotoxin that it elaborates have been implicated. The most important factor in its initiation is intestinal ischemia. Treatment is directed at repopulating the bowel surface with normal enteric flora.

The antibiotics commonly used for antibiotic bowel preparations are kanamycin, neomycin, erythromycin, and metronidazole (Table 10–8). Kanamycin is the best single agent, and is also useful for irrigating the peritoneal cavity when peritoneal contamination occurs. Neomycin and erythromycin when combined are particularly effective (Nichols et al., 1973). One gram of neomycin plus one gram of erythromycin base is given at 1, 2, and 11 PM on the day before surgery. A third regimen involves neomycin and metronidazole. One gram of neomycin and 750 mg of metronidazole are administered orally three times a day for two days before the operation (Dion et al., 1980).

Perioperative antibiotics have also been suggested to lessen the incidence of complications arising from violating the integrity of the bowel. Intravenous ampicillin given preoperatively and postoperatively with a mechanical preparation was as successful as oral antibiotics in reducing wound infection. Indeed, in a series utilizing this regimen, the wound infection rate was only 6% (Menaker et al., 1981). Others, however, have not achieved the same success rate and note a 30% incidence of wound infection as opposed to a 16% incidence with an oral antibiotic and mechanical bowel preparation regimen. More recent evidence indicates that prophylaxis with an antimicrobial agent that acts solely against aerobic microorganisms does not lower wound infection rates, whereas prophylaxis with antibiotic agents that are effective against anaerobes significantly reduces infectious wound complications (Eykyn et al., 1979). A mechanical bowel preparation combined with the IV administration of a single dose of tinidazole (a metronidazole derivative) and doxycycline results in the development of infectious complications in only 5% of patients undergoing colon and rectal operations (Giercksky et al., 1982).

SHOCK

An all-encompassing definition of shock is impairment of cellular function due either to a reduction in

TABLE 10–8.

Antibiotic Bowel Preparation

PREOPERATIVE DAY	KANAMYCIN	NEOMYCIN PLUS ERYTHROMYCIN	NEOMYCIN PLUS METRONIDAZOLE	TINIDAZOLE PLUS DOXYCYCLINE
3	1 gm kanamycin p.o. q 1 hr × 4, then q.i.d.	—	—	—
2	1 gm kanamycin p.o. q.i.d.	—	1 gm neomycin q.i.d. plus 750 mg metronidazole q.i.d.	—
1	1 gm kanamycin p.o. q.i.d.	1 gm erythromycin plus 1 gm neomycin at 1 P.M., 2 P.M., 11 P.M.	1 gm neomycin q.i.d. plus 750 mg metronidazole q.i.d.	2 hr preoperatively: 1,600 mg tinidazole plus 400 mg doxycycline infused IV over 2 hr

effective delivery to the cell of oxygen and nutrients or inability of the cell to utilize these substrates. Shock may be divided into hypovolemic, neurogenic, endocrine, anaphylactic, cardiogenic, and septic.

Hypovolemic Shock

When loss of blood, loss of fluid, or sequestration of fluid is etiologic, cardiac output and blood pressure fall while plasma renin and antidiuretic hormone levels rise. The sympathetic nervous system is stimulated, which results in the constriction of arterioles and major veins, an increase in heart rate and strength of contraction, a reduction in microcirculatory flow and sludging, and mobilization of glucose for energy production. If hypotension and peripheral vasoconstriction continue, a fall in transmembrane potential occurs, membrane permeability increases, and fluid is sequestered, particularly in muscle cells. If the perfusion defect continues, cellular death supervenes.

Hypovolemic shock is generally divided into three phases. The first phase occurs with volume deficits between 0% and 10% and begins with a sequestration of fluid resulting in cellular edema. During this phase, the blood volume deficit is counteracted by stimulation of the sympathetic nervous system, which results in peripheral vasoconstriction and increased heart rate in an attempt to maintain blood pressure. Antidiuretic hormone, renin, and aldosterone also contribute to the vasoconstriction and reduced urinary excretion in an attempt to restore extracellular fluid volume. The second stage occurs with a blood volume deficit of 15%–20%. There is a marked reduction in cardiac output and arterial pressure despite the compensatory mechanisms described above. The low blood pressure and the intense adrenergic discharge result in tachycardia, tachypnea, cutaneous vasoconstriction, pallor, diaphoresis, piloerection, apprehension, and restlessness. The third stage occurs when the deficit exceeds 25%. During this stage, tissue perfusion is inadequate, and if it persists for long periods, cellular death occurs. Vital organ function is severely impaired, capillary membrane integrity is lost, and disseminated intravascular coagulation may occur. It is at this point that irreversible shock follows.

TREATMENT.—Hypovolemic shock due to third space fluid loss as occurs in burns and retroperitoneal dissections is treated with isotonic fluid replacement. Many advocate partial replacement with 5% albumin, suggesting that the total fluid requirement is less with a smaller amount extravasating into the third space and pulmonary interstitium. Other studies suggest that crystalloid is more likely to produce pulmonary edema than solutions of 5% albumin, hetastarch, or dextran. When blood loss is the cause, whole blood is adminis-

tered which is appropriately typed and cross matched. Dextran may be used as a blood substitute for volume expansion; however, it carries the risk of creating clotting abnormalities through antigen–antibody reactions. Prolonged use of dextran may also engender renal failure. Hetastarch is a polysaccharide which is available in a 6% solution. It has replaced dextran as the drug of choice for volume expansion when blood and albumin are unavailable since renal and bleeding complications are less frequent.

Attention must also be paid to the ventilatory status of the patient. Hypoventilation with hypoxemia severely augments the deleterious effects of underperfusion.

In the successfully treated patient, there are three phases of fluid dynamics. The first phase follows the insult and is characterized by sequestration of fluid and cellular edema. This accounts for the major initial pathophysiologic problems encountered in hypovolemic shock. The sequestration generally ceases between 24 and 36 hours, provided the patient has been adequately resuscitated. In the ensuing 2–3 days, a homeostatic period occurs in which the sequestered fluid remains unavailable to the intravascular space. At approximately 4–7 days postinsult, the third phase is initiated in which the fluid is mobilized. It is during this phase that patients with limited cardiac reserve may develop congestive heart failure as the fluid is mobilized from the third space into the intravascular space. The judicious use of diuretics may be necessary to rapidly restore intravascular volume to normal.

Neurogenic, Endocrine, and Anaphylactic Shock

Neurogenic shock results from a sudden loss of vasomotor tone due to a neural injury. This commonly occurs in traumatic injuries to the CNS.

Endocrine deficiencies due to lack of function of the adrenals or pituitary result in circulatory failure. Uncontrollable diabetes may result in an osmotic diuresis with resultant hypotension.

Anaphylactic shock is a result of an acute systemic autoimmune reaction that causes a marked increase in vascular permeability with sequestration of edema fluid, bronchospasm, and laryngospasm. This sequestered fluid leaves the intravascular space, resulting in hypotension.

TREATMENT.—Neurogenic shock is treated with isotonic fluid administration until blood pressure returns to normal levels. Patients with adrenal insufficiency or pituitary insufficiency are treated with replacement corticosteroids and sodium-containing fluids. Anaphylactic shock is treated with epinephrine, fluids, antihistamines and corticosteroids.

Cardiogenic Shock

Cardiogenic shock occurs when the heart fails to act effectively as a pump. This may be brought about by a myocardial infarction, cardiac arrythmia, or depression of myocardial contractility due to a metabolic aberration. Rare causes include mechanical blockage which may occur in a tension pneumothorax, vena caval obstruction, and cardiac tamponade. The most common cause of the disorder is clearly myocardial infarction. Approximately 10%–15% of patients with an acute myocardial infarction will develop cardiogenic shock. The mortality in this group is exceedingly high, approaching 75% (Geddes et al., 1980). These patients present with cold, clammy skin, hypotension, tachycardia, anuria, and oliguria, and are often confused and agitated. The pathophysiologic process includes a decreased cardiac output, decreased ejection fraction, a normal or increased CVP, increased pulmonary capillary wedge pressure and an increased left ventricular end-diastolic pressure. Since the hypotension results in decreased coronary artery perfusion with a propensity toward limiting the amount of oxygen delivered, one of the first hallmarks of therapy is to provide the patient with oxygen. Decreased oxygenation of the injured myocardium will result in progression of the infarction. Oxygen is administered by face mask, nasal prongs, or rebreathing bag. Patients with chronic lung disease need to be monitored carefully if their respiratory drive is secondary to hypercapnia. These patients may require intubation. It is during the initial postmyocardial infarction period that life-threatening cardiac arrhythmias frequently occur. Therefore, these patients need to be continuously monitored, and a large-bore IV line must be available to administer antiarrhythmogenic drugs. Often these patients will have a slight metabolic acidosis, which must be corrected with bicarbonate administration. The placement of a pulmonary artery catheter is generally indicated, since the diagnosis can be conveniently made utilizing this device, and treatment is often dictated by measurements performed. With the pulmonary artery catheter, cardiac output, preload and afterload can be determined and will verify the diagnosis by revealing a decreased cardiac output. Preload is determined by pulmonary capillary wedge pressure measurement and afterload by peripheral resistance calculation. When the peripheral resistance is normal, an agent is chosen which will improve the cardiac output by increasing myocardial contractility. Dobutamine or isoproterenol is the drug of choice. When there is an increase in peripheral resistance, therapy is directed at increasing cardiac output and decreasing afterload. The afterload may be reduced with nitroprusside and cardiac output increased with either dobutamine or isoproterenol. The dose of nitroprusside must be monitored carefully, since it may result in thiocyanate intoxication. When there is a decreased peripheral resistance and a decreased cardiac performance, the use of vasopressors such as dopamine and epinephrine is useful (Houston, 1984).

SEPTIC SHOCK

Septic shock, unlike hypovolemic shock, does not necessarily result in a perfusion deficit to the tissues. Rather, it results in an inability of the body cell mass to effectively use substrate delivered to it. Septic shock is an important cause of morbidity and mortality among hospitalized patients. About 25%–40% of patients who sustain a bacteremia develop septic shock (Parker and Parrillo, 1983). In most cases it originates from an infection with gram-negative enteric bacilli. *E. coli* is the most common pathogen followed by *Klebsiella, Enterobacter, Serratia,* and *Pseudomonas.* Gram-positive organisms on occasion cause sepsis, and rarely viruses, fungi, rickettsia, and protozoa have been implicated. The genitourinary tract is the most common source, followed in decreasing order of frequency by the GI tract, the respiratory tract, wound infections, infected IV catheters, and pelvic infections. Patients at increased risk for the development of septic shock include those who have cirrhosis, diabetes mellitus, and neoplastic diseases (Mizock, 1984). Radiotherapy and/or chemotherapy particularly predisposes the latter group of patients to sepsis. The mortality ranges between 40% and 90%.

Endotoxin, a lipopolysaccharide protein complex, which is part of the cell wall of gram-negative bacteria, has been implicated as the initiating agent in the pathophysiologic process. Since other than gram-negative organisms also cause the syndrome, it is clear that if endotoxin is an initiating agent, it is not the only one. The infecting organism activates certain protein systems, resulting in the formation of active mediators from inactive precursors. The complement, kinin, and clotting systems are stimulated and the release of histamine and prostaglandins facilitated. There are two phases of septic shock, the hyperdynamic and the hypodynamic. In the first phase, patients are hypotensive and have warm, dry skin. Arterial vasodilatation, an increase in body temperature, hyperventilation with respiratory alkalosis, and an increase in pulse pressure and cardiac output occur. Although the absolute value of cardiac output in patients with septic shock is generally increased, it cannot achieve adequate perfusion of essential organs. These events may be heralded by a shaking chill, which is followed by a rapid rise in temperature. Lactic acidosis may supervene, indicating the cell's inability to utilize or obtain oxygen and substrate.

The degree of lactic acidemia, however, does not correlate with the lack of oxygenation in septic shock, nor is it of any prognostic significance (Rosenberg and Rush, 1966). The vasoactive kinins, histamine and prostaglandins, result in vasodilatation and perhaps some vasoconstriction in various vascular beds. As sepsis continues, myocardial depression occurs, with a reduced ejection fraction and left ventricular dilatation. The activation of the kinin system results in the release of bradykinin, an agent known to produce vasodilatation of arterioles. Activation of the complement system results in depletion of the C3 component (Parker and Parrillo, 1983). The WBC count is elevated with a shift to the left; however, rarely sepsis may result in a depressed WBC count, a poor prognostic sign. Blood glucose is elevated due to the elevated levels of glucagon, growth hormone, and catecholamines. Progression of the disease leads to leukocyte aggregation with development of disseminated intravascular coagulation, vascular endothelial damage producing capillary leak with interstitial edema, and depression of myocardial function. Platelet aggregation results in the release of vasoactive substances. Pulmonary platelet aggregation has been suggested as the initiating event in the adult respiratory distress syndrome (ARDS). Myocardial depression increases capillary permeability, and impaired cellular function leads to the second phase of septic shock, a hypodynamic state in which unresponsive hypotension and death occur.

TREATMENT.—Treatment is directed at supporting the patient and defining the source of infection so that it can be eliminated. Initially, blood, urine, sputum, and wound drainage, if present, are sent for culture. Intravenous bactericidal antibiotics are begun immediately. The choice of antibiotics is dictated by the suspected source of the sepsis. An aminoglycoside at full dosage is administered initially. Dose level subsequently is adjusted to the renal function. If an intra-abdominal source is suspected, anaerobic coverage is added: clindamycin, chloramphenicol, or metronidazole. If a pulmonary source seems likely, a cephalosporin is added; and if the urinary tract is suspect, ampicillin is given. Blood pressure is supported with crystalloid and colloid infusions, the rate dictated by the blood pressure, CVP, or right atrial filling pressure and urine output. Colloid resuscitation is preferred as it appears that less capillary leakage occurs with less subsequent interstitial edema formation. Corticosteroid administration is controversial. Shumer administered corticosteroids to a group of septic shock patients and noted a substantial decrease in the mortality (Shumer, 1976). Lucas and Ledgerwood (1954), however, were unable to substantiate the reduction in mortality, al-

though steroids seemed to increase systolic blood pressure and CVP pressure. In experimental models, steroids are particularly effective if they are given very early in the course. The earlier they are given, the more likely they are to prevent capillary leakage and myocardial depression. The theoretical advantage for steroid administration includes stabilization of lysosomal membranes, supporting the blood pressure and prevention of the formation of β-lipoprotein. β-Lipoprotein, the precursor of β-endorphin, a vasodilator, originates from the same molecule as ACTH. Thus, steroid administration supresses ACTH and therefore endorphin production (Sheagren, 1981). Although the issue of corticosteroid administration is unsettled, it seems apparent that if they are administered late in the course of the disease, they are ineffective; to be effective at all, they must be given early. Methylprednisolone and dexamethasone have both been used with success. They are given in two doses six hours apart.

The use of antiserum has improved survival in some studies. Gram-negative bacilli share a common core lipopolysaccharide. Antiserum to this antigen from human volunteers when given to septic patients in one series reduced the mortality by 50% (Ziegler, 1982). In selected cases, continuous intraperitoneal lavage may also be helpful. This therapy is particularly amenable to patients who are septic as a result of pelvic inflammation and abscesses. Dialysis catheters placed into the depths of the pelvis and irrigated with saline and antibiotic solutions are particularly helpful in these patients (Moukhtar and Romney, 1980).

If hypotension occurs after adequate fluid resuscitation or if it persists, blood pressure may be supported with dopamine. The advantages of dopamine over other agents are that in addition to its ionotrophic effect, it increases renal blood flow. However, dopamine does increase intrapulmonary shunting, and therefore dobutamine may be preferable in patients with pulmonary complications. Although dobutamine does support blood pressure, it does not improve renal blood flow. β-Endorphins released in response to the stress of shock may be partially responsible for the hypotension. It is clear that pituitary endorphins play a role in the pathophysiology of shock, but their exact role is somewhat controversial. The cardiodepressant effect of endorphins is mediated by opiate receptors in the CNS (Faden, 1984). Naloxone, an inhibitor of β-endorphin binding, may result in significant improvement in the blood pressure and reversal of myocardial depression (Peters et al., 1981). Thyrotropin-releasing hormone has also been utilized effectively in patients with persistent hypotension. This drug apparently acts centrally as does naloxone and, like naloxone, requires an intact sympathoadrenomedullary axis to be effective. Thyrotropin-releasing hormone and naloxone have an additive effect (Faden, 1984). The prostaglandins PGI and thromboxane A_2 may mediate some of the cardiovascular changes in shock but are not solely responsible for myocardial depression (Carmona et al., 1984). They contribute locally to vasodilatation, and it has been found experimentally that blocking the vasoconstrictor thromboxane A_2 in animals improves survival (Cook et al., 1980).

CARDIAC DYSFUNCTION

When venous filling pressure is increased and hypotension persists, a low cardiac output accounts for the poor tissue perfusion. The normal resting cardiac output is 3.5 $L/m^2/min$; an output less than 2.0 $L/m^2/min$ requires treatment. Digitalis preparations, catecholamines, and calcium increase myocardial contractility, whereas hypoxia, acidosis, hypercapnia, elevated serum potassium, barbiturates, anesthetic agents, propranolol, procainamide, quinidine, and lidocaine reduce myocardial contractility. Metabolic abnormalities that impair contractility should be corrected and drugs eliminated when possible. Ionotropic drugs are useful for increasing myocardial contractility and thereby increasing cardiac output. The catecholamines are most appropriate in the acute situation. Norepinephrine is a peripheral vasoconstrictor and increases afterload to the heart. It is appropriate for cardiopulmonary resuscitation but not for prolonged use, since vital organ perfusion is not optimized. Dobutamine and isoproterenol are β-adrenergic agents that increase myocardial contractility but at high doses tend to cause cardiac irritability and increase myocardial oxygen consumption. Epinephrine is a bronchodilator and peripheral vasoconstrictor and is useful in patients with pulmonary bronchoconstriction and associated cardiac dysfunction. Dopamine is the most useful, since it increases cardiac output while improving vital organ perfusion.

When afterload or ventricular wall tension is increased, cardiac output may be limited. Afterload can be reduced by selective use of vasodilators. Phentolamine (Regitine), an α-blocker, dilates both veins and arteries. It is particularly useful in patients with increased sympathetic tone. Nitroprusside is a short-acting vasodilator whose predominant effect is on the smooth muscle of the arteries. It is an excellent agent for titrating blood pressure in the acute situation. Nitroglycerin is particularly useful in patients with an associated ischemic myocardium, since it improves collateral flow to the heart while dilating the peripheral vasculature.

Sinus rhythm is the most effective rhythm for optimal cardiac output. When supraventricular and ventricular arrhythmias occur in association with a diminished cardiac output, restoration of normal sinus rhythm may be all that is required.

VENOUS THROMBOEMBOLISM

Pelvic operations have a high propensity for thromboembolic disease. Its prevention in the urologic patient is of particular importance, since many operations are performed in the pelvic area. Urologic patients are often placed in lithotomy position, which impairs venous return. Indeed, since most thrombi develop in the venous plexus along the calf and usually begin during the operation, it is easy to appreciate the significance of impaired venous return which is promoted by positioning. The prevention of thromboembolic disease has been a subject of considerable controversy. Elevation of the foot of the bed, avoidance of extremity compression, early ambulation, physical therapy, and leg wraps have all been utilized in an effort to prevent the sequelae of thromboembolic disease. Unfortunately, none of these devices has been shown to be of particular efficacy in the prevention of the disorder. More recently, the development of an alternating pressure cuff on the lower extremities which facilitates venous return appears promising; however, due to the positioning of urologic patients, it is not easily accomplished.

The use of mini-dose heparin, i.e., 5,000 units subcutaneously every 12 hours, has been suggested in several studies to decrease the incidence of pulmonary emboli. One study found a marked reduction in pulmonary emboli as diagnosed by routine postoperative scans (Lahnborg et al., 1974). Other studies have failed to show any significant difference with and without the use of mini-dose heparin. It is clear, however, that in patients undergoing pelvic lymphadenectomy, administration of preoperative and postoperative mini-dose heparin results in a markedly increased incidence of lymphocele formation (Koonce et al., 1986). Therefore in urologic practice, prophylactic mini-dose heparinization in the preoperative and postoperative period has not enjoyed success and therefore has rarely been used.

Pulmonary emboli may occur silently and be an incidental finding on a chest roentgenogram or they may suggest their presence by causing dyspnea, chest pain, hemoptysis, and rarely, when massive, circulatory collapse. On physical examination, the pulmonic portion of the second heart sound may be increased, a parasternal heave may occur, on occasion a friction rub can be heard, and the ECG often shows right heart strain as evidenced by right axis deviation. The chest x-ray, when positive, reveals a lucent area that lacks vascular markings. Later a wedge-shaped infiltrate develops. Pulmonary scans may be used to support the diagnosis, but the definitive study is a pulmonary angiogram. Therapy is directed at identifying the source and treating it while anticoagulating the patient, initially with a continuous heparin infusion. If the pulmonary embolus is large or saddle-type and is causing circulatory collapse that is unresponsive to supportive measures, a pulmonary embolectomy is indicated. In selected circumstances, the use of streptolysin is most effective in dissolving emboli. It should be noted, however, that in postoperative patients, vascular suture lines and puncture sites are likely to bleed with the utilization of this drug.

RESPIRATORY DYSFUNCTION

Respiratory Insufficiency

Inadequate ventilation in the posttraumatic and postoperative periods results in hypercapnia or hypoxemia or both. The primary goal of the therapy is to provide the patient with the capability of maintaining an arterial oxygen partial pressure of at least 60 mm Hg on an inspired oxygen content as close to room air as possible. To achieve this goal, oxygen delivered by nasal prongs, rebreathing mask, a face mask, or an endotracheal tube with respiratory support may be required. There are, however, constraints to the amount of oxygen which can be delivered. Limitation of the amount may be a consequence of the device used for delivery; however, more commonly the amount that can be safely delivered is limited by the fact that inhalation of high oxygen concentrations results in pulmonary toxicity. The hazards of high concentrations include suppression of the respiratory drive in patients with chronic lung disease, retrolental fibroplasia (primarily a disease of the newborn but described in adults), segmental atelectasis due to the greater solubility of oxygen compared to nitrogen, impairment of respiratory ciliary function, decrease in pulmonary surfactant, and direct injury to capillary endothelial cells. The arterial carbon dioxide partial pressure, which is normally 40 mm Hg, is a primary indication of the adequacy of ventilation. Common causes of hypercapnia include obstructive lung disease, ARDS, metabolic alkalosis, and respiratory depression due to sedation or CNS trauma. Hypocapnia may be a result of hypoxia, anxiety, pulmonary embolism, sepsis, and pulmonary insufficiency. Although an indicator of ventilatory adequacy, alteration of PCO_2 itself is rarely an indication for respiratory support. Of more importance is the PO_2, which should be maintained above 60 mm Hg. A PO_2 less than 60 mm Hg requires a change in respiratory management. A normal PO_2 for a particular patient breathing room air prior to injury may be estimated by subtracting one half the individual's age from 100. If impending airway obstruc-

tion is not a problem, initial support of the PO_2 may be obtained by the use of nasal prongs or face masks. Humidified oxygen should be used when possible to prevent drying the nasotracheal mucosa. Oxygen delivered by nasal prongs generally cannot provide an inspired concentration much above 50%. Even though humidified high flows have a drying effect on the mucosa, Venturi masks provide constant flows of oxygen ranging between 24% and 40%, depending on the mask. Partial rebreathing masks can deliver in excess of 80% oxygen; however, humidity cannot be added to the system. On occasion, posttraumatic patients require endotracheal intubation, preferably by the nasotracheal route with a prestretched low-pressure cuff and respiratory support (Table 10–9). The indications for intubation include (1) the facilitation of pulmonary toilet, (2) the prevention of upper airway occlusion, (3) as protection against aspiration, and (4) the need for mechanical ventilation. The requirement for mechanical ventilation is assessed by vital capacity, inspiratory force, respiratory rate, arterial oxygen content, and work of breathing. Vital capacity, or the volume of a maximal inspiration following a maximal expiration, is normally 60–70 ml/kg of body weight. If it is less than 15 ml/kg, ventilatory support is indicated. The inspiratory force, or amount of pressure that the patient is able to generate against a closed airway, is normally −75 to −100 cm H_2O. Patients who can achieve no more than −25 cm H_2O require mechanical support. The normal respiratory rate is 12–20 breaths per minute. A rate that exceeds 35 breaths per minute suggests the need for ventilatory assistance. The arterial oxygen partial pressure, PO_2, should exceed 60 mm Hg. If this cannot be accomplished by raising the oxygen content of inspired air through the use of face masks and nasal prongs, intubation should be performed. Severe intercostal retractions and a tracheal tug indicate an increased work of breathing and are forerunners of respiratory insufficiency. Initially, the respirator is adjusted to deliver

---

TABLE 10–9.

Indications for Endotracheal Intubation

Facilitation of pulmonary toilet
Prevention of upper airway occlusion
Protection against aspiration
Need for mechanical ventilation as determined by:
1. VC <15 ml/kg of body weight
2. Inspiratory force < −25 cm H_2O
3. Respiratory rate >35/min
4. PO_2 <60 mm Hg despite high ambient O_2 concentration
5. Excessive, prolonged increase in the work of breathing

12–15 ml/kg body weight at a frequency of 8–14 times per minute for the adult and 15–30 times per minute for the child. Inspired oxygen content, or FIO_2, should be the lowest needed to maintain the PO_2 above 60 mm Hg (an FIO_2 of 40% is a good level to begin with, adjusting it upward or downward as required). Not only must blood gases be monitored adjusting the respirator accordingly, but also the circulatory status must be carefully followed, for on occasion institution of mechanical ventilation will cause a fall in the cardiac output with lowering of the blood pressure. When the PO_2 cannot be maintained by an acceptable FIO_2 (<60%), the addition of positive end-expiratory pressure (PEEP) may be helpful. This technique maintains a specified pressure at the end of each respiration rather than allowing end-expiratory pressure to fall to zero. It is particularly useful in ARDS. Initially, 5 cm H_2O pressure is employed. If the desired response is not achieved, it is increased in increments of 5 cm H_2O, carefully monitoring the blood pressure for signs of a significant reduction in cardiac output. Usually no more than 15 cm H_2O is required. However, on rare occasions, as much as 25 cm H_2O pressure may be needed. With the use of PEEP, PO_2 can be maintained at acceptable levels using reduced FIO_2. Other advantages include a decrease in pulmonary shunting and an increase in functional residual capacity (FRC). A proposed advantage is that it drives pulmonary edema fluid from the alveoli and the interstitium into the pulmonary capillaries. Its major disadvantages are a reduction in cardiac output and a diminished urine output. The latter effect is perhaps the result of an increased release of antidiuretic hormone. One method of anticipating future respiratory difficulties as well as determining how the patient is progressing on the respirator is by sequentially determining the arterial oxygen gradient, $P(A-a)O_2$. This gradient is a sensitive indicator of early resiratory impairment. To calculate the $P(A-a)O_2$, the patient receives 100% oxygen for 20–30 min, arterial blood gases are drawn, and the barometric pressure is recorded. The calculation is as follows: barometric pressure − water vapor pressure (47 mm Hg) − the partial pressure of alveolar CO_2. Because alveolar CO_2 rapidly equilibrates with arterial CO_2, the PCO_2 obtained from the blood gas analysis may be substituted. This quantity minus the PO_2 is equal to the arterial alveolar oxygen gradient:

$$P(A-a)O_2 = \text{barometric pressure} - 47 - PCO_2 - PO_2$$

A normal value lies between 25 and 65. A value exceeding 450 suggests failure. Ventilatory support is continued until the indications for its use no longer apply. If mechanical ventilation was the reason for intubation, the patient may be weaned from the respirator when the chest roentgenogram reveals no deterioration, the spontaneous respiratory rate is less than 30/minute, PEEP is no longer required, the inspiratory force exceeds −25 cm H_2O, the vital capacity exceeds 15 ml/ kg, and the PO_2 can be maintained over 60 mm Hg on an FIO_2 below 50%. The patient is weaned by placing him on a T piece with humidified oxygen for 10 minutes of each hour. If the T piece is tolerated, the time on it is gradually increased until the respirator is no longer required. The blood gases and patient's respiratory effort must be carefully monitored both during weaning and following extubation.

Adult Respiratory Distress Syndrome

Acute posttraumatic pulmonary insufficiency (ARDS) occurs following major trauma, burns, hypoproteinemia, or inadequate fluid resuscitation during shock, severe sepsis, pancreatitis, or transplant rejection crisis (antigen–antibody reaction). Following the initiating event, platelet microaggregates form in the pulmonary capillaries and injure the alveolar capillary endothelium. Vasoactive substances are released, resulting in increased capillary permeability (Blaisdell, 1974). Peribronchiolar edema follows, which causes an increase in small airway resistance and a reduction in lung compliance, making aeration of the lungs difficult. Pulmonary shunting also occurs. The PO_2 falls and the PCO_2 rises, often despite increases in the FIO_2. Clinically the patient becomes dyspneic, tachypneic, and hypoxemic. There is a reduced FRC, reduced lung compliance, and often bilateral pulmonary infiltrates are present on the chest film. The syndrome should be suspected in the septic or traumatized patient when the PO_2 falls despite efforts to increase the FIO_2. Treatment involves nasotracheal intubation and mechanical ventilation. PEEP is often necessary. If PEEP results in a reduced cardiac output, inotropic agents may be required to return blood pressure to acceptable levels. The use of colloid to increase intravascular oncotic pressure and thereby draw fluid from the pulmonary perivascular space into the capillaries is controversial, as is the use of steroids. Prophylactic antibiotics administered either by the parenteral route or by inhalation have little to recommend them. Infections are treated when they occur with the antibiotic to which the bacteria are sensitive.

REFERENCES

1. Abel RM, Beck CH Jr, Abbott WM, et al: Improved survival from acute renal failure after treatment with intravenous essential L-amino acids and glucose. *N Engl J Med* 1973; 288:695.
2. Anderson CF, Wochos DN: The utility of serum albumin values in the nutritional assessment of hospitalized patients. *Mayo Clin Proc* 1982; 57:181.

3. Blaisdell FW: Pathophysiology of the respiratory distress syndrome. *Arch Surg* 1974; 108:44.

4. Boyce WH, Vest SA: The role of ammonia reabsorption in the acid base imbalance following ureterosigmoidostomy. *J Urol* 1952; 67:169.

5. Boyd JD: Chronic acidosis secondary to ureteral transplantation. *Am J Dis Child* 1931; 42:366.

6. Calloway DH, Spector H: Nitrogen balance as related to calorie and protein intake in active young men. *Am J Clin Nutr* 1957; 2:405.

7. Carmona RH, Tsao TC, Trunkey DD: The role of prostacyclin and thromboxane in sepsis and septic shock. *Arch Surg* 1984; 119:189.

8. Ching N, Grossi CE, Angers J, et al: The outcome of surgical treatment as related to the response of the serum albumin level to nutritional support. *Surg Gynecol Obstet* 1980; 151:199.

9. Christensen PB, Kronborg O: Whole-gut irrigation versus enema in elective colorectal surgery: A prospective, randomized study. *Dis Colon Rectum* 1981; 24:592.

10. Clark SS: Electrolyte disturbance associated with jejunal conduit. *J Urol* 1974; 112:42.

11. Clarke JS, Condon RE, Barlett JG, et al: Preoperative oral antibiotics reduce septic complications of colon operations. *Ann Surg* 1977; 186:251.

12. Cook JA, Wise WC, Halushka PV: Elevated thromboxane levels in the rat during endotoxin shock: Protective effects of imidazole, 13-azaprostanoic acid, or essential fatty acid deficiency. *J Clin Invest* 1980; 65:227.

13. Copeland EM, MacFadyen BV, Dudrick SJ: Effect of intravenous hyperalimentation on established delayed hypersensitivity in the cancer patient. *Ann Surg* 1976; 184:60.

14. Creevy CD: Renal complications after ileal diversion of urine in non-neoplastic disorders. *J Urol* 1960; 83:394.

15. Creevy CD, Webb EA: A fatal hemolytic reaction following transurethral resection of the prostate gland. *Surgery* 1947; 21:56.

16. D'Agostino A, Leadbetter WF, Schwartz WB: Alterations in the ionic composition of isotonic saline solution instilled into the colon. *J Clin Invest* 1953; 32:444.

17. Daly JM, Dudrick SJ, Copeland EM: Intravenous hyperalimentation: Effect in delayed cutaneous hypersensitivity in cancer patients. *Ann Surg* 1980; 192:587.

18. Dion YM, Richards GK, Prentis JJ, et al: The influence of oral versus parenteral preoperative metronidazole on sepsis following colon surgery. *Ann Surg* 1980; 192:221.

19. Dudrick SJ, MacFadyen BV, Van Buren CT, et al: Parenteral hyperalimentation: Metabolic problems and solutions. *Ann Surg* 1972; 176:259.

20. Eykyn SJ, Jackson BT, Lockhart-Mummery HE, et al: Prophylactic preoperative intravenous metronidazole in elective colorectal surgery. *Lancet* 1979; 2:761.

21. Faden AI: Opiate antagonists and thyrotropin-releasing hormone. *JAMA* 1984; 252:1177.

22. Fairfull-Smith R, Abunassar R, Freeman JB, et al: Rational use of elemental and nonelemental diets in hospitalized patients. *Ann Surg* 1980, 192:600.

23. Farris DO, Odel HM: Electrolyte pattern of the blood after bilateral ureterosigmoidostomy. *JAMA* 1950; 142:634.

24. Feigal D, Blaisdell W: The estimation of surgical risk. *Med Clin North Am* 1979; 63:1131.

25. Freund H, Atarmain S, Fischer J: Comparative study of parenteral nutrition in renal failure using essential and nonessential amino acid containing solutions. *Surg Gynecol Obstet* 1980; 151:652.

26. Freund H, Dienstag J, Lehrich J, et al: Infusion of branched-chain enriched amino acid solution in patients with hepatic encephalopathy. *Ann Surg* 1982; 196:209.

27. Geddes JS, Adgey AAJ, Pantridge JF: Prevention of cardiogenic shock. *Am Heart J* 1980; 99:243.

28. Giercksky KE, Danielson S, Garberg O, et al: A single dose tinidazole and doxycycline prophylaxis in elective surgery of colon and rectum. *Ann Surg* 1982; 195:227.

29. Glass RL, Winship DH, Rogers WA: Comparison of intragastric infusion with conventional mechanical bowel preparation. *Dis Colon Rectum* 1981; 24:589.

30. Goldman L, Caldera DL, Nussbaum SR, et al: Multifactoral index of cardiac risk in noncardiac surgical procedures. *N Engl J Med* 1977; 297:845.

31. Goldman L, Caldera DL, Southwick FS, et al: Cardiac risk factors and complications in noncardiac surgery. *Medicine* 1978; 57:357.

32. Hares MM, Nevak E, Minervini S, et al: An attempt to reduce the side effects of mannitol bowel preparation by intravenous infusion. *Dis Colon Rectum* 1982; 25:289.

33. Houston MC, Thompson WL, Robertson D: Shock diagnosis and management. *Arch Intern Med* 1984; 144:1433.

34. Johnson LR: *Physiology of the Gastrointestinal Tract.* New York, Raven Press, 1981, vol 2.

35. Jolinbur M, Morales P: Jejunal conduits: Technique and complications. *J Urol* 1975; 113:787.

36. Kinney JM: The metabolic response to injury, in Richards JR, Kinney JM (eds): *Nutritional Aspects of Care in the Critically Ill.* London, Churchill-Livingstone, 1977.

37. Koch MO, McDougal WS: Chlorpromazine: Adjuvant therapy for the metabolic derangements created by urinary diversion through intestinal segments. *J Urol* 1985a; 134:165.

38. Koch MO, McDougal WS: Nicotinic acid treatment for the hyperchloremic acidosis following urinary diversion through intestinal segments. *J Urol* 1985b; 134:162.

39. Koch MO, McDougal WS:The pathophysiology of hyperchloremic metabolic acidosis following urinary diversion through intestinal segments. *Surgery* 1985; 98:561.

39a. Koonce J, Selikowitz S, McDougal WS: Complications of low-dose heparin prophylaxis following pelvic lymphadenectomy. *Urology* 1986; 28:21.

40. Lahnborg G, Friman L, Bergstrom K, et al: Effect of low-dose heparin on incidence of post-operative pulmonary embolism detected by photoscanning. *Lancet* 1974; 1:329.

41. Lapides J: Mechanism of electrolyte imbalance following ureterosigmoid transplantation. *Surg Gynecol Obstet* 1951; 93:691.

42. Law DK, Dudrick SJ, Abdon NI: Immunocompetence of

patients with protein-calorie malnutrition: The effects of nutritional repletion. *Ann Intern Med* 1973; 79:545.

43. Long CL, Spencer JL, Kinney JM, et al: Carbohydrate metabolism in normal man and effect of glucose infusion. *J Appl Physiol* 1971; 31:102.

44. Lucas CE, Ledgerwood AM: The cardiopulmonary response to massive doses of steroids in patients with septic shock. *Arch Surg* 1984; 119:537.

45. McDougal WS, Koch MO: Accurate determination of renal function in patients with intestinal urinary diversions. *J Urol* 1986; 135:1175.

46. McDougal WS, Wilmore DW, Pruitt BA Jr: Effect of near isosmotic intravenous nutrient infusions on nitrogen balance in critically ill injured patients. *Surg Gynecol Obstet* 1977a; 145:408.

47. McDougal WS, Wilmore DW, Pruitt BA Jr: Glucose dependent hepatic membrane transport in non-bacteremic and bacteremic thermally injured patients. *J Surg Res* 1977b; 22:697.

48. McDougal WS, Wright FS: Defect in proximal and distal sodium transport in post-obstructive diuresis. *Kidney Int* 1972; 2:304.

49. Menaker GJ, Litvak S, Bendix R, et al: Operations on the colon without preoperative oral antibiotic therapy. *Surg Gynecol Obstet* 1981; 152:36.

50. Mizock B: Septic shock: A metabolic perspective. *Arch Intern Med* 1984; 144:579.

51. Montemurro DG, Stevenson JA: Survival and body composition of normal and hypothalamic obese rats in acute starvation. *Am J Physiol* 1960; 198:757.

52. Moukhtar M, Romney S: Continuous intraperitoneal antibiotic lavage in the management of purulent sepsis of the pelvis. *Surg Gynecol Obstet* 1980; 150:548.

53. Nichols RL, Broido P, Condon RE, et al: Effect of preoperative neomycin-erythromycin intestinal preparation on the incidence of infectious complications following colon surgery. *Ann Surg* 1973; 178:453.

54. Parker MM, Parrillo JE: Septic shock: Hemodynamics and pathogenesis. *JAMA* 1983; 250:3324.

55. Passmore R: Recommended intakes of nutrients for the United Kingdom. Reports on Public Health and Medical Subjects, no. 120:34. London, Her Majesty's Stationery Office, 1969.

56. Peters WP, Johnson MW, Friedman PA, et al: Pressor effect of naloxone in septic shock. *Lancet* 1981; 1:529.

57. Rosenberg JC, Rush BF: Lethal endotoxin shock. *JAMA* 1966; 196:767.

58. Rosenberg ML: The physiology of hyperchloremic acidosis following ureterosigmoidostomy: A study of urinary reabsorption with radioactive isotopes. *J Urol* 1953; 70:569.

59. Sheagren JN: Septic shock and corticosteroids. *N Engl J Med* 1981; 305:456.

60. Shumer W: Steroids in the treatment of clinical septic shock. *Ann Surg* 1976; 184:333.

61. Stamey TA: The pathogenesis and implications of electrolyte imbalance in ureterosigmoidostomy. *Surg Gynecol Obstet* 1956; 103:736.

62. Steiger E, Daly JM, Allen JR, et al: Postoperative intravenous nutrition: Effects on body weight, protein, regeneration, wound healing and liver morphology. *Surgery* 1973; 73:686.

63. Tarhan S, Moffitt EA, Taylor WF, et al: Myocardial infarction after general anesthesia. *JAMA* 1972; 220:1451.

64. Wilmore DW, Long JM, Mason AD Jr, et al: Catecholamines mediator of the hypermetabolic response to thermal injury. *Ann Surg* 1974; 180:653.

65. Wilmore DW, Mason AD Jr, Pruitt BA Jr: Insulin response to glucose in hypermetabolic burn patients. *Ann Surg* 1976; 183:314.

66. Wolfe BM, Culebras JM, Sim AJW, et al: Substrate interaction in intravenous feeding: Comparative effects of carbohydrate and fat on amino acid utilization in fasting man. *Ann Surg* 1977; 186:518.

67. Ziegler EJ: Treatment of gram-negative bacteremia and shock with human antiserum to a mutant Escherichia coli. *N Engl J Med* 1982; 307:1125.

Chapter 11

Renal Injuries

JACK W. MCANINCH, M.D.

Trauma is the leading cause of death between the ages of 1 and 44 years. In addition, millions of dollars are spent each year to rehabilitate trauma victims. The team approach to management provides the highest level of expertise in reducing morbidity and preventing mortality. The urologic surgeon is relied on by the trauma surgeon to deal with the complex injuries of the urogenital system. He may not become involved in the initial resuscitation phases of trauma care, because most injuries to the urogenital system are not life-threatening, but may become the most important member of the team two weeks later, when urinary extravasation, abscess formation, and septic complications develop secondarily.

The kidney is the most commonly injured organ in the urogenital system (McAninch, 1985), and renal injuries continue to create controversy in diagnosis and management. In spite of our current refined approach, many questions remain unanswered about when and for which patient a study should be done to evaluate a renal injury. Even when the diagnosis has been established, there is great controversy over operative vs. nonoperative management. The urologic and surgical literature is replete with articles debating the merits of each (Cass and Luxenberg, 1983; Peterson, 1977; Evins et al., 1980). No simple answers exist. One must review the literature critically and familiarize oneself with the variety of approaches in renal trauma care to be able to apply this information to an individual patient.

The following quotation from Erickson's 1860 *Textbook of Surgery* indicates the importance of careful clinical evaluation in patients with major renal injuries:

If the kidneys are injured, the patient will commonly experience a frequent desire to pass water, and this will be tinged with blood, often to a considerable extent. The absence of blood in the urine must not, however, be taken as an indication that the kidney is not injured; it may be so disorganized as to be totally incapable of secreting, and subsequently no bloody urine finds its way into the bladder. A man was admitted into the hospital under my care for a buffer in-

jury of the back; he passed water untinged with blood, but after death his right kidney was found completely smashed by the blow, with an extensive extravasation of blood in the celluloadipose tissue around it. Here it was evident that the disorganization was so sudden and complete that no bloody urine had found its way into the bladder.

This clinical picture continues to be reported today.

Approximately 8%–10% of blunt and penetrating abdominal injuries will involve the kidneys. In rural settings, blunt trauma will account for the largest percentage of renal injuries (90%–95%) (Krieger et al., 1984); in urban settings, the percentage of penetrating renal injuries increases (20%) (Sagalowsky et al., 1983).

This chapter is intended to provide a logical diagnostic approach to the patient with renal injury and to establish a rationale for management that will ultimately preserve the largest amount of functioning renal tissue.

MODE OF INJURY AND PRESENTATION

The mechanism of injury to the kidney is broadly classified as blunt or penetrating. Blunt trauma is the more common, accounting for 80%–90% of injuries, and results from automobile accidents, falls, contact sports, assaults, and personal violence (Ahmed and Morris, 1982; Banowsky et al., 1970; Cass, 1983). Gunshot and stab wounds cause 10%–20% of renal injuries and represent the most common causes of penetrating injury. In 1968, Carlton and co-workers reported associated intra-abdominal injuries in 80% of their patients with penetrating renal injury. This observation was supported by Sagalowsky et al. (1983): of 122 patients with gunshot wounds, all had associated intra-abdominal injury. Liver, small intestine, stomach, and colon were the most commonly injured organs. Stab wounds less frequently have associated intra-abdominal injury, with reports ranging from 30% to 70% (Sagalowsky et al., 1983; Heyns et al., 1983). This wide variation may be based on the location of the stab wound. Bernath and others (1983) have noted that stab wounds posterior to the anterior axillary line were associated with intra-abdominal injury in fewer than 12% of cases.

Renal injuries from blunt trauma occur consequent to upper abdominal injury and rapid deceleration. Gross or microscopic hematuria is usually present (Bright et al., 1978; Cockett et al., 1975). These patients will often have profuse abdominal tenderness, lower rib fractures, vertebral body fractures, and flank contusions. A palpable abdominal mass with associated shock may be indicative of a rapidly developing retroperitoneal hematoma from a major renal parenchymal or renal vascular injury. Rapid deceleration injuries usually involve multiple organ systems, and patients are often in shock and unconscious. Head-on automobile collisions and falls from great heights account for the majority of such injuries. Multiple bony fractures are usually present as well as injuries to the abdominal viscera, vascular system, chest, and head. The renal injury seen in such cases is often a renal pedicle avulsion or acute thrombosis of the main renal artery or one of the segmental arterial branches. Hematuria may not be present, and the diagnosis must be established by radiographic imaging prompted by a high index of suspicion (Guerriero et al., 1971; Stables et al., 1976). Excretory urography (IVP) is therefore indicated in all patients with rapid deceleration injury, even though hematuria may not be present.

Stab wounds to the kidneys generally have their entrance points in the lower thorax, flank area, or upper abdomen. The size of the entrance wound has little correlation with the extent of injury and the depth of penetration. Hematuria, most often gross, is usually present with major parenchymal injuries. The incidence of associated intra-abdominal injuries varies greatly and may be related to the entrance site (Bernath et al., 1983). Careful abdominal examination may reveal marked tenderness and generalized rigidity, indicating bowel perforation. Peritoneal lavage is useful for evaluating intra-abdominal injury after stab wounds to the torso (Danto, 1982). Hemorrhagic shock is a common presenting sign, and reestablishment of circulatory volume is of prime importance in the initial treatment. Renal imaging should be done in all patients with stab wounds in the upper abdomen, flank, back, and lower chest, whether hematuria is present or not.

Patients with gunshot wounds to the torso that penetrate the kidney often present in hemorrhagic shock with multiple organ injury. Rapid resuscitation is of prime importance, and immediate surgery is often required before diagnostic studies can be performed. The type of weapon must be ascertained. The damage a missile can inflict is related to the kinetic energy expended, determined from the formula:

$$KE = MV^2/2$$

(where KE = kinetic energy, M = mass of the missile, V = muzzle velocity of the weapon). The higher the muzzle velocity, the greater the tissue damage. Figure 11–1 demonstrates the method by which bullets of high velocity cause extensive tissue damage: on entering the soft tissue of the body, the bullet creates a temporary pulsatile cavity and vaporizes the surrounding tissue. This temporary cavity and vaporization can cause extensive damage (the "blast effect"), which may not be appreciated at operation. Intensive debridement may be required to remove the nonviable tissue created by such injuries. Most common handguns are low-velocity weapons (muzzle velocity <2,000 ft/sec), and less extensive debridement may be required. Rifles and specialized handguns will commonly cause extensive damage, as can bullet fragmentation (Fackler et al., 1984). Table 11–1 lists some muzzle velocities of common weapons.

HEMATURIA

The presence of blood in the urine is usually the first indicator of renal injury. In most reported series, over 95% of patients have had microscopic or gross hematuria (Nicolaisen and McAninch, 1985; Bright et al., 1978; Radwin et al., 1976; Wein et al., 1977). However, Bright and co-workers (1978) have noted that the degree of hematuria does not correlate with the severity of injury. For example, in a patient with gross hematuria after blunt abdominal trauma but normal results on IVP, a renal contusion is diagnosed. A patient with microscopic hematuria after a rapid deceleration injury

TABLE 11–1.

Muzzle Velocity of Common Weapons*

WEAPON	VELOCITY (FT/SEC)
.22 Short	1,045
.22 Magnum	2,000
.38 Caliber	1,330
.45 Caliber	1,320
Carbine (.30 caliber)	1,970
7.62 mm (M-14)	2,400–2,800
5.56 mm (M-16)	3,250

*From McAninch JW: Renal injuries, in McAninch JW (guest ed): *Urogenital Trauma*, vol 2 of Blaisdell WF, Trunkey DD (series eds): *Trauma Management.* New York, Thieme Stratton, 1985, pp 27–49. Used by permission.

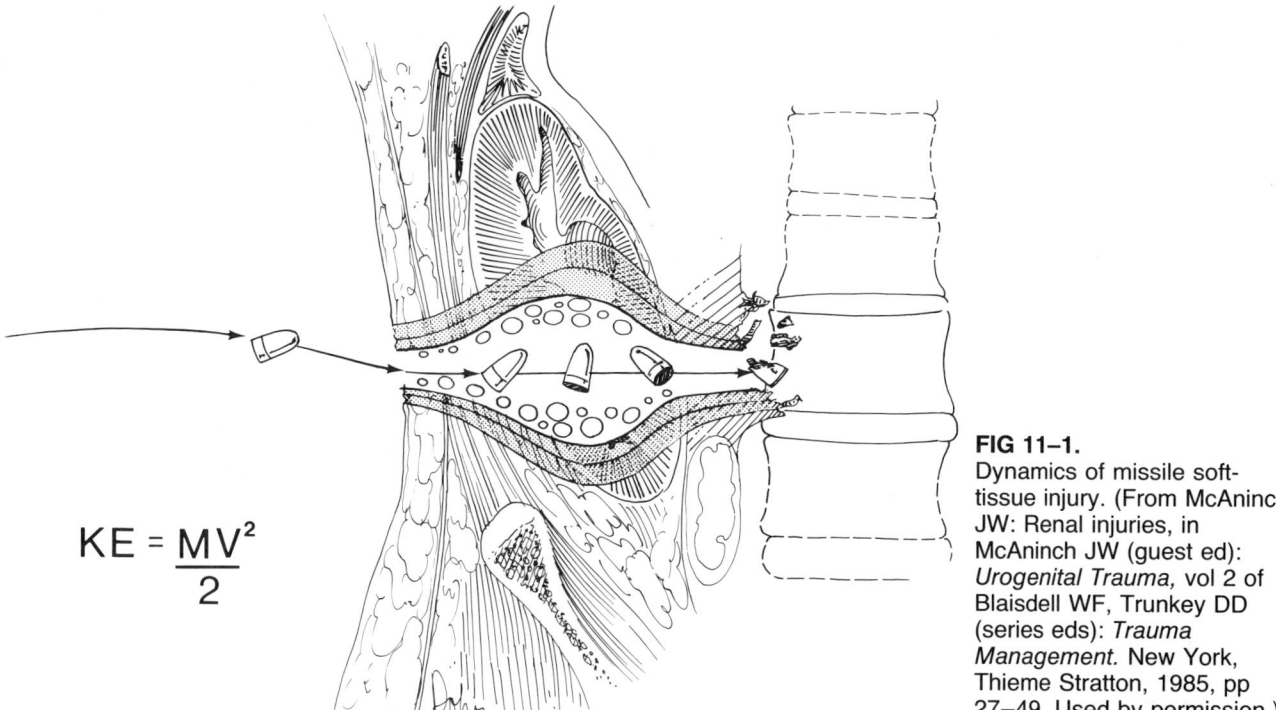

$$KE = \frac{MV^2}{2}$$

FIG 11–1.
Dynamics of missile soft-tissue injury. (From McAninch JW: Renal injuries, in McAninch JW (guest ed): *Urogenital Trauma,* vol 2 of Blaisdell WF, Trunkey DD (series eds): *Trauma Management.* New York, Thieme Stratton, 1985, pp 27–49. Used by permission.)

may well demonstrate a major renal vascular injury. Guerriero and associates (1971) noted gross hematuria in only 10 of 33 patients with renal vascular injuries, and Stables et al. (1976) found no hematuria in 24% of patients with traumatic renal artery occlusion. All series note the importance of hematuria as an indicator in injury, but emphasize that it is a nonspecific finding and does not correlate with the seriousness of the renal damage.

The patient should be evaluated in the emergency room where urine should be obtained for study. Unconscious patients with serious injuries should be catheterized for dipstick urinalysis. If positive, a finding of >10 RBCs/HPF would be expected, and the urinary system should be promptly evaluated. When time permits, microscopic urinary evaluation should be done, and >5 RBCs/HPF indicate potential injury. Guice and associates (1983) suggest that gross hematuria is the best indicator of significant renal injury.

Nicolaisen and McAninch (1985), in an attempt to refine the indications for radiographic assessment based on the presence of hematuria, noted that 221 patients with microscopic hematuria after blunt trauma who were free of hemorrhagic shock had renal contusions. This prospective study also noted that 85 patients with gross hematuria or shock (blood pressure <90 mm Hg) associated with microscopic hematuria after blunt trauma constituted the group demonstrating significant renal injuries. In 53 patients with penetrating

trauma, no combination of parameters could predict the severity of injury. If this study can be corroborated by other centers, it may prove useful in selecting patients with blunt trauma in whom imaging studies are necessary.

CLASSIFICATION OF RENAL INJURIES

Categorizing renal injuries according to severity helps to select appropriate therapy and predict results of treatment. Renal injuries can be classified under four large groups: (1) renal contusions—bruises or subcapsular hematomas associated with an intact renal capsule and collecting system; (2) minor lacerations—superficial cortical disruptions that do not involve the deep renal medulla or the collecting system; (3) major lacerations—parenchymal disruptions extending through the cortex and into the deep medulla, which may involve the collecting system (fractured kidneys are also classified as major parenchymal lacerations); (4) vascular injuries—occlusions or tears of the renal artery or vein or their segmental branches (Fig 11–2).

Vascular injuries and major and minor lacerations constitute significant trauma. These patients should have complete radiographic assessment to determine the full extent of injury and select appropriate management. Categorization does not mandate operation, but it will aid the surgeon in directing the care of these patients.

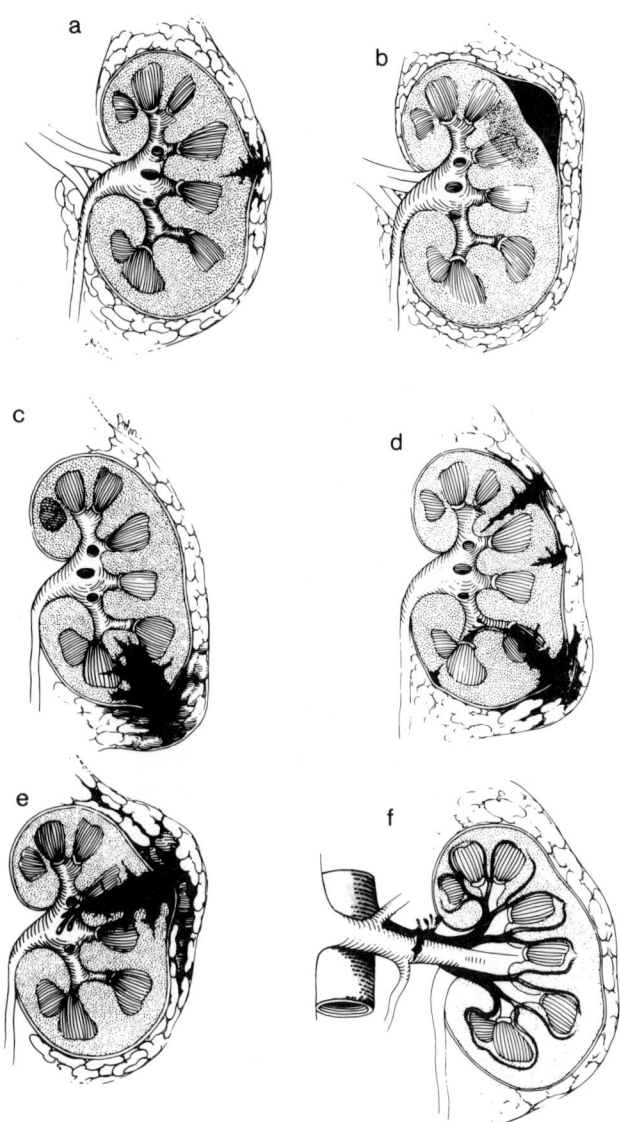

FIG 11–2.
Renal injuries classified by severity: *(a)* minor laceration; *(b)* renal contusion; *(c)* major laceration (deep medullary lacerations); *(d)* major laceration (fractured kidney); *(e)* major laceration (laceration into the collecting system); and *(f)* vascular injury. (From Nicolaisen GS, McAninch JW: Evaluation and management of traumatic renal injuries, In *AUA Update Series.* Houston, American Urological Association Office of Education, 1985, vol IV, lesson 37. Used by permission.)

STAGING AND ASSESSMENT OF INJURY

Staging is the orderly process by which a renal injury is completely defined by history, physical examination, and radiographic or other imaging techniques. For instance: A patient with blunt trauma from an automobile accident has gross hematuria; an IVP is obtained and the kidneys are found to be normal. This patient's in-jury is categorized as a renal contusion, and no additional studies are necessary. However, if the results of IVP had been indeterminate or abnormal, additional information would be needed to complete the staging and document the full extent of the injury.

History, physical examination, and the determination of hematuria are the initial evaluative measures. The presence of >5 RBCs/HPF continues to be the best indicator of renal injury, and radiographic assessment should be done. All patients with rapid deceleration injury should undergo excretory urography, whether hematuria is present or not, to avoid missing acute renal artery thrombosis or other vascular injury. Figure 11–3 depicts a staging algorithm that is a useful systematic approach to radiographic imaging.

Once the presence of hematuria has been documented, the initial recommended study is the high-dose IVP (in most circumstances, 2.0 ml/kg radiographic contrast material). If this is combined with nephrotomography a significant percentage of patients will be adequately staged (87% of patients with blunt trauma and approximately 68% of those with penetrating injuries [Nicolaisen and McAninch, 1985; Cass, 1982]). The IVP should establish the presence or absence of both kidneys, clearly define the renal parenchyma, and outline the collecting system and ureters. The addition of nephrotomography will aid in detecting renal lacerations, intrarenal hematomas, and decreased vascular perfusion to segmental areas of the kidney (Lang, 1975; Mahoney and Persky, 1968). In the patient with critical injuries, the IVP should be initiated in the trauma suite, with the contrast material injected directly into the IV lines during the initial resuscitation phase. The first plain film of the abdomen, which can then be taken, will indicate the presence of the kidneys and the extent of renal injury. This is a time-saving step in these serious injuries, where in most cases only a single film of the abdomen can be obtained before the patient is moved to the operating suite. The excreted contrast material seen on such films does not obscure other pathologic conditions, such as bony fractures, usually noted on the initial plane radiograph of the abdomen.

When the results of excretory urography are abnormal or indeterminate (i.e., incomplete or poor visualization), additional studies should be obtained, time permitting. Computed tomography (CT) is being used extensively (McAninch and Federle, 1982; Berger et al., 1980; Sandler and Toombs, 1981) and is immediately available in most trauma centers 24 hours per day. For staging renal injuries, CT has several advantages: noninvasiveness; clear delineation of parenchymal lacerations; sensitive detection of urinary extravasation; outline of nonviable tissue; definition of the extent and

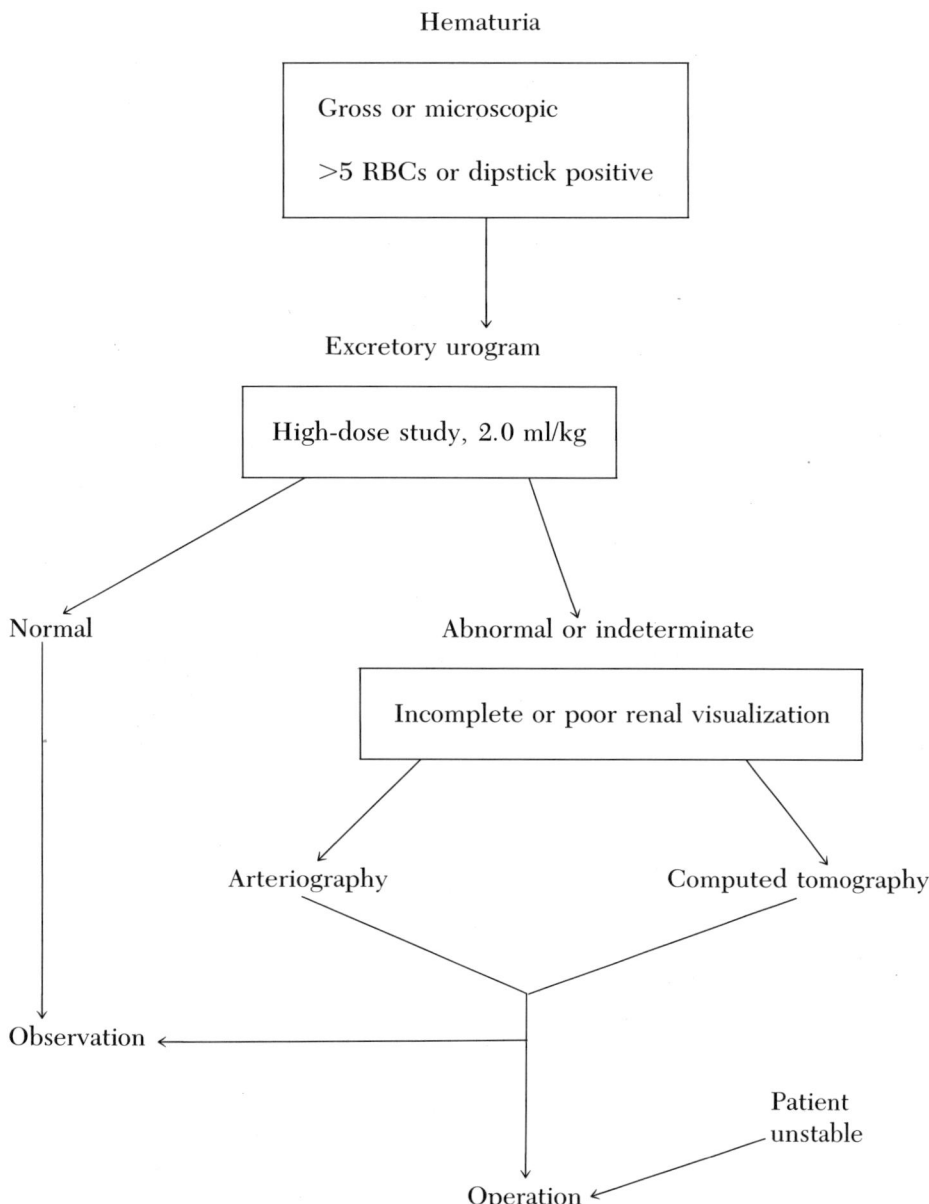

Hematuria

Gross or microscopic

>5 RBCs or dipstick positive

Excretory urogram

High-dose study, 2.0 ml/kg

Normal

Abnormal or indeterminate

Incomplete or poor renal visualization

Arteriography

Computed tomography

Observation

Patient unstable

Operation

FIG 11–3.
Staging algorithm for systematic approach to radiographic imaging.

size of the surrounding hematoma; detection of associated injury; three-dimensional views of the kidney and retroperitoneum. CT has also been useful in detecting arterial injury to the kidney (Reagan et al., 1984; Steinberg et al., 1984).

In 24 patients in whom incomplete visualization in IVP or nephrotomography (Table 11–2) prompted suspicion of major renal injury, CT clearly differentiated major lacerations and detected extravasation more sensitively than the IVP (Figs 11–4 and 11–5) (McAninch and Federle, 1982). In five patients, CT detected extravasation of opacified urine not visible on IVP. Renal

lacerations and perirenal and intrarenal hematomas were well defined by CT. As a result, CT enabled proper management in all instances. Eighteen patients were managed nonoperatively and six underwent surgery. The renal findings at operation confirmed the observations on CT in all surgical cases. In addition, CT detected major injuries to the liver, spleen, and pancreas in four patients.

The physiologic enhancement of the kidney by contrast material makes it ideal for CT imaging and allows improved definition of lacerations, hematomas, and nonviable tissue (Erturk et al., 1985). By clearly defin-

TABLE 11–2.

Computed Tomography in 24 Patients With Suspected Major Renal Trauma*

	MINOR LACERATION OR CONTUSION	MAJOR LACERATION WITHOUT EXTRAVASATION	MAJOR LACERATION WITH EXTRAVASATION	PATIENTS
Trauma				24
Blunt	11	4	6	
Penetrating		2	1	
		No. of occurrences		
IVP or nephrotomography				24
Poor visualization	11	6	6	
Extravasation	0	(?) 1	(?) 2	
Computed tomography				24
Perirenal hematoma	11	6	7	
Intrarenal hematoma	2	6	6	
Laceration	6	6	7	
Extravasation	0	0	7	
Operations	0	2	4	6

*From McAninch JW: Renal trauma, in Resnick MI (ed): *Current Trends in Urology.* Baltimore, Williams & Wilkins Co, 1985, vol. 3. Used by permission.

ing the depth of laceration and extent of injury into the collecting system, CT ensures more appropriate management decisions. The three-dimensional image provides a great deal of collective information. CT can indirectly detect major vascular injuries as well as segmental artery injuries. In the patient shown in Figure 11–5, in whom the IVP showed nonvisualization of the kidney after a rapid deceleration injury, CT demonstrated a nonenhancing soft-tissue shadow of a normal-sized kidney on the involved side. One can use angiography to confirm the diagnosis.

The major limitation of CT is the lack of detection of venous injuries to the main renal vein or its segmental branches.

Arteriography is more widely available and, in the past, had been the definitive study in staging major renal injuries. With the advent of CT, arteriography has been supplanted. However, it can still provide adequate information for management (Lang, 1975; Wynn et al., 1978). It will define parenchymal lacerations and vascular injuries and is recommended when CT is unavailable. The single most common indication for arteriography is nonvisualization of a kidney on excretory urography after major blunt abdominal trauma. Several causes for nonvisualization exist and should be considered: total avulsion of the renal artery and vein; renal artery thrombosis; absence of the kidney, either congenital or from surgical removal; and severe contusion causing major vascular spasm. Either digital subtraction arteriography or conventional arteriography may be used to evaluate vascular injuries. CT can detect vascular injuries, but arteriography gives more detailed in-

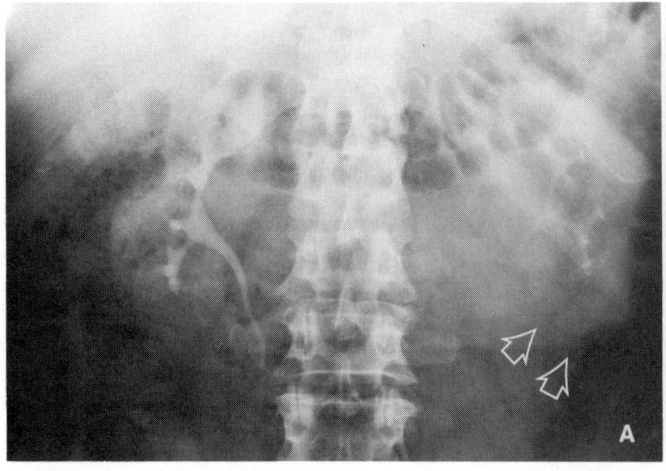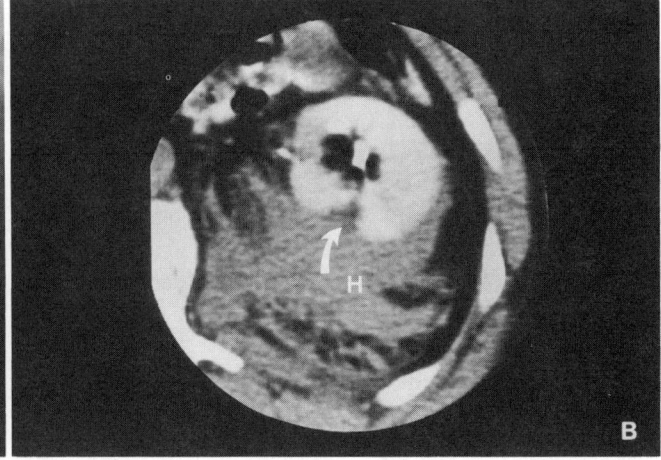

FIG 11–4.

A, excretory urogram in a young man with abdominal pain and gross hematuria after blunt trauma reveals poor visualization of the lower pole of the left kidney and lateral deviation *(arrows).* **B,** CT showed retroperitoneal hematoma *(H)* and a minor laceration of the renal parenchyma *(arrow).* Nonoperative management was successful.

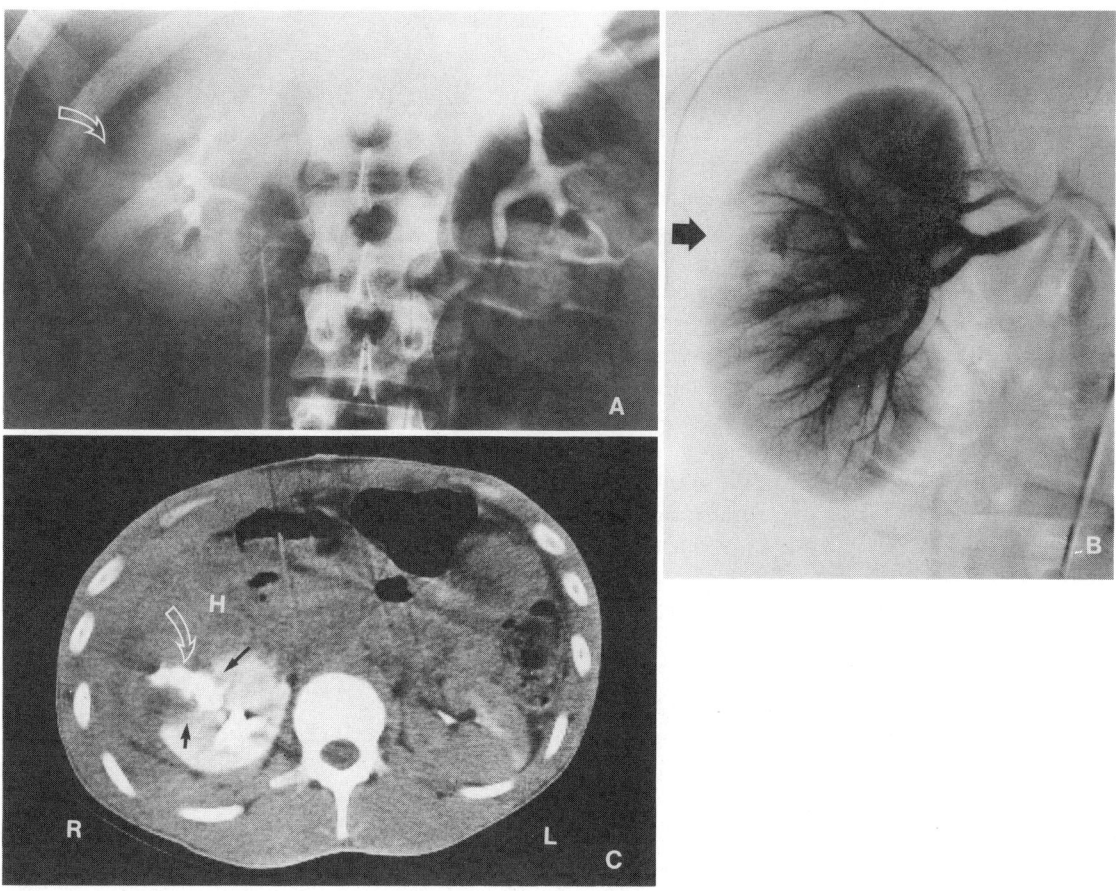

FIG 11–5.
A, excretory urogram in a patient with a right flank stab wound and microscopic hematuria suggests a defect in the upper lateral border of the right renal parenchyma *(arrow).* **B,** arteriography shows a minimal defect in the renal cortex *(arrow)* and no extravasation. **C,** CT scan reveals a large right renal laceration *(black arrows)* with extensive extravasation of opacified urine *(white arrow).* A large retroperitoneal hematoma is also noted *(H).* Operative repair resulted in renal salvage.

formation and defines the exact anatomical area of vascular injury.

When injuries to the segmental veins, main renal vein, and vena cava are suspected, venography must be used if the patient is stable enough (Peterson et al., 1985). In these injuries, immediate operative intervention may be required to control bleeding and maintain the patient's hemodynamic stability.

Although sonography has been used to evaluate and stage renal injuries (Berger et al., 1980; Kay et al., 1980; Schmoller et al., 1981), it provides less information than CT or arteriography. It can detect renal lacerations, but will not definitively assess their depth and extent. It cannot accurately detect vascular injuries. Sonographic techniques are improving, however, and its use in the future may have some merit.

Radionuclide scanning has been limited. In 24 patients, Chopp et al. (1980) combined this method with high-dose excretory urography and found that the number of arteriograms required for further staging was sig-

nificantly reduced. This technique appears to provide less information than arteriography or CT (Woodruff et al., 1967; Federle et al., 1981).

Retrograde pyelography has little benefit in evaluating renal injuries but is most useful in detecting associated ureteral or renal pelvic disruptions and perforations (Mendez, 1977).

With an orderly approach to the staging of renal trauma, the full extent of the injury can be defined to allow intelligent and accurate management decisions. The ultimate goal of complete staging is to provide sufficient information for management that will result in the preservation of renal parenchyma and salvage of injured kidneys.

INDICATIONS FOR OPERATION

The indications for operative intervention after renal injury vary greatly from one center to another. Blunt traumatic injuries create the most controversy. Contusions represent 85%–90% of blunt renal injuries and, in general, all series agree that they can be managed nonoperatively. The remaining 10%–15% of blunt injuries comprise minor and major lacerations and vascular injuries. Most vascular injuries, when recognized early, should have operation and reconstruction when possible. The management of minor and major lacerations, however, is highly variable. Peterson (1977) would avoid renal operation unless bleeding is life-threatening; in most cases, when an operation is required a nephrectomy would be performed. Cass (1983) has taken a completely opposite view and recommends immediate surgical management of major lacerations with or without extravasation. It is difficult to assess individual series, because no group directly compares operative and nonoperative management in a controlled fashion at one institution. The lack of a uniform classification system also makes comparison between series extremely difficult.

Selecting the patient in whom complications will not develop with nonoperative management is difficult. Delayed bleeding, persistent extravasation with hematoma, and the potential for infection cause concern. In the 10%–15% of patients with blunt injuries more serious than contusions, Carlton (1978) reported that, in his experience, 90% of the complications from expectant treatment occurred in this group. Others have noted that, when delayed operation is required to manage a complication, total nephrectomy frequently results (Holcroft et al., 1975).

We take an aggressive approach to staging the injury and, by so doing, obtain adequate information to select the appropriate management, be it operative or nonoperative. We use the following specific indications for renal exploration: expanding or uncontained hematoma; pulsatile hematoma; extensive urinary extravasation; nonviable renal parenchyma; and vascular injury (McAninch and Carroll, 1982). When the degree of urinary extravasation is minor, operation is not required—assuming, of course, that no other indication necessitates surgical intervention. In a series of 154 patients with blunt renal trauma (McAninch and Carroll, 1982), the above indications mandated renal exploration in 14. All 14 kidneys were repaired and no total nephrectomies were required (Table 11–3). This experience indicates the high renal salvage rate that can be achieved in patients who require renal exploration after blunt trauma. With these indications, and improved staging techniques, only approximately 3% of patients with blunt renal trauma now require renal exploration at our trauma center.

Penetrating renal injuries should have operative exploration. Only when preoperative staging clearly indicates that the extent of injury is minor can a nonoperative approach be used successfully. Approximately 70% of patients with penetrating renal injuries at San Francisco General Hospital now require operative intervention (see Table 11–3). The indications for operative exploration are the same as those listed above when careful preoperative staging has been accomplished. Recently, Carroll and McAninch (1985) reported that CT provided accurate preoperative assessment in 11 patients with penetrating renal injury, which allowed nonoperative management in 8. Associated intra-abdominal injuries occur in 80% of patients with penetrating renal injury, and these often require immediate surgical intervention and leave no time for careful preoperative staging. In such circumstances, bleeding and life-threatening conditions should be controlled in the operating room, and an IVP should be obtained on the operating table to be certain that at least one normal functioning kidney is present and to gain information regarding the potentially injured kidney. If findings on IVP are abnormal, exploration of the ipsilateral kidney should be carried out. This careful, selective approach to penetrating renal injuries has not resulted in delayed renal operation at our institution.

Bernath et al. (1983) and Heyns et al. (1983) advocated a nonoperative approach for stab wounds. From their data, it appears that when the entrance site is dorsal to the posterior axillary line, the incidence of associated abdominal injury requiring renal exploration is low.

Vascular injury of major renal vessels has been reported in 1%–3% of patients with blunt renal injuries (Guerriero et al., 1971; Lohse et al., 1982). Total avulsion of the renal artery and vein, seen after rapid deceleration, is the most serious injury because of acute hemorrhage. Acute renal artery thrombosis is also seen in rapid deceleration injuries and is difficult to diagnose. The degree of hematuria is often insignificant or, in some cases, absent (Stables et al., 1976). All patients known to be involved in rapid deceleration accidents should undergo excretory urography whether hematuria is present or not. Nonvisualization of the kidney on IVP demands immediate arteriography or CT. The free

TABLE 11–3.

Renal Injuries at San Francisco General Hospital*†

TYPE OF INJURY	NO. OF OPERATIONS (%)		
Blunt (154 kidneys)	14 (9)	Repair	14
		Nephrectomy	0
Penetrating (36 kidneys)	25 (70)	Repair	18
		Nephrectomy	7

*185 patients.
†From McAninch JW: Renal injuries, in McAninch JW (guest ed): *Urogenital Trauma*, vol 2 of Blaisdell WF, Trunkey DD (series eds): *Trauma Management*. New York, Thieme Stratton, 1985, pp 27–49. Used by permission.

movement of the kidneys in the retroperitoneum results in sudden stretch on the renal artery. The arterial intima, having little elasticity, tears, which produces thrombosis within the vessel lumen (Figs 11–6 and 11–7). These thrombi quickly reduce blood flow to the kidney, which may then be viable for only a limited time. Rapid diagnosis and immediate operation are needed to salvage the kidney (McAninch, 1975).

Venous injuries of the main renal vein or segmental renal branches constitute a serious, at times lethal, condition. These injuries can result from blunt or penetrating trauma and massive blood loss can be expected. Preoperative staging for an accurate diagnosis is difficult because excretory urography, nephrotomography, CT, and arteriography do not adequately image venous injuries. Often, particularly on the right side, a renal and vena caval injury will coexist (Peterson et al., 1985), from which the mortality rate can be as high as 50%.

RETROPERITONEAL HEMATOMA

The general surgeon is often confronted with the unexpected finding of a large retroperitoneal hematoma during exploration for an abdominal injury. These hematomas may be found in blunt or penetrating injuries. Exploring the kidney that is surrounded by hematoma has traditionally been regarded as unnecessary, and complete nephrectomy has often resulted (Wein et al., 1977; Thompson et al., 1977). The urologic surgeon who is called in for consultation should have a systematic approach to evaluating these hematomas. High-dose excretory urography should be performed on the operating room table to evaluate the status of the po-

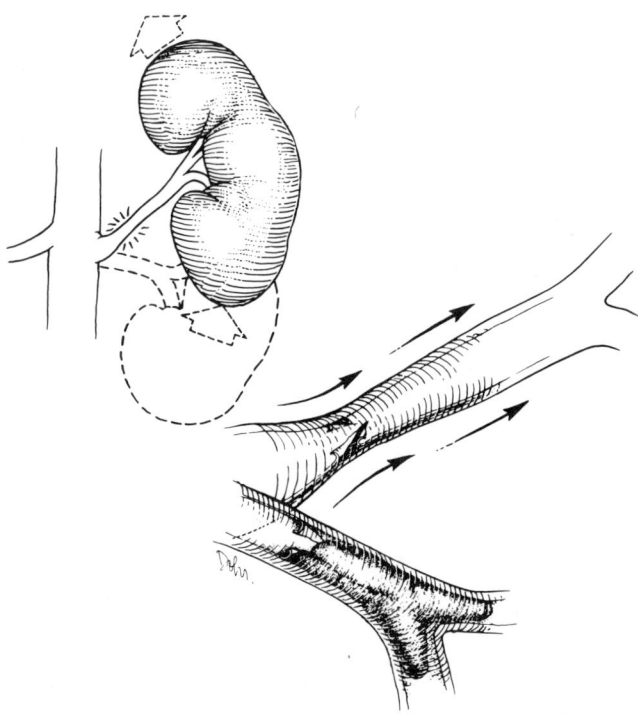

FIG 11–7.
The mechanism of arterial thrombosis from blunt trauma. (From McAninch JW: Renal injuries, in McAninch JW (guest ed): *Urogenital Trauma,* vol 2 of Blaisdell WF, Trunkey DD (series eds): *Trauma Management.* New York, Thieme Stratton, 1985, pp 27–49. Used by permission.)

tentially injured kidney and to confirm the presence of a functioning contralateral renal unit. If the IVP appears normal and no continued expansion of the hematoma is noted, surgical exploration can be avoided. However, when the IVP is indeterminate or abnormal, surgical exploration of the hematoma should be carried out. It is imperative to isolate the renal artery and vein before entering the hematoma to control the heavy bleeding that may develop during exploration. In cases of bilateral retroperitoneal hematomas requiring exploration, we would choose to explore first the kidney suspected of having the lesser injury, assuming the patient's hemodynamic stability is maintained.

OPERATIVE EXPLORATION AND RENAL EXPOSURE

Once the decision for operative exploration is made, the preferred approach is a midline, transabdominal incision (Scott and Selzman, 1966; Carlton et al., 1968; McAninch and Carroll, 1982). This allows assessment of other intra-abdominal visceral organs and major abdominal vessels. Repair of major vascular, spleen, liver, and bowel injuries should generally be carried out before renal exploration and repair. However, should renal

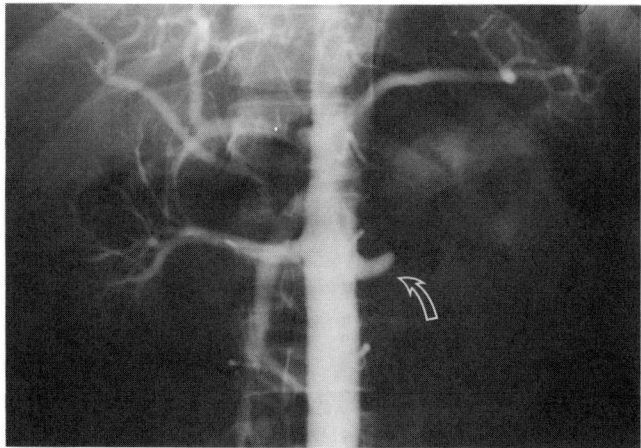

FIG 11–6.
In a 29-year-old man with blunt trauma, arteriography demonstrates acute left renal arterial thrombosis. (From McAninch JW: Renal injuries, in McAninch JW (guest ed): *Urogenital Trauma,* vol 2 of Blaisdell WF, Trunkey DD (series eds): *Trauma Management.* New York, Thieme Stratton, 1985, pp 27–49. Used by permission.)

bleeding be massive and persistent, renal exploration takes precedence.

To control massive bleeding before renal exploration, it is important to isolate the renal artery and vein individually (Fig 11–8, A). The surgeon must be careful to examine the anatomic relationships within the posterior abdomen and posterior parietal peritoneum before beginning vascular isolation. In many cases, the surgical landmarks have been distorted by urinary extravasation and massive hematomas. The transverse colon is lifted from the abdomen and placed on the anterior chest. This allows the small bowel to be lifted freely from the abdomen superiorly to the right, to expose the small bowel mesentery and the posterior parietal peritoneum. Anatomical landmarks to be identified at this point are the inferior mesenteric vein and the aorta. Should the aorta be covered by large hematoma, an incision can be made in the retroperitoneum just medial to the inferior mesenteric vein; by palpation through the hematoma, the aorta will be found. At this level, the aorta is free of major branches on its anterior surface and is usually easily and safely identified. The aorta is dissected superiorly on the anterior surface up to the area of the ligament of Treitz, where the left renal vein

will be found crossing anterior to the aorta. This is a major anatomical landmark, since the left renal artery originates from the aorta just lateral and superior to the left renal vein, and the right renal artery lies medial and superior as it originates from the aorta (Fig 11–8, B). Vessel loops of soft silicone can be placed around the individual vessels for retraction as well as occlusion. In most circumstances, it is unnecessary to occlude the vessels at the time of initial isolation. The right renal vein ordinarily can be isolated through the retroperitoneal incision; however, if it is difficult, mobilization of the second portion of the duodenum will readily expose the right renal vein and vena cava to make the vessel more accessible. Only vessels to the injured kidney need to be isolated.

Once vessel isolation is complete, an incision is made in the peritoneum just lateral to the colon and the colon is reflected medially to expose the retroperitoneal hematoma in its entirety (Fig 11–8, C). The kidney should then be totally exposed and mobilized for complete inspection. This can be done quickly and safely without concern for great blood loss because of the vascular control that has been obtained. Should heavy bleeding be encountered, vascular clamps or vessel clamps can

FIG 11–8.
The operative approach to renal vessels. **A,** the bowel is retracted superiorly to expose the retroperitoneum, where an incision is made medial to the inferior mesenteric vein over the aorta. **B,** the renal vessels are exposed and vessel loops are placed. **C,** the hematoma is then entered from a lateral approach. (From McAninch JW, Carroll PR: Renal trauma: Kidney preservation through improved vascular control—a refined approach. *J Trauma* 1982; 22:285–290. Used by permission.)

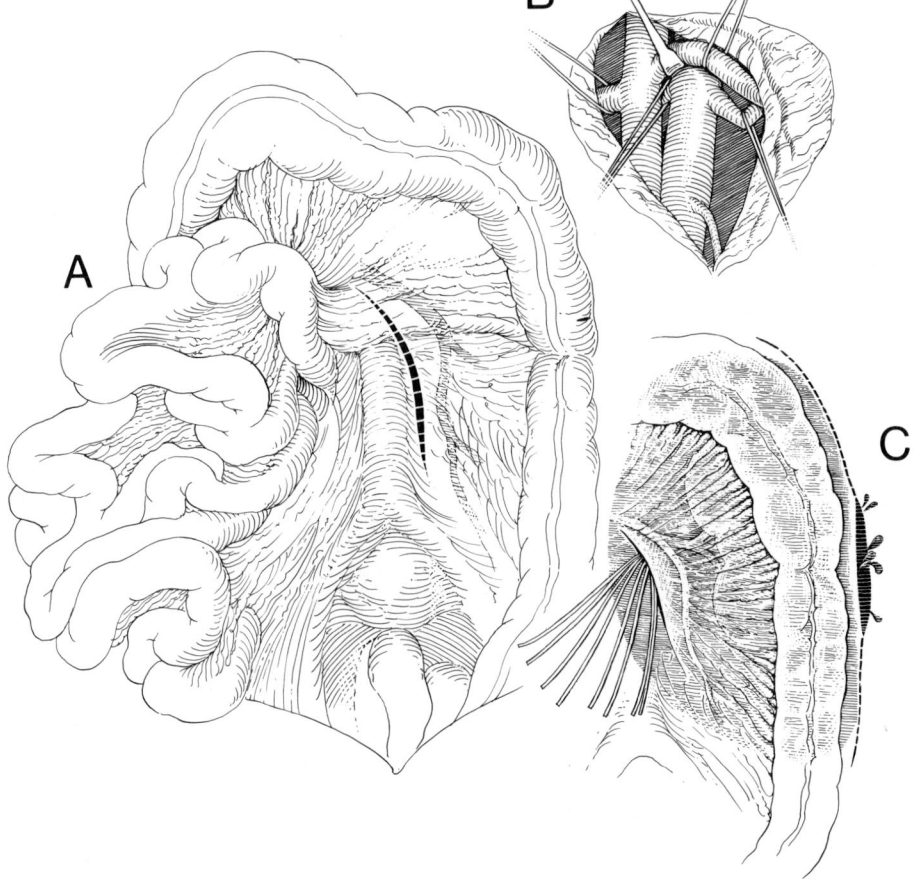

be used. In most circumstances, occlusion of the renal artery beyond 30 minutes is not required, and the kidney tolerates this amount of warm ischemia time well. Should the time extend beyond this limit, renal cooling is advised during the continued reconstruction process.

Scott and Selzman (1966) originally described this technique of early vascular control when exposing a traumatized kidney. Renal bleeding can be prevented and nephrectomy rates reduced. McAninch and Carroll (1982) compared a series of patients in whom early vascular control was achieved with another group in whom vascular control was inconsistent (Table 11–4). The nephrectomy rate in the former group was 18%, and all patients who required nephrectomy had sustained penetrating injuries. In the group with poor vascular control, the nephrectomy rate was 56%—a statistically significant difference. Clearly, when nephrectomy due to hemorrhage is prevented, as it can be by this technique, the renal salvage rate is greatly improved.

OPERATIVE FINDINGS AND RENAL RECONSTRUCTION

Complete renal exposure is of primary importance (Fig 11–9). The kidney is often surrounded by large hematoma, which should be completely swept away so that the entire surface area of the kidney, as well as the hilar vessels, is completely in view for inspection. Should massive bleeding be encountered from a vascular injury or a parenchymal laceration, temporary occlusion of the renal artery may be necessary to control hemorrhage. Large intrarenal hematomas that reside in lacerations should be completely evaluated and the margins of the laceration inspected. All nonviable tissue should be completely removed. Hemostasis should be obtained on the laceration margins with 4–0 chromic sutures on a fine-tapered needle placed in a figure-of-8 over individual bleeding points. Chromic suture is preferred, because its monofilament characteristics allow it to slide through the tissue without tearing or cutting the renal parenchyma. Larger sutures often cause increasing amounts of tissue ischemia and are unnecessary.

TABLE 11–4.

Nephrectomy Rates in Patient Series With and Without Early Vascular Control*

PROCEDURE		EARLY VASCULAR CONTROL (SERIES II; 39 PATIENTS), NO. (%)	POOR VASCULAR CONTROL (SERIES I; 34 PATIENTS), NO. (%)
Nephrectomy	(N = 26)	7 (18)†	19 (56)†
Repair	(N = 47)	32 (82)	15 (44)

*Data from McAninch JW, Carroll PR: Renal trauma: Kidney preservation through improved vascular control—a refined approach. *J Trauma* 1982; 22:285–290.
†P<.001.

As bleeding within the laceration comes under control, inspection should be directed to the collecting system in the depth of the renal laceration. It may be very obvious that the collecting system is open; if so, interrupted sutures of 4–0 chromic or 4–0 PDS should be used. Running sutures of the same material may be used to ensure a water-tight closure if the collecting system is wide open. One should then attempt to approximate the margins of the laceration. When the renal capsule has been spared and is available, it offers sufficient strength to approximate and hold the renal margins in place. A running suture approximating the capsule over the laceration will help to maintain hemostasis and provide a template for wound healing at the margins of the laceration. Very often the capsule will have been destroyed by the traumatic injury and is not available to be used in closure. In such cases, we prefer a pedicle flap of omentum to cover the defect and to aid in hemostasis and prevention of urinary extravasation. The omentum has the advantage of being viable tissue, rich in lymphatics and blood supply, which promotes healing of the injured area. When, as is often the case, omentum is not available and fragments of the capsule remain, interrupted sutures should catch the margins of the remaining capsule (without extending into the parenchyma) and these can be tied over a Gelfoam bolster. This technique provides excellent hemostasis and will be an aid in any delayed urinary extravasation or delayed bleeding. The Gelfoam is reabsorbed within three weeks; the risk of future infection or calculus formation from its use is minimal (McAninch et al., 1979). Multiple lacerations may coexist, and each laceration can be reconstructed similarly. Occasionally, after reconstruction and control of hemorrhage within each laceration, the entire kidney can be wrapped in omentum to protect it against future problems and to promote wound healing.

Deep lacerations through the upper or lower pole of the kidney may devascularize a large segment of tissue, and a partial nephrectomy may be indicated. The capsule should be preserved and the segmental artery supplying the involved area may require ligation. In most cases, we are able to spare the segmental vessel, which often supplies additional surrounding tissue, and remove only the nonviable area. Hemostasis on the margins of the parenchyma should be achieved with the interrupted suture technique described above and the collecting system should be carefully closed. To be certain that closure is watertight, we often occlude the ureter digitally just below the level of the renal pelvis and inject 2–3 ml of indigo carmine into the renal pelvis. Any extravasation will become obvious. Coverage of the renal parenchyma after partial nephrectomy is important. We use any remaining capsule or, if unavail-

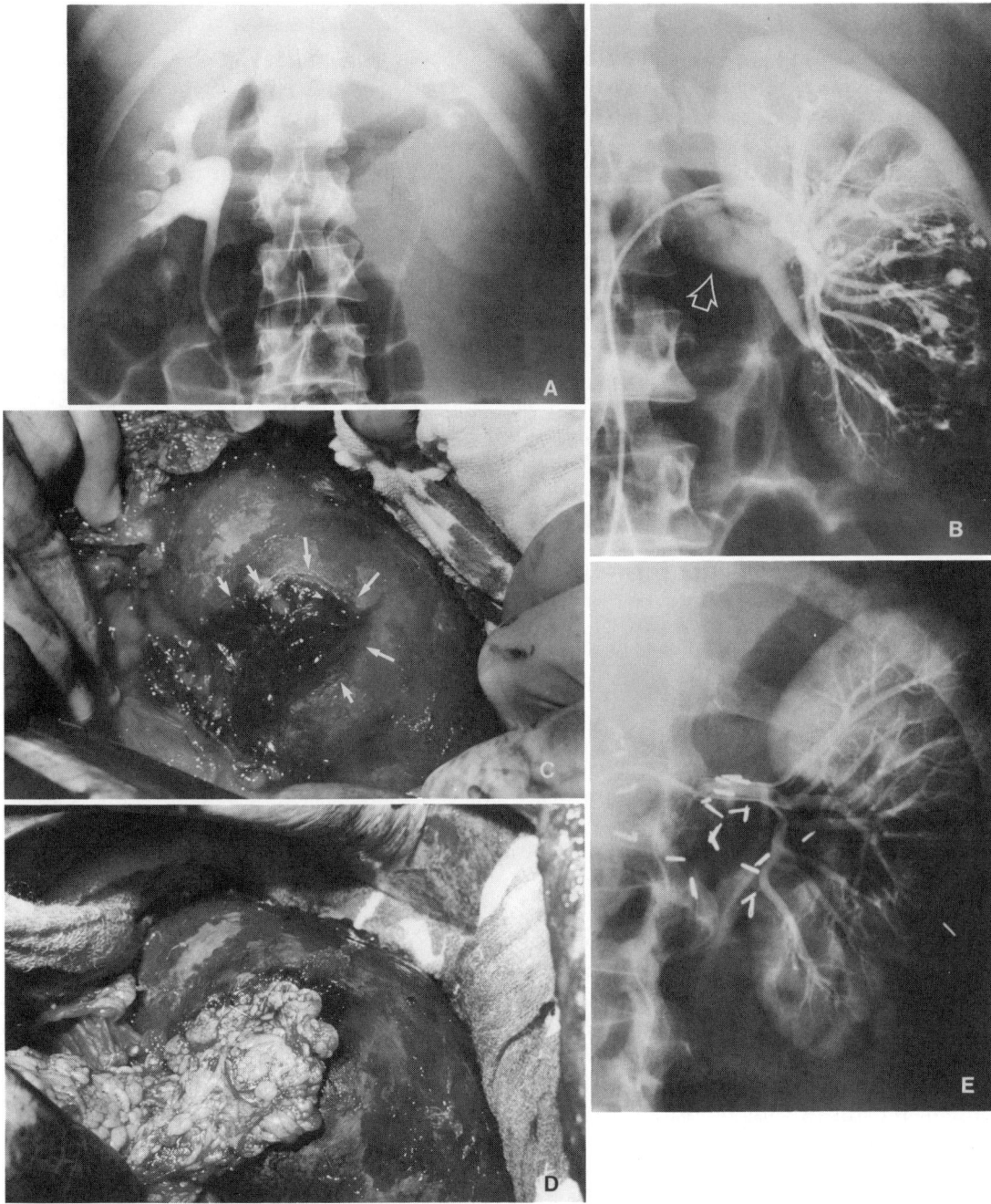

FIG 11–9.
A, excretory urogram in a 19-year-old patient with blunt abdominal trauma and gross hematuria demonstrates poor visualization of the left kidney. **B,** arteriogram shows numerous areas of vascular extravasation in the mid and lower left kidney. The prompt filling of the renal vein indicates massive arteriovenous shunting *(arrow).* **C,** at operation, a deep parenchymal laceration *(arrows)* was noted on the medial aspect of the kidney near the renal hilum. **D,** the laceration was debrided and the bleeding vessels suture-ligated. The defect was covered with an omental pedicle flap. **E,** selective left renal arteriography six months after injury demonstrates complete resolution of arteriovenous shunting and healing of the renal laceration. (From McAninch JW, Carroll PR: Renal trauma: Kidney preservation through improved vascular control—a refined approach. *J Trauma* 1982; 22:285–290. Used by permission.)

able, a pedicle flap of viable omentum. On the occasions when these are not available, a free peritoneal graft can be sutured into place over the defect.

Vascular injuries to the kidney create important decision-making situations. Acute renal artery thrombosis requires rapid diagnosis and immediate surgical intervention if total renal loss is to be avoided. The amount of time in which such a kidney remains salvageable is dependent upon the degree of renal arterial obstruction from the clot that forms within the lumen. This may range from one to several hours. Diagnostic studies are not likely to define viability, and only by immediate surgical intervention can renal salvage be expected (Stables et al., 1976). These injuries are best managed by excising the segment of the renal artery that is involved in the injury and perfusing the kidney with cold hypertonic solution to improve viability. Several options can be exercised at this point: primary arterial reanastomosis; hypogastric artery graft; synthetic graft; or autotransplantation to the pelvis. Autotransplantation appears to offer the best salvage rate (Fay et al., 1974; Guttman et al., 1978). Lacerations of the renal artery should be repaired with 5–0 interrupted vascular sutures. Segmental renal artery lacerations can be repaired similarly; extensive segmental artery injury may require ligation. Should necrosis involve more than 15% of the kidney, active debridement and excision of the nonviable tissue should be done.

Venous injuries cause massive bleeding, and repair on the right is complicated by the shortness of the renal vein and the frequent involvement of the vena cava. Complete occlusion of the vena cava above and below the area of injury will temporarily control bleeding until vascular clamps can be applied to the exact areas of damage. Fine 5–0 vascular sutures should be used to close these defects. Segmental renal vein injuries are best managed by ligation of the vessel. This does not cause ischemic damage because of the inner communication of the veins within the renal parenchyma.

Total renal pedicle avulsion, which involves complete laceration of the renal artery and vein from their attachments, requires immediate surgical intervention and does not allow time for diagnostic studies. The kidney is often viable at the time of exploration; with cold hypertonic solution perfused into the renal artery, one has ample time to make intraoperative management decisions. Autotransplantation appears to be the most successful approach (Guttman et al., 1978). Massive internal injuries may mandate nephrectomy, assuming that the patient has a normal contralateral kidney.

Retroperitoneal drains are left in place when there is a question of urinary leakage, but careful management is obligatory to avoid infection that may ascend along a drain tract. In situations where urinary extravasation will not occur, drains are unnecessary. They should not be left in place in an attempt to drain retroperitoneal hematoma, since they are not effective and only provide an avenue for potential infection.

POSTOPERATIVE CARE AND FOLLOW-UP

The postoperative care of the patient with renal trauma is similar to that for any major transabdominal surgical procedure, with nasogastric suction or gastrostomy for bowel decompression and urethral catheter drainage until the patient is stable enough to void. Antibiotics should be given for bowel perforation or severely contaminated wounds. If the urine is infected, preoperative and postoperative antibiotics should be continued through a full ten-day course.

In most circumstances, the urine becomes free of clots within the first 12–24 hours and is free of gross blood within 48 hours. Serial hematocrit readings should be obtained to be certain that continued bleeding does not occur. When drains have been left in place, significant drainage is often noted, although this is commonly intraperitoneal fluid and not urine. One should check the creatinine content of the fluid, which will be many times the serum concentration if urine is present. Intravenous injection of methylene blue may also be used to evaluate possible urinary extravasation. Blood pressure should be followed closely in the early postoperative period as well as later on.

The patient can generally be discharged 8–10 days after a reconstructive renal procedure. At the time of discharge, patients are allowed free ambulation without restriction and should be encouraged to return to normal physical activity as soon as the incisional pain has subsided.

The patient should be seen in follow-up every one to two weeks for the first six weeks to monitor blood pressure and evaluate the urine for the presence of blood. Hypertension is uncommon, and microscopic hematuria should gradually subside. Approximately three weeks after injury, the patient should have a follow-up IVP to evaluate the anatomical configuration of the kidney and to be certain that no obstruction has resulted from perirenal scarring. Renal radionuclide scans are often helpful to evaluate the functional status of the injured kidney.

COMPLICATIONS

Early complications occur within the first four weeks of injury and include delayed bleeding, abscess, sepsis, urinary fistula, urinary extravasation and urinoma, and hypertension (Spark and Berg, 1976; Von Knorring et al., 1981; Jakse et al., 1984). Delayed bleeding can occur from the immediate postoperative period until several weeks later. The greatest risk of delayed heavy retroperitoneal bleeding occurs within the first two weeks of injury. Abscess may develop in the perinephric space, and in most circumstances will be noted within the first seven days. This may be associated with sepsis and will be manifested by increasing fevers that may reach 41° C. Prompt surgical exploration and drainage are usually required; however, after appropriate diagnostic procedures, localized abscesses may be drained percutaneously. Symptoms of urinary extravasation in the first four weeks of injury may be manifested by a low-grade fever and continued pain in the area of the kidney. Excretory urography and CT can be an aid in establishing the diagnosis and the extent of extravasation. Percutaneous drainage has successfully managed extensive extravasation, but small amounts of urine in the retroperitoneal space appear to be of no particular consequence if they remain uninfected. These often resolve spontaneously without intervention. Hypertension in the postoperative period is uncommon; however, its presence and duration are extremely variable (Von Knorring et al., 1981). In most circumstances, the hypertension does not require treatment; when indicated, medical therapy usually controls the problem. The hypertension appears to be renin-mediated in most cases and is usually transient when it occurs in the early postinjury period.

Late complications include arteriovenous fistula, hydronephrosis, hypertension, calculus formation, and chronic pyelonephritis. Delayed hypertension has been noted by several authors. In a group of carefully followed up patients, Jakse and colleagues (1984) noted the onset of hypertension some 15 years after renal trauma. Long-standing hypertension persists because of partial renal ischemia, resulting in a renin-mediated type of hypertension (Spark and Berg, 1976). Most cases can be managed medically, but surgical intervention may be necessary in unresponsive patients. Arteriovenous fistulas are caused by both blunt and penetrating injuries, but mainly by stab wounds (Cosgrove et al., 1973). Delayed urinary bleeding is the usual presenting symptom, and many patients have associated hypertension. Large fistulas require surgical correction and at times nephrectomy. Hydronephrosis can develop in the late postoperative period because of surrounding fibrosis and obstruction to the upper ureter or ureteropelvic junction. This condition may lead to calculus formation or recurrent pyelonephritis or both.

To detect many of these developing delayed complications, an IVP is strongly recommended within three months of major renal injury. Additional follow-up should be done in these patients when persistent problems or suspicion of abnormalities exist.

REFERENCES

1. Ahmed S, Morris LL: Renal parenchymal injuries secondary to blunt abdominal trauma in childhood: A 10-year review. *Br J Urol* 1982; 54:470–477.
2. Banowsky LH, Wolfel DA, Lackner LH: Considerations in diagnosis and management of renal trauma. *J Trauma* 1970; 10:587–597.
3. Berger PE, Munschauer RW, Kuhn JP: Computed tomography and ultrasound of renal and perirenal diseases in infants and children: Relationship to excretory urography in renal cystic disease, trauma and neoplasm. *Pediatr Radiol* 1980; 9:91–99.
4. Bernath AS, Schutte H, Fernandes RRD, et al: Stab wounds of the kidney: Conservative management in flank penetration. *J Urol* 1983; 129:468–470.
5. Bright TC, White K, Peters PC: Significance of hematuria after trauma. *J Urol* 1978; 120:455–456.
6. Carlton CE Jr: Injuries of the kidney and ureter, in Harrison JH, et al (eds): *Campbell's Urology*. Philadelphia, WB Saunders Co, 1978, pp 881–905.
7. Carlton CE Jr, Scott R Jr, Goldman M: The management of penetrating injuries of the kidney. *J Trauma* 1968; 8:1071–1083.
8. Carroll PR, McAninch JW: Operative indications in penetrating renal trauma. *J Trauma* 1985; 25:587–593.
9. Cass AS: Blunt renal trauma in children. *J Trauma* 1983; 23:123–127.
10. Cass, AS: Immediate radiologic and surgical management of renal injuries. *J Trauma* 1982; 22:361–363.
11. Cass AS, Cass BP: Immediate surgical management of severe renal injuries in multiple-injured patients. *Urology* 1983; 21:140–145.
12. Cass AS, Luxenberg M.: Conservative or immediate surgical management of blunt renal injuries. *J Urol* 1983; 130:11–16.
13. Chopp RT, Hekmat-Ravan H, Mendez R: Technetium-99m glucoheptonate renal scan in diagnosis of acute renal injury. *Urology* 1980; 15:201–206.
14. Cockett ATK, Frank IN, Davis RS, et al: Recent advances in the diagnosis and management of blunt renal trauma. *J Urol* 1965; 113:750–754.
15. Cosgrove MD, Mendez R, Morrow JW: Traumatic renal arteriovenous fistula: Report of 12 cases. *J Urol* 1973; 110:627–631.
16. Danto LA: Paracentesis and diagnostic peritoneal lavage, in Blaisdell WF, Trunkey DD (eds): *Trauma Management*. New York, Thieme Stratton, 1982, vol 1.
17. Erturk E, Sheinfeld J, DiMarco PL, et al: Renal trauma: Evaluation by computerized tomography. *J Urol* 1985, 133:946–949.
18. Evins SC, Thomason WB, Rosenblum R: Non-operative management of severe renal lacerations. *J Urol* 1980; 123:247–249.
19. Fackler ML, Surinchak JS, Malinowski JA, et al: Bullet fragmentation: A major cause of tissue disruption. *J Trauma* 1984; 24:35–39.
20. Fay R, Brosman S, Lindstrom R, et al: Renal artery

thrombosis: A successful revascularization by autotransplantation. *J Urol* 1974; 111:572–577.

21. Federle MP, Kaiser JA, McAninch JW, et al: The role of computed tomography in renal trauma. *Radiology* 1981; 141:455–460.

22. Free AH, Free HM: Urinalysis, critical discipline of clinical science. *CRC Crit Rev Clin Lab Sci* 1972; 3:481–531.

23. Guerriero WG, Carlton CE Jr, Scott R Jr, et al: Renal pedicle injuries. *J Trauma* 1971; 11:53–62.

24. Guice K, Oldham K, Eide B, et al: Hematuria after blunt trauma: When is pyelography useful? *J Trauma* 1983; 23:305–311.

25. Guttman FM, Homsy Y, Schmidt E: Avulsion injury to the renal pedicle: Successful autotransplantation after 'bench surgery.' *J Trauma* 1978; 18:469–471.

26. Heyns CF, Klerk DP de, Kock MLS de: Stab wounds associated with hematuria—a review of 67 cases. *J Urol* 1983; 130:228–231.

27. Holcroft JW, Trunkey DD, Minagi H, et al: Renal trauma and retroperitoneal hematomas—indications for exploration. *J Trauma* 1975; 15:1045–1052.

28. Jakse G, Putz A, Gassner I, et al: Early surgery in the management of pediatric blunt renal trauma. *J Urol* 1984; 131:920–924.

29. Kay CJ, Rosenfield AT, Armm M: Gray-scale ultrasonography in the evaluation of renal trauma. *Radiology* 1980; 134:461–466.

30. Knorring J von, Fyhrquist F, Ahonen J: Varying course of hypertension following renal trauma. *J Urol* 1981; 126:798–801.

31. Krieger JN, Algood CB, Mason JT, et al: Urological trauma in Pacific Northwest: Etiology, distribution, management and outcome. *J Urol* 1984, 132:70–73.

32. Lang EK: Arteriography in the assessment of renal trauma: The impact of arteriographic diagnosis on preservation of renal function and parenchyma. *J Trauma* 1975; 15:553–556.

33. Lohse JR, Botham RJ, Waters RF: Traumatic bilateral renal artery thrombosis: Case report and review of literature. *J Urol* 1982; 127:522–525.

34. McAninch JW: Acute renal artery thrombosis following blunt trauma. *Urology* 1975; 6:74–77.

35. McAninch JW: Renal injuries, in McAninch JW (guest ed): *Urogenital Trauma*, vol 2 of Blaisdell WF, Trunkey DD (series eds): *Trauma Management.* New York, Thieme Stratton, 1985, pp 27–49.

36. McAninch JW, Carroll PR: Renal trauma: Kidney preservation through improved vascular control—a refined approach. *J Trauma* 1982; 22:285–290.

37. McAninch JW, Federle MP: Evaluation of renal injuries with computerized tomography. *J Urol* 1982; 128:456–460.

38. McAninch JW, Rodkey WG, Stutzman RE, et al: Experimental penetrating renal trauma: A comparison of bench and in situ repair. *Invest Urol* 1979; 17:33–36.

39. Mahoney SA, Persky L: Intravenous drip nephrotomography as an adjunct in the evaluation of renal injury. *J Urol* 1968; 99:513–516.

40. Mendez R: Renal trauma. *J Urol* 1977; 118:698–703.

41. Nicolaisen GS, McAninch JW, Marshall GA, et al: Renal trauma: Re-evaluation of the indications for radiographic assessment. *J Urol* 1985; 133:183–187.

42. Peterson NE: Intermediate-degree blunt renal trauma. *J Trauma* 1977; 17:425–435.

43. Peterson NE, Millikan JS, Moore EE: Combined renal and vena caval trauma: A review of personal and recorded experience. *J Urol* 1985; 133:567–574.

44. Radwin HM, Fitch WP, Robison JR: A unified concept of renal trauma. *J Urol* 1976; 116:20–22.

45. Reagan K, Beckmann CF, Larsen CR, et al: Renal infarction: Computerized tomographic appearance with angiographic correlation. *J Urol* 1984; 132:331–334.

46. Sagalowsky AI, McConnell JD, Peters PC: Renal trauma requiring surgery: An analysis of 185 cases. *J Trauma* 1983; 23:128–131.

47. Sandler CM, Toombs BD: Computed tomographic evaluation of blunt renal injuries. *Radiology* 1981; 141:461–466.

48. Schmoller H, Kunit G, Frick J: Sonography in blunt renal trauma. *Eur Urol* 1981; 7:11–15.

49. Scott RF Jr, Selzman HM: Complications of nephrectomy: Review of 450 patients and a description of a modification of the transperitoneal approach. *J Urol* 1966; 95:307–312.

50. Spark RF, Berg S: Renal trauma and hypertension. The role of renin. *Arch Intern Med* 1976; 136:1097–1100.

51. Stables DP, Fouche RF, Villiers van Niekerk JP de, et al: Traumatic renal artery occlusion: 21 cases. *J Urol* 1976; 115:229–233.

52. Steinberg DL, Jeffrey RB, Federle MP, et al: The computerized tomography appearance of renal pedicle injury. *J Urol* 1984; 132:1163–1164.

53. Thompson IM, Latourette H, Montie JE, et al: Results of non-operative management of blunt renal trauma. *J Urol* 1977; 118:522–524.

54. Wein AJ, Arger PH, Murphy JJ: Controversial aspects of blunt renal trauma. *J Trauma* 1977; 17:662–666.

55. Wein AJ, Murphy JJ, Mulholland SG, et al: A conservative approach to the management of blunt renal trauma. *J Urol* 1977; 117:425–427.

56. Woodruff JH Jr, Cockett ATK, Cannon R, et al: Radiologic aspects of renal trauma with the emphasis on arteriography and renal isotope scanning. *J Urol* 1967; 97:184–188.

57. Wynn WW, Ricketts HJ, McRoberts JW, et al: Comparison of arteriography, venography, and pyelography in experimental renal trauma. *Invest Urol* 1978; 16:62–66.

Chapter 12

Ureteral Injuries

JOSEPH N. CORRIERE, JR., M.D.

The sole function of the ureter is to transport urine from the kidney to the bladder. When the ureter is injured, it may become obstructed, creating hydroureteronephrosis and loss of renal function, or a fistula may occur, leading to urinary extravasation into the retroperitoneum or peritoneal cavity. If injury to another structure has occurred at the time of the ureteral injury, a fistula may develop. Most fistulas are to the vagina, skin, or bowel.

Extravasated urine can induce an intense inflammatory reaction, resulting in secondary fibrosis and ureteral obstruction. If the urine is infected, a phlegmon and possibly life-threatening sepsis may develop, necessitating emergency surgical intervention.

In the past few years, the use of percutaneous and endourologic techniques has decreased the complications from ureteral injuries and prevented many patients from having to undergo surgical procedures that once were standard therapy for these distressing problems.

ETIOLOGY

There are two major types of ureteral injuries—those due to *external violence*, usually penetrating missiles, and the more common injuries due to *surgical misadventure*. The late complications of *radiotherapy* or *migrating foreign bodies* can also cause injury to the ureter. Table 12–1 lists the various etiologic agents that can injure the ureter.

INJURY DUE TO EXTERNAL VIOLENCE

The most common cause of ureteral injury due to external violence is gunshot wounds. They account for over 95% of the lesions (Bright and Peters, 1977; Eickenberg and Amin, 1976; Fisher et al., 1972; Holden et al., 1976; Liroff et al., 1977; Pitts and Peterson, 1981; Rusche, 1948; Steers et al., 1985; Stone and Jones, 1962; Walker, 1969). Knife wounds are the next most common agent and, very rarely, patients fall and become impaled on a spike. Uncommonly, a crushing blow that usually damages bone involves the ureter in the wound. Finally, a well-described but rare injury occurs when the ureter is avulsed from the renal pelvis (Fig 12–1) (Ambiavagar and Nambiar, 1979; LaBerge et al., 1970; Palmer and Drago, 1981). These injuries are usually seen in children, who have hyperextensible spinal columns. The child is usually struck from behind and the ureter tenses and snaps against the 12th rib and transverse processes of the upper lumbar vertebrae.

SURGICAL INJURY

Ureteral injury may complicate 0.5%–1.0% of all pelvic operations. Over one half of these are usually gynecologic procedures, while urinary tract procedures commonly account for 30%. The most common procedures are hysterectomy, salpingo-oophorectomy, vesicourethral suspension, and ureterolithotomy (Bright and Peters, 1977; Carlton et al., 1971; Dowling et al., 1986; Gangai et al., 1976; Higgins, 1967; Ihse et al., 1975; Zinman et al., 1978).

Surgical procedures on the great vessels and colon as well as retroperitoneal tumor excision are the next most frequent procedures that lead to ureteral injury (Andersson and Bergdahl, 1976; Beahrs et al., 1978; Leff et al., 1982; Schapira et al., 1981; Selman et al., 1981). As listed in Table 12–1, many other procedures have been implicated but are infrequently seen (Altebarmakian et al., 1981; Eisenkop et al., 1982; Irvin et al., 1975; Kuzmorov et al., 1980; Selman et al., 1981).

RADIATION INJURY

Radiation injury to the ureter is very rare. Although often considered when a patient with a previously treated pelvic tumor is found to have ureteral obstruction, the incidence of radiation damage to the ureter is 0.04%, while the incidence of ureteral obstruction due

TABLE 12–1.
Etiology of Ureteral Injuries

External violence
  Penetrating injuries
    Gunshot wound
    Knife wound
    Impalement on spike
  Blunt injuries
    Avulsion
    Crushing injury
Surgical injuries
  Gynecologic procedures
    Wertheim's hysterectomy
    Abdominal hysterectomy
    Vaginal hysterectomy
    Salpingo-oophorectomy
    Vesicovaginal fistula repair
    Dilatation and curettage
    Excision of cervical stump
    Cystocele repair
    Colpocleisis
    Laparoscopy
    Endometrioma resection
  Obstetrical procedures
    Forceps delivery
    Precipitous delivery
    Cesarean section
    Therapeutic abortion
  Urinary tract procedures
    Retrograde pyelogram
    Ureteroscopy
    Ureterolithotomy
    Renal pelvic surgery
    Vesicourethral suspension
    Vesicocolic fistula repair
    Suprapubic excision of bladder tumor
    Bladder diverticulectomy
    Stone basket manipulation
    Transurethral resection of a bladder tumor
    Transurethral resection of the prostate
    Radical prostatectomy
  Vascular surgery
    Vena cava ligation
    Aortic aneurysmectomy
    Bypass procedures
    Lumbar sympathectomy
  Abdominal procedures
    Colectomy
    Colostomy
    Colostomy closure
    Abdominal-perineal resection
    Appendectomy
    Exploratory laparotomy
    Enterolysis
    Duodenal resection
    Pancreatic surgery
    Herniorrhaphy
    Biliary surgery
  Retroperitoneal procedures
    Retroperitoneal fibrosis surgery
    Retroperitoneal lymphadenectomy
    Retroperitoneal tumor resection
    Neurosurgical procedures
      Laminectomy
      Paravertebral nerve block
  Radiation injury
  Migrating foreign bodies
    Urinary calculi
    Bullets
    Swallowed objects

to recurrent tumor in these patients is over 95% (Corriere, 1978; Underwood et al., 1977).

MIGRATORY FOREIGN BODIES

Finally, migratory foreign bodies, most commonly urinary calculi, bullets, and swallowed objects, can perforate or obstruct the ureter.

DIAGNOSIS

EXTERNAL VIOLENCE INJURY

When a patient has had a penetrating injury of the abdomen, retroperitoneum, or pelvis in the area of the

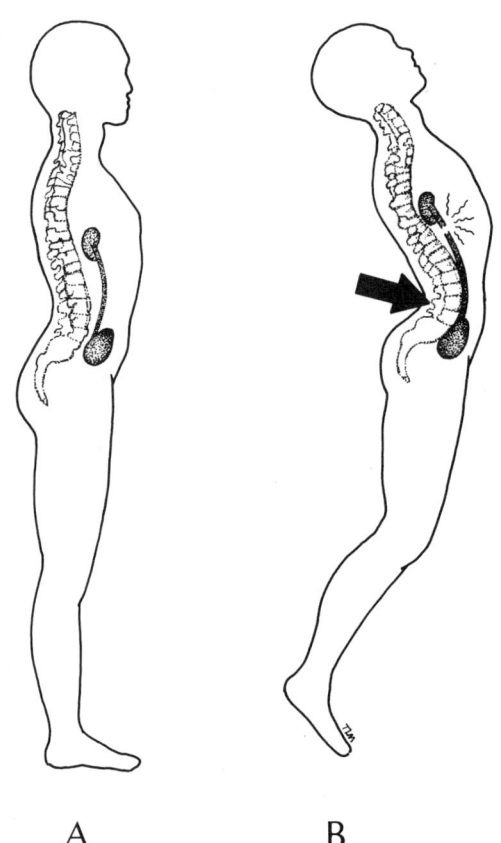

A   B

FIG 12–1.
A, normal relation of urinary tract to spine in child. **B,** with sudden blow to back, ureter tenses against hyperextended vertebral column and avulses at ureteropelvic junction.

urinary tract, an excretory urogram must be performed. If a fracture of the 11th or 12th rib, a transverse lumbar process, or the bony pelvis is present, an excretory urogram must be performed. If hematuria is present in a patient who has had significant abdominal or pelvic trauma, an excretory urogram must be performed.

The excretory urogram is the best method of diagnosing a ureteral injury. Urinary extravasation will be seen on the study as well as some decrease in collecting system visualization (Fig 12–2).

If the patient is to undergo surgical exploration, the ureter should be dissected from its bed and examined where it lies in proximity to the missile tract. If it cannot be positively determined if an injury is present, one vial (5 ml) of indigo carmine should be injected IV. Within 7–10 minutes, the dye should leak into the periureteral tissues if the ureter has been injured.

If the patient is not going to undergo exploration and there remains a question as to the presence of an injury, the most definitive study would be a retrograde ureterogram. Often this is not feasible in the multiply injured trauma patient. In this instance, an ultrasound examination or, preferably, a CT scan may demonstrate the presence of extravasation.

SURGICAL INJURY

If confronted in the operating room with a possible ureteral injury, IV indigo carmine as described above will help the urologist decide if urinary extravasation is present. Unfortunately, there is no good way to determine if ureteral devascularization from the surgical procedure or the blast effect of a high-velocity missile has occurred other than by cutting the ureter and seeing whether it bleeds. Some surgeons advocate the use of IV fluorescein and a Wood's lamp. If there is fluorescence of the ureter, the vasculature is said to be intact. The presence of ureteral peristalsis is not helpful, for peristaltic movement may continue in the ureter for hours after it has been removed from the body.

Most of the time, the diagnosis is not made until many days after the injury has occurred. Table 12–2 lists the most common signs and symptoms seen in these patients. If a ureteral injury is suspected, an excretory urogram is mandatory. If the urogram shows extravasation, delayed function, or hydroureteronephrosis, a retrograde ureterogram should be done to confirm the type of injury, if it is not well delineated on the urogram. This study should be done just prior to institution of treatment to prevent sepsis from instrumenting a closed space. In the critically ill patient a bedside ultrasound can be done if there is a question on the urogram.

In some patients it may be possible to determine that a percutaneous nephrostomy and possibly antegrade ureteral stent should be placed as therapy (i.e., an ileal diversion patient). In this instance, an antegrade rather than retrograde ureterogram should be done. Occasionally, a CT scan will be needed to complete the evaluation.

The proper retrograde study is performed using a Braasch bulb or a cone-tipped catheter and the contrast material injected at the level of the ureteral orifice. Passing a whistle-tip catheter into the renal pelvis does not rule out an obstructed ureter and is to be discouraged as a diagnostic study. As will be discussed later, this may, however, become part of the therapy.

CLASSIFICATION

A useful classification of ureteral injuries is presented in Table 12–3.

EXTERNAL VIOLENCE INJURY

If a missile passes close to but does not penetrate a ureter, a *contusion* is present. If the ureter is penetrated, either a *partial laceration* or *complete laceration* will be present. Rarely, a ureter will be *crushed*, usu-

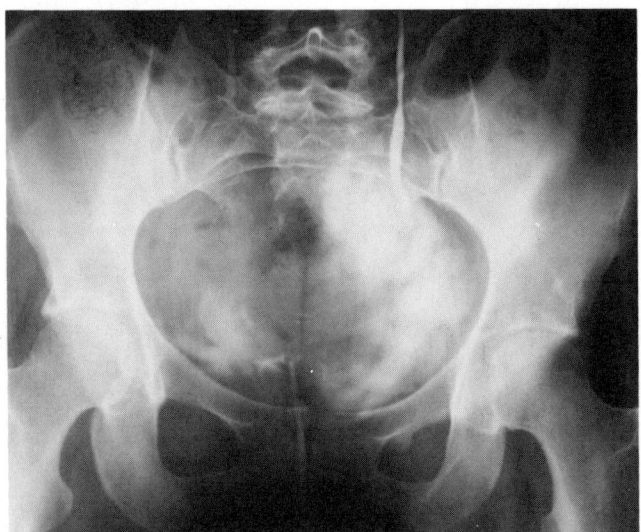

FIG 12–2.
Woman, 26 years old, with lacerated left ureter secondary to stab wound of the abdomen. IVP the day of injury shows contrast extravasation in left lower retroperitoneal area.

TABLE 12–2.
Signs and Symptoms of Ureteral Injuries

Flank pain	1–21 days
Fever	>100° F
Anuria	Bilateral only
Ureterovaginal fistula	1–30 days
Ureterocutaneous fistula	1–30 days

TABLE 12–3.

Classification of Ureteral Injuries

External violence
    Contusion
    Partial laceration
    Complete laceration
    Crush
    Avulsion
Surgical injury
    Crush
    Avulsion
    Transection
    Ligation
    Devascularization
    Fistula formation
Radiation injury

ally in association with a nearby bony injury of the same type, or *avulsed* from the ureteropelvic junction by a hyperextension injury.

SURGICAL INJURY

A surgeon may *crush* a ureter with a clamp, *avulse* a ureter with a retractor, *transect* the ureter with a knife or scissors, or *ligate* the ureter inadvertently. If the ureter is stripped of its adventitia and, therefore, blood vessels, it becomes *devascularized* and necrosis will usually occur in about 10–14 days. This may lead to *fistula* formation, as can any of the other injuries above.

RADIATION INJURY

Uncommonly, ureteral injury will occur secondary to irradiation of the organ. These lesions may not be seen for months to years after therapy and usually result in ureteral obstruction.

THERAPY

EXTERNAL VIOLENCE INJURY

CONTUSION.—This injury is discovered during exploration in a patient who has had a missile pass close to the ureter, but the structure has remained intact. No therapy is necessary in these patients. If a high-velocity bullet (over 2,500 fps) is implicated, there is always the danger of late necrosis of the ureter. In this instance, placement of an internal stent and drain in the area of the injury should be considered. This problem is seen more often in military conflicts than in civilian life (Cass, 1978; Christenson et al., 1983; Rohner, 1971; Stutzman, 1977).

LACERATION.—If a partial laceration is present and the ureter that is still in continuity is viable, placement of an indwelling double-J stent and closure of the

wound with interrupted 4–0 or 5–0 absorbable sutures will give the best results (Steers et al., 1985; Parker, 1971; Sieben et al., 1978). Some authors advocate running closure of the wound and elimination of all stents (Carlton et al., 1971). Prior to the advent of the totally indwelling stent, this was clearly a better way to handle minor lesions, for the placement of a transcutaneous stent and formal nephrostomy may increase the complication rate and extend the scope of the procedure (Weaver, 1956). With the use of the indwelling stent, drainage is minimal, and the patient can be discharged at an early date. Of course, all of the devitalized tissue must be debrided and a Penrose drain placed at the site of the repair and brought out through a separate stab wound.

However, if the remaining intact ureter is of questionable viability or if there is a complete laceration of the ureter, all devitalized tissue must be excised before deciding on a repair.

Clearly, the procedure with the lowest complication rate is the *ureteroneocystostomy* (Fig 12–3). This repair can be performed only on the patient with an injury below the level of the iliac vessels. The kidney can usually be mobilized and lowered so an additional few centimeters of the gap between the ureter and bladder can be decreased. The use of a *bladder flap* can also be used to bring the bladder closer to the ureter (Fig 12–4). Sometimes merely suturing the bladder to the psoas fascia *(psoas hitch)* can ensure that there will not be tension on the repair. A nonrefluxing reimplantation is most desirable but cannot always be performed. As adults, especially women, usually have little trouble with vesicoureteral reflux, this should not result in major problems later in life.

However, if the injury is too high to perform a ureteroneocystostomy, a *ureteroureterostomy* should be

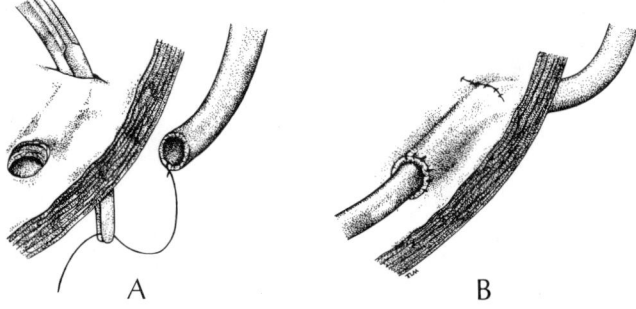

FIG 12–3.
Ureteroneocystostomy. **A,** clamp through bladder wall, mucosa to serosa where ureter will enter bladder. Neo-orifice created in mucosa of bladder. Submucosal tunnel created from neo-orifice to entrance of ureter into bladder. **B,** ureter enters bladder, runs in submucosal tunnel to neo-orifice and sewn in place. Stent is in ureter.

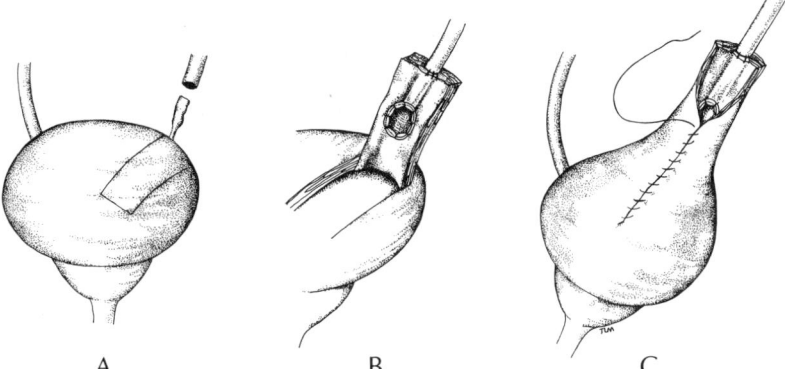

Fig 12–4.
A, bladder flap to be created outlined on bladder. **B,** flap created, ureter sewn in place via submucosal tunnel. **C,** flap sewn into tube to close bladder defect.

done (Fig 12–5). Traditionally, this was performed with a running suture of 4–0 or 5–0 absorbable material without stenting. The use of a stent and nephrostomy added major surgical time and complications to the procedure. If the wound was contaminated by bowel contents, stenting was mandatory.

With the introduction of the double-J stent, using an indwelling stent and interrupted sutures has become popular. The surgical time is shorter and the margin of safety increased. No matter which technique is used, a Penrose drain must be placed as described above. If a major length of ureter is lost, consideration should be given to a *transureteroureterostomy* or merely bringing the cut end of the ureter to the skin as a *cutaneous ureterostomy* for later definitive repair. *Autotransplantation* of the kidney to the hypogastric vessels plus ureteroneocystostomy should also be considered. Of course, this adds major operative time and risk to the patient, but in the patient with a solitary kidney it can be lifesaving.

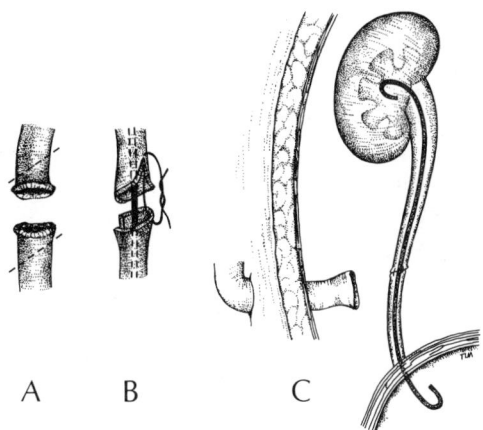

FIG 12–5.
A, traumatized and severed ureter. *Dotted lines* show where spatulated edges will be trimmed. **B,** ureter spatulated; suturing begun; stent in place. **C,** anastomosis complete; double-J catheter in ureter; Penrose drain via stab wound to area of anastomosis.

CRUSH INJURY.—When the ureter has been crushed along with other adjacent tissues, debridement and usually ureteroureterostomy must be performed. All of the above described techniques should be considered, and whichever seems appropriate should be employed.

AVULSION INJURY.—This injury is essentially a complete laceration and requires debridement and definitive repair with stenting as described above.

SURGICAL INJURY

Prior to a discussion of the types of repairs that should be employed when faced with a surgical injury of the ureter, some comments should be made about the prophylactic *preoperative placement of ureteral catheters* to identify the ureters at the time of exploration. Before performing a surgical procedure, especially when it is clear that the dissection will be difficult because of prior surgery, inflammation, or an inflammatory disease process such as endometriosis, many physicians will place retrograde ureteral catheters into the ureters. It is difficult to document that this technique actually decreases the incidence of ureteral injury in these patients. Perhaps the best comment that can be made is that it helps to identify an injured ureter when the catheter is seen in the operative field.

CRUSH INJURY.—During a surgical procedure, a surgeon may inadvertently crush a ureter by placing a clamp on the structure or ligating it and then removing the clamp or ligature. Now a decision must be made. Has the crushed ureteral segment been devascularized enough so it will eventually necrose and either develop a stricture or, more likely, a fistula? Unfortunately, there is no good intraoperative test to solve this dilemma.

Obviously, if there is good evidence that major injury has occurred, the ureteral segment should be excised and either a ureteroneocystostomy or ureteroureterostomy, with internal stenting, performed as described previously. If the surgeon chooses not to resect the segment, placement of an indwelling stent by either opening the bladder or transurethrally at the end of the procedure is a good safety measure. Placement of a drain, in this instance, is probably not mandatory.

LACERATION INJURY.—When the ureter has been completely severed, either by avulsion or transection, and recognized intraoperatively, repair by any of the above techniques outlined in the section on lacerations associated with external violence should be employed. However, if a laceration is not recognized until after the surgery has been completed, a variety of therapeutic decisions will have to be made. Initially, retrograde ureteral catheterization should be attempted. Unfortunately, most of the time the catheter cannot be negotiated past the laceration into the proximal ureter. Obviously, this technique is only applicable with a partial laceration. However, if successful, a double-J stent should be placed and the patient observed for resolution of the extravasated urine. At the first sign of deterioration, the patient should have a percutaneous or formally placed drain inserted to remove the extravasation.

If the retrograde catheterization is unsuccessful, a percutaneous nephrostomy should be placed and an antegrade stent passed into the bladder. After the extravasation resolves, a double-J stent can be placed for long-term drainage (Dowling et al., 1986; Lang et al., 1979; Persky et al., 1981; Stables et al., 1978).

If drainage of the kidney cannot be established by the percutaneous route, surgical exploration must be performed (Hoch et al., 1975; Mendez and McGinty, 1978). In the first few postoperative days primary repair may be considered as discussed above, but if discovered later in the postoperative period, formal nephrostomy tube placement and drainage of the extravasation should be done, with delayed repair planned many months in the future. The use of an *ileal ureter* when large segments of the ureter have been lost can be performed when primary repair can be planned and is not done in an emergency situation. However, since autotransplantation has gained popularity, this procedure is used less today. Often in the seriously ill patient with a normal opposite renal unit, *nephrectomy* is the best choice.

LIGATION INJURY.—When complete obstruction of the ureter is discovered, it is always tempting to return to the operating room and deligate the ureter. Except for ureters ligated at the time of vesicourethropexy, this may be a hazardous procedure and lead to delayed necrosis and fistula formation. Probably the reason it works so well with vesicourethral suspension is that during this procedure, large amounts of tissue are caught in the suture, so devascularization is less of a

problem than is mechanical obstruction from angulation. Perhaps, if the ureter has been loosely ligated and a stent is placed after deligation, this procedure has some merit (Gurin et al., 1982).

The more conservative approach is to place a percutaneous nephrostomy tube in the kidney and attempt to pass a stent antegrade past the obstruction (Dowling et al., 1986; Harshman et al., 1982; Lang et al., 1979; Persky et al., 1981; Stables et al., 1978). Retrograde stenting fails almost all of the time. If stenting is successful, balloon dilatation, in an effort to disrupt the suture, may be tried but is unnecessary (Kaplan et al., 1982). If the ureter has been ligated with chromic suture material, the obstruction will usually resolve in 3–4 weeks (Harshman et al., 1982). If ligated with polyglycolic acid suture, it may take 6–8 weeks to resolve (Fig 12–6). If it has not resolved in 4–6 months, formal repair will be necessary by one of the above described techniques. Once again, often the patient who cannot withstand the

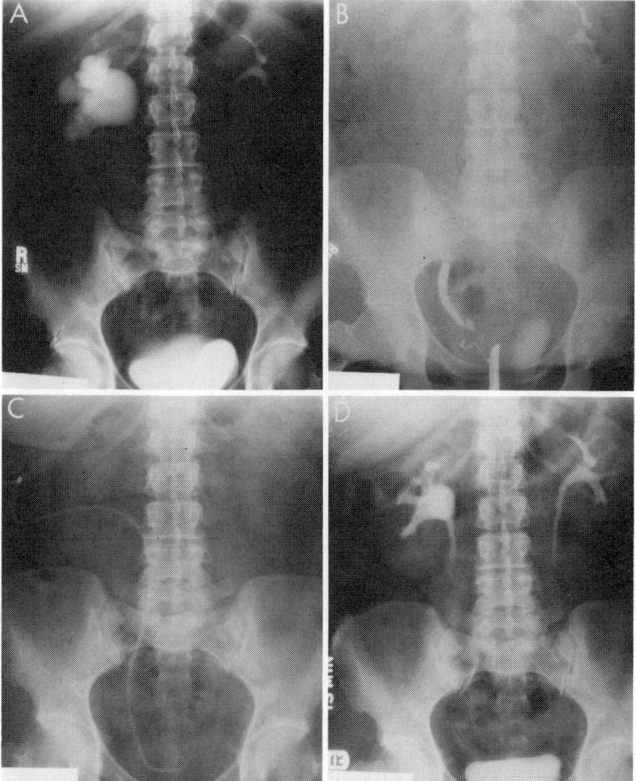

FIG 12–6.
Woman, 46 years old, who suffered a ureterovaginal fistula following an abdominal hysterectomy. **A,** IVP done on the tenth postoperative day demonstrates obstruction of the right ureter. **B,** retrograde ureterogram demonstrates the area of obstruction and some extravasation. **C,** an antegrade ureteral stent was placed beyond the fistula the following day and left in place nine weeks. **D,** an IVP done six months later demonstrates resolution of the hydronephrosis and a normal collecting system.

complications associated with reconstructive procedures may best be handled by nephrectomy.

FISTULA FORMATION.—If necrosis occurs from any of the injuries outlined in Table 12–3 and the urine either collects in the retroperitoneum or abdomen or tracts to the skin, bowel, or vagina, a percutaneous nephrostomy and ureteral stent should be placed. As with ureteral lacerations, if successful, in time the ureter will heal and the fistula will usually close. If a stent cannot be passed, the ureter will usually stricture at the site of the fistula, and formal repair must be performed 4–6 months after the injury using one of the previously described procedures (Dowling et al., 1986; Lang, 1981; Lang et al., 1979).

RADIATION INJURY

This uncommon injury is usually discovered months to years after the therapy has been completed. Ureteral stricture formation is usually present. Repair is very difficult, for irradiated tissue heals poorly. Permanent internal stent diversion is one approach as is nephrectomy or diversion of the urine into an isolated bowel conduit using both ureter and bowel outside the field of treatment. Occasionally, reconstructive procedures using irradiated tissue wrapped in omentum will be successful (Underwood et al., 1977).

POSTOPERATIVE CARE AND COMPLICATIONS

Indwelling ureteral stents can be left in place for up to two months without fear of complications. Although many have been in place with little difficulty for over a year, after two months, some of them will develop calculi on the stent and may then cause obstruction of the ureter (Dowling et al., 1986). If a drain has been placed, it can usually be removed in two weeks, even if the stent is left for a longer period. Most patients can be promptly discharged from the hospital and have these items removed in the outpatient setting.

Patients with percutaneous nephrostomy tubes should be taught to irrigate the tubes and to change dressings. A nephrostogram should be done at three-week intervals to see if the obstruction has been relieved. When the ureter is again draining, the tube can be removed on an outpatient basis.

Stents placed across fistulas should be left in place for at least 4–6 weeks and a pull-out ureterogram performed to ensure that the fistula has closed. If the fistula site can be seen (skin or vagina), it should be inspected every two weeks until the site has sealed. The stent should not be removed until the opening is completely closed and the overlying skin or mucosa is intact.

Once all foreign materials have been removed, the patient should be treated with antibiotics to ensure a sterile urinary tract. Repeated cultures are mandatory.

An excretory urogram should be performed 3–6 months after the repair and again a year later. Delayed obstruction is rare but can occur.

With formal surgical repairs, return to full activity will take 4–6 weeks. The use of indwelling stents or percutaneous tubes will not delay recovery longer than that usually seen with the primary surgical procedure.

The major complications from indwelling tubes, either totally indwelling or exiting from the kidney, are infection, tube obstruction, and calculus formation (Dowling et al., 1986). If patients are instructed in how to irrigate the tubes at home, many trips to the emergency room or office can be avoided. Tubes may have to be changed if they obstruct, and pyelonephritis must be treated with appropriate antibiotics. Calculi must be handled at the time of tube removal.

REFERENCES

1. Altebarmakian VK, Davis RS, Khuri FJ: Ureteral injury associated with lumbar disk surgery. *Urology* 1981; 17:462.
2. Ambiavagar R, Nambiar R: Traumatic closed avulsion of the upper ureter. *Injury* 1979; 11:71.
3. Andersson A, Bergdahl L: Urologic complications following abdominoperineal resection of the rectum. *Arch Surg* 1976; 111:969.
4. Beahrs JR, Beahrs OH, Beahrs MM, et al: Urinary tract complications with rectal surgery. *Ann Surg* 1978; 187:542.
5. Bright TC III, Peters PC: Ureteral injuries due to external violence: 10 years' experience with 59 cases. *J Trauma* 1977; 17:616.
6. Bright TC III, Peters PC: Ureteral injuries secondary to operative procedures. *Urology* 1977; 9:22.
7. Carlton CE Jr, Scott R Jr, Guthrie AG: The initial management of ureteral injuries: A report of 78 cases. *J Urol* 1971; 105:335.
8. Cass AS: Ureteral contusion and delayed necrosis from gunshot injury. *Urology* 1978; 12:195.
9. Christenson PJ, O'Connell KJ, Clark M, et al: Ballistic ureteral trauma: A comparison of high and low velocity weapons. *Contemp Surg* 1983; 23:45.
10. Corriere JN Jr: The obstructed ureter in the patient with cancer, in Miller TJ, Dudrick SJ (eds): *The Management of Difficult Surgical Problems*. Austin, Tex, The University of Texas Press, 1978, pp 109–115.
11. Dowling RA, Corriere JN Jr, Sandler CM: Iatrogenic ureteral injury. *J Urol* 1986; 135:912.
12. Eickenberg H, Amin M: Gunshot wounds to the ureter. *J Trauma* 1976; 16:562.
13. Eisenkop SM, Richman R, Platt LD, et al: Urinary tract injury during ceasarean section. *Obstet Gynecol* 1982; 60:591.

14. Fisher S, Young DA, Malin JM Jr, et al: Ureteral gunshot wounds. *J Urol* 1972; 108:238.
15. Gangai MP, Agee RE, Spence CR: Surgical injury to the ureter. *Urology* 1976; 8:22.
16. Gurin JI, Garcia RL, Melman A, et al: The pathologic effect of ureteral ligation, with clinical implications. *J Urol* 1982; 128:1404.
17. Harshman MW, Pollack HM, Banner MP, et al: Conservative management of ureteral obstruction secondary to suture entrapment. *J Urol* 1982; 127:121.
18. Higgins CC: Ureteral injury during surgery. *JAMA* 1967; 199:118.
19. Hoch WH, Kursh ED, Persky L: Early aggressive management of intraoperative ureteral injuries. *J Urol* 1975; 114:530.
20. Holden S, Hicks CC, O'Brien DP III, et al: Gunshot wounds of the ureter: A 15 year review of 63 consecutive cases. *J Urol* 1976; 116:562.
21. Ihse I, Arnesjo B, Jonsson G: Surgical injuries of the ureter. *Scand J Urol Nephrol* 1975; 9:39.
22. Irvin TT, Goligher JC, Scott JS: Injury to the ureter during laparoscopic tubal sterilization. *Arch Surg* 1975; 110:1501.
23. Kaplan JO, Winslow OP Jr, Sneider SE, et al: Dilatation of a surgically ligated ureter through a percutaneous nephrostomy. *AJR* 1982; 139:188.
24. Kuzmorov IW, MacIsaac SG, Sioufi J, et al: Iatrogenic ureteral injury secondary to lumbar sympathetic ganglion blockade. *Urology* 1980; 16:617.
25. LaBerge I, Homsy YL, Dadour G, et al: Avulsion of ureter by blunt trauma. *Urology* 1979; 13:172.
26. Lang EK: Diagnosis and management of ureteral fistulas by percutaneous nephrostomy and antegrade stent catheter. *Radiology* 1981; 138:311.
27. Lang EK, LaNasa JA, Garrett J, et al: The management of urinary fistulas and strictures with percutaneous ureteral stent catheters. *J Urol* 1979; 122:736.
28. Leff EI, Groff W, Rubin RJ, et al: Use of ureteral catheters in colonic and rectal surgery. *Dis Colon Rectum* 1982; 25:457.
29. Liroff SA, Pontes JES, Pierce JM Jr: Gunshot wounds of the ureter: 5 years of experience. *J Urol* 1977; 118:551.
30. Mendez R, McGinty DM: The management of delayed recognized ureteral injuries. *J Urol* 1978; 119:192.
31. Palmer JM, Drago JR: Ureteral avulsion from non-penetrating trauma. *J Urol* 1981; 125:108.
32. Parker JM: Re-emphasizing the importance of urinary tract diversion and splinting in injuries of the upper third of the ureter. *J Urol* 1971; 106:368.
33. Persky L, Hampel N, Kedia K: Percutaneous nephrostomy and ureteral injury. *J Urol* 1981; 125:298.
34. Pitts JC III, Peterson NE: Penetrating injuries of the ureter. *J Trauma* 1981; 21:978.
35. Rohner TJ Jr: Delayed ureteral fistula from high velocity missiles: Report of 3 cases. *J Urol* 1971; 105:63.
36. Rusche CF: Injury of the ureter due to gunshot wounds. *J Urol* 1948; 60:63.
37. Schapira HE, Li R, Gribetz M, et al: Ureteral injury: Complication of partial excision of aortofemoral vascular prosthesis. *J Urol* 1981; 126:817.
39. Sieben DM, Howerton L, Amin M, et al: The role of ureteral stenting in the management of surgical injury of the ureter. *J Urol* 1978; 119:330.
40. Stables DP, Ginsburg NJ, Johnson ML: Percutaneous nephrostomy: A series and review of the literature. *AJR* 1978; 130:75.
41. Steers WD, Corriere JN Jr, Benson GS, et al: The use of indwelling stents in managing ureteral injuries due to external violence. *J Trauma* 1985; 25:1001.
42. Stengel JN, Felderman ES, Zamora D: Ureteral injury: Complication of laparoscopic sterilization. *Urology* 1974; 4:341.
43. Stone HH, Jones HQ: Penetrating and nonpenetrating injuries to the ureter. *Surg Gynecol Obstet* 1962; 114:52.
44. Stutzman RE: Ballistics and the management of ureteral injuries from high velocity missiles. *J Urol* 1977; 118:947.
45. Underwood PB Jr, Lutz MH, Smoak DL: Ureteral injury following irradiation therapy for carcinoma of the cervix. *Obstet Gynecol* 1977; 49:663.
46. Walker JH: Injuries of the ureter due to external violence. *J Urol* 1969; 102:410.
47. Weaver RG: The effect of large caliber splints on ureteral healing. *Surg Gynecol Obstet* 1956; 103:590.
48. Zinman LM, Libertino JA, Roth RA: Management of operative ureteral injury. *Urology* 1978; 12:290.

Chapter 13

Trauma to the Lower Urinary Tract

JOSEPH N. CORRIERE, Jr., M.D.

BLADDER INJURIES

In the child the bladder is an abdominal organ and as such is vulnerable to external trauma. As the bony pelvis grows, the bladder becomes protected from injury, especially if it is empty of urine. The bladder is located extraperitoneally in the space of Retzius. Laterally it is bound by the internal obturator muscles and the lateral umbilical ligaments. Its base is attached to the urogenital diaphragm, and Denonvilliers' fascia, or the rectovesical fascia, binds it loosely posteriorly. However, unlike the rest of the organ, the dome of the bladder is mobile and distensible.

When the bladder is distended or the pelvis is fractured, the normal protective influence of the intact pelvic ring is lost and, in fact, the shearing force of a pelvic fracture commonly tears the bladder at its moorings. A spicule of bone may lacerate the organ, or it may rupture at the dome by a direct blow to the abdomen without bony injury. Missiles, on the other hand, from an outside force, from internal migration, or in the hands of a well-meaning surgeon, may find the bladder despite its position.

ETIOLOGY

Penetrating Injuries

In usual civilian practice, the most common penetrating injuries of the bladder will occur from surgical misadventure. Table 13–1 lists the types of procedures and instruments that have been associated with operative injury to the bladder (Corriere et al., 1986; Dmochowski et al., 1986; Loffer and Pent, 1975; Mattingly and Borkowf, 1978). Trauma from external violence is most commonly due to gunshot wounds (Carroll and McAninch, 1984; Corriere and Harris, 1981; Corriere et al., 1986; Johnson, 1971; McConnell et al., 1982; Waterhouse and Gross, 1969).

Rarely, injury may occur from migration and erosion of internally placed foreign materials, most commonly surgical drains, intrauterine devices, or hip prostheses (Burnette, 1982; Cohen et al., 1977; Hubbard et al., 1979; Puranen and Koivisto, 1978; Zakin, 1984). Finally, long-term Foley catheters have been known to erode through the bladder, usually at the dome, where it sits at the tip of the catheter (Spees et al., 1981).

Blunt Injuries

The most common etiology of bladder injuries due to external violence is blunt trauma to the abdomen, mostly from motor vehicle accidents but also from falls, crushing injuries to the bony pelvis or blows to the abdomen (Table 13–2). The full bladder is especially vulnerable to a deceleration injury (Bonavita and Pollack, 1983; Carroll and McAninch, 1984).

In motor vehicle accidents, injuries are seen in passengers wearing seatbelts, where the force of the collision may focus on the abdomen and thus the full bladder; in the unrestrained child (or, less commonly, the adult) who is thrown by the impact against an unyielding object; or secondary to a pelvic fracture (Cass, 1976, 1984).

In our experience with 111 bladder injuries over a seven-year period, 86% were due to blunt trauma, and 90% of the blunt injuries were secondary to motor vehicle accidents (Corriere et al., 1986). A total of 89% of the blunt injuries was associated with pelvic fractures. On the other hand, 9% of patients with pelvic fractures have a concomitant injury to the bladder.

Spontaneous Rupture

It is difficult to understand how a bladder can rupture "spontaneously" as is reported in most large series of these injuries. It is usually seen in the patient with preexisting bladder pathology, usually in chronic retention, and is most likely associated with minor blunt trauma.

TABLE 13–1.
Etiology of Penetrating Injuries
of the Bladder

Operative injury
  Transurethral procedures
    Resectoscope
    Lithotrite
    Cystoscope
    Urethral instrumentation
  Gynecologic procedures
    Abdominal hysterectomy
    Vaginal hysterectomy
    Removal of cervical stump
    Salpingo-oophorectomy
    Caesarean section
    Laparoscopy—Veress needle, trochar
    Dilatation and curettage
    Suction curettage
    Neovaginal construction
  Abdominal procedures
    Herniorrhaphy
    Abdominal-perineal resection
    Anterior colon resection
    Neonatal umbilical artery catheterization
External violence
  Gunshot wound
  Knife wound
  Spike impalement
Internal migration
  Surgical drains
    Penrose
    Saratoga sump
    Foley catheter
  Intrauterine devices
    Lippes loop
    Dalkon shield
    Copper-7
    Copper-T
  Hip prosthesis
    Pins
    Trochanteric plate
  Neurosurgical shunts
  Long-dwelling Foley catheter

DIAGNOSIS

Signs and Symptoms

The signs and symptoms of rupture of the bladder are usually nonspecific. The patient may complain of suprapubic pain or relate that he attempted to urinate and could not. Commonly, the discomfort of a concomitant

TABLE 13–2.
Etiology of Blunt Injuries of the Bladder

      Motor vehicle accident
      Fall
      Crush of bony pelvis
      Abdominal blow

fractured pelvis or other organ system injury overshadows the pain from the damaged urinary tract.

Tenderness is present in the suprapubic area and bowel sounds absent, especially if it is an intraperitoneal rupture. Shock is rarely due to an isolated bladder rupture. When it is present, another cause for the hypotension should be sought.

Bladder perforation during a transurethral surgical procedure with the patient under spinal anesthesia is commonly associated with acute symptoms on the operating table. Extraperitoneal injuries will cause lower abdominal pain, and the patient's blood pressure may begin to rise. Intraperitoneal injuries with extravasation of large quantities of fluid lead to abdominal distention and referred pain to the tip of the shoulder if the fluid irritates the diaphragm.

If recognition of an intraperitoneal injury is delayed, uroascites may develop and cause marked abdominal distension. This may cause respiratory distress and even lower limb venous occlusion, especially in the neonate (Dmochowski et al., 1986). Peritoneal signs of tenderness and rebound will develop and, if the urine is infected, frank peritonitis may eventually be seen (Culp, 1942).

Hematuria is a hallmark finding with bladder injuries. In our experience and that of others, gross hematuria occurs over 95% of the time, with microscopic hematuria present in the remaining cases (Carroll and McAninch, 1984; Hayes et al., 1983; McConnell et al., 1982).

Radiographic Examination

The static cystogram is the only study that will definitely diagnose a ruptured bladder (Bonavita and Pollack, 1983; Carroll and McAninch, 1983; Del Villar et al., 1972; Prather and Kaiser, 1950; Sandler et al., 1981a). If a urethral injury is suspected because of a pelvic fracture, the presence of blood at the urethral meatus, a high-riding prostate on rectal examination, or marked ecchymosis and edema of the perineum, scrotum, and/or penis, a retrograde urethrogram must be done before attempted urethral catheterization.

If a ruptured urethra is found, urethral catheterization may be contraindicated and a suprapubic cystostomy performed. If the tube is placed percutaneously, a static cystogram must still be done to rule out a concomitant bladder injury. If the cystotomy tube is placed surgically, the bladder can be inspected at the time of the exploration and the cystogram eliminated.

It is best to perform this examination using fixed equipment in the radiology suite. A satisfactory study can be done in the emergency room using portable equipment and grid cassettes if absolutely necessary. A

Foley catheter is placed in the bladder and the bladder emptied of urine. A 300-ml bottle of standard infusion contrast material (25%–30%) and a similar amount of saline solution are attached to a Y connector and then to the catheter to obtain a 50-50 mixture.

A scout radiograph is taken and then 100 ml of the mixture infused. A second film is exposed to check for gross extravasation. If a bladder rupture is seen, the catheter is immediately placed to straight drainage. If extravasation is not seen, the remainder of the solution is instilled and films obtained in the AP, oblique, and lateral projections (Buxton, 1978).

The bladder is then drained of all contrast material and an additional film taken. This is especially important in patients in whom the oblique and lateral films may have been omitted because of concomitant injuries or the patient's clinical state. Small amounts of extravasation may be present behind a contrast-filled bladder and may only be seen on this view. The bladder should also be drained before proceeding with studies of the upper urinary tract. An excretory urogram should then be performed.

It cannot be overemphasized that a normal cystogram on a urogram is not sufficient evidence to rule out bladder injury. Frequently, a blood clot or omentum temporarily seals a small rent and the bladder will appear intact. Only the static cystogram with full vesical distention and a postdrainage film can verify that a bladder rupture is not present.

CLASSIFICATION

As can be seen in Table 13–3, bladder injuries secondary to blunt trauma are subclassified as to the extent and location of the injury, while penetrating injuries are usually grouped together. However, as discussed in the section on therapy, when considering the treatment of iatrogenic penetrating injuries, extent and location become critical and must be differentiated.

Bladder Contusion

A bladder contusion results from damage to the bladder mucosa or muscularis without loss of wall continu-

TABLE 13–3.
Classification of Bladder Injuries

Blunt trauma
    Contusion
    Interstitial rupture
    Intraperitoneal rupture
    Extraperitoneal rupture
    Intraperitoneal and extraperitoneal rupture
Penetrating trauma

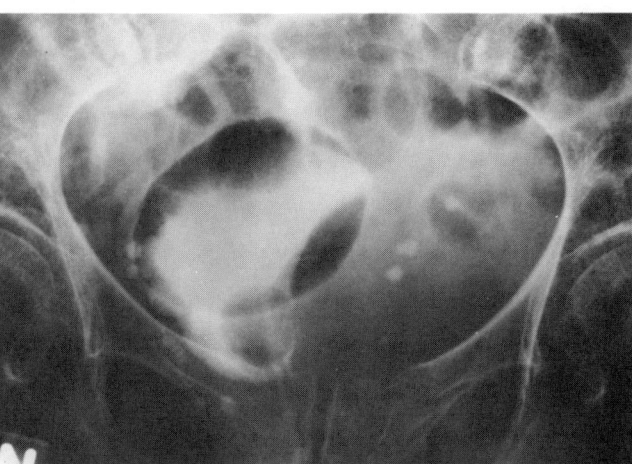

FIG 13–1.
Bladder contusion secondary to a pelvic hematoma and pelvic fracture. (From Sandler CM, Phillips JM, Harris JD, et al: Radiology of the bladder and urethra in blunt pelvic trauma. *Radiol Clin North Am* 1981; 19:195. Used with permission.)

ity. Extravasation is not seen on cystogram, but the bladder outline may be distorted (Fig 13–1). The exact incidence of this injury is difficult to determine, because often the diagnosis is made by exclusion in the patient with trauma to the lower abdomen and hematuria. The best overall estimate is that this injury accounts for about one third of all bladder injuries.

Interstitial Rupture

Occasionally, an incomplete (non–full-thickness) tear of the bladder wall is seen secondary to blunt trauma. As with the bladder contusion, no extravasation is seen on the cystogram (Fig 13–2). It is important to distinguish this injury from a bladder contusion, for the therapy requires a longer period of catheterization, since it may represent a full-thickness injury that has sealed with clots or at least needs a longer time to heal due to the extent of the damage to the bladder wall.

Intraperitoneal Rupture

As previously mentioned, intraperitoneal rupture of the bladder occurs when there is a sudden rise in intravesical pressure secondary to a blow to the pelvis or lower abdomen. This increased pressure results in rupture of the dome, the weakest and most mobile part of the bladder. Contrast material will fill the cul-de-sac, outline loops of bowel, and eventually extend into the paracolic gutter (Fig 13–3). This injury is common in children because of the intraperitoneal position of the bladder. Intraperitoneal bladder ruptures probably account for one third of all bladder injuries and are about equal in incidence to extraperitoneal ruptures.

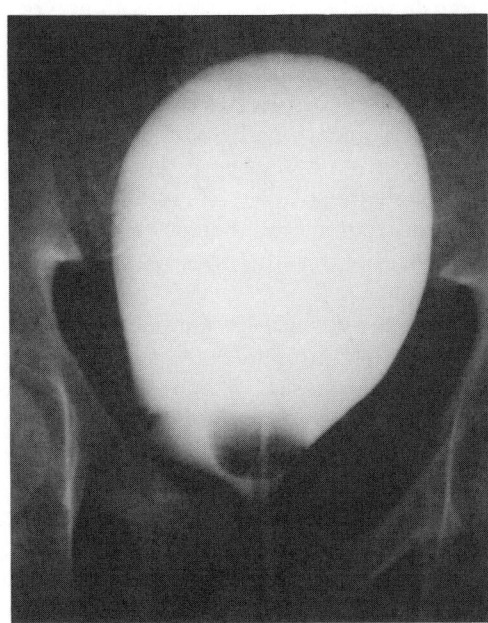

FIG 13–2.
Interstitial bladder rupture secondary to a pelvic fracture.

Extraperitoneal Rupture

Extraperitoneal bladder ruptures are almost exclusively seen with pelvic fractures. Most of the time, the bladder is sheared on the anterior lateral wall near the bladder base by the distortion of the pelvic ring disrup-

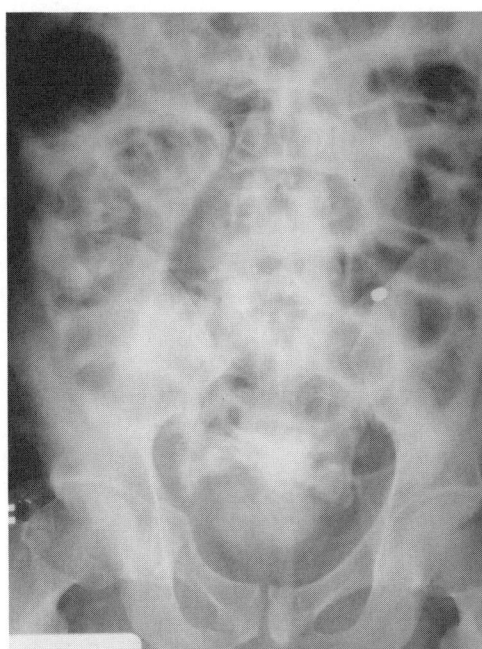

FIG 13–3.
Intraperitoneal bladder rupture secondary to blunt abdominal trauma.

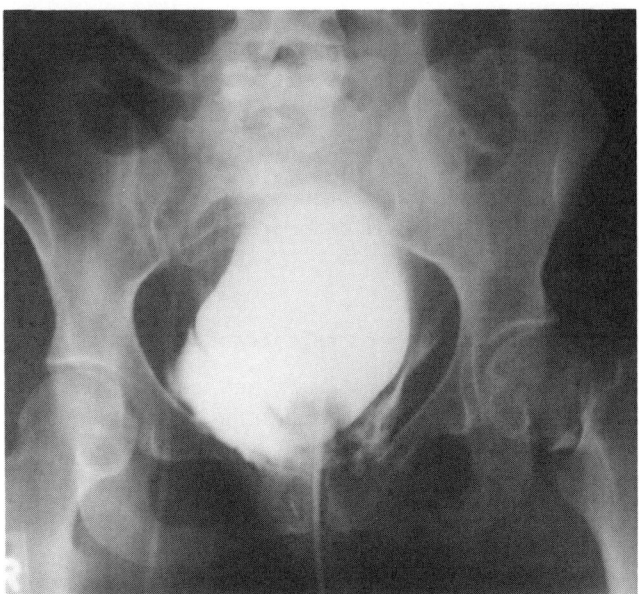

FIG 13–4.
Extraperitoneal bladder rupture secondary to pelvic fracture.

tion. Occasionally, the bladder will be lacerated by a sharp, bony spicule. On cystography, flame-shaped areas of extravasation that are usually confined to the perivesical soft tissues are visualized (Fig 13–4). If there is a large pelvic hematoma, the bladder will often be compressed into the "teardrop deformity" (Prather and Kaiser, 1950). Urinary extravasation may extend to the thigh via the obturator foramen to the scrotum via the inguinal canal, up the anterior abdominal wall, or retroperitoneally as high as the kidneys (Fig 13–5).

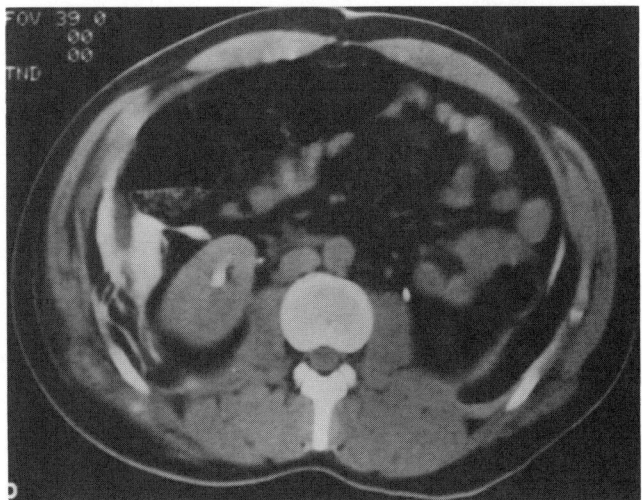

FIG 13–5.
CT scan of patient with an extravesical bladder rupture. Note extravasation as high as kidneys.

Intraperitoneal and Extraperitoneal Bladder Rupture

Occasionally, the bladder will be ruptured both intraperitoneally and extraperitoneally. These injuries are due to penetrating trauma or pelvic fractures. The radiologic findings are a mixture of the above descriptions of the single injuries.

THERAPY

Penetrating Injuries From External Violence

All patients with penetrating injuries from external violence should undergo exploration of the abdomen and of the track the missile followed from its entrance wound to its exit wound. The peritoneal cavity should be opened even if the injury is thought to be entirely extraperitoneal and the intra-abdominal viscera and major vasculature examined for damage. All devitalized tissue and debris (bullets, bone spicules, clothing, etc.) should be removed from the bladder and abdomen (Carroll and McAninch, 1984; McConnell et al., 1982).

If the ureteral orifices are involved in the injury or there is concern about the integrity of the ureters, 5 ml of indigo carmine should be injected IV. In 7–10 minutes, the blue dye should appear in the bladder. A search for extravasation should be made and the ureters intubated with 5 F whistle-tip catheters if there is any concern that they have been damaged.

If a large pelvic hematoma is present, it is best left undisturbed if one can be sure by radiography (plain film, cystogram, urogram, and/or arteriogram) that there is no major disruption of the bladder neck, ureters, or vasculature. The bladder is then entered through its peritoneal surface at the dome and thoroughly inspected.

After debridement, the extraperitoneal vesical defect should be closed with a one-layer running suture of 3-0 chromic or polyglycolic suture through the interior of the bladder. Extensive mobilization of the bladder to ensure a water-tight closure or to place the knots on the outside of the bladder will usually only increase bleeding. If it is impossible to close the extraperitoneal defects, they should not be disturbed. With adequate bladder drainage, they will eventually heal without difficulty.

A suprapubic cystotomy tube is then inserted in the bladder through a separate stab wound. At least a 24 F size should be used to ensure egress of blood clots. Either a Malecot or mushroom tube is recommended. The mushroom style has a firmer flange and is less likely to become dislodged, but the side holes are small. If the tip of the catheter is excised prior to insertion, this drawback is overcome.

The tube should be sewn in place with an O chromic or polyglycolic purse-string suture, which is then used to fix the bladder to the wall at the site where it crosses the abdomen, again through a separate stab wound. This will ensure a controlled fistula if leakage occurs at the time of tube removal. The catheter should not be brought through the bladder wound, as removal may disrupt the suture line; nor should it come through the abdominal incision, for this increases the chance for wound infection.

The intraperitoneal bladder incision is closed with a double layer of 3–0 chromic or polyglycolic suture in a running water-tight fashion. This is best done after the suprapubic tube is sewn into the bladder but before the tube is brought through and fixed to the abdominal wall.

One-inch Penrose drains are placed near the suture lines and brought through separate stab wounds. The drains and suprapubic tubes are sutured to the skin with 3–0 nonabsorbable suture material. The wound is closed in a standard fashion and a sterile dressing is applied.

Iatrogenic Penetrating Injuries

When the bladder is inadvertently injured during a surgical procedure, prompt repair with a double layer of 3–0 absorbable suture material and tube drainage with a Foley catheter or suprapubic tube will usually ensure an excellent result. The most common mistake made is not thoroughly inspecting the entire bladder when the injury is discovered, thereby overlooking a second rent in the organ. This is especially disturbing during gynecologic surgery when a lesion of the dome is seen and repaired, but one at the base that also involves the vagina is missed.

Delayed recognition of operative injuries need individualized attention. If recognized in the first few days after surgery, immediate correction will usually be successful. However, after a week or two, tissue edema will impede proper wound healing, and repair is best delayed for months (Mattingly and Borkowf, 1978). Only intraperitoneal injuries and uroascites demand prompt repair whenever they are discovered.

Injury to the bladder by an endoscope (cystoscope, resectoscope, or laparoscope)—especially if the injury is extraperitoneal and properly recognized, the procedure immediately terminated, and the rent small—can be handled with large-caliber urethral catheter drainage and expectant therapy (Culp, 1942; Loffer and Pent, 1975). These patients must be observed very closely and, at the least sign of deterioration, abdominal exploration and repair of the bladder performed. If there is any evidence of uroascites or if the urine is infected, formal repair is mandatory.

Internal Migrating Objects

These are rare injuries that necessitate removal of the foreign material and urethral catheter or suprapubic tube drainage. Formal repair performed as exploration is usually necessary to remove most of these objects and must be done if the injury is intraperitoneal (Cohen et al., 1977).

Bladder Contusions

These injuries necessitate Foley catheter drainage for a few days or, if minor, no therapy at all. If there is a large pelvic hematoma and marked bladder neck distortion, the patient may have difficulty voiding. These cases may require prolonged catheter drainage. If there is a major injury to the sacrum and the patient cannot urinate, a cystometrogram should be performed to be sure there has not been damage to the sacral nerve roots that innervate the bladder. If no detrusor contraction is seen and the patient continues to be unable to void after multiple trials, intermittent self-catheterization should be instituted. Most of these problems are temporary unless accompanied by major neurologic deficit.

Interstitial Ruptures

As stated in the section on classification, these are incomplete bladder wall ruptures or may represent a small full-thickness rupture that has sealed itself with clot or possibly omentum. They should probably be treated by ten days of catheter drainage, just as the complete but unrepaired rupture, and followed up closely for a change in clinical status. A cystogram should be performed prior to catheter removal.

Intraperitoneal Ruptures

All intraperitoneal bladder ruptures due to blunt abdominal trauma should undergo formal repair. The peritoneal cavity should be opened, all urine and blood evacuated, the viscera and vasculature inspected for injury, and appropriate therapy instituted. The bladder will usually have a large, 5-cm or greater, rent at the dome. It should be widened, if necessary, thoroughly to inspect the interior of the organ. Any concomitant extraperitoneal rents should be closed with a single running 3–0 chromic or polyglycolic suture from inside the bladder. Devitalized tissue should be excised and, after a suprapubic tube has been placed as previously described, the dome wound closed with absorbable suture material. The suprapubic tube is brought through a separate stab wound and a peritoneal drain placed and brought out through a second stab wound. The peritoneal cavity cannot and should not be drained. Closure of the abdomen is as previously described (Bacon, 1943;

Carroll and McAninch, 1984; Culp, 1942; Hayes et al., 1983; McConnell et al., 1982).

There are a few scattered reports that intraperitoneal bladder injuries can be treated with simple Foley catheter drainage (Mulkey and Witherington, 1974; Richardson and Leadbetter, 1975; Robards et al., 1976). However, when the literature is carefully reviewed, it is clear that most of these authors are discussing iatrogenic transurethral bladder perforations and not wounds due to external violence. Most patients with intraperitoneal bladder ruptures due to abdominal blows or a fractured pelvis have large gaping rents and marked uroascites when first seen. These patients must undergo prompt surgical repair because they rapidly deteriorate if not treated in a timely fashion (Bacon, 1943; Carroll and McAninch, 1984; Culp, 1942; Rieser and Nicholas, 1963).

Extraperitoneal Ruptures

Isolated extraperitoneal bladder ruptures can be easily handled by ten days of Foley catheter drainage (Brosman and Paul, 1976; Corriere et al., 1986; Mulkey and Witherington, 1974; Richardson and Leadbetter, 1975; Robards et al., 1976). One cannot decide, as some authors state, to only treat small extraperitoneal ruptures with catheter drainage and formally close large ruptures, for it is difficult to relate the amount of contrast extravasation to the extent of the injury (Brosman and Paul, 1976; Carroll and McAninch, 1984; Cass et al., 1983; McConnell et al., 1982). Extravasation is related to the amount of contrast instilled as well as to the size of the injury. However, in our experience, extravasation into the pelvis, down the inguinal canal to the scrotum, and up the retroperitoneum as high as the kidneys can be successfully treated with catheter drainage. If the patient has uninfected urine and appropriate catheter care is used, the urine will quickly absorb and the bladder rent will heal.

If the patient with an extraperitoneal bladder rupture is to be explored for associated injuries and is not gravely ill, it is best to open the dome of the bladder, not disturb the pelvic hematoma, repair the rupture intravesically, close the bladder, and insert a suprapubic tube as previously described (Palomar et al., 1980). If the pelvic hematoma is opened for another reason, a drain should be placed. If it is not opened, no drain is necessary.

Intraperitoneal and Extraperitoneal Ruptures

All of these injuries need to be formally repaired as described above. Most of these patients have major pelvic fractures and often have injured their urethra, bladder neck, or, in the female, vagina as well. Prompt reconstruction, even in the face of a marked pelvic disruption, will usually be necessary for a good long-term result. These cases all need to be individualized, especially those with major tissue destruction.

POSTOPERATIVE CARE AND FOLLOW-UP

Postoperatively, if no other injuries are present, oral alimentation may be resumed when GI peristalsis returns to normal—usually within 4–7 days of the injury. The bladder heals remarkably quickly and, if a good repair was achieved, the suprapubic tube or Foley catheter may be removed within a week. If there is any question about the closure, a cystogram should be performed before tube removal. If extravasation is present, the catheter should be left in place for further drainage and the cystogram repeated on the tenth postoperative day.

When the patient is voiding normally and there is no leakage of urine from the drainage site for 24 hours, the drain may be removed. Routine antibiotics are not necessary, but once the catheter has been removed, the urine should be cultured and appropriate antibiotics given if bacteriuria is present. If the urine is infected at the time of the injury and subsequent extravasation, antibiotics should be begun preoperatively and continued for at least a week. Frequent urine cultures should be obtained to avoid serious complications.

If the patient had an extraperitoneal injury that was treated with Foley catheter drainage alone, a cystogram should be done on the tenth postoperative day. In our experience, over 85% of the bladders will be healed by that time and the catheter can then be removed. Virtually all injuries treated in this fashion will be healed with less than three weeks of catheter drainage. In the rare male patient who has persistent extravasation, a punch suprapubic tube should be placed in the bladder to prevent urethral complications.

Full activity may be resumed within 3–4 weeks of surgery unless other injuries or complications dictate further therapy or rest. Other than urine cultures if indicated, no long-term follow-up is necessary.

COMPLICATIONS

The most serious complications for bladder ruptures are secondary to delay in diagnosis. When urine leaks into the peritoneal cavity, it will equilibrate with serum, so peritoneal fluid analysis for creatinine and urea will not be helpful in diagnosis. If uroascites becomes marked, respiratory difficulty will develop, especially in infants. Emergency paracentesis may be life-saving.

Sepsis from infected urine is a major threat. In the unrecognized intraperitoneal rupture, generalized peritonitis or loculated abscesses may develop. The bladder rent must be closed, the urine diverted, and all purulent collections drained surgically. Appropriate antibiotics must be given.

The mortality rate of patients with bladder ruptures is about 12% (Carroll and McAninch, 1984; Corriere et al., 1986; McConnell et al., 1982). The cause of death in these patients should never be secondary to the bladder wound if it is properly diagnosed and treated, but will be due to associated visceral or vascular injuries.

Injuries to the bladder neck, urethra, and vagina, if not promptly and properly repaired at the time of the injury, may result in incontinence, fistula, or stricture formation (Merchant et al., 1984). In these cases, reconstruction will have to be delayed for months to allow edema, infection, and induration to disappear. In some patients, a neuropathic bladder may accompany the injury and voiding may be impossible. Intermittent self-catheterization is a good way to overcome this problem.

URETHRAL INJURIES

Anatomically, the male urethra can be divided into: (a) the prostatic urethra, (b) the membranous urethra, (c) the bulbous urethra, and (d) the penile, or pendulous, urethra. When considering injuries to the male urethra, a better classification that helps to determine appropriate therapy is: (a) *posterior urethral injuries*—those of the prostatic and membranous urethra, above and including the urogenital diaphragm, and (b) *anterior urethral injuries*—those of the bulbous and penile, or pendulous, urethra, below the urogenital diaphragm.

The female urethra is very rarely injured (Bolgar et al., 1977; Bredael et al., 1979; Casselman and Schillinger, 1977; Netto et al., 1983). When it is damaged, however, it is usually accompanied by a severe bony pelvic disruption with concomitant injury to the bladder neck and vagina. It is usually more common in children than adults (Buxton, 1978; Merchant et al., 1984; Parkhurst et al., 1981; Persky, 1978; Williams, 1975).

ETIOLOGY

Posterior Urethral Injuries

Almost all injuries of the posterior urethra in the male occur in conjunction with fracture of the bony pelvis (Devine and Devine, 1982; Mitchell, 1968, 1975; Palmer et al., 1983). In modern civilian society, 90% of these injuries are due to motor vehicle accidents involving automobiles, motorcycle riders, or pedestrians. Falls from a height, industrial crushing injuries, and sporting accidents make up the other 10% of the pelvic fracture patients (Cass and Godec, 1978).

The pathophysiology of this injury is most commonly due to the shearing force of the bone disruption, with the prostate, attached by the puboprostatic ligaments, being pulled in one direction while the membranous urethra, attached to the urogenital diaphragm, is pulled in another direction.

Penetrating wounds of the posterior urethra from external violence are uncommon but do occur. Urethral instrumentation with perforation of the prostatic urethra, on the other hand, is more frequently seen (Table 13–4). There are two cases in the literature where a lateral blow to the pelvis and thigh resulted in rupture of the posterior urethra at the junction of the prostatic and membranous urethra with probable concomitant rupture of the puboprostatic ligaments without fracture of the pelvis (Das, 1977; Redman and O'Donnell, 1980).

Anterior Urethral Injuries

Similarly, most injuries to the anterior urethra due to external violence are due to blunt trauma to the perineum (Blandy, 1975; Blumberg, 1969; Cass and Godec, 1978; Kiracofe et al., 1975; Macleod, 1976; Mitchell,

TABLE 13–4.

Etiology of Urethral Injuries

Posterior urethral injuries
  Fracture of the pelvis
    Motor vehicle accidents
    Fall, crush
    Sporting accidents
  Penetrating injuries
    External violence
      Gunshot
      Stab
    Urethral instrumentation
      Resectoscope
      Sounds
      Filiforms, followers
  Lateral pelvic blow
Anterior urethral injuries
  Straddle injury
    Fall
    Fence, ladder
    Kick
    Bicycle
  Penetrating injury
    Gunshot
    Machine injury
    Knife wound
  Urethral instrumentation
    Catheter
    Cystoscope
    Sounds
    Filiforms, followers
    Self-instrumentation
  Penile surgery
    Prosthesis placement, erosion
    Circumcision
  Sexual intercourse
    Urethral laceration
    Fracture of the penis

1968; Pontes and Pierce, 1978; Witherington and McKinney, 1983). The bulbous urethra is usually crushed against the pelvic arch, usually when the patient falls astride an object. This may be while falling from a height, straddling a fence, or having a foot slip from the rung of a ladder. It is sometimes due to a kick to the perineum or hitting a bump in the road while riding a bicycle and coming down hard on the seat or the crossbar.

Penetrating injuries of the anterior urethra due to external objects are less common, but iatrogenic damage due to urethral instrumentation, especially inflation of a Foley catheter balloon in the bulbous urethra, occurs frequently (see Table 13–4) (Sellett, 1971). Surgery of the penis, most notably for penile prosthesis placement—or, later, prosthesis erosion—and circumcision can inadvertently damage the urethra.

Accidents secondary to sexual activity, either intercourse with urethral laceration or fracture of the penis, masturbation by urethral instrumentation, or genital mutilation by a mentally disturbed patient, have also been reported.

DIAGNOSIS

Signs and Symptoms

Patients with a history of trauma to the perineum or who have a fracture of the bony pelvis should be suspected of having a urethral injury. If the patient has attempted to void, he may find he cannot or he may relate that he had the sensation of voiding but no urine came out of his urethra. This patient may have voided into his tissues.

The majority of patients with a ruptured urethra will have blood at the urethral meatus, and many of them will have swelling and ecchymosis of the penis, scrotum, and/or perineum. This is due to urine and/or blood leaking into these structures. If the edema is only in the penis, it is probably contained within Buck's fascia. If it extends to the scrotum, perineum, or anterior abdominal wall, it will be contained by Colles' fascia.

In the patient with a fractured pelvis, rectal examination may reveal the prostate to be in a higher position than usual. This "high-riding prostate" is due to disruption of the urethra with the prostate elevated from its normal position by a large pelvic hematoma. If the puboprostatic ligaments did not rupture, the prostate may have been lifted from its bed attached to a comminuted bone fragment. The soft, boggy hematoma will be felt where the prostate is normally found.

If the patient can urinate, he will usually have gross hematuria. It may be only grossly bloody at the beginning of the stream and/or at the end of the stream if the injury is a partial rupture.

Radiographic Examination

Any patient with a suspected urethral injury must have a retrograde urethrogram performed (Corriere and Harris, 1981; Del Villar et al., 1972; Sandler et al., 1981a). Under no circumstances should an attempt be made to catheterize the urethra until this study delineates the anatomy and any damage done to that organ. Injudicious catheterization of the injured urethra carries the risk of converting a partial urethral rupture into a complete urethral rupture as well as possibly infecting a sterile periurethral or pelvic hematoma.

Trauma surgeons will occasionally comment that placement of a urethral catheter is one of the first maneuvers that should be done in a multiple trauma patient, especially one who is hypotensive. They argue that urine output monitoring is critical in determining organ perfusion and proper fluid infusion rates. The urologic specialist should first counter with the point that blood pressure measurements also give the physician an idea of the level of organ perfusion. Second, he should comment that virtually all multiple trauma patients will be sent to the radiology department to undergo appropriate studies within minutes of emergency room evaluation. At that time, a urethrogram can be performed along with other indicated x-rays.

Some desperately ill patients who may have urethral injuries must be taken directly to the operating room for abdominal exploration prior to radiographic examination. At the time of surgery in these patients, a suprapubic tube should be inserted into the dome of the bladder for drainage. The lower tract may then be studied in the postoperative period and indicated therapy instituted. If the tube is not needed, it may easily be removed. As will be discussed under complications of urethral ruptures, the long-term problems of incontinence, stricture formation, and impotence, which seem very unimportant at the time of injury, loom large in the postoperative period of these unfortunate men. The less urethral damage done, the more the patient will recover free of these complications.

Ideally, the urethrogram should be performed under fluoroscopic monitoring. If necessary, it may be obtained with fixed or even portable equipment in the emergency room. An easy technique that does not require special equipment is to insert a number 14 or 16 F Foley catheter into the urethra so the balloon of the catheter is just 2–3 cm proximal to the meatus. One to two millimeters of saline solution is injected into the balloon to seat it in the fossa navicularis. No lubricant should be used or the catheter may slide out of the urethra.

The patient is then moved to 25–30° oblique, and about 25 ml of 25%–30% contrast media is injected with a Toomey syringe into the urethra. An exposure is taken during the active injection of the contrast medium to distend the urethra and produce a dynamic urethrogram (Fig 13–6).

A penile clamp (i.e., Brodney, Knudson) may also be used but is cumbersome and may not always be available. Inserting the tip of the syringe directly into the urethra should not be done, since the examiner's hand will be exposed to the x-ray beam.

The oblique position is best for demonstrating the entire anterior and posterior urethra. In the AP position, the bulbous urethra is foreshortened and the urethra, as well as areas of extravasation, will overlap, making the study uninterpretable.

If the urethra is normal, the balloon should be deflated, the catheter advanced into the bladder, the balloon reinflated, and a cystogram performed as previously discussed to rule out a bladder rupture. If a catheter has already been inserted into the bladder, it should not be removed. If a urethral injury is suspected, a urethrogram should be done around the indwelling catheter to rule out urethral damage. This can be done by inserting a 16-gauge intracath alongside the Foley and compressing the urethra while injecting contrast material.

Finally, an excretory urogram should be performed to evaluate the kidneys and ureters. If a large pelvic hematoma is present, the bladder will have a "teardrop" appearance and ride high out of the pelvis (Fig 13–7). This is commonly called a "pie-in-the-sky" bladder.

POSTERIOR URETHRAL INJURIES

CLASSIFICATION

As previously stated, virtually all posterior urethral injuries are secondary to fracture of the bony pelvis. Table 13–5 lists the classification of these injuries, which is helpful in planning their management (Colapinto and McCallum, 1977; Sandler et al., 1981b).

Type I Injury

The posterior urethra and proximal bulbous urethra are stretched because the moorings of the prostate to the urogenital diaphragm have been ruptured and a hematoma has collected in the perivesical space (Fig 13–8). Although stretched, the urethra is not ruptured. This injury accounts for 17% of posterior urethral ruptures.

Type II Injury

Until recently, this was thought to be the most common type of posterior urethral injury. The classic description is rupture of the prostatomembranous urethra

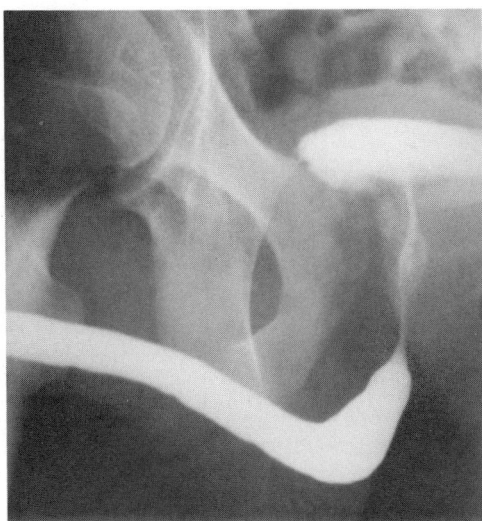

FIG 13–6.
Normal retrograde urethrogram.

TABLE 13–5.

Classification of Urethral Injuries

Posterior urethral injuries
  Blunt trauma
    Type I
      Urethra stretched by pelvic hematoma
    Type II
      Prostatomembranous urethral rupture
      Urogenital diaphragm intact
      Partial or complete rupture
    Type III
      Prostatomembranous urethral rupture
      Urogenital diaphragm ruptured
      Bulbous urethra injured
      Partial or complete rupture
  Penetrating injury
Anterior urethral injuries
  Urethral contusion
  Partial urethral rupture
  Complete urethral rupture
  Penetrating urethral injury

at the apex of the prostate above the urogenital diaphragm. Extravasation of contrast occurs superiorly into the pelvis but is limited inferiorly by an intact urogenital diaphragm (Figs 13–9 and 13–10). The rupture may be complete or incomplete. Recent work has shown that this injury is present only 17% of the time.

Type III Injury

This is the most common injury seen. It is present 66% of the time. The prostatomembranous urethra is either partially or completely ruptured and contrast extends into the pelvis. However, in this injury, the urogenital diaphragm and bulbous urethra are also injured and contrast material will be seen extending into the perineum and into the bulous urethra (Fig 13–11).

Penetrating Injuries

When the posterior urethra has been damaged by an external missile or intraurethral instrument, extravasation will follow the tract of the injury. The extent and location of the injury are variable and, when due to a gunshot wound, the bladder neck is also commonly damaged.

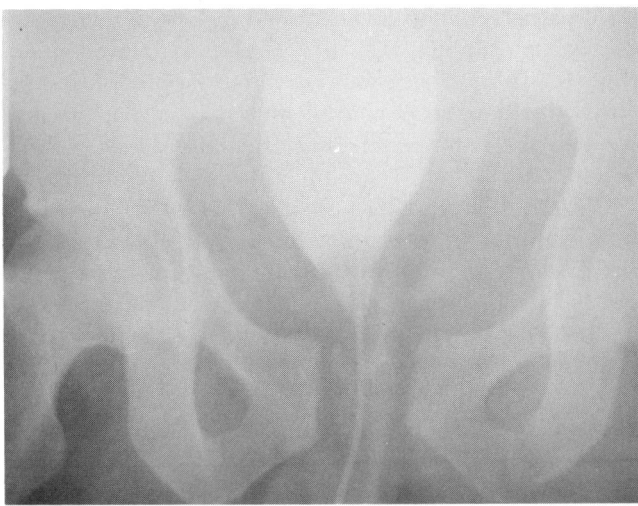

FIG 13–7.
"Teardrop" or "pie-in-the-sky" bladder from a pelvic hematoma in a patient with a pelvic fracture.

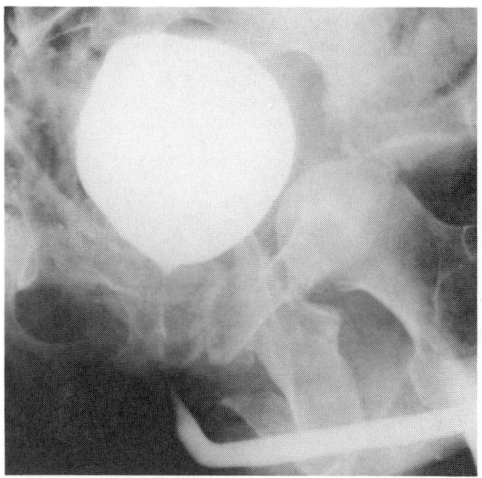

FIG 13–8.
Type I urethral injury. Posterior urethra is compressed by hematoma. (From Sandler CM, Phillips JM, Harris JD, et al: Radiology of the bladder and urethra in blunt pelvic trauma. *Radiol Clin North Am* 1981; 19:195. Used with permission.)

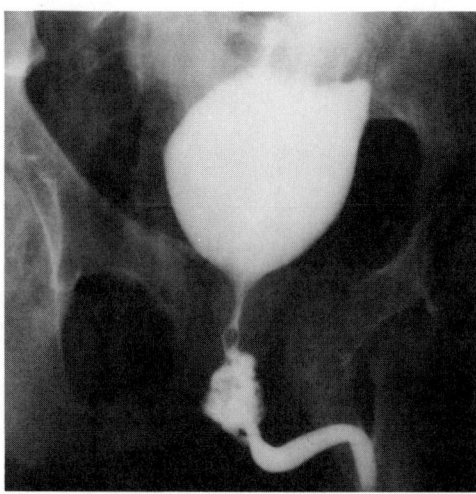

FIG 13–9.
Type II partial urethral injury. All extravasation is in pelvis. Also note intraperitoneal bladder rupture. (From Sandler CM, Harris JH Jr, Corriere JN Jr, et al: Posterior urethral injuries after pelvic fracture. *AJR* 1981; 137:1233. Used with permission.)

THERAPY

Type I Injury

The patient with a pelvic hematoma compressing the urethra will many times have difficulty voiding, so a urethral catheter should be left indwelling for a few days. However, once he has recovered sufficiently from his injuries to discontinue urine output monitoring and is alert enough to attempt to void, the catheter should be removed. Patients with a severe pelvic disruption,

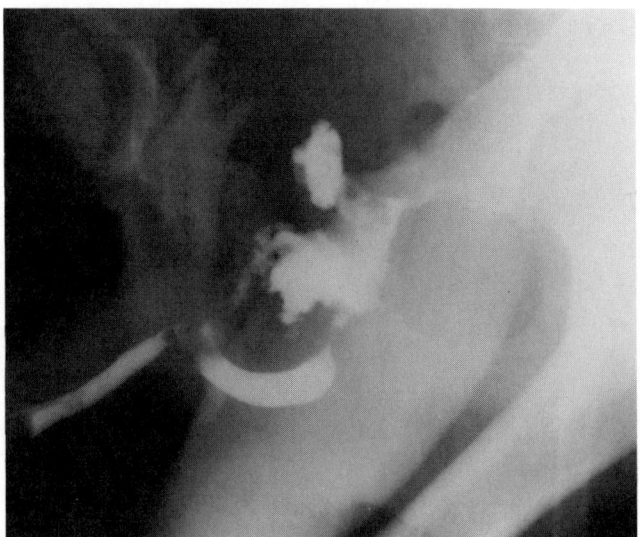

FIG 13–10.
Type II complete urethral injury. All extravasation is in pelvis.

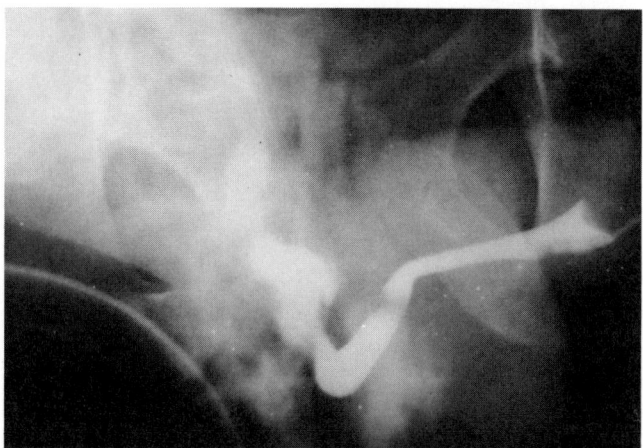

FIG 13–11.
Type III complete urethral injury. Extravasation extends into perineum as urogenital diaphragm is ruptured. (From Sandler CM, Harris JH Jr, Corriere JN Jr, et al: Posterior urethral injuries after pelvic fracture. *AJR* 1981; 137:1233. Used with permission.)

especially if the sacrum is damaged, may have some neurologic deficit and be unable to urinate. A cystometrogram should help to determine the status of the bladder. If multiple trials of voiding fail, the patient should be taught clean intermittent self-catheterization while awaiting return of detrusor function.

Partial Urethral Rupture

The patient with minimal partial urethral rupture may be treated with urethral catheter drainage for 14–21 days, followed by a voiding cystourethrogram to ensure healing of the injury. If the catheter does not pass easily into the bladder or if the injury is extensive, urethral catheterization should not be done for fear that the partial rupture may be converted to a complete rupture. The patient should undergo suprapubic cystotomy by either the trochar technique or formal surgical placement. A cystogram must then be done in these patients to rule out a concomitant bladder rupture.

Clearly, the most conservative way to handle all of these injuries is to place a suprapubic cystotomy and not attempt urethral instrumentation. A voiding cystourethrogram through the cystotomy tube should be performed 14–21 days after the injury. If extravasation is no longer present and the urethra is of normal caliber or there is only a minimal stricture at the site of the injury, the tube should be removed and the patient allowed to void (Corriere and Harris, 1981).

If there is a marked narrowing or total occlusion of the urethra at the area of previous extravasation, panendoscopy and a few days of urethral catheter drainage should be carried out. The visual urethrotome should be available to incise any strictures that may be present.

If the strictured area cannot be successfully negotiated with the panendoscope, the patient should be left on suprapubic drainage for six months and delayed repair by one of the reconstructive procedures discussed below performed. Partial urethral ruptures account for about one third of the posterior urethral ruptures.

Complete Urethral Rupture

There are two ways to treat a complete rupture of the posterior urethra: immediate surgical realignment, or suprapubic cystotomy and delayed surgical repair. *Immediate surgical realignment* is the procedure of choice in the stable patient who: (1) is going to have immediate pelvic exploration for a concomitant vascular or rectal injury; (2) has a severe prostatourethral dislocation with perhaps fixation of a "pie-in-the-sky" bladder and prostate to a displaced comminuted bone fragment by the puboprostatic ligaments; or (3) has major bladder neck lacerations or prostatic fragmentation (seen frequently in children) (Al-ali and Husain, 1983; Banks, 1926; Crassweller et al., 1977; DeWeerd, 1977; Glass et al., 1978; Janknegt, 1975; Janosz et al., 1975; Meyers and DeWeerd, 1972; Patterson et al., 1983; Pierce, 1972; Radge and McInn, 1969; Turner-Warwick, 1977; Webster and Selli, 1983; Wilkinson, 1961).

The procedure is performed through a lower abdominal midline incision. The hematoma is evacuated and a regular or fenestrated 16 or 18 F catheter passed per urethra through the urogenital diaphragm into the prevesical space. The catheter is identified by sight or feel and brought into the surgical field. The anterior bladder wall is opened and the interior of the organ inspected. Lacerations are repaired with 3-0 absorbable suture. Another catheter is then passed out the bladder neck, through the prostatic urethra, and brought into the surgical field by sight. A single 0 nylon suture is used to tie the tips of the catheters together through their distalmost eyes, and the bladder catheter is used to guide the urethral catheter into the bladder (Fig 13–12, A). The bladder catheter is removed, a second 0 nylon catheter is placed through the eye of the urethral catheter, and the suture eventually is brought out through the bladder and abdominal wall as a lock stitch to fix this catheter in place. Some authors have used interlocking sounds to effect this alignment, but the high incidence of false passages with this technique has led to its virtual abandonment.

It is now important to verify whether the puboprostatic ligaments have ruptured. If not, they must be transected, freeing the bladder and prostate from their bony attachments. If this is not done and there is severe distortion of the anterior pelvic arch, the bladder

and prostate will never be able to be brought down to their normal position alongside the urogenital diaphragm (Fig 13–12, B).

One of the older techniques used to attempt realignment of the prostate was to insert a Foley catheter into the urethra and place it on traction. The idea was to pull the catheter's balloon snugly against the bladder neck and prostate to bring the organ into position. Unfortunately, to effect such a maneuver, if the prostate was still attached to the pubis, the Foley catheter would have to realign the pelvic bone as well. The procedure failed to accomplish this most of the time and, unfortunately, many times the force of the balloon on the base of the bladder caused pressure necrosis of the bladder neck, and an irreparable injury followed.

The prostate and bladder should now easily and without tension be repositioned against the urogenital diaphragm. Although some surgeons would attempt directly to anastomose the severed ends of the urethra, an easier and equally effective repair can be effected by placing 0 nylon (Vest) traction sutures through the distal prostate, through the urogenital diaphragm, onto the perineum, where they are tied snugly over a bolster (Fig 13–12, C). A suprapubic tube is placed in the bladder, a drain in the prevesical space, and the wounds closed.

The Vest sutures and drain can be removed in 14 days, but the urethral catheter should be left in place for three weeks. Then a retrograde urethrogram should be done around the tube, and if there is no extravasation it can be removed. A voiding cystourethrogram is then performed, and if the patient voids normally the suprapubic tube is removed.

Delayed surgical repair is the procedure of choice if (1) the patient is medically unstable or (2) the surgeon is unskilled in performing major urethral reconstructive surgery. In fact, in the past few years, more and more urologists have made the delayed repair their procedure of choice for almost all patients with posterior urethral ruptures as the long-term results and complications of the various techniques have become available. As will be discussed below, there is now evidence that the incidence of stricture, incontinence, and impotence is lower with the delayed approach to the repair of this injury (Coffield and Weems, 1977; Lucey et al., 1972; Morehouse and MacKinnon, 1977).

Once the diagnosis of a complete posterior urethral disruption has been made, a suprapubic tube is placed into the bladder either by trocar or formal cystotomy. A cystogram is then performed to ensure that there is no bladder rupture, and nothing more is done about the urethral injury at that time. A cystotomy tube is changed at monthly intervals and the patient followed up clinically until the pelvic hematoma has completely

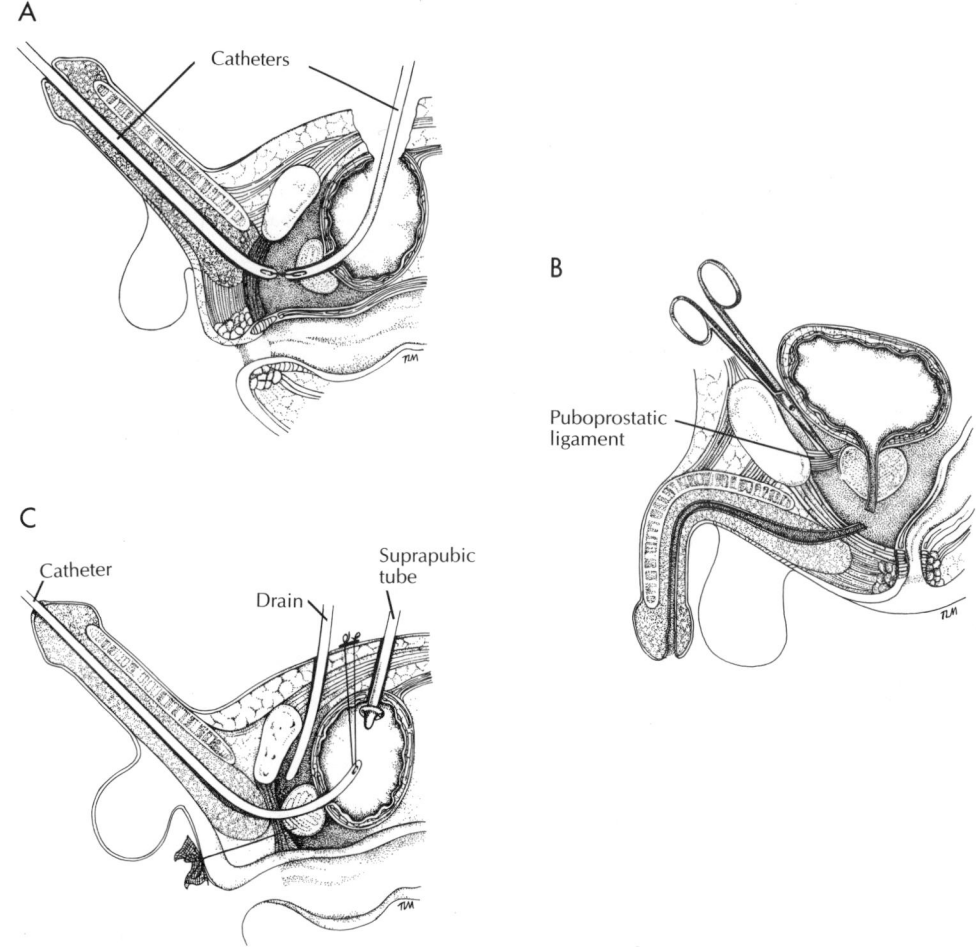

FIG 13–12.
A, catheter introduced into urethral meatus and, via surgically opened bladder, into bladder neck, then tied together in prevesical space. **B,** unruptured puboprostatic ligament cut before repositioning of bladder and prostate. **C,** urethral catheter lock-stitched to abdominal wall. Prostate kept in position against urogenital diaphragm by vest sutures.

reabsorbed, all scar tissue has softened, the pelvic injury has healed, and the patient has been otherwise rehabilitated. This usually takes from six to nine months after the accident.

Before the repair, a cystometrogram is performed to see whether the patient has normal bladder function, as well as a combined cystogram/retrograde urethrogram x-ray to determine the length of the urethral defect (Fig 13–13). Most high-riding prostates and bladders will spontaneously return to the pelvis during the delay as the pelvic hematoma reabsorbs. With this information, a procedure for definitive repair can be made. Broadly, the repairs that have been described are: (1) a two-stage reconstruction; (2) a one-stage reconstruction; and (3) endoscopic reestablishment.

The *two-stage urethroplasty* has been used since the early 1950s as a way to repair posterior urethral strictures (Coffield and Weems, 1977; Johanson, 1953; Ko-

ratim, 1985; Morehouse et al., 1972; Morehouse and MacKinnon, 1977, 1980). At the first stage, either through a perineal midline or pedicled flap incision, the urethra is incised ventrally through the entire stricture as proximal as the verumontanum. A scrotal or perineal skin inlay flap is then sutured to the cut edges of the urethra with interrupted absorbable suture material (Fig 13–14). These techniques, in essence, marsupialize the strictured urethra. A catheter is placed through the proximal ostium into the bladder and a pressure dressing applied to the wound. The previously placed suprapubic tube is removed on the day of surgery and the Foley catheter and dressing a few days later. The patient should void spontaneously.

Between the first and second stages, the ostia, both proximally and distally, must be periodically calibrated. Once all the scar has softened and the wounds healed, the second stage may be performed. The ostia must re-

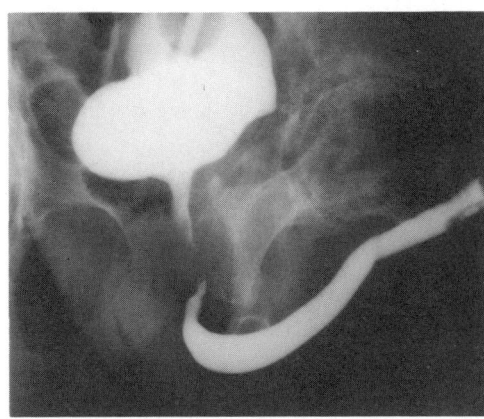

FIG 13–13.
Combined cystogram and retrograde urethrogram in patient with a complete urethral rupture six months postinjury.

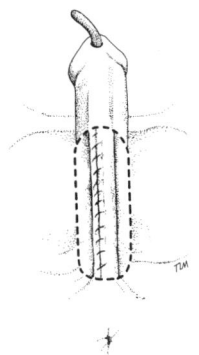

FIG 13–15.
Second stage of a two-stage urethroplasty. Inlay circumcised (*dotted line*) and neourethra formed and closed. Bulbocavernosus muscles, skin, and subcutaneous tissue will then be closed in layers.

main at least a 26 F size. It takes 4–6 months to be sure that the urethra is ready for closure.

During the second stage, the inlay is circumcised along with the underlying urethra to form an even tube of at least a 24 F caliber. The urethra is closed with interrupted or running absorbable sutures, as are the bulbocavernous muscles, subcutaneous tissues, and skin (Fig 13–15). Some surgeons prefer nonabsorbable running pullout sutures for closure.

Urethral catheter and/or suprapubic tube drainage is used for 10–14 days and a voiding cystourethrogram performed when the tube is removed. Catheter drainage should be continued for another week if extravasation is present.

There are basically three techniques described for *one-stage reconstruction* of posterior urethral strictures. The original technique was an intussusception of the

distal normal urethra into the proximal scarred urethra and prostate (Badenoch, 1950; Dobrowolski, 1982; Netto, 1985; Pierce, 1979). More recently, this procedure has been refined by performing a direct urethroprostatic anastomosis by either bypassing or excising the scarred area (Allen, 1975; Diokno, 1980; Khan and Furlow, 1976; Kishev, 1976; Koraitim, 1985; Malloy et al., 1980; Netto, 1985; Paine and Coombes, 1968; Pierce, 1979; Strong and Hodges, 1977; Turner-Warwick, 1968, 1977; Waterhouse et al., 1974, 1980; Webster et al., 1983; Zincke and Furlow, 1985). Resection of the symphysis pubis to attain access to the distal prostate has been very helpful (Pierce, 1962; Waterhouse et al., 1973).

With all of these procedures, a midline perineal incision is made and carried through the bulbocavernosus muscle to expose the urethra. The urethra with its corpus spongiosum is then dissected free in both directions to obtain adequate length. It is then divided just distal to the stricture, which means almost flush with the urogenital diaphragm.

If the intussusception technique is to be used, a curved sound is now passed via the suprapubic sinus tract, where the suprapubic tube has been in place, and guided into the internal urethral meatus. The tip of the sound is felt in the perineal wound at the site of the stricture. An incision is made with a scalpel onto the sound, which is then forced through into the wound. This tract is dilated to 30 F (Badenoch, 1950; Netto, 1985; Pierce, 1979).

A small sound is now placed into the suprapubic sinus and out the distal tract and its tip firmly invaginated into the open end of a 16 F red rubber catheter. The sound and catheter are withdrawn into the bladder and out the suprapubic sinus, and the free end of the catheter is inserted into the distal urethra for a distance

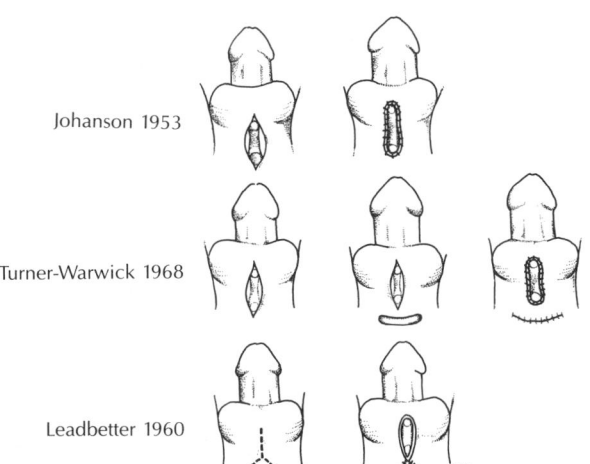

Johanson 1953

Turner-Warwick 1968

Leadbetter 1960

FIG 13–14.
Various types of first-stage procedures of a two-stage urethroplasty.

of 5–8 cm (Fig 13–16). The edges of the urethra are sewn to the catheter with 3-0 chromic catgut and the catheter, with the urethra attached, pulled through the opening in the prostate into the bladder (Fig 13–17). The cut end of the urethra should be made to rest about 1 cm proximal to the verumontanum. A marking suture on the urethra prior to invagination is helpful to ensure this position. The outer wall of the urethra is now sewn with a few absorbable sutures to the area where it enters the tract.

The catheter, which is now exiting the suprapubic sinus, is fixed with tension to the abdominal wall. A suprapubic tube is inserted into the bladder and placed to straight drainage. The bulbocavernous muscles, subcutaneous tissues, and skin are closed in layers and a pressure dressing applied to the wound.

The dressing is removed on the third day. The red rubber catheter will loosen and can be removed in about a week. A voiding cystourethrogram should be done on the 14th day and the patient allowed to void. A Foley catheter should be placed into the urethra if voiding is difficult and voiding trials begun. The suprapubic tube can be removed. These patients may need periodic urethral dilatation for a few months if voiding becomes difficult.

If an *end-to-end prostatourethral anastomosis* is to be performed after the urethra has been transected distal to the stricture, it should be spatulated (Diokno, 1980; Khan and Furlow, 1976; Kishev, 1976; Koraitim, 1985; Malloy et al., 1980; Netto, 1985; Pierce, 1979; Strong and Hodges, 1977; Turner-Warwick, 1977; Webster et al., 1983; Zincke and Furlow, 1985). A curved sound is now passed through the suprapubic sinus into the prostatic urethra. If negotiation of the internal urethral meatus is difficult, a panendoscope can be passed through the suprapubic sinus into the prostatic urethra. Palpation of the tip of the sound or panendoscope in the wound acts as a guide to identify the prostate. Sim-

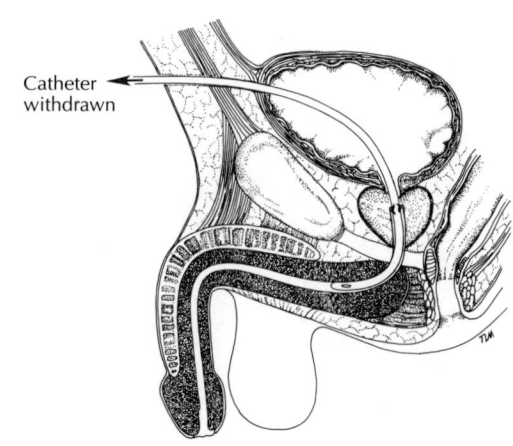

FIG 13–17.
Catheter previously sewed to cut end of urethra drawn into bladder, intussuscepting bulbous urethra into prostatic urethra.

ilarly, the light in the panendoscope may shine through the tissues in the perineal wound and help identify the prostate by sight. Now the scar is excised with a scalpel and the prostate beveled from the verumontanum posteriorly to the midprostate anteriorly.

A 16 F fenestrated catheter is placed via the urethral meatus and drawn into the wound. An 0 nylon suture is tied to the distal eye of the catheter and the suture drawn by the sound or panendoscope out the suprapubic sinus onto the abdominal wall. The catheter is now pulled into the bladder by the suture to be eventually anchored with the suture to the abdominal wall over a button. The spatulated end of the urethra is now sutured with four 3-0 absorbable quadrant sutures to the beveled-cut end of the prostate. The wound is closed in layers and a suprapubic tube inserted and placed to straight drainage. A pressure dressing is applied to the wound (Fig 13–18).

On the third postoperative day, the dressing is removed. The urethral stent is removed in three weeks and a voiding cystourethrogram performed. If the urethra is well healed, the suprapubic tube is removed. If extravasation is present, the patient is kept on suprapubic drainage for another week and the voiding cystourethrogram repeated. By this time, the urethra should be healed, the patient should void normally, and the suprapubic catheter can be removed.

Occasionally, the prostate and bladder will not descend far enough to allow the anastomosis to be made perineally. If this occurs, a midline abdominal incision should be made and the bladder and prostate freed from above. Many times, gouges will have to be used to remove the underside of the pubis to obtain adequate mobilization. The anastomosis, stenting, and closure are still carried out as described above in these

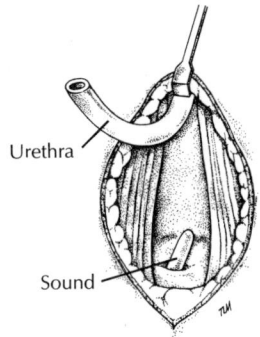

FIG 13–16.
Urethra severed distal to stricture and retracted. Sound passed via bladder, into bladder neck and prostatic urethra. Incision over tip of sound allows extrusion of sound into wound.

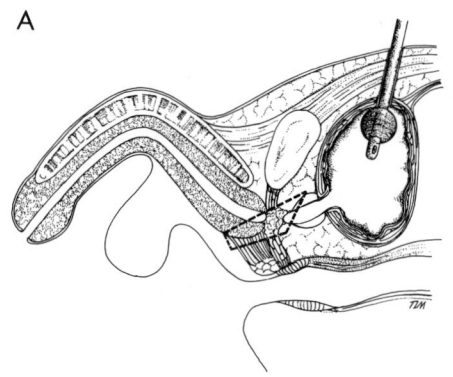 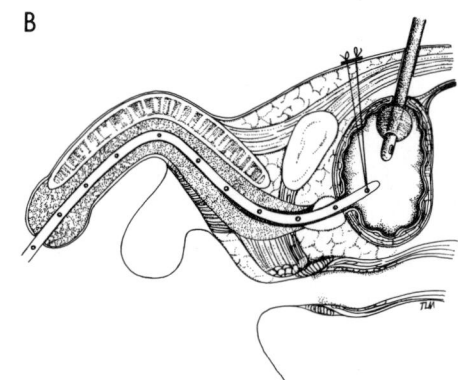

FIG 13–18.
A, preoperative status. Suprapubic tube in place. Area bounded by *dotted line* will be excised (includes scar). Prostate and urethra beveled. **B,** urethra sewed to prostate end-to-end. Fenestrated urethral stent and suprapubic tube in place.

cases, and a prevesical drain is placed. Because of the increased amount of dead space created when the prostate is dissected from above, the omentum should be freed, brought into the pelvis, and wrapped around the anastomosis to fill in this area. Postoperative management is similar to that described above, except the patient cannot be fed orally until his GI tract recovers from the abdominal exploration. The drain can be removed in one week.

Transpubic urethroprostatic anastomosis without resection of the scarred area is performed through a lower midline abdominal incision (Waterhouse et al., 1974, 1980). A Gigli saw is then passed beneath the pubis on either side of the midline and a trapezoidal piece of pubic bone removed. This gives access to the anterior surface of the prostate, which is entered. The previously perineally mobilized and spatulated distal urethra is brought through the crura and anastomosed to the prostatic incision with 3-0 quadrant absorbable sutures over a 22 F Foley catheter. With this procedure, the prostate is not mobilized nor is the stricture excised. A suprapubic tube and prevesical drain are brought out through the abdomen. The wounds are closed as described previously (see Fig 13–18). The timing of urethral catheter, suprapubic tube, and drain removal are essentially identical to the end-to-end anastomosis procedure.

Finally, some urologists have advocated merely *re-establishing urethral continuity endoscopically* (Gonzalez et al., 1983; Islam, 1978; Lieberman and Barry, 1982). With this technique, a panendoscope is passed via the suprapubic sinus into the proximal urethra by one surgeon, and either a visual urethrotome or a resectoscope fitted with a Colling's knife is passed into the distal urethra by a second surgeon. Using the light from the panendoscope as a guide, the distal operator cuts his way through the scar and into the prostatic urethra. A 22 F urethral catheter is then placed into the bladder, removed one week later, and the patient allowed to void. Periodic dilatations are usually necessary in these patients.

ANTERIOR URETHRAL INJURIES

CLASSIFICATION

The majority of these injuries due to blunt trauma are secondary to straddle injuries (Blumberg, 1969; Kiracofe et al., 1975; Macleod, 1976; Pontes and Pierce, 1978; Witherington and McKinney, 1983). In these instances, the pelvis is usually not fractured, and the overlying skin remains intact. Perhaps today there is a greater volume of iatrogenic injuries to the anterior urethra due to urethral instrumentation. A classification of these injuries is listed in Table 13–5 and is useful when planning appropriate therapy.

Urethral Contusion

When a patient has undergone a straddle injury and has initial or terminal hematuria but has a normal urethrogram, he has sustained a urethral contusion.

Partial Urethral Rupture

When a patient has had a urethral injury either due to external blunt trauma or urethral instrumentation and on a retrograde urethrogram demonstrates extravasation of contrast material, but the urethra is in continuity and contrast material goes freely into the bladder, he is said to have a partial urethral rupture (Fig 13–19).

Complete Urethral Rupture

If, after blunt trauma, extravasation is demonstrated on a retrograde urethrogram and urethral continuity is lost, a complete urethral rupture is present (Fig 13–20).

Penetrating Urethral Injury

If the urethra has been injured by an external missile, urethral instrumentation, or migration of a penile prosthesis, a partial or complete disruption of the urethra may result. Only major lacerations or those associated with extensive tissue destruction are handled in a unique way from blunt injuries. However, because of

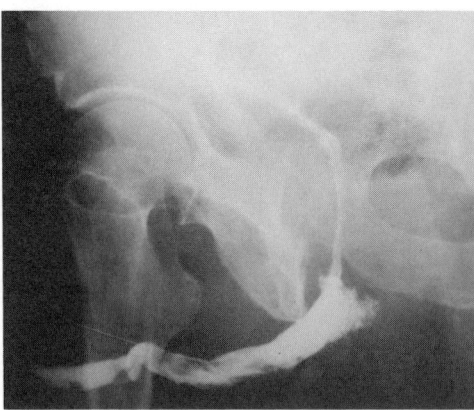

FIG 13–19.
Partial anterior urethral rupture secondary to Foley catheter balloon blown up in bulbous urethra.

these therapeutic decisions, penetrating injuries are best placed into a separate category.

THERAPY

Urethral Contusions

No special therapy is necessary for patients with this injury. They usually are able to void normally, and their hematuria promptly clears. In the absence of extravasation, there are probably no long-term sequelae of this injury.

Partial Urethral Rupture

If the extravasation on urethrogram is minimal, contained by Buck's fascia, and there is good urethral con-

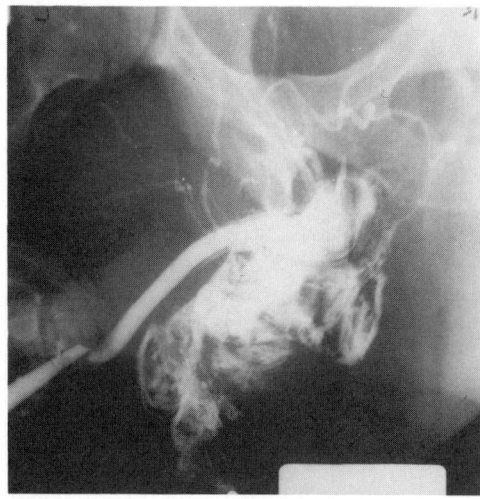

FIG 13–20.
Complete anterior urethral rupture secondary to a straddle injury. Note the venous extravasation. (From Sandler CM, Phillips JM, Harris JD, et al: Radiology of the bladder and urethra in blunt pelvic trauma. *Radiol Clin North Am* 1981; 19:195. Used with permission.)

tinuity, these patients may be allowed to void, or a urethral catheter can be placed into their bladder for a few days. If the injury is extensive and extends outside Buck's fascia, a suprapubic tube should be placed into the bladder and a voiding cystourethrogram repeated in 10–14 days. If the urethra is normal or there is only a large-caliber stricture, the tube can be removed and the patient allowed to void. Periodic urinary flow rates and urethrograms should be done to be sure a stricture does not develop at the site of the injury.

If a marked narrowing has developed, the patient should undergo panendoscopy and visual urethrotomy of the strictured area. A urethral catheter should be left in the urethra for 24 hours and the suprapubic tube removed. When the urethral catheter is removed, the patient should be allowed to void. Periodic urinary flow rates and urethrograms are critical follow-up measures in these patients.

Complete Urethral Rupture

These patients should all have a suprapubic tube placed into their bladders and, if extensive perineal extravasation of blood and urine is present, followed closely. If the urine is infected, they should receive antibiotics. Occasionally, these patients with extensive extravasation delay in presenting to the physician, and erythema, purulence, and frank necrosis may be present in the penis, scrotum, or perineum. These patients will need subcutaneous drains placed, debridement and suprapubic urinary diversion performed, and antibiotics prescribed.

When the skin of the genitalia and perineum is normal and at least 14 days have elapsed from the injury, a voiding cystourethrogram and possibly a combined voiding cystourethrogram and retrograde urethrogram should be done to delineate the injury. A stricture of some magnitude will probably have developed. If urethral continuity is intact, panendoscopy and a *visual urethrotomy* may be all that is needed to incise the stricture (Katz and Waterhouse, 1971; Waterhouse and Selli, 1978). If this is successful, a urethral catheter should be placed for 24 hours and the suprapubic tube removed. These patients must be carefully followed up with voiding flow rates and retrograde urethrograms for recurrence of their stricture.

However, most of these patients will have developed complete occlusion of their urethra. If infection did not supervene on the extravasation of blood and urine, it will commonly be very short. The patient should be left on suprapubic drainage until the perineum is well healed prior to reconstruction. This will usually take 4–6 months. A combined voiding cystourethrogram and retrograde urethrogram will delineate the extent of the stricture to be repaired as was described in posterior urethral ruptures.

The *two-stage urethroplasties* that are discussed above and illustrated in Figures 13–14 and 13–15 can be used to repair strictures of the anterior urethra as well as the posterior urethra (Johananson, 1953; Leadbetter, 1960; Swinney, 1952, 1957). Because the stricture area is usually at the penoscrotal junction, a simple midline incision rather than the scrotal or perineal inlay flaps is usually adequate for the first stage.

However, if a short stricture is present, a *one-stage end-to-end urethroplasty* with a ventral wall skin patch can be performed (Devine et al., 1976). In this procedure, a midline incision is made over the area of the stricture, which is identified with a urethral sound, and the urethra is mobilized in both directions. The strictured area is then excised and the dorsal wall of the urethra reanastomosed. A full-thickness skin graft is then taken from the shaft of the penis and defatted. This graft is used to widen the anastomosis and is sutured in place with 4-0 absorbable sutures (Fig 13–21).

The bulbocavernous muscles, subcutaneous tissue, and skin are closed in layers. A urethral catheter is left in place for 3–5 days and the suprapubic tube continued on drainage for 14 days. A voiding cystourethrogram is performed at that time and, if no extravasation is seen, the suprapubic tube is removed. If extravasation is seen, suprapubic drainage is continued for another week and the study repeated.

If the stricture is long, a *one-stage patch graft urethroplasty* or *pedicled urethroplasty* can be performed (Blandy et al., 1960; Devine et al., 1976; Orandi, 1968). In these procedures, the stricture is identified as described above for the end-to-end urethroplasty, but the urethra is not mobilized. Instead, the stricture is incised ventrally in the manner of a first-stage procedure of the two-stage technique, and then a free graft taken from the skin of the penis as described above and the entire incised urethra "patched" open (Fig 13–22). In the pedicled technique, instead of a free graft, a suitable length of skin is freed next to the urethral incision but left attached to its underlying blood supply. This pedicled graft is then inverted and sewn into the urethral defect (Fig 13–23). Postoperative care is similar to that of the end-to-end technique.

Penetrating Urethral Injury

The most common penetrating injuries of the anterior urethra are secondary to urethral instrumentation. Most of these injuries are minor, and patients are usually allowed to void, or a Foley catheter is placed in their bladder for a few days. Occasionally, a sound or filiform and follower is forced through the urethral wall and into the rectum. These patients need a suprapubic tube placed in their bladders for a few weeks, and their wounds will usually heal without sequelae. A retrograde urethrogram and/or voiding cystourethrogram should be done to ensure that the injury has healed before the patient is allowed to void.

During dilatation of the corpus cavernosum for placement of a penile prosthesis, the urethra may be ruptured. If the prosthesis is not inserted and a Foley catheter is placed into the bladder for a few days, the perforation will heal. Sometimes a prosthesis will spontaneously erode through the urethra, especially in patients with decreased sensation from neurologic disease or if a Foley catheter is left in the urethra for a long time in patients with penile prostheses. Merely removing the prosthesis will allow the urethral lacerations to heal. A Foley catheter may be placed for a few days to facilitate the process.

Penetrating injuries that need prompt attention are those due to external violence. Clean knife wounds merely need minimal debridement, closure of the defect with absorbable sutures, and suprapubic catheter drainage for 2–3 weeks. A voiding cystourethrogram

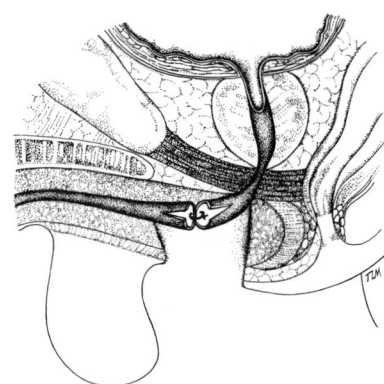

FIG 13–21.
Strictured area has been removed. Cut ends of urethra spatulated ventrally. Dorsal walls anastomosed. Ventral defect will be "patched open" with a full-thickness skin graft (see Fig 13–22).

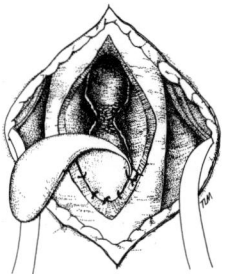

FIG 13–22.
Free full-thickness skin graft sewed into incised urethra at area of stricture to widen stenotic section.

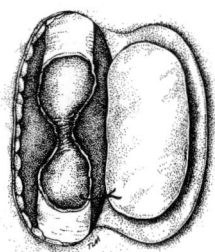

FIG 13–23.
Pedicled skin graft sutured into area of stricture to widen stenotic section.

and/or retrograde urethrogram should be done prior to removal of the suprapubic tube.

Dirty wounds with extensive tissue destruction and foreign material in the wound (metal pellets, oil, grease, hair, clothing, etc.) need to be thoroughly cleansed with antiseptic solutions and copious irrigation. Although debridement of devitalized tissue is important, it must be stressed that contused corpus spongiosum tissue is hemorrhagic and ecchymotic and may appear necrotic when it is only badly bruised. If debridement is vigorous, more urethra than is necessary may be removed and discarded, making eventual repair a formidable task. However, if a large section of urethra is lost, the ends of the urethra that are left should be sutured to the skin in the manner of a first-stage urethroplasty. Skin will have to be brought between the ostia to close the wound (Fig 13–24).

Suprapubic diversion should be done until the perineum has healed. This may take weeks to months. Radiographic studies will demonstrate residual stricture, which can be handled by one of the above-described urethroplasty techniques.

POSTOPERATIVE CARE AND FOLLOW-UP

A discussion of when the stents, drains, and catheters should be removed has been listed under each procedure above. Antibiotics should be withheld in these patients until they have all tubes removed from their urinary tracts. At that time, they should receive appropriate antibiotics and urine cultures should be performed to ensure sterilization has been accomplished.

Periodic voiding flow rates should be done at three-month intervals for at least a year. If there is a decrease in flow, a retrograde urethrogram should be performed. Patients with neurologic damage may not be able to void and may have to be taught intermittent self-catheterization. Those with bladder neck damage may be incontinent and need α-adrenergic drugs to become dry. Complete neurologic evaluation is imperative in these patients.

RESULTS AND COMPLICATIONS

When the literature is reviewed for the results and complications of these various techniques, it is well established that the primary realignment patients have a much higher long-term complication rate than do patients treated with delayed repair, despite the technique employed (Jackson and Williams, 1974; McAninch, 1981; Whitehead and Morales, 1972). The restricture rate with primary repair is about 70% (30%–100%), the impotence rate is 44% (20%–50%) and the incontinence rate 20% (2%–44%).

With the delayed procedures, it should be noted that although virtually all partial ruptures heal without stricture, almost all patients with complete ruptures initially develop strictures. However, following definitive repair, it is the rare patient who has a long-term stricture problem. Moreover, the impotence rate is only 11% (0%–56%) and the incontinence rate, 2% (0%–5%) in well reported series. It appears to be well worth the trade-off of wearing a suprapubic catheter for six or more months when compared to the long-term problems associated with early repair.

There is considerable concern that patients who undergo a transphincteric urethroplasty and, in essence, have their internal urethral mechanism destroyed by the procedure, are later at risk for incontinence if a transurethral resection of the prostate is performed. This risk is certainly real. However, when one considers that these are usually young men with 40–50 years of life ahead of them and that only 10% of males will need a prostatectomy as they get older, the argument becomes thin for withholding the procedure.

Finally, when faced with an elderly patient with complete obstruction due to a posterior urethral rupture who has previously had a transurethral prostatectomy and has therefore no bladder neck continence mechanism, the use of the intussusception procedure should be chosen. With this technique, the stricture and sphincter (if it is still intact) are not resected but merely dilated. If the sphincter is intact and the stric-

FIG 13–24.
Damaged urethra has been discarded. Freshened edges of severed urethra sewed to skin. Normal skin brought between ostia to cover defect.

ture can be kept open with periodic dilatations, the patient should be continent and able to void. If the sphincter has been irreparably damaged, the stricture, if not dilated, should merely recur.

INJURIES IN FEMALES

There are very few reports of rupture of the female urethra in the literature (Bolgar et al., 1977; Bredael et al., 1979; Casselman and Schillinger, 1977; Netto et al., 1983). Most of them are in children and most of them also involve the bladder neck and vagina (Buxton, 1978; Merchant et al., 1984; Parkhurst et al., 1981; Persky, 1978; Williams, 1975). These injuries need immediate reconstruction of the bladder neck and repair of all vaginal lacerations to ensure that the continence mechanism will be intact postinjury as well as to prevent the formation of vesicovaginal fistulas. Retropubic repair over a stenting urethral catheter and use of a suprapubic tube will usually produce a good result. Occasionally, a urethrovaginal fistula will develop and require secondary closure.

INJURIES IN CHILDREN

As mentioned above, most urethral injuries in females are in children and need immediate repair to prevent long-term complications (Buxton, 1978; Merchant et al., 1984; Parkhurst et al., 1981; Persky, 1978; Williams, 1975). Males are treated by any of the above described procedures but seem to do best when treated with the delayed technique (Aubert and Court, 1975; Devereaux and Williams, 1972; Garrett, 1975; Gibbons et al., 1979; Glassberg et al., 1979; Haller et al., 1979; Leadbetter and Leadbetter, 1962; Malek et al., 1977; McGuire and Weiss, 1973). If the patient has an anterior stricture which can be incised with a visual urethrotome, this is preferable. If restricturing occurs, he should be taught intermittent self-catheterization to keep the stricture dilated. Eventually, a formal urethroplasty should be performed (Brock et al., 1981; Kramer et al., 1981; McGuire and Weiss, 1973; Waterhouse, 1976).

FRACTURE OF THE BONY PELVIS

Most patients that have fractures of the bony pelvis have multisystem injuries secondary to high-speed motor vehicle accidents (Antoci and Schiff, 1982; Belis et al., 1979; Cass and Ireland, 1973; Clark and Prudencio, 1972; Conolly and Hedberg, 1969; Fallon et al., 1984; Flaherty et al., 1968; Gibson, 1974; Glass et al., 1978; Iversen and Jessing, 1973; Kaiser and Farrow, 1965; Levine and Crampton, 1963; Looser and Crombie, 1976; McCague and Semans, 1944; Morehouse and MacKinnon, 1969; Pokorny et al., 1979; Reynolds et al., 1973; Trunkey et al., 1974; Weems, 1979). When an anterior arch fracture is present, the probability of a urinary tract injury is high. With a single break of the pelvic ring, 12.5% of patients will have an injury to the lower urinary tract, and with two breaks in the ring, 22% of patients will have such an injury (Palmer et al., 1983).

Bladder ruptures, mostly extraperitoneal in nature, will be seen 9% of the time, urethral injuries 3.5% of the time, and in 1% of the patients both a bladder and urethral injury will be present. Aside from the urinary tract complications of stricture and incontinence, in the male, impotence can be seen in up to 50% of these patients (Gibson, 1970). As discussed above, there is a wide range of variation in the long-term complication rate depending on the initial care and the reconstructive procedure chosen for the repair of the injury. It must be stressed to emergency room and trauma team personnel that injudicious instrumentation of the urethra in the patient with a fractured pelvis and possible urethral injury may lead to a lifetime of chronic debilitation because of a momentary lapse in the proper management of this injury (Corriere and Harris, 1981).

REFERENCES

1. Al-Ali IH, Husain I: Disrupting injuries of the membranous urethra—the case for early surgery and catheter splinting. *Br J Urol* 1983; 55:716.
2. Allen TD: The transpubic approach for strictures of membranous urethra. *J Urol* 1975; 114:63.
3. Antoci JP, Schiff J Jr: Bladder and urethral injuries in patients with pelvic fracture. *J Urol* 1982; 128:25.
4. Aubert J, Court B: Traumatic ruptures of the urethra in children. *Eur Urol* 1975; 1:122.
5. Badenoch AW: A pull-through operation for impassable traumatic stricture of the urethra. *Br J Urol* 1950; 22:404.
6. Bacon SK: Rupture of the urinary bladder: Clinical analysis of 147 cases in the past ten years. *J Urol* 1943; 49:432.
7. Banks H: Ruptured urethra: A new method of treatment. *Br J Surg* 1927; 15:262.
8. Belis JA, Recht KA, Milam DF: Simultaneous traumatic bladder perforation and disruption of the prostatomembranous urethra. *J Urol* 1979; 122:412.
9. Blandy J: Injuries of the urethra in the male. *Injury* 1975; 7:77.
10. Blandy JP, Singh M, Tresidder GC: Urethroplasty by scrotal flap for large urethral strictures. *Br J Urol* 1960; 40:261.
11. Blumberg N: Anterior urethral injuries. *J Urol* 1969; 102:210.
12. Bolgar GC, Duncan RE, Evans AT: Primary repair of completely transected female urethra by advancement. *J Urol* 1977; 118:118.
13. Bonavita JA, Pollack HM: Trauma of the adult bladder and urethra. *Semin Roentgenol* 1983; 18:299.

14. Bredael JJ, Kramer SA, Cleeve LK, et al: Traumatic rupture of the female urethra. *J Urol* 1979; 122:560.

15. Brock WA, Kaplan GW: Use of the transpubic approach for urethroplasty in children. *J Urol* 1981; 125:496.

16. Brosman SA, Paul JG: Trauma of the bladder. *Surg Gynecol Obstet* 1976; 143:605.

17. Burnette DG Jr: Bladder perforation and urethral catheter extrusion: An unusual complication of cerebrospinal fluid—peritoneal shunting. *J Urol* 1982; 127:543.

18. Buxton RA: Rupture of the urethra in a female child with a fractured pelvis: A case report. *Injury* 1978; 9:209.

19. Carroll PR, McAninch JW: Major bladder trauma: The accuracy of cystography. *J Urol* 1983; 130:887.

20. Carroll PR, McAninch JW: Major bladder trauma: Mechanisms of injury and a unified method of diagnosis and repair. *J Urol* 1984; 132:254.

21. Cass AS: Bladder trauma in the multiply injured patient. *J Urol* 1976; 115:667.

22. Cass AS: The multiply injured patient with bladder trauma. *J Trauma* 1984; 24:731.

23. Cass AS, Ireland GW: Bladder trauma associated with pelvic fractures in severely injured patients. *J Trauma* 1973; 13:205.

24. Cass AS, Johnson CP, Khan AU, et al: Nonoperative management of bladder rupture from external trauma. *Urology* 1983; 22:27.

25. Cass AS, Godec CJ: Urethral injury due to external trauma. *Urology* 1978; 11:607.

26. Casselman RC, Schillinger JF: Fractured pelvis with avulsion of the female urethra. *J Urol* 1977; 117:385.

27. Clark SS, Prudencio RF: Lower urinary tract injury associated with pelvic fractures: Diagnosis and management. *Surg Clin North Am* 1972; 52:183.

28. Coffield KS, Weems WL: Experience with management of posterior urethral injury associated with pelvic fracture. *J Urol* 1977; 117:722.

29. Cohen MS, Warner RS, Fish L, et al: Bladder perforation after orthopaedic hip surgery. *Urology* 1977; 9:291.

30. Colapinto V, McCallum RW: Injury to the male posterior urethra in fractured pelvis: A new classification. *J Urol* 1977; 118:575.

31. Colodny AH: Bladder injury during herniorrhaphy. *Urology* 1974; 3:89.

32. Conolly WB, Hedberg EA: Observations on fractures of the pelvis. *J Trauma* 1969; 9:104.

33. Corriere JN Jr, Harris JD: The management of urologic injuries in blunt pelvic trauma. *Radiol Clin North Am* 1981; 19:187.

34. Corriere JN Jr, Sandler CM: Management of the ruptured bladder: 7 years experience with 111 cases. *J Trauma* 1986; 26:830.

35. Crassweller PO, Farrow GA, Robson CJ, et al: Traumatic rupture of the supramembranous urethra. *J Urol* 1977; 118:770.

36. Culp OS: Treatment of ruptured bladder and urethra: Analysis of 86 cases of urinary extravasation. *J Urol* 1942; 48:266.

37. Das S: Complete rupture of the posterior urethra without fractured pelvis. *J Urol* 1977; 118:116.

38. Del Villar RG, Ireland GW, Cass AS: Management of bladder and urethral injury in conjunction with the immediate surgical treatment of the acute severe trauma patient. *J Urol* 1972; 108:581.

39. Devereux MH, Williams DI: The treatment of urethral stricture in boys. *J Urol* 1972; 108:489.

40. Devine PC, Devine CJ Jr: Posterior urethral injuries associated with pelvic fractures. *Urology* 1982; 20:467.

41. Devine PC, Fallon B, Devine CJ Jr: Free full thickness skin graft urethroplasty. *J Urol* 1976; 116:444.

42. DeWeerd JH: Immediate realignment of posterior urethral injury. *Urol Clin North Am* 1977; 4:75.

43. Diokno AC: Late genitourinary tract complications associated with severe pelvic injury. *Surg Gynecol Obstet* 1980; 150:150.

44. Dmochowski RR, Crandell SS, Corriere JN Jr: Bladder injury and uroascites from umbilical artery catheterization. *Pediatrics* 1986; 77:421.

45. Dobrowolski Z: Long-term results of surgical treatment of posttraumatic posterior urethral strictures by Solovov's method. *J Urol* 1982; 128:700.

46. Fallon B, Wendt JC, Hawtrey CE: Urological injury and assessment in patients with fractured pelvis. *J Urol* 1984; 131:712.

47. Flaherty JJ, Kelly R, Bradford B, et al: Relationship of pelvic bone fracture patterns to injuries of urethra and bladder. *J Urol* 1968; 99:297.

48. Garrett RA: Pediatric urethral and perineal injuries. *Pediatr Clin North Am* 1975; 22:401.

49. Gibbons MD, Koontz WW Jr, Smith MJV: Urethral strictures in boys. *J Urol* 1979; 121:217.

50. Gibson GR: Impotence following fractured pelvis and ruptured urethra. *Br J Urol* 1970; 42:86.

51. Gibson GR: Urological management and complications of fractured pelvis and ruptured urethra. *J Urol* 1974; 111:353.

52. Glass RE, Flynn JT, King JB, et al: Urethral injury and fractured pelvis. *Br J Urol* 1978; 50:578.

53. Glassberg KI, Talete-Velcek F, Ashley R, et al: Partial tears of prostatomembranous urethra in children. *Urology* 1979; 13:500.

54. Gonzalez R, Chiou R, Hekmat K, et al: Endoscopic reestablishment of urethral continuity after traumatic disruption of the membranous urethra. *J Urol* 1983; 130:785.

55. Haller JC, Kassner EG, Waterhouse K, et al: Traumatic strictures of the prostatomembranous urethra in children: Radiologic evaluation before and after urethral reconstruction. *Urol Radiol* 1979; 1:43.

56. Hayes EE, Sandler CM, Corriere JN Jr: Management of the ruptured bladder secondary to blunt abdominal trauma. *J Urol* 1983; 129:946.

57. Hubbard JG, Amin M, Polk HC Jr: Bladder perforations secondary to surgical drains. *J Urol* 1979; 121:521.

58. Islam M: Posterior urethral trauma and strictures: An attempt to solve a controversy. *J Urol* 1978; 119:418.

59. Iversen HG, Jessing P: Urinary tract lesions associated with fractures of the pelvis. *Acta Chir Scand* 1973; 139:201.

60. Jackson DH, Williams JL: Urethral injury: A retrospective study. *Br J Urol* 1974; 46:665.

61. Janknegt RA: Management of complete disruption of the posterior urethra. *Br J Urol* 1975; 47:305.

62. Janosz F, Zielinski J, Szkodny A, et al: Surgical technique and results of primary repair in recent urethral injuries. *Eur Urol* 1975; 1:278.

63. Johanson B: Reconstruction of the male urethra in strictures: Application of the buried intact epithelium technic. *Acta Chir Scand [Suppl]* 1953; 176:1.

64. Johnson PA: Rectal impalement with peforation of the bladder. *Br Med J* 1971; 2:748.

65. Kaiser TF, Farrow FC: Injury of the bladder and prostatomembranous urethra associated with fracture of the bony pelvis. *Surg Gynecol Obstet* 1965; 120:99.

66. Katz AS, Waterhouse K: Treatment of urethral strictures in men by internal urethrotomy: A study of 61 patients. *J Urol* 1971; 105:807.

67. Khan AU, Furlow WL: Transpubic urethroplasty. *J Urol* 1976; 116:447.

68. Kiracofe HL, Pfister RR, Peterson NE: Management of nonpenetrating distal urethral trauma. *J Urol* 1975; 114:57.

69. Kishev SV: Excision of the urogenital diaphragm: A method of repair of the completely obstructed membranous urethra. *J Urol* 1976; 115:548.

70. Koraitim M: Experience with 170 cases of posterior urethral strictures during 7 years. *J Urol* 1985; 133:408.

71. Kramer SA, Furlow WL, Barrett DM, et al: Transpubic urethroplasty in children. *J Urol* 1981; 126:767.

72. Leadbetter GW Jr: A simplified urethroplasty for strictures of the bulbous urethra. *J Urol* 1960; 83:54.

73. Leadbetter GW Jr, Leadbetter WF: Urethral strictures in male children. *J Urol* 1962; 87:409.

74. Levine JI, Crampton RS: Major abdominal injuries associated with pelvic fractures. *Surg Gynecol Obstet* 1963; 116:223.

75. Lieberman SF, Barry JM: Retreat from transpubic urethroplasty for obliterated membranous urethral strictures. *J Urol* 1982; 128:379.

76. Loffer FD, Pent D: Indications, contraindications and complications of laparoscopy. *Obstet Gynecol Surv* 1975; 30:407.

77. Looser KG, Crombie HD Jr: Pelvic fractures: An anatomic guide to severity of injury. *Am J Surg* 1976; 132:638.

78. Lucey DT, Smith MJV, Koontz WW Jr: Modern trends in the management of urologic trauma. *J Urol* 1972; 107:641.

79. Macleod DAD: Anterior urethral injuries. *Injury* 1976; 8:25.

80. Malek RS, O'Dea MJ, Kelalis PP: Management of ruptured posterior urethra in childhood. *J Urol* 1977; 117:105.

81. Malloy TR, Wein AJ, Carpiniello VL: Transpubic urethroplasty for prostatomembranous urethral disruption. *J Urol* 1980; 124:359.

82. Mattingly RF, Borkowf HI: Acute operative injury to the lower urinary tract. *Clin Obstet Gynecol* 1978; 5:123.

83. McAninch JW: Traumatic injuries to the urethra. *J Trauma* 1981; 21:291.

84. McCague EJ, Semans JH: The management of traumatic ruptures of the urethra and bladder complicating fracture of the pelvis. *J Urol* 1944; 52:36.

85. McConnell JD, Wilkerson MD, Peters PC: Rupture of the bladder. *Urol Clin North Am* 1982; 9:293.

86. McGuire EJ, Weiss RM: Scrotal flap urethroplasty for strictures of the deep urethra in infants and children. *J Urol* 1973; 110:599.

87. Merchant WC, Gibbons MD, Gonzales ET: Trauma to the bladder neck, trigone and vagina in children. *J Urol* 1984; 131:747.

88. Mitchell JP: Injuries to the urethra. *Br J Urol* 1968; 40:649.

89. Mitchell JP: Trauma to the urethra. *Injury* 1975; 7:84.

90. Morehouse DD, Belitsky P, MacKinnon KJ: Rupture of posterior urethra. *J Urol* 1972; 107:255.

91. Morehouse DD, MacKinnon KJ: Urologic injuries associated with pelvic fractures. *J Trauma* 1969; 9:479.

92. Morehouse DD, MacKinnon KJ: Posterior urethral injury: Etiology, diagnosis, initial management. *Urol Clin North Am* 1977; 4:69.

93. Morehouse DD, MacKinnon KJ: Management of prostatomembranous urethral disruption: 13-Year experience. *J Urol* 1980; 123:173.

94. Mulkey AP Jr, Witherington R: Conservative management of vesical rupture. *Urology* 1974; 4:426.

95. Myers RP, DeWeerd JH: Incidence of stricture following primary realignment of the disrupted proximal urethra. *J Urol* 1972; 107:265.

96. Netto NR Jr: The surgical repair of posterior urethral strictures by the transpubic urethroplasty or pull-through technique. *J Urol* 1985; 133:411.

97. Netto NR Jr, Ikari O, Zuppo VP: Traumatic rupture of female urethra. *Urology* 1983; 22:601.

98. Orandi A: One-stage urethroplasty. *Br J Urol* 1968; 40:717.

99. Paine D, Coombes W: Transpubic reconstruction of the urethra. *Br J Urol* 1968; 40:78.

100. Palmer JK, Benson GS, Corriere JN Jr: Diagnosis and initial mangement of urological injuries associated with 200 consecutive pelvic fractures. *J Urol* 1983; 130:712.

101. Palomar J, Polanco E, Frentz G: Rupture of the bladder following blunt trauma: A plea for routine peritoneotomy in patients with extraperitoneal rupture. *J Trauma* 1980; 20:239.

102. Parkhurst JD, Coker JE, Halverstadt DB: Traumatic avulsion of the lower urinary tract in the female child. *J Urol* 1981; 126:265.

103. Patterson DE, Barrett DM, Myers RD, et al: Primary realignment of posterior urethral injuries. *J Urol* 1983; 129:513.

104. Persky L: Childhood urethral trauma. *Urology* 1978; 9:603.

105. Pierce JM Jr: Exposure of the membranous and posterior urethra by total pubectomy. *J Urol* 1962; 88:256.

106. Pierce JM: Management of dismemberment of the prostatic membranous urethra and ensuing stricture disease. *J Urol* 1972; 107:259.

107. Pierce JM Jr: Posterior urethral stricture repair. *J Urol* 1979; 121:739.

108. Pokorny M, Pontes JE, Pierce JM Jr: Urological injuries associated with pelvic trauma. *J Urol* 1979; 121:455.

109. Pontes JE, Pierce JM Jr: Anterior urethral injuries: Four years of experience at the Detroit General Hospital. *J Urol* 1978; 120:563.

110. Prather GC, Kaiser TF: The bladder in fracture of the bony pelvis: The significance of a "tear drop bladder" as shown by cystogram. *J Urol* 1950; 63:1019.

111. Puranen J, Koivisto E: Perforation of the urinary bladder and small intestine caused by a trochanteric plate. *Acta Orthop Scand* 1978; 49:65.

112. Radge H, McInnn GF: Transpubic repair of the severed prostatomembranous urethra. *J Urol* 1969; 101:335.

113. Redman JF, O'Donnell PD: Traumatic severance of prostatomembranous urethra without associated fractured pelvis. *Urology* 1980; 16:292.

114. Reynolds BM, Balsano NA, Reynolds FX: Pelvic fractures. *J Trauma* 1973; 13:1011.

115. Richardson JR Jr, Leadbetter GW Jr: Nonoperative treatment of the ruptured bladder. *J Urol* 1975; 114:213.

116. Rieser C, Nicholas E: Rupture of the bladder: Unusual features. *J Urol* 1963; 90:53.

117. Robards VL Jr, Haglund RV, Lubin EN et al: Treatment of rupture of the bladder. *J Urol* 1976; 116:178.

118. Sandler CM, Phillips JM, Harris JD, et al: Radiology of the bladder and urethra in blunt pelvic trauma. *Radiol Clin North Am* 1981; 19:195.

119. Sandler CM, Harris JH Jr, Corriere JN Jr, et al: Posterior urethral injuries after pelvic fracture. *Am J Radiol* 1981b; 137:1233.

120. Sellett T: Iatrogenic urethral injury due to preinflation of a Foley catheter. *JAMA* 1971; 217:1548.

121. Spees EK, O'Mara C, Murphy JB, et al: Unsuspected intraperitoneal perforation of the urinary bladder as an iatrogenic disorder. *Surgery* 1981; 89:224.

122. Strong DW, Hodges CV: Transpubic urethroplasty for membranous urethral strictures. *Urology* 1977; 9:27.

123. Swinney J: Reconstruction of the urethra in the male. *Br J Urol* 1952; 24:229.

124. Swinney J: Urethroplasty: An assessment after seven years experience. *Br J Urol* 1957; 29:293.

125. Trunkey DD, Chapman MW, Lim RC, et al: Management of pelvic fractures in blunt trauma. *J Trauma* 1974; 14:912.

126. Turner-Warwick R: The repair of urethral strictures in the region of the membranous urethra. *J Urol* 1968; 100:303.

127. Turner-Warwick R: Complex traumatic posterior urethral injuries. *J Urol* 1977; 118:564.

128. Turner-Warwick R: A personal view of the management of traumatic posterior urethral strictures. *Urol Clin North Am* 1977; 4:111.

129. Waterhouse K: The surgical repair of membranous urethral strictures in children. *J Urol* 1976; 116:363.

130. Waterhouse K, Abrahams JI, Caponegro P, et al: The transpubic repair of membranous urethral strictures. *J Urol* 1974; 111:188.

131. Waterhouse K, Abrahams JI, Gruber H, et al: The transpubic approach to the lower urinary tract. *J Urol* 1973; 109:486.

132. Waterhouse K, Gross M: Trauma to the genitourinary tract: A 5 year experience with 251 cases. *J Urol* 1969; 101:241.

133. Waterhouse K, Laungani G, Patel U: The surgical repair of membranous urethral strictures: Experience with 105 consecutive cases. *J Urol* 1980; 123:500.

134. Waterhouse K, Selli C: Technique of optical internal urethrotomy. *Urology* 1978; 11:407.

135. Webster GD, Mathes GL, Selli C: Prostatomembranous urethral injuries: A review of the literature and a rational approach to their management. *J Urol* 1983; 130:898.

136. Webster GD, Selli C: Management of traumatic posterior urethral stricture by one stage perineal repair. *Surg Gynecol Obstet* 1983; 156:620.

137. Weems WL: Management of genitourinary injuries in patients with pelvic fractures. *Ann Surg* 1979; 189:717.

138. Whitehead ED, Morales PA: Complications of urethroplasty for stricture. *J Urol* 1972; 107:412.

139. Wilkinson FOW: Rupture of the posterior urethra. *Lancet* 1961; 1:1125.

140. Williams DI: Rupture of the female urethra in childhood. *Eur Urol* 1975; 1:129.

141. Witherington R, McKinney JE: An unusual case of anterior urethral injury. *J Urol* 1983; 130:564.

142. Zakin D: Perforation of the bladder by the intrauterine device. *Obstet Gynecol Surv* 1984; 39:59.

143. Zincke H, Furlow WL: Long-term results with transpubic urethroplasty. *J Urol* 1985; 133:605.

Chapter 14

The Adrenals

STUART S. HOWARDS, M.D.
ROBERT CAREY, M.D.

ANATOMY

The adrenal glands are a pair of retroperitoneal organs embedded in perirenal adipose tissue. They lie superior and medial to the upper poles of the kidneys. Their lowest extent, particularly on the left, is very close to the renal vessel; thus, care must be taken to avoid injury to the renal blood supply during adrenalectomy. CT scanning has made it clear that the adrenal glands are anterior to the kidney. The cortex has a characteristic yellow color, which makes it easy to recognize during surgery. The medulla, which is usually not visualized, is brown or red. The glands are flattened with distinct edges and measure about 5 cm × 3 cm × 1 cm. The normal human gland weighs 4–5 gm in vivo and 6 gm in death (Nelville and O'Hare, 1979). The terminal weight increase is due to ACTH release during stress. The shape and size of the glands are variable.

The cortex has a mesodermal origin, arising in utero from the dorsal mesentery near the cranial pole of the mesonephros. The gland differentiates into an outer zone, which will form the adult cortex, and a much larger inner fetal zone, which will be 80% of the gland at birth but rapidly degenerates while the outer portion grows (Fig 14–1). The net result is that the gland at birth is twice as large as it is a few weeks later. The medulla arises from ectodermal neural crest tissue, which also generates sympathetic ganglion cells. Strands of this chromaffin tissue migrate to the adrenal cortex on its medial side and become centrally located in the gland.

Each adrenal gland is supplied with blood from three arteries that are branches of the aorta, renal, and inferior phrenic arteries (Fig 14–2). These arteries form a plexus in the capsule which gives rise to cortical arteries, which in turn supply sinusoids surrounding the cords of cells in the cortex. There are no veins in the cortex. The medulla has a dual blood supply from major capsular arteries that pass directly through the cortex

without branching and from the cortical sinusoids that connect to medullary capillaries. Thus, the medullary cells are exposed to the cortical effluent with high concentrations of cortical steroids. The medulla and cortex drain into a large central medullary vein, the adrenal vein, which inserts directly into the vena cava on the right and into the left renal vein on the left.

The cortical cells have sparse, possibly adrenergic innervation (Long, 1983). The medullary cells are innervated by preganglionic sympathetic nerve fibers arising from the intermediolateral column of the lower thoracic spinal cord. These fibers travel with the splanchnic nerves and synapse with a group of medullary cells to form a functional unit. The medullary pheochromocytes are the equivalents of sympathetic ganglion cells.

The adrenal cortex contains three concentric zones: a very thin outer zona glomerulosa, a thick zona fasciculata, and a thin zona reticularis (Fig 14–3). The zona glomerulosa is composed of small (12–15 μ) columnar cells with sparse cytoplasm. The cells are packed in clusters and arcades. In man, the zona glomerulosa may be absent in some areas of the cortex. Aldosterone synthesis occurs in steps in the smooth endoplasmic reticulum and mitochondria where the essential enzymes are located (Junqueira and Carneiro, 1983, 1984).

The zona fasciculata consists of polyhedral large (20-μ) cells arranged in straight cords one to two cells thick and filled with lipid droplets. The zona reticularis consists of smaller cells with few lipid droplets. The cells are arranged in now parallel anastomosing cords (see Fig 14–3).

The adrenal medulla consists of chromaffin cells generously supplied with nerves and blood vessels. They are polyhedral and arranged in cords with close association to the vascular spaces. Epinephrine- and norepinephrine-secreting cells are distinct; each contains a specific type of granule (Junqueira and Carneiro, 1984).

Ectopic adrenal tissue can be found in many locations, particularly near the adrenal glands, kidneys, ce-

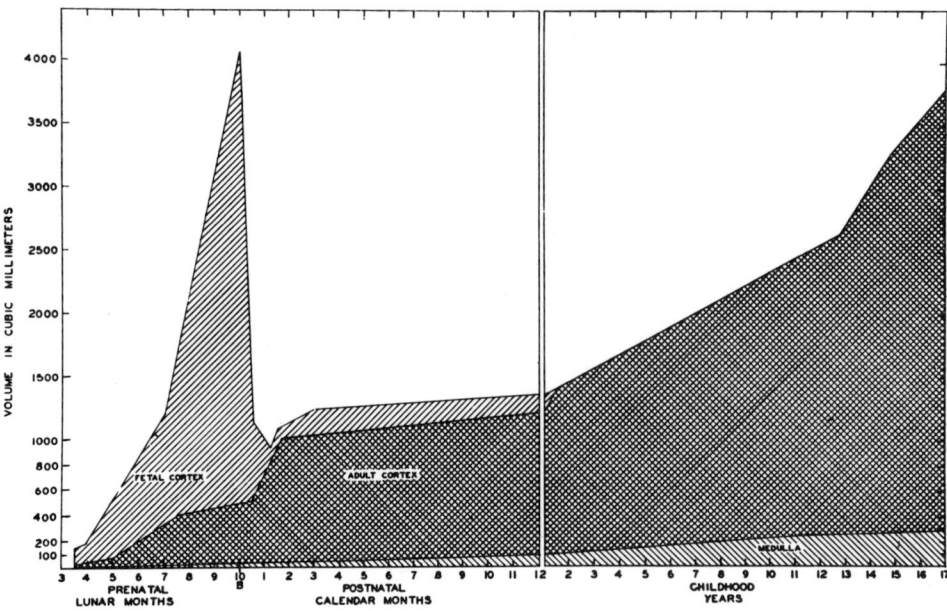

FIG 14–1.
Growth of the adrenal cortex including the fetal cortex *in utero* and after birth. There is a striking decrease in the size of the fetal cortex after birth and a gradual increase in the adult cortex with aging. (From Bethune JE: The adrenal cortex, A Scope monograph. Kalamazoo, Mich, The Upjohn Co. Used with permission.)

liac axis, or testis and spermatic cord. Unilateral congenital absence of an adrenal gland is a rare anomaly. When the kidney is absent, the adrenal is present in its normal position but is more disk-like than triangular.

PHYSIOLOGY

The adrenal cortex synthesizes cholesterol and also takes it up from the circulation. The first and rate-limiting process in the synthesis of adrenal steroids is the conversion of cholesterol to pregnenolone (Fig 14–4).

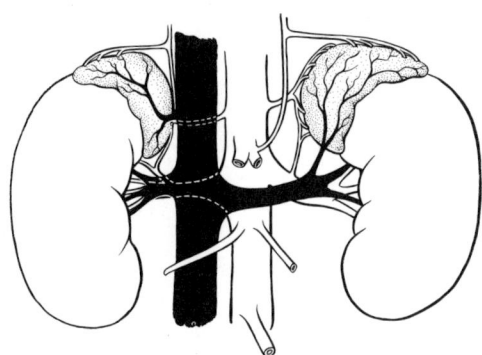

FIG 14–2.
Blood supply to the adrenal gland. The gland is supplied by three major arteries, the superior middle and inferior adrenal arteries, which are branches of the inferior phrenic artery, the aorta, and the renal artery, respectively. One major vein drains into the renal vein on the left and vena cava on the right.

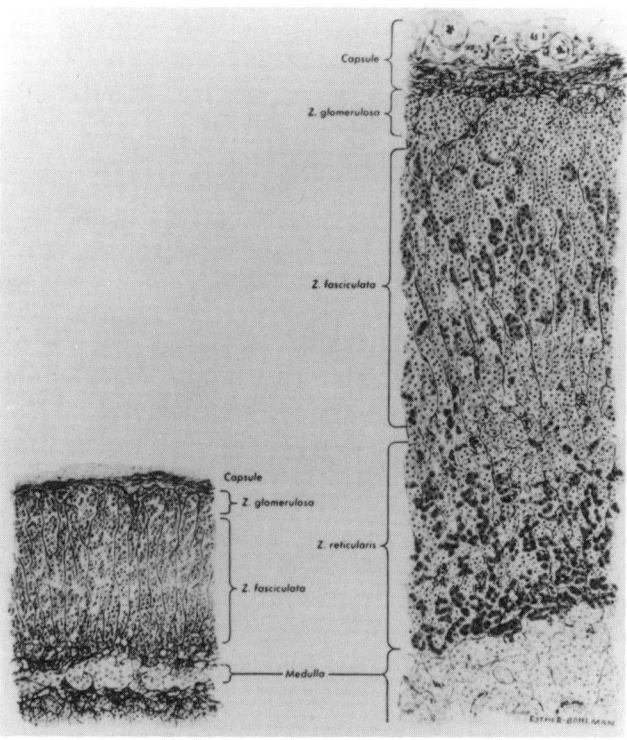

FIG 14–3.
Histology of the adrenal gland. Note the cortex has three major components, the zona glomerulosis, zona fasciculata, and zona reticularis.

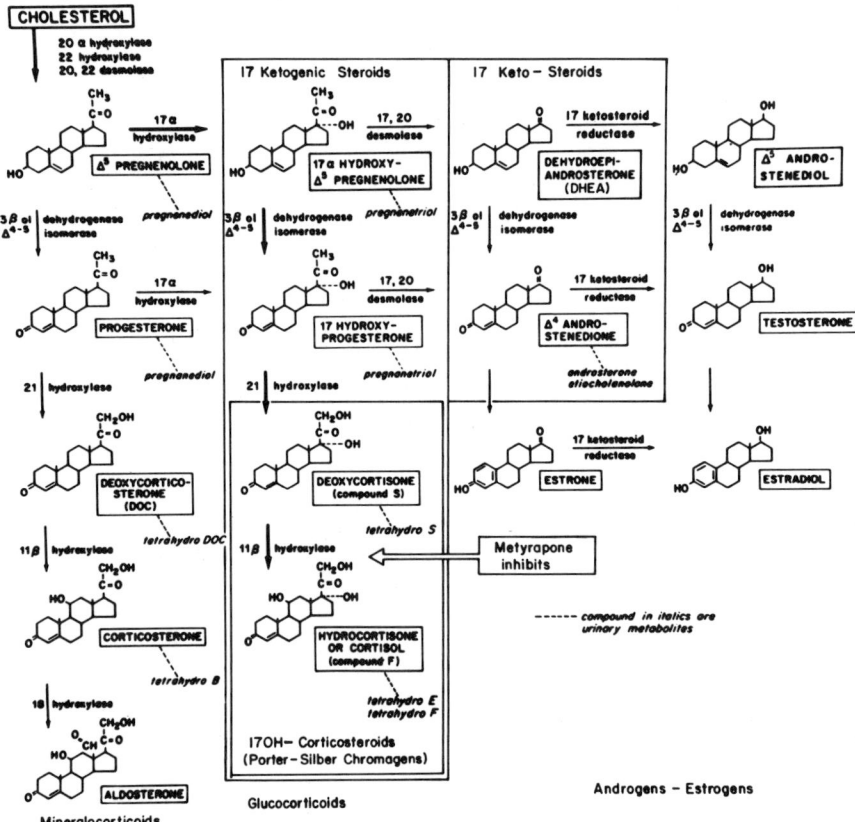

FIG 14–4.
Biosynthetic pathways for the synthesis of adrenal cortical steroids. Also the site of action of metyrapone, urinary metabolites, and the Porter Silver chromagens.

The hormones critical to life produced by the cortex are: (1) the glucocorticoids, particularly cortisol, which affect carbohydrate and protein metabolism, and (2) a mineralocorticoid, aldosterone, which regulates sodium and potassium balance. These hormones must be replaced after bilateral adrenalectomy. The adrenal cortex also synthesizes androgens and estrogens, which in normal individuals are not as important in sexual development and function as their counterparts produced by the testis and ovaries. In certain pathologic states (see below) these sex steroids have dramatic virilizing or feminizing effects.

STEROIDOGENESIS

All adrenal steroid hormones have the same steroid nucleus (Fig 14–5). The presence of a ketone at the 3 position and hydroxyl groups at the 11 and 21 positions is required for potent glucocorticoid activity. An oxygenated carbon at the 18 position results in powerful mineralocorticoid activity. Elimination of the C_{20-21} side chain and incorporation of an oxygenated carbon at the 18 position creates a potent androgen. Aromatization of the A ring generates an estrogen.

The synthesis of glucocorticoids and sex steroids occurs primarily, although not exclusively, in the zona fasciculata and reticularis, respectively. Aldosterone is synthesized exclusively in the zona glomerulosa. The synthesis pathways for production of these hormones are presented in Figure 14–4. In certain situations, al-

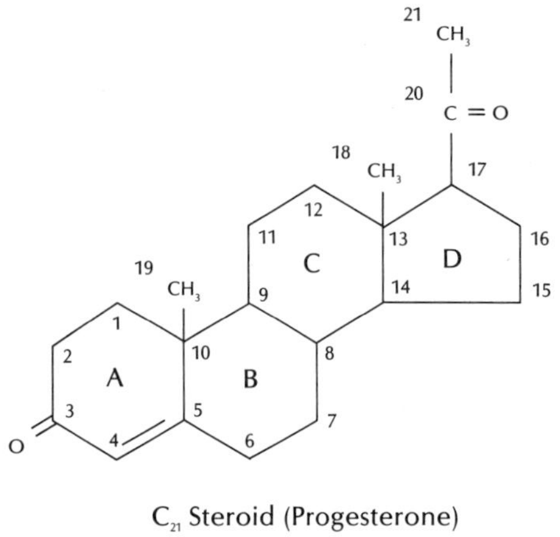

C_{21} Steroid (Progesterone)

FIG 14–5.
Common steroid nucleus.

ternative pathways may become important; for example, if cortisol synthesis is specifically blocked, corticosterone synthesis may be increased to provide the necessary glucocorticoid.

Iatrogenic interruption of these pathways may be useful in diagnosis and treatment. Metyrapone decreases the conversion of cholesterol to pregnenolone. Thus when the drug is used preoperatively the levels of all adrenal steroids may be reduced. Metyrapone also inhibits 11-hydroxylation, the last step in cortisol synthesis (see Fig 14–4), resulting in an increase in ACTH and thus cortisol percursor production in the normal individual. Therefore, metyrapone can be administered to test the hypothalamic-pituitary-ACTH axis (see below) and in the treatment of Cushing's syndrome, especially prior to surgical therapy. Aminoglutethimide inhibits the desmolase and the aromatase reactions, thus decreasing steroid synthesis. This property has been utilized in the treatment of hypersecretion of adrenal steroids, prostate cancer, and breast cancer (see below).

METABOLISM

Circulating cortisol is 75%–80% bound to an α_2 globulin, transcortin, 15% bound to albumin, and only 5%–10% unbound. The biologically active factor is the free hormone; thus, alterations in the level of transcortin can have physiologic significance. The half-life of cortisol is approximately 70 minutes. Most of the hormone is metabolized in the liver, conjugated, and excreted as glucuronides in the urine. A small fraction of the cortisol (about 50 µg) is excreted unaltered in urine. Measurement of the urinary metabolites and/or free cortisol can be used to evaluate adrenal function (see below).

Plasma aldosterone is weakly bound to a specific binding globulin and albumin. Ninety percent of the circulating aldosterone is cleared in one pass through the liver, reduced, and excreted in the urine as the 3 and 18 glucuronides. Less than 1% of the aldosterone is excreted as the free hormone. Because of its rapid metabolism, the half-life of plasma aldosterone is only 20 minutes.

Adrenal androgens are excreted in the urine as DHEA-S and two reduced isomers, androsterone and etiocholanolone (see Fig 14–4). These compounds make most of the urinary 17-ketosteroids. Two thirds of the 17-ketosteroids come from the adrenal and one third from the gonad in normal individuals. Elevation of the urinary 17-ketosteroids almost always indicates adrenal hyperfunction.

REGULATION

ACTH is the regulator of glucocorticoid secretion and is also the primary determinate of the secretion of adre-

nal sex steroids. It is derived from a 31,000-dalton glycoprotein, pro-opiomelancortin, which is cleaved in corticotropic cells into ACTH, B-lipotropin, and a glycopeptide (Wilson and Foster, 1985). ACTH is a single-chain polypeptide containing 39 amino acids. The active component of the molecule is the initial segment of 23 amino acids. The hormone is secreted in irregular bursts throughout the day, but the most active secretion occurs in the early morning, thus causing the diurnal secretion of cortisol. Cortisol secretion increases within minutes of an elevation in plasma ACTH levels. ACTH binds to an adrenal plasma membrane receptor and activates adenylate cyclase, which in turn raises tissue cyclic AMP (cAMP) levels. The cAMP activates critical protein kinases, which effect the phosphorylation of proteins that increase steroidogenesis.

The secretion of ACTH is controlled by corticotropin-releasing factor (CRF), a protein with 41 amino acid residues, which is synthesized in the median eminence of the hypothalamus and travels to the pituitary via the portal-hypophyseal vessels (Fig 14–6). CRF secretion in turn is regulated by three main factors—circadian rhythm, glucocorticoid feedback, and stress. Stress increases the secretion of CRF and ACTH, whereas there is a negative feedback relationship between plasma glucocorticoid levels and ACTH secretion. This inhibition of ACTH secretion is probably mediated at both the hypothalamus and the pituitary.

Three clinically relevant points related to ACTH secretion are: (1) some neoplasms secrete ACTH-like substances that can stimulate the adrenal to release glucocorticoids and thus cause Cushing's syndrome; (2) after treatment with large doses of exogenous glucocorticoids, not only is the adrenal gland unresponsive, but

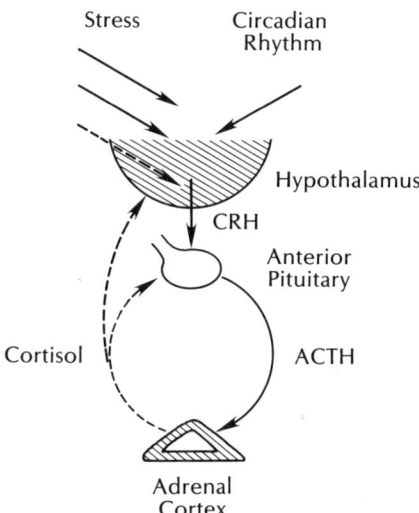

FIG 14–6.
Hypothalamic pituitary adrenal axis including the long- and short-loop feedback circuits.

the hypothalamic-pituitary axis may be unable to secrete normal quantities of ACTH; and (3) ACTH is not required for aldosterone synthesis and secretion; therefore, a patient with hypopituitarism does not require mineralocorticoid replacement.

ACTIONS OF GLUCOCORTICOIDS

Glucocorticoids are essential for life. An adrenalectomized man will die even if provided with mineralocorticoids. Adrenal steroids and, indeed, all steroid hormones easily cross the cell membrane to bind with cytoplasmic receptor proteins that are present in target tissues. The steroid may be metabolized within the cell to a more or less active form. The steroid-receptor complex migrates to the cell nucleus, where it attaches to a specific group of genes. This causes the production of new RNA, which in turn effects the synthesis of proteins that serve as structural building blocks of enzymes that regulate cellular function in various tissues.

Glucocorticoids are so named because they have major effects on carbohydrate metabolism including the promotion of liver glycogen deposits and gluconeogenesis. In the starving patient, they allow survival by enhancing proteolysis, preventing death from hypoglycemia. They also have an anti-insulin effect. Thus they are diabetogenic, causing hyperglycemia during stress or when present in pharmacologic quantities.

Glucocorticoids have many permissive actions in that they facilitate processes that they do not initiate. Several additional effects of glucocorticoids are listed in Table 14–1 along with their clinical implications.

ACTIONS OF ADRENAL ANDROGENS AND ESTROGENS

The mechanism of action of the major adrenal androgens DHEA, DHEA-S, and androstenedione is similar to that described above for the glucocorticoids. These compounds are very weak androgens that have little effect in physiologic quantities. However, in pathologic states they may virilize a female fetus (adrenogenital syndrome—see Chapter 58) causing pseudohermaphroditism, and virilize either prepubertal children or adult females with Cushing's syndrome. The effects of excess adrenal androgens are not clinically evident in the adult male, thus delaying the diagnosis of Cushing's syndrome in some men. Excess adrenal estrogens may cause breast enlargement in children and men.

REGULATION OF GLOMERULOSA FUNCTION

The primary regulator of aldosterone secretion is the renin-angiotensin system. *Renin* is an enzyme synthesized in the juxtaglomerular apparatus of the nephron. When released into the circulation, renin cleaves *renin*

TABLE 14–1.

Effects and Implications of Glucocorticoids

EFFECTS OF GLUCOCORTICOIDS	CLINICAL IMPLICATIONS
Enhance skeletal and cardiac muscle contraction	Absence results in weakness
Cause protein catabolism	Excess results in wastage and weakness
Inhibit bone formation	Excess decreases bone mass
Inhibit collagen synthesis	Excess causes thin skin and fragile capillaries
Increase vascular contractility and decrease permeability	Absence makes it difficult to maintain blood pressure
Have anti-inflammatory activity	Exogenous steroid useful in treating inflammatory diseases
Have anti-immune system activity	Exogenous steroids useful in treating transplantation and various immune diseases
Maintain normal glomerular filtration	Absence reduces glomerular filtration

substrate, a globulin secreted from the liver, releasing the decapeptide *angiotensin I*. Angiotensin I is hydrolyzed to *angiotensin II*, an octapeptide, by *converting enzyme*, which is found primarily in the lung. Angiotensin II is a potent stimulator of aldosterone secretion. It is rapidly destroyed in the plasma by *angiotensinases*.

Renin secretion is regulated by a complex intrarenal mechanism (see Chapter 25). Decreased perfusion pressure in the renal artery and decreased chloride absorption at the macula densa cause increased renin release and thus ultimately increased circulating levels of aldosterone. Angiotensin II and aldosterone can inhibit renin secretion through short and long loop feedback systems, respectively. Catecholamines may increase renin release, but dopamine inhibits aldosterone secretion in sodium-depleted individuals. ACTH increases the sensitivity of the zona glomerulosa to angiotensin II and stimulates aldosterone secretion, but this latter effect is not long lasting. Potassium also causes an increase in aldosterone secretion. Thus, aldosterone secretion is increased in the following clinical settings: (1) stress; (2) hemorrhage; (3) sodium depletion; (4) dehydration; (5) hyperkalemia; (6) congestive heart failure; (7) hepatic cirrhosis; (8) nephrotic syndrome, (9) estrogen administration, (10) renal artery stenosis.

EFFECTS OF MINERALOCORTICOIDS

Mineralocorticoids are steroid hormones that effect ion transport in the epithelial cells of the kidney, GI tract, sweat glands, and salivary glands, causing sodium absorption and loss of potassium. Their mechanism of action is to increase production of a yet uncharacterized protein. Since protein synthesis requires time, the effect of mineralocorticoids is not seen until one or two hours after the tissue is exposed to the steroid. The most important physiologic effects of these compounds are to increase reabsorption of sodium and secretion of potassium and hydrogen ions in the distal tubule of the kidney. Mineralocorticoids do not cause excessive potassium excretion in sodium-depleted subjects. This fact is clinically important, since sodium-depleted patients with hyperaldosteronism do not demonstrate a significant kaluresis. Excess aldosterone causes weight gain, increased blood pressure, hypokalemia, and mild metabolic acidosis as a result of the physiologic effects described above. However, normal subjects "escape" from the effects of excessive mineralocorticoids after about 14 days, and eventually blood volume returns to baseline levels. The mechanisms of the "escape phenomenon," which has been intensely studied for years, remain uncertain. It appears that the rise in blood pressure may produce a pressure natriuresis, and also the arterial natriuretic factor (ANF) may play a role. Mineralocorticoid deficiency results in sodium loss, and if uncorrected, death from hypovolemic shock. The naturally occurring mineralocorticoids are, in order of potency, aldosterone, deoxycorticosterone (DOC), 18-hydroxy-DOC, corticosterone, and cortisol. Table 14–2 lists the relative mineralocorticoid and glucocorticoid activity of several naturally occurring and synthetic steroids.

THE ADRENAL MEDULLA

The adrenal medulla secretes catecholamines into the circulation. The primary compound secreted by the medulla is epinephrine. Small quantities of norepinephrine and trace amounts of dopamine are also released from the gland. The adrenal medulla usually, but not invariably, acts in concert with the rest of the sympathetic nervous system. The hormones from the medulla are not essential for life.

SYNTHESIS AND METABOLISM OF CATECHOLAMINES

Catecholamines are synthesized from tyrosine in the adrenal medulla. The pathway of catecholamine synthesis is illustrated in Figure 14–7. In approximately 15% of the granules in the normal medulla, the last step in biosynthesis is the conversion of dopamine to norepinephrine, thus accounting for the secretion of norepinephrine by the normal adrenal and by tumors of the gland. The majority of the granules contain phenylethanolamine-N-methyltransferase (PNMT), which converts norepinephrine to epinephrine, the major hormone of the adrenal medulla. In contrast, sympathetic nerves and other extraadrenal chromaffin tissue do not contain PNMT and thus do not secrete epinephrine.

TABLE 14–2.
Relative Activities of Glucocorticoid and Mineralocorticoid

STEROID	GLUCOCORTICOID ACTIVITY	MINERALOCORTICOID ACTIVITY
Cortisol	1.0	1.0
Corticosterone	0.3	15.0
Aldosterone	0.3	3,000.0
Deoxycorticosterone	0.2	100.0
Cortisone	0.7	1.0
Prednisolone	4.0	0.8
9-α Fluorocortisol	10.0	125.0
Dexamethasone	25.0	0

This point is clinically useful when attempting to localize a catecholamine-secreting tumor.

Norepinephrine (NE) and epinephrine (E) have very short half-lives in plasma, ranging from 1 to 3 minutes. They are degraded by two principal enzymes, catechol-O-methyl transferase (CCMT) and monoamine oxidase (MAO). Figure 14–8 illustrates the major metabolic pathways for circulating catecholamines. CCMT converts norepinephrine and epinephrine, respectively, to normetanephrine (NMN) and metanephrine (MN). Determination of NMN or MN produces vanillylmandelic acid (VMA), a major metabolic product of catecholamine degradation. Less than 5% of the circulating NE and E are secreted intact in the urine. In a normal individual, NE from sympathetic nerve terminals makes up most of the intact urinary catecholamines, while E

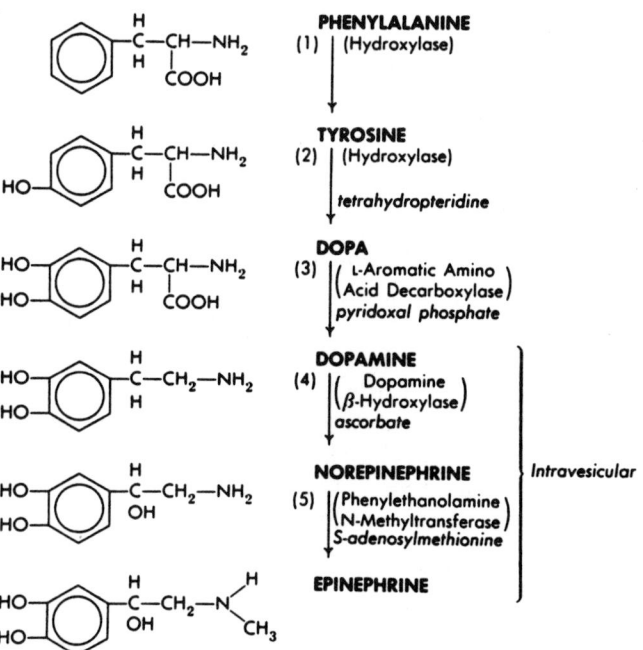

FIG 14–7.
Biochemical pathway for the synthesis of norepinephrine and epinephrine.

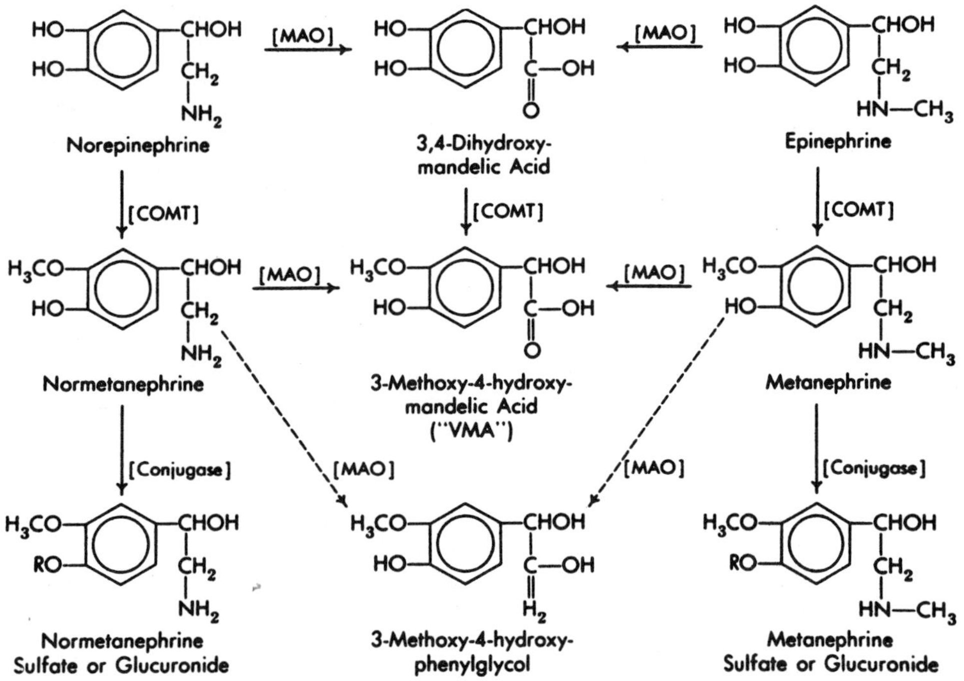

FIG 14–8.
Metabolic pathways for the degradation of norepinephrine and epinephrine. The urinary metabolites are important in the diagnostic evaluation of patients with suspected pheochromocytomas.

composes 20% of the total. The vast majority of the catecholamines are secreted in the urine as NMN, MN, VMA, and other metabolites shown in Figure 14–6. Measurement of these various products is important in the diagnosis of pheochromocytoma. The average daily excretion rates in normal men are listed here for the compounds measured in clinical laboratories: (1) NE and E, 50 μg; (2) NMN and MN, 300 μg; and (3) VMA, 400 μg.

REGULATION OF ADRENAL MEDULLARY CATECHOLAMINE SYNTHESIS AND SECRETION

Stimulation of the sympathetic nervous system during stress caused by stimuli such as fear, pain, or hemorrhage results in increased secretion of catecholamines from the adrenal medulla. Hypoglycemia is also a potent stimulator of adrenal catecholamine secretion. In most situations, the ratio of NE to E remains stable, although exceptions do occur. The basal plasma levels of epinephrine are 25–50 pg/ml. A total of approximately 150 μg is secreted daily. Norepinephrine release at sympathetic nerve endings may spill over into the intracellular fluid or be retaken up by the tissues. The plasma level of norepinephrine is determined by a balance between the amount released into the circulation and the quantities metabolized and retaken up by the tissues.

ACTIONS OF CATECHOLAMINES

Epinephrine and norepinephrine are potent hormones which activate α_1, α_2, β_1, and β_2 plasma membrane receptors. Epinephrine primarily stimulates β-receptors but also has some effect on α-receptors, particularly when the plasma levels of the hormone are high. Conversely, norepinephrine primarily affects α-receptors. Because these catecholamines can have alpha and beta action, the effects of IV-administered hormones vary both quantitatively and qualitatively with the dose.

For example, at low doses epinephrine causes vasodilatation (a beta effect), which may result in hypotension, whereas at higher doses there is a net increase in vascular resistance causing hypertension with a larger increment in systolic than diastolic pressure, resulting in an increased pulse pressure. Both compounds have a positive inotropic effect on the heart and therefore increase cardiac output. Intravenously administered epinephrine causes tachycardia, whereas norepinephrine results in bradycardia because of the vagal reflex response to the increase in blood pressure.

Catecholamines help restore plasma glucose during exercise or stress by stimulating glycogenolysis in the liver, inhibiting insulin secretion, and increasing glucagon secretion. They also facilitate the reuse of lactate by exercising muscle and increase in the release of free fatty acids into the circulation.

Excessive concentrations of catecholamines from exogenous sources or endogenous secretion in patients with pheochromocytoma may cause symptoms relating to the above-described effects (see "Pheochromocytoma"). The symptoms in patients with pheochromocytoma are due to epinephrine and norepinephrine. Norepinephrine elevates the blood pressure and thus may cause headaches.

CUSHING'S SYNDROME

INTRODUCTION

Cushing's syndrome is a complex of symptoms and signs caused by excess circulating glucocorticoids. The term is used to describe all patients with the clinical syndrome regardless of the etiology. It is important to understand that Cushing's *disease* refers to the form of Cushing's *syndrome* caused by pituitary hypersecretion of ACTH. The most common etiology of Cushing's syndrome is exogenously administered glucocorticoids. Twenty-five percent of endogenous Cushing's syndrome is due to primary adrenal diseases, i.e., adenoma or carcinomas. Seventy-five percent of the endogenous disease is excessive ACTH secretion, usually from the pituitary gland but occasionally from an ectopic source. The vast majority of patients with pituitary hypersecretion of ACTH (Cushing's disease) have microadenomas, although they may not be obvious during surgery. It is easier to understand the evaluation and treatment of patients with Cushing's syndrome if one keeps in mind that the *first* task of the clinician is to determine whether or not the patient has Cushing's syndrome. The *second* goal is to determine the etiology of the syndrome, and the *final* assignment is to formulate a treatment plan.

SYMPTOMS AND SIGNS

Cushing's syndrome occurs in men, women, and children of all races but is most commonly diagnosed in women between the ages of 20 and 60 years. It is usually characterized by plethoric "moon" facies and central or "buffalo" obesity in the nuchal, truncal, and girdle areas (Fig 14–9,A and B). Protuberance of superclavicular fat pads is the cardinal physical finding that distinguishes Cushing's syndrome from obesity. Serial photographs of the patient are very helpful in making the diagnosis, because these pictures always document a shift in fat distribution, even if the patient is not obese. The protein wasting secondary to excess glucocorticoids causes easy bruising and thick skin, which results in pink or purple striae. Muscle wasting may cause severe weakness. The weakness is almost al-

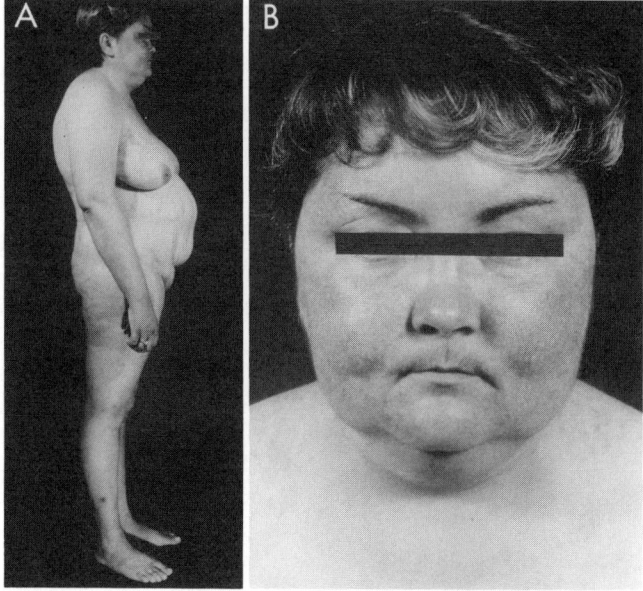

FIG 14–9.
Typical appearance of a patient with Cushing's syndrome. **A,** the central obesity striae, buffalo hump, and moon facies. **B,** the protruberance of the supraclavicular fat pads, hirsutism, facial pigmentation, and the fact that one cannot see the ears, which is typical of patients with Cushing's syndrome. There are also acne fold lesions on the skin.

ways proximal and can be brought out by asking the patient to do deep knee bends. Emotional symptoms and headaches are common. Adrenal androgens frequently cause hirsutism in women and prepubertal boys (Lee et al., 1985). Oligomenorrhea in women and acne are also common symptoms related to adrenal androgens. Fifteen percent of the patients have urinary stones due to hypercalciuria. The full-fledged syndrome is easy to recognize, but the differential diagnosis may be difficult, particularly in men and also in women with hirsutism unrelated to Cushing's syndrome. The most prevalent symptoms and signs of the syndrome are listed in Table 14–3.

Common physical findings in patients with Cushing's syndrome are hypertension and edema secondary to the mineralocorticoid activity of the adrenal steroids and, as mentioned above, plethoric moon facies, central obesity, striae, acne, and hirsutism.

DIFFERENTIAL DIAGNOSIS

Cushing's syndrome is caused by the actions of glucocorticoids. Therefore, anything that causes excess circulating levels of these hormones can evoke the syndrome. The most common etiology is iatrogenic administration of glucocorticoids. Pituitary Cushing's syndrome or Cushing's disease accounts for the majority of the noniatrogenic cases. Although approximately

TABLE 14–3.

Symptoms and Signs of Cushing's Syndrome in Order of Frequency

SYMPTOMS	SIGNS
Central obesity	Central obesity
Hirsutism	Hypertension
Oligomenorrhea	Hirsutism
Purple striae	Purple striae
Plethoric facies	Plethoric facies
Easy bruisability	Acne
Personality change	Edema
Acne	Muscle weakness
Edema	
Muscle weakness	
Poor wound healing	
Backache	
Polyuria	
Polydipsia	
Impotence	
Growth arrest (children)	

95% of the patients with pituitary Cushing's syndrome have detectable pituitary tumors, it is not clear whether these patients suffer from primary hypersecretion of ACTH or the pituitary is stimulated by excessive CRF from the hypothalamus. Ectopic secretion of ACTH or CRF (Carey et al., 1984) from tumors can also cause the syndrome. Finally, primary adrenal adenomas or carcinomas may secrete enough hormone to produce the syndrome. Adrenal tumors tend to be autonomous in their secretion of glucocorticoids, a characteristic that assists in the differential diagnosis of Cushing's syndrome. Most of these tumors are unilateral, but bilateral tumors do occur. The majority of children with Cushing's syndrome have adrenal neoplasms.

In patients under 15 years old, adrenal carcinoma is the most common cause of Cushing's syndrome. Since most patients with adrenal cortical carcinoma present with endocrine symptoms or a retroperitoneum, the diagnosis is rarely suspected because of distant metastasis, although pulmonary and liver metastases are not uncommon, and skeletal, brain, pleura, and mediastinal metastases do occur (Hutter et al., 1966). Patients with ectopic or pituitary ACTH-dependent Cushing's syndrome may have hyperpigmentation. Galactorrhea occurs occasionally and only in individuals with pituitary hypersecretion of ACTH. Patients with adrenal carcinoma tend to present with hirsutism and virilism due to androgenic adrenal steroids, and individuals with ectopic ACTH tend to have severe manifestations of the syndrome due to the high levels of ACTH. The presence of a decreased serum potassium in patients with this finding is highly suggestive of ectopic ACTH secretion, and the diagnosis should be pursued even if there is also suspicion of a pituitary etiology.

The diagnosis of Cushing's syndrome is most frequently entertained when a physician is consulted by an overweight woman concerned about hirsutism. The vast majority of these patients suffer from excessive secretion of androgens from the ovary (Kirschner et al., 1976) rather than hyperadrenocorticism. However, late onset adrenal hyperplasia has been reported in 24 of 400 women with hirsutism (Kutten et al., 1985).

LABORATORY DIAGNOSIS OF CUSHING'S SYNDROME

There are two basic steps in the laboratory evaluation of individuals with suspected Cushing's syndrome: first, whether they have the syndrome, and second, the etiology.

Many tests have been recommended to make the diagnosis of Cushing's syndrome. We prefer to screen patients with a determination of the free cortisol in a 24-hour urine specimen. In our laboratory, the normal value is lower than 80 μ/24 hr. Values vary with age and from laboratory to laboratory. Following are examples of such results (Frants, 1973):

URINARY FREE CORTISOL	ADULTS (N-11)	CHILDREN (N-30)
μg/24 hr	96 ± 35	41 ± 23
μg/gm creatinine/24 hr	68 ± 15	75 ± 18

A normal finding rules out Cushing's syndrome unless the clinical level of suspicion is very high. If the urinary free cortisol is elevated, we then do a low-dose dexamethasone test. Dexamethasone is a biologically active glucocorticoid that suppresses the secretion of ACTH and thus endogenous adrenal glucocorticoids but does not affect the measurement of the endogenous hormones in the serum and their metabolism in the urine. The patient is given dexamethasone 0.5 mg, orally, every six hours, four times daily. Then at 9 A.M. following the last 3 A.M. dosage, a blood sample is obtained. The normal individual will have a suppressed serum cortisol of less than 2–5 μg/dl. Twenty-four-hour urine levels of free cortisol should also be depressed to less than 20 μg/L on the second and third days. Porter-Silber 17-OH-corticoids and ketogenic steroids in the urine should be less than 2.0 and 5.0 mg/gm of creatinine, respectively. The best criterion to use for the dexamethasone test is the value of 17-OH corticoids/gm of creatinine. The creatinine corrects for surface area. If the serum and urinary values are normal, the patient does not have Cushing's syndrome; if elevated, Cushing's syndrome is likely. However, there are exceptions, including endogenous depression and alcoholic pseudo-Cushing's syndrome.

In other institutions, the A.M. and P.M. serum cortisol levels and/or overnight dexamethasone test are used to make the diagnosis of Cushing's syndrome. In the normal individual, there is a diurnal rhythm in serum cortisol concentration, with a peak of 8–25 µg/dl at 6–9 A.M. and a nadir of less than 7–10 µg/dl in the afternoon (Fig 14–10). Therefore, the test is done by drawing blood samples at 8 A.M. and 4 P.M. Normal persons should conform to the above-quoted values and display at least a 50% fall between A.M. and P.M. determinations. This diurnal rhythm is not present in very young patients (see Fig 14–10). Patients with Cushing's syndrome usually have elevated values, particularly in the afternoon, and also do not exhibit the normal diurnal rhythm. The overnight dexamethasone suppression test is done by giving the patient 1.0 mg of dexamethasone at 11 P.M. and obtaining a blood sample the next morning at 8 A.M. The normal individual will suppress his serum cortisol to less than 5.0 µg/dl, whereas the patient with Cushing's syndrome will have a serum cortisol greater than 5.0 µg/dl (usually greater than 20 µg/dl).

LABORATORY DETERMINATIONS OF THE ETIOLOGY OF CUSHING'S SYNDROME

There are many tests available to facilitate the differential diagnosis of Cushing's syndrome. None of these tests is always correct, and the diagnosis can be very difficult. Disagreement among test results is the rule rather than the exception. Among the many tests available, we prefer three: plasma ACTH, high-dose dexamethasone suppression, and the metyrapone test. The normal plasma ACTH level in an adult is 10–80 pg/ml

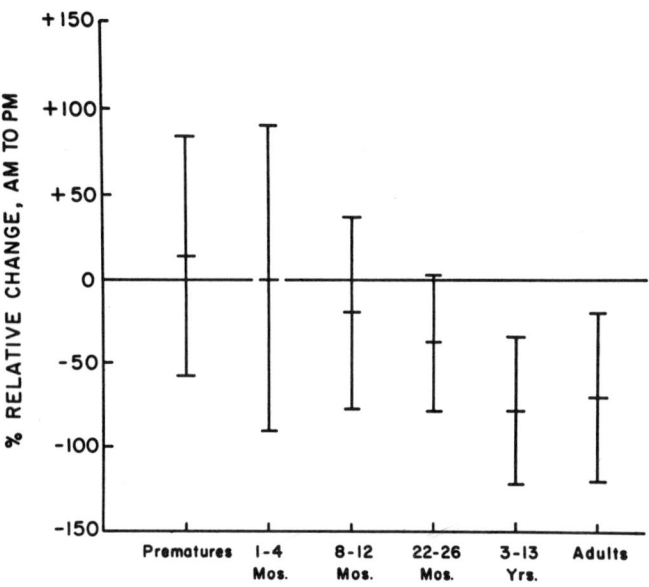

FIG 14–10.
Diurnal rhythm in serum cortisol levels and its variation with age.

in the morning and at least 50% less in the evening (Besser, 1973; Ratcliffe, 1972). The A.M. plasma ACTH is suppressed to undetectable levels in patients with adrenal adenomas or cancers. In patients with ectopic secretion of ACTH, the value is elevated (200–1,000 pg/ml), and in patients with Cushing's disease it is normal or high (40–100 pg/ml) but elevated inappropriately for the level of cortisol. The high-dose dexamethasone test is done in the same manner as the low-dose test described above, except that each dose of the drug is 2.0 mg. Patients with ectopic ACTH secretion or adrenal tumors do not suppress but those with Cushing's disease do.

Metyrapone inhibits the enzyme 11-β-hydroxylase, thus preventing 11-β-hydroxylation during steroidogenesis (see Fig 14–4). This eliminates the production of cortisol, which in turn increases the secretion of ACTH (since the negative feedback has been removed) and the secretion of adrenal steroids such as 11-deoxycortisol. In patients with ectopic ACTH secretion or adrenal tumors, metyrapone has no effect. Patients with Cushing's disease have an exaggerated response. There is a potential danger that metyrapone will cause adrenal insufficiency and adrenal crisis.

LOCALIZATION OF ADRENAL CAUSES OF CUSHING'S SYNDROME

The techniques that are currently utilized to localize adrenal tumors are listed in Table 14–4.

Intravenous pyelography, arteriography, and adrenal vein venography are no longer recommended for localizing adrenal lesions in patients with Cushing's syndrome.

Most tumors of the adrenal causing Cushing's syndrome are larger than 2 cm and therefore are easily visualized with CT scanning (Figs 14–11 and 14–12). Therefore, the CT scan is the single most useful radiologic test (Adams et al., 1983; Scott et al., 1985). It should be noted that the adrenal gland on the affected side in patients with renal agenesis or inferior ectopy is a paraspinal disk-shaped organ that has a linear appearance on CT scan (Kenney, 1985). An IVP, especially with tomography, may also reveal an adrenal mass, although it is less sensitive. Adrenal tumors typically displace the upper pole of the kidney laterally (Fig 14–13), changing the axis of the kidney and the collecting system. They rarely displace the kidney caudally without

TABLE 14–4.

Localization Test for Cushing's Syndrome

1. CT or MRI imaging
2. Scintiscan: 6-β-iodomethyl norcholesterol (NP-59)
3. Ultrasonography

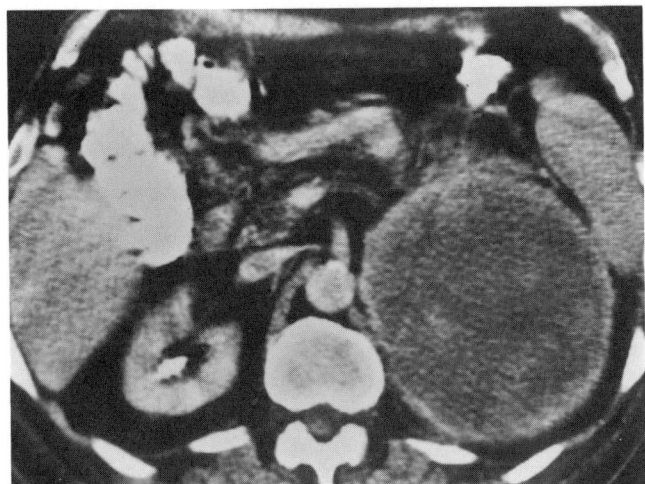

FIG 14–11.
CT scan of a patient with a large adrenal hematoma. CT scan of a patient with a tumor causing Cushing's syndrome might be quite similar in configuration, although the intratumor densities would vary.

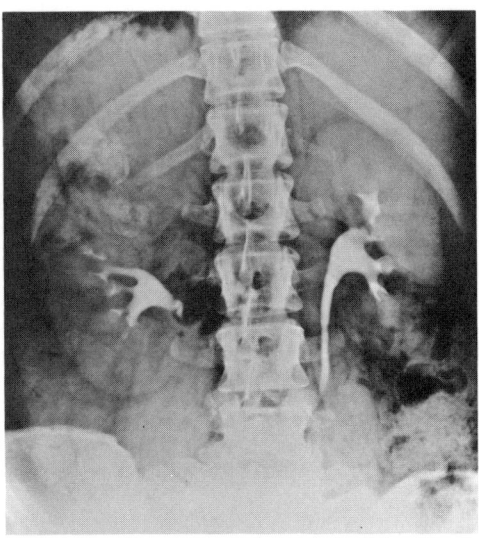

FIG 14–13.
IVP of a patient with an adrenal mass. Note the renal axis and the axis of the collecting system are deviated with the upper pole pushed outward. This is quite typical of an adrenal tumor. The tumors of the adrenal usually push the upper pole outward rather than pushing the kidney down.

shifting the axis. Sonography can be a useful screening technique, particularly for large tumors, and is very useful in determining whether a retroperitoneal mass extends from or into the kidney or is separated from the kidney and/or the liver. If it is unclear whether a retroperitoneal mass is of adrenal or renal origin, arteriography may be useful, since it is often pathognomonic of renal cell carcinoma and may also be useful in planning surgery. However, we do not routinely perform arteriograms in patients with suspected adrenal or renal tumors.

Venography with catheterization of the adrenal veins

is an extremely precise technique for making the diagnosis of endocrinologically active adrenal tumors and localization of same. However, it is difficult to catheterize the adrenal veins, particularly on the right, and even in experienced hands this method may fail. Also there is danger of damaging the adrenal gland if undue pressure is exerted in the adrenal vein. Adrenal masses can also be identified with scintillation scanning techniques using NP-59 (Fig 14–14). We have found this technique useful in the evaluation of patients with Conn's syndrome. In the future, MRI imaging will probably play an important role in the evaluation of adrenal masses, giving functional as well as anatomical information.

TREATMENT OF CUSHING'S SYNDROME

Ectopic ACTH

Cushing's syndrome secondary to ectopic ACTH carries an ominous prognosis. A malignant tumor that secretes an ACTH-like compound is usually the cause of this syndrome. Many different tumors have been associated with the ectopic ACTH syndrome. The most common in order of frequency are: (1) pulmonary oat cell carcinoma; (2) carcinoid tumor; (3) epithelial carcinoma of the thymus; (4) islet cell tumor of the pancreas; (5) carcinoma of the thyroid; and (6) pheochromocytoma (Davies et al., 1982). Treatment of these patients obviously must be individualized. Very often the primary disease is too advanced for anything but palliative therapy. However, if the cancer has been diagnosed early

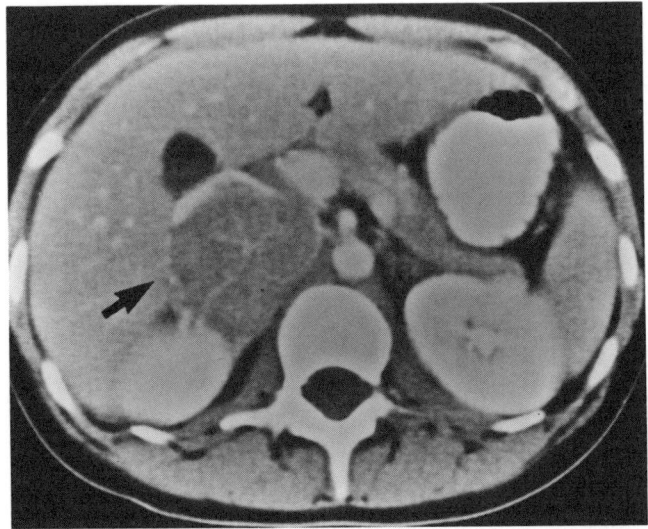

FIG 14–12.
CT scan of a patient with a pheochromocytoma *(arrow)*.

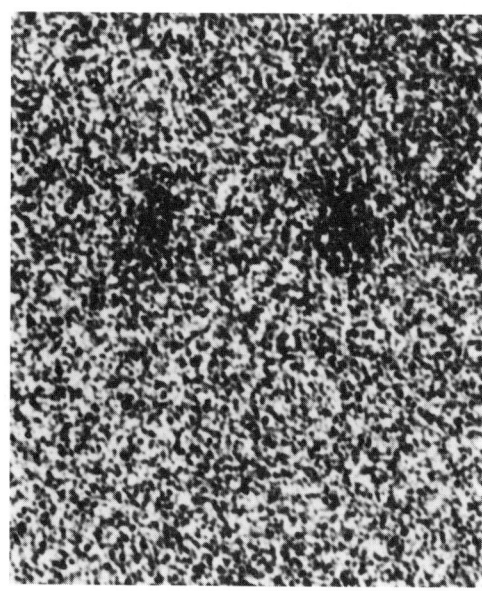

FIG 14–14.
IP59 scintillation scan of a patient with bilateral adrenal hyperplasia. Note that both adrenal glands are seen on this scan and are of about equal density. If the patient had an adrenal adenoma, the side with the adenoma would be highlighted, whereas the other side would not be visualized.

or if the tumor is benign, excision of the primary tumor may be curative (Davies et al., 1982). Even in advanced cases, the endocrine symptoms may respond to one or another of the medical or surgical treatments discussed below.

Cushing's Disease

Transsphenoidal hypophysectomy, pituitary irradiation, bilateral adrenalectomy, and medical therapy are all used to treat Cushing's disease. The current treatment of choice in most centers is transsphenoidal removal of the microadenoma (Boggan et al., 1983; Styne et al., 1984; Ludecke and Niedworok, 1985). The success rate has been reported between 70% and 95%. This treatment is less morbid than bilateral adrenalectomy and avoids the complication of Nelson's syndrome (see below). Pituitary irradiation (cobalt 60) with 4,000–5,000 rad has a reasonable success rate in children, approximately 80%, but only about a 20% cure rate in adults. Pituitary irradiation with a cyclotron (proton beam) may have a higher cure rate in adults, but this technique is available in only a few locations in the United States. We reserve bilateral adrenalectomy for those patients who have had failed pituitary surgery. The technique of adrenalectomy is discussed below. Bilateral adrenalectomy may be complicated in 10%–20% of the cases by rapid postoperative growth of pituitary tumors and hyperpigmentation (Nelson's syndrome).

Preoperative pituitary irradiation decreases the incidence of Nelson's syndrome.

Medical treatment with cyproheptadine 6 mg, orally, four times daily, has been reported to cause remission in 60%–65% of patients with Cushing's disease after 6–8 weeks. Mitotane, o,p-DDD, 6.0 gm/day, or aminoglutethimide has also been used to treat patients with Cushing's disease, but because of significant side effects we do not recommend these compounds except for inoperable cases. We prefer metyrapone, 1.0 gm/day. This drug has a very short half-life and must be given every four hours. It is monitored by following serum and not urinary cortisol. Hypertension may result from accumulation of DOC. This can be treated with spironolactone. In general, in Cushing's disease medical therapy is used only when indicated to prepare patients for surgical treatment. The major use of medical treatment is in palliation of patients with Cushing's syndrome due to malignancies.

Adrenal Tumors

The treatment of Cushing's syndrome secondary to adrenal tumors is surgical removal unless the tumor is unresectable. Bilateral adrenalectomy is the treatment of choice for Cushing's disease that has not responded to pituitary surgery. Adrenal excision has also been used to treat breast and prostatic cancer. Preoperative preparation requires specific measures to correct metabolic abnormalities caused by endocrinologically active tumors such as those associated with Cushing's syndrome, primary aldosteronism, and pheochromocytoma. The details of preoperative management of patients with the latter two diseases will be discussed below. Patients with Cushing's syndrome may be prepared for surgery by treatment with metyrapone (250–500 mg, orally, every four hours) while awake. To the extent that these drugs reverse Cushing's syndrome, they will decrease the operative morbidity and the mortality. Malignant adrenal tumors causing Cushing's syndrome in the adult do not carry a good prognosis. In the M. D. Anderson series, 13/18 (72%) had recurrent disease (Shirkhoda, 1985). Eight of the 13 had local recurrence. The prognosis is better in children.

BILATERAL POSTERIOR ADRENALECTOMY.—The bilateral posterior approach has several advantages. The glands can be exposed simultaneously, the peritoneal cavity need not be entered, obesity presents less of a problem, and the postoperative morbidity is decreased. The major disadvantages of this approach are that exposure is limited and the abdominal cavity cannot be thoroughly explored. The patient is placed in the prone position with the table flexed approximately 35° (Fig 14–15). In addition, the kidney rest or a sandbag may

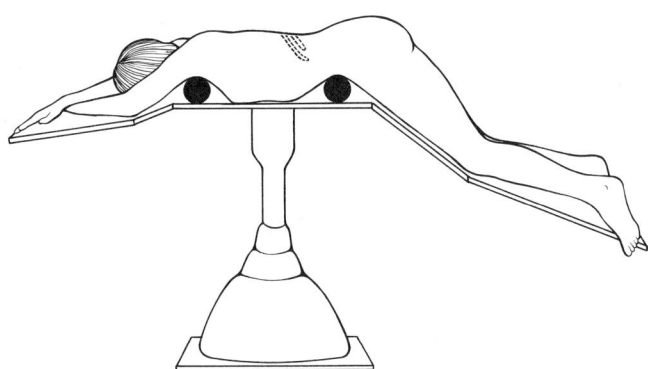

FIG 14–15.
Appropriate positioning for a patient for a bilateral posterior approach to the adrenal gland, with the possible incisions.

be used. Either a straight or hockey-stick incision is made over the 11th or 12th rib or in the 11th intercostal space. A dorsal lumbotomy incision may also be made, but in obese or muscular patients it may be difficult to gain good exposure. We prefer the 11th rib incision, particularly on the right (see Fig 14–15). The rib is resected subperiosteally and the rib bed is incised in the usual fashion, taking care not to injure the subcostal nerve and vessel or enter the pleura. The incision is extended laterally to the rib bed and the medial attachments of the diaphragm are cut, allowing the pleura to retract upward. At this point Gerota's fascia is entered and the kidney is easily identified. Some authorities recommend freeing the upper pole of the kidney at this point, but we prefer to retract the kidney inferiorly and free the upper edge of the adrenal gland initially (Fig 14–16). The adrenal gland is identified by its golden-

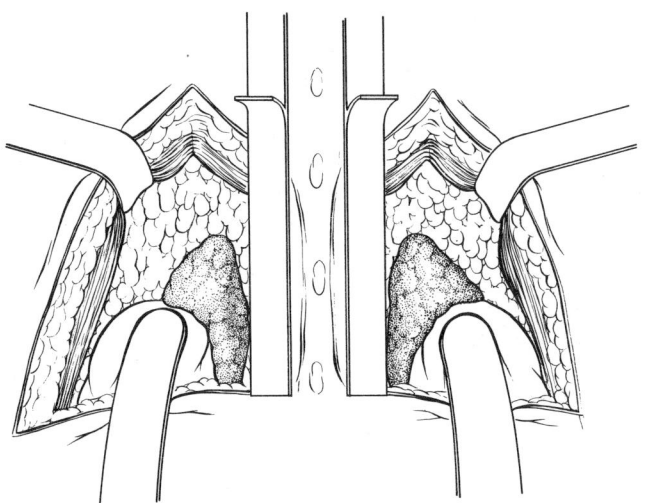

FIG 14–16.
Posterior surgical approach of the adrenal gland. The upper edge of the adrenal gland is freed initially.

yellow color, its triangular shape, and its sharp edges. In Cushing's disease, these characteristics are more difficult to observe. The numerous small arteries and veins can be ligated, clipped, or cauterized. On the left, the gland is usually rather medial with its lower margin just above the renal vessels. The adrenal vein that drains into the left renal vein is carefully ligated (see Fig 14–16). Extreme care should be taken not to ligate the renal artery or a branch to the upper pole of the kidney, which may be very close to the adrenal. On the right, the gland is usually at least in part retrocaval, and great care is necessary to dissect the adrenal away from the vena cava and ligate the short adrenal vein, which enters the cava directly (see Fig 14–2). If bleeding occurs from a laceration of the vena cava or renal vein, the area should be packed while the surgical team gets adequate exposure, prepares the suction, and obtains a 4-0 traumatic vascular silk to repair the injury. Sutures placed without adequate exposure may injure adjacent structures.

There is nothing unusual about the closure. No drains are used. We prefer permanent suture material. Obviously, the patient will require replacement glucocorticoids and mineralocorticoids indefinitely. We usually give hydrocortisone sodium succinate (Solu-Cortef), 100 mg IV every eight hours for the first few days and taper to an oral dose of 25–37.5 mg of cortisone acetate and 0.1 mg of fludrocortisone acetate (Florinef) daily.

UNILATERAL APPROACH.—We prefer the unilateral flank approach for adrenal tumors causing Cushing's syndrome and also since the advent of modern localizing techniques use a unilateral approach in most cases of pheochromocytoma and primary aldosteronism secondary to an adrenal adenoma. The flank may be entered through a standard incision, resecting a rib subperiosteally, a supracostal incision, or a thoracoabdominal incision. A thoracoabdominal incision is often advantageous for malignant tumors, but is usually not necessary for benign lesions (Scott et al., 1985). For a description of these approaches, see Chapter 1. In most instances, we approach the right adrenal gland either through the bed or just below the 10th or 11th rib without entering the chest or dividing the diaphragm. The attachments of the diaphragm are released to allow the pleura to retract cephalad. If the tumor is large, we do not hesitate to make a thoracoabdominal incision. The left adrenal is exposed by reflecting the peritoneum and descending the colon medially and retracting the kidney inferiorly (Fig 14–17). Care must be taken not to injure the spleen, the renal vessels, and the pancreas. Gerota's fascia is entered, and, as described above, the upper edge of the adrenal gland is mobilized initially, if possible. Another option is to divide the

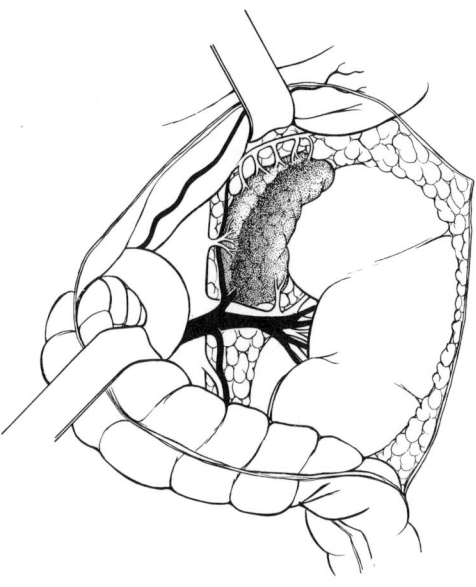

FIG 14–17.
Flank approach to a left adrenalectomy. Note the colon is retracted medially and the spleen superiorly and the adrenal lies in part medial to the kidney.

adrenal vein early in the procedure and use it as a "handle" during the dissection.

The right adrenal is exposed by reflecting the hepatic flexure of the colon and the duodenum medially and retracting the kidney inferiorly (Fig 14–18). A right adrenalectomy can be difficult, particularly in patients with Cushing's syndrome. We believe it is very important to release the upper margin of the gland from the liver before dissecting out the remaining adrenal. Almost invariably, the vena cava must be retracted medially to completely expose the gland and divide the

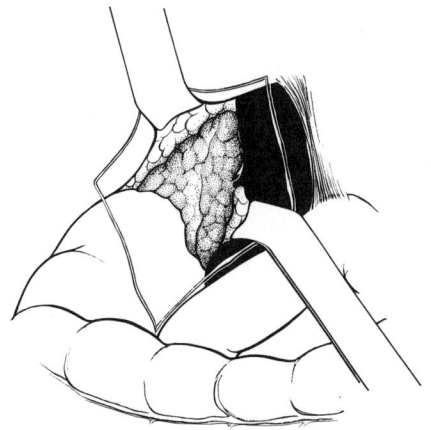

FIG 14–18.
Approach to the right adrenal gland. Note the colon is retracted medially and the duodenum is also retracted. The adrenal lies in part medial to the kidney and the adrenal vein is very short.

short adrenal vein as it enters the vena cava. Care must be exercised to avoid injury to the duodenum, renal vessels, and the vena cava. If a significant vascular injury occurs, it should be handled as described above in the section on the bilateral posterior approach.

ANTERIOR APPROACH.—The anterior approach has the advantages of exposure of both adrenal glands and the entire abdominal cavity. This is particularly important in patients with pheochromocytoma or primary aldosteronism when the tumor(s) cannot be definitely localized preoperatively. The anterior approach may also provide better exposure for bilateral adrenalectomy in patients with Cushing's disease. The disadvantages of this approach include the increased possibility of postoperative intestinal obstruction, more difficulty with obesity, and less than optimal exposure of the right adrenal gland.

The incision may be a vertical midline or a chevron. We prefer the latter. We usually expose the adrenal glands exactly as described above (see Figs 14–17 and 14–18) by mobilizing the hepatic and splenic flexures of the colon and retracting the kidneys inferiorly. An alternative is to detach the omentum from the inferior portion of the transverse colon and enter the lesser sac before reflecting the splenic flexure. Occasionally for ectopic adrenal tumors, one has to reflect the stomach and/or pancreas upward to gain access to the lesion. The left adrenal also can be exposed by opening the gastrocolic ligament and incising the peritoneum on the inferior border of the pancreas. The pancreas is retracted up, and the renal and adrenal veins are exposed. The abdominal closure should be very secure, particularly in patients with Cushing's syndrome. We do not use drains, but do recommend a bowel preparation and a postoperative nasogastric tube.

PRIMARY HYPERALDOSTERONISM

Deming and Luetscher described a sodium retaining substance in the urine in 1950. Three years later, Simpson and associates chemically identified the compound as the 18-aldehyde of corticosterone, aldosterone. Within a year, aldosterone had been synthesized, and, remarkably, Conn described the clinical syndrome of primary hyperaldosteronism, or Conn's syndrome. Never before in the history of medicine had an important clinical advance followed so closely on the heels of a basic scientific discovery.

Conn's syndrome is defined as the adrenal hypersecretion of aldosterone in a hypertensive, nonedematous patient. The exact incidence of the disease is not known, but it accounts for approximately 1% of hypertensive patients. It should be pointed out that 1% of 35 million hypertensive patients in the United States is

350,000 people. Women outnumber men about 2.5 to 1, and 75% of the patients are between 30 and 50 years old.

Primary hyperaldosteronism should be suspected in any hypertensive patient with hypokalemia. The patients typically present with hypertension, muscle weakness, polyuria, hypokalemia, and mild metabolic alkalosis. The incidence of the common symptoms of primary hyperaldosteronism reported in Conn's original description of 103 cases (1964) is listed in Table 14–5.

In recent years the diagnosis of Conn's syndrome has been confirmed in many individuals with no obvious *symptoms* of hypokalemia. Indeed, there are rare patients with primary hyperaldosteronism who do not have unprovoked hypokalemia.

The headaches, of course, are due to the hypertension, whereas the muscle weakness, polyuria, and paresthesias relate to the effect of hypokalemia on skeletal muscle, the renal concentrating mechanism, and peripheral nerves, respectively.

The major physical finding in patients with primary hyperaldosteronism is hypertension without edema. Mild retinopathy may be present. Routine laboratory evaluation reveals (1) a persistently dilute urine with a pH of 6.5 or higher, (2) a plasma potassium below 3.5 to 4.0 mEq/L off all diuretic medication, and (3) mild metabolic alkalosis (elevated serum HCO_3^-). The serum sodium concentration may be slightly elevated, and mild proteinuria is often present. The ECG often reveals premature ventricular contractions, depression of the ST segments, T waves, and the presence of U waves. When the diagnosis of primary hyperaldosteronism is entertained, three questions must be sequentially answered: (1) Does the patient have primary hyperaldosteronism? (2) If so, what is the etiology? (3) If the disease is due to an adenoma, what is the location of the tumor?

DOES THE PATIENT HAVE PRIMARY HYPERALDOSTERONISM?

By far the most common cause of hypertension and hypokalemia is essential hypertension treated with diuretics. Also, hypertension, hypokalemia, and hypersecretion of aldosterone are more often due to secondary hyperaldosteronism than to primary hyperaldosteron-

ism. Any stimulus that compromises renal blood flow may increase renin secretion and thus cause secondary hyperaldosteronism. Common clinical situations which evoke secondary hyperaldosteronism are listed in Table 14–6.

Rare causes of secondary hyperaldosteronism include renin-secreting renal tumors and Bartter's syndrome. Bartter's syndrome is characterized by elevations in plasma renin and aldosterone, hypokalemia, and hyperplasia of the juxtaglomerular cells in the kidney.

It is not always easy to determine whether a patient has primary hyperaldosteronism. Many different approaches have been recommended. We make the diagnosis of primary hyperaldosteronism if the patient has: (1) spontaneous hypokalemia (less than 4.0 mEq/L) and inappropriate kaliuresis (more than 3.0 mEq/day); (2) furosemide-stimulated plasma renin activity of 2 ng/ml/hr; and (3) increased plasma concentration and urinary excretion of aldosterone after potassium repletion to more than 4.0 mEq/L (Herf et al., 1979).

WHAT IS THE ETIOLOGY OF PRIMARY HYPERALDOSTERONISM?

Primary hyperaldosteronism may be due to an adrenal tumor (usually a unilateral adenoma), adrenal hyperplasia ("idiopathic"), deoxycorticosterone acetate (DOCA)-suppressible adrenal hyperplasia (indeterminate hyperplasia), or glucocorticoid-remediable family hyperplasia. The last two forms of the disease occur in adolescents and children, respectively, and are rare. The major differential diagnosis is between adenoma and hyperplasia. We make the diagnosis of adenoma if: (1) there is no elevation of plasma aldosterone with upright posture; (2) the blood pressure is normalized on spironolactone (400 mg/L for six weeks); (3) the serum aldosterone-stimulating factor is *not* elevated (Carey et al., 1984); (4) imaging tests are consistent with the diagnosis of adenoma (see below). Biglieri and associates (1979) have noted that the plasma 18-hydroxycorticosterone level after overnight recumbency is much higher in patients with an adenoma than in individuals with hyperplasia. In spite of these tests, the differential diagnosis may be difficult. Once the diagnosis of adenoma is made, we attempt to localize the tumor.

TABLE 14–5.

Symptoms of Conn's Syndrome

Muscle weakness	73%
Polyuria (nocturia)	72%
Headache	51%
Polydipsia	46%
Paresthesia	24%
No symptoms	6%

TABLE 14–6.

Causes of Secondary Hyperaldosteronism

Shock
Dehydration
Renal artery stenosis
Cardiac failure
Hepatic cirrhosis
Pregnancy

WHERE IS THE TUMOR?

Methods used to localize adrenal adenomas have included IVP, aortography, phlebography, adrenal vein catheterization, radionuclide scan, CT, and NMR imaging. We have found the NP-59 scan very useful (see Fig 14–4). We prepare the patient for three days with Lugol's solution, 5 drops twice daily, and oral dexamethasone, 0.5 mg every six hours. The Lugol's solution is continued for two weeks. Imaging is done after the three-day preparation and again in four more days. CT scanning will often identify an adenoma (Vetter et al., 1985), but false negative results are not rare, because these adenomas are typically rather small; therefore, we do not find CT scanning useful. Adrenal vein catheterization with sampling and analysis of the effluent is a precise way to localize tumor, but as pointed out in the discussion of Cushing's syndrome, it is technically difficult.

TREATMENT

Adrenal hyperplasia causing primary hyperaldosteronism should be treated medically since both glands are involved and adrenalectomy will not eliminate the hypertension. Medical therapy involves using one or more of amiloride, spironolactone, or triamterene. Amiloride is the best potassium-sparing diuretic. Adrenal adenomas are treated surgically. We remove the gland via a unilateral flank incision. The surgical technique is described above.

ADRENAL INSUFFICIENCY

Adrenal insufficiency was described in 1855 by Thomas Addison; thus the eponym Addison's disease. Addison's disease is a rare entity, usually secondary to tuberculosis or autoimmune adrenal cortical atrophy. Other etiologies include amyloidosis, histoplasmosis, blastomycosis, and metastatic carcinoma. Of course, iatrogenic adrenal insufficiency secondary to high-dose adrenal steroid therapy or surgical adrenalectomy is much more common.

The symptoms of adrenal insufficiency include muscular weakness, fatigability, weight loss, hyperpigmentation, anorexia, nausea, vomiting, and diarrhea. Cardiac tampanode is a rare, often fatal complication of adrenal insufficiency. Hypotension is the cardinal sign. Hyponatremia and hyperkalemia are common. Plasma cortisol levels are low, as are levels of the various urinary metabolites of the endogenous adrenal steroids. The diagnosis can be confirmed by finding an elevated serum ACTH and a lack of significant response in serum cortisol after administration of ACTH (Fig 14–19).

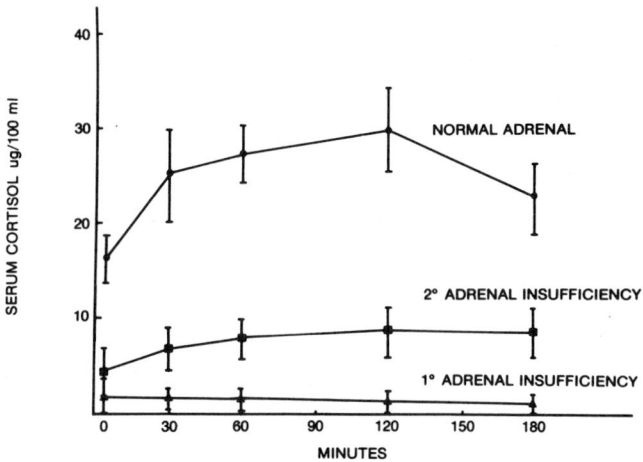

FIG 14–19.
Serum cortisol response and to ACTH infusion in normal persons and in patients with primary and secondary adrenal insufficiency and adrenal insufficiency.

The initial treatment is 100 mg of hydrocortisone every eight hours. All patients maintained on adrenal steroid therapy should receive this dose for several days following any surgical procedure. The dose is tapered to a maintenance dose of 30 mg/day, given as 20 mg in the morning and 10 mg in the evening, to mimic the physiologic diurnal variations in serum levels. In patients with adrenal insufficiency secondary to pituitary disease, it is usually not necessary to provide mineralocorticoid replacement. In patients with no functioning adrenal tissue, fluorocortisol 0.05–0.10 mg/day will provide adequate mineralocorticoid replacement. The adequacy of the replacement therapy can be determined by monitoring blood pressure, serum electrolytes, renal function, and plasma ACTH and renin activity.

PHEOCHROMOCYTOMAS

Pheochromocytomas are fascinating tumors that secrete catecholamines that cause a variety of symptoms, usually including hypertension. Most pheochromocytomas secrete norepinephrine and epinephrine. Occasional tumors release only norepinephrine, and rare lesions secrete dopa, dopamine, serotonin, somatostatin, ACTH, or epinephrine. The tumors are composed of chromaffin cells and may arise anywhere in the body where chromaffin tissue derived from primitive neuroectoderm is located. Approximately 95% of the tumors are located in the adrenal gland, but pheochromocytomas have been found in the bladder, organ of Zuckerkandl, and, indeed, anywhere in the body that sympathetic nervous tissue is located. Three percent of pheochromocytomas are extra-abdominal. Although only 0.1%–0.2% of hypertensive patients have a pheochromocytoma, the diagnosis should always be suspected, because the tumors are curable in 90% of the patients, and untreated pheochromocytomas frequently cause fatal complications such as cardiac arrhythmias, congestive heart failure, myocardial infarct, cerebral vascular accidents, and hemorrhage.

Pheochromocytomas may be familial. Patients with familial pheochromocytomas often have multiple tumors. These Sturge-Weber syndrome cases also are often associated with one of the variants of the multiple endocrine neoplasia syndrome (MEN), the most common of which is type 2A, characterized by carcinoma of the thyroid, hyperparathyroidism, and pheochromocytoma. Pheochromocytoma is also linked with von Recklinghausen's disease (café-au-lait spots) and von Hippel-Lindau disease (angiomatosis of the retina and hemangioblastoma of the cerebellum). Patients with von Hippel-Lindau disease may also have neurologic symptoms including seizures and mental retardation. Patients with a pheochromocytoma usually present with hypertension, although 15%–20% are normotensive, and rare patients are hypotensive. Approximately 50% of the hypertensive patients have sustained elevations in blood pressure, while the remainder have the characteristic paroxysmal hypertension. Hypertensive patients with sweating, tachycardia, and headaches have over a 90% chance of having a pheochromocytoma, whereas individuals with none of these characteristics have less than a 1% incidence of pheochromocytoma (Plouin et al., 1981). The common symptoms and signs are listed in Table 14–7.

Patients, even those with sustained hypertension, often experience paroxysmal symptoms that may occur very dramatically and suddenly. The frequency of these attacks is rather variable and the duration is usually less than an hour. The attacks may be precipitated by emotional, physical, or pharmacologic stimuli. Neurocutaneous lesions, progressive diabetes associated with hypertension, and unexplained intraoperative hypertension all should suggest the presence of pheochromocytoma.

The hypertension is due to the release of catecholamines from the tumor *and* increased sympathetic neural tone. Bravo and associates (1982) have shown that clonidine, a drug that decreases sympathetic tone, reduces blood pressure and heart rate in patients with pheochromocytoma without affecting plasma catecholamines or renin concentrations. Tumors that produce only norepinephrine are associated with fewer symptoms than those that release norepinephrine and epinephrine. The latter compound has a more pronounced affect on β-receptors, thus causing symptoms such as diaphoresis and palpitations.

LABORATORY EVALUATION

The diagnosis of pheochromocytoma is made by documenting elevated levels of catecholamines in the blood or urine. The assays are rather precise, but false positive results can occur secondary to drugs, stress, and other diseases that increase the concentrations of catecholamines (Table 14–8).

We screen patients for pheochromocytoma by analyzing a 24-hour urine sample for combined metanephrine (normal 0.2–1.3 mg/24 hr) and VMA. Normal values for our laboratory are shown in Table 14–9. Urinary metanephrine is the best single test. Urinary-free catechol-

TABLE 14–7.

Symptoms and Signs of Pheochromocytoma*

	FREQUENCY, %
Symptom	
Headache	80–85
Weakness	75–80
Diaphoresis	65–70
Palpitations	60–65
Orthostatic hypotension	50–60
Nausea and vomiting	35–60
Nervousness	35–40
Constipation	30–35
Sign	
Hypertension	80–85
Retinopathy	50–70
Decreased body weight	40–70
Fasting hyperglycemia	40–50

*Adapted from Atuk NO: Pheochromocytomas: Diagnosis, localization, and treatment. *Hosp Pract*, April 1983, p 187.

TABLE 14–8.

Some Causes of Increased Catecholamines

DRUGS	DISEASES
Methyldopa	Guillain-Barré syndrome
Theophylline	Neuroblastoma
Ephedrine	Porphyria
Levodopa	Brain tumor
Tricyclic antidepressants	Carcinoid syndrome
Clonidine (withdrawal)	Intrahepatic cholelithiasis

TABLE 14–9.

Normal Values for Catecholamines

	ADULT (μG)	INFANT AND CHILD (RANGE, μG/KG)
Norepinephrine	80	0.4–1.6
Epinephrine	25	0.02–1.6
VMA	8	83 ± 26
Normetanephrine	450	4.9–20.8
Metanephrine	300	3.1–15.6

*Catecholamine values for 24-hour urine collection.

amines can also be useful; however, false negative results are more frequent with this test. An advantage of measuring urinary-free catecholamines is that if epinephrine is present and norepinephrine is absent, the tumor is extremely likely to be located in an adrenal gland. These tests in combination are over 95% accurate.

Some authorities have recommended screening patients with an assay of plasma catecholamines and point out that 24-hour urine collections are difficult to obtain. Bravo (1984) has had excellent results with the plasma assay. It should be pointed out that the blood test also must be very carefully done, with a fasting, supine, and relaxed patient, and the laboratory must be experienced with the assay. False negative results do occur, especially if the sample is taken while the patient is normotensive and asymptomatic. In summary, the plasma test is very useful if properly done, but for most clinicians the urinary screening tests are more practical. In rare situations when the above mentioned tests are equivocal, a provocative test is indicated. Glucagon should be used rather than histamine or tyramine because glucagon is safer. An IV dose of 0.5–2.0 mg of glucagon is given. Patients with pheochromocytoma have a threefold increase in plasma catecholamines and at least a 35/35 mm Hg rise in blood pressure within a few minutes, whereas normal subjects do not. This test should only be done under controlled conditions with phentolamine and/or nitroprusside available. Bravo (1984) has described a clonidine suppression test that he thinks is safer than the glucagon test and very reliable.

LOCALIZATION OF PHEOCHROMOCYTOMAS

Once the diagnosis of pheochromocytoma has been made, it is necessary to localize the tumor. CT scanning will usually identify the lesion because the vast majority of pheochromocytomas are large enough to be identified by this technique (see Fig 14–12). Arteriograms and venograms are no longer indicated. The ^{131}I-meta-iodobenzylguanidine (MIBG) scan is also a very accurate method to localize a pheochromocytoma (Sisson et al., 1981; Shapiro et al., 1985) (Fig 14–20). We often use both methods so that we can make a unilateral surgical approach to solitary tumors.

In rare situations, venous sampling along the vena cava may be necessary to find an occult pheochromocytoma.

TREATMENT

The treatment of pheochromocytoma is surgical removal unless there is a contraindication to operation. In the latter situations, medical therapy is indicated. In

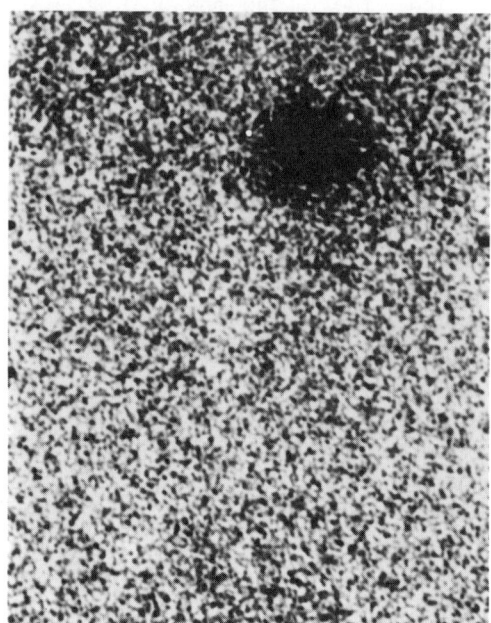

FIG 14–20.
MIBG scan of a pheochromocytoma.

the 10% of patients who have malignant pheochromocytomas, treatment should be governed by the basic principles of cancer management. The diagnosis is not frequently made preoperatively and, indeed, may be difficult to confirm histologically. Tumors which invade adjacent structures or metastasize are considered malignant. Alpha- and beta-adrenergic blocking agents and α-methyl-L-tyrosine (metyrosine, which blocks synthesis of norepinephrine by inhibiting tyrosine hydroxylase) can all be used to alleviate symptoms in patients with inoperable malignant pheochromocytomas.

Unfortunately, there is no good chemotherapy for this disease, and radiotherapy is only marginally effective. These patients should never receive β-blockers unless they have full α-blockade, because β-blockade with inadequate α-blockade can cause a hypertensive crisis.

We routinely provide α-blockade to patients with pheochromocytomas preoperatively with oral phenoxybenzamine, 10–100 mg/day for 7–14 days, to normalize blood pressure and expand intravascular volume. We do not give preoperative blood transfusions as has been recommended by some authorities. We use β-blockade only if cardiac arrhythmias or tachycardias create problems. Again, β-blockade without α-blockade can cause a hypertensive crisis. We have prepared some patients for surgery with metyrosine 2.0–2.5 mg/day for 5–10 days. This approach has the advantage of reducing the catecholamine levels to normal but the disadvantage of causing sleepiness and diarrhea.

As mentioned above, because of improved localization of the tumor(s) with CT and MIBG scanning, we no longer routinely explore the abdomen in patients with pheochromocytoma. The unilateral approach has also been advocated by others (Cullen et al., 1985). The surgical techniques are similar to those described above for Cushing's syndrome. Intraoperative arterial lines and ready access to nitroprusside and phentolamine are essential to manage hypertension related to manipulation of the tumor. After the tumor is removed, hypotension is common and should be treated with large volumes of crystalloid-reserving pressors for those instances when volume expansion will not correct the hypotension. Pressors are necessary in some patients probably because of previous down-regulation of α-adrenergic receptors. Hypotonic volume expansion is inadequate.

NONFUNCTIONING ADRENAL MASSES

In the past, nonfunctioning adrenal masses were seldom diagnosed except for very large cysts and those secondary to adrenal hemorrhage in the newborn period. However, with the advent of CT scanning and particularly since its emergence as a routine diagnostic study, nonfunctioning adrenal masses are being detected with increasing frequency (see Fig. 14–11). In patients with known cancer, the vast majority of these lesions are metastasis to the adrenal gland.

Asymptomatic adrenal masses in patients without known malignancies create a clinical problem. In the past, almost all adrenal masses were surgically explored. However, current practice in our institution is initially to rule out a functioning adrenal tumor as described. In this setting, if there is no evidence of endocrine function and the tumor is less than 4 cm in diameter, several studies have shown that it is almost certainly a benign lesion that does not require treatment (Geelhoed and Druy, 1982; Prinz et al., 1982). However, nonfunctioning malignant adrenal tumors do occur (Seddon et al., 1985); therefore, good clinical judgment and close follow-up are necessary. This is not surprising, since the autopsy incidence of benign adrenal tumors ranges from 1.4% to 8.7% (Copeland, 1982), and silent primary adrenal cancers are exceedingly rare. The most common benign tumors of the adrenal are adrenal cortical adenomas, myelolipomas (Muller et al., 1985), and cysts. If the lesion is less than 4 cm in diameter, our policy is to repeat a sonogram or CT scan in three months and then every six months for two years. If the tumor is larger, we base our treatment on the results of the fine-needle aspiration biopsy. Obviously, clinical judgment is required in the management of these patients, and the physician must be flexible.

REFERENCES

1. Atuk NO: Pheochromocytoma: Diagnosis, localization, and treatment. *Hosp Pract*, April 1983, p 187.
2. Besser GM: ACTH and MSH assays and their clinical application. *Clin Endocrinol* 1973; 2:175.
3. Bethune JE: The adrenal cortex: A Scope monograph. Kalamazoo, Mich, The Upjohn Co.
4. Biglieri EG, Schambelan M, et al: The significance of elevated levels of plasma 18-hydroxycorticosterone in patients with primary aldosteronism. *J Clin Endocrinol Metab* 1979; 49:87.
5. Boggan JE, Tyrrell JB, Wilson CB: Transsphenoidal microsurgical management of Cushing's disease: Report of 100 cases. *J Neurosurg* 1983; 59:195.
6. Bravo EL, Gifford RW Jr: Current concepts: Pheochromocytoma: Diagnosis, localization, and management. *N Engl J Med* 1984; 311:1298.
7. Bravo EL, et al: Blood pressure regulation in pheochromocytoma. *Hypertension* 1982; 4(suppl II):193.
8. Carey RM, Sen S, Dolan IM, et al: Idiopathic hyperaldosteronism: A possible role for aldosterone-stimulating factor. *N Engl J Med* 1984; 311:94.
9. Conn JW, Knopf RF, Nesbit RM: Clinical characteristics of primary aldosteronism from an analysis of 145 cases. *Am J Surg* 1964; 107:159.
10. Copeland PM: The incidentally discovered adrenal mass. *Ann Surg* 1984; 199:116.
11. Cullen ML, Staren ED, Straus AK, et al: Pheochromocytoma: Operative strategy. *Surgery* 1985; 98:927.
12. Davies CJ, Joplin GF, Welbourn RB: Surgical management of the ectopic ACTH syndrome. *Ann Surg* 1982; 196:246.
13. Franks RC: Urinary 17-hydroxycorticosteroid and cortisol excretion in childhood. *J Clin Endocrinol Metab* 1973; 36:702.
14. Geelhoed GW, Druy EM: Management of adrenal "incidentaloma." *Surgery* 1982; 92:866.
15. Herf SM, Teates CD, Tegtmeyer CJ, et al: Identification and differentiation of surgically correctable hypertension due to primary aldosteronism. *Am J Med* 1979; 67:397.
16. Junqueira LC, Carneiro J: Adrenal islets of Langerhans, thyroid, parathyroids, and pineal body, in *Basic Histology*, ed 4. Los Altos, Calif, Lange Medical Publications, 1984, pp 428–436.
17. Kenney PJ, Robbins GL, Ellis DA, et al: Adrenal glands in patients with congenital renal anomalies: CT appearance. *Radiology* 1985; 155:181.
18. Kirschner MA, Zucker IR, Jespersen D: Idiopathic hirsutism—an ovarian abnormality. *N Engl J Med* 1976; 294:637.
19. Kuttenn F, Couillin P, Girard F, et al: Late-onset adrenal hyperplasia in hirsutism. *N Engl J Med* 1985; 313:224.
20. Lieberman LM, Beierwaltes WH, Conn JW, et al: Diagnosis of adrenal disease by visualization of human adrenal glands with ^{131}I-19-iodocholesterol. *N Engl J Med* 1971; 285:1387.
21. Ludecke DK, Niedworok G: Results of microsurgery in

Cushing's disease and effect on hypertension. *Cardiology* 1985; 72(1):91–96.

22. Muller SC, Schreyer T, Rumpelt H-J: Myelolipoma of the adrenal gland. *Urol Int* 1985; 40:132.

23. Plouin PF, Degoulet P, Tugaye A, et al: Le depistage du pheochromocytoma: Chez quels hypertendus? Étude semiologique chez 2585 hypertendus dont 11 ayant un pheochromocytome. *Nouv Presse Med* 1981; 10:869.

24. Prinz RA, Brooks MH, Churchill R, et al: Incidental asymptomatic adrenal masses detected by computed tomographic scanning: Is operation required? *JAMA* 1982; 248:701.

25. Ratcliffe JG, Knight RA, Besser GM, et al: Tumour and plasma ACTH concentrations in patients with and without the ectopic ACTH syndrome. *Clin Endocrinol* 1972; 1:27.

26. Scott HW Jr, Abumrad NN, Orth DN: Tumors of the adrenal cortex and Cushing's syndrome. *Ann Surg* 1985; 201:586.

27. Seddon JM, Baranetsky N, Boxel PJV: Adrenal "incidentalomas"—need for surgery. *Urology* 1985; 25:1.

28. Shapiro B, Sisson JC, Eyre P, et al: ^{131}I-MIBG—new agent in diagnosis and treatment of pheochomocytoma. *Cardiology* 1985; 72(1):137–142.

29. Shirkhoda A: Computed tomography after adrenalectomy in adrenal cortical carcinoma. *Urol Radiol* 1985; 7:132.

30. Sisson JC, Frager MS, Valk TW, et al: Scintigraphic localization of pheochromocytoma. *N Engl J Med* 1981; 305:12.

31. Styne DM, Grumbach MM, Kaplan SL, et al: Treatment of Cushing's disease in childhood and adolescence by transsphenoidal microadenomectomy. *N Engl J Med* 1984; 310:889.

32. Vetter H. Fischer M, Galanski M, et al: Primary aldosteronism: Diagnosis and noninvasive lateralization procedures. *Cardiology* 1985; 71(1):57–63.

33. Wilson JD, Foster DW: *Williams Textbook of Endocrinology,* ed 7. Philadelphia, WB Saunders Co, 1985.

Chapter 15

Kidney and Ureter

W. Scott McDougal, M.D.

RENAL ANATOMY

The renal parenchyma is divided into an outer cortical zone and an inner medullary area. The medulla is subdivided into inner and outer portions, the former containing the 4–12 renal papillae. Each papilla is surrounded by transitional cell epithelium, the complex referred to as a calyx. It drains into an infundibulum, and the infundibulae join one to another and thence to the renal pelvis. A detailed knowledge of the calyceal anatomy is important, since endoscopic location of lesions within the kidney requires the ability to identify the location of the calyx in which the lesion is present on the two-dimensional roentgenogram and to correlate it with the gross anatomy of the kidney when visualizing the infundibulae from the renal pelvis. The calyces located in the upper and lower poles are often compound and project directly to their respective poles. The remainder are arranged into an anterior and posterior row (Fig 15–1). The anterior row of calyces forms an angle 70° to the frontal plane; these are the calyces visualized laterally on an AP view of an IV urogram. The posterior calyces form an angle of 20° with the frontal plane and are those visualized medially on the urogram (Fig 15–2).

Segments of the cortex and medulla are served by end arteries, i.e., arteries that do not anastomose with other arteries and therefore have little potential for the development of collateral vessels. The segments of the kidneys are the two polar, the anterosuperior, the anteroinferior and the posterior (Fig 15–3). The first main branch off the renal artery serving the renal parenchyma supplies the posterior segment. Thus, location of this artery allows for identification of the subsegmental anatomy of the kidney. It allows identification of the avascular plane, adjacent to Brödel's line between the anterior and posterior segments along which the kidney may be split and calyces entered through the parenchyma without violation of the blood supply to segments of parenchyma (see Fig 15–2). The arteries

branch from the main renal artery into, successively, the interlobar, arcuate, interlobular and afferent, glomerular, and efferent arterioles. The efferent arterioles supply the proximal tubule from the glomerulus in the cortex from which it arose and the proximal tubule and the loop of Henle when it arises from a juxtamedullary glomerulus. The venous drainage follows the arterial supply; however, unlike the arteries, veins intercommunicate between the segments.

Lymphatic vessels in the kidney follow the arterial vessels from interlobular arteries proximally. The exact course of the lymphatics distal to the interlobular arteries is not known. It has been suggested that the medulla contains few if any lymphatics, whereas the cortex may be drained by regional areas from lymphatics that actually do not penetrate very deeply into the parenchyma (Rouiller and Muller, 1969). The renal capsule is supplied richly by lymphatics. Capsule lymphatics and parenchyma lymphatics drain to the renal hilum and from there to the great vessel closest to the respective kidney.

The kidneys are supplied by sympathetic nerves from T4 to L4, which course to the celiac, superior, and inferior mesenteric and aorticorenal ganglia. The postganglionic fibers course to the renal hilum and thence to the renal parenchyma. Once in the renal parenchyma, they travel with the vessels, making contact eventually with the juxtaglomerular apparatus and the basement membranes of the proximal and distal tubule cells. Innervation of the glomerulus is controversial. Stimulation of these nerves causes vasoconstriction and an acute reduction in renal blood flow. Denervation of the kidney results in an increase in electrolyte excretion primarily due to a diminished proximal tubule fractional reabsorption.

The hilum of the kidney contains the renal vessels, renal nerves, lymphatics, lymph node tissue, and renal pelvis. The renal capsule extends to the renal sinus and incorporates it into the kidney proper.

There are two types of nephrons: cortical nephrons

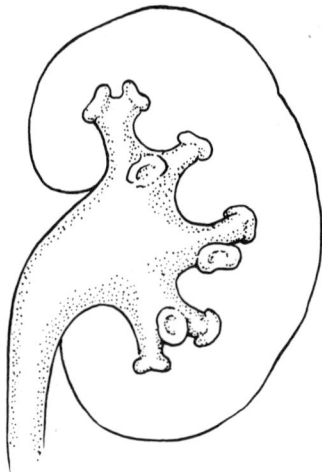

FIG 15–1.
Calyceal anatomy. The polar calyces project to their respective poles while the middle calyces form an anterior and posterior row.

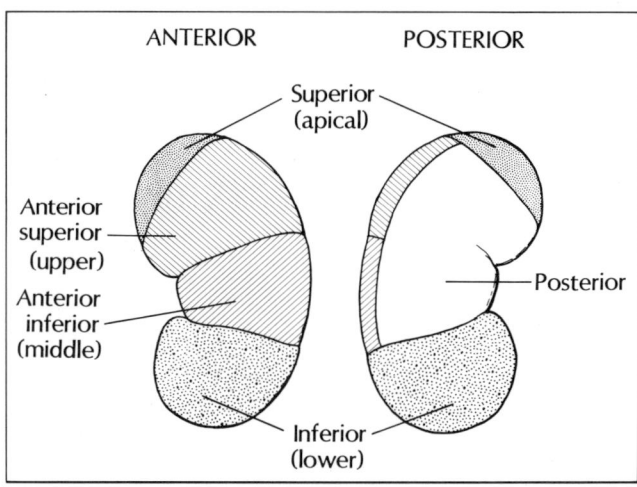

FIG 15–3.
Segmental anatomy of the kidney. There are five segments: superior, inferior, posterior, anterosuperior, and anteroinferior.

with short loops of Henle and juxtamedullary nephrons with long loops of Henle (Fig 15–4). There are approximately 2 million nephrons per kidney, seven eighths of which are cortical and one eighth juxtamedullary. The arterial supply to the cortical nephrons differs from that of the juxtamedullary nephrons. In the former, the efferent arteriole courses along the proximal tubule belonging to the glomerulus from which it came. This allows that arteriole to reabsorb filtrate that it left behind in the glomerulus. Since these nephrons have short loops of Henle, they play a lesser role in maximal urinary concentration than do the juxtamedullary nephrons with long loops of Henle. The efferent arterioles from the juxtamedullary nephrons not only course along the proximal tubule from its respective glomerulus but

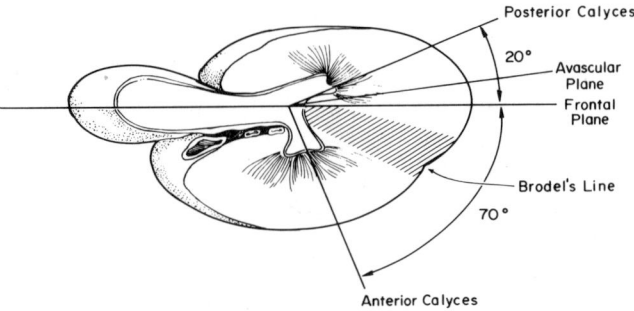

FIG 15–2.
Location of anterior and posterior calyces. Notice that the anterior calyces form an angle of 70° with the frontal plane and the posterior calyces a 20° angle with that plane. The avascular plane adjacent to Brödel's line through which the calyces may be entered without violating segmental vasculature is depicted.

also dive deep into the medulla, coursing along their respective loops of Henle. The vessels which follow the loops of Henle are called vasa recta.

The components of the nephron include the glomerulus, the proximal tubule, loop of Henle, distal tubule, and collecting duct. The glomerulus consists of an afferent arteriole, capillary and efferent arteriole, a juxtaglomerular apparatus, and Bowman's space and capsule (Fig 15–5). The afferent arteriole, and to a lesser extent the efferent arteriole, are in intimate contact with the first portion of the distal tubule belonging to that glomerulus. This association is called the juxtaglomerular apparatus and consists of two parts: a macula densa, or the dark cells of the distal tubule, and the juxtaglomerular cells, or the endothelial cells of the afferent and efferent arterioles that contain renin granules. It has been postulated that this structure is responsible for the regulation of sodium conservation through the stimulation of aldosterone production as well as a feedback mechanism for the intrarenal regulation of each nephron's blood flow and therefore GFR (vide infra). The glomerular capillary network is the area through which plasma is filtered. The capillaries contain an endothelium, which lies on a basement membrane. On the other side of the basement membrane is Bowman's space. It is on this side of the basement membrane that the epithelial cells, or foot processes (podocytes), are found (Fig 15–6, A). Finally, within the glomerular tuft are found mesangial cells that support the structures of the glomerulus. These cells may play a role in causing arteriole constriction as well as providing support for the structures named above. Bowman's space empties

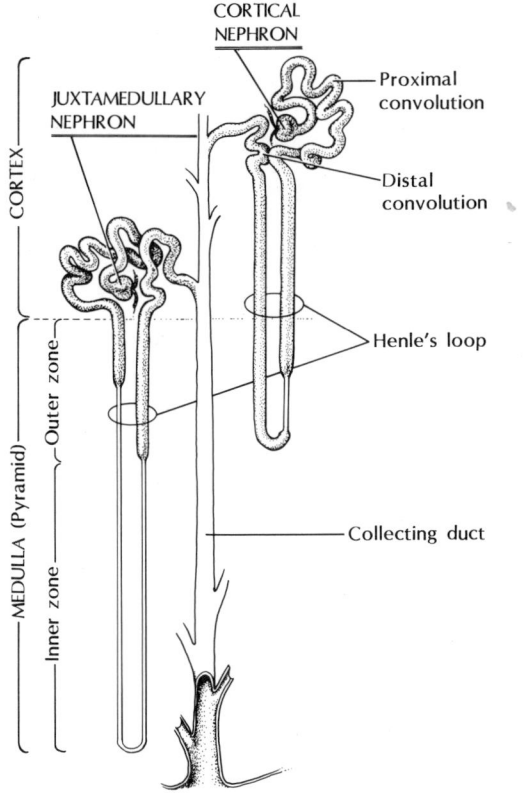

FIG 15–4.
The renal parenchyma is divided into cortex and medulla and contains two types of nephrons: cortical and juxtamedullary.

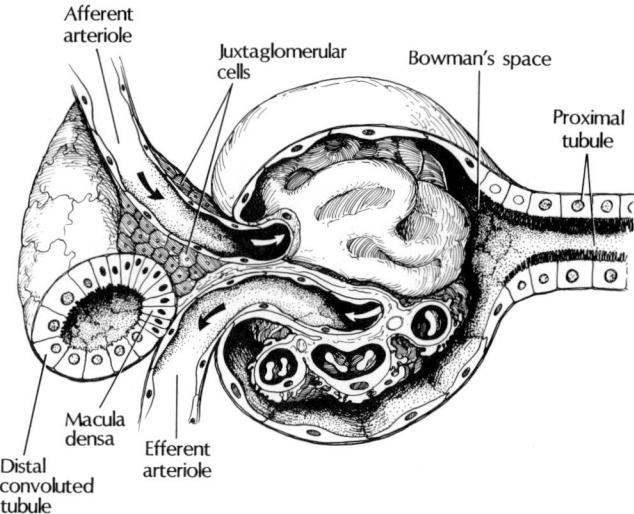

FIG 15–5.
Glomerulus and juxtaglomerular apparatus.

RENAL PHYSIOLOGY

GLOMERULAR FILTRATION

By a process called ultrafiltration, plasma, water, and nonprotein crystalloids are separated from blood cells and protein within the glomerulus. The forces involved in ultrafiltration are the hydrostatic pressure within the glomerular capillary, the permeability of the glomerular membrane, the oncotic pressure of the plasma, and the hydrostatic pressure in Bowman's space. The hydrostatic pressure in the glomerular capillary remains relatively constant throughout its length, while the oncotic pressure rises along the course of the capillary as filtrate leaves. The net filtration pressure is the hydrostatic pressure minus the oncotic pressure minus the pressure in Bowman's space. Net filtration is determined by this pressure and the permeability (reflection coefficient) of the glomerular membrane. As blood courses along the length of the capillary, the net ultrafiltration pressure declines, because oncotic pressure rises due to the increased protein concentration as fluid is removed from the capillary lumen (Fig 15–7). This serves to keep the amount of filtrate constant despite capillary length and thus prevents loss of excessive fluid into Bowman's space. The permeability of the membrane is a function of the permeability of its component parts: the endothelium, the basement membrane, and the epithelial cells (podocytes or foot processes). The podocytes are separated by filtration slits that contain pores having dimensions of 40–140 A. Negatively charged glycoproteins are attached to the surfaces of the endothelial cells, basement membrane, and podocytes. Thus the membrane acts as a barrier by discriminating according

into the convoluted proximal tubule. The cells of this portion of the nephron have a dense brush border on the luminal side (Fig 15–6, B) and are attached to each other by impermeable tight junctions. At the end of the proximal tubule, the nephron dives toward the medulla, where it attaches to the thin limb of the loop of Henle. This portion of the proximal tubule is called the pars recta. Its brush border is less dense, and this area perhaps is less active in filtrate reabsorption than the other portion of the proximal tubule. The thin limb of the loop of Henle descends and then turns back on itself and ascends toward its respective glomerulus. As it ascends it changes to the thick limb of the loop of Henle, which has notably different permeability as well as transport characteristics than the thin limb. The thick limb joins the first portion of the distal tubule whose cells are called the macula densa—a component of the juxtaglomerular apparatus. The distal convoluted tubule joins with many others to enter a collecting duct. The collecting duct traverses the parenchyma to the renal papilla (Rouiller and Muller, 1969).

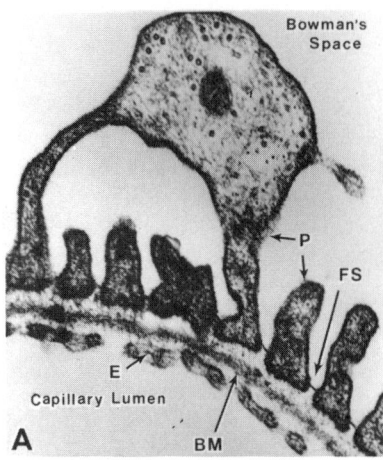
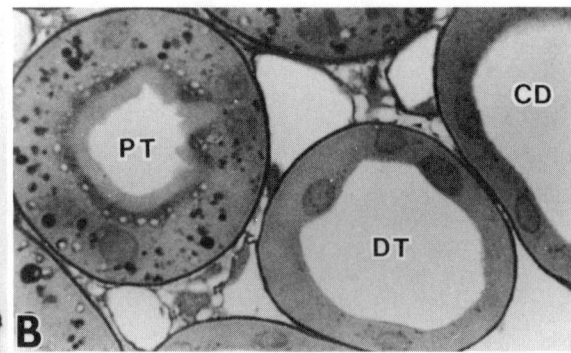

FIG 15–6.
A, electron photomicrograph of the glomerulus. *E,* endothelial cell; *BM,* basement membrane; *P,* podocyte; *FS,* filtration slit. **B,** light photomicrograph of renal tubules. *PT,* proximal tubule (notice dense brush border); *DT,* distal tubule; *CD,* collecting duct. Notice the dense staining basement membrane surrounding each tubule.

to both the molecule's size and its electric charge (an electrostatic barrier).

The measurement of the quantity of plasma filtered by the glomeruli can be determined by a clearance technique. A clearance is defined as the volume of plasma from which the kidney removes a substance per unit time. To measure the amount of fluid arriving in Bowman's space or the GFR, a substance is chosen that is freely filtered at the glomerulus but neither secreted nor reabsorbed by the tubule. Inulin, a starchlike polymer of fructose with a molecular weight of about 5,000, is such a substance, the standard against which other substances used to measure GFR are compared. Inulin clearance is independent of serum concentration and urine flow. The clearance is calculated from the formula: UV/P

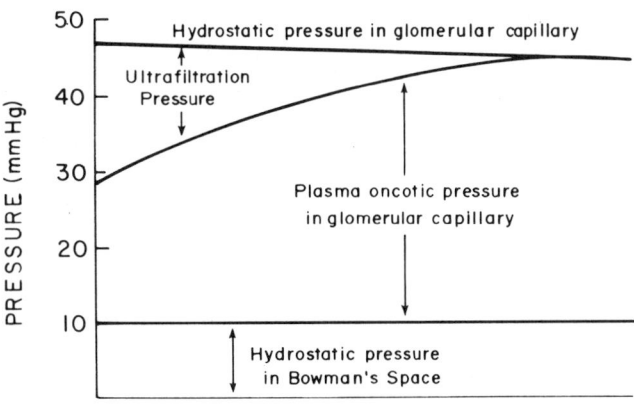

FIG 15–7.
Forces of glomerular ultrafiltration. As fluid traverses the capillary and as some is filtered into Bowman's space, oncotic pressure within the capillary increases until it completely counteracts the hydrostatic pressure gradient. This limits the amount of fluid filtered into the proximal tubule.

where U is the urine concentration, V the urine volume per unit per period of time, and P the plasma concentration (vide infra).

The amount of filtrate removed from the plasma remains relatively constant due to the forces described above. The fraction of fluid in the glomerular capillary bed entering Bowman's space is called the filtration fraction and is about 20%; i.e., 20% of the plasma arriving in the glomerulus leaves as filtrate into Bowman's space.

TUBULE REABSORPTION

Reabsorption by the tubule epithelium of substances essential to normal body function such as water, sugars, amino acids, and electrolytes is critical for homeostasis. Reabsorption of a substance may be determined by comparing its clearance to that of inulin: if the clearance of the filtered substance is less than that of inulin, it must be reabsorbed. Many substances that are reabsorbed are actively transported, thus having the potential to saturate the carrier mechanism, at which time a transport maximum (Tm) is achieved. The Tm is the maximum amount of substance that can be actively transported when all carriers are saturated. Glucose, other sugars, sulfate, amino acids, phosphate, uric acid, and albumin have a Tm. The transport of a number of compounds can be facilitated by sodium transport. Glucose, uric acid, amino acids and phosphate movement are enhanced in the presence of sodium, a process called cotransport.

Not all substances are reabsorbed by an active process. Passive reabsorption accounts for urea movement across the tubule. This type of transport is strongly affected by urine flow. At low flow rates, there is an increased reabsorption of urea, and at high flow rates urea is flushed from the tubule, thereby lessening its absorption. The flow of urea after filtration is not unidirectional. In certain portions of the tubule urea is reabsorbed and in others secreted. Potassium and weak acids and bases (including drugs) also are secreted and reabsorbed in different portions of the nephron. The process is called bidirectional transport.

TUBULE SECRETION

Substances secreted are usually either weak acids or weak bases, are foreign to the body, and are not metabolized or are metabolized slowly or incompletely. Secretion is confirmed when a filtered substance's clearance exceeds that of the inulin clearance. Examples of substances secreted include drugs (diuretics, antibiotics, salicylates), para-aminohippuric acid, and thiamine.

RENAL HEMODYNAMICS

Approximately 20%–25% of the cardiac output flows to the kidneys, more than 90% of which perfuses the cortical region. Even though the medullary region gets but a small fraction of the total renal blood flow, in absolute amounts, it is perfused with about the same amount of blood as is resting muscle. The blood flow is very sensitive to hemorrhage, the cortex being primarily affected. Antidiuretic hormone, the prostaglandins, and renin as well as sympathetic nerve stimulation also affect renal blood flow. Antidiuretic hormone causes vasoconstriction and therefore prevents washout of osmotically active particles from the medulla. Prostaglandins are both vasodilatory and vasoconstrictive. Their vasodilatory action exerts a protective role in diseases causing vasoconstriction (Dunn, 1980). Thromboxane, a potent prostaglandin vasoconstrictor, may play a role in altering renal blood flow during obstructive uropathy (Morrison et al., 1977). Others are not convinced that it alters renal hemodynamics. Renin, through the release of angiotensin, also causes vasoconstriction, and its release may be stimulated by an alteration of distal tubule sodium concentration and/or afferent arteriole pressure. Finally, stimulation of the sympathetic nerves supplying the kidney, through their innervation of the arterioles, causes vasoconstriction by stimulating contraction of endothelial muscle cells.

Under most conditions renal blood flow remains relatively constant by a phenomenon known as autoregulation. Over a range of systolic blood pressures from 80 to 180 mm Hg, renal blood flow and GFR remain constant. It has been suggested that the renin-angiotensin system, prostaglandins, and the catecholamines, kinins or other substances regulate renal blood flow by altering vascular resistance in the afferent and efferent arterioles. Through this combined effect, intraglomerular hydrostatic pressure is maintained relatively constant over wide variations in systemic pressure.

Renal blood flow is usually measured by a clearance technique (see below). Para-aminohippurate (PAH) is used, since it is both filtered and secreted by the kidney. Thus, the amount of plasma that delivers the amount of PAH found in the urine per unit period of time is the amount of plasma that passed through the kidney or the renal plasma flow. The calculation is simple. However, from a practical point of view, its determination is difficult, since both the renal vein and renal artery concentrations must be known to calculate the amount of plasma delivered to the kidney as the arteriovenous difference is the amount removed in one pass. The renal artery concentration is easily determined from a peripheral venous sample; however, an invasive technique must be used whereby a catheter is passed to sample the renal vein directly. This difficulty can be obviated if the renal vein concentration approaches zero. If the PAH concentration is kept low enough in the systemic circulation, essentially all PAH entering the renal artery is cleared in one pass. Thus the calculation simplifies to a clearance: UV/P of PAH. Since about 10% of the arterial concentration remains in the renal vein at low systemic PAH concentrations, and since this may vary by as much as 20% under varying experimental conditions, the calculation of renal plasma flow determined without actual measurement of renal venous concentration is termed "effective renal plasma flow" to indicate that it is not actual renal plasma flow but rather an approximation thereof, since renal vein concentration is assumed to be zero, which, in fact, may not be the case. If the renal vein concentration were known, then actual renal plasma flow could be calculated by the clearance technique described above. Renal blood flow may be calculated by dividing the plasma flow by the quantity: $(1 - \text{hematocrit})$.

SODIUM AND WATER BALANCE

Sodium is actively transported from the luminal contents to the interstitium. The bulk of energy expended by the kidney is involved in this active process. In the proximal tubule, approximately 60%–70% of the filtered sodium and fluid is reabsorbed. Since water follows sodium passively and since the proximal tubule is freely permeable to water, the fluid in this portion of the tubule is reabsorbed isosmotically. A constant fraction of the filtered sodium is reabsorbed in the proximal tubule (glomerulotubular balance), normally about 70% despite variations in the GFR. Glomerulotubular balance appears to be brought about by two processes. The first involves changes in filtration fraction, perhaps brought about by nervous or humoral factors. The second involves the process of cotransport. Since sodium reabsorption is linked to the reabsorption of various substances that are almost completely reabsorbed by the proximal tubule (such as glucose), when increases in filtered load of the substance occur, an increase in sodium reabsorption is stimulated and vice versa.

In the thin limb of the loop of Henle, water moves according to its concentration gradient. Thus in the proximal portion water moves out, while in the distal portion it moves back in. Sodium follows passively. In the thick ascending limb, sodium and chloride are actively pumped from the lumen; however, this portion is impermeable to water, and thus the fluid becomes hypotonic. In the distal tubule sodium is actively removed under the influence of aldosterone. Water moves according to the movement of sodium and its concentration gradient. In the collecting duct, sodium is reabsorbed and water movement (collecting duct permeability to water) is influenced by the concentration of antidiuretic hormone.

Sodium balance is regulated by three factors. First factor, or GFR, alters sodium reabsorption through the mechanism of glomerular tubule balance. Occasionally, glomerular tubule balance may be disrupted under circumstances of excessive sodium loads. Second factor, or aldosterone, regulates sodium reabsorption in the distal tubule. Aldosterone is released either as a direct action of hyperkalemia on the adrenal or as a result of the release of renin. Renin is released from the juxtaglomerular cells either as a consequence of a low afferent arteriolar pressure (secondary to hypotension) or decreased distal tubule sodium concentration. Renin acts on a protein substrate to cleave a decapeptide, angiotensin I. Angiotensin I is converted by converting enzyme to a vasoactive octapeptide, angiotensin II. Angiotensin II results in vasoconstriction and stimulation of the adrenal to release aldosterone. Third factor or factors are nervous and/or humoral influences, which alter sodium transport. They are not well-defined; however, their effects are known. For example, volume expansion with sodium results in natriuresis due to a decreased fractional reabsorption in the proximal tubule from 70% to 40% and a decreased sodium reabsorption in the distal tubule. This is not dependent on either factors I or II. A substance recently described that may be responsible for some of these effects is atrial natriuretic factor, a 28–amino acid polypeptide that has been identified in the atrium. It has potent diuretic, natriuretic, and vasorelaxant properties. It inhibits the synthesis and release of aldosterone and the release of antidiuretic hormone. Atrial natriuretic factor's release is stimulated by stretch of the atrium, volume expansion, sodium concentration, osmolality, and certain vasopressor agents. Finally, prostaglandins play a minor role in the modulation of sodium and water excretion. Prostaglandins synthesized in the renal medulla act locally to enhance water and sodium excretion by: (1) increasing renal blood flow, (2) inhibiting sodium trans-port from the thick ascending limb of the loop of Henle, (3) antagonizing the action of vasopressin on the collecting duct, and (4) inhibiting urea and sodium reabsorption from the collecting duct (Dunn, 1983).

CONCENTRATION AND DILUTION

The ascending limb of the loop of Henle is impermeable to water, and since sodium and chloride are actively reabsorbed, electrolyte transport occurs without water following. This establishes a hyperosmotic medullary interstitium, which is the primary determinant of the kidney's ability to concentrate. The loop of Henle acts as a countercurrent multiplier increasing the concentration of solutes, whereas the vasa recta coursing along with the loop of Henle preserve medullary tonicity by serving as countercurrent exchangers. This is particularly important since plasma flow to the medulla is about ten times greater than is tubule fluid flow. This unique relationship preserves medullary tonicity and thus concentration capabilities.

Urea recycling helps maintain osmolality in the medulla, both in the medullary interstitium as well as the tubule lumen. Urea concentration in the loop of Henle increases as water leaves. Under the influence of antidiuretic hormone, urea reabsorption in the collecting duct is facilitated as water is reabsorbed. This maintains a high medullary tonicity. Thus, during antidiuresis urea provides 40% of the medullary osmotically active particles, whereas in diuretic states urea constitutes only 10% of the medullary solutes.

Antidiuretic hormone (ADH) adjusts the amount of water that is reabsorbed from the late distal tubule and collecting ducts. ADH increases tubule permeability, thus allowing water to travel according to its concentration gradient. In the collecting duct, where medullary tonicity is high, water is reabsorbed. ADH also causes vasoconstriction of the vasa recta, preventing the removal of solute from the medulla, thereby maintaining a high medullary tonicity. Prostaglandins may also play a role in this regulation by opposing the action of ADH. ADH is released as a result of stimuli from volume receptors or osmoreceptors. Osmoreceptors located in the hypothalamus, when exposed to increased osmolality, stimulate the posterior pituitary to release ADH. Similarly, volume receptors located in the left atrium or pulmonary veins also result in the release of ADH.

ACID-BASE BALANCE

The daily production of 40–70 mmole of inorganic and organic acids (the fixed acids) and 13,000 mmole of CO_2, which momentarily generates hydrogen (the volatile acid), requires elimination by the body. This acid load requires buffering so that major pH shifts do not occur locally. The buffers include hemoglobin, protein, inorganic phosphate, organic phosphate, and bicarbonate. Organic phosphate is the major intracellular buffer, while the main extracellular buffer is the carbonic acid bicarbonate system. The latter is particularly effective since the concentration of one component of the pair, CO_2, can be rapidly altered by the lungs.

Perhaps one of the most effective initial methods of buffering the volatile acid load is the reaction hydrogen has with hemoglobin. Oxygenated hemoglobin is more acidic than nonoxygenated hemoglobin. Thus, when hemoglobin gives up its oxygen, it can take up a great deal of hydrogen without any overall change in local pH. The carbon dioxide arising from metabolism is hydrated to carbonic acid and dissociates into hydrogen and bicarbonate. The hydrogen is taken up by hemoglobin, and the CO_2 travels to the lungs as bicarbonate. The reverse reaction occurs in the lungs where hydrogen is given up, CO_2 is eliminated, and hemoglobin is oxygenated. Fixed acids are initially buffered by bicarbonate, phosphate, and proteins, thus lessening their effect on systemic pH. In the process of buffering, fixed acids consume bicarbonate. The volatile acid is excreted by the lungs and the fixed acids by the kidney, thus restoring buffer capacity.

The kidney reclaims bicarbonate by two mechanisms. Filtered bicarbonate is reclaimed in the proximal tubule, where 80%–90% of that which is filtered is reabsorbed. This occurs by the secretion of hydrogen ion by the proximal tubule cell, combination of the hydrogen with filtered bicarbonate to form carbonic acid, dehydration of carbonic acid to CO_2 and water within the tubule lumen, diffusion of CO_2 into the proximal tubule cell, formation of carbonic acid, removal of hydrogen to be secreted into the lumen and thereby, in effect, reabsorbing bicarbonate. Filtered bicarbonate is reclaimed in this manner; however, bicarbonate consumed in the process of buffering fixed acid is restored and the hydrogen eliminated by hydrogen secretion in the distal tubule. The hydrogen arises from carbonic acid and thus results in bicarbonate generation within the tubule cell. The bicarbonate stores are thus replenished. The hydrogen ion secreted in the distal tubule is fixed to phosphate, creatinine, and urate. These act as buffers and are excreted as weak acids. Hydrogen excreted in this manner is called titratable acid and, as such, lowers urinary pH. Since an intraluminal hydrogen concentration cannot exceed a concentration difference of more than 1,000:1, any amount excreted in excess of this back-diffuses into serum. The kidney is therefore incapable of lowering urinary pH values to less than a pH of 4.4. This mechanism limits the amount of hydrogen ion capable of being secreted. Fortunately, the ammonium system allows for further elimination of hydrogen ion not at the expense of lowering urinary pH. Ammonia is generated from glutamine in the distal tubule. Ammonia then diffuses across the membrane into the tubule lumen and combines with the hydrogen ion to form ammonium. Ammonium is trapped and combines with an available anion—usually chloride, sulfate, or phosphate—to form a neutral salt. The neutral salts do not influence pH and are excreted as such. Distal hydrogen secretion is promoted by increasing the severity of the acidosis, the presence of aldosterone, potassium depletion, and increased delivery of sodium to the distal nephron. Factors that alter bicarbonate reabsorption in the proximal tubule include the serum concentration of CO_2, potassium, chloride, phosphate, calcium, and parathormone, and the volume status of the patient. Increased bicarbonate reabsorption occurs with elevated CO_2, decreased potassium, diminished chloride, and elevated phosphate and decreased volume. Increased parathormone, hyperkalemia, and reduced CO_2 decrease bicarbonate reabsorption.

POTASSIUM

Only about 10%–20% of the filtered load of potassium is excreted. It is reabsorbed in the proximal tubule and may also be reabsorbed in the distal tubule and collecting duct under certain experimental conditions. It is secreted in the distal tubule, and it is by this mechanism that the majority of potassium is excreted in the urine. The rate of secretion is influenced by the presence of mineralocorticoids, urine flow, acid-base balance, and sodium intake. Mineralocorticoids stimulate the secretion of potassium mainly in the cortical collecting duct. Their presence, of course, results in sodium reabsorption, but there is not a 1:1 ratio of sodium for potassium. Alkalosis results in increased potassium secretion, whereas acidosis results in hydrogen secretion at the expense of potassium secretion. Potassium and hydrogen ion transport are linked in some fashion, explaining this phenomenon. An increased flow rate keeps the concentration gradient high between tubule cell and lumen and therefore promotes secretion. Sodium, by increasing flow rates in the distal tubule or by its effect on the Na/K cellular exchange pump, also increases potassium secretion.

CALCIUM

The bulk of filtered calcium is reabsorbed in the proximal tubule. However, the remainder is reabsorbed in the thick ascending limb of the loop of Henle and the distal tubule. Its reabsorption depends on phosphate, magnesium, and parathormone. Hypophosphatemia decreases calcium reabsorption as does an increase in magnesium concentration. Parathormone increases calcium reabsorption and reduces phosphate reabsorption. Parathormone also stimulates the kidney to produce 1,25-dihydroxyvitamin D_3, which increases gut absorption and bone reabsorption for calcium. Magnesium, calcium, and sodium may share some of the same carriers, since if there is an excess of one, others tend to be excreted, i.e., the common pump is saturated by the ion in excess.

MAGNESIUM

Magnesium is reabsorbed in the proximal tubule, thick ascending limb, and distal tubule. Increased sodium, calcium, and mineralocorticoid reduce reabsorption. Osmotic agents, renal vasodilatation, glucose, and hyperthyroidism increase excretion (Valtin, 1983).

URETERAL ANATOMY

The ureter is a thick-walled tube approximately 20–30 cm long. The proximal portion of the ureter is attached to the renal pelvis, and it is this structure into which the calyces and infundibula drain. The ureter travels through the retroperitoneum adjacent to the gonadal vessels and courses over the iliac vessels to enter the bladder posteriorly at the ureteral vesical junction. The ureter has three layers: an external adventitia, a smooth muscle coat, and a mucous membrane. In the adventitia, blood vessels and unmyelinated nerve fibers run longitudinally. The nerves are mostly of sympathetic origin; however, some are of parasympathetic derivation. The muscular coat consists of an intermingling mass of fiber bundles laid down in a spiral or helical fashion. There is no continuity of contractile elements and therefore no syncytium between cells. The muscle cells have close contact, and it is thought that in these areas electrical conduction occurs from one cell to another. The areas of close contact between cells are termed nexus (Burr et al., 1968). The innermost layer is composed of transitional epithelial cells. Between the muscular layer and the mucosal layer is the lamina propria, composed of loose connective tissue in which collagen fibers and blood vessels reside. The mucosa consists of 4–5 layers of transitional epithelial cells that are not supported by a basement membrane.

The ureter's blood supply comes from multiple sources. The upper portion and renal pelvis are supplied by branches of the renal artery. The midportion receives numerous branches from the gonadal vessels as well as from the common iliac vessels. The lowermost ureter receives its blood supply from the superior and inferior vesical arteries. The veins and lymphatics of the ureter accompany the arteries. In the cephalad portion of the ureter they follow the renal artery, in the midportion they course to the gonadal and internal iliac vessels, and in the caudal portion they drain with the bladder veins and lymphatics. Lymph node drainage,

therefore, occurs to the hilum of the kidney, to the periaortic, pericaval, retroaortic, retrocaval, perivesical and iliac nodes, depending on the segment of ureter. The nerve supply is derived from the 7th thoracic to the 1st lumbar sympathetic ganglia. There is perhaps some parasympathetic innervation from fibers coursing through the hypogastric plexus.

URETERAL PHYSIOLOGY

Ureteral peristalsis occurs from renal pelvis to bladder. The mechanism by which it is stimulated is not now well defined. It is clear that initial contraction occurs in the renal pelvis, perhaps initiated by passive stretching of its wall. The pelvis serves as a low-pressure pump funneling urine into the ureter (Djurhuus et al., 1976). It does not appear that innervation plays any role in the propagation of the peristalsis. Since the ureteral smooth muscle cells have numerous close contacts between them, the so-called nexus, muscular contraction arising in the renal pelvis is in all likelihood propagated along the ureter by these muscle cell connections. Transection and reanastomosis of the ureter does not permanently impair peristaltic waves across the suture line. Indeed, when healed, a peristaltic wave passes through the anastomosis and propagates distally in a coordinated fashion as though no anastomosis were present. Transection of the proximal ureter, however, may result in retrograde peristaltic waves, a phenomenon less likely with transections of the distal ureter.

Under conditions of normal urine flow, renal pelvic pressure remains relatively low, perhaps 1–2 mm Hg. Both renal pelvic pressure and ureteral pressure are unaffected at low flow rates by bladder pressure. However, during periods of diuresis, ureteral pressures are increased as bladder filling occurs. Indeed, with prolonged diuresis, ureteral dilatation may occur (Chung and Mantel, 1952). Moreover, during obstruction, maximal renal pelvic pressures can be obtained in excess of 150 cm H_2O (Kill and Aukland, 1961). Both α- and β-adrenergic receptors have been found in the ureter, and it is thought that the α-receptors are excitatory and the β-receptors inhibitory to ureteric contraction (Malin et al., 1970). Infection has also been demonstrated to inhibit ureteral peristalsis.

DETERMINATION OF RENAL FUNCTION

Because of the complexity and wide range of functions the kidney performs, the assessment of renal function requires measurement of individual processes occurring in various portions of the nephron. It is not practical to measure each one; therefore, selected functions are determined and, unless otherwise indicated, it is assumed that these reflect the general function of the entire kidney. For convenience we divide these studies into five groups: (1) glomerular filtration, (2) renal blood flow, (3) tubule electrolyte transport, (4) concentration and dilution, and (5) protein conservation.

GLOMERULAR FILTRATION RATE

The GFR is the amount of filtrate arriving in the proximal tubule per unit time. It assesses the normalcy of renal blood flow, glomerular integrity and proximal tubule pressure. Its measurement is based on the concept of a clearance (see above). The ideal substance for this measurement is freely filtered at the glomerulus and is not metabolized, secreted, or reabsorbed by the tubule. The material used to perform the clearance must be maintained at a constant concentration in the serum. For inulin this requires maintaining a constant IV infusion, since the substance is not endogenous to the body. Once a constant serum level is obtained the amount of inulin excreted in the urine is measured per unit time (urine concentration, U, multiplied by the volume, V, of urine excreted: U·V). To find the amount of serum from which this amount of inulin was extracted, the product is divided by the serum concentration, P. Thus U·V/P gives the amount of serum completely cleared of the substance per unit time or the GFR.

A normal GFR (measured by inulin clearance) for a young adult is 130 ml/min for males and 120 ml/min for females per 1.73 m^2 body surface area (BSA). Exercise lowers GFR while pregnancy may increase it by as much as 50%. Hydration, either extreme overhydration or dehydration, may also affect filtration rate. In children over the age of 1 year, when the GFR is corrected to 1.73 m^2 BSA, it is the same as for adults. Below the age of 1 year, it is less—only about 50% of the corrected adult value in the neonatal period. Similarly, with age the GFR falls, until by the age of 60 it is again only 50% of the young adult value.

Creatinine clearance is perhaps the most widely used method of clinically determining GFR. Its advantage over inulin lies in the fact that it is endogenously produced by muscle at a relatively constant rate and therefore need not be infused. Its disadvantages include the fact that it is secreted by the tubule and therefore is not an ideal substance for measurement of GFR. This results in a significant overprediction of GFR when the filtration is less than 20 ml/min—often by as much as 50%. At more normal levels of renal function, the amount secreted is offset by the concomitant measurement of serum chromogens during creatinine analysis. The inclusion of chromogens in the creatinine determination thus falsely elevates the serum value. The two errors cancel themselves at relatively normal levels of

renal function. Muscle wasting may reduce production of creatinine and thereby alter the constant production. Other causes of spurious serum creatinine values include diabetic ketoacidosis (Nanji and Campbell, 1981) and cephalosporin, cimetidine, trimethoprim-sulfamethoxazole, or propranolol administration (Muther, 1983). The bowel may secrete and metabolize creatinine in some disease states, thus altering the serum values. Although there are many difficulties with creatinine clearance, it serves as a useful clinical tool as long as its shortcomings are kept in mind and one of the interfering substances mentioned above is not present. It is particularly useful when assessing changes in renal function over the long term. The clearance is performed by obtaining a 24-hour urine collection, recording the volume and creatinine content and obtaining a serum creatinine concentration. The completeness of the 24-hour urine collection may be ascertained in the normal individual by the following calculation: the normal individual excretes between 0.5 and 1.5 mg creatinine/hr/kg of body weight. Thus, if the collection does not have that amount of creatinine in it, it is incomplete.

Several formulas have been developed to assess GFR from the measurement of serum creatinine alone without the need for a urine collection (Hull et al., 1981). Perhaps the most widely used is:

$$\frac{(140 - age) \cdot body\ weight\ (kg)}{72 \cdot Cr_s}$$

where Cr_s is the serum creatinine concentration. This gives the GFR for males; for females the value is multiplied by 0.85. This formula is useful for calculating GFR when the need to adjust drug doses arises. It is reasonably accurate when the GFR exceeds 40 ml/min. This formula is inaccurate in patients with spinal cord injury (muscle wasting), those with bowel interpositions in the urinary tract and those in whom one of the interfering substances is present (see above). It must be remembered that serum creatinine concentrations vary with muscle mass (more muscular individuals have a higher serum creatinine), the daily protein consumption and the metabolic state of the patient. Alterations in serum creatinine are a less accurate method of determining renal function.

Because of the inaccuracies of creatinine, other substances and methods have been developed in an attempt to measure GFR accurately without the difficulties encountered with the inulin clearance. Vitamin B_{12}, edetate (EDTA), sodium iothalamate, and sodium diatrizoate all have been shown to have clearance rates similar to inulin. They appear to have no advantages, however, over endogenous creatinine clearances except in selected circumstances.

Radiopharmaceuticals have enjoyed recent popularity as a less cumbersome method of determining renal function. The radiopharmaceutical technetium labeled pentetic acid (diethylenetriaminepentaacetic acid; DTPA) is used for GFR measurements and iodohippurate sodium I 131 (Hippuran) for renal blood flow (RBF) determination. There are two commonly used methods to determine GFR and RBF. The GFR or RBF can be calculated from an empiric formula which requires knowledge of the amount of isotope taken up by the kidneys over a six-minute period (Gates, 1982). The second method involves injecting the isotope and withdrawing serial blood samples over several hours. The blood samples are counted for isotope and a disappearance curve plotted. The GFR or RBF is calculated from the half-life of the disappearance curve (Britton and Maisey, 1983). In patients with relatively normal renal function who are in good health, these methods correlate well with inulin clearances. However, they are inaccurate in patients with significant edema or ascites, those with intestine interposed in the urinary tract and in patients with poor renal function.

RENAL BLOOD FLOW

The kidneys receive about one fifth of the cardiac output. The measurement of RBF allows an assessment of the vascular integrity of the kidney and, to a lesser extent, the viability of the renal parenchyma. A clearance technique or isotope washout method is usually employed. Clinically, the clearance methodology is most convenient, since it requires no special instrumentation and is based on the Fick principle. The amount excreted in the urine (urine concentration, U, multiplied by urine volume, V) divided by the renal artery minus the renal vein concentration difference gives the amount of plasma required to deliver the substance. Since it is from renal plasma that the substance is extracted, renal plasma flow is determined ($RPF = UV/A - RV$; where U is urine concentration; V, urine volume per unit time; A, renal artery concentration; and RV, renal vein concentration). Since the determination of renal vein concentration is cumbersome, a substance which is completely extracted in one pass is chosen, so that renal vein concentration is zero. Thus, the calculation simplifies to UV/P. Since no known substance is completely extracted in one pass, the measurement is called "effective" renal plasma flow. Para-aminohippurate (PAH), iodopyracet (Diodrast), and Hippuran, when administered at low-dose levels, are almost completely removed and are the agents employed to measure renal plasma flow. The renal extraction is about 90%; thus there is about a 10% error. Moreover, not all blood to the kidney is exposed to functioning tubules and glomeruli, so this technique

may be significantly inaccurate in disease. Normal values are 600–700 ml/min/1.73 m². Renal blood flow (RBF) may be calculated from renal plasma flow (RPF) by $RBF = RPF/(1 - Hct)$ where Hct is the patient's hematocrit.

Washout methods involve injecting a bolus of isotope (xenon or krypton) in the renal artery and then measuring the speed with which the isotope leaves renal parenchyma by recording the rate at which radioactivity diminishes over the kidney by conventional scanning techniques. The concentration of isotope in the kidney as a function of time is plotted. The equation that describes the disappearance curve is of the fourth order or greater. By stripping the curve, i.e., generating the equation for the curve, RBF to various segments (cortex and medulla) may be determined (Rosen et al., 1968).

Finally, the disappearance of isotopically labeled Hippuran from the blood or timed accumulation by the kidney may be used as a gross estimate of RBF (see above).

TUBULE ELECTROLYTE TRANSPORT

Integrity of transport processes can be determined by assessing the transport of various electrolytes. This is important in certain disease entities such as renal tubular acidosis in which there is a defect in hydrogen ion secretion. That defect may be brought to light by stressing the kidney's hydrogen ion transport process. Other defects such as distal tubule ammonium dysfunction, as occurs in certain patients who form uric acid stones, and salt-losing nephropathy may similarly be assessed.

Sodium excretion rate may be used as a crude index of proximal and distal tubule function. The normal kidney under conditions of severe salt restriction should be able to excrete less than 0.1% of the filtered load of sodium in a urine concentration of less than 1 mEq/L.

Distal tubule function may be assessed by the administration of a mineralocorticoid and observation of a decrease in urine sodium concentration and a rise in urine potassium concentration. Distal tubule hydrogen secretion may also be assessed by administering hydrochloric acid, ammonium chloride, or one of the cationic amino acids (lysine, arginine, or histidine) and observing a fall in the urine pH to less than 5 and/or a lack of ammonia generation.

CONCENTRATION AND DILUTION

Alterations in concentration and dilution indicate distal tubule and collecting duct dysfunction. Many entities affect these functions (see Table 15–3), most often pyelonephritis and urinary obstruction. The normal kidney can dilute to an osmolality of 40 mOsm/kg and concentrate to 1,200 mOsm/kg. Concentrating ability is determined either by water deprivation or by the administration of vasopressin. Diluting ability is determined by having the patient drink 1 L of water over a short time. The patient will excrete more than half the volume over three hours with a urine osmolality of 80 mOsm/kg or less.

PROTEIN CONSERVATION

The normal excretion of protein in the urine is 75–150 mg/24 hours. The bulk of the protein is albumin and 60%–70% of that excreted is of renal origin. Abnormal excretion is usually a reflection of glomerular dysfunction or a systemic nonrenal disease. In the former, it generally indicates significant renal deterioration. Total protein excretion is measured per 24 hours. If abnormal, orthostatic proteinuria that does not imply disease is eliminated by dividing the collections into an overnight specimen and a daytime specimen. Protein in the upright or daytime specimen but not in the supine or overnight specimen confirms the orthostatic nature of the proteinuria. Finally, certain disease entities such as myeloma produce abnormal proteins. These are determined by subjecting the urine to electrophoresis.

THE DETERMINATION OF RENAL FUNCTION IN THE PRESENCE OF INTESTINAL SEGMENTS

The assessment of renal function in patients in whom a segment of intestine is interposed in the urinary tract requires modifying the techniques normally employed to determine renal function in the intact collecting system. This is necessary, since the intestine, which is used as a conduit, functions both as a vehicle of urine transport and as an absorptive and secretory surface. Thus, urine may be considerably modified after transversing intestine, making it difficult to determine whether the measured substance has been altered by bowel absorption or secretion. Many of the substances used to assess renal glomerular function and RBF are absorbed by the intestinal segment. Urea is perhaps the most permeable, but creatinine, inulin, and PAH are also significantly absorbed (Koch and McDougal, 1985). Their absorption depends on their concentration and the amount of time they remain in contact with the intestine, which is a function of the surface area to which they are exposed and the urine flow rate. Figures 15–8 to 15–10 illustrate the dependency of urine flow rate on absorption of urea, creatinine, and inulin in patients with ileal and colon conduits. For clarity, the figures compare the percent maximal clearance achieved as a function of flow rate in normal collecting systems and ileal and colon interpositions. The illustrations demonstrate that when urine flow rates exceed 250 ml/hr, in-

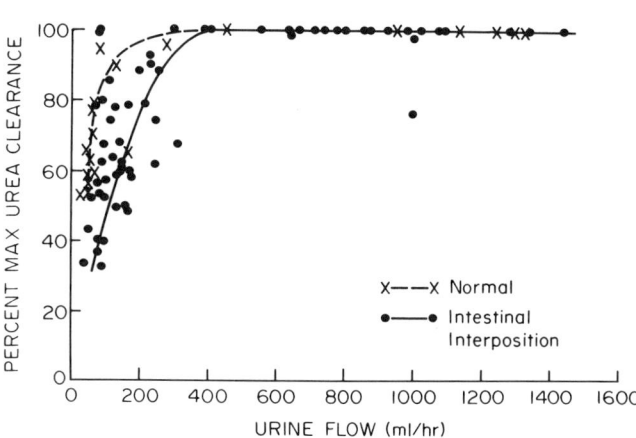

FIG 15–8.
Flow dependency of urea clearance in patients with normal urinary tracts compared with those with ileal and colon interpositions.

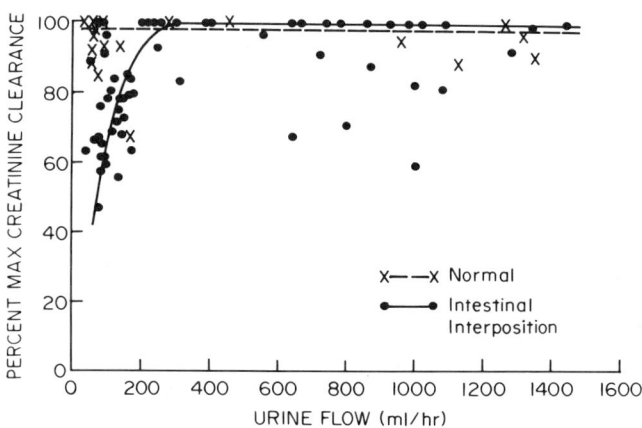

FIG 15–9.
Flow dependency of creatinine clearance in patients with normal urinary tracts compared with those with ileal and colon interpositions.

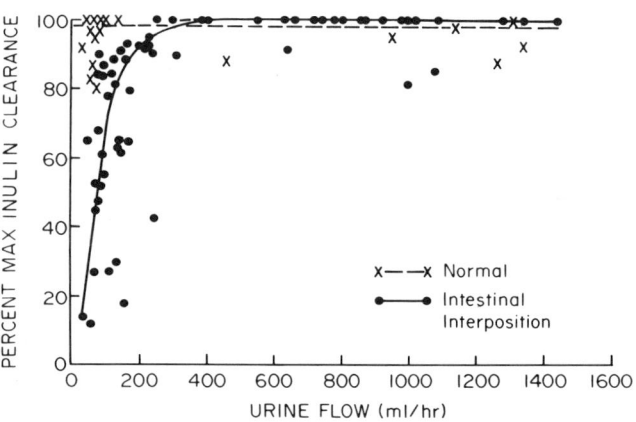

FIG 15–10.
Flow dependency of inulin clearance in patients with normal urinary tracts compared with those with ileal and colon interpositions.

testinal absorption for creatinine and inulin is negligible, and at these flow rates the substances more closely measure the true GFR and renal plasma flow (PAH shows a similar flow dependency). These curves were derived from a group of patients with ileal conduits, colon conduits and intact urinary tracts. Since renal function varied from poor to normal, the patients were compared by factoring each measured clearance by the maximal clearance obtained for that patient. At very high flow rates, patients with stomal stenosis tend to diminish their clearance, indicating that at these flows, the intestine serves as a functional obstruction (McDougal and Koch, 1986).

MEASUREMENT OF GFR

Since clearance is flow dependent, it is necessary to establish a diuresis in patients with ileal and colonic segments at urine flows between 300 and 700 ml/hour. These flows may be achieved by fluid administration alone. Rarely a diuretic such as furosemide may also be required. The urine is collected over one hour. The volume is measured and the substance used for the clearance measured in the serum and urine. Creatinine is perfectly adequate in these patients, provided the serum does not contain excessive ketoacids, cephalosporins, cimetidine, trimethoprim-sulfamethoxazole, or propranolol. Since diuretic creatinine clearance parallels diuretic inulin clearance, the former is the clinically expedient method of determining GFR. Creatinine clearance measured in this manner is about 20% less than inulin clearance at GFRs that exceed 20 ml/min. If the GFR is less than 20 ml/min, the creatinine clearance may overestimate the inulin clearance due to secretion by the renal tubule. At very low GFRs (i.e., less than 10 ml/min), the error may be as great as 50% (Fig 15–11).

Renal isotope determination of GFRs in these patients is not useful, since there is little correlation between GFRs obtained by this method and those obtained by inulin clearance. This is true at all levels of renal function (McDougal and Koch, 1986).

Since the bowel absorbs, secretes, and metabolizes creatinine, the calculation of GFR from serum creatinine, age, weight, and sex by empiric formulas is totally unreliable and not useful in these patients (McDougal and Koch, 1986).

RENAL BLOOD FLOW

PAH clearance obtained when urine flows range between 300 and 700 ml/hr accurately reflects effective renal plasma flow. Isotope scans are not accurate in determining renal blood flow in these patients.

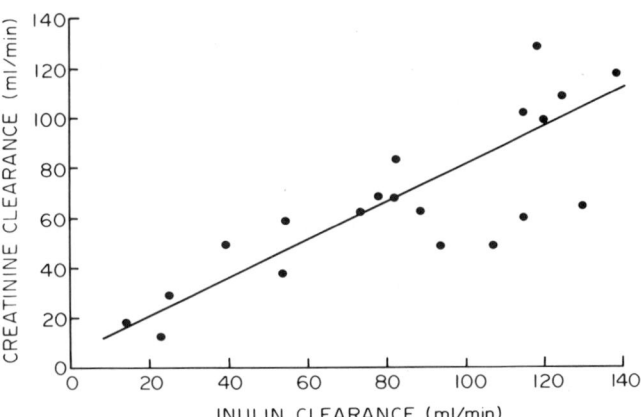

FIG 15–11.
Diuretic inulin and creatinine clearances in patients with normal urinary tracts compared with those with ileal and colon interpositions. Notice the excellent correlation and that creatinine clearance is consistently 20% less than inulin clearance.

TUBULE FUNCTION

The determination of tubule function by the measurement of sodium excretion and hydrogen ion secretion under conditions of sodium deprivation or hydrogen loading (see above) cannot be accurately performed in these patients. Although some qualitative indication of the way electrolytes are handled can be obtained, the maximal capabilities of the kidneys with respect to electrolyte transport cannot be determined when urine is analyzed after it has traversed an intestinal segment.

CONCENTRATION AND DILUTION

Since the bowel cannot concentrate much above 50 mOsm/kg greater than serum, urine traversing an intestinal segment, if more concentrated than serum, is diluted. Thus, in patients with ileal and colon conduits, maximal urinary concentration does not usually exceed 400 mOsm/kg. The determination of maximal urinary concentrating ability of the kidney cannot be performed in patients with intestinal segments (McDougal and Koch, 1986). Although the kidney's ability to dilute can be assessed utilizing the techniques described above, maximal dilutional capabilities cannot be determined due to absorption by the intestine. Therefore statements about distal tubule and collecting duct function based on maximal concentrating ability in patients with colon or ileal segments are meaningless.

It is clear that patients who have colon or ileal conduits, those with intestine interposed in the urinary tract, and those in whom intestine is used as a urinary reservoir require special manipulations to measure renal function. The GFR and renal plasma flow can be accurately assessed if, while they are being measured, urine flows range between 300 and 700 ml/hr. They must be measured by classic clearance techniques, not by either isotopic scans or empiric calculation from serum creatinine. Tubule function is most difficult to assess in these patients, and at present there is no good way of determining the kidney's maximal capacity for conserving sodium, excreting hydrogen, or reabsorbing water.

ACUTE RENAL FAILURE

Acute renal failure (ARF) in surgical patients has numerous etiologies (Table 15–1). Whatever the primary cause, it is associated with significant morbidity and mortality and presents as a progressive rise in BUN and creatinine, often with a fall in urine output. ARF is said to be oliguric when the urine output is less than 400 ml/24 hr and nonoliguric when the urine output exceeds this amount. The initial therapy and prognosis for ARF are not only determined by examining the serum chemistries and state of fluid balance but depend as well on the expeditious assignment of the patients to one of the three subdivisions of this disease: prerenal, intrarenal, or postrenal. It is necessary, therefore, to begin diagnostic procedures which will determine the type of ARF simultaneously with therapy directed at correcting fluid imbalance and electrolyte abnormalities.

IMMEDIATE MANAGEMENT OF ACUTE RENAL FAILURE

A careful history and physical examination will often suggest the type of ARF; however, the immediate management of this disease is dictated by the extent of serum chemical aberrations and fluid imbalance. Thus initial steps are directed at defining these abnormalities. An ECG, serum electrolytes, and urine for microscopic and chemical analysis are obtained. If the serum potassium is elevated, the ECG changes will indicate the rapidity with which the hyperkalemia must be corrected and will serve not only as a baseline against which the success of therapy may be measured but also as an immediate approximation of potassium concentration during the patient's course. Hyperkalemia results in progressive peaking of the T wave, prolongation of the QRS complex, and, at high concentrations, absence of the P wave. When the serum potassium is greater than 7 mEq/L or when prominent alterations of the ECG occur (particularly when the P wave is absent), immediate reduction in serum potassium is accomplished by infusing hypertonic sodium bicarbonate and/or administration of insulin and glucose in a ratio of 1

TABLE 15–1.

Etiology of Acute Renal Failure

Prerenal failure
  Excessive nitrogen load
  Myocardial pump failure
  Hypovolemia
Intrarenal failure
  Occlusion of renal arteries or renal veins
  Arteriolar damage
    Malignant hypertension
    Polyarteritis
    Hypersensitivity angiitis
  Disseminated intravascular coagulation
  Glomerulonephritis
    Lupus erythematosus
    Poststreptococcal glomerulonephritis
  Parenchymal damage
    Acute interstitial nephritis
    Acute pyelonephritis
    Papillary necrosis
      Diabetes mellitus
      Nephrosclerosis
      Sickle cell anemia
      Analgesic medications
    Cortical necrosis
    End-stage renal disease
  Hepatorenal syndrome
  Hepatic artery ligation
    Vasomotor nephropathy
      Shock
      Sepsis
      Transfusion reactions
      Crush injury
  Tubule toxins
    Drugs
    Myoglobin
    Poisons
Postrenal failure
  Obstruction of the collecting system
    Tumors
    Calculi
    Infection and inflammatory lesions
    Fibrosis
    Blood clots
    Renal papillae
    Increased intra-abdominal pressure
    Retroperitoneal hemorrhage

unit/5 gm. Calcium is administered IV to stabilize myocardial conductivity. Acidosis and hyperkalemia may be partially corrected by giving bicarbonate, which results in a shift of potassium into the cell with preservation of electroneutrality by concomitant movement of hydrogen ion out of the cell. If some renal function is preserved, potassium secretion is increased in the distal tubule, since bicarbonate reduces the amount of hydrogen ion competing for the common hydrogen-potassium secretory mechanism, thus allowing it increased access to potassium. Glucose and insulin result in movement of potassium into the cell, either by binding it during glucose transport or during glycolytic phosphorylation. These mechanisms result in a transitory fall in serum potassium, since it moves out of the cell in the former when acidosis recurs and in the latter when substrate has been metabolized. It is important, therefore, to lower the total body serum potassium by simultaneously employing an ion exchange resin such as polystyrene sulfonate (Kayexalate), 10–20 gm orally or 50 gm by enema, both given with sorbitol or by peritoneal or hemodialysis. Usually 2–3 hours are required before ion exchange resins show an effect or before dialysis can be instituted. Most commonly, alkalinization, ion exchange resins, and occasionally dialysis are all that are required, with glucose and insulin being reserved only for those cases in which hyperkalemia is an immediate threat to life. The state of fluid balance is determined by analysis of intake and output, weight, venous or left atrial filling pressure, physical examination, and history. Inappropriate intake for the amount of output usually accounts for the imbalance; however, therapeutic maneuvers used to treat electrolyte abnormalities may be contributory as well. Kayexalate, by exchanging sodium for potassium and sodium bicarbonate, can cause an excessive sodium load and fluid retention. When either overhydration or hypernatremia becomes an urgent problem, dialysis is most effective in correcting the disorder. Usually time is not of the essence, and sodium and fluid restriction will suffice. Low venous or pulmonary wedge pressure, clinical signs of dehydration, a history of blood loss, and hypotension indicate volume depletion and are treated by appropriate fluid replacement.

During the acute treatment of hyperkalemia and fluid imbalance, an attempt should be made to define the type of ARF involved. A careful history and physical examination can often be diagnostic. The aseptic and atraumatic passage of a Foley catheter is helpful from both a diagnostic and occasionally a therapeutic point of view. If the patient is capable of voiding spontaneously and the residual is less than 30 ml, the catheter is withdrawn. If the diagnosis remains unclear after these simple maneuvers, a more sophisticated work-up is in order and will include chemical analysis of the urine and serum for sodium, potassium, urea, creatinine, and osmolality; central venous or right atrial pressure determinations; and ultrasonograms of the kidneys. Intravenous pyelography is rarely, if ever, indicated. The degree of renal deterioration is usually of such an extent that visualization does not occur. Indeed, the contrast material may cause further deterioration of the severely compromised kidney and, therefore, indiscriminate use of contrast material is to be condemned. If the diagnosis is still in doubt, retrograde or antegrade pyelography or sonography is indicated. The characteristics and therapy for each type of ARF are described in detail.

PRERENAL FAILURE

Prerenal failure occurs when there is an excessive nitrogen load or reduced blood supply to the kidney. The former may result from increased muscle catabolism, blood breakdown within the GI tract, or excessive protein alimentation. The latter appears in the presence of volume depletion, congestive heart failure, in valvular heart disease, or any disease that causes myocardial pump failure.

PATHOPHYSIOLOGY.—In prerenal oliguria there is a reduction in renal blood flow. When the blood flow is reduced to a level such that the afferent arteriolar pressure falls below 60 mm Hg, reduced filtration pressure results in a decrease in the amount of filtrate delivered to the proximal tubule, glomerulotubular balance is disrupted, and sodium, water, and urea are reabsorbed in increased amounts. A BUN/creatinine ratio of greater than 10:1 occurs because creatinine, when filtered, is not reabsorbed by the tubule, whereas urea is reabsorbed in increased amounts. The proportional excretion of BUN and creatinine which occurs in health is disrupted and the urea recirculates, thus resulting in the abnormally high BUN/creatinine ratio.

DIAGNOSIS.—The serum BUN/creatinine ratio is greater than 10:1, and the central venous pressure (CVP) and right arterial pressure in volume depletion are low, providing coexisting myocardial disease is not present. The small volume of urine excreted is highly concentrated with a low sodium content (<15 mEq/L). The urine urea concentration divided by the plasma urea concentration (U/P urea) is greater than 20:1 and the U/P osmolality greater than 1.5. The fractional excretion of sodium or the ratio of U/P sodium to U/P creatinine ($U/P_{Na} : U/P_{Cr}$) is less than 1%. The urine sediment may reveal hyaline casts but generally will be free of casts and red and white blood cells (Table 15–2).

TREATMENT.—The treatment of prerenal oliguria is directed at the primary disease: (1) correction of the lesion which has caused the increased nitrogen load, (2) improvement of the failing myocardium, or (3) volume repletion.

INTRARENAL FAILURE

Acute intrarenal failure may follow an ischemic or a nephrotoxic injury. Examples of the former include sepsis, hemorrhagic shock, aortic cross clamping, and surgical or obstetric misadventures; examples of the latter include drugs (aminoglycosides), crush injuries

TABLE 15–2.
Differential Diagnosis of Prerenal, Intrarenal, and Postrenal Failure

MEASUREMENT	NORMAL	PRERENAL	INTRARENAL	POSTRENAL
Blood				
CVP, cm H_2O	5–8	Low to normal	Normal to elevated	Normal to elevated
BUN/creatinine ratio	10:1	>10:1	10:1	10:1+
Urine				
Sodium, mEq/L	15–40	<15	>40	>40
Potassium, mEq/L	15–40	Variable	Variable	Variable
Osmolality, mOsm/L	400–600	>450	<300	<300
Volume, ml	800–1200	Low	Variable	Variable; initially low
Urine/blood ratio				
Urea	20:1	>20:1	<10:1	<5:1
Osmolality	1.5–2.0	>1.5:1	<1.2:1	<1.0:1
Creatinine	20:1	>40:1	<20:1	<20:1
Fractional sodium excretion, %	Variable	<1.0	>1.0	>1.0
Urine—microscopic analysis	0–1 RBC 0–1 WBC Occasional hyaline cast No cellular casts	Occasional hyaline cast	Tubule epithelial casts RBCs, free heme or myoglobin	RBCs and WBCs Malignant cells Crystals

(myoglobin), and diagnostic agents (IV contrast). One half to two thirds of the patients will be oliguric, while the remainder will have urine outputs exceeding 400 ml/24 hr. Patients who present with nonoliguric ARF have higher sodium concentrations, greater fractional sodium excretions, shorter hospital stays, a lower mortality, and require fewer dialyses than those who present with oliguria. Indeed, oliguric renal failure carries with it a mortality of 50%–70%, while the mortality is approximately 25% in nonoliguric patients. The importance of sodium excretion rate is emphasized by the finding that the less fatal nonoliguric renal failure patient has a higher sodium excretion rate than does the oliguric patient. Moreover, in the oliguric acute renal failure patient, a persistent urine sodium concentration less than 40 mEq/L is associated with only a 37% survival, while those with a urinary concentration greater than 40 mEq/L have a survival of 56% (Baek and Makabali, 1975). It has been suggested that a low fractional excretion for sodium suggests that the etiology of the injury involves tubular obstruction or alterations in renal hemodynamics rather than direct tubular nephrotoxicity (Corwin et al., 1984). The recovery is also somewhat dependent on the cause of the acute renal failure. Approximately 80% of patients with acute cortical necrosis and 66% of those with thrombotic thrombocytopenic purpura or the hemolytic uremic syndrome do not recover renal function adequate to support life. This is in contrast to patients with acute tubular necrosis and acute interstitial nephritis in whom 62% recover normal renal function. Thirty-one percent show a par-

tial recovery and only 6% have no recovery whatsoever (Bonomini et al., 1984). Another indication of recoverability is correlated with the length of time anuria occurs and the degree to which the kidney takes up nuclide on a renal scan. In one study, infants who remained anuric for at least four days following the acute insult and revealed no uptake of radionuclide on scan invariably died of their ARF, whereas neonates with ischemic ARF who were nonoliguric and whose kidneys took up the nuclide had a more favorable prognosis (Chevalier et al., 1984).

PATHOPHYSIOLOGY.—A great deal of clinical and experimental data have been accumulated trying to define a pathophysiology for ARF. Unfortunately, most of the information comes from experimental animals, and correlation with the clinical situation is often contrived. There are four proposed mechanisms: (1) vasomotor, (2) increased nephron permeability, (3) tubule obstruction, and (4) decreased ultrafiltration (Hermreck, 1982). Since the initiating events are multiple and varied, it is likely that in each case the acute renal failure has been caused by a combination of these factors. It is important to understand each of the mechanisms, since it is on these hypotheses that therapy of established ARF is based.

The vasomotor theory proposes that there is not only a fall in total renal blood flow but also a redistribution of intrarenal blood flow. Xenon washout and microsphere injection studies have demonstrated a shift of blood flow away from the cortical nephrons toward the juxtamedullary nephrons (Siegel et al., 1977). The renin angiotensin system may play a role in the renal blood flow shift. Moreover, prostaglandins may also be contributory. If the initial injury causes an increased release of renin, cortical afferent arteriolar constriction would result in reduction of cortical blood flow. Since stimulation of the autonomic nervous system releases renin, its role in the etiology is unclear but may partially explain the occurrence of acute renal failure in patients whose aortas have been cross clamped. Further evidence for the importance of renin is demonstrated by depleting it in experimental models by either chronic salt loading or antirenin antibodies. A renin-depleted animal has not only a greater chance of recovery but also a more rapid return of function. Moreover, in many patients, renin activity is elevated early in the course of acute renal failure (Stein et al., 1978). Unfortunately, specific antagonists of renin or angiotensin do not alter the course of these patients once the ARF is established.

Additional evidence for the significance of hormonal regulation of intrarenal blood flow during ARF comes from reports of indomethacin-induced acute renal failure. Indomethacin blocks the synthesis of prostaglandins. PGE_2, a vasodilator, appears to play a significant role in maintaining renal blood flow in the face of pathologic forces that tend to compromise it. Thus in the proper setting, eliminating PGE_2 results in the development or exacerbation of ARF (Torres et al., 1975). Although renal vasoconstriction and low renal blood flow may play key roles in the early stages of intrinsic renal failure, later in the course spontaneous increases in renal blood flow often occur without a concomitant increase in GFR.

The hypothesis of increased tubule permeability proposes that though filtration pressure may or may not be reduced when the filtrate arrives in the proximal tubule, it leaks out, resulting in an effective reduction in filtration. Conflicting data supporting this theory have been reported. Some investigators have demonstrated leakage of labeled inulin and mannitol from the tubule. Moreover, the dye lisamine green has been observed to leak out of the tubule. Others, however, using micropuncture techniques have failed to show increased permeability in split drop experiments.

Tubule blockade may play a major role in ARF after surgery. Myoglobin can precipitate in the tubule and cause mechanical blockage of the tubule lumen. In models of renal artery cross clamping, up to 90% of the proximal tubules have been found to be occluded by swollen cells and desquamated proximal tubule microvillae; however, other experimental models that simulate medical causes of acute renal failure are less convincing.

A decrease in ultrafiltration probably plays a major role in many causes of ARF. Numerous studies have demonstrated alterations in the glomerular basement membrane and supporting structures. These alterations result in a decrease in the membrane permeability by as much as 60%. This, coupled with decreased blood flow, results in a lessened hydrostatic pressure gradient and a marked decrease in the delivery of filtrate to the proximal tubule. Perhaps the changes which occur in glomerular membrane properties may explain the dissociation between renal blood flow and GFR, since a decreased glomerular capillary permeability would continue to cause a decreased GFR even in the face of increased renal blood flow (Kon and Ichikawa, 1984).

DIAGNOSIS.—The diagnosis is based on historical, chemical, and radiologic determinations. The BUN/creatinine ratio is 10:1, and the CVP or left arterial filling pressure is normal or elevated. The urinary sodium concentration exceeds 40 mEq/L with a variable potassium secretion—usually less than 20 mEq/L. The U/P osmolality is less than 1.2 and the U/P urea less than 10. The fractional sodium excretion exceeds 1%. Tubule epithelial cells and tubule epithelial cell casts may be observed in the urine (see Table 15–2). A radioimmunoassay of tubule antigens has been developed which successfully diagnoses 80% of patients with acute tubule necrosis (Zager et al., 1980). This test has not gained popularity, because the diagnosis of intrarenal failure may be made with this degree of accuracy by conventional techniques.

Ultrasonography or roentgenographic studies of the abdomen without the administration of pyelographic contrast should be used in an effort to determine whether the acute problem is superimposed on prior renal disease. Preexisting renal disease may be demonstrated by unilateral or bilateral small kidneys indicative of a vascular or infectious etiology. An enlarged renal outline may imply either acute renal vein thrombosis or an infiltrative lesion such as myeloma or lymphoma or, when associated with dilated calyces, suggests postrenal failure as the cause. Radionuclide renal scans are occasionally helpful in cases of bilateral renal artery thrombosis and may suggest that arteriography should be performed.

The BUN rises between 10 and 20 mg/dl/day, while creatinine increases 0.5–1.0 mg/dl/day. Plasma potassium increases 0.5 mEq/L/day; and since 50–100 mEq of fixed acid is retained, the plasma bicarbonate falls by 1–2 mEq/L/day. In the posttraumatic state, the hypercatabolic response may result in a more rapid rise in the BUN, potassium, and fixed acid accumulation. A crush injury involving muscle necrosis and myoglobin nephropathy can result in rises in serum creatinine up to 2 mg/dl/day and rapid increases in serum potassium and uric acid. Hypocalcemia is often noted and may have significant cardiac consequences, particularly if the serum potassium is elevated.

TREATMENT.—Careful fluid balance, the judicious use of ion exchange resins, dialysis, and administration of a potent diuretic when oliguria occurs are the hallmarks of therapy. The use of furosemide in the treatment of acute oliguric intrarenal failure is controversial. Several studies have failed to demonstrate more rapid recovery, improved GFR, or reduction in the number of dialyses required, except when cardiac decompensation is present (Brown et al., 1981). On the other hand, converting oliguria to nonoliguria makes subsequent fluid management less cumbersome. If it is given, a dose of 80 mg is tried; if unsuccessful, it is doubled and the doubling repeated until finally a bolus of 1 gm is achieved. One of three responses occurs: (1) Oliguria persists. The patients are treated with replacement of net water requirements (insensible loss − water of metabolism = 10 ml/kg/24 hr). If a return in urine output does not occur within 21 days, the chance of recovery of renal function is poor. (2) A diuresis ensues, GFR increases, and an immediate reversal in the rise of BUN and creatinine occurs. It is important to rehydrate the patient in this setting, since this response suggests prerenal oliguria. These patients recover—often obtaining normal renal function. (3) A diuresis occurs, GFR remains low and BUN and creatinine continue to rise. These patients are treated with replacement of net water loss (10 ml/kg/24 hr) plus urine output, GI, and tube losses. Large sodium losses occur during this phase and require replacement. Potassium losses are small, although rarely they may be excessive and require replacement. Appropriate serum potassium concentrations are maintained by restriction of potassium and the use of ion exchange resins and dialysis.

Low dose dopamine (1 to 2 μg/kg/min) has not been shown to alter the course of ARF. There is some evidence, however, to indicate that it is helpful in the hepatorenal syndrome and in preventing ARF if used during the early prodromal period (Tiller and Mudge, 1980). The administration of dopamine to these patients results in a significant increase in the diuresis and natriuresis and when combined with furosemide may improve renal function in at least some patients with ARF. Indeed, experimental evidence shows a synergistic protective effect with the combination of dopamine and furosemide in several models of ARF. Dopamine causes marked vasodilatation of both afferent and efferent arterioles (Graziani et al., 1984). Mannitol has been shown to be helpful in preventing ARF if given before interrupting renal artery flow and in lessening the toxic effects of myoglobin.

Beta-blockers have produced variable results, and angiotensin and renin inhibitors have not proved use-

ful. Alkalizing agents, by solubilizing organic acids (drugs) when they are etiologic, and diuresis, by diluting toxic substances (cisplatin and myoglobin), are helpful prophylactically. Finally, if given early in the course of the disease, calcium channel blockers may be helpful by reducing intracellular calcium in tubule cells that have suffered an ischemic injury and are deprived of ATP.

Perhaps the one drug regimen that all agree is most helpful, particularly in surgical patients, is nutritional support. Hyperalimentation with essential L-amino acids has been shown to improve recovery and reduce the number of dialyses required (Freund et al., 1980).

Dialysis plays an important role in the management of these patients. If the BUN is greater than 100 and the creatinine exceeds 12, dialysis is mandatory. Recent evidence suggests that if the BUN is kept below 70, the incidence of sepsis is reduced from 88% to 63%, bleeding from 60% to 36%, and mortality from 80% to 36% (Conger, 1975). Others have shown consistently that the mortality rate is lowered, extrarenal complications are less frequent, and the clinical course is better in any type of ARF when dialysis is employed before the clinical signs of uremia occur. Dialysis may be accomplished either peritoneally or hemically.

Plasma exchange may be useful in treating patients whose renal failure is due to nondialyzable substances such as dextran. Slow continuous ultrafiltration using a filter placed between two limbs of a Scribner shunt has been useful in selected critically ill patients. Its advantages include hemodynamic stability, ability to remove large fluid volumes—thus making treatment modalities such as IV nutrition feasible—and no requirement for systemic anticoagulation. It has many disadvantages, among them electrolyte abnormalities, bleeding, clotting of the shunt filter, excessive drug removal, azotemia, hypotension, and bacteremia (Synhaivsky et al., 1983).

The mortality in this group is high; however, between 25% and 70% of patients surviving an acute tubule injury eventually recover sufficient renal function to support life without dialysis. Of those recovering, 20%–40% will have a reduced GRF for one year or longer. Abnormalities of tubule function including renal glucosuria and decreased concentrating ability may persist indefinitely.

The most frequent cause of death is infection (Schrier, 1979). It causes 36% of the deaths and is usually pulmonary in origin. The next most common cause of death is GI hemorrhage. Delayed wound healing, poor generation of granulation tissue, anorexia, anemia, and bleeding abnormalities due to platelet malfunction all contribute to the high level of morbidity in this disease.

Because of the high mortality of ARF, perhaps the

most effective treatment is prophylaxis. Adequate hydration and selective use of dopamine, mannitol, furosemide, and alkalinization in patients undergoing surgery who are known to have a high risk for the development of postoperative ARF should lessen the incidence of the disease. Risk factors that carry with them an increased incidence of postoperative ARF include advanced age, cardiac or hepatic failure, sepsis, jaundice, rhabdomyolysis, and massive blood transfusions.

MYOGLOBIN NEPHROTOXICITY

Rhabdomyolysis has been reported to be the cause of ARF in approximately 5% of all patients suffering from acute renal insufficiency. The prognosis for recovery is excellent, provided rhabdomyolysis is recognized promptly and therapy is begun immediately. There are traumatic and nontraumatic causes of rhabdomyolysis (Koffler et al., 1976). The former include crush injuries, electric burns, arterial occlusion, and surgery. Nontraumatic causes include increased muscle exertion, inflammatory muscle diseases, infection, toxic drugs, and metabolic abnormalities (particularly hypokalemia).

Following the precipitating event, the patient becomes acutely ill with fever, weakness, and pain. The urine is brownish and is dipstick-positive for heme. Since the molecular weight of myoglobin is much lower than that of hemoglobin, it is rapidly cleared from the serum. Thus, a spun blood sample will reveal a clear serum in contradistinction to hemolysis, where the serum will be red due to retained hemoglobin. A high serum CPK level establishes the diagnosis. Serum potassium and phosphate are often elevated, and hyperuricemia may be severe. The mechanism of ARF is unclear but is probably due to a combination of events including low flow, prolonged contact of the tubule lumen with myoglobin, and high uric acid concentration. Therapy is directed at establishing a diuresis with either a loop diuretic or by volume expansion, with saline combined with either mannitol or a loop diuretic. The diuresis dilutes the toxic myoglobin, removes it from the kidney, reduces the serum uric acid concentration, and modestly alkalinizes the urine. Alkalinization is advisable initially, since an acid medium promotes myoglobin dissociation into ferrihemate and globin. It is the former which is toxic (Eneas et al., 1979).

RADIOCONTRAST-INDUCED ACUTE RENAL FAILURE

The incidence of radiocontrast-induced ARF has been reported to be as high as 12% of all patients receiving contrast material. The incidence of contrast nephropathy in patients with normal renal function is lower than 2% after angiography and between 1.5% and 15% after IV urography. In contrast, among patients with diabetes and coexisting renal insufficiency, more than half develop significant renal damage after IV urography (Berkseth and Kjellstrand, 1984). The occurrence of clinically significant ARF in the general population is exceedingly low; it is most commonly found following angiography or IV urography in patients with significant risk factors for the development of the disease. Almost all patients who develop clinically significant ARF have either previously existing renal disease with compromised renal function or diabetes with moderate renal impairment. Adult onset diabetic patients with normal renal function do not, however, appear to be at increased risk for development of ARF. Patients with juvenile onset diabetes are more likely to develop renal insufficiency even though they do not demonstrate significant renal impairment prior to the injection of contrast medium. Other factors that have been less well correlated with the development of renal insufficiency following the injection of radiocontrast include advanced age, dehydration, multiple contrast exposures within 24 hours, hyperuricemia, proteinuria, hypoalbuminemia, multiple myeloma, and impaired liver function. Dehydration per se does not appear to be a specific risk factor unless the patient has preexisting renal disease. Conversely, in patients with prior renal disease, establishing a diuresis following the injection of contrast appears to lessen the incidence of renal deterioration. In patients with multiple myeloma, there is no compelling evidence of any increased risk for contrast-induced renal failure unless there is preexisting renal insufficiency. However, contrast-induced renal failure in patients with multiple myeloma often has catastrophic implications because it is irreversible.

In summary, existing renal insufficiency is the single most important predisposing factor. Moreover, diabetes does not increase the risk of contrast-induced renal failure if renal function is normal and if the patient does not have juvenile onset diabetes. When radiocontrast-induced renal failure does occur, 75% of patients are oliguric for the first 24 hours following contrast exposure. Prolonged oliguria results in only partial return of renal function. The serum creatinine generally peaks by the seventh postinsult day, and renal function returns to normal in three quarters of the cases. Dialysis is rarely required, and almost all patients have an adequate return of function and do not require long-term dialysis (Mudge, 1980). The pathogenesis of the disorder has been hypothesized to be due to direct tubule toxicity, to renal ischemia due to alterations in blood flow with shunting of blood from cortex to medulla, and to intratubular obstruction. Because contrast agents are uricosuric, tubule obstruction by crystals of uric acid has been proposed as the pathogenic mechanism. Finally, immunologic factors have been proposed but the evidence is not compelling. It appears that the renal toxicity after contrast exposure is probably the result of direct tubule toxicity and renal ischemia (Berkseth and Kjellstrand, 1984).

POSTRENAL FAILURE

The causes of postrenal failure are divided into lower and upper urinary tract obstruction. The lower urinary tract is evaluated in the preliminary treatment, as indicated above, by the passage of the Foley catheter. Obstruction of the upper urinary tract may be due to ureteral calculi, blood clots, sloughed papillae, ureterovesical or ureteropelvic junction obstruction, ureteral tumors, ureteral stricture, extrinsic compression by retroperitoneal tumors, hemorrhage, fibrosis, tumor, or an inflammatory lesion. Recently, increased intra-abdominal pressure has been reported as a cause of postrenal failure.

PATHOPHYSIOLOGY.—In the patient with postrenal failure, renal blood flow is markedly reduced with a reduction in GFR to 1/10 normal or less, causing a decrease in filtration with decreased delivery of filtrate to the proximal tubule (McDougal and Wright, 1972). This may be mediated through the renin angiotensin system (McDougal et al., 1976) and/or the prostaglandin system. Elevated concentrations of renin as well as the vasodilator PDE_2 and the vasoconstrictor thromboxane A_2 have been reported in obstructive uropathy (Morrison et al., 1977). The significance of the prostaglandins in the pathophysiology of obstructive uropathy has recently been questioned. Specific inhibitors of thromboxane synthesis have failed to show any major effect on renal hemodynamics (Loo et al., 1985). Moreover, inhibition of prostaglandin synthesis during the obstructive phase has little influence on return of renal function, whereas inhibition of renin during this phase results in a major improvement in the return of renal function (McDougal, 1982). Even though complete obstruction does occur, GFR does not cease but remains modest, and the fluid that is filtered is reabsorbed by the renal tubules, the peripelvic lymphatics and veins, and the perirenal tissues when the urine extravasates about the fornices of the calyx. With the increased hydrostatic pressure within the ureter, destruction of tubule tight junctions occurs, causing increased tubule permeability (McDougal et al., 1976). Permeability is increased throughout the entire nephron, proximal tubule, loop, distal tubule, and collecting duct, and there is a reduction in the net sodium and water reabsorption with net addition of sodium by the collecting duct. Increased potassium secretion by the distal tubule occurs in the postobstructed phase, and during recovery concentrating ability remains impaired for days (McDougal and Wright, 1972). Renal blood flow, although reduced,

persists and removes solute from the medulla. Thus, following the relief of such obstructions, there is often an excessive volume output with large sodium losses and inability of the kidney to concentrate and excrete an acid load. The loss of medullary tonicity and the rapidity with which it is built up determines the degree and length of time urinary concentration is impaired. Rarely, concentrating ability may take as long as 6–9 months to return to normal. Should obstruction occur in the presence of infection or protein depletion and exist for prolonged periods, parenchymal destruction occurs. Under these circumstances, complete return of renal function does not occur. The longest report of complete obstruction with the return of renal function to support life is 90 days.

DIAGNOSIS.—The diagnosis of upper tract obstruction is established by retrograde pyelography or, on occasion, by percutaneous puncture and antegrade pyelography. However, it should be remembered that indiscriminate manipulation of the urinary tract must be avoided, since sepsis, the most common cause of death in patients with intrarenal failure, all too often is a consequence of genitourinary manipulation. Complete anuria after lower urinary tract obstruction has been ruled out by the introduction of a Foley catheter demands retrograde or antegrade pyelography. In addition to obstruction, total anuria may be due to bilateral renal artery thrombosis, aortic vascular catastrophes, acute glomerulonephritis, or cortical necrosis, and it may be present during the first 12 to 24 hours of acute tubular injury.

Abdominal roentgenograms, CT scans, and renal ultrasonograms are useful. The diagnosis may be suggested by the presence of calcification in the course of the ureters, osteoblastic lesions of the bones implying carcinoma of the prostate that may have invaded the bladder, or large pelves on ultrasonography implying obstruction. Rarely, the ultrasonogram will show a nondilated system even in the presence of bilateral obstruction (Rascoff et al., 1983). Therefore, if one's index of suspicion is high, retrograde or antegrade pyelography should be performed even though the ultrasonogram does not show hydronephrosis.

These patients are well hydrated, have normal or slightly elevated CVP, a BUN/creatinine ratio of 10:1 or greater, and a urine microscopic examination that may reveal red cells, white cells, crystals, or malignant cells. Urine sodium concentration is greater than 40 mEq/L, and potassium concentration is variable, usually ranging between 20 and 40 mEq/L. That the renal concentrating ability is severely impaired is reflected by a U/P urea of less than 5 and a U/P osmolality of less than 1.0. The fractional sodium excretion is in excess of 1%.

TREATMENT.—Therapy is directed at the site of obstruction requiring a urethra catheter, a suprapubic cystotomy, a nephrostomy, a cutaneous ureterostomy, or indwelling ureteral catheters; or double-J stents may be left temporarily until the metabolic status of the patient is stable enough to permit the appropriate surgical procedure. During the postobstructed period, large quantities of urine are excreted with a low osmolality and a high sodium concentration (50–70 mEq/L). These defects are unresponsive to antidiuretic hormone and aldosterone administration. Volume and sodium should be replaced as lost. Dextrose 5% in 0.5 N saline solution is usually the appropriate infusion. Potassium losses are therapeutic initially; but if prolonged and excessive, replacement of potassium may be necessary later. Sodium and potassium conservation returns to normal within 48–72 hours; however, the concentrating defect may persist for 7–12 days, making dehydration a potential danger should fluid be inappropriately restricted. The majority of these patients, if they respond with a diuresis, will go on to recovery of renal function. If oliguria persists, the obstruction has caused destruction of parenchyma, and such patients are managed as described for intrarenal ARF.

POLYURIC STATES

Polyuria is defined as a urine volume greater than 2,000–2,500 ml/24 hr. For the purpose of differential diagnosis, acute polyuric states may be divided into three general subclassifications: postrenal, prerenal, and intrinsic renal polyuria.

POSTRENAL POLYURIA

Postrenal polyuria occurs during partial chronic urinary obstruction or following the release of complete bilateral ureteral occlusion (postobstructive diuresis) provided that significant parenchymal damage has not occurred. As described above, basic defects involve a washout of the medullary concentration gradient and increased tubule permeability. Such patients lose excessive volumes of fluid that generally contain 40–80 mEq/L of sodium and variable concentrations of potassium. They are incapable of both concentrating and acidifying their urine. Since there is no medullary osmotic gradient, ADH is ineffective. The management of these patients is as described above and involves careful intake and output records, daily weights, frequent serum osmolalities, and careful physical examination to determine the presence of dehydration or edema.

INTRARENAL POLYURIA

Intrarenal polyuria results from an intrinsic impairment of the renal concentrating mechanism. It may be due to anatomical disruption from disease, metabolic aberrations affecting the ADH collecting duct interaction, or functional renal derangements (Table 15–3). A reduction in renal medullary tonicity may be due to anatomical disruptions such as found in sickle cell anemia. In such patients, portions of the renal medulla are infarcted due to low oxygen tonicity and sickling of the erythrocytes. Other diseases such as medullary cystic disease, pyelonephritis, nephrocalcinosis and amyloidosis may similarly disrupt the medulla. Metabolic aberrations of the ADH-collecting duct interaction include hypercalcemia, prolonged hypokalemia, and renal tubular acidosis (RTA). These disorders are not uncommon in postsurgical and posttrauma patients.

Hypercalcemia initially interferes with the action of ADH on the collecting duct epithelium, and if it is corrected early, there is an immediate reversal of the concentrating defect. Prolonged hypercalcemia with calcium deposition in the thick ascending limb of the loop of Henle, distal tubule and collecting duct may result in permanent loss of renal concentrating ability. Prolonged hypokalemia may exert its effect by inhibition of the generation of adenyl cyclase, cyclic adenosine monophosphate (cAMP), or protein kinase and their interaction between ADH and the collecting duct. Restoration of the appropriate potassium level corrects the concentrating disorder. RTA may also cause lack of concentrating ability. Since RTA often results in low serum potassium concentrations, it may be hypokalemia, not the RTA as such, that is responsible for the concentrating defect. In any event, restoration of normal serum

TABLE 15–3.

Differential Diagnosis of Polyuria

Postrenal polyuria
  Obstruction
Intrarenal polyuria
  Anatomical disruption
    Sickle cell anemia
    Pyelonephritis
    Amyloidosis
    Nephrocalcinosis
  Metabolic abnormalities of ADH-collecting duct
    interaction
    Hypercalcemia
    Hypokalemia
    Renal tubular acidosis
  Functional abnormalities
    Nephrogenic diabetes insipidus
    Medullary washout
Prerenal polyuria
  Lack of ADH
    Suppressed:expanded extracellular fluid
    Diabetes insipidus
    Trauma
    Diuretic agents

acid base and electrolyte balance corrects the problem provided nephrocalcinosis has not complicated the disease.

Finally, functional derangements such as nephrogenic diabetes insipidus in which the collecting duct appears to be insensitive to circulating levels or ADH and lack of medullary tonicity due to persistent excessive urine output are other causes of polyuria. The management of intrarenal polyuric states involves daily monitoring of body weights and serum osmolality and fluid replacement volume for volume. Specific metabolic disorders such as hypercalcemia, hypokalemia, and systemic acidosis are corrected. Anatomical abnormalities such as sickle cell disease, pyelonephritis, polycystic disease, myeloma, amyloidosis, and polyarteritis are not amenable to any specific treatment with respect to polyuria, and therefore these patients are best treated in a supportive manner. Exogenous ADH, of course, is not effective in these disorders unless they are of metabolic etiology, and then only when the abnormality has been corrected.

PRERENAL POLYURIA

Prerenal polyuria is due to insufficient ADH or to exogenous or endogenous diuretic agents. The intrinsic ability of the concentrating mechanism to function is maintained; however, if the polyuria is prolonged, it may result in washout of the medullary osmotic gradient, in which case an intrinsic renal defect would be superimposed on a prerenal defect (DeWardener and Herxheimer, 1957). Insufficient ADH activity may result from suppression of ADH release in the pituitary, either as a normal consequence of volume overload or as a consequence of iatrogenic-induced reduction in serum osmolality. The common causes in volume expansion include overadministration of IV fluids, compulsive water drinking, and pathologic stimulation of the thirst center. Dilution of plasma osmolality occurs when sodium loss exceeds that replaced or when water is replaced in an electrolyte-poor fluid. Absence of or insufficient ADH production occurs in cerebral trauma and diabetes insipidus, which may be an inherited or acquired abnormality. Since the intrinsic concentrating mechanism is unaffected in these states, it is of paramount importance to determine the primary etiology. If a lack of ADH is responsible, administration of this hormone is therapeutic. If, however, volume abnormalities have caused this disorder, their correction will be curative.

Prolonged, persistent diuresis can cause dilatation of the collecting system. After the prolonged diuresis, IV urography may reveal bilateral hydronephrosis. This hydronephrosis occurs on the basis of excessive diuresis, not on the basis of any mechanical obstruction. Since diuresis causes an increase in intraureteral pressure as bladder filling occurs, it is easy to understand why ureteral dilatation occurs (see Ureteral Physiology).

CHRONIC RENAL FAILURE

Chronic renal failure occurs most commonly as a result of glomerulonephritis due to immunopathogenic mechanisms; diabetes, infection, toxins, obstruction, congenital disorders, and hereditary nephropathies account for the majority of the remainder. The loss of renal function is usually insidious, not manifesting itself until late in the disease. The initial symptoms that herald a significant lack of renal function are protean and include fatigue, lassitude, and weakness. As the renal function continues to deteriorate, the patient may complain of a metallic taste, anorexia, nausea, vomiting, abdominal pain, hiccups, and diarrhea. With progression, there is an inability to concentrate, and somnolence, twitching, bone pain, pericarditis, and hypertension (and its sequelae) supervene. All organ systems are affected at various stages in the disease. The detrimental effects of renal failure can be ameliorated by proper recognition and anticipation of the various complications with appropriate prophylactic therapy instituted before the abnormalities become clinically manifest.

PATHOPHYSIOLOGY

As renal function decreases, those nephrons which continue to function increase in size as do their respective glomeruli. This concept was originally suggested by Bricker and co-workers (1960) and is known as the intact nephron hypothesis. Although this hypothesis has been questioned since micropuncture data suggest a heterogeneity of tubule function in disease, it appears that within the constraints of biologic variability, the intact nephron hypothesis has the most evidence to support it and thus best explains the observed deterioration of renal function in chronic renal failure. GFR and renal plasma flow decrease and functional mass decreases in parallel, while each individual nephron that continues to function increases its single nephron filtration rate. Since fewer nephrons are functioning and those that are have increased their activity, if solute excretion is to be maintained, the fractional excretions for the respective solutes must increase. Sodium homeostasis is maintained until late in the disease by the mechanism of increasing single nephron fractional excretion from less than 1% in health to 10%–20% in disease. Late in the disease, sodium excretion becomes fixed, and the nephron is incapable of varying its fractional excretion according to the needs of the patient, thus resulting in a salt-losing nephropathy. Salt restriction in such patients will result in a significant hyponatremia. This occurs late in the disease process. Similarly, water conservation and excretion are maintained until the GFR falls below 20 ml/min, at which time the kidney is incapable of compensating for varying water loads or significant water deprivation. The kidney retains a remarkable ability to excrete potassium until very late in the process. The increased flow rate obtained by the remaining functional nephrons facilitates potassium secretion. Acidosis, which is associated with increasing renal failure, also promotes potassium secretion. Mild hyperkalemia occurs when the GFR falls below 10 ml/min. Significant hyperkalemia is uncommon until oliguria supervenes with a urine output less than 500 ml/24 hr.

Acid-base balance is reasonably well maintained until the GFR falls below 20 ml/min; then acid-base imbalance is a result of the failure of the distal tubule to maintain ammonia excretion (Kurtzman, 1982). Indeed, even in advanced renal failure, not much alkali is excreted. Bicarbonate reabsorption is strongly influenced by the extracellular volume, so that late in the disease, when the kidney is not capable of responding to fluid loads, excessive fluid intake often results in bicarbonate wasting.

The determination of renal function using creatinine presents certain difficulties in patients with chronic renal failure. Serum creatinine may be deceptively low due to increased metabolism by the gut and decreased muscle mass as a result of muscle wasting due to the uremia. When the GFR is less than 20 ml/min, the creatinine clearance will overestimate the true GFR (inulin clearance) due to tubule secretion of creatinine.

Unfortunately, some aspects of renal function are maintained at the expense of others. Phosphate homeostasis is an excellent example of this "trade-off" phenomenon. As the need to excrete phosphate continues in the face of decreased renal function, the intact nephrons are not capable of increasing their fractional excretion for phosphate. Serum phosphate rises, which results in a fall in serum calcium and the subsequent release of parathormone. Parathormone acts on tubules to increase phosphate excretion, thereby returning serum phosphate to normal levels. The other effect of parathormone on bone reabsorption also occurs, resulting in bone destruction. Bone destruction is the price, or trade-off, paid for maintaining a normal level of serum phosphate.

The increased filtration that each nephron must maintain appears to be responsible for its ultimate demise. It has been shown experimentally that increasing single nephron GFR results in stripping off of the glomerular endothelial cells, thus destroying the charge

barrier that prevents deposition of circulating proteins in the glomerular basement membrane (Olson et al., 1982). The circulating serum proteins are then deposited and accumulate in the mesangium, resulting in the histologic appearance of glomerulosclerosis. Thus, an initial insult that reduces renal mass, such as segmental injuries due to obstruction, reflux, or dysplasia, may have occurred in the distant past, with renal failure occurring slowly over the ensuing years due to increased workload of the remaining functional nephrons, protein deposition causing their ultimate failure. In chronic renal failure patients in the Christchurch series, 12% of cases were due to reflux nephropathy, very few of whom had been infected. These children began dialysis at about age 18 years, suggesting that the initial insult caused progressive renal deterioration as each remaining nephron was required to do more work (Bailey and Lynn, 1984).

The fact that protein is detrimental to renal function was suggested many years ago (Addis, 1948) and led to the development of the protein-restricted "renal failure diet." When patients eat these diets, many of the symptoms of uremia are alleviated or reduced. Administration of essential amino acids and their keto analogs has been successful in slowing the rise in BUN; however the substances are not generally very palatable. More recently, it has been suggested that dietary protein may indeed cause renal failure in patients who have sustained a renal insult (Brenner et al., 1982). Thus, dietary protein restriction may not only alleviate symptoms of uremia in the patients with severely compromised renal function, but it may also prevent the progression of the disease in patients who have a modest decrease in renal function, of insufficient degree to manifest signs or symptoms. Protein induces an acute increase in GFR through renal vasodilation and glomerular hyperperfusion. The elevated intracapillary pressures and transcapillary flux of ultrafiltrate eventually disrupt the integrity of the glomerular capillary membrane. Proteinuria ensues, and the increased flux of proteins exacerbates the glomerular capillary injury and causes progressive accumulation of mesangial deposits—the forerunner of focal glomerular sclerosis. Others have suggested that it may not be protein per se which causes the renal injury, but rather the high phosphate intake from the protein. The phosphorus-induced renal injury is mediated through calcium-phosphate deposition in the kidney due to increased filtered load of phosphate per nephron (Haut et al., 1980).

COMPLICATIONS

Chronic renal failure is a systemic disease and, as such, affects every organ system in the body. The more common complications most relevant to the urologist include bladder and renal/perirenal infections, infertility, impotence, and gynecomastia.

RENAL/PERIRENAL AND BLADDER INFECTIONS.—Urinary tract infection is an important cause of morbidity and mortality in chronic renal failure and has been reported to be responsible for up to 20% of all deaths in patients with end-stage renal disease (Montgomerie et al., 1968; Keane et al., 1977). Indeed, as many as 10% of patients receiving hemodialysis have been found at any one time to have a bacteremia secondary to an infection (Nsouli et al., 1979). Pyocystis, pyonephrosis, perinephric abscesses, and infected polycystic kidneys are not uncommon (Lees et al., 1985). The frequent absence of leukocytosis and fever, coupled with a low index of suspicion, since a poorly functioning urinary tract is often dismissed, results in a delayed or missed diagnosis. A history of stone disease, congenital abnormalities of the collecting system, and pus from the urethra, coupled with urethral catheterization, renal ultrasonography, or CT will usually lead to the diagnosis.

INFERTILITY.—Both men and women who have end-stage renal disease are infertile. Even dialysis does not seem to restore their fertility. Serum hormone studies reveal an elevated LH and FSH and a depressed testosterone in males. It appears that the failure occurs at the level of the testes.

IMPOTENCE.—Gonadotrophins are elevated, prolactin is elevated, and testosterone is depressed. Some patients respond to dialysis and others respond to androgen supplementation. A trial of bromocriptine to inhibit the elevated prolactin has been used with success in some patients, but the side effects of the drug usually prevent patients from continuing therapy.

GYNECOMASTIA.—Gynecomastia is not unusual, particularly early in dialysis. There is no satisfactory treatment; however, certain drugs aggravate the situation and should be discontinued if possible. The drugs include digitalis preparations, methyldopa (Aldomet), and spironolactone (Aldactone).

GASTROINTESTINAL TRACT.—Mucosal ulcerations, ascites, pancreatitis, pruritus, and peptic ulceration are not uncommon. More common, however, are the symptoms of anorexia, nausea, vomiting, diarrhea, uremic breath, and metallic taste.

CARDIOVASCULAR DISORDERS.—Hypertension occurs in more than 80% of patients with end-stage renal disease. Three quarters of these patients are hypertensive due to volume overload. In order for hypertension to occur, the patient must be overloaded by at least 5%

of his body weight. This can be completely corrected by dialysis and the selected use of potent diuretics. The other 25% of uremic hypertensive patients are volume independent. These patients are very labile when dialyzed and often require potent antihypertensives for blood pressure control. Even so, some patients ultimately require nephrectomy for control of their hypertension. Congestive heart failure is not infrequent in these patients. Uremic pericarditis continues to be a frequent complication even in the face of dialysis. These patients often present with fever, chest pain, leukocytosis, and sometimes a pericardial friction rub. Often, they will respond to dialysis alone, but indomethacin or corticosteroid administration may be necessary. Rarely, cardiac tamponade occurs, signaled by an enlarging cardiac silhouette on chest x-ray, an elevated venous pressure, and hypotension. Surgical drainage is the treatment of choice. Echocardiography has been used with great success to follow the size of pericardial effusions.

NEUROMUSCULAR SYSTEM.—Peripheral neuropathy is usually sensory, with patients complaining of burning sensation in the feet and legs, particularly at night. Its incidence has decreased with the advent of early dialysis. Left untreated, the neuropathy will progress to loss of sensation, muscle atrophy, and motor nerve involvement. Therapy consists of dialysis and replacement of nutritional deficits. Renal transplantation reverses the disorder. CNS disorders include loss of attention, short memory, and disturbed sleep. Without treatment, agitation, seizures, and psychotic behavior with eventual coma supervene. Aluminum intoxication has been implicated, but its role remains controversial.

HEMATOLOGIC DISORDERS.—The major cause of anemia in uremic patients is the lack of bone marrow production. Other contributing factors include GI bleeding; vitamin deficiency states, particularly folate, pyridoxine, and iron; and hemolysis. Therapy consists of dialysis, replacement of nutritional deficits, minimizing blood loss, and occasionally the use of androgens. Transfusion should be reserved for situations in which active bleeding occurs or in cases where the anemia is severe, i.e., hematocrit less than 14%. These patients also have a bleeding diathesis, since platelet function appears to be disturbed. It is a qualitative rather than a quantitative defect. This abnormality will reverse with dialysis alone.

SKELETAL SYSTEM.—Osteomalacia and hyperparathyroidism are often seen together in uremia. Renal osteodystrophy includes osteitis fibrosa cystica, osteoporosis, osteomalacia, and osteosclerosis. Hyperparathyroidism results in subpareosteal resorption of bone and, rarely, brown tumor formation. Growth retardation in children and deforming skeletal rickets also occur. Chronic acidosis leads to the osteopenia and depression of skeletal growth in children. The osteomalacia may be a result of aluminum intoxication. The complications are fractures, aseptic necrosis of the hips, bone pain, and deformities. Treatment involves correction of serum phosphorus, usually with aluminum-containing antacids, the administration of calcium supplements and dihydroxyvitamin D_3. (Care must be taken not to give calcium or vitamin D until the serum phosphorus is normal, since metastatic calcification will occur and damage many organs.) Parathyroidectomy is indicated if roentgenographic changes of hyperparathyroidism or metastatic calcification occur.

Arthropathy or joint pain is usually secondary to gout when there is hyperuricemia or to pseudogout when the serum uric acid is relatively normal but the serum calcium and phosphorus are both high. Both disorders respond to correction of the metabolic disturbances.

METABOLIC DISTURBANCES.—Metabolic acidosis occurs late in the disease. Hypercalcemia occasionally occurs, usually a result of secondary hyperparathyroidism or excessive calcium or vitamin D supplementation. Hyperuricemia is extremely common and is not generally associated with gout. Hypermagnesemia is very common in patients who are not on dialysis. This can be significantly aggravated by magnesium-containing compounds (antacids). It is very important in patients' with end-stage renal disease not to give magnesium-containing drugs, since these patients are prone to magnesium intoxication.

TREATMENT

Currently, treatment of end-stage renal disease involves four basic approaches. Early in the disease, dietary manipulation with the use of selected drugs to maintain serum electrolytes is all that is necessary. As the GFR falls below 10 ml/min, generally either peritoneal dialysis or hemodialysis will be required. Renal transplantation is the ultimate therapy for the disorder.

DRUG ADMINISTRATION

The kidneys are the major route of drug metabolism and secretion. Renal failure necessitates altered handling of most drugs, and giving them in the usual dosages to uremic patients will lead to elevated blood levels, toxic symptoms, and perhaps even death. Antibiotics, sedatives, narcotics, and cardiovascular drugs often must be adjusted to an appropriate dose level. Certain drugs should be totally avoided in uremic patients (Table 15–4). Drugs that commonly require altered dosage schedules are listed in Table 15–5.

TABLE 15–4.

Drugs to Avoid in Uremic Patients

Antimicrobials	Diuretics
Methenamine mandelate	Acetazolamide
Nalidixic acid	Mercurials
Nitrofurantoin	Triamterene
Tetracyclines	Spironolactone
(other than doxycycline or	Thiazides
minocycline)	Arthritis-gout
Neomycin	Gold salts
Analgesics-narcotics	Probenecid
Salicylates	Phenylbutazone
Acetaminophen	Antineoplastics
Phenacetin	Nitrosourea
Phenazopyridine	Cisplatin
Meperidine	Neuromuscular agents
Morphine	Gallamine
Sedatives-tranquilizers	Pancuronium
Barbiturates	Succinylcholine
Glutethimide	Antiarrhythmics
Lithium	Bretylium
Ethchlorvynol	Others
Methaqualone	Terbutaline
Phenothiazines	Acetohexamide
Antacids-laxatives	Chlorpropamide
Magnesium-containing	
Phosphate-containing	
Mineral oil	
Antihypertensives	
Methyldopa	
Guanethidine	
Reserpine	

TABLE 15–5.

Drugs Requiring Major Dosage Revision in Uremic Patients

Antimicrobials	Antihypertensives
Aminoglycosides*	Clonidine
Cephalosporins*	Diazoxide*
Penicillins*	Nitroprusside*
Polymyxins	Cardiac glycosides
Vancomycin	Digoxin
Flucytosine*	Digitoxin
Trimethoprim-	Arthritis-gout
sulfamethoxazole*	Allopurinol
Amantadine	Anti-neoplastics
Analgesics-narcotics	Bleomycin
Salicylates*	Cyclophosphamide*
Meperidine	Methotrexate
Morphine	Mithramycin
Antiarrhythmics	Neuromuscular agents
Procainamide*	Neostigmine
Quinidine*	Miscellaneous
N-acetylprocainamide*	Insulin
	Cimetidine*
	Clofibrate
	Methimazole
	Nicotinic acid
	Propylthiouracil

*Dialyzed significantly.

REFERENCES

1. Addis T: *Glomerular Nephritis: Diagnosis and Treatment.* New York, Macmillan Co, 1948, p 222.
2. Baek S, Makabali G: Clinical determinants of survival from postoperative renal failure. *Surg Gynecol Obstet* 1975, 140:685.
3. Bailey RR, Lynn KL: End-stage reflux nephropathy. *Contrib Nephrol* 1984, 39:102.
4. Berkseth RO, Kjellstrand CM: Radiologic contrast-induced nephropathy. *Med Clin North Am* 1984; 68:351.
5. Bonomini V, Stefoni S, Vangelista A: Long-term patient and renal prognosis in acute renal failure. *Nephron* 1984; 36:169.
6. Brenner BM, Meyer TW, Hostetter TH: Dietary protein intake and the progressive nature of renal disease: The role of hemodynamically mediated injury in the pathogenesis of progressive glomerulosclerosis in aging, renal ablation and intrinsic renal disease. *N Engl J Med* 1982; 307:652.
7. Bricker NS, Morris PAD, Kime JW Jr: The pathologic physiology of chronic Bright's disease: An exploration of the intact nephron hypothesis. *Am J Med* 1960; 28:77.
8. Britton KE, Maisey MN: Renal radionuclide studies, in Maisy MN, Britton KE, Gilday DL (eds): *Clinical Nuclear Medicine.* London, Chapman & Hall, 1983, p 99.
9. Brown CB, Ogg CS, Cameron JS: High dose frusemide in acute renal failure: A controlled trial. *Clin Nephrol* 1981; 15:90.
10. Burr L, Berger W, Dewey MM: Electrical transmissions at the nexus between smooth muscle cells. *J Gen Physiol* 1968; 51:347.
11. Chevalier RL, Campbell F, Brenbridge ANAG: Prognostic factors in neonatal acute renal failure. *Pediatrics* 1984; 74:265.
12. Chung RCH, Mantel LK: Urographic changes in diabetes insipidus: Report of case. *JAMA* 1952; 150:1307.
13. Conger JP: A controlled evaluation of prophylactic dialysis in post traumatic acute renal failure. *J Trauma* 1975; 15:1056.
14. Corwin HL, Schreiber MJ, Fang LST: Low fractional excretion of sodium: Occurrence with hemoglobinuric- and myoglobinuric-induced acute renal failure. *Arch Intern Med* 1984; 144:981.
15. DeWardener HE, Herxheimer AW: The effect of a high water intake in the kidney's ability to concentrate the urine in man. *J Physiol* (Lond) 1957; 139:42.
16. Djurhuus JC, Nerstrom B, Rask-Anderson H: Dynamics of the upper urinary tract in man. *Acta Clin Scand* 1976; 472:49.
17. Dunn MJ: Renal prostaglandins, in Dunn MJ (Ed): *Renal Endocrinology.* Baltimore, Williams & Wilkins Co, 1983.
18. Dunn MJ, Zambraski EJ: Renal effects of drugs that inhibit prostaglandin synthesis. *Kidney Int* 1980; 18:609.
19. Eneas JF, Schoenfeld PY, Humphreys MH: The effect of infusion of mannitol-sodium bicarbonate on the clinical course of myoglobinuria. *Arch Intern Med* 1979; 139:801.
20. Freund H, Atarmain S, Fischer J: Comparative study of parenteral nutrition in renal failure using essential and nonessential amino acid containing solutions. *Surg Gynecol Obstet* 1980; 151:652.
21. Gates GF: Glomerular filtration rate: Estimation from fractional renal accumulation of ^{99m}Tc-DTPA (stannous). *AJR* 1982; 138:565.
22. Graziani G, Cantaluppi A, Casati S, et al: Dopamine and frusemide in oliguric acute renal failure. *Nephron* 1984; 37:39.
23. Haut LL, Alfrey AC, Guggenheim S, et al: Renal toxicity of phosphate in rats. *Kidney Int* 1980; 17:722.
24. Hermreck AS: The pathophysiology of acute renal failure. *Am J Surg* 1982; 144:605.
25. Hull JH, Hals LJ, Koch GG, et al: Influence of range of renal functions and liver disease in predictability of creatinine clearance. *Clin Pharmacol Ther* 1981; 29:516.
26. Keane WF, Shapiro FL, Raij L: Incidence and type of infections occurring in 445 chronic hemodialysis patients. *Trans Am Soc Artif Intern Organs* 1977; 23:41.
27. Kiil F, Aukland K: Renal concentration mechanism and haemodynamics at increased ureteral pressure during osmotic and saline diuresis. *Scand J Clin Lab Invest* 1961; 13:276.
28. Koch ML, McDougal WS: Determination of renal function following urinary diversion through intestinal segments. *J Urol* 1985; 133:517.
29. Koffler A, Friedler R, Massry S: Acute renal failure due to nontraumatic rhabdomyolysis. *Arch Intern Med* 1976; 85:23.
30. Kon V, Ichikawa I: Physiology of acute renal failure. *J Pediatr* 1984; 105:351.
31. Kurtzman NA: Chronic renal failure: Metabolic and clinical consequences. *Hosp Pract* 1982; 7:114.
32. Lees JA, Falk RM, Stone WJ et al: Pyocystis, pyonephrosis, and perinephric abscess in end-stage renal disease. *J Urol* 1985; 134:716.
33. Loo MH, Marion DN, Vaughn ED: Failure of thromboxane inhibition to improve renal blood flow in dogs with complete unilateral ureteral obstruction. *Abstr Am Urol Assoc* 1985; 63:129.
34. Malin JM, Deane RF, Boyarsky S: Characterisation of adrenergic receptors in human ureter. *Br J Urol* 1970; 42:171.
35. McDougal WS: Pharmacologic preservation of renal mass and function in obstructive uropathy. *J Urol* 1982; 128:418.
36. McDougal WS, Koch MO: Accurate determination of renal function in patients with ileal and colon interpositions in the urinary tract. *J Urol* 1986; 135:1175.
37. McDougal WS, Rhodes RS, Persky L: A histochemical and morphologic study of postobstructive diuresis in the rat. *Invest Urol* 1976; 14:169.
38. McDougal WS, Wright FS: Defect in proximal and distal sodium transport in post-obstructive diuresis. *Kidney Int* 1972; 2:304.
39. Montgomerie JZ, Kalmanson GM, Guze LB: Renal failure and infection. *Medicine* 1968; 47:1.
40. Morrison AR, Nishikawa K, Needleman P: Unmasking of thromboxane A_2 synthesis by ureteral obstruction in the rabbit kidney. *Nature* 1977; 267:259.

41. Mudge G: Nephrotoxicity of urographic radiocontrast drugs. *Kidney Int* 1980; 19:540.

42. Muther RS: Drug interference with renal function tests. *Am J Kidney Dis* 1983; 3:118.

43. Nanji AA, Campbell DJ: Falsely-elevated serum creatinine values in diabetic ketoacidosis—clinical implications. *Clin Biochem* 1981; 14:91.

44. Nsouli KA, Lazarus JM, Shoenbaum SC, et al: Bacteremic infection in hemodialysis. *Arch Intern Med* 1979; 139:1255.

45. Olson JL, Hostetter TH, Rennke HG, et al: Altered glomerular permeability and progressive sclerosis following ablation of renal mass. *Kidney Int* 1982; 22:112.

46. Rascoff JH, Golden RA, Spinowitz BS, et al: Nondilated obstructive nephropathy. *Arch Intern Med* 1983; 143:696.

47. Rosen SM, Hollenberg NK, Dealy JB, et al: Measurement of the distribution of blood flow in the human kidney using the intra-arterial injection of [133]Xe: Relationship to function in the normal and transplanted kidney. *Clin Sci* 1968; 34:287.

48. Rouiller C., Muller AF: *The Kidney*. New York, Academic Press, 1969, vol 1.

49. Schrier RW: Acute renal failure. *Kidney Int* 1979; 15:205.

50. Siegel NJ, Gunstream SK, Handler RI, et al: Renal function and cortical blood flow during the recovery phase of acute renal failure. *Kidney Int* 1977; 12:199.

51. Stein JH, Lifschitz MD, Barnes LD: Current concepts in the pathophysiology of acute renal failure. *Am J Physiol* 1978; 234:F171.

52. Synhaivsky A, Kurtz SB, Wochos DN, et al: Acute renal failure treated by slow-continuous ultrafiltration. *Mayo Clin Proc* 1983; 58:729.

53. Tiller DJ, Mudge GH: Pharmacologic agents used in the management of acute renal failure. *Kidney Int* 1980; 18:700.

54. Torres VE, Strong CG, Romero JC, et al: Indomethacin enhancement of glycerol-induced acute renal failure in rabbits. *Kidney Int* 1975; 7:170.

55. Valtin H: *Renal Function*. Boston, Little, Brown & Co, 1983.

56. Zager R, Rubin N, Ebert T, et al: Rapid radioimmunoassay for diagnosing acute tubular necrosis. *Nephron* 1980; 26:7.

Chapter 16

Renal, Perirenal, and Ureteral Neoplasms

RICHARD D. WILLIAMS, M.D.

Primary neoplasms of the kidney including those of the parenchyma, collecting and drainage systems, and immediate surrounding structures, are varied in histologic appearance but are generally uniform in clinical presentation. For the most part, symptoms are absent until late in the course of the disease although hematuria, either microscopic or gross, is an almost universal (often early) sign, particularly in renal parenchymal and collecting system malignancies. In adults, renal cell cancer is by far the most common primary renal neoplasm (85%–90%), while upper tract urothelial cancer is second (7%–8%) and the various sarcomas third (2%–3%) in incidence. Neoplasms metastatic to the kidney and surrounding structures, while more common than primary neoplasms, rarely are encountered prior to the patient's death and therefore are of limited clinical significance. Diagnosis of renal neoplasms has undergone some evolution within the last 5–10 years, primarily owing to the advent of computed tomographic (CT) scanning; in fact, detection of incidental tumors (those discovered serendipitously during screening examinations) is increasing. Magnetic resonance (MR) imaging and immunoscanning are expected to provide useful information as well. Treatment of localized renal, perirenal, and ureteral tumors, whether benign or malignant, remains primarily surgical, although the results of the treatment of advanced malignant disease continue to be dismal.

CLASSIFICATION

The diverse nature of upper urinary tract neoplasms defies a simplified yet comprehensive classification system. In the past, long and detailed classification schema have been presented that, although complete, are not practically useful. The schema presented here (Table 16–1) represents not only an operational guide to this chapter, but also one that is rapidly committed to memory and thus recall. This chapter and the schema presented focus on adult neoplasms only.

Renal and perirenal tumors include renal parenchymal, urothelial (renal pelvis), and connective tissue neoplasms. Ureteral tumors are primarily urothelial, but include connective tissue and metastatic tumors as well. Benign renal and perirenal tumors include cortical adenomas, vascular and connective tissue tumors, and hamartomas. Primary malignant tumors in the kidney and surrounding tissues include renal cell carcinoma, transitional cell carcinoma of the renal pelvis, and the various sarcomas. Malignant renal tumors originating from distant sites include: (1) neoplasms involving the kidney by direct extension (e.g., adrenal cortical cancer, pancreatic and colonic cancer, and various sarcomas); (2) the common hematogenous metastases from lung, stomach, and breast primary tumors; and (3) the hematologic tumors, which may be primary in the kidney but most commonly are manifestations of a systemic neoplasm such as lymphoma, leukemia, or myeloma. Ureteral tumors, similar to upper tract neoplasms, may be benign or malignant, but the malignant urothelial tumors are most common. Secondary malignant tumors of the ureter are most often metastases from breast, prostate cancers, or lymphomas, but direct extension from sarcomas or intestinal tumors may also occur.

EVALUATION OF RENAL MASSES

The initial discovery of a renal or perirenal mass in the past relied on intravenous urography with tomography. Renal angiography supplied further diagnostic information, particularly in a mass with characteristic neovascularity, but frequently renal exploration and biopsy were required to provide the ultimate diagnostic denouement. Improved ultrasonographic techniques, radiologic modalities such as CT, and percutaneous fine-needle aspiration cytologic methods have increased diagnostic accuracy to the extent that the diagnosis is secure prior to (and often obviates) surgical exploration

TABLE 16-1.

Classification of Upper Urinary Tract and Retroperitoneal Neoplasms

Renal tumors
  Benign
    Adenoma
    Oncocytoma
    Angiomyolipoma-hamartoma
    Fibroma
    Leiomyoma
    Lipoma
    Hemangioma
    Juxtaglomerular tumor
  Primary malignant
    Renal cell carcinoma
    Urothelial-renal collecting system and pelvis-transitional cell, squamous cell, and adenocarcinoma
    Sarcoma
  Secondary malignant
    Adrenal carcinoma
    Retroperitoneal sarcoma, pancreas, colon (direct extension)
    Lung, stomach, breast, prostate (hematogenous metastases)
    Hematologic tumors-lymphoma, leukemia,
    Multiple myeloma (primary or metastatic)
Primary retroperitoneal tumors
  Benign—lipoma
  Malignant—lymphoma, sarcoma
Ureteral tumors
  Benign
    Papilloma
    Hemangioma
    Leiomyoma
    Fibroepithelial polyps
  Primary malignant
    Urothelial
    Sarcoma
  Secondary malignant
    Retroperitoneal sarcoma, colon, lymphoma, testicular (direct extension)
    Breast, prostate, renal (lymphatic or hematogenous metastases)

in more than 90% of masses detected on intravenous urography.

Because modern diagnostic technology offers an array of choices for the study of renal and perirenal masses, the optimal workup requires an algorithm to provide a systematic approach to the definition of renal masses. Use of this approach should decrease to less than 10% the number of masses remaining indeterminate prior to definitive therapy (Fig 16–1).

Intravenous urography (IVU) continues to be the primary urinary tract screening modality owing to its familiarity and ease of interpretation. The impetus for more cost-effective and less invasive measures, coupled with marked technical improvements, has caused ultrasonography (US) to rival IVU as the initial screening technique (Smith-Harrison et al., 1983), particularly for patients with chronic renal failure, diabetes mellitus,

pregnancy, or other conditions for which the use of x-ray and/or contrast materials may be contraindicated. The addition of nephrotomography to IVU has increased diagnostic accuracy considerably, but renal masses still can be accurately defined in only approximately 75% of cases. US can quite accurately define whether the mass detected on IVU is cystic or solid (Figs 16–2 and 16–3). If the mass fulfills strict criteria for a simple cyst, including through transmission without internal echoes and a strong posterior wall—true in nearly 65% of renal masses detected by IVU—there is little need for further workup in the asymptomatic patient. In the past, needle puncture of cysts with aspiration for cytology and chemical analyses as well as injection of contrast medium for cystography, were considered important screening techniques. Now, however, the 98% accuracy rate of ultrasound precludes routine use of these studies. In the symptomatic patient (flank pain or hematuria) further diagnostic steps such as cyst aspiration and/or CT should be considered. The presence of abnormal cytology or blood in the aspirate indicates a need for CT and, probably, surgical exploration.

In 15% of cases a renal mass detected on IVU will not be confirmed on US. Although the mass may be an artifact, the clinician should conduct further studies, particularly in symptomatic patients. Some mass lesions are not detected by US because of overlying bone, a similar echographic pattern to renal parenchyma, or operator error. In this situation a nuclear renal scan is recommended. Cortical scanning agents such as ^{99m}Tc DMSA will show the homogenous appearance of normal renal parenchyma if no mass is present; an area of increased density if the mass is a pseudotumor or hypertrophied column of Bertin (Fig 16–4); and decreased density if the mass is cystic or a solid tumor. In the latter case, CT is then indicated.

Twenty percent of renal masses identified by IVU show multiple internal echoes characteristic of a solid mass on US. This finding alone suggests that the mass is a neoplasm, but further evaluation by CT is required. In the past renal angiography was the next step after detection of a renal mass on IVU or US but at present CT is the preferred modality because it is as accurate (approximately 95%) as angiography without the potential morbidity. In addition, CT provides accurate local staging information sufficient in most cases to allow definitive surgical management. CT, with and without contrast injection, is recommended to take full advantage of the contrast enhancement characteristic of the highly vascular renal parenchymal tumors (Fig 16–5). Solid masses that also have substantial areas of negative CT attenuation numbers indicative of fat are diagnostic of angiomyolipoma; when visualized in an asymptomatic

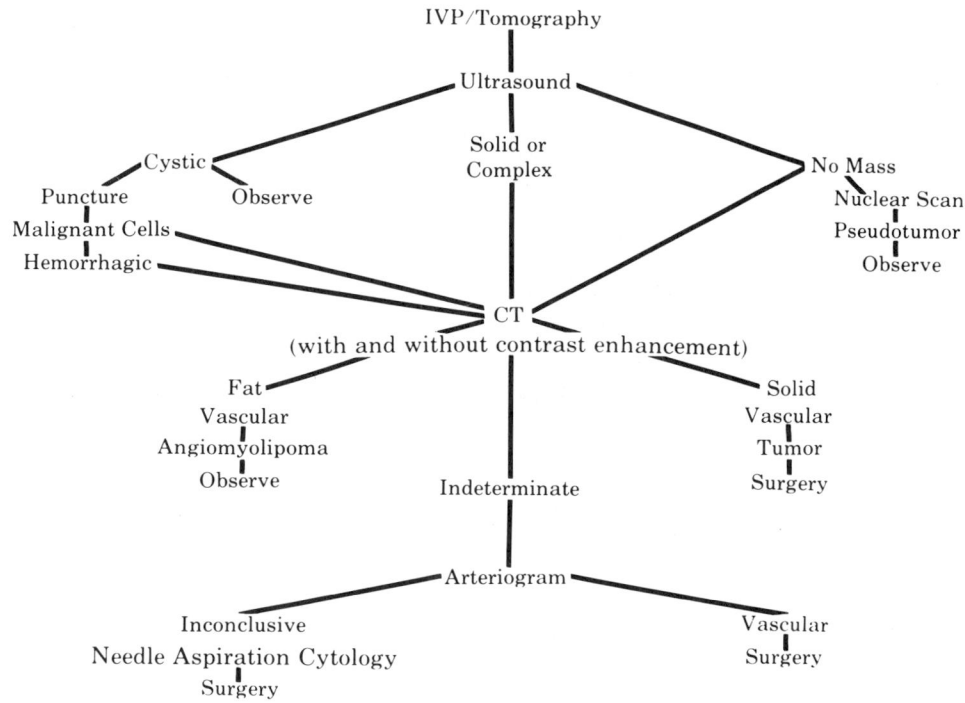

FIG 16–1.
Algorithm for evaluation of a renal mass.

patient they require no further study (Fig 16–6). In 10% of cases or less, CT will be indeterminate and a renal arteriogram or needle aspiration cytology or both may be required to further define the mass. In a very few cases (less than 5%) surgical exploration

may be needed to provide the definitive diagnosis.

The majority of renal parenchymal masses can be accurately defined following the principles of this algorithm.

Another diagnostic advance is magnetic resonance imaging (MRI): initial studies indicate that MRI is

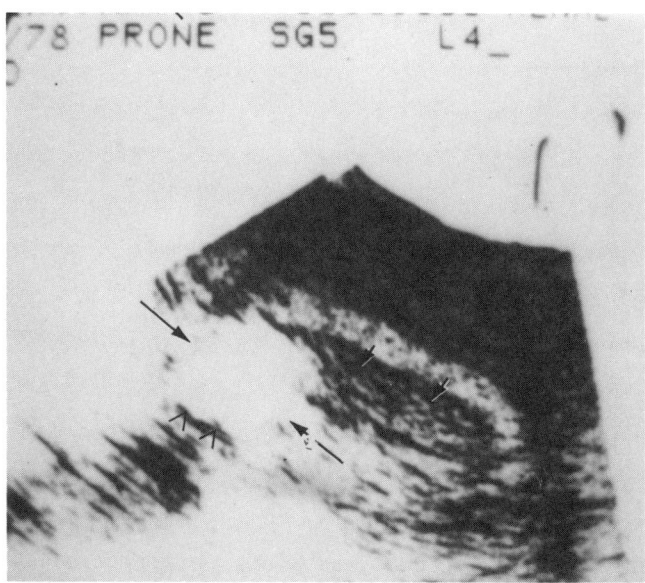

FIG 16–2.
Ultrasound of a simple renal cyst showing renal parenchyma *(short arrows)*, cyst wall *(long arrows)*, and strong posterior wall *(arrowheads)*.

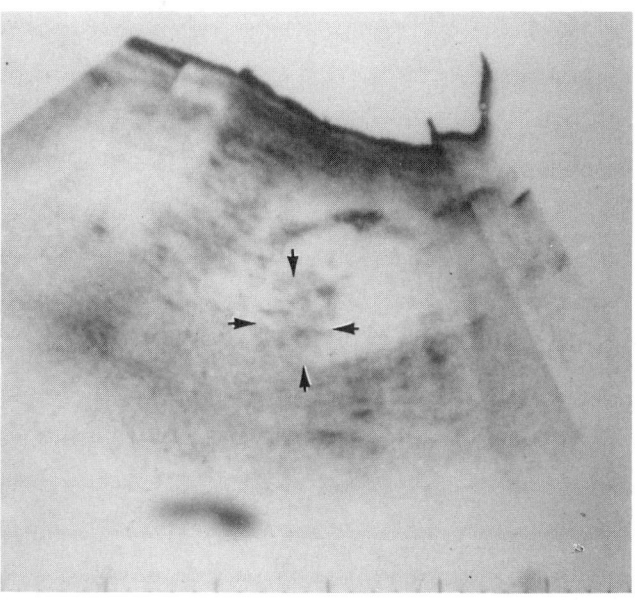

FIG 16–3.
Ultrasound examination of a solid renal mass *(arrows)*.

FIG 16–4.
DMSA scan of a renal pseudotumor (hypertrophied column of Bertin) shown by *arrows.*

equivalent to CT in the diagnosis of renal masses (Fig 16–7) but is more accurate in local tumor staging due to the precise display of tumor extension into perinephric fat, renal vein, and vena cava (Hricak et al., 1985). The recent advent of paramagnetic contrast enhancement with MRI may permit even better differentiation in the future (see Chapter 3C).

BENIGN RENAL TUMORS

Benign tumors of the kidney and its surroundings are not uncommon but because they do not tend to cause significant symptoms or morbidity they have been infrequently detected in the past. The liberal use of di-

agnostic studies, particularly CT, and the emergence of new syndromes, such as neoplasia associated with acquired renal cystic disease, have combined to increase the discovery of these masses and thus require their differentiation from malignant renal masses.

ADENOMA

The adenoma is the most common benign solid renal parenchymal lesion (Williams, 1985). True adenomas are small cortical lesions with a papillary histology (Fig 16–8). For the most part these lesions are asymptomatic and found only incidentally at autopsy or nephrectomy for unrelated causes. There is substantial controversy in the literature, however, regarding whether a solid renal mass under 3 cm in size should be designated an adenoma or a carcinoma. The controversy emanates from an often misquoted autopsy study (Bell, 1950) in which over 60 renal tumors less than 3 cm in size were shown to have metastasized in only three cases, whereas tumors larger than 3 cm had a much higher rate of metastasis. Although Bell cautioned that size was not an absolute criteria, many urologists and pathologists have embraced the "3 cm rule" and base their diagnosis on it arbitrarily. The size of the tumor, however, is not sufficient for diagnosis and thus adenomas cannot be clearly differentiated preoperatively from low-grade carcinomas. Murphy and Mostofi (1970) reviewed 180 cases of renal adenoma of which 37 were greater than 3 cm in diameter and only one had metastasized further, indicating the difficulty in preoperative delineation. Bennington and Beckwith (1975), in an elegant review, suggested that adenomas are not histo-

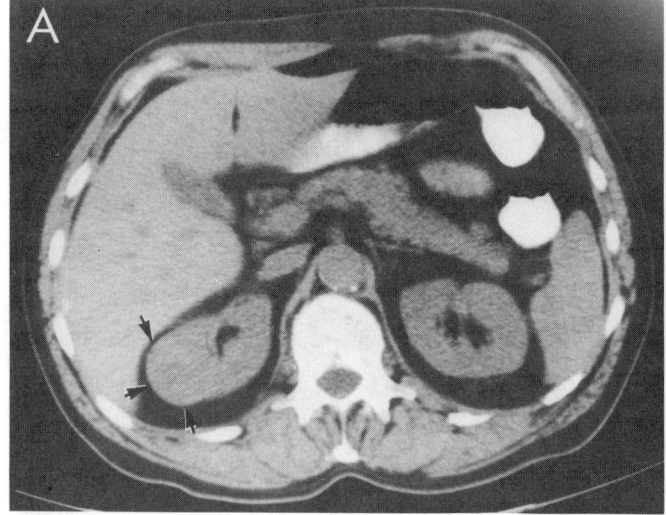
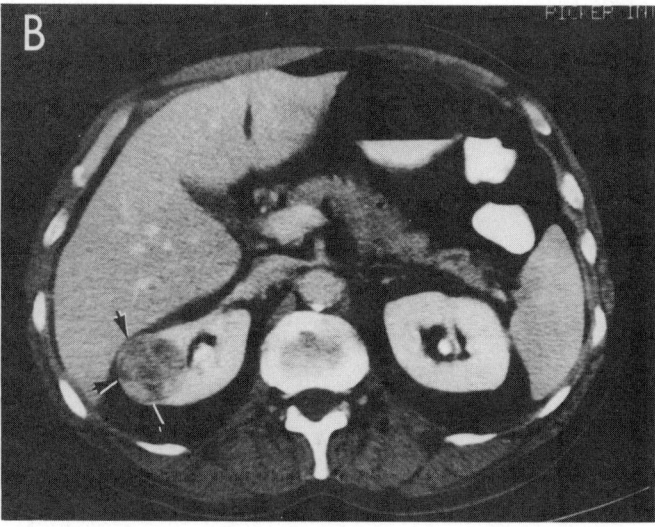

FIG 16–5.
Computed tomograph of a renal cell carcinoma *(arrows).* **A,** without contrast enhancement. **B,** with contrast enhancement. (From Williams RD, Tanagho EA: *Current Surgical* *Diagnosis and Treatment.* Los Altos, Calif., Lange Medical Publications, 1985, p 860. Used with permission.)

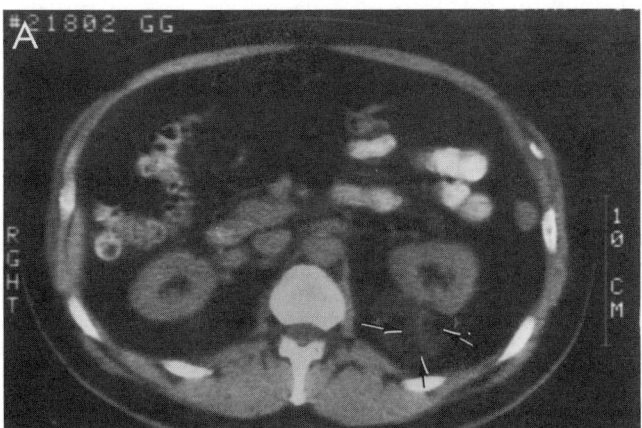

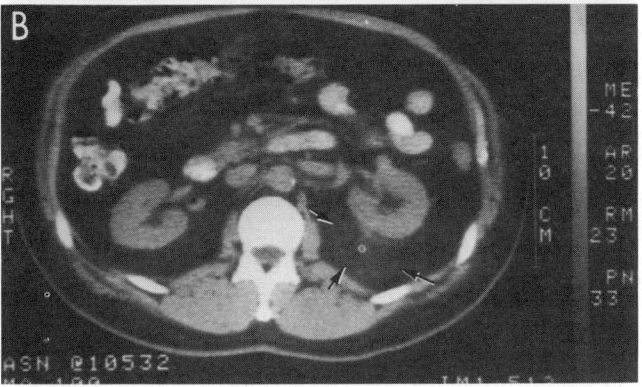

FIG 16–6.
Computed tomograph of a renal angiomyolipoma. **A,** typical mass lesion with internal areas of fat *(arrows).* **B,** same lesions as in **A** showing abundant fat measuring −42.7 Hounsfield units *(arrows).*

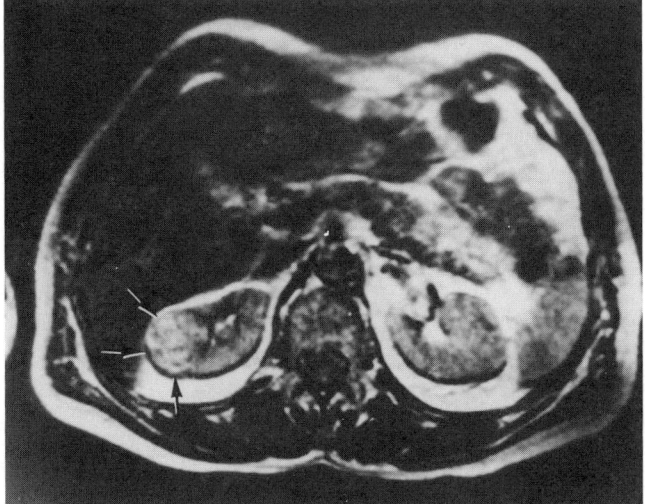

FIG 16–7.
Transaxial magnetic resonance image of the same renal cell carcinoma *(arrows)* shown in Figures 16–5, A and B. (From Williams RD, Tanagho EA: *Current Surgical Diagnosis and Treatment.* Los Altos, Calif, Lange Medical Publications, 1985, p 860. Used with permission.)

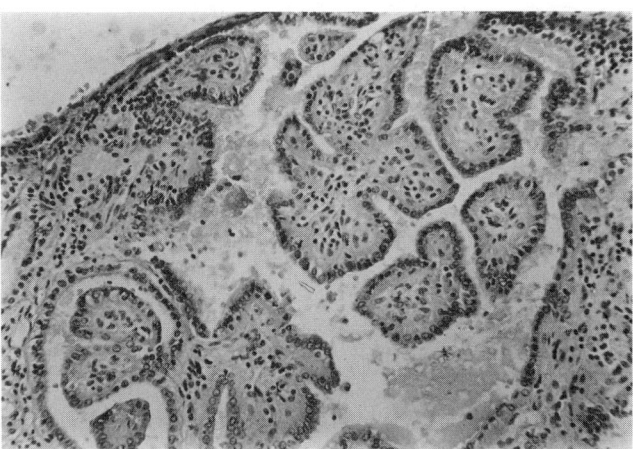

FIG 16–8.
Histologic section of true renal adenoma (original magnification, × 100).

logically distinguishable from carcinomas. Because these small tumors occur in the same 2:1 male-female ratio, are rarely found adjacent to a renal cancer (except in patients with von Hippel-Lindau disease or acquired renal cystic disease associated with renal neoplasia), and are histologically indistinguishable from larger carcinomas, they should be considered renal cancers.

The etiology and clinical incidence of true renal adenomas is unknown although 7%–22% of patients will have them at autopsy (Bonsib, 1985). Adenomas are rarely symptomatic and paraneoplastic syndromes are not encountered as a rule. Because the biologic potential of a small renal mass cannot be predicted preoperatively, urologic oncologists consider them malignant and continue to recommend standard radical nephrectomy as treatment for appropriate surgical candidates. At the present time, nephron-sparing surgery is appropriate in selected cases because detection of incidental (and often small) tumors is increasing; the prognosis of patients with either renal adenocarcinoma in a solitary kidney or bilateral renal cancers treated by subtotal nephrectomy is high (approximately 70% five-year survival); and results obtained in patients treated with radical nephrectomy are comparable. Symptomatic patients with paraneoplastic syndromes or calcifications within small tumors should, however, be treated by radical nephrectomy if they have adequate renal function and a normal contralateral kidney.

ACQUIRED RENAL CYSTIC DISEASE

Approximately 45% of patients with end-stage renal disease develop multiple renal cortical (ARCD) cysts (Fig 16–9). The incidence of renal tumor development in these patients is 9%—2,500 times that of patients with normal renal function. The majority of ARCD pa-

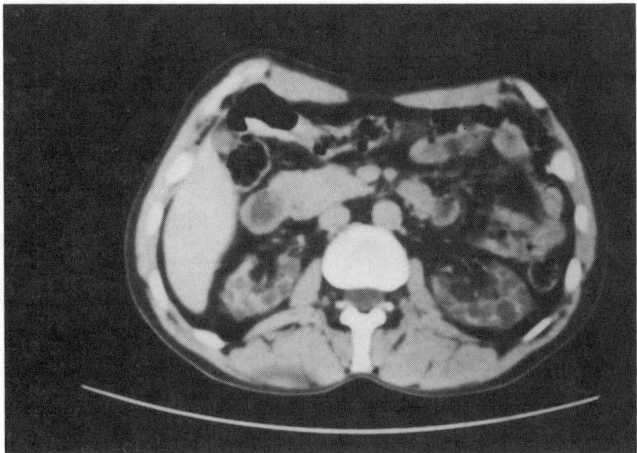

FIG 16–9.
Computed tomograph of a patient on chronic hemodialysis for end-stage renal disease showing multiple bilateral cysts of acquired renal cystic disease.

tients with tumor present with microscopic or gross hematuria and exhibit enhancing solid renal masses on contrast-enhanced CT studies (Fig 16–10). The pathologic findings include multiple, bilateral tumors with histologic characteristics varying from epithelial proliferation in cyst walls to frank carcinoma. The majority of the lesions are small benign-appearing tumors although there is a 6% rate of metastases in ARCD patients with renal tumors (0.5% of all ARCD patients). The etiology of these tumors is unknown, although the rate of incidence appears to increase directly with the duration of chronic hemodialysis (43.5% less than three years; 79.3% greater than three years). Patients on peritoneal

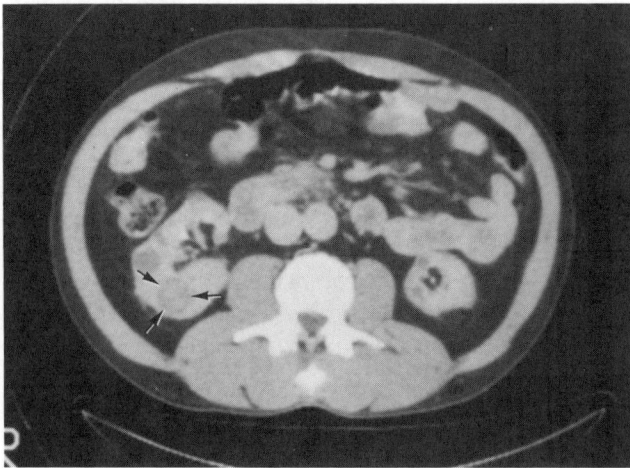

FIG 16–10.
Computed tomograph of a patient with acquired renal cystic disease and renal neoplasia *(arrows).* (From Bretan PN, et al: Chronic renal failure: A significant risk factor in the development of acquired renal cysts and renal cell carcinoma. *Cancer* 1986; 57:1874. Used by permission.)

dialysis do not seem to have the same propensity to form tumors, although definitive studies are lacking. There is some suggestion that dialysis only allows patients to live long enough for ARCD and associated neoplasia to develop, since a high incidence of renal cysts has also been reported in patients with chronic renal failure not requiring dialysis (Bretan et al., 1986). Currently it is thought that a poorly excreted metabolite that is not dialyzable may be responsible for the development of ARCD and subsequent renal tumors. It is recommended that end-stage renal disease patients on dialysis receive a baseline US at the beginning of dialysis with yearly follow-up to screen for renal cysts. If the patient develops hematuria and/or cysts, abdominal CT with contrast should be done (Levine et al., 1984). Solid renal masses are most assuredly renal tumors requiring surgical resection. Because these tumors tend to be multiple and bilateral, both kidneys must be carefully scrutinized for the presence of tumor.

ONCOCYTOMA

Oncocytoma is a benign renal tumor that has only recently been identified (Klein and Valensi, 1976; Lieber et al., 1981). Prognostic reports of patients with renal cancer prior to 1976 may have included these tumors and thus presented overly optimistic survival statistics. It is estimated that 5%–7% of renal tumors are oncocytomas accounting for approximately 750 cases per year (Lieber, 1984a). These tumors are asymptomatic and are most often found incidentally even though they may become very large. Males are affected twice as often as females. Oncocytomas occur most commonly in the fifth decade, as is the case with renal cell cancer. It is felt that although oncocytomas have only recently been discovered, the apparent rising incidence is due to improvement in diagnostic methods of visualizing smaller asymptomatic lesions.

The diagnosis of oncocytoma is primarily based on pathologic considerations, as there are no reliable distinguishing clinical characteristics. Hematuria is present in fewer than 10% of patients and paraneoplastic syndromes and symptoms are routinely absent. There are no characteristic features of the tumors on IVU, US, or CT. Weiner and Bernstein (1977) have described several angiographic findings including a "spoke-wheel" pattern of interlobar arteries as typical of oncocytomas. These findings are not consistently present, however, and occasionally similar findings are seen in patients with renal cancer (Maatman et al., 1984). Prior to treatment, only a high index of suspicion in asymptomatic small lesions may suggest an oncocytoma.

Pathologically, oncocytomas are tan or light brown renal masses that are well encapsulated and rarely invade adjacent tissue. A central stellate scar is often present, but necrosis typical of renal adenocarcinoma is not. The tumors are usually solitary and unilateral, although several bilateral cases and even multiple sites within one kidney have been reported. Histologically the cellular pattern is monotonously uniform, consisting of large well-differentiated cells with an intensely eosinophilic cytoplasm (Fig 16–11), which on ultrastructural studies is packed with mitochondria. The oncocytes rarely exhibit mitosis or invade other structures. Cellular origin is unknown, although the oncocytes resemble proximal convoluted tubular cells (Merino and Librelsi, 1982). Oncocytomas are actually more common in other sites, including salivary, thyroid, parathyroid, and adrenal glands.

Considering only those patients with tumors consisting of pure oncocytoma cells (grade I), there have been over 100 cases reported in the literature with no metastases and no deaths due to tumor. Previous reference to grade II tumors where approximately 15% of patients developed metastases and died is thought to represent mixed tumors with significant elements of renal adenocarcinoma cells present.

It is now well recognized that pure oncocytomas are benign, and thus if recognition of their histologic characteristics could be ascertained preoperatively, renal-sparing surgery (subtotal nephrectomy) would be in order. Although attempts have been made to define oncocytomas by needle biopsy (Rodriguez et al., 1980) and flow cytometry (Rainwater et al., 1986) the results are equivocal and thus the recommended treatment remains radical nephrectomy.

HAMARTOMA—ANGIOMYOLIPOMA

Renal hamartomas are benign tumorous lesions that are associated in over 50% of patients with tuberous sclerosis. Tuberous sclerosis is congenital (1/150,000 births) and familial, and characterized by brain gliosis, mental retardation, epilepsy, adenoma sebaceum of the face, submucosal fibromas, and hamartomas of retina, lungs, liver, pancreas, bone, and kidneys. Approximately 80% of tuberous sclerosis patients will have angiomyolipomas of the kidney, most often bilateral and asymptomatic. Those patients *without* tuberous sclerosis commonly have unilateral lesions and often present with symptoms caused by spontaneous rupture and hemorrhage into the retroperitoneum.

Angiomyolipomas are unencapsulated yellow to gray lesions of multicentric origin detected in middle-aged females four times more often than in men. As the name implies, the lesions are characterized by three major histologic components in varying proportions: mature fat cells, smooth muscle, and abnormal vessels (Fig 16–12). Mitotic figures are rare but multinucleated giant cells may be seen. Renal hamartomas have been seen in renal hilar lymph nodes (Busch et al., 1976). There have been no documented distant metastases and thus these extrarenal deposits are considered to be further evidence of multicentric origin rather than histologic aggressiveness (Mostofi and Davis, 1984).

The diagnosis of angiomyolipomas relies on the same algorithim previously presented (see Fig 16–1). As described, many patients will present with spontaneous hemorrhage requiring emergent treatment. Previous to CT and current generation US, renal hamartomas were difficult to distinguish from renal cancer. On IVU with tomography a solid mass is seen that cannot be differentiated from malignancy. Arteriography may reveal neovascularity similar to renal cancer and is thus not helpful in differential diagnosis. Computed tomography and ultrasonography, however, are often diagnostic due

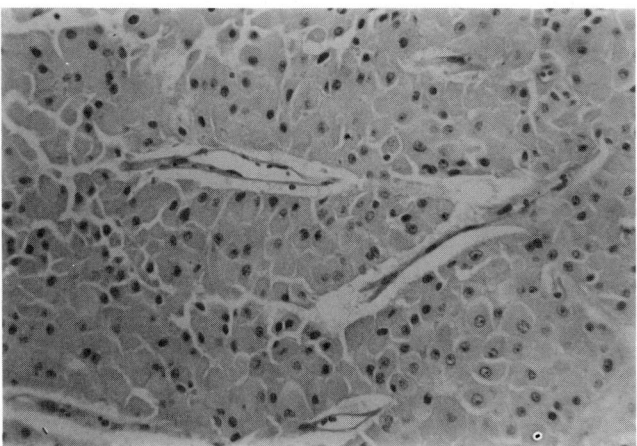

FIG 16–11.
Histologic section of a benign renal oncocytoma (original magnification, × 200).

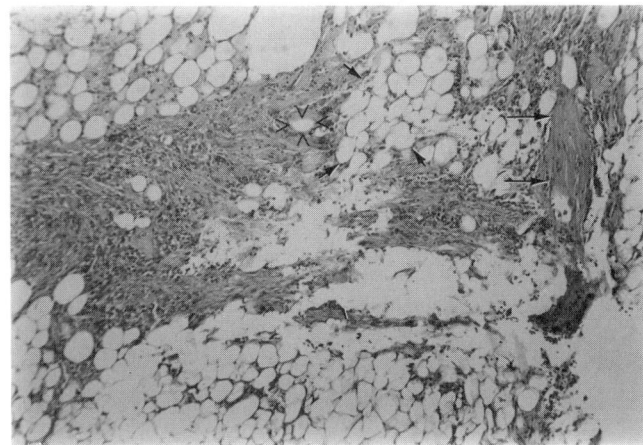

FIG 16–12.
Histologic section of a renal angiomyolipoma showing typical fat *(short arrows),* muscle *(long arrows),* and vessels *(arrowheads)* (original magnification, × 63).

to the large amounts of fat present within the tumors. Fat on CT has a negative density (-20 to -80 Hounsfield units), which is pathognomonic for angiomyolipomas when observed in the kidney (see Fig 16–6) (Pitts et al., 1980). Ultrasonographically, high-intensity internal echoes may suggest fat, but confirmation by CT is recommended.

When the CT scan is unequivocal and there are no symptoms, surgical treatment of an angiomyolipoma is unnecessary (Lingeman et al., 1982; Oesterling et al., 1986) although annual monitoring by CT or US is recommended. In the symptomatic patient or in cases in which the diagnosis is uncertain, exploration, biopsy, and/or complete excision may be required. In selected patients with hemorrhage, percutaneous angioinfarction may be suitable (Rosen et al., 1984). Because the tumor is benign, if surgical intervention is required, subtotal nephrectomy guided by intraoperative frozen section should be considered.

OTHER BENIGN TUMORS

There are a variety of other benign—albeit rare—renal tumors, which include fibromas, leiomyomas, lipomas, hemangiomas, and juxtaglomerular cell tumors. The *fibroma* is a variably sized benign tumor found primarily in females. Although cortical fibromas are described in the literature, recent studies have documented only those arising in the medulla. In this location they represent the most common benign mesenchymal tumor and are found in an average of 35% of autopsy cases (Bonsib, 1985). Few cases are diagnosed during an individual's lifetime, as most fibromas are small and cause no symptoms. Pathologically they are gray-white, without a capsule, and are bilateral in 50% of the cases. *Leiomyomas* may arise from the renal capsule or renal vessels, tend to be less than 1.0 cm in size, and are very infrequently diagnosed other than at autopsy. The kidney is the most frequent urinary tract site although the lesion is more common in the uterus and gastrointestinal (GI) tract (Belis et al., 1979). The lesions are gray-white and no mitoses are evident. Treatment is rarely required. The *lipoma* is not common in the kidney in its pure form (i.e., not associated with angiomyolipoma) (Xipell, 1971). Renal lipomas are seen primarily in middle-aged females. They arise from renal capsule or perirenal tissue and consist of mature adipose cells without evident mitoses. Diagnosis may be made by CT when the tumor is large due to the characteristic CT differentiation of fat. *Hemangiomas* are uncommon but occasionally will be responsible for hematuria with an elusive cause. They tend to be less than 2 cm in diameter and are sometimes multiple and bilateral. They can be found within the renal medulla or in the submucosa of the renal collecting system. Di-

agnosis may be determined by angiography or direct vision of the lesions in the renal collecting system by endoscopy (Ekelund and Gothlin, 1975). The lesions are dark red and composed of endothelial lined spaces. The most important of this subgroup of benign tumors is the *juxtaglomerular cell tumor* because it causes profound hypertension, which can be cured by surgical treatment. The tumor is quite rare, with less than 20 cases reported in the literature. Patients are usually in their early adult years (over 25 years) and present with marked diastolic hypertension and hypokalemia. The tumors arise from epithelial cells of the juxtaglomerular apparatus and can be shown to contain renin secretory granules by specific stains (Bowie's). Pathologically the tumors resemble hemangiopericytoma, but metastases have not been identified. Diagnosis is suspected by secondary hyperaldosteronism and confirmed by selective renal vein sampling for renin. If the tumor can be localized by CT or angiography a subtotal nephrectomy might be possible; however, in most cases a complete nephrectomy has been curative (Bonsib, 1985).

In general, the majority of benign tumors are not a therapeutic challenge since they are rarely encountered during the patient's lifetime. Those found incidentally would best be handled by close surveillance and those causing symptoms would best be treated by renal conserving surgery. Because there are no features to unequivocally establish the diagnosis prior to surgery (excepting juxtaglomerular tumors), the pathologist most often provides the diagnosis following total nephrectomy.

PRIMARY MALIGNANT RENAL TUMORS

Primary malignant neoplasms of the renal and perirenal tissues comprise approximately 6% of all malignancies in adults. The various forms include the more common epithelial tumors of the parenchyma, such as renal adenocarcinoma, and urothelial tumors, including transitional cell cancer of the renal pelvis. Less common types are adult nephroblastoma and the mesenchymal tumors such as liposarcoma and leiomyosarcoma.

RENAL CARCINOMA

Renal cell carcinoma (RCC) accounts for 85% of all renal parenchymal cancers, while uroepithelial cancer accounts for 7%–8%, and adult Wilms' and sarcomas 3%–4% of all kidney cancers. RCC represents 2%–3% of all adult malignancies and is the ninth most common malignant tumor in white men and the 13th most common in women. It has been thought that RCC is increasing in incidence but recent reports show that the 1977 incidence in the United States was essentially unchanged from 1969 in both white men and women, 9.4 per 10^5 and 4.6 per 10^5, respectively (Page and Asire, 1985). The incidence has increased for blacks during the same time period, however, and is now exactly the same as for whites. It is estimated that between 15,000 and 18,000 new cases will be seen and over 7,500 persons will die this year from RCC.

Etiology

The etiology of RCC is completely unknown, although reports have implicated tobacco use (particularly pipes and cigars), environmental or industrial contaminants such as cadmium (Kalonel, 1976) and other trace elements, asbestos, petroleum by-products, and exposure to herpes simplex virus as having a possible role (Kalonel, 1976; Karcioglu, 1978; Cocchiora et al., 1980; Fallon, 1985). Further risk factors include obesity and urban environments. A familial incidence has been described, and, in fact, a specific translocation of chromosomes 3 and 11 in tumor cells in one family has been observed (Pathak et al., 1982). Recent evidence also suggests that oncogene localization to the short arm of chromosome 3 may have etiologic implications in renal cancer (Carroll et al., 1986). In addition patients with HLA antigen types Bw44 and DR8 appear to be more prone to development of RCC (Cohen, 1979). There is an association of RCC with several clinical syndromes, which include von Hippel-Lindau disease (cerebellar hemangioblastoma, retinal angiomata, and bilateral RCC), adult polycystic kidney disease, horseshoe kidneys, and acquired renal cystic disease from chronic renal failure. Carcinogen studies in animals have revealed a number of possible etiologic agents, such as dimethyl nitrosamines (present in tobacco smoke) and estrogen, but a direct link to causation in humans from either of these is lacking. There are no precise models of human RCC in animals, although several groups have reproduced human tumors in immune-deficient animals (Williams et al., 1984b; Clayman et al., 1985) and recently streptozotocin has been shown in mice to induce renal tumors that are very similar to those found in primary and metastatic sites (Hand, 1985).

RCC occurs most commonly in the fifth to sixth decade, although it has been described in children and young adults as well. Males are affected twice as often as females. The cell of origin of RCC has been shown to be the proximal convoluted tubule. Substantiating evidence is the similar ultrastructural characteristics of tumor cells and proximal tubular cells (Gondos, 1981) and cell surface antigen homology (Wallace and Nairn, 1972). Renal adenomas have similar findings, further suggesting that there is little difference between adenomas and carcinomas.

Diagnosis and Staging

RCC continues to be an enigmatic cancer with an extremely variable mode of presentation. An increasing number of tumors are being detected incidentally because of increased use of CT: prior statistical data involving presenting characteristics thus may no longer be accurate. Previously a triad of flank pain, palpable mass, and hematuria was stressed as being suggestive of RCC (Table 16–2). However, less than 10% of patients will exhibit this combination of findings, and both pain and a palpable mass are late events occurring only with very large tumors (those that invade surrounding structures or those that have undergone spontaneous hemorrhage). Sixty percent of patients will present with gross or total painless hematuria. Symptoms due to bone or brain metastases may in fact be the initial complaint, as 30% of patients will have distant metastatic extension when first seen. Physical findings indicative of renal cancer are limited: a palpable mass in either upper quadrant that can be moved anteriorly by costophrenic angle palpation; a varicocele of recent onset usually—but not always—on the left that does not recede in the supine position (due to renal vein or vena caval tumor thrombus); anemia; and cachexia. There are no laboratory findings or tumor markers specific for the diagnosis of RCC. Urine cytologic analysis is of little benefit (Piscioli et al., 1985). Recent studies have suggested plasma transcobalamin II (Jensen et al., 1983) or serum haptoglobin levels (Babaian and Swanson, 1982) might be helpful in the diagnosis of RCC. Although neither was shown to be diagnostically specific they may be useful to detect recurrence or to follow re-

TABLE 16–2.

Presenting Findings in Renal Cell Carcinoma Patients*

FINDING	OCCURRENCE, %
Hematuria	50–60
Elevated erythrocyte sedimentation rate	50–60
Abdominal mass	24–45
Anemia	21–41
Flank pain	35–40
Hypertension	22–38
Weight loss	28–36
Pyrexia	7–17
Hepatic dysfunction	10–15
Classic triad (hematuria, abdominal mass, flank pain)	7–10
Hypercalcemia	3–6
Erythrocytosis	3–4
Varicocele	2–3

*Adapted from Skinner et al., 1971; Chisholm, 1974; and Fallon, 1985.

sponse to therapy. Erythrocyte sedimentation rate (ESR) is elevated in more than 55% of patients with RCC, but this too is nonspecific and of no major diagnostic benefit.

RCC has been dubbed the "internist's tumor" because of its protean manifestations (see Table 16–2). Paraneoplastic syndromes are quite common, occurring in 10%–40% of patients, and occasionally represent the initial finding that leads to diagnosis (Cronin et al., 1976). In general, these paraneoplastic syndromes are unrelated to, and do not imply, the presence of metastases. Anemia occurs in 20%–40% of patients, but its exact etiology is unknown. Because blood loss is not usually sufficient to account for the anemia, it is considered to be caused by a chronic disease process. Erythrocytosis (not polycythemia, which implies an increase in all formed elements of the blood) occurs in 3%–4% of patients and is presumed to be caused by an overproduction of erythropoietin by tumor cells. Indeed RCC tumor cells in vitro have been shown to produce erythropoietin (Okabe et al., 1985) and serum erythropoietin levels have been shown to be elevated in selected RCC patients (Toyama et al., 1979). Pyrexia occurs in approximately 20% of patients; it is presumed that the tumor produces a pyrogenic factor, but thus far none has been isolated. Twenty percent to 40% of patients have hypertension. While the hypertension may predate tumor-related factors, renin production by RCC has been documented (Hollifield et al., 1975). Hypercalcemia is exhibited by RCC in 3%–6% of patients. Much speculation currently attends the etiology of the hypercalcemia of malignancy (Mundy and Martin, 1982). In RCC, parathyroid hormone, prostaglandins, or osteoclast-activating factor serum levels have not generally been found to be elevated in patients with hypercalcemia. Recent studies indicate an adenylate cyclase stimulating factor that reacts with parathyroid hormone receptors to cause bone resorption and/or interaction with growth factors (epithelial growth factor and transforming growth factor)—the more likely genesis of the hypercalcemia in renal cancer (Strewler et al., 1983). Hepatic dysfunction (Stauffer's) syndrome occurs in 10%–15% of patients. The etiology is unknown but is suspected to be caused by a hepatotoxic product of the tumor. This syndrome is manifested by an elevated alkaline phosphatase, increased prothrombin time, and an elevated α_2-globulin. Similar to other paraneoplastic syndromes, Stauffer's is not related to the presence of metastases; however, unlike them, the hepatic dysfunction syndrome portends a poor prognosis (Boxer et al., 1978). In general, the presence of one of these syndromes may raise suspicion of RCC, but they do not suggest metastases and are not prognostic, because re-

moval of the primary tumor in the absence of metastases will usually eliminate the syndrome. They may, however, be useful as markers of tumor should they persist or recur following definitive treatment.

The initial diagnosis of renal cell carcinoma relies heavily on the algorithm for workup of a renal mass presented previously (see Fig 16–1). Patients with unexplained constitutional symptoms, microscopic or gross hematuria, a paraneoplastic syndrome, or typical metastases with an undetermined primary tumor, most often cause the physician to suspect a renal malignancy and to begin a diagnostic workup. In those patients with microhematuria, a voiding three-glass urine test may help to define the upper tract as the source of red cells if they are equally present in all three specimens. Intravenous urography with nephrotomography is still recommended for initial examination. IVU alone is only 75% accurate and thus further tests are required for confirmation, particularly in symptomatic patients. A simple benign renal cyst is the most common renal mass and on IVU exhibits a radiolucent center with a thin wall and a sharp interface between the mass and the renal cortex, i.e., the typical "beak sign" of a cortical cyst (Fig 16–13). All renal masses seen on IVU should have further definition by US to determine if the mass is cystic or solid. Ultrasound is approximately 98% accurate in defining simple cysts and thus needle aspiration of typical cysts is unnecessary. Equivocal lesions may be subjected to needle puncture, but further studies with CT or angiography should be performed first. Should cyst fluid be analyzed, cytologic analysis is the only reliable test as cyst chemistries are not specific for or against the presence of tumor (Juul et al., 1985).

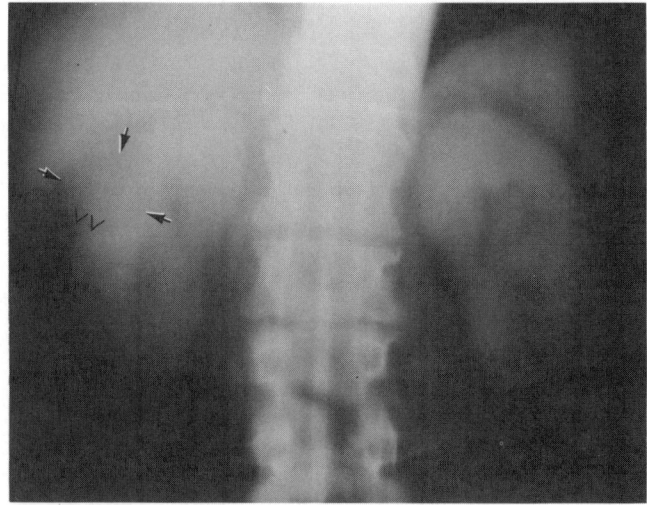

FIG 16–13.
Nephrotomogram of a patient with a simple renal cyst *(arrows)* showing the typical "beak sign" *(arrowheads)*.

Bloody aspirates are not diagnostic but are associated with malignancy in approximately 25% of patients (Lang et al., 1972). There has been considerable controversy in the literature concerning the possibility of needle-tract seeding of RCC (Gibbons et al., 1977); however, the incidence is extremely low and, thus, of little concern in well-selected patients (VonSchreeb et al., 1967).

Calcification of renal masses is an important diagnostic finding as its presence significantly increases the probability of cancer. Daniel and others (1972) found that 87% of renal masses with central calcifications and 20% of those with peripheral calcifications were malignant. Recently Weyman and others (1982) have shown that calcifications with soft tissue overlying them are highly associated with malignancy, whereas those without surrounding soft tissue at the external surface are less likely to be neoplasms. Simple benign renal cysts contain calcium in less than 1% and, thus, calcium-containing masses require at least further diagnostic study and probably surgical definition as well. Lesions seen on IVU but unconfirmed on US or equivocal lesions on US can be studied by radioisotope scanning. This modality can detect normal from abnormal renal parenchyma but cannot determine cyst vs. tumor and thus is not useful as an isolated test.

Renal angiography of solid renal masses has been supplanted by the equally reliable but less morbid and less costly CT scan. Prior to the advent of CT, selective renal angiography was used almost exclusively for the final diagnostic step before surgical treatment of solid renal masses. Angiography demonstrates neovascularity, arteriovenous fistulae, venous pooling on contrast, and accentuation of capsular vessels singly or in combination in 85%–95% of renal tumors (Fig 16–14). Unfortunately, approximately 10% of tumors are not hypervascular and thus present some difficulty in their definition. Angiography is also capable of defining renal vein and vena caval involvement with tumor. Angiography carries a small but real risk of complications including hemorrhage or pseudoaneurysm formation at the puncture site, arterial emboli, and contrast impairment of renal function. In addition, because angiography generally requires hospitalization, there are significant cost factors favoring its replacement by CT in the routine case (Zimmer et al., 1984).

CT imaging is now the diagnostic procedure of choice when a solid mass is seen on ultrasound or in many instances when a mass appears solid on an initial IVP. There have been numerous studies to document that the CT scan is as accurate as renal angiography in diagnosis of renal cancer (Cronan et al., 1982; Mauro et al., 1982). Weyman and associates (1980), for example,

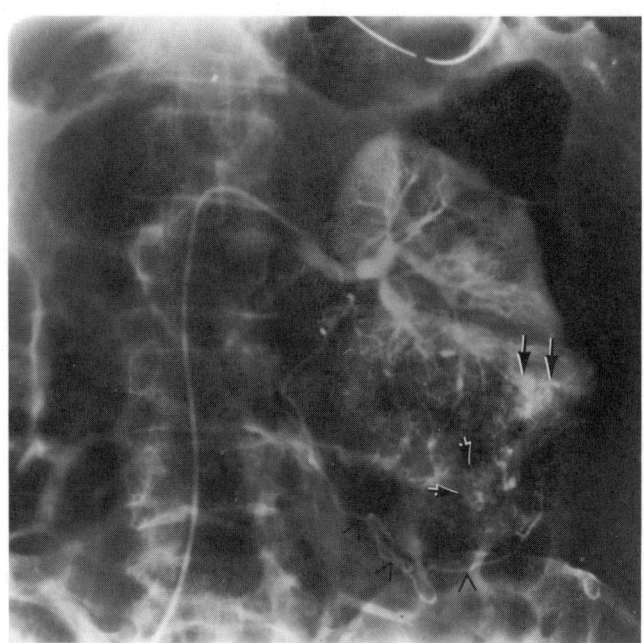

FIG 16–14.
Selective left renal angiogram of typical renal cell carcinoma showing neovascularity *(short arrows)*, vascular puddling *(long arrows)*, and parasitic capsular vessels *(arrowheads)*.

studied 49 RCC patients who had both CT and angiography and found the diagnostic accuracy to be 95% with CT and 89% with angiography.

A typical finding of RCC on CT is a mass that "enhances" with the use of contrast media. In general, RCC exhibits an overall decreased density in Hounsfield units as compared to normal renal parenchyma but will show a heterogeneous pattern of enhancement or increased attenuation (slightly decreased from surrounding parenchyma) when contrast is used (Kosko et al., 1984). In some smaller tumors differentiation with contrast studies alone may be confusing and thus studies with and without contrast are recommended (Fig 16–15). Further suspicious diagnostic findings include an indistinct interface between the mass and surrounding parenchyma and secondary findings of obvious regional node enlargement, perinephric fat invasion, or involvement of the renal vein or vena cava. In addition, larger tumors may exhibit central areas with a marked decrease in density corresponding to areas of necrosis within the tumor. Approximately 10% of renal masses will be indeterminate by CT due to lack of contrast enhancement and other distinguishing criteria. Balfe and others (1982) evaluated 60 such masses and found that in only 16% was the addition of angiography helpful. In most cases motion artifacts on slow scanners and obesity were thought to be the limiting factors, and repeated

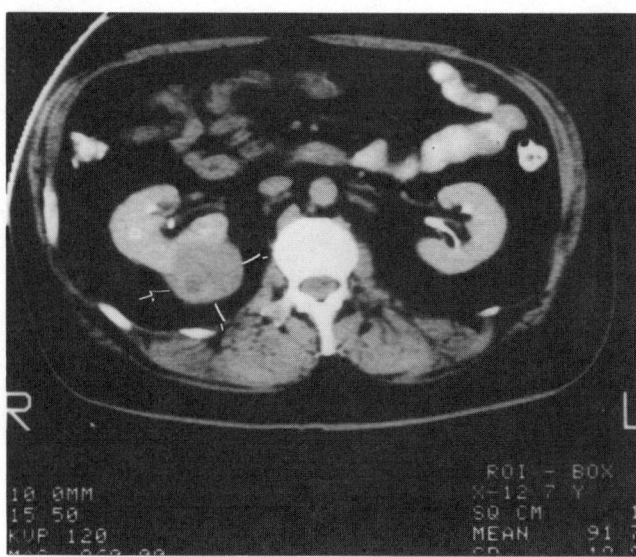

FIG 16–15.
Computed tomograph of a contrast enhanced right renal cell carcinoma *(arrows).* (From Williams RD: *Cecil Textbook of Medicine.* Philadelphia, WB Saunders, 1985, p 642. Used with permission.)

CT was recommended; when the diagnosis was still in question ultrasound-guided needle aspiration was diagnostic in 84% of these patients.

Staging

CT has also been shown to be better suited for local staging of renal cancer than any other modality except perhaps MRI (Lang, 1984; Weyman et al., 1980; Cronan et al., 1982; Probst et al., 1981; Levine et al., 1980). Cronan et al. (1982) studied 23 patients and found CT to be accurate in staging RCC in 91% while US and angiography demonstrated accuracy in 70% and 61%, respectively. Weyman et al. (1980) compared CT to angiography in 62 cases: perinephric extension was accurately predicted in 83% by CT and 68% by angiography; renal vein involvement in 82% by CT and 79% by angiography; vena caval involvement in 93% by CT and 100% by angiography; and lymph node extension in 80% by CT and 85% by angiography. Probst and others (1981) had similar results in 40 patients but found CT staging more accurate than angiography for lymph node involvement. Lang (1984) studied 22 patients using several techniques and found dynamic CT (rapid-sequence CT immediately following bolus contrast infusion) to be more accurate for staging than conventional CT, angiography, US, or radionuclide studies. Thus, at present CT is the preferred diagnostic and local staging method for renal cancer. There is still a place in the diagnostic armamentarium for angiography for patients with equivocal lesions; for guiding the operative approach to nephron-conserving surgery (e.g., those with a solitary kidney or renal insufficiency); and for those who are suspected of having bilateral lesions that were not detected on CT (von Hippel-Lindau disease).

Magnetic resonance imaging (MRI) also has the ability to diagnose and locally stage renal cancer although there are to date very few studies that use pathologic confirmation to compare the accuracy of MRI to other screening modalities. In general, MRI has the advantages of being noninvasive, harmless, and able to provide images in multiple planes. Initial studies of renal masses by Hricak and others (1985) in 27 patients with renal cancer showed MRI to be equivalent to CT in the diagnosis of RCC but more accurate than CT in staging renal cancer (96% vs. 70%), particularly with respect to perinephric extension, adjacent organ involvement, and the detection and extent of renal vein and vena caval involvement. The advent of paramagnetic contrast agents is expected to improve the accuracy of RCC diagnosis by MRI, perhaps allowing it to become the preferred technique for both diagnosis and staging of renal cancer (Ramchandani et al., 1986).

Following definition of a renal cancer by clinical means determination of therapy is based on local and distant extent of disease (Table 16–3). Local clinical staging relies primarily on CT, as discussed previously. Local renal vein extension, which occurs in 10%–20% of patients, has previously been diagnosed most often by delayed angiographic films or venacavography to delineate the precise extent of thrombi. CT has been shown to be quite accurate in identifying tumor thrombi, but determination of extent in the cava may still require further studies. Pedal injection of contrast during CT may be helpful in selected patients but abdominal ultrasound has been shown to be quite accurate in determining the presence (5%–10%) and extent of caval thrombi (Kosko et al., 1984; Aziri et al., 1979;

TABLE 16–3.

Common Sites of Metastases in Renal Cancer*

SITE	%
Lung	50–60
Bone	30–40
Regional nodes	15–30
Main renal vein	15–20
Perirenal fat	10–20
Adrenal (ipsilateral)	10–15
Vena cava	8–15
Brain	10–13
Adjacent organs (colon, pancreas)	10
Kidney (contralateral)	2

*Adapted from Fallon (1984), Richie and Garnick (1982), and Johnson et al. (1983).

Levine et al., 1980). Currently when a tumor thrombus is suspected in the vena cava, either US or MRI is recommended for confirmation and determination of the cephalad extension.

Involvement of renal hilar and paraortic lymph nodes (15%–30% of patients) is currently best assessed by CT, although it must be noted that only nodes greater than 1.5 cm in diameter are visualized. Because intranodal architecture is not determined, enlarged nodes are not necessarily tumorous and microscopic tumor in normal-sized nodes will not be detected. The prognosis in patients with nodal metastasis is poor, and thus large nodal masses demonstrated on CT could be further staged by CT- or US-guided needle aspiration cytologic analysis; however, minimal enlargement of regional nodes should not deter surgical staging as nodal detection by CT or US does not provide specific criteria to determine malignant involvement.

Assessment of distant spread to intra-abdominal organs, such as the pancreas, spleen, and particularly the liver, can also be adequately assessed by CT. Radionuclide scans are rarely indicated for this purpose. In fact, Lindner and others (1983) have shown that patients with a nonpalpable liver and normal serum liver studies rarely have hepatic metastases on such scans and thus do not recommend its use. Liver lesions suspected on CT can be further examined by CT-guided needle aspiration (Fig 16–16), thus obviating surgical diagnosis in selected patients.

Determination of metastases to bones is most accurate by radionuclide bone scan, although the study is nonspecific and bone x-rays of identified abnormalities

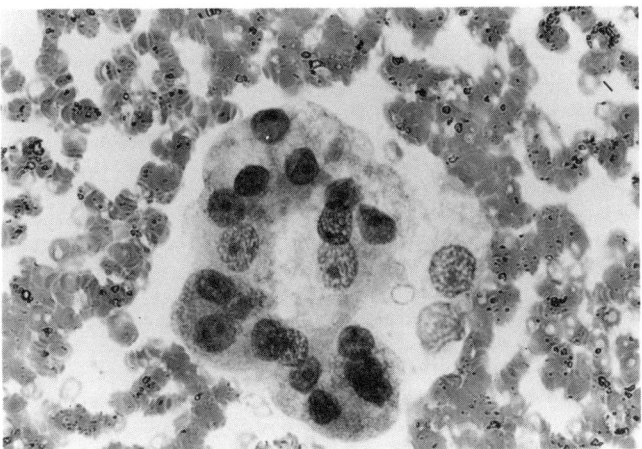

FIG 16–16.
Needle aspiration cytology of renal cell carcinoma metastatic to the liver. The malignant cell clusters are arranged in three-dimensional glandular formation. The cells have abundant finely vacuolated cytoplasm and the nuclei are hyperchromatic, with prominent central nucleoli and moderate anisonucleosis (original magnification, × 300).

are required to confirm the presence of the typical osteolytic lesions. Bone metastases are most common in the pelvis, spine, skull, and long bones (femur and humerus), although spread to other osseous sites is not infrequent. Occasionally, abnormal areas on bone scan not confirmed by x-ray may require CT scan, MRI, or bone biopsy to establish the presence or absence of tumor (Baker et al., 1985). There is controversy, however, concerning whether a bone scan is indicated in the clinical staging of renal cancer, since in two studies patients without bone pain or abnormal serum alkaline phosphatase values did not exhibit metastases on scans (Lindner et al., 1983; Campbell et al., 1985). Despite the cost savings by eliminating the bone scan, the prognosis for patients with osseous metastases is so poor that the discovery of bone tumors would contraindicate surgical intervention; thus, use of bone scanning is still recommended.

The lung is the most common site of metastases from RCC, occurring in 50%–60% of patients at autopsy (deKernion et al., 1978). A posteroanterior and lateral chest x-ray is an adequate screen for pulmonary lesions, but chest tomography or chest CT is more sensitive. Chest CT has the advantages of less radiation dosage, better delineation of mediastinal structures, and determination of lesions as small as 3–5 mm. The controversy over a cost-effective staging workup would suggest that chest x-ray may be sufficient; however, if the presence of lung metastases would alter subsequent therapy, chest CT is recommended. Further staging studies such as brain CT, celiac angiography, supraclavicular node biopsy, and mediastinoscopy as suggested in the past are unnecessary in routine cases.

Although clinical stage is the key to determining appropriate therapy, prognosis is based on the combination of clinical and pathologic staging. There are a variety of staging systems in use but all are based primarily on the early study of Flocks and Kadesky (1958). Modification to include vascular involvement was proposed by Robson et al. (1969) and this system has been used universally in the United States since that time (Fig 16–17). Although the Robson system is easy to use it does not relate directly to prognosis (particularly with respect to stage IIIa) and thus the TNM system has been proposed as the superior approach.

Table 16–4 compares the Robson staging classification with the TNM scheme (Robson et al., 1968; Holland, 1973; Wallace et al., 1975; Beahrs and Myers, 1983). Obviously the TNM system allows more precise distinction between subdiagnostic and prognostic categories and would provide a more accurate method of comparing treatment results from the United States and abroad. It is also obvious that the TNM system is lengthy and cumbersome and thus not amenable to

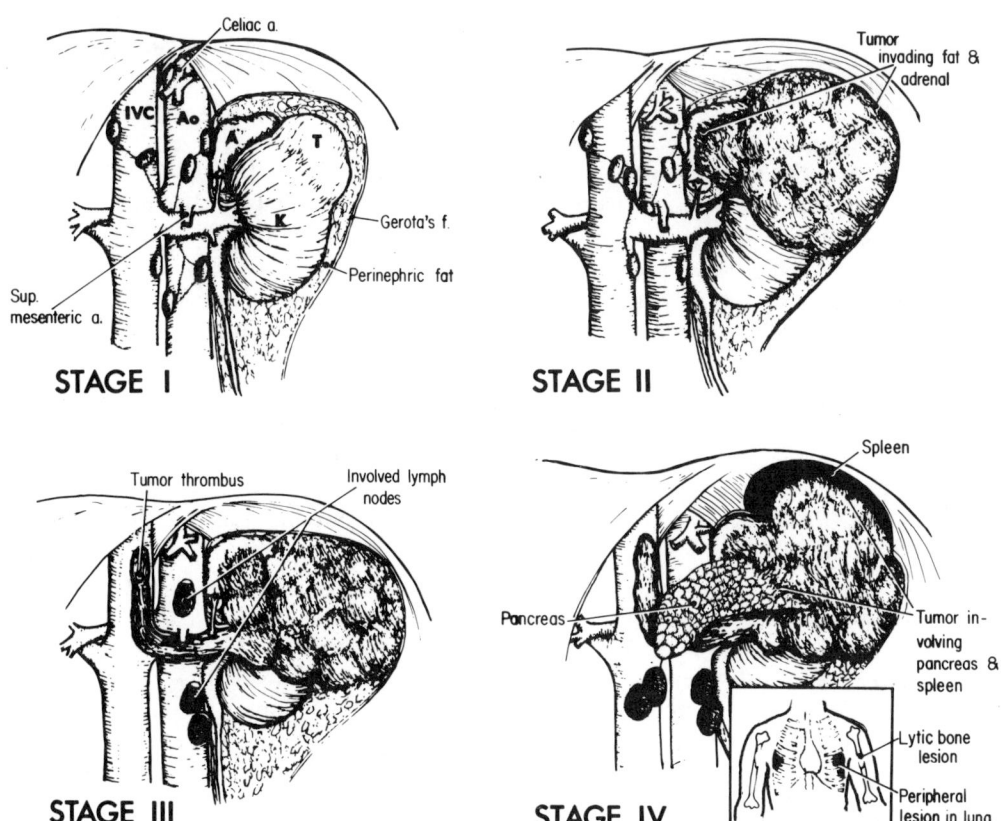

FIG 16–17.
Robson staging system for renal cell carcinoma (see Table 16–4). (Adapted from Robson J, et al: The results of radical nephrectomy for renal cell carcinoma. *J Urol* 1969; 101:297.)

practical use. Even though the TNM system has distinct advantages it is doubtful that it will become used routinely in the United States.

Pathology

As stated previously in this chapter, renal cell carcinoma is an epithelial malignancy that has been designated by many terms, including hypernephroma, Grawitz' tumor, renal adenocarcinoma, and nephrocarcinoma; however, the term *renal cell carcinoma* (RCC) is universally accepted and preferred. RCC occurs with equal frequency in either kidney and has no predilection for upper or lower poles. Due to its cortical genesis, RCC tends to grow out from the surface of the kidney, causing the characteristic bulge or mass effect critical to its detection on diagnostic imaging studies (Fig 16–18). RCC is often quite large when detected, averaging 7–8 cm in diameter. Grossly, the tumor is characteristically yellow to orange due to the abundance of lipids present, particularly in the clear cell variety (25%). The granular type, which makes up another 25%, tends to be more gray to white. The rest of the tumors are mixed cell types with only 2% being a sar-

comatoid variety. Small tumors are homogeneous on cut surface but larger tumors exhibit hemorrhage, necrosis with secondary cystic areas, and, occasionally, calcification. A definite capsule is not demonstrable in most cases, although a pseudocapsule of compressed renal parenchyma is often observed. RCC has a tendency to invade vascular spaces and extend into the main renal vein in 15%–20% of cases and the inferior vena cava in 8%–15% (see Table 16–3). Tumor thrombi that reach the right atrium have been noted. These thrombi do not commonly invade the cava but can occlude it, leading to such manifestations as an acute varicocele that does not empty in the supine position, or lower limb edema.

Microscopically RCC is most often a mixed adenocarcinoma containing clear cells, granular cells, and, occasionally, sarcomatoid-appearing cells. The clear cell variety is most often predominant. These cells are usually round with a distinct cell border and abundant clear or vacuolated cytoplasm (Fig 16–19). The cytoplasm contains cholesterol, triglycerides, glycogen, and lipids, the latter two of which are removed during histologic preparation and account for the clear cytoplasm ob-

TABLE 16–4.

Comparison of Conventional and TNM Staging Classification of Renal Cell Cancer

ROBSON STAGE	T	N	M
I.—Tumor confined by renal capsule	T_1 (Small tumor with minimal calyceal distortion T_2 (large tumor with calyceal deformity)		
II.—Tumor extension to perirenal fat or ipsilateral adrenal but confined by Gerota's fascia	T_{3a}		
IIIa.—Renal vein or IVC involvement	T_{3b} (renal vein involvement) T_{3c} (renal vein and caval involvement below the diaphragm) T_{4b} (caval involvement above the diaphragm)	N_0 (nodes negative)	M_0 (lack of distant metastases)
IIIb.—Lymphatic involvement	T_{1-3}	N_1 (single homolateral regional node involved) N_2 (multiple regional, contralateral or bilateral nodes involved) N_3 (fixed regional nodes) N_4 (juxtaregional nodes involved)	
IIIc.—Combination of IIIa and IIIb	T_{3-4}	N_{1-4}	
IVa.—Spread to contiguous organs except ipsilateral adrenal	T_{4a}	N_{0-4}	
IVb.—Distant metastases	T_{1-4}	N_{0-4}	M_1

served by light microscopy. The nuclei are small, round, and dark, and mitotic figures and giant cells are infrequent. Mucin, although common to other adenocarcinomas, is not present within any of the primary renal adenocarcinoma cell types. Although the clear cell type tends to form acini or tubules they may be solid or papillary as well. Commonly the tumors will contain distinct areas of more than one pattern. Ultrastructure

examination reveals the glycogen and lipids in the cytoplasm as well as epithelial characteristics (microvilli and tight junctions), a poorly developed Golgi apparatus, and sparse mitochondria and cytosomes (Bonsib, 1985; Mostofi and Davis, 1984).

The granular cell variety is often present in these tumors, even though the clear cell component may predominate. Typically the cells are arranged in sheets and

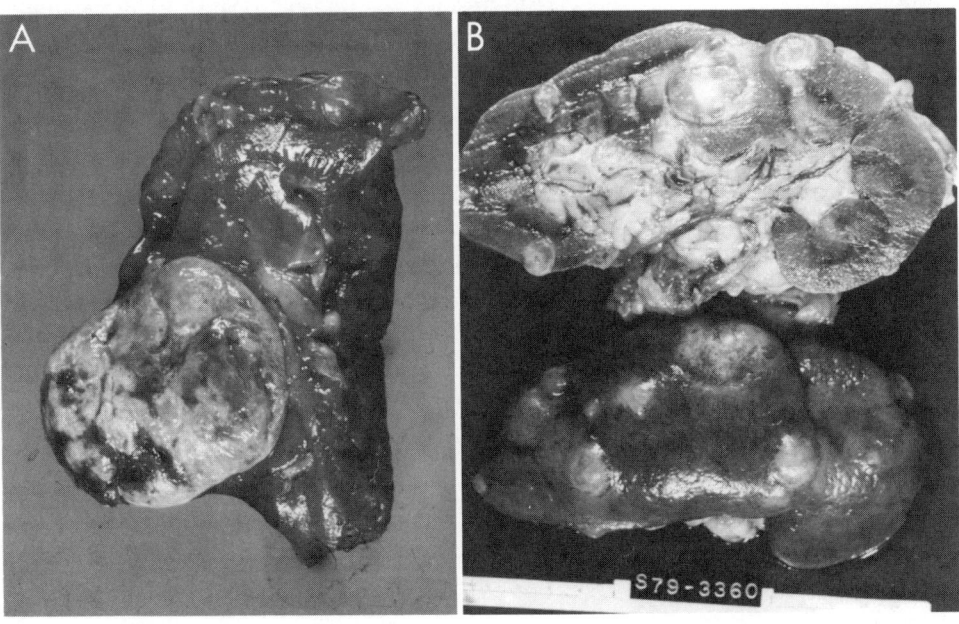

FIG 16–18.
A, gross pathology of the same right renal cell cancer shown in Figure 16–15. **B,** photograph of the gross pathology of multiple renal cell carcinomas in a patient with von Hippel-Lindau disease.

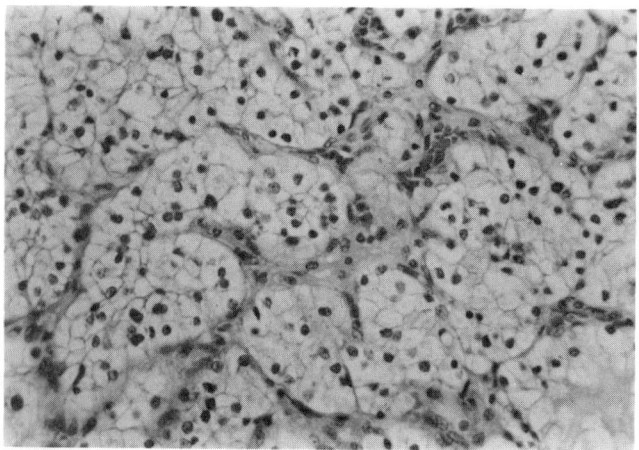

FIG 16–19.
Photomicrograph of clear cell renal adenocarcinoma (original magnification, × 250).

exhibit a homogeneous eosinophilic cytoplasm and large nuclei (Fig 16–20). Electron microscopy reveals the granular cytoplasm to contain large numbers of mitochondria and cytosomes; the Golgi apparatus is visible and there is less lipid and glycogen than in the clear cell type.

The sarcomatoid cell type is rare in its pure form and is more commonly found to be a small component of either the granular or clear cell variety (or both). The cells are spindle-shaped, form sheets or bundles, and are often hard to differentiate from fibrosarcoma cells. Electron microscopic features, however, can clearly identify them as epithelial (Deitchman and Sidhur, 1980).

There is considerable controversy as to whether grading of RCC bears prognostic significance and

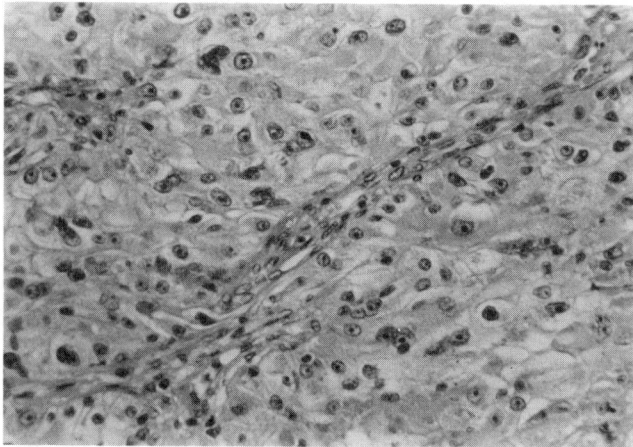

FIG 16–20.
Photomicrograph of granular cell renal adenocarcinoma (original magnification, × 250).

whether the cell types have a differing prognosis. In general, although grading is possible (as attested to by the many grading systems available), there are no universally agreed upon criteria, and since high-grade lesions tend to occur in patients with high-stage disease the importance of grading seems moot. Those espousing their own grading scale would disagree with this point of view, yet in practical use grade does not alter treatment of the patient and thus is not critical clinically. When high-grade, predominantly granular tumors are corrected for grade and stage, there does not appear to be a difference between clear cell and granular cell tumor prognosis (McNichols et al., 1981). The sarcomatoid variant, however, does tend to have a poorer prognosis, although few such tumors are available at any one institution to study the biologic correlates (Colvin and Dickersin, 1978).

Treatment

Appropriate treatment of renal cell cancer relies almost entirely on the stage of the tumor at presentation and thus requires attention to the details of proper staging described previously. The prognosis of patients with stages I, II, and IIIa (renal vein only) are very similar if complete surgical removal of the kidney and its enveloping fascia is accomplished (Table 16–5). Patients with stages IIIb and IIIc have a poorer prognosis but unless multiple large nodes are confirmed as tumorous preoperatively, surgical therapy is still recommended. Indeed, the diagnosis of stage IIIb is most often made by the pathologist on postoperative examination of permanent histologic sections.

Radical nephrectomy has been recommended as the definitive surgical treatment for localized RCC since the classic articles by Robson (1963, 1969). Although not controlled in the strict sense, his study showed a 5-year survival advantage of radical nephrectomy over simple nephrectomy. There continue to be no properly controlled statistical data available to favor radical nephrectomy, but because the operation as described is relatively simple to perform and adheres to the surgical

TABLE 16–5.
Prognosis of Surgically Treated
Patients With Renal Cancer*

STAGE	5-YR SURVIVAL, %
I	60–70
II	50–65
IIIa (Renal vein)	50–60
IIIa (Vena cava)	25–35
IIIb and IIIc	15–35
IVa and IVb	0–5

*Adapted from Robson et al. (1969), Skinner et al. (1971), and Johnson et al. (1984).

oncology principle of "a wide margin beyond the malignancy" it has become universally accepted. Radical nephrectomy entails en bloc removal of the kidney and its enveloping fascia (Gerota's), including the ipsilateral adrenal, the proximal half of the ureter, and lymph nodes up to the extent of the area of transection of the renal vessels (Fig 16–21). A key element in performance of a radical nephrectomy is early ligature of the renal vessels. Much is made in the literature about decreasing potential hematogenous tumor spread by observance of this practice; however, in the absence of supporting data the fact that it will very effectively limit potential blood loss during the surgery is a more compelling reason. Ligation of the renal artery prior to the vein is also considered an important step to decrease the pooling of blood that would eventuate from early vein ligation; however, this is of minor consideration and the vein can be ligated first when necessary.

There are a variety of incisions that will provide optimal access for a radical nephrectomy and the urologic surgeon should be familiar with more than one. I prefer a unilateral chevron (anterior subcostal) incision from the tip of the 12th rib to just beyond the midline. The patient is positioned with a sandbag under the ipsilateral chest at 30° and the table is slightly broken (Fig 16–22). This approach can be extraperitoneal but there is little reason to avoid entering the abdomen and, in fact, intraabdominal examination of the liver and paraortic nodes is helpful. A large Balfour retractor (or two smaller ones) is quite helpful to atraumatically retract the abdominal wall. Following incision of the line of Toldt (peritoneal reflection) the colon on either side can be displaced medially. This allows rapid location of the renal vessels in the middle of the field and easy access for isolation. One should recall that the left renal vein (and sometimes the right) often has an ascending lumbar branch entering directly posterior, which should be found and ligated to prevent unnecessary hemorrhage. Following ligation of the major renal vessels, much of the dissection of Gerota's fascia can be done by the fingers as the plane is easily developed. The adrenal should be taken with the specimen and care must be taken to be certain that all arterial and venous branches are ligated (or silver-clipped). The conjoint Gerota's fascia, lymphatics, and sympathetic tissue on the midline can then be easily ligated (clipped) and divided and the specimen removed. Alternative incisions include the thoracoabdominal, through the bed of the 9th, 10th, or 11th ribs with diaphragmatic incision and midline extension. This incision is best used for large upper pole tumors or when excision of an ipsilateral solitary lung metastasis is planned. In some instances a classic flank incision may be used but compromise of the exposure, particularly of the adrenal, is common. A midline incision is also appropriate in some cases.

Regional lymphadenectomy has been advocated by some (Peters and Brown, 1980; Marshall and Powell, 1982; Skinner et al., 1971) yet there is no proof that it is more than prognostic. Certainly the finding of positive nodes portends a poorer prognosis, but there is evidence that when microscopic tumor is present in 1–2 nodes the five-year survival is 30% as compared to near 0% with larger volume node involvement. Removal of all lymph nodes potentially involved in the variable lymphatics of the retroperitoneum is not routinely recommended because (1) extensive dissection would be required (Fig 16–23); (2) the few positive nodes would most often be encompassed by the radical nephrectomy; and (3) there is no real evidence that extended lymphadenectomy is beneficial.

Preoperative renal artery embolization has been used as a surgical adjunct in the past in the hope that it would decrease operative blood loss both by allowing the main renal vein to be taken first and by decreasing collateral venous flow. In addition, infarction is thought to facilitate the operation by causing edema in tissue planes (McLean and Meranze, 1985). Many substances

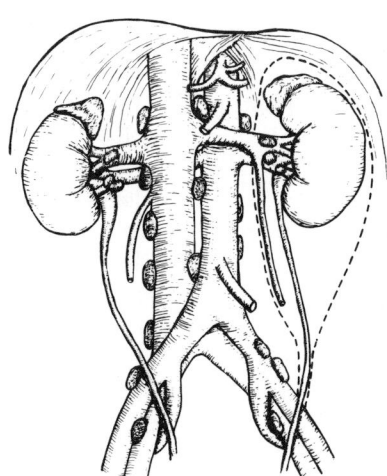

FIG 16–21.
Boundaries of a left radical nephrectomy. Dotted line represents both the surgical margin and Gerota's fascia.

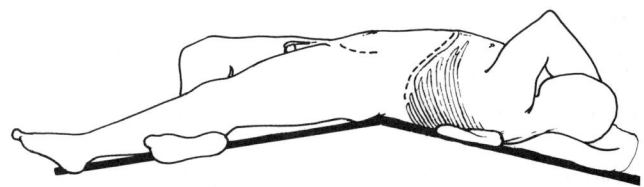

FIG 16–22.
Patient positioning for a radical nephrectomy through an anterior subcostal incision (dotted line at costal margin). The lower Gibson incision is used for completion of a radical nephroureterectomy.

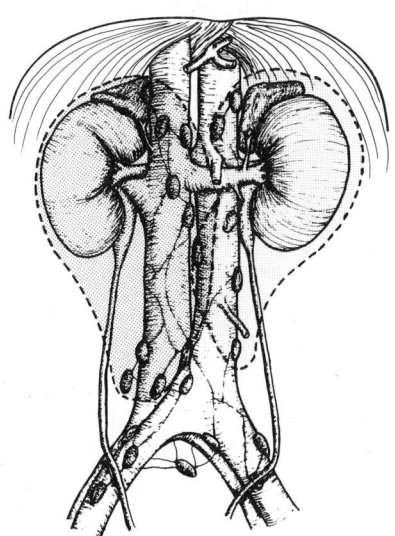

FIG 16–23.
Surgical extent of a radical nephrectomy and lymphadenectomy for renal cell carcinoma on either side.

have been used including particulate Gelfoam, polyvinyl alcohol sponge, detachable balloons, and steel coils; however, absolute ethanol has become the agent recommended because it causes superior occlusion in less time and minimizes the resultant "postinfarction syndrome" (Leionen, 1985). The renal artery is well-suited for embolization as it is an end artery and easily accessible by modern angiographic techniques. Possible complications of the embolization procedure include risk of damage to the contralateral kidney caused by excessive use of contrast materials; risk of overflow of the embolization agent causing ischemic injury to lower limbs or bowel; and migration of coils or balloons to distant sites during surgery (Cox et al., 1982; McLean and Meranze, 1985). The "postinfarction syndrome," marked by flank pain, fever, leukocytosis, and occasional hypertension, is decreased by the use of ethanol, but symptoms may persist for up to 72 hours. Thus, if embolization is used, it should be performed immediately prior to surgery. To date there is no conclusive evidence that preoperative embolization actually decreases blood loss or facilitates the surgery. In most cases in which large venous collaterals are present, the tumor has a parasitic blood supply from a source other than the renal artery. Occasionally arteriovenous fistulae are present, which contraindicate the use of embolization due to the real risk of embolizing particles or ethanol to the general circulation and lungs. Further, the tissue planes encountered during a radical nephrectomy are usually easily defined even with very large tumors. The possible immunotherapeutic advantages of preoperative embolization proposed by Swanson and

colleagues (1980) seemed plausible but subsequent studies by Swanson et al. (1983) and Gottesman et al. (1985) have not shown a salutary effect in patients with metastatic disease. Taken together, these facts suggest that preoperative renal infarction should not be used routinely, except perhaps in those patients with very large tumors in whom the renal artery may be difficult to reach early in the operation. The technique is, however, well-suited for palliation in patients with nonresectable tumors and significant local symptoms such as hemorrhage, flank pain, and, perhaps, some paraneoplastic syndromes.

Radiation therapy has also been proposed as a preoperative or postoperative surgical adjunct. The noncontrolled studies completed prior to widespread improvement in surgical techniques during the late 1970s suggested improved survival (Cox et al., 1970; Riches, 1966); however, a subsequent controlled study showed no real difference in survival between patients randomized to receive radiation vs. those who received none (van der Wurf-Messing, 1973). There is no evidence that postsurgical radiation therapy to the renal bed, whether residual tumor is present or not, has any benefit. In fact, although radiation has a proved palliative effect in patients with bone metastasis, tumor is not eliminated and thus survival is unaffected.

BILATERAL RCC AND TUMOR IN A SOLITARY KIDNEY

Renal cell carcinoma in a solitary kidney (congenital absence or prior removal of the contralateral organ), bilateral RCC, and RCC with renal insufficiency are all special situations that may require departure from the radical nephrectomy dictum. Obviously if the tumor and all (or a majority) of the functioning renal tissue were to be removed in these situations, chronic dialysis and/or transplantation would be required. While this approach is advocated by some to provide optimum tumor treatment, it is not justified in most cases because of the morbidity and mortality of chronic dialysis and renal transplantation. Initial reports suggested that patients who had RCC in the previously removed kidney had a survival rate/duration only half that of patients who never had a contralateral kidney or had it removed for nonmalignant causes (Wickham, 1975; Malek et al., 1976; Marberger et al., 1981). In most of these studies, however, the patient groups were not comparable in that the treatment (surgical or not) and follow-up were not similar. Schiff and others (1979) reviewed their cases and those in the literature (62 patients) and compared only patients who had complete tumor removal (with maintenance of renal function). They found that at 46 month follow-up, the average survival was not substantially different (78%) for patients with bilateral

RCC (whether synchronous or asynchronous) from that of patients with a prior nephrectomy for benign disease. Topley et al. (1984) reported an overall 70% five-year survival rate for 27 RCC patients, of whom 11 had bilateral RCC and 12 had no contralateral renal malignancy. These statistics taken together suggest that the prognosis for patients with bilateral RCC (whether synchronous or asynchronous) and patients with a solitary kidney due to absence or removal of an opposite benign kidney is the same and depends primarily on the adequacy of the surgical procedure and the pathologic stage of the tumor. Further, because these data are similar to those for stage I and II renal cancers treated by radical nephrectomy, they suggest that a partial nephrectomy, which removes all of the cancer in low-stage RCC, is adequate treatment in patients with a solitary kidney (no matter what the antecedent history), chronic renal insufficiency, and perhaps those with small incidental tumors as well. Although most urologic surgeons would treat patients with synchronous bilateral tumors by performing a radical nephrectomy on the side with the larger tumor and a partial nephrectomy on the opposite side, these same data could be used to argue for bilateral partial nephrectomies in appropriately staged patients.

Staging of these patients should follow the same guidelines mentioned previously for RCC patients with a normal contralateral kidney. The only exception is that angiography is required to properly assess the extent of the tumor within the kidney and the renal artery anatomy. The surgical approach is no different from that described previously, except that the surgeon should consider whether removal of the kidney with "bench surgery" followed by autotransplantation will be required to completely excise the tumor. In the past, Novick and others (1980) have recommended this approach for large tumors but over the last few years enthusiasm for "bench surgery" has declined so that it is only deemed necessary in large, centrally located tumors. Several authors have recently advocated enucleation of renal cancers in patients with RCC in a solitary kidney or bilateral RCC (Graham and Glenn, 1979; Novick et al., 1986). Marshall et al. (1986) studied the feasibility of enucleation in 16 kidneys first removed by radical nephrectomy, with tumor enucleation completed in the pathology department. The results showed that 43% of the enucleated tumor beds had residual cancer and that preoperative determination of tumors amenable to successful enucleation was not accurate. The risk of inadequate excision and subsequent recurrence does not favor enucleation; thus, partial nephrectomy with an adequate parenchymal margin is the preferred treatment. Patients with multiple small RCC such as those with von Hippel-Lindau disease

(Pearson et al., 1980) and those with chronic renal insufficiency may be candidates for enucleation, but every attempt should be made to remove all of the tumor and a margin of kidney if possible.

The intraoperative approach for partial nephrectomy is modified so that with large tumors the kidney is cooled with saline slush and a vascular clamp is used to occlude the renal artery after an intravenous infusion of 20% mannitol has been started. The renal artery is then dissected to the branches supplying the cancer, where they are ligated and divided. The renal fascia is dissected from the normal portion of the kidney to at least 1 cm away from the tumor and a standard partial nephrectomy is performed, removing the tumor and all of the renal fascia and leaving the adrenal intact. The renal pelvis is then closed and the open vessels ligated. In patients with small tumors in whom rapid excision is possible, renal cooling may not be necessary. The adrenal is left despite the fact that adrenal metastases occur in 10%–15% of patients because it may have been removed previously (or will be removed) on the contralateral side. A recent retrospective review by Robey and Schellhammer (1986) has suggested that routine resection of the ipsilateral adrenal may not be required in standard radical nephrectomy because occurrence/recurrence in the adrenal is very rare. While this may be true particularly in low-stage, lower pole tumors, further study is required before this concept should be universally accepted in the routine RCC case. In the situation described above, leaving the adrenal in place seems appropriate.

Renal Vein and Vena Caval Tumor Thrombi

RCC has an unusual propensity to invade renal vascular spaces and to produce tumor thrombi in renal veins within the kidney, the main renal vein, inferior vena cava, and, occasionally, extending into the right atrium. Prior to 1970 it was presumed that patients with renal vein thrombi, and particularly caval extension from RCC, had such a dire prognosis that surgical intervention would be pointless. Indeed, Robson's staging classification placed these patients in the same category as those with regional node involvement (see Fig 6–17). Beginning with the classic article by Skinner et al. (1971), it became increasingly clear that renal vein involvement and caval involvement below the hepatic veins in patients without evidence of regional or distant metastases had a better prognosis than other stage III patients and a similar prognosis to those with stage II when treated by complete excision. Subsequently, experience with successful removal of caval thrombi even into the right atrium has been described (Kearney et al., 1981; Schefft et al., 1978; Cherrie et al., 1982; Cummings et al., 1979; Novick and Cosgrove, 1980).

Extension of RCC into the main renal vein occurs in 15%–20% of patients whereas caval thrombi occur in only 8%–15% of patients (see Fig 16–3). Tumor in the main renal vein is more common on the right and is directly related in incidence to the size of the primary tumor (Gancharenko et al., 1979). In general, tumor in the renal vein does not pose a more difficult surgical challenge than a routine radical nephrectomy except that care must be used to not dislodge tumor thrombi during the operation. Most often if the thrombus is near the caval orifice of the vein it can be cautiously milked back into the vein and a vascular clamp applied at the caval junction. The tumor and the vein containing the thrombus can then be removed in toto and the caval incision oversewn with minimal blood loss. Ligation of the renal artery prior to dealing with the vein is recommended except when excessive manipulation of the vein is required with the attendant risk of dislodging tumor emboli.

Caval thrombi on the other hand present a formidable surgical challenge depending primarily on their cephalad extent within the cava. As previously mentioned, when a caval thrombus is suspected, further diagnostic studies are required. CT can often diagnose the thrombus but not infrequently the cephalad extent of the tumor will be difficult to assess by CT alone. Abdominal ultrasound can often show a distended cava without the normal pulsations and abnormal echoes within the cava up to the proximal end of the thrombus, particularly using longitudinal scans. Venacavography has been used in the past and is highly reliable; however, false positive results can occur from extrinsic compression or anterior displacement (Fig 16–24). If the thrombus is above the hepatic veins superior venacavography may be required to completely delineate the extent of the thrombus. MRI is particularly accurate in delineating caval thrombi and can be used in place of any of the techniques previously described.

The surgical approach to the removal of caval thrombi depends entirely on the cephalad extent. Skinner et al. (1972), Lieskovsky et al. (1984), and Pritchett et al. (1986) described three groups of caval thrombi with regard to level within the cava. Group I lesions, comprising approximately 50% of the total thrombi described, are subhepatic (below the hepatic veins) and are the easiest to resect. In general, these thrombi do not invade the cava and thus can be removed without resection of the caval wall. Although many investigators suggest a thoracoabdominal approach, an anterior subcostal incision is optimal for a right-sided group I tumor, while a bilateral subcostal incision (chevron) can be used for a left-sided tumor. Early attention must be placed on isolating the cava above the thrombus and securing all of the tributary veins including: the cava below the thrombus, the opposite renal vein, gonadal veins, the adrenal veins, and, in most cases, the lumbar veins. The renal artery is ligated at its emergence from the aorta and then the kidney is mobilized. An L-shaped incision can then be made from the tumor containing renal vein into the cava. Most often the thrombus can be extracted using a balloon catheter such as a Fogarty and then a vascular clamp placed across the cava. Following removal of the thrombus the kidney is mobilized and removed. If the caval wall is infiltrated by tumor the prognosis is dismal, but usually the surgeon is already committed to removal of the tumor by the time this is discovered. If the cava is totally obstructed it can be resected because sufficient venous collaterals will have formed. If not totally obstructed, continuity between the remaining renal vein and the cava must be preserved. It is possible on the left to tie off the renal vein at its insertion into the cava and preserve renal function by maintaining normal collateral circulation via lumbar, adrenal, and gonadal vessels, but these patients will often require short-term hemodialysis postoperatively. In approximately 40% of cases (group II), thrombi extend above the intrahepatic portion of the cava up to the insertion of the cava on the right atrium. Group III thrombi that actually penetrate into the atrium account for only 12% of tumors. A right thoracoabdominal incision is best used for group II and III lesions. The ligament attaching the liver to the diaphragm can be severed, the liver displaced medially, and the intrahepatic tumor thrombus adequately visualized. The inferior vena cava can be secured with a tourniquet just above the diaphragm if the thrombus does not extend into the atrium. The operation can proceed using the same principles observed for removal of group I thrombi. In addition, a vascular clamp across the superior mesenteric artery will decrease the hepatic venous flow as will a vascular clamp across the porta hepatis (Pringle maneuver). The tumor thrombus can then be removed with a Fogarty catheter. If the thrombus has reached the atrium, consultation with a cardiovascular surgeon will be required to perform cardiopulmonary bypass, open the atrium, and push the tumor back into the inferior vena cava where a vascular clamp can be placed on the cava above the thrombus. The operation then proceeds as previously described. Probably the best anatomic description of the technical and anatomic principles involved in caval thrombectomy is by Clayman et al. (1980). Although the surgical approach to caval thrombi from RCC is straightforward, a good prognosis depends on the removal of the entire tumor in the absence of regional or distant metastases. If the tumor thrombus is not invading the cava, the prognosis for five-year survival is nearly 50%. However, those with regional or distant metastases concom-

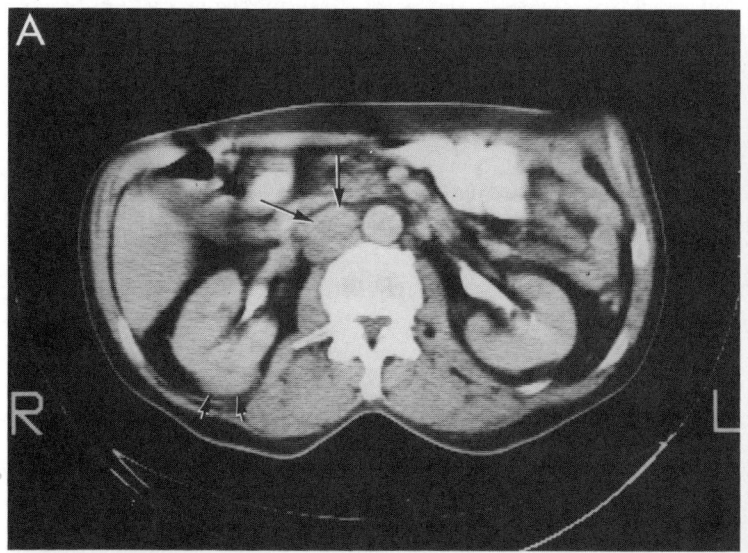

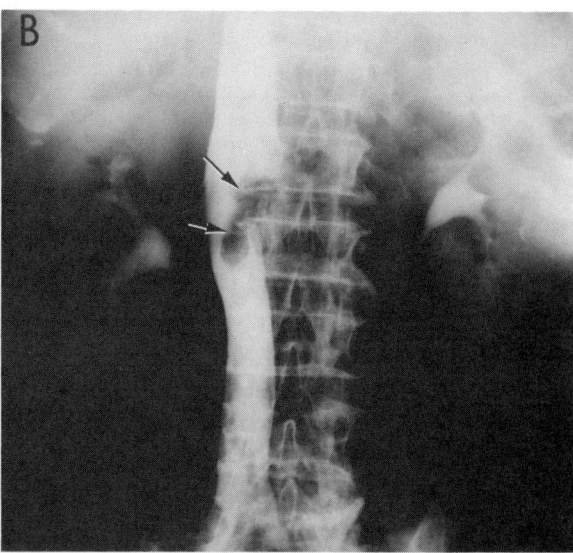

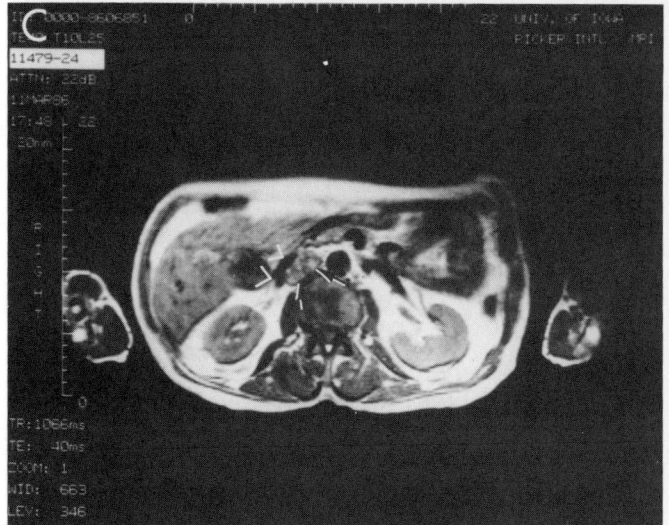

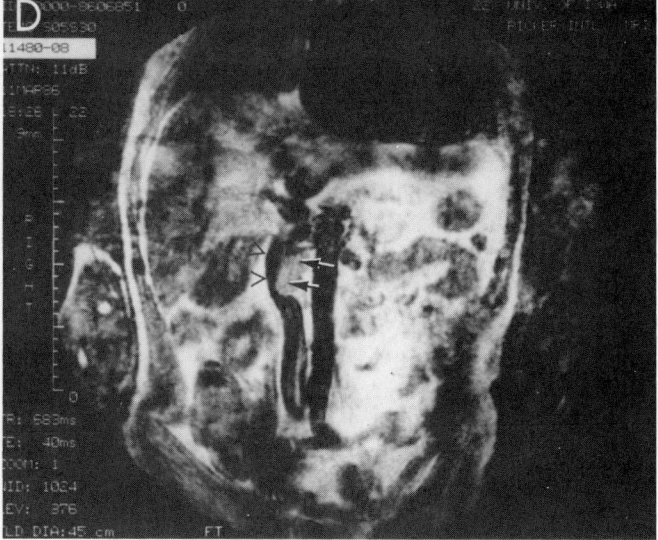

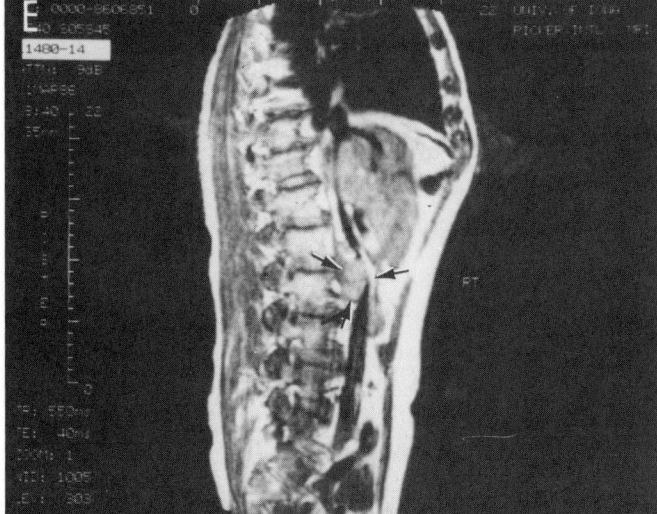

FIG 16–24.
A patient with a renal cell carcinoma suspected to be invading the inferior vena cava. **A,** computed tomograph showing renal mass *(short arrows)* and suspected vena cava thrombus *(long arrows).* **B,** inferior vena cavagram showing suspected intraluminal mass *(arrows).* **C,** transaxial magnetic resonance image (MRI) showing tumorous retroperitoneal node *(arrows)* displacing vena cava anterior *(arrowheads).* **D,** coronal MRI showing tumorous lymph node *(arrows)* displacing vena cava laterally *(arrowheads).* **E,** sagittal MRI showing tumorous lymph node *(arrows)* displacing vena cava anteriorly *(arrowheads).*

itant with a caval thrombus have a 0% five-year survival (Cherrie et al., 1982). Thus aggressive surgical management is justified in those patients without metastases or known tumor invading the wall of the cava.

Adjacent Organ Involvement

RCC tends to involve organs or structures adjacent to the tumor in approximately 10% of patients. In general, this occurs with very large tumors and occasionally will cause symptoms that lead to the initial diagnosis. Organs affected may include the ipsilateral adrenal (in 10%–15% of patients) and other retroperitoneal structures, such as the second portion of the duodenum on the right, or the descending colon and the tail of the pancreas and/or spleen on the left, and the liver and the diaphragm on either side. In the case of adrenal involvement the stage is classified as II; however, if other organs or structures are involved these tumors are classified as stage IVa. Because the adrenal is usually removed in concert with a classical radical nephrectomy, no special adjustment during surgery is required except with a fairly large lesion, in which case a thoracoabdominal approach may be indicated. The prognosis of a patient with ipsilateral adrenal involvement is no different than that of the usual stage II patients even though these metastases may be by direct extension or vascular means. Involvement of the contralateral adrenal, although uncommon, does occur. We have encountered patients with an adrenal adenoma on the side opposite the RCC and thus the assumption should not be made that a contralateral adrenal mass is necessarily a metastasis. Stage IVa disease is usually discovered during surgery; careful preoperative scrutiny of the CT or MRI may be very valuable in predicting adjacent organ involvement. In many cases surgical intervention involving partial resection of the colon, tail of the pancreas and/or spleen, or partial hepatectomy (including a portion of the diaphragm) is technically feasible; however, the prognosis is usually less than 5% five-year survival (deKernion et al., 1978). In patients with symptoms or signs of local extension such as pain, hematochezia, or intra-abdominal hemorrhage surgical resection of the primary tumor with total or partial resection of the secondary organ involved is justified. Asymptomatic patients with known extension have such a poor prognosis that total resection may not be indicated. Obviously, the management of each case must take into account the individual patient and the experience and technical ability of the surgeon.

Advanced Disease

Distant metastases from renal cell cancer are present in approximately 30% of patients at the time of diag-

nosis. The distribution of these metastases is highly variable (see Table 16–3) but tends to involve primarily lungs and bones. Many urologic surgeons justify adjunctive nephrectomy in these patients, citing the morbidity of the primary tumor, the lack of other therapeutic options, and reports of spontaneous regression. There is, however, no properly controlled series to validate this approach. Recent clinical series comparing nonrandomized patients have shown no survival advantage to adjunctive nephrectomy except in patients with solitary metastases that can be resected simultaneously. Montie et al. (1977) reviewed 22 patients who had adjunctive nephrectomy and 48 patients who did not. All of the patients in the first group had died at two years, while 12.5% from the latter group were alive at two years. Middleton (1967) reported that of 29 patients managed by nephrectomy, only two were alive beyond one year. Johnson et al. (1975) reported an 11.3 month median survival in 43 patients treated by nephrectomy as opposed to 7.9 months for 50 patients treated without surgery. A subgroup of patients with metastases to bone only, curiously, had a 16.1 month median survival. DeKernion and others (1978) reported an average survival of patients presenting with metastatic disease of four months with only 10% alive at one year. These results, although not from randomized studies, and thus consisting of noncomparable groupings, do not support routine adjunctive nephrectomy in patients with multiple metastases.

Spontaneous regression has been reported in the past, but usually the original diagnosis was not histologically documented and when it was, most patients suffered recurrence and died of RCC within a short period of time. Recent reviews have shown that fewer than 1% of patients (primarily only those with lung metastases) can expect a spontaneous remission (deKernion and Berry, 1980). Because surgical mortality in this group averages 2%–10%, adjunctive nephrectomy based on the expectation of spontaneous remission is not reasonable.

Patients with solitary metastases to lung or perhaps liver accessible to surgical removal have been shown to have a five-year survival near 30% (O'Dea et al., 1978; Toho and Whitmore, 1975) and thus should be considered candidates for combined nephrectomy and resection of metastases if all of the tumor can be removed. The patient and the surgeon should be aware, however, that micrometastases to other sites are extremely likely and the prognosis for each individual patient is thus very limited. Palliative nephrectomy in patients with intractable symptoms such as persistent gross hematuria or flank pain or incapacitating paraneoplastic syndromes is justifiable, although angiographic infarction is a reasonable alternative. Finally, if an innovative therapeutic protocol is being tested, adjunctive nephrectomy may be an advantage, although to date there are no sufficient modalities available to recommend such an approach. Swanson and others (1980) suggested that angioinfarction followed by nephrectomy may be more effective than nephrectomy alone. Initially, they felt that the embolization procedure might cause exposure of tumor antigens to the circulation so that an immunologic response to tumor cells remaining after nephrectomy might improve patient survival. Subsequent studies by Swanson et al (1983) showed minor improvement and only in those patients with parenchymal pulmonary metastases. Others have not been able to corroborate these findings: Gottesman et al. (1985) studied 30 patients with infarction-nephrectomy and achieved only one partial remission. They concluded, as have other urologic oncologists, that infarction-nephrectomy is not an effective modality for the treatment of metastatic renal cancer.

The role of radiotherapy in patients with metastatic renal cancer has been limited previously to treatment of painful osseous metastases. Although this approach has been shown by experience to palliate pain it has no effect on ultimate survival. There is little evidence that radiation therapy is beneficial in the treatment of locally advanced disease or metastases other than that mentioned above. Lang and deKernion (1981) described the use of ^{125}I radioactive seeds infused into the renal artery for patients with locally advanced but unresectable lesions. They asserted that survival was increased and local palliation was improved; however, data from appropriately controlled studies are lacking.

Chemotherapy

Because metastatic renal cell cancer has a dismal prognosis, an almost desperate search for an effective chemotherapeutic agent has been undertaken. The data can be summarized simply by stating that no cytotoxic agent used either alone or in combination has shown more than a 10%–16% response rate (Table 16–6). Vinblastine alone or in combination with lomustine (CCNU) has been the most effective drug, but even then responses tend to be partial and of very limited duration. The use of human renal cancer cells grown in cell culture or immunodeficient animals for chemotherapy assays was engendered by the early work of Salmon et al. (1978) in ovarian tumors, but success has not been achieved in renal cancer. Studies to date have only shown agents that are not efficacious, and little information exists to assist in the selection of effective agents even when using the patient's own tumor cells for testing (Lieber, 1984b).

TABLE 16–6.

Chemotherapy for Treatment of Metastatic Renal Cell Carcinoma*

AGENT(S)	NO.	% CR/PR†
Single		
Vinblastine	296	16
Hydroxyurea	140	11
Lomustine (CCNU)	79	10
Cyclophosphamide	132	9
Methyl-GAG	76	9
Fluorouracil	201	5
Doxorubicin	65	0
Cisplatinum	60	0
Combination		
Vinblastine, cyclophosphamide, hydroxyurea, progesterone, and prednisone	45	16
Vinblastine and CCNU	93	13
Vinblastine and progesterone	38	8

*Adapted from Richie and Garnick (1982); and Olver and Leavitt (1984).

†Indicates total percent of complete and partial remissions.

Hormonal Therapy

Treatment of metastatic renal cell cancer with hormones is based on studies showing an enhanced rate of renal cancer formation in hamsters exposed to estrogen and the prevention of tumors in animals treated simultaneously with progesterone (Kirkman and Bacon, 1949). The initial clinical studies by Bloom (1973) led to the approval of medroxyprogesterone acetate by the Food and Drug Administration for treatment of metastatic renal cancer; to date, this is still the only drug approved for such therapy. Subsequent studies in a large number of patients, however, have shown at best only a 5%–10% response rate (Table 16–7). Various other hormonal agents have proved no more efficacious than progesterone. There are no controlled studies of hormonal agents in the treatment of metastatic renal cancer, primarily because the response rates are so poor; however, there has recently been an excellent prospective randomized study of nephrectomy with and

TABLE 16–7.

Hormonal Therapy for Metastatic Renal Cell Carcinoma*

AGENT	NO.	% CR/PR†
Progesterone	695	5
Androgen	190	3
Tamoxifen	106	3
Nafoxidine	39	10

*Adapted from Bodey (1979) and deKernion (1983).

†Indicates percent of complete and partial remissions.

without adjuvant progesterone in patients with up to stage III tumor (Pizzocaro et al., 1986). A total of 136 patients were studied and the results showed no difference in recurrence rate between those treated with or without progesterone (25.8% and 23.8%, respectively). These data taken together do not support significant activity for any of the hormones tested in the treatment of metastatic renal cancer.

Immunotherapy

Immunotherapy of renal cell carcinoma has great appeal based on the supposed role of immunity in the natural history of the disease. Although no specific evidence exists that renal cell cancer is more responsive to immunomodulation than other cancers, the observed spontaneous regression rate as well as the variable rate of growth of tumors (exhibited by appearance of metastases from primary tumors resected 15–20 years earlier) suggest that immune factors may be responsible. A variety of clinical approaches to immunotherapy of renal cell cancer have been tested, which include: nonspecific immunostimulation by agents such as BCG (Morales et al., 1982; Montie et al., 1982) or *C. parvum*; specific adoptive immunotherapy with agents such as transfer factor (Montie et al., 1983) and immune RNA (deKernion and Ramming, 1980; Steele et al., 1981); and biologic response modifiers such as thymosin fraction 5 (Dimitrov et al., 1985) and the various forms of interferon (Quesada et al., 1983, 1985; deKernion et al., 1983; Kirkwood et al., 1985; Niedhart et al., 1984; Marumo et al., 1984; Vugrin et al., 1985). The initial results with each of these agents were encouraging, but larger-scale studies did not show significant sustained remissions with the exception of the interferons nor did they demonstrate any documented effect on survival. The interferons (particularly α-interferon), on the other hand, consistently show an approximate 16%–30% response rate of which a few are sustained complete remissions (Niedhart, 1986). The early studies with interferon used leukocyte-derived α-interferon. For the most part, the availability of the drug was extremely limited due to its production from human material; thus, dose was also limited. Quesada and others (1983) reported a study of 50 patients with 3×10^6 units daily by intramuscular injection which revealed a 26% complete and partial remission (CR/PR) rate. DeKernion and associates (1983) reported a similar series in which 16.5% of 43 patients achieved a CR or PR. Molecular biologic techniques rapidly advanced such that subsequent studies have been accomplished using interferon produced by recombinant DNA methods, which have both high purity and specific activity. Quesada and others (1985) recently reported treatment of 41 patients with high doses (20×10^6 units/m^2 daily by intramuscular injection) of recombinant α-interferon and showed a 29% response rate. Williams et al. (1984a) treated 39 patients randomized to a high intermittent bolus dose, intravenously, or a lower continuous dose, subcutaneously, of recombinant α-interferon and showed an overall 23% response rate, including two complete remissions. In general, these studies provide the most reproducible and best results of any agent used to date to treat metastatic renal cancer. Characteristics of patients responding to interferon tend to be those with minimal tumor burden (i.e., primary kidney tumor removed), lung metastases only, and excellent performance status. Trials with β-interferon have been recently reported but it is not generally as effective as α-interferon (Marumo et al., 1984; Vugrin et al., 1985). γ-Interferon trials are currently in progress; however, early unpublished results do not appear to be as good as with α-interferon. Toxic effects of the interferons are primarily limited to flu-like symptoms, which include pyrexia, nausea, vomiting, and fatigue, although in most patients the symptoms are tolerable. There is no doubt that the interferons (particularly α-interferon) have unsurpassed activity in metastatic renal cell cancer; however, remissions are not durable (averaging 4–12 months) and, thus, interferon alone is unlikely to provide substantial improvement in patient survival. There are in vitro cell culture and animal tumor data to support additive and synergistic effects of combinations of interferons with cytotoxic chemotherapeutic agents; however, one initial study using leukocyte interferon and vinblastine (Figlin et al., 1985) in 24 patients showed only a 13% response rate. Future studies will be directed toward optimizing the schedule and dose of interferon, combinations of recombinant interferons (α-interferon plus γ-interferon, etc.) and futher trials of interferon in combination with cytotoxic agents.

Currently there are a variety of other lymphokines being studied that may have clinical activity in renal cell cancer, which include interleukin 2 and tumor necrosis factors. The initial unpublished studies in renal cell cancer using interleukin 2 injections alone are not particularly promising; however, recent data from Rosenberg and Williams (1986) at the National Cancer Institute using reinfusion of autologous lymphocytes from renal cancer patients exposed to interleukin 2 in vitro suggest that this approach to activating killer cells may be particularly efficacious. Large-scale studies in several institutions in the United States are just beginning to determine the potential of this approach. It would seem that other methods similar to this that could educate the immune system to actively and specifically destroy renal cancer cells may very well be effective.

Future Directions in Renal Cell Cancer

Despite what is presented here, relatively little continues to be known concerning the etiology, methods of early diagnosis, and highly effective treatment regimens of locally advanced or disseminated renal cancer. Further studies aimed at discovering the probable multifactorial sequence of events leading to the development of renal cancer are of the utmost importance. Likely avenues of research include detailed environmental and familial studies of renal cancer patients, molecular biologic probes into the interaction of DNA structure and function (oncogenes), and, perhaps, detailed studies of specific patient conditions in which renal cancer is common, such as von Hippel-Lindau disease and acquired renal cystic disease associated with renal neoplasms. Research on improved diagnostic methods will very likely include magnetic resonance imaging and spectroscopy, and monoclonal antibodies to renal cancer-associated antigens (or perhaps to renal cancer–specific antigens in serum or urine) attached to radioactive tracers, positron emitters, or paramagnetic materials for specific imaging studies. Antibodies to RCC cells have already been used to detect and treat human RCC cells in animals and may well be used in the near future to treat renal cancer, either alone or tagged to radiation-emitting substances (Vessella et al., 1985; Lange et al., 1985). Finally, because treatment with various immune modulators appears to be the most productive approach thus far in patients with metastatic disease, investigations into lymphokine production, function, and interaction appear most likely to produce treatments with a favorable impact on patient survival in the foreseeable future.

RENAL UROTHELIAL CANCER: RENAL COLLECTING SYSTEM AND PELVIS

Cancer of the mucosa of the upper urinary tract accounts for 4.5%–9% of all renal tumors but only 5% of all urothelial tumors. Bladder cancer remains the most common urothelial cancer, comprising more than 90% of such lesions. The incidence of renal urothelial cancer is said to be increasing, although there is some question as to whether better diagnostic methods have caused an increase in the rate of detection rather than a real rise in incidence (Fraley, 1978). Renal urothelial cancer has a peak incidence in the sixth to seventh decade and occurs three times more commonly in men than women. The cancer appears in each kidney with equal frequency but occurs in both kidneys in only 2%–4% of patients. Exceptions to the lower frequency in women and the infrequency of bilateral tumors are analgesic-associated tumors and those found in residents of Balkan countries, respectively.

The etiology of renal urothelial cancer is unknown; however, it is likely that it is similar to that of bladder cancer (McLaughlin et al., 1983). A variety of environmental risk factors including cigarette smoking, ingestion of caffeine, and exposure to chemicals used in rubber and textile industries as well as intrinsic risk factors, such as chronic urinary tract infection and urolithiasis, have been implicated. Previously, a viral etiology was proposed, but there is little evidence to support this contention. The artificial sweeteners (saccharin and cyclamates) previously implicated as possible bladder carcinogens now appear to have no relationship to human urothelial cancer (Morrisson and Buring, 1980).

Patients with long-term exposure to phenacetin have been shown to develop a characteristic nephropathy, and in nearly 70% urothelial cancer of the renal pelvis also is found (Johansson and Wahlquist, 1979; Mahoney et al., 1977). Because the primary metabolite of phenacetin (4-acetoaminophenol) is similar to that of acetaminophen, concern has been raised that the substitution of acetaminophen for phenacetin currently practiced may not change the risk of urothelial cancer.

Inhabitants of the Balkan countries (Bulgaria, Greece, Rumania, and Yugoslavia) tend to develop a curious nephropathy that is associated with cancer of the renal pelvis and accounts for more than 40% of the renal cancers in these countries (Markovic, 1972). These tumors are commonly bilateral and tend not to be biologically agressive.

Use of thorium dioxide (Thorotrast) as a contrast agent for retrograde pyelography in the 1930s and 1940s has also been associated with the development of renal urothelial cancer. The agent is thought to induce tumors due to its α-ray emittance. The tumors are delayed an average of 19 years from the time of injection of the agent to the time of detection and are highly malignant epidermoid cancers (Verhood et al., 1974).

Recent studies have implicated the cellular oncogene C-HA-*ras* to be related to formation of urothelial cancers. The oncogene has been shown to be expressed in a few bladder cancers; however, deletion of the short arm of chromosome 11, which is the site of the C-HA-*ras* oncogene, has also been noted in some bladder tumors (Carroll et al., 1986). Whether this or other oncogenes are related to urothelial cancer genesis is unknown, but they are unlikely to be solely responsible.

Renal urothelial cancers occur in association with prior bladder cancer in only 2%–4% of patients, whereas patients with an initial renal urothelial cancer develop similar tumors within the bladder in 50%–75% of cases. These data imply that urothelial cancer is a field disease that may occur simultaneously or subsequently anywhere along the urinary tract. A common etiology is suspected; however, the reason that upper tract tumors occur less frequently when the tumor is initially seen in the bladder is not entirely understood. Perhaps interconnecting factors such as a carcinogen plus a promoter (inflammation, etc.) must be present to initiate an upper tract tumor. Supporting data for this concept are currently only circumstantial.

The diagnosis of renal urothelial cancer relies routinely on radiographic studies, as there are no characteristic clinical symptoms. Up to 15% of patients will be asymptomatic and the diagnosis established serendipitously as a result of studies initiated for other reasons. Microscopic or gross hematuria occurs in 60%–75% of patients. A three-glass urine test may be helpful in implicating the upper urinary tract when red blood cells are present in equal numbers throughout urination (total hematuria). Flank pain occurs in 30%–40% of patients and is usually a dull ache, although acute renal colic with passage of vermiform clots indicative of upper tract bleeding occasionally can occur. A palpable flank mass is present in less than 15% of patients and usually indicates an extensive lesion. Constitutional symptoms (including weight loss and anorexia) are late events, occurring in patients with widely disseminated disease (7%–10%). There are no tumor markers specific for renal urothelial cancer although hypercalcemia (Bourne et al., 1964), elevated levels of serum gonadotrophins (Golde et al., 1974), and carcinoembryonic antigen (CEA) (Glashan et al., 1980) have been reported. Elevated urine levels of fibrinogen degradation products and acute-phase reactive proteins have been reported, but also are nonspecific (Messing, 1986).

Intravenous urography reveals an abnormal filling defect suspicious for an intrinsic lesion in 50–75% of renal urothelial cancers (Fig 16–25). Characteristically, an irregular lesion contiguous with the wall of the renal pelvis or incomplete visualization of an infundibulum or calyx is suggestive of a urothelial lesion. Complete renal pelvis obstruction with nonvisualization occurs in approximately 10% of cases. A large variety of other lesions may produce similar filling defects in the collecting system, such as nonopaque calculi, blood clots, sloughed renal papillae, fungus balls, extrinsic compression by a renal vessel, or—more rarely—a cholesteatoma or a hemangioma. Ultrasound, although not of general use in the diagnosis of renal pelvic tumors, can detect a large stone by the exhibition of a characteristic acoustic shadow.

Cystoscopy, particularly during active gross hematuria, can be quite helpful in determining the bleeding site and is necessary to eliminate the presence of synchronous bladder tumors. Retrograde ureteropyelography at the same time can be diagnostic (75% of cases) and is required in those patients for whom visualization of the collecting system is not possible due to complete obstruction (Fig 16–26). Occasionally, injection of a

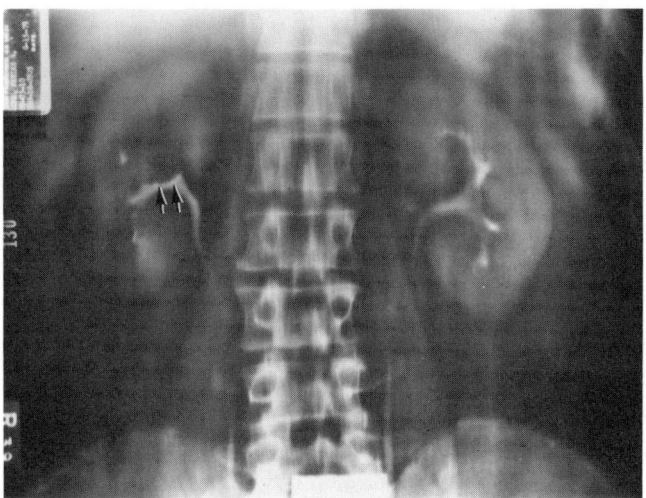

FIG 16–25.
Nephrotomogram from an intravenous urogram of a patient with a right renal urothelial cancer. The renal pelvis filling defect is marked by *arrows*.

combination of air and contrast will be helpful (Fig 16–27). Ureteral urine cytologic analysis by saline barbotage or direct brushing of the lesion (Brown et al., 1973) at the time of retrograde studies can be diagnostic. Antegrade pyelography using a percutaneous approach has been advocated in especially enigmatic lesions; however, the risk of spreading tumor cells with this method obviates its routine use.

Renal angiography is of little value in the diagnosis of

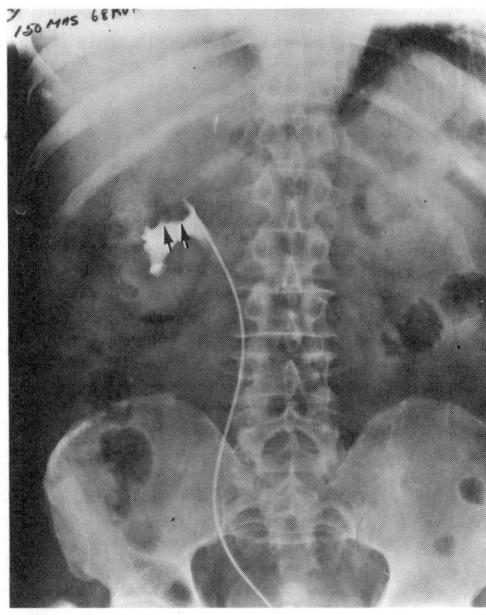

FIG 16–26.
Retrograde pyelogram of the same patient as in Figure 16–25. (From Williams RD, Tanagho EA: *Current Surgical Diagnosis and Treatment.* Los Altos, Calif, Lange Medical Publications, 1985, p 863. Used with permission.)

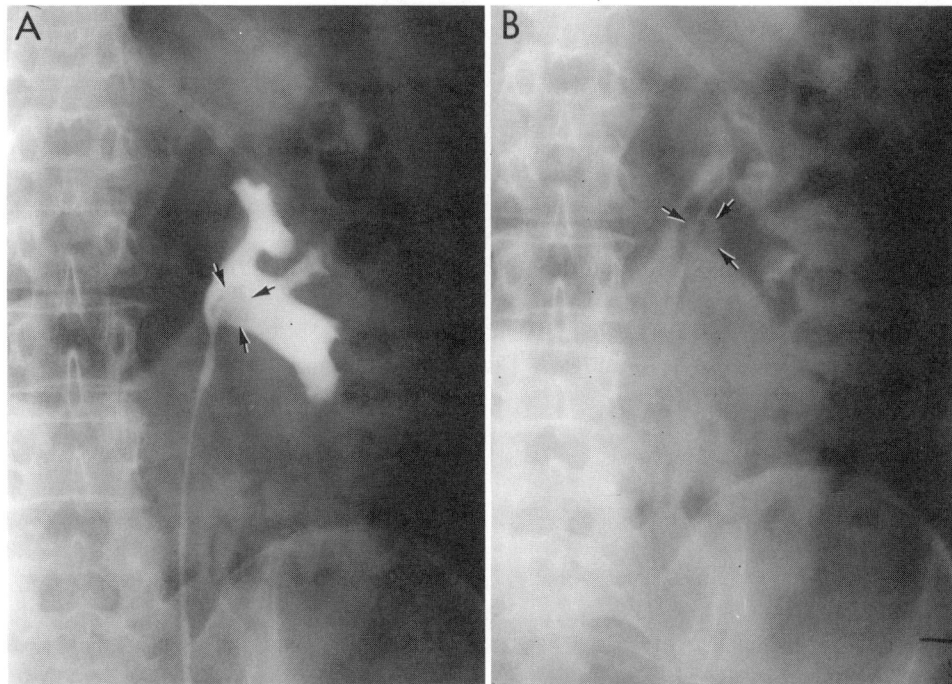

FIG 16–27.
Retrograde pyelogram in a patient with a right renal pelvis urothelial tumor. **A,** contrast study. **B,** air study. The tumor is marked by *arrows.*

upper tract urothelial tumors, since there are no findings pathognomonic for the disease. In selected cases, however, angiography may help differentiate the prominent neovascularity of RCC or a vascular impression on an infundibulum or calyx. If segmental resection of a low-grade superficial lesion is contemplated, angiography may be useful to identify the distribution of the renal vessels.

Computed tomography (CT) can be diagnostic in renal urothelial lesions (Gatewood et al., 1982; Baron et al., 1982) but is rarely indicated for diagnosis alone. In equivocal cases, CT may distinguish RCC from urothelial lesions. CT is most useful in determining the presence of local involvement of renal parenchyma or perirenal tissues, local lymph nodes, or intraabdominal organs. In low-grade superficial lesions, CT is unnecessary but with large, high-grade tumors a staging abdominal CT would be appropriate.

Magnetic resonance imaging (MRI) has proved valuable in diagnosis and staging of RCC, but too few patients have been studied by MRI to determine its usefulness in renal urothelial cancer. It is expected that (as with RCC) staging may be improved by MRI, whereas initial diagnosis will not. MRI will require further improvements in technique and paramagnetic contrast agents before the modality can be recommended for routine use in renal urothelial cancer management.

The advent of rigid ureteroscopes has led to their use in direct vision and biopsy of urothelial lesions in the upper urinary tract (Fig 16–28) (Huffman, 1986). The current scopes are well-suited to visualization of the ureter but are difficult to manipulate into the renal pelvis and selectively into other than the upper pole calyces. Flexible ureteroscopes with operating channels are currently being developed and are expected to be optimal for diagnostic purposes. Percutaneous nephroscopy has also been advocated; however, as with antegrade studies, the risk of tumor cell spill contraindicates this approach except in highly selected situations such as in a solitary kidney, when transrenal percutaneous resection of low-grade tumors may be indicated.

Urine cytologic analysis can be very useful in determining whether an upper tract filling defect is a urothelial tumor. Specimens can be obtained from voided urine, by ureteral catheter (using simple collection or saline barbotage), or by the use of a special brush that can be directed toward the lesion with an angiographic catheter under fluoroscopic guidance. Voided urine is the least accurate, while direct collections from the upper tract have been shown to markedly improve accuracy (Leistenschneider and Nagel, 1980). There are important limitations to upper tract cytologic analysis, which include (1) urolithiasis or inflammation, which can cause exfoliated cells to appear to be low-grade tumor cells (false positive); (2) shedding of only normal-appearing cells in low-grade tumors (false negative); and (3) wide difference in interpretation of urothelial cytologic characteristics from one pathologist to another. False positive interpretations occur in approximately 10% of specimens and false negatives range from 30%–60% (Sarnacki et al., 1971). The false negatives are higher with low-grade lesions than with high-grade lesions and can be improved by saline barbotage using a ureteral catheter adjacent to the lesion. Flow cytome-

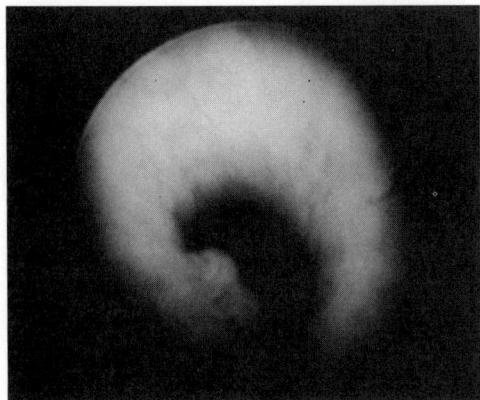

FIG 16–28.
Ureteroscopic view of a renal urothelial tumor. (Courtesy of Jeffrey Huffman, M.D., Division of Urology, University of California, San Diego.)

try, perhaps in combination with determination of cell surface antigens on exfoliated urothelial cells, has considerable promise in improving the accuracy of diagnosis and perhaps will provide prognostic information as well (Czerniak and Koss, 1985; King et al., 1983).

Staging

Following clinical diagnosis of a renal urothelial lesion, therapy is based on determination of the extent of local and distant disease (clinical stage). In low-grade and/or low-volume lesions, local extension within the renal parenchyma or perinephric fat would not change

the surgical approach; thus, local staging preoperatively may be unnecessary. If, however, a renal conserving approach is contemplated (segmental resection, or transureteral or percutaneous resection), an abdominal CT is recommended to determine the local extent of tumor. Patients with low-grade and/or low-volume tumors should have a chest roentgenogram to eliminate the possibility of pulmonary metastases. Consideration should be given to a bone scan, although patients without bony symptoms will rarely have bone metastases. Patients with high-grade and/or high-volume renal urothelial tumors have a much higher incidence of metastases, primarily to local lymph nodes, lung, bones and liver in decreasing order of frequency. Staging studies in these patients should include: (1) abdominal CT; (2) chest roentgenogram and CT; and (3) bone scan. Lesions detected on these studies should be biopsied if necessary to prove the presence of metastases; fine-needle aspiration cytologic analysis is often quite helpful in this regard. If metastases are documented, the prognosis is extremely poor, and nonsurgical treatment may then be appropriate.

Pathologic staging of renal urothelial cancer provides the most accurate data to predict prognosis. There is no completely satisfactory staging system available in the literature, although most reports rely on the system proposed by Grabstald and others (1971), which was modified to correlate more readily to the bladder cancer system by Cummings (1980). A staging system adapted from Grabstald and Cummings is shown in

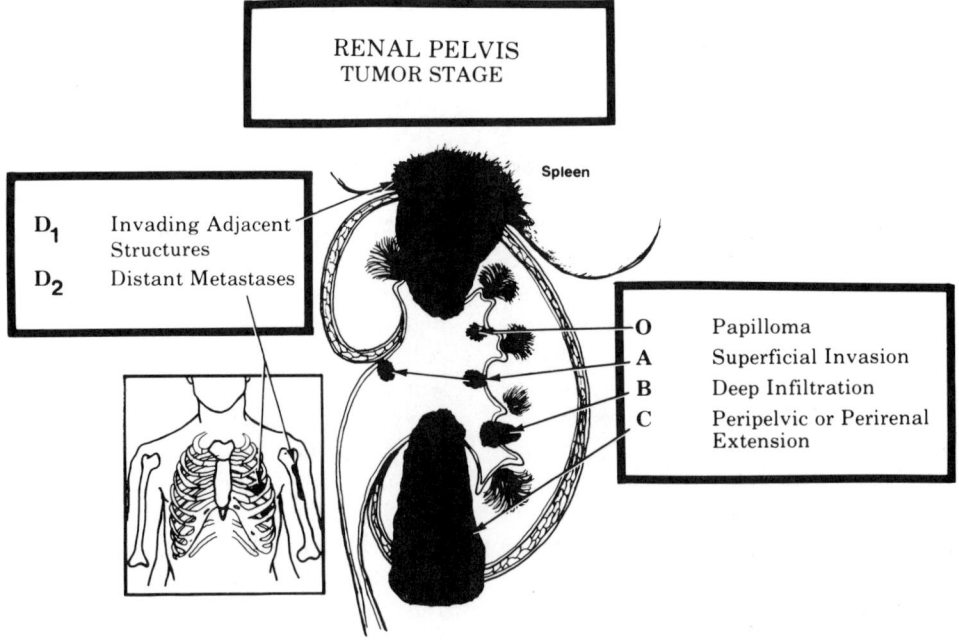

FIG 16–29.
Staging system for renal urothelial cancer.

Figure 16–29. Stage O includes low-grade tumors or carcinoma in situ without deeper invasion. Stage A includes tumors that involve the submucosa but do not extend deeper. Stage B refers to tumors deeply invasive into pelvis musculature or renal parenchyma. Stage C includes tumors involving peripelvic or perirenal extension. Stage D_1 tumors involve adjacent organs or local lymph nodes and D_2 tumors are metastatic to distant sites. The relative frequency of occurrence of the various stages at presentation are stages O and A, 40%; stage B, 30%; and stages C and D, 30% (Grabstald et al., 1971; Cummings, 1980; Wagle et al., 1975). Prognosis for patients with these various stages is difficult to determine from the literature for the following reasons: a variety of staging systems are used, the grade of the tumors is not uniformly described, and various surgical approaches have been used. In general, high-grade tumors tend to be high stage as well, but this correlation is not absolute. Prognosis for low-grade and/or low-stage tumors (O and A) tends to be excellent (approaching 100%) as long as complete tumor excision is accomplished. The presence of multiple urothelial sites of tumor apparently has a negative effect on the prognosis, but confirmatory data are scarce. Prognosis for patients with stage B disease is an approximate 75% five-year survival, but varies with the grade of the lesion and the extent of surgical therapy. Stages C and D disease tend to occur only in patients with high-grade lesions; the five-year survival in these patients is poor (5%–10%).

Pathology

Over 90% of tumors involving the urothelium of the upper urinary tract are epithelial. Of these, 90% or more are transitional cell cancers (TCC), most being papillary in configuration (Figs 16–30 and 16–31). So-called benign papillomas are really quite rare and their observance depends more on the philosophy of the pathologist examining the lesion than on any established criteria. In general, papillomas are difficult to distinguish clinically from carcinomas and thus are generally treated as malignant lesions. Squamous cell carcinoma of the renal urothelium occurs in 7%–9% of cases and is usually associated with chronic urolithiasis and/or chronic inflammation. These lesions tend to be extensive when detected; thus, the prognosis is poor, with few survivals despite radical surgical extirpation. Adenocarcinoma of the renal urothelium occurs in less than 1% of cases, with only approximately 60 reported cases. These tumors tend to occur in females and also are associated with urolithiasis, hydronephrosis, and chronic pyelonephritis (Fraley, 1978). Adenocarcinomas invariably are found in high stages and patients have a uniformly poor prognosis.

Histologic grading of renal urothelial tumors is similar to that of bladder cancer. High-grade tumors (grades

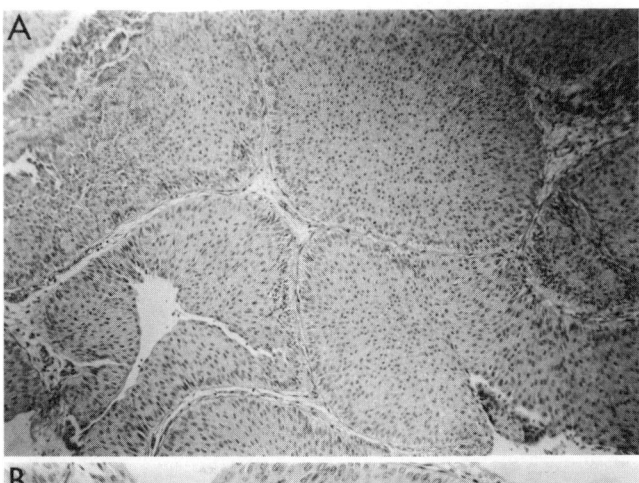

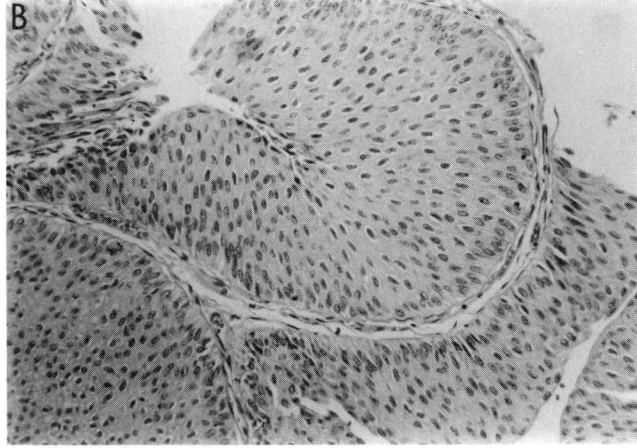

FIG 16–30.
A, histologic section of a transitional cell cancer of the renal pelvis (original magnification, × 80). **B,** same section at higher magnification (original magnification, × 157).

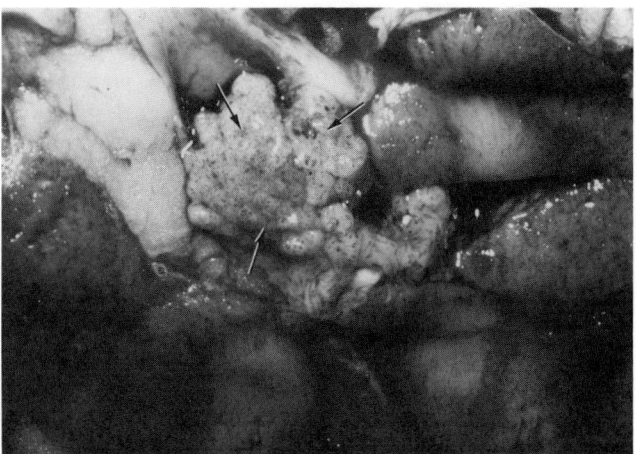

FIG 16–31.
Gross pathology of a renal pelvis transitional cell carcinoma *(arrows).*

3 and 4) have a negative impact on survival. Attempts to correlate prognosis with the presence or absence of cell-surface blood group antigens in both urine cytologic specimens and tumor tissue have been equivocal and are generally of little clinical significance (Messing, 1986). It is possible that the retention of these antigens by malignant cells could predict a favorable prognosis and perhaps allow the consideration of local excision or transureteral resection. There are, however, no data at present to substantiate this approach. It is more likely that flow cytometry with the use of monoclonal antibodies to TCC-associated antigens or karyotypic analyses of exfoliated cytologic specimens will be helpful in the future.

Treatment

There are no appropriately controlled and randomized series providing the necessary information to determine optimum therapy for renal urothelial cancer. Recommended treatment for patients without metastases currently entails a radical nephroureterectomy including Gerota's fascia, the ipsilateral adrenal, the entire ureter, and a cuff of bladder surrounding the distal ureter. Previous results have shown that a simple nephrectomy with subtotal ureterectomy results in substantial local recurrences in the renal bed (30%–40%) and distal ureteral stump (30%–60%) or in the bladder (50%–60%) (Johansson and Wahlquist, 1979; Cummings, 1980; Strang and Pearse, 1976). Convincing data from Johansson and Wahlquist (1979) showed that a radical nephroureterectomy with retroperitoneal lymphadenectomy resulted in an 84% five-year survival, whereas simple nephroureterectomy resulted in only a 51% five-year survival figure. This study was not a randomized series; therefore, inherent biases are possible. Summary of the available data suggests that the prognosis for low-stage and/or low-grade tumors is excellent, using almost any approach from local excision to radical nephroureterectomy as long as all tumor is removed. There is evidence, however, that local recurrence is higher with the more conservative approach. Tumors of high stage and/or grade have a much better prognosis when treated by the radical approach. Because it is difficult to accurately determine the stage of the tumor preoperatively, a radical nephroureterectomy is the recommended procedure. Patients with biopsy-proved low-stage and/or low-grade unifocal disease who also have a solitary kidney or renal insufficiency may be candidates for more conservative surgery; however, the risks of recurrence should be kept in mind.

The role of a formal retroperitoneal lymphadenectomy is undetermined. It is unecessary in patients with low-stage and/or low-grade disease and is probably only diagnostic in those of higher stage and grade. Because a limited node dissection adds very little in time and complexity to the operation, it seems reasonable to include it in healthy patients with normal vessels, but no therapeutic benefit should be ascribed to it.

There are a variety of surgical incisions advocated for a radical nephroureterectomy; a two-incision technique using an anterior subcostal approach for the radical nephrectomy (as described previously) and a Gibson incision for the distal ureterectomy (see Fig 16–22) is optimal. The ureter can be dissected in continuity down to the bladder mucosa from behind the bladder and an adequate portion of vesical mucosa removed without an anterior cystotomy. A ligature on the proximal ureter early in the case may prevent tumor cell extrusion into the bladder and subsequent implantation on the fresh surgical margin.

The recent advent of ureteroscopes for diagnostic purposes has caused several investigators to attempt to treat superficial low-grade renal urothelial lesions by transurethral ureteroscopic resection/fulguration (Huffman, 1986). Few patients have been treated, and longterm results are as yet unknown. Patients with special conditions, which might include a solitary kidney or renal insufficiency, are potential candidates for transurethral ureteroscopic treatment but conventional treatment is most prudent in all others. Percutaneous nephroscopic treatment through the flank has also been advocated but is not advisable due to the risks of tumor seeding.

In addition, laser approaches to urothelial cancer are receiving some attention (Rosenberg and Williams, 1986). Transurethral endoscopic laser of superficial lesions has theoretic merit but little information is available for lesions outside of the bladder. The Nd:YAG laser, argon laser, and carbon dioxide laser all could be useful for this purpose if appropriate adaptations for ureteral endoscopes become available. Treatment of upper tract lesions will likely be enhanced when flexible ureteroscopes with operating ports are perfected. If a conservative surgical approach (segmental resection or endourologic resection) is chosen, meticulous follow-up with periodic retrograde utereropyelography, upper tract barbotage cytologic analysis, and, possibly, ureteropyeloscopy is mandatory.

Because 30%–50% of patients with renal urothelial cancer will have bladder occurrence and/or recurrence, periodic surveillance by cystoscopy and bladder wash cytologic analysis is necessary following definitive treatment (Grabstald et al., 1971). A useful regimen is to repeat both examinations quarterly the first year, semiannually the second year, and annually thereafter. Should a recurrence be found, the regimen should start over again.

In the past, adjuvant therapy in the form of radiation and chemotherapy has not been shown to be of any major benefit. Recent advances in chemotherapy for metastatic bladder TCC, however (discussed later in this chapter), suggest that adjuvant chemotherapy in selected cases with a high risk of recurrence or metastases might be beneficial.

Advanced Disease

Renal urothelial cancer with local extension into adjacent organs or distant metastases has a dismal prognosis (0%–10%) despite treatment. Selected patients with intractable symptoms from hematuria or ureteral obstruction may benefit from a nonextensive nephrectomy, although angioinfarction of the kidney is a reasonable alternative. There is no evidence that removal of the primary lesion affects survival in any way; spontaneous regression is virtually nonexistent in this disease. Further, there are no data that support removal of the primary tumor and a solitary metastasis should they coexist.

Radiation therapy as a surgical adjunct in patients with locally extensive disease has no proved benefit but is useful in palliation of symptomatic metastases, particularly of bone or liver.

A variety of chemotherapeutic agents has been used for renal urothelial tumors (primarily TCC) without significant durable remissions. Agents used singly that have shown occasional responses in renal TCC are similar to those used for bladder TCC and include methotrexate, cyclophosphamide, fluorouracil, doxorubicin, and cisplatin; however, none of these has produced more than a 10%–20% partial response rate with virtually no complete responders. Recently two regimens of chemotherapy combining cisplatin, methotrexate, and vinblastine alone (CMV) or with the addition of doxorubicin (M-VAC) have shown a 50%–70% combined complete and partial remission rate in patients with uroepithelial tumors (Meyers et al., 1986; Sternberg et al., 1985). The majority of these patients had TCC of the bladder but a few patients with renal TCC were treated. Meyers et al. (1986) reported seven patients with renal TCC of which 71% responded (two CR, three PR). Patients with other cell types do not appear to respond well to either regimen, although one recent patient at the University of Iowa with bladder adenocarcinoma had a 50% reduction of tumor volume with CMV. Despite some evidence that the durability of complete remission may be less than one year with either regimen, these response data represent a marked improvement. Thus, patients with metastatic renal urothelial cancer who have adequate renal function (which may often be the limiting factor) may benefit from either of these protocols.

There have been very few studies of immunotherapy of renal urothelial cancer. Recently Herr (1985) reported successful treatment of a superficial tumor in a solitary kidney by BCG instillation and Hawtrey and Williams (1986) have successfully treated a similar patient with α-interferon. Systemic leukocyte interferon has been used in only a very few patients with urothelial tumors with only limited success (Shortliffe, 1986).

Future Perspectives

Perhaps the greatest impact on renal urothelial cancer would be the effective means of detecting early disease. Similar to bladder cancer, a large portion of patients with invasive disease present in an advanced state without prior symptomatology. Patients at risk eventually could be screened by urine cytologic analysis, flow cytometric analysis, or urothelial-associated or specific antigen techniques. Elimination of risk factors (primarily tobacco use) may be helpful but is unlikely to occur. Determination of the nuclear events leading to tumor formation may allow reversion or a block in the process as molecular biology advances. More accurate staging techniques using immunoscanning with urothelial-associated antigens or MRI may be useful. Finally, improvement of the already effective chemotherapeutic regimens for bladder cancer may prove to be effective for renal urothelial cancer as well.

SARCOMAS

Primary renal sarcomas account for only 3% of malignancies within the kidney. Diagnosis is often delayed until the tumors are quite large and symptomatic. There are no particular distinguishing features; differentiation from renal adenocarcinoma is therefore difficult. The majority of these tumors are thought to arise from the renal capsule, although local renal vasculature has also been implicated. Fibrosarcoma was once considered the most common type, but recent data indicate that these sarcomas may have been misdiagnosed primary parenchymal sarcomatoid renal adenocarcinomas suggesting that fibrosarcoma is quite rare (Bonsib, 1985). Leiomyosarcoma is now considered the most common (approximately 60%), with over 85 cases reported in the literature (Niceta et al., 1974). These tumors tend to occur primarily in females with peak incidence in the fourth decade. Liposarcoma is next in incidence (Evans, 1979) with hemangiopericytoma, rhabdomyosarcoma (Berk et al., 1960) and osteogenic sarcoma occurring in decreasing order of frequency (Farrow et al., 1968). For the most part, the renal sarcomas are large circumscribed tumors that are not encapsulated. They tend to compress renal cortex and may invade the collecting system and renal veins. Leiomyosarcomas are fibrous masses with a whorled grey to white appearance with few mitoses on histologic section (see Fig 16–25). Liposarcomas are bulky soft masses that are lobulated and yellow to grey on cut surface. They are distinguished by lipoblasts or pleomorphic fat cells on histologic section.

The diagnosis of a renal sarcoma is most frequently made by the pathologist following surgical removal of a suspected primary renal carcinoma. Angiography is not diagnostic, although sarcomas generally do not exhibit neovascularity or arteriovenous fistulae within the tumor. CT may be helpful in identifying a capsular origin of these tumors, but because they tend to be hypovascular (as are 10% of renal adenocarcinomas), differentiation is problematic.

In most cases, treatment of renal sarcomas involves radical nephrectomy. Despite this approach, local recurrence is common; thus, surgery alone is seldom curative. Adjuvant therapy has been recommended although radiation does not appear to be of benefit and no specific chemotherapeutic regimen has proved uniformly beneficial. Minor success has been achieved with regimens containing doxorubicin; however, durable remissions have not been reported (Williams, 1985). Because so few sarcomas have been treated, the true prognosis is unknown; overall there appears to be no greater than a 10% five-year expected survival (Richie and Garnick, 1982).

SECONDARY MALIGNANT TUMORS

Malignant tumors secondarily involving the kidney are not uncommon, but in the past they have rarely been discovered during the patient's lifetime. The current trend for complete tumor staging using highly sensitive studies, such as abdominal CT, prior to and/or during therapy, and the emergence of more effective therapy in many cancers has resulted in more frequent clinical detection of secondary renal malignancies. In general, nonprimary malignancies within the kidney arise either by direct extension (adrenal, retroperitoneal sarcomas, pancreatic, or colonic cancer) or by hematogenous spread (melanoma, lung, breast, stomach). Lymphoma, leukemia, or myeloma may occur primarily in the kidney, but they are much more likely to represent a systemic manifestation.

Direct Extension

Reliable figures describing the incidence of direct involvement of the kidney by malignancy from retroperitoneal or intra-abdominal organs are not available. Virtually any tumor of the retroperitoneum, duodenum, pancreas, or colon could invade the kidney, but the occurrence is very infrequent. The various sarcomas have been previously mentioned. Urothelial tumors of the renal pelvis and collecting system also can invade the

renal parenchyma. In general, direct extension by any of the tumors mentioned above is a late event in the course of the primary tumor, and when detected predicts an extremely poor prognosis.

Hematogenous Extension of Solid Tumors

Metastatic spread of nonhematologic neoplasms to the kidney is not uncommon, occurring in 7.7% of patients in combined autopsy series (Mayer, 1982). These metastases also occur as a late event and very infrequently cause symptoms that lead to their detection before the patient's death. Results of compiled autopsy series of the relative frequency of solid tumor metastases to the kidneys from various primary sites are shown in Table 16–8 (Bracken et al., 1979; Olsson et al., 1971; Klinger, 1951; Wagle et al., 1975). These data may be useful in determining the likely primary site of a metastasis detected in the kidney. Mayer (1982) has suggested alternatively that the probability of renal metastasis can be predicted from the histology of a nonrenal primary tumor: melanoma, 38%; contralateral kidney, 20%; lung, 17%; pancreas, 10%; and breast, 8%. Mayer's study, together with data from Table 16–8, indicates that lung, breast, and pancreas are the most common primary tumors that produce renal metastases.

In general, solid tumors metastatic to the kidney do not exhibit typical findings on either renal angiography or CT (Fig 16–32). Metastatic lesions usually are not highly vascular and, thus, neovascularity and contrast enhancement are not often encountered. Ten percent of primary RCC lesions are also not highly vascular; therefore, the lack of tumor vessels and/or contrast en-

TABLE 16–8.

Solid Tumors Metastatic
to the Kidney*

PRIMARY SITE	%
Lung	27
Breast	14
Stomach	12
Pancreas	7
Colon	6
Cervix	5
Esophagus	4
Prostate	4
Gallbladder, testis, thyroid, bladder, or melanoma	2–3
Endometrium, kidney, ovary, head and neck, or bone	1–2

*Adapted from Mayer RJ: Infiltrative and metastatic disease of the kidney, in Rieselback RE, (Garnick MB (eds): *Cancer and the Kidney.* Philadelphia, Lea & Febiger, 1982, p 707.

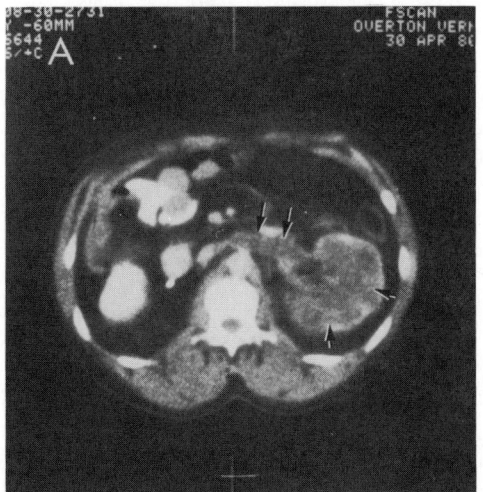

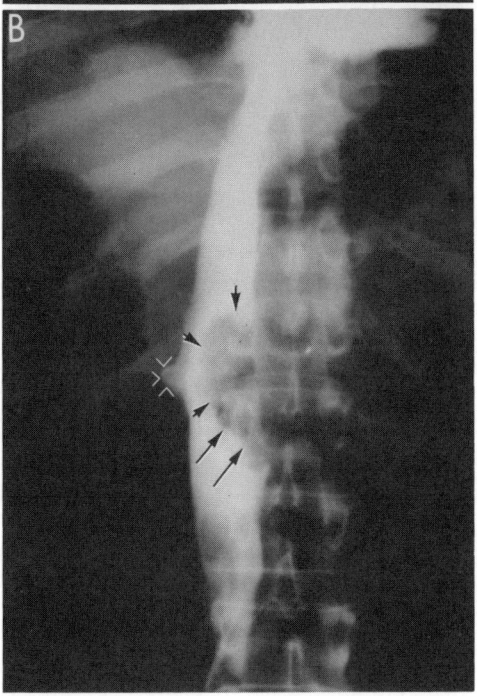

FIG 16–32.
Patient with colon carcinoma metastatic to the kidney with atypical findings. **A,** computed tomograph showing enhancing tumor diffusely invading the left kidney *(short arrows)* and left renal vein *(long arrows)*. **B,** inferior venocavogram of same patient showing tumor occluding the left renal vein and involving the vena cava *(short arrows)*. The tumor was invading the wall of the vena cava predicted by inferior growth of the tumor below the left renal vein *(long arrows)*. Backflow of contrast into the normal right renal vein is seen *(arrowheads)*. (From Carroll PR, et al: Microscopic hematuria, left renal mass with renal vein obstruction and elevated serum level of carcinoembryonic antigen in a 56-year-old man. *J Urol* 1983; 129:569–570. Used with permission.)

hancement is by no means diagnostic. Fine-needle aspiration cytologic analysis of lesions suspected to be metastases may be very helpful in establishing the diagnosis.

Despite surgical removal of the primary and secondary lesion, survival statistics for patients with renal metastases from solid tumors are uniformly poor. For optimum management of the primary neoplasm, diagnosis of solid tumors metastatic to the kidney should be established nonsurgically whenever possible. More effective chemotherapy of primary malignancies will likely cause an increase in the premortem incidence and detection of secondary renal malignancies.

Hematogenous Extension of Hematologic Tumors

Tumors of hematologic origin frequently involve the kidney, but similar to other secondary tumors they infrequently cause symptoms indicating their presence. Common examples include lymphoma, leukemia, and multiple myeloma. Occasionally (less than 1%) these tumors may be primary in the kidney; however, most often they represent extension of the systemic neoplasm (Mayer, 1982).

Prior to the modern era of chemotherapy, leukemias were found to infiltrate the kidneys in nearly 70% of autopsied cases (acute lymphocytic most common) (Kirschbaum and Preuss, 1943). The involvement tends to be bilateral and is diffuse in the renal cortex. Although renal enlargement on intravenous urogram (IVU) is common, the excretory system may be normal. Renal failure may occur but is most likely due to dehydration and urate nephropathy. Treatment of a renal leukemic infiltrate is directed at the primary disease and usually entails chemotherapy, although small doses of radiation therapy may be helpful. Prognosis depends on the course of the primary disease, but with modern cytotoxic therapy renal involvement may be eliminated.

Involvement of the kidneys by lymphoma has been reported in 36% of patients prior to effective chemotherapy (Mayer, 1982). The incidence was highest in lymphosarcoma (46%), with histiocytic lymphoma (45%) and Hodgkin's disease (15%) seen in decreasing order of frequency. The disease is found in both kidneys in more than 60% of patients. Involvement may occur in several patterns, including multiple nodules, diffuse infiltration, extension from contiguous lymphatics, and—rarely—solitary masses (Heiken et al., 1983). CT has been described as the most accurate method of detection but there are no characteristic findings to help in the differential diagnosis (Fig 16–33). The lesions are usually devoid of neovascularity on angiographic studies. Surgical treatment is not recommended unless the lesion is solitary and no other lesions suspicious of lymphoma are detected. Fine-needle aspiration cytologic

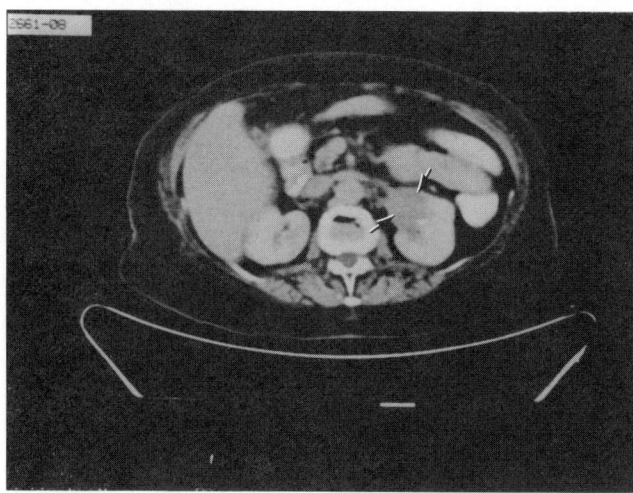

FIG 16–33.
Computed tomograph of a patient with metastatic lymphoma to the left kidney *(arrows)*.

analysis may be helpful in establishing the diagnosis in equivocal cases, but definition of the subtype of the disease (and thus optimum treatment) may require a piece of tissue for immunospecific studies.

Multiple myeloma involvement of the kidneys occurs in only approximately 11% of patients in autopsy series. More commonly, myeloma causes renal disease secondary to tubular precipitation of myeloma proteins rather than by plasma cell infiltration. Treatment is directed at hydration, elimination of hyperuricemia, and decreasing proteinuria by chemotherapy.

PRIMARY RETROPERITONEAL TUMORS

Eighty-five percent of primary retroperitoneal tumors are malignant, while 15% include such benign masses as lipoma and congenital cystic lesions (Wheatley, 1983). Seventy percent of the malignant tumors are lymphomas or sarcomas. The lymphomas may be lymphosarcomas, reticulum cell sarcomas, and variations of Hodgkin's disease. Sarcomas in decreasing order of incidence are liposarcomas (21%), leiomyosarcomas (15%), rhabdomyosarcomas, and fibrosarcomas. The latter are often mistaken for angiomyoplipoma, xanthogranulomas, or sarcomatoid renal adenocarcinoma. They are associated in 12% of patients with von Recklinghausen's disease. These retroperitoneal malignancies as a group often invade contiguous abdominal organs such as kidney, colon, and pancreas.

Prior to CT, diagnosis of retroperitoneal tumors was usually established at autopsy; despite technical advances there are no characteristic radiographic findings. MRI, because of its superior ability to visualize anatomy in multiple planes and distinguish tissue signals more clearly, is advantageous for diagnosis of retroperitoneal masses (Bretan et al., 1986).

Treatment of malignant retroperitoneal tumors remains primarily surgical, although local recurrence is common and patients are rarely cured (Cody et al., 1981; Wile et al., 1981).

URETERAL TUMORS

Ureteral tumors are rare, accounting for fewer than 1% of all urinary tract neoplasms. Approximately 75% of these tumors are malignant. As with other urinary tract tumors the incidence of ureteral tumors appears to be increasing, but these figures may reflect an increase in detection due to improved diagnostic modalities. Transitional cell carcinoma (TCC) is the most common ureteral tumor and is frequently diagnosed in an advanced stage. Treatment of all ureteral tumors remains surgical, although recently endoscopic diagnosis and treatment of low-grade and/or low-stage tumors is proving effective. Prognosis of advanced malignant lesions continues to be poor, except in cases in which combination chemotherapy is beginning to provide meaningful responses.

BENIGN URETERAL TUMORS

Benign neoplasms of the ureter are uncommon, accounting for only 20%–25% of all ureteral tumors. Hematuria or ureteral obstruction with flank pain (often intermittent) are common presenting findings. Benign papillomas have been described that are very similar to low-grade TCCs in gross appearance and histologic characteristics. Clinically they are difficult to distinguish from carcinomas. They tend to occur, as does TCC of the ureter, in the fifth to seventh decade and in males three times more commonly than females. Radiographically the benign papilloma may be more smooth and better defined than a ureteral TCC. Diagnosis is best established by ureteroscopic biopsy. If the biopsy confirms low-grade tumor and radiographic studies do not show evidence of invasion of the ureteral wall, treatment may be by segmental ureteral resection or transurethral ureteroscopic resection. Because there is controversy whether papillomas are truly benign, larger lesions are treated similar to renal urothelial tumors with a radical nephroureretectomy. A variety of other, albeit rare, ureteral tumors have been described including leiomyomas of the ureteral musculature and neurolemmomas and angiomas (Richie, 1978).

The fibroepithelial polyp is the most common benign ureteral tumor. This tumor presents in a younger age group than malignant ureteral tumors and tends to occur more often in males (Richie, 1978). They often arise from the upper third of the ureter and tend to be solitary. Diagnosis is often made on IVU, although differentiation from a primary ureteral malignancy may be difficult without a ureteropyelogram or direct ureteroscopy. Typically the polyp is smooth and thin, arising from a long stalk. Treatment (when these lesions are suspected preoperatively) is by local excision and fulguration of the base, although most often the lesion is treated as a suspected carcinoma. Endoscopic resection may be possible in selected cases.

PRIMARY MALIGNANT URETERAL TUMORS

Cancer of the ureter accounts for only 1% of upper urinary tract malignancies. The incidence of ureteral cancers is low, with a ratio of 1:50 in relation to bladder cancers in just over 1,500 reported cases (Richie, 1978). Primary ureteral cancer occurs twice as often in males as in females and has a peak incidence in the seventh decade. These tumors are in the distal third of the ureter in approximately 70% of cases and are rarely (less than 1%) bilateral.

The etiology of ureteral cancer is unknown but is thought to be similar to urothelial cancer in other sites of the urinary tract. Schmouz and Cole (1974) have specifically implicated cigarette smoking, coffee drinking, and leather working as risk factors. Chronic inflammation in patients with vesical schistosomiasis also is antecedent to ureteral cancer (Makar, 1948). Primary ureteral cancer occurs in association with prior bladder cancer in up to 50% of patients (Batata and Grabstald, 1976).

There are no pathognomonic clinical signs or symptoms indicative of ureteral cancer. Gross, painless hematuria is the frequent initial event (60%–90%), although microscopic hematuria is also frequently observed. Flank pain occurs with passage of clots and with rapid onset of ureteral obstruction (anaplastic tumors). Nearly 50% of patients will describe lower tract irritative symptoms (Richie, 1982). The ureteral tumor is rarely palpable, although resultant hydronephrosis may be. Tumor markers and paraneoplastic syndromes are not evident.

Diagnosis of ureteral cancer relies almost entirely on radiographic modalities. IVU continues to be the primary screening study, and when properly conducted will routinely reveal the abnormality (the entire ureter must be visualized bilaterally). Typical findings on IVU include an intraluminal filling defect (19%), hydronephrosis, (34%), or nonvisualization of the kidney (46%) (Batata et al., 1975). Similar defects in the ureter may be caused by blood clots, calculi, sloughed renal papillae, ureteritis cystica, ureteral varices, or extrinsic masses such as endometriosis or tumorous retroperitoneal lymph nodes.

Cystoscopy is important in determining the presence of synchronous bladder tumors, defining the site of hematuria and visualizing a ureteral tumor exiting the ureteral orifice. Simultaneously retrograde ureterograms can be confirmatory and diagnostic and thus are required for proper definition of a ureteral lesion. Typical findings on retrograde studies are either an intraluminal mass that is displaced by contrast, revealing a "wine goblet" sign (Fig 16–34) or a more diffuse strictured area (Fig 16–35). Urine cytologic analysis by saline barbotage or brushing at the time of retrograde studies, as previously described, may be diagnostic. Percutaneous antegrade pyelography has its proponents but is rarely necessary except in patients with inaccessible ureters. CT and MRI are not useful as primary diagnostic techniques, although both may be quite useful as staging modalities, particularly in high-grade tumors. Because coronal scans can visualize the ureter in longitudinal section, MRI may be the more valuable study but neither CT nor MRI has been studied extensively for this purpose. CT is useful to differentiate ureteral calculi in enigmatic cases. Angiography is of little benefit in the diagnosis or staging of primary ureteral tumors.

Direct visualization of ureteral lesions with biopsy has become a very accurate means of diagnosis. Simultaneous treatment of small low-grade superficial lesions

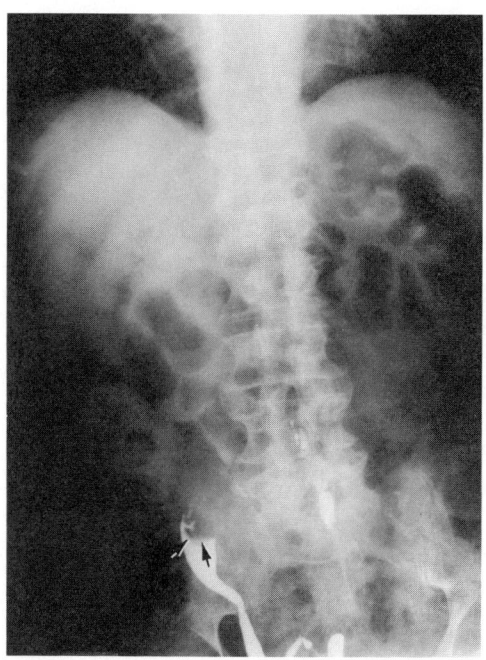

FIG 16–34.
Retrograde ureterogram showing a "wine goblet" sign of a typical low-grade transitional cell carcinoma *(arrows)* of the right ureter. (From Williams RD, Tanagho EA: *Current Surgical Diagnosis and Treatment.* Los Altos, Calif, Lange Medical Publications, 1985, p 864. Used with permission.)

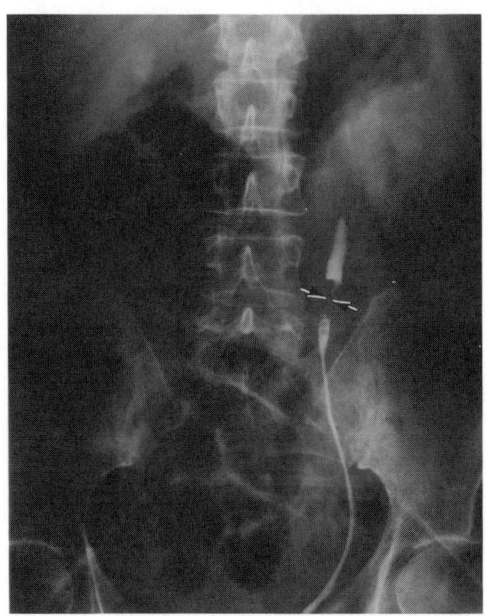

FIG 16–35.
Retrograde ureterogram showing a strictured area *(arrows)* in a patient with a high-grade undifferentiated primary ureteral cancer. The filling defects below the tumor are air bubbles.

is currently being tested for efficacy in a few centers (Huffman, 1986). Urine cytologic analysis, preferably by retrograde catheterization, can be diagnostic. Similar to renal urothelial lesions, false negative results occur in approximately 30% of cases.

STAGING

Staging of ureteral cancer is similar to that of bladder cancer and relates to depth of penetration into the ureteral wall (Batata et al., 1975). Stages O and A tumors are those that are confined to the mucosa or submucosa, respectively (55%); stage B tumors involve the ureteral musculature (10%); stage C penetrate the entire ureteral wall and invade surrounding fat (20%); and stage D tumors invade adjacent organs or are metastatic to distant sites (15%), primarily lung, bone, and liver. Distant metastases are rare in patients with clinical stages O and A, but are seen in 40% of patients with stage B and 75% with stage C lesions.

Prognosis of ureteral cancer depends on stage and grade more than on extent of treatment. As with renal urothelial cancer, high-stage tumors tend to be high grade as well, whereas low-stage tumors commonly are also low grade. Five-year survival related to grade ranges from 50%–80% for low-grade tumors (grades 1 and 2) and 0%–16% for high-grade tumors (grades 3 and 4) (Batata et al., 1975). Five-year survival figures by stage have been: stages O and A, 60%–90%; stage

B, 40%–50%; stage C, 20%–30%; and stage D, 0% (Bloom, 1973; Batata et al., 1975; McCarron et al., 1983).

PATHOLOGY

Over 90% of primary ureteral malignancies are TCC. Mixed TCC with squamous or adenomatous elements accounts for 20% of these (Bloom, 1970). Pure squamous cell cancers account for 8%, usually in association with bilharziasis. Both adenocarcinoma and undifferentiated tumors constitute less than 1%. Nearly 80% of ureteral cancers are papillary while the rest are solid.

TREATMENT

Classic therapy of primary ureteral cancer is similar to that for renal urothelial cancer: radical nephroureterectomy using the surgical approach previously described (see Fig 16–22). There are, however, no adequately controlled randomized studies comparing conservative (segmental resection) with radical surgery. Proximal ureteral tumors have a higher risk of distal recurrence than distal tumors and thus conservative resection is not recommended. If it is possible to precisely determine tumor grade and stage preoperatively, those of low grade and/or stage could be treated adequately by segmental resection but all others should be managed by standard radical surgery. Distal ureteral lesions are less likely to involve proximal sites; therefore, low-grade and/or low-stage lesions can be managed successfully by distal ureterectomy and ureteroneocystostomy (Johnson et al., 1984). In high-grade and/or high-stage lesions radical nephroureterectomy is optimal, although the prognosis depends on the degree of anaplasia and depth of penetration rather than the extent of surgical treatment. There is little evidence that addition of an extended lymphadenectomy is more than diagnostic and thus it is not mandatory. As mentioned before, the low-grade and/or low-stage lesions may be optimally managed by ureteral endoscopic resection and/or fulguration, but recurrences appear to be common (Huffman, 1986).

Adjuvant therapy of locally advanced lesions would be advantageous if proven therapy existed. Radiation therapy has been advocated but substantive efficacy data are unavailable. Intraureteral BCG or interferon may be useful, but durability of responses has not been reported.

Patients with primary ureteral cancer have an approximate 30% chance of recurrence in a retained ureter and up to 50% in the bladder. Bladder and ureteral cytologic analysis, retrograde ureterography, ureteroscopy, and cystoscopy should be routine, as discussed, for primary renal urothelial cancers.

ADVANCED DISEASE

There is no evidence that surgical treatment of the primary lesion in the face of metastatic disease has any merit. Palliation can be achieved by local radiation (Johnson et al., 1984). Chemotherapy of metastatic disease has been uniformly disappointing, although cisplatin achieves short-term palliation in some patients. MCV or M-VAC regimens are expected to be effective in patients with metastatic ureteral TCC as they are in bladder TCC, but too few patients have been treated to make a definitive statement.

SARCOMAS

Leiomyosarcomas of the ureteral musculature are rare, but are the most common ureteral sarcoma. Treatment by any means is rarely curative, and although radical nephroureterectomy is the standard approach, local recurrence is the rule.

SECONDARY MALIGNANT URETERAL TUMORS

Malignancies metastatic to the ureter are uncommon, with few cases reported. They can occur by direct extension (cervix, colon, retroperitoneal lymphoma) or by hematogenous spread (breast, stomach, prostate, lung, kidney [RCC], cervix, colon, or lymphoma). The breast is the most common site involving metastases to the ureter as reported in 18.3% of patients at autopsy (Richie and Garnick, 1969). The lesions are infrequently symptomatic (Cohen et al., 1979). All levels of the ureter may be involved although most are in the lower third. Bilateral involvement occurs in 30%. Generally, the ureteral mucosa is not involved; thus, hematuria is not frequent. There are no characteristic radiographic findings. Treatment must be predicated on the state of the primary tumor. Rarely, if ever, is surgical removal indicated or beneficial. As the primary tumor is not always evident preoperatively, on occasion the diagnosis will be made pathologically only after the tumor is removed. Palliation by ureteral intubation or percutaneous nephrostomy may be indicated to relieve symptoms or to provide adequate renal function for chemotherapy optimum for the primary lesion.

REFERENCES

1. Aziri F, Morangola JP, Kirch KH: Ultrasonic diagnosis of tumor thrombosis of right renal vein. *Urology* 1979; 12:106.
2. Babaian RJ, Swanson DA: Serum haptoglobin: A nonspecific tumor marker for RCC. *South Med J* 1982; 75:1345–1348.
3. Baker HL Jr, Berquist TH, Kispert DB, et al: Magnetic resonance imaging in a routine clinical setting. *Mayo Clin Proc* 1985; 60:75–90.
4. Balfe DM, McClennan BL, Stanley RJ, et al: Evaluation of renal masses considered indeterminate on computed tomography. *Radiology* 1982; 142:421–428.
5. Baron RL, McClennan RL, Lee JKT, et al: Computed tomography of transitional cell carcinoma of the renal pelvis and ureter. *Radiology* 1982; 144:125.
6. Batata M, Grabstald H: Upper urinary tract urothelial tumors. *Urol Clin North Am* 1976; 3:79.
7. Batata M, Whitmore WF Jr, Hilaris BS, et al: Primary carcinoma of the ureter: A prognostic study. *Cancer* 1975; 35:1629.
8. Beahrs OH, Myers MH (eds): *American Joint Committee on Cancer: Manual for Staging of Cancer.* ed 2. Philadelphia, JB Lippincott, 1983, p 178.
9. Belis JH, Post GJ, Rochman SC, et al: Genitourinary leiomyomas. *Urology* 1979; 13:424.
10. Bell ET: *Renal Diseases,* ed 2. Philadelphia, Lea & Febiger, 1950, p 435.
11. Bennington JL, Beckwith JR: *Tumors of the Kidney, Renal Pelvis, and Ureter,* fascicle 12, *Atlas of Tumor Pathology,* series 2. Washington, DC, Armed Forces Institute of Pathology, 1975.
12. Berk LE, Erinc AI, McManus RG: Hemangiopericytoma involving the kidney. *N Engl J Med* 1960; 263:1185.
13. Bloom HJG: Hormone-induced and spontaneous regression of metastatic renal cancer. *Cancer* 1973; 32:1066.
14. Bodey GP: Current status of chemotherapy in metastatic renal carcinoma, in Johnson DE, Samuels ML (eds): *Cancer of the Genitourinary Tract.* New York, Raven Press, 1979, p 67.
15. Bonsib SM: Pathologic features of renal parenchymal tumors, in Culp DA, Loening SA (eds): *Genitourinary Oncology.* 1985, p 185.
16. Bourne HE, Trembloy RE, Ansell JS: Stupor, hypercalcemia and carcinoma of the renal pelvis. *N Engl J Med* 1964; 271:1005.
17. Boxer RJ, Waismer J, Lieber MM: Non-metastatic hepatic dysfunction associated with renal carcinoma. *J Urol* 1978; 119:468–471.
18. Bracken RB, et al: Secondary renal neoplasms: An autopsy study. *South Med J* 1979; 72:806.
19. Bretan PN, Busch MP, Hricak H, et al: Chronic renal failure: A significant risk factor in the development of acquired renal cysts and renal cell carcinoma. *Cancer* 1986; 57:1871.
20. Bretan PN Jr, Williams RD, Hricak H: Preoperative assessment of retroperitoneal pathology by MRI: Primary leiomyosarcoma of inferior vena cava. *Urology* 1986; 27:251–255.
21. Brown RC, Hawtrey CE, Pixley EE: Brush biopsy of the renal pelvis. *AJR* 1973; 119:779.
22. Busch FM, Bark CJ, Clyde HR: Benign renal angiomyolipoma with regional node involvement. *J Urol* 1976; 116:715.
23. Campbell RJ, Broaddus SB, Leadbetter GW: Staging of

renal cell carcinoma: Cost effectiveness of routine preoperative bone scans. *Urology* 1985; 25:326–329.

24. Carroll PR, Fair WR, Chaganti RSK: Oncogene expression and chromosome rearrangements in urologic tumors, in Williams RD (ed): *Advances in Urologic Oncology.* New York, MacMillan Co, 1987.

25. Carroll PR, Pellegrini C, Hedgcock MW, et al: Microscopic hematuria, left renal mass with renal vein obstruction and elevated serum level of carcinoembryonic
antigen in a 56-year-old man. *J Urol* 1983; 129:568.

26. Cherrie RJ, Goldman DG, Lindner A, et al: Prognostic
implications of vena caval extension of renal cell carcinoma. *J Urol* 1982; 128:910–912.

27. Chisholm GD: Nephrogenic ridge tumors and their syndromes. *Ann NY Acad Sci* 1974; 230:402.

28. Clayman RV, Figenshou RS, Bear A, et al: Transplantation of human RCC into athymic mice. *Cancer Res*
1985; 45:2650.

29. Clayman RV, Gonzalez R, Fraley EE: Renal cell cancer
invading the inferior vena cava: Clinical review and anatomical approach. *J Urol* 1980; 123:157.

30. Cocchiora R, Torro G, Flaminio G: Purification of
herpes simplex virus tumor associated antigen from human kidney cancer. *Cancer* 1980; 46:1594.

31. Cody HS, Turnbull AD, Fortner JG, et al: The continuing challenge of retroperitoneal sarcomas. *Cancer*
1981; 47:2147.

32. Cohen AJ, Li FP, Berg S, et al: Hereditary RCC associated with chromosomal translocation. *N Engl J Med*
1979; 301:592–595.

33. Colvin RV, Dickersin GR: Pathology of renal tumors, in
Skinner DG, DeKernion JB (eds): *Genitourinary Cancer.* Philadelphia, WB Saunders, 1978.

34. Cox CE, Lacy SS, Montgomery WG, et al: Renal adenocarcinoma: A 28 year review with emphasis on rationale
and feasibility of preoperative radiotherapy. *J Urol* 1970;
104:51.

35. Cox GG, Lee KR, Price HE, et al: Colonic infarction
following ethanol embolization of renal cell carcinoma.
Radiology 1982; 145:343.

36. Cronan JJ, Zeman RK, Rosenfield AT: Comparison of
computerized tomography, ultrasound and angiography
in staging renal cell cancer. *J Urol* 1982; 127:712.

37. Cronin RE, Kaehy WD, Miller PD, et al: Renal cell
carcinoma: Unusual systemic manifestations. *Medicine*
1976; 55:291.

38. Cummings KB: Nephroureterectomy: Rationale in the
management of transitional cell carcinoma of the upper
urinary tract. *Urol Clin North Am* 1980; 7:569.

39. Cummings KB, Li W, Ryan JA, et al: Intraoperative
management of renal cell carcinoma with supradiaphrag-
matic caval extension. *J Urol* 1979; 122:829.

40. Czerniak B, Koss LG: Expression of Ca antigen on human urinary bladder tumors. *Cancer* 1985; 55:2380.

41. Daniel WW, Hartman GW, Witten DM, et al: Calcified
renal masses: A review of 10 years' experience at the
Mayo Clinic. *Radiology* 1972; 103:503.

42. Deitchman B, Sidhur GS: Ultrastructural study of a sarcomatoid variant of renal cell carcinoma. *Cancer* 1980;
46:1152.

42a. deKernion JB: Treatment of advanced renal cell cancer—traditional methods and innovative approaches. *J
Urol* 1983; 130:2.

43. deKernion JB, Berry D: The diagnosis and treatment of
renal cell carcinoma. *Cancer* 1980; 44:(suppl):1947.

44. deKernion JB, Ramming JB, Smith RB: The natural history of metastatic renal cell carcinoma: A computer analysis. *J Urol* 1978; 120:148.

45. deKernion JB, Ramming KP: The therapy of renal adenocarcinoma with immune RNA. *Invest Urol* 1980;
17:378.

46. deKernion JB, Sarna G, Figlin R: The treatment of renal
cell carcinoma with human leukocyte alpha interferon. *J
Urol* 1983; 130:1063.

47. Dimitrov NV, Arnold D, Munson J, et al: Phase II study
of thymosin fraction 5 in the treatment of metastatic
renal cell carcinoma. *Cancer Treat Rep* 1985; 69:137.

48. Ekelund L, Gothlin J: Renal hemangiomas: An analysis
of 13 cases diagnosed by angiography. *AJR* 1975;
125:788.

49. Evans HL: Liposarcoma: A study of 55 cases with reassessment of its classification. *Am J Surg Pathol* 1979;
3:301.

50. Fallon B: Renal parenchymal tumors B: Clinical and diagnostic features, in Culp DA, Loening SA (eds): *Genitourinary Oncology.* Philadelphia, Lea & Febiger, 1985,
p 202.

51. Farrow GM, Harrison EG Jr, Utz DC et al: Sarcomas
and sarcomatoid and mixed malignant tumors of the kidney in adults. *Cancer* 1968; 22:545.

52. Figlin RA, deKernion JB, Maldozys J, et al: Treatment
of renal cell carcinoma with alpha (human leukocyte) interferon and vinblastine in combination: A phase I-II
trial. *Cancer Treat Rep* 1985; 69:263.

53. Flocks RH, Kadesky MC: Malignant neoplasms of the
kidney: An analysis of 353 patients followed five years or
more. *J Urol* 1958; 79:196.

54. Fraley EE: Cancer of the renal pelvis, in Skinner DG,
deKernion JB (eds): *Genitourinary Cancer.* Philadelphia, WB Saunders, 1978, pp 134–149.

55. Gancharenko V, Gerlock JA Jr, Kadir S, et al: Incidence
and distribution of venous extension in 70 hypernephromas. *AJR* 1979; 133:263.

56. Gatewood OMB, et al: Computerized tomography in the
diagnosis of transitional cell carcinoma of the kidney. *J
Urol* 1982; 127:876.

57. Gibbons RP, Bush WH, Burnett LL: Needle tract seeding following aspiration of renal cell carcinoma. *J Urol*
1977; 118:865.

58. Glashan RW, Higgens E, Nevill AM: The clinical value
of plasma and urinary carcinoembryonic antigen (CEA)
assays in patients with hematuria and urothelial carcinoma. *Eur Urol* 1980; 6:344.

59. Golde DW, Schanhelan M, Weintraub BD, et al: Go-
nadotropin-secreting renal carcinoma. *Cancer* 1974;
33:1048.

60. Gondos B: Diagnosis of tumors of the kidney: Ultrastructural classification. *Ann Clin Lab Sci* 1981; 11:308.

61. Gottesman JE, Crawford ED, Grossman HB et al: Infarction-nephrectomy for metastatic renal carcinoma: Southwest Oncology Group Study. *Urology* 1985; 25:248.

62. Grabstald H, Whitmore WG Jr, Melamed M: Renal pelvic tumors. *JAMA* 1971; 218:845.

63. Graham SD Jr, Glenn JF: Enucleative surgery for renal malignancy. *J Urol* 1979; 122:546.

64. Hand GC: Identification of a high-frequency model for renal carcinoma by the induction of renal tumors in the mouse with a single injection of streptozotocin. *Cancer Res* 1985; 45:703–708.

65. Hawtrey CE, Williams RD: Personal communication, 1986.

66. Heiken JP, Gold RP, Schnur MJ, et al: Computed tomography of renal lymphoma with ultrasound correlation. *J Comput Assist Tomogr* 1983; 7:245.

67. Herr HW: Durable response of a carcinoma in situ of the renal pelvis to topical BCG. *J Urol* 1985; 134:531.

68. Holland JM: Cancer of the kidney: Natural history and staging. *Cancer* 1973; 32:1030.

69. Hollifield JW, Page DL, Smith C, et al: Renin-secreting clear cell carcinoma of the kidney. *Arch Intern Med* 1975; 135:859.

70. Hricak H, Demas BE, Williams RD, et al: MRI in the diagnosis and staging of renal and perirenal neoplasms. *Radiology* 1985; 154:709.

71. Huffman JL: Endourologic diagnosis and treatment of upper tract urothelial tumors, in Williams RD (ed): *Advances In Urologic Oncology*. New York, MacMillan Publishing Co. 1987.

72. Jensen HS, Gimsing P, Pedersen F, et al: Transcobalamin II as an indicator of activity in metastatic RCC. *Cancer* 1983; 52:1700–1704.

73. Johannson S, Wahlquist L: A prognostic study of urothelial renal pelvic tumors: Comparison between the prognosis of patients treated with intrafascial nephrectomy and perifascial nephroureterectomy. *Cancer* 1979; 43:2525.

74. Johnson DE, Swanson DA, VonEschenbach AC: Tumors of the genitourinary tract, in Smith DR (ed): *General Urology*. Los Altos, Calif, Lange Medical Publications, 1984.

75. Johnson DE, Kaesler KE, Samuels ML: Is nephrectomy justified in patients with metastatic renal cell carcinoma? *J Urol* 1975; 114:27.

76. Juul N, Torp-Pedersen S, Growall S, et al: Ultrasonically guided fine needle aspiration biopsy of renal masses. *J Urol* 1985; 133:579.

77. Kalonel LN: Association of cadmium with renal cancer. *Cancer* 1976; 37:1782.

78. Karcioglu ZA, Sorper RM, VanReinsvelt HA, et al: Trace element concentrations in RCC. *Cancer* 1978; 42:1330.

79. Kearney GP, Waters WB, Klein LA, et al: Results of inferior vena cava resection for renal cell carcinoma. *J Urol* 1981; 125:769.

80. King CT, Clark TD, Lovett J, et al: A comparison of clinical course with blood group antigen testing by specific red cell adherence and immunoperoxidase in ureteral and renal pelvic tumors. *J Urol* 1983; 130:871.

81. Kirkman H, Bacon RL: Renal adenomas and carcinomas in diethylstilbestrol treated male golden hamsters. *Anat Rec* 1949; 103:475.

82. Kirkwood JM, Harris JE, Vera R, et al: Randomized trial of low and high doses of leukocyte interferon in metastatic renal cell carcinoma. *Cancer Res* 1985; 45:863–871.

83. Kirschbaum JD, Preuss GG: Leukemia: A clinical and pathologic study of 123 fatal cases in a series of 14,000 necropsies. *Arch Intern Med* 1943; 71:777.

84. Klein MJ, Valensi QJ: Proximal tubular adenomas of the kidney with so called oncocytic features: A clinicopathologic study of 13 cases of a rarely reported neoplasm. *Cancer* 1976; 38:906.

85. Klinger ME: Secondary tumors of the genitourinary tract. *J Urol* 1951; 65:144.

86. Kosko JW, Lipuma JP, Resnick MI: Radiological evaluation of renal masses, in Javadpour N (ed): *Cancer of the Kidney*. New York, Thieme-Stratton Inc, 1984.

87. Lang EK: Angiocomputed tomography and dynamic computed tomography in staging of renal cell carcinoma. *Radiology* 1984; 151:149.

88. Lang E, deKernion JB: Transcatheter embolization of advanced renal cell carcinoma with radioactive seeds. *J Urol* 1981; 126:581.

89. Lang EK, Johnson B, Chance HL, et al: Assessment of avascular renal mass lesions: The use of nephrotomography, arteriography, cyst puncture, double contrast study, and histochemical and histopathologic examinations of the aspirate. *South Med J* 1972; 65:1.

90. Lange PH, Vessella RL, Diou RK, et al: Monoclonal antibodies in human renal cell carcinoma and their use in radioimmune localization and therapy of tumor xenografts. *Surgery* 1985; 98:143.

91. Leionen A: Embolization of renal carcinoma. *Ann Clin Res* 1985; 17:299–305.

92. Leistenschneider W, Nagel R: Lavage cytology of the renal pelvis and ureter with special reference to tumors. *J Urol* 1980; 124:597.

93. Levine E, Grantham JJ, Slusher SL, et al: CT of acquired cystic kidney disease and renal tumors in long-term dialysis patients. *AJR* 1984; 142:125–131.

94. Levine E, Macklace N, Rosenthal SJ, et al: Comparison of computed tomography and utlrasound in abdominal staging of renal cancer. *Urology* 1980; 16:317–322.

95. Lieber MM: Renal oncocytoma, in Javadpour N (ed): *Cancer of the Kidney*. New York, Thieme-Stratton Inc, 1984a, p 139.

96. Lieber MM, Tomera KM, Farrow GM: Renal oncocytoma. *J Urol* 1981; 125:481.

97. Lieber MM: Soft agar colony formation assay for in vitro chemotherapy sensitivity testing of human renal cell car-

cinoma: Mayo Clinic experience. *J Urol* 1984b; 131:391.

98. Lieskovsky G, Pritabett R, Skinner DG: Surgical management of renal cell carcinoma. *Monogr Urol* 1984; 5:98.

99. Lindner A, Goldman DG, deKernion JB: Cost-effective analysis of prenephrectomy radiosiotope scans in renal cell carcinoma. *Urology* 1983; 22:127.

100. Lingeman JE, Donohue JP, Madrua JA, et al: Angiomyolipoma: Emerging concepts in management. *Urology* 1982; 20:566.

101. Maatman TJ, Novick AC, Jancino BF, et al: Renal oncocytoma, a diagnostic and therapeutic dilemma. *J Urol* 1984; 132:878.

102. Mahoney JF, Storey BG, Ibanez RC, et al: Analgesic abuse: Renal parenchymal disease and carcinoma of the kidney or ureter. *Aust NZ J Med* 1977; 7:463.

103. Makar N: Bilharzial ureter: Some observations on surgical pathology and surgical treatment. *Br J Surg* 1948; 36:148.

104. Malek RS, Utz DC, Culp OS: Hypernephroma in the solitary kidney: Experience with 20 cases and review of the literature. *J Urol* 1976; 116:553.

105. Marberger M, Pugh RCB, Auvert J, et al: Conservative surgery of renal carcinoma: The EIRSS experience. *Br J Urol* 1981; 53:528.

106. Markovic B: Endemic nephritis and urinary tract cancer in Yugoslavia, Bulgaria and Rumania. *J Urol* 1972; 107:212–219.

107. Marshall FF, Taxy JB, Fishman EK, et al: The feasibility of surgical enucleation for renal cell carcinoma. *J Urol* 1986; 135:231.

108. Marshall F, Powell KC: Lymphadenectomy for renal cell carcinoma: Anatomical and therapeutic considerations. *J Urol* 1982; 128:677.

109. Marumo K, Murai M, Hayakama M, et al: Human lymphoblastoid interferon therapy for advanced renal cell carcinoma. *Urology* 1984; 24:567.

110. Mauro MA, Wadsworth DE, Stanley RJ, et al: Renal cell carcinoma: Angiography in the CT era. *AJR* 1982; 139:1135.

111. Mayer RJ: Infiltrative and metastatic disease of the kidney, in Rieselback RE, Garnick MB (ed): *Cancer and the Kidney.* Philadelphia, Lea & Febiger, 1982, p 707.

112. McCarron JP Jr, Mullis D, Vaughn ED Jr: Tumors of the renal pelvis and ureter: Current concepts and management. *Semin Urol* 1983; 1:75.

113. McLaughlin JK, et al: Etiology of Ca of renal pelvis. *J Natl Cancer Inst* 1983; 71:287.

114. McLean GK, Meranze SG: Embolization techniques in the urinary tract. *Urol Clin North Am* 1985; 12:743.

115. McNichols DW, Segura JW, DeWeerd JH: Renal cell carcinoma: Long term survival and late recurrence. *J Urol* 1981; 126:17.

116. Merino MJ, Librelsi VA: Oncocytomas of the kidney. *Cancer* 1982; 50:1952.

117. Messing E: Tumor antigens in the diagnosis, staging and prognosis of urologic cancer, in Williams RD (ed): *Advances In Urologic Cancer.* New York, MacMillan Publishing Co, 1987.

118. Meyers F, Palmer J, Hannigan J: Chemotherapy of disseminated transitional cell carcinoma, in Williams RD (ed): *Advances in Urologic Oncology.* New York, MacMillan and Co, 1987.

119. Middleton RG: Surgery for metastatic renal cell carcinoma. *J Urol* 1967; 97:973.

120. Montie JE, Stewart BH, Straffon RA, et al: The role of adjunctive nephrectomy in patients with metastatic renal cell carcinoma. *J Urol* 1977; 117:272.

121. Montie JE, Bukowski RM, James RE, et al: A critical review of immunotherapy of disseminated renal adenocarcinoma. *J Surg Oncol* 1982; 21:5.

122. Morales A, Wilson JL, Pates JL, et al: Cytoreductive surgery and systemic bacillus Calmette-Guerin therapy in metastatic renal cancer: A phase II trial. *J Urol* 1982; 127:230.

123. Morrisson AS, Buring JE: Artificial sweeteners and cancer of the lower urinary tract. *N Engl J Med* 1980; 302:437.

124. Mostofi FK, Davis CJ Jr: Pathology of tumors of the kidney, in Javadpour J (ed): *Cancer of the Kidney.* New York, Thieme-Stratton Inc, 1984, pp 15–32.

125. Mundy GR, Martin TJ: The hypercalcemia of malignancy: Pathogenesis and management. *Metabolism* 1982; 31:1247.

126. Murphy GP, Mostofi FK: Histologic assessment and clinical prognosis of renal adenoma. *J Urol* 1970; 103:31.

127. Niceta P, Lavengood RW, Fernandes M, et al: Leiomyosarcoma of kidney: Review of the literature. *Urology* 1974; 3:270.

128. Niedhart JA, Gagen MM, Young D, et al: Interferon alpha therapy of renal cancer. *Cancer Res* 1984; 44:4140.

129. Niedhart JA: Interferon therapy for the treatment of renal cancer *Cancer* 1986; 57(suppl):1696.

130. Novick AC, Stewart BH, Straffon RA: Extracorporeal renal surgery and autotransplantation: Indications, techniques and results. *J Urol* 1980; 123:806.

131. Novick AC, Zincke H, Neves RJ, et al: Surgical enucleation for renal cell carcinoma. *J Urol* 1986; 135:235.

132. Novick AC, Cosgrove DM: Surgical approach for removal of renal cell carcinoma extending into the vena cava and the right atrium. *J Urol* 1980; 123:947.

133. O'Dea MJ, et al: The treatment of renal cell carcinoma with solitary metastasis. *J Urol* 1978; 120:540.

134. Oesterling JE, Fishman EK, Goldman SM et al: The management of renal angiomyolipoma. *J Urol* 1986; 135:1121.

135. Okabe T, Urobe A, Kato T: Production of erythropoietin-like activity by human renal and hepatic carcinomas in cell culture. *Cancer* 1985; 55:1918.

136. Olsson CA, Mozen JD, Laferte RO: Pulmonary cancer metastatic to the kidney—a common renal neoplasm. *J Urol* 1971; 105:492.

137. Olver IN, Leavitt RD: Chemotherapy and immunotherapy of disseminated renal cancer, in Javadpour N (ed): *Cancer of the Kidney.* New York, Thieme-Stratton Inc, 1984.

138. Page HS, Asire JH (eds): *Cancer Rates and Risks.*

Washington, DC, NIH publication no. 85–691, April 1985.

139. Pathak S, Strong LC, Ferrell R, et al: A familial renal cell carcinoma with a 3:11 chromosome translocation limited to tumor cells. *Science* 1982; 217:939.

140. Pearson JC, Weiss J, Tanagho EA: A plea for conservation of kidney in renal adenocarcinoma associated with Von Hippel-Lindau disease. *J Urol* 1980; 124:910.

141. Peters PC, Brown GL: The role of lymphadenectomy in the management of renal cell carcinoma. *Urol Clin North Am* 1980; 7:705.

142. Piscioli F, Pusiol T, Scappini P, et al: Urine cytology in the detection of renal adenocarcinoma. *Cancer* 1985; 56:2251.

143. Pitts WR, Kazam E, Gray G, et al: Ultrasonography, computed tomography and pathology of angiomyolipoma of the kidney: Solution to a diagnostic dilemma. *J Urol* 1980; 124:907.

144. Pizzocaro G, Piva L, Salvioni R, et al: Adjuvant medroxyprogesterone acetate and steroid hormone receptors in category M_o renal cell carcinoma: An interim report of a prospective randomized study. *J Urol* 1986; 135:18.

145. Pritchett TR, Lieskovsky G, Skinner DG: Extension of renal cell carcinoma into the vena cava: Clinical review and surgical approach. *J Urol* 1986; 135:460.

146. Probst P, Hoogewand HM, Haertel M, et al: Computerized tomography versus angiography in the staging of malignant renal neoplasm. *Br J Radiol* 1981; 54:744.

147. Quesada JR, Swanson DA, Trinidade A, et al: Renal cell carcinoma: Antitumor effects of leukocyte interferon. *Cancer Res* 1983; 43:940.

148. Quesada JR, Rios A, Swanson DA, et al: Antitumor activity of recombinant-derived interferon alpha in metastatic renal cell carcinoma. *J Clin Oncol* 1985; 3:1522.

149. Rainwater LM, Farrow GM, Lieber MM: Flow cytometry of renal oncocytoma: Common occurrence of DNA polyploidy and aneuploidy. *J Urol* 1986; 135:1167.

150. Ramchandani P, Soulen RL, Schmall RL, et al: Impact of magnetic resonance on staging of renal carcinoma. *Urology* 1986; 27:664.

151. Riches EW: The place of radiotherapy in the management of parenchymal carcinoma. *J Urol* 1966; 95:313.

152. Richie JP: Management of ureteral tumors, in Skinner DG, deKernion JB (eds): *Genitourinary Cancer.* Philadelphia, WB Saunders, 1978.

153. Richie JP, Garnick MB: Primary renal and ureteral cancer, in Rieselback RE, Garnick MB (eds): *Cancer and the Kidney.* Philadelphia, Lea & Febiger, 1982, pp 683–685.

154. Robey EL, Schellhammer PF: The adrenal gland and renal cell carcinoma: Is ipsilateral adrenalectomy a necessary component of radical nephrectomy? *J Urol* 1986; 135:453.

155. Robson CJ: Radical nephrectomy for renal cell carcinoma. *J Urol* 1963; 89:37.

156. Robson CJ, Churchill BM, Anderson W: The results of radical nephrectomy for renal cell carcinoma. *J Urol* 1969; 101:297.

157. Robson CJ, Churchill BM, Anderson W: The results of radical nephrectomy for renal cell carcinoma. *Trans Am Assoc Genitourinary Surg* 1968; 60:122.

158. Rodriguez CA, Buskop A, Johnson J, et al: Renal oncocytoma: Preop diagnosis by aspiration biopsy. *Acta Cytol* 1980; 24:355.

159. Rosen RJ, Schlossberg P, Raven SJ, et al: Management of symptomatic renal angiomyolipomas by embolization. *Urol Radiol* 1984; 6:196.

160. Rosenberg SA: The adoptive immunotherapy of cancer using the transfer of activated lymphocytic cells and interleukin-2. *Semin Oncol* 1986; 13:200.

161. Rosenberg SJ, Williams RD: Photodynamic therapy of bladder carcinoma. *Urol Clin North Am* 1986; 13:435.

162. Salmon SE, Hamburger AW, Soebulen B, et al: Quantitation of differential of sensitivity of human tumor stem cells to anti-cancer drugs. *N Engl J Med* 1978; 198:1321.

163. Sarnacki CT, et al: Urinary cytology and the clinical diagnosis of urinary tract malignancy: A clinicopathologic study of 1,400 patients. *J Urol* 1971; 106:761.

164. Schefft P, et al: Surgery for renal cell carcinoma extending into the inferior vena cava. *J Urol* 1978; 120:128.

165. Schiff MD, Bagley DH, Lytton B: Treatment of solitary and bilateral renal carcinomas. *J Urol* 1979; 121:581.

166. Schmouz R, Cole P: Epidemiology of cancer of the renal pelvis and ureter. *J Natl Cancer Inst* 1974; 52:1431.

167. Shortliffe L: Immune modifiers in genitourinary cancers, in Williams RD (ed): *Advances in Urologic Oncology.* New York, MacMillan Publishing Co, 1987.

168. Skinner DG, Pfister RF, Colvin R: Extension of renal cell carcinoma into the vena cava: The rationale for aggressive surgical management. *J Urol* 1972; 107:711.

169. Skinner DG, Colvin RB, Vermillion CD, et al: Diagnosis and management of renal cell carcinoma. *Cancer* 1971; 28:1165.

170. Smith-Harrison LI, Laing FC, Jeffrey RB, et al: Ultrasonography versus excretory urography prior to transurethral prostatectomy—A prospective study. *Welcome Trends in Urology* 1983; 5:11.

171. Steele G Jr, Wang BS, Richie JP, et al: Results of oncogenic 1-RNA therapy in patients with metastatic renal cell carcinoma. *Cancer* 1981; 47:1286–1288.

172. Sternberg CN, Yagoda A, Scher HI, et al: Preliminary results of M-VAC (methotrexate, vinblastine, doxorubicin, and cisplatin) for transitional cell carcinoma of the urothelium. *J Urol* 1985; 133:403.

173. Strang DW, Pearse HD: Recurrent urothelial tumors following surgery for transitional cell carcinoma of the upper urinary tract. *Cancer* 1976; 38:2173.

174. Strewler GJ, Williams RD, Nissenson RA: Human renal carcinoma cells produce hypercalcemia in the nude mouse and a novel protein recognized by parathyroid hormone receptors. *J Clin Invest* 1983; 71:769.

175. Swanson DA, Wallace S, Johnson, DE: The role of embolization and nephrectomy in the treatment of metastatic renal carcinoma. *Urol Clin North Am* 1980; 7:719.

176. Swanson DA, et al: Angioinfarction plus nephrectomy for metastatic renal cell carcinoma, an update. *J Urol* 1983; 130:449.

177. Toho BM, Whitmore WE Jr: Solitary metastases from renal cell carcinoma. *J Urol* 1975; 114:836.

178. Topley M, Novick AC, Montie JE: Long term results following partial nephrectomy for localized renal adenocarcinoma. *J Urol* 1984; 131:1050.

179. Toyama K, Fujiyand N, Suzuki H, et al: Erythopoietin levels in the course of a patient with eythropoietin producing renal cell carcinoma and transplantation of this tumor in nude mice. *Blood* 1979; 54:245.

180. Van der Wurf-Messing B: Carcinoma of the kidney. *Cancer* 1973; 32:1056.

181. Verhood R, Harmsen A, Kunik A: On the frequency of tumor induction in a Thorotrast kidney. *Cancer* 1974; 34:2061.

182. Vessella RL, Chiou RK, Lange PH: Monoclonal antibodies in urology: Review of reactivities and applications in diagnosis, staging and therapy. *Semin Urol* 1985; 3:158–167.

183. VonSchreeb T, Arner O, Skausted G, et al: Renal adenocarcinoma: Is there a risk of spreading tumor cells in diagnostic puncture? *Scand J Urol Nephrol* 1967; 1:270.

184. Vugrin D, Hood L, Taylor W, et al: Phase II study of human lymphoblastoid interferon in patients with advanced renal cell carcinoma. *Cancer Treat Rep* 1985; 69:817.

185. Wagle DG, Moore RH, Murphy GP: Secondary carcinomas of the kidney. *J Urol* 1975; 114:30.

186. Wallace AC, Nairn RC: Renal tubular antigens in kidney tumors. *Cancer* 1972; 29:977.

187. Wallace DM, Chisholm GD, Hendry WF: TNM classification for urological tumors (UICC-1974). *Br J Urol* 1975; 47:1.

188. Weiner SN, Bernstein RG: Renal oncocytoma: Angiographic features of two cases. *Radiology* 1977; 125:633.

189. Weyman PJ, McClennan BL, Stanley RJ, et al: Comparison of computed tomography and angiography in the evaluation of renal cell carcinoma. *Radiology* 1980; 137:417.

190. Weyman PJ, McClennan BL, Lee JKT, et al: CT of calcified renal masses. *AJR* 1982; 138:1095.

191. Wheatley JK: Retroperitoneal fibrosis, tumors, and cysts, in Glenn JF (ed): *Urologic Surgery.* Philadelphia, JB Lippincott, 1983, pp 391–403.

192. Wickham TEA: Conservative renal surgery for adenocarcinoma: The place of bench surgery. *Br J Urol* 1975; 47:25.

193. Wile AG, Evans HL, Romsdahl MM: Leiomyosarcoma of soft tissue: A clinicopathologic study. *Cancer* 1981; 48:1022.

194. Williams RD: Tumors of the kidney, ureter and bladder, in Wyngaarden JB, Smith LH (eds): *Cecil's Textbook of Medicine,* Philadelphia, WB Saunders, 1985, p 639.

195. Williams RD, Jensen B, Higgins M, et al: Alpha$_2$ interferon therapy of disseminated renal cell cancer. *J Urol* 1984a; 131:178.

196. Williams RD, Nissenson RA, Strewler GJ: Human malignancy-associated hypercalcemia model in nude mice. Proceedings of the Fourth International Workshop on Immune-Deficient Animals in Experimental Research. *Exp Cell Res* 1984b; 52:298.

197. Xipell JM: The incidence of benign renal nodule (a clinico-pathologic study). *J Urol* 1971; 100:503.

198. Zimmer WD, Williamson B, Hartman GW, et al: Changing patterns in the evaluation of renal masses: Economic implication. *AJR* 1984; 143:285.

Chapter 17

Kidney Stone Surgery

J. Patrick Spirnak, M.D.
Martin I. Resnick, M.D.

With recent attention directed toward the newer techniques of stone removal such as percutaneous nephrostolithotomy and extracorporeal shock-wave lithotripsy (ESWL), there has been a tendency to forget the significant technological advances in open surgical stone procedures over the past decade. The development and acceptance of these new surgical techniques predictably have been accompanied by significant diminution in the number of open surgical stone procedures performed. While it is clear that the role of open stone surgery will not be as prominent in the future, it is imperative that residency training programs continue to train future urologists in a variety of these techniques. These traditional procedures will continue to be required for removal of large complex or staghorn calculi, particularly when associated with intrarenal scarring and infundibular and/or calyceal stenosis. Additionally, when stone disease is associated with a correctable anatomical abnormality (e.g., ureteropelvic junction obstruction), an open procedure will likely be required. In these particular instances, reconstructive procedures are necessary, and for this reason it is essential that the urologic surgeon understand the intrarenal vascular anatomy and remain proficient in open surgical stone removal.

HISTORICAL OVERVIEW

Archeological studies have shown urinary stone disease to have been an affliction of man dating before 4800 B.C. A review of ancient records also shows stone disease to have been present in the Greek and Roman civilizations (Shattock, 1905). Ancient physicians recorded the symptoms and treatment associated with urinary stone disease but directed little attention toward the localization of the stone or to the cause of its formation. Hippocrates, in addition to being credited with having performed the first renal operation, observed that many patients with kidney or bladder stones had sandy urinary sediment and theorized the cause of stones to be due to the ingestion of muddy river water or water containing lime (Butt, 1956).

The operative removal of renal calculi has been documented in both the Greek and Roman literature and usually occurred incidental to the drainage of a renal or perirenal abscess. Celsus, Galen, and other early physicians mentioned kidney stones in their writings, but none believed surgical extraction to be safe or feasible. In 1510, Cardan of Milan reportedly extracted 18 renal stones while draining a renal abscess. Other isolated reports of stone removal performed either through the collecting system or via incidental nephrotomies appeared during the 1500s and 1600s. Hevin, who is credited with coining the word "nethrolithotomy," concluded in 1775 that renal operations for the removal of stone should be limited solely to infected kidneys (Hevin, 1778). European surgeons during that time heeded his advice and limited the use of nephrotomies to kidneys that had become infected and distended with purulent matter. If the stones could not be readily removed at the time of surgery, they were left to pass spontaneously through urinary fistuli that often developed.

Dr. William Ingalls on Oct 8, 1872, at the Boston City Hospital performed what is believed to be the first planned nephrotomy in the United States (Bundy and Ingalls, 1882). The 31-year-old female patient, who earlier had a perirenal abscess drained, recovered after a postoperative course complicated by sepsis and was in good health eight years later. Sir Henry Morris, an English surgeon, in 1880 performed the first nephrolithotomy on a healthy kidney free of abscesses (Morris, 1881). In 1898 he published the results of 34 nephrolithotomies and reported only one death (Morris, 1913). With these successful results the dangers of nephrolithotomy were disproved and the feasibility of removing renal stones before the kidney was converted into an abscess sac was clearly shown. Also in 1880, Vincenz Czerny (Gil-Vernet, 1983) is credited with performing the first pyelolithotomy and in 1887 was the first to use catgut to reapproximate the nephrotomy incision.

Controversy developed over which method of stone removal was preferable. Nephrolithotomy offered the major advantage of visualizing the entire collecting system directly and it evolved into the conservative procedure of choice. With the development of useful clinical radiography in the early 1900s, accurate stone localization became possible and resulted in a renewed interest in pyelotomy for stone removal. Lower in 1913 advocated pyelolithotomy as the procedure of choice for simple pelvic stones (Stewart and Straffon, 1975).

Surgeons practicing in the 1800s had an excellent understanding of the gross renal anatomy and the renal blood supply; however, an accurate understanding of the intrarenal anatomy and vasculature was lacking. Josef Hyrtl, a 19th century Viennese anatomist, was the first to study the intrarenal vasculature (Hyrtl, 1872) and in 1882 described an avascular plane between the anterior and posterior vascular segments. In 1901, Max Brödel, a medical artist at the Johns Hopkins Medical School, published the classic work "The intrinsic blood vessels of the kidney and their significance in nephrotomy" (Brödel, 1901). In this report he accurately described the branching of the main renal artery, emphasized the lack of free anastomosis between segmental branches, and confirmed Hyrtl's initial observation. Graves (1954) confirmed the accuracy of Brödel's observations and described the existence of specific anatomical renal segments based on the intrarenal arterial distribution.

Progress in the field of renal stone surgery continued into the 1900s. With improved understanding of the intrarenal anatomy and blood supply, surgeons experimented with a number of different nephrotomy incisions in an attempt to identify an approach to the intrarenal collecting system that would result in minimal parenchymal damage. Zuckerkandl (1908) described the extension of a pyelotomy incision into the medial surface of the lower renal pole. About the same time, Marwedel (1907) recommended transverse parenchymal incisions across the convex border of the kidney as a means of removing calyceal stones. Cullen and Derge (1909) recommended the nephrotomy incision be made with a thin silver wire passed between arterial branches from within outward. They further recommended the use of an "L"-shaped nephrotomy incision, with the upper limb of the "L" following the avascular plane described by Hyrtl and Brödel with extension of the lower limb across the inferior renal pole parallel to the posterior segmental artery. Marion (1922) and Eisendrath (1923) advocated extension of a pyelotomy incision onto the midposterior surface of the kidney after first ligating the posterior segmental artery. Kerr (1970)

modified this incision slightly by placing it in the intersegmental line located between the posterior and inferior vascular segments and thus obviated ligating the posterior segmental artery.

In the late 1920s Deming (1928a, 1928b), after studying the effects of the various renal incisions on the intrarenal circulation, concluded that the approach that best minimized parenchymal damage was a longitudinal incision through the avascular plane initially described by Hyrtl and Brödel. Smith and Boyce (1967) have combined this incision with reconstruction of the intrarenal collecting system to remove staghorn calculi and improve renal drainage.

INDICATIONS FOR STONE REMOVAL

The objectives of renal stone surgery include: (1) removal of all calculi; (2) repair of all correctable anatomic abnormalities including obstructed or strictured infundibula and calyces; (3) complete eradication of associated urinary tract infections; (4) preservation of functioning renal tissue; and (5) prevention of recurrent stone formation.

The indications for surgical removal of a renal calculus include intractable urinary infection, progressive renal damage, urinary obstruction, persistent pain, or gross hematuria. In most instances, surgery can be delayed until a preliminary metabolic evaluation is completed. However, if severe obstruction or sepsis occurs, prompt temporizing urinary drainage is indicated. This can be accomplished by retrograde catheter placement, percutaneous nephrostomy drainage, or at times by formal open stone removal.

PREOPERATIVE EVALUATION

The preoperative evaluation of the stone patient should be directed toward defining the nature and configuration of the calculus, the degree of function of the affected as well as the contralateral kidney, and identification of the cause of stone formation.

Prior to surgical stone removal, a preliminary metabolic evaluation should be performed in all clinically stable patients. Laboratory data including serum calcium, phosphorus, uric acid, creatinine, urea nitrogen, and electrolytes are routinely obtained. These studies identify patients with severe metabolic derangements such as hypercalcemia, hyperuricemia, systemic acidosis, and/or renal failure and may lead to the diagnosis of hyperparathyroidism or renal tubular acidosis as the underlying cause of stone formation. If hyperparathyroidism is suggested and confirmed by an elevation in the serum parathormone level, surgical removal of the parathyroid adenoma should precede surgical stone removal (Carlson et al., 1960; Thomas et al., 1958). Individuals in whom a possibility for cystinuria exists (as suggested by family history or early onset of stone disease) should be screened by the qualitative sodium nitroprusside test. A screening 24-hour urine quantitative collection for creatinine clearance determination, urinary calcium, phosphorus, uric acid, and cystine is also obtained. Depending on the results of the above studies, a more detailed metabolic evaluation may be carried out in the postoperative period (Menon and Krishnan, 1984; Pak, 1980; Pitts and Resnick, 1980). In the past, controversy has existed over the optimal time to obtain quantitative urine studies in the patient who has recently undergone stone removal. Recent evidence has shown the ideal time for such studies may be as long as 2–3 months after complete resolution of an acute episode of renal colic (Norman et al., 1984). By waiting until the patient has returned to his normal environment and diet, a more accurate metabolic assessment is possible.

A midstream clean-catch urine sample is obtained in every patient. Urinalysis should include urinary pH determination and the documentation of crystals. By determining the pH one may be able to identify a cause of stone formation and the type of stone present. The presence of an acid urine suggests the presence of uric acid, cystine, or calcium oxalate stones, whereas stone formation in the presence of an alkaline urine should lead one to consider the presence of either calcium phosphate stones in association with renal tubular acidosis or struvite stones. The presence of the pathognomonic hexagonal cystine crystals confirms the diagnosis of cystinuria, while the presence of oxalate or uric acid crystals suggest the possibility, but is not diagnostic, of the presence of those respective calculi.

Urine culture and sensitivity studies are routinely performed when urinalysis, clinical course, or stone configuration suggests the possibility of a struvite calculus or the presence of a concomitant urinary tract infection. Infectious stones occur in the presence of urea-splitting organisms. These organisms produce the enzyme urease which splits urinary urea into ammonium and bicarbonate ions thereby causing an abnormally alkaline urinary pH and the subsequent precipitation of carbonate apatite and magnesium ammonium phosphate. Common urea-splitting organisms include *Proteus, Pseudomonas, Klebsiella*, and *Staphylococcus. Escherichia coli* does not produce urease and, when associated with this type of stone, most likely represents a superinfection (Spirnak and Resnick, 1984). Patients with suspected struvite stones should receive appropriate antibiotics at least 48 hours prior to surgical intervention in an attempt to sterilize the urine.

A thorough radiologic evaluation of the urinary tract is essential prior to stone removal. Excretory urography is routinely performed. An adequate study must identify the size and location of all stones in addition to visualizing the entire collecting system including both ureters and bladder. Plain nephrotomograms may be helpful in detecting small fragments and caliceal extensions. Retrograde pyelography is indicated when the kidney is poorly functioning and when the complete collecting system, including the ureters, is not well visualized.

In the presence of suspected radiolucent stones, computerized axial tomography (CAT) performed without contrast injection will prove diagnostic and delineate the extent of renal involvement (Resnick et al., 1984). Renal angiography is not routinely obtained but may be helpful in providing a "renal road map" in the patient with a staghorn calculus who has undergone prior renal surgery and in whom an anatrophic nephrolithotomy is planned. Individuals with an anomalous urinary tract including the presence of renal fusion anomalies such as horseshoe kidneys or crossed fused ectopia may likewise benefit from presurgical renal angiography.

The use of radionuclide renal scans has proved helpful in assessing baseline renal function prior to parenchymal splitting procedures and in assessing the degree of function and salvageability of the kidney damaged by chronic stone disease. Voiding cystourethrography and urodynamic evaluation may be helpful in diagnosing subtle lower urinary tract disorders that could predispose the patient to chronic urinary infection and recurrent stone formation.

INTRARENAL ANATOMY

A basic knowledge and appreciation of the intrarenal vasculature and collecting system is essential for the successful performance of renal stone surgery. The division of the renal pelvis and the number and distribution of the calyces are the most variable components of the renal anatomy. The renal pelvis is usually saccular and funnels directly into the ureter. In most instances the pelvis is well defined and divides within the substance of the kidney into two or three primary divisions referred to as major calyces or infundibula (Fig 17–1). In some kidneys the renal pelvis is practically nonexistent, and it appears that the ureter is continuous with each of the major pelvic divisions. Although there is much variation in size and contour, no one form can be considered normal. From the infundibula secondary and sometimes even tertiary divisions arise, which are called minor calyces. The renal calyces drain and are in immediate apposition to the papillae and may be classified as anterior, posterior, superior, inferior, or medial.

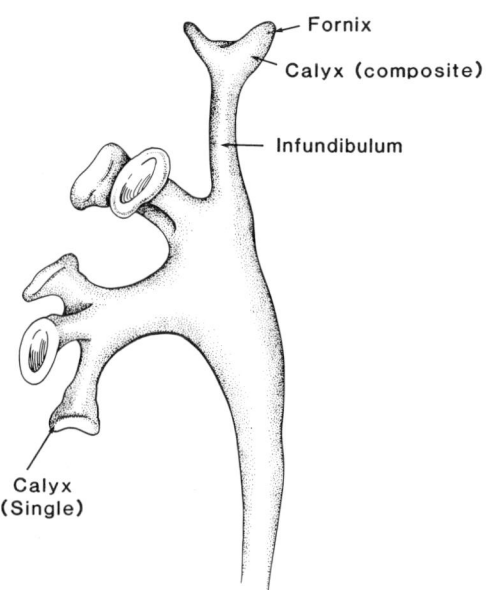

FIG 17-1.
Anatomy of the intrarenal collecting system.

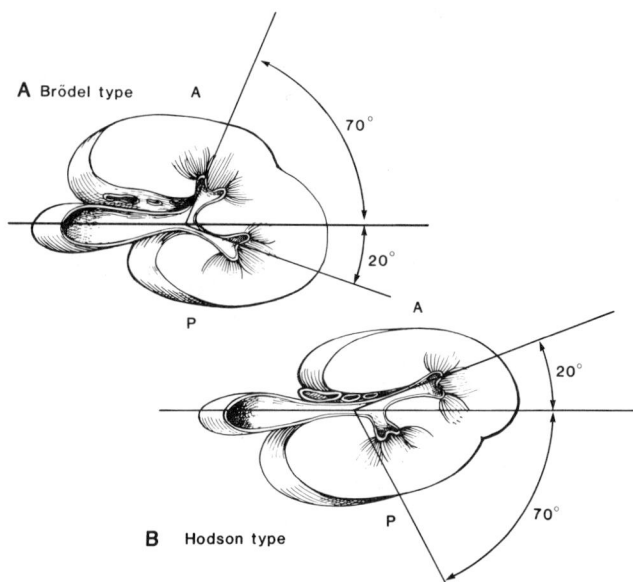

FIG 17-2.
Diagram showing the two types of kidneys first described by Brödel **(A)** and Hodson **(B)**.

The calyces are well-defined anatomical structures that vary from the single calyx draining one papilla to the composite calyx draining multiple papillae. The polar calyces generally demonstrate the greatest variation in size and shape and are usually compound. The total number of calyces in a single kidney is usually 8, but may vary from 4 to 12. There are usually 2–3 infundibula that drain the calyces and empty into the renal pelvis.

Recent development in percutaneous stone removal techniques have rekindled an interest in the anatomy of the intrarenal collecting system. The calyx of the uppermost infundibulum is usually single but receives multiple papillae and is therefore referred to as a compound or composite calyx. The remainder of the calyces are paired and arranged anterior or posterior to an imaginary line that divides the kidney longitudinally into a posterior and anterior half. Previous conflicting reports by Brödel (1901) and Hodson (1972) have led to confusion and a poor understanding of the intrarenal calyceal anatomy. In Brödel's classic work, he depicted the anterior calyces as lying at about a 70° angle to the midfrontal plane, with the posterior row calyces projecting at about a 20° angle. In a conflicting report, Hodson (1972) described the typical kidney as being nearly a mirror image of the Brödel type, with the anterior row calyces projecting at about a 20° angle and the posterior calyces at a 70° angle to the midfrontal plane. A study by Kaye and Reinke (1984) recognized the presence of both kidney types. They found the majority of right kidneys to be of the Brödel type (69%), while about 80% of left kidneys demonstrated the Hodson calyceal arrangement (Fig 17–2).

The main renal arteries arise from the lateral surface of the aorta below the level of the superior mesenteric artery at the level of the second lumbar vertebra. About 70% of all kidneys are supplied by a single renal artery that divides into the anterior and posterior segmental vessels (Boijsen, 1959). This primary division usually occurs at the level of the renal hilum.

Although Brödel was the first to recognize the segmental distribution of the renal artery, it was not until 1954 that Graves described the constant pattern of the intrarenal arterial distribution. He described five anatomical segments—apical, lower, posterior, upper, and middle—and noted an absence of collateral circulation.

In approximately 50% of all kidneys, the posterior division is the first branch of the renal artery, and in about one third of cases the first branch will be the vessel to the lower pole segment. In the remaining instances there is about an equal division between those in which the upper segmental artery arises as the first branch of the main renal artery and those in which the posterior artery arises simultaneously with the anterior segmental artery (Elkins and Resnick, 1982).

The anterior division represents the main continuation of the renal artery and typically divides into superior, anterior, and inferior segmental branches as well as the apical artery of Graves. This vessel usually branches either immediately before or within the superior aspect of the renal hilum (Fig 17–3). The five segmental branches divide further into the interlobar or paracalyceal arteries. The interlobar arteries, usually numbering 8–12, ascend in the columns of Bertin along opposite margins of the infundibula. Prior to reaching

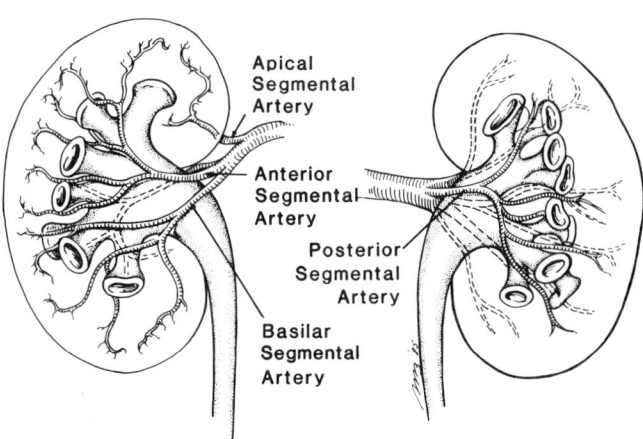

FIG 17–3.
Intrarenal vasculature.

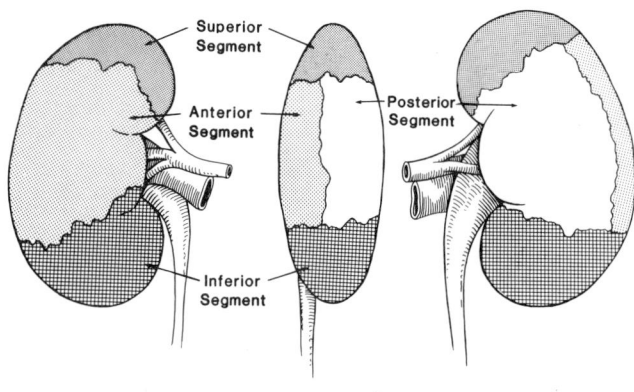

FIG 17–5.
Vascular renal segments.

the corticomedullary junction, these vessels divide into the arcuate arteries. The arcuate arteries arch over the bases of the medullary pyramids and give off the interlobular arteries. Branches of the interlobular arteries become the afferent arterioles supplying the glomeruli. Renal arteriography will identify vessels to the arcuate level (Straffon, 1982) (Fig 17–4).

Based on the segmental arterial distribution, the kidney is divided into four surgical segments: superior, inferior, anterior and posterior (Fig 17–5). The apical artery and corresponding vascular segment is so variable that it is included as part of the superior segment. The upper and lower polar segments include the single po-

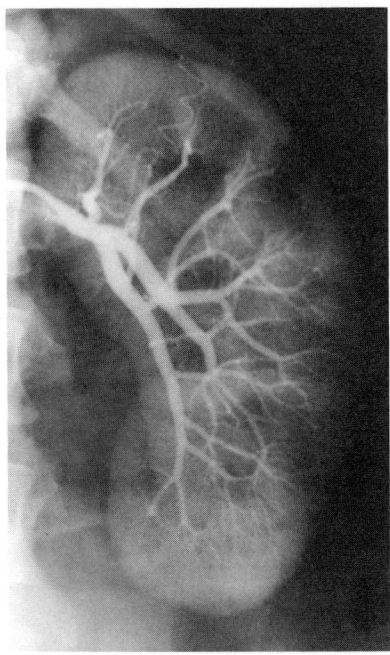

FIG 17–4.
Arteriogram demonstrating the normal intrarenal vasculature.

lar calyceal system. The double row of pyramids between these polar segments comprise the anterior and posterior renal segments.

The artery to the superior or upper pole segment has an extrahilar source of origin in nearly all instances. It courses upward and medially to the pelvis toward the upper pole calyx and is usually easily isolated at the time of surgery. It routinely supplies the anterior surface of the superior renal segment and in about 50% of all kidneys will supply the posterior surface of the upper pole as well. In the remaining kidneys the posterior surface will be supplied by a superior polar extension of the posterior segmental artery.

The arterial branch to the inferior renal segment also has an extrahilar source of origin, and like the artery to the superior renal segment, may be easily isolated at the time of surgery. The artery descends obliquely in front of the renal pelvis and after crossing the lower pelvic border divides into an anterior and posterior branch.

The anterior renal segment is the largest of the four surgical segments and occupies not only the entire anterior surface of the midkidney but extends laterally to include the convex border and a variable amount of posterior parenchyma as well. Included in this segment are the anterior and posterior surfaces of the anterior row calyces as well as the anterior surface of the posterior calyces. The anterior segmental artery arises as the direct continuation of the anterior division of the renal artery. In about half of all kidneys a single artery will supply the entire anterior segment. Variations, depending on the blood supply to the superior and inferior polar segments, may also occur.

The posterior branch of the renal artery supplies the entire posterior surgical segment. It is usually extrahilar in origin and runs posterior to the renal pelvis in the superior renal hilum. The size of the posterior segment is highly variable and cannot be predicted solely on the basis of external topography. In more than 50% of all kidneys the posterior segment is large and confined to the middle or upper half of the posterior renal surface without extending to the convex lateral border of the kidney. In these kidneys, the posterior segmental artery vascularizes only the parenchyma of the posterior calyces. In 30% of kidneys the posterior segment extends to the lateral renal border or even beyond and onto the anterior surface. In those instances, the posterior artery provides more of the blood supply to the posterior row of calyces; however, it is rare for this vessel to supply both the anterior and posterior surfaces of these calyces.

Brödel's white line is the linear whitish depression located on the posterior surface of the kidney that overlies the highly vascular longitudinal columns of Bertin. This landmark should not be confused with the relatively avascular plane, or Brödel's line, which is the proper site of incision for anatrophic nephrolithotomy and overlies the division between the anterior and posterior renal segments (Myers, 1971) (Fig 17–6).

The renal venous system does not follow the corresponding arterial segmental distribution. The venous system demonstrates free anastomosis between the intrarenal veins, which allows the occlusion of venous channels without jeopardizing the renal parenchyma. In general, the main renal vein may arise from as many as four trunks that leave the hilum and join prior to emptying into the vena cava.

CLINICAL APPLICATIONS OF INTRARENAL ANATOMY TO INTRARENAL STONE SURGERY

In more than 80% of kidneys, the renal artery divides into the primary branches before reaching the hilum, and in most kidneys this primary branching provides easy access to the posterior segmental artery. Approaching the mobilized kidney posteriorly allows one easily to isolate the main renal artery as well as its primary branches.

In more than half of all kidneys, the posterior segmental artery arises as the first branch of the main renal artery. Occlusion of this vessel followed by IV administration of methylene blue or indigo carmine will provide complete delineation of this vascular segment. In more than 50% of all kidneys, this posterior segment is confined to the middle or upper half of the posterior renal surface and does not reach the lateral renal border. Therefore, two or three intersegmental planes with obvious surgical implication are discernible.

The largest plane lies between the anterior and posterior segments on the posterior surface of the kidney and conforms closely to Brödel's classic description of the posterior intersegment plane (Brödel, 1901). The

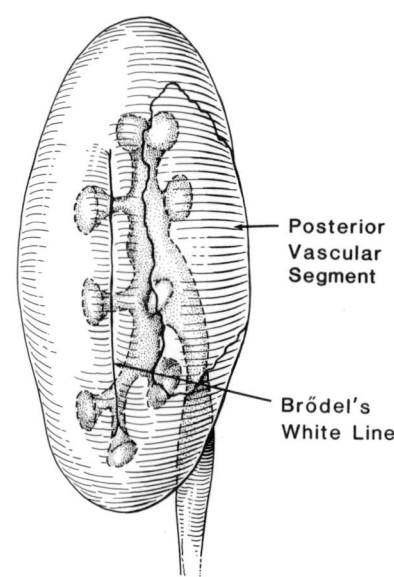

FIG 17–6.
The posterior vascular segment and Brödel's white line.

length of this avascular plane varies from kidney to kidney and depends largely on the size of the inferior and superior vascular segments. Utilization of the intersegmental plane requires entry into the pelvis through a posterior calyx in order to avoid injury to the anterior circulation as it supplies the anterior surface of the posterior row of calyces. Smith and Boyce (1967) have utilized this plane for their intersegmental anatrophic nephrolithotomy.

The second intersegmental plane with surgical importance lies between the posterior and inferior segments on the posterior surface of the kidney. In more than 80% of kidneys, the lower segmental artery supplies the anterior and posterior surface of the inferior segment. Consequently, the artery to the posterior surface of the inferior segment is of anterior origin and allows for placement of renal incisions on the posterior renal surface between these two segments without injury to either circulation.

A third intersegmental plane may also be demonstrated on the posterior surface of the kidney lying between the posterior segment and the superior segment. This plane is present only in those kidneys where the superior segmental artery, usually of anterior origin, supplies the posterior surface of the superior segment.

It is important to remember that while each kidney possesses four surgical segments, the extent of each is highly variable and no segment is constant enough to permit blind parenchymal incisions without risking segmental devascularization. The posterior vascular distribution is most highly variable and no two kidneys have been found to have exactly the same parenchymal distribution. Although one can easily demonstrate the intrarenal segmental planes, it must be emphasized that the kidney does not possess a surgically useful completely avascular plane. Therefore, the indiscriminate use of nonintersegmental renal incisions, such as the bivalve nephrotomy, needlessly sacrifices functioning renal parenchyma and is to be condemned.

SURGICAL APPROACHES

A number of different surgical approaches to the kidney have been described (Bodner and Briskin, 1950; Kropp, 1983). Gustav Simon in 1869 performed the first nephrectomy utilizing what he believed to be the most direct and atraumatic route to the kidney—the vertical lumbar incision (Gil-Vernet, 1983). As the art of renal surgery progressed, surgeons often approached the kidney through an anterior transperitoneal approach. This incision was useful in the days prior to the development of current radiographic techniques in that it allowed for the complete exploration of both kidneys. With the development of sophisticated radiographic techniques the need to grossly inspect both kidneys became obsolete, and the anterior transabdominal approach was replaced by the lumbar or flank incision. By surgically exploring the kidney through an extraperitoneal incision, the morbidity associated with the transabdominal approach was avoided. We currently utilize all three surgical approaches to the kidney and believe it essential for the stone surgeon to be well versed in all techniques.

POSTERIOR VERTICAL LUMBOTOMY

The posterior approach to the kidney remains the most direct route to the renal fossa and avoids the trauma and postoperative pain associated with other muscle-transecting incisions (Bensiman, 1982; Gil-Vernet, 1983; Kropp, 1983; Novick, 1980; Pansadoro, 1983). When properly performed, it also avoids entry into either the peritoneal or pleural cavities, avoiding the potential associated complications when these areas are entered. We currently use this approach for all simple renal pelvic stones not amenable to removal by percutaneous techniques.

After anesthetizing and intubating are accomplished, the patient is placed prone on the operating table. To allow effective ventilation and promote venous return the abdomen and anterior chest are elevated off of the operating table by placing the patient on two laterally placed, rolled bed sheets. The table is flexed slightly to increase the distance between the 12th rib and posterior iliac crest.

To maximize the exposure the incision is started at the lower margin of the 12th rib, directly over the sacrospinalis muscle, and is extended caudally to the level of the posterior iliac crest in a gentle lateral curve. In

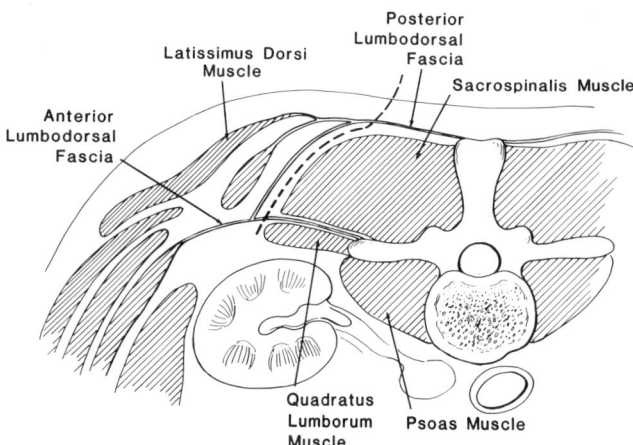

FIG 17–7.
Cross-section of the lumbar region demonstrating the relationship of the back muscles to the muscle-splitting lumbotomy incision (dotted line).

the obese patient or in a high-positioned kidney, additional exposure can be achieved by removing a small segment of the 12th rib. The incision is deepened through the posterior lumbodorsal fascia, revealing the sacrospinalis muscle group (Fig 17–7). This muscle is retracted medially using broad retractors and the anterior lumbodorsal fascia is incised. The quadratus lumborum muscle is identified and is likewise medially retracted, exposing the retroperitoneal space. To further facilitate exposure the costovertebral ligamentous attachment of the 12th rib is incised. A self-retaining retractor is then placed. While making the incision, care should be taken to avoid traumatizing the iliohypogastric nerve. Using sponge sticks, the kidney contained within Gerota's fascia is mobilized downward and laterally. Gerota's fascia is incised and the kidney, pelvis, and upper ureter dissected. After removing the stone and closing the pelvis the fascial layers are closed.

FLANK APPROACH

The lateral flank incision is probably the most frequently utilized approach to the kidney. Depending on the position of the kidney a subcostal incision, 12th or 11th rib resection, or intercostal incision can be effectively employed. This incision when properly performed avoids entry into both the peritoneal and pleural cavities. Unfortunately, unlike the posterior approach, exposure gained through the flank depends on transection of the latissimus dorsi, external and internal oblique muscles, as well as the transversus abdominus muscle. Nerves to these muscles may also be traumatized, resulting in an unsightly postoperative flank bulge.

After obtaining a suitable level of anesthesia, the patient is placed on the operating table in the flank posi-

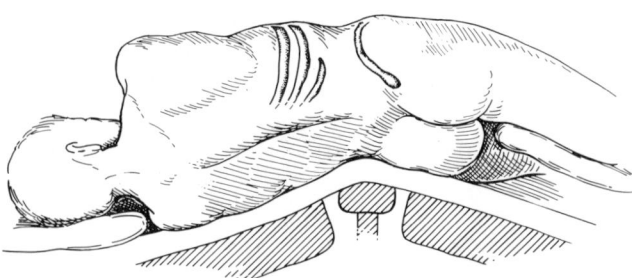

FIG 17–8.
Patient in standard flank position with the kidney rest elevated.

tion, with the iliac crest positioned over the kidney rest (Fig 17–8). The bottom leg is flexed at the hip and knee, and the upper leg remains straight and is supported on a pillow. The patient's back should parallel the edge of the operating table and be as close to the side as possible. The kidney rest is slowly elevated while the blood pressure is carefully monitored. After completely elevating the kidney rest, the table is flexed to further increase the distance between the lower rib and the iliac crest. A rolled sheet is placed beneath the lower axilla to remove pressure from the brachial plexus and avoid postoperative neuropathy. The head is likewise supported. The uppermost extremity is extended and supported on an armboard. The patient is then taped to the table with 3-in. adhesive tape. A standard flank incision is then performed.

We use this approach in all previously operated kidneys and when complete mobilization of the kidney is required, as in anatrophic nephrolithotomy.

ANTERIOR APPROACH

The anterior transperitoneal approach to the kidney is reserved for those patients with physical deformities who could not tolerate the flank position. It is also useful in patients with renal fusion anomalies such as a pelvic or horseshoe kidney. We utilize either a standard midline or chevron incision depending on the position of the kidney.

RENAL HYPOTHERMIA

The development of practical and efficacious in situ renal hypothermia has provided the renal stone surgeon with the ability to occlude the renal artery for long periods without impairing renal function. Many studies have been conducted over the past 100 years not only to investigate the effects of renal ischemia on function but also to develop additional techniques to prevent renal damage during the period of renal circulatory arrest.

In 1880 Litten (Stueber et al., 1958) performed a se-

ries of experiments assessing the histologic effects of varying periods of ischemia on the kidney. Since that time other investigators have carried out similar studies on a variety of mammalian kidneys (Badenoch and Darmady, 1947; Eisendrath and Strauss, 1910; Koletsky, 1954; Wickham et al., 1967). These investigators noted damage to proximal tubular epithelial cells during periods of ischemia in excess of 30 minutes. Histologically, while the proximal tubular cells showed extensive ischemic injury, the glomeruli and blood vessels were generally spared. The mechanism of damage to the proximal tubule is related to the availability of energy-rich adenosine triphosphate (ATP). During periods of warm ischemia, ATP is utilized as the primary energy source to maintain the cellular membrane transport system (Collins et al., 1977). With the depletion of available ATP, intracellular influx of salt and water occurs, followed by cellular swelling, resulting ultimately in cell death (Novick, 1983).

Investigators studying the effects of ischemia on total renal function have demonstrated an immediate impairment of function following any length of complete renal ischemia (Mayer et al., 1957; Selkurt, 1945). The overall effect is directly related to the period of warm ischemia, and after 60 minutes the deleterious effects are only partially reversible.

It has been suggested that the solitary kidney is more resistant to ischemic changes and can safely tolerate ischemic periods of greater than 30 minutes. Askari and associates (1982) have performed renal revascularization procedures in patients with a solitary kidney and noted no postoperative renal failure in spite of unprotected warm ischemia times ranging from 14 to 59 minutes. These findings confirm the belief that the solitary kidney can better withstand prolonged renal ischemia. The exact mechanism is as yet unknown.

The method by which renal ischemia occurs also appears to have some role in affecting the extent of injury. It has been shown that simultaneous clamping of both the renal artery and vein is more deleterious to the kidney than is clamping of the renal artery alone (Neely and Turner, 1959). The proposed reason for this is the loss of retrograde flow to the kidney via the renal vein. Intermittent clamping of the renal artery has also been shown to be more deleterious to renal function than continuous arterial occlusion, presumably due to the intrarenal trapping of vasoconstricting agents (Wilson et al., 1971).

It is now known that renal cooling during periods of ischemia inhibits the metabolic activity of the renal tubular cells, decreases oxygen consumption, and preserves renal function (Harvey, 1959; Levy, 1959). Lowering the renal core temperature to 15–20° C provides maximal protective benefit. All methods of producing hypothermia appear to offer comparable protection.

To cool the kidney we routinely use a physiologic iced slush solution. The completely mobilized kidney is encircled with a rubber sheet and surrounded in iced slush. Although it has been suggested that this technique unreliably cools the kidney to the desired level, no loss of renal function occurring as a direct result of the technique has been reported, and it is excellent in preserving renal function in patients with a solitary kidney and a staghorn calculus (Stubbs et al., 1978).

The application of external cooling appliances and immersion techniques, also successful in reducing the renal temperature, have the added inconvenience of providing only temporary hypothermia (Resnick et al., 1984) and in cases requiring prolonged renal ischemia must be intermittently employed (Shattock, 1905). Perfusion cooling via the collecting system is also possible but produces inhomogeneous hypothermia and is difficult to maintain (Marshall and Blandy, 1974). Transvenous perfusion with an iced solution has also been attempted but has resulted in venous congestion and a deleterious effect on postischemic renal function (Wilhelm et al., 1978).

Transarterial perfusion cooling is probably the most successful technique currently available in terms of providing a rapid and homogeneous cooling of the renal parenchyma while allowing the efflux of blood and metabolic by-products from the vascular tree. Various techniques have been developed, including perfusion with cannulation of the renal artery (Farcon et al., 1974) and percutaneous transcatheter perfusion (Marberger et al., 1978). Both techniques are being used with increasing success in achieving hypothermia at the time of renal stone surgery.

Ongoing studies continue in an effort to develop pharmaceutical agents that will either improve the kidney's ischemic tolerance during periods of hypothermia or obviate the use of renal cooling altogether. Wickham and co-workers (1979) successfully used the purine nucleotide inosine to preserve renal function for periods of ischemia up to 90 minutes. Although the exact mechanism of its function is as yet unknown, inosine is believed to act by maintaining a high intrarenal level of ATP precursors (IMP, ADP, AMP). Once renal blood flow has been restored, these precursors are rapidly converted back to ATP (Fernando et al., 1976). Other agents including vasopressin, allopurinol, cortisone, heparin, and ethacrynic acid have all been experimentally shown to improve the ischemic tolerance of the kidney; however, none of these agents when used alone provides sufficient protection even for short periods of complete renal ischemia (Marberger and Stackl, 1981).

Now no method is as successful as hypothermia in preserving renal function for periods of ischemia of two hours or more.

INTRAOPERATIVE LOCALIZATION OF RENAL STONES

Perhaps nothing is more distressing to the renal stone surgeon than an inability to locate easily and entirely remove the complete renal calculus. Various authors have reported a retained stone rate ranging from 5%–30% following nephrolithotomy (Boyce and Elkins, 1974; Marshall et al., 1975; Singh et al., 1973; Sutherland, 1981; Wickham et al., 1974). In the past, the surgeon has relied on various indirect methods to identify and remove residual stone fragments, including direct palpation, multiple parenchymal needle sticks, and various stone forceps or metal probes. These adjunctive procedures frequently resulted in significant renal trauma and often left the surgeon frustrated and the patient with the stone. Since Braasch and Carmen (1919) first reported the use of intraoperative fluoroscopy to facilitate open stone removal, a number of technological advances have occurred to aid in the identification and removal of small stone fragments. The use of intraoperative radiography and ultrasonography as well as the development of rigid and flexible nephroscopes have greatly enhanced our ability to remove all renal calculi. This technology should be readily available to the stone surgeon (Berte and Resnick, 1984).

With recent improvements in intraoperative x-ray techniques utilizing portable x-ray units and sterile x-ray film, it is now possible to intraoperatively identify calculi as small as 1–2 mm in diameter (Fig 17–9). Although a relatively safe imaging technique, disadvantages include the need to completely dissect the kidney, the ability to visualize the kidney in only two dimensions, prolongation of operative time, and the possibility of trauma to the vascular pedicle (Marshall, 1983). Recently Gil-Vernet and Culla (1984) have developed a technique for obtaining three-dimensional intraoperative radiographs which may further aid in identifying and localizing small residual fragments.

Intraoperative ultrasound is able to localize radiolucent calculi that otherwise might not have been identified using conventional operative radiography. It is also less time consuming and offers the additional advantage of depth determination and AP orientation. Recently, Marshall and associates have reported success using intraoperative ultrasonography to localize stone fragments 2–3 mm in diameter (Marshall et al., 1981). Personnel experienced in the use of ultrasound techniques and in the interpretation of the sonogram are essential if this method is to be employed successfully.

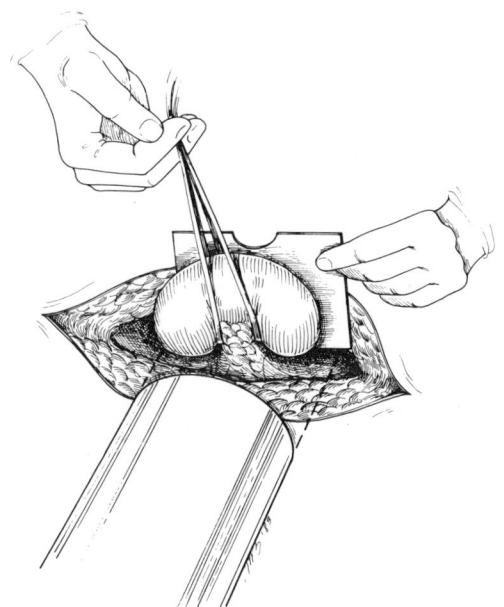

FIG 17–9.
Technique of intraoperative radiography.

The use of intraoperative nephroscopy to visualize the collecting system has also proved helpful in identifying small residual stone fragments; however, even the experienced nephroscopist may accurately visualize only 60% of the renal calyces (Zingg and Futterlieb, 1980). With further experience in the use of these instruments, operative nephroscopy can be expected to become a valuable tool in both identifying and removing small stone fragments.

PYELOLITHOTOMY

Czerny in 1880 is credited with performing the first pyelotomy. Until recently, with the advent and widespread acceptance of percutaneous and extracorporeal lithotripsy, this approach remained the surgical procedure of choice for the removal of moderate-sized unbranched calculi confined to the renal pelvis. Its use for removal of simple renal pelvic stones will certainly diminish and perhaps disappear. As initially described by Czerny the renal pelvic incision was made vertically with total disregard for the anatomical, vascular, and functional capacity of the renal pelvis. In the 1960s, Gil-Vernet, contrary to earlier teaching, demonstrated the renal pelvic musculature to be arranged in circular fashion (Gil-Vernet, 1983). Based on his studies the vertical pyelotomy with all of its potential drawbacks, including pelvic hypotonia, inadvertent extension with damage to the ureteropelvic junction, and persistent urinary drainage, was abandoned in favor of the transverse pyelotomy incision.

The surgical approach to the kidney depends on multiple factors, including the anatomical position of the kidney, the body habitus of the patient, the history of prior renal surgical procedures, the size and location of the stone, and, perhaps most important, on the skill and preference of the operating surgeon. Most urologic surgeons have been trained in the flank approach, and it probably remains the preferred incision for pelvic stone removal. Recently, a renewed interest has developed in the use of the posterior or lumbotomy approach for pelvic stone removal. The lumbotomy incision, with all of its inherent benefits, has in some centers all but replaced the standard flank incision for the removal of the uncomplicated pelvic stone (Gittes and Belldegrun, 1983; Kropp, 1983).

After adequately exposing the perinephric space a posterior longitudinal incision is made in Gerota's fascia and the kidney identified. We attempt to preserve Gerota's vascia with the perinephric fat and reapproximate it at the end of the procedure to prevent the formation of adhesions to the psoas muscle. The upper ureter is identified as it lies on the psoas muscle and a small Penrose drian or vascular loop placed around it. Tightening the Penrose drain and obstructing the ureter prior to stone manipulation ensures against downward migration of stone fragments. In the presence of a solitary pelvic stone it is usually not necessary or advantageous to completely dissect the entire kidney. Instead, the dissection is limited to identifying the proximal ureter and posterior renal pelvis. After identifying the stone, two 4-0 chromic stay sutures are placed in the posterior pelvic wall on either side of the planned transverse pyelotomy incision. The incision is initiated with a curved scalpel blade and may be extended with vascular or Pott's scissors. The incision should be large enough to allow stone removal without tearing the renal pelvis (Fig 17–10). In the presence of a very large stone it is possible to extend the incision in a "visor" fashion onto the anterior pelvic wall. Nerve hooks, spatulae, or vascular forceps are used to free completely the stone prior to its removal. A 10 F straight catheter is passed into the bladder to demonstrate the patency of the ureteropelvic junction and ureter. The renal pelvis is irrigated with saline. The pyelotomy is closed in watertight fashion with 5-0 absorbable suture. A Penrose or suction type drain is placed near the pyelotomy incision and brought out dependently through a separate stab incision.

EXTENDED PYELOLITHOTOMY

In 1965 Gil-Vernet described a modification of the simple pyelotomy incision which allowed additional exposure of the intrarenal collecting system. This incision

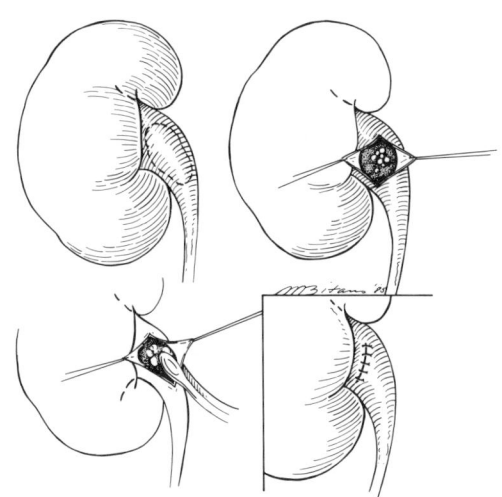

FIG 17–10.
Technique of simple pyelolithotomy.

facilitates the removal of branched calculi without the added inherent risks or blood loss associated with parenchymal incisions (Gil-Vernet, 1965).

Gil-Vernet, in his initial report, utilized a posterior vertical lumbotomy incision. However, in patients with large staghorn calculi associated with chronic urinary

and parenchymal infections, the kidney is frequently embedded in thick scar tissue, and adequate mobilization may be difficult through this limited incision. We therefore routinely use a flank incision through the bed of the 11th or 12th rib.

After entering the retroperitoneal space, Gerota's fascia is opened and the posterior renal pelvis dissected free of all adherent peripelvic fat. We routinely identify and dissect the renal pedicle sufficient to allow rapid clamping or manual compression of the vessels if necessary. By sharply dividing the adventitia that connects the renal parenchyma to the collecting system, access to the renal sinus is achieved. The plane between the posterior renal surface and the renal pelvis is further developed using small vein retractors to rotate bluntly the posterior parenchyma anteriorly and medially (Fig 17–11). A Küttner or "peanut" dissector is helpful in further dissecting the parenchyma from the intrarenal collecting system. Using this technique it is possible to expose all of the infundibula. The only blood vessel that may be jeopardized is the branch to the posterior renal segment, which is usually easily retracted and preserved when the parenchyma is dissected free of the pelvis.

A curved pyelotomy incision conforming to the intra-

FIG 17–11.
Technique of extended pyelolithotomy.

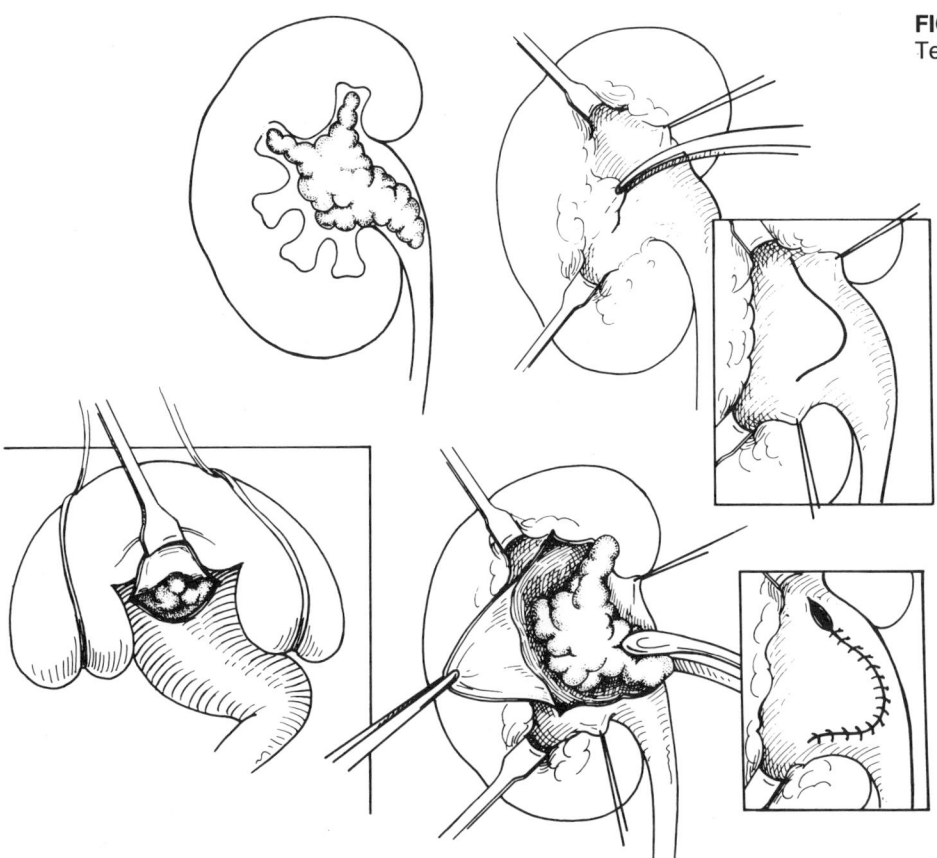

renal collecting system is made with a hooked scalpel. The incisional apex is directed toward the ureteropelvic junction thereby avoiding the tearing of this structure. To promote maximum stone exposure the incision is extended longitudinally onto the upper and lower infundibula. The stone is freed by gentle manipulation under direct vision using a blunt probe or nerve hook and is then removed intact. In cases of very large staghorn calculi, Gil-Vernet has found it helpful to free the pelvic vertex of the stone first followed by removal of the shortest and most easily mobilized infundibular extension (Gil-Vernet, 1983).

When possible, all calyces are explored under direct vision using a nasal speculum and malleable probe. Intraoperative x-rays are routinely obtained to confirm complete stone removal. Small remaining impacted calyceal bell-shaped stones may be removed by adjunctive radial nephrotomy. The use of the flexible nephroscope or intraoperative ultrasound may further aid in identifying stone fragments noted on x-ray. In patients with very large stones requiring extensive intrarenal exposure or fragments that are difficult to localize using routine techniques, clamping the renal artery and cooling the kidney produces a softening of the renal parenchyma that may facilitate stone identification and removal (Wulfson, 1981).

Prior to closure, the collecting system is vigorously irrigated with saline to remove all mucosal calcifications and fragments. The patency of the ureteropelvic junction and ureter is demonstrated by the passage of a 10 F straight catheter into the bladder. The pyelotomy incision is closed with interrupted or running 4-0 or 5-0 absorbable suture. It is not necessary to close the infundibular incision as the renal parenchyma will cover the infundibula. The pelvis is drained either with a Penrose or suction-type drain and the incision closed in routine fashion.

PYELONEPHROLITHOTOMY

An appreciation and accurate understanding of the anatomical relationship of the intrarenal vasculature to the collecting system is useful not only in the performance of anatrophic nephrolithotomy but is applicable to other stone removal procedures as well (Resnick et al., 1981). An extension of the routine pyelotomy incision onto and through the posterior parenchyma utilizing the avascular intersegmental plane between the posterior and inferior renal segments has been well described (Resnick, 1981). It is particularly useful in removing large intrapelvic renal stones with extension into the lower pole collecting system. Because of the anatomical distribution of the inferior and posterior segmental arteries, it is possible to perform a lower pole

nephrotomy without endangering either vessel and without first clamping the renal artery and cooling the kidney.

After completely exposing the kidney and renal pelvis, it is suspended with 1-in. umbilical tapes. The renal pedicle is carefully dissected and the renal artery identified and controlled with a vessel loop. The kidney is then positioned to bring the entire pelvis and posterior surface into the operative field. A preliminary x-ray is obtained to identify any small calculi or lines of stone cleavage which may not be apparent on the routine plain films (Fig 17–12). This initial film also allows comparison with later x-rays and is helpful in deciding whether or not residual stone fragments exist.

If one desires, it is possible to identify the exact location of the intersegmental plane between the posterior and inferior renal segments. This is performed by first clamping the posterior division of the main renal artery with a small vascular clamp and then infusing 20 ml of methylene blue or indigo carmine. The posterior segment will blanch, while the remainder of the parenchyma stains intensely blue. Although this step is not essential, hemostasis will be easier to maintain and parenchymal damage minimized if the incision is directed through this avascular plane.

The posterior renal pelvis is incised with a hooked scalpel, and the pyelotomy extended toward the lower pole infundibulum (Fig 17–13). We attempt to prevent the inadvertent extension of the pyelotomy incision through the ureteropelvic junction by directing the distal part of the incision away from the proximal ureter. The lower pole infundibulum is then identified and the capsule overlying the parenchyma sharply incised. The renal parenchyma is separated by blunt dissection using the back of the scalpel handle. Small vessels are carefully ligated with 5-0 chromic catgut. The pyelotomy incision is extended along the infundibulum but not through the minor calyx. After exposing the entire stone, it is carefully freed from the mucosa using delicate blunt dissection and removed. After irrigating with saline and prior to closing, a repeated x-ray is obtained to ensure complete stone removal. Should a fragment

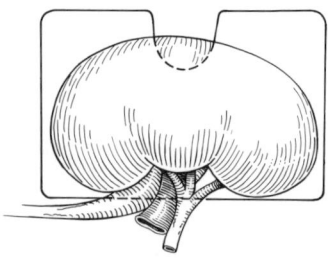

FIG 17–12.
Preliminary x-rays are obtained using a portable x-ray film. Note the notch is placed opposite the renal hilum.

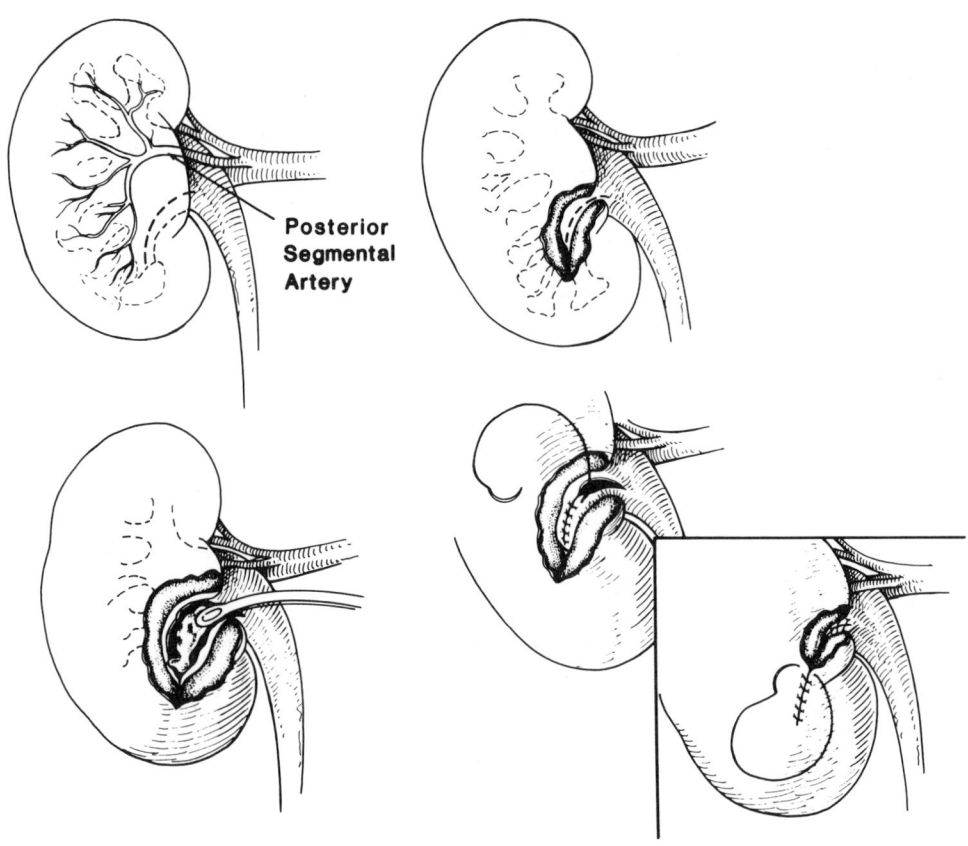

Posterior
Segmental
Artery

FIG 17–13.
Technique of extended pyelonephrolithotomy. Note the position of the posterior segmental artery paralleling but not crossing the lower pole infundibulum.

be identified the flexible nephroscope or intraoperative ultrasound is useful in identifying its location. To promote good drainage and limit urinary extravasation a Silastic ureteral catheter is routinely used.

The infundibulum and renal pelvis are closed with a running 5-0 or 6-0 absorbable suture. The renal capsule is closed with a running 4-0 or 5-0 absorbable suture. No mattress-type parenchymal sutures are used. The kidney is returned to the renal fossa and the lower pole and ureter covered with perinephric fat. The area is well drained and the incision closed.

ANATROPHIC NEPHROLITHOTOMY

The presence of an untreated staghorn calculus is associated with significant patient morbidity resulting from persistent urinary tract infection and obstruction. Patient survival is also reduced if the staghorn is left in situ, with reported mortality rates ranging from 28% to 50% over a 10-year period (Blandy and Singh, 1976; Singh et al., 1973). In approximately 50% of patients with unoperated staghorn calculi, renal function will deteriorate even in the absence of progressive radiographic changes in the size of the stone.

Anatrophic nephrolithotomy is indicated when the surgeon must remove branched or staghorn calculi as-

sociated with infundibular stenosis or when prior stone surgery has been performed, making dissection of the renal sinus difficult, if not impossible. In addition, the procedure is useful in those situations in which a pyelolithotomy is not feasible, such as in a kidney with a small intrarenal pelvis. The term "anatrophic" was coined by Boyce and relates to the segmental distribution of the intrarenal arteries. Division of these vessels results in atrophy of the parenchyma supplied by that vessel; incisions in the renal parenchyma that do not transect these vessels spare this tissue and are therefore not atrophic but anatrophic. The objectives of anatrophic nephrolithotomy include: (1) removal of all stones and stone fragments; (2) repair of obstructed or strictured infundibula and calyces; (3) complete eradication of urinary infection; (4) preservation of a maximal amount of functioning renal tissue; and (5) prevention of recurrent infection and stone formation (Spirnak and Resnick, 1983).

Anatrophic nephrolithotomy with reconstructive calyorrhaphy, as first described by Boyce, is an operative procedure whose principle is based on the segmental arterial blood supply (Smith and Boyce, 1967). When properly performed, the procedure provides for maximal exposure of the intrarenal collecting system and allows for the complete removal of all stones. When indicated, intrarenal reconstructive procedures to maximize urinary drainage can also be performed.

Once the patient has been anesthetized, a Foley catheter is placed in the bladder to allow for the continued monitoring of the renal output. The patient is placed in the lateral flank position, the kidney rest elevated, and the table flexed to provide maximum separation of the ribs from the iliac crest. Three-in. wide adhesive tape is used to fix the patient to the operating table.

Depending on the position of the kidney, the incision is made through the bed of either the 11th or 12th rib. If previous renal surgery has been performed, it is preferable to approach the kidney through an incision placed above the area of prior surgery. This allows for initial renal access to be made through unscarred tissue. Gerota's fascia is identified and opened in a cephalad-caudad direction. By preserving Gerota's fascia the kidney may be replaced in its fatty envelope after completing the procedure. The kidney is mobilized and all perinephric fat carefully dissected off of the renal capsule. The capsule is carefully preserved, and if inadvertently opened, closed with chromic catgut sutures. Following complete mobilization, the kidney is suspended in the operative field with two 1-in. umbilical tapes, and a preliminary portable radiograph obtained (see Fig 17–9).

The main renal artery and its branches are dissected. The relatively avascular plane as described by Hyrtl and Brödel, located between the anterior and posterior renal segments, is identified by clamping the posterior segmental artery and injecting 20 ml of methylene blue through a peripheral vein (Fig 17–14). All of the kidney will stain blue with the exception of the posterior renal segment. In an effort to avoid the complete dissection of the renal artery and all of its inherent risks, Bryniak and Chesley (1981) reported success in identifying the avascular plane by using the Doppler stethoscope. They found the intensity of blood flow to be minimized along the avascular line and corroborated their findings by identifying the anterior division of the renal artery, clamping it, and injecting methylene blue in the usual manner. In all cases there was close correlation between the Doppler findings and the more conventional means of demarcating the avascular line of incision.

A modification of the initial technique described by Smith and Boyce has also been reported by Redman and associates (Redman et al., 1979). Instead of completely dissecting the renal artery and its major divisions, they rely on the relatively constant segmental renal vascular arrangement and perform the anatrophic incision at the expected avascular site after first clamping the renal pedicle with a Satinsky clamp. Although these modifications may prove useful in selected instances, specifically in the reoperated kidney where pedicle dissection may be difficult, we continue to identify the avascular plane between the anterior and posterior renal segments and believe parenchymal loss to be minimized.

The main renal artery is occluded with a noncrushing bulldog vascular clamp. A rubber drape is affixed around the kidney, and the kidney is cooled in an iced saline slush solution. Dry abdominal packs placed on the peritoneum prevent cooling of the intra-abdominal viscera.

Prior to clamping the main renal artery, 25 gm of mannitol is injected IV. In addition to promoting a postischemic diuresis, the mannitol increases the osmolarity of the glomerular filtrate and intratubular fluid, thereby preventing the formation of ice crystals.

The renal capsule is incised sharply over the previously identified avascular plane, and the renal paren-

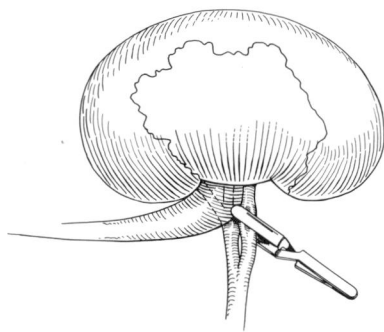

FIG 17–14.
Clamping of the posterior segmental artery followed by the IV injection of methylene blue will identify the appropriate avascular incision site.

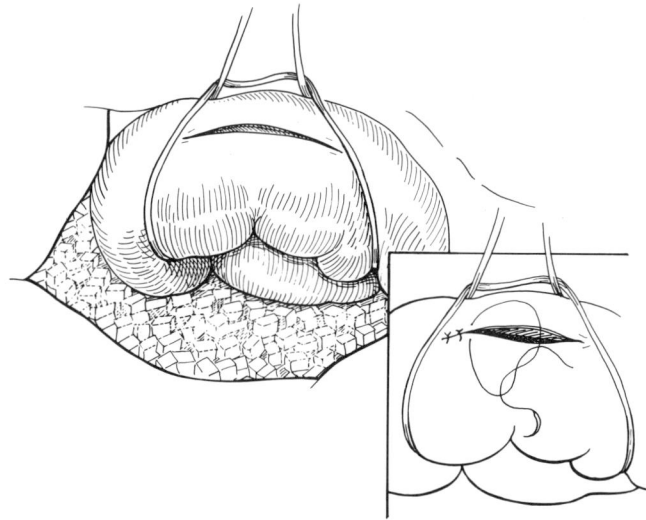

FIG 17–15.
The kidney is suspended in umbilical tapes and cooled in saline slush. The nephrotomy incision is made with the back of a scalpel handle.

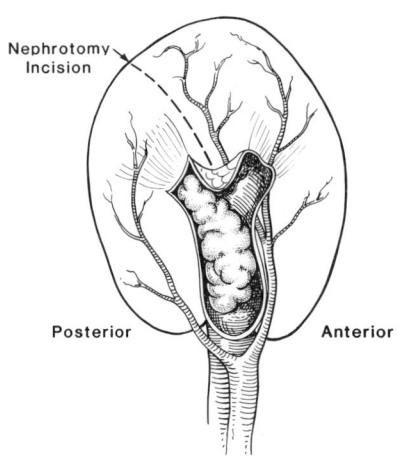

FIG 17–16.
Renal cross-section showing the proper placement of the nephrotomy incision in the avascular plane and the desired site of entry into the collecting system.

chyma is bluntly dissected using the back of a scalpel handle (Fig 17–15). By using blunt dissection, the intrarenal arteries are minimally traumatized. Small bleeding vessels are readily identified and can be controlled with 5-0 or 6-0 chromic catgut suture. If renal back bleeding proves to be a problem, the renal vein may be clamped.

By angling the incision toward the midportion of the renal hilum, it is often possible to remain completely within the avascular plane. Ideally, the collecting system is entered at the base of the posterior infundibula (Fig 17–16).

In the presence of large posterior calyceal stones, a dilated calyx may initially be entered. Using a probe or stone forceps, the remaining collecting system can then be identified and opened.

Recent studies have revealed a dual blood supply to the renal papillae; therefore, when encountered, it is less traumatic to bisect the papillae rather than at-

tempting to circumvent them (Boyce et al., 1979) (Fig 17–17). To minimize stone fractures and reduce the incidence of retained calculi, attempts to remove the stone should not be made until all of the calyceal and infundibular extensions have been identified and the stone mobilized from those structures. After the stone has been removed, the pelvis and calyces are vigorously irrigated and carefully inspected for retained fragments.

Intrarenal reconstruction of the collecting system is performed when scarred or narrowed infundibula or calyces have been identified. Calyorrhaphy refers to the repair of a single scarred calyx and is performed by incising the calyx along its appropriate margin and suturing the walls to the renal pelvis (Fig 17–18). Calycoplasty refers to the repair of adjacent strictured calyces and is performed by suturing the adjacent walls of two separate calyces, thus forming a single structure (Fig 17–19). Posterior calyces are opened along anterior margins and anterior ones are opened posteriorly. Polar calyces are opened along their appropriate lateral margins. These reconstructive procedures effectively act to shorten and widen the lumen and provide free drainage of urine. Size 5-0 or 6-0 chromic catgut is used for all intrarenal repairs. Care should be taken to include only the epithelium and muscularis in the repair, otherwise the underlying interlobular arteries might be inadvertently damaged. The ultimate goal of all intrarenal reconstructive procedures is to provide for a nonobstructed mucosal-lined passageway from the renal papillae into the renal pelvis.

Silastic stents are used routinely and are passed from the renal pelvis into the bladder prior to manipulating the calculus. In addition to providing good drainage

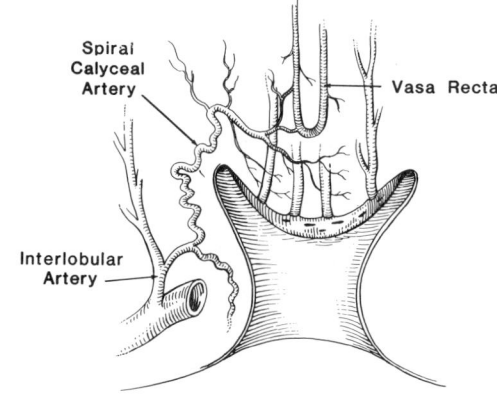

FIG 17–17.
Diagram depicting the dual blood supply to a renal papilla.

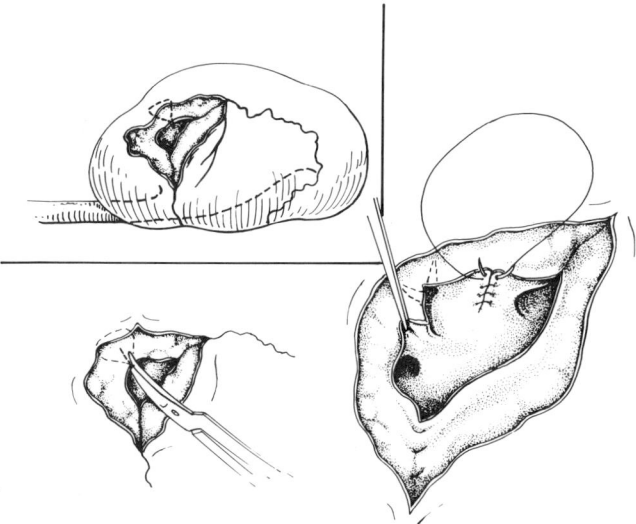

FIG 17–18.
Technique of calyorrhaphy useful in improving the drainage of an obstructed calyx.

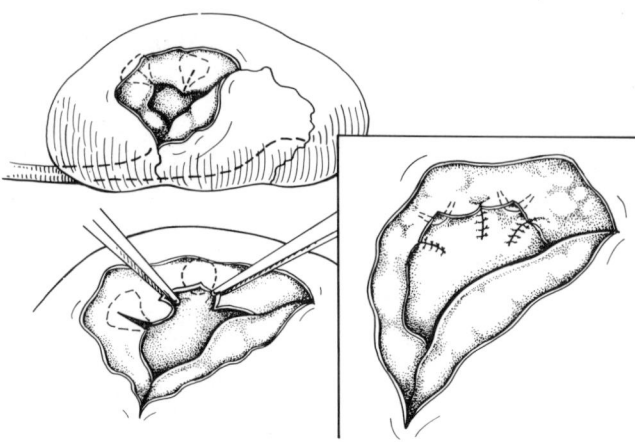

FIG 17–19.
Technique of calycoplasty useful in repairing adjacent strictured calyces.

with a minimal amount of postoperative urinary extravasation, the stent also prevents the migration of small calculi into the ureter. After the complete removal of all identifiable stone fragments, an intraoperative portable x-ray of the kidney is obtained prior to closure. This x-ray will identify any residual stone fragments that should be searched for and removed. The renal pelvis is closed with a running 6-0 chromic catgut suture, and the renal capsule is closed with a running 4-0 chromic suture (Fig 17–20). The use of mattress-type parenchymal sutures leads to tissue ischemia and should be avoided.

Once the renal capsule has been closed and adequate hemostasis obtained, the slush is removed from the kidney and the vascular clamp removed from the renal artery. The kidney is returned to Gerota's fascia. In an attempt to decrease the amount of postoperative scarring and adhesion formation, the kidney and upper ureter are covered with perirenal fat. If Gerota's fascia has

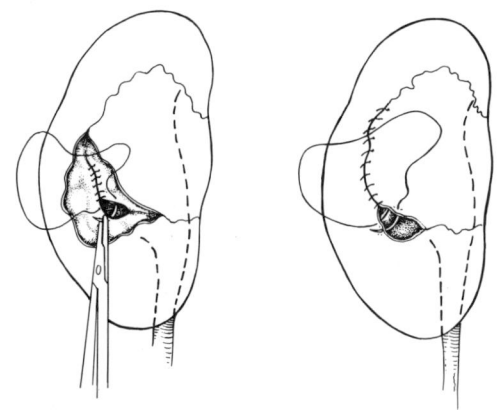

FIG 17–20.
The collecting system and renal capsule are closed in separate layers with fine chromic catgut.

been obliterated due to previous surgical procedures or extensive scarring, the use of a mobilized omental flap will serve the same purpose.

A small Penrose or suction-type drain is placed within Gerota's fascia to drain any fluid collection; this can usually be removed postoperatively within the first 72 hours. The incision is closed in layers with absorbable suture material. Nephrostomy tubes, which may cause further renal damage and act as a nidus for continued infection, are not routinely used.

COAGULUM PYELOLITHOTOMY

Coagulum pyelolithotomy, described by Dees (1943), did not gain widespread urologic acceptance until being rediscovered by Patel 30 years later (Patel, 1973). Since then, in an effort to simplify the technical aspects of the procedure, decrease the associated morbidity, and improve the surgical results, the coagulum recipe has undergone a number of modifications (Broeker and Hackler, 1979; Burns and Finlayson, 1982; Fisher et al., 1980; Kalaoh et al., 1983; Marshall et al., 1978; Roth, 1984; Sherer, 1980). Currently coagulum pyelolithotomy has gained an accepted role in the removal of multiple renal stones.

The clot or coagulum on which the procedure is based is formed by a reaction between local fibrinogen and thrombin. Fibrin monomers organize to form stable fibrin in the presence of both calcium and thrombin. Cryoprecipitate, readily available in most blood banks, is routinely used as the source of fibrinogen and when combined with calcium and thrombin will form a clot of great tensile strength. Calcium chloride serves to neutralize the anticoagulant (citrate) present in the cryoprecipitate and acts as a cofactor to facilitate the conversion of prothrombin to thrombin (Marshall, 1983).

The ideal candidate for coagulum pyelolithotomy has a large extrarenal pelvis with multiple small calculi contained in the intrarenal collecting system. It may also be helpful in the patient with multiple calyceal stones undergoing anatrophic nephrolithotomy or pyelonephrolithotomy.

Prior to performing the coagulum procedure, it is essential that all materials be readily available in the operating room. The kidney is exposed in the usual fashion, the upper ureter identified and occluded with a soft vessel loop. Topical thrombin (5,000 units) is reconstituted with 5 ml of normal saline. Ten ml of a 10% calcium chloride solution is placed in a small sterile basin and to it is added 0.25 ml of the reconstituted thrombin solution. Four units of cryoprecipitate (40–60 ml) and a few drops of methylene blue, useful in facilitating future identification and removal of the entire clot, are placed in a separate sterile container. Heating the solution does not increase the tensile strength;

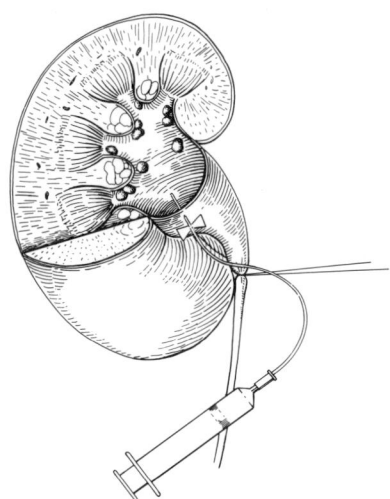

FIG 17–21.
The desired coagulum recipe is injected via a butterfly needle placed in the renal pelvis. Note the vessel loop tightened around the proximal ureter.

therefore, the entire procedure is performed at room temperature.

A 19-gauge butterfly or angiocatheter is positioned in the renal pelvis and all urine aspirated and measured (Fig 17–21). The cryoprecipitate is aspirated in a 50-ml syringe and to this is added 1.0 ml of the thrombin-$CaCl_2$ solution. If the capacity of the renal pelvis is larger than 50 ml, 2.0 ml of the thrombin-$CaCl_2$ solution is used. Further addition of calcium chloride acts only to reduce the set time without increasing the tensile strength and is not recommended (Burns and Finlayson, 1983). The solution is then briefly mixed and the volume corresponding to the predetermined pelvic capacity is injected into the renal pelvis. Overdistention resulting in pyelovenous backflow should be avoided. Clotting typically occurs 30–45 seconds after mixing the solution. After 5 minutes, the renal pelvis is incised, and the coagulum with the entrapped stones is removed. Burns and Finlayson (1982) showed that waiting an additional 5 minutes results in no further significant increase in tensile strength and is therefore not warranted. X-rays are routinely obtained to confirm complete stone removal.

In an effort to further simplify the procedure Kalash and associates (1982) rely on tissue levels of thromboplastin to provide the necessary thrombin and obtain excellent results by injecting only cryoprecipitate and calcium chloride. Marshall (1983) has further revised their technique and performs the procedure by first injecting the predetermined amount of cryoprecipitate into the renal pelvis and then adding 1 ml of 10% calcium chloride. Pyelotomy is performed 5–7 minutes later.

Coagulum pyelolithotomy has generally been considered to be a safe procedure. However, at least one case of a fatal pulmonary embolism directly related to the procedure has been reported (Pence et al., 1982) The pulmonary embolism is believed to have occurred as a result of overdistending the renal pelvis, with subsequent pyelovenous backflow and dissemination of the thrombogenic material into the systemic circulation. Studies on dogs have confirmed this theory (Pence et al., 1982). To avoid this potentially lethal complication, one must carefully measure the pelvic capacity and avoid overdistention of the renal pelvis. Deleting thrombin from the coagulum recipe might further enhance the safety of this procedure (Pence et al., 1982). The risk of hepatitis and allergic reactions has been minimized with the use of cryoprecipitate. Retained clots pose little threat to the patient and will usually be dissolved within 24–48 hours by the urokinase/plasmin-mediated mechanism of clot lysis (Burns and Finlayson, 1982).

RADIAL NEPHROTOMY

The use of single or multiple radial nephrotomy incisions may be extremely useful as a primary procedure to remove peripherally located stones or, as is more often the case, as a complementary or adjunctive procedure useful in combination with any of the already described surgical techniques. The segmental division of the main renal artery as well as their branches progress to the periphery of the kidney in radial-like fashion (Graves, 1954) (Fig 17–22). This radial distribution allows for parenchymal incisions to be made parallel to the renal vessels with a minimal loss of functioning parenchyma.

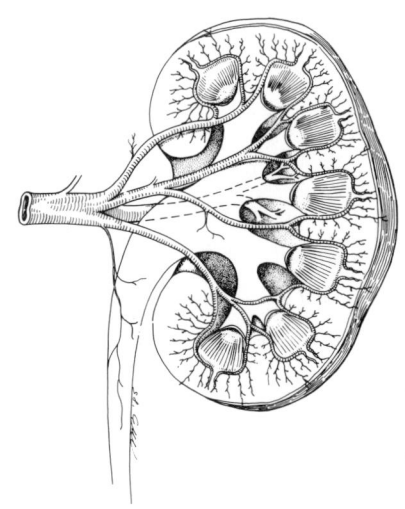

FIG 17–22.
Radial distribution of the renal vasculature.

The kidney is exposed through a flank incision and completely mobilized as in anatrophic nephrolithotomy. The renal artery is identified and controlled with a vessel loop. An intraoperative x-ray is obtained and the stone accurately localized. It is not unusual for peripheral calculi to be easily palpated through areas of thinned-out renal cortex. If the stone is not easily palpated, the use of ultrasound may be helpful in localizing its exact position. Stones associated with normal parenchymal thickness are oftentimes not readily palpable; therefore, clamping the renal artery will soften and shrink the kidney and may facilitate stone removal. The kidney can tolerate warm ischemic times of 10–15 minutes without permanent loss of function (Wickham et al., 1967); however, if multiple nephrotomies or a prolonged ischemic time is anticipated, the kidney should be cooled. Clamping the renal artery not only facilitates stone localization but also allows parenchymal dissection to proceed in a bloodless field.

As with anatrophic nephrolithotomy the renal capsule is sharply incised and the parenchyma bluntly dissected either with the back of a scalpel handle or with small brain or malleable retractors (Fig 17–23). Often the intralobar arteries will be identified, and these may be spared by gentle retraction as the parenchymal incision is deepened. The blunt dissection is continued until the pericalyceal fat is identified. The calyx is sharply incised directly over the stone. The stone is carefully mobilized and removed intact.

To avoid the major renal vasculature the parenchymal incision is made on the convexity of the posterior renal surface. Most calyces can be approached in this manner, and large venous branches crossing on the anterior surface will be avoided (Resnick, 1983). The use of the intraoperative Doppler stethoscope may further decrease the chance of ischemic injury.

The calyx is irrigated and then directly visualized to

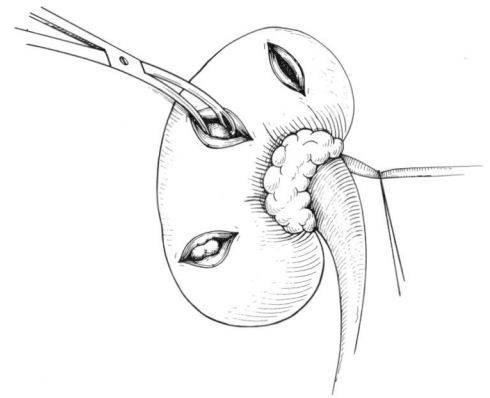

FIG 17–23.
Radial distribution of the multiple nephrotomies and the vessel loop placed around the renal artery.

ensure complete stone removal. X-rays are obtained to identify residual fragments. If no stones remain, the calyx is closed with fine absorbable sutures. If this is not technically possible, simple closure of the capsule including 1–2 mm of parenchyma with fine absorbable suture will suffice. Again, deep mattress sutures are to be avoided. The area is drained and the incision closed in the usual fashion. Utilizing the vascular sparing principles just described, it is possible to perform multiple radial nephrotomies with very little trauma to the renal parenchyma (Wickham, 1983).

PARTIAL NEPHRECTOMY

Partial nephrectomy is a therapeutic procedure for many localized renal conditions including vascular lesions, localized carcinoma in a solitary kidney, trauma, segmental obstruction or infection, and renal stone disease localized to either pole (Coleman and Witherington, 1979). This procedure, when first introduced, was embraced with great enthusiasm, and stone disease became one of the major indications for its use. Advocates believed stone recurrence rates to be reduced when the stone and associated abnormal renal parenchyma were surgically removed. We now know that although partial nephrectomy meets many of the proposed criteria for a suitable stone procedure, in most instances it needlessly sacrifices functioning renal parenchyma and does not necessarily reduce the incidence of future stone formation.

Wells in 1884 is credited with performing the first inadvertent partial nephrectomy in a patient with a perirenal lipoma (Kim, 1969). It was not until 1889 that Kümmell performed the first deliberate partial nephrectomy for stone disease (Kim, 1969) At the turn of the century, surgeons frustrated by the frequent complications of hemorrhage and urinary fistula all but abandoned the procedure (Redman, 1983). Stewart (1952) resurrected interest in the procedure when he demonstrated that about 75% of all stones were formed in either radiographically or histologically abnormal lower pole calyces. In 1960, he published the first large series of over 200 patients treated by partial nephrectomy and noted a 6.4% incidence of ipsilateral stone recurrence in patients followed up from 1 to 8 years (Stewart, 1960). Other studies with similarly short follow-up periods appeared in the literature corroborating Stewart's low recurrence rate (Papathanassadis and Swinney, 1976; Puigvert, 1966). These results, when compared to the 58% ipsilateral recurrence rate reported by Williams, appear remarkable at best and suggest a significant place for partial nephrectomy in the treatment of stone disease (Williams, 1963).

It was not until 1971, when Myrvold and Fritjoffson

reported a 57% recurrence rate following partial nephrectomy in 65 patients followed up for an average of
nine years, that the procedure began to fall into disfavor (Myrvold and Fritjofsson, 1971). Rose and Fellows
(1977) reexamined the initial patient population reported by Stewart, only now with a 20-year follow-up
period, and found essentially no difference in the recurrence rate for the operated and nonoperated kidneys. Marshall and associates (1975) also found no difference in the stone recurrence rate when they
compared the various surgical procedures including pyelolithotomy, nephrolithotomy, and partial nephrectomy for the single small polar stone.

In general, we think that partial nephrectomy most
often needlessly sacrifices salvageable renal tissue and
should be reserved for those patients with severe obstruction associated with parenchymal damage in whom
recovery of renal function is expected to be negligible
or the rare individual with calyceal stones associated
with a polar diverticulum. In our experience it is the
rare stone patient who benefits from partial nephrectomy.

Although several different techniques have been advocated for the performance of partial nephrectomy
(Goldstein and Abeshouse, 1937; Semb, 1955; Wein et
al., 1978; Poutasse, 1962; Kim, 1969; Williams et al.,
1967), no procedure is clearly superior to any of the
others. The kidney is approached through a standard
flank incision and mobilized sufficiently to allow complete access to the involved pole and hilar vessels. The
proximal ureter is identified and traced to the renal pelvis. It is important to mobilize the pelvis sufficiently so
as not to leave diseased tissue behind. The vascular
pedicle is dissected and the branch to the involved pole
identified. Ligating the branch to the involved segment
not only reduces the blood loss but also facilitates tissue
resection by identifying the margins of excision. The
capsule over the involved pole is incised and bluntly
peeled back beyond the proposed margins of excision.
Preserving the capsule will later facilitate closure. The
parenchyma is bluntly transected with the back of a
scalpel handle and encountered vessels sequentially ligated. The collecting system is sharply transected and
the diseased segment removed. Manual compression of
the remaining parenchyma will allow identification and
ligation of bleeding vessels with 5-0 chromic catgut.
The collecting system is closed in a watertight fashion
with a running 4-0 absorbable suture. The capsule is
trimmed and the edges closed over the remaining parenchyma with a 4-0 absorbable suture. If after ligating
all obvious bleeding vessels, the cut parenchymal surface continues to bleed, it is helpful to place a piece of
Gelfoam on the raw surface and reapproximate the cut
edges with horizontal mattress sutures of 2-0 chromic

catgut. These sutures are placed only a few millimeters
from the edge of the surface in an attempt to limit ischemic necrosis of the remaining parenchyma. Prior to
closing, an x-ray is obtained to confirm complete stone
removal. The area is well drained and the incision
closed.

NEPHRECTOMY

Nephrectomy, although meeting many of the objectives of stone surgery, cannot be endorsed as routine
treatment for renal stone disease. Nephrectomy needlessly sacrifices renal tissue, and in as many as one third
of stone patients, calculi will develop in the contralateral kidney (Rose and Fellows, 1977). We believe
nephrectomy should be reserved for the rare patient
with chronic obstruction and a nonfunctioning kidney
as determined by radionuclide renal scan using one of
the glomerular isotopes. If the involved kidney demonstrates a clearance of greater than 20 ml/min, recognizing the fact that renal scanning techniques frequently underestimate renal function in the obstructed
kidney, we usually attempt to preserve that kidney by
removing the stone if the patient is a reasonable surgical candidate. Additionally, the patient with a solitary
kidney and less than a 10 ml/min clearance deserves an
attempt at stone removal, as it is well known that relief
of obstruction will frequently result in a dramatic improvement of renal function (Stubbs et al., 1978; Witherow and Wickham, 1980). Nephrectomy is indicated in
the elderly or debilitated patient with a complex renal
stone, recurrent infection, and a normally functioning
contralateral kidney who is thought to be a poor candidate for a difficult and possibly prolonged operative
procedure.

POSTOPERATIVE CARE

Postoperative management of the stone patient is
similar to that for any patient following a major operation. Intravenous fluids are maintained until the patient
has fully recovered from the effects of anesthesia and is
able to tolerate a clear liquid diet. It is desirable to
maintain a urine output of at least 50 ml/hour to help
clear any debris or small clots from the collecting system. Intravenous antibiotics are selected on the basis of
preoperative urine culture and sensitivity results and,
in the presence of infection or struvite stones, maintained for at least 3–5 days. Appropriate oral antibiotics
are selected based on repeated urine culture results.
The patient who has had an infectious type of stone removed is discharged receiving prophylactic antibiotic
therapy and requires close urologic follow-up. Ambulation is begun on the first postoperative day. Coughing
and deep breathing are encouraged. The use of incen-

tive spirometry and respiratory therapy is routinely used in the patient with pulmonary disease.

Intraoperatively placed drains are removed when all drainage has stopped, usually by the fifth postoperative day. If a plain Silastic ureteral stent has been used, it will usually pass spontaneously prior to discharge. If by the seventh postoperative day it has failed to pass, it is removed cystoscopically. Routine nephrotograms obtained without contrast and repeated IV urography are obtained three months postoperatively. Repeated 24-hour urine collection for calcium, phosphorus, creatinine, and uric acid is obtained at least six weeks from the time of surgery and after the patient has resumed his normal life-style. If metabolic abnormalities are detected, appropriate therapy is then instituted.

COMPLICATIONS

Atelectasis is the most common complication of stone surgery and has been reported to occur in as many as 37% of patients undergoing anatrophic nephrolithotomy (Zuckerkandl, 1908). Since most stone procedures are performed with the patient in the flank position, one would expect a greater incidence of pulmonary complications to occur in the dependent lung; however, Birch and Mims (1975) have reported the incidence of atelectasis to be equally distributed between the two lungs. Clinical experience has shown that postoperative atelectasis may be avoided by a brief period of hyperventilation with vigorous reexpansion of the lungs after returning the patient to the supine position and just prior to removing the endotracheal tube.

Patients with a history of pulmonary disease routinely undergo preoperative pulmonary function testing and instruction in proper breathing techniques. In the postoperative period all patients are encouraged to cough and breathe deeply. The use of intermittent positive-pressure breathing devices, as well as ultrasonic aerosols, has proved helpful in preventing pulmonary complications in high-risk patients.

Pneumothorax occurs in fewer than 5% of patients undergoing stone surgery. Previous surgical procedures and the presence of chronic urinary tract infections with perinephritis can cause adherence of the upper pole of the kidney to the diaphragm and overlying pleura and increase the likelihood of an intraoperative pneumothorax. Additionally, 11th rib or interspace incisions tend to be associated with a higher incidence of pleural injury than 12th rib or subcostal incisions. Inadvertent pleural injury is immediately repaired with a running chromic catgut suture. Prior to the placement of the final stitch, the lung is hyperinflated. If the integrity of the pleural closure is in doubt, a chest tube is placed. A routine chest x-ray is obtained in the recovery room.

Pulmonary embolism is a complication that can follow any surgical procedure. In an effort to decrease the incidence of deep vein thrombosis, all patients are fitted with elastic support hose prior to surgery, and early ambulation is encouraged. In the high-risk patient the use of sequential venous compression stockings during surgery may further decrease the incidence of deep vein thrombosis.

Postoperative renal hemorrhage will occur in fewer than 10% of patients undergoing a renal parenchymal incision (Gil-Vernet, 1965). Properly placed intersegmental nephrotomy incisions with meticulous closure of the collecting system and capsule will usually avoid bleeding complications. When bleeding occurs, it is usually seen on the seventh to 14th postoperative day and is most likely due to failure of the sutures used to close the intrarenal structures. These patients can usually be treated expectantly with fluids and blood transfusions as necessary. Ureteral stents are useful in promoting drainage and relieving associated urinary obstruction. We have found the antifibrinolytic agent aminocaproic acid (Amicar) to be of value in treating postoperative urinary tract hemorrhage. It is given IV until the bleeding stops. Orally administered Amicar is then continued for several days, and the drug is gradually discontinued. Its usage is contraindicated in the patient with suspected disseminated intravascular coagulation (DIC).

In the rare instance where the bleeding fails to respond to conservative measures, arteriography is performed further to define the cause and localize the site of bleeding. If open vessels or arteriovenous fistula are present, embolization of the involved vessel will frequently prove therapeutic. If the bleeding fails to respond to these measures, renal exploration with repair of the damaged vessel is indicated.

In spite of the many intraoperative adjunctive techniques available to avoid ischemic renal injury during stone surgery, one can expect a few patients to demonstrate a postoperative decrease in renal function. Boyce reported a 2% incidence of postoperative renal deterioration following anatrophic nephrolithotomy (Boyce and Elkins, 1974). Traumatic injury to the main renal artery with subsequent stenosis and hypertension has not been a significant problem in patients undergoing stone surgery. Boyce and Elkins (1974) reviewed 100 cases of nephrolithotomy and found that only two patients became hypertensive postoperatively. Neither patient required medication for blood pressure control. By following the previously described intersegmental planes, significant vascular injury can be avoided.

The routine utilization of internal Silastic double-J ureteral catheters following difficult or complicated stone surgery has significantly decreased the incidence of prolonged urinary drainage. If a stent is not used, the flank drain should remain in place until the drainage has stopped. If the drainage persists longer than three weeks, a retrograde urogram is performed to rule out distal obstruction. Passage of a ureteral stent will generally promote drainage and facilitate healing.

Perhaps the most distressing complication associated with stone surgery is the identification of retained stone fragments. The reported incidence varies from 5% to 30% (Boyce and Elkins, 1974; Sutherland, 1981). With recent improvements in intraoperative stone localization techniques, fragments as small as 1–2 mm in diameter can effectively be localized and removed. If, at the time of surgery, retained stone fragments are identified and are unable to be located, we routinely leave a nephrostomy tube. Postoperatively, either irrigation and chemolysis or attempted percutaneous removal may then be performed.

The management of retained stones depends on the composition of the fragment. A variety of solutions and techniques have been used for chemolysis (Sheldon and Smith, 1982). Highly alkaline irrigating solutions such as sodium bicarbonate and tromethamine (THAM) used in conjunction with systemic alkalinization have been successful in dissolving cystine and uric acid stones (Crissey and Gittes, 1979; Rodman et al., 1984).

Struvite or infection-induced stones are soluble in acid solution. Because of the presence of associated infection, urinary irrigation with an appropriate solution must be performed in conjunction with the appropriate systemic antibiotic therapy. Renacidin is a commercially available solution of citric acid, D-gluconic acid, magnesium hydroxycarbonate, magnesium acid citrate, calcium carbonate, and water at a pH of 4.0. Although it has not yet been approved by the FDA for renal pelvic irrigation, it is useful in dissolving retained struvite stones.

Stone irrigation can begin one to two days postoperatively provided a nephrostogram or retrograde pyelogram demonstrates free drainage of contrast medium down the ureter with no extravasation. A variety of techniques are currently available to introduce the irrigants into the renal pelvis. If a nephrostomy tube is in place, we utilize it to deliver the irrigant and allow drainage to occur down the unobstructed ureter. During pelvic irrigation with Renacidin it is important to monitor and maintain an intrapelvic pressure below 25 cm H₂O. An overflow safety apparatus utilizing a CVP manometer can easily be fashioned and will avoid the systemic complications associated with absorption of the irrigant (Cato and Tulloch, 1974; Sheldon and Smith, 1982) (Fig 17–24).

Prior to instituting Renacidin irrigation, urine cultures must be sterile. Appropriate antibiotics are con-

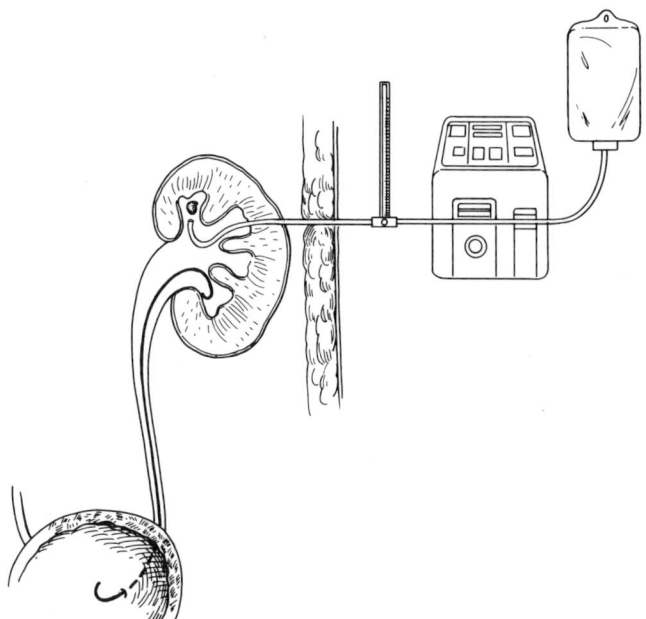

FIG 17–24.
Set-up for the postoperative irrigation of residual stone fragments. Note the CVP monitor in place, helpful in preventing excessive intrarenal pelvic pressures and the excessive systemic absorption of the irrigant.

tinued throughout the duration of irrigant therapy and daily urine cultures obtained. Serum magnesium levels are also monitored. The renal pelvis is irrigated with sterile saline for 24 hours prior to initiating stone dissolution. If the patient complains of pain, becomes febrile, or develops signs of magnesium toxicity, the irrigant is immediately stopped. The rate of infusion should generally not exceed 120 ml/hour. Plain radiographs and/or nephrotograms are obtained to monitor the dissolution process. The procedure is terminated when all stone fragments have been dissolved.

REFERENCES

1. Askari A, Novick AC, Stewart BH, et al: Surgical treatment of renovascular disease in the solitary kidney: Results in 43 cases. *J Urol* 1982; 127:20.
2. Badenoch AW, Darmady EM: The effects of temporary occlusion of the renal artery in rabbits and its relationship to traumatic uraemia. *J Pathol* 1947; 59:79.
3. Bensimon H, Bresette JF, Maxted WC, et al: Misconceptions about posterior approach for renoureteral surgery. *Urology* 1982; 19:462.
4. Berte M, Resnick MI: Intraoperative imaging in renal calculus surgery. *Urol Radiol* 1984; 6:144.
5. Birch AA, Mims GR: Anesthesia considerations during nephrolithotomy with slush. *J Urol* 1975; 113:433–435.
6. Blandy JP, Singh M: The cases for a more aggressive approach to staghorn stones. *J Urol* 1976; 115:505.
7. Bodner H, Briskin H: Subdiaphragmatic renal exposure

by resection of the eleventh rib. *Urol Cutan Rev* 1950; 54:272.

8. Boijsen E: Angiographic studies of the anatomy of single and multiple renal arteries. *Acta Radiol [Suppl]* 1959; 183:1.

9. Boyce WH, Elkins IB: Reconstructive renal surgery following anatrophic nephrolithotomy: Follow-up of 100 consecutive cases. *J Urol* 1974; 111:307.

10. Boyce WH, Russell JM, Webb R: Management of the papillae during intrarenal surgery. *Trans Am Assoc Genitourinary Surg* 1979; 71:76.

11. Braasch WF, Carmen RD: Renal fluoroscopy at the operating table. *JAMA* 1919; 73:1751.

12. Brödel M: The intrinsic blood vessels of the kidney and their significance in nephrotomy. *Johns Hopkins Med J* 1901; 12:10.

13. Broecker BH, Hackler RH: Simplified coagulum pyelolithotomy using cryoprecipitate. *Urology* 1979; 14:143.

14. Bryniak SR, Chesley AE: The use of the Doppler stethoscope in anatrophic nephrotomy. *J Urol* 1981; 126:295.

15. Bundy FE, Ingalls W: Nephrolithotomy. *Boston Med Surg J* 1982; 106:483.

16. Burns JR, Finlayson B: Coagulum pyelolithotomy: Tensile strength of coagula as a function of variables. *Urology* 1982; 19:381.

17. Burns JR, Finlayson B: Coagulum pyelolithotomy, in Resnick MI (ed): *Current Trends in Urology*. Baltimore, Williams & Wilkins Co, 1982, vol 2, pp 31–44.

18. Butt AJ: Historical survey, in Butt AJ (ed): *Etiologic Factors in Renal Lithiasis*. Springfield, Ill, Charles C Thomas Publishers, 1956, pp 3–47.

19. Carlson KP, Bates HB, Boyce WH: Death due to parathyroid crisis. *J Urol* 1960; 84:219.

20. Cato AR, Tulloch AGS: Hypermagnesemia in a uremic patient during renal pelvis irrigation with renacidin. *J Urol* 1974; 111:313.

21. Coleman CH, Witherington R: A review of 117 partial nephrectomies. *J Urol* 1979; 122:11.

22. Collins GM, Taft P, Green RD, et al: Adenine nucleotide levels in preserved and ischemically injured canine kidneys. *World J Surg* 1977; 1:237.

23. Crissey MM, Gittes RF: Dissolution of cystine ureteral calculus by irrigation with tromethamine. *J Urol* 1979; 121:811.

24. Cullen EK, Derge HF: The use of silver wire in opening the kidney: Preliminary report. *Johns Hopkins Med J* 1909; 20:350.

25. Dees JE: The use of an intrapelvic coagulum in pyelolithotomy: Preliminary report. *South Med J* 1943; 49:503.

26. Deming CL: Renal circulation following various types of nephrotomies. *Am J Surg* 1928; 4:424.

27. Deming CL: Renal circulation following various types of elongations of pyelotomy incisions. *J Urol* 1928; 20:713.

28. Eisendrath DN: Technique of enlarged pyelotomy for renal calculi. *Surg Gynecol Obstet* 1923; 36:715.

29. Eisendrath DN, Strauss DC: The effect on the kidney of temporary compression of its vessels. *JAMA* 1910; 55:2286.

30. Elkins IB, Resnick MI: Intrarenal anatomy, in Resnick MI, Parker MD (eds): *Surgical Anatomy of the Kidney*. Mt Kisco, Futura Publishing Co, 1982, pp 17–34.

31. Farcon EM, Morales P, Al-Askari S: In vivo hypothermic perfusion during renal surgery. *Urology* 1974; 3:414.

32. Fernando AR, Armstrong DMG, Griffiths JR, et al: Enhanced preservation of the ischemic kidney with inosine. *Lancet* 1976; 1:555.

33. Fischer CP, Sonda CP III, Diokno AC: Use of cryoprecipitate coagulum in extracting renal calculi. *Urology* 1980; 15:6.

34. Gil-Vernet JM: New surgical concepts in removing renal calculi. *Urol Int* 1965; 20:255.

35. Gil-Vernet JM: Pyelolithotomy, in Roth RA, Finlayson B (eds): *Stones Clinical Management of Urolithiasis*. Baltimore, Williams & Wilkins Co, 1983, pp 297–331.

36. Gil-Vernet JM, Culla A: Intraoperative 3-dimensional radiography of kidney: Modified technique. *J Urol* 1984; 132:872.

37. Gittes RF, Belldegrun A: Posterior lumbotomy: Surgery for upper tract calculi. *Urol Clin North Am* 1983; 10:625.

38. Goldstein AE, Abeshouse BS: Partial resections of the kidney: A report of 6 cases and a review of the literature. *J Urol* 1937; 38:15.

39. Graves FT: The anatomy of the intrarenal arteries and its application to segmental resection of the kidney. *Br J Surg* 1954; 42:132.

40. Harvey RB: Effect of temperature on function of isolated dog kidney. *Am J Physiol* 1959; 197:181.

41. Hevin, P: Recherches historiques et critiques sur la nephrotomie. *OU Tarle Du Rein Mem Acad R Chir* 1778; 3:238.

42. Hodson J: The lobar structure of the kidney. *Br J Urol* 1972; 44:246.

43. Hyrtl J: Das Nierenbecken der Saugethiere und des Menschen. *Naturwissenschaften* 1872; 31:107.

44. Kalash SS, Campbell EM Jr, Young JD: Further simplification of cryoprecipitate coagulum pyelolithotomy without thrombin. *Urology* 1982; 22:483.

45. Kalash SS, Young JD, Harne G: Modification of cryoprecipitate coagulum pyelolithotomy technique. *Urology* 1982; 19:467.

46. Kaye KW, Reinke DB: Detailed clinical anatomy for endourology. *J Urol* 1984; 132:1085.

47. Kerr WS Jr: Surgical management of renal stones with emphasis on infundibulotomy. *J Urol* 1970; 103:130.

48. Kim SK: New techniques of partial nephrectomy. *J Urol* 1969; 102:165.

49. Koletsky S: Effects of temporary interruption of renal circulation in rats. *Arch Pathol* 1954; 58:592.

50. Kropp KA: Surgical approaches to renal and ureteral calculi. *Urol Clin North Am* 1983; 10:617.

51. Levy M: Oxygen consumption and blood flow in the hypothermic perfused kidney. *Am J Physiol* 1959; 197:1111.

52. Marberger M, Georgi M, Guenther R, et al: Simulta-

neous balloon occlusion of the renal artery and hypo-thermic perfusion in in-situ surgery of the kidney. *J Urol* 1978; 119:463.

53. Marberger M, Stackl W: Renal hypothermia in situ, in Resnick MI (ed): *Current Trends in Urology.* Baltimore, Williams & Wilkins Co, 1981, vol 1, pp 70–89.

54. Marion G: La pyelotomie elargie. *J Urol Med Chir* 1922; 13:1.

55. Marshall FF: Intraoperative localization of renal calculi. *Urol Clin North Am* 1983; 10:629.

56. Marshall FF, Smith NA, Murphy JB, et al: A compari-son of ultrasonography and radiography in the localiza-tion of renal calculi: Experimental and operative expe-rience. *J Urol* 1981; 126:576.

57. Marshall S: Coagulum pyelolithotomy. *Urol Clin North Am* 1983; 10:659.

58. Marshall S, Lyon RP, Scott MP Jr: Further simplifica-tions for coagulum pyelolithotomy. *J Urol* 1978; 119:588.

59. Marshall V, Blandy J: Simple renal hypothermia. *Br J Urol* 1974; 46:253.

60. Marshall VF, Lavengood RW, Kelly D: Complete lon-gitudinal nephrolithotomy and the Shorr regimen in the management of staghorn calculi. *Ann Surg* 1975; 162:366.

61. Marshall VR, Singh M, Tresidder GC, et al: The place of partial nephrectomy in the management of renal ca-lyceal calculi. *Br J Urol* 1975; 47:759.

62. Marweder G: Querer Nierensteinschinilt. *Zentralbl Chir* 1907; 34:875.

63. Menon M, Kirshnan CS: Evaluation and medical man-agement of the patient with calcium stone disease. *Urol Clin North Am* 1984; 10:595.

64. Morris H: A case of nephro-lithotomy in the extraction of a calculus from an undilated kidney. *Trans Clin Soc London* 1881; 14:31.

65. Morris H: Quoted in *History of Urology.* Baltimore, Williams & Wilkins Co, 1913, vol 2, p 17.

66. Mayer JH, Heider C, Morris GC Jr, et al: Hypothermia: III. The effect of hypothermia on renal damage resulting from ischemia. *Ann Surg* 1957; 146:152.

67. Myers RP: Brödel's line. *Surg Gynecol Obstet* 1971; 132:424.

68. Myrvold H, Fritjofsson JA: Late results of partial neph-rectomy for renal lithiasis. *Scand J Urol Nephrol* 1971; 5:57.

69. Neely WA, Turner MD: The effect of arterial, venous and arteriovenous occlusion on renal blood flow. *Surg Gynecol Obstet* 1959; 108:669.

70. Norman RW, Bath SS, Robertson WG, et al: When should patients with symptomatic urinary stone disease be evaluated metabolically? *J Urol* 1984; 132:1137.

71. Novick AC: Posterior surgical approach to the kidney and ureter. *J Urol* 1980; 124:192.

72. Novick AC: Renal hypothermia: In vivo and ex vivo. *Urol Clin North Am* 1983; 10:637.

73. Pak CYC: Ambulatory evaluation of nephrolithiasis: Classification, clinical presentation, and diagnostic cri-teria. *Am J Med* 1980; 69:19.

74. Pansadoro V: The posterior lumbotomy. *Urol Clin North Am* 1983; 10:573.

75. Papathanassadis S, Swinney J: Results of partial neph-rectomy compared with pyelolithotomy and nephrolith-otomy. *Br J Urol* 1966; 38:403.

76. Patel JJ: The coagulum pyelolithotomy. *Br J Surg* 1973; 60:230.

77. Pence JR II, Airhart RA, Novicki DE, et al: Pulmonary emboli associated with coagulum pyelolithotomy. *J Urol* 1982; 127:572.

78. Pitts GW, Resnick MI: Urinary stone formation: Patient evaluation and management. *Urol Clin North Am* 1980; 7:45.

79. Poutasse EF: Partial nephrectomy; new techniques, ap-proach, operative indication and review of 51 cases. *J Urol* 1962; 88:153.

80. Puigvert A: Partial nephrectomy for renal lithiasis: Ex-perience with 208 cases. *Int Surg* 1966; 46:555.

81. Redman JF, Bissada NK, Harper DL: Anatrophic nephrolithotomy: Experience with a simplification of the Smith and Boyce technique. *J Urol* 1979; 122:595.

82. Redman JF: Partial nephrectomy *Urol Clin North Am* 1983; 10:677.

83. Resnick MI: Surgery of renal calculi. *AUA Update Series* 1983; 2:lesson 29.

84. Resnick MI: Pyelonephrolithotomy for removal of calculi from the inferior renal pole. *Urol Clin North Am* 1981; 8:585.

85. Resnick MI, Kursh ED, Cohen AM: Computed tomog-raphy in the delineation of uric acid calculi. *J Urol* 1984; 131:9.

86. Resnick MI, Prends DM, Boyce WH: Surgical anatomy of the human kidney and its applications. *Urology* 1981; 17:367.

87. Rodman JS, Williams JJ, Peterson CM: Dissolution of uric acid calculi. *J Urol* 1984; 131:1039.

88. Rose MB, Fellows OJ: Partial nephrectomy for stone disease. *Br J Urol* 1977; 49:605.

89. Roth RA: Surgery of renal calculi. *Semin Urol* 1984; 2:45.

90. Selkurt EE: The changes in renal clearance following complete ischemia of the kidney. *Am J Physiol* 1945; 144:395.

91. Semb C: Partial resection of the kidney; operative tech-nique. *Acta Chir Scand* 1955; 109:360.

92. Shattock SG: A prehistoric or predynastic Egyptian cal-culus. *Trans Pathol Soc London* 1905; 61:275.

93. Sheldon CA, Smith AD: Chemolysis of calculi. *Urol Clin North Am* 1982; 9:121.

94. Sherer JF: Cryoprecipitate coagulum pyelolithotomy. *J Urol* 1980; 123:621.

95. Singh M, Chapman R, Tresidder GC, et al: The fate of the unoperated staghorn calculus. *Br J Urol* 1973; 45:581.

96. Smith MJ, Boyce WH: Anatrophic nephrotomy and plastic calyorrhaphy. *Trans Am Assoc Genitourinary Surg* 1967; 59:18.

97. Spirnak JP, Resnick MI: Anatrophic nephrolithotomy. *Urol Clin North Am* 1983; 10:665.

98. Spirnak JP, Resnick MI: Urinary stones, in Smith DR (ed): *General Urology.* Los Altos, Calif, Lange Medical Publications, 1984, pp 253–278.

99. Stewart BH, Straffon RA: Lithotomy procedure, in Stewart GH (ed): *Operative Urology.* Baltimore, Williams & Wilkins Co, 1975, pp 168–179.

100. Stewart HH: Partial nephrectomy in the treatment of renal calculi. *Ann R Coll Surg* 1952; 11:32.

101. Stewart HH: The surgery of the kidney in the treatment of renal stones. *Br J Urol* 1960; 32:392.

102. Straffon RA: The surgical management of staghorn calculi, in Novick AC, Straffon RA (eds): *Vascular Problems in Urologic Surgery.* Philadelphia, WB Saunders, 1982, pp 99–108.

103. Stubbs AJ, Resnick MI, Boyce WH: Anatrophic nephrolithotomy in the solitary kidney. *J Urol* 1978; 119:457.

104. Stueber PJ Jr, Kovacs S, Koletsky S, et al: Regional hypothermia in acute renal ischemia. *J Urol* 1958; 79:793.

105. Sutherland JW: Residual postoperative upper urinary tract stone. *J Urol* 1981; 126:573.

106. Thomas WC, Wiswell JG, Conner TB, et al: Hypercalcemic crisis due to hyperparathyroidism. *Am J Med* 1958; 24:229.

107. Wein AJ, Carpiniello VL, Murphy J: A simple technique for partial nephrectomy. *Surg Gynecol Obstet* 1978; 146:621.

108. Wickham JEA: Paravascular multiple nephrotomy, in Roth RA, Finlayson B (eds): *Stones Clinical Manage-* *ment of Urolithiasis.* Baltimore, Williams & Wilkins Co, 1983, pp 333–347.

109. Wickham JEA, Coe N, Ward JP: One hundred cases of nephrolithotomy under hypothermia. *J Urol* 1974; 112:702.

110. Wickham JEA, Fernando AR, Hendry WF, et al: Intravenous inosine for ischaemic renal surgery. *Br J Urol* 1979; 51:437.

111. Wickham JEA, Hanley HG, Joekes AM: Regional renal hypothermia. *Br J Urol* 1967; 39:727.

112. Wilhelm E, Schrott KM, Krönert E, et al: Transvenous perfusion cooling of the kidney: A new technique of local renal hypothermia. *Invest Urol* 1978; 16:87.

113. Williams DF, Schapiro AE, Arconti JS, et al: A new technique of partial nephrectomy. *J Urol* 1967; 97:955.

114. Williams RE: Long term survey of 538 patients with upper tract stones. *Br J Urol* 1963; 35:416.

115. Wilson DH, Barton BB, Parry WL, et al: Effects of intermittent versus continuous renal arterial occlusion on hemodynamics and function of the kidney. *Invest Urol* 1971; 8:507.

116. Witherow R, Wickham JE: Nephrolithotomy in chronic renal failure: Saved from dialysis! *Br J Urol* 1980; 52:419.

117. Wulfson MA: Extended pyelolithotomy: The use of renal artery clamping and regional hypothermia. *J Urol* 1981; 125:467.

118. Zingg EJ, Futterlieb A: Nephroscopy in stone surgery. *Br J Urol* 1980; 52:33.

119. Zuckerkandl O: Ueber der Diagnose und Operation und Nierensteinen. *Arch Klin Chir* 1908; 87:481.

Chapter 18

Percutaneous Stone Removal

JOHN C. HULBERT, M.D.
PAUL H. LANGE, M.D.

Urinary calculi have been a major affliction since prehistoric times, and attempts to alleviate the resultant suffering gave birth to the specialty of urology. Initial attempts at removal of bladder stones without anesthesia were not only a credit to the speed of the medieval lithotomists but also to the fortitude of the long-suffering patients. The development of anesthesia and antiseptic surgical techniques, however, made it possible to remove bladder stones with less haste and fewer complications, and subsequently allowed techniques to be developed for the removal of stones in the ureter and renal pelvis. The refinement of these techniques in this century, the use of potent antibacterial agents, a more precise knowledge of renal arterial anatomy, and use of methods of localized renal hypothermia have permitted the safe surgical removal of most renal and ureteral calculi. Simultaneous developments in the techniques of endoscopy have allowed successful manipulation of calculi in the bladder and the lower ureter by the transurethral route.

Until recently the accepted criteria for the surgical management of renal and ureteral calculi have been to use open surgical techniques for calculi in the renal pelvis and upper two thirds of the ureter and to use retrograde transurethral techniques for most calculi in the lower third of the ureter. Recent advances in the management of urinary calculi have so revolutionized the approach to these problems that many securely established techniques are becoming obsolete.

The first published account of percutaneous instrumentation of the renal pelvis was in 1941 (Rupel and Brown, 1941). A patient had required the surgical placement of a nephrostomy tube in a solitary kidney to alleviate calculus anuria; several days later, a cystoscope was gently introduced down the nephrostomy tube tract and a stone visualized and removed from the area of the ureteropelvic junction. It is significant that comment was made at the time of the surprising lack of bleeding and the mature appearance of the nephrostomy tract. However, such a technique could not become widely used until less invasive techniques for inserting the nephrostomy tubes were developed. Initial experience with percutaneous drainage of the upper urinary tract (Goodwin et al., 1955) was in poor-risk patients with large hydronephrotic kidneys. It was not until more sophisticated radiologic imaging techniques such as fluoroscopy and ultrasound were developed that percutaneous puncture became more routine, so that now even undilated renal pelves can be punctured with relative impunity.

Percutaneous puncture of the collecting system requires a clear understanding of the surface anatomy of the kidney, of the intrarenal arterial anatomy, and of the relationships of the calyces to one another (Kaye, 1983). The flank should be punctured posterolaterally to avoid possible injury to surrounding viscera (Fig 18–1) and the collecting system entered through the lateral aspect of the renal parenchyma, rather than through the renal pelvis, to puncture through the avascular plane and facilitate passage of a guide wire into the ureter. Opacification of the collecting system may be achieved by the IV infusion of radio contrast or by prior insertion of a retrograde catheter to allow retrograde infusion. Ultrasound is also useful for localizing the collecting system, particularly in the neonatal period, when it can be used at the bedside or in a bassinet (Stanley et al., 1983).

Passage of a guide wire into the collecting system is a basic principle of any form of intrarenal manipulation (Fig 18–2); in addition, a second wire should be placed alongside in the collecting system as a safety wire in case access to the kidney is lost (Fig 18–3). A nephrostomy tube may be inserted over the working wire (Fig 18–4), and nephrostomy tubes changed merely by removing the existing tube over such a wire and passing the replacement tube over the same wire. Insertion of small drainage tubes in this fashion is routine; in addition, this small conduit will allow for techniques such

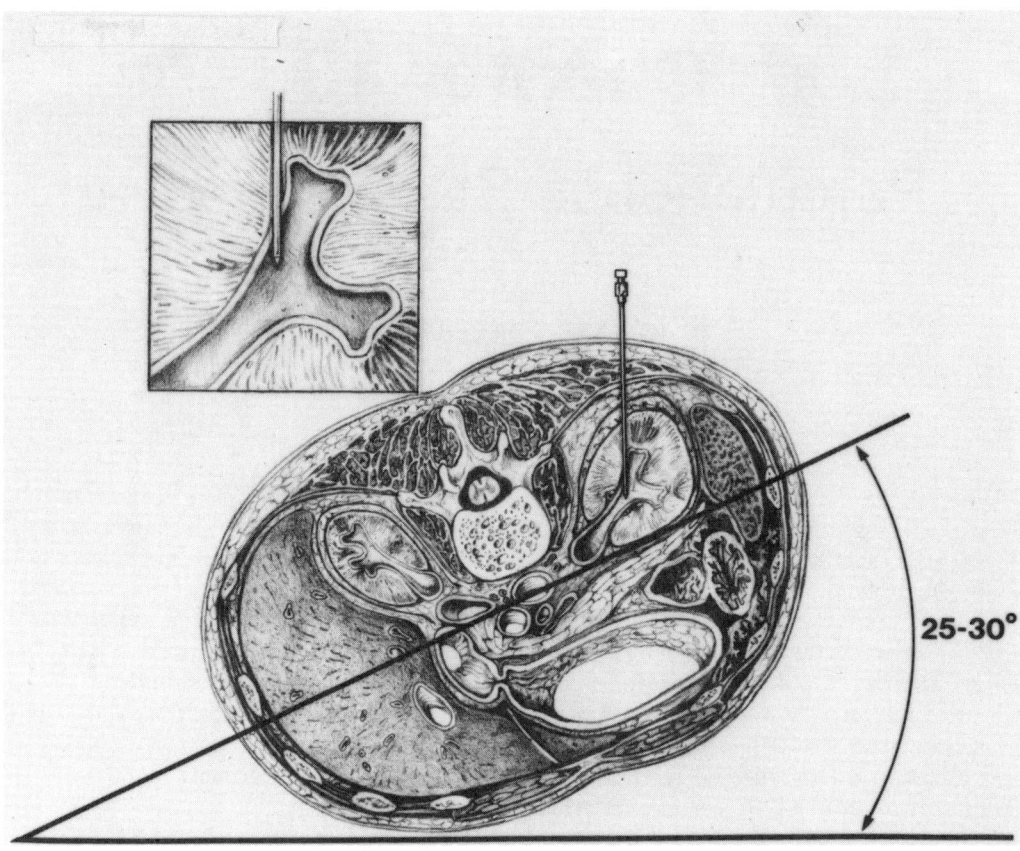

FIG 18–1.
The cross-sectional anatomy of the kidney to illustrate the safest site for percutaneous puncture of the collecting system. (From Coleman CC, Castaneda-Zuniga WR, Hunter DW, et al: A logical approach to renal stone removal. *AJR* 1984; 143:609–615. Used with permission.)

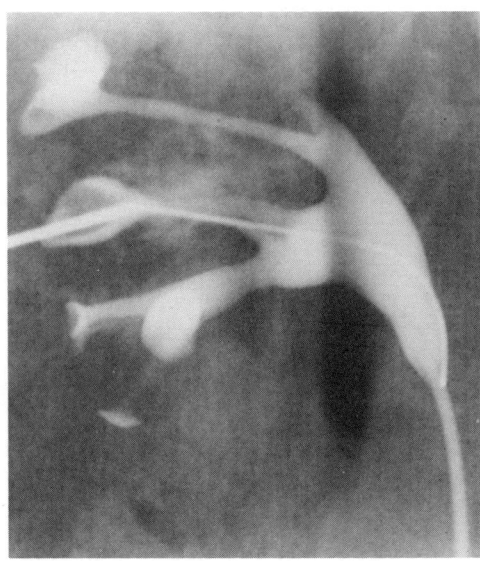

FIG 18–2.
Puncture into the calyx of the kidney with guide wire passed into the renal pelvis.

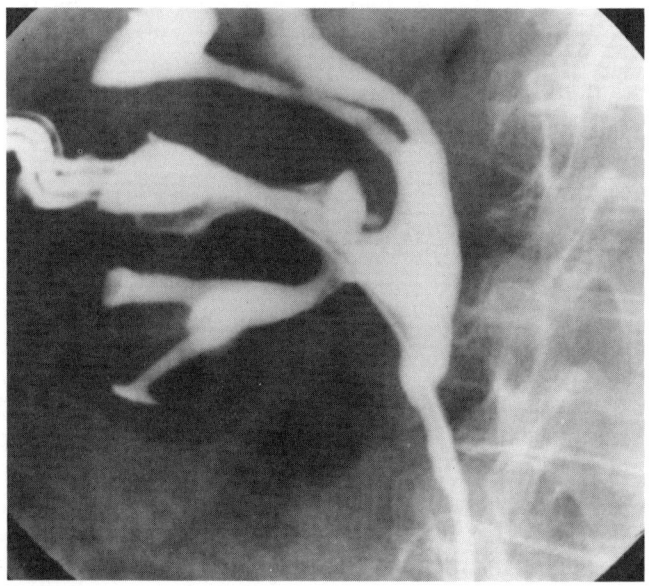

FIG 18–3.
Nephrostomy tube in place with safety wire positioned down ureter.

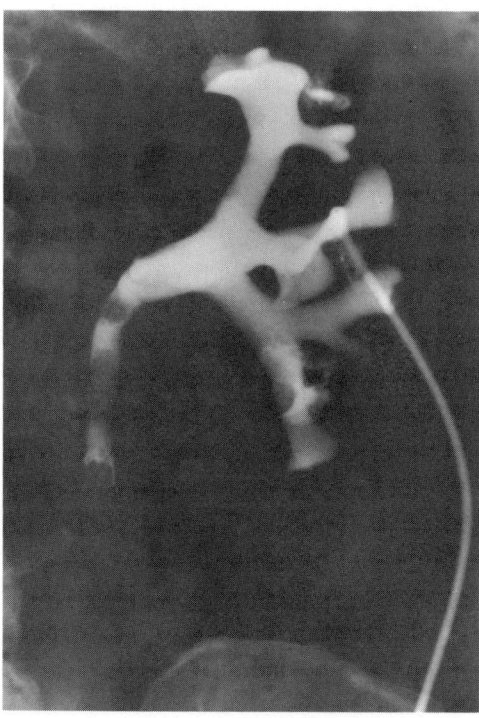

FIG 18–4.
Small nephrostomy tube in collecting system of obstructed kidney.

FIG 18–5.
Complete set of Teflon dilators required for tract dilatation.

as pressure perfusion studies to evaluate equivocal upper urinary tract obstruction (Whitaker, 1979) and percutaneous irrigation with medication for the management of some types of renal calculi (Sheldon and Smith, 1982) or certain upper urinary tract infections (Mazer and Bartone, 1982).

TRACT DILATATION

The development of safe techniques of percutaneous puncture and drainage of the upper urinary tract created the possibility of managing renal calculi by this route. Initial attempts through undilated tracts utilizing stone baskets in an attempt to engage small ureteral calculi met with some success (Smith et al., 1979), but it became clear that the tract would have to be dilated sufficiently to allow insertion of instruments more easily. The idea of dilating a tract to the caliber of 28–30 F through the renal parenchyma was considered with some trepidation at first. Initial efforts at dilatation were performed over several days with modified urethral dilators, but soon it became apparent that dilatation could be performed more rapidly. Teflon dilators were designed as a coaxial system enabling dilatation of the nephrostomy tract with dilators of increasing size up to 30 F in a single session (Fig 18–5). More recently

the adaptation of angioplasty balloon dilators for this purpose has allowed an even more rapid method for tract dilatation (Clayman et al., 1983a).

MAINTENANCE OF THE TRACT

Maintenance of the tract following dilatation can be achieved by three methods. First is the insertion of a nephrostomy tube and awaiting maturation of the tract before instrumentation (usually 5–6 days); the tract will then have an appearance similar to that of the urethra, with a surprising lack of bleeding. This is favored by the less experienced, as the absence of hemorrhage and blood clot optimizes visibility within the renal pelvis. Secondly, insertion of the nephroscope sheath over a guide wire immediately after dilatation, without a Teflon sheath, is preferred by some. It is a suitable alternative for the more experienced but has a disadvantage in that repeated passage of the instrument in and out of the kidney, such as for the removal of multiple small calculi, cannot be made easily; it requires tedious reinsertion of the instrument over a guide wire each time. A more favored technique is the insertion of a Teflon sheath over the largest of the dilators, providing a smooth conduit between the interior of the collecting system and the skin surface (Figs 18–6 and 18–7), that additionally achieves hemostasis by tamponading the nephrostomy tract and allows repeated nontraumatic passes with different instruments in and out of the kidney. This technique permits the whole procedure to be completed in one stage and is the method of choice at our institution.

PATIENT PREPARATION AND ANESTHESIA

The patient is admitted on the morning of surgery after abstinence from fluids or solids since the preceding night. After a history and physical examination, rou-

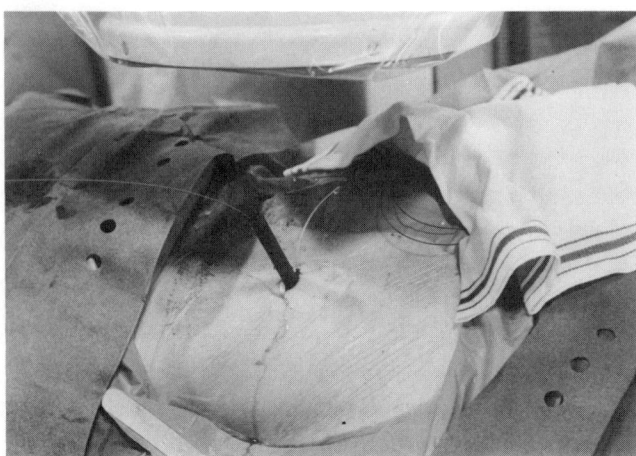

FIG 18–6.
External view of Teflon sheath after dilatation and insertion within the collecting system. Note the safety catheter alongside the sheath and the presence of a guide wire through the sheath.

tine laboratory studies include CBC, serum electrolytes, BUN, creatinine, coagulation studies, and urinalysis and urine culture. Blood is routinely crossmatched. After appropriate premedication and administration of IV antibiotics, a retrograde catheter is placed in the renal pelvis by either rigid or flexible endoscopy using local anesthesia; this will allow retrograde infusion of contrast material that will both outline

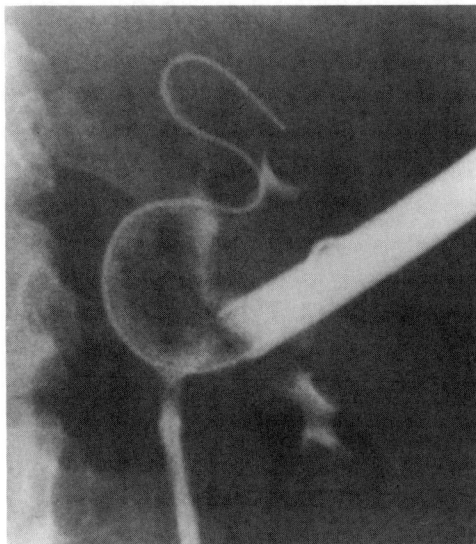

FIG 18–7.
Radiologic view of sheath into collecting system outlined with contrast. Note a safety wire coiled in upper pole calyx of kidney. (From Hulbert JC: Percutaneous nephrolithotomy: Endoscopic access and manipulation, in Amplatz K, Lange PH: *Atlas of Endourology.* Chicago, Year Book Medical Publishers, 1986, pp 219–226. Used with permission.)

and distend the collecting system. The patient is then transferred to the operating room. Types of anesthesia for percutaneous stone removal vary, but it has been our practice to perform most using local anesthesia with IV sedation. Local anesthetic in the form of 0.25% bupivacaine is infiltrated subcutaneously down to the renal capsule. This is combined with IV administration of sedatives and narcotics by an anesthesiologist. Exceptions are children and young men who are routinely given a general or epidural anesthetic. Many institutions, however, prefer general or epidural anesthesia for all cases, and the method can depend on the surgeons' and anesthesiologists' preference. The patient is positioned prone on the table with the appropriate flank prepared and draped so that only the puncture site is exposed; a drainage bag placed beneath this site and attached to the drapes will catch any irrigation fluid or blood and prevent it from leaking onto the table or the floor (Fig 18–8). A sterile table with all catheters, nee-

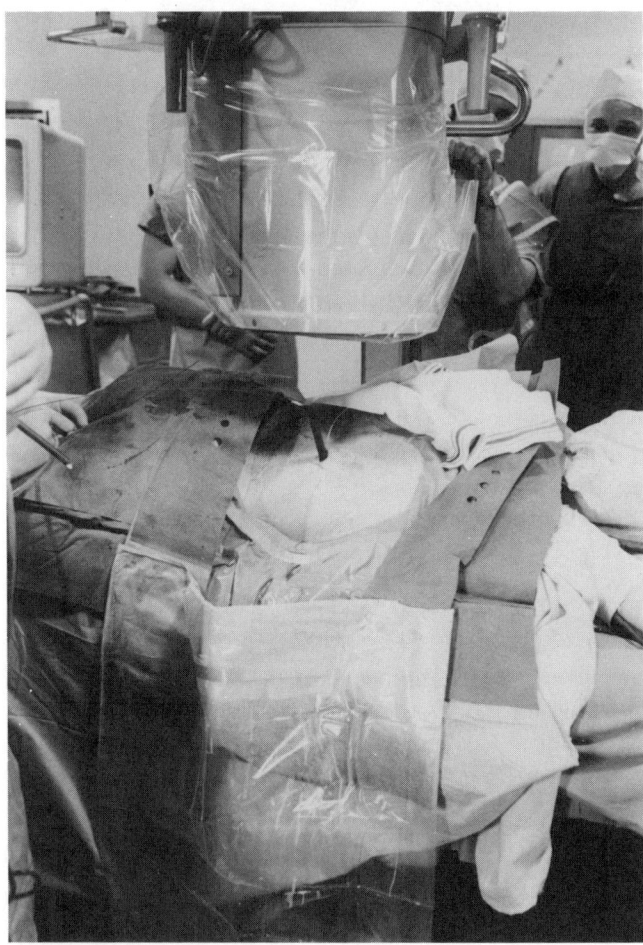

FIG 18–8.
Operative field for percutaneous stone removal showing position of fluoroscopic C-arm and waterproof drapes and drainage bag adjacent to percutaneous puncture site.

dles, and dilators should be on hand for tract establishment, and a second cart with all endoscopic instruments, light sources, and ultrasonic generators should also be prepared.

INSTRUMENTATION

Calculi can be manipulated under fluoroscopic control, endoscopic control, or by a combination. Fluoroscopic techniques include the use of baskets or forceps (Randall's or Mazzariello-Caprini) (Clayman et al., 1983b). However, these techniques have been less used since the refinement of percutaneous endoscopic instruments. Initially we performed endoscopy with a cystoscope, using grasping forceps to remove small calculi. Subsequently, specialized endoscopic instruments with triradiate grasping attachments allowed slightly larger stones to be removed intact under direct vision (Miller and Wickham, 1984), but it was clear that larger stones still could not be removed unless they were broken up. At first we used electrohydraulic shock wave lithotripsy to shatter stones into fragments by direct application of the electrohydraulic probe to the stone, but the disadvantage of this technique was that the fragments had to be removed piecemeal (there was no method of suction to remove the fragments) and on occasion the fragments would move to inaccessible parts of the kidney, where removal would be very difficult. Ultrasonic lithotripsy, which has largely superseded electrohydraulic lithotripsy, is used with specifically designed nephroscopes with offset eyepieces to allow the insertion of the rigid ultrasonic probes (Figs 18–9 and 18–10). It permits more gentle pulverization of the stone and allows fragments to be sucked out through the center of the probe as lithotripsy is taking place (Marberger, 1983). The rigid instruments, however, may not be able to reach all the kidney through the nephrostomy tract. Subsequent advances in fiberoptics resulted in the development of the flexible nephroscope, which in skilled hands can be used to examine all areas of the kidney (Fig 18–11). Small calculi may be grasped with grasping forceps smaller than 5 F cir-

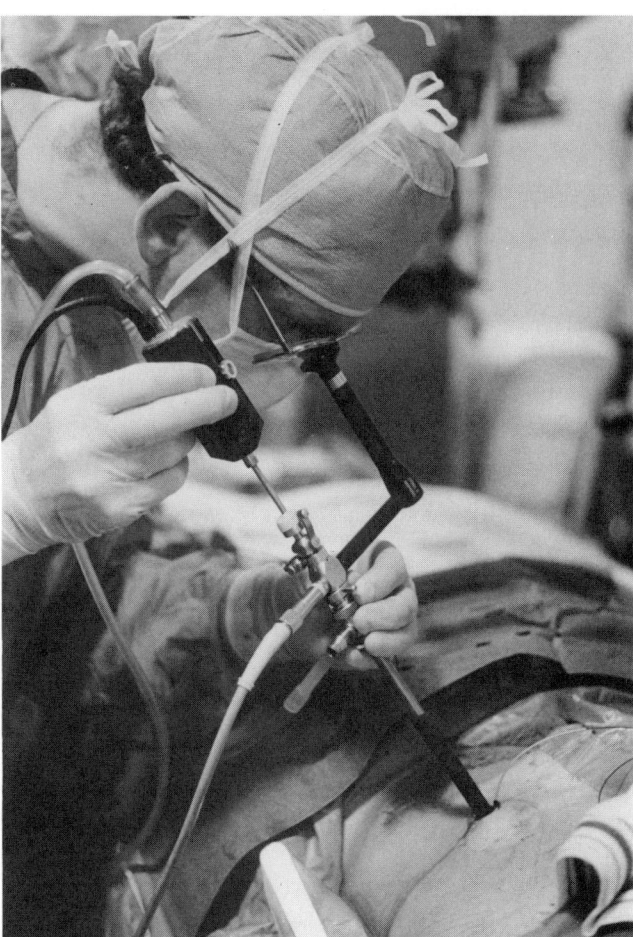

FIG 18–10.
Nephroscope with ultrasound in use intraoperatively.

cumference, and snares and baskets may be passed to engage stones or at least move them to areas of the kidney where they may be reached more easily with rigid instruments. Frequently a pressure bag must be placed around the irrigation bag when using the flexible nephroscope with an instrument through the instru-

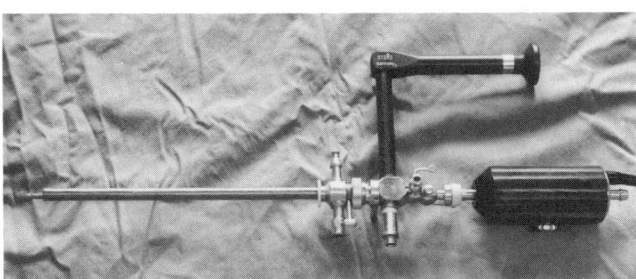

FIG 18–9.
Nephroscope with ultrasonic sonotrode in place.

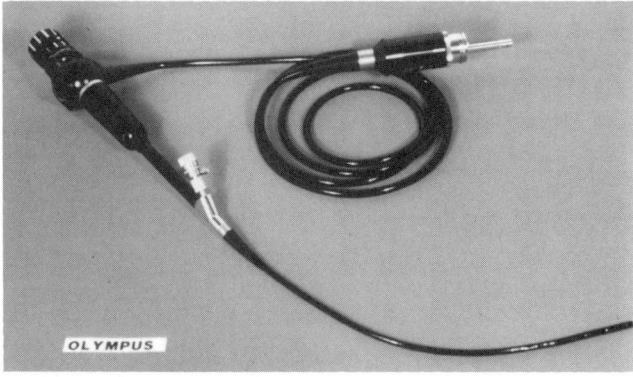

FIG 18–11.
Flexible nephroscope.

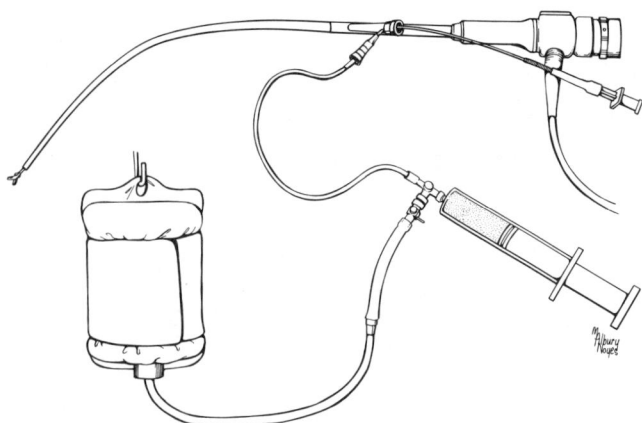

FIG 18–12.
Assembly used to improve flow through flexible nephroscope during instrumentation. (From Lange PH, Reddy PK, Hulbert JC, et al: Percutaneous removal of caliceal and other "inaccessible" stones: Instruments and techniques. *J Urol* 1984; 132:439–442. Used with permission.)

mentation channel, as the presence of an instrument in the channel will reduce the flow of irrigation to almost zero unless pressure is applied (Fig 18–12 and Table 18–1).

Before embarking on percutaneous stone removal it is wise to gain the confidence of a radiologist or, alternatively, to develop the appropriate skills oneself to become familiar with guide wires and catheters. No renal manipulation should take place without the presence of a safety wire alongside the working wire in the renal pelvis or down the ureter, and one's initial cases should be straightforward stones less than 2 cm in diameter situated in the renal pelvis. Indeed, it may be most appropriate initially to perform the technique in two stages to allow optimal vision within the renal pelvis. Frustration, disappointment, and disaffection with the technique may result if the novice endourologist prematurely attempts removing more difficult stones.

CALYCEAL CALCULI

Calyceal calculi are frequently asymptomatic and may not require surgical removal unless they are enlarging, result in infection, or are causing pain by obstructing

TABLE 18–1.
Comparative Flow Rates Through Flexible Nephroscope With and Without Graspers

	FLOW RATE, ML/MIN	
PRESSURE, MM HG	WITHOUT GRASPERS	WITH GRASPERS
0	125	17
150	375	43
300	575	83

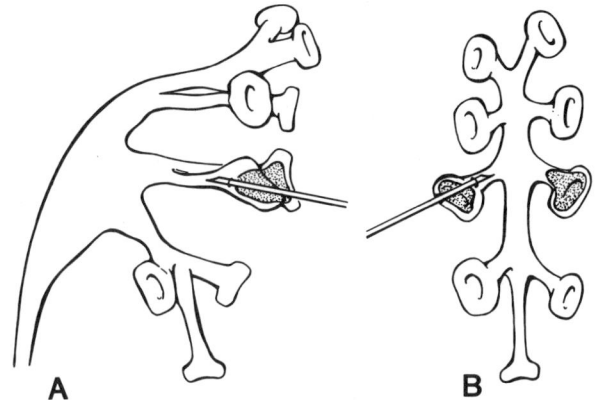

FIG 18–13.
Optimal site for puncture of a calyceal stone is immediately medial to the stone in the calyceal infundibulum.

the calyceal infundibulum or intermittently obstructing the ureteropelvic junction. An important exception are airline or military personnel who may be grounded from flying because of the roentgenographic appearance of a stone in the kidney. Successful management of calyceal calculi requires precise puncture; it is preferable to puncture into the calyceal infundibulum immediately medial to the stone (Fig 18–13), so that following dilatation, the Teflon sheath will trap the stone in the calyx and prevent its migration into the renal parenchyma or into another less accessible part of the kidney (Lange et al., 1984). Calculi located in anterior calyxes require puncture into the common infundibulum of that calyx, and then the stone can be reached with either a rigid or flexible nephroscope. If the calyx is punctured directly, it is very difficult to negotiate a wire back into the pelvis because of the steep angulation encountered, and it may prove impossible to dilate adequately with-

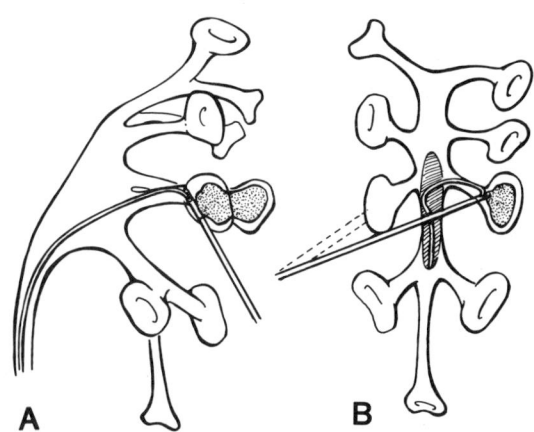

FIG 18–14.
When the stone is in an anterior calyx, direct puncture into the calyx itself may make it impossible to manipulate a guide wire into the main part of the renal pelvis because of acute angulation.

out pushing the stone into the renal parenchyma (Fig 18–14) (Coleman et al., 1984). Visualization within a freshly dilated tract into a calyx requires patience as the narrow confines of the calyx demand familiarity with intrarenal anatomy and great gentleness during instrumentation; even minor trauma to the wall of the calyx may make it impossible to engage the stone successfully. Calculi in the middle and upper calyx were initially approached indirectly—that is, by puncture into the lower calyx—and then using the flexible nephroscope to visualize the stone. However, the flexible nephroscope requires considerable experience, and only small instruments can be used to engage stones; furthermore, it is often difficult to open graspers or baskets in a small calyx sufficiently to engage the stone; a loop snare may be useful in this instance. To puncture directly into the upper calyxes, the puncture may have to be established from between the 11th and 12th rib. This was avoided at first because of fears of intrapleural complications; however, it is now performed routinely with few complications (see below) (Young et al., 1985a). More than one puncture into the kidney may be necessary if stones occupy more than one calyx (Fig 18–15,A); this may involve a completely fresh puncture, or if the targeted calyx is close to the existing tract but inaccessible, a Y puncture off the existing tract is useful (Fig 18–15,B) (Lange et al., 1984).

Calculi in calyceal diverticula present a special problem; they can be approached directly or indirectly (Fig 18–16). The direct approach involves puncture into the diverticulum; however, it may be necessary to dilate the tract with a guide wire coiled precariously in the diverticulum only, if the narrow opening of the diverticulum cannot be negotiated with a guide wire (Fig 18–17). The chance of losing access during this manip-

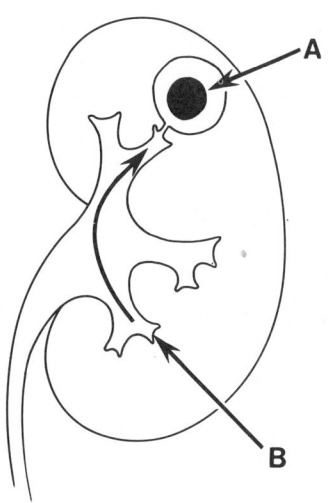

FIG 18–16.
A, the direct approach to calyceal diverticula. **B,** the indirect approach to calyceal diverticula. (From Hulbert JC, Reddy PK, Hunter DW, et al: Percutaneous techniques for the management of caliceal diverticula containing calculi. *J Urol* 1986; 135:225–227. Used with permission.)

ulation is therefore significant, and it should be performed only by those with considerable experience. A wire may be passed into the renal pelvis under direct vision once the tract has been dilated; the stone can then be removed and the small opening of the diverticulum can then be dilated or incised and stented with a wide caliber nephrostomy tube. Following dilatation and stenting, it appears that the diverticulum then shrinks and disappears after removal of the nephrostomy tube. The indirect approach to a calyceal divertic-

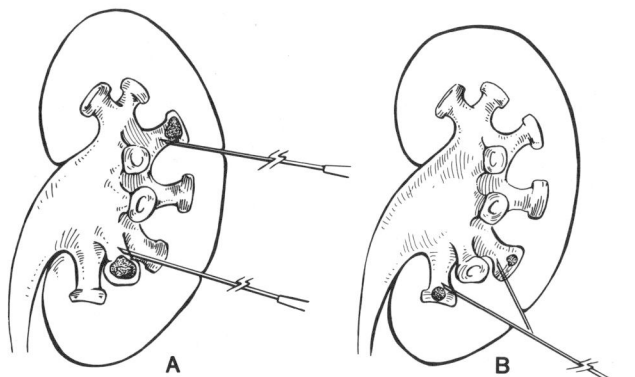

FIG 18–15.
A, principle of double puncture. **B,** principle of Y cutaneous puncture. (From Lange PH, Reddy PK, Julbert JC, et al: Percutaneous removal of caliceal and other "inaccessible" stones: Instruments and techniques. *J Urol* 1984; 132:439–442. Used with permission.)

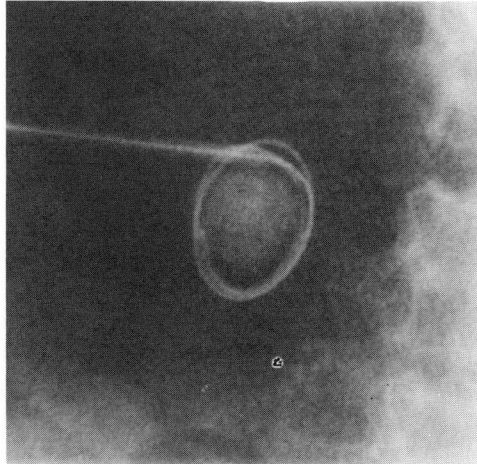

FIG 18–17.
Direct puncture onto a stone in the calyceal diverticulum with wire coiled in the diverticulum only. (From Hulbert JC, Reddy PK, Hunter DW, et al: Percutaneous techniques for the management of caliceal diverticula containing calculi. *J Urol* 1986; 135:225–227. Used with permission.)

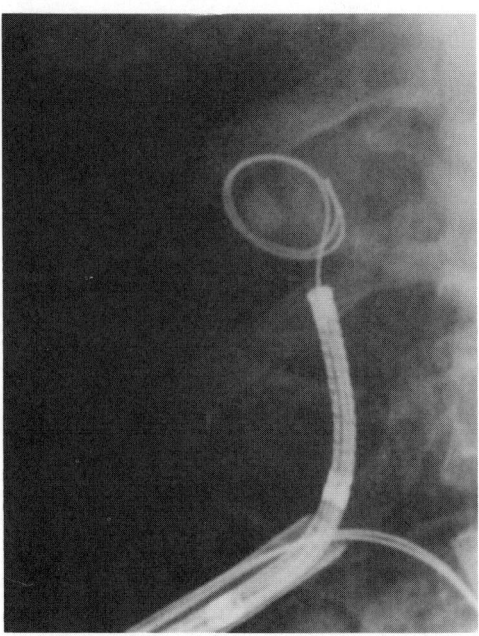

FIG 18–18.
Indirect approach to calyceal divertuculum utilizing flexible nephroscope and passing wire through ostium of calyceal diverticulum. (From Hulbert JC, Reddy PK, Hunter DW, et al: Percutaneous techniques for the management of caliceal diverticula containing calculi. *J Urol* 1986; 135:225–227. Used with permission.)

ulum involves puncture through a calyx remote from the site of the diverticulum and then locating the small opening of the diverticulum with a flexible nephroscope (Fig 18–18). It can be incised with a cutting electrode or dilated with a balloon enough to allow the diverticulum to be entered and the stone engaged with grasping forceps or a basket (Hulbert et al., 1986).

STAGHORN CALCULI

Staghorn calculi present a major challenge (Fig 18–19) and should not be attempted percutaneously without considerable experience of endourologic techniques generated by experience with less difficult cases (Hulbert et al., 1985a). These calculi tend to be infected, and prophylaxis with IV antibiotics is essential. The optimal site for puncture of the collecting system is determined by careful inspection of the stone and its relations to the collecting system by using C-arm fluoroscopy. The puncture onto the stone itself is straightforward, as the stone is usually easy to see. However, great difficulty may be encountered attempting to pass a guide wire beyond the stone into the renal pelvis and down the ureter. Retrograde distention of the collecting system with contrast medium may help, but if this is not possible, the wire can be buckled up

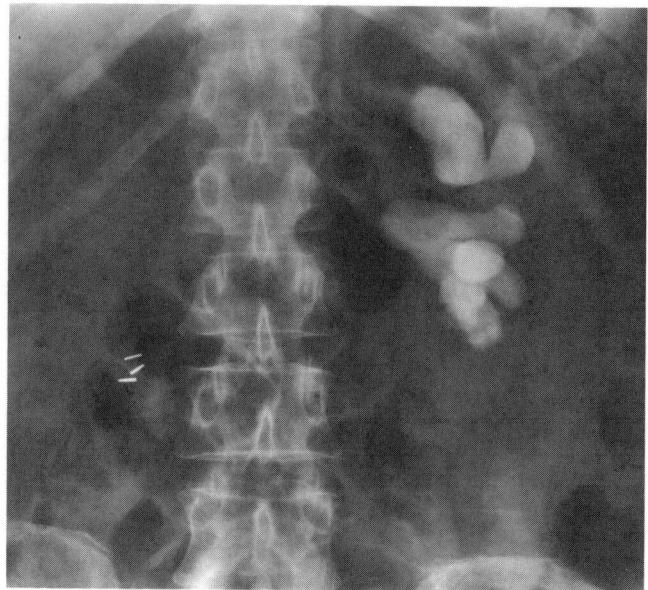

FIG 18–19.
Large staghorn calculus successfully managed percutaneously.

beside the stone, a second wire passed to maintain access, and the initial wire manipulated past the stone with appropriately curved catheters. Occasionally this procedure fails, in which instances the wire can be coiled within the calyx and the surrounding parenchyma and the tract very carefully established; the collecting system is then penetrated endoscopically and enough stone disintegrated to allow the wire to be passed into the collecting system more adequately (Young et al., 1985b).

The use of double and Y punctures is often appropriate to remove all of the calculus, particularly since chronic inflammation may have resulted in sequestration of large stone fragments within the parenchyma within pseudodiverticula. These large stones may require more than one session for their complete removal. The flexible nephroscope is very useful in these instances because it can be used to inspect the interior of the collecting system carefully to locate and pinpoint residual fragments so that the approach can be planned more precisely. Patients may be discharged between sessions with nephrostomy tubes in place and readmitted on the morning of subsequent procedures. Percutaneous irrigation with hemiacidrin may also be used in attempts to dissolve small fragments, although its use in shrinking large stones is questionable; it is wise in these instances to have two tubes in the collecting system to allow through-and-through drainage. This reduces the risk of severe electrolyte imbalance, which occasionally ensues with this type of treatment.

URETERAL CALCULI

Stones in the ureter are common and may well pass spontaneously; intervention is reserved for those that are associated with infection, severe pain, and obvious impaction. In the lower ureter retrograde techniques are considered more appropriate, particularly with the development of ureterorenoscopy. However, even in the most experienced hands, only a 50% success rate can be expected with stones in the middle and upper ureter when approached with the ureterorenoscope (Huffman et al., 1983); percutaneous techniques, therefore, are highly suitable for these stones. However, difficulties may be encountered when attempting to manipulate ureteral calculi in antegrade fashion. Wires and baskets may not pass down the ureter past the stone because of edema, and baskets and graspers may not open sufficiently to engage the stone. Significant trauma to the ureter may result in perforation and extravasation, especially in cases where the stone has been impacted for some time. A technique that has proved effective is controlled retrograde flushing through a 7 F retrograde ureteral catheter with carbon dioxide or liquid contrast (Fig 18–20) (Hulbert et al., 1985b). This has the effect of dislodging and propelling the stone back into the renal pelvis, where it can be grasped easily (Fig 18–21). (Of course, no attempt at

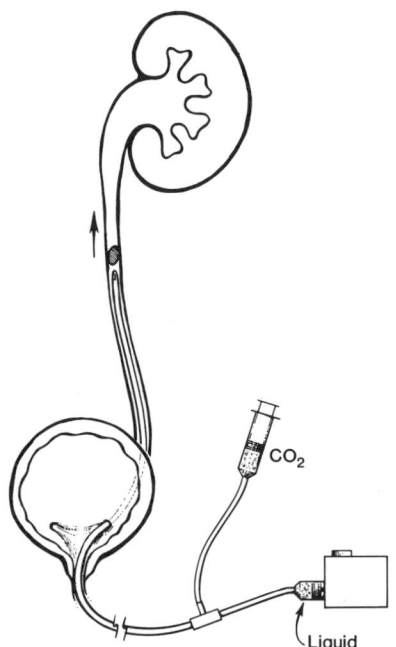

FIG 18–20.
Apparatus for retrograde flushing of ureteral stones. (From Hulbert JC, Reddy PK, Hunter DW, et al: Percutaneous management of ureteral calculi facilitated by retrograde flushing with carbon dioxide or diluted radiopaque dye. *J Urol* 1985; 134:19–32. Used with permission.)

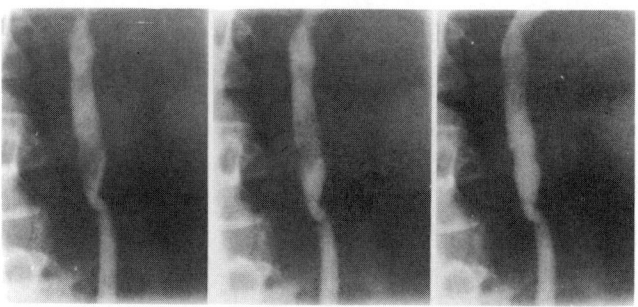

FIG 18–21.
Sequential cinefluoroscopic films of ureteral stones being flushed up the ureter into the renal pelvis. (From Hulbert JC, Reddy PK, Hunter DW, et al: Percutaneous management of ureteral calculi facilitated by retrograde flushing with carbon dioxide or diluted radiopaque dye. *J Urol* 1985; 134:29–32. Used with permission.)

flushing is made until adequate percutaneous access is established because of the risk of blowing out calyces.) This technique has been successful at our institution and it is routinely used for upper and middle ureteral calculi.

OTHER SPECIAL SITUATIONS
CALCULI IN TRANSPLANTED KIDNEYS

Calculi in transplanted kidneys are a rare but significant cause of diminution in renal allograft function (Hulbert et al., 1985c). Open surgical techniques are associated with a marked increase in morbidity because of the delay in healing associated with immunosuppression. Percutaneous techniques are well suited to transplant kidneys because of the superficial location of the graft. Antegrade puncture, pressure perfusion studies, and stenting have become accepted techniques for the management of complications of renal transplantation (Hunter et al., 1983) and stone removal is clearly a natural extension of these.

CALCULI IN RENAL PAPILLAE

Papillary calculi are usually associated with medullary sponge kidney or distal renal tubular acidosis. Many are asymptomatic; intervention should be reserved for those with recurrent ureteric colic. Calculi within papillae may be plucked out by careful use of the flexible nephroscope and grasping forceps (Fig 18–22); however, in many instances so many collecting ducts contain small calculi that attempts to remove all are counterproductive (Fig 18–23).

CALCULI IN PEDIATRIC PATIENTS

Calculi in the upper urinary tract in the pediatric age group in the United States are rare and are usually as-

FIG 18–22.
Management of papillary calculi may involve plucking them out of the collecting tubules with the aid of a flexible nephroscope. (From Reddy PK, Lange PH, Hulbert JC, et al: Percutaneous removal of caliceal and other "inaccessible" stones: Results. *J Urol* 1984; 132:443–447. Used with permission.)

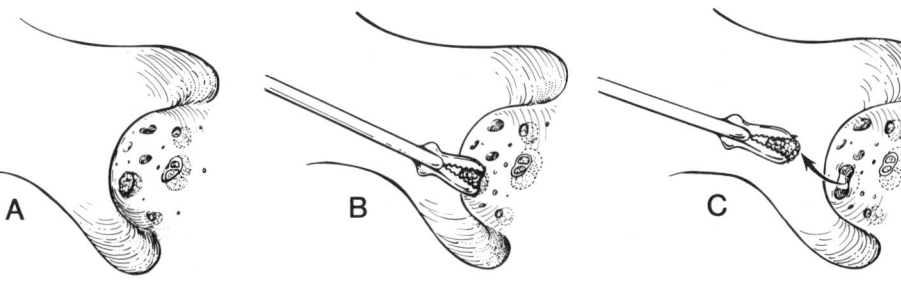

sociated with a hereditary metabolic defect. However, percutaneous techniques can be readily applied to children above the age of about 7 or 8 years (Hulbert et al., 1985d). Younger children have kidneys of significantly smaller size, and concern exists that establishing tracts of 24–30 F through kidneys of this size can damage an appreciable portion of the kidney. However, the development of smaller-caliber instruments and smaller-caliber tracts into the kidney should make this a more suitable technique for children of a much younger age group in the future.

INTRARENAL ELECTROSURGERY

Intrarenal electrosurgery involves techniques of percutaneous incision of strictures within the collecting system (Clayman et al., 1984) and can be helpful when

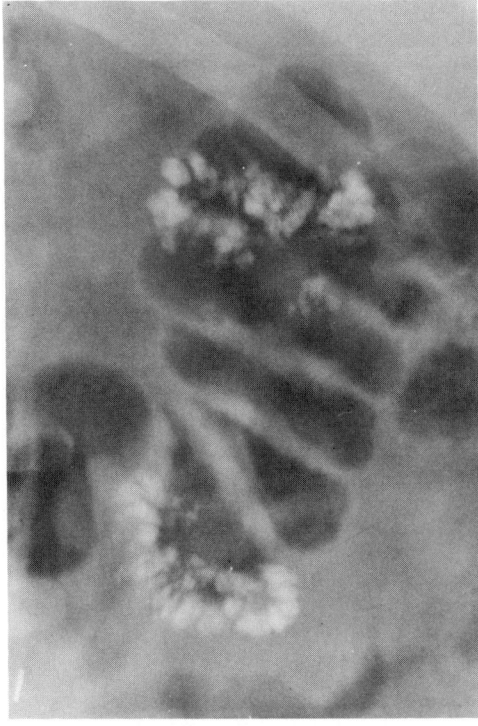

FIG 18–23.
Advanced papillary calculus is not amenable to percutaneous techniques.

calculi are trapped behind stenosed infundibula within calyceal diverticula or actually embedded within the renal parenchyma (Figs 18–24 and 18–25). Incision should be performed with great care using a fine electrode, the fibrous bands being incised fiber by fiber with a low cutting current. Knowledge of and attention to the distribution of the intrarenal vessels with relation to infundibula is essential to prevent significant hemorrhage. These techniques can be useful aids to stone removal and can also be applied to the successful management of strictures in the absence of calculous disease located in the ureter, at the ureteropelvic junction or at the calyceal infundibulum.

COMPLICATIONS OF PERCUTANEOUS NEPHROLITHOTOMY

Despite the lower morbidity from percutaneous stone removal than from open stone removal, complications do occur. The most significant is hemorrhage. Hemorrhage at the time of the procedure is not infrequent, but adherence to basic principles will prevent major problems (Clayman et al., 1984). A safety wire should always be present alongside the working wire to allow the rapid insertion of a nephrostomy tube of wide caliber should bleeding occur. This will have the effect of tamponading any bleeding. If this fails, the nephrostomy tube should be removed and a balloon catheter inserted and inflated until hemorrhage ceases; it should be left inflated for several minutes and then deflated again. Usually hemorrhage will have ceased and a ne-

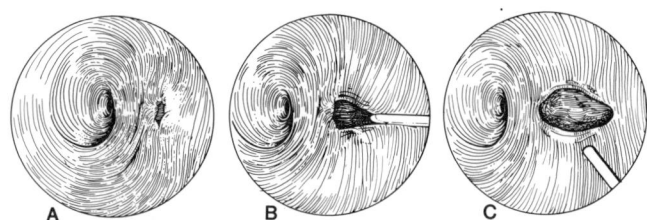

FIG 18–24.
Stenosed infundibulum being incised to reveal calculus beneath. (From Hulbert JC, Reddy PK, Young AT, et al: Reappraisal of the management of staghorn stones. *World Urol Update Series,* vol 2, lesson 23, 1985. Used with permission.)

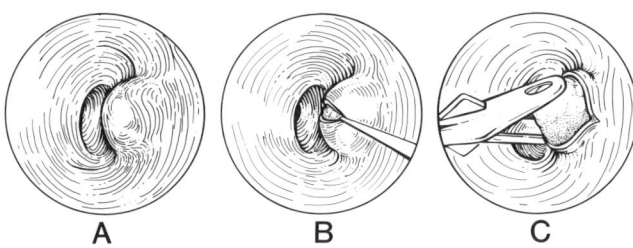

FIG 18–25.
Stone embedded in substance of kidney being released by intrarenal electrosurgery. (From Reddy PK, Lange PH, Hulbert JC, et al: Percutaneous removal of caliceal and other "inaccessible" stones: Results. *J Urol* 1984; 132:443–447. Used with permission.)

phrostomy tube can be reinserted. If bleeding persists, the balloon can be reinflated and left in place overnight on the ward. When significant bleeding does occur, the procedure should be terminated and plans made to continue only after the tract has matured for several days. It is fortunately very rare that drastic measures such as emergency embolization or nephrectomy must be employed in the presence of acute hemorrhage.

Delayed hemorrhage—that is, days or weeks after the procedure—is usually due to the formation of arteriovenous aneurysms, which are rare (no more than 0.5–1%) (Patterson et al., 1985). In such instances ther-

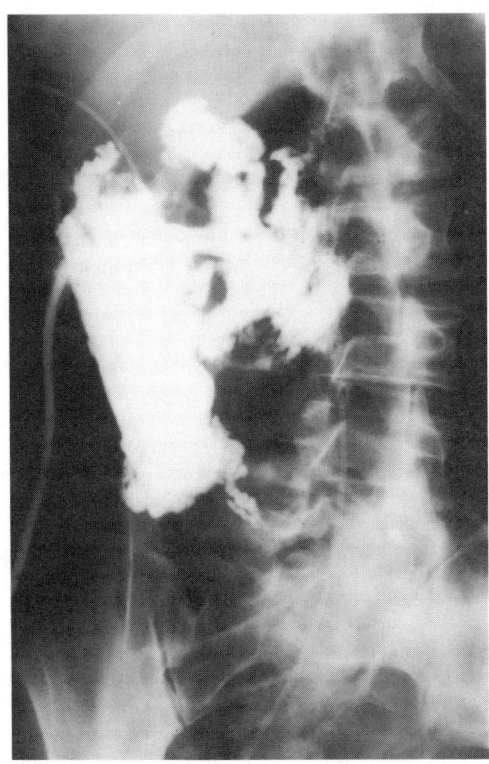

FIG 18–26.
Severe extravasation during percutaneous manipulation.

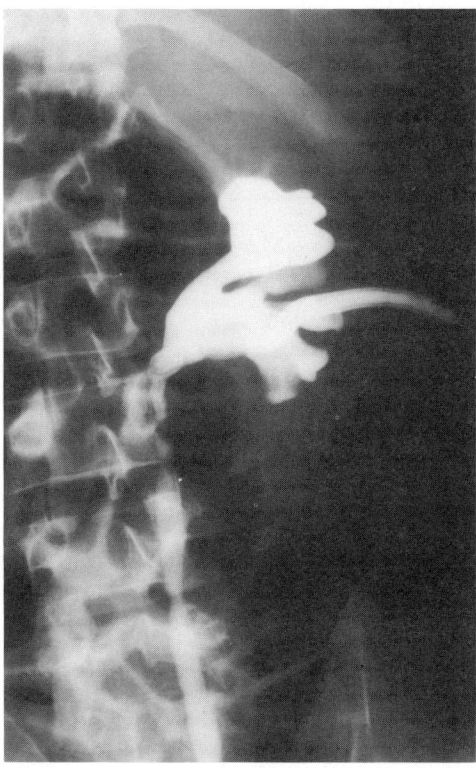

FIG 18–27.
Nephrostogram after four days and nephrostomy drainage showing complete resolution.

apeutic embolization will control bleeding adequately; partial or total nephrectomy need not be employed unless this fails.

Extravasation is a common feature and can appear alarming, but should not be cause for alarm so long as adequate drainage of the collecting system is achieved; if trauma to the ureter is sustained then a stent should be passed down the ureter. Healing is very rapid under these circumstances (Figs 18–26 and 18–27); if drainage is not adequate perirenal and periureteral fibrosis with scarring and stricture formation may occur.

Pleural effusion is seen as a complication of punctures between the 11th and 12th ribs. It is not as common as originally anticipated (about 5% of cases) and almost always subsides spontaneously. The use of a Teflon sheath in both fresh and mature tracts should prevent leakage of irrigation fluid into the pleural space during stone manipulation. The necessity for thoracocentesis is very rare, and in our experience no case of pneumothorax has occurred (Young et al., 1985a). A chest x-ray in the recovery room following the procedure is important to determine that neither hydrothorax nor pneumothorax has developed. Other potential complications include damage to adjacent viscera, injuries to the spleen, liver, and colon having been recorded. Fluid overhydration may occur, but can be obviated by using normal saline as irrigation fluid, with sterile water only used for cases where intrarenal electrocautery is required. Paralytic ileus is rare provided the collecting system is adequately drained, and the potential for other complications, such as pulmonary embolism, myocardial infarction, and deep vein thrombosis, exists as in any other type of surgical procedure. Care and experience will limit the effects of complications, and it is essential to have the necessary equipment to deal with sudden complications.

ADVANTAGES OF PERCUTANEOUS STONE REMOVAL

The advantages of percutaneous nephrostolithotomy over open surgical techniques are clear. First, the procedure may be performed using local anesthesia (although in many cases it is performed using general anesthesia). Second, no major flank incision is required, and the convalescence as a result is very much shorter. Indeed, many of our patients have returned to work almost immediately after being discharged from the hospital. Third, the ability to place a nephrostomy tube in the kidney and return at another date allows the surgeon the luxury of being able to reinspect the collecting system at his leisure to remove all remaining stone fragments. In experienced hands, the hospital stay may be as little as two days, and even for the larger stones, hospital stays can be kept to a minimum by discharging patients with nephrostomy tubes in place and performing subsequent procedures on an outpatient basis.

THE FUTURE OF PERCUTANEOUS STONE REMOVAL

It is clear that the percutaneous approach to renal calculi has revolutionized the management of this disease. Hard on its heels, however, has come the development of extracorporeal shock wave lithotripsy (ESWL) (Chaussy and Schmiedt, 1983), and initially some thought that percutaneous stone removal would rapidly become obsolete. This, however, is very far from the truth. It is becoming clear that many cases require either a combination of both techniques or percutaneous stone removal alone (Eisenberger et al., 1985). In particular, staghorn calculi seem to require debulking percutaneously prior to ESWL, as the many fragments from stones above 2.5 cm in diameter will tend to totally obstruct the ureter after ESWL. Stones in obstructed kidneys, in calyceal diverticula, radiolucent stones, cystine stones, and stones impacted in the ureter also may not be suitable for ESWL. Finally, and perhaps more important, the scope of the percutaneous approach to the kidney is not confined to calculous disease alone, and techniques for the management of strictures at the ureteropelvic junction (Badlani et al., 1986), the calyceal infundibulum, and within the ureter, and even techniques of reconstruction of the upper urinary tracts are now being developed (Hulbert, 1985).

In the future the urologist will not be able to ignore these new techniques, and a sound working knowledge of both percutaneous stone removal and ESWL will be essential for those wishing to be involved in the management of ureteral and renal calculi.

REFERENCES

1. Badlani B, Eshghi M, Smith AD: Percutaneous surgery for ureteropelvic junction obstruction (endopyelotomy). *J Urol* 1986; 135:26–28.
2. Chaussy C, Schmiedt E: Shock wave treatment for stones in the upper urinary tract. *Urol Clin North Am* 1983; 10:743–748.
3. Clayman RV, Castaneda-Zuniga WR, Hunter DW, et al: Rapid balloon dilatation of a nephrostomy track for nephrostolithotomy. *Radiology* 1983a; 147:884–887.
4. Clayman RV, Hunter DW, Surya V, et al: Percutaneous intrarenal electrosurgery. *J Urol* 1984; 131:864–867.
5. Clayman RV, Surya V, Castaneda-Zuniga WR, et al: Percutaneous nephrostolithotomy with Mazzariello-Caprini forceps. *J Urol* 1983b; 129:12123–1215.
6. Coleman CC, Castaneda-Zuniga WR, Miller RP, et al: A logical approach to renal stone removal. *AJR* 1984; 143:609–616.
7. Eisenberger F, Fuchs G, Miller K, et al: Extracorporeal shockwave lithotripsy (ESWL) and endourology: an ideal combination for the treatment of kidney stones. *World J Urol* 1985; 1(3):41–47.
8. Goodwin WE, Casey WC, Woolf W: Percutaneous trocar (needle) nephrostomy in hydronephrosis. *JAMA* 1955; 157:891–897.
9. Huffman JL, Bagley DH, Lyon ES: Extending cystoscopic techniques into the ureter and renal pelvis: Experience with ureteroscopy and pyeloscopy. *JAMA* 1983; 250:2002–2006.
10. Hulbert, JC, Reddy PK, Young AT, et al: Reappraisal of the management of staghorn stones. *World Urol Update Series* vol 2, lesson 23, 1985a.
11. Hulbert JC, Reddy PK, Hunter DW, et al: Percutaneous management of ureteral calculi facilitated by retrograde flushing with carbon dioxide or diluted radiopaque dye. *J Urol* 1985b; 134:29–32.
12. Hulbert JC, Reddy PK, Young AT, et al: The percutaneous removal of calculi from transplanted kidneys. *J Urol* 1985c; 134:324–326.
13. Hulbert JC, Reddy PK, Hunter DW, et al.: Percutaneous techniques for the management of caliceal diverticula containing calculi. *J Urol* 1986; 135:225–227.
14. Hulbert JC: Percutaneous intrarenal endoscopy. Presented at the American Urological Association annual meeting, Atlanta, May 12–16, 1985.
14a. Hulbert JC, Reddy PK, Gonzalez R, et al: Percutaneous

nephrostolithotomy: An alternative approach to the management of pediatric calculus disease. *Pediatrics* 1985d; 176:610–612.

15. Hunter DW, Castaneda-Zuniga WR, Coleman CC, et al: Percutaneous techniques in the management of urological complications in renal transplant patients. *Radiology* 1983; 148:407–415.

16. Kaye KW: Renal anatomy for endourologic stone removal. *J Urol* 1983; 130:647–649.

17. Lange PH, Reddy PK, Hulbert JC, et al: Percutaneous removal of caliceal and other "inaccessible" stones: Instruments and techniques. *J Urol* 1984; 132:439–442.

18. Marberger M: Disintegration of renal and ureteral calculi with ultrasound. *Urol Clin North Am* 1983; 10:729–735.

19. Mazer MJ, Bartone FF: Percutaneous antegrade diagnosis and management of candidiasis of the upper urinary tract. *Urol Clin North Am* 1982; 9:157–161.

20. Miller RA, Wickham JEA: Optical triradiate nephroscope: New concept in percutaneous renal surgery. *Urology* 1984; 23:20–23.

21. Patterson DE, Segura JW, LeRoy AJ, et al: The etiology and treatment of delayed bleeding following percutaneous lithotripsy. *J Urol* 1985; 133:447–451.

22. Rupel E, Brown R: Nephroscopy with removal of stone following nephrostomy for obstructive calculous anuria. *J Urol* 1941; 46:177–182.

23. Sheldon CA, Smith AD: Chemolysis of calculi. *Urol Clin North Am* 1982; 9:121–144.

24. Smith AD, Reinke DB, Miller RP, et al: Percutaneous nephrostomy in the management of ureteral and renal calculi. *Radiology* 1979; 133:49–54.

25. Stanley P, Bear JW, Reid BS: Percutaneous nephrostomy in infants and children. *AJR* 1983; 144:73–76.

26. Whitaker R: Clinical application of upper urinary tract dynamics. *Urol Clin North Am* 1979; 6:137–141.

27. Young AT, Hunter DW, Castaneda-Zuniga WR, et al: Percutaneous extraction of urinary calculi: Use of the intercostal approach. *Radiology* 1985a; 154:633–638.

28. Young AT, Hulbert JC, Cardella JF, et al: Percutaneous nephrostolithotomy: Application to staghorn calculi. *AJR* 1985b; 145:1265–1269.

Chapter 19

Ureteroscopy for Diagnosis and Treatment in the Upper Urinary Tract

DEMETRIUS H. BAGLEY, M.D.

Techniques of ureteropyeloscopy have rendered the entire urinary tract available to endoscopic inspection and manipulation. Although flexible ureteroscopes were first placed into the smaller ureteral lumen for diagnostic purposes, the development of small-diameter rigid ureteroscopes and the adaptation of techniques for dilation of the ureterovesical junction for ureteroscopy opened the ureter to therapeutic endoscopic procedures.

Changes in instrumentation have extended the range of the rigid instrument to the upper ureter and renal pelvis. The flexible ureteropyeloscope has extended the range even further into the intrarenal infundibula and calyces. However, the therapeutic capabilities of these instruments are much more limited.

Increasingly widespread use of the techniques of ureteropyeloscopy has seen their rapid acceptance by urologists and an expansion of the application of these techniques. Calculi throughout the entire ureter and renal pelvis can be approached and treated endoscopically. The high success rates seen for the treatment of middle and lower ureteral calculi may support ureteroscopy as a technique of choice for stones in these locations. Filling defects or tumors throughout the ureter and the pelvis can also be inspected, biopsied, and possibly treated. The rapid development of even more versatile instruments holds the promise of even greater applications in the future.

HISTORY OF URETEROPYELOSCOPY

The first reported endoscopy of a ureter was anecdotally noted by Young, who passed a cystoscope into the dilated ureter in a child with posterior urethral valves in 1929 (Young and McKay, 1929). Routine ureteroscopy was not considered until Goodman (Goodman, 1977) and Lyon (Lyon et al., 1978) independently reported techniques for dilation of the ureterovesical junction and distal ureteral endoscopy. Lyon initially used pediatric instruments in a female patient but soon used juvenile cystoscopes to enter the distal ureter in males (Lyon et al., 1979). In 1980 Perez Castro-Ellendt reported on a long (40-cm) rigid ureteropyeloscope that could be passed transurethrally and transureterally to the level of the renal pelvis (Perez Castro-Ellendt and Martinez-Pineiro, 1980).

Since these instruments had a small working channel, inspection, tissue sampling, and stone retrieval could be performed throughout the ureter and into the renal pelvis. Removal of calculi was limited by the size of the calculus that could be withdrawn through the distal nondistended ureter, until techniques for calculus fragmentation were adapted for ureteroscopy. Ultrasonic lithotripsy initially gained wide acceptance for the percutaneous fragmentation and removal of renal calculi. As appropriately sized ultrasound probes were designed for use with the ureteropyeloscope, fragmentation and removal of large calculi from the upper urinary tract became possible, as reported by Huffman et al. (1983).

RIGID URETEROPYELOSCOPY

The capabilities and thus the indications and applications of rigid and flexible ureteropyeloscopy differ from those of the rigid endoscopes, so they are considered separately.

INDICATIONS

The indications for rigid ureteropyeloscopy have grown as experience with the technique has increased and as a wider range of instruments has become available. The indications can be grouped by their intended use into diagnostic or therapeutic procedures (Table 19–1).

TABLE 19–1.

Indications for Ureteropyeloscopy

Diagnostic
    Radiologic filling defects
    Unilateral hematuria
    Unilateral abnormal urinary cytology
    Strictures
Therapeutic
    Calculi
    Tumors
    Stricture or other ureteral obstruction
    Foreign bodies

Diagnostic

The diagnostic indications for ureteropyeloscopy are generally to define or to eliminate the presence of a malignancy. The diagnostic procedures consist of passing the endoscope to a specific area within the upper urinary tract to visualize an abnormality and to pass a working instrument, often to sample the tissue under direct vision.

Radiologic filling defects in the upper urinary tract seen on excretory urography constitute a major indication for ureteropyeloscopy (Fig 19–1). Visual inspection and possibly biopsy can generally provide the diagnosis. These techniques have been particularly useful in differentiating radiolucent calculi from blood clot or from tumors. Ureteral obstruction can often be localized and visualized ureteroscopically. Sampling of tissue in the area under direct vision by brushing or cold cup biopsy can provide specific histologic evaluation of the area. The presence of an obvious urothelial malignancy may be diagnosed by visual inspection while the presence of an intact, smooth urothelium may eliminate such a diagnosis. Thus, endoscopy is used to confirm the diagnosis of radiologic abnormalities.

Gross painless hematuria or abnormal urinary cytology localized to one ureterorenal unit has been an indication for ureteroscopy. Although the flexible ureteroscope can provide more thorough examination of the intrarenal collecting system and should be used in all of these patients who undergo ureteroscopy, the rigid instrument offers finer optical resolution and may provide some benefit in visualization of the ureter and renal pelvis. Uncommon ureteral lesions may be identified and lesions within the renal pelvis may also be seen. Inspection of the renal pelvis and infundibula through the lateral telescope may indicate a source of hematuria within the pelvis or from a specific infundibulum. Carcinoma in situ may also be seen or suspected, and selected brushings or biopsy specimen can be obtained as indicated.

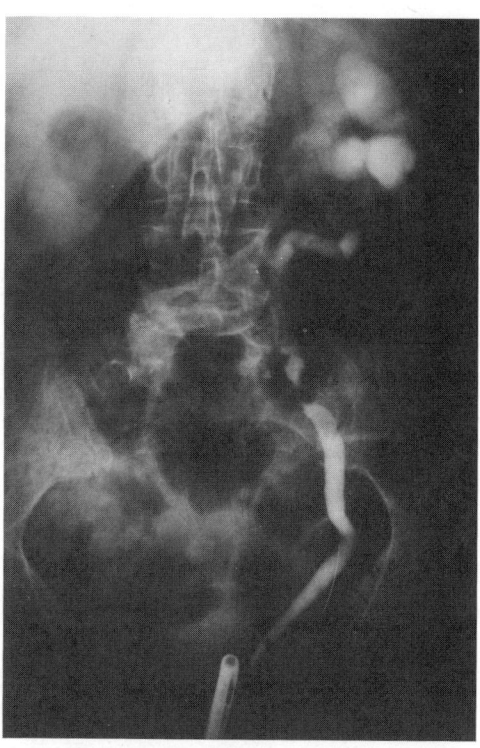

FIG 19–1.
Retrograde ureteropyelogram demonstrates a mid ureteral filling defect in an 81-year-old woman who had gross hematuria and poor visualization of the left kidney on excretory urography. Ureteroscopy confirmed the presence of a high-grade urothelial tumor.

Therapeutic Indications

Rigid ureteropyeloscopy has achieved its widest application for therapeutic procedures, most frequently for the treatment of urinary calculi. Small calculi that can be withdrawn through the lumen of the ureter can be grasped through the ureteroscope and removed intact. Larger calculi that will not pass freely through the ureter can be fragmented within the lumen and these smaller fragments then removed (Fig 19–2). Success of these techniques has resulted in its widespread application.

Ureteral obstruction can be visualized through the ureteroscope and may be bypassed under vision when cystoscopic manipulation is unsuccessful. Strictures can also be dilated under vision with a balloon catheter placed directly into the lumen.

Ureteropyeloscopy holds a promising role in the treatment of tumors in the upper urinary tract. Urothelial tumors can certainly be diagnosed, biopsied, and in some cases even resected. Ureteroscopy can then be used diagnostically to maintain surveillance of the treated area to detect recurrences. Although low-grade

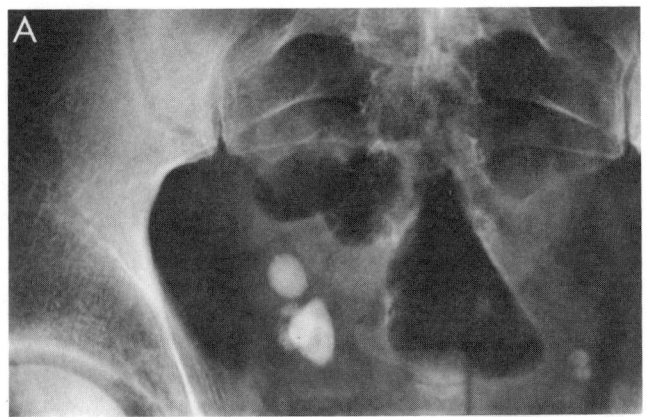

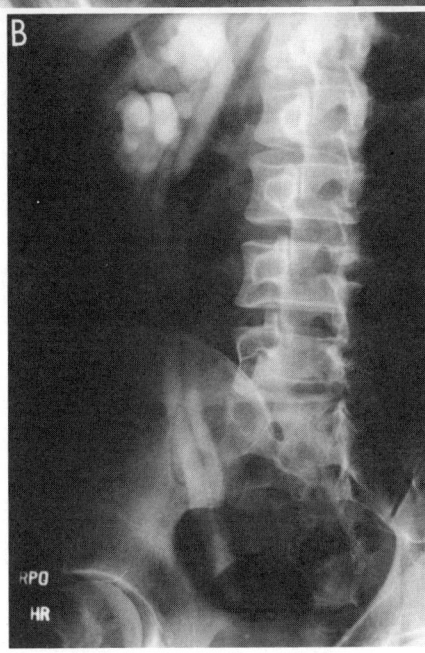

FIG 19–2.
A and **B,** four ureteral calculi are located in the distal ureter in a duplicated collecting system. The size of the calculi is shown on the KUB film **(A)** and the position confirmed on a contrast study **(B).** They could be removed by ureteroscopy and ultrasonic lithotripsy.

transitional cell carcinomas of the ureter have been cured by local resection, the appropriate role of this procedure is currently being evaluated.

INSTRUMENTS

Rigid ureteropyeloscopes are available from several companies. The earliest design, which remains a very practical and useful instrument, is the standard sheathed instrument with interchangeable telescopes (Fig 19–3). Generally, a smaller diagnostic sheath, approximately 10 F or larger working sheath (approximately 12 F) with a 5 F working channel, will accept interchangeable telescopes with forward or lateral view-

ing lenses. The larger sheath has been particularly useful, since the working channel will accept an instrument for manipulation within the ureter. The smaller diagnostic sheath does not have this capability but its smaller diameter has particular benefits. Often it can be passed into the undilated ureter and through areas impassable by the larger sheath without dilation. Recent additions to the group of sheathed instruments include those that will accept an offset telescope. With this combination, an ultrasound probe can be passed through the instrument and positioned under direct vision. These instruments offer the greatest range of instrumentation for rigid ureteropyeloscopy.

The first operating ureteropyeloscopes with an offset ocular eyepiece were designed with an integral telescope, incorporated within the instrument itself (Fig 19–4). The eyepiece is offset from the working channel at approximately 30°, allowing for a straight working channel and use of an ultrasonic probe under direct vision. The integral design allows use of a larger optical system with a relatively smaller outer sheath. The offset design, however, may make rotation of the instrument to change the field of view difficult, since the endoscopist must follow the changing position of the eyepiece.

An endoscope has been designed that combines rigid and flexible features to overcome the problem of the operator's position. A fiberoptic viewing bundle is incorporated into a rigid sheath with a flexible, offset eyepiece (Fig 19–5). Thus, for insertion and passage of the instrument, the eyepiece can be parallel to the rigid portion of the instrument while it can be offset as necessary and allow ultrasound probe under vision. This design also permits use of a small integral optical system, which minimizes the outer circumference of the instrument.

Since the majority of ureteroscopic procedures are limited to the distal ureter, two companies have marketed short ureteroscopes that will pass only to the level of the iliac vessels. Although these instruments offer no particular advantages over the longer instruments in terms of working capabilities, many urologists find them easier and less cumbersome to use in the lower ureter.

LITHOTRIPTERS

Ultrasonic lithotripters are made by several manufacturers and are basically similar. A generator is used to power a transducer consisting of piezoceramic crystals and an acoustical horn with a cylindrical metal probe (see Figs 19–4 and 19–5). These probes are necessarily narrow and long because of the limitations of working through a ureteroscope. The largest probe that will fit through the ureteroscope sheath alone and be used with the tactile ultrasonic lithotripsy technique is 2.7

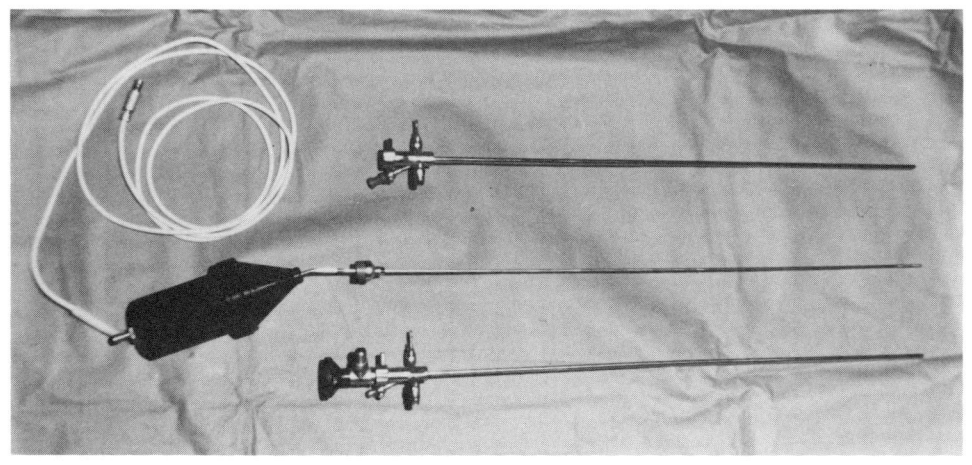

FIG 19–3.
A rigid ureteropyeloscope with interchangeable forward and lateral viewing lenses and both a working and diagnostic sheath.

mm in diameter. Smaller probes are required for visual ultrasonic lithotripsy through the operating ureteropyeloscopes. These are either 1.5 or 1.8 mm in diameter.

A solid wire probe has been introduced for use with a special generator (TUUL system). This instrument provides rapid, powerful lithotripsy and also has the advantage of fragmenting calculi when the side of the probe is in contact. However, it does not have a provision for suction or removal of fragments, which must be basketed or allowed to pass.

Electrohydraulic lithotripters consist of a generator producing an electrical discharge that passes across the tip of a coaxial cable probe. This discharge sets up a shock wave in a fluid medium, which can impinge on the rigid calculus. The probe must be 5 F or smaller for use within the ureteroscope, and the generator should have provision for single-impulse discharge rather than solely multiple discharge. Thus, more con-

trolled fragmentation of the calculus can be induced, and heat build-up can be limited.

TECHNIQUE

PREPARATION FOR THE PROCEDURE

As with all surgical and endoscopic procedures, successful completion depends to a large extent on careful preoperative planning and preparation. Selection of an appropriate patient is of utmost importance. On first attempting ureteroscopy, it is best for the urologist to choose a patient with a stone that is relatively easy to approach ureteroscopically. An ideal candidate would be a female patient with a calculus lodged a few centimeters above the bladder and only partially obstructing the lumen.

The patient should be informed of the plan for endoscopic treatment, its advantages as well as its risks, possible complications, and also the possible alternative techniques. As with other endoscopic procedures, infected urine should be treated with the appropriate an-

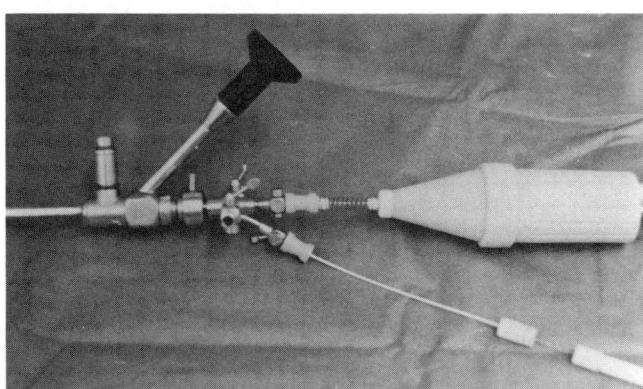

FIG 19–4.
An operating ureteropyeloscope contains a straight channel for introduction of the ultrasound probe and the side working channel for flexible instruments. The ultrasound transducer contains piezoceramic crystals, which, when excited, vibrate at an ultrasonic rate to activate the hollow metal probe. Suction applied through the probe can remove fragments of calculus.

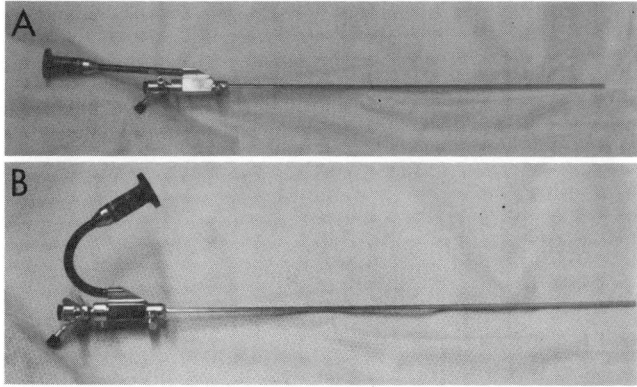

FIG 19–5.
A working ureteropyeloscope **(A)** with a flexible offset eyepiece **(B)** combines fiberoptic imaging and illumination with a rigid working sheath.

tibiotics before endoscopy. We have generally administered antibiotics preoperatively even when the urine is sterile, since we consider these patients to be at a high risk for infection.

ANESTHESIA

General anesthesia has usually been preferred for ureteropyeloscopy. It can offer good relaxation of the pelvic musculature and can allow complete control of respiration by the anesthesiologist. Thus, one can avoid the sudden cough or movement of patients who have had spinal or epidural anesthesia. Regional anesthesia will provide satisfactory relaxation and has been used successfully in removing distal ureteral lesions. Local intraluminal anesthesia with IV sedation has been advocated by some urologists. It should be stressed, however, that appropriate patient selection is essential for success with these techniques.

POSITIONING

The patient is placed in the dorsal lithotomy position, with the leg opposite the involved ureter slightly abducted to provide room for the operator as he positions the long ureteropyeloscope into the distal ureter. As the calculus is approached ureteroscopically, the patient should be placed in a reverse Trendelenburg position to allow the calculus to move freely toward the distal ureter rather than in a retrograde fashion toward the kidney.

RADIOLOGIC CONTROL

Fluoroscopy is extremely valuable for monitoring the position of the calculus, for ureteral dilation, and even during passage of the ureteropyeloscope. The position of guide wires can be readily determined with a fluoroscope, and the appropriate position of a catheter to be placed postoperatively can be immediately confirmed. As discussed later, fluoroscopy is essential for use of the flexible ureteropyeloscope.

IRRIGATION

Normal saline at body temperature should be used for irrigation during ureteropyeloscopy. Since there are some risks of absorption of the fluid through pyelolymphatic or pyelovenous backflow or even from ureteral perforation, a nonhemolyzing solution satisfactory for IV administration should be employed. The pressure of irrigation should be maintained at 30 cm H_2O or lower to minimize the chance of backflow.

CYSTOSCOPY

The bladder should be inspected with a cystoscope or other endoscope prior to ureteroscopy. A cone-tipped retrograde ureteropyelogram is then performed in all patients except those with a very distal ureteral calculus. This indicates the distensibility of the ureter and defines any strictures, tortuosities, or nondistensible bands that might provide an obstruction to passage of the instrument.

DILATION OF THE URETER

Dilation of the orifice and the intramural portion of the ureter has been an essential step for passage of the rigid ureteropyeloscope in most patients. Although the smaller instruments (10 F or less) can often be passed without ureteral dilation, distention of the lumen has been essential to accommodate the larger working sheaths.

The safety of ureteral dilation has been studied in animals and confirmed in clinical studies in patients. Although temporary hypertonicity and diminished peristalsis were documented in dogs, long-term effects of ureteral dilation to 14 F were not apparent (Green, 1944; Ford et al., 1984). Among patients treated with ureteroscopy, the lack of ureterovesical junction stricture formation and the lack of reflux appear to support the safety of this procedure.

Several techniques are available for ureteral dilation. Each can provide satisfactory results and each has its own disadvantages and risks. These techniques can be grouped into the nonguided and those that use a guide wire for placement (Table 19–2).

Nonguided

Nonguided ureteral catheters can be employed for dilating the ureter. Successively larger catheters can be placed cystoscopically. The risk of perforation is high, and catheters often lack the strength to ensure adequate dilation. A ureteral catheter can also be passed and left in place for 24 or more hours to provide gradual dilation of the ureter. This often gives excellent dilation, even of long, narrow areas, but prolongs the hospital stay.

Cone-tipped metal bougie dilators can be placed cystoscopically to dilate the ureteral lumen. These instruments give full dilation but carry some risk of ureteral

TABLE 19–2.

Ureteral Dilation

Nonguided dilators
  Ureteral catheters
  Metal cone-tipped bougies
Guided dilators
  Ureteral catheters and self-retaining stents
  Olive-tipped metal bougies
  Graduated flexible dilators
  Balloon dilating catheters

perforation. They are particularly useful for the impacted calculus that will not permit passage of a guide wire (Fig 19–6).

Wire-Guided Dilation

Dilators can be placed more accurately with certainty of their entering and remaining within the ureteral lumen by first placing a radiographic guide wire into the lumen. Several different types of dilators can then be passed over the guide wire. Ureteral catheters that will accept a guide wire can be used. Self-retaining stents can also be used and are particularly effective for chronic dilation of long, narrow ureteral segments. Olive-shaped metal bougie dilators have the advantage of full dilation and reusability, but the larger dilators that cannot pass through the working part of the cystoscope must be backloaded into the instrument to be placed under direct vision into the ureteral orifice. They are readily evident on fluoroscopy. Graduated plastic dilators can also pass over a guide wire, but the larger sizes that will not fit through the sidearm of the cystoscope must be placed with fluoroscopic control alone.

Balloon dilating catheters are easy to place and can provide full dilation. Inflation of the balloon should be monitored fluoroscopically by filling the balloon with diluted radiographic contrast medium (Fig 19–7). High-pressure balloons capable of maintaining pressures greater than 12 atm should be employed, since as many as 20% of normal ureters require such pressures for adequate dilation. These balloons can also be used without a guide wire, and a 4.5 F version is available that can pass through the ureteroscope. The major disadvantages of the balloon catheters are their expense and their design for one-time use.

URETERAL ENDOSCOPY

After completion of dilation, if a guide wire has been used, it can be left in place throughout the procedure to provide an immediate guide to the ureteral orifice, even if there is bleeding, and to aid in placement of a drainage catheter at termination. The ureteropyeloscope is then placed and advanced under direct vision through the urethra and through the bladder to enter

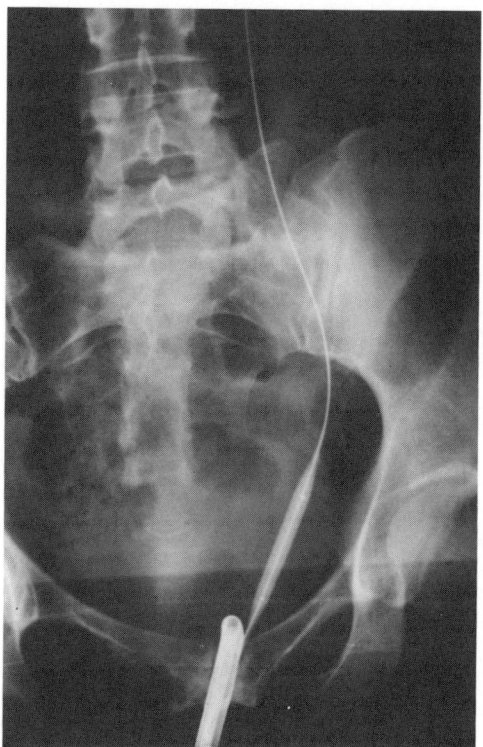

FIG 19–7.
A 5-mm × 4-cm ureteral dilating balloon catheter has been inflated with contrast to ensure full dilation of the ureteral orifice and intramural portion of the ureter.

the ureteral orifice. It is passed along the ureteral lumen under direct vision. Irrigation pressure should be minimized and should be maintained at less than 30 cm H_2O to avoid dislodging a calculus. If there is difficulty in passing the instrument at the level of the iliac vessel, in a tortuous area, or at the ureteropelvic junction, a ureteral catheter or guide wire can be passed through the working channel of the instrument to straighten that part of the ureter.

URETERAL CALCULI

Calculi located within the ureter can be removed ureteroscopically. Stones small enough to pass through the distal ureter after dilation can be removed intact. Calculi that are too large to remove in this way must be fragmented either by ultrasonic lithotripsy or electrohydraulic lithotripsy (Table 19–3).

Retrieval

Calculi smaller than approximately 6 mm in diameter may be retrieved through the ureter after adequate dilation and ureteroscopy. The calculus can be grasped with various instruments, including helical stone baskets, snares, double snares, and wire-pronged graspers. A calculus of any size can be approached with an at-

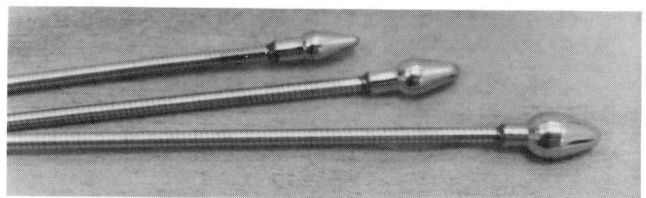

FIG 19–6.
Cone-tipped metal dilators are particularly useful for dilating the orifice when the ureter is totally obstructed.

TABLE 19–3.
Endoscopic Treatment of Ureteral Calculi

Retrieval
Fragmentation
    Ultrasonic lithotripsy
        Nonvisual or tactile
        Visual
    Electrohydraulic lithotripsy

tempt to extract it, since the operator can observe the stone, the working instrument, and the ureter directly and see whether the stone moves freely within the ureter. If the ureteral mucosa closes on the calculus and folds with the instruments as they advance, the calculus cannot be withdrawn and should be fragmented in situ.

LITHOTRIPSY

The ultrasonic lithotripter has been preferred for fragmenting ureteral calculi, since it provides a very controlled fragmentation and also removes fragments as they are formed. Both tactile and visual techniques for ultrasonic lithotripsy have been described. Both techniques require firm contact between the probe and the calculus, which should be secured in a basket for fragmentation.

TACTILE OR NONVISUAL LITHOTRIPSY

When utilizing the tactile technique, the operator first secures a calculus in a basket and then positions it at the tip of the ureteroscope to fill the entire visual field. The telescope is then removed, and the long ultrasound probe is passed through the lumen of the ureteropyeloscope until it makes contact with the calculus. This can be confirmed by feeling the pressure of the probe against the calculus within the basket. It can also be confirmed fluoroscopically. Suction is applied through the probe, and irrigant is passed into the sheath through the normal inflow channel to provide a cooling stream of saline through the instrument around the calculus and back into the probe. As the ultrasonic probe is activated, it is maintained in firm contact with the calculus to disintegrate a portion. As the stone is fragmented, the probe will pass the stone, and the pressure contact between the stone and probe will be lost. At this point the probe is removed and the telescope replaced to allow repositioning of the calculus within the basket. The ultrasonic fragmentation is then repeated as necessary.

VISUAL ULTRASONIC LITHOTRIPSY

The development of the direct-vision operating ureteropyeloscope, when combined with narrow ultrasound probes, has made possible intraureteral ultrasonic lithotripsy under direct vision. This technique provides the advantages of constant visual monitoring for positioning of the ultrasound probe. It is also more rapid, since the probe can be constantly repositioned under vision to fracture the calculus most efficiently. Although it is best to engage the calculus in a basket for the procedure, it is not essential. The ureter can be obstructed proximally with a basket or a balloon. It is particularly useful for the impacted calculus that cannot be engaged in a basket. The major disadvantage is the small diameter of the probe, which frequently becomes obstructed with fragments of calculus and must be cleared to allow flow of the irrigant for cooling.

ELECTROHYDRAULIC LITHOTRIPSY

Electrohydraulic lithotripsy has been used in the ureter as well. Although early experimental reports indicated damage to the ureter with this technique, at that time it was not performed under visual monitoring, and damage resulted from contact of the probe with the ureteral mucosa (Raney, 1978; Goodfriend et al., 1984). Electrohydraulic lithotripsy provides less control of fragmentation, and the calculus may migrate proximally within the ureter. However, it provides more rapid fragmentation and has produced satisfactory results (Green and Lytton, 1985).

The calculus is first visualized with the endoscope as with other techniques. However, a basket is not placed above the stone because of the possibility of sparking from the lithotripter probe to the wire of the basket. The lithotripter probe is passed and placed adjacent to, but not in contact with, the calculus. The lithotripter is then activated with single impulses. As the calculus is broken, it is reduced to a size that can be removed through the ureter or the individual fragments are removed.

RESULTS

Ureteroscopic stone removal has generally been quite successful (Table 19–4). Among several series the success rate has been approximately 80%. Patients have tolerated the procedure well and the complication rate has been acceptably low. The success rate varies with the location of the calulus within the ureter. Lower ureteral stones are removed in 90%–97% of patients as reported in several series (Bagley, 1986). For stones located more proximally within the middle third of the ureter, this success rate falls to 68%–83%. The success rate decreases even further for proximal ureteral stones, to approximately 50%. As expected, the success rate increases when all modalities of instrumentation including ultrasonic lithotripsy are available. Only when techniques for fragmentation are available can larger calculi be removed. This procedure has resulted

TABLE 19–4.

Success of Ureteroscopic
Removal of Ureteral Calculi*

LOCATION	% REMOVAL
Upper ureter	33–60
Middle ureter	66–83
Lower ureter	90–97
All sites	78–89

*With ultrasonic lithotripsy available.
Modified from Bagley DH: Ureteropyeloscopy for the treatment of stone
disease, in Harrison L, Kandel L
(eds): *Techniques in Urologic Stone
Surgery.* Mt Kisco, NY, Futura Publishing Co, 1986, p 47.

in shorter hospital stays and less narcotic use than in open ureterolithotomy (Seeger et al., 1985).

TUMOR

Ureteroscopy has introduced a new concept into the treatment of urothelial neoplasms. Radiologic filling defects can be diagnosed by endoscopic visualization and biopsy. Tumors can be resected endoscopically, and accurate treatment can be planned based on pathologic findings. Surveillance for recurrence can be performed more accurately with endoscopic monitoring utilizing the ureteropyeloscope than with radiologic studies alone.

DIAGNOSIS

Findings that may suggest the presence of urothelial tumors and that therefore may indicate ureteroscopy include the presence of abnormalities such as radiographic filling defects, unilateral hematuria, or unilateral abnormal cytologic findings. Endoscopic surveillance of the upper tract after previous treatment of a urothelial lesion can also be added to this list.

The diagnostic approach to a radiologic filling defect may be altered as experience with the ureteropyeloscope increases. The standard urologic evaluation of gross hematuria includes an excretory urogram and cystoscopic evaluation of the bladder. Discovery of a filling defect on the excretory urogram suggestive of a urothelial neoplasm has usually prompted further radiologic evaluation with a retrograde pyelogram, ultrasound, or CT scan. The high accuracy of endoscopic inspection and biopsy may obviate these further radiologic studies. If a urologist can provide a diagnosis endoscopically without an increase in morbidity, this study may appropriately replace the other, less direct measures.

Ureteral lesions can be studied directly. By using the standard techniques for placement of the ureteroscope, the operator can visualize the entire ureteral mucosa. Even tiny papillary lesions that cannot be detected ra-

diographically can be appreciated with endoscopic inspection (Huffman et al., 1985). In performing the procedure, the ureterovesical junction should be dilated with great caution to avoid damage to papillary lesions before they can be visualized endoscopically.

Exophytic urothelial tumors of the upper tract can be diagnosed visually by their characteristic papillary appearance just as similar lesions in the bladder can be diagnosed. Papillary inflammatory lesions may be confused, and high-grade lesions may often be confused with urothelial edema. Therefore, tissue should be obtained for diagnosis in all patients. A cup biopsy forceps of 5 F will fit through the working channel of most rigid ureteroscopes. Although this instrument obtains a very small piece of tissue, it can provide a sample to indicate the cytologic and architectural nature of the tissue. Alternatively, a biopsy brush, usually with a 5-F sheath, can be passed through the working channel of the rigid instrument to brush the suspicious area under direct vision. Deflectable brushes can be used for lesions located laterally within the renal pelvis.

When hematuria has been localized to a specific renal collecting system but there are no other radiologic indications of abnormality, endoscopy can be an appropriate diagnostic step. Careful, thorough inspection is again the key to endoscopic diagnosis. The ureteroscope is passed through the length of the ureteral lumen until it reaches the renal pelvis, at which time the forward viewing telescope is replaced with the lateral telescope to inspect the pelvis and the infundibula. Even greater endoscopic range can be obtained by passing a deflectable, flexible ureteroscope through the rigid sheath to use it for inspection of the renal pelvis and collecting system. If both instruments, the rigid and the flexible, are available, they should be combined for the greatest resolution and endoscopic range. The rigid instrument can inspect the ureter and the flexible instrument the pelvis and infundibula.

A similar approach can be taken when abnormal cytologic findings can be lateralized to one collecting system that is also radiologically normal. Again, as the instrument passes through the ureter into the renal pelvis, any suspicious lesion should be brushed or biopsied. The flexible instrument may add a greater range in this situation as well. If the entire endoscopic inspection reveals no gross lesion, selected brushings or biopsy specimens should be taken. Accessible major infundibula should be biopsied individually so that the intrarenal collecting system can be sampled and mapped cytologically.

RESECTION OF UPPER URINARY UROTHELIAL TUMORS

Urothelial lesions can also be resected with the ureteroresectoscope (Fig 19–8). The technique is similar to

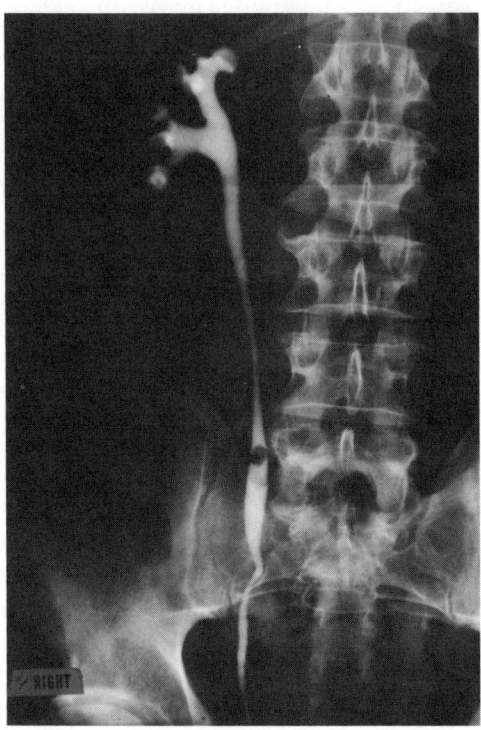

FIG 19–8.
A midureteral filling defect was found to be a low-grade urothelial tumor. It was resected ureteroscopically and in follow-up over two years had no recurrences.

that used with a pediatric resectoscope, since these instruments are very similar except for the added length in the ureteral model. A nonconducting irrigant such as glycine or water is used as well as an electrosurgical unit with the lowest settings of both the coagulation and cutting currents that can provide the most effective resection with the least tissue injury.

The actual technique of resection is different from that used in the bladder or prostate. Only tumor directly within the lumen is resected, and no deep bites, such as those used in the bladder or in the prostate, are attempted, since this can result in immediate perforation of the thin ureteral wall. The resectoscope loop is extended beyond the tumor and then withdrawn again into the instrument to cut the tissue. This prevents injury to the surrounding ureteral wall. The base of the lesion can be fulgurated with the loop and low current.

Smaller lesions can be obliterated completely with a cup forceps and can also be fulgurated with 5-F Bugby electrode. Continuous irrigation is necessary to disperse the bubbles produced during fulguration.

RESULTS

In a combined series from the University of Chicago Hospital and Memorial Sloan-Kettering Cancer Center,

Huffman and associates (1985) reviewed 59 patients who were evaluated or treated ureteroscopically for urothelial neoplasms. By ureteropyeloscopic evaluation of patients with some suspicion of neoplasm, the procedure was successfully completed in 54 patients but had to be discontinued in five. Among the 54 patients evaluated, 27 urothelial tumors were diagnosed, 1 metastatic squamous cell carcinoma, 3 uric acid calculi; 1 patient had bleeding localized to the lower pole but otherwise undefined, and in 22 patients no abnormality could be detected.

Among the patients with urothelial tumors, 16 had primary ureteroscopic fulguration or resection, while six underwent nephroureterectomy and three had segmental ureterectomy. The two patients with metastatic lesions underwent systemic chemotherapy.

Three patients underwent segmental ureterectomy and were followed up ureteroscopically to search for recurrences. Each tumor was found to have been staged accurately with the endoscopic procedure. Three low-grade recurrences were identified during the course of follow-up in one patient, all located near the site of the anastomosis. These lesions were also treated endoscopically.

Ureteroscopic staging was compared with pathologic mapping in the six patients who underwent nephroureterectomy. Two had accurate clinical staging, while four had understaging by endoscopy. All patients had extensive tumors extending into the pelvis and calyces, and not all areas of tumor were accessible endoscopically.

Sixteen patients were treated ureteroscopically for apparently localized low-grade transitional cell tumors. At an average follow-up of 16 months, 14 low-grade recurrences had been biopsied and treated endoscopically in five patients. Of significance among these patients was the observation that local recurrences could be detected endoscopically before they could be seen radiographically.

Thus, ureteroscopy clearly has a valuable role in the diagnosis of urothelial neoplasms. However, its value in therapy must be studied in longer-lasting series. In these early reports, one can be encouraged by the potential for staging of ureteral lesions in planning for segmental resection and also for the possible role of endoscopic resection as a therapeutic modality.

Certainly, patients treated in this way require continued surveillance. The bladder should be examined cystoscopically and the upper collecting system ureteroscopically at three-month intervals after tumor resection. Other techniques that are available including excretory urography and urine cytology should also be employed to increase the sensitivity of detection of recurrences.

FLEXIBLE URETEROPYELOSCOPY

Flexible instrumentation for ureteropyeloscopy offers many specific differences from the features provided by rigid instrumentation. The major advantage is specifically the flexibility of the instruments. They can pass tortuosities within the ureter and enter portions of the collecting system totally inaccessible to rigid instruments. Since these instruments must pass through the ureter, they are usually of a small diameter and, thus, small enough to fit into the renal infundibula.

The same characteristics may be liabilities in the use of the endoscopes. Flexibility makes it very difficult to pass the instrument through the urethra and bladder into the ureter. Some type of guide system must be used to facilitate this passage. The small diameter of the instrument will accept, at best, a very small working channel.

Flexible instruments were the first to be used as ureteropyeloscopes. The benefits of a small flexible instrument that could be passed into the ureter just as a ureteral catheter were obvious. However, the possibility of rigid ureteropyeloscopy and the therapeutic superiority of those instruments were not considered at that time. Marshall (Marshall, 1964) described his associate's use of a 9 F ureteroscope to visualize a calculus within the distal ureter. Takagi et al. (1971) reported on a 2-mm "pyeloureteroscope" that could be passed transurethrally through the ureter and into the intrarenal collecting system. It did not have an irrigating system and could be passed into the ureter only with difficulty, until Takayasu and Aso (1974) described a Teflon guide tube that could be passed through a cystoscope into the distal ureter, through which the flexible instrument could be introduced.

These earlier studies with flexible ureteropyeloscopes indicated several features that constitute major advantages (Table 19–5). A channel for irrigation is extremely valuable in all patients, even those in whom a diuresis has been induced, and it is essential whenever there is any bleeding within the lumen.

Deflectability is also extremely valuable. Although nondeflectable instruments can be passed through the ureter and into the renal pelvis, their course within the intrarenal collecting system is dictated by the anatomical configuration. An instrument the tip of which can be deflected at least 150° can be passed into the infundibula in the upper pole and the midportion of the kidney. If the flexible tip of the instrument is too short, it may not be able to pass fully into the lower pole calyx, but it should be able to enter the infundibulum.

The fiberoptic image of the flexible instruments lacks the precise resolution of solid-lens instruments, yet it is adequate for most endoscopic procedures. As urologists become familiar with the use of the flexible nephroscope and the cystoscope, they have little difficulty in accepting the images provided by the small flexible ureteropyeloscope.

INSTRUMENTS

Several flexible ureteropyeloscopes are now available as prototypes or as early production models. Nondeflectable instruments are made in sizes from 6 F to 9 F. They are available both with and without irrigating channels as well as with and without focusable eyepieces.

Instruments are also available with manipulable tips, with diameters from 2.7–3.6 mm. The 3.2- and 3.6-mm instruments have working channels of 0.5 and 1.2 mm, respectively. Each instrument is deflectable to 90° or 160° in opposite directions on the same plane (Table 19–6).

TECHNIQUES

INTRODUCTION OF INSTRUMENTS

Because of the flexibility of the instruments, they must be introduced through the urethra and bladder into the ureter with the benefit of some internal or external assistance. As noted above, Takayasu and Aso (1974) described a guide tube that could be passed into the ureter initially. A cystoscope sheath or rigid ureteroscope sheath can be used in a similar fashion. Instru-

TABLE 19–5.

Desirable Features of Flexible Ureteropyeloscope

Deflectable tip (to 150°)
Channel for working instruments and irrigation
Small diameter (<4 mm)

TABLE 19–6.

Flexible Ureteropyeloscopes

COMPANY	DIAMETER, F	CHANNEL, F	DEFLECTION,°
ACMI	6.5–7.5	None	None
	13*	4.2	160 + 90
Olympus	8.1*		160 + 90
	9.6*	1.5	160 + 90
	10.8*	3.6	160 + 90
	14.0	5.0	160 + 90
Reichert	7.0	1.5	None
	9.0	3.5	None
Van Tec	6.0	1.0 + 2.0	None†
	8.5	1.5 + 3.0	None†

*Prototype.
†"Disposable" shaft.

ments with an adequate working channel can be passed over a guide wire into the ureter.

It is extremely helpful to use an irrigant containing radiographic contrast material when passing the instrument within the intrarenal collecting system so that it can be visualized fluoroscopically (Bagley et al., 1983). It is often difficult to remain visually oriented within the kidney, since infundibula enter the pelvis in positions that may be difficult to appreciate in two dimensions.

The collecting system should be inspected in a systematic fashion to be certain of full visualization. Specific anatomical configurations such as compound calyces and infundibula in certain positions can be recognized as landmarks during the procedure. The most reliable landmarks, however, are the ureteropelvic junction and any catheter or wire passing from the ureter and lying within the pelvis.

APPLICATION OF FLEXIBLE URETEROPYELOSCOPY

Flexible ureteropyeloscopes have been particularly useful for endoscopic visualization of the proximal urinary tract when the rigid instrument cannot be employed. The flexible instrument can pass tortuous portions of the ureter. It can pass laterally, anteriorly, and posteriorly into portions of the intrarenal collecting system inaccessible to the rigid instrument.

The major application of this technique has been for the diagnostic evaluation of patients with radiologic filling defects in the upper urinary tract and for gross unilateral hematuria for which no cause has been demonstrated radiologically (Fig 19–9). A very high success rate (21 of 21 patients studied) has been observed in evaluation of filling defects, while evaluation of hematuria has been less successful, with approximately 50% of patients having some specific abnormality detected. This may be partially based on the inability to enter the lower infundibulum in many patients.

In use, the deflectable instruments have proved to be much more valuable than the nondeflectable design. The latter often is unable to visualize the entire ureter if there are any curves within the lumen (Bagley and Rittenberg, 1986). The nondeflectable instruments cannot be directed within the intrarenal collecting system, and when used in a retrograde fashion, pass only to the upper infundibulum toward which the ureter directs the instrument.

The availability of working instruments has made it possible to perform manipulations and procedures through the working channel of the larger flexible ureteropyeloscopes. Snares, double snares, three-pronged graspers, and electrodes can be passed through these channels and perform tasks such as retrieval or fulguration within even the most lateral portions of the intrarenal collecting system. In the future, procedures that might employ electrohydraulic lithotripsy or lasers can be expected to render these instruments more useful.

FIG 19–9.
A, a cone-tipped retrograde ureteropyelogram outlines an irregularly filling infundibulum and calyx. **B,** the 3.6-mm Olympus ureteropyeloscope has been passed to the adjoining infundibulum, and a papillary transitional cell carcinoma was visually evident.

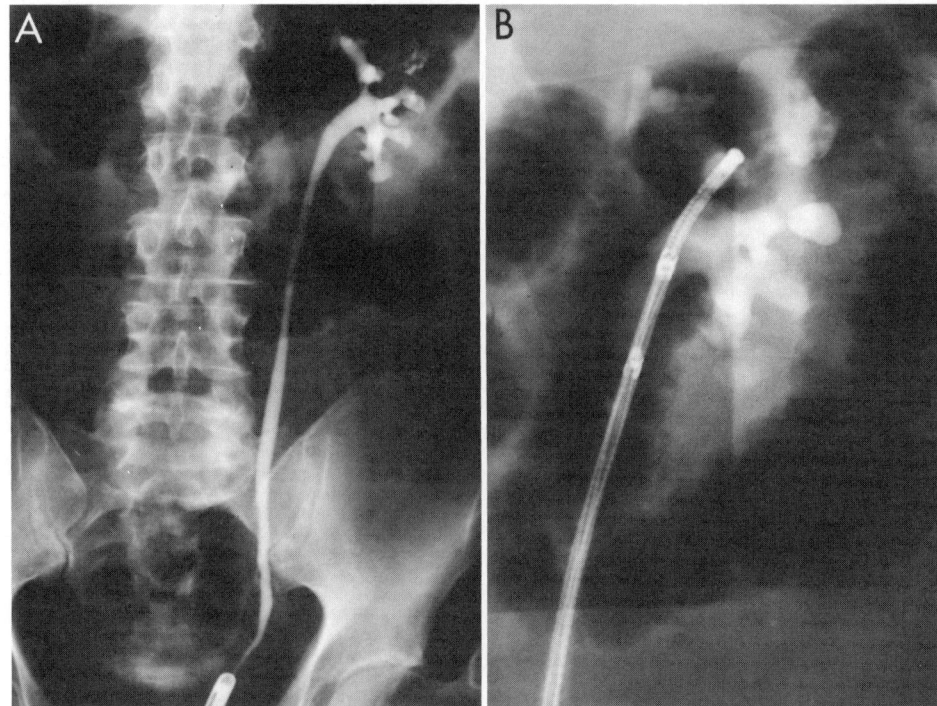

POSTURETEROSCOPY CARE

URETERAL CATHETER

Most urologists routinely leave a ureteral catheter in place in the ureter to drain urine from the renal pelvis and decrease any possibility of extravasation or obstruction by ureteral edema. The ureteral catheter is placed into the ureter and renal pelvis at the completion of the ureteroscopy. A Foley catheter is also placed into the bladder, to which the ureteral catheter is secured. Both catheters are attached to urinary collection devices and are left in place until any hematuria has cleared.

SELF-RETAINING STENTS

Self-retaining ureteral stents have also been used for drainage following ureteroscopy. These can be removed by cystoscopy later or removed manually, if a nylon suture has been passed through the stent and left passing through the urethra. One potential disadvantage to the use of these stents is that any perforation of the ureter may be subject to intravesical pressure, since urine can reflux as well as drain along the stent.

PAIN

Pain is not a prominent symptom after ureteroscopy. Although patients may experience some aching in the ipsilateral lower quadrant or flank pain, severe flank pain should be considered unusual and a cause for investigation. The most frequent cause of such pain is obstruction of the ureteral catheter by blood clot or by tight ligation to the Foley catheter.

The postoperative narcotic analgesic requirements were compared in two groups of patients. The first had standard open surgical ureterolithotomy, while the second had ureteral calculi removed by ureteroscopy. Patients in the surgical group required an average of 11 doses of analgesics postoperatively, and the patients treated endoscopically needed only three doses (Seeger et al., 1985). This comparison, although imprecise, offers some relative indication of the pain induced by each procedure for the removal of ureteral calculi. It is clear that patients can tolerate the endoscopic procedure with less discomfort.

PREVENTIVE ANTIBIOTICS

Since many of these procedures are long and subject to potential contamination, and also since stones are often present with obstruction, we use broad-spectrum antibiotic coverage started preoperatively and continued into the postoperative period. If the urine is infected preoperatively, ureteropyeloscopy is delayed until effective treatment has been started. This treatment is continued through the intraoperative and postoperative periods.

COMPLICATIONS OF URETEROSCOPY

HEMATURIA

Mild hematuria is routinely seen after ureteroscopy. Severe bleeding during ureteroscopy has been distinctly unusual and would be a basis for termination of the procedure if visibility is impossible. Prolonged postoperative hematuria is also uncommon. It usually clears within one to three days. Persistence of gross hematuria should be the basis for further evaluation.

URETERAL INJURY

Severe ureteral injury is distinctly uncommon in experienced hands. Mild injury to the mucosa is much more common and is probably the basis for the frequently observed hematuria. Minor perforation evidenced by extravasation of contrast occurs in 10%–15% of patients undergoing ureteroscopy. Such perforations are usually undetectable endoscopically. Larger perforations in which periureteral fat can be seen endoscopically may also occur. Depending on the position, they may be the basis for termination of the procedure. Treatment of any type of injury is similar, with placement of a ureteral catheter for drainage.

INFECTION

Infection after ureteroscopy has been the second most common complication in the combined series reviewed by Sosa and Huffman (1986) (Table 19–7). At the rate observed, however, it remains relatively infrequent (1.3%). The use of antibiotics on a preventive basis may have minimized the infection rate in our series, although there was no controlled series to support this.

TABLE 19–7.

Complications*

TYPE OF COMPLICATION	NO.	%
Mechanical		
Mucosal injury	13	1.5
Perforation	7	0.8
Hemorrhage	4	0.5
Extravasation (major)	2	0.2
Stricture	1	0.1
Infections	11	1.3

*Adapted from Sosa E, Huffman JL: Complications of ureteropyeloscopy, in Huffman JL, Bagley DH, Lyon ES (eds): *Ureteroscopy.* Philadelphia, WB Saunders Co., 1987.

STRICTURE

Ureteral stricture following ureteroscopy has been unusual. Since it has been seen after passage or removal of calculi, it may be difficult to differentiate the cause. At one meeting on endoscopy at which 838 ureteropyeloscopies were reported by several authors, only a single stricture was reported (Sosa and Huffman, 1986). In the 12-month follow-up of patients after ureteroscopy, no strictures were noted (Stoehl and Marberger, 1986). Ureteral stricture can occur, however. Gillenwater has cited patients sent to his institution for treatment of stricture following ureteroscopy (Gillenwater, 1985). Long-term studies of larger numbers of patients will be necessary before the risk of stricture can be well defined.

BREAKAGE OF ENDOSCOPES

In a few instances, rigid ureteropyeloscopes have been bent and even broken within the patient. Removal has required an open surgical operation. Such breakage may occur in the long rigid instrument after repeated flexing of the sheath during ureteroscopy. Therefore, any sheaths should be replaced rather than merely straightened before continuing the procedure.

REFERENCES

1. Aso Y, Ohtawara Y, Suzuki K, et al: Usefulness of fiberoptic pyeloureteroscope in the diagnosis of the upper urinary tract lesions. *Urol Int* 1984; 39:355–357.
2. Bagley DH: Ureteropyeloscopy for the treatment of stone disease, in Harrison L, Kandel L (eds): *Techniques in Urologic Stone Surgery.* Mt Kisco, NY, Futura Publishing Co., 1986, pp 33–51.
3. Bagley DH, Huffman JL, Lyon ES: Combined rigid and flexible ureteropyeloscopy. *J Urol* 1983; 130:243–244.
4. Bagley DH, Rittenberg MH: Percutaneous antegrade flexible ureteroscopy. *Urology* 1986; 27:331–334.
5. Ford TF, Parkinson MD, Wickham JEA: Clinical and experimental evaluation of ureteric dilatation. *Br J Urol* 1984; 56:460–463.
6. Gillenwater J: *Ureteroscopy*, in Gillenwater J, Howards S (eds): *Yearbook of Urology.* Chicago, Year Book Medical Publishers, 1985, p 122.
7. Goodfriend R: Ultrasonic and electrohydraulic lithotripsy of ureteral calculi. *Urology* 1984; 23:5–8.
8. Goodman TM: Ureteroscopy with a pediatric cystoscope in adults. *Urology* 1977; 9:394.
9. Green DF, Lytton B: Early experience with electrohydraulic lithotripsy of ureteral calculi using direct vision ureteroscopy. *J Urol* 1985; 133:767.
10. Greene LF: Effects of ureteral dilatation on ureter and kidney. *J Urol* 1944; 52:505–512.
11. Huffman JL, Bagley DH, Lyon ES: Abnormal ureter and intrarenal collecting system, in Bagley, DH, Huffman JL, Lyon ES (eds): *Urologic Endoscopy: A Manual and Atlas.* Boston, Little, Brown & Co, 1985, p 69.
12. Huffman JL, Bagley DH, Schoenberg HW, et al: Transurethral removal of large ureteral and renal pelvic calculi using ureteroscopic ultrasonic lithotripsy. *J Urol* 1983; 130:31.
13. Huffman JL, Morse MJ, Bagley DH, et al: Endoscopic diagnosis and treatment of upper tract urothelial tumors—a preliminary report. *Cancer* 1985; 55:1422–1428.
14. Lyon ES, Banno JJ, Schoenberg HW: Transurethral ureteroscopy in men using juvenile cystoscopy equipment. *J Urol* 1979; 122:151.
15. Lyon ES, Kyker JS, Schoenberg HW: Transurethral ureteroscopy in women: A ready addition to urologic armamentarium. *J Urol* 1978; 119:35.
16. Marshall VF: Fiberoptics in urology. *J Urol* 1964; 91:110–114.
17. Perez-Castro Ellendt E, Martinez-Pineiro JA: Transurethral ureteroscopy—a current urological procedure. *Arch Esp Urol* 1980; 33:445.
18. Raney AM: Electrohydraulic ureterolithotripsy. *Urology* 1978; 12:284–285.
19. Seeger AR, Rittenberg MH, Bagley DH: Ureteral calculi: ureteroscopic removal vs. open ureterolithotomy. Presented at the 43rd Annual Meeting of the Mid-Atlantic Section AUA, October 3–6, 1985, Philadelphia.
20. Sosa E, Huffman JL: Complications of ureteropyeloscopy, in Huffman JL, Bagley DH, Lyon ES (eds): *Ureteroscopy.* Philadelphia, WB Saunders Co., 1987.
20a. Stoehl W, Marberger M: Late sequelae of the management of ureteral calculi with the ureterorenoscope. *J Urol* 1986; 136:386–389.
21. Takagi T, Go T, Takayasu H, et al: Fiberoptic pyeloureteroscope. *Surgery* 1971; 70:661–666.
22. Takayasu H, Aso Y: Recent development for pyeloureteroscopy: Guide tube method for its introduction into the ureter. *J Urol* 1974; 112:176–178.
23. Young HH, McKay RW: Congenital valvular obstruction of the prostatic urethra. *Surg Gynecol Obstet* 1929; 48:409.

Chapter 20

Extracorporeal Shock Wave Lithotripsy (ESWL) for the Treatment of Upper Urinary Stones

CHRISTIAN G. CHAUSSY, M.D.
GERHARD J. FUCHS, M.D.

Extracorporeal shock wave lithotripsy (ESWL) is a noninvasive method for the treatment of renal and upper ureteral stones. The contact-free destruction of kidney stones by high-energy shock waves was made possible by basic research in acoustic physics. Pitting seen on aircraft wings was found to be due to shock waves generated on collision of the aircraft with micrometeorites or raindrops. Shock waves were found to have a destructive effect when traversing the interface of materials with different acoustical properties, such as fluids and brittle solids. The reaction at the interface creates a strong tensile force in the solid. Mechanical breakdown on the surface of the solid takes place when the tensile force exceeds the solid's comprehensive strength (Forssmann et al., 1977).

Use of shock waves to treat human medical conditions was hampered by great expense and finding a safe means of generating shock waves of sufficient strength. The problem was solved by developing a means of producing shock waves by an underwater spark discharge (Chaussy et al., 1976, 1977, 1978, 1979; Eisenberger et al., 1977a, 1977b).

PHYSICAL PROPERTIES OF SHOCK WAVES

The basic technical principles of shock waves are often confused with those of ultrasound waves. Although both shock waves and ultrasonic waves are governed by the same laws of acoustics, they are fundamentally different. Ultrasound consists of a sinusoidal wave of defined wave length with alternating positive and negative deflections. Shock waves consist of a single positive-pressure front of multiple frequencies with a steep onset and gradual decline (Fig 20–1). Shock waves undergo substantially lower attenuation than ultrasound waves when propagated through water or body tissue. Thus, shock waves can be transmitted through water and into the body without major loss of energy or damage to the tissue.

The use of shock waves in the medical field for the destruction of urinary stones is based on the following properties, which were evaluated during a six-year investigational phase:

1. Shock waves lead to disintegration of brittle material such as human kidney stones.

2. Shock waves generated by the underwater discharge of a capacitor can be reliably reproduced.

3. Shock wave energy can be propagated to the stone without energy loss or damage to the tissue.

4. Shock waves can be precisely focused by integrating the energy source into a suitable reflecting system (Chaussey et al., 1980, 1982).

EXPERIMENTAL INVESTIGATIONS

MECHANISM OF STONE DISINTEGRATION

Shock waves give rise to mechanical stresses in brittle materials, such as kidney stones, which eventually exceed the comprehensive strength of the material and lead to disintegration after repeated application.

Once the focused shock wave reaches the stone, the pressure front is partially reflected at the front surface of the stone, thus being split into compressive and tensile components, which leads to the build-up of a high pressure gradient. This force eventually exceeds the comprehensive strength of the stone in this particular region and causes disintegration of the front surface of the stone (Fig 20–2, A). A portion of the wave contin-

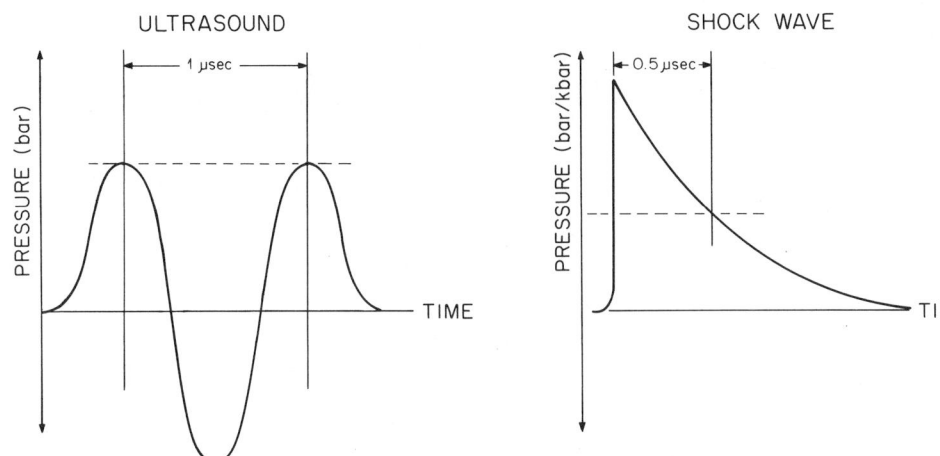

FIG 20–1.
Comparison of time-pressure profile of ultrasound and shock wave.

ues through the stone and is reflected at the rear surface where the same effect takes place (Fig 20–2, B). The disintegration of the outer layers exposes new surfaces which, in turn, are broken into fragments once the comprehensive strength is exceeded. This finally results in complete disintegration of all stone parts (Fig 20–2, C).

It was further found that repeated exposure to low-energy shock waves in the pressure range of 700–900 bar resulted in considerably smaller particle size than shock waves in the pressure range of 3,000–4,000 bar, which were used in the beginning.

Human kidney stones of different chemical compositions can be disintegrated by multiple exposures to shock waves (Chaussey et al., 1978). This finding is independent of the chemical composition of the stone for all but cystine stones. Most cystine stones require a significantly higher energy level and more exposures.

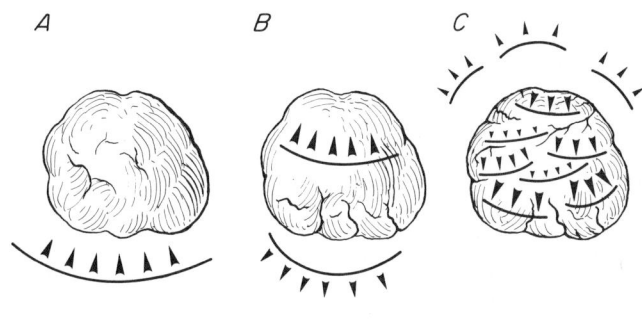

$\blacktriangle$ = Compressive wave

$\blacktriangledown$ = Tensile wave

FIG 20–2,A–C.
Principle of stone disintegration: shock waves create areas of shear and tear forces with build-up of high-pressure gradients.

EFFECT ON CELL POPULATIONS

Red blood cells subjected to shock waves showed an increased degree of hemolysis. The amount of free hemoglobin produced during a session of ESWL treatment, however, has no clinical relevance, as the relatively small volume of blood involved at the target area is only a minimal fraction of the entire blood volume.

When human lymphocyte cultures were exposed to shock waves, no cytolysis was found. Stimulation capability of human lymphocytes cultured after shock wave exposure also showed no difference in mitogenic stimulation or in the proliferation of mixed lymphocyte cultures. As indicated by these in vitro experiments, detrimental effects on proliferative activity of normal cells can be ruled out (Chaussy et al., 1976, 1980; Eisenberger et al., 1977).

EFFECTS ON IN VIVO EXPERIMENTAL SYSTEMS

Shock waves can be freely transmitted and propagated through the body without major loss of energy if an appropriate transmission medium with similar acoustic impedances, such as water, is used. The effects of shock wave exposure on rats are listed in Table 20–1. These experiments proved the safety of shock waves for all tissues but lung tissue. However, this is of little clinical significance, for as the shock wave is focused exactly on the stone, the lung region is outside the focal area.

EXPERIMENTAL STONE MODEL

Human kidney stones implanted in the renal pelvis of dogs were disintegrated by repeated exposures to extracorporeally induced shock waves to a size allowing their spontaneous passage. Using a semiellipsoidal reflector, extracorporeally induced shock waves can be fo-

TABLE 20–1.

Results of In Vivo Experiments Exposing Rats to Shock Waves

EXPOSURE SITE	CLINICAL RESULTS	PATHOLOGIC CHANGES			
		(24 HR AFTER EXPERIMENT)		(14 DAYS AFTER EXPERIMENT)	
		MACROSCOPIC	MICROSCOPIC	MACROSCOPIC	MICROSCOPIC
Thorax (n = 20)	Massive hemoptysis	+ + +	+ + +	– – –	– – –
With a sheet of Styrofoam (n = 20)	None	0	0	0	0
Abdominal cavity (n = 20)	None	0	0	0	0
Liver (n = 20)	None	(+)	(+)	0	0
Colon	None	(+)	(+)	0	0

cused and thus brought to bear on specific stones (Fig 20–3). The exact localization of the stone in the F_2 focal point was achieved by using a biplane x-ray system. First trials with ultrasound showed that a stone could be localized with ultrasound (Fig 20–4). However, the degree of stone disintegration and the progressive fragmentation could not be assessed properly with ultrasound. With a biplane x-ray system the stone can be

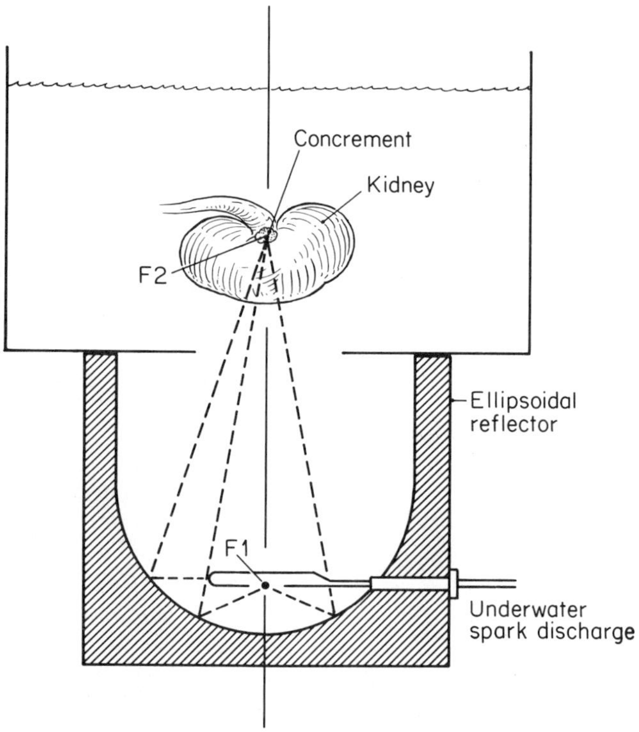

FIG 20–3.
Schematic drawing of arrangement of spark electrode (shock wave source in focal point F1), ellipsoidal reflector, and target (kidney stone in focal area F2).

detected fluoroscopically and the degree of stone disintegration assessed from the frame pictures.

All animal experiments proved that the shock waves used in the experimental system could be reliably reproduced, were safe, and led to successful disintegration of the stones. The shock wave energy needed for in vivo disintegration of human kidney stones implanted in the renal pelvis of mongrel dogs is within a pressure range that does not lead to damage of body tissues. Thus, the method was ready for clinical use in man (see Table 20–1; Chaussy et al., 1978, 1979, 1980, 1982).

CLINICAL EXPERIENCE WITH THE LITHOTRIPTER

Following these successful and encouraging experimental results, an apparatus for clinical use in humans was constructed (Chaussy et al., 1980, 1981). The major features of this machine are: the water tub with the integrated ellipsoid to focus the shock waves, the biplane x-ray system, the patient support and the control panel (Fig 20–5). Between the first human applications in February of 1980 and the installation at the second center in Stuttgart by the end of 1983, two refined models were constructed based on the experience of the Munich group with more than 1,000 patients. The third-generation model of the lithotripter (Human model 3) was the first commercially available model, and beginning in 1984 it was distributed worldwide (Fig 20–6).

TREATMENT

PATIENT PREPARATION

The preparation for ESWL is the same as for any conventional surgical stone removal (ECG, chest x-ray,

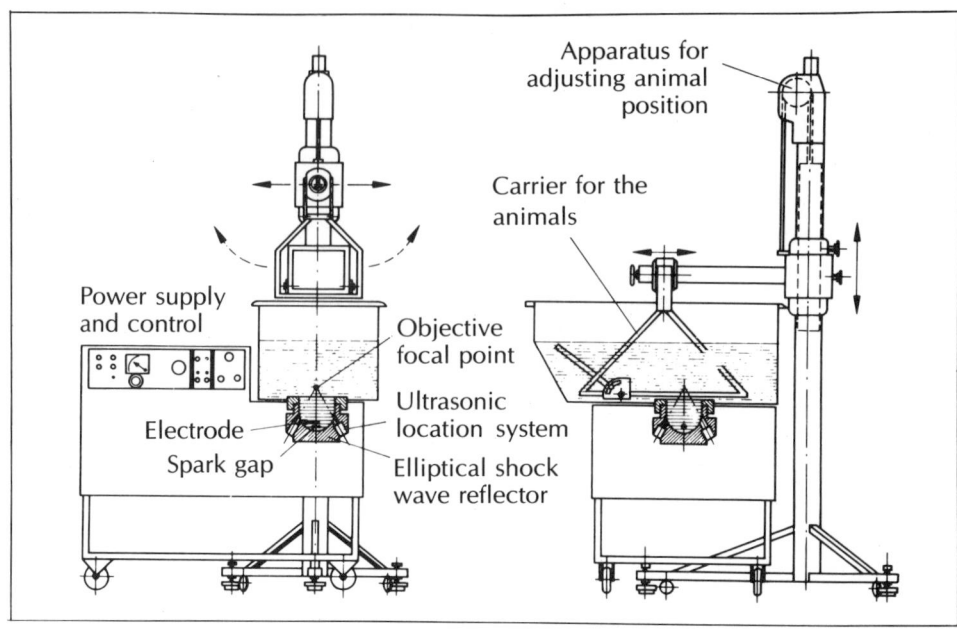

FIG 20–4.
Device for animal experiments with integrated ultrasound transducers for stone localization.

laboratory data). But there is no need to cross-match blood. No special measures are necessary aside from the administration of laxatives the day before treatment to have the bowels free of air on the day of treatment. On the day of treatment, before induction of anesthesia, a plain x-ray (KUB) is obtained to verify the exact position of the stone and to exclude strong shadowing by intestinal gas.

Experimental studies and clinical experience have both shown that shock wave treatment in the presence of intestinal gas does not cause any traumatization of the intestines. However, localization of small stones and assessing the degree of stone disintegration are more difficult when bowel gas is present.

ANESTHESIA

Anesthesia is mandatory, since shock wave therapy is painful. Although the pain caused by individual shock waves can be tolerated, studies with volunteers have shown that the pain caused by a series of shock waves

FIG 20–5.
First clinical device for ESWL (Dornier human model no. 1, used between February 1980, and May 1982).

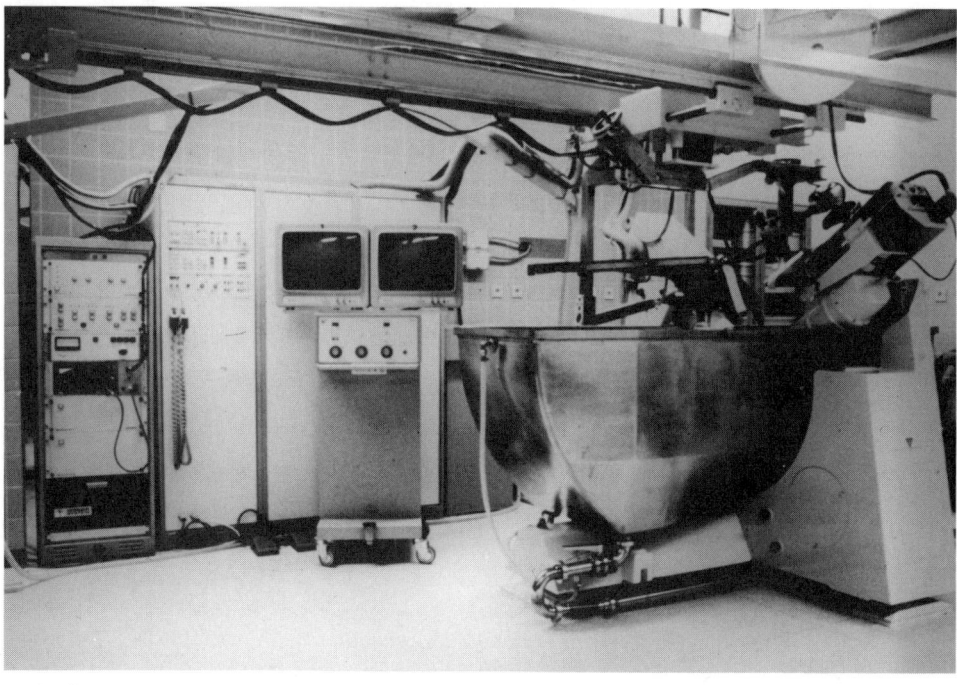

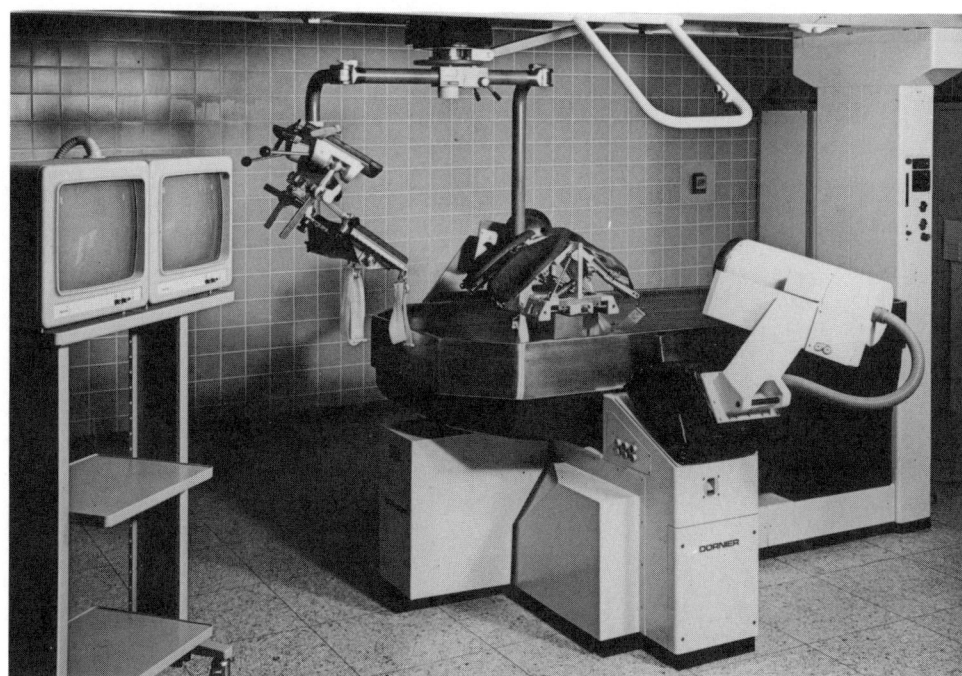

FIG 20–6.
First commercially available kidney lithotripter (Dornier human model no. 3, used from October 1983 to present).

cannot be tolerated without some form of anesthesia.

In most centers continuous epidural catheter anesthesia was found to be the method of choice. Epidural anesthesia is being used worldwide in 75%–95% of all patients. The catheter entry site is sealed watertight with a surgical incision drape. The indwelling epidural catheter offers the possibility of easily giving more anesthesia should the procedure be delayed or prolonged.

In our center, general anesthesia is only performed if there are contraindications for regional anesthesia or if requested by the patient. In high-risk patients general anesthesia is preferred, since it permits more extensive monitoring (CVP, PWP, AP) and it is preferred in children to prevent uncontrolled movement.

COURSE OF A ROUTINE TREATMENT

On satisfactory induction of anesthesia the patient is brought to the lithotripter unit. He is then placed on the patient support and fixed with straps to prevent involuntary movements due to buoyancy once he is immersed in the water. The patient is then suspended in the water bath, and the stones are localized using the x-ray system.

Two independent x-ray image conversion systems are used. Their axes intersect at the second focal point of the semiellipsoid. To achieve stone disintegration the patient has to be moved in such a way that the targeted stone is located in the center of the hairpins on both x-ray screens. Once the stone can be detected in the center of the crosshairs on both monitors, it is exactly lo-

cated in the second focal area of the shock wave front and treatment commences (Fig 20–7).

Depending on the stone size and hardness and the number of stones, treatment consists of 500–2,000 shock wave applications and lasts for 20–60 minutes. Fluoroscopy and/or frame pictures are taken after every 100 shock wave exposures to assess the degree of stone disintegration and to guide the repositioning of the stone(s) until complete disintegration of all stone parts is achieved (Table 20–2 and Fig 20–8). Average fluoroscopy time during the treatment is 100 sec ± 80 sec, including cases of multiple stone and staghorn stones. In these cases each stone or stone part has to be localized separately, which understandably adds to the overall radiation incurred by the patient.

AFTERCARE

Immediately after the treatment a KUB film is obtained to assess the final degree of stone disintegration. If the KUB reveals remaining particles larger than 4 mm, treatment can be continued if the patient is still under epidural anesthesia. Once treatment is definitely terminated, the patient is brought to the recovery

TABLE 20–2.

Lithotripter Treatment Data

ASPECT	RANGE	MEAN
Duration of treatment, min	20–60	45
No. of shock waves	300–2,000	1,350
Fluoroscopy time, sec	40–160	100

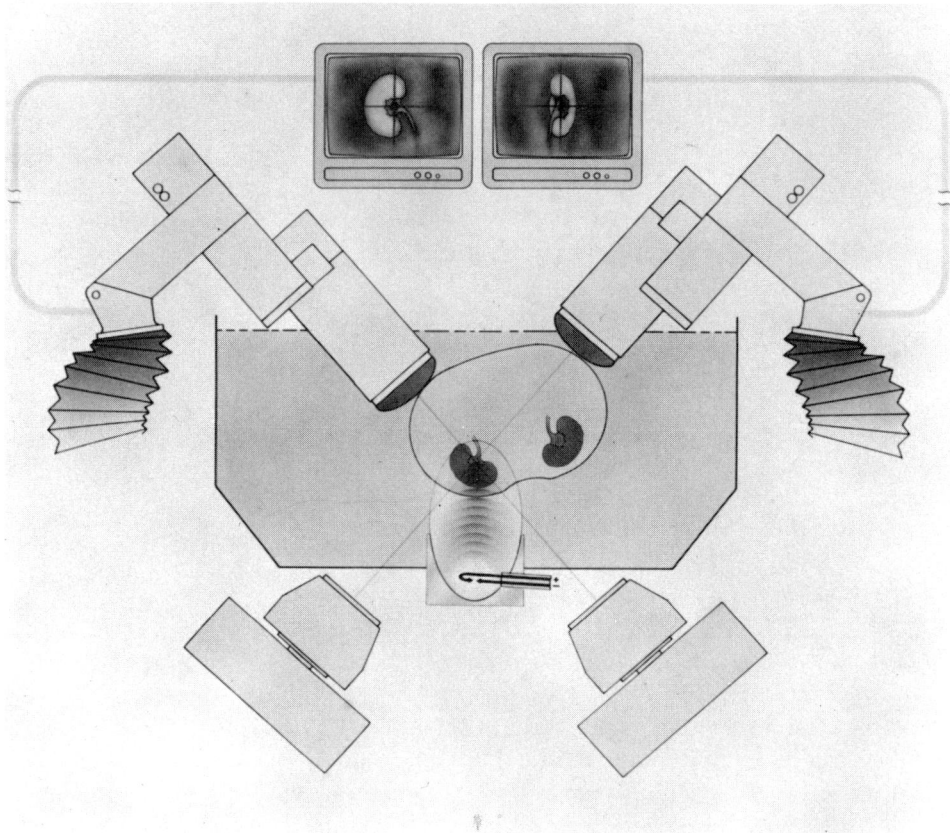

FIG 20–7.
Schematic drawing of localization system: ellipsoid with spark electrode in F1 and stone in F2 *(center)* and biplane fluoroscopic x-ray system with intersection of the central beams in F2.

room, and after he recovers from the anesthesia he is taken back to the ward. The IV line is disconnected as soon as the patient is ambulatory. He is then asked to move and exercise as much as possible, and a daily oral fluid intake of 2–3 L is recommended.

Pain medication is given on request. Although approximately 70% of all patients ask for an injection during the first 12 hours because of muscle discomfort in the treated area or at the skin entry site of the epidural catheter, the incidence of pain or colic is relatively low

FIG 20–8.
Arrangement of patient positioning with the patient fixed onto the support, allowing transmission of shock waves from the back.

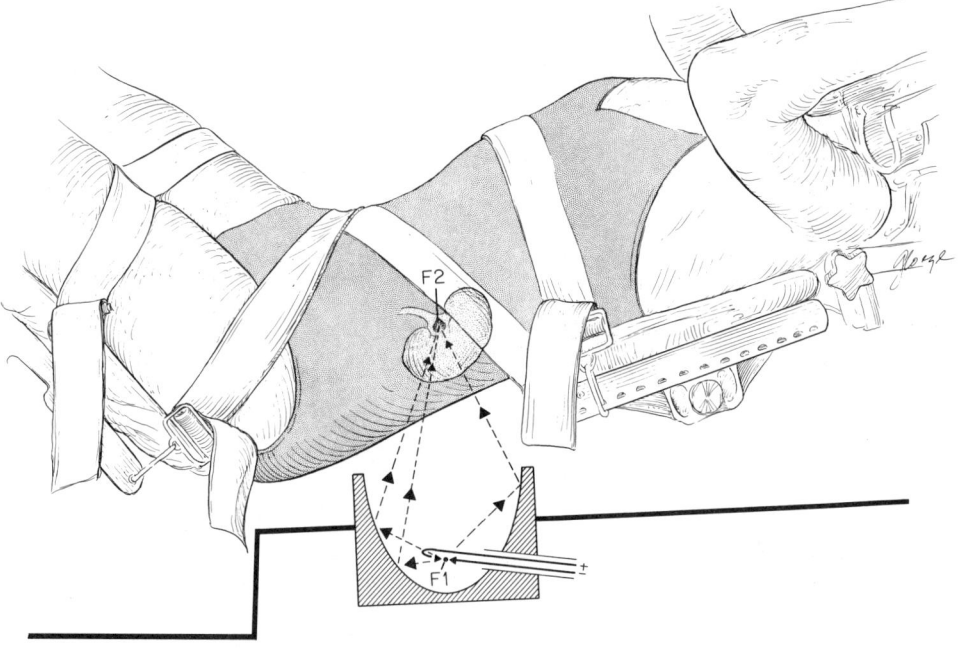

(30%). Fifty percent of these patients get relief of their symptoms from either oral pain medication or suppositories. Fifteen percent of the entire collective require IV or IM administration of narcotics.

KUB films are taken daily during the hospital stay to assess the degree of stone disintegration and to verify the location of the passing fragments. Ultrasound is performed to rule out perirenal fluid and blood collection and to monitor the degree of hydronephrosis, which is temporarily seen in more than 60% of the patients.

Thirty to forty percent of all stone patients leave the hospital stone-free after an average stay of 2.5 days posttreatment. In principle, patients are discharged once the KUB confirms complete disintegration as well as the beginning of the passage of the fragments, and ultrasound reveals no evidence of complications such as perirenal hematoma or gross hydronephrosis. The outpatient follow-up examination of the asymptomatic patient consists of an examination after day 14 (KUB, sonogram, blood pressure, urinalysis) and a three-month follow-up including the same examinations.

ESWL MANAGEMENT OF URINARY STONES

CONTRAINDICATIONS

Excluded from treatment with the lithotripter are patients with untreated or untreatable bleeding disorders and patients with decompensated heart deficiency (Table 20–3). Technical problems in positioning the patient on the support and/or difficulty in localizing the stone exist in grossly obese patients weighing over 150 kg, patients larger than 200 cm tall, and children smaller than 120 cm. Anatomical and/or functional alterations of the upper urinary tract can be urologic contraindications when adversely influencing drainage from the kidney.

When deciding whether a patient is eligible for ESWL, the impact of anatomical or functional anomalies is best assessed by KUB and IVP films. Compression of the abdomen during the IVP is helpful to obtain better contrast shadowing of the structures of the collecting system. If, with this, the intrarenal anatomy cannot be fully understood, a retrograde filling study under fluoroscopy is undertaken. In cases of severe narrowing at any level (calyceal neck ureteropelvic junction [UPJ], ureter, ureteral orifice, BPH, urethral stricture) with secondary distension of the collecting system that is not attributable to the stone itself, ESWL should not be considered as primary therapy because of possible problems with the discharge of the fragmented particles. Relative obstruction not attributable to the stone does not, however, necessarily preclude the successful use of ESWL. If the patient complains of intermittent pain at the time of diuresis with quick resolution thereafter, ESWL is the preferred method. Patients who are elective candidates for ESWL presenting with BPH and a residual urine volume in excess of 100 ml should undergo prostatic surgery prior to ESWL. Irregularly urodynamic patterns with functional impairment of the ureter need careful evaluation as to whether they will be amenable to ESWL (Chaussy et al., 1982, 1984; Eisenberger et al., 1985; Fuchs et al., 1985; Schmiedt and Chaussy, 1984).

Pregnancy is a firm contraindication for shock wave treatment. In the rare cases where interventional stone treatment has to be considered (frequent and untreatable pain, imminent urosepsis) placement of a nephrostomy tube using local anesthesia is preferable, and ESWL can be performed as early as seven days after delivery.

INDICATIONS

Pelvic and Calyceal Stones

Although the first clinical experience with ESWL only dates back a few years, the method has already become a routine procedure with well-established indications. As with every method being introduced into clinical use, the initial indications were restricted to have parameter-free conditions. Initially only those patients in good general condition with pelvic stones smaller than 1 cm or single calyceal stones were accepted. This confined the applicability of ESWL to 20% of a nonselected stone patient group (Fig 20–9). Success with these patients and lack of complications allowed the range of indications to be extended carefully (Table 20–4). Patients with pelvic stones larger than 1 cm and those with multiple stones have been included in the range of routine indications (Fig 20–10). Treatment of patients with urinary tract infections and/or treatment of infected stones showed no adverse effects when done under antibiotic cover. Experience has shown that approximately 75% of kidney stones can be treated by ESWL monotherapy (Fig 20–11).

TABLE 20–3.
ESWL Contraindications

General	Untreated bleeding disorders
	Decompensated heart insufficiency
	Pregnancy
Technical	Obesity (>150 kg)
	Height (adults >200 cm; children <120 cm)
Urologic	Obstruction distal to the stone (calyceal neck, UPJ, ureteral, BPH, urethral stenosis)

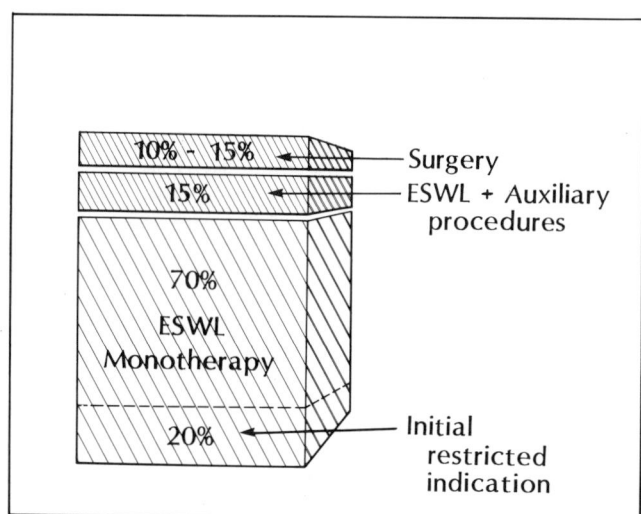

FIG 20–9.
Treatment-strategy trends.

10% – 15% — Surgery
15% — ESWL + Auxiliary procedures
70% ESWL Monotherapy
20% — Initial restricted indication

TABLE 20–4,

Present Range of Indications

Single and multiple stones in the kidney
   and the ureter
Partial and complete staghorn stones
Infected stones
Stones in solitary kidneys
High-risk patients
Children (>120 cm)
Radiolucent stones

TABLE 20–5.
Distribution of Treatment Modalities for Renal Stones, %*

CHARACTERISTIC	MONO-ESWL	MULTIPLE ESWL	ESWL AND PCN	SURGERY
Stones <2.5 cm	93	1	4	2
Partial staghorn stones	50	35	10	5
Complete staghorn stones	3	20	63	15

*Munich study results.

Partial and Complete Staghorn Stones

In the treatment of partial and complete staghorn stones, different strategies have evolved. Due to the large overall stone burden, success with one treatment in large partial or complete staghorn stones is the exception. This is only feasible in calculi in nondilated collecting systems (Fig 20–12). Cases of staghorns with substantially larger stone mass, i.e., when they are filling a dilated collecting system, can be managed in a two-stage procedure, which is usually performed within 2–4 days (Fig 20–13, A and B).

In principle, fractionated disintegration might be feasible in most cases. Because of the large stone burden, however, the period until the patient is stone free is considerably prolonged, since no obstructing particles should be remaining in the upper tract before a consecutive session is undertaken. Generally, multiple sessions are required to completely disintegrate those stones, and there is a considerably higher incidence for the need of post-ESWL auxiliary procedures during stone passage. Between sessions close patient surveillance is necessary to prevent the possible hazards of urosepsis secondary to prolonged obstructive pyelonephritis. This requires good patient compliance, especially when the patient will be followed up on an outpatient basis for a long time.

Because of the large stone fragment burden, in stones of partial staghorn size with extremely large mass and in complete staghorn calculi, a percutaneous procedure is performed first to debulk the stone. In a second session, usually after two to four days, ESWL is employed for the remaining calyceal stone parts. In some cases three or more sessions, either percutaneous nephrostomy or additional ESWL treatment under epidural anesthesia, have to be undertaken due to the size of the original stone. This combined approach is now routinely used for the treatment of excessively large stones (Table 20–5).

Primary indications for open surgery are stone cases requiring repair of UPJ stenosis. In these cases all stone parts that can be reached from the renal pelvis are removed at the time of correction of the underlying obstruction. Approximately 5–8 days postoperatively, the UPJ anastomosis has healed, and when a nephrostogram reveals no extravasation, ESWL can be performed for the remaining calyceal stone parts (Fig 20–14). This combined approach allows removal of complicated staghorn stones with minimal trauma to the parenchyma, since calyceal stone fragments formerly accessible only by nephrostomies can be readily disintegrated by ESWL (Chaussy and Schmiedt, 1983, 1984; Eisenberger et al., 1985; Schmiedt and Chaussy, 1984). Another indication for open surgery is in staghorn stones with a small central (pelvic) and a large peripheral (calyceal) stone mass, especially when associated

FIG 20–10.
Course of typical ESWL case. **A,** KUB before treatment. **B–E,** x-ray control before ESWL and after 100, 400, and 600 shock wave applications, showing increasing degrees of stone disintegration. **F,** KUB after termination of treatment (800 shocks), showing complete stone disintegration.

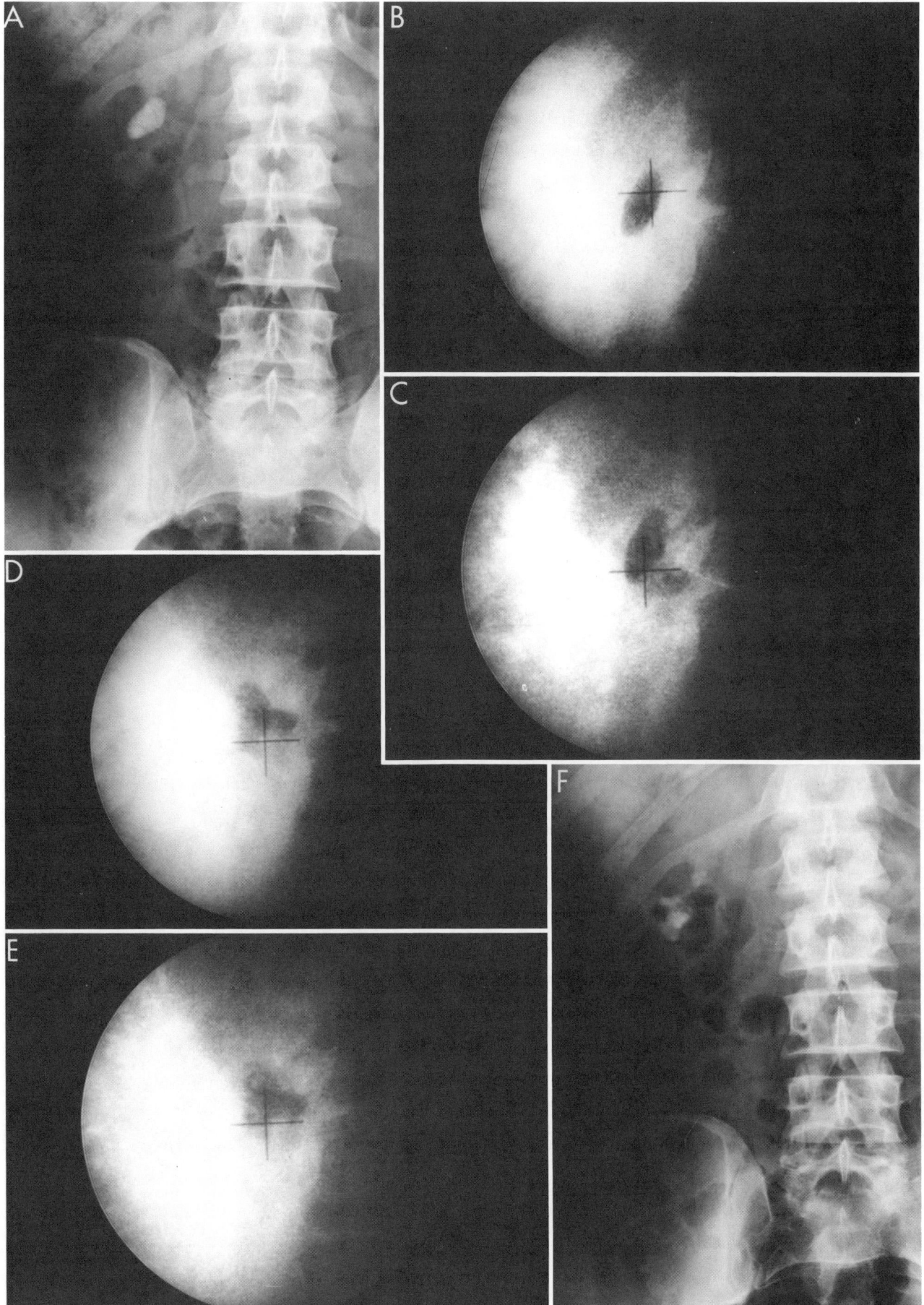

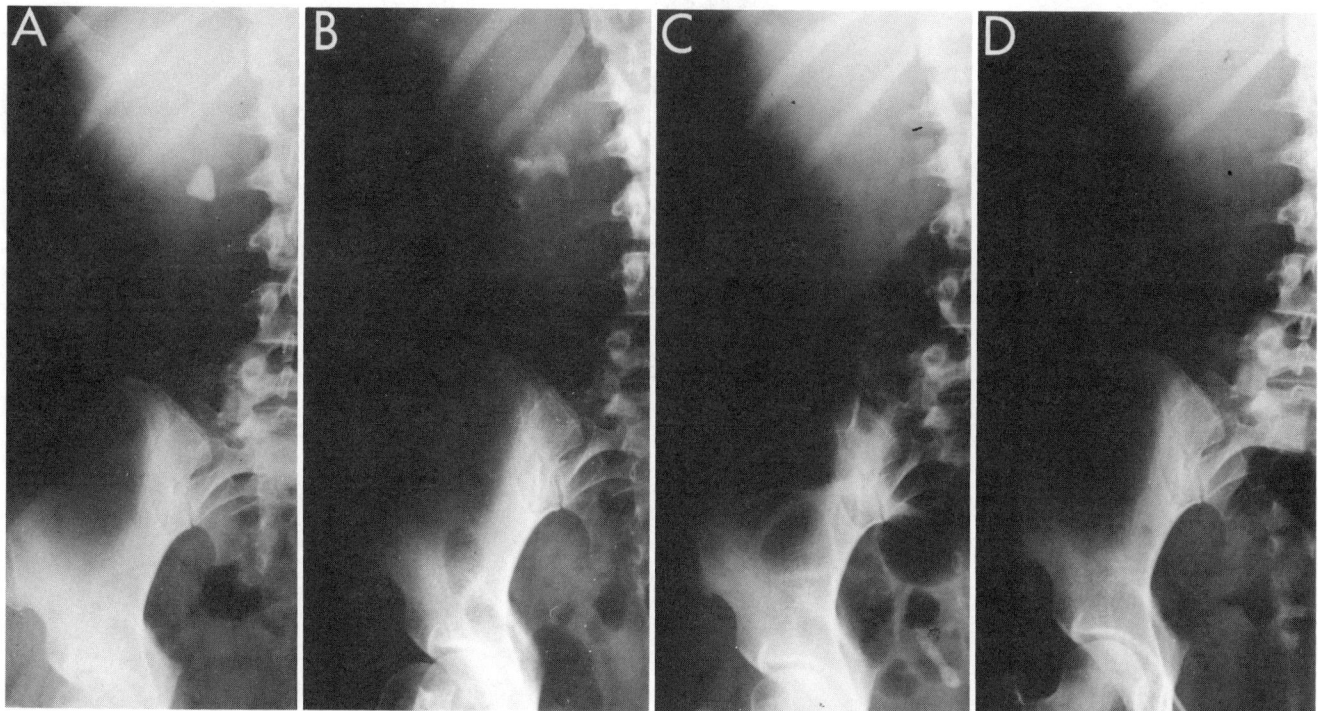

FIG 20–11.
ESWL of "easy" kidney stone. **A,** KUB before treatment. **B,** KUB immediately after ESWL showing complete disintegration; stone parts spread all over collecting system. **C,** after day 3: "steinstrasse" in pelvic ureter. **D,** after day 7: stone-free.

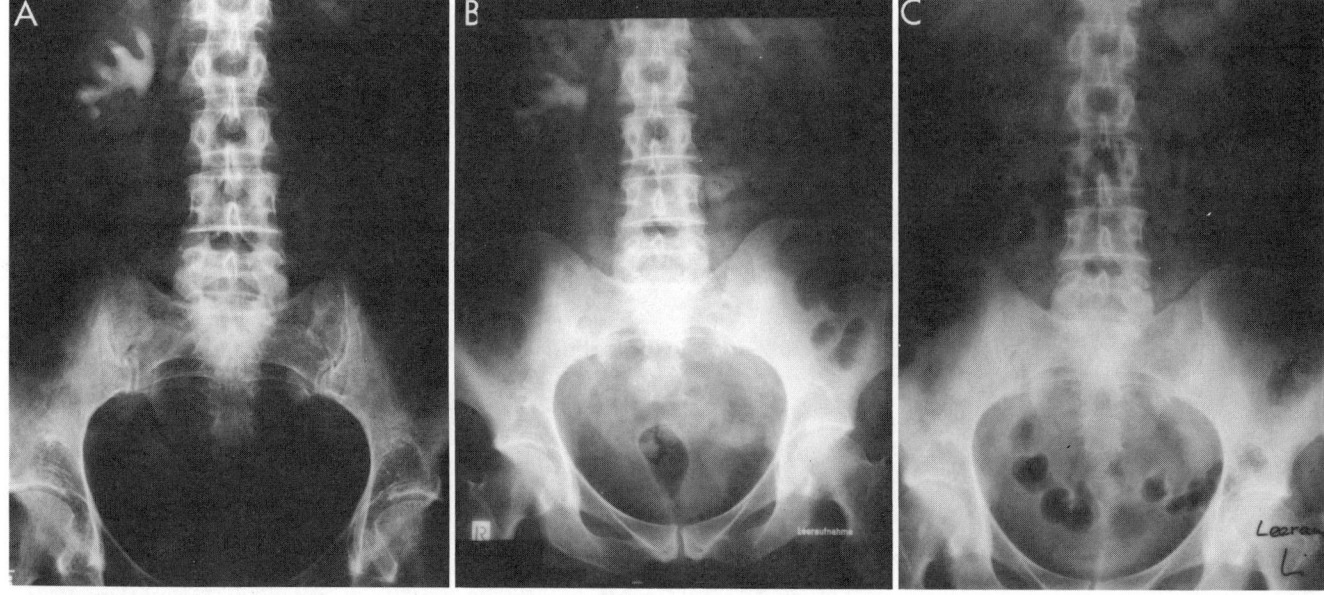

FIG 20–12.
ESWL of complete staghorn stone. **A,** KUB before treatment. **B,** KUB immediately after ESWL showing disintegration and stone particles in proximal ureter. **C,** follow-up KUB two weeks after ESWL, showing no more fragments in the upper urinary tract.

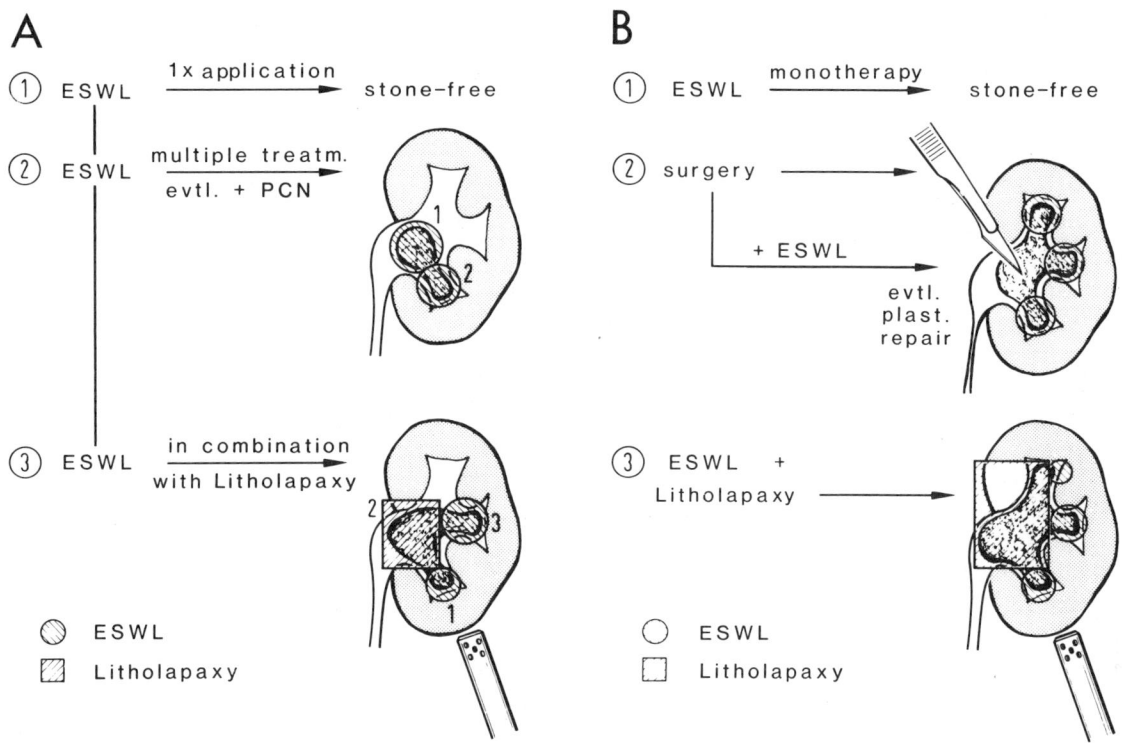

FIG 20–13.
Strategies for the treatment of **(A)** partial and **(B)** complete staghorn stones.

with long, narrow infundibuli. In these cases nephrostomies are mandatory, since the disintegrated particles would not pass through the narrow infundibuli.

Ureteral Stones

Ureteral stones located above the ilial crest that cause symptoms and are too large to be passed sponta-

neously are primarily eligible for ESWL treatment.

Initially ureteral stones were treated in situ with a reported success rate no higher than 50%. At the time of open surgery following ESWL it was found that most of the stones had been disintegrated, but the pieces were being held together (Chaussy et al., 1981). They could not pass because they were wedged in by the

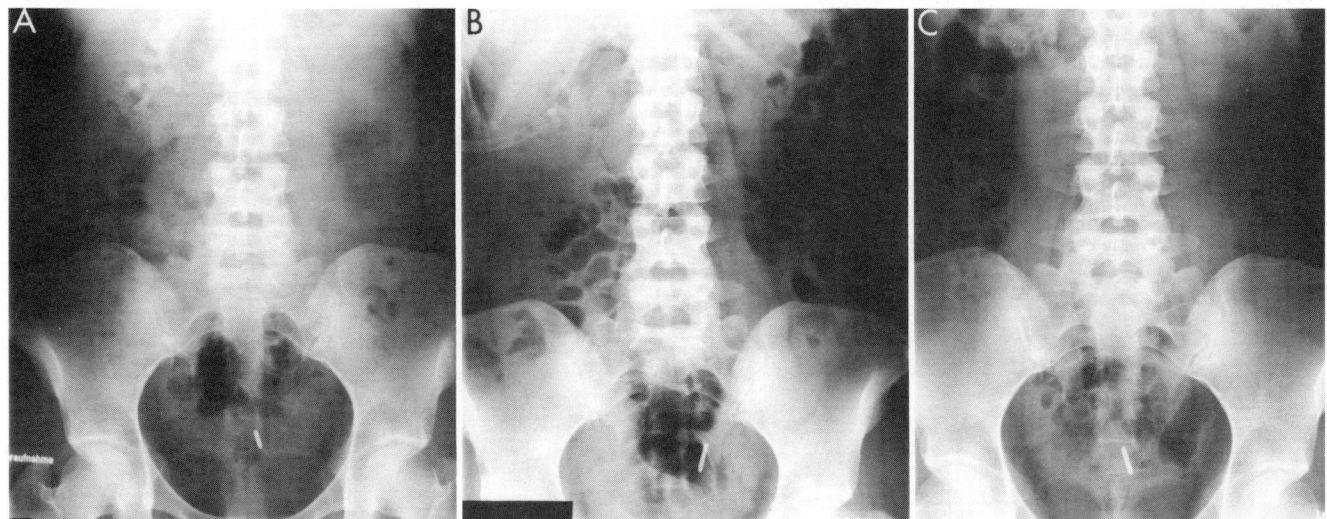

FIG 20–14.
Complete staghorn treated with a combination of open surgery and ESWL. **A,** KUB before treatment. **B,** after open surgery with remaining stone parts in lower calyceal group and nephrostomy tube. **C,** after ESWL (postoperative day 5).

edematous ureteral wall. This was noted mainly in stones totally obstructing the ureter which were situated in the same place longer than six weeks (Fig 20–15).

Thus, only stones with a shorter history were then accepted for primary ESWL treatment. The remainder were manipulated back into the renal collecting system, followed by ESWL under the same anesthesia. For ureteral manipulation, ureteral catheters (4 or 6 F) are used. If this fails, a ureteroscope is employed either to reposition the stone or if possible to retrieve it instantly. With this the overall success of the initial treatment varied between 75% and 95%, the discrepancy being explained by the difficulty in selecting patients for either approach, since the exact length of the history cannot be obtained in most cases (Miller et al., 1985).

The most recent policy in the management of upper ureteral stones is therefore primarily to plan for a combined procedure. First an attempt at pushing the stone back is made. If this is not possible, an attempt is made to pass a 4 F stent around the stone, which gives the stone at least some additional space for successful disintegration (Table 20–6; Fuchs et al., 1985; Miller et al., 1985).

Radiolucent and Semiopaque Stones

Nonobstructive radiolucent and semiopaque stones, such as uric acid stones and certain cystine stones, should have initial medical therapy. If the symptoms include obstruction with urinary tract infection, the primary approach is to relieve obstruction by insertion of a percutaneous nephrostomy accompanied by appropriate antibiotic coverage. X-ray visualization of radiolucent and slightly opaque stones is only possible after administration of contrast medium. The stone can then

TABLE 20–6.

Success Rate of ESWL Treatment of Ureteral Stones in Relation to the Stone Location at Treatment (UCLA Results)

LOCATION	NO.	(%)	SUCCESSFUL DISINTEGRATION, %	STONE FREE, % AT 2 WK	STONE FREE, % AT 3 MO
Ureteral stones repositioned into renal collateral system	71	(59)	100	85	98
Upper ureter plus stent	33	(27)	94.6	92	99
Midureter plus stent	11	(7.5)	91.2	89	99
Upper ureter, no stent	6	(4.4)	33	25	33*
Midureter, no stent	2	(2.1)	50	50	50*

*The remainder was rendered stone free after auxiliary procedures.

be localized by using the filling defect as a target area or by focusing the area in which the stone was previously determined to be. Assessment of the amount of disintegration is difficult because the fragments cannot be detected radiographically. Thus, a clear-cut decision as to when fragmentation of the stone is sufficient cannot be obtained, and treatment is arbitrarily terminated based on experience with previous cases with similar stones.

High-risk Patients

High-risk patients, such as those with hypertension or a heart condition, can also receive ESWL without major problems. These patients are usually treated using general anesthesia, which allows for close monitoring of the cardiocirculatory and lung parameters (ECG, BP, CVP, PWP). Analysis of the Munich patient group

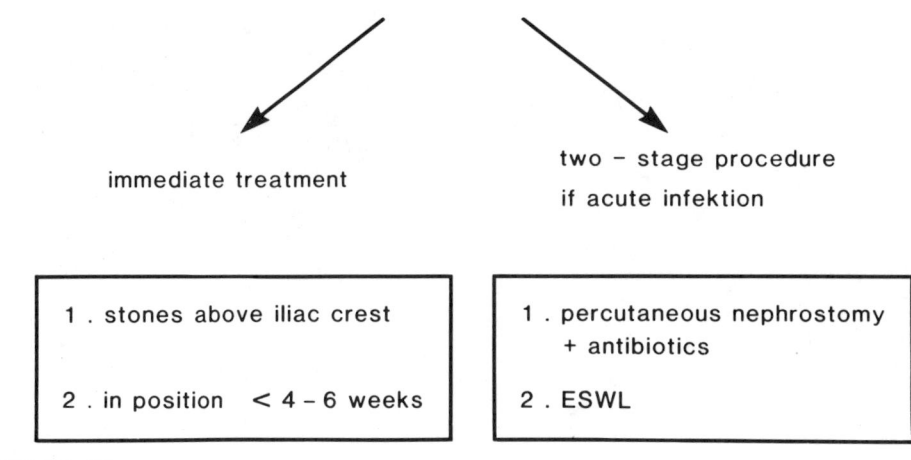

FIG 20–15.
Strategy for the treatment of ureteral stones with ESWL.

according to the risk group grading of the American Association of Anesthesiologists shows that 4.9% of the patients who were successfully treated belonged to ASA group IV. That means they were cardiopulmonary high-risk patients (Table 20–7).

Treatment of Children

Although nephrolithiasis is relatively rare in the pediatric age group, the advent of a noninvasive treatment modality for children was especially appreciated. ESWL is particularly advantageous because it can be repeated without detrimental adverse effects. With the Dornier lithotripter currently available, children no smaller than 120 cm can receive ESWL treatment (Fig 20–16). Smaller children cannot be positioned on the patient support (Chaussy et al., 1984; Fuchs et al., 1985).

To prevent the lungs from being hit by the shock waves, a Styrofoam layer is positioned on the back of the little patient to cover the lower region of the lungs. Treatment is performed with general anesthesia to ensure that the child does not move during the procedure and to keep the respiration movements low—a further prophylaxis of lung exposure. In general children get rid of the disintegrated stone material with fewer problems than adults with stones of similar size.

CLINICAL ASSESSMENT OF ESWL

The success rate of ESWL depends on various factors, such as the size of the original stone(s), the stone

TABLE 20–7.
Distribution of ASA-Risk Group Patients Treated With ESWL (Munich Results)

RISK GROUP	% OF PATIENTS
I	40.91
II	37.80
III	17.00
IV	4.9
V	0

location, and the presence of anatomical and/or function alterations of the upper urinary tract. Taking these into account, the overall results are broken down into different groups to obtain comparable results. During the past five years, three main groups of ESWL stones have evolved. Accordingly, "easy" kidney stones, "complicated" kidney stones, and upper ureteral stones are evaluated separately.

In all kidney stones the rate of successful stone disintegration is in the 97% to 99% range (Table 20–8). Only a small minority of all renal stones—namely, cystine stones and a few stones of different chemical composition with a very long history—cannot be successfully treated with ESWL. It can by no means be predicted, however, which of these stones will not respond sufficiently. Experience has shown that solitary or multiple kidney stones of an overall size of up to 2 cm without obstruction distal to the stone can be considered easy stones for ESWL treatment, with a 99%

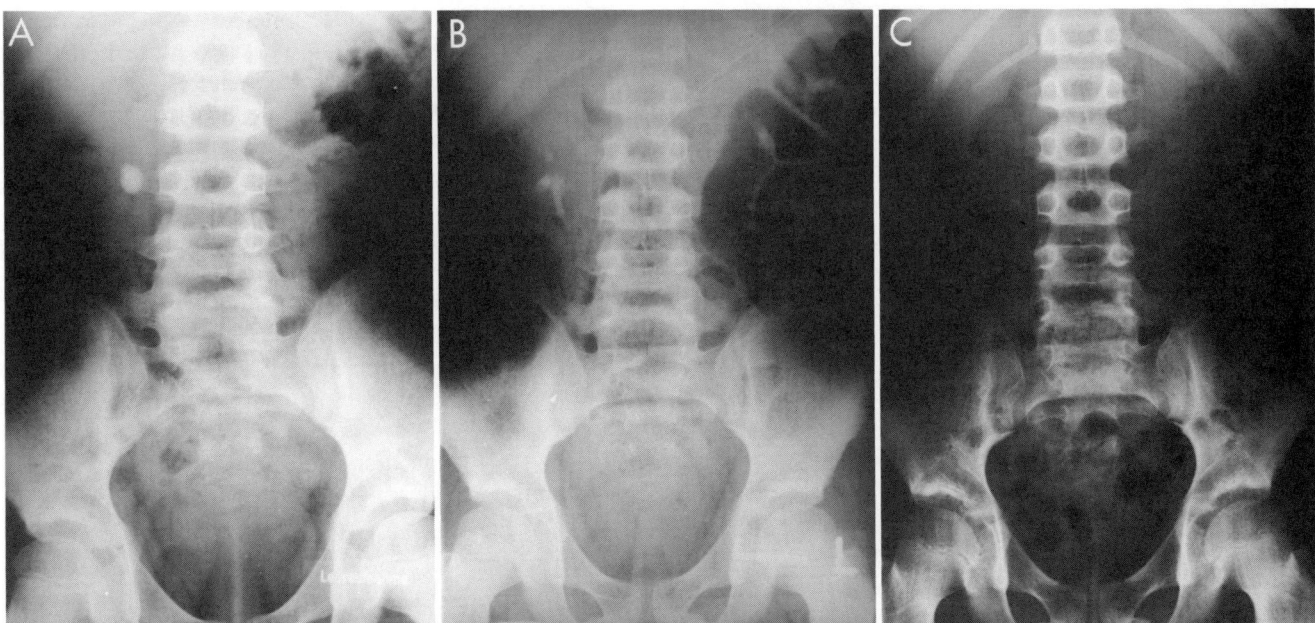

FIG 20–16.
ESWL of 6-year-old boy. **A,** KUB before treatment. **B,** KUB immediately after ESWL confirming success of treatment. **C,** postoperative day 5: stone-free after asymptomatic postprocedural course.

TABLE 20–8.

World Experience With ESWL

FEATURES OF STUDY	MUNICH	STUTTGART	SAPPORO	UCLA
Date of opening	2/80	10/83	9/84	3/85
No. of patients	2,200	1,800	500	600
Success rate, %	99.0	99.1	97.0	99.5
Stone free, %	85	75	77	80
Spontaneously passable, %	11	20	19	16
Fragments > 4 mm, %	2	4	3	4
Open surgery, %	1	0.2	0.5	0
Complications				
Fever, %	6	3	5	4
Pain/colic, %	25	28	32	22
Auxiliary measures, %	16.4	14.6	15.8	17.0
Pre-ESWL, %	5.1	8.3	4.4	12.2
Post-ESWL, %	11.3	6.3	11.4	4.8

TABLE 20–9.

Complications After ESWL (Munich and UCLA Results)

EASY, %	STONES	PROBLEM, %
24	Pain/colic	34
5	Fever	36
6	Auxiliary measures	28

TABLE 20–10.

Auxiliary Measures After ESWL (Munich and UCLA Results)

EASY, %	STONES	PROBLEM, %
5	Ureteral manipulation (UC, loop, URS)	22
1	Percutaneous nephrostomy	4
—	Open operation	2

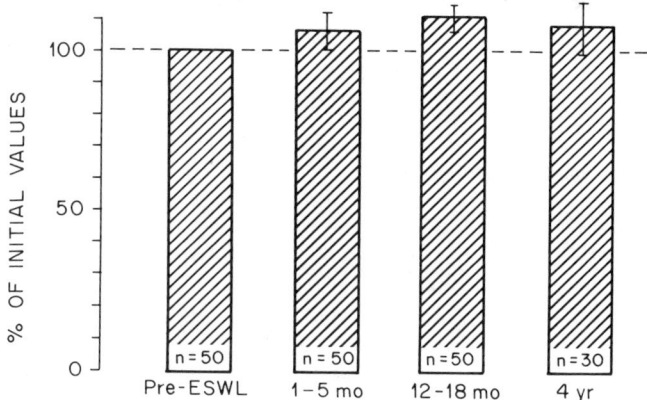

FIG 20–17.
Short- and long-term follow-up control of renal function after ESWL (in percent of initial value).

success rate and a low rate of postprocedural complications (Tables 20–9 and 20–10; Fig 20–17). Coincidentally, stones smaller than 2 cm are also considered ideal for percutaneous removal. When planning a percutaneous removal the actual stone location needs special attention, however, since stones located in the middle and upper calyces are more difficult to approach percutaneously. ESWL offers the advantage that all radiopaque renal stones can be easily localized and treated irrespective of their actual location in the collecting system. Since 25%–35% of renal stones are located in or extend into middle and upper calyces, the advantage ESWL has in the treatment of these patients is easy to see. Noninvasiveness, the simplicity of treatment, and the lower rate of postprocedural complications explain why percutaneous lithotripsy is being increasingly superseded by ESWL in the treatment of the so-called easy stones.

When is it best to use percutaneous lithotripsy in the treatment of small renal calculi? In renal stones with stenosis hampering proper discharge of the disinte-

grated fragments, or in stones presenting with severe functional alterations of the motility of the upper urinary tract, percutaneous lithotripsy has the advantage that the underlying anatomical anomaly can be corrected at the time of percutaneous stone surgery. Furthermore, in cases of disturbed upper tract motility, it is advantageous to remove the stone rather than disintegrate it. So stone location and the configuration of the upper tract mainly influence the choice of the appropriate treatment modality for the removal of small- to medium-sized kidney stones. Table 20–5 shows that approximately 70% of the nonselected stone patients belong by size in the group of easy stones.

What is the role of ESWL monotherapy in the treatment of solitary or multiple stones of a size between 2.5 and 4 cm? The common euphoria over the belief that all stones, irrespective of size, could and should be managed by ESWL monotherapy, which was prevalent after the first successful series, has faded away. ESWL experience with larger stones showed a significant increase in the rate of complications during fragment discharge after successful stone disintegration. This finding is consistent with a relatively high rate of auxiliary procedures needed and a prolongation of the hospitalization time (see Tables 20–9 and 20–10).

The choice between ESWL and a percutaneous procedure as a primary treatment strategy must be thoroughly weighed. These considerations also apply in determining the indications for the treatment of partial and complete staghorn stones.

REFERENCES

1. Alken P, Hutschenreiter G, Gunther R, et al: Percutaneous stone manipulation. *J Urol* 1981; 125:463–466.
2. Chaussy C, Eisenberger F, Wanner K, et al: The use of shock waves for the destruction of renal calculi without direct contact. *Urol Res* 1976; 4:175.

3. Chaussy C, Eisenberger F, Wanner K: Die Implantation humaner Nierensteine- ein einfaches experimentelles Steinmodell. *Urologe Ausg* 1977; 16:35.

4. Chaussy C, Eisenberger F, Wanner K, et al: Extrakorporale Anwendung von Stosswellen bei der Behandlung des Harnsteinleidens; Teil 11. *Aktuel Urol* 1978; 9:95.

5. Chaussy C, Schmiedt E, Forssmann B, et al: Contact-free renal stone destruction by means of shock waves. *Eur Surg Res* 1979; 11:36.

6. Chaussy C, Brendel W, Schmiedt E: Extracorporeally induced destruction of kidney stones by shock waves. *Lancet* 1980; 1:1265–1268.

7. Chaussy C, Schmiedt E, Jocham D, et al: First clinical experience with extracorporeally induced destruction of kidney stones by shock waves. *J Urol* 1981; 127:417–420.

8. Chaussy C, Schmiedt E, Jocham D, et al: in Chaussy C (ed): *Extracorporeal Shock Wave Lithotripsy*. Munich, Karger Verlag, 1982.

9. Chaussy C, Schmiedt E, Jocham D: Nonsurgical Treatment of Renal Calculi with Shock Waves, in Roth RA, Finlayson B, (eds): *Stones, Clinical Management of Urolithiasis*. Baltimore, Williams & Wilkins, 1982.

10. Chaussy C, Schmiedt E: Shock wave treatment for stones in upper urinary tract. *Urol Clin North Am* 1983; 10:743–750.

11. Eisenberger F, Schmiedt E, Chaussy C, et al: Aspekte zum derzeitigen Stand der beruehrungsfreien Harnsteinzertruemmerung. *Dtsch Aerzteblatt* 1977; 17:1145.

12. Eisenberger F, Chaussy C, Wanner K: Extracorporale Anwendung von hochenergetischen Stosswellen—ein neuer Aspekt in der Behandlung des Harnsteinleidens. *Aktuel Urol* 1977; 8:3–15.

13. Eisenberger F, Fuchs G, Miller K, et al: Non-invasive renal stone therapy with extracorporeal shockwave lithotripsy (ESWL), in *Proceedings of the 3rd Multinational Postgraduate Course: Radiology Today*. Berlin, Springer, 1984.

14. Eisenberger F, Fuchs G, Miller K: Extracorporeal shock wave lithotripsy and endourology: An ideal combination for the treatment of kidney stones. *World J Urol* 1985; 3:41–47.

15. Fernstroem I, Johannson B: Percutaneous pyelolithotomy: A new extraction technique. *Scand J Urol* 1976; 4:257–259.

16. Forssmann B, Hepp W, Chaussy C, et al: Eine Methode zur beruhrungsfreien Zertrummerung von Nierensteinen durch Stosswellen. *Biomed Tech (Berlin)* 1977; 22:164.

17. Fuchs G, Miller K, Rassweiler J: Alternatives to open surgery for renal calculi: Percutaneous nephrolithotomy and extracorporeal shockwave lithotripsy, in Shilling A (ed): *Klinische und experimentelle Urologie*. Munchen, Zuckschwerdt, 1984.

18. Fuchs G, Miller K, Rassweiler J, et al: Extracorporeal shock wave lithotripsy: One-year experience with the Dornier lithotripter. *Eur Urol* 1985; 11:145.

19. Gumpinger R, Miller K, Fuchs G, et al: Antegrade ureteroscopy for stone removal. *Eur Urol* 1985; 11:199–202.

20. Huffman JL, Bagley DH, Lyon ES: Transurethral removal of large ureteral and renal pelvic calculi using ureteroscopic ultrasonic lithotripsy. *J Urol* 1983; 130:31.

21. Korth K: *Percutaneous Renal Stone Surgery*. Berlin, Springer, 1984.

22. Kurth KH, Hohenfellner R, Altwein JE: Ultrasound litholapaxy of staghorn calculus. *J Urol* 1977; 117:242–243.

23. Marberger M, Stackl W, Hruby W: Percutaneous litholapaxy of renal stones. *Eur Urol* 1982; 8:236–242.

24. Miller K, Fuchs G, Bub P, et al: Financial analysis, personnel planning and organizational requirements for the installation of a kidney lithotripter at a urologic department. *Eur Urol* 1984; 2:210–217.

25. Miller K, Fuchs G, Rassweiler J, et al: Treatment of ureteral stone disease: The role of ESWL and endourology. *World J Urol* 1985; 3:445.

26. Perez-Castro Ellendt E, Martinez Pineiro JA: Transurethral ureteroscopy—a current urological procedure. *Arch Exp Urol* 1980; 33:445.

27. Schmiedt E, Chaussy C: Extracorporeal shock-wave lithotripsy of kidney and ureteric stones. *Urol Int* 1984; 193–198.

28. Segura JW, Patterson DE, Le Roy AJ, et al: Percutaneous lithotripsy. *J Urol* 1983; 130:1051.

29. Wickham J, Miller R: *Percutaneous Surgery of Renal Calculi*. London, Churchill-Livingstone, 1983.

Chapter 21

Renal Cystic Disease

MARGUERITE C. LIPPERT, M.D.

Renal cysts are abnormal dilations of either renal tubules, ducts, or glomerular capsules or are diverticulum-like structures possibly in continuity with the nephron. Renal cystic disease can involve both kidneys diffusely as autosomal dominant polycystic kidney disease, one particular area of both kidneys as medullary sponge kidneys, or just one kidney or part of it as multicystic kidney. Alternatively, a cystic tumor can replace normal renal parenchyma as in multilocular cyst. Finally, simple cysts can occur alone or together anywhere in the kidney.

Classification of cysts for the purpose of simplifying concepts does not necessarily provide a practical, clinical framework for such diverse cystic diseases. Hence, although the classification of Osathanondh and Potter (1964), based on microdissection studies, provided a basis for future research on the pathogenesis of renal cystic disease, the clinician was not aided in making his diagnosis from history, physical examination and radiographic studies. Meaningful clinical classifications of renal cystic disease based on clinical and radiographic features, genetic investigation, and morphology, as that of Bernstein (1976), honestly portray the diversity of renal cystic diseases, but do not lend themselves to a simple conceptual framework. Table 21–1 gives a classification from Glassberg and Filmer (1985) that offers a compromise of conceptually broad genetic categories, but still displays adequately the diversity of renal cystic diseases. The advent of ultrasonography, magnetic resonance imaging (MRI), and computed tomography has drastically improved the clinician's ability to diagnose and differentiate between the various renal cystic diseases. However, the importance of a careful family history is emphasized by the classification of renal cystic diseases into genetic categories.

AUTOSOMAL DOMINANT POLYCYSTIC KIDNEY DISEASE (ADULT TYPE)

Autosomal dominant polycystic kidney disease (DPK) is the most common form of cystic disease of the kidney. It is the third most common cause of end-stage renal disease (ESRD); 6% to 12% of patients receiving chronic dialytic therapy have DPK. Patients typically develop chronic renal failure in the fifth decade of life. The highest incidence rate of ESRD from DPK is 1.5 per 100,000 population, which occurs between the ages of 45 and 64 years (Eggers et al., 1984). DPK occurs in about one of every 1,250 live births and is characterized by bilateral cystic disease in enlarged kidneys with a retained reniform shape. There is an association with cysts in other organs, the most common of which is the liver.

CLINICAL FEATURES

Typically, for the first 10 years of life, the kidneys are normal in function and anatomy. From age 10 to 30 years, ultrasound may show the presence of cysts, although the patient may be asymptomatic. However, from age 30 to 40 years, a diagnosis may be made because of symptomatic presentation such as palpable kidneys, microscopic or gross hematuria, urinary tract infection, flank pain, and renal colic from passing clots. Although elevation of creatinine may begin between 40

TABLE 21–1.

Cystic Disease of the Kidney

Genetic
  Autosomal dominant (adult) polycystic kidneys (DPK)
  Autosomal recessive (infantile) polycystic kidneys (RPK)
  Juvenile nephronophthisis—medullary cystic disease complex (NMCD)
    Juvenile nephronophthisis (autosomal recessive)
    Medullary cystic disease (autosomal dominant)
  Congenital nephrosis (autosomal recessive)
  Cysts associated with multiple malformation syndromes
Nongenetic
  Multicystic kidney (multicystic dysplasia)
  Multilocular cyst (multilocular cystic nephroma)
  Simple cysts
  Medullary sponge kidneys (MSK) (<5% inherited)
  Glomerulocystic kidneys
  Acquired renal cystic disease in chronic dialysis patients (ARCD)

and 50 years of age, dialytic or transplant therapy for ESRD does not typically become necessary until after age 50 years, and the mean age at death is 50 years (Dalgaard, 1957). Without dialytic or transplant therapy, patients die of uremia (59%), heart failure, and cerebral hemorrhage. However, more recent studies suggest that many patients today are being diagnosed as having DPK without symptomatic presentation and that these patients may remain free of ESRD until their 70s (Delaney et al., 1985; Churchill et al., 1984).

Half of DPK patients also have hepatic cysts, but these are not pathologic (Levine et al., 1985a). There are also scattered reports of cysts in the pancreas, testes, ovaries, thyroid, and spleen, which may or may not be coincidental findings. Affected individuals may present with rupture of an intracerebral aneurysm. Approximately 30% of DPK patients have occult intracranial aneurysms at autopsy with brain examination (Suter, 1949; Brown, 1951; Bigelow, 1953). However, a prospective angiographic study of DPK patients found asymptomatic, unruptured cranial aneurysms in 41% of the patients (Wakabayashi et al., 1983). Possible other disease associations include valvular and aortic abnormalities (Leier et al., 1984).

DPK patients may have a positive family history, flank pain, hypertension, or flank masses. Pain is the most frequent presenting symptom of DPK. It antedates renal palpability and occurs in 59% of affected individuals (Dalgaard, 1957). Pain usually occurs in the flank or lateral abdomen but may radiate to the epigastrium or suprapubic area. Although pain increases as the disease progresses and cysts enlarge, no etiology has been confirmed. Perhaps pain results from the stretching of the renal capsule, pressure on adjacent organs, or traction on the renal pedicle. Distinguishing the pain from the pain of a simultaneous renal disease as obstruction, hemorrhage, infection, or tumor is difficult but of great importance for the preservation of future renal function.

Mild to moderate hypertension occurs in about 46% of DPK patients (Dalgaard, 1957), and antedates the onset of renal dysfunction in many cases. The degree of hypertension does not correlate with renal impairment or patient age. Sodium and fluid retention may play a greater role than increased renin production in the etiology of this mild to moderate hypertension (Reubi, 1985), which usually responds well to diuretics or diuretic-containing drug combinations.

Hematuria (gross or microscopic) is common, occurring in up to 64% of affected individuals (Delaney et al., 1985). Hematuria can result from spontaneous or traumatic cyst rupture or concomitant calculi, infection, or neoplasm. An evaluation of hematuria in DPK patients should include all of these possibilities. Clots can cause renal colic and obstruction of the urinary tract.

Approximately 60% of DPK patients have palpable flank masses, of which one third are bilateral and two thirds are unilateral. However, it is a less common finding (15% of patients) at initial presentation.

Fewer than 10% of cases of DPK present in the first decade of life (Kissane and Smith, 1975). Children presenting at an early age with symptoms or complications leading to the diagnosis of DPK are a distinctly different group than those children in whom the diagnosis is made by screening the family of an affected patient. Children with symptomatic presentation of DPK early in life do poorly although some children survive several years with azotemia. The prognosis of children diagnosed with DPK from family screening is uncertain given the paucity of data. Prenatal diagnosis has been accomplished sonographically (Zerres et al., 1982).

MORPHOLOGY

On gross inspection the DPK kidneys are huge because they are filled with hundreds of fluid-filled cysts (Fig 21–1). Their surfaces are distorted by innumerable large cysts. However, they retain their reniform shape unlike dysplastic kidneys. At autopsy, mean combined kidney weights of clinically asymptomatic DPK patients was 512 gm; symptomatic patients, 930 gm (Hatfield and Pfister, 1972); and ESRD DPK patients who came to renal dialysis, 2,600 gm (Bennett et al., 1973). The cut surface shows extensive parenchymal replacement of cortical and medullary cysts, which vary from a few millimeters to a few centimeters in diameter (Fig 21–2). Cyst fluid varies from clear yellow to chocolate brown and from watery to gelatinous.

On microscopic study, islands of normal renal parenchyma are found, although most parenchyma has been replaced with cysts. The cysts are lined by a single layer of flattened cuboidal or columnar epithelium, but glomerular and tubular cysts may also be recognized. Cysts involve all elements of the nephron and collecting ducts.

PATHOGENESIS

Unlike single solitary cysts, cysts in DPK patients are connected to functional renal units (Bricker and Patton, 1955). Cysts can be categorized into two basic types according to solute composition. Cysts of proximal tubule origin have concentrations of sodium, chloride, potassium, urea, hydrogen ion, and creatinine similar to those of plasma. Cysts of distal tubule origin have concentrations of sodium and chloride lower than those in plasma while concentrations of potassium, hydrogen ion, creatinine, and urea are greater than those of serum (Huseman et al., 1980).

Theories have been expressed suggesting why these DPK kidneys form cysts and how that relates to development of renal failure. Obstruction has long been sug-

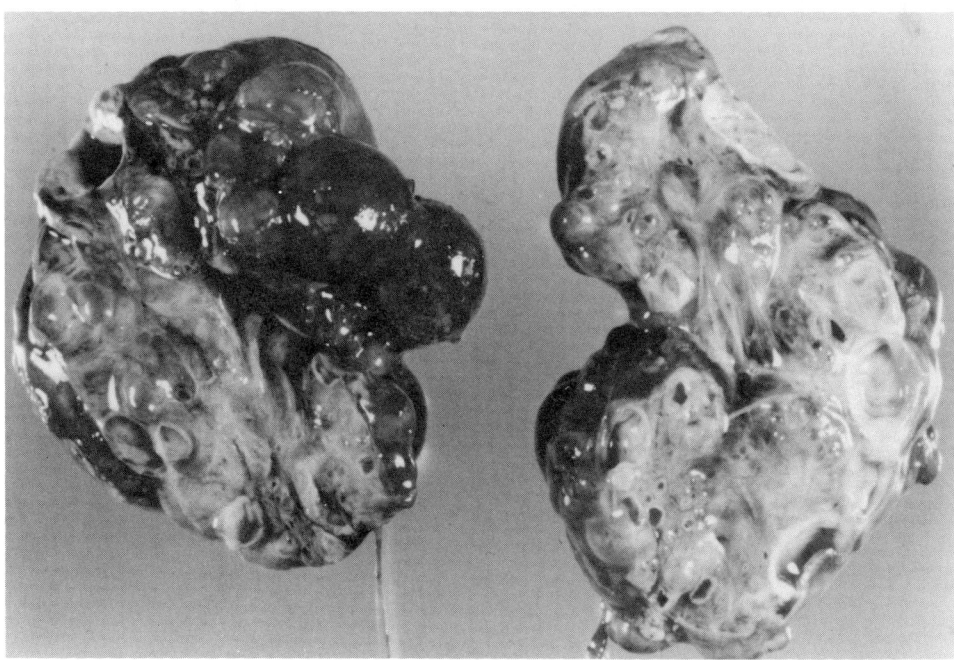

FIG 21–1.
DPK. Innumerable cysts distort the surfaces of these bilaterally enlarged kidneys. (Courtesy of Dr. S.E. Mills, Department of Pathology, University of Virginia Medical School, Charlottesville.)

gested to cause cyst development. Tubular cyst formation has been thought to be due to a prolonged increase in intratubular pressure secondary to partial obstruction from hyperplastic cells (Evan et al., 1979). An alternative theory has been that the tubular basement membrane is structurally altered with resultant decreased compliance of the tubular wall and resultant cyst formation. (Kanwar and Carrone, 1984). Animal studies show that 2-amino-4,5-diphenyl thiazole induces cystic disease in rat kidneys due to this tubular basement membrane defect. Electron microscopic studies in DPK

human renal cysts also reveal thickened tubular basement membranes (Cuppage et al., 1980), raising the possibility that DPK involves an altered synthesis or metabolism of tubular basement membrane.

Cysts enlarge progressively and may eventually cause compression of normal renal parenchyma with resultant atrophy (Franz and Reubi, 1983). Such compression might account for the late development of renal failure following many years of progressive renal enlargement from growing cysts. Of course, hypertension or urinary tract infection may decrease glomerular filtration rate in some patients.

GENETICS

DPK has an autosomal dominant pattern of inheritance. Penetrance is complete by age 80 years, but may be less than 100% prior to that time. Although up to 25% of patients have no family history for DPK (Hatfield and Pfister, 1972), family history might have been missed because (1) relatives might have died of other causes before DPK became symptomatic or was diagnosed, (2) other family members might not have been aware of the diagnosis, (3) paternity may not have been clear, and (4) spontaneous new mutations may have occurred.

Reeders et al. have shown that the DPK locus is closely linked to the α-globulin locus on the short arm of chromosome 16 (Reeders et al., 1985). If further studies confirm that this locus for the common DPK phenotype is also the locus for the more atypical manifestations of this disease, this locus discovery should facilitate presymptomatic detection and prenatal diagno-

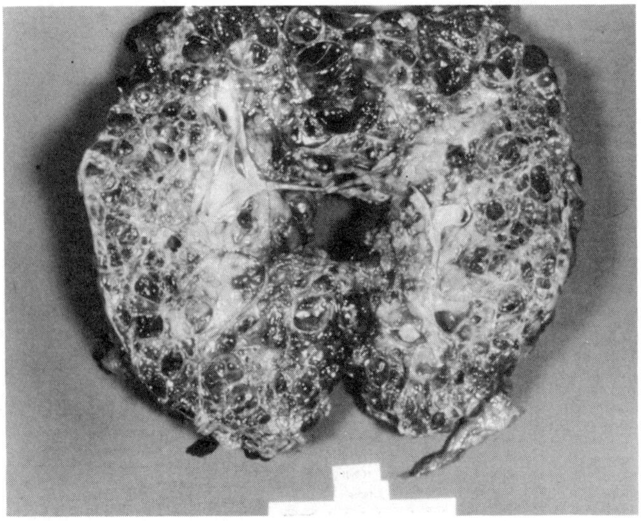

FIG 21–2.
DPK. Cut section reveals extensive parenchymal replacement by cortical and medullary cysts of varying sizes. (Courtesy of Dr. S.E. Mills.)

sis of DPK. The early diagnosis of DPK facilitates improved genetic counseling and rational decision-making about reproduction.

Dalgaard (1957) demonstrated an adherence to the mean age at onset of the symptoms of DPK among affected family members. This adherence may not be true for other groups of DPK patients (Torres et al., 1986).

RADIOLOGIC FINDINGS

Ultrasonography is an excellent screening test for identification of patients with DPK (Fig 21–3). Recent improvements in ultrasound techniques make it possible to detect cysts smaller than 5 mm in size (Melson et al., 1983). This technique is more sensitive than excretory urography and avoids radiation exposure (Gabow et al., 1984). Further, the technique is so sensitive that even in those aged 15 to 25 years, there is only a 7.2% chance that cysts will be missed by ultrasound examination in someone at risk for developing DPK (Mellins, 1985). In those over age 30 years, ultrasound should pick up 100% of DPK because of the larger cyst size. Further, ultrasound has the advantage over intravenous urography of performing a concomitant examination of the liver for cysts.

Both computed tomography and renal angiography

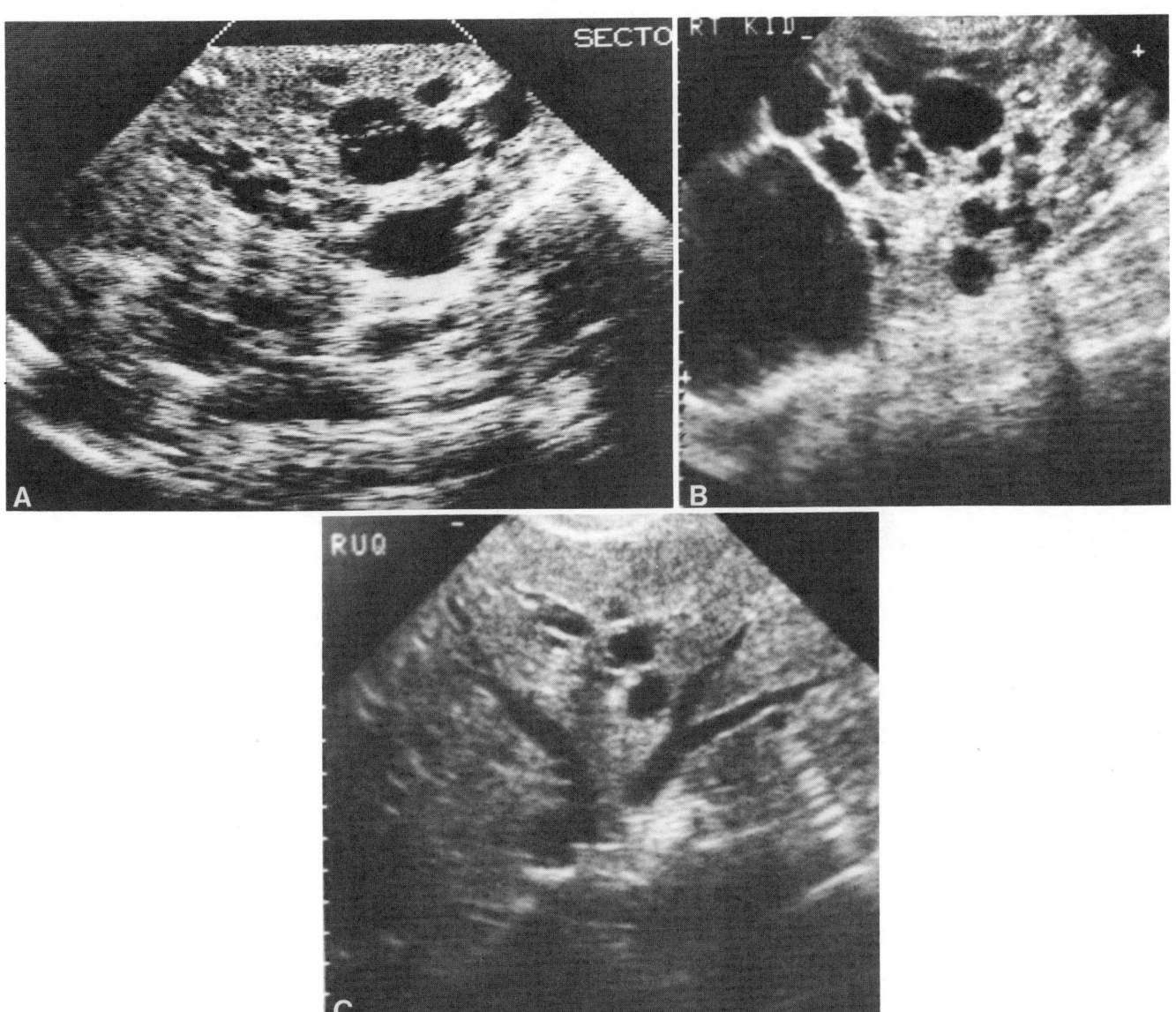

FIG 21–3.
DPK in two adults. **A,** gray-scale renal ultrasonography reveals multiple cysts of varying sizes in an enlarged right kidney of a 60-year-old woman. **B,** gray-scale ultrasonography reveals markedly enlarged kidneys with numerous cysts of varying sizes in same patient. **C,** two small cysts within the liver in a 37-year-old woman.

are highly sensitive and can reveal renal cystic disease missed by ultrasound and intravenous urography (Figs 21–4 and 21–5). If ultrasound examination is equivocal in a patient at risk for DPK, computed tomography is probably the procedure of choice as a second screening test or even as a first screening test in an institution with ultrasound methodology that is not adequate. Computed tomography is less invasive than renal angiography while providing simultaneous examination of the liver for cysts. The potential genetic marker on chromosome 16 (Reeders et al., 1985) is not yet clinically available. However, ultrasound and/or computed tomography affords an early diagnosis of DPK before the patient has become symptomatic. Hence, screening family members at risk for DPK allows improved genetic counseling.

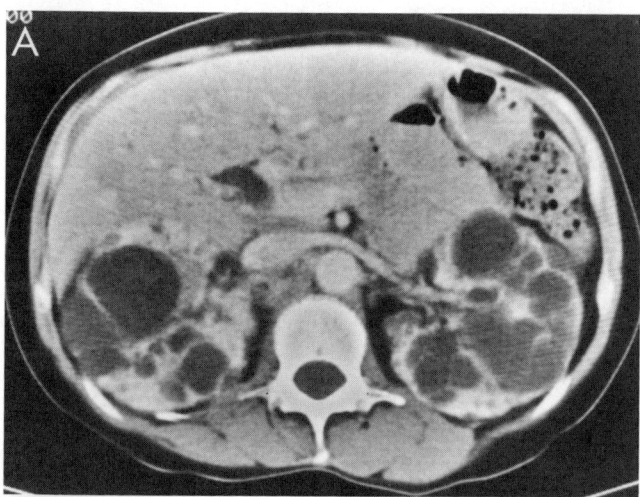

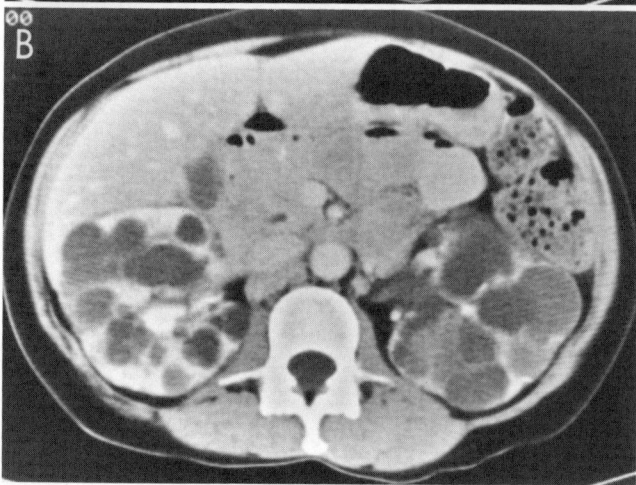

FIG 21–4.
DPK in a 51-year-old woman. Enhanced computed tomography reveals bilaterally markedly enlarged kidneys with multiple cysts of varying sizes, shapes, and attenuation values. There is some enhancing renal cortical tissue remaining, but excretion into the collecting system is identified only with difficulty.

Computed tomography is an excellent technique for evaluating possible complications in DPK, such as cyst or parenchymal infection, cyst hemorrhage, parenchymal calcifications, and renal calculi. Renal angiography has been the procedure of choice for demonstrating renal cell carcinoma in patients with DPK and can evaluate suspected renal hemorrhage and renovascular lesions. Intravenous urography with nephrotomography will detect most but not all cases of DPK diagnosed by ultrasound (Gabow et al., 1984). Intravenous urography without nephrotomography will detect advanced cases of DPK and is helpful in evaluating urinary tract obstruction in DPK. If intravenous contrast is contraindicated in a DPK patient, computed tomography without contrast is the most useful test to evaluate urinary tract obstruction, since ultrasonography may not distinguish between parenchymal cysts and dilated calyces. Scanning with radiolabeled gallium can help localize the cyst or parenchymal infection in DPK, but false positive and false negative results may reduce the value of this test in individual patients. Retrograde pyelography is contraindicated because of the high risk in DPK patients of introducing infection. Magnetic resonance imaging can help differentiate a complicated cyst from a neoplasm in DPK patients in whom sonography and computed tomography are indeterminant (Hilpert et al., 1986).

MANAGEMENT

Therapy of DPK must be directed toward the management of the inevitable end-stage renal disease (ESRD) and toward the management of the complications of DPK that occur before and after renal failure.

Renal function is well-preserved until late in DPK, at which time it decreases rapidly. The first sign of kidney failure in DPK is the inability to maximally concentrate urine. Other signs of renal failure soon follow. Hence, general medical therapy for treatment of chronic renal failure such as low-protein, low-phosphorus (Maschio et al., 1985), and low-sodium diets (Reubi, 1985) are appropriate, as are slowing the development of osteodystrophy with aluminum hydroxide gel and treating metabolic acidosis.

However, DPK patients will eventually require dialysis or renal transplantation. Effective peritoneal dialysis has, surprisingly, not been hindered by the large cystic renal and hepatic masses. DPK patients do as well as other ESRD patients when receiving hemodialysis, taking into account their older age at the time of dialysis. Long-term renal dialysis also markedly reduces blood pressure in hypertensive DPK patients (Reubi, 1985). Renal transplantation results for DPK patients have been as good or better than those in non-DPK patients (Sanfilippo et al., 1983), despite the fact that DPK patients represent an older group of cadaver graft

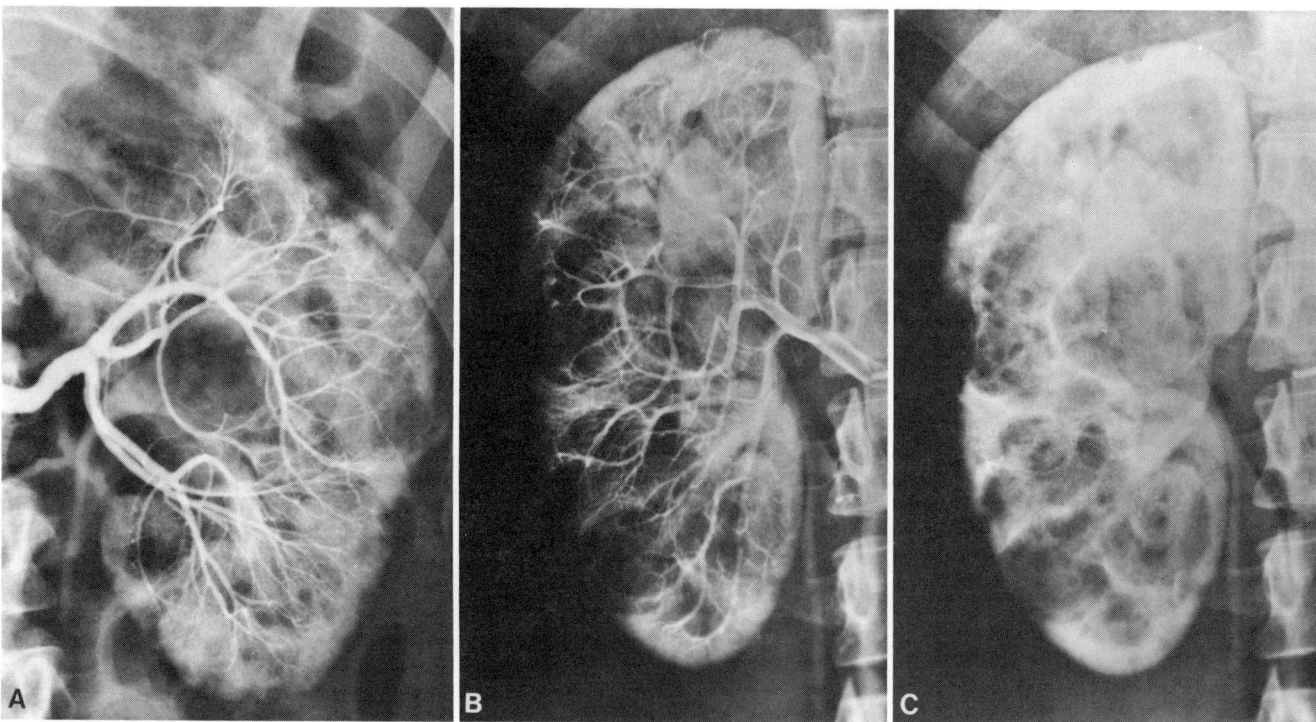

FIG 21–5.
DPK in two adults. **A,** arterial phase of renal arteriography in a 23-year-old woman reveals stretching of the vessels around avascular masses in enlarged kidneys. **B,** arterial phase of renal arteriography in a 30-year-old woman also reveals the intrarenal arterial branches to be stretched, elongated, and displaced by many cysts. **C,** nephrogram phase of same arteriogram reveals mottled or Swiss-cheese appearance in both enlarged kidneys.

recipients. Despite superior results with transplants from living relatives in non-DPK patients, the use of such transplants has been somewhat limited in DPK patients, since the disease is familial. However, if the ability to detect the potential gene marker in the short arm of chromosome 16 (Reeders et al., 1985) reaches clinical availability, living-related donors can be found at any age. Until then, living-related donors for DPK patients are identified by negative radiographic findings at age 30 years or older. With improved imaging techniques, perhaps the age of identification of potential living-related donors for DPK patients may be lowered to 20 years (Mellins, 1985).

Routine urologic disease such as urinary infections, calculi, and obstruction can be very difficult problems in DPK patients and require special therapeutic consideration. Symptomatic urinary tract infections are common with an overall incidence of 53% and recurrence rate of 61% in DPK patients. Complicated urinary infections are less common but difficult to treat. Of 23 female patients who had symptomatic urinary tract infections, 12 (52%) had pyelonephritis, of which three developed perinephric abscess (Delaney et al., 1985). Modern antibiotic therapy, even if prolonged, can be unsuccessful. In infected patients a urine culture may or may not be sterile since the infection may be present in noncystic parenchyma, in cysts of proximal tubular origin, or in cysts of distal tubular origin. Computed tomography or gallium scan may help localize the infection to cysts, but few antibiotics penetrate the cysts, and of these, and fewer still are effective against gram-negative organisms. Antibiotics that penetrate cysts well are clindamycin or metronidazole (active against anaerobes), prolonged treatment with trimethoprim-sulfamethoxazole or ampicillin, vancomycin, cefotaxime (Bennetts et al., 1985), and chloramphenicol (Schwab, 1985). Aminoglycosides do not reach detectable levels in both distal and proximal cyst fluids (Bennetts et al., 1985). Successful treatment with trimethoprim-sulfamethoxazole requires that organisms be sensitive to each drug of the combination individually (Schwab and Weaver, 1986). Failure of prolonged systemic antibiotics to treat recalcitrant infection may make nephrectomy and surgical drainage necessary because of the attendant mortality.

Urinary tract instrumentation (cystoscopy, retrograde pyelograms, and bladder catheterizations) should be avoided when possible in DPK patients because of the high association of infection and mortality despite sterile urine cultures prior to instrumentation. Of 14 DPK patients who underwent urinary tract instrumentation, six (43%) developed symptomatic urinary tract infections, of whom three had pyelonephritis, which led to death in two despite parenteral antibiotic therapy (Delaney et al., 1985). Most urinary tract infections in DPK patients are felt to be ascending in origin.

Pain occurs in 61% of DPK patients (Gabow et al., 1984) and must be evaluated to rule out obstruction, infection, neoplasms, and hemorrhage. If these possible causes of pain are eliminated, pain from DPK can be treated conservatively with analgesics such as acetaminophen. However, many patients must live with a fair amount of chronic pain for which pain clinics may be helpful.

Obstruction of the urinary tract can occur from clots, calculi, or cysts and is difficult to diagnose. Computed tomography without contrast may be necessary for diagnosis of obstruction due to the confusion of differentiating cysts from dilated calyces on ultrasound (Levine and Grantham, 1981). Computed tomography revealed that 11.6% of 43 DPK patients had renal calculi, whereas 44.2% had parenchymal calcifications (Levine and Grantham, 1985b). Treatment of obstructing calculi with extracorporeal shock wave lithotripsy in DPK has been done with good results and without complications (Fig 21–6). Alternatively, appropriate stone surgery can be done,

Obstruction of the urinary tract in DPK patients can also be caused by renal cyst compression, which can be treated by surgical or percutaneous decompression. Blood clots can also cause obstruction and can usually be managed conservatively. Bleeding in polycystic kidneys can usually be diagnosed by computed tomography and treated with bed rest and observation. However, more severe hemorrhage can be treated by segmental renal artery embolization or percutaneous nephroscopic balloon occlusion of infundibuli to preserve some renal function in DPK patients. Alternatives to control hemorrhage include complete or partial nephrectomy.

Mild to moderate hypertension occurs in 46% to 62% of DPK patients (Dalgaard, 1957; Gabow et al., 1984) but responds well to diuretics or diuretic-containing drug combinations as well as to hemodialysis (Reubi, 1985).

The incidence of bilateral renal cell carcinoma in DPK is 4 to 14 times greater than normal kidneys although unilateral hypernephroma may or may not be of higher incidence in DPK patients (Gardner and Evan, 1984). Von Hippel-Lindau disease (the association of bilateral renal cell carcinoma and polycystic kidneys) must be ruled out. Renal arteriography is the diagnostic test of choice to rule out malignancy in DPK although magnetic resonance imaging can help differentiate tumor from cyst (Hilpert et al., 1986).

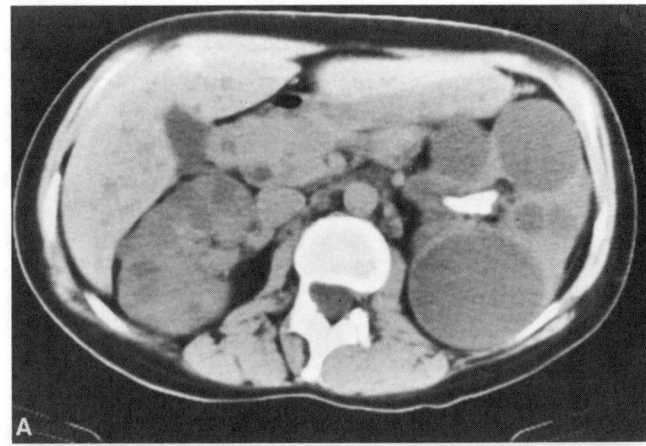

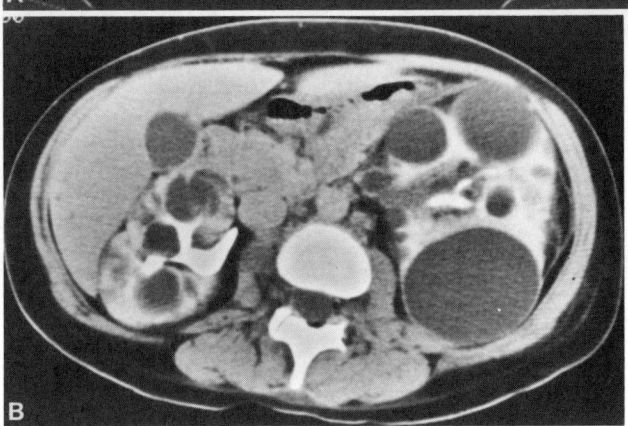

FIG 21–6.
DPK with left staghorn calculus treated by extracorporeal shock wave lithotripsy (ESWL) in a 31-year-old woman. **A,** before ESWL, unenhanced computed tomography reveals a left staghorn calculus, bilaterally huge kidneys with multiple cysts of many sizes, and hepatic cysts. **B,** one day after ESWL, enhanced computed tomography reveals no change in the renal cysts from ESWL. The residual gravel cannot be seen because of the contrast.

AUTOSOMAL RECESSIVE POLYCYSTIC KIDNEY DISEASE (INFANTILE TYPE)

Autosomal recessive polycystic kidney disease (RPK) includes a spectrum of phenotypic appearances ranging from renal failure in infants to hepatic disease in older children. Additionally, newborns with the severe form of the disease die of respiratory disorders within a few days after birth. RPK occurs in about one in 10,000 births (Cussen and Stephens, 1983) and is characterized by bilateral cystic disease in enlarged kidneys with retained reniform shape. RPK is transmitted in an autosomal recessive pattern only.

CLINICAL FEATURES

The clinical picture of RPK covers a continuum of presentations of renal failure and hepatic disease developing at different ages. The basis of this continuum is the coexistence of renal and hepatic abnormalities that are each progressing independently and to different degrees. Children presenting in the newborn and infantile period suffer more from renal failure. Children presenting after the infantile period are more likely to have liver disease, although a continuum exists.

The most common RPK presentation is in the newborn period, at which time the child has tremendously enlarged kidneys (up to ten times normal size) (Cussen and Stephens, 1983) and pulmonary hypoplasia. Vaginal delivery may be impeded by the massive kidneys, and oligohydramnios is commonly noted. Respiratory distress develops partly from the pulmonary hypoplasia. Neonatal mortality results from respiratory insufficiency despite resuscitative measures, which frequently lead to pneumomediastinum and pneumothorax. Oliguria and Potter's facies may be present, but renal failure is not the cause of death. Should the child survive the newborn period, he most likely will have eventual renal failure and systemic hypertension. After the first month, growth failure and congestive heart failure are usually the major clinical problems (Landing et al., 1976).

After the first year of life, the clinical presentation of RPK is more variable because of the spectrum of renal failure and hepatic disease. As the children with RPK grow, the renal cystic dilation can regress as the kidneys become smaller. These children can have either slowly progressive renal insufficiency or simply a mild renal functional impairment as a mild inability to concentrate urine. Systemic hypertension is common. However, these older children with RPK usually have more severe hepatic involvement. The hepatic disease in RPK is called congenital hepatic fibrosis (CHF) because the liver has an increase in the number of portal bile ducts and fibrosis of the periportal spaces. CHF presents clinically as portal hypertension, which can result in massive upper gastrointestinal hemorrhage from ruptured esophageal varices. There is no hepatocellular disease. In this older age group (over 6 years), CHF, not the renal cystic disease, is usually responsible for death. However, the continuum of renal failure and hepatic disease in older children with RPK also includes children symptomatic from both kidney and liver disease (McClean et al., 1978). The clinical cause and pathological expression of renal disease can be entirely dissimilar within the same family (Gang and Herrin, 1986).

MORPHOLOGY, PATHOGENESIS, AND GENETICS

The gross and microscopic characteristics of RPK kidneys vary with the age of the child at the time of presentation, but the disease is always bilateral. In newborns with the severe form of RPK, gross inspection

reveals the kidneys to be adult size and reniform. The kidney surface is smooth and studded wtih 1- to 2-mm cysts. The cut surface shows striking radial arrangement of thin-walled channels extending from the pelvis to the cortex. Microscopically, these channels are dilated collecting tubules and ducts, which are the fusiform cysts of RPK. Interstitial edema is severe.

In those children surviving the neonatal period, the kidneys become smaller, the collecting ducts become less dilated, and the cysts that still remain are spherical as opposed to saccular (Lieberman et al., 1971). These findings are also true for those children who present with RPK after infancy. However, these children are more likely to develop CHF. Grossly, the liver in CHF is firm and enlarged with a granular surface. Microscopically, there is an excessive number of interlobular bile ducts. The liver is subdivided by bands of fibrous tissue in the periportal and interlobular spaces.

The pathogenesis is unknown except that it is genetically transmitted as autosomal recessive. Hence, a very careful family history covering at least three generations should be taken to exclude autosomal dominant polycystic kidney disease. Recently described has been a mouse model for congenital polycystic kidney disease, which is inherited as an autosomal recessive allele. Cystic change was found in kidney collecting ducts and proximal tubules and to a lesser extent in the liver and pancreas (Fry et al., 1985). With the availability of this model, perhaps information leading to the pathogenesis of RPK can be discovered.

Because of the continuum of clinical presentation of RPK, four types were described by Blyth and Ockenden with different ages at onset. Each type was to be transmitted as an autosomal recessive allele and each type was to have occurred in all of the affected children of a family (Blyth and Ockenden, 1971). However, further work has shown that not only do different clinical and pathologic courses occur within one family (Gang and Herrin, 1986), but that RPK has a spectrum of phenotypic expressions and not just four genetically determined rigidly defined subgroups (Gang and Herrin, 1986; Lieberman et al., 1971).

RADIOLOGIC FINDINGS AND CLINICAL MANAGEMENT

In infants with bilateral palpable abdominal masses, the initial study should be renal and abdominal sonography (Fig 21–7). In newborns with the severe form of RPK, the kidneys are very large with diffusely increased parenchymal echogenicity. This increased echogenicity is caused by the sound beam bouncing off the innumerable dilated tubules. Using modern scanners with 5-MHz transducers, some investigators recently have shown a more heterogeneous pattern with either "striped" or "pepper and salt" echoes (Wernecke et al.,

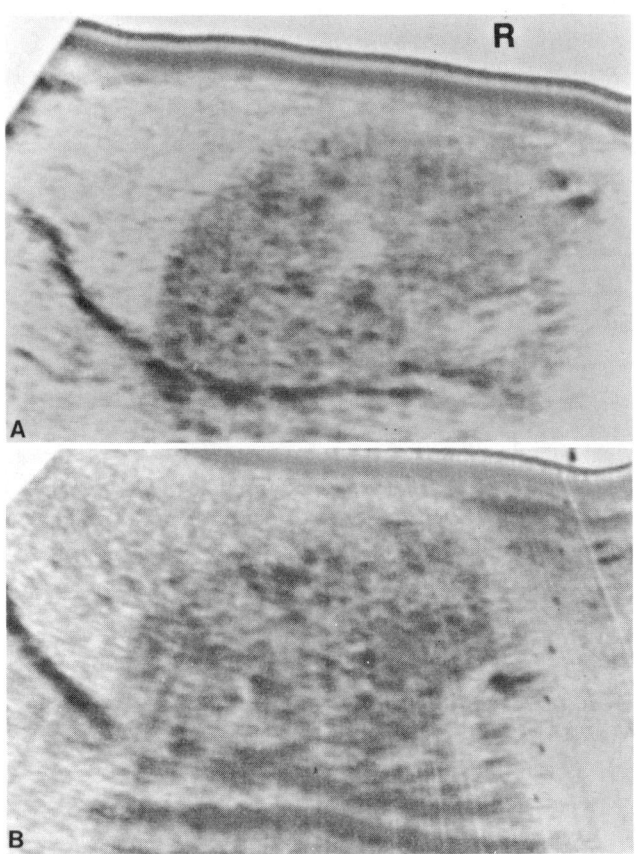

FIG 21–7.
RPK in newborn girl. **A,** gray-scale renal ultrasonography at age 4 days reveals massively enlarged kidneys with increased echogenicity. **B,** ultrasonography in same patient at age 9 days confirms the hyperechogenicity from the echo-reflective surfaces of the innumerable tiny cysts.

1985), or a peripheral zone of normally echogenic cortex in some newborns (Melson et al., 1985). Occasionally, small cysts can be identified in the hyperechogenic medullary areas.

Intravenous urography in infants with RPK may show only progressively dense nephrograms of these massive kidneys or may show the classic radial streaking of contrast in dilated collecting tubules. The collecting system is usually not seen although contrast in the bladder may be seen on delayed films.

In the older child with RPK, the excretory urogram pattern is less diagnostic since the kidneys are only somewhat enlarged and the renal pelvis and calyces are better visualized. The calyces may be blunted or the pyramids may be opacified with a brush-like pattern. Ultrasonography in the older child with RPK can demonstrate diffuse hyperechogenicity of the parenchyma or reveal a prominent rim of renal cortex with normal echogenicity and a moderately echogenic medulla (Melson et al., 1985).

Since there is no known amniotic fluid protein pattern to diagnose individuals affected with RPK prenatally, attempts have been made at prenatal ultrasound diagnosis. Unfortunately, the gestational age at which renal sonographic changes were noted within the same family varied from 20 to 34 weeks and resulted in both false positive and false negative diagnoses (Luthy and Hirsch, 1985).

There is no known cure for RPK, but genetic counseling can inform parents that siblings of an affected child have a 25% chance of being affected and a 50% chance of being heterozygous carriers. Otherwise, treatment is supportive. Respiratory resuscitation and support may be necessary for newborns diagnosed with RPK. Older children may need antihypertensive medications for hypertension, digitalization for congestive heart failure, and alkali for metabolic acidosis to permit normal growth, as well as other supportive measures for treating renal failure in growing children. Chronic dialysis and renal transplantation may be necessary eventually, but liver involvement should be evaluated first. Portal hypertension may result from CHF in RPK and can be managed by portacaval shunting to treat esophageal varices if present.

JUVENILE NEPHRONOPHTHISIS–RENAL MEDULLARY CYSTIC DISEASE COMPLEX

Juvenile nephronophthisis–renal medullary cystic disease complex (NMCD) comprises a group of hereditary diseases characterized by progressive renal failure. These diseases have different hereditary patterns but share similar clinical features as well as similar pathologic features. Both kidneys are small and reniform, with small cysts at the corticomedullary junction. NMCD is uncommon, but is the most common cause of renal failure in the adolescent.

CLINICAL FEATURES AND GENETICS

Despite the genetic variants of NMCD, the pathologic and clinical features are the same except for the age at onset. The autosomal recessive variant, juvenile nephronophthisis, affects mostly children and adolescents. The autosomal dominant variant, medullary cystic disease, affects mostly young adults. Finally, NMCD can rarely occur sporadically with no documented family history.

While adults with progressive idiopathic renal failure in NMCD may present to the physician with fatigue or anemia, children may present with growth retardation or skeletal deformity. Polyuria, nocturia, and polydipsia occur in over 80% of patients (Gardner, 1976) because of a severe defect in tubular function resulting in an inability to concentrate urine. As a result, these patients may waste salt. The salt wasting occurs late in the disease and can be heralded by a decrease of already present arterial hypertension (Gardner, 1979). However, the polyuria and nocturia can be present for more than ten years before ESRD occurs (Chamberlin et al., 1977; Burke et al., 1982). Both sexes are affected equally. Urinary infections are not common.

The autosomal recessive variant is the most common hereditary pattern of NMCD and results in death or ESRD about age 14 years (Gardner, 1976). This genetic variant can occur with or without retinal degeneration such as retinitis pigmentosa. Renal-retinal syndrome refers to the autosomal recessive form of NMCD with retinal degeneration, and it is far less common than juvenile nephronophthisis, the autosomal recessive form of NMCD without retinal degeneration. The autosomal dominantly inherited disease, medullary cystic disease, is less common than juvenile nephronophthisis and results in death or ESRD at about 32 years (Gardner, 1976). A negative family history does not exclude NMCD, since there are a few sporadic cases recognized. However since these different hereditary patterns have similar pathologic and clinical features, and since there are exceptions to the aforementioned coupling of age at onset to mode of inheritance, categorization of juvenile nephronophthisis from medullary cystic disease on the basis of age at onset may not be possible (Burke et al., 1982).

MORPHOLOGY AND PATHOGENESIS

On gross inspection, both kidneys in NMCD are contracted and pale with granular subcapsular surfaces. On cut section, there is poor demarcation between the medulla and the attenuated cortex. The cysts are actually diverticulae of the distal convoluted tubules. When the cysts are macroscopic, they are typically located on cut section at the corticomedullary junction. They can range from microscopic to 1 cm in size. Cysts are usually present, but occasionally are not.

Microscopic examination reveals glomerular sclerosis, tubular atrophy, and interstitial fibrosis with chronic inflammation. The tubular basement membrane is markedly thickened. Without the presence of cysts, the pathologic findings resemble those of interstitial nephritis. Indeed, NMCD has been considered a primary interstitial nephritis by some (Brown et al., 1981). A renal biopsy can, therefore, be misleading if the corticomedullary cysts are missed in the biopsy specimen. Such biopsy specimens have been diagnosed as chronic interstitial nephritis initially, but were changed because a later examination of the entire kidney from a subsequent nephrectomy or autopsy may show the medullary cysts (Brown et al., 1981; Proesmans et al., 1975).

The pathogenesis of NMCD is unknown except that it can be genetically transmitted. It is possible that a nephrotoxic substance—such as the product of an inborn enzyme defect—leads to tubular dysfunction, an inability to concentrate urine, and then to progressive renal failure. When fed to rats, diphenylalamine and diphenyl thiazide produce cysts in the medullary collecting ducts, and an inability to concentrate urine maximally is present before the cysts are discernible as gross pathologic findings (Martinez-Maldonado, 1985). Since renal transplants do not undergo medullary cystic changes in NMCD patients, a circulating cystogenetic metabolite would seem unlikely. However, in patients with cystinosis, the transplanted kidney does not demonstrate either the functional or histologic findings of cystinosis.

RADIOLOGIC FINDINGS AND MANAGEMENT

Radiographic findings vary with the stage of progression of NMCD. Intravenous urography would reveal normal findings early in the disease, but would show bilaterally small kidneys later in the disease, at which time high-dose nephrotomography could reveal renal cysts (Rosenfield et al., 1977). However, excretory urography is of little value very late in the disease when renal function is greatly reduced. Ultrasonography with a 5-MHz short-focus probe can demonstrate a few small medullary cysts when patients have terminal uremia, but not in patients with mild uremia. Such ultrasonography reveals disappearance of the corticomedullary differentiation in all but the early stages of NMCD, as well as small kidneys and increased parenchymal echogenicity (Rosenfield et al., 1977; Garel et al., 1984). Ultrasonography and/or computed tomography could complement renal biopsy in the diagnosis of NMCD if the cysts were missed on biopsy but revealed by ultrasonography or computed tomography.

Treatment of NMCD is directed toward general therapy for renal failure. If salt wasting is present, it can be treated with salt supplementation. Dialysis and transplantation are successful. However, recognition of genetic variants is vital to avoid selecting a living donor for transplant who is himself at risk of developing the disease (Avasthi et al., 1976).

RENAL CYSTS ASSOCIATED WITH MULTIPLE MALFORMATION SYNDROMES

Renal cystic disease, macroscopic to microscopic, has been found in some hereditary disorders. Knowledge of these disorders is important for genetic counseling.

CHROMOSOME DISORDERS

Patients with trisomy 21 (Down's syndrome), trisomy 18 (Edward's syndrome), trisomy 13 (Patau's syndrome), and trisomy C can develop microscopic cysts, usually of the glomerular spaces of the renal cortex.

AUTOSOMAL RECESSIVE SYNDROMES

Renal cystic disease is commonly found in the Meckel syndrome (encephalocele, polydactyly, microcephaly) (Miller and Gang, 1983), in the Jeune syndrome (asphyxiating thoracic dystrophy: small chest, renal dysplasia), and in Zellweger's syndrome (cerebrohepatorenal syndrome: high forehead, hepatomegaly). Renal cystic changes have been less commonly reported in a few other autosomal recessive syndromes, such as the Goldston syndrome (cerebral malformations), the Majewsky and Saldino-Noonan types of short rib polydactyly, lissencephaly (microcephaly, smoothness of the brain, and cystic renal dysplasia), and the Elijalde syndrome (acrocephalopolydactylous dysplasia: gigantism, renal dysplasia).

X-LINKED DOMINANT SYNDROME

Renal cystic disease occurs later in life in the orofaciodigital syndrome, type I, and somewhat resembles autosomal dominant polycystic kidney disease.

AUTOSOMAL DOMINANT SYNDROMES

Tuberous sclerosis and von Hippel-Lindau disease are the most likely to require the attention of a urologist of all the multiple malformation syndromes. Hence, both autosomal dominant syndromes are described in detail.

Von Hippel-Lindau Disease

Von Hippel-Lindau disease consists of renal cysts and renal adenocarcinomas as well as retinal hemangiomas; cerebellar, medullary, and spinal hemangioblastomas; pheochromocytomas; pancreatic cysts; and epididymal cysts. The disease is inherited as an autosomal dominant disorder with an incomplete penetrance. Most manifestations of von Hippel-Lindau disease do not become apparent until after the patient has reached the end of his second decade. However, retinal angiomatosis frequently is the earliest manifestation and has been found in an asymptomatic patient as young as 9 years (Levine et al., 1982). Cerebellar hemangioblastomas were once the most common cause of death in von Hippel-Lindau disease. As more patients survive the cerebellar lesions because of improved treatment methods, the incidence of renal cysts and tumors in live patients is increasing (Pearson et al., 1980). Computed tomography of the abdomen and brain and indirect ophthalmoscopy were done as screening examinations on family members at risk. Of those who were diagnosed as having von Hippel-Lindau disease, 35% had renal cell carcinoma and 76% had renal cysts (Levine et al., 1982). Renal cell carcinoma was only found in those patients who had renal cysts (Frimodt-Moller et al., 1981; Levine et al., 1982). The renal cysts of von Hippel-Lindau disease are lined by variably hyperplastic epithelium. Nodular hyperplasia within the cyst walls have been observed to progress to clear cell carcinoma (Bernstein, 1976). Unlike renal cell carcinoma in patients who do not have von Hippel-Lindau disease, renal cell carcinoma in patients with the disease occurs in younger patients, has an equal sex distribution, and is usually bilateral and multicentric (Pearson et al., 1980).

Radiographic evaluation of renal lesions in von Hippel-Lindau disease is useful for screening family members and for following patients once they have been diagnosed as having the disease. Intravenous urography can reveal several masses, but cannot differentiate cysts from tumors. Computed tomography can reveal solid-mass lesions and many reveal renal cysts, but cannot differentiate between a small cyst and a small tumor, or find a small tumor in the wall of the cyst. Sonography and, sometimes, angiography can fail to detect these small masses (Levine et al., 1982). Therefore, sequential computed tomographic studies are most sensitive in finding tumors before they become large enough to metastasize. Annual studies are recommended once a diagnosis of von Hippel-Lindau disease is made. Renal cell carcinoma has been diagnosed in the early part of the third decade in patients with von Hippel-Lindau disease. Hence, early screening with computed tomography of family members provides diagnosis of von Hippel-Lindau disease by detection of renal cysts as the earliest sign of the disease (Levine et al., 1982), and provides early diagnosis of renal cell carcinoma. Diagnosis of the disease is made if the patient has more than one manifestation of von Hippel-Lindau disease or has one of its manifestations along with a family history of the disease.

Conservative surgical procedures to resect renal cell carcinoma in von Hippel-Lindau disease are appropriate because of the multifocal and recurring nature of the tumors. Hence, renal function can be preserved despite the development of new tumors over several years (Pearson et al., 1980). Further, annual computed tomographic studies allow assessment of the success or failure of the local resection.

Tuberous Sclerosis

Tuberous sclerosis is an autosomal dominant disorder which is characterized by the presence of adenoma sebaceum, epilepsy, and mental retardation. However, these findings begin at different times and in different sequences in individual patients; therefore, the diagnosis may be delayed until many years after the first symptoms begin. Renal angiomyolipomata (hamartomas) may occur in 40% to 80% of tuberous sclerosis patients (Chonko et al., 1974) and renal cysts in about 10% (Elkin and Bernstein, 1969). Renal hamartomas are rare before the sixth year of life in tuberous sclerosis patients (Nickel et al., 1974; Compton et al., 1976), whereas renal cysts have been found more frequently in younger patients than older patients (Compton et al., 1976). Children with tuberous sclerosis have been misdiagnosed as having autosomal dominant polycystic kidney disease (DPK) because of bilateral renal enlargement from renal cysts (Engstrom et al., 1962; Wenzl et al., 1970). The correct diagnosis is made later when additional features of tuberous sclerosis present as adenoma sebaceum and renal hamartomas. Confusion also arises when a tuberous sclerosis patient with renal hamartomas is misdiagnosed as DPK because hamartomas on intravenous urogram are misinterpreted as renal cysts and no further studies are done. Renal cysts can occur without the presence of renal hamartomas (Rosenberg et al., 1975; Okada et al., 1982) or with hamartomas (Compton et al., 1976) in tuberous sclerosis.

Renal failure is rare in tuberous sclerosis, but may be more likely to occur when renal cystic changes are present (with or without hamartomas) than when renal hamartomas are present alone (Okada et al., 1982). Perhaps, renal failure had been rare in tuberous sclerosis patients because traditionally 75% of these patients were dead by 20 years of age and 40% were dead by 5 years of age because of neurological problems (O'Callaghan et al., 1975). However, with the development of more effective anticonvulsant drugs, patients are living longer and perhaps, thereby, are developing renal insufficiency from replacement of renal parenchyma by growing hamartomas and from compression of renal parenchyma by cystic development.

The pathologic findings vary with the degree of cystic involvement in tuberous sclerosis. Grossly, when cysts are present in tuberous sclerosis the kidneys can be enlarged with cysts projecting throughout the renal cortex bilaterally. On cut section, multiple cysts (microscopic to 5.0 cm) are found throughout the cortex and medulla. Microscopically, renal cysts of tuberous sclerosis can be easily differentiated from other renal cysts because of their hyperplastic epithelium of eosinophilic cells (Stapleton et al., 1980). Microdissection shows that cysts develop from all parts of the nephron.

The pathogenesis of cystic kidneys in tuberous sclerosis is unknown, but tuberous sclerosis occurs one in every 150,000 births (Chonko et al., 1974). Although tuberous sclerosis is inherited as an autosomal dominant trait, variable expressivity may explain why only 50% of the tuberous sclerosis patients have a family member with any feature of tuberous sclerosis complex (Stapleton et al., 1980).

Excretory urography can reveal calyceal distortion and renal enlargement resulting from either the renal cysts of tuberous sclerosis or from the renal hamartomas (Compton et al., 1976). Ultrasonography can differentiate the cysts from the hamartomas (Fig 21–8). However, ultrasonography cannot differentiate the renal cysts of tuberous sclerosis from the cysts of DPK (unless multiple hepatic cysts are demonstrated that would confirm the diagnosis of DPK). Computed tomography might demonstrate renal hamartomas in addition to renal cysts and thereby confirm the diagnosis of tuberous sclerosis. The renal angiographic manifestations of tuberous sclerosis consist of multiple hamartomas, multiple renal artery aneurysms, renal cortical cysts, or some combination of these. The multiple renal lesions do not present a diagnostic problem on angiography (Compton et al., 1976).

MULTICYSTIC DYSPLASIA (CONGENITAL MULTICYSTIC KIDNEY)

Multicystic dysplasia is the most common cause of an abdominal mass in the newborn. Although present at birth, unilateral multicystic dysplasia can be clinically silent throughout adulthood. However, bilateral multicystic dysplasia is fatal at birth because of insufficient renal function. The involved kidneys do not maintain a reniform shape, but are replaced by clusters of cysts with fibrotic tissue. Multicystic dysplasia is a form of renal dysplasia and is rarely a genetically transmitted disorder (Al Saadi et al., 1984).

CLINICAL FEATURES

Multicystic dysplasia is present from birth, but may be detected by prenatal ultrasound (Legarth et al., 1981). Its most common presentation in newborns is almost always that of an asymptomatic abdominal mass that is mobile, irregular, located in the flank, and may be transilluminated (Walker et al., 1978). Of those children diagnosed as having multicystic dysplasia, 60% were diagnosed when younger than 1 year (Bloom and Brosman, 1978). Multicystic dysplasia is considered to be the most common or the second most common abdominal mass in the newborn with congenital hydronephrosis as the alternative diagnosis (Griscom, 1965).

In older children and adults, unilateral multicystic dysplasia can be found incidentally during radiographic study being done for unrelated reasons (Kyaw and

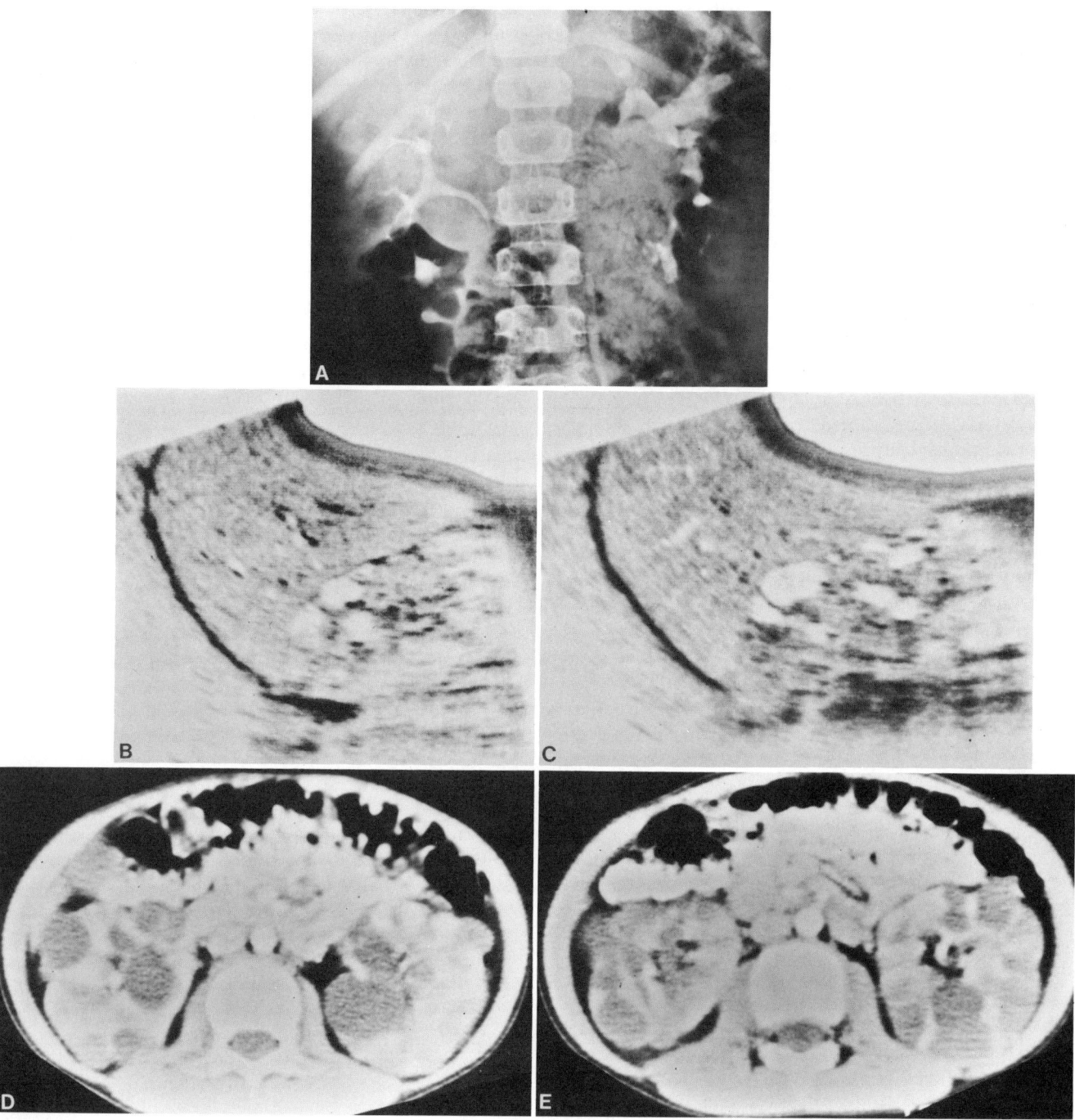

FIG 21–8.
Tuberous sclerosis in a 10-year-old boy whose two brothers and mother were also affected. **A,** intravenous urography reveals distortion of the collecting systems bilaterally. The splaying of the calyces could be secondary to either cysts or hamartomas. **B** and **C,** gray-scale renal ultrasonography reveals multiple cysts in both kidneys. **D** and **E,** enhanced computed tomography reveals bilateral multiple renal cysts, but no hamartomas. The presence of cysts can precede the presence of hamartomas. From the radiographic findings, this boy would have been misdiagnosed as having autosomal dominant polycystic kidney disease had it not been for his seizure disorder, hypertension, and adenoma sebaceum.

Koehler, 1976). These older patients with unilateral multicystic dysplasia can be entirely asymptomatic or can sometimes have abdominal pain (Spence, 1955; Ambrose et al., 1982). Hypertension was documented in a 6-year-old child with multicystic dysplasia in whom hypertension disappeared following nephrectomy (Javadpour et al., 1970). However, multicystic dysplasia is not felt to cause hypertension or urinary infections ordinarily. Contralateral compensatory hypertrophy always develops in older patients whether the patient has had a nephrectomy or not (Griscom et al., 1975).

Bilateral multicystic dysplasia is uncommon and incompatible with life (Griscom et al., 1975). Bilateral disease can be associated with Potter's facies, pulmonary hypoplasia, and oligohydramnios.

The prognosis of patients with unilateral multicystic dysplasia is dependent on the status of the contralateral urinary tract irrespective of the removal of the dysplastic kidney (DeKlerk et al., 1977). Autopsy series show a higher frequency of contralateral abnormalities (30%–100%) than do series of live patients who are surgical candidates where only a 14% incidence of contralateral abnormalities is demonstrated (Taxy, 1985). DeKlerk et al. found a higher incidence of contralateral renal disease in patients with lower ureteral atresias and, therefore, a worse prognosis than in patients with higher ureteral or renal pelvic atresias (DeKlerk et al., 1977). Therefore, careful study of the contralateral kidney is most important in evaluating a patient with multicystic dysplasia since the prognosis is otherwise excellent.

MORPHOLOGY

Multicystic dysplasia can appear grossly as a disorganized patternless mass of variably sized cysts (Fig 21–9). The calyces and pelvis are not recognizably present grossly and the reniform shape may not be evident. Usually, part or all of the ureter is atretic. Microscopically, immature ducts, ductules, and glomeruli along with cysts and islands of cartilage reside within fibrous stroma or cellular mesenchyme. Mature glomeruli may be present, although rare. The cysts are lined by low cuboidal epithelium. Contrast injection into the cysts reveals communication between the cystic spaces via tubular structures (Saxton et al., 1981).

PATHOGENESIS AND RENAL DYSPLASIA

Multicystic dysplasia is a category of renal dysplasia in which the kidney is totally affected and usually cystic. Another category of renal dysplasia is segmental or partial renal dysplasia in which a partially involved kidney is associated with an obstructive anomaly of the urinary tract. Additionally, renal dysplasia can occur in individuals with multiple anomaly syndromes as Meckel's syndrome (Miller and Gang, 1983), prune-belly syn-

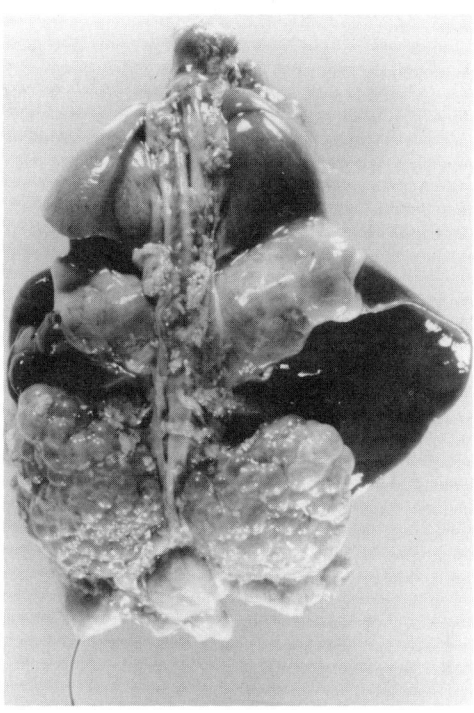

FIG 21–9.
Bilateral multicystic dysplasia. Bilateral multicystic kidneys are seen at the inferior margin of this posterior view of the viscera of a newborn infant. Both kidneys appear as disorganized, patternless masses of cysts of varying sizes. Bilateral multicystic dysplasia is fatal at birth. (Courtesy of Dr. S.E. Mills.)

drome (Taxy, 1985), Jeune's syndrome (asphyxiating thoracic dysplasia), and Zellweger's syndrome (cerebrohepatorenal syndrome (Bernstein, 1979). The abovementioned forms of renal dysplasia can be considered as a heterogeneous continuum of totally or segmentally involved kidneys that may be cystic, solid, or both.

The diagnosis of renal dysplasia is made from recognizing histological criteria thought to represent the aborted remnants of the ureteral bud or its poorly induced progeny (Taxy, 1985). Such histologic findings include primitive glomeruli, ductules, tubules, cartilage, and primitive ducts. However, all of these features, except for the primitive ducts, can also be found near renal scars or in renal inflammation. Hence, primitive ducts are the only noncontroversial histologic evidence of embryonic renal maldevelopment (Filmer at al., 1974).

The pathogenesis of dysplasia has attracted many hypotheses. An association exists between renal dysplasia and urinary tract obstruction. Controversy exists concerning whether the urinary tract obstruction causes renal dysplasia in utero (Bernstein, 1976) or whether the cause of the obstruction (such as a malfunctioning ureteral bud or its branch) is also the cause of dysplasia in the associated kidney or segment (Taxy, 1985). The association of unilateral renal dysplasia with unilateral ureteral obstruction, bilateral renal dysplasia with posterior urethral valves, and segmental renal dysplasia with segmental obstruction actually could support both theories. Furthermore, although the severity of the renal dysplasia relates to the severity of obstruction (Bernstein, 1979), the severity of both could relate to the degree of malfunction of the ureteral bud. Furthermore, although early fetal ureteral obstruction in lambs resulted in hydronephrosis and cortical cysts (Beck, 1971), no typical primitive ducts diagnostic of dysplasia were demonstrated (Taxy, 1985). Mackie and Stephens (1975) relate the degree of lateral ectopia of lower pole orifices in duplicated systems to the degree of dysplasia of the lower pole.

The multicystic dysplasia category of renal dysplasia includes a large, cystic mass associated commonly with either an ipsilateral ureteropelvic occlusion or ureteral atresia or absence. When the dysplasia is less severe and the kidney is smaller and not cystic, the lesion is described as a "solid" cystic kidney or an aplastic kidney, which is somewhat less likely to have an obstructed ureter. These lesions are opposite ends of a spectrum in which dysplasia involves an entire kidney.

RADIOLOGIC FINDINGS

Multicystic dysplasia appears as a nonfunctional mass on intravenous urography. Contralateral compensatory hypertrophy is a common finding in older children and adults. Since differentiation between multicystic dysplasia and congenital hydronephrosis is necessary in evaluating an infant with a palpable abdominal mass, ultrasound features unique to multicystic dysplasia have been described (Stuck et al., 1982). The most useful criteria are (1) the presence of interfaces between cysts, (2) nonmedial location of the largest cyst, and (3) absence of an identifiable renal sinus (Fig 21–10). Other confirmatory but not necessarily unique features on ultrasonography are multiplicity of cysts that do not communicate and an absence of parenchymal tissue. Renal angiography may show an absent or hypoplastic renal artery with no nephrogram (Fig 21–11). A retrograde ureterogram might reveal an atretic or absent ureter. Computed tomography is not commonly done in infants because of the requirement of prolonged periods of time without movement. However, when computed tomography of multicystic dysplasia in adults is performed, a small cystic kidney with or without calcifications can be revealed since the multicystic dysplastic kidneys identified in adults tend to be smaller. Percutaneous injection of contrast material into cysts of multicystic dysplasia in children has revealed communication between cysts (Saxton et al., 1981).

MANAGEMENT

In the past, multicystic kidneys were routinely removed for pathologic diagnosis. However, the availability of ultrasonography, computed tomography, and other studies deemed appropriate for a particular patient can confirm the diagnosis of multicystic dysplasia preoperatively. Since the development of carcinoma and hypertension is unlikely, routine nephrectomy is not mandatory (Glassberg and Filmer, 1985). In patients found to have pain from a retained multicystic kidney, nephrectomy alleviated the pain (Ambrose et al., 1982). However, whether routine nephrectomy in children is necessary to relieve possible pain later in adulthood is uncertain. Evaluation of the contralateral kidney is essential.

Diagnosis of multicystic dysplasia can be made by prenatal ultrasound (Legarth et al., 1981). However, the diagnosis of presumed obstructive uropathy has been given erroneously to multicystic kidneys on fetal ultrasonography (Kramer, 1983). Histologic changes of renal dysplasia occur so early in gestation that in utero intervention to relieve distal obstruction is technically not feasible. Indeed, surgical in utero intervention done inadvertently on multicystic kidneys thought to be obstructive uropathy by prenatal ultrasound did not improve the status of the multicystic kidney (Kramer, 1983).

MULTILOCULAR CYSTS (MULTILOCULAR CYSTIC NEPHROMA)

Multilocular cysts can usually be considered as a benign cystic tumor of the kidney that is usually unilateral and compresses the adjacent normal renal tissue. Hence, although multilocular cysts can be considered a cystic disease of the kidney, multilocular cysts do not diffusely replace normal renal parenchyma bilaterally, but instead compress normal renal tissue unilaterally. Multilocular cyst of the kidney is rare, with fewer than 100 cases reported (Cussen and Stephens, 1983). Children and adults are affected with equal frequency. The cysts are limited to the area of the kidney involved with the mass and vary in size from a few millimeters to 2.5 cm.

CLINICAL FEATURES

Although children and adults are affected equally with multilocular cysts, 73% of cases occurring in children under 4 years of age are in boys; in contrast, 80% of cases occurring in all patients over the age of 4 years are in females (Madewell et al., 1983). Children present with a nonpainful abdominal mass, whereas adults present with a symptom such as abdominal pain or hema-

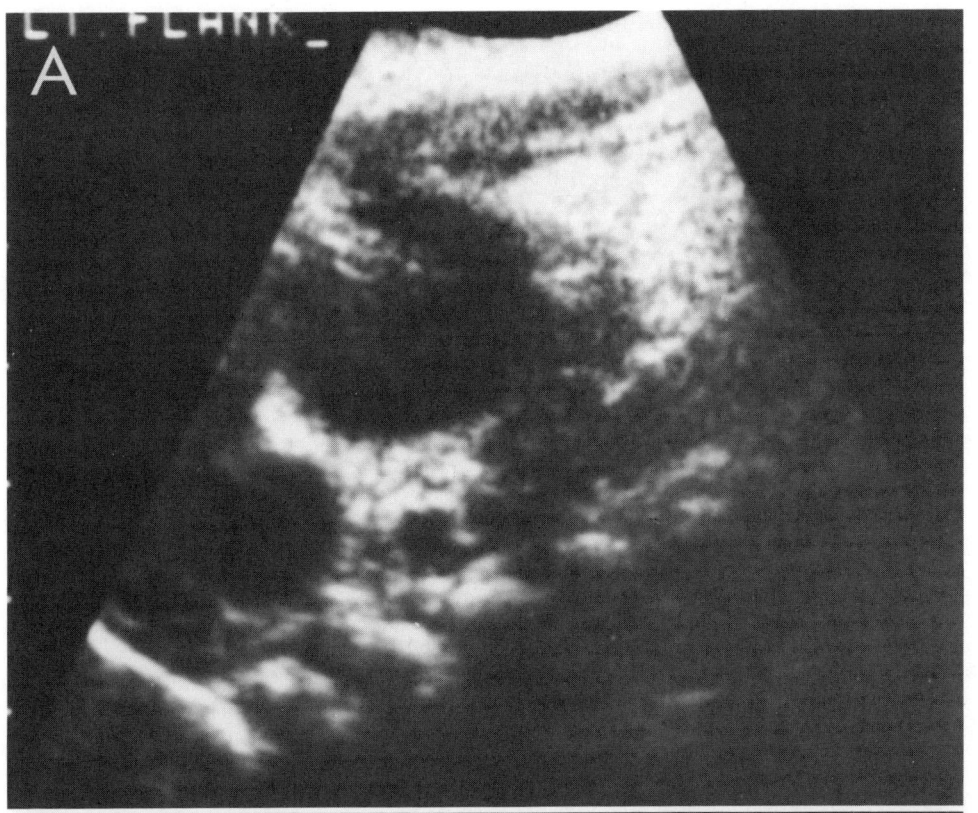

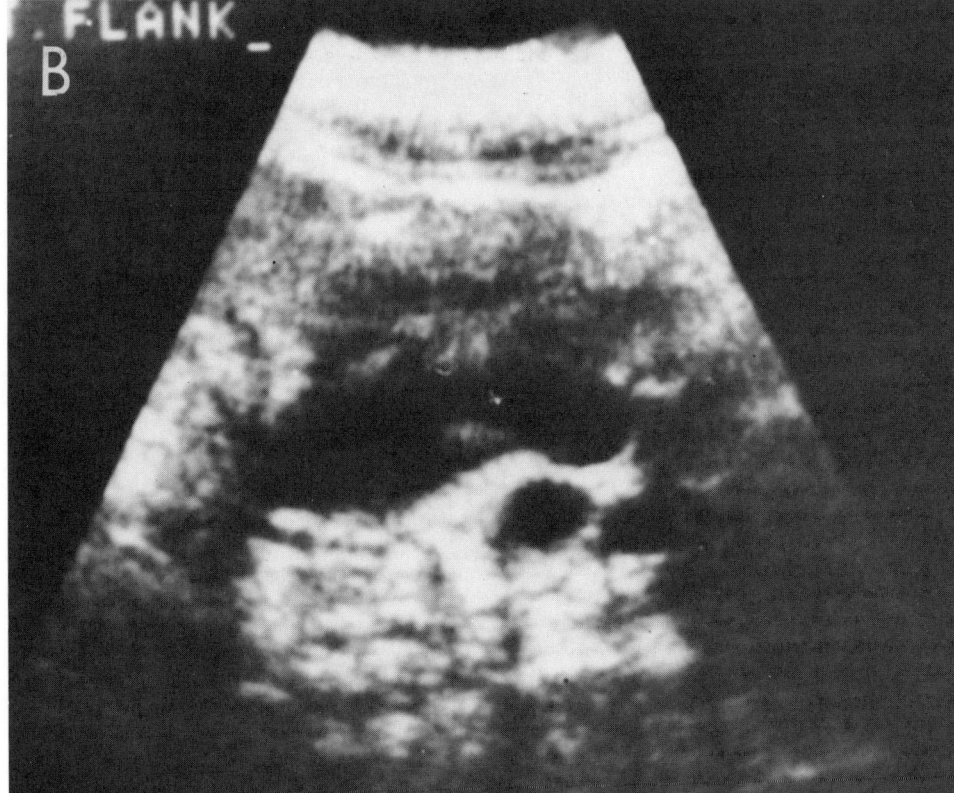

FIG 21–10.
Multicystic dysplasia in female newborn. **A** and **B,** gray-scale renal ultrasonography at age 1 week reveals multiple cysts in the left flank with septation between the cysts, but no normal renal parenchyma. No communication was identified between the cysts as would be present with hydronephrosis. The right kidney (not shown) was identified as normal.

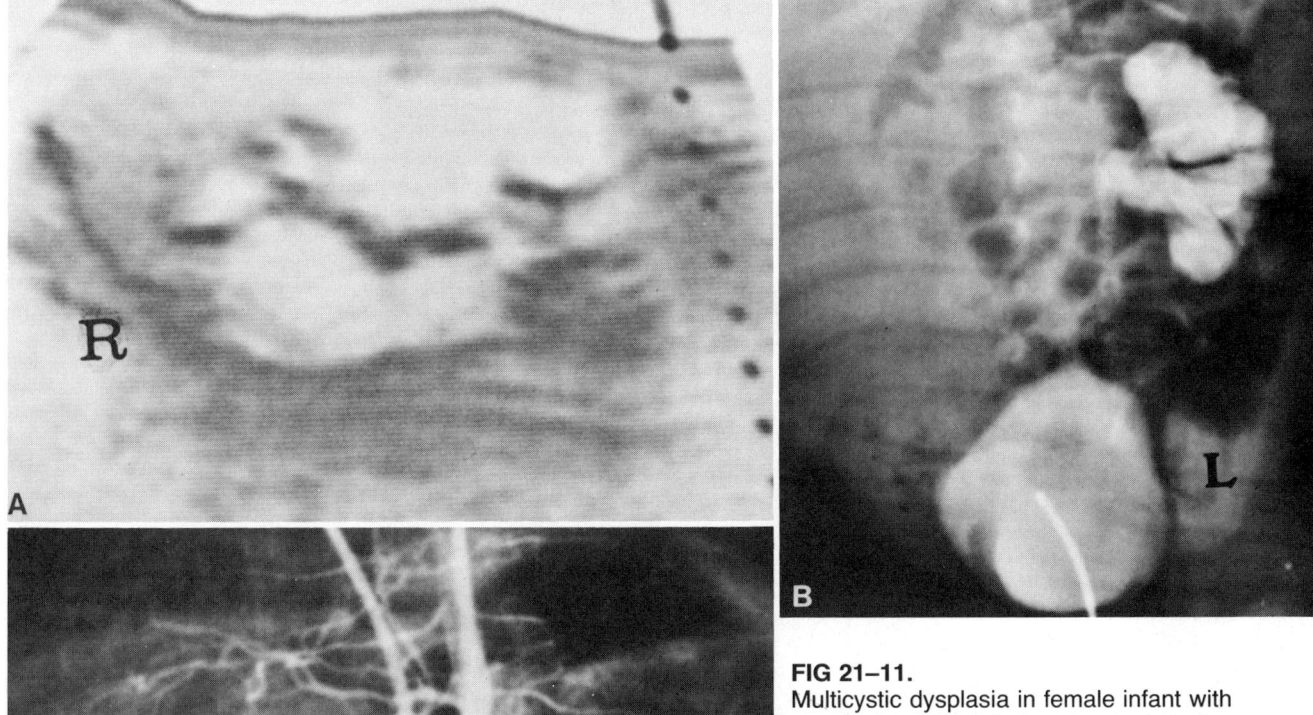

FIG 21–11.
Multicystic dysplasia in female infant with ventriculoseptal defect and patent ductus. Although this 2-month-old had a right multicystic kidney and a left congenital ureteropelvic junction obstruction, only a right abdominal mass was palpated. **A,** gray-scale renal ultrasonography reveals a large mass in the right flank with multiple cysts which are not communicating. The left kidney (not shown) was imaged as hydronephrotic. **B,** intravenous urography reveals a hydronephrotic left kidney and nonfunction on the right. **C,** midstream aortogram reveals no right renal artery.

turia. When hematuria does occur, it is usually associated with herniation of the multilocular cysts into the renal pelvis. The cystic mass has been observed to grow slowly in some patients (three years) and rapidly in others (two months) according to sequential physical examinations (Madewell et al., 1983). Although rare, metastases can occur.

MORPHOLOGY AND PATHOGENESIS

The pathologic criteria for establishing a diagnosis of multilocular cysts have not changed since proposed in 1956 by Boggs and Kimmelstiel. The criteria include: (1) multilocularity; (2) no communication between cysts; (3) cysts lined by epithelium; (4) no communication between the cyst and pelvis; (5) remaining kidney essentially normal; and (6) no normal nephrons in the septae of the cysts. Grossly, the multilocular cystic mass is well circumscribed by a thick capsule and compresses

normal adjacent renal tissue. Bilateral lesions are rare. Cut surface reveals multiple noncommunicating cysts varying from a few millimeters to 2.5 cm in diameter.

Microscopically, the cysts are lined by flattened or cuboidal epithelium. In children, the stroma of the septae is usually edematous or myxomatous. Microscopic foci of nephroblastoma were found in the stroma of the septae in six of 33 children studied by Madewell et al. (1983), whereas in adults, stromal septae had pronounced cellularity, with a spindle-cell pattern sometimes seen. Sarcoma was found in the septal stroma in four out of 21 adults, with three of these adults developing metastasis containing the stromal element (Madewell et al., 1983). The chemical content of the cyst fluid is similar to that of serum (Abt et al., 1979). The pathogenesis of multilocular cyst is not known except that there is no familial tendency and it behaves like a tumor in its growth, local recurrence, and ability to metastasize. The lack of normal renal tissue in its septae distinguishes the multilocular cyst from contiguous simple cysts and from congenital renal cystic disease with multiple cysts.

RADIOLOGIC FINDINGS AND MANAGEMENT

Intravenous urography of multilocular cystic kidney usually demonstrates an intrarenal mass in a normally functioning kidney. Ultrasound reveals a cluster of fluid-filled masses but can only clearly image the larger cysts (Banner et al., 1981). Computed tomography reveals a cluster of cysts with thick walls and calcifications. However when the cysts are small, it is more difficult for computed tomography to confirm the cystic component of the mass (Parienty et al., 1981). The septae on computed tomography enhance with intravenous contrast medium. Renal angiography usually reveals a hypovascular mass often with a tumor blush and neovascularity (Madewell et al., 1983). Hence, although ultrasonography, computed tomography, and arteriography can differentiate a multicystic kidney from a multilocular cyst, these tests cannot reliably differentiate a benign multilocular cyst from a multilocular cyst with foci of nephroblastoma or sarcoma, cystic Wilm's tumor, or a multiloculated renal cell carcinoma.

Treatment is nephrectomy after assessment of the status of the contralateral kidney, although bilateral cases are rare. Inadequate local resection has resulted in recurrence. Since radiographic methods cannot rule out malignancy, a partial nephrectomy is adequate only after the surgeon has confirmed that there is no malignancy in the surgical specimen pathologically.

SIMPLE CYSTS

Although simple cysts are an entity in their own right, concern regarding simple renal cysts is usually directed toward their differentiation from neoplastic renal masses because they are so frequently discovered as an incidental finding on radiographic studies. Indeed, 24% of routine abdominal computed tomographic scans in one study (Laucks and McLachlan, 1981) and 20% in another (Tada et al., 1983) revealed unexpected simple renal cysts. Analyzing the data separately in both studies, 0% to 6% of patients under age 40 years were found to have simple renal cysts as opposed to 18% to 19% of patients aged 40 to 60 years and about 30% of patients over age 60 years. However, at autopsy, at least 50% of patients older than 50 years have grossly recognizable cysts of the renal parenchyma (Kissane, 1976). By computed tomography, the number of cysts per patient and the cyst diameter increased with

the age of the patient (Laucks and McLachlan, 1981). Such data suggest that simple renal cysts are acquired abnormalities perhaps related to the aging process, but the pathogenesis is unknown. They are not familial.

Simple renal cysts are rare in children. Before the routine use of intravenous urography, ultrasonography, and computed tomography in children, the most common presentation of renal cysts in a child was of an asymptomatic abdominal mass. However, in recent series of pediatric patients, simple renal cysts were incidental findings on radiographic studies (Gordon et al., 1979; Bartholomew et al., 1980). Hypertension can be a symptomatic but rare presentation of a simple renal cyst in children (Babka et al., 1974; Hoard and O'Brien, 1976).

Simple renal cysts by definition are unilocular, do not communicate with the collecting system, occur in a kidney that is otherwise normal, and have an epithelial lining that contains no renal elements. Grossly, cysts are tense and thin-walled and vary in diameter from millimeters to 7 cm (Figs 21–12 and 21–13). Microscopically, the cysts are lined with low cuboidal to attenuated flattened epithelium. Fluid of simple cysts has chemical features of an ultrafiltrate of plasma (Kissane, 1976). A dynamic equilibrium exists between production and absorption of fluid in simple renal cysts such that the cyst fluid has levels of tritiated water that reach 73% of the serum level within 2–5 hours after an intravenous injection of tritiated water (Jacobsson et al., 1977). Perhaps cyst growth is dependent on the ratio of fluid production to absorption. When followed by serial ultrasound examinations, 42% of patients with simple cysts had a change in diameter of 1 cm or more over a

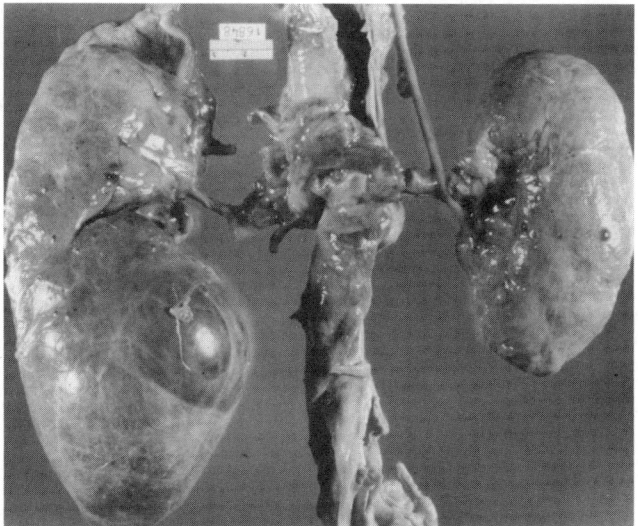

FIG 21–12.
Simple cyst. A tense, thin-walled cyst is seen in the left upper pole. (Courtesy of Dr. S.E. Mills.)

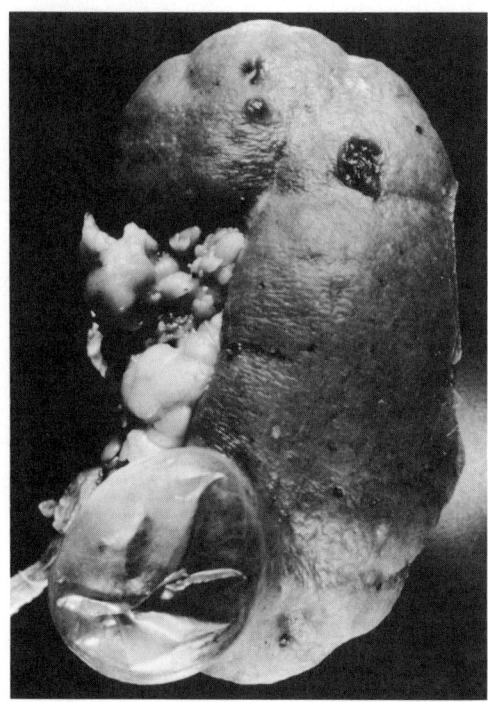

FIG 21–13.
Simple cyst. A simple cyst is seen in the lower pole of this left kidney. Cyst was collapsed inadvertently at surgery. (Courtesy of Dr. S.E. Mills.)

two- to three-year period (Dalton et al., 1986). However, only 64% of the cyst diameter changes resulted from an increase in size and all of these had a change of only 1 to 2 cm, which Dalton felt was not significant when evaluating cysts by serial ultrasonography. However, 29% of these patients with simple renal cysts did develop another cyst during this two- to three-year observation period. However when decades of years of patients' ages are compared, computed tomography studies reveal that as the patients' ages increase, both the number of cysts per patient and the diameter of the cysts increase (Laucks and McLachlan, 1981).

Radiographic methods to diagnose a simple renal cyst have improved tremendously over the last decade. Intravenous urography cannot differentiate between solid renal masses and a cystic renal mass, both of which distort the pyelocalyceal system (Fig 21–14). If the ultrasound requirements for a simple renal cyst are met on a study, investigation of a simple cyst is complete (Bartholomew et al., 1980). The ultrasound appearance of a simple renal cyst is an anechoic mas with a smooth posterior wall and posterior wall enhancement (Fig 21–15). However, if ultrasonography is equivocal, computed tomography is then useful because of its high accuracy in differentiating solid from cystic masses. Ultrasonography is preferable as an initial study because of lower cost, less radiation, and less of a requirement for the pediatric patient to remain motionless for extended time periods. Renal angiography is less important in the diagnosis of a simple renal cyst, but would show an avascular mass with renal cortex compressed at the margins. Percutaneous cyst puncture for cytology and injection of contrast is not now as commonly needed because of the high quality of ultrasonography and computed tomography available. Magnetic resonance imaging can help differentiate between a simple cyst and a hemorrhagic cyst (Fig 21–16).

After the diagnosis of simple renal cyst has been made, no treatment is necessary except for treatment of symptoms. Even though simple renal cysts are rare in the pediatric age group, once malignancy has been excluded appropriately by radiographic methods, no surgical exploration to confirm the diagnosis is necessary (Gordon et al., 1979; Ravden et al., 1980; Siegel and McAlister, 1980). Rarely, cysts may require surgical resection or percutaneous decompression because of pain. Cyst fluid can reaccumulate to the original dimensions after percutaneous aspiration as documented by ultrasonography (Bartholomew et al., 1980).

MEDULLARY SPONGE KIDNEY (MSK)

Medullary sponge kidney (MSK) is a relatively benign entity that is diagnosed by excretory urography. The diagnosis is usually made in adults who present for excretory urography for evaluation of calculi, hematuria, or infection. The kidneys are more commonly bilaterally involved with 1- to 5-mm cysts at the papillary tips. In MSK, the kidneys retain their reniform shape and the incidence varies from 1 in 5,000 to 1 in 20,000 (Kuiper, 1976). Although described histologically to some extent in 1908 (Ekstrom et al., 1959), it was the onset of intravenous urography that made possible the description and classification of MSK as a disease by Cacchi and Ricci (1948). Even today, intravenous urography is the mainstay of diagnosis.

CLINICAL FEATURES

MSK is most commonly diagnosed in patients from 20 to 50 years of age when an intravenous urogram has been performed to evaluate their presenting symptom, which is either renal colic (50% to 60%), gross hematuria (10% to 18%), or urinary infection (20% to 33%) (Kuiper, 1976). Of those who have this radiographic diagnosis made, 60% will pass calculi at some time (Kuiper, 1976). However, in an uncertain number the condition is undiagnosed because patients are asymptomatic. A small number of patients with MSK have received a diagnosis when they were evaluated by intravenous urography for other conditions, i.e., hyper-

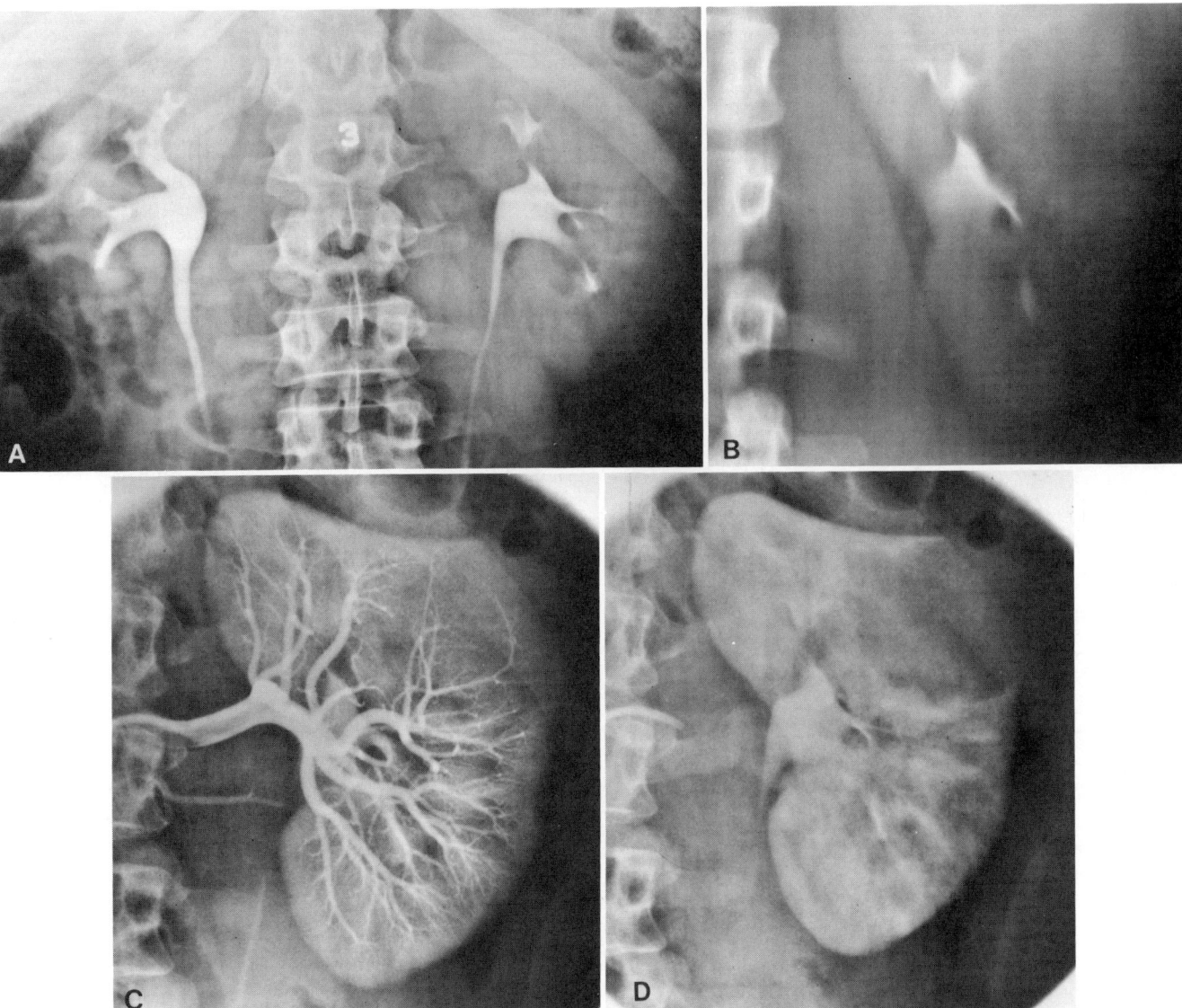

FIG 21–14.
Simple cyst in 36-year-old man. **A,** intravenous urography reveals bilaterally normal kidneys except for a spherical mass arising from the superolateral margin of the left kidney. **B,** nephrotomography reveals the mass to be relatively lucent and to have smooth margins. **C** and **D,** renal arteriography confirms the presence of a simple cyst by the draping of vessels around the mass, the lack of neovascularity, and its thin rim and "beaks" at its margin with the renal cortex.

tensive screening, abdominal tumors. Hence, although 60% of patients diagnosed with MSK will pass calculi, it is uncertain what percentage of all patients with MSK will pass stones, have infections, etc. Nephrocalcinosis in MSK is clinically benign except that it leads to stone formation and passage of calculi. Fifty-eight of 70 kidneys examined in affected patients had intracavitary calculi ranging in number from 1 to infinity (Ekstrom et al., 1959). Stone analysis reveals calcium phosphate stones are the most common, with calcium oxalate stones making up most of the remainder.

In MSK, the medullary cysts are dilated collecting ducts in the pyramids and papillae. Urinary stagnation in these dilated or cystic tubules may be the cause of the calcifications found in these dilated ducts. Alternatively, perhaps the biochemical composition of the urine is altered in patients with MSK who have calculi (Yendt, 1982). Although recent studies show no association between MSK and hypercalciuria, 15% to 21% of patients with calcium stones were diagnosed as having MSK by intravenous urography (Parks et al., 1982; Yendt, 1982).

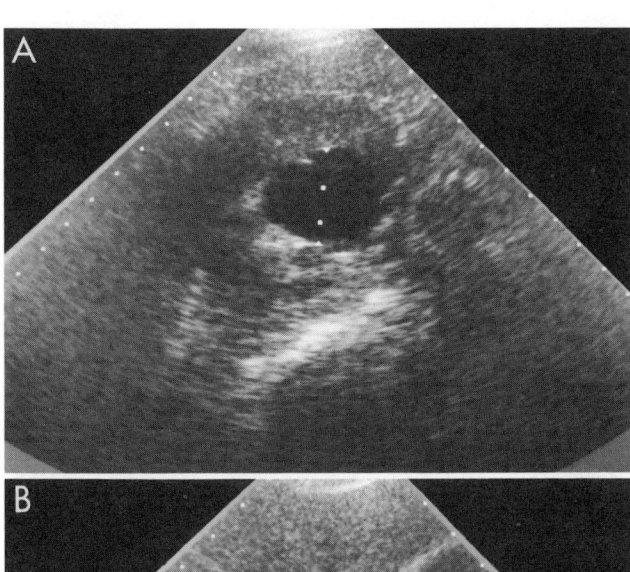

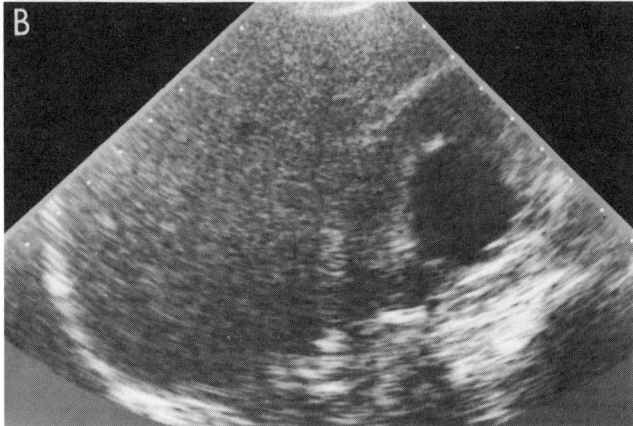

FIG 21–15.
Simple cyst in 53-year-old woman. **A** and **B**, gray-scale renal ultrasonography reveals an anechoic renal mass with a smooth posterior wall and posterior wall enhancement.

Progression of MSK as evidenced by the size or location of the dilated collecting tubules on intravenous urography is uncommon (Kuiper, 1976). However, nephrocalcinosis in MSK is acquired (Kuiper, 1976). Segmental ultrasounds of six children with MSK have revealed increasing echogenicity of the renal pyramids. This increasing echogenicity proved to be calcifications in computed tomography, although the calcifications were too small to be seen on intravenous urography (Patriquin and O'Regan, 1985). Three of these six children had hematuria, which was probably associated with their microlithiasis. Microlithiasis may be the cause of most of the hematuria seen in those MSK patients who have no demonstrable calcifications on intravenous urography.

About 20% to 30% of patients with MSK have a urinary tract infection as their presenting symptom (Kuiper, 1976). Urinary infection does not play a causal role in stone formation in most patients with MSK, most of whom have sterile urine (Yendt, 1982). However, when

present, infection has the potential of accelerating stone growth by alkalinizing the urine. Fortunately, infection in affected patients responds well to antibiotics, unlike that in patients with autosomal dominant polycystic kidney disease. Additionally, patients affected with MSK have an incidence of hypertension similar to that found in the normal population. Occasionally, a mild urine-concentrating defect can be found in patients with MSK who otherwise have normal renal function.

MORPHOLOGY, PATHOGENESIS, AND GENETICS

Grossly, the kidneys of MSK are of normal size or slightly larger and have kept their reniform shape. Cut surface reveals the larger of the dilated collecting duct cysts, which are present at the papillary tips of the renal pyramids. The smaller of the dilated collecting duct cysts may only be visible microscopically since they vary from 1 to 5 mm in diameter. Microscopically, the collecting ducts are lined by flattened epithelium and the columns of Bertin are normal. As few as one or as many as all renal pyramids can have MSK. Calcifications can be seen in the cysts clustered toward the papillary tips.

The pathogenesis of MSK is unknown, but it is probably congenital. Successive ultrasound studies in children with MSK show the progressive development of calcifications as the children grow as well as the presence of the disease in children (Patriquin and O'Regan, 1985). MSK has some basis for at least occasional ge-

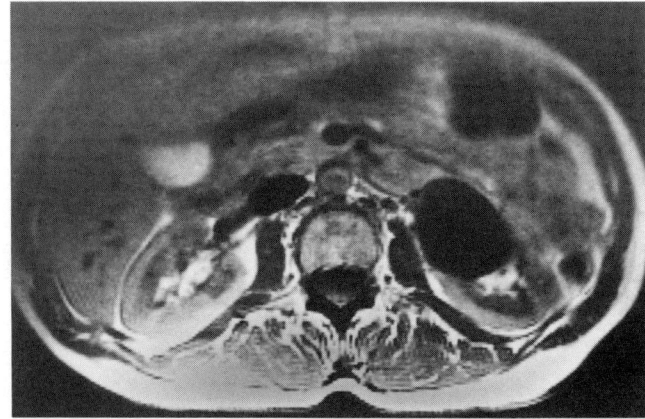

FIG 21–16.
Simple cyst in a 51-year-old woman. Magnetic resonance imaging reveals a T_1 weighted image diagnostic of a simple left renal cyst because of the distinct smooth margin and because of the low signal intensity (e.g., the cyst contents are very black). To confirm the diagnosis of a simple cyst, a T_2 weighted image (not shown) would reveal the cyst contents to have a high signal intensity (e.g., to be very white), which would be consistent with water.

netic transmission because of documented family history in a few families (Kuiper, 1976) and the association with other congenital manifestations.

RADIOLOGIC FINDINGS AND MANAGEMENT

The diagnosis of MSK is made by excretory urography (Fig 21–17). The contrast medium stagnates in the dilated or cystic linear tubules in one or more renal papillae. These appear as linear radiations from the calyces and vary in shape from beads to strands. Whether a "pyramidal blush" is an early sign of MSK on excretory urography or is a normal finding is undecided. Calculi are located at papillary tips on a plain abdominal radiograph. Ultrasonography only shows the medullary cysts if they are large. However in children, the renal medulla is better seen on ultrasonography because there is less renal sinus fat and overlying muscle. As a result, hyperechoic areas of the pyramids are visualized

and either represent ductal calcifications or dilated collecting ducts (Patriquin and O'Regan, 1985). On retrograde pyelography, the cysts or ectatic ducts either fail to fill with contrast or fill less prominently.

Some patients with MSK will have an asymptomatic course and require no medical intervention while 10% of the symptomatic patients have a poor prognosis because of recurrent renal calculi and infection (Kuiper, 1976). Infections in patients with MSK respond well to antibiotics. Prevention of stone formation can be attempted with increased fluid intake, administration of thiazides, and inorganic phosphates (Yendt, 1982). Once calculi are formed, extracorporeal shock wave lithotripsy can be used to disintegrate stones in the collecting system and, to some extent, fragment the ductal calcifications. Appropriate antibiotic prophylaxis during the perioperative period should be used because of the high risk of obstruction in these patients from passage of more calculi.

GLOMERULOCYSTIC DISEASE

Glomerulocystic disease is a rare congenital bilateral cystic disease in which Bowman's space is dilated and, thus, appears as a glomerular cyst. These multiple cysts are confined to the renal cortex in these reniform but very large kidneys. However, a morphologically heterogeneous group of patients have the microscopic findings of glomerular cysts in their kidneys, including patients in whom this is the only anomaly, patients with Zellweger's syndrome, patients with other defined syndromes such as trisomy 13, and patients with autosomal dominant polycystic kidney disease presenting in infancy. Whether patients with glomerulocystic disease as the only anomaly are a separate entity is undecided (McAlister et al., 1979).

Patients present in infancy or early childhood with bilaterally palpable flank masses (McAlister et al., 1979; Taxy et al., 1976; Sellers et al., 1978). Renal function can deteriorate (McAlister et al., 1979) or remain stable for an unknown amount of time (Sellers and Richie, 1978). There is no known genetic transmission.

The cortical cysts have varied in diameter from a few millimeters to 7 cm (Sellers and Richie, 1978). Intravenous urography demonstrates huge kidneys. Ultrasonography demonstrates large kidneys and may show numerous small cysts depending on cyst size (McAlister et al., 1979). Selective renal arteriography can show intrarenal vessels stretched around numerous cortical cysts (Sellers and Richie, 1978).

Treatment is appropriate management of renal failure. Both dialysis and renal transplantation have been successful (McAlister et al., 1979).

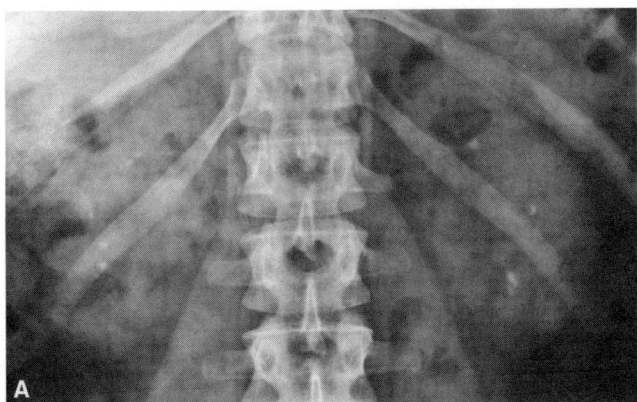
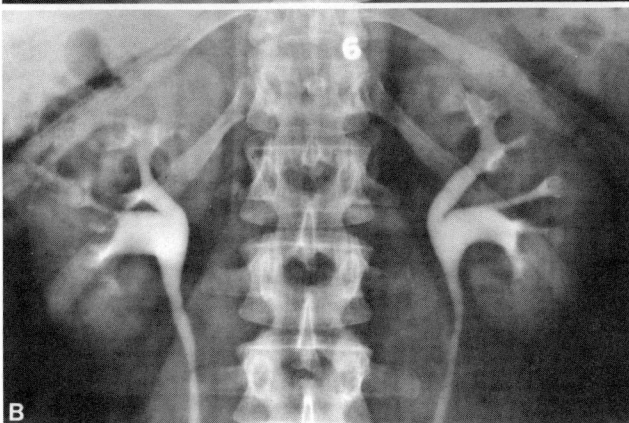

FIG 21–17.
Medullary sponge kidney. **A,** plain abdominal radiograph reveals bilateral nephrocalcinosis. The calculi are located at papillary tips. **B,** intravenous urogram in same patient reveals linear radiations from calyces. These radiations result from stagnation of contrast medium in cystic or dilated collecting ducts.

ACQUIRED RENAL CYSTIC DISEASE

Long-term maintenance dialysis therapy is now accepted therapy for patients with end-stage renal disease (ESRD). However, dialytic therapy has ushered in a new class of renal cystic disease called acquired renal cystic disease (ARCD), so named in 1977 by Dunnill et al. when they observed that 47% of 30 long-term dialysis patients had bilateral renal cystic disease at autopsy. None of these patients had renal cystic disease before beginning hemodialysis. Since then, multiple reports (Levine and Grantham, 1985c; Ishikawa et al., 1980) have confirmed the presence of ARCD in patients receiving hemodialysis or peritoneal dialysis and to a lesser extent in patients who have longstanding renal insufficiency but who have never undergone dialysis.

The incidence of ARCD in dialysis patients increased with the length of time the patient received dialysis (Levine and Grantham, 1985c). Ishikawa et al. reported a 43.5% incidence of ARCD in patients who had undergone less than three years of hemodialysis as opposed to 79.3% incidence in patients who had more than three years (Ishikawa et al., 1980). Patients with ARCD are usually asymptomatic and are, therefore, not usually diagnosed unless screened radiologically. Hemorrhage into the cysts is not uncommon. Rarely, ARCD patients have retroperitoneal hemorrhage due to rupture of the hemorrhagic cyst into the retroperitoneal space (Kassirer and Gang, 1982). Such hemorrhage occurs usually during anticoagulation therapy for hemodialysis, at which time patients may have such symptoms as sudden flank pain, hypotension, and decrease in hematocrit (Ishikawa, 1985). Patients with ARCD may also develop renal abscess.

Successful renal transplantation in dialysis patients leads to regression of established ARCD and reduction in size of the affected native kidneys (Levine and Grantham, 1985c; Ishikawa et al., 1983). Conversely, prolongation of the dialysis period not only increases the incidence of ARCD but also the number and size of acquired cysts. A sudden rise in hematocrit after years of stable anemia while on hemodialysis is unusual, but has been observed in a few cases of ARCD (Ishikawa, 1985).

Of concern is the association of ARCD with benign and malignant renal tumors. Reviewing the literature, ARCD has an overall incidence of 43.6% of all dialysis patients surveyed. Renal tumors, usually multiple, are found in 7.1% of chronic dialysis patients, but in 16.4% of ARCD patients. Many of the tumors were small and were therefore considered adenomas, but 4.7% of the renal tumors had metastasized by the time of presentation (Grantham and Levine, 1985).

MORPHOLOGY AND PATHOGENESIS

Grossly, renal cortex in ARCD is replaced by multiple, bilateral cysts as is the medulla to some extent. The kidneys are reniform and small. Most cysts are 0.5 to 2 cm in diameter but can be 4 cm in diameter. Cyst fluid is clear to hemorrhagic.

Microscopically, few glomeruli remain. Most cysts are lined by low cuboidal or columnar epithelium. However, some cysts have atypical proliferations of columnar, multilayered living cells in papillary formation. The tumors found in ARCD can be categorized histologically into three groups: (1) papillary tumors that project into the cyst lumen without invading the cyst wall; (2) parenchymal tumors that reside adjacent to but not within the cyst; and (3) solid tumors that are sheets of less differentiated cells with either foamy or acidophilic cytoplasm. Central areas may have hemorrhage (Dunnill, 1985). Controversy exists over whether tumors less than 3 cm in size in ARCD patients should be called adenomas or adenocarcinomas, since there is no histologic difference and since tumors greater than 3 cm had to once be less than 3 cm. However, observation of the natural history of these tumors in ARCD patients should soon reveal the malignant potential of these adenomas.

The pathogenesis of ARCD is unknown but is thought to involve the accumulation of biologically active substances that accumulate during dialysis. Certain chemicals, such as diphenylthiazole, cause acquired cystic disease in animals, which regresses when diphenylthiazole is taken out of their diet (Kanwar and Carrone, 1984). Such biologically active substances may cause kidneys to develop ARCD and renal tumors, if these kidneys are exposed to the biologically active substance long enough. Support for this hypothesis lies in the fact that successful renal transplantation results in regression of established ARCD. Since ARCD does occur, although less commonly, in patients who have longstanding renal insufficiency before beginning dialysis, dialysate-leached chemicals are not felt to be the initiators of ARCD. Hence, this suggests that although hemodialysis and peritoneal dialysis prolong life, they are not complete kidney substitutes since they may allow some biologically active substance to accumulate (Grantham and Levine, 1985).

RADIOLOGIC FINDINGS AND MANAGEMENT

Because the incidence of ARCD increases with the length of time a patient has been on dialysis, asymptomatic dialysis patients should be screened after 3–4 years of dialysis with either ultrasound or computed tomography. If ultrasound detects the presence of cysts, then computed tomography should be done to assess the presence of a renal tumor, since ARCD patients are at increased risk for renal tumors. Unfortunately, it can be very difficult for ultrasonography to detect the small cysts of ARCD in these sonographically abnormal kidneys of ESRD patients (Grantham and Levine, 1985) (Fig 21–18). Hence, computed tomography would be good as a first test to diagnose ARCD as well to diagnose tumor (Fig 21–19). Alternatively, if an ARCD patient becomes symptomatic with flank pain, computed tomography should be obtained to diagnose and localize retroperitoneal hemorrhage.

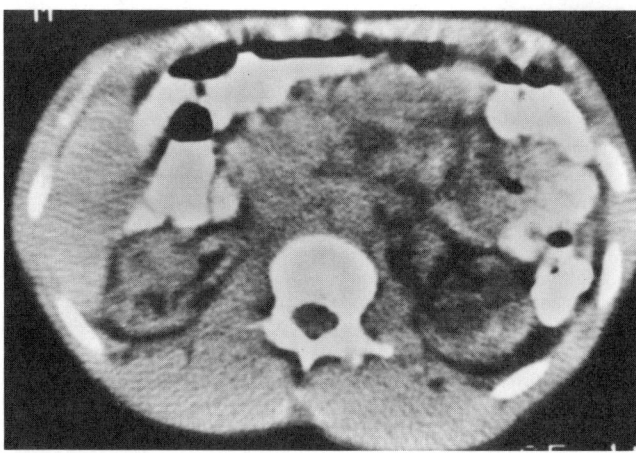

FIG 21–19.
ARCD in 25-year-old man on chronic dialysis. Enhanced computed tomography reveals bilaterally small kidneys, which do not enhance with intravenous contrast. Multiple cysts are visualized in both kidneys.

If a tumor is diagnosed radiographically and either the tumor is greater than 3 cm or the patient is a candidate for transplantation, the patient should undergo nephrectomy. If the tumor is less than 3 cm in diameter, the patient can either have a nephrectomy or undergo observation by serial computed tomography or ultrasonography to assess tumor growth (Levine and Grantham, 1985c). Aggressive tumor growth should warrant a nephrectomy.

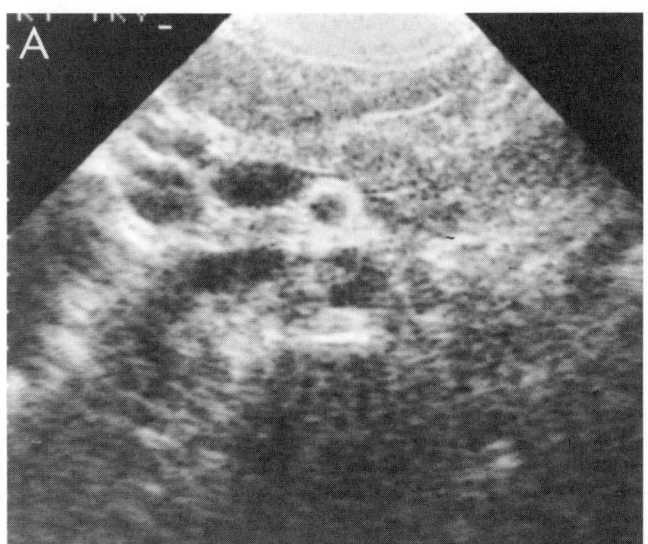

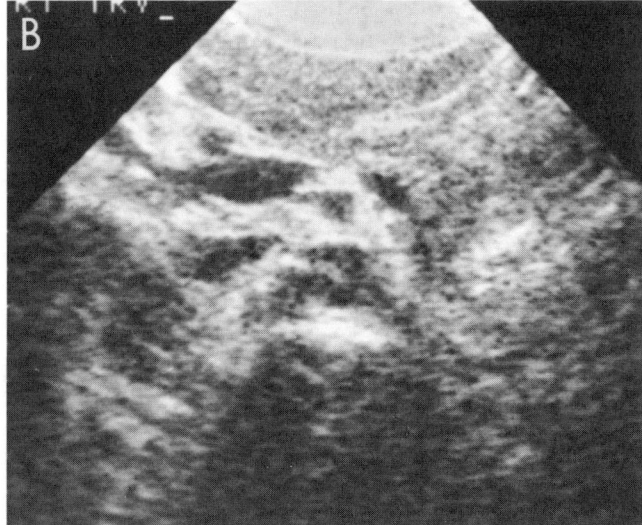

FIG 21–18.
ARCD in a 52-year-old woman on chronic dialysis. **A** and **B,** gray-scale renal ultrasonography reveals small kidneys with increased cortical echoes as found in chronic renal disease. Multiple cysts are present and are of varying sizes.

REFERENCES

1. Abt AB, Demers LM, Shochat SJ: Cystic nephroma: An ultrastructural biochemical study. *J Urol* 1979; 122:539.
2. Al Saadi AA, Yoshimoto M, Bree R, et al: A family study of renal dysplasia. *Am J Med Genet* 1984; 19:669.
3. Ambrose SS, Gould RA, Trulock TS, et al: Unilateral multicystic renal disease in adults. *J Urol* 1982; 128:366.
4. Avasthi PS, Erickson DG, Gardner KD Jr: Hereditary renal-retinal dysplasia and the medullary cystic disease-nephronophthisis complex. *Ann Intern Med* 1976; 84:157.
5. Babka JC, Cohen MS, Sode J: Solitary intrarenal cyst causing hypertension. *N Engl J Med* 1974; 291:343.
6. Banner MP, Pollack HM, Chatten J, et al: Multilocular renal cysts: Radiologic pathologic correlation. *AJR* 1981; 136:239.
7. Bartholomew TH, Slovis TL, Kroovand LR, et al: The sonographic evaluation and management of simple renal cysts in children. *J Urol* 1980; 123:732.
8. Beck AD: The effect of intrauterine urinary obstruction upon the development of the fetal kidney. *J Urol* 1971; 105:784.
9. Bennett AH, Stewart W, Lazarus JM: Bilateral nephrec-

tomy in patients with polycystic renal disease. *Surg Gynecol Obstet* 1973; 137:819.

10. Bennetts WM, Elzinga L, Pulliam JP, et al: Cyst fluid antibiotic concentrations in autosomal-dominant polycystic kidney disease. *Am J Kidney Dis* 1985; 6:400.

11. Bernstein J: Congenital malformations of the kidney, in Early LE, Gottschalk CW (eds): *Strauss Welt's Diseases of the Kidney*, ed 3. Boston, Little, Brown & Co, 1979, p 1989.

12. Bernstein J: A classification of renal cysts, in Gardner KD Jr (ed): *Cystic Diseases of the Kidney*. New York, John Wiley & Sons, 1976, p 7.

13. Bigelow NH: The association of polycystic kidneys with intracranial aneurysms and other related disorders. *Am J Med Sci* 1953; 225:485.

14. Bloom DA, Brosman S: The multicystic kidney. *J Urol* 1978; 120:211.

15. Blyth H, Ockenden BG: Polycystic disease of kidneys and children presenting in childhood. *J Med Genet* 1971; 8:257.

16. Boggs LK, Kimmelstiel P: Benign multilocular cyst nephroma: Report of two cases of so-called multilocular cyst of the kidney. *J Urol* 1956; 76:530.

17. Bricker NS, Patton JF: Cystic disease of the kidneys: A study of dynamics and chemical composition of cyst fluid. *Am J Med* 1955; 18:20.

18. Brown RAP: Polycystic disease of the kidneys and intracranial aneurysms: The etiology and interrelationship of these conditions: Review of recent literature and report of seven cases in which both conditions coexisted. *Glasgow Med J* 1951; 32:333.

19. Brown RS, McClusky RT, Gang DL: Weekly clinicopathological exercises: Case records of the Massachusetts General Hospital, case 48–1981. *N Engl J Med* 1981; 305:1334.

20. Burke JR, Inglis JA, Craswell PW, et al: Juvenile nephronophthisis and medullary cystic disease—the same disease (report of a large family with medullary cystic disease associated with gout and epilepsy). *Clin Nephrol* 1982; 18:1.

21. Cacchi R, Ricci V: Sopra una rara e forse ancora non descritta affezione cristica delle piramidi renali ('rene a spugna'). *Atti Soc Ital Urol* 1948; 5:59.

22. Chamberlin BC, Hagge WW, Stickler GB: Juvenile nephronophthisis and medullary cystic disease. *Mayo Clinic Proc* 1977; 52:485.

23. Chonko AM, Weiss SM, Stein JH, et al: Renal involvement in tuberous sclerosis. *Am J Med* 1974; 56:124.

24. Churchill DN, Bear JB, Morgan J, et al: Prognosis of adult onset polycystic kidney disease re-evaluated. *Kidney Int* 1984; 26:190.

25. Compton WR, Lester PD, Kyaw MM, et al: The abdominal angiographic spectrum of tuberous sclerosis. *AJR* 1976; 126:807.

26. Cuppage FE, Huseman RA, Chapman A, et al: Ultrastructure and function of cysts from human adult polycystic kidneys. *Kidney Int* 1980; 17:372.

27. Cussen LG, Stephens FD: Renal dysgenesis: A 'urologic' classification, in Stephens FD (ed): *Congenital Malformations of the Urinary Tract*. New York, Praeger Publishing, 1983, p 463.

28. Dalgaard OZ: Bilateral polycystic disease of kidneys: A follow-up of 284 patients and their families. *Acta Med Scand* 1957; 328 (suppl):10.

29. Dalton D, Neiman H, Grayhack JT: The natural history of simple renal cysts: A preliminary study. *J Urol* 1986; 136:905.

30. DeKlerk DP, Marshall FM, Jeffs RD: Multicystic dysplastic kidney. *J Urol* 1977; 118:306.

31. Delaney VB, Adler S, Bruns FJ, et al: Autosomal dominant polycystic kidney disease: Presentation, complications, and prognosis. *Am J Kidney Dis* 1985; 5:104.

32. Dunnill MS, Millard PR, Oliver D: Acquired cystic disease of the kidneys: A hazard of long-term intermittent maintenance hemodialysis. *J Clin Pathol* 1977; 30:868.

33. Dunnill MS: Acquired cystic disease, in Grantham JJ, Gardner KD Jr (eds): *Problems in Diagnosis and Management of Polycystic Kidney Disease*. Kansas City, PKR Foundation, 1985, p 211.

34. Eggers PW, Connerton R, McMullan M: The Medicare experience with end-stage renal disease: Trends in incidence, prevalence, and survival. *Health Care Financing Rev* 1984; 5:69.

35. Ekstrom T, Engfeldt B, Lagergren C, et al: *Medullary Sponge Kidney*. Stockholm, Almqvist & Wiksell, 1959.

36. Elkin M, Bernstein J: Cystic diseases of the kidney: Radiological and pathological considerations. *Clin Radiol* 1969; 20:65.

37. Engstrom N, Ljungqvist A, Persson B, et al: Tuberous sclerosis with a localized angiomatous malformation in the ileum and excessive albumin loss into the lower intestinal tract. *Pediatrics* 1962; 39:681.

38. Evan AP, Gardner KD Jr, Bernstein J: Polypoid and papillary epithelium hyperplasia: A potential cause of ductal obstruction in adult polycystic disease. *Kidney Int* 1979; 16:743.

39. Filmer RB, Taxy JB, King LR: Renal dysplasia: Clinicopathological study. *Trans Am Assoc Genitourinary Surg* 1974; 66:18.

40. Franz KA, Reubi FC: The rate of functional deterioration in polycystic kidney disease. *Kidney Int* 1983; 23:526.

41. Frimodt-Moller PC, Nissen HM, Dyreborg U: Polycystic kidneys as the renal lesion in Lindau's disease. *J Urol* 1981; 125:868.

42. Fry JL Jr, Koch WE, Jennette JC, et al: A genetically determined murine model of infantile polycystic kidney disease. *J Urol* 1985; 134:828.

43. Gabow PA, Ikle DW, Holmes JH: Polycystic kidney disease, prospective analysis of nonazotemic patients and family members. *Ann Intern Med* 1984; 101:238.

44. Gang DL, Herrin JT: Infantile polycystic disease of the liver and kidneys. *Clin Nephrol* 1986; 25:28.

45. Gardner KD Jr: Juvenile nephronophthisis and renal medullary cystic disease, in Gardner KD Jr (ed): *Cystic Diseases of the Kidney*. New York, John Wiley & Sons, 1976, p 173.

46. Gardner KD Jr: The medullary cystic disease: The ne-

phronophthisis-cystic renal medulla complex and medullary sponge kidney, in Early LE, Gottschalk CW (eds): *Strauss Welt's Diseases of the Kidney*, ed 3. Boston, Little, Brown & Co, 1979, p 1147.

47. Gardner KD Jr, Evan AP: Cystic kidneys: An enigma evolves. *Am J Kidney Dis* 1984; 3:403.

48. Garel LA, Habib R, Pariente D, et al: Juvenile nephronophthisis: Sonographic appearance in children with severe uremia. *Radiology* 1984; 151:93.

49. Glassberg K, Filmer RB: Renal dysplasia, renal hypoplasia and cystic disease of the kidney, in Kelatis PO, King LR, Belman AB (eds): *Clinical Pediatric Urology*. Philadelphia, WB Saunders Co, 1985, vol 2, p 941.

50. Gordon RL, Pollack HM, Popky GL, et al: Simple serous cysts of the kidney in children. *Radiology* 1979; 131:357.

51. Grantham JJ, Levine E: Acquired cystic disease: Replacing one kidney disease with another. *Kidney Int* 1985; 28:99.

52. Griscom NT: The roentgenology of neonatal abdominal masses. *AJR* 1965; 93:447.

53. Griscom NT, Vawter GF, Fellers FX: Pelvoinfundibular atresia: The usual form of multicystic kidney: 44 unilateral and 2 bilateral cases. *Semin Roentgenol* 1975; 10:125.

54. Hoard TD, O'Brien DP III: Simple renal cyst and high renin hypertension cured by cyst decompression. *J Urol* 1976; 115:326.

55. Hatfield PM, Pfister RC: Adult polycystic disease of the kidneys (Potter type 3). *JAMA* 1972; 222:1527.

56. Hilpert PL, Friedman AC, Radecki PD, et al: MRI of hemorrhagic renal cysts in polycystic kidney disease. *AJR* 1986; 146:1167.

57. Huseman R, Grady A, Welling D, et al: Macropuncture study of polycystic disease in adult human kidneys. *Kidney Int* 1980; 18:375.

58. Ishikawa I, Saito Y, Onouchi A, et al: Development of acquired cystic disease and adenocarcinoma of the kidney in glomerulonephritis chronic hemodialysis patients. *Clin Nephrol* 1980; 14:1.

59. Ishikawa I, Yuri T, Kitada H, et al: Regression of acquired cystic disease of the kidney after successful renal transplantation. *Am J Nephrol* 1983; 3:310.

60. Ishikawa I: Uremic acquired cystic disease of the kidney. *Urology* 1985; 26:101.

61. Jacobsson L, Lindqvist B, Michaelson G, et al: Fluid turnover in renal cysts. *Acta Med Scand* 1977; 202:327.

62. Javadpour N, Chelouhy E, Moncada L, et al: Hypertension in a child caused by a multicystic kidney. *J Urol* 1970; 104:918.

63. Kanwar YS, Carrone FA: Reversible changes of tubular wall and basement membrane in drug-induced renal cystic disease. *Kidney Int* 1984; 26:35.

64. Kassirer JP, Gang DL: Weekly clinicopathological exercises, case records of the Massachusetts General Hospital, case 16–1982. *N Engl J Med* 1982; 306:975.

65. Kissane JM, Smith MG: *Pathology in Infancy and Childhood*, ed 2. St Louis, CV Mosby Co, 1975, p 587.

66. Kissane JM: The morphology of renal cystic disease, in Gardner KD Jr (ed): *Cystic Diseases of the Kidney*. New York, John Wiley & Sons, 1976, p 31.

67. Kramer SA: Current status of fetal intervention for congenital hydronephrosis. *J Urol* 1983; 130:641.

68. Kuiper JJ: Medullary sponge kidney, in Gardner KD Jr (ed): *Cystic Diseases of the Kidney*. New York, John Wiley & Sons, 1976, p 151.

69. Kyaw MM, Koehler PR: Congenital multicystic kidney, in Gardner KD Jr (ed): *Cystic Diseases of the Kidney*. New York, John Wiley & Sons, 1976, p 115.

70. Landing BH, Gwinn JL, Lieberman E: Cystic diseases of the kidney in children, in Gardner KD Jr (ed): *Cystic Diseases of the Kidney*. New York, John Wiley & Sons, 1976, p 187.

71. Laucks SP, McLachlan SF: Aging and simple cysts of the kidney. *Br J Radiol* 1981; 54:12.

72. Leier CV, Baker PB, Kilman JW, et al: Cardiovascular abnormalities associated with adult polycystic kidney disease. *Ann Intern Med* 1984; 100:683.

73. Legarth J, Verder H, Gronvall S: Prenatal diagnosis of multicystic kidney by ultrasound. *Acta Obstet Gynecol Scand* 1981; 60:523.

74. Levine E, Grantham JJ: The role of computed tomography in the evaluation of adult polycystic kidney disease. *Am J Kidney Dis* 1981; 1:99.

75. Levine E, Collins DL, Horton WA, et al: CT screening of the abdomen in von Hippel-Lindau disease. *AJR* 1982; 139:505.

76. Levine E, Cook LT, Grantham JJ: Liver cysts in autosomal-dominant polycystic kidney disease: Clinical and computed tomography. *AJR* 1985a; 145:229.

77. Levine E, Grantham JJ: High density renal cysts in autosomal dominant polycystic kidney disease demonstrated by CT. *Radiology* 1985b; 154:477.

78. Levine E, Grantham JJ: Clinical diagnosis, in Grantham JJ, Gardner KD Jr (eds): *Problems in Diagnosis and Management of Polycystic Kidney Disease*. Kansas City, PKR Foundation, 1985c, p 235.

79. Lieberman E, Salinas-Madrigal L, Gwinn JL, et al: Infantile polycystic disease of the kidneys and liver: Clinical, pathological and radiologic correlations and comparisons with congenital hepatic fibrosis. *Medicine* 1971; 50:277.

80. Luthy DA, Hirsch JH: Infantile polycystic kidney disease: Observations from attempts at prenatal diagnosis. *Am J Med Genet* 1985; 20:505.

81. Mackie CG, Stephens FD: Duplex kidneys: A correlation of renal dysplasia with position of the ureteral orifice. *J Urol* 1975; 114:274.

82. Madewell JE, Goldman SM, Davis CJ Jr, et al: Multilocular cystic nephroma: A radiographic-pathologic correlation of 58 patients. *Radiology* 1983; 146:309.

83. Martinez-Maldonado M: Functional aspects: Electrolyte and uric acid excretion, in Grantham JJ, Gardner KD Jr (eds): *Problems in the Management and Diagnosis of Polycystic Kidney Disease*. Kansas City, PKR Foundation, 1985, p 70.

84. Maschio G, Oldrizzi L, Rugio C: The effects of dietary protein restriction on the course of early chronic renal

failure, in Mitch WE (ed): *The Progressive Nature of Renal Disease.* New York, Churchill Livingstone, 1985, p 203.

85. McAlister WH, Siegel MJ, Shackelford G, et al: Glomerulocystic kidney. *AJR* 1979; 133:536.

86. McClean RH, Gang DL, Herrin JT: Weekly clinicopathological exercises: Case records of the Massachusetts General Hospital, case 48–1978. *N Engl J Med* 1978; 299:1294.

87. Mellins HZ: Radiologic diagnosis, in Grantham JJ, Gardner KD Jr (eds): *Problems in the Management and Diagnosis of Polycystic Kidney Disease.* Kansas City, PKR Foundation, 1985, p 19.

88. Melson GL, Biello DR, Lee JKT: Comparative imaging, in Lee KT, Sagel SS, Stanley RI (eds): *Computed Body Tomography.* New York, Raven Press, 1983, p 535.

89. Melson GL, Shackelford GD, Cole BR, et al: The spectrum of sonographic findings in infantile polycystic kidney disease with urographic and clinical correlations. *J Clin Ultrasound* 1985; 13:113.

90. Miller WA, Gang DL: Weekly clinicopathological exercises. Case records of the Massachusetts Hospital, case 11–1983. *N Engl J Med* 1983; 308:642.

91. Nickel WR, Reed WB, Lachman R: *The Kidney in Tuberous Sclerosis,* The Fifth Conference of the Clinical Delineation of Birth Defects, part XVI, Original Article Series, vol 10. Baltimore, Williams & Wilkins Co, 1974, p 80.

92. O'Callaghan TJ, Edwards JA, Tobin M, et al: Tuberous sclerosis with striking renal involvement in a family. *Arch Intern Med* 1975; 135:1082.

93. Okada RD, Platt MA, Fleishman J: Chronic renal failure in patients with tuberous sclerosis association with renal cysts. *Nephron* 1982; 30:85.

94. Ostathanondh V, Potter EL: Pathogenesis of polycystic kidneys: Historical survey. *Arch Pathol* 1964; 77:459.

95. Parienty RA, Pradel J, Imbert M, et al: Computed tomography of multilocular cystic nephroma. *Radiology* 1981; 140:135.

96. Parks JH, Coe FL, Strauss AL: Calcium nephrolithiasis and medullary sponge kidney in women. *N Engl J Med* 1982; 306:1088.

97. Patriquin HB, O'Regan S: Medullary sponge kidney in childhood. *AJR* 1985; 145:315.

98. Pearson JC, Weiss J, Tanagho EA: A plea for conservation of kidney in renal adenocarcinoma associated with von Hippel-Lindau disease. *J Urol* 1980; 124:910.

99. Proesmans W, Van Damme B, Macken J: Nephronophthisis and tapetoretinal degeneration associated with liver fibrosis. *Clin Nephrol* 1975; 3:160.

100. Ravden MI, Zuckerman HL, Kay CJ, et al: Evaluation of solitary simple renal cysts in children. *J Urol* 1980; 124:904.

101. Reeders ST, Breuning MH, Davies KE, et al: A highly polymorphic DNA marker linked to adult polycystic disease on chromosome 16. *Nature* 1985; 317:542.

102. Reubi FC: Hypertension, in Grantham JJ, Gardner KD Jr (eds): *Problems in the Management and Diagnosis of Polycystic Kidney Disease.* Kansas City, PKR Foundation, 1985, p 121.

103. Rosenberg JC, Bernstein J, Rosenberg B: Renal cystic disease associated with tuberous sclerosis complex: Renal failure treated by cadaveric kidney transplantation. *Clin Nephrol* 1975; 4:109.

104. Rosenfield AT, Siegel NJ, Kappelman NB, et al: Gray scale ultrasonography in medullary cystic disease of the kidney and congenital hepatic fibrosis with tubular ectasia: New observations. *AJR* 1977; 129:297.

105. Sanfilippo FP, Vaughn WK, Peters TG, et al: Transplantation for polycystic kidney disease. *Transplantation* 1983; 36:54.

106. Saxton HM, Golding SJ, Chantler C, et al: Diagnostic puncture in renal cystic dysplasia (multicystic kidney). Evidence on the etiology of the cysts. *Br J Radiol* 1981; 54:555.

107. Schwab SJ: Efficacy of chloramphenicol in refractory cyst infections in autosomal dominant polycystic kidney disease. *Am J Kidney Dis* 1985; 5:258.

108. Schwab SJ, Weaver ME: Penetration of trimethoprim and sulfamethoxazole into cysts in a patient with autosomal-dominant polycystic kidney disease. *Am J Kidney Dis* 1986; 7:434.

109. Sellers B, Richie JP: Glomerulocystic kidney: Proposed etiology and pathogenesis. *J Urol* 1978; 119:678.

110. Siegel MJ, McAlister WH: Simple cysts of the kidney in children. *J Urol* 1980; 123:75.

111. Spence HM: Congenital unilateral multicystic kidney: An entity to be distinguished from polycystic kidney disease and other cystic disorders. *J Urol* 1955; 74:693.

112. Stapleton FB, Johnson D, Kaplan GW, et al: The cystic renal lesion in tuberous sclerosis. *J Pediatr* 1980; 97:574.

113. Stuck KJ, Koff SA, Silver TM: Ultrasonic features of multicystic dysplastic kidney: Expanded diagnostic criteria. *Radiology* 1982; 143:217.

114. Suter W: Das kongenitale aneurysma der basalen gehirnarterien und cystennieren. *Schweiz Med Wochenschr* 1949; 79:471.

115. Torres VE, Holley KE, Offord KP: Epidemiology, in Grantham JJ, Gardner KD Jr (eds): *Problems in the Management and Diagnosis of Polycystic Kidney Disease.* Kansas City, PKR Foundation, 1986, p 49.

116. Tada S, Yamagishi J, Kobayashi H, et al: The incidence of simple renal cyst by computed tomography. *Clin Radiol* 1983; 34:437.

117. Taxy JB, Filmer RB: Glomerulocystic kidney. *Arch Pathol Lab Med* 1976; 100:186.

118. Taxy JB: Renal dysplasia: A review. *Pathol Annu* 1985; 20(pt 2):139.

119. Wakabayashi T, Fujita S, Ohbora Y, et al: Polycystic kidney disease and intracranial aneurysms: Early angiographic diagnosis and early operation for the unruptured aneurysm. *J Neurosurg* 1983; 58:488.

120. Walker D, Fennell R, Garin E, et al: Spectrum of multicystic dysplasia. *Urology* 1978; 11:433.

121. Wenzl JE, Lagos JC, Albers D: Tuberous sclerosis presenting as polycystic kidneys and seizures in an infant. *J Pediatr* 1970; 77:673.

122. Wernecke K, Heckemann R, Bachman H, et al: Sonography of infantile polycystic kidney disease. *Urol Radiol* 1985; 7:138.

123. Yendt ER: Medullary sponge kidney and nephrolithiasis. *N Engl J Med* 1982; 306:1106.

124. Zerres K, Weiss H, Bulla M, et al: Prenatal diagnosis of an early manifestation of autosomal dominant adult-type polycystic kidney disease. *Lancet* 1982; 2:988.

Chapter 22

Renal Infection

ANTHONY J. SCHAEFFER, M.D.

Although renal infection is less prevalent than bladder infection, it frequently is a more difficult problem for the patient and his physician because of its often varied and morbid presentation and course, the difficulty in establishing a firm microbiologic and pathologic diagnosis, and its potential for significantly impairing renal function. Although the classic symptoms of acute onset of fever, chills, and flank pain are usually indicative of renal infection, some patients with these symptoms do not have pyelonephritis. Conversely, significant renal infection may be associated with an insidious onset of nonspecific local or systemic symptoms or be entirely asymptomatic. Therefore, a high clinical index of suspicion and appropriate radiologic and laboratory studies are required to establish the diagnosis of renal infection.

Unfortunately, the relationship between laboratory findings and the presence of renal infection often is poor. Bacteriuria and pyuria, the hallmarks of urinary tract infection, are not necessarily indicative of renal infection. Conversely, patients with significant renal infection may present with sterile urine.

The pathologic and radiologic criteria for diagnosing renal infection may also be misleading. Interstitial renal inflammation, once thought to be predominantly caused by pyelonephritis, is now recognized as a nonspecific histopathologic change associated with a variety of immunologic, congenital, or chemical lesions that usually develop in the absence of bacterial infection. Infectious granulomatous diseases of the kidney frequently have either radiologic or pathologic characteristics that mimic renal cystic disease, neoplasia, or other renal inflammatory disease.

The impact of renal infection on renal function is varied. In the absence of obstruction, acute or chronic pyelonephritis may transiently or permanently alter renal function, but nonobstructive pyelonephritis is no longer recognized as a major cause of renal failure. However, pyelonephritis, when associated with urinary tract obstruction, or granulomatous renal infection may lead

rapidly to significant inflammatory complications, renal failure, or even death.

This chapter will review the pathology and pathogenesis of acute and chronic bacterial and granulomatous infectious diseases of the kidney. The clinical presentation, appropriate diagnostic steps and management for each disease will be discussed.

BACTERIAL NEPHRITIS

The majority of patients with urinary tract infection are asymptomatic or have symptoms of dysuria, frequency, and urgency judged to indicate infection of the urinary bladder. If urinary tract infection is associated with fever, chills, and flank pain, the infection is judged to be more severe and usually involving the kidneys. Bacterial nephritis, whether isolated or recurrent, may cause acute or chronic renal parenchymal damage. The kidney itself may become colonized with bacteria and act as the source for recurrent episodes of renal or lower urinary tract infection.

This section will review the pathology and pathogenesis of bacterial nephritis. Bacterial virulence and host risk factors will be elaborated. The clinical presentation, evaluation, and management of acute pyelonephritis will be discussed. The diagnostic and therapeutic steps indicated for patients with complicated acute bacterial nephritis such as focal nephritis, emphysematous pyelonephritis, renal abscess, infected hydronephrosis, pyonephrosis, or perinephric abscess will also be outlined. Last, the question of how the urologist can best diagnose and manage chronic pyelonephritis will be considered.

Interstitial renal inflammation is a nonspecific cellular response of the renal interstitium that may or may not be complicated by fibrosis and varying degrees of tubular or glomerular damage. It has been generally believed that bacterial infection of the kidney, such as pyelonephritis, was the most frequent cause of interstitial renal inflammation and subsequent development of serious renal disease. More recently, however, the non-specific nature of the histopathologic changes of interstitial renal inflammation has been appreciated. It is now recognized that interstitial renal inflammation is associated with immunologic reactions, congenital lesions, or papillary damage in the absence of bacterial infection and that bacterial infection is frequently a secondary event as a result of urologic evaluations of patients with chronic preexisting interstitial renal inflammation. Thus, histologic evidence alone is too often assumed indicative of bacterial nephritis and is not sufficient to establish whether interstitial changes in the kidney are either primary or secondary to bacterial infection or of noninfectious etiologies.

The term "bacterial nephritis" should be reserved for interstitial renal inflammation primarily due to the immediate or late effects of bacterial infection in the renal parenchyma. Pyelonephritis refers to bacterial nephritis involving the renal parenchyma and collecting system. Acute pyelonephritis refers to a clinical symptom complex or pathologic lesion that is always associated with urinary tract infection. It may, however, have no morphological or functional components detectable by routine clinical modalities. Chronic pyelonephritis can only be diagnosed when there is postinfectious morphological, radiologic, or functional evidence of renal disease but need not be associated with urinary tract infection at the time of study.

PATHOLOGY

The histologic changes in acute and chronic interstitial inflammation due to bacterial infection are not pathognomonic of bacterial infection but rather are common to a variety of diseases resulting in interstitial inflammation.

The opportunity for pathologic confirmation of acute bacterial nephritis is rare. The kidney may be edematous. Focal acute suppurative bacterial nephritis caused by hematogenous dissemination of bacteria to the renal cortex is characterized by multiple focal areas of suppuration on the surface of the kidney (Fig 22–1). Histologic examination of the renal cortex shows focal suppurative destruction of glomeruli and tubules. Adjacent cortical structures and the medulla are not involved in the inflammatory reaction. Acute ascending pyelonephritis is characterized by linear bands of inflammation extending from the medulla to the renal capsule (Fig 22–2). Histologic examination usually reveals a focal wedge-shaped area of acute interstitial inflammation with the apex of the wedge in the renal medulla. Polymorphonuclear leukocytes or a predominantly lymphocytic and plasma cell response are seen. Bacteria may also be present.

The changes that appear to be most specific for

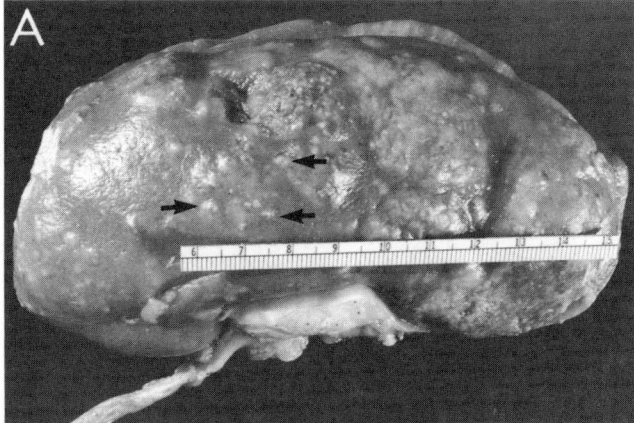

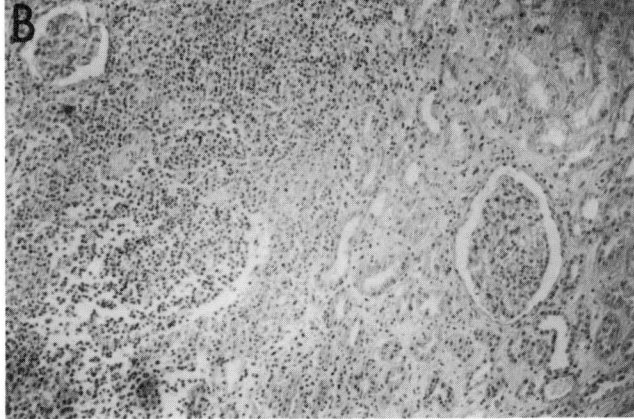

FIG 22–1.
Acute focal suppurative bacterial nephritis. **A,** surface of kidney. *Arrows* indicate focal areas of suppuration. **B,** renal cortex showing focal suppurative destruction of glomeruli and tubules.

chronic pyelonephritis are evident on careful gross examination of the kidney and consist of a cortical scar associated with retraction of the corresponding renal papilla (Hodson and Wilson, 1965a; Hodson, 1965b, 1967; Freedman, 1979; Heptinstall, 1974). The kidney shows evidence of patchy involvement with numerous chronic inflammatory foci, mainly confined to the cortex but also involving the medulla (Fig 22–3). The scars may be separated by intervening zones of normal parenchyma, causing a grossly irregular renal outline. The microscopic appearance, as with most chronic interstitial disease, includes the presence of lymphocytes and plasma cells. Although glomeruli within scars may be surrounded by a cuff of fibrosis or be partially or completely hyalinized, glomeruli outside these severely scarred zones are relatively normal. Vascular involvement is variable but in patients with hypertension, nephrosclerosis may be found. Papillary abnormalities include deformity, sclerosis, and sometimes necrosis. Studies in animals have clearly indicated the critical

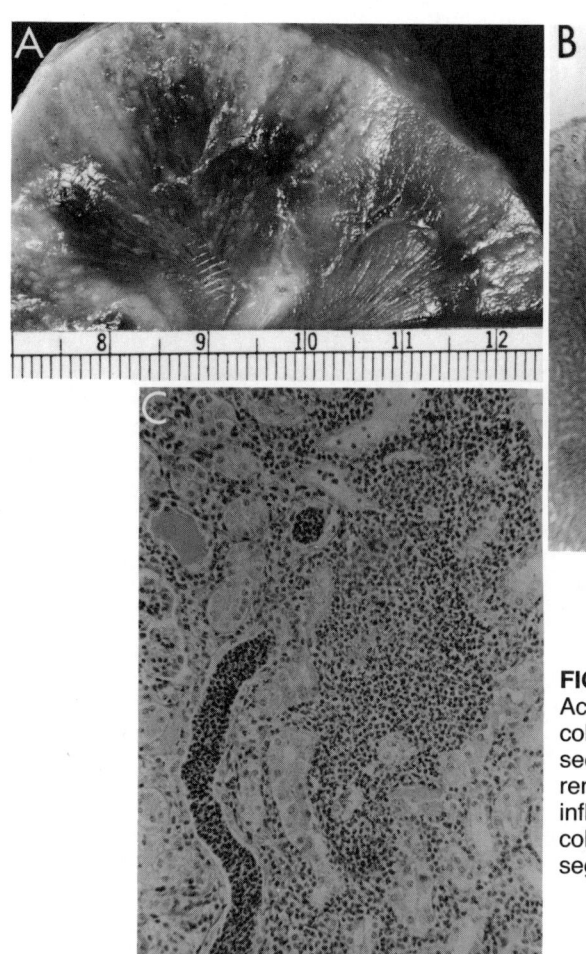

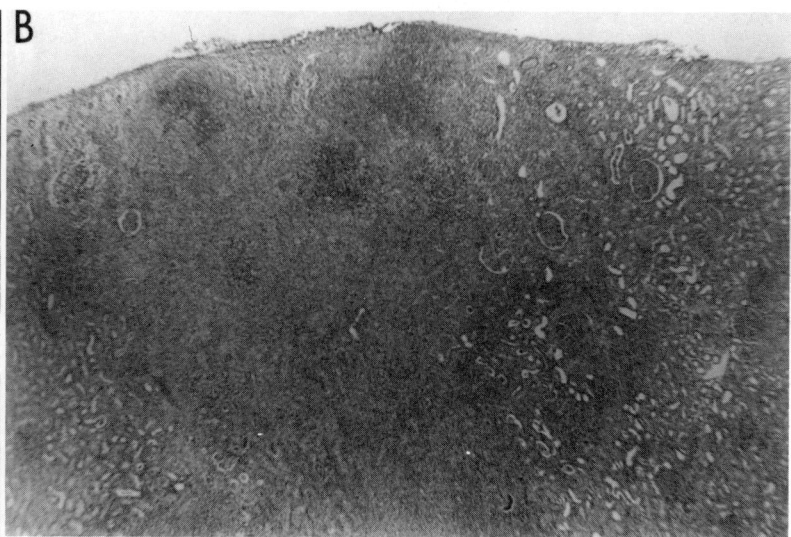

FIG 22–2.
Acute ascending pyelonephritis. **A,** cortical structures, tubules, and collecting ducts diffusely infiltrated with inflammatory cells. **B,** section of the renal cortex showing wedge-shaped destruction of renal cortical structures due to ascending infiltration with inflammatory cells. **C,** thickened and inflamed tissue surrounding the collecting ducts in the medulla. A polymorphonuclear cast of segmented neutrophils is clearly visible.

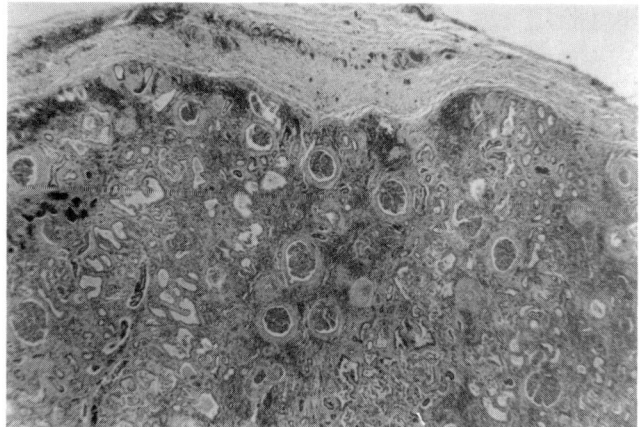

FIG 22–3.
Chronic pyelonephritis. The renal cortex shows thickened fibrous capsule and focal retracted scar on surface of kidney. Focal destruction of tubules in center of picture is accompanied by periglomerular fibrosis and scarring.

role of the papilla in the initiation of pyelonephritis (Freedman and Beeson, 1958). However, these changes are not necessarily specific for bacterial infection, and in fact may occur in the absence of infection owing to other disorders such as analgesic abuse, diabetes, and sickle cell disease.

The classic pathologic description of chronic pyelonephritis has traditionally been that of Weiss and Parker (1939). Their autopsy studies, however, included late stages of the disease, which are often complicated by hypertension and vascular changes and are best referred to as end-stage kidneys. They repeatedly emphasized that patients with this form of renal disease did not always have clinical evidence of bacterial infections of the urinary tract sufficient to explain the severe loss of renal tissue. Stamey and Pfau (1963a) presented a case of pure symptomatic pyelonephritis, incurable with drug therapy and uncomplicated by vascular hypertension. The microscopic sections together with Heptinstall's comments represent an unusual opportunity to study the pathologic characteristics of this disease in its purest form.

PATHOGENESIS

Pathways of Renal Infection

Three possible pathways for renal infection have been cited: hematogenous, lymphatic, and urinary. The weight of clinical and experimental evidence strongly suggests that most episodes of pyelonephritis are caused by retrograde ascent of bacteria from the bladder, through the ureter to the renal pelvis and parenchyma. Although cystitis is frequently restricted to the bladder, in approximately 50% of instances there is further extension of the infection into the upper urinary tract (Stamey, 1980). Although reflux of urine is probably not required for ascending infections, edema associated with cystitis may cause sufficient changes in the ureterovesical junction to permit reflux. Once the bacteria are introduced into the ureter they may ascend to the kidney unaided. However, this ascent would be greatly increased by any process that interferes with the normal ureteral peristaltic function. Gram-negative bacteria and their endotoxins as well as pregnancy and ureteral obstruction have a marked antiperistaltic effect.

Once the bacteria reach the renal pelvis, they can enter the renal parenchyma via the collecting ducts at the papillary tips and then ascend upward within the collecting tubules. This process is hastened and exacerbated by ureteral obstruction or vesicoureteral reflux, particularly when it is associated with intrarenal reflux. The hematogenous route plays a limited role in initiation of pyelonephritis; an exception is the cortical abscess that occurs following gram-positive bacteremia. Clinical evidence to support an extramural lymphatic route is very limited.

Bacterial Virulence

A number of bacterial factors have been proposed to be associated with uropathogenicity. These include production of hemolysin, K antigen, O antigen, and bacterial adhesin. The ability of bacteria to bind to epithelial cells by specific adhesins may be the most relevant one.

Bacterial adherence to surface structures of host epithelial cells is of fundamental importance in the initiation of many infections. The virulence of *E. coli* strains isolated from patients with urinary tract infection have been correlated with their ability to adhere to human uroepithelial cells. Bacteria capable of adhering avidly are frequently associated with upper urinary tract infections (Svanborg-Eden et al., 1976). Adhesin is frequently associated with the presence on the bacteria of proteinaceous filamentous organelles named fimbriae or pili (Duguid and Gillies, 1957), which appear to recognize specific receptors on the host epithelial cells. An epithelial cell receptor that seems to bind specifically most pyelonephritogenic *E. coli* has recently been identified. Kallenius et al. (1980a) and Leffler and Svanborg-Eden (1980) independently reported that pyelonephritogenic *E. coli* possess specific P-fimbriae that adhere to glycolipid receptors in the cell membrane of uroepithelial cells as well as to human red cells that contain the P-antigen. Further work by Kallenius et al. (1980b) has shown that a disaccharide (α-D-Galp-(1-4)-β-D-Galp) is the active minimal carbohydrate receptor. Clinical studies of patients with urinary tract infection have strongly associated P-fimbriated *E. coli* with acute pyelonephritis. Ninety percent of children with acute pyelonephritis were infected with P-fimbriated strains. On the other hand, only 19% of strains causing cystitis, 14% of strains associated with asymptomatic bacteriuria, and 7% of fecal *E. coli* isolates from healthy controls carried P-fimbriae (Kallenius et al., 1981). P-fimbriated strains of *E. coli* were isolated significantly more often from adult women with clinical signs and symptoms of pyelonephritis than from women with cystitis or asymptomatic bacteriuria (Latham and Stamm, 1984). Thus, P-fimbriae appear to be a virulence factor of *E. coli* causing acute pyelonephritis.

Although *E. coli* bearing P-fimbriae are associated with clinical pyelonephritis, factors other than P-fimbriae must play an important role in the pathogenesis of upper urinary tract infections. Studies have shown that more than 40% of adult women with acute pyelonephritis were infected with *E. coli* strains not bearing P-fimbriae as were most of the women without symptoms of pyelonephritis but with infections localized in the upper urinary tract (Latham and Stamm, 1984; Hagberg et al., 1983). Using a mouse model, Hagberg et al. (1983) noted that avirulent fecal strains transformed with plasmids that coded for P-fimbriae were recovered from kidneys of infected animals in smaller quantities than were mutants of pyelonephritogenic strains with the same adhesin. Moreover, the presence of vesicoureteral reflux, which is frequently associated with development of pyelonephritis and renal scarring, has been correlated with a decreased prevalence of P-fimbriated strains of *E. coli* among children with acute pyelonephritis (Lomberg et al., 1983).

Host Factors

The main host determinants of renal damage in bacterial nephritis are obstruction, vesicoureteral reflux, age, underlying disease, pregnancy, susceptibility to recurrent urinary tract infection, and delay in institution of therapy.

OBSTRUCTION.—The fact that obstruction increases renal damage due to infection is well established in the literature and has received much support from animal experiments. Although a normal kidney exhibits little susceptibility to bacterial infection, the obstructed kidney is readily infected (Guze and Beeson, 1956). The mechanism by which interference with urine flow increases susceptibility of the kidney to infection is not clear. Presumably stagnation of the urine enhances the growth of bacteria both in suspension and on the urothelial surface. The effect of residual urine on facilitating bacterial multiplication in the bladder cavity is well known clinically and experimentally (Freedman, 1967a). In addition, ischemia caused by increased tissue pressure in the kidney and ureter may impair delivery of leukocytes and antibacterial serum factors and thus impair clearance of bacteria from the kidney.

VESICOURETERAL REFLUX.—Hodson and Edwards (1960) first described the association of vesicoureteral reflux, urinary tract infection, and renal clubbing and scarring. Winberg et al. (1974) in a study of children 2 months to 16 years of age with their first symptomatic urinary tract infection found that approximately half of the infants had reflux and that the incidence decreased to 18% with increasing age. These incidences are similar to those observed in neonates and schoolchildren who had screening bacteriuria (Abbott, 1972; Kunin et al., 1965; Savage et al., 1973; Asscher et al., 1973).

In children the prognostic significance of reflux varies with its degree as determined by voiding cystourethrography. Children with gross reflux and urinary tract infections usually develop progressive renal damage manifested by renal scarring, proteinuria, and renal failure. Those with a lesser degree of reflux usually improve or completely recover spontaneously or after treatment of the urinary tract infection. Minimal and moderate reflux is seldom if ever a cause of new scar formation, progression of scarring, or reduced renal growth. It is important to distinguish cortical atrophy caused by coarse, focal renal scarring from the more uniform back pressure type of atrophy. Focal renal scarring, even when severe, is undoubtedly caused by infection and vesicoureteral reflux. But thin rim or symmetric back pressure atrophy of the renal cortex may be a congenital process essentially complete at birth (Mackie and Stephens, 1975). It often accounts for renal failure and remains basically unalterable in terms of influencing renal function by subsequent correction of associated vesicoureteral reflux. In adults, the presence of reflux does not appear to decrease renal function unless there is stasis and concurrent urinary tract infections.

Both Hodson et al. (1975) and Ransley and Risdon (1978) showed in a porcine model with obstructed bladder neck that pyelotubular backflow would occur only in areas of the kidney drained by complex renal papillae

that were relatively flat rather than convex. Bacteria caused pyelonephritis only in these areas of the kidney. Subsequently children have been shown to have some of the same complex papillae, especially in the polar regions (Tamminen and Kaprio, 1977). This helps to explain the predilection for renal scars from infections in these areas.

AGE.—Younger patients with pyelonephritis appear to be most susceptible to renal scarring. As the patient becomes older, the kidney is more resistant to infection. The age and renal scarring in 116 males at first infection and follow-up are shown in Table 22–1. Among the boys with their first infection during the first year of life, only one had a scar at the first IVP, whereas six (about 5%) developed a scar later. Among those who were older than 1 year of age at the so-called first infection, a focal scar was already present at the first IVP in one fourth. The data suggest that when first infections are diagnosed during the first year and treated, the infection carries a 5%–10% risk for scar formation. When urinary tract infection is diagnosed after the first year of life, the patient probably had a previous unrecognized, untreated infection that caused scarring in a higher percentage. Subsequent infection in these individuals carries a small risk, as none had a new scar on follow-up studies. McLachlan et al. (1975) found that the frequency of scarring was similar at 5 and 12 years in girls with asymptomatic bacteriuria. These findings emphasize that attempts to prevent renal scarring should be directed at very young children.

This concept was further supported by an analysis of the sex distribution of the first urinary tract infection and focal renal scarring (Winberg et al., 1975). The male-to-female ratio for pyelonephritic scars among 0–16-year-olds was close to 1. The male-to-female ratio among infected 0–12-month-olds was 0.8 and among 0–6-month-olds was 1.1. In older children and in adults urinary tract infections are about ten times more common in females. Thus, the similar male-to-female ratio

TABLE 22–1.

Renal Scarring in Males at "First" Infection and at Follow-up*

AGE AT "FIRST" URINARY TRACT INFECTION	NO. OF PATIENTS	% WITH SCARS	
		AT "FIRST" INTRAVENOUS PYELOGRAM	AT FOLLOW-UP†
1–30 days	54	0	4
2–12 mo	62	1	6
1–16 years	44	25	0

*From Winberg J, Bollgren I, Kallenius G, et al: Clinical pyelonephritis and focal renal scarring: A selected review of pathogenesis, prevention and prognosis. *Pediatr Clin North Am* 1982; 29:805. Used by permission.
†Follow-up of earlier undamaged kidneys.

for scars in children and adults is probably due to the initial similarity in the infection rate in neonates. The increased incidence of infection in older girls and women apparently does not have a significant impact on the incidence of renal pyelonephritic scars.

UNDERLYING DISEASE.—There is a high incidence of renal scarring in patients with underlying conditions that cause chronic interstitial nephritis, virtually all of which produce primary renal papillary damage. These include: diabetes mellitus, sickle cell disorders, adult nephrocalcinosis, hyperphosphatemia, hypokalemia, analgesic abuse, sulfonamide nephropathy, gout, heavy-metal poisoning, and aging (Freedman, 1979).

An increased incidence of clinical overt urinary tract infections appears to occur in women with diabetes mellitus, but there is no substantial increase among diabetic men (Forland et al., 1977; Vejlsgaard, 1973; Ooi and Chen, 1974). Autopsy studies have shown the incidence of pyelonephritis to be four- to five-fold higher in diabetic than in nondiabetic individuals (Robbins and Tucker, 1944). However, such studies may be misleading, because it is difficult to distinguish renal parenchymal changes due to pyelonephritis from the interstitial inflammatory changes of diabetic nephropathy.

Although the majority of urinary tract infections in diabetic patients are asymptomatic, diabetes appears to predispose to more severe infections. One study using antibody-coated bacteria techniques to localize the site of infection showed the upper urinary tract to be involved in nearly 80% of diabetic patients with urinary tract infections (Forland et al., 1977). This evidence of increasing immunologic response in diabetic patients who acquire bacteriuria suggests renal parenchymal involvement and a potential increase in morbidity. Once established, upper urinary tract infections are frequently complicated by emphysematous pyelonephritis, papillary necrosis, perinephric abscess, or metastatic infection (Wheat, 1980).

Other conditions that may increase the susceptibility of the kidney to infection include hypertension and vascular obstruction (Freedman, 1979). Association of renal infection with several other renal diseases including glomerulonephritis, atherosclerosis, and tubular necrosis, which are not associated with papillary necrosis, does not lead to pyelonephritis and scarring.

PREGNANCY.—The prevalence of bacteriuria in pregnant women varies from 4% to 7%, and the incidence of acute clinical pyelonephritis ranges from 25% to 35% in untreated bacteriuric women (Stamey, 1980). This is probably the result of dilatation of the ureters and pelvis of the kidney secondary to pregnancy-related hormonal alterations. It is not surprising that untreated bacteriuria in the first trimester is accompanied by a substantial incidence of acute pyelonephritis, since Fairley et al. (1966) documented that half of these women have upper tract bacteriuria. Untreated bacteriuria involving these dilated upper tracts would be expected to produce a significant number of abnormalities that should be radiologically apparent. Kincaid-Smith and Bullen (1965) cultured 4,000 women at their first antenatal visit. Of 240 bacteriuric women, 148 returned for IV urography six weeks after delivery. Approximately 40% of these patients had radiologic abnormalities consistent with pyelonephritis or analgesic nephritis. Brumfitt et al. (1967) showed that the incidence of radiologic abnormalities in bacteriuria of pregnancy was proportional to the difficulty in clearing the infection. Patients who responded promptly to a single course of therapy had a 23% incidence of radiologic abnormalities, but those who remained bacteriuric despite repeated therapeutic efforts had a 65% incidence of radiologic changes. Thus, prolonged bacteriuria and pyelonephritis of pregnancy appear to be associated with significant radiologic abnormalities. However, there is very little evidence to suggest that bacteriuria of pregnancy or acute pyelonephritis of pregnancy cause these renal radiologic abnormalities.

SUSCEPTIBILITY TO URINARY TRACT INFECTION.—Patients with an increased susceptibility to lower urinary tract infection probably are more likely to develop pyelonephritis than those who rarely have bacteriuria. Women and girls with increased colonization of the vaginal introitus, men with chronic bacterial prostatitis, and patients with an indwelling catheter are predisposed to urinary tract infections and subsequent development of pyelonephritis. There is no evidence, however, that in the absence of underlying abnormalities such as obstruction, diabetes mellitus, or analgesic abuse, recurring episodes of symptomatic or asymptomatic urinary tract infections in adults lead to functional or morphological renal damage or hypertension. In a follow-up study of young schoolgirls with persistent bacteriuria, however, Gillenwater et al. (1979) observed that although no subjects showed end-stage renal disease or hypertension, there was a significantly higher incidence of renal scars and bacteriuria during pregnancy.

DELAY IN ONSET OF THERAPY.—An important determinant of the extent of renal damage is the duration of infection before treatment begins. Winberg et al. (1975) identified 41 girls in whom the first known infection was inadequately treated. The incidence of renal damage was four times higher in this group with therapeutic delay than in 440 girls in whom diagnosis and treatment were prompt and adequate. The impact of therapeutic delay was also demonstrated in a recent study by Miller and Phillips (1981). They induced pyelonephritis in rats and then delayed therapy for varying periods. With each prolongation of the interval from eight hours to seven days, the renal damage became more and more severe.

ACUTE PYELONEPHRITIS

Clinical Findings

The onset of acute pyelonephritis is usually abrupt. The classic clinical features are chills, fever (100°F +), and costovertebral angle or flank pain accompanied by symptoms of cystitis.

Although some authors regard loin pain and fever in combination with significant bacteriuria as diagnostic of acute pyelonephritis, it is clear from localization studies utilizing ureteral catheterization (Stamey and Pfau, 1963b) or the bladder washout technique (Fairley et al., 1967) that clinical symptoms correlate poorly with the site of infection (Fairley, 1972; Stamey et al., 1965; Eykyn et al., 1972; Smeets and Gower, 1973). In a large study of 201 women and 12 male patients with recurrent urinary tract infection, Busch and Huland (1984) showed that fever and flank pain are no more diagnostic of pyelonephritis than they are of cystitis. Of patients with flank pain and/or fever, over 50% had lower tract bacteriuria. Conversely, patients with bladder symptoms or no symptoms frequently had upper tract bacteriuria. Approximately 75% of patients give a history of previous lower urinary tract infections.

On physical examination there frequently is tenderness to deep palpation in the costovertebral angle. Variations of this clinical presentation have been recognized. Infants and young children may show no indication of pain or tenderness in the region of the kidney, and their sole clinical manifestation may be pyrexia. Acute pyelonephritis may also simulate GI tract abnormalities with abdominal pain, nausea, vomiting, and diarrhea. Asymptomatic progression of acute pyelonephritis to chronic pyelonephritis, particularly in compromised hosts, may occur in the absence of overt symptoms.

Laboratory Findings

The patient may have leukocytosis with predominance of neutrophils. Urinalysis usually reveals numerous WBCs, frequently in clumps, and bacteria. Leukocytes exhibiting brownian motion in the cytoplasm (glitter cells) may be present if the urine is hypotonic but are not in themselves diagnostic of pyelonephritis. The presence of large amounts of granular or leukocyte casts in the urinary sediment is suggestive of acute pyelonephritis. A specific type of urinary cast characterized by the presence of bacteria in its matrix has been demonstrated in the urine of patients who have had acute pyelonephritis (Fig 22–4) (Lindner et al., 1980). Bacteria in the casts were not easily distinguished by simple bright-field microscopy without special staining of the sediment. Staining the sediment with a basic dye such as dilute toluidine blue or KOVA (I.C.L. Scientific, Fountain Valley, Calif.) stain demonstrated the bacteria in casts without difficulty. Urine cultures are invariably positive. Most often the causative microorganism is *E. coli*. However, more resistant species, such as *Proteus, Klebsiella, Pseudomonas,* or *Serratia,* should be suspected in patients who have recurrent urinary tract infections, are hospitalized, have indwelling catheters, or required recent urinary tract instrumentation. Except for *Streptococcus faecalis* and *Staphylococcus epidermidis,* gram-positive bacteria rarely cause pyelonephritis. Blood cultures should be obtained because bacteremia and sepsis are common.

Differential Diagnosis

Acute appendicitis, diverticulitis, and pancreatitis can all cause a similar degree of pain, but the location of the pain often is different. The urine examination is usually normal. Herpes zoster can cause superficial pain in the region of the kidney; the diagnosis will be apparent when shingles appear.

Initial Management

Hospitalization, initially with complete bed rest, IV fluids, and antipyretics, is usually required. Bladder outlet obstruction and associated urinary retention

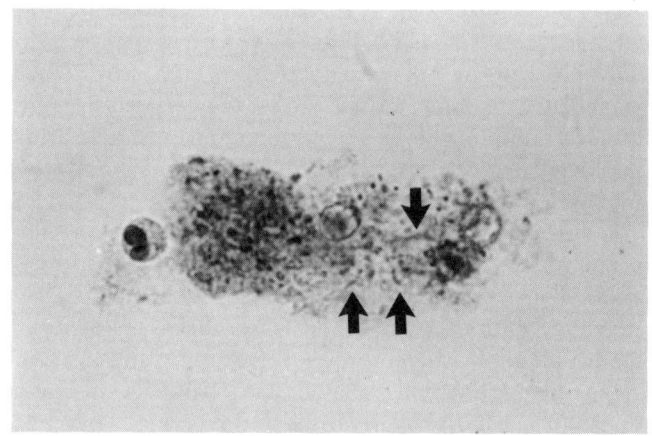

FIG 22–4.
Bright-field micrograph of a mixed bacterial leukocyte cast from patient with acute pyelonephritis. Only the bacteria and the nucleus of a leukocyte stain strongly. *Arrows* show some bacteria. Many more are clearly demonstrated by through-focusing (toluidine blue O, ×640). (From Lindner LE, Jones RN, Haber MH: A specific urinary cast in acute pyelonephritis. *Am J Clin Pathol* 1980; 73:810. Used with permission.)

should be relieved by an indwelling urethral or suprapubic catheter. If upper tract obstruction is suspected, an IV urogram should be obtained. An obstructed kidney has difficulty concentrating and excreting antimicrobial agents. In addition, obstruction in effect creates a potential abscess, pyohydronephrosis, which can rapidly destroy the renal parenchyma and endanger the patient's life. Any significant obstruction must be relieved expediently by the safest and simplest means.

Until the results of the culture and sensitivities are available, broad-spectrum antimicrobial therapy should be instituted. A Gram stain of the urine sediment is helpful to guide selection of the initial empiric antimicrobial therapy. Single-drug parenteral therapy with a third-generation cephalosporin or uriedo penicillin is usually effective for most patients with domiciliary infections. An aminoglycoside plus ampicillin may be required for compromised hosts with nosocomial infections.

Subsequent Management

Even though the urine usually becomes sterile within a few hours of starting antimicrobial therapy, patients with acute pyelonephritis may continue to have fever, chills, and flank pain for several more days.

Alterations in antimicrobial therapy may be made depending on the patient's clinical response and the results of the culture and sensitivity tests. Sensitivity tests should also be utilized to replace potentially toxic drugs such as aminoglycoside with less toxic drugs such as the cephalosporins. In patients with bacteremia, parenteral therapy should be continued for 7–10 days. If the blood cultures are negative, parenteral therapy can be discontinued after 3–5 days. In either case an appropriate oral antimicrobial drug should be continued in full dosage for an additional 10–14 days.

Excretory urography is usually performed following institution of adequate therapy and resolution of the patient's symptoms, and therefore it is not surprising that the majority of patients with pyelonephritis have a normal excretory urogram (Silver et al., 1976; Wicks and Thornbury, 1979). The most common radiologic abnormality is renal enlargement, which occurs from generalized renal edema as a consequence of the inflammatory process (Fig 22–5,A). An overall length of 15 cm or a length 1½ cm greater than the unaffected side have been established as criteria for the diagnosis of renal enlargement in acute pyelonephritis (Silver et al., 1976; Corriere and Sandler, 1982; Wicks and Thornbury, 1979). The inflammatory response may also cause cortical vasoconstriction, which is presumably responsible for the diminished nephrogram and delayed appearance of the pyelogram as well as compression of the collecting structures so that the calyces have an attenuated or

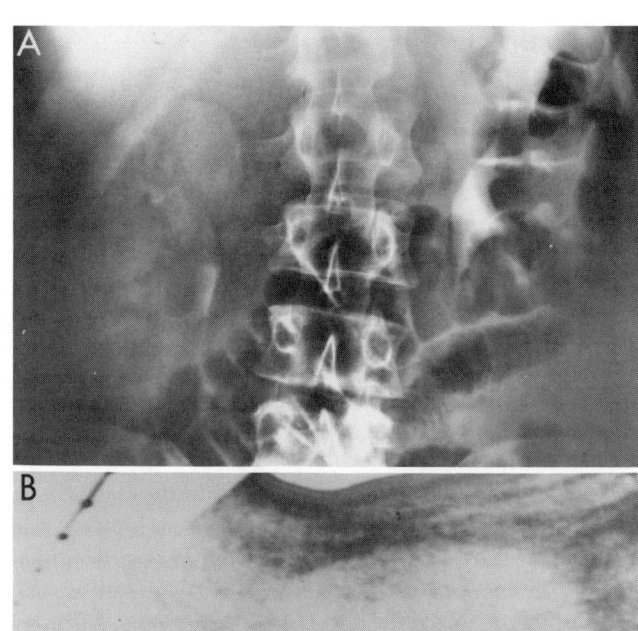

FIG 22–5.
Acute pyelonephritis. **A,** excretory urogram. Ten-minute film demonstrates enlarged right kidney with minimal function. Findings are consistent with edema. **B,** ultrasound of the right kidney demonstrates renal enlargement, hypoechoic parenchyma, and compressed central collecting complex *(arrows)*.

spidery appearance. In addition to these abnormalities, calyceal and ureteral dilation have occasionally been reported (Harrison and Shaffer, 1979). This finding has been commonly attributed to decrease in ureteral peristalsis caused by bacterial endotoxin. Although ureteral dilatation may occur with infection, this diagnosis should not be made until obstruction, either past or present, has been excluded.

Ultrasound (Fig 22–5,B) and computed tomography (CT) show renal enlargement, hypoechoic or attenuated parenchyma, and a compressed collecting system. These studies are most helpful for ruling out obstruction and identifying complicated renal and perirenal infections.

Unfavorable Response to Therapy

When the response to therapy is slow or the urine continues to show infection, an immediate reevaluation is mandatory. Urine and blood cultures must be repeated and appropriate alterations in antimicrobial therapy made on the basis of sensitivity testing. Radiologic investigation is indicated to attempt to identify unsuspected obstructive uropathy, urolithiasis, or underlying anatomical abnormalities that may have predisposed the patient to infection, prevented a rapid therapeutic response, or caused complications of the infectious process such as renal or perinephric abscess. When the clinical picture is not characteristic of acute pyelonephritis, radionuclide imaging may be useful to demonstrate functional changes associated with acute pyelonephritis (decrease in renal blood flow, delay in peak function, and a delay in excretion of the radionuclide (Fischman and Roberts, 1982).

Follow-up

Repeat urine cultures should be performed 5–7 days during therapy and 10–14 days and 4–6 weeks after discontinuing antimicrobial therapy to ensure that the urinary tract remains free of infection. Depending on the clinical presentation and response and initial urologic evaluation, some patients may require additional evaluation (e.g., voiding cystourethrogram, cystoscopy) and correction of an underlying abnormality of the urinary tract.

ACUTE FOCAL OR MULTIFOCAL BACTERIAL NEPHRITIS

This is an uncommon, severe form of acute renal infection in which a heavy leukocyte infiltrate is confined to a single renal lobe (focal) or multiple lobes (multifocal).

Clinical Findings

The clinical presentation of patients with acute bacterial nephritis is similar to those with acute pyelonephritis but usually much more severe. About half of the patients are diabetic, and sepsis is common. Generally, leukocytosis and urinary tract infection due to gramnegative organisms are found, and over 50% of the patients are bacteremic (Wicks and Thornbury, 1979).

Radiologic Findings

The diagnosis must be made by radiologic examination. The urographic findings are those of a mass, most commonly poorly marginated, and suggestive of renal abscess or tumor (Fig 22–6,A). The mass has slightly less nephrographic density than the surrounding normal renal parenchyma.

Ultrasonography and CT aid in establishing the diagnosis. On ultrasonography the lesion is typically poorly marginated and relatively sonolucent, with occasional low-amplitude echoes that disrupt the cortical medullary junction (Corriere and Sandler, 1982) (Fig 22–6,B). Contrast enhancement is necessary with CT studies, since the lesion is difficult to visualize on the unenhanced study (Fig 22–6,C). Wedge-shaped areas of decreased enhancement are seen. No definite wall is evident, and frank liquefaction is absent. Abscesses, on the other hand, tend to have liquid centers, are usually round, and are present both prior to and following contrast enhancement. More chronic abscesses may also show a ring-shaped area of increased enhancement surrounding the lesion (Corriere and Sandler, 1982). Gallium scanning reveals uptake that is in the region of and larger than the previously demonstrated mass (Rosenfeld et al., 1979). In patients with multifocal disease the findings are similar but multiple lobes are involved.

Management

Acute bacterial nephritis probably represents a relatively early phase of frank abscess formation. In a series of cases reported by Lee et al. (1980), a patient with acute focal bacterial nephritis did progress to abscess formation. Treatment includes hydration and IV antibiotics for at least seven days. Patients with bacterial nephritis typically respond to medical therapy, and follow-up studies will show resolution of the wedge-shaped zones of diminished attenuation. Failure to respond to antimicrobial therapy is an indication for appropriate studies to rule out obstructive uropathy, renal or perirenal abscess, renal carcinoma, or acute renal vein thrombosis. Long-term follow-up studies performed in a few patients with multifocal disease have demonstrated decrease in renal size and focal calyceal deformities suggestive of papillary necrosis (Davidson and Talner, 1978).

EMPHYSEMATOUS PYELONEPHRITIS

Emphysematous pyelonephritis is an acute necrotizing parenchymal and perirenal infection caused by gasforming uropathogens. The pathogenesis is poorly understood. Since the condition almost always occurs in diabetic patients, it has been postulated that the high tissue glucose levels provide the substrate for microorganisms such as E. coli, which are able to produce carbon dioxide by the fermentation of sugar (Schainuck et al., 1968). Although glucose fermentation may be a factor, the explanation does not account for the rarity of emphysematous pyelonephritis despite the high frequency of gram-negative urinary tract infection in diabetic patients, nor does it explain the rare occurrence of the condition in nondiabetics. In addition to diabetes, many patients have urinary tract obstruction associated with urinary calculi or papillary necrosis and significant renal functional impairment.

It seems more reasonable to postulate that impaired

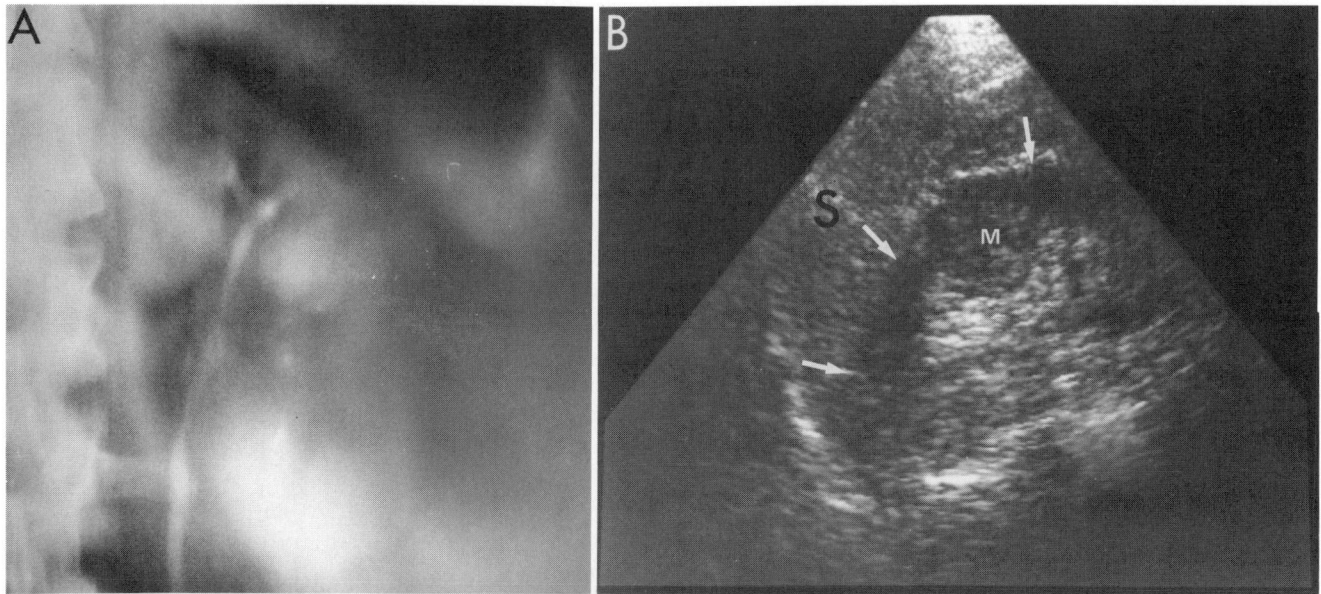

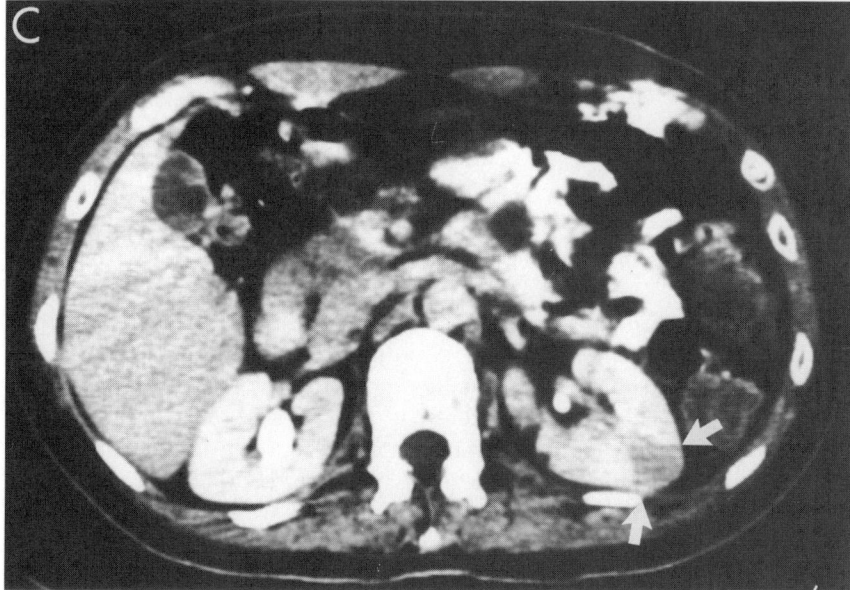

FIG 22–6.
Acute focal bacterial nephritis. **A,** excretory urogram. Five-minute tomogram demonstrates normally functioning upper and lower poles and a poorly marginated midrenal mass with poor function and absent collecting system visualization. **B,** ultrasound; longitudinal view of the left kidney demonstrates spleen *(S)* and left kidney *(arrows)*. Note irregular midpole mass *(M)* of slightly higher echotexture than surrounding normal renal parenchyma. **C,** contrast-enhanced CT scan demonstrates a wedge-shaped area of low density *(arrows)* in the midportion of the left kidney. The findings resolved after antibiotic therapy.

host response caused by local factors such as obstruction or a systemic condition such as diabetes allows organisms with the capability of producing carbon dioxide to utilize necrotic tissue as a substrate to generate gas in vivo. Thus, emphysematous pyelonephritis should be considered a complication of severe pyelonephritis rather than a distinct entity.

Clinical Findings

All of the documented cases of emphysematous pyelonephritis have been adults (Hawes et al., 1983). Juvenile diabetics do not appear to be at risk. Women are affected more often than men.

The usual clinical presentation is severe, acute pyelonephritis, although in some instances a chronic infec-

tion precedes the acute attack. Almost all patients display the classic triad of fever, vomiting, and flank pain (Schainuck et al., 1968). Pneumaturia is absent unless the infection involves the collecting system. Urine cultures are invariably positive. *E. coli* is most frequently identified; *Klebsiella* and *Proteus* are less common.

Radiologic Findings

The diagnosis is established radiographically. Tissue gas that is distributed in the parenchyma may appear on abdominal x-rays as mottled gas shadows over the involved kidney (Fig 22–7). This finding is often mistaken for bowel gas. A crescentric collection of gas over the upper pole of the kidney is more distinctive. As the infection progresses, gas will extend to the perinephric

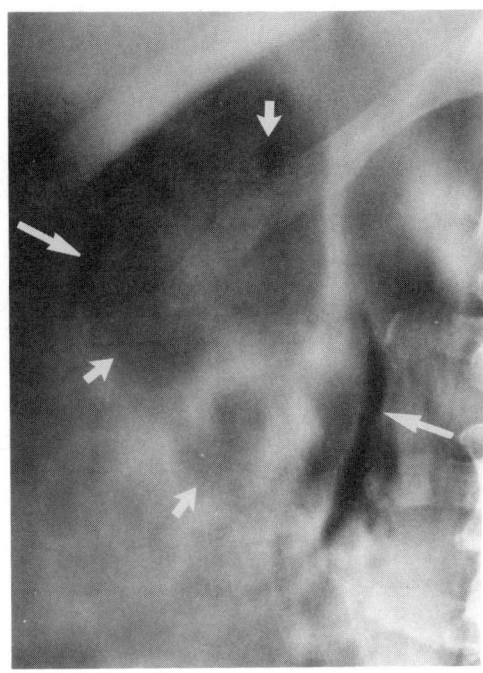

FIG 22–7.
Emphysematous pyelonephritis; plain film. Extensive perinephric *(large arrows)* and intraparenchymal *(small arrows)* gas secondary to acute bacterial pyelonephritis.

space and retroperitoneum. This distribution of gas should not be confused with cases of pyelonephritis in which air is in the collecting system of the kidney. This condition is secondary to a gas-forming bacterial urinary tract infection, frequently occurs in nondiabetics, is less serious, and usually responds to antimicrobial therapy.

Excretory urography is rarely of value, since the affected kidney usually is nonfunctioning or poorly functioning. Because of the significant risk of contrast nephropathy in critically ill, dehydrated diabetics with abnormal renal function, retrograde pyelography rather than excretory urography is advisable to demonstrate obstruction. Obstruction is demonstrated in approximately 25% of the cases. Ultrasonography usually demonstrates strong focal echoes suggesting the presence of intraparenchymal gas (Conrad et al., 1979; Brenbridge et al., 1979). CT is also helpful to demonstrate and define the extent of the emphysematous process (Kim et al., 1979). A renal scan should be performed to assess the degree of renal function impairment in the involved kidney as well as the status of the contralateral kidney.

Management

Emphysematous pyelonephritis is a surgical emergency. Most patients are septic. Fluid resuscitation and broad-spectrum antimicrobial therapy are essential. If the kidney is functioning, medical therapy can be con-

sidered. Nephrectomy is recommended for patients who do not improve after a few days of therapy (Elder, 1984). If the affected kidney is nonfunctioning and not obstructed, nephrectomy should be performed, since medical treatment alone is usually lethal. If a kidney is obstructed, catheter drainage must be instituted. If the patient's condition improves, nephrectomy may be deferred pending a complete urologic evaluation. Although there are isolated case reports of retention of renal function after medical therapy combined with relief of obstruction, most patients require nephrectomy.

RENAL ABSCESS

Renal abscess or carbuncle is a collection of purulent material confined to the renal parenchyma. Prior to the antibiotic era, 80% of renal abscesses were attributed to hematogenous seeding by staphylococci (Campbell, 1930). While experimental and clinical data document the facility for abscess formation in normal kidneys after hematogenous inoculation with staphylococci, widespread use of antibiotics in the past 25 years appears to have diminished the propensity for gram-positive abscess formation (DeNavasquez, 1950; Cotran, 1969).

During the past two decades gram-negative organisms have been implicated in the majority of adults and children with renal abscesses. Hematogenous renal seeding by gram-negative organisms may occur, but this is not likely to be the primary pathway for gram-negative abscess formation. Clinically there is no evidence that gram-negative septicemia antedates most lesions. Further, gram-negative hematogenous pyelonephritis is virtually impossible to produce in animals unless the kidney is traumatized or completely obstructed (Cotran, 1969; Timmons and Perlmutter, 1976). The partially obstructed kidney rejects bloodborne gram-negative inocula as well as a normal kidney. Thus, ascending infection associated with tubular obstruction from prior infections or calculi appears to be the primary pathway for establishment of gram-negative abscesses. Two thirds of gram-negative abscesses in adults are associated with renal calculi or damaged kidneys (Salvatierra et al., 1967). While the association of pyelonephritis with vesicoureteral reflux is well established, the association of renal abscess with vesicoureteral reflux has been infrequently noted (Segura and Kelalis, 1973). However, recent observations indicate that reflux is frequently associated with renal abscessses and persists long after sterilization of the urinary tract (Timmons and Perlmutter, 1976).

Clinical Findings

The patient may present with fever, chills, abdominal or flank pain, and occasionally weight loss and malaise. Lower urinary tract infections including cystitis usually occur as well. Occasionally these symptoms may be vague and delay diagnosis until surgical exploration or, in more severe cases, at autopsy (Anderson and Mc-Aninch, 1980). A thorough history may reveal a gram-positive source of infection 1–8 weeks prior to the onset of urinary tract symptoms. The infection may have occurred in any area of the body. Multiple skin carbuncles and IV drug abuse introduce gram-positive organisms into the bloodstream. Other common sites are the mouth, lungs, and bladder (Lyons et al., 1972). Complicated urinary tract infections associated with stasis, calculi, pregnancy, neurogenic bladder, and diabetes mellitus also appear to predispose to abscess formation (Anderson and McAninch, 1980).

Laboratory Findings

The patient typically has marked leukocytosis. The blood cultures are usually positive. Pyuria and bacteriuria may not be evident unless the abscess communicates with the collecting system. Since gram-positive organisms are most commonly blood-borne, urine cultures in these cases will typically show no growth or microorganism different from that isolated from the abscess. When the abscess contains gram-negative organisms, the urine culture will usually demonstrate the same organism isolated from the abscess.

Radiologic Findings

The urographic findings depend on both the nature and duration of the infection. In patients in whom abscess formation has progressed from an episode of acute bacterial nephritis or those in whom the kidney has been seeded by an outside infection, radiologic examination may demonstrate generalized renal enlargement with distortion of the renal contour on the affected side. There also may be renal fixation evident on inspiratory and expiratory films and obliteration of the corresponding psoas shadow. Scoliosis is often present, with a concavity of the curve facing the affected kidney. If renal involvement is diffuse the nephrogram will be delayed or even absent. When an abscess is more localized, the findings may be similar to those of acute focal bacterial nephritis.

In a more chronic abscess, the predominant urographic abnormalities are those of a renal mass lesion (Fig 22–8). The calyceal system may be poorly defined or show distortion or even amputation (Resnick and Older, 1982). Nephrotomography usually reveals a relative radiolucency in the involved area. Occasionally the excretory urogram will appear normal despite the presence of a renal abscess, particularly if the abscess involves the anterior or posterior portion of the kidney without impinging on the parenchyma or collecting system.

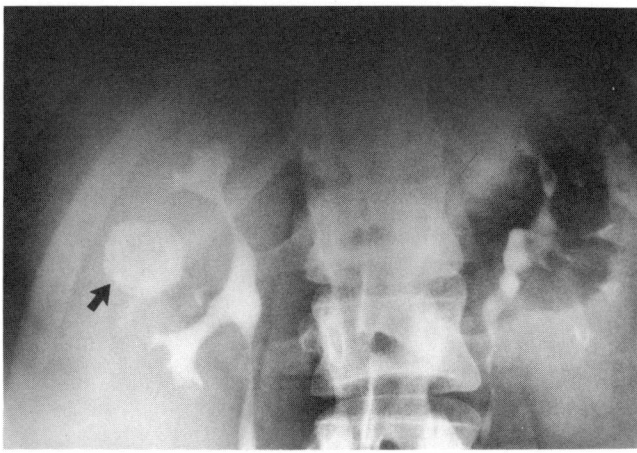

FIG 22–8.
Renal abscess associated with infection stone. Excretory urogram. Twenty-minute film demonstrates a large right midpole calculus *(arrow)*. Abscess associated with this infection stone causes displacement of adjacent collecting system.

CT and ultrasonography are very helpful to distinguish abscess from other inflammatory renal diseases. CT appears to be the diagnostic procedure of choice for renal abscesses, since it provides excellent delineation of the tissue. On CT, abscesses are characteristically well defined both prior to and following contrast enhancement. Initially CT shows renal enlargement and focal, rounded areas of decreased attenuation (Fig 22–9,A). After several days of the onset of the infection, a thick fibrotic wall begins to form around the abscess. CT of a chronic abscess shows obliteration of adjacent tissue planes, thickening of Gerota's fascia, a round or oval parenchymal mass of low attenuation, and a surrounding inflammatory wall of slightly higher attenuation that forms a ring when the scan is enhanced with contrast material (Fig 22–10,A). The ring sign is due to the increased vascularity of the abscess wall (Gerzof and Gale, 1982; Callen, 1979).

Ultrasonography is the quickest and least expensive method to demonstrate a renal abscess. An echo-free or low-echo density space-occupying lesion with increased transmission is found on the sonogram. The margins of an abscess are indistinguishable in the acute phase, but the structure contains a few echoes, and the surrounding renal parenchyma is edematous (Fiegler, 1983) (Fig 22–9,B). Subsequently the appearance tends to be that of a well-defined mass. However, the internal appearance may vary from a virtually solid lucent mass to that with large numbers of low-level internal echoes (Schneider et al., 1976) (Fig 22–10,B). The number of echoes depends on the amount of cellular debris within the abscess. Presence of air results in a strong echo with a shadow. Differentiation between abscess and a

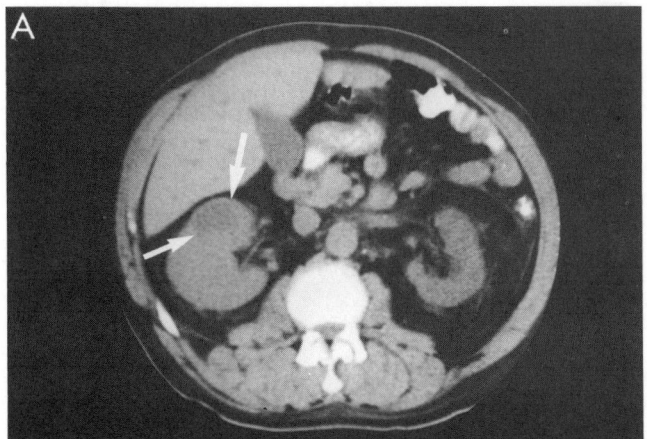

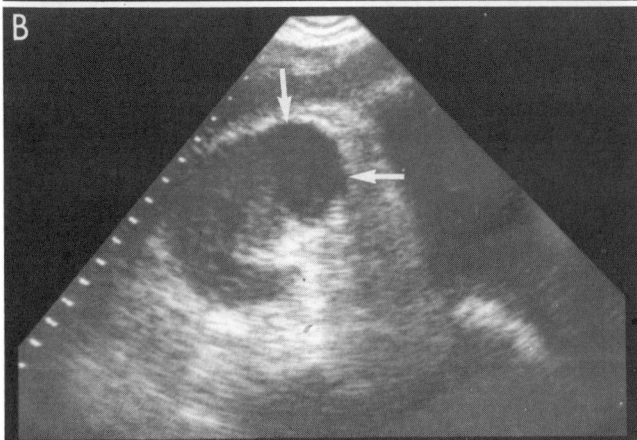

FIG 22–9.
Renal abscess, acute. **A,** nonenhanced CT scan through the midpole of the right kidney demonstrates right renal enlargement and an area of decreased attenuation *(arrows).* After antimicrobial therapy, follow-up scan showed complete regression of these findings. **B,** ultrasound. Transverse scan of the right kidney demonstrates a poorly marginated rounded focal hypoechoic mass *(arrows)* in the anterior portion of the kidney.

chronic tumor is impossible in many cases. Arteriography is used infrequently to demonstrate abscesses. The center of the mass tends to be hypovascular or avascular, with increased vascularity at the cortical margins and lack of vascular displacement and neovascularity.

Radionuclide imaging with gallium or indium is sometimes useful in evaluating patients with renal abscesses. The exact mode of ^{67}Ga localization in tissues is not clear. Suggested possible mechanisms include concentration within labeled polymorphonuclear leukocytes, leakage of protein-bound gallium through capillaries and increased vascularity of the lesion. Delayed imaging is often necessary. Gallium remains a nonspecific method of identifying an inflammatory lesion, and more importantly it lacks anatomical detail. In addition, gallium is excreted into the colon as well as sequestered

in postsurgical beds, inflammatory sites, and tumors; the interpretation of gallium scans therefore can be extremely difficult (Hampel et al., 1980).

Indium 111-labeled leukocyte scanning has recently been introduced as a clinically effective method for detecting inflammatory diseases and abscesses. Indium 111-labeled leukocytes accumulate only in sites of inflammation and not in normal kidneys nor in tumors. Thus, their presence appears highly specific for inflammation. However, the indium scan has limitations. Hyperalimentation and hyperglycemia can prevent the accumulation at the site of inflammation, and the distribution of leukocytes is altered in patients who have had splenectomy or bone marrow radiation. Lastly, the necessity of high doses of radiation with indium scanning may make it unsuitable for pediatric patients (Fawcett et al., 1981; Godec et al., 1981).

Management

Although the classic treatment for an abscess has been incision and drainage, in the past few years there

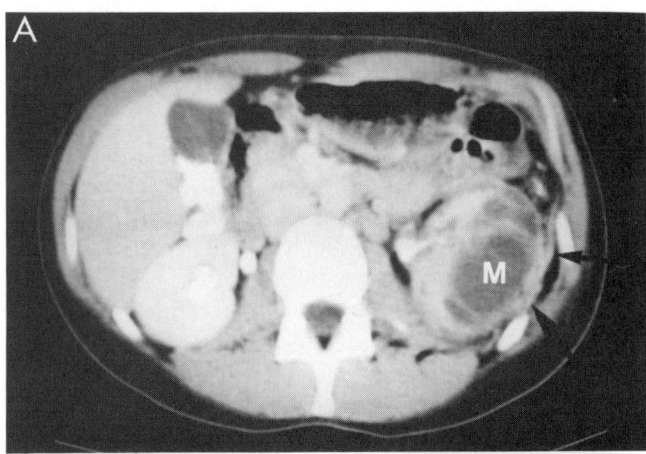

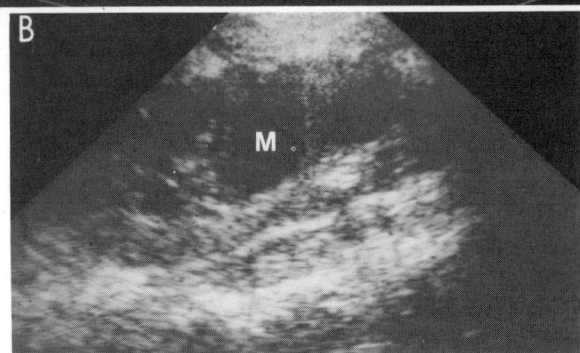

FIG 22–10.
Renal abscess, chronic. **A,** CT. Enhanced scan shows irregular septated low-density mass *(M)* extensively involving the left kidney. Note thickening of perinephric fascia *(arrows)* and extensive compression of the renal collecting system. Findings are typical of renal abscess. **B,** ultrasound. Longitudinal scan demonstrates septated hypoechoic mass *(M)* occupying much of the renal parenchymal volume.

has been good evidence that the use of IV antimicrobials and careful observation, if begun early enough in the course of the process, may obviate surgical procedures (Hoverman et al., 1980; Levin et al., 1984).

When hematogenous dissemination is suspected, the pathogenic organism most frequently is penicillin-resistant *Staphylococcus*, and the antibiotic of choice therefore is a penicillinase-resistant penicillin (Schiff et al., 1977). If a history of penicillin hypersensitivity is present, the recommended drugs are either cephalosporin or vancomycin. Cortical abscesses that occur in the abnormal urinary tract are associated with more typical gram-negative pathogens and should be treated empirically with IV third-generation cephalosporins, uriedo penicillins, or aminoglycosides until specific therapy can be instituted. These individuals can be followed by ultrasonography or CT until the abscess resolves. A clinical course contrary to this should lead to the suspicion of misdiagnosis or an uncontrolled infection with development of perinephric abscess or infection with an organism resistant to the antibiotics used in therapy. In these instances drainage is usually necessary.

CT- or ultrasound-guided needle aspiration may be necessary to differentiate an abscess from a hypovascular tumor. Aspirated material can be cultured and appropriate antimicrobial therapy based on the findings.

If antibiotic therapy is not effective, some patients with renal abscesses may be treated successfully by percutaneous drainage (Finn et al., 1982; Fernandez et al., 1985). These preliminary reports are encouraging and support the use of percutaneous drainage of renal abscess in selected patients. However, surgical drainage currently remains the procedure of choice for most renal abscesses.

INFECTED HYDRONEPHROSIS AND PYONEPHROSIS

If bacterial infection occurs in a hydronephrotic kidney, a purulent exudate will collect in the renal collecting system. The term "pyonephrosis" refers to infected hydronephrosis associated with suppurative destruction of the parenchyma of the kidney, with total or near-total loss of renal function (Fig 22–11). The point at which infected hydronephrosis ends and pyonephrosis begins is difficult to determine prospectively.

Clinical Findings

The patient usually will be very ill with high fever, chills, and flank pain and tenderness. A previous history of urinary tract calculi, infection, or surgery is common. Bacteriuria may not be present if the ureter is completely obstructed.

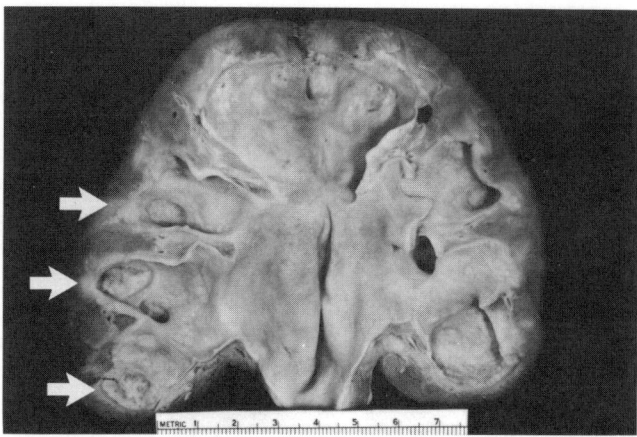

FIG 22–11.
Pyonephrosis; gross specimen. Kidney shows marked thinning of renal cortex and medulla, suppurative destruction of the parenchyma *(arrows),* and distension of the pelvis and calyces. Previous incision released large quantity of purulent material. Ureter showed obstruction distal to point of section.

Radiologic Findings

The urographic findings are those of urinary tract obstruction and depend on the degree and duration of obstruction. Typically, the obstruction is long-standing, and excretory urography shows a poorly functioning or nonfunctioning hydronephrotic kidney. Ultrasound demonstrates hydronephrosis and fluid debris levels within the dilated collecting system (Corriere and Sandler, 1982) (Fig 22–12, A). The diagnosis of pyonephrosis is suggested if focal areas of decreased echogenicity are seen within the hydronephrotic parenchyma.

Management

Appropriate IV antimicrobial therapy is important but immediate drainage of the kidney is mandatory. Percutaneous nephrostomy (Fig 22–12, B) or retrograde ureteral catheterization, or if necessary open nephrostomy must be performed. When the patient's condition has stabilized, appropriate corrective surgery or nephrectomy can be performed depending on the patient's age, the cause and type of obstruction, and the degree of renal function impairment.

PERINEPHRIC ABSCESS

Perinephric abscess usually results from rupture of an acute cortical abscess into the perinephric space. Patients with pyonephrosis, particularly when a calculus is present in the kidney, are very susceptible to perinephric abscess formation. Diabetes mellitus is present in about one third of the patients with perinephric abscess (Thorley et al., 1974). In about one third of the cases, perinephric abscess is caused by hematogenous

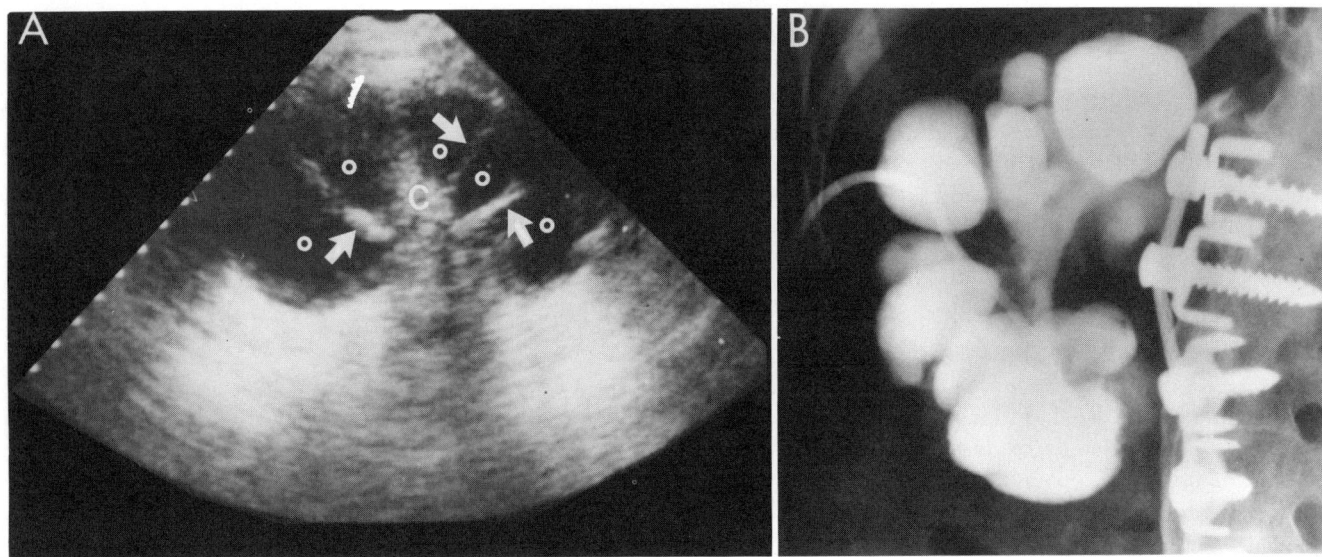

FIG 22–12.
Pyonephrosis. **A,** ultrasound. Longitudinal scan of the right kidney demonstrates echogenic central collecting complex *(C)* with radiating echogenic septae *(arrows)* and thinned hypoechoic parenchyma. Multiple dilated calyces *(O)* with diffuse low-level echoes are seen. **B,** antegrade pyelogram performed through a percutaneous nephrostomy catheter correlates well with the ultrasound image. Dilated pus filled calyces are demonstrated. The renal pelvis is obliterated by chronic scarring and stone disease. The kidney did not regain function.

spread, usually from sites of skin infection. A perirenal hematoma can become secondarily infected by the hematogenous route or direct extension of a primary renal infection. Rarely, perinephric abscess may be the result of bowel perforation or spread of osteomyelitis from the thoracolumbar spine. *E. coli, Proteus,* and *Staphylococcus aureus* account for the majority of infections.

Clinical Findings

The onset of symptoms is typically insidious. Symptoms have been present for more than five days in the majority of patients with perinephric abscess compared with only about 10% of patients with pyelonephritis. The clinical presentation is otherwise similar to that of pyelonephritis. An abdominal or flank mass can be felt in about half the cases. Laboratory features include leukocytosis, elevated serum creatinine, and pyuria in over 75% of cases. Pyelonephritis usually responds within 4–5 days of appropriate antibiotic therapy; perinephritic abscess does not. Thus, perinephric abscess should be suspected in a patient with urinary tract infection and abdominal or flank mass or persistent fever after four days of antimicrobial therapy.

Radiologic Findings

Excretory urography is abnormal in 80% of cases. However, the abnormalities are not specific. Classically the radiographic features of perinephric abscess have been the absence of psoas shadow, a mass in the perirenal area, often associated with indistinct renal outlines, and an elevated or immobile diaphragm. With large abscessses, the soft tissue density may extend to the pelvis following the renal fascia. In patients with perinephric abscess secondary to gas-forming organisms, bubbled collections of extraluminal gas are seen surrounding the kidney (Love et al., 1973). CT is particularly valuable for demonstrating the primary abscess (Fig 22–13). In some cases the abscess is confined to the perinephric space; however, extension to the flank or psoas muscle may occur. CT is able to show with exquisite anatomical detail the route of spread of infection into the surrounding tissues. This information may be helpful in planning the approach for surgical drainage. Ultrasound will demonstrate a diverse sonographic appearance ranging from a nearly anechoic mass displacing the kidney to an echogenic collection that tends to blend with normally echogenic fat within Gerota's fascia (Corriere and Sandler, 1982).

Occasionally, a retroperitoneal or subdiaphragmatic infection may spread to the paranephric fat that is outside Gerota's fascia. The clinical symptoms of insidious onset of fever, flank mass, and tenderness are indistinguishable from those associated with perinephric abscess. However, urinary tract infection is absent. Ultrasonography and CT can usually delineate the abscess outside Gerota's fascia (Fig 22–14).

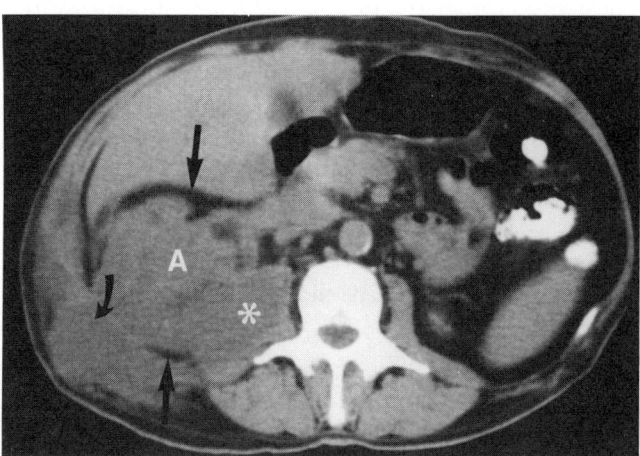

FIG 22–13.
Extensive perinephric abscess. Nonenhanced CT scan through the lower pole of the right kidney (previous left nephrectomy). Extensive abscess *(A)* distorts and enlarges renal contour, infiltrates perinephric fat *(arrows),* and also extends into the psoas muscle *(asterisk)* and the soft tissues of the flank *(curved arrow).* Also note that normal renal collecting system fat has been obliterated by the process.

Management

Once the diagnosis of perinephric or paranephric abscess is established, the primary treatment is surgical drainage. Antibiotics alone cannot be relied on for the treatment of a frank perinephric abscess. However, antimicrobial therapy with an aminoglycoside together with an antistaphylococcal agent such as methicillin or oxacillin should be started prior to surgery. If there is penicillin hypersensitivity, cephalothin or vancomycin

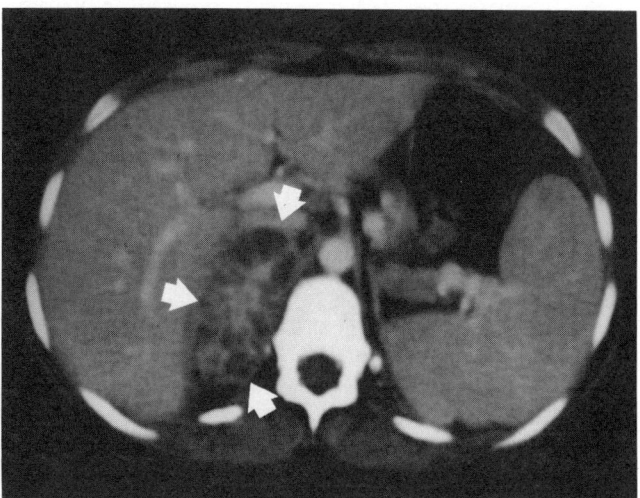

FIG 22–14.
Paranephric abscess involving the right adrenal gland. CT scan. Large right pararenal mass *(arrows)* with multiple low-density areas within. At surgery a large pararenal abscess with extensive involvement of the right adrenal was found.

may be used. If surgical intervention is not immediately possible such as in debilitated patients, percutaneous tube placement, aspiration of pus, and antibiotic irrigation may be effective temporarily.

Once the perinephric abscess has been incised and drained through a retroperitoneal incision, the underlying problem must be dealt with. Some conditions such as renal cortical abscess or enteric communication require prompt attention. Nephrectomy for pyonephrosis may be performed concurrent with drainage of the perinephric abscess if the patient's condition is good. In other instances it is best to drain the perinephric abscess first and correct the underlying problem or perform a nephrectomy when the patient's condition has improved.

CHRONIC PYELONEPHRITIS

Although it has been generally accepted that repeated bacterial infections of the kidney cause chronic renal insufficiency, it is now apparent that in the absence of underlying renal or urinary tract disease, chronic pyelonephritis is rare. However, in patients with underlying functional or structural urinary tract abnormalities, chronic renal infection can cause significant renal impairment. Hence, it is essential that appropriate studies are utilized to diagnose, localize, and treat chronic renal infection.

Prevalence

Despite the difficulty in establishing the bacterial etiology of chronic interstitial renal inflammation by gross and microscopic examination of the kidney at autopsy, it was thought initially that chronic interstitial nephritis primarily was caused by chronic or healed pyelonephritis. In the past 20 years, however, realization that the bacterial nature of chronic interstitial renal inflammation cannot be inferred solely on the basis of morphological examination of the kidneys has led to a reassessment of the prevalence of chronic pyelonephritis in patients with end-stage renal disease. The extent to which nonobstructive chronic pyelonephritis is overdiagnosed is illustrated in a classic study of 15 patients who died of uremia in a series of over 4,500 autopsies and who were diagnosed at autopsy as having nonobstructive pyelonephritis (Freedman, 1967b). Of these 15 patients, none had convincing evidence that bacterial infection was a significant factor in the production of renal disease. Many findings, however, suggested a role of familial, vascular, or toxic disorders known to be capable of producing chronic interstitial renal inflammation. Similar studies have also indicated that it is very difficult to identify urinary tract infection as a pathogenic factor of primary importance in the production of significant renal disease (Kleeman and

Freedman, 1960; Pawlowski et al., 1963; Farmer and Heptinstall, 1970).

The prevalence of chronic pyelonephritis has also been assessed in patients undergoing dialysis for end-stage renal disease. Despite a 2%–5% prevalence of bacteriuria in women, only 1,000–2,000 women have end-stage renal disease as a result of pyelonephritis. Schechter et al. (1971) analyzed the cause for renal failure in 170 patients referred to them for dialysis. Chronic pyelonephritis was the primary cause of end-stage renal disease in 22 (13%) but was almost always associated with an underlying structural defect. Unequivocal nonobstructive chronic pyelonephritis was not found. The authors also observed that symptomatic infections tended to occur prior to the onset of azotemia in most patients with chronic pyelonephritis. Similarly, Huland and Busch (1982) evaluated 161 patients with end-stage renal disease and found that 42 had chronic pyelonephritis. However, in addition to a history of urinary tract infection, these 42 patients had complicating defects such as vesicoureteral reflux, analgesic abuse, nephrolithiasis or obstruction. Nonobstructive uncomplicated urinary tract infection alone was never found to be the cause of renal insufficiency. Thus, starting from end-stage renal disease seen at autopsy or at the dialysis clinic, the prevalence of uncomplicated chronic bacterial pyelonephritis is very rare.

In addition, the role of bacterial infection in development of chronic renal disease can be assessed prospectively in patients with renal interstitial and tubular damage similar to that which has classically been termed chronic pyelonephritis. The frequency with which various potential causes of interstitial damage are operative in patients with interstitial nephritis was assessed by Murray and Goldberg (1975). They began with a group of 320 patients who were newly diagnosed as having renal disease (serum creatinine level greater than 1.3 mg/100 ml) and who did not have evidence over a four-year period of primary glomerular disease. Of the original 320 patients, 101 fulfilled the criteria for interstitial nephritis. Thirty-one patients had anatomical obstructive disease of the urinary tract, 20 had analgesic nephropathy, 11 had hyperuricemia, 10 had nephrosclerosis, and 9 had obstructive renal calculi. The cause of renal disease was indeterminate in 11, and 7 had multiple causes. Bacterial infection, although documented in 27 patients, was never the primary cause of renal damage. These investigators not only concluded that urinary tract infection is rarely the sole cause of chronic renal disease in the adult, but they also observed that 89% of their azotemic patients had a readily identifiable primary cause of their interstitial nephritis. Thus, when patients with clinical diagnosis of chronic interstitial nephritis are selected as the starting point,

it is easy to associate many factors with this disease, but urinary tract infection does not seem to be one of them.

These studies were conducted on adults. The combination of infection and reflux appears to play the major role in development of chronic pyelonephritis in children. Although infection in a neonate may result in renal impairment in the absence of other predisposing factors (see above), substantial renal impairment is usually associated with ureterovesical reflux. In addition, it is clear in most studies of children that infection occurring after the first year of life seems to carry a small risk of future scar development, even in patients who have had established history of infection and renal scarring.

Clinical Findings

There are no symptoms of chronic pyelonephritis until it produces renal insufficiency, and then the symptoms are similar to those of any other form of chronic renal failure. If chronic pyelonephritis is an end result of many episodes of acute pyelonephritis, intermittent symptoms of fever, flank pain, and dysuria may be exhibited by a patient with chronic pyelonephritis. Similarly, physical findings in chronic pyelonephritis are nonexistent.

Laboratory Findings

Microscopic study of the urine is nonspecific. There may be increased numbers of leukocytes, but their presence in urine does not prove the existence of upper urinary tract infection per se. Leukocyte casts are suggestive but not diagnostic of chronic pyelonephritis. Bacteriuria alone is not diagnostic of chronic pyelonephritis. Asscher (1980) has tabulated 8 long-term follow-up studies from the literature on kidneys of adults with urinary tract infections. The data from these reports on 901 patients show that bacteriuria present in otherwise healthy adults for long periods may be associated with nonexistent or extremely minimal evidence of kidney damage. Conversely, patients who have chronic pyelonephritis may have negative urine cultures.

Radiologic Findings

What measure then should the clinician use as evidence of chronic progressive pyelonephritis? The diagnosis of chronic pyelonephritis can be made with the greatest confidence on the basis of pyelographic findings. The essential features are asymmetry and irregularity of the kidney outlines, blunting and dilatation of one or more calyces, and cortical scars at the corresponding site (Fig 22–15). In the absence of stones, obstruction and tuberculosis and with the single exception of analgesic nephritis with papillary necrosis (which can be readily excluded by history) chronic pyelonephritis is virtually the only disease that will produce a localized

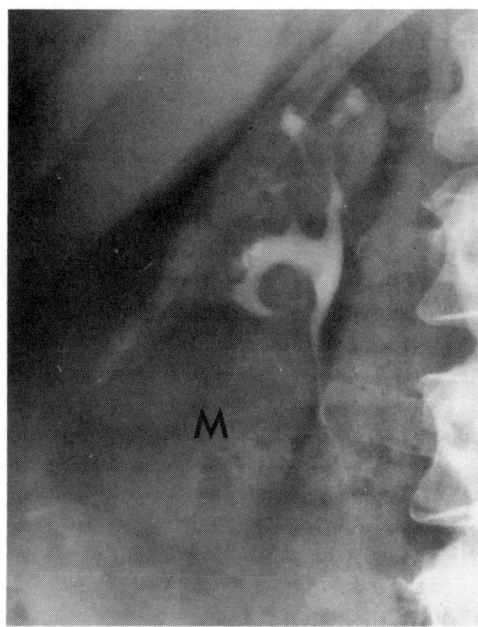

FIG 22–15.
Chronic pyelonephritis. Excretory urogram. Ten-minute film demonstrates irregular renal outline with upper pole parenchymal atrophy. Note significant loss of renal cortical thickness over blunted and dilated calyces. Lower pole mass *(M)* is a simple cyst.

scar over a deformed calyx (Stamey, 1980). In advanced pyelonephritis, calyceal distortion and irregularity together with cortical scars complete the picture. Hodson (1965b) pointed out that renal infarction, an extremely rare condition, may closely resemble pyelonephritic scars, but the renal pyramid remains in renal infarction in contradistinction to pyelonephritis.

Management

Management of radiographic evidence of pyelonephritis should be directed at treating infection if present, preventing future infections, and monitoring and preserving renal function. The treatment of existing infection must be based on careful antimicrobial sensitivity tests and selection of drugs that can achieve bactericidal concentrations in the urine and yet are not nephrotoxic. Achievement of acceptable bactericidal levels of a drug in the urine of a patient with chronic pyelonephritis may be difficult because the diminished concentrating ability of pyelonephritis may impair excretion and concentration of the antimicrobial agent. The duration of antimicrobial therapy is frequently prolonged to maximize the chance of cure. In patients in whom renal damage develops or progresses in the presence of urinary tract infection, the working hypothesis should be that there is an underlying renal, usually papillary, lesion or underlying urologic condition such

as obstruction or calculus that has increased susceptibility to renal damage. Appropriate nephrologic and urologic evaluation should be undertaken to identify and if possible correct these abnormalities.

EVALUATION AND MANAGEMENT OF RECURRENT RENAL INFECTION

It has been proposed that the site of urinary tract infection influences the probability for recurrent bacteriuria. Renal infections are thought to represent a deep-tissue infection in which a significant immunologic response occurs, and delivery of effective concentrations of antimicrobial agents may be difficult. In contrast, bladder infections primarily involve the superficial mucosa, which is constantly bathed by urine containing exceedingly high concentrations of effective antimicrobial agents.

Thus, it has been postulated that renal infections require more intensive and longer duration of antimicrobial therapy and that failure to use adequate therapy initially may cause a recurrent infection due to "relapse" from bacterial persistence in the kidney. Patients with recurrent infections with the same strain within a 1–2-week period have been judged to have "relapse" from bacterial persistence in the kidney. Recurrent infections at longer intervals, even if caused by the same strain, were judged to be "reinfections"; that is, new infections from an outside source such as the bowel flora.

To assess more accurately the validity of these concepts and to determine whether recurrent urinary tract infections were due to "relapse" or to "reinfection," investigators developed direct and indirect techniques to differentiate renal from bladder infections.

LOCALIZATION OF URINARY TRACT INFECTION

Direct Techniques

URETERAL CATHETERIZATION.—To differentiate between bladder bacteriuria and renal bacteriuria, Stamey and Pfau (1963b) developed a technique of cystoscopy based on extensive washings of the bladder, a controlled culture of the ureteral catheters before entering the ureteral orifices, and multiple serial cultures from each kidney that distinguish between bladder and renal pelvic bacteriuria.

Localization patterns in patients with lower urinary tract infections present few diagnostic difficulties. Fortunately, all of the bacteria from the bladder can be removed by washing with 2–3 L of sterile irrigating solution. In patients with bladder bacteriuria, the first ureteral collections may contain a few contaminating bacteria carried by the catheter into the ureter. Subse-

TABLE 22–2.
Localization Patterns in Patients Without Renal Bacteriuria

PATIENT	SOURCE*	BACTERIA/ML	ORGANISM
1	CB	>100,000	Escherichia coli
	WB	200	E. coli
	RK$_1$	0	
	LK$_1$	0	
	RK$_2$	0	
	LK$_2$	0	
	RK$_3$	0	
	LK$_3$	0	
2	CB	70,000	Proteus mirabilis
	WB	50	P. mirabilis
	RK$_1$	10	P. mirabilis
	LK$_1$	0	
	RK$_2$	0	
	LK$_2$	0	
	RK$_3$	0	
	LK$_3$	0	
3	CB	>100,000	E. coli
	WB	0	0
	RK$_1$	0	0
	LK$_1$	0	0
	RK$_2$	0	0
	LK$_2$	0	0
	RK$_3$	0	0
	LK$_3$	0	0

*CB indicates catheterized bladder urine; WB, washed bladder (after 2–3 L of sterile water irrigation); RK$_1$ to RK$_3$, serial right renal urines; and LK$_1$ to LK$_3$, serial left renal urines.

TABLE 22–3.
Localization Patterns in Patients With Unilateral Renal Bacteriuria

PATIENT	SOURCE*	BACTERIA/ML	ORGANISM
1	CB	>100,000	Proteus mirabilis
	WB	2,500	P. mirabilis
	RK$_1$	15,000	P. mirabilis
	LK$_1$	100	P. mirabilis
	RK$_2$	10,000	P. mirabilis
	LK$_2$	0	0
	RK$_3$	8,000	P. mirabilis
	LK$_3$	0	0
2	CB	>100,000	Escherichia coli
	WB	0	0
	RK$_1$	0	0
	LK$_1$	900	E. coli
	RK$_2$	0	0
	LK$_2$	700	E. coli
	RK$_3$	0	0
	LK$_3$	500	E. coli
3	CB	17,000	P. mirabilis
	WB	0	0
	RK$_1$	0	0
	LK$_1$	7,000	P. mirabilis
	RK$_2$	0	0
	LK$_2$	6,000	P. mirabilis
	RK$_3$	0	0
	LK$_3$	6,000	P. mirabilis

*CB indicates catheterized bladder urine; WB, washed bladder (after 2–3 L of sterile water irrigation); RK$_1$ to RK$_3$, serial right renal urines; and LK$_1$ to LK$_3$, serial left renal urines.

quent specimens will demonstrate sterility of the renal pelvic urine (Table 22–2).

The localization patterns in patients with unilateral and bilateral renal bacteriuria are presented in Tables 22–3 and 22–4. Note that bilateral renal bacteriuria is nearly always associated with higher bacterial counts in renal urine. The determination of unilateral renal bacteriuria is important, because it is frequently associated with a congenital or acquired abnormality. If the abnormality can be surgically removed or corrected, the focus of bacterial persistence in the kidney and the cause of recurrent bacteriuria usually can be eradicated.

Approximately 50% of the adult men and women with urinary tract infections studied by Stamey and co-workers (1965) had renal bacteriuria. These findings were initially thought to be diagnostic of pyelonephritis. However, most of these patients had IV urograms that were nonsurgical; that is, only rarely was there obstruction, stones, or gross loss of renal cortex. Reflux was rare in those patients who had cystograms. If evidence of destruction of renal tissue by bacteria is an acceptable definition of pyelonephritis, most of the patients who had renal bacteriuria did not have pyelonephritis.

Thus, ureteral catheterization studies are not useful for diagnosing pyelonephritis. However, this test is es-sential for determining a renal focus of bacterial persistence in patients with recurrent urinary tract infections and is the benchmark for evaluation of other techniques utilized to identify renal bacteriologic involvement. Subsequent studies have confirmed the accuracy of this technique (Fairley et al., 1966; Ronald et al., 1969, Eykyn et al., 1972).

BLADDER WASHOUT TEST.—This test is based on the premise that infections localized to the bladder can be washed free of bacteria by irrigating via a Foley catheter and that subsequent serial cultures essentially represent ureteral urine. As described by Fairley et al. (1971), after collecting the initial specimen, the bladder is emptied through the urethral catheter and 40 ml of 0.2% neomycin together with one ampule of Elase (a combination of two lytic enzymes, fibrinolysin and desoxyribonuclease) is introduced into the bladder. After 10 minutes, the bladder is distended with 0.2% neomycin to reduce the folds in the mucosa, and the catheter is occluded for 20 minutes. The bladder is then emptied and washed out with 2 L of sterile saline solution. Some of the saline in the final washout is collected for culture, and after emptying the bladder, three timed specimens are collected at 10-minute intervals. Bacterial counts are done on all of the specimens.

TABLE 22–4.
Localization Patterns in Patients With Bilateral
Renal Bacteriuria

PATIENT	SOURCE*	BACTERIA/ML	ORGANISM
1	CB	>100,000	*Escherichia coli*
	WB	3,000	*E. coli*
	RK$_1$	50,000	*E. coli*
	LK$_1$	60,000	*E. coli*
	RK$_2$	80,000	*E. coli*
	LK$_2$	60,000	*E. coli*
	RK$_3$	>100,000	*E. coli*
	LK$_3$	>100,000	*E. coli*
2	CB	>100,000	*Proteus mirabilis*
	WB	2,000	*P. mirabilis*
	RK$_1$	25,000	*P. mirabilis*
	LK$_1$	2,000	*P. mirabilis*
	RK$_2$	30,000	*P. mirabilis*
	LK$_2$	9,000	*P. mirabilis*
	RK$_3$	30,000	*P. mirabilis*
	LK$_3$	18,000	*P. mirabilis*
3	CB	>100,000	*E. coli*
	WB	1,000	*E. coli*
	RK$_1$	>100,000	*E. coli*
	LK$_1$	>100,000	*E. coli*
	RK$_2$	>100,000	*E. coli*
	LK$_2$	>100,000	*E. coli*
	RK$_3$	>100,000	*E. coli*
	LK$_3$	>100,000	*E. coli*

*CB indicates catheterized bladder urine; WB, washed bladder (after 2–3 L of sterile water irrigation); RK$_1$ to RK$_3$, serial right renal urines; and LK$_1$ to LK$_3$, serial left renal urines.

Renal infection was assumed to be present when the timed specimen collected 20–30 minutes after bladder washout contained more than 3,000 bacteria/ml, and this 20–30-minute specimen also contained more than five times as many bacteria as were present in the final bladder washout. Bladder infection was assumed to be present when the final timed specimen was sterile. These criteria, however, have never been evaluated with ureteral catheterization as a standard. Hooton et al. (1984) reassessed the criteria for interpretation of the bladder washout test and suggested that a test be considered positive if a given organism was present in quantities greater than or equal to 1,000 bacteria/ml in at least two specimens after washing the bladder; other results were considered negative. These criteria appear reasonable for patients with intact urinary tracts. Correlations of ureteral catheterization studies with bladder washout results have been disappointing in a small number of patients with neurogenic bladders studied to date and suggest that bladder washout tests may not be reliable in this population.

Indirect Techniques

ANTIBODY-COATED BACTERIA.—Ureteral catheterization and the bladder washout are not practical for routine clinical use. The antibody-coated bacteria test described by Thomas et al. (1974) and Jones et al. (1974) is rapid, noninvasive, and inexpensive. The antibody-coating technique utilizes a simple and direct immunofluorescent procedure to detect antibody-coated bacteria in urine sediments of adults with urinary tract infections. Antibody-coated bacteria are usually found in urine samples from patients with kidney infections but not from adults with uncomplicated bladder infections. This test is not useful in children (Akerlund et al., 1979).

The reliability of the urine fluorescent antibody test as the means of localizing upper urinary tract infection in adults has been established by comparing it with direct localization studies using ureteral catheterization or bladder washout catheterization (Thomas and Forland, 1982). Combined results of eight groups of investigators indicate that for those patients with upper tract infection there was agreement in 88%, and for those infections confined to the bladder there was agreement in 76%.

Variation in methodology may explain some of the discrepancy among the studies. The method of interpreting the fluorescent antibody test as positive differed among investigators. Thomas et al. (1974) interpreted the test as positive when greater than or equal to 20% or 25% of the organisms in each microscopic field fluoresced. Other investigators used much smaller numbers of fluorescent organisms to indicate a positive test (Jones et al., 1974; Harding et al., 1978). Unfortunately, many of the patients in these studies were on urologic services and had chronic urinary tract infections. Further experience comparing the antibody-coated bacteria test with direct localization procedures in acutely symptomatic younger women is still needed.

False negative and false positive results occur with the antibody-coated bacteria test. False negative tests usually result from delayed or inadequate antibody response. Most women with a first infection will receive treatment before the development of antibodies against their infecting strain. In a study by Ruben et al. (1980) the average duration of symptoms in antibody-coated bacteria-negative patients was two days compared with six days in antibody-coated bacteria-positive patients. Thus, it appears that sufficient time for an immune response to develop is necessary before the antibody-coated bacteria test becomes positive. False negative results with the antibody-coated bacteria test may also be seen in patients with infections characterized by fewer than 100,000 bacteria/ml, presumably because there are insufficient bacteria to bind antibody.

False positive fluorescent antibody tests may result from overinterpretation of small numbers of fluorescing antibodies in patients with lower urinary tract infections. Proteinuria may also give a false positive fluorescent antibody test in about 30% of patients (Thomas et al., 1975). Vaginal or fecal strains of *E. coli* may be antibody-coated and can contaminate urine specimens. Other tissue infections, primarily prostatitis and hemorrhagic cystitis as well as ileal conduits and bladder tumors, may result in antibody production (Thomas and Forland, 1982). Lastly, certain microorganisms (*Pseudomonas*, staphylococci, and yeast) often produce false positive results (Jones et al., 1974; Harding and Merz, 1975; Marrie et al., 1979).

It appears that a positive test for antibody-coated bacteria suggests that the patient has more than simple urinary tract infection; either kidney or prostatic involvement is present, or an "invasive" bladder infection has occurred related to the presence of stones, malignancy, recent surgery, or hemorrhagic cystitis. A negative antibody test in the absence of the clinical picture of acute pyelonephritis provides reasonable assurance that renal, prostatic or invasive bladder infection is absent. Therapeutic guidelines predicted by these findings suggest that the patient with a negative antibody-coated test will respond well to a short course of therapy, even of 1–3 days' duration (Ruben et al., 1980). Unfortunately, the limited availability and variability in the sensitivity and specificity of the test, as well as its poor performance in children and women with acute urinary tract infections, has led most investigators to recommend that the antibody-coated bacteria test be reserved for epidemiologic and investigative purposes and not be used clinically at this time.

RENAL CONCENTRATING ABILITY.—A defect in concentrating ability has long been recognized in pyelonephritis. Since upper tract bacteriuria in women is unilateral as often as it is bilateral, effects on renal concentrating mechanisms are best studied at the time of ureteral catheterization studies. Ronald et al. (1969) performed bilateral ureteral catheterization studies on 66 women with documented bacteriuria. Most of these patients were asymptomatic at the time of the study, and none had symptoms of acute pyelonephritis. Nevertheless, patients with renal bacteria concentrated less well than those whose infection was limited to the bladder, and the uninfected side of patients with unilateral bacteriuria concentrated better than the contralateral, infected kidney. This defect seemed to disappear with adequate antimicrobial therapy. Thus, the presence of renal pelvic bacteriuria is accompanied by a functional defect in renal concentrating ability. However, this does not necessarily mean that the patient has pyelonephritis. Stamey (1982) thinks that minimal edema at the papillary tip, well within the renal pelvis, could influence greatly the countercurrent concentrating mech-

anism. Thus, impaired concentrating ability may mean nothing more than the presence of pelvic bacteriuria. Efforts to distinguish between bladder and renal bacteriuria by measuring the maximal urinary concentration in bladder urine have not been effective (Clark et al., 1969).

C-REACTIVE PROTEIN.—C-reactive protein is an acute reaction found in the serum of girls with acute clinical pyelonephritis but only infrequently in those with clinical cystitis. However, it is nonspecific for acute pyelonephritis and will reflect any parenchymal infection (Jodal et al., 1975). Further, the measurement of serum C-reactive protein does not help determine the site of infection in girls with significant bacteriuria who do not have clinical signs of acute pyelonephritis (Hellerstein et al., 1982).

Of the techniques outlined above, only bilateral ureteral catheterization can demonstrate and localize upper tract infection with certainty. The bladder washout test gives reasonable insurance that the upper tracts are infected, but it cannot localize the focus of infection. The indirect techniques may be helpful in research for characterizing groups of patients, but they are not specific in individual patients and should not be applied routinely in the management of patients with recurrent bacteriuria.

BACTERIAL "RELAPSE" FROM A NORMAL KIDNEY

The concept that bacteria persist in the renal parenchyma between bacteriuric episodes and cause "relapsing" urinary tract infections was based on a study by Turck et al. (1968). Using ureteral catheterization to localize the site of infection, they reported that 80% of posttreatment recurrences in 29 patients with renal bacteriuria were caused by relapse with the same organism that produced the initial infection; in contrast, 71% of posttreatment recurrences in 25 patients with bladder bacteria were caused by reinfection with new organisms. Thus, the clinical implication was made that bacterial persistence could be recognized by simply identifying two consecutive recurrent infections with the same organism. Unfortunately, this study did not indicate whether the urine was cultured during therapy to ensure that the original infection had in fact been eradicated. It is possible that some of these so-called relapses were in fact unresolved initial infections and that ureteral edema associated with catheterization may have impeded clearance of the initial infecting strain.

Subsequent studies summarized by Stamey (1980) have shown that in a normal urinary tract, recurrent infections are not due to relapse from bacterial persistence in the kidney. Using ureteral catheterization techniques, Cattell et al. (1973) localized the site of bacteriuria in 42 patients who were followed up for six months after therapy. They analyzed the response to antimicrobial therapy of two weeks' duration. Sixteen of the 26 patients who were cured of their initial infection had recurrence with the same organism. Eight had upper tract infections, and eight had bladder bacteriuria.

Using fluorescent antibody-coated bacteria in adults to distinguish renal from bladder infection, Forland et al. (1977) observed the pattern of recurrent urinary tract infections in diabetics. Ten of fifteen patients who initially had antibody-coated bacteriuria had reinfections with a different species or serotype, 8 of which were antibody-coated negative; 5 recurred with the same organism, 4 of which were antibody-coated. Harris et al. (1976) performed antibody-coated bacterial localization studies on pregnant patients with bacteriuria. There were 14 recurrences among 35 patients who initially had antibody-negative bacteriuria; 10 of the 14 recurrences were reinfections and not "relapses."

Last, using C-reactive protein and other parameters to separate acute pyelonephritis from lower tract infection in children, Wientzen et al. (1979) observed that ten days of therapy in both groups produced a similar number of reinfections, 25% and 32%, respectively, and relapses, 6% and 4%. Thus, these authors all concluded that localization studies were of no predictive value in planning the management of patients with recurrent bacteriuria.

Other investigators have examined the relationship between radiologic abnormalities such as pyelonephritic scarring and reflux and the pattern of bacterial recurrence (Bergstrom et al., 1967; Guttman, 1973). Kidney abnormalities such as stones, congenital anomalies, and papillary necrosis were excluded. Reinfections were just as common as relapse in patients with reflux as in those without. Further, Kunin et al. (1964) in children and Guttman (1973) in adults could identify no relationship between the frequency of bacteriuric recurrence after therapy and the presence of radiologic abnormalities.

Last, if bacterial persistence in the kidney is a major problem after therapy, one would expect that patients who have more recurrent infections would also have more relapses than those who have less frequent recurrences. Mabeck (1972) analyzed this, however, and found that with increasing number of recurrences, the relationship among treatment failure, relapse, and reinfection remained unchanged.

Thus, bacteria do not persist in normal kidneys between recurrent urinary tract infections (Stamey, 1980), and recurrences with the same strain are not due to "relapse" from the kidney. In women and girls, recurrences with the same strain are usually due to reinfections by the same strain that caused the initial infection and subsequently persisted on the vaginal mucosa. In men, recurrences are usually caused by bacterial persistence in prostatic secretions. Less often, recurrent urinary tract infections with the same strain are caused by bacterial persistence in a congenital or acquired abnormality (urachal cyst, ureteral stump). Therefore, thorough urologic evaluation of the lower urinary tract is warranted in patients with normal upper urinary tracts and recurrent bacteriuria.

BACTERIAL PERSISTENCE IN AN ABNORMAL KIDNEY

Bacterial persistence in a renal abnormality should be suspected in patients with recurrent bacteriuria with the same strain and at close intervals. Appropriate radiologic studies such as excretory urography, plain film tomography, retrograde pyelography, and CT scans must be performed. If an abnormality such as a stone, calyceal diverticulum, atrophic pyelonephritis, or necrotic papillae from papillary necrosis is identified, it is highly likely that the cause of recurrence is a persistent focus of bacteria within the kidney. Ureteral catheterization studies are required to localize the site of infection to the abnormality. The only effective cure for patients with a persistent focus of bacteria within the kidney is surgical correction or removal of the abnormality. Alternatively, if the patient is not a candidate for surgery, long-term suppressive antimicrobial therapy can be utilized to maintain sterile urine and prevent acute infection. Failure to localize bacteria to the kidneys indicates that another focus of bacterial persistence (prostate, ureteral stump, urachal cyst) or reinfection (entero- or vaginal-urinary tract fistula) may be present, identified, and corrected by appropriate microbiologic, radiologic, and urologic procedures.

INFECTIOUS GRANULOMATOUS NEPHRITIS

RENAL TUBERCULOSIS

Tuberculosis is an acute or chronic infectious disease which in the United States is almost always due to *Mycobacterium tuberculosis*. *M. bovis*, *M. kansasii*, and *M. intracellulare* are rare causes of tuberculosis.

Mycobacteria customarily gain access to the human body by inhalation, although the bovine organisms may be acquired by ingestion of unpasteurized milk. Following initiation of the tuberculous infection, a primary pathologic focus develops, which usually heals spontaneously. In addition, the primary infection frequently results in an initial silent bacillemia that is responsible for systemic spread of *Mycobacterium* with latent infection of many organs. These latent foci of tuberculous infection may break down and result in overt tuberculosis of the kidney or other organs many years later. Bacillemia and seeding of the kidneys or other organs may also occur from a focus of progressive primary or reactivation tuberculosis in the lung or from clinically evident secondary tuberculosis in other organs. Therefore, any individual with previous tuberculous infection is at risk for development of renal involvement.

Renal infection is among the most common sites for extrapulmonary tuberculosis. About 5% of the estimated 250,000 patients with active tuberculosis in the United States have cavitary tuberculosis of the genitourinary tract (Narayana et al., 1982; Wechsler et al., 1960; Smith and Lattimer, 1973). Thirty percent to 50% of patients with renal tuberculosis have had previous tuberculous pulmonary disease documented by history or implied from chest roentgenograms (Simon et al., 1977). However, it is uncommon for pulmonary tuberculosis to be active at the time of diagnosis of renal tuberculosis. Although effective chemotherapy has resulted in a significant decrease in the prevalence of pulmonary tuberculosis, the frequency of renal tuberculosis has not declined significantly in recent decades (Narayana et al., 1982; Lester, 1980).

Pathology and Pathogenesis

In the hematogenous phase that takes place after the primary infection, both kidneys are seeded with tubercle bacilli in 90% of cases. However, clinically apparent renal tuberculosis is usually unilateral. The initial lesions involve the renal cortex with multiple small granulomas in the glomeruli and in the juxtaglomerular regions. In patients in whom acquired cellular immunity develops, there is inhibition of bacterial multiplication and containment of the disease process to the renal cortex. Microscopic examination reveals central caseation necrosis surrounded by pink staining epithelial histiocytes, Langhans' giant cells and, more peripherally, lymphocytes and plasma cells (Hartman, 1985a). Most patients are asymptomatic and have normal findings on radiologic examination. The asymptomatic cortical disease may be stable for many years and be an incidental finding at nephrectomy or autopsy. These early lesions may resolve completely either spontaneously or as a result of treatment.

In untreated patients who fail to heal spontaneously, the lesions may progress slowly and remain asymptomatic for variable periods. In most individuals the latent period between initial exposure and reactivation of renal disease is 10–40 years (Simon et al., 1977) and may be increased by appropriate therapy (Narayana, 1982). The cortical areas of infection may seed the glomerular filtrate, creating lesions in the tubules and

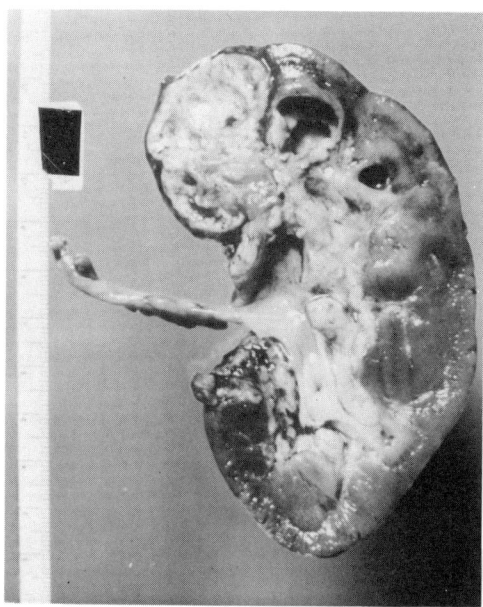

FIG 22–16.
Tuberculoma; gross specimen. A destructive necrotic mass present in the upper pole. (From Hartman DS: Radiologic pathologic correlation of the infectious granulomatous diseases of the kidney: I and II. *Monogr Urol* 1985a; 6:3. Used with permission.)

loops of Henle, resulting in additional foci in the renal pyramid. As the lesions progress they produce areas of caseous necrosis, chronic interstitial nephritis with papillary necrosis, and parenchymal cavitation (Fig 22–16). Large tumorlike parenchymal lesions or tuberculomas frequently have a fibrous wall and can resemble a solid mass lesion. Their content may vary from caseous to calcified material (Hartman, 1985a). Larger blood vessels may show obliterative arteritis.

Once cavities form, spontaneous healing is rare, and destructive lesions result. With necrotizing papillitis, there is spread of the infection to the renal pelvis, ureter, and bladder. The inflammation and edema produce obstruction of the infundibula and ureter leading to caliectasis, calyceal clubbing, and ureteral strictures with consequent pelvic and ureteral dilatation. Extensive peripelvic fibrosis may cause a marked decrease in the pelvic capacity (Fig 22–17). With extensive renal tuberculosis, parenchymal calcification is often present, varying from faint punctate foci to a complete cast of the kidney. Total destruction of the kidney may occur, resulting in autonephrectomy (Fig 22–18).

Seeding of the urine may also result in involvement of the bladder and male genital organs. Tuberculous inflammation in the bladder produces necrosis, ulceration, and fibrosis, which results in a thick-walled sac of small capacity.

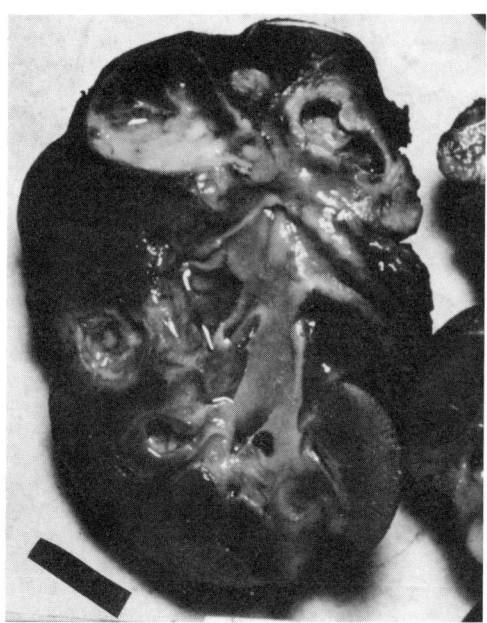

FIG 22–17.
Tuberculosis with pelvic fibrosis. Gross specimen shows several large necrotic cavities. The peripelvic fibrosis has caused marked diminution in the pelvic volume. (From Hartman DS: Radiologic pathologic correlation of the infectious granulomatous diseases of the kidney: I and II. *Monogr Urol* 1985a; 6:3. Used with permission.)

Clinical Findings

Renal tuberculosis is predominantly a disease of young to middle-aged adults. Approximately 50% of the patients are in the 20–40-year age group, and approximately 75% are under age 50 (O'Flynn et al., 1970; Wechsler et al., 1960).

Because of the slow progression and variable course of the disease, there is no classic presentation. Approximately 20% of patients subsequently found to have tuberculosis will be symptom free. Tuberculous nephropathy is an insidious process that often goes unrecognized for long periods. Up to 70% of patients with even advanced cavitary tuberculosis of the kidney may have few diagnostic renal symptoms (Lattimer, 1965; Borthiwick, 1956). Gross hematuria, dull vague flank discomfort, and ureteral colic secondary to passage of clots, debris, or calculi are the most frequent renal symptoms. Constitutional complaints such as fevers, chills, night sweats, weight loss, and malaise are uncommon. It is only when the bladder is involved that the patients become severely symptomatic. Frequency is the most common presenting symptom and is often progressive and occurs both during the day and at night. Pain, urgency, and dysuria are also common with bladder involvement.

The physical examination is usually not helpful diagnostically. A chronic draining fistula tract from previous renal surgery or palpably enlarged, firm seminal vesicles on rectal examination should arouse suspicion. Patients with chronic epididymitis that is unresponsive to therapy should also be evaluated for tuberculosis.

Patients may also present with complications of renal tuberculosis including draining sinus, hypertension, renal failure, secondary amyloidosis and adenocarcinoma of the renal pelvis (Bhargava 1982; Flechner and Gow, 1980; Wesson, 1982; Studer and Weidmann, 1984; Wisnia et al., 1978; Kulkarni et al., 1981). A draining sinus or abscess cavity can occur many years after completion of chemotherapy even in the presence of sterile organs. Pyelocutaneous fistula is often associated with calculus obstruction of the ureteropelvic junction.

Hypertension is present in approximately 5%–10% of patients with renal tuberculosis; the incidence in patients with unilateral nonfunctioning or poorly functioning kidneys is approximately 25%. Although some of these individuals have evidence of renal ischemia as determined by renal vein renin studies and normalization

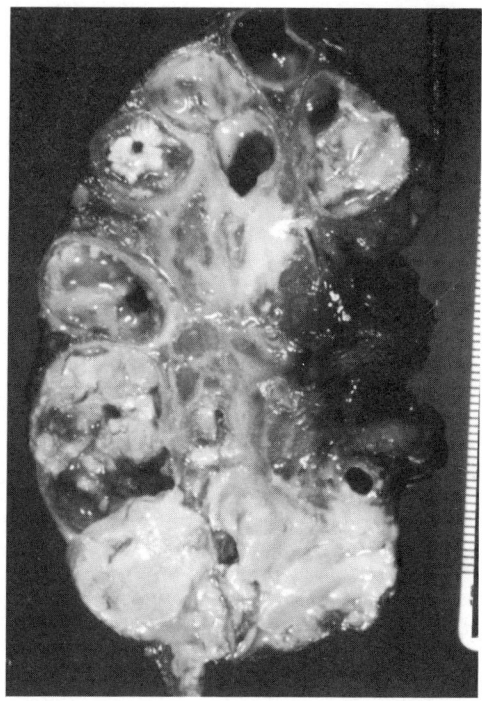

FIG 22–18.
Tuberculosis, autonephrectomy. Gross specimen shows complete parenchymal destruction by extensive caseous necrosis. (From Hartman DS: Radiologic pathologic correlation of the infectious granulomatous diseases of the kidney: I and II. *Monogr Urol* 1985a; 6:3. Used with permission.)

of blood pressure following nephrectomy, most patients appear to have hypertension that is not mediated by the renin-angiotension system and is not cured by nephrectomy (Ocon et al., 1984).

Laboratory Findings

The urinalysis is abnormal in 90% of the patients. The most common finding is sterile, acid pyuria, frequently accompanied by hematuria and proteinuria. Pyuria in the absence of a positive culture for the usual uropathogens should always warrant consideration of diagnosis of renal tuberculosis. Acid-fast smears of urinary concentrates are usually negative and are not totally reliable if positive, because the saprophytic organism *Mycobacterium smegmatis* may contaminate the urine and is morphologically indistinguishable from *M. tuberculosis* on acid-fast smear.

The most important laboratory test is urine culture for *M. tuberculosis*. Cultures are positive in approximately 90% of affected individuals (Corigliano and Leedom, 1983). First morning urine specimens are more reliable than 24-hour collections because they are easier to collect and there is much less chance of bacterial contamination. Since discharge of *M. tuberculosis* into the urine may be intermittent, it is advisable to collect morning urine specimens on a total of at least three separate days (Narayana, 1982). Cultures should be obtained whether or not pus is present in the urine, because positive cultures have been found in otherwise normal specimens (Innes, 1981). Sensitivity testing should be performed on isolated organisms.

Although the classic finding of renal tuberculosis is sterile pyuria, many patients will have a concurrent positive culture for uropathogenic organisms at presentation initially. This finding should not preclude consideration of diagnosis of renal tuberculosis in the proper clinical setting. Such patients will continue to demonstrate pyuria after the uropathogenic organism is eradicated by appropriate antimicrobial therapy.

Rarely, the sealing off of cavitary lesions in chronic infections may result in persistently negative findings on urine culture for tuberculosis and may mean that the disease in some patients remains undiagnosed despite repeated cultures (Lowe et al., 1983). The tuberculin test, while not diagnostic, is of value only if positive.

Tuberculosis depresses renal function but the damage has to be widespread before the serum creatinine level is elevated. Biochemical evidence of renal functional impairment is seen in less than 10% of patients with renal tuberculosis. Renal failure may be accentuated by obstructive uropathy that results from ureteral stricture or by secondary amyloidosis (Mallinson et al., 1981). Estimations of renal function are essential to preliminary assessment of patients and determining

the dose levels of antituberculous drugs. Wisnia et al. (1978) showed that 58% of patients had chronic renal failure at the time of diagnosis; 43% had subclinical impairment of creatinine clearance; 10% had compensated renal failure; and 5% had uremia.

It is apparent that the clinical presentation may be subtle, and therefore a high index of suspicion of renal tuberculosis is indicated in patients with a history of past or present tuberculosis, with chronic cystitis that fails to respond to adequate antibiotics, or with sterile pyuria accompanied by gross or microscopic hematuria.

Radiologic Findings

Approximately 90% of patients with renal tuberculosis will have abnormal excretory urograms. The roentgenographic findings vary according to the severity of the destructive process and the duration of the disease. Early changes in the radiologic appearance of the tuberculous kidney may be subtle and difficult to find, and hence a normal excretory urogram does not rule out renal tuberculosis.

The most suggestive urographic features of renal parenchymal tuberculosis are the presence of cavities that communicate with the collecting system and fill with contrast medium and dilatation of part or all of the calyceal system (Narayana, 1982; Hartman, 1985a). Initially the cavities are small, appear as a slight irregularity of a minor calyx, and show a moth-eaten, feathery, irregular appearance (Fig 22–19,A). Fibrous stenosis may cause amputation of one or more calyces (Fig 22–19,B). As the calyceal system becomes eroded, the parenchyma is destroyed by cavitation and the picture may closely resemble pyelonephritis (Fig 22–20,A). The kidney will be either enlarged if caseous sacs are present or atrophic if there is long-standing infection. Parenchymal curvilinear or confluent calcifications in areas of caseous necrosis are common (Fig 22–21). Autonephrectomy with complete nonvisualization of kidney may result from complete parenchymal destruction or hydronephrosis due to stenosis of the renal pelvis or ureter (Fig 22–22).

Fibrosis may cause the volume of the renal pelvis to be markedly reduced (Fig 22–23). Tuberculous ureteritis is common and results in rigidity and a beading or corkscrew appearance (Fig 22–24). Ureterovesical junction obstruction is caused by tuberculous cystitis or strictures of the distal third of the ureter. Although not common, mural calcification in the wall of the calyces, pelvis, ureter, or bladder is extremely suggestive of renal tuberculosis (Hartman, 1985a).

Retrograde urography should be employed in cases of nonvisualized or poorly functioning kidneys. A voiding cystogram may show a severely contracted bladder with vesicoureteral reflux in one or both sides.

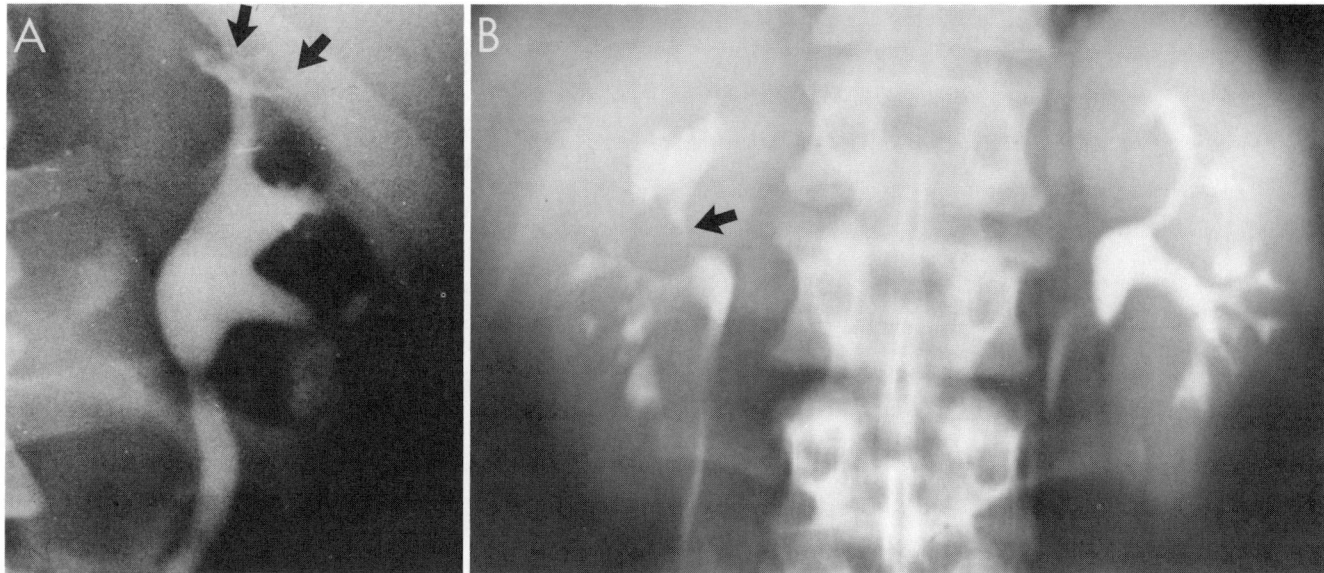

FIG 22–19.
Tuberculosis. Early disease. Excretory urograms. **A,** the calyceal surface is irregular with linear rays of contrast material extending into the medulla *(arrows).* **B,** tomogram demonstrates upper pole infundibular stenosis *(arrow).* No ureteral involvement is seen. (From Hartman DS: Radiologic pathologic correlation of the infectious granulomatous diseases of the kidney: I and II. *Monogr Urol* 1985a; 6:3. Used with permission.)

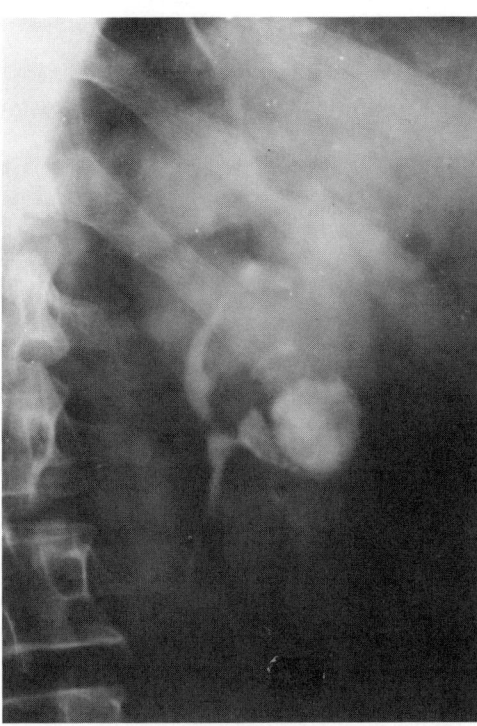

FIG 22–20.
Tuberculosis, advanced disease. Excretory urogram shows a large cavity communicating with the calyx in the middle portion of the kidney. Note the irregularity and narrowing of the remainder of the collecting system. (From Hartman DS: Radiologic pathologic correlation of the infectious granulomatous diseases of the kidney: I and II. *Monogr Urol* 1985a; 6:3. Used with permission.)

Sonographic examination and CT have limited diagnostic value. However, these modalities are valuable for monitoring patients for evidence of obstruction and perinephric extension or perinephric hemorrhage that may complicate renal tuberculosis. CT is also the best technique for demonstrating pelvic and ureteral mural thickening and calcification. Angiography is of limited usefulness in diagnosing renal tuberculosis. Arterial fibrosis results in narrowing or amputation of the small intrarenal arteries (Hartman, 1985a; Gajaraj and Victor, 1981).

Differential Diagnosis

If all of the radiographic findings are typical of renal tuberculosis, the diagnosis is straightforward. However, all the classic findings are frequently absent and the radiographs may suggest medullary sponge kidney, papillary necrosis, invasive transitional cell carcinoma, chronic pyelonephritis, or schistosomiasis. The pertinent radiologic discriminators have been summarized by Hartman (1985a).

Medullary sponge kidney and papillary necrosis both demonstrate extracalyceal accumulations of contrast material and medullary nephrocalcinosis but cortical calcification and parenchymal masses are rare. Ring calcification near a necrotic or sloughed papillus is extremely suggestive of papillary necrosis. Chronic urinary tract infection and pyelonephritis will show similar distortion of the renal cortex and collecting system, but calyceal, pelvic, and ureteral strictures are not com-

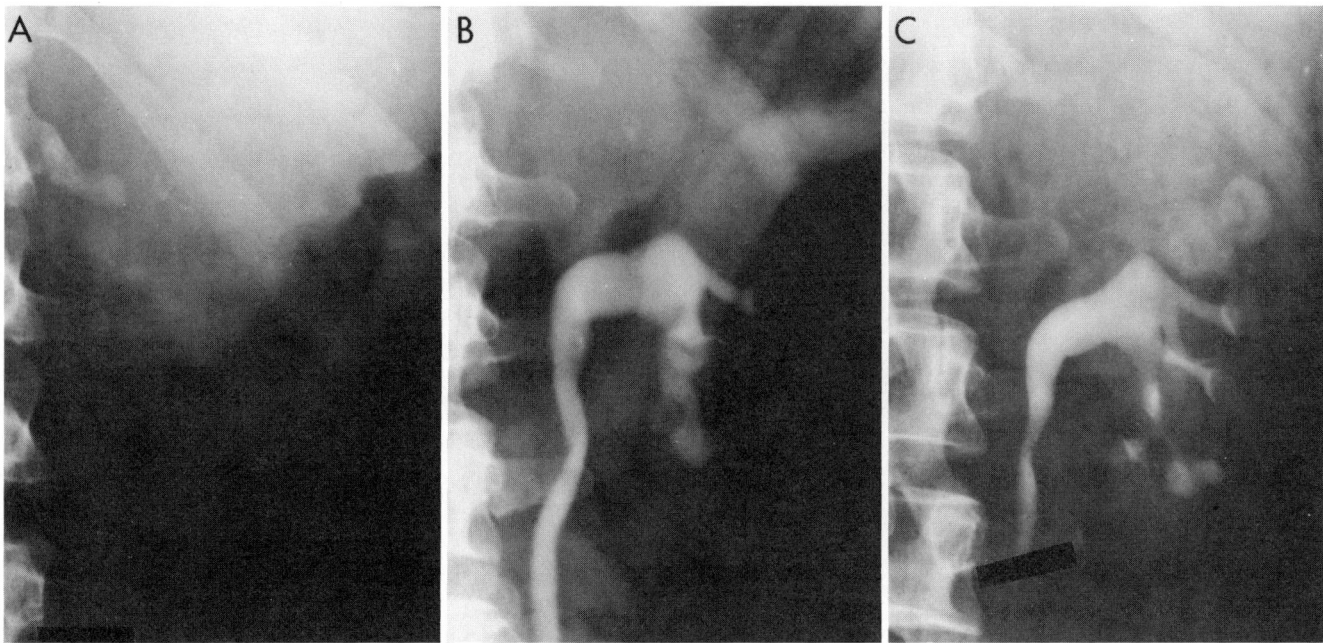

FIG 22–21.
Tuberculoma. **A,** KUB. Faint calcification over the renal fossa. **B,** 15-minute film from an excretory urogram shows "amputation" of the upper pole calyx in the region of calcification. **C,** delayed film reveals contrast material filling the mass, indicating its communication with the collecting system. (From Hartman DS: Radiologic pathologic correlation of the infectious granulomatous diseases of the kidney: I and II. *Monogr Urol* 1985a; 6:3. Used with permission.)

monly associated with typical urinary tract infections. Invasive transitional cell carcinoma may show infundibular narrowing or calyceal amputation and may be difficult to differentiate radiographically from tuberculosis. Transitional cell carcinoma is not associated with communicating cavities or parenchymal calcification. Infun-

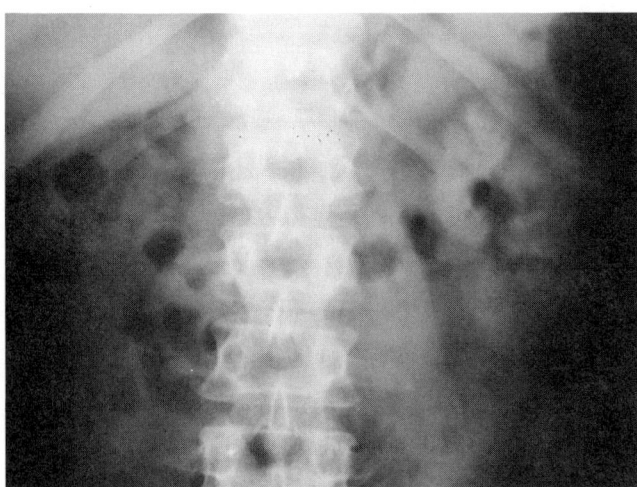

FIG 22–22.
Tuberculosis, autonephrectomy. KUB. Extensive confluent calcifications in the renal fossa. The excretory urogram indicated no function. (From Hartman DS: Radiologic pathologic correlation of the infectious granulomatous diseases of the kidney: I and II. *Monogr Urol* 1985a; 6:3. Used with permission.)

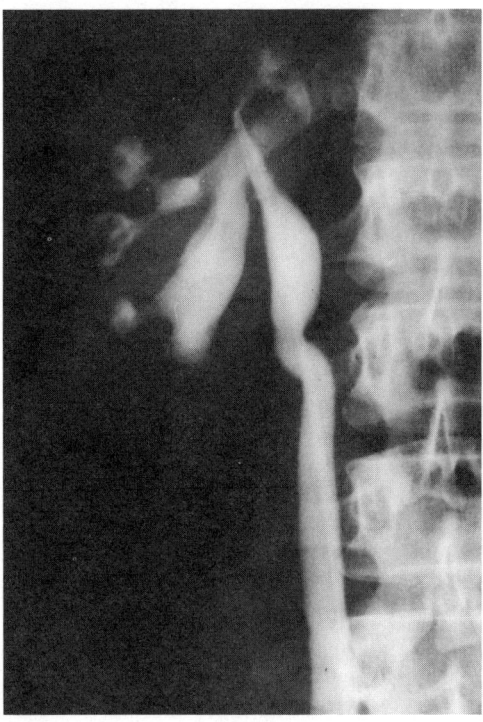

FIG 22–23.
Tuberculosis with pelvic fibrosis. Retrograde pyelogram shows marked calyceal irregularity. The calyces come together and empty into the proximal ureter. (From Hartman DS: Radiologic pathologic correlation of the infectious granulomatous diseases of the kidney: I and II. *Monogr Urol* 1985a; 6:3. Used with permission.)

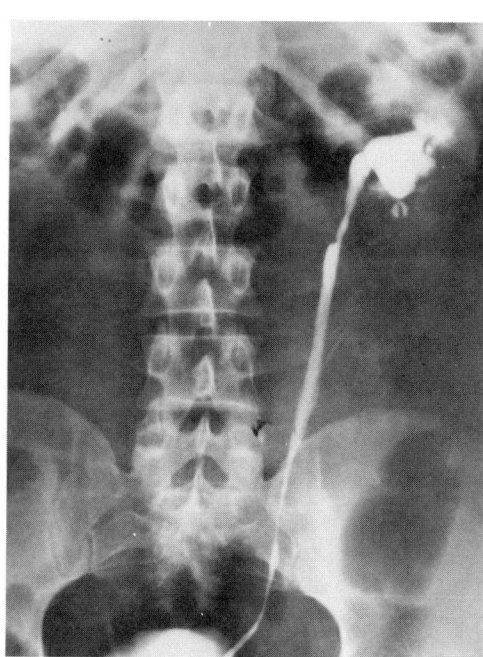

FIG 22–24.
Tuberculosis, advanced disease. Retrograde pyelogram shows extensive ureteral changes including irregularity and rigidity, giving "pipestem" appearance. Note extensive distortion, irregularity, and amputation of the upper pole calyx.

dibular narrowing and submucosal calcifications are seen in amyloidosis of the kidney, but amyloidosis is not associated with destructive communicating cavities or parenchymal calcification. *Schistosoma haematobium* infection of the urinary tract may cause ureteral and bladder calcifications. However, schistosomiasis does not involve the renal parenchyma primarily and is not associated with renal calcification. Schistosomiasis only affects the distal ureter, while tuberculosis often involves the entire ureteral length.

Cystoscopy

In early cases the bladder is diffusely red and extremely sensitive. Bladder wall ulcerations, severe contracture of the bladder, and golf hole ureteral orifices are seen in more advanced disease. Ulcerated areas in the bladder strongly suggest tuberculosis. Tuberculous ulcers are irregular and shallow with undermined edges, often well-circumscribed from adjoining normal-appearing mucosa. Neoplasms may have a similar appearance and must be differentiated by biopsy.

Management

Appropriate management must be based on an accurate bacteriologic diagnosis and initial assessment of the extent of the disease, the level of renal function, and the nature and severity of ureteric obstruction. Renal tuberculosis in the absence of active pulmonary tuber-

culosis does not represent a significant infectious risk, and therefore isolation is not required. Management always includes appropriate chemotherapy and surgery when indicated. With close and long-term follow-up, the overall mortality for renal tuberculosis has dropped from approximately 50% to 2%.

Table 22–5 gives an overview of agents available for treatment of tuberculosis. There is no uniformly recommended chemotherapeutic program. Patients with positive urine cultures but negative findings on urinalysis and normal urograms usually are treated with isoniazid and rifampin for one year. Patients with clinically manifest renal tuberculosis are usually treated with three antituberculous drugs such as isoniazid, ethambutol, and rifampin for two years (Wechsler et al., 1960). If the patient's initial urinary isolate is *M. tuberculosis* resistant to isoniazid or is one of the nontuberculous (atypical) mycobacteria, isoniazid therapy should be continued, but two other drugs are chosen on the basis of in vitro susceptibilities.

A short-term regimen of only four months' duration has been recommended by Gow (1981) and is summarized in Table 22–6. The initial results with this regimen appear promising, but longer follow-up is required.

Usually tuberculous bacilli disappear from urine immediately after initiating chemotherapy, but to monitor the drug's effect, repeated cultures should be done every 3–4 months. All the antituberculous drugs have toxic effects when used on a continuous basis. Renal and hepatic function should be checked during the treatment. Rifampin and isoniazid appear to be the safest drugs in the presence of impaired renal function (Reidenberg et al., 1973).

Surgery was once commonly used in the treatment of renal tuberculosis, but since the advent of effective antituberculous chemotherapy, it is reserved primarily for management of local complications such as ureteral stricture or for treatment of nonfunctioning kidneys. If surgery is warranted, it is wise to precede the operation with at least three weeks and preferably three months of triple-drug chemotherapy.

The incidence of ureteral strictures has been reported as high as 10% in centers that treat large numbers of patients with urinary tuberculosis (O'Flynn, 1979). The stricture may be present at the time of diagnosis of renal tuberculosis, but it often develops or progresses during otherwise effective treatment with chemotherapeutic agents (Claridge, 1970). Recommended treatment of strictures includes ureteroneocystostomy, construction of a Boari flap, or transluminal balloon ureteral dilatation (Badenock, 1961; Waller et al., 1983; Murphy et al., 1982). Success rates with these procedures have been reported at 60%–90%.

Horne and Tulloch (1975) added prednisolone, 5 mg,

TABLE 22–5.
Commonly Used Agents in the Treatment of Tuberculosis*

DRUG	USUAL DOSE	ROUTE	FREQUENCY	ADVERSE EFFECTS
Isoniazid	300 mg/day	po†	Once daily	Hepatotoxicity, peripheral neuritis, asymptomatic transaminase rise
Ethambutol	25 mg/kg/day	po	Once daily for 60 days	Retrobulbar neuritis seen at 25 mg/kg dose, less frequent at 15 mg/kg dose; completely reversible if drug stopped immediately
	15 mg/kg/day	po	Once daily for remainder of treatment period	
Rifampin	600 mg/day	po	Once daily	Hepatotoxicity, thrombocytopenia
Pyrazinamide	1,000 mg/day	po	Once daily	Hepatoxicity: dose- and duration-related
Ethionamide	750–1,000 mg/day	po	Divided doses, 250 mg each	Nausea, vomiting, anorexia, abdominal pain
Cycloserine	750–1,000 mg/day	po	Divided doses, 250 mg each	Neurotoxicity: headache, drowsiness, convulsions, psychotic disturbance
Para-amino salicyclic acid	10–12 gm/day	po	2–3 Divided doses	GI irritation, rash and fever; hemolysis in glucose-6-phosphate dehydrogenase deficiency
Streptomycin	750–1,000 mg/day	im	Once daily for 30 days and twice weekly thereafter	Ototoxicity, especially in older patients; rash and fever
Capreomycin	500–1,000 mg/day	im	Once daily for 90 days and twice weekly thereafter	Renal toxicity and ototoxicity, especially in older patients

*Compiled from Kucers AM, Bennett N: *The Use of Antibiotics*, ed. 3. London, William Heinemann Medical Books Ltd, 1979, by Corigliano and Leedom (1983).
†po, per os.

4 times a day, to the standard chemotherapy regimen in all patients with renal tuberculosis who had evidence of stricture formation at presentation or in whom it developed during treatment. No important side effects were noted. Of 29 patients so treated, 72% were relieved of obstruction, and only two patients showed recurrence after withdrawal of steroids; both individuals responded to further courses of steroids. However, success with steroid therapy has not been universal (Claridge, 1970).

Removal of a nonfunctioning kidney is usually indicated for advanced unilateral disease complicated by sepsis, hemorrhage, intractable pain, newly developed severe hypertension, suspicion of malignancy, inability to sterilize the urine with drugs alone, abscess formation with development of fistula or inability to have ap-

propriate follow-up (Narayana, 1982; Lorin et al., 1983; Lattimer and Wechsler, 1980). Prophylactic removal of a nonfunctioning kidney to prevent complications, remove a potential source of viable organisms, and shorten the duration of convalescence and requirement for chemotherapy is advocated by some authors (Osterhage et al., 1980; Flechner and Gow, 1980; Wong and Lau, 1980).

Others who have followed up a large series of patients treated with medical therapy alone have concluded that, since the frequency of late complications is only 6%, routine nephrectomy should not be performed for every nonfunctioning kidney (Wechsler and Lattimer, 1975; Lattimer, 1980). All of these authors treated patients for two years. The merits of short-term therapy and prophylactic nephrectomy vs. long-term, two year chemotherapy and selected nephrectomy warrant further study.

TABLE 22–6.
Gow's Short-Course Regimen for Treatment of Renal Tuberculosis*

PHASE	DRUG	DOSE, MG/DAY	FREQUENCY
Intensive phase (2 mo)	Isoniazid	300	Daily
	Rifampin	450	Daily
	Pyrazinamide	1,000	Daily
Continuation phase (2 mo)	Isoniazid	600	3 times/wk at night
	Rifampin	900	3 times/wk at night

*From Gow JG: The management of genitourinary tuberculosis. *J Antimicrob Chemother* 1981; 7:590–591. Used by permission.

XANTHOGRANULOMATOUS PYELONEPHRITIS

Xanthogranulomatous pyelonephritis is a chronic inflammatory disease characterized by accumulation of lipid-laden foamy macrophages. It begins within the pelvis and calyces and subsequently extends into and destroys renal parenchymal and adjacent tissues. In most cases, xanthogranulomatous pyelonephritis is unilateral and results in a nonfunctioning, enlarged kidney associated with obstructive nephropathy secondary to nephrolithiasis. Although uncommon, approximately 500 cases have been reported since the disorder was first described by Schlagenhaufer in 1916. It has been known to imitate virtually every other inflammatory disease of the kidney as well as renal cell carcinoma on radiographic examination. In addition, the microscopic appearance of xanthogranulomatous pyelonephritis has been confused with clear cell adenocarcinoma of the kidney on frozen section and led to radical nephrectomy (Butnick, 1971).

Pathology and Pathogenesis

The kidney is usually massively enlarged and has a normal contour. Xanthogranulomatous pyelonephritis may be diffuse, as in approximately 80% of the patients, or segmental. In the diffuse form of the disease the entire kidney is involved whereas in segmental xanthogranulomatous pyelonephritis only the parenchyma surrounding one or more calyces or one pole of a duplicated collecting system is involved. Upon sectioning, the kidney usually demonstrates nephrolithiasis and peripelvic fibrosis. The calyces are dilated and filled with purulent material, but fibrosis surrounding the pelvis usually prevents dilatation. The papillae are frequently destroyed by papillary necrosis (Goodman et al., 1979). In advanced stages of the disease multiple

parenchymal abscesses are filled with viscous pus and lined by yellowish tissue (Fig 22–25,A). The cortex is often thin and is frequently replaced by xanthogranulomatous tissue. The capsule is frequently thickened, and extension of the inflammatory process into the perinephric or paranephric space is common (McDonald, 1981; Goodman et al., 1979).

On microscopic examination, the yellowish nodules that line the calyces and surround the parenchymal abscesses contain dark sheets of lipid-laden macrophages (foamy histiocytes with small, dark nuclei and clear cytoplasm) intermixed with lymphocytes, giant cells, and plasma cells (Fig 22–25,B). Xanthogranulomatous cells are not specific to xanthogranulatomous pyelonephritis but may be present anywhere inflammation or obstruction coexists. The origin of the fatty substance is disputed. Cholesterol esters that make up a part of the lipid might be derived from lysis of erythrocytes following hemorrhage (Saedd and Fine, 1963).

The primary factors involved in the pathogenesis of xanthogranulomatous pyelonephritis are obstruction and infection. It has been proposed clinically and demonstrated experimentally that primary obstruction followed by infection with *E. coli* can lead to tissue destruction and collections of lipid material by histiocytes (Povysil and Konickova, 1972). Other possible interrelated factors include venous occlusion and hemorrhage, abnormal lipid metabolism, lymphatic blockage, failure

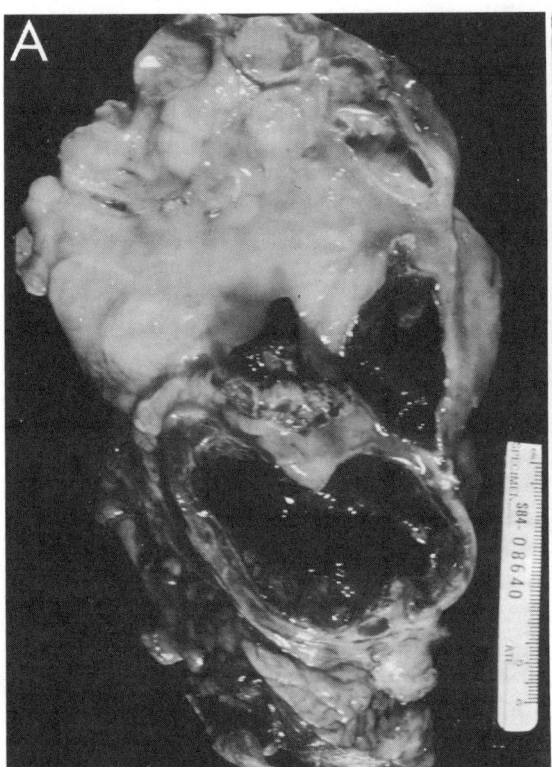
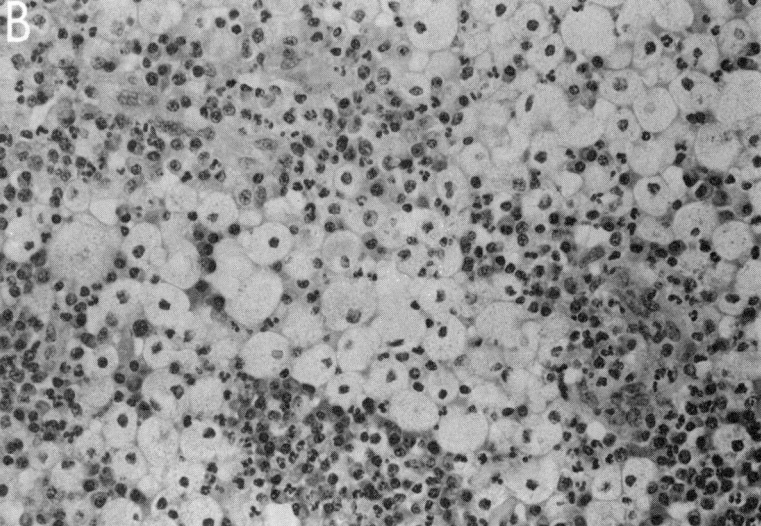

FIG 22–25.
Xanthogranulomatous pyelonephritis. **A,** gross specimen. Kidney is massively enlarged, measures 23 × 12 cm; the normal architecture replaced by a shaggy yellow upper pole mass corresponding to xanthogranulomatous inflammation and numerous distorted and dilated calyces. **B,** microscopically, the shaggy yellow tissue is composed primarily of lipid-laden histiocytes mixed with other inflammatory cells.

of antibiotic therapy in urinary tract infection, altered immunologic competence, and renal ischemia (McDonald, 1981; Goodman et al., 1979; Tolia et al., 1981; Friedenberg and Spjut, 1963; Mering and Kaplan, 1973). The concept that xanthogranulomatous pyelonephritis is related to incomplete bacterial degradation and altered host response has received mixed support (Khalyl-Mawad et al., 1982; Overgaard Nielsen and Lorentzen, 1981). Thus, it appears that there is probably no single factor instrumental in the pathogenesis of this disease. Rather, there is an inadequate host acute inflammatory response within an obstructed, ischemic, or necrotic kidney.

Clinical Findings

Xanthogranulomatous pyelonephritis should be suspected in patients presenting with urinary tract infections and a unilateral enlarged nonfunctioning or poorly functioning kidney with a stone or a mass lesion indistinguishable from malignant tumor. Although it may present at any age, the peak incidence of xanthogranulomatous pyelonephritis is in the fifth to the seventh decade. The clinical findings of focal and diffuse xanthogranulomatous pyelonephritis in adults and diffuse xanthogranulomatous pyelonephritis in children are similar (Elder and Marshall, 1980; Elder, 1984). The patients are in poor general medical condition, and children suffer from failure to thrive (Schulman and Denis, 1977). Both sexes are affected relatively equally, and there is no predilection for either kidney. Most children with focal xanthogranulomatous pyelonephritis have been girls who were in good medical condition, and the left kidney was nearly always involved.

Most patients have multiple symptoms which are variable and nonspecific. Patients usually have flank pain, fever or chills, malaise, weight loss, symptoms of cystitis, calculi or palpable mass, and a history of recurrent urinary tract infection. Many patients are admitted to the hospital for urosepsis. Less commonly, hypertension, hematuria, or hepatomegaly is the presenting complaint. Past medical history is often positive for urinary tract infections and urologic instrumentation (Malek and Elder, 1978; Goodman et al., 1979; Grainger et al., 1982; Flynn et al., 1979; Tolia et al., 1963; Yazaki et al., 1982).

Laboratory Findings

Hematologic evaluation commonly shows anemia and leukocytosis. Diabetes mellitus is present in approximately 15% of the patients. Urinalysis reveals proteinuria and pyuria in nearly all cases. Urine cultures are positive in approximately 70% of patients. Although earlier reports indicated that *Proteus mirabilis* was the primary pathogen, in recent studies *E. coli* was cultured in 40% of specimens and *P. mirabilis* was cultured in approximately 30%. Other pathogens include *Klebsiella, Pseudomonas,* and *Bacteroides* (Elder, 1984). Approximately 10% of the patients will have mixed cultures. It is noteworthy that about one third of the patients have sterile urine, probably because many patients receive long-term antibiotics or are taking antibiotics when the culture is obtained. The infecting organism may be revealed only by tissue cultures obtained during surgery.

Recently Ballesteros and associates (1980) reported accurate preoperative diagnosis of xanthogranulomatous pyelonephritis by serial urinary cytology in 80% of their cases.

Xanthogranulomatous pyelonephritis is almost always unilateral, and therefore azotemia or frank renal failure is uncommon (Goodman et al., 1979). Reversible hepatic dysfunction has been observed in 20%–40% of patients with diffuse xanthogranulomatous pyelonephritis (Tolia et al., 1981; Malek and Elder, 1978). Liver enzymes return to normal after nephrectomy.

Radiologic Findings

Approximately 50%–80% of patients will show the classic triad of unilateral renal enlargement with little or no function and a large calculus in the renal pelvis (Elder, 1984). At times the enlargement may be localized and resemble a renal mass. Less commonly excretory urography demonstrates delayed function and hydronephrosis, which may be massive. Smaller calcifications within the mass are not uncommon but are much less specific (Fig 22–26,A). Although there is abundant intracellular fat, the plane almost never demonstrates significant lucency (Hartman, 1985b). Retrograde pyelography may show the point of obstruction and dilatation of the renal pelvis and calyces. If there is extensive parenchymal damage, contrast studies may demonstrate an ulcerated pyelocalyceal system with multiple irregular filling defects.

Sonography usually demonstrates global enlargement of the kidney. The normal renal architecture is replaced by multiple hypoechoic fluid-filled masses that correspond to debris-filled dilated calyces or foci of parenchymal destruction (Fagerholm, 1983; Hartman et al., 1984). With focal involvement, a solid mass involving a segment of the kidney is demonstrated with an associated calculus in the collecting system or ureter. Renal cell carcinoma and other solid renal lesions must be considered in the differential diagnosis (Elder, 1984).

CT is probably the most useful radiologic technique in evaluating patients with xanthogranulomatous pyelonephritis. CT usually demonstrates a large, reniform

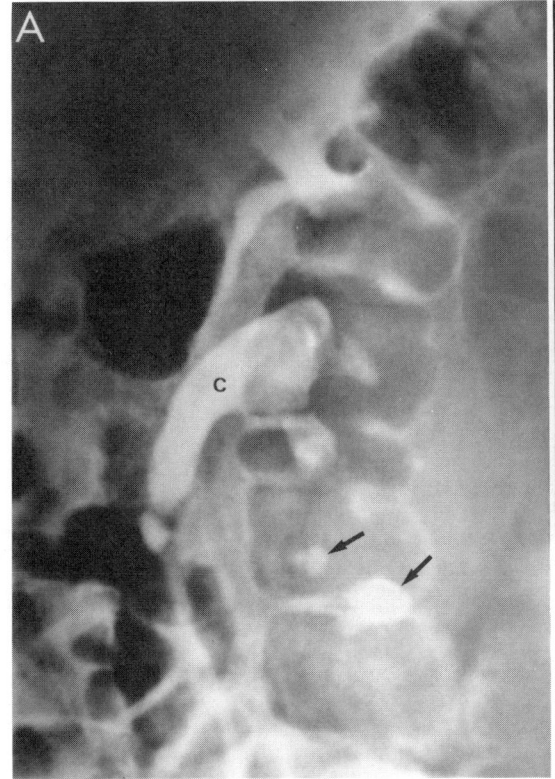
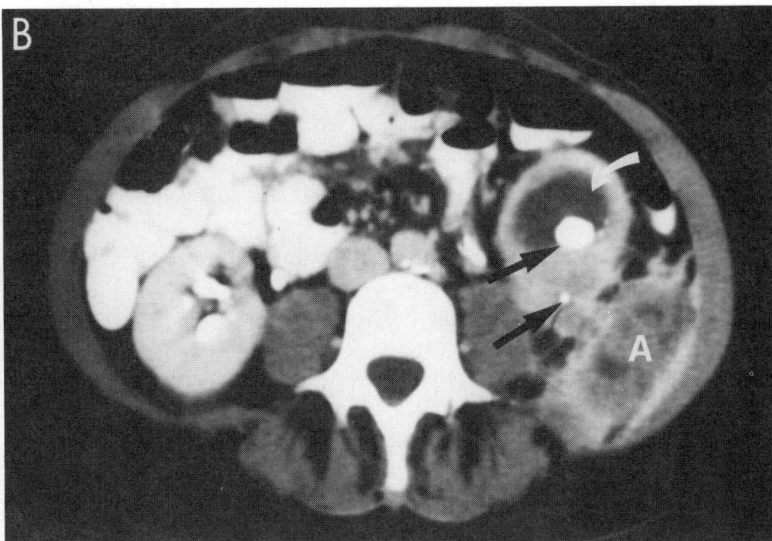

FIG 22–26.
Xanthogranulomatous pyelonephritis. Excretory urogram. **A,** ten-minute film shows lower pole enlargement and nonfunction. Note a large pelvic calculus *(C)* and several smaller parenchymal calculi *(arrows).* The upper pole of the bifid collecting system functions normally. **B,** enhanced CT scan shows collecting system and parenchymal calculi *(arrows)* with lower pole pyonephrosis *(curved arrow)* and an irregular, predominantly low-density perinephric abscess *(A)* extending into the soft tissues of the flank.

mass with the renal pelvis tightly surrounding a central calcification without pelvic dilatation (Hartman, 1985b; Goldman et al., 1984; Solomon et al., 1983) (Fig 22–26,B). Renal parenchyma is replaced by multiple water density masses representing dilated calyces and abscess cavities filled with varying amounts of pus and debris. On enhanced scans, the walls of these cavities demonstrate a prominent blush owing to the abundant vascularity within the granulation tissue. The cavities themselves, however, fail to enhance, whereas tumors and other inflammatory lesions usually do. The CT scan is particularly helpful in demonstrating the extent of renal involvement and may indicate whether adjacent organs or the abdominal wall are involved by xanthogranulomatous pyelonephritis.

Angiography is seldom required for diagnosing xanthogranulomatous pyelonephritis. Most commonly there is stretching and attenuation of vessels around avascular masses with an irregular nephrogram (Elder, 1984). Benign neovascularity, absence of irregular vessels, and AV shunting are also characteristic.

Differential Diagnosis

Diagnosis of segmental xanthogranulomatous pyelonephritis or xanthogranulomatous pyelonephritis without calculi may be very difficult. Xanthogranulomatous

pyelonephritis in association with massive pelvic dilatation cannot be distinguished from pyonephrosis. When xanthogranulomatous pyelonephritis occurs within a small contracted kidney the radiographic findings are nonspecific and nondiagnostic. Renal parenchymal malakoplakia may show renal enlargement and multiple inflammatory masses replacing the normal renal parenchyma, but calculi are usually not present. Renal lymphoma may be associated with multiple hypoechoic masses surrounding the contracted, nondilated pelvis, but lymphoma is usually clinically obvious, and renal involvement is usually bilateral and not associated with calculi (Hartman, 1985b).

Management

The management of xanthogranulomatous pyelonephritis is surgical. Antimicrobial therapy may be necessary to stabilize the patient preoperatively, but long-term antimicrobial therapy will not eradicate the infection or restore renal function. Nephrectomy is required for diffuse xanthogranulomatous pyelonephritis. Partial nephrectomy is the preferred treatment for focal xanthogranulomatous pyelonephritis if the diagnosis can be established preoperatively or if necessary by frozen section at the time of surgery. However, the lipid-laden macrophages associated with xanthogranulomatous pyelonephritis closely resemble clear cell adenocarcinoma and may be difficult to distinguish solely on the basis of frozen section. Further, xanthogranulomatous pyelonephritis has been associated with renal cell carcinoma, papillary transitional cell carcinoma of the pelvis or bladder, and infiltrating squamous cell carcinoma of the pelvis (Tolia et al., 1981; Pitts et al., 1981; Schoborg et al., 1980). Therefore, if malignant renal tumor cannot be excluded, nephrectomy should be performed.

It is important to remove the entire inflammatory mass, because in nearly three fourths of patients xanthogranulomatous tissue is infected. If incision and drainage alone are performed rather than nephrectomy, the patient may continue to suffer from protracted debilitating illness and may develop a renocutaneous fistula, and an even more difficult nephrectomy will be necessary (Elder, 1984). Extensive xanthogranulomatous pyelonephritis may make surgical dissection and ligation of the friable vascular pedicle particularly difficult. Extension of the inflammatory reaction to adjacent bowel, diaphragm, and aorta will require tedious dissection and at times resection of the involved tissues.

RENAL PARENCHYMAL MALAKOPLAKIA

Malakoplakia, from the Greek words meaning "soft plaque," is an uncommon inflammatory lesion described originally by Michaelis and Gutmann (1902). It was characterized by Von Hansemann (1903) as soft, yellow-brown plaques with granulomatous lesions in which the histiocytes contain distinct basophilic inclusions, or Michaelis-Gutmann bodies. While its exact pathogenesis is unknown, malakoplakia probably results from abnormal macrophage function in response to a bacterial infection, which is most often *E. coli*. The inclusions probably represent calcification around incompletely digested bacteria (Abdou et al., 1977; McClurg et al., 1973).

The disease usually affects the lower urinary tract; only about 50 patients with renal parenchymal malakoplakia have been reported (McClure, 1983). Clinically, radiologically, and at surgery, malakoplakia frequently mimics a neoplastic growth. Mortality can exceed 50%, and the morbidity can be substantial (Stanton and Maxted, 1981). Extension of renal parenchymal malakoplakia into the perirenal space is uncommon (Angell and Smith, 1968). Renal parenchymal malakoplakia may also be complicated by renal vein thrombosis and inferior vena cava thrombosis (Hartman et al., 1980; McClure, 1983).

Concomitant nonrenal foci of malakoplakia are seen infrequently in patients with renal parenchymal malakoplakia (Stanton and Maxted, 1981). The most common locations of other foci are the bladder and ureter. Less common sites that have been reported include the retroperitoneum, abdominal wall, colon, testis and epididymis, lungs, scrotum, prostate, adrenal gland, lymph nodes, and diaphragm (McClure, 1983).

Pathology and Pathogenesis

Two basic patterns of renal parenchymal malakoplakia have been described: (1) multifocal and (2) unifocal. The multifocal pattern accounts for 75% of the reported cases and is bilateral in about half of the patients (Hartman et al., 1980). The kidney is usually enlarged and contains multiple masses varying from several millimeters to several centimeters. They are usually yellow, well-demarcated, and may have a focus of hemorrhage or suppuration (Fig 22–27,A and B). The masses often coalesce to form larger nodules. These nodules often project beyond the cortical margin, resulting in an irregular contour. Less commonly the nodules are limited to the papilla or the medulla and occasionally mimic necrotizing papillitis, in which cases the renal contour is normal (Hartman, 1985a; Bowers and Cathey, 1971; Ravel, 1967; Dridder et al., 1977).

Unifocal disease usually appears as a large yellow-gray mass, 2.5–8 cm in diameter (Hartman et al., 1980). The mass is usually smooth and well-marginated, and central necrosis or cyst formation may be present. Calcification of the mass is unusual (Scullin and Hardy, 1972).

Microscopically, malakoplakia is characterized by the Von Hansemann histiocyte, a large polygonal cell with foamy eosinophilic cytoplasm and compact, dark nucleus admixed with intracellular and extracellular inclusions known as Michaelis-Gutmann bodies. Calcium phosphate crystals and iron are precipitated on these membranes so that both periodic acid-Schiff and von Kossa calcium stains and iron stains are useful in demonstrating the inclusions (Lou and Teplitz, 1974). They are slightly smaller than a red blood cell and are recognized by concentric laminations that impart a targetoid or owl's-eye appearance (Garrett and McClure, 1982) (Fig 22–27,C). Histochemical and ultrastructural studies have shown that these Michaelis-Gutmann bodies represent incompletely destroyed bacteria surrounded by concentric lipoprotein membranes. Extracytoplasmic Michaelis-Gutmann bodies probably represent debris released from dead cells (McClure, 1983).

Recently it has been shown that macrophages in malakoplakia involving the kidney and bladder contain large amounts of immunoreactive α_1-antitrypsin (Callea et al., 1982). The amount of α_1-antitrypsin remains unchanged during the morphogenetic stages of the pathologic process. Macrophages from other pathologic processes, closely resembling malakoplakia but without Michaelis-Gutmann bodies, do not contain α_1-antitryp-

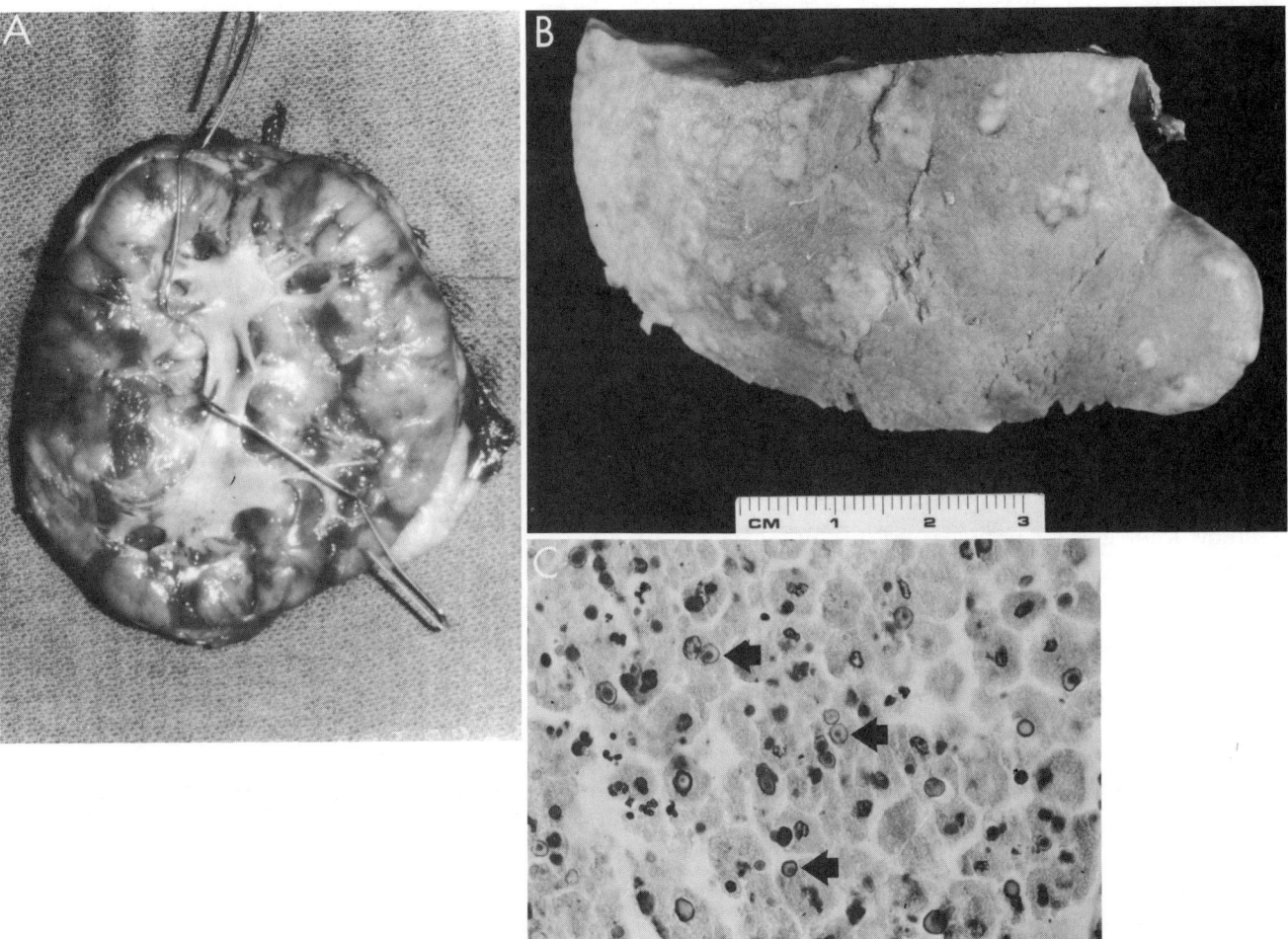

FIG 22–27.
Renal parenchymal malakoplakia. **A,** the cut surface demonstrates extensive cortical and upper medullary replacement by multifocal, confluent, tumorlike masses. **B,** the cortical surface exhibits multiple, firm, plaquelike lesions. **C,** hallmark of malakoplakia is demonstration of the Michaelis-Gutmann body *(arrows),* which represents incompletely destroyed bacteria surrounded by lipoprotein membrane (hematoxylin-eosin). (**A** and **B** from Hartman DS: Radiologic pathologic correlation of the infectious granulomatous diseases of the kidney: I and II. *Monogr Urol* 1985a; 6:3. Used with permission.)

sin except for a few macrophages in tuberculosis and xanthogranulomatous pyelonephritis. Therefore, immunohistochemical staining for α_1-antitrypsin may be a useful test for an early and accurate differential diagnosis of malakoplakia.

Megalocytic interstitial nephritis shows histologic changes that are similar to those of renal parenchymal malakoplakia, but the lesions are usually confined to the renal cortex, and Michaelis-Gutmann bodies are less prevalent. Some authors believe that megalocytic interstitial nephritis and renal parenchymal malakoplakia represent two ends of the spectrum of a similar process (Hartman, 1985a; Ravel, 1967; Garrett and McClure, 1982).

The pathogenesis of malakoplakia is unknown. Almost all reported cases are associated with gram-negative urinary tract infections, and one of the most popular theories is that malakoplakia results from incomplete resolution of the bacterial infection. Hematogenous dissemination of *E. coli* and subsequent foci of renal parenchymal malakoplakia may explain the finding of lesions limited to the cortex and medulla without renal pelvic or lower tract disease. Patients with renal malakoplakia limited to the renal papilla or associated with contiguous renal pelvic involvement or obstructive uropathy probably have had an ascending infection (Hartman, 1985a).

The demonstration of bacteria within phagocytic vacuoles of histiocytes suggests an inability of the histiocyte to digest the bacteria. It is not clear, however, why a histiocytic response is found instead of the usual polymorphonuclear leukocyte response. The association of malakoplakia with debilitating diseases such as sarcoidosis, diabetes mellitus, and tuberculosis suggests an immunologic defect as a prerequisite for its development. Defective monocyte function has been demonstrated in one patient with malakoplakia (Abdou et al., 1977). In this case phagocytosis was normal, but complete degradation of the bacteria was impossible because of low levels of cyclic guanine monophosphate (cGMP), which may have resulted in decreased lysosomal degradation and inability of the cell to release the lysosomal enzymes. This defect appears to be reversible by cholinergic agonists that cause accumulation of cGMP in monocytes and enhance chemotaxis, but this finding has yet to be confirmed. At present malakoplakia should be considered an unusual inflammatory lesion resulting from altered host macrophage or histiocytic response.

Clinical Findings

The diagnosis of renal parenchymal malakoplakia is difficult but should be suspected if there is marked enlargement of the kidney in the presence of urinary tract infection. Renal parenchymal malakoplakia usually affects middle-aged women with recurrent urinary tract infections. The most common presenting signs and symptoms are fever, flank pain, or a palpable flank mass (Stanton and Maxted, 1981; Hartman, 1985a). The most common pathogen is *E. coli*, which accounts for about 75% of the cases. *Aerobacter aerogenes*, *Klebsiella pneumonae*, or *Proteus* are cultured less frequently (Garrett and McClure, 1982; McClure, 1983). In addition, *E. coli* cultures have been obtained from the resected kidney, perinephric abscess, or blood or CSF of affected patients. Renal failure without evidence of obstruction is not uncommon when malakoplakia is bilateral or occurs in a solitary kidney.

The association of some cases of malakoplakia with altered immune states such as transplantation, sarcoidosis, hemolytic uremic syndrome, tuberculosis, steroid therapy, lymphoma, leukemia, alcoholism, or emaciation is common (Hartman, 1985a; Cadnapaphornchai et al., 1978). Nevertheless, many patients in whom malakoplakia occurs have no recognized defects.

Radiologic Findings

Multifocal malakoplakia on excretory urography typically presents as enlarged kidneys with multiple filling defects (Fig 22–28,A). Renal calcification, lithiasis, and hydronephrosis are absent. The multifocal nature is best appreciated using ultrasonography, CT, or arteri-

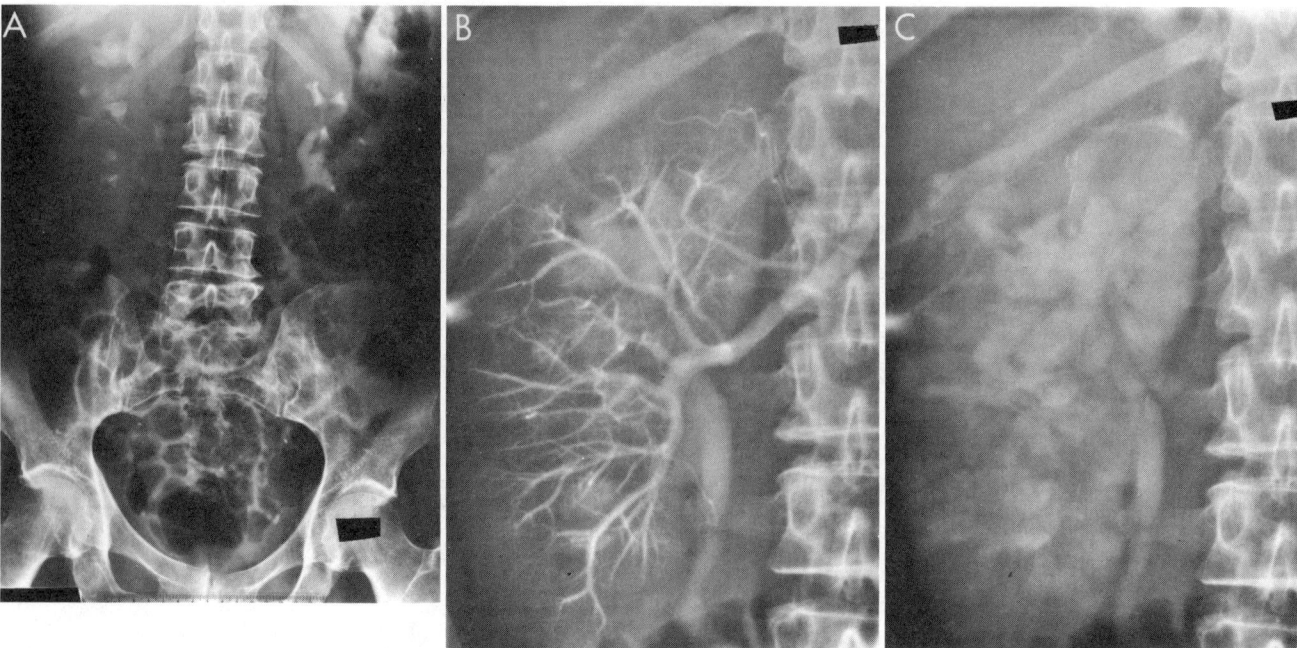

FIG 22–28.
Multifocal renal parenchymal malakoplakia. **A,** excretory urogram. The right kidney is enlarged (16.5 cm) with dilatation of the upper pole calyces and poor filling of the renal pelvis. **B,** early angiogram. Separation of the intrarenal vessels without neovascularity. **C,** angiographic nephrogram. Multiple irregular filling defects located primarily within the cortex. (Courtesy of Charles E. Bickham Jr, MD, Bethesda, Md. From Hartman DS, Davis CJ, Lichtenstein JE, et al: Renal parenchymal malakoplakia. *Radiology* 1980; 136:33–42. Used with permission.)

ography. Sonographic examination may demonstrate renal enlargement and distortion of the central echo complex. The masses are often confluent, resulting in an overall increase in the echogenicity of the renal parenchyma (Hartman et al., 1980). On CT, the foci of malakoplakia are less dense than the surrounding enhanced parenchyma (Hartman, 1985a). Arteriography typically reveals a hypovascular mass with peripheral neovascularity (Fig 22–28,B and C) (Cavins and Goldstein, 1977; Trillo et al., 1977).

Unifocal malakoplakia on IV urography presents as a noncalcified mass that is indistinguishable from other inflammatory or neoplastic lesions. Ultrasound and CT may demonstrate a solid or cystic structure depending on the degree of internal necrosis. Angiography may demonstrate neovascularity (Trillo et al., 1977). Extension beyond the kidney, which can occur with either multifocal or uniform malakoplakia, is best demonstrated by CT.

Differential Diagnosis

The differential diagnosis includes renal cystic disease, neoplasia, and renal inflammatory disease (Hartman, 1985a). Malakoplakia should be considered when one or more renal masses are observed, particularly in females with recurrent urinary tract infections with E. coli, altered immune response syndromes or cystoscopic evidence of malakoplakia or filling defects in the collecting system (Charboneau et al., 1980). Malakoplakia should also be suspected when these radiographic findings occur in a renal transplant patient who has persistent urinary tract infection despite appropriate antimicrobial therapy. Cystic diseases can be generally excluded by careful sonographic and CT evaluations. Renal involvement with metastatic disease or lymphomas usually occurs late in the course of the disease, which is well established. Multifocal renal cell carcinoma is most often seen in the context of von Hippel-Lindau syndrome with its other clinical manifestations. Patients with xanthogranulomatous pyelonephritis usually present with signs and symptoms of urinary tract infection. As with malakoplakia, the involved kidney is enlarged, but renal calculi and obstruction are frequent. Multiple renal abscesses are frequently associated with hematogenous dissemination owing to cardiac disease.

Management

Initial management is directed at control of the urinary tract infection, correction of the immunologic defect, and improvement of renal function if failure is present (Hartman, 1985a). Nephrectomy is usually performed for unilateral disease.

Long-term antibiotic therapy, such as rifampin, trimethoprim-sulfamethoxazole, and doxycycline, has been used successfully in approximately 10%–15% of patients with renal parenchymal malakoplakia established by biopsy (Stanton and Maxted, 1981). Dramatic improvement has also been reported in patients with intra-abdominal malakoplakia treated with cholinergic agents and ascorbic acid (Abdou et al., 1977).

The long-term prognosis appears to be related most directly with the extent of disease. When parenchymal renal malakoplakia is bilateral or occurs in the transplanted kidney, death usually occurs within six months (Deridder et al., 1977; Bowers and Cathey, 1971; Ho et al., 1979). Patients with unilateral disease usually have a long-term survival following nephrectomy.

RENAL ECHINOCOCCOSIS

Echinococcosis is a parasitic infection caused by the larval stage of the tapeworm *Echinococcus granulosus*. The disease is prevalent in dogs, sheep, cattle, and humans in South Africa, Australia, New Zealand, Mediterranean countries (especially Greece), and some parts of the Soviet Union. In the United States the disease is rare, but it is found in immigrants from Eastern Europe or other foreign endemic areas or as an indigenous infection among native Americans in the southwest and Eskimos (Plorde, 1977).

Pathogenesis and Pathology

Echinococcosis is produced by the larval form of the tapeworm, which in its adult form resides in the intestine of the dog, the definitive host. The adult worm is 3–9 mm long. The ova in the feces of the dog contaminate grass and farmlands and are ingested by sheep, pigs, or man, the intermediate hosts. Larva hatch, penetrate venules in the wall of the duodenum, and are carried by the bloodstream to the liver. Those larva that escape the liver are next filtered by the lungs. Approximately 3% of the organisms that escape entrapment in the liver and lungs may then enter the systemic circulation and infect the kidneys. The larva undergo vesiculation, and the resultant hydatid cyst gradually develops at a rate of about 1 cm per year. Thus, the cyst may take 5–10 years to reach pathologic size.

Echinococcal cysts of the kidney are usually single and located in the cortex (Nabizadeh et al., 1983). The wall of the hydatid cyst has three zones: a peripheral zone of fibroblasts derived from tissues of the host becomes the adventitia and may calcify, an intermediate laminated layer becomes hyalinized, and a single inner layer is composed of nucleated epithelium and is called the germinal layer. The germinal layer gives rise to brood capsules which increase in number, become vacuolated and remain attached to the germinal membrane by a pedicle. New larva (scolices) develop in large numbers from the germinal layer within the brood capsule (Fig 22–29). The hydatid cyst is also filled with fluid. When brood capsules detach they enlarge and move freely in the fluid and are then called daughter cysts. Hydatid sand is composed of free larva and daughter cysts.

Clinical Findings

The symptoms of echinococcosis are those of a slowly growing tumor. Most patients are asymptomatic or have flank mass, dull pain, or hematuria (Nabizadeh et al., 1983; Gilsanz et al., 1980). Because the cyst is focal, it rarely affects renal function. Rarely the cyst ruptures into the collecting system, and the patient may experience severe colic and passage of debris resembling grape skins in the urine (hydatiduria). The cyst may also rupture into an adjacent viscus or the peritoneal cavity. The fluid is extremely antigenic (Hartman, 1985b).

Laboratory Findings

If cyst rupture occurs, the definitive diagnosis can be established by identifying daughter cysts in the urine or by identifying the laminated wall of the cyst (Sparks et al., 1976).

Fewer than half of the patients have eosinophilia. The most reliable diagnostic test uses partially purified hydatid *arc* 5 antigens in a double diffusion test (Coltorti and Varela-Diaz, 1978). Complement fixation, hemagglutination, and the Casoni intradermal skin tests are less reliable but when combined are positive in about 90% of patients (Sparks et al., 1976).

Radiologic Findings

Excretory urography typically shows a thick-walled cystic mass, occasionally calcified (Buckley et al., 1985). If the cyst ruptures into the collecting system, daughter cysts may be outlined in the pelvis as an irregular mass or as multiple solitary lesions (Gilsanz et al., 1980). Occasionally direct filling of the cyst with contrast medium occurs.

Ultrasonography and CT are useful in characterizing the mass. Ultrasonography usually demonstrates a multicystic or multiloculated mass. A sudden change in position may demonstrate bright falling echoes corresponding to hydatid sand, which can be observed during real-time evaluation of a hydatid cyst (Saint Martin and Chiesa, 1984).

On CT several patterns of renal echinococcosis may be recognized. The most specific is a cystic mass with discrete round daughter cysts and a well-defined enhancing membrane (Martorana et al., 1981). The less specific pattern is that of a thick-walled multiloculated cystic mass (Gilsanz et al., 1980). The presence of daughter cysts within the mother cyst differentiates the lesion from simple renal cyst and from renal abscesses, infected cysts, and necrotic neoplasm.

Both CT and ultrasound are useful in evaluating the liver. Angiography is seldom required. Diagnostic aspiration should not be performed because of the danger of rupture and spillage of the highly antigenic cyst contents and risk of fatal anaphylaxis (Roylance et al., 1973).

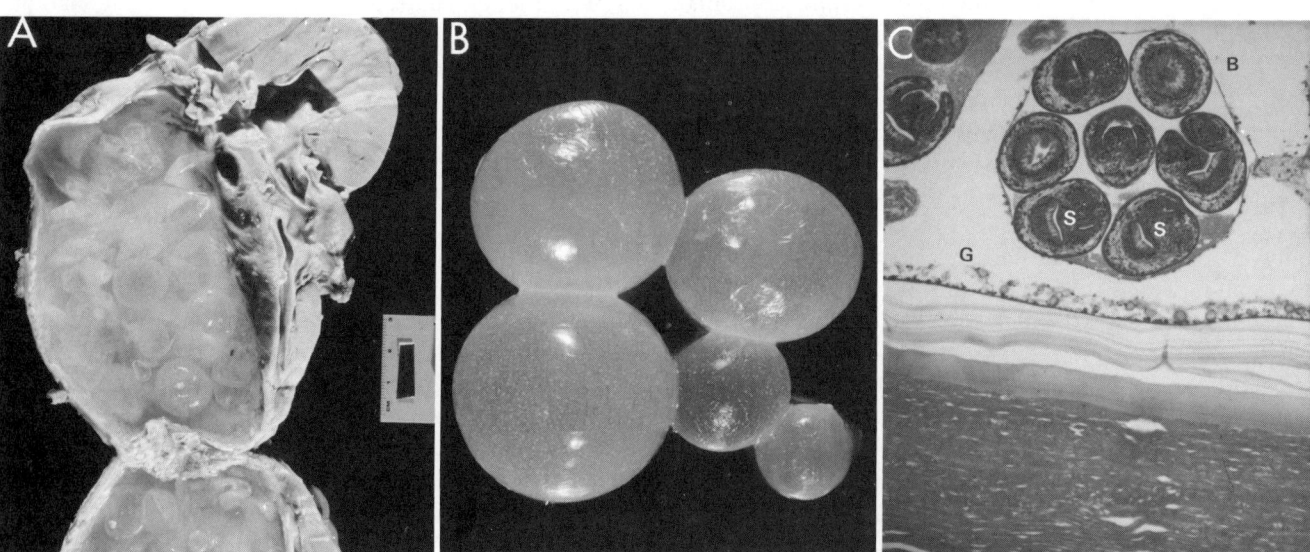

FIG 22–29.
Echinococcosis. **A,** gross specimen. A cystic mass 7 × 11 cm in lower pole. Smaller daughter cysts are identified within the larger cystic mass. **B,** gross specimen. Daughter cysts represent brood capsules that have detached and move freely. **C,** photomicrograph. Brood capsules *(B)* arising from the germinal layer *(G)* contain viable and degenerating scolices *(S)*. (From Hartman DS: Radiologic pathologic correlation of the infectious granulomatous diseases of the kidney: III and IV. *Monogr Urol* 1985b; 6:26. Used with permission.)

Management

The prognosis of echinococcosis is good but depends on the site and size of the cysts. Although there have been preliminary reports on the use of mebendazole in the treatment of hydatid disease, the results have been less than satisfactory (editorial, *Br Med J*, 1979).

Surgery remains the mainstay of treatment of renal echinococcosis. The cyst should be removed without rupture to reduce the chance of seeding and recurrence. If the cyst wall is calcified, the larva are probably dead, and the risk of seeding is low, although a daughter cyst may be viable. If the cyst may rupture or cannot be removed and marsupialization is required, the contents of the cyst initially should be aspirated and filled with a scolecidal agent such as 30% sodium chloride, 2% formalin, or 1% iodine for approximately five minutes to kill the germinal portions (Sparks et al., 1976; Nabizadeh et al., 1983).

ACKNOWLEDGMENT

Appreciation for assistance in obtaining and preparing the figures is gratefully extended to Robert Vogelzang, M.D., Assistant Professor of Radiology, Northwestern University Medical School and Northwestern Memorial Hospital; David S. Hartman, M.D., C.D.R., M.C., U.S.N., Chairman and Registrar, Department of Radiologic Pathology, Armed Forces Institute of Pathology; Herbert Sommers, M.D., Professor of Pathology, Northwestern University Medical School; and Tariq Murad, M.D., Professor of Pathology, Northwestern University Medical School and Director of Surgical Pathology, Northwestern Memorial Hospital.

REFERENCES

1. Abbott CD: Neonatal bacteriuria; a prospective study in 1,460 infants. *Br Med J* 1972; 1:267.
2. Abdou NI, NaPombejara C, Sagawa A, et al: Malakoplakia: Evidence for monocytic lysosomal abnormality correctable by cholinergic agonist in vitro and in vivo. *N Engl J Med* 1977; 297:1413.
3. Akerlund AS, Ahlstedt S, Hanson LA, et al: Antibody response in urine serum against *Escherichia coli* O antigen in childhood urinary tract infection. *Acta Pathol Microbiol Scand* 1979; 87:29.
4. Anderson KA, McAninch JW: Renal abscesses: Classification and review of 40 cases. *Urol* 1980; 16:333.
5. Angell JC, Smith I: Renal malakoplakia with perinephric extension. *Br J Urol* 1968; 40:429.
6. Asscher AW: *The Challenge of Urinary Tract Infection.* New York, Academic Press, 1980.
7. Asscher AW, McLachlan MSF, Vierrier Jones R, et al: Screening for asymptomatic urinary-tract infection in schoolgirls. *Lancet* 1973; 2:1.
8. Badenock AW: Reparative surgery for tuberculous stricture of the ureter. *X Congress Society International D Urology* 1961; 1:163.
9. Ballesteros JJ, Faus R, Gironella J: Preoperative diagnosis of renal xanthogranulomatosis by serial urine cytology: Preliminary report. *J Urol* 1980; 124:9.
10. Bergstrom T, Lincoln K, Orskov F, et al: Studies of urinary tract infections in infancy and childhood: VIII. Reinfection vs. relapse in recurrent urinary tract infections: Evaluation by means of identification of infecting organisms. *J Pediatr* 1967; 71:13.
11. Borthiwick WM: Genitourinary tuberculosis. *Tubercle* 1956; 37:120.
12. Bowers JH, Cathey WJ: Malakoplakia of the kidney with renal failure. *Am J Clin Pathol* 1971; 55:765.
13. Brenbridge AN, Buschi AJ, Cochrane JA, et al: Renal emphysema of the transplanted kidney: Sonographic appearance. *AJR* 1979; 132:656.
14. Brumfitt W, Gruneberg RN, Leigh DA: Bacteriuria in pregnancy, with reference to prematurity and long-term effects on the mother, in *Symposium on Pyelonephritis.* Edinburgh, E & S Livingstone, 1967, p 20.
15. Buckley RJ, Smith S, Herschorn S, et al: Echinococcal disease of the kidney presenting as a renal filling defect. *J Urol* 1985; 133:660.
16. Busch R, Huland H: Correlation of symptoms and results of direct bacterial localization in patients with urinary tract infections. *J Urol* 1984; 132:282.
17. Butnick R: Xanthogranulomatous pyelonephritis: An unusual case. *J Urol* 1971; 106:815.
18. Cadnapaphornchai P, Rosenberg BF, Taher S, et al: Renal parenchymal malakoplakia: An unusual cause of renal failure. *N Engl J Med* 1978; 299:1110.
19. Callea F, Van Damme B, Desmet VJ: Alpha-1-antitrypsin in malakoplakia. *Virchows Arch (Pathol Anat)* 1982; 395:1.
20. Callen PW: Computed tomographic evaluation of abdominal and pelvic abscesses. *Radiol* 1979; 131:171.
21. Campbell MF: Perinephric abscess. *Surg Gynecol Obstet* 1930; 51:654.
22. Cattell WR, Charlton CAC, McSherry A, et al: *The Localization of Urinary Tract Infection and Its Relationship to Relapse, Reinfection and Treatment.* London, Oxford University Press, 1973.
23. Cavins JA, Goldstein AMB: Renal malakoplakia. *Urology* 1977; 10:155.
24. Charboneau JW, Hattery RR, Williamson B, et al: Malakoplakia of the urinary tract with renal parenchymal involvement. *Urol Radiol* 1980; 2:89.
25. Claridge M: Ureteric obstruction in tuberculosis. *Urology* 1970; 42:688.
26. Clark H, Ronald AR, Cutler RE, et al: The correlation between site of infection and maximal concentrating ability in bacteriuria. *J Infect Dis* 1969; 120:47.
27. Coltorti EA, Varela-Diez VM: Detection of antibodies against Echinococcus granulosus arc 5 antigens by double diffusion test. *Trans R Soc Trop Med Hyg* 1978; 72:226.
28. Conrad MR, Bregman R, Kilman WJ: Ultrasonic recognition of parenchymal gas. *AJR* 1979; 132:395–399.

29. Corigliano B, Leedom JM: Renal tuberculosis: Part 2, in Massry SG, Glassock RJ (eds): *Textbook of Nephrology.* Baltimore, Williams & Wilkins Co, 1983, p 6.73.

30. Corriere JN, Sandler CM: The diagnosis and immediate therapy of acute renal and perirenal infections. *Urol Clin North Am* 1982; 9:219.

31. Cotran RS: Experimental pyelonephritis, in Rouiller C, Muller AF (eds): *The Kidney.* New York, Academic Press, 1969, vol 2, pp 269–361.

32. Davidson AG, Talner LB: Late sequelae of adult onset bacterial nephritis. *Radiology* 1978; 127:367.

33. DeNavasquez S: Experimental pyelonephritis in the rabbit produced by staphylococcal infection. *J Pathol* 1950; 62:429.

34. Deridder PA, Koff SA, Gikas PW, et al: Renal malakoplakia. *J Urol* 1977; 117:428.

35. Duguid JP, Gillies RR: Fimbriae and adhesive properties in dysentery bacilli. *J Pathol* 1957; 74:397.

36. Editorial: Medical treatment for hydatid disease. *Br Med J* 1979; 2:563.

37. Elder JS: Xanthogranulomatous pyelonephritis and gas forming infections of the urinary tract. *AUA Update* 1984; lesson 31, vol III, p 2.

38. Elder JS, Marshall FF: Focal xanthogranulomatous pyelonephritis in adulthood. *Johns Hopkins Med J* 1980; 146:141.

39. Eykyn S, Lloyd-Davies RW, Shuttleworth KED, et al: The localization of urinary tract infection by ureteric catheterization. *Invest Urol* 1972; 9:271.

40. Fagerholm M: Case of the autumn season. *Semin Ultrasound* 1983; 4:145.

41. Fairley KF: The routine determination of the site of infection in the investigation of patients with urinary tract infection, in Kincaid-Smith P, Fairley KF (eds): *Renal Infection and Renal Scarring.* Melbourne, Mercedes Publishing Co, 1972, p 107.

42. Fairley KF, Bond AG, Adey FD: The site of infection in pregnancy bacteriuria. *Lancet* 1966; 1:939.

43. Fairley KF, Bond AG, Brown RB, et al: Simple test to determine the site of urinary tract infection. *Lancet* 1967; 2:427.

44. Fairley KF, Grounds AD, Carson NE, et al: Site of infection in acute urinary-tract infection in general practice. *Lancet* 1971; 2:615.

45. Farmer ER, Heptinstall RH: Chronic non-obstructive pyelonephritis: A reappraisal, in Kincaid-Smith P, Fairley KF (eds): *Renal Infection and Renal Scarring.* Melbourne, Mercedes Publishing Co, 1972, pp 233–236.

46. Fawcett HD, Goodwin DA, Lantieri RL: In-111-leukocyte scanning in inflammatory renal disease. *Clin Nucl Med* 1981; 6:237.

47. Fernandez JA, Miles BJ, Buck AS, et al: Renal carbuncle: Comparison between surgical open drainage and closed percutaneous drainage. *Urology* 1985; 25:142.

48. Fiegler W: Ultrasound in acute renal inflammatory lesions. *Eur J Radiol* 1983; 3:354.

49. Finn DJ, Palestraint AM, DeWolf WC: Successful percutaneous management of renal abscess. *J Urol* 1982; 127:425.

50. Fischman NH, Roberts JA: Clinical studies in acute pyelonephritis: Is there a place for renal quantitative camera study? *J Urol* 1982; 128:452.

51. Flechner SM, Gow JG: Role of nephrectomy in the treatment of non-functioning or very poorly functioning unilateral tuberculous kidney. *J Urol* 1980; 123:822.

52. Flynn JT, Molland EA, Paris AMI, et al: The underestimated hazards of xanthogranulomatous pyelonephritis. *Br J Urol* 1979; 51:443.

53. Forland M, Thomas V, Shelokov A: Urinary tract infections in patients with diabetes mellitus: Studies on antibody coating of bacteria. *JAMA* 1977; 238:1924.

54. Freedman LR: Experimental pyelonephritis: XIII. The ability of water diuresis to increase susceptibility of *E. coli* bacteriuria in normal rat. *Yale J Biol Med* 1967a; 39:255.

55. Freedman LR: Chronic pyelonephritis at autopsy. *Ann Intern Med* 1967b; 66:697.

56. Freedman LR: Interstitial renal inflammation, including pyelonephritis and urinary tract infection, in Earley LE, Gottschalk CW (eds): *Strauss and Welt's Diseases of the Kidney,* ed 3. Boston, Little, Brown & Co, 1979, vol 2, pp 817–876.

57. Freedman LR, Beeson PB: Experimental pyelonephritis: IV. Observations on infections resulting from direct inoculation of bacteria in different zones of the kidney. *Yale J Biol Med* 1958; 30:406.

58. Friedenberg MJ, Spjut HJ: Xanthogranulomatous pyelonephritis. *AJR* 1963; 90:97.

59. Gajaraj A, Victor S: Tuberculous aortoarteritis. *Clin Radiol* 1981; 32:461.

60. Garrett IR, McClure J: Renal malakoplakia: Experimental production and evidence of a link with interstitial megalocytic nephritis. *J Pathol* 1982; 136:111.

61. Gerzof SG, Gale ME: Computed tomography and ultrasonography for diagnosis and treatment of renal and retroperitoneal abscesses. *Urol Clin North Am* 1982; 9:185.

62. Gillenwater JY, Harrison RB, Kunin CM: Natural history of bacteriuria in school girls: A long-term case-control study. *N Engl J Med* 1979; 301:396.

63. Gilsanz V, Lozano G, Jimenez J: Renal hydatid cysts: Communicating with collecting system. *AJR* 1980; 135:357.

64. Godec CJ, Tsai SH, Smith SJ, et al: Diagnostic strategy in evaluation of renal abscess. *Urology* 1981; 18:535.

65. Goldman SM, Hartman DS, Fishman EK, et al: CT of xanthogranulomatous pyelonephritis: Radiologic-pathologic correlation. *AJR* 1984; 141:963.

66. Goodman M, Curry T, Russell T: Xanthogranulomatous pyelonephritis (XGP): A local disease with systemic manifestations: Report of 23 patients and review of the literature. *Medicine* 1979; 58:171.

67. Gow JG: Results of treatment in a large series of cases of genitourinary tuberculosis and the changing pattern of the disease. *Br J Urol* 1970; 42:647.

68. Gow JG: The management of genitourinary tuberculosis. *J Antimicrob Chemother* 1981; 7:590.

69. Grainger RG, Longstaff AJ, Parsons MA: Xanthogranulomatous pyelonephritis: A reappraisal. *Lancet* 1982; 1:1398.

70. Guttmann D: Follow-up of urinary tract infections in

domiciliary patients, in *Urinary Tract Infection*. Proceedings of the Second National Symposium, London, 1972. Oxford University Press, 1973, chap 8, p 62.

71. Guze LB, Beeson PB: Experimental pyelonephritis: I. Effect of ureteral ligation on the course of bacterial infection in the kidney of the rat. *J Exp Med* 1956; 104:803.

72. Hagberg L, Hull R, Hull S, et al: Contribution of adhesion to bacterial persistence in the mouse urinary tract. *Infect Immun* 1983; 40:265.

73. Hampel N, Class RN, Persky L: Value of ^{67}gallium scintigraphy in the diagnosis of localized and renal and perirenal inflammation. *J Urol* 1980; 124:311.

74. Harding A, Merz WG: Evaluation of antibody coating of yeast in urine as indicator of the site of urinary tract infection. *J Clin Microbiol* 1975, 2:222.

75. Harding GKM, Marrie TJ, Ronald AR, et al: Urinary tract infection localization in women. *JAMA* 1978; 240:1147.

76. Harris RE, Thomas VL, Shelokov A: Asymptomatic bacteriuria in pregnancy: Antibody-coated bacteria, renal function, and intrauterine growth retardation. *Am J Obstet Gynecol* 1976; 126:20.

77. Harrison RB, Shaffer HA Jr: The roentgenographic findings in acute pyelonephritis. *JAMA* 1979; 241:1718.

78. Hartman DS: Radiologic pathologic correlation of the infectious granulomatous diseases of the kidney: Parts I and II. *Monogr Urol* 1985a; 6:3.

79. Hartman DS: Radiologic pathologic correlation of the infectious granulomatous diseases of the kidney: Parts III and IV. *Monogr Urol* 1985b; 6:26.

80. Hartman DS, Davis CJ, Lichtenstein JE, et al: Renal parenchymal malakoplakia. *Radiology* 1980; 136:33.

81. Hartman DS, Sanders RC, Davis CJ: Xanthogranulomatous pyelonephritis: Sonographic-pathologic correlation of 16 cases. *J Ultrasound Med* 1984; 3:481.

82. Hawes S, Whigham T, Ehrmann S, et al: Emphysematous pyelonephritis. *Infect Surg* 1983; 2:191.

83. Hellerstein S, Duggan E, Welchert E, et al: Serum C-reactive protein and the site of urinary tract infections. *J Pediatr* 1982; 100:21.

84. Heptinstall RH: *Pathology of the Kidney*, ed 2. Boston, Little, Brown & Co, 1974.

85. Ho Khang-Loon, Rassekh ZS, Nam SH: Bilateral renal malakoplakia. *Urology* 1979; 13:321.

86. Hodson CJ, Wilson S: Natural history of chronic pyelonephritis scarring. *Br Med J* 1965a; 2:191.

87. Hodson CJ: Coarse pyelonephritis scarring or "atrophic pyelonephritis" *Proc R Soc Med* 1965b; 58:785.

88. Hodson CJ: The radiological contribution toward the diagnosis of chronic pyelonephritis. *Radiology* 1967; 88:857.

89. Hodson CJ, Edwards D: Chronic pyelonephritis and vesicoureteral reflux. *Clin Radiol* 1960; 11:219.

90. Hodson CJ, Maling TMJ, McManamon PJ, et al: The pathogenesis of reflux nephropathy (chronic atrophic pyelonephritis). *Br J Radiol* 1975; 13(suppl):1–26.

91. Hooton TM, O'Shaughnessy EJ, Clowers D, et al: Localization of urinary tract infection in patients with spinal cord injury. *J Infect Dis* 1984; 150:85.

92. Horne NW, Tulloch WS: Conservative management of renal tuberculosis. *Br J Urol* 1975; 47:481.

93. Hoverman IV, Gentry LO, Jones DW, et al: Intrarenal abscess: Report of 14 cases. *Arch Intern Med* 1980; 140:914.

94. Huland H, Busch R: Chronic pyelonephritis as a cause of end stage renal disease. *J Urol* 1982; 127:642.

95. Innes JA: Non respiratory tuberculosis. *J R Coll Physiol (Lond)* 1981; 15:227.

96. Jodal U, Lindberg U, Lincoln K: Level diagnosis of symptomatic urinary tract infections in children. *Acta Paediatr Scand* 1975; 64:201.

97. Jones SR, Smith JW, Sanford JP: Localization of urinary tract infections by determination of antibody-coated bacteria in urine sediment. *N Engl J Med* 1974; 290:591.

98. Jones SR: Prostatitis as cause of antibody-coated bacteria in urine. *N Engl J Med* 1974; 291:365.

99. Jones SR: Antibody-coated bacteria in urine (letter). *N Engl J Med* 1976; 295:1380.

100. Kallenius G, Mollby R, Svenson SB, et al: The P^k antigen as receptor for the haemagglutinin of pyelonephritic *Escherichia coli*. *FEMS Microbiol Lett* 1980a; 7:297.

101. Kallenius G, Mollby R, Svenson SB, et al: Identification of a carbohydrate receptor recognized by uropathogenic *Escherichia coli*. *Infection* 1980b; 8:298.

102. Kallenius G, Mollby R, Svenson SB, et al: Occurrence of P-fimbriated *Escherichia coli* in urinary tract infections. *Lancet* 1981; 2:1369.

103. Kerr WK: Tuberculosis of the genitourinary tract (abstract). *Br J Urol* 1977; 49:237.

104. Khalyl-Mawad J, Greco MA, Schinella RA: Ultrastructural demonstration of intracellular bacteria in xanthogranulomatous pyelonephritis. *Hum Pathol* 1982; 13:41.

105. Kim DS, Woesner ME, Howard TF, et al: Emphysematous pyelonephritis demonstrated by computed tomography. *Am Roentgen Ray Soc* 1979; 132:287–288.

106. Kincaid-Smith P, Bullen M: Bacteriuria in pregnancy. *Lancet* 1965; 1:395.

107. Kleeman SEJ, Freedman LB: The finding of chronic pyelonephritis in males and females at autopsy. *N Engl J Med* 1960; 263:988.

108. Kucers AM, Bennett N: *The Use of Antibiotics*, ed 3. London, William Heinemann Medical Books Ltd, 1979, p 77.

109. Kulkarni SH, Kolhatkar RK, Ranabhise AM, et al: Adenocarcinoma of the renal pelvis with associated tuberculosis of the kidney. *Indian J Cancer* 1981; 18:229.

110. Kunin CM, Deutscher R, Paquin AJ: Urinary tract infection in school children: An epidemiologic, clinical and laboratory study. *Medicine* 1964; 43:91.

111. Latham RH, Stamm WE: Role of fimbriated *Escherichia coli* in urinary tract infections in adult women: Correlation with localization studies. *J Infect Dis* 1984; 149:835.

112. Lattimer JK: Current concepts of renal tuberculosis. *N Engl J Med* 1965; 273:208.

113. Lattimer JK, Wechsler MW: Editorial comment: Surgical management of nonfunctioning tuberculous kidneys. *J Urol* 1980; 124:191.

114. Lee JKT, McClennan BL, Melson GL, et al: Acute focal

bacterial nephritis: Emphasis on gray scale sonography and computed tomography. *AJR* 1980; 135:87.

115. Leffler H, Svanborg-Eden C: Chemical identification of a glycosphingolipid receptor for *Escherichia coli* attaching to human urinary tract epithelial cells and agglutinating erythrocytes. *FEMS Microbiol Lett* 1980; 8:127.

116. Lester TW: Extrapulmonary tuberculosis. *Clin Chest Med* 1980; 1:219.

117. Levin R, Burbige KA, Abramson S, et al: The diagnosis and management of renal inflammatory processes in children. *J Urol* 1984; 132:718.

118. Lindner LE, Jones RN, Haber MH: A specific urinary cast in acute pyelonephritis. *Am J Clin Pathol* 1980; 73:809.

119. Lomberg H, Hanson LA, Jacobsson B, et al: Correlation of P blood group, vesicoureteral reflux, and bacterial attachment in patients with recurrent pyelonephritis. *N Engl J Med* 1983; 308:1189.

120. Lorin MI, Hsu KHF, Jacob SC: Treatment of tuberculosis in children. *Pediatr Clin North Am* 1983; 30:333.

121. Lou TY, Teplitz C: Malakoplakia: Pathogenesis and ultrastructural morphogenesis: A problem of altered macrophage (phagolysosomal) response. *Hum Pathol* 1974; 5:191.

122. Love L, Baker D, Ramsey R: Gas-producing perinephric abscess. *AJR* 1973; 119:783.

123. Lowe J, Pfau A, Stein H: Reactivated musculoskeletal tuberculosis with concomitant asymptomatic genitourinary infection. *Isr J Med Sci* 1983; 19:262.

124. Lyons RW, Long JM, Litton B, et al: Arteriography and antibiotic therapy of a renal carbuncle. *J Urol* 1972; 107:524.

125. Mabeck CE: Treatment of uncomplicated urinary tract infection in nonpregnant women. *Postgrad Med J* 1972; 48:69.

126. Mackie GG, Stephens FD: Duplex kidneys: A correlation of renal dysplasia with position of the ureteral orifice. *J Urol* 1975; 114:274.

127. Malek RS, Elder JS: Xanthogranulomatous pyelonephritis: A critical analysis of 26 cases and of the literature. *J Urol* 1978; 119:589.

128. Mallinson JW, Fuller RW, Levison DA, et al: Diffuse interstitial renal tuberculosis—an unusual cause of renal failure. *Br J Med* 1981; 198:137.

129. Marrie TJ, Harding GKM, Ronald AR, et al: Influence of mucoidy or antibody-coating of Pseudomonas aeruginosa. *J Infect Dis* 1979; 139:357.

130. Martorana G, Gilberti C, Pescatore D: Giant echinococcal cyst of the kidney associated with hypertension evaluated by computerized tomography. *J Urol* 1981; 126:99.

131. McClure J: Malakoplakia. *J Pathol* 1983; 140:275.

132. McClurg FV, D'Agostino AN, Martin JH, et al: Ultrastructural demonstration of intracellular bacteria in three cases of malakoplakia of the bladder. *Am J Clin Pathol* 1973; 60:780.

133. McDonald GS: Xanthogranulomatous pyelonephritis. *J Pathol* 1981; 133:203.

134. McLachlan MSF, Meller S, Verrier-Jones ER, et al: The urinary tract in school girls and covert bacteriuria. *Arch Dis Child* 1975; 50:253.

135. Mering JH, Kaplan GW, McLaughlin AP: Xanthogranulomatous pyelonephritis. *Urology* 1973; 1:338.

136. Michaelis L, Gutmann C: Ueber Einschlusse in Blasentumoren. *Z Klin Med* 1902; 47:208–215.

137. Miller T, Phillips S: Pyelonephritis: The relationship between infection, renal scarring and antimicrobial therapy. *Kidney Int* 1981; 19:654.

138. Murphy DM, Fallon B, Lane V, et al: Tuberculous stricture of ureter. *Urology* 1982; 20:382.

139. Murray T, Goldberg M: Chronic interstitial nephritis: Etiologic factors. *Ann Intern Med* 1975; 82:453.

140. Nabizadeh I, Morehouse HT, Freed SZ: Hydatid disease of the kidney. *Urology* 1983; 22:176.

141. Narayana AS: Overview of renal tuberculosis. *Urology* 1982; 3:231.

142. Bhargava BN, Mehta FS, et al: Pyeloduodenal fistula secondary to renal tuberculosis. *J R Coll Surg Edinb* 1982; 27:242.

143. Ocon J, Novillo R, Villavicencio H, et al: Renal tuberculosis and hypertension: Value of the renal vein renin ratio. *Eur Urol* 1984; 10:114.

144. O'Flynn JD: Genitourinary tuberculosis. *Urol Dig* 1979; 18:25.

145. O'Flynn JD: Surgical treatment of genitourinary tuberculosis. *Br J Urol* 1970; 42:667.

146. Ooi BS, Chen NU: Prevalence and site of bacteriuria in diabetes mellitus. *Postgrad Med J* 1974; 50:497.

147. Osterhage HR, Fischer V, Haubensak K: Positive histological tuberculous findings despite stable sterility of the urine on culture: Results of 111 nephrectomies and partial nephrectomies. *Eur Urol* 1980; 6:116.

148. Overgaard Nielsen H, Lorentzen M: Xanthogranulomatous pyelonephritis: An immunohistochemical and ultrastructual study on the occurrence of bacteria and bacterial antigen. *Urol Int* 1981; 36:335.

149. Pawlowski JM, et al: Chronic pyelonephritis in morphologic and bacteriologic study. *N Engl J Med* 1963; 268:965.

150. Pitts JC, Peterson NE, Conley MC: Calcified functionless kidney in a 51 year old man. *J Urol* 1981; 125:398.

151. Plorde LL: Echinococciasis, in *Harrison's Principles of Internal Medicine*, ed 8. New York, McGraw-Hill, 1977, pp 1117–1118.

152. Povysil C, Konickova L: Experimental xanthogranulomatous pyelonephritis. *Invest Urol* 1972; 9:313.

153. Ransley PG, Risdon RA: Reflux and renal scarring. *Br J Radiol* 1978; 14(suppl):1–35.

154. Ravel R: Megalocytic interstitial nephritis: An entity probably related to malakoplakia. *Am J Clin Pathol* 1967; 47:781.

155. Reidenberg MM, Shear L, Cohen RV: Estimation of isoniazid in patients with impaired renal function. *Am Rev Respir Dis* 1973; 108:1426.

156. Resnick MI, Older RA: *Diagnosis of Genitourinary Disease.* New York, Thieme-Stratton, 1982.

157. Robbins SL, Tucker AW: The cause of death in diabetes. *N Engl J Med* 1944; 231:865.

158. Ronald AR, Cutler RE, Turck M: Effect of bacteriuria on renal concentrating mechanisms. *Ann Intern Med* 1969; 70:723.

159. Rosenfeld AT, Glickman MG, Taylor KJW, et al: Acute focal bacterial nephritis (acute lobar nephronia). *Radiol* 1979; 132:553.

160. Roylance J, Davies ER, Alexander WD: Translumbar puncture of a renal hydatid cyst. *Br J Radiol* 1973; 46:960.

161. Ruben RH, Fein LST, Jones SR, et al: Single dose amoxacillin therapy for urinary tract infection, multicenter trial using antibody-coated bacteria localization technique. *JAMA* 1980; 244:561.

162. Saedd SM, Fine G: Xanthogranulomatous pyelonephritis. *Am J Clin Pathol* 1963; 39:616.

163. Saint Martin G, Chiesa JC: 'Falling snowflakes': An ultrasound sign of hydatid sand. *J Ultrasound Med* 1984; 3:257.

164. Salvatierra O Jr, Bucklew WB, Morrow JW: Perinephric abscess: A report of 71 cases. *J Urol* 1967; 98:296.

165. Savage DCL, Wilson MI, McHardy M, et al: Covert bacteriuria of childhood: A clinical and epidemiological study. *Arch Dis Child* 1973; 48:8.

166. Schainuck LT, Fouty R, Cutler RE: Emphysematous pyelonephritis: A new case and review of previous observations. *Am J Med* 1968; 44:134.

167. Schechter H, Leonard CD, Scribner BH: Chronic pyelonephritis as a cause of renal failure in dialysis candidates: Analysis of 173 patients. *JAMA* 1971; 216:514.

168. Schiff M Jr, Glickman M, Weiss RM: Antibiotic treatment of renal carbuncle. *Ann Intern Med* 1977; 87:305.

169. Schlagenhaufer F: Uber eigentumliche Staphylomykosen der Nieven und des pararenalen Bindegewebes. *Frankfurt Z Pathol* 1916; 19:139.

170. Schneider M, Becker JA, Staiano S, et al: Sonographic-radiographic correlation of renal and perirenal infections. *AJR* 1976; 127:1007.

171. Schoborg TW, Saffos RO, Urdaneta L, et al: Xanthogranulomatous pyelonephritis associated with renal carcinoma. *J Urol* 1980; 124:125.

172. Scullin DR, Hardy R: Malakoplakia of the urinary tract with spread of the abdominal wall. *J Urol* 1972; 107:908.

173. Schulman CC, Denis R: Xanthogranulomatous pyelonephritis in childhood (letter.) *J Urol* 1977; 117:398.

174. Segura JW, Kelalis PP: Localized renal parenchymal infections in children. *J Urol* 1973; 109:1029.

175. Silver TM, Kass EJ, Thornbury JR, et al: The radiological spectrum of acute pyelonephritis in adults and adolescence. *Radiology* 1976; 118:65.

176. Simon MB, Weinstein AJ, Pasternak MN, et al: Genitourinary tuberculosis. Clinical features in a general hospital population. *Am J Med* 1977; 63:410.

177. Smeets F, Gower PE: The site of infection in 123 patients with bacteriuria. *Clin Nephol* 1973; 1:290.

178. Smith A, Lattimer JK: Genitourinary tract involvement in children with tuberculosis. *NY State J Med* 1973; 73:2325.

179. Solomon A, Braf Z, Papo J, et al: Computerized tomog-raphy in xanthogranulomatous pyelonephritis. *J Urol* 1983; 130:323.

180. Sparks AK, Connor DH, Neafie RC: Echinococcosis, in Binford CH, Connor DH (eds): *Pathology of Tropical and Extraordinary Diseases*. Washington, DC, The Armed Forces Institute of Pathology, 1976, pp 530–533.

181. Stamey TA: *Pathogenesis and Treatment of Urinary Tract Infections*. Baltimore, Williams & Wilkins, 1980.

182. Stamey TA, Govan DE, Palmer JM: The localization and treatment of urinary tract infections: The role of bactericidal urine levels as opposed to serum level. *Medicine* 1965; 44:1.

183. Stamey TA, Pfau A: Some functional, pathologic, bacteriologic, and chemotherapeutic characteristics of unilateral pyelonephritis in man: I. Fundamental and pathologic characteristics. *Invest Urol* 1963a; 1:134.

184. Stamey TA, Pfau A: Some functional, pathologic, bacteriologic, and chemotherapeutic characteristics of unilateral pyelonephritis in man: II. Bacteriologic and chemotherapeutic characteristics. *Invest Urol* 1963b; 1:162.

185. Stanton MJ, Maxted W: Malakoplakia: A study of the literature and current concepts of pathogenesis. *J Urol* 1981; 125:139.

186. Studer UE, Weidmann P: Pathogenesis and treatment in renal tuberculosis. *Eur Urol* 1984; 10:164.

187. Svanborg-Eden C, Hanson LA, Jodal U, et al: Variable adherence to normal human urinary tract epithelial cells of *Escherichia coli* strains associated with various forms of urinary tract infections. *Lancet* 1976; 2:490.

188. Tamminen TE, Kaprio EA: The relation of the shape of renal papillae and of collecting duct openings to intra-renal reflux. *Br J Urol* 1977; 49:345.

189. Thomas V, Forland M: Antibody-coated bacteria in urinary tract infections. *Kidney Int* 1982; 21:1.

190. Thomas V, Forland M, Shelokov A: Immunoglobulin levels and antibody-coated bacteria in urines from patients with urinary tract infections. *Proc Soc Exp Biol Med* 1975; 148:1198.

191. Thomas V, Shelokov A, Forland M: Antibody-coated bacteria in the urine and the site of urinary tract infection. *N Engl J Med* 1974; 290:588.

192. Thorley JD, Jones SR, Sanford JP: Perinephric abscess. *Medicine* 1974; 53:441.

193. Timmons JW, Perlmutter AD: Renal abscess: A changing concept. *J Urol* 1976; 115:299.

194. Tolia BM, Friedenberg MJ, Spjut HJ: Xanthogranulomatous pyelonephritis. *AJR* 1963; 90:97.

195. Tolia BM, Iloreta A, Freed SZ, et al: Xanthogranulomatous pyelonephritis: Detailed analysis of 29 cases and a brief discussion of atypical presentations. *J Urol* 1981; 126:437.

196. Trillo A, Lorentz WB, Whitley NO: Malakoplakia of kidneys simulating renal neoplasm. *Urology* 1977; 10:472.

197. Turck M, Ronald AR, Petersdorf RG: Relapse and reinfection in chronic bacteriuria: II. The correlation between site of infection and pattern of recurrence in chronic bacteriuria. *N Engl J Med* 1968; 278:422.

198. Vejlsgaard R: Studies on urinary infections in diabetes: IV. Significant bacteriuria in pregnancy in relation to

age of onset, duration of diabetes, angiopathy and urological symptoms. *Acta Med Scand* 1973; 193:337.

199. Von Hansemann D: Uber Malakoplakie Der Harnblase. *Arch Pathol Anat* 1903; 173:302.

200. Waller RM, Finnerty DP, Casarella WJ: Transluminal balloon dilation of a tuberculous ureteral stricture. *J Urol* 1983; 129:1225.

201. Wechsler H, Lattimer JK: An evaluation of the current therapeutic regimen for renal tuberculosis. *J Urol* 1975; 113:760.

202. Wechsler H, Westfall M, Lattimer JK: The earliest signs and symptoms of 127 male patients with genitourinary tuberculosis. *J Urol* 1960; 83:801.

203. Weiss S, Parker F Jr: Pyelonephritis: Its relation to vascular lesions and to arterial hypertension. *Medicine* 1939; 18:221.

204. Wesson LG: Unilateral renal disease and hypertension. Nephron 1982; 31:2.

205. Wheat LJ: Infection and diabetes mellitus. *Diabetes Care* 1980; 3:187.

206. Wicks JD, Thornbury JR: Acute renal infections in adults. *Radiol Clin North Am* 1979; 17:245.

207. Wientzen RL, McCracken GH Jr, Petruska ML, et al: Localization and therapy of urinary tract infections of childhood. *Pediatrics* 1979; 63:467.

208. Winberg J, Andersen HJ, Bergstrom T, et al: Epidemiology of symptomatic urinary tract infection in childhood. *Acta Paediatr Scand [Suppl]* 1974; 252.

209. Winberg J, Bergstrom T, Jacobsson B: Morbidity, age, and sex distribution, recurrences and renal scarring in symptomatic urinary tract infection in childhood. *Kidney Int* 1975; 8(suppl):101.

210. Winberg J, Bollgren I, Kallenius R, et al: Clinical pyelonephritis and focal renal scarring: A selected review of pathogenesis, prevention and prognosis. *Pediatr Clin North Am* 1982; 29:805.

211. Winberg J, Miller T, Phillips S: Pyelonephritis: The relationship between infection, renal scarring and antimicrobial therapy. *Kidney Int* 1981; 19:654.

212. Wisnia LG, Kukol JS, DeSanta Maria JL, et al: Renal function damage in 131 cases of urogenital tuberculosis. *Urology* 1978; 11:457.

213. Wong SH, Lau WY: The surgical management of nonfunctioning tuberculous kidney. *J Urol* 1980; 124:187.

214. Yazaki T, Ishikawa A, Ogawa Y, et al: Xanthogranulomatous pyelonephritis in childhood: Case report and review of English and Japanese literature. *J Urol* 1982; 127:80.

Chapter 23

Hydronephrosis

JAY Y. GILLENWATER, M.D.

Hydronephrosis is the distention of the renal calyces and pelvis with urine due to obstruction of the outflow of urine distal to the renal pelvis. Prolonged hydronephrosis results in renal parenchymal atrophy. Elevated renal pelvic pressures and decreased renal blood flow are postulated to be mechanisms of injury and cellular atrophy. Obstructive uropathy progressively impairs all renal functions except urinary dilution. The longer and more complete the obstruction, the more severe the pathophysiologic changes become.

This chapter summarizes both the pathophysiologic changes and the operative approaches to correcting hydronephrosis. Excellent reviews of obstructive uropathy have been written by Bell (1946), Bricker (1967), Wilson (1977), Wright and Howards (1979), Klahr et al. (1977), and Gillenwater (1985). Excellent reviews of surgical techniques for hydronephrosis have been written by Blandy (1978), DeWeerd (1983), and Schaeffer and Grayhack (1985).

PATHOPHYSIOLOGY OF OBSTRUCTIVE UROPATHY

HISTORICAL ASPECTS OF HYDRONEPHROSIS

Hinman first studied experimental hydronephrosis systematically (Hinman, 1919, 1926, 1934). These studies showed that infection plus obstruction caused severe and rapid renal damage (Hinman, 1919). Histologic changes were noted after one week of obstruction, with some histologic recovery after release of 60 days of obstruction in rats (Hinman, 1919). Denervation of the kidney did not alter the course of hydronephrosis (Hinman and Hepler, 1925c). Renal arterial or venous obstruction accentuated the damage in hydronephrosis (Hinman and Hepler, 1925a, 1925b, 1926). Release of two weeks' complete ureteral obstruction with contralateral nephrectomy resulted in the return of most function to the previously obstructed kidney. The same experiment with three weeks of complete obstruction resulted in only a 50% return of function. The animals could not survive release of obstruction lasting longer than two or three weeks if the opposite kidney was removed at the time of release of the ureteral obstruction. The animals could survive release of 30–60 days of unilateral hydronephrosis if the opposite normal kidney was damaged gradually after release of the ureteral obstruction.

PATHOLOGY

Urinary tract obstruction causes proximal dilation with anatomical changes in the proximal ureter, renal pelvis, and renal parenchyma. Initially the proximal ureter and renal pelvis react with muscle hypertrophy and hyperplasia. Production of connective tissue consisting of collagen and elastic tissue occurs later and is thought to impair myogenic impulse transmission and cause disturbance of peristalsis (Cussen and Tymms, 1972; Ladofoged and Djurhuus, 1976; Djurhuus, 1976; Gee and Kiviat, 1975; Gosling and Dixon, 1974). An increase in collagen, which acts as an inelastic collar preventing distention, has been found in the obstructing segments of the ureteropelvic junction and in megaureters (Notfrey, 1971, 1972; Hanna et al., 1976).

Hydronephrosis eventually causes tubular dilation with cellular atrophy. Within seven days atrophy is seen in the distal nephron. By 14 days there is progressive dilation of the distal tubules and atrophy of the proximal tubular epithelial cells. At 28 days there is loss of 50% of the medulla with marked atrophy of the proximal tubules and thinning of the cortex. Glomerular changes are not noted before 28 days of obstruction. There is no evidence of microscopic changes in the arteries to account for the markedly reduced blood flow observed in hydronephrosis. Venous drainage is impaired, causing some of the renal damage (Hinman and Morison, 1924; Deming, 1951; Hinman, 1945; Sheehan and Davis, 1959; Strong, 1940; Shimamura et al., 1966; Altschul and Fedor, 1953; Rao and Heptinstall, 1968).

FLUID TURNOVER IN HYDRONEPHROSIS

Urine exits the renal pelvis in complete ureteral obstruction by (1) extravasation, (2) pyelolymphatic backflow, and (3) pyelovenous backflow. Replacement glomerular filtration maintains the hydronephrosis. Thus, there is an active turnover of urine in the hydronephrotic renal pelvis despite complete ureteral obstruction. Substances injected into the hydronephrotic renal pelvis have been shown to appear in the systemic circulation, confirming the dynamic state of the hydronephrosis (strychnine [Tuffier, 1894], phenolsulfonphthalein [Burns and Schwartz, 1918], dye [Morison, 1929], and indigo carmine [Hinman and Lee-Brown, 1924]). Extravasation of urine first occurs through rupture of the fornix (Hinman and Lee-Brown, 1924; Bird and Moise, 1926). Narath (1940) and Olsson (1948) studied urine backflow during ureteral obstruction. Initially, ureteral obstruction produces pyelocanalicular and pyelosinus backflow. Higher pressures lead to traumatic rupture of the calyceal fornix and egress of urine into both the lymphatic and the venous sytems. With low pressures, most of the fluid exits into the lymphatics. In chronic hydronephrosis, most of the urine exits into the renal venous system.

The quantity of urine exiting the renal pelvis after acute complete ureteral obstruction was measured at 0.06 ml/min. In chronic hydronephrosis of 6–34 days, the quantity of urine exiting the renal pelvis ranged from 0.04 to 0.16 ml/min. Replacement glomerular filtration was calculated to be 1.74 ml/min in complete ureteral obstruction of one week's duration and 0.4 ml/min after five weeks of complete ureteral obstruction (Naber and Madsen, 1974).

SUMMARY

With acute obstruction and high pressures, as during retrogrades or with ureteral calculi, most of the fluid exits the renal pelvis by extravasation at the calyceal fornix. With low-pressure obstruction, much of the fluid exits into the lymphatics. In chronic hydronephrosis, most of the fluid exits into the renal venous system.

RENAL COUNTERBALANCE, COMPENSATORY RENAL HYPERTROPHY, AND RENAL HYPOTROPHY

The concepts of renal counterbalance, compensatory renal hypertrophy and renal hypotrophy (disuse atrophy) were first introduced by Hinman in 1922. The theory of renal counterbalance is based on the premise that there is a mechanism to monitor total renal function and that the function of each kidney can be modulated up or down as appropriate. Thus, if one kidney is removed or rendered nonfunctioning by obstruction, the opposite kidney would undergo compensatory hypertrophy. If additional renal tissue is added by release of unilateral obstruction or transplantation of additional kidneys, some mechanism would modulate total renal function downward by renal hypotrophy. Renal hypertrophy and hypotrophy involve changes in function. Some misunderstandings and controversy have resulted from efforts to understand renal counterbalance in terms of renal mass instead of function.

Recent studies have provided new information about compensatory renal hypertrophy. Two forms of renal growth have been postulated: obligatory growth thought to be under the stimulus of growth hormone and compensatory growth under an unknown humoral stimulus. Compensatory renal growth includes both hypertrophy and hyperplasia (Preuss, 1983) and is not as great in older animals (Hayslett, 1983). Following unilateral ureteral obstruction, there is a bilateral increase in renal mass the first week, followed by atrophy in the obstructed kidney and continued hypertrophy in the opposite unobstructed kidney. The increased mass in the obstructed kidney may be a local response to injury (Zelman, 1983) or a response to the unknown humoral factor. In most models, the renotrophic factor is present in anephric animals. Renotrophic factor stimulates three forms of growth: embryonic growth, wound repair compensatory growth, and neoplastic growth (Patt and Houk, 1983). The renotrophic factor is thought to be humorally mediated and must be continually present to maintain compensatory growth (Malt, 1983). Similar renotrophic factors are present in both urine and serum (Harris et al., 1983). During compensatory hypertrophy, glomeruli increase in size but not in number.

Renal hypotrophy (atrophy) is most easily understood in terms of function. When a hypertrophied kidney is experimentally transplanted in a recipient with a hypertrophied kidney, both kidneys return to their previous size (Silber, 1974, 1975). When two additional kidneys were transplanted into male rats with normal kidneys, there was no change in the size of either the two transplanted kidneys or the two normal kidneys. However, total renal blood flow and glomerular filtration showed no increase over normal conditions (Rist et al., 1975). Silber (1974) found that, after transplanting an additional kidney, there was an increase in total renal mass and a 50% increase in total renal blood flow and GFR.

Studies in our laboratory showed that with release of unilateral complete obstruction the obstructed kidney regains function over the next four months. Compensatory hypertrophy persists in the contralateral kidney during that time. Total renal function is not fully recovered to control values at four months.

RETURN OF FUNCTION AFTER COMPLETE URETERAL OBSTRUCTION

Recovery of renal function after release of complete ureteral obstruction varies with the species, total time of obstruction, presence or absence of infection, degree of pyelovenous and pyelolymphatic backflow and whether or not the renal pelvis is intrarenal or extrarenal. In the dog with a normal contralateral kidney after release of two weeks' obstruction, the GFR rate can recover up to 46% of control function in 3–4 months. After release of four weeks of total ureteral obstruction, there was recovery of the GFR to 35% of control by five months. No return of function was noted after release of six weeks of total ureteral obstruction in the dog (Vaughan et al., 1971).

How long the human kidney can be completely obstructed and still regain function after release is not known for certain. Reports in the literature have been difficult to evaluate because the length and completeness of obstruction and the return of renal function are difficult to document adequately. With greater use of renal scanning and better awareness of possible ureteral obstruction after pelvic surgery, better documentation should be possible. The cases with the longest period of complete ureteral obstruction meeting the above criteria and showing some recovery of function are 56 and 69 days (Lewis and Pierce, 1962; Graham, 1962; Reisman et al., 1957).

PREDICTION OF RECOVERY POTENTIAL PRIOR TO RELEASE OF THE OBSTRUCTION

For the urologist to make the correct decision about nephrectomy or correcting an obstruction he needs to know whether the kidney will regain function after the obstruction is released. The two best means of assessing recoverability preoperatively are the placement of temporary nephrostomy tubes or the use of renal scans with sophisticated analysis. The simplest method is to place percutaneous nephrostomy tubes to relieve the obstruction and monitor creatinine clearances. If the previously obstructed kidney has not regained at least a 6–10-ml/minute creatinine clearance by 2–3 months, I do not think it merits repair if the other kidney is normal.

Prediction of recoverability using the renal scans is more difficult and has not been successful in many studies. However, evaluating cortical zones of interest by arbitrary mathematical analysis has been successful (^{99}Tc pentetic acid; DTPA [Lome et al., 1979], ^{99m}Tc DMSA [McDougal and Flanigan, 1981], iodohippurate sodium I131; ^{131}I Hippuran [Kalika et al., 1981; Gillenwater et al., 1979]). That the different scanning agents measure different renal functions has not seemed to be as important as the method of quantitative analysis of the results. ^{131}I Hippuran measures renal blood flow and correlates with the GFR. ^{99m}Tc DMSA accumulates in the cytoplasm of proximal tubular cells and is used to assess functioning cortical tissue. ^{99}Tc DTPA is eliminated by glomerular filtration and correlates with cortical renal blood flow.

INTRAPELVIC AND RENAL TUBULAR PRESSURES

NORMAL.—Normal renal pelvis pressure is 6.5 mm Hg, which slightly exceeds the intraperitoneal, bladder, and ureteral pressures. Normal proximal tubular pressure is 14 mm Hg (Michaelson, 1974).

UNILATERAL URETERAL OBSTRUCTION.—Ureteral and renal pelvic pressures after complete unilateral ureteral obstruction rise abruptly, then decrease within 24 hours to 50% of the peak values (Vaughan et al., 1970). Ureteral pressures continue to decline over the next eight weeks to 15 mm Hg despite the continued completed obstruction (Vaughan et al., 1970). Thus, static ureteral pressure measurements have never been of help in determining the degree of ureteral obstruction.

The peak ureteral pressures with acute complete ureteral obstruction vary with the hydration, osmotic diuresis, degree of ureteral contractions and amount of reabsorption. The maximal pressure from filtration is the stop-flow pressure of 15–20 mm Hg (glomerular capillary pressure [60 mm Hg] minus capillary oncotic pressure [25 to 30 mm Hg] minus hydrostatic pressure in Bowman's capsule [15 mm Hg]). The higher pressures measured during acute obstruction are a result of both filtration pressure and active muscle contractions in the ureter and renal pelvis. Ureteral pressures of 50–70 mm Hg have been measured after acute ureteral obstruction in man (Michaelson, 1974; Obniski, 1907; Backlund and Nordgren, 1966; Hinman and Morison, 1924; Murphy and Scott, 1966; Taylor and Ullman, 1961; McDonald et al., 1937). Ureteral pressure after obstruction can be further increased to 100 mm Hg by saline or mannitol diuresis (Papadopoulou et al., 1969; Vaughan et al., 1971).

Pressure transmitted back to the renal pelvis varies with the conditions of the obstruction. In acute unilateral obstruction in nondiuretic rats the ureteral and proximal tubular pressure rises to 14 mm Hg, which is normal proximal tubule pressure (Gottschalk and Mylle, 1956, 1957). In diuretic rats the maximal pressure in the ureter and proximal renal tubule rises to 40 mm Hg (Gottschalk and Mylle, 1956, 1957). Further elevation of ureteral pressure by injecting fluid to 80 mm Hg fails to transmit and does not raise proximal tubular pressure above 40 mm Hg, presumably due to compression of papillary foramina (Gottschalk and Mylle, 1956).

Within 24 hours the elevated proximal tubular pressure is below normal due to afferent arteriole constriction (Dominguez and Adams, 1958; Gottschalk and Mylle, 1956). Another consequence of elevation of ureteral pressure is that the proximal and distal tubules become leaky to creatinine, mannitol, and sucrose. Studies have shown that the aqueous junctional complexes of membranes of adjacent cells are disrupted. The permeability is restored when tubular pressures return to normal (Lorentz et al., 1972; Bulger et al., 1974).

When unilateral ureteral obstruction persists ureteral pressure slowly declines to reach normal levels in three to four weeks (Vaughan et al., 1970). Michaelson (1974) found an average decrease of intrapelvic pressures of 6.6 mm Hg after following up six patients with partial obstruction for eight weeks. While ureteral and pelvis pressures do decrease toward normal ranges with chronic obstruction, they must continue to be slightly elevated, since something is sending signals maintaining the alterations seen in renal function during obstruction. Since relief of the obstruction promptly reverses the process, another possible explanation would be that the distention and increased tension are sending the signals initiating the physiologic response.

ALTERATIONS OF RENAL FUNCTION DURING UNILATERAL OBSTRUCTION: ACUTE EFFECTS

GLOMERULAR FILTRATION RATE.—As previously described, within minutes of ureteral obstruction the hydraulic pressure of the fluid proximal to the obstruction rises. As the proximal tubular pressure rises, the GFR falls. The decrease in the GFR is due to both the rise in tubular pressure and the decrease in the area of filtering membrane when the afferent arteriole constriction begins after five hours of obstruction. Glomerular filtration rates in rats with complete obstruction were 52% at four hours, 23% at 12 hours, 4% at 24 hours and 2% at 48 hours (Harris and Gill, 1981; Provoost and Molenaar, 1981).

RENAL BLOOD FLOW.—Ipsilateral renal blood flow and ureteral pressure have a triphasic relationship during the first 24 hours of acute unilateral ureteral obstruction (Fig 23–1) (Moody et al., 1975). The initial transient (1½-hour) response is an increase in renal blood flow and ureteral pressure, indicating a preglomerular vasodilatation. Pretreatment with prostaglandin inhibitors prevents this vasodilatation, indicating that

PGE_2 and PGI_2, which dilate vessels and are produced in the kidneys, may be responsible (Allen, 1978; Schramm and Carlson, 1975; Gaudio, 1980; Ichikawa and Brenner, 1979). Nishikawa et al. (1977) and Needleman et al. (1978) measured increased production of the vasodilating prostaglandins during acute ureteral obstruction. The increase in renal blood flow caused by acute ureteral obstruction appears to be limited to cortical blood flow, with the inner cortex having the greatest increase (Solez et al., 1976). The transient increase in renal blood flow is thought to be due to a vasodilating prostaglandin release.

The second vascular phase, which consists of a decrease in renal blood flow with continuation of rising ureteral pressures, occurs 1½—5 hours after the obstruction. The postulated mechanism is a rise in postglomerular vascular resistance. During the third and chronic phase (5–18 hours), both renal blood flow and ureteral pressure decrease. This fall in renal blood flow and ureteral pressure continues chronically (Vaughan et al., 1970). Preglomerular vasoconstriction causes the decrease in renal blood flow and lower ureteral pressure. Similar preglomerular vasoconstriction is seen in a single nephron with tubular obstruction (Arendshorst et al., 1974).

The mechanism of the chronic preglomerular vasoconstriction in hydronephrosis is not known. Recent studies by Chevalier and Peach (1985) and Chevalier and Jones (1986) with experimental neonatal chronic partial ureteral obstruction in guinea pigs showed that angiotensin II may play a role in the preglomerular vasoconstriction. Enalapril inhibits both intrarenal and extrarenal converting enzymes, inhibiting formation of angiotensin II. Enalapril increased renal blood flow 83% in the partially obstructed kidney and increased the number of perfused glomeruli by 26%. It is postulated that Enalapril does not have the agonist (constrictor activity) of some of the other previously used angiotensin blockers. Earlier studies suggested that angiotensin was not involved, since renin depleted animals (Vaughan et al., 1971; Huguenin et al., 1976) and infusion of the competitive antagonist, saralasin, had no effect (Moody et al., 1977). Preglomerular vasoconstriction has been observed despite renal denervation and adrenergic blockage with both α- and β-antagonists (Vaughan et al., 1971). Thromboxane A_2 is another prostaglandin vasoconstrictor found in hydronephrotic rat kidney (Morrison et al., 1977, 1978). Imidazole, a thromboxane inhibitor, has been reported to restore renal blood flow to normal levels (Yarger et al., 1980). However, other studies by Vaughan (personal communication, 1987) have not been able to verify thromboxane A_2 as the preglomerular vasoconstrictor by chemi-

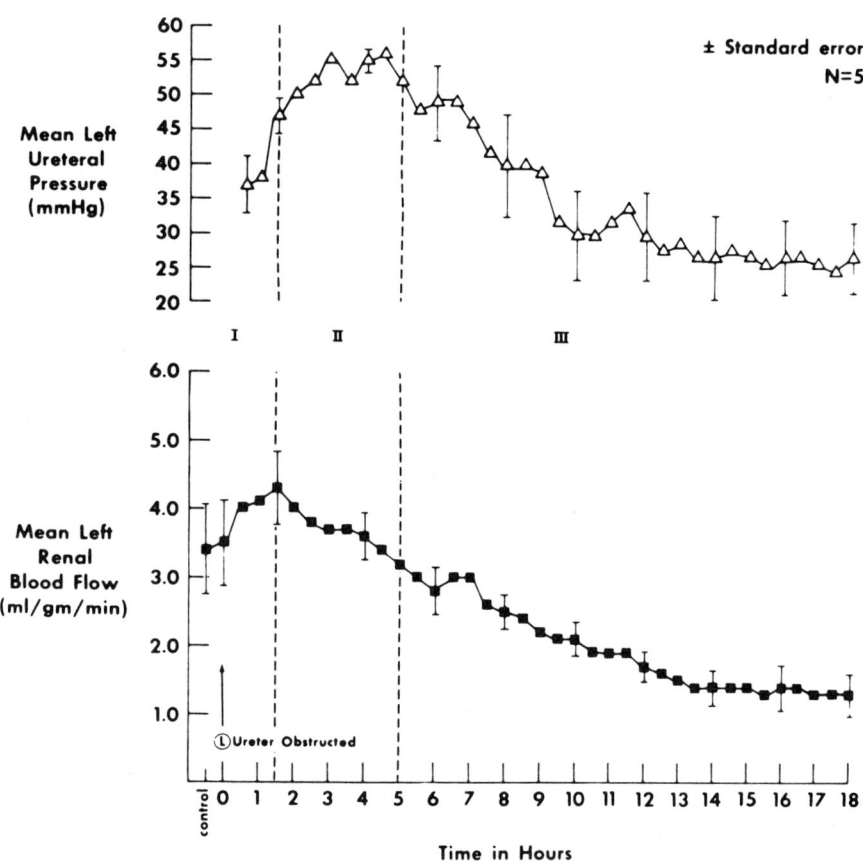

FIG 23–1.
Triphasic relationship between ipsilateral renal blood flow and ureteral pressure during 18 hours of left complete ureteral obstruction in the dog. In phase I both left renal blood flow and ureteral pressure increase. In phase II left renal blood flow decreases, while ureteral pressure continues to increase. In phase III both left renal blood flow and ureteral pressure decline together. (From Vaughan ED Jr, Sorenson EJ, Gillenwater JY: The renal hemodynamic response to chronic unilateral complete ureteral occlusion. *Invest Urol* 1970; 8:78. Used with permission.)

cal measurements or competitive blockades. The mechanism of the gradual preglomerular vasoconstriction of hydronephrosis is not known at this time.

TUBULAR FUNCTION.—In *partial* acute ureteral obstruction, urine volume decreases, osmolality increases, and urinary sodium concentration is reduced (Hermann, 1859; Winton, 1931; Abbrecht and Malvin, 1960; Suki et al., 1966). These changes result from slower rate of tubular flow due to a decreased GFR and higher pressures. If the ureteral pressure is elevated to 70 mm Hg, there is a 50% reduction in Tm glucose and Tm PAH (Malvin et al., 1964). After acute *complete* ureteral obstruction, there is an additional decrease in the glomerular filtration rate and decreased sodium concentration in the distal tubule. After *release* of acute ureteral obstruction of 5–60 minutes at 75–120 mm Hg pressure there is a temporary concentrating defect (Finkle et al., 1968; Jaenike and Bray, 1960; Kessler, 1960).

SUMMARY

After acute ureteral obstruction, ureteral and pelvic pressures rise as high as 50–70 mm Hg, depending on the diuretic state. This is a higher pressure than the 20-mm Hg net filtration pressure, indicating a component of muscle contraction. Renal blood flow and ureteral pressure relationships respond in a triphasic pattern to acute ureteral obstruction. The first phase (lasting 1½ hours) is renal vasodilatation from vasodilating prostaglandins associated with a rise in ureteral pressure. The second phase (1½–5 hours) consists of a continued increase in ureteral pressure associated with a decrease in renal blood flow, indicating postglomerular vasoconstriction mediated in some unknown manner. The third and chronic phase exhibits both decreasing ureteral pressure and renal blood flow, as a result of preglomerular vasoconstriction mediated by an unknown mechanism. During acute complete ureteral obstruction, the GFR decreases, and tubular function becomes impaired. With acute partial obstruction, tubular pressure rises and the GFR decreases, with a resultant decrease in urine volume because of better reabsorption, increased osmolality, and lowered urine sodium concentration.

CHRONIC EFFECTS—PARTIAL OBSTRUCTION

Studies have been performed during chronic partial obstruction in patients and experimental animal models. Evaluation of renal function during the obstruction is important, since there is a different environment after release of the obstruction. Studies by Suki et al (1966), Olesen and Madsen (1968), Stecker and Gillenwater (1971), and Wilson (1972) show significant impairment in renal functions during mild degrees of obstruction. These studies during chronic unilateral partial ureteral obstruction show reductions in renal blood flow, GFR, concentrating ability, hydrogen excretion, and sodium reabsorption. Since tubular transit time is increased, sodium reabsorption must be impaired more than the filtration rate and tubule flow rate to account for the increased urinary sodium concentrations.

Most patient studies during partial obstruction have been with bilateral ureteral obstructions. All studies have shown impairment of urinary concentration (Dorhout-Mees, 1960; Earley, 1956; Holliday et al., 1967; Winberg, 1958, 1959; McCrory et al., 1971; Zetterstrom et al., 1958; Berlyne, 1961; Better et al., 1973; Gillenwater et al., 1975; Muldowney et al., 1966; Platts and Williams, 1963; Vaughan et al., 1973). Impairments of all phases of acidification (ammonia excretion, titratable acidity and bicarbonate absorption) have been reported (Earley, 1956; McCrory et al., 1971; Winberg, 1958, 1959; Zetterstrom et al., 1958; Berlyne, 1961; Better et al., 1973; Gillenwater et al., 1975).

CHRONIC EFFECTS—COMPLETE OBSTRUCTION

URETERAL AND TUBULAR PRESSURES.—Ureteral pressures peak 3–5 hours after complete unilateral ureteral obstruction and within 24 hours decrease to 50% of peak values. Ureteral pressures continue to decline over the next 6–8 weeks to approximately 15 mm Hg (Vaughan et al., 1970). Proximal tubular pressure may return to normal (Arendshorst et al., 1974) or to 70% below normal (Jaenike, 1970; Yarger et al., 1972). Numerous collapsed tubules are observed on the kidney surface (Jaenike, 1970; Harris and Yarger, 1974). Glomerular capillary pressure is reduced (Dal Canton et al., 1970).

RENAL BLOOD FLOW.—Renal blood flow measured by flow probes during continued complete unilateral ureteral obstruction shows progressive decreases that are due to afferent arteriole constriction. Measurement showed renal blood flow to be 70% of control at 24 hours, 50% at 72 hours, 30% by 6 days, 20% by 2 weeks, 18% at 4 and 6 weeks, and 12% at 8 weeks (Moody et al., 1975; Vaughan et al., 1970) (Fig 23–2). Blood flow is decreased most in the outer cortex and inner medulla (Solez et al., 1976; Yarger and Griffith, 1974; Siegel et al., 1977). The mechanism of the afferent arteriole vasoconstriction is not known.

GLOMERULAR FILTRATION RATE.—Measurements of the GFR decrease progressively during complete ureteral obstruction in the dog, to 1.74 ml/min at one week and 0.4 ml/min at five weeks of occlusion (Naber and Madsen, 1974). The fluid exiting by pyelolymphatic, pyelovenous, and pyelotubular backflow is replaced by the continuing glomerular filtration. Immediately after ureteral obstruction is released, there is minimal urine flow. One week after release of two weeks' obstruction, the GFR is 15% of control. Maximal recovery of the GFR after release from two weeks' obstruction is 46% of control (Vaughan et al., 1973). After release of four weeks of ureteral obstruction, the GFR was 3% and recovered to 35% of control at five months (Vaughan et al., 1971). No return of glomerular filtration was noted in dogs after six weeks of obstruction with a normal contralateral kidney.

TUBULAR FUNCTION.—All tubular functions studied, with the exception of urinary dilution, are progressively impaired by complete ureteral obstruction. Perhaps urinary dilution is unimpaired because it is not an energy-requiring system. Immediately after release, concentrating ability is severely impaired. This tubular function is one of the first to be injured. Recovery of concentrating ability can be complete after release of two weeks of complete obstruction. After release of four weeks of obstruction there is permanent impairment of concentrating ability.

During short partial ureteral obstruction urine osmolality is higher due to slow tubular transit times in the obstructed kidney. When the partial obstruction is released, urine osmolality falls (Jaenike et al., 1960).

Other tubular functions that have been shown to be impaired by chronic unilateral ureteral obstruction are maximal tubular excretory capacity (Tm) of PAH, Tm glucose, potassium excretion, sodium reabsorption, and urinary acidification (Berlyne and Macken, 1962; Chisholm, 1964; Kerr, 1954, 1956; Vaughan et al., 1971).

The main tubular effect is in concentrating ability. After release of up to 24 hours of obstruction, there is a normal flow rate of dilute urine with no sodium-losing tendency (Wilson, 1974).

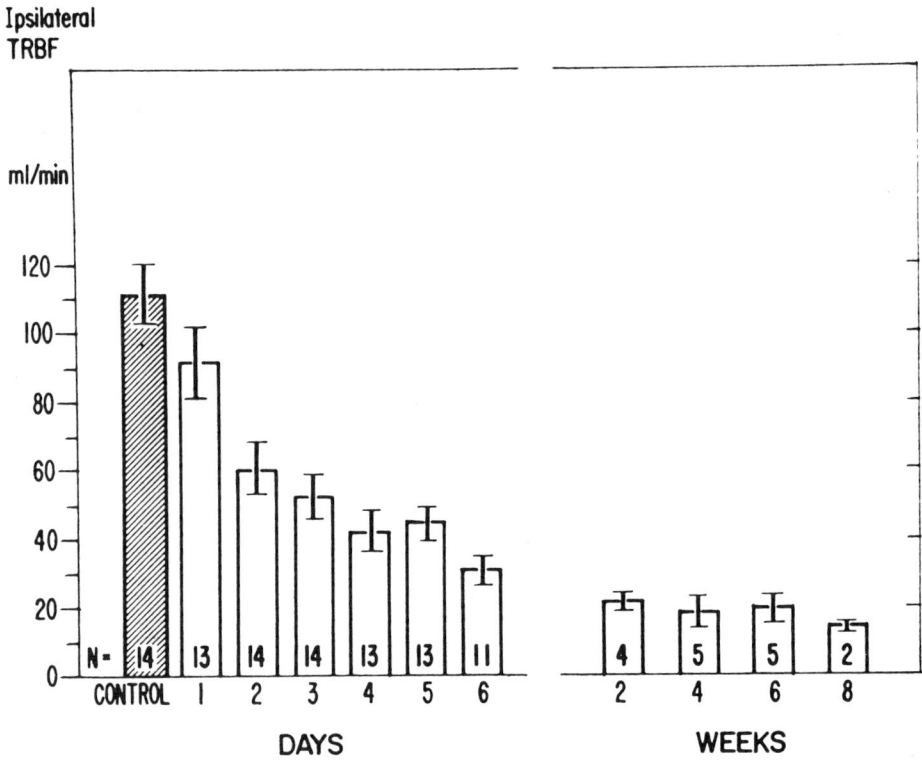

IPSILATERAL RENAL BLOOD FLOW WITH STANDARD ERROR
OF MEAN DURING CHRONIC TOTAL URETERAL OCCLUSION

FIG 23–2.
Changes in left renal blood flow with chronic total left ureteral occlusion in 14 dogs with chronic indwelling blood flow probes. TRBF indicates total renal blood flow. (From Moody TE, Vaughan ED Jr, Gillenwater JY: Relationship between renal blood flow and ureteral pressure during 18 hours of total unilateral occlusion. *Invest Urol* 1975; 13:246. Used with permission.)

CONTRASTING CONDITIONS DURING URETERAL OBSTRUCTION

Physiologic changes are different depending on whether the ureteral obstruction is unilateral or bilateral. With bilateral ureteral obstruction, the uremic state starts with retention of substances that are normally excreted, and one or more of these substances apparently affects renal hemodynamics and tubular function. Better understanding of clinical problems such as postobstructive diuresis has resulted from the study of this situation.

The surface tubules in rats look normal after 24 hours of bilateral ureteral obstruction, in contrast to the poorly perfused and filtering nephrons with collapsed tubules observed in unilateral ureteral obstruction. Proximal and distal tubular pressures are elevated in bilateral ureteral obstruction, but are lower than normal in unilateral ureteral obstruction. Afferent arteriole pressure is increased in bilateral ureteral obstruction and decreased in unilateral ureteral obstruction. Glomerular capillary pressure increases to higher levels when both ureters are obstructed than when ureteral obstruction is unilateral. Renal blood flow is reduced to one third of the control value after release of both bilateral and unilateral ureteral obstruction (24 hours). The single nephron GFR is 40% of normal in bilateral ureteral obstruction. The reason for the decreased single nephron GFR is the elevated tubular pressure, since the glomerular capillary pressure is normal.

The explanation for the differences between the renal vascular responses in unilateral and bilateral ureteral obstruction is not known (Harris and Yarger, 1974; Jaenike, 1972; McDougal and Wright, 1972; Thirakomen et al., 1976; Walls et al., 1975; Wilson, 1972, 1974; Yarger and Griffith, 1974; Dal Canton et al., 1980).

CONTRASTING STUDIES AFTER RELEASE OF URETERAL OBSTRUCTION

The contrast in the physiologic effects of chronic (24 hours or greater) unilateral and bilateral ureteral obstruction has provided a better understanding of postobstructive diuresis. The major difference is the significant natriuresis and diuresis occurring after release of bilateral ureteral obstruction. Through different mechanisms, renal blood and GFR are reduced to 33% of control in both unilateral and bilateral ureteral obstruction after release of 24 hours' obstruction. In unilateral ureteral obstruction there is afferent arteriole constriction with abnormal distribution between the cortex and the medulla, with a shift of blood flow from the outer cortex to the inner cortex and medulla (Jaenike, 1970, 1972). In bilateral ureteral obstruction there is efferent arteriole constriction and normal distribution of blood flow. The reduced GFR in unilateral ureteral obstruction is due to the afferent arteriole constriction, and the reduced GFR in bilateral ureteral obstruction is due to the increased proximal tubule pressure. After release of bilateral obstruction, the tubular pressure returns toward normal and the GFR remains low because of new afferent arteriole constriction (Schnermann et al., 1970; Wright and Briggs, 1977). Wright (1974) postulated that the new afferent arteriole constriction after release of bilateral ureteral obstruction is due to the macula densa feedback mechanism responding to increased distal delivery of tubule fluid.

Urine flow is increased up to ten times that of control after release of bilateral ureteral obstruction, in contrast to the low rates of flow and solute excretion after release of unilateral ureteral obstruction. The excretion rates of potassium, phosphate, and urea are significantly increased. After release of bilateral ureteral obstruction, diuresis occurs despite the withholding of fluid and food during the period of obstruction. The diuresis persists for several days until sodium balance is restored. The concentrating defect persists several days longer than the salt loss. The impaired sodium reabsorption occurs in both the proximal and distal tubule (McDougal and Wright, 1972; Buerkert et al., 1976, 1977). The mechanism of the postobstructive diuresis is not known, but cross-perfusion studies have shown a build-up of a natriuretic factor in the plasma of animals with bilateral ureteral obstruction (Wilson and Honrath, 1976). Wright and Howards (1979) postulated that postobstructive diuresis results from two factors: (1) distention and damage to the collecting ducts by increased luminal pressure and (2) inhibition of proximal tubular sodium reabsorption by an unidentified factor that is normally excreted in the urine.

After release of 24 hours of unilateral ureteral obstruction, the previously obstructed kidney has a normal urine flow rate of dilute urine with no natriuresis. These conditions result from reduced renal blood flow and GFR, slightly impaired sodium reabsorption, and severely impaired concentrating ability.

CLINICAL POSTOBSTRUCTIVE DIURESIS

Patients rarely have a severe, life-threatening diuresis after release of urinary tract obstruction. Almost always, diuresis occurs only if all nephron units were obstructed in a way similar to the animal studies described previously. Excellent reviews have been published by Goldsmith (1968), Howards (1973), Vaughan and Gillenwater (1973), and Wright (1979). A diuresis almost always occurs after release of bilateral obstruction. The diuresis is physiologic, usually mild, and self-limiting. Patients with bilateral obstruction have a retention of sodium and water, and the diuresis is just restoring normal fluid and electrolyte balance (Muldowney et al., 1966).

The postulated mechanisms for the rare pathologically significant postobstructive diuresis are: (1) impaired urine concentrating ability, (2) impaired sodium reabsorption, and (3) solute diuresis due to retained urea or administered glucose. Transient peak urine flow rates of up to 69 ml/min, with an average of 30 ml/min, have been reported (Eiseman et al., 1955). Cases of diabetes insipidus unresponsive to vasopressin have been reported (Earley, 1956; Roussak and Oleesky, 1954).

In the clinical situation, a patient could develop major fluid and electrolyte problems if he had the rare pathologic sodium or water diuresis and it went unrecognized. Our plan of management after release of obstruction is to have the patient weighed and blood pressures recorded in the upright and supine positions and ask that the physician be notified if urine volume exceeds 200 ml/hr. The thirst mechanism will correct any abnormal water loss in the conscious and alert patient. Pathologic sodium loss with resultant contraction of extracellular fluid volumes can be recognized by orthostatic hypotension. Pathologic sodium loss can be replaced by 0.5N saline, initially at 50% of output to avoid perpetuating a possible salt and water overload.

RELATIONSHIPS OF HYPERTENSION AND HYDRONEPHROSIS

Acute unilateral ureteral obstruction is frequently associated with increased renin output and resultant hypertension in animal and patient studies (Vaughan et al., 1971; Moody et al., 1975). Animal studies show that elevated peripheral renin returns to normal with acute unilateral ureteral obstruction in the dog (Vaughan et al., 1970). Chronic unilateral hydronephrosis is not infrequently associated with hypertension but rarely has increased peripheral renin levels (Vaughan et al., 1974). If the hypertension is of renal origin, the lesion may be an associated renal artery stenosis. Hypertension with elevated renin secretion has been documented in some cases of unilateral hydronephrosis (Riehle and Vaughan, 1981; Weidmann, 1977). Surgical correction of the hydronephrosis is justified to improve renal function. If the hypertension also improves, that is an added bonus. Correction of unilateral chronic hydronephrosis will cure associated hypertension only 20%–30% of the time.

SURGICAL MANAGEMENT OF HYDRONEPHROSIS

All operative techniques for correction of ureteropelvic junction obstructions strive to achieve dependent drainage with an unobstructive anastomosis of the ureter to the renal pelvis. Attempts must be made to avoid parenchymal loss or ligation of any of the renal vessels. It must be remembered that any anastomosis heals by scar tissue, which will contract by about one third of its original diameter as it matures. End-to-end or end-to-side anastomoses without spatulation of the ureter are doomed to failure.

DETERMINING WHETHER A DILATED UPPER URINARY TRACT IS SIGNIFICANTLY OBSTRUCTED

In the past decade several tests were developed for determining whether a dilated upper urinary tract has a significant obstruction that will benefit from surgery. The most popular tests are diuresis renography (O'Reilly et al., 1978) and the perfusion-pressure flow test (Whitaker, 1973). Other proposed tests are radionuclide parenchymal transit time estimations (Whitfield et al., 1978) and renal pelvic morphology (Gosling and Dixon, 1978; Lupton et al., 1981; English et al., 1982). Historically, washout of contrast instilled through ureteral catheters was used to determine whether an obstruction was significant. Diuresis renography is most often used in our institution because it is noninvasive. Excellent reviews of these techniques have recently been published (Lupton, 1985; O'Reilly, 1986).

Upper urinary tract dilatation is usually initially diagnosed by IV urography or ultrasonography. The dilated upper tracts are sometimes not significantly dilated and will not benefit from surgery. Such cases are seen in reflux, prune belly syndrome, idiopathic megaureter with a nondilating segment, megacalicosis, or postsurgical repair.

DESCRIPTION OF TECHNIQUES

In the diuretic renogram an isotope such as ^{123}I-Hippuran is given and at a given time a diuretic is administered IV. The results are followed on a time activity curve (Fig 23–3). A normal curve excludes obstruction.

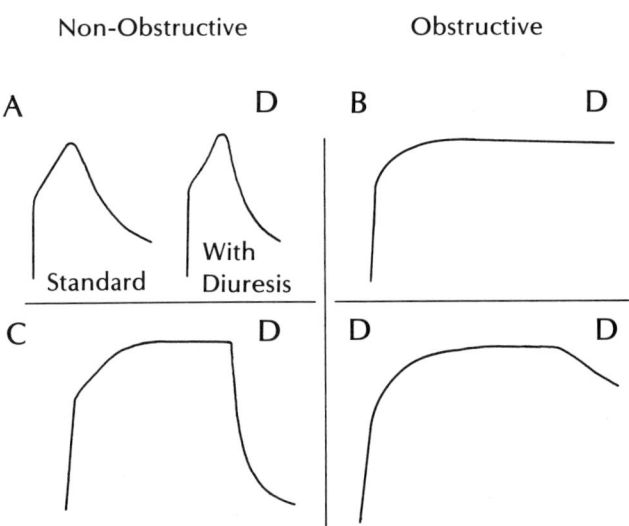

FIG 23–3.
Typical diagrammatic diuresis renograms. *D* is where diuretics were administered; *A* is nonobstructive; *B* is obstructive with no washout; *C* is nonobstructive after diuresis; *D* is obstructive with no washout after diuresis. (Adapted from Lupton EW, Richards D, Testa HJ, et al: A comparison of diuresis renography, the Whitaker test, and renal pelvic morphology in idiopathic hydronephrosis. *Br J Urol* 1985; 57:119.)

A rising curve unaffected by a diuretic is indicative of obstruction. A rising curve that is converted into a normal third phase by the diuretic suggests a dilated but unobstructed system (O'Reilly et al., 1978).

Parenchymal mean transit time estimation involves obtaining activity curves over areas of interest in the whole kidney, renal pelvis, bladder, and a vascular region after the IV injection of ^{99m}Tc-DTPA or ^{123}I-Hippuran. Through a complex mathematical analysis, the mean transit times of the tracer through the renal parenchyma can be calculated. Delay in transit time in a dilated upper tract is said to confirm obstruction (Whitfield et al., 1978).

In the perfusion-pressure flow studies a percutaneous nephrostomy catheter is passed, allowing perfusion at 10 ml/min with simultaneous recording of pressure. A pressure rise of greater than 22 cm H_2O is said to indicate obstruction. A rise of pressure less than 15 cm H_2O excludes obstruction while the intervening range is equivocal (Whitaker, 1973).

Morphological changes in the obstructing segment and renal pelvis in hydronephrosis show an increase in collagen and elastic tissue in the muscle bundles of the renal pelvis. Increased collagen in the narrow ureteral segment prevents distention. Electron microscopy of the dilated segments shows a reduction in the number of myofilaments and the development of large quantities of granular reticulum and Golgi membranes (Gosling and Dixon, 1978; Lupton et al., 1981; Notfrey, 1971, 1972; Hanna et al., 1976).

ACCURACY OF THE VARIOUS TESTS

The available literature regarding the accuracy of these tests to determine whether an upper urinary tract dilatation represents a significant obstruction was recently reviewed by Lupton et al. (1985) and O'Reilly (1986). In general there is good correlation of the tests when there is clear-cut severe obstruction or clear-cut nonobstruction. On the equivocal cases, however, there is much disagreement.

Two studies (Cosgriff and Berry, 1982; Lupton et al., 1985) showed excellent correlation between diuresis renography and mean transit time through the renal parenchyma. Studies comparing diuresis renography with perfusion pressure flow studies have shown variable results: correlations of 86% (Senac et al., 1985), 84% (Hay et al., 1984), 67% (Lupton et al., 1985), 53% (Whitaker and Buxton-Thomas, 1984), and 54% (Gonzalez and Chiou, 1985).

Studies comparing diuresis renography with renal pelvic morphology report an 87.5% correlation between the two tests (Gosling and Dixon, 1978; Lupton et al., 1981; English et al., 1982). Israel (1982) reported a 100% correlation between diuresis renography and synchronous intrapelvic pressure measurements.

Equivocal washout studies present clinical problems. If the kidney has good function and the patient is well-hydrated, then O'Reilly (1986) thinks an equivocal response indicates partial obstruction. When renal function is impaired, it is difficult to distinguish whether an equivocal response is due to obstruction or to the renal impairment itself. By O'Reilly's analysis (1986), 15 of 188 cases (8%) had obstruction on the diuresis renograms but normal perfusion-pressure studies. O'Reilly stated that two cases had intermittent obstruction and the other 13 had gross hydronephrosis and/or poor renal function. Forty-eight of the 188 cases (25%) had normal diuresis renograms but abnormal perfusion-pressure studies (32 came from Hay et al., 1984). A possible explanation put forward by O'Reilly for the discrepancies is that the systems tolerated lower flow rates but not the 10 ml/min used in the perfusion-pressure studies.

INDICATIONS FOR SURGERY

The major indications for surgical repair of hydronephrosis are relief of pain or relief of significant obstruction that will destroy renal function. Intermittent hydronephrosis classically occurs during a marked diuresis and is best demonstrated by radiologic studies during the symptomatic episodes (Nesbit, 1956). Determina-

tion whether a mild partial obstruction is significant or not is harder. The best studies are the IV urogram to demonstrate calyceal clubbing and dilation, diuretic renograms, Whitaker renal perfusion studies, retrograde pyelograms with washout studies, or longitudinal follow-up urograms showing progressive dilation of the renal pelvis and calyces. When an individual case is first seen with mild to moderate dilation of the renal pelvis, it is not always possible to forecast its natural history. Investigational studies of the natural history of hydronephrosis are being done in at least one medical center by following cases of hydronephrosis without surgical intervention unless there is deterioration of function or worsening of the hydronephrosis.

I have seen patients whose previous radiologic studies showed a very mild hydronephrosis, presenting with significant hydronephrosis that had progressed over periods of one to ten years. So ureteropelvic junction obstruction is progressive in some patients.

SELECTION OF OPERATIVE PROCEDURES

All the operative techniques provide a dependent and progressive funneling of the ureteropelvic junction. The two basic techniques are the use of some type of pelvic flap (Foley, 1937; Culp and DeWeerd, 1951; Scardino and Prince, 1953) and the dismembered pyeloplasty (Foley, 1937; Anderson and Hynes, 1949). The most frequently used procedure is the dismembered pyeloplasty. All the operations work, and one cannot be dogmatic in declaring that one or another of the operations is best. The various techniques are described and illustrated below.

Meticulous care and delicate handling of the renal pelvis and ureter are essential for the success of the pyeloplasty. Small, sharp scissors and fine vascular forceps are essential to prevent crush damage of the tissues. Tissues that are frequently lifted or held should have sutures placed for traction, or skin hooks should be used. Tissue must be approximated with fine sutures (I prefer chromic catgut 4–0 or 5–0 swedged on an atraumatic needle). Knots should be placed on the outside. I prefer to use interrupted sutures beginning at the apex to avoid a "dog-ear" in this location.

I agree with Blandy (1978) that too much time has been wasted arguing over whether to use stents or nephrostomies. Any urologist should know how to use both. Stents should be used if the anastomosis appears to want to kink or is obstructed by edema during the first few days. Nephrostomies are needed if renal function is unclear or if there is a high likelihood of leakage or obstruction. I usually favor using both stents and nephrostomies.

In bilateral hydronephrosis, if one side is infected, that side should be repaired first. In uninfected hydronephrosis, if one side has poorer function, that side should be repaired first.

Ureteral strictures in association with ureteropelvic junction obstruction are uncommon in my experience. Ureteral strictures can be corrected during the pyeloplasty. The Davis intubated ureterotomy (Davis, 1943) or end-to-end anastomosis can be used.

FOLEY Y-PLASTY

The Foley Y-plasty (Foley, 1937) was designed for the correction of an obstructive, congenital high insertion of the ureter into the renal pelvis (Fig 23–4). The pelvis and ureter need to be dissected free. Careful planning of the pelvis incisions is essential. The anterior and posterior pelvic incisions and the ureteral incision all should be approximately the same length. I have found it helpful to mark the end of each planned incision with a 4–0 silk suture. The anterior pelvic incision is started at the ureteropelvic junction and extended laterally and downward toward the hilum of the kidney. The posterior incision is then similarly made, giving the "V" portion of the "Y" incision. The ureteral incision is then made down the lateral margin (the side facing the renal pelvis). The sharp tip of the pelvic flap is rounded off. Prior to closure, the nephrostomy tube and stent are placed if they are being used. Interrupted 4–0 or 5–0 chromic catgut is used for closure, starting at the tip to leave any dog-ear in the renal pelvis. I drain the

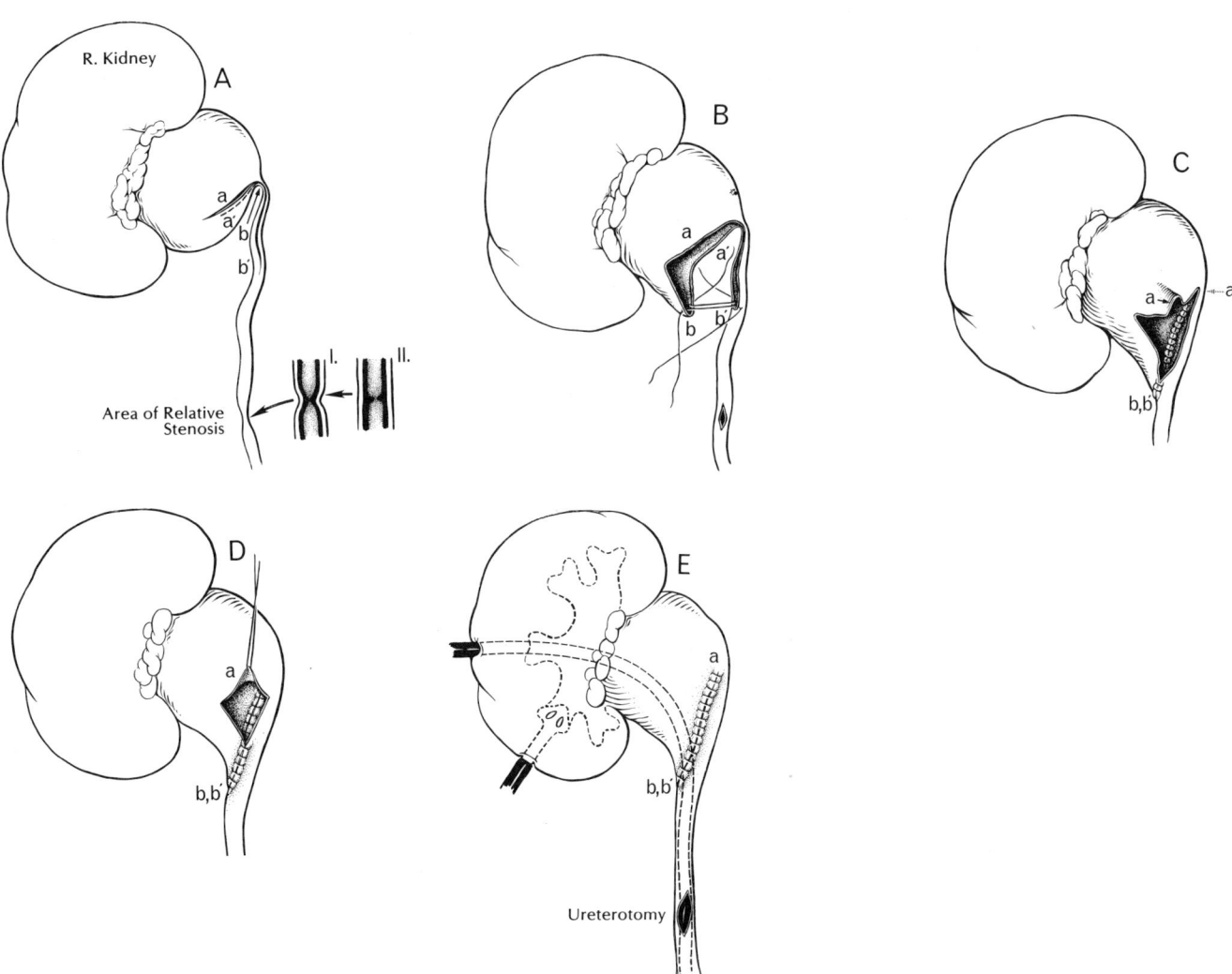

FIG 23–4.
Classic Foley Y-plasty. **A,** the anterior and posterior pelvic incisions should be the same length as the ureteral incision. **B,** the tip of the flap and the lower end of the ureterotomy should be approximated first. **C** and **D,** the closure is started at *(b,b')* and extended upward to *(a,a')*. A dog-ear may form at point *a.* *I* and *II* illustrate the infrequent relative narrowing of the ureter, which is occasionally seen lower in the ureter. **E,** if there is a narrow segment, it can be treated by a ureterotomy or dilatation. (Adapted from Smart WR: Surgical correction of hydronephrosis, in Harrison JH, et al (eds): *Campbell's Urology,* ed 4. Philadelphia, WB Saunders, 1979.)

area with two drains. All tubes and drains are brought out the posterior part of the wound, since it has less sensation than the anterior portion. The Foley Y-type ureteropelvioplasty can be combined with a Davis intubated ureterotomy (Davis, 1943) when the ureteral structure is longer than can be corrected with the Foley Y-plasty alone (Fig 23–5). Use of stents is essential in the Davis intubated ureterotomy to provide the scaffolding for the new growth. I usually loosely place several sutures from the edges across the catheter to ensure that the sides lie flat and do not curl up. Davis states that the healing by secondary intention requires six weeks. When exposing the ureter, there should be minimal dissection to preserve the blood supply. Careful nontraumatic handling of the tissues is essential.

CULP-SCARDINO URETEROPELVIOPLASTY

Turning down a pelvic flap to widen the narrowed area in the upper ureter is common to both the Culp (Culp and DeWeerd, 1951) and Scardino (Scardino and Prince, 1953) techniques. Both of these operations work well when there is low insertion of the ureter and are not indicated with high insertion of the ureter. Culp uses a spiral type of flap and Scardino employs the vertical flap (Figs 23–6 and 23–7). These procedures can only be used when there is an external pelvis present. The spiral flap of the Culp operation is longer than the Scardino vertical flap. The Culp spiral flap has been recommended when there is an obtuse angle between the renal pelvis and the ureter, and the Scardino flap has been recommended when the ureteropelvic angle is approximately 90° (Smart, 1963). After freeing the kidney and ureter, the pyelotomy incisions are marked with stay sutures or a marking pen before opening the

pelvis. The necessary length of the flaps is determined by the length of the ureterotomy. The ureteral incision should extend 1 cm below the obstructive segment. If the ureteral incision must be longer than the pelvic flap, the distal portion can be left open as a Davis intubated ureterotomy. If ureteral stents and nephrostomy tubes are used, they should be placed prior to closure. Closure is usually begun on the posterior surface with 4–0 chromic interrupted sutures. It is essential when planning the flaps to study the vessels of the renal pelvis to preserve as much of the blood supply as possible.

DISMEMBERED URETEROPYELOSTOMY

The operation first described by Foley in 1937 and Anderson and Hynes in 1949 is the most frequently used pyeloplasty and works well in most situations. The operation consists of suturing a spatulated ureter to a generous V-shaped pelvic flap. The pelvic flap is essential to provide the funneling. The kidney and ureter are dissected free, noting carefully whether there are any lower pole vessels. It is important to know how long the narrow upper ureteral segment is to plan the pelvic flap. If needed, additional length can be gained by mobilizing the kidney. Preoperative evaluation should have provided information about the length of the narrowed ureteral segment. If additional information is needed, one can open the renal pelvis and pass down a calibrating ureteral catheter or bougie à boule.

Before opening the pelvis, I plan and map out the incision, the pelvic flap, and any excess renal pelvis I am going to remove, placing marking sutures or using a marking pen. I always mark the lateral edge of the ureter where I plan to do the spatulation to prevent

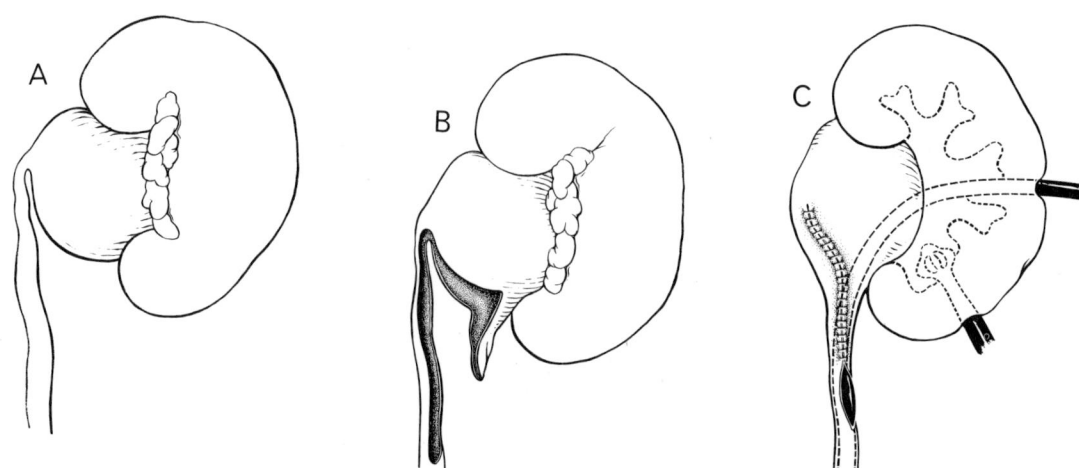

FIG 23–5.
The classic Foley Y-plasty for high ureteral insertion with a long obstructing ureteral defect. **A,** the high insertion of the ureter into the renal pelvis. **B,** the long ureterotomy and fun-

neling flap in the renal pelvis. **C,** closure of the renal pelvis with the lower portion of the ureteral incision left open as a Davis intubated ureterotomy.

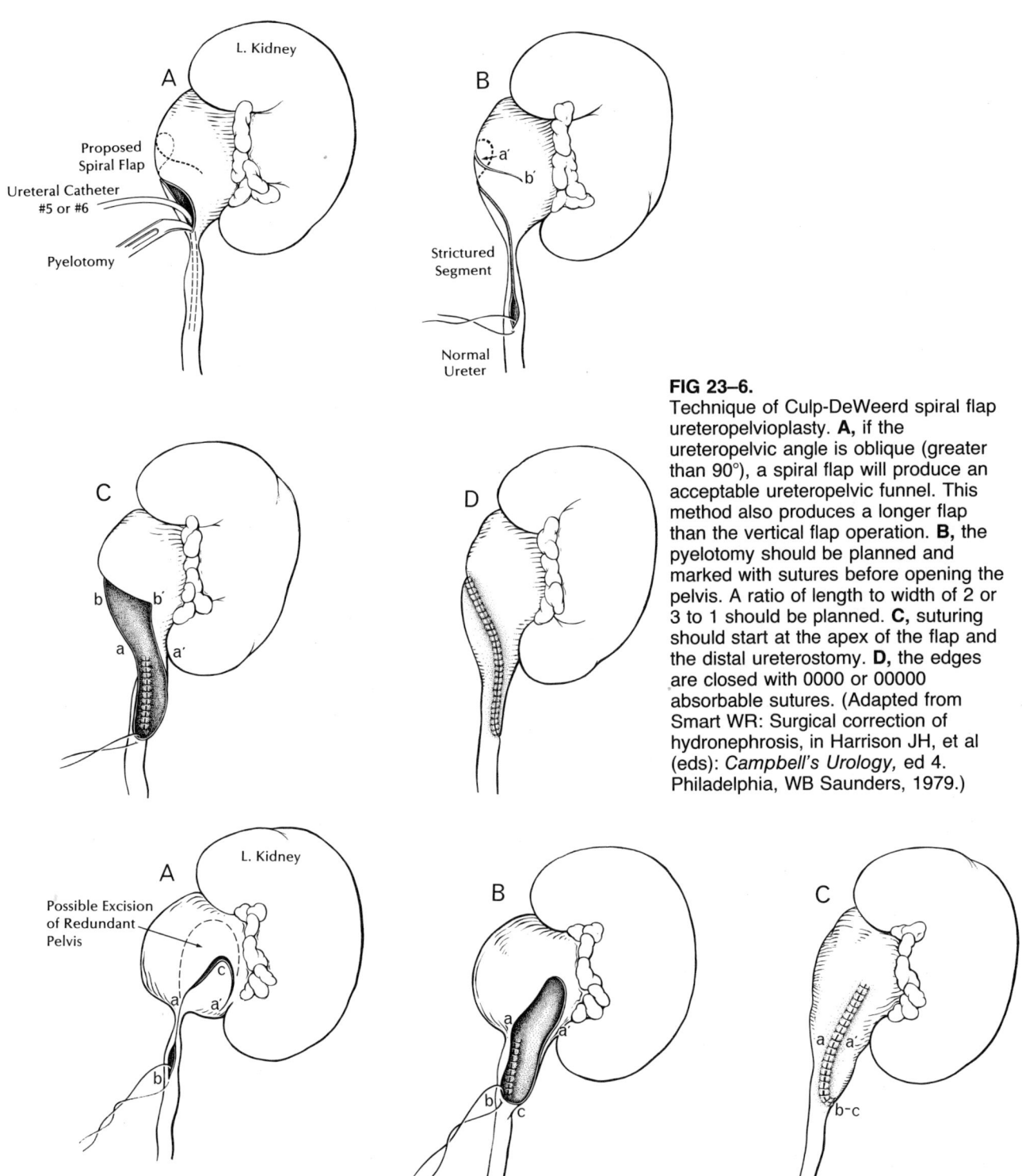

FIG 23–6.
Technique of Culp-DeWeerd spiral flap ureteropelvioplasty. **A,** if the ureteropelvic angle is oblique (greater than 90°), a spiral flap will produce an acceptable ureteropelvic funnel. This method also produces a longer flap than the vertical flap operation. **B,** the pyelotomy should be planned and marked with sutures before opening the pelvis. A ratio of length to width of 2 or 3 to 1 should be planned. **C,** suturing should start at the apex of the flap and the distal ureterostomy. **D,** the edges are closed with 0000 or 00000 absorbable sutures. (Adapted from Smart WR: Surgical correction of hydronephrosis, in Harrison JH, et al (eds): *Campbell's Urology,* ed 4. Philadelphia, WB Saunders, 1979.)

FIG 23–7.
Technique of Scardino-Prince vertical flap ureteropelvioplasty. **A,** this technique is best used when the ureteropelvic angle is approximately 90°. The flap should have a 2 or 3 to 1 length to width ratio and be the same length as the ureterotomy. **B,** a pyelotomy is made in the lower pelvis and carried down the ureter. **C,** the flap is cut and redundant pelvic tissue excised using sharp plastic surgical scissors. The apex of the flap is sewn to the end of the ureterotomy incision. 0000 or 00000 absorbable sutures are used, starting at the apex so any dog-ears will be in the renal pelvis. (Adapted from Smart WR: Surgical correction of hydronephrosis, in Harrison JH, et al (eds): *Campbell's Urology,* ed 4. Philadelphia, WB Saunders, 1979.)

rotation and cutting the ureter in the incorrect place. Proper orientation is essential. It is easy to misplace the anastomosis and cause rotations or angulation if one does not pay attention, stay oriented, and properly mark the tissues. A nice aspect of this operation is that it can be used with either an intrarenal or an extrarenal pelvis. There are several ways one can fashion the renal pelvic flap (Figs 23–8 to 23–11) (Blandy, 1978).

If there is a lower pole vessel the ureter needs to be brought anterior to the vessel. If one is going to use a stent and nephrostomy tube, these must be placed prior to closure. I do not think it matters whether one uses interrupted or continuous running sutures. I usually use interrupted 4–0 chromic catgut sutures. I usually close the posterior margins first, starting at the tip. In some cases the renal pelvis is very thick. It is important that the mucosal edges of the renal pelvis be approximated to the ureteral mucosa. Suture knots should be placed on the outside.

URETEROCALYCEAL ANASTOMOSIS

The ureterocalyceal anastomosis can be used when there seems to be no other option for anastomosing the kidney. This anastomosis is most useful when the renal

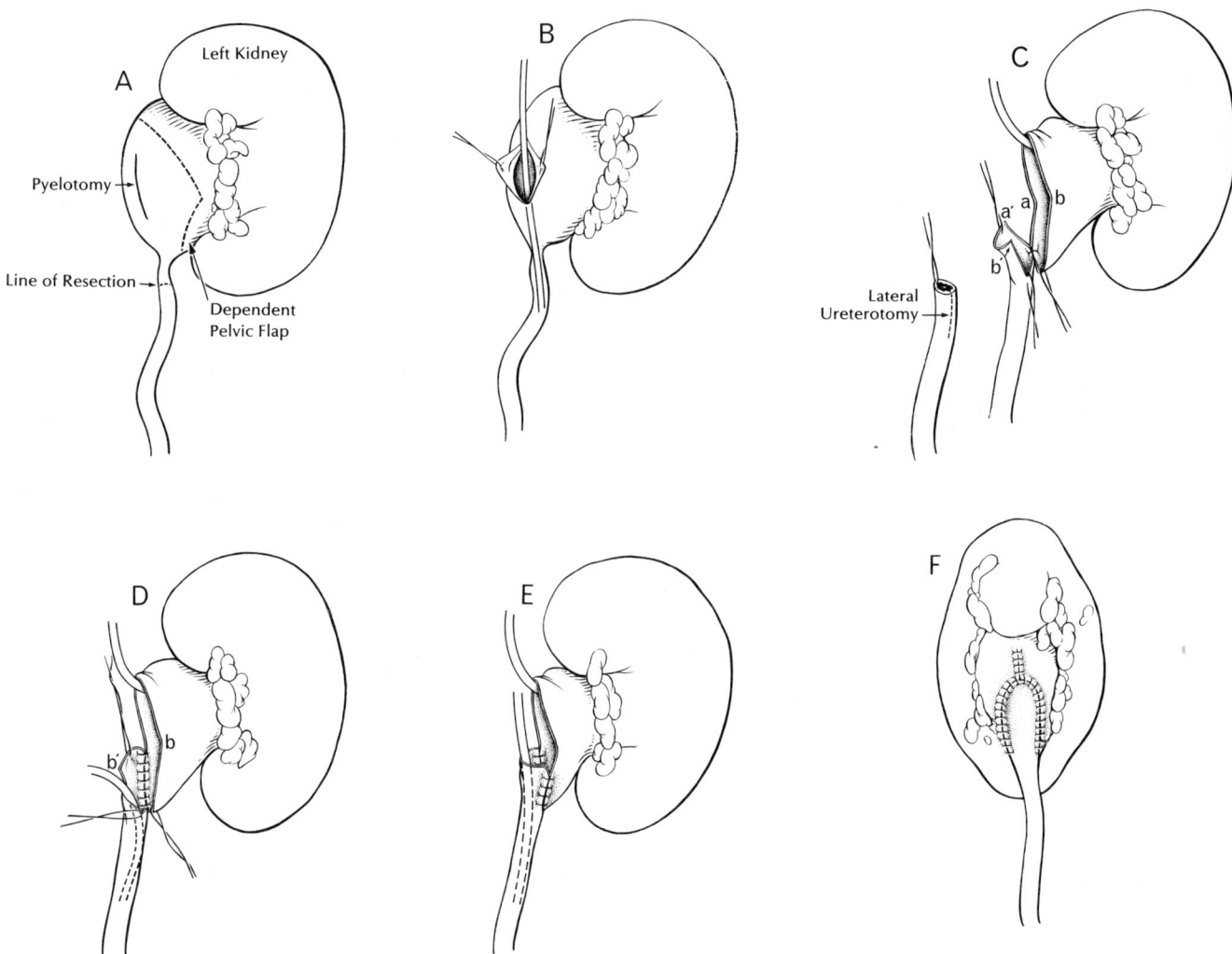

FIG 23–8.
Dismembered Foley Y-plasty operation. In this operation the obstructing segment is excised and a dependent funneling of the pelvis is achieved. **A,** the pyelotomy and ureterotomy incisions are planned and marked. **B,** the pyelotomy incision is started. **C,** the lateral ureterotomy is done and ureteral stent passed after excising the ureteral segment. **D,** suturing with 0000 or 00000 absorbable sutures is started on the posterior surface at the apex. **E,** traction sutures are used to approximate the edges. It is important not to crush the tissue with heavy forceps. The tip of the flap is sewn to the ureterotomy. It is important to start at the apex. **F,** the funnel is completed and any redundant pelvis can be resected. The pelvis is reconstituted. (Adapted from Smart WR: Surgical correction of hydronephrosis, in Harrison JH, et al (eds): *Campbell's Urology,* ed 4. Philadelphia, WB Saunders, 1979.)

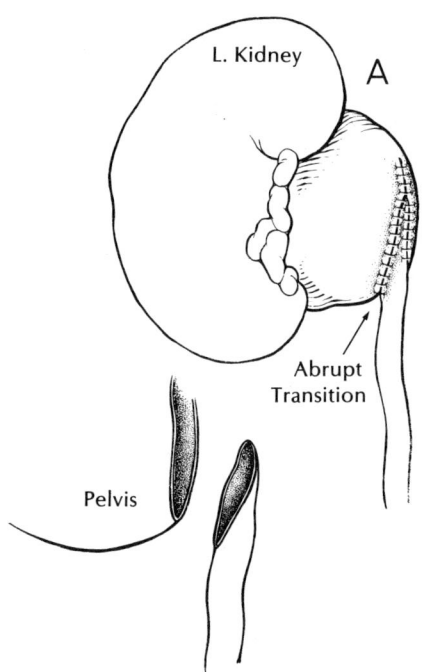

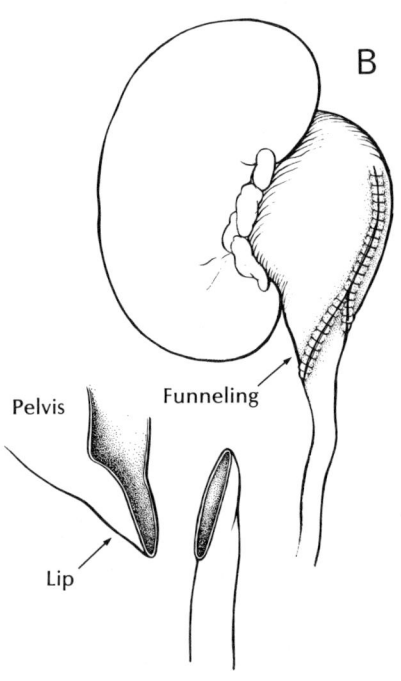

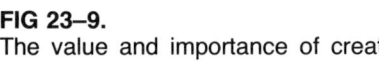

FIG 23–9.
The value and importance of creating a funnel **(B)** in the pelvis as opposed to creating an abrupt transition between the pelvis and the ureter **(A)**. Proper funneling gives dependent drainage with a nice transition between the pelvis and ureter. (Adapted from Smart WR: Surgical correction of hydronephrosis, in Harrison JH, et al (eds): *Campbell's Urology, ed 4.* Philadelphia, WB Saunders, 1979.)

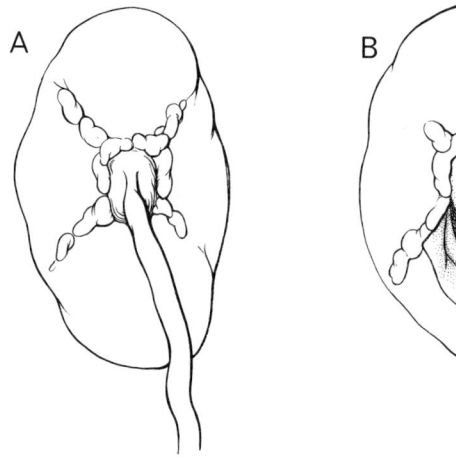

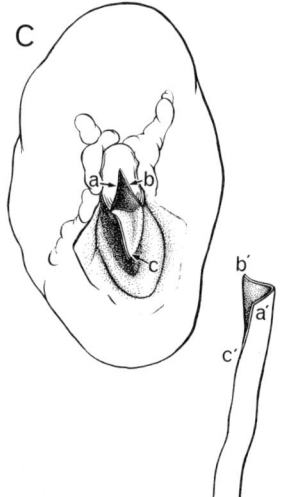

FIG 23–10.
Difficult high-insertion repair. The Foley Y-plasty can be adapted to patients with a small extrarenal pelvis. **A,** area of narrowing in upper ureter with small extrarenal pelvis. **B,** resection of renal parenchyma to expose the large intrarenal pelvis. **C,** ureter is cut distal to the obstruction and spatulated. Pelvis flap is also formed after resection of stenotic segment. Approximate *a-a', b-b',* and *c-c'.* **D,** anastomosis completed with a funneled and dependent pelvis. (Adapted from Smart WR: Surgical correction of hydronephrosis, in Harrison JH, et al (eds): *Campbell's Urology,* ed 4. Philadelphia, WB Saunders, 1979.)

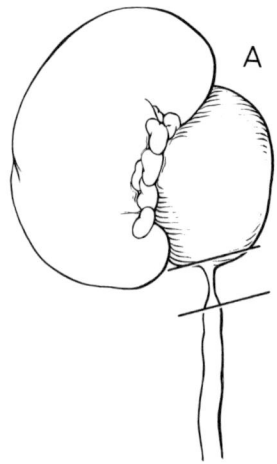
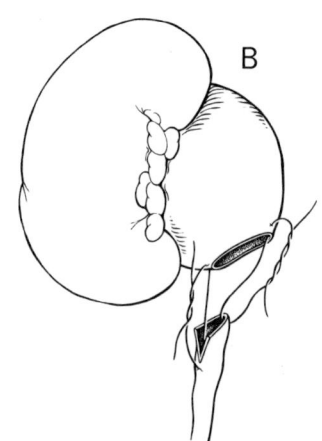
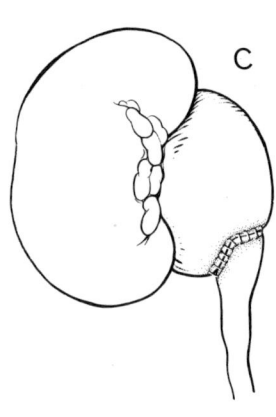

FIG 23–11.
Simple pyeloplasties in children have been successful. The narrow segment **(A)** is excised, leaving the pelvis tunneled, and the ureter is spatulated **(B).** Closure is with 0000 or 00000 absorbable sutures **(C).** (Adapted from Zincke H, et al: Ureteropelvic obstruction in children. *Surg Gynecol Obstet* 1974; 139:876.

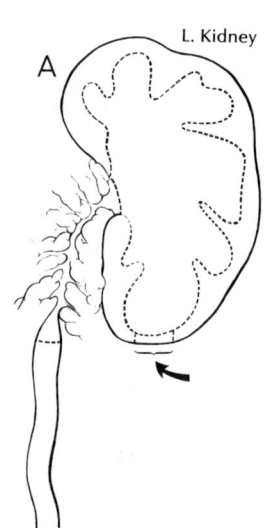
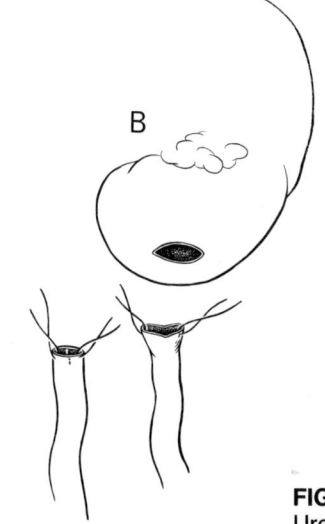
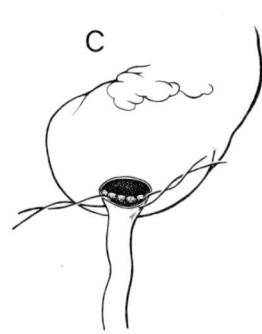
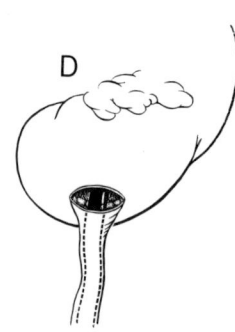
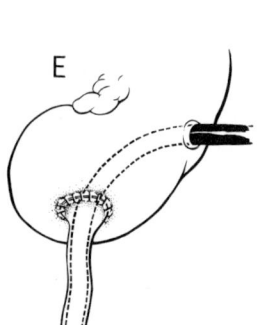
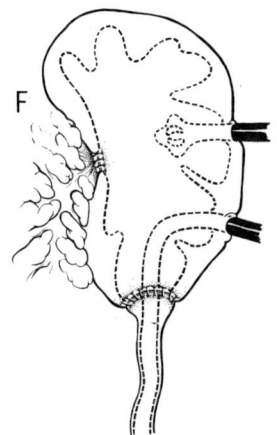

FIG 23–12.
Ureterocalyceal anastomosis for difficult situations in which anastomosis to the pelvis is impossible. **A,** the scar tissue is excised. The renal pelvis is closed after placing a stent and nephrostomy tubes. **B,** ureterotomies are done after orientation sutures are placed in the ureter. The ureteral opening is enlarged, ready for the anastomosis. An elliptical segment of renal parenchyma is excised. The dimensions of the opening should be the same as those of the ureter to be anastomosed. **C,** ureter is sutured to calyceal mucosa with interrupted 0000 or 00000 absorbable sutures. **D,** ureteral stent is passed after posterior anastomosis is complete. **E,** the anastomosis is completed. **F,** diagram of funneling. (Adapted from Smart WR: Surgical correction of hydronephrosis, in Harrison JH, et al (eds): *Campbell's Urology,* ed 4. Philadelphia, WB Saunders, 1979.)

pelvis cannot be dissected out or used. The procedure is more difficult and will have a higher failure rate than other procedures. The two different methods are removing a button of renal parenchyma over the lower calyx (Fig 23–12) or opening the lower calyx by incising down to the calyx in the medial portion of the lower pole of the kidney (Fig 23–13).

In ureterocalyceal anastomosis one usually has not been able to adequately dissect out the renal pelvis due to scar tissue. Usually there is a longer narrowed seg-

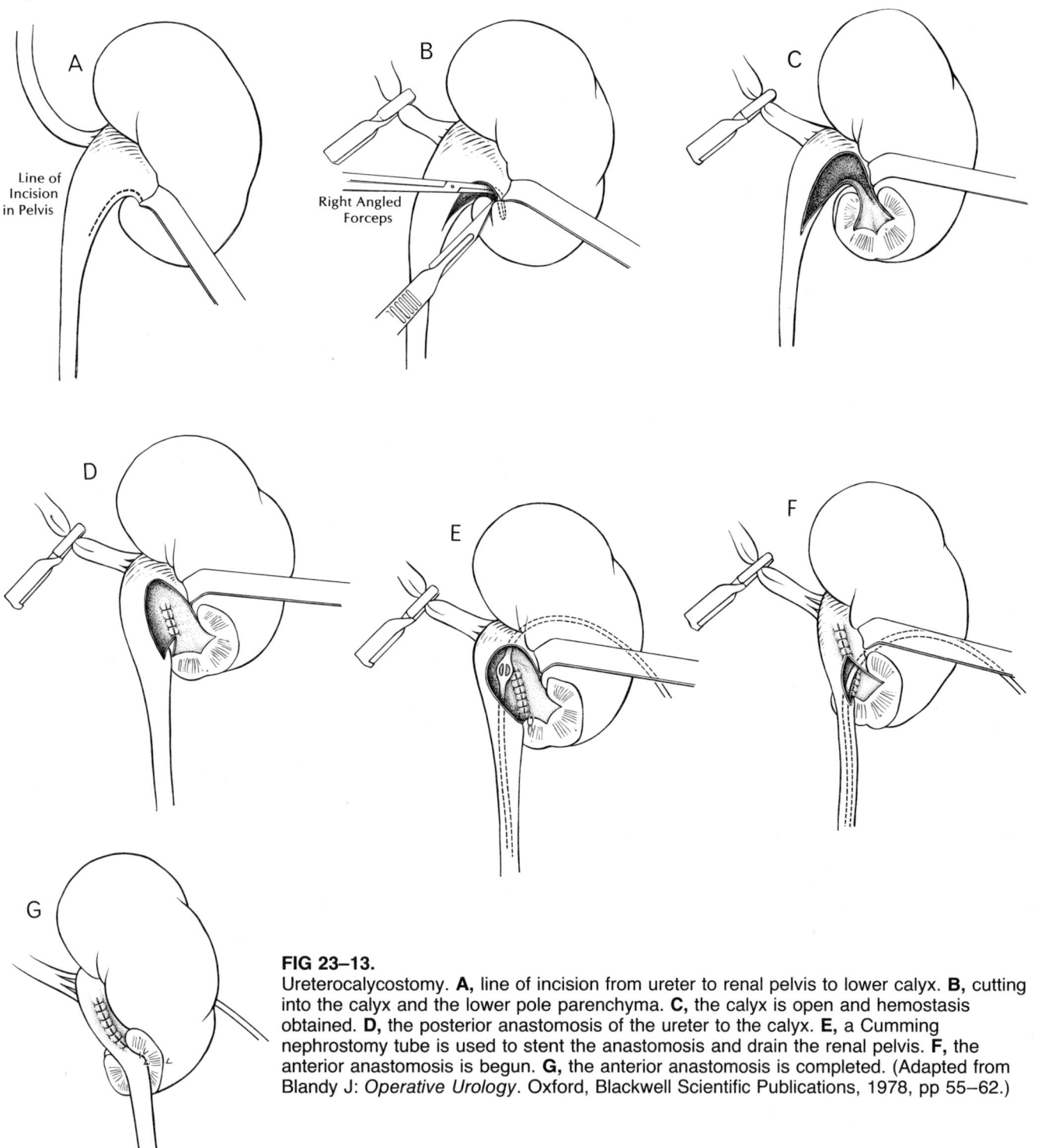

FIG 23–13.
Ureterocalycostomy. **A,** line of incision from ureter to renal pelvis to lower calyx. **B,** cutting into the calyx and the lower pole parenchyma. **C,** the calyx is open and hemostasis obtained. **D,** the posterior anastomosis of the ureter to the calyx. **E,** a Cumming nephrostomy tube is used to stent the anastomosis and drain the renal pelvis. **F,** the anterior anastomosis is begun. **G,** the anterior anastomosis is completed. (Adapted from Blandy J: *Operative Urology.* Oxford, Blackwell Scientific Publications, 1978, pp 55–62.)

ment of ureter. The ureter is cut back to normal tissue and spatulated on the lateral border. If the renal pelvis can be entered, I pass a sound down to the lower calyx and cut out an adequate button of renal parenchyma, marking the mucosal edges of the calyx for later anastomosis. If the renal pelvis cannot be entered, a guillotine type of procedure will provide access to the lower calyx.

The other type of procedure that can be used in some circumstances is the ureterocalicostomy. This procedure was devised to correct a narrow infundibulum. The renal pelvis is opened and a right-angle clamp is passed into the lower calyx. Before making the cut, it is advisable to place a bulldog clamp around the renal artery, since bleeding from large veins can be tremendous. The parenchyma is cut, including the mucosa of the lower calyx. The ureter is cut on its lateral border and sutured with 3–0 or 4–0 chromic sutures to the caliceal mucosa. The sutures should be interrupted at the apex. Nephrostomy tubes and stents should always be used, since these operations have a higher risk of failure.

EXTRINSIC VASCULAR OBSTRUCTION

The Hellstrom vascular relocation procedure (Hellstrom, 1949) is a simple technique for correcting renal outlet obstruction from a crossing vessel (Fig 23–14). This procedure should be used solely when there is extrinsic obstruction only. One has to be very careful that there is no intrinsic obstruction. I have operated on at least six hydronephrotic kidneys with intrinsic obstruction that had been previously thought to have only extrinsic obstruction. Intrinsic obstruction cannot be ruled out by passing a catheter through the area in question. Some of the obstructions are due to impaired function, not intrinsic or extrinsic scarring. If there is any doubt about an area of intrinsic obstruction, I think it should be corrected at the time of surgery. In the Hellstrom procedure the anterior herniation of the renal pelvis is placed behind the vessels, and the vessel is fixed to the renal pelvis to relieve the extrinsic obstruction.

HYDRONEPHROSIS IN A HORSESHOE KIDNEY

Hydronephrosis can occur in horseshoe kidneys. The obstruction is usually ureteropelvic junction obstruction, not from pressure where the ureter crosses the kidney. The isthmus is usually medial to the obstruction, not causing the obstruction. The problem with routine cutting of the isthmus is that one can injure major renal vessels or enter the collecting system. In my experience most pyeloplasties can be accomplished without having to divide the isthmus. If the isthmus is thin, it should be divided with lateral placement of the lower poles.

The pyeloplasty can be done like any other pyeloplasty. The dismembered pyeloplasty is usually the preferred procedure. To expose the horseshoe kidney, make a midline or paramedian incision. The bowel is mobilized medially. The kidney usually lies in the region of the bifurcation of the aorta. Renal vessels come from the aorta and iliac arteries.

HYDRONEPHROSIS FROM RETROCAVAL URETER

The retrocaval ureter is a congenital anomaly in which the right ureter passes behind the vena cava,

FIG 23–14.
Hellstrom technique. The obstructing vessel is moved to a nonobstructing position. An intrinsic defect should be carefully ruled out by perfusion studies on the operating table. **A,** the anterior herniation of the renal pelvis over an obstructing renal vessel. **B,** fixation of the renal vessels in a nonobstructing position. (Adapted from Smart WR: Surgical correction of hydronephrosis, in Harrison JH, et al (eds): *Campbell's Urology,* ed 4. Philadelphia, WB Saunders, 1979.)

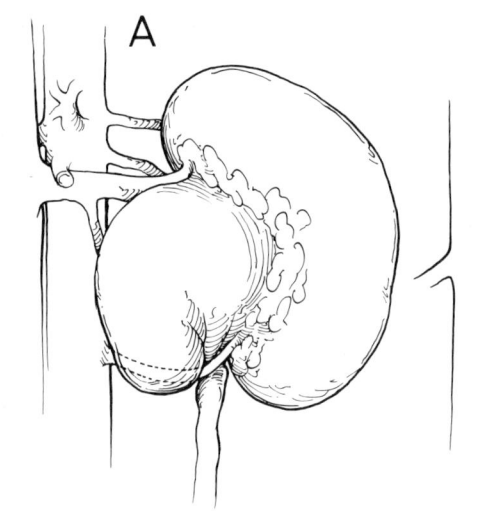
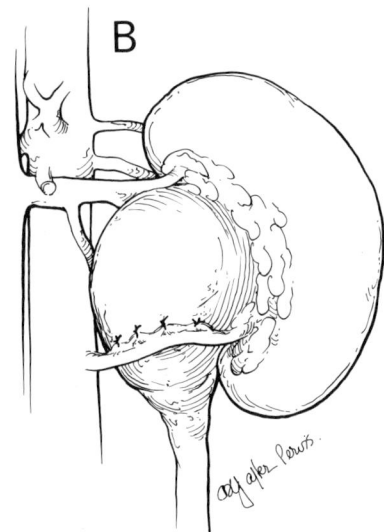

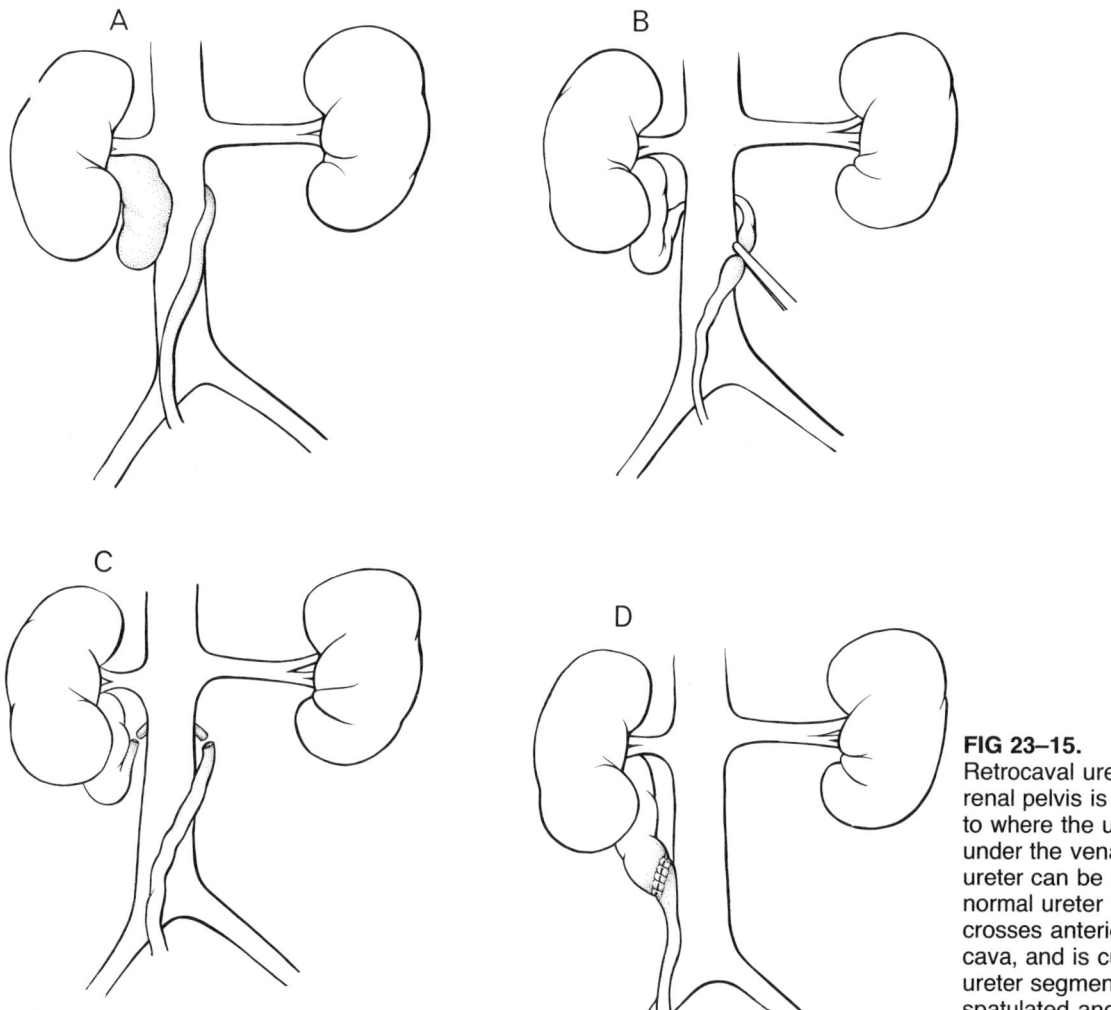

FIG 23–15.
Retrocaval ureter **(A)**. The renal pelvis is dissected down to where the ureter passes under the vena cava **(B)**. The ureter can be cut here **(C)**; the normal ureter is located as it crosses anterior to the vena cava, and is cut. The distal ureter segment is then spatulated and an anastomosis with a U-shaped flap of the renal pelvis is performed **(D)**.

causing obstruction (Fig 23–15). The abnormality is due to persistence of the posterior cardinal vein as the major portion of the infrarenal inferior vena cava. The kidney is approached through a flank or abdominal incision. The renal pelvis is dissected down to where the ureter passes under the vena cava. The ureter can be cut here. The normal ureter is located as it crosses anterior to the vena cava and is cut. One then spatulates the distal ureter segments and does an anastomosis with a U-shaped flap of the renal pelvis.

REFERENCES

1. Abbrecht PH, Malvin RL: Flow rate of urine as a determinant of renal countercurrent multiplier system. *Am J Physiol* 1960; 199:919.
2. Allen JT, Vaughan ED Jr, Gillenwater JY: The effect of indomethacin on renal blood flow and ureteral pressure in unilateral obstruction in awake dogs. *Invest Urol* 1978; 15:324.
3. Altschul R, Fedor S: Vascular changes in hydronephrosis. *Am Heart J* 1953; 46:291.
4. Anderson JC, Hynes W: Retrocaval ureter; case diagnosed preoperatively and treated successfully by plastic operation. *Br J Urol* 1949; 21:209.
5. Arendshorst WJ, Finn WF, Gottschalk CW: Nephron stop-flow pressure response to obstruction for 24 hours in the rat kidney. *J Clin Invest* 1974; 53:1497.
6. Backlund L, Nordgren L: Pressure variations in the upper urinary tract and kidney at total ureteric occlusion. *Acta Soc Med Ups* 1966; 71:285.
7. Bell ET: *Renal Diseases.* Philadelphia, Lea & Febiger, 1946.
8. Berlyne GM: Distal tubular function in chronic hydronephrosis. *Q J Med* 1961; 30:339.
9. Berlyne GM, Macken A: On the mechanism of renal inability to produce a concentrated urine in chronic hydronephrosis. *Clin Sci* 1962; 22:315.
10. Better OS, Arieff AI, Massry SG, et al: Studies on renal function after relief of complete unilateral ureteral ob-

struction of three months' duration in man. *Am J Med* 1973; 54:234.

11. Bird CE, Moise TS: Pyelovenous backflow. *JAMA* 1926; 86:661.

12. Blandy J: Emergency situations: Acute retention of urine. *Br J Hosp Med* 1978; 19:109.

13. Blandy J: *Operative Urology.* Oxford, Blackwell Scientific Publications, 1978, pp 55–62.

14. Bricker NS: Obstructive nephropathy, in Black DAK (ed): *Renal Disease.* Philadelphia, FA Davis Co, 1967.

15. Buerkert J, Alexander E, Purkerson ML, et al: On the site of decreased fluid reabsorption after release of ureteral obstruction in the rat. *J Lab Clin Med* 1976; 87:397.

16. Buerkert J, Head M, Klahr S: Effects of acute bilateral ureteral obstruction on deep nephron and terminal collecting duct function in the young rat. *J Clin Invest* 1977; 59:1055.

17. Bulger RE, Lorentz WB Jr, Colindres RE, et al: Morphologic changes in rat renal proximal tubules and their tight junctions with increased intraluminal pressure. *Lab Invest* 1974; 30:136.

18. Burns JE, Schwartz EO: Absorption from the renal pelvis in hydronephrosis due to permanent and complete occlusion of the ureter. *J Urol* 1918; 2:445.

19. Chevalier RL, Peach MJ: Hemodynamic effects of enalapril on neonatal chronic partial ureteral obstruction. *Kidney Int* 1985; 28:891.

19a. Chevalier RL, Jones CE: Contribution of endogenous vasoactive compounds to renal vascular resistance in neonatal chronic partial ureteral obstruction. *J Urol* 1986; 136:532.

20. Chisholm GD: Bilateral renal clearance studies in experimental obstructive uropathy. *Proc R Soc Med* 1964; 57:571.

21. Cosgriff DF, Berry JM: A comparative assessment of deconvolution and diuresis renography in equivocal upper urinary tract obstruction. *Nucl Med Commun* 1982; 53:377.

22. Culp OS, DeWeerd JH: Pelvic flap operation for certain types of ureteropelvic obstruction: Preliminary report. *Mayo Clin Proc* 1951; 26:483.

23. Cussen LJ, Tymms A: Hyperplasia of ureteral muscle in response to acute obstruction of the ureter. *Invest Urol* 1972; 9:504.

24. Dal Canton A, Corradi A, Stanziale R, et al: Effects of 24 hour unilateral ureteral obstruction on glomerular hemodynamics in rat kidney. *Kidney Int* 1979; 15:457.

25. Dal Canton A, Corradi A, Stanziale R, et al: Glomerular hemodynamics before and after release of 24-hour bilateral ureteral obstruction. *Kidney Int* 1980; 17:491.

26. Davis DM: Intubated ureterotomy; new operation for ureteral and ureteropelvic stricture. *Surg Gynecol Obstet* 1943; 76:513.

27. Deming CL: The effects of intrarenal hydronephrosis on the components of the renal cortex. *J Urol* 1951; 65:478.

28. DeWeerd JH: Ureteropelvioplasty, in Glenn JF (ed): *Urologic Surgery.* Philadelphia, JB Lippincott, 1983, pp 227–252.

29. Djurhuus JC, Nerstrom B, Gyrd-Hansen N, et al: Experimental hydronephrosis *Acta Chir Scand* 1976; 472:17.

30. Dominguez R, Adams RB: Renal function during and after acute hydronephrosis in the dog. *Lab Invest* 1958; 7:292.

31. Dorhout-Mees EJ: Reversible water losing state, caused by incomplete ureteric obstruction. *Acta Med Scand* 1960; 168:193.

32. Earley LE: Extreme polyuria in obstructive nephropathy: Report of a case of "water-losing nephritis" in an infant with a discussion of polyuria. *N Engl J Med* 1956; 255:600.

33. Eiseman B, Vivion C, Vivian J: Fluid and electrolyte changes following relief of urinary obstruction. *J Urol* 1955; 74:222.

34. English PJ, Testa HJ, Gosling JA, et al: Idiopathic hydronephrosis in childhood—a comparison between diuresis renography and upper urinary tract morphology. *Br J Urol* 1982; 54:603.

35. Finkle AL, Karg SJ, Smith DR: Parameters of renal functional capacity in reversible hydroureteronephrosis in dogs: II. Effects of one hour of ureteral obstruction upon urinary volume, osmolality, TcH_2O, C_{PAH}, RBF_{kr} and pUO_2. *Invest Urol* 1968; 6:26.

36. Foley FEB: New plastic operation for strictures at ureteropelvic junction; report of 20 operations. *J Urol* 1937; 38:643.

37. Gaudio KM, Siegel NJ, Hayslett J, et al: Renal perfusion and intratubular pressure during ureteral occlusion in the rat. *Am J Physiol* 1980; 238:F205.

38. Gee WE, Kiviat MD: Ureteral response to partial obstruction: Smooth muscle hyperplasia and connective tissue proliferation. *Invest Urol* 1975; 12:309.

39. Gillenwater JY, Westervelt FB Jr, Vaughan ED, et al: Renal function after release of chronic unilateral hydronephrosis in man. *Kidney Int* 1975; 7:179.

40. Gillenwater JY, Teates D, Marion DN: Prediction of recoverability in hydronephrosis with ^{131}I hippuran renograms. Presented at Annual Meeting, American Urological Association, New York, May 13–17, 1979.

40a. Gillenwater JY: The pathophysiology of urinary obstruction, in Walsh PC, Gittes RF, Perlmutter AD, et al (eds): *Campbell's Urology,* ed 5. Philadelphia, WB Saunders, 1986, pp 542–571.

41. Goldsmith C: Postobstructive diuresis. *Kidney* 1968; 2:1.

42. Gonzalez R, Chiou RK: Diagnosis of upper urinary tract obstruction in children: Comparison of diuresis renography and perfusion pressure flow studies. *J Urol* 1985; 133:646.

43. Gosling JA, Dixon JS: Species variation in the location of upper urinary tract pacemaker cells. *Invest Urol* 1974; 11:418.

44. Gosling JA, Dixon JS: Functional obstruction of the ureter and renal pelvis: A histological and electron microscopic study. *Br J Urol* 1978; 50:145.

45. Gottschalk CW, Mylle M: Micropuncture study of pressures in proximal tubules and peritubular capillaries of

the rat kidney and their relation to ureteral and renal venous pressures. *Am J Physiol* 1956; 185:430.

46. Gottschalk CW, Mylle M: Micropuncture study of pressures in proximal and distal tubules and peritubular capillaries of the rat kidney during osmotic diuresis. *Am J Physiol* 1957; 189:323.

47. Graham JB: Recovery of kidney after ureteral obstruction. *JAMA* 1962; 181:993.

48. Hanna MK, Jetts RD, Sturgess JM, et al: Ureteral structure and ultrastructure: II. Congenital ureteropelvic junction obstruction and primary obstructive megaureter. *J Urol* 1976; 116:725.

49. Harris RH, Gill JM: Changes in glomerular filtration rate during complete ureteral obstruction in rats. *Kidney Int* 1981; 19:603.

50. Harris RH, Hise MK, Best CF: Renotrophic factors in urine. *Kidney Int* 1983; 23:616.

51. Harris RH, Yarger WE: Renal function after release of unilateral ureteral obstruction in rats. *Am J Physiol* 1974; 227:806.

52. Hay AW, Norman WJ, Price ML, et al: A comparison between diuresis renography and the Whitaker test in 64 kidneys. *Br J Urol* 1984; 56:561.

53. Hayslett JP: Effect of age on compensatory renal growth. *Kidney Int* 1983; 23:599.

53a. Hellstrom J, Giertz G, Lindblom K: Pathogenesis and treatment of hydronephrosis, in *VIII Congreso de la Sociedad Internacional de Urologia*. Paris, Libraire Gaston Doin, 1949.

54. Hermann W: Sitzungsberichte d. k. Akad, der Wissensch. zu Wein. *Math-Naturwiss* 1859; 36:349. Cited in Cushny AR: *The Secretion of the Urine*. London, Longmans, Green and Co, 1917.

54a. Hinman F: Experimental hydronephrosis: Repair following ureterocystoneostomy in white rats with complete ureteral obstruction. *J Urol* 1919; 3:147.

55. Hinman F: Renal counterbalance: An experimental and clinical study with reference to significance of disuse atrophy. *Trans Am Soc Genitourin Surg* 1922; 15:241.

56. Hinman F: Renal counterbalance. *Arch Surg* 1926; 12:1105.

57. Hinman F: Pathogenesis of hydronephrosis. *Surg Gynecol Obstet* 1934; 58:356.

58. Hinman F: Hydronephrosis: I. The structural change. II. The functional change. III. Hydronephrosis and hypertension. *Surgery* 1945; 17:816.

59. Hinman F, Hepler AB: Experimental hydronephrosis: The effect of changes in blood pressure and in blood flow on its rate of development: I. Splanchnotomy: Increased intrarenal blood pressure and flow; diuresis. *Arch Surg* 1925a; 11:578.

60. Hinman F, Hepler AB: Experimental hydronephrosis: The effect of changes in blood pressure and in blood flow on its rate of development: II. Partial obstruction of the renal artery: Diminished blood flow, diminished intrarenal pressure and oliguria. *Arch Surg* 1925b; 11:649.

61. Hinman F, Hepler AB: Experimental hydronephrosis; The effect of changes in blood pressure and in blood flow on its rate of development, and the significance of the venous collateral system: III. Partial obstruction of the renal vein without and with ligation of all collateral veins. *Arch Surg* 1925c; 11:917.

62. Hinman F, Hepler AB: Experimental hydronephrosis: The effect of ligature of one branch of the renal artery on its rate of development: IV. Simultaneous ligation of the posterior branch of the renal artery and the ureter on the same side. *Arch Surg* 1926; 12:830.

63. Hinman F, Lee-Brown RK: Pyelovenous outflow, its relation to pelvic reabsorption, to hydronephrosis and to accidents of pyelography. *JAMA* 1924; 82:607.

64. Hinman F, Morison DM: An experimental study of the circulatory changes in hydronephrosis. *J Urol* 1924; 21:435.

65. Howards SS: Postobstructive diuresis: A misunderstood phenomenon. *J Urol* 1973; 110:537.

66. Huguenin M, Ott CE, Romero JC, et al: Influence of renin depletion on renal function after release of 24 hours ureteral obstruction. *J Lab Clin Med* 1976; 87:58.

67. Ichikawa I, Brenner BM: Role of local intrarenal vasoconstrictor-vasodilator interactions in mild partial ureteral obstruction. *Am J Physiol* 1979; 236:F131.

68. Israel AR: Validation of the diuretic renogram with simultaneous intrapelvic pressure monitoring. Presentation to the 77th Annual Meeting of the American Urological Society, Kansas City, 1982.

69. Jaenike JR: The renal response to ureteral obstruction: A model for the study of factors which influence glomerular filtration pressure. *J Lab Clin Med* 1970; 76:373.

70. Jaenike JR: The renal functional defect of postobstructive nephropathy: The effects of bilateral ureteral obstruction in the rat. *J Clin Invest* 1972; 51:2999.

71. Jaenike JR, Bray GA: Effects of acute transitory urinary obstruction in the dog. *Am J Physiol* 1960; 199:1219.

72. Kalika V, Bard RH, Iloreta A, et al: Prediction of renal functional recovery after relief of upper urinary tract obstruction. *J Urol* 1981; 126:301.

73. Kerr WS Jr: Effect of complete ureteral obstruction for one week on kidney function. *J Appl Physiol* 1954; 6:762.

74. Kerr WS Jr: Effects of complete ureteral obstruction in dogs on kidney function. *Am J Physiol* 1956; 184:521.

75. Kessler RH: Acute effects of brief ureteral stasis on urinary and renal papillary chloride concentration. *Am J Physiol* 1960; 199:1215.

76. Klahr S, Buerkert J, Purkerson JL: The kidney in obstructive nephropathy. *Contrib Nephrol* 1977; 7:220.

77. Ladefoged O, Djurhuus JC: Morphology of the upper urinary tract in experimental hydronephrosis in pigs. *Acta Chir Scand* 1976; 472:29.

78. Lewis HY, Pierce JM: Return of function after relief of complete ureteral obstruction of 69 days' duration. *J Urol* 1962; 88:377.

79. Lome LG, Pinsky S, Levy L: Dynamic renal scan in the nonvisualizing kidney. *J Urol* 1979; 121:148.

80. Lorentz WB Jr, Lassiter WE, Gottschalk CW: Renal tubular permeability during increased intrarenal pressure. *J Clin Invest* 1972; 51:484.

81. Lupton EW, O'Reilly PH, Testa HJ, et al: Diuresis re-

nography and morphology in urinary tract obstruction. *Br J Urol* 1981; 51:449.

82. Lupton EW, Richards D, Testa HJ, et al: A comparison of diuresis renography, the Whitaker test and renal pelvic morphology in idiopathic hydronephrosis. *Br J Urol* 1985; 57:119.

83. Malt RA: Humoral factors in regulation of compensatory renal hypertrophy. *Kidney Int* 1983; 23:611.

84. Malvin RL, Kutchai H, Ostermann F: Decreased nephron population resulting from increased ureteral pressure. *Am J Physiol* 1964; 207:835.

85. McCrory WW, Shibuya M, Leumann E, et al: Studies of renal function in children with chronic hydronephrosis. *Pediatr Clin North Am* 1971; 18:445.

86. McDonald JR, Mann FC, Priestly JT: The maximum intrapelvic pressure (secretory) of the kidney of the dog. *J Urol* 1937; 37:326.

87. McDougal WS, Flanigan RC: Renal functional recovery of the hydronephrotic kidney predicted before relief of the obstruction. *Invest Urol* 1981; 18:440.

88. McDougal WS, Wright FS: Defect in proximal and distal sodium transport in postobstructive diuresis. *Kidney Int* 1972; 2:304.

89. Michaelson G: Percutaneous puncture of the renal pelvis, intrapelvic pressure, and the concentrating capacity of the kidney in hydronephrosis. *Acta Med Scand (Suppl)* 1974; 559:1.

90. Moody TE, Vaughan ED Jr, Gillenwater JY: Relationship between renal blood flow and ureteral pressure during 18 hours of total unilateral occlusion. *Invest Urol* 1975; 13:246.

91. Moody TE, Vaughan ED Jr, Gillenwater JY: Comparison of the renal hemodynamic response to unilateral and bilateral ureteral obstruction. *Invest Urol* 1977; 14:455.

92. Morison DM: Routes of absorption in hydronephrosis: Experimentation with dyes in the totally obstructed ureter. *Br J Urol* 1929; 1:30.

93. Morrison AR, Nishikawa K, Needleman P: Unmasking of thromboxane A$_2$ synthesis by ureteral obstruction in the rabbit kidney. *Nature* 1977; 267:259.

94. Morrison AR, Nishikawa K, Neddleman P: Thromboxane A$_2$ biosynthesis in the ureter-obstructed isolated perfused kidney of the rabbit. *J Pharmacol Exp Ther* 1978; 205:1.

95. Muldowney FP, Duffy GJ, Kelly DG et al: Sodium diuresis after relief of obstructive uropathy. *N Engl J Med* 1966; 274:1294.

96. Murphy GP, Scott WW: The renal hemodynamic response to acute and chronic ureteral occlusions. *J Urol* 1966; 95:636.

97. Naber KG, Madsen PO: Renal function in chronic hydronephrosis with and without infection and the role of lymphatics: An experimental study in dogs. *Urol Res* 1974; 2:1.

98. Narath PA: The hydromechanics of the calix renalis. *J Urol* 1940; 43:145.

99. Needleman P, Bronson SD, Wyche A, et al: Cardiac and renal prostaglandin I$_2$ biosynthesis and biological effects in isolated perfused rabbit tissues. *J Clin Invest* 1978; 61:839.

100. Nesbit RM: Diagnosis of intermittent hydronephrosis: Importance of pyelography during episodes of pain. *J Urol* 1956; 67:787.

101. Nishikawa K, Morrison A, Needleman P: Exaggerated prostaglandin biosynthesis and its influence on renal resistance in the isolated hydronephrotic rabbit kidney. *J Clin Invest* 1977; 59:1143.

102. Notfrey RG: The structural basis for normal and abnormal ureteral motility. *Ann R Coll Surg Engl* 1971; 49:250.

103. Notfrey RG: Electron microscopy of the primary obstructive megaureter. *Br J Urol* 1972; 44:229.

104. Obniski: *Centralbl Physiol* 1907; 21:548. Cited by Cushny A: *The Secretion of the Urine*. London, Longmans, Green and Co, 1917.

105. Olesen S, Madsen PO: Renal function during experimental hydronephrosis: Function during partial obstruction following contralateral nephrectomy in the dog. *J Urol* 1968; 99:692.

106. Olsson O: Studies on back-flow in excretion urography. *Acta Radiol (Suppl)* 1948; 70.

107. O'Reilly PH, Testa HJ, Lawson RS, et al: Diuresis renography in equivocal urinary tract obstruction. *Br J Urol* 1978; 50:76.

108. O'Reilly PH: Diuresis renography eight years later: A critical update. *J Urol* 1986; 136:993.

109. Papadopoulou ZL, Slotkoff LM, Eisner GM, et al: Glomerular filtration during stop-flow. *Proc Soc Exp Biol Med* 1969; 130:1206.

110. Patt LM, Houck JC: Role of polypeptide growth factors in normal and abnormal growth. *Kidney Int* 1983; 23:603.

111. Preuss HG: Compensatory renal growth symposium— an introduction. *Kidney Int* 1983; 23:571.

112. Provoost AP, Molenaar JC: Renal function during and after a temporary complete unilateral ureter obstruction in rats. *Invest Urol* 1981; 18:242.

113. Rao NR, Heptinstall RH: Experimental hydronephrosis: A microangiographic study. *Invest Urol* 1968; 6:183.

114. Reisman DD, Kamholz JH, Kantor HI: Early deligation of the ureter. *J Urol* 1957; 78:363.

115. Riehle RA, Vaughan ED Jr: Renin participation in hypertension associated with unilateral hydronephrosis. *J Urol* 1981; 126:243.

116. Rist M, Lee S, Gittes RF: Glomerular filtration rate and effective renal plasma flow in four-kidney rats. *Surg Forum* 1975; 26:577.

117. Roussak NJ, Oleesky S: Water-losing nephritis, a syndrome simulating diabetes insipidus. *Q J Med* 1954; 23:147.

118. Scardino PL, Prince CL: Verical flap ureteropelvioplasty: Preliminary report. *South Med J* 1953; 46:325.

119. Schaeffer T, Grayhack JT: Surgical management of ureteropelvic junction obstruction, in Walsh PC, Gittes

RF, Perlmutter AD et al. (eds): *Campbell's Urology*, ed. 5. Philadelphia, WB Saunders Co, 1985, pp 2505–2531.

120. Schnermann J, Wright FS, Davis JM, et al: Regulation of superficial nephron filtration rate by tubulo-glomerular feedback. *Pflugers Arch* 1970; 318:147.

121. Schramm LP, Carlson DE: Inhibition of renal vasoconstriction by elevated ureteral pressure. *Am J Physiol* 1975; 228:1126.

122. Senac MU Jr, Miller JH, Stanley P: Evaluation of obstructive uropathy in children; radionuclide renography versus the Whitaker test. *Am J Radiol* 1985; 1:11.

123. Sheehan HL, Davis JC: Experimental hydronephrosis. *Arch Pathol* 1959; 68:185.

124. Shimamura T, Kissane JM, Gyorki F: Experimental hydronephrosis: Nephron dissection and electron microscopy of the kidney following obstruction of the ureter and in recovery from obstruction. *Lab Invest* 1966; 15:629.

125. Siegel NJ, Feldman RA, Lytton B, et al: Renal cortical blood flow distribution in obstructive nephropathy. *Circ Res* 1977; 40:379.

126. Silber S: Compensatory and obligatory renal growth in babies and adults. *Aust NZ J Surg* 1974; 44:421.

127. Silber S: Growth of baby kidneys transplanted into adults. *Surg Forum* 1975; 26:579.

128. Smith DR, Schulte JW, Smart WR: Surgery of the kidney, in Campbell MF (ed): *Urology*, ed 2. Philadelphia, WB Saunders Co, 1963, chap 53, pp 2325–2415.

128a. Smart WR: Surgical correction of hydronephrosis, in Harrison JH, et al (eds): *Campbell's Urology*, ed 4. Philadelphia, WB Saunders, 1979, pp 2047–2114.

129. Solez K, Pouchak S, Buono RA, et al: Inner medullary plasma flow in the kidney with ureteral obstruction. *Am J Physiol* 1976; 231:1315.

130. Stecker JR Jr, Gillenwater JY: Experimental partial ureteral obstruction: I. Alteration in renal function. *Invest Urol* 1971; 8:377.

131. Strong KC: Plastic studies in abnormal renal architecture: The parenchymal alterations in experimental hydronephrosis. *Arch Pathol* 1940; 29:77.

132. Suki W, Eknoyan G, Rector FC Jr, et al: Patterns of nephron perfusion in acute and chronic hydronephrosis. *J Clin Invest* 1966; 45:122.

133. Taylor MJ, Ullmann E: Glomerular filtration after obstruction of the ureter. *J Physiol* 1961; 157:38.

134. Thirakomen K, Kozlov N, Arruda J, et al: Renal hydrogen ion secretion following the release of unilateral ureteral obstruction. *Am J Physiol* 1976; 231:1233.

135. Tuffier M: Etude clinique et experimentale sur l'hydronephrose. *Ann Mal Org Genitourin* 1894; 12:14. Cited by Cushny A: *The Secretion of the Urine*. London, Longmans, Green and Co, 1917.

136. Vaughan ED Jr, Buhler FR, Laragh JH: Normal renin secretion in hypertensive patients with primarily unilateral chronic hydronephrosis. *J Urol* 1974; 112:153.

137. Vaughan ED Jr, Gillenwater JY: Diagnosis, characterization and management of postobstructive diuresis. *J Urol* 1973; 109:286.

138. Vaughan ED Jr, Shenasky JH II, Gillenwater JY: Mechanism of acute hemodynamic response to ureteral occlusion. *Invest Urol* 1971; 9:109.

139. Vaughan ED Jr, Sorenson EJ, Gillenwater JY: The renal hemodynamic response to chronic unilateral complete ureteral occlusion. *Invest Urol* 1970; 8:78.

140. Vaughan ED Jr, Sweet RE, Gillenwater JY: Unilateral ureteral occlusion: Pattern of nephron repair and compensatory response. *J Urol* 1973; 109:979.

141. Walls J, Buerkert JE, Purkerson JL, et al: Nature of the acidifying defect after relief of ureteral obstruction. *Kidney Int* 1975; 7:304.

142. Weidmann P, Beretta-Piccoli C, Hirsch D, et al: Curable hypertension with unilateral hydronephrosis: Studies on the role of circulating renin. *Ann Intern Med* 1977; 87:437.

143. Whitaker RH: Methods of assessing obstruction in dilated ureters. *Br J Urol* 1973; 45:15.

144. Whitaker RH, Buxton-Thomas MS: A comparison of pressure-flow studies and renography in equivocal upper urinary tract obstruction. *J Urol* 1984; 131:446.

145. Whitfield HN, Britton KE, Hendry WF, et al: The distinction between obstructive nephropathy and uropathy by radionuclide transit times. *Br J Urol* 1978; 50:433.

146. Wilson DR: Micropuncture study of chronic obstructive nephropathy before and after release of obstruction. *Kidney Int* 1972; 2:119.

147. Wilson DR: The influence of volume expansion on renal function after relief of chronic unilateral ureteral obstruction. *Kidney Int* 1974; 5:402.

148. Wilson DR: Renal function during and following obstruction. *Ann Rev Med* 1977; 28:329.

149. Wilson DR, Honrath U: Cross-circulation study of natriuretic factors in postobstructive diuresis. *J Clin Invest* 1976; 57:380.

150. Winberg J: Renal function in congenital bladder neck obstruction. *Acta Chir Scand* 1958; 116:332.

151. Winberg J: Renal function in water-losing syndrome due to lower urinary tract obstruction before and after treatment. *Acta Paediatr* 1959; 48:149.

152. Winton FR: Influence of increase of ureteral pressure on the isolated mammalian kidney. *J Physiol* 1931; 71:381.

153. Wright FS: Intrarenal regulation of glomerular filtration rate. *N Engl J Med* 1974; 291:135.

154. Wright FH, Howards SS: Obstructive injury, in Brenner BM, Rector FC (eds): *The Kidney*. Philadelphia, WB Saunders Co, 1979.

155. Wright FS, Briggs JP: Feedback regulation of glomerular filtration rate. *Am J Physiol* 1977; 233:F1.

156. Yarger WE, Aynedjian HS, Bank N: A micropuncture study of postobstructive diuresis in the rat. *J Clin Invest* 1972; 51:625.

157. Yarger WE, Griffith LD: Intrarenal hemodynamics following chronic unilateral obstruction in the dog. *Am J Physiol* 1974; 227:806.

158. Yarger WE, Schocken DD, Harris RH: Obstructive uropathy in the rat. Possible roles for the renin-angiotensin system, prostaglandins, and thromboxanes in postobstructive renal function. *J Clin Invest* 1980; 65:400.

159. Zelman SJ, Zenser TV, Davis BB: Renal growth in response to unilateral ureteral obstruction. *Kidney Int* 1983; 23:594.

160. Zetterstrom R, Ericsson NO, Winberg J: Separate renal function studies in predominantly unilateral hydronephrosis. *Acta Paediatr* 1958; 47:540.

161. Zincke H, Kelalis PP, Culp OS: Ureteropelvic obstruction in children. *Surg Gynecol Obstet* 1974; 139:873.

Chapter 24

Renal Transplantation

Arthur I. Sagalowsky, M.D.
J. Harold Helderman, M.D.
Paul C. Peters, M.D.

The ability to sustain life in patients with end-stage renal disease (ESRD) has become a commonplace clinical reality during the past 30 years owing to the development of dialysis and renal transplantation. Criteria for patient selection for transplantation are discussed below. Briefly, successful renal transplantation offers a higher quality of life and greater rehabilitation than dialysis in appropriate patients.

Review of recently released statistics of the U.S. ESRD program for 1983 places the magnitude of renal failure and its management in perspective (U.S. ESRD Program, 1984). Through 1983, approximately 70,000

patients in the United States were on dialysis (Fig 24–1). Assuming continued growth in the ESRD program, more than 80,000 patients will be on dialysis by 1986. Patient characteristics in the ESRD program by sex, race, age, and primary diagnosis through 1983 are shown in Table 24–1. These data are based on 49,039 patients reported from 25 of 32 networks in the ESRD program. The percentage of male and female patients is nearly equal (53.0% and 47.0%, respectively). The greater proportion of whites to blacks (62.1% vs. 34.6%, respectively) is unexplained. Nearly 39% of the ESRD patients are less than 50 years old. Hypertensive

TABLE 24–1.

Characteristics of Patients in U.S. End-Stage Renal Disease Program to 1983*†

TOTAL PATIENTS BY PRIMARY DIAGNOSIS	PRIMARY DIAGNOSIS‡					ALL OTHERS	NOT REPORTED	TOTAL, %	% OF TOTAL YEAR-END DIALYSIS PATIENTS
	1	2	3	4	5				
Race									
White	15.0	22.6	16.0	9.1	4.8	23.5	9.0	100	62.1
Black	35.8	13.8	15.2	2.1	4.0	14.6	14.5	100	34.6
Other	18.4	27.1	20.8	5.8	4.4	18.8	4.7	100	3.3
Sex									
Male	23.7	21.9	14.1	5.8	4.3	19.5	10.7	100	53.0
Female	20.8	17.4	17.8	7.5	4.5	21.2	10.8	100	47.0
Age, yr									
0–9	5.1	19.4	2.6	5.9	1.8	58.6	6.6	100	0.6
10–19	3.3	33.4	0.8	2.5	3.4	47.7	8.9	100	2.4
20–29	8.6	33.5	10.1	2.0	4.2	29.1	12.5	100	7.9
30–39	15.5	28.4	17.6	3.1	3.3	19.4	12.7	100	12.9
40–49	22.0	20.8	15.4	9.1	3.6	16.6	12.5	100	15.0
50–59	22.6	16.4	18.9	9.9	5.1	16.0	11.1	100	21.3
60–69	26.7	15.4	18.3	7.6	4.8	17.8	9.4	100	24.2
70+	31.1	13.2	12.1	4.9	6.8	23.4	8.5	100	15.7
% of total year-end dialysis patients	22.3	19.7	15.8	6.6	4.5	20.3	10.8	100	100.0

*From the U.S. ESRD Program: Selected 1983 statistics. *Contemp Dialysis Nephrol* 1984; 5:51–58. Used with permission.
†Based on a sample containing 1983 year-end statistics from 25 of 32 networks (total patients: 49,039).
‡1 indicates hypertensive disease; 2, glomerulonephritis; 3, diabetic nephropathy; 4, polycystic kidney disease; and 5, etiology unknown.

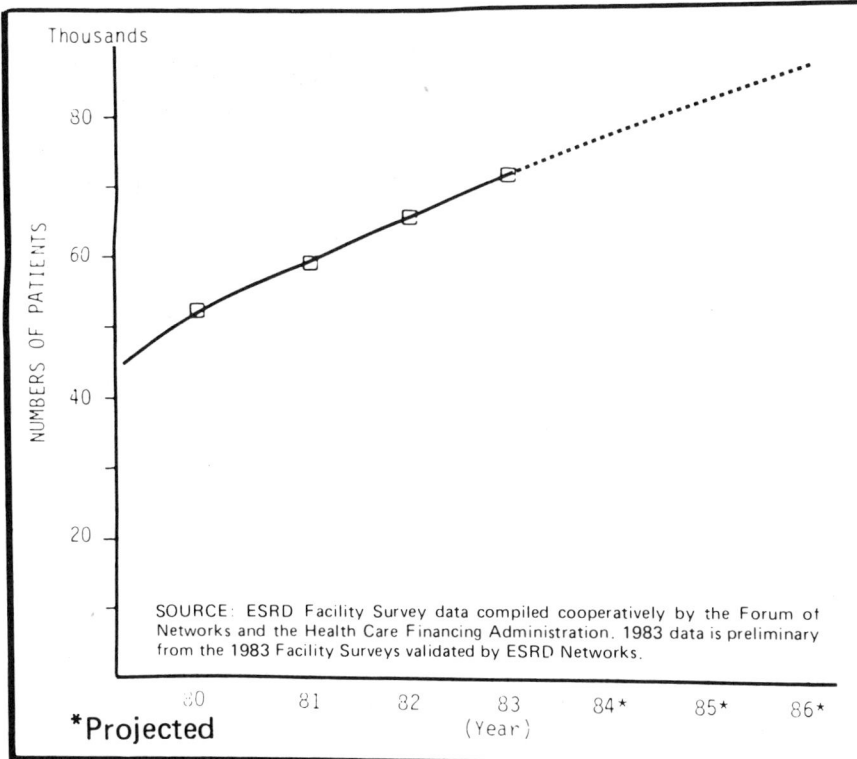

FIG 24–1.
Actual and projected number of patients on dialysis 1980–1986. (From The U.S. ESRD program: Selected 1983 statistics. *Contemp Dialysis Nephrol* 1984; 5:51–58. Used with permission.)

disease currently accounts for the greatest number of patients with renal failure, followed by glomerulonephritis and diabetes mellitus (22.3%, 19.7%, and 15.8%, respectively). Characteristics of patients newly entered into the ESRD program during 1983 are shown in Table 24–2. The continuing increase in hypertension or diabetes mellitus as the cause of renal failure is noted (24.9% and 25.4%, respectively). The number of new patients beginning long-term dialysis in the United States during 1983 varied from 60 to 140 per million population (Fig 24–2). Over 6,000 renal transplants were performed in the United States during 1983 (Fig 24–3). Approximately two thirds to three fourths of the transplants were from cadaver donors.

IMMUNOBIOLOGY OF TRANSPLANTATION

Transplantation immunity and rejection follow the basic laws of genetics. A brief definition of terms is essential to their discussion. A *locus* is the location of a gene on a chromosome. *Alleles* are alternate forms of a gene at a given locus on homologous chromosomes. A *haplotype* is the sum of genetic material or unit of inheritance contributed by each parent. Each individual receives a maternal and a paternal haplotype. The *phenotype* is the total histocompatibility antigens expressed by an individual without distinguishing which antigens are maternally or paternally derived. The *genotype* is the total of histocompatibility antigens expressed by an individual, designating the maternal and paternal contributions.

Understanding of the conceptual organization of transplantation immunology occurred parallel to discoveries regarding tumor immunology and the immune process in general. Landsteiner (1901) first described histocompatibility antigens specific for blood groups. Some 30 years later, Haldane (1933) predicted that tumors and other tissues were rejected owing to differences in alloantigens analogous to differences in blood group antigens. In 1937 Gorer proved the preceding hypothesis and developed a genetic theory of transplantation that simply stated: "Normal and neoplastic tissues contain isoantigenic factors which are genetically determined. Isoantigenic factors present in grafted tissue and absent in the host are capable of eliciting a response which results in destruction of the graft." Medawar (1944) proved that the same rules determine rejection or acceptance of skin allografts and isografts in rabbits. Subsequently, Medawar proved that the same results also apply to skin grafts in human burn victims. Billingham and associates (1956) demonstrated the phenomenon of accelerated or second set rejection in previously sensitized recipients.

HUMAN MAJOR HISTOCOMPATIBILITY COMPLEX (MHC)

In 1958 Dausset described an antibody detecting a leukocyte antigen Mac(HLA-A$_2$). Subsequently, the hu-

TABLE 24–2.

Characteristics of Patients Entered Into U.S. End-Stage Renal Disease (ESRD) Program During 1983*†

TOTAL NEWLY DIAGNOSED PATIENTS BY PRIMARY DIAGNOSIS	PRIMARY DIAGNOSIS‡					ALL OTHERS	NOT REPORTED	TOTAL, %	% OF TOTAL NEW ESRD PATIENTS
	1	2	3	4	5				
Race									
White	19.5	17.5	25.4	6.5	5.0	24.5	1.6	100	66.8
Black	39.0	12.7	23.5	2.2	3.8	16.9	1.9	100	28.9
Other	13.6	20.5	33.4	3.2	7.6	17.9	3.8	100	4.3
Sex									
Male	26.6	17.9	22.4	4.9	4.7	21.8	1.7	100	54.6
Female	23.3	15.0	27.9	5.3	4.1	22.5	1.9	100	45.4
Age, yr									
0–9	7.7	13.5	3.2	9.6	3.2	60.3	2.5	100	1.0
10–19	5.0	31.3	1.0	3.1	4.4	51.8	3.4	100	2.5
20–29	10.5	29.2	21.9	2.2	4.8	30.0	1.4	100	6.9
30–39	15.6	26.9	30.0	3.4	4.1	18.6	1.4	100	11.1
40–49	21.5	16.8	28.8	7.8	3.9	19.2	2.0	100	12.5
50–59	23.8	14.2	31.0	8.2	4.0	17.6	1.2	100	18.7
60–69	32.1	14.1	27.5	4.2	4.2	16.5	1.4	100	24.3
70+	33.9	11.6	20.7	2.5	6.3	23.0	2.0	100	23.0
% of total new ESRD patients	24.9	16.3	25.4	5.1	4.8	22.0	1.5	100	100.0

*From the U.S. ESRD Program: Selected 1983 statistics. *Contemp Dialysis Nephrol* 1984; 5:51–58.
†Based on a sample containing 1984 statistics from 25 of 32 networks (total patients: 16,705).
‡1 indicates hypertensive disease; 2, glomerulonephritis; 3, diabetic retinopathy; 4, polycystic kidney disease; and 5, etiology unknown.

man MHC was designated HLA for "human leukocyte antigen." HLA antigens are present in serum, saliva, and on the surface of all mature nucleated cells. The HLA region in man is located on the short arm of the sixth chromosome (Fig 24–4). (DeWolf, 1981). HLA-A, -B, and -C alleles encode for class 1 antigens, which are glycoproteins composed of two noncovalently bound polypeptide chains. There is an allospecific heavy chain (44,000 daltons molecular weight) and a common light chain (12,000 daltons mol wt), which is identical to β_2

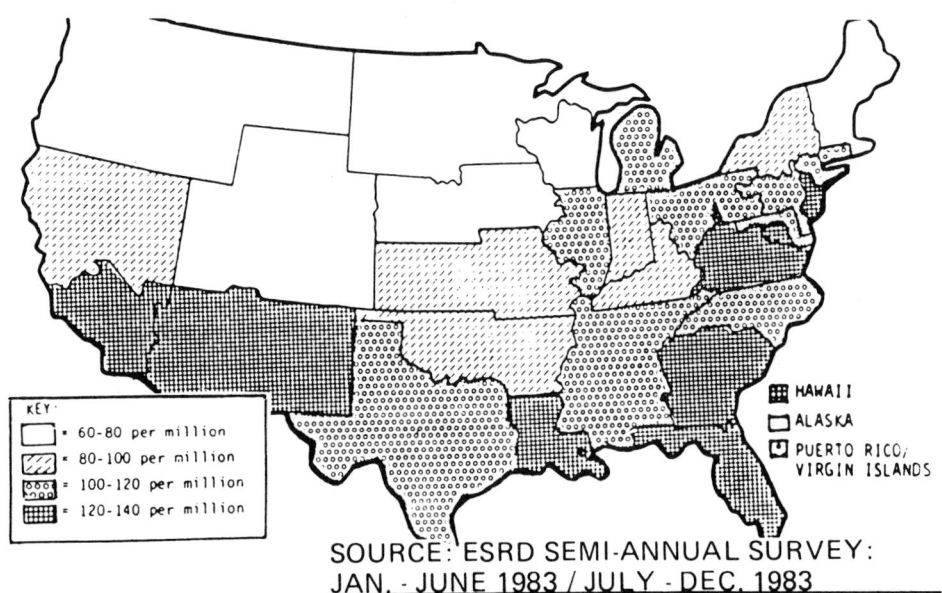

FIG 24–2.
New patients on dialysis in United States during 1983. (From the U.S. ESRD program: Selected 1983 statistics. *Contemp Dialysis Nephrol* 1984; 5:51–58. Used with permission.)

KEY
- 60-80 per million
- 80-100 per million
- 100-120 per million
- 120-140 per million

HAWAII
ALASKA
PUERTO RICO/ VIRGIN ISLANDS

SOURCE: ESRD SEMI-ANNUAL SURVEY: JAN. - JUNE 1983 / JULY - DEC. 1983

NOTE: This figure is computed based on location of the facility where patients began chronic maintenance dialysis, not by the residence of the patient. Also, 1982 population estimates have been used for computing rates.

Thousands

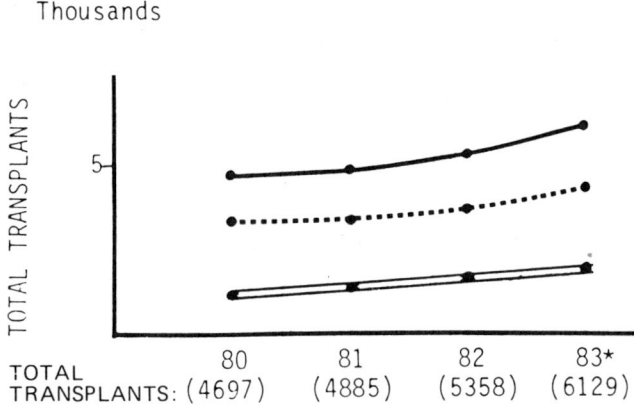

TOTAL TRANSPLANTS: (4697) 80 (4885) 81 (5358) 82 (6129) 83*

KEY: All Transplants: ▬▬ Cadaver: ••••• Live-Related: ═══

FIG 24–3.
Number of renal transplants in United States during 1983. (From The U.S. ESRD program: Selected 1983 statistics. *Contemp Dialysis Nephrol* 1984; 5:51–58. Used with permission.)

microglobulin encoded on another chromosome. These class 1 antigens are spatially located predominantly on the outer surface of (80%) and partially within (20%) the cell membrane. Among the class 1 molecules the HLA-B, followed by HLA-A antigens, are the dominant histocompatibility determinants, and HLA-C antigens are weak. Class 1 antigens stimulate predominantly cytotoxic T-lymphocytes and B-lymphocytes (see section on mechanisms of rejection).

HLA alleles of the "D region" (DR, DQ, DP) encode for class 2 antigens (Fig 24–5). Class 2 molecules contain two chains of approximately 34,000 and 28,000 dal-

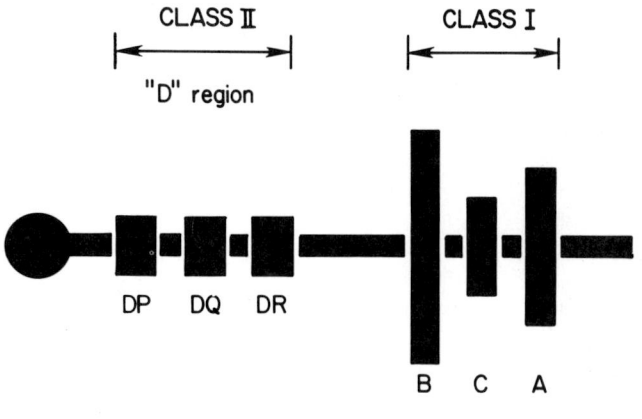

6th CHROMOSOME

FIG 24–4.
Location of the major human histocompatibility complex. (From Helderman JH: Renal transplantation, in *Internal Medicine Grand Rounds.* Dallas, University of Texas Health Science Center at Dallas, Oct 25, 1984. Used with permission.)

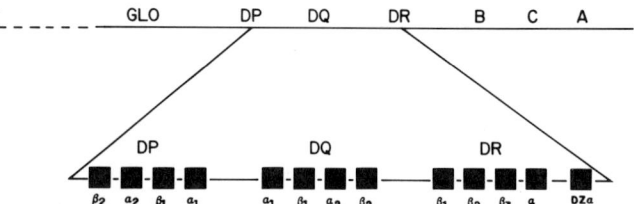

FIG 24–5.
Components of the histocompatibility D region. (From Helderman JH: Renal transplantation, in *Internal Medicine Grand Rounds.* Dallas, University of Texas Health Science Center at Dallas, Oct 25, 1984. Used with permission.)

tons mol wt each. Class 2 antigens stimulate proliferation of helper T-lymphocytes and B-lymphocytes and are present predominantly on macrophages and on B cells.

The most recent list of HLA specificities recognized at the Ninth International Histocompatibility Workshop held in Munich in 1984 is shown in Table 24–3. The enormous polymorphism of the HLA system is readily apparent, with 23 A, 48 B, 8 C, 19 D, 16 DR, 6 DP,

TABLE 24–3.

HLA Specificities—1984*

HLA-A	HLA-B	HLA-B	HLA-C	HLA-DR	HLA-D
A1	B5	Bw48	Cw1	DR1	Dw1
A2	B7	B49(21)	Cw2	DR2	Dw2
A3	B8	Bw50(21)	Cw3	DR3	Dw3
A9	B12	B51(5)	Cw4	DR4	Dw4
A10	B13	Bw52(5)	Cw5	DR5	Dw5
A11	B14	Bw53	Cw6	DRw6	Dw6
Aw19	B15	Bw54(w22)	Cw7	DR7	Dw7
A23 (9)	B16	Bw55(w22)	Cw8	DRw8	Dw8
A24 (9)	B17	Bw56(w22)		DRw9	Dw9
A25(10)	B21	Bw57(17)		DRw10	Dw10
A26(10)	Bw22	Bw58(17)		DRw11(5)	Dw11(w7)
A28	B27	Bw59		DRw12(5)	Dw12
A29(w19)	B35	Bw60(40)		DRw13(w6)	Dw13
A30(w19)	B37	Bw61(40)		DRw14(w6)	Dw14
A31(w19)	B38(16)	Bw62(15)			Dw15
A32(w19)	B39(16)	Bw63(15)		DRw52	Dw16
Aw33(w19)	B40	Bw64(14)		DRw53	Dw17(w7)
Aw34(10)	Bw41	Bw65(14)			Dw18(w6)
Aw36	Bw42	Bw67		HLA-DQ	Dw19(w6)
Aw43	B4412	Bw70			
Aw66(10)	B45(12)	Bw71(w70)		DQw1	
Aw68(28)	Bw46	BW72(w70)		DQw2	
Aw69(28)	Bw47	Bw73		DQw3	
		Bw4			
		Bw6		HLA-DP	
				DPw1	
				DPw2	
				DPw3	
				DPw4	
				DPw5	
				DPw6	

*From Nomenclature of factors of the HLA system, 1984, in Albert ED, et al (eds): *Histocompatibility Testing, 1984.* Berlin, Springer-Verlag, 1984. Used with permission.

and 3 DQ alleles identified to date. More than 10^{10} genotypes are possible. The evolutionary purpose of the extreme polymorphism in the immune system is unclear. The currently favored primary and secondary theoretical purposes of the immune system are the fighting of infection and tumor surveillance. Transplant rejection is an iatrogenic immune function. Nevertheless, the study of transplant rejection may lead to increased understanding of tumor immunology and of the fundamental immune process.

HISTOCOMPATIBILITY TESTING

Blood group compatibility (A, B, and O) between donor and recipient is the first immunologic requirement for transplantation (Table 24–4). The crossmatch is a serologic test for preformed recipient cytotoxic antibody against donor antigens, primarily class 1 HLA-A and -B antigens (Braun, 1976). Transplantation with a positive crossmatch risks a high probability of hyperacute rejection immediately following surgery. Incubation of donor lymphocytes with recipient serum is fundamental to all crossmatch techniques. The microcytotoxicity test originally developed by Terasaki remains in greatest use and involves incubation of donor lymphocytes with recipient serum and standard typing serums having known HLA specificities, standard rabbit complement C1, and eosin as a vital dye (Mittal et al., 1968). Many other crossmatch techniques have been used in attempts to detect lower levels of anti-HLA antibody and to detect non-HLA antibody, thereby increasing the sensitivity of the test (Table 24–5). Currently the antiglobulin crossmatch for anti-IgG, IgM, and IgA antibodies is believed to be highly sensitive and specific. Not all of the antibodies detected by the various tests listed in Table 24–3 are relevant to transplantation. Warm antibodies (ie, those detected at 37° C) are directed against T-lymphocytes bearing class I HLA antigens and are highly significant for transplantation. Autoantibodies are not significant, and cold antibodies (ie, those detected at 4° C) directed against B-lymphocytes are unlikely to influence graft survival. Prior transplantation is the most common cause of sensitization among patients awaiting transplantation; also im-

TABLE 24–4.
Blood Group Antigen Histocompatibility Between Donor and Recipient

RECIPIENT BLOOD GROUP	DONOR BLOOD GROUP			
	A	B	AB	O
A	Yes	No	No	Yes
B	No	Yes	No	Yes
AB	Yes	Yes	Yes	Yes
O	No	No	No	Yes

TABLE 24–5.
Crossmatching Techniques*

Cytotoxicity for HLA—modifications and recommendations
  Test recipient serum obtained <48 hr before transplantation
  Test all positive recipient sera
  Use dilutions of sera
  Prolong incubation time
  Add subcytotoxic ALG or ATG
  Trypsinize lymphocytes
  Consider weakly positive readings as a positive crossmatch
  Construct a "safe donor antigen profile"
  Test organ perfusates for cadaver organ transplants
Antiglobulin (anti-IgG, IgM, IgA)
Antibody-mediated cell-dependent immune lympholysis (ABCIL)†
Antibody-dependent cell-mediated cytotoxicity (ADCC)†
Lymphocyte-antibody lymphocytolytic interaction (LALI)†
Lymphocyte-dependent antibody (LDA)
Direct lymphocyte-mediated cytotoxicity (LMC)†
Blocking of MLC
B-cell antibody testing‡
Capillary agglutination for lymphocytes
Capillary agglutination for neutrophils
Inhibition of macrophage migration‡

*From Braun WE: The new serology of histocompatibility testing and its significance in human renal transplantation. *Urol Clin North Am* 1976; 3:503–525. Used with permission.
†These techniques are based on the same concept of detecting complement-independent antibody-dependent cytotoxicity, but have some distinguishing features.
‡Rejections have been reported in HLA-identical, MLC-compatible siblings who had positive crossmatches with these techniques.

portant are blood transfusions and pregnancy. Patients are characterized as high or low responders in terms of antibody formation. Patient serums are collected at regular intervals and tested against a random panel of known HLA antigens. Previously, all available recipient serums were tested against the cells of a given potential donor, and a positive crossmatch with any recipient serum precluded transplantation. With the availability of the more effective immunosuppressive agent cyclosporine (see below), transplantation is possible with a negative crossmatch on current serum despite a positive crossmatch with historical serums. The explanation for this clinical fact, which defies immunologic theory of recall antibody formation against prior antigen exposure, is unclear.

Serologic testing for class 2 HLA-D antigens is not possible. However, DR antigens may be identified serologically in the same manner described above for class 1 HLA-A and -B antigens. Thus, donor and recipient DR matching (see below) is feasible for cadaver renal transplantation. Compatibility for class 2 HLA-D antigens is measured by the mixed lymphocyte reaction (MLR). Donor and recipient lymphocytes are mixed and cultured together for 3–5 days. The results of mixed lymphocyte culture (MLC) may be quantified as a stimulation index. A low or nonstimulating MLC correlates with high graft survival in living donor transplants. Owing to the time required, the MLC is not applicable to cadaver transplants. The sequence of events in MLR and antibody formation are described in the next section.

MECHANISMS OF IMMUNOLOGIC RESPONSE—ANTIGEN RECOGNITION, CELLULAR AND HUMORAL RESPONSE

Lymphocytes are the primary immunocompetent cells and consist of different subpopulations with separate origins and specific functions (Fig 24–6) (Hamburger, 1981). Thymus-dependent lymphocytes become cytotoxic, suppressor, or helper T cells. Bone marrow-derived cells become plasmocytes capable of antibody production for humoral immunity. Cytotoxic T cells are capable of destroying allogeneic target cells. Suppressor T cells inhibit both the cellular and humoral immune responses. Helper T cells mediate the simultaneous and interrelated T-cell and humoral B-cell responses. In addition, normal peripheral blood contains a class of spontaneous cytolytic cells termed "natural killer" (NK) cells, which may function primarily for tumor surveillance but may also have a role in allograft rejection. Another group of killer cells (K) that are neither ordinary B- nor T-lymphocytes—thus the term "null" cells—produce antibody-dependent cytotoxicity.

The immune response that culminates in rejection of vascularized grafts is a cascade of cellular and humoral events (Fig 24–7) (Strom and Carpenter, 1983). Each arm of the process will be described separately. However, all levels of the process occur simultaneously in vivo and are interrelated.

The immune response is initiated by recognition of foreign class 2 antigens by T-lymphocytes, predominantly T helper cells. Strong experimental evidence exists in animals that donor blood-borne cells, so-called passenger leukocytes, present the stimulating class 2 antigens to the recipient. In humans class 2 antigens are present on endothelial cells in the kidney and may be the sensitizing cells. During inflammation, activated lymphocytes release γ-interferon, which may in turn result in expression of class 2 antigens on cell surfaces. Thus, the role, if any, of passenger leukocytes in human alloantigen presentation is unclear.

Activated precursor T helper cells in the kidney and in draining lymph nodes undergo postantigenic differentiation. A variety of cell surface receptors and products is synthesized by these activated lymphocytes. The most important is the lymphokine interleukin-2 and its receptor, which enhances lymphocyte activation and leads to clonal expansion of the T helper subset.

Simultaneously precursor T helper cell activation results in release of the lymphokine macrophage-stimulating factor (MSF). Activated macrophages then produce interleukin-1, which further enhances T helper cell activation.

Presentation of class 1 alloantigens to activated lymphocytes undergoing T helper clonal expansion results in production of effector cytotoxic T-lymphocytes. Effector T cells then mediate B-lymphocyte differentiation and antibody production against the allograft and generation of cytotoxic T cells that may directly attack the allograft. Rejected allografts contain both antigen-specific cytotoxic T cells, B cells, and specific antibodies; and NK cells and macrophages from nonspecific inflammation.

MODIFICATION OF THE IMMUNE RESPONSE—PREVENTION OF REJECTION

Blood Transfusion Effect

Blood transfusion presents foreign antigens and may result in recipient sensitization with antibody forma-

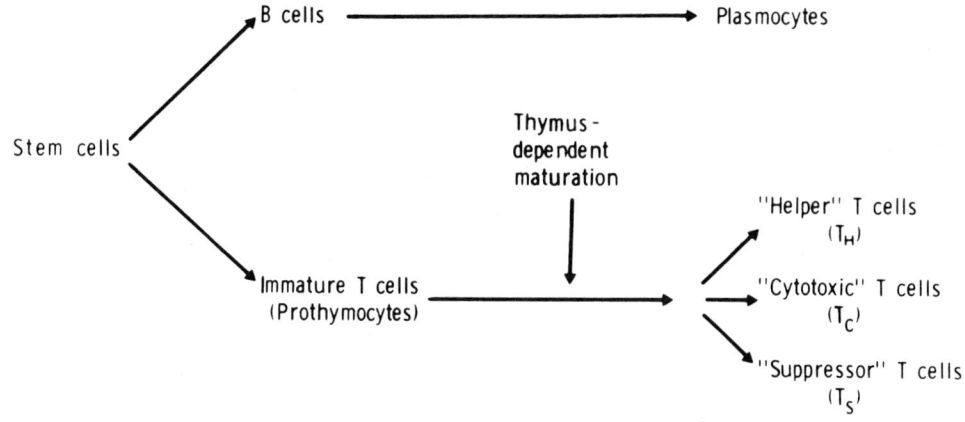

FIG 24–6.
Lymphocyte subset development from bone marrow-derived and thymus-dependent precursors. (From Hamburger J: Transplantation immunology, in Hamburger J, Crosnier J, Bach JF, et al (eds): *Renal Transplantation Theory and Practice.* Baltimore, Williams & Wilkins, 1981, pp 1–22. Used with permission.)

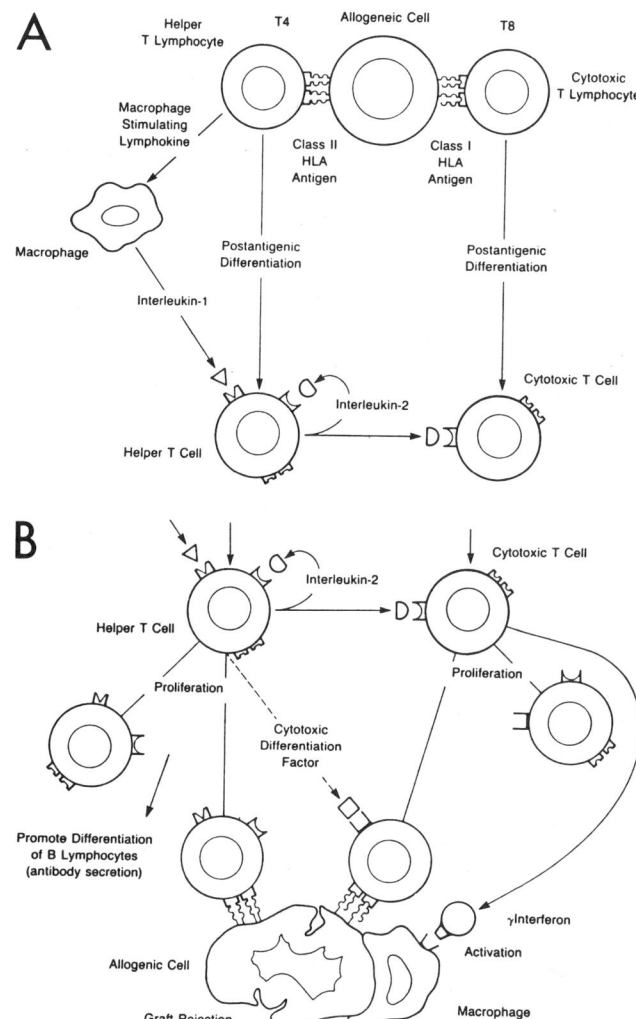

FIG 24–7.
The normal host response to alloantigen presentation. Antigen activates helper T-lymphocytes, which produce macrophage stimulation and interleukin-2 (IL-2) receptor formation on helper T- and on cytotoxic T-lymphocytes. Stimulated macrophages release interleukin-1 (IL-1), which in turn stimulates IL-2 release. IL-2 causes proliferation of B- and T-lymphocytes and release of γ-interferon. The final common path of antigen-directed antibodies and effector cytotoxic T cells produces allograft rejection. (Adapted from Helderman JH: Renal transplantation, in *Internal Medicine Grand Rounds.* Dallas, University of Texas Health Science Center at Dallas, Oct 25, 1984; and Strom TB, Carpenter CB: Transplantation: Immunogenetic and clinical aspects, part 2. *Hosp Pract* 1983; 18:135–150.

tion. Therefore, the conventional early wisdom in clinical transplantation was to avoid transfusion in transplant candidates whenever possible. In 1973 Opelz and colleagues first reported that, in a review of more than 1,000 renal transplants, blood transfusions significantly improved graft survival. Subsequently, this observation has been substantiated by many reports throughout the world (Figs 24–8 and 24–9). The optimal number, tim-

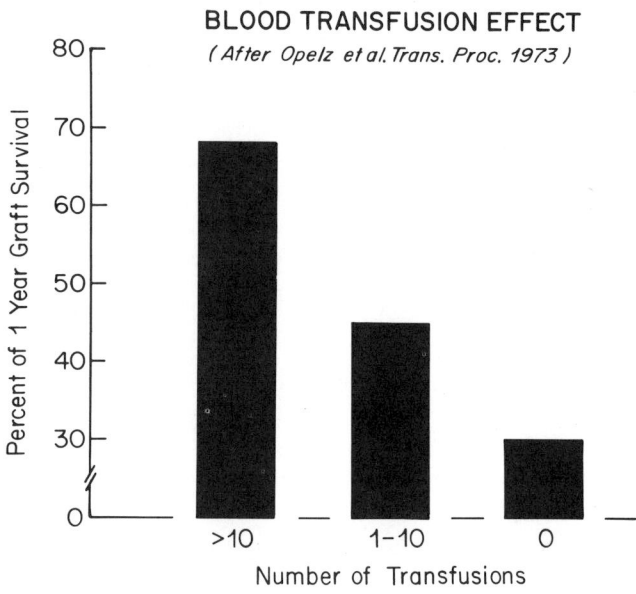

FIG 24–8.
The effect of blood transfusions on one-year graft survival from collected series prior to 1973. (From Helderman JH: Renal transplantation, in *Internal Medicine Grand Rounds.* Dallas, University of Texas Health Science Center at Dallas, Oct 25, 1984. Used with permission.)

ing, frequency, and type (whole blood vs. blood components and fresh vs. stored) of pretransplant transfusion is unsettled.

The mechanism of the transfusion effect on graft survival is unclear. The first hypothesis was that transfu-

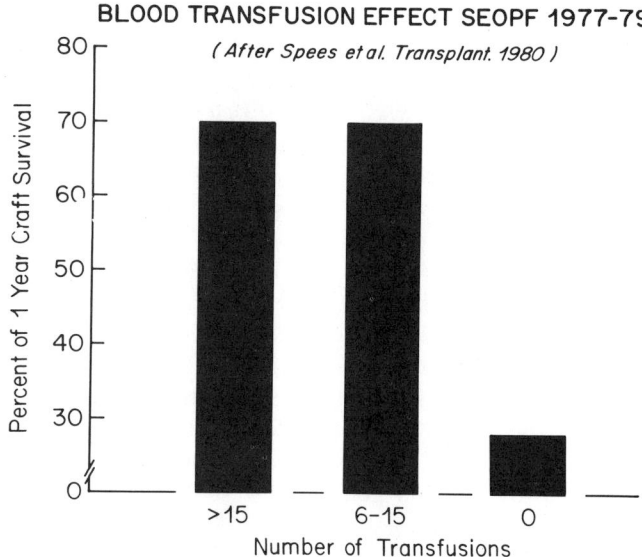

FIG 24–9.
The effect of blood transfusions on one-year graft survival from collected series, 1977–78. (From Helderman JH: Renal transplantation, in *Internal Medicine Grand Rounds.* Dallas, University of Texas Health Science Center at Dallas, Oct 25, 1984. Used with permission.)

sions primarily select favorable patients for transplantation by sensitizing high responder patients, who are then unlikely to receive a transplant (Van Rood and Balner, 1978). High graft survival is obtained in the low responder patients who were not sensitized and who are more likely to receive a transplant. There are two important observations against the selection hypothesis of the transfusion effect. First, transfusions only sensitize a small fraction of the patients. Second, the transfusion effect on graft survival is present in both highly sensitized patients and in nonresponders.

The second hypothesis is that transfusions enhance a state of recipient immunologic tolerance (Van Rood and Balner, 1978). Observations supporting this theory are that transfusions increase suppressor T-cell number and function, produce anti-idiotypic antibodies, and reduce measurable in vitro cellular immunity. However, immunosuppressive medication is still required following transplantation.

The third hypothesis for the transfusion effect is the clonal deletion theory of Terasaki (1984). Terasaki proposed that transfusions routinely sensitize the recipient as in other forms of immunization. This sensitization causes lymphoblast formation. The addition of high-dose immunosuppression following transplantation and alloantigen presentation results in cytotoxicity of these activated lymphocytes, leading to specific clonal deletion of allosensitized cells.

The preceding remarks pertain to random third-party blood transfusions. Salvatierra and associates (1980) reported a dramatic increase in graft survival among living related donor (LRD) kidney recipients who received a series of three donor-specific transfusions (DST) from haploidentical donors. Heretofore, one-year LRD graft survival with azathioprine and prednisone immunosuppression was 60%–80% for high and low mixed lymphocyte culture stimulation pairs, respectively. In the original DST protocol approximately one third of patients developed a positive crossmatch to the intended donor and could not receive a kidney from that donor. However, the sensitization was narrow and did not preclude other LRD or cadaver transplantation. The remaining two thirds of DST patients were crossmatch negative and underwent LRD transplantation. The one-year graft survival in these patients was 94%, equal to the results in HLA-identical LRD transplants. These results have been confirmed in hundreds of patients worldwide, and graft survival and renal function show only slight decrease annually at 3–5 years of follow-up. Newer protocols including recipient immunosuppression during DST reduce the sensitization rate to approximately 15% while maintaining excellent graft survival. Although the mechanism of DST as with random transfusions is incompletely understood, specific recipient immune alteration, rather than simple favorable selection, occurs.

Irradiation

Whole body irradiation represented the first attempt at immunosuppression in clinical transplantation. The technique quickly fell out of favor owing to unacceptably high morbidity and mortality rates, primarily from sepsis, and to the development of immunosuppressive drugs (see below). A modern technique of total lymphoid irradiation (TLI) was successfully applied to the treatment of Hodgkin's disease and later shown to provide a brief interval during which organ transplants in animals resulted in permanent acceptance (chimerism) when accompanied by donor bone marrow transplantation. The technique of TLI both with and without donor marrow transplantation has been applied to limited human trials of renal transplantation in recipients at high risk for rejection. Despite resulting in moderately good graft survival, TLI continues to be associated with frequent and severe complications.

Azathioprine

Azathioprine inhibits DNA and RNA synthesis by blocking enzymes involved in purine metabolism (Fig 24–10). Thus, production of rapidly growing cells such as activated clones of sensitized lymphocytes is impeded. Azathioprine is a nonspecific immunosuppressant that acts proximally in the chain of lymphocyte activation and is not very effective against activated lymphocytes. For LRD transplants, azathioprine is begun at a dose of 2.5–3 mg/kg/day several days before transplantation. Azathioprine is started at doses up to 5

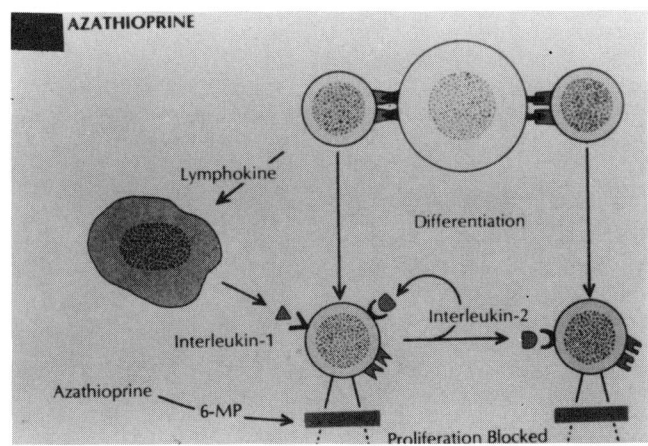

FIG 24–10.
The azathioprine metabolite 6-mercaptopurine blocks lymphocyte proliferation by inhibiting DNA replication. (From Strom TB, Carpenter CB: Transplantation: Immunogenetic and clinical aspects: II. *Hosp Pract* 1983; 18:135–150. Used with permission.)

mg/kg/day just prior to cadaver transplantation and tapered to a maintenance dosage.

Azathioprine is metabolized primarily by the liver and is minimally excreted in the urine. The major toxicity of azathioprine is myelosuppression with marrow elements affected in the following order: activated lymphocytes greater than polymorphonuclear leukocytes greater than platelets. Dose reductions are required when the peripheral WBC count is less than 5,000/mm^3 to avoid potentially lethal sepsis and other complications. The second major toxicity is an infrequent dose-related cholestatic jaundice, which must be distinguished from viral hepatitis. Chemically induced hepatitis occasionally is severe enough to require conversion from azathioprine to other cytotoxic drugs.

Cyclophosphamide

Cyclophosphamide interferes with DNA synthesis and causes death of activated lymphocytes and other rapidly dividing cells. This agent may be substituted for azathioprine at doses of 1 mg/kg/day. The therapeutic-to-toxic ratio is narrower for cyclophosphamide than for azathioprine.

Cyclosporine

The most important recent advance in clinical organ transplantation has been the discovery of the profoundly immunosuppressive drug cyclosporine. This agent is a metabolite of the soil fungi *Trichoderma polysporum rifai* and *Cylindrocarpum pelucidum*. Cyclosporine A has a molecular weight of 1,200 daltons and is composed of 11 amino acids, one of which is hydrophobic and makes the agent soluble only in lipids or or-

ganic solvents (Fig 24–11). Cyclosporine inhibits interleukin-2 production and/or release from activated T helper cells and reduces interleukin-1 release from macrophages (Fig 24–12). The net effect of cyclosporine is to block the activation of mature cytotoxic T-lymphocytes without altering suppressor T-cell function. Mature T cells remain capable of cytotoxicity following exposure to interleukin-2 even in the presence of cyclosporine.

Corticosteroids

Corticosteroids and cytotoxic agents have been the mainstay of immunosuppression since the beginning of clinical transplantation. Increased doses of corticosteroids remain the most common initial antirejection therapy. The initial and maintenance dosages of corticosteroids for immunosuppression vary widely. The mechanism of action of corticosteroids as immunosuppressants can be understood by referring to the sequence of antigen-induced lymphocyte activation, leading to the elaboration of lymphokines in the entire cellular and humoral cascade described earlier (Figs 24–7 and 24–13). Corticosteroids inhibit the release of interleukin-1 by monocytes and block the progression of lymphocyte activation and effector cell production. Corticosteroids greatly reduce the number of circulating K cells. The nonspecific effects of corticosteroids as anti-inflammatory agents that stabilize lysosomal membranes and diminish swelling also play a role.

Antilymphocyte Preparations

A variety of polyclonal animal-derived antilymphocyte preparations has been used in clinical transplanta-

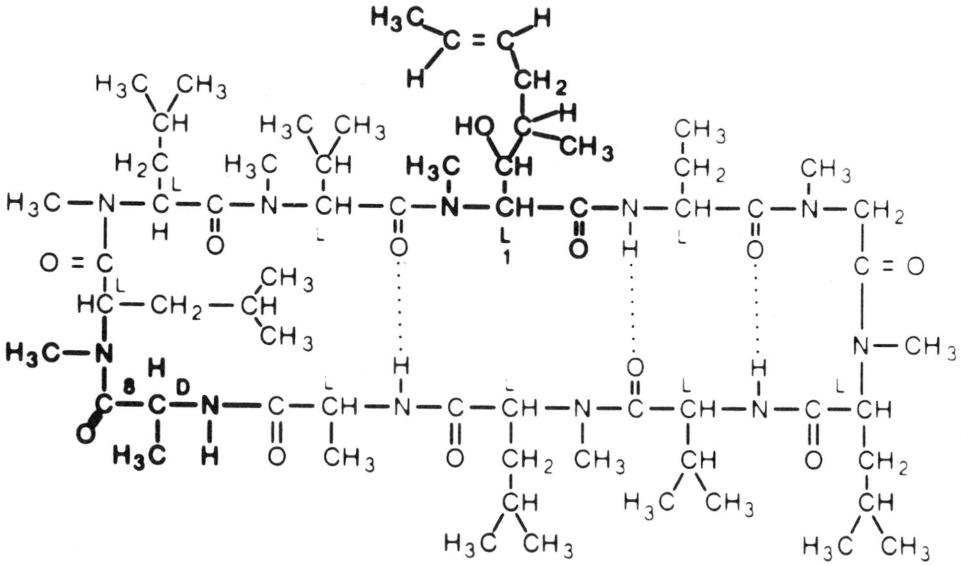

FIG 24–11.
The chemical structure of cyclosporine A.

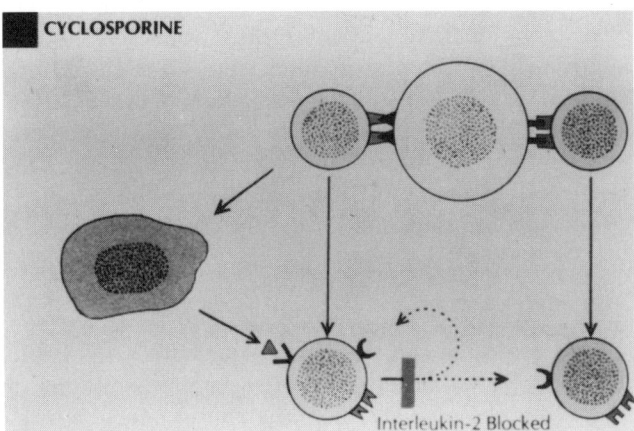

FIG 24–12.
The major effect of cyclosporine is blockade of IL-2 release, limiting cytotoxic T-cell activation and proliferation. (From Strom TB, Carpenter CB: Transplantation: Immunogenetic and clinical aspects: II. *Hosp Pract* 1983; 18:135–150. Used with permission.)

tion for 20 years. These biologic serums have variable potency and purity from one batch to the next. Owing to the infusion of large quantities of animal protein, patients receiving antilymphocyte globulin (ATG) or antilymphocyte serum (ALS) must be pretreated with antihistamines and increased doses of corticosteroids. These preparations are potent inhibitors of cellular rejection and may be used initially as an adjunct to prevent rejection or to treat established rejection. Some transplant physicians routinely administer ATG or ALS for 2–4 weeks following transplantation to prevent early rejection episodes. While these agents are extremely effective at preventing rejection, they are expensive and are associated with risks of overimmunosuppression

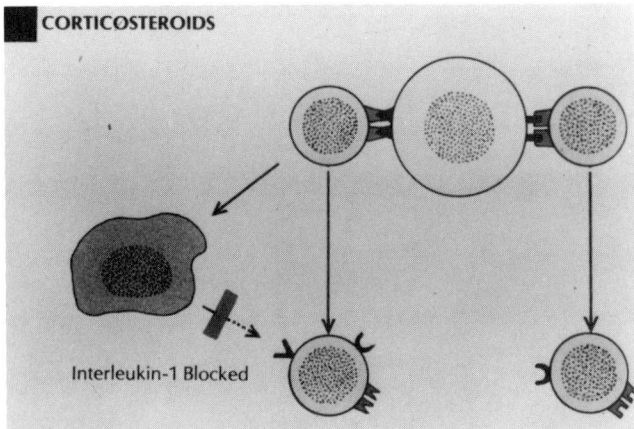

FIG 24–13.
Corticosteroids prevent IL-1 release and IL-2 formation. (From Strom TB, Carpenter CB: Transplantation: Immunogenetic and clinical aspects: II. *Hosp Pract* 1983; 18:135–150. Used with permission.)

and sepsis. Therefore, an altenative approach is to begin ATG at 14 mg/kg/day for 10–14 days as an adjunct to treat rejection crisis (Filo et al., 1981; Nowygrod et al., 1981). During ATG therapy, the maintenance cytotoxic agents are withheld to prevent excessive immunosuppression. Patients may develop a serum sickness syndrome requiring discontinuation of ATG.

Monoclonal Anti-T-cell Antibody Therapy

Specific monoclonal antibodies against lymphocyte subsets can be produced using mouse hybridoma technology. These preparations, unlike ATG or ALS, have constant potency and purity and require infusion of much lower doses of animal protein. Preliminary clinical trials of monoclonal antibody therapy for the treatment of acute rejection episodes have been completed recently (see Antirejection Therapy). In the future initial monoclonal antibody therapy directed against antigen recognition or interleukin-2 receptor expression may block the effector limb of the immune response, resulting in permanent graft tolerance.

CLINICAL RENAL TRANSPLANTATION
RECIPIENT SELECTION AND PREPARATION

Although theoretically anyone suffering from fatal irreversible renal disease with a GFR in the range of 5–8 ml/min or less is a candidate for renal transplantation, the prime candidates are those whom renal transplantation will return to the full productive capability that they enjoyed before becoming ill. Recipients have been documented from 1 month of age (Najarian et al., 1971) to greater than 74 years of age (Hakala, 1985). With the availability of the autoanalyzer, diagnosis of azotemia is often made years in advance of ESRD status, and careful medical management may postpone the need for dialysis or transplantation.

In the United States, diabetes mellitus is rapidly rising as a major cause of ESRD, requiring transplantation. Many centers are now performing transplantation on such patients who have a modicum of renal function (GFR 8–10 ml/min) remaining before the ravages of advanced diabetic neuropathy coupled with ESRD have permanently precluded rehabilitation, i.e., blindness or advanced neuropathy with loss of bowel or bladder control. Although figures for long-term success of renal allografts are low in groups of recipients less than 1 year of age and older than 50 years of age (Table 24–6), recent advances in recipient preparation—i.e., donor-specific transfusions (DST), cortisone reduction with cyclosporine A (CSA), availability of potent ALG and ATG preparations, and donor and recipient pretreatment—have made worthy individual consideration regardless of age (Helderman, 1984; Briggs, 1984).

TABLE 24–6.

Transplantation in Infants Less Than 1 Year of Age*†

AUTHOR(S)	AGE, MO	PRIMARY RENAL DISEASE	DONOR	FUNCTION, MO	OUTCOME
Goodwin et al. (1963)	6 Days	Prune-belly syndrome	C, An	NF	Died
Cerilli et al. (1972)	5	HUS	C	13	Alive
De Shazo et al. (1974)	12	—	C	—	Died
	5	—	C	—	Died
	4	—	C	—	Died
	1½	—	C	—	Died
Lawson et al. (1975)	¼	—	C	NF	Died
	9	—	LD	—	Died
	12	—	C	—	Died
Kwun et al. (1978)	1	PKD	C, An	NF	Died
Moel and Butt (1982)	3	CN	C, An	6R	Died
	8	OU	C	42R	Died
Miller et al. (1982)	11	OU	C	44	Alive

*From Fine RN: Renal transplantation in children, in Morris PJ (ed): *Kidney Transplantation: Principles and Practice*, ed 2. London, Grune & Stratton, 1984, p 514. Used with permission.
†HUS indicates hemolytic-uremia syndrome; PKD, polycystic kidney disease; CN, cortical necrosis; OU, obstructive uropathy; LD, living donor; C, cadaver; An, anencephalic; NF, nonfunction; and R, rejection.

The incidence of ESRD in children is estimated by Alexander (1984) to be 1.5–4.5 children per million to total population. The incidence in infants less than 1 year of age is in the order of 0.2/1 million total population (Fine, 1984).

Although there are many conflicting reports concerning the use of newborns and infants less than 1 year of age as donors or recipients, results are improving, and each case must be decided on its own merits. One can conclude that a kidney from a 1½-year-old child can provide satisfactory function for an adult recipient (Fine, 1984).

Age

Age is no longer a sole consideration for transplantation. Certain groups have done poorly historically to date (i.e., those less than 1 year of age and those greater than 50 years of age), but the skills of medical and surgical management might be applied to permit renal transplantation in any given case. As immunosuppression becomes more specific and less toxic, the age range will be extended. Improved intensive pediatric care and improved harvesting and matching techniques will increase the number of pediatric kidneys available. Peritoneal dialysis has aided in making the newborn and infant under 6 months of age possible candidates.

General Medical Problems

Diabetes mellitus is a common associated condition in U.S. renal transplant recipients and also has been found not uncommonly to exclude willing but unsuspecting donors. Many patients in the United States have been candidates for simultaneous pancreas and renal transplantation, and currently diabetes and hy-

pertensive renal disease exceed glomerulonephritis as a cause of renal failure. Control of diabetes is mandatory, and regular insulin administration is preferred for blood glucose control in the perioperative period. Patients with active GI symptoms or history of proved ulcer within the preceding two years should receive antacids and cimetidine and should undergo endoscopy (gastroduodenoscopy) before transplantation. In patients with irritative GI symptoms, cimetidine may be used with caution because there is an increased incidence of renal damage in transplantation patients.

Voiding cystourethrography should be carried out before transplantation. This allows assessment of bladder capacity and rules out obstructive urethral lesions, e.g., stricture. Any voiding dysfunction problems detected can have more detailed urodynamic workup with three-channel studies including EMG. Periodic hydrostatic dilation can be carried out if an unused bladder has a very small capacity (less than 100 ml), but this is seldom necessary, since the volume of urine put out by the newly transplanted kidney will accomplish this end within a few days. Vesicoureteral reflux per se is not an indication for pretransplant nephrectomy if there is just a wisp of reflux into nondilated ureters and there is no infection in the urine. Pretransplant nephrectomy is rarely needed for hypertension, since in most patients it can be controlled by medication on long-term hemodialysis. A few patients require major antihypertensive medication. If the patient does not have a salvageable urinary bladder but does have a continence mechanism, formation of a bladder from large or small bowel can be considered, or augmentation of a small bladder with an antireflux ileocecal augmentation, or ileac or colon augmentation can be considered. If no satisfactory urethral continence mechanism remains, an abdominal reservoir with a continent stoma may be constructed, or diversion into an ileac segment may still be considered. We prefer to perform the bowel surgery before giving any immunosuppression therapy, i.e., while the patient is on long-term dialysis, and to implant the transplanted ureter into the new "bladder," which has had a chance to heal with no immunosuppression prior to transplantation.

In patients with urinary tract infection and stones, pretransplant bilateral nephrectomy is preferred. If infection alone is present and no reflux is present, localization studies may be done (Stamey et al., 1965). Patients with cystitis, no reflux, and a normal-appearing bladder can usually be spared bilateral pretransplant nephrectomy by appropriate antimicrobial therapy. If infection is present in a cystic or dysplastic kidney, it is best removed, usually before the transplantation, through a posterior approach rather than at the time of transplantation.

Patients with polycystic kidney who are not infected and who are normotensive, either on dialysis or with mild antihypertensive therapy, usually do not require pretransplant nephrectomy. Usually with adequate pretransplant dialysis and particularly after transplantation, there is a decrease in the size of the polycystic kidneys and adequate space for the transplanted kidney extraperitoneally in the groin.

Patients who develop hematuria on long-term dialysis should be screened for renal cell carcinoma, and if a lesion is demonstrated, bilateral nephrectomy should be performed. Although more severe anemia will result and transfusion requirement will increase on dialysis (not necessarily bad unless marked sensitization occurs), one should wait at least one year before considering transplantation to reduce markedly, although not eliminate, the chance that metastasis has occurred.

Secondary hyperparathyroidism should be treated with phosphorus binders, i.e., cellulose phosphate and dialysis three or four times a week. Skeletal monitoring via bone densitometry should be carried out to detect a need for parathyroidectomy to prevent vertebral collapse and extremity fractures (particularly of tibia and fibia).

Patients who have had a previous transplant and who are being maintained on dialysis present unique problems. If the kidney has been left in and is causing no problems without immunosuppression, one must check carefully antigens of any proposed new graft to be certain that the patient has not been sensitized to antigens that the previous donor kidney had but the recipient lacked. Also, screening for non-A, non-B hepatitis is important, since its presence may preclude grafting or greatly modify the immunosuppressive reigmen.

Low serum complement in the recipient, although perhaps indicative of an active process in the recipient's own kidneys, is not a contraindication to transplantation but should be monitored, since these patients are at risk for recurrent disease in the allograft.

Our own experience in performing transplants in patients with Fabry's disease, anti-GBM antibody disorders such as Wegener's granulomatosis, and Goodpasture's syndrome has been satisfactory. One patient with oxalosis soon had his transplant filled with oxalate crystals and had a large ureteral calculus. Although 1–3 year successes have been reported (Welchel et al., 1983), the kidneys were filled with oxalate crystals. Oxalosis and a positive crossmatch remain contraindications to transplantation in our view. Patients having a simultaneous successful liver and kidney transplant would be spared the ravages of systemic oxalosis.

PREPARATION FOR OPERATION

LRD Recipient

These patients are usually established on dialysis. Donor specific transfusion of 300 ml of whole blood q 3 wk × 3 is usually carried out on 1-haplotype parent-to-child or child-to-parent transfers. Perioperative antibiotics are used. Typed blood is available. WBC-poor packed RBCs are given if needed (seldom). A Swan-Ganz catheter is not used unless there is a question of cardiac status, but CVP monitoring often is used. All skin preparation is done in the operating room immediately before surgery. Immunosuppression is begun on the morning of surgery. There is no theoretical reason to begin immunosuppression before, since azathioprine (Imuran) or cyclosporine is not effective until there is antigenic challenge of the recipient's immune system.

Cadaver Recipient

A donor-recipient lymphocyte cytotoxicity crossmatch is done and the latest serum results used. Dialysis is carried out unless the transplant is done within 24 hours of previous dialysis. Immediate function is the goal; this avoids early postoperative dialysis with bleeding problems secondary to heparinization. Electrolytes are checked. We correct serum potassium greater than 5.5 mEq/L with sodium polystyrene sulfonate (Kayexalate) enema (20 gm). The abdomen is shaved in the operating room. Four units of packed RBCs are available (type and hold). Central venous pressure monitoring is used frequently but Swan-Ganz monitoring only for cardiac problems. Fluids are given IV on call to surgery and for 48 hours postoperatively. The patient receives a transfusion if the hematocrit reading is less than 20.

ORGAN PROCUREMENT

Living Donor Evaluation

The renal donor enjoys an enviable success rate among individuals undergoing a major surgical procedure. Certainly, this is partly because every effort is made in the screening process to ascertain normality, but also it is a tribute to the attention given preoperatively, during surgery, and postoperatively by those attending the individual willing to risk himself to help another less fortunate. Anesthesia mortality to date has been less than 0.01% and operative mortality 0.001% in more than 20,000 cases.

General evaluation consists of CBC, SMA-12, liver profile, check for lues and hepatitis, ECG, and chest x-ray (lateral and apical lordotic views). Typing for HLA and DR antigens (MTB ant) is carried out. DST is requested for 1-haplotype HLA matches (Salvatierra,

1985). We perform abdominal CT with contrast in lieu of IVP as a screen for renal mass lesions. A KUB examination taken at the conclusion of the CT study demonstrates the renal collecting system. We prefer renal arteriography to IV injection with CT scanning or subtraction technique because of the better definition of arterial anatomy.

During arteriography, oblique views are taken on occasion to rule out overlap of renal arteries. If both arteries are single, we prefer the left kidney because of its longer renal vein. Carrel patches are not taken from LRDs (see Operative Technique). The kidney with a single renal artery is used if one kidney has multiple renal arteries. A donor WBC crossmatch (lymphocyte microcytotoxicity) with recipient serum is performed just before transplantation.

Many donors are excluded because of unsuspected disease. Diabetes, atherosclerosis, solitary kidney, silent myocardial infarction, asymptomatic renal calculi, unsuspected prostate cancer, and unsuspected cancer of the cervix all have been discovered during a thorough donor workup. Hume and associates (1963) found it necessary to exclude 57 of the first 83 live donors considered in their program. In our program, most of the studies usually can be completed on an outpatient basis by cooperation with the referring physician.

EVALUATION OF POTENTIAL CADAVER DONOR

Determination of Death

Legally, a patient is dead when the physician declares him dead. The death should be pronounced by the attending physician, preferably not a member of the transplant team. Death is brain stem death. All the key functions that define a live human being as a nondependent biologic unit are brain stem functions. These include, most prominently, the ability to be conscious of the surrounding environment and the ability to breathe. That portion of the brain stem particularly concerned with regulating these vital functions is the tegmentum of the mesencephalon and the rostral pons (Pallis, 1984). Structural brain damage in this area sufficient to cause an individual to be declared dead is an irreversible change. In one group of collected cases reported by Pallis, no patient survived who met the criteria of brain stem death in over 1,100 patients so evaluated (Pallis, 1984). To declare a patient brain-dead, the patient should be comatose and should have been on a ventilator long enough to ascertain that brain damage is not remediable. Drug ingestion, particularly of alcohol and barbiturates, and recently administered neuromuscular blocking agents should be ruled out as having been administered (Table 24–7). Hypothermia

TABLE 24–7.
Coma: Duration of Drug Effects*†

DRUG	PLASMA HALF-LIVES, HR
Aspirin	0.25–3.0
Pentazocine	2
Paracetamol	2–4
Diphenhydramine	4–10
Imipramine	8–24
Morphine	10–60
Nortriptyline	15–93
Methadone	18–97
Pentobarbitone	20–35
Chlorpromazine‡	24 (mean)
Carbamazepine	24–48
Diazepam and active metabolites	24–96
Phenobarbitone	50–140

*From Pallis C: Brainstem death: The evolution of a concept, in Morris PJ (ed): *Kidney Transplantation: Principles and Practice*, ed 2. London, Grune & Stratton, 1984, p 119. Used by permission.
†Alcohol metabolism (zero-order kinetics): 10 ml H^{-1}.
‡Prolonged at high concentration.

and severe metabolic derangements, as seen in advanced uremia or diabetic acidosis or sepsis, should be ruled out. Bedside tests should confirm that the patient is apneic (no breathing response to a PCO_2 of greater than 50 Hg reached in a 10-minute period off the ventilator). The patient should have no brain stem reflex activity. This means no blinking, no grimacing with supraorbital nerve compression, no gagging or coughing when stimulated by a catheter introduced into the pharynx, and no eye movements or pupillary response to light. One must comply with the laws of his state, but neither an EEG nor a cerebral angiogram is necessary for the diagnosis of brain stem death.

Maintenance of the Cadaver

After the individual has been declared dead, maintenance of the heart, lung, and renal function is paramount to have quality organs for transplantation. Ventilation is continued. Plasma expanders are given as needed to maintain a high urine output (greater than 100 ml/hr) and to avoid obstruction of the individual nephron. Diuretics are used to ensure an adequate urine volume. Ringer's lactate or one half strength normal saline is given as needed to maintain a systolic blood pressure at 90 mm Hg or above. This will help to minimize the incidence of ATN, which has been reported to be 5% or less (if warm ischemia time can be diminished—less than 5 minutes) (Miller, 1985). Considering the cellular machinery, cooling (6°–10° C) slows the engine (mitochondria); ischemia depletes the fuel (ATP). Warm ischemia time must be minimized to avoid ATP depletion and ADP accumulation in the cell.

Although vasopressors are not recommended, occasionally it is necessary to use one in addition to volume expansion to maintain systolic pressure at 90 mm Hg or greater and urine output at greater than 100 ml/hr. Dopamine is preferred by our group. Chlorpromazine, steroids, verapamil (particularly if the patient has received catecholamines), and furosemide have been recommended and are used to minimize renal ischemia. If at all possible, multiple organ harvest should be performed, details of which are described later in this chapter. The importance of having an immediately functioning organ as a result of careful preservation of the cadaver donor has been made even more meaningful by the introduction of cyclosporine, an effective immunosuppressant but cytotoxic (especially nephrotoxic) agent.

Organ Preservation

Renal preservation time is increasing steadily. In the early days of clinical renal transplantation, donor hypothermia was used to cool the kidneys. Semb and associates (1960), working with normal dog kidneys, suggested that if renal arterial occlusion or the time that the kidneys were without blood is to exceed 23 minutes, the preservation of renal function could be enhanced markedly by cooling. Currently, cold slush storage and extracorporeal perfusion systems are used to deliver cold fluid to the kidneys. Cold storage is not as satisfactory as extracorporeal perfusion for renal preservation much beyond 48 hours. Preservation should maintain the kidney in a functional state and not increase the acute tubular necrosis (ATN) rate over harvest conditions. With the introduction of cyclosporine as an immunosuppressive agent, immediate function is highly desirable. Also, it is important to avoid dialysis in the first few hours following transplantation, since a higher incidence of wound hematomas seems to occur in such patients.

As cadaver organs become increasingly the source for grafting, the importance of preservation increases. Of considerable importance in the ultimate preservation is the condition of the dying donor. The individual who has been in shock but yet not excluded for donation because there was no sepsis or systemic disorder is the one who particularly needs maintenance of blood and extracellular fluid volume during the harvesting procedure. The administration of large volumes of cooled perfusate is helpful, as is the use of adjunctive agents to maintain diuresis to prevent precipitation of protein in the distal nephron, which is so commonly associated with ATN. Such agents include mannitol and renal arterial spasmolytic agents such as dibenzyline. Measures should be taken to correct acidosis and to minimize infection in the cadaver donor. Heparinization of the do-

nor to minimize the risk of clotting is also important in the periharvest period. The use of diuretics such as furosemide may be needed for circulatory support while preparing for donor harvest.

The effect of a high potassium hyperosmolar solution on preservation of kidneys has been studied thoroughly by Collins and Sachs. The flushing and storage solutions currently popular are EuroCollins (Collins C_2 with Mg and PO_4 removed) and Sacks solution, which has a potassium content mimicking that of intracellular potassium (155 mEq/L[3]). The successful perfusate solutions should be cool, prevent cell swelling, and if possible stimulate ATP synthesis at 6°–10° C. The high levels of potassium in the Sacks perfusate are thought to prevent cellular potassium loss during storage and preservation and, therefore, to contribute to the stability of the cell membrane. Table 24–8 shows the ingredients in preservation solutions commonly used today. The Belzer machine in 1967 represented the first clinically applicable model used widely, the principles of which are in use today (Belzer, 1967). Extracorporeal continuous hypothermic perfusion provides effective preservation of kidneys requiring preservation for more than 48 hours and certainly has been the best method of achieving satisfactory cooling and preservation beyond 72 hours. With today's modern means of communication and transportation, preservation times shorter than 72 hours usually suffice. However, development of a method to store organs indefinitely or for months would have immediate worldwide application.

Continuous hypothermic perfusion at 6°–10° C is associated with less measurable metabolic (mitochondrial) activity of the kidney in contrast to the very major activity at 36°C. Evidence for utilization of substances by

TABLE 24–8.

Composition of Cold Storage Solutions

SUBSTANCE	SOLUTION, mM*			
	C_2	EC	S_2	HOC
Na^+	10	9.3	15	80
K^+	115	107	143	80
Mg^{2+}	30	0	16	72
Ca^{2+}	0	0	0	0
Cl^-	15	14	15	0
HCO_3^-	10	9.3	38	0
SO_4^{2-}	60	0		70
PO_4^{2-}	57.5	93	120	0
Glucose	126	182	206	0
Mannitol	. . .	. . .	. . .	150
Citrate	. . .	. . .	. . .	162
mOsm/L	320	325	430	400
pH	7.0	7.0	7.0	7.1

*C_2 indicates Collins solution; EC, EuroCollins; S_2, Sacks; and HOC, hypertonic citrate.

the kidney at 0°–10° C is present, but pales in comparison to the requirements at 36° C, at which temperature free fatty acids, amino acids, ketone bodies, and glucose are all consumed. At 6°–10° C, anaerobic glycolysis is predominant. Pegg and associates (1982) suggested that caproate provides the major fuel when oxygen tension is less than 150 mm Hg and the temperature is 0°–10° C. Evidence for the need for metabolites at 0°–10° C for periods of less than 72 hours is indeed scanty. After three days of simple storage, virtually all stored ATP has been broken down, and only about 20% of total adenine nucleotides remain, yet Southard and Belzer (1985) have shown that such kidneys may function immediately. These authors advocated the use of adenosine to prevent utilization of ATP and other nucleotides. Acidosis, which often occurs with anaerobic glycolysis at lower temperatures, of course, can be abated by washing out lactate and hydrogen ions formed as a result of this process, and perfusion may have a benefit in this area contrasted with simple cold storage. Oncotic agents have been added to most perfusates to prevent kidney cell swelling. We have shown to our satisfaction that the dead kidney may still respond to osmotic pressures and will swell or shrink on an extracorporeal apparatus, depending on the osmolality of the perfusate. At this writing, there is little that one can rely on to predict whether a kidney ultimately will function after a period of prolonged extracorporeal perfusion. Pressures, resistance to flow, and measurable metabolic products have all been cited but are often unreliable in predicting ultimate function. The condition of the dying donor, the milieu of the kidney during the terminal phase of the donor's life, the care after brain death, particularly volume expansion, and the time on the preservation unit seem to us still to be our major parameters in predicting ultimate function in this crude state of extracorporeal maintenance of viable organs.

OPERATIVE AND POSTOPERATIVE MANAGEMENT

Careful attention to details of intraoperative management is the first step toward a smooth course following transplantation. Only small to moderate doses of muscle relaxants are required during renal transplantation. If depolarizing muscle relaxants that require renal excretion are used, the patient must be carefully observed for adequate ventilation prior to extubation. The nondepolarizing agent atracurium (Tracrium) undergoes spontaneous hydrolysis over 30–45 minutes and is the preferred muscle relaxant in patients with renal failure.

Intraoperative urine flow is the rule in living related donor kidney transplants and in the majority of optimally procured and preserved cadaver allografts.

Therefore, the recipient should receive liberal intravascular compartment fluid expansion to stimulate diuresis. Although intraoperative blood loss usually is low, transfusion with 2 units of packed RBCs usually is indicated due to the anemia associated with renal failure. Crystalloid solutions devoid of potassium (0.9 N saline and 0.45 N saline) are delivered at 10–15 ml/kg/hr along with colloids (12.5% albumin, 0.5 gm/kg). Osmotic diuretics are used routinely (12.5–25 gm of mannitol), and furosemide is occasionally added for cadaver allograft recipients. Despite aggressive fluid administration, the majority of patients do not require CVP lines or Swan-Ganz monitoring. These adjuncts may be reserved for pediatric recipients, in whom the margin for change in intravascular volume is small, and for patients with known cardiac disease.

The insulin-dependent diabetic recipient poses special intraoperative challenges. These patients have an increased incidence of cardiovascular disease. Wide fluctuations in blood glucose may occur during anesthesia and while patients are unable to take oral feedings. To avoid this problem, we adhere to the recommendations of Rasking and Rosenstock (personal communication) as follows: Via a central venous catheter a constant glucose infusion of D50W at 17 ml/hr is maintained. The concentration of the glucose solution minimizes the volume required, and the rate was calculated to deliver slightly more than the daily hepatic glucose production of an average-sized adult over 24 hours. A separate insulin infusion is mixed by adding 250 units of regular insulin in 500 ml of 0.45 N saline. The insulin infusion rate is adjusted by the following sliding scale:

GLUCOSE, MG/DL	RATE, ML/HR
70	1
71–100	2
101–150	4
151–200	6
201–250	8
251–300	10
301–350	12
351–400	14
400	16

The glucose-insulin infusions are maintained until the patient resumes a regular diet and maintenance insulin replacement. Hourly blood glucose determinations must be performed to detect any hypoglycemia while the infusions continue. This is especially important while the patient is anesthetized. In our experience this regimen allows satisfactory preoperative blood glucose control even in patients with brittle diabetes.

After obtaining maximal patient hydration, we replace postoperative urine output with 0.45 N saline at a rate equal to hourly urine output up to 350 ml/hr and 75% of output for urine output exceeding 350 ml/hr. Patients experiencing heavy diuresis require serum and urinary potassium monitoring and may require potassium replacement.

Immunosuppressive regimens vary considerably among transplant centers. The following protocols represent our current policy. Living related donor recipients begin therapy with azathioprine, 2.5 mg/kg/day, and prednisolone, 50 mg BID three days before surgery and receive methylprednisolone (Solu-Medrol), 375 mg IV on the day of surgery. Azathioprine therapy remains at the same dose as tolerated, and prednisolone is tapered to 30 mg/day by one month postoperatively. Prednisolone dosage is further reduced to 10–15 mg/day by one year. Formerly, cadaver allograft reicipients received azathioprine 5 mg/kg on the day of surgery and the same postoperative doses of azathioprine and corticosteroids as described above. During the past year, we have substituted cyclosporine at an initial dose of 14 mg/kg/day in place of azathioprine for all cadaver allograft recipients. Patients receiving cyclosporine undergo a more rapid taper in the steroid dose. Whole blood cyclosporine levels are monitored by high-pressure liquid chromatography technique three times per week, aiming for a therapeutic trough level of 100–200 ng/dl. Clinical signs of cyclosporine toxicity include hyperkalemia, hypertension, tremor, and occasionally seizures. Cyclosporine-induced hypertension responds well to calcium channel blockers. Other objective signs of cyclosporine toxicity include increases in serum creatinine and liver enzymes.

The optimal cyclosporine protocol in cadaver renal transplantation to minimize nephrotoxicity in kidneys already at risk of acute renal failure is not yet clear. One approach is to maintain standard cyclosporine doses during anuria and make adjustments based on drug levels determined by radioimmunoassay (Kahan et al., 1984). An alternative approach is to initiate cyclosporine therapy preoperatively and then substitute ALS or ATG if there is anuria. A third approach is to begin all cadaver recipients on therapy with corticosteroids and antilymphocyte preparations and convert to cyclosporine after 10–14 days, when there is renal function. Finally, triple-drug therapy with low doses of cyclosporine, azathioprine, and corticosteroids is being studied.

Careful daily measurement of patient weight, fluid intake, and urinary output is essential. Daily hematocrit, serum electrolyte, and creatinine determinations are recommended. Patients are usually ambulatory and often are able to resume a diet on the first postoperative day. Intravenous lines and urinary catheters are re-

moved as soon as possible to minimize the risk of sepsis. We remove the bladder catheter within 36 hours of surgery in most cases.

DIAGNOSTIC AIDS

The diagnosis of a rejection episode may be obvious or subtle. Increase in serum creatinine, decrease in GFR as measured by ^{125}iodine isothalamate (Glofil) clearance, weight gain, oliguria, proteinuria, malaise, fever, pain and swelling over the allograft, and hypertension all may be signs of rejection. The decrement in serum creatinine toward normal plotted against time follows an asymptotic curve. If the serum creatinine falls off this curve, one can pronounce a decline in renal function, even if the current creatinine level is lower than the preceding value. Dehydration, ATN, obstruction, urine leak, nephrotoxicity, and rejection all must be considered as possible causes of decreasing renal function.

Sonography

Sonography is extremely helpful in the early diagnosis of renal transplant obstruction, lymphocele, or urinary leakage (Figs 24–14 and 24–15). Retrograde pyelography may be difficult to perform in the early postoperative period but occasionally is required to exclude obstruction.

Renal Scans

Serial radionuclide renal scans are of great value in identifying early ATN and subsequent rejection (McConnell et al., 1984). We obtain a ^{99m}Tc diethylenetriamine pentaacetic acid (^{99m}Tc DTPA) renal

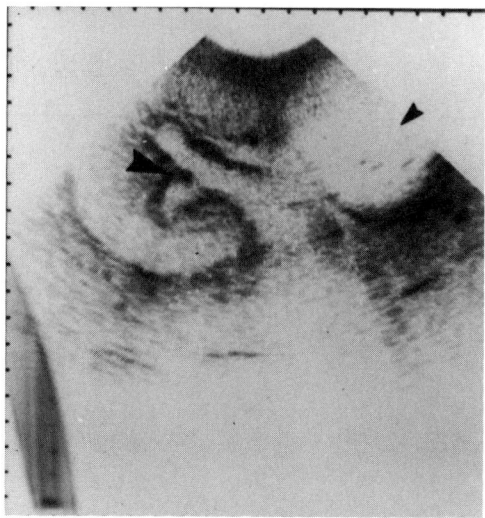

FIG 24–14.
Sonogram revealing hydronephrosis *(large arrowhead)* and lymphocele inferior to kidney *(small arrowhead)*.

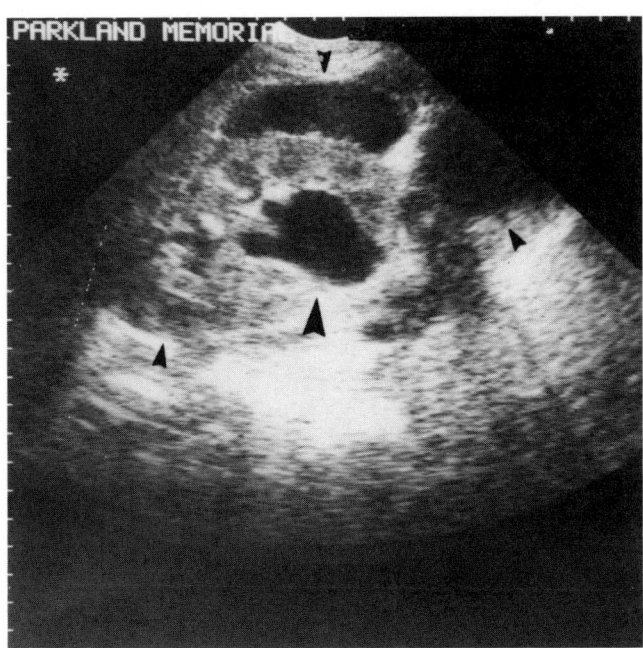

FIG 24–15.
Sonogram revealing hydronephrosis *(large arrowhead)* and fluid around kidney secondary to urinary leak *(small arrowheads)*.

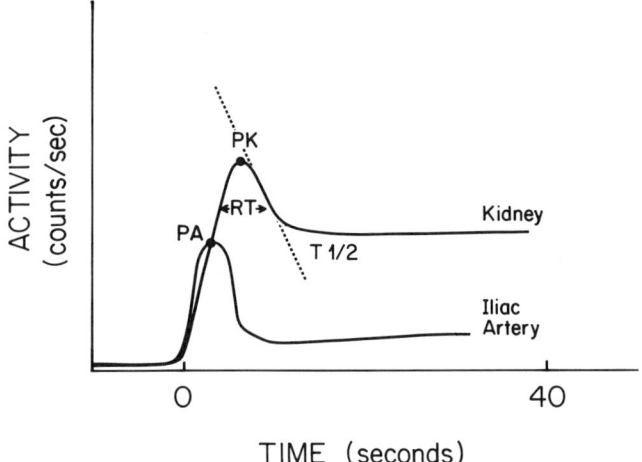

FIG 24–16.
Normal schematic ^{99m}Tc DTPA renal allograft and iliac artery blood flow curves. *P,* time (sec) between iliac artery and allograft peak activities (PK-PA). T½, washout parameter corresponding to time from peak allograft activity to one half the peak value. *RT,* intrarenal transit time. (From McConnell JD, Sagalowsky AI, Lewis SE, et al: Prospective evaluation of renal allograft dysfunction with ^{99m}Technetium diethylenetriaminepentaacetic acid renal scans. *J Urol* 1984; 131:875–879. Used with permission.)

scan in all patients during the first 24 hours postoperatively and thereafter as clinically indicated. Scans reveal both renal blood flow and excretion. Quantitative computer-generated flow curves are analyzed for interval to peak activity for the allograft compared with the iliac artery, intrarenal transit time, and washout (Fig 24–16). The excretory portion of the study depicts tracer accumulation in the renal collecting system and bladder. Normal renal function, as in a living related donor recipient, is shown by superimposition of renal and iliac artery blood flow curves and excellent excretion (Fig 24–17). During ATN, one finds slight delay in renal blood flow and poor excretion on the initial scan, with subsequent improvement in both parameters on later studies (Figs 24–18 and 24–19). Acute tubular necrosis alone should not be progressive. Thus, deterioration of renal blood flow on serial scans strongly suggests rejection or rare technical complications (Figs 24–20 to 24–22). In a prospective study at our institution the sensitivity and positive predictive value of ^{99m}Tc DTPA renal scans in identifying rejection episodes were 90.4% and 88.7%, respectively (McConnell et al., 1984).

Allograft Biopsy

Allograft needle biopsy provides a core of tissue for histologic evaluation in difficult cases of transplant dysfunction. Complications including pain, bleeding, and urinary extravasation occur infrequently but are potentially serious. Recently, fine-needle (25-gauge) aspiration cytology of the allograft has emerged as a potentially useful tool (Hayry and Von Willebrand, 1981). Daily aspiration may be performed with minimal pain and essentially no morbidity. The aspirate provides renal tubular cells and the entire spectrum of inflammatory cells (lymphoblasts, plasmablasts, activated lymphocytes, macrophages, and monocytes). Inflammatory cell activity in peripheral blood is measured at the same time as allograft aspiration, so that specific immunologic activity within the allograft may be indexed. The presence of increasing numbers of lymphoblasts coincides with acute rejection episodes, and large numbers of mononuclear cells in the aspirate despite antirejection therapy may portend loss of the allograft. There are limited data to suggest that this technique may distinguish tubular cell dysfunction owing to cyclosporine toxicity from rejection. Greater experience with fine-needle aspiration in more centers is required to confirm these early findings.

ANTIREJECTION THERAPY

Bolus Steroids

Increased doses of corticosteroids are the mainstay of therapy for initial acute rejection episodes. Our policy is to give 4 gm of IV methylprednisolone over a six-day period. Two thirds of patients with initial rejection ep-

FIG 24–17.
Normal day 1 scan in living related donor recipient with good function. The allograft blood flow curve is superimposed on the iliac artery curve, and the excretory phase reveals prompt calyceal visualization and early drainage into the bladder. (From McConnell JD, Sagalowsky AI, Lewis SE, et al: Prospective evaluation of renal allograft dysfunction with [99m]Technetium diethylenetriaminepentaacetic acid renal scans. *J Urol* 1984; 131:875–879. Used with permission.)

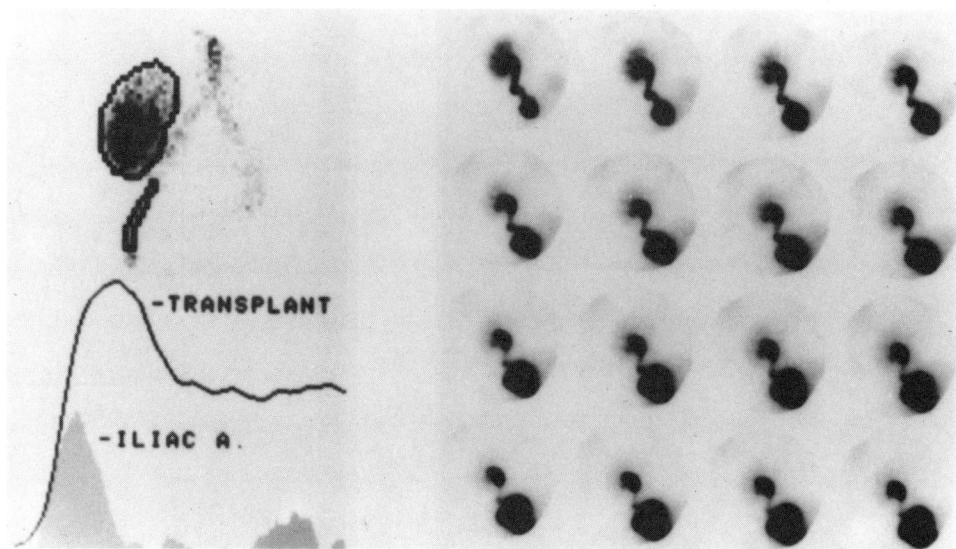

FIG 24–18.
Baseline renal scan in cadaver renal transplant recipient with oliguric leak tubular necrosis (ATN). The blood flow curves show mild impairment of allograft perfusion and excretion markedly decreased. (From McConnell JD, Sagalowsky AI, Lewis SE, et al: Prospective evaluation of renal allograft dysfunction with [99m]Technetium diethylenetriaminepentaacetic acid renal scans. *J Urol* 1984; 131:875–879. Used with permission.)

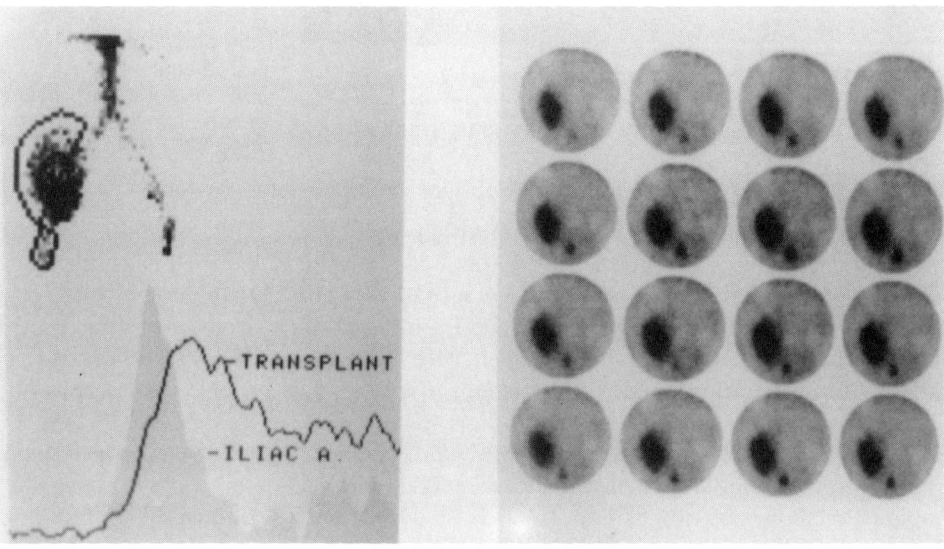

FIG 24–19.
Renal scan one week later in same patient as Figure 24–18 shows resolution of ATN with improvement in allograft blood flow and excretion. (From McConnell JD, Sagalowsky AI, Lewis SE, et al: Prospective evaluation of renal allograft dysfunction with [99m]Technetium diethylenetriaminepentaacetic acid renal scans. *J Urol* 1984; 131:875–879. Used with permission.)

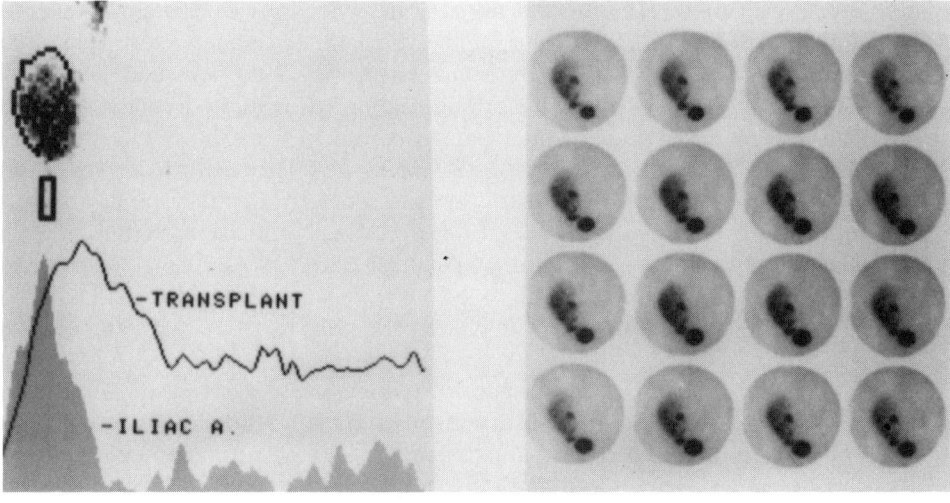

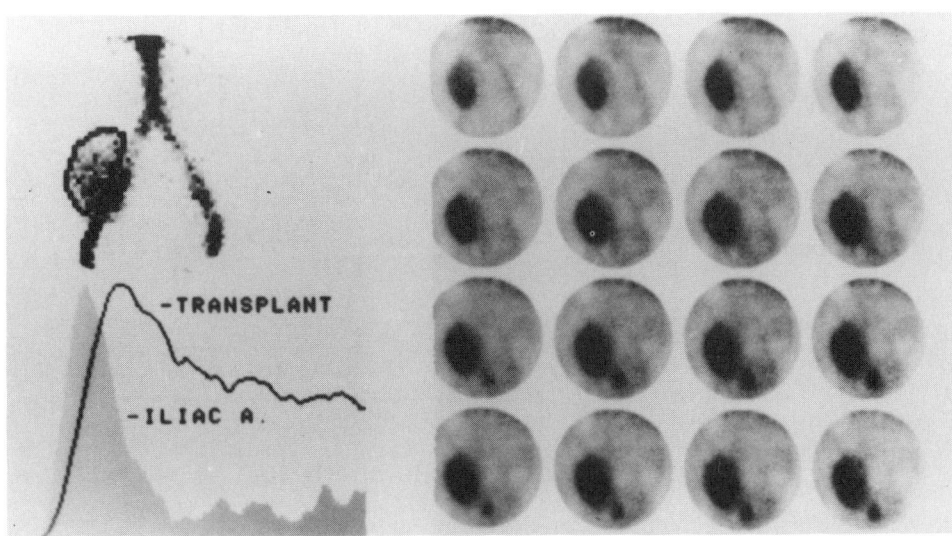

FIG 24–20.
Baseline renal scan in patient with nonoliguric ATN. (From McConnell JD, Sagalowsky AI, Lewis SE, et al: Prospective evaluation of renal allograft dysfunction with ^{99m}Technetium diethylenetriaminepentaacetic acid renal scans. *J Urol* 1984; 131:875–879. Used with permission.)

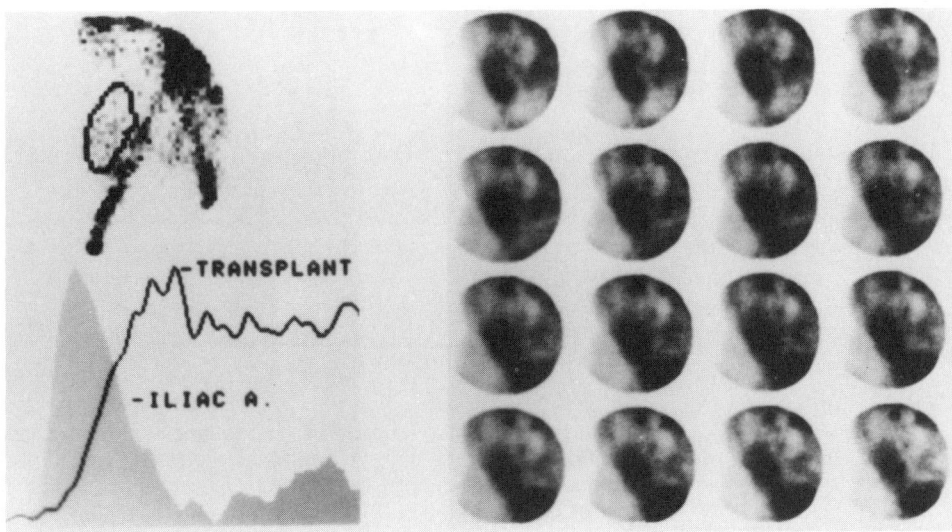

FIG 24–21.
Renal scan one week later in same patient as Figure 24–20 during clinical acute rejection manifested by fever, allograft pain and swelling, and decreased urine output. Both allograft blood flow and excretion are decreased compared with baseline. (From McConnell JD, Sagalowsky AI, Lewis SE, et al: Prospective evaluation of renal allograft dysfunction with ^{99m}Technetium diethylenetriaminepentaacetic acid renal scans. *J Urol* 1984; 131:875–879. Used with permission.)

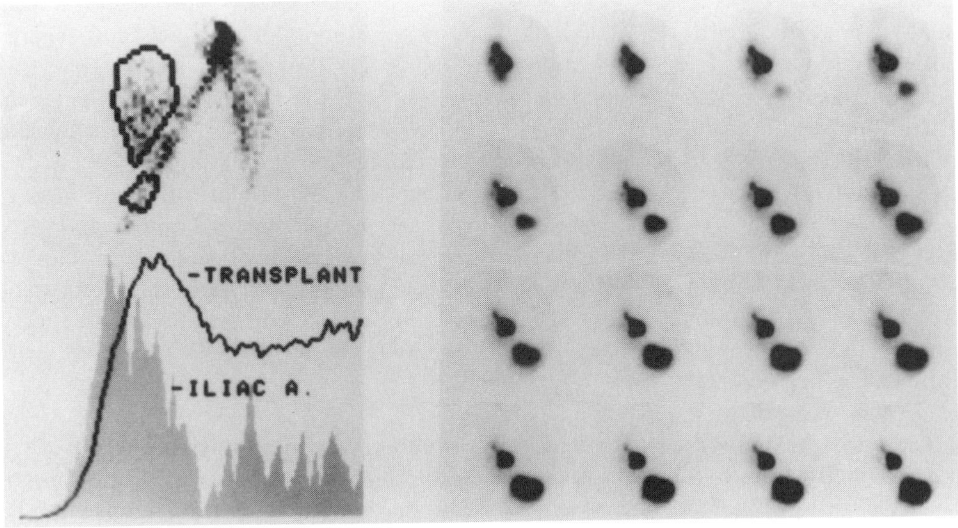

FIG 24–22.
Renal scan in same patient as Figures 24–20 and 24–21 following treatment with bolus steroids. Renal blood flow and excretion are markedly improved and are consistent with resolution of rejection and ATN. Clinically the patient was well and had a serum creatinine of 1.5 mg/dl and sodium isothalamate (Glofil) clearance of 50 ml/min. (From McConnell JD, Sagalowsky AI, Lewis SE, et al: Prospective evaluation of renal allograft dysfunction with ^{99m}Technetium diethylenetriaminepentaacetic acid renal scans. *J Urol* 1984; 131:875–879. Used with permission.)

isodes respond to bolus steroid therapy. Large cumulative doses of corticosteroids are associated with increased infectious complications, aseptic necrosis of the hip, cataracts, and GI bleeding. Therefore, patients who fail to respond to a first series of bolus steroids or who have repeated rejection episodes are best managed by switching to another form of therapy.

Antithymocyte Globulin and Antilymphocyte Serum (ATG, ALS)

Polyclonal antilymphocyte preparations are highly effective as initial therapy for acute rejection episodes, with more than 90% of patients responding (Filo et al., 1981). In our experience, 87% of patients with steroid-resistant initial rejection episodes and the majority of patients with recurrent rejection episodes responded to a 10–14-day course of ATG. Cyclosporine or azathioprine must be withheld or greatly reduced during ATG therapy to avoid overimmunosuppression.

Monoclonal Antibody Therapy

In clinical trials the pan–T-cell monoclonal antibody OKT_3 produces rapid disappearance of circulating T-lymphocytes following an initial IV dose and produces reversal of acute cellular rejection in over 90% of cases during a 14-day course of treatment (Cosimi et al., 1981). Nearly all patients experience fever and chills during OKT_3 infusion for the first few days. Rare anaphylactoid reactions have occurred in patients who had volume overload.

During 1983, our transplant center participated in a multicenter trial comparing PAN OKT_3 monoclonal antibody with bolus steroid therapy for acute allograft rejection occurring after the fifth postoperative day in 123 patients. There were no statistical differences between the two groups of patients. The treatment protocols of the two groups are shown in Fig 24–23 (Ortho Multicenter Transplant Study Group, 1985). Patients in the steroid therapy group received either 4 gm of IV methylprednisolone or 3 mg/kg/day of oral prednisone for five days. Patients in the OKT_3 group received 5 mg/day of OKT_3 IV for 14 days and had their azathioprine dose reduced. Our center randomized 12 patients into the multicenter trial and entered ten patients with rejection occurring before the fifth postoperative day. All seven patients with rejection beyond day 5 responded to OKT_3, three of whom suffered no further rejection episodes (Table 24–9). Only two of the five steroid-treated patients had reversal of their rejection episode. OKT_3 reversed the rejection in only one of five patients in the early rejection group. The 71% one-year allograft survival iin our OKT_3-treated patients was the same as in the larger multicenter trial and was statistically better than the graft survival in the steroid group (Fig 24–

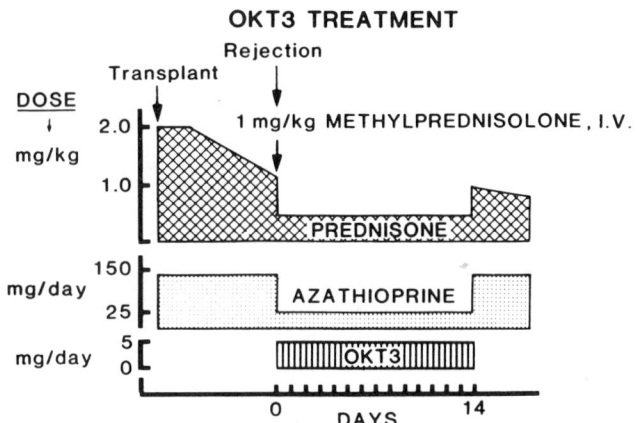

FIG 24–23.
Immunosuppressive regimens for patients randomly assigned to OKT_3 or steroid treatment groups. (From Ortho Multicenter Transplant Study Group: A randomized clinical trial of OKT_3 monoclonal antibody for acute rejection of cadaver renal transplants. *N Engl J Med* 1985; 313:337–342. Used with permission.)

24) (Ortho trial). Experiments on the cells and serums of patients receiving OKT_3 therapy reveal that this monoclonal antibody is cytotoxic but not mitogenic to activated lymphocytes and removes donor-responsive cells from the circulation (Bowen et al., 1984).

Additional monoclonal antibodies directed against specific T-cell subsets of the helper-inducer and suppressor-cytotoxic lines are being readied for clinical testing. Theoretically, deletion of helper-inducer T cells without alteration of other T-cell functions is an ideal therapy against cellular rejection.

Cyclosporine Rescue

Because the predominant mechanism of action of cyclosporine is inhibition of T-cell activation, one might predict that this agent would be ineffective as therapy

TABLE 24–9.
Reversal of Acute Rejection

	REJECTION AFTER DAY 5		REJECTION BEFORE DAY 5	
	OKT₃	STEROIDS	OKT₃	STEROIDS
Reversal of rejection	7/7	2/5	1/5	2/5
Rerejection	4/7	5/5	0/1	1/2

Adapted from Helderman JH: Renal transplantation, in *Internal Medicine Grand Rounds.* Dallas, The University of Texas Health Science Center at Dallas, Oct 25, 1984.

for an established rejection episode. To test this hypothesis, we treated 20 allograft recipients suffering from steroid- and ATG-resistant rejection episodes with 14 mg/kg/day of cyclosporine as rescue therapy (Sagalowsky et al., 1985). Sixteen of twenty (80%) patients responded and had a mean cumulative serum creatinine of 1.8 mg/dl at 1–9 months of follow-up. Cyclosporine rescue therapy also was successful in five patients who required cessation of ATG therapy owing to serum sickness or thrombocytopenia. The four nonresponding patients showed extensive humoral rejection in the allograft.

Plasmapheresis

All of the antirejection therapies described above are ineffective against antibody-mediated humoral rejec-

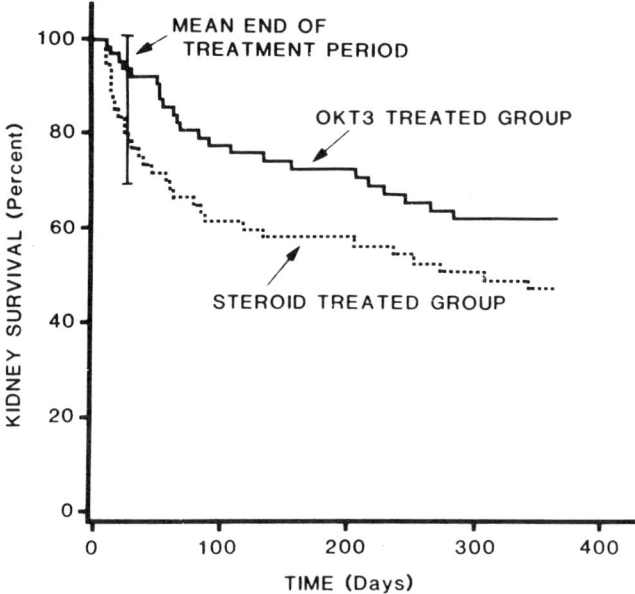

FIG 24–24.
Life-table analysis of kidney survival in patients randomly assigned to OKT₃ or steroid treatment for acute renal allograft rejection. (From Ortho Multicenter Transplant Study Group: A randomized clinical trial of OKT₃ monoclonal antibody for acute rejection of cadaver renal transplants. *N Engl J Med* 1985; 313:337–342. Used with permission.)

tion. This form of rejection is characterized on biopsy by platelet aggregates and microthrombi in vascular spaces and polymorphonuclear leukocytes along vessel walls. The technique of plasmapheresis for the removal of pathogenic substances from the circulation has been available for more than a decade. Development of continuous- and intermittent-flow centrifuges permits removal of plasma constituents "on line" with the patient circulation. Plasmapheresis is beneficial in a number of autoimmune- and antibody-complex-mediated diseases (AMA Council on Scientific Affairs, 1985). The attempt to treat humoral allograft rejection with this technique was a logical step. In several small, uncontrolled series plasmapheresis or lymphocytoplasmapheresis was effective in treating rejection (Slapak et al., 1971; Cordella et al., 1977; Lyngaard et al., 1980; Allen et al., 1982; Darr et al., 1982; Kleinman et al., 1982). In three controlled series, including our own trial, an intensive short course of plasmapheresis was not of benefit in reversing steroid-resistant humoral rejection (Power et al., 1981; Kirkubakaran et al., 1981; Helderman et al., 1982). We treated nine patients with oliguric rejection that revealed a humoral pattern on renal biopsy with five consecutive daily two-body volume plasma exchanges. Although eight of nine patients obtained an increase in urinary output, only two patients retained allograft function by three months. To date the role, if any, of plasmapheresis as antirejection therapy is speculative. Each of these trials must be interpreted in terms of the time course when plasmapheresis was initiated and the number and frequency of treatments. The possibility remains that circulating allospecific antibody may be removed and then kept at a low level until host immune tolerance occurs, mediated by suppressor cells or other mechanisms.

COMPLICATIONS

The diseases that produce ESRF and the systemic consequences of chronic uremia make renal transplant recipients a challenging group of patients who are at risk of a wide range of complications. Nevertheless, as long-term dialysis has become more efficient, surgical technique refined, and immunosuppression made safer and more effective, the incidence of all of the complications to be described has decreased.

Surgical

Surgical complications may be divided into vascular, urologic, and wound-related problems. Renal artery thrombosis occurs in fewer than 1% of cases. Patients receiving cyclosporine may be at increased risk of renal artery thrombosis (Canadian Multicentre Transplant Study Group, 1983). Early arterial suture line bleeding invariably is due to technical problems, and rare arterial bleeding beyond 48 hours postoperatively is usually due to infection and pseudoaneurysm formation at the anastomosis (Sagalowsky and Peters, 1982). The incidence of significant transplant renal artery stenosis is approximately 10% and may be due to fibrous dysplasia accompanying chronic rejection, technical error, or kinking or axial rotation of the vessels (Sagalowsky and Peters, 1984) (Figs 24–25 and 24–26).

Bleeding from the renal vein anastomosis is very rare. Renal vein obstruction may result from extrinsic compression by the renal artery if the kidney is improperly positioned. Pediatric recipients of adult allografts are at increased risk of iliac or renal vein compression by the kidney itself if the space created for the kidney is too tight following fascial closure. Acute renal vein obstruction may result in hemorrhagic infarction or rupture of the allograft. Renal vein thrombosis occurs in only 1%–3% of transplants and is suggested by sudden pain and swelling over the allograft, oliguria, hematuria, proteinuria, and ipsilateral leg edema. Venography is necessary to confirm renal vein thrombosis, and anticoagulation with heparin is the usual therapy.

Major urologic complications after renal transplantation require prompt diagnosis and therapy to avoid high patient morbidity and mortality. Ureteral leak is the most serious problem and usually results from ureteral ischemia. This problem presents during the first six weeks postoperatively. The renal hilar fat and blood vessels must be preserved during organ procurement to minimize the occurrence of ureteral leak (Salvatierra et al., 1974). Retrograde pyelography is difficult to per-

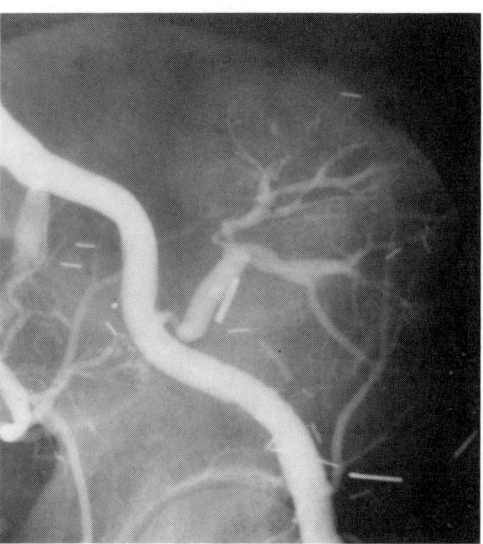

FIG 24–26.
Stenosis of end-to-side renal artery anastomosis may be due to turbulent flow as well as to rejection or kinking.

form during the early postoperative period but may establish the diagnosis. More often ureteral intubation is not possible, and a perirenal fluid collection is identified by ultrasonography (see Fig 24–15). Needle aspiration of the fluid confirms urinary extravasation by the presence of urea and potassium in concentrations above serum levels. The most important principle in the treatment of ureteral leak following transplantation is to divert the urine away from the fresh vascular anastomoses (Sagalowsky et al., 1983a). Placement of a retrograde or percutaneous ureteral stent or nephrostomy allows the patient to be stabilized. Occasionally such procedures provide definitive management. More often, however, they allow for semielective open surgical repair of the ureteral injury. Ureteral debridement and closure over a stent for minimal perforations and ureteral reimplantation or resection with nephrostomy diversion for major ureteral slough are the procedures of choice (Figs 24–27 and 24–28). Definitive reconstruction is accomplished 4–6 months later by ipsilateral native nephrectomy and ureteropyelostomy of native ureter to transplant renal pelvis (Fig 24–29).

Posttransplant ureteral obstruction is rare in the early postoperative period. Late ureteral obstruction develops in a small percentage of cases and usually occurs near the ureterovesical junction. This problem most likely is due to distal ureteral ischemia and subsequent fibrosis. Ureteral obstruction is best treated initially by an internalized stent. Patient and allograft status as well as ease of maintaining the stent determine the choice for definitive management between direct surgical repair and continued stenting or percutaneous nephrostomy.

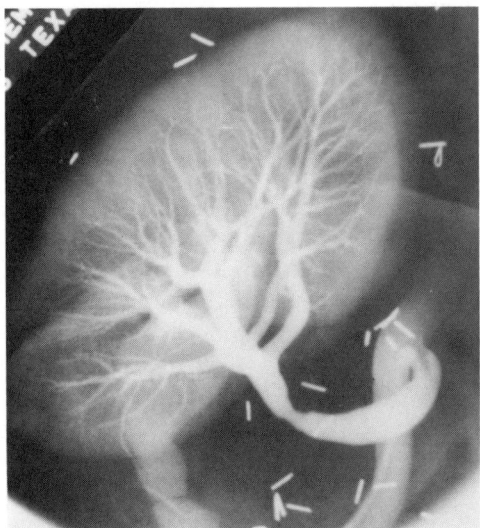

FIG 24–25.
Transplant renal artery stenosis following end-to-end anastomosis to hypogastric artery. Preanastomotic stenosis may be due to perfusion injury. Anastomotic stenosis may result from technical error. Postanastomotic stenosis is most often due to fibrous dysplasia of rejection.

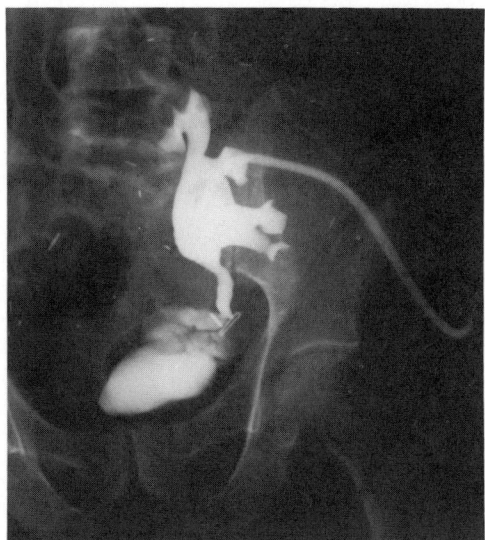

FIG 24–27.
Following the diagnosis of urine leak by sonography and aspiration in a cadaver transplant recipient as nephrostomy diverts the urine and a nephrostogram confirms leakage is from the ureter.

The incidence of bladder leak following transplantation is very low if a meticulous three-layer closure is employed. Although most authors leave the bladder catheter in for 3–7 days (Simmons et al., 1971; Smith and Erlich, 1976; Salvatierra et al., 1977), the catheter may be removed safely within 36 hours in the majority of cases (Sagalowsky et al., 1983a).

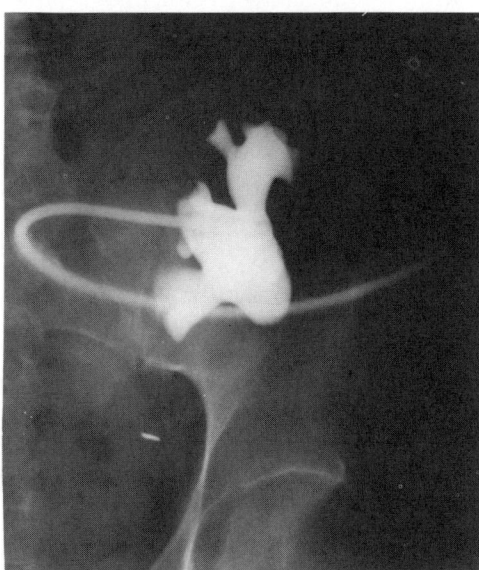

FIG 24–28.
The usual finding at exploration is a necrotic distal ureter and infected urinoma. Primary repair usually is not indicated and debridement, nephrostomy drainage, and closure of the renal pelvis allow for safe initial management.

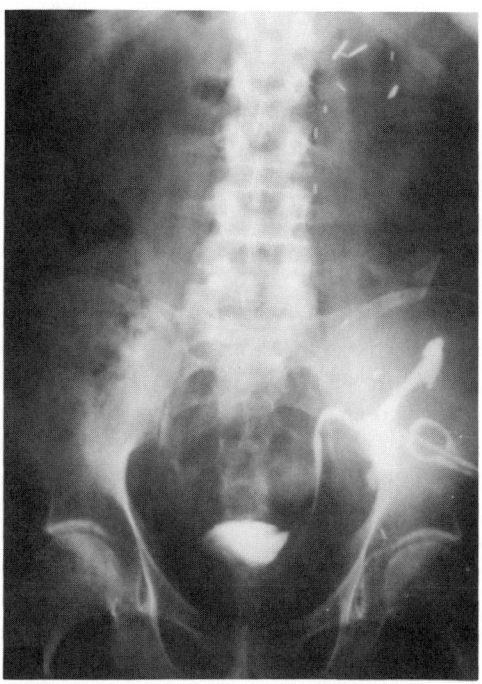

FIG 24–29.
Four to six months later, when the patient is on lower doses of immunosuppression and has stable allograft function, ipsilateral native nephrectomy and ureteropyelostomy to the allograft provide definitive repair.

Lymphoceles may produce ureteral obstruction and decreased renal function (Ehrlich and Smith, 1977). Leakage of lymph from unligated lymphatics that are disrupted during dissection of the iliac vessels is the major cause of lymphocele formation. The first signs of a lymphocele are the edema of the ipsilateral lower extremity and rise in serum creatinine. The diagnosis is established by ultrasonography (see Fig 24–14). Percutaneous aspiration of the fluid collection is the recommended initial therapy for symptomatic lymphoceles. Intraperitonealization or external drainage is recommended for recurrent lymphoceles.

Spontaneous rupture of the allograft may occur in the first days following transplantation owing to graft swelling. The kidney swells during ATN and rejection. The role of capsulotomy at the time of transplantation in preventing graft rupture is unclear and is not practiced by the authors. Patients with allograft rupture present with sudden onset of severe pain over the allograft and have a firm, swollen mass and rigid abdomen on palpation accompanied by decrease in hematocrit. These clinical findings are so constant that sudden decrease in hematocrit and appearance of fluid around the kidney in a patient who is free of pain must suggest anastomotic bleeding. Surgical exploration is required to control bleeding and often results in nephrectomy. Occasionally the allograft appears savageable, and the parenchymal bleeding is controllable. However, the decision to leave the allograft in is a decided risk.

The incidence of wound infections in renal transplant recipients can be reduced to 1%–2% by the routine use of perioperative antibiotics, vigorous wound irrigation, and avoidance of drains (Banowsky et al., 1974; Novick, 1981; Muakassa et al., 1983). Closed suction drains are preferable to Penrose drains when external drainage is necessary.

Medical

Medical complications following renal transplantation may involve every system of the body. Major complications are directly related to the metabolic consequences of corticosteroids. The occurrence of cataracts and aseptic necrosis of the femoral head correlates with the cumulative dose of corticosteroids. Delayed wound healing and easy bruisability are well-recognized consequences of the hypercorticoid state. Pancreatitis and peptic ulcer disease are two GI problems that are exacerbated by corticosteroids. Most renal failure patients have asymptomatic increases in serum amylase levels owing to its decreased renal clearance. All transplant recipients receive antacids as ulcer prophylaxis. Formerly, potential recipients with documented episodes of peptic ulcer disease underwent vagotomy and a gastric drainage procedure before transplantation (Prompt et al., 1977). Currently, the need for such preventive surgery is being reevaluated owing to the use of lower doses of steroids and the availability of histamine receptor blockers. One must be aware that histamine receptor blockers cause a rise in serum creatinine and may diminish renal function.

Infections

The immunosuppressed transplant patient is at increased risk of opportunistic infections that rarely affect an intact host. Sepsis is the leading cause of death in transplant patients. Pneumonia occurs in 10%–25% of renal transplant recipients and has a reported mortality of 2%–80% (Moore et al., 1983). *Legionella, Candida, Pneumocystis, Aspergillus, Nocardia, Cryptococcus,* herpesvirus, and cytomegalovirus, along with common bacterial causes, all must be considered in the differential diagnosis of pneumonia in the immunocompromised host (Chatterjee et al., 1978a, 1978b; Taylor et al., 1981; Ludmerer and Kissane, 1982; Moore et al., 1983). The key to high patient survival in these cases is prompt diagnosis and specific therapy. Sputum samples are usually contaminated with upper airway flora and are nondiagnostic. Transtracheal aspiration, bronchoscopy, and percutaneous or open lung biopsy all have their proponents, depending on the morbidity rate and diagnostic yield for each test in a given institution (Fulkerson, 1984).

Viral hepatitis occurs in 6%–16% of renal transplant recipients and is a frequent cause of morbidity (Ware et al., 1979). The majority of cases are due to non-A and non-B transfusion-acquired virus (Strom, 1982). The risk of transplantation and immunosuppression in patients who are serologically positive for hepatitis B surface antigen (HBsAg) has been unclear. In a prospective study renal transplant recipients who were HBsAg positive had a much higher risk of chronic liver disease (cirrhosis, 32%; chronic persistent hepatitis, 23%; death due to liver failure, 14%; death due to hepatoma, 9%) than HBsAg-positive patients who remain on hemodialysis (persistent hepatitis, 10%) (Parfrey et al., 1984). In the same study, 40% of the dialysis patients reverted to HBsAg-negative status, while none of the transplant recipients reverted. Immunosuppression per se may facilitate hepatitis viral replication. Azathioprine, cyclophosphamide, and cyclosporine are direct hepatotoxins. Reduction or cessation of immunosuppression in transplant recipients with progressive liver disease must be individualized based on clinical course, enzyme changes, and liver biopsy results.

Cytomegalovirus (CMV) is ubiquitous in the environment, and clinical infection may cause profound immunosuppression leading to opportunistic infections (Glenn, 1981; Schooley et al., 1983). Active CMV infection is uncommon and occurs most often in transplant recipients who were seronegative before transplantation, i.e., those suffering primary infection (Chatterjee et al., 1978b). Seronegative recipients of kidneys from seropositive donors have the highest rate of CMV disease and major complications (Smiley et al., 1985). Fever owing to reactivated CMV is a diagnosis of exclusion in a clinically well patient found to have rising CMV titers in the blood. New CMV vaccines are now undergoing initial clinical trials in transplant recipients. Infection with herpes simplex virus also is common among transplant recipients.

The recently described human T-cell lymphotropic virus, type III (HTLV-III), which produces the acquired immune deficiency syndrome (AIDS), may become a frequent and highly fatal pathogen in transplant recipients. During the past 15 years, blood transfusions have been frequent among patients awaiting renal transplant owing to the salutary effect on allograft survival. However, only recently has a screening antibody test for HTLV-III been available to safeguard the blood supply. It is reasonable to speculate that a substantial number of current transplant recipients are already HTLV-III antibody-positive and that an unknown proportion of these will develop AIDS.

Hypertension

The majority of dialysis patients and renal transplant recipients have hypertension (Prompt et al., 1977; Sagalowsky and Peters, 1984). Hypertension following transplantation is multifactoral. Corticosteroids contribute to volume overload. Cyclosporine frequently produces hypertension owing to increased renal vascular resistance. Native kidneys, rejection, and transplant renal artery stenosis all may produce renin-mediated hypertension in the transplant recipient. Satisfactory blood pressure control is essential to preservation of renal allograft function. Native bilateral nephrectomy is indicated when optimal antihypertensive medications do not control blood pressure in an allograft recipient who does not have severe rejection or transplant renal artery stenosis.

Hyperparathyroidism

Secondary hyperparathyroidism and osteodystrophy are natural consequences of ESRD. Improved medical management with phosphate binders and vitamin D metabolites before transplantation reduce the metabolic consequences. Successful renal transplantation allows the return to normal of calcium and phosphorus metabolism and involution of parathyroid hyperplasia in most patients. Nevertheless, a fraction of patients remain hypercalcemic, have clinical sequelae, and require parathyroidectomy (Prompt et al., 1977; Garvin et al., 1985).

Malignancy

There is an increased incidence of malignancy in renal transplant recipients compared with the general population owing to the effects of immunosuppression on reducing host immune surveillance, direct oncogenesis by immunosuppressants (e.g., cyclophosphamide), and increased susceptibility to oncogenic viruses (Prompt et al., 1977; Penn, 1981). Squamous carcinomas of the skin and mucous membranes are the most common tumors and are usually amenable to simple excision. Lymphomas are the second most common tumors in transplant patients, have a predilection for CNS involvement, and frequently are fatal. The development of lymphoma in three renal transplant recipients treated in one of the early cyclosporine trials raised fears about this agent (Thiru et al., 1981). Now that several thousand transplant recipients have been treated with cyclosporine, an increased incidence of lymphoma similar to that seen in patients treated with other immunosuppressants has been noted (Penn, 1984). Further experience is required to exclude an exaggerated risk for lymphoma following cyclosporine treatment.

However, it appears likely that the tendency for lymphomas in cyclosporine-treated patients is dose related and correlates with the degree of immunosuppression.

SURGERY OF RENAL TRANSPLANTATION

Cadaver Donor Harvest

Today the kidneys are frequently harvested as a part of multiple-organ retrieval from a single donor. Properly done, the heart, liver, and kidneys may be removed in sequence with other organs (intrathoracic and intra-abdominal), with excellent preservation in a functional state. In situ perfusion and cooling of the organs of the beating heart cadaver with exposure through a jugular notch to pubis incision has made this possible. The reader is referred to the following articles for lucid diagrams and detailed descriptions in color of the technique: Starzl et al., 1984; Simmons and Najarian, 1984). Salient features are covered in the description of transplantation techniques later in this chapter.

LRD Nephrectomy

We currently prefer the patient to be placed on the operating table in the supine position. Large-bore, 16-gauge needles or angiocatheters are placed in the forearm. The patient is in the supine position, with the left side slightly elevated on a folded sheet. The incision usually begins about the tip of the 12th rib and is angled superiorly toward the xiphoid, reaching a peak at the level of the xiphoid and then angling inferiorly again across the rectus sheath on the opposite side. We have found it helpful to cut into the opposite rectus sheath a distance of 2.5–3 cm on occasion. If there is an unusually wide costal margin, the extension across to the opposite rectus is not necessary. The skin incision is often made with the Bovie unit. The incision is carried down through the muscle layers, with the rectus being divided as described above and the external-internal oblique and transversus being divided. Hemostasis is achieved by cautery and 3-0 chromic gut ligatures. The peritoneal cavity is opened, and the colon is reflected medially by an incision lateral to the mesentery of the colon along the line of Toldt. We prefer to carry this incision from the flexure of the colon, splenic or hepatic, inferiorly to a point where the incision in the posterior parietal peritoneum crosses the common iliac artery, either lateral to the cecum or to the descending colon, depending on the side from which the kidney is being removed. One can then mobilize the peritoneum bluntly by placing the index and middle finger in the retroperitoneum, and then a curved large Deaver retractor can be used to elevate the posterior parietal peritoneum, thus exposing the ureter where it crosses the common iliac artery and enters the pelvis.

The ureter is dissected distally as near to the bladder as is feasible. A large hemoclip is placed distally on the ureter, or 2-0 silk ligature may used, and the ureter is divided proximal to the tie or clip. It is then dissected superiorly, with care being taken to preserve periadventitial areolar tissue and blood vessels (Fig 24–30). When the ureter has been dissected to the lower pole of the kidney, Gerota's fascia is removed by making an incision down to the kidney capsule and dissecting the areolar tissue adhesions of Gerota's capsule free from the underlying renal capsule. When mobilization of the kidney has thus been completed, attention is directed to the pedicle of the kidney. Any major tributaries to the renal vein are ligated with 2-0 and 3-0 silk suture ligatures, and the vessels are divided between these ligatures. On the left side, we have found it necessary to divide the gonadal and adrenal tributaries to obtain proper length of the renal vein. On the right side, occasionally the gonadal tributary drains into the renal vein, and occasionally one finds a large lumbodorsal tributary, which must be carefully divided and secured with a suture ligature on either side. This vessel is also commonly found on the left side. The renal artery is identified and may be exposed easily by retraction of the renal vein after it has been mobilized completely. Superior or inferior retraction of the renal vein usually aids dissection and identification of the renal artery. The artery is dissected down to its periampullary por-

tion, at which point it is usually seen to be slightly wider than the rest of the vessel. We prefer, when removing a kidney, to ligate the renal artery as close to the aorta as possible with a 2-0 or 0 silk ligature and then to place, immediately distal to this point, a 3-0 arterial silk suture ligature for additional hemostasis.

After all of the lymphatics and the areolar tisssue around the pedicle have been carefully ligated and divided between ligatures, only the renal artery and the renal vein hold the kidney. At this point, we usually give heparin, 100 units/kg. A 70-kg man would receive 7,000 units of heparin. An effort is made to maintain a high rate of urine output during the prenephrectomy dissection, which may require 1–1½ hours, the time being spent in achieving meticulous hemostasis. The urine output usually is in the range of 60–100 ml/hr during this phase of the surgery, and 12.5 gm of mannitol is usually given in addition to the heparin.

After ten minutes, the kidney is removed by ligating the renal artery and suture-ligating it as described above. The renal vein is divided, usually distal to the tributary of the adrenal vein to permit easy anastomosis in the recipient. After the kidney has been removed, the effect of the heparin is reversed by giving 1 mg of protamine for each 100 units of heparin; thus, a 70-kg man would receive 70 mg of protamine. Closure of the venous stump is completed with running 4-0 vascular silk. Protamine is infused slowly by the anesthesiologist to minimize bradycardia and cardiac arrhythmias. The removed kidney is flushed with a Ringer's lactate-heparin-procaine mixture containing 1,000 units of heparin and procaine, 100 mg/L cooled to between 4° and 7°C. A #18 angiocatheter is inserted into the renal artery and held gently between thumb and index finger to permit flushing of the kidney. When the venous effluent is clear, the kidney is taken to the adjacent room for implantation into the proposed recipient.

When multiple renal arteries are present in the living donor, the arteries are divided separately, and no attempt is made to take a Carrel patch from the donor site in the living donor transfer. The artery to the inferior pole of the kidney, which is usually smaller, is anastomosed end-to-side to the main renal artery with 6-0 Prolene, essentially making it a branch of the main renal artery. When the arteries are of equal size and length, one may incise them and produce a common stoma of the two renal arteries, incising the superior one on its inferior border and the inferior one on its superior border and closing the two as a single vessel with a running 6-0 Prolene suture. This gives a single artery to anastomose in the recipient, a technique we occasionally use in renovascular surgical procedures. The donor then has a layered closure with no drain, after the hemostasis is carefully checked. A counterac-

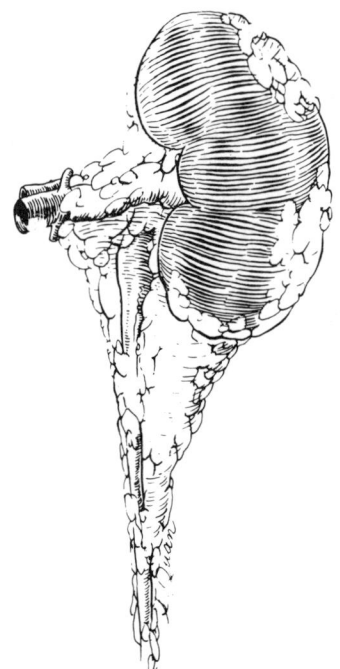

FIG 24–30.
LRD nephrectomy. Note preservation of perihilar fat and triangle of fatty and areolar tissue extending from lower pole of kidney along course of ureter.

tion of the heparin by the use of protamine is discussed with the anesthesiologist, and closure is begun. A running 2-0 chromic gut is used to approximate the posterior parietal peritoneum along the line of Toldt to the medial border of the colonic peritoneum, after the renal fossa has been carefully checked to make sure there are no adrenal venous bleeders. If present, these may be controlled easily with a hemoclip or ligature, which will make certain the renal fossa is dry and that there is no bleeding from the distal ureteral stump. Next, the closure of the posterior parietal peritoneum is completed with a running 2-0 chromic gut suture; then the anterior abdominal wall is closed in layers. The fascial layers are closed with interrupted 0 Prolene, the subcutaneous tissue with 3-0 Dexon or 3-0 chromic gut, and the skin with clips. Usually no blood is required for the donor, although we have 6 units of blood available (type and hold) for the donor nephrectomy. The correct position of the nasogastric tube is ascertained before closure and is usually left in until bowel sounds are vigorous or until the patient has passed gas during the postoperative period. An effort is made to maintain urine output at a high level until the kidney has been removed, and then fluids are administered to maintain an adequate urine output of more than 30 ml/hr.

Complications of Living Donor Nephrectomy

This operation has the lowest morbidity and mortality of any major surgical procedure performed today. It compares very favorably with the morbidity and mortality of circumcision with local anesthesia and is an example of what can be accomplished with increased sensitivity of the surgeons and anesthesiologists to the care of a normal individual who is volunteering to undergo a major surgical procedure. Most complications in living donor nephrectomy can be prevented by attention to detail. Careful intraoperative and postoperative fluid therapy will minimize the possibility of tubular necrosis from insufficient fluid replacement or oliguria secondary to extracellular fluid depletion. Nasogastric suction losses in the early postoperative period may be replaced with one half strength normal saline, since there has been no tendency to unusual electrolyte loss from the surgery. Postoperative atelectasis may be prevented by vigorous pulmonary toilet, encouraging the patient to cough and making certain that the anesthesiologist has expanded the lung regularly. A chest x-ray is made in the recovery room immediately after the surgery to ascertain that there has been no evidence of pneumothorax from rupture of a pulmonary bleb during anesthesia or from a small nick in the parietal pleura, which might have occurred when the kidney was being dissected from the attachments to the adrenal and to the body wall laterally.

Thrombophlebitis has not been a problem in our experience. We do not routinely wrap the legs preoperatively. The supine position also favorably influences postoperative morbidity. Early ambulation is encouraged, and the donor is often out of bed and walking about his room on the evening of surgery or certainly by the next morning. The preoperative workup often screens out individuals with any detectable bleeding abnormality. A study done at the University of Texas at Galveston revealed only 11 cases of unusual postoperative bleeding in more than 25,000 patients who had a normal PTT, prothrombin time, and platelet count before surgery. These are routinely done on transplant donors. In the Galveston experience only 11 patients undergoing major surgical procedures had a bleeding tendency if these studies were normal, and of those 11, 4 had hemophilia A, 4 had hemophilia B, and 3 had von Willebrand's disease, which could have been detected by a routine bleeding time determination. The stress of the donor nephrectomy operation is borne out by the fact that peptic ulcer occasionally develops postoperatively in donors, as was the case in one of our donors, who had massive hemorrhage from a bleeding duodenal ulcer six weeks after a successful transplantation and required emergency surgery.

Pneumothorax, if it does occur, is treated simply by underwater sealed-catheter drainage. The donor is encouraged to breathe vigorously postoperatively; if he seems reticent, blow bottles can be used to encourage deeper respiration. Wound infections have not been a major problem, occurring in 1%–2% of our donors postoperatively. Meticulous attention to hemostasis has been the major factor in prevention of postoperative wound infections in donors. The majority of the infections have been due to *Staphylococcus epidermidis* or, occasionally, *S. aureus*.

Multiple Organ Procurement from the Cadaver

The cadaver donor is taken to the operating room after brain death has been declared and appropriate forms within the institution have been signed. Terminal care and harvest care include the judicious administration of large volumes of IV fluid to restore or maintain blood volume, and diuretics, furosemide (Lasix), 20–40 mg IV every 2–3 hours and mannitol, 12.5 gm every hour are administered to promote diuresis. Multiple organ harvest is preferred and will be described as will the deviations when only the kidneys are being harvested. The patient is prepared and draped in the supine position with endotracheal respiratory support, and the body is opened through a midline incision from the jugular notch to the symphysis (Fig 24–31). The great vessels are rapidly encircled, particularly the distal aorta and vena cava and the proximal aorta just

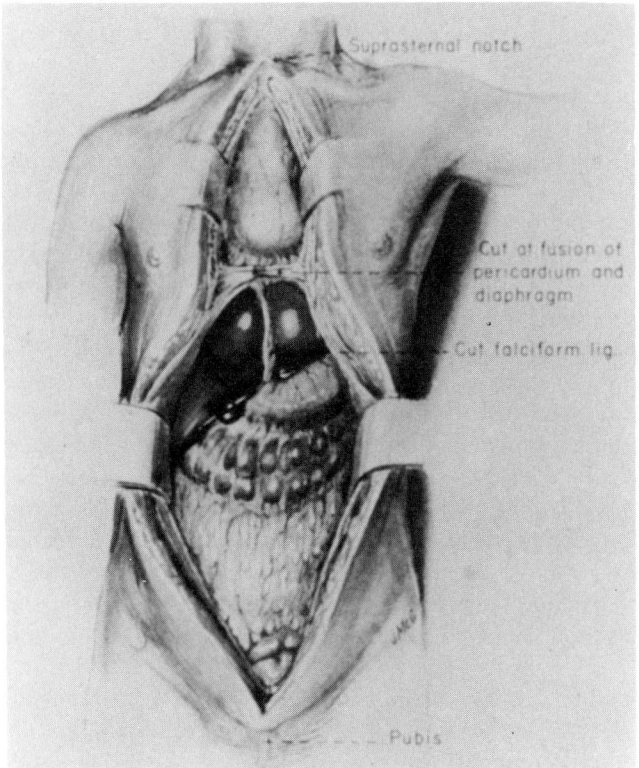

FIG 24–31.
Incision of cadaver harvest extending from suprasternal notch to pubis. Falciform ligament has been divided as have the diaphragmatic attachments of pericardium anteriorly. (From Starzl TE, Hakala TR, Shaw BW, et al: A flexible procedure for multiple cadaveric organ procurement. *Surg Gynecol Obstet* 1984; 158:223–230. Used with permission.)

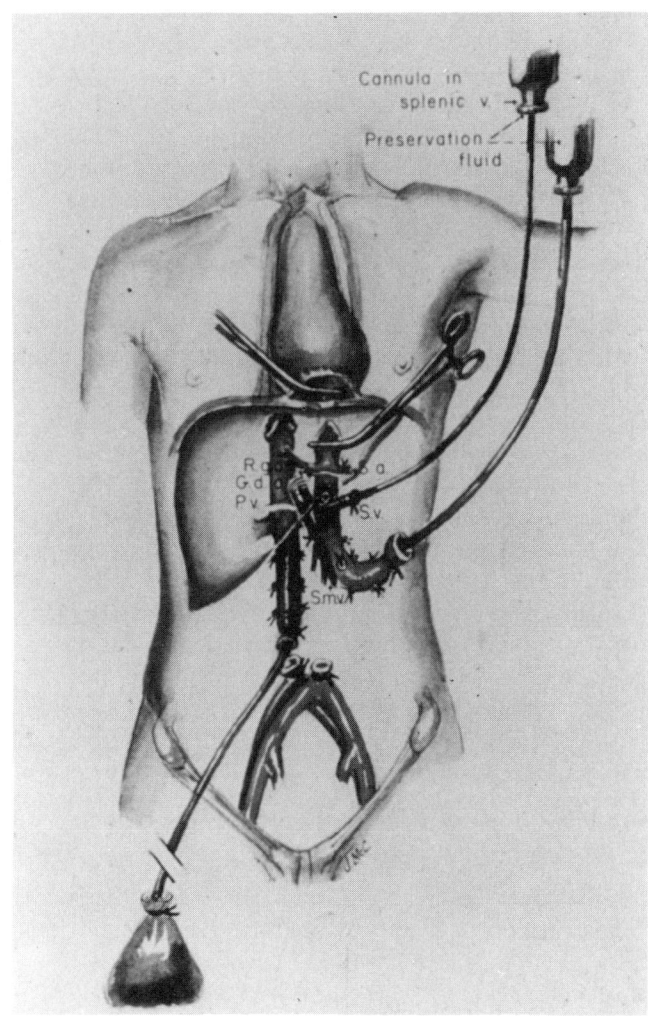

FIG 24–32.
Aorta and vena cava have been divided and tributaries and branches ligated. Splenic vein has been cannulated and liver is being perfused with preservation fluid as are systemic organs through the aortic stump. Note cannula in vena cava to remove effluent. Preservation of common and external and internal iliac arteries and veins that may be harvested subsequently. (From Starzl TE, Hakala TR, Shaw BW, et al: A flexible procedure for multiple cadaveric organ procurement. *Surg Gynecol Obstet* 1984; 158:223–230. Used with permission.)

above and below the level of the diaphragm. Catheters are inserted into the distal aorta and vena cava and portal vein (for liver harvest) for perfusion and for removal of blood (Figs 24–32 and 24–33). Perfusion is carried out using chilled Ringer's lactate solution, which contains heparin, 20,000 units/L, and mannitol, 18–20 gm/L. The dissection proceeds from the pelvis superiorly. Care is taken to preserve the distal iliac arteries and veins, since they may be harvested as donor vessels, particularly if liver implantation is to be subsequently carried out. The dissection is carried superiorly to the diaphragm, where the aorta has been encircled as part of obtaining vascular control. The dorsal branches of the aorta and dorsal tributaries of the vena cava are ligated. The inferior mesenteric artery is ligated, and the superior mesenteric artery is isolated and hepatic artery and celiac axis dissected, allowing exposure of the entire retroperitoneum by evisceration of the bowel. The kidneys are removed en bloc with the aorta and vena cava, after ligation of necessary venous tributaries such as gonadal and lumbodorsal. If the liver and heart are to be

removed, the heart is procured first to avoid an effect of hyperkalemia on the heart. Except for the cardioplegic, a solution of 20% KCl is used at the time of removal. The kidneys are usually cooled quickly by the infusion of 0.5–1 L of perfusate, and the perfusion is continued at a slower rate if multiple organs are to be removed. The heart is removed first, then the liver, then the kidneys.

Once the distal aorta and vena cava are freed and the vena cava and aorta secured superiorly, the kidneys can be mobilized using Gerota's capsule as a plane of dis-

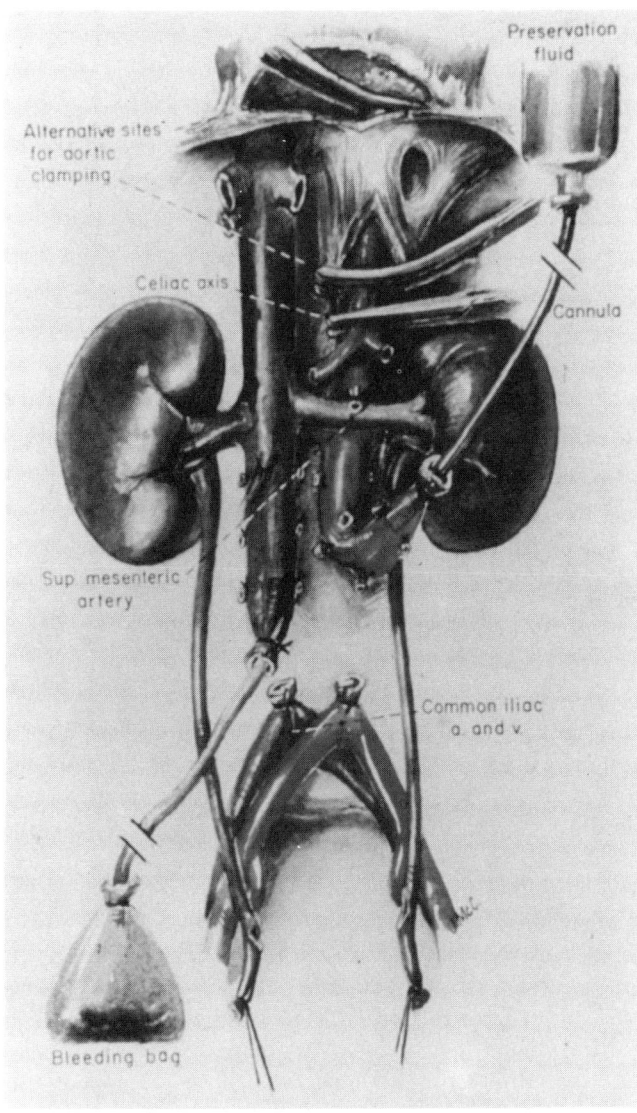

FIG 24–33.
Technique for renal harvest showing alternative sites for aortic clamping, preservation fluid being instilled through aortic stump, and effluent egressing through cannula in proximal vena cava. Note division of ureters just below iliac artery crossing. (From Starzl TE, Hakala TR, Shaw BW, et al.: A flexible procedure for multiple cadaveric organ procurement. *Surg Gynecol Obstet* 1984; 158:223–230. Used with permission.)

section. The distal aorta, vena cava, kidneys, and proximal aorta and vena cava are removed en bloc, after ligation of the lumbar vessels posteriorly, and are placed in a large washbasin of ice slush, where a complete dissection and assessment of the anatomy, particularly of the renal arterial circulation, is undertaken. The excess perinephric fat is removed, taking care to preserve the perihilar fat, which contains the blood supply for the most distal portion of the ureter, i.e., the

renal artery collaterals to the ureter. The kidneys may be separated by the division of aorta and vena cava in the midline, although if a preservation technique using a machine is to be carried out, we think it is best to close the aorta superiorly and to perfuse inferiorly by a single cannula inserted into the aorta. The multiple organ donor must, of course, have available lymph nodes removed so that subsequent histocompatibility and other immunologic studies can be performed. Donor nodes are easily obtained from the abdominal mesenteries. The spleen is harvested as well to obtain lymphocytes for histocompatibility studies.

RENAL TRANSPLANTATION

Consideration of Technique

The recipient is usually prepared and draped on the operating table in the supine position. Sterile drapes are applied for an oblique lower abdominal incision, usually extending from the tip of the 12th rib to the pubic tubercle (Fig 24–34). The incision is lateral to the rectus muscle and curves medially to the edge of the rectus as one approaches the pubic tubercle inferiorly. The external oblique aponeurosis is identified and separated in the direction of its fibers. The internal oblique and transversus abdominis are divided perpendicular to their fibers. Hemostasis is achieved by cautery and by 3-0 gut ligature. The parietal peritoneum is identified, peritoneal attachments to the spermatic cord are sharply divided, and the areolar fascia over the psoas muscle is easily separated bluntly from the peritoneum. The peritoneum is then mobilized medially and superiorly, and retractors are placed. The recipient bed involving the lower vena cava and iliac veins and the common internal and external iliac arteries is now ready for necessary sharp dissection and exposure as needed. One may use a Balfour retractor or, if short of assistants, the Bookwalter retractor will permit one blade to retract the bladder out of the way, thus facilitating the exposure and operation with one assistant. If a single donor renal artery is present, we prefer to utilize the recipient hypogastric artery for anastomosis, unless it contains a large plaque. In this situation, the recipient common or external iliac artery is used. If multiple donor arteries are present with Carrel patches, or if the hypogastric artery has large plaques and is not suitable for endarterectomy, the external iliac artery of the recipient is used (Fig 24–35). The site of implantation is chosen to avoid kinking of donor artery or vein, the limitations therefore being the length of the artery and vein and the length of the ureter for implantation into the bladder. We prefer to use the right side of the recipient for implantation, since the venous system is more accessible on the right side. The hypogastric ar-

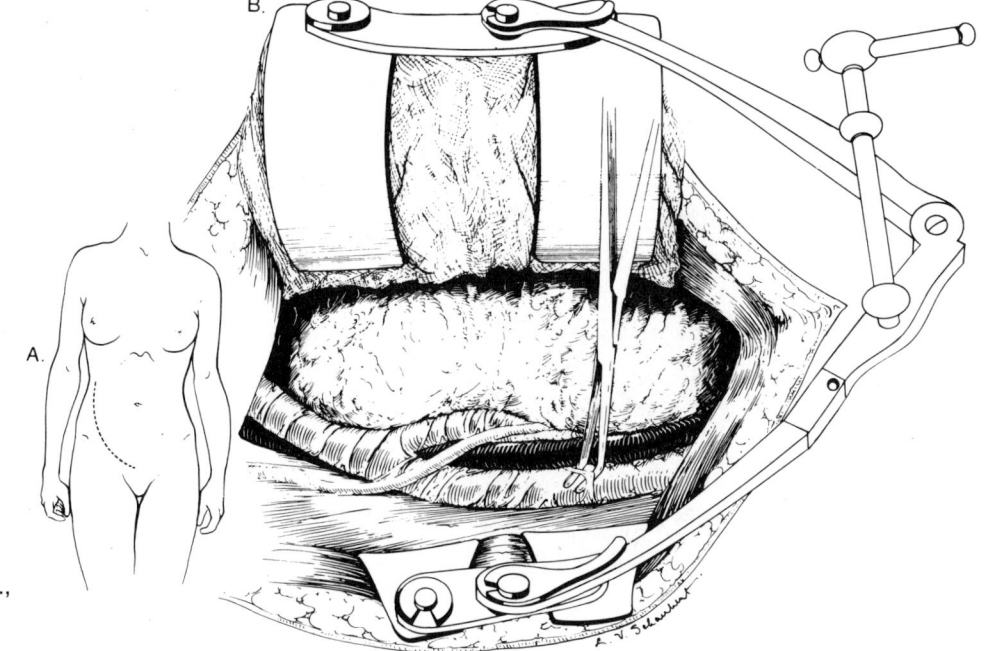

FIG 24–34.
Incision for renal transplantation. Peritoneum mobilized medially showing external and internal iliac artery and external iliac vein. **A,** line of incision in skin and muscle layers. **B,** view into iliac fossa after medial mobilization of peritoneum. (From Salvatierra O Jr: Renal transplantation, in Glenn JF (ed): *Urologic Surgery.* Philadelphia, JB Lippincott Co., 1983, p 360. Used with permission.)

tery is dissected by a combination of blunt and sharp dissection. Meticulous attention must be paid to ligation of lymphatics encountered during the venous and arterial dissection to avoid postoperative lymphocele. Only as much external iliac vein should be cleared as is necessary for implantation of the renal vein. One should be able to place a Satinsky clamp comfortably on the external iliac vein to create an adequate site for implantation of the renal vein. We do not heparinize the

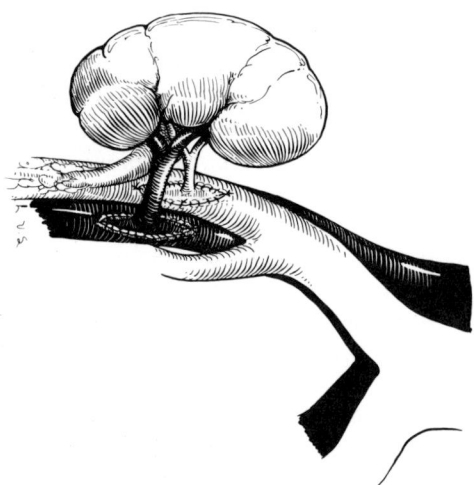

FIG 24–35.
Use of Carrel patch on donor artery and vein end-to-side to external iliac artery and vein. (From Salvatierra O Jr: Renal transplantation, in Glenn JF (ed): *Urologic Surgery.* Philadelphia, JB Lippincott Co, 1983, p 364. Used with permission.)

recipient in routine implantations. When we are forced to use the external or common iliac artery as a site of implantation, the recipient is heparinized with 100 units/kg ten minutes before clamping the artery for implantation. In these instances, the vein is sewed in first and then the artery to minimize the time required for occlusion of the artery to the lower extremity. We prefer 6-0 Prolene suture and always spatulate the arterial anastomosis. This is mandatory if vascular disparity exists, and it permits gradual transition in luminal caliber with laminar flow, even if no disparity exists. It also increases the cross-sectional area of the lumen at the point of anastomosis. The anastomosis is usually completed with a running 6-0 Prolene suture when the vessels are large, i.e., greater than 6 mm in diameter (Fig 24–36). With smaller vessels or when children are receiving a transplant, we prefer to use an interrupted suture technique to permit growth and expansion of the spatulated vessels. Quadrant sutures may be placed in the venous system to prevent axial torsion at the time of anastomosis.

The arterial techniques vary (Fig 24–37), but we prefer, if possible, to place a suture directly anteriorly and directly posteriorly, so that to complete the arterial anastomosis, rotation of the vessel through only 90° is required rather than 180°. Meticulous attention is given to technique at this point to preclude reclamping once circulation is restored. The patient is given 12.5 gm of mannitol about ten minutes before removal of the vascular clamps and 125 mg of Solu-Medrol 2–3 minutes before removal of the vascular clamps. At the time of

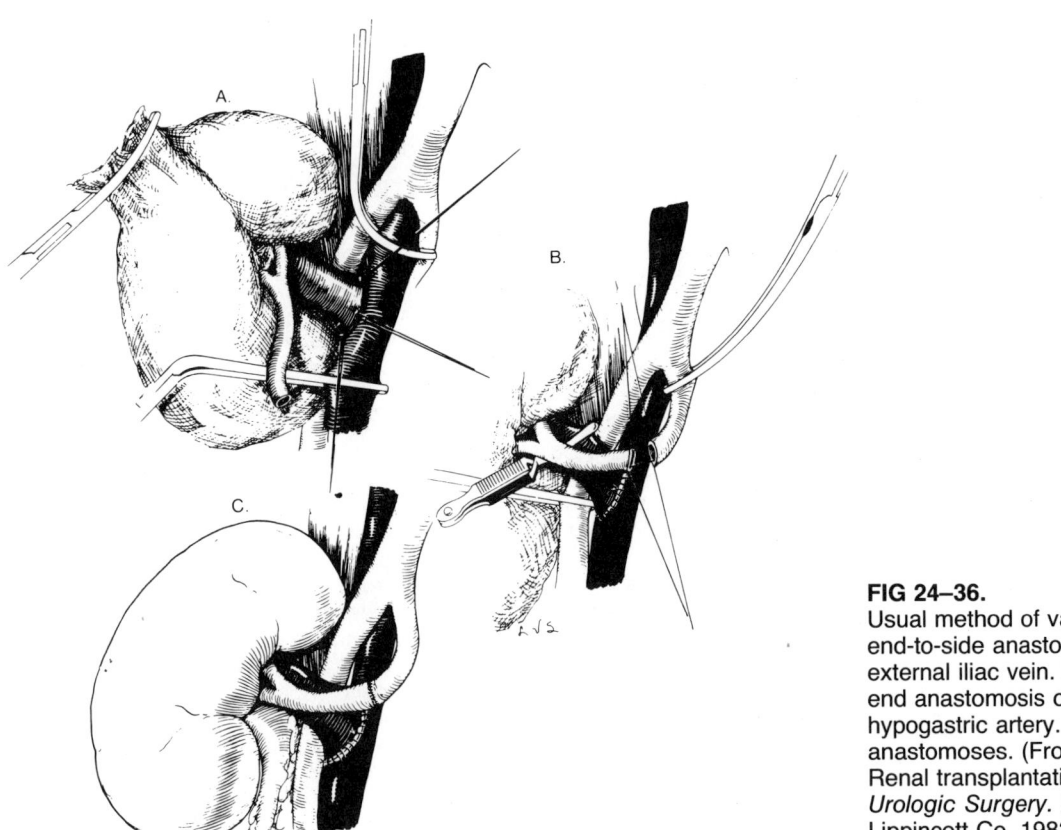

FIG 24–36.
Usual method of vascular anastomosis. **A,** end-to-side anastomosis of renal vein to external iliac vein. **B,** beginning of end-to-end anastomosis of renal artery to hypogastric artery. **C,** completed vascular anastomoses. (From Salvatierra O Jr: Renal transplantation, in Glenn JF (ed): *Urologic Surgery.* Philadelphia, JB Lippincott Co, 1983, p 362. Used with permission.)

removal of the clamps, we prefer the cadaver kidney recipient to have received about 2,500 ml of IV fluids. The kidney is an excellent monitor of the extracellular fluid volume, and if it does not expand promptly and become firm, the anesthesiologist is urged to increase

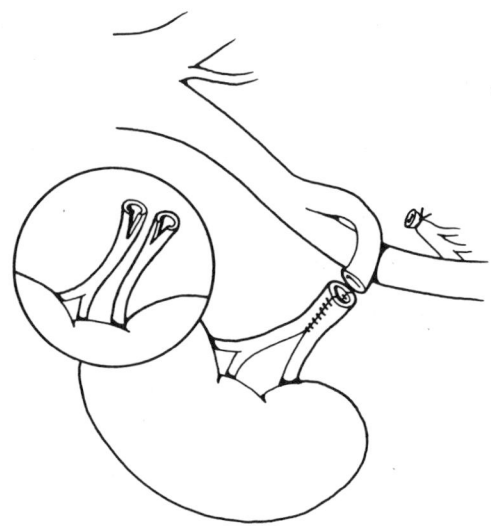

FIG 24–37.
Technique for construction of a common stoma to have single artery for implantation.

the rate of fluid administration. In cases in which the kidneys function promptly and the urine output is greater than 1 ml/min, fluids may be administered at a rate consistent with maintaining a CVP between 7 and 12 cm. The patient who makes no urine must be administered fluids judiciously to maximally safe hydration and then given 20–200 mg of Lasix IV to initiate urine flow. Once circulation is restored, additional arterial sutures are occasionally required, for which we use 6-0 Prolene. If Ethibond sutures are used, it is important to run the Ethibond sutures through mineral oil to minimize their tendency to pull as they transverse the tissue planes. A catheter is passed through the urethra into the bladder, 100–150 ml of triple-antibiotic solution is used to distend the bladder slightly, and the catheter is clamped and connected to sterile closed drainage. Filling the bladder facilitates its identification in the abdomen and enlarges its capacity to facilitate the reimplantation of the ureter, particularly if performing a Politano-Leadbetter tunnel technique through an open cystostomy (Politano and Leadbetter, 1958). When using an extravesical tunnel technique, as described and applied recently by Barry and Hatch (1985) of Oregon, filling the bladder before implantation facilitates identification and exposure of the bladder in preparation of the tunnel. The ureter is implanted into the bladder through an open cystotomy using a suburothelial tunnel (Politano and Leadbetter, 1958) (Fig 24–38). The ureteral mucosa is anchored to the bladder mucosa with a 4-0 chromic gut anchoring suture along with interrupted sutures of 5-0 chromic gut, and the cystotomy is closed anteriorly in three layers with a running 5-0 to the mucosa, 3-0 to the muscle, and 3-0 to the adventitia. Alternatively, one may choose an extravesical approach as described by Lich et al. (1961) (Fig 24–39), or as applied by Barry and Hatch (1985), in which parallel incisions are made in the adventitia and muscle of the bladder down to the mucosa. The seromuscular or adventitiomuscular tunnel is carefully dissected off the underlying mucosa, and distally a small opening is made in the mucosa. A running anastomosis of the ureteral mucosa to the bladder mucosa is accomplished after the ureter has been pulled through the seromuscular tunnel, care being taken not to perforate the mucosa except at the point where the anas-

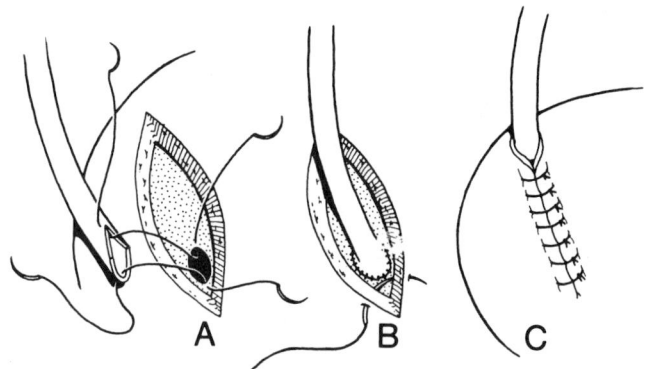

FIG 24–39.
Lich-Howerton-Gregoir extravesical technique, which may be used as an alternative to Politano-Leadbetter tunnel. (Barry technique differs in that two parallel slits are made instead of oblique extravesical incision and ureter is pulled through tunnel from superior to inferior incision and then sewed to mucosa.) **A,** extravesical dissection of bladder seromuscular layer and opening into bladder lumen at inferior aspect. **B,** ureter is anastomosed to bladder mucosa and then seromuscular layer is closed over ureter. **C,** completed ureteral anastomosis. (Adapted from Simmons RL, Najarian JS: Kidney transplantation, in Simmons RL, et al (eds): *Manual of Vascular Access, Organ Donation, and Transplantation.* New York, Springer-Verlag, 1984, p 314.)

tomosis of ureteral and bladder mucosa is to occur. Once this watertight anastomosis of ureteral mucosa to bladder mucosa has been accomplished with 5-0 chromic gut sutures, the seromuscular incision distally is then closed over the ureter with interrupted 3-0 chromic gut sutures, and the proximal incision is closed as tightly as necessary about the point of entrance of the ureter to the tunnel. We do not drain the transplant wound. A layered closure with no drain is accomplished.

Hemostasis, despite qualitative coagulation defects in the recipient, is usually complete, and meticulous care must be taken to accomplish hemostasis to avert the complication of postoperative hematoma requiring reexploration. This especially is true in cadaver recipients, since they may often require dialysis within a few hours of transplantation, and the heparinization necessary for dialysis, even if regional techniques are used, may result in hematoma formation in the fresh operative wound. A layered closure with interrupted 0 Prolene to the muscle fascia of the internal and transversus as one layer is usually used. The aponeurosis of the external oblique is then closed with a buried running 0 Prolene, the subcutaneous tissue with 3-0 chromic gut or 3-0 Dexon, and the skin with clips. Urine output is usually classified as greater than or less than 1 ml/min. The catheter is unclamped at the time of opening of the cystotomy to minimize the escape of fluid into the op-

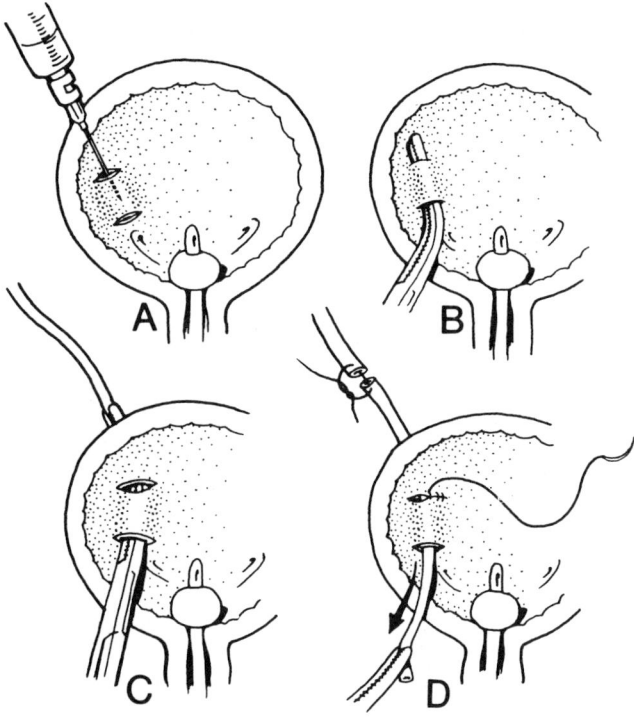

FIG 24–38.
Tunneled technique for reimplantation similar to Politano-Leadbetter. **A,** two parallel incisions in bladder mucosa. May distend submucosal space with saline injection as shown. **B,** submucosal dissection to create tunnel for ureter. **C,** ureter is pulled into bladder. **D,** ureter may be pulled directly through the submucosal tunnel or by suturing a catheter to the ureter, as shown, to aid in positioning. (Adapted from Simmons RL, Najarian JS: Kidney transplantation, in Simmons RL, et al (eds): *Manual of Vascular Access, Organ Donation and Transplantation.* New York, Springer-Verlag, 1984, p 311. Used with permission.)

erative wound and to permit the anesthesiologist to assess urine output after the bladder closure is completed.

Complications of the renal transplant operation include those of any major lower abdominal procedure. One must be careful to separate the peritoneum, particularly at the superior portion of the incision, from the underlying transversalis fascia, so that the peritoneum and adjacent colon are not encompassed by Prolene sutures during the layered muscular closure. Such a complication will undoubtedly result in infection, may actually threaten the life of a patient, and may force modification of his immunosuppressive regimen when infection occurs during the postoperative course. Graft and life loss are obvious complications of this technical problem. Urine leaks, of course, may be detected, and when a mass is felt in the early postoperative period, one must make the differential diagnosis between hematoma, particularly if the patient has required dialysis within a few hours of surgery, and urinoma. Lymphoceles usually appear later, although we have seen lymphoceles as early as three weeks following the implantation. We believe that lymphoceles are related to escape of fluid not controlled by meticulous lymphatic ligation during the dissection to expose a portion of the external iliac vein and external iliac artery or internal iliac artery as needed. A small lymphocele may be treated by aspiration under sonographic control. Lymphoceles larger than 5 cm may obstruct the course of the transplanted ureter and may require repeated aspiration or, occasionally, exteriorization of the lymphocele or interiorization of the lymphocele into the peritoneal cavity. Occasionally, a thick-walled lymphocele discovered late after transplantation has been treated by exteriorization to the skin and drainage until the cavity collapsed and sealed off. Usually, in those lymphoceles requiring surgery internal anastomosis of the lymphocele wall to the parietal peritoneum, making the lymphocele essentially a wide-mouthed diverticulum of the peritoneal cavity, has sufficed. Omentum has occasionally been placed into the lymphocele to prevent premature closure of the neck of the lymphocele and re-formation. We have seen lymphocele formation in cases in which the recipient had not been heparinized, i.e., retroperitoneal lymphadenectomy, and believe that meticulous attention to lymphatic closure is the best prophylactic measure. Certainly, with the advent of sonography and CT scanning, lymphoceles are found commonly following a large variety of retroperitoneal dissections.

Infection or wound abscess, of course, may occur following renal transplantation. Sources of fungal infections must be sought in the environment. There was a case of three consecutive *Aspergillus* infections in transplant recipients at the Johns Hopkins Hospital that were attributed to a pigeon nest being built on the intake valve of the air-conditioning system for the operating room. Removal of this nest resulted in the immediate reduction of the incidence of *Aspergillus* infections in the recipient population in that institution. The principle is that if such an infection occurs, one must make a careful search of the environment for the fomite.

Urine leaks, which may occur at any time, are often due to inadequate vascularization, with subsequent necrosis of the distal ureter, or to an opening in the urinary bladder. The use of IVP and cystography and sonography is of great help in the diagnosis of extravasation of urine. Repair and drainage are mandatory and must be done promptly. Immunosuppressed patients who have had renal transplantation tolerate extravasation poorly, and sepsis soon intervenes. Corrective surgery may involve the use of an indwelling nephrostomy and a stent. The general principles in reconstitution of the integrity of the collecting system of the urinary tract including reimplantation are carried out in the transplant patient as in any other patient with such a defect. Wound abscesses, of course, must be drained promptly, and when exploring an infected transplant wound, general principles are followed, leaving the skin open and closing it later with Steri-Strips, as in any contaminated abdominal wound.

Vascular abnormalities are rare. Seventeen renal artery stenoses have occurred in more than 500 transplants in our experience. All but one were distal to the anastomosis, associated with the arteritis of rejection, and were not due to problems of technique.

RESULTS OF RENAL TRANSPLANTATION

In general, in the past decade (1975–85) both patient survival and graft survival have improved compared with those of previous decades. Patient survival has been reported by Salvatierra at 96%–99% for two years following transplantation, and graft survival 87%–92% for the same period in living related donors. Miller (1985) reported a 93% living related donor graft survival, with a 91% cadaver graft survival, for one year. Using machine preservation exclusively, Miller reported a 5% ATN rate for three years (January 1983 to January 1986) for kidneys harvested at his own institution and placed on the extracorporeal pump immediately (Miller, 1985). Barry and Hatch (1985) reported an ATN approximating 70% in kidneys stored longer than 24 hours in simple cold solution. The establishment of patients on high-quality hemodialysis, the improvement in typing serums, the addition of ALG and

TABLE 24–10.

Current Results Attainable: First Grafts

DATA SOURCE (YR)	NO.	MATCH	IMMUNOSUPPRESSION*	SURVIVAL, %			
				PATIENT		GRAFT	
				1 YR	3 YR	1 YR	3 YR
Living related donor							
Salvatierra (1985)	186	HLA identical	AZA-steroids	97 ± 1	94 ± 2	90 ± 2	85 ± 3
Salvatierra (1985)	221	0 and 1-Haplotype	DST-AZA-steroids	97	95 ± 2	94 ± 2	85 ± 3
Cadaver							
Opelz et al. (1985)	161	HLA-B, DR	CSA-steroids			86 ± 3	
Opelz et al. (1985)	181	4HLA-DR Mismatch	CSA-steroids			67 ± 4	
Opelz et al. (1985)	381	HLA-B, DR	Non-CSA			75 ± 2	
Opelz et al. (1985)	378	4HLA-DR Mismatch	Non-CSA			57 ± 3	
Opelz et al. (1985)	2,198	All groups†	CSA-steroids			76 ± 1	
Opelz et al. (1985)	6,392	All groups	Non-CSA			64 ± 1	

*DST, donor-specific transfusion; CSA, cyclosporine A; AZA, azathioprine.
†HLA-DR matched grafts did 10% better than mismatched ones regardless of whether CSA was used. Data of Salvatierra suggest that effect of CSA is not lost up to three years. Neoplasms, particularly lymphomas, seem to appear earlier posttransplant in patients on CSA as opposed to conventional AZA-steroid immunosuppression, but the incidence does not appear increased. It should be noted that Opelz data are from many centers (multicenter data), and the CSA effect has yet to be confirmed in single centers.

ATG, and the introduction of cyclosporine A have all contributed to the increased patient survival and graft survival in the past three years. The reduction in steroid dosage permissible with cyclosporine and the introduction of donor specific transfusions as advocated by Salvatierra have further increased the survival in the 1-haplotype match, e.g., child-to-parent or parent-to-child transfer. In 1975, one-year survival in 1-haplotype transfer was 65%. In 1985, 93%–94% one- and two-year survival was reported by Salvatierra (1985). The development of serologic tests for D-related (DR) antigens on the short arm of the 6th chromosome and the tendency to give multiple blood transfusions to patients on long-term hemodialysis have contributed to better results in cadaver transplantation. A value of 40% one-year cadaver graft survival was anticipated by most major centers in 1970. This has improved steadily throughout the past 15 years to a current value of 86% ± 3% one-year survival in transfused 2-DR matched cadaver recipients (Opelz, 1985). Survival rates in recipients over 60 years of age and under 1 year of age are lower (Table 24–10), but the study numbers are small. We anticipate increasing success in these groups with the advent of safer immunosuppressive agents permitting steroid dose reduction and of improvements in the techniques of surgical and medical management in the newborn and infant groups.

Survival in the diabetic patient has steadily improved, until many centers are willing to perform transplantation without asking the patient to suffer the ravages of advanced uremia. Cyclosporine with the concomitant ability to reduce the steroid dosage is especially useful in the diabetic recipient. Superior figures have been reported from the Minnesota group in the management of such diabetic patients (Table 24–11).

If one defines rehabilitation after renal transplantation as the ability to work full or part time, living related donor recipients are clearly superior to recipients of cadaver kidneys. According to Morris (Morris, 1984), 78% of patients who received a living donor transplant were fully employed, whereas of those who received cadaver transplantation, only 64.7% were able to resume full-time work, identical to that in the British experience of the working patients on home dialysis, i.e., 64.6% (Morris, 1984). The longer the patient survives transplantation, the more apt he is to have a successful rehabilitation. In Morris's experience, 37.6% of the patients were employed 4–6 months after transplantation, whereas 71% were fully employed two years after transplantation.

SECOND TRANSPLANTS

The fate of the second graft is especially related to the source of the first graft and also to the source of the

TABLE 24–11.

Patient and Graft Survival in 305 Uremic and Diabetic Patients in Minneapolis

TYPE OF DONOR	TWO-YEAR, %	
	PATIENT SURVIVAL	GRAFT SURVIVAL
HLA-identical sibling	90	90
Other related	73	67
Cadaver	68	55

From Najarian JS, Sutherland DR, Simmons RL, et al: Ten year experience with renal transplantation in juvenile onset diabetes. *Ann Surg* 1979; 190:487–500. Used by permission.

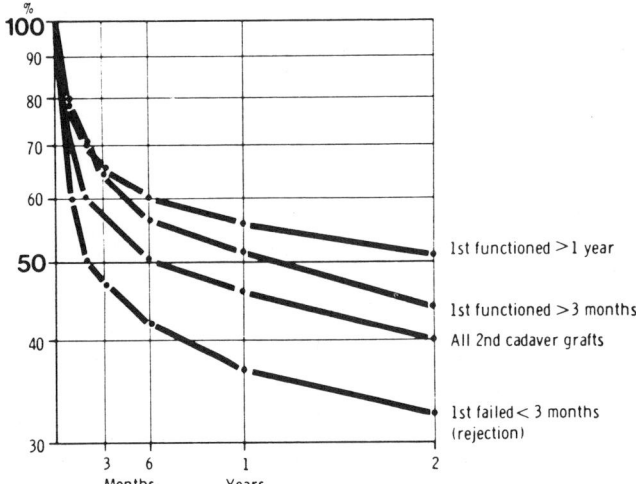

FIG 24–40.
Second cadaver graft survival related to the fate of the first graft in Europe. (From Morris PJ (ed): Results of renal transplantation, in *Kidney Transplantation: Principles and Practice*, ed 2. London, Grune & Stratton, 1984, p 555. Used with permission.)

second graft. A second LRD graft generally does better than a second consecutive cadaver graft. The second cadaver graft does especially poorly if the first cadaver graft was rejected in less than three months (Fig 24–40).

SUMMARY

Renal transplantation offers suitable patients with end-stage renal disease the highest patient survival and the best quality of life and rehabilitation. Current histocompatibility matching and immunosuppressive regimens provide one-year allograft survival of greater than 90% for living related donor recipients and graft survival of 70%–80% in first-time cadaver recipients. Efficient multiorgan procurement is more important than at any previous time owing to the increased availability of organ transplantation and greater patient and physician acceptance of these procedures because of lower morbidity and improved graft survival. Concepts of preservation, organ sharing, histocompatibility matching, and drug regimens to provide optimal graft survival will continue to evolve. More specific antirejection therapy that will be more effective and less toxic will be developed.

REFERENCES

1. Allen N, Dyer P, Harris K, et al: Effects of plasma exchange on immunoglobulins, complement, and immune complexes in renal transplant recipients. *Plasma Ther Transfusion Technol* 1982; 3:157–162.

2. Alexander S: Treatment of infants with ESRD, in Fine RN, Gruskin A (eds): *ESRD in Children*. Philadelphia, WB Saunders, 1984, pp 17–29.

3. Banowsky LH, Montie JE, Braun WE, et al: Renal transplantation—prevention of wound infections. *Urology* 1974; 4:656–659.

4. Barry J, Hatch DA: Parallel incision, unstented extravesical ureteroneocystostomy: Followup of 203 kidney transplants. *J Urol* 1985; 134:249–251.

4a. Belzer FO, Ashby BS, Dunphy JE: Twenty-four-hour and 72-hour presentation of canine kidney. *Lancet* 1967; 2:536.

5. Billingham RE, Brent L, Medawar PB: The antigenic stimulus in transplantation immunity. *Nature* 1956; 178:514–519.

6. Bowen A, Edwards LC, Gailiunas P, et al: Lymphocyte function in patients treated with monoclonal anti-T$_3$ antibody for acute cadaveric renal allograft rejection. *Transplantation* 1984; 38:489–493.

7. Braun WE: The new serology of histocompatibility testing and its significance in human renal transplantation. *Urol Clin North Am* 1976; 3:503–525.

8. Briggs JD: The recipient of a renal transplant, in *Kidney Transplantation: Principles and Practice*, ed 2. Morris PJ (ed): London, Grune & Stratton, 1984, pp 59–79.

9. The Canadian Multicentre Transplant Study Group: A randomized clinical trial of cyclosporine in cadaveric renal transplantation. *N Engl J Med* 1983; 309:809–815.

10. Chatteerjee SN, Gottlieb L, Berne TV: Fulminant pulmonary infections in renal transplant recipients. *Surg Gynecol Obstet* 1978a; 147:583–587.

11. Chatterjee S, Fiala M, Weiner J, et al: Primary cytomegalovirus and opportunistic infections: Incidence in renal transplant recipients. *JAMA* 1978b; 240:2446–2449.

12. Cordella CJ, Sutton DM, Falk JA, et al: Intensive plasma exchange, complement dependent microcytotoxicity and renal transplant rejection. *Proc Eur Dial Transplant Assoc* 1977; 14:328–335.

13. Cordella CJ, Sutton DM, Falk JA, et al: A controlled trial evaluating intensive plasma exchange in renal transplant recipients. *Proc Eur Dial Transplant Assoc* 1980; 17:429–434.

14. Cosimi AB, Wortis HH, Delmonico FL, et al: Randomized clinical trial of antithymocyte globulin in cadaver renal allograft recipients: Importance of T cell monitoring. *Surgery* 1976; 80:155–163.

15. Cosimi AB, Burton RC, Colvin RB, et al: Treatment of acute renal allograft rejection with OKT$_3$ monoclonal antibody. *Transplantation* 1981; 32:535–540.

16. Council on Scientific Affairs of the AMA: Current status of therapeutic plasmapheresis and related techniques. *JAMA* 1985; 253:819–825.

17. Darr FW, McCurdy PR, Helfrick GB, et al: Treatment of steroid resistant renal allograft rejection with lymphocytopheresis. *Plasma Ther Transfusion Technol* 1982; 3:423–427.

18. Dawidson I, Dausset J: Iso-leuco-anticorps. *Acta Haematol (Basel)* 1958; 20:156–166.

19. De Wolf WG: Transplantation immunology. *J Urol* 1981; 18:317–321.

20. Ehrlich RM, Smith RB: Surgical complications of renal transplantation. *Urology* 1977; 10(Suppl):43–56.

21. Filo RS, Smith EJ, Leapman SB: Reversal of acute renal allograft rejection with adjunctive ATG therapy. *Transplantation Proc* 1981; 13:482–491.

22. Fine RN: Renal transplantation in children, in Morris PJ (ed): *Kidney Transplantation: Principles and Practice.* London, Grune & Stratton, 1984, pp 509–546.

23. Fulkerson WJ: Fiberoptic bronchoscopy. *N Engl J Med* 1984; 311:511–515.

24. Garvin PJ, Castaneda M, Linderer R, et al: Management of hypercalcemic hyperparathyroidism after renal transplantation. *Arch Surg* 1985; 120:578–583.

25. Glenn J: Cytomegalovirus infections following renal transplantation. *Rev Infect Dis* 1981; 3:1151–1178.

26. Gorer PA: The genetic and antigenic basis of tumor transplantation. *J Pathol Bacteriol* 1937; 44:691–697.

27. Hakala T: Unpublished data, 1985.

28. Haldane JBS: The genetics of cancer. *Nature* 1933; 132:265.

29. Hamburger J: Transplantation immunology, in Hamburger J, Crossnier J, Bach JF, et al (eds): *Renal Transplantation Theory and Practice.* Baltimore, Williams & Wilkins, 1981, pp 1–22.

30. Hayry P, Von Willebrand E: Monitoring of human renal allograft rejection with fine needle aspiration cytology. *Scand J Immunol* 1981; 13:87–97.

31. Helderman JH, Gailiunas P Jr, Silva F: Plasmapheresis and renal transplant rejection, in Tindall RSA (ed): *Therapeutic Apheresis and Plasma Perfusion.* New York, Alan R Liss Inc, 1982, pp 271–282.

32. Helderman JH: Renal transplantation, in *Internal Medicine Grand Rounds.* Dallas, The University of Texas Health Science Center at Dallas, Oct 25, 1984.

33. Hume DM, Magee JH, Kauffman MH, et al: Renal homotransplantation in man in modified recipients. *Ann Surg* 1963; 158:608–644.

34. Kahan BD, Van Buren CT, Flechner SM, et al: Cyclosporine immunosuppression mitigates immunologic risk factors in renal allotransplantation in Cyclosporine. Kahan BD (ed): Orlando, Fla, Grune & Stratton Inc, 1984, pp 253–262.

35. Kleinman S, Nichols M, Strauss F, et al: Use of lymphoplasmapheresis or plasmapheresis in the management of acute renal allograft rejection. *J Clin Apheresis* 1982; 1:14–17.

36. Kirkubakaran MG, Disney APS, Norman J, et al: A controlled trial of plasmapheresis in the treatment of renal allograft rejection. *Transplantation* 1981; 32:164–165.

37. Landsteiner K: Uber Agglutination serscheinungen normalen menschliachen Blutes Vivien. *Klin Wochenschr* 1901; 14:1132.

38. Lich R, Howerton LW, Davis LA: Recurrent urosepsis in children. *J Urol* 1961; 86:554.

39. Ludmerer KM, Kissane JM (eds): Clinicopathologic conference cavitary lung disease following renal transplantation. *Am J Med* 1982; 72:145–155.

40. Lyngaard F, Ladeford J, Jans H: Intensive plasma exchange prior to renal transplantation to prevent rejection. *Plasma Ther* 1980; 1:55–57.

41. McConnell JD, Sagalowsky AI, Lewis SE, et al: Prospective evaluation of renal allograft dysfunction with [99m]technetium diethylenetriaminepentaacetic acid renal scans. *J Urol* 1984; 131:875–879.

42. Medawar PB: The behavior and fate of skin autografts and skin homografts in rabbits. *J Anat* 1944; 78:176–199.

43. Miller J: Machine preservation and ATN rate. *Sandoz Pharm Proc* November 1985.

44. Mittal KK, Mickey MR, Singal DP, et al: Serotyping for homotransplantation: XVIII. Refinement of microdroplet lymphocyte cytotoxicity test. *Transplantation* 1968; 6:913–927.

45. Moore FD Jr, Kohler TR, Strom TB, et al: The declining mortality from pneumonia in renal transplant recipients. *Infect Surg* 1983; 2:13–19.

46. Muakkassa WF, Goldman MH, Mendez-Picon G, et al: Wound infections in renal transplant patients. *J Urol* 1983; 130:17–19.

47. Najarian JS, Tallent B, Kjellstrand C, et al: Renal transplantation in infants and children. *Ann Surg* 1971; 174:583.

48. Najarian JS, Sutherland DR, Simmons RL, et al: Ten-year experience with renal transplantation in juvenile onset diabetes. *Ann Surg* 1979; 190:487–500.

48a. Nomenclature of factors of the HLA system, 1984, in Albert ED, Baur MP, Mayr WR (eds): *Histocompatibility Testing; 1984.* Berlin, Springer-Verlag, 1984, pp 4–8.

49. Novick AC: The value of intraoperative antibiotics in preventing renal transplant wound infections. *J Urol* 1981; 125:151–152.

50. Nowygrod R, Appel G, Hardy MA: Use of ATG for reversal of acute allograft rejection. *Transplant Proc* 1981; 13:469–472.

51. Opelz G, Sengar DPS, Mickey, et al: Effect of blood transfusions on subsequent kidney transplants. *Transplant Proc* 1973; 5:253–259.

51a. Opelz G: Correlation of HLA matching with kidney graft survival in patients with or without cyclosporine treatment. *Transplantation* 1985; 40:240–243.

52. Ortho Multicenter Transplant Study Group: A randomized clinical trial of OKT$_3$ monoclonal antibody for acute rejection of cadaver renal transpalnts. *N Engl J Med* 1985; 313:337–342.

53. Pallis C: Brainstem death: The evolution of a concept, in Morris PJ (ed): *Kidney Transplantation: Principles and Practice.* London, Grune & Stratton, 1984, pp

54. Parfrey PS, Forbes RDC, Hutchinson TA, et al: The clinical and pathological course of hepatitis B liver disease in renal transplant recipients. *Transplantation* 1984; 37:461–466.

55. Pegg DE, Jacobsen IA, Halasz NA: *Organ Preservation: Basic and Applied Aspects.* MTP Press, Lancaster, PA, 1982.

56. Penn I: The price of immunotherapy. *Current Probl Surg* 1981; 18:689–751.

57. Penn I: Lymphomas complicating organ transplantation, in Kahan BD (ed): *Cyclosporine.* Orlando, Fla, Grune & Stratton, 1984, pp 574–581.

58. Politano V, Leadbetter WF: An operative technique for

the correction of vesicoureteral reflux. *J Urol* 1958; 79:932.

59. Power D, Nicholls A, Muirhead N, et al: Plasma exchange in acute renal allograft rejection: Is a controlled trial really necessary? *Transplantation* 1981; 32:162–163.

60. Prompt CA, Lee DBN, Upham AT: Medical complications of renal transplantation. *Urology* 1977; 9(suppl):32–48.

61. Sagalowsky AI, Helderman JH, Dawidson I, et al: Cyclosporine rescue therapy for renal transplant rejection. Presented at American Urological Association Meeting, Atlanta, May 15, 1985.

62. Sagalowsky AI, Ransler CW, Peters PC, et al: Urologic complications in 505 renal transplants with early catheter removal. *J Urol* 1983a; 129:929–932.

63. Sagalowsky AI, Gailiunas P, Helderman JH, et al: Renal transplantation in diabetic patients: The end result does justify the means. *J Urol* 1983b; 129:253–255.

64. Sagalowsky AI, Peters PC: Emergencies in renal transplantation. *Urol Clin North Am* 1982; 9:215–218.

65. Sagalowsky AI, Peters PC: Renovascular hypertension following renal transplantation. *Urol Clin North Am* 1984; 11:491–502.

66. Salvatierra O Jr, Vincenti F, Amend W, et al: Deliberate donor-specific blood transfusions prior to living related renal transplantation. *Ann Surg* 1980; 192:543–552.

67. Salvatierra O Jr, Kountz SL, Belzer FO: Prevention of ureteral fistula after renal transplantation. *J Urol* 1974; 112:445–448.

68. Salvatierra O Jr, Olcott C IV, Amend WJ Jr, et al: Urological complications of renal transplantation can be prevented or controlled. *J Urol* 1977; 117:421–424.

68a. Salvatierra O Jr: Renal transplantation, in Glenn JF (ed): *Urologic Surgery*, ed 3. Philadelphia, JB Lippincott, 1983, pp 359–367.

69. Salvatierra O Jr: Seven year experience with donor-specific blood transfusions (DSTs): Results and considerations for maximum efficacy. *Transplantation* 1985; 40:654–659.

70. Schooley RT, Hirsch MS, Colvin RB, et al: Association of herpesvirus infections with T-lymphocyte subset alterations, glomerulopathy, and opportunistic infections after renal transplantation. *N Engl J Med* 1983; 308:307–313.

71. Semb C, Krog J, Johansen K: Renal metabolism and blood flow during local hypothermia, studied by means of renal perfusion in situ. *Acta Chir Scand* 1960; 253:196.

72. Simmons RL, Tallent MB, Kjellstrand CM, et al: Kidney transplantation from living donors with bilateral double renal arteries. *Surgery* 1971; 69:201.

73. Simmons RL, Finch ME, Ascher NL, et al (eds): *Manual of Vascular Access, Organ Donation and Transplantation*. New York, Springer-Verlag, 1984.

74. Slapak M, Naik RB, Lee HA: Renal transplant in a patient with major donor-recipient blood group incompatibility: Reversal of acute rejection by the use of modified plasmapheresis. *Transplantation* 1971; 31:4–7.

75. Smiley ML, Wlodaver CG, Grossman RA, et al: The role of pretransplant immunity in protection from cytomegalovirus disease following renal transplantation. *Transplantation* 1985; 40:157–161.

76. Smith RB, Ehrlich RM: The surgical complications of renal transplantation. *Urol Clin North Am* 1976; 3:621–646.

77. Southard JH, Rice MJ, Belzer FO: Preservation of renal function by adenosine stimulated synthesis in hypothermically perfused dog kidneys. *Cryobiology* 1985; 22:237–242.

78. Stamey TA, Govan DE, Palmer JM: The localization and treatment of urinary tract infections: The role of bactericidal urine levels as opposed to serum levels. *Medicine* 1965; 44:1.

79. Standards Committee of the American Society of Transplant Surgeons: Current results and expectations of renal transplantation. *JAMA* 1981; 246:1330–1331.

80. Starzl TE, Hakala TR, Shaw BW, et al: A flexible procedure for multiple cadaveric organ procurement. *Surg Gynecol Obstet* 1984; 158:223–230.

81. Strom TB, Carpenter CB: Transplantation: Immunogenetic and clinical aspects, part 2. *Hosp Pract* 1983; 18:135–150.

82. Strom TB: Hepatitis B, transfusions and renal transplantation—five years later. *N Engl J Med* 1982; 307:1141–1142.

83. Taylor RJ, Schwentker FN, Hakala TR: Opportunistic lung infections in renal transplant patients: A comparison of Pittsburgh pneumonia agent and Legionnaires' disease. *J Urol* 1981; 125:289–292.

84. Terasaki PI: The beneficial transfusion effect on kidney graft survival attributed to clonal deletion. *Transplantation* 1984; 37:119–125.

85. Thiru S, Calne RY, Nagington J: Lymphoma in renal allograft patients treated with cyclosporin-A as one of the immunosuppressive agents. *Transplant Proc* 1981; 13:359–364.

86. The U.S. ESRD Program: Selected 1983 statistics. *Contemp Dialysis Nephrol* 1984; 5:51–58.

87. Van Rood JJ, Balner H: Blood transfusion and transplantation. *Transplantation* 1978; 26:275–276.

88. Ware AJ, Juby JP, Hollinger B, et al: Etiology of liver disease in renal transplant recipients. *Ann Intern Med* 1979; 91:364–371.

89. Welchel JD, Alison DV, Luke RJ, et al: Successful renal transplantation in hyperoxaluria, a report of two cases. *Transplantation* 1983; 35:161–164.

Chapter 25

Renovascular Hypertension

R. Ernest Sosa, M.D.
E. Darracott Vaughan, Jr., M.D.

Sustained hypertension is a leading risk factor for premature illness and death. Poorly controlled or uncontrolled pressure elevation leads to small-vessel disease and end-organ damage, principally affecting the heart, kidneys, and CNS. Hypertension is the leading cause of congestive heart failure in the United States (Laragh, 1981). The kidney, as a result of hypertensive vascular disease, suffers progressive deterioration of filtering capacity at the glomerular level, expressed as progressive renal insufficiency. Moreover, renin-dependent hypertension results in more severe vascular damage and hence in a higher incidence of myocardial infarctions and strokes at a younger age than in patients with normal or low renin hypertension, who are generally older (Brunner et al., 1972).

Sixty million Americans suffer from high blood pressure (Rowland and Roberts, 1982). Hypertension secondary to renovascular disease is potentially surgically treatable. The true incidence of renovascular hypertension (RVH) has not been accurately established. It has been estimated that 5%–10% of all hypertensive patients have RVH. Recent advances in pharmacology and in interventional radiology have increased our understanding of the pathophysiology of the renin-angiotensin-aldosterone system, leading to the development of sensitive and specific diagnostic screening tests for RVH. With these more reliable means available it is becoming apparent that the true incidence of RVH may have been underestimated.

The importance of diagnosing RVH is multifold. First, RVH is difficult to manage medically (Hunt et al., 1974). Second, even if adequate blood pressure control is maintained by pharmacologic means, progression of the renal artery disease is not prevented (Schreiber et al., 1984), and renal ischemia may actually worsen when the pressure is lowered to clinically desirable levels (Textor et al., 1985). Third, RVH is potentially curable (Novick et al., 1984), and renal function can be stabilized or enhanced by renal revascularization (Ying et al., 1984).

This chapter presents current understanding of the natural history and pathophysiology of RVH. A strategy for a cost-effective and reliable diagnostic evaluation for RVH, utilizing recently developed pharmacologic probes and radiologic maneuvers, is described. Finally, a critical comparison of the various available treatment options is made.

DEFINITIONS

It is acknowledged that 60 million individuals in the United States suffer from hypertension (Rowland and Roberts, 1982). A patient with a marked, persistent pressure elevation, such as 220/110 mm Hg, is said to be hypertensive without disagreement. However, to call an elderly patient with a blood pressure of 140/90 mm Hg hypertensive would arouse both objection and agreement. It is indeed difficult to define a precise cutoff that marks the upper limit of acceptable blood pressure, but it is generally acknowledged that the higher the blood pressure, the worse the resultant morbidity and mortality (Pickering, 1973). As a working definition, the 1984 Joint National Committee on Detection, Evaluation, and Treatment of High Blood Pressure recommended that a sustained blood pressure of 140/90 mm Hg be considered the cutoff point between normal blood pressure and mild hypertension for all patients over 18 years of age. Children with a sustained increase in arterial pressure greater than or equal to the 90th percentile for age are considered hypertensive. Whatever the patient's age, an elevation in blood pressure should be documented on at least three separate occasions, when measured with an appropriate-sized blood pressure cuff, to justify the diagnosis of hypertension. It is not unusual to find that a patient who is hypertensive at an initial evaluation remains normotensive at subsequent measurements (Carey et al., 1976).

RENOVASCULAR DISEASE

Renal disease has been recognized in association with hypertension since the early 19th century (Bright, 1836). In 1898 Tigerstadt and Bergmann demonstrated that a water-soluble extract that they termed renin, derived from the renal cortex of a healthy rabbit, could produce a marked and sustained hypertension when injected IV into a second rabbit (Tigerstadt and Bergmann, 1898). Interest in the relation between renal disease and hypertension did not flourish, however, until the classic experiments by Goldblatt and associates (1934) in the dog demonstrated that reversible elevation in the systemic arterial pressure could be produced by clamping the main renal artery of one of two healthy kidneys. The blood pressure returned to normal on removal of the kidney or the clamp. The development and widespread use of arteriography in the 1950s focused attention on renal arterial disease in hypertensive patients and led to advances in renovascular surgery (Freeman et al., 1954). However, it quickly became apparent that lesions of the renal artery could be demonstrated angiographically in normotensive patients. Moreover, the results of the national Cooperative Study on renovascular surgery for treatment of hypertension revealed a sobering 34% failure rate in hypertensive patients subjected to renal revascularization (Foster et al., 1975). It became clear that angiographic documentation of renal artery disease in a hypertensive patient was not sufficient to justify surgical correction of the arterial lesion. More reliable diagnostic screening tests were needed to predict which patients would benefit from renal revascularization.

Greater insight into the pathophysiology of renal artery disease was gained by experiments in vessel hemodynamics. It was determined that the internal diameter of an artery must be reduced by greater than 70% for a significant fall in blood flow to occur (Mann et al., 1938) and that a pressure gradient of >40 mm Hg across a stenosis in the renal artery was necessary to produce a significant decrease in the renal plasma flow (RPF), GFR, urinary sodium excretion (UNaV), and urine flow rate (Selkurt, 1951).

The discovery of angiotensin (Page and Helmer, 1940; Braun-Menendez et al., 1940) and determination of its sequence (Skeggs et al., 1956) led to the eventual development of accurate radioimmunoassays to quantify the activity of the renin-angiotensin system (RAS). Researchers established that a significant decrease in blood flow to a kidney results in the activation of the renin-angiotensin-aldosterone cascade, establishing a hypertensive state. Correction of the stenosis or removal of the ischemic kidney eliminates the hyperreninemic state, allowing the blood pressure to return to normal levels.

Accordingly, RVH can be defined as a sustained blood pressure elevation secondary to a physiologically significant renal artery stenosis that is correctable by repair of the lesion or by removal of the kidney. From a clinical point of view, however, it is imperative to be able to diagnose RVH prospectively. Fortunately, the availability of pharmacologic agents that block different steps in the renin-angiotensin-aldosterone cascade, such as converting enzyme inhibitors (CEI), angiotensin II analogues, and renin inhibitors, as well as technical advances in interventional radiology, have contributed to our understanding of the pathophysiology of RVH. Based on this knowledge, sensitive and specific tests that prospectively identify patients with RVH are currently being developed and utilized.

PATHOLOGY AND NATURAL HISTORY OF RENAL ARTERY DISEASE

Atherosclerosis and fibromuscular disease (FMD) of the renal artery account for most cases of RVH. Two thirds of patients with RVH are found to have atheromatous lesions of the renal artery. Atherosclerotic plaques are typically located in the proximal 2 cm of the main renal artery but may also involve the distal artery and its branches. The lesions begin as a proliferation of myointimal cells. Subsequent lipid deposition occurs, with necrosis, inflammation, and formation of atherosclerotic intimal plaques that protrude into the lumen. Calcification, surface erosion with thrombus formation, or dissection of the vessel wall may ensue (Ratliff, 1985).

Atheromatous renal arterial disease predominantly afflicts males in the older age groups. The disease is often diffuse, affecting the aorta and its major branches as well as the coronary and cerebral arteries. The renal arteries are involved bilaterally in up to 40% of cases. In the National Cooperative Study on RVH, patients with bilateral disease had the lowest cure rate and the highest morbidity (Foster et al., 1975). A recent review by Schrieber et al. (1984) of 85 patients with atheromatous renovascular disease treated medically and followed up for a mean of 52 months revealed that the disease was progressive in over 44% of patients, with progression to complete occlusion of the renal artery in 16% of renal units. There was no correlation between the adequacy of blood pressure control and progression of the arterial lesion.

Fibromuscular diseases of the renal artery are responsible for one third of cases of RVH. Four pathologically different types of renal artery dysplasia have been described (Stewart, 1981).

Intimal fibroplasia accounts for approximately 10% of the fibromuscular diseases. This disorder primarily involves the intima by the circumferential accumulation of collagen, compromising the arterial lumen. The disease is progressive, and dissecting hematomas may form. Children and young male adults are principally afflicted. Vessels other than the renals may be involved. Angiographically, a smooth focal stenosis is typically seen at the midrenal artery or its branches. Dissection may alter the appearance of the stenosis and of the vessel. Owing to the progressive nature of this disease, prompt repair of the lesion is advised.

Fibromuscular hyperplasia is the rarest of the fibrous dysplasias. It is characterized by hyperplasia of the smooth muscle and fibrous tissue, producing a concentric thickening of the renal artery wall. This disease afflicts children and young adults and is progressive.

Medial fibroplasia accounts for 75%–80% of the fibromuscular dysplasias. On angiogram the diseased artery has the appearance of a string of beads. This angiographic pattern is caused by a series of collagenous rings alternating with aneurysmal dilatations, involving the media of the main renal artery, often extending into its branches. Women in their 30s and 40s are usually afflicted. This lesion does not dissect, and complete occlusion has not been reported. Schreiber and co-workers (1984) followed up a group of 75 patients with this disease for a mean of 65 months and noted that progression of the lesion occurred in 33% of patients regardless of their age. Correction of the lesion by angioplasty is the treatment of choice.

In perimedial fibroplasia, a collar of dense collagen envelops the renal artery just beneath the adventitia. The lesions are tightly stenotic, and therefore extensive collateral vessels are commonly identified on angiography. Young women 15–30 years old are most commonly afflicted. The lesion may be progressive, and repair is recommended.

PHYSIOLOGY OF THE RENIN-ANGIOTENSIN SYSTEM

The renin-angiotensin system (RAS) plays an important role in the regulation of blood pressure and sodium-volume homeostasis in health and disease states. First, the RAS is involved in the maintenance of a constant arterial blood pressure despite extremes of sodium intake. A decrease in sodium intake increases the formation and secretion of angiotensin II by mechanisms that will be discussed below. In addition, the number of angiotensin II receptors in the adrenal cortex increases, while the density of angiotensin II receptors in the smooth muscle cells decreases in response to sodium depletion (Aguilera and Catt, 1981). Angiotensin II enhances aldosterone biosynthesis by the glomerulosa cells of the adrenal cortex and increases sodium reabsorption at the rat proximal tubule (Jackson et al., 1985). Angiotensin II increases the ultrafiltration coefficient of the glomerular capillary, decreasing the GFR. Angiotensin II also stimulates thirst (Fitzsimons, 1978) and has an antidiuretic effect (Levens et al., 1981). In states of increased sodium intake, the activity of the RAS is depressed, and excess sodium is excreted to maintain balance.

Second, the RAS protects the organism from the potential catastrophic effects of rapid-onset hypotension by responding to a sudden pressure decrease with an immediate increase in renin release. As will be discussed below, renin initiates a cascade of enzymatic reactions resulting in the formation of angiotensin II. Angiotensin II is a potent vasopressor that helps to restore the systemic blood pressure. On a molar basis, angiotensin II is 10–40 times more potent than norepinephrine (NE). Angiotensin II is particularly effective in constricting the precapillary and postcapillary resistance vessels. Its effect on capacitance vessels is less pronounced, and vessels in the skeletal muscles and lungs tend to be more resistant to the direct actions of angiotensin II. In the CNS, angiotensin II stimulates receptors in the area postrema, which is not protected by a blood-brain barrier, increasing the discharge rate of the sympathetic nerves. In addition, in the peripheral nerves angiotensin II potentiates the effects of NE at the noradrenergic neuroeffector junctions by increasing NE release, decreasing NE uptake, and increasing vascular sensitivity to NE (Jackson et al., 1985). In summary, the RAS buffers a fall in blood pressure by direct constriction of resistance vessels, by interaction with the noradrenergic receptors in the vascular smooth muscles, and by stimulation of the area postrema to increase sympathetic activity.

Renin is a proteolytic enzyme formed in modified smooth muscle cells known as the juxtaglomerular cells of the afferent arteriole found in intimate proximity to the macula densa of the distal tubule. Together, these microstructures are known as the juxtaglomerular apparatus. Renin is primarily released in response to: (1) a low renal perfusion pressure in the afferent arteriole baroreceptor mechanism) due either to renal artery stenosis or to a decrease in the mean systemic arterial pressure. This response is independent of renal innervation or of the GFR; (2) a low chloride (or sodium) concentration in the filtrate reaching the distal convoluted tubule (macula densa mechanism). The initiating stimulus to renin release appears to be Cl^- reabsorption (Kotchen et al., 1978); (3) adrenergic stimulation of the β_1-receptors in the juxtaglomerular cells. In the intact animal, decreases in mean arterial pressure and central blood volume activate baroreceptor and volume reflexes, eliciting adrenergic efferent signals that stimulate renin release. In addition, studies with β-adrenoreceptor antagonists have suggested that sympathetic tone is important in maintaining renin secretion under basal conditions and increasing renin release in response to upright posture and sodium depletion (Frolich, 1977).

The enzymatic action of renin splits angiotensin I off the $α_2$-globulin angiotensinogen. Angiotensin I is generally considered to be an inactive molecule. Angiotensin-converting enzyme (ACE) then cleaves a dipeptide off angiotensin I to produce angiotensin II. Aminopeptidase A may then cleave the amino residue off angiotensin II to form the heptapeptide angiotensin III. Both angiotensin II and III have a short half-life of two minutes. Angiotensin III activates the zona glomerulosa of the adrenal gland to increase the production of aldosterone in a similar manner to angiotensin II, but it is a much less potent vasoconstrictor.

Angiotensin II has a wide range of biologic activities. It is the protagonist through which the RAS regulates blood pressure and fluid-volume homeostasis. It is believed that angiotensin II plays an important role in renal autoregulation at low renal perfusion pressures. Normal dogs with an intact RAS subjected to graded decreases in mean renal perfusion pressure are able to maintain normal GFR. However, if the activity of the RAS is blocked by high salt intake or by converting enzyme inhibition, decrease of mean renal perfusion pressure to the same levels is associated with a decrease in GFR (Hall et al., 1977, 1978, 1981). It is hypothesized that at low renal perfusion pressures, angiotensin II produces a selective efferent arteriolar vasoconstriction to increase the glomerular capillary hydrostatic pressure and thus maintain GFR (Fumio et al., 1983). Clinically, it has been shown that use of CEIs in patients with bilateral renal artery stenosis or stenosis of a renal artery to a solitary kidney is associated with azotemia that reverses when the drug is suspended. However, lowering of the pressure to the same level with antihypertensives that do not interfere with the RAS does not appear to compromise renal function (Blythe, 1983; Curtis et al., 1983; Textor et al., 1984). Use of CEI in patients with unilateral renal artery stenosis has also been shown to decrease the GFR and renal blood flow to the ipsilateral kidney but not of the contralateral normal kidney (Wenting et al., 1984).

PATHOPHYSIOLOGY OF RENOVASCULAR HYPERTENSION

Our current understanding of the pathophysiology of RVH has been derived to a large extent from work in experimental models of hypertension. Goldblatt and associates (1934) first demonstrated that clamping one renal artery in a normal dog produced hypertension that was correctable by removing the clamp or the ischemic kidney. Currently, two models of RVH have been identified and characterized. The two-kidney, one-clip (2K,1C) model is prepared by clipping the renal artery to one of the two kidneys. The ischemic kidney is designated the ipsilateral kidney and the untouched kidney is the contralateral kidney. In the one-kidney, one-clip model (1K,1C), the renal artery to one kidney is clipped, and the contralateral kidney is removed. The degree of hypertension is similar for these two hypertensive models, but the mechanisms of hypertension differ in some respects.

The decrease in perfusion pressure and blood flow to the ipsilateral kidney in the 2K,1C model is associated with a decrease in the ipsilateral GFR, filtered sodium load, and urine sodium excretion (Stamey et al., 1961). An increase in the reabsorption of water and electrolytes is dictated by glomerular-tubular balance in the ipsilateral kidney, increasing the urinary concentration of unreabsorbable solutes and hence of urine osmolality. The decrease in mean renal perfusion pressure stimulates the baroreceptor mechanism, and the decrease in the filtered load of sodium (or chloride) activates the macula densa mechanism to increase renin secretion from the ischemic kidney.

The increase in the plasma renin activity (PRA) invokes an angiotensin II–induced increase in mean systemic arterial blood pressure. The contralateral GFR does not change significantly, but a pressure-induced natriuresis and diuresis can be observed (Sosa and Vaughan, 1983). Contralateral renin secretion is totally suppressed as the baroreceptor and macula densa mechanisms are turned off by the described physiologic changes. In addition, angiotensin II could directly suppress renin release by negative feedback. Tubular absorption of electrolytes and water is less marked than in the ipsilateral kidney as GFR is maintained. The contralateral urine osmolality is accordingly lower than that of the ipsilateral kidney.

The increase in arterial blood pressure in the 1K,1C model of hypertension is initially maintained by the vasoconstrictive properties of angiotensin II. Removal of the source of renin by reestablishing perfusion to the ischemic kidney, by removing the ischemic kidney, or by blocking the effects of RAS with the administration of an angiotensin II analogue, a CEI, or a renin inhibitor (RI) will lower the blood pressure at this stage.

If the renal artery clip is maintained for several weeks in the dog or rat, PRA is noted to return to normal or below normal levels. In this "chronic phase" of RVH, the degree of pressure elevation is equal to that observed in the acute phase, but salt and volume retention primarily account for the hypertension. The acute administration of RAS blockers will not have a significant depressor effect. Sodium deprivation at this point will increase the PRA and reestablish sensitivity to the depressor effects of RAS blockers (Gavras et al., 1973).

In the 1K,1C hypertensive model PRA is only briefly increased. The decreased capacity to excrete sodium and water inherent to this renodeprived model leads to hypertension that is chiefly maintained by sodium and water retention (Liard et al., 1974). The PRA is greatly reduced, similar to the chronic phase of the 2K,1C model, and RAS blockers are likewise ineffective in lowering the pressure. Salt deprivation increases the role of renin in maintaining the hypertension, and the addition of RAS blockers similarly elicits a depressor response (Bengis et al., 1978).

PHARMACOLOGIC PROBES FOR THE RENIN-ANGIOTENSIN SYSTEM

Further insight into the physiologic activities of the RAS has been gained from the development of pharmacologic compounds that inhibit the release of renin, block the conversion of angiotensin I to angiotensin II, or are specific angiotensin II receptor analogues (Miller et al., 1972; Brunner et al., 1971) (Fig 25–1).

It has been recognized that electrical stimulation of the renal nerves or of the vasomotor center in the brain stem produces an increase in renin release (Johnson et al., 1971; Richardson et al., 1974). Infusion of β-adrenergic blocking agents but not of α-adrenergic blockers abolishes adrenergic-stimulated increases in renin release. More recently, it has been determined that the β_1-receptors mediate the adrenergic stimulated renin release (Dibona, 1982).

The next group of drugs that interfere with the RAS are the angiotensin II analogues. Synthesis of these drugs was possible once the structure of angiotensin II was known. Saralasin is an angiotensin II analogue with affinity for the angiotensin II receptor but with partial agonist activity (Pals et al., 1971). Administration of saralasin to 2K,1C hypertensive models or to patients with renovascular hypertension produces a depressor response. However, it can produce a brief but dangerous elevation of the blood pressure in patients with normal or low renin hypertension, a feature that is undesirable in patients who are already hypertensive (Case et al., 1979). Saralasin has been used to screen for renovascular hypertension, but has recently lost favor to other drugs owing to its agonist activity, which led to an underestimation of the contribution of angiotensin II to the maintenance of hypertension (Horn et al., 1979).

Converting enzyme inhibitors (CEI) prevent the conversion of angiotensin I to angiotensin II. Naturally oc-

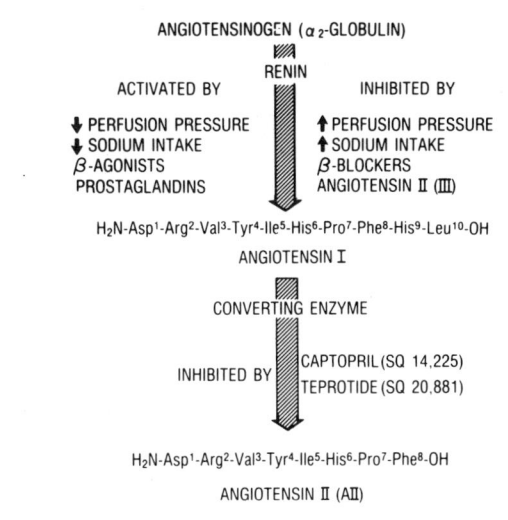

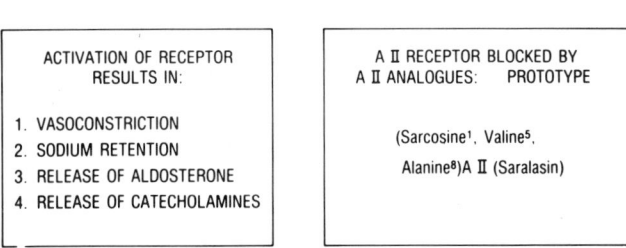

FIG 25–1.
The renin-angiotensin-aldosterone system. (From Felsen D, Vaughan ED Jr, in Rajfer J (ed): *Urologic Endocrinology.* Philadelphia, WB Saunders Co. In press. Used with permission.)

curring CEIs were initially found in snake venom (Bakhle, 1968). Teprotide was the first synthetic CEI to be used in humans (Cheung and Cushman, 1973) but was inconvenient as a screening test owing to a lack of oral activity. Captopril, an orally active CEI with a rapid onset of action, was subsequently developed following the promising studies with teprotide. Captopril is currently utilized in a screening role for renin-dependent hypertension (see Single-dose Captopril Test below) as well transiently to treat RVH in selected cases.

More recently, enalapril, an orally active CEI, has been developed and clinically tested. Enalapril does not have the sulfhydryl group of captopril, which is thought to be responsible for rashes and taste loss associated with the latter drug. Enalapril displays a greater affinity for converting enzyme than captopril, but has a slow onset of action, requiring two hours to block angiotensin I conversion to angiotensin II (Brunner et al., 1981) in contrast to the 15 minutes needed by captopril. Enalapril is, however, more potent and has a longer-lasting effect. Plasma converting enzyme activity does not return to control levels until 72 hours after a single oral dose. Blood pressure has been measured to be significantly decreased four hours after oral administration.

Enalaprilat, or enalaprilic acid, is a more rapidly acting, equally potent CEI related to enalapril. Enalaprilat requires IV administration and is currently being evaluated as a screening agent for RVH in a role similar to that of captopril.

The depressor action of these pharmacologic probes has confirmed the view that the level of PRA closely reflects the degree of associated vasoconstriction in the arteriolar bed (Pratchett, 1983). In addition, angiotensin-converting enzyme, also known as kininase II, inhibits the breakdown of bradykinin. Therefore, no matter how specific a CEI is for the converting enzyme, it will not be specific with regard to the RAS and may influence the metabolism of peptide substrates participating in other metabolic processes. Highly specific renin inhibitors that prevent the formation of angiotensin I have been shown to produce a depressor effect very similar to that produced by other RAS-blocking agents when administered to animals with experimental, renin-dependent hypertension (Brunner et al., 1983). These studies serve further to confirm the correlation between the level of PRA and the degree of arteriolar vasoconstriction.

CLINICAL EVALUATION OF RENOVASCULAR HYPERTENSION

Renovascular hypertension is potentially reversible by renal revascularization. It is now clear that the angiographic finding of a stenosed renal artery in a hypertensive patient is insufficient evidence on which to diagnose RVH. In this regard, hypertensive patients with a renal artery stenosis may not improve following renal revascularization (Foster et al., 1975). Moreover, radiologic (Eyler et al., 1962) and autopsy series (Holley et al., 1964) have shown that normotensive patients can have severely diseased renal arteries. Thus, the emphasis in screening for RVH should not be on the anatomical demonstration of a renal artery lesion. Rather, the diagnosis of RVH is predicated on the demonstration of a physiologically significant decrease in renal blood flow to one or both kidneys associated with the hypersecretion of renin from the more ischemic kidney, the absence of renin secretion from the contralateral kidney, and lowering of the blood pressure on removal of the source of hyperreninemia, or on blockage of the renin-angiotensin system. It is the clinician's task to identify potentially curable patients in a safe, cost-effective, and reliable manner.

Until recent times means by which to reliably identify patients with a physiologically significant renal artery stenosis have eluded the clinician. There are no pathognomonic clinical characteristics that reliably lead to the diagnosis (Simon et al., 1972), but there are several clinical features that should arouse suspicion that RVH may be present (Table 25–1).

The lack of reliable clinical clues stimulated further study in laboratory animals in pursuit of understanding the physiologic profiles that characterize laboratory models of Goldblatt hypertension. These endeavors have resulted in the delineation of a variety of approaches to screen for patients with RVH. The observation in animals that partial occlusion of the renal artery results in increased fractional reabsorption of sodium and water (Blake et al., 1950) and increased urinary concentration of nonreabsorbable solute led to the development of differential renal function studies (Howard et al., 1954). The initial criteria for a positive test was a 50% decrease in urine flow and a 15% of greater decrease in sodium concentration from the affected kidney. This test underwent various modifications (Stamey et al., 1963), but for the most part has been eliminated in favor of less clinically complex procedures that accurately measure the activity of the RAS. At the present time, differential renal function studies are performed in the setting of unilateral renal parenchymal disease or when renin studies are equivocal. In the former case, renal function tests determine the residual GFR and the sodium- and water-excreting capacity of a diseased kidney in consideration for nephrectomy.

The rapid sequence IV urogram (Maxwell and Lupu, 1968) and the renal scan (Maxwell et al., 1968) have been utilized as screening tests for RVH. These studies provide anatomical information about renal size and in addition indirectly assess renal blood flow and function. Compared with renal arteriograms, the urogram had a specificity of 86% and sensitivity of 75%, and the scan a specificity of 77% and sensitivity of 74% (Harvey et al., 1985) in identifying a renal artery stenosis of 50% or more. It must be remembered that demonstration of a renal artery stenosis on an arteriogram is not sufficient evidence to determine whether the ipsilateral kidney is producing the hypertensive state. Therefore, the sensitivity and specificity of these two tests for identifying RVH would be even lower.

Thornbury and associates (1982) reexamined the database of the National Cooperative Study to assess the reliability of the hypertensive urogram as a screening test for RVH. Recalculation of the data showed a higher (41.6%) false negative rate than the authors of the study found, noting that 35% of patients cured and 44% of patients improved by renal revascularization had normal screening urograms. Based on these findings, the authors concluded that the excretory urogram failed to achieve a suitable sensitivity and specificity to serve as a screening test for RVH.

Table 25–2 compares the sensitivities and specifici-

TABLE 25–1.
Clinical Clues Suggestive of Renovascular Hypertension (RVH)*

CLUES	COMMENT
Historical	
Hypertension in the absence of any family history of hypertensive disease	Suspect if family history negative; however, about 1/3 of patients with RVH will have a positive family history.
Age of onset of hypertension <25 or >45 year	The average age of onset for essential hypertension is 31 ± 10 (SD) yr. Children and young adults usually have fibromuscular disease; adults over 45 years are more likely to have atherosclerotic narrowing of arteries.
Abrupt onset of moderate to severe hypertension	Essential hypertension usually begins with a "labile" phase before mild hypertension becomes established; usually has a more telescoped natural history, often first appearing as moderate hypertension of recent onset.
Development of severe or malignant hypertension	RVH often becomes moderately severe and is prone to produce acceleration or malignant-phase hypertension; both forms of hypertension involve markedly increased renin release.
Headaches	Essential hypertension is usually asymptomatic. There seem to be more headaches with RVH, possibly related to its severity or to high levels of angiotensin II, a potent cerebrovascular vasoconstrictor.
Cigarette smoking	In a recent survey, 74% of patients with fibromuscular renal artery stenosis were smokers; 88% of those with atherosclerotic disease smoke (Nicholson et al., 1983).
White race	RVH is uncommon in the black population.
Resistance to or escape from blood pressure control with standard diuretic therapy or antiadrenergic	Probably the most typical feature of RVH is that it responds poorly to diuretics and often only transiently to antiadrenergic drugs.
Excellent antihypertensive response to converting enzyme inhibitors eg, captopril	Converting enzyme inhibitors block the renin-angiotensin system most effectively and are, therefore, highly specific agents.
Examination and routine laboratory results	
Retinopathy	Hemorrhages, exudates, or papilledema indicate acceleration or malignant phase.
Abdominal or flank bruit	A helpful clue, but commonly present in elderly individuals, and occasionally present in younger patients who have no apparent vascular stenosis
Carotid bruits or other evidence of large-vessel disease	Commonly the vascular pathology is not limited to the renal bed.
Hypokalemia—in the untreated state or in response to a thiazide diuretic	Increased aldosterone stimulation by the renin-angiotensin system tends to reduce the serum potassium level. In untreated essential hypertension this does not occur. Thiazide diuretics accentuate this phenomenon in RVH.

*From Vaughan ED Jr, Case DB, Pickering TG, et al: Clinical evaluation of renovascular hypertension and therapeutic decisions. *Urol Clin North Am* 1984; 11:393–407. Used by permission.

TABLE 25–2.

Predictive Value of Screening Tests for Renovascular Hypertension*

	IVP†	DIVA†	PRA‡/§	SINGLE-DOSE CAPTOPRIL‖
Sensitivity (%)	75	88	80	100
Specificity (%)	86	89	84	95
False positive (%)	14	11	16	5
False negative (%)	25	12	20	0
Predictive value¶ (%)				
Prevalence				
2	9.9	14.6	9.3	29
5	22.1	30.5	20.8	51.3
10	37.5	48.1	35.7	69
Exclusion value (%)				
Prevalence				
2	99.4	99.7	99.5	100
5	98.5	99.3	98.8	100
10	96.7	98.5	97.4	100

*Calculations:

Predictive value

$$= \frac{\text{sensitivity} \times \text{prevalence}}{(\text{sensitivity} \times \text{prevalence}) + \text{false positive rate} \times (100 - \text{prevalence})} \times 100$$

Exclusion value

$$= \frac{\text{specificity} \times (100 - \text{prevalence})}{\text{specificity} \times (100 - \text{prevalence}) + \text{false negative rate} \times \text{prevalence}} \times 100$$

Abbreviations: IVP, intravenous pyelogram; DIVA, digital angiography; PRA, plasma renin activity.

†From Harvey RJ, Krumlovsky F, delGroco F, et al: Screening for renovascular hypertension. *JAMA* 1985; 254:388. Used by permission.

‡From Pickering TG, Sos TA, Vaughan ED Jr, et al: Predictive value and changes of renin secretion in hypertensive patients with unilateral renovascular disease undergoing successful renal angioplasty. *Am J Med* 1984; 76:398–404. Used by permission.

§From Brunner HR, Gavras C, Laragh JH, et al: Angiotensin II blockade in man by SAR¹-Ala⁸-angiotensin II for understanding and treatment of high blood pressure. *Lancet* 1973; 2:1045. Used by permission.

‖From Muller FB, Sealey JE, Case DB, et al: The captopril test for identifying renovascular disease in hypertensive patients. *Am J Med* 1986; 80:633. Used by permission.

ties of the more commonly used screening tests for RVH and lists the predictive value of each test for three different prevalence rates. The single-dose captopril test has the highest predictive and exclusion values, confirming its reliability as a screening test for RVH.

The recent emergence of methods to reliably assay the activity of the renin-angiotensin system and the development of pharmacologic probes that block specific steps in the renin-angiotensin cascade have led to the use of renin determinations to identify RVH. Goldblatt's initial animal work led to the assumption that the underlying derangement in RVH is excess renin secretion resulting in angiotensin II formation. The hypertensive animal model most analogous to human RVH is the 2K,1C Goldblatt preparation. In this model the hypertension is initially dependent on the increased renin secretion from the clipped kidney. The administration of CEIs (Weed et al., 1979) or competitive angiotensin II analogues (Brunner et al., 1971) can prevent or reverse the hypertension. This early phase of 2K,1C hypertension exhibits four characteristics: increased secretion of renin from the clipped kidney; absence of renin secretion from the opposite kidney; decreased renal blood flow to the clipped kidney; and elevated pressure secondary to angiotensin II–induced vasoconstriction. The identification of these characteristics permitted the development of a rational approach to the use of PRA determinations and angiotensin blockade in the diagnosis of RVH (Vaughan et al., 1973; Pickering et al., 1984) (Fig 25–2).

Unfortunately, the use of peripheral PRA was challenged by the finding that it can be normal in a fraction of patients with RVH (Marks and Maxwell, 1975) as well as in Goldblatt animal models (Ayers et al., 1969). With respect to the former observations, careful review of the protocols used to measure peripheral PRA revealed that samples were obtained under conditions that are now recognized to alter PRA. Several antihypertensive agents reduce PRA, while salt depletion by dietary restraint or diuretic consumption increases PRA. Now, all peripheral ambulatory PRA samples are

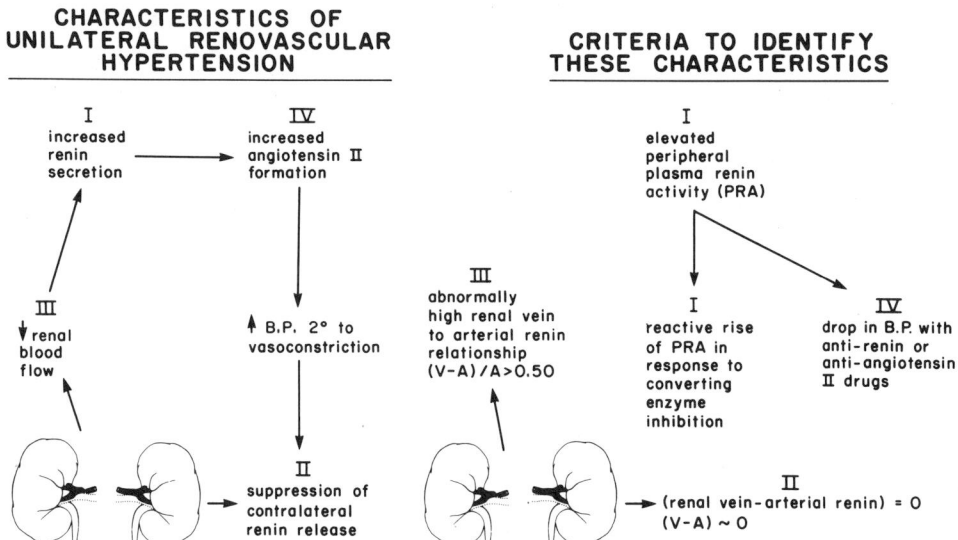

CHARACTERISTICS OF UNILATERAL RENOVASCULAR HYPERTENSION

CRITERIA TO IDENTIFY THESE CHARACTERISTICS

FIG 25–2.
Characteristics of the early phase of two-kidney, one-clip Goldblatt hypertension in the rat *(left)* and the criteria derived from the animal model that identify correctable renal hypertension. (From Vaughan ED Jr, Case DB, Pickering TG, et al: Clinical evaluation of renovascular hypertension and therapeutic decisions. *Urol Clin North Am* 1984; 11(3):393–407. Used by permission.)

collected in a standardized setting that requires that the patient be salt replete and off all antihypertensive medicines that influence PRA for at least two weeks. In addition, the PRA is sampled following four hours of ambulation and is indexed against a 24-hour urine sodium determination.

The rationale for emphasizing the peripheral PRA determination is that it is an index of renin secretion (Sealey et al., 1973). In a recent study of hypertensive patients who had successful angioplasty, the peripheral PRA was elevated in 80% (Pickering et al., 1984) prior to angioplasty. The PRA always decreased and often returned to normal following successful angioplasty (Fig 25–3).

The 20% false negative rate places definite limitations on the use of peripheral PRA as a screening test for RVH. In addition, many patients with proved RVH have severe, life-threatening hypertension that precludes cessation of antihypertensive medications before blood sampling. Performance of PRA determinations while the patient is taking medication invalidates the accuracy of the test and eliminates the quantification of peripheral PRA as a practical screening tool. A third factor is the finding that 16% of patients with essential hypertension also have high PRA when indexed against a 24-hour urine sodium determination (Brunner et al., 1972).

ENHANCED ACCURACY OF PERIPHERAL PLASMA RENIN ACTIVITY BY STIMULATION WITH ANGIOTENSIN-BLOCKING AGENTS

The first angiotensin-blocking agent used for testing in human hypertension was saralasin. The initial results

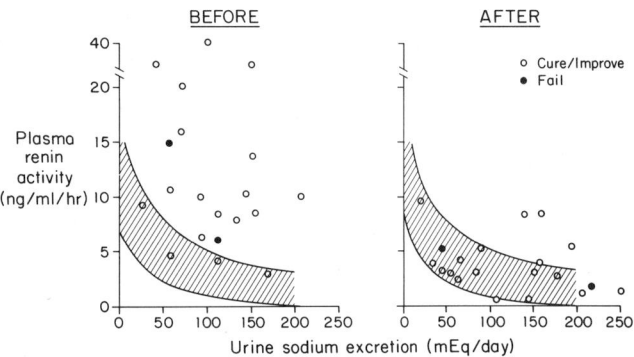

FIG 25–3.
Effect of angioplasty on peripheral plasma renin activity indexed against 24-hour sodium excretion. *Left panel:* before angioplasty. *Right panel:* 6 months after angioplasty. *Hatched area* shows normal range. (From Pickering TG, Sos TA, Vaughan ED Jr, et al: Predictive value and changes of renin secretion in hypertensive patients with unilateral renovascular disease undergoing successful renal angioplasty. *Am J Med* 1984; 76:398–404. Used with permission.)

demonstrated that the compound did, as predicted, lower blood pressure in high-renin forms of hypertension (Brunner et al., 1973). However, the partial agonist activity of the drug in clinical settings where PRA is low represented an undesirable feature (Case et al., 1976a; Carey et al., 1978). Accordingly, although reasonably accurate in identifying RVH (Horn et al., 1979), the usefulness of the test is limited because of the need of IV infusion, the agonist effect of the drug, and the influence of sodium balance on the blood pressure response to the agent.

A second approach to the use of angiotensin blockade to expose RVH came from experience following the development of CEIs that block angiotensin II formation. One of the peptides, teprotide, was shown to block the vasopressor effect of angiotensin I, and it was possible to demonstrate a close direct correlation between the pretreatment level of PRA and the magnitude of the depressor response (Case et al., 1976b). Moreover, in untreated hypertensive patients on the same normal sodium intake, IV teprotide lowered blood pressure to the greatest extent in the group profiled as high-renin by the renin-sodium index, but it did not lower blood pressure significantly in the low-renin subgroup. From these studies, it was possible to predict that inhibition of converting enzyme might be a useful diagnostic approach for detecting angiotensin-dependent forms of hypertension.

The success of teprotide was a potent stimulus to the development of the orally active CEI captopril. Captopril has the potential for use as a diagnostic probe, like teprotide, since it has a rapid onset of action (within 10–15 minutes), reaching a peak effect by 90 minutes (Case et al., 1978). Initial studies showed a close relationship between the pretreatment peripheral PRA and the magnitude of the depressor response induced after the first dose of captopril (Case et al., 1978).

During the early studies of the effect of these agents on blood pressure in hypertensive patients, it was noted that angiotensin blockade resulted in a marked rise in PRA in selected patients. Accordingly, the induction of marked rises in PRA in 31 of 32 renovascular patients but only 2 of 64 with essential hypertension appeared to be a more specific test for RVH than the induction of depressor responses. These studies also revealed that prior sodium depletion, either with diuretics or by low-sodium diet, abolished the specificity of this test (Case et al., 1979a).

With the availability of the oral CEI captopril, a single oral test dose was utilized. Results very similar to those obtained using the IV competitive antagonist saralasin or the CEI teprotide were found (Case et al., 1979b). An additional advantage was the observation that renovascular hypertensive patients being treated with a β-adrenergic blocker still responded to oral administration of captopril, with a fall in blood pressure and a rise in PRA (Fig 25–4) (Case et al., 1982).

SINGLE-DOSE CAPTOPRIL TEST

The "captopril test" is a highly reliable screening test well suited for outpatient use (Table 25–3). Diuretics and preferably all other antihypertensive medications are discontinued at least two weeks before testing. Patients are advised not to restrict their salt intake. Twenty to thirty minutes before testing, the patient is seated comfortably in a quiet room, during which time an IV catheter is placed for blood sampling, and baseline blood pressure measurements are performed. Venous blood is drawn for a baseline PRA determination prior to the oral administration of 25 mg of captopril. Blood pressure measurements are then taken every 15 minutes for one hour. Venous blood sampling is repeated at 30 and 60 minutes after captopril administration. A depressor response proportional to the pretreatment PRA level (Pickering et al., 1984) will be seen in patients with renin-dependent hypertension (diastolic pressure decrease of 15% or more). However, the change in blood pressure has low specificity, since patients with essential hypertension may also have a depressor response.

Formation of angiotensin II is greatly decreased by captopril. The resulting decrease in systemic blood pressure in patients with renin-dependent hypertension further decreases the blood flow to the ischemic kidney. The lower perfusion pressure produces a decrease in the glomerular capillary hydrostatic pressure, diminishing the GFR. In addition, the efferent arteriole, which undergoes angiotensin II–induced vasoconstriction at low perfusion pressures to maintain glomerular filtration, relaxes as angiotensin II formation is blocked, thus decreasing the ipsilateral GFR further. The macula densa and baroreceptor mechanisms are activated, and more renin is released from the ischemic kidney. In renin-dependent hypertension, but not in essential hypertension, inhibition of the converting enzyme is associated with a rise in PRA to 12 ng/ml/hr or greater, an increment in PRA of 10 ng/ml/hr or more above baseline levels, and a percent increase in PRA of 170% or more (or 400% if the baseline PRA was <3 ng/ml/hr). If these three criteria are present in a patient with normal renal function off diuretics, RVH can be distinguished from essential hypertension with a specificity of 100% and a sensitivity of 95% (Muller et al., 1986). Prior sodium depletion by diuretics or by dietary restraint increases PRA and abolishes the specificity of this test. Patients in treatment with β-blockers remain responsive to captopril as described above unless their

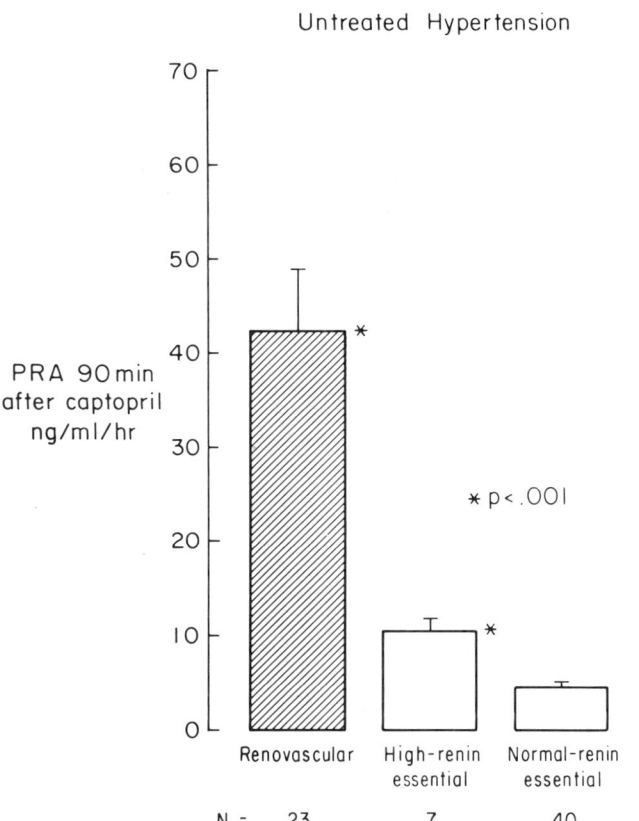

FIG 25–4.
Levels of plasma renin activity in renovascular and essential hypertension 90 minutes after a single dose of captopril. A marked reactive hyperreninemia was found in group with renovascular hypertension whether or not already receiving β-blocker therapy. (From Case DB, Atlas SA, Laragh JH: Physiologic effects and blockade, in Laragh JH, Buhler FR, Seldin DW (eds): *Frontiers in Hypertension Research.* New York, Springer-Verlag, 1982, pp 541–550. Used with permission.)

TABLE 25–3.

Single-Dose Captopril Test*

Drugs

    Discontinue all antihypertensive medicines for at least 2 wks, if possible; otherwise, maintain β-blocker but avoid diuretics, converting enzyme inhibitors, and nonsteroidal anti-inflammatory drugs for at least 1 wk, ideally 2 wks

Diet

    A diet with normal or high sodium content is necessary. Too low a sodium intake will produce false positive results. If there is a question about diet, a 24-hr urine collection for sodium will closely reflect the intake.

Procedure

    The patient is seated comfortably for 20–30 min before testing and maintained in this position for the duration of the test.

    Blood pressure is measured at 20, 25, and 30 min (obtain three stable baseline measurements) then blood is sampled for PRA (in a lavender top Vacutainer kept at room temperature).

    A 25-mg captopril tablet is crushed (to ensure that it dissolves) and 30 ml of water added to prepare a suspension. The patient is instructed to drink the suspension, wash the contents out twice, and drink those also.

    Blood pressure and PRA are remeasured after 30 and 60 min.

*From Vaughan ED Jr, Case DB, Pickering TG, et al: Clinical evaluation of renovascular hypertension and therapeutic decisions. *Urol Clin North Am* 1984; 11(3):393–407. Used by permission.

baseline PRA is less than 2.5, in which case the test may be unreliable (Table 25–4).

In summary, the single-dose captopril test appears to accurately separate patients with RVH from those with essential hypertension. In addition, a 24-hour urine collection is not necessary, and the patients can remain on β-blockade. Hence, the test is complementary to the renin-sodium index to identify renin hypersecretion.

CONTRALATERAL SUPPRESSION OF RENIN SECRETION AND AN ELEVATED RENAL VEIN-TO-ARTERIAL RENIN RELATIONSHIP—USE OF DIFFERENTIAL RENAL VEIN RENIN DETERMINATIONS

Differential renal vein renin determinations have emerged as a most useful tool in identifying correctable RVH (Judson and Helmer, 1965). The most common approach has been to calculate the renal vein renin ra-

TABLE 25–4.

Criteria to Distinguish Renovascular
Hypertension From Essential Hypertension*

Stimulated PRA ≥12 ng/ml/hr
Absolute increase in PRA of ≥10 ng/ml/hr
% Increase in PRA ≥150%, or 400% if baseline
    PRA <3 ng/ml/hr

*From Muller FB, Sealey JE, Case DB, et al: The captopril test for identifying renovascular disease in hypertensive patients. *Am J Med* 1986; 80:633. Used by permission.

tio, that is, stenotic divided by normal side PRA values with some arbitrary "positive" ratio (usually 1.5:1). The major limitation in this method is in selecting a ratio with the accuracy to divide precisely patients with RVH from those with essential hypertension (Marks and Maxwell, 1975). Generally, a positive renal vein ratio predicts a fall in blood pressure following correction of the vascular lesion in over 90% of cases. However, a negative ratio does not preclude a successful response to revascularization. In the review of Marks and Maxwell (1975) 49% of patients with negative ratio were cured by appropriate surgical intervention. However, these patients were a minority of the total patients studied, and the actual false negative rate was 15% (62/412 patients).

In addition, the collection of renal vein blood alone for PRA is subject to the risk of a sampling error due to either incorrect catheter placement or sampling from a renal vein which does not subtend the renal area supplied by the stenotic artery. Sampling from the inferior vena cava below the renal veins is a safeguard against this source of error, and the absence of a 50% renin increment from both kidneys together identifies an inadequate differential renal vein renin study (Vaughan et al., 1973).

Finally, a positive renal vein ratio does not exclude bilateral renin secretion, albeit asymmetric, which could indicate bilateral renal disease. In this setting correction of a unilateral lesion might not totally correct the underlying pathology, with subsequent unmasking of contralateral renin secretion producing failure of total blood pressure control.

In view of these limitations of the traditional renal vein ratio analysis, another method for analysis of renin values has been devised and is based on the characteristics of experimental 2K,1C Goldblatt hypertension (Table 25–5 and Figure 25–2).

Hypersecretion of renin, as determined by the renin-sodium index or captopril stimulation, serves as the primary criterion for the diagnosis of RVH. A second criterion is the demonstration of the absence of renin secretion from the contralateral (or noninvolved) kidney. Suppression of renin secretion from this kidney can be determined by subtracting the arterial plasma renin activity (A) from the renal venous renin activity (V). Since the inferior vena caval (IVC) renin and aortic renin are the same, the IVC renin value can be substituted for (A) in this equation (Fig 25–5) (Sealey et al., 1973). Hence, patients with curable RVH exhibit an absence of renin secretion from the opposite kidney; i.e., (V − A) = 0, also termed "contralateral suppression" of renin (Stockigt et al., 1972; Vaughan et al, 1973). Contralateral suppression of renin indicates that the noninvolved kidney is responding in an appropriate, "normal" fash-

TABLE 25–5.

Renin Values for Predicting Curability of Renovascular Hypertension

Collection of Samples (moderate sodium intake ± 100 mEq/day)
 Ambulatory peripheral renin and 24-hr urine sodium excretion under steady state conditions (i.e., not on
  day of arteriography)
 Collection of blood for PRA before and after converting enzyme blockade
 Collection of supine:
  Renal vein renin from suspect kidney (V1) and inferior vena caval renin (A1)
  Renal vein from contralateral kidney (V2) and inferior vena caval renin (A2)
 Enhancement of renin secretion by converting enzyme blockade if initial renin sampling is inconclusive

Criteria for Predicting Cure	
High PRA in relation to UNaV	Measurement of hypersecretion of renin
Contralateral kidney: (V2 − A2) = 0	An indicator of absent renin secretion from the contralateral kidney
Suspect kidney: (V1 − A1)/A1 = 0.50	Measurement of reduced renal blood flow

$$\frac{(V-A)}{A} + \frac{(V-A)}{A} = 0.50 \text{ in patients with high PRA means}$$

Incorrect sampling	
Segmental disease	Repeat with segmental sampling

‖ From Vaughan ED Jr, Sos TA, Sniderman KW, et al: Renal vein renin secretory patterns before and after transluminal angioplasty in patients with renovascular hypertension: Verification of analytic criteria, in Laragh JH (ed): *Frontiers in Hypertension Research*. New York, Springer-Verlag, 1981. Used by permission.

ion to the elevated blood pressure, increased circulatory angiotensin II levels, or increased sodium chloride at the macula densa by shutting off renin secretion. This phenomenon is at times present not only in patients with unilateral renal arterial lesions but also in patients with bilateral disease demonstrated by arteriograms who have a dominant lesion on one side (Pickering et al., 1983).

A third criterion is based on studies of renal vein and arterial vein relationships in patients with essential hypertension. The mean renal venous renin level has been determined to be about 25% higher than arterial PRA (Fig 25–6) (Sealey et al., 1973). Hence, a total renin increment (both kidneys) of approximately 50% is necessary to maintain a given peripheral renin level, (V − A)/A = 50%. However, a reduction in renal blood flow also influences the renal venous renin level. In this setting, the renal venous renin concentration will be misleadingly high, shifting the renal vein-to-arterial renin relationships upward. Hence, the elevation of the increment above approximately 50% becomes an index of the severity of the reduction in blood flow consequent to the obstructing vascular lesion (see Fig 25–6). The combination of these criteria found in a group of patients managed by renal revascularization is shown in Figure 25–7 (Vaughan et al., 1979).

In a group of 46 hypertensive patients with arteriographically proved unilateral renal artery stenosis who underwent successful angioplasty, 35 patients had technically successful sampling (Pickering et al., 1984). Twelve patients had studies that were judged to be un-

successful; 9 sets of values were rejected because the combined increment from the two sides was less than 0.50. Six of nine patients were taking captopril over a long term and exhibited high systemic levels of PRA (between 14 and 99 ng/ml/hr) at the time of sampling. These data suggest that long-term converting enzyme blockade with captopril may invalidate renal vein renin analysis by a mechanism that remains unclear. At the other extreme, three patients had extremely low peripheral PRA (less than 1 ng/ml/hr) at the time of sampling, two of whom were taking β-blockers. This latter situation is potentially clarified by captopril stimulation (see below).

Of the 34 patients with technically acceptable values, there were 23 true positives, no false positives, 3 true negatives, and 8 false negatives. Of the 8 false negatives, 4 had a (V − A)/A between 0.40 and 0.48, and 6 showed contralateral suppression. Only two patients had symmetric renal vein renins (Pickering et al., 1984). Hence, the present approach gave a sensitivity of 74% and specificity of 100%, better indices than found by renin ratio analysis.

An additional aid to renal vein sampling is the utilization of segmental renal venous sampling (Schambelan et al., 1974), especially in cases where sampling of blood from the major renal veins fails to demonstrate a combined renin increment of 50% from both kidneys, suggesting either a technical error or segmental disease. This approach may be particularly helpful in children with segmental parenchymal disease (Parrott et al., 1984).

ESSENTIAL HYPERTENSION
RENIN ACTIVITY IN PLASMA
FROM AORTA AND VENA CAVA

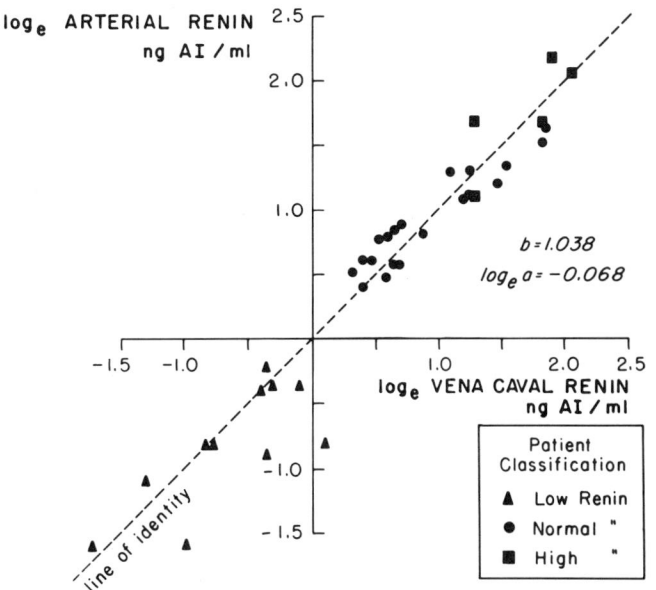

FIG 25–5.
Relationship of renin activity in blood collected from the aorta to that found in vena caval blood. The two values are not different in patients with essential hypertension. (From Sealey JE, Buhler FR, Laragh JE, et al: The physiology of renin secretion in essential hypertension: Estimation of renin secretion rate and renal plasma flow from peripheral and renal vein levels. *Am J Med* 1973; 55:391–401. Used with permission.)

The patterns of renal vein renin activity in bilateral RVH are less consistent than in unilateral disease. Pickering et al. (1985) contrasted renal vein renin activity in hypertensive patients with unilateral vs. bilateral RVH and found that most patients in both groups had a $(V-A)/A \geq 0.48$ on the more ischemic side. They noted, however, that contralateral suppression was demonstrable in only 13/25 (54%) patients with bilateral disease. Most patients with bilateral RVH present with varying levels of azotemia. Pickering found that contralateral suppression was less likely to occur in patients with a serum creatinine of >2 mg/dl. Although contralateral suppression is less pronounced in patients with bilateral stenoses, the asymmetry of renal vein renin activity is still present. Captopril challenge stimulates renin secretion from the more ischemic side in bilateral RVH as it does in unilateral disease.

CONVERTING ENZYME INHIBITION TO ENHANCE THE ACCURACY OF RENAL VEIN RENIN ANALYSIS

Following the initial report of renin stimulation by angiotensin blockade (Re et al., 1978), several groups of

RENAL VEIN RENIN DIAGNOSTIC PATTERNS

ESSENTIAL HYPERTENSION

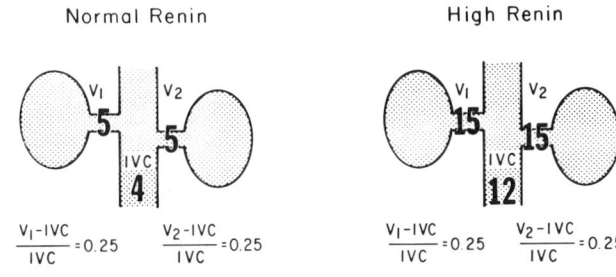

RENOVASCULAR HYPERTENSION

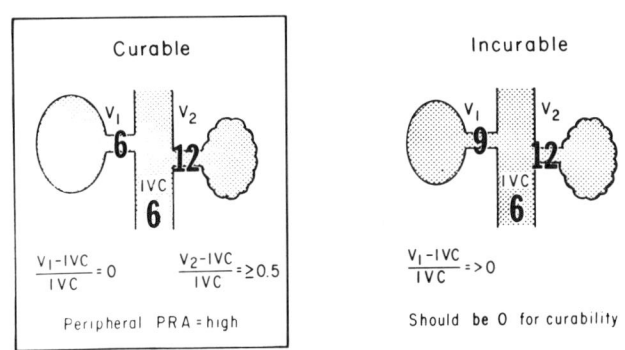

FIG 25–6.
Renal vein renin diagnostic pattern. In essential hypertension *(top)* at all levels of renin secretion the renin level of each renal vein is about 25% greater than either the peripheral arterial or venous levels. In unilateral renin secretion (curable renovascular hypertension) the active kidney is solely responsible for maintaining the peripheral renin levels. Hence, the increment is 50% (0.5) and becomes progressively greater as renal blood flow is reduced. Unequal bilateral renin secretion *(bottom right)* indicates bilateral disease and decreases the chance of cure following corrective unilateral surgery. *(From Laragh JH, Sealey JE: Cardiovasc Med* 1977; 2:1053. Used with permission.)

investigators reported increased renin release from the ischemic kidney with converting enzyme inhibition (Lyons et al., 1983; Thibonnier et al., 1984). The magnitude of these induced changes is shown in Figure 25–8. These data were determined from 26 patients with unilateral RVH. Not shown in this illustration is the observation that the inferior vena caval levels (systemic or peripheral levels) were comparable to those values measured from the normal side, revealing the continued suppression of renin secretion from that kidney even after stimulation. Although more information is required to distinguish the subtleties (i.e., bilateral renovascular disease, branch lesions, total occlusions, coexistent renal parenchymal disease), it is clear that stimulation adds to the analysis in certain specific situations: (1) when patients are already taking drug ther-

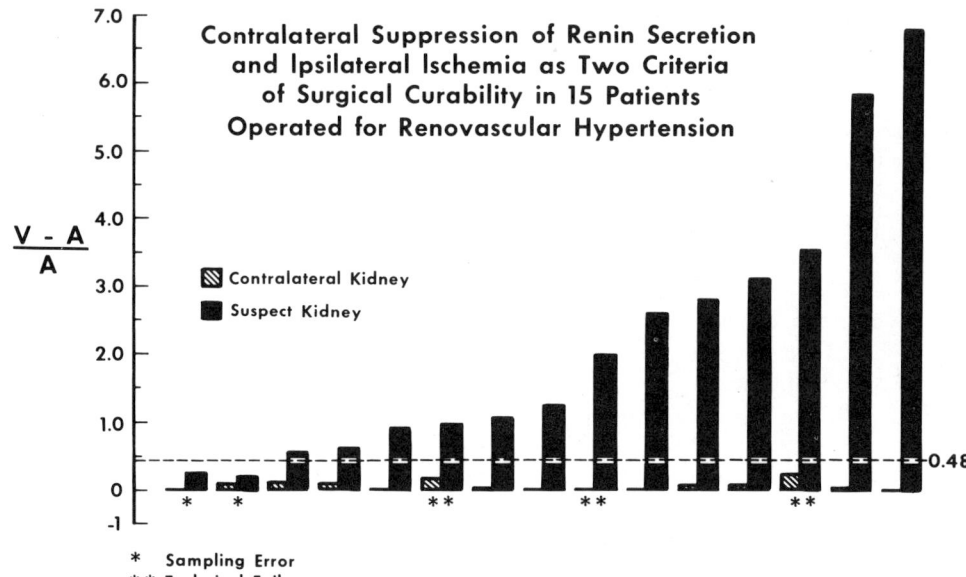

Contralateral Suppression of Renin Secretion and Ipsilateral Ischemia as Two Criteria of Surgical Curability in 15 Patients Operated for Renovascular Hypertension

$\frac{V - A}{A}$

Contralateral Kidney
Suspect Kidney

* Sampling Error
** Technical Failure

FIG 25–7.
Of 15 patients with renovascular hypertension, 13 exhibited (V − A)/A in excess of 48% from the suspect kidney and a suppressed value from the contralateral kidney (asterisk). V is renal venous PRA; A is arterial or infrarenal inferior vena cava PRA. Double asterisks denote three patients whose values suggested surgical curability yet had residual or recurrent hypertension due to technical failure. (From Vaughan ED Jr, Carey RM, Ayers CR, et al: A physiologic definition of blood pressure response to renal revascularization in patients with renovascular hypertension. *Kidney Int* 1979; 15:S83. Used with permission.)

apy, e.g., β-blockers, and the renin levels are generally reduced; (2) when there is a question about the reliability of the PRA measurements, particularly of low levels; (3) where experience performing renal vein renins is limited and subject to sampling errors; and (4) where there already exist equivocal values.

In summary, renal vein renin sampling after captopril stimulation accentuates renin release from ischemic renal tissue, which is particularly useful when values are equivocal, branch stenoses are present, or when renovascular disease is superimposed on coexisting hypertension or renal disease. We anticipate that the addition of converting enzyme inhibition during renal venous sampling for renin will virtually eliminate the false negative analyses that have previously limited the accuracy of the studies.

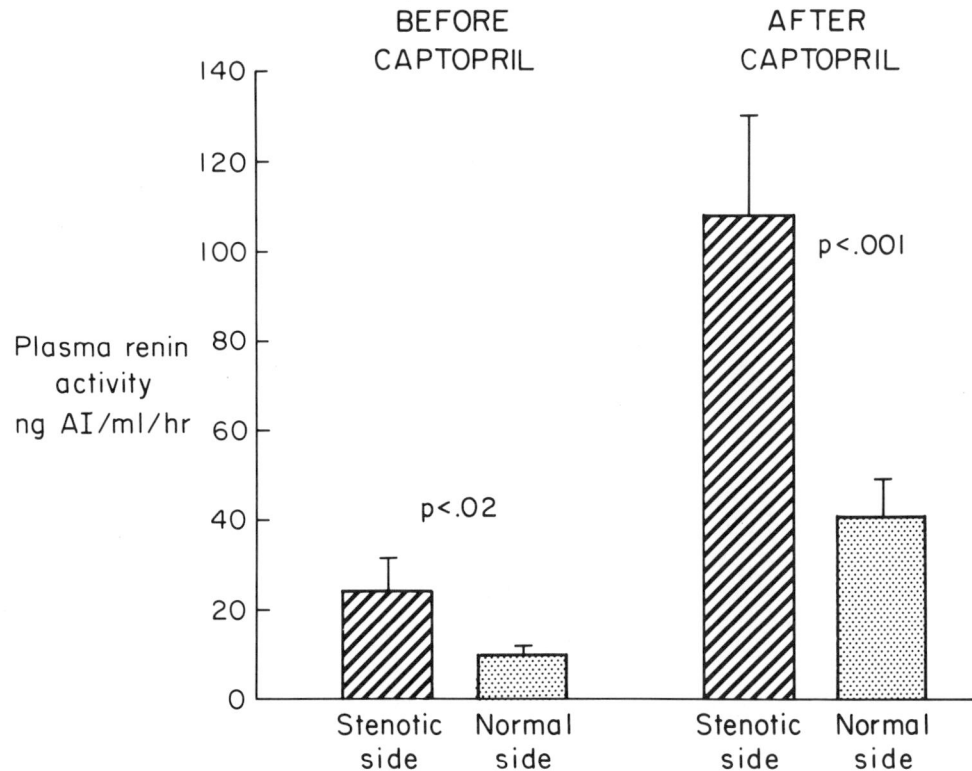

FIG 25–8.
Renal vein determinations (renal vein levels only) in patients with documented renovascular hypertension before and after captopril stimulation. Captopril accentuates renin secretion from the ischemic kidney. Not shown are the inferior vena caval levels, which were the same as those measured from the normal side both before and after captopril stimulation. (From Vaughan ED Jr, Case DB, Pickering TG, et al: Clinical evaluation of renovascular hypertension and therapeutic decisions. *Urol Clin North Am* 1984; 11:393–407. Used with permission.)

ANATOMICAL CONFIRMATION OF THE FOUR CRITERIA

It should be noted that virtually the entire evaluation of the hypertensive patient can now be accomplished without hospitalization, thus avoiding much of the expense previously thought to be too excessive to warrant evaluation (McNeil, 1975). We have performed over 300 renal vein renin samplings in outpatients, 100 of them accompanied by digital angiography (DIVA), which is about 90% accurate in identifying renal arterial lesions (Osborne et al., 1981). We utilize a central injection for the DIVA, which gives better definition of the renal vasculature. Major drawbacks to the use of DIVA include the limited resolution in branch and segmental renal artery disease and impaired image quality in obese or uncooperative patients, as well as in the setting of cardiac insufficiency and renal failure (Harvey et al., 1985).

The concept of IV angiography is not new and was first performed in the 1930s (Robb and Steinberg, 1939). In fact, the use of the early technique in urology was reviewed in 1961 (Steinberg and Marshall, 1961). However, the technique was further developed by refinements in catheter angiography. Recently, the combination of equipment designed for image enhancement and computer analysis has revolutionized the field and led to the use of DIVA as a major tool in the management of RVH (Hillman et al., 1983).

Accordingly, DIVA can be utilized both to accompany renal vein sampling for renin to give anatomical definition of the renal vasculature and to follow up patients after correction of a renovascular lesion (Sos et al., 1983). The only caveat to remember is that DIVA demonstrates only anatomical vascular disease, and the functional significance of the lesion must still be demonstrated.

Altogether, screening criteria have been identified that provide great insurance that the patient with renovascular disease will be both physiologically and anatomically characterized and the success of a corrective procedure can be predicted. When these criteria have been identified, the patient is admitted to the hospital for definitive management of the offending lesion.

VALIDATION OF THE FOUR CRITERIA

In addition to a favorable clinical response to renal angioplasty, we have also had the unique opportunity to study the effect of restoration of blood flow on renal vein renin concentration and renin secretion (Vaughan et al., 1981; Pickering et al., 1984). To accomplish this goal, we have monitored the immediate effect of successful angioplasty on renal renin secretion. Thirty minutes following angioplasty, there was a marked reduc-

tion in the renal vein renin from the previously stenotic side (Fig 25–9). The residual ipsilateral increment of renal vein renin was about 50% above the peripheral level, while contralateral renin suppression persisted. This 50% increment has been predicted previously to occur in the setting of unilateral renin secretion and normal renal blood flow (Sealey et al., 1973).

Several months after angioplasty, the peripheral PRA had returned to normal levels in most patients, indicating a reduction in renin secretion (see Fig 25–3). Of equal interest was the reestablishment of bilateral renin increments of about 25% above the inferior vena caval renin level (see Fig 25–9). Hence, contralateral renin suppression reversed following successful angioplasty. This 25% increment from both kidneys is characteristic of the renin secretory pattern found in patients with essential hypertension (Sealey et al., 1973).

The finding that the renal renin secretory characteristics of RVH reverse following successful angioplasty with correction of the hypertension is strong support that they truly reflect the abnormal secretory behavior of renin in curable RVH.

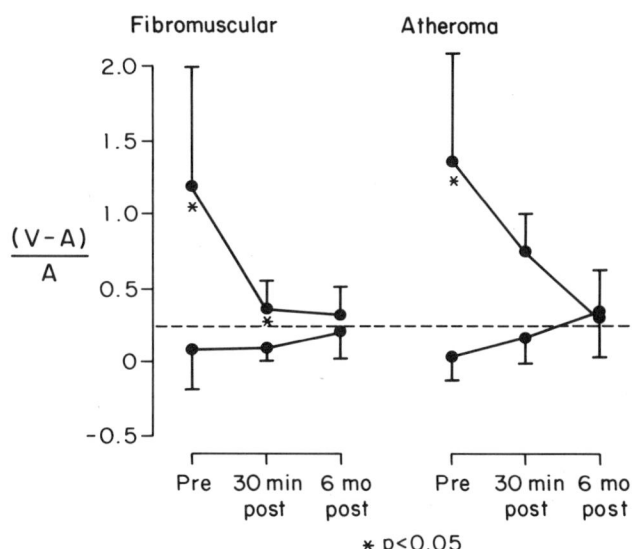

EFFECT OF ANGIOPLASTY ON RENAL VEIN RENINS

FIG 25–9.
Effect of angioplasty on renal vein renins. Samples were taken immediately before angioplasty, 30 minutes after, and 6 months after. The higher values for the ischemic kidney, the lower for the contralateral kidney. *Asterisk* indicates significant difference between the 2 kidneys; *dotted line* is the normal level of $(V - A)/A$ (0.24). (From Pickering TG, Sos TA, Vaughan ED Jr, et al: Predictive value and changes of renin secretion in hypertensive patients with unilateral renovascular disease undergoing successful renal angioplasty. *Am J Med* 1984; 76:398–404. Used with permission.)

IDENTIFYING THE POTENTIALLY CURABLE PATIENT: A COST-EFFECTIVE APPROACH

Our current approach is outlined in Figure 25–10. All patients with fixed hypertension are potential candidates for this protocol, since we believe that nearly all patients with identified RVH can be best managed by angioplasty or surgical revascularization. With respect to this empirical protocol, it could be argued that highly suspected patients could be screened initially with a digital angiogram. However, the functional significance of an anatomical lesion with respect to ischemia-induced renin release still must be established, so we routinely begin with a peripheral PRA determination.

In our experience, a low PRA is rarely found in untreated patients with nonazotemic renal arterial disease, and we therefore do not usually continue this evaluation unless these patients demonstrate refractoriness to treatment. Patients with high or normal PRA undergo a peripheral captopril test. The test cannot be performed if the patient has been taking captopril over a long term. If the test is positive, the differential renal vein sampling and digital angiography are performed together. The sampling procedure is ideally done first, and then the catheter is advanced into the superior vena cava for injection of contrast material. This combined study is performed in the radiology suite, following which the patient lies quietly in the hypertension unit for 2–4 hours before returning home.

After the diagnostic criteria have been established, the patient is briefly hospitalized for selective arteriography and PTA or surgical revascularization.

We continue to believe that the functional significance of a renal artery lesion must be identified before the therapeutic decision is directed toward intervention.

TREATMENT OF RENOVASCULAR HYPERTENSION

The therapeutic options for the treatment of RVH include percutaneous transluminal angioplasty (PTA), surgical revascularization, autotransplantation, pharmacologic interference with the renin-angiotensin system, and removal of the ischemic kidney.

RENAL ARTERY ANGIOPLASTY FOR RENOVASCULAR HYPERTENSION

Percutaneous transluminal angioplasty (PTA) was first introduced by Dotter and Judkins in 1964 for the treatment of peripheral vascular stenoses. The introduction of flexible, double-lumen balloon catheters by Gruntzig and Hopff (1974) permitted the development of percutaneous balloon angioplasty of renal artery stenoses. The fibromuscular dysplasias and unilateral, nonostial, nonoccluded atherosclerotic renal artery stenoses are the most suitable lesions for treatment with PTA. The ability to avoid a general anesthetic and the relatively low morbidity of this procedure have allowed renal revascularization in patients who might have been deemed unsuitable for surgery secondary to concurrently present diseases. However, PTA does have inherent limitations and is not suitable for dilation of ostial lesions or arterial occlusions. Therefore, careful patient selection is necessary. Additionally, PTA is not universally available, as it can be performed only by well-trained and skilled interventional radiologists with a vascular surgical team back-up.

Tables 25–6 and 25–7 summarize a number of the larger series reporting the technical feasibility of angioplasty and the blood pressure benefit derived in patients with RVH of atherosclerotic and fibromuscular disease etiologies.

Technical success is defined as complete if the residual stenosis is 50% or less and as partial if the residual stenosis is 50%–70% (Sos et al., 1983). The criteria to

TABLE 25–6.

Renal Angioplasty for Atherosclerotic Disease

AUTHOR(S)	NO. OF PATIENTS DILATED	NO. SUCCESSFULLY DILATED	BLOOD PRESSURE BENEFIT		LENGTH OF FOLLOW-UP, MO (MEAN)
			CURED	IMPROVED	
Tegtmeyer	20	19	10	6	0–10 (3.5)
Sos et al.	51	19	7	10	4–40 (16)
Martin et al.	60	52	10	27	3–39 (16)
Colapinto et al.	51	44	8	29	1–36
Schwarten et al.	54	52	23	25	1–18
Total, no.(%)	236	186(79)	58(31)	97(52)	
Blood pressure benefit, no.(%)			155/186	(83)*	
			155/236	(66)†	

*Of those successfully dilated.
†Of the total group.

TABLE 25–7.

Renal Angioplasty for Fibromuscular Dysplasia

AUTHOR(S)	NO. OF DILATED	NO. SUCCESSFULLY DILATED	BLOOD PRESSURE BENEFIT		LENGTH OF FOLLOW-UP, MO (MEAN)
			CURED	IMPROVED	
Tegtmeyer	31	31	10	21	4–64 (31)
Sos et al.	31	27	16	9	4–40 (16)
Martin et al.	20	18	5	12	3–39 (16)
Colapinto et al.	11	9	5	5	1–36
Total, no.(%)	93	85(91)	35(41)	47(55)	
Blood pressure benefit, no.(%)			82/85	(96)*	
			82/93	(88)†	

*Of those successfully dilated.
†Of the total group.

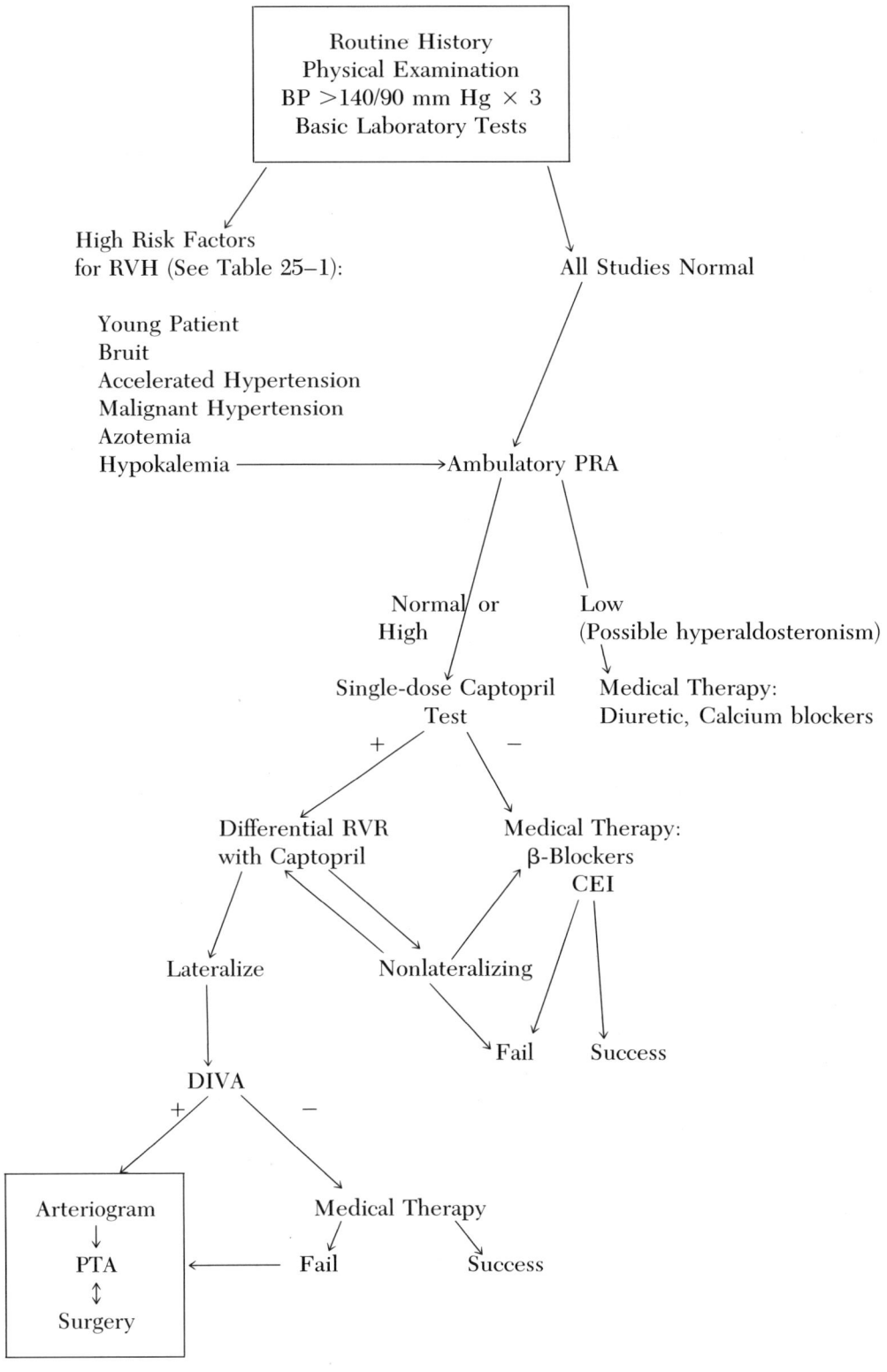

FIG 25–10.
Evaluation plan to identify renovascular hypertension. (Modified from Sosa RE, Vaughan ED Jr: Evaluation of surgically curable hypertension. *Am Urol Assoc Update* 1983; 2(32). Used with permission.)

determine blood pressure are those of the National Cooperative Study (Foster et al., 1975). A decrease in the diastolic blood pressure to <90 mm Hg in the absence of antihypertensive medication is considered a cure, and a decrease in diastolic blood pressure ≥15% with or without antihypertensive medication is considered an improvement. A decrement in diastolic blood pressure of <15% is considered a failure.

Renal artery angioplasty was attempted in 236 patients with atherosclerotic RVH (see Table 25–6) and was technically successful in 186 (79%). The blood pressure benefit enjoyed by the subgroup that was successfully dilated was 83% (31% cured, 52% improved). Nonostial, nonoccluded unilateral lesions were noted to be most amenable to angioplasty. Ostial lesions, total renal artery occlusions, and bilateral lesions, which often tend to be ostial and occluded, were not found to respond satisfactorily to angioplasty. The technical success rate in these patients as a group was 12% (Sos et al., 1983). Ostial lesions result from aortic plaques that impinge on the renal artery orifice. Inflation of the balloon catheter displaces the plaque, with recoil and assumption of the original position occurring at deflation of the balloon. Martin et al. (1985) reported a 25% blood pressure benefit with no cures for a group of 20 patients with ostial lesions. In patients with bilateral RVH the cure rate was a disappointing 5% and the improved rate 41% in contrast to a 25% cure and a 47% improved rate for patients with unilateral disease.

Follow-up reports on most renal angioplasty series have been short term and limited in the number of patients. Sos et al. (1983) reported that follow-up angiography of 16 renal arteries in 15 patients performed an average of 21.8 months after angioplasty revealed no evidence of recurrence of the stenoses. In addition, there was an average increase in kidney area of 12%. The other series summarized here have reported restenosis rates of about 5%–10%. However, redilation is often possible and beneficial to the patient.

Renal artery angioplasty has been found to be more successful in the treatment of the fibromuscular dysplasias (FMD). Table 25–7 summarizes four published series involving 93 patients with FMD. In 91% of patients the stenosed arteries were dilated successfully. Eighty-eight percent of this subgroup enjoyed a blood pressure benefit as a result of the treatment (38% cured and 50% improved). Long-term follow-up of this group is also limited. Tegtmeyer et al. (1984) reviewed eight published series plus his own, totaling 154 patients, and noted a long-term patency rate of 83%–100% for follow-ups ranging 4–64 months. Restenosis following dilation occurred in fewer than 5% of cases.

Complications of renal PTA are reported in 10% of cases and include trauma to the access vessels (femoral or axillary) with hemorrhage. Renal artery damage by dissection, aneurysm formation, perforation, and balloon malfunction or rupture have been reported (Flechner, 1984; Cohn et al., 1984; Tegtmeyer et al., 1984). Cholesterol embolization of the kidney with resulting loss in renal function or to the lower extremities resulting in distal vascular compromise have been reported. Complications of angioplasty resulting in the loss of renal unit have been infrequent, and associated deaths have been rare.

SURGICAL TREATMENT FOR RENOVASCULAR HYPERTENSION

The first described surgical treatment for RVH was nephrectomy (Leadbetter and Burklund, 1938). During the 1940s and 1950s, small kidneys were removed from hypertensive patients in hopes of controlling the blood pressure. In 1956, a careful review of the effects of nephrectomy on the treatment of hypertension revealed a dismal 19% success rate. However, it was appreciated that patients with unilateral renovascular disease fared somewhat better than patients with unilateral renal parenchymal disease (Thompson and Smithwick, 1957). This observation plus the development and widespread use of angiography in the 1950s redirected the focus of surgical treatment of renovascular disease to revascularization of ischemic kidneys. Freeman et al. (1954) pioneered revascularization of ischemic kidneys by performing an aortic and bilateral renal artery thromboendarterectomy to treat hypertension. Aortorenal bypass quickly became the preferred method of renal revascularization, and angiography was relied on as the definitive test to predict blood pressure response to surgical correction.

The National Cooperative Study (Foster et al., 1975) reviewed the efficacy of surgically correcting renal artery lesions to cure hypertension. Despite a 51% cure rate, the results revealed a disappointing 34% failure rate and an unacceptable operative mortality approaching 10%. The patients at highest risk for operative morbidity and mortality were those with concurrent coronary or cerebrovascular disease or with bilateral renovascular disease often associated with azotemia.

In the past decade the results of surgical renal revascularization have improved owing to better patient selection, which can be attributed to a better understanding of the natural history of renal artery disease, to a better understanding of the physiology and pathophysiology of the renin-angiotensin-aldosterone system, and to the development of pharmacologic probes that serve as highly reliable screening tests. There is growing recognition that patients with nonfunctioning kidneys and total occlusion of the renal artery need not be subjected to nephrectomy for control of the hypertension (Libertino et al., 1980). These patients are likely to have recovery of renal function and a blood pressure benefit following renal revascularization if patent distal vessels have been continually fed by a rich network of hilar collateral vessels. In these instances, a delayed nephrogram can be demonstrated angiographically despite complete main renal artery occlusion, and, at surgery, renal biopsy will frequently reveal an intact glomerular morphology, predicting a successful revascularization (Libertino et al., 1980).

The previously unacceptable operative morbidity and mortality associated with renal revascularization has been substantially improved by careful patient selection and preparation. Novick and colleagues (1981) evaluated their RVH patients preoperatively for the often-found coexistence of coronary or cerebral vascular disease. Cardiac and cerebral revascularizations are performed, when appropriate, before renal revascularization. Utilizing this protocol, operative morbidity and mortality have been greatly reduced. Novick et al. reported a 2% operative mortality, while achieving a 91% blood pressure benefit in 100 consecutive renal revascularizations for atherosclerotic renovascular hypertension.

Aortorenal bypass with an autogenous vascular graft is the surgical treatment of choice for renal revascularization. The saphenous vein is most commonly employed. Patients with branch renal artery stenoses can be treated with multiple branch grafts employing microsurgical techniques in vivo or with ex vivo bench surgery (Novick, 1984). Branch intraparenchymal stenoses may also be amenable to intraoperative dilation with rigid dilators (Fig 25–11) or angiographic-type balloon catheters introduced through the main renal artery. Most patients with bilateral disease require repair of only one side (Novick, 1984; Textor et al., 1985). If a bilateral repair is planned, it is most safely carried out as a staged procedure.

In patients with severe aortoiliac disease, it is possible to revascularize the kidneys without performing a complete aortic replacement. Libertino et al. (1980) developed and described hepatorenal and splenorenal bypass operations to revascularize the right and left kidneys, respectively (Chibaro et al., 1984). The splenorenal bypass requires patent celiac and splenic arteries. A singular end-to-end vascular anastomosis is fashioned (Fig 25–12). The spleen is preserved and is nourished by collateral short gastric arteries.

The hepatic artery is ideal for bypass to the right kidney, as this vessel is rarely involved by atherosclerotic disease. The most common method of hepatorenal bypass is interposition of a saphenous vein graft, end-to-side to the common hepatic artery, just beyond the gastroduodenal artery (Fig 25–13). If the right hepatic ar-

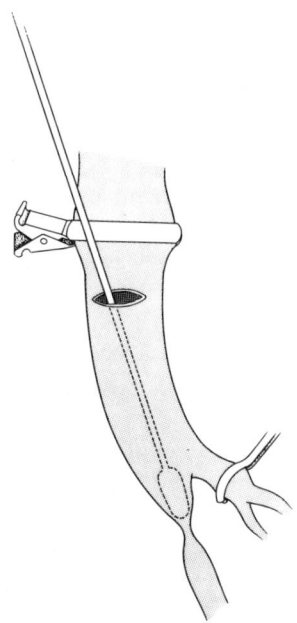

FIG 25–11.
Intraoperative dilatation of a stenosed renal artery branch with a rigid dilator.

tery is utilized, the gallbladder becomes ischemic, and an adjunctive cholecystectomy is necessary (Fig 25–14). An end-to-end gastroduodenal-renal anastomosis with saphenous vein interposition is also feasible (Fig 25–15), although the gastroduodenal artery is difficult to mobilize. Follow-up studies of 1–8 years (mean, four years) after hepatorenal bypass are very encouraging (Chibaro et al., 1984). The operation was successful in 33/36 patients (92%). There was no evidence of permanent hepatic impairment, graft stenosis, or late complications within this period of follow-up.

Renal autotransplantation and iliorenal bypass with a long saphenous vein graft are useful alternative tech-

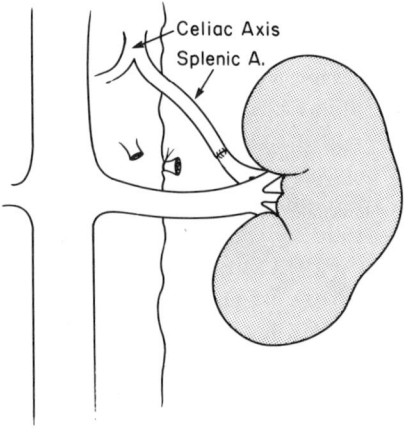

FIG 25–12.
End-to-end splenorenal bypass to the left kidney.

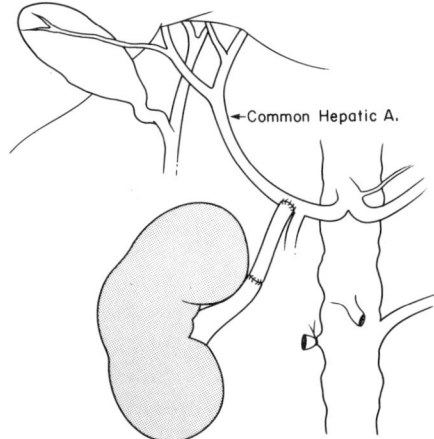

FIG 25–13.
End-to-side hepatorenal bypass from the common hepatic artery to the right renal artery with a saphenous vein graft interposition.

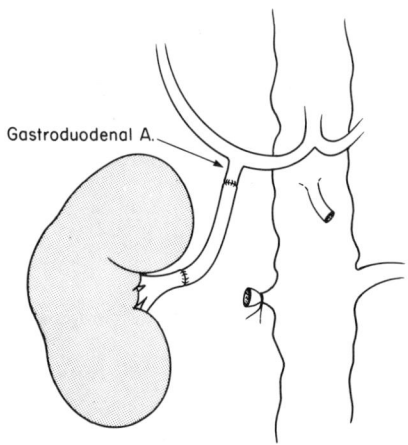

FIG 25–15.
End-to-end gastroduodenal-renal artery anastomosis with a saphenous vein interposition.

niques for renal revascularization in patients with severe aortic atherosclerosis and absence of severe iliac disease (Figs 25–16 and 25–17). However, the atherosclerotic process may progress to involve the iliac vessels and compromise the blood flow of the revascularized kidney. Accordingly, Novick et al. (1984) recommend that renal autotransplantation be considered only when a splenorenal or hepatorenal bypass cannot be performed. Early results following renal revascularization certainly are very favorable. However, data on the durability of vascular anastomoses and the long-term blood pressure benefit are more limited. Dean (1984) reviewed the 20-year experience at Vanderbilt University with aortorenal bypass. He found that blood pressure was cured or improved in 96% of

patients with fibromuscular renal artery disease and in 78% of patients operated on for atheromatous renal artery disease. The beneficial effect of operative treatment was maintained in this group over a follow-up period of up to 23 years. Dean noted that the fate of a bypass graft is determined by several factors. These include the graft material, its length, and the flow through the conduit, as well as the surgical precision of the anastomosis. A short graft length and high flow rate are favorable for long-term patency. An autogenous artery would seem the ideal graft to use.

Wylie and colleagues utilized the hypogastric artery for renal revascularization (Wylie, 1975). They reported a 98% patency rate and no late degenerative changes at

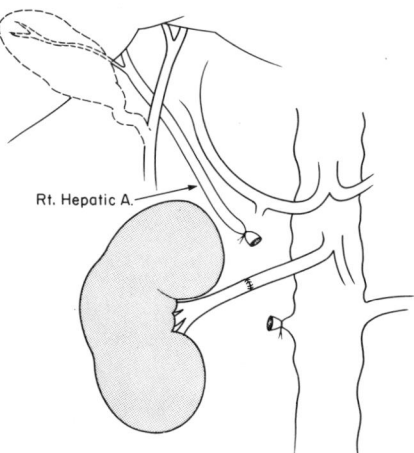

FIG 25–14.
End-to-end hepatorenal bypass from the right hepatic artery to the right renal artery. Note an adjunctive cholecystectomy is necessary as the gallbladder loses its nutrient artery.

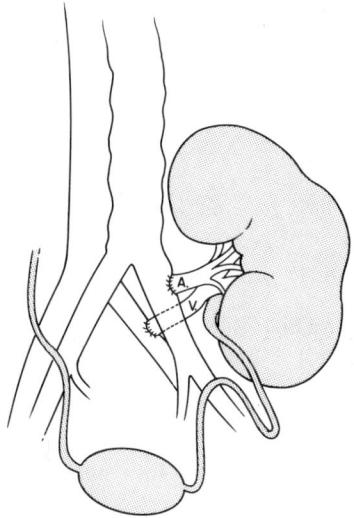

FIG 25–16.
Renal revascularization by autotransplantation leaving the ureter intact.

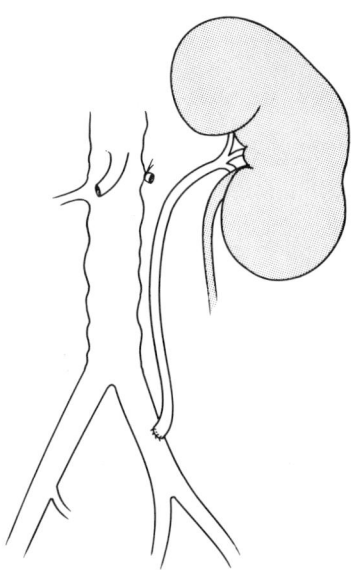

FIG 25–17.
Renal revascularization by ileorenal bypass via a long saphenous vein graft.

a five-year follow-up. The hypogastric autograft has particular appeal in the pediatric patient, since the graft enlarges with age in proportion to the increased demand for flow. However, the iliac arteries are not immune from the fibromuscular and atherosclerotic diseases involving the renal arteries and may succumb to these same disease processes, or may already be involved at the time that renal revascularization is planned.

Venous autografts are structurally weaker than arterial autografts and represent an increased risk for aneurysmal degeneration. The saphenous vein has been frequently used in aortorenal revascularization. There is a 3%–7% degeneration in follow-up of at least one year (Ernst et al., 1973; Dean, 1984). Dacron was previously employed as a synthetic graft, but recently polytetrafluoroethylene (Gore-Tex) grafts are preferred owing to greater ease in suturing. Synthetic grafts also have the advantage of being available in a variety of sizes. However, they are subject to anastomotic aneurysmal degeneration and stenosis.

MEDICAL MANAGEMENT OF RENOVASCULAR HYPERTENSION

Medical management of RVH has been unsatisfactory in the past. Hunt et al. (1974) followed up 214 patients with RVH for 7–14 years. One hundred patients were selected for surgical therapy after failing three months of medical therapy. Patients treated medically had a higher mortality (40% vs. 16%) and overall less effective control of their blood pressure than their surgically treated counterparts. Dean (1981) treated 41 patients

with RVH medically. He found that the arterial stenosis progressed in 41% of patients and that progression to complete occlusion occurred in 12%. In all fairness to the medical management of RVH, it must be pointed out that the drugs used in these studies were far less specific to control a hyperactive renin-angiotensin system than the β-blockers and CEIs available today. These newer drugs have been found to be more effective in controlling blood pressure in patients with fibromuscular dysplasia than in those with atherosclerotic renovascular disease (Vidt, 1984). Patients in the latter group tend to be older, have more target organ damage, and require more drugs for blood pressure control than the younger patients with fibromuscular disease.

Despite the greater specificity of these drugs for the renin-angiotensin system, their safety and efficacy in the treatment of RVH remain in doubt. Many reports have documented the onset of reversible decrease in renal function associated with use of a CEI in patients with bilateral renal artery stenosis or stenosis of the renal artery to a solitary kidney (Curtis et al., 1984; Blythe, 1984). The decrement in renal function does not appear to be entirely due to the depressor effect of these drugs, since lowering the blood pressure to the same degree with a drug that does not interfere with the RAS does not usually lead to a comparable decrease in renal function. Micropuncture studies suggest that at low perfusion pressures the glomerular capillary hydrostatic pressure is increased by an angiotensin II-dependent constriction of the efferent arteriole (Hall et al., 1981). Inhibition of angiotensin II formation in the setting of a decreased perfusion pressure lowers glomerular filtration. Discontinuation of the CEI restores renal function to baseline.

Unfortunately, the drug-induced decrease in renal function is not limited to patients with bilateral renal artery disease. Experimental administration of CEIs in 2K,1C animals reveals that the ischemic kidney suffers a severe decline in GFR, while the contralateral kidney remains unaffected. Clinically, Wenting et al. (1984) demonstrated by use of nuclear renal scans significant impairment of renal function in the ischemic kidney of patients with unilateral renal artery stenosis treated with captopril.

Textor et al. (1985) argued against medical management for RVH beyond the use of CEIs, possibly to implicate a wide range of antihypertensive medications. In eight patients with bilateral renal artery disease in which both renal arteries had luminal stenoses of 75% or greater, decreasing the blood pressure from a baseline of approximately 200/100 mm Hg to 140/85 mm Hg by graded increases in IV nitroprusside produced significant decreases in total renal plasma flow and GFR. Nitroprusside is known to increase renin secretion, and therefore interference with the autoregulatory actions of AII cannot be implicated. It appears that renal units whose blood flow is already compromised cannot tolerate a reduction in blood pressure to clinically relevant levels.

Four of those eight patients underwent unilateral renal revascularization of the more ischemic kidney. Although renal function per se was not significantly improved by surgery, repeated challenge with graded nitroprusside infusions, attaining the same lower levels of blood pressure, did not lower renal plasma flow or GFR. This observation strongly suggests that revascularization may protect the kidneys from loss of renal function when the blood pressure is lowered to clinically desirable levels by antihypertensive medicines (Textor et al., 1984).

Finally, it must be borne in mind that RVH is a vascular disease whose manifestations can include loss of renal function and renin-dependent hypertension. Pharmacologic treatment of the hypertension does not address the cause of the problem. However, despite the superior results generally achieved by the revascularization of a renal artery stenosis, there is a role for medical management in the treatment of RVH. Medical management must be employed in patients who have not benefited from angioplasty or surgical revascularization, in patients who are not candidates for intervention because of the presence of other serious diseases, and in those who have refused intervention. Whatever the case, all patients with RVH who are treated pharmacologically must be carefully and routinely evaluated for changes in renal function and size.

REFERENCES

1. Aguilera G, Catt K: Regulation of vascular angiotensin II receptors in the rat during altered sodium intake. *Circ Res* 1981; 49:751–758.
2. Ayers CR, Harris RH Lefer LG: Control of renin release in experimental hypertension. *Cir Res* 1969; 24/25 (suppl 1):103.
3. Bakhle YS: Conversion of angiotensin I to angiotensin II by cell-free extracts of dog lung. *Nature* 1968; 220:919.
4. Bengis RG, Coleman TG, Young DB, et al: Long term blockade of angiotensin formation in various normotensive and hypertensive rat models using converting enzyme inhibitor (SE 14225). *Circ Res* 1978; 43(suppl 1):145.
5. Blake WD, Wegria R, Ward HP, et al: Effect of renal arterial constriction on excretion of sodium and water. *Am J Physiol* 1950; 163:422.
6. Blythe WB: Captopril and renal autoregulation. *N Engl J Med* 1983; 308:390.
7. Braun-Menendez E, Fasciolo JC, Leloir LR, et al: The substance causing renal hypertension. *J Physiol* 1940; 98:283.
8. Bright: Causes and observations, illustrative of renal disease accompanied with the secretion of albuminous urine. *Surg Hosp Rep* 1836; 1:338.
9. Brunner HR, Kirshmann JD, Sealey JE, et al: Hypertension of renal orgin: Evidence for two different mechanisms. *Science* 1971; 174:1344.
10. Brunner HR, Laragh JH, Baer L, et al: Essential hypertension: Renin and aldosterone, heart attack, and stroke. *N Engl J Med* 1972; 286:441–449.
11. Brunner HR, Gavras H, Laragh JH, et al: Angiotensin II blockade in man by SAR¹-Ala⁸-angiotensin II for understanding and treatment of high blood pressure. *Lancet* 1973; 2:1045.
12. Brunner DB, Desponds G, Biollaz J, et al: Effect of a new angiotensin converting enzyme inhibitor MK421 and its lysine analogue on the components of the renin-angiotensin-system in healthy subjects. *Br J Clin Pharmacol* 1981; 11:461–467.
13. Brunner HR, Biollaz J, Waeber B, et al: Experience with enalapril in normotensive volunteers and in patients with hypertension and congestive heart failure, in Doyle AE, Bearn AG (eds): *Hypertension and the Angiotensin System.* New York, Raven Press, 1983, pp 193–210.
14. Carey RM, Reid RA, Ayers CR, et al: The Charlottesville blood pressure survey: Value of repeat blood pressure measurement to define the prevalence of labile and sustained hypertension. *JAMA* 1976; 236:847.
15. Carey RM, Vaughan ED Jr, Ackerly JA, et al: The immediate pressor effect of saralasin in man. *J Clin Endocrinol Metab* 1978; 46:36.
16. Case DB, Wallace JM, Keim HJK, et al: Usefulness and limitations of saralasin, a weak competitive agonist for angiotensin II, for evaluating the renin and sodium factor in hypertensive patients. *Am J Med* 1976a; 60:825.
17. Case DB, Wallace JM, Keim HJ, et al: Estimating renin participation in hypertension: Superiority of converting enzyme inhibitor over saralasin. *Am J Med* 1976b; 61:790.
18. Case DB, Atlas SA, Laragh JH, et al: Clinical experience with blockade of the renin-angiotensin-aldosterone system by an oral converting enzyme inhibitor (SQ 14225, captopril) in hypertensive patients. *Prog Cardiovasc Dis* 1978; 21:195.
19. Case DB, Laragh JH: Reactive hyperreninemia following angiotensin blockade with either saralasin or converting enzyme inhibitor: A new approach to screen for renovascular hypertension. *Ann Intern Med* 1979a; 91:153.
20. Case DB, Atlas SA, Laragh JH: Reactive hyperreninemia to angiotensin blockade identified renovascular hypertension. *Clin Sci* 1979b; 57(suppl):313.
21. Case DB, Atlas SA, Laragh JH: Physiologic effects and blockade, in Laragh JH, Bjhler FR, Seldin DW (eds): *Frontiers in Hypertension Research.* New York, Springer-Verlag, 1982, pp 541–550.
22. Cheung HS, Cushman DW: Inhibition of homogenous angiotensin-converting enzyme of rabbit lung by synthetic venom peptides of Bothrops jararaca. *Biochem Biophys Acta* 1973; 293:451.
23. Chibaro EA, Libertino JA, Novick AC: Use of hepatic circulation for renal revascularization. *Ann Surg* 1984; 199:406.

24. Cohn DJ, Sos TA, Saddekni S, et al.: Transluminal angioplasty for atherosclerotic renal artery stenosis. *Semin Intervent Radiol* 1984; 1:279.

25. Colapinto RF, Stronell RD, Harries-Jones EP, et al: Percutaneous transluminal dilatation of the renal artery: Follow-up studies on renovascular hypertension. *AJR* 1982; 139:727–732.

26. Curtis JJ, Luke RG, Whelchel JD, et al: Inhibition of angiotensin-converting enzyme in renal-transplant recipients with hypertension. *N Engl J Med* 1983; 308:377.

27. Dean RH, Kieffer RW, Smith BM, et al: Renovascular hypertension: Anatomic and functional changes during drug therapy. *Arch Surg* 1981; 116:1408.

28. Dean RH: Renovascular hypertension: An overview, in Rutherford RB (ed): *Vascular Surgery.* Philadelphia, WB Saunders Co, 1984.

29. Dibona GF: The functions of the renal nerves. *Rev Physiol Biochem Pharmacol* 1982; 94:76–91.

30. Ernst CB, Stanley JC, Marshall FF, et al: Renal revascularization for atherosclerotic renovascular hypertension: Prognostic implications of focal renal arterial versus overt generalized arteriosclerosis. *Surgery* 1973; 73:859.

31. Eyler WR, Clark MD, Garman JE, et al: Angiography of the renal areas including a comparative study of renal arterial stenosis in patients with and without hypertension. *Radiology* 1962; 78:879.

32. Fitzsimons JT: Angiotensin, thirst, and sodium appetite: Retrospect and prospect. *Fed Proc* 1978; 37:2669–2675.

33. Flechner SM: Percutaneous transluminal dilatation: A realistic appraisal in patients with stenosing lesions of the renal artery. *Urol Clin North Am* 1984; 11:515.

34. Foster JH, Maxwell MH, Franklin SS, et al: Renovascular occlusive disease: Results of operative treatment. *JAMA* 1975; 231:1043.

35. Freeman NE, Leeds FM, Elliot W: Thromboendarterectomy for hypertension due to renal artery occlusion. *JAMA* 1954; 156:1077.

36. Frolich ED: The adrenergic nervous system and hypertensive state of the art. *Mayo Clin Proc* 1977; 52(6):361–368.

37. Fumio G, Jackson EK, Branch RA, et al: The effects of renal perfusion pressure on angiotensin II-induced changes in glomerular filtration rate. *Pharmacologist* 1983; 25:344.

38. Gavras H, Brunner HR, Vaughan ED Jr, et al: Angiotensin-sodium interaction in blood pressure maintenance of renal hypertensive and normotensive rats. *Science* 1973; 180:1369.

39. Goldblatt H, Lynch J, Hanzal RF, et al: Studies on experimental hypertension: I. The production of persistent elevation of systolic blood pressure by means of renal ischemia. *J Exp Med* 1934; 59:347.

40. Gruntzig A, Hopff H: Perkutane rekanalisation chronischer arterieller Verschlusse mit einem neuen Dilatationskatheter: Modifikation der Dotter-Technik. *Deutsch Med Wochenschr* 1974; 99:2502–2511.

41. Hall JE, Guyton AC, Cowley AW Jr: Dissociation of renal blood flow and filtration rate autoregulation by renin depletion. *Am J Physiol* 1977; 232:F215–F221.

42. Hall JE, Guyton AC, Jackson TE, et al: Control of glomerular filtration rate by renin-angiotensin system. *Am J Physiol* 1977b; 233:F366–F372.

43. Hall JE, Guyton AC, Salgado HC, et al: Renal hemodynamics in acute and chronic angiotensin II hypertension. *Am J Physiol* 1978; 235:F174–F179.

44. Hall JE, Coleman TG, Guyton AC, et al: Control of glomerular filtration rate by circulating angiotensin II. *Am J Physiol* 1981; 241:R190–R197.

45. Harvey RJ, Krumlovsky F, delGroco F, et al: Screening for renovascular hypertension. *JAMA* 1985; 254:388.

46. Hillman BJ, Smith JRL, Pond GD, et al: Photoelectric radiology, in Hanafec WN, Wilson GH (eds): *Radiology.* New York, John Wiley & Sons, 1983.

47. Holley KE, Hunt JC, Brown AL, et al: Renal artery stenosis: A clinical-pathologic study in normotensive patients. *Am J Med* 1964; 37:14.

48. Horn ML, Conklin VM, Keenan RE, et al: Angiotensin II profiling with saralasin: Summary of the Eaton Collaborative Study. *Kidney Int* 1979; 9(suppl):S115.

49. Howard JE, Conner TB: Use of differential renal function studies in the diagnosis of renovascular hypertension. *Am J Surg* 1954; 107:58.

50. Howard JE, Berthrong N, Gould D, et al: Hypertension resulting from unilateral renovascular disease and its relief by nephrectomy. *Johns Hopkins Med J* 1954; 94:51.

51. Hunt JC, Sheps SG, Harrison EG Jr, et al: Renal and renovascular hypertension: A reasoned approach to diagnosis and management. *Arch Intern Med* 1974; 133:988–999.

52. Jackson EK, Branch RA, Margolius HS, et al: Physiological functions of the renal prostaglandin, renin, and kallikrein systems, in Seldin DW, Giebisch G (eds): *The Kidney: Physiology and Pathophysiology.* New York, Raven Press, 1985, p 613.

53. Johnson JA, Davis JO, Witty RT: Effects of catecholamines and renal nerve stimulation on renin release in the non-filtering kidney. *Circ Res* 1971; 29:646.

54. Judson WE, Helmer OM: Diagnostic and prognostic values of renin activity in renal venous plasma in renovascular hypertension. *Hypertension* 1965; 13:79–89.

55. Kotchen TA, Galla JH, Luke RG: Contribution of chloride to the inhibition of plasma renin by sodium chloride in the rat. *Kidney Int* 1978; 13:201.

56. Laragh JH: Basic principles for the office evaluation and treatment of high blood pressure: Part I. *Cardiol Rev Rep* 1981; 2:1318.

57. Leadbetter WF, Burkland CF: Hypertension in unilateral renal disease. *J Urol* 1938; 39:611.

58. Levens NR, Peach MJ, Carey RM: Role of the intrarenal renin-angiotensin system in the control of renal function. *Circ Res* 1981; 49:157.

59. Liard JF, Cowley AW, McCaa RE, et al: Renin, aldosterone, body fluid volumes, and the baroreceptor reflex in the development and reversal of Goldblatt hypertension in conscious dogs. *Circ Res* 1974; 34:549.

60. Libertino JA, Zinman L, Breslin DJ, et al: Renal artery revascularization: Restoration of renal function. *JAMA* 1980; 244:1340.

61. Lyons DF, Streck WF, Kem DC, et al: Captopril stimulation of differential renins in renovascular hypertension. *Hypertension* 1983; 5:65.

62. Mann FC, Herrick JF, Essex HE, et al: The effect of the blood flow of decreasing lumen of a blood vessel. *Surgery* 1938; 4:249.

63. Marks LS, Maxwell MH: Renal vein renin value and limitations in the prediction of operative results. *Urol Clin North Am* 1975; 2:311.

64. Martin LG, Price RB, Casarella WJ, et al: Percutaneous angioplasty in clinical management of renovascular hypertension: Initial and long-term results. *Radiology* 1985; 155:629–633.

65. Maxwell MH, Lupu AN: Excretory urogram in renal arterial hypertension. *J Urol* 1968; 100:395–406.

66. Maxwell MH, Lupu AN, Taplin GV: Radioisotope renogram in renal arterial hypertension. *J Urol* 1968; 100:376.

67. McNeil BJ, Adelstein SJ: Measures of clinical efficacy: The value of case finding in hypertensive renovascular disease. *N Engl J Med* 1975; 293:211.

68. McNeil BJ, Varady PD, Burrows BA: Measures of clinical efficacy: Cost-effectiveness calculations in the diagnosis and treatment of hypertensive renovascular disease. *N Engl J Med* 1975; 293:216.

69. Miller ED Jr, Samuels AI, Haber E: Inhibition of angiotensin conversion in experimental renovascular hypertension. *Science* 1972; 177:1108.

70. Muller FB, Sealey JE, Case DB, et al: The captopril test for identifying renovascular disease in hypertensive patients. *Am J Med* 1986; 80:633.

71. Novick AC, Straffon RA, Stewart BH, et al: Diminished operative morbidity and mortality in renal revascularization. *JAMA* 1981; 256:749.

72. Novick AC, Khauli RB, Vidt DG: Diminished operative risk and improved results following revascularization for atherosclerotic renovascular disease. *Urol Clin North Am* 1984; 11(3):435–449.

73. Novick AC: Microvascular reconstruction of complex branch artery disease. *Urol Clin North Am* 1984; 11:465.

74. Novick AC, Textor SC, Bodie B, et al: Revascularization to preserve renal function in patients with atherosclerotic renovascular disease. *Urol Clin North Am* 1984; 11:477.

75. Osborne RW Jr, Goldstone J, Hillman BJ, et al: Digital video subtraction angiography: Screening technique for renovascular hypertension. *Surgery* 1981; 90:932.

76. Page IH, Helmer OM: A crystalline pressor substance (angiotonin) resulting from the reaction between renin and renin activator. *J Exp Med* 1940; 71:29.

77. Pals DT, Masucci FD, Denning GS Jr, et al: Role of the pressor action of angiotensin II in experimental hypertension. *Circ Res* 1971; 29:673.

78. Parrott TS, Woodard JR, Trulock TS, et al: Segmental renal vein renins and partial nephrectomy for hypertension. *J Urol* 1984; 131:736.

79. Pickering G: Hypertension definitions, natural histories and consequences, in Laragh JH (ed): *Hypertension Manual.* New York, Yorke Medical Books, 1973.

80. Pickering TG, Case DB, Sos TA, et al: Unilateral suppression of renin secretion in patients with bilateral renal artery stenosis. *Clin Res* 1983.

81. Pickering TG, Sos TA, Vaughan ED Jr, et al: Predictive value and changes of renin secretion in hypertensive patients with unilateral renovascular disease undergoing successful renal angioplasty. *Am J Med* 1984; 76:398–404.

82. Pickering TG, Sos TA, James GD, et al: Comparison of renal vein renin activity in hypertensive patients with stenosis of one or both renal arteries. *J Hypertension* 1985; 3:1–3.

83. Pratchett AA: Design of enalapril, in Doyle AE, Bearn AG (eds): *Hypertension and the Angiotensin System.* New York, Raven Press, 1983, pp 155–165.

84. Ratliff NB: Renal vascular disease: Pathology of large blood vessel disease. *Am J Kidney Dis* 1985; 5:893.

85. Re R, Noveline R, Escourrou M-T, et al: Inhibition of angiotensin-converting enzyme for diagnosis of renal-artery stenosis. *N Engl J Med* 1978; 298:582–586.

86. Richardson D, Stella A, Geonetti G, et al: Mechanisms of renal release of renin by electrical stimulation of the brainstem in the cat. *Circ Res* 1974; 34:425–434.

87. Robb GP, Steinberg I: Visualization of the chambers of the heart, the pulmonary circulation and the great vessels in man. *AJR* 1939; 41:1.

88. Rowland M, Roberts J: Blood pressure levels and hypertension in person 6–74 years: United States, 1976–80. Advance Data from Vital and Health Statistics, No. 84. Hyattsville, Md, US Department of Health and Human Services, Public Health Service, National Center for Health Statistics. Oct 8, 1982, DHHS publication no. (PHS) 82-1250.

89. Schambelan M, Glickman M, Stockigt JR, et al: The selective renal vein renin sampling in hypertensive patients with segmental renal lesions. *N Engl J Med* 1974; 290:1153.

90. Schreiber MJ, Pohl MA, Novick AC: The natural history of atherosclerotic and fibrous renal artery disease. *Urol Clin North Am* 1984; 11(3):383–392.

91. Schwarten DE, Yune HY, Klatte EC, et al: Clinical experience with percutaneous transluminal angioplasty (PTA) of stenotic renal arteries. *Radiology* 1980; 135:601–604.

92. Sealey JE, Buhler FR, Laragh JH, et al: The physiology of renin secretion in essential hypertension: Estimation of renin secretion rate and renal plasma flow from peripheral and renal vein renin levels. *Am J Med* 1973; 55:391–401.

93. Selkurt EE: The effect of pulse pressure and mean arterial pressure modification of renal hemodynamics and electrolyte and water excretion. *Circulation* 1951; 4:541.

94. Simon N, Franklin SS, Bleifer KH, et al: Clinical characteristics of renovascular hypertension. *JAMA* 1972; 220:1209.

95. Skeggs LT Jr, Lentz KE, Kahn JR, et al: Amino acid sequence of hypertension II. *J Exp Med* 1956; 104:193.

96. Sos TA, Pickering TG, Phil D, et al: Percutaneous transluminal renal angioplasty in renovascular hypertension

due to atheroma or fibromuscular dysplasia. *N Engl J Med* 1983; 309:274–279.

97. Sosa RE, Vaughan ED Jr: Evaluation of surgically curable hypertension. *Am Urol Assoc Update* 1983; 2(32).

98. Stamey TA, Nudelman IJ, Good TH, et al: Functional characteristics of renovascular hypertension. *Medicine* 1961; 40:347.

99. Stamey TA: *Renovascular Hypertension*. Baltimore, Williams & Wilkins Co, 1963.

100. Stewart BH: Renovascular hypertension: Surgical treatment. *Monogr Urol* 1981; 5:3–36.

101. Steinberg I, Marshall VG: Intravenous abdominal aortography in urologic diagnosis. *J Urol* 1961; 86:456.

102. Stockigt JR, Noakes CA, Collins RD, et al: Renal-vein renin in various forms of renal hypertension. *Lancet* 1972; 1:1194.

103. Tegtmeyer CJ, Dyer R, Teates CD, et al: Percutaneous transluminal dilatation of the renal arteries. *Radiology* 1980; 135:589–599.

104. Tegtmeyer CJ, Tegtmeyer VL, Kellum CD, et al: Percutaneous transluminal angioplasty: The treatment of choice for vascular lesions caused by fibromuscular dysplasia. *Semin Intervent Radiol* 1984; 1:289.

105. Textor SC, Tarazi RC, Novick AC, et al: Regulation of renal hemodynamics and glomerular filtration in patients with renovascular hypertension during converting enzyme inhibition with captopril. *Am J Med* 1984; 76:29–37.

106. Textor SC, Novick AC, Tarazi RC, et al: Critical perfusion pressure for renal function in patients with bilateral atherosclerotic renal vascular disease. *Ann Intern Med* 1985; 102:308–314.

107. Thibonnier M, Joseph A, Sassano P, et al: Improved diagnosis of unilateral renal artery lesions after captopril administration. *JAMA* 1984; 251:56–60.

108. Thompson JE, Smithwick RH: Human hypertension due to unilateral renal disease, with special reference to renal artery lesions. *Angiology* 1957; 3:493.

109. Thornbury JR, Stanley JC, Fryback DG: Hypertensive urogram: A nondiscriminatory test of renovascular hypertension. *AJR* 1982; 138:43–49.

110. Tigerstedt R, Bergmann TG: Niere und kreislauf. *Skand Arch Physiol* 1898; 8:233.

111. Vaughan ED Jr: Renovascular hypertension, in Harrison JH (ed): *Campbell's Urology* ed 5. Philadelphia, WB Saunders Co, 1986, chap 58.

112. Vaughan ED Jr, Buhler FR, Laragh JH, et al: Renovascular hypertension; renin measurements to indicate hypersecretion and contralateral suppression, estimate renal plasma flow and score for surgical curability. *Am J Med* 1973; 55:402.

113. Vaughan ED Jr, Carey RM, Ayers CR, et al: A physiologic definition of blood pressure response to renal revascularization in patients with renovascular hypertension. *Kidney Int* 1979; 15:S83.

114. Vaughan ED Jr, Sos TA, Sniderman KW, et al: Renal vein renin secretory patterns before and after transluminal angioplasty in patients with renovascular hypertension: Verification of analytic criteria, in Laragh JH (ed.): *Frontiers in Hypertension Research*. New York, Springer-Verlag, 1981.

115. Vaughan ED Jr, Case DB, Pickering TG, et al: Clinical evaluation of renovascular hypertension and therapeutic decisions. *Urol Clin North Am* 1984; 11(3):393–407.

116. Wallace ECH, Morton JM: Chronic captopril infusion in two-kidney, one clip rats with normal plasma renin concentration. *J Hypertension* 1984; 2:285–289.

117. Weed WC, Vaughan ED Jr, Peach MJ: Prolongation of the saralasin responsive state of two-kidney, one-clip hypertension in the rat by the orally administered converting enzyme inhibitor captopril (SQ 14225). *Hypertension* 1979; 1:8.

118. Wenting GJ, Tan-tijong HL, Derkx FHM, et al: Split renal function after captopril in unilateral renal artery stenosis. *Br Med J* 1984; 288:886–890.

119. Wylie EJ: Endarterectomy and autogenous arterial grafts in the surgical treatment of stenosing lesions of the renal artery. *Urol Clin North Am* 1975; 2:351.

120. Ying CY, Tifft CP, Gavras H, et al: Renal revascularization in the azotemic hypertensive patient resistant to therapy. *N Engl J Med* 1984; 311:1070–1075.

Chapter 26

The Ureter*

Robert M. Weiss, M.D.
Bo L.R.A. Coolsaet, M.D.

The function of the upper urinary tract is to transport urine from the minor calyces toward the bladder and to protect the renal parenchyma and cranial portions of the urinary tract from distally generated backflow and back pressure. The upper urinary tract manifests peristaltic activity and can adapt its mechanical activity to changes in diuresis and alterations in the state of the bladder. Peristaltic activity begins with the origin of electrical activity at pacemaker sites situated in the proximal portion of the renal collecting system (Bozler, 1942; Weiss et al., 1967; Gosling and Dixon, 1974; Constantinou, 1974; Tsuchida and Yamaguchi, 1977). The electrical activity is transmitted distally from cell to cell and gives rise to the ureteral contraction wave, which propels the bolus of urine distally. Efficient collection and propulsion of urine is dependent on the passive and active properties of the ureter and on the ability of the ureter to coapt its walls completely (Woodburne and Lapides, 1972). Urine passes into the bladder via the ureterovesical junction (UVJ), which permits antegrade transport of urine from the ureter into the bladder and prevents retrograde passage of urine from the bladder into the ureter.

ANATOMY

GROSS ANATOMY

The ureter is a 25–30-cm tube extending from the renal pelvis to the bladder. The three narrowest areas of the ureter are (1) at the ureteropelvic junction, (2) at the site where the ureter crosses the iliac vessels, and (3) at the intramural portion of the distal ureter. It is at these sites that calculi most frequently become impacted.

As the ureter descends in a medial direction from the

kidney, it lies on the psoas muscle in close apposition to the peritoneum to which it is attached. In its abdominal course, both ureters are crossed anteriorly by the gonadal vessels, and they in turn cross anterior to the genitofemoral nerve (Fig 26–1,A). On the right side the descending portion of the duodenum usually lies in front of the ureter, and more distally the right ureter is crossed anteriorly by the right colic and ileocolic vessels in the root of the mesentery (Fig 26–1,B). The appendix may overlie the right ureter. On the left side the ureter is crossed anteriorly by the left colic vessels, and as it passes over the pelvic brim it passes behind the sigmoid colon.

As the ureters descend into the true pelvis they pass over the terminal portions of the common or the first portion of the external iliac arteries (Fig 26–1,C). At this point, the two ureters are only about 5 cm apart. After crossing into the pelvis, the ureters first diverge laterally and then converge medially toward the trigone. In the male, the pelvic ureter passes anterior to the internal iliac (hypogastric) artery and then just before it enters the bladder it is crossed anteriorly by the vas deferens (Fig 26–1,D). In the female the pelvic ureter passes anterior to the internal iliac artery, below the root of the broad ligament of the uterus, and then runs along the lateral aspect of the cervix, passing under the uterine artery. The ureter is crossed by the obliterated umbilical vessels in both males and females.

BLOOD SUPPLY

The ureter is supplied by a variable number of segmental arteries that arise from the aorta and from a variety of its branches or subbranches, including the renal, gonadal, adrenal, common iliac, internal iliac, external iliac, superior vesical, inferior vesical, vesiculodeferential, uterine, obturator, gluteal, vaginal, and middle hemorrhoidal arteries (Daniel and Shackman, 1952). As these segmental vessels reach the ureter,

---

*The original work was supported in part by Public Health Service grant AG 00112 from the National Institute on Aging.

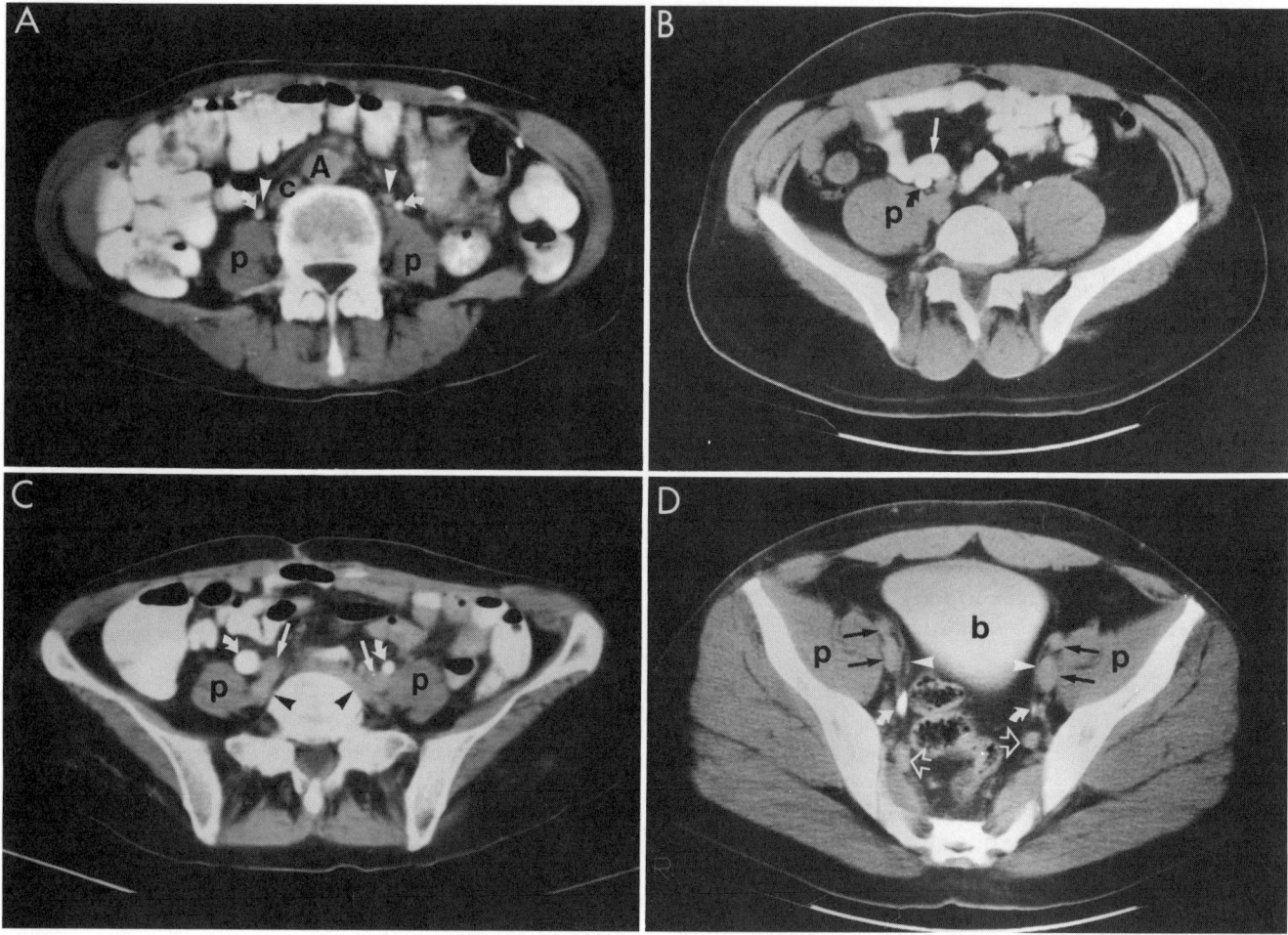

FIG 26–1.
CAT scans: **A,** transverse section showing ureters *(curved arrows)* lying on top of psoas muscle *(p)*. The gonadal vessels *(arrowhead)* are located anterior and medial to the ureter. *A* indicates aorta; *c,* vena cava. **B,** close relationship between right ureter *(curved arrow)* and bowel *(straight arrow)*. *p* indicates psoas muscle. Right ureter is dilated. **C,** ureters *(curved arrows)* passing over common iliac arteries *(straight arrows)* and veins *(arrow heads)*. *p* indicates psoas muscle. Ureters are dilated. **D,** ureters *(curved arrows)* are crossed anteromedially by vas deferens *(arrowhead)*. *Straight arrows* show external iliac arteries and veins; *open arrows* show hypogastric vessels. *p* indicates iliopsoas muscles; *b,* bladder.

they divide into ascending and descending branches that run in the adventitial layer of the ureter. The descending branches of proximally located segmental arteries anastomose with the ascending branches of more distally located segmental vessels, thus forming long, longitudinally running vascular channels. These anastomosing arteries may give off secondary branches that form arterial plexuses in the adventitial layer of the ureter. Tributaries from the adventitial plexus pierce the muscular coat and form delicate plexuses in the submucosal layer.

LYMPHATIC SUPPLY

Ureteral lymph vessels begin in communicating submucosal, intramuscular, and adventitial plexuses (Williams and Warwick, 1980). Lymphatics from the proximal ureter may join lymphatics of the kidney, which follow the course of the renal vein to end in the lateral aortic nodes, or they may drain directly into the lateral aortic nodes near the origin of the gonadal arteries. Lymphatics from the midureter drain into the lumbar nodes along the aorta and inferior vena cava and lymphatics from the lower ureter terminate in the common, external, and internal iliac glands.

NERVE SUPPLY

In a syncytial-type smooth muscle such as the ureter, there is a diffuse release of transmitter from nerve bundles, with the subsequent spread of excitation from one muscle cell to another (Burnstock, 1970). The lack of discrete neuromuscular junctions in such a system accounts for the difficulty in anatomically demonstrating the presence of ureteral innervation. Although nerve supply to the ureter has been described, the above anatomical configuration has left open the question as to whether these nerves supply the blood vessels of the ureter or the ureteral muscle itself.

The nerves to the ureter arise from the celiac, aorticorenal, and mesenteric ganglia and also from the aortic, superior hypogastric, and inferior hypogastric plexuses. The sympathetic fibers are from T11–12 and L1. The parasympathetic fibers to the upper ureter are derived from the vagus and those to the lower ureter are from S2–4.

CROSS-SECTIONAL ANATOMY

Histologically the ureter is composed of three layers: (1) an inner mucosal layer consisting of urothelium and its supporting lamina propria, (2) a muscular layer, and (3) an outer adventitial layer. The inner lining of the ureter or urothelium is composed of transitional-cell epithelium. In the contracted state the urothelium assumes a stellate appearance, with the lumen of the ureter being completely occluded. Beneath the urothelium and separating it from the muscular coat is the lamina propria, which contains both elastic and collagenous fibers.

Typical spindle-shaped smooth muscle cells, grouped together in compact bundles and rich in nonspecific cholinesterase, are observed to originate in the distal part of each minor calyx in man and to spread distally (Dixon and Gosling, 1982). These cells, which compose the outer muscle layer of the calyces and renal pelvis, are continuous with the muscular coat of the ureter.

The arrangement of the fibers in the muscular coat of the ureter has been variously described. A classic description is that the muscular coat consists of an inner and outer longitudinal layer separated by a circular layer (Copenhaver and Johnson, 1958). Satani (1919) noted that the musculature of the upper ureter ran haphazardly in all directions, while that of the lower ureter consisted of inner longitudinal and outer circular fibers. The most popular description is that the muscle bundles assume a helical or spiral configuration (Tanagho, 1971). Murnaghan (1958a) described the muscle bundles as a long spiral beginning as longitudinal strands in the outer region of the musculature, forming a middle circular layer as the spiral turn is made and terminating in a longitudinal inner layer. More recent studies have not confirmed this spiral arrangement. Gosling and Dixon (1985) noted that individual muscle bundles do not spiral around the ureter but rather form a complex meshwork of interweaving and interconnecting smooth muscle bundles without distinct longitudinal and circular layers. A recent scanning electron microscopic study of the guinea pig ureter showed a primarily circular arrangement in the upper ureter with a few outer longitudinal and obliquely oriented fibers, a highly irregular orientation in the midureter of interlacing muscle bundles, and predominantly longitudinally oriented muscle bundles with an underlying circular muscle coat in the lower ureter (Tachibana et al., 1985). The adventitia of the ureter is composed of areolar and fibroelastic connective tissue, which contains the blood vessels, lymphatics, and nerve fibers that subdivide, ramify, and enter the ureter proper.

CELLULAR ANATOMY

The primary functional anatomical unit of the ureter is the smooth muscle cell whose main function is to contract. The cell is extremely small, approximately 250–400 μm long and 5–7 μm in diameter, which permits a significant proportion of the calcium (Ca^{++}) involved in the contractile process to enter the cell from extracellular sources at the time of excitation. The nucleus of the cell is ellipsoid and contains a darkly staining body, the nucleolus, and the genetic material of the cell. Surrounding the nucleus is the cytoplasm or sarcoplasm, which contains the structures involved in cell function. In the cytoplasm, frequently in close approximation to the nucleus, are mitochondria that perform many of the nutritive functions of the cell (Fig 26–2). Also within the cytoplasm are lattice-shaped structures called the endoplasmic or sarcoplasmic reticulum. These structures, along with the mitochondria and certain sites on the internal surface of the cell membrane, serve as the site for internal storage of calcium, which plays an important role in the contraction of the smooth muscle cell.

Dispersed in the sarcoplasm are the contractile proteins, actin and myosin, that interact—depending on the local Ca^{++} concentration—to result in contraction or relaxation. The actin is dispersed throughout the sarcoplasm in hexagonal clumps and is interspersed with the less numerous clumps of the more deeply staining myosin. Attachment plaques are dark bands along the cell surface that serve as attachment devices for the actin. Any process that leads to an increase in Ca^{++} concentration in the region of the contractile proteins results in contraction, and, conversely, any process that leads to a decrease in Ca^{++} concentration in the region of the contractile proteins results in relaxation.

Along the periphery of the cell are numerous cavitary structures, some of which open to the outside of the cell, referred to as caveolae or pinocytic vesicles. Their exact function is not known, although they may serve a role in the nutritive functions of the cell or in the transport of ions across the cell membrane. Surrounding the

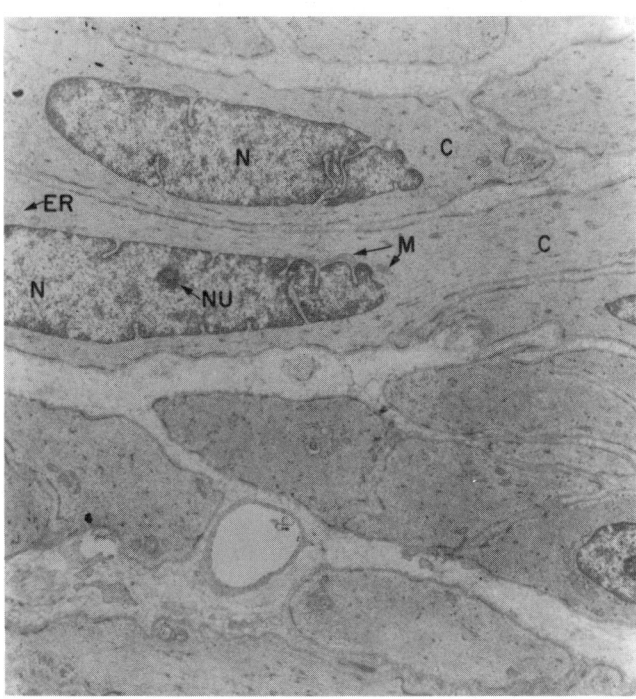

FIG 26–2.
Electron micrograph of human ureter. *N* indicates nucleus; *N*, nucleolus; *ER*, endoplasmic or sarcoplasmic reticulum; *M*, mitochondria; and *C*, cytoplasm or sarcoplasm. (Adapted from Weiss RM: *Urology* 1978; 12:114. Used with permission.)

cell is a doubled layer of cell membrane. The inner plasma membrane surrounds the entire cell, whereas the outer basement membrane is absent at areas of close cell-to-cell contact, referred to as intermediate junctions (Fig 26–3).

PHYSIOLOGY

ELECTRICAL ACTIVITY

The electrical properties of excitable tissues depend on the distribution of ions on the inside and outside of the cell membrane and on the relative permeability of the cell membrane to these ions (Hodgkin, 1958). When a ureteral muscle cell is in a nonexcited or resting state, the electrical potential difference across the cell membrane, or transmembrane potential, is referred to as the resting membrane potential (RMP). The RMP is primarily determined by the distribution of potassium ions (K^+) across the cell membrane and by the preferential permeability of the resting membrane to potassium (Washizu, 1966; Hendrickx et al., 1975). In the resting state, the tendency for the positively charged K^+ ions to diffuse from the inside of the cell, where they are more concentrated, to the outside of the cell, where they are less concentrated, creates an elec-

trical gradient in which the inside of the cell membrane is more negative than the outside of the cell membrane. The electrical gradient that is formed tends to oppose the further movement of K^+ ions outward across the cell membrane along its concentration gradient, and an equilibrium is reached with a greater concentration of K^+ on the inside of the membrane and with the inside of the cell membrane being negatively charged with respect to the outside of the cell membrane.

The RMP in smooth muscle is lower than that which would be expected if the resting cell membrane were exclusively permeable to potassium. The RMP in the ureter is -33 to -70 mV (Washizu, 1966; Kobayashi, 1969). In the resting state, sodium concentration on the outside of the cell membrane is greater than that on the inside. If the resting membrane were somewhat permeable to Na^+, both the concentration and electrical gradient would support an inward movement of sodium across the cell membrane, with a resultant decrease in the electronegativity of the inner surface of the cell membrane. Such a process could be a factor in the maintenance of a low RMP in smooth muscle. The RMP also may be maintained in part by an active mechanism capable of extruding Na^+ from within the cell against a concentration and electrochemical gradient and by the relative permeability and distribution

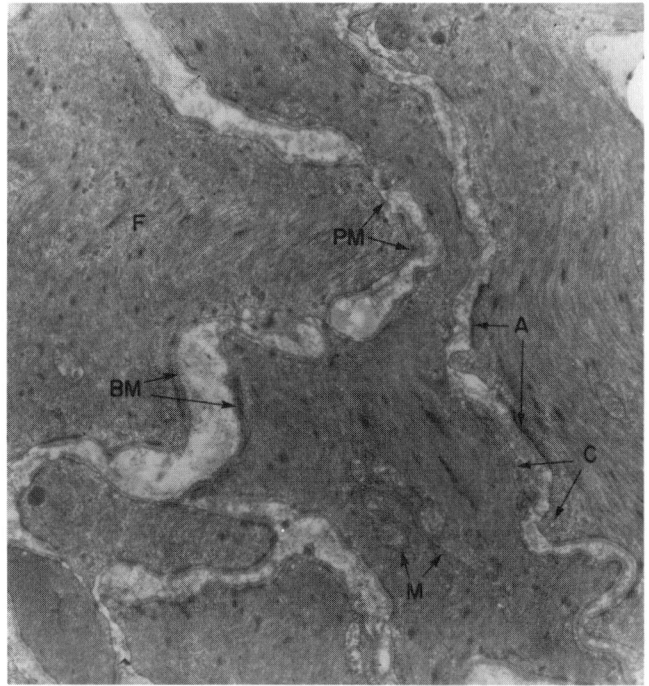

FIG 26–3.
Electron micrograph of human ureter. *A* indicates attachment plaques; *C*, caveolae or pinocytic vesicles; *M*, mitochondria; *BM*, basement membrane; *PM*, plasma membrane; and *F*, actin and myosin filaments. (From Weiss RM: *Urology* 1978; 12:114. Used with permission.)

of chloride (Cl⁻) ions across the cell membrane (Kuriyama, 1963).

The transmembrane potential of an inactive or resting ureteral cell remains stable until it is excited by an external stimulus, whether electrical, mechanical (stretch), or chemical, or by conduction of electrical activity from an already excited adjacent cell. When a ureteral cell is stimulated, depolarization occurs, the inside of the cell membrane becoming less negative than it was before stimulation. If a sufficient area of the cell membrane is depolarized rapidly enough to reach a critical level of transmembrane potential, referred to as the threshold potential, a regenerative depolarization, or action potential, is initiated (Fig 26–4). The action potential has the capability to act as the stimulus for excitation of adjacent quiescent cells and through a complicated chain of events gives rise to the ureteral contraction.

When the ureteral smooth muscle cell is excited, its membrane loses its preferential permeability to K^+ and becomes more permeable to Na^+ and Ca^{++} ions, which move inward across the cell membrane and give rise to the upstroke of the action potential (Kobayashi, 1965; Vereecken et al., 1975a, 1975b). As the positively charged Na^+ and Ca^{++} ions move inward across the cell membrane, the inside of the membrane becomes less negative with respect to the outside and may even become positive at the peak of the action potential, a state referred to as overshoot. The rate of rise of the upstroke of the ureteral action potential is relatively slow, 1.2 ± 0.06 V/sec in the cat (Kobayashi, 1969), and accounts in part for the slow conduction velocity in the ureter.

After reaching the peak of its action potential, the ureter maintains its potential for a time (plateau of the action potential) before the transmembrane potential returns to its resting level (repolarization). The plateau phase appears to depend on the persistence of a high inward Na^+ conductance, and the repolarization phase appears at least in part to be due to an outward movement of K^+ across the cell membrane (Kuriyama and Tomita, 1970; Kuriyama et al., 1967). The duration of

the action potential in the cat ranges from 259 to 405 msec (Kobayashi and Irisawa, 1964).

In summary, the transmembrane potential of the resting ureteral cell (RMP) is approximately -33 to -70 mV and is primarily determined by the distribution of K^+ ions across the cell membrane and the relative selective permeability of the resting cell membrane to potassium. When excited by a suprathreshold stimulus, the membrane becomes less permeable to K^+ and more permeable to Na^+ and Ca^{++}, which move inward across the cell membrane and provide the ionic mechanism for the development of the upstroke of the action potential. Calcium also plays a prominent role in the contractile mechanism. After reaching the peak of its action potential, the membrane maintains a depolarized state, plateau of the action potential, for a time before the membrane potential of the activated cell returns to its resting level, repolarization. The plateau appears to be related to a persisting inward Na^+ current, and repolarization of the membrane probably is related to a decrease in the membrane permeability to Na^+ and Ca^{++} and a renewed increase in permeability to potassium.

Pacemaker Potentials and Pacemaker Activity

Cells that develop electrical activity spontaneously are referred to as pacemaker cells. Pacemaker cells differ from nonpacemaker cells in that their resting transmembrane potential does not remain constant but rather undergoes a slow spontaneous depolarization. If the spontaneously changing membrane potential reaches the threshold potential, the upstroke of an action potential occurs.

Dixon and Gosling (1970, 1971, 1973, 1974, 1982) provided morphological evidence of specialized pacemaker tissue in the proximal portion of the urinary collecting system. In man, Dixon and Gosling (1982) identified atypical smooth muscle cells in the region of attachment of each minor calyx to the renal parenchyma that are devoid of nonspecific cholinesterase. These distinctive cells, loosely arranged and separated from each other by connective tissue, form a thin sheet of muscle that covers each minor calyceal fornix. The atypical cells run across the renal parenchyma that lies between the renal attachments of the minor calyces and thus serve as a connector between the minor calyces. Atypical cells are arranged longitudinally as an inner layer of the muscle coat of the minor and major calyces and of the renal pelvis. They appear to be closely applied to and to interconnect with typical muscle cells in these structures but do not extend beyond the ureteropelvic junction into the ureter. The atypical cells may act as pacemaker cells or as a preferential conduction pathway.

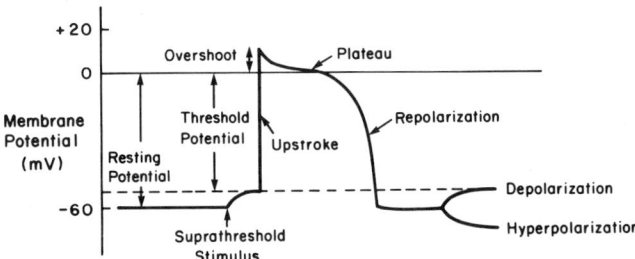

FIG 26–4.
Schematic diagram of ureteral action potential.

In the multicalyceal kidney, Morita and associates (1981, 1986), using extracellular electrodes, recorded low voltage potentials, which appear to be pacemaker potentials, from the border of the minor calyces and the major calyx. They noted that the contraction rhythm varied in each calyx. Calyceal contractions, with resultant coaptation of the walls of the calyces, facilitate outflow of urine from the papillae and protect the renal parenchyma from pressure increases transmitted from the renal pelvis. At normal urine flow rates in the multicalyceal kidney, pacemaker contractions of the calyces occur at a rate of approximately 6 per minute. At normal rates of flow, the contraction waves are frequently blocked in the renal pelvis or at the ureteropelvic junction (UPJ) (Morita et al., 1981). With increasing flow, there is a cessation of this block, and a 1:1 relationship is observed between pacemaker and ureteral contractions (Constantinou and Yamaguchi, 1981). In other words, at high flows, ureteric contractions occur at the same frequency as that of the calyces, whereas at low flows calyceal contraction frequencies are greater than that of the ureter.

Under normal conditions, electrical activity arises proximally and is conducted distally from one muscle cell to another across areas of close cellular apposition referred to as intermediate junctions (Uehara and Burnstock, 1970; Libertino and Weiss, 1972). These close cellular contacts are similar to nexuses, which have been shown in other smooth muscles to be low-resistance pathways for cell-to-cell conduction (Barr et al., 1968). Although the primary pacemaker for ureteral peristalsis is located in the proximal portion of the collecting system, other areas of the ureter may act as latent pacemakers. Under normal conditions, these latent pacemaker regions, such as the ureterovesical junction, are dominated by activity arising at the primary pacemaker sites. When the latent pacemaker site is freed of its domination by the primary pacemaker, it in turn may act as a pacemaker.

CONTRACTILE ACTIVITY

The contractile event is dependent on the concentration of free sarcoplasmic Ca^{++} in the region of the contractile proteins actin and myosin. Any process that results in an increase in Ca^{++} in the region of the contractile proteins favors the development of a contraction; any process that results in a decrease in Ca^{++} in the region of the contractile proteins favors relaxation.

Contractile Proteins

In smooth muscle, the most widely accepted theory suggests that phosphorylation of myosin is involved in the contractile process. With excitation, there is a transient increase in the sarcoplasmic Ca^{++} concentration

from a steady state concentration of $10^{-8}M$ to $10^{-7}M$ to a concentration of $10^{-6}M$ or higher. At this higher concentration, Ca^{++} forms an active complex with the calcium-binding protein calmodulin (Watterson et al, 1976) (Fig 26–5). Calmodulin without Ca^{++} is inactive. The calcium-calmodulin complex activates a calmodulin-dependent enzyme, myosin light chain kinase, which in turn catalyzes the phosphorylation of the 20,000-dalton light chain of myosin (Fig 26–6). Phosphorylation of the myosin light chain is a prerequisite for activation by actin of myosin Mg^{++}-ATPase activity, with resultant hydrolysis of ATP and the development of smooth muscle tension or shortening (Fig 26–7).

When the Ca^{++} concentration in the region of the contractile proteins is low, myosin light chain kinase is not activated, since calmodulin requires Ca^{++} to activate the enzyme. This prevents activation of the contractile apparatus, since the myosin light chain cannot be phosphorylated, a process that must precede tension development. Furthermore, a phosphatase dephosphorylates the myosin light chain, preventing actin activation of myosin ATPase activity, and relaxation results. Recently, there has been evidence that phosphorylation of the enzyme myosin light chain kinase by a cAMP-dependent protein kinase decreases myosin light chain kinase activity by decreasing the affinity of this enzyme for calmodulin (Adelstein et al., 1981). Such a process may in part account for cAMP-dependent relaxation of smooth muscle.

The calcium involved in the ureteral contraction is derived from two main sources (Fig 26–8). Since smooth muscle cells have a small diameter, the inward movement of extracellular Ca^{++} into the cell during

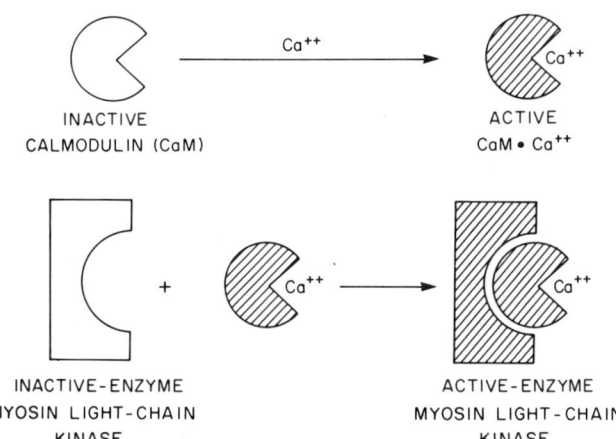

FIG 26–5.
Schematic representation of contractile process in smooth muscle. Calmodulin is activated by Ca^{++}. The activated calcium-calmodulin complex activates the enzyme myosin light chain kinase. (Adapted from Weiss RM: in Harrison JH, et al (eds): *Campbell's Urology,* ed 5. Philadelphia, WB Saunders Co, 1986, pp 94–128. Used with permission.)

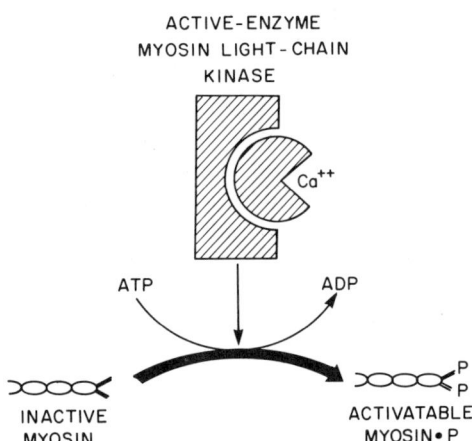

FIG 26–6.
Schematic representation of contractile process in smooth muscle. The activated enzyme, myosin light chain kinase, catalyzes the phosphorylation of myosin. Myosin must be phosphorylated for actin to activate myosin ATPase. (Adapted from Weiss RM: in Harrison JH, et al (eds): *Campbell's Urology,* ed 5. Philadelphia, WB Saunders Co, 1986, pp 94–128. Used with permission.)

the upstroke of the action potential provides a significant source of sarcoplasmic calcium. Calcium release from more tightly bound storage sites, i.e., the endoplasmic reticulum, mitochondria, and membrane-binding sites, also provides a source of sarcoplasmic calcium (Vereecken et al., 1975a). Relaxation results from a decrease in the concentration of free sarcoplasmic Ca^{++} in the region of the contractile proteins. The decrease in sarcoplasmic Ca^{++} can result from uptake of Ca^{++} into intracellular storage sites or from extrusion of Ca^{++} from the cell.

ROLE OF NERVOUS SYSTEM IN URETERAL FUNCTION

That the ureter is a syncytial type of smooth muscle without discrete neuromuscular junctions has caused confusion in determining the role of the autonomic ner-

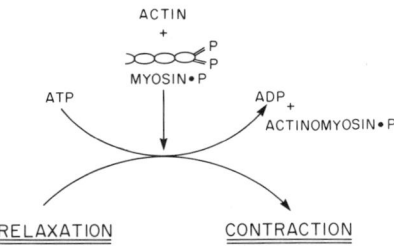

FIG 26–7.
Schematic representation of contractile process in smooth muscle. Actin activates ATPase activity of phosphorylated myosin with the resultant development of contraction. (Adapted from Weiss RM: in Harrison JH, et al (eds): *Campbell's Urology,* ed 5. Philadelphia, WB Saunders Co, 1986, pp 94–128. Used with permission.)

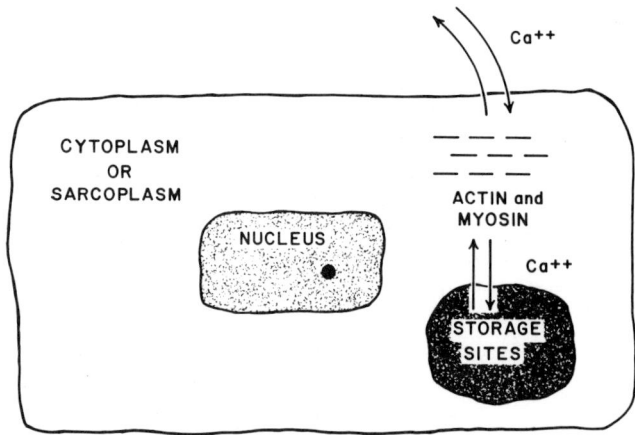

FIG 26–8.
Schematic representation of calcium movements involved in ureteral contraction. Ca^{++} involved in contractile process may be derived from extracellular sources or from tightly bound intracellular storage sites.

vous system in its function. Since peristalsis may persist after transplantation (O'Conor and Dawson-Edwards, 1969) or denervation (Wharton, 1932), and since normal antegrade peristalsis persists after reversal of a segment of ureter in situ (Melick et al., 1961), it is apparent that ureteral peristalsis can occur without innervation. It is, however, also apparent that the nervous system plays at least a modulating role in ureteral peristalsis. There is strong evidence to support the presence of excitatory α-adrenergic and inhibitory β-adrenergic receptors in the ureter (McLeod et al., 1973; Rose and Gillenwater, 1974; Weiss et al., 1978). Norepinephrine, which is primarily an α-adrenergic agonist although it also is able to stimulate β-adrenergic receptors, increases the force of electrically induced ureteral contractions (Weiss et al., 1978). When administered in the presence of phentolamine (Regitine), an α-adrenergic blocking agent, norepinephrine decreases the force of ureteral contractions (Weiss et al., 1978). This reversal of action can be explained by norepinephrine's acting primarily on inhibitory β-adrenergic receptors when the excitatory α-adrenergic receptors are blocked. Propranolol (Inderal), a β-adrenergic antagonist, potentiates the increase in contractile force induced by norepinephrine (Weiss et al., 1978) by permitting norepinephrine to act more exclusively on excitatory α-adrenergic receptors when the inhibitory β-adrenergic receptors are blocked. Furthermore, isoproterenol, a β-adrenergic agonist, depresses contractility (Weiss et al., 1978).

Other support for the presence of excitatory α-adrenergic and inhibitory β-adrenergic receptors in the ureter includes the demonstration of adenylate cyclase activity in the ureter (Weiss et al., 1977), the demonstration of α- and β-adrenergic receptors in the ureter using receptor binding techniques (Latifpour et al., 1987), and the finding that the application of a given intraluminal pressure results in a greater degree of ureteral deformation in rabbits depleted of catecholamines by the administration of reserpine than results from the application of the same pressure load to a ureter of a normal nonreserpine-treated animal (Weiss et al., 1974). Last, electrical stimulation with high intensity, high frequency, and short duration pulses has been shown to release neurotransmitter, presumably from neural tissue intrinsic within the wall of the ureter (Weiss et al., 1978) and renal calyx (Longrigg, 1975).

Available data suggest that cholinergic (parasympathetic) agonists potentiate ureteral contractility by directly stimulating cholinergic receptors (Vereecken, 1973) or by indirectly causing the release of catecholamines (Rose and Gillenwater, 1974). Cholinergic (muscarinic) receptors have been demonstrated in the ureter (Latifpour et al., 1987). Last, Del Tacca's demonstration (1978) of acetylcholine release from isolated guinea pig, rabbit, and human ureters during field stimulation and the inhibition of this release by the neural poison, tetrodotoxin, provide further evidence for a role of the parasympathetic nervous system in the control of ureteral activity.

Physiology of the Ureteropelvic Junction

Griffiths and Notschaele (1983) recently described the mechanics of urine transport within the ureter. As the renal pelvis fills, there is a rise in renal pelvic pressure, and urine is ultimately extruded into the upper ureter, which is initially in a collapsed state. After transporting the urine into the upper ureter, renal pelvic pressure declines to its baseline value, and the cycle of pelvic filling, increase in pressure, and launching of urine into the ureter occurs again. The closed ureteropelvic junction (UPJ) may be protective to the kidney in dissipating back pressure from the ureter, since ureteral contractile pressures are higher than renal pelvic pressures.

The mechanism for urine launching into the upper ureter is not completely understood, and abnormalities in a rather complicated regulatory mechanism may cause the hydronephrosis associated with functional ureteropelvic junction obstructions in which urine transport is impaired, even though large-caliber catheters readily can be passed through the UPJ. While in the normal system, peristalsis is controlled by pacemaker cells that generate high-frequency action potentials in the proximal pelvicalyceal region, in the chronically dilated system, the frequency gradient in the renal pelvis is lost. This can be associated with latent pacemakers initiating dystropic orthograde peristaltic activity or retrograde contractions. Obstruction alters the hierarchical organization of the multiple coupled pacemakers that normally coordinate peristaltic activity (Constantinou and Djurhuus, 1981; Djurhuus and Constantinou, 1982). Such disruption causes uncoordinated pelvic contractions that may result in hypertrophy of the renal pelvic smooth muscle and impaired transport of urine into the ureter. Whether these functional changes are secondary to or the cause of the dilatation is not certain.

Murnaghan (1958b) related the functional abnormality to an alteration in the configuration of the muscle bundles at the UPJ, and Foote and associates (1970) observed a decrease in musculature at the UPJ. Hanna (1978) noted in severe UPJ obstruction abnormalities in the musculature of the renal pelvis and disruption of intercellular relationships at the UPJ itself. Gosling and Dixon (1978) also observed histologic abnormalities in the dilated renal pelvis. In some instances, there may be areas of actual narrowing or valvelike processes at the UPJ. Furthermore, a vessel or adhesive band crossing the UPJ may potentiate the degree of dilatation in any of the forms of UPJ obstruction.

Propulsion of Urinary Bolus

Following extrusion into the ureter, the urine forms a bolus owing to a ureteral contraction ring that completely coapts the ureteral walls (Griffiths and Notschaele, 1983). The contraction ring at the rear end of the bolus progresses distally at a constant velocity, while the velocity of the leading edge of the bolus varies along the ureter (Durbin et al., 1980). Therefore, the width and length of the bolus is not uniform as it moves from the renal pelvis to the bladder. The bolus of urine that is pushed distally in front of the contraction ring lies almost entirely in a passive, noncontracting part of the ureter (Weinberg, 1974). Contraction waves normally occur 2–6 times per minute in the normal human ureter (Edmond et al., 1970) and are conducted at a velocity of approximately 2–5 cm/sec (Butcher and Sleator, 1955). Baseline (resting) ureteral pressure is approximately 0–5 cm H_2O, and contractile pressures may range from 20–60 cm H_2O (Edmond et al., 1970). The major component of the recorded pressure is derived from the contraction wave, with only a small component derived from the bolus pressure. The urine bolus is forced into the bladder by the advancing contraction wave, which dissipates at the ureterovesical junction (Fig 26–9). When functioning properly the ureterovesical junction (UVJ) ensures one-way transport of urine.

As with any tubular structure, the ureter can transport a set maximum amount of fluid per unit time. Under normal flows, in which bolus formation occurs, the amount of urine transported per unit time is significantly less than the maximum transport capacity of the ureter. At high flows, the ureteral walls do not coapt,

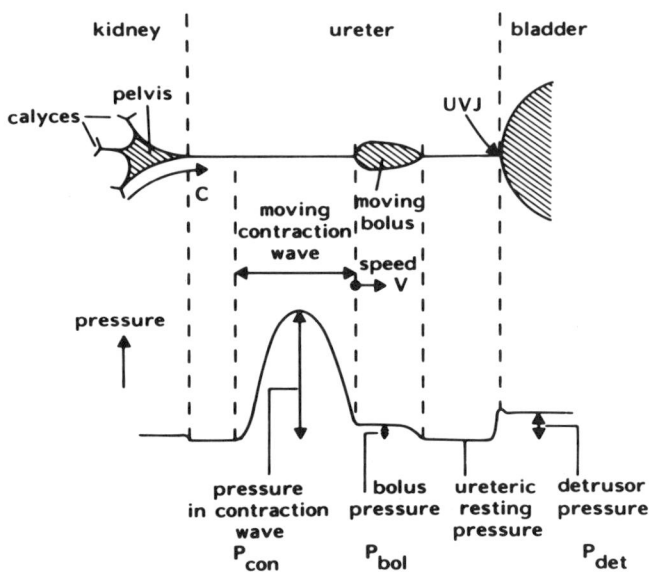

FIG 26–9.
Schematic representation of a single bolus in the ureter moving from the renal pelvis to the bladder. Corresponding distribution of pressure within the urinary tract is shown *(lower tracing).* (From Griffiths DJ, Notschaele C: The mechanics of urine transport in the upper urinary tract: I. The dynamics of the isolated bolus. *Neurol Urodynam* 1983; 2:155. Used with permission.)

and fluid is transported in a continuous column rather than in a series of boluses.

When transport becomes inadequate, there is backup of urine with resultant ureteral dilatation. Inadequate transport can result from either too much fluid entering the ureter per unit time or from too little fluid exiting the ureter per unit time. Both input and output, as well as the compliance of the system, must be considered when predicting whether ureteral dilatation will occur. For example, a minor degree of obstruction to outflow will cause more dilatation at high flows than at low flows.

Changes in the dimensions of the ureter that occur in pathologic states may in themselves account for inefficient urine transport, even if the contractile force of the individual fibers remains unchanged (Griffiths, 1983; Biancani et al., 1982; Weiss and Biancani, 1982). The Laplace equation expresses the relationship between the variables that affect intraluminal pressure:

$$\text{pressure} = \frac{\text{stress} \times \text{wall thickness}}{\text{radius}}$$

Thus, an increase in ureteral diameter in itself can cause a decrease in intraluminal pressure and result in inefficient urine transport.

The Laplace relationship also may provide a rationale for ureteral tapering of the dilated ureter. With ureteral tapering, muscle thickness and the ability of the ureter to contract remains unchanged. The decrease in radius occurring with ureteral tapering may in itself account for higher intraluminal pressures, with resultant improved urine transport. The tapered ureter can coapt its walls more readily and generate a higher intraluminal pressure, even though the material itself has not been changed (Weiss and Biancani, 1982). Diagrammatically, with ureteral tapering, force or the number of blocks remains unchanged, but intraluminal pressure increases as the load is supported over a smaller area; thus, pressure or the height of the pile of blocks increases (Fig 26–10).

EFFECT OF DIURESIS ON URETERAL FUNCTION

The upper urinary tract alters its characteristics according to the amount of urine to be transported. Smooth muscle function is affected by the amount of stretch and the rate with which the stretch is applied. The initial response of the ureter to an increase in urine flow is an increase in the frequency of peristalsis. At relatively low flow rates, small increases in flow result in large increases in peristaltic frequency. At higher flow rates, relatively large increases in flow result in only small increases in peristaltic frequency. With increasing urine flow, peristaltic frequency increases to a maximum and the further increases in urine transport occur by means of increases in bolus volume (Constantinou et al., 1974). As the volume of the bolus increases, the leading edge of the bolus approximates the contraction ring of the preceding bolus. The pressure in the bolus increases owing to increased resistance at the leading edge of the bolus, where it touches the preceding contraction wave. At very high flows, the pressure within the bolus exceeds the contraction pressure in the ring, and the contraction pressure is no longer sufficient to coapt the ureteral wall. The boluses then coalesce, and the ureter becomes filled with a column of fluid and dilates. At high flows, urine is transported through an open tube by columnar flow rather than by a series of boluses.

PHYSIOLOGY OF THE URETEROVESICAL JUNCTION

Griffiths (1983) analyzed the factors involved in urine transport across the UVJ. Under normal conditions and at normal flow rates, the contraction wave that occludes the ureteral lumen propagates distally with the urine bolus in front of it. When the bolus reaches the UVJ, the pressure within the bolus must exceed intravesical pressure for the bolus of urine to pass across the UVJ into the bladder. For the contraction wave to coapt the ureter walls and move the urine bolus distally, the

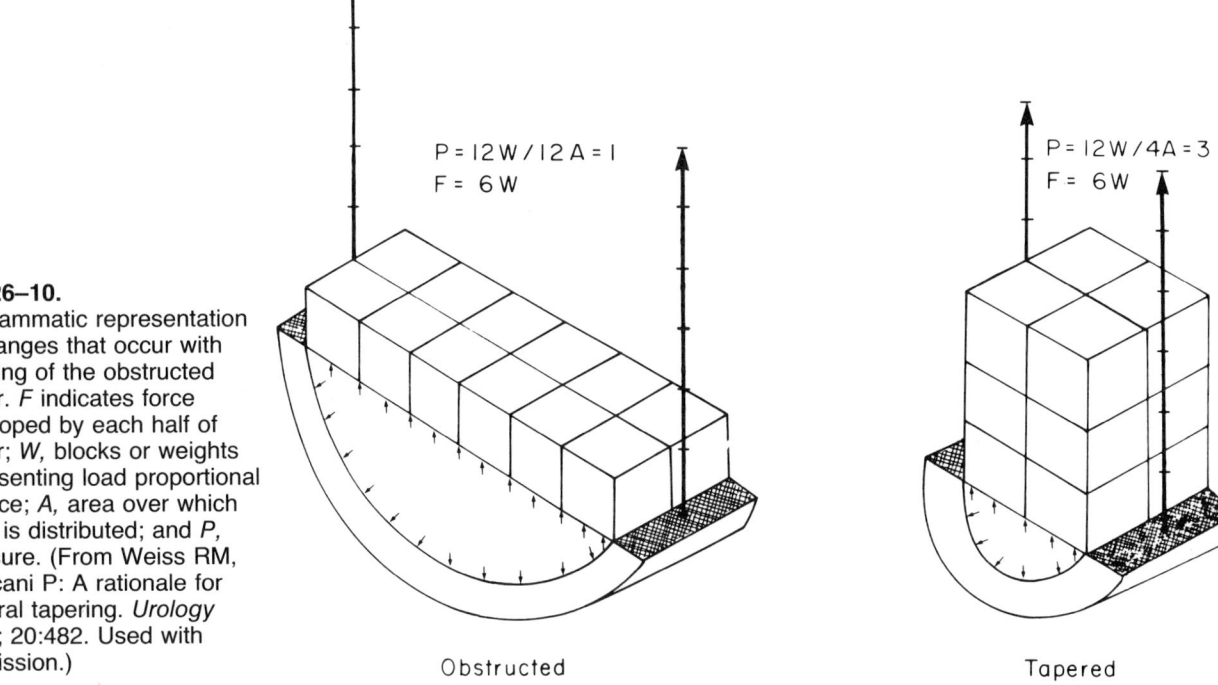

FIG 26–10.
Diagrammatic representation of changes that occur with tapering of the obstructed ureter. *F* indicates force developed by each half of ureter; *W,* blocks or weights representing load proportional to force; *A,* area over which force is distributed; and *P,* pressure. (From Weiss RM, Biancani P: A rationale for ureteral tapering. *Urology* 1982; 20:482. Used with permission.)

Obstructed

$P = 12W/12A = 1$
$F = 6W$

Tapered

$P = 12W/4A = 3$
$F = 6W$

pressure generated by the contraction wave must exceed the pressure within the urinary bolus. The UVJ does not relax (Weiss and Biancani, 1983).

Blok et al. (1985) demonstrated fast and slow pressure waves at the UVJ. The fast pressure waves originate from intrinsic ureteral contractions and are related to retraction of the distal ureter within its sheaths and are accompanied by bolus ejection of fluid into the bladder. This telescoping, dynamic event decreases the UVJ resistance to flow and thus facilitates urine passage into the bladder. The slow pressure waves originate from the surrounding detrusor muscle and represent the influence of detrusor activity on flow through the UVJ. Coolsaet and associates (1982) demonstrated that resistance to flow through the UVJ is due to (1) stretch of the bladder base and UVJ, (2) intravesical pressure transmitted to the submucosal ureteral segment, (3) activity of the detrusor muscle surrounding the UVJ, and (4) dynamics of the ureteral segment composing the UVJ.

The relationship between upper urinary tract function and resistance to outflow at the UVJ is of major clinical importance. The pressure within the blader during the storage phase is of paramount importance in determining the efficacy of urine transport across the UVJ. This is the pressure that the ureter needs to work against for the greatest time. During filling of the normal bladder, sympathetic impulses and the viscoelastic properties of bladder wall inhibit the magnitude of the intravesical pressure rise. With filling, the normal blad-

der maintains a relatively low intravesical pressure (McGuire, 1983). The low intravesical pressure facilitates urine transport across the UVJ and prevents ureteral dilatation.

Renal function and the functional integrity of the upper urinary tract are at risk in individuals with noncompliant or hyperactive bladders. In the noncompliant fibrotic bladder and in some forms of neurogenic vesical dysfunction, the bladder is autonomous, and relatively small increases in bladder volume result in large increases in intravesical pressure with impairment of ureteral emptying. Regular emptying of these bladders may not be sufficient to protect the upper tracts, and intravesical pressure may need to be lowered by reduction in detrusor tonus, i.e., relaxation. Furthermore, outflow resistance at the UVJ may be high in the overdistended bladder even at low intravesical pressures. This results from stretch of the UVJ and decreased retractability of the submucosal ureteral segment. Intravesical volume should thus be kept at reasonable levels with intermittent catheterization. Increased resistance to bolus outflow across the UVJ can occur when there is obstruction at the UVJ, when intravesical pressure or volume is excessive, or when flow rates are so high as to exceed the transport capacity of the UVJ. Under such conditions, in which the bolus of urine cannot freely pass into the bladder, the pressure within the bolus, propelled by the ureteral contraction ring, increases and may exceed the pressure within the contraction ring. Under these conditions the contraction wave will be unable to coapt the ureteral wall and intraureteric reflux will occur. This impaired bolus transport will cause secondary widening of the ureter and weaker ureteral contractions, with only a fraction of the bolus volume passing distally. Griffiths (1983) presented theoretical evidence to show that a similar situation of impaired bolus transport across the UVJ would be expected if the ureter was wide and/or weakly contracting, even if the UVJ was perfectly normal. Under these conditions, a similar breakdown of bolus discharge into the bladder can occur in the wide and/or weakly contracting ureter at high flow rates, even if the UVJ is normal. Such a situation of impaired urine transport might go undetected with a Whitaker test (Whitaker, 1973, 1979).

There is some evidence that gravity may assist urine transport and that the erect position, by enhancing hydrostatic loading of the UVJ, may facilitate urine transport across the UVJ, especially in individuals with wide upper tracts (Schick and Tanagho, 1973). From a clinical view, some workers have suggested that bed rest may be deleterious to renal function in the individual who presents with urinary retention and dilated upper urinary tracts (George et al., 1984).

Relationship Between Vesicoureteral Reflux and Ureteral Function

The intravesical ureter is approximately 1.5 cm long and takes an oblique course through the bladder wall. It is composed of an intramural segment, which is surrounded by detrusor muscle, and of a submucosal segment, which lies directly under the bladder urothelium (Tanagho et al., 1968). The relationship between the length and diameter of the intravesical segment of ureter appears to be a factor in the prevention of vesicoureteral reflux (Paquin, 1959). Trigonal function also may be a factor in the prevention of vesicoureteral reflux (Tanagho et al., 1965). Furthermore, the development of vesicoureteral reflux in individuals with bladder outlet obstruction and neurogenic vesical dysfunction provides evidence that increased intravesical pressures also may be a factor in certain instances of reflux.

Although an abnormality of the UVJ is the primary etiologic factor in most cases of reflux, there is evidence to suggest that decreased ureteral peristaltic activity may be a contributory factor (Weiss and Biancani, 1983; Kirkland et al., 1971). Support for this contention may be derived from the findings that a normal ureter may not reflux, even when reimplanted into a bladder without a submucosal tunnel (Debruyne et al., 1978), and that a defunctionalized refluxing ureter may cease to reflux when a proximal diversion is taken down and urine flow through the ureter is reinstated (Teele et al.,

1976; Weiss, 1979). The observation that vesicoureteral reflux may temporarily cease following electrical stimulation of the ureter (Melick et al., 1966) provides further support for the role of ureteral peristalsis in the prevention of vesicoureteral reflux. Last, the success rate of antireflux procedures is less with poorly functioning dilated ureters and although this may be related to technical factors, decreased peristaltic activity may be a reason for failure in many instances.

Studies in normal and mildy refluxing systems have shown that there is a high-pressure zone in the distal ureter with a resultant pressure gradient across the UVJ (Weiss and Biancani, 1983). With bladder filling, the resultant UVJ-bladder pressure gradient increases in nonrefluxing systems, while it decreases and may disappear in refluxing systems (Weiss and Biancani, 1983) (Fig 26–11). This decrease in the pressure gradient may be related to lateralization of the ureteral orifice and shortening of the intravesical tunnel and may correspond to the time when reflux occurs.

EFFECT OF INFECTION ON URETERAL FUNCTION

Infection within the upper urinary tract may impair urine transport. Bacteria and endotoxins have been shown to inhibit ureteral activity (Primbs, 1913; Teague and Boyarsky, 1968; King and Cox, 1972), and pyelonephritis in the monkey has been associated with decreased peristaltic activity (Roberts, 1975). Further-

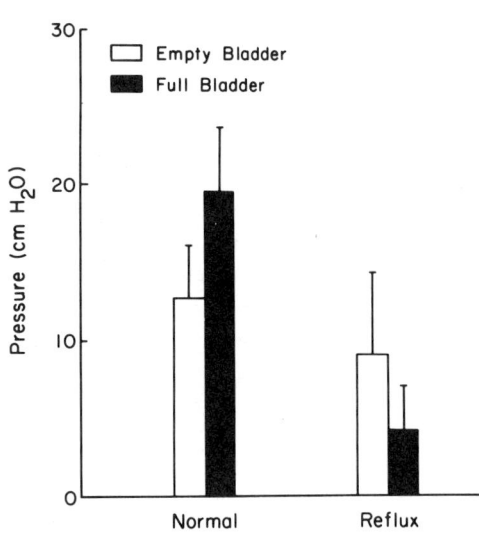

FIG 26–11.
Pressure gradient across the ureterovesical junction in normal and mildly refluxing (grades II–III) systems. (From Weiss RM, Biancani P: Characteristics of normal and refluxing ureterovesical junctions. *J Urol* 1983; 129:858. Used with permission.)

more, Rose and Gillenwater (1973) have shown that infection can potentiate the deleterious effects of obstruction on ureteral function.

In man, irregular peristaltic contractions, often with a decreased amplitude, have been recorded with infection. In more severe cases absence of activity has been noted (Ross et al., 1972). Furthermore, ureteral dilatation may occur with retroperitoneal inflammatory processes secondary to appendicitis, regional enteritis, ulcerative colitis, or peritonitis (Makker et al., 1972). Infection also may reduce the compliance of the intravesical ureter and permit reflux to occur in situations in which the UVJ is intrinsically of marginal competence (Cook and King, 1979).

EFFECT OF CALCULI ON URETERAL FUNCTION

Factors that can affect the spontaneous passage of calculi include: (1) the size and shape of the stone (Ueno et al., 1977), (2) areas of narrowing within the ureteral lumen, (3) ureteral peristalsis, (4) the hydrostatic pressure of the column of urine proximal to the calculus (Sivula and Lehtonen, 1967), and (5) edema, inflammation, and spasm of the ureter at the site of the stone (Holmlund and Hassler, 1965).

Two factors that appear to be most useful in facilitating stone passage are an increase in hydrostatic pressure proximal to a calculus and relaxation of the ureter in the vicinity of the stone. In support of the theory that increased hydrostatic pressure facilitates stone passage, it has been shown in the rabbit and dog ureter that artificial concretions with holes move more slowly than those without holes (Sivula and Lehtonen, 1967). Furthermore, ureteral ligation proximal to a concretion has been shown to decrease hydrostatic pressure and peristaltic activity proximal to a stone, with resultant inhibition of stone passage (Sivula and Lehtonen, 1967). Theoretically, high fluid intake will increase hydrostatic pressure proximal to a stone and, thus, may aid in its passage; however, the elevated hydrostatic pressure, if prolonged, is potentially deleterious to the kidney.

With respect to the potential facilitative effect of ureteral relaxation on stone passage, Peters and Eckstein (1975) showed that the spasmolytic agents phentolamine, an α-adrenergic antagonist, and orciprenaline, a β-adrenergic agonist, dilated the canine ureter at the level of an artificial concretion and thus permitted increased fluid flow beyond a partially obstructive concretion. Norepinephrine, on the other hand, although increasing the force and frequency of peristalsis proximal to the concretion, led to a decreased flow beyond the concretion. It thus appears that the resistance to urine flow in this experimental model was caused in part by the artificial concretion and in part by local spasm and

that the resistance could be decreased by spasmolytic drugs, such as phentolamine and orciprenaline. Although it has not been determined whether this spasmolytic effect would aid in stone passage, the same principle has been used to float upper ureteral stones proximally into the renal pelvis as a prelude to percutaneous nephrostolithotomy. Obstruction of the ureter with a balloon catheter leads to dilatation of the ureter, with the impacted stone then being able to migrate proximally (Beckmann and Roth, 1985).

In a study in man, renal colic was relieved by meperidine in 83% of patients, by the α-adrenergic antagonist phentolamine in 63% and by propranolol, a β-adrenergic antagonist that presumably would interfere with the β-adrenergic inhibitory actions of catecholamines, in 0% (Kubacz and Catchpole, 1972).

Although the above pharmacologic data can be interpreted to imply that ureteral relaxation in the region of a concretion would aid in stone passage, a controlled study currently is not available. Such a study with an agent known to have strong relaxant effects on the ureter, such as theophylline (Weiss et al., 1977), would be of value, but the interpretation of the data obtained might be difficult because of the marked variability of spontaneous stone passage in the clinical setting.

Ureterolithotomy

The management of urinary tract calculi has undergone significant changes in the past decade with the innovation of ureteroscopy, percutaneous nephrostolithotomy, and extracorporeal lithotripsy. Expectant observation with the hope of spontaneous passage still remains the hallmark in the management of the great majority of small ureteral calculi. The newer technological advances have improved our ability to manage these calculi, but should not significantly affect the decision as to which calculi need such management.

Ureteroscopy, with its auxiliary techniques of direct-vision stone extraction, ultrasonographic disintegration, and electrohydraulic lithotripsy, has revolutionized management of middle and lower ureteral calculi. Some of the endoscopic procedures also have a role in the management of upper ureteral calculi. Percutaneous techniques with or without floating the upper ureteral calculus back into the renal pelvis and extracorporeal lithotripsy for the nonimpacted stone also can aid in the management of the upper ureteral stone.

As these newer techniques are discussed in detail elsewhere, this section will confine itself to a less frequently used procedure, i.e., the standard ureterolithotomy. For upper ureteral calculi, a standard flank incision below or through the bed of the 12th rib has been the standard approach, although a posterior lumbotomy has a role in the management of the larger and/

or well-impacted calculi. During the procedure, avoidance of stone migration, especially proximally, should be ensured by control of the ureter above and below the calculus with vessel loops or Babcock forceps. A longitudinal incision in the ureter is made directly over the stone and the stone extracted. A small catheter is passed proximally into the renal pelvis for irrigation and to be sure that the system is unobstructed. Although distal passage of the catheter into the bladder can be used to assess distal obstruction, it can cause edema at the UVJ, with prolongation of flank drainage postoperatively. For this reason, the use of distal catheter passage can be individualized depending on the clinical aspects and the preoperative assessment of the ureter distal to the stone.

Some authors favor a loose closure of the ureter with only one or two fine sutures, and others seek a watertight closure with either interrupted or continuous fine sutures. Obviously, both methods work. Our preference is multiple interrupted 5–0 chromic adventitial sutures, which avoid excessive prolonged leakage yet provide a means for urine egress, if necessary. One or two Penrose drains brought to the exterior through a stab wound provide for drainage.

The technical details for removal of midureteral stones are essentially the same as for upper ureteral calculi. The approach can either be via a subcostal flank incision or via an anterior extraperitoneal approach, using a horizontal incision beginning below the tip of the 12th rib and extending anteriorly, with the patient in the supine position with a roll of towels beneath the shoulder and buttock. This approach, classically employed for sympathectomy, can involve muscle-splitting or the muscle layers can be divided.

Lower ureteral calculi can be approached through a modified Gibson's incision as used for renal transplantation or through a vertical suprapubic incision with retraction of the bladder toward the contralateral side. At times with periureteral inflammatory reaction or scarring from previous surgery, the ureter and stone may be difficult to identify. Identification of the ureter proximally, where it crosses the bifurcation of the common iliac artery, may facilitate its localization, and opening the peritoneum with visualization of the ureter through the posterior parietal peritoneum can be helpful. Identification and division of the obliterated umbilical vessels also can aid in localizing the ureter and facilitate removal of the calculus. The actual technique for stone removal is similar to that employed for higher stones, and early control of the ureter proximal to the calculus is imperative. A vessel loop should be placed around the ureter wherever it is first identified proximal to the stone.

For calculi in the most distal portions of the ureter, such as in the intramural portion of the UVJ, a transvesical approach can be used. If the stone can be palpated with the bladder open, an incision is made in the vesical mucosa directly over the calculus. If the stone is somewhat more proximal, proximal ureteral control with a vessel loop is obtained at any level extravesically, and the incision in the ureter can be extended proximally from within the bladder. If there is concern about distal ureteral damage in cases in which there is significant periureteral inflammation, a ureteroneocystotomy using a short submucosal tunnel may be required and would be preferable to the development of a ureteral obstruction necessitating a secondary procedure. In women a transvaginal approach can also be used for distal ureteral stones.

EFFECT OF PREGNANCY ON URETERAL FUNCTION

Hydroureteronephrosis of pregnancy begins in the second trimester of gestation and subsides within the first month after parturition. It is more severe on the right than on the left side, and ureteral dilatation does not occur below the pelvic brim. Roberts (1976) presented a strong argument in favor of obstruction as the primary etiologic factor in the development of hydroureteronephrosis of pregnancy. Others have suggested a hormonal mechanism for the ureteral dilatation of pregnancy (van Wagenen and Jenkins, 1939). As emphasized by Roberts (1976): (1) elevated baseline (resting) ureteral pressures consistent with obstruction have been recorded above the pelvic brim in pregnant women, which decrease when positional changes permit the uterus to fall away from the ureters (Sala and Rubi, 1967); (2) normal ureteral contractile pressures have been recorded in pregnant women, suggesting that hormonally induced ureteral atony is not the prime factor in ureteral dilatation of pregnancy; (3) women whose ureters do not cross the pelvic brim, i.e., those with pelvic kidneys or ileal conduits, do not develop ureteral dilatation of pregnancy; (4) hydronephrosis of pregnancy usually does not occur in quadripeds whose uterus hangs away from the ureters (Traut and Kuder, 1938); and (5) elevated ureteral pressures in the pregnant monkey return to normal when the uterus is elevated from the ureters at laparotomy or when the fetus and placenta are removed from the uterus.

Studies of the effects of hormones of pregnancy on ureteral function have been conflicting. Several studies have shown an inhibitory effect of progesterone on ureteral function (Lubin et al., 1941; Hundley et al., 1942; Kumar, 1962): however, this has not been a universal finding (Payne and Hodes, 1939; McNellis and Sherline, 1967; Schneider et al., 1953). Progesterone has been noted to increase the degree of ureteral dilatation during pregnancy and to retard the rate of disappearance of hydroureter in postpartum women (Lubin et al., 1941). Furthermore, hydronephrosis has been reported in women taking oral contraceptives (Guyer and Delaney, 1970; Marshall et al., 1966). Others, however, failed to induce changes in ureteral activity in women by the administration of estrogen, progesterone, or a mixture of these drugs (Marchant, 1972; Clayton and Roberts, 1973). Last, Raz and associates (1972) suggested that hormones may alter ureteral responses to catecholamines. Thus, although obstruction appears to be the primary factor in the development of hydronephrosis of pregnancy, it is possible that a combination of hormonal and obstructive factors is involved (Fainstat, 1963).

EFFECT OF AGE ON URETERAL FUNCTION

Aging affects the structure and function of the ureter. Cussen (1967) noted in a human autopsy study of subjects, ranging in age from 12 weeks of gestation to 12 years of age, a progressive increase in the population of smooth muscle cells and a small increase in the overall size of the individual smooth muscle cells with age. This is accompanied by an irregular increase in the number of elastic fibers. A progressive increase in ureteral cross-sectional muscle area is also observed in the ureter of the guinea pig between 3 weeks and 3 years of age, accompanied by an increase in developed force (Hong et al., 1980) (Fig 26–12). The increase in force developed between 3 weeks and 3 months of age seems to be attributable to an increase in contractility, since

there is an associated increase in active stress, or force per unit area of muscle. The increase in force developed between 3 months and 3 years of age can be explained by an increase in muscle mass alone, since there is no change in active stress between these two age groups (Fig 26–13). Although changes in the force-length relationships of guinea pig ureter occur with age, the force-velocity relationships do not change with age (Biancani et al., 1984).

The response of the ureter to pathologic insults depends not only on the magnitude and duration of the pathologic insult but also on the age of the individual affected. The neonatal rabbit ureter undergoes greater degrees of deformation in response to an applied intraluminal pressure than does the adult rabbit ureter (Akimoto et al., 1977). This decrease in compliance with age is also noted clinically where more marked degrees of ureteral dilatation occur in the neonate and young child in response to obstruction than in the adult. Recent experimental data suggest that aging has effects on the mechanical and biochemical properties of the ureter (Hong et al., 1980; Akimoto et al., 1977; Wheeler et al., 1986), and these changes may affect the response of the ureter to pathologic insult.

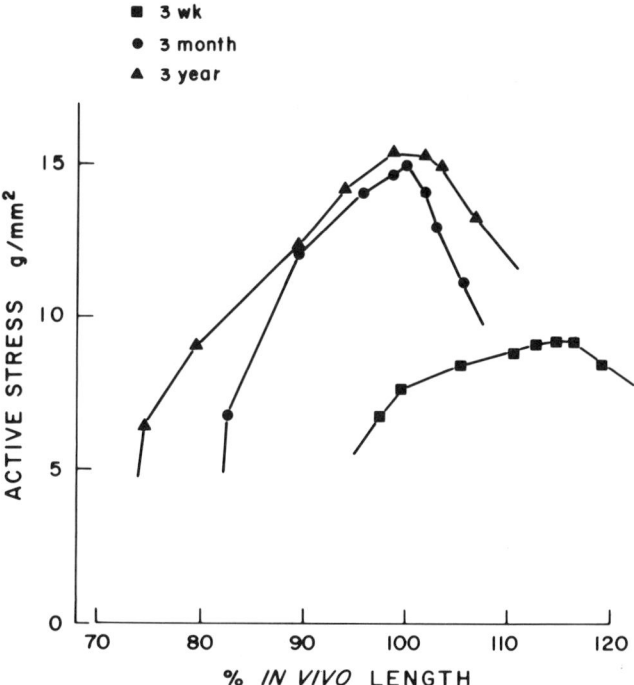

FIG 26–13.
Active stress (force/unit area)-length curves of isolated guinea pig ureteral segments as a function of age. Stress increases between the 3-week and 3-month age groups and then remains constant with further aging. (Adapted from Hong KW, Biancani P, Weiss RM: Effect of age on contractility of guinea pig ureter. *J Urol* 1980; 17:459. Used with permission.)

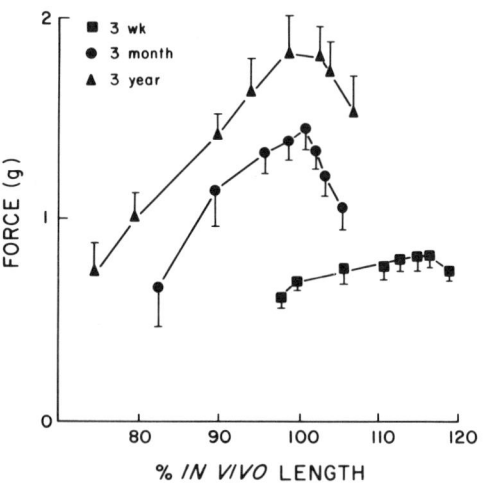

FIG 26–12.
Active force-length curves of isolated guinea pig ureteral segments as a function of age. Developed force increases with age. Data are shown as mean ± SEM. (Adapted from Hong KW, Biancani P, Weiss RM: Effect of age on contractility of guinea pig ureter. *J Urol* 1980; 17:459. Used with permission.)

URETERAL PHARMACOLOGY

The ureter is affected by both the parasympathetic (cholinergic) and sympathetic (adrenergic) branches of the autonomic nervous system.

PARASYMPATHETIC (CHOLINERGIC) SYSTEM

Acetylcholine, the prototype cholinergic agonist, serves as the neurotransmitter at (1) neuromuscular junctions of somatic motor nerves (nicotinic sites), (2) preganglionic parasympathetic and sympathetic neuroeffector junctions (nicotinic sites), and (3) postganglionic parasympathetic neuroeffector sites (muscarinic sites). The latter represents the type of receptor found on the smooth muscle cells. Acetylcholine is synthesized in the nerve terminals from acetyl CoA and choline by the enzyme choline acetyltransferase, is released into the synaptic cleft, and interacts with postganglionic receptor sites to elicit a functional response. Acetylcholine subsequently is hydrolyzed (degraded) by the enzyme acetylcholinesterase (AchE). The muscarinic effects of cholinergic agonists can be blocked by the parasympatholytic agent, atropine. The effects of nicotinic agonists can be blocked by nondepolarizing ganglionic blocking agents or by high concentrations of the nicotinic agonist itself, which may cause ganglionic blockade by desensitization of receptor sites after an initial period of ganglionic stimulation.

Acetylcholine and other cholinergic agonists such as methacholine, carbamylcholine (carbachol), and bethanechol (Urecholine) have in general been observed to have an excitatory effect on ureteral function, i.e., increase the frequency and force of contractions (Rose and Gillenwater, 1974; Vereecken, 1973). The excitatory effects of cholinergic agonists may be related to an indirect release of catecholamines (Rose and Gillenwater, 1974; Labay et al., 1968) or to a direct effect on muscarinic receptors (Vereecken, 1973). The excitatory responses to carbachol and Ach appear to be more marked in ureters from younger than older guinea pigs (Weiss et al., 1985).

Anticholinesterases (anti-ChEs) prevent hydrolysis of Ach by cholinesterases and thus potentiate the actions of acetylcholine. The effects of anticholinesterases such as physostigmine and neostigmine parallel the excitatory effects of Ach and other parasympathomimetics on the ureter (Macht, 1916). Although atropine, a competitive antagonist of the muscarinic effects of acetylcholine, has been shown to inhibit the excitatory effects of parasympathomimetic agents (Vereecken, 1973; Macht, 1916) and physostigmine (Macht, 1916) on a variety of ureteral and calyceal preparations, the majority of studies have shown that atropine itself has little direct effect on ureteral activity in many species, including humans (Reid et al., 1976; Kiil, 1957). Even when atropine has been observed to inhibit ureteral activity, its effects are frequently minimal and inconsistent (Ross et al., 1967), providing little rationale for its use in the treatment of ureteral colic.

SYMPATHETIC (ADRENERGIC) SYSTEM

The majority of investigators have noted that α-adrenergic agonists such as norepinephrine and phenylephrine stimulate ureteral activity (McLeod et al., 1973; Rose and Gillenwater, 1974; Weiss et al., 1978; Vereecken, 1973; Hannappel and Golenhofen, 1974). Norepinephrine, the chemical mediator responsible for adrenergic transmission, is synthesized in the neuron from tyrosine. Once released from the nerve terminal, some of the norepinephrine combines with postsynaptic receptors on the effector organs, such as smooth muscle, leading to a physiologic response. There are both α1- and α2-postsynaptic receptors in smooth muscle, and stimulation of either results in a contractile response. Presynaptic α2-adrenergic receptors on the neuron inhibit the release of norepinephrine from the nerve terminal, and thus excitation of these receptors are inhibitory to smooth muscle function (Fig 26–14). Reuptake or neuronal uptake of norepinephrine into the neuron limits the amount of time that norepinephrine is in contact with the innervated tissue and, thus, regulates the magnitude and duration of the catecholamine-induced response. Agents such as cocaine and imipramine (Tofranil), which inhibit neuronal uptake, potentiate the physiologic response to norepinephrine.

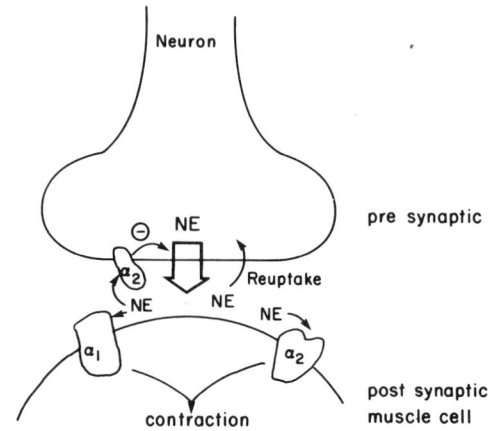

FIG 26–14.
Diagrammatic representation of the role of presynaptic and postsynaptic α receptors. Norepinephrine *(NE)* released from nerve terminals may combine with postsynaptic α1- and α2-receptors on smooth muscle to cause contraction. Norepinephrine combining with presynaptic α2-receptors on the nerve terminal prevents further release of NE from the neuron.

The enzymes monoamine oxidase (MAO) and catechol-o-methyltransferase (COMT) provide degradation pathways for norepinephrine.

β-adrenergic agonists such as isoproterenol inhibit ureteral activity (McLeod et al., 1973; Rose and Gillenwater, 1974; Weiss et al., 1978; Vereecken, 1973). The relaxant effects of isoproterenol are greater in ureters from young than from old guinea pigs (Weiss et al., 1985). Tyramine, whose adrenergic agonist effects are due primarily to the release of norepinephrine from adrenergic terminals, has a stimulatory effect on the upper urinary tract (Finberg and Peart, 1970; Longrigg, 1974).

The α-adrenergic antagonists phentolamine (Regitine) and phenoxybenzamine (Dibenzyline) have been shown to inhibit the stimulatory effects of norepinephrine and other α-adrenergic agonists in a variety of ureteral preparations (McLeod et al., 1973; Rose and Gillenwater, 1974; Weiss et al., 1978; Vereecken, 1973; Hannappel and Golenhoffen, 1974; Finberg and Peart, 1970; Longrigg, 1974). The β-adrenergic antagonist propranolol (Inderal) has been shown to block the inhibitory effects of β-adrenergic agonists such as isoproterenol in a variety of ureteral preparations. The prototype α_1-adrenergic antagonist is prazosin, and the prototype α_2-adrenergic antagonist is yohimbine.

SECOND MESSENGERS

Adrenergic and cholinergic agonists regulate physiologic processes via their interaction with a variety of specific membrane-bound receptors (Alquist, 1948; Furchgott, 1964). There is increasing evidence that the responses to these agonists are mediated via "second messengers," such as cAMP, cGMP, Ca^{++}, inositol 1,4,5-triphosphate (IP_3), and diacylglycerol (DG).

Cyclic adenosine 3′,5′-monophosphate (cAMP) is believed to mediate the relaxing effects of β-adrenergic agonists in a variety of smooth muscles including the ureter (Weiss et al., 1977; Triner et al., 1971; Andersson, 1972; Kroeger and Marshall, 1974). According to this concept, the β-adrenergic agonist, such as isoproterenol, combines with a receptor on the outer surface of the cell membrane (Fig 26–15). The β-adrenergic agonist itself does not enter the cell. The agonist-receptor complex in turn activates the enzyme adenylate cyclase on the inner surface of the cell membrane with the conversion of adenosine triphosphate (ATP) to cAMP. A guanine nucleotide-regulatory protein (G protein) acts as a functional communication between the hormone-occupied receptor and the catalytic or active unit of the adenylate cyclase. Age-dependent changes in the ability of isoproterenol to activate adenylate cyclase play a role in the effect of age on isoproterenol-induced ureteral relaxation (Wheeler et al., 1986). Cyclic AMP acts as a

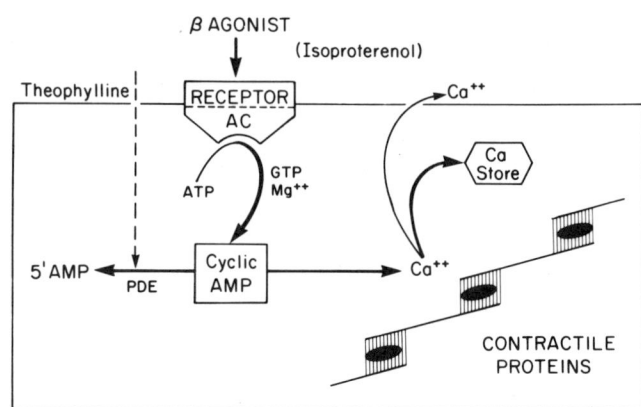

FIG 26–15.
Diagrammatic representation of the role of cAMP in β-adrenergic agonist–induced relaxation of smooth muscle. Agonist combines with receptor on the outer surface of the cell membrane. The receptor-agonist complex, in turn, activates the enzyme adenylate cyclase on the inner surface of the cell membrane, which in the presence of magnesium (Mg^{++}) and guanosine triphosphate (GTP) results in the conversion of ATP to cAMP. Cyclic AMP is postulated to cause an increased uptake of Ca^{++} into intracellular storage sites with a resultant decrease in Ca^{++} in the region of the contractile proteins, resulting in relaxation. Cyclic AMP may also act directly on the contractile proteins to inhibit the contractile process. The enzyme phosphodiesterase (PDE) degrades cAMP to 5′AMP. Theophylline is a PDE inhibitor and can thus also increase cAMP with resultant smooth muscle relaxation.

"second or internal messenger" for the response elicited by the β-adrenergic agonist. Cyclic AMP through activation of an enzyme, protein kinase, and phosphorylation of proteins has been suggested to lead to the uptake of Ca^{++} into intracellular storage sites such as the endoplasmic reticulum with a resultant decrease in free sarcoplasmic Ca^{++} and the development of relaxation (Andersson and Nilsson, 1972). The role of cAMP in relaxation of smooth muscle also may be related to phosphorylation of the enzyme myosin light chain kinase by a cAMP-dependent protein kinase (Adelstein et al., 1981).

Cyclic AMP levels may be increased by increasing its synthesis or by decreasing its degradation. Synthesis of cAMP involves activation of the enzyme adenylate cyclase, and degradation of AMP involves activation of the enzyme phosphodiesterase. Two agents that relax ureteral smooth muscle increase cAMP levels: isoproterenol by increasing synthesis and theophylline by decreasing degradation (Weiss et al., 1977). Further support for a role of cAMP in smooth muscle relaxation can be derived from the finding that dibutyryl cAMP, which more readily diffuses into the intact cell and is more resistant to breakdown by phosphodiesterase than cAMP, can relax smooth muscle (Takago et al., 1971).

Some of the actions of α_2-adrenergic agonists appear to be related to inhibition of adenylate cyclase by stimulation of an inhibitory G-protein, G_i (Fig 26–16). Some of the actions of α_1-adrenergic agonists and other hormones or neurotransmitters whose actions are associated with an increase in intracellular Ca^{++} appear to be related to changes in inositol lipid metabolism. Interaction of these agents with a receptor leads to the hydrolysis of polyphosphatidylinositol 4,5 biphosphate (PIP_2) by a phosphodiesterase with the formation of IP_3 and DG (Fig 26–17). IP_3 is involved in Ca^{++} mobilization from intracellular stores, i.e., endoplasmic reticulum (Streb et al., 1983). DG activates protein kinase C with the resultant physiologic response being dependent on protein phosphorylation (Nishizuka, 1984). DG also serves as a source of arachidonic acid (AA), the substrate for prostaglandin synthesis (Mahadevappa and Holub, 1983). Arachidonic acid may be involved in the stimulation of guanylate cyclase with the formation of cGMP (Berridge, 1984). This would explain both the calcium-dependent increase in cGMP associated with cholinergic agonist—carbachol—and the α_1-adrenergic agonist, norepinephrine-induced smooth muscle contraction. These increases in cGMP follow the onset of the contractile event. Cyclic GMP itself appears to be inhibitory to smooth muscle function. 8-Bromo cGMP relaxes the ureter and other smooth muscles, and sodium nitroprusside-induced smooth muscle relaxation is associated with an increase in cGMP (Schultz et al., 1979; Cho et al., 1984).

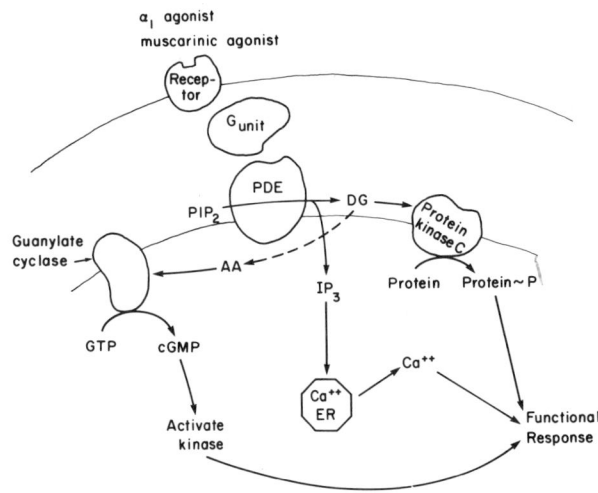

FIG 26–17.
Diagrammatic representation of the role of inositol lipid metabolism and smooth muscle function. Agonists such as α_1-adrenergic agonists and some muscarinic agonists interact with a receptor, with the resultant hydrolysis of polyphosphatidylinositol 4,5 biphosphate *(PIP₂)* by a phosphodiesterase *(PDE)* with the formation of 1,4,5-trisphosphate *(IP₃)* and diacylglycerol (DG). The functional response resulting from IP_3 is related to calcium (Ca^{++}) mobilization. DG activates *protein kinase C*, with the resultant functional response being dependent on protein phosphorylation. There is some evidence, less clear, that DG may cause an increase in cyclic GMP *(cGMP)*, which may explain the increase in cGMP observed in association with the contractile response to some autonomic agonists. *AA* indicates arachidonic acid.

Other Pharmacologic Agents

Although there are numerous reports of the excitatory effects of morphine and meperidine (Demerol) (Weiss, 1982) on the ureter, these findings have not been universal (Kiil, 1957; Ross et al., 1967). Both morphine and Demerol thus may have ureteral spasmogenic effects that theoretically would detract from their value in the management of ureteral colic. The efficacy of these agents in treating ureteral colic depends on their CNS actions, which decrease the perception of pain.

Prostaglandins are derived from fatty acids and have a variety of actions in various systems in the body. The primary prostaglandins, PGE_1, PGE_2, and $PGF_{2\alpha}$, are synthesized from the fatty acid AA by enzymatic reactions that can be inhibited by indomethacin and aspirin. The functional response to prostaglandins varies with the specific smooth muscle. In the ureter, prostaglandin E_1 (PGE_1) inhibits activity (Abrams and Feneley, 1976), whereas prostaglandin $F_{2\alpha}$ $(PGF_{2\alpha})$ is excitatory (Boyarsky and Labay, 1969).

Indomethacin has been employed in the management of ureteral colic (Holmlund and Sjöden, 1978). The beneficial effects of indomethacin probably are due

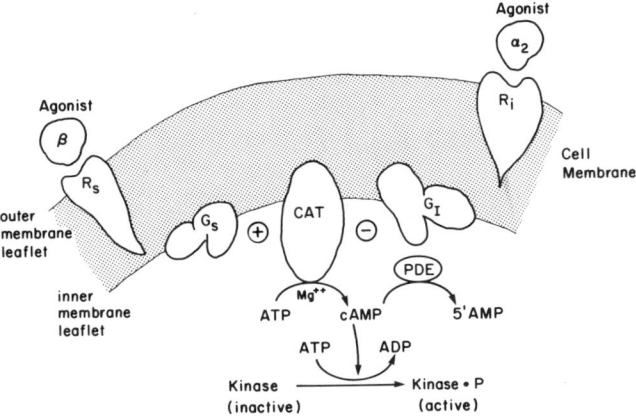

FIG 26–16.
Diagrammatic representation of the relationship between the functional response to α_2-adrenergic agonist and the cyclic nucleotide system. β-agonists activate the enzyme adenylate cyclase and increase cAMP production by working through a stimulatory receptor, R_s and a stimulatory G-protein, G_s. α_2-Adrenergic agonists inhibit adenylate cyclase activity by working through an inhibitory receptor, R_i, and an inhibitory G-protein, G_i. CAT indicates catalytic subunit of adenylate cyclase. The enzyme phosphodiesterase (PDE) degrades cAMP to 5'AMP.

to inhibition of the prostaglandin-mediated vasodilatation that occurs subsequent to obstruction (Allen et al., 1978; Sjöden et al., 1982). The vasodilatation theoretically would result in an increase in glomerular capillary pressure and subsequent increase in pelvoureteral pressure. Indomethacin, by reducing pelvoureteral pressure and thus pelvoureteral wall tension, might eliminate some of the pain of renal colic that is dependent on distention of the upper urinary tract.

Since calcium is necessary for the development of the action potential and contraction of the ureter, agents that block the movement of Ca^{++} into the cell would be expected to depress ureteral function. Indeed, calcium channel blockers such as verapamil have been shown to inhibit ureteral activity (Vereecken et al., 1975; Golenhofen and Lammel, 1973).

In the clinical situation the ureter's relatively sparce blood supply limits the distribution of drug to the ureter. In addition, many drugs that have theoretical potential use in the management of ureteral pathology have untoward side effects when used in the necessary concentrations. Although many drugs can affect ureteral function, their current clinical usefulness is limited.

URETERAL REPLACEMENT

A multitude of clinical conditions has provided an impetus for the use of ureteral substitutes. Indications for ureteral replacement include recurrent calculi, ureteral trauma, hydronephrosis, ureteral vaginal fistulae, ureteral obstruction, inflammatory disease, ureteral tumors, retroperitoneal fibrosis and a variety of undiversions. Methods for ureteral replacement can be classified as synthetic prostheses, free grafts and pedicle grafts. The use of free grafts and prosthetics has primarily been experimental, with most of the success being found with pedicle grafts.

A variety of synthetic materials has been used as ureteral substitutes including vitallium (Lord and Eckel, 1942), tantalum (Lubash, 1947), polyvinylchloride (Ulm and Lo, 1959), silicone (Schreiber et al., 1979), Teflon (Ulm and Krauss, 1960), Dacron (Block et al., 1977), polyethylene tubing (Hardin, 1957), and silicone rubber (Blum et al., 1963). These materials have not gained general acceptance because of the occurrence of salt deposits and the lack of peristalsis resulting in a functional obstruction. Studies continue to find a suitable biocompatible synthetic material. Recently, a tube of collagen sponge has been tried in an effort to capitalize on the regenerative growth potential of the ureter (Tachibana et al., 1985). The collagen sponge acts as a biodegradable scaffold for the regenerative activity of the ureter.

Free grafts have included bladder mucosa (Hovnan-

ian and Kingsley, 1966), peritoneum (Eposti, 1956), vessels (Sewell, 1955), stomach (Morelle, 1963), and fetal umbilical vessels (Klippel and Hohenfellner, 1979). Disruption and the lack of peristalsis have interfered with the clinical use of these materials.

In clinical practice, ureteral replacement has in general involved the use of pedicle grafts. Bladder flaps (Boari flaps) with or without hitching the bladder to the psoas muscle have been a highly effective means of replacing the lower ureter (Ockerblad, 1947). Flaps of the renal capsule have been used to enlarge the UPJ (Thompson et al., 1963); however, the renal capsule has not been used to replace the upper ureter. Although pedical flaps of the fallopian tube and appendix have been tried, they have not provided the required peristaltic function, and hydronephrosis has resulted (Melnikoff, 1912; Schein et al., 1956).

Ileum has been the most widely used pedicle graft and the most successful (Goodwin et al., 1959; Bower et al., 1978; Skinner and Goodwin, 1975). Proponents of the use of the ileum for replacement of the entire ureter have cautioned against its use in individuals with a serum creatinine level ≥ 2 mg/dl because of the ensuant electrolyte abnormalities (Bower et al., 1979). Middletown (1977) suggested that the absorptive surface of the ileal ureter could be minimized by tapering the segment. This procedure could also aid in the construction of an antirefluxing ileovesical anastomosis. When using the ileum to replace the ureter, the ileum should be used in an isoperistaltic manner, and there should be no evidence of distal obstruction. Immediately following the procedure, there may be considerable mucus in the urine, but this tends to clear with time. 3N Saline or acetylcysteine (Mucomyst) may aid in preventing mucous clogging of the urinary tract. Colon has also been used as an ureteral substitute, but less widely (Struthers and Scott, 1974).

Last, Lytton and Schiff (1981) used a short, isolated segment of ileum to replace a segment of ureter, with the distal anastomosis of the ileum being to the ureter. This interposition rather than replacement gives one the opportunity to take advantage of a normal, nonrefluxing UVJ and diminishes the risk of electrolyte disturbances, especially in the patient with mild renal insufficiency. The ilial segment is tapered on its distal antimesenteric surface and, if possible, placed retroperitoneally by passing it through a window in the mesentery.

REFERENCES

1. Abrams PH, Feneley RCL: The actions of prostaglandins on the smooth muscle of the human urinary tract in vitro. Br J Urol 1976; 47:909.
2. Adelstein RS, Pato MD, Conti MA: The role of phos-

phorylation in regulating contractile proteins. *Adv Cyclic Nucl Res* 1981; 14:361.

3. Akimoto M, Biancani P, Weiss RM: Comparative pressure-length diameter relationships of neonatal and adult rabbit ureters. *Invest Urol* 1977; 14:297.

4. Allen JT, Vaughan ED Jr, Gillenwater JY: The effect of indomethacin on renal blood flow and ureteral pressure in unilateral ureteral obstruction in awake dogs. *Invest Urol* 1978; 15:324.

5. Alquist RP: Study of adrenotropic receptors. *Am J Physiol* 1948; 153:586.

6. Andersson RGG: Cyclic AMP and calcium ions in mechanical and metabolic responses of smooth muscle: Influence of some hormones and drugs. *Acta Physiol Scand [Suppl]* 1972; 382:1.

7. Andersson R, Nilsson K: Cyclic AMP and calcium in relaxation in intestinal smooth muscle. *Nature* 1972; 238:119.

8. Barr L, Berger W, Dewey MM: Electrical transmission at the nexus between smooth muscle cells. *J Gen Physiol* 1968; 51:347.

9. Beckmann CF, Roth RA: Use of retrograde occlusion balloon catheters in percutaneous removal of renal calculi. *Urology* 1985; 25:277.

10. Berridge MJ: Inositol trisphosphate and diacylglycerol as second messengers. *Biochem J* 1984; 220:345.

11. Biancani P, Hausman M, Weiss RM: Effect of obstruction on ureteral circumferential force-length relations. *Am J Physiol* 1982; 243:F204.

12. Biancani P, Onyski JH, Zabinski MP, et al: Force velocity relationships of the guinea pig ureter. *J Urol* 1984; 131:988.

13. Block NL, Stover E, Politano VA: A prosthetic ureter in the dog. *Trans Am Soc Artif Intern Organs* 1977; 23:367.

14. Blok C, van Venrooij GEPM, Coolsaet BLRA: Dynamics of the ureterovesical junction; A qualitative analysis of the ureterovesical pressure profile in the pig. *J Urol* 1985; 134:818.

15. Blum J, Skemp C, Reisner M: Silicone rubber ureteral prosthesis. *J Urol* 1963; 90:276.

16. Boxer RJ, Fritzsche P, Skinner DG, et al: Replacement of the ureter by small intestine: Clinical application and results of the ileal ureter in 89 patients. *J Urol* 1979; 121:728.

17. Boxer RJ, Johnson SF, Ehrlich RM: Ureteral substitution. *Urology* 1978; 12:269.

18. Boyarsky S, Labay P: Ureteral motility. *Annu Rev Med* 1969; 20:383.

19. Bozler E: The activity of the pacemaker previous to the discharge of a muscular impulse. *Am J Physiol* 1942; 136:543.

20. Burnstock G: Structure of smooth muscle and its innervation, in Bulbring E, Brading AF, Jones AW, et al: (eds): *Smooth Muscle.* Baltimore, Williams & Wilkins Co, 1970, pp 1–69.

21. Butcher HR Jr, Sleator W Jr: A study of the electrical activity of intact and partially mobilized human ureters. *J Urol* 1955; 73:970.

22. Cho YH, Biancani P, Weiss RM: Adenyl and guanyl nu-

cleotide induced relaxation of ureteral smooth muscle. *Fed Proc* 1984; 43:353.

23. Clayton JD, Roberts JA: Radionuclide postpartum evaluation of the urinary tract during anovular therapy. *Surg Gynecol Obstet* 1973; 137:215.

24. Constantinou CE: Renal pelvic pacemaker control of ureteral peristaltic rate. *Am J Physiol* 1974; 226:1413.

25. Constantinou CE, Djurhuus JC: Pyeloureteral dynamics in the intact and chronically-obstructed multicalyceal kidney. *Am J Physiol* 1981; 241:R398.

26. Constantinou CE, Grenato JJ Jr, Govan DE: Dynamics of the upper urinary tract: Accommodation in the rate and stroke volume of ureteral peristalsis as a response to transient alteration in urine flow rate. *Urol Int* 1974; 29:249.

27. Constantinou CE, Yamaguchi O: Multiple-coupled pacemaker system in renal pelvis of the unicalyceal kidney. *Am J Physiol* 1981; 241:R412.

28. Cook WA, King LR: Vesicoureteral reflux, in Harrison JH, Gittes RF, Perlmutter AD, et al (eds): *Campbell's Urology.* Philadelphia, WB Saunders Co, 1979, pp 1596–1634.

29. Coolsaet BLRA, van Venrooij GEPM, Blok C: Detrusor pressure versus wall stress in relation to ureterovesical resistance. *Neurourol Urodynam* 1982; 1:105.

30. Copenhaver WM, Johnson DD (eds): *Bailey's Textbook of Histology,* ed 14. Baltimore, Williams & Wilkins Co, 1958.

31. Cussen LJ: The structure of the normal human ureter in infancy and childhood: A quantitative study of the muscular and elastic tissue. *Invest Urol* 1967; 5:179.

32. Daniel O, Shackman R: The blood supply of the human ureter in relation to ureterocolic anastomoses. *Br J Urol* 1952; 24:334.

33. Debruyne FMJ, Wijdeveld PGAB, Koene RAP, et al: Uretero-neo-cystomy in renal transplantation: Is an antireflux mechanism mandatory? *Br J Urol* 1978; 50:378.

34. DelTacca M: Acetylcholine content of and release from isolated pelviureteral tract. *Nauyn-Schmeidebergs Arch Pharmacol* 1978; 302:293.

35. Dixon JS, Gosling JA: The fine structure of pacemaker cells in the pig renal calices. *Anat Rec* 1973; 175:139.

36. Dixon JS, Gosling JA: The musculature of the human renal calices, pelvis and upper ureter. *J Anat* 1982; 135:129.

37. Djurhuus JC, Constantinou CE: Chronic ureteric obstruction and its impact on the coordinating mechanisms of peristalsis (pyeloureteric pacemaker system). *Urol Res* 1982; 10:267.

38. Durbin G, Gerlach R, Eichhorn G, et al: The time distance diagram: A new method to analyze ureteral peristalsis by cineradiography. *J Urol* 1980; 18:207.

39. Edmond P, Ross JA, Kirkland IS: Human ureteral peristalsis. *J Urol* 1970; 104:670.

40. Eposti PL: Regeneration of smooth muscle fibers of the ureter following plastic surgery with free flaps of the parietal peritoneum. *Minerva Chir* 1956; 11:1208.

41. Fainstat T: Ureteral dilatation in pregnancy: A review. *Obstet Gynecol Surv* 1963; 18:845.

42. Finberg JPM, Peart WS: Function of smooth muscle of the rat renal pelvis—response of the isolated pelvis muscle to angiotensin and some other substances. *Br J Pharmacol* 1970; 39:373.

43. Foote JW, Blennerhassett JB, Wiglesworth FW, et al: Observations on the ureteropelvic junction. *J Urol* 1970; 104:252.

44. Furchgott RF: Receptor mechanisms. *Annu Rev Pharmacol Toxicol* 1964; 4:21.

45. George NJR, O'Reilly PH Jr, Barnard RJ, et al: Practical management of patients with dilated upper tracts and chronic retention of urine. *Br J Urol* 1984; 56:9.

46. Golenhofen K, Lammel E: Selective suppression of some components of spontaneous activity in various types of smooth muscle by iproveratril (verapamil). *Pflugers Arch* 1973; 331:233.

47. Goodwin WE, Winter CC, Turner RD: Replacement of the ureter by small intestine: Clinical application and results of the "ileal ureter." *J Urol* 1959; 81:406.

48. Gosling JA: Atypical muscle cells in the wall of the renal calix and pelvis with a note on their possible significance. *Experientia* 1970; 26:769.

49. Gosling JA, Dixon JS: Morphologic evidence that the renal calyx and pelvis control ureteric activity in the rabbit. *Am J Anat* 1971; 130:393.

50. Gosling JA, Dixon JS: Species variation in the location of upper urinary tract pacemaker cells. *Invest Urol* 1974; 11:418.

51. Gosling JA, Dixon JS: Functional obstruction of the ureter and renal pelvis: A histological and electron microscopic study. *Br J Urol* 1978; 50:145.

52. Gosling JA, Dixon JS: Upper urinary tract: Structure, in *Textbook of Genito-urinary Surgery*. Whitfield HN, Hendry WF (eds): Edinburgh, Churchill Livingstone, 1985, pp 279–285.

53. Griffiths DJ: The mechanics of urine transport in the upper urinary tract: 2. The discharge of the bolus into the bladder and dynamics at high rates of flow. *Neurourol Urodynam* 1983; 2:167.

54. Griffiths DJ, Notschaele C: The mechanics of urine transport in the upper urinary tract: I. The dynamics of the isolated bolus. *Neurourol Urodynam* 1983; 2:155.

55. Guyer PB, Delaney D: Urinary tract dilatation and oral contraceptives. *Br Med J* 1970; 4:588.

56. Hanna MK: Some observations on congenital ureteropelvic junction obstruction. *Urology* 1978; 12:151.

57. Hannappel J, Golenhofen K: The effect of catecholamines on ureteral peristalsis in different species (dog, guinea pig and rat). *Pflugers Arch* 1974; 55:350.

58. Hardin CA: Experimental repair of ureters by polyethylene tubing and ureteral and vessel grafts. *Arch Surg* 1957; 68:57.

59. Hendrickx H, Vereecken RL, Casteels R: The influence of potassium on the electrical and mechanical activity of the guinea pig ureter. *Urol Res* 1975; 3:155.

60. Hodgkin AL: Ionic movements and electrical activity in giant nerve fibers. *Proc R Soc Lond (Biol)* 1958; 148:1.

61. Holmlund D, Hassler O: A method of studying the ureteral reaction to artificial concrements. *Acta Chir Scand* 1965; 130:335.

62. Holmlund D, Sjöden JG: Treatment of ureteral colic with intravenous indomethacin. *J Urol* 1978; 120:676.

63. Hong KW, Biancani P, Weiss RM: Effect of age on contractility of guinea pig ureter. *Invest Urol* 1980; 17:459.

64. Hovnanian AP, Kingsley IA: Reconstruction of the ureter by free autologous bladder mucosa graft. *J Urol* 1966; 96:167.

65. Hundley JM Jr, Diehl WK, Diggs ES: Hormonal influences upon the ureter. *Am J Obstet Gynecol* 1942; 44:858.

66. Kiil F: *The Function of the Ureter and Renal Pelvis*. Philadelphia, WB Saunders Co, 1957.

67. Kirkland IS, Ross JA, Edmond P, et al: Ureteral function in vesicoureteral reflux. *Br J Urol* 1971; 43:289.

68. King WW, Cox CE: Bacterial inhibition of ureteral smooth muscle contractility: I. The effect of common urinary pathogens and endotoxin in an in vitro system. *J Urol* 1972; 108:700.

69. Klippel KF, Hohenfellner R: Umbilical vein as ureteral replacement. *Invest Urol* 1979; 16:447.

70. Kobayashi M: Effects of Na and Ca on the generation and conduction of excitation in the ureter. *Am J Physiol* 1965; 208:715.

71. Kobayashi M: Effect of calcium on electrical activity in smooth muscle cells of cat ureter. *Am J Physiol* 1969; 216:1279.

72. Kobayashi M, Irisawa H: Effect of sodium deficiency on the action potential of the smooth muscle of ureter. *Am J Physiol* 1964; 206:205.

73. Kroeger EA, Marshall JM: Beta-adrenergic effects on rat myometrium: Role of cyclic AMP. *Am J Physiol* 1974; 226:1298.

74. Kubacz GJ, Catchpole BN: The role of adrenergic blockade in the treatment of ureteral colic. *J Urol* 1972; 107:949.

75. Kumar D: In vitro inhibitory effect of progesterone on extrauterine smooth muscle. *Am J Obstet Gynecol* 1962; 84:1300.

76. Kuriyama H: The influence of potassium, sodium and chloride on the membrane potential of the smooth muscle of *Taenia coli*. *J Physiol (Lond)* 1963; 166:15.

77. Kuriyama H, Tomita T: The action potential in the smooth muscle of the guinea pig *Taenia coli* and ureter studied by the double sucrose-gap method. *J Gen Physiol* 1970; 55:147.

78. Kuriyama H, Osa T, Toida N: Membrane properties of the smooth muscle of guinea-pig ureter. *J Physiol (Lond)* 1967; 191:225.

79. Labay PC, Boyarsky S, Herlong JH: Relation of adrenal to ureteral function. *Fed Proc* 1968; 27:444.

80. Latifpour J, O'Hollaren B, Weiss RM: Unpublished data, 1987.

81. Libertino JA, Weiss RM: Ultrastructure of human ureter. *J Urol* 1972; 108:71.

82. Longrigg N: Autonomic innervation of the renal calyx. *Br J Urol* 1974; 46:357.

83. Longrigg N: Minor calyces as primary pacemaker sites for ureteral activity in man. *Lancet* 1975; 1:253.

84. Lord JW Jr, Eckel JH: The use of Vitallium tubes in the urinary tract of the dog. *J Urol* 1942; 48:412.

85. Lubash S: Experiences with tantalum tubes in the reimplantation of the ureters into the sigmoid in dogs and humans. *J Urol* 1947; 57:1010.

86. Lubin S, Drexler LS, Bilotta WA: Post-partum pyeloureteral changes following hormone administration. *Surg Gynecol Obstet* 1941; 73:391.

87. Lytton B, Schiff M: Interposition of an ileal segment for repair of ureteral injuries. *J Urol* 1981; 125:739.

88. Macht DI: On the pharmacology of the ureter: II. Actions of drugs affecting the sacral autonomics. *J Pharmacol Exp Ther* 1916; 8:261.

89. Mahadevappa VG, Holub BJ: Degradation of different molecular species of phosphatidylinositol in thrombin-stimulated human platelets. *J Biol Chem* 1983; 258:5337.

90. Makker SP, Tucker AS, Izant RJ Jr, et al: Nonobstructive hydronephrosis and hydroureter associated with peritonitis. *N Engl J Med* 1972; 287:535.

91. Marchant DJ: Effects of pregnancy and progestational agents on the urinary tract. *Am J Obstet Gynecol* 1972; 112:487.

92. Marshall S, Lyon RP, Minkler D: Ureteral dilatation following use of oral contraceptives. *JAMA* 1966; 198:206.

93. McGuire EJ: Physiology of the lower urinary tract. *Am J Kid Dis* 1983; 2:402.

94. McLeod DG, Reynolds DG, Swan KG: Adrenergic mechanisms in the canine ureter. *Am J Physiol* 1973; 224:1054.

95. McNellis D, Sherline DM: The rabbit ureter in pregnancy and after norethynodrel-mestranol administration: A radiographic and histologic study. *Obstet Gynecol* 1967; 30:336.

96. Melick WF, Brodeur AE, Herbig F, et al: Use of a ureteral pacemaker in the treatment of ureteral reflux. *J Urol* 1966; 95:184.

97. Melick WF, Naryka JJ, Schmidt JH: Experimental studies of ureteral peristaltic patterns in the pig: II. Myogenic activity of the pig ureter. *J Urol* 1961; 86:46.

98. Melnikoff AE: Sur le replacement de l'uretere par une anse isolee de l'intestin Grele. *Rev Clin J Urol* 1912; 1:601.

99. Middleton AW Jr: Tapered ileum as ureter substitute in severe renal damage. *Urology* 1977; 9:509.

100. Morelle VR: Replacement of the ureter by a segment of the stomach in pigs and dogs. *Arch Chir Neerl* 1963; 15:293.

101. Morita T, Ishizuka G, Tsuchida S: Initiation and propagation of stimulus from the renal pelvic pacemaker in pig kidney. *Invest Urol* 1981; 19:157.

102. Morita T, Kondo S, Suzuki T, et al: Effect of calyceal resection on pelviureteral peristalsis in isolated pig kidney. *J Urol* 1986; 135:151.

103. Murnaghan GF: The dynamics of the renal pelvis and ureter with reference to congenital hydronephrosis. *Br J Urol* 1958b; 30:321.

104. Murnaghan GF: Mechanisms of congenital hydronephrosis with reference to factors influencing surgical treatment. *Ann R Coll Surg Engl* 1958a; 23:25.

105. Nishizuka Y: The role of protein kinase C in cell surface signal transduction and tumor production. *Nature* 1984; 308:693.

106. O'Conor VJ Jr, Dawson-Edwards P: Role of the ureter in renal transplantation: I. Studies of denervated ureter with particular reference to ureteroureteral anastomosis. *J Urol* 1969; 82:566.

107. Ockerblad NF: Reimplantation of the ureter into the bladder by a flap method. *J Urol* 1947; 57:845.

108. Paquin AJ Jr: Ureterovesical anastomosis: The description and evaluation of a technique. *J Urol* 1959; 82:573.

109. Payne FL, Hodes PJ: The effect of the female hormones and of pregnancy upon the ureters of lower animals as demonstrated by intravenous urography. *Am J Obstet Gynecol* 1939; 37:1024.

110. Peters HJ, Eckstein W: Possible pharmacological means of treating renal colic. *Urol Res* 1975; 3:55.

111. Primbs K: Untersuchungen uber die Einwirkung von Bakterientoxinen auf der uberlebenden Meerschweinchenureter. *Z Urol Chir* 1913; 1:600.

112. Raz, S, Ziegler M, Caine M: Hormonal influence on the adrenergic receptors of the ureter. *Br J Urol* 1972; 44:405.

113. Reid RE, Herman R, Teng C: Attempts at altering ureteral activity in the unanesthetized, conditioned dog with commonly employed drugs. *Invest Urol* 1976; 12:74.

114. Roberts JA: Experimental pyelonephritis in the monkey: III. Pathophysiology of ureteral malfunction induced by bacteria. *Invest Urol* 1975; 13:117.

115. Roberts JA: Hydronephrosis of pregnancy. *Urology* 1976; 8:1.

116. Rose JG, Gillenwater JY: Pathophysiology of ureteral obstruction. *Am J Physiol* 1973; 225:830.

117. Rose JG, Gillenwater JY: The effect of adrenergic and cholinergic agents and their blockers upon ureteral activity. *Invest Urol* 1974; 11:439.

118. Ross JA, Edmond P, Griffiths JM: The action of drugs on the intact human ureter. *Br J Urol* 1967; 39:26.

119. Ross JA, Edmond P, Kirkland IS: *Behaviour of the Human Ureter in Health and Disease.* Edinburgh, Churchill Livingstone, 1972.

120. Sala NL, Rubi RA: Ureteral function in pregnant women: II. Ureteral contractility during normal pregnancy. *Am J Obstet Gynecol* 1967; 99:228.

121. Satani Y: Histological study of the ureter. *J Urol* 1919; 3:247.

122. Schein CJ, Sanders AR, Hurwitt ES: Experimental reconstruction of the ureters. *Arch Surg* 1956; 73:47.

123. Schick E, Tanagho EA: The effect of gravity on ureteral peristalsis. *J Urol* 1973; 109:187.

124. Schneider DH, Eichner E, Gordon MB: An attempt at production of hydronephrosis of pregnancy, artificially induced. *Am J Obstet Gynecol* 1953; 65:660.

125. Schreiber B, Homann W, Mlynek M, et al: Ureteral re-

placement with a new prosthesis. *Trans Am Soc Artif Intern Organs* 1979; 25:61.

126. Schultz K, Bohme E, Volker AWK, et al: Relaxation of hormonally-stimulated smooth muscular tissues by the 8-bromo derivative of cyclic GMP. *Naunyn Schmiedebergs Arch Pharmacol* 1979; 306:1.

127. Sewell WH: Failure of freeze-dried homologous arteries used as arterial grafts. *J Urol* 1955; 74:600.

128. Sivula A, Lehtonen T: Spontaneous passage of artificial concretions applied in the rabbit ureter. *Scand J Urol Nephrol* 1967; 1:259.

129. Sjöden JG, Wahlberg J, Persson AEG: The effect of indomethacin on glomerular capillary pressure and pelvic pressure during ureteral obstruction. *J Urol* 1982; 127:1017.

130. Skinner DG, Goodwin WE: Indications for the use of intestinal segments in management of nephrocalcinosis. *J Urol* 1975; 113:436.

131. Streb H, Irvine RF, Berridge MJ, et al: Release of Ca^{++} from a non-mitochondrial store in pancreatic acinar cell by inositol, 1,4,5-triphosphate. *Nature* 1983; 306:67.

132. Struthers NW, Scott R: Reconstruction of the upper ureter with colon. *J Urol* 1974; 112:179.

133. Tachibana M, Nagamatsu GR, Addonizio JC: Ureteral replacement using collagen sponge tube grafts. *J Urol* 1985; 133:866.

134. Tachibana S, Takeuchi M, Uehara Y: The architecture of the musculature of the guinea-pig ureter as examined by scanning electron microscopy. *J Urol* 1985; 134:582.

135. Takago K, Takayanagi I, Tomiyama A: Actions of dibutyryl cyclic adenosine monophosphate, papaverine and isoprenaline on intestinal smooth muscle. *Jpn J Pharmacol* 1971; 21:477.

136. Tanagho EA: Ureteral embryology, developmental anatomy, and myology, in Boyarsky S, Gottshalk GW, Tanagho EA, et al (eds): *Urodynamics: Hydrodynamics of the Ureter and Renal Pelvis.* New York, Academic Press, 1971, pp 3–27.

137. Tanagho EA, Hutch JA, Meyers FH, et al: Primary vesicoureteral reflux: Experimental studies of its etiology. *J Urol* 1965; 93:165.

138. Tanagho EA, Meyers FH, Smith DE: The trigone: Anatomical and physiological considerations: 1. In relation to the ureterovesical junction. *J Urol* 1968; 100:623.

139. Teague N, Boyarsky S: The effect of coliform bacilli upon ureteral peristalsis. *Invest Urol* 1968; 5:423.

140. Teele RL, Lebowitz RL, Colodny AH: Reflux into the unused ureter. *J Urol* 1976; 115:310.

141. Thompson IM, Kovacsi L, Portersfield J: Reconstruction of the ureteropelvic junction with pedicle grafts and renal capsule. *J Urol* 1963; 89:573.

142. Traut HF, Kuder A: Inflammation of the upper urinary tract complicating the reproductive period of woman: Collective review. *Int Abst Surg* 1938; 67:568.

143. Triner L, Nahas GG, Vulliemoz Y, et al: Cyclic AMP and smooth muscle function. *Ann NY Acad Sci* 1971; 185:458.

144. Tsuchida S, Yamaguchi O: A constant electrical activity of the renal pelvis correlated to ureteral peristalsis. *Tohoku J Exp Med* 1977; 121:133.

145. Uehara Y, Burnstock G: Demonstration of "gap junctions" between smooth muscle cells. *J Cell Biol* 1970; 44:215.

146. Ueno A, Kawamura T, Ogawa A, et al: Relation of spontaneous passage of ureteral calculi to size. *Urology* 1977; 10:544

147. Ulm AH, Krauss L: Total unilateral Teflon ureteral substitutes in the dog. *J Urol* 1960; 83:575.

148. Ulm AH, Lo MC: Total bilateral polyvinyl ureteral substitutes in the dog. *Surgery* 1959; 45:313.

149. van Wagenen G, Jenkins RH: An experimental examination of factors causing ureteral dilatation of pregnancy. *J Urol* 1939; 42:1010.

150. Vereecken RL: *Dynamical Aspects of Urine Transport in the Ureter.* Louvain, Acco, 1973.

151. Vereecken RL, Hendrickx H, Casteels R: The influence of calcium on the electrical and mechanical activity of the guinea pig ureter. *Urol Res* 1975a; 3:149.

152. Vereecken RL, Hendrickx H, Casteels R: The influence of sodium on the electrical and mechanical activity of the guinea pig ureter. *Urol Res* 1975b; 3:159.

153. Washizu Y: Grouped discharges in ureter muscle. *Comp Biochem Physiol* 1966; 19:713.

154. Watterson DM, Harrelson WG Jr, Keller PM, et al: Structural similarities between the Ca^{++}-dependent regulatory proteins of $3':5'$-cyclic nucleotide phosphodiesterase and actomyosin ATPase. *J Biol Chem* 1976; 251:4501.

155. Weinberg SL: Ureteral function: I. Simultaneous monitoring of ureteral peristalsis. *Invest Urol* 1974; 12:103.

156. Weiss RM: Clinical implications of ureteral physiology. *J Urol* 1979; 121:401.

157. Weiss RM: Pharmacology of the ureter, in Finkbeiner AE, Barbour GL, Bissada NK (eds): *Pharmacology of the Urinary Tract and Male Reproductive System.* New York, Appleton-Century-Crofts, 1982, pp 137–173.

158. Weiss RM, Bassett AL, Hoffman BF: Adrenergic innervation of the ureter. *Invest Urol* 1978; 16:123.

159. Weiss RM, Biancani P: A rationale for ureteral tapering. *Urology* 1982; 20:482.

160. Weiss RM, Biancani P: Characteristics of normal and refluxing ureterovesical junctions. *J Urol* 1983; 129:858.

161. Weiss RM, Biancani P, Zabinski MP: Adrenergic control of the ureteral tonus. *Invest Urol* 1974; 12:30.

162. Weiss RM, Vulliemoz Y, Verosky M, et al: Adenylate cyclase and phosphodiesterase activity in rabbit ureter. *Invest Urol* 1977; 15:15.

163. Weiss RM, Wagner ML, Hoffman BF: Localization of pacemaker for peristalsis in the intact canine ureter. *J Urol* 1967; 5:42.

164. Weiss RM, Wheeler MA, Biancani P: Age related changes in the response of ureteral smooth muscle to autonomic agonists. *Fed Proc* 1985; 44:506.

165. Wharton LR: The innervation of the ureter, with respect to denervation. *J Urol* 1932; 28:639.

166. Wheeler MA, Housman A, Weiss RM: Age dependence of adenylate cyclase activity in guinea pig ureter homogenate. *J Pharmacol Exp Ther* 1986; 99:239.

167. Whitaker RH: Methods of assessing obstruction in dilated ureters. *Br J Urol* 1973; 45:15.

168. Whitaker RH: The Whitaker test. *Urol Clin North Am* 1979; 6:529.

169. Williams PL, Warwick R (eds): *Gray's Anatomy*, Philadelphia, WB Saunders Co, 1980.

170. Woodburne RT, Lapides J: The ureteral lumen during peristalsis. *Am J Anat* 1972; 133:255.

Chapter 27

Voiding Function: Relevant Anatomy, Physiology, and Pharmacology

ALAN J. WEIN, M.D.
ROBERT M. LEVIN, Ph.D
DAVID M. BARRETT, M.D.

Despite the fact that there have been an extraordinary number of good recent reviews on the subject (Tanagho, 1978; Wein and Raezer, 1979; Gosling, 1979; Andersson and Sjogren, 1982; Bradley and Sundin, 1982; Bhatia and Bradley, 1983; McGuire, 1983, 1984a; DeGroat and Booth, 1984; Mundy, 1984a), it is obvious that the anatomy, physiology, and pharmacology of the central neural and peripheral neural and muscular structures controlling micturition have not been completely elucidated. These reviews have been written from various orientations—those of anatomist, neurophysiologist, pharmacologist, and, sometimes, clinician. However, there exists in and between many of these reviews imprecise and noncomparable terminology; major conflicts regarding fundamental concepts; and very basic disagreements on fundamental facts of anatomy, physiology, and pharmacology; either by virtue of different findings and their citation, or simply by different interpretations of fundamental data. In this chapter, we hope to at least summarize many currently accepted views on the structure and function of the lower urinary tract. In many instances we will hopefully be able to present tenable hypotheses that link clinical observations to anatomic, physiologic, and pharmacologic facts.

At the outset, it might be well to briefly summarize exactly what we are trying to explain. The lower urinary tract functions as a group of interrelated structures whose joint function is to bring about efficient bladder filling and urine storage and its voluntary expulsion. During normal filling there is a very slow rise in bladder pressure, despite a large increase in volume. There is a gradual increase in proximal urethral resistance, and there is no involuntary bladder contractile activity. Increases in intra-abdominal pressure do not cause urine leakage. At a certain intravesical pressure, the sensation of bladder distention is perceived and micturition is initiated voluntarily, proceeding to completion and involving a decrease in bladder outlet resistance and contraction of the bladder smooth musculature, with some reshaping of the proximal urethra. To explain these relatively simple-sounding phenomena, an impressive literature has been generated. It will be our goal to summarize and synthesize this literature into consistent working hypotheses. It will become readily apparent that no one prior author's viewpoint will be constantly supported. This is simply because scientists are not consistent, and no two groups of people agree about everything. We have tried to find viewpoints expressed by persons whom we respect that agree with our own particular concepts, but it will be obvious that although we may agree with and endorse one author's viewpoint in one area, this may not be the case in another area.

No matter what type of organization one uses for this subject, it is impossible to list all data in an order that seems entirely logical on first reading. We therefore suggest an initial more rapid review of this material followed by an in-depth reading. Figures are used, especially in the anatomy section, to illustrate particularly relevant points, but we have tried to avoid graphic illustration of mere laboratory data. In some cases inclusive review articles, rather than original ones, are cited; we apologize to those whose "classic" references may have been omitted.

To provide the reader with a basis for criticizing our conclusions, as well as those of others, it is well to keep in mind at the outset the following observations:

1. The bewildering array of experimental models must be carefully considered. A small change in one experimental model may make results totally noncomparable to those obtained in another.

2. Great differences may exist between species; results in certain animal models do not necessarily imply exact similarity to humans.

3. Results obtained in one type of tissue may not equal those obtained in another, even if the tissues are from the same species, seem similar, and are adjacent anatomically.

4. Significant changes in anatomy, physiology, and pharmacology may be secondary to alterations induced by or related to sex, age, hormonal status, infection, denervation or decentralization, stretch, and the effects of certain drugs, including laboratory and clinical anesthetics.

5. In vitro experimental results are not necessarily the same as in vivo experimental results, and neither of these are necessarily equivalent to "normal."

6. Anatomy (structure) does not necessarily imply function or physiology, and vice versa.

7. Receptors do not necessarily imply innervation or function.

8. The effect produced by an agonist or antagonist does not necessarily impart physiologic significance to a particular neurotransmitter, or, by inference, physiologic significance to a component of the nervous system that secretes that neurotransmitter.

9. Pressure or tension generation by a whole organ model or muscle strip is not necessarily the same as contractility, and neither of these is necessarily translatable into bladder-emptying ability. The same caveat applies to assumptions made about outlet resistance.

10. The altered physiology of an abnormal clinical state can usually be explained according to at least a part of one theory of micturition. However, this does not necessarily mean that that theory is entirely correct.

11. Authors, ourselves included, tend to be somewhat selective in their citation of references. Generally, references that support a particular viewpoint are cited more frequently, more conspicuously, and more favorably than those that do not.

RELEVANT ANATOMY

THE BLADDER

The urinary bladder wall consists of three layers: (1) an outer adventitia composed primarily of connective tissue; (2) a smooth muscle layer; and (3) an inner mucous membrane that totally lines the interior. Although the mucosa has recently been demonstrated to contain irregularly arranged smooth muscle bundles (Dixon and Gosling, 1983), their function is uncertain, and they will not be considered further. The main smooth muscle layer of the bladder is classically divided into detrusor and trigone. Anatomically, the trigonal area is defined as that region of the posterior bladder wall between the ureteral orifices and the vesicourethral junction. This area is best divided into a superficial and a deep trigone. The deep trigone is actually continuous with, and an integral part of, the detrusor smooth muscle (El Badawi, 1982; Tanagho, 1982; Gosling and Chilton, 1984). Tanagho and Pugh (1963) seem to have been among the first to recognize the subdivision of the trigonal area, and originally described three separate layers. The superficial layer of the trigonal musculature is wholly ureteral in origin and, as such, is mesodermal in origin, while the origin of the remainder of the bladder is endodermal (Tanagho and Pugh, 1963; Woodburne, 1967). In his most recent description, Tanagho (1982) still considered the actual trigone to be divided into superficial and deep layers, both of ureteral origin (the former from the longitudinal musculature of the intravesical ureter and the latter from Waldeyer's sheath), with a separate layer of detrusor muscle below this.

The detrusor can best be described as a mesh of smooth muscle bundles that are, to a greater (Donker et al., 1982) or lesser extent (Bro-Rasmussen et al., 1965; Woodburne, 1968; Hald and Bradley, 1982; Gosling and Chilton, 1984) loosely organized in the area of the bladder base into an outer longitudinal, a middle circular, and an inner longitudinal layer. There is significant disagreement (see below) regarding the continuity of a smooth muscle layer of the bladder (including the trigone) into the urethra.

The terms "bladder body" and "bladder base" refer to a functional rather than anatomic division of the bladder smooth muscle, which is based on distinct differences in neuromorphology and neuropharmacology between the smooth muscle lying circumferentially above and below the level of the ureterovesical junction. This concept was first introduced by El Badawi and Schenk (1966), and subsequently confirmed by them and others (El Badawi and Schenk, 1968; Raezer et al., 1973; El Badawi, 1982).

THE URETHRA

Female Urethra

The adult female urethra is approximately 4 cm in length and 6 mm in diameter. It extends from the vesicourethral junction behind the symphysis pubis and is embedded in the anterior wall of the vagina. The wall of the female urethra comprises an outer muscular layer and inner epithelium, which is generally thrown into longitudinal folds and apposed to itself except during the act of voiding. The smooth muscle portion of the female urethra extends throughout its length. There is general agreement on the existence of an inner longitudinal layer (Tanagho, 1978; Gosling and Chilton, 1984), but disagreement over the extent of an outer circular or semicircular layer. Tanagho (1978) feels that the outer semicircular coat is substantial and a direct continuation of the detrusor outer longitudinal layer, but Gosling and Chilton (1984) describe only a few circularly arranged muscle fibers in the outer aspect of the smooth muscle layer. This distinction is important when considering whether or not an active urethra smooth muscle sphincteric mechanism is involved in maintaining continence. Urethral smooth muscle is embedded in a matrix of collagen, and actually collagen seems to be the major structural component of the human female urethra (Hickey et al., 1982). These collagen fibers have an orientation similar to the smooth muscle fibers.

Male Urethra

The anterior urethra (distal to the membranous portion) in the male seems to function purely as a conduit and has no role in either maintaining continence or in facilitating micturition. Gosling and Chilton (1984) described the preprostatic urethra as approximately 1–1.5 cm in length and containing smooth muscle bundles oriented as a distinct circular collar and becoming continuous distally with the prostatic capsule. These muscle bundles are separated by much connective tissue containing elastin and collagen. The prostatic urethra is approximately 3–4 cm in length. Tanagho (1986) seemed to imply that the entire male posterior urethra contains a representation of an inner longitudinal smooth muscle layer and an outer circular layer, least developed in the midline dorsally. He noted that the urethral musculature in this area is very compact and contains abundant collagen and elastic fibers. This type of arrangement also seems to be implied by Donker et al. (1982), i.e., the existence to some extent of both longitudinally and circularly oriented smooth muscle bundles in the male preprostatic and prostatic urethra. The membranous portion of the male posterior urethra extends for a distance of at least 2.5 cm distal to the prostate, to the bulb of the penis. The presence of a smooth muscular element in the wall of the membranous urethra has been described by Donker et al. (1982) and by Tanagho (1986), and it would seem from their descriptions that some such longitudinal and some circular fibers are present. Gosling (1984, 1985) describes a relatively small smooth muscle component in this area, but one which consists of outer circular and inner longitudinal components. Proximally this smooth muscle is in continuity with the fibromuscular trabecu-

lae of the prostatic apex and distally is continuous into the bulbous part of the urethra.

Anatomic Continuity Between Bladder and Urethra

Whether bladder smooth muscle bundles are continuous into the urethra is a matter of dispute. The concept of the continuation of the inner longitudinal layer of bladder smooth muscle directly into the urethra is endorsed by Woodburne (1967, 1968), Lapides (1958, 1970), Lapides et al. (1960), Tanagho and Smith (1966, 1968), Bradley and Sundin (1982), and El Badawi (1982, 1983). Tanagho and Smith (1968) described continuation of the superficial trigonal layer into the dorsal urethral wall up to the verumontanum in the male and into the proximal two-thirds of the female urethra. Donker et al. (1982) confirm this finding relative to the superficial layer of the trigone, an idea also attributed to Hutch (1967) and to Hutch and Rambo (1967). Such an anatomical arrangement is obviously consistent with the hypothesis that opening of the bladder neck and proximal urethra during active bladder contraction is based at least in part on anatomic factors. This view, however, is disputed by McNeal (1972), Gosling (1979, 1984), Donker et al. (1982), and Gilpin and Gosling (1983). These authors regard the urethral smooth muscle system as entirely independent from that of the bladder, and developmentally unrelated to it. This would imply that the adaptive changes that occur in the bladder outlet during voiding are mediated by factors other than those related to muscular continuity and arrangement.

THE STRIATED MUSCLE COMPONENT

The classic view of the external sphincter is that of a striated muscle within the leaves of the urogenital diaphragm, which, on contraction, is capable of stopping the urinary system (Wein and Raezer, 1979). Such a structure was described by Woodburne (1967) as completely annular in the male, though decreased in bulk posteriorly, and blending with the fibers of the prostatic capsule. In the female, the muscle is classically described as tapering considerably and being deficient posteriorly.

A contemporary discussion of the striated muscle component includes both an intramural or intrinsic (to the urethra) and an extramural or extrinsic portion. The intrinsic portion, termed the "rhabdosphincter" by El Badawi (1982), denotes skeletal muscle intimately associated with the proximal (prediaphragmatic) urethra and consisting of several bundles oriented in an oblique vertical direction along the urethra, freely interdigitating with and forming an integral part of its outer muscular layer. In the male, Gosling (1979) and Gosling and Chilton (1984) described this component as being more circularly oriented and completely encircling the urethra from the verumontanum to the striated musculature of the pelvic floor. Tanagho (1986) basically endorses this description in the male, but feels that this intrinsic rhabdosphincter is deficient in the midline posteriorly. He describes some striated muscle fibers as extending over the apex of the prostate in the male and a few as being distributed up to the level of the bladder neck. In the female the intrinsic striated component is thickest in the middle third of the urethra, but was described by Gosling (1984) as being relatively deficient posteriorly in this region. Tanagho (1986), while basically agreeing with this description, simply refers to a raphe in the posterior midline.

In both sexes, the caudal end of the intramural or intrinsic striated component is classically described as adjacent to a bulky skeletal muscle structure oriented in the horizontal plane of the pelvic floor, encircling the membranous urethra, and corresponding to the so-called "external urethral sphincter" of standard anatomy textbooks. This "periurethral muscle" (Gosling, 1979, 1984; Gosling and Chilton, 1984) is related to, but structurally separate from, the urethral wall, and corresponds to Gosling's description of the medial parts of the levator ani muscles. In reviewing older literature, it is quite apparent that at least the anatomic concept of an intrinsic (intramural) vs. an extrinsic (extramural) striated component was clearly implied in the works of Woodburne (1967, 1968), Hutch (1967), Gil Vernet (1968), Tanagho and Smith (1968), and Sant (1962). Evidence presented later will show that this intrinsic striated component is also functionally different from the extrinsic component and particularly suited for the maintenance of urinary continence.

Oelrich has presented a very scholarly description of his concept of the striated muscle component in both the male (1980) and the female (1983). In the male, he described the striated urethral sphincter as being derived from a single primordium, which originally extended from the bladder to the perineal membrane. Development of the prostate occurs from the urethra and into the overlying striated muscle, thinning the overlying parts and causing a reduction or atrophy of some of the muscle in this area, while overlying it and incorporating it within the prostatic perimeter. This accounts for what appear to be isolated segments of striated sphincter muscle distributed around and in the prostate. Further, the fascia of this sphincter muscle was described as being inseparable from the prostatic sheath, oriented vertically, and passing through the urogenital hiatus to unite with the fascia of the pudendal canals at the level of the ischiopubic rami. The concept of a urogenital diaphragm comprising a transverse plane of muscle spanning between the two ischiopubic

rami between a superior and inferior fascial plane, and surrounding the membranous urethra (implying the existence of an extrinsic portion of a striated urethral sphincter) was not confirmed. In the female, Oelrich described the striated component as consisting of two parts, one surrounding the urethra and the other surrounding the urethra and vagina. Both of these are infiltrated with smooth muscle, which somewhat obscures their visibility on gross dissection. The orientation of the portion that surrounds the urethra is predominantly circular, and thins out posteriorly. A separate bundle of fibers is attached to the ischiopubic rami and constitutes a compressor muscle, arching across the ventral side of the urethra. Additionally, there is a urethrovaginal sphincter surrounding the urethra and vaginal vestibule collectively. These designated muscles are in continuity with one another and are located within the pelvic cavity in the urogenital hiatus of the pelvic diaphragm.

THE SPHINCTER CONCEPT

The Smooth Sphincter

There is no anatomic sphincter at the bladder neck. Rather, there is a physiologic internal sphincter that consists of the bladder neck and proximal urethra. Both of these areas are rich in elastic fibers, and the inherent tension exerted by their walls on the lumen is dependent on these fibers and on the tone of the smooth musculature. This "smooth sphincter" tone rises during normal urine storage so that urethral pressure exceeds intravesical pressure. Normal micturition implies a number of events, one being a reduction in resistance in this "smooth sphincter" area. It is important to remember that a functional obstruction can occur at the level of the smooth sphincter either because of a failure of the usual adaptive changes that occur in this area during detrusor contraction, or because of actual simultaneous active contraction (smooth sphincter dyssynergia).

The Striated Sphincter

The striated sphincter comprises both the intramural and extrinsic striated component surrounding the urethra in the female and the posterior urethra in the male. Lapides' classical experiments involving progressive shortening of the urethra proximal to the urogenital diaphragm to within 2 cm of the bladder neck showed that the striated sphincter is not necessary for urinary continence (1957, 1958). Curarization of this striated sphincter alone does not normally result in incontinence. Incontinence may result under these circumstances, however, if the smooth sphincter apparatus has been destroyed, as with a prostatectomy.

Pathologic hyperreactivity (striated sphincter dyssynergia) can produce severe functional obstruction during bladder contraction.

ANATOMIC ILLUSTRATIONS

The following figures demonstrate and will hopefully reinforce the previous discussion of relevant anatomy of the bladder and the bladder outlet. Figure 27–1 shows arching smooth muscle bundles in the area of the bladder neck. This is the only area in which there exists even a loose arrangement of the bladder smooth muscle into three vaguely distinct layers. Woodburne's concept (1967) was that the fibers at the neck are arranged in such a way that their coordinated contraction pulls the neck open rather than constricts it. This would at least seem to imply that some longitudinal component of bladder smooth muscle extends into the proximal urethra. Also, it might be inferred that bladder distention without active contraction might actually serve to constrict the vesicourethral junction. Figure 27–2 shows Bradley's concept of why the bladder neck opens during bladder contraction. Randomly distributed interdigitating smooth muscle fibers from the bladder dome form arcades as the bladder base is approached. The trigonal arcade intersects the detrusor arcade, which faces in an opposite direction. When normal detrusor reflex excitation occurs, there is contraction of the two smooth muscle arcades, producing separation at the

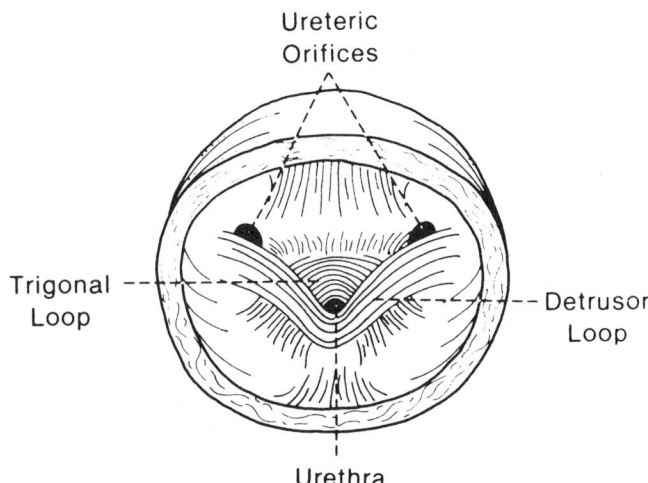

FIG 27–2.
Dorsal view of the bladder neck with schematic representation of the trigonal and detrusor smooth muscle loops. (From Hald T, Bradley W: *The Urinary Bladder: Neurology and Dynamics.* Baltimore, Williams & Wilkins, 1982. Used with permission.)

bladder neck with expulsion of urine into the proximal urethra. Figure 27–3 diagrams Hutch's concept of the female bladder neck and proximal urethra. He pictured an "internal sphincter" as being located in the bladder base and consisting anteriorly and laterally of the fundus ring and posteriorly of the trigonal muscle. His urethral sphincter consists of the smooth muscle layers of the urethra. Both the outer circular layer and inner lon-

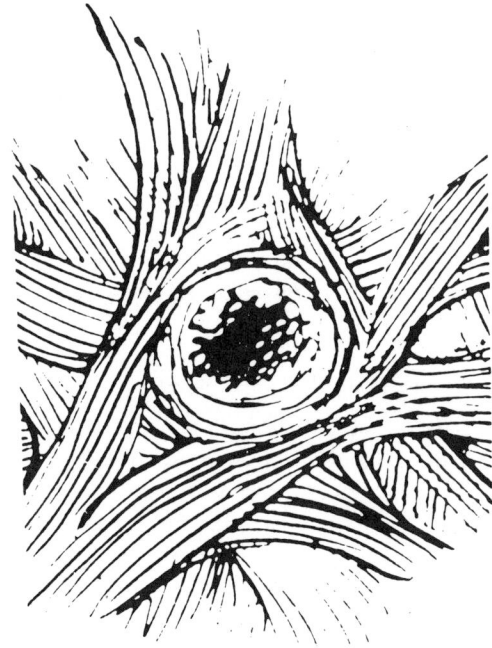

FIG 27–1.
Smooth muscle bundles in the area of the bladder neck. (From Woodburne RT: Anatomy of the bladder, in Boyarsky S (ed): *The Neurogenic Bladder.* Baltimore, Williams & Wilkins, 1967. Used with permission.)

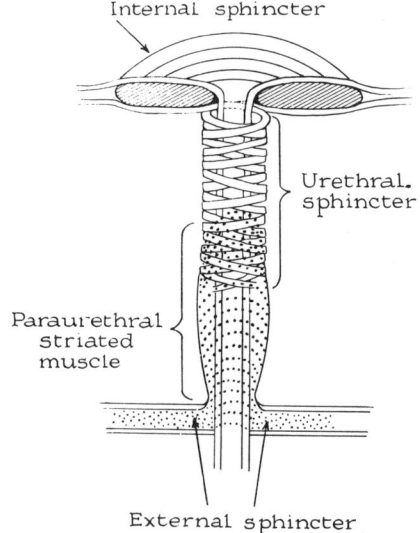

FIG 27–3.
The female bladder neck and proximal urethra according to Hutch. (From Hutch JA: A new theory of the anatomy of the internal urinary sphincter and the physiology of micturition: IV. The urinary sphincteric mechanism. *J Urol* 1967; 97:705. Used with permission.)

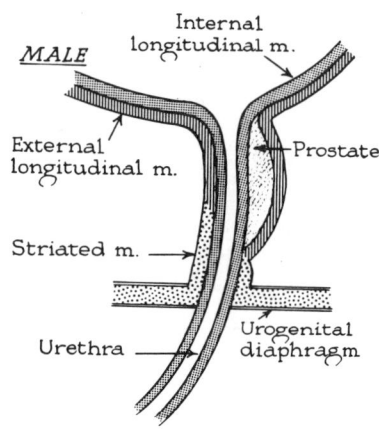

FIG 27–4.
Sagittal section of the male urethra according to Hutch and Rambo. (From Hutch JA, Rambo OA Jr: A new theory of the anatomy of the internal urinary sphincter and the physiology of micturition: III. Anatomy of the urethra. *J Urol* 1967; 97:696. Used with permission.)

gitudinal layer of the urethra were described as continuous with bladder smooth muscle layers, the outer circular layer fusing caudally with the striated sphincter and the inner longitudinal layer extending down the entire urethral length. Hutch's striated sphincter consists of two parts: (1) a true "external" sphincter, consisting of striated muscle within the leaves of the urogenital diaphragm; and (2) the periurethral striated muscles, which he conceived as arising from the urogenital diaphragm, advancing upward around the lower urethra, and inserting cranially into the urethral sphincter. Figure 27–4 shows Hutch's concept of the male bladder neck and proximal urethra. The inner longitudinal muscle of the bladder is pictured as extending down the entire length of the posterior urethra. The outer muscular layer is similarly continuous, but only through the upper half of the posterior urethra. Figure 27–5 shows Woodburne's representation of the "classical" external sphincter concept in the male. Figure 27–6 shows Gosling's diagramatic representation of the male lower urinary tract. The lumens of the bladder and urethra are dilated and the right half of the trigone

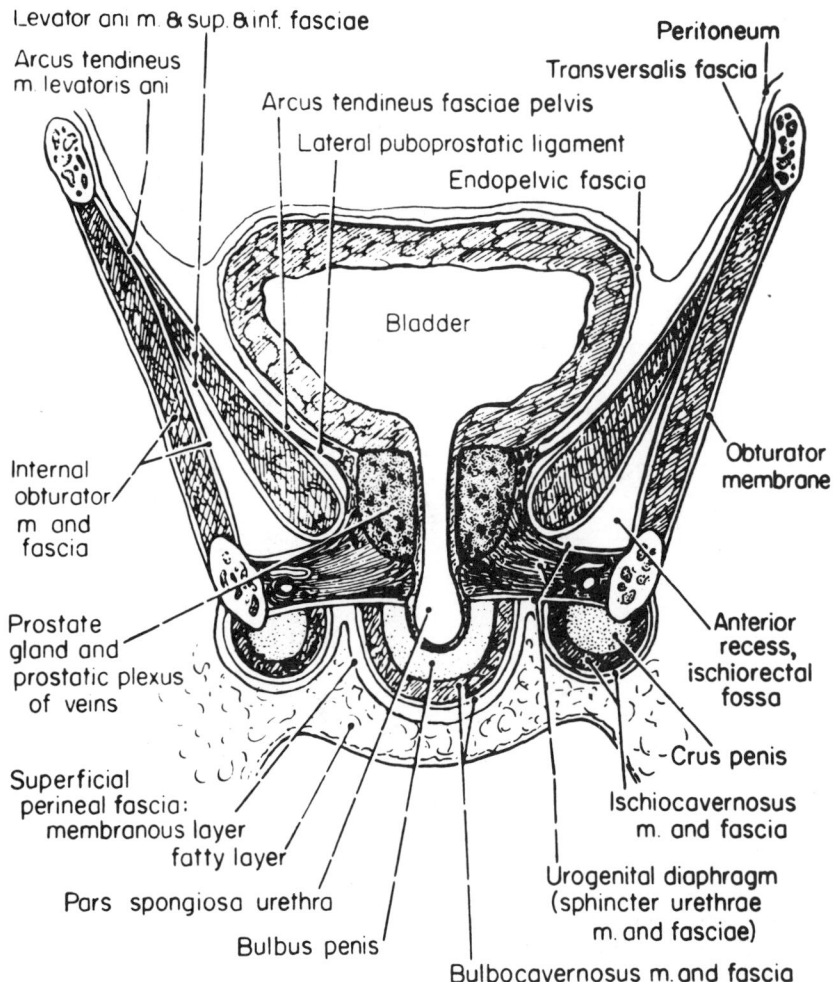

FIG 27–5.
The external sphincter and its relationships in the male. (From Woodburne RT: Anatomy of the bladder, in Boyarsky S (ed): *The Neurogenic Bladder.* Baltimore, Williams & Wilkins, 1967. Used with permission.)

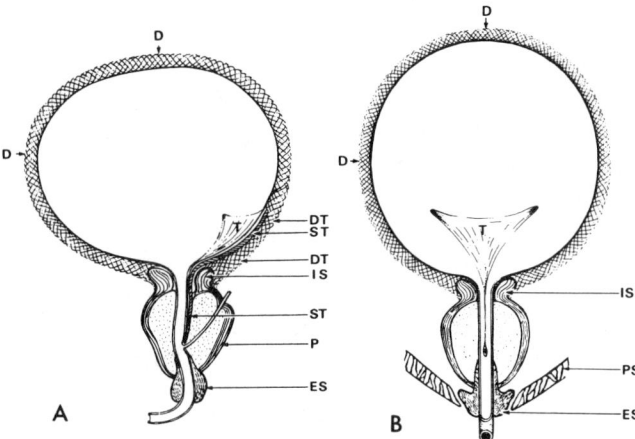

FIG 27–6.
A, sagittal section through the adult male lower urinary tract. **B,** corresponding coronal section. See text for more detailed descriptions. (From Gosling J: The structure of the bladder and urethra in relation to function. *Urol Clin North Am* 1979; 6(Feb):31. Used with permission.)

(T) in Figure 27–6,A, and the entire trigone in Figure 27–6, B are shown as surface features. The detrusor muscle (D) is in direct continuity with the deep trigone (DT), with which it is morphologically and histochemically identical. The extension of the ureteric muscle formed by the trigonal muscle is represented as a thin layer (ST) extending inferiorly as far as the verumontanum. The smooth muscle of the bladder neck forms the internal urethral sphincter (IS) and is continuous with the prostatic capsule (P). The two components of the striated sphincter are shown, ES representing the intrinsic component and PS representing the extrinsic or periurethral component. Figure 27–7 shows Gosling's concept of corresponding sections (to Figure 27–6) through the female lower urinary tract. The trigone (T) is represented as a surface feature. The detrusor smooth muscle (D) continues posterior to the superficial trigone and at the bladder neck region is replaced by a thin layer of smooth muscle (SM), which extends throughout the length of the urethra. Gosling does not, however, consider the smooth muscle of the urethra actually continuous with the smooth muscle of the detrusor. Both components of the striated sphincter are pictured. The intramural intrinsic striated component

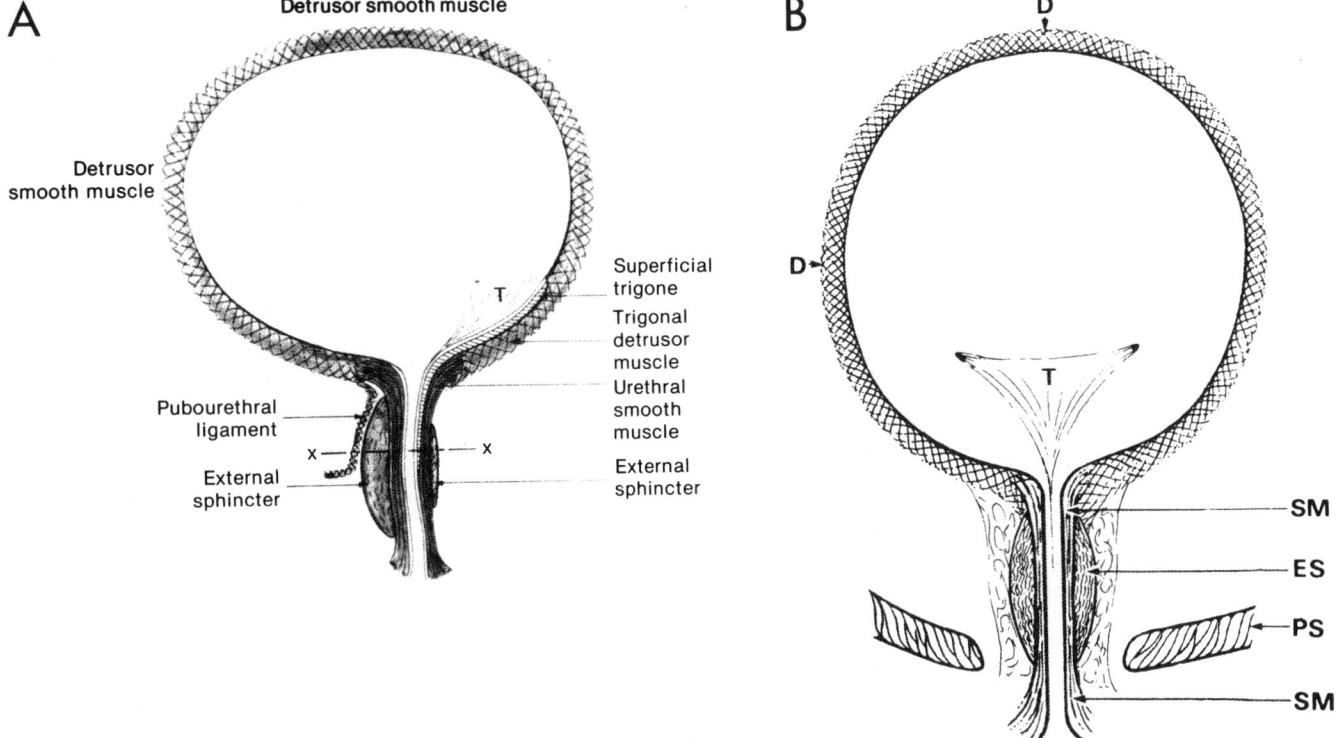

FIG 27–7.
A, sagittal section through the adult female lower urinary tract. **B,** corresponding coronal section. See text for more detailed description. (From Gosling J: Anatomy, in Stanton SL (ed): *Clinical Gynecologic Urology*. St Louis, CV Mosby, 1984. From Gosling J: The structure of the bladder and urethra in relation to function. *Urol Clin North Am* 1979; 6(Feb):31. Used with permission.)

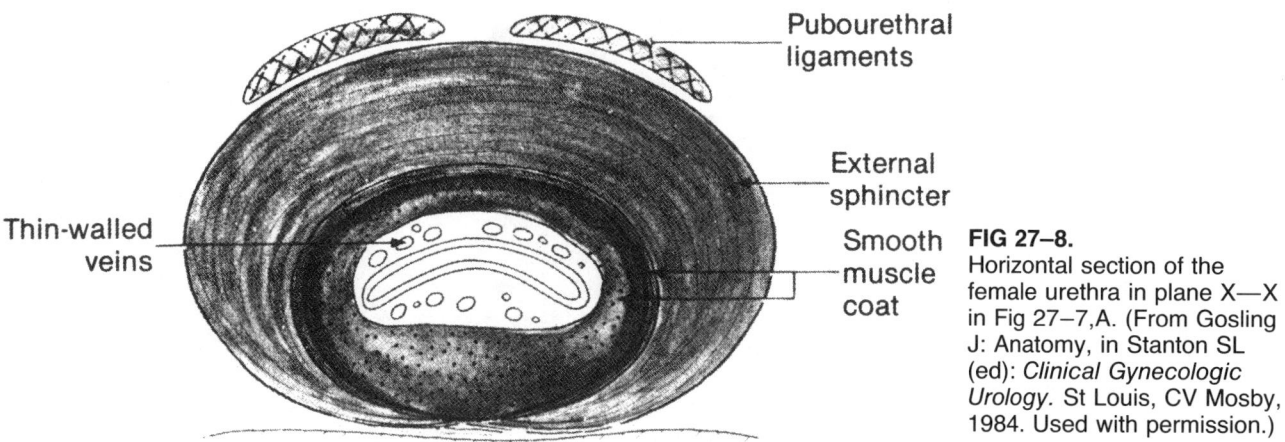

FIG 27–8.
Horizontal section of the female urethra in plane X—X in Fig 27–7,A. (From Gosling J: Anatomy, in Stanton SL (ed): *Clinical Gynecologic Urology.* St Louis, CV Mosby, 1984. Used with permission.)

(external sphincter [ES]) is thickest along the middle third of the urethra and relatively deficient posteriorly. This is separate from the extramural or extrinsic component, designated periurethral striated muscle (PS). The pubourethral ligament extends toward the pubis and lies anterior to the intrinsic component of the striated sphincter. Figure 27–8 is a cross-section of the female urethra at the plane denoted by the X in Figure 27–7, A. The intramural or intrinsic portion of the striated sphincter is again designated "external sphincter," an admittedly confusing term here. The smooth muscle coat consists of a relatively minor outer circular part and a thicker inner longitudinal component. Figure 27–9 shows El Badawi's concept of the muscular structures involved in micturition. BBD denotes the bladder body detrusor, and UT denotes the ureterotrigonal muscle unit, which extends into the proximal urethra. His "lissosphincter" consists of the periureteral sheaths, bladder base detrusor muscle, and proximal urethral smooth muscle, all of which are described as being in direct anatomic continuity. His "rhabdosphincter" corresponds to the intrinsic component of the striated sphincter and is distinct from the striated muscle of the pelvic diaphragm (PD), which encircles the urethra proximal to its pelvic subdiaphragmatic portion (SDU), comprising the extrinsic component of the striated sphincter.

The synthesis of vesicourethral structure by Khanna et al. (1981, 1983) is depicted in Figure 27–10. They considered the predominant muscle element at the vesicourethral junction to be the superficial bladder base, the circular layer of which sweeps around to form the circular vesical neck, and the longitudinal layer of which extends to form the posterior vesical neck and urethra. They thought that this muscle element participates in active (neurally induced) closure of the outlet during bladder filling. During voiding, they thought this muscular component actively relaxed, again in re-

sponse to neural stimulation. The longitudinal layer of the detrusor muscle extends to form the anterolateral and posterolateral bladder neck and urethra, providing a structural basis for adaptive funneling of the outlet and some shortening of the proximal urethra during the emptying phase of micturition. The prostatic capsule, in their schema, was felt to be simply a continuation of the smooth muscle layers of the proximal urethra in the male, constituting a single morphological unit with the bladder neck and proximal urethra.

Figure 27–11 shows Oelrich's illustration of a median

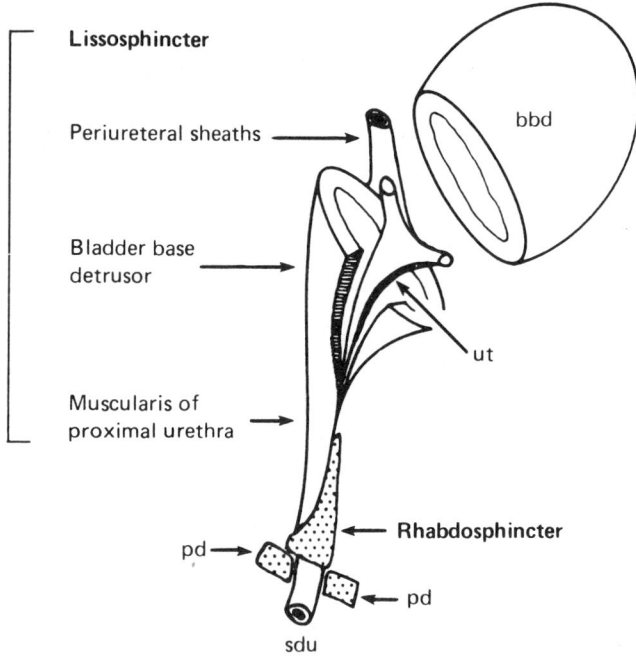

FIG 27–9.
Units of muscular apparatus of micturition. See text for more detailed description. (From El Badawi A: Autonomic muscular innervation of the vesical outlet and its role in micturition, in Hinman F Jr (ed): *Benign Prostatic Hypertrophy.* Berlin, Springer-Verlag, 1983. Used with permission.)

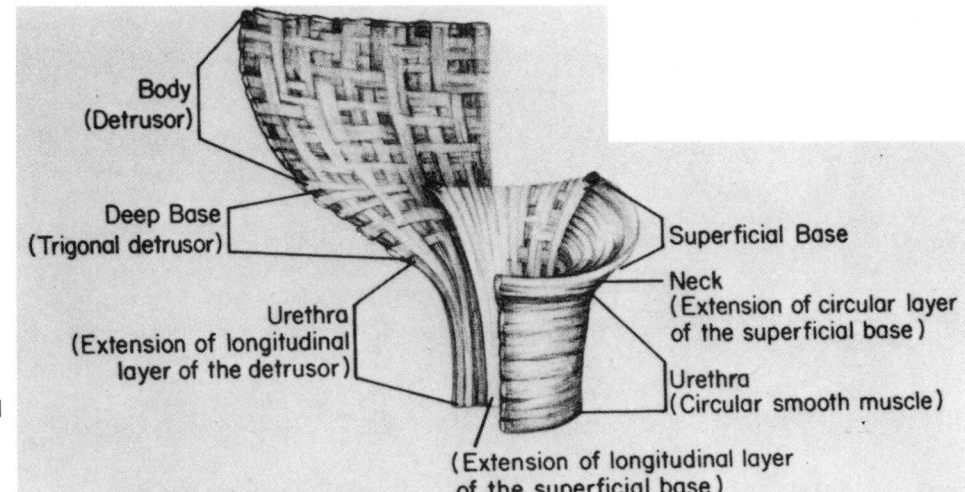

FIG 27–10.
Vesical neck smooth musculature. (From Khanna OP, et al: Vesicourethral smooth muscle: Function and relation to structure. *Urology* 1981; 18:211. Used with permission.)

section of the lower urinary tract of a 21-year-old man, demonstrating the extent of contact between the urethra (U) and the striated sphincter (SU, called "sphincter urethrae"). B denotes bladder and PR denotes prostate. Figure 27–12 shows Oelrich's diagram of a frontal

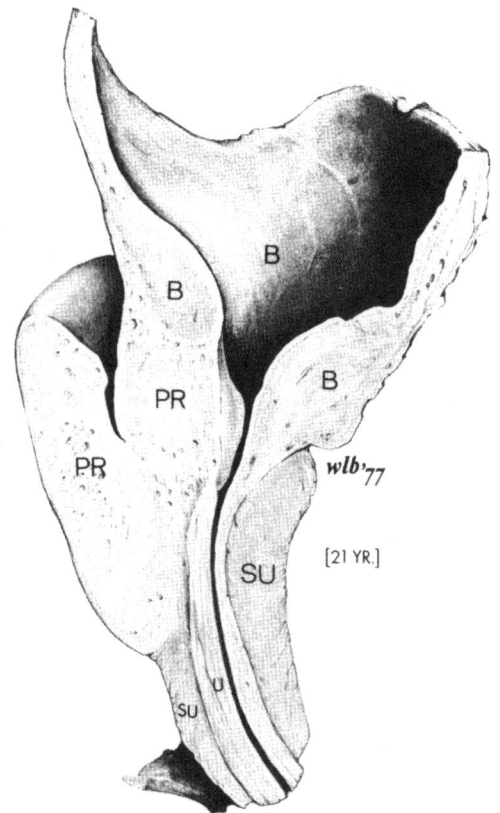

FIG 27–11.
Median section of bladder, urethra, prostate, and urethral sphincter in a 21-year-old man. See text for more detailed description. (From Oelrich TM: The urethral sphincter muscle in the male. *Am J Anat* 1980; 158:229. Used with permission.)

section of a male pelvis taken at right angles to the perineal membrane and through the membranous urethra. The relationships of the fascial planes in the pelvis, urogenital hiatus, and perineum are demonstrated. Note the lack of a "classical" horizontally oriented urogenital diaphragm surrounded by two distinct parallel fascial layers. Figure 27–13 shows Oelrich's concept of an oblique view of the muscular relationships of the lower urinary tract and vagina in a 27-year-old woman. US represents the striated urethral sphincter. Distal to and directly continuous with the lower border of this is a bundle of fibers that begins as a small tendon attached to the ischiopubic ramus (IR) and extends forward toward the anterior surface of the urethra, widening to a band of approximately 6 mm, then becoming continuous with the corresponding fibers of the opposite side. This is called the "compressor urethrae muscle" (CU). The urethrovaginal sphincter (UVS) is a thin, flat muscle that blends ventrally with the compressor urethrae. From the ventral side of the urethra its fibers extend dorsally along the lateral wall of the urethra and vagina. An additional layer of striated muscle, the transverse vaginal muscle (TV), parallels the compressor urethrae and radiates from it to attach to the anterior portion of the lateral wall of the vagina, attaching just above the urethrovaginal sphincter. SM denotes smooth muscle of the urethrovaginal compartment; VW, vaginal wall (arrow indicates continuity of this beneath the muscle); U, urethra; and B, bladder smooth muscle.

RELEVANT SMOOTH AND STRIATED MUSCLE STRUCTURE AND FUNCTION

SMOOTH MUSCLE

The physiology of mammalian smooth muscle has been studied far less extensively than that of skeletal or cardiac muscle, yet an extensive literature exists, in-

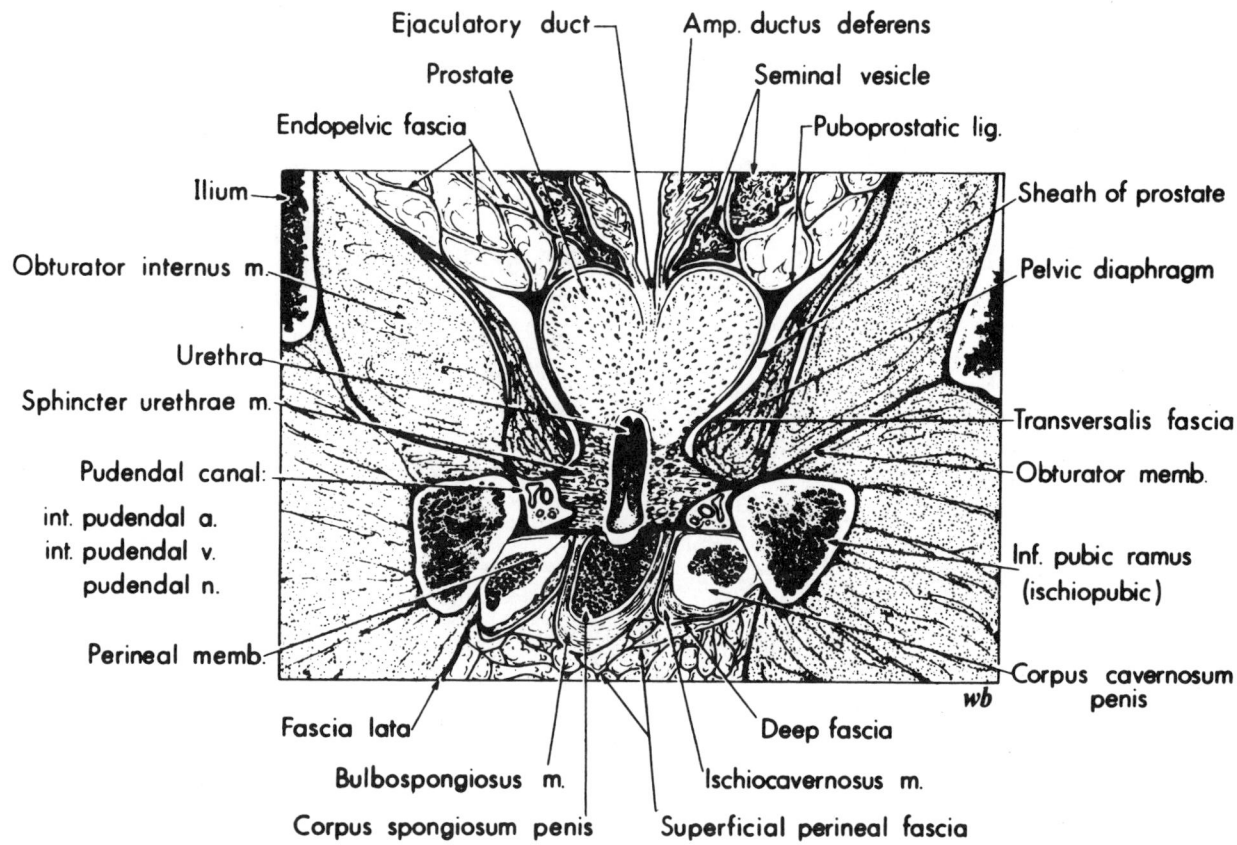

FIG 27–12.
Frontal section of pelvis in a 28-year-old man. (From Oelrich TM: The urethral sphincter muscle in the male. *Am J Anat* 1980; 158:229. Used with permission.)

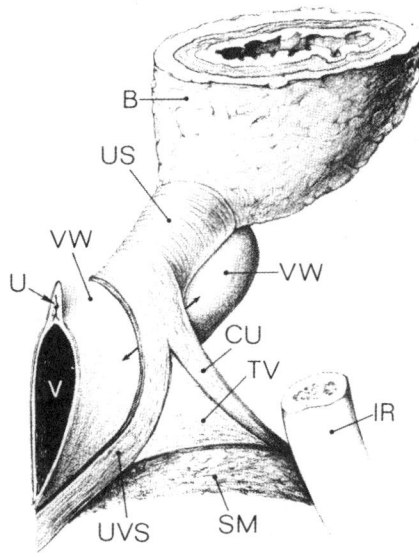

FIG 27–13.
Oblique view of urogenital sphincter muscles, bladder, and vagina in a 27-year-old woman. See text for more detailed description. (From Oelrich TM: The striated urogenital sphincter muscle in the female. *Anat Rec* 1983; 205:223. Used with permission.)

cluding contemporary reference volumes (Bulbring et al., 1981). However, the majority of information concerning smooth muscle anatomy and physiology is derived from studies on vascular smooth muscle, with a lesser body of information available on the gut and on other smooth muscle organs. One must be careful in directly applying information and conclusions derived from one smooth muscle system to another. With this in mind, we shall briefly review only those general principles related to the structure, function, and control of the smooth muscle of the lower urinary tract.

Structure

Smooth muscle is composed of far smaller fibers than skeletal muscle—usually 2–5 μ in diameter and 50–200 μ in length, although ranging up to nearly 1 mm in some organs (Gabella, 1981; Guyton, 1986). Cell length is an uncertain parameter, as cells can shorten and elongate over a wide range. The cell membrane is similar to that of other cell types, but presents numerous in-pocketings known as "caveolae," whose significance remains obscure, but which seem to be closely associated with portions of the abundant sarcoplasmic reticulum. Smooth muscle contains both actin and myosin filaments, which have chemical characteristics similar to but not exactly like those of the filaments in skeletal muscle. Smooth muscle also contains tropomyosin, but probably not troponin or a troponin-like substance. However, as in skeletal muscle, a chemical interaction between actin and myosin forms the basis for the contractile process and, as in skeletal muscle, although the details differ, this contractile process is activated by calcium ions, with ATP ultimately providing the energy source for this process. The actin and myosin filaments lack the periodic spatial arrangement found in striated muscle, and thus no cross-striations are visible microscopically—hence the term "smooth" muscle. However, smooth muscle does contain large numbers of thin actin filaments attached to so-called dense bodies (Figure 27–14). Some of these dense bodies are attached to the cell membrane, whereas others seem held in place by a scaffold of structural protein forming attachments from one dense body to another. Thick filaments about 2.5 times the diameter of the thin actin filaments, and assumed to be myosin, are interspersed throughout.

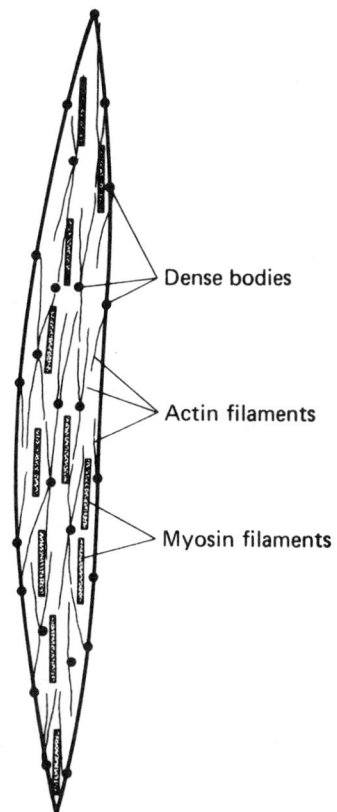

FIG 27–14.
Arrangement of actin and myosin filaments in smooth muscle cell, showing dense bodies. (From Guyton AC: *Textbook of Medical Physiology.* Philadelphia, WB Saunders, 1986. Used with permission.)

Despite the fact that these are only $\frac{1}{12}$–$\frac{1}{15}$ as numerous as the actin filaments, it is presumed that there are sufficient cross-bridges to interact with the actin filaments and cause contraction by a sliding filament mechanism similar to that of skeletal muscle. Although the contractile process seems relatively similar, contraction in smooth muscle develops much less rapidly.

Fiber Arrangement and Spread of the Contractile Impulse

Bozler (1941) originally proposed two categories of smooth muscle—multiunit and unitary—on the basis of general physiologic properties. Multiunit smooth muscle receives an extensive innervation, often having a 1:1 ratio between nerve endings and muscle fibers. Motor coordination is effected primarily through the nervous system, and fibers appear to be organized in some sort of a motor plan. Spontaneous activity generally does not occur. The outer surfaces of these fibers are covered by a thin layer of glycoprotein that seems to help insulate the separate fibers from each other (Guyton, 1986). Typical examples of multiunit smooth muscle include the fibers of the ciliary muscle and iris of the eye, piloerector muscles, and the muscle of many larger blood vessels. Unitary smooth muscle, often called visceral smooth muscle, receives a much less extensive innervation, and the muscle fibers are usually arranged in sheets or bundles such that the cell membranes contact each other at multiple points to form "gap junctions" through which ions can flow from one cell to the next. These fibers usually form a functional syncytium whose motor coordination is primarily dependent on myogenic conduction from muscle fiber to fiber. Classical unitary smooth muscle is found in the wall of the gut, bile duct, ureter, and uterus.

The smooth muscle of the bladder seems to share characteristics of both multiunit and unitary type. Most authors agree on an abundant cholinergic innervation of the bladder smooth muscle, but there was originally some disagreement as to the degree of innervation, some authors arguing for a nearly 1:1 nerve-muscle cell basis (El Badawi and Schenk, 1966, 1968; Schulman et al., 1972; Taira, 1972), while others argued for a nerve-muscle cell ratio of considerably less than 1:1 (Raezer et al., 1973). El Badawi's ultrastructural studies, however, clearly indicate that no part of the vesicourethral smooth muscle is supplied by nerves on a one-to-one basis (El Badawi, 1982). Daniel et al. (1983), studying adult human female detrusor muscle, found many nerve profiles (about one for each eight cells cut in cross-section) and felt that the muscle cells did not possess any large gap junctions such as would provide typical low-resistance contacts between cells. There were observed, however, some regions of contact that might represent small gap junctions, and there were close apposition contacts and intermediate contacts between cells. Fletcher and Bradley (1978) also summarized morphologic evidence of close contact between adjacent bladder smooth muscle cells. However, they felt that neurotransmitter was delivered to a "domain" within a bladder muscle fascicle. From the site of each varicosity, they hypothesized neurotransmitter molecules diffusing in all directions, exciting whichever smooth muscle cells the transmitter reached in sufficient quantity. This theory necessitates bladder smooth muscle having an extensive coating with receptor sites.

Thus, the bladder does not seem to fit exclusively in either category of smooth muscle, but seems to possess properties more of a multiunit than a unitary smooth muscle, with some morphologic substrates compatible with cell coupling and the capability of impulse propagation from cell to cell. Few data are available regarding the path of spread of the contractile sequence during an emptying bladder contraction. Experimentally, it has been shown by Conway and Bradley (1969) that bladder contraction during reflex micturition begins in the dorsal urethrovesical junction area and spreads to the fundus, probably ventrally.

Innervation

Burnstock and Bell (1974) and Burnstock (1977, 1985, 1986) has extensively reviewed and updated current concepts of the structural relationships of peripheral autonomic nerves and smooth muscle. The neuroeffector junction differs significantly from the synapses of skeletal muscle and those of ganglia, both of which are elaborate, with separation of specialized presynaptic and postsynaptic membranes by about 20–50 nm. Transmitters are released from presynaptic sites, diffuse across a narrow cleft, and occupy receptors located on the postsynaptic or postjunctional membrane. Peripheral autonomic nerve fibers have extensive terminal varicose regions, which are free of Schwann cell envelopments. In these areas vesicle-filled varicosities, 1–2 μ in diameter, release transmitter *en passage*. Prejunctional varicosity membranes sometimes show thickenings, but there are rarely postjunctional specializations, and there is a variable and often wide (20–120 nm) cleft, allowing the potential for significant modulation (Fig 27–15). Vesicles are the most characteristic component of peripheral autonomic nerve endings. Since these contain neurotransmitters and are believed to be involved in their release, attempts to classify nerve endings rest mainly on the identification of the neurotransmitters at the EM level. Endings containing a large majority of small agranular (clear) vesicles are considered to be cholinergic, those containing numerous

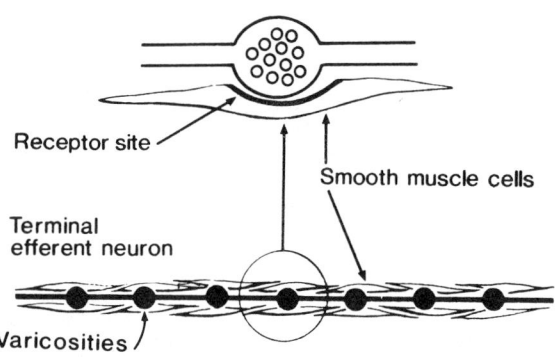

FIG 27–15.
Peripheral autonomic neuroeffector junction. (From Mundy AR: Clinical physiology of the bladder, urethra and pelvic floor, in Mundy AR, Stephenson TP, Wein AJ (eds): *Urodynamics: Principles, Practice, and Application.* New York, Churchill Livingstone, 1984. Used with permission.)

small granular vesicles are considered to be adrenergic (Gabella, 1981). Varicosities are also seen that contain large granular vesicles, and these have been considered to be purinergic (containing ATP).

Excitation, Excitation-Contraction Coupling, and Contraction

Excitation of smooth muscle cells can occur in a number of different ways (Guyton, 1986). The intracellular calcium concentration in smooth muscle is maintained at less than 10^{-7}M via active extrusion against an electrochemical gradient across an essentially impermeable plasma membrane. With excitation, whether by direct membrane depolarization or receptor activation, calcium acts as the predominant messenger linking stimuli from extracellular origin to the intracellular environment (Kuriyama, 1981; Movsesian, 1982; Droogmans et al., 1985). The threshold for mechanical activation of the contractile process occurs at a calcium concentration of about 10^{-7}M and full activation occurs at a free calcium concentration in the cytosol of about 10^{-5}M. The increase in intracellular calcium occurs as a result of an increased permeability of the membrane to extracellular calcium with resultant influx, and from a secondary release of calcium from intracellular stores. The importance of this latter pathway varies from tissue to tissue, but the location of these stores seems to be the sarcoplasmic reticulum and storage sites on the inner surface of the plasma membrane. In general, smooth muscle is considerably more dependent on extracellular calcium translocation than is skeletal muscle. Excitation, with increase in cytosol calcium concentration, can occur by a number of different mechanisms, which can be briefly summarized as follows (Kuriyama, 1981;

Movsesian, 1982; Droogmans et al., 1985; Guyton, 1986): (1) action potentials can produce associated calcium influx through potential dependent channels (voltage-regulated channels); (2) agonist-induced receptor activation can induce calcium influx by causing depolarization, but such excitation may also occur with no depolarization at all, through so-called receptor-dependent channels.

Membrane potential varies from one type of smooth muscle to another, but is generally of the order of -50 to -60 mV, about 30 mV less negative than in skeletal muscle. Action potentials in visceral (unitary) smooth muscles are qualitatively similar to those in skeletal muscle. However, it is said that action potentials probably do not occur in pure multiunit smooth muscle (Guyton, 1986). Some types of visceral smooth muscle generate a slow-wave rhythmic fluctuation of the membrane potential that is similar to that seen in true pacemaker cells. These waves cannot cause muscle contraction, but when they rise above the approximately -35 mV threshold potential, an action potential develops and spreads over the visceral muscle mass with resultant contraction. When visceral smooth muscle is stretched, spontaneous action potentials are usually generated, and these are thought to result from a combination of a decrease in the membrane potential caused by the stretch itself plus the normal slow-wave potentials. The question of whether bladder smooth muscle exhibits "spontaneous activity" has not been settled. The presence of "slow-wave fluctuations" is not the same as spontaneous activity. Although spontaneous activity is a property of some smooth muscle, it is not a property of all. The function of most smooth muscle organs is to slowly move contents through their length via peristalsis (such as the gastrointestinal tract and the renal collecting system and ureter). Thus, spontaneous activity would be a mechanism for initiation and control of peristalsis. Unlike most other smooth muscle organs, the urinary bladder functions to passively store its contents at low pressure until micturition occurs, at which time the bladder undergoes a coordinated contraction of sufficient magnitude to fully empty its contents. The presence of spontaneous activity in this system would be counterproductive. This topic will be discussed in greater detail later.

The actual contractile process in smooth muscle fibers is different from that in skeletal and cardiac muscle (Perry and Grand, 1979; Movsesian, 1982; Guyton, 1985). Smooth muscle probably does not have an effective tropinin complex, which regulates arrangement of the contractile proteins of skeletal muscle. The sequence of excitation-contraction coupling is as follows: (1) excitation of the muscle leads to the depolarization of the cell membrane and translocation of extracellular calcium into the cell; (2) this extracellular calcium can trigger the release of intracellular calcium, which increases the cytosol calcium concentration above 10^{-7}M; (3) the intracellular calcium binds to calmodulin (a calcium-binding protein) and the combination activates myosin kinase; (4) the activated myosin kinase phosphorylates myosin via adenosine triphosphate (ATP) hydrolysis and results in actin-myosin interaction; (5) the actin and myosin filaments slide past each other, resulting in contraction; (6) relaxation occurs following repolarization of the cell membrane and the resultant decrease in intracellular free calcium (mediated by both sequestration within intracellular storage sites and active calcium translocation back across the cell membrane). In the absence of sufficient free cytosol calcium, myosin kinase inactivates, myosin dephosphorylates, and the muscle relaxes.

What actually causes the signal transduction between excitation and the intracellular processes that cause contraction or relaxation is a field that is of great interest at this time, and a few relevant facts bear mentioning. The signaling system using cyclic adenosine monophosphate (cAMP) as a second messenger may be best known. There is a reasonable rationale for the hypothesis that cAMP can induce smooth muscle relaxation by stimulating calcium uptake into the sarcoplasmic reticulum or perhaps by stimulating calcium extrusion from the cytosol (Hardman, 1981). Cyclic guanosine monophosphate (GMP) may mediate smooth muscle relaxation, at least in vascular smooth muscle, by likewise somehow antagonizing the accumulation of free cytosolic calcium (Ignarro and Kadowitz, 1985). Most recently, inositol triphosphate has been proposed as a link between the plasma membrane and the increase in cytosol calcium (which occurs with smooth muscle contraction), especially in pharmacomechanical coupling, the type that occurs with the various physiological neurotransmitters (Berridge and Irvine, 1984; Somlyo, 1985).

Tonus

Tonus can be considered as a state of continuous stress in muscle maintained by activity of the contractile elements. Such activity may be contributed to by the intrinsic characteristics of smooth muscle, which are responsible for its response to stretch and for its spontaneous activity (where this exists), by prolonged direct smooth muscle excitation via local tissue factors or circulating hormones, and by tonic activity initiated extrinsically in the autonomic nervous system (Wein and Raezer, 1979; Guyton, 1986). The status of the intravesical pressure during bladder filling—which, up to a certain point, normally rises very little as a function of volume—is thought to be a result of this inherent tonus, the presence or absence of phasic excitatory or inhibitory neuronal impulses acting at the cell membrane or ganglia, and those important characteristics of smooth muscle that describe its ability to change length greatly without marked changes in tension, its elastic properties.

The urinary bladder appears to have intrinsic tone (in the absence of neuronal influences). Using an isolated whole bladder model, Levin et al. (1984) demonstrated that, in the absence of any extrinsic influences, the cystometric curve could be shifted to the right (more compliant) by diltiazem (calcium-channel blocker), EGTA (calcium chelator), or isoproterenol (β-adrenergic agonist).

Elasticity and Viscoelasticity

The bladder wall is composed of passive elements and active elements. The passive elements consist of collagen, elastic, and those elements of smooth muscle that do not require energy utilization for their function. The passive properties attributable to these elements account for much of the reason why a very gradual increase in bladder volume, with consequent stretch on the bladder wall elements, results in only a very low increase in pressure or wall tension, whereas a rapid stretch results in a large increase in pressure or tension, but one that begins to disappear immediately and that gradually returns almost to the prestretch level, even though the muscle is still lengthened. Similarly, if a segment of smooth muscle is rapidly shortened, all tension will be immediately lost, but over a period of time much of the tension again returns (Guyton, 1986). These properties collectively describe the elastic and viscoelastic properties of the bladder wall. They constitute the so-called passive characteristics that are extremely important in determining the bladder response to filling, and may be important as well in determining contractility characteristics during the emptying phase of micturition. These properties, and their measurements, have been described by a number of authors. Note, however, that the spherical bladder and its pressure volume relationships during storage and active contraction may not be directly related to those of a longitudinal length-tension muscle system. Further, unfortunately, terminology is not consistent, and makes comparison of one author's viewpoint to that of another quite difficult (Kondo et al., 1972; Van Mastrigt et al., 1978; Zinner et al., 1983; Coolsaet, 1984; Griffiths, 1984; Van Duyl, 1985).

STRIATED MUSCLE

Skeletal muscle is generally divided into slow-type fibers and twitch-type fibers. Slow-type fibers, not to be confused with slow-twitch–type fibers, are widely distributed in animals, but in mammals are found only in the extrinsic muscles of the ear and in the extraocular muscles, and will not be considered further (Kirchberger and Schwartz, 1985).

Twitch-type fibers have been further classified into subtypes on the basis of their metabolic and functional characteristics. Such classification was initiated about 15 years ago, and now eight such classification systems are listed by Kirchberger and Schwartz (1985). Speed of contraction seems to correlate with histochemical reaction for adenosine triphosphatase (ATPase) at an alkaline pH. Fibers strongly reactive are fast-twitch, whereas those that are weakly reactive are slow-twitch. Resistance to fatigue is directly related to the intensity of oxidative enzyme staining in the same fibers. Slow-twitch fibers are high in oxidative enzyme activity and are relatively fatigue-resistant. Fast-twitch fibers may be fatigable or relatively fatigue-resistant. Gosling et al. (1981) stated that the entire intramural (intrinsic) striated sphincter is composed of slow-twitch type fibers, whereas the periurethral levator ani (the extramural or extrinsic component) consist of both slow-twitch and fast-twitch fibers. Teleologically, this would be convenient because the intramural striated component would then consist of specialized fibers functionally capable of maintaining tension over prolonged time periods without fatigue. Their data indicated that the extramural component of the striated sphincter is also composed of a majority of slow-twitch type fibers rather than fast-twitch. They postulated this might be related to a role played by this muscle in actively supporting the pelvic viscera, and that the slow-twitch fibers are responsible for "background activity" during electromyographic (EMG) recordings. The fast-twitch population of the extramural component of the striated sphincter is functionally associated with rapid, forceful muscle contraction. It is these fibers, then, that are recruited to increase the force and speed of contraction of the levator ani during those events that might otherwise cause stress incontinence by raising intra-abdominal pressure.

Bazeed et al. (1982), studying the dog, reported a different fiber distribution in the intramural component of the striated sphincter. They found that only 35% of the intramural striated sphincter consisted of slow-twitch fatigue-resistant fibers. Like Gosling, they felt that it was these fibers that were responsible for the contribution of the striated sphincter to continence at rest. Of the fast-twitch fibers, they reported that fatigue-resistant ones constituted 20%, with fatigable fibers constituting the remainder. The intramural striated sphincter in the rabbit was found by Tokunaka et al. (1984) to consist solely of fast-twitch fatigue-resistant striated muscle. Bowen et al. (1976) stated that the intramural striated sphincter of the cat was fast-twitch fatigable muscle, but El Badawi and Atta (1985) showed ultrastructural evidence that it consists of both fast- and slow-twitch fibers.

It is tempting to look upon the physiology of the striated sphincter as changeable (Schmidt, 1983), change potentially occurring in response to a change in the type of stimuli or innervation. Such a change causing an increase in the proportion of slow-twitch fibers or nonfatigable fast-twitch fibers might augment sustained outlet resistance during urine storage.

NEURAL CONTROL OF MUSCLE FUNCTION

AUTONOMIC VS. SOMATIC NERVOUS SYSTEM

The physiology of smooth muscle cannot be separated from that of the autonomic nervous system, which includes all efferent pathways having ganglionic synapses outside the central nervous system, and therefore includes the efferent innervation of all smooth muscle cells, cardiac muscle cells, and several types of secretory cells (see references cited by Wein and Raezer, 1979; Appenzeller, 1982; Guyton, 1986). There are no efferent somatic nerve cell body synapses outside of the central nervous system. Other differences between autonomic and somatic nerves include the following: (1) autonomic nerves may form extensive peripheral plexuses, while somatic nerves generally do not; (2) the most distal autonomic motor neurons are generally non-myelinated, while somatic neurons are myelinated; (3) neurotrophic atrophy occurs when the somatic nerve supply to a skeletal muscle is sectioned, whereas such changes do not generally occur in a smooth muscle structure whose autonomic nerve supply has been destroyed.

CLASSICAL ANATOMY AND FUNCTION OF THE AUTONOMIC NERVOUS SYSTEM

The terms sympathetic and parasympathetic refer only to anatomic divisions of the autonomic nervous system (Wein and Raezer, 1979; Appenzeller, 1982; Guyton, 1986). The sympathetic division consists of fibers originating in the thoracic and lumbar regions of the spinal cord, while the parasympathetic division consists of fibers originating in the cranial and sacral spinal nerves. The classical view of the peripheral autonomic nervous system, necessary for a full understanding of contemporary modifications, was described by us in a previous chapter (Wein and Raezer, 1979), based on the schema outlined by Jacobowitz (1974) (Fig 27–16). Each division was thought to generally involve a two-neuron system: "preganglionic" neurons emanating from the central nervous system and making synaptic contact with cells within ganglia, from which "postganglionic" neurons emerge to innervate peripheral organs. Three classes of ganglia were described in the sympathetic division: (1) those located adjacent to the spinal cord (paraganglia); (2) those located between the paravertebral ganglia and the end organ (preganglia); and (3) those located adjacent to or within the end organ (peripheral ganglia). In the parasympathetic system, ganglia were described as generally located in or near the innervated organs. Peripheral sympathetic ganglia were described that were close to peripheral parasympathetic cell bodies, providing an anatomic substrate for sympathetic influences on parasympathetic ganglion cell transmission. Afferent or sensory fibers from any autonomically innervated organ pass uninterruptedly through peripheral (infraspinal) autonomic ganglia to end in a neurosynapse only in the spinal cord. According to the classical scheme then, spinal neural control of an autonomically innervated organ is

FIG 27–16.
Classical view of the autonomic nervous system. Black circles represent cholinergic cell bodies; white circles, adrenergic cell bodies; R, receptor organ; BV, blood vessel; SG, sweat gland; and SAN, sinoatrial node. *Small dotted arrow* indicates intraganglionic adrenergic system of fibers. (From Jacobowitz DM: The peripheral autonomic system, in Hubbard JI (ed): *The Peripheral Nervous System.* New York, Plenum Press, 1974. Used with permission.)

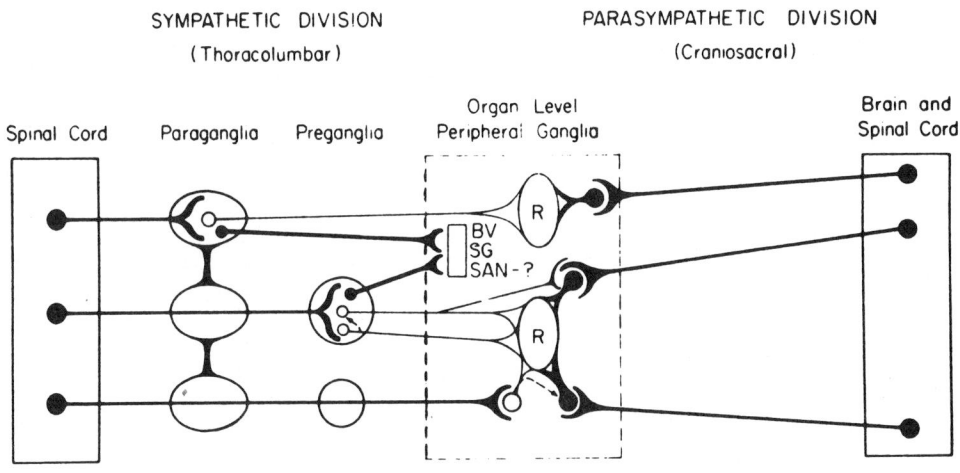

achieved by a "reflex arc" type of arrangement involving three or perhaps four neurons establishing an organ–spinal cord–peripheral autonomic ganglion–organ neural circuit.

CONTEMPORARY MODIFICATION OF CLASSICAL AUTONOMIC NERVOUS SYSTEM SCHEME

Some of these traditional concepts of the autonomic nervous system require modification to be able to understand the extensive literature that exists about innervation and neural control of the lower urinary tract. El Badawi (1982, 1983, 1984, 1985) has very nicely summarized the anatomic aspect of these modifications and the following represents what seems to us to be a synthesis of his conclusions (Fig 27–17). The muscular innervation of the lower urinary tract is derived exclusively from postganglionic neurons of what is termed "the urogenital short neuron system." Although paraganglia and preganglia exist, actual innervation predominantly (or exclusively) emanates from peripheral ganglia that are at a short distance from, adjacent to, or within the organs they innervate, thus the name "short" (as opposed to the "long" postganglionic fibers from cell

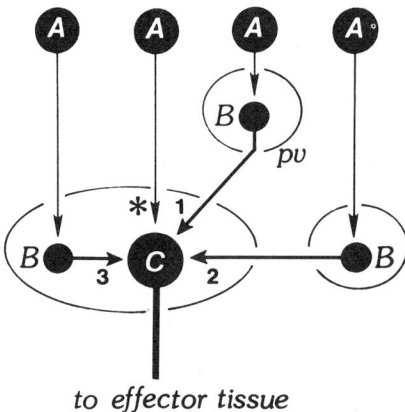

FIG 27–17.
Basic and simplified scheme of neural control of vesicourethral smooth muscle. **A,** preganglionic spinal cord–derived neurons; **B,** postganglionic modulator neurons; **C,** postganglionic effector neurons innervating vesicourethral muscularis and other effector tissues. Whether cholinergic or adrenergic, effector neuron **C** can be influenced by multiple neurosynapses established by its corresponding preganglionic neuron *(asterisk),* a modulator sympathetic adrenergic neuron derived from paravertebral ganglia (pv; terminal 1), a modulator sympathetic adrenergic or parasympathetic cholinergic or SIF cell catecholaminergic neuron that is located in the same ganglion of the urogenital short neuron system (terminal 2), or in a more proximal ganglion of the same system (terminal 3). (From El Badawi A: Autonomic muscular innervation of the vesical outlet and its role in micturition, in Hinman F Jr (ed): *Benign Prostatic Hypertrophy.* Berlin, Springer-Verlag, 1983. Used with permission.)

bodies that lie away from the target organ in paraganglia or preganglia). The ganglia themselves are composed of three cell types: cholinergic principal neurons, adrenergic principal neurons, and small intensely fluorescent (SIF) cells (see also Jacobowitz, 1974). The latter cells are thought to play an important role in the modulation of interganglionic vasomotor function and ganglionic transmission and are identifiable both histochemically and ultrastructurally. Urogenital ganglia also include a complex intraganglionic network of cholinergic and adrenergic fibers that either pass uninterruptedly through the ganglion or terminate in it as neuron synapses with one or more of its cell populations. Thus, the efferent autonomic pathways of the lower urinary tract do not necessarily conform to the classical autonomic bineuronal efferent model, as they may be interrupted by more than one synaptic relay. Postganglionic neurons innervate effector tissues, but also may terminate within ganglia of the same system, or in relation to its individually distributed neurons, similar to the depiction in the "classical" scheme of postganglionic sympathetic terminations on parasympathetic ganglion cells. Finally, some urogenital nerve fibers may establish infraspinal sensory synapses in the urogenital short neuron system or central ganglia to form organ-to-ganglion-to-organ neural circuits that bypass and function independently of the spinal cord.

Another relatively contemporary concept requires mention for subsequent understanding of studies involving autonomic stiumulation and blockade. For many years the only autonomic neurotransmitters recognized were acetylcholine and norepinephrine. However, it has become obvious that other transmitters might indeed be present in components of the autonomic nervous system. Burnstock (summarized in Burnstock, 1985, 1986) has repeatedly emphasized this concept and the concept of modulatory transmitter mechanisms, including prejunctional inhibition or enhancement of transmitter release, postjunctional modulation of transmitter action, and the secondary involvement of locally synthesized hormones and prostaglandins. Cotransmitters may be released along with a classical neurotransmitter in response to nervous activation and may interact at the level of the receptor and/or second messenger before evoking a functional response. The cotransmitter may have a direct action on postjunctional cells, or may facilitate the action of the classical neurotransmitter or/and act as an inhibitor of its release. Cotransmitter and classical transmitter may be stored in the same vesicle, and therefore be released in a parallel fashion at all impulse frequencies, or be stored in separate vesicle types, in which case differential release at different impulse frequencies might be possible. This concept may explain why pharmacologic blockade of classical receptors may only partially abolish the effects of neural stimulation. These somewhat complicated concepts and possibilities are nicely summarized in a diagram by Lundberg and Hökfelt (1983) (Fig 27–18).

NEUROTRANSMISSION AND RECEPTORS

Although smooth muscle cell contraction can occur via excitation in a variety of different ways (see previous discussion), the classical model involves synaptic re-

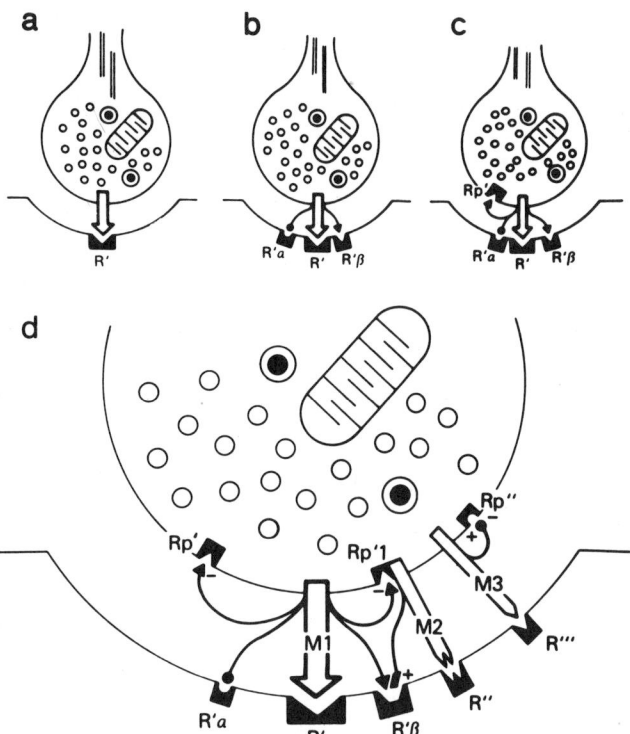

FIG 27–18.
Development of the concept of chemical neurotransmission. Schematic illustration of the development of the concept of chemical transmission. One transmitter *(a)* acts on one postsynaptic receptor (R'). One transmitter *(b)* acts on multiple types of postsynaptic receptors (R'α,R'β). The transmitter *(c)* acts in addition on a presynaptic receptor (Rp'). Multiple compounds (M1–3), possibly differentially stored in small vesicles (classical transmitter) and in large dense-core vesicles (classical transmitter plus peptide), are released from the same nerve ending *(d)*. The main possible interactions indicated are: (1) inhibition of release of the second messenger (peptide M2) by the classical transmitter (M1) via presynaptic action (Rp'1); (2) interaction at the postsynaptic receptor (R'β) level between M1 and M2; (3) facilitation or inhibition of release of the classical transmitter by the peptide (M3) via action on a presynaptic receptor (Rp''); and (4) activation by the peptide (M3) of electrical activity in the presynaptic neuron via action on a presynaptic receptor (Rp''). (From Lundberg JM, Hokfelt T: Coexistence of peptides and classical neurotransmitters. *Trends Neurol Sci* 1983; 6:325. Used with permission.)

lease of a neurotransmitter in response to neural stimulation, with subsequent conbination of the transmitter agent with a recognition site, or receptor, on the postsynaptic smooth muscle cell membrane. The transmitter receptor combination then initiates changes in the postsynaptic effector cell, which ultimately result in what we consider to be the "characteristic" effect of that particular neurotransmitter. Although other mechanisms exist, classical synaptic neurotransmission involves a vesicular-exocytotic mechanism (see Cooper and Meyer, [1984] for review). According to this theory, an action potential initiates an influx of calcium ions into the presynaptic nerve terminal, resulting in a process of exocytosis of the neurotransmitter in the presynaptic nerve terminal. The molecular mechanism for this remains obscure.

Clinicians are often confused because they assume that the terms "sympathetic" and "parasympathetic" necessarily imply particular neurotransmitters. As previously stated, these terms imply strictly anatomic origin, whereas in both subsystems of the autonomic nervous system other adjectives are used to describe the nature of the neurotransmitter involved in the chemical transmission of impulses from nerve to smooth muscle.

Cholinergic Receptor Sites

The term "cholinergic" refers to those receptor sites at which acetylcholine is the neurotransmitter. Peripheral cholinergic fibers are thought to include somatic motor fibers, all preganglionic autonomic fibers, and all postganglionic parasympathetic fibers. Classically, cholinergic receptor sites have been subdivided into two major classes: muscarinic and nicotinic receptors. These divisions are based on the original observations that the alkaloid muscarine mimics the effects of acetylcholine at some cholinergic sites, while low doses of nicotine mimic the effects of acetylcholine at other sites. Muscarinic sites include all autonomic effector cells (postsynaptic autonomic cholinergic receptors), whereas nicotinic sites are located on autonomic ganglia and motor end plates of skeletal muscle. These nicotinic sites are not identical. Atropine competitively inhibits muscarinic receptors, whereas high doses of nicotine inhibit nicotinic sites. Large concentrations of atropine may, in addition to blocking muscarinic receptors, exert some inhibitory effect on nicotinic receptors as well.

During the last ten years there have been major technical advances in receptor physiology and pharmacology, including the development of highly specific agonists and antagonists, along with specific radioligands to label and quantitate these, and more exact characterization of the intracellular changes that occur as a result of receptor-effector coupling. Although the identification of receptor subtypes may be confusing at times, the concept does provide a basis to explain the differing actions sometimes exhibited by a neurotransmitter at different effector junctions, and also offers the opportunity for the development of more selective pharmacologic tools for the therapy of effector dysfunctions. The distinction between nicotinic and muscarinic cholinergic receptors has already been mentioned. There are at least two subtypes of muscarinic receptors: M_1 and M_2 (Hammer and Giachetti, 1982; Eglen and Whiting, 1985). The M_1 class was originally defined as exhibiting a high affinity toward the antagonist pirenzepine, and was thought to be located primarily in the central nervous system, while M_2 receptors were described as exhibiting a low affinity toward this antagonist and thought to be located mainly on peripheral effector organs. Further subclassification (an M_3 receptor has already been proposed) will doubtless occur (Nilvebrant, 1986).

Events at the cholinergic neuromuscular junction were described by Burnstock (1974) and Burnstock and Bell (1977) and summarized by Benson (1983) (Fig 27–19). Acetylcholine is synthesized from acetyl coenzyme A and choline by the enzyme choline acetyltransferase (*a* in Fig 27–19). It is then stored in agranular vesicles (*b*) and is released from the nerve terminal into the synaptic cleft with nerve stimulation (*c*). It then diffuses to receptor sites on the smooth muscle membrane (*d*). Inactivation is due primarily to local hydrolysis by postjunctional acetylcholinesterase (*e*). A small fraction may diffuse away to be hydrolyzed elsewhere in the effector or in the general circulation. Active reuptake of choline by the nerve terminal does occur (*f*) but active reuptake of acetylcholine does not.

Adrenergic Receptor Sites

Receptor sites at which a catecholamine is the neurotransmitter are termed adrenergic and include most postganglionic sympathetic fibers (including those to the lower urinary tract smooth muscle), where the catecholamine responsible for neurotransmission is norepinephrine. Adrenergic receptor sites are further classified as α or β on the basis of the differential effects elicited by a series of catecholamines (see Wein and Raezer, [1979] for references). Classically, α-adrenergic effects consist of vasoconstriction and contraction of the smooth musculature. These effects are stimulated most potently by norepinephrine and methoxamine and cannot be elicited by isoproterenol. α-Adrenergic effects were described originally as being inhibited by phentolamine and phenoxybenzamine, though a number of other selective inhibitors now exist. β-Adrenergic effects are classically considered to include cardiac stimulation, vasodilation, bronchodilatation, and other types of smooth muscle relaxation. These effects are stimu-

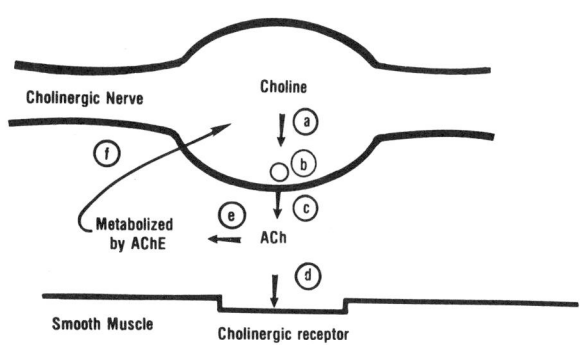

ACh - acetylcholine
AChE - acetylcholinesterase

FIG 27–19.
Events at the peripheral cholinergic nerve terminal. See text for more detailed description. (From Benson GS: Mechanisms of autonomic drug action on the bladder outlet, in Hinman F Jr (ed): *Benign Prostatic Hypertrophy.* Berlin, Springer-Verlag, 1983. Used with permission.)

lated most potently by isoproterenol, less so by epinephrine, and least by norepinephrine.

Events at the peripheral adrenergic neuromuscular junction have been described by Burnstock (1974) and Burnstock and Bell (1977) and summarized by Benson (1983) (Fig 27–20). Norepinephrine is synthesized from tyrosine through steps involving the synthesis of dopa and dopamine. The rate-limiting step is the tyrosine hydroxylase–induced transformation of tyrosine to dopa (*a* in Fig 27–20). The norepinephrine is then stored in granular vesicles (*b*), and in response to a nerve stimulus, the transmitter is released from the nerve terminal (*c*) and diffuses to a receptor site on the smooth muscle membrane (*d*). Inactivation occurs largely through active reuptake into peripheral nerve terminals (*e*), where

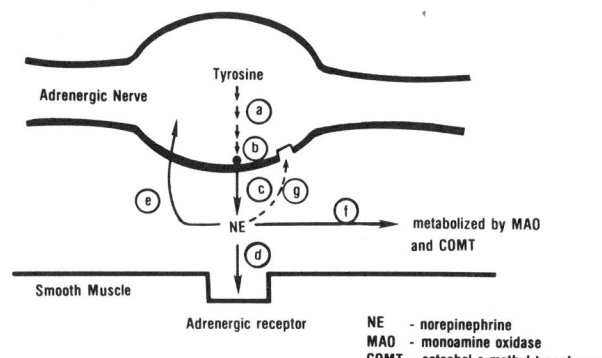

NE - norepinephrine
MAO - monoamine oxidase
COMT - catechol-o-methyl-transferase

FIG 27–20.
Events at the peripheral adrenergic nerve terminal. See text for more detailed description. (From Benson GS: Mechanisms of autonomic drug action on the bladder outlet, in Hinman F Jr (ed): *Benign Prostatic Hypertrophy.* Berlin, Springer-Verlag, 1983. Used with permission.)

granular reincorporation and/or degradation by monamine oxidase and catechol-O-methyltransferase occurs. Inactivation also occurs to a lesser extent by uptake into smooth muscle cells, where enzymatic inactivation occurs, and by similar enzymatic inactivation in other tissues, primarily the liver and kidneys. Site g in Figure 27–20 depicts the ability of norepinephrine to interact with presynaptic receptors and thereby regulate its own rate of release from the nerve terminal (see below). In the adrenal medulla, an additional synthetic change occurs—the synthesis of epinephrine from norepinephrine by the enzyme phenylethanolamine-N-methyltransferase.

Much attention has been paid to the identification and subclassification of adrenergic receptor subtypes. α-Adrenergic receptors were originally classified into two subtypes, defined by their anatomic location on presynaptic and postsynaptic sites (Berthelson and Pettinger, 1977; Lavin et al., 1981; Kalsner, 1984; Ruffolo, 1984; Wein, 1986 [and references contained therein]). Presynaptic α-receptors were thought to mediate feedback inhibition of the release of norepinephrine from nerve terminals (site g in Fig 27–20), whereas postsynaptic receptors were thought to mediate the typical responses exhibited by effector organs, such as smooth muscle contraction. Originally, postsynaptic receptors were termed α_1, and presynaptic autoreceptors termed α_2. Phentolamine and phenoxybenzamine are both nonspecific α-adrenergic antagonists and, like dihydroergocryptine, bind to both α_1 and α_2 sites. Prazosin is a relatively selective antagonist of α_1 receptor sites, whereas rauwolscine and yohimbine are relatively selective antagonists of α_2 sites. The agonist clonidine appears to selectively bind to α_2 receptor sites. Although the anatomic subclassification of presynaptic as equivalent to α_2 and postsynaptic as equivalent to α_1 subtype is generally true, recent studies have indicated that not all α_1-adrenergic receptors are located postsynaptically, and not all α_2 adrenergic receptors are located presynaptically.

β-Adrenergic receptors are generally likewise agreed to exist in two forms (see references in Minneman and Molinoff, 1980; Lefkowitz et al., 1984). β_1-Adrenergic receptors are postsynaptic, produce an increase in the rate and force of cardiac contraction, and stimulate lipolysis and amylase secretion by salivary glands. Norepinephrine is considered to have a selectivity for β_1 receptors, and epinephrine for β_2, whereas isoproterenol has no distinct selectivity but is a more potent stimulator (effective at lower concentrations) of both subtypes. Propranolol and alprenolol antagonize both β_1 and β_2 effects, but β_1 effects are preferentially inhibited by metoprolol and betaxolol. β_2-Adrenergic receptors are present both presynaptically and postsynapti-cally. They mediate smooth muscle relaxation in bronchi, blood vessels, and the genitourinary and gastrointestinal tracts; facilitate norepinephrine release from presynaptic nerve terminals; increase glycogenolysis in liver and muscle; increase gluconeogenesis in liver; increase insulin and glucagon secretion by pancreatic cells; and stimulate renin release by juxtaglomerular cells in the kidney. β_2-Adrenergic receptors are also selectively stimulated by salbutamol and soterenol.

Signal transduction from receptor to effector mechanism is thought to depend on adenyl cyclase stimulation in the case of β receptors, on inhibition of adenyl cyclase or an increased calcium influx through the cell membrane in the case of α_2 effects, and on alteration of cellular calcium ion fluxes for α_1 (perhaps an increased release of intracellular calcium) (Andersson, 1985) and muscarinic cholinergic receptors. In dog bladder body muscle strips Rohner and Hannigan (1980) have shown that β-adrenergic–induced relaxation (and that induced by theophylline) increases cAMP content. Contractions in response to cholinergic and α-adrenergic stimulation were not associated with changes in cAMP.

Other Receptors

Excitatory and inhibitory effects are produced by various endogenous compounds in the central and peripheral nervous system in addition to acetylcholine and norepinephrine. Some of these substances may serve as modulators of transmission rather than as actual excitatory or inhibitory neurotransmitters. It should be noted, however, that when a proposed transmitter substance causes a tissue to respond, this does not necessarily imply that the tissue is innervated under normal circumstances by nerves releasing that substance. Specific criteria must be satisfied before a substance can be established as a neurotransmitter in a particular tissue (Burnstock, 1977, 1985, 1986; Wein and Raezer, 1979). This mistake is often made in the urologic literature, i.e., ascribing neurotransmitter status to a substance that is simply active in an in vitro or in vivo preparation. Substances besides acetylcholine and norepinephrine that have been proposed as neurotransmitters are listed in Table 27–1. Those that are particularly relevant to lower urinary tract function are discussed in subsequent sections.

DENERVATION

When smooth muscle is denervated its sensitivity to neurohumoral transmitters increases, and the increase is greater when postganglionic rather than preganglionic fibers are destroyed (Wein and Raezer, 1979). The resultant supersensitivity of smooth muscle so produced has been elegantly reviewed by Westfall (1981),

TABLE 27–1.

Proposed Neurotransmitters and Neuromodulators in Addition to Acetylcholine and Norepinephrine*

	CENTRAL	PERIPHERAL
Adenosine triphosphate (ATP)		X
Prostaglandins (F$_2$, E, E$_2$)		X
Peptides		
Opioids	X	X
Vasoactive intestinal polypeptide (VIP)	X	X
Substance P	X	X
Neuropeptide Y		X
Somatostatin		X
Bradykinin		X
Amines and amino acids		
Dopamine	X	X
Serotonin	X	X
Histamine	X	X
γ-Amino butyric acid (GABA)	X	X
Glycine	X	X
Glutamate	X	X
Taurine	X	
Proline		X
Carnosine		X
Octopamine		X

*Data from Wein and Raezer, 1979; Anderson and Sjogren, 1982; Hald and Bradley, 1982; Burnstock, 1985.

and has direct potential application to the lower urinary tract. Supersensitivity may be experimentally manifested by one of three states: (1) the maximal response of a tissue may increase with no change in the dose required to produce a response that is 50% of maximum (ED$_{50}$); (2) there may be a decrease in the ED$_{50}$ with no change in the maximum response; and (3) there may be both an increase in the maximum response and a decrease in the ED$_{50}$. Westfall divides supersensitivity into two broad categories. Deviation supersensitivity implies an increased percentage of the administered drug or neurotransmitter reaching the receptors. It does not involve a change in the actual responsiveness of target cells. This type of change is termed "prejunctional" in some of the older (and current) literature. This type of change usually develops relatively rapidly, and is specific for a given agonist or neurotransmitter. The mechanisms involved are either a loss of neuronal uptake or inactivation mechanisms or loss of extraneuronal uptake or inactivation mechanisms. Nondeviation supersensitivity, sometimes called postjunctional or postsynaptic, involves an actual change in the responsiveness of the effector (smooth muscle) cells. This type of response generally requires time to develop, several days up to weeks. It is a nonspecific type response in that drugs besides the specific neurotransmitter originally involved may likewise produce a supersensitive response. The mechanism seems to be related to a chronic interruption, pharmacologic or surgical, of the

normal pattern of neurotransmission to effector cells. In other words, it is a consequence of a chronic decrease in the normal contact between an excitatory neurotransmitter and its effector cells. It is intriguing to speculate that some receptor alteration may be involved, but studies have shown no change in receptor affinity, and thus far no increase in receptor density in smooth muscle, though such an increase does occur in skeletal muscle. There appears to be some electrophysiologic alteration that, in some smooth muscles, involves a change in the state of depolarization—the resting potential is decreased but the threshold potential is not changed so that the magnitude of depolarization necessary to move the membrane potential from resting to threshold is decreased. Sensitivity of the effector would then be expected to increase to any agonist that induces contraction by cell membrane depolarization. If this is so, the ED$_{50}$ should decrease. Another possibility is that the threshold potential itself may increase (become less negative and require less depolarization to achieve). An increase in maximum response may also occur because of what is termed an improvement in the electrical coupling between cells, leading to enhanced synchronization of agonist-induced contractions. Alterations in calcium homeostasis may also play a part in the development of nondeviation supersensitivity at the molecular level.

It should be remembered that when we speak of "denervation" regarding the lower urinary tract, that we are more often talking about "decentralization." Denervation implies a loss of the postsynaptic neuron that innervates the muscle, whereas decentralization refers to the loss of intermediate neuron(s) between the central nervous system and the final postsynaptic neuron. Because of the nature of the short neuron system innervating the smooth muscle of the bladder and urethra, surgical "denervation" distal to the spinal cord generally affects only nerves that are proximal to the ganglia from which originate the final postganglionic efferent neurons.

PERIPHERAL INNERVATION OF THE LOWER URINARY TRACT

GROSS NEUROANATOMY

The pelvic and hypogastric nerves supply the bladder and urethra with efferent parasympathetic and sympathetic neurons, and both convey afferent (sensory) neurons from these organs to the spinal cord (Fig 27–21) (Bors and Comarr, 1971; Bradley et al., 1974; Fletcher and Bradley, 1978; Bradley and Sundin, 1982; El Badawi, 1982, 1983). The parasympathetic efferent supply is classically described as originating in the intermediolateral region of the gray matter of sacral spinal cord

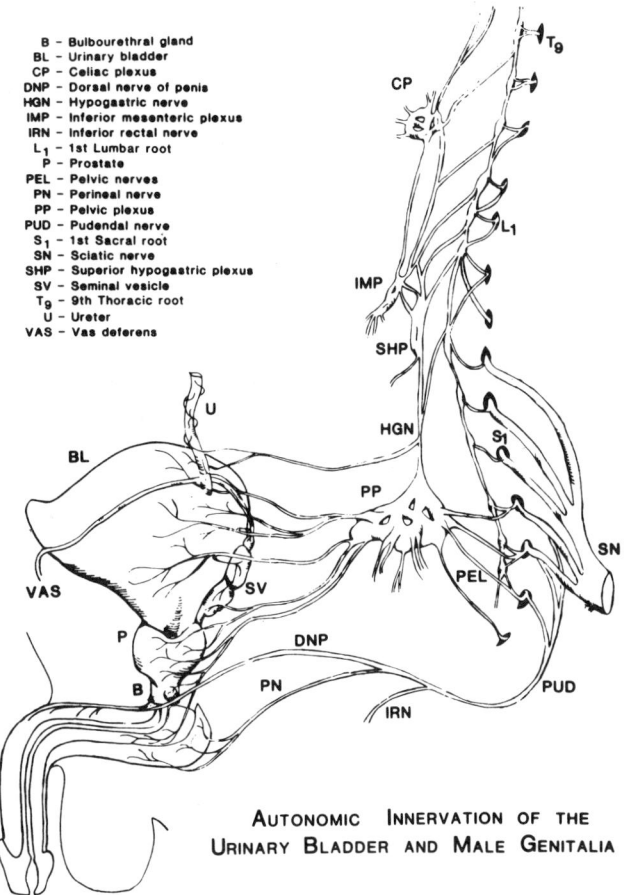

B - Bulbourethral gland
BL - Urinary bladder
CP - Celiac plexus
DNP - Dorsal nerve of penis
HGN - Hypogastric nerve
IMP - Inferior mesenteric plexus
IRN - Inferior rectal nerve
L₁ - 1st Lumbar root
P - Prostate
PEL - Pelvic nerves
PN - Perineal nerve
PP - Pelvic plexus
PUD - Pudendal nerve
S₁ - 1st Sacral root
SN - Sciatic nerve
SHP - Superior hypogastric plexus
SV - Seminal vesicle
T₉ - 9th Thoracic root
U - Ureter
VAS - Vas deferens

AUTONOMIC INNERVATION OF THE
URINARY BLADDER AND MALE GENITALIA

FIG 27–21.
Autonomic innervation of the lower urinary tract (in the male). (From DeGroat WC, Booth AM: Autonomic systems to the urinary bladder and sexual organs, in Dyck PJ, Thomas PK, Lambert EH (eds): *Peripheral Neuropathy.* Philadelphia, WB Saunders, 1984. Used with permission.)

segments S2–4, and emerges as preganglionic fibers in the ventral nerve roots. This parasympathetic preganglionic supply is ultimately conveyed by the pelvic nerve, which courses deep in the pelvis on each side of the rectum as three or four trunks in man. Efferent sympathetic nerves to the bladder and urethra are thought to originate in the intermediolateral nuclei of spinal cord segments T11–L2 (Norlen, 1982), although El Badawi (1982) mentions only an upper and midlumbar origin. These nerves traverse lumbar sympathetic (paravertebral) ganglia and branches of these ganglia join the presacral nerve (superior hypogastric plexus), which is a plexiform nerve arrangement in the lumbosacral area anterior to the aorta. This divides into the right and left hypogastric nerves, each of these actually being an elongated plexus. Bilaterally, at a variable distance from the bladder and urethra the hypogastric and pelvic nerves meet and branch to form the pelvic

plexus, sometimes known as the inferior hypogastric plexus or plexus of Frankenhauser. This is a plexus of freely interconnected nerves in the pelvic fascia that is lateral to the rectum, internal genitalia, and lower urinary organs. Divergent branches of this plexus innervate these pelvic organs. Afferent neurons are conveyed by both the hypogastric and pelvic nerves to the dorsal columns of the lumbosacral spinal cord. Most of the afferent fibers reach the spinal cord through dorsal nerve roots, but some through ventral roots (El Badawi, 1982, 1983). As would be expected, the main trunks of both nerves contain nonmyelinated (preganglionic efferent) and myelinated (sensory) nerve fibers.

As has been previously mentioned, the actual neuronal innervation of bladder and urethral smooth muscle emanates from ganglia, which theoretically may be located anywhere between the spinal cord and organ of innervation. Ganglia that innervate the bladder and urethral smooth musculature generally lie close to or within these structures, and give rise to "short neurons," which comprise the bulk of the nerves that actually innervate these structures. All peripheral ganglia are composed of cholinergic, adrenergic, and small intensely fluorescent cells, and represent one morphological substrate in which modulation of one neural component by another can occur. Dixon et al. (1983) feel that the intramural autonomic ganglion cells of the bladder (at least those in the dome and lateral walls) possess only cholinergic neurons, and distinguish these from those in the pelvic plexus. Other authors have not made this distinction.

Efferent innervation of the striated sphincter is likewise classically thought to emanate from the sacral spinal cord, via the pudendal nerve (Fletcher and Bradley, 1978; El Badawi, 1982, 1983). However, some think that the striated sphincter is innervated primarily by the autonomic nervous system through the hypogastric plexus (Donker et al., 1982), while Gosling thinks that the slow-twitch fibers in both the intramural and extramural component of the striated sphincter are innervated by somatic fibers within the pelvic nerve (1979). In the cat at least, El Badawi and others (El Badawi and Schenk, 1974; El Badawi, 1982, 1983; El Badawi and Atta, 1985) think that the intramural component of the striated sphincter possesses somatic and autonomic (both cholinergic and adrenergic) innervation. It is interesting to note that using studies employing retrograde axonal transport of horseradish peroxidase, Tanagho et al. (1982) and Morita et al. (1984) came to different conclusions regarding the innervation of the striated sphincter in the dog. Tanagho thought that the pudendal nucleus was the only motor center, whereas Morita et al. thought that the pelvic nerve might contain some somatic fibers innervating the striated sphincter. Electrophysiologic studies in the human by Vodusek and Light (1983) seemed to show that the intramural component of the striated sphincter is innervated by the pudendal nerve, but the authors did not exclude the possibility that the autonomic nervous system also innervates this muscle. These somewhat contradictory (and therefore confusing) views of the neural supply of the striated sphincter are summarized in Table 27–2.

BLADDER AND URETHRAL SMOOTH MUSCLE

Abundant cholinergic innervation of the bladder has been demonstrated by histochemical and ultrastructural technology in a variety of experimental animals (El Badawi and Schenk, 1968, 1974; Raezer et al., 1973; Alm and Elmer, 1975; Fletcher and Bradley, 1978; El Badawi, 1982). There is some disagreement as to regional variation of the degree of innervation. At this time there does not seem to be any clear consensus regarding this (El Badawi, 1982), and functionally these subtleties would seem unimportant, because of the overall relative abundance of the cholinergic innervation. These observations regarding cholinergic innervation in animals seem to be applicable to and to extend at least into the proximal urethra (El Badawi, 1982; Ekstrom and Malmberg, 1984). Adrenergic innervation of the bladder and urethral smooth musculature has been likewise extensively studied in animals using similar techniques (El Badawi and Schenk, 1966 and 1968; Oman et al., 1971; Raezer et al., 1973; Wein et al., 1974; Alm and Elmer, 1975; Benson et al., 1976; Fletcher and Bradley, 1978; El Badawi, 1982). These

TABLE 27–2.
Nerve Supply of the Striated Sphincter (Intramural and Extramural)

VIEW	SELECTED REFERENCES
Pudendal nerve only (both components, somatic fibers only)	Traditional (see text) Fletcher and Bradley (1978) Kluck (1980) Tanagho et al. (1982)
Pelvic plexus only (both components)	Gil Vernet (1968) Donker et al. (1982)
Pudendal nerve—fast-twitch fibers (extramural only) Pelvic nerve—slow-twitch fibers (both components, somatic fibers only)	Gosling (1979)
Somatic via pudendal nerve Autonomic cholinergic via pelvic nerve Autonomic adrenergic via hypogastric, perhaps pelvic and pudendal nerve	El Badawi (1982, 1983) El Badawi and Atta (1985)

animal studies show that the smooth musculature of the bladder base and proximal urethra has a rich adrenergic innervation, while the bladder body has sparse but definite such innervation. The density of innervation seems in all areas to be less than that of the cholinergic system.

Studies of cholinergic and adrenergic nerve endings in the smooth muscle of the human bladder and urethra have produced inconsistent conclusions. Ek et al. (1977) described an abundance of cholinergic nerves in all parts of the bladder musculature of both sexes, and a similar less extensive cholinergic innervation of the urethra. Kluck (1980), Daniel et al. (1983) and Gosling (1984) supported this view of an extensive cholinergic innervation of the human bladder body smooth musculature. Gosling stated that this type of cholinergic innervation extended throughout the human bladder base, except for what he termed the "superficial trigonal muscle," an area that he considers to be of little functional significance. Gosling made a distinction between the bladder neck and what others term the "bladder base," and he noted that bladder neck smooth muscle is histologically, histochemically, and pharmacologically distinct from the remainder of the bladder. He described the male bladder neck as having a very sparse cholinergic innervation, but the female bladder neck as being supplied with cholinergic fibers (1979, 1985). Kluck (1980) described the trigonal musculature (of at least the male) as having a relatively sparse distribution of cholinergic nerves, but described cholinergic innervation of the remainder of the bladder base and of at least the ventral part of the urethra, with no differences between male and female.

Mobley et al. (1966) described adrenergic fibers in the anterior wall of the human bladder; Sundin et al. (1977) and Benson et al. (1979) showed clear evidence of sparse but definite innervation of the human bladder body. Kluck (1980), however, did not find adrenergic innervation in the human female bladder body, while Daniel et al. (1983) found that small granular vesicles (presumably adrenergic) were the predominant vesicle type in approximately 15% of the nerve profile seen before 5-hydroxydopamine treatment (which generally makes adrenergic fibers more prominent), and 23% afterward. Both Ek et al. (1977) and Sundin et al. (1977) described a dense adrenergic innervation of the human trigone. Nordling (1983) stated that superficial trigonal musculature possesses a moderately dense adrenergic innervation (though ascribes little anatomic mass or significant function to this area), and that an otherwise sparse supply of adrenergic nerves exists in the bladder base. Norlen (1982) agreed with this. The male bladder neck was described by Gosling (1985) and Kaneko et al. (1980) as possessing a rich adrenergic innervation, but

Kluck (1980), in specimens that he described as being from the dorsal part of the male bladder neck, failed to find adrenergic nerve fibers. Gosling (1985) found human female bladder neck to contain relatively few adrenergic nerves; no other authors have made definite statements about the morphology of this area in the female. Kluck (1980) reported no adrenergic innervation of the ventral part of the female urethra, but evidence of a sparse adrenergic innervation of his only biopsy sample of anterior female urethra. Gosling stated that the smooth muscle of the female urethra is associated with relatively few adrenergic nerves, but that the preprostatic urethra in the male possesses a rich adrenergic innervation (1984).

El Badawi (1985) wisely commented on the extreme heterogeneity of findings related to cholinergic and adrenergic innervation of the human lower urinary tract. He stated that all human material thus far studied morphologically cannot be considered as representative of the adult norm in either sex for a variety of reasons, and cautioned likewise against the unqualified acceptance of the existence of sex differences in vesicourethral innervation, without further confirmation of standardly processed truly "normal" specimens. The functional aspects of adrenergic and cholinergic stimulation will be discussed subsequently, but an obvious question to ask is, how much of a particular type of innervation is actually necessary to subserve a physiologic function, and the answer, of course, is unknown.

Other neuroactive substances have been identified in the autonomic nerves supplying the lower urinary tract. Leucine enkephalin was found present in preganglionic terminals in cat bladder ganglia (deGroat et al., 1983). ATP and a number of peptides have been shown to exist in bladder nerves in a variety of animals and in man (El Badawi, 1982; Gu et al., 1984; Mundy, 1984b). These other peptides include vasoactive intestinal polypeptide (VIP), substance P, and neuropeptide Y and somatostatin. An independent purinergic (Burnstock et al., 1972) and peptidergic (Alm et al., 1977) population of efferent autonomic nerves has been postulated, but El Badawi (1982) stated that no unequivocal morphologic evidence exists that such nerves exist as a separate and independent population in the bladder or urethra smooth musculature. He pointed out that although these substances may function as neurotransmitters or neuromodulators, the vesicles containing them may simply be stored in an otherwise classical cholinergic or adrenergic nerve (1982).

STRIATED SPHINCTER

Most authors agree that the striated sphincter, including both components—the intramural (intrinsic) and extramural (extrinsic)—is innervated only through

motor end plates, implying purely somatic innervation, though there may be differences in opinion regarding the actual nerve trunks carrying these fibers (Fletcher and Bradley, 1978; Gosling, 1979, 1984). El Badawi reviewed his evidence for triple innervation (somatic plus cholinergic and adrenergic autonomic) of the intramural striated sphincter of the male cat (1982) and updated this with ultrastructural evidence of such innervation in the same species (El Badawi and Atta, 1985). Additionally, his laboratory (Atta and El Badawi, 1985) studied the effects of bilateral ventral sacral rhizotomy and found that autonomic cholinergic and adrenergic innervation was preserved (and exhibited sprouting), whereas somatic axons totally degenerated. Morphologic evidence of autonomic innervation of the striated sphincter has not yet been demonstrated in other species or in man (Fletcher and Bradley, 1978; Gosling, 1979, 1984; Wein et al., 1979; Kluck, 1980). El Badawi (1982) pointed out that his conclusions are applicable only to the intramural portion of the striated sphincter, and that other authors' conclusions regarding the intramural component may have been erroneously drawn from specimens from the adjacent extramural component. Further studies will obviously be necessary to settle this question.

SENSORY INNERVATION

Afferent nerve fibers have been demonstrated in the pelvic, pudendal, and hypogastric nerves (Kuru, 1965). Fletcher and Bradley (1978) concluded that most vesical afferent axons terminate as free nerve endings, and that no specialized receptors are identifiable, except for sparse pacinian corpuscles. Afferent terminals are found in the submucosa and muscularis; Fletcher and Bradley stated that, in the cat at least, a greater terminal density of afferent fibers to the pelvic nerve exists in the muscularis, while a greater density of afferents to the hypogastric nerve exists in the submucosa (1978). deGroat and Booth (1984) and deGroat and Kawatani (1985) state that afferent fibers subserving the sensation of bladder distention originate in tension receptors in the bladder wall and travel in the pelvic nerves, while mechanoreceptor afferents are present in the hypogastric nerve. Further, according to them, both pelvic and hypogastric afferent pathways carry nociceptive afferents, while afferent pathways from the striated sphincter and from the urethra transmit sensations of temperature, pain, wall distention (urethra), and urine passage, and travel in the pudendal nerve. Fletcher and Bradley (1978) summarized studies in the cat that they interpreted as consistent with the fact that afferents subserving the sensation of distention (and active therefore in evoking micturition), are more prominent in the muscle layer than in the submucosa and are distributed evenly to all regions of the bladder, whereas afferents traveling in the hypogastric nerve are most dense in the trigone and anterior bladder neck region, and subserve the sensations of pain, conscious touch, and distention.

PERIPHERAL, NEURAL, AND HUMORAL INFLUENCES ON LOWER URINARY TRACT FUNCTION

CHOLINERGIC RECEPTOR DISTRIBUTION, STIMULATION, AND BLOCKADE

Cholinergic contractile receptor sites that can be blocked by atropine (muscarinic) have been demonstrated in the bladder body and bladder base in the rat (Elmer, 1974), cat (Nergardh and Boreus, 1973; Nergardh, 1975), dog (Leoni et al., 1973; Raezer et al., 1973), rabbit (Levin et al., 1980), and man (Todd and Mack, 1969; Benson et al., 1976; Levin et al., 1983a). Khanna et al. (1981) studied the differential effects of cholinergic stimulation on the different layers of the rabbit bladder body and base. They found that acetylcholine had a marked stimulatory effect on both longitudinal and circular bladder body smooth musculature, a moderate contractile effect on the longitudinal and circular (less so) layers of the bladder base, a weak contractile response from the longitudinal layer of the urethra, and none from the urethral circular layer. In vitro studies of human urethral strips by Ek et al. (1977b) demonstrated a very low contractile response to acetylcholine, and the responses that did occur could be blocked with atropine. Similar data were presented by Persson and Andersson (1976) for the cat and guinea pig. Abdel-Rahman et al. (1983) showed a very minor contractile effect of cholinergic stimulation on the longitudinal and circular smooth muscle of the cat proximal urethra, contrasted with a moderate contractile effect in the bladder. Ekstrom and Malmberg (1984), however, recorded rapid and marked contractile responses of the rat urethra to cholinergic stimulation and felt that these effects were exerted by a direct action on muscarinic receptors.

Levin et al. (1980) found that the response to cholinergic stimulation in the rabbit bladder was not uniform, and was substantially greater in the bladder body than in the bladder base. They were the first to use radioligand-binding technology to quantitate the muscarinic cholinergic receptor density in animal and man (Levin et al., 1980, 1982, 1983a), and found, using tritiated (^{3}H-QNB) as a binding agent, that the distribution of receptor sites was qualitatively similar. Nilvebrant et al. (1983, 1985) confirmed, in the bladder body of the rabbit and human, the presence of a single population of noninteracting binding sites with pharmacologic specificity characteristic of muscarinic receptors. Lepro and

Kuhar reported similar findings in the rabbit bladder (1984). Anderson and Marks (1982) reported considerable "spare" cholinergic receptors in the bladder, implying that one could inhibit a significant proportion of muscarinic receptors without affecting cholinergically induced contractile function. Levin et al. (1983c) found no such evidence of spare muscarinic receptors in the rabbit, a finding supported by Nilvebrant et al. (1985) in human bladder. Adami et al. (1985) demonstrated the muscarinic receptors in rat bladder are of the M_2 variety, a finding recently confirmed in our laboratory in both rabbit and human bladder body (Ruggieri et al., 1986).

PARASYMPATHETIC NERVOUS SYSTEM FUNCTION

Stimulation and Blockade

A sustained bladder contraction is produced by stimulation of the pelvic nerves (Ursillo and Clark, 1956; Ingersoll et al., 1957; Gyermek, 1961; Kuru, 1965). It is generally agreed that reflex activation of this pelvic nerve excitatory outflow tract is responsible for the emptying bladder contraction of normal micturition. Whether acetylcholine is the sole neurotransmitter released during such stimulation is highly controversial. Muscarinic blockade only partially antagonized the bladder response to pelvic nerve stimulation (Ambache, 1955; Ursillo and Clark, 1956; Ambache and Zar, 1970; Burnstock et al., 1972; Elmer, 1975). Some urethral responses to pelvic nerve stimulation and cholinergic stimulation have also shown atropine resistance (Tanagho et al., 1969a, 1969b; Nergardh and Boreus, 1973; Khanna et al., 1975). These authors found increased urethral resistance to flow in response to cholinergic stimulation, attributed by Khanna et al. and Nergardh and Boreus to be mediated by intramural ganglion cells belonging to the short adrenergic system. McGuire (1978) and McGuire and Herlihy (1978) presented compelling evidence, however, that urethral smooth muscle relaxation occurs during detrusor activity, and they attributed those apparent increases in urethral "resistance" previously described in response to pelvic nerve stimulation to simple transmission of an intravesical pressure head to an open urethral chamber. They explained that this occurred normally after an initial prevoiding urethral relaxation, following which the bladder and urethra become essentially isobaric. McGuire (McGuire, 1977, 1978; McGuire and Herlihy, 1978, 1979) obviated this experimental problem of pressure transmission by diverting the urinary stream. He then reported urethral smooth muscle relaxation in response to pelvic nerve stimulation, which was blocked after propranolol administration, implying that this is a β-adrenergic response to pelvic nerve and presumably short adrenergic stimulation. This is an important observation, as it forms the basis for one theory of the mechanism of relaxation of the bladder neck and proximal urethra at the onset of voiding.

Field stimulation of isolated bladder strips has been used to reproduce the effects of nerve-mediated activity. Hindmarsh et al. (1977) showed that such electrical stimulation produced contractions in human bladder strips that were only partially blocked by atropine. The variable blockade achieved by atropine of pelvic nerve–induced and field stimulation–induced bladder contractile activity has given rise to an exceptionally large literature having to do with "atropine resistance" of the urinary bladder. In many ways, atropine resistance is a clinically convenient concept, as it can be invoked to explain our clinical difficulty in abolishing involuntary bladder contractions with anticholinergic agents, and it can be invoked also to support the rationale of the treatment of such types of bladder activity with more than one pharmacologic agent with different mechanisms of action (Wein, 1986). Typical examples of such atropine resistance in field-stimulated specimens include data reported by Krell et al. (1981), showing a maximum inhibitory ability of 55%–60% for guinea pig bladder strips, by Nergardh (1981) showing less than 20% inhibition in the cat, by Adami et al. (1985) showing 50% inhibition in the rat, and by Cowan and Daniel (1983) showing 50% inhibition in human female bladder body smooth muscle. Atropine resistance in human bladder muscle, however, is by no means agreed upon, and there is some evidence that it may not in fact occur in normal bladder. Sibley (1984) studied bladder strips from rabbit, pig, and man in the same experimental model. In the rabbit, he found that atropine was at most 56%–64% effective in blocking the response to field stimulation (at higher frequencies only—the response at low frequency was only 14%), and that this atropine-resistant response was completely abolished by the addition of tetrodotoxin, indicating that it was nerve-mediated. In the pig, he found that up to 83% of the contractile response was abolished at higher frequencies of stimulation. In the human, atropine produced an effective blockade at all frequencies of stimulation, with only 1%–7% of the response persisting, and this persistent response was not significantly reduced by tetrodotoxin, indicating that the nerve-mediated contractile responses were purely cholinergic. These human strips were obtained from patients undergoing surgery for lower urinary tract disorders or from patients undergoing donor nephrectomy; no difference was found between specimens from the two sources. Kinder and Mundy (1985a) demonstrated an almost total inhibition by atropine of the response of human

bladder body strips to field stimulation. In those strips in which there was some resistance to inhibition by atropine, the residual contractile response to stimulation was only minimally affected by tetradotoxin. The 13 patients from whom the samples were taken all had normal bladder function on urodynamic study. Sjogren et al. (1982) demonstrated marked atropine inhibition (more than 95% at higher concentrations) of field stimulation responses in bladder preparations from young patients and from those undergoing cystectomy for cancer. However, in bladder strips from patients who either had a diagnosis of bladder instability or who exhibited bladder hypertrophy, atropine was markedly less effective, a maximum inhibitory effect of approximately 50% being obtained. The reasons for this were not obvious, but it is tempting to speculate that noncholinergic nonadrenergic activation of the contractile apparatus may be more prominent in patients with bladder dysfunction than in those with normal bladder function.

Atropine Resistance and the Role of Adenosine Triphosphate

Many theories have been proposed to explain those instances of the relative insensitivity of bladder contractility, elicited by pelvic nerve or field strip stimulation, to atropine. Historically these hypotheses have included the following: (1) that at least some of the cholinergic receptors in the bladder are nicotinic rather than muscarinic (Gyermek, 1961); (2) that the nerve impinges so closely on the muscarinic receptor complex that atropine cannot penetrate this sufficiently to produce complete inhibition (Ursillo and Clark, 1956; Elmer, 1975; Carpenter, 1977); and (3) that a major portion of the parasympathetic transmission is noncholinergic (and nonadrenergic) (Ambache and Zar, 1970; Burnstock, 1972; Johns and Paton, 1976). Most contemporary literature points to hypothesis no. 3 as the most probable explanation of this phenomenon. In this regard, ATP (Burnstock et al., 1972; Burnstock, 1985, 1986) and prostaglandins (Kuru, 1965; Johns and Paton, 1976) have been the substances most frequently proposed as noncholinergic nonadrenergic parasympathetic neurotransmitters. Burnstock (1972, 1977, 1985, 1986) has provided evidence that ATP is an excitatory neurotransmitter involved in the bladder response to pelvic nerve stimulation. Further, he has proposed that this concept explains atropine resistance in those species where it in fact does occur. In 1978 he and coworkers presented evidence (quinacrine fluorescence histochemistry) of what were presumed to be purinergic nerves in the guinea pig urinary bladder and evidence that field stimulation produced a bladder contraction that was predominantly noncholinergic, associated with ATP release, and, further, that the release could be blocked by tetrodotoxin but not by adrenergic nerve destruction (Burnstock et al., 1978a, 1978b). Downie and Dean (1977) and Dean and Downie (1978) showed that a noncholinergic excitatory substance was involved in motor neurotransmission in rabbit bladder and concluded that this probably was ATP. Further work from Burnstock's laboratory (Kasakov and Burnstock, 1983) showed that mechanical responses to ATP and field stimulation in guinea pig bladder strips were simultaneously abolished after desensitization of these strips by a slowly degradable analogue of ATP, α, β-methylene ATP. Recent studies have shown that desensitization with this compound blocks not only the atropine-resistant contractile response to field stimulation but the atropine-resistant excitatory junction potential as well (Hoyle and Burnstock, 1985). Arylazido aminopropionyl ATP (ANAPP$_3$) is a photoaffinity analogue of ATP that antagonizes contractile responses to it. Theobald (1983) reported that this compound blocks ATP-induced contractions in the cat urinary bladder. Most recently, he has shown this antagonism is restricted to one purinergic receptor subtype (P$_2$ as opposed to P$_1$; see Theobald [1986] and Burnstock [1986] for discussion). Theobald has looked quite carefully at the bladder response in this animal to parasympathetic nerve–induced contraction. Two distinct components could be observed. The initial or first response was a sharp transient spiked rise in pressure (phase 1), as opposed to a second phase that consisted of a tonic maintenance of contraction during stimulation. This second phase was maintained at a certain level or fell slowly to a lower pressure level as long as the pelvic nerves were stimulated. He showed that ANAPP$_3$ partially blocked the first phase of the response to parasympathetic nerve–induced contraction. Atropine had no effect on phase 1 response or on ATP-induced contraction, but blocked the second phase of the pelvic nerve contraction, changing the character of this response so that it resembled that of an ATP-induced contraction. These data indicate that the second phase of the pelvic nerve–induced contraction is atropine-sensitive and produced by the release of acetylcholine. These data also suggest that acetylcholine is not solely responsible for the first phase of the nerve-induced contraction, but that ATP, released by pelvic nerve stimulation, is at least partially responsible for this event. The fact that the first phase of pelvic nerve–induced contraction is only partially antagonized by ANAPP$_3$ while ATP-induced contractions are completely antagonized suggests further that there may be another transmitter involved in the first phase of these contractions. This other transmitter may be acetylcholine or another neurotransmitter released on pelvic nerve stimulation (Theobald, 1983). Maggi et al. (1985c)

came to similar conclusions regarding two components of the twitch response to field stimulation of rat bladder strips. They found that the inhibition of atropine of the second or tonic component was much greater (over 60%) than of the early or phasic component (less than 10%). Both were almost completely inhibited by tetrodotoxin.

Levin et al. (1986) likewise found that the response of their whole bladder model responded to field stimulation with a biphasic response: an initial rapid rise in tension followed by a plateau period of increased intravesical pressure. Functionally, virtually all bladder emptying occurred during the plateau period. Although purinergic stimulation accounted for approximately 50% of the initial phase of bladder contraction, because of the phasic nature of the purinergic response, it mediated virtually no bladder emptying. These authors concluded that whereas purinergic stimulation may play a role in initiating micturition, bladder emptying appears to be primarily a cholinergic phenomenon. Further, work from this laboratory has shown that β,γ-methylene ATP is approximately 100 times more potent than ATP itself in stimulating this initial phasic contractile activity. Studies on ATPase activity have demonstrated that this activity is too slow to account for this difference in potency, and studies on enzyme kinetics have demonstrated that this analogue of ATP does not bind to ATPase, indicating that purinergic response is not mediated by activation of a surface membrane ATPase (Levin et al., 1983b, 1986). Although high-affinity binding of ^{3}H-ATP to membrane fractions of rabbit bladder has been demonstrated, distribution of these binding sites is not consistent with those of neurotransmitter receptor sites, and probably represents other specific binding sites for ATP.

Nicotine induces an increase in intravesical pressure in cats that is atropine resistant and not mediated either through adrenergic or cholinergic mechanisms. However, Koley et al. (1984) felt that a purinergic mechanism was involved, and in fact quinidine (which is felt under some circumstances to block ATP-induced contractions) and ATP desensitization (which also inhibits the response to subsequent exogenous ATP) abolish the response to nicotine.

PROSTAGLANDINS

In the lower urinary tract, prostaglandins have been prominently mentioned as perhaps having a potential role in excitatory neurotransmission, contractile activity, and, perhaps, an involvement in the response to ATP. Pharmacologic evidence for the role of prostaglandins in bladder contractility has been succinctly and well summarized by Andersson and Sjogren (1982). Evidence that various prostaglandins of the E and F types produce contraction of bladder muscle strips is summarized, and the point made that the contractions were slower in development and with a longer duration than those produced by acetylcholine. These contractions likewise were reported as not influenced by tetrodotoxin, atropine, or phenoxybenzamine (Bultitude et al., 1976; Abrams and Feneley, 1976; Andersson et al., 1977). Khanna et al. (1978) suggested the existence of a specific prostaglandin receptor in the bladder.

Khalaf et al. (1981) showed that distention of the dog bladder and pelvic nerve stimulation both caused a release of prostaglandins. Bultitude et al. (1976) summarized evidence that prostaglandin synthetase inhibitors decreased spontaneous activity and basal tone of the rabbit bladder, effects that were antagonized by prostaglandin $F_{2\alpha}$, suggesting a connection between contractile effects mediated by acetylcholine and by prostaglandins. Johns and Paton (1976) suggested that prostaglandins are liberated during neural excitation of bladder contraction and contribute to the atropine-resistant portion of this response. Andersson and Sjogren (1982) summarized evidence that indicates that it is unlikely that these agents are directly involved in bladder emptying, but more probable that they contribute to tone or spontaneous activity in the bladder muscle, or that they may be modulators of transmission, increasing the release of or the effectiveness of excitatory neurotransmitters.

There is some evidence that prostaglandins may be involved in the contractile response of bladder smooth muscle to ATP. Andersson et al. (1980) showed that the response to ATP in rabbit bladder strips consisted of an initial sharp phasic component, then a tonic maintained component, whereas in the guinea pig, there was only the phasic component. Prostaglandin synthetase inhibitors abolished only the tonic portion in the rabbit and had no effect in the guinea pig. Downie and Larson (1981) found a similar response to ATP in the rabbit, and found further that prostaglandin synthetase inhibition decreased the second tonic portion of the response to 30% of control but did not affect the initial phasic response. Husted et al. (1983) found that indomethacin abolished the second tonic phase of the response of rabbit and human bladder muscle strips to ATP and reduced the first phasic response by about 30%. They pointed out, however, that the ATP response was heavily dependent on extracellular calcium, and therefore that the inhibitory action of indomethacin might be attributable to a calcium antagonist effect rather than to an exclusive effect on prostaglandin synthesis. They also noted parenthetically that hypertrophic detrusor seemed to be more sensitive to the action of ATP than normal detrusor.

Downie and Karmazyn (1984) have presented interesting findings that suggest a different type of contractile influence of prostaglandins on detrusor muscle. They found that mechanical irritation of the epithelium of rabbit bladder increased basal tension and spontaneous activity in response to electrical stimulation of bladder muscle, and that these responses were related to the intensity of the irritative trauma. The effects were mimicked by prostaglandins of the E, F, and I series, and the transfer effect was significantly reduced by pretreatment of the epithelium, but not the muscle, with prostaglandin synthetase inhibitors. Certainly this suggests a potential role of prostaglandins in the pathophysiology of hyperactivity secondary to traumatic changes of the bladder urothelium. These prostaglandin products would be capable of causing muscle contraction per se and enhancing neurotransmission in the muscle layer of the bladder. In a similar vein, Maggi et al. (1984a) found that the ability of various prostaglandin synthetase inhibitors to block rhythmic contractions induced by distention in the rat bladder was proportional to their anti-inflammatory effectiveness. Although prostaglandin release during bladder distention has been classically described as mediated from the muscle layer, it may be that this occurs from the epithelial layer.

It is interesting that at least some prostaglandins have been shown to cause relaxation of urethral preparations as well as increased contractility of bladder. Andersson et al. (1977) showed that prostaglandin E_1 (PGE_1) and PGE_2 relaxed urethral preparations that had been contracted by norepinephrine, epinephrine, or PGF_2. Khalaf et al. (1981) similarly found that PGE_2 produced a decrease in resting urethral pressure in the dog, and Finkbeiner and Bissada (1981) found that a similar effect was produced in guinea pig by PGE_2, PGE_1, PGF_1, and PGF_2. It should be noted that Finkbeiner and Bissada are the only authors who note relaxation of urethral smooth muscle with PGF_2; all others have noted contraction.

Prostaglandins, then, have been potentially implicated in bladder contractility or tension development that occurs during bladder filling and also in the emptying contractile response of bladder smooth muscle to neural stimulation. It has also been suggested that they are involved in the maintenance of urethral tone during the storage phase of micturition, and also in the release of this tone during bladder contraction. As has been implied, the mechanisms surrounding the maintenance of continence during the filling phase of micturition have not been established, and, moreso, there is no agreement regarding what causes opening of the bladder outlet with reflex-induced bladder contraction during the emptying phase of micturition. Ito and Kimoto (1985) have suggested the involvement of prostaglandins in maintenance of tension in the rabbit proximal urethra and in its reduction during micturition. They found that indomethacin and a specific prostaglandin antagonist markedly reduced the resting tension of urethral muscle strips and that activation of intrinsic inhibitory nerve fibers failed to evoke muscle relaxation at this reduced muscle tone (after treatment with indomethacin or the specific inhibitor). In other words, the amplitude of muscle relaxation invoked by field stimulation was dependent on the pre-existent level of muscle tone of the urethral strips, and the data indicate that endogenous prostaglandins play a role in maintaining this muscle tone. Electrical field stimulation evoked a phasic contraction followed by relaxation, and guanethidine, phentolamine, and atropine each reduced the amplitude of the electrical changes so produced during the phasic contraction portion of the response as well as of the mechanical responses. The combined application of guanethidine and atropine further reduced the amplitude of these changes. The muscle relaxation phase was not affected by propranolol, phentolamine, guanethidine, or atropine. These results taken together indicate that proximal urethral smooth muscle cells in the rabbit are innervated by adrenergic and cholinergic excitatory fibers, and by noncholinergic, nonadrenergic inhibitory fibers. Thus, the muscle tone of the proximal urethra may be maintained by endogenous prostaglandins and by adrenergic and cholinergic excitatory neural stimuli, and reduction in the muscle tone may occur by the activation of noncholinergic, nonadrenergic inhibitory fibers.

It is interesting that a similar concept of the role of nonadrenergic, noncholinergic inhibitory fibers in the opening of the bladder outlet during micturition was expressed by Andersson et al. (1983) and Klarskov et al. (1983) in closely related articles that appeared within a few months of one another. Andersson et al. showed that field stimulation caused a contraction of isolated rabbit and human urethral smooth muscle when the resting tension was low, and that this response was abolished by tetrodotoxin. The contractions were markedly but not completely reduced by α-adrenergic blockade. When the preparations were contracted by norepinephrine, however, which might mimic the state of the proximal urethra during bladder filling, electrical stimulation produced relaxation, and this response was inhibited by tetrodotoxin, suggesting the involvement of nerves and the release of a relaxation-producing transmitter. The electrically induced relaxation was not blocked by indomethacin, propranolol, acetylcholine, atropine, or metoclopramide. The authors speculated that, during transmural electrical stimulation of the urethra, both excitatory and inhibitory transmitters may be released, and the ultimate effect might depend on initial baseline tension. When the tension is low, the excitatory effect predominates, probably through α-adrenergic neurotransmission and effect. When the tension is increased, which may mimic the physiologic situation during bladder filling, the inhibitory effect of the neurotransmitter is demonstrable. Klarskov et al. (1983) exposed human and pig urethral, bladder neck, and trigonal smooth muscle to field stimulation. Responses were composed of different combinations of relaxation and contraction, but there were no significant consistent differences in the responses from various areas. The configuration of the response was slightly shifted in favor of contraction by β-adrenergic blockade and prostaglandin synthetase inhibition. The reverse effect was seen, with augmentation of the relaxation phase after α-adrenergic blockade and muscarinic cholinergic blockade. Cholinergic blockade had the most pronounced effect and, in some strips, changed a pure contraction into a relaxation. α-Adrenergic blockade diminished the contraction and β-adrenergic blockade inhibited it. Relaxation was slightly diminished, and contractions slightly enhanced, by prostaglandin synthetase inhibition. However, after blockade of all transmitter systems, the major part of the relaxation was still present during electrical field stimulation, and this relaxation phase was totally abolished by tetrodotoxin. It was concluded then that this relaxation phase was nerve mediated through noncholinergic, nonadrenergic, nonprostaglandin transmitter or modulator systems.

ADRENERGIC RECEPTOR DISTRIBUTION, STIMULATION, AND BLOCKADE

Regardless of controversies surrounding the presence or absence of adrenergic innervation, there is general agreement that the smooth muscle of the bladder body, bladder base, and proximal urethra in a variety of animals and in man contains both α-adrenergic receptors and β-adrenergic receptors. Originally, functional studies done on isolated bladder strips showed that responses to β-adrenergic stimulation were greatest in the bladder body, whereas α-adrenergic responses predominated in the bladder base and proximal urethra (Edvardsen, 1968a, 1968b, 1968c, 1968d; Edvardsen and Setekleiv, 1968; Todd and Mack, 1969; Anderson et al., 1971; Rohner et al., 1971; Donker et al., 1972; Leoni et al., 1973; Raezer et al., 1973; Awad et al., 1974; Elmer, 1974; Benson et al., 1976). Some investigators found no smooth muscle response in the bladder body to α-adrenergic stimulation (Nergardh and Boreus, 1972; Sundin et al., 1977). In all other studies, the β-adrenergic effect in this area simply predominated over the α-adrenergic effect. At the bladder neck area, β-adrenergic responses have been found to be weak and inconsistent (Nergardh and Boreus, 1972;

Raezer et al., 1973; Caine et al., 1975; Benson et al., 1976). Urethral strips in both humans and animals show a predominance of α-adrenergic responses over β-adrenergic responses with no gradation from the bladder neck, at least through the proximal urethra (Raz et al., 1972a; Ek et al., 1977). Studies have demonstrated that the distribution of α- and β-receptors parallels the functional responses to adrenergic stimulation. β-Adrenergic receptors predominate in the bladder body and α-adrenergic receptors predominate in the bladder base of man, dog, and rabbit (Wein and Levin, 1979; Levin and Wein, 1979a, 1979b; Levin et al., 1980). Attempts to subcategorize the β-adrenergic receptors in bladder body have not produced a great degree of consistency. Nergardh et al. (1977) found the cat bladder body to contain β_1 receptors, whereas they found a third type of β-adrenergic receptor in human bladder, corresponding neither to β_1 nor β_2. Larsen (1979) found β_2-adrenergic receptors in the detrusor muscle of the pig, but also found β-adrenergic receptors in the human bladder with neither β_1 nor β_2 characteristics. Maggi and Meli (1982) found that the rat bladder contains β_2-adrenergic receptors, while Anderson and Marks (1984) found the rabbit bladder body to contain primarily β_2-adrenergic receptors but with a minor population (13%–23%) of classical β_1-adrenergic receptors.

Andersson et al. (1984) used radioligand-binding techniques to evaluate the proportion of α_1- and α_2-adrenergic receptors in female rabbit bladder base and urethra. In membrane preparations, they found 25% α_1- and 75% α_2-adrenergic receptors. Functional experiments of the urethral smooth muscle determined that both α_1 and α_2 subtypes were present postjunctionally, since norepinephrine, phenylephrine, and clonidine all caused contraction; prazosin inhibited norepinephrine-induced contraction by 84%, and rauwolscine by 52%. Larsson et al. (1986) reported that the number of α-adrenergic receptors in the female bladder base and urethra increased toward the urethral meatus, but that this increase was due primarily to a selective increase in the number of α_2-adrenergic receptors. The distal urethra was more responsive to clonidine than the proximal portion (contractile response—indicating postjunctional α_2-adrenergic receptors), while no such difference in responses to norepinephrine occurred. If, as some have suggested (Langer and Shepperson, 1982), the α_2 postsynaptic receptor responds to humoral stimulation, while the α_1 postsynaptic receptor responds to neural stimulation, the two subtypes in the urethra may serve different functions, i.e., a greater tonic-type response in the distal urethra (Larsson et al., 1986). Yablonsky et al. (1986) reported a sex-related difference in regional α subtypes in the rabbit urethra. They found the relative proportion of α_1- and α_2-adrenergic recep-

tors corresponded to 20% and 80% of the total α-adrenergic receptor number in the female (about the same as reported by Andersson et al., 1984), whereas it was 65% and 35% in the male. Additionally, they reported that α_2-adrenergic receptor-binding capacity and affinity were 6.6 times higher and 2 times lower, respectively, in the female than in the male. Mattiasson (1984) found that in the human, however, the urethral contractile response was mediated by postjunctional α_1-adrenergic receptors. He found α_2-adrenergic receptors as well, but these were of the presynaptic autoregulatory type. Similarly, Kunisawa et al. (1985), Honda et al. (1985), and Tsujimoto et al. (1986) reported that postjunctional α_1-adrenergic receptors mediate contractile responses in the human and rabbit bladder base.

SYMPATHETIC NERVOUS SYSTEM FUNCTION

The role of the sympathetic nervous system in lower urinary tract function has been a fertile issue for research, discussion, and debate since the early 1900s. Those who advocate a major role for the sympathetic nervous system in lower urinary tract function generally do so along the lines of McGuire's figure (Fig 27–22), the implications of which have been recently summarized by multiple authors (Wein and Raezer, 1979; Norlen, 1982; El Badawi, 1982; McGuire, 1984a; deGroat and Booth, 1984). According to evidence cited by these

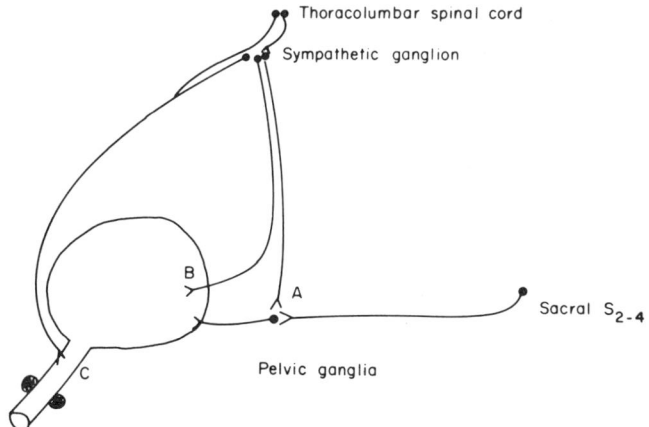

FIG 27–22.
Summary of potential effects of the sympathetic nervous system on the lower urinary tract. **A,** inhibition of parasympathetic ganglionic transmission. **B,** increased accommodation during filling/storage by stimulation of the predominantly β-adrenergic receptors of the bladder body. **C,** increased outlet resistance by stimulation of the predominantly α-adrenergic receptors of the bladder base and proximal urethra. All tend to facilitate the filling/storage phase of micturition. (From McGuire EJ: Clinical evaluation and treatment of neurogenic vesical dysfunction, in Libertino J (ed): *International Perspectives in Urology.* Baltimore, Williams & Wilkins, 1984. Used by permission.)

authors (which will be summarized below), the sympathetic nervous system acts primarily to facilitate the filling/storage phase of micturition, and does so by three mechanisms: (1) increasing accommodation, by stimulation of the β-adrenergic receptors in the bladder body; (2) increasing outlet resistance, by stimulation of the predominantly α-adrenergic receptors in the bladder base and proximal urethra; and (3) inhibiting bladder contractility via a blocking effect on parasympathetic ganglionic transmission. Evidence to support this view comes mainly from animal experimentation. Inferences are made about human physiology primarily on the basis of (1) similar effects of adrenergic stimulation and blockade on isolated bladder muscle strips; (2) urodynamic studies of bladder and outlet function during filling/storage and emptying in normal individuals, and in individuals with various voiding dysfunctions; and (3) the pathology produced by various lesions of the neural axis thought to result in relatively pure sympathetic neural deficits. There is considerable disagreement, however, regarding the role of the sympathetic nervous system in human lower urinary tract function. Many are of the opinion that it plays a very minor role, and this side of the controversy is well summarized by Tanagho (1978), Gosling (1979), Nordling (1983), and Mundy (1984a).

The results of hypogastric nerve stimulation in different experimental animals are sometimes contradictory, probably because of (1) species differences in the relative densities of α- and β-adrenergic receptors in different areas of the bladders; (2) stimulation of different populations of α- and β-adrenergic receptors according to different stimulus characteristics used by different investigators; (3) varying effects, through ganglionic interactions, on postganglionic cholinergic and nonadrenergic noncholinergic neurons; and (4) the presence of nonadrenergic postganglionic fibers in the hypogastric nerves of some animals. Kuru (1965) summarized experiments in the cat that show that hypogastric stimulation produces an initial transient rise of intravesical pressure, thought to be primarily due to trigonal contraction, followed by inhibition of spontaneous or evoked bladder contractions. deGroat and Saum (1971, 1972, 1976) showed that this inhibition is mediated by two distinct mechanisms: (1) a direct depression of bladder smooth muscle brought about by activation of primarily β-adrenergic receptors and (2) a depression of transmission in parasympathetic ganglionic cells mediated by α-adrenergic receptors. deGroat and Theobald (1976) showed that this inhibitory, sympathetic efferent pathway is activated only by afferent impulses in the pelvic nerves, thought to occur as a response to bladder filling. This is thus a spinal reflex whose efferents were originally thought to emanate from the lower

thoracic and upper lumbar spinal cord segments (deGroat, 1975). Recent anatomic and physiologic studies by deGroat's group (Kuo et al., 1984) have shown that the sympathetic pathways to the lower urinary tract of the cat originate in the sacral sympathetic chain as well as in the upper lumbar and inferior mesenteric ganglia. Edvardson (1968a, 1968b) likewise postulated a spinal reflex in the cat—with afferents in the pelvic nerves and efferents in the hypogastric nerves—causing bladder relaxation during filling and, therefore, an increased volume threshold for micturition. Consistent with this hypothesis is the fact that, in the cat, β-adrenergic blockade or surgical sympathectomy has been reported to increase bladder activity, decrease bladder capacity, and produce a shift to the left of the accommodation limb of the cystometric curve (Gjone, 1965; Edvardson, 1968a, 1968b, 1968c, 1968d; deGroat and Saum, 1972; deGroat and Theobald, 1976). Chemical and surgical sympathectomy have been found to produce similar results in the dog (Wein et al., 1974; Nishizawa et al., 1985). Fall et al. (1977) showed that intravaginal electrical stimulation at a low intravesical pressure induced detrusor inhibition in cats via the hypogastric nerve and that this inhibition was eliminated by β-adrenergic blockade. Pudendal nerve afferent stimulation was noted by Sundin et al. (1974) to similarly produce detrusor inhibition in the cat. Creed (1979) showed that hypogastric nerve stimulation in the dog increased urethral pressure, but she also found that this increased bladder pressure as well. The urethral response was attributed to direct adrenergic innervation. The bladder response was attributed to an indirect action via ganglia in the pelvic plexus but also a probable direct action as well. Ohtsuka et al. (1980) stimulated the hypogastric nerve of dogs and observed a rapid transient increase in bladder pressure but an inhibition of bladder contraction induced by stimulation of the pelvic nerve proximal to the pelvic plexus. Two mechanisms of inhibition were thought to exist, one at the level of the ganglia in the pelvic plexus, mediated by α-adrenergic receptors, and one peripherally at or in the bladder wall, mediated by β-adrenergic receptors.

Elliot (1907) reported that the bladder neck of the monkey contracted in response to sympathetic stimulation. Kleeman (1970) reported that electrical stimulation of the presacral nerves and the administration of α-adrenergic agonists caused bladder neck closure in dogs. He also made what seems to be the first observation that sympatholytic drugs were capable of facilitating bladder emptying in certain patients. Krane and Olsson (1973a, 1973b) described the concept of a physiologic internal sphincter, controlled partially via tonic stimulation from the sympathetic nervous system of contractile α-adrenergic receptors in the smooth muscle of the bladder neck and proximal urethra. Awad and Downie (1976a, 1976b, 1977) likewise endorsed the concept of sympathetic nervous system involvement in the continence mechanism at the level of the bladder outlet. McGuire (1984a, 1984b) not only supported the concept of a sympathetically induced β-adrenergically mediated moderating effect on bladder tone during filling, but also endorsed the concept of a major influence of the sympathetic nervous system on proximal urethral activity. His observations were a synthesis of data from the literature plus experimental observations on urethral closure pressure and pressure profiles made in his laboratory on the cat (McGuire, 1977; McGuire and Herlihy, 1978, 1979), and from clinical observations made by him and his associates in patients with nonfunctional bladder outlets (Woodside and McGuire, 1979; McGuire, 1984a, 1984b). He postulated that during bladder filling there was an excitatory influence on urethral closing pressure mediated by the sympathetic nervous system through α-adrenergic receptors. Urethral smooth muscle closure function was felt to be totally independent of preganglionic parasympathetic influences and totally lost when the thoracolumbar sympathetic outflow was destroyed in man. He argued strongly that bladder filling clearly results in a sympathetic motor response, normally resulting in vasomotor changes that are best seen when profoundly augmented in spinal cord injury patients with autonomic hyperreflexia. Further, he stated that smooth sphincter activity increases with bladder filling as a result of this sympathetic efferent activity, and that this rise can be blocked with α-adrenergic blocking agents. Additionally, he hypothesized an involvement of the adrenergic component in the opening of the bladder outlet that occurs at the initiation of the emptying phase of micturition. This was felt to be a result of β-adrenergic stimulation of the outlet smooth musculature, but through parasympathetic stimulation. In other words, this theory involves preganglionic parasympathetic stimulation not only as the prime factor in bladder contraction (through postganglionic primarily cholinergic fibers), but also as a major contributing factor to the decrease in outlet resistance seen during the empyting phase of micturition (through postganglionic adrenergic fibers—the "short adrenergic system"). This short adrenergic component activates β-adrenergic receptors (not α-adrenergic) in the outlet.

El Badawi (1982), on reviewing the available physiologic and pharmacologic data, totally endorsed the view that "the storage phase of micturition is controlled primarily by sympathetic, and the voiding phase by parasympathetic vesicourethral innervation." Further, he endorsed the concept of β-adrenergic–induced relaxa-

tion of the smooth sphincter at the outset of voiding, and synthesized this with his own view that unitary contraction of the body detrusor and smooth sphincter occurred during micturition and was essential for the maintenance of the voiding phase. As he put it, "This proposal does not contradict McGuire-Herlihy's thesis of lissosphincter relaxation at the outset of voiding, which the present author endorses unreservedly, believing that relaxation of a continent lissosphincter, maintained under sympathetic α-adrenergic excitation during the storage phase, must be replaced by parasympathogenic contraction, to sustain a rigidly open vesicourethral outlet in the face of the pressure force of the urine stream." Our interpretation is that this proposal involves a sympathetic α-adrenergic contribution to smooth sphincter tone during storage, replaced by a parasympathetic short adrenergic β-receptor–induced relaxation at the onset of emptying. During emptying, parasympathetic cholinergic stimulation of this also keeps it rigid (in the "open" position) to prevent ballooning and loss of the pressure force.

An equally impressive body of evidence and opinions by equally impressive investigators, however, argues against a significant role of the sympathetic nervous system during either the filling/storage or the emptying phase of micturition. Klevmark (1977) failed to demonstrate a significant bladder pressure rise in anesthetized cats with an intact neural axis at physiological (very low) filling rates (one hourly diuresis [HD] equilvalent to 1.1 ml/kg/hr). At rates of 3–4 HD the intravesical pressure gradually increased. No change occurred after sympathectomy, whether at high or low volume, leading Klevmark to conclude that no significant influence on bladder activity during filling was exerted by the sympathetic nervous system. Clinically significant alteration of bladder capacity with β-adrenergic antagonists does not occur in normal individuals, though there is some suggestion that some change might occur in selected individuals with a decreased bladder capacity (Wein, 1986). Tanagho (1978) has very nicely summarized some of the impressive array of his work on the dog, which has led him to raise significant questions regarding the reality of a sympathetic effect on the intrinsic smooth musculature of the urethra. His group (Tanagho and Myers, 1969; Tanagho et al., 1969a, 1969b; Tanagho and Miller, 1970) did not find a consistent and lasting response of the urethra in response to sympathetic stimulation, but did find that this smooth musculature responded uniformly to parasympathetic and cholinergic stimulation, and that its activity was obliterated by muscarinic antagonists. The effects of adrenergic stimulation and blockade were attributed to receptors in the vasculature of the urethra rather than on the smooth muscle itself. A definite adrenergic influence was postulated for a genital sphincter in the male that prevents retrograde ejaculation. This genital sphincter consists of smooth muscle in the supramontanal male urethra, a view that is similar to that of Gosling, who postulated a similar role for adrenergic innervation in the lower urinary tract of the male, but restricted this area more precisely to the area of the bladder neck (Bruschini et al., 1978; Gosling, 1979; Gosling and Chilton, 1984).

Finally, Nordling (1983) summarized an impressive amount of data relative to the role of the sympathetic nervous system in lower urinary tract function and concluded that, at least in the normal individual, the role is indeed a minor one. He pointed out that "sympathectomy" and operations that affect sympathetic ganglia, such as retroperitoneal lymphadenectomy, rarely if ever cause a lasting effect on the micturition of a normal individual. Studying urethral pressure changes in response to α-adrenergic blocking agents in a group of patients with localized neurologic lesions of the spinal cord, cauda equina, or peripheral nerves, he concluded that only lesions of the peripheral sympathetic nervous system caused a decreased effect of α-adrenergic blockade on maximum urethral pressure. He felt that this did confirm the existence of urethral α-adrenergic innervation-induced tonus at rest, but found no evidence for a significant influence of the sympathetic system on bladder function at rest or during voiding. Additionally, he questioned the classic localization of the origin of the sympathetic nervous supply to the lower urinary tract by pointing out that in the patients whom he studied, no specific level of the spinal cord could be demonstrated to act as a center for sympathetic innervation of the lower urinary tract. He added that this might be due to an extension of the sympathetic origin over so many spinal segments that damage to only a few would not significantly alter sympathetic influence on the urethra. Attempts to correlate the appearance of an open bladder neck with the bladder at rest and the level of spinal cord or peripheral neurologic lesion produced the conclusion that an intact parasympathetic rather than sympathetic innervation was the prerequisite for keeping the bladder neck closed at rest. Severe stress on the bladder neck produced by strong detrusor contractions in the presence of sphincter dyssynergia was also noted as a possible cause of the "open bladder neck." On the basis of experiments done in normal females, Nordling suggests that urethral pressure is partly maintained through peripheral α-adrenergic receptors in the urethral vessels rather than the smooth muscle itself, and by α-adrenergic receptors in the central nervous system that are more involved in the regulation of striated sphincter tone than smooth sphincter tone.

α-ADRENERGIC EFFECTS ON THE STRIATED SPHINCTER.—Much of the confusion relative to whether or not the sympathetic nervous system has an effect on the striated sphincter relates to the interpretation of clinical observations and experimental data referable to the effect of α-adrenergic blocking agents on urethral pressure in the region of the urogenital diaphragm and on electromyographic (EMG) activity in the periurethral striated muscle of this area. Nanninga et al. (1977) found that the EMG activity recorded in the "external sphincter" decreased after phentolamine administration in three paraplegic patients. They attributed the effect to a peripheral blockade of a direct sympathetic action on the striated sphincter (through sympathetic innervation). Nordling et al. (1981) demonstrated that clonidine and phenoxybenzamine (both of which pass the blood-brain barrier) decreased urethral pressure and EMG activity from the area of the striated sphincter in five normal women, but that phentolamine (which does not pass the blood-brain barrier) also decreased urethral pressure in this area, while it had no effect on the EMG activity. The authors concluded that phentolamine's effect was due purely to smooth muscle relaxation, while the effect of clonidine (and, possibly, phenoxybenzamine) was elicited mostly through centrally induced changes in striated urethral sphincter tonus, but that these agents also had an effect on the smooth muscle component of urethral pressure. These pressure effects were measured as effects on the maximum urethral pressure during urethral profilometry. It should be noted, however, that none of the drugs affected the reflex rise in urethral pressure or the reflex increase in EMG activity seen during bladder filling. Likewise, there was no drug-induced decrease in the urethral pressure or EMG activity response to voluntary contraction of the pelvic floor striated musculature. Pedersen et al. (1980) studied the effects of thymoxamine on the urethral pressure profile in ten patients with spastic paraplegia and on striated sphincter EMG activity in five. This drug is an α-adrenergic blocking agent that passes the blood-brain barrier. Peak urethral pressure was reduced in all patients studied, and EMG activity was likewise decreased, leading the authors to speculate that the drug had its action on the striated sphincter on a central basis.

Gajewski et al. (1984) conducted an elegant series of experiments on cats to explore the effect of α-adrenergic blocking agents on the striated sphincter. They found that hypogastric and pudendal nerve stimulation produced rises in urethral perfusion pressure, and that prazosin and phentolamine decreased the response only to hypogastric nerve stimulation. However, when the striated sphincter response was evoked indirectly, by stimulating the central cut end of the contralateral pudendal nerve, the urethral pressure response was significantly decreased by α-adrenergic blockade, moreso by prazosin than by phentolamine. The EMG potentials elicited by the indirect stimulation were completely abolished by prazosin, while the potentials elicited by direct stimulation remained unchanged, but did disappear after gallamine, a striated muscle blocker, was administered. Hypogastric nerve stimulation produced no EMG activity. The calculated reflex characteristics that were elicited by the contralateral pudendal cut-end stimulation were determined to be characteristic of a multisynaptic pathway in a spinal reflex. The authors concluded that α-adrenergic blocking agents do not influence the pudendal nerve–dependent urethral response through a peripheral action, but that at least prazosin can significantly inhibit this response at a central level.

PERIPHERAL PEPTIDES

The word "peptide" loosely refers to any member of a class of compounds of low molecular weight that yields two or more amino acids on hydrolysis. Those mentioned as putative neurotransmitters affecting the lower urinary tract are listed in Table 27–1. Of these, one large family is composed of the opioids. This term refers to any directly acting compound with effects that are stereospecifically antagonized by naloxone. This definition does not exclude the possibility that an opioid or part of it may have additional actions not antagonized by naloxone and therefore not mediated by opioid receptors. The term "opioid peptide" is usually applied to a naturally occurring peptide with opiate-like biological properties, i.e., at least some of the characteristic properties of morphine, which effects should be reversible by naloxone (Hughes and Kosterlitz, 1983; Morley, 1983; Mundy, 1984b). The terminology in this field is just becoming established and is still somewhat confusing. There are three distinct families of endogenous opioid peptides—enkephalins, endorphins, and dymorhins. These families are distinguished by genetically distinct precursor polypeptides and by characteristic anatomical distributions. To further complicate matters, there is reasonably firm evidence for at least four major categories of opioid receptors, designated μ, κ, δ, and σ. There may in fact be as many as eight different types of opioid receptors. A full discussion of the general pharmacology of this area, still developing, is beyond the scope of this chapter, and the interested reader is referred to the symposium edited by Hughes (see Hughes and Kosterlitz, 1983) and to Jaffee and Martin (1985), though both these reference sources will probably be out of date at the time of, or shortly after, the publication of this text.

SUBSTANCE P.—Substance P is an undecapeptide originally detected in extracts of gut and brain, and prepared as a powder, hence the designation "P." The pharmacologic effects of substance P include vasodilation; stimulation of intestinal, bronchial, and other smooth muscles; stimulation of salivary secretion; diuresis and natriuresis; and a variety of effects on peripheral and central neurons apparently attributable to depolarization (Douglas, 1985). It is found in small amounts in the human bladder, mainly in a suburothelial position, and in the dorsal root ganglia and dorsal horn of the spinal cord, particularly in the lumbosacral cord (Alm et al., 1978; Mundy, 1984b). In vitro, substance P produces a contractile effect on the canine bladder (Norlen et al., 1983), which is atropine-resistant (Mundy, 1984b). Lack of effect of the substance P antagonist on contractile responses to electrical field stimulation in rabbit and guinea pig muscle strips, and an increase in response in rat bladder strips, suggests that substance P is not, however, a mediator of excitatory noncholinergic activation (Husted et al., 1981), though a role as a local neuromodulator cannot be excluded. deGroat and Kawatani (1985) have identified substance P in 22% of bladder afferents identified in feline sacral dorsal root ganglia. They speculated that substance P may be a sensory neurotransmitter, and this is consistent with experiments that indicate that the systemic administration of capsaicin, an agent that depletes substance P and other peptides in small-diameter afferents, depressed the micturition reflex in rats and caused urinary retention (Maggi et al., 1984b).

VASOACTIVE INTESTINAL POLYPEPTIDE.—Vasoactive intestinal polypeptide (VIP) is a peptide that contains 28 amino acid residues and was first isolated from the small intestine, hence its name. Its range of effects is very broad and includes potent vasodilation, stimulation of cardiac contractility, stimulation of glycogenolysis, and relaxation of many smooth muscles. It also stimulates secretion of several pituitary and hypothalamic hormones and of pancreatic, salivary, intestinal, and other exocrine secretions (Douglas, 1985). Of the peptide-containing nerves in the bladder, those containing VIP are the most abundant. VIP is found in all bladder layers, but especially beneath the transitional epithelium, around blood vessels, and in the muscle layers (Polak and Bloom, 1984; Mundy, 1984b). Although Finkbeiner (1983) found no significant effect of VIP on the spontaneous activity of guinea pig bladder muscle or that induced by acetylcholine or electrical stimulation, others (Levin and Wein, 1981; Larsen et al., 1981; Klarskov and Gerstenberg, 1984; Kinder and Mundy, 1984b) have shown an inhibitory effect on muscle strip spontaneous contractility and on that evoked by electrical field stimulation or stimulation with various pharmacologic agonists. This effect, also seen in urethral smooth muscle, was not affected by tetrodotoxin (Klarskov and Gerstenberg, 1984), β-adrenergic blockade (Levin and Wein, 1981; Klarskov and Gerstenberg, 1984), or prostaglandin synthesis blockade (Klarskov and Gerstenberg, 1984). Sjogren et al. (1985) reported similar inhibition effects on rabbit bladder strip preparations but not on human specimens. They also noted an inhibitory effect of VIP on the contractile response of the rabbit urethra to electrical stimulation and norepinephrine, but a less marked effect on human urethral response to norepinephrine and no consistent effect on the response of this tissue to field stimulation.

Gu et al. (1983) have reported a markedly reduced concentration of VIP in the "unstable bladder" as compared with the normal. This observation, with the results of some of the experiments just cited, has led to speculation regarding a possible role of VIP as a modifier of bladder activity during filling. There is by no means agreement on this hypothesis (Sjogren et al., 1985). In the pharmacologic literature the emerging view of VIP seems to be that of a neuroeffector agent involved in the control of flow in local vascular beds and in the modulation of cholinergic transmission (Haynes, 1985). deGroat and Kawatani (1985) showed that 24% of bladder afferents in sacral dorsal root ganglia contained VIP. Sixty-two percent of the VIP bladder afferent neurons also contained substance P. This raises speculation as to a potential afferent neurotransmitter role for this substance as well, especially since VIP-positive nerves have also been found in the spinal cord (Mundy, 1984b).

NEUROPEPTIDE Y.—Neuropeptide Y is a 36 amino acid peptide found to occur in various peripheral locations, including the bladder (Lundberg et al., 1984). Mundy (1984b) describes its bladder distribution as similar to that of VIP, and describes it a a potent contractor of smooth muscle. Varied potential functions in the mammalian nervous system have been hypothesized. Although it does cause generalized vasoconstriction, it also inhibits contractile responses to electrical stimulation in vas deferens and cervix smooth muscle (Gray and Morley, 1986). Lundberg et al. (1984) found NP-Y to induce a reversible reduction of the noncholinergic, nonadrenergic contractile response to field stimulation of the guinea pig urinary bladder. Gu et al. (1984) report numerous neuropeptide Y–containing nerves in the human bladder, primarily in the muscle layer, and particularly in the trigonal region.

PERIPHERAL OPIOIDS.—Most discussions on the potential role of the opioids as a major influence on lower urinary tract function concern a central inhibitory role, but there is some suggestion that enkephalins may function as an inhibitory neuromodulator at the peripheral level as well (deGroat and Kawatani, 1985), at the ganglionic level. A full discussion of the potential role of opioid mechanisms influencing lower urinary tract activity will be presented in the section on central mechanisms.

OTHER SUBSTANCES

The effects of histamine are mediated through specific H_1 and H_2 receptors, so classified by their susceptibility to blockade by pyrilamine and burimamide (Khanna, 1983; Caine, 1984). Contractile H_1 receptors have been demonstrated in rabbit (Fredericks, 1975), guinea pig (Khanna et al., 1977), and dogs and humans (Van Buren and Anderson, 1979; Khanna, 1983). No H_2 receptors have been identified in bladder smooth muscle. Atropine partially blocks the response to histamine, and so muscarinic receptors may be involved in this response (Fredericks, 1975).

Serotonin (5-hydroxytryptamine [5-HT]) produces a contractile response in guinea pig, dog, and cat urinary bladder (Gyermek, 1962; Erspamer et al., 1973; Saum and deGroat, 1973). The bladder effect is biphasic, consisting of an initial transient contraction due to stimulation of the autonomic ganglia, followed by a more prolonged contraction due to a direct effect on the smooth muscle itself (Caine, 1984). There is an additional inhibitory effect on parasympathetic ganglia produced by serotonin as well (Saum and deGroat, 1973). Two types of excitatory serotonin receptors have been identified in the cat urinary bladder, the so-called M and $5\text{-}HT_2$ type (Saxena et al., 1985). The physiologic significance of the action of serotonin on urinary bladder is open to speculation. Vaidyanathan et al. (1981a) have shown that administration of clomipramine, an inhibitor of serotonin uptake, decreased bladder capacity, and cite other work reporting the occurrence of urinary frequency with this drug, suggesting that serotonin might have a potentiating effect on excitatory bladder reflex activity.

γ-Aminobutyric acid (GABA) is thought to be primarily a central inhibitory neurotransmitter. However, GABA receptors have been shown to be present in guinea pig bladder dome strips. Maggi et al. (1985a, 1985b) have shown that GABA may transiently enhance excitatory neurotransmission early in the phase of neurogenic activation, and then exhibit a more sustained inhibition of contractility, mediated by both $GABA_A$ and $GABA_B$ receptors. In the mouse bladder, Santicioli et al. (1986) postulated that prejunctional $GABA_B$ receptors are present in the postganglionic excitatory innervation and act to reduce neurotransmitter release when stimulated.

Involvement of a nonadrenergic, noncholinergic neurotransmitter in the urethral relaxation that occurs during bladder emptying has already been somewhat considered in the section on prostaglandins. Additionally, Mattiasson et al. (1985) have reported that the electrically induced contraction of urethral smooth muscle from rabbit and man consists of two main components, one adrenergically mediated and one nonadrenergic and atropine-resistant. This suggests an unknown contractile neurotransmitter released by stimulation of known nerves or by stimulation of another type of nerve yet to be established.

CENTRAL NERVOUS SYSTEM INFLUENCES ON THE LOWER URINARY TRACT

Micturition is basically a function of the peripheral autonomic nervous system. However, the ultimate control of lower urinary tract function, including that of the periurethral striated musculature, obviously resides at higher neurologic levels. At any of these levels, and within the pathways that connect them, facilitation and inhibition of various aspects of bladder and outlet activity during the filling/storage and emptying phases of micturition can occur. There is considerable disagreement regarding details of the location and relative importance of central nervous system structures regulating micturition, and as well about ascending and descending pathways connecting these structures to each other and to the thoracic, lumbar, and sacral spinal cord. In this section we shall attempt to summarize the most widely quoted views that are consistent with our present working knowledge of micturition and its disorders.

NEURAL PATHWAYS

Spinal Cord

The micturition center in the spinal cord is primarily localized to three spinal nerve segments, S2–4, with its major part focused at S3 (Bradley et al., 1974; Fletcher and Bradley, 1978; deGroat and Booth, 1980; Hald and Bradley, 1982; Bradley and Sundin, 1982; Bhatia and Bradley, 1983; Schmidt, 1983). This is anatomically located at vertebral levels T12–L1. Rockswold et al. (1980a) reported that the origin of vesical pelvic nerve fibers in the rhesus monkey was S1–2 and in the chimpanzee S1–4. They also noted that the spinal origin of the pudendal nerve was one segment rostral (Rockswold et al., 1980b). Schmidt (1983) nicely summarized the neuroanatomy of the conus medullaris in the dog as follows. The gray matter is divided into a series of zones, and the parasympathetic preganglionic motor neurons lie within an intermediolateral gray column. The cells extend in a vertical plane along the border

between the white and gray matters, and in a horizontal plane at the base of the dorsal horn at lamina 7 and lamina 5, the former serving as the center for bladder control, and the latter more medial group of parasympathetic cells serving as the center for rectal control. Between the rectal center and the bladder center lies a zone of interneurons believed to be a site of convergence for afferent input to these two visceral centers. Somatic nerve fibers originate from the nucleus of Onuf, in the midventral spinal gray matter at the level of S1 (in animals), drifting laterally to the edge of the spinal white and gray matters at S2. This area corresponds to human levels S2–3. The innervation to most striated muscles of the pelvic floor, including those of the periurethral and anal sphincters, originates within the nucleus of Onuf. Figure 27–23 conceptualizes Schmidt's description. Hald and Bradley's cross-sectional diagrams at the level of the sacral cord (Fig 27–24) and above (Fig 27–25) are likewise helpful in understanding the efferent innervation of the lower urinary tract at the spinal cord level.

Ascending sensory stimuli from the lower urinary tract include exteroceptive (pain, temperature and touch—generated in the urothelium) and proprioceptive sensory impulses (initiated in the bladder muscle and periurethral striated muscle). Exteroceptive impulses ascend in the spinothalamic tract, while proprioceptive impulses travel in the posterior columns (Hald and

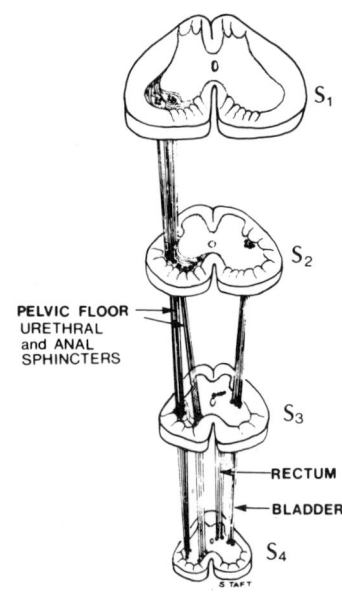

FIG 27–23.
Micturition center (sacral spinal cord) in the human. Based on horseradish peroxidase transport studies in animals and the results of sacral root stimulation studies in humans (see text). (From Schmidt RA: Urethrovesical reflexes and their inhibition, in Hinman F Jr (ed): *Benign Prostatic Hypertrophy.* Berlin, Springer-Verlag, 1983. Used with permission.)

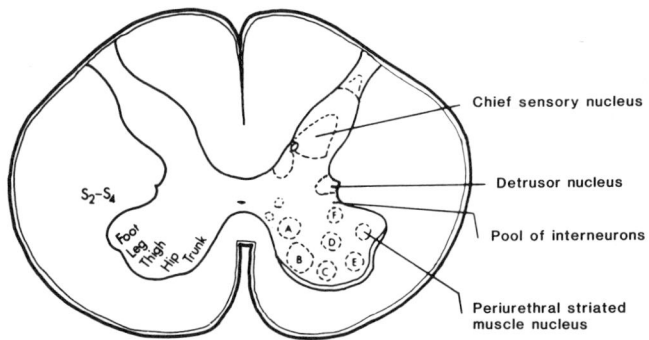

FIG 27–24.
Spinal cord at level S2–4. Location of detrusor and pudendal motor nuclei. Bilateral representation, though not pictured, exists. (From Hald T, Bradley W: *The Urinary Bladder: Neurology and Dynamics.* Baltimore, Williams & Wilkins, 1982. Used with permission.)

Bradley, 1982; Bradley and Sundin, 1982; Bhatia and Bradley, 1983) (Figs 27–26 and 27–27). Proprioceptive axons from the detrusor muscle enter the dorsal portion of the gray matter, and turn to travel rostrally to synapse in the area of the pontine-mesencephalic reticular formation designated as the nucleus tegmentolateralis dorsalis tegmentalis of the pons (Hald and Bradley, 1982; Bradley and Sundin, 1982; Bhatia and Bradley, 1983). This exclusive rostral passage of propriospinal afferents is termed "long routing," a concept originally implied by Barrington (1921). This phenomenon certainly occurs in the experimental animal, but awaits confirmation in man (Hald and Bradley, 1982; Bradley and Sundin, 1982; Bhatia and Bradley, 1983). Exteroceptive sensory impulses travel in the spinothalamic tracts and synapse in the thalamus, which projects to the sensorimotor cortex. Whether propriospinal affer-

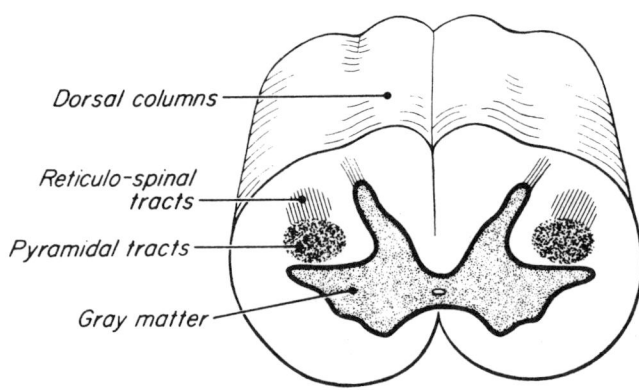

FIG 27–25.
Spinal tracts innervating the bladder (reticulospinal tracts) and striated periurethral musculature (pyramidal or corticospinal tracts.) (From Hald T, Bradley W: *The Urinary Bladder: Neurology and Dynamics.* Baltimore, Williams & Wilkins, 1982. Used with permission.)

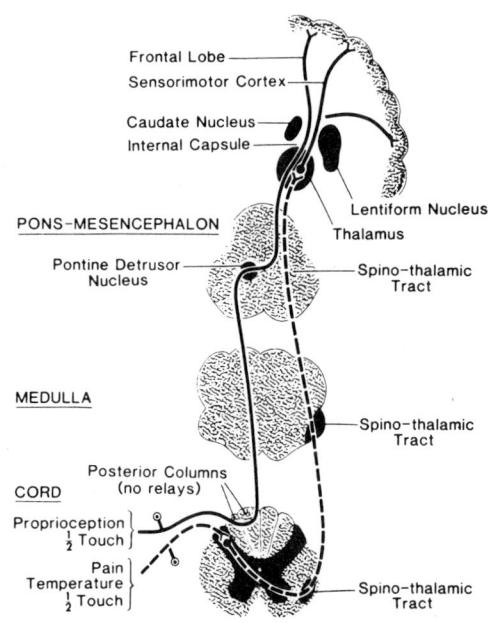

FIG 27–26.
Afferent neural pathways of exteroceptive and proprioceptive sensory neurons (see text). (From Hald T, Bradley W: *The Urinary Bladder: Neurology and Dynamics.* Baltimore, Williams & Wilkins, 1982. Used with permission.)

ents are exclusively "long routed," or whether it simply seems that way with an intact neural axis remains a question. If exclusive long routing occurs, then there can be no true spinal micturition center under any circumstances. If pathways do exist, but are simply vestigial in the normal animal, then there at least exists the potential for spinal reflex pathways to emerge under certain circumstances (such as after spinal cord transection—see below).

It is agreed, however, that pudendal reflex pathways are organized at a spinal level, although subject to supraspinal influences (Bradley and Sundin, 1982; Hald and Bradley, 1982; Bhatia and Bradley, 1983). Afferent proprioceptive axons that originate in the striated musculature of the pelvic floor enter the sacral spinal cord and divide. One group synapses on pudendal motor neurons in the ventral gray matter of the sacral spinal cord, while another group travels rostrally in the dorsal columns to synapse on neurons in the cerebellum, as well as ascending in the medial lemniscus to synapse on neurons in the nucleus ventralis posterolateralis (thalamus), ending eventually in the sensorimotor cortex (Fig 27–28).

Brain Stem

Evidence that indicates that the brain stem, specifically the neurons of the pontine-mesencephalic gray matter, contains the nuclei that are the origin of the

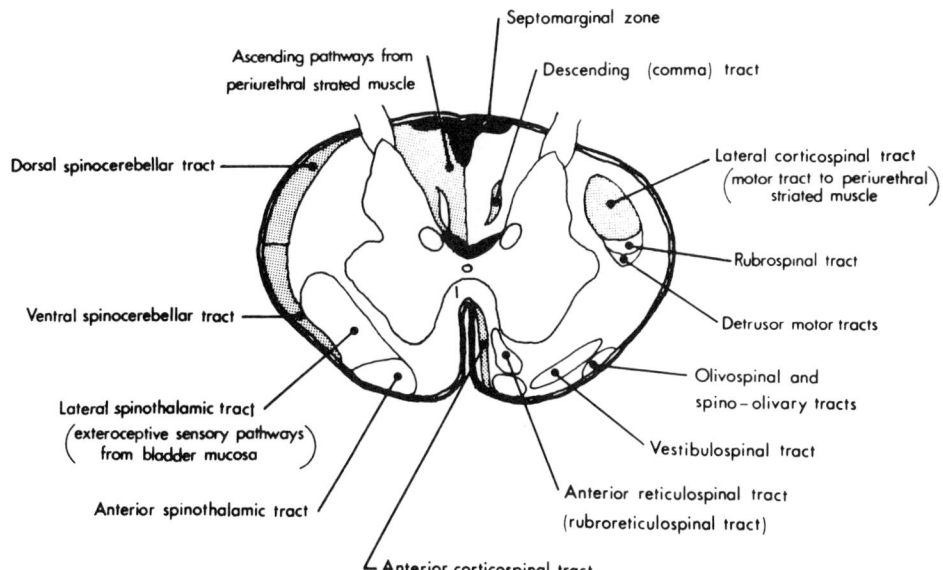

FIG 27–27.
Spinal cord cross-section (above Fig 27–25) showing relationship of efferent and afferent tracts concerned with lower urinary tract function (see text). (From Bhatia NN, Bradley WE: Neuroanatomy and physiology: Innervation of the urinary tract, in Raz S (ed): *Female Urology.* Philadelphia, WB Saunders, 1983. Used with permission.)

final common pathway to bladder motor neurons has been summarized by Kuru (1965), Bradley and Teague (1968), deGroat and Ryall (1969), Bradley et al. (1974), deGroat (1975), Carlsson (1978), Bradley and Sundin (1982), Hald and Bradley (1982), and Bhatia and Bradley (1983). The area called Barrington's center (Barring-

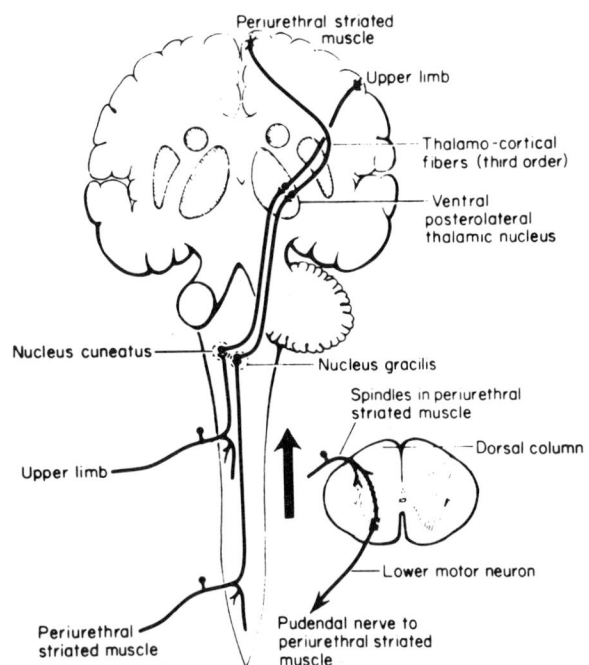

FIG 27–28.
Spinal and supraspinal organization of pudendal reflex pathways (see text). Connections to cerebellum not pictured. (From Bhatia NN, Bradley WE: Neuroanatomy and physiology: Innervation of the urinary tract, in Raz S (ed): *Female Urology.* Philadelphia, WB Saunders, 1983. Used with permission.)

ton, 1921) in the anterior pontine region is considered to be the origin of facilitory impulses to the bladder. Destruction of this center or transsection just below it causes a permanent disturbance of bladder emptying (Barrington, 1925; Bradley and Conway, 1966; Ruch, 1974). Transsection above this center at the intracollicular level results in detrusor hyperactivity (Ruch, 1974; Carlsson, 1978). Originally, these neurons were described as located in the nucleus locus ceruleus, but subsequent studies have suggested that the location is in the nucleus lateralis dorsalis tegmentalis, which is rostral to the nucleus locus ceruleus (Hald and Bradley, 1982; Bhatia and Bradley, 1983). Input to this area is derived from the cerebellum, basal ganglia, thalamus and hypothalamus, and cerebral cortex (Fig 27–29). Carlsson (1978) added that experimental stimulation of Barrington's center causes not only bladder contraction, but also a marked decrease in EMG discharge in the periurethral striated musculature.

Cerebellum

The cerebellum serves as a modulator of neural activities that are initiated in other parts of the central nervous system. Insofar as the lower urinary tract is concerned, it receives sensory input from the bladder and from the pelvic floor musculature. Its ultimately efferent functions are felt to include maintenance of tone in the pelvic floor striated musculature; an influence on the rate, range, and force of bladder and pelvic floor striated muscle contraction; an influence on the coordination between emptying bladder contraction and periurethral striated muscle relaxation; and a direct effect on the brain stem. Electrical stimulation of the fastigial nucleus in the experimental animal produces

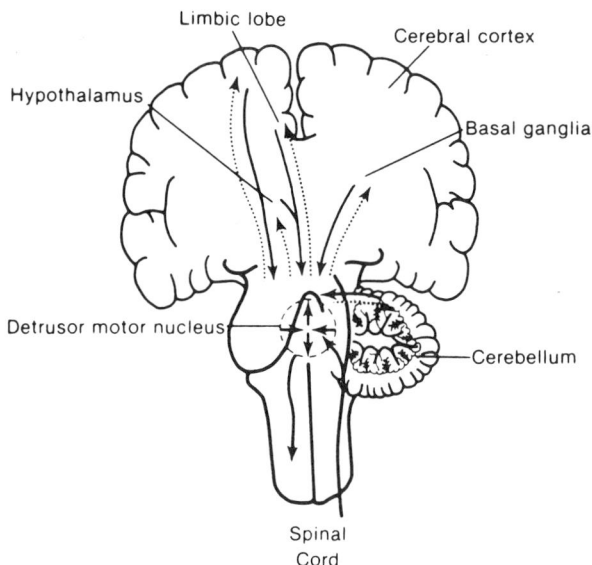

FIG 27–29.
Pontine detrusor nucleus and its connections (see text). (From Hald T, Bradley W: *The Urinary Bladder: Neurology and Dynamics.* Baltimore, Williams & Wilkins, 1982. Used with permission.)

suppression of detrusor reflex activity through an inhibitory influence on Barrington's center; ablation of the anterior vermis of the cerebellum produces detrusor hyperactivity (Bradley et al., 1974; Bradley and Sundin, 1982; Hald and Bradley, 1982; Bhatia and Bradley, 1983).

Basal Ganglia

These subcortical nuclei comprise the caudate nucleus, red nucleus, putamen and globus pallidus, and cells of the substantia nigra located in the midbrain. Animal experiments have suggested an inhibitory effect of these nuclei on spontaneous detrusor reflex contractions (Carlsson, 1978; Hald and Bradley, 1982; Bhatia and Bradley, 1983). This correlates with the detrusor hyperactivity often seen in patients with Parkinson's disease with basal ganglia dysfunction.

Thalamus, Hypothalamus, and Limbic System

The thalamus consists of a collection of nuclei that collectively represent the principal relay nucleus for sensory axons ascending to the cerebral cortex (Bradley and Sundin, 1982; Hald and Bradley, 1982). Splanchnic afferent stimulation has implicated the dorsomedial thalamic nucleus as a relay station, but the precise routing of afferents from the pelvic and pudendal nerves is unknown. The hypothalamus consists of a collection of nuclei at the base of the thalamus, and has a major influence on neuroendocrine function and body water balance. Its function, insofar as the lower urinary tract

is concerned, has been little investigated, but bladder distention in the experimental animal has been reported to cause neuronal firing in a posterior group of nuclei (Hald and Bradley, 1982; Bhatia and Bradley, 1983). Carlsson (1978) cited studies that show that the hypothalamus may be involved in the organization of that portion of micturition or defecation mediated by the autonomic nervous system. The limbic system comprises areas that are principally located in the temporal lobe and associated subcortical nuclei considered to be the rostral extensions of the autonomic nervous system. This includes the amygdala, the hippocampal formation, and the neurons of the cingulate gyrus. Electrical stimulation of specific nuclei in the experimental animal has been shown to both facilitate and depress bladder activity, depending on the specific nucleus stimulated (Bradley and Sundin, 1982), but a typical type of bladder dysfunction does not seem to occur in patients with temporal lobe dysfunction (Hald and Bradley, 1982; Bhatia and Bradley, 1983).

Cerebral Cortex

The region of the cerebral hemispheres concerned with bladder muscle innervation consists of the superomedial portion of the frontal lobes and the genu of the corpus callosum (Fig 27–30) (Andrew and Nathan, 1965; Bradley et al., 1974; Hald and Bradley, 1982; Bhatia and Bradley, 1983). Stimulation of specific areas of the cerebral cortex in the experimental animal can demonstrate either facilitation or inhibition of detrusor muscle contraction (Gjone and Setekleiv, 1963), but transsection experiments seem to indicate that the net effect of these areas is inhibitory (Tang, 1955; Tang and Ruch, 1956; Kuru, 1965; Ruch, 1974; Nathan, 1976). Voiding dysfunction in patients with functional ablation of cerebral cortical areas is generally characterized by a hyperactive detrusor reflex. The cerebrocortical areas concerned with innervation of the periurethral striated musculature are distinct and geographically separate, located on the medial aspect of the sensorimotor cortex (see Fig 27–30) (Hald and Bradley, 1982; Bhatia and Bradley, 1983). Axons that originate from neurons in this region pass caudally through the internal capsule and cerebral peduncles in the brain stem, continuing ultimately in the lateral columns of the spinal cord as the corticospinal tracts. The axons from the detrusor motor area in the frontal lobes traverse the basal ganglia to terminate in the pontine mesencephalic reticular formation. Ascending axons from sensory receptors located in the detrusor muscle and in the periurethral striated muscle terminate in both the detrusor portion of the frontal lobes and the pudendal portion of the sensory motor cortex.

Thus, the prevailing concept regarding central ner-

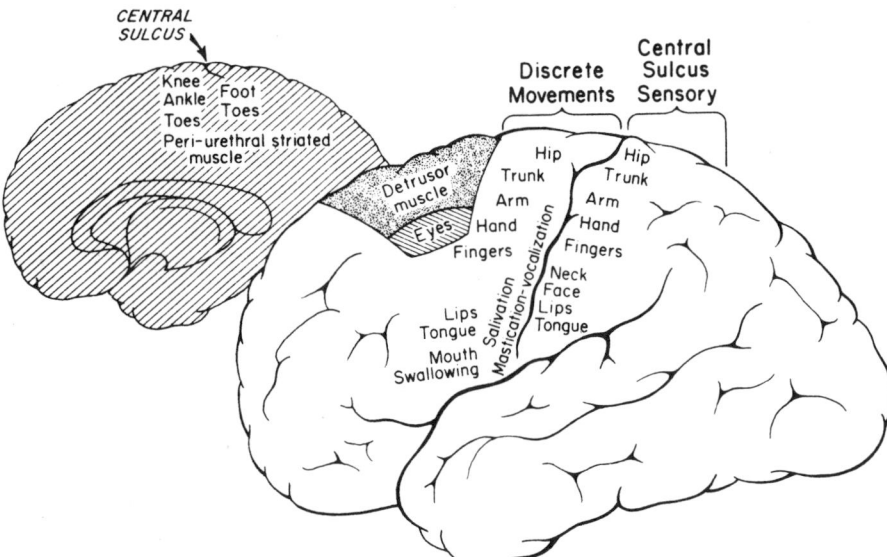

FIG 27–30.
Areas in the cerebral hemispheres concerned with innervation of the bladder (lateral surface) and periurethral striated muscle (medial surface) (see text). (From Bhatia NN, Bradley WE: Neuroanatomy and physiology: Innervation of the urinary tract, in Raz S (ed): *Female Urology.* Philadelphia, WB Saunders, 1983. Used with permission.)

vous system influences is that of an anterior pontine micturition center, which is considered to be the origin of efferent facilitory impulses in the normal animal or individual. This conclusion, however, has been reached by transsection, destruction, and stimulation experiments only in the laboratory animal, and almost exclusively by inference in the human. Additionally, such experiments have disclosed, at least in cats, alternating inhibitory and facilitory areas rostral to this, each level being capable of functioning in the absence of the more rostral one. The inhibitory areas are found in structures above the transhypothalamic level and in the midbrain; the facilitory areas in the posterior hypothalamus and in the anterior pons (Carlsson, 1978).

CENTRAL TRANSMITTERS AND THE ROLE OF THE ENDOGENOUS OPIOIDS

Although data exist on the location of the central nervous system structures involved in the central regulation of lower urinary tract function, the neurotransmitter mechanisms involved have received little study and are largely unknown. Potential central neurotransmitters are listed in Table 27–1, and, of these, GABA and glycine have been shown to be inhibitory neurotransmitters. Sillen (1980) has summarized a series of experiments on anesthetized rats, providing certain observations regarding central neurotransmitter mechanisms involved in the control of bladder function in this experimental model. Central adrenergic stimulation was achieved by injection of the amine precursor levodopa, after blockade of its peripheral metabolism, resulting in central accumulation of the metabolites dopamine and noradrenalin, thought responsible for the functional ef-

fects of levodopa. Activation of central catecholamine neurons in this fashion resulted in a hyperactive bladder response. This bladder hyperactivity to levodopa seemed to be elicited mainly via stimulation of central dopamine receptors in the pontine-mesencephalic brain region. This hyperactive bladder response occurred even in the presence of maximally inhibited catecholamine neurons, suggesting that impulses to the bladder from these dopamine receptors were propagated via neurons of a noncatecholamine type. Acetylcholine was also reported to serve as a neurotransmitter in the central nervous system as well as in peripheral tissues. Muscarinic receptors predominate in the brain, while nicotinic receptors predominate in the spinal cord. Activation of central muscarinic receptors also (like levodopa) produced a hyperactive bladder response, likewise generated in pontine-mesencephalic structures. Further, there was felt to be an interaction in some instances between dopamine and muscarinic receptors in the generation of the hyperactive bladder response to levodopa. Central GABA receptor activation inhibited the bladder hyperactivity that occurred in response to levodopa. This inhibition was felt to occur at a site postsynaptic to the dopaminergic neurons. Thus, Sillen concluded that excitatory dopaminergic and muscarinic receptors, and inhibitory GABAergic receptors, were involved in the supraspinal modulation of lower urinary tract function, probably by an action on the basic micturition reflex pathways originating in the pontine-mesencephalic area. Further data from other experimental systems are awaited.

There is strong evidence that endogenous opioid peptides influence micturition by a tonic inhibitory effect on detrusor reflex pathways. The evidence for this has been well summarized by deGroat and Kawatani (1986) and by Booth et al. (1985). In cats, systemic administration of naloxone increased the force of bladder contractions at constant bladder volumes, decreased bladder capacity, increased the frequency and magnitude of pressure increases during bladder filling and during micturition, promoted more effective emptying of the bladder, and increased multiunit firing in parasympathetic postganglionic nerves on the surface of the bladder (Roppolo et al., 1983; Hisamitsu and deGroat, 1984; Booth et al., 1985). The inhibitory effects could be mediated at several possible levels, including the peripheral bladder ganglia, the sacral spinal cord, or the brain-stem micturition center (Booth et al., 1985). At each of these sites dense collections of enkephalinergic terminals have been identified (Glazer and Basbaum, 1980; Kawatani et al., 1983; Leger et al., 1983). Also, at each of these sites, local administration of exogenous opioid peptides has produced an inhibition of the micturition reflex (Simonds et al., 1983; Roppolo et al., 1983; Hisamitsu and deGroat, 1984; deGroat and Kawatani, 1985; Booth et al., 1985). In the rat, both intracerebroventricular and systemic morphine inhibited spontaneous bladder contractions, the former effect being reversed by similarly administered naloxone, and the latter by intrathecal naloxone, most effectively in the region of the lumbosacral spinal cord (Dray and Metsch, 1984a, 1984b). Sillen and Rubenson (1986) likewise reported inhibition of the bladder response in rats to central levodopa administration by intracerebroventricular or systemic morphine. Regional (intraarterial) administration produced an initial weak excitatory effect followed by a slight depression of the bladder reactivity to receptor agonists and peripheral nerve stimulation. Jubelin et al. (1984) showed that the synthetic opiate [D-Ala^2,Me-Phe^4] leu-enkephalin (DAMLE) inhibited bladder contractions and voiding in acutely decerebrate cats (intercollicular transsection of the brain stem). However, there was no effect on the bladder contraction of acute spinal cats (C5–6) or on electrically induced contraction of in vitro bladder strips. Naloxone antagonized this inhibitory effect and per se induced a bladder contraction. The conclusion from this set of experiments was that the inhibitory effect was exerted either at the level of ascending spinal pathways (spinobulbar or spinopontine) or in the pontine micturition center, probably the latter. Murray and Feneley (1982), using cystometric recordings, showed that naloxone increased intravesical pressure throughout filling in a group of normal men and women, and reduced bladder capacity. Maximum urethral closure pressure was reduced in females, but not in males. Vai-

dyanathan et al. (1981b) showed that intravenous naloxone resulted in a rapid and evanescent enhancement of detrusor reflex activity in five patients with neurogenic bladder dysfunction secondary to incomplete suprasacral spinal cord lesions. No effect was seen in two patients with complete suprasacral lesions. Thor et al. (1983) noted that naloxone did not alter bladder function in acute spinal cats (before the development of reflex bladder contractions) but, in large doses, did enhance reflex bladder activity and promote micturition in chronic spinal cats after reflex bladder activity had developed.

Different mechanisms and different opioid receptors seem to be involved in mediating these inhibitory effects at each level of the nervous system. In bladder ganglia, agents that selectively activate δ receptors are effective in blocking transmission, whereas μ and κ agonists are relatively ineffective, indicating that the δ receptors present on preganglionic nerve terminals are primarily responsible for enkephalinergic inhibition at this level (deGroat and Kawatani, 1985). These authors proposed that the mechanism of inhibition in bladder parasympathetic ganglia by endogenous enkephalins is by depression of the release of acetylcholine from the preganglionic nerve. They hypothesized that enkephalins are co-released with acetylcholine at this level and likely mediate a negative feedback or presynaptic inhibitory action in these ganglia. Summarizing a large amount of data from their laboratory, deGroat and Kawatani (1985) and Booth et al. (1985) theorized that μ receptors in the brain stem primarily are concerned with the alteration of bladder capacity, while δ receptors in the spinal cord are primarily involved in the modulation of the magnitude and duration of bladder contraction. The μ brain-stem receptors are thought also to be possibly involved in the control of bladder stability during urine storage, since small doses of naloxone in the cat also increased the frequency and magnitude of uninhibited pressure rises during the tonus limb of the cystometrogram. Equally intriguing is the fact that sphincter motor neurons in the nucleus of Onuf (supplying the periurethral striated muscle) are also densely innervated by enkephalingeric terminals (Glazer and Basbaum, 1980; deGroat et al., 1983). Booth et al. (1985) cited further work in their laboratory that shows that sphincter reflexes in the cat can be inhibited by κ opiate agonists. The κ receptors seem to have little effect on bladder reflex activity at any level.

Thus, enkephalinergic mechanisms may have an important influence on bladder capacity and stability during filling, on bladder contractility during emptying, and on reflex-induced sphincter activity. These mechanisms may be predominantly mediated at different sites by different opiate receptors, raising the very interesting possibility of the development of specific opioid drugs to selectively alter different components of the lower urinary tract during the micturition cycle (deGroat et al., 1985; Booth et al., 1985).

ORGANIZATION OF THE MICTURITION REFLEX

Sacral vs. Supraspinal Micturition Center

The pelvic nerves contain major afferent pathways from the lower urinary tract and also the preganglionic parasympathetic fibers responsible for the bladder contraction of normal micturition. It has long been recognized that the sacral spinal cord, the origin of this nerve, contains neural pathways involved in the micturition reflex (Langley and Anderson, 1895, 1896; Learmoth, 1931; Gruber, 1933). Originally, controversy existed regarding the location of the primary neurologic organizational center for the micturition reflex. Denny-Brown and Robertson (1933), McClellan (1939), and Lapides (1970) all have implied that the micturition reflex is primarily a sacral reflex mediated by the pelvic and pudendal nerves, and subject to facilitation or inhibition by descending central nervous system pathways (Fig 27–31). Much evidence has been presented in this chapter that in fact the actual organizational center for micturition is localized in the pontine-mesencephalic reticular formation. A full synthesis of data supporting the existence of a pontine-mesencephalic micturition center was more or less concurrently published by Bradley (1974) and by deGroat (1975), but the concept of a supraspinal organization of the basic reflex arc responsible for micturition was originally implied by Barrington in 1914. The theories popularized by Bradley and by deGroat, both of which incorporate the concept of the pontine-mesencephalic micturition center, differ somewhat, and the differences will be pointed out. Over the years, Bradley's theory has evolved into his defining several neurophysiologic "loops" (described below), which now constitute an overall theory of micturition and its dysfunction.

In the writings of deGroat (deGroat, 1975; deGroat and Booth, 1984; deGroat and Kawatani, 1985), one finds a continuously evolving synthesis of the sacral vs. supraspinal organizational theories, and finds further that these are not mutually exclusive, but complementary. According to deGroat's initial studies of cats with an intact neural axis, bladder distention caused a reflex firing of the preganglionic parasympathetic fibers with a subsequent micturition contraction. The electrophysiologic characteristics of the entire reflex pathway, however, were those of a supraspinal reflex rather than a sacral reflex. In chronic animals (cats) with transected

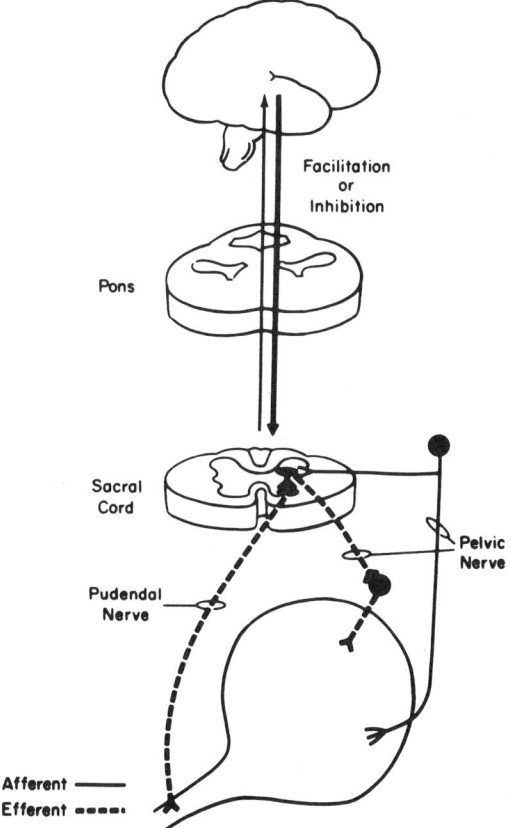

FIG 27–31.
Sacral micturition reflex theory of bladder function. Bladder distention results in afferent pelvic nerve stimuli, which enter the sacral cord through the dorsal roots and synapse with pelvic and pudendal nerve nuclei, resulting in efferent discharges mediating bladder contraction and striated sphincter relaxation. Central nervous system pathways, and local influences as well, mediate facilitation and inhibition of this sacral reflex. (From Blaivas JG: The neurophysiology of micturition: A clinical study of 550 patients. *J Urol* 1982; 127:958. Used with permission.)

spinal cords, most ultimately develop reflex contractions of the bladder in response to bladder distention. The electrophysiologic properties of this neural circuit were those of a spinal reflex, i.e., a reflex organized in the area of the sacral spinal cord. Figure 27–32 shows deGroat's initial conceptualization of these data. The so-called reflex neurogenic bladder that develops in patients with spinal cords that have been transected or severely injured above the sacral level provides an analogy to the bladder contractions during filling that develop in the chronic spinal cat. Whether the appearance of this reflex voiding pattern after spinal transsection results from the formation of new neural pathways by "collateral sprouting" (Liu and Chambers, 1958; Guth, 1974), or from the removal of an inhibitory influence on an existing pathway, or by the emergence

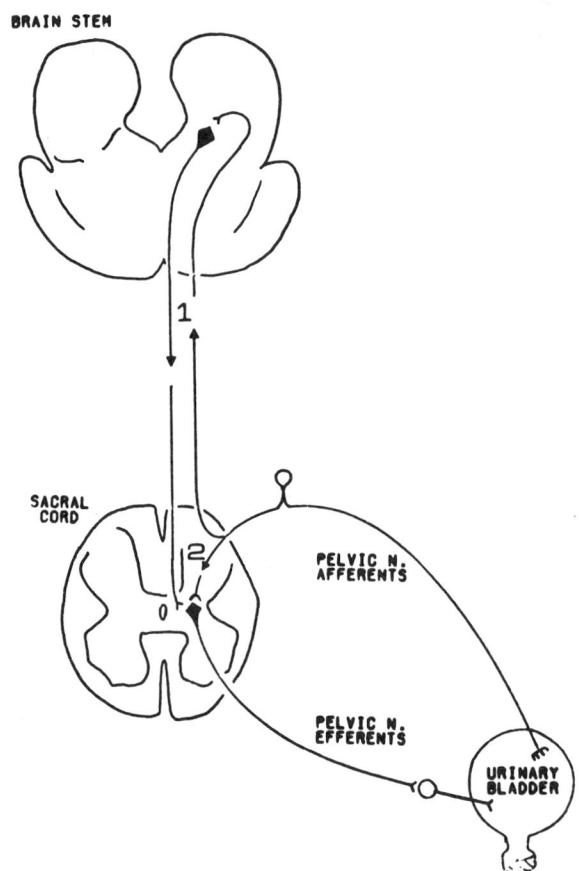

FIG 27–32.
Reflex micturition pathways involved in micturition in normal (1-supraspinal reflex pathway) and chronic spinal (2-spinal reflex pathway) cats. (From deGroat WC: Nervous control of the urinary bladder of the cat. *Brain Res* 1975; 87:201. Used with permission.)

of a vestigial reflex pathway is still speculative, although deGroat and co-workers (deGroat et al., 1981; deGroat and Booth, 1984; deGroat and Kawatani, 1985) have provided compelling evidence to support the latter theory. Their later electrophysiologic studies not only have confirmed the existence of a supraspinal reflex pathway to the bladder in cats, but have provided evidence for the existence of a spinal reflex pathway as well. In animals with an intact neural axis the supraspinal reflex mechanism was the most prominent. Recordings from postganglionic nerves on the surface of the urinary bladder showed that reflex firing occurred with a long latency (mean of 100 msec) following stimulation of myelinated afferent fibers in the pelvic nerve from the bladder. The supraspinal reflex was present in decerebrate animals, but absent in acute or chronic spinal animals. A second reflex, less prominent and noted in only 60% of animals with an intact neural axis, occurred with a considerably longer latency (180–200

msec) following stimulation of vesical afferents. This second reflex occurred only at intensities of stimulation that were sufficient to activate unmyelinated afferent fibers. This reflex was abolished by transsection of the spinal cord at lower thoracic or upper lumbar levels, but was again detectable 3–7 days after transsection in chronic spinal animals. For several weeks this second reflex continued to increase in magnitude in conjunction with the development of reflex micturition. Thus, deGroat's data synthesize the two theories regarding micturitional reflex organization. Both a spinal and supraspinal central pathway appear to exist, each with different afferent limbs for the mediation of detrusor to detrusor reflexes. The supraspinal reflex pathway appears to have the major role in the initiation of bladder contractions in animals with an intact neural axis. The function of the spinal reflex pathway in normal animals is unknown, but appears to be responsible for the development of reflex bladder contractions in the chronic spinal animals, which it may, by inference, in the human clinical counterpart as well. Figure 27–33 illus-

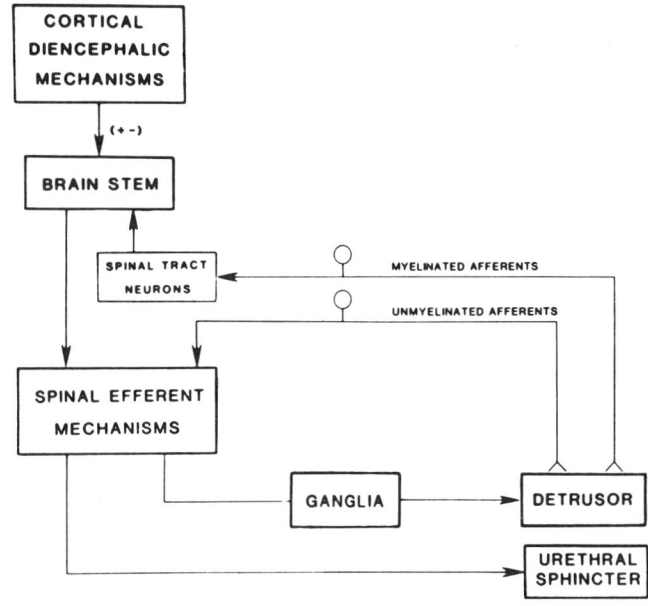

FIG 27–33.
Central micturition reflex pathways in the cat. With an intact neural axis, micturition is initiated by a supraspinal reflex pathway centered in the pontine-mesencephalic reticular formation. The reflex is triggered by myelinated afferents (A-δ) subserving tension receptors in the bladder wall. In chronic spinal animals, micturition is initially blocked as connections between the brain stem and sacral cord are interrupted. A spinal reflex emerges that is triggered by unmyelinated C-fiber vesical afferents. This reflex is usually weak or undetectable in animals with an intact neural axis. (From deGroat WC, Kawatani M: Neural control of the urinary bladder: Possible relationship between peptidergic inhibitory mechanisms and detrusor instability. *Neurourol Urodynamics* 1985; 4:285. Used with permission.)

trates this concept. deGroat's theory of the reflex pathways involved in micturition (Fig 27–34) (deGroat and Booth, 1980; deGroat and Booth, 1984; deGroat and Kawatani, 1985) differs from that of Bradley (Bradley et al., 1974; Bradley and Sundin, 1982; Hald and Bradley, 1982; Bhatia and Bradley, 1983) in a few respects. deGroat and associates stressed the importance of the sympathetic nervous system in modulating lower urinary tract function to facilitate urine storage, while Bradley and associates ascribed minimal significance to this. deGroat postulated a synapse between efferent fibers from the brain-stem center and the nuclei of the preganglionic parasympathetic fibers in the sacral spinal cord; Bradley and associates did not originally picture such a synapse (1974), although subsequent conceptualizations have included an efferent synapse in the sacral spinal cord (Bradley and Sundin, 1982; Hald and Bradley, 1982; Bhatia and Bradley, 1983). deGroat implied that striated sphincter integration is primarily a sacral spinal cord phenomenon, while Bradley and associates have provided data that separate neural pathways govern voluntary control of the striated sphincter and the bladder (see below).

The Loop Concept

Bradley has extended his original concept of a pontine-mesencephalic micturition center to include a conceptualization of central nervous system control of the lower urinary tract that identifies four neurologic "loops." This concept has evolved over the years (Bradley et al., 1974; Bradley and Sundin, 1982; Hald and Bradley, 1982; Bhatia and Bradley, 1983) and represents a very coherent theory of micturition, consistent with which Bradley has developed one of the popular classification systems of voiding dysfunction. Close interaction exists among all four loops and successful function of one loop depends on the integrity of the remaining loops. Successful integration of the micturition reflex requires a balanced contribution by all loops.

Loop 1 (Fig 27–35) consists of neuronal connections between the specific areas of the frontal lobes of the cerebral cortex and the micturition center in the pontine-mesencephalic reticular formation. The cerebellum and basal ganglia provide additional input. Loop 1 coordinates voluntary control of the detrusor reflex. Maturation of loop 1 may account for the attainment of vol-

FIG 27–34.
deGroat's diagram of the reflex pathways involved in micturition. Plus and minus signs indicate excitatory and inhibitory synaptic actions. Voluntary control is accomplished by connection (not pictured) between the frontal cortex and pontine micturition center. Note the prominent role of the sympathetic nervous system and the synapse between the descending fibers from the micturition center (excitatory) and the preganglionic parasympathetic fibers in the sacral spinal cord. (From deGroat WC, Booth AM: Physiology of the urinary bladder and urethra. *Ann Intern Med* 1980; 92(pt 2):321. Used with permission.)

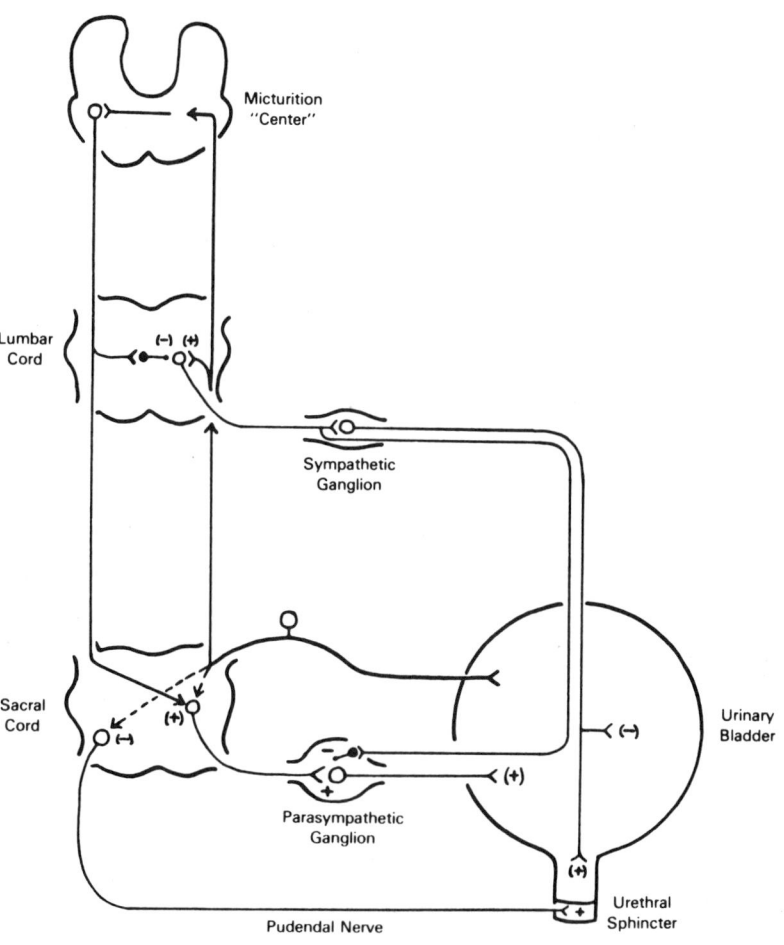

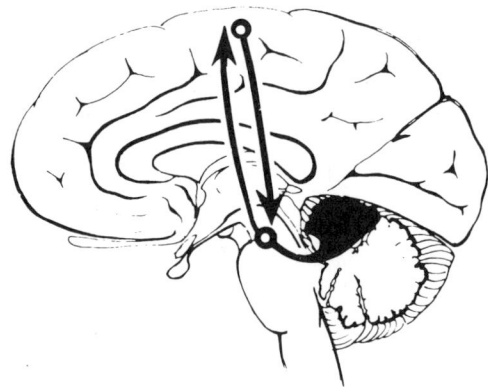

FIG 27–35.
Bradley's loop 1. See text for description. (From Bhatia NN, Bradley WE: Neuroanatomy and physiology: Innervation of the urinary tract, in Raz S (ed): *Female Urology*. Philadelphia, WB Saunders, 1983. Used with permission.)

untary control over the micturition reflex during infancy, and interruption of this loop may cause involuntary bladder contractions. Loop 2 (Fig 27–36) consists of the routing of bladder sensory afferents in the spinal cord and brain stem and descending pathways from the micturition center to the detrusor motor neurons in the sacral gray matter. Loop 2 includes the intraspinal pathway of detrusor muscle afferents to the brain-stem micturition center and the motor impulses from this center to the sacral spinal cord. Strictly speaking, it does not include peripheral sensory or motor pathways (Hald and Bradley, 1982), and so is not a true "loop," but does include the long routed portion of the sensory afferent fibers that originate in the bladder musculature and synapse in the brain stem. Loop 2 is thought to coordinate and provide for a detrusor reflex of adequate temporal duration to allow total evacuation of intravesical contents. Partial interruption of this loop by spinal cord injury is felt to result in a detrusor reflex of low threshold and in poor emptying, as manifested by residual urine. Spinal cord transsection of this circuit produces detrusor arreflexia and urinary retention, the urinary tract correlates of "spinal shock," which refers to the loss of detrusor reflex activity response of the detrusor and periurethral striated muscle seen following acute spinal cord transsection. The development of reflex bladder activity following suprasacral spinal cord transsection, according to this theory, occurs subsequent to the development of neuroma formation and "ephaptic interaction between neurites and axons at the level of spinal cord transsection" (Hald and Bradley, 1982)—conceptually similar to but not exactly the same as collateral sprouting.

Loop 3 consists of the peripheral detrusor afferent axons and their pathway in the spinal cord, which ter-

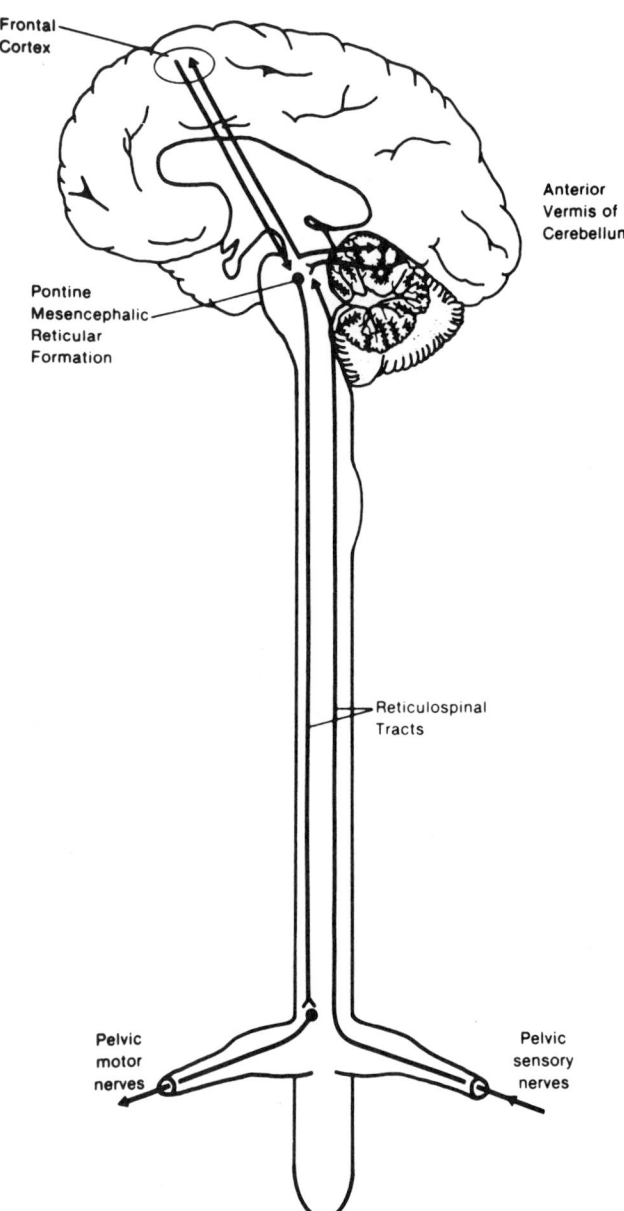

FIG 27–36.
Bradley's loops 1 and 2. See text for description. (From Bhatia NN, Bradley WE: Neuroanatomy and physiology: Innervation of the urinary tract, in Raz S (ed): *Female Urology*. Philadelphia, WB Saunders, 1983. Used with permission.)

minates by synapsing on pudendal motor neurons that ultimately innervate the periurethral striated muscle (Fig 27–37). This loop provides the neurological substrate whereby impulses from detrusor afferents can inhibit pudendal motor efferents, and thus allow the contracted striated sphincter (during bladder filling and urine storage) to relax in synchrony with bladder contraction during the emptying phase of micturition. Dys-

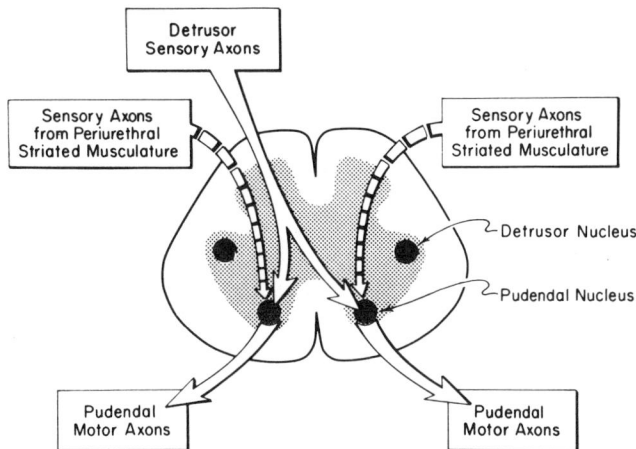

FIG 27–37.
Bradley's loops 3 and 4B. See text for description. (From Hald T, Bradley W: *The Urinary Bladder: Neurology and Dynamics.* Baltimore, Williams & Wilkins, 1982. Used with permission.)

function of this loop may be responsible for detrusor striated sphincter dyssynergia or for voluntary sphincter relaxation. Loop 4 (Fig 27–38) consists of two components. Loop 4A consists of the suprasacral afferent and efferent innervation of the pudendal motor neurons to

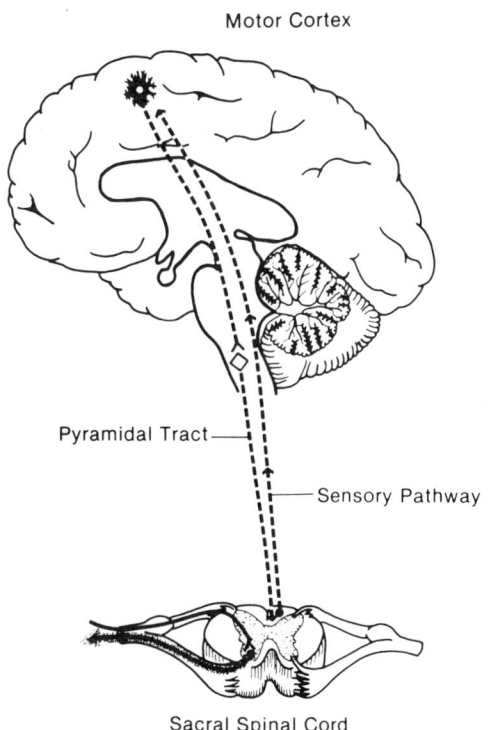

FIG 27–38.
Bradley's loop 4A. See text for description. (From Hald T, Bradley W: *The Urinary Bladder: Neurology and Dynamics.* Baltimore, Williams & Wilkins, 1982. Used with permission.)

the periurethral striated musculature. Loop 4B (see Fig 27–37) consists of afferent fibers from the periurethral striated musculature, which synapse on pudendal motor neurons, i.e., the segmental innervation of the periurethral striated muscle. In contrast to stimulation of detrusor afferent fibers, which produce inhibitory postsynaptic potentials in pudendal motor neurons through loop 3, pudendal nerve afferents produce excitatory postsynaptic potentials in these motor neurons through loop 4B. This provides for contraction of the periurethral striated muscle during bladder filling and urine storage. The sensory impulses concerned arise from muscle spindles and tendon organs in the pelvic floor musculature. Loop 4 provides for volitional control of the striated sphincter. Abnormalities of the suprasacral portion result in abnormal responses of the pudendal motor neurons to bladder filling and emptying, which may be manifested as detrusor striated sphincter dyssynergia or/and loss of voluntary contraction of the striated sphincter on command.

Injury or Disease of the Neural Axis

Blaivas (1982) has made a nice attempt to correlate the clinical and urodynamic data from 550 patients with voiding dysfunction with current theories of micturition. He has made the following observations. In 155 patients with suprasacral neurologic lesions (complete and incomplete), physiologically normal voiding was seen in only 41%. Detrusor-striated sphincter dyssynergia was demonstrated in 34% and, surprisingly (and paradoxically), detrusor areflexia was noted in 25%. These data suggest that coordinated voiding is not a simple segmental sacral reflex, but requires higher neurologic integration. Detrusor striated sphincter dyssynergia was found in 45% of 119 patients with suprasacral spinal cord lesions. However, none of 36 patients with supraspinal neurologic lesions (nor none of 213 patients without neurologic disease) had striated sphincter dyssynergia. These data support the view that coordinated voiding is regulated by neurologic centers above the spinal cord, and suggest that a clinical diagnosis of striated sphincter dyssynergia implies a neurologic lesion that interrupts the neural axis between the pontine-mesencephalic reticular formation and the sacral spinal cord. All 27 patients with neurologic lesions above the pons and who were able to void, did so synergistically (with relaxation of the striated sphincter preceding detrusor contraction). Twenty of these patients had involuntary bladder contractions, but 12 of these 20 had voluntary control of the striated sphincter. This supports Bradley's concept of separate neural pathways governing voluntary control of the bladder and the periurethral striated musculature. Most of Blaivas' patients with involuntary bladder contractions secondary to neurologic lesions above the pons were able to voluntarily contract their striated sphincter without abolishing bladder contraction. This would seem to suggest that the inhibition of bladder contraction by pudendal motor activity is not merely a simple sacral reflex, but that such inhibition is a complex neurologic event. Twenty-two of Blaivas' patients had evidence of sacral or infrasacral neurologic impairment of bladder function but had suprasacral control of striated sphincter function, or vice versa, providing a clinical correlate to the distinct anatomical locations of the parasympathetic motor nucleus and the pudendal nucleus of the sacral spinal cord.

Component Reflexes of Micturition

Preceding discussions give some idea of the complexity and diversity of the experimental data and resulting hypotheses relating to the component functions thought to be important for normal filling/storage and emptying of the lower urinary tract. Historically, based on what information was available at the time, authors have been tempted to divide micturition into various "reflexes" related to its initiation, continuation and completion, and cessation. In a series of experiments in the cat reported between 1914 and 1941 Barrington (1914 and cited by Kuru, 1965) originally described seven component reflexes of micturition (Table 27–3). Kuru (1965) classified all "reflexes" related to micturition according to whether they were concerned primarily with the storage of urine or the initiation, continuation, and cessation of micturition. Kuru's initial scheme has been used by others (Mahoney et al., 1977; Wein and Raezer, 1979; Schmidt, 1983) to include additional "reflexes" affecting micturition (Tables 27–4 through 27–7). Although it is impressive to list these, and quite interesting from a historical standpoint, close inspection will readily reveal that some of these reflexes are contradictory, some are without much experimental or any clinical correlation, and some seem to be based on an attempt to explain a particular event in simple terms rather than to physiologically investigate it. In other words, some are consistent with neuromorphologic, neurophysiologic, neuropharmacologic, and clinical data and make sense in this respect, and others are inconsistent with such data, and therefore do not.

IMPORTANT FUNCTIONAL QUESTIONS

Based on the preceding experimental and clinical data, and the various theories and hypotheses that have been proposed relative to these, we can now at least address, and in some cases answer, certain important questions that relate to the activity of the bladder and the outlet during the filling/storage and emptying phases of micturition. Further, we can now also look at at least some of the alterations of lower urinary tract function produced by a variety of experimental and clinical conditions and try to provide neuromorpho-

TABLE 27–3.

Component Reflexes of Micturition According to Barrington*

REFLEX	AFFERENT PATHWAY	EFFERENT PATHWAY	CENTER
Strong vesical contraction following bladder distention (I)	Pelvic nerve	Pelvic nerve	Supraspinal
Strong vesical contraction evoked by running liquid in the urethra (II)	Pudendal nerve	Pelvic nerve	Supraspinal
Transient, slight vesical contraction following distention of proximal urethra (III)	Hypogastric nerve	Hypogastric nerve	Spinal
Relaxation of urethra following running of liquid through the urethra (IV)	Pudendal nerve	Pudendal nerve	Spinal
External sphincter relaxation evoked by bladder distention (V)	Pelvic nerve	Pudendal nerve	Spinal
Relaxation of smooth musculature of proximal urethra following bladder distention (VI)	Pelvic nerve	Pelvic nerve; hypogastric nerve	Spinal
Transient, slight vesical contraction evoked by running liquid through the urethra (VII)	Pelvic nerve	Pelvic nerve	Spinal

*Adapted from Kuru M: Nervous control of micturition. *Physiol Rev* 1965; 45:425.

TABLE 27–4.

Reflexes Related to Urine Storage*

REFLEXES	AFFERENT PATHWAY	EFFERENT PATHWAY	CENTER
Inhibition of vesical contraction during bladder filling	Pelvic nerve; dorsal funiculus; lateral funiculus	Ventral reticulospinal tract (?), pelvic nerve	Bulbar
Increased external sphincter (sacral) contraction evoked by bladder filling	Pelvic nerve	Pudendal nerve	Spinal
Increased external sphincter (sacral) contraction evoked by urine involuntarily entering the proximal urethra	Pudendal nerve	Pudendal nerve	Spinal
Inhibition of vesical contraction during bladder filling†	Pelvic nerve	Hypogastric nerve	Spinal (thoracolumbar)
Increased contraction (closure) of bladder neck and proximal urethra during bladder filling	Pelvic nerve	Hypogastric nerve	Spinal (thoracolumbar)

*After Kuru M: Nervous control of micturition. *Physiol Rev* 1965; 45:425. Original references cited by Kuru (1965) and Mahoney et al. (1977).
†Directly by an inhibitory effect on bladder muscle cells, and indirectly by an inhibitory effect on bladder parasympathetic ganglia (see text).

TABLE 27–5.

Reflexes Related to Initiation of Micturition*

REFLEXES	AFFERENT PATHWAY	EFFERENT PATHWAY	CENTER
Vesical contraction and proximal urethral relaxation evoked by bladder distention	Pelvic nerve; hypogastric nerve	Pelvic nerve; hypogastric nerve	Spinal
Vesical contraction evoked by bladder distention†	Pelvic nerve; dorsal funiculus; lateral funiculus	Lateral reticulospinal tract (?); pelvic nerve	Bulbar
Relaxation of pelvic musculature evoked by bladder distention and subsequent contraction	Pelvic nerve; lateral funiculus	Corticospinal tract (?); pudendal nerve	Bulbar

*After Kuru M: Nervous control of micturition. *Physiol Rev* 1965; 45:425. Original references cited by Kuru (1965) and Mahoney et al. (1977).
†Also central inhibition of bulbar relaxer center and thereby of the first reflex listed in Table 27–4.

logic, neurophysiologic, and neuropharmacologic correlates, based on these data, theories, and hypotheses. Most of the material in this section is strictly based on references previously cited in this chapter. However, some of the viewpoints expressed, though based on selected data, reflect our own personal prejudices, consistent with our own experimental and clinical experience.

WHAT DETERMINES BLADDER RESPONSE DURING FILLING?

It is clear that the normal bladder response to filling at a physiological rate is an almost imperceptible change in intravesical pressure. During at least the initial stages of bladder filling, there is little question that this very high compliance is due to purely passive properties of the bladder wall. These passive properties have been described by various authorities (see Zinner et al. [1983], Coolsaet [1984], and Griffiths [1984] for summaries and references). The definition employed and the graphs and formulas used by these authors may be somewhat different, but the implied conclusions are the same. The elastic properties of the bladder wall allow it to stretch to a certain degree without any increase in tension, thus allowing intravesical pressure to remain virtually unchanged within certain limits. The viscoelastic properties of the bladder wall allow for an increase in intravesical pressure when the rate of stretch (filling volume) exceeds the rate of stress relaxation. They also allow for a decrease in this tension when this rapid filling slows or stops. Although the

TABLE 27–6.

Reflexes Related to Continuation and Completion of Micturition*

REFLEX	AFFERENT PATHWAY	EFFERENT PATHWAY	CENTER
Continued vesical contraction during voiding† evoked by receptors in detrusor muscle	Pelvic nerve; lateral funiculus; juxtasolitariothalamic tract	Lateral reticulospinal tract (?); pelvic nerve	Pontine
Promotion of vesical contraction during voiding‡ evoked by receptors in detrusor muscle	Pelvic nerve; lateral funiculus; juxtasolitariothalamic tract	Tectobulbar, mesencephalobulbospinal, lateral reticulospinal tract (?); pelvic nerve	Bulbar
Continued vesical contraction during voiding evoked by running liquid through the urethra	Pudendal nerve; lateral funiculus; juxtasolitariothalamic tract	Lateral reticulospinal tract (?); pelvic nerve	Pontine; bulbar; mesencephalic
Continued vesical contraction during voiding evoked by running liquid through the urethra	Pelvic nerve	Pelvic nerve	Spinal
Continued vesical contraction during voiding evoked by receptors in detrusor muscle	Pelvic nerve	Pelvic nerve	Spinal

*After Kuru M: Nervous control of micturition. *Physiol Rev* 1965; 45:425. Original references cited by Kuru (1965) and Mahoney et al. (1977).
†Also centrally based (probably pontine) reflex inhibition of external urethral sphincter tone.
‡By inhibition of bulbar relaxer center.

TABLE 27–7.

Reflexes Related to Cessation of Micturition*

REFLEX	AFFERENT PATHWAY	EFFERENT PATHWAY	CENTER
Vesical relaxation following cessation of stream in urethra†	Pelvic and/or pudendal nerve; dorsal funiculus; lateral funiculus	Ventral reticulospinal tract (?); pelvic nerve	Bulbar‡
Recovery of external sphincter tone following cessation of stream in urethra	Pelvic and/or pudendal nerve	Pudendal nerve	Spinal§

*After Kuru M: Nervous control of micturition. *Physiol Rev* 1965; 45:425. Original references cited by Kuru (1965) and Mahoney et al. (1977).
†Similar reflex probably exists with afferent limb from cessation of stimuli on receptors in series with detrusor muscle fibers.
‡Similar reflex probably exists at spinal level.
§Similar reflex probably exists at central (most likely pontine) level.

mathematical descriptions and terminologies may be relatively new, the concept of nonneurogenic tonus during filling is not, and was clearly implied and supported by Nesbit and Lapides (1947, 1948) and by Tang and Ruch (1955). Filling cystometry, in the usual clinical setting, seems to show a slight increase in intravesical pressure during infusion of the medium employed. However, Klevmark (1974, 1977, 1980) has shown, in an elegant series of experiments on cats, that this pressure rise during filling is a function of the fact that the filling is carried out at a greater than physiologic rate, and that, at physiologic filling rates, there is essentially no rise in bladder pressure until bladder capacity is reached. During filling, if an acute increase in volume

introduction occurs, the bladder wall will respond with an increase in intravesical pressure, which is a combination of the elastic and viscoelastic properties of the bladder wall and the inherent response of smooth muscle to stretch. Clinically, loss of the elastic fibers of the bladder wall and their replacement by fibrous tissue is most commonly observed in patients with chronic pancystitis, especially those who have had an indwelling catheter present for some time. The effect of the loss of elastic and viscoelastic properties of the bladder wall is clearly seen in these patients, whose filling cystometrograms exhibit a decrease in compliance at low filling rates. Such decreased compliance, secondary to loss of the passive properties of the bladder wall, is generally unresponsive to pharmacologic manipulation, hydro-distention, or nerve section, and most often requires augmentation cystoplasty to achieve satisfactory bladder reservoir function.

A small fraction of our previous discussions has mentioned the possibility of substances released from the bladder muscle or epithelium during filling (primarily prostaglandins), which contribute to the small rise in bladder wall tension that occurs. A much larger fraction of our discussion has considered the question of inhibitory neural mechanisms that are operative during the filling/storage phase of micturition. It is certainly clear that in animals a spinal sympathetic reflex, which is evoked by bladder filling, exists. One arm of this seems to involve direct neurally mediated stimulation of β-adrenergic receptors located predominantly in the bladder body smooth musculature. Evidence for a major such inhibitory influence in humans is sparse, but a minor such role could conceivably exist. It is quite clear that the "normal" bladder response during filling is most vividly disrupted by efferent activity in the pelvic nerve excitatory tract, causing involuntary bladder contraction. Most discussions of neural reflex mechanisms causing an inhibitory influence on bladder activity during the filling/storage phase of micturition have centered on indirect inhibitory mechanisms of this parasympathetic efferent activity. Such an inhibitory role at the level of the bladder parasympathetic ganglia has clearly been confirmed by deGroat (deGroat and Saum, 1971 and 1972; deGroat and Theoblad, 1976; deGroat and Booth, 1984) as a part of the spinal sympathetic reflex that occurs during bladder filling in cats. Compelling, although indirect, evidence exists to support such a role in humans. The afferent stimuli that initiate this reflex are mediated primarily through the pelvic nerves, but there is evidence to support inhibitory reflex mechanisms that are activated through pudendal nerve stimuli as well. McGuire (1978, 1983, 1984a) implies a direct inhibition of the detrusor motor neurons in the sacral spinal cord due to increased afferent pudendal nerve activity generated by receptors in the striated sphincter. Such a reflex was first described in man by Kock and Pompeius (1963), who showed an inhibition of bladder contractile activity initiated by anal distention. Sundin et al. (1974) have proposed a similar pudendal nerve–mediated inhibition of bladder activity during stimulation, but one that is mediated through hypogastric nerve activity.

Good evidence also seems to exist to support a strong tonic inhibitory effect of endogenous opioids on bladder activity at at least two levels, the spinal cord and the brain stem (Hisamitsu and deGroat, 1984; deGroat and Booth, 1984; deGroat and Kawatani, 1984). Such activity likely affects the bladder response (or lack of it) during filling, and may affect contractility during emptying as well.

The presence and functional significance (if any) of spontaneous activity during filling in the normal urinary bladder is a very controversial subject that certainly bears on the question of what determines the bladder response during filling. In the anesthetized cat, intravesical pressure recordings have revealed marked "spontaneous" rhythmic bladder activity (Klevmark, 1974, 1977, 1980). Rhythmic activity, however, was unable to be elicited in the awake state in this animal. In the deeply relaxed or anesthetized animals, when rhythmic bladder contractions were demonstrated during filling, they were demonstrated at physiologic filling rates, as opposed to other experiments (Plum and Colfelt, 1960), where more rapid and unphysiologic rates of filling were utilized. Klevmark (1980) found that these spontaneous bladder contractions in deeply asleep or anesthetized cats during filling were not affected by spinal cord transsection above the T10 level, but were either abolished or converted to an irregular, dissimilar pattern by partial or complete peripheral denervation, supporting the notion that such spontaneous rhythmic activity in response to distention was dependent on intact bladder innervation from the spinal cord. Plum and Colfelt (1960) maintained that the rhythmic contractions of the cat urinary bladder were nonneurogenic and, further, that the rhythmic waves were a fundamental and necessary component of normal micturition, a micturition contraction developing as a rapid summation and fusion of smaller waves. They suggested that an increase in such contractions followed sacral rhizotomy or distal cordectomy, implying that the sacral spinal cord segments tonically inhibited bladder rhythmicity.

Older reports (Plum, 1960; Kock and Pompeius, 1963) cite the occurrence of rhythmic bladder activity in awake humans with neurologically intact bladders. It is difficult to correlate this information with, and integrate it into, current urodynamic doctrine, which suggests that phasic increases in intravesical pressure that occur during bladder filling are always abnormal, regardless of whether there is any sensory perception of these increases. Any contemporary description of filling cystometry (see Abrams [1984]) will be found to state quite conclusively that "in the normal bladder no detrusor activity is evident during the filling phase of cystometry; therefore, strictly speaking, any detrusor contractions occurring during the filling phase should be regarded as abnormal." In the anesthetized rabbit, bladder activity during filling is absent at low intravesical volumes and pressures, and develops only on reaching a critical pressure, this activity being completely inhibited by ganglionic blockade. In the presence of ganglionic blockade, the in vivo bladder is devoid of activity at any volume or pressure, suggesting that this activity is not really "spontaneous," but occurs in response to bladder filling and is in fact a neuronal reflex phenomenon (Levin et al., 1986). In the rabbit, the in vitro whole bladder preparation is totally devoid of any spontaneous phasic contractile activity at any volume or pressure during filling unless longitudinal tension is applied to the preparation (Levin et al., 1986). This study further suggested that the "spontaneous" activity commonly observed in isolated urinary bladder strip preparations is in fact a simple myogenic response to stretch, and has little physiologic significance under normal conditions. Klevmark (1980) wisely commented on the term "spontaneous," pointing out that this term generally applies only to contractions due to membrane potential variations that occur either spontaneously in pacemaker regions or following chemical or mechanical stimulation of smooth muscle. Thus, the type of "spontaneous" activity seen in bladder smooth muscle strips (invariably under some stretch) would seem to have little functional significance insofar as the emptying contraction of a normal micturition is concerned. It is possible, however, that localized contractile activity in response to various types of pathology might be sufficient to generate a rise in intravesical pressure, causing pathologic symptomatology during the filling phase of micturition (urgency, frequency, and, perhaps, urge incontinence), irrespective of the emptying ability of such activity.

WHAT DETERMINES OUTLET RESPONSE DURING FILLING?

Virtually all authorities would agree that there is a gradual increase in urethral pressure during bladder filling (see references at the beginning of this chapter, and also Tanagho and Smith, 1966; Jonas and Tanagho, 1975; Ghonheim et al., 1975; Khalaf et al., 1979; and Abdel-Rahman et al., 1981, 1983). Urethral pressure is an ill-defined phenomenon, measurement of which is far from standardized. It is clear that, whatever definition one accepts, this is not equivalent to urethral resistance, nor are increases or decreases in either "pressure" or "resistance" reproducibly correlated with urinary continence and incontinence. Experiments and discussions that have attempted to delineate the various components of urethral "pressure," "resistance," and continence/incontinence are flawed by the lack of a definitive method to separate even active from passive characteristics of the urethral wall. Moreover, in our opinion, there is no truly reliable way to separate the smooth from the striated muscular contribution to urethral pressure by strictly pharmacologic means, as long as significant disagreement exists regarding the inner-

vation and receptor content of the smooth muscle of the wall of the urethra, of the intramural striated muscle, and of the extramural striated muscle. Various estimates of the contribution of components of the urethral wall to urethral "pressure" are given in Table 27–8. It must be recognized that these estimates were derived from different experimental animals (and man), under totally different experimental and clinical conditions and with totally different rationales. However, it is clear that most authors at least recognize the presence of relatively passive elements in the urethral wall, as opposed to active elements, in which the tension can change in response to neural or humoral mechanisms.

The rise in urethral pressure that is seen during the filling phase of micturition can be correlated with an increase in efferent pudendal nerve and hypogastric nerve impulse frequencies, at least in cats (deGroat and Booth, 1980, 1984). This, and the well-known phenomenon of a gradual increase in pelvic EMG activity during bladder filling, make it certain that a reflex increase in striated sphincter activity contributes significantly to this urethral response. Neurophysiologic, neuromorphologic, and neuropharmacologic data are all consistent with this. However, it should be recognized that most of these data are based on the "classic" model of a striated sphincter innervated by the pudendal nerve, and do not include the possibility of the contribution of efferent pelvic nerve impulses. Furthermore, if the possibility of a different innervation of the intramural vs. the extramural striated sphincter is considered, it is not quite clear to us exactly how one goes about discretely separating the contribution of the intramural vs. the extramural striated component to the urethral response during bladder filling.

Although it seems logical (and compatible with neuropharmacologic, neurophysiologic, and neuromorphologic data) to assume that the urethral smooth muscle component also contributes to the change in urethral response during bladder filling, it is extremely difficult

TABLE 27–8.

Estimates of Urethral Pressure Components (Percentage)

AUTHOR(S)	STRIATED MUSCLE*	SMOOTH MUSCLE	CONNECTIVE TISSUE†	VASCULATURE
Tanagho et al., (1969a, 1969b, 1978)	50	50	. . .	. . .
Donker et al. (1972)	Minimal	. . .	. . .	. . .
Raz et al. (1972b)	. . .	. . .	. . .	30
Awad and Downie (1976b)	16(F) 31(M)	40 45	41 24	
Koff (1977)	43	. . .	. . .	. . .
Rud et al. (1980)	33	33		33
Rossier et al. (1982)	60	. . . 30	. . .	. . .

*Includes (and does not differentiate between) extramural and intramural.
†Includes elastic, collagenous, and fibrous tissue within or outside urethral wall.

to prove this either experimentally or clinically. Those, like ourselves, who intuitively believe that an increase in smooth sphincter activity contributes to urethral response during bladder filling should not ignore significant questions that inevitably arise during consideration of this subject. To say that the urethral smooth muscle possesses a tonus, and that this tonus itself is sufficient to maintain continence, is certainly not equivalent to saying that this tonus is neurologically mediated or that it changes during bladder filling. Abdel-Rahman et al. (1981) failed to demonstrate a change in proximal urethral smooth muscle activity in the cat during bladder filling, but did, however, feel that the intrinsic tone of this area was what maintained urinary continence. It is difficult to prove that the increases in hypogastric nerve activity seen during bladder filling in experimental animals affect urethral smooth sphincter tone. In other words, these efferent impulses may not terminate on the smooth musculature of the urethra or on ganglion cells affecting urethral motility, but may simply be terminating only on parasympathetic ganglion cells relaying to the smooth musculature of the bladder.

There is also a potential for substances, such as prostaglandins, released either from the bladder mucosa or bladder musculature during filling, to raise urethral resistance by a humoral effect on the urethral wall, and without any neurologic mediation.

Finally, it is certainly possible, if one believes that there is a continuity of smooth musculature from the bladder base into the proximal urethra, that bladder filling increases mural tension in the bladder neck area, and that this may be transmitted to the urethra and reflected by tension changes in its wall.

The passive properties of the urethral wall certainly deserve mention, as these undoubtedly play a large role in the maintenance of continence (see Zinner et al. [1983] and references contained therein). To maintain urinary continence, the urethral lumen must be completely sealed. Urethral wall tension develops within the outer layers of the urethra and is a product not only of the active characteristics of smooth and striated muscle, but of the passive characteristics of the elastic and collagenous tissue as well. This tension must be exerted on a soft or plastic inner layer capable of being compressed to a closed configuration. This region of compression full of "filler material" (Zinner et al., 1980, 1983) represents the submucosal portion of the urethra. The softer and more plastic this area is, the less pressure is required by the tension-producing area to produce continence. Finally, whatever the compressive forces, the lumen must be capable of being obliterated by a watertight seal. This mucosal seal mechanism (the term seems to have been originated by Caine and Raz [1973]) explains why a very thin-walled rubber tube requires less pressure to close an open end when the inner layer is coated with a fine layer of grease than when it is not, the latter case being much like a scarred or atrophic urethral mucosa. Also, this explains why one can pass a pair of small clamped catheters through a normal urethra into the bladder with no leakage around the catheters (Zinner et al., 1983).

WHY DOES URINARY LEAKAGE NOT OCCUR WITH INCREASES IN INTRA-ABDOMINAL PRESSURE?

Outlet resistance to the flow of urine from the bladder is maintained by the passive and active factors mentioned above. During voluntarily initiated micturition, the bladder pressure becomes higher than the outlet pressure, certain adaptive changes occur in the shape of the bladder outlet, and urine pases into and through the proximal urethra. One could reasonably ask why such changes do not occur with increases in intravesical pressure, which are similar in magnitude, but which are produced only by changes in intra-abdominal pressure, such as straining and coughing. First of all, a coordinated bladder contraction does not occur in response to such stimuli, and this fact clearly points out that increases in intravesical pressure are by no means equivalent to emptying ability. For urine to flow into the proximal urethra, not only must an increase in intravesical pressure occur, but it must be a product of a coordinated bladder contraction, occurring through a neurally mediated reflex mechanism, and associated with characteristic changes in the bladder neck and proximal urethral area, themselves either directly or indirectly a part of this reflex mechanism. A decrease in tension in the striated sphincter is reflexly coordinated with this type of bladder contraction, and so is, most probably, a decrease in smooth sphincter tension as well.

A major factor in the prevention of urinary leakage during increasees in intra-abdominal pressure is the location of the sphincter unit (the bladder neck and proximal urethra). Normally, this is situated within the abdominal cavity to allow at least equal pressure transmission of any increase in intra-abdominal pressure to the bladder neck and proximal urethra. This phenomenon was first extensively described by Enhorning (1961), has been confirmed in virtually every urodynamic laboratory that has studied stress urinary incontinence, and is illustrated in Figure 27–39. The structures that normally support the female bladder neck and proximal urethra, and that prevent the descent of this sphincteric area into a position where intra-abdominal pressure increases are not transmitted at least equally to it, are the pubourethral ligaments, the pubocervical fascia, and the levator ani muscles (Stan-

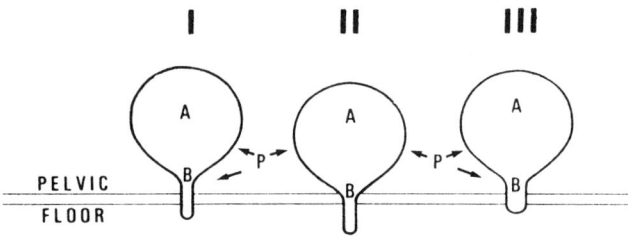

FIG 27–39.
Transmission of increases in intra-abdominal pressure to bladder neck and proximal urethra. "I" represents the normal state. Intra-abdominal pressure increases are transmitted at least equally to the bladder neck/proximal urethral area (P + B ≥ P + A). "II" illustrates the genesis of stress urinary incontinence. During increases in abdominal pressure, the bladder neck/proximal urethra area exhibits abnormal descent to a position in which a positive pressure gradient no longer occurs (P + B < P + A). In "III" the support of the bladder neck/proximal urethra is adequate, but the bladder neck and proximal urethra are nonfunctional. They exhibit essentially no closure pressure and are isobaric with the bladder. P + B is never greater than P + A. Restoring a support mechanism to check descent will correct the incontinence resulting from "II" but not from "III." Correction of "III" requires, assuming the support mechanism is satisfactory, reconstruction or compression of the bladder neck/proximal urethra (urethral sling, artificial sphincter, periurethral polytef injection). (From Stanton SL: Urethral sphincter incompetence (stress incontinence), in Mundy AR, et al (eds): *Urodynamics: Principles, Practice, Application.* New York, Churchill Livingstone, 1984. Used with permission.)

ton, 1984). Alteration of these structures such that pathologic descent does occur with abdominal straining occurs secondary to many etiologic factors, and stress incontinence is the result. To correct this, it is necessary only to substitute a mechanism to prevent this descent and to restore the normal resting position of the bladder neck and proximal urethra. This is accomplished by any of the standard vesicourethral suspension procedures.

There is evidence (Tanagho, 1978) to show that the increase in urethral closure pressure seen with increments in extrinsic pressure applied to the abdomen actually exceeds the extrinsic pressure increase, indicating that active muscular function, in addition to simple transmission of pressure, is involved. Tanagho (1978) cogently summarized evidence that the increment in urethral closure pressure involves a reflex increase in sphincter activity, primarily of the striated sphincter, but involving the smooth sphincter as well.

WHY DOES VOIDING ENSUE WITH A NORMAL BLADDER CONTRACTION?

As previously indicated, there is a reflex coordination between a voluntarily induced bladder contraction of

adequate magnitude and active responses of the proximal urethra. Most would agree with the observations cited and reconfirmed by Tanagho (1978) and McGuire (1983, 1984a) that urethral pressure actually decreases prior to bladder contraction, although some (Khalaf et al., 1979) would disagree. The decrease in pelvic floor striated muscle EMG activity prior to voluntary bladder contraction would strongly suggest that at least a portion of this decrease in urethral response is secondary to a reflex mechanism involving the striated sphincter and mediated through the pudendal nerves (Tanagho, 1978; McGuire, 1983, 1984a; deGroat, 1984). It is certainly tempting to speculate that a similar coordination of smooth sphincter activity exists, reflected by a decrease in efferent hypogastric nerve activity that occurs with the onset of bladder contraction (deGroat and Booth, 1980, 1984). Although there is much indirect evidence to support a reflex inhibition of smooth sphincter activity during bladder contraction, it is very difficult to accrue direct evidence to confirm this. McGuire (1984a) seems to agree that there is indeed an active mechanism that occurs that decreases urethral smooth muscle activity during bladder contraction, but postulates, on the basis of data previously cited, that this is an active mechanism involving stimulation of β-adrenergic receptors, a concept supported by El Badawi (1982, 1983) and Khanna (1983).

Tanagho (see 1978 for summary) originated the ingenious bipartite animal bladder, where the detrusor compartment was separated from the sphincter compartment. This effectively excludes vesicourethral smooth muscle continuity as a mechanism contributing to bladder neck and proximal urethral pressure changes during bladder filling and bladder contraction. The bipartite model in the dog clearly showed that there was indeed a rise in urethral pressure with bladder filling, and active contraction of the bladder was still preceded by a drop in urethral closure pressure. However, this by no means excludes a contributory mechanistic effect on the bladder neck and proximal urethra by virtue of smooth muscle continuity between the bladder base and proximal urethra in the intact animal or in man. There is clearly a difference between the type of tension developed in the bladder wall smooth musculature during filling and that seen during active contraction. Some contribution to the architectural changes seen in the area of the bladder neck and proximal urethra during bladder emptying may very likely be a result of mural continuity.

Finally, consistent with the observation of Anderson et al. (1983) and Klarskov et al. (1983), we cannot eliminate the possibility that some of the changes that occur in the bladder neck/proximal urethral area during active bladder contraction are wholly or in part neurologically mediated, but through a noncholinergic and nonadrenergic postganglionic neurotransmitter mechanism.

NORMAL LOWER URINARY TRACT FILLING/STORAGE AND EMPTYING: SIMPLE OVERVIEW, EXTRAPOLATION, AND APPLICATION OF THE TWO-PHASE CONCEPT

It is obvious that a discussion of any magnitude regarding the central and peripheral factors involved in micturition is bound to engender at least some disagreement. What follows is a concise and simplified description of the normal processes involved in lower urinary tract function, which seems to us to be consistent with the data and opinions presented.

The two functions of the lower urinary tract are the storage and the active expulsion of the urine. During bladder filling at physiological rates, intravesical pressure initially rises slowly despite large increases in volume. This phenomenon is due primarily to the passive elastic and viscoelastic properties of the smooth muscle and connective tissue of the bladder wall. There is little neural efferent activity to the lower urinary tract until a certain critical intravesical pressure is reached, following which there is a gradual reflex increase in somatic nerve efferent activity through the pudendal nerve, and perhaps the pelvic nerve, which causes increased activity in the striated sphincter. In animals, and perhaps in humans as well, an additional spinal sympathetic reflex also results from pelvic nerve afferents. The efferent limb of this reflex is through the hypogastric nerve, resulting in inhibition of bladder contractile activity by an inhibitory effect on parasympathetic ganglionic transmission, and also in an increase in outlet resistance because of active stimulation of the predominantly α-adrenergic receptors of the smooth muscle of the bladder neck and proximal urethra. A decrease in bladder body contractility, because of active stimulation through this reflex of the predominantly β-adrenergic receptors in that area, may also play a role. The net effect of all these actions is filling of the bladder cavity, but with a low increase in intravesical pressure relative to volume, and without any evidence of active coordinated bladder contractile activity. A tonic inhibitory influence exerted by endogenous opioids very likely plays a contributory role in maintaining this stability.

Intraurethral pressure is always greater than intravesical pressure, and this relationship is maintained during increases in intra-abdominal pressure because of the anatomic localization of the "sphincter unit" and the positive pressure transmission ratio to this area with respect to intravesical contents.

Although many factors are involved in the micturition reflex, it is intravesical pressure producing the sensation of distention that is primarily responsible for the initiation of voluntarily induced emptying of the lower urinary tract. The origin of the parasympathetic neural outflow to the bladder, the pelvic nerve, is in the sacral spinal cord. However, the actual organizational center for the micturition reflex in an intact neural axis is in the brain stem, and the complete neural circuit for normal micturition includes the ascending and descending spinal cord pathways to and from this area and the facilitory and inhibitory influences from other parts of the brain.

The final step in voluntarily induced micturition involves inhibition of the somatic neural efferent activity to the striated sphincter and an inhibition of all aspects of the spinal sympathetic reflex evoked during filling. Efferent parasympathetic pelvic nerve activity is ultimately what is responsible for a highly coordinated contraction of the bulk of the bladder smooth musculature. A decrease in outlet resistance occurs, with adaptive shaping or funneling of the relaxed bladder outlet. This may involve reflex inhibition of both the striated sphincter and of α-adrenergically mediated smooth sphincter tone. It may also involve active β-adrenergic–induced smooth muscle relaxation or smooth muscle relaxation through a noncholinergic, nonadrenergic mechanism. The adaptive changes are also due at least in part to the anatomic interrelationships of the smooth muscle of the bladder base and proximal urethra. Other reflexes that are elicited by bladder contraction and by the passage of urine through the urethra may reinforce and facilitate complete bladder emptying. Superimposed on these autonomic and somatic reflexes are complex modifying supraspinal inputs from other central neuronal networks. These facilitory and inhibitory impulses, which originate from several levels of the nervous system, including the brain stem, cerebellum, and cerebral cortex, allow for the full conscious control of micturition.

Whatever disagreements exist, it is extremely important to note that all "experts" would agree, however, on certain points. The micturition cycle involves two relatively discrete processes: (1) bladder filling and urine storage, and (2) bladder emptying. Whatever the neuromorphologic, neurophysiologic, neuropharmacologic, and mechanical details involved, the following summary must apply for these processes to occur normally.

BLADDER FILLING AND URINE STORAGE

1. Accommodation of increasing volumes of urine at a low intravesical pressure and with appropriate sensation.
2. The bladder outlet must be closed at rest and re-

main so during increases in intra-abdominal pressure.

3. There must be no involuntary bladder contractions.

BLADDER EMPTYING

1. There must be a coordinated contraction of the bladder smooth musculature of adequate magnitude.
2. A concomitant lowering of resistance must occur at the level of the smooth sphincter and the striated sphincter.
3. There must be no anatomic obstruction.

Further extrapolating this concept, one recognizes that any type of voiding dysfunction *must* result from an abnormality of one of the factors listed above. Thus, any abnormality of bladder filling and/or urine storage must result from a problem related to one of three simple factors or from some combination thereof. Any problem with bladder emptying must occur on a similar basis. One may reasonably argue about the anatomic, physiologic, and pharmacologic factors involved in these two relatively discrete phases of micturition, but one cannot argue about the conceptual simplicity of component contributions involved in each phase. This division, with its implied subdivisions under each category into causes related to the bladder and causes related to the outlet, provides a perfect rationale for the classification of all types of voiding dysfunction into disorders related to bladder filling or urine storage, and disorders related to bladder emptying. There are indeed some types of voiding dysfunction that represent combinations of filling/storage and emptying disorders. Within this scheme, these become readily understandable. Further, one may take all aspects of urodynamic and videourodynamic evaluation and classify the individual component studies as to exactly what they evaluate, in terms of either bladder or outlet (smooth and striated sphincter) activity during bladder filling/storage, or during bladder emptying. Urodynamics thus become readily understandable. Finally, one can then easily classify all known treatments for voiding dysfunction under the broad categories of whether they facilitate bladder filling/storage or bladder emptying, and whether they do so by an action primarily on the bladder or on the components of the bladder outlet. These clinical considerations are fully expanded in the next chapter.

ACKNOWLEDGMENT

The authors wish to thank Karl Erik Andersson and David M. Jacobowitz for reviewing the manuscript and for providing valuable advice for its final form.

REFERENCES

1. Abdel-Hakim A, Hassouna M, Rioux F, et al: Response of urethral smooth muscles to pharmacological agents. II. Non-cholinergic non-adrenergic agonists and antagonists. *J Urol* 1983; 130:988.
2. Abdel-Rahman M, Galeano C, Elhilali M: New approach to study of voiding cycle in cat: Preliminary report on pharmacologic studies. *Urology* 1983; 22:91.
3. Abdel-Rahman M, Galeano C, Lamarch J, et al: A new approach to the study of the voiding cycle in the cat. *Invest Urol* 1981; 18:475.
4. Abrams P: The practice of urodynamics, in Mundy AR, Stephenson TP, Wein AJ (eds): *Urodynamics: Principles, Practice and Applications.* New York, Churchill Livingstone, 1984, pp 76–92.
5. Abrams PH, Feneley RCL: The actions of prostaglandins on the smooth muscle of the human urinary tract in vitro. *Br J Urol* 1976; 47:909.
6. Adami M, Bertaccini G, Coruzzi C, et al: Characterization of cholinoreceptors in the rat urinary bladder by the use of agonists and antagonists of the cholinergic system. *J Auton Pharmacol* 1985; 5:197.
7. Alm P, Alumets J, Hakonson P, et al: Peptidergic nerves in the genitourinary tract. *Neuroscience* 1977; 2:751.
8. Alm P, Alumets J, Broden E, et al: Peptidergic (substance P) nerves in the genitourinary tract. *Neuroscience* 1978; 3:419.
9. Alm P, Elmer M: Adrenergic and cholinergic innervation of the rat urinary bladder. *Acta Physiol Scand* 1975; 94:36.
10. Ambache N: The use and limitation of atropine for pharmacological studies on autonomic effectors. *Pharmacol Rev* 1955; 7:467.
11. Ambache N, Zar MA: Non-cholinergic transmission by post-ganglionic motor neurones in the mammalian bladder. *J Physiol Lond* 1970; 210:761.
12. Anderson GF, Marks BH: Spare cholinergic receptors in the urinary bladder. *J Pharmacol Exp Ther* 1982; 221:598.
13. Anderson GF, Marks BH: Beta adrenoceptors in the rabbit detrusor muscle. *J Pharmacol Exp Ther* 1984; 228:283.
14. Anderson GF, Pierce JM, Fredericks CM, et al: A pharmacologic evaluation of the adrenergic receptors of the rabbit detrusor muscle. *Arch Int Pharmacodyn Ther* 1971; 191:220.
15. Andersson KE: New trends in lower urinary tract pharmacology. *Trends Pharmacol Sci* 1984; 5:521.
15a. Andersson KE: Classification and function of peripheral vascular α-adrenoreceptors, in Refsum H, Mjøs O (eds): *α-Adrenoreceptor Blockers in Cardiovascular Disease.* New York, Churchill Livingstone, 1985, pp 3–18.
16. Andersson KE, Ek A, Persson CGA: Effects of prostaglandins on the isolated human bladder and urethra. *Acta Physiol Scand* 1977; 100:165.
17. Andersson KE, Husted S, Sjogren C: Contribution of prostaglandins to the adenosine triphosphate induced

contraction of rabbit urinary bladder. *Br J Pharmacol* 1980; 70:443.

18. Andersson KE, Larsson B, Sjogren C: Characterization of the adrenoceptors in the female rabbit urethra. *Br J Pharmacol* 1984; 81:293.

19. Andersson KE, Mattiasson A, Sjogren C: Electrically induced relaxation of the noradrenaline contracted isolated urethra from rabbit and man. *J Urol* 1983; 129:210.

20. Andersson KE, Sjogren C: Aspects on the physiology and pharmacology of the bladder and urethra. *Prog Neurobiol* 1982; 19:71.

21. Andrew J, Nathan PW: The cerebral control of micturition. *Proc R Soc Med* 1965; 58:553.

22. Appenzeller O: Anatomy and histology; Autonomic hyperreflexia, in Appenzeller O (ed): *The Autonomic Nervous System.* Elsevier Biomedical Press, 1982, pp 1–30, 406–407.

23. Atta MA, El Badawi A: Intrinsic neuromuscular defects in the neurogenic bladder: IV. Loss of somato-motor and preservation of autonomic innervation of the male feline rhabdosphincter following bilateral sacral ventral rhizotomy. *Neurourol Urodynamics* 1985; 4:219.

24. Awad SA, Bruce AE, Carro-Ciampi G, et al: Distribution of alpha and beta adrenoreceptors in human urinary bladder. *Br J Pharmacol* 1974; 50:525.

25. Awad SA, Downie JW: The effect of adrenergic drugs and hypogastric nerve stimulation on the canine urethra. *Invest Urol* 1976a; 13:298.

26. Awad SA, Downie JW: Relative contribution of smooth and striated muscles to the canine urethral pressure profile. *Br J Urol* 1976b; 48:347.

27. Awad SA, Downie JW: The adrenergic component in the proximal urethra. *Urol Int* 1977; 32:192.

28. Barrington FJF: The nervous mechanism of micturition. *Q J Exp Physiol* 1914; 8:33.

29. Barrington FJF: The relation of the hindbrain to micturition. *Brain* 1921; 44:23.

30. Barrington FJF: The effect of lesions of the hind- and mid-brain on micturition in the cat. *Q J Exp Physiol* 1925; 15:81.

31. Bazeed MA, Thuroff JM, Schmidt RA, et al: Histochemical study of urethral striated musculature in the dog. *J Urol* 1982; 128:406.

32. Benson GS: Mechanisms of autonomic drug action on the bladder outlet, in Hinman F Jr (ed): *Benign Prostatic Hypertrophy.* Berlin, Springer-Verlag, 1983, pp 373–382.

33. Benson GS, Jacobowitz D, Raezer DM, et al: Adrenergic innervation and stimulation of canine urethra. *Urology* 1976; 7:337.

34. Benson GS, McConnell JA, Wood JG: Adrenergic innervation of the human bladder body. *J Urol* 1979; 122:189.

35. Benson GS, Wein AJ, Raezer DM, et al: Adrenergic and cholinergic stimulation and blockade of the human bladder base. *J Urol* 1976; 116:174.

36. Berridge MJ, Irvine RF: Inositol triphosphate, a novel second messenger in cellular signal transduction. *Nature* 1984; 312:315.

37. Berthelson S, Pettinger WA: A functional basis for classification of alpha adrenergic receptors. *Life Sci* 1977; 21:595.

38. Bhatia NN, Bradley WE: Neuroanatomy and physiology: Innervation of the urinary tract, in Raz S (ed): *Female Urology.* Philadelphia, WB Saunders Co, 1983, pp 12–32.

39. Blaivas JG: The neurophysiology of micturition: A clinical study of 550 patients. *J Urol* 1982; 127:958.

40. Booth AM, Hisamitsu T, Kawatani M, et al: Regulation of urinary bladder capacity by endogenous opioid peptides. *J Urol* 1985; 133:339.

41. Bors E, Comarr AE: *Neurological Urology.* Baltimore, University Park Press, 1971.

42. Bowen JM, Timm GW, Bradley WE: Some contractile and electrophysiologic properties of the periurethral striated muscle of the cat. *Invest Urol* 1976; 13:327.

43. Bozler E: Action potentials and conduction of excitation in smooth muscle. *Biol Symp* 1941; 3:95.

44. Bradley WE, Conway CJ: Bladder representation in the pontine-mesencephalic reticular formation. *Exp Neurol* 1966; 16:237.

45. Bradley WE, Sundin T: The physiology and pharmacology of urinary tract dysfunction. *Clin Neuropharmacol* 1982; 5:131.

46. Bradley WE, Teague CT: Spinal cord organization of micturition reflex afferents. *Exp Neurol* 1968; 22:504.

47. Bradley WE, Timm GW, Scott FB: Innervation of the detrusor muscle and urethra. *Urol Clin North Am* 1974; 1:3.

48. Bro-Rasmussen F, Sorensen AH, Bredahl E, et al: The structure and function of the urinary bladder. *Urol Int* 1965; 19:280.

49. Bruschini H, Schmidt RA, Tanagho EA: The male genitourinary sphincter mechanism in the dog. *Invest Urol* 1978; 15:284.

50. Bulbring E, Brading AF, Jones AW, et al: *Smooth Muscle: An Assessment of Current Knowledge.* Austin, University of Texas Press, 1981.

51. Bultitude MI, Hills NH, Shuttleworth KED: Clinical and experimental studies on the action of prostaglandins and their synthesis inhibitors on detrusor muscle in vitro and in vivo. *Br J Urol* 1976; 48:631.

52. Burnstock G: Purinergic nerves. *Pharmacol Rev* 1972; 24:509.

53. Burnstock G: Cholinergic, adrenergic, and purinergic transmission. *Fed Proc* 1977; 36:2434.

54. Burnstock G: Nervous control of smooth muscle by transmitters, co-transmitters and modulators. *Experientia* 1985; 41:869.

55. Burnstock G: The changing face of autonomic neurotransmission. *Acta Physiol Scand* 1986; 126:67.

56. Burnstock G, Bell C: Peripheral autonomic transmission, in Hubbard JI (ed): *The Peripheral Nervous System.* New York, Plenum, 1974, pp 277–327.

57. Burnstock G, Cocks T, Grave R, et al: Purinergic innervation of the guinea pig urinary bladder. *Br J Pharmacol* 1978a; 63:125.

58. Burnstock G, Cocks T, Kasakov L, et al: Direct evi-

dence for ATP release from non-adrenergic non-cholinergic ('purinergic') nerves in the guinea pig *Taenia coli* and bladder. *Eur J Pharmacol* 1978b; 49:145.

59. Burnstock G, Dumsday B, Smythe A; Atropine resistant excitation of the urinary bladder: The possibility of transmission via nerves releasing a purine nucleotide. *Br J Pharmacol* 1972; 44:451.

60. Caine M: *The Pharmacology of the Urinary Tract.* Berlin, Springer-Verlag, 1984, pp 5–29, 31–47.

61. Caine M, Raz S: The role of female hormones in stress incontinence. Presentation at Societe Internationale d'Urologie, 16th Congress, Amsterdam, 1973.

62. Caine M, Raz S, Zeigler M: Adrenergic and cholinergic receptors in the human prostate, prostatic capsule and bladder neck. *Br J Urol* 1975; 47:193.

63. Carlsson CA: The supraspinal control of the urinary bladder. *Acta Pharmacol Toxicol* 1978; 43:8.

64. Carpenter FG: Atropine resistance and muscarinic receptors in the rat urinary bladder. *Br J Pharmacol* 1977; 59:43.

65. Conway CJ, Bradley WE: Measurement of the spread of excitation in the urinary detrusor muscle during reflex induction. *J Urol* 1969; 101:533.

66. Coolsaet BLRA: Cystometry, in Stanton SL (ed): *Clinical Gynecologic Urology.* St Louis, CV Mosby Co, 1984, pp 59–81.

67. Cooper JR, Meyer EM: Possible mechanisms involved in the release and modulation of release of neuroactive agents. *Neurochem Int* 1984; 6:419.

67a. Cowan WD, Daniel EE: Human female bladder and its noncholinergic contractile function. *Can J Physiol Pharmacol* 1983; 61:1236.

68. Creed K: The role of the hypogastric nerve in bladder and urethral activity of the dog. *Br J Pharmacol* 1979; 65:367.

69. Daniel EE, Cowan W, Daniel VP: Structural basis for neural and myogenic control of human detrusor muscle. *Can J Physiol Pharmacol* 1983; 61:1247.

70. Dean DM, Downie JW: Contribution of adrenergic and 'purinergic' neurotransmission to contraction in the rabbit detrusor. *J Pharmacol Exp Ther* 1978; 207:431.

71. deGroat WC: Nervous control of the urinary bladder of the cat. *Brain Res.* 1975; 87:201.

72. deGroat WC, Booth AM: Autonomic systems to the urinary bladder and sexual organs, in Dyck PJ, Thomas PK, Lambert EH, et al (eds): *Peripheral Neuropathy.* Philadelphia, WB Saunders Co, 1984, pp 285–299.

73. deGroat WC, Booth AM: Physiology of the urinary bladder and urethra. *Ann Intern Med* 1980; 92(pt 2): 321.

74. deGroat WC, Kawatani M: Neural control of the urinary bladder: Possible relationship between peptidergic inhibitory mechanisms and detrusor instability. *Neurourol Urodynamics* 1985; 4:285.

75. deGroat WC, Kawatani M, Hisamitsu T, et al: The role of neuropeptides in the sacral autonomic reflex pathways of the cat. *J Auton Nerv Syst* 1983; 7:339.

76. deGroat WC, Nadelhaft I, Milner J, et al: Organization of the sacral parasympathetic reflex pathways to the urinary bladder and large intestine. *J Auton Nerv Syst* 1981; 3:135.

77. deGroat WC, Ryall RW: Reflexes to sacral parasympathetic neurons concerned with micturition in the cat. *J Physiol* 1969; 200:87.

78. deGroat WC, Saum WR: Adrenergic inhibition in mammalian parasympathetic ganglia. *Nature* 1971; 231:188.

79. deGroat WC, Saum WR: Sympathetic inhibition of the urinary bladder and of pelvic ganglionic transmission in the cat. *J Physiol Lond* 1972; 220:297.

80. deGroat WC, Saum WR: Synaptic transmission in parasympathetic ganglia in the urinary bladder of the cat. *J Physiol Lond* 1976; 256:137.

81. deGroat WC, Theobald RJ: Reflex activation of sympathetic pathways to vesical smooth muscle and parasympathetic ganglia by electrical stimulation of vesical afferents. *J Physiol Lond* 1976; 259:223.

82. Denny-Brown D, Robertson EG: On the physiology of micturition. *Brain* 1933; 56:149.

83. Dixon JS, Gilpin SA, Gilpin CT, et al: Intramural ganglia of the human urinary bladder. *Br J Urol* 1983; 55:195.

84. Dixon JS, Gosling JA: Histology and fine structure of the muscularis mucosae of the human urinary bladder. *J Anat* 1983; 136:265.

85. Donker PJ, Droes JTPM, Van Ulder BM: Anatomy of the musculature and innervation of the bladder and urethra, in Chisholm GD, Williams DI (eds): *Scientific Foundations of Urology.* Chicago, Year Book Medical Publishers, 1982, pp 404–411.

86. Donker PJ, Ivanovici F, Noach EL: Analyses of the urethra pressure profile by means of electromyography and the administration of drugs. *Br J Urol* 1972; 44:180.

87. Douglas WW: Polypeptides—angiotensin, plasma kinins, and others, in Gilman AG, Goodman LS, Rall TW, (eds): *The Pharmacological Basis of Therapeutics.* New York, Macmillan Publishing Co, 1985, pp 639–659.

88. Downie JW, Dean DM: The contribution of cholinergic postganglionic neurotransmission to contractions of rabbit detrusor. *J Pharmacol Exp Ther* 1977; 203:417.

89. Downie JW, Karmazyn M: Mechanical trauma to bladder epithelium liberates prostanoids which modulate neurotransmission in rabbit detrusor muscle. *J Pharmacol Exp Ther* 1984; 230:445.

90. Downie JW, Larsson C: Prostaglandin involvement in contractions evoked in rabbit detrusor by field stimulation and by adenosine 5-triphosphate. *Can J Physiol Pharmacol* 1981; 59:253.

91. Dray A, Metsch R: Opioid receptor subtypes involved in the central inhibition of bladder motility. *Eur J Pharmacol* 1984a; 104:47.

92. Dray A, Metsch R: Inhibition of urinary bladder contractions by a special action of morphine and other opioids. *J Pharmacol Exp Ther* 1984b; 231:254.

93. Droogmans G, Himpens B, Casteels R: Co-exchange, co-channels and co-antagonists. *Experientia* 1985; 41:895.

94. Edvardsen P: Nervous control of the urinary bladder in

cats: I. The collecting phase. *Acta Physiol Scand* 1968a; 72:157.

95. Edvardsen P: Nervous control of the urinary bladder in cats: II. The expulsion phase. *Acta Physiol Scand* 1968b; 72:172.

96. Edvardsen P: Nervous control of the urinary bladder in cats: III. Effects of autonomic blocking agents in the intact animal. *Acta Physiol Scand* 1968c; 72:183.

97. Edvardsen P: Nervous control of the urinary bladder in cats: IV. Effects of autonomic blocking agents on responses to peripheral nerve stimulation. *Acta Physiol Scand* 1968d; 72:234.

98. Edvardsen P, Setekleiv J: Distribution of adrenergic receptors in the urinary bladder of cats, rabbit, and guinea pigs. *Acta Pharmacol Toxicol* 1968; 26:437.

99. Ek A, Alm P, Andersson KE, et al: Adrenergic and cholinergic nerves of the human urethra and urinary bladder: A histochemical study. *Acta Physiol Scand* 1977a; 99:345.

100. Ek A, Alm P, Andersson KE, et al: Adrenoreceptor and cholinoceptor mediated responses of the isolated human urethra. *Scand J Urol Nephrol* 1977b; 11:97.

101. Eglen RM, Whiting RL: Muscarinic receptors' subtypes: Problems of classification. *Trends Pharmacol Sci* 1985; 6:357.

102. Ekstrom J, Malmberg L: On a cholinergic motor innervation of the rat urethra. *Acta Physiol Scand* 1984; 120:237.

103. El Badawi A: Neuromorphologic basis of vesicourethral function. I. Histochemistry, ultrastructure, and function of intrinsic nerves of the bladder and urethra. *Neurourol Urodynamics* 1982; 1:3.

104. El Badawi A: Autonomic muscular innervation of the vesical outlet and its role in micturition, in Hinman F Jr (ed): *Benign Prostatic Hypertrophy.* Berlin, Springer-Verlag, 1983, pp 330–348.

105. El Badawi A: Ultrastructure of vesicourethral innervation: II. Postganglionic axoaxonal synapses in intrinsic innervation of the vesicourethral lissosphincter: A new structural and functional concept in micturition. *J Urol* 1984; 131:781.

106. El Badawi A: Ultrastructure of vesicourethral innervation: III. Anoaxonal synapses between postganglionic cholinergic axons and probably SIF-cell derived processes in the feline lissosphincter. *J Urol* 1985; 133:524.

107. El Badawi A, Atta MA: Ultrastructure of vesicourethral innervation: IV. Evidence for somatomotor plus autonomic innervation of the male feline rhabdosphincter. *Neurourol Urodynamics* 1985; 4:23.

108. El Badawi A, Schenk EA: Dual innervation of the mammalian urinary bladder: A histochemical study of the distribution of cholinergic and adrenergic nerves. *Am J Anat* 1966; 119:405.

109. El Badawi A, Schenk EA: A new theory of the innervation of bladder musculature: I. Morphology of the intrinsic vesical innervation apparatus. *J Urol* 1968; 99:585.

110. El Badawi A, Schenk EA: A new theory of the innerva-

tion of bladder musculature: III. Innervation of the vesicourethral junction and external urethral sphincter. *J Urol* 1974; 111:613.

111. Elliot TR: The innervation of the bladder and urethra. *J Physiol Lond* 1907; 35:367.

112. Elmer M: Action of drugs on the innervated and denervated urinary bladder of the rat. *Acta Physiol Scand* 1974; 91:289.

113. Elmer M: Atropine sensitivity of the rat urinary bladder during nerve degeneration. *Acta Physiol Scand* 1975; 93:202.

114. Enhorning G: Simultaneous recording of the intravesical and intraurethral pressures. *Acta Chir Scand* 1961; 276(suppl):1.

115. Erspamer GF, Negri L, Piccinelli D: The use of preparations of urinary bladder smooth muscle for bioassay of and discrimination between polypeptides. *Naunyn Schmiedebergs Arch Pharmacol* 1973; 276:61.

116. Fall M, Erlandson BE, Carlsson CA, et al: The effect of intravaginal stimulation on the feline urethra and urinary bladder neuronal mechanisms. *Scand J Urol Nephrol* 1977; 44(suppl):19.

117. Finkbeiner AE: In vitro effects of vasoactive intestinal polypeptide on guinea pig urinary bladder. *Urology* 1983; 22:275.

118. Finkbeiner AS, Bissada NK: In vitro effects of prostaglandins on the guinea pig detrusor and bladder outlet. *Urol Res* 1981; 9:281.

119. Fletcher TF, Bradley WE: Neuroanatomy of the bladder-urethra. *J Urol* 1978; 119:153.

120. Fredericks CM: Characterization of the rabbit detrusor response to histamine through pharmacologic antagonism. *Pharmacology* 1975; 13:5.

121. Gabella G: Structure of smooth muscles, in Bulbring F, Brading AF, Jones AW, et al (eds): *Smooth Muscle.* Austin, University of Texas Press, 1981, p 1–46.

122. Gajewski J, Downie J, Awad S: Experimental evidence for a central nervous system site of action in the effect of alpha adrenergic blockers on the external urethral sphincter. *J Urol* 1984; 133:403.

123. Ghonheim MA, Fretin JA, Gagnon DJ, et al: The influence of vesical distention on urethral resistance to flow: The collecting phase. *Br J Urol* 1975; 47:657.

124. Gilpin SA, Gosling JA: Smooth muscle in the wall of the developing human bladder and urethra. *J Anat* 1983; 137:503.

125. Gil Vernet S: *Morphology and Function of Vesico-Prostato-Urethral Musculature.* Treviso, Italy, Canora, 1968, pp 245–269.

126. Gjone R: A dual peripheral and supraspinal autonomic influence on the urinary bladder. *J Oslo City Hosp* 1965; 15:173.

127. Gjone R, Setekleiv J: Excitatory and inhibitory responses to stimulation of the cerebral cortex in the cat. *Acta Physiol Scand* 1963; 59:337.

128. Glazer E, Basbaum A: Leucine enkephalin: Localization in and axoplasmic transport by sacral parasympathetic preganglionic neurons. *Science* 1980; 208:1479.

129. Gosling J: The structure of the bladder and urethra in

relation to function. *Urol Clin North Am* 1979; 6(Feb):31.

130. Gosling J: Anatomy, in Stanton SL (ed): *Clinical Gynecologic Urology.* St. Louis, CV Mosby Co, 1984, pp 3–12.

131. Gosling JA: Personal communication (letter), June 12, 1985.

132. Gosling JA, Chilton CP: The anatomy of the bladder, urethra and pelvic floor, in Mundy AR, Stephenson TP, Wein AJ (eds): *Urodynamics: Principles, Practice and Applications.* London, Churchill Livingstone, 1984, pp 3–13.

133. Gosling JA, Dixon JS, Critchley HOD, et al: A comparative study of the human external sphincter and periurethral levator ani muscles. *Br J Urol* 1981; 53:35.

133a. Gray TS, Morley JE: Neuropeptide Y: Anatomical distribution and possible function in mammalian nervous system. *Life Sci* 1986; 38:389.

134. Griffiths DJ: Hydrodynamics and mechanisms of the bladder and urethra, in Mundy AR, Stephenson TP, Wein AJ (eds): *Urodynamics: Principles, Practice and Applications.* London, Churchill Livingstone, 1984, pp 42–49.

135. Gruber CM: The autonomic innervation of the genitourinary system. *Physiol Rev* 1933; 13:497.

136. Gu J, Blank MA, Huang WM, et al: Peptide containing nerves in human urinary bladder. *Urology* 1984; 24:353.

137. Gu J, Restorich JM, Blank MA, et al: Vasoactive intestinal polypeptide in the normal and unstable bladder. *Br J Urol* 1983; 55:645.

138. Guth L: Axonal regeneration and functional plasticity in the central nervous system. *Exp Neurol* 1974; 45:606.

139. Guyton AC: *Textbook of Medical Physiology.* Philadelphia, WB Saunders Co, 1986, pp 120–135, 136–149, 686–696.

140. Gyermek L: Cholinergic stimulation and blockade on urinary bladder. *Am J Physiol* 1961; 201:325.

141. Gyermek L: Action of 5-hydroxytryptamine on the urinary bladder of the dog. *Arch Int Pharmacodyn Ther* 1962; 137:137.

142. Hald T, Bradley W: *The Urinary Bladder: Neurology and Dynamics.* Baltimore, Williams & Wilkins, 1982, pp 5–21, 22–36, 82–88.

143. Hammer R, Giachetti A: Muscarinic receptor subtypes: M1 and M2 biochemical and functional characterization. *Life Sci* 1982; 31:2991.

144. Hardman JG: Cyclic nucleotides and smooth muscle contraction: Some conceptual and experimental considerations, in Bulbring E, Brading AF, Jones AW (eds): *Smooth Muscle.* Austin, University of Texas Press, 1981, pp 249–262.

145. Hassouna M, Abdel-Hakim A, Abdel-Rahman M, et al: Response of the urethral smooth muscles to pharmacological agents: I. Cholinergic and adrenergic agonists and antagonists. *J Urol* 1983; 191:1262.

146. Haynes LW: Modulation of cholinergic transmission by VIP, a peptide co-transmitter. *Trends Pharmacol Sci* 1985; 6:427.

147. Hickey DS, Phillips JI, Hukins DWL: Arrangement of collagen fibers and muscle fibers in the female urethra and their implications for the control of micturition. *Br J Urol* 1982; 54:556.

148. Hindmarsh JR, Idowu OA, Yeates WK, et al: Pharmacology of electrically evoked contractions of human bladder. *Br J Pharmacol* 1977; 61:115P.

149. Hisamitsu T, deGroat WC: The effect of opioid peptides and morphine applied intrathecally and intracerebroventricularly on the micturition reflex in the cat. *Brain Res* 1984; 298:51.

150. Honda K, Osawa AM, Takenaka T: α_1-Adrenoceptor subtype mediating contraction of the smooth muscle in the lower urinary tract and prostate of rabbits. *Naunyn Schmiedebergs Arch Pharmacol* 1985; 330:16.

151. Hoyle CHV, Burnstock G: Atropine resistant excitatory junction potentials in rabbit bladder are blocked by α, β-methylene ATP. *Eur J Pharmacol* 1985; 114:239.

152. Hughes J, Kosterlitz HW: Introduction: Opioid peptides. *Br Med Bull* 1983; 39:1.

153. Husted S, Sjogren C, Andersson KE: Direct effects of adenosine and adenine nucleotides on isolated human urinary bladder and their influence on electrically induced contractions. *J Urol* 1983; 130:392.

154. Husted S, Sjogren C, Andersson KE: Substance P and somatostatin and excitatory neurotransmission in rabbit urinary bladder. *Arch Int Pharmacodyn* 1981; 252:72.

155. Hutch JA: A new theory of the anatomy of the internal urinary sphincter and the physiology of micturition: IV. The urinary sphincteric mechanism. *J Urol* 1967; 97:705.

156. Hutch JA, Rambo OA Jr: A new theory of the anatomy of the internal urinary sphincter and the physiology of micturition: III. Anatomy of the urethra. *J Urol* 1967; 97:696.

157. Ignarro LJ, Kadowitz PJ: The pharmacological and physiological role of cyclic GMP in vascular smooth muscle relaxation. *Ann Rev Pharmacol Toxicol* 1985; 25:171.

158. Ingersoll EH, Jones LL, Hegre ES: Effect on urinary bladder of unilateral stimulation of pelvic nerves in the dog. *Am J Physiol* 1957; 189:167.

159. Ito Y, Kimoto Y: The neural and non-neural mechanisms involved in urethral activity in rabbits. *J Physiol* 1985; 367:57.

160. Jacobowitz DM: The peripheral autonomic system, in Hubbard JI (ed): *The Peripheral Nervous System,* New York, Plenum, 1974, pp 87–110.

161. Jaffee JH, Martin WR: Opioid analgesics and antagonists, in Gilman AG, Goodman LS, Rall TW, et al (eds): *The Pharmacological Basis of Therapeutics.* New York, Macmillan Publishing Co, 1985, pp 491–531.

162. John A, Paton DM: Evidence for a role of prostaglandins in atropine-resistant transmission in the mammalian urinary bladder. *Prostaglandins* 1976; 11:595.

163. Jonas U, Tanagho E: Studies on vesicourethral reflexes. *Invest Urol* 1975; 12:357.

164. Jubelin B, Galeano C, Ladouceus D, et al: Effect of enkephalin on the micturition cycle of the cat. *Life Sci* 1984; 34:2015.

165. Kalsnro S (chairman): The noradrenergic presynaptic re-

ceptor controversy (symposium). *Fed Proc* 1984; 43:1351.

166. Kaneko S, Minami K, Yachiku S, et al: Bladder neck dysfunction: The effect of the adrenergic agent phentolamine on bladder neck dysfunction and a fluorescent histochemical study of bladder neck smooth muscle. *Invest Urol* 1980; 18:212.

167. Kasakov L, Burnstock G: The use of the slowly degradable analog, α, β-methylene ATP to produce desensitization of the P-2 purinoceptor effect on nonadrenergic noncholinergic responses of guinea pig urinary bladder. *Eur J Pharmacol* 1983; 86:292.

168. Kawatani M, Lowe IP, Booth AM, et al: The presence of leucine-enkephalin in the sacral preganglionic pathway to the urinary bladder of the cat. *Neurosci Lett* 1983; 39:143.

169. Khalaf I, Elshawarby LA, Lehoux JG, et al: Release of prostaglandins into the pelvic venous blood of dogs in response to vesical distention and pelvic nerve stimulation. *Invest Urol* 1979; 17:244.

170. Khalaf IM, Ghonheim MA, Elhilali MM: The effect of exogenous prostaglandins F2 and E2 and indomethacin on micturition. *Br J Urol* 1981; 43:21.

171. Khalaf I, Toppercer A, Elhilali M: Urethral pressure changes on reflex micturition. *Invest Urol* 1979; 17:141.

172. Khanna OP: Effect of nonautonomic drugs on the vesical neck, in Hinman F Jr (ed): *Benign Prostatic Hypertrophy.* Berlin, Springer-Verlag, 1983, pp 384–404.

173. Khanna OP, Barbieri EJ, McMichael R: Effects of prostaglandins on vesicourethral smooth muscle of rabbit. *Urology* 1978; 12:674.

174. Khanna OP, Barbieri EJ, Altamura M, et al: Vesicourethral smooth muscle: Function and relation to structure. *Urology* 1981; 18:211.

175. Khanna OP, DeGregorio GJ, Sample RG, et al: Histamine receptors in urethrovesical smooth muscle. *Urology* 1977; 10:375.

176. Khanna OP, Heber D, Gonick P: Cholinergic and adrenergic neuroreceptors in urinary tract of female dogs: Evaluation of function with pharmacodynamics. *Urology* 1975; 5:616.

177. Kinder RB, Mundy AR: Atropine blockade of nerve mediated stimulation of the human detrusor. *Br J Urol* 1985a; 57:418.

178. Kinder RB, Mundy AR: Inhibition of spontaneous contractile activity in isolated human detrusor muscle strips by vasoactive intestinal polypeptide. *Br J Urol* 1985b; 57:20.

179. Kirchberger MA, Schwartz IL: Excitation and contraction of skeletal muscle, in West JB (ed): *Best and Taylor's Physiological Basis of Medical Practice.* Baltimore, Williams & Wilkins Co, 1985, pp 58–107.

180. Klarskov P, Gerstenberg T: Vasoactive intestinal polypeptide influence on lower urinary tract smooth muscle from human and pig. *J Urol* 1984; 131:1000.

181. Klarskov P, Gerstenberg TC, Ramirez D, et al: Noncholinergic, non-adrenergic nerve mediated relaxation of trigone, bladder neck and urethral smooth muscle in vitro. *J Urol* 1983; 129:848.

182. Kleeman FJ: The physiology of the internal urinary sphincter. *J Urol* 1970; 104:549.

183. Klevmark B: Motility of the urinary bladder in cats during filling at physiologic rates: I. Intravesical pressure patterns studied by a new method of cystometry. *Acta Physiol Scand* 1974; 90:565.

184. Klevmark B: Motility of the urinary bladder in cats during filling at physiological rates: II. Effects of extrinsic bladder denervation on intramural tension and on intravesical pressure patterns. *Acta Physiol Scand* 1977; 101:176.

185. Klevmark B: Motility of the urinary bladder in cats during filling at physiological rates: III. Spontaneous rhythmic contractions in the conscious and anesthetized animal: Influence of distention and innervation. *Scand J Urol Nephrol* 1980; 14:219.

185a. Kluck P: The autonomic innervation of the human urinary bladder, bladder neck, and urethra: A histochemical study. *Anat Rec* 1980; 198:439.

186. Kock N, Pompeius R: Inhibition of vesical motor activity induced by anal stimulation. *Acta Clin Scand* 1963; 126:244.

187. Koff SA: Striated muscle determinants of intraurethral resistance. *Invest Urol* 1977; 15:147.

188. Koley B, Koley J, Saka JK: The effect of nicotine on spontaneous contractions of cat urinary bladder in situ. *Br J Pharmacol* 1984; 83:347.

189. Kondo A, Susset JG, Lefaivre J: Viscoelastic properties of bladder: I. Mechanical model and its mathematical analysis. *Invest Urol* 1972; 10:154.

190. Krane RJ, Olsson CA: Phenoxybenzamine in neurogenic bladder dysfunction: I. A theory of micturition. *J Urol* 1973a; 110:650.

191. Krane RJ, Olsson CA: Phenoxybenzamine in neurogenic bladder dysfunction: II. Clinical considerations. *J Urol* 1973b; 110:653.

192. Krell RD, McCoy JL, Ridley PT: Pharmacological characterization of the excitatory innervation to the guinea pig urinary bladder in vitro: Evidence for both cholinergic and non-adrenergic non-cholinergic neurotransmission. *Br J Pharmacol* 1981; 74:15.

193. Kunisawa Y, Kawabe K, Nijima T, et al: A pharmacological study of alpha adrenergic receptor subtypes in smooth muscle of human urinary bladder base and prostatic urethra. *J Urol* 1985; 134:296.

194. Kuo DC, Hisamitsu T, deGroat WC: A sympathetic projection from sacral paravertebral ganglia to the pelvic nerve and to post-ganglionic nerves on the surface of the urinary bladder and large intestine of the cat. *J Compr Neurol* 1984; 226:76.

195. Kuriyama H: Excitation-contraction coupling in various visceral smooth muscles, in Bulbring E, Brading AF, Jones AW, et al (eds): *Smooth Muscle.* Austin, University of Texas Press, 1981, pp 171–197.

196. Kuru M: Nervous control of micturition. *Physiol Rev* 1965; 45:425.

197. Langer SZ, Shepperson NB: Recent developments in vascular smooth muscle pharmacology: the postsynaptic 2 adrenoceptor. *Trends Pharmacol Sci* 1982; 11:440.

198. Langley JN, Anderson HK: The innervation of the pelvic and adjoining viscera: II. The bladder. *J Physiol* 1895; 19:71.

199. Langley JN, Anderson HK: The innervation of the pelvic and adjoining viscera: VII. Anatomical observations. *J Physiol* 1896; 20:372.

200. Lapides J: Neuromuscular vesical and ureteral dysfunction, in Campbell MF, Harrison JH (eds): *Campbell's Urology*. Philadelphia, WB Saunders Co, 1970, p 1343.

201. Lapides J: Structure and function of the internal vesical sphincter. *J Urol* 1958; 80:341.

202. Lapides J, Ajemian EP, Stewart BH, et al: Further observations on the kinetics of the urethrovesical sphincter. *J Urol* 1960; 84:86.

203. Lapides J, Sweet RB, Lewis LW: Role of striated muscle in urination. J Urol 1957; 77:247.

204. Larsen JJ: α and β Adrenoceptors in the detrusor muscle and bladder base of the pig and β adrenoceptors in the detrusor muscle of man. *Br J Pharmacol* 1979; 65:215.

205. Larsen JJ, Ottesen B, Fahrenkrug J, et al: Vasoactive intestinal polypeptide in the male genitourinary tract: Concentration and motor effect. *Invest Urol* 1981; 19:211.

206. Larsson B, Sjogren C, Andersson KE: Regional distribution of α-adrenoceptor subtypes in the female rabbit urethra. *Acta Physiol Scand* 1986; 126:39.

207. Lavin TN, Hoffman BB, Lefkowitz RJ: Determination of subtype selectivity of alpha adrenergic antagonists. *Mol Pharmacol* 1981; 20:28.

208. Learmoth JR: A contribution to the neurophysiology of the urinary bladder in man. *Brain* 1931; 54:147.

209. Lefkowitz RJ, Caron MG, Stiles GL: Mechanisms of membrane receptor regulation. *N Engl J Med* 1984; 24:1570.

210. Leger L, Charmay Y, Chayvialle JA, et al: Localization of substance P and enkephalin like immunoreactivity in relation to catecholamine containing cell bodies in the cat dorsolateral pontine tegmentum: Immunofluorescence study. *Neuroscience* 1983; 8:525.

211. Leoni JV, Wein AJ, Raezer DM, et al: The effect of β-adrenergic stimulation on the contractile response of canine detrusor muscle. *Invest Urol* 1973; 10:419.

212. Lepor H, Kuhar M: Characterization of muscarinic cholinergic receptor binding in the vas deferens, bladder, prostate and penis of the rabbit. *J Urol* 1984; 132:392.

213. Levin RM, Goldman M, Wein AJ: Effect of isoproterenol and EGTA on the volume-pressure relationship of the in-vitro whole bladder preparation. *Neurourol Urodynamics* 1984; 3:133.

214. Levin RM, High J, Wein AJ: Evidence against the presence of spare muscarinic receptors in the rabbit urinary bladder. *Neurourol Urodynamics* 1983c; 2:317.

215. Levin RM, Jacoby R, Wein AJ: Effect of adenosine triphosphate on contractility and adenosine triphosphatase activity of the rabbit urinary bladder. *Mol Pharmacol* 1981; 19:525.

216. Levin RM, Jacoby R, Wein AJ: High affinity divalent ion specific binding of ^{3}H-ATP to homogenates derived from rabbit urinary bladder: Comparison with divalent ion ATP-ase activity. *Mol Pharmacol* 1983b; 23:1.

217. Levin R, Ruggieri M, Velagapudi S, et al: Relevance of spontaneous activity in urinary bladder function: An invitro and in-vivo study. *J Urol* 1986; 136:517.

218. Levin RM, Ruggieri MR, Wein AJ: Functional effects of the purinergic innervation of the rabbit urinary bladder. *J Pharmacol Exp Ther* 1986; 236:452.

219. Levin RM, Shofer F, Wein AJ: Cholinergic, adrenergic and purinergic response of sequential strips of rabbit urinary bladder. *J Pharmacol Exp Ther* 1980; 212:536.

220. Levin RM, Staskin DR, Wein AJ: The muscarinic cholinergic kinetics of the human urinary bladder. *Neurourol Urodynamics* 1983a; 2:221.

221. Levin RM, Wein AJ: Direct measurement of the anticholinergic activity of a series of pharmacologic compounds on the canine and rabbit urinary bladder. *J Urol* 1982; 128:396.

222. Levin RM, Wein AJ: Distribution and function of adrenergic receptors in the urinary bladder of the rabbit. *Mol Pharmacol* 1979a; 16:441.

223. Levin RM, Wein AJ: Effect of vasoactive intestinal peptide on the contractility of the rabbit urinary bladder. *Urol Res* 1981; 9:217.

224. Levin RM, Wein AJ: Quantitative analysis of alpha and beta adrenergic receptor densities in the lower urinary tract of the dog and the rabbit. *Invest Urol* 1979b; 17:75.

225. Liu CN, Chambers WW: Intraspinal sprouting of dorsal root axons: Development of new collaterals and preterminals following partial denervation of the spinal cord in the cat. *Arch Neurol Psychiatry* 1958; 79:46.

226. Lundberg JM, Hökfelt T: Coexistence of peptides and clasical neurotransmitters. *Trends Neurol Sci* 1983; 6:325.

227. Lundberg JM, Hua XY, FrancoCereceda A: Effects of neuropeptide Y on mechanical activity and neurotransmission in the heart, vas deferens, and urinary bladder of the guinea pig. *Acta Physiol Scand* 1984; 121:325.

228. Mahoney DT, Laferte RO, Blais DJ: Integral storage and voiding reflexes: Neurophysiologic concept of continence and micturition. *Urology* 1977; 9:95.

229. Maggi C, Evangelista S, Grimaldi G, et al: Evidence for the involvement of arachidonic and metabolites on spontaneous and drug induced contractions of rat urinary bladder. *J Pharmacol Exp Ther* 1984a; 230:500.

230. Maggi CA, Meli A: Modulation by β-adrenoceptors of spontaneous contractions of rat urinary bladder. *J Auton Pharmacol* 1982; 2:255.

231. Maggi CA, Santicioli P, Meli A: The effects of topical capsaicin on rat urinary bladder motility in vivo. *Eur J Pharmacol* 1984b; 103:41.

232. Maggi CA, Santicioli P, Meli A: GABA$_A$ and GABA$_B$ receptors in detrusor strips from guinea pig bladder dome. *J Auton Pharmacol* 1985a; 5:55.

233. Maggi CA, Santicioli P, Meli A: Dual effect of GABA on the contractile activity of the guinea pig isolated urinary bladder. *J Auton Pharmacol* 1985b; 5:131.

234. Maggi CA, Santicioli P, Meli A: Pharmacological evidence for the existence of two components in the twitch

response to field stimulation of detrusor strips from the rat urinary bladder. *J Auton Pharmacol* 1985c; 5:221.

235. Mattiasson A: *On the Peripheral Nervous Control of the Lower Urinary Tract*, thesis. University of Lund, Sweden, 1984.

236. Mattiasson A, Andersson KE, Sjogren C: Adrenergic and non-adrenergic contraction of isolated urethral muscle from rabbit and man. *J Urol* 1985; 133:298.

237. McGuire EJ: Experimental observations on the integration of bladder and urethral function. *Trans Am Assoc Genitourinary Surgeons* 1977; 68:38.

238. McGuire EJ: Experimental observations on the integration of bladder and urethral function. *Invest Urol* 1978; 15:303.

239. McGuire EJ: Physiology of the lower urinary tract. *Am J Kidney Dis* 1983; 2:402.

240. McGuire EJ: Clinical evaluation and treatment of neurogenic vesical dysfunction, in Libertino J (ed): *International Perspectives in Urology*. Baltimore, Williams & Wilkins, 1984a, vol 2, pp 1–15.

241. McGuire EJ: Mechanisms of urethral continence and their clinical application. *World J Urol* 1984b; 2:272.

242. McGuire EJ, Herlihy E: Bladder and urethral responses to isolated sacral motor root stimulation. *Invest Urol* 1978; 16:219.

243. McGuire EJ, Herlihy E: Bladder and urethral responses to sympathetic stimulation. *Invest Urol* 1979; 17:9.

244. McLellan FC: *The Neurogenic Bladder*. Springfield, Ill, Charles C Thomas, 1939, pp 57, 116.

245. McNeal JE: The prostate and prostatic urethra: A morphologic synthesis. *J Urol* 1972; 107:1008.

246. Minneman KP, Molinoff PB: Classification and quantitation of beta adrenergic receptor subtypes. *Biochem Pharmacol* 1980; 29:1317.

247. Mobley TL, El Badawi A, McDonald DF, et al: Innervation of the human urinary bladder. *Surg Forum* 1966; 17:505.

248. Morita T, Nishizawa O, Noto H, et al: Pelvic nerve innervation of the external sphincter of urethra as suggested by urodynamic and horseradish studies. *J Urol* 1984; 131:591.

249. Morley JS: Chemistry of opioid peptides. *Br Med Bull* 1983; 39:5.

250. Movsesian AM: Calcium physiology in smooth muscle. *Prog Cardiovasc Dis* 1982; 25:211.

251. Mundy AR: Clinical physiology of the bladder, urethra and pelvic floor, in Mundy AR, Stephenson TP, Wein AJ (eds): *Urodynamics: Principles, Practice and Application*. New York, Churchill Livingstone, 1984a, pp 14–25.

252. Mundy AR: Neuropeptides in lower urinary tract function. *World J Urol* 1984b; 2:211.

253. Murray KHA, Geneley RCL: Endorphins—a role in urinary tract function? The effect of opioid blockade on the detrusor and urethral sphincter mechanisms. *Br J Urol* 1982; 54:638.

254. Nanninga J, Kaplan P, Lal S: Effect of phentolamine on peripheral muscle EMG activity in paraplegia. *Br J Urol* 1977; 49:537.

255. Nathan PW: The central nervous connections of the bladder, in Williams DJ, Chisholm GD (eds): *Scientific Foundations of Urology*. Chicago, Year Book Medical Publishers, 1976, vol 2, pp 51–58.

256. Nergardh A: Autonomic receptor functions in lower urinary tract: A survey of recent experimental results. *J Urol* 1975; 113:180.

257. Nergardh A: Neuromuscular transmission in the corpus fundus of the urinary bladder. *Scand J Urol Nephrol* 1981; 15:103.

258. Nergardh A, Boreus LD: Autonomic receptor function in the lower urinary tract of man and cat. *Scand J Urol Nephrol* 1972; 6:32.

259. Nergardh A, Boreus LO: The functional role of cholinergic receptors in the outlet region of the urinary bladder: An in vitro study in the cat. *Acta Pharmacol Toxicol* 1973; 32:467.

260. Nergardh A, Boreus LO, Haglo AS: Characterization of the adrenergic beta receptor in the urinary bladder of man and cat. *Acta Pharmacol Toxicol* 1977; 40:14.

261. Nesbit R, Lapides J: Tonus of the bladder during spinal shock. *Arch Surg* 1948; 56:138.

262. Nesbit R, Lapides J, Volk W, et al: The effects of blockade of the autonomic ganglia on the urinary bladder in man. *J Urol* 1947; 57:242.

263. Nilvebrant L, Andersson KE, Mattiasson A: Characterization of the muscarinic cholinoceptors in the human detrusor. *J Urol* 1985; 134:418.

264. Nilvebrant L, Sparf B: Muscarinic receptor binding in the guinea pig urinary bladder. *Acta Pharmacol Toxicol* 1983; 52:30.

264a. Nilvebrant L: On the muscarinic receptors in the urinary bladder and the putative subclassification of muscarinic receptors. *Acta Pharmacol Toxicol* 1986; 59(suppl 1):1.

265. Nishizawa O, Fukada T, Matsuzaki A, et al: Role of the sympathetic nerve in bladder and urethral sphincter function during the micturition cycle in the dog evaluated by pressure flow EMG study. *J Urol* 1985; 134:1259.

266. Nordling J: Influence of the sympathetic nervous system on lower urinary tract in man. *Neurourol Urodynamics* 1983; 2:3.

267. Nordling J, Meyhoff H, Hald T: Sympatholytic effect on striated urethral sphincter. A peripheral or central nervous system effect? *Scand J Urol Nephrol* 1981; 15:173.

268. Norlen L: Influence of the sympathetic nervous system on the lower urinary tract and its clinical implications. *Neurourol Urodynamics* 1982; 1:129.

269. Norlen LJ, Blaivas JG, Groden W, et al: Contractile effect of substance P on the canine urinary bladder in vivo. *Neurourol Urodynamics* 1983; 2:323.

270. Oelrich TM: The urethral sphincter muscle in the male. *Am J Anat* 1980; 158:229.

271. Oelrich TM: The striated urogenital sphincter muscle in the female. *Anat Rec* 1983; 205:223.

272. Ohtsuka M, Mori J, Tsujioka K, et al: Pharmacological study on sympathetic inhibition of the urinary bladder in dogs. *Jpn J Pharmacol* 1980; 30:181.

273. Owman C, Owman T, Sjoberg NO: Short adrenergic neurons innervating the female urethra of the cat. *Experientia* 1971; 27:313.

274. Pedersen E, Torring J, Kleman B: Effect of the alpha adrenergic blocking agent thymoxamine on the neurogenic bladder and urethra. *Acta Neurol Scand* 1980; 61:107.

275. Perry SV, Grand RJA: Mechanisms of contraction and the specialized protein components of smooth muscle. *Br Med Bull* 1979; 35:219.

276. Persson CGA, Anderson KE: Adrenoceptor and cholinoceptor mediated effects in the isolated urethra of the cat and guinea pig. *Clin Exp Pharmacol Physiol* 1976; 3:415.

277. Plum F: Autonomous urinary bladder activity in normal man. *AMA Arch Neurol* 1960; 2:497.

278. Plum F, Colfelt R: The genesis of vesical rhythmicity. *AMA Arch Neurol* 1960; 2:487.

279. Polak JM, Bloom SR: Localization and measurement of VIP in the genitourinary system of man and animals. *Peptides* 1984; 5:225.

280. Raezer DM, Greenberg SH, Jacobowitz DM, et al: The innervation of the trigonal area of canine urinary bladder. *Urology* 1976; 7:369.

281. Raezer DM, Wein AJ, Jacobowitz D, et al: Autonomic innervation of canine urinary bladder: Cholinergic and adrenergic contributions and interaction of sympathetic and parasympathetic systems in bladder function. *Urology* 1973; 2:211.

282. Raz S, Caine M, Zeigler M: The vascular component in the production of intraurethral pressure. *J Urol* 1972b; 108:93.

283. Raz S, Zeigler M, Caine M: Isometric studies on canine urethral musculature. *Invest Urol* 1972a; 9:443.

284. Rockswold GI, Bradley WE, Chou SN: Innervation of the urinary bladder in higher primates. *J Compr Neurol* 1980a; 193:509.

285. Rockswold GI, Bradley WE, Chou SN: Innervation of the external urethral and external anal sphincters in higher primates. *J Compr Neurol* 1980b; 193:521.

286. Rohner TJ, Hannigan J: Effect of norepinephrine and isoproterenol on in vitro detrusor muscle contractility and cyclic AMP content. *Invest Urol* 1980; 17:324.

287. Rohner TJ, Raezer DM, Wein AJ, et al: Contractile responses of dog bladder neck muscle to adrenergic drugs. *J Urol* 1971; 105:657.

288. Roppolo JR, Booth AM, deGroat WC: The effects of naloxone on the neural control of the urinary bladder of the cat. *Brain Res* 1983; 264:335.

289. Rossier A, Fam B, Lee I, et al: Role of striated and smooth muscle components in the urethral pressure profile in traumatic neurogenic bladders: A neuropharmacologic and urodynamic study. Preliminary report. *J Urol* 1982; 128:529.

290. Ruch TC: The urinary bladder, in Ruch TC, Patton HD (eds): *Circulation, Respiration, and Fluid Balance.* Philadelphia, WB Saunders, 1974, pp 525–546.

291. Rud T, Andersson KE, Asmussen M, et al: Factors maintaining the urethral pressure in women. *Invest Urol* 1980; 17:343.

292. Ruffolo RR: Interactions of agonists with peripheral alpha adrenergic receptors. *Fed Proc* 1984; 43:2910.

293. Ruggieri M, Levin RM, Wein AJ: Muscarinic receptor subtypes in rabbit and human bladder smooth muscle. *Neurourol Urodynamics*, in press.

294. Sant GR: The anatomy of the external striated urethral sphincter. *Paraplegia* 1972; 10:153.

295. Santicioli P, Maggi CA, Meli A: The postganglionic excitatory innervation of the mouse urinary bladder and its modulation by prejunctional GABA$_B$ receptors. *J Auton Pharmacol* 1986; 6:53.

296. Saum WR, deGroat WC: The actions of 5-hydroxytryptamine on the urinary bladder and on vesical autonomic ganglia in the cat. *J Pharmacol Exp Ther* 1973; 185:70.

297. Saxena PR, Herligers J, Mylecharane EJ, et al: Excitatory 5-hydroxytryptamine receptors in the cat urinary bladder are the M and 5 HT$_2$ type. *J Auton Pharmacol* 1985; 5:101.

298. Schmidt RA: Urethrovesical reflexes and their inhibition, in Hinman F Jr (ed): *Benign Prostatic Hypertrophy.* Berlin, Springer-Verlag, 1983, pp 361–372.

299. Schulman CC, Duarte-Escalante O, Boyarsky S: The ureterovesical innervation: A new concept based on a histochemical study. *Br J Urol* 1972; 44:698.

300. Sibley GNA: A comparison of spontaneous and nerve mediated activity in bladder muscle from man, pig, and rabbit. *J Physiol* 1984; 354:431.

301. Sillen U: Central neurotransmitter mechanisms involved in the control of urinary bladder function. *Scand J Urol Nephrol*, suppl 58, 1980, pp 1–45.

302. Sillen U, Rubenson A: Central and peripheral motor effects of morphine on the rat urinary bladder. *Acta Physiol Scand* 1986; 126:181.

302a. Simonds WF, Booth AM, Thor KB, et al: Parasympathetic ganglia: Naloxone antagonizes inhibition by leucine-enkephalin and GABA. *Brain Res* 1983; 271:365.

303. Sjogren C, Andersson KE, Husted S, et al: Atropine resistance of transmurally stimulated isolated human bladder muscle. *J Urol* 1982; 128:1368.

304. Sjogren C, Andersson KE, Mattiasson A: Effects of vasoactive intestinal polypeptide on isolated urethral and urinary bladder smooth muscle from rabbit and man. *J Urol* 1985; 133:136.

305. Somlyo AP: The messenger across the gap. *Nature* 1985; 316:298.

306. Stanton SL: Urethral sphincter incompetence (stress incontinence), in Mundy AR, Stephenson TD, Wein AJ (eds): *Urodynamics: Principles, Practice, Application.* New York, Churchill Livingstone, 1984, pp 230–241.

307. Sundin T, Carlsson CA, Kock NA: Detrusor inhibition induced from mechanical stimulation of the anal region and from electrical stimulation of pudendal nerve afferents: An experimental study in cat. *Invest Urol* 1974; 11:374.

308. Sundin T, Dahlstrom A, Norlen L, et al: The sympathetic innervation and adrenoreceptor function of the human lower urinary tract in the normal state and after

parasympathetic denervation. *Invest Urol* 1977; 14:322.

309. Taira N: The autonomic pharmacology of the bladder. *Annu Rev Pharmacol* 1972; 12:197.

310. Tanagho EA: The anatomy and physiology of micturition. *Clin Obstet Gynecol* 1978; 5 (April):3.

311. Tanagho EA: The ureterovesical junction: Anatomy and physiology, in Chisholm GD, Williams D (eds): *Scientific Foundations of Urology*. Chicago, Year Book Medical Publishers, 1982, pp 295–404.

312. Tanagho EA: Anatomy of the lower urinary tract, in Walsh PC, Gillen RF, Perlmutter AD, et al (eds): *Campbell's Urology*. Philadelphia, WB Saunders, 1986, pp 46–74.

313. Tanagho EA, Meyers FH: The 'internal sphincter': Is it under sympathetic control? *Invest Urol* 1969; 7:79.

314. Tanagho EA, Meyers FH, Smith DR: Urethral resistance: Its components and implications: I. Smooth muscle component. *Invest Urol* 1969a; 7:136.

315. Tanagho EA, Meyers FH, Smith DR: Urethral resistance: Its components and implications: II. Striated muscle component. *Invest Urol* 1969b; 7:195.

316. Tanagho EA, Miller ER: Initiation of voiding. *Br J Urol* 1970; 42:175.

317. Tanagho EA, Pugh RCB: The anatomy of the ureterovesical junction. *Br J Urol* 1963; 35:151.

318. Tanagho EA, Smith DR: The anatomy and function of the bladder neck. *Br J Urol* 1966; 38:54.

319. Tanagho EA, Smith DR: Mechanism of urinary continence: I. Embryologic, anatomic, and pathologic considerations. *J Urol* 1968; 100:640.

320. Tanagho EA, Schmidt RA, Gomes deAranjo C: Urinary striated sphincter: What is its nerve supply? *Urology* 1982; 20:415.

321. Tang PC: Levels of brain stem and diencephalon controlling micturition reflex. *J Neurophysiol* 1955; 18:583.

322. Tang P, Ruch T: Non-neurogenic basis of bladder tonus. *Am J Physiol* 1955; 181:249.

323. Tang PC, Ruch TC: Localization of brain stem and diencephalic areas controlling the micturition reflex. *J Compr Neurol* 1956; 106:213.

324. Theobald RJ: The effect of arylazido aminoproprionyl ATP on atropine resistant contractions of the cat urinary bladder. *Life Sci* 1983; 32:2479.

325. Theobald RJ: The effect of ANAPP₃ on inhibition of pelvic nerve induced contractions of the cat urinary bladder. *Eur J Pharmacol* 1986; 120:351.

326. Thor KB, Roppolo JR, deGroat WC: Naloxone induced micturition in unanesthetized paraplegic cats. *J Urol* 1983; 129:202.

327. Todd JK, Mack AJ: A study of human bladder detrusor muscle. *Br J Urol* 1969; 41:448.

328. Tokunaka S, Murakami U, Hashi K, et al: Electrophoretic and ultrastructural analysis of the rabbit's striated external urethral sphincter. *J Urol* 1984; 132:1040.

329. Tsujimoto G, Timmins PV, Hoffman BB: Alpha adrenergic receptors in the rabbit bladder base smooth muscle: Alpha-1 adrenergic receptors mediate contractile responses. *J Pharmacol Exp Ther* 1986; 236:384.

330. Ursillo RC, Clark BB: The action of atropine on the urinary bladder of the dog and on the isolated nerve-bladder strip preparation of the rabbit. *J Pharmacol Exp Ther* 1956; 118:338.

331. Vaidyanathan S, Rao MS, Chary KSN, et al: Clinical impact of serotonin activity in the bladder and urethra. *J Urol* 1981a; 125:42.

332. Vaidyanathan S, Rao M, Chary K, et al: Enhancement of detrusor reflex activity by naloxone in patients with chronic neurogenic bladder dysfunction: Preliminary report. *J Urol* 1981b; 126:500.

333. Van Buren GA, Anderson GF: Comparison of the urinary bladder base and detrusor to cholinergic and histaminergic receptor activation in the rabbit. *Pharmacology* 1979; 18:136.

334. Van Duyl WA: A model for both the passive and active properties of urinary bladder tissue related to bladder function. *Neurourol Urodynamics* 1985; 4:275.

335. Van Mastrigt R, Coolsaet BLRA, Van Duyl WA: The passive properties of the urinary bladder in the collection phase. *Urol Int* 1978; 33:14.

336. Vodusek DB, Light JK: The motor nerve supply of the external urethral sphincter muscles: An electrophysiologic study. *Neurourol Urodynamics* 1983; 2:193.

337. Wein AJ: Pharmacology of the bladder and urethra, in Stanton SL, Tanagho EA (eds): *Surgery of Female Incontinence*. Berlin, Springer-Verlag, 1986, pp 229–250.

338. Wein AJ, Benson GS, Jacobowitz DM: Lack of evidence of adrenergic innervation of the external urethral sphincter. *J Urol* 1979; 121:324.

339. Wein AJ, Gregory JG, Cromie WJ, et al: Sympathetic innervation and chemical sympathectomy of canine bladder. *Urology* 1974; 4:27.

340. Wein AJ, Levin RM: Comparison of adrenergic receptor density in the urinary bladder in man, dog, and rabbit. *Surg Forum* 1979; 30:576.

341. Wein AJ, Raezer DM: Physiology of micturition, in Krane RJ, Siroky MB (eds): *Clinical Neurourology*. Boston, Little Brown & Co, 1979, pp 1–33.

342. Westfall DP: Supersensitivity of smooth muscle, in Bulbring E, Brading AF, Jones AM, et al (eds): *Smooth Muscle*. Austin, University of Texas Press, 1981, pp 285–309.

343. Woodburne RT: Anatomy of the bladder, in Boyarsky S (ed): *The Neurogenic Bladder*. Baltimore, Williams & Wilkins, 1967, pp 3–17.

344. Woodburne RT: Anatomy of the bladder and bladder outlet. *J Urol* 1968; 100:474.

345. Woodside JR, McGuire EJ: Urethral hypotonicity after suprasacral spinal cord injury. *J Urol* 1979; 121:783.

346. Yablonsky F, Riffaud JP, Lacolle JY, et al: β-1 and β-2 adrenoceptors in the smooth muscle of male and female rabbit urethra. *Eur J Pharmacol* 1986; 121:1.

347. Zinner NR, Sterling AR, Ritter RC: Role of inner urethral softness in urinary continence. *Urology* 1980; 16:115.

348. Zinner NR, Sterling AM, Ritter RC: Structure and forces of continence, in Raz S (ed): Female Urology. Philadelphia, WB Saunders Co, 1983, pp 33–41.

Chapter 28

Voiding Dysfunction: Diagnosis, Classification, and Management

DAVID M. BARRETT, M.D.
ALAN J. WEIN, M.D.

OVERVIEW

Abnormalities of the micturition cycle are responsible for a large number of patient visits to the urologist. The symptoms and problems that result from these disorders can range from simply annoying and not serious to completely disabling and hazardous. These abnormalities may occur as a result of neurologic injury or disease, inflammatory or infectious disease, bladder outlet obstruction, structural changes in the bladder and urethra or loss of their supporting structures because of surgical or nonsurgical trauma or aging, or strictly psychogenic factors. For the purposes of description and teaching, micturition is best divided into two relatively discrete phases, one of bladder filling and urine storage, and one of bladder emptying. The previous chapter has considered in great detail the anatomic, neuromorphologic, neurophysiologic, neuropharmacologic, and mechanical details involved in both the storage and expulsion of urine by the lower urinary tract. It was emphasized and will be re-emphasized that, despite disagreements by "experts" on various details, one can succinctly summarize these processes from a conceptual point of view. Bladder filling and urine storage require:

1. Accommodation of increasing volumes of urine at a low intravesical pressure and with appropriate sensation.
2. A closed bladder outlet at rest that remains so during increases in intra-abdominal pressure.
3. Absence of involuntary bladder contractions.

Bladder emptying requires:

1. A coordinated contraction by the bladder smooth musculature of adequate magnitude.
2. Concomitant lowering of resistance at the level of the smooth sphincter and of the striated sphincter.
3. Absence of anatomic obstruction.

Any type of voiding dysfunction must by definition result from an abnormality of one or more of the factors involved in the two phases of the micturition cycle. This type of division, with its implied subdivisions under each category into causes related to the bladder and causes related to the outlet, seems to provide a reasonable rationale for consideration of all types of voiding dysfunction as disorders either related primarily to (1) bladder filling or urine storage or (2) bladder emptying. There are, indeed, some types of dysfunction that represent a combination of filling/storage and emptying disorders. However, within this type of scheme, these combined disorders become readily understandable and a clear rationale for treatment—or choices of treatments—is generally evident. In our description of various types of lower urinary tract dysfunction, we will try to emphasize this type of view of the clinical pathophysiology involved, because we believe that this provides the best rationale for the explanation of symptoms, urodynamic findings, and choices for therapy.

This chapter will begin with a very brief consideration of the pathophysiology, symptomatology, and general outline of treatment for disorders that fall primarily under the category of "failure to empty" and of "failure of filling/storage." We will then consider all aspects of the neurourologic evaluation, and the integration of the data so obtained within various classification systems for both neurogenic and nonneurogenic voiding dysfunction. We will then consider the treatment of all types of voiding dysfunction, again using primarily a simple-minded functional division of specific modes of therapy as to whether they primarily facilitate lower urinary tract emptying by increasing intravesical pressure or decreasing outlet resistance, or facilitate bladder filling and urine storage by inhibiting bladder contractility or increasing outlet resistance. Specific voiding dysfunctions associated with neurologic disease will be considered next, as will other types of voiding dysfunction, some of which are relatively clear-cut and some of which defy the usual type of pathophysiologic explanation.

FAILURE TO EMPTY

Absolute or relative failure to empty results from decreased bladder contractility, increased outlet resistance, or both. Absolute or relative failure of adequate bladder contractility may result from temporary or permanent alteration in any one of the neuromuscular mechanisms necessary for initiating and maintaining a normal detrusor contraction. Inhibition of the micturition reflex in a neurologically normal individual may occur via a reflex mechanism secondary to painful stimuli, especially from the pelvic and perineal areas. Nonneurogenic causes include impairment of bladder smooth muscle function, which may result from overdistention, severe infection, or fibrosis. Increased outlet resistance is generally secondary to anatomic obstruction, but may be secondary to a failure of coordination of the smooth or striated sphincter during bladder contraction. Treatment for failure to empty generally consists of attempts to increase intravesical pressure or facilitate the micturition reflex, to decrease outlet resistance, or both.

FAILURE TO STORE

The pathophysiology of failure of the lower urinary tract to fill with or store urine adequately may be secondary to reasons related to the bladder, the outlet, or both. Hyperactivity of the bladder during filling can be expressed as discrete involuntary contractions or as low compliance (Δ volume/Δ pressure) without a discrete and separable contraction. Involuntary contractions are most commonly seen in association with neurological disease or following neurological injury, but may also be associated with inflammatory or irritative processes in the bladder wall, with bladder outlet obstruction, or they may be idiopathic. Decreased compliance during filling may be secondary to the sequelae of neurological injury or disease, but may also result from any process that destroys the viscoelastic properties of the bladder wall. Decreased outlet resistance may result from damage to the innervation or the structural elements of the smooth or striated sphincter, neurologic disease or injury, or surgical or other mechanical trauma; aging may also be responsible for such pathology. Classical stress urinary incontinence in the female seems primarily to be a sphincter dysfunction caused by a failure of the normal transmission of increases in intra-abdominal pressure to the area of the bladder neck and proximal urethra. This is felt to be due mainly to a change in the anatomic position of the vesicourethral junction and proximal urethra during increases in intra-abdominal pressure, and is felt to accompany pelvic floor weakness

or relaxation, which can be due to a number of causes. The treatment of abnormalities related to the filling/storage phase of micturition is directed toward inhibiting bladder contractility or decreasing sensory input during filling, or toward increasing outlet resistance, either continuously or just during abdominal straining.

THE NEUROUROLOGIC EVALUATION

An orderly approach to the workup of the patient with voiding dysfunction includes the modalities as outlined in Table 28–1. As with all clinical problems, the degree of disease complexity, patient cooperation, time and equipment available, and economic resources will modify the progression through these diagnostic stages. In general, one should proceed using the simplest, least-invasive, least-expensive methods first, as these will often yield adequate information to support a reasonable therapeutic trial. When unsatisfactory results are achieved or a definite answer is elusive, then the more complex tests are introduced.

HISTORY

Assessment of the clinical history of individuals with voiding dysfunction should be designed to exclude previous neurologic disease, trauma, and surgery. A list of medications should be compiled, as these are frequently the sole cause for a change in voiding habits. Symptoms related to other organ systems that are compositely innervated by the somatic and autonomic nerves in the pelvis are to be included; this would include bowel and sexual function. Fecal incontinence, constipation, lower abdominal distention, and cramping may suggest neuropathy affecting intestinal or rectal sphincteric function. Changes in sexual function such as the frequency, duration, and rigidity of erections or a decrease in the volume of semen and force of ejaculation similarly would imply altered innervation of the pelvic organs.

The functional classification of voiding dysfunction, i.e., failure to empty and failure to store, can provide a basis for collecting historical data from the patient. A "wet" patient with an empty bladder is most character-

TABLE 28–1.

Neurourologic Workup

History
Physical examination
Neurologic examination
Urine bacteriologic studies
Renal function studies
Intravenous urogram
Voiding cystourethrogram
Endoscopic examination
Videourodynamic studies

istic of a storage failure abnormality, while a "dry" patient with a bladder that cannot be emptied is typical of evacuation failure. Primary symptoms such as these are a useful guide to an ultimate diagnosis. However, the development of secondary symptoms or the combination of symptoms confuses the picture and makes the implementation of therapeutic measures based on historical factors alone fraught with error. Table 28–2 attempts to categorize the common presenting symptoms of a patient with the voiding dysfunction.

Incontinence (failure to store) may result from ureteral ectopy, congenital or acquired fistulae, or neuromuscular defects of the detrusor or urethral sphincter. Patients with wetness and ureteral ectopy or congenital fistulae are almost all female and generally present as a consequence of failed toilet training. Not infrequently, these defects are associated with other abnormalities of the ureteral bud and/or urogenital sinus, and further examination should be undertaken. Deficiencies in the smooth or striated muscle or fibroelastic tissue that make up the bladder neck and proximal urethra and the surrounding structures can give rise to varying degrees of urinary incontinence. The characteristics of incontinence may be further divided into the following categories: stress, dripping, precipitous, or total. Varying in severity, leakage may occur only with associated increases in intra-abdominal pressure as occurs with stress urinary incontinence. A scarred, fibrotic, or noncompliant urethra and bladder neck, which may be seen after trauma, multiple surgical procedures, or radiation therapy, may also be associated with incontinence; this, however, generally results in a constant dripping type of leakage. Incontinence may also occur as a result of involuntary detrusor contractions or detrusor hyperreflexia. This type of incontinence is generally associated with some type of neurologic disease and is characterized by precipitous voiding with a more sustained type of leakage than is seen with pure sphincteric insufficiency. Total incontinence implies that little

if any voiding occurs, and that urine essentially passes through the bladder and out, the continence mechanism being completely incompetent.

Incontinence is not always a symptom of primary failure to store urine adequately. Overflow or paradoxic incontinence may develop in a patient with insidious detrusor decompensation who has a failure-to-empty defect. The leakage in this case is generally associated with sudden increases in intra-abdominal pressure. Detrusor hyperreflexia can be seen as a result of outlet obstruction and failure to empty. The exact changes that result in increased detrusor contractility are unknown, but incontinence is the resultant symptom.

Urgency is defined as the extreme desire to void, either because of pain or because of a fear of leaking. The urgency that is associated only with pain is generally secondary to inflammatory disease of the lower urinary tract. Urodynamic studies may be normal except for hypersensitivity. The urgency that is associated with a fear of leaking, or a history of doing so, is usually associated with detrusor instability or hyperreflexia. Thus, urgency may either be a symptom of a primary failure to fill or store normally, or a manifestation of the detrusor hyperreflexia that develops secondary to bladder outlet obstruction. When using the term "urgency," it is important to at least try to distinguish between the impending sensation of micturition and the simple fear of leaking urine. The former is usually associated with detrusor hyperreflexia, while the latter may be associated with leakage from simple sphincteric incontinence.

An increase in daytime urinary frequency may simply be psychogenic in origin. However, such an increase may represent a genuine need to void, and in this case it may result from either pain on low-volume bladder distention (usually indicative of inflammatory disease of the lower urinary tract) or primary detrusor hyperreflexia. Increased frequency of urination may result secondarily from a failure to empty adequately, either because of a decreased functional bladder capacity (due to a substantial residual urine volume) or in association with outlet obstruction–induced hyperreflexia. Nocturia usually accompanies nonpsychogenic urinary frequency and may be associated, on the same basis, with either a failure of adequate urine storage or a failure of adequate urine emptying. Patients with true nocturnal enuresis often have detrusor hyperreflexia; therefore, this symptom may be seen in patients who exhibit a primary failure of storage or in patients who have developed hyperreflexia because of a failure to empty secondary to outlet obstruction.

The symptom of "pressure" defies exact definition. It is not quite the urge to void; rather, in many persons there is a feeling that the bladder is full or that the urge to void will occur shortly. Often no discernible voiding

TABLE 28–2.

Symptoms of Voiding Dysfunction

	FAILURE TO FILL OR STORE	FAILURE TO EMPTY
Incontinence	P*	S*
Urgency	P	S
Frequency	P	S
Nocturia	P	S
Enuresis	P	S
Pressure	P	P
Dysuria	?	?
Hesitancy	. . .	P
Straining to void	. . .	P

*P indicates primary symptom; S, secondary symptom (may develop consequent to the primary abnormality).

dysfunction is discovered in patients who complain of this symptom. However, it can be due to an intravesical pressure during filling that is pathologically elevated but below the level necessary to elicit the sensation of distention or urgency. This symptom also may be representative of the patient's accurate perception of the fact that the bladder does not in fact empty completely, and that a modest or large residual urine volume is present.

Hesitancy and straining to void, when associated with true voiding dysfunction, generally reflect a failure to empty adequately.

Finally, in considering the symptoms of voiding dysfunction, it must be remembered that combination abnormalities can occur. Outlet obstruction may exist in combination with detrusor hyperreflexia, and this combination may give rise to hesitancy, straining to void, frequency, nocturia, and urgency with urge incontinence. Likewise, elements of both sphincteric and detrusor incontinence can exist in the woman who complains of stress incontinence that is accompanied by frequency, nocturia, and—especially—urgency incontinence. An orderly review of these symptoms with the patient, noting the temporal aspects of each, will enhance the clinician's ability to tailor subsequent tests that will be needed to confirm any presumptive diagnoses.

The preceding section clearly describes the value of a symptomatic history in suggesting a voiding dysfunction that represents an abnormality of storage, of emptying, or of both. *Physical examination* may provide visual confirmation of the presence and degree of urinary incontinence in the form of cutaneous excoriation of the genitalia. A distended bladder suggests a failure to empty. Cutaneous scars may provide evidence of previous neurosurgical, orthopedic, or pelvic surgical procedures that may be related to the problem under evaluation, either because they were performed to correct a similar problem or because the problem under evaluation developed subsequent to a surgical procedure.

A careful neurologic examination is especially important as it may provide evidence of a neurologic lesion or disease that is presenting initially with symptoms of voiding dysfunction. Review should include inspection of the skull and vertebral column, assessment of mentation, sensory examination, motor function, and coordination evaluation. Concentrating on the pelvis, evaluation of rectal sphincter tone and the bulbocavernous reflex is essential to help establish the integrity of sacral reflex arcs. Acknowledgement of the presence of one or another type of neurologic disease may considerably alter the philosophy of treatment for a given patient.

Abnormal *renal function studies* may reflect an unsuspected failure of emptying or may reflect intrinsic renal or prerenal disease, thereby changing some of the considerations for treatment. *Urine bacteriologic studies* are needed as urinary tract infection can cause or worsen all of the symptoms of a failure to store and may coexist with other causes of storage failure. Persistent or recurrent infection may be a reflection of inadequate emptying with persistent residual urine.

Many urologists feel that *intravenous urography* is not necessary in patients with lower tract dysfunction. They point out that a significant abnormality is detected only in a small percentage of the patients studied. There is no argument, however, that upper tract deterioration can suggest emptying failure that has occurred on the basis of either bladder distention, pathologically elevated voiding pressures, or both. In addition, urography is useful in assessing the upper urinary tracts prior to the institution of any treatments designed to alter voiding function. Silent deterioration of the upper tracts can occur after surgical or pharmacologic therapy designed to decrease bladder contractility or increase outlet resistance, and it is therefore useful to have some sort of a baseline for later comparison.

Voiding cystourethrography as an isolated study may provide direct confirmation of leakage with stress and important details of bladder and urethral configuration and position. An alert uroradiologist can detect, in some cases, leakage secondary to an involuntary detrusor contraction without the aid of differential bladder and abdominal pressures. An abnormality of emptying might be suggested by secondary signs of outlet obstruction, such as trabeculation, cellules, diverticula, or vesicoureteral reflux. If obstruction has been documented, a VCUG carefully done during bladder contraction can generally pinpoint the site. *Endoscopic evaluation* may be helpful in detecting or suggesting previously unsuspected inflammatory or neoplastic disease or outlet obstruction.

URODYNAMICS*

Urodynamics is a neurourologic diagnostic tool concerned with the identification and measurement of physiologic and pathologic factors involved in the storage, transportation, and evacuation of urine. The purpose of urodynamic testing is to identify and quantitate the etiologic factors that contribute to voiding dysfunction, whether it is a problem of storage or emptying. Since the bladder is a notoriously poor witness and will respond to a variety of pathologies with the same symptoms, the need for urodynamic testing becomes important before directing therapy. A number of studies have

*Assistance in preparing this section was provided by Dr. Benad Goldwasser, Special Fellow in Neurourology and Prosthetics, Mayo Clinic, 1985–1986.

documented the poor correlation between the patient's symptoms and the findings of urodynamic testing. This is true both for patients without (Farrar et al., 1975; Powell et al., 1981) and those with (Blaivas et al., 1979; Blaivas, 1980) neurourologic disorders. In a study done in patients with multiple sclerosis, it was found that when treatment was instituted based on symptoms and signs, it was effective in 27% of patients. When treatment was based on urodynamic testing, it was effective in 83% of patients (Blaivas et al., 1979; Blaivas, 1980).

Urodynamic tests are not without their limitations and potential for misinterpretation. Ideally, the clinical symptom should be reproduced during the urodynamic testing sequence. If the symptom is not reproduced during routine testing, then the study must be repeated in as close an approximation as possible to the situation in which the patient's symptoms actually occur, be it rising, standing, coughing, laughing, or jogging. Patients should be asked to evaluate the similarity of their usual voiding to the voiding sequence evaluated urodynamically. This helps establish the credibility of the urodynamic data. On the other hand, there may be certain symptoms produced during the testing sequence that have urodynamic correlates, but that do not constitute a part of the patient's original complaints. These findings may represent artifact or be early subclinical abnormalities not generally manifested during normal voiding, but brought to light by the challenge of urodynamic testing.

Interpreting sensory data during urodynamic testing is subjective at best. This is unfortunate as many patients with voiding dysfunction have only sensation abnormalities and have no demonstrable urodynamic abnormalities. It is relatively easy to assess whether the sensation of bladder filling is normal, decreased, or absent. However, there are a number of sensations that may occur during filling that are difficult to classify. The most troublesome of these is the symptom of urgency or the dire need to void. Although this can correlate with involuntary bladder contractions or detrusor hyperreflexia, it is not uncommon to reproduce this symptom during urodynamic testing and not be able to demonstrate a urodynamic abnormality. The explanation for this rests within the limitations of the test to be sensitive enough to record the subtle changes that occur in the bladder or outlet or the inhibitory factors that come into play in the laboratory setting.

Operator expertise is essential for successful completion of most studies and physician participation is recommended in the more complex tests.

There are four basic urodynamic modalities: (1) cystometry, (2) uroflowmetry (with residual volume determination), (3) urethral pressure profilometry, and (4) combined studies (with or without fluoroscopy). The exact sequence in which these tests are administered to the patient is entirely dependent on the presenting symptoms and the presumptive diagnosis based on other neurourologic tests. Table 28–3 lists the urodynamic modalities as they relate to evaluation of the bladder or the bladder outlet, and, in general, the simplest, least invasive test should be used first.

General urologic practices will derive most benefit from simple uroflowmetry and cystometry. Sphincter electromyography is important should neurogenic bladder dysfunction make up a significant proportion of the study population. In urogynecologic practices, cystometry, uroflowmetry, and urethral pressure studies will be of most value. Complex cases that cannot be evaluated adequately in this manner are best referred to a urodynamic center capable of performing multifunction studies—with the option of combined videocystourethrography—which is staffed by a urodynamic technician with knowledge and expertise in this specialty.

It is important to keep in mind constantly that urodynamic testing is but one part of the overall evaluation. It is imperative that all portions of the neurourologic evaluation and the conclusions drawn from them be carefully reviewed and, when discrepancy exists between the symptoms and the urodynamic findings, these studies should be questioned and repeated, especially before irreversible therapy is undertaken.

CYSTOMETRY

Cystometry is the method by which changes in bladder pressure are measured with progressive increase in bladder volume. The test is basically designed to evaluate the filling/storage phase of detrusor function. The presence or absence of a detrusor contraction, although an important observation, is not necessary to glean important information from this test.

TABLE 28–3.
Urodynamics Made Easy

	BLADDER	OUTLET
Filling/storage	$P_{det(FCMG)}$[1]	P_{ureth}[2]
		FLUORO[3]
Emptying	$P_{det(VCMG)}$[4]	P_{ureth}[2]
		FLUORO[3]
		EMG[5]
	FLOW[6]	
	RU[7]	

1. Filling cystometrogram (recording of detrusor pressure during filling).
2. Urethral pressure(s).
3. Fluoroscopy.
4. Voiding cystometry (recording of detrusor pressure during voluntary or involuntary emptying).
5. EMG of periurethral striated muscle.
6. Flowmetry.
7. Residual urine determination.

The normal adult bladder capacity averages 400–750 ml. Within this capacity, bladder pressure should not rise above 15 cm H_2O, the mean rise in normal bladders being 6 cm H_2O. Schematically, the normal cystometrogram may be divided into four phases (Fig 28–1). There is an initial rise in pressure to achieve resting bladder pressure. The second phase—the tonus limb—reflects the vesicoelastic properties of the smooth muscle, collagen, and mucopolysaccharides of the bladder wall. During this filling/storage phase, bladder pressure should increase by very little as the normal bladder is able to accommodate to increasing urine volumes. Rapid filling rates may generate a steeper tonus limb. Bladder wall fibrosis of various etiologies (infection, radiation, etc.) and detrusor hypertrophy may also cause compromised accommodation with a steeper tonus limb. At peak capacity, the detrusor muscle and other elastic tissue have stretched to their limit and any additional increase in volume will be accompanied by a rise in intravesical pressure. During this third stage, the patient is still able to suppress voluntary voiding contractions. Ruch and Tang (1969) have demonstrated that the characteristic shape of the cystometry curve is independent of neural control and is an inherent property of the structural elements of the bladder wall. The fourth phase of the normal cystometrogram is the generation of a voluntary voiding reflex that is dependent on intact neural pathways to the micturition center located in the brain stem. The normal patient should be able to suppress voiding even at capacity. It is not uncommon during a study to find that the patient is unable to generate a micturition reflex at command. This is more common in females and is related to psychological inhibition resulting from the unnatural circumstances of the study. Performance of this phase of the cystometrogram with the male patient in an erect posture and the female on a commode may facilitate generating a micturition reflex. Alternatively, voiding may be stimulated by administering 2.5 mg of bethanechol subcutaneously. If voiding does not occur in response to bethanechol administration, the bladder should be emptied and filled again while measuring bladder pressures to test for evidence of denervation supersensitivity (see below).

Both water and gas (carbon dioxide) are used for cystometry. Controversy exists as to the relative merits of each. From a practical point of view, there seems not to be a statistically significant difference between the two. The advantage of gas (carbon dioxide) is that it is clean and efficient, allowing rapid rate of bladder filling, and thereby takes little time to perform. On the other hand, when the bladder is filled at a rapid rate, accommodation is impaired, with resultant low values of total capacity. Carbon dioxide is irritating to the urothelium and some patients complain of discomfort and dysuria. Because gas is compressible, phasic bladder contractions of small amplitude may escape recording and significant high-pressured contractions may appear as of low amplitude. Gas may also leak unobtrusively at instrument connections and also from the bladder around the catheter, particularly if outflow resistance is low.

The cystometric variables that may be observed during the study are those of compliance, contractility, sensation, and capacity. Although the stable bladder should remain so even at an unphysiologic rate of filling of 100 ml/min, certain patients require a slower rate of bladder filling. These include patients with a known neurological condition suspected of having a hyperreflexic bladder, those with bladders with decreased compliance, and children. In these situations the bladder should be filled at a rate of 25–50 ml/min. During filling, a record is made of the bladder volumes at which first sensation of filling, sensation of urgency to void, and the sensation of maximal capacity occur. During the filling phase, provocative measures such as coughing and the Valsalva maneuver should be utilized to observe for increased detrusor contractility. The pressure measured within the bladder is composed of pressure induced by the detrusor itself and also by intra-abdominal pressure. Therefore, pressure increments recorded on a simple cystometrogram may at least partially reflect intra-abdominal pressure rises. To eliminate such artifactual interferences, it may be necessary to measure simultaneously intra-abdominal pressure by means of a rectal catheter. Cystometers are available that will electronically subtract the rectal pressure from the total bladder pressure, thus giving the subtracted bladder pressure or detrusor pressure. This measurement is important both for provocative cystometry and for voiding studies to determine the efficiency of the voiding contraction.

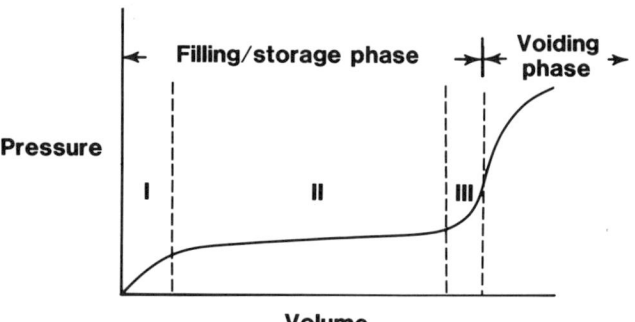

FIG 28–1.
Characteristics of the normal cystometrogram.

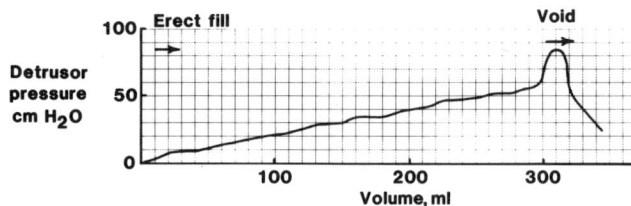

FIG 28–2.
Cystometrogram depicting decreased detrusor compliance.

Abnormal Cystometric Patterns

Abnormalities of bladder function that may be detected by cystometry include decreased detrusor compliance, increased detrusor contractility or detrusor hyperreflexia, involuntary detrusor contraction, increased bladder compliance, and decreased detrusor contractility or detrusor areflexia.

Bladder compliance refers to the change in detrusor pressure that accompanies an increase in bladder volume during filling. Normally, this should not exceed 15 cm H_2O. A bladder with decreased compliance is one in which the pressure rises steeply with filling (Fig 28–2). Reduced compliance may result from detrusor hypertrophy, fibrotic changes in the bladder wall, bladder wall inflammation, and possibly neurologic lesions.

Involuntary detrusor contraction refers to a phasic rise in bladder pressure. This may occur in response to provocation such as a cough, stress, or posture change, or may occur spontaneously (Fig 28–3). States of increased detrusor contractility have been referred to as "detrusor instability" or "detrusor hyperreflexia." In general, the term detrusor hyperreflexia refers to detrusor function that is a direct result of associated neurologic disease, while detrusor instability is seen in the absence of neurologic disease (Hald, 1984). Detrusor hyperreflexia commonly occurs due to suprapontine cerebral disorders, such as cerebrovascular accidents or parkinsonism. This may also occur in patients with suprasacral spinal cord disease processes, such as multiple sclerosis or trauma with or without concomitant detrusor-sphincter dyssynergia.

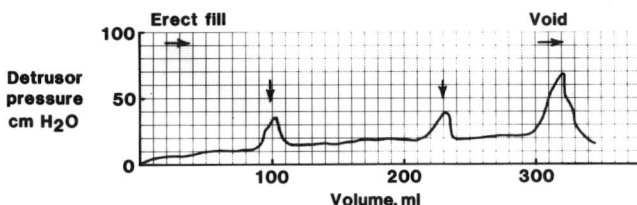

FIG 28–3.
Cystometrogram depicting increased detrusor contractility.

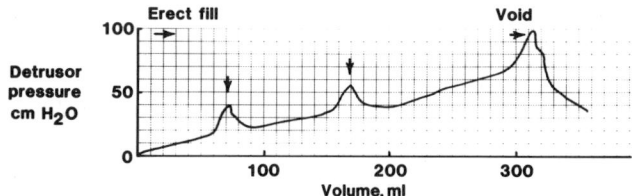

FIG 28–4.
Cystometrogram showing decreased compliance and increased detrusor contractility.

With marked detrusor instability (or hyperreflexia), the compliance may also be reduced, probably secondary to detrusor muscle hypertrophy (Fig 28–4). At the most severe end of the spectrum of increased detrusor contractility is the bladder with reduced capacity, in which detrusor hyperreflexia occurs at a low volume (Fig 28–5). Steepness of the curve is due to muscle contraction as well as to reduced compliance. This may be due to a neurogenic etiology, though it may be seen in patients with severe outlet obstruction. Since both decreased compliance and detrusor hyperreflexia result in a precipitous bladder pressure elevation, differentiation of the two can be difficult.

A large-capacity bladder of normal or increased compliance (Fig 28–6) may result from decreased sensation as commonly seen in the diabetic patient or from chronic outlet obstruction. It may also be a behavioral phenomenon in patients who learned to voluntarily inhibit voiding for long periods. Weir and Jaques (1974) found that 30% of patients with bladder capacities in excess of 800 ml were urodynamically normal, so that increased bladder capacity in itself is not necessarily an indication of pathology, especially in patients who can generate a normal detrusor contraction and void to completeness.

Decreased bladder capacity may be purely sensory in origin (normal compliance and stability) and is com-

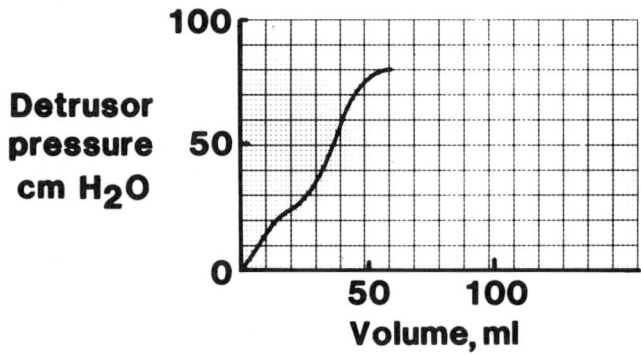

FIG 28–5.
Cystometrogram showing detrusor hyperreflexia.

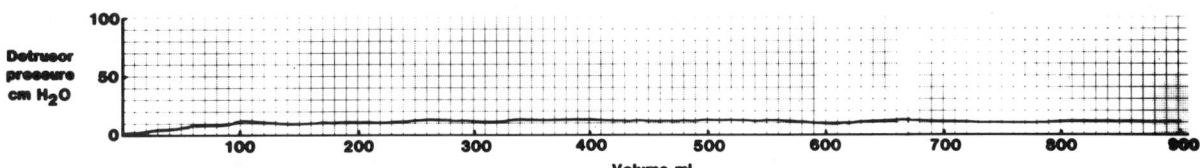

FIG 28–6.
Cystometrogram demonstrating increased detrusor compliance.

monly seen in females with the idiopathic frequency syndrome and in those with an inflamed bladder. These patients are usually able to produce a voluntary detrusor contraction (Fig 28–7).

BETHANECHOL SUPERSENSITIVITY TEST

Bethanechol chloride is an acetylcholine-like parasympathomimetic agent that exhibits a relatively selective action on the urinary bladder and gut with little or no action at therapeutic dosages on ganglia or on the cardiovascular system (Koelle, 1975; Ursillo, 1967). It is cholinesterase-resistant and causes a contraction in vitro of smooth muscle from all areas of the bladder (Raezer et al., 1973). It has minimal effect on the normal bladder, decreasing the capacity slightly, increasing detrusor tone, and increasing the maximum voluntary micturition pressure. It will not cause the normal bladder to become unstable.

The bethanechol supersensitivity test is based on Cannon's law of denervation, which states that when an organ is deprived of its nerve supply it will develop hypersensitivity to its own neurotransmitter substance. Lapides reported this to be quite accurate in patients with denervated bladder (Lapides et al., 1962). However, others (Blaivas et al., 1980) have found significant rates of false negative and false positive responses.

The test requires the subcutaneous injection of 2.5 mg of bethanechol chloride followed in 15–30 minutes by repeated cystometric examination. In patients with areflexic bladders of neurogenic etiology, a rise in pressure of at least 15 cm H_2O at 100 ml filling, in excess

of the pretreated cystometric pressure, will result (Fig 28–8). This does not occur in normal patients and in those with myogenically decompensated bladders. A positive test should not imply a potential benefit from the therapeutic use of oral bethanechol chloride (Barrett, 1981).

The use of bethanechol chloride is contraindicated in patients with bronchial asthma, peptic ulcer, hyperthyroidism, enteritis, bowel obstruction, bladder outlet obstruction, cardiac disease, or a history of recent gastrointestinal surgery (Wein, 1979, 1980).

PROPANTHELINE (PRO-BANTHINE) STIMULATION TEST

Propantheline blocks acetylcholine at muscarinic receptor sites. The test, which is of limited clinical value, is based on the ability of propantheline bromide (as well as of other antimuscarinic agents, such as glycopyrrolate, isopropamide, hyoscyamine, and anisotropine methylbromide) to abolish or cause a decrease in amplitude and frequency of involuntary bladder contractions and to increase the bladder capacity in patients with hyperreflexia of neurologic etiology (Fig 28–9).

The production of propantheline bromide has been discontinued and glycopyrrolate, 0.1 mg, is currently being used for this test. It is of particular value in the evaluation of the unstable bladder in the man with suspected bladder outlet obstruction where it is important to know whether the instability is of neurologic etiology or secondary to obstruction. This of course has an implication on the therapy that will be recommended to the patient. Whereas in the patient with obstructive unstable bladder, the instability may resolve following relief of obstruction, the patient with a neurologic etiology will most likely become incontinent if surgical treatment is instituted to reduce outlet obstruction.

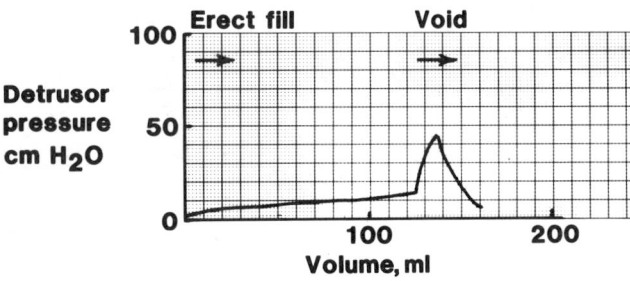

FIG 28–7.
Cystometrogram depicting normal compliance and contractility but small capacity.

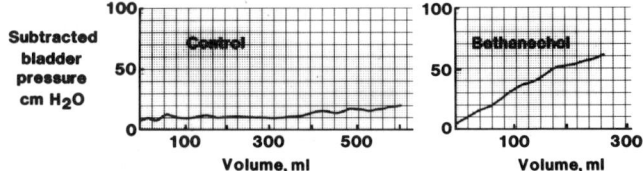

FIG 28–8.
Cystometrogram with positive bethanechol stimulation test.

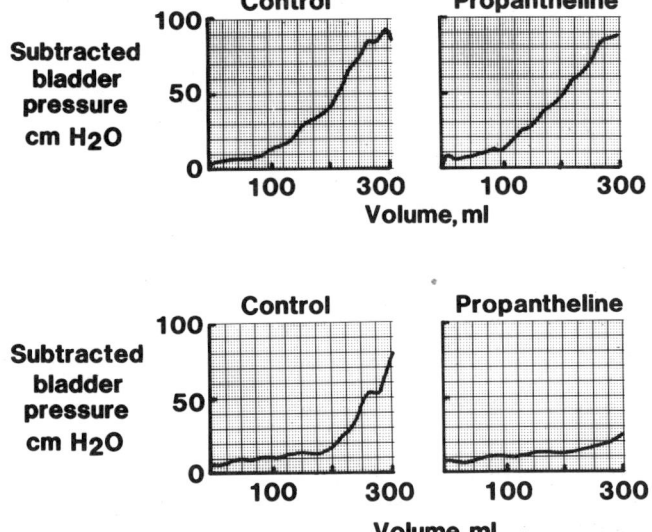

FIG 28–9.
Cystometrogram before and after administration of 15 mg parenteral propantheline. A positive response is indicated by flattening of tonus limb *(bottom panels)*. No change indicates positive study *(top panels)*.

Contraindications to glycopyrrolate include patients with glaucoma and those with obstructive gastrointestinal disease.

UROFLOWMETRY

Flow rate is defined as the volume of fluid expelled from the urethra per unit time and is expressed in milliliters per second. Modern instruments allow us to record not only the overall rate but also a flow pattern (Fig 28–10). The urine flow rate is an expression of the combined activity of the detrusor and urethra. A normal flow rate will usually indicate a good function of both.

An obstruction in the urethra (benign prostatic hyperplasia, urethral stricture, etc.) may be overcome by a more forceful detrusor contraction, which would bring higher bladder pressures during the stage of micturi-

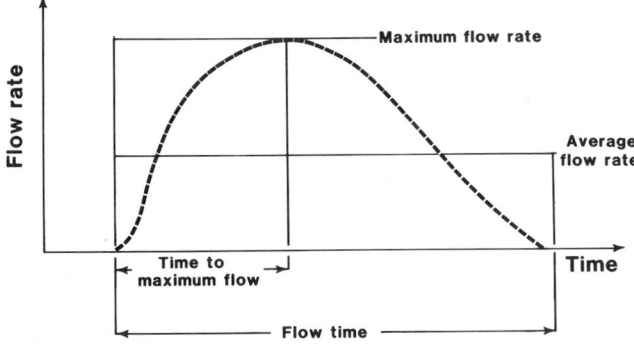

FIG 28–10.
Characteristics of normal uroflow.

tion. This may result in a normal peak flow rate—at least during the early stages of the obstruction—but a reduced average flow rate. Therefore, for a full definition of lower tract function, simultaneous pressure (cystometry) and flow studies during voiding may be indicated. However, a urine flow study even by itself has value as a screening test for lower urinary tract dysfunction, for preoperative and postoperative assessment of lower urinary tract surgery, and to study the effects of pharmacologic agents on urethral resistance and voiding efficiency.

Urine Flow Rate Parameters

To the clinician, the urine flow rate parameters (Table 28–4) of most importance are the maximum flow rate, voided volume, and flow pattern. Our measurement of urine flow rate is dependent mostly on the patient's comfort with the environment in which the test is conducted and lack of feelings of intimidation in the urodynamics laboratory. Additionally, the patient must have an adequately full bladder, as urine flow rate is dependent on voided volume. Ideally, urine flow interpretation should be for volumes of at least 150 ml. Interpretation of flow rates with smaller voided volumes should be done with caution. Volumes of over 500 ml are also accompanied by reduced flow rates, perhaps due to overstretching of the detrusor fibers. However, flow rates are also related to the sex and age of the patient. Most data in the literature relate to measurements of flow in men younger than 55 years and cite norms of 15 and 25 ml/second for mean and maximum rates. Abrams (1984) states correctly that normality should be defined in terms of age, and that the use of nomograms relating to young men when assessing the older male patient is a doubtful practice. Females have significantly higher flow rates than males matched for age and voided volumes. The values presented in Table 28–5 are quoted from Abrams (1984) and represent the minimal flow rate for a given sex, age, and voided volume.

The normal flow pattern exhibits a rapid rise to maximum flow rate, attaining this level within one third of the ultimate voiding time, and in addition at least 45%

TABLE 28–4.

Urine Flow Rate Parameters
_____

1. *Flow time.*—Time over which measurable flow occurs
2. *Time to maximum flow.*—Time elapsed from the onset of flow to maximum rate
3. *Maximum flow rate.*—The maximal rate of flow
4. *Voided volume.*—The total volume expelled by way of the urethra
5. *Average flow rate.*—The voided volume divided by flow time
6. *Voiding time.*—The total duration of micturition, including interruptions
7. *Flow pattern.*—May be continuous, interrupted, or specifically described
_____

TABLE 28–5.

Minimum Acceptable Urine Flow Rates*

| AGE, YR | MINIMUM VOIDED VOLUME, ML | FLOW RATES, ML/SEC | |
		MALES	FEMALES
4–7	100	10	10
8–13	100	12	15
14–45	200	21	18
46–55	200	22	15
56–80	200	9	10

*From Abrams P: The urethral pressure profile measurement, in Mundy AR, et al (eds): *Urodynamics: Principles, Practice, and Application.* Edinburgh, Churchill-Livingstone, 1984.

of the total volume voided should be evacuated within this same period and before maximum flow is achieved. After achieving maximum rate, flow decreases more slowly; hence, a true bell-shaped curve is not achieved. In normal patients, average flow rate should be approximately 50% of the maximum urine flow rate.

If abdominal straining is used to augment voiding, the stream is interrupted (Fig 28–11). Other causes of an interrupted stream may be voluntary sphincter contraction during voiding by the patient with detrusor sphincter dyssynergia or, more commonly, by an anxious patient, and artifactual recording, which most commonly occurs when a male patient directs the urinary stream across the collecting funnel (Fig 28–12).

In outlet obstruction, a flat, elongated curve with a low maximum flow rate is seen characteristically, which is reached in the initial part of the trace (Fig 28–13). In the patient with detrusor underactivity, the flow pattern will be intermittent, as abdominal muscles are used to augment voiding. The urine flow rate is reduced and the maximum flow rate is seen in the middle part of the tracing.

Considerable care must be taken not to overinterpret flow study results, and considerable clinical judgment is necessary in their interpretation. Abnormal findings should be confirmed by the use of more elaborate urodynamic studies.

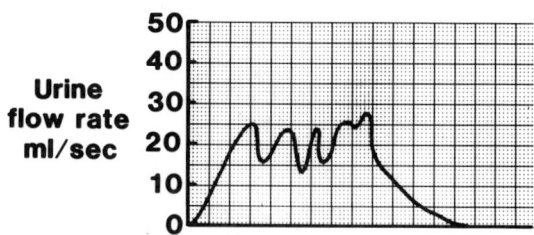

FIG 28–12.
Uroflow with interrupted pattern due to intermittent sphincter activity.

RESIDUAL URINE VOLUME

Residual urine volume is another parameter that integrates the activity of the bladder and outlet during the emptying phase of micturition. A consistently increased residual urine volume generally indicates increased outlet resistance, decreased bladder contractility, or both. Absent residual urine is compatible with normal lower urinary tract function during emptying but may also exist in the presence of significant disorders of filling/storage (incontinence) and disorders of emptying in which the intravesical pressure is sufficient to overcome increases in outlet resistance up to a certain point. Typical of this is the patient with outlet obstruction due to benign prostatic obstruction. Initially, despite significant obstruction, the detrusor is capable of emptying the bladder by contracting with a greater force, leading to hypertrophy and increased intravesical pressure. However, with time, the detrusor will fail, leading to increased residual volumes and decrease in intravesical pressures produced during voiding. This is an important example of the fact that it is extremely rare to be able to make an adequate diagnosis on the basis of any single urodynamic study. All of the study results must fit together and these must ultimately be compatible with the symptoms and with the results of the rest of the neurourologic evaluation.

ELECTROMYOGRAPHY

Electromyography (EMG) is the study of the bioelectric potentials generated by depolarization of skeletal

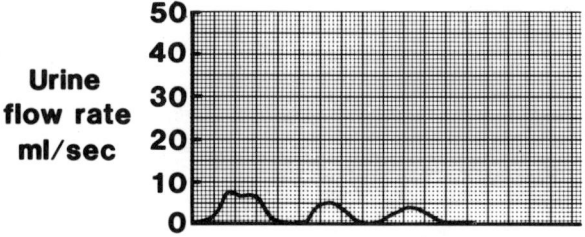

FIG 28–11.
Uroflow with interrupted stream during straining to void.

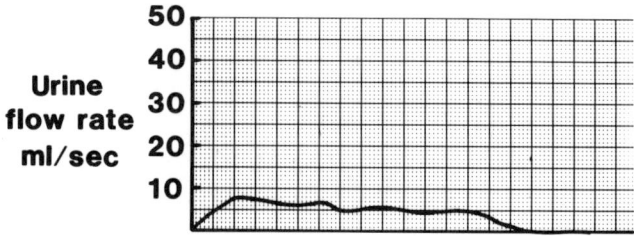

FIG 28–13.
Uroflow with abnormal flow rate characteristic of detrusor outlet obstruction.

muscle. Skeletal muscle is innervated by neurons whose cell bodies lie in the anterior horn of the spinal cord. The anterior horn cell in the grey matter of the spinal cord, its axon, and all of the muscle fibers (their number may vary) that it innervates is called a "motor unit." An excitatory impulse from an anterior horn cell causes contraction of all the muscle fibers in that motor unit, which is brought about by membrane depolarization. The electrical discharge produced on contraction of the muscle fibers of the motor unit by their depolarization is termed "motor unit action potential." This may be detected by electrodes and displayed on an oscilloscopic screen or strip chart, or it may be converted to an audible sound. Individually recorded on an oscilloscope, the motor unit action potentials may take a variety of configurations, being biphasic, triphasic, or, rarely, polyphasic. In the relaxed state, the normal striated muscle is almost electrically quiescent, and only infrequent action potentials are recorded. However, with progressive muscle contraction, increasing numbers of motor units are recruited and each motor unit fires at a more rapid rate. These firings can be individually recorded electromyographically, and the configuration of the action potentials is of significance in diagnosis. At the point of maximal contraction, motor unit action potentials are so frequent that total overlap occurs and EMG separation cannot be achieved, resulting in an interference pattern. It takes considerable EMG experience to interpret the various parameters recorded on an oscilloscope during sphincter EMG. Identification of these variations is important in making an accurate neurologic diagnosis. Individual motor unit action potentials may be detected by needle electrodes placed directly into or near the muscle to be studied. When surface electrodes are used, individual motor unit action potentials are not visualized; rather, an overall gross recording of the activity of the muscle is detected (Barrett, 1980). If one is not interested in motor unit action potentials, surface electrodes are adequate. They will detect whether the pelvic floor muscles are contracting or relaxing at any given instant. However, being unable to detect individual motor unit action potential, surface electrodes cannot help in assessing the integrity of these muscles and their nerve supply.

During cystometric bladder filling there should be incremental increase in EMG activity as more motor units are recruited. This activity will reach a maximum when peak bladder capacity is reached and at the command to void there should be sudden cessation of sphincter activity, which should persist throughout voiding (Blaivas et al., 1977). On completion of bladder emptying, resumption of baseline sphincter activity occurs. In assessing external sphincter activity, the patient is asked to interrupt voiding in the middle of the stream, at which point there should be an abrupt increase in sphincter activity that should be sufficient to stop the flow. Resumption of voiding should occur thereafter; however, if the holding pattern is maintained, the detrusor reflex should ideally be lost in approximately 10 seconds (Webster, 1982).

Abnormal EMG patterns may be detected in a number of situations. Detrusor sphincter dyssynergia describes sphincter activity that is inappropriate to the activity of the detrusor. Three varieties of such discoordination (Fig 28–14) were described (McGuire, 1979). One pattern involves an appropriate increase in EMG activity with bladder filling, which is followed by inappropriate involuntary increase in activity at the onset of detrusor contraction. Thus, the detrusor contracts against a closed sphincter.

A second type of discoordination involves failure to develop a proper reflex detrusor contraction because of increased EMG activity during voiding, which causes inhibition of the detrusor motor nucleus in the sacral spinal cord with resultant inhibition of detrusor contraction. This type of discoordination may be seen in patients with suprasacral spinal cord injury.

The third type involves contraction and relaxation of the sphincter during bladder filling. This amounts to periods of uninhibited sphincter relaxation, which is associated with reflex detrusor contraction leading to urgency and urge incontinence.

Simultaneous EMG activity with an increase in intravesical pressure is not always sphincter dyssynergia. Detrusor striated sphincter dyssynergia is undoubtedly the most overdiagnosed entity in the field of voiding dysfunction. Patients suspected of having this diagnosis should always be further investigated with urodynamic and/or radiologic evaluation to study the activity of the bladder and the outlet during the emptying phase of micturition. It must be remembered that true detrusor

DETRUSOR EXTERNAL SPHINCTER DYSSYNERGIA

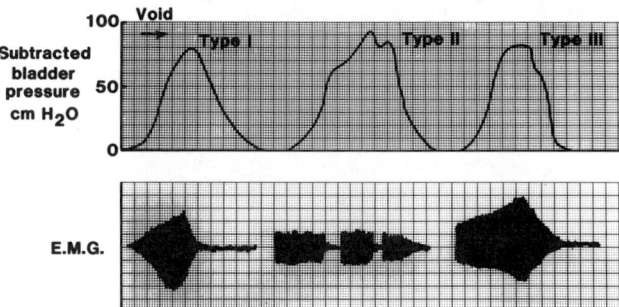

FIG 28–14.
Varieties of external sphincter discoordination. (Adapted from McGuire EJ: Electromyographic evaluation of sphincter function and dysfunction. *Urol Clin North Am* 1979; 6:121.)

striated sphincter dyssynergia is extremely uncommon (or does not exist) in patients without neurologic disease. Such a diagnosis in such a patient deserves exhaustive study before it is in fact confirmed.

Sphincter electromyographic studies have gained popularity for the investigation of children with voiding dysfunction (Maizels and Firlit, 1979). It is suggested that learned dysfunctional sphincter habits may be responsible for voiding abnormalities, with resultant recurrent urinary infections, reflux, and upper tract deterioration. An attempt has been made to retrain these patients by the use of biofeedback techniques utilizing sphincter electromyography during voiding (Maizels et al., 1979). The patient can appreciate from an EMG audio monitor whether or not sphincter relaxation is being accomplished; hence, positive reinforcement may correct the problem.

EVOKED RESPONSES

Evoked responses are potential changes in neural tissue, recorded using averaging techniques, resulting from distant stimulation, usually electrical (*Sixth Report on the Standardization of Terminology of Lower Urinary Tract Function*, 1985). Evoked responses may be used to test the integrity of peripheral, spinal, and central nervous pathways. As with nerve conduction studies, the conduction time (latency) may be measured. In addition, information may be gained from the amplitude and configuration of these responses. An example of such procedures is the sacral evoked response, which is measured by the latency of the bulbocavernosus reflex (Nordling and Meyhoff, 1979; Blaivas et al., 1981; Blaivas, 1984). The bulbocavernosus reflex arc is mediated by afferent and efferent pudendal nerve fibers. The reflex is seemingly polysynaptic, traversing at least several spinal cord segments. Clinically the bulbocavernosus reflex is elicited by briskly squeezing the glans penis or clitoris and observing or feeling a reflex contractility response of the external anal sphincter or bulbocavernosus muscle. Alternatively, the reflex may be stimulated by pulling the balloon of a Foley catheter against the bladder neck. The bulbocavernosus reflex is present in almost all normal men and in approximately 70% of normal women. Absence of this reflex in a man is strongly suggestive of a sacral neurological lesion, but it is present in approximately 50% of patients with an incomplete lower motor neuron lesion. Since it is difficult to grade the reflex clinically, measurement of the bulbocavernosus reflex latency offers a more quantitative means of evaluating the sacral reflex arcs. When one side of the penis is stimulated electrically, as by a surface stimulating electrode, there is a bilateral contractile response in the bulbocavernosus muscle that

may be detected by needle electrodes placed bilaterally into this muscle (Blaivas et al., 1981; Krane and Siroky, 1980). Considerable EMG expertise is required to determine the onset of the evoked response because of the possibility of interference from units that are firing either randomly or as a result of the patient's anxiety. Although some authors have therefore recommended that evoked responses of 20 to 30 stimulations be electronically averaged to determine the latency, Blaivas (1984) claims that the shortest latency, rather than the average, is a more specific means of evaluating the sacral segments, since the bulbocavernosus reflex is polysynaptic.

The bulbocavernosus reflex is a crossed response and it is therefore possible to stimulate on one side and record from both the ipsilateral and contralateral bulbocavernosus muscles. This is useful in detecting subtle abnormalities that affect only a single afferent or efferent pathway. Thus, when stimulating one side of the penis, recordings may be made from both sides of the bulbocavernosus muscle. The other side of the penis is then stimulated and the latencies from the bulbocavernosus muscle again determined. In this fashion, it is possible to evaluate the right and left afferent and efferent pathways individually. The normal bulbocavernosus reflex latency varies from approximately 30–40 msec, but the exact values vary slightly from one laboratory to another. Any neurological process that interferes with the integrity of the reflex arc will result in a prolonged latency. Common disorders that result in prolonged latencies include diabetes mellitus, alcoholic neuropathy, and prolapsed disks. Less commonly, a prolonged latency may be a very early manifestation of a spinal cord tumor or multiple sclerosis. Bulbocavernosus reflex latencies may also be obtained by stimulating the proximal urethra or bladder and neck instead of the penis. However, limited experience has been gained with these techniques. A word of caution is warranted in the interpretation of the result of sacral evoked responses. This test quantitates the integrity of innervation of the striated pelvic floor and perineal muscles and the supraspinal neurological pathways involved in lower urinary tract function. It does not give information concerning the status of the smooth muscle of the detrusor, bladder neck, and proximal urethra. Although in general it is reasonable to assume that when a neurological lesion is found to affect the striated perineal floor muscles, it is likely that the same process also involves the detrusor, since the pudendal and parasympathetic nuclei in the sacral spinal cord are practically adjacent to one another. Nevertheless, certain neurological disorders may involve one portion of the nervous system and spare another.

URETHRAL PRESSURE PROFILE

The urethral pressure profile (UPP) is a graphic recording of the pressure within the urethra at each point along its length. In 1969, Brown and Wickham described a technique that is the basis for modern perfusion urethral profilometry. They utilized a specially designed catheter with multiple side holes and an occluded tip. Fluid infused along the catheter escaped through the side holes and the UPP measured the resistance of the urethral walls to distention by this escaping fluid. This resistance is expressed in terms of the pressure necessary to maintain a steady flow of fluid through the catheter system. The UPP recording commences in the bladder and constant withdrawal of the catheter is accomplished through the entire length of the urethra. Being a measure of the urethra's response to distention, various factors affecting urethral compliance will alter the appearance of the profile curve. Contributing to the normal urethral compliance are smooth muscle activity, striated muscle activity, fibroelastic component of the urethral wall, vascular tension due to the rich spongy network around the urethra, and an extrinsic compression component of varying degrees. Obviously, alteration in any of these variables may significantly alter the appearance of the curve. However, the diagnostic value of the static UPP is limited because it is a study that is carried out neither during filling/storage nor during emptying. In other words, it is a study done at rest and it is difficult to prove that the events recorded bear any relationship to what goes on in the bladder or in the outlet during filling/storage or emptying.

Despite these shortcomings, static infusion profilometry does have a certain value in evaluating artificial sphincter function, gel prosthesis function, and the success of medical or surgical therapy for sphincteric incontinence. Alterations in the static profile certainly correlate with some disease entities, such as stress incontinence, an enlarged prostate, and striated sphincter dyssynergia. However, the overlap is invariably large and, because of this, the utility of infusion profilometry as a specific diagnostic study is limited. More recently, other methods of UPP measurement have been developed. These methods include dynamic UPP and stress UPP. Before we elaborate on these more sophisticated techniques, let us review some aspects of static UPP.

STATIC URETHRAL PRESSURE PROFILE

Although this technique may be modified by altering the position of the patient and the degree of bladder filling, the static UPP measures urethral pressure with the patient lying down, the bladder at rest, and the urethra closed. Although perfusion profilometry may be the method most commonly used, the other two techniques used to measure static UPP are the balloon catheter technique and the catheter tip transducers.

PERFUSION URETHRAL PRESSURE PROFILE

This technique uses a catheter with side holes through which saline or gas is perfused with a motorized syringe pump. The pressure is measured via a side arm from the catheter. The pressure registered by the transducer represents the resistance to flow from the catheter side hole, as previously mentioned. Consequently, when the catheter is in the bladder, only bladder pressure opposes the outflow of saline and the measured pressure is low. The catheter is then withdrawn at a constant rate through the entire length of the urethra. The initial pressure recorded on the chart strip recording is the intravesical pressure, followed by a positive deflection at the bladder neck and progressive increase in urethral pressure to the midportion of the urethra in the female (Fig 28–15) and the membranous urethra in the male (Fig 28–16). Beyond this point, pressure progressively decreases again until the external meatus is reached. Recommended nomenclature for UPP has been proposed by the International Continence Society. The most frequently measured parameters are: (1) maximum urethral pressure, which is the maximum pressure of the profile; (2) maximum urethral closing pressure, which is the difference between the maximum urethral pressure and bladder pressure; and (3) functional profile length, which is the length of the urethra along which the pressure exceeds bladder pressure. Perfusion profilometry may be performed using gas instead of saline. However, it has been demonstrated that carbon dioxide perfusion rate of 120 ml/min is needed to achieve a satisfactory reading of urethral

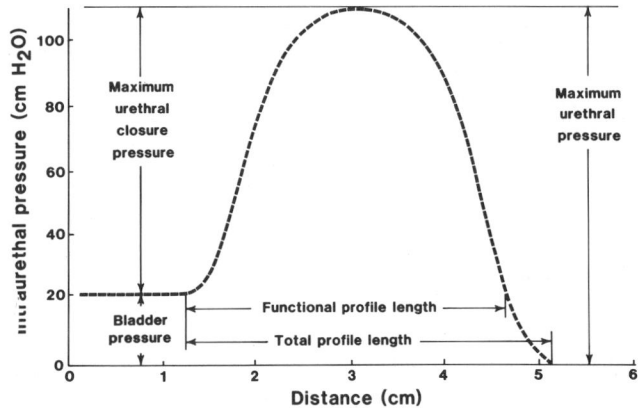

FIG 28–15.
Normal perfusion urethral pressure profile in a woman.

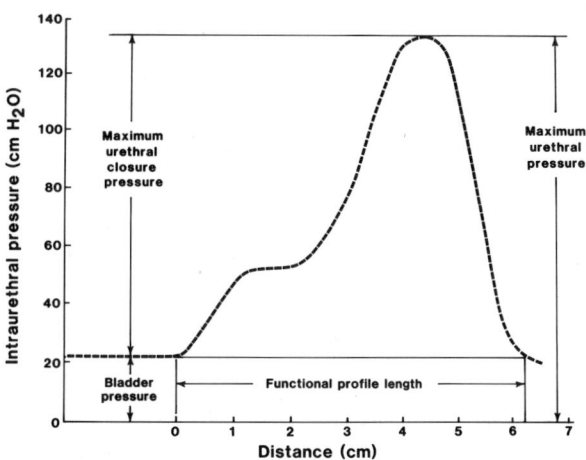

FIG 28–16.
Normal perfusion urethral pressure profile in a man.

pressure. The use of such high flow rates makes the practical recording of urethral pressure very difficult in that the bladder is rapidly filled and the investigator is measuring urethral pressure whilst the bladder is being distended swiftly. In addition, the speed of gas flow may theoretically lead to the recording of artificially high pressures due to the gas pushing the urethral walls apart. In this instance, the gas profile may be measuring urethral elasticity rather than urethral pressure.

BALLOON CATHETER SYSTEM

With this system, the eye holes of the catheter are covered by a fine plastic balloon. This technique relies on a closed system that must be free of all bubbles or leaks. Although accurate, frequent calibration and frequent replacement of balloon catheters is required due to a gradual change in compliance of the thin-walled balloons.

CATHETER TIP TRANSDUCER

This technique utilizes a catheter on which one or two transducers are mounted. The transducer is at the site of recording and obviates the problems inherent in recording at a distance from the organ being monitored. It allows for urethral pressure measurement during voiding and double transducer tip catheters allow for the simultaneous measurement of intravesical and urethral pressure. The disadvantages of the Mickel transducer catheter are its fragility and expense.

Whatever technique is used to measure UPP, it is probably essential to simultaneously record external sphincter activity. This will ensure that unintentional external sphincter contraction does not cause artifactual increase in maximum closing pressure. Many patients find it impossible to inhibit the external sphincter during the perfusion withdrawal process, and this is even

more true of those with hypersensitive urethras and neurogenic bladder dysfunction. Simultaneous measurement of rectal pressure is advocated as a means to identify unintentional abdominal straining, which might also alter the normal profile curve.

In the normal female patient, there is a gradual tendency for the maximum urethral pressure to decrease with age. This occurs mainly after the menopause. The shape of the UPP curve is symmetrical. In male patients, there is no decrease in maximum urethral pressure with age. However, there is a tendency for the length of the prostatic urethra to increase, particularly over the age of 45 years. The shape of the UPP curve in the male is asymmetrical. In the proximal part of the profile, there is a variable plateau due to the bladder neck and prostatic tissues. In the distal part of the profile, the pressure recorded is higher than from the distal female urethra due to the length and configuration of the male urethra. However, the distal urethra is rarely of clinical significance except on the rare occasion when the profile is recorded in a patient with a urethral stricture.

STRESS URETHRAL PRESSURE PROFILE

The stress UPP is best performed using a dual-sensor catheter tip transducer. The sensors should be separated by a 5–10 cm interval. The catheter is introduced into the bladder and then slowly withdrawn through the urethra while the patient is asked to cough at regular intervals. In order that coughs may be recorded for each 0.2 cm of urethral length, a slow withdrawal speed of 0.1 cm/sec is used (Abrams, 1984).

In normal patients, the increased pressure seen during coughing on the intravesical pressure trace is also seen superimposed on the urethral pressure trace. In normal patients, the raised intra-abdominal pressure is transmitted to the proximal two-thirds of the female urethra. In patients with genuine stress incontinence, there is a failure of pressure transmission. The findings on stress UPP have a far better correlation with the presence of genuine stress incontinence than does the static UPP. Stress urinary incontinence in women is the main indication for stress UPP.

DYNAMIC PRESSURE PROFILE

Static infusion profilometry is criticized mainly because it provides limited information as to the pressure in the outlet at rest. It supplies no information as to the outlet's behavior during filling, storage, and voiding.

The dynamic pressure profile is intended to show variation in sphincteric closure pressure under various physiological events, as well as under various stresses and commands. This dynamic pressure profile is obtained easily by the membrane catheter or microtransducer technique, but is practically impossible to obtain by perfusion techniques (Tanagho, 1984). The pressure profile will rise normally when the bladder is filled. In addition, postural change will affect the urethral pressure. The lowest pressure will be recorded when the patient is in the supine position. Sitting up will cause an increase in pressure, and further increase in functional length and magnitude of closing pressure will occur when the upright position is assumed. Sharp increase in intra-abdominal pressure, as by coughing, and low sustained increase in intra-abdominal pressure, as from bearing down, should result in simultaneous increase of pressure in the urethra, which is normally much higher than the increase in intra-abdominal pressure.

A drop in urethral pressure just before normal voiding occurs has been reported (Tanagho, 1979). Whether meaningful information can be gleaned from urethral measurement during voiding is debatable and the technique is unlikely to replace the conventional ways of assessing dynamic urethral function by fluoroscopy. However, where fluoroscopic evaluation is unavailable, dynamic profilometry may be applied to diagnose such uncommon causes of obstruction as proximal sphincter dyssynergia. The establishment of a pressure gradient in an area of the urethra that is normally wide open with a detrusor contraction establishes the diagnosis of obstruction. An elevated bladder pressure that falls off sharply at some point between the bladder neck and the bulbomembranous urethral junction is sufficient to diagnose proximal sphincter dyssynergia (McGuire, 1984b).

Dynamic UPP measurement may prove to be useful in the assessment of pharmacotherapy, but this field has been little explored as yet.

MULTIFUNCTION STUDIES

As mentioned earlier in this chapter, the clinical applicability of urodynamics depends to a large extent on the reenactment of the patient's symptom complex during the urodynamic assessment. Which urodynamic study to perform will depend on the nature of the clinical problem, the available electronic equipment, the ease with which the studies can be performed, and the interest and expertise of the urodynamic technician. In most clinical settings, it is usually practical and cost-effective to screen patients by performing uroflowmetry, measurement of postvoid residual volume, and cystometry (Blaivas, 1984). These three simple tests will help understand the patient's symptoms in the vast majority of cases. However, occasionally a more sophisticated study is called for, the indications for which include (Blaivas et al., 1981): (1) when simple diagnostic procedures have proved inconclusive, (2) when the pa-

tient has persistent symptoms despite what appears to be appropriate treatment, (3) when the patient has a preexisting condition known to be associated with complex urodynamic abnormalities, and (4) when history and physical examination suggest neurologic pathology that requires more elaborate investigation for diagnosis.

Despite what has previously been stated, the patient with an overt neurological disease (such as spinal cord injury or multiple sclerosis) in whom a urodynamic evaluation is required should probably routinely undergo a multifunction study, preferably with simultaneous cystourethrography. This is because such conditions are often accompanied by unpredictable pathophysiology.

Ideally, abdominal pressure should be measured so that subtracted bladder pressure can be monitored during voiding studies, as it is this entity that will alter in obstruction or detrusor dysfunction. This is particularly important when the detrusor contractions are of small magnitude and when voiding is accompanied by abdominal straining.

A poor urinary flow rate may be caused by outflow obstruction (Fig 28–17), impaired detrusor contractility (Fig 28–18), or a combination of both. From a practical standpoint, bladder outlet obstruction may be defined as a poor urinary flow rate in the presence of an "adequate" detrusor contraction (Blaivas, 1984). For an individual patient, however, it may be difficult to determine whether or not the detrusor pressure is "adequate" with respect to the flow rate, and for this reason many investigators have attempted to formulate a mathematical definition of bladder outlet obstruction by constructing a "resistance coefficient." The hydrodynamic principles from which these resistance coefficients are derived assume that the bladder behaves as

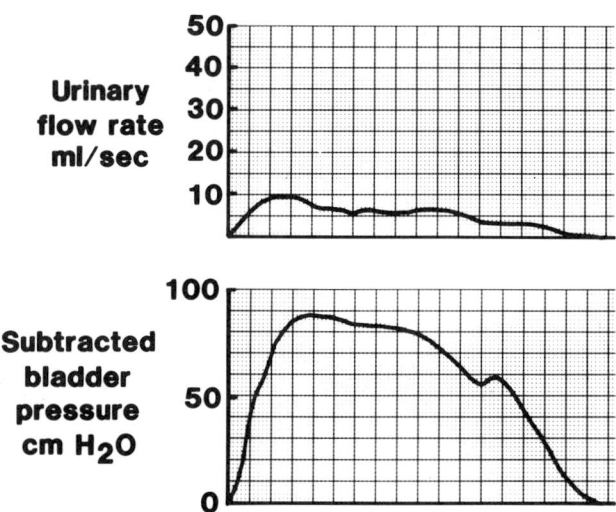

FIG 28–17.
Abnormal uroflow consistent with outlet obstruction.

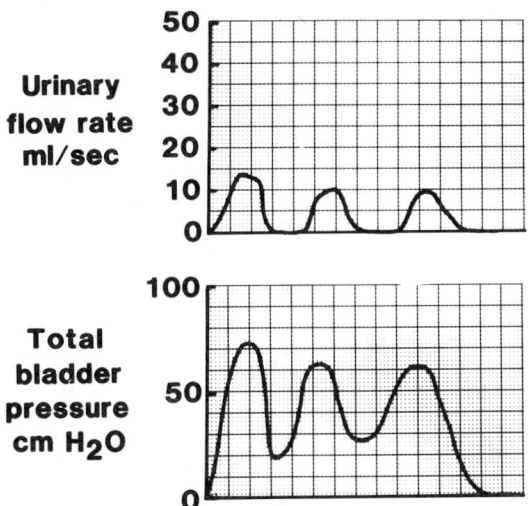

FIG 28–18.
Variable flow rate and bladder pressure. Pattern may be seen in patients with detrusor-sphincter dyssynergia or abdominal straining.

a geometric sphere and that the urethra behaves as a rigid tube. This is clearly not the case, and this limits the usefulness of these formulas. If a bladder has normal contractility, the pressure generated during contraction is high, and a low flow rate is produced (see Fig 28–17), then a formula is not needed to verify that an obstruction exists. On the other hand, if the bladder is decompensated and incapable of producing an adequate rise in intravesical pressure, then there is no way that obstruction (or resistance) can be quantitated by these formulas. These formulas assume that if obstruction exists the bladder is capable of monitoring a proportionate contraction in response to the obstruction. This is not always the case. If the bladder cannot contract, or contracts poorly, then obstruction cannot be defined. Thus, it appears that when these formulas are most necessary they are least accurate, and are therefore of little clinical use at present.

Synchronous cystometry and sphincter EMG is a most useful urodynamic technique for assessing neurologic function (Blaivas, 1977, 1984). This examination helps define the interrelationship between the striated external urethral sphincter and the detrusor during the storage and voiding phases of micturition.

The most important information to be gained from simultaneous cystometry and EMG relates to the presence or absence of detrusor external sphincter dyssynergia (DSD). In normal patients or those with detrusor instability or detrusor hyperreflexia due to neurologic lesions above the brain stem, the external urethral sphincter relaxes completely immediately prior to the onset of the rise in detrusor pressure. The relaxation continues throughout the detrusor contractions unless

the patient voluntarily attempts to interrupt micturition. Detrusor external sphincter dyssynergia is characterized by an involuntary contraction of the external sphincter coincident with or immediately preceding the rise in detrusor pressure. Blaivas (1981a) has classified DSD into three types (see Fig 28–14). Type 1 is characterized by an abrupt increase in EMG activity whose onset is approximately coincident with the onset of the measurable detrusor contraction. At the peak of the detrusor contraction there is a sudden complete relaxation of the external sphincter, and widening ensues as the detrusor pressures fall. In type 2 there are sporadic sphincter contractions throughout the detrusor contraction. Type 3 is characterized by a crescendo-decrescendo pattern of external sphincter contraction, which parallels the detrusor contraction. Detrusor sphincter dyssynergia is an abnormal reflex that occurs only when the neurologic pathways between the pons and sacral micturition center are interrupted. In the absence of such a neurologic lesion, extreme caution should be exercised in diagnosing DSD.

Pseudodyssynergia is believed to be an abnormality of learned behavior in which the patient subconsciously or consciously attempts to inhibit micturition as it progresses (Wein and Barrett, 1982). This results in "voluntary" contraction of the external sphincter throughout micturition. These pseudodyssynergic syndromes are encountered most frequently in children with persistent voiding symptoms, in men with prostatitis, and in women with the urethral syndrome.

It is not possible to distinguish between true dyssynergia and pseudodyssynergia by simultaneous cystometry and EMG unless intra-abdominal pressure is measured concurrently. In pseudodyssynergia there is usually some elevation of intra-abdominal pressure as the patient tightens the abdominal and pelvic floor musculature.

Another, possibly more simple, way to screen for detrusor-sphincter dyssynergia is by performing the combined test of uroflow and EMG (Barrett and Wein, 1981). When uroflow is accompanied by no external sphincter activity, dyssynergia may be excluded.

STOP FLOW TEST

This is a combined measurement of uroflow and intravesical pressure, which is performed to evaluate the isometric detrusor pressure. The patient is asked to void. After the stream is well initiated and intravesical pressure is being recorded, the stream is abruptly stopped. In normal detrusor function the detrusor pressure usually increases to a level well above voiding pressure. Theoretically, this test reflects on the integrity of detrusor tone and contractility.

VIDEOURODYNAMICS

Videourodynamics is a technique utilizing synchronously recorded urodynamic studies and cystourethrography for the evaluation of complex lower urinary tract problems. It comprises the performance of pressure/flow/external-sphincter EMG studies during filling, storage, and voiding phases of the micturition cycle, together with the periodic screening of the synchronous cystourethrographic appearance of the bladder and outflow tract.

Radiographic contrast material is used as the infusant for cystometry, permitting radiographic visualization of the lower urinary tract. Urodynamic parameters are transduced and displayed on a storage oscilloscope. A television camera scans the oscilloscope screen and the resulting image is displayed on a television monitor. The fluoroscopic image of the bladder and urethra is electronically mixed with that of the urodynamic data and displayed on the same monitor. The advantage of these studies is that they combine the objectivity of urodynamics with the visual radiographic image of the part being studied, making for far more logical interpretation of results. The financial investment required to establish such a study center limits its use to the medical center setting where a large enough patient population in whom such a study is indicated is seen, making this investment justified.

Videourodynamics has proved particularly valuable in the identification of complex bladder outlet obstruction problems. If bladder outlet obstruction has been diagnosed or is suspected, but the site of obstruction is not clear, a micturitional static urethral pressure profile determination will usually provide definitive diagnostic criteria (Blaivas, 1981). The demonstration of a drop in pressure between the bladder and the membranous urethra during voiding, at which time the bladder and proximal urethra are normally isobaric, establishes the diagnosis of an obstruction between the bladder and the site of the decreased pressure. Fluoroscopy will also allow visualization of the site of obstruction (Yalla et al., 1980).

Detrusor external sphincter dyssynergia is characterized by involuntary contraction of the external urethral sphincter during involuntary detrusor contractions (Fig 28–19). This may be readily identified by fluoroscopy during voiding. If a patient was treated by external sphincterotomy and a question arises as to the efficacy of surgery, this may be best evaluated by a videourodynamic study.

Videourodynamics is also valuable in the evaluation of some women with stress incontinence. The symptoms of stress incontinence may be caused not only by urethral and bladder hypermobility, but also by dener-

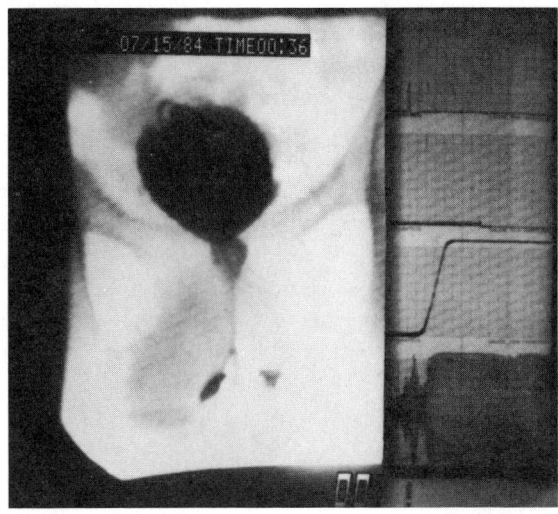

FIG 28–19.
Videourodynamic study in a 15-year-old paraplegic (T5) boy who had high residual urine volumes and recurrent urinary infection. Intravesical pressure exceeded 100 cm H_2O (scale, 0–100 cm H_2O) and EMG activity was excessive during the detrusor contraction. Although the bladder neck opened, the prostatic urethra was dilated to a level of the external sphincter. There was no urinary flow *(top strip recording)* or increase in abdominal pressure *(second strip recording from top)*. This is classic detrusor-external sphincter dyssynergia.

vation of the smooth muscle of the proximal urethra, as following radical pelvic surgery or in myelodysplasia. The latter condition maybe diagnosed by demonstrating an open bladder neck during bladder filling and a proximal urethra that is isobaric with the bladder. The practical implication of this is the choice of surgical repair, which is quite different for the two.

CLASSIFICATION OF VOIDING DYSFUNCTION

The purpose of any classification system is to facilitate understanding and management and to avoid confusion among those who are concerned with the problem for which the system was designed. A classification system should be able to serve as a sort of an intellectual shorthand and to convey, in a few key words or phrases, the essence of a clinical situation. An ideal classification system for voiding dysfunction should include or imply a number of factors. Foremost among these should be the conclusions reached from urodynamic testing during the filling/storage and the emptying phases of micturition. Expected clinical symptoms should be able to be inferred from a given category within an ideal system, as should the knowledge of whether the individual has normal sensory appreciation. The approximate site and type of a neurologic lesion or a lack of one may also be implied. If the various categories in the classification system accurately portray pathophysiology, treatment options should then be obvious, and a treatment "menu" evident. Most current systems of classification for voiding dysfunction were formulated primarily to describe those types of dysfunction secondary to neurologic disease or injury. The ideal classification system should be applicable to all types of voiding dysfunction. Besides the primarily neurologic systems, urodynamic and functional classification schemes also exist. We will review the major systems or types of systems currently used for the classification of neurogenic and nonneurogenic voiding dysfunction and comment about the advantages, disadvantages, and applicability of each. In so doing, we will describe many of the common forms of micturitional abnormalities.

THE BORS-COMARR CLASSIFICATION

Bors and Comarr (Bors, 1957; Comarr, 1967; Bors and Comarr, 1971) made a remarkable contribution with their classification system, one which was deduced by pure logic from clinical observation of patients with traumatic spinal cord injury (Table 28–6). This system is applicable only to patients with neuropathic dysfunction and considers (1) the anatomic localization of the lesion, (2) the neurologic completeness or incompleteness of the lesion, and (3) a designation as to whether the lower urinary tract is "balanced" or "unbalanced." The latter terms are based simply on the percentage of residual urine relative to bladder capacity. Unbalanced refers to the presence of greater than 20% residual

TABLE 28–6.

The Bors-Comarr Classification

Sensory neuron lesion
  Incomplete, balanced
  Complete, imbalanced
Motor neuron lesion
  Balanced
  Imbalanced
Sensory motor neuron lesion
  Upper motor neuron lesion
    Complete, balanced
    Complete, imbalanced
    Incomplete, balanced
    Incomplete, imbalanced
  Lower motor neuron lesion
    Complete, balanced
    Complete, imbalanced
    Incomplete, balanced
    Incomplete, imbalanced
  Mixed lesion
    Upper somatomotor neuron, lower
      visceromotor neuron
    Lower somatomotor neuron, upper
      visceromotor neuron
    Normal somatomotor neuron, lower
      visceromotor neuron

urine in a patient with an upper motor neuron lesion or 10% in a patient with a lower motor neuron lesion. This relative residual urine volume would ideally imply coordination (synergy) or lack of coordination (dyssynergy) between the smooth and striated sphincters of the outlet and the bladder during bladder contraction or attempted micturition by abdominal straining or the Credé maneuver.

The determination of whether a lesion is complete or incomplete is made on the basis of a thorough neurologic examination. This system assumes that the sacral spinal cord is the primary reflex center for micturition. Lower motor neuron in this system is a term used to describe collectively the preganglionic and postganglionic parasympathetic autonomic fibers that innervate the bladder and outlet and originate as preganglionic fibers in the sacral spinal cord. This term is used in an analogy to efferent somatic nerve fibers, such as those of the pudendal nerve, which originate in the same sacral cord segment but terminate directly on pelvic floor striated musculature without the interposition of ganglia. Upper motor neuron is used in this classification system to refer to those descending autonomic pathways above the sacral spinal cord (the origin of the parasympathetic nerve supply to the bladder and outlet).

In the Bors-Comarr system, an upper motor neuron bladder refers to the pattern of micturition that results from an injury to the suprasacral spinal cord, after the period of spinal shock has passed, assuming that the sacral spinal cord and the sacral nerve roots are intact and that the pelvic and pudendal nerve reflexes are intact. A lower motor neuron bladder refers to the pattern that results if the sacral spinal cord or sacral roots are damaged and the reflex pattern through the autonomic and somatic nerves that emanate from these segments is absent. In essence, this classification system implies that if skeletal muscle spasticity exists below the level of the lesion, the lesion is above the sacral spinal cord and is by definition an upper motor neuron lesion. This type of lesion is characterized by detrusor hyperreflexia during the filling/storage phase of micturition. If there is flaccidity of the skeletal muscle below the level of a lesion, a lower motor neuron lesion is said to exist, implying associated detrusor areflexia. Exceptions do occur and are classified in the "mixed lesion" group, characterized either by detrusor hyperreflexia in association with a flaccid skeletal muscle paralysis below the level of the lesion or by detrusor areflexia with skeletal muscle spasticity or normal skeletal muscle tone below the level of the lesion.

The use of this classification system is best illustrated by example. An upper motor neuron lesion, complete, imbalanced, is a neurologically complete lesion above the level of the sacral spinal cord that results in skeletal muscle spasticity below the level of the injury. Detrusor hyperreflexia exists, but with a residual urine volume of greater than 20% of bladder capacity, which implies that there is obstruction in the area of the bladder outlet (generally striated sphincter dyssynergia) during the hyperreflexic detrusor contraction. This is typically seen in a complete T10 paraplegic. A lower motor neuron lesion, complete, imbalanced, is typified by a neurologically complete lesion at the level of the sacral spinal cord or of the sacral roots that results in skeletal muscle flaccidity below that level. Detrusor areflexia results, and whatever maneuvers the patient may use to increase intravesical pressure are not sufficient to decrease residual urine to less than 10% of bladder capacity.

The Bors-Comarr classification system applies best to post-spinal shock, spinal cord–injured patients with complete neurologic lesions. It is difficult to use in patients with multicentric neurologic diseases and cannot be used at all for voiding dysfunction from nonneurologic causes. Additionally, it should be recognized that the terms upper and lower motor neuron do not, strictly speaking, apply to bladder innervation. The autonomic nervous system pathways from the spinal cord to the bladder contain not just one neuron but both preganglionic and postganglionic fibers, and the neural circuit from the spinal cord to the bladder is not truly analogous to the lower motor neuron of the somatic nervous sytem. Additionally, a neurologic description does not always imply a certain clinical or urodynamic picture. The period of spinal shock that immediately follows severe cord injuries is associated with bladder areflexia, whatever the status of the sacral somatic reflexes.

Temporary or permanent changes in bladder and/or bladder outlet activity during filling/storage or emptying may occur secondary to a number of factors such as chronic overdistention, urinary infection, or reinnervation phenomenon, making it impossible to always predict correctly lower urinary tract activity solely on the basis of the level of a neurologic lesion. Although the terms balanced and imbalanced are helpful in that they describe the presence or absence of a certain relative percentage of residual urine, these do not necessarily imply the true functional significance of the lesion, which depends on the potential for damage to the lower or upper urinary tract and also on the social and vocational disability that results.

THE GIBBON AND HALD-BRADLEY CLASSIFICATION

Gibbon (1976) pointed out that the complexity of the Bors-Comarr system could be lessened considerably by eliminating reference to the completeness of the lesion and to its balanced or unbalanced nature (Table 28–7). He also suggested that a rearrangement into suprasa-

TABLE 28–7.

The Gibbon Classification

Suprasacral lesion
Sacral lesion
Motor
Sensory
Motor and sensory
Mixed lesion

cral, sacral, and mixed categories would preserve the neurologic basis of the scheme, while simplifying it even further for everyday use by urologists. However, he noted that, for reasons similar to those just mentioned, the utility of such a system is limited. Hald and Bradley (1982) further modified the Gibbon system into what they described as a simple neurotopographical classification (Table 28–8). Within this classification system, a supraspinal lesion is characterized by synergy between detrusor contraction and the smooth and striated sphincters, but a defective inhibition of the voiding reflex exists. Detrusor hyperreflexia is the most constant finding. Sensation is usually preserved. However, depending on the site of the lesion, detrusor areflexia and defective sensation may be seen. A suprasacral spinal lesion is generally equivalent to what is described as an upper motor neuron lesion in the Bors-Comarr classification. An infrasacral lesion is equivalent to a lower motor neuron lesion. Peripheral autonomic neuropathy is most frequently encountered in the diabetic and is characterized by a deficient bladder sensation, gradually increasing residual urine, and ultimate decompensation, with loss of detrusor contractility. A muscular lesion is described as involving the detrusor itself, the smooth sphincter, or any portion or all of the striated sphincter. The resultant dysfunction is dependent on which structure is affected. Detrusor dysfunction is the most common, and generally results from decompensation following long-standing bladder outlet obstruction.

THE BRADLEY CLASSIFICATION

Bradley's "loop system" of classification is another primarily neurologic system based on his concept of central nervous system control of the lower urinary tract as consisting of four neurologic loops (see Chapter 27 for full description).

TABLE 28–8.

The Hald-Bradley Classification

Supraspinal lesion
Suprasacral spinal lesion
Infrasacral lesion
Peripheral autonomic neuropathy
Muscular lesion

Loop 1 consists of neuronal connections between the cerebral cortex and the micturition center in the pontine-mesencephalic reticular formation. Loop 1 coordinates voluntary control of the detrusor reflex. Interruption of this loop may be seen in conditions such as brain tumor, cerebrovascular accident or disease, and cerebral atrophy with dementia. The final result is usually detrusor hyperreflexia.

Loop 2 includes the intraspinal pathway of detrusor muscle afferents to the brain-stem micturition center and the motor impulses from this center to the sacral spinal cord. Strictly speaking, it does not include peripheral sensory or motor pathways and so is not a true "loop," but does include the long routed portion of the sensory afferent fibers that originate in the bladder musculature and synapse in the brain stem. Loop 2 is thought to coordinate and provide for a detrusor reflex of adequate temporal duration to allow total evacuation of intravesical contents. Partial interruption of this loop by spinal cord injury is felt to result in a detrusor reflex of low threshold and in poor emptying, as manifested by residual urine. Spinal cord transsection of this circuit acutely produces detrusor areflexia and urinary retention, the urinary tract correlates of spinal shock. After spinal shock has passed, detrusor hyperreflexia results. According to the loop theory, this reflex bladder activity occurs subsequent to a process similar to collateral sprouting from afferent to efferent pathways at the level of the transsection.

Loop 3 consists of the peripheral detrusor afferent axons and their pathway in the spinal cord, which terminates by synapsing on pudendal motor neurons that ultimately innervate the periurethral striated muscle. This loop provides the neurological substrate for coordinated reciprocal action of the bladder and striated sphincter. Dysfunction of this loop may be responsible for detrusor striated sphincter dyssynergia or for involuntary sphincter relaxation.

Loop 4 consists of two components. Loop 4A consists of the suprasacral afferent and efferent innervation of the pudendal motor neurons to the periurethral striated musculature. Loop 4B consists of afferent fibers from the periurethral striated musculature, which synapse on pudendal motor neurons, i.e., the segmental innervation of the periurethral striated muscle. In contrast to stimulation of detrusor afferent fibers, which produce inhibitory postsynaptic potentials in pudendal motor neurons through loop 3, pudendal nerve afferents produce excitatory postsynaptic potentials in these motor neurons through loop 4B. This provides for contraction of the periurethral striated muscle during bladder filling and urine storage. The sensory impulses concerned arise from muscle spindles and tendon organs in the pelvic floor musculature. Loop 4 provides for volitional control of the striated sphincter. Abnormalities of the suprasacral portion result in abnormal responses of the pudendal motor neurons to bladder filling and emptying, which may be manifested as detrusor striated sphincter dyssynergia or/and loss of voluntary contraction of the striated sphincter on command.

This classification system is very sophisticated and is a tribute to the ingenuity and neurophysiological expertise of its originator. For the neurologist this method of classification may be an excellent way to conceptualize the relevant neurophysiology, assuming that there is agreement on the existence and significance of all four loops, and the types and locations of the lesion produced in these loops by various injuries and disease processes. Many urologists, however, find this classification difficult to use for all types of neurogenic voiding dysfunction, and not at all applicable to voiding dysfunction of other etiology.

Cystometry and sphincter EMG are the primary tools for categorizing the different loop lesions; however, it may be extremely difficult to test the intactness of each loop system, even using extremely sophisticated neuromuscular testing. Multicentric and partial lesions are difficult to describe in this system and the type and functional significance of voiding dysfunction they produce are not necessarily implied by their anatomic localization.

THE LAPIDES CLASSIFICATION

Lapides (1967, 1970) contributed significantly to the classification and care of the patient with neuropathic voiding dysfunction by popularizing a modification of a scheme originally proposed by McLellan in 1939 (Table 28–9). The systems are virtually identical except that Lapides further divided McLellan's category of atonic bladder into motor neurogenic and sensory neurogenic bladder. The Lapides classification is the scheme most familiar to urologists and describes in recognizable shorthand the clinical and cystometric condition in most types of neurogenic voiding dysfunction. In the uninhibited neurogenic bladder and reflex neurogenic bladder groups, the exact categorization further implies whether the striated sphincter is dyssynergic (reflex neurogenic bladder) or synergic (uninhibited neurogenic bladder) during bladder contraction.

TABLE 28–9.

The Lapides Classification

Sensory neurogenic bladder
Motor paralytic bladder
Uninhibited neurogenic bladder
Reflex neurogenic bladder
Autonomous neurogenic bladder

A sensory neurogenic bladder results from any disease that selectively interrupts the sensory fibers between the bladder and spinal cord or the afferent tracts to the brain. Most commonly, this is seen with diabetes mellitus, tabes dorsalis, and pernicious anemia. The first clinical changes consist only of impaired sensation of bladder distention. Unless voiding is initiated out of habit or on a timed basis, varying degrees of bladder overdistention often result, with resultant hypotonicity. With bladder decompensation significant amounts of residual urine usually are found and, at this time, the cystometric curve generally demonstrates a large bladder capacity with a flat low-pressure filling curve (high compliance). The bethanechol supersensitivity test is usually positive in the early stages, but may later become negative as decompensation of the bladder smooth muscle occurs.

A motor paralytic bladder results from disease processes that destroy the parasympathetic motor innervation of the bladder. Herpes zoster and extensive pelvic surgery or trauma can produce a motor paralytic bladder. The early symptoms may vary from painful urinary retention to a relative inability to initiate and maintain normal micturition. In the early stages the filling limb of the cystometrogram is normal with normal sensation, but without a voluntary bladder contraction at bladder capacity. Later, chronic overdistention and bladder decompensation may occur and a large-capacity bladder with a flat low-pressure filling limb and, generally, a large residual urine volume will result. The bethanechol test is positive and remains so unless severe decompensation occurs.

The uninhibited neurogenic bladder was described originally as resulting from injury or disease in the "corticoregulatory tract." The sacral spinal cord was presumed to be the reflex center for micturition and this corticoregulatory tract was believed normally to exert an inhibitory influence on the primary micturition reflex. A destructive lesion in this tract would then result in overfacilitation of the micturition reflex. Cerebrovascular accident, brain or spinal cord tumor, and demyelinating disease are the most common causes of this type of lesion. This generally results in a voiding dysfunction characterized clinically by frequency, urgency, and incontinence and cystometrically by normal sensation with detrusor hyperreflexia at low filling volumes. Residual urine is characteristically small or absent unless anatomic outlet obstruction or true involuntary smooth or striated sphincter dyssynergia occurs. The patient usually can initiate a bladder contraction voluntarily but is often unable to do so during cystometry because sufficient urine storage cannot occur before detrusor hyperreflexia is stimulated.

The term reflex neurogenic bladder is most commonly used to describe the post–spinal shock condition existing after complete interruption of the sensory and motor pathways between the sacral spinal cord and the brain stem. This occurs most commonly in traumatic spinal cord injury and transverse myelitis, but may occur with extensive demyelinating disease or tumor as well. Typically, the patient has no bladder sensation and is unable to initiate micturition voluntarily. Incontinence results because of low-volume detrusor hyperreflexia, which generally occurs with striated sphincter dyssynergia. This type of lesion is equivalent to a complete upper motor neuron lesion in the Bors-Comarr system.

An autonomous neurogenic bladder results from complete motor and sensory separation of the bladder from the sacral spinal cord. Any disease process that destroys the sacral spinal cord or causes extensive damage to the sacral roots or pelvic nerves may result in this state of affairs. The patient is unable to initiate micturition voluntarily, has no bladder reflex activity, and no specific bladder sensation. This type of bladder is equivalent to the complete lower motor neuron lesion in the Bors-Comarr system. This is also the type of dysfunction seen in patients with spinal shock. The characteristic cystometric pattern is initially similar to the late stages of the motor or sensory paralytic bladder with a marked shift to the right of the filling curve and a large bladder capacity at low intravesical pressure. However, secondary changes in the filling limb may occur that cause an increase in slope (decreased compliance). This may be secondary to chronic inflammatory change or to the effects of the denervation with secondary neuromorphologic and neuropharmacologic changes. Emptying capacity in an autonomous neurogenic bladder may vary from nil to a large percentage of bladder capacity, depending on the ability of the patient to increase intravesical pressure and on the resistance offered during this increase by the smooth and striated muscle of the bladder outlet.

These five classic cystometric categories and their usual clinical settings are easily understood and remembered, and this sytem therefore provides an excellent framework for teaching the fundamentals of neurogenic voiding dysfunction to students and nonurologists. Unfortunately, many patients do not exactly fit into one or another of these classic descriptions. Gradations of sensory, motor, and mixed lesions occur, and the patterns produced after different types of peripheral denervation may vary widely from those classically described. The detrusor hyperreflexia associated with the classic uninhibited neurogenic bladder may be associated with dyssynergic striated sphincter activity as well as synergic activity. Secondary changes caused by infection, chronic distention, and reinnervation can be superimposed on and alter any of these classic patterns.

THE URODYNAMIC CLASSIFICATION

With the development of more sophisticated urodynamic equipment and techniques to categorize the activity of the bladder and the outlet during the filling/storage and emptying phase of micturition, systems of classification have been formulated based solely on objective urodynamic data. Evolution of this type of classification system has paralleled urodynamic expertise and, in the United States, was pioneered by Krane and Siroky (1979, 1984) whose initial urodynamic classification system is shown in Table 28–10. When exact urodynamic classification is possible, this system provides a truly exact description of the particular voiding dysfunction under consideration. If a normal or hyperreflexic detrusor exists with coordinated smooth and striated sphincters and without anatomic obstruction, total bladder emptying should be produced. Striated sphincter dyssynergia is most commonly seen in the patient with a complete suprasacral spinal cord injury after the period of spinal shock has passed. Smooth sphincter dyssynergia is seen most classically in autonomic dysreflexia, when it is characteristically associated with detrusor hyperreflexia and striated sphincter dyssynergia. Detrusor hyperreflexia is most commonly associated with neurologic lesions above the sacral spinal cord, but may be associated with inflammatory or infectious disease or may be idiopathic. Detrusor areflexia may be secondary to bladder muscle decompensation or to various other conditions that produce inhibition at either the level of the brain-stem micturition center, sacral spinal cord, bladder ganglia, or bladder smooth muscle. Patients with a voiding dysfunction that falls into this category generally attempt bladder emptying by abdominal straining or the Credé maneuver. Their continence status and the efficiency of their emptying efforts are determined by the status of the smooth and striated sphincters of the outlet.

This classification system is easiest to use when detrusor hyperreflexia or normoreflexia exists, since sufficiently sophisticated and reproducible urodynamic techniques exist to describe the activity of the smooth and striated sphincters during bladder contraction. Thus, a typical T10 paraplegic exhibits detrusor hyperreflexia, smooth sphincter synergia, and striated sphincter dyssynergia. When a voluntary or hyperreflexic bladder contraction cannot be elicited, this system is more difficult to use since it is not appropriate to speak of true dyssynergia in the absence of an opposing bladder contraction.

Urodynamic classification systems assume total urodynamic agreement among the classifiers. Often this is easy to obtain, but one has only to attend the case presentation sections of various neurourology symposia to realize that there are many voiding dysfunctions that do not fit neatly into such a classification system. This does not mean that such a system is not the ideal system, but simply that the "experts" sometimes do not agree about the significance of the data generated using a given urodynamic evaluation. As sophisticated urodynamic technology and understanding improve, this type of classification system of voiding dysfunction may supplant others in general use.

THE FUNCTIONAL SYSTEM

Classification of voiding dysfunction can also be formulated on a functional basis, describing the dysfunction simply in terms of whether the deficit produced is primarily one of the filling/storage phase of micturition or of the emptying phase (Wein, 1981, 1984; Quesada et al., 1968; Bors and Comarr, 1971; Wein et al., 1976) (Table 28–11). This type of classification system is an excellent alternative when a particular dysfunction does not lend itself to a generally agreed-on classification in one of the other systems. Our groups have adopted and promoted this system primarily because of dissatisfaction with attempts to exactly classify urodynamically some types of voiding dysfunction.

This simple-minded scheme assumes only that, whatever their differences, all "experts" would agree on certain general principles concerning the micturition cycle. Bladder filling and urine storage require: (1) accommodation of increasing volumes of urine at a low intravesical pressure and with normal and appropriate sensation, (2) absence of involuntary bladder contractions, and (3) a bladder outlet that is closed at rest and

TABLE 28–10.

The Krane-Siroky Urodynamic Classification

Detrusor hyperreflexia (or normoreflexia)
    Coordinated sphincters
    Striated sphincter dyssynergia
    Smooth sphincter dyssynergia
    Nonrelaxing smooth sphincter
Detrusor areflexia
    Coordinated sphincters
    Nonrelaxing striated sphincter
    Denervated striated sphincter
    Nonrelaxing smooth sphincter

TABLE 28–11.

The Functional Classification

Failure to store
    Because of the bladder
    Because of the outlet
Failure to empty
    Because of the bladder
    Because of the outlet

---

remains so with stress. Storage failure can then result because of problems related to bladder hyperreflexia or/ and low compliance and because of a permanent or intermittent decrease in outlet resistance. Bladder emptying requires: (1) a coordinated bladder contraction of adequate magnitude, and (2) lack of anatomic obstruction and concomitant lowering of resistance at the level of the smooth and striated sphincter. Failure to empty can then result from inadequate bladder contractility or increased outlet resistance. Failure in either category generally is not absolute, but is more often relative.

Such a system can easily be "expanded" and made more complicated to include etiologic or urodynamic connotations (Table 28–12). However, the simplified system avoids argument in those complex situations in which the exact etiology or urodynamic mechanism for a voiding dysfunction cannot be agreed on. A different type of logical extension of this system functions especially well as a "menu" for the treatment of voiding dysfunction (see Tables 28–14 through 28–18).

Proper use of this system for a given voiding dysfunction obviously requires a reasonably accurate notion of what the urodynamic data show. However, an exact diagnosis is not required for treatment. Some patients do not have just a discrete storage or emptying failure, and the existence of combination deficits must be recognized to properly utilize this system of classification. For example, the T10 paraplegic generally exhibits a relative failure to store because of detrusor hyperreflexia and a relative failure to empty because of striated sphincter dyssynergia. In such a combination deficit, to utilize this classification system one must assume that either one deficit is primary and that significant improvement will result from its treatment alone or that the voiding dysfunction can be converted primarily to a disorder either of storage or emptying by means of surgical or pharmacologic therapy. The resultant deficit can then be treated or circumvented. Using the same example, the combined deficit in a T10 paraplegic can be converted primarily to a failure to store by surgical procedures directed at the dyssynergic striated sphincter and the resultant incontinence (secondary to detrusor hyperreflexia) circumvented (in a male) with an external collecting device. Alternatively, the deficit can be converted to primarily a failure to empty by surgical or pharmacologic measures designed to abolish or reduce the detrusor hyperreflexia, and the resultant failure to empty can then be circumvented with clean intermittent catheterization. Other examples of combination deficits include impaired bladder contractility with sphincter dysfunction, bladder outlet obstruction with detrusor hyperactivity, and bladder outlet obstruction with sphincter malfunction.

The major problem with this classification system is that not every voiding dysfunction can in fact be reduced or converted primarily to a failure of storage or emptying. Additionally, although the functional classification of therapy that is a correlate of this scheme seems logical, there is a danger of accepting an easy therapeutic solution and of thereby overlooking an etiology for a voiding dysfunction that is reversible at the primary level of causation. Nonneurogenic voiding dysfunction can, however, be classified within this system, with the possible exception of those dysfunctions involving primarily the sensory aspect of micturition. Unfortunately, those disorders are virtually ignored by all classification systems, unless one broadly includes such disorders as the urgency-frequency syndrome without urodynamic abnormalities under the category of "failure to store."

THE INTERNATIONAL CONTINENCE SOCIETY CLASSIFICATION

Finally, the International Continence Society (ICS) has proposed a classification based on the functional state of the bladder and urethra, which operates with a "three by three" system (ICS Standardization Committee, 1981) (Table 28–13). Overactive detrusor function

TABLE 28–12.

The Expanded Functional Classification

Failure to store
  Because of the bladder
    Detrusor hyperactivity
      Involuntary contractions
        Suprasacral neurologic disease
        Bladder outlet obstruction
        Idiopathic
      Decreased compliance
        Fibrosis
        Idiopathic
    Sensory urgency
      Inflammatory
      Infectious
      Neurologic
      Psychologic
      Idiopathic
  Because of the outlet
    Stress incontinence
    Nonfunctional bladder neck/proximal urethra
Failure to empty
  Because of the bladder
    Neurologic
    Myogenic
    Psychogenic
    Idiopathic
  Because of the outlet
    Anatomic
      Prostatic obstruction
      Bladder neck contracture obstruction
      Urethral stricture
    Functional
      Smooth sphincter dyssynergia
      Striated sphincter dyssynergia

TABLE 28–13.
The International Continence
Society Classification

Detrusor
  Normal
  Overactive
  Underactive
Urethra
  Normal
  Overactive
  Incompetent
Sensation
  Normal
  Hypersensitive
  Hyposensitive

is indicated when, during the filling phase, there are involuntary detrusor contractions that the patient cannot suppress. Involuntary bladder contractions that are associated with known neurologic disease fall under the category of detrusor hyperreflexia, while the term "unstable detrusor" or "bladder instability" is applied to such activities that occur in a patient without known neurologic disease. The term "compliance" refers to the volume-pressure relationship in a bladder, and the term "low compliance" represents another form of detrusor overactivity, consisting of a tonically high pressure rise in response to a given increment in bladder volume. The underactive detrusor may be described as noncontractile (no contraction under any circumstances), areflexic (the complete absence of centrally coordinated contractions secondary to neurologic disease); a patient with an underactive detrusor may be described as having a high-compliance bladder (a capacious bladder that shows little change in pressure in response to a given volume increment). An overactive urethral closure mechanism contracts involuntarily against the detrusor contraction (dyssynergia) or fails to relax during attempted micturition. An incompetent urethral closure mechanism allows urinary incontinence. This may occur only during a rise in intra-abdominal pressure (stress incontinence) or it may be persistent, in which case continuous leakage will occur. It may also be due to an involuntary fall in urethral pressure in the absence of detrusor activity (the unstable urethra).

Within the ICS system, the condition of a classical T10 paraplegic would be classified as follows: overactive detrusor, overactive urethra, hyposensitive; the condition of a stroke patient with urgency incontinence would most likely be classified as follows: overactive detrusor, normal urethra, normal sensation; the condition of a patient with stress incontinence would be classified as follows: normal detrusor, incompetent urethra, normal sensation.

No type of classification system for voiding dysfunction is perfect. Some systems clearly apply only to neuropathic disorders. Others are more widely applicable, but each has its obvious shortcomings. Each offers something to every clinician, the amount being dependent on his level and type of training, interest, experience, and prejudices regarding the accuracy and interpretation of urodynamic data. The ideal approach for a given patient remains a thorough neurourologic evaluation and an attempt at categorization using each classification system described. If the clinician can classify a given patient's condition in each system, or understand why he cannot, he has a working knowledge of the situation sufficient to proceed with treatment. However, if the clinician is familiar with only one or two of these classification systems and a given patient does not fit these and the reason is uncertain, the patient should be studied further and the voiding dysfunction better characterized, certainly before irreversible therapy is undertaken.

TREATMENT OF VOIDING DYSFUNCTION

This section will consider the various therapies available for the treatment of voiding dysfunction. There are only a discrete number of such therapies available, and these are easily categorized on a functional basis according to their effect on the bladder or on the outlet (Tables 28–14 and 28–15). This basic outline will be followed in discussing each therapy. Some of the surgical forms of therapy are well discussed elsewhere in this text; therefore, an attempt has been made to restrict detailed descriptions of surgical technique or illustrations of some of those procedures that are important but do not receive extensive coverage in other chapters.

The initial choice of a surgical vs. a nonsurgical mode of management for a given problem is based on a great many factors. Although many urologists who lecture and write about the management of voiding dysfunction are associated primarily with one approach or procedure or another, all would doubtless agree on certain goals of management satisfied by the ideal treatment for voiding dysfunction (Table 28–16). One of the most important concepts to be put forth in recent years is a urodynamic clarification of exactly what "adequate storage at low intravesical pressure" means. McGuire (1984) and co-workers (1981, 1983) have clearly shown that upper tract deterioration is apt to occur when storage, even though adequate in terms of continence, occurs at sustained intravesical pressures of greater than 40 cm H_2O. Application of this concept to patients with storage problems because of decreased bladder compliance has resulted in the concept of the "leak-point" as

TABLE 28–14.

Therapy to Facilitate Bladder Emptying

Increasing intravesical pressure/bladder contractility
    External compression, Valsalva
    Promotion or initiation of reflex contractions
        Trigger zones or maneuvers
        Bladder training, tidal drainage
    Pharmacologic therapy
        Parasympathomimetic agents
        Prostaglandins
        Blockers of inhibition
           α-Adrenergic antagonists
        Opioid antagonists
    Electrical stimulation
        Directly to the bladder
        To the spinal cord or nerve roots
    Reduction cystoplasty
Decreasing outlet resistance
    At a site of anatomic obstruction
        Prostatectomy
        Urethral stricture repair/dilation
    At the level of the smooth sphincter
        Transurethral resection or incision of the
           bladder neck
        Y-V plasty of the bladder neck
        Pharmacologic therapy
           α-Adrenergic antagonists
           β-Adrenergic agonists
    At the level of the striated sphincter
        External sphincterotomy
        Urethral overdilation
        Pudendal nerve block or interruption
        Pharmacologic therapy
           Skeletal muscle relaxants
               Centrally acting relaxants
               Dantrolene
               Baclofen
           α-Adrenergic antagonists
        Biofeedback, psychotherapy
Circumventing problems
    Intermittent catheterization
    Continuous catheterization
    Urinary diversion

TABLE 28–15.

Therapy to Facilitate Urine Storage

Inhibiting bladder contractility/decreasing sensory
    input/increasing bladder capacity
    Timed bladder emptying
    Pharmacologic therapy
        Anticholinergic agents
        Musculotropic relaxants
        Polysynaptic inhibitors
        Calcium antagonists
        β-Adrenergic agonists
        α-Adrenergic antagonists
        Prostaglandin inhibitors
        Tricyclic antidepressants
        Dimethyl sulfoxide
        Bromocriptine
    Biofeedback, bladder retraining
    Bladder overdistention
    Electrical stimulation (reflex inhibition)
    Interruption of innervation
        Central (subarachnoid block)
        Peripheral (sacral rhizotomy, selective sacral
           rhizotomy)
        Perivesical (peripheral bladder denervation)
    Augmentation cystoplasty
Increasing outlet resistance
    Physiotherapy
    Electrical stimulation of the pelvic floor
    Pharmacologic therapy
        α-Adrenergic agonists
        Tricyclic antidepressants
        β-Adrenergic antagonists
        Estrogen
    Vesicourethral suspension (SUI)
    Bladder outlet reconstruction
    Surgical mechanical compression
    Nonsurgical mechanical compression
Circumventing problem
    Antidiuretic hormone–like agents
    External collecting device
    Intermittent catheterization
    Continuous catheterization
    Urinary diversion

TABLE 28–16.

Voiding Dysfunction: Goals of Management

Upper urinary tract preservation or improvement
Absence or control of infection
Adequate storage at low intravesical pressure
Adequate emptying at low intravesical pressure
Adequate control
No catheter or stoma
Social acceptability/adaptability
Vocational acceptability/adaptability

a very significant piece of urodynamic data. Patients with leakage pressures less than 40 cm H_2O, or those who can store (or whose bladders can be pharmacologically or surgically manipulated to store) at a pressure less than that, will not experience upper tract deterioration in the absence of other complicating factors, such as significant vesicoureteral reflux with urinary tract infection.

In treating patients with voiding dysfunction, a perfect result is rarely, if ever, achieved, and a very flexible approach must therefore be adopted in choosing therapy. This approach must take into account the individual wishes of each patient and the practicality of each proposed solution in each case, especially in those patients with neurologic disease (Table 28–17). The decision as to whether initial therapy is to be surgical or nonsurgical is best made with the patient and with his or her family. In every case, the patient and family must be informed of all possible methods of management, including reversibility and side effects that occur with some regularity, ultimate best and worst possible

TABLE 28–17.

Patient Factors to Consider in Choosing Therapy

Prognosis of underlying disease, especially if progressive
    or malignant disease
Limiting factors: inability to perform certain tasks (hand
    dexterity, ability to transfer)
Mental status
Motivation
Desire to remain catheter- or appliance-free
Desire to avoid surgery
Sexual activity status
Reliability
Educability
Psychosocial environment; interest, reliability, and
    cooperation of family
Economic resources
Age

result, and frequency and extent of follow-up. It has always been our subjective bias that the simplest, least destructive, and most reversible form of therapy that has a chance of satisfying the goals of treatment should be tried first. It is important to remember that a combination of therapeutic maneuvers, drugs, and surgery can sometimes be used to achieve a particular end, especially if they act through different mechanisms and their side effects are not synergistic. There are circumstances and locales in which hospital resources and bed usage efficiency must also be considered, especially in this day and age when some health care systems and third-party payers have placed considerable financial restraints on health care.

Absolute or relative indications for changing or augmenting a particular regimen exist, and, likewise, these are generally agreed on, although the relative importance of the indication for change might be disputed (Table 28–18). Remember that here, too, the term "inadequate," when applied to storage or emptying, does not imply only completeness, but includes performance of the function at unacceptably high intravesical pressures.

TABLE 28–18.

Reasons to Change or Augment a Given
Regimen

Upper urinary tract deterioration
Recurrent sepsis or fever of urinary tract
    origin
Lower urinary tract deterioration
Inadequate storage
Inadequate emptying
Inadequate control
Unacceptable side effects
Skin changes secondary to incontinence
    or collecting device

THERAPY TO FACILITATE BLADDER EMPTYING

Increasing Intravesical Pressure/Bladder Contractility

EXTERNAL COMPRESSION: VALSALVA.—Manual compression of the bladder (Credé maneuver) is most effective in those patients with decreased bladder tonicity who can generate a pressure greater than 50 cm H_2O with this maneuver and whose outlet resistance is borderline or decreased (Glahn, 1974; Wein et al., 1976). The technique of voiding by the open-hand Credé method involves the placement of the thumb of each hand over the area of the left and right anterior superior iliac spine and of the digits over the suprapubic area, with slight overlap at the finger tips (Opitz, 1984). The slightly overlapped digits are then pressed into the abdomen and, when they have gotten behind the symphysis, downward, to compress the fundus of the bladder. Both hands are then pressed as deeply as possible downward into the real pelvic cavity. At times, the compression of the bladder can be accomplished more efficiently by the use of the fist of one hand (closed-hand method) or a rolled-up towel. A similar increase in intravesical pressure may be achieved by abdominal straining. The technique of straining (Valsalva) involves sitting and resting the abdomen forward on the thighs in both men and women. During straining in this position, hugging of the knees and legs may be advantageous to prevent any bulging of the abdomen. To increase intravesical pressure in this manner requires voluntary control of the abdominal wall and diaphragmatic muscles, or, in the case of the Credé maneuver, adequate hand control. Straining at the time the Credé maneuver is applied should be avoided because this increases the intra-abdominal pressure and causes bulging of the abdominal wall, which then tends to lift the compressing hands off the fundus of the bladder. The Credé maneuver is much easier in a patient with a lax, lean abdominal wall than with a taut or obese one, and is more readily performed in a child than in an adult.

Voiding by these maneuvers is unphysiologic and is resisted by the same forces that normally resist stress urinary incontinence. In our experience, adaptive change (funneling) of the bladder neck and proximal urethra does not generally occur with external compression maneuvers of any kind. In patients with intact pelvic floor striated muscle reflexes, increases in outlet resistance may actually occur. If adequate emptying does not occur, other types of therapy to decrease outlet resistance may be considered, but these may adversely affect continence. Vesicoureteral reflex is a relative contraindication to Credé or Valsalva maneuvers, especially in patients who are capable of generating a high intravesical pressure by so doing. The greatest likelihood of success with this mode of therapy is in the patient with an areflexic and hypotonic or atonic bladder, and some outlet denervation (striated or smooth sphincter or both). Such a patient not uncommonly has stress incontinence as well. The continued use of external compression or Valsalva maneuver implies that the intravesical pressure between attempted voidings is consistently below that necessary to cause upper tract deterioration. This may be an erroneous assumption, and close follow-up is necessary, especially in those patients with close to normal outlet resistance, to avoid this complication. The most flagrant misuse of this form of management is in the patient with a decentralized or denervated bladder in whom decreased compliance during filling has developed. Such a patient may silently develop intravesical pressures greater than those necessary to cause upper tract deterioration with minimal filling.

PROMOTION OR INITIATION OF REFLEX CONTRACTION.—In those types of spinal cord injury or disease characterized by detrusor hyperreflexia, manual stimulation of areas within the sacral and lumbar dermatomes may sometimes provoke a reflex bladder contraction (Bors, 1957; Comarr, 1959; Gibbon, 1974). Pulling the skin or hair of the pubis, scrotum, or thigh, squeezing the clitoris, or digital rectal stimulation are examples of the type of activity that sometimes induces "trigger voiding" in these patients. According to Glahn (1974), the most effective method of inducing such a contraction is rhythmic suprapubic manual pressure (seven or eight pushes every 3 seconds). This is thought to produce a summation effect on the tension receptors in the bladder wall, resulting in an afferent neural discharge, which activates the bladder reflex arc. Ideally, the contraction so produced will be sustained and of adequate magnitude. Patients who are potentially able to induce bladder contractions by such a maneuver should be encouraged to find their own optimal "trigger points" and position for urination. Manual dexterity and either the ability to transfer to a commode or an external collecting device are required. Unfortunately, this type of patient often has sphincter dyssynergia, and such maneuvers may have to be combined with measures to decrease outlet resistance at the level of the striated and/or smooth sphincter.

Occasionally, this form of induced bladder contraction is possible and desirable in patients with supraspinal disease and involuntary bladder contractions. If induced emptying can be carried out frequently enough in these patients so as to keep bladder volume below the threshold for activation of the micturition reflex, in-

continence can be "controlled." This actually amounts to a form of timed voiding. If regular frequent attempts at voiding are unsuccessful, these maneuvers may be helpful as an initial sensory cue (Opitz, 1984).

Some clinicians still feel that the establishment of a rhythmic pattern of bladder filling and emptying by maintaining a copious fluid intake and by periodically clamping and unclamping an indwelling catheter or with intermittent catheterization can "condition" or "train" the micturition reflex. Others, with whom we would agree, feel that there is no cause-and-effect relationship between the attainment of balanced lower urinary tract function and such programs, and that such regimens benefit patients primarily by focusing their attention on the urinary tract and by ensuring an adequate fluid intake.

Pharmacologic Manipulation

PARASYMPATHOMIMETIC AGENTS.—Since at least a major portion of the final common pathway in physiologic bladder contraction is stimulation of the muscarinic cholinergic receptor sites at the postganglionic parasympathetic neuromuscular junction, agents that imitate the actions of acetylcholine (ACh) might be expected to be useful in the management of patients who cannot empty because of inadequate bladder contractility. Acetylcholine itself cannot be used for therapeutic purposes because of actions at central and ganglionic levels and because of its rapid hydrolysis by acetylcholinesterase and by nonspecific cholinesterase (Taylor, 1980). Many acetylcholine-like drugs exist. However, only bethanechol chloride exhibits a relatively selective action on the urinary bladder and gut with little or no action at therapeutic dosages on ganglia or on the cardiovascular system (Ursillo, 1967; Taylor 1980). It is cholinesterase-resistant and causes a contraction in vitro of smooth muscle from all areas of the bladder (Raezer et al., 1973).

Bethanechol has been recommended for the treatment of postoperative or postpartum urinary retention in a subcutaneous dose of 5–10 mg. In this instance, it should be used only if the patient is awake and alert and if there is no outlet obstruction (Wein, 1986). For over 30 years it has been used in the treatment of the atonic or hypotonic bladder (Lee, 1949) and it has been reported to be effective in achieving "rehabilitation" of the chronically atonic or hypotonic detrusor (Lapides, 1964; Diokno and Koppenhoeffer, 1976; Sonda et al., 1979). When so used, it is recommended that the drug be initially administered subcutaneously in a dose of 5–10 mg (usually 7.5 mg) every 4–6 hours. This is done preferably with an intermittent bladder decompression regimen. The patient is asked to try to void 20–30 minutes after each dose. When the residual urine volume has decreased to an acceptable level, the dose is gradually decreased and ultimately changed to an oral dose of 50 mg four times daily.

In cases of partial bladder emptying, a therapeutic trial with an oral dose of 25–100 mg, four times daily, may be utilized in conjunction with attempted voiding every four hours. Bethanechol has also been used to stimulate or facilitate the development of reflex bladder contractions in patients with suprasacral spinal cord injury (Perkash, 1975). Twiddy et al. 1980) have, in fact, concluded that—in spinal cats at least—intact pelvic reflex pathways are required for bethanechol to produce what they described as a brisk and sustained increase in intravesical pressure during bladder filling.

Although anecdotal success in specific patients seems to occur occasionally, attempts to facilitate bladder emptying in series of patients where bethanechol was the only variable have been generally disappointing. A pharmacologically active subcutaneous dose (5 mg) did not demonstrate significant changes in flow parameters or residual urine volume in women with a residual urine volume equal to or greater than 20% of bladder capacity, but no evidence of neurologic disease or outlet obstruction, or in a group of 27 "normal" women of approximately the same age (Wein et al., 1980a). A similar dose likewise failed to produce urodynamic evidence of improved emptying in patients with a positive bethanechol supersensitivity test (Wein et al., 1980b). This dosage did increase the intravesical pressure at all points along the filling limb of the cystometrogram, and also decreased the bladder capacity threshold, findings previously described by others (Lapides et al., 1963; Sonda et al., 1979). In adequate doses, bethanechol is capable of eliciting an increase in tension in bladder smooth muscle such as would be expected from in vitro studies, but its ability to stimulate or facilitate a physiologic-like bladder contraction in the vast majority of patients with voiding dysfunction has been unimpressive. Similar sentiments, at least with respect to its use in neuropathic dysfunction, have been expressed by Gibbon (1975), Merrill and Rotta (1977), Yalla et al. (1977), and Light and Scott (1982). In fact, it is difficult to find reproducible urodynamic data that support recommendations for the usage of bethanechol in any specific category of patients. Most, if not all, long-term studies in such patients are neither prospective nor double-blind, and do not exclude the effects of other simultaneous regimens, such as treatment of urinary infection, bladder decompression by continuous or intermittent catheterization, timed voiding with the Credé maneuver and other types of treatment affecting the bladder or outlet, an important lesson when considering the design of drug protocols for treatment of voiding dysfunction. Short-term studies in which the drug was

the only variable have generally failed to demonstrate significant efficacy in terms of flow and residual urine volume data. Whether repeated doses of any cholinergic agonist can achieve a clinical effect that a single dose cannot is speculative. If this is not the case, the long-term response to therapy could be predicted by a urodynamic assessment before and after a short trial. It is generally agreed that, at least in a denervated bladder, an oral dose of 200 mg is required to produce the same effect as a subcutaneous dose of 5 mg (Diokno et al., 1976; Philp et al., 1980).

Much of the emotion surrounding the question of whether bethanechol is efficacious can be dispelled by a brief urodynamically controlled trial in which institution of therapy is the only variable (Blaivas, 1984). In the laboratory, a functioning micturition reflex is an absolute requirement for the production of a sustained bladder contraction by a subcutaneous injection of drug (Downie, 1984). Clinically, there is little logic for its use in patients with detrusor hyperreflexia who already have bladder contractions, though the contractions are uncontrollable. Patients with incomplete lower motor neuron lesions seem to constitute the most reasonable group for a trial of bethanechol (Awad et al., 1984), though caution is advised in drawing conclusions regarding efficacy of a single modality when other modes of treatment (such as bladder decompression) are employed simultaneously.

No agreement exists as to whether cholinergic stimulation produces a increase in urethral resistance (Wein, 1980b). It would appear that pharmacologically active doses do in fact increase urethral closure pressure, at least in patients with neurogenic bladder dysfunction with detrusor hyperreflexia (Sporer et al., 1978). If so, a reasonable question would seem to be whether cholinergic agonists could be combined with agents to decrease outlet resistance to facilitate emptying. Khanna (1976) reported that a combination of a total daily oral dose of bethanechol of 50–100 mg with 20–30 mg of oral phenoxybenzamine produced satisfactory results in a group of patients with an atonic bladder and functional outlet obstruction. Our own experience with such therapy, utilizing even 200 mg of oral bethanechol daily, has been extremely disappointing. Certainly, most clinicians would agree that a daily dose of 50–100 mg rarely affects any urodynamic parameter at all.

The potential side effects of the cholinergic agonists and the anticholinesterase agents are similar and include flushing, nausea, vomiting, diarrhea, gastrointestinal cramps, bronchospasm, headache, salivation, sweating, and difficulty with visual accommodation (Taylor, 1980). Intramuscular or intravenous use is contraindicated, as such use can precipitate acute and severe side effects, resulting in acute circulatory failure and cardiac arrest. Contraindications to the use of this general category of drug include bronchial asthma, peptic ulcer, bowel obstruction, enteritis, history of recent gastrointestinal surgery, cardiac arrhythmia, hyperthyroidism, and any type of bladder outlet obstruction.

PROSTAGLANDINS.—It has been hypothesized (see previous chapter) that prostaglandins are intimately involved in and necessary for bladder contractile activity. Using this rationale, Bultitude et al. (1976) reported that instillation of 0.5 mg of prostaglandin E_2 (PGE_2) into the bladders of females with varying degrees of urinary retention resulted in acute emptying and in improvement of long-term emptying in two thirds of the patients studied. Desmond et al. (1980) reported further results with intravesical use of this agent in patients whose bladders exhibited no contractile activity or in whom bladder contractility was relatively impaired. Twenty of 36 patients showed a strongly positive and six showed a weakly positive immediate response. Fourteen patients were reported to show prolonged beneficial effects, all but one of whom had shown a strongly positive immediate response. Stratification of the data revealed that an intact sacral reflex arc was a prerequisite for any type of positive response. The authors noted additionally that the effects of PGE_2 appeared to be additive or synergistic with cholinergic stimulation in some patients. Vaidyanathan et al. (1981) reported that intravesical instillation of 7.5 mg of 15(S)-15-methyl prostaglandin F_2 produced reflex voiding in some patients with incomplete suprasacral spinal cord lesions. The favorable response to a single dose of drug, when present, lasted from 1.0–2.5 months. Other investigators including Stanton (1978) and Delaere et al. (1981) have reported absolutely no success with this type of treatment. Prostaglandins have a relatively short half-life and it is difficult to understand how any effects after a single application can last up to even several months. If such does occur, it must be the result of a "triggering effect" on some as yet unknown physiologic or metabolic mechanism. Potential side effects of prostaglandin usage include bronchospasm, chills, hypotension, tachycardia, cardiac arrhythmia, convulsions, hypocalcemia, and diarrhea (Moncada et al., 1980).

BLOCKERS OF INHIBITION.—As discussed in the preceding chapter, a sympathetic reflex exists, at least in the cat, which promotes urine storage by exerting an inhibitory effect on pelvic parasympathetic ganglionic transmission. This effect, although inhibitory, is α-adrenergic in nature. Some have suggested on this basis that α-adrenergic blockade, in addition to decreasing outlet resistance, may in fact facilitate transmission through these ganglia and thereby enhance bladder

contractility. Methyldopa (Raz et al., 1977) has been used with this rationale with at least some good initial results. Likewise, on this basis, Raz and Smith (1976) have advocated a trial of an α-adrenergic blocking agent for the treatment of nonobstructive urinary retention. Although we and others have had occasional anecdotal success using this or another exotic approach, we would caution against the assumption that improvement during administration of a drug occurs solely because of it.

The recent knowledge "explosion" in the field of neuropeptide physiology has also provided potentially new insights into lower urinary tract function and its pharmacologic alteration. Endogenous enkephalins are thought to exert a tonic inhibitory effect on the micturition reflex (see Chapter 27), and agents such as narcotic antagonists offer a new possibility for stimulating reflex bladder activity (Vaidyanathan et al., 1981; Thor et al., 1981). Comprehensive clinical trials will doubtless be done in the future.

ELECTRICAL STIMULATION.—Clinical trials of direct electrical stimulation of the bladder originated in 1940 (Dees, 1965), but have met with only partial success and intermittent enthusiasm since that time. Merrill (1974, 1975), Merrill and Conway (1974), and Halverstadt and Parry (1975) reviewed the literature through the mid-1970s and reported the results of their clinical trials. Direct electrical stimulation was most effective in patients with hypotonic and areflexic bladders. Initial success, defined as a low postvoid residual with sterile urine, was achieved in only 50%–60% of patients, and secondary failure, usually related to equipment malfunction, often supervened. The spread of current to other pelvic structures whose stimulus thresholds were lower than that of the bladder often resulted in abdominal, pelvic, and perineal pain; a desire to defecate; contraction of the pelvic and leg muscles; and erection and ejaculation in males. It was also noted that the increase in intravesical pressure was not generally coordinated with bladder neck opening or with pelvic floor relaxation, and that other measures to accomplish these ends might be necessary. Grimes et al. (1973, 1975) applied electrical stimulation directly to the sacral spinal cord, attempting to take advantage of the remaining intact motor pathways to initiate micturition. Although some short-term success was noted, many of the side effects seen with direct bladder stimulation occurred, since the stimulus, so applied, was also unphysiologic. Enthusiasm for both of these approaches has waned considerably, and resurrection seems unlikely.

Tanagho (1986) and his associates have logically pursued neurostimulation for both emptying and storage problems, and have recently described the evolution of their work (Schmidt, 1983; Tanagho, 1986). They too tried direct spinal cord stimulation but found that a very high rise in outlet resistance resulted. Although some emptying occurred, it was with a high intravesical pressure, a high outlet pressure, and occurred only at the end of stimulation, when the outlet resistance decreased. They found that it was not possible, at the spinal cord level, to separate a bladder center from a striated sphincter center. However, they have pursued, with some success, stimulation of individual sacral nerve roots. In an ingenious series of experiments, they found that, to achieve maximal specific detrusor stimulation with minimal sphincter activation, it was necessary to separate the dorsal component from the ventral component of the sacral root involved and to isolate and selectively section the somatic fibers of the sacral root to be stimulated.

REDUCTION CYSTOPLASTY.—On the surface, a reduction cystoplasty seems an attractive alternative for patients with chronic urinary retention who have large decompensated bladders. Using techniques of reduction cystoplasty involving either fundus invagination or detrusor duplication (but not simple excision of bladder tissue only), symptomatic successes with lower bladder capacity and lower residual urine volumes have been achieved (Roberts et al., 1979; Hanna, 1982; Kinn, 1985). However, it is puzzling that, in spite of such improvement, at least Roberts et al. and Kinn report absolutely no change in bladder or sphincter activity. In patients with no bladder contractility, flow rates seemed to change only when outlet reduction is also performed. However, there may in fact be merit in Hanna's suggestion (1982) that such procedures are rational in selected patients with very large bladder capacities who exhibit poor emptying because of only a partial deficiency of detrusor contractility.

Decreasing Outlet Resistance at a Site of Anatomic Obstruction

Prostatic enlargement and urethral stricture are two of the more common causes of bladder outlet obstruction in males. The pathophysiology of their development and their surgical and nonsurgical correction are very well dealt with elsewhere in this volume.

Decreasing Outlet Resistance at a Level of the Smooth Sphincter

TRANSURETHRAL RESECTION OR INCISION OF THE BLADDER NECK.—Emmett (1945) performed the first transurethral bladder neck resection in a patient with neurogenic bladder dysfunction in 1937, and for years this procedure represented the first line of surgical attack in cases of poorly balanced bladder function. Orig-

inally, the operation was performed primarily in two groups of patients: (1) those with weak or absent detrusor contractions, and (2) those with anatomic or functional obstruction at the level of the bladder neck and proximal urethra, which prevents emptying even with a sustained detrusor contraction (Gibbon et al., 1965; Bunce, 1967; Wein et al., 1976). Some urologists still prefer to resect the bladder neck whenever signs of outlet obstruction are associated with a neuropathic bladder, and treat other areas of outlet resistance later, if this proves necessary (Glahn, 1974; Gibbon, 1974a, 1974b).

More refined urodynamic techniques have resulted in the realization that dyssynergia at the level of the bladder neck or proximal urethra is uncommon, both in patients with neurologic disease and in patients with obstruction but without neurologic disease. We think that the prime indication for this procedure is the demonstration of true obstruction at the bladder neck or proximal urethra level by combined detrusor pressure-flow studies with either fluoroscopic demonstration of a failure of opening of the smooth sphincter area or a micturitional urethral profile showing that the bladder pressure falls off sharply at some point between the bladder neck and the area of the striated sphincter (McGuire, 1984). In the past, it was felt that another category of patient for whom this procedure would be useful was one with a sacral spinal cord lesion and an areflexic bladder who could achieve a measurable increase in intravesical pressure by straining and/or use of the Credé maneuver, but who could not empty the bladder adequately by these methods (Wein et al., 1976). Currently, we feel that the procedure simply creates a form of "graded" stress incontinence in these patients, and that other alternatives for adequate emptying should be sought first. Although documented instances of bladder neck obstruction in women have been reported (Diokno et al., 1984), extreme caution must be exercised in surgically treating this entity in females because of the risk of incontinence.

Techniques of resection include: (1) a thorough circumferential resection of all tissue between the internal orifice and the verumontanum (in males); (2) a limited resection of dorsal tissue from the 3 o'clock through the 9 o'clock position; (3) a resection further limited to the posterior lip; and (4) transurethral incision of the bladder neck at the 5 o'clock and/or 7 o'clock position. Turner-Warwick (1984) recommends a single full-thickness incision in one position in the male, extending from the bladder base down to the level of the verumontanum. He recommends deepening the incision until pinpoints of reflected light reveal minute interstitial fat globules between the latticework of the residual prostatic capsule fibers. Hemostasis is achieved by electrocoagulation with the flat surface of the knife electrode. Turner-Warwick (1984) reports the incidence of diminished volume of ejaculate to be only 10%, and in fewer than 5% of his patients has it been absent altogether. Our experience in this regard falls somewhere between the reported incidences of Moisey et al. (1982) (16%) and of Blaivas and Norlen (1984) (50%).

Y-V Plasty of the bladder neck.—A simple Y-V plasty of the bladder neck can accomplish the same effect as a transurethral resection or incision. However, especially in females, an "overcorrection" can result, and it would seem logical to recommend this only when an open surgical procedure is already required to correct a concomitant disorder. When outlet reduction is required to render a refractory, poorly emptying bladder totally incontinent prior to inserting an artificial sphincter, an open Y-V plasty—though originally mentioned as a possibility (Scott et al., 1974)—certainly makes subsequent placement of a bladder neck cuff extremely difficult.

Pharmacologic therapy.—Krane and Olsson (1973a, 1973b) endorsed the concept of a physiologic internal sphincter partially controlled by tonic sympathetic stimulation of contractile α-adrenergic receptors in the smooth musculature of the bladder neck and proximal urethra. Further, they hypothesized that some obstructions that occur at this level during detrusor contraction are a result of inadequate opening of the bladder neck and/or of inadequate decrease in resistance in the area of the proximal urethra. They also theorized and presented evidence that α-adrenergic blockade could be useful in promoting bladder emptying in such a patient with an adequate detrusor contraction but without anatomic obstruction or detrusor–striated sphincter dyssynergia. Abel et al. (1974) called attention to the fact that such a functional obstruction, which they, too, presumed to be mediated by the sympathetic nervous system, could be maximal in the urethra rather than at the bladder neck, and coined the term "neuropathic urethra." The implication that α-adrenergic blockade could be useful in certain patients with a failure to empty despite an adequate increase in intravesical pressure was first made by Kleeman in 1970 and was subsequently supported by others (Johnston and Farkas, 1975; Stockamp, 1975; Whitfield et al., 1976). Successful results, usually defined as an increase in flow rate, decrease in residual urine, and improvement in upper tract appearance (where pathological), could often be correlated with an objective decrease in urethral profile closure pressures. Although one would expect success with such therapy to be most evident in patients without detrusor–striated sphincter dyssynergia as reported by Hachen (1980), Mobley (1976) reported a startling 86% success rate in 21 patients with a reflex neurogenic bladder, with a corresponding success rate of 66% in what was termed "flaccid" and 57% in what was termed "autonomous" neurogenic bladder dysfunction, success being defined as postvoid residual urine volume consistently less than 100 ml. Such effects could be explained by a decrease in perineal striated muscle activity induced by α-adrenergic blockade (Nanninga et al., 1976), an effect that is probably mediated by a central mechanism, as evidence of adrenergic innervation of the human striated sphincter is lacking (Wein et al., 1979; Rossier et al., 1982). Scott and Morrow (1978), on the other hand, noted excellent results with phenoxybenzamine therapy in 9 of 10 patients with a flaccid bladder and a flaccid external sphincter, a single patient with an upper motor neuron bladder with intact sympathetic innervation, but only 8 of 21 patients with hyperreflexia and autonomic dysreflexia, and 0 of 6 patients with an upper motor neuron bladder and sympathetic denervation (lesion between T10 and L2).

There is also a rationale for the addition of α-adrenolytic therapy in the patient with inadequate emptying secondary to neuropathic voiding dysfunction after conventional pharmacologic treatment has failed. Parasympathetic decentralization has been reported to lead to a marked increase in adrenergic innervation of the bladder, with a resultant conversion of the usual β (relaxant) response of the bladder body in response to sympathetic stimulation to an α (contractile) effect (Norlen et al., 1976; Sundin et al., 1977). Although the alterations in innervation have been disputed (Nordling et al., 1980), the alteration in receptor function has not. Koyanagi (1979) showed supersensitivity of the urethra to α-adrenergic stimulation in a group of patients with autonomous neurogenic bladders. This implies a similar change in adrenergic receptor function in the urethra following parasympathetic decentralization. Parsons and Turton (1980) observed a similar phenomenon but provided a different pharmacologic explanation. They ascribed the cause to adrenergic supersensitivity of the urethral smooth muscle caused by sympathetic decentralization, resulting in an inappropriately high sensitivity to alterations in circulating catecholamine levels brought about as a part of normal cardiovascular homeostasis. Nordling et al. (1981) described a similar phenomenon in females after radical hysterectomy and ascribed this change to damage to the sympathetic innervation of the lower urinary tract.

The pentolamine stimulation test (Olsson et al., 1977) can usually predict the effectiveness of α-adrenolytic therapy in a given situation. Flow rates are measured before and after an intravenous dose of 5 mg, and the values are plotted on a nomogram that relates flow rate to volume voided. An increase of 0.8 units on the nomogram predicts improved voiding with oral α-adrenergic blocker therapy.

Phenoxybenzamine is the α-adrenolytic agent that was most commonly used in the treatment of voiding dysfunction (Wein, 1986). The initial adult dosage of this agent is 10 mg/day. The dose may be increased by 10 mg every 4–5 days to a recommended maximum of 60 mg/day. Daily doses larger than 10 mg are generally divided and given every 8–12 hours. The maximum effect of a particular dose usually becomes apparent only after a week following the initiation of or a change in therapy. After discontinuation, the effects of daily administration persist for about the same period of time. In our experience, patients who responded favorably to this agent generally did so at doses less than 30 mg and did not respond to dose increases with incremental improvement. Some patients who responded to 10 mg daily could be maintained on an even lower dose. Potential side effects include orthostatic hypotension, reflex tachycardia, nasal congestion, diarrhea, miosis, sedation, nausea, and vomiting (secondary to local irritation). Ejaculatory failure, due to lack of seminal emission and not to retrograde ejaculation, frequently occurs, but without any adverse effect on erection. Those who still use phenoxybenzamine for long-term therapy should be aware of the recently reported adverse in vitro and in vivo mutagenicity studies with this agent (see latest *Physicians' Desk Reference*). Further, the manufacturer has indicated a dose-related incidence of gastrointestinal tumors in rats. Although this agent has been in use for some 30 years in humans without any adverse epidemiologic associations, it is obvious that one must now carefully consider the potential medical-legal ramifications of long-term phenoxybenzamine therapy, especially in younger persons.

Prazosin hydrochloride is one of a class of new antihypertensive agents with an affinity for postsynaptic α_1 receptors, at least in vascular smooth muscle (Weiner, 1980), in contrast to classical α-adrenergic blocking agents such as phentolamine and phenoxybenzamine, which have blocking properties at both α_1 and α_2 receptor sites. Prazosin has been shown to cause α_1 blockade in the smooth muscle of the canine and human urethra (MacGregor and Diokno, 1981) and, in addition, it has been safely used clinically to lower outlet resistance in some patients (Andersson et al., 1981). Therapy with prazosin is generally begun in daily divided doses of 2–3 mg. The dose may be very gradually increased to a maximum of 20 mg daily, although we seldom use more than 9–12 mg daily in three divided doses. The potential side effects of prazosin are consequent to its α_1 blockade. Additionally, there occasionally occurs the "first-dose phenomenon," a symptom complex of faintness, dizziness, palpitation, and, infrequently, syncope,

thought to be due to acute postural hypotension. The incidence of this phenomenon can be minimized by restricting the initial dose of the drug to 1 mg and by administering this at bedtime. Other side effects associated with chronic prazosin therapy are generally mild and rarely necessitate withdrawal of the drug. Thus far, reports of adverse effects on seminal emission have been sparse.

Other agents with some α-adrenergic blocking properties at various levels of neural organization have urologic side effects that may be therapeutically useful in certain circumstances. Methyldopa is an antihypertensive agent that is converted to α-methyl norepinephrine, and that functions as a false neurotransmitter at a central and perhaps peripheral level, with the end result being a decreased peripheral sympathetic effect. Raz et al. (1977) have reported improved emptying in patients with neurogenic bladder dysfunction with this agent. Clonidine is another antihypertensive agent whose net action results in a decreased peripheral sympathetic effect, reflected in the urinary tract by a decrease in the urethral closure pressure profile (Nordling and Meyhoff, 1979). Other commonly used pharmacologic agents with significant α-adrenergic blocking properties include chlorpromazine and haloperidol (Weiner, 1980).

These and other agents with α-adrenergic blocking properties at various levels of neural organization have been utilized in patients with various types of voiding dysfunction. Our own experience would suggest that a trial of such an agent in such patients may certainly be worthwhile, as the effect or noneffect will become obvious in a matter of days, and the pharmacologic side effects are of course reversible. However, our results with such therapy in these patients have been considerably less spectacular than those of at least some other investigators.

β-Adrenergic stimulation has been shown experimentally to decrease the urethral pressure profile and by inference urethral resistance (Raz and Caine, 1971). Vaidyanathan et al. (1980) reported a decrease in urethral closure pressure after administration of terbutaline, a relatively specific β_2 agonist. Whether this drug or other pharmacologically similar agents will prove useful to facilitate bladder emptying by decreasing outlet resistance remains to be investigated.

Decreasing Outlet Resistance at the Level of the Striated Sphincter

EXTERNAL SPHINCTEROTOMY.—Therapeutic destruction of the external urethral sphincter was first performed in 1936 (Watkins) in a patient with nonspastic obstruction. Although this procedure was mentioned again by Baumrucker in 1943, the first large clinical series was reported in 1958 by Ross et al. The primary indication for this procedure is the failure of the bladder to empty in a male patient with a suprasacral lesion, and good but involuntary detrusor contractions when prostatic obstruction and obstructive inflammatory lesions at the bladder neck have been ruled out and when other types of management have been unsuccessful or are not possible.

A substantial improvement in bladder emptying occurs in 70%–90% of cases (Wein et al., 1976). Upper tract deterioration is rare after successful sphincterotomy, and vesicoureteral reflux—if present previously—often disappears because of lower bladder pressures and the reduced incidence of infection in a catheterless patient with a low residual urine volume. Although an external collecting device is generally worn postoperatively, total dripping incontinence or severe stress incontinence should be unusual unless the proximal sphincter mechanism (bladder neck and proximal urethra) has been compromised, either by prior surgical therapy or as a secondary effect of the striated sphincter dyssynergia.

Different techniques have historically been used for external sphincterotomy, including incisions at the 5 and 7 o'clock positions, the 3 and 9 o'clock positions, the 2, 4, 8, and 10 o'clock positions, and the 11 and 1 o'clock positions (Wein et al, 1976). The 12 o'clock sphincterotomy, originally proposed by Madersbacher and Scott (1975), is our method of choice for a number of reasons, which have been confirmed and commented on by others (Madersbacher and Scott, 1975; Yalla et al., 1977; Hachen, 1977; Madersbacher, 1986). First of all, the anatomy of the striated sphincter is such that its main bulk is anteromedial. The blood supply, on the other hand, is primarily lateral; thus there is less of a chance of significant hemorrhage with an incision at the 12 o'clock position. There is disagreement about the rate of postoperative erectile impotence in those individuals who have erections in series that have utilized the 3 and 9 o'clock technique. Estimates vary from 5%–30%, and whatever the true figure is, it is clear that most would agree that this complication is far less common (approximately 5%) with incision in the anteromedial position (Madersbacher, 1986). Other complications of external sphincterotomy may include significant hemorrhage and urinary extravasation. Urethral stricture rarely occurs at the site of sphincterotomy.

External sphincterotomy can be performed using a knife electrode or by resection with a loop electrode. The incision must extend from the level of the verumontanum at least to the bulbomembranous junction. A gradually deepening incision allows good visual control and minimizes the chances of significant hemorrhage and extravasation.

Sphincterotomy failure generally indicates an inadequate incision or resection of tissue. Failure may also be due to the presence of another unsuspected obstruction (e.g., at the bladder neck), or to inadequate bladder contractions, either those of insufficient magnitude or those that are poorly sustained. Residual urine does not always indicate "failure," as low pressure but incomplete emptying in an individual may be quite acceptable.

URETHRAL OVERDILATION.—Urethral overdilation to 40–50 F usually achieves the same objective as external sphincterotomy in females (Gibbon, 1974a). External sphincterotomy is rarely carried out, however, in females because of the lack of a suitable external collecting device. In young boys, when sphincterotomy is contemplated, a similar stretching of the posterior urethra can be accomplished through a perineal urethrostomy, obviating or postponing the need for formal sphincterotomy (Johnston and Farkas, 1975).

PUDENDAL NERVE BLOCK OR INTERRUPTION.—Relief of obstruction at the level of the striated sphincter can also be achieved by a pudendal neurectomy, which was first described in 1899 by Rocket (Stark, 1969). Historically, this method enjoyed the same popularity in patients with fixed neurologic lesions and striated sphincter dyssynergia, but is seldom used today (Wein et al., 1976). In the rare instance of usage, therapeutic assessment of the results of a pudendal block should precede the formal procedure, which should be carried out only unilaterally. Bilateral nerve section results in an unacceptably high rate of impotence in the male, and may result in fecal and severe stress urinary incontinence.

PHARMACOLOGIC THERAPY.—There is no class of pharmacologic agents that will selectively relax the striated musculature of the pelvic floor. Chlordiazepoxide, methocarbamol, orphenadrine, and diazepam belong to a group of agents classified as centrally acting muscle relaxants (Bianchine, 1980). The primary side effect of all members of this group is sedation, which many feel is primarily responsible for their muscle-relaxing effect when administered orally. In general, we have not found the recommended oral doses to be effective in controlling the classical types of striated sphincter dyssynergia seen secondary to neurological disease. If the etiology of incomplete emptying in a neurologically normal patient is obscure, and the patient has what appears to be inadequate relaxation of the pelvic floor striated musculature, a trial of an agent such as diazepam may be worthwhile. The theoretical rationale in such circumstances is either that of relaxation of the pelvic floor striated musculature during bladder contraction or of such relaxation removing a stimulus that is inhibitory to bladder activity. Improvement under such circumstances may simply be due, however, to the drug's antianxiety effect or to the intensive explanation, encouragement, and modified biofeedback therapy that usually accompanies this type of treatment in such patients.

Dantrolene sodium is a skeletal muscle relaxant that has been shown to dissociate excitation-contraction coupling at a site distal to the neuromuscular end-plate—in the sarcoplasmic reticulum (Bianchine, 1980). It has been shown to have therapeutic benefits for chronic spasticity associated with CNS disorders. The drug has been used in patients with classical detrusor striated sphincter dyssynergia, and was initially reported as being very successful in improving voiding function in these patients (Murdock et al., 1976). Therapy in adults is begun at a dose of 25 mg twice daily, increasing the dose weekly by 50- to 100-mg increments up to a recommended daily maximum of 400 mg given in divided doses. Hackler et al. (1980) achieved improvement in voiding function in approximately half of their patients treated with dantrolene, but found that such improvement required doses of 600 mg/day. Although the drug has no autonomic side effects, it may induce a generalized weakness severe enough to compromise its therapeutic benefits, especially at higher doses. Other potential side effects include euphoria, dizziness, diarrhea, and hepatotoxicity. The toxicity to liver is seemingly related to high-dose long-term usage, and has apparently resulted in some fatalities.

Baclofen is an agent that is thought to hyperpolarize primary afferent fiber terminals in the spinal cord (Bianchine, 1980) and to inhibit monosynaptic and polysynaptic spinal reflex activity (Duncan et al., 1976). It has been found useful in the treatment of skeletal spasticity due to a variety of causes. Treatment is started at an initial dose of 5 mg, three times daily, and the dose is doubled every three days until a daily total dose level of 60 mg is reached. The manufacturer recommends that the total daily dose not exceed 20 mg, four times daily. Florante et al. (1980) reported that 73% of their patients with voiding dysfunction secondary to acute and chronic spinal cord injury showed lower striated sphincter responses and decreased residual urine volume following treatment, but only with an average daily oral dose of 120 mg. Hachen and Krucker (1977) found a 75 mg daily oral dose to be ineffective in patients with striated sphincter dyssynergia and traumatic paraplegia, whereas they found a daily intravenous dose of 20 mg to be highly effective. Potential side effects of baclofen include drowsiness, insomnia, rash, pruritus, dizziness, and weakness. Hallucinations may sometimes occur after abrupt withdrawal.

BIOFEEDBACK, PSYCHOTHERAPY.—Biofeedback is a useful adjunct in the improvement of voluntary control of neural, visceral, and skeletal responses (Miller, 1975; Miller and Dworkin, 1977; Schwartz, 1979). Understanding the concept of biofeedback is much easier than formulating an all-inclusive unequivocal definition of the term. Literally, the feedback is information (data) about biologic processes that is used in a training program whose goal is the acquisition of some voluntary control over these processes. The data utilized in biofeedback may be as sophisticated as visual or auditory information produced by modern electronic equipment that is monitoring a physiologic process, or it may be as simple as a self-kept record of symptoms and biologic events (such as a chart for recording the time and volume of voiding and the presence of urgency and incontinence). The list of physiologic processes subject to some self-control includes blood pressure, heart rate, blood flow, sweat gland activity, skin temperature, body temperature, respiratory function, genital responses, bowel and bladder motility, and skeletal muscle control.

Schwartz (1979) points out that the clearest evidence for the efficacy of biofeedback is in the regulation of skeletal muscle activity. As he further states, this should not be surprising since, of all bodily systems, the skeletal muscles are under the most voluntary control. Hinman (1974, 1980) and Allen (1972, 1980) were perhaps the first to recognize that nonneurogenic striated sphincter dyssynergia (or a disorder qualitatively similar to it) can occur on a psychogenic basis and cause gross upper tract damage. Each used a combination of psychologic retraining, suggestion, medication, and parent cooperation in a treatment program. Wear and colleagues (1979) described the use of a urodynamic display of external sphincter EMG activity to facilitate this type of treatment. They reported significant clinical improvement in four of eight patients so treated. Maizels and co-workers (1979) further simplified this technique by using anal skin electrodes for the repeated display of striated sphincter activity without pain or significant inconvenience to the patient. Successful therapy in this type of situation obviously requires a strongly motivated patient who is capable of understanding the instruction for biofeedback training.

Circumventing Problems

INTERMITTENT CATHETERIZATION.—Intermittent catheterization (IC) has proved to be the most effective means of attaining a catheter-free state in the majority of patients with acute spinal cord lesions (Frankel, 1974). It is also an extremely effective method of treating the adult or child whose bladder fails to empty, especially when efforts to increase intravesical pressure and/or decrease outlet resistance have been unsuccessful. In those patients who have inadequate urine storage due to involuntary bladder contractions, decreased compliance, or stress incontinence with adequate or inadequate emptying, it may also be used if the dysfunction can be converted solely or primarily to one of emptying by pharmacologic or surgical means (Wein et al., 1976). Lapides (Lapides et al., 1974, 1976) deserves enormous credit for first applying the concept of self-IC to large groups of outpatients with voiding dysfunction. Subsequently, he and his co-workers and many others have demonstrated the long-term efficacy and safety of such a regimen. IC requires a cooperative, well-motivated patient or family. The patient must have adequate hand control (or a family member willing to catheterize), and adequate urethral exposure must be obtained. It is advantageous to have a special urologic nurse who instructs the patients and/or families in the regimen; provides them with understandable written instructions to refresh their memory regarding technique, precautions, and danger signals; and who provides continuing support for patients who call with questions or problems referable to their regimen. Teaching self-IC requires an approach that communicates acceptance of the procedure by the instructor. Many patients are initially extremely reluctant to perform any procedure on themselves that involves the genitalia. Patients need a thorough explanation of the advantages of IC along with assurance that it is simple and will not tie them to their homes nor to an absolute time schedule. Additionally, proper selection of equipment for the patient's intelligence and financial level will increase patient acceptance of and compliance with the self-IC program. Initially reticent patients are continually amazed by the ease with which such a regimen is established.

For adult male patients, 14 or 16 F red rubber catheters are generally used. In some cases (for example, in patients with impaired fine motor skills), stiffer plastic catheters may be easier for the patient to insert. These are also commercially available in a disposable form. A notable advantage to the red rubber catheters is their longevity. They can be reused indefinitely and boiled for sterilization if desired. For patients without insurance it is almost necessary to use them. Disposable plastic female catheters, now manufactured by several companies, are recommended for female patients. They are inexpensive, very convenient, and can be obtained under most insurance plans. These 14 F 6-in. catheters are easy to handle. Female patients may catheterize themselves on the toilet without a mirror, making IC less confusing, cleaner, and quicker. Red rubber or Robinson catheters may also be used for female patients if desired. For the patient's convenience, liquid cleansing agents are usually easier to handle in the form of cotton balls soaked in the agent. These can be easily carried

in a small jar. Any water-soluble lubricant is suitable for lubricating the catheter prior to insertion.

Continuous catheterization.—Indwelling urethral catheters may be used for short-term bladder drainage, and careful such use of a small-bore catheter (for a short time) does not, in our opinion, seem to adversely affect the ultimate outcome. Occasionally, an indwelling catheter is a last resort type of therapy for long-term bladder drainage. Virtually all patients with an indwelling urethral catheter have bacteriuria after a certain period of time. A contracted, fibrotic bladder may be the ultimate result. Bladder calculi may form on the catheter or on the retention balloon. Urethral complications are relatively uncommon in females, but bladder spasm may occur, producing urinary incontinence. The temptation to use a large retention balloon should be resisted, as the continuous use of such a balloon combined with some pressure on the catheter may cause erosion of the bladder neck. A suprapubic trochar catheter may be initially more comfortable than a urethral catheter and obviates urethral complications in the male, the main advantage of this type of continuous drainage over longer periods. Bladder spasm with incontinence may, however, be more of a problem, and when blockage occurs, nursing personnel are more reluctant to change this type of catheter without physician assistance. Finally, it should be mentioned that Kaufman et al. (1977) reported development of squamous cell carcinoma of the bladder in six of 59 patients with spinal cord injuries who had long-term indwelling catheters. Four of these had no obvious tumor visible on endoscopy and the diagnosis was made by bladder biopsy. Five of these patients also had transitional-cell elements in their tumor. Broecker et al. (1981) surveyed 81 consecutive spinal cord injury patients with an indwelling urinary catheter for greater than ten years and, although they did not find frank carcinoma in any, found squamous metaplasia of the bladder in 11, and leukoplakia in one. Thus, periodic urinary cytology and cystoscopy, perhaps with random biopsy, especially with the new onset of gross hematuria, seems necessary in patients on long-term indwelling catheter drainage.

Urinary Diversion

Supravesical diversion.—Although commonly employed in the past for treatment of neurogenic voiding dysfunction, supravesical diversion is rarely indicated in any patient with only voiding dysfunction. Indications for performing supravesical diversion may include (1) progressive hydronephrosis and intractable upper urinary tract dilatation (which may be due to blockade of the ureterovesical junction caused by a trabeculated, thick bladder, or to vesicoureteral reflux that does not respond to conservative measures), and (2) recurrent upper urinary tract infections. Although the various techniques of supravesical urinary diversion are covered in other chapters of this book, it is important to note that urinary diversion cannot always prevent upper urinary tract deterioration. The initial results of the ileal conduit seemed encouraging in terms of upper tract preservation (Bricker, 1950). Unfortunately, long-term results have been disappointing (Schwartz and Jeffs, 1975; Pitts and Muecke, 1979; Dunn et al., 1979; Orr et al., 1981). The colonic conduit, which allows a reflux-preventing implantation of the ureters, was expected to produce better long-term results. Although earlier reports seemed to support this supposition (Mogg and Syme, 1969; Altwein et al., 1977; Althausen et al., 1978), long-term results seem to be equally disappointing (Elder et al., 1979).

Continent forms of diversion that may be applicable to patients with voiding dysfunction involve the application of a continent abdominal wall stoma (Koch pouch, ileocecal reservoir, or Mainz pouch) (Goldwasser and Webster, 1985). Severe scoliosis and kyphosis are relative contraindications to conduit stomal diversion because of difficulty in fitting a stomal bag. However, this problem does not exist with continent stomal diversion (Ashken, 1982). On the other hand, continent stomal diversion requires that the patient be able to perform intermittent catheterization or that dependable assistance is available for catheterizing the stoma. Currently, there are no reports of long-term results in patients with these forms of continent diversion, and little is known about their long-term effects on the upper tracts. The subject of bladder reconstruction and substitution is covered more completely elsewhere in this book.

THERAPY TO FACILITATE URINE STORAGE

Inhibiting Bladder Contractility/Decreasing Sensory Input/Increasing Bladder Capacity

TIMED BLADDER EMPTYING.—This is actually a form of bladder conditioning (Opitz, 1984). Although it is not a sophisticated concept, like many we use it seems to work quite well in some circumstances. The idea is to completely or partially empty the bladder on a frequent enough schedule so as to keep the intravesical volume below that at which storage failure results. This may involve only frequent voiding with a limited fluid intake in the patient with detrusor hyperactivity. It may involve a more complicated coordination of an IC schedule with bladder volume in the patient with a detrusor hyperreflexia and sphincter dyssynergia whose status has been converted to that of relative emptying failure, but only if the bladder is emptied periodically

(before the threshold for hyperreflexia or decreased compliance causing storage failure or upper tract damage is reached).

Pharmacologic Therapy

ANTICHOLINERGIC AGENTS.—Because at least a major portion of the neurohumoral stimulus for physiologic bladder contraction is acetylcholine-induced stimulation of postganglionic parasympathetic cholinergic receptor sites on bladder smooth muscle, atropine and atropine-like agents will depress true detrusor hyperreflexia of any etiology (Pedersen and Grynderup, 1966; Blaivas et al., 1980; Jensen, 1981a). In such patients the volume to the first involuntary contraction will generally be increased, the amplitude of the contractions decreased, and the total bladder capacity increased, with a proportionate reduction in symptomatology. Interestingly, bladder compliance in normal individuals and in those with detrusor hyperreflexia, where the initial slope of the filling curve on cystometry is normal prior to the involuntary contraction, does not seem to be significantly altered (Jensen, 1981a). The effect of these agents on intravesical pressure during filling in those patients who exhibit only decreased compliance has not been well studied. In patients with detrusor hyperreflexia, although significant clinical improvement may be achieved, only partial inhibition generally results, because of the phenomenon of atropine resistance. (See Chapter 27 for a complete discussion.) Atropine only partially antagonizes the bladder response to either pelvic nerve stimulation or direct electrical stimulation. Atropine does, however, completely inhibit the response of the bladder smooth muscle to exogenously administered acetylcholine. At present, there is no completely accepted explanation for this relative atropine resistance, but the most attractive theory is that a major portion of the neurotransmission mediated by parasympathetic (pelvic) nerve stimulation is noncholinergic and nonadrenergic. In other words, part of the excitation caused by pelvic nerve stimulation is secondary to release of a transmitter other than acetylcholine or norepinephrine. Most of the drugs discussed under the heading of inhibiting bladder contractility are agents that inhibit muscarinic cholinergic receptor sites. These agents, though helpful, leave much to be desired in terms of efficacy in certain situations, and there certainly exists a potential pharmacologic basis for attempting to further inhibit bladder contractility by inhibiting the various other types of receptor sites thought to be responsible for the noncholinergic portion of the contractile response elicited through pelvic nerve efferent impulses. Unfortunately, clinically useful agents of this type are either unavailable (as in the case of ATP) or have not been shown to produce consistently useful clinical effects, at least at this time.

The potential side effects of all antimuscarinic agents include inhibition of salivary secretion (dry mouth), blockade of the iris sphincter muscle (pupillary dilation), blockade of the ciliary muscle of the lens to cholinergic stimulation (blurred vision for near objects), tachycardia, drowsiness, and inhibition of gut motility (Weiner, 1980). Those agents that possess some ganglionic blocking activity may also cause orthostatic hypotension at high doses. Antimuscarinic agents are generally contraindicated in patients with narrow-angle glaucoma and should be used with caution in patients with significant bladder outlet obstruciton, as complete urinary retention may be precipitated.

Although atropine sulfate is available as a 0.5-mg tablet form, and it and all related anticholinergics are well absorbed from the gastrointestinal tract, propantheline bromide is the oral agent that is most commonly used to produce an antimuscarinic effect in the lower urinary tract. The usual adult oral dosage is 15 to 30 mg every 4–6 hours. Oral administration in the fasting state rather than with or after meals is preferable from the standpoint of bioavailability (Gibaldi and Grundhofer, 1975). The clinical efficacy can generally be predicted by the observation of the effect of a parenteral dose on the cystometrogram (Blaivas et al., 1980). The parenteral preparation is no longer available in the United States, and atropine or glycopyrrolate may be used instead. There seems to be little difference between the antimuscarinic effects on bladder smooth muscle of propantheline and those of other antimuscarinic agents. Although there are obviously many other considerations that account for the activity of a given dose of a drug at its site of action, there is no oral drug available whose direct antimuscarinic binding potential, at least in vitro, approximates that of atropine better than the long available and relatively inexpensive propantheline bromide (Levin et al., 1982).

It would seem that an agent with a significant ganglionic blocking effect as well as such an action at the peripheral receptor level might be more effective in suppressing bladder contractility. Methantheline has a higher ratio of ganglionic blocking to antimuscarinic activity than does propantheline, but the latter drug, clinical dose per dose, seems to be at least as potent in each respect (Weiner, 1980). Methantheline does have similar effects on the lower urinary tract, however, and some clinicians still prefer it—in doses of 50–100 mg four times a day—over other anticholinergic agents (Lapides and Dodson, 1953; Hebjorn, 1977).

MUSCULOTROPIC RELAXANTS (ANTISPASMODICS).—These agents fall under the general heading of direct-acting smooth muscle depressants, whose "antispasmodic" activity reportedly is directly on smooth muscle at a site that is metabolically distal to the cholinergic receptor mechanism (Finkbeiner et al., 1978). Although all three of the agents to be discussed do relax smooth muscle in vitro by a papavarine-like activity, all have been found in addition to possess variable antimuscarinic and local anesthetic properties. There is still a question as to how much of their clinical efficacy is due simply to their atropine-like effect. If in fact any of these agents do exert a clinically significant inhibitory effect that is independent of an antimuscarinic action, there exists a therapeutic rationale for combining their use with that of a relatively pure antimuscarinic agent.

Oxybutynin chloride has been described as a moderately potent anticholinergic agent with a strong independent musculotropic relaxant activity, and local anesthetic activity as well (Lish et al., 1965; Finkbeiner et al., 1978; Fredericks et al., 1978). This agent has been used successfully to depress detrusor hyperreflexia in patients with neurogenic bladder dysfunction (Thompson and Lauvetz, 1976). A randomized double-blind control study in 30 patients with detrusor instability comparing oxybutynin, 5 mg three times daily, and placebo was carried out by Moisey et al. (1980). Of 23 patients who completed the study with oxybutynin, 17 had symptomatic improvement and nine had evidence of urodynamic improvement—mainly an increase in bladder volume at first contraction and an increase in total bladder capacity. The recommended adult dose of oxybutynin is 5 mg three to four times daily. The potential side effects are the same as those of propantheline.

Dicyclomine hydrochloride is also reported to possess a direct relaxant effect on smooth muscle in addition to an antimuscarinic action (Johns et al., 1976; Downie et al., 1977; Khanna et al., 1979). An oral dose of 20 mg, three times daily, in adults has been reported to increase bladder capacity in patients with detrusor hyperreflexia (Fischer et al., 1978). Our own experience suggests that the individual dose must often be raised to at least 30 mg to achieve a good clinical effect. The potential side effects are antimuscarinic ones. Flavoxate hydrochloride is another compound that has been reported to have a direct inhibitory action on smooth muscle in addition to anticholinergic and local analgesic properties (Kohler and Morales, 1968). Favorable clinical effects have been noted in patients with urodynamically documented detrusor hyperreflexia. However, Briggs et al. (1980) reported essentially no effect of this agent on detrusor hyperreflexia in an elderly population, an experience that would coincide with our own subjective impression of limited clinical efficacy in situations in which other less expensive agents have failed (Benson et al., 1977). However, as with all agents in this group, a short clinical trial may be worthwhile. The recommended adult dose is 100–200 mg three to four times daily. Reported side effects are rare and are primarily antimuscarinic in nature.

POLYSYNAPTIC INHIBITORS.—Baclofen has been discussed previously in the section about agents that decrease outlet resistance secondary to striated sphincter dyssynergia. It has also been shown to depress detrusor hyperreflexia secondary to a spinal cord lesion (Roussan et al., 1975; Kiesswetter and Schober, 1975). Taylor and Bates (1979), in a double-blind crossover study, reported it to be very effective also in decreasing daytime and nighttime urinary frequency and incontinence in patients with idiopathic detrusor hyperreflexia. Cystometric changes were not recorded, however, and it should be noted that considerable improvement was also obtained in this study with placebo.

CALCIUM ANTAGONISTS.—The rationale underlying the potential use of calcium antagonists for the inhibition of bladder contractility has been described previously (see Chapter 27). The calcium antagonist nifedipine has been shown to be an effective inhibitor of contraction induced by several mechanisms in human and guinea pig bladder muscle (Forman et al., 1978; Sjogren and Andersson, 1979). It is also capable of completely blocking the noncholinergic portion of the contraction produced by electrical field stimulation in rabbit bladder (Husted, Sjogren, and Andersson, cited by Husted et al., 1980). Nifedipine more effectively inhibited potassium-induced than carbachol-induced contractions in bladder strips, whereas terodiline, an agent with both calcium antagonistic and anticholinergic properties, had the opposite effect (Husted et al., 1980). However, terodiline caused a complete inhibition of the response of rabbit bladder to electrical field stimulation. This agent in low concentrations seemed to have mainly an antimuscarinic action, whereas at higher concentrations a calcium antagonistic effect became evident. In vitro experiments appeared to show that these two effects are at least additive with regard to bladder contractility. Whether the calcium antagonistic properties of terodiline contribute to its clinical effectiveness in vivo and whether the drug is actually more effective than standard antimuscarinic agents alone remain to be established. Rud et al. (1980) reported that in oral dosages of 12.5 mg 2 or 3 times daily it produced a marked decrease in the number of hyperreflexic contractions in a group of seven women with urgency incontinence and two with nocturnal enuresis. Bladder capacity was approximately doubled, and the amplitude of the contractions was decreased. In a double-blind crossover study in 12 women with motor urge incontinence, Ekman et al. (1980) reported an increase in bladder capacity and in the volume at which the sensation of urgency was experienced in all but one of the patients treated with terodiline, whereas placebo treatment had no effect on either objective or subjective parameters.

Palmer et al. (1981) reported a double-blind placebo trial with a single 20-mg daily dose of flunarizine in 14 females with urinary frequency, incontinence, and urodynamically proved detrusor instability. A statistically significant decrease in urgency was produced in the flunarizine-treated group, but there was no change in the frequency of micturition. Although there was a trend toward improvement of cystometric parameters, this was not statistically significant at the $P \le .05$ level. The side effects produced in patients who have been treated with calcium antagonists for voiding dysfunction have been small in number, but it should be noted that the potential side effects of these agents can be considerable and consist of hypotension, facial flushing, headache, dizziness, abdominal discomfort, constipation, nausea, rash, weakness, and palpitations. Although work in this area is obviously in its infancy, this class of agents may yet prove to be a promising alternative or addition to existing treatment for the inhibition of bladder contractility.

β-ADRENERGIC AGONISTS.—The presence of β-adrenergic receptors in human bladder muscle has prompted attempts to increase bladder capacity with β-adrenergic stimulation. Such stimulation can cause significant increases in the capacity of animal bladders, which contain a moderate density of β-adrenergic receptors (Larsen and Mortensen, 1978). In vitro studies show a strong dose-related relaxant effect of β_2-agonists on the bladder body of rabbits, but little effect on the bladder base or proximal urethra (Khanna et al., 1981). Terbutaline (Bricanyl), in oral dosages of 5 mg three times daily, has been reported to have a "good clinical effect" in some patients with urgency and urgency incontinence, but no significant effect on the bladders of neurologically normal humans without voiding difficulty (Norlen et al., 1978). Although these results are compatible with those in other organ systems (β-adrenergic stimulation causes no acute change in total lung capacity in normal humans while it does favorably affect patients with bronchial asthma), few if any adequate studies are available on the effects of β-adrenergic stimulation in patients with detrusor hyperactivity.

α-ADRENERGIC ANTAGONISTS.—Jensen has reported an increased α-adrenergic effect in the bladders of patients he characterized as "uninhibited" (1981b). Short-term and long-term administration of prazosin have been reported by him to increase bladder capacity and decrease the amplitude of contractions in this category of patients (Jensen, 1981b, 1981c). Rohner et al. (1978) found a change in the normal β response of canine bladder body to an α response after bladder outlet obstruction. Perlberg and Caine (1982) studied the in vitro response of bladder dome muscle from patients with obstructive prostatic hypertrophy and found an α-adrenergic response to noradrenaline (instead of the usual β-adrenergic response) in 23% of 47 patients. They speculated as to a potential relationship between irritative symptoms in these patients and this altered adrenergic response. They further theorized that at least some of the symptomatic improvement in irritative symptoms seen in patients with benign prostatic hypertrophy treated with α-adrenergic blocking agents may be due to a direct effect of these agents on bladder muscle, rather than their effect on outflow resistance. Although highly speculative, this potential avenue of therapy seems worthy of further study, though it should be noted that considerable improvement was obtained also with a placebo.

PROSTAGLANDIN INHIBITORS.—As mentioned previously, prostaglandins are one class of compound with a potentially important role in excitatory neurotransmission in the lower urinary tract. Thus, there exist multiple mechanisms whereby inhibitors of prostaglandin synthesis might decrease bladder contractility.

Cardozo et al. (1980) reported such effects in a double-blind placebo study of 30 women with detrusor instability, with which they used the prostaglandin synthetase inhibitor flurbiprofen at a dosage of 50 mg three times daily. Abnormal bladder activity, however, was not abolished in significantly more drug-treated than placebo-treated patients, and actual bladder capacity likewise showed no change. It was concluded that the drug did not abolish detrusor hyperreflexia, but delayed the intravesical pressure rise to a greater level of distention. Forty-three per cent of the patients experienced side effects from the drug, primarily nausea, vomiting, headache, indigestion, gastric distress, constipation, and rash. Cardozo and Stanton (1980) reported symptomatic improvement in patients with detrusor instability given indomethacin, another prostaglandin synthetase inhibitor, in dosages of 50–200 mg/day. This was a short-term study with no cystometric data, and the drug was compared only with bromocriptine. The incidence of side effects was high, although no patient had to stop treatment because of these. It is interesting that this category of agents has proved to be useful in primary dysmenorrhea, a condition felt to be related to a high level of menstrual endometrial prostaglandin synthesis (Chan et al., 1981). Numerous prostaglandin inhibitors exist, most of which fall under the heading of nonsteroidal anti-inflammatory

drugs. The most common such agent, aspirin, is only a relatively weak inhibitor of prostaglandin synthesis.

TRICYCLIC ANTIDEPRESSANTS.—Some authors have found tricyclic antidepressants, particularly imipramine hydrochloride, to be particularly useful for facilitating urinary storage (Milner and Hills, 1968; Petersen et al., 1974; Raezer et al., 1977; Castleden et al., 1981). These agents have been the subject of a voluminous amount of highly sophisticated pharmacological investigation to determine the mechanisms of action responsible for their varied effects (Hollister, 1978; Baldessarini, 1980; Richelson, 1983). Most of the investigations have had the aim of explaining their antidepressant properties, and consequently have been carried out primarily on central nervous system tissue. The results, and the conclusions and speculations inferred from them, are extremely interesting, but it must be emphasized that it is essentially unknown whether they apply to or have relevance for the lower urinary tract.

All of these agents possess varying degrees of at least three major pharmacological actions: they have central and peripheral anticholinergic actions at some—but not all—sites; they block the active transport system in the presynaptic nerve ending that is responsible for the reuptake of the released amine neurotransmitters serotonin and noradrenaline; and they are sedatives, an action that occurs presumably on a central basis but is perhaps related to their antihistaminic properties. There is also evidence that they desensitize α_2 receptors on central noradrenergic neurons (Spyraki and Fibiger, 1980). Paradoxically, they have also been shown to block α-adrenergic receptors and serotonin receptors. Many such compounds are available and have been categorized as to their relative potency insofar as their multiple effects are concerned.

Imipramine has prominent systemic anticholinergic effects (Baldessarini, 1980). However, it appears to have only a weak antimuscarinic effect on bladder smooth muscle (Dhattiwala, 1976; Olubadewo, 1980; Levin et al., 1983). A strong direct inhibitory effect on bladder smooth muscle does exist, however, which is neither anticholinergic nor adrenergic (Dhattiwala, 1976; Benson et al., 1977; Fredericks et al., 1978; Olubadewo, 1980). This may be due to a local anesthetic–like action at the nerve terminals in the adjacent effector membrane, an effect that seems to occur also in cardiac muscle (Bigger et al., 1977), or to an inhibition of the participation of calcium in the excitation-contraction coupling process (Olubadewo, 1980). Clinically, the drug seems effective in decreasing bladder contractility and in increasing outlet resistance (Cole and Fried, 1972; Mahony et al., 1973; Raezer et al., 1977; Castleden et al., 1981). In trying to correlate mechanism of action with clinical effect, one might postulate that the increase in outlet resistance is due to a peripheral blockade of noradrenaline reuptake, which would tend to produce or enhance an α-adrenergic effect in the smooth muscle of the bladder base and proximal urethra. Theoretically at least, the latter action, if indeed it occurs in the lower urinary tract as it does centrally, might also tend to stimulate predominantly β-adrenergic receptors in bladder body smooth musculature, an action that would further facilitate urine storage by decreasing the excitability of smooth muscle in that area.

Castleden et al. (1981) began therapy in elderly patients with detrusor instability with a single 25-mg nighttime dose, which was increased every third day by 25 mg until either the patient was continent, had side effects, or a dose of 150 mg was reached. Six of ten patients became continent and, in those who underwent repeated cystometry, bladder capacity increased by a mean of 105 ml and bladder pressure at capacity decreased by a mean of 18 cm H_2O. Maximal urethral pressure increased by a mean of 30 cm H_2O. Although our subjective impression (Raezer et al., 1977) was that such effects became evident only after days of treatment, some patients in this series became continent after only 3–5 days of treatment. Our usual adult dosage of imipramine for voiding dysfunction is 25 mg, four times daily; half that dose is given in elderly patients. In our experience the effects of imipramine on the lower urinary tract are often additive to those of the atropine-like agents. Consequently, a combination of imipramine and propantheline is sometimes especially useful for decreasing bladder contractility (Raezer et al., 1977). If imipramine is to be used in conjunction with an atropine-like agent, it should be noted that the anticholinergic side effects of the drugs may be additive.

When used in the generally larger doses employed for antidepressant effects, the most frequent side effects of this group of agents are those attributable to their systemic anticholinergic activity. Allergic phenomena, including rash, abnormal liver function, obstructive jaundice, and agranulocytosis may also rarely occur. Side effects on the central nervous system may include weakness, fatigue, a parkinsonian effect (a fine tremor most noted in the upper extremities), and a manic or schizophrenic behavioral picture. Sedation may also result from an antihistaminic effect. Postural hypotension may be seen, presumably on the basis of α_1 receptor blockade.

Antidepressants can be shown electrophysiologically and hemodynamically to have a depressant effect on the myocardium shortly after their institution (Burgess and Turner, 1980; Muller and Schultze, 1980). However, Veith et al. (1982) have pointed out that in the animal

studies showing a direct myocardial depressant effect, this occurs at plasma concentrations that are in the toxic range for humans. This group of investigators followed 24 depressed patients with heart disease treated for four weeks and found that tricyclic antidepressants had no effect on left ventricular ejection fraction at rest or during maximal exercise. The mean daily dose of imipramine in their series was 129 mg and they suggest that such antidepressant doses can be used in depressed patients with pre-existing heart disease, except for those with severely impaired myocardial performance, without an adverse effect on ventricular rhythm or hemodynamic function. Whether cardiotoxic effects will prove to be a legitimate concern in patients receiving somewhat smaller doses for lower urinary tract dysfunction remains to be seen, and is a potential matter of concern.

If imipramine or any of the tricyclic antidepressants is to be prescribed for the treatment of voiding dysfunction, the patient should be thoroughly informed of the usual indications, the potential side effects, and, in the United States, of the fact that this drug is not approved by the FDA for this purpose. The onset of significant side effects (severe abdominal distress, nausea, vomiting, headache, lethargy, and irritability) following abrupt cessation of high doses of imipramine in children (Petti and Law, 1981) would suggest that the drug should be discontinued gradually, especially in patients receiving high doses. Tricyclic antidepressants can also cause excess sweating of obscure etiology and a delay of orgasm and orgasmic impotence, whose cause is likewise unclear (Baldessarini, 1980). The use of imipramine is contraindicated in patients receiving monoamine oxidase inhibitors, as severe central nervous system toxic effects (hyperpyrexia, seizures, and coma) can be precipitated.

DIMETHYL SULFOXIDE.—Dimethyl sulfoxide (DMSO) is a relatively simple, naturally occurring organic compound that has been used as an industrial solvent for many years. Among its properties is its ability to penetrate the intact skin, transporting chemicals along with it for absorption into the bloodstream. It has multiple pharmacological actions (membrane penetrant, anti-inflammatory, local analgesic, bacteriostatic, diuretic, cholinesterase inhibitor, collagen solvent, vasodilatory) and is used for the treatment of arthritis and other musculoskeletal disorders [*Medical Letter*, 1980; Council on Scientific Affairs, 1982]. The only formulation approved for use on human bladder is a 50% solution; a 70% solution is generally used topically for musculoskeletal disorders.

Stewart and Shirley (1976) reported symptomatic improvement in 75% of patients with interstitial cystitis following a course of treatment with this agent and an improvement in bladder capacity in 80% of these patients. Generally, one 50-ml intravesical instillation is carried out every other week for a total of six treatments. The solution is allowed to remain in the bladder for 15 minutes, after which it is expelled by spontaneous voiding. Fowler (1981) reported results with dimethyl sulfoxide in 20 patients with early interstitial cystitis; three complete and 16 partial symptomatic remissions were achieved. However, functional bladder capacity, measured at the termination of therapy in 18 cases, was observed to increase by more than 25% in only four patients. In ten patients with severe detrusor hyperreflexia, Andersen et al. (1981) were able to demonstrate no subjective or objective effects of the drug. The unpleasant garlic-like odor of dimethyl sulfoxide is its primary side effect and makes double-blind placebo studies impossible. Cataracts have been reported in experimental animals but not in humans; eye evaluation is, however, recommended with chronic therapy. As a last resort, this drug has been used by some clinicians in the frustrating patient with the urgency-frequency syndrome but without objective evidence of true interstitial cystitis or detrusor hyperreflexia. Although anecdotal improvement is sometimes reported, few—if any—formal reports of the results of such therapy exist. Currently, we use this drug only for the treatment of interstitial cystitis.

BIOFEEDBACK, BLADDER RETRAINING.—Cardozo and associates (1978, 1980) used biofeedback-assisted training in the treatment of a group of patients with idiopathic detrusor hyperreflexia and consequent symptoms of urinary incontinence, urgency, and frequency. With a detailed explanation of the problem and of the procedure, subtracted detrusor pressure data were converted to auditory and visual stimuli. The patients were asked to try to "control" the rise in intravesical pressure by any maneuver they found helpful. In between weekly sessions, they kept a diary to record the occurrence of urgency, frequency, and incontinence. Twenty-seven of 32 female patients completed the study. According to subjective data, 11 were cured, 11 improved, and 5 were considered failures of therapy. Objectively, the corresponding numbers were 12, 4, and 11. In this series, the treatment was most effective in patients with mild to moderate detrusor instability and was not helpful in severe cases, defined as a large detrusor contraction simultaneous with the first sensation of the desire to void.

Frewen (1978, 1980) used "bladder drill retraining" in patients with cystometrically documented idiopathic detrusor instability and urgency incontinence. Each patient was given a detailed explanation of the presumed causation of the symptomatology along with a written description. A micturition chart, on which was recorded the times of and intervals between urination, the volume produced, and the degree of incontinence, served as the biofeedback data. In this study, three quarters of the patients were hospitalized for an average stay of ten days. Supportive therapy with sedatives and anticholinergics was used in all patients. The objective cure rate was 82.5%. In a later study, 30 consecutive patients with urgency and/or frequency, ten of whom had urgency incontinence, were treated on an outpatient basis. Provocative cystometry was carried out in 22 of these patients, only eight of whom showed detrusor instability. At the conclusion of 12 weeks of treatment, 24 of the 30 patients were subjectively free of urinary symptoms. However, of the eight patients with bladder instability, only three showed objective disappearance of detrusor hyperreflexia.

Pengelly and Booth (1980) compared the results of bladder training in 25 patients (of 28 who completed their protocol) with a group of historical controls. All patients had cystometrically proved detrusor instability without obvious cause. Drug therapy was either stopped before admission to the study or remained constant throughout its duration. The overall symptomatic assessment of the patients in the treatment group showed that 7 were cured, 12 improved, 5 showed no change, and 1 was worse. The corresponding figures in the control group were 1, 7, 16, and 1. According to cystometric data, 8 patients in the treatment group were cured, 3 were improved, 12 showed no change, and 2 were worse. In the control group the corresponding figures were 0, 4, 17, and 4. Poor prognostic factors were a history of nocturnal enuresis after the age of 10 years, isometric bladder contraction above 100 cm H_2O during voluntary interruption of micturition, and failure to show any improvement within the first two weeks of treatment.

Elder and Stephenson (1980) reported the results of a Frewen-type regimen in the treatment of 21 of 27 patients who completed their study. All patients initially had symptoms of increased frequency, urgency, and urgency incontinence. Cystometric studies showed that 7 had detrusor instability, 11 had "reduced compliance," and 3 had normal filling curves. In addition to the bladder training program, 18 of 21 patients were admitted to the hospital for their initial therapy, which in all cases included an anticholinergic (emepronium bromide) and a tricyclic antidepressant (a combination of nortriptyline and fluphenazine). Of the three patients with normal cystometric filling curves, two were symptomatically cured and one was improved. In 5 of the 11 patients with "reduced compliance" there was a reversion of the cystometrogram to normal; all five of these women were symptomatically cured. In six patients the filling curve was unchanged; symptomatically, two of these were cured, three were improved, and one was unchanged. None of the seven patients with detrusor instability showed any change in the bladder filling curves. However, symptomatically, three were cured, while two were improved.

Although it is impossible to isolate the various factors in each of these studies, the value of some type of biofeedback data as an adjunct to the patients' training and conditioning seems obvious. Although in the United States hospitalization for such a regimen is apt to be disallowed because of objections to the cost by third-party insurance carriers, the combination of thorough explanation with continued support and perhaps adjunctive pharmacologic therapy, along with feedback data, certainly seems justified as a trial in these often extremely difficult and frustrating cases.

BLADDER OVERDISTENTION.—Cystodistention involves prolonged stretching of the bladder wall using a hydrostatic pressure equal to systolic blood pressure. Smith (1981) has summarized the experience at his center and detailed the technique used. The procedure is a modification of the original cystodistention procedure described by Helmstein. The bladder is distended to systolic blood pressure for four successive periods of 30 minutes. After each distention, the bladder is emptied and its capacity measured. An indwelling catheter is left overnight. For repeat distentions, longer periods are used. Distention is usually carried out under continuous regional (epidural) anesthesia. Improvement, when it occurs, is generally attributable to ischemic changes in the nerve endings or terminals in the bladder wall (Sehn, 1978). Thus, the procedure can be grouped under the general heading of those procedures designed to produce bladder denervation. Potential complications include bladder rupture (5%–10%), hematuria, and retention. We have been unimpressed with the results of this procedure in patients with storage failure secondary to neuropathic detrusor hyperactivity. Pengelly et al. (1978) studied 46 patients with urinary symptoms associated with detrusor hyperreflexia and came to similar conclusions. Of 43 patients fully reevaluated urodynamically, none showed a conversion from detrusor hyperreflexia to normal bladder behavior. Only four patients were symptomatically improved, while five patients reported that their symptoms worsened.

ELECTRICAL STIMULATION.—Electrical stimulation, applied through removable anal and vaginal devices and peripherally through patch electrodes, has been used to facilitate urine storage by an inhibition of bladder contractility (Glen, 1975; Godec and Krolg, 1976; Erlandson and Fall, 1977; Merrill, 1979; Plevnik and Janež, 1979; Godec et al., 1981). When effective, this is primarily by an inhibitory pudendal to pelvic nerve reflex. In cats, this depression of bladder contractility also involves a pudendal to hypogastric reflex with further inhibition mediated through a peripheral β-adrenergic effect on the bladder smooth muscle (Erlandson and Fall, 1977). There is much disagreement in the literature regarding the mechanisms involved, the optimal parameters for electrical stimulation, the necessity for chronic vs. acute stimulation, the criteria for patient selection, and the actual cure or improvement rates. Optimal criteria for successful use in a cooperative patient include the following: (1) preservation of the morphology of the urinary tract, (2) preservation of the sacral reflex center, (3) a low degree of peripheral denervation of the pelvic floor musculature, (4) positive urodynamic and neurophysiologic responses to test stimuli, and (5) ability to empty satisfactorily when the stimulus is turned off or with intermittent catheterization. In what authors term "properly selected cases" of detrusor incontinence, initial rates of cure or improvement as high as 50% are seen. McGuire (1983) reported 16 patients with involuntary contractions of varying etiologies treated with common perineal or posterior tibial nerve stimulation. Twelve patients were dry, three improved, and one "possibly improved." The parameters used were 5–8 V, 2–10 cycles per second, and a pulse width of 5–20 msec. It is very important when using this therapeutic option to make sure that the stimulation delivered by the device selected coincides with that desired. Even so, this modality has been much less successful in our hands than in others.

Interruption of Innervation

CENTRAL.—Historically, central (subarachnoid) block was not used solely for urologic indications but was used to convert a state of severe somatic spasticity to flaccidity and to abolish autonomic dysreflexia (Bors, 1957; Comarr, 1959; Bors and Comarr, 1971). As a byproduct, a reflex neurogenic bladder was converted acutely to an areflexic one that could be emptied by methods described previously or by IC. In 8 of 29 patients in Gibbon's series (1966) and in 7 of 11 patients in that of Misak et al. (1962), a flaccid bladder developed that emptied without further surgical or medical treatment. Additional therapy, when necessary, was directed at increasing intravesical pressure and decreasing outlet resistance. The obvious disadvantage of this type of procedure is a lack of selectivity, with unintended motor or sensory loss other than related to the urinary bladder. Impotence was very common, and in those patients with some residual motor or sensory function, these were often significantly altered or lost. Additionally, as will be further discussed in a subsequent section, the conceptually simple result of an areflexic bladder, although it may be produced acutely, very often is not maintained on a long-term basis after decentralization. Such patients often develop decreased compliance, resulting in significant storage problems.

PERIPHERAL (SACRAL RHIZOTOMY OR SELECTIVE SACRAL RHIZOTOMY).—In most cases bilateral anterior and posterior sacral rhizotomy or conusectomy will also convert a hyperreflexic, hypertonic bladder to an autonomous one (Meirowsky et al., 1950; Misak et al., 1962). Adequate emptying was achieved in 23 of 28 such patients treated by Misak et al. (1962), but 13 required subsequent catheterization because of persistent vesicoureteral reflux or hypersensitivity to or technical difficulty with the required collecting device. Erections were lost or impaired in 85% of patients who were potent preoperatively. A temporary impairment of bowel function, lasting 6–12 weeks, also occurred postoperatively.

Selective sacral nerve section was introduced as a treatment to increase bladder capacity by abolishing only the motor supply responsible for uninhibited contractions, leaving sphincter and sexual function intact. The initial use of this procedure followed the observation by Heimburger et al. in 1948 that the third sacral anterior root provides the dominant motor innervation of the human bladder. To enhance the clinical response and minimize side effects, differential sacral rhizotomy should be preceded by stimulation and blockade of the individual sacral roots with cystometric and sphincterometric control. Rockswold, Bradley, and Chou (1973) and Torrens and Griffith (1974) described their experiences with differential sacral rhizotomy. Of Torrens' nine cases, two were unequivocal failures, two were successes, and five were "improved." No significant postoperative disturbance of bowel or sexual function occurred.

Although technique refinements, such as percutaneous radiofrequency selective sacral rhizotomy (Mulcahy and Young, 1978), have occurred, there is still much argument as to the place of these procedures within a plan of treatment for detrusor hyperactivity. Torrens (1985) quotes successful results in collected groups of patients that range from 48% in patients with idiopathic instability (presumably urodynamically documentable involuntary bladder hyperactivity without neurologic disease) to 81% in patients classified as having a paraplegic bladder. However, as he points out, what is meant by "success" varies from one series to another and from one patient to another. These procedures are seldom used today, and certainly should be preceded by urodynamic and urologic evaluation of the effects of selective nerve blocks before their performance, especially in patients without fixed neurologic disease or injury. Even then, unintended effects on pelvic and lower extremity sensory or motor functions may occur, with disastrous sequelae. Needless to say, the informed consent before such a procedure should be the responsibility of an experienced neurosurgeon.

PERIVESICAL (PERIPHERAL BLADDER DENERVATION).—There are a number of procedures that fall into the category of those intending to achieve a peripheral parasympathetic decentralization of the bladder. As will be evident from neuroanatomic considerations discussed in the previous chapter, such attempts achieve, at best, partial peripheral denervation. The reported success rates for many of these procedures are high enough to make one wonder why they are not more commonly utilized. There seem to be a number of reasons for this. First, many clinicians feel that standard and more conservative methods are often successful in managing bladder hyperactivity in these patients. In their hands the success rates of peripheral bladder denervation in patients who fail vigorous but nonsurgical attempts at therapy are much lower than those reported. Second, there is almost a total lack of long-term follow-up in any of these procedures, and "postoperative assessment" usually means within a few months of the procedure.

Transvaginal partial denervation of the bladder was originally described by Ingelmann-Sundberg in 1959. This procedure has been utilized mostly for the treatment of refractory urge incontinence, and in Ingelmann-Sundberg's hands, success rates of up to 80% have been achieved (Ingelmann-Sundberg, 1980). Mundy (1985) cites reports of the successes of others in 50%–65% of cases.

Cystolysis is a term used to describe extensive perivesical dissection and mobilization with division of the superior vesical pedicle and the ascending branches of the inferior vesical pedicle (Leach et al., 1983). At the conclusion of this procedure, the bladder is completely mobilized from all surrounding attachments except for the ureters, a portion of the inferior vesical pedicle that contributes blood supply to the ureter, and the completely mobilized urethra. The intent originally was to provide relief of bladder pain in patients with refractory interstitial cystitis. The procedure was originally reported by Worth and Turner-Warwick (1973) and was used by Freiha and Stamey (1980) in a small heterogenous group of patients with symptomatic bladder hyperactivity. Leach et al. (1983) originally reported that 28 of 32 patients so treated experienced partial or complete relief of their preoperative pain. Permanent retention occurred in seven of their 32 cases, and in six patients, intractable frequency continued.

Supratrigonal transsection of the bladder (cystocystoplasty) was first reported by Turner-Warwick and Ashken (1967) as a method of achieving a similar peripheral neural decentralization of the bladder body. Mundy (1985) has cited the encouraging reports of others, and reviewed his larger experience. Of 104 patients with a follow-up of 1–5 years, 74% were cured, 14% were improved, and 12% were failures. Between 20–32 months, 10% of the group initially judged to have a satisfactory response suffered relapse, giving a long-term subjective success rate of 65%. Thirty-five percent of those who claimed to be symptomatically cured had reverted to stable detrusor behavior, prompting Mundy to comment that a symptomatic cure did not necessarily mean a urodynamic one, and a symptomatic failure did not necessarily mean a urodynamic failure.

AUGMENTATION CYSTOPLASTY.—Enterocystoplasty is the most major of surgical procedures for the treatment of storage failure secondary to bladder dysfunction. Originally, this procedure was used primarily for the treatment of bladder contraction secondary to tuberculosis or other types of fibrosis (Smith et al., 1977). The original UCLA experience included some patients with neurogenic bladder dysfunction, and the indications of the procedure were subsequently widened to produce very favorable results in this group of patients who had failed more conservative management (Linder et al., 1983). Although controversy exists regarding the optimal method of bladder reconstruction, and regarding the amount of bladder to be removed, successful results have been reported by a number of clinicians, as long as the technique employed avoids making a diverticulum of the augmenting bowel patch (Bramble, 1982; Mundy and Stephenson, 1985). Goldwasser and Webster (1986) have nicely reviewed the subject and categorized success rates by dysfunction etiology. Positive results have been obtained in 80%–90% of patients with tuberculous cystitis, interstitial cystitis, and neurogenic bladder dysfunction, while the success rate in a small number of patients with radiation injury to the bladder has been in the neighborhood of 50%. Many of these patients will experience emptying failure afterward, and in most of these, preoperative prediction of this phenomenon is possible by careful urodynamic evaluation. Sphincter weakening procedures may then be employed, or IC may be used to obviate this problem. Storage failure secondary to bladder dysfunction and a significant sphincter component may be safely managed with augmentation cystoplasty in combination with the use of an artificial urinary sphincter. IC can even be performed safely under these circumstances.

Increasing Outlet Resistance

PHYSIOTHERAPY.—By periodically carrying out various pelvic floor exercises for several months, it is possible for at least females to increase pelvic striated muscle tone, and to control minor degrees of prolapse and sphincteric incontinence (Jeffcoate, 1975; Greenhill, 1979). Kegel's exercises are designed to improve the functioning of the pubococcygeus muscles. These exercises involve voluntary tensing of the perineal muscles approximately 20 times in the morning, afternoon, and evening to strengthen the perineal muscular, fascial, and elastic tissues. Patients are instructed to contract their perineal muscles as though interrupting the flow of urine and to squeeze the vaginal muscles while drawing up the perineum and rectum as though interrupting a bowel movement. Every effort should be made to motivate the patient to learn the exercise routine correctly and to practice it conscientiously and on a daily basis. In uncomplicated urinary stress incontinence, consistent exercise is said to bring symptomatic improvement within six weeks, although several months are required to bring all supportive and sphincteric structures to optimal capacity. In our experience, only the most minor degrees of sphincteric incontinence can be significantly improved in this way.

ELECTRICAL STIMULATION.—Intravaginal and anal electrical stimulation have been utilized to treat storage failure by increasing outlet resistance as well as by decreasing bladder contractility (see previous text). The mechanism involves indirect stimulation of the striated pelvic floor musculature through branches of the pudendal nerve (Erlandson and Fall, 1977; Merrill et al., 1975; Teague and Merrill, 1977). In cats, additional urethral closure is provided by a pudendal to hypogastric reflex that stimulates the smooth muscle of the bladder neck and proximal urethra via an α-adrenergic effect. In "properly selected cases," initial cure or improvement rates range as high as 50%–80%, figures that we have not been able to reproduce. One must remember to select the proper unit, as parameters for this type of stimulation differ from those employed to inhibit bladder contractility.

Pharmacologic Therapy

α-ADRENERGIC AGONISTS.—The bladder neck and proximal urethra contain a preponderance of α-adrenergic receptor sites that, when stimulated, produce smooth muscle contraction. Such stimulation alters the urethral pressure profile by increasing maximum urethral pressure (MUP) and maximum urethral closure pressure (MUCP) (Ek et al., 1978). Various orally administered pharmacologic agents are available to produce α-adrenergic sitmulation with relatively mild side effects. Potential side effects of all of the agents that produce a peripheral α-adrenergic sympathetic effect include blood pressure elevation, anxiety, and insomnia due to central nervous system stimulation. They may also cause headache, tremor, weakness, dizziness, respiratory difficulties, palpitations, and cardiac arrhythmias. All of these agents should be used with caution in patients with hypertension, cardiovascular disease, and hyperthyroidism (Weiner, 1980).

Ephedrine is a noncatecholamine sympathomimetic agent that owes part of its peripheral action to the release of norepinephrine, but that also directly stimulates both α- and β-adrenergic receptors (Weiner, 1980). The oral adult dosage is 25–50 mg, four times daily. Some tachyphylaxis develops to its peripheral actions, probably as a result of depletion of norepinephrine stores. Pseudoephedrine is a stereoisomer of ephedrine that is used for similar indications with similar precautions (Wein, 1986). The adult dose is 30–60 mg, four times daily, and the 30-mg dose form is available in the United States without a prescription. Norephedrine chloride in a dose of 75–100 mg was shown to acutely increase MUP and MUCP in women with urinary stress incontinence (Ek et al., 1978). At a 300-ml bladder volume, the MUP rose from 82 to 100 cm H_2O and the MUCP rose from 63 to 95 cm H_2O. The functional profile length did not change significantly. A 14-day double-blind crossover study comparing the effects of norephedrine to placebo showed that reduction of urinary leakage with the drug as compared with placebo occurred in 12 of 22 patients. Diokno and Taub (1975) reported a good to excellent response in 27 of 38 patients with sphincteric incontinence treated with ephedrine sulfate. They noted that beneficial effects were most often achieved in those with minimal to moderate wetting, and that little benefit was achieved in patients with severe stress incontinence. Obrink and Bunne (1978), however, noted that 100 mg of norephedrine chloride twice daily did not improve severe stress incontinence sufficiently to offer an alternative to surgical treatment. They further noted in their group of ten such patients that the MUCP was not influenced at rest or with stress at low or moderate bladder volumes.

Phenylpropanolamine hydrochloride shares the pharmacologic properties of ephedrine and is approximately equal in peripheral potency while causing less central stimulation (Weiner, 1980). Utilizing doses of 50 mg, three times daily, Awad et al. (1978) found that, after four weeks of therapy, 11 of 13 females and six of seven males with stress incontinence (severity not noted) were significantly improved. MUCP was increased from a mean of 47 to 72 cm H_2O in the empty bladder and from 43 to 58 cm H_2O with a full bladder. Fifty milligrams of phenylpropanolamine was combined with 8 mg of chlorpheniramine (an antihistamine) and 2.5 mg of isopropamide (an antimuscarinic) as a sustained-release capsule used primarily for relief of symptoms of allergic rhinitis (Ornade). Using one capsule twice daily, Stewart et al. (1976) found that, of 77 women with urinary stress incontinence, 18 were completely cured, 28 were much better, 6 were slightly better, and 25 were no better. In 11 men with postprostatectomy stress incontinence, the numbers in the corresponding categories were 1, 2, 1, and 7. Subsequently, Montague and Stewart (1979) carried out urethral profilometry in 12 women with moderate to marked stress incontinence and six women with no history of incontinence. The MUP increased more than 20% in 11 of the incontinent women and only 1 in the continent group. Rees and Ransley (1980) reported on the use of Ornade in 83 children with daytime wetting from a variety of causes. Of 24 children with bladder neck incompetence, 41% were cured, 29% had minimal symptoms, and 12% were improved. Interestingly, of 31 patients with bladder instability, the corresponding improved percentages were 24%, 24%, and 20%. The latter beneficial effect may have been due to the anticholinergic agent in the preparation. Treatment had to be discontinued in eight children because of side effects; 23 other children had mild and transient side effects. Major surgery was avoided by drug treatment in four children. The formulation of Ornade has now been changed, such that each capsule contains only 75 mg of phenylpropanolamine and 12 mg of chlorpheniramine. It is also available as a liquid.

Phenylpropanolamine is no longer a drug that can be obtained only by prescription in the United States. It is readily available as a component of many appetite suppressants, and a readily available and inexpensive source of this agent is available simply by asking patients to obtain a particular brand that contains 50–75 mg of phenylpropanolamine with no caffeine and with a minimum of extra ingredients.

Although some clinicians report spectacular cure and improvement rates with α-adrenergic agonists in patients with sphincteric urinary incontinence, our own experience coincides with those who report that treatment with such agents often produces improvement but rarely total dryness in cases of severe or even moderate stress incontinence. A clinical trial is certainly worthwhile, however, and will at the least assure the patient that the possibility of one type of nonsurgical therapy has been explored.

β-ADRENERGIC ANTAGONISTS.—Theoretically, β-adrenergic blocking agents might be expected to "unmask" or potentiate an α-adrenergic effect, thereby increasing outlet resistance. Gleason et al. (1974) reported success in treating certain patients with stress urinary incontinence with propranolol, a β-adrenergic blocking agent, with oral doses of 10 mg four times daily. The beneficial effect, however, became manifest only after 4–10 weeks of treatment, a difficult fact to explain on a pharmacologic basis. Although such treatment has been suggested as an alternative method of drug therapy in patients with hypertension and sphincteric incontinence, few—if any—subsequent reports of such efficacy have appeared, and others have reported no significant changes in urethral profile pressures in normal women after β-adrenergic blockade (Donker and Van der Sluis, 1976). Though 10 mg four times a day is a relatively small dosage of propranolol, it should be recalled that the major potential side effects of the drug are related to the drug's therapeutic β-blocking effects. Heart failure may develop as well as an increase in airway resistance, and asthma is a contraindication to its use. Abrupt discontinuation may precipitate an exacerbation of anginal attacks and rebound hypertension (Weiner, 1980).

ESTROGEN.—Salmon et al. first reported the use of estrogen in the treatment of stress urinary incontinence in 1941. Raz et al. (1973) found that a daily dose of 2.5 mg of Premarin (conjugated estrogens) improved stress incontinence and increased urethral pressures in postmenopausal patients, effects that they attributed to mucosal proliferation with a consequently improved "mucosal seal effect" and to enhancement of the α-adrenergic contractile response of urethral smooth musculature to endogenous catecholamines. Schreiter et al. (1976) reported similar benefits after ten days of treatment with daily divided doses of 6 mg of estriol. They showed also that the effects of estrogen and of exogenous α-adrenergic stimulation were additive. Hodgson et al. (1978) reported that the sensitivity of the rabbit urethra to α-adrenergic stimulation was estrogen dependent, as castration caused a decreased sensitivity, and treatment with low levels of estrogen reversed the defect.

Rud (1980a) studied the effects of 4-mg daily doses of estradiol and 8-mg daily doses of estriol on 30 women with an average age of 61 years, 24 of whom had stress urinary incontinence. Profilometry parameters were recorded at a bladder volume of 200 ml using microtransducer technique. Small but statistically significant changes occurred in the MUP (59–63 cm H_2O), functional urethral length (25–28 mm), and actual urethral length (33–37 mm). No statistically significant change occurred in urethral closure pressure (37–39 cm H_2O). Eight of the 24 incontinent patients experienced subjective and objective improvement, 9 experienced subjective improvement only, and 7 experienced neither

subjective nor objective improvement. There was no correlation between subjective or objective improvement and the previously mentioned urodynamic measurements. However, in 18 patients pressure transmission to the urethra was recorded during cough, and in seven of these this improved. All of these had subjective improvement and five were shown to be objectively dry. Rud pointed out that it is hard to believe that the small changes in urodynamic measurements, even though statistically significant, are directly related to resumption of continence. He noted also that the increased pressure transmission ratio might be due to factors outside the urethra—either in the striated musculature of the pelvic floor or in the periurethral vasculature or supporting tissues. Interestingly, he found no changes in urodynamic measurements in five continent and three stress-incontinent females with cystic glandular hyperplasia treated with a single injection of 1,000 mg of intramuscular progesterone, except that the pressure transmission ratio was lower in the three patients in whom this was measured.

Rud (1980b) also studied profilometry during the menstrual cycle in six females. There was no change in any profilometric values during the menstrual cycle and no correlation between estrogen levels and MUP. It may be, as he suggested, that at physiological levels estrogens have little influence on urodynamic measurements related to continence, and that only pharmacological doses cause urodynamically significant changes; further, pharmacological doses may alter responses to other exogenous autonomic stimulation, particularly α-adrenergic, as the previously described laboratory experiments by Hodgson et al. (1978) would suggest.

Beisland et al. (1981) administered 80 mg of estriol intramuscularly every four weeks in combination with an oral dosage of phenylpropanolamine of 50 mg twice daily to a group of 13 patients with what they described as an incompetent urethral closure mechanism, that is, proximal sphincter hypofunction. Patients with genuine stress incontinence were excluded. These patients had a poor urethral pressure profile and a low MUCP. All patients, however, had adequate transmission of increased abdominal pressure to the bladder and to the urethra. Eight of these patients became continent, and four showed improvement. The average MUCP increased from 24.6 to 43.6 cm H_2O ($P<.005$). In the one patient in whom MUCP did not change, the clinical condition likewise was unchanged.

Levin et al. (1981) have shown that parenteral estrogen administration can change the α-adrenergic receptor content and the autonomic innervation of the lower urinary tract of immature female rabbits. Whether these experiments have any clinical significance is unknown. Certainly, estrogen therapy seems capable of facilitating urinary storage in some female patients by increasing outlet resistance, and there is evidence of an augmentative or perhaps additive effect with α-adrenergic therapy in this regard. Whether the levels achieved by the commonly used oral or parenteral estrogen preparations or by estrogen vaginal creams (which simply provide a convenient vehicle for systemic absorption) actually increase the α-adrenergic receptor content of the smooth muscle of the bladder outlet or the "mucosal seal effect" is still a matter for speculation and is currently under study.

The potential long-term effects of such treatment must be carefully considered, however, in the light of the current controversy over whether estrogen therapy predisposes to the development of endometrial carcinoma. At the very least, if a beneficial effect on the lower urinary tract is achieved by a regimen that uses estrogenic therapy alone or in combination with other therapy, the lowest maintenance dose of estrogen therapy possible should ultimately be used, one which it is to be hoped is below that used for replacement therapy in menopausal and postmenopausal women.

VESICOURETHRAL SUSPENSION.—Fixation of the vesicourethral junction in a physiologic position has been observed to correct genuine stress urinary incontinence in the female in 85%–90% of cases. Although the origins of this procedure are steeped in uromythology, an approach employing a retropubic exposure seems to have been reported first by Hepburn (1920), who anchored the bladder neck of a young girl to the periosteum of the pubis for the treatment of urethral prolapse. Bonney (1923) seems to have been one of the first clinicians to correlate correction of stress urinary incontinence with elevation and fixation of the vesicourethral junction. As far as the timing of the first true description of a retropubic vesicourethral suspension procedure for stress urinary incontinence, there is some disagreement. Williams (1974) described an operation involving retropubic mobilization of the vesicourethral junction with paired catgut sutures placed through the bladder wall and the pubic periosteum. The Marshall-Marchetti-Krantz operation was not reported until 1949, but was actually first reported as being done initially in 1944 (Marshall et al., 1949).

Innumerable modifications of the Marshall-Marchetti-Krantz procedure and innumerable other procedures designed to fix the vesicourethral junction in a physiologic position have been described, enough to convince these authors that: (1) there is not a "best" procedure for correcting genuine stress urinary incontinence; (2) a good surgeon who understands the pathophysiology of incontinence and who intelligently arrives at a diagnosis of genuine stress urinary

incontinence in a given patient will have surgical success with any one of a number of methods with which he or she is comfortable; (3) there is a substantial lack of agreement as to what these procedures are actually designed to do; and (4) substantial disagreement exists as to the urodynamic effects of these various procedures. All of these operations seem to have in common the fact that they do indeed restore the physiologic position of the bladder neck-proximal urethral area, allowing transmission of intra-abdominal pressure increases to urethral as well as bladder pressure. This seems to us to be the common denominator shared by all these procedures, although other authors have cited other factors that they feel are important in correcting stress urinary incontinence, notably, increasing urethral closure pressure, increasing functional urethral length, restoring bladder neck competence, and increasing support to the bladder neck. A brief description of some of these procedures follows.

The Marshall-Marchetti-Krantz operation (Fig 28–20) most commonly is carried out by approximating the periurethral and vaginal fascia to the back of the cartilaginous portion of the pubic symphysis. Although the original procedure was done through a horizontal suprapubic incision, a vertical incision may yield better exposure in patients who represent failures of previous pelvic surgical procedures. It should be noted that the original procedure (Marshall et al., 1949) involved mobilization of the urethra to within 1 cm of the meatus, and employed three no. 1 doubled chromic sutures inserted equidistant from each other on either side of the urethra, the sutures catching the deep vaginal wall and the lateral wall of the urethra. A similar suture was placed lateral to the vesicourethral junction, and additional sutures placed in the vaginal wall to bolster the repair. Additional sutures were also placed between the musculature of the lower and lateral bladder and the posterior rectus sheath to further close the space of Retzius. Most surgeons who perform this operation do so in a fashion similar to that pictured here, and perform the surgery utilizing absorbable sutures. Some surgeons prefer to open the bladder to facilitate suture placement, some use nonabsorbable sutures, and some, who consider urethral lengthening to be the primary therapeutic goal, place sutures only between the bladder and rectus fascia (Lapides, 1971).

Retropubic colposuspension was first described by Burch in 1961, and was designed to accomplish repositioning and fixation of the vesicourethral junction without the need for retaining sutures in the pubis (Fig 28–21). The suture placement generally commences distally at the level of the vesicourethral junction, although some surgeons prefer a midurethral level. Fixation is to Cooper's ligament. Some feel that this

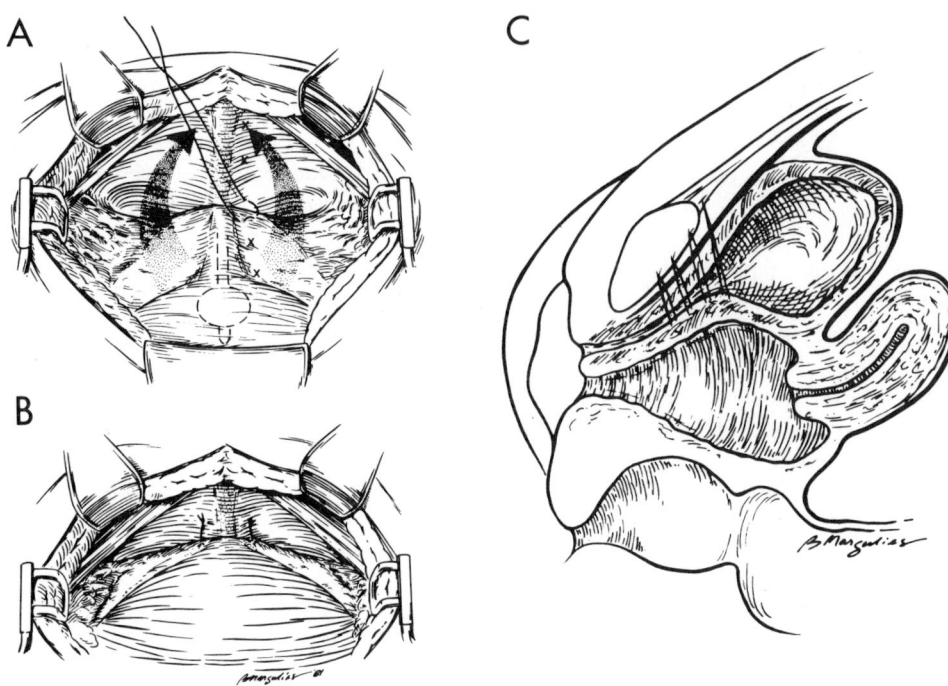

FIG 28–20.
The Marshall-Marchetti-Krantz (MMK) procedure. This procedure corrects stress urinary incontinence by approximating the periurethral and vaginal fascia to the underside of the pubic symphysis with three or four pairs of sutures. **A,** this shows the placement of the sutures in the periurethral and vaginal fascia and at the level of the vesicourethral junction. The *large, stippled, upward-curved arrows* point out the sites of suture attachment on the underside of the pubic symphysis. The orientation of this view is retropubical, as the surgeon would see it. Accurate placement of the sutures is important to achieve satisfactory suspension, and in order to ensure good fixation, it is helpful to take a double bite through the periurethral fascia and anterior vaginal wall, taking care not to put the suture through the vaginal mucosa. Insertion of the sutures is facilitated by elevation of the anterior vaginal wall. It is usually helpful for the surgeon to use his left hand to do this while placing the sutures with the right hand. **B,** this is the retropubic view after the sutures have been tied individually, commencing with the more distal pair. While the surgeon is tying the sutures it is helpful to have the assistant elevate the vaginal wall. The most proximal suture may need to be passed through the insertion of the rectus abdominal muscle. **C,** lateral view showing correction of the anatomic abnormality contributing to stress urinary incontinence by the MMK procedure. Here, the most proximal sutures are shown passing through the insertion of the rectus abdominus muscle. (From Webster GD: The urethra, in Paulson DF (ed): *Genitourinary Surgery*. New York, Churchill Livingstone, 1984. Used with permission.)

operation gives broader support to the urethra and bladder base, and avoids the risk of urethral compression. Potential enteroceles in the pouch of Douglas must be obliterated by successive purse-string sutures in this area (Stanton, 1986).

The needle suspension procedure was first introduced by Pereyra in 1959. Numerous modifications of this procedure have evolved, and currently there are two modifications in common use by urologists in the United States, both of which involve endoscopic control to make sure that the suspending sutures are not passed through the lumen of the bladder or urethra. The Stamey procedure (Fig 28–22) utilizes the intact tissues lateral to the vesicourethral junction as the primary source of support for the suspension of the bladder neck. The Raz needle suspension (Fig 28–23) perforates the retropubic space between the pubic bone and the endopelvic fascia, freeing the endopelvic fascia from its lateral pubic attachments. The suspension sutures are placed through the edge of the lateral vaginal wall, incorporating the endopelvic fascia in this area (Hadley et al., 1985). Each of these modified Pereyra procedures has a large number of ardent supporters, and the results in the treatment of stress incontinence, when performed by experienced operators, seem to be similar to the best results achieved by modifications of the Marshall-Marchetti-Krantz procedure or the Burch procedure. However a comparison of long-term success rates, reports of which are few in number even concerning the older procedures, will be necessary before any valid statements regarding relative merit can be made.

Bladder Outlet Reconstruction

Reconstruction of the bladder neck is one method of restoring sphincteric continence in patients with a

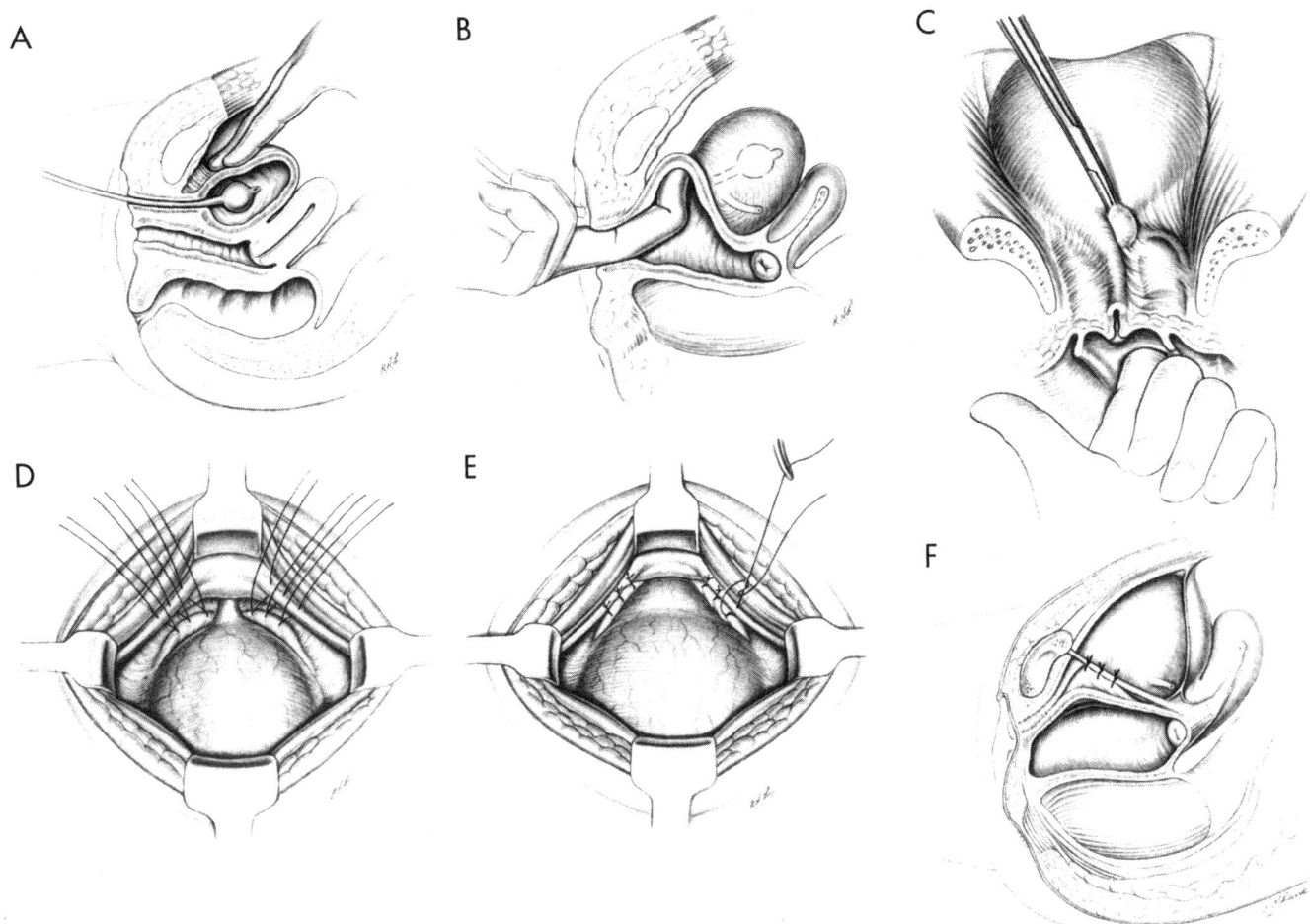

FIG 28–21.

The Burch colposuspension. **A,** sagittal section showing blunt finger dissection in the retropubic space of Retzius, between symphysis pubis and anterior surface of bladder and urethra. **B,** sagittal section showing upward pressure in the lateral vaginal fornix, to aid the surgeon in the abdominal section. **C,** retropubic view, from area of pubis, but with pubis removed. With the operator's finger still in the vagina and elevating the lateral vaginal fornix, the bladder base is dissected medially off the paravaginal fascia. **D,** retropubic surgeon's view showing three pairs of sutures in place but not tied. Some surgeons use heavy absorbable sutures, while others use nonabsorbable sutures. The sutures are inserted into the paravaginal fascia and then to the nearest point on the ipsilateral iliopectineal ligament. The most distal (caudal) sutures are inserted opposite the bladder neck and not lower. The next two sutures are inserted more proximally (cephalad) alongside the bladder base. The lateral vaginal fornix should be elevated toward the iliopectineal ligament as the corresponding suture is being placed through the ligament, so as to position the suture accurately. **E,** this illustrates tying of the sutures, showing approximation of the paravaginal tissue to the iliopectineal ligaments. **F,** sagittal section showing elevation of the bladder neck and bladder base on a shelf of paravaginal fascia, sutured to the iliopectineal ligament. (From Stanton SL: Colposuspension, in Stanton SL, Tanagho EA (eds): *Surgery of Female Incontinence.* Berlin, Springer Verlag, 1986. Used with permission.)

fixed, open bladder outlet. Other alternatives involve mechanical compression of the bladder outlet and include sling procedures, periurethral polytef injections, and the use of an artificial sphincter. Bladder neck reconstruction for the treatment of urinary incontinence was introduced by Young in 1907, and was subsequently modified by Dees (1949) and Leadbetter (1964). Procedures utilizing the Young-Dees principle involve construction of a neourethra from the posterior surface of the bladder wall and trigone. In the male,

the prostatic urethra affords additional substance for closure and increase in outlet resistance. The Leadbetter modification involves proximal reimplantation of the ureters to allow more extensive tubularization of the trigone. Long-term success rates of between 60% and 70% have been achieved (Leadbetter, 1985). Tanagho (1969, 1981) has described a procedure based on a similar concept but using the anterior bladder neck to create a functioning neourethral sphincter. A success rate of 70% has been reported.

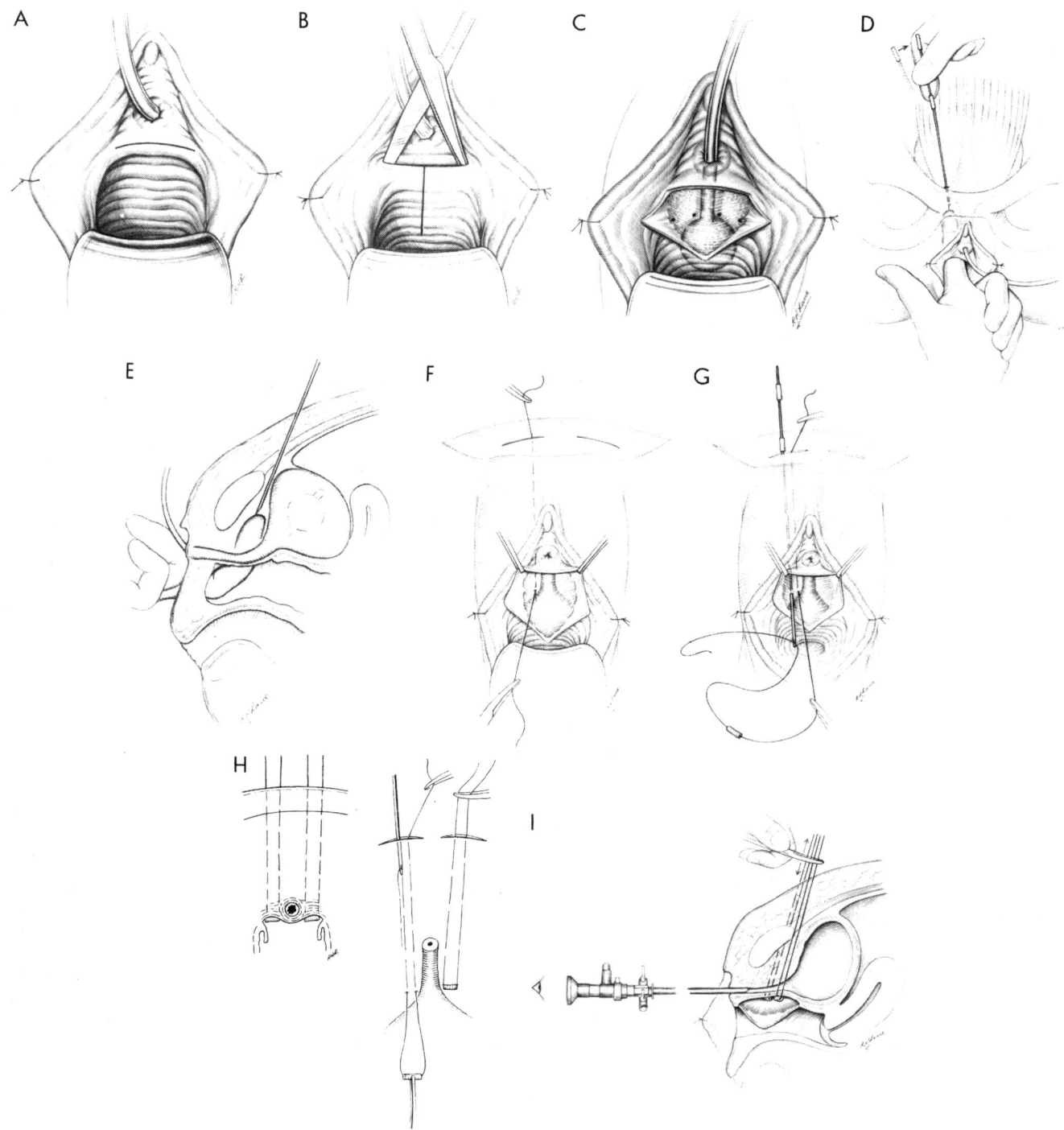

FIG 28–22.

The Stamey procedure. **A,** a transverse incision is made below the urethral meatus at the level of the vaginal wall. If the urethra is 2.5 cm or less, the incision is begun just below the meatus. If the urethra is longer, the incision can be made closer to the bladder neck. There is generally a convenient fold of tissue at least 2 cm from the bladder neck, and the incision may often be made at this level. **B,** the vaginal wall is carefully separated by dissection from the urethra and trigone. The tips of the scissors must be directed posteriorly toward the floor at all times to avoid en-

tering the bladder. Adequate exposure is generally achieved by simply spreading the Metzenbaum scissors laterally in a posterior direction, and then incising this anterior flap in the midline. The Foley catheter balloon should be easily palpable within the incision, thereby determining exactly the location of the bladder neck. **C,** the two dots on either side of the urethrovesical junction indicate the sites of suture placement. Dissection of the fascia between the vaginal wall and the overlying urethra and trigone should be avoided. This pubovesicocervical fascia, together with the overlying en-

Surgical Mechanical Compression

Periurethral polytef injection.—The injection of polytef paste periurethrally to increase urethral resistance was first reported by Berg (1973), and its use has been promoted and developed primarily by Politano and co-workers (Politano et al., 1974; Politano, 1982; Vorstman et al., 1985). Most clinicians inject the polytef paste transurethrally, and an ingrowth of fibroblasts is stimulated, which produces a bolstering effect, but does not stricture the urethra. Good results have been reported by Politano's group in up to 70% of patients. Schulman et al. (1983) reported a 70% cure rate in 56 women with stress urinary incontinence, with an additional 16% showing marked improvement. Lim et al. (1983) described cure in 21% and benefit in an additional 54% of 28 females so treated. Deane et al. (1985) described four cures and two improvements in six females with genuine stress incontinence, and five cures and six improvements in 22 females with bladder neck incompetence at rest. They were able to achieve only one improvement in eight males with postprostatectomy incontinence. Malizia et al. (1984) have expressed concern regarding the potential of distal embolization of the polytef particles, such as they observed in laboratory animals. They recommended against its use in children until long-term effects were further investigated. However, clinically significant instances of such embolization have not been reported in patients so treated up to this point.

Sling procedures.—The suburethral sling procedure, as cited by Ullery (1953), was first reported by vonGiordono in 1907. Since that time, a considerable number of procedures utilizing this principle have been described, employing either autologous or alloplastic material. These procedures have been used for patients with all types of stress incontinence, and they are quite effective in patients with recurrent stress incontinence and pelvic floor weakness, especially after repeated unsuccessful surgeries. Among urologists, McGuire (1978, 1981, 1985) has written most extensively about this procedure and feels that its use should probably be restricted to patients with poorly functional or nonfunctional urethral sphincter mechanisms, as opposed to those with stress incontinence because of urethral hypermobility. The procedure does involve elevation and fixation of the urethra in a normal retropubic position, and thus can be used as a treatment for genuine stress incontinence. However, the noncircumferential compression afforded by the sling is optimal treatment for patients with no urethral closing function of any kind and those with poor urethral smooth muscle function. One of three general types of urethral dysfunction generally forms the indication for this procedure: (1) total failure, usually resulting from direct injury to the urethra or sloughing of the urethral musculature; (2) severe weakness of the urethral smooth musculature, usually as a result of neural injury or following radical pelvic surgery; and (3) fixation and rigidity of the urethra, usually due to prior operative procedures. Al-

---

dopelvic fascia, is the primary source of support of the bladder neck in this type suspension. **D,** two suprapubic incisions, 3–5 cm long, are made to the left and right of the midline, at the upper border of the pubic symphysis. These are carried down to the rectus fascia. This is generally done at the beginning of the procedure, and the incisions are packed with antibiotic soaked sponges. After the vaginal dissection, this coronal view shows a 15° angled Stamey needle passed through the right anterior rectus fascia and turned toward the undersurface of the pubic symphysis. This avoids any intestinal perforation and allows guidance alongside the urethrovesical junction. After the needle perforates the rectus fascia, it is not long before its tip will be felt by the finger shown in the vaginal incision at the point of the bladder neck. **E,** sagittal view showing location of the needle point on the tip of the index finger. Once the needle is in place, the right-angle cystoscope is used to make sure that the needle has not entered the bladder and that it has not been placed submucosally or intramuscularly. Vertical motion of the needle will tell the surgeon endoscopically whether the needle is exactly positioned at the level of the bladder neck. If the needle is placed improperly, it should be repositioned. **F,** a heavy monofilament nylon suture is threaded through the eye of the needle and the needle withdrawn suprapubically. Both ends of the suture are clamped. **G,** the Foley catheter is reintroduced, and the special needle once again passed 1 cm lateral to the first passage, and

a Dacron or Gore-Tex sleeve is threaded on. It is important to maintain the position of the second passage 1 cm lateral to the first at the level of the bladder neck. After the suture is threaded through the Dacron or Gore-Tex bolster, the needle is then withdrawn suprapubically, pulling the second pass of the suture with it, thereby establishing the suspended loop with the buttress vaginally, reinforcing the pubocervical and endopelvic fascia. The bolsters should settle easily into the vaginal space at the side of the bladder neck. **H,** the identical procedure is then repeated for the incision on the opposite side, so that the two suspending sutures, each with its buttress, are located on either side of the bladder neck. The two loops, as pictured, support a shelf of pubovesicocervical fascia on either side of the bladder neck. **I,** the panendoscope is then introduced and used to observe the point of urethral closure when each suspending loop is pulled up suprapubically. Closure generally occurs exactly at the bladder neck with the suspending loops symmetrically placed. The vaginal incision must be closed first before tying the suspending sutures. If the rectus fascia is weak, each suture is passed through a small oval of Silastic, and tied over this, which then rests directly on the rectus fascia. (From Stamey TA: Endoscopic suspension of the bladder neck, in Stanton SL, Tanagho EA (eds): *Surgery of Female Incontinence.* New York, Springer-Verlag, 1986. Used with permission.)

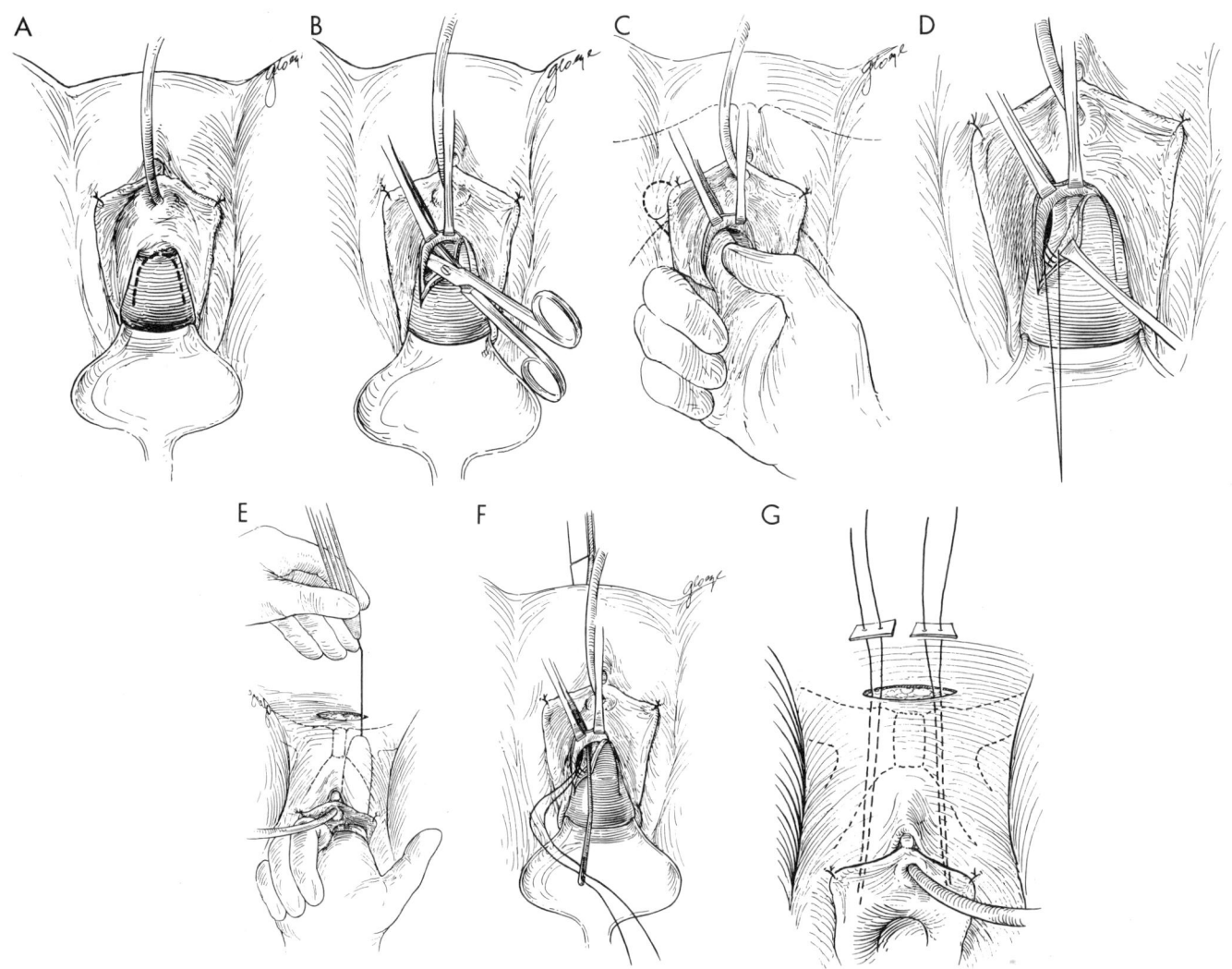

FIG 28–23.
The Raz bladder neck suspension. **A,** in the dorsal lithotomy position, a Foley catheter is inserted into the bladder and a posterior vaginal retractor is used to expose the anterior vaginal wall. Traction on the catheter facilitates identification of the level of the bladder neck. An inverted-U incision is made in the anterior surface of the vaginal wall. The apex should cross the level of the midurethra. **B,** a plane is created underneath the vaginal wall and advanced to the endopelvic fascia. Dissection proceeds laterally toward the pubic bone. The retropubic space is entered either bluntly or sharply between the pubic bone and the endopelvic fascia. **C,** the index finger is used to enter the retropubic space and to mobilize the urethra and the bladder neck. The endopelvic fascia is gently freed from its lateral attachments to the pubic bone. If the retropubic space is entered in the proper plane, bleeding should not be troublesome. The mobilization should be extended down to the level of the ischial tuberosity to adequately mobilize the vaginal wall, bladder neck, and urethra. With scar tissue, achievement of sufficient mobility may require sharp dissection of the paraurethral and perivaginal tissues. **D,** a no. 1 Prolene helical suture is secured to the vaginal wall (excluding the epithelium) and endopelvic fascia. The suture should include at least three separate passes. After each pass, traction should be placed on the suture to test the tissue integrity and the holding power of the suture. **E,** through a transverse suprapubic incision, the anterior rectus fascia is exposed. Through this fascia the suspension needle is guided into the retropubic space and out of the vaginal introitus under finger guidance. We pass the needle twice on each side, the needle enters into the rectus fascia betin about 1–2 cm apart. Through the area in which the endopelvic fascia has been penetrated, the finger should be able to contact the needle immediately after its passage through the rectus fascia. Penetration by the bladder should therefore not occur. **F,** the two ends of each Prolene suture are threaded through the eye of the needle and withdrawn through the suprapubic incision. The identical procedure is repeated on the contralateral side; 4–5 cm of rectus fascia should lie between the Prolene sutures in the suprapubic incision. **G,** the two ends of each Prolene suture then exit through the rectus fascia approximately 1 cm apart, with 4–5 cm, of rectus fascia separating the sutures on the two sides. Endoscopic examination is then carried out to make sure there has been no perforation of the bladder or urethra. Indigo carmine and furosemide (Lasix) is given, and efflux should be seen from both ure-

though originally described through a retropubic approach, a combined vaginal and abdominal approach (Fig 28–24) often facilitates the procedure. The vaginal access to and dissection of the retropubic space for sling placement is similar to that provided by the Raz-type suspension procedure. Success rates of up to 90% with the pubovaginal sling operation have been reported. Many patients have difficulty resuming volitional voiding after surgery, and permanent self-intermittent catheterization may be required. Additionally, detrusor instability may occur after the procedure or be exacerbated by it, if pre-existent. Adjustment of the proper sling tension requires considerable operator experience or necessitates intraoperative urethral pressure profilometry. An increase in urethral pressure in the area of the urethra subjacent to the sling of 10 cm H_2O has given consistently good postoperative results in McGuire's experience.

Artificial Sphincter

Control of sphincteric urinary incontinence with implantable prosthetic devices has evolved rapidly over the past 15 years (Berry, 1961; Kaufman, 1970; Foley, 1947; Rosen, 1976). Clearly the most significant contribution was the introduction in 1973 by Scott et al. of a totally implantable artificial sphincter mechanism that could be used in adults and children of both sexes. Through biomechanical evolution, device improvements have led to the currently used AS 800 device (manufactured by American Medical Systems, Minnetonka, Minn.), which is composed of an inflatable snap-on cuff, a pressure-balloon reservoir, and a control assembly (Fig 28–25). Constructed of silicone rubber, this hydraulic device is filled with iso-osmotic contrast medium. Cuff closing pressure is regulated by the elasticity in the wall of the balloon and is partially dependent on the volume of fluid within the balloon. Fluid flow through the system is mediated by the pump and a series of unidirectional valves within the control assembly. On the side of the control assembly is a small button that allows for manual activation and deactivation of the sphincter mechanism.

Balloon pressures have been manufactured that, under normal clinical situations, will provide for enough cuff closing pressure to ensure urinary continence but

similarly will not damage the underlying tissue of the urethra or bladder neck.

Too much pressure will result in tissue ischemia and ultimate pressure necrosis with erosion of the cuff into the lumen of the underlying urethra or bladder neck. This generally is manifested by recurrent incontinence and concurrent infection around the implanted sphincter. Device removal is required for adequate healing; however, another device may be implanted later.

Identification of candidates for implantation requires a thorough neurourologic workup to properly establish the unique aspects of the patient's incontinence. Specific criteria for implantation have been established (Barrett, 1986), with most patients having true sphincteric incontinence with normal detrusor contractility and compliance. Ideal candidates are individuals with incontinence related to radical prostatectomy or transurethral resection of the prostate (TURP), neurogenic etiologies such as myelomeningocele, previous pelvic trauma, and females with failed conventional anti-incontinence procedures.

Surgical implantation of the device involves the implantation of the cuff around the bulbous urethra of adult males (Fig 28–26) and around the bladder neck in females and children (Fig 28–27). Bladder neck cuff placement may be technically challenging in patients with previous bladder neck or pelvic surgery. Dissection of the vesical rectal plane or the vesicovaginal plane, which is required for cuff placement, is carried out inferior to the point of entry of the ureters into the posterior aspect of the bladder. The pressure balloon reservoir is placed in a pocket underneath the fascia of the anterior abdominal wall and the pump control assembly is positioned in a convenient location in the subcutaneous tissues of the scrotum or labia.

The device remains in a deactivated configuration for 6–8 weeks to allow for complete healing around the pump and cuff. Device activation is carried out manually by merely compressing the pump chamber firmly. The device may again be deactivated by compressing the button on the site of the control assembly.

Results of implantation of 221 patients with the AS 800 device followed for up to three years would indicate 95% of the patients are acceptably dry. Mechanical reliability based on life-table analysis projects a 97% sur-

---

ters. With the endoscopic instrument withdrawn to the level of the midurethra, the bladder neck should elevate and close during upward traction on the Prolene sutures. Considerable force should not be necessary to accomplish this. If it is, this generally means that further mobilization is necessary. The vaginal incisions are then closed. The two ends of each suture on each side are then tied to one another. If

the rectus fascia is weak these can be first threaded through a small piece of Silastic mesh, as pictured. After each suture is secured individually, they are tied to one another across the midline. (From Hadley HR, Zimmern PE, Staskin DR et al: Transvaginal needle bladder neck suspension. *Urol Clin North Am* 1985; 12:291. Used with permission.)

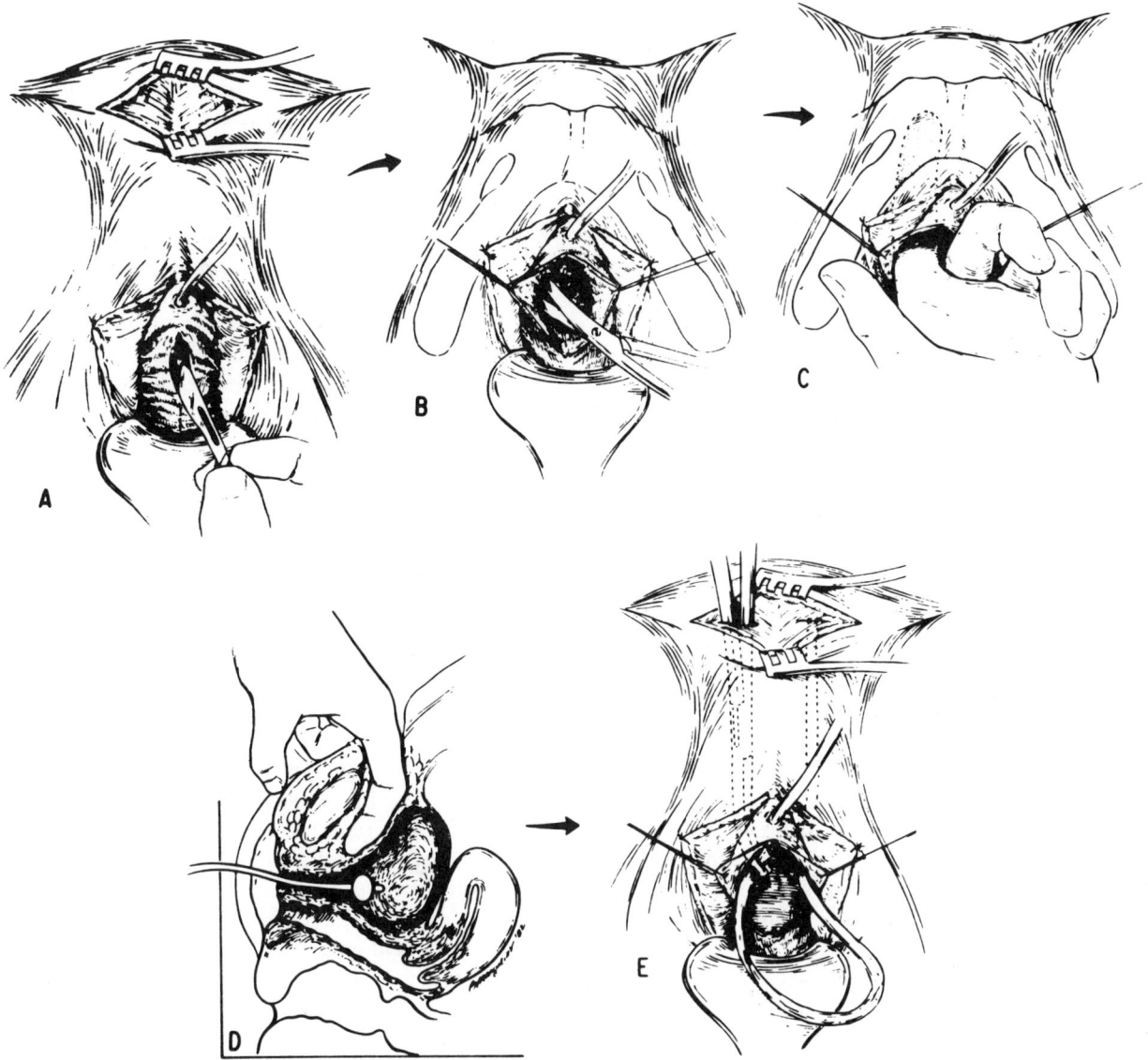

FIG 28–24.
Fascial sling procedure. **A,** in this variation, a combined vaginal and abdominal approach is used. **B,** after reflecting the vaginal wall from the urethra, the paraurethral musculo-fascial tissue is dissected into the retropubic space at the level of the bladder neck. This approach is very similar to that used in the Raz suspension procedure. **C,** this space is further digitally dissected, as in the Raz procedure. **D,** rarely, a retropubic dissection is also necessary, as pictured. **E,** the fascial sling, which may be obtained from rec-

tus or external oblique fascia, or fascia lata, is looped around the urethrovesical junction, and fixed to the rectus fascia suprapubically. As mentioned in the text, the tension of the sling must be carefully adjusted to provide the minimum necessary to achieve the desired goal of elevation or compression. (From Webster GD: The urethra, in Paulson DF (ed): *Genitourinary Surgery.* New York, Churchill-Livingstone, 1984. Used with permission.)

vivorship at three years. Although the initial results appear reasonable, the long-term results will help establish more accurately the exact role of the artificial sphincter in treating severe incontinence.

Nonsurgical Mechanical Compression

Nonsurgical occlusive devices have been used for the control of female incontinence for years. The status of

these devices is somewhat confusing. Published reports are rare, and usually include a description of the particular device with "success" rates ranging from 70%–90%. Yet, there seems to be little enthusiasm for their use in general, or for any particular device, on the part of any clinician when questioned in person. Historically available devices are well described by Edwards (1975) and by Stanton (1978). "New" ones periodically appear

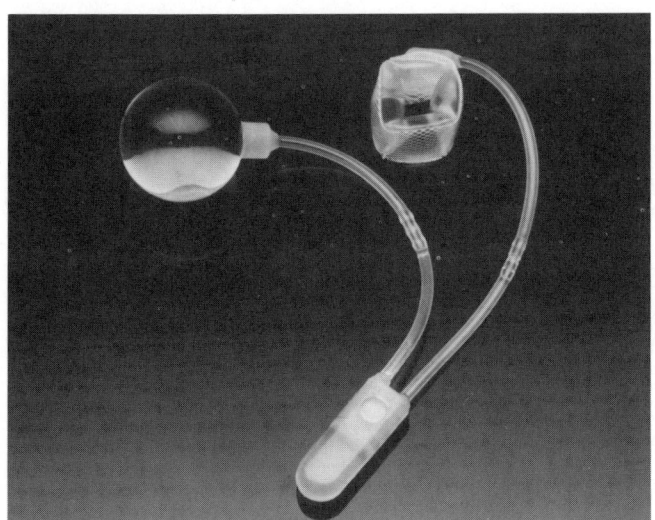

FIG 28–25.
The AS 800 artificial sphincter mechanism.

and then characteristically disappear after adequate clinical trials. The idea is to maintain light but firm compression of the urethra with an intravaginal device. Although the concept seems simple, its execution apparently is not. Were a compression device available that was easy to apply, stayed in position for extended periods of time, was reasonably comfortable to wear, was extremely and continuously effective in preventing urinary leakage, and was easy to remove or deflate when micturition was necessary, it would obviously constitute a distinct advance in the treatment of female urinary incontinence.

Nonsurgical occlusive devices for the male are exter-

nal and compress the penile urethra by squeezing the penis from two sides. Their main disadvantages include bulk, discomfort (in those with sensation) and, potentially, pressure necrosis of the urethra, especially in those patients with impaired sensation.

Continence achieved with urethral compression is applicable only to those patients with low-pressure bladders. Although any pressure system can theoretically be occluded, continued obstruction of a high-pressure system can result in disastrous lower and upper urinary tract sequelae.

Circumventing Storage Failure

ANTIDIURETIC HORMONE–LIKE AGENTS.—A novel approach to the treatment of urinary frequency occurring within a specific period of time has been the use of synthetic antidiuretic hormone (vasopressin) analogues. Desmopressin acetate (DDAVP) has been used effectively in patients with central diabetes insipidus to increase the osmolality of urine and decrease its volume (intranasal spray administration of 5–20 μg produced 8–20 hours of antidiuresis in eight patients with central diabetes insipidus) (Robinson, 1976). The drug is well tolerated in both adults and children (Kosman, 1978). For diabetes insipidus, the usual adult dosage is 0.01-0.04 mg daily as a single or divided dose. (Desmopressin has a much greater antidiuretic potency than lysine or arginine vasopressin, a more prolonged duration of action [generally 8–10 hours], and virtually no pressor activity.) For children, the usual dosage range is from 0.005 to 0.03 mg daily. Large doses may cause transient headaches, nausea, and a slight increase in blood pressure. Fluid intake should be adjusted during

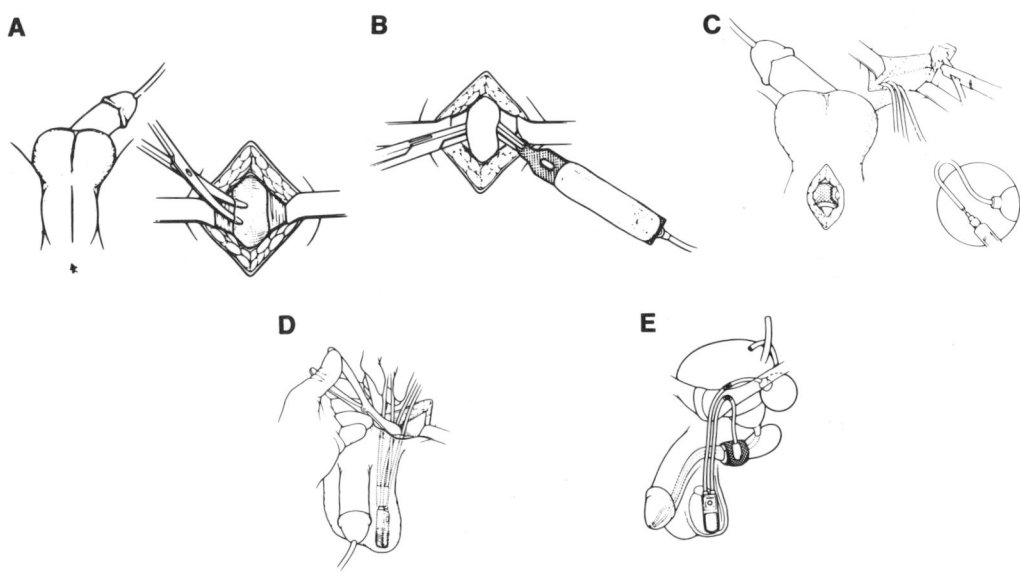

FIG 28–26.
Implantation of the artificial sphincter cuff around the bulbous urethra.

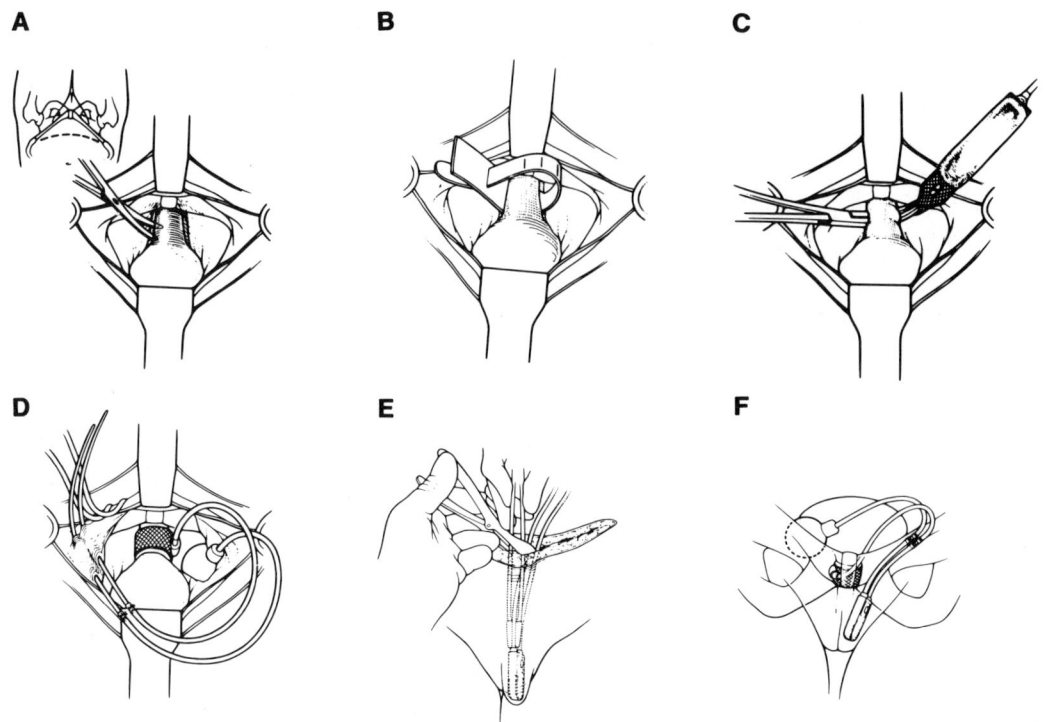

FIG 28–27.
Implantation of the artificial sphincter cuff around the bladder neck.

this therapy to avoid hyponatremia and water intoxication.

The drug has been used to decrease the frequency of nocturnal enuresis. In one study of 22 children, the number of wet nights per two weeks for desmopressin-treated children was 4.2 compared with 10.9 for those given a placebo. In 19 of 22 patients the response during desmopressin therapy was better than that during placebo (Birkasova et al., 1978). A single nighttime dose of 10–40 µg was used. Ramsden et al. (1982) completed a double-blind crossover trial of desmopressin and a placebo in 21 adult patients with nocturnal enuresis who were wet more than 20 nights per month. The total number of nights wet on the active drug was statistically less than that on placebo (91 vs. 167). Sixteen patients had fewer nights wet on treatment with active drug, two showed improvement with placebo, and three showed no difference. Eight patients on the active drug became entirely dry or had only one night wet, compared with only one patient on placebo. The authors noted that three patients who had been taking active drug remained entirely dry. The dosage used was 20 mg intranasally at bedtime. Side effects were few and minor.

Hilton and Stanton (1982) used a nighttime dose of 20 mg intranasally in a double-blind placebo study of 25 female patients with nocturnal urinary frequency. The number of mean episodes per night decreased from 3.17 in the pretreatment group to 1.94 in the group treated with active drug, as compared with 2.61 in the placebo-treated group. The nocturnal urinary output, as expected, also decreased significantly in the drug-treated group. One patient already receiving treatment for hypertension with a diuretic became hypertensive with a diastolic pressure of 110 mm Hg (entrance pressure was 80 mm Hg). The authors emphasize that hypertension, ischemic heart disease, and congestive failure should all be considered contraindications to the use of this type of medication. A double-blind placebo-controlled study was also carried out in a group of 21 males with benign prostatic hypertrophy and significant nocturia (Mansson et al., 1980). Active treatment consisted of 20 mg of drug administered intranasally. This produced a mean fall in nocturia from 2.60 episodes during the control period to 1.93, statistically significantly better than the response to placebo (2.31), but hardly a significant clinical change.

Although the overall results with this type of agent in patients other than those with central diabetes insipidus seem more statistically than clinically significant, this class of treatment may prove to be useful on a long-term basis in the occasional case of nocturia or enuresis that has proved refractory to all other forms of therapy or in acute situations in which a decrease in intravesical volume and consequent urinary frequency is desired for a limited period of time.

External Collection Devices

Unfortunately, no satisfactory external collecting device has been devised for the female, primarily because of the difficulties of fixation and leakproof collection. Absorptive padding, a collection device of sorts, is a last resort for many patients. The ideal substance is one that is highly permeable and absorbent. Immediately next to the patient is generally a layer of hydrophobic material, through which the urine passes into the absorbent pad, in turn surrounded by a waterproof material to keep the clothing dry. The hydrophobic material next to the patient will ideally keep her relatively dry and reduce chafing as much as possible.

External collecting devices for the male (Texas or condom catheter) are generally successful insofar as urine collection is concerned, but are unacceptable to many patients because of the visible equipment required, the incontinence, and the "leaks" of often foul-smelling urine that can result. They can also cause severe pressure necrosis of the penis, down to and including the urethra, especially in a patient with impaired sensation (Golz, 1981). A collecting device without a single discrete roller band or application ring would seem to offer at least a theoretical advantage for this reason.

SPECIFIC VOIDING DYSFUNCTIONS ASSOCIATED WITH NEUROLOGIC DISEASE

Discrete neurologic lesions generally affect bladder filling and urine storage and bladder emptying in a relatively consistent fashion, dependent on the area affected. Neurologic lesions above the brain stem that affect micturition generally result in involuntary bladder contractions with changes in the smooth sphincter and striated sphincter areas analogous to those that occur in normal micturition. Patients with complete lesions of the spinal cord above spinal cord level S2, after they recover from spinal shock, generally exhibit involuntary bladder contraction with smooth sphincter synergia but striated sphincter dyssynergia. Patients with significant spinal cord trauma below that level generally do not manifest involuntary bladder contractions per se. Detrusor areflexia is the rule initially, and, depending on the extent of neurologic injury, various forms of decreased compliance during filling may occur. Damage to the neurologic outflow from the lower thoracic and upper lumbar levels may result in an open smooth sphincter area, while the area of the striated sphincter generally retains a residual resting sphincter tone, but is not under voluntary control. Spotty lesions of the spinal cord and brain stem area generally result in involuntary bladder contractions with variable degrees of striated sphincter dyssynergia or synergia, depending on the severity of the disease and the intactness of the cortical and pontine-mesencephalic regulatory tracts. In this section the most common types of voiding dysfunction that occur with specific neurologic diseases or trauma are discussed.

AT OR ABOVE THE BRAIN STEM (SUPRASPINAL)

Cerebrovascular Disease

Cerebrovascular disease is about the third most common cause of death and the most common cause of disability in the United States and Europe. Arterial thrombosis, occlusion, and hemorrhage are the most common causes, leading to ischemia and infarction of variably sized areas in the brain, usually around the internal capsule. Transient ischemia has not been associated with any particular type of bladder dysfunction, but following an occlusive episode or a hemorrhagic episode of a cerebral vessel, focal neurologic deficits often occur and voiding dysfunction of a particular type may result (Kahn, 1981; Hald and Bradley, 1982; McGuire, 1984; Mundy and Blaivas, 1984). After the initial acute episode, acute urinary retention may occur. Following the variable degree of recovery of the neurological lesion, a fixed deficit becomes apparent over a few weeks or months. During this time the symptoms of bladder dysfunction generally become apparent. The most common long-term expression of lower urinary tract dysfunction after cerebrovascular accident is detrusor hyperreflexia. Sensation is variable, but generally intact, and thus the patient has urgency. The appropriate response is to try to inhibit the involuntary bladder contraction by forceful contraction of the striated sphincter. If this can be accomplished, only urgency results; if not, urgency with incontinence results. The increase in striated sphincter EMG activity during bladder filling is usually normal until just before the involuntary contraction occurs. A burst of striated sphincter activity will be seen when the patient voluntarily attempts to inhibit this voiding contraction. If one pays attention just to the tracings, and not to the patient, this is indistinguishable from true detrusor striated sphincter dyssynergia, when in reality it is an appropriate response for a patient who does not wish to soil himself, the urodynamic table, or the urodynamic technician ("pseudodyssynergia"—Wein and Barrett, 1982). Smooth sphincter synergia is the rule, and thus the disorder is one solely of filling and storage, occasioned by detrusor hyperactivity.

Mundy and Blaivas (1984) point out some important modifying factors that must be considered in the care of such patients. This problem generally occurs in an elderly population, some of whom have pre-existent involuntary bladder activity during filling and/or out et obstruction and/or stress incontinence. These individuals may not have sought medical attention because these problems were manageable, but the additional difficulty of a physical disability makes the situation intolerable. In other words, the bladder dysfunction may be primarily due to a pre-existent problem, and may exist with outlet pathology as well, and the entire voiding dysfunction may be adversely or beneficially affected by treatment regimens that concentrate on the detrusor hyperactivity alone. Another problem relates to the effect of the confusion and disorientation that often results after such a neurologic catastrophe and the fact that many such patients become depressed—especially if they have lost their independence and then require nursing home care. Vigorous pharmacologic treatment of the detrusor hyperactivity may make these associated problems worse. Finally, some patients with dementia will lose the awareness of the desirability of voluntary urinary control. These individuals may either have involuntary bladder contractions or not. In any case, they will simply void where and when they want, because they see no reason why they shouldn't. This is generally a reflection of dementia, and there really is no treatment for this, other than trying to circumvent this problem with a collection device or pads.

Concussion

Generally, when voiding dysfunction occurs following a closed head injury, there is an initial period of detrusor areflexia followed by recovery of unstable reflex detrusor dysfunction (McGuire, 1984). This is not a very common sequela of concussion, but when it occurs, the same considerations apply as for voiding dysfunction secondary to cerebrovascular disease.

Brain Tumor

Both primary and metastatic brain tumors have been reported to be associated with disturbances of bladder function (Hald and Bradley, 1982; McGuire, 1984). When voiding dysfunction results, it is related to the localized area involved. The most frequently involved area that results in bladder dysfunction is the superior aspect of the frontal lobe (Blaivas, 1985a). Resultant voiding dysfunction consists of detrusor hyperreflexia and urinary incontinence. These individuals may have a markedly diminished awareness of all vesical events and, if so, are totally unable to even attempt suppression of the micturition reflex. Smooth and striated sphincter activity is generally synergic in the absence of attempts to suppress the involuntary detrusor reflex.

Parkinson's Disease

This degenerative disorder primarily affects the pigmented neurons of the substantia nigra. This results in a relative dopamine deficiency and a predominance of cholinergic activity in the corpus striatum. Classical neurologic symptoms include bradykinesia, tremor, and skeletal rigidity. The annual prevalence is estimated at between 100 and 150 cases per 100,000 population, with onset between the ages of 45 and 65 years, and affecting men and women equally (Mundy and Blaivas, 1984; Blaivas, 1985a). Voiding dysfunction occurs in 25%–75% of patients, but—as with cerebrovascular disease—there may be pre-existing detrusor or outlet abnormalities, and the symptomatology may be affected by the various treatments that are instituted.

Symptoms generally consist of urgency, frequency, nocturia, and urge incontinence. The most common urodynamic correlate is detrusor hyperreflexia (Murnaghan, 1961; Andersen et al., 1976; Pavlokis et al., 1983). Striated sphincter dyssynergia with detrusor hyperreflexia generally does not occur, but it is less clear whether these patients retain voluntary striated sphincter control (Blaivas, 1985a).

One of the significant problems that we and others (McGuire, 1984; Mundy and Blaivas, 1984) have had in dealing with these patients is deciding whether or not male patients have anatomic outlet obstruction secondary to prostatic enlargement and whether prostatectomy is indicated in these patients. In our experience, the considerations here are slightly different than they are for a patient with voiding dysfunction following a stroke who also has outlet obstruction. Generally, detrusor contractility seems to be unimpaired in stroke patients with involuntary bladder hyperactivity, and, in the absence of dyssynergia, poor flow rates and high residual urine volumes in the male generally indicate prostatic obstruction. Outlet reduction in these patients is generally beneficial, because the functional bladder capacity improves as a result of a decrease in the residual urine volume, and one feels more comfortable in treating these patients with agents to decrease bladder contractility if they do not have concomitant anatomic bladder outlet obstruction. Male patients with Parkinson's disease and the identical symptoms and urodynamic findings do not seem to fare as well after prostatectomy. Poorly sustained bladder contractions, sometimes with slow sphincter relaxation, may occur as a result of the neurologic disease, and in these individuals prostatectomy may result in no change or in a worsening of the voiding symptoms. Acontractile bladders are unusual in Parkinson's disease, and are generally seen only in women, in which case the situation is probably representative of a pre-existent problem and unrelated to the Parkinson's disease per se.

Shy-Drager Syndrome

This uncommon degenerative disorder results in atrophy of areas in the cerebellum, brain stem, peripheral autonomic ganglia, and thoracolumbar preganglionic sympathetic neurons (Blaivas, 1985a). Patients generally present with parkinsonian symptomatology coupled with orthostatic hypotension and anhidrosis. Voiding dysfunction generally consists of detrusor hyperreflexia, a bladder neck that is open at rest, and denervation of the striated sphincter. Areflexia or poorly sustained bladder contractions may occur, however, and with diminished smooth and striated sphincter tone, may make management of the voiding dysfunction quite difficult, especially if prostatectomy seems indicated.

SPINAL CORD LESIONS

Suprasacral Spinal Cord Injury

Traumatic spinal cord injury may occur as a consequence of injury by a high-velocity missile, fracture or fracture dislocation of the spinal column, or after sudden or severe hyperextension. Spinal cord bony segments are numbered by the vertebral level, and these have a differing relationship to spinal cord segmental level at different locations. The sacral spinal cord begins at about the spinal column level of T12–L1. The spinal cord terminates in the cauda equina at the spinal column level of approximately L2. In addition to trauma, spinal cord damage may be produced by vascular disease, arteriovenous malformations, myelopathy, arachnoiditis, or myelitis.

Over the past 25 years or so, the survival of the patient with spinal cord injury has improved dramatically. This has resulted from the development of specialized centers for the care of such patients resulting in a decreased number of potentially significant complications of all types. Urologic care and urologic surveillance have undergone decided improvements, but renal failure is still one of the most common late causes of death, at least in male patients. Although amyloid disease of the kidneys accounts for some of these, bladder and outlet dysfunction and urinary tract infection with urosepsis are still significant problems. Controlled and coordinated bladder emptying depends on an intact neural axis. Bladder contractility and the occurrence of reflex bladder contractions depends on an intact conus medullaris (sacral spinal cord segments) and its afferent and efferent connections. Complete lesions above this area but below the area of the sympathetic outflow generally result in detrusor hyperreflexia, absent sensation below the level of the lesion, smooth sphincter synergia, and striated sphincter dyssynergia. Lesions above the spinal column level of T6 (spinal cord level of T7 or T8) may result in smooth sphincter dyssynergia as well.

Following a significant spinal cord injury, "spinal shock" occurs. This refers to a decreased excitability of spinal cord segments below the level of the lesion (Hald and Bradley, 1982; Thomas, 1984; McGuire, 1984). This is synonymous with absent somatic reflex activity and a state of flaccid muscle paralysis below this level. Although classic teaching refers to generalized areflexia, Thomas (1984) points out that the most peripheral somatic reflexes of the sacral cord segments (the anal and bulbocavernous reflexes) may never disappear or, if they do, return within minutes or hours of the injury. Spinal shock includes a suppression of autonomic activity, and the bladder is acontractile and areflexic. The smooth sphincter mechanism generally functions, however, and EMG activity can generally be recorded from the striated sphincter (McGuire, 1984; Awad et al., 1977; Yalla et al., 1978). Because sphincter tone exists, urinary incontinence does not result, urinary retention is the rule, and intermittent catheterization (or continuous catheterization in some instances) is necessary to circumvent this problem. If the distal spinal cord is intact but simply isolated from higher centers, there will eventually be a return of detrusor contractility. At first such reflex activity is poorly sustained and produces only low-pressure changes, but the strength and duration of such involuntary contractions increases, producing involuntary voiding, usually with incomplete bladder emptying. The return of reflex bladder activity in such patients is generally manifested by involuntary voiding between intermittent catheterizations, and occurs along with recovery of lower extremity deep tendon reflexes. In evolving lesions every attempt should be made to preserve as low bladder pressure storage as possible, and to avoid any measures that might impair this, such as indwelling catheter drainage and any operations designed to decrease outlet resistance.

The characteristic fixed dysfunction that results when a patient has a complete lesion above the sacral spinal cord is detrusor hyperreflexia, smooth sphincter synergia (with lesions below T6), and striated sphincter dyssynergia. Neurologic examination shows spasticity of skeletal muscle distal to the lesion, hyperreflexic deep tendon reflex, and extensor plantar responses. There is impairment of superficial and deep sensation. Incomplete bladder emptying generally results, usually because of striated sphincter dyssynergia. The neurological center responsible for coordinating bladder and striated sphincter activity is in the pontine mesencephalic formation (see Chapter 27) and any lesion between this area and the sacral spinal cord may interfere with this coordination. True dyssynergia, either continuous or intermittent, generally occurs, but in some instances there is simply a failure of relaxation without an increase in sphincter activity over baseline. Occasionally, incomplete bladder emptying may result from poorly sustained detrusor contractions. This seems to occur more commonly in lesions close to the conus medullaris than with lesions higher up (Hald and Bradley, 1984). Poorly sustained contractions may also be due to locally functioning reflex arcs, which result in detrusor inhibition from strong striated pelvic floor muscle contraction or are due to a loss of higher center mediated detrusor facilitation, which normally occurs after the initial increase in intravesical pressure during a bladder contraction (Thomas, 1984).

The involuntary bladder contractions may occur spontaneously or in response to external somatic stimulation of certain areas. Detrusor hyperreflexia with true striated sphincter dyssynergia is generally easily demonstrable urodynamically, and fluoroscopic studies generally demonstrate that, with the bladder contraction, the smooth sphincter area opens normally, and there is obstruction at the level of the striated sphincter, usually most pronounced in the region of the urogenital diaphragm. The primary urologic risk factor is high intravesical pressure. It is essential to know what the bladder pressures are prior to contraction during filling, to know the volume tolerated by the bladder at a safe intravesical pressure, and to recognize what the "leak point" is. To avoid upper tract sequelae, the pressures at volumes recovered by intermittent catheterization must be less than 40 cm H_2O, or pressures at the time of urinary leakage must be less than 40 cm H_2O. If bladder pressures are suitably low, or can be made low with nonsurgical or surgical treatment, intermittent catheterization can be continued as a safe and effective way of satisfying many of the goals of management. Alternatively, external sphincterotomy can be used in males to lower the leak point to an acceptable level.

Autonomic dysreflexia refers to the syndrome of exaggerated sympathetic activity in response to stimuli below the level of a spinal cord lesion (McGuire, 1984; Thomas, 1984; Hald and Bradley, 1982). It occurs most commonly in patients with lesions of the cervical spinal cord, but may occur with any lesion above T6. The syndrome consists of hypertension with headache and profuse sweating in the skin of the face and neck, and reflex bradycardia. The stimulus for this exaggerated response commonly arises from the rectum or bladder. This syndrome may be precipitated by instrumentation or by a change of a tube or catheter, and, in such cases, resolves quickly if the stimulus is withdrawn. Striated sphincter dyssynergia invariably occurs, and smooth sphincter dyssynergia is generally a part of the syndrome as well. Acutely, the hemodynamic effects of the syndrome may be managed with parenteral ganglionic or α-adrenergic blockade or with parenteral chlorpromazine (Thorazine) (McGuire, 1984). On a chronic basis, such episodes can be treated with α-adrenergic blocking agents or with chlorpromazine. To maintain volumes and pressures below which such changes will not

occur may require more frequent intermittent catheterization or measures to decrease bladder hyperactivity or increase bladder capacity or measures to decrease outlet resistance so that leakage occurs at a lower pressure and lower volume.

Sacral Spinal Cord Injury

Following recovery from spinal shock, there is generally a depression of deep tendon reflexes below the level of the lesion with varying degrees of flaccid paralysis. Sensation is absent below the level of the lesion. Detrusor areflexia is the common initial result, and is generally accompanied by a competent but nonrelaxing smooth sphincter and a striated sphincter that retains some tone, but is generally not under voluntary control, and that shows absent or diminished EMG activity (Hald and Bradley, 1982; McGuire, 1984; Thomas, 1984). Although the classical description of this problem is that of a decent capacity bladder with high compliance, it is obvious that in many of these patients decreased compliance develops, and because such decreased compliance also develops in neurological lesions distal to the sacral spinal cord, this type of change most likely represents a response to decentralization. At a point at which bladder pressure becomes greater than urethral pressure, leakage results, and if this leak point is high enough, upper tract deterioration may occur. It is not uncommon for incomplete lesions to result at this level, and so varying degrees of bladder and sphincter(s) dysfunctions and combinations may occur. Neurourologic evaluation should be able to exactly characterize each of these patients, and management modalities are generally obvious, consistent with the goals and practicalities previously mentioned.

Multiple Sclerosis

Multiple sclerosis is one of the most common neurologic diseases causing voiding dysfunction (Blaivas, 1985b; Mundy and Blaivas, 1984; Hald and Bradley, 1982; Blaivas et al., 1979; McGuire and Savastano, 1984; Andersen and Bradley, 1976; Schoenberg and Gutrich, 1980). The disease is due to focal neural demyelination, which causes impairment of nerve conduction. The slowing of nerve conduction results in varying neurologic abnormalities that tend to be subject to exacerbation and remission over time. The demyelinating process most commonly involves the posterior and lateral columns of the cervical spinal cord, the site of pathways that subserve bladder and outlet function. Fifty percent to 80% of patients with this disease complain of voiding symptoms at some time. Bladder involvement is part of the presenting symptom complex in approximately 10% of patients, and may constitute the sole initial complaint, either in the form of acute urinary retention of unknown etiology or in the acute onset of involuntary bladder contractions with urgency and frequency. Detrusor hyperreflexia is the most common urodynamic abnormality detected, and occurs in 50%–90% of cases (McGuire, 1984; Mundy and Blaivas, 1984; Blaivas, 1985b). Of patients with detrusor hyperreflexia, 30%–65% also have striated sphincter dyssynergia. Bladder areflexia may also occur, and in fact has been reported in 1–40% of cases, but, of these, a substantial proportion progress to detrusor hyperreflexia. Smooth sphincter synergia is the rule.

Blaivas and Barbalias (1984) have identified certain risk factors referable to voiding dysfunction in patients with multiple sclerosis. These include an indwelling catheter, detrusor striated sphincter dyssynergia in men, and decreased compliance, resulting in sustained intravesical pressures greater than 40 cm H_2O. Because sensation is frequently intact, one must be careful to separate pseudodyssynergia from true striated sphincter dyssynergia. Also, there are some varieties of striated sphincter dyssynergia that are more worrisome than others (Blaivas et al., 1981), and it is those that are sustained—resulting in high bladder pressures of long duration—that are most associated with urologic complications.

PERIPHERAL NERVE LESIONS

Diabetes Mellitus

Peripheral and autonomic neuropathies are common in diabetes. The etiology is a metabolic derangement of the Schwann cell that results in segmental demyelinization and impairment of nerve conduction (Blaivas, 1985b; Mundy and Blaivas, 1984). Neuropathy tends to develop in middle-aged and elderly patients with longstanding or poorly controlled diabetes. The exact incidence is uncertain, as unselected patients generally do not complain of bladder symptoms. If specifically questioned, 5%–50% report symptoms of voiding dysfunction. Frimodt-Moller (1976a, 1976b) coined the term "diabetic cystopathy" to refer to involvement of the lower urinary tract by this disease. The insidious onset of impaired bladder sensation is usually the first manifestation of such involvement. There results a gradual increase in the time interval between voiding, which may progress to a point at which the patient voids only once or twice a day without ever sensing any real urgency. Ultimately, detrusor decompensation occurs; detrusor contractility is therefore diminished, and abdominal straining is necessary to initiate and maintain the voided stream, the strength and force of which are impaired. The typical urodynamic findings include impaired bladder sensation, increased cystometric capacity, decreased bladder contractility, impaired uroflow, and residual urine. The main differential diagnosis is generally from outlet obstruction, as both conditions produce a low flow rate. However, the flow pattern in diabetic patients is more commonly one of abdominal straining, and, of course, pressure/flow studies are easily able to differentiate the two. What one generally sees in these patients is actually the secondary manifestations of resultant bladder decompensation, and this could be obviated by early awareness of the problem and the institution of strictly timed voiding.

Tabes Dorsalis

Luetic involvement of the posterior sacral roots and the dorsal columns of the spinal cord may result in the loss of bladder sensation and the same sequelae as seen in the patient with diabetes and resultant voiding dysfunction (Hald and Bradley, 1982). Pernicious anemia may also result in a similar type of "sensory neurogenic bladder."

Herpes Zoster

Herpes zoster may affect the urinary tract by causing urgency and frequency, followed by urinary retention (Hald and Bradley, 1982). Invasion of the sacral spinal ganglia with the causative virus results in the typical skin eruption within the corresponding dermatomes. Endoscopic examination may reveal a similar type of vesicles within the vesical mucosa. This condition is generally self-limited and resolves spontaneously within a month or two.

Disk Disease

Most disk protrusions compress the spinal roots in the L4–5 or L5–S1 disk interspaces. Voiding dysfunction may occur as a result of a prolapsed disk, and, when present, generally occurs with the usual clinical manifestations of low back pain radiating in a girdle-like fashion along the involved spinal root areas (Hald and Bradley, 1982). The most consistent urodynamic finding is detrusor areflexia. The striated sphincter may be normal or show evidence of denervation. Patients with voiding dysfunction generally present with the onset of difficulty voiding, straining, or urinary retention. Laminectomy may not improve bladder function, and it may be difficult in these cases to separate causation as a result of the disk sequelae itself from changes secondary to the surgery.

Radical Pelvic Surgery

Voiding dysfunction is unfortunately relatively common after pelvic plexus injury. This most commonly occurs after abdominoperineal resection and radical hysterectomy. Neurologic dysfunction after these procedures is reported in 10%–60% of patients, and in 15%–20%, voiding dysfunction is permanent (McGuire, 1984; Mundy, 1984; Hald and Bradley, 1982). The type of voiding dysfunction that occurs is dependent on the specific nerves involved and the degree of injury, and on the pattern of reinnervation or altered innervation that results over time. Urinary retention, with varying degrees of sensory preservation, is generally the initial manifestation of such voiding dysfunction. The permanent pattern is generally a failure of voluntary bladder contraction, or impaired bladder contractility, with obstruction by residual striated sphincteric tone, which is not subject to voluntarily induced relaxation. Often, the smooth sphincter area is open and nonfunctional, a result—some think—of destruction of the terminal sympathetic nerve supply to this area. Alternatively, such an appearance in a patient whose bladder neck has not been operated on may result from increased intravesical pressure and obstruction at the level of the striated sphincter. Decreased compliance is common in these patients, and this, with the obstruction caused by fixed residual striated sphincter tone, results in the paradoxical occurrence of both storage and emptying failure. These patients often leak across the distal sphincter area and, in addition, are unable to empty their bladder, because although they have an increased intravesical pressure, they have nothing that approximates a true bladder contraction. Thus, they represent a combined problem of filling/storage and emptying. They often present with urinary incontinence, which is characteristically most manifest with increases in intra-abdominal pressure. This is usually most obvious in females, as the prostatic bulk in males often masks a deficit in urethral closure function. Alternatively, patients may present with variable degrees of urinary retention. Urodynamic studies may show decreased compliance, poor proximal urethral closure function, loss of voluntary control of the striated sphincter, and a positive bethanechol supersensitivity test. Upper tract risk factors are related to the "leak point," and the therapeutic goal is low-pressure bladder storage with periodic emptying. The temptation to perform a prostatectomy should be avoided unless a clear demonstration of outlet obstruction is possible. Prostatectomy will often simply decrease urethral sphincteric function and thereby result in occurrence, or worsening, of stress urinary incontinence.

Sacral Agenesis, Tethered Cord Syndrome, and Myelodysplasia

These are all discussed in the section Voiding Dysfunction in Children.

OTHER VOIDING DYSFUNCTIONS AND RELATED TOPICS
BLADDER OUTLET OBSTRUCTION
General Pathophysiology

Bladder outlet obstruction is one of the most commonly encountered disorders in urology. Surgical treatment of the most common variety, benign enlargement of the prostate, constitutes one of the most frequently performed operative procedures within the specialty. Generally, bladder outlet obstruction can be divided by etiology into fixed anatomic obstructions and functional obstructions. "Functional" in this case does not mean psychogenic, but simply indicates increased urethral resistance to the forces of bladder emptying by neuromuscular phenomena, which may be involuntary or voluntary. Anatomic causes of bladder outlet obstruction include prostatic enlargement (benign or malignant), bladder neck contracture, urethral stricture, and meatal stricture. Functional causes of bladder outlet obstruction include dyssynergia at the level of the smooth sphincter and at the level of the striated sphincter.

Bladder Response to Obstruction

The bladder responds to obstruction in a somewhat unpredictable fashion, but the dysfunction usually involves either a form of hyperactivity or a form of decompensation. Abrams (1985), although he does not entirely endorse the hypothesis that outlet obstruction causes involuntary bladder contractions, summarizes data that strongly suggest that outlet obstruction does cause detrusor instability. Gathering information from eight sources, he cites evidence that the incidence of detrusor instability in patients with outlet obstruction secondary to prostatism ranges from 53%–80%, with a mean of 62% in the series that he cites. Postoperatively, the incidence ranged from 0%–55% in these same series, with a mean of 24%. The fact that outlet obstruction in the form of prostatic enlargement is a major cause of bladder instability in the male is certainly supported by the high rate of instability preoperatively in such patients and the high reversal rate of this phenomenon following prostatectomy. This concept is widely endorsed (Hald and Bradley, 1982; Turner-Warwick, 1984; Jones and Schoenberg, 1983; McGuire, 1984b). Although the proof of cause and effect generally consists of retrospective correlation of the disappearance of this phenomenon after the relief of outlet obstruction, this nevertheless seems to be a reasonable argument, even though the disappearance of the phenomena generally takes between one and six months (and reversion to a normal cystometric accommodation limb and disappearance of all irritative symptoms may take as long as 12 months). In those instances in which the involuntary contractions seem to result from outlet obstruction, the exact pathophysiology is uncertain. Abrams (1985) points out that no series has demonstrated an association between the incidence of instability and the severity of obstruction. He also points out an increasing incidence of detrusor instability in "asymptomatic" elderly men. Chalfin and Bradley (1982) have presented evidence that there may be an increased input of stimuli that trigger detrusor reflex activity in the area of obstruction. In prostatic enlargement, this could be a result of anatomical distortion and/or compression of the abundant nerve endings in this area due to prostatic enlargement. Reversion to normal after prostatectomy could then be explained by a decompression phenomenon. Involuntary contractions occurring in response to outlet obstruction are looked on by some as a compensatory or protective mechanism to keep bladder and residual urine volumes small, thus preventing elongation of detrusor fibers by overdistention to a point at which they cannot contract efficiently (Jones and Schoenberg, 1983). These involuntary contractions may also occur on an irritative or neurologic basis. Perhaps the sensory tension receptors in the bladder wall are more prone to early activation when the bladder wall is under stress or when it has undergone the morphological changes of trabeculation or thickening. The most reasonable view of this phenomenon seems to be that although it can occur in response to outlet obstruction alone, its occurrence in most cases probably reflects a multifactorial origin—a response to a combination of neurologic, architectural, and strictly urodynamic factors related to pressure/volume changes within the bladder. All would agree that the phenomenon is much less common in response to urethral stricture disease and, although it may occur in response to smooth or striated sphincter dyssynergia, is much less common in response to these abnormalities as well. Regardless of arguments to the contrary, it is very important to keep in mind that the phenomenon of the production of involuntary bladder contractions in response to outlet obstruction is reproducible experimentally, as it has been clearly shown that in pigs, infravesical obstruction results in involuntary bladder contractions or decreased compliance in 80%–90% of animals (Jorgenson et al., 1983; Sibley, 1985).

Decreased bladder compliance during filling without discrete involuntary contractions is another interesting phenomenon that can likewise be caused by outlet obstruction (or simply coexist with it). In either case, its significance is the same as that of involuntary contractions, insofar as its detection prior to a procedure designed to correct the outlet obstruction is concerned. The possible relationship of this phenomenon to outlet obstruction has been the subject of much investigation. Benson et al. (1975) showed that increasing the degree of stretch on muscle strips from the bladder body changed the response to norepinephrine from the usual β or relaxation response to an α or contractile one. Rohner et al. (1978) recorded the same type of change of response to adrenergic stimulation after producing chronic bladder outlet obstruction in dogs. Perlberg and Caine (1982) showed that muscle strips from the bladder dome of 23% of patients with prostatic enlargement requiring surgery showed an α-adrenergic response instead of the normal β-adrenergic response. Decreased compliance during filling may result also from destruction of the viscoelastic properties of the bladder wall (see Chapter 27), generally resulting from infection, inflammation, and fibrosis. The clinical significance of all these forms of detrusor hyperactivity that occur in response to outlet obstruction is that they—and not the simple phenomenon of residual urine with a decreased functional bladder capacity—are often responsible for the clinical correlates of frequency, nocturia, urgency, and urge incontinence.

Detrusor hypoactivity, as opposed to hyperactivity, may also occur in response to significant infravesical obstruction (Hald and Bradley, 1982; McGuire, 1984; Turner-Warwick, 1984). The finding of residual urine is more a sign of detrusor decompensation than of the outlet obstruction that caused it. For this reason, the relief of outlet obstruction more commonly leads to satisfactory urodynamic resolution in patients with a low, as opposed to a large, residual urine volume (Hald and Bradley, 1982). Patients with outlet obstruction and decompensation of the detrusor muscle generally find it necessary to void with the aid of abdominal straining, easily detectable urodynamically.

Morphologically, the clinically detectable changes to bladder outlet obstruction are bladder muscle hypertrophy followed by connective tissue replacement with increased collagen deposits within the bladder wall (McGuire, 1984b). Smooth muscle hypertrophy and hyperplasia occur in response to outlet obstruction along with collagen deposition. Although the absolute amount of collagen increases, the collagen concentration decreases (Uvelius and Mattiasson, 1984). What regulates the degree of cellular hypertrophy in response to obstruction is unknown, but Gilpin et al. (1985) reported that the largest increases in cell size were observed only in patients with concomitant connective tissue infiltration. Whether the endoscopic appearance of trabeculation is more related to collagen deposition or muscular hypertrophy is a matter of discussion (McGuire, 1984b). Experimentally, models of bladder outlet obstruction vary with respect to the species employed and the degree or duration of the obstruction.

Even so, general response patterns exist (Brent and Stephens, 1975; Uvelius et al., 1984; Levin et al., 1984). An initial dilation of the bladder is followed by concentric bladder wall thickening with variable degrees of smooth muscle hypertrophy/hyperplasia and collagen deposition. After one week of partial bladder outlet obstruction, the rabbit bladder shows a ninefold increase in mass. Although clinically this is felt to represent a compensatory response that maintains normal emptying (as defined by low residual urine volume, but with high intravesical pressures), experimentally the response of the obstructed bladder to cholinergic stimulation drops to 50% of control values, and the ability of such a bladder to expel urine in response to such stimulation drops to 28% of control values. These rapidly induced contractile and metabolic changes are partially reversible, as a two-week recovery period (following one week of obstruction) results in a decrease in bladder mass to only twice control levels, a return of the contractile and pressure responses to cholinergic stimulation to control levels, and a recovery of the ability of the bladder to expel its contents to 75% of control levels (Levin et al., 1984, 1985). Although it is difficult to compare experimental and clinical situations, this would seem to grossly correlate with the fact that after relief of outlet obstruction, some patients experience total resolution of their symptoms and exhibit normal pressure flow/emptying characteristics, while others experience only partial resolution of some or all of these parameters, suggesting that the reversibility of changes in the coordination of contraction and in pressure development may not occur in the same proportion.

Outlet Obstruction in the Male

Prostatic obstruction may be due to hyperplasia, carcinoma, or inflammation. Obstruction is defined by an elevated intravesical pressure and a subnormal flow rate during voiding. Many asymptomatic elderly males satisfy this definition, but without the usual symptom complex known as "prostatism" (Hald and Bradley, 1982; Andersen et al., 1978). When symptoms are associated with prostatic obstruction, they fall into two general categories: obstructive and irritative. The obstructive symptoms consist of hesitancy, a slow stream, prolonged urination, and a feeling of incomplete emptying. These are variable among individuals, and do not occur in any direct relationship to either pressure or flow parameters. However, they are clearly distinct from the irritative symptoms, which consist of frequency, nocturia, urgency, and urge incontinence. These findings are statistically related to detrusor hyperactivity, which, as we have just discussed, occurs either in response to outlet obstruction or along with it

in a substantial percentage of cases. Frequency and nocturia can also occur on the basis of a substantial residual urine volume, with a resultant decrease in the functional bladder capacity. Bladder trabeculation can occur in response to either detrusor hyperactivity or obstruction without such hyperactivity (Andersen and Nordling, 1979, 1980; McGuire, 1984; Turner-Warwick, 1984; Hald and Bradley, 1982). The optimal neurourologic evaluation for the patient with suspected outlet obstruction secondary to prostatic enlargement differs markedly from center to center, and may vary from a history, physical examination, and endoscopic examination to sophisticated combined pressure/flow studies. It should be noted that significant outlet obstruction can exist in the presence of relatively normal flow rates, and that diagnoses can be made in this subgroup of patients only by pressure/flow studies that show grossly elevated intravesical pressures (Gerstenberg et al., 1982).

Other anatomic causes of bladder outlet obstruction in the male include sclerosis or fibrosis of the bladder neck, urethral stricture disease, and urethral valves. Functional causes of outlet obstruction in the male include smooth sphincter dyssynergia (sometimes called bladder neck dyssynergia or dyskinesia) and striated sphincter dyssynergia. Each of these can be secondary to neurologic or nonneurologic causes. We have previously discussed the neurologic causes of sphincter dyssynergia. Turner-Warwick deserves credit for defining the condition of dyssynergic bladder neck obstruction (Turner-Warwick and Whiteside, 1970; Turner-Warwick et al., 1973; Turner-Warwick, 1984). This syndrome characteristically occurs in males between the ages of 30 and 55 years who complain of long-standing obstructive and irritative symptoms. Many of these patients have been to see many urologists and have been diagnosed as having psychogenic voiding dysfunction because of a normal prostate on rectal examination, a negligible residual urine volume, and a normal endoscopic bladder appearance. In this group of patients, objective evidence of outlet obstruction is easily obtainable by uroflowmetry or/and pressure-flow studies. Secondary bladder changes may occur, manifested by detrusor hyperactivity or decompensation with residual urine volumes. Once obstruction is diagnosed, it is localized at the level of the bladder neck in these individuals either by micturitional urethral profilometry or by a formal videourodynamic study or cystourethrography during a bladder contraction that shows obstruction at the level of the bladder neck, prostatic enlargement having already been excluded endoscopically. The diagnosis may also be made indirectly by the finding of outlet obstruction in such a patient in the absence of a distal urethral stricture, prostatic enlargement, or evidence of striated sphincter dyssynergia. The etiology of this problem is unknown (Turner-Warwick, 1984). Bates et al. (1975) proposed that these patients had an abnormal arrangement of musculature in the bladder neck region, such that coordinated detrusor contraction caused bladder neck narrowing instead of the normal funneling. The occurrence of this problem in young, anxious, and high-strung individuals and its partial relief by α-adrenergic blocking agents have prompted some to speculate that it may in some way be related to sympathetic hyperactivity. Neither of these theories has ever been substantiated. When individuals with this problem develop prostatic enlargement, a double obstruction results, and Turner-Warwick has applied the term "trapped prostate" to this entity (Turner-Warwick, 1984). A patient so afflicted generally has a lifelong history of voiding dysfunction that has gone relatively unnoticed because he has accepted this as normal. Exacerbation of these symptoms occurs during a relatively short and early period of prostatic enlargement, and marked relief in these cases is generally effected by a "small" prostatic resection. As Turner-Warwick (1984) points out, such patients often note after outlet resection that they have "always" had some voiding problem, and that they have "never" voided so well as after their treatment.

Functional obstruction at the level of the striated sphincter in males and females occurs most commonly in association with neurologic disease, and such dysfunction has previously been discussed. It is easy to voluntarily produce obstruction in this area, and we do so routinely when obeying the command to stop voiding. This is, in fact, one form of pseudodyssynergia (Wein and Barrett, 1982), which must be distinguished from true involuntary functional obstruction at this level. Many believe that true detrusor striated sphincter dyssynergia, implying involuntary contraction of the striated sphincter during the emptying phase of micturition, occurs only in patients with neurologic disease. The existence of this entity in individuals without neurologic disease seems to be reported most commonly in children, but is reported in some adults as well, in whom the entity has been postulated to cause problems ranging from obstructive or/and irritative symptoms of obscure etiology to recurrent urinary tract infection (Van Gool and Tanagho, 1977; Raz and Smith, 1976; Kaplan et al., 1980; Hinman, 1974; Allen, 1977). It should be noted that it is very difficult to prove urodynamically that an individual has such an entity, and it should further be noted that diagnoses in most of the patients reported have been made on the basis of history, isolated flowmetry, isolated measurements of total

intravesical pressure, and pelvic floor EMG activity. Unequivocal demonstration of this entity, in our opinion, requires simultaneous pressure-flow-EMG evidence of bladder emptying occurring simultaneously with involuntary striated sphincter contraction in the absence of any element of an abdominal straining component, either in an attempt to augment bladder contraction or as a response to discomfort during urination. Such reports do exist (Jorgensen et al., 1982; Gerstenberg et al., 1983), and seem to confirm the existence of the syndrome. The etiology is uncertain, and may represent a persistent transitional phase in the development of micturitional control or persistence of a reaction response to the stimulus of urethral pain during voiding long after the problem has disappeared (Jorgensen et al., 1982).

Outlet Obstruction in the Female

The frequency of infravesical obstruction in the female is rare when compared to incidence in the male. Farrar and Osborne (1984) cite a figure of 10% of females referred to urodynamic centers as having outlet obstruction. They report that the most common site of urodynamically proved outlet obstruction in the female is in the distal urethra, and that this condition usually results from a failure of this area to fully attain its potential caliber during voiding. The exact cause of this is uncertain, and authors note the apparent discrepancy of their conclusion with the fact that the distal urethra is rarely smaller than the 10–12 F size that seems necessary to produce significant obstruction. Nonneurologic striated sphincter dyssynergia is possible as an etiology, but as we have just discussed, this diagnosis is quite difficult to make. Whether findings such as these justify empiric urethral dilation in females with obscure voiding symptoms is doubtful, but it may explain the small number of patients that seems to obtain reproducible clinical and urodynamic benefit in response to such treatment.

Bladder neck obstruction in females is quite rare. Diokno et al. (1984) have clearly defined this entity in a small number of patients on the basis of urodynamic video-pressure-flow criteria. They point out that the diagnosis of obstruction in this situation must be based on the demonstration of a relationship between pressure and flow, and should not be made on the basis of a single isolated parameter or any study that does not include simultaneous measurements of such parameters. Once the diagnosis of obstruction has been made, the site of obstruction must be localized further, either by simultaneous (during flow) periurethral EMG recording and fluoroscopic monitoring or by micturitional urethral pressure profilometry. Even then, surgical treatment of this entity should obviously be approached with caution, as incontinence presents a significant risk.

URINARY INCONTINENCE

Definitions

Urinary incontinence is not a disease, but rather a symptom that can result from many different types of lower urinary tract dysfunction. The International Continence Society has defined urinary incontinence as a condition in which involuntary loss of urine is a social or hygienic problem and is objectively demonstrable (International Continence Society Committee on Standardization of Terminology). Such loss of urine can first of all be divided into loss that occurs through the urethra and through channels other than the urethra. We are concerned here only with urethral incontinence. The terminology used to describe various types of urethral urinary incontinence is somewhat confusing because symptoms, signs, physical conditions, urodynamic findings, and anatomic factors are sometimes mixed together to describe the situation. There are a number of terms, however, that seem to have general acceptance and are understood by all to mean certain things. The International Continence Society's Committee on the Standardization of Terminology is responsible for most of these definitions, and this committee now has input from the American-based Urodynamics Society. Stress incontinence can either denote a symptom, a sign, or a condition. The symptom of stress incontinence indicates the patient's statement of involuntary loss of urine when exercising physically. The sign denotes the observation of involuntary loss of urine from the urethra immediately on an increase in abdominal pressure. The condition of "genuine stress incontinence" (many criticize this term, but none have been able to come up with a better one) denotes involuntary loss of urine when the total intravesical pressure exceeds the maximum urethral pressure, but in the absence of detrusor activity ("subtracted bladder pressure": total intravesical pressure minus intra-abdominal pressure). Urgency incontinence refers to the involuntary loss of urine associated with a strong desire to void. Urgency incontinence may be subdivided into motor urge incontinence, which is associated with involuntary bladder contractions, and sensory urge incontinence, which is not associated with involuntary bladder hyperactivity. Reflex incontinence applies to the involuntary loss of urine due to abnormal reflex activity in the spinal cord in the absence of sensation. Overflow incontinence is the involuntary loss of urine when intravesical pressure exceeds the maximum urethral pressure

due to an elevation of intravesical pressure associated with distention of the bladder, but in the absence of detrusor activity.

Incidence

True prevalence figures for urinary incontinence vary widely. There are reports that cite the occasional leakage of small quantities of urine in over half of healthy females when exercising (Nemir and Middleton, 1954; Wolin, 1969). However, because this is not a social or hygienic problem, this does not strictly qualify for inclusion. To cite another example, most urologists would agree that the incidence of urinary incontinence following radical prostatectomy is somewhere between 1% and 15%. Rudy et al. (1984) reported an incidence of 87%. This was based on detailed evaluation, and basically counted any leakage of urine as incontinence. In fact, only 13% of their patients had incontinence severe enough to warrant treatment. Three patients complained of incontinence in whom it could not be demonstrated by sophisticated urodynamic techniques, and two others had easily demonstrable incontinence despite a negative clinical history. Thus, in almost one-third of their patients, either the clinical or urodynamic assessment would have been incorrect had they been the only means of defining incontinence. Thomas et al. (1980) reported the prevalence of urinary incontinence in Great Britain, determining this by the number of incontinent patients under the care of various health and social service agencies in two London boroughs ("recognized" incontinence), and by a postal survey of 22,430 persons on the practice lists of 12 general practitioners in different parts of the country ("unrecognized" incontinence). Recognized incontinence occurred in 0.2% of women and 0.1% of men aged 15–64 years and in 2.5% of women and 1.3% of men aged 65 years and over. Unrecognized incontinence showed a prevalence of 8.5% in women and 1.6% in men aged 15–64 years and 11.6% in women and 6.9% in men aged 65 years and over. Feneley (1979) found the prevalence of recognized incontinence to be approximately 1% of a community population, and found the prevalence of unrecognized incontinence to be 2.0% in the male population aged 15–64 years and 5.7% in the male population aged 65 years and over. For females, the corresponding rates were 7.6% and 13.9%. Yarnell et al. (1981) carried out a small survey in women in South Wales and found that 45% admitted to some degree of incontinence. Some degree of stress incontinence was reported by 22% of women, urge incontinence by 10%, and combined incontinence by 14%. In most women the urinary loss was small and infrequent, but 5% experienced a loss sufficient to necessitate a change of clothes, and in 2.6% such loss occurred daily. Over 3% reported urinary incontinence that interfered with their social or domestic life, but only half of these had sought medical advice.

Pathophysiology

Very simply, discounting those individuals who, most commonly because of cerebral disease and dementia, do not recognize a need to urinate in a socially acceptable manner or place, the pathophysiology of urinary incontinence can be thought of as either primarily related to the bladder, to the outlet, or to both. Bladder hyperactivity may be manifested either as discrete involuntary contractions or as a more tonic decrease in the volume-pressure curve during filling (decreased compliance). Bladder-related incontinence may also occur in an individual without hyperactivity but with such hypersensitivity on filling that urinary incontinence results from an uncontrollable need to rid oneself of a painful bladder sensation that has occurred in response to filling. Incontinence primarily related to the outlet most commonly results from intermittent decreases in urethral pressure below that of bladder pressure, which occur during abdominal straining and are unrelated to bladder hyperactivity (another way of defining "genuine stress incontinence"). Episodic decreases in outlet pressure that are unrelated to increases in bladder pressure may also be responsible for urinary incontinence (urethral instability). Finally, the bladder outlet may simply have lost its closure potential and be nonfunctional. Although the incontinence in individuals so affected increases greatly with abdominal straining, the pathophysiology, and therefore the correction, is quite different. Virtually any combination of bladder- and urethra-related causes may exist, compounding the complexity of finding an adequate solution.

Outlet-Related Incontinence in the Female

Under this category we will consider genuine stress incontinence, the nonfunctional bladder neck and proximal urethra, and urethral instability. Urethral instability refers to a seemingly reflex-caused total loss of urethral closure pressure provoked by bladder filling without a corresponding increase in intravesical pressure (McGuire, 1978). Although others have subsequently reported significant urethral pressure variations during bladder filling, some in association with involuntary bladder contractions and some in association with stress incontinence (Ulmsten et al., 1982; Hindmarsh et al., 1983; Vereecken and Das, 1985), it is not clear whether this entity is really the same as that reported by McGuire. In McGuire's series it was clearly only the urethral instability that was responsible for the

urinary incontinence that occurred, and it may have been that these patients were experiencing a reflex phenomenon related to their lower urinary tract without any component of increased bladder pressure because of bladder hypotonicity. The etiology of urethral instability is unknown, and it is unknown whether it is related to bladder instability. This is an entity that cannot be diagnosed by conventional urodynamic study, but obviously requires the performance of urethral pressure recordings. Additionally, there have been no systematic investigations regarding the treatment of this entity, but conceptually it would seem to make sense to try to raise the urethral pressure well above the threshold at which a precipitous decrease results from whatever stimulus is involved. Fossberg et al. (1981) have reported some success with this approach in individuals with sensory urge incontinence without bladder hyperactivity.

Many of the factors related to maintaining outlet continence have been discussed in Chapter 27. These fall under three general categories: those related to urethral closure pressure, those related to urethral length, and those related to support of the urethral trigonal anatomy. Urethral length has been considered an important factor by some in the prevention of stress urinary incontinence. Lapides (1958) considered the normal functional urethral length in the female to be 3–4.5 cm, and felt that stress incontinence occurred when and because the functional length of the urethra was less than 3 cm in the erect position. Although this may be correct much of the time, there is considerable disagreement about whether there is more than a minimal change in the resting urethral pressure profile and functional urethral length following surgical correction of stress urinary incontinence. Also, there seems to be a minimal change in functional urethral length in the elderly between those patients who have stress urinary incontinence and age-matched controls. Factors such as these have led most people to believe that reduction of urethral length is rarely a factor per se in promoting stress urinary incontinence. Urethral closure pressure certainly contributes to a cough competent outlet, and any decreases in factors that contribute to urethral closure pressure, though they may not cause stress incontinence by themselves, in combination can certainly either result in stress incontinence or make a preexisting problem much worse. Alternatively, augmenting these factors may result in enough of an increase in urethral closure pressure to significantly ameliorate the problem. The smooth muscle of the outlet, as previously discussed, is estimated to account for 30%–50% of urethral closure pressure. The smooth muscle has inherent tonus and is responsive to pharmacologic stimulation, particularly α-adrenergic stimulation. Those factors that decrease smooth muscle tone and responsiveness include overstretching, trauma, α-adrenergic blockade, β-adrenergic stimulation, progesterone administration, and age. Factors that increase smooth muscle tone and responsiveness include stretch (up to a certain point), α-adrenergic stimulation, β-adrenergic blockade, and estrogen administration. The striated muscle of the outlet likewise has been estimated to account for 30%–50% of urethral closure pressure. As we have discussed, this striated muscle effect is not just confined to the urethra in the area of the urogenital diaphragm, as the striated muscle continues along the urethra from this level for a variable distance to the bladder neck. Striated muscle also possesses an inherent tonus and is capable of voluntary or reflex activity. The factors that decrease striated muscle tone and responsiveness include a section of the pudendal nerve (and perhaps the pelvic nerve), skeletal muscle blockade, age, trauma, and overstretching. Factors that increase striated muscle tone and responsiveness include pudendal nerve stimulation, and perhaps pelvic nerve stimulation as well. The vascular cushion effect refers to the turgor of the blood vessels in the urethral submucosa, and is estimated by Raz et al. (1972) to account for 20%–30% of urethral closure pressure, although others think its contribution is minor. Factors that decrease this effect include estrogen deficiency, trauma, and age. Factors that increase this effect include estrogen administration. This "inner urethral softness" (Zinner et al., 1980) acts as a framework of spongy tissue that supplements the pliability and elasticity of the urethral mucosa. The "mucosal seal effect" is due to the infolding of the urethral epithelium surfaces and the complete apposition of the surface folds for which the mucus secretion is necessary. Although the exact contribution of this mucosal seal effect to urethral closure pressure is unknown, the lack of it certainly increases the amount of compression pressure required to form a complete seal. Factors that decrease this effect include estrogen deficiency, trauma, and age, and factors that increase this effect include estrogen administration in an estrogen-deficient female. Staskin et al. (1985) have concisely summarized current concepts of the anatomic support and configuration of the urethral trigonal region. Pelvic support of the vesicourethral segment is contributed mostly by the pubourethral ligaments, which are condensations of the endopelvic fascia. The main support is from the posterior pubourethral ligaments, which extend from the inferior portion of the pubis to the mid-urethra. These prevent downward and posterior rotational displacement of the urethra. A "hammock" of levator ani muscle, covered by endopelvic fascia, supports the bladder and urethra in their intra-abdominal position. The pubourethralis, a division of this muscle, forms a sling around the proximal urethra as it passes through the pelvic diaphragm, and aids in preventing posterior displacement of the proximal urethra and bladder neck. The perineal musculature and fasciae provide additional inferior support, especially during abdominal straining. The fascia of the bladder and anterior vaginal wall fuse to form the pubocervical fascia. At the level of the bladder neck and proximal urethra, and at the level of the bladder and the fascial ring of the cervix, these two fasciae are densely adherent. Along the base of the bladder, the vesicovaginal space can be developed between these two fascial planes. The pubocervical fascia prevents herniation of the bladder and urethra into the vagina. The vagina provides a potential space for both anterior and posterior vaginal prolapse. Anterior vaginal prolapse may occur through the upper (cystocele), middle (trigonocele), or lower (urethrocele) third of the pubocervical fascia. The sacrouterine and cardinal ligaments support the uterus and cervix laterally and posteriorly, respectively, and provide direct and indirect supports to the bladder and urethra through these attachments. The primary function of these suspensory mechanisms is to support the upper urethra and urethrovesical junction and to provide a stabilizing effect to check the descent of these areas, which might otherwise be produced by an increase in intra-abdominal pressure.

The one factor that seems to be common to patients with genuine stress urinary incontinence is a failure of transmission of intra-abdominal pressure increases such that these increases do not act to occlude the proximal urethra. Enhorning (1961) was the first to urodynamically demonstrate this failure of abdominal pressure transfer to the upper urethra. Anatomically, the pathophysiology seems to involve downward and posterior rotational movement of the bladder neck and proximal urethra during intra-abdominal straining such that the bladder neck and proximal urethra descend to a position in which concomitant transmission of intra-abdominal pressure to these areas does not occur. The urethra is in a dependent position with respect to the bladder base, and the tensile forces of the full bladder open the bladder neck with stress and permit flow into the proximal urethra (Staskin et al., 1985). Whatever other factors contribute to decreases in urethral closure pressure, this abnormal descent of the bladder base and proximal urethra seems to be the only factor regarding which a consensus has been reached as an etiology common to all patients with genuine stress incontinence. All of the true vesicourethral suspension procedures previously discussed correct this problem by anchoring the vesicourethral junction in a physiologic position and preventing its descent during abdominal straining.

Many gynecologists, though not all, are fond of the "angle concept" in discussing stress incontinence and recommending various types of therapy. At rest with a full bladder, the bladder base lies parallel to or 1–2 cm above a line joining the lower sacral border to the lower border of the pubic symphysis. The posterior urethrovesical angle is equal to or less than 100°. With straining, a downward and backward displacement of 1–1.5 cm of the bladder base and proximal urethra occurs, but the posterior urethrovesical angle stays the same. The angle of inclination of the upper urethral axis to the vertical is approximately 35° or at least less than 45°. The classic gynecologic definition and classification of stress incontinence into type 1 and 2 was made on the basis of changes in these angles (Green, 1962, 1975, 1982; Hodgkinson, 1970). These diagnoses were made on the basis of a lateral cystogram, sometimes using a chain to visualize the path of the urethra. Type 1 incontinence referred to complete or nearly complete loss of the posterior urethrovesical angle, but with maintenance of the angle of inclination of the urethral axis. Type 2 incontinence referred to those patients who had loss of the posterior urethrovesical angle and an abnormal angle of inclination of the urethral axis—greater than 45°—and often even completely reversed—greater than 90°. The angle concept held that type 1 incontinence could be corrected by an anterior vaginal repair, which supported the urethra and the bladder neck, while a true suspension procedure was necessary for type 2, which actually comprised downward and backward rotational descent of the urethral axis. Although the angle concept provides some quantitative determination of the degree of failure of support of the vesicourethral segment, it is obvious to any who have studied women with voiding dysfunction that some who are continent have abnormal angles and some who have stress incontinence have normal angles. Stanton (1985) comments further that the angle concept is "unphysiological and mechanically unsound," implying support of the view that angle changes per se are not the cause of stress urinary incontinence, but simply are often a correlate of the changes that occur that cause failure of intra-abdominal pressure increases to the bladder neck/proximal urethra. We have found the diagrammatic conceptualization of Constantinou et al. (1981) to be an excellent teaching aid in explaining both the deficit in stress incontinence and just what it is that vesicourethral suspension procedures actually do (Fig 28–28).

The pathophysiology of classical stress urinary incontinence is corrected by any of the standard vesicourethral suspension procedures, all of which, when properly done, restore the anatomic and physiological position of the proximal urethra in such a fashion as to allow transmission of intra-abdominal pressure so that ure-

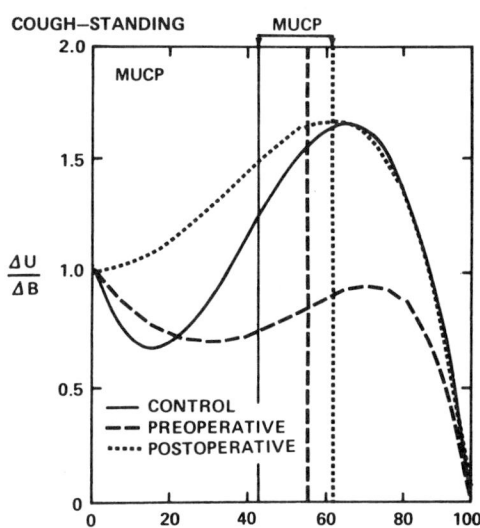

FIG 28–28.
Transmission of intra-abdominal pressure along urethra in normal females and in those with stress incontinence. Data shown were recorded in standing position with a soft perfusion urethral catheter. The vertical lines show position of the maximum urethral closure pressure (MUCP) expressed as a percentage of urethral length for comparison. Note that this maximum is most proximal in controls. Vertical axis shows the ratio of change in urethral pressure to change in bladder pressure. Note that in controls this is greater than 1 for most of urethral length. Note that in preoperative patients with stress urinary incontinence, this transmission ratio is less than 1 along the urethral length, making it easy to understand conceptually why stress incontinence occurs. Note that following correction, a transmission ratio of > 1 is restored for entire urethral length. (From Constantinou CE, et al: The impact of bladder neck suspension on the mode of distribution of abdominal pressure along the female urethra, in Zinner NR, Sterling AM (eds): *Female Incontinence: Progress in Clinical and Biological Research.* New York, Alan R Liss, 1981, pp 121–132. Used with permission.)

thral closure pressure is enhanced at least to a similar degree, and sometimes to a greater degree. In doing so, proper support is provided to prevent rotational descent of the bladder neck and proximal urethra. Such correction assumes that the factors responsible for urethral closure pressure in the first place are physiologically intact and capable of translating these intra-abdominal pressure increases into increased urethral resistance. There are some patients who have gross incontinence during abdominal straining in whom a vesicourethral suspension may not be effective for correction. These patients, classified by McGuire as having type 3 stress incontinence (McGuire, 1984), have a nonfunctional bladder neck and proximal urethra, in the sense that there is no or poor closure pressure in this area. On an x-ray film with the patient erect and the bladder full, this area appears fixed in an open or relatively open position. The incontinence that is seen in patients with this finding is generally quite severe, although they often complain primarily of stress incontinence, i.e., incontinence with any increases in intraabdominal pressure. They may state that they "leak" constantly, and they very well may, in the sense that almost any slight increase in intravesical pressure will cause urinary leakage. Often, they complain particularly of gravitational incontinence that occurs simply on assuming the erect position. These patients often have a history of previous pelvic surgery, and the characteristic history includes three or more failed procedures for what was diagnosed as genuine stress incontinence. These patients have a particularly low urethral closure pressure in the first 1–2 cm distal to the bladder neck, and radiographically they exhibit a bladder neck and proximal urethra that never quite closes in the erect position with the bladder full, but such patients do not exhibit an increase in subtracted bladder pressure. Hypermobility of the vesicourethral segment is generally not a problem in these patients, as this segment is relatively fixed because of previous scarring and fibrosis.

Bladder-Related Incontinence in the Female

Bladder-related incontinence in the female may be related to phasic rises in detrusor pressure, either of neurologic (detrusor hyperreflexia) or nonneurologic (detrusor instability) causation. Such incontinence may also be related to an abnormal tonic rise in intravesical pressure (decreased compliance). The neurologic causes of involuntary bladder contraction have already been reviewed. The etiology of detrusor instability is obscure, especially in women, in whom outlet obstruction (a common cause in males) is very uncommon and more uncommon still as a cause of detrusor instability (Mundy and Stevenson, 1984; Mundy, 1985). Although some work has begun to elucidate neurophysiologic and neuromorphologic differences in the detrusor muscle in patients with bladder instability from those of normal subjects, the etiology of this troublesome and common clinical problem remains obscure. One subject that should be mentioned is the simultaneous occurrence of bladder instability and stress urinary incontinence. Detrusor instability per se is a common cause of urinary incontinence in females, and has been cited as one of the most common causes of failure of a suspension operation for stress urinary incontinence (Stanton et al., 1978; Webster et al., 1984). Whether these urodynamic findings really do indicate that such an operation is likely to fail is an important question, as an affirmative answer would mandate a very careful urodynamic evaluation in all such patients, and might well decrease the enthusiasm for corrective surgery in such patients who also have genuine stress incontinence. Many, ourselves included, have noted the occurrence of detrusor insta-

bility in patients with stress incontinence, but have repaired the stress incontinence anyway, hoping to ameliorate the involuntary bladder contraction component of the incontinence by nonsurgical means. More often than not, however, this does not constitute a problem, and McGuire and Savastano (1985) have documented their long-stated position that only a small number of women with both stress incontinence and detrusor instability and/or urge incontinence will still have symptomatic urinary loss as a result of persistent detrusor abnormalities at eight weeks after corrective surgery for the stress incontinence. They reviewed 603 women with urodynamically and radiographically diagnosed stress incontinence, 180 of whom had associated detrusor instability and/or urge incontinence. The overall incidence of detrusor instability was 30%. By eight weeks after surgery, only 25 patients, or 4% of the entire group, and only 14 of those with detrusor instability and/or urge incontinence at any time, were still troubled with incontinence due to that condition. Thus, in McGuire's experience, women with both stress incontinence and detrusor instability fare worse after operation with respect to total freedom from incontinence than those with only stress incontinence. However, 90% of women with stress incontinence and detrusor instability and/or urge incontinence were totally relieved of urinary incontinence at six months by an operation designed to treat only the stress incontinence. These data also suggest that, at least in some patients, stress incontinence and detrusor instability are causally related. Patients with stress incontinence who also have bladder-related incontinence because of decreased compliance do not seem to fare so well after surgery to correct just the stress incontinence. The decreased compliance generally does not change, and if it was refractory to nonsurgical therapy before a suspension operation, it will remain that way, at least in our experience. There are few published reports regarding the fate of such patients after only an operation for stress incontinence, probably because experience elsewhere has been the same as ours.

Urgency-Frequency-Incontinence in the Female Without Detrusor Hyperactivity or Outlet Incompetence

There are a number of patients who have symptoms that suggest involuntary bladder contractions, but who in fact have no detectable consistent pathology during the filling phase of micturition. Most of these patients report symptoms of urgency, frequency, dysuria, pressure, and incontinence in varying proportions. The treatment of this symptom complex is, initially at least, usually empirical and consists of one or a combination of the agents to decrease detrusor hyperactivity (Wein,

1985). The fact that most of these patients do not have involuntary contractions most likely accounts for the generally less than optimal results achieved only with this type of treatment (Rees et al., 1976; Ulmstein et al., 1977). Urodynamic evaluation of these patients may be entirely normal, may show only hypersensitivity during filling, or may show variations in urethral pressures. Fossberg et al. (1981) did report successful subjective results in 22 of 34 female patients with sensory urge incontinence, many of whom had wide urethral pressure variations during filling, with phenylpropranolamine (50 mg, twice daily). The ideal pharmacologic treatment for at least the urgency, frequency, and dysuria component of this symptom complex, which seems to be more sensory than motor in origin, would be an agent that produces topical anesthesia or hypoesthesia of the bladder and urethral mucosa. Although there are agents that are reported to have this action as their primary effect, the clinical results that have been obtained with such compounds in these circumstances, at least by us, have been poor. There is no doubt that, in at least some of these patients, the symptoms are psychosomatic. However, more likely reasons for the relatively poor treatment results in this group are (1) that current "state of the art" techniques and practices fail to detect pathophysiology from the myriad possibilities (Table 28–19) and (2) optimal treatment for some of these has not yet been defined.

Bladder-Related Incontinence in the Male

Detrusor hyperreflexia and detrusor instability in the male have previously been discussed, and the relationship between outlet obstruction secondary to prostatic enlargement and detrusor instability has received particular mention. Likewise, decreased compliance as a cause of urinary incontinence in the male has also been covered. Overflow incontinence has not yet been discussed and does represent a form of bladder-related incontinence that seems to be more common in the male than in the female, though it occurs in both. Overflow incontinence, although everyone "knows what it is," is in reality very poorly understood. The primary pathophysiology is actually a failure of emptying, leading to urinary retention with "overflow" incontinence, which results from either continuous or episodic elevation of the intravesical pressure over the urethral pressure. Overflow incontinence is said to result from outlet obstruction or from detrusor inactivity, either neurological or pharmacological in origin, or may be secondary to inadvertent overdistention of the bladder. Our subjective impression is that it generally occurs in patients whose bladder neck and proximal urethral resistance is somewhat decreased to begin with, or it occurs in patients in whom the detrusor is not totally decompensated. The treatment is obviously directed at the primary cause of the failure to empty, or at circumvention of this if correction of the primary cause is not possible or if irreversible bladder decompensation has in fact occurred.

Outlet-Related Incontinence in the Male

In theory at least, categories of outlet-related incontinence should exist in the male that are similar to those in the female, i.e., urethral instability, stress urinary incontinence, and a nonfunctional bladder neck and proximal urethra. In reality, there is little if any information regarding the topic of urethral instability in the male, and we personally have never seen a case analogous to that reported in the female. Stress-related incontinence does exist following neurological injury (this is discussed in the sections Voiding Dysfunction Following Radical Pelvic Surgery, and Myelodysplasia). The best known type of outlet-related incontinence in the male is post prostatectomy incontinence, best known perhaps because it is iatrogenic. In discussing this entity, we must first recognize that post-prostatectomy incontinence is not solely due to outlet-related reasons. A substantial proportion of these patients have bladder dysfunction (hyperactivity) that accounts for all or part of their symptoms. The incidence of bladder hyperactivity in patients with postprostatectomy incontinence has been variously reported as 55% (Mayo and Ansell, 1979), 40% (Abrams, 1980), 66% (Reid et al., 1980), and 38% (Yalla et al., 1982). Although many of these patients had not been studied urodynamically prior to surgery, it can be assumed that—because of

TABLE 28–19.

Possible Etiologies for Urgency-Frequency-Incontinence Syndrome

Lower urinary tract or genital infection, bacterial or nonbacterial
Meatal obstruction
Bladder outlet or urethral obstruction
Detrusor hyperactivity
Smooth sphincter dyssynergia
Striated sphincter dyssynergia
Urethral diverticulum
Urethral caruncle
Urethral carcinoma
Bladder carcinoma or carcinoma in situ
Bladder stone
Estrogen deficiency
Interstitial cystitis
Radiation cystitis
Tuberculous cystitis
Emotional factors
Decreased central or peripheral opioid/peptide influence
Nonlaminar turbulent distal urethral flow

the high incidence of bladder instability secondary to outlet obstruction in the male—a substantial proportion of them actually had the detrusor hyperactivity preoperatively. Thus, in most cases, it did not develop in response to the prostatectomy. Recognition of the abnormal bladder function is of utmost importance in the management of these patients, as satisfactory continence may most often be obtained with pharmacologic therapy alone when the patient has only a minimal amount or no element of sphincteric insufficiency.

Postprostatectomy incontinence is one of the most distressing complications that can befall the patient and the urologist. Caine and Edwards (1958) demonstrated quite conclusively that the continence mechanism after a nonradical prostatectomy is located between the verumontanum and the urogenital diaphragm. One can characterize the various types of prostatectomy, including radical prostatectomy, as to the potential anti-incontinence mechanisms left intact, and therefore as to the probability of urinary incontinence (related to the outlet) after prostatectomy. All types of prostatectomy essentially ablate the proximal sphincter mechanism (smooth muscle of the bladder neck and proximal prostatic urethra, as well as the nonmuscular elements of the wall of these structures). These hopefully leave intact enough of the distal sphincter mechanism (the smooth muscle of the urethra from the level of the verumontanum to the urogenital diaphragm, the striated muscle between the leaves of the urogenital diaphragm, the striated muscle that is applied along the outer wall of the urethra for a variable distance from the urogenital diaphragm to and beyond the apex of the prostate, and the nonmuscular elements of the urethral wall in that area) to produce continence in the absence of bladder hyperactivity and sensory urgency. Radical prostatectomy has the greatest potential for significant distal sphincter mechanism destruction, while transurethral prostatectomy generally results in the least disturbance of this mechanism. If one considers the potential damage to proximal or/and distal sphincter mechanisms with various types of prostatectomy, a correlation with the commonly quoted figures for incontinence after different types of prostatectomy is evident: transurethral prostatectomy for benign prostatic hypertrophy, 0.2%–1.4%; transurethral prostatectomy for carcinoma, 10%–20% (presumably because the distal sphincter mechanism is in some cases compromised by the carcinoma itself); suprapubic prostatectomy, 1%–6%; nonradical retropubic prostatectomy, 0.2%–2.6%; radical retropubic prostatectomy, 1%–87% (depending on the definition of incontinence); simple perineal prostatectomy, 1.9%–10.5%; and radical perineal prostatectomy, 2%–15%.

The evaluation of postprostatectomy incontinence should be carried out in a nonpanicky fashion. Incontinence associated with detrusor hyperactivity or sensory urgency will generally resolve within three months. Incontinence without urgency is more worrisome and less apt to resolve spontaneously. There is little excuse for a less-than-expert urodynamic evaluation of patients with these problems to ascertain the exact contribution of outlet and bladder-related factors. Although by no means foolproof, the following suggestions, which we have found useful, might help to avoid postprostatectomy incontinence:

1. When evaluating these patients, first remove all tags and tissue remnants protruding into the lumen, especially those at the level of the resected bladder neck and distal sphincter mechanism.
2. With a transurethral resection avoid capsular perforation, as it is very possible that extravasation of irrigating fluid through a capsular perforation can result in fibrosis and/or fixation with resultant dysfunction of the distal sphincteric mechanism.
3. It is sometimes necessary to resect prostatic tissue distal to the verumontanum. Be careful when doing this. The classical teaching to stop a resection at the level of the verumontanum is not because the so-called external sphincter is at this level, but because this generally gives a safe margin of sphincteric mechanism distal to this point (between the verumontanum and the urogenital diaphragm).
4. With an open prostatectomy for benign disease, try to divide the apical urethra at the verumontanum or close to it.
5. With the radical prostatectomy, try to maintain some sphincteric mechanism distal to the prostate, but do not sacrifice total removal of the prostate for this (leaving a "button" of prostatic tissue is not a radical prostatectomy). If in doubt, create a pseudo–bladder neck.
6. Be careful of instrumenting the urethra after any type of prostatectomy in such a fashion as to potentially injure what is left of the distal sphincteric mechanism.
7. Remember that any patient whose proximal or distal sphincteric mechanism is already compromised is at greater risk from incontinence when another procedure in the area of the bladder neck and anterior urethra is carried out. Remember that any neuromuscular disease affecting the tone of the smooth or striated muscle of the distal sphincteric mechanism will predispose an individual patient to a greater chance of urinary incontinence after even a well-done transurethral resection. This may include any neuromuscular

disease, diabetic neuropathy, or alcoholic neuropathy. Remember also that a patient who has had a transsphincteric urethroplasty or who has a rigid urethral stricture in this area is at greater risk for urinary incontinence following a prostatectomy.

Incontinence in the Elderly

This is a common and distressing problem for elderly patients, their families and friends, and for the health professionals who care for them. This subject, of immense social and economic importance, has recently been reviewed by Ouslander (1981), Williams and Pannill (1982), Farrar (1984), and Resnick and Yalla (1985). The following represents a summation of salient points from these excellent reviews. Studies in Great Britain and Europe have fixed the prevalence of urinary incontinence in the hospitalized elderly at between 18% and 40% in men and at between 24% and 46% in women. In elderly persons living in the community, the corresponding figures are 5% to 25% in men and 7% to 42% in women. In the United States it is estimated that 28% to 50% of elderly nursing home–bound patients are incontinent of urine. It is estimated that over $8 billion per year in additional care is required for the 15% of incontinent elderly who are institutionalized. As with all statistics regarding incontinence, it is difficult to arrive at an exact, agreed-on prevalence in this population, simply because the definition of incontinence varies so widely from clinician to clinician, and may or may not include the implication of social or vocational disability. The subject of incontinence in the elderly requires somewhat different considerations than incontinence in other age groups. First of all, there is a large segment of the elderly who may be classified as having "transient incontinence." Transient incontinence is most often associated with acute medical conditions, psychological responses, or iatrogenic factors, particularly associated with the administration of various types of pharmacologic agents. Restricted mobility and impaction of stool are also common causes of transient incontinence. Appropriate early management can often prevent this type of incontinence from becoming an established problem. Such transient incontinence may represent as much as 50% of the total number of cases of incontinence in the elderly. Another type of problem, not particular to the elderly, but certainly more common in this group, consists of "functional" incontinence. This refers to the inability of a normally continent person to reach the toilet in time to avoid an accident. The reason may be an illness—commonly a musculoskeletal limitation—that prevents an appropriately quick response to a need to void, or it may be a failure—either on an organic, cerebral or psychological basis—of the individual to appreciate the social necessity for voiding in the right place at the right time, or it may represent spiteful behavior associated with depression, suppressed hostility, or anger. Bladder hyperactivity, usually with sensation (urge incontinence), is the most common of urodynamically definable conditions contributing to incontinence in the elderly, and is estimated to occur in between 40% and 70% of elderly incontinent patients. Involuntary voiding without stress or sensation (reflex incontinence) may occur as well, either because of interruption of sensory pathways, spinal cord disease, or cortical damage that can both cause detrusor hyperreflexia and also diminish the awareness of bladder filling. Stress incontinence in the elderly is much more common in the female than in the male who has not undergone prostatectomy. Nonsurgical management of these patients is often beneficial and extremely gratifying. Overflow incontinence represents an easily diagnosable problem in these patients. If this is secondary to outlet obstruction with an intact detrusor, surgical treatment is often successful. If the detrusor is decompensated, intermittent catheterization, if practical or available, will totally obviate the problem.

THE PAINFUL BLADDER AND URETHRA

Evaluation and management of patients afflicted with purely sensory disorders of the lower urinary tract is a frustrating and challenging experience for the urologist. These disorders, as such, defy precise definition, and have been aptly characterized by the term "a hole in the air," attributed to Tage Hald (George, 1986). These patients often complain bitterly of symptoms suggestive of lower urinary tract dysfunction, without an obvious infectious cause or urodynamic correlate, but occasionally with endoscopic abnormalities; however, these abnormalities are by no means constant. As George (1986) so concisely puts it, "these patients constitute the very substance of the 'hole in the air'; the boundary is formed by those who have been identified and abstracted by due clinical process."

Interstitial Cystitis and the Urethral Syndrome in the Female

Hunner (1915) first described a painful bladder condition manifested by frequency, nocturia, urgency and suprapubic pain in a group of young female patients whose endoscopic examination revealed ulcers on the vesical mucosa. Subsequently, it has become evident that the classical finding of a scarred, contracted bladder that splits and bleeds on distention is present only in about 10% of cases of interstitial cystitis. The onset of the disease is commonly subacute and full development of the symptom complex takes place over a relatively short time (Hald and Holm-Bentzen, 1986). Many women can pinpoint exactly the time and the circumstances under which the symptoms first developed. Many times, the symptoms initially seem to be similar to those of urinary tract infection, of which many of these women have a repeated history. However, the symptom complex, though initially thought to be due simply to another infection, "never disappears." The most common symptoms of the patient with "classic" interstitial cystitis are pain, frequency, nocturia, and urgency because of suprapubic discomfort. In most cases, voiding relieves the pain somewhat, but the pain recurs with bladder filling. The patients are usually females between 30 and 70 years old, although men and children of either sex constitute approximately 10% of cases (Hald and Holm-Bentzen, 1986). The incidence of the disease varies with the investigator. Oravisto (1975) found the occurrence in Helsinki to be 18 per 100,000 women. Kinder and Smith (1958) found only 27 cases in more than 17,000 urological registrations in London.

One of the problems in gathering data about this disease entity is the total lack of a precise or agreed-on definition. The symptom complex is certainly not specific for whatever this entity is. Urodynamic investigation is generally unrevealing except for the presence of hypersensitivity during filling. Occasionally, patients will exhibit decreased compliance during filling, and in our experience, these are usually "classic" cases, with a long duration of disease, and with fibrosis of the bladder wall. Endoscopic findings vary widely. Classical Hunner's ulcers are rarely seen, but bleeding on refilling the bladder following cystoscopic distention is common. The bleeding may result from petechial hemorrhage, ecchymoses, or free bleeding from vessels, and is estimated by Hald and Holm-Bentzen (1986) to occur in up to 90% of patients. Messing and Stamey (1978) described glomerulation (discrete rounded petechial mucosal hemorrhage) as a manifestation of early interstitial cystitis. It should be noted, however, that such endoscopic findings are seen after cystoscopic filling and emptying of the bladder in patients with other inflammatory bladder conditions but few symptoms, and in patients undergoing endoscopic examination for periodic surveillance (e.g., bladder tumors) without either definable disease or symptoms. The etiology of the disease has variously been considered to be infectious (bacterial and nonbacterial), traumatic, and autoimmune. Dixon and Hald (1986) have nicely summarized the varied pathological findings, and also the recent attempts to implicate a defective bladder urothelium in the pathogenesis of this disease, an idea originally suggested by Gordon et al. (1973). They summarize the available pathologic evidence, which suggests that the disease is associated with a pancystitis. The bladder wall is frequently thickened and the perivesical tissues are infiltrated. There is frequent thinning of the mu-

cosal lining, and in some areas the mucosa may be denuded. The areas of ulceration are frequently covered by a layer of fibrin. An intense reaction occurs in the subepithelial connective tissue beneath the areas of mucosal thinning. The lamina propria is edematous, congested, and contains dilated capillaries and perivascular hemorrhages. A diffuse cellular infiltration of all areas of the bladder wall by lymphocytes exists, but this is most evident in the lamina propria. Mast cell infiltration in the muscle coat ("detrusor mastocytosis") has been proposed as a diagnostic feature by Larsson et al. (1982), and this has been adopted as a diagnostic feature by others. However, there is no agreement as to this criterion among many other clinicians (Messing and Stamey, 1978), particularly in "early" cases. Parsons et al. (1983) have proposed a theory of pathophysiology involving a defect in the protective layer of glycosaminoglycans (GAG) lining the mucosa of the urinary bladder, allowing a marked increase in bladder urothelial permeability, allowing the "leakage of urine" into the bladder wall, thereby inducing a local inflammatory response. Dixon and Hald (1986) could not support such a theory implicating a defective glycocalyx, as they detected no structural differences in the GAG layer in patients with interstitial cystitis from that in controls. However, they mention the possibility that abnormalities of bladder urothelial permeability may occur because of other reasons, perhaps defective "urothelial tight junctions," with resultant urine "leakage" into the bladder wall. Attempts to formulate any sort of a meaningful review of this disease are hampered by the fact that there is absolutely no agreement on a definition of the disease, or any sort of way to grade its severity. The disease is often defined on the basis of symptoms alone without endoscopic or pathologic findings; it may be defined on the basis of symptoms with endoscopic findings but without specific pathology, or on the basis of pathologic findings with symptoms but with or without typical endoscopic findings. Doubtless, responses to treatment differ among patients in these various categories, probably accounting for the wide variations and disagreements regarding successful results with hydrodistention, intravesical instillations of dimethyl sulfoxide and sodium oxychlorosene (Clorpactin), surgically or laser-induced "denervation," synthetic GAG therapy, and biofeedback.

The urethral syndrome in the female is another imprecisely defined entity that to most urologists is a diagnosis of exclusion for those patients with frequency and/or urgency and/or dysuria without evidence of any pathological process in the urinary or genital tract. George (1986b) has nicely summarized the clinical features. Frequency by day is the primary complaint of most patients who describe a constant, unpleasant sensation in the area of the urethra that is only temporarily relieved by voiding. Nocturia is generally of mild degree, and never reaches the levels experienced by patients with true interstitial cystitis. Patients complain that their voiding is typically slow and hesitant, and in fact the measured flow rate, even with an adequate volume, may indeed be low.

However, the diagnosis of obstruction should certainly not be made only on the basis of flowmetry, but the diagnosis of obstruction in this syndrome must include pressure-flow documentation. Any positive finding uncovered during the course of a search for infectious, inflammatory, or urodynamically definable pathologic characteristics should be treated specifically. Unfortunately, this leaves a large number of patients who must be treated either by biofeedback, reassurance and encouragement (really a form of biofeedback, but simply practiced by the urologist in the course of his or her office hours), empiric pharmacologic therapy, and a host of other treatments, generally directed toward the urethra or the surrounding structures. Schmidt (1985) and Tanagho have come up with a very interesting theory of striated sphincter dysfunction as a cause for the urethral syndrome. Treatment of such dysfunctional striated sphincter activity, in their hands, has been successful in ameliorating the clinical condition, and such treatment has ranged from simple reeducation—with or without pharmacologic therapy—to various forms of neurostimulation, to fatigue, to presumed erratic behavior of the striated sphincter.

Prostatodynia in the Male

Prostatodynia is a term that was formulated to describe a syndrome in the male that is somewhat similar to the urethral syndrome in the female (Drach et al., 1978). This describes a patient with ill-defined discomfort in the prostatic area, but with no evidence of true acute or chronic prostatitis, and with no evidence of infectious or inflammatory findings in the expressed prostatic secretions. This syndrome of chronic genital discomfort, which often includes symptoms of irritative voiding dysfunction, typically occurs in tense and anxious persons (Osborn, 1986). In a series of 37 such patients in whom the diagnosis had been established by clinical, bacteriologic, and urodynamic screening (with no identifiable pathology), Osborn (1986) reports the somatic symptoms as consisting of perineal pain (76%), pain after ejaculation (56%), suprapubic discomfort (55%), penile pain (44%), orchalgia (35%), and loin pain (17%), with voiding symptoms consisting of a postmicturition dribble (35%), hesitancy (26%), a persistent nagging urethral sensation (23%), and a necessity of straining to void (17%). The pathophysiology is obscure, and etiologic considerations have ranged from

undiagnosed bacterial, trichomonal, or chlamydial infections to what is described as tension myalgia of the pelvic floor striated musculature, occurring primarily in persons with a tendency to a tense, neurotic personality (Segura et al., 1979). Specific disorders that are uncovered in the evaluation of these patients, whether infectious, inflammatory, or urodynamic, are treated with some rationale. In the remainder of patients, which is unfortunately quite large, empiric anti-infective therapy, often long-term, is often employed, sometimes with equally empiric therapy to either decrease bladder contractility or decrease outlet resistance.

PEDIATRIC VOIDING DYSFUNCTION*

A full spectrum of voiding abnormalities is seen in pediatric patients (Allen and Bright, 1978). Urologists, pediatricians, and primary care clinicians are charged with the responsibility of defining the characteristics of voiding problems in children early so as to prevent subtle but potentially significant deterioration of the bladder and upper urinary tract. Certain congenital abnormalities that are obvious at birth, such as myelodysplasia or urogenital sinus abnormalities, signal the potential for underlying voiding dysfunction (Bauer et al. 1982). In such instances, it has been shown that an early assessment of the neurourologic status of the patient allows for placement of the patient into either a therapeutic or a regular surveillance group (McGuire et al., 1981; Bauer et al., 1984). Such a program greatly decreases the likelihood of insidious vesical deterioration and upper tract disease as the child matures (Raezer et al., 1977).

Unfortunately, not all children with the potential for voiding problems can be identified early. Many patients have no apparent problem with urination. Thus, parents and physicians are unable to note subtle changes in voiding activity until retention, marked incontinence, or urosepsis develops. Typically, children are poor historians and are unlikely to communicate a subtle voiding problem to their parents or physicians. Also, because they do not perceive voiding changes to be abnormal, significant micturitional abnormalities may go undetected for some time. This does not mean that all children with the most mundane alterations in urination require extensive neurourologic evaluation. However, awareness of the potential for serious underlying disease does mandate the need for a thorough understanding of the peculiarities of voiding dysfunction in children. Knowledge of normal and abnormal micturition physiology, the indications and reliability of diagnostic

---

*Prepared with assistance from Dr. Jeffrey Woodside, Professor of Urology, University of New Mexico.

testing modalities, and a current concept of therapeutic alternatives is necessary for all practitioners caring for children with voiding disturbances. The symptoms and conditions indicating further investigations of a child's neurourologic status are listed in Table 28–20. However, clinical experience and intuition must also be included as important ingredients in the decision-making process as to which child should undergo further investigation of vesicourethral function.

It is the goal of this presentation to outline a practical approach to the problem of vesicourethral dysfunction in children. Diagnostic modalities will be presented, with references to their relative importance and usefulness in the pediatric age group. Certain conditions common to vesicourethral dysfunction in children will be discussed and the nuances relative to their diagnosis and management will be reviewed.

NORMAL VESICOURETHRAL MATURATION

In the newborn child, urination is a reflex event. It is regulated by a relatively simple spinal reflex (Bradley et al., 1974). When the detrusor becomes distended by urine, stretch receptors in the wall of the detrusor evoke a sensory impulse conducted to the spinal cord by autonomic afferent fibers, which in turn evoke an autonomic efferent stimulus that causes contraction of the detrusor smooth muscle. This spinal reflex mechanism is coordinated by a simultaneous relaxation in the periurethral striated skeletal muscle sphincter during the voiding reflex sequence, and probably requires an intact brain stem (deGroat, 1975; deGroat and Saum, 1971; deGroat and Theobald, 1976; El Badawi and Schenk, 1968, 1971, 1974). This mechanism has been described in detail and is outlined elsewhere in this text; however, in very young children this reflex mechanism lacks the inhibitory influence from central nervous system areas above the brain stem. Although there is little volitional control of voiding, the functions of urine storage and continence are maintained and it is

TABLE 28–20.

Indications for Pediatric Neurourologic Investigation

Urosepsis
Retention
Hematuria
Dripping incontinence
Overt neurologic disease (including trauma)
Painful urination
Straining or squatting during urination
Secondary enuresis
Frequency and urgency incontinence
Encopresis
Polydipsia
Urogenital sinus abnormality

the relative state of bladder fullness and bladder wall distention that triggers the detrusor contraction and subsequent bladder emptying.

Goellner and co-workers (1981) studied the various aspects of micturition during the first 3 years of life. They pointed out that most children sleep approximately two-thirds of the day and that almost half of the voidings occur during the sleep period. During the first 12 months, the number of voidings averages 20 per day. As the child matures, the frequency of voiding decreases but the volume increases to approximately four times that of the newborn infant. It has been suggested that the increase in voided volume and the decrease in voiding frequency are a function of a subtle assumption of the central (above brain stem) neurologic control of urination (Yeates, 1973). However, physiologic studies indicate that the change in voided volume–frequency relationship during the first 3 years of life is related more to fluid volume ingested for body weight and the changing functional volume of the child's bladder.

It would seem reasonable to conclude that both factors influence enhanced urinary control. By the end of the third year, most children are able to have some conscious control over urination. However, enuresis and daytime accidents do occur. At the age of 4 years, bladder volume is adequate and central urinary control is complete in most individuals. The degree of neuromuscular maturation that occurs is most dramatic on the assumption of voluntary control of the external (skeletal muscle) urethral sphincter and voluntary control of detrusor contractility. Both of these functions are related to neurologic organization of the centers in the midbrain and cerebral cortex. There is no question that there are some neurologically normal children who have a temporary delay in the maturation of these functions and may well exhibit significant voiding dysfunction at the ages of 4, 5, or 6 years and who eventually "outgrow" their problem. As with adults, for normal coordinated voiding to occur in the child, the brain stem must be intact, allowing for adequate contraction of the detrusor smooth muscle and vesicourethral segment and a simultaneous relaxation of the skeletal muscle sphincter at the bladder outlet. During the period of maturation, if one area of maturation lags behind the other, discoordinated voiding patterns may be seen.

PEDIATRIC NEUROUROLOGIC EVALUATION

History and Physical Examination

For all practical purposes, neurourologic tests should not be ordered for a child with suspected vesicourethral dysfunction without first pursuing a complete history and physical examination. A history suggestive of subtle central nervous system disease, such as a traumatic birth, neonatal anoxia, convulsive disorder, and head or back trauma, may explain many voiding abnormalities despite the absence of apparent outward neurologic disease.

During the initial interview with the child and parents, an assessment should be made of the fluid ingested in a 24-hour period. Historical approximations of this volume may be misleading, however, and when clinically related to a voiding problem such as diabetes insipidus, the child or parents should bring a documented list of volumes ingested and voided.

The voiding history itself may well hold the clue as to which of the various available tests will offer the best opportunity for an accurate and cost-effective diagnosis. In the very young child, parents should be questioned about their observations of diaper wetness, number of diapers used, diaper rash, dripping loss of urine, leakage while crying, stream characteristics, straining during urination, and apparent pain or discomfort when voiding. The older child may be asked questions about painful urination, urgency, frequency, and unusual voiding positions, although corroboration from parents is often necessary for information related to enuresis, soiling, and the problem of malodor.

Aside from the potential damage to the bladder and upper tracts, vesicourethral dysfunction in children, especially when incontinence is involved, may have a dramatic impact on the psychologic and sociologic adjustment of the child or family as a unit. During the initial interview, an assessment should be made regarding the impact of the "voiding problem" on the child and parents. This information is important not only diagnostically, because psychogenic voiding dysfunction must be excluded, but also therapeutically, because eventual selection of therapeutic alternatives depends greatly on the psychosocial status of the patient and family.

The physical examination should be directed to those conditions that may foretell underlying neurologic diseases or suggest the more immediate problem of bladder overdistention and hydronephrosis. In the absence of overt neurologic problems, the examination should first be directed toward the integrity of the spinal column by looking for clues to vertebral dysrhaphism. Hairy patches, depressions, or dimples over the spine or sacral area may indicate underlying dysrhaphism or sacral agenesis. A rough estimate of intact reflex arc activity can be evaluated by evoking deep tendon reflexes in the lower extremities, and also a rough estimate can be made of sensory neurologic integrity of the abdomen, perineum, and lower extremities. Further neurologic assessment should be left to a pediatric neurologist, and any clues that may be uncovered during the initial physical examination may lead to a consultation that would include more specificity.

After the initial history and physical examination have been completed and a urinalysis is obtained, a determination must be made as to the severity of the "voiding problem." Does it require further evaluation or can the child and parents be reassured that the symptoms are normal variants and there is no threat to the unseen portions of the urinary tract? Clinical judgment becomes the most important indicator at this time. Children with those conditions listed earlier will be evaluated further. However, in those children who cannot be placed in a particular category, one must weigh the relative costs and degree of invasiveness involved in additional tests against the risk of failing to diagnose a potentially serious vesicourethral abnormality.

Urodynamics

In many children, a satisfactory evaluation for the cause of a voiding abnormality can be made with various combinations of uroradiographic studies and cystoscopy. The voiding cystourethrogram is most efficient in evaluating the detrusor and bladder outlet. Vesicoureteral reflux, ureterocele, urethral valves, ureteral ectopy, and fistula all may be related to voiding difficulties and must be excluded before an abnormality of intrinsic vesicourethral function can be implicated as the cause of a particular voiding problem. It is also not unusual to have a combination of abnormalities (e.g., vesicoureteral reflux and detrusor hyperreflexia) that simultaneously produce an abnormal voiding symptom (Blaivas et al., 1977).

Once structural abnormalities have been excluded, a urodynamic evaluation should be considered to more accurately define voiding dynamics (Bauer, 1984). The goals of urodynamics in children are to define the nature and cause of voiding symptoms, help determine the best mode of therapy, and provide a means to assess therapeutic progress (Gierup, 1970). The urodynamic parameters available to most clinicians include (1) uroflow determination, (2) cystometry, (3) pelvic floor skeletal muscle EMG, (4) urethral pressure profilometry, and (5) combined studies—which include simultaneous measurements of various combinations of the aforementioned studies, with or without abdominal pressure recordings. For all practical purposes, videourodynamics, in which multiple voiding parameters are recorded in sequence with voiding cystourethrography, are only occasionally indicated in children, and the availability of such urodynamic equipment is rarely found outside the medical center environment (Mundy et al., 1982). For that reason, only the most complex voiding disorders in children would require videourodynamic assessment.

As the urodynamic test becomes more complex, the

degree of invasiveness tends to increase. With children, invasiveness increases the possibility of alienating the patient and thereby introducing an unacceptable degree of artifact into the study. Therefore, it is generally best to begin the urodynamic testing sequence with a minimally invasive examination, yet one that will produce reliable data relative to the abnormal voiding symptom. At the beginning, the examiner should ask the question, "What do we want to know?" Do the symptoms suggest a detrusor abnormality or an outlet abnormality, or could it be a combination of both problems?

In some patients, the use of sedation, such as intramuscular meperidine (1 mg/kg), has been proposed to decrease discomfort and anxiety during urodynamic testing (Evans et al., 1979). Others report reliable results in children by using nitrous oxide anesthesia (Koff et al., 1980).

Uroflowmetry

When dealing with children, patient cooperation is essential for the recording of any urodynamic parameter, and because uroflowmetry is the least invasive of all urodynamic tests, many clinicians prefer to use this procedure as an initial screening tool. This, of course, requires that the child be of an age to follow instructions and void in a fashion that is typical for that particular individual. For all practical purposes, this eliminates children much younger than the age of 4 or 5 years. If the uroflow characteristics that are recorded are normal, more invasive urodynamic testing may not be necessary subsequently. On the other hand, when uroflow is abnormal, the additional tests may be tailored to concentrate on the specific voiding abnormality.

Not infrequently, the volume voided may be insufficient to accurately assess the flow rate and pattern. Although some degree of invasiveness is introduced into the study, an accurate test may be obtained by filling the child's bladder through a small catheter with body-temperature sterile water to a point of comfortable bladder fullness (Barrett, 1984). The catheter is removed and the child is instructed to void. An accurate residual volume measurement may also be obtained by calculating the difference between the volume voided and the volume instilled·in the bladder.

Many clinicians prefer to record pelvic floor or abdominal skeletal muscle EMG activity during uroflowmetry (Barrett and Wein, 1981; Barrett, 1980; Maizels and Firlit, 1979). This technique requires the placement of electrodes on or in the muscles of the perineum and abdominal wall (Koff and Kass, 1982). Normal micturition occurring in the absence of pelvic floor EMG activity confirms the relaxation of the external

urethral sphincter during a detrusor contraction (Fig 28–29). A lack of coordination between detrusor contractility and simultaneous external sphincter relaxation is manifested by an interrupted flow pattern and corresponding bursts of EMG activity. Straining to void produces concomitant contractions in the abdominal and perineal skeletal muscles that may or may not produce significant outflow obstruction. When obstruction is evident, the flow rate will be decreased and EMG activity will be increased.

Regardless of the exact methodology used, which may be modified to suit the needs of the clinician and child, uroflowmetry offers a cost-effective means of assessment for significant underlying vesicourethral dysfunction.

Cystometry

Although moderately invasive and dependent on a cooperative child, cystometry in children can provide enlightening information regarding detrusor contractility and detrusor compliance (Merrill et al., 1972). Every investigator has preferences regarding the age of

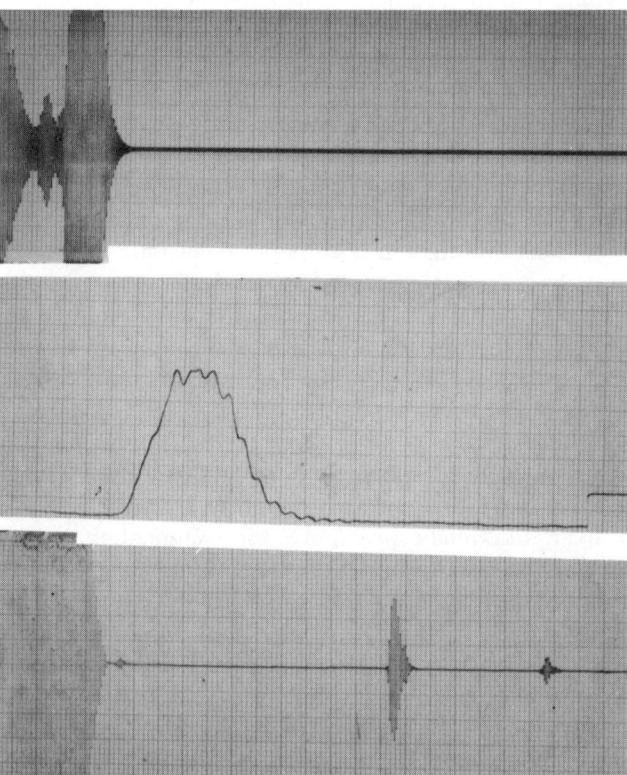

FIG 28–29.
Uroflow rate with abdominal wall EMG *(top)* and urethral sphincter EMG *(bottom)*. The flow rate *(middle)* is normal, with a peak rate of 30 ml/sec (flow rate scale, 0–50 ml/sec). Both the abdominal wall musculature and the urethral sphincter appropriately relax during voiding.

the child to be investigated, catheter size, infusion medium, rate of filling, and the need for simultaneous EMG recording (Gleason et al., 1977). The amount of information obtained from childhood cystometry is directly related to the degree of interest and care directed to the test by the clinician. Moreso than with adults, cystometry in children requires a gentle hand and the ability to abate the fears of the child during the testing sequence. Extremely young children who undergo cystometry may be surprisingly cooperative, especially if there are associated neurologic deficits; however, others may require sedation or even light general anesthesia (Fig 28–30).

Many younger children, especially males, require the use of small intraurethral catheters (5–8 F) to infuse the bladder during cystometry. When water is used as a testing medium, larger tubes are preferred so as to decrease the erroneous pressure increases that occur during moderate infusion rates (5–100 ml of water per minute). Alternatively, carbon dioxide may be conveniently used without fear of false pressure increases, even at high infusion rates. However, one should be aware of the alteration in detrusor compliance as the infusion rate is increased. Infusion rates higher than 100 ml of water or carbon dioxide per minute are unphysiologic and will produce a recording that may reflect a significant decrease in detrusor compliance. When testing for the possibility of detrusor instability, higher infusion rates may be used advantageously to provoke what may otherwise be an undetected hyperreflexic detrusor contraction. On the other hand, infusion rates that are too high and extremely unphysiologic may induce a detrusor contraction that ordinarily would not be seen during the child's normal voiding sequence. This is especially true in children under the age of 4 years in whom complete maturation of the inhibitory mechanism of micturition has not occurred. Therefore, the interpretation of cystometry data must take into account all such variables and care must be taken not to produce extreme unphysiologic conditions that may produce data that are overinterpreted and thus lead to erroneous conclusions regarding vesicourethral dysfunction (Grossman et al., 1977).

At times during cystometry it may be unclear as to whether an increase in intravesical pressure is related to a "true" detrusor contraction or is transferred from increases in intra-abdominal pressure during straining. The measurement of intra-abdominal pressure through a closed-catheter system placed in the rectum may clarify this source of artifact. Again, more invasiveness is required and additional pressure recording equipment is necessary. However, most modern urodynamic recording devices will provide the capability of simultaneously measuring rectal pressure, bladder pressure, and subtracted detrusor pressure (bladder pressure minus rectal pressure).

It is clear that some children are at risk of developing upper tract decompensation secondary to ureterovesical obstruction and/or reflux that is a result of excessively high resting or voiding pressure. In bladders that have decreased contractility and decreased compliance, the pressure at which the urine leaks out of the urethra has been shown to be of predictive value in defining the potential for upper tract decompensation. A measurement of "leak pressure" may be undertaken in children of all ages by merely infusing fluid through a small catheter and then noting the intravesical pressure at which the fluid leaks out of the urethra around the catheter. The critical pressure is at or near 40 cm H_2O (McGuire et al., 1983).

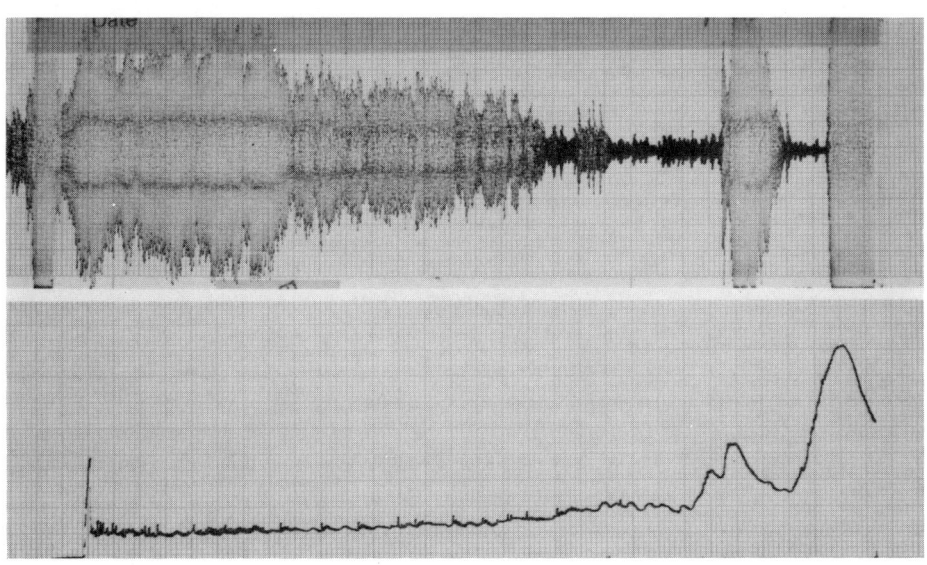

FIG 28–30.
A normal CMG-EMG in a 12-year-old girl. The tonus limb reflects normal compliance *(bottom)*, and EMG activity *(top)* is noted to decrease during detrusor contractions that are under voluntary control. Electrocardiographic patch electrodes were used to record EMG activity.

Urethral Pressure Profile

Static urethral pressure profilometry has only limited use in pediatric urodynamic testing. It was devised to assess the status of the bladder outlet and offer information relative to the inherent resistance of the continence mechanism. Its drawback centers on evaluation of tissues when they are static and equating this information to their function during the dynamics of voiding.

In most situations, information relative to the continence mechanism (combined smooth and skeletal muscle sphincters) can be obtained more accurately from other urodynamic tests or combinations thereof.

Urethral pressure profile measurement has occasionally been useful in following the function of artificial sphincter occlusive cuff pressures; however, radiographic studies are usually more diagnostic when there is a sphincter malfunction.

Sphincter Electromyography

In pediatric urodynamics, accurate recording of EMG activity in the pelvic floor skeletal muscles has been reliably produced by using surface electrodes such as the patch ECG electrodes that may be pasted on the skin overlying the perineal skeletal muscles. Not only is invasiveness reduced, but also the reliability of recording electrical activity in the muscles has been shown to be accurate. This type of electrical activity measurement can easily be recorded on a strip-chart recorder and combined with other urodynamic testing modalities such as cystometry or uroflow.

Combined Studies

During the initial screening urodynamic investigation, it may not be clear whether an obvious voiding defect is a result of a detrusor contractility abnormality or an outlet abnormality. A combination study of simultaneous cystometry (with or without rectal pressure measurement), uroflowmetry, and EMG is ideally suited to better define the voiding abnormality (Firlit et al., 1978). Children who are candidates for this relatively invasive test are those that fail to completely empty or those with interrupted voiding thought to be related to sphincter-detrusor dyssynergia. From the information gained from this study, a rough calculation of outflow resistance can be made when voiding pressure is compared to peak flow rate.

Cooperation from the child is necessary during combined studies, as a considerable amount of manipulation is necessary to set up for this test (Cook et al., 1977). EMG electrodes must be placed on the perineum or abdominal musculature, rectal pressure catheters must be inserted, and one or two urethral catheters must be placed for the infusion of fluid and the recording of pressure. For this study one has the option of inserting one catheter through which the water medium is infused and intravesical pressure is recorded. However, experience has shown that it is very convenient to insert into the bladder a small 4–5 F polyethylene feeding tube through which pressure is measured, followed by an 8 F catheter through which the water medium is instilled into the bladder. Prior to the recording of voiding pressures, the 8 F catheter is removed, allowing for easy voiding around the small catheter. There is no question that the catheter, especially in younger children with small-diameter urethras, will introduce some element of outflow obstruction and thereby decrease the peak flow rate. However, this appears to be negligible in most clinical settings.

Children with high voiding pressures and low flow rates may have outflow obstruction. When there is increased EMG activity during the voiding sequence, one may suspect an uncoordinated external sphincter giving rise to this degree of obstruction. On the other hand, if EMG activity is coordinated, other causes of outflow obstruction must be excluded. When voiding pressures are low and flow rates are also low, it is reasonable to conclude that detrusor contractility is lacking. This may be a result of neurologic impairment to the detrusor or myogenic detrusor decompensation. One must also take into account the possibility of poor patient cooperation during the testing sequence. In such a situation, a patient may volitionally fail to initiate a detrusor contraction and augment voiding by straining or subtle increases in intravesical pressure. This generally results in low detrusor pressures and also low flow rates.

In most instances, an experienced urodynamic technician can define the degree of patient compliance during urodynamic testing and thereby guard against misinterpretation of this type of study.

Videourodynamics

The true essence of pediatric urodynamics is achieved when the parameters of the combined study are coupled with voiding cystourethrography. The bladder is filled with contrast medium and the voiding parameters are recorded while fluoroscopic monitoring of the sequence is carried out. Limited use of the fluoroscope during bladder filling keeps the radiation exposure to a minimum (Webster and Older, 1980).

Webster et al. (1984) have defined other categories of micturitional abnormalities in neurologically normal children that may be observed with videourodynamics. In general, those conditions that fail to result in a detrusor contraction, such as detrusor areflexia and increased detrusor compliance, can be evaluated with tests that are less expensive and invasive. As before, patient cooperation is extremely important during this testing sequence, as actual voiding must occur for an adequate study to be obtained.

ENURESIS

Enuresis may be defined as the nocturnal or diurnal incontinence of urine in a child more than age 4 years. Primary enuresis is present from birth, and secondary enuresis is preceded by a dry interval of 6 to 12 months. The incidence of enuresis has been estimated at 12% of children aged 4 to 5 years (Forsythe and Redmond, 1974) and it is known to have a familial predisposition (Whiteside and Arnold, 1975). Theories regarding the cause of enuresis include such factors as delayed level of central nervous system maturation, psychogenic and behavioral components, environmental influences, deep sleep, allergies, small bladder capacity, uninhibited neurogenic bladder, and structural abnormalities of the urinary tract (Firlit et al., 1978; Kass et al., 1979; Lennert and Mowad, 1979; Marshall et al., 1973). The incidence of underlying organic problems is higher in the older patient population. This discussion focuses on those children with enuresis who have no overt neurologic or urologic disorder.

The extent of the urologic evaluation of enuretic children has recently undergone reevaluation. In the past, most children underwent intravenous urography (IVP), voiding cystourethrography (VCUG), and endoscopy. However, many investigators currently recommend a much more selective evaluation that is tailored to the individual child and family. The importance of the history, physical examination, and results of urinalysis and urine culture in this evaluation should be stressed. One should determine the pattern of incontinence—whether nocturnal, diurnal, or both; associated urinary frequency or urgency; fecal soiling or constipation; and a history of urinary tract infection and obstructive voiding symptoms. Careful examination of the spine, abdomen, genitalia, and sacral spinal cord segments is necessary and the voided stream should be observed. A urinalysis and urine culture, if indicated, are performed. A history or laboratory findings suggestive of urinary tract infection or evidence suggestive of a hyperactive external urethral sphincter, urinary tract obstruction, or a neurologic problem mandates further evaluation. Determination of the uroflow rate, possibly combined with surface perineal EMG, may be indicated (Barrett, 1980). If a screening IVP is indicated, additional information can be obtained from voiding urethral and postvoiding bladder films.

Many investigators have performed urodynamic evaluations of enuretic children in an attempt to identify any specific abnormality of vesicourethral function responsible for their incontinence. Whiteside and Arnold (1975) studied 50 patients who had normal urinary flow rates and complete bladder emptying. In 13 patients with nocturnal enuresis only, 15% had detrusor instability. In contrast, 97% of 37 patients with both diurnal and nocturnal enuresis had detrusor instability. Blaivas et al. (1977) evaluated 13 children with diurnal and nocturnal enuresis and found two with neurogenic bladder dysfunction. Seven of the remaining 11 patients had decreased bladder capacity. Firlit et al. (1978) performed pressure, flow, and EMG urodynamic studies on 34 children with diurnal and nocturnal enuresis and found 13 patients with uninhibited bladders, 17 with hyperactive external sphincters, and four with normal results. Kass et al. (1979) urodynamically evaluated 65 children with primary enuresis. Forty-two of these patients had either a history or evidence of a urinary tract infection. Among these, 30 had uninhibited detrusor contractions, 10 were normal, and 2 had mixed types of neurogenic bladder dysfunction. Of the remaining 23 patients, 5 had uninhibited detrusor contractions and 18 were normal. Smey et al. (1978) reported a subgroup of enuretic children who had hyperactive external urethral sphincters as a contributing factor to incontinence. Diurnal enuresis, urinary frequency and urgency, urinary tract infection, and fecal soiling or constipation appear to correlate with the presence of uninhibited detrusor contractions, hyperactive external urethral sphincters, or both.

The role of radiographic studies of enuretic children has also undergone reevaluation. Redman and Seibert (1979) obtained an IVP or VCUG or both in 138 enuretic children. Twenty-one children had abnormal results and each had a history of urinary tract infection, signs and symptoms of lower urinary tract obstruction, or both. No significant radiographic abnormalities were found in children with diurnal enuresis unless also accompanied by symptoms or findings suggestive of urinary tract infection or obstruction. Kass et al. (1979) similarly reported that an IVP revealed a significant abnormality only in children with a history of urinary tract infection.

Various treatment modalities have been used for the enuretic child, including reassurance, psychotherapy, behavior modification techniques, hypnotherapy, pharmacologic manipulation, and various surgical techniques. The preponderance of evidence would currently suggest that the history, physical examination, and basic laboratory findings will indicate the appropriate extent of evaluation and initial management of most enuretic children. Children with past or present urinary tract infection, evidence of lower urinary tract obstruction, or abnormalities on physical or neurologic examination should undergo thorough urologic evaluation followed by appropriate treatment. Meatotomy may be helpful in the male child with documented meatal stenosis but may be associated with a high relapse rate. The routine use of meatotomy or urethral dilations is to be condemned. Children with none of the above findings or those who have had a negative urologic evaluation may be managed in a more conservative manner.

When deciding whether or not to initiate therapy, it is important to assess the effect of the problem on the patient and the family as a whole. When some parents and children are informed that no significant organic or psychologic problem is causing enuresis, reassurance and fluid restriction before bedtime may be the only treatment required. Many are quite willing to wait for the spontaneous resolution of enuresis that occurs at a rate of 15% per year in patients between the ages of 5 and 19 years (Forsythe and Redmond, 1974). Should treatment be desired, a conditioning device such as a pad and bell bed alarm is successful in up to 60% of patients, and behavior modification therapy, aimed at making the patient responsible for changing his behavior and complemented by positive reinforcement, is helpful in up to 70% of patients (Marshall et al., 1973).

For patients who are unable to accept or who fail these approaches, pharmacologic therapy may be effective. The results of the thorough study by Kass et al. (1979) suggest that successful empiric pharmacologic treatment of these patients can be instituted on the basis of historical information, physical examination, and urinalysis. Eighty-five percent of patients with nocturnal enuresis and daytime urgency and frequency will respond well to anticholingeric medication. Seventy percent of patients with nocturnal enuresis only will be successfully treated with imipramine hydrochloride. It is the authors' experience that nortriptyline hydrochloride may also be efficacious in those patients who do not respond to imipramine hydrochloride. It has the additional benefit of a more rapid onset of action and shortened half-life should undesirable side effects be encountered. Some of those who fail imipramine hydrochloride therapy will respond to an anticholineric drug. A high relapse rate can be expected if drug therapy is withdrawn early, particularly in younger children. Pharmacologic therapy is more effective if it is suggested in a positive manner and combined with a program of behavior modification and positive reinforcement techniques. Patients with hyperactive external urethral sphincters may be successfully treated with diazepam (Smey et al., 1978) or biofeedback training (Libo et al., 1983; Sugar and Firlit, 1982). Further urologic and radiologic evaluation of these patients is frequently indicated if pharmacologic therapy fails, particularly in the older child.

MYELODYSPLASIA

Myelodysplasia (spinal dysrhaphism) is a serious congenital anomaly occurring in 1–2 per 1,000 births in the United States. Multifactorial genetic and environmental influences are thought to be responsible for the unfused neural arches and associated spinal cord defects (Emery and Lendon, 1973). Although early neurosurgical intervention has decreased the incidence of meningitis and hydrocephalus in infancy (Laurence, 1964), urologic complications resulting from neurogenic vesical dysfunction (NGVD) remain the leading cause of morbidity and mortality in the survivors (Smith, 1965). For most, urinary incontinence is a formidable social handicap and, in some, vesicoureteral reflux, hydroureteronephrosis, chronic pyelonephritis, and urolithiasis ultimately result in renal failure (Light and Van Blerk, 1977). In recent years, sophisticated urodynamic evaluation of these children has permitted more precise and rational treatment. Pharmacologic therapy—often in conjunction with intermittent catheterization—and application of new surgical techniques have rendered many of these patients continent and promoted the preservation of renal function.

The term "spinal dysrhaphism" encompasses a spectrum of anomalies from the more occult dermal sinus, lipoma of the spinal cord, aberrant roots or fibrous traction bands, and abnormal filum terminale to the more apparent and generally more serious lesions of the meningocele and myelomeningocele (Andersen, 1974). Although the occult forms may be associated with significant NGVD, this discussion will primarily be concerned with the more common form encountered by the urologist, myelomeningocele. The location of myelomeningoceles is most frequently lumbosacral (42%), thoracolumbar (27%), and sacral (21%); only 10% are in a thoracic or cervical location (Kaplan, 1978). Some investigators maintain that the level of vertebral defect largely determines the degree of neurologic damage (Barson, 1970), but most studies suggest no correlation between the level and the NGVD (Thomas et al., 1974; Bauer et al., 1977). Indeed, significant NGVD has been associated with minimal bony defects (Andersen, 1975). These disparities can be explained by the various possible neuropathologic lesions of the spinal cord, which include hydromyelia, syringomyelia, failure of upward migration or tethering of the cord, and dysplasia of the cord itself. Further, these changes may extend several segments above and below the actual site of the myelomeningocele. The importance of urodynamic evaluation to determine the precise vesicourethral dysfunction present in an individual patient is thus apparent.

Clinical Presentation

Nearly all forms of spinal dysrhaphism are associated with a cutaneous abnormality of the back such as hypertrichosis, skin pigmentation, lipoma, skin dimple, or the more obvious meningocele (Andersen, 1975). Many patients have lower extremity orthopedic deformities, paresis, or paralysis. Inability to void, urinary incontinence, and fecal incontinence unfortunately are common problems.

Urologic Evaluation

As soon as practical after repair of the myelomeningocele and shunting of the cerebrospinal fluid, if necessary, several urologic studies are done. These include IVP, VCUG, urine culture, and serum blood urea nitrogen and creatinine determinations. These are obtained both for the purpose of identifying those patients requiring immediate treatment and as a baseline in others. Approximately 20% of patients have an abnormal IVP demonstrating hydroureteronephrosis, caliectasis, or hydroureter (Gaum et al., 1982). Of particular importance, nearly two-thirds of patients with a normal IVP have an abnormal VCUG showing vesicoureteral reflux, a thickened bladder wall, diverticula, or abnormalities of the bladder shape and outlet (Gaum et al., 1982). Thus, these radiographic examinations are complementary and we think that a VCUG should be obtained even when the IVP is normal. Urodynamic evaluation of neonates has been infrequently performed, but recently Bauer et al. (1984) have obtained important data suggesting that neonates should routinely undergo urodynamic investigation.

Urodynamic evaluation of children with myelodysplasia is essential in order to render rational therapy for urinary incontinence and to preserve renal function. It is important to ascertain several urodynamic parameters, including the bladder capacity, intravesical filling pressure, intravesical pressure at the moment of urethral urinary loss, presence or absence of reflex detrusor activity, and the competence of the internal and external sphincteric mechanisms. The degree of coordination of the detrusor and sphincter mechanisms and the postvoiding residual urine volume should also be determined.

Several investigators have reported similar results of urodynamic studies in myelodysplastic children (McGuire et al., 1981; Bauer et al., 1977; Mayo et al., 1979). Reflex detrusor activity in response to bladder filling is present in 17%–40% of patients. The reported incidence of detrusor-sphincter dyssynergia depends on its definition. If it is defined as an actual increase in the EMG activity and urethral pressure at the area of the external sphincter during a reflex detrusor contraction, then the incidence is 7% (McGuire et al., 1981). If it is defined as failure of sustained sphincter relaxation during reflex detrusor contraction or voiding maneuvers such as Valsalva or Credé, then the incidence is 50% (Bauer et al., 1977). From 60%–80% of patients have detrusor areflexia. In many, this is the result of the neural injury, but in others it may represent myogenic failure resulting from chronic distention or recurrent urinary tract infection with bladder wall fibrosis. Detrusor hypertonia (decreased detrusor compliance) during bladder filling occurs in up to 85% of patients with detrusor areflexia and is manifested as a progressive increase in intravesical pressure with increasing volume, culminating in urethral urinary loss when the intravesical and intraurethral pressures are equal. A striking relationship exists between intravesical pressure at the time of urethral urinary loss and the incidence of ureteral complications. In one report, if urethral urinary loss occurred at an intravesical pressure of less than 40 cm H_2O, no patient had vesicoureteral reflux and only 10% had ureteral dilatation (McGuire et al., 1981). On the other hand, if urethral urinary loss occurred at an intravesical pressure of more than 40 cm H_2O, 68% of patients had vesicoureteral reflux and 81% had ureteral dilatation. The majority of patients had open bladder necks, determined radiographically, and these patients tend to have a lower maximum urethral closure pressure than patients with closed bladder necks. The major clinical problem in these patients tends to be urinary incontinence. In those with higher urethral closure pressures, upper urinary tract deterioration and vesicoureteral reflux tend to occur.

The results of these investigations have important prognostic and therapeutic implications for children with myelodysplasia. McGuire et al. (1983) described six patients with detrusor hypertonia on initial urodynamic evaluation, but they had no upper tract complications. Subsequently, upper urinary tract deterioration or vesicoureteral reflux developed in all patients and persisting detrusor hypertonia was demonstrated on follow-up urodynamic testing. This upper tract deterioration may occur after many years of normalcy and as late as the midteenage years. Similar urodynamic abnormalities have been demonstrated in neonates, including upper tract deterioration developing in patients who initially had detrusor-sphincter dyssynergia and normal upper tracts (Bauer et al., 1984).

Bladder wall fibrosis may cause decreased detrusor compliance in some patients. However, the fact that bilateral surgical transection of all sacral roots often results in a hypertonic, areflexic detrusor response to bladder filling suggests that this response is usually intrinsic to the vesical musculature freed from central neural regulation (McGuire and Wagner, 1977). Experimental data indicate that an increase in adrenergic innervation and α-receptor activity occurs in the bladder after parasympathetic denervation (Sundin and Dahlstrom, 1973). Histochemical fluorescent studies of excised detrusor muscle from normal and parasympathetically denervated human bladders show that the adrenergic nerve terminals of the denervated detrusors are thicker, more strongly fluorescent, and more densely distributed than in normal detrusors (Sundin et al., 1977). Interestingly, the patients with denervated bladders had filling curves typical of decreased compliance on cystometry. Although the precise mechanism responsible for increased adrenergic activity is unknown, unmasking of already existing α-receptors or an induction of α-receptors by parasympathetic denervation seems probable (Sundin et al., 1977).

Treatment is based on vesicourethral function as previously outlined and also the degree of social and functional disability.

CAUDAL REGRESSION SYNDROME

The syndrome of caudal regression was described by Duhamel (1961); in its most severe form it results in sacral agenesis, imperforate anus, fusion of the lower extremities, and absence of the genitourinary tract except for the gonads. The disordered embryogenesis appears to occur at the fourth to fifth week of fetal life when the lumbosacral spine and nephrogenic masses are forming and the cloacal septum is separating the genitourinary sinus from the rectum. Fortunately, the most severe form is rare, but imperforate anus and sacral agenesis are relatively common, both as separate and combined congenital anomalies.

Imperforate Anus

Several classifications of imperforate anus have been proposed, and that of Santulli et al. (1970) is widely used. They classified imperforate anus into four major groups based on the level of the imperforation in relation to the puborectalis muscle or the pelvic floor: low or translevator lesions, intermediate lesions (which are uncommon), high or supralevator lesions, and miscellaneous lesions. In nearly all males with supralevator lesions, a fistulous communication exists between the prostatic urethra and the blind end of the rectal pouch; in females, a rectovaginal fistula is common. Patients with translevator lesions do not have fistulas between the intestinal and urinary tracts; rather the rectal fistula opens near the posterior scrotal raphe in males and the distal vagina or posterior fourchette in females.

Patients with supralevator lesions are managed with a diverting colostomy shortly after birth, with the defin-

itive rectal pull-through and repair of rectourethral fistula performed when the child has grown to approximately 25 pounds. Those who have infralevator lesions are treated with a primary anoplasty. All patients should have uroradiographic studies to detect associated genitourinary anomalies. Patients with supralevator lesions have a 52%–67% incidence of genitourinary anomalies, but those with infralevator lesions have a 16%–18% incidence (Belman and King, 1972; Wiener and Kiesewetter, 1973).

Imperforate anus may be associated with many urologic problems, including a broad spectrum of congenital upper tract anomalies. Additionally, complications may result from the definitive surgical correction of the lesion. Excessive surgical dissection of the fistula has resulted in urethral stricture or sphincteric damage and urinary incontinence (Persky et al., 1974; Resnick and King, 1976), and failure to adequately excise the fistula has resulted in pseudodiverticulum formation (Persky et al., 1974).

Although it is difficult to ascertain the precise cause, neurogenic vesical dysfunction is also common among patients with imperforate anus. Approximately 25% of patients have abnormalities of the lumbosacral spine and most of these are supralevator lesions, which alone could be responsible for voiding dysfunction (see Sacral Agenesis, below) (Tank, 1979). In addition, the pelvic plexuses can be injured by the surgical dissection of the rectal pull-through procedure. These plexuses lie deep in the pelvis on either side of the rectum and contain preganglionic and postganglionic parasympathetic and sympathetic neurons. Thus, abnormalities of innervation of the bladder or urethra or both may occur. These include partial or complete denervation of the bladder manifested as poor detrusor contractility or detrusor areflexia or decreased detrusor compliance with increased intravesical filling pressure. Varying degrees of urethral denervation may also occur that result in a low maximum urethral closure pressure and urinary incontinence. A similar clinical situation exists in patients who have undergone radical or Wertheim's hysterectomy or abdominoperineal resection of the rectum (Woodside and McGuire, 1982; McGuire, 1975). Because of the spectrum of potential abnormalities in patients with imperforate anus, urodynamic investigation is important in order to precisely identify the disordered vesicourethral function and thus to direct specific treatment. Most commonly encountered are patients with decreased detrusor compliance, usually detrusor areflexia, as the cause of incomplete bladder emptying and as a contributing factor to urinary continence. Some patients have also had sphincteric insufficiency, with a bladder neck that appears open radiographically, and severe urinary incontinence.

Sacral Agenesis

Sacral agenesis results from the failure of ossification of the sacral segments early in fetal development and it is manifested as a complete or partial absence of one or more sacral vertebral segments. Several authors have noted an increased incidence of sacral agenesis in the offspring of mothers with diabetes mellitus (Koff and Deridder, 1977; Mariani et al., 1979; White and Klauber, 1976). The neuropathologic abnormalities of the spinal cord, spinal roots, and spinal canal have been ascertained from postmortem anatomic dissections and from the findings encountered at neurosurgical procedures. In some patients, the sacral roots caudad to the lowest formed vertebral segment fail to develop, but in others, sacral roots are present despite the absence of the corresponding vertebral bodies (Smith, 1959). There may be dense fibrous tissue within the spinal canal as well as an intradural lipoma, a dermoid cyst, or tethering of the cord (Cumes, 1977). Sacral agenesis may also coexist with forms of spinal dysrhaphism such as myelomeningocele and diastematomyelia (White and Klauber, 1976). The resulting neurologic deficits may result from absence of spinal roots, compression or entrapment of the spinal cord or spinal roots, and traction on the spinal cord during growth of the vertebral column due to tethering.

Many patients with sacral agenesis have either no apparent or only very subtle neurologic symptoms and the condition may elude diagnosis for many years (Braren and Jones, 1979). Unfortunately, some patients with neurogenic vesical dysfunction have undergone multiple urologic surgical procedures before the underlying condition was diagnosed (Cumes, 1977; Koontz and Prout, 1968). The condition may be misdiagnosed if only anteroposterior radiographs of the lower spine are obtained, because the sacrum may be poorly visualized if an exaggerated lumbosacral angle exists. Therefore, it is important to obtain lateral sacral radiographs (Mariani et al., 1979). Patients with neurogenic vesical dysfunction may initially present with hydronephrosis, urinary tract infection, vesicoureteral reflux, difficulty in voiding, urinary incontinence, or bladder trabeculation (Cumes, 1977). Several investigators have determined that neurogenic vesical dysfunction will occur in patients with agenesis of one or two sacral segments (Mariani et al., 1979; Williams and Nixon, 1957). The findings on physical examination may include a palpable defect in the sacrum, sacral dimple, a flattened gluteal region, foot deformities, decrease in perineal sensation and resting external anal sphincter tone, and flaccid paralysis of the muscles innervated by the second through fifth sacral roots (White and Klauber, 1976; Cumes, 1977). Motor deficits are generally more pronounced than sensory deficits (Mariani et al., 1979; Sarnat et al., 1976) and in many patients these deficits do not correlate with those expected from the bony deficiency (Koff and Deridder, 1977; Smith, 1959).

Similarly, the type of neurogenic vesical dysfunction does not correlate with the bony lesion (Koff and Deridder, 1977; Williams and Nixon, 1957). Because the clinical neurologic evaluation is also not helpful in identifying the specific type of neurogenic vesical dysfunction in a given patient, urodynamic evaluation is extremely important. Koff and Deridder (1977) performed sophisticated urodynamic evaluations in 11 patients with sacral agenesis. Three patients had uninhibited detrusor dysfunction, four had detrusor areflexia, and four had a mixed type of detrusor dysfunction. Perineal EMG was performed in 12 patients and seven showed evidence of denervation, but some had normal external anal sphincter activity in the presence of detrusor dysfunction. The pattern of neurogenic vesical dysfunction may change with time, probably because of traction of the spinal cord due to tethering.

NONNEUROGENIC VOIDING DYSFUNCTION

The true challenge to the pediatric neurourologist is the definition of vesicourethral dysfunction in the child who has no outward neurologic deficits. These children present to the clinician only after voiding difficulties have led to either urinary tract infection or significant social disability. The spectrum of voiding abnormalities seen is not unlike that of patients with true neurologic disease (Allen, 1977). Thus, to label this type of vesicourethral dysfunction as nonneurogenic speaks only to the general neurologic state of the child but not to the problem of vesicourethral dysfunction. Because this group of patients has no outward indicators for the potential of underlying vesicourethral problems, it becomes important for all clinicians to assess carefully the subtle changes in voiding mentioned by the child or his parents. When in doubt as to the significance of a particular symptom, a simple noninvasive test such as uroflowmetry may better define its importance.

When voiding dysfunction is identified without outward neurologic disease or evidence of spinal dysrhaphism, confusion abounds as to its cause. Some investigators have suggested an underlying psychological defect and others emphasize an occult neuropathy (Hinman, 1974; Dorfman et al., 1969; Johnston and Farkas, 1975; Martin et al., 1971; Williams et al., 1974). Regardless of origin, many of the voiding abnormalities are thought to be related to a disorder of coordination between the detrusor contraction and external sphincter relaxation. Support for a nonneurologic acquired abnormality comes from Bauer and associates (Bauer et al., 1980), who failed to show denervation potentials on raw EMG data obtained from the external sphincter in such children. In addition, many such voiding abnormalities are associated with acquired bowel defects such as encopresis. Also supporting an acquired cause is the fact that many such voiding abnormalities can be corrected by a period of bladder retraining, which may include biofeedback methods. The full spectrum of voiding dysfunction may be seen and the diagnostic characteristics are the same as those seen with true neurologic disease (Wein, 1984). Many such patients have subtly sustained irreversible damage to the detrusor, resulting in myogenic decompensation (Fig 28–31). In others, vesicoureteral reflux may have developed and surgical repair will be necessary to ensure adequate bladder emptying (Snyder et al., 1982). In all such patients, therapy should be designed to promote adequate bladder emptying and an acceptably low voiding pressure to protect the upper urinary tracts.

Those with urgency, frequency, and wetting related to vesicourethral incoordination are ideally suited to bladder retraining with the biofeedback methods. However, those with significant failure-to-empty defects may be better managed with conventional techniques of intermittent catheterization, with or without the use of adjunctive pharmacologic agents. The length of therapy in both categories of patients will be dependent on the degree of detrusor decompensation, upper tract disease, and the likelihood of relapse into a vesicourethral incoordination state. Regular follow-up is necessary in these patients to ensure that relapse has not occurred and that the upper tracts are, in fact, being protected.

PSYCHOGENIC VOIDING DYSFUNCTION

Voiding dysfunction in the absence of organic disease has been classified as psychogenic in origin (Krane and Siroky, 1979a, 1979b). Voiding problems of all types will be associated with psychological manifestations. However, which problem arrived first, the voiding problem or the psychological problem, is not always clear. Considering all voiding problems, some are obviously organic and others psychogenic, but numerous patients with significant voiding dysfunction fall in between, resulting in diagnoses that are indeterminate. Voiding symptoms related to psychological disturbances range from frank urinary retention to marked urgency, frequency, and enuresis (Rowan, 1975).

Psychogenic Retention

For those patients with defects related to a "failure to empty," the symptoms may range from acute retention to those suggesting a more chronic course, such as urinary infection or overflow type incontinence (Allen, 1972; Barrett, 1976; Doran and Roberts, 1976; Larson et al., 1963; Norden and Friedman, 1961). Acute retention is often temporally related to some catastrophic psychological stress in the patient's life. These patients are often young or middle-aged women. However, select male patients may also demonstrate this phenomenon (Chapman, 1959; Fox et al., 1976; Khan, 1971;

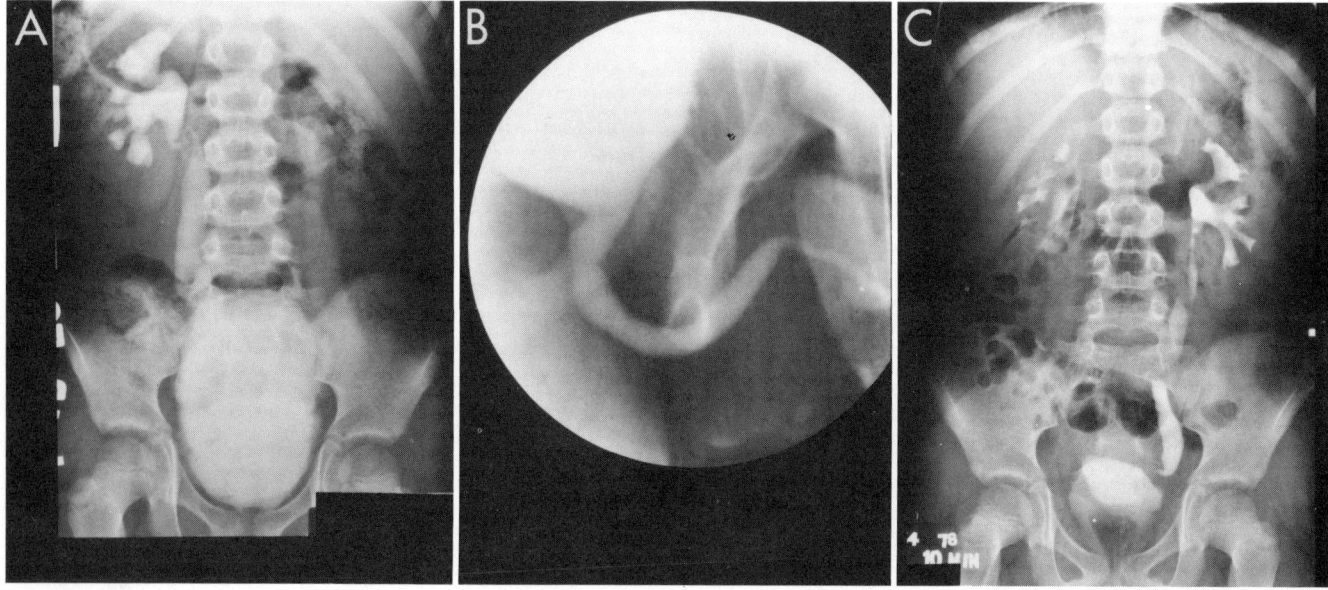

FIG 28–31.
A, excretory urogram in a 10-year-old boy with wetting. Neurologic examination reflected no neurologic disease. **B,** voiding cystourethrogram is normal, with exception of excessive postvoided residual volume. **C,** excretory urogram after six-month interval following combined bladder retraining and intermittent self-catheterization.

Krugman et al., 1971). Such patients may exhibit a propensity for repeated episodes of acute retention with essentially normal voiding in between.

Other patients with a more chronic pattern will continuously maintain considerable residual volumes. A propensity toward urinary infection may be seen and, depending on the degree of retention and the integrity of the ureterovesical angle, upper tract disease may develop. The patient may give a history of having been toilet-trained at an early age and exhibiting excellent urinary control. Many will be life-long infrequent voiders and notably require voiding only 1–4 times daily. Micturition in such patients may require straining, pushing, or a Credé maneuver to augment emptying. The sensation to void will become impaired and the desire to void is noticed only when large volumes occupy the bladder. Periodic "treatments" such as urethral dilation or internal urethrotomy may transiently benefit the patient; however, excessive residual urine generally returns.

The pathophysiology of psychogenic retention is open to controversy but is much easier to accept in the patient with acute retention. Theoretically, excessive stimuli, either excitatory or inhibitory, originate in centers above the brain stem and influence coordinated micturition. Speculation exists as to the point at which impairment—leading to retention—occurs in the cascade of events that take place during normal micturition. The two most plausible explanations are: (1) excessive stimulation of the centers of the pyramidal tracts that directly innervate the skeletal muscle urethral sphincter, resulting in a failure of the sphincter to relax in preparation for voiding (Lewin et al., 1967), or (2) inhibition of the autonomic motor nucleus in the brain stem that innervates the bladder body, resulting in decreased detrusor contractility (Bradley and Conway, 1966; Bradley and Teague, 1968).

Urodynamic parameters in patients with acute urinary retention generally reflect decreased detrusor contractility, normal compliance, and slightly elevated urethral closing pressure. After the acute phase resolves, the parameters—including uroflow rate—may be normal. Those with a chronic history of urinary retention will have decreased detrusor contractility, increased compliance, increased intra-abdominal pressure during voiding, and reduced average flow rate.

The diagnosis of psychogenically induced urinary retention is one of exclusion, and subtle neurologic or metabolic disease must be excluded (Barrett, 1978). Thus, consultations from appropriate specialists are indicated and often extensive neurologic testing, including scans of the brain and spinal cord, may be necessary. Since the symptoms that lead to medical scrutiny are urologic in origin, it becomes the responsibility of the urologist to coordinate this evaluation. Also, management of the voiding problem must be undertaken along the lines outlined earlier for patients with failure to empty. Psychologic consultation is required when patients have overt psychological disease or when they fail to understand the potential for a psychological disturbance as a cause for voiding abnormalities.

Psychogenic Urgency-Frequency Syndrome

Much of the urologic practice deals with patients exhibiting symptoms of urinary urgency, frequency, and urge incontinence (Smith, 1962). Lower abdominal and perineal discomfort may also accompany this syndrome. These patients, who are predominantly female, must undergo a thorough neurologic evaluation to exclude underlying disease (Rees et al., 1976). Similar symptomatology will lead to the diagnosis of unsuspected urinary infections in many patients, and those without a psychogenic relationship will improve with antibiotic treatment (Gallagher et al., 1965). Those of psychogenic etiology again will be suspected when all tests are essentially normal. Even males with typical "prostatitis" symptoms may fall into this category (Fritjofsson et al., 1973).

Uroradiography, endoscopy, and urodynamics are normal in most of these patients. However, in a select few, micturition may be impaired by functional outflow obstruction (Farrar et al., 1976). Cystometry will reflect normal compliance but a decreased functional capacity (Rees et al., 1976). Occasionally, detrusor instability will be seen. Uroflow will reflect a small voided volume and the average rate will be decreased. Combined studies including fluoroscopy are best for detecting the functional outlet obstruction syndrome. Voiding pressure will be high and flow rate decreased. There will be little if any bladder neck funneling during the detrusor contraction. Thus, the condition suggests dyssynergia of the bladder neck and proximal urethra.

It is unclear whether this extreme version of the urgency-frequency syndrome is truly psychogenic in origin, but the lack of associated neurologic disease in most of these patients makes the diagnosis likely (Mulholland et al., 1974). Many of these patients respond, at least temporarily, to measures designed to reduce outflow resistance (Raz and Smith, 1976). Internal urethrotomy or bladder neck resection is a last resort and should be performed only in the presence of data substantiating outlet obstruction (Emmett et al., 1950). The goal of such therapy is to reduce voiding pressures and reverse the stimulus that provokes the sensation of urgency.

The psychodynamics of the psychogenic urgency-frequency abnormality are similar to the retention syndrome in that inhibitory or excitatory stimuli from the central nervous system above the brain alter the micturition sequence in such a way that either sensation or contraction is impaired.

REFERENCES

1. Abel B, Gibbon N, Jameson R, et al: The neuropathic urethra. *Lancet* 1974; 2:1229.

2. Abrams PH: Investigation of post-prostatectomy problems. *Urology* 1980; 15:209.

3. Abrams P: The practice of urodynamics, in Mundy AR, Stephenson TP, Wein AJ (eds): *Urodynamics: Principles, Practice and Application.* Edinburgh, Churchill-Livingstone, 1984a, p 76.

4. Abrams P: The urethral pressure profile measurement, in Mundy AR, Stephenson TP, Wein AJ (eds): *Urodynamics: Principles, Practice and Application.* Edinburgh, Churchill-Livingstone, 1984b, p 127.

5. Abrams P: Detrusor instability and bladder outlet obstruction. *Neurourol Urodynamics* 1985; 4:317.

6. Allen TD: Psychogenic urinary retention. *South Med J* 1972; 65:302.

7. Allen TD: The non-neurogenic neurogenic bladder. *J Urol* 1977; 117:232–238.

8. Allen TD: Commentary on dysfunctional abnormalities of the urinary tract. *Urol Clin North Am* 1980; 7:357.

9. Allen TD, Bright TC III: Urodynamic patterns in children with dysfunctional voiding problems. *J Urol* 1978; 119:247–249.

10. Althausen AF, Hagen-Cook K, Hendren WH III: Nonrefluxing colon conduit: Experience with 70 cases. *J Urol* 1978; 120:35.

11. Altwein JE, Jonas U, Hohenfellner R: Long-term follow-up of children with colon conduit urinary diversion and ureterosigmoidostomy. *J Urol* 1977; 118:832.

12. Andersen JT, Nordling J: Prostatism: I. The correlation between symptomatic cystometric and urodynamic findings. *Scand J Urol Nephrol* 1979; 13:229.

13. Andersen JT, Nordling J: Prostatism: II. The correlation between cystourethroscopic, cystometric, and urodynamic findings. *Scand J Urol Nephrol* 1980; 14:23.

14. Andersen JT, Walter S, Vejesgaaard R: A clinical and bacteriological trial with DMSO in the treatment of severe detrusor hyperreflexia. *Scand J Urol Nephrol* 1981; 60(suppl):63.

15. Andersen JT, Bradley WE: Abnormalities of detrusor and sphincter dysfunction in multiple scleroma. *Br J Urol* 1976; 48:193.

16. Anderson FM: Occult spinal dysrhaphism: A series of 73 cases. *Pediatrics* 1975; 55:826–835.

17. Andersen JT, Hebjorn S, Frimodt-Moller C, et al: Disturbances of micturition in Parkinson's disease. *Acta Neurol Scand* 1976; 53:161.

18. Andersen JT, Jacobson O, Worm-Petersen J, et al: Bladder function in healthy elderly males. *Scand J Urol Nephrol* 1978; 12:123.

19. Andersson K, Ek A, Hedlung H, et al: Effects of prazosin on isolated human urethra and in patients with lower neuron lesions. *Invest Urol* 1981; 19:39.

20. Ashken MH: Urinary reservoirs, in Ashken MH (ed): *Urinary Diversion.* New York, Springer-Verlag, 1982.

21. Awad SA, Bryniak SR, Downie JN, et al: Urethral pressure profile during the spinal shock stage in man: A preliminary report. *J Urol* 1977; 117:91.

22. Awad S, Downie J, Kiruluta H: Alpha adrenergic agents in urinary disorders of the proximal urethra: I. Stress incontinence. *Br J Urol* 1978; 50:332.

23. Awad S, McGinnis R, Downie J: The effectiveness of bethanechol chloride in lower motor neuron lesions: The importance of mode of administration. *Neurourol Urodynamics* 1984; 3:173.

24. Baldessarini R: Drugs and the treatment of psychiatric disorders, in Gilman AG, Goodman LS, Gilman A (eds): *The Pharmacological Basis of Therapeutics.* New York, Macmillan, 1980, pp 391–447.

25. Barrett DM: Psychogenic urinary retention in women. *Mayo Clin Proc* 1976; 51:351.

26. Barrett DM: Evaluation of psychogenic urinary retention. *J Urol* 1978; 120:191.

27. Barrett DM: Disposable (infant) surface electrocardiogram electrodes in urodynamics: A simultaneous comparative study of electrodes. *J Urol* 1980; 124:663.

28. Barrett DM: The effect of oral bethanechol chloride on voiding in female patients with excessive residual urine: A randomized double-blind study. *J Urol* 1981; 126:640.

29. Barrett DM: Urinary flow rate: Theory and methods, in Barrett DM, Wein AJ (eds): *Controversies in Neuro-urology.* New York, Churchill-Livingstone, 1984, pp 133–143.

30. Barrett DM: *AUA Update: The Artificial Sphincter,* lesson 32 Houston, AUA, 1986.

31. Barrett DM, Wein AJ: Flow evaluation and simultaneous external sphincter electromyography in clinical urodynamics. *J Urol* 1981; 125:538–541.

32. Barson AJ: Spina bifida: The significance of the level and extent of the defect to the morphogenesis. *Dev Med Child Neurol* 1970; 12:129–144.

33. Bates CP, Arnold EP, Griffiths DJ: The nature of the abnormality in bladder neck obstruction. *Br J Urol* 1975; 47:651.

34. Bauer SB: Urodynamics in children: Indications and methods, in Barrett DM, Wein AJ (eds): *Controversies in Neuro-urology.* New York, Churchill-Livingstone, 1984, pp 193–202.

35. Bauer SB, Colodny AH, Retik AB: The management of vesicoureteral reflux in children with myelodysplasia. *J Urol* 1982; 128:102–105.

36. Bauer SB, Hallett M, Khoshbin S, et al: Predictive value of urodynamic evaluation in newborns with myelodysplasia. *JAMA* 1984; 252:650–652.

37. Bauer SB, Labib KB, Dieppa RA, et al: Urodynamic evaluation of boy with myelodysplasia and incontinence. *Urology* 1977; 10:354–362.

38. Bauer SB, Retik AB, Colodny AH, et al: The unstable bladder of childhood. *Urol Clin North Am* 1980; 7:321–336.

39. Baumrucker GO: Management of the paralyzed bladder. *Arch Surg* 1948; 56:484.

40. Beisland HO, Fossberg E, Sander S: On incompetent urethral closure mechanism: Treatment with estriol and phenylpropanolamine. *Scand J Urol Nephrol* 1981; 60(suppl):67.

41. Belman AB, King LR: Urinary tract abnormalities associated with imperforate anus. *J Urol* 1972; 108:823–824.

42. Benson G, Sarshik S, Raezer D, et al: Comparative effects and mechanisms of action of atropine, propantheline, flavoxate, and imipramine on bladder muscle contractility. *Urology* 1977; 9:31.

43. Benson GS, Rogers DM, Wein AJ, et al: Effect of muscle length on adrenergic stimulation of canine detrusor. *Urology* 1975; 5:769.

44. Berg S: Polytef augmentation urethroplasty. *Arch Surg* 1973; 107:379.

45. Berry JL: A new procedure for correction of urinary incontinence: Preliminary report. *J Urol* 1961; 85:771–775.

46. Bianchine J: Drugs for Parkinson's disease: Centrally acting muscle relaxants, in Gilman AG, Goodman LS, Gilman A (eds): *The Pharmacological Basis of Therapeutics.* New York, Macmillan, 1980, pp 475–493.

47. Bigger J, Giardino E, Perel J, et al: Cardiac antiarrhythmic effect of imipramine hydrochloride. *N Engl J Med* 1977; 296:206.

48. Birkasova M, Birkas O, Flynn MJ, et al: Desmopressin in the management of nocturnal enuresis in children: A double-blind study. *Pediatrics* 1978; 62:970.

49. Blaivas JG: Management of bladder dysfunction in multiple sclerosis. *Neurology* 1980; 30:12.

50. Blaivas J: If you currently prescribe bethanechol chloride for urinary retention, please raise your hand. *Neurourol Urodynamics* 1984; 3:209.

51. Blaivas JG: Non-traumatic neurogenic voiding dysfunction in the adult: I. Physiology and approach to therapy. AUA Update Series, lesson 11, vol 4, 1985a.

52. Blaivas JG: Non-traumatic neurogenic voiding dysfunction in the adult: II. Multiple sclerosis and diabetes mellitus. AUA Update Series, lesson 12, vol 4, 1985b.

53. Blaivas JG: Electromyography and sacral evoked responses, in Mundy AR, Stephenson TP, Wein AJ (eds): *Urodynamics: Principles, Practice and Application.* Edinburgh, Churchill-Livingstone, 1984a, p 139.

54. Blaivas JG: Multichannel urodynamic studies. *Urology* 1984b; 23:421.

55. Blaivas JG, et al: Failure of bethanechol denervation supersensitivity as a diagnostic aid. *J Urol* 1980; 123:199.

56. Blaivas JG, et al: Detrusor-external sphincter dyssynergia: A detailed EMG study. *J Urol* 1981; 125:545.

57. Blaivas JG, Barbalias GA: Detrusor external sphincter dyssynergia in men with multiple sclerosis: An ominous urologic condition. *J Urol* 1984; 131:94.

58. Blaivas JG, Bhimani G, Labib KB: Vesicourethral dysfunction in multiple sclerosis. *J Urol* 1979; 122:342.

59. Blaivas J, Labib K, Michalik S, et al: Cystometric response to propantheline in detrusor hyperreflexia: Therapeutic implications. *J Urol* 1980; 124:259.

60. Blaivas JG, Labib KL, Bauer SB, et al: A new approach to electromyography of the external urethral sphincter. *J Urol* 1977; 117:773.

61. Blaivas JG, Labib KL, Bauer SB, et al: Changing concepts in the urodynamic evaluation of children. *J Urol* 1977; 117:778–781.

62. Blaivas JG, Norlen LJ: Primary bladder neck obstruction. *World J Urol* 1984; 2:195.

63. Blaivas JG, Sinha HP, Zayed AA, et al: Detrusor-external sphincter dyssynergia. *J Urol* 1981a; 125:541.

64. Blaivas JG, Zayed AAH, Labib KB: The bulbocavernosus reflex in urology: A prospective study of 299 patients. *J Urol* 1981b; 126:197.

65. Bonney V: On diurnal incontinence of urine in women. *Br J Obstet Gynecol* 1923; 30:358.

66. Bors E: Neurogenic bladder. *Urol Survey* 1957; 7:177.

67. Bors E, Comarr AE: *Neurological Urology.* Baltimore, University Park Press, 1971.

68. Bradley WE, Chou S, Markland C: Classifying neurologic dysfunction of the urinary bladder, in Boyarsky S (ed): *The Neurogenic Bladder.* Baltimore, Williams & Wilkins Co, 1967, pp 139–146.

69. Bradley WE, Conway CJ: Bladder representation in the pontine-mesencephalic reticular formation. *Exp Neurol* 1966; 16:251.

70. Bradley WE, Teague CT: Spinal cord organization of micturitional reflex afferents. *Exp Neurol* 1968; 22:504.

71. Bradley WE, Timm GW, Scott FB: Innervation of the detrusor muscle and urethra. *Urol Clin North Am* 1974; 1:3–27.

72. Bramble FJ: The treatment of adult enuresis and urge incontinence by enterocystoplasty. *Br J Urol* 1982; 54:693.

73. Braren V, Jones WB: Sacral agenesis: Diagnosis, treatment and follow up of urological complications. *J Urol* 1979; 121:543–544.

74. Brent L, Stephens FD: The response of smooth muscle cells in the rabbit urinary bladder to outflow obstruction. *Invest Urol* 1975; 12:494.

75. Bricker EM: Bladder substitution after pelvic evisceration. *Surg Clin North Am* 1950; 30:1511.

76. Briggs R, Castleden C, Asher M: The effect of flavoxate on uninhibited detrusor contractions and urinary incontinence in the elderly. *J Urol* 1980; 123:665.

77. Broecker BH, Klein FA, Hackler RH: Cancer of the bladder in spinal cord injury patients. *J Urol* 1981; 125:196.

78. Brown M, Wickham JEA: The urethral pressure profile. *Br J Urol* 1969; 51:211.

79. Bultitude M, Hills N, Shuttleworth K: Clinical and experimental studies on the action of prostaglandins and their synthesis inhibitors on detrusor muscle in vitro and in vivo. *Br J Urol* 1976; 48:631.

80. Bunce PL: Transurethral resection and Y-V plasty for neurogenic bladder, in Boyarsky S (ed): *The Neurogenic Bladder.* Baltimore, Williams & Wilkins Co, 1967, pp 196–199.

81. Burch JC: Urethrovaginal fixation to Cooper's ligament for correction of stress incontinence, cystocele and prolapse. *Am J Obstet Gynecol* 1961; 81:281.

82. Burgess C, Turner T: Cardiotoxicity of antidepressant drugs. *Neuropharmacology* 1980; 19:1195.

83. Caine M, Edwards D: The peripheral control of micturition: A cineradiological study. *Br J Urol* 1958; 30:34.

84. Cardozo L, Stanton S: A comparison between bromocriptine and indomethacin in the treatment of detrusor instability. *J Urol* 1980; 123:399.

85. Cardozo L, Stanton S, Robinson H, et al: Evaluation of flurbiprofen in detrusor instability. *Br Med J* 1980; 280:281.

86. Cardozo LD, Abrams PD, Stanton SL, et al: Idiopathic bladder instability treated by biofeedback. *Br J Urol* 1978; 50:521.

87. Castleden C, George C, Redwick A, et al: Imipramine—a possible alternative to current therapy for urinary incontinence in the elderly. *J Urol* 1981; 125:218.

88. Chalfin SA, Bradley WE: The etiology of detrusor hyperreflexia in patients with infravesical obstruction. *J Urol* 1982; 127:938.

89. Chan WY, Dawood MY, Fuchs F: Prostaglandins in primary dysmenorrhea. *Am J Med* 1981; 70:535.

90. Chapman AH: Psychogenic urinary retention in women. *Psychosom Med* 1959; 21:119.

91. Cole A, Fried F: Favorable experiences with imipramine in the treatment of neurogenic bladder. *J Urol* 1972; 107:44.

92. Comarr AE: The practical urologic management of the patient with spinal cord injury. *Br J Urol* 1959; 31:1.

93. Comarr AE: Diagnosis of the traumatic cord bladder, in Boyarsky S (ed): *The Neurogenic Bladder.* Baltimore, Williams & Wilkins Co, 1967, pp 147–152.

94. Constantinou CE, Faysal MH, Rother L, et al: The impact of bladder neck suspension on the mode of distribution of abdominal pressure along the female urethra, in Zinner NR, Sterling AM (eds): *Female Incontinence: Progress in Clinical and Biological Research.* New York, Alan R Liss Inc, vol 78, 1981, pp 121–132.

95. Cook WA, Firlit CF, Stephens FD, et al: Techniques and results of urodynamic evaluation of children. *J Urol* 1977; 117:346–349.

96. Council on Scientific Affairs: DMSO. *JAMA* 1982; 248:1369.

97. Cumes D: Sacral dysgenesis associated with occult spinal dysraphism causing neurogenic bladder dysfunction. *J Urol* 1977; 117:127–128.

98. Deane AM, English P, Hehn M, et al: Teflon injection in stress incontinence. *Br J Urol* 1985; 57:78.

99. Dees JE: Congenital epispadias with incontinence. *J Urol* 1949; 612:513.

100. Dees JE: Contraction of the urinary bladder produced by electric stimulation: Preliminary report. *Invest Urol* 1965; 2:539.

101. deGroat WC: Nervous control of the urinary bladder of the cat. *Brain Res* 1975; 87:201–211.

102. deGroat WC, Saum WR: Adrenergic inhibition in mammalian parasympathetic ganglia. *Nature* 1971; 231:188–189.

103. deGroat WC, Theobald RJ: Reflex activation of sympathetic pathways to vesical smooth muscle and parasympathetic ganglia by stimulation of vesical afferents. *J Physiol* 1976; 259:223–237.

104. Delaere TC, Moonen T, et al: The value of prostaglandin E_2 and $F_{2\alpha}$ in women with abnormalities of bladder emptying. *Br J Urol* 1981; 53:306.

105. Desmond A, Bultitude M, Hills N, et al: Clinical experience with intravesical prostaglandin E_2: A prospective study of 36 patients. *Br J Urol* 1980; 53:357.

106. Dhattiwala A: The effect of imipramine on isolated innervated guinea pig and rat urinary bladder preparations. *J Pharmacol* 1976; 28:453.

107. Diokno A, Koppenhoeffer R: Bethanechol chloride in neurogenic bladder dysfunction. *Urology* 1976; 8:455.

108. Diokno A, Taub M: Ephedrine in treatment of urinary incontinence. *Urology* 1975; 5:624.

109. Diokno AC, Hollander JB, Bennett CJ: Bladder neck obstruction in females: A real entity. *J Urol* 1984; 132:294.

110. Dixon JS, Hald T: Morphological studies of the bladder wall in interstitial cystitis, in George NJR, Gosling JA (eds): *Sensory Disorders of the Bladder and Urethra.* New York, Springer-Verlag, 1986, pp 63–71.

111. Donker P, Van der Sluis C: Action of beta adrenergic blocking agents on the urethral pressure profile. *Urol Int* 1976; 31:6.

112. Doran J, Roberts M: Acute urinary retention in the female. *Br J Urol* 1976; 47:793.

113. Dorfman LE, Bailey J, Smith JP: Subclinical neurogenic bladder in children. *J Urol* 1969; 101:48–54.

114. Downie J: Bethanechol chloride in urology—a discussion of issues. *Neurourol Urodynamics* 1984; 3:211.

115. Downie J, Twiddy D, Awad S: Antimuscarinic and noncompetitive antagonist properties of dicyclomine hydrochloride in isolated human and rabbit bladder muscle. *J Pharmacol Exp Ther* 1977; 201:662.

116. Drach GW, Fair WR, Meares EM, et al: Classification of benign diseases associated with prostatic pain: Prostatitis or prostatodynia? *J Urol* 1978; 120:286.

117. Duhamel B: From the mermaid to anal imperforation: The syndrome of caudal regression. *Arch Dis Child* 1961; 36:152–155.

118. Duncan G, Shahani B, Young R: An evaluation of baclofen treatment for certain symptoms in patients with spinal cord lesions. *Neurology* 1976; 26:441.

119. Dunn M, Roberts JBM, Smith PJB, et al: The long-term results of ileal conduit urinary diversion in children. *Br J Urol* 1979; 51:458.

120. Edwards L: Mechanical and other devices, in Caldwell KPS (ed): *Urinary Incontinence.* New York, Grune & Stratton, 1975, pp 115–127.

121. Ek A, Andersson KE, Gullberg B, et al: The effects of long-term treatment with norephedrine on stress incontinence and urethral closure pressure profile. *Scand J Urol Nephrol* 1978; 12:105.

122. Ekman G, Andersson KE, Rud T, et al: A double blind crossover study of the effects of xerodilene in women with unstable bladder. *Acta Pharmacol Toxicol* 1980; 46(suppl):39.

123. El Badawi A, Schenk EA: A new theory of the innerva-

tion of the bladder musculature: Morphology of the intrinsic vesical innervation apparatus. *J Urol* 1968; 99:585–592.

124. El Badawi A, Schenk EA: A new theory of the innervation of bladder musculature: II. The innervation of bladder musculature: 2. The innervation apparatus of the ureterovesical junction. *J Urol* 1971; 105:368–371.

125. El Badawi A, Schenk EA: A new theory of the innervation of bladder musculature: IV. Innervation of the vesicourethral junction and external urethral sphincter. *J Urol* 1974; 111:613–615.

126. Elder DD, Moisey CU, Rees RWM: A long-term follow-up of the colonic conduit operation in children. *Br J Urol* 1979; 51:462.

127. Elder DD, Stephenson TP: An assessment of the Frewen regime in the treatment of detrusor dysfunction in females. *Br J Urol* 1980; 52:467.

128. Emmett JL: Transurethral resection in treatment of true and pseudo cord bladder. *J Urol* 1945; 53:4.

129. Emery JL, Lendon RG: The local cord lesion in neurospinal dysrhaphism (meningomyelocele). *J Pathol* 1973; 110:83–96.

130. Emmett JL, Hutchins SPR, McDonald JR: The treatment of urinary retention in women by transurethral resection. *J Urol* 1950; 63:1031.

131. Enhorning G: Simultaneous recording of intravesical and intraurethral pressure. *Acta Chir Scand* 1961; 276(suppl):1.

132. Erlandson BE, Fall M: Intravaginal electrical stimulation in urinary incontinence. *Scand J Urol Nephrol* 1977; 44(suppl):5.

133. Evans AT, Felker JR, Shank RA III, et al: Pitfalls of urodynamics. *J Urol* 1979; 122:220–222.

134. Farrar DJ: Urodynamics and the elderly, in Mundy AR, Stephenson TP, Wein AJ (eds): *Urodynamics: Principles, Practice, and Application.* New York, Churchill-Livingstone, 1984, pp 249–255.

135. Farrar DJ, Osborne JL, Stephenson TP, et al: A urodynamic view of bladder outflow obstruction in the female: Factors influencing the results of treatment. *Br J Urol* 1976; 47:815.

136. Farrar DJ, Whiteside CG, Osborne JL, et al: A urodynamic evaluation of micturition symptoms in the female. *Surg Gynecol Obstet* 1975; 141:875.

137. Feneley RCL, Shepherd AM, Powell PH, et al: Urinary incontinence: Prevalence and needs. *Br J Urol* 1979; 51:493.

138. Finkbeiner A, Welch L, Bissada N: Uropharmacology: IX. Direct acting smooth muscle stimulants and depressants. *Urology* 1978; 12:231.

139. Firlit CF, Smey P, King LD: Micturition urodynamic flow studies in children. *J Urol* 1978; 119:250–253.

140. Fischer C, Diokno A, Lapides J: The anticholinergic effects of dicyclomine hydrochloride in uninhibited neurogenic bladder dysfunction. *J Urol* 1978; 120:328.

141. Florante J, Leyson J, Martin B, et al: Baclofen in the treatment of detrusor-sphincter dyssynergy in spinal cord injury patients. *J Urol* 1980; 124:82.

142. Foley FB: An artificial sphincter: A new device and operation for control of enuresis and urinary incontinence. *J Urol* 1947; 58:250–259.

143. Forman A, Andersson K, Henriksson L, et al: Effects of nifedipine on the smooth muscle of the human urinary tract in vitro and in vivo. *Acta Pharmacol Toxicol* 1978; 43:111.

144. Forsythe WI, Redmond A: Enuresis and spontaneous cure rate: Study of 1,129 enuretics. *Arch Dis Child* 1974; 49:259–263.

145. Fossberg E, Bersland HO, Sander S: Sensory urgency in females: Treatment with phenylpropanolamine. *Eur Urol* 1981; 7:157.

146. Fowler JE Jr: Prospective study of intravesical dimethylsulfoxide in the treatment of suspected early interstitial cystitis. *Urology* 1981; 18:21.

147. Fox M, Jarvis GJ, Henry L: Idiopathic chronic urinary retention in the female. *Br J Urol* 1976; 47:797.

148. Frankel HL: Intermittent catheterization. *Urol Clin North Am* 1974; 1:115.

149. Fredericks C, Green R, Anderson G: Comparative in vitro effects of imipramine, oxybutynin and flavoxate on rabbit detrusor. *Urology* 1978; 12:487.

150. Freiha FS, Stamey TA: Cystolysis: A procedure for the selective denervation of the bladder. *J Urol* 1980; 123:360.

151. Frewen WK: An objective assessment of the unstable bladder of psychosomatic origin. *Br J Urol* 1978; 50:246.

152. Frewen WK: The management of urgency and frequency of micturition. *Br J Urol* 1980; 512:367.

153. Frimodt-Moller C: Diabetes cystopathy: I. A clinical study on the frequency of bladder dysfunction in diabetics. *Danish Med Bull* 1976; 23:267.

154. Frimodt-Moller C: Cystopathy: II. Relationship to some late diabetic manifestations. *Danish Med Bull* 1976b; 23:279.

155. Fritjofsson A, Kihl B, Danielsson G: Chronic prostatovesiculitis: Incidence and significance of bacterial findings. *Scand J Urol Nephrol* 1973; 8:173.

156. Gallagher DJA, Montgomerie JZ, North JDK: Acute infections of the urinary tract and the urethral syndrome in general practice. *Br Med J* 1965; 1:622.

157. Gaum LD, Wese FX, Alton DJ, et al: Radiologic investigation of the urinary tract in the neonate with myelomeningocele. *J Urol* 1982; 127:510–511.

158. George NJR: Preface in George NJR, Gosling JA (eds): *Sensory Disorders of the Bladder and Urethra.* New York, Springer-Verlag, 1986a, p vii.

159. George NJR: Urethral syndrome—clinical features, in George NJR, Gosling JA (eds): *Sensory Disorders of the Bladder and Urethra.* New York, Springer-Verlag, 1986b, pp 91–102.

160. Gerstenberg TC, Andersen JT, Klarshov P, et al: High flow infravesical obstruction in the male. *J Urol* 1982; 127:943.

161. Gerstenberg TC, Lykkegaard, Nielsen M, et al: Spastic striated external sphincter syndrome initiating recurrent urinary tract infection in females. *Eur Urol* 1983; 9:87.

162. Gibaldi M, Grundhofer B: Biopharmaceutic influences

on the anticholinergic effects of propantheline. *Clin Pharmacol Ther* 1975; 18:457.

163. Gibbon N: Management of the bladder in acute and chronic disorders of the nervous system. *Acta Neurol Scand* 1966; 42(suppl):133.

164. Gibbon N: Urinary incontinence in disorders of the nervous system. *Br J Urol* 1975; 37:624.

165. Gibbon NOK: Later management of the paraplegic bladder. *Paraplegia* 1974a; 12:87.

166. Gibbon NOK: Neurogenic bladder in spinal cord injury: Management of patients in Liverpool, England. *Urol Clin North Am* 1974b; 1:147.

167. Gibbon NOK: Nomenclature of neurogenic bladder. *Urology* 1976; 8:423.

168. Gibbon NOK, Ross JC, Damanski M: Bladder neck resection in the paraplegic: Report of over 100 cases. *Paraplegia* 1965; 2:264.

169. Gierup J: Micturition studies in infants and children: Intravesical pressure, urinary flow and urethral resistance in boys without infravesical obstruction. *Scand J Urol Nephrol* 1970; 4:217–230.

170. Gilpin SA, Gosling JA, Barnard RJ: Morphological and morphometric studies of the human obstructed trabeculated urinary bladder. *Br J Urol* 1985; 57:525.

171. Glahn BE: Neurogenic bladder in spinal cord injury: Management of patients. *Urol Clin North Am* 1974; 1:163.

172. Gleason D, Reilly R, Bottaccini M, et al: The urethral continence zone and its relation to stress incontinence. *J Urol* 1974; 112:81.

173. Gleason DM, Bottaccini MR, Reilly RJ: Comparison of cystometrograms and urethral profiles with gas and water media. *Urology* 1977; 9:155–160.

174. Glen E: Control of incontinence by electrical devices, in Caldwell KBS (ed): *Urinary Incontinence.* New York, Grune & Stratton, 1975, pp 89–114.

175. Godec C, Krolg B: Selection of patients with urinary incontinence for application of functional electrical stimulation. *Urol Int* 1976; 31:124.

176. Godec CJ, Fravel R, Cass AS: Optimal parameters of electrical stimulation in the treatment of urinary incontinence. *Invest Urol* 1981; 18:239.

177. Goellner MH, Ziegler EE, Fomon SJ: Urination during the first three years of life. *Nephron* 1981; 28:174–178.

178. Goldwasser B, Webster GD: Continent urinary diversion. *J Urol* 1985; 134:227.

179. Goldwasser B, Webster GD: Augmentation and substitution enterocystoplasty. *J Urol* 1986; 135:215.

180. Golz H: Complications of external condom drainage. *Paraplegia* 1981; 19:189.

181. Gordon HL, Rossen RD, Hersh EM, et al: Immunologic aspects of interstitial cystitis. *J Urol* 1983; 109:228.

182. Green T: The problem of urinary stress incontinence in the female: An appraisal of its current status. *Obstet Gynecol Surv* 1962; 23:603.

183. Green T: Urinary stress incontinence: Differential diagnosis, pathophysiology and management. *Am J Obstet Gynecol* 1975; 122:368.

184. Green T: Urinary stress incontinence: Diagnosis and classification, in Buchsbaum H, Schmidt J (eds): *Gynecologic and Obstetric Urology.* Philadelphia, WB Saunders, 1982, pp 199–223.

185. Greenhill JP: The nonsurgical therapy of stress incontinence associated with vaginal relaxation, in Cantor EB (ed): *Female Urinary Stress Incontinence.* Springfield, Ill, Charles C Thomas, 1979, pp 195–206.

186. Grimes JH, Nashold BS, Anderson EE: Clinical application of electronic bladder stimulation in paraplegics. *J Urol* 1975; 113:338.

187. Grimes JH, Nashold BS, Currie D: Chronic electrical stimulation of the paraplegic bladder. *J Urol* 1973; 109:242.

188. Grossman HB, Koff SA, Doikno AC: Cystometry in children. *J Urol* 1977; 117:646–648.

189. Hachen H: Clinical and urodynamic assessment of alpha adrenolytic therapy in patients with neurogenic bladder function. *Paraplegia* 1980; 18:229.

190. Hachen H, Krucker V: Clinical and laboratory assessment of the efficacy of baclofen on urethral sphincter spasticity in patients with traumatic paraplegia. *Eur Urol* 1977; 3:237.

191. Hachen H: Sexual impotence: A complication of external sphincterotomy. *Urol Int* 1977; 32:336.

192. Hackler R, Broecker B, Klein F, et al: A clinical experience with dantrolene sodium for external urinary sphincter hypertonicity in spinal cord injured patients. *J Urol* 1980; 124:78.

193. Hadley HR, Zimmern PE, Staskin D, et al: Transvaginal needle bladder neck suspension. *Urol Clin North Am* 1985; 12:291.

194. Hald T: *The Committee on Standardization of Terminology of Lower Urinary Tract Function.* Copenhagen, International Continence Society, 1984.

195. Hald T, Bradley WE: *The Urinary Bladder: Neurology and Dynamics.* Baltimore, Williams & Wilkins, 1982.

196. Hald T, Holm-Bentzen M: Clinical symptom complex, in George NJR, Gosling JA (eds): *Sensory Disorders of the Bladder and Urethra.* New York, Springer-Verlag, 1986, pp 49–62.

197. Halverstadt DP, Parry WL: Electronic stimulation of the human bladder: Nine years later. *J Urol* 1975; 113:341.

198. Hanna M: New concept in bladder remodeling. *Urology* 1982; 19:6.

199. Hebjorn S: Treatment of detrusor hyperreflexia in multiple sclerosis. *Urol Int* 1977; 32:209.

200. Heimburger RF, Freeman LW, Wilde NJ: Sacral nerve innervation of the human bladder. *Int J Neurosurg* 1948; 5:154.

201. Hepburn TN: Prolapse of the female urethra. *Surg Gynecol Obstet* 1920; 31:83.

202. Hilton P, Stanton SL: The use of desmopressin (DDAVP) in nocturnal urinary frequency in the female. *Br J Urol* 1982; 54:252.

203. Hindmarsh JR, Gosling PT, Deane AM: Bladder instability: Is the primary defect in the urethra? *Br J Urol* 1983; 55:648.

204. Hinman F Jr: Urinary tract damage in children who wet. *Pediatrics* 1974; 54:142–150.

205. Hinman F Jr: Syndromes of vesical incoordination. *Urol Clin North Am* 1980; 7:311.

206. Hodgkinson JC: Stress urinary incontinence. *Am J Obstet Gynecol* 1970; 108:1141.

207. Hodgson BT, Dumas S, Bolling DR, et al: Effect of estrogen on sensitivity of rabbit bladder and urethra to phenylephrine. *Invest Urol* 1978; 16:67.

208. Hollister L: Tricyclic antidepressants. *N Engl J Med* 1978; 299:1106.

209. Hunner GL: A rare type of bladder ulcer in women: Report of cases. *Boston Med Surg J* 1915; 172:660.

210. Husted S, Andersson KE, Sommer L, et al: Anticholinergic and calcium antagonistic effects of etrodilene in rabbit urinary bladder. *Acta Pharmacol Toxicol* 1980; 46:20.

211. Merillo C, Conway CJ: Clinical experience with the Mentor bladder stimulator: III. Patients with urinary vesical hypotonia. *J Urol* 1975; 113:335.

212. Ingelmann-Sundberg A: Partial denervation of the bladder: A new operation for the treatment of urge incontinence and similar conditions in women. *Acta Obstet Gynecol Scand* 1959; 38:487.

213. Ingelmann-Sundberg A: Denervation of the bladder, in Stanton SL, Tanagho EA (eds): *Surgery of Female Incontinence.* New York, Springer-Verlag, 1980, pp 93–97.

214. International Continence Society Standardization Committee: Fourth report on the standardization of terminology of lower urinary tract function. *Br J Urol* 1981; 53:333.

215. Jeffcoate N: *Principles of Gynecology.* London, Butterworths, 1975, pp 701–708.

216. Jensen D Jr: Pharmacological studies of the uninhibited neurogenic bladder. *Acta Neurol Scand* 1981a; 64:175.

217. Jensen D Jr: Altered adrenergic innervation in the uninhibited neurogenic bladder. *Scand J Urol Nephrol* 1981b; 60:61.

218. Jensen D Jr: Uninhibited neurogenic bladder treated with prazosin. *Scand J Urol Nephrol* 1981c; 15:229.

219. Johns A, Tasker J, Johnson C, et al: The mechanism of action of dicyclomine hydrochloride on rabbit detrusor muscle and vas deferens. *Arch Int Pharmacodyn Ther* 1976; 224:109.

220. Johnston JH, Farkas A: Congenital neuropathic bladder: Practicalities and possibilities of conservative management. *Urology* 1975; 5:719–727.

221. Jones KW, Schoenberg HW: Obstruction and the uninhibitable detrusor, in Hinman F Jr (ed): *Benign Prostatic Hypertrophy.* New York, Springer-Verlag, 1983, pp 706–710.

222. Jorgensen TM, Djurhuus JC, Jorgenson HS, et al: Experimental bladder hyperreflexia in pigs. *Urol Res* 1983; 11:239.

223. Jorgensen TM, Djurhuus JC, Schroder HD: Idiopathic detrusor sphincter dyssynergia in neurologically normal patients with voiding abnormalities. *Eur Urol* 1982; 8:107.

224. Kahn Z: Predictive correlation of urodynamic dysfunction after cerebrovascular accident. *J Urol* 1981; 126:88.

225. Kaplan GW: Myelomeningocele and related disorders. *J Cont Educ Urol* 1978, pp 15–28.

226. Kaplan WE, Firlit CF, Schoenberg HW: The female urethral syndrome: External sphincter spasm as etiology. *J Urol* 1980; 124:48.

227. Kass EJ, Diokno AC, Montealegre A: Enuresis: Principles of management and result of treatment. *J Urol* 1979; 121:794–796.

228. Kaufman JJ: A new operation for male incontinence. *Surg Gynecol Obstet* 1970; 131:295–299.

229. Kaufman JM, Fam B, Jacobs SC, et al: Bladder cancer and squamous metaplasia in spinal cord injury patient. *J Urol* 1977; 118:967.

230. Khan A: Psychogenic urinary retention in a boy. *J Urol* 1971; 106:432.

231. Khanna O: Disorders of micturition: Neuropharmacologic basis and results of drug therapy. *Urology* 1976; 8:316.

232. Khanna O, Barbieri E, McMichael R: The effects of adrenergic agonists and antagonists on vesicourethral smooth muscle of rabbits. *J Pharmacol Exp Ther* 1981; 216:95.

233. Khanna O, DiGregorio C, Barbieri E, et al: In vitro study of antispasmodic effects of dicyclomine hydrochloride on vesicourethral smooth muscle of guinea pig and rabbit. *Urology* 1979; 13:457.

234. Kiesswetter H, Schober W: Lioresal in the treatment of neurogenic bladder dysfunction. *Urol Int* 1975; 30:63.

235. Kinder CH, Smith RD: Hunner's ulcer. *Br J Urol* 1958; 30:338.

236. Kinn AC: The lazy bladder—appraisal of surgical reduction. *Scand J Urol Nephrol* 1985; 19:93.

237. Koelle G: Parasympathomimetic agents, in Goodman LS, Gilman A (eds): *The Pharmacologic Basis of Therapeutics.* New York, Macmillan Publishing Co, 1975, p 467.

238. Koff SA, Deridder PA: Patterns of neurogenic bladder dysfunction in sacral agenesis. *J Urol* 1977; 118:87–89.

239. Koff SA, Kass EJ: Abdominal wall electromyography: A noninvasive technique to improve pediatric urodynamic accuracy. *J Urol* 1982; 127:736–739.

240. Koff SA, Solomon MH, Lane GA, et al: Urodynamic studies in anesthetized children. *J Urol* 1980; 123:61–63.

241. Kohler R, Morales P: Cystometric evaluation of flavoxate hydrochloride in normal and neurogenic bladder. *J Urol* 1968; 100:729.

242. Koontz WW Jr, Prout GR Jr: Agenesis of the sacrum and the neurogenic bladder. *JAMA* 1968; 203:481–486.

243. Kosman ME: Evaluation of a new antidiuretic agent, desmopressin acetate (DDAVP). *JAMA* 1978; 240:1896.

244. Koyanagi T: Further observation on the denervation supersensitivity of the urethra in patients with chronic neurogenic bladders. *J Urol* 1979; 122:348.

245. Krane R, Olsson C: Phenoxybenzamine in neurogenic bladder dysfunction: I. A theory of micturition. *J Urol* 1973a; 110:650.

246. Krane R, Olsson C: Phenoxybenzamine in neurogenic

bladder dysfunction: II. Clinical considerations. *J Urol* 1973b; 110:653.

247. Krane RJ, Siroky MB: Classification of neuro-urologic disorders, in Krane RJ, Siroky MB (eds): *Clinical Neuro-Urology.* Boston, Little Brown & Co, 1979a, pp 143–158.

248. Krane RJ, Siroky MB: Psychogenic voiding dysfunction, in Krane RJ, Siroky MB (eds): *Clinical Neuro-Urology.* Boston, Little Brown & Co, 1979b, chap 15.

249. Krane RJ, Siroky MB: Studies on sacral evoked potentials. *J Urol* 1980; 123:872.

250. Krane RJ, Siroky MB: Classification of voiding dysfunction: Value of classification systems, in Barrett DM, Wein AJ (eds): *Controversies in Neuro-Urology.* New York, Churchill-Livingstone, 1984, pp 223–238.

251. Krugman A, Boyarsky S, Glenn JF Jr, et al: Psychological considerations in bladder function disturbances, in Hinman F Jr (ed): *Hydrodynamics of Micturition.* Springfield, Ill, Charles C Thomas, 1971.

252. Lapides J: Urecholine regimen for rehabilitating the atonic bladder. *J Urol* 1964; 91:648.

253. Lapides J: Cystometry. *JAMA* 1967; 201:618.

254. Lapides J: Neuromuscular, vesical and ureteral dysfunction, in Campbell MF, Harrison JH (eds): *Urology.* Philadelphia, WB Saunders Co, 1970, pp 1343–1379.

255. Lapides J: Simplified operation for stress incontinence. *J Urol* 1971; 105:262.

256. Lapides J, et al: Denervation supersensitivity as a test for neurogenic bladder. *Surg Gynecol Obstet* 1962; 114:241.

257. Lapides J, Diokno AC, Gould FR, et al: Further observations on self-catheterization. *J Urol* 1976; 116:169.

258. Lapides J, Dodson A: Observations on effect of methantheline bromide in urological disturbances. *Arch Surg* 1953; 66:1.

259. Lapides J, Friend C, Ajemian E, et al: Comparison of action of oral and parenteral bethanechol chloride upon the urinary bladder. *Invest Urol* 1963; 1:94.

260. Larsen J, Mortensen S: Effect of ritodrine on the bladder capacity in unanesthetized pigs. *Acta Pharmacol Toxicol* 1978; 43:405.

261. Larsen JW, Swenson WM, Utz DC, et al: Psychogenic urinary retention in women. *JAMA* 1963; 184:111.

262. Larsen S, Thompson SA, Hald T, et al: Mast cells in interstitial cystitis. *Br J Urol* 1982; 54:283.

263. Laurence KM: The natural history of spina bifida cystica: Detailed analysis of 407 cases. *Arch Dis Child* 1964; 39:41–57.

264. Leach GE, Goldman D, Raz S: Surgical treatment of detrusor hyperreflexia, in Raz S (ed): *Female Urology.* Philadelphia, WB Saunders Co, 1983, pp 326–334.

265. Leadbetter GW Jr: Surgical correction of total urinary incontinence. *J Urol* 1964; 91:261.

266. Leadbetter GW Jr: Surgical reconstruction for complete urinary incontinence: A 10- to 22-year followup. *J Urol* 1985; 133:205.

267. Lee L: The clinical use of urocholine in dysfunctions of the bladder. *J Urol* 1949; 62:300.

268. Lennert JB, Mowad JJ: Enuresis: evaluation of perplexing symptom. *Urology* 1979; 13:27–29.

269. Levin RM, High J, Wein AJ: The effect of short term obstruction on urinary bladder function in the rabbit. *J Urol* 1984; 132:789.

270. Levin RM, Malkowicz SB, Wein AJ, et al: Recovery from short term obstruction of the rabbit urinary bladder. *J Urol* 1985; 134:388.

271. Levin R, Staskin D, Wein A: The muscarinic cholinergic binding kinetics of the human urinary bladder. *Neurourol Urodynamics* 1982; 1:221.

272. Levin RM, Jacobson TZD, Wein AJ: Autonomic innervation of rabbit urinary bladder following estrogen administration. *Urology* 1981; 17:449.

273. Levin RM, Staskin DR, Wein AJ: Analysis of the anticholinergic and musculotropic effects of desmethylimipramine on the rabbit urinary bladder. *Urol Res* 1983; 11:259.

274. Lewin RJ, Dilland GU, Porter RW: Extrapyramidal inhibition of the urinary bladder. *Brain Res* 1967; 4:301.

275. Libo LM, Arnold GE, Woodside JR, et al: EMG biofeedback for functional bladder-sphincter dyssynergia: A case study. *Biofeedback Self Regul* 1983; 8:243–253.

276. Light J, Scott F: Bethanechol chloride and the traumatic cord bladder. *J Urol* 1982; 128:85.

277. Light K, Van Blerk PJP: Causes of renal deterioration in patients with meningomyeloceles. *Br J Urol* 1977; 49:257–260.

278. Lim K, Ball AJ, Fenehey RCL: Periurethral Teflon injection: A simple treatment for urinary incontinence. *Br J Urol* 1983; 55:208.

279. Linder A, Leach GE, Raz S: Augmentation cystoplasty in the treatment of neurogenic bladder dysfunction. *J Urol* 1983; 129:491.

280. Lish P, Labbude J, Peters E, et al: Oxybutynin: A musculotropic antispasmodic drug with moderate anticholinergic action. *Arch Int Pharmacodyn* 1965; 156:467.

281. MacGregor R, Diokno A: The alpha adrenergic blocking action of prazosin hydrochloride on the canine urethra. *Invest Urol* 1981; 18:426.

282. Madersbacher H: Striated sphincter dyssynergia. *Neurourol Urodynamics* 1986; 5:307.

283. Madersbacher H, Scott FB: Twelve o'clock sphincterotomy: Technique, indications, results. *Urol Int* 1975; 30:75.

284. Mahony D, Laferte F, Majoney J: Observations on sphincter augmenting effect of imipramine in children with urinary incontinence. *Urology* 1973; 2:317.

285. Maizels M, Firlit CF: Pediatric urodynamics: A clinical comparison of surface versus needle pelvic floor/external sphincter electromyography. *J Urol* 1979; 122:518–522.

286. Maizels M, King LR, Firlit CF: Urodynamic biofeedback: A new approach to treat vesical sphincter dyssynergia. *J Urol* 1979; 122:205.

287. Malizia AA Jr, Reiman HM, Meyers RP, et al: Migration and granulomatous reaction after periurethral injection of polytef. *JAMA* 1984; 251:3277.

288. Mansson W, Sundin T, Gullberg B: Evaluation of a synthetic vasopressin analogue for treatment of nocturia in benign prostatic hypertrophy. *Scand J Urol Nephrol* 1980; 14:139.

289. Mariani AJ, Stern J, Khan AU, et al: Sacral agenesis: An

analysis of 11 cases and review of the literature. *J Urol* 1979; 122:684–686.

290. Marshall S, Marshall HH, Lyon RP: Enuresis: An analysis of various therapeutic approaches. *Pediatrics* 1973; 52:813–817.

291. Marshall VF, Marchetti AA, Krantz KE: The correction of stress incontinence by simple vesicourethral suspension. *Surg Gynecol Obstet* 1949; 88:509.

292. Martin DC, Datta NS, Schweitz B: The occult neurological bladder. *J Urol* 1971; 105:733–738.

293. Mayo ME, Ansell JS: Urodynamic assessment of incontinence after prostatectomy. *J Urol* 1979; 122:60.

294. Mayo ME, Chapman WH, Shurtleff DB: Bladder function in children with meningomyelocele: Comparison of cine-fluoroscopy and urodynamics. *J Urol* 1979; 121:458–461.

295. McGuire E: Reflex urethral instability. *Br J Urol* 1978; 50:200.

296. McGuire E: Electromyographic evaluation of sphincter function and dysfunction. *Urol Clin North Am* 1979; 6:121.

297. McGuire EJ: Urodynamic evaluation after abdominalperineal resection and lumbar intervertebral disk herniation. *Urology* 1975; 6:63–70.

298. McGuire EJ: *Urinary Incontinence.* New York, Grune & Stratton, 1981.

299. McGuire EJ: Clinical evaluation and treatment of neurogenic vesical dysfunction, in Libertino J (ed): *International Perspectives in Urology.* Baltimore, Williams & Wilkins, 1984a, vol 11.

300. McGuire EJ: Detrusor response to outlet obstruction. *World J Urol* 1984b; 2:208.

301. McGuire EJ: The neuropathic urethra, in Mundy AR, Stephenson TP, Wein AJ (eds): *Urodynamics: Principles, Practice and Application.* Edinburgh, Churchill-Livingstone, 1984c, pp 288–296.

302. McGuire EJ: Abdominal procedures for stress incontinence. *Urol Clin North Am* 1985; 12:285.

303. McGuire EJ, Lytton B: The pubovaginal sling in stress urinary incontinence. *J Urol* 1978; 119:82.

304. McGuire EJ, Savastano JA: Urodynamic findings and long term outcome management of patients with multiple sclerosis induced lower urinary tract dysfunction. *J Urol* 1984; 132:713.

305. McGuire EJ, Savastano JA: Stress incontinence and detrusor instability/urge incontinence. *Neurourol Urodynamics* 1985; 4:313.

306. McGuire EJ, Wagner FC Jr: The effects of sacral denervation on bladder and urethral function. *Surg Gynecol Obstet* 1977; 144:343–346.

307. McGuire EJ, Shi-Chun Z, Horwinski ER, et al: Treatment for motor and sensory detrusor instability by electrical stimulation. *J Urol* 1983; 129:78.

308. McGuire EJ, Woodside JR, Borden TA: Upper urinary tract deterioration in patients with myelodysplasia and detrusor hypertonia: A follow-up study. *J Urol* 1983a; 129:823–826.

309. McGuire EJ, Woodside JR, Borden TA, et al: Prognostic value urodynamic testing in myelodysplastic patients. *J Urol* 1983b; 126:205.

310. McGuire EJ, Woodside JR, Borden TA, et al: The prognostic value of urodynamic testing in myelodysplastic patients. *J Urol* 1981; 126:205–209.

311. McLellan FC: *The Neurogenic Bladder.* Springfield, Ill, Charles C Thomas, 1939, pp 57–70, 116–185.

312. Dimethylsulfoxide (DMSO). *Medical Letter* 1980; 22:94.

313. Meirowsky AM, Scheibert CD, Hinchey TR: Studies on the reflex arc in paraplegia. *Int J Neurosurg* 1950; 7:33.

314. Merrill DC: Clinical experience with the Mentor bladder stimulator: II. Meningomyelocele patients. *J Urol* 1974; 112:823.

315. Merrill DC: The treatment of detrusor incontinence by electrical stimulation. *J Urol* 1979; 122:515.

316. Merrill DC, Conway CJ: Clinical experience with the Mentor bladder stimulator: I. Patients with upper motor neuron lesions. *J Urol* 1974; 112:52.

317. Merrill D, Rotta J: A clinical evaluation of detrusor denervation supersensitivity using air cystometry. *J Urol* 1977; 111:27.

318. Merrill DC, Bradley WE, Markland C: Air cystometry: II. A clinical evaluation of normal adults. *J Urol* 1972; 108:85–88.

319. Merrill DC, Conway C, DeWolf W: Urinary incontinence treatment with electrical stimulation of the pelvic floor. *Urology* 1975; 5:67.

320. Messing EM, Stamey TA: Interstitial cystitis: Early diagnosis, pathology and treatment. *Urology* 1978; 12:381.

321. Miller NE: Applications of learning and biofeedback to psychiatry and medicine, in Freedman AM, Kaplan HI, Sadock BJ (eds): *Comprehensive Textbook of Psychiatry.* Baltimore, Williams & Wilkins, 1975, pp 349–365.

322. Miller NE, Dworkin BR: Effects of learning on visceral functions—biofeedback. *N Engl J Med* 1977; 296:1274.

323. Milner G, Hills N: A double-blind assessment of antidepressants in the treatment of 212 enuretic patients. *Med J Aust* 1968; 1:943.

324. Misak SJ, Bunts RC, Ulmer JL, et al: Nerve interruption procedures in the urologic management of paraplegic patients. *J Urol* 1962; 88:392.

325. Mobley D: Phenoxybenzamine in the management of neurogenic vesical dysfunction. *J Urol* 1976; 116:737.

326. Mogg RA, Syme RRA: The results of urinary diversion using the colonic conduit. *Br J Urol* 1969; 41:434.

327. Moisey C, Stephenson T, Brendler C: The urodynamic and subjective results of treatment of detrusor instability with oxybutynin chloride. *Br J Urol* 1980; 52:472.

328. Moisey CU, Stephensen TP, Evans C: A subjective and urodynamic assessment of unilateral bladder neck incision for bladder neck obstruction. *Br J Urol* 1982; 54:114.

329. Moncada S, Flower R, Vane J: Prostaglandins, prostacyclin and thromboxane A$_2$, in Gilman AG, Goodman LS, Gilman A (eds): *The Pharmacological Basis of Therapeutics.* New York, Macmillan, 1980, pp 668–681.

330. Montague D, Stewart B: Urethral pressure profiles before and after Ornade administration in patients with stress incontinence. *J Urol* 1979; 122:198.

331. Mulcahy JJ, Young AB: Percutaneous radiofrequency sacral rhizotomy in treatment of hyperreflexic bladder. *J Urol* 1978; 120:557.

332. Mulholland SG, Yalla SV, Raezer DM, et al: Primary external urethral sphincter hyperkinesia in a boy. *Urology* 1974; 4:577.

333. Muller J, Schulze S: Imipramine cardiotoxicity: An electrocardiographic and hemodynamic study in rabbits. *Acta Pharmacol Toxicol* 1980; 46:191.

334. Mundy AR: Pelvic plexus injury, in Mundy AR, Stephenson TP, Wein AJ (eds): *Urodynamics: Principles, Practice, and Application.* New York, Churchill-Livingstone, 1984, pp 273–277.

335. Mundy AR: The surgical treatment of detrusor instability. *Neurourol Urodynamics* 1985a; 4:357.

336. Mundy AR: The unstable bladder. *Clin Obstet Gynecol* 1985b; 12:431.

337. Mundy AR, Blaivas JG: Non-traumatic neurological disorders, in Mundy AR, Stephenson TP, Wein AJ (eds): *Urodynamics: Principles, Practice and Application.* New York, Churchill-Livingstone, 1984, pp 278–287.

338. Mundy AR, Stephenson TP: The urge syndrome, in Mundy AR, Stephenson TP, Wein AJ (eds): *Urodynamics: Principles, Practice, and Application.* New York: Churchill-Livingstone, 1984, pp 212–228.

339. Mundy AR, Stephenson TP: Clam ileocecocystoplasty for the treatment of refractory urge incontinence. *Br J Urol* 1986; 58.

340. Mundy AR, Borzykowsky M, Saxton HM: Videourodynamic evaluation of neuropathic vesicourethral dysfunction in children. *Br J Urol* 1982; 54:645–649.

341. Murdock M, Sax D, Krane R: Use of dantrolene sodium in external sphincter spasm. *Urology* 1976; 8:133.

342. Murnaghan GF: Neurogenic disorders of the bladder in Parkinsonism. *Br J Urol* 1961; 33:403.

343. Nanninga J, Kaplan P, Lal S: Effect of phentolamine on perineal sphincter spasm. *Urology* 1976; 8:133.

344. Nemir A, Middleton RP: Stress incontinence in young nulliparous women. *Am J Obstet Gynecol* 1954; 68:1166.

345. Norden CW, Friedman EA: Psychogenic urinary retention. *N Engl J Med* 1961; 264:1096.

346. Nordling J, Meyhoff HH: Dissociation of urethral and anal sphincter activity in neurologic bladder dysfunction. *J Urol* 1979; 122:352.

347. Nordling J, Christensen B, Gosling J: Noradrenergic innervation of the human bladder in neurogenic dysfunction. *Urol Int* 1980; 35:188.

348. Nordling J, Meyhoff H, Hald T, et al: Urethral denervation supersensitivity to noradrenaline after radical hysterectomy. *Scand J Urol Nephrol* 1981; 15:21.

349. Nordling J, Meyhoff HH, Christensen NJ: Effects of clonidine on urethral pressure. *Invest Urol* 1979; 16:289.

350. Norlen L, Dahlstrom A, Sundin T, et al: The adrenergic innervation and adrenergic receptor activity of the feline urinary bladder and urethra in the normal state and after hypogastric and/or parasympathetic denervation. *Scand J Urol Nephrol* 1976; 10:177.

351. Norlen L, Sundin T, Waagstein F: Beta-adrenoceptor stimulation of the human urinary bladder in vivo. *Acta Pharmacol Toxicol* 1978; 43:5.

352. Obrink A, Bunne G: The effect of alpha-adrenergic stimulation in stress incontinence. *Scand J Urol Nephrol* 1978; 12:205.

353. Olsson C, Siroky M, Krane R: The phentolamine test in neurogenic bladder dysfunction. *J Urol* 1977; 117:481.

354. Olubadewo J: The effect of imipramine on rat detrusor muscle contractility. *Arch Int Pharmacodyn Ther* 1980; 245:84.

355. Opitz JL: Treatment of voiding dysfunction in spinal cord injured patients: Bladder retraining, in Barrett DM, Wein AJ (eds): *Controversies in Neuro-urology.* London, Churchill-Livingstone, 1984, pp 437–451.

356. Oravisto KJ: Epidemiology of interstitial cystitis. *Ann Chir Gynaecol* 1975; 64:75.

357. Orr JD, Shand JEG, Watters DAK, et al: Ileal conduit urinary diversion in children: An assessment of the long-term results. *Br J Urol* 1981; 53:424.

358. Osborn DE: Prostatodynia: Clinical aspects, in George NJR, Gosling JA (eds): *Sensory Disorders of the Bladder and Urethra.* New York, Springer-Verlag, 1986, pp 139–147.

359. Ouslander IG: Urinary incontinence in the elderly. *West J Med* 1981; 135:482.

360. Palmer J, Worth P, Exton-Smith A: Flunarizine: A once daily therapy for urinary incontinence. *Lancet* 1981; 2:279.

361. Parsons CL, Schmidt JP, Pollen JJ: Successful treatment of interstitial cystitis with sodium pentosonpolysulfate. *J Urol* 1983; 130:51.

362. Parsons K, Turton M: Urethral supersensitivity and occult urethral neuropathy. *Br J Urol* 1980; 52:131.

363. Pavlakis AJ, Siroky MB, Goldstein I, et al: Neurourologic findings in Parkinson's disease. *J Urol* 1983; 129:80.

364. Pederson E, Grynderup V: Clinical pharmacology of the neurogenic bladder. *Acta Neurol Scand* 1966; 42(suppl):111.

365. Pengelly AW, Booth CM: A prospective trial of bladder training as treatment for detrusor instability. *Br J Urol* 1980; 52:463.

366. Pengelly AW, Stephenson TP, Mulroy EJG, et al: Results of prolonged bladder distention as treatment for detrusor instability. *Br J Urol* 1978; 50:243.

367. Pereyra AJ: A surgical procedure for the correction of stress incontinence in women. *West J Surg Obstet Gynecol* 1959; 67:223.

368. Perkash I: Intermittent catheterization and bladder rehabilitation in spinal cord injury patients. *J Urol* 1975; 114:230.

369. Perlberg S, Caine M: Adrenergic response of bladder muscle in prostatic obstruction. *Urology* 1982; 20:524.

370. Persky L, Tucker A, Izant RJ Jr: Urological complications of correction of imperforate anus. *J Urol* 1974; 111:415–418.

371. Petersen KE, Andersen OO, Hansen T: Mode of action and relative value of imipramine and similar drugs in the treatment of nocturnal enuresis. *Eur J Clin Pharmacol* 1974; 7:187.

372. Petti T, Law W: Abrupt cessation of high dose imipramine treatment in children. *JAMA* 1981; 246:768.

373. Philp N, Thomas D, Clarke S: Drug effects on the voiding cystometrogram: A comparison of oral bethanechol and carbachol. *Br J Urol* 1980; 52:484.

374. Pitts WR Jr, Muecke EC: A 20-year experience with ileal conduits: The fate of the kidneys. *J Urol* 1979; 122:154.

375. Plevnik S, Janež J: Maximal electrical stimulation for urinary incontinence: Report of 98 cases. *Urology* 1979; 14:638.

376. Politano VA: Periurethral polytetrafluoroethylene injection for urinary incontinence. *J Urol* 1982; 127:439.

377. Politano VA, Small MP, Harper JM, et al: Periurethral Teflon injection for urinary incontinence. *J Urol* 1974; 111:180.

378. Powell PH, Shepard AM, Lewis P, et al: The accuracy of clinical diagnosis assessed urodynamically, in Zinner NR, Sterling AM (eds): *Female Incontinence.* New York, Alan R Liss Inc, 1981, p 201.

379. Quesada EM, Scott FB, Cardus D: Functional classification of neurogenic bladder dysfunction. *Arch Phys Med Rehabil* 1968; 49:692.

380. Raezer D, Wein A, Jacobowitz D, et al: Autonomic innervation of canine urinary bladder: Cholinergic and adrenergic contributions and interaction of sympathetic and parasympathetic systems in bladder function. *Urology* 1973; 2:211.

381. Raezer DM, Benson GS, Wein AJ, et al: The functional approach to the management of the pediatric neuropathic bladder: A clinical study. *J Urol* 1977; 117:649.

382. Ramsden PD, Hindmarsh JR, Price DA, et al: DDAVP for adult enuresis—a preliminary report. *Br J Urol* 1982; 54:256.

383. Raz S, Caine M: Adrenergic receptors in the female canine urethra. *Invest Urol* 1971; 9:319.

384. Raz S, Smith R: External sphincter spasticity syndrome in female patients. *J Urol* 1976; 115:443.

385. Raz S, Caine M, Ziegler M: The vascular component in the production of urethral pressure. *J Urol* 1972; 108:93.

386. Raz S, Kaufman J, Ellison G, et al: Methyldopa in treatment of neurogenic bladder disorders. *Urology* 1977; 9:188.

387. Raz S, Ziegler M, Caine M: The role of female hormones in stress incontinence. Proceedings of 16th Congress of Societe International d'Urologie, Amsterdam. Paris, Doin, vol 2, 1973, pp 397–402.

388. Redman JF, Seibert JJ: The uroradiographic evaluation of the enuretic child. *J Urol* 1979; 122:799–801.

389. Rees D, Ransley P: Eskornade in the treatment of diurnal incontinence in children. *Br J Urol* 1980; 52:476.

390. Rees DLP, Whitfield HN, Islam AKM, et al: Urodynamic findings in adult females with frequency and dysuria. *Br J Urol* 1967; 47:853.

391. Reid GF, Fitzpatrick JM, Worth PHL: The treatment of patients with urinary incontinence post prostatectomy. *Br J Urol* 1980; 52:532.

392. Resnick MI, King LR: Use of the Leadbetter anti-incontinence procedure in children. *J Urol* 1976; 116:366–368.

393. Resnick NM, Yalla SV: Management of urinary incontinence in the elderly. *N Engl J Med* 1985; 313:800.

394. Richelson E: Antimuscarinic and other receptor blocking properties of antidepressants. *Mayo Clin Proc* 1983; 58:40.

395. Roberts JBM, Smith PJB, Dunn M, et al: Vesicoplication in the management of chronic urinary retention. *Br J Urol* 1979; 51:532.

396. Robinson AG: DDAVP in the treatment of central diabetes insipidus. *N Engl J Med* 1976; 294:507.

397. Rohner T, Hannigan J, Sanford E: Altered in vitro adrenergic responses of dog detrusor muscle after chronic bladder outlet obstruction. *Urology* 1978; 11:357.

398. Rosen M: A simple artificial implantable sphincter. *Br J Urol* 1976; 48:675–680.

399. Ross JC, Gibbon NOK, Damanski M: Division of the external urethral sphincter in the treatment of the paraplegic bladder: A preliminary report on a new procedure. *Br J Urol* 1958; 30:204.

400. Rossier A, Fam B, Lee I, et al: Role of striated and smooth muscle components in the urethral pressure profile in traumatic neurogenic bladders: A neuropharmacological and urodynamic study: A preliminary report. *J Urol* 1982; 128:529.

401. Roussan M, Abramson A, Levine S: Bladder training: Its role in evaluating the effect of an antispasticity drug on voiding in patients with neurogenic bladder. *Arch Phys Med Rehabil* 1975; 56:463.

402. Rowan EL: Psychophysiologic disorders of micturition. *Coll Health* 1975; 23:251.

403. Ruch TC, Tang PC: The higher control of the bladder, in Boyarsky S (ed): *The Neurogenic Bladder.* Baltimore, Williams & Wilkins, 1969, pp 34–45.

404. Rud T: The effects of estrogens and gestagens on the urethral pressure profile in urinary continent and stress incontinent women. *Acta Obstet Gynecol Scand* 1980a; 59:265.

405. Rud T: Urethral pressure profile in continent women from childhood to old age. *Acta Obstet Gynecol Scand* 1980b; 59:331.

406. Rud T, Andersson K, Boye N, et al: Terodiline inhibition of human bladder contraction: Effects in vitro and in women with unstable bladder. *Acta Pharmacol Toxicol* 1980; 46:31.

407. Rudy DC, Woodside JR, Crawford ED: Urodynamic evaluation of incontinence in patients undergoing modified Campbell radical prostatectomy: A prospective study. *J Urol* 1984; 132:708.

408. Salmon UJ, Walter RI, Geist SH: The use of estrogen in the treatment of dysuria and incontinence in postmenopausal women. *Am J Obstet Gynecol* 1941; 42:845.

409. Santulli TV, Kiesewetter WB, Bill AH Jr: Anorectal anomalies: A suggested international classification. *J Pediatr Surg* 1970; 5:281–287.

410. Sarnat HB, Case ME, Graviss R: Sacral agenesis: Neurologic and neuropathologic features. *Neurology* 1976; 26:1124–1129.

411. Schmidt RA: Neural prostheses and bladder control. *Eng Med Biol* 1983; 2:31.

412. Schmidt RA: The urethral syndrome. *Clin Obstet Gynecol* 1985; 12:477.

413. Schoenberg HC, Gutrich JM: Management of vesical dysfunction in multiple sclerosis. *Urology* 1980; 16:444.

414. Schreiter F, Fuchs P, Stockamp K: Estrogenic sensitiv-

ity of alpha receptors in the urethral musculature. *Urol Int* 1976; 31:13.

415. Schulman CC, Simon J, Wesper E, et al: Endoscopic injection of Teflon for female urinary incontinence. *Eur Urol* 1983; 9:246.

416. Schwartz GE: Biofeedback and the behavioral treatment of disorders of disregulation. *Yale J Biol Med* 1979; 52:581.

417. Schwartz GR, Jeffs RD: Ileal conduit urinary diversion in children: Computer analysis of follow up from 2 to 16 years. *J Urol* 1975; 114:285.

418. Scott FB, Bradley W, Timm GW: Treatment of urinary incontinence by an implantable prosthetic urinary sphincter. *J Urol* 1974; 112:75–80.

419. Scott M, Morrow J: Phenoxybenzamine in neurogenic bladder dysfunction after spinal cord injury: I. Voiding dysfunction. *J Urol* 1978; 119:480.

420. Segura JW, Opitz JL, Greene CF: Prostatosis, prostatitis, or pelvic floor tension myalgia? *J Urol* 1979; 122:168.

421. Sehn JT: Anatomic effect of distention therapy in unstable bladder: New approach. *Urology* 1978; 11:581.

422. Sibley GNA: An experimental model of detrusor instability in the obstructed pig. *Br J Urol* 1985; 57:292.

423. *Sixth Report on the Standardization of Terminology of Lower Urinary Tract Function.* New York, The International Continence Society Committee on Standardization of Terminology, May 1985.

424. Sjogren C, Andersson K: Effects of cholinoceptor blocking drugs, adrenoceptor stimulants and calcium antagonists on the transmurally stimulated guinea pig urinary bladder in vitro and in vivo. *Acta Pharmacol Toxicol* 1979; 44:228.

425. Smey P, Firlit CF, King LR: Voiding pattern abnormalities in normal children: Results of pharmacologic manipulation. *J Urol* 1978; 120:574–577.

426. Smith DR: Psychosomatic 'cystitis.' *J Urol* 1962; 87:359.

427. Smith ED: Congenital sacral anomalies in children. *Aust NZ J Surg* 1959; 29:165–176.

428. Smith ED: *Spina Bifida and the Total Care of Spinal Myelomeningocele.* Springfield, Ill, Charles C Thomas Publisher, 1965, pp 92–123.

429. Smith JC: The place of prolonged bladder distention in the treatment of bladder instability and other disorders. *Br J Urol* 1981; 53:283.

430. Smith RB, Van Cangh P, Skinner DG, et al: Augmentation enterocystoplasty: A critical review. *J Urol* 1977; 118:35.

431. Snyder HM, Caldamone AA, Wein AJ, et al: The Hinman syndrome—alternatives for treatment. Presented at the Annual Meeting of the American Urological Association. Kansas City, May 16–20, 1982.

432. Sonda L, Gershon C, Diokno A: Further observations on the cystometric and uroflowmetric effects of bethanechol chloride on the human bladder. *J Urol* 1979; 122:775.

433. Sporer A, Leyson J, Martin B: Effects of bethanechol chloride on the external urethral sphincter in spinal cord injury patients. *J Urol* 1978; 120:62.

434. Spyraki C, Fibiger H: Functional evidence for subsensitivity of noradrenergic α_2 receptors after chronic desipramine treatment. *Life Sci* 1980; 27:1863.

435. Stamey TA: Endoscopic suspension of the vesical neck, in Starlon SL, Tanagho EA (eds): *Surgery of Female Incontinence.* New York, Springer-Verlag, 1986, pp 115–132.

436. Stanton S: Disease of the urinary system: Drugs acting on the bladder and urethra. *Br Med J* 1978; 1:1607.

437. Stanton SL: Stress incontinence: Why and how operations work. *Clin Obstet Gynecol* 1985; 12:369.

438. Stanton SL: Colposuspension, in Stanton SL, Tanagho EA (eds): *Surgery of Female Incontinence.* New York, Springer-Verlag, 1986, pp 95–103.

439. Stanton SL, Cardozo L, Williams JE, et al: Clinical and urodynamic features of failed incontinence surgery in the female. *J Obstet Gynecol* 1978; 51:515.

440. Stark G: Pudendal neurectomy in management of neurogenic bladder in myelomeningocele. *Arch Dis Child* 1969; 44:698.

441. Staskin DR, Zimmern PC, Hadley HR, et al: Pathophysiology of stress incontinence. *Clin Obstet Gynecol* 1985; 12:357.

442. Stewart B, Banowsky L, Montague D: Stress incontinence: Conservative therapy with sympathomimetic drugs. *J Urol* 1976; 115:558.

443. Stewart BH, Shirley SW: Further experience with intravesical DMSO in the treatment of interstitial cystitis. *J Urol* 1976; 116:36.

444. Stockamp K: Treatment with phenoxybenzamine of upper urinary tract complications caused by intravesical obstruction. *J Urol* 1975; 113:128.

445. Sugar EC, Firlit CF: Urodynamic biofeedback: A new therapeutic approach for childhood incontinence/infection (vesical voluntary sphincter dyssynergia). *J Urol* 1982; 128:1253–1257.

446. Sundin T, Dahlstrom A: The sympathetic innervation of the urinary bladder and urethra in the normal state and after parasympathetic denervation at the spinal root level: An experimental study in cats. *Scand J Urol Nephrol* 1973; 7:131–149.

447. Sundin T, Dahlstrom A, Norlen L, et al: The sympathetic innervation and adrenoreceptor function of the human lower urinary tract in the normal state and after parasympathetic denervation. *Invest Urol* 1977; 14:322–328.

448. Tanagho EA: Membrane and microtransducer catheters: Their effectiveness for profilometry of the lower urinary tracts. *Urol Clin North Am* 1979; 6:110.

449. Tanagho EA: Bladder neck reconstruction for total urinary incontinence: Ten years of experience. *J Urol* 1981; 1215:321.

450. Tanagho EA: Urethral pressure profile: Membrane catheter, in Barrett DM, Wein AJ (eds): *Controversies in Neuro-urology.* New York, Churchill-Livingstone, 1984, p 55.

451. Tanagho EA: Electrostimulation, in Stanton SL, Tanagho EA (eds): *Surgery of Female Incontinence.* New York, Springer-Verlag, 1986, pp 251–257.

452. Tanagho EA, Smith DR, Meyers FH, et al: Mechanism

of urinary continence: II. Technique for surgical correction of incontinence. *J Urol* 1969; 101:305.

453. Tank ES: The urologic complications of imperforate anus and cloacal dysgenesis, in Harrison JH, Gittes RF, Perlmutter AD, et al: *Campbell's Urology*, ed 4. Philadelphia, WB Saunders Co, 1979, pp 1889–1900.

454. Taylor MC, Bates CP: A double-blind crossover trial of baclofen: A new treatment for the unstable bladder syndrome. *Br J Urol* 1979; 51:505.

455. Taylor P: Cholinergic agonists, in Gilman AG, Goodman LS, Gilman A (eds): *The Pharmacological Basis of Therapeutics*. New York, Macmillan, 1980, pp 91–99.

456. Teague CT, Merrill DC: Electric pelvic floor stimulation: Mechanisms of action. *Invest Urol* 1977; 15:65.

457. Thomas DG: Spinal cord injury, in Munday AR, Stephenson TP, Wein AJ (eds): *Urodynamics: Principles, Practice and Application*. New York, Churchill-Livingstone, 1984, pp 260–272.

458. Thomas GG, Zachary RB, Lister J: Serial follow-up studies of bladder pressure in spina bifida infants. *J Pediatr Surg* 1974; 9:471–476.

459. Thomas TM, Plymat KR, Blannen J, et al: Prevalence of urinary incontinence. *Br Med J* 1980; 281:1243.

460. Thompson I, Lauvetz R: Oxybutynin in bladder spasm, neurogenic bladder and enuresis. *Urology* 1976; 8:452.

461. Thor K, Rappolo J, deGroat W: Naloxone-induced micturition in unanesthetized paraplegic cats. *J Urol* 1981; 129:202.

462. Torrens MJ: The role of denervation in the treatment of detrusor instability. *Neurourol Urodynamics* 1985; 4:353.

463. Turner-Warwick R: Bladder outflow obstruction in the male, in Mundy AR, Stephenson TP, Wein AJ (eds): *Urodynamics: Principles, Practice and Application*. Edinburgh, Churchill-Livingstone, 1984, pp 183–204.

464. Turner-Warwick RT, Ashken MH: The functional results of partial, subtotal and total cystoplasty with special reference to ureterocecocystoplasty, selective sphincterotomy and cystocystoplasty. *Br J Urol* 1967; 39:3.

465. Twiddy D, Downie J, Awad S: Response of the bladder to bethanechol after acute spinal cord transection in cats. *J Pharmacol Exp Ther* 1980; 215:500.

466. Ullery JC: *Stress Incontinence in the Female*. New York, Grune & Stratton Inc, 1953.

467. Ulmsten U, Henrickson L, Iosif S: The unstable female urethra. *Am J Obstet Gynecol* 1982; 144:93.

468. Ursillo RC: Rationale for drug therapy in bladder dysfunction, in Boyarsky S (ed): *The Neurogenic Bladder*. Baltimore, Williams & Wilkins, 1967, pp 187–190.

469. Uvelius B, Mattiasson A: Collagen content in the rat urinary bladder subjected to infravesical outflow obstruction. *J Urol* 1984; 132:587.

470. Uvelius B, Persson L, Matthiasson A: Smooth muscle cell hypertrophy and hyperplasia in the rat detrusor after short term infravesical outflow obstruction. *J Urol* 1984; 131:173.

471. Vaidyanathan S, Rao M, Bapna B, et al: Beta adrenergic activity in human proximal urethra: A study with terbutaline. *J Urol* 1980; 124:869.

472. Vaidyanathan S, Rao M, Chary K, et al: Enhancement of detrusor reflex activity by naloxone in patients with chronic neurogenic bladder dysfunction. Preliminary report. *J Urol* 1981; 126:500.

473. Vaidyanathan S, Rao M, Mapa M, et al: Study of instillation of 15(S)–15-methyl prostaglandin $F_{2\alpha}$ in patients with neurogenic bladder dysfunction. *J Urol* 1981; 126:81.

474. Van Gool J, Tanagho EA: External sphincter activity and recurrent urinary tract infection in girls. *Urology* 1977; 10:348.

475. Vereecken RL, Das J: Urethral instability: Related to stress incontinence? *J Urol* 1985; 134:698.

476. Veith RC, Rashkind MA, Caldwell JH, et al: Cardiovascular effects of tricyclic antidepressants in depressed patients with chronic heart disease. *N Engl J Med* 1982; 306:954.

477. Vorstman B, Lockhart J, Kaufman MR, et al: Polytetrafluoroethylene injection for urinary incontinence in children. *J Urol* 1985; 133:248.

478. Watkins RH: The bladder function in spinal injury. *Br J Surg* 1936; 23:734.

479. Wear JB Jr, Wear RB, Cleeland C: Biofeedback in urology using urodynamics: Preliminary observations. *J Urol* 1979; 121:464.

480. Webster GD: Urodynamic studies, in Resnick MI, Older RA (eds): *Diagnosis of Genitourinary Disease*. New York, Thieme-Stratton, 1982, p 173.

481. Webster GD, Koefoot RB Jr, Sihelnik S: Urodynamic abnormalities in neurologically normal children with micturition dysfunction. *J Urol* 1984; 132:74–77.

482. Webster GD, Older RA: Video urodynamics. *Urology* 1980; 16:106–114.

483. Webster GD, Sihelnik SA, Stone AR: Female urinary incontinence: The incidence, identification and characteristics of detrusor instability. *Neurourol Urodynamics* 1984; 3:235.

484. Wein A, Benson G, Jacobowitz D: Lack of evidence for adrenergic innervation of the external urethral sphincter. *J Urol* 1979; 121:324.

485. Wein A, Malloy T, Shofer F, et al: The effects of bethanechol chloride on urodynamic parameters in normal women and in women with significant residual urine volumes. *J Urol* 1980a; 124:397.

486. Wein A, Raezer D, Malloy T: Failure of the bethanechol supersensitivity test to predict improved voiding after subcutaneous bethanechol administration. *J Urol* 1980b; 123:202.

487. Wein AJ: Pharmacologic approaches to the management of neurogenic bladder dysfunction. *Urology* 1979; 18:17.

488. Wein AJ: Pharmacology of the bladder and urethra, in Stanton SL, Tanagho EA (eds): *Surgery of Female Incontinence*. New York, Springer-Verlag, 1980, p 185.

489. Wein AJ: Classification of neurogenic voiding dysfunction. *J Urol* 1981; 125:605.

490. Wein AJ: Classification of voiding dysfunction: A simple approach, in Barrett DM, Wein AJ (eds): *Controversies in Neuro-urology*. New York, Churchill-Livingstone, 1984, pp 239–250.

491. Wein AJ: Specific methods of pharmacologic treatment, in Stanton SL, Tanagho EA (eds): *Surgery of Female Incontinence*. New York, Springer-Verlag, 1986, pp 229–250.

492. Wein AJ, Barrett DM: Etiologic possibilities for increased pelvic floor electromyography activity during bladder filling. *J Urol* 1982; 127:949.

493. Wein AJ, Raezer DM, Benson GS: Management of neurogenic bladder dysfunction in the adult. *Urology* 1976; 8:432.

494. Weiner N: Atropine, scopolamine and related antimuscarinic drugs, in Gilman AG, Goodman LS, Gilman A (eds): *The Pharmacological Basis of Therapeutics*. New York, Macmillan, 1980a, pp 120–137.

495. Weiner N: Drugs that inhibit adrenergic nerves and block adrenergic receptors, in Gilman AG, Goodman LS, Gilman A (eds): *The Pharmacological Basis of Therapeutics*. New York, Macmillan, 1980b, pp 176–210.

496. Weiner N: Norepinephrine, epinephrine and the sympathomimetic amines, in Gilman A, Goodman LS, Gilman A (eds): *The Pharmacological Basis of Therapeutics*. New York, Macmillan, 1980, pp 138–175.

497. White RI, Klauber GT: Sacral agenesis: Analysis of 22 cases. *Urology* 1976; 8:521–525.

498. Whiteside CG, Arnold EP: Persistent primary enuresis: A urodynamic assessment. *Br Med J* 1975; 1:364–367.

499. Whitfield H, Doyle P, Mayo M, et al: The effect of adrenergic blocking drugs on outflow resistance. *Br J Urol* 1976; 47:823.

500. Wiener ES, Kiesewetter WB: Urologic abnormalities associated with imperforate anus. *J Pediatr Surg* 1973; 8:151–157.

501. Weir J, Jaques PF: The large capacity bladder: A urodynamic study. *Urology* 1974; 5:544.

502. Williams DI, Nixon HH: Agenesis of the sacrum. *Surg Gynecol Obstet* 1957; 105:84–88.

503. Williams DI, Hirst G, Doyle D: The occult neuropathic bladder. *J Pediatr Surg* 1974; 9:35–41.

504. Williams E: Discussion on stress incontinence. *Proc R Soc Med* 1947; 40:361.

505. Williams ME, Panvill FC: Urinary incontinence in the elderly: Physiology, pathophysiology, diagnosis and treatment. *Ann Intern Med* 1982; 97:895.

506. Wolin LH: Stress incontinence in young healthy nulliparous female subjects. *J Urol* 1969; 101:545.

507. Woodside JR, McGuire EJ: Detrusor hypertonicity as a late complication of a Wertheim hysterectomy. *J Urol* 1982; 127:1143–1145.

508. Worth PHL, Turner-Warwick R: The treatment of interstitial cystitis by cystolysis with observations on cystoplasty. *Br J Urol* 1973; 95:65.

509. Yalla S, Blunt K, Fam B, et al: Detrusor-sphincter dyssynergia. *J Urol* 1977; 118:1026.

510. Yalla SV, DiBenedello M, Blunt KJ, et al: Urethral striated sphincter responses to electro-bulbocavernosus stimulation. *J Urol* 1978; 119:406.

511. Yalla SV, Fam BA, Gagilondo FB, et al: Anteromedian external urethral sphincterotomy: Technique, rationale and results in 32 patients. *Paraplegia* 1977; 13:247.

512. Yalla SV, Kirsth L, Kearney G, et al: Postprostatectomy incontinence: Urodynamic assessment. *Neurourol Urodynamics* 1982; 1:77.

513. Yalla SV, Sharma GV, Barsamian EM: Micturitional static urethral pressure profile: A method of recording urethral pressure profile during voiding and the implications. *J Urol* 1980; 124:649.

514. Yarnell JWG, Voyle GJ, Richards CJ, et al: The prevalence and severity of urinary incontinence in women. *J Epidemiol Community Health* 1981; 35:71.

515. Yeates WK: Bladder function in normal micturition. *Clin Dev Med* 1973; 48/49:28–36.

516. Young HH: Suture of the urethral and vesical sphincters for the cure of incontinence of urine with a report of a case. *Trans South Surg Gynecol Assoc* 1907; 20:210.

517. Zinner NR, Sterling AM, Ritter RC: Role of inner urethral softness in urinary continence. *Urology* 1980; 16:115.

Chapter 29

Inflammatory Diseases of the Bladder

DAVID T. UEHLING, M.D.

Inflammatory diseases of the bladder are among the most common conditions afflicting the genitourinary tract. These inflammatory conditions may have a specific infectious etiology or be due to chemical, radiation, irritative, or unknown factors. Symptoms of frequency, urgency, and dysuria are the hallmarks of bladder inflammation, but other conditions such as urethritis and prostatitis may share these symptoms. The differential diagnosis of dysuria relates to many conditions. A useful algorithim for the diagnosis and treatment of dysuria has been described by Bruskewitz (1983). Diagnosis and optimal treatment of inflammatory diseases of the bladder require an understanding of the underlying pathophysiology and an ability to differentiate the specific infectious and noninfectious causes. Since bacterial inflammations of the bladder are so common in clinical practice, a review of current concepts in the pathophysiology of bacterial cystitis is a logical starting point for the understanding of bladder inflammations.

The pathology of bladder inflammation follows that of inflammation in general, with some special features being characteristic of the specific types (Pugh, 1985). Acute cystitis is characterized by an inflammatory cell infiltrate in the lamina propria of the bladder and mucosal congestion and edema. In more severe cases, the epithelium becomes ulcerated or hyperplastic, the inflammatory response extends into muscle, and the blood vessels become thick walled. In chronic conditions, extensive bladder fibrosis occurs. Acute cystitis often is characterized as hemorrhagic because of the vascularity of the bladder wall and the hematuria that so commonly is a symptom. Diabetics with infections due to gas-forming organisms may have a more severe type of emphysematous inflammation in the bladder. This emphysematous cystitis manifests as a gas collection within the bladder lumen and is particularly serious in the presence of outlet obstruction (LeDuc et al., 1985).

PATHOPHYSIOLOGY OF BACTERIAL INFLAMMATIONS OF THE BLADDER

The bladder in some patients is extremely susceptible to bacterial infections. This most typically occurs when the storage and emptying function is deranged, but it also occurs without a recognizable functional or anatomical abnormality. One obvious factor for the frequency of bladder infection is proximity to the normally colonized mucosal surfaces of the urethra so that frequent infections might be expected as they are in other similar areas (lungs, mouth, eyes). However, in nonsusceptible individuals, it is difficult to establish a bacterial bladder infection. Evidence for this is the rapid clearing, or washout, of *Escherichia coli* placed in the bladder of experimental subjects (Cox and Hinman, 1961). This rapid clearing of experimentally induced infection without antibiotic treatment and similar observations in patients with rapid resolution of naturally occurring bladder infections has been termed the "bladder defense mechanism" and may be thought of as several factors working together in the bladder. The most important factor is the coordinated and efficient muscle contractions that normally empty the bladder. Urine itself acts variably to inhibit or promote growth of bacteria in the bladder; urine can have strong antibacterial activity when its pH, osmolality, etc., are nonphysiologic for bacteria growth (Uehling and Balish, 1977). The transitional epithelium of the bladder is covered by a mucin layer of a mucopolysaccharide termed "glycosaminoglycans." When the mucin is experimentally denuded, bacteria adhere in greater numbers to the bladder epithelium (Parsons, 1982). Whether a deficient mucin layer plays an important clinical role in susceptibility to inflammation remains an open question. Urinary antibody against infecting bacteria is known to increase after urinary tract infection (UTI) (Uehling and Steihm, 1971). In vitro studies suggest that this antibody, which is mainly IgA and IgG, exerts a protective

963

effect by coating the infecting bacteria and blocking their binding sites for epithelium (Svanborg-Eden et al., 1976). Locally secreted immunoglobulins in the bladder can be increased experimentally, and this increase is associated with faster clearance of induced infection (Jensen et al., 1984; Uehling et al., 1982). Some patients may be susceptible to bladder infections because bacteria are able to attach more avidly to their bladder epithelium. This susceptibility for bacteria adherence is thought to be controlled in part by genetic factors including postulated differences in density of receptors for bacterial attachment that the individual carries on epithelial cells (Källenius and Winberg, 1978; Leffler and Svanborg-Eden, 1981; Schaeffer et al., 1981). Evidence for this is the seemingly greater incidence of UTI in persons with certain blood group antigens. In one group of 27 infection-prone women, 85% were found to have the P_2 blood group antigens, whereas only 21% would be expected in the general population (Mulholland et al., 1984). Bacterial "soiling," or colonization of the bladder, probably occurs by retrograde flow within the colonized urethra or by expression of urethral bacteria into the bladder during coitus (Bran et al., 1972; Chenoweth and Clawater, 1960). Whether these bacteria attach firmly to the bladder wall or are washed out determines whether tissue infection occurs. Use of a vaginal diaphragm contraceptive has been labeled an "accomplice" in recurrent UTI, purportedly by elevating the bladder neck and altering urinary flow (Gillespie, 1984).

In addition to the host's bladder defenses, initiation of bladder inflammation also depends on the virulence or pathogenicity of the bacteria. The virulence of *E. coli* in the urinary tract is related to the ability of the organisms to adhere to the epithelial surfaces. This is mediated most importantly by protein filaments on bacterial surfaces called fimbriae or pili (Leffler and Svanborg-Eden, 1981). Just as the O, H, and K antigen classification of *E. coli* is complex (Kaijser et al., 1977), so too is the fimbrial system. Type 1 fimbriae are thought to bind to mannoside residues on bladder epithelial cells or mucin; thus, *E. coli* carrying type 1 fimbriae commonly are associated with bladder infections (Kunin and McCormack, 1966; Orskov et al., 1980). *E. coli* which are likely to cause nonobstructive pyelonephritis have another type of fimbriae termed "P-fimbriae." P-fimbriae were present in 91% of urinary strains cultured from one series of children with acute pyelonephritis (Huovinen, 1984). In a group of adult women with *E. coli* UTIs, 57% with pyelonephritis but only 19% with cystitis were infected with P-fimbriated *E. coli* (Latham and Stamm, 1984). P-fimbriae derive their virulence from specific recognition and binding to urothelial cell glycosphingolipids, which correlate with

patients' carrying of certain P blood group antigens (Svensen et al., 1982).

In the past decade, investigations into the inception and progression of UTIs have focused on the process of bacterial adherence to urothelial surfaces as the initiating, disease-producing event. What is now known about the interactions between bacterial virulence factors and host factors has been learned by studies of bacteria interacting with patients' voided epithelial or scraped periurethral cells (in vitro tests) or by in vivo animal experiments (rat, rabbit, nonhuman primates) (Källenius and Winberg, 1978; Schaeffer et al., 1981; Svanborg-Eden et al., 1976; Uehling et al., 1982). Both in vitro and in vivo studies have conceptual advantages and limitations, and both have contributed to the current understanding of the role of bacterial adherence in the pathophysiology of inflammatory diseases of the bladder. Both also have the potential for leading to new methods of treatment for UTI, since they may reveal ways to prevent bacterial adherence to the mucosal surfaces of the urinary tract.

Another study used broadly in biology to study epithelial surfaces is scanning electron microscopy. From such studies of the bladder after experimentally induced *E. coli* infections in the rat, a hypothesis relating to the pathophysiology of bladder inflammation has been developed (Balish et al., 1982). The sequence of early changes is as follows: (1) adherence of bacteria to the surface mucin layer; (2) desquamation, or lifting off, of the surface mucin; and (3) swelling and desquamation of the superficial cell layer (Fig 29–1). The next changes seen are bacteria adhering in greater numbers to the subsurface epithelial cell layers after the desquamation of the superficial cells and mucin layer (Fig 29–2). By this process, bladder colonization by pathogenic bacteria may progress to a tissue infection. The bladder mucin layer appears to be a relative but not absolute barrier to bacterial adherence. Bacteria later appear to become enmeshed in the desquamating mucin strands. This process may facilitate washing out of bacteria from the bladder and may be another way in which the surface mucin layer exerts a protective effect. During the recovery phase, a reversal of these pathologic changes can also be seen. Scanning electron microscopic studies are subjective, and this hypothesis for the pathophysiology of bladder inflammation will require further objective studies for confirmation.

BACTERIAL CYSTITIS

The bacteria that cause cystitis characteristically come from the Enterobacteriaceae family (Stamey, 1980). *E. coli* is the predominant organism in patients with uncomplicated cystitis. Among one study of 200 adult women with uncomplicated cystitis, 182 (87%)

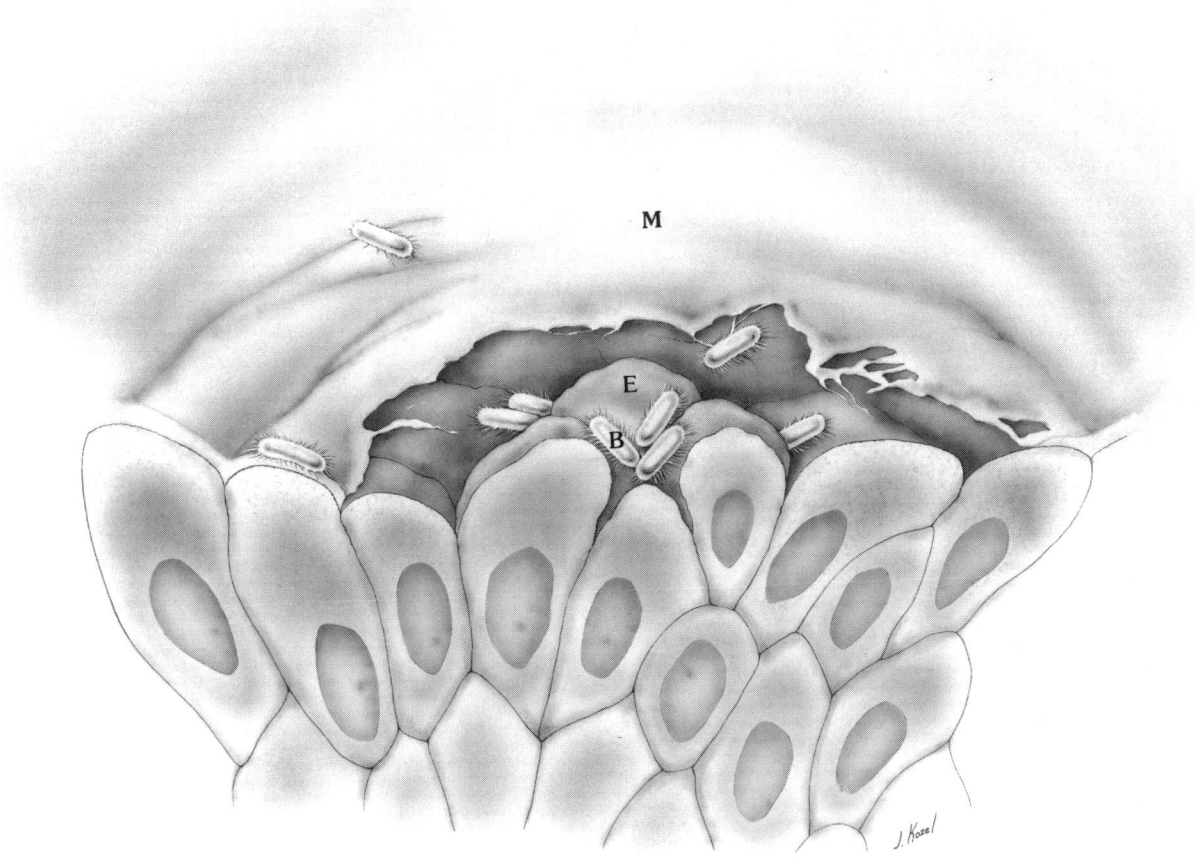

FIG 29–1.
Adherence of bacteria to bladder epithelium. Artist's rendering based on scanning electron microscopic studies of experimentally induced *E. coli* cystitis in the rat. Pathogenic bacteria *(B)* first adhere to surface mucin layer *(M)* and cause swelling and desquamation of underlying transitional epithelial cells *(E)*.

were due to *E. coli*, 9 (4%) to *Staphylococcus saprophyticus*, 7 (3%) to *Proteus mirabilis*, and 6 (3%) to *Klebsiella* (Ronald, 1984). In children with asymptomatic bacteriuria, *E. coli* is usually the infecting organism (Lindberg et al., 1975). Strict anaerobes seldom cause cystitis despite their overwhelming predominance in the feces. *Staphylococcus epidermidis* is being recognized more frequently as a cause of nosocomial UTI, especially in men requiring urethral catheters where the organisms may be resistant to many antibiotics and where the UTI may lead to bacteremia despite being present in the urinary tract in relatively low numbers (Arpi and Renneberg, 1984). Mixed bacterial cultures of urine usually represent improper collection of the urine specimen, and the culture should be repeated instead of treating all of the mixed flora (Bartlett and Treiber, 1984).

While it has long been recognized that *E. coli* of a few common O antigen serotypes are the causative organisms in most UTIs (Kaijser et al., 1977), recent interest has focused on other bacterial virulence factors, especially fimbriae. Bacterial agglutination of mammalian erythrocytes is used in vitro as a marker for what is thought to be occurring in vivo. Bacterial agglutination of guinea pig erythrocytes that can be inhibited by D-mannose has been termed "mannose-sensitive" (MS) and is caused by fimbriae called type 1 (Källenius et al., 1969). Agglutination not inhibited by D-mannose is described as mannose-resistant (MR), and *E. coli* with P-fimbriae are characteristically MR. Patients with uncomplicated cystitis typically are infected with MS *E. coli*, possibly because type 1 fimbriae may promote attachment of the bacteria to bladder mucin or allow trapping by uromucoid (Latham and Stamm, 1984; Orskov et al., 1980). Even low urinary counts of P-fimbriated *E. coli* should raise the suspicions for the presence of pyelonephritis, even though the symptoms are primarily those of lower tract infection (Bollgren et al., 1984). In the current literature, the MS or MR characterization of *E. coli* has largely replaced the previously used O antigen typing in describing the *E. coli* that cause UTIs.

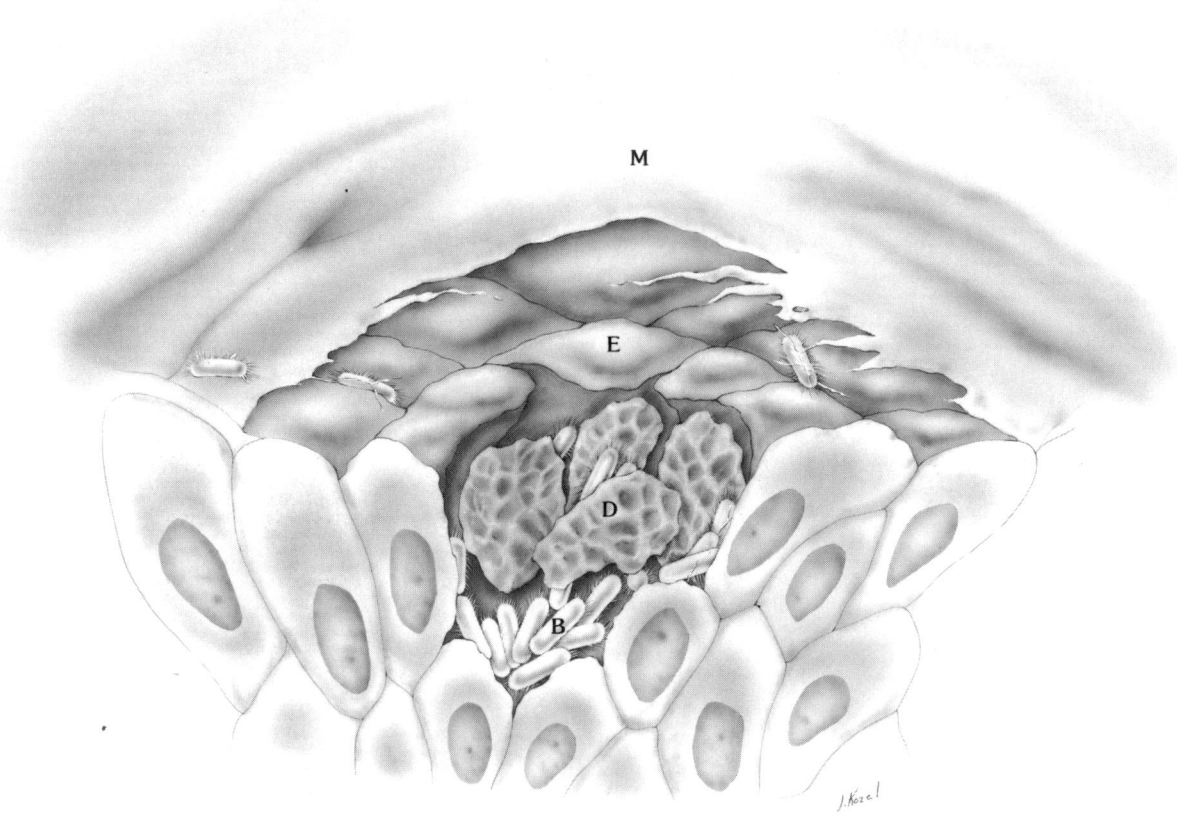

FIG 29–2.
Adherence of bacteria to bladder epithelium. Artist's rendering based on scanning electron microscopic studies of experimentally induced *E. coli* cystitis in the rat. Bacteria *(B)* can then adhere in greater numbers to subsurface epithelial cells *(E)* and cause a tissue infection after sloughing of surface mucin *(M)* and desquamated epithelial cells *(D).*

Since the bladder is in continuity with normally colonized mucosal services, urine sampling for the diagnosis of UTI must be interpreted carefully. The criterion of $>10^5$ organisms per milliliter in a clean voided midstream urine specimen as representing "significant bacteriuria" is based on the urinary colony counts in patients with full-blown pyelonephritis (Kass and Finland, 1956). However, lower colony counts also may be significant in certain clinical situations. In comparing clean midstream voided urine specimens to specimens obtained by urethral catheterization or suprapubic aspiration, Stamm et al. (1982) found the best criterion for significant infection in acutely dysuric adult women to be $\geq 10^2$ bacteria per milliliter of urine obtained by the clean voided method. In patients with indwelling bladder catheters, lower colony counts from a urine culture also are clinically significant, and the low-level bacteriuria usually becomes high-level within three days (Stark and Maki, 1984). The methods used by clinical laboratories to identify the type of bacteriuria and determine antibiotic sensitivity have been extensively described (Edwards and Ewing, 1972; Stamey, 1980). Because of the time and expense involved in the standard laboratory culture and sensitivity, simpler methods to diagnose bacteriuria have been sought. One such way of identifying bacteriuria in office practice is the use of the methylene blue stained smear of the urinary sediment. Bacteria seen on the stained smear of the urinary sediment correlate well with a significant urinary bacterial colony count and allow a treatment decision to be made on the spot (Wear, 1966). Nitrite indicator strips can be used to detect gram-negative bacteriuria and have been particularly useful for home screening programs (Kunin et al., 1976). Detection of urinary leukocytes by a strip test for esterase correlates with urinary leukocyte counts and so can be used to accurately identify bladder inflammation (Gillenwater, 1981). The combined use of esterase and nitrate test strips has been used successfully as a cost-cutting method to detect bacteriuria (Juchau and Nauschuetz,

1985). Obtaining a clean midstream voided urine for the evaluation of bacteriuria remains the critical step in all of these diagnostic steps. While it is not easy or infallible, its careful use is essential in accurately identifying UTI. It also prevents catheterization solely for the purpose of obtaining of a urinary specimen, which has been reported to result in iatrogenic UTI in 1%–20% of such cases (Turck et al., 1962).

In addition to urinalysis and culture, other diagnostic tests may be indicated in patients with bladder inflammation caused by bacteria. In adult women, the diagnosis of bacterial cystitis commonly involves differentiating it from nonbacterial urethritis (urethral syndrome), which also is very common but is characterized by frequency and dysuria without bacteriuria (Gallagher et al., 1965; O'Dowd et al., 1984). In uncomplicated bacterial cystitis in adult women, diagnostic evaluation for contributing anatomical problems has been disappointing. In one study of 153 women with recurrent UTI, 89% had a normal IVP and only one had a cystoscopic finding (colorectal fistula) that required treatment (Engel et al., 1980). In women having flank pain, fever or chills, or other signs or symptoms suggestive of upper UTIs, further evaluation should be used to rule out an upper tract involvement, which may warrant a longer duration of antibiotic therapy (Fang et al., 1978). In urologic practice, an IVP has traditionally been the usual screening test for recurrent UTI. However, there are more accurate and cost-effective ways of localizing the sites of these infections. The gold standard in UTI localization tests is by urine cultures obtained through ureteral catheterization. Stamey reported that up to 40% of adult women who otherwise would have been thought to have uncomplicated cystitis actually had upper urinary tract infection when studied by ureteral catheterization (Stamey, 1980). In another localization test designed to obviate ureteral catheterization and still provide urine for culture and sensitivity from the upper urinary tract, Fairley et al. (1971) utilized proteolytic enzymes to wash out bladder bacteria and to allow culture of the urine coming down from the kidneys into the bladder. Among a group of adult women with a variety of symptoms suggestive of acute UTI, one half were found to have upper tract infection by this test. Using the Fairley test, Busch and Huland (1984) reported that 18% of patients with documented upper UTIs had only symptoms of lower tract infection. Despite being laborious and expensive, the Fairley test remains the localization test with which others are most often compared. Thomas developed an immunofluorescent test for the detection of antibody coating of bacteria (ACB) in the urine based on the premise that upper but not lower tract UTIs cause an increase in urinary antibody (Thomas et al., 1974). Due

to a high incidence of false positive tests, especially in children's bladder infections, the usefulness of the ACB test has been questioned, and it has not come into wide use (Lorentz and Resnick, 1979). Other laboratory tests for UTI localization are urinary concentrating ability (Uehling, 1971), serum quantitative CRP assay (Jodal et al., 1975), gallium scan (McKerrow et al., 1984; Schardjin et al., 1984), urinary NAG assay (Principi et al., 1982) urinary LDH (Lorentz and Resnick, 1979), and β-microglobulin excretion (Hall and Vasiljevic, 1973). P-fimbriated *E. coli* in the urine cannot be used as a localization test because of the false negative and false positive findings (Latham and Stamm, 1984). The clinician who is frequently involved in the management of UTIs will benefit from having one of these localization tests available to aid in diagnosis and treatment.

Other diagnostic tests are useful in inflammatory bladder conditions. While a radiologic workup is usually not indicated in adult women with uncomplicated bladder inflammation, radiologic workup has been recommended for children. Saxena and associates (1975) reported that, after a first UTI, the IVP was abnormal in 15% of girls and 33% of boys and that the voiding cystourethrogram was abnormal in 34% of girls and 39% of boys. Bladder saccules, diverticula, residual urine, and vesicoureteral reflux were the important clinical entities identified by the voiding cystourethrogram. McKerrow and associates (1984) reported that 45% of children with one or more UTIs had a positive radiographic finding requiring treatment, that those referred after one infection had a higher yield of positive findings, and that 90% of children younger than 2 years old had such an abnormality. Currently, ultrasonography of the genitourinary tract is replacing some radiographic evaluation and has found wide use for screening children with UTI (Ben Ami, 1984; Redman and Seibert, 1984). In addition to diagnosing dilatations of the urinary tract, ultrasound can delineate thickenings and irregularities of the bladder suggestive of inflammation (Fig 29–3). While ordinarily not indicated in the diagnostic evaluation of bladder inflammation, CT can also diagnose these changes (Fig 29–4). In addition to their lower cost, ultrasound images may be more informative than CT for bladder inflammation (Fig 29–5). Cystoscopic evaluation of bladder inflammation is commonly utilized to seek predisposing causes and to rule out or to biopsy vesical neoplasms that may present as chronic cystitis. Cystoscopy probably is not generally informative or indicated in children with undocumented UTI or in those with normal x-rays and only occasional infections while not taking antibiotics (Walther and Kaplan, 1979). In chronic bladder inflammation, the inflammatory response becomes endoscopically evident as a characteristic pebbled appearance on the trigone or dif-

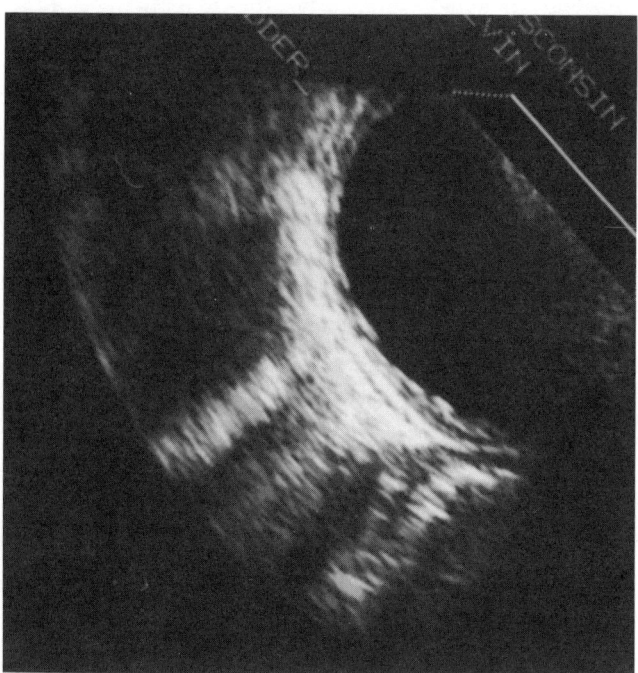

FIG 29–3.
Ultrasound of the bladder in a 5-year-old boy with frequency, urgency, and microscopic hematuria showing bladder wall thickening and mucosal irregularity. Findings resolved in six weeks. The presumed diagnosis was viral cystitis.

fusely throughout the bladder which is usually called cystitis cystica. Histologically, cystitis cystica is characterized by lymphoid follicles or cysts that usually indicate a long-standing infectious process (Figs 29–6 and 29–7). Increased urinary secretory IgA has been re-

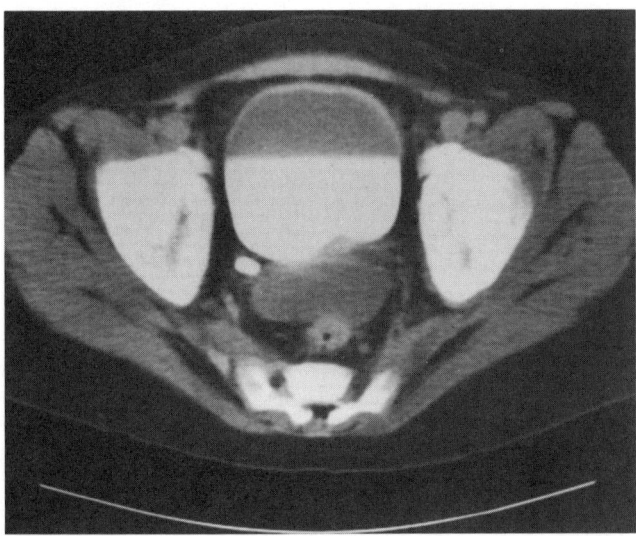

FIG 29–4.
Computed tomogram of the bladder in a 15-year-old girl with a fungus ball on the posterior wall of the bladder and severe fungal cystitis.

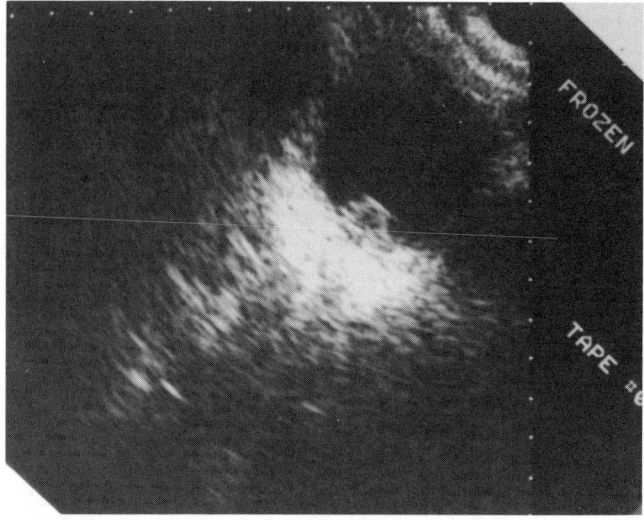

FIG 29–5.
Same child as in Figure 29–4. Fungus ball and posterior bladder wall thickening and inflammation as seen on ultrasound of the bladder.

ported in association with the lymphoid follicles found in cystitis cystica (Uehling and King, 1973). In some cases of cystitis cystica, large inflammatory, cystic, and lymphoid lesions can be present in the bladder in the absence of documented UTI (Fig 29–8) (Smey, 1986). Cystitis cystica, cystitis glandularis, and Brunn's nests are related conditions commonly seen in chronic inflammation. They are also seen around the stalk of papillary tumors and in the margins of invasive tumors. Brunn's nests are invaginations of transitional epithe-

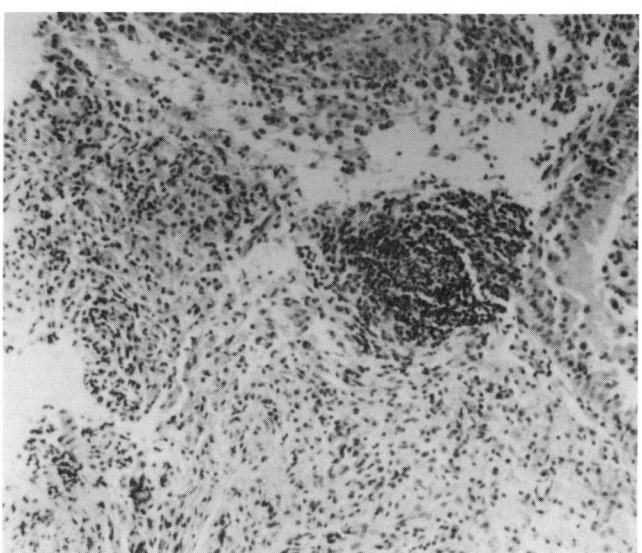

FIG 29–6.
Cystitis follicularis. Bladder biopsy in a 67-year-old woman with recurrent cystitis and cystoscopic appearance of cystitis cystica. Sections of a trigonal biopsy show inflammation, fibrosis, and lymphoid follicles in the lamina propria.

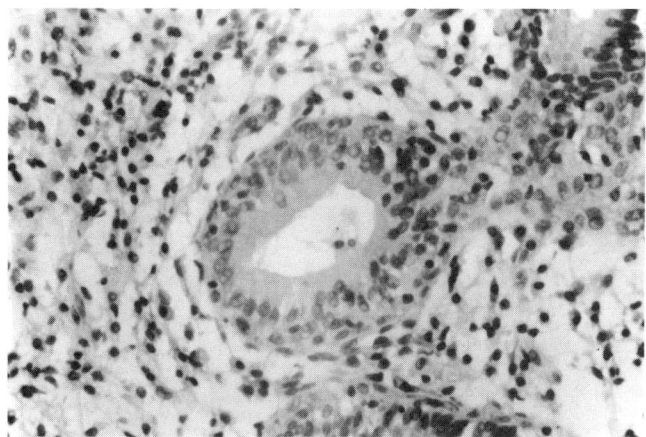

FIG 29–7.
Cystitis glandularis. Bladder biopsy in same patient as in Figure 29–6. Sections of lateral wall biopsy show transitional epithelium overlying lamina propria within which are inflammation and cystitis glandularis.

lium growing down from bladder surface and eventually becoming separated by connective tissue. Central cavitation of the infolded mucosa follows because the nests are obstructed when they are no longer in continuity with the surface and then appear as cystitis cystica. The usual mucosal lining of the cysts in cystitis cystica is by low cuboidal cells, but the mucosa may differentiate into columnar epithelium in which case the cysts are termed "cystitis glandularis" (see Figs 29–6 and 29–7). These lesions can be confused with but only uncommonly convert to adenocarcinoma (Edwards et al., 1972). Pugh (1985) believes that these lesions are

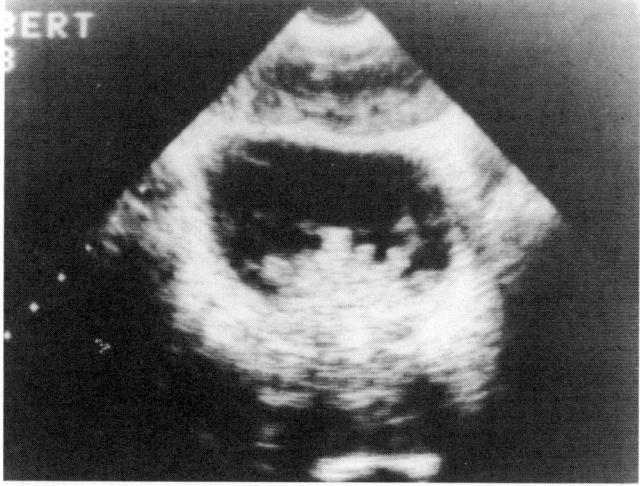

FIG 29–8.
Proliferative polypoid cystitis. A 15-year-old boy with frequency and urgency and no urinary tract infections. Ultrasound showed large filling defect on the posterior wall of the bladder. Biopsy showed proliferative polypoid cystitis and cystitis cystica.

not strongly associated with neoplasia and more properly should be regarded as indicating mucosal instability.

On cystoscopy, the trigone of adult women with and without recurrent infections frequently reveals whitish patches with clear-cut edges which, if biopsied, show squamous metaplasia. These patches are seen in normal women (such as those undergoing cystoscopy as part of a staging evaluation for cervical cancer), and biopsy is not ordinarily indicated. Leukoplakia also manifests as a whitish patch on cystoscopy, but leukoplakia seldom is on the trigone and is more commonly found in men with recurrent infections. Bladder lesions suspicious for leukoplakia should be biopsied. Histologically, it is distinguished from squamous metaplasia by its surface keratination, parakeratosis, and the presence of a stratum granulosum with intracellular bridges (Pugh, 1985). Squamous cell carcinoma of the bladder occurs eventually in about one fourth of cases of leukoplakia. In addition to histologic examination, bladder biopsy can be used for estimation of ABO isoantigen expression in the tissue, which is characteristically retained in benign and lost in malignant lesions. Using a specific red cell adherence assay, Gahzizadeh and associates (1985) reported that all of 14 cases of cystitis cystica, cystitis glandularis, and chronic cystitis retained their antigen expression, and 13 of 18 lesions of squamous metaplasia retained antigen expression. Srinivas et al., (1985) reported on a semiquantitative red cell adherence assay that yielded ABO antigen counts significantly higher for benign than for malignant lesions.

In severe cases of cystitis, and especially in patients with pneumaturia, an enteric fistula into the bladder should be suspected. Cystoscopically, enterovesical fistulas will appear as reddened, edematous areas that may have a shaggy necrotic center and bowel contents emanating into the bladder. Sigmoid diverticulitis is the usual bowel pathology causing fistulas into the bladder. Lippert et al. (1984) described an oral isotope-labeled sodium chromate test to diagnose enterovesical fistulas and reported that this test is more reliable than the oral charcoal test, cystography, or barium enema. Vesicovaginal fistulas occur primarily after gynecologic surgery, and vesicouterine fistulas have been reported after cesarean section (Pawar, 1985). Both of the latter types of fistulas cause bladder inflammation that is less severe than that from enterovesical fistulas, because the urine ordinarily drains freely from the bladder rather than being sequestered in the bladder.

Treatment of uncomplicated bacterial inflammations of the bladder can usually be accomplished with commonly used antibacterial agents that give high urinary concentration (Stamey et al., 1974). For conventional 7–10-day treatment regimens, trimethoprim-sulfamethoxazole, amoxicillin, nitrofurantoin, and nalidixic acid are usually considered safe and effective (Keys and Edson, 1983). With acute uncomplicated first episodes of cystitis, demonstration of bacteriuria on the stained smear of the urinary sediment usually can be used to institute a course of therapy, since the offending urinary pathogen can be presumed to be E. coli, and it will be sensitive to most antibiotics. In recurrent, difficult, or inpatient UTIs, treatment decisions must be based on the results of the urine culture and sensitivity. In bacterial bladder infections there are few indications for treatment with the potentially nephrotoxic antibiotics (Smith et al., 1980). In enterococcal UTIs, in vitro sensitivities to trimethoprim-sulfamethoxazole may be misleading, since in vivo the organisms can incorporate exogenous folates and thus escape the antibacterial action of trimethoprim-sulfamethoxazole (Goodhart, 1984). In chronic infections, long-term oral therapy may be attempted, but low success rates may be anticipated because the bladder abnormalities that first made the patient susceptible often cannot be corrected (Turck et al., 1962). The widespread use of nitrofurantoin and trimethoprim-sulfamethoxazole for UTI has been remarkably well tolerated (Keys and Edson, 1983). Nitrofurantoin has the theoretical advantage of complete absorption in the upper gastrointestinal tract with less resultant selection for antibiotic resistant forms in rectal flora (Stamey, 1980; Winberg et al., 1973). The uncommon adverse reactions to full-dose nitrofurantoin are largely pulmonary or allergic in their manifestations (Holmberg et al., 1980). The widespread use of trimethoprim-sulfamethoxazole has led to concern about selection for resistant bacteria in the fecal, introital, and periurethral flora. The current literature is divided on the significance and threat posed by trimethoprim-sulfamethoxazole-resistant bacteria. Lacey and Brumfitt reported little rise in trimethoprim-sulfamethoxazole resistant organisms causing UTIs (Brumfitt et al., 1983; Lacey, 1982). Schaeffer et al. (1981) reported a 13%–20% incidence of anal and vaginal colonization with Enterobacteriaceae resistant to trimethoprim-sulfamethoxazole. Trimethoprim resistance may be particularly high in catheter-associated UTIs (Kraft et al., 1984). Additionally, Kraft et al. (1984) reported a 10% increase in trimethoprim-resistant E. coli urinary isolates between 1980 and 1982 and presented evidence that the resistance was chromosomally as well as plasmid mediated. Future research on use of antibiotics for UTI will need to focus on the changing mechanisms of bacterial resistance as well as on the efficacy of the drug (Neu, 1984).

Partly from concern for the problems of antibiotic resistance and partly from concern for cost and side effects, single-dose therapy for cystitis has received much attention. Most studies based single-dose therapy on tests localizing the UTI to the lower tract and reported equal efficacy with single-dose therapy (1 gm of sulfisoxazole, one single- or double-strength trimethoprim-sulfamethoxazole, 2 gm of cefaclor, 3 gm of amoxicillin, etc.) as with 7–10 days of therapy in adult women with uncomplicated cystitis (Buckwold et al., 1982; Fang et al., 1978; Schultz et al., 1984; Sheehan et al., 1984; Tolkoff-Rubin et al., 1984). Single dose therapy in children with uncomplicated UTIs has not been as effective as a 10-day course (McCracken et al., 1981).

Since recurrent bacterial infections of the bladder are a source of discomfort and aggravation to so many women and female children, prophylaxis has received broad acceptance and is usually accomplished with low-dose nitrofurantoin or trimethoprim-sulfamethoxazole (Lohr et al., 1977; Stamey et al., 1977). In adult women, the beneficial effect of trimethoprim-sulfamethoxazole may be related to lesser vaginal colonization with Enterobacteriaceae and lesser urethral contamination following coitus (Stamey et al., 1977; Vosti, 1975). Prophylaxis for 6–12 months against UTIs appears to benefit children for 1–2 years after stopping the prophylaxis (Smellie, 1978). In endoscopic surgery on the bladder and prostate, prophylactic antibiotics are gaining in use. The efficacy of antibiotics in surgical prophylaxis relates in an important way to the duration of need for an indwelling bladder catheter (Nielsen et al., 1981). As with full-dose therapy, single-dose therapy, long-term prophylaxis, and surgical prophylaxis must be assessed for selection of antibiotic-resistant organisms. For example, cinoxacin has been recommended as an antibacterial agent with a low incidence of posttreatment antibiotic resistance in vaginal and anal flora (Schaeffer et al., 1982).

Treatment of catheter-associated UTIs is important because of their high morbidity and expense. They account for 40% of all nosocomial infections, occur in 7%–16% of patients requiring indwelling catheterization and commonly lead to bacteremia (Fowler, 1983). Avoidance of unnecessary catheterization and prevention of catheter-induced UTI by closed drainage remain the basic tenets of good catheter care (Kunin and McCormack, 1966). Prophylactic antibiotics, meatal care regimens, and instillation of hydrogen peroxide into drainage bags have not further reduced morbidity from catheter use (Breitenbucher, 1984; Burke et al., 1981; Platt et al., 1983; Thompson et al., 1984). Chronic use of an indwelling catheter will result in bullous or polypoid cystitis, which can take months to subside and can be confused cystoscopically with bladder tumor (Fig 29–9).

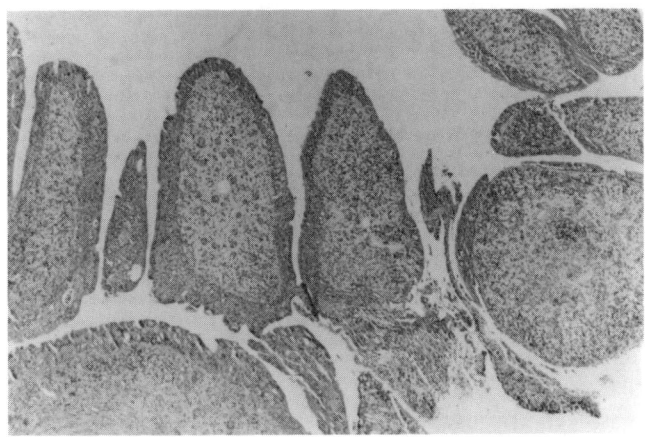

FIG 29–9.
Bullous cystitis. Elderly male with long-standing indwelling urethral catheter. Bladder biopsy of raised edematous lesion on dome showed hyperplastic urothelium and a chronic submucosal inflammatory response.

TUBERCULOSIS

Genitourinary tuberculosis is secondary to hematogenous spread from pulmonary tuberculosis and remains a problem in areas of the world where health conditions are poor and adequate chemotherapy is not available (Latham and Stamm, 1984). Bladder tuberculosis is the result of downward seeding from renal tuberculosis. Accordingly, patients' symptoms may follow at widely varying time intervals after the onset of their pulmonary or renal TB. Women and children are thought to present earlier with the frequency, dysuria, and hematuria characteristic of implantation of tubercle bacilli on the bladder epithelium, and men more commonly present with epididymitis.

Cystoscopically, tuberculosis appears as patchy erythematous ulcerations from which an exudate may arise. Since the bladder lesions may resemble carcinoma, cup biopsy should be considered. The ureteral orifices and the bladder around them appear most reddened in the acute phase. Even with adequate treatment, these areas later take on a pink-and-white speckled appearance. In untreated cases, the dome of the bladder eventually undergoes more fibrosis and contracture than the base. Since the trigone contracts less, the ureteral orifices assume a position in the angle at the top of the bladder. This makes the ureteral orifices hard to catheterize should ureteral dilatation become necessary for the hydronephrosis that occurs as the tuberculous ureteritis heals with treatment.

The diagnosis of genitourinary TB depends on positive acid-fast smears or cultures of urine for *Mycobacterium tuberculosis*. Acid-fast smears and guinea pig inoculations are not now performed by most laboratories, so diagnosis depends more on positive urine cultures. First morning urine specimens yield as many positive cultures as 24-hour collections. At least three successive morning urine specimens should be obtained when TB is suspected. Drug-resistant TB bacilli are known to occur, so sensitivities to all of the antituberculous chemotherapeutic agents should be carried out. This process may require the sending of specimens to specialized public health laboratories.

Treatment of tuberculous cystitis is based on the premise that pulmonary and renal involvement also are present. Treatment with three drugs is preferred and all can usually be given orally (Wechsler, 1980). Isoniazid (10–30 mg/kg/day) is the most effective and important antituberculous agent, and its use requires pyridoxine supplementation. Rifampin is a newer oral drug whose use has helped eliminate the need for parenteral therapy and that should be given at a 10–20 mg/kg/day dosage. Ethambutol (15–20 mg/kg/day) has largely replaced the use of para-aminosalicylic acid but is not recommended for children because of risk of optic neuritis. Patients receiving ethambutol should have monthly visual examinations. Streptomycin and other drugs are occasionally needed when the patient has problems with the preferred drugs or in treating drug-resistant cases. Two years of therapy is still preferred, although nine months of therapy has also been advocated as being less costly and equally effective (Gow, 1979). Bladder fibrosis and contracture are generally preventable by medical therapy. Bladder augmentation or urinary diversion uncommonly are necessary.

VIRAL INFLAMMATIONS

Viral inflammations in the bladder cause symptoms similar to those of bacterial infections except that the viral lesions have more hemorrhagic manifestations (Hashida et al., 1976). The usual causative agents are adenovirus types 11 and 21 and papovavirus (Mininberg et al., 1982). Affected patients are typically children and immunosuppressed adults (Lecatsas et al., 1973; Mufson et al., 1971). The urinary sediment contains epithelial cells with intranuclear inclusion bodies which, when inoculated into fetal glial cell cultures, cause cytopathic changes on electron microscopy which are characteristic for the different types of virus (Padgett et al., 1983). Viral cultures are not generally available, so the diagnosis of viral cystitis is often made on clinical grounds alone in a patient with hemorrhagic cystitis, a negative culture for bacteria, and a self-limited course. The presence of hematuria and its proper evaluation is a concern, since the viral etiology of the inflammation often cannot be confirmed with viral cultures. Ultrasound examination of the bladder can be used to document the characteristic thickening and irregularity of the bladder wall in viral cystitis (see Fig 29–3). The symptoms and the hematuria typically resolve in 2–3 weeks. If the ultrasound changes also resolve, other diagnostic modalities such as cystoscopy usually employed for the evaluation of hematuria may be selectively omitted (Walther and Kaplan, 1979). If the symptoms and ultrasound changes persist or if there is a palpable bladder mass on rectal or abdominal examination, further evaluation is needed. Voiding cystourethrogram ordinarily will not delineate the bladder lesion as well as ultrasound but may show a transitory type of reflux (Mininberg et al., 1982). Since viral cystitis is presumed to be hematogenous, a diagnostic evaluation for obstruction or other predisposing anatomical abnormalities usually is not warranted.

FUNGAL INFLAMMATIONS

Fungal inflammations in the bladder are increasing in frequency due to increased use of broad-spectrum antibiotics, chemotherapy for malignancy, and immunosuppressive agents (Michigan, 1976). Bladder infections due to fungi are considered ascending in origin, whereas renal fungal infections are from hematogenous spread. Genitourinary fungal infections are most commonly caused by *Candida albicans* (Fig 29–10) but may also be caused by *Aspergillus fumigatus, Cryptococcus neoformans, Torulopsis glabrata,* and others (Fisher et al., 1982). Symptoms range widely from the asymptomatic patient with a fungus-colonized bladder catheter to the patient with severe urgency, frequency, and dysuria. Considerations similar to those for bacterial infections apply to appropriate evaluation and localization. The number of organisms in the urine required to constitute a "significant infection" differs for fungal cystitis. Colony counts of 10,000–15,000/ml of urine on a clean voided midstream urine specimen or any fungal growth from a catheterized specimen should be considered significant and require investigation (Wise et al., 1980). No growth on a bacterial culture of the urine in a patient with severe frequency and dysuria also can be the harbinger of a fungal UTI. Methylene blue–stained smear of the urinary sediment is a good method to show the hyphae and budding forms (Wear, 1966). Ultrasound of the bladder can be helpful in making the diagnosis, since it can delineate the thickened inflamed bladder wall and an intravesical aggregation of hyphae, or so-called fungus ball (see Fig 29–5). CT of the bladder may also delineate a filling defect and bladder wall thickening but is less cost-effective (see Fig 29–4). In candida cystitis, the mucosa is hemorrhagic, covered with a whitish pseudomembrane, or covered with whitish elevated plaques similar to those seen in fungal stomatitis (Michigan, 1976). Cystoscopy may also show an edematous, shaggy, raised, or ulcerated mucosa. Bladder biopsy may be required to differentiate the fungal lesion from a necrotic tumor. Fungus balls appear as gray, shaggy aggregations in the bladder and may be adherent to the bladder wall. In a patient with a fungal bladder infection, the presence or absence of an upper urinary tract fungus ball must be ascertained by an imaging modality such as ultrasound, since such a renal or ureteral fungus ball can be intermittently obstructing. Candida precipitin titers, if available, correlate well with severity of involvement and response to treatment (Michigan, 1976).

Treatment of genitourinary fungal UTIs requires parenteral or intravesical amphotericin B and oral flucytosine (Fisher et al., 1982). Iodides given intravesically are too irritating, and nystatin or miconazole are insoluble in the usual vehicles (Utz, 1980). Treatment with ketoconazole has been reported to have a poor success rate when administered for genitourinary infections (Dismukes et al., 1983; Wise et al., 1985). Parenteral administration of amphotericin B requires attention to prescribing guidelines because of its toxicity (Michigan, 1976; Utz, 1980). Flucytosine is not available for parenteral administration but reaches high urinary levels after oral administration. The recommended dosage of flucytosine is 150 mg/kg/day in divided doses, with reduced dosages for patients with reduced renal function. Therapy should be based on laboratory sensitivities, since resistant strains are encountered or are seen to emerge during a course of therapy. Treatment of fungal cystitis with intravesical irrigations of amphotericin B may be preferable to systemic therapy if renal or systemic involvement is absent. Such instillations are 90% effective and obviate most of the problems with sys-

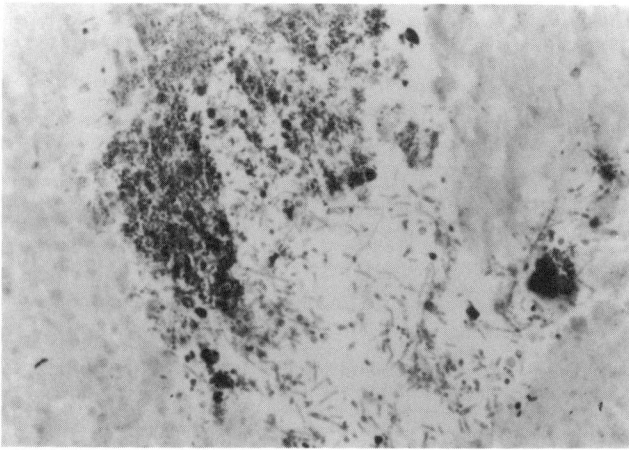

FIG 29–10.
Fungal cystitis. Bladder biopsy showing inflammatory reaction in the lamina propria, necrotic amorphous material, calcification, and *Candida albicans* hyphae.

temic use (Wise et al., 1982). Amphotericin irrigations also may be effective when given as supplements to oral flucytosine (Rohner and Tuliszewski, 1980). In the presence of bloodstream involvement from a fungal UTI, both amphotericin B and flucytosine should be given if renal function is adequate for the former and the infecting organism is sensitive to the latter (Solomkin et al., 1982).

OTHER INFLAMMATIONS

Trichomonas vaginalis, a flagellate protozoa, may infect the urinary tract but usually does so as urethritis or prostatitis. Since the symptoms of frequency and urgency may be the same as cystitis, microscopic examination and culture of urine and genital secretions should include this possibility (Krieger, 1981). Similarly, *Chlamydia trachomatis* is primarily a causative agent for urethritis, epididymitis, and prostatitis. Chlamydial infections are transmitted by sexual exposure and diagnosed through inoculation into cell cultures (Kramer, 1985; Rettig and Nelson, 1981). *Ureaplasma urealyticum* is recovered infrequently in patients with UTI but must be considered in the differential diagnosis of patients with frequency and dysuria in whom the bacterial urine culture is negative (Taylor-Robinson and McCormack, 1980).

Parasitic infestations cause intensive bladder inflammation and symptoms when involving the genitourinary tract. *Schistosoma haematobium* is endemic in Egypt, and the ova in patients infected with this parasite enter the bladder muscle from veins in the pelvis and are embedded in the bladder wall. This causes an eosinophilic inflammatory response, which becomes chronic and eventually results in fibrosis and calcification (Pugh, 1985). As the ova are excreted in the urine, an epithelial reaction occurs, with severe mucosal hyperplasia and subsequent dysplasia, bladder stones, and squamous cell carcinoma (Kambal, 1981). Children with schistosomal cystitis may have nodular filling defects in the bladder, vesicoureteral reflux, and obstructive uropathy, but all of these have been reported to improve markedly with praziquantel therapy (Farid et al., 1984). *Echinococcus granulosus* infestation of the urinary tract typically causes calcified cysts of the kidney but may also infiltrate the bladder and cause hematuria, cystitis, bladder cysts, and hydronephrosis (Fuloria et al., 1975; Hansman and Brown, 1974).

NONINFECTIOUS INFLAMMATIONS

INTERSTITIAL CYSTITIS

Interstitial cystitis is an inflammatory condition of the bladder characterized by severe irritative voiding symptoms; sterile, cytologically normal urine; and char-

acteristic cystoscopic findings. Interstitial cystitis typically affects middle-aged women, who experience severe frequency, urgency, and nocturia. Suprapubic pain is commonly part of the symptom complex and is variably relieved by voiding. In the past decade, the term "interstitial cystitis" has largely replaced the previous nomenclature of cystitis parenchymatosa (Nitze, 1907), Hunner's ulcer (Hunner, 1914), elusive ulcer (Howard, 1944), and others. Interstitial cystitis is being diagnosed with increasing frequency because of an increased awareness of the disease, an improved ability to rule out infectious causes of similar symptoms, and more interest in active treatment. The condition occurs 6 to 11 times more commonly in women than men, has been reported to occur in children, and has been diagnosed in elderly patients (Chenoweth and Clawater, 1960; Geist and Antolak, 1970; Brock et al., 1979; Oravisto, 1975). In an epidemiologic study of interstitial cystitis, Oravisto (1975) reported that 1.8/10,000 adult women in Helsinki had the condition, that symptoms usually had been present for 3–5 years before diagnosis, and that symptoms in most cases stabilized at some point after a rapid initial onset (Oliver, 1976).

The etiology of interstitial cystitis is unknown. Hunner and others earlier proposed that interstitial cystitis resulted from chronic bacterial infection (Hunner, 1914). In more recent reports, culture and histologic studies have not shown an infectious etiology (Collan et al., 1976; Hanash and Pool, 1970). Moreover, it is the lack of response to treatment with antibiotics that frequently suggests the diagnosis initially. The histologic findings of submucosal edema and fibrosis have suggested lymphatic obstruction, but lack of a specific diagnostic test or any therapy to improve lymphatic drainage has precluded verification of this etiologic theory. An immunologic etiology has been suggested by an electron microscopy study of bladder wall blood vessels in 20 patients with interstitial cystitis in which 70% had severe endothelial injury and 50% had severe smooth muscle injury (Matilla et al., 1983). A neurogenic etiology has been proposed based on the focal inflammation seen around perivesical nerve bundles in patients and on monkey experiments in which sacral nerve root stimulation caused pathologic changes in the bladder similar to interstitial cystitis (Franksson, 1957; Meirowsky, 1969). In recent years, the immunologic theory of etiology has aroused the most interest. Patients with interstitial cystitis have been reported to have positive serum fluorescent antinuclear antibody tests, immunoglobulin deposition in the bladder wall, and serum antibladder antibodies (Gordon et al., 1973; Jokinen et al., 1972; Mattila, 1982; Oravisto, 1982; Silk, 1970). However, all of these findings also have been disputed (Messing and Stamey, 1978). A deficiency in bladder mucin has been proposed as another possible etiology for interstitial cystitis but has not been verified pathologically (Parsons et al., 1983). Endocrine, psychoneurotic, and other causes also have been postulated.

The differential diagnosis of interstitial cystitis includes all of the inflammatory conditions in the bladder which cause irritative symptoms. Urethritis, or the "urethral syndrome," in women may be confused with interstitial cystitis, but the latter causes less dysuria and more nocturia and suprapubic pain. Prostatitis may have similar symptoms in men but prostatitis is more common and has the characteristic findings in the expressed prostatic secretions (Meares, 1973). Urine cultures should be negative to invoke the diagnosis of interstitial cystitis, but the urinalysis may show microscopic hematuria and pyuria (Messing and Stamey, 1978). Carcinoma in situ can present with symptoms and cystoscopic findings similar to interstitial cystitis, so urinary cytology also should be unequivocally negative (Utz and Zincke, 1974). Cystoscopic examination is best accomplished using regional or general anesthesia to allow hydrodistention and subsequent verification of the characteristic round submucosal hemorrhages or "glomerulations" that are indicators of the disease. Ulceration and reduced capacity are less common cystoscopic findings and are not necessary to make the diagnosis. During the first filling of the bladder (to under 70 cm H$_2$O), the bladder mucosa will appear normal but the terminal portion of the drained fluid will become blood tinged. In patients with reduced bladder capacity (less than 400 ml under anesthesia), the glomerulations will appear earlier in the hydrodistention process. While the diagnosis is substantiated in this way, the severity of symptoms rarely correlates with the severity of the cystoscopic findings (Messing and Stamey, 1978). The value of bladder wall biopsy remains controversial. Histologic findings typically include pancystitis, submucosal edema, and fibrosis. In addition, specific pathologic findings that have been reported to be present in interstitial cystitis are eosinophilic infiltrates, perineural inflammation cells, and mast cell infiltrates (Larsen et al., 1982; Walsh, 1978). Some authors believe that no pathognomonic findings are obtained on bladder biopsy (Messing and Stamey, 1978). In addition to cystoscopy, other genitourinary evaluations can be based on the differential diagnosis, e.g., IVP, renal ultrasound, cystogram, or urodynamics. Characteristic findings on ultrasound or CT have not been reported to our knowledge.

Treatment of interstitial cystitis is mandated by the highly symptomatic patient but is difficult because of lack of objective findings and an uncertain natural course. Explaining the condition may be reassuring enough to allow the patient to live with the symptoms. If not, hydrodistention with anesthesia is the next step and will improve symptoms in up to 30% of patients. Hydrodistention done once or at intervals may sufficiently relieve the symptoms. Use of systemic medications has been tried extensively but without consistent results. Corticosteroids have been reported to effect improvement in 70% of patients in one series but were ineffective in others (Hoyt, 1952; Messing and Stamey, 1978). Azathioprine and other immunosuppressive agents have been given on the theory that interstitial cystitis is an immunologically mediated disease and have been reported to show a 50%–60% success rate in relieving pain (Oravisto and Alfthan, 1976). There have not been recent or corroborating reports on the use of immunosuppression, which should be used with the usual concerns about infectious complications. Based on the deficient bladder mucin theory, Parsons (1982) treated interstitial cystitis with an experimental oral heparin analogue, sodium pentosanpolysulfate, and reported an 83% improvement in symptoms. This optimistic report also awaits further confirmation.

Local treatment for interstitial cystitis is currently in broader use than systemic therapy. Intravesical instillation of dimethyl sulfoxide (DMSO) has been used extensively since 1967 (DeJuana and Everett, 1977; Stewart et al., 1967). In one review, 15% of patients had complete relief and 65% had mild improvement that lasted at least four months (Fowler, 1981). Therapy must be continued monthly, and in some patients improvement will not continue. DMSO remains in general use because of relative lack of side effects, moderate cost, the ability to do instillations as an office procedure, and FDA approval for intravesical use. Another agent for intravesical instillation is sodium oxychlorosene, which has had varying popularity since 1957 (O'Conor, 1955). Messing and Freiha (1979) reported that 72% of patients treated with 2–4 instillations with general anesthesia experienced improvement. Patients undergoing oxychlorosene instillations require anesthesia for the painful bladder instillation. Most patients have worsened symptoms temporarily after treatment. Long-lasting improvement may result from 1–2 instillations. Oxychlorosene should not be given in the presence of vesicoureteral reflux. Other reports do not corroborate this improvement following intravesical instillations of oxychlorosene (DeJuana and Everett, 1977). Marberger et al. (1974) recommended multiple injections of the metalloprotein anti-inflammatory agent orgotein around the involved areas of the bladder wall, but this agent is not now available in the United States.

The surgical treatment of interstitial cystitis is reserved for the most severe cases and those that have not responded to medical treatment. Surgical treatment by cystolysis attempts to provide bladder denervation by dissecting blood vessels, nerves, and adventitia off the bladder down to the ureteral sheaths and trigone (Worth and Turner-Warwick, 1973). Short-term results have been reported to be favorable, but longer-term evaluations have not been reported (Freiha and Stamey, 1980; Leach and Raz, 1983). Cystolysis has largely replaced surgical treatment by presacral neurectomy or rhizotomy (Pearl and Strauss, 1938). Augmentation of bladder capacity through the use of bowel segments has been utilized in severely afflicted patients, but there remain questions on whether the interstitial cystitis then affects the bowel segment itself and how much of the bladder base should be retained (Freiha et al., 1980; McGuire et al., 1973; Seddon et al., 1977; Sidh et al., 1980; Smith et al., 1977). Urinary diversion offers relief for the most severely afflicted patients from the intractable symptoms but should be rarely needed, since other treatment options are available that involve less morbidity.

MALAKOPLAKIA

Malakoplakia is an uncommon granulomatous disease that primarily involves the genitourinary tract. The bladder is involved in 40% of reported cases, the kidney in 16%, the testes in 12%, the ureters in 11%, and the prostate in 10% (Stanton and Maxted, 1981). Malakoplakia of the bladder causes severe irritative symptoms and hematuria, so that it must be distinguished by biopsy from other benign and malignant conditions. The term itself derives from the Greek "malakos" (soft) and "plakas" (plaque). Grossly, malakoplakia consists of yellow-brown granulomas in the bladder. When viewed cystoscopically, the granulomas appear as brown flat nodules with intact mucosa or ulcerated yellow plaques with surrounding hyperemia. The cystoscopic lesion may appear similar to amyloidosis, which rarely involves the bladder (Malek et al., 1971). Sarcoidosis is not thought to involve the bladder, and endometriomas would ordinarily not be confused with malacoplakia because they appear bluish. Microscopically, the lesions of malakoplakia consist of infiltrates of large macrophages mixed with lymphocytes and plasma cells (Fuloria et al., 1975). The large macrophages contain abundant cystoplasm within which are inclusions named Michaelis-Gutmann bodies (Michaelis and Gutmann, 1902). The Michaelis-Gutmann bodies are rounded, 5–10-μm inclusions with a concentric "owl-eye" or "bird's-eye" appearance (Fig 29–11). By x-ray microprobe analysis, the inclusions contain varying amounts of calcium (Kuthy and Ormos, 1978). By electron microscopy they arise within phagolysosomes (Lewin et al., 1974). Bacterial fragments have been found in Mi-

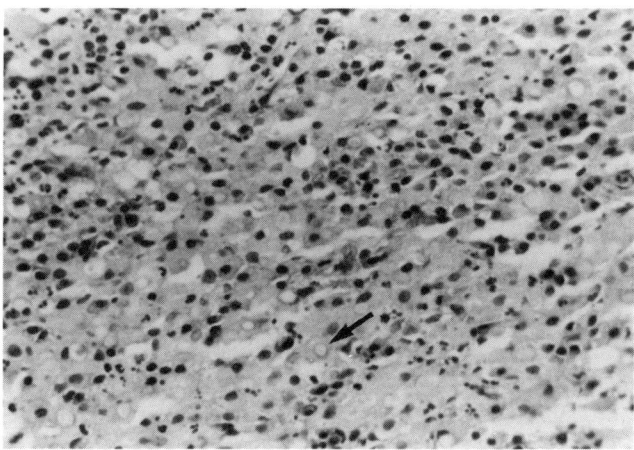

FIG 29–11.
Malakoplakia of bladder. Bladder biopsy in a 2½-year-old boy with an *E. coli* urinary tract infection and hematuria; inflammatory reaction with Michaelis-Gutmann body *(arrow)* seen in lamina propria. (From Witherington R, Branan WJ, Wray BB, et al: Malacoplakia associated with vesicoureteral reflux and selective immunoglobulin A deficiency. *J Urol* 1984; 132:975. Used with permission.)

chaelis-Gutmann bodies, so it is believed that malakoplakia is due to an abnormality of bacterial digestion by macrophages (McClory et al., 1973). Abdou et al. (1977) reported that blood mononuclear cells from a patient with malakoplakia contained intracytoplasmic lysosomal inclusions with low levels of cyclic 3′,5′-guanosine monophosphate (cGMP), which resulted in decreased release of lysosomal enzymes. Oliver (1976) had shown previously that a similar cGMP defect in the Chédiak-Higashi syndrome could be reversed in vitro by cholinergic agents. This led to the successful treatment of malakoplakia with the cholinergic agent bethanechol chloride (Abdor et al., 1977; McKay, 1978).

Because malakoplakia is associated with recurrent UTI, an immunologic deficiency has been suspected. McKay (1978) reported normal humoral immunity but deficient T cells in patients with malakoplakia. Witherington et al. (1984) reported a child with bladder malakoplakia, vesicoureteral reflux, selective absence by serum IgA, depressed skin test reactivity, and decreased bacterial killing by the patient's monocytes. The child's altered immunologic markers improved after long-term trimethoprim-sulfamethoxazole and bethanechol chloride treatment.

Clinically, malakoplakia commonly presents with an *E. coli* or other coliform UTI (Stanton and Maxted, 1981). Additionally, 40% of malakoplakia patients will have another systemic disease such as a carcinoma, an immunodeficiency, or an autoimmune disease. Malakoplakia has been reported in close proximity to a transitional cell carcinoma of the bladder (Skudowitz and

Weintraub, 1977). Following renal transplantation, malakoplakia has particularly serious implications and may be associated with graft loss and mortality rates of 80% and 60%, respectively. Streem (1984) has advocated stopping azathioprine in cases of bladder involvement following transplantation and consideration of early transplant nephrectomy in cases of renal parenchymal (transplant) involvement. Even without renal transplantation, mortality from malakoplakia may be 50% owing to involvement of other vital organs in addition to bladder (Stanton and Maxted, 1981).

Treatment of malakoplakia is based on the current theories of pathogenesis. Long-term antibiotic treatment of the associated UTI is primary. Trimethoprim-sulfamethoxazole may play a special role because of its lipophilic quality, postulated cell membrane binding, and cell entry capacity, all of which facilitate macrophage phagocytosis (Maderazo et al., 1979, Stanton and Maxted, 1981). Bethanechol chloride treatment has had a consistent therapeutic benefit in reported cases. Because of the effective medical regimens, total endoscopic removal of the bladder lesions of malakoplakia may not be necessary (Zornow et al., 1979). Since in vitro assessment of macrophage functions is not generally available, careful clinical follow-up is necessary to evaluate the effect of bethanechol chloride and to be sure the observed bladder lesions are not carcinomas. In the patient who has had renal transplantation, cessation of azathioprine therapy should be considered in the management of malakoplakia involving either the kidney or bladder (Arnesen et al., 1977; Streem, 1984).

CYCLOPHOSPHAMIDE CYSTITIS

Cyclophosphamide (Cytoxan) has been widely used since 1958 for the treatment of malignant and autoimmune disease (Plotz et al., 1979). Since 1961 it has been recognized that cyclophosphamide administration results in bladder toxicity, which is manifested in the early stages by a hemorrhagic cystitis and later by severe bladder fibrosis and serious urinary bleeding (Phillips et al., 1961). A metabolic product of cyclophosphamide, acrolien, has been shown to be the cause of the bladder toxicity (Brock et al., 1979). Other alkylating agents such as ifosfamide also cause a toxic hemorrhagic cystitis (Watson, 1984). Approximately 7% of patients treated with cyclophosphamide will develop the characteristic clinical picture of frequency, dysuria, and gross or microscopic hematuria (Duckett et al., 1973; Pyeritz et al., 1978). The severity of the cystitis correlates with the total dosage given and the rapidity of treatment. Cystoscopically, the bladder initially appears edematous and has multiple punctate hemorrhagic areas. Later, diffuse telangiectasia, reduced capacity, and trabeculation are present. Histologically, the early

lesion shows edema, ulceration, and hemorrhage in the mucosa and lamina propria. The late histologic appearance is chronic inflammation and fibrosis (Pugh, 1985).

Treatment of cyclophosphamide cystitis should begin with prevention. The injury depends on the time of contact and on the concentration of the acrolien in the bladder, which can be reduced by hydration and frequent voiding or by indwelling catheterization. Droller et al. (1982) reported that only one of 198 patients experienced hemorrhagic cystitis with such a regimen, whereas eight of 97 previously treated patients without this preventive regimen had the toxicity, and six of the eight died as a direct result of the hemorrhages. Once the bladder complications have occurred and bladder hemorrhage is present, the cyclophosphamide therapy should be stopped if at all possible, which usually allows the toxic cystitis and hematuria to subside in most cases. The next step might be the use of epsilon aminocaproic acid, an agent which inhibits clot lysis by urinary urokinase. Epsilon aminocaproic acid has been used for urinary tract bleeding from a wide variety of urinary conditions including cyclophosphamide cystitis, but the success rate and complications of this drug have not been reported for cyclophosphamide cystitis (Persky and Selman, 1980). In patients experiencing severe hematuria, clot retention may ensue. At that time, cystoscopy with electrocoagulation of any major visible bleeding points in the bladder should be carried out. This also accomplishes evacuation of clots from the bladder, which promotes cessation of the bleeding. If this is not effective, bladder irrigations with 1% silver nitrate delivered through a three-way indwelling Foley catheter will exert a temporary astringent effect and be successful in some cases (Kumar, 1976). Closed continuous irrigation with a 1% alum solution may control massive bladder hemorrhage and has the advantages of being nonpainful, nontoxic, and inexpensive (Ostroff and Chenault, 1982). Open surgical exposure of the bladder and application of phenol to actively bleeding points has been described as successful treatment in a child with cyclophosphamide cystitis (Duckett et al., 1973). Used as a closed-bladder instillation, phenol treatment resulted in a fatal methemoglobinemia reaction in an infant (Lebowitz, 1980). Because cyclophosphamide treatment may have to be continued in some patients even after hematuria has appeared, agents to prevent acrolien contact with vesical mucosa have been employed to prevent toxicity. One such agent, 2-mercaptoethane sodium sulfonate, is thought to form a nontoxic thioether with acrolien and prevent bladder toxicity without affecting the chemotherapeutic efficacy of the cyclophosphamide (Fowler, 1981; Tolley, 1977). How this agent fits in with other preventive and therapeutic measures has not been described.

An accepted treatment modality for severe intractable hemorrhage following cyclophosphamide therapy is intravesical instillation of formalin. As described by Fair (1974), 1% formalin given by a careful and precise protocol will be generally successful. The important points of the protocol are: (1) use of general anesthesia; (2) complete removal of all bladder clots by irrigation; (3) active electrocoagulation of actively bleeding vessels; (4) passive instillation of 1% formalin without pressure in and out of the bladder for ten minutes; and (5) copious saline irrigation of the bladder to follow formalin treatment. Higher concentrations of formalin result in high rates of perforation, bladder fibrosis, and hydronephrosis. Reflux must be recognized by cystography or prevented by ureteral occlusion or Fowler position to prevent renal necrosis and ureteritis cystica (Mahboubi et al., 1976). In cases where the above methods have been unsuccessful, therapeutic embolization of pelvic blood vessels also has been unsuccessful (Benson et al., 1980). Intravenous vasopressin has not had wide use but seemed helpful in one reported case in which 4% formalin was not effective (Pyeritz et al., 1978). Urinary diversion may become necessary as a lifesaving desperation maneuver (Golin and Benson, 1977). A final point about cyclophosphamide therapy is that bladder cancer has been reported to occur with a higher incidence after its use (Fairchild et al., 1979; Wall and Clausen, 1975). Such patients should have appropriate urologic surveillance by periodic urinalysis, cytology, or cystoscopy.

RADIATION CYSTITIS

Radiation cystitis is similar to cyclophosphamide cystitis in that it is given for treatment of pelvic malignancy but has severe adverse as well as therapeutic sequelae for the bladder. Radiation cystitis of clinical significance is thought to occur in 2%–12% of patients receiving therapeutic radiation (Bloedorn et al., 1962). Pool (1959) reported on 50 women with radiation cystitis seen at the Mayo Clinic over a five-year period. Acute radiation cystitis occurs either during or shortly after the radiation therapy. Histologically, it is characterized by edema and inflammation of the mucosa and lamina propria (Persky and Austen, 1953). The majority of cases of acute radiation cystitis subside spontaneously during the 12–18 months after completion of therapy. The late sequelae of irradiation may occur up to 8–10 years after radiation therapy and result from an endarteritis, which leads to inflammatory cell infiltrates, fibrosis throughout the bladder wall, and increased mucosal vascularity. Before rotating ports and other contemporary techniques were used for radiation therapy, encrusted ulcers and other serious complications were not uncommon (Persky and Austen, 1953). In

1962, Bloedorn and co-workers described the factors that contributed to radiation cystitis as: (1) obstruction to ureters or bladder outlet; (2) urinary tract infection; (3) previous radiation therapy or surgical procedures; (4) presence of large ulcerated tumors; and (5) excessive radiation dosage, which at that time was thought to be dosages exceeding 10,000 rad.

Diagnostic evaluation of radiation cystitis depends on the severity and chronicity. There is little reason to document early acute radiation cystitis cystoscopically. Treatment is directed at prevention of UTI and relief of bladder spasms. In chronic and severe cases, cystoscopy will show the characteristic diffuse telangiectasia and reduced capacity. Cystoscopy may also be important in ruling out persistent or newly recurring bladder tumors. Symptomatic treatment of radiation cystitis consists of anticholinergic and antispasmodic agents for the associated frequency and urgency. As in the case of troublesome bleeding from cyclophosphamide, a progressive treatment regimen of epsilon aminocaproic acid, bladder fulguration, formalin instillations, and diversion or cystectomy may be needed (DeKernion, 1976). ACTH therapy has been described but is of principally historic interest (Persky and Austen, 1953).

EOSINOPHILIC CYSTITIS

Eosinophilic cystitis is a rare type of chronic cystitis occurring primarily in allergic patients and manifesting an eosinophilic infiltrate in the lamina propria and muscle layers of the bladder (Pugh, 1985). Forty cases have been reported since 1959, but many more have likely occurred but not been reported (Brown, 1960). Most patients with eosinophilic cystitis present with frequency, dysuria, urgency, and hematuria. Recurrent UTI is less of a clinical manifestation than in other bladder lesions. A history of allergic problems and eosinophilia are each present in one quarter of patients (Sidh et al., 1980). Cystoscopically, the characteristic findings are yellow raised plaques, necrotic ulcerated lesions resembling bladder tumors, and generalized edema and erythema (Kessler et al., 1975; Littleton et al., 1982; Sutphin and Middleton, 1984). Because of the clinical presentation and cystoscopic findings, the disease has been confused with malacoplakia, tuberculosis, interstitial cystitis, and bladder neoplasms. Histologically, the biopsy will show a diffuse infiltration with eosinophils and other inflammatory cells. Electron microscopy and special staining techniques have not provided other specific findings (Brown, 1960; Littleton et al., 1982).

An allergic etiology for eosinophilic cystitis is suggested by the strong allergic histories and the bladder histology. This theory has been questioned by Kessler and associates (1975), who reported newborn children with this disease and with elevated serum IgE levels but who would be expected to have been protected from allergens while in utero. Also noted has been the high proportion of black children among all of the children reported with this condition (Kessler, 1971). The diagnosis is made by a history of irritative bladder symptoms (with or without infections) and bladder biopsy. Related diagnostic findings on ultrasound may show a thickened and irregular bladder wall, and vesicoureteral reflux is an associated finding in approximately one third of cases.

Because of the supposed allergic etiology, therapy has usually been with ACTH or corticosteroids (Brown, 1960). The natural course of the inflammatory process is self-limited in most reported cases, so that improvement after medication may not be due to it. The average duration of symptoms in treated or untreated cases is difficult to discern from the available case reports. Serious sequelae such as hydronephrosis have been reported, so surveillance of kidney function or upper tract dilatation is important in addition to monitoring of symptoms (Kessler et al., 1975; Littleton et al., 1982). In mildly symptomatic cases without upper tract dilatation, antihistamines alone have seemed beneficial (Keeton, 1978). In another report, a one-week course of prednisone and antihistamines seemed effective (Walsh, 1978).

Serious sequelae from eosinophilic cystitis appear to be rare. One case of renal failure due to hydronephrosis with a documented eosinophilic infiltration of the kidneys has been reported (Horowitz et al., 1972). In rare cases, the symptoms, bleeding, and upper tract dilatation have warranted the use of radiation to the bladder, extensive cystoscopic fulguration, and partial or total cystectomy (Brown, 1960; Sidh et al., 1980). Bladder biopsies may be needed to rule out a neoplastic lesion, but cystectomy and urinary diversion seldom are indicated. The need for augmentation of a bladder with severe fibrosis and reduced capacity due to eosinophilic cystitis has not been reported. Recurrence of severe symptoms and bleeding following partial cystectomy have led to one report cautioning against such bladder augmentation (Brown, 1960).

SUMMARY

Inflammatory conditions of the bladder are among the most common and symptomatic of conditions affecting the genitourinary tract. New concepts in the pathophysiology of bacterial cystitis offer therapeutic possibilities in addition to the prevalent current use of antibiotics. While endoscopic examination of the bladder remains the mainstay in diagnosis, a wide spectrum of localization tests and imaging modalities can add to diagnostic ability and can facilitate therapy. With the use of cultures for common and uncommon infecting agents, endoscopy, cytology, and bladder biopsy, the noninfectious inflammatory conditions can be diagnosed and often treated. Bladder neoplasm frequently must be carefully excluded from diagnostic consideration before appropriate therapy can be instituted for these highly symptomatic inflammations.

REFERENCES

1. Abdou NI, NaPombejara C, Sagawa A, et al: Malacoplakia: Evidence for monocyte lysosomal abnormality correctable by cholinergic agonist in vitro and in vivo. *N Engl J Med* 1977; 297:1413.
2. Arnesen E, Halvorsen S, Skjorten F: Malacoplakia in a renal transplant. *Scand J Urol Nephrol* 1977; 11:93.
3. Arpi M, Renneberg J: The clinical significance of *Staphylococcus aureus* bacteriuria. *J Urol* 1984; 132:697.
4. Balish MJ, Jensen J, Uehling DT: Bladder mucin: A scanning electron microscopy study in experimental cystitis. *J Urol* 1982; 128:1060.
5. Bartlett RC, Treiber N: Clinical significance of mixed bacterial cultures of urine. *Am J Clin Pathol* 1984; 82:319.
6. Ben Ami T: The sonographic evaluation of urinary tract infections in children. *Semin Ultrasound CT MR* 1984; 5:19.
7. Benson RC, Marquis WE, Crummy AB, et al: Embolization for genitourinary disorders. *Urology* 1980; 16:587.
8. Bloedorn FG, Young JD, Cuccia CA, et al: Radiotherapy in carcinoma of the bladder: Possible complications and their prevention. *Radiology* 1962; 79:576.
9. Bollgren I, Endstrom CF, Hammarlind M, et al: Low urinary counts of P-fimbriated Escherichia coli in presumed acute pyelonephritis. *Arch Dis Child* 1984; 59:102.
10. Bran JL, Levison ME, Kaye D: Entrance of bacteria into the female urinary bladder. *N Engl J Med* 1972; 286:626.
11. Breitenbucher RB: Bacterial changes in the urine samples of patients with long-term indwelling catheters. *Arch Intern Med* 1984; 144:1585.
12. Brock N, Stebar J, Pahl J, et al: Acrolein, the causative factor of urotoxic side effects of cyclophosphamide, isosfamide, trofosfamide, and sufosfamide. *Drug Res* 1979; 29:659.
13. Brown EW: Eosinophilic granuloma of the bladder. *J Urol* 1960; 83:665.
14. Brumfitt W, Hamilton-Miller JMT, Wood A: Evidence for showing in trimethoprim resistance during 1981—a comparison with earlier years. *J Antimicrob Chemother* 1983; 11:503.
15. Bruskewitz RC: Dysuria, in Peckham BM, Shapiro SS (eds): *Signs and Symptoms in Gynecology*. Philadelphia, JB Lippincott Co, 1983, pp 245–253.
16. Buckwold FJ, Ludwig P, Harding GKM, et al: Therapy for acute cystitis in adult women: Randomized comparizon of single-dose sulfisoxazole vs trimethoprim-sulfamethoxazole. *JAMA* 1982; 247:1839.

17. Burke JP, Garibaldi RA, Britt MR, et al: Prevention of catheter-associated urinary tract infections: Efficacy of daily meatal care regimens. *Am J Med* 1981, 70:655.

18. Busch R, Huland H: Correlation of symptoms and results of direct bacterial localization in patients with urinary tract infections. *J Urol* 1984; 132:282.

19. Chenoweth CV, Clawater EW: Interstitial cystitis in children. *J Urol* 1960; 83:150.

20. Collan Y, Alfthan O, Kivilaakso E, et al: Electronic microscopic and histological findings in urinary bladder epithelium in interstitial cystitis. *Eur Urol* 1976; 2:242.

21. Cox CE, Hinman F: Experiments with induced bacteriuria, vesical emptying and bacterial growth on the mechanism of bladder defense to infection. *J Urol* 1961; 86:739.

22. DeJuana CP, Everett JC: Interstitial cystitis: Experience and review of recent literature. *Urology* 1977; 10:325.

23. DeKernion JB: Complications of adjunct cancer therapy, in Smith RR, Skinner DG (eds): *Complications of Urologic Surgery.* Philadelphia, WB Saunders, 1976, pp 436–458.

24. Dismukes WE, Stamm AM, Graybill JR, et al: Treatment of systemic mycoses with ketoconazole: Emphasis on toxicity and clinical response in 52 patients. *Ann Intern Med* 1983, 98:13.

25. Droller MJ, Saral R, Santos G: Prevention of cyclophosphamide-induced hemorrhagic cystitis. *Urology* 1982; 20:256.

26. Duckett JW, Peters PC, Donaldson MH: Severe cyclophosphamide hemorrhagic cystitis controlled with phenol. *J Pediatr Surg* 1973; 8:55.

27. Edwards PD, Hurm RA, Jaeschke WH: Conversion of cystitis glandularis to adenocarcinoma. *J Urol* 1972; 108:568.

28. Edwards PR, Ewing WH: *Identification of the Enterobacteriaceae,* ed 3. Minneapolis, Burgess Publishing Co, 1972, pp 7–47.

29. Engel G, Schaeffer AJ, Grayhark JT, et al: The role of excretory urography and cystoscopy in the evaluation and management of women with recurrent urinary tract infections. *J Urol* 1980; 123:190.

30. Fair WR: Formalin in the treatment of massive bladder hemorrhage: Techniques, results, and complications. *Urology* 1974; 3:573.

31. Fairchild WV, Spence CR, Solomon HD, et al: The incidence of bladder cancer after cyclophosphamide therapy. *J Urol* 1979; 122:163.

32. Fairley KF, Grounds AD, Carson NE, et al: Site of infection in acute urinary-tract infection in general practice. *Lancet* 1971; 2:615.

33. Fang LST, Tolkoff-Rubin NE, Rubin R: Efficacy of single-dose and conventional amoxicillin therapy in urinary-tract infection localized by the antibody-coated bacteria technic. *N Engl J Med* 1978; 298:413.

34. Farid Z, El-Masry NA, Bassily S, et al: Treatment of biharzial obstructive uropathy with Praziquantel. *J Infect Dis* 1984; 150:307.

35. Fisher JF, Chew WH, Shadomy S, et al: Urinary tract infections due to *Candida albicans. Rev Infect Dis* 1982; 4:1107.

36. Fowler JE: Nosocomial catheter-associated urinary tract infection. *Infect Surg* 1983; 2:43.

37. Fowler JE: Prospective study of intravesical dimethyl sulfoxide in treatment of suspected early interstitial cystitis. *Urology* 1981; 18:21.

38. Franksson C: Interstitial cystitis: A clinical study of 59 cases. *Acta Chir Scand* 1957; 113:51.

39. Freedman A, Ehrlich RM, Ljung BM: Prevention of cyclophosphamide cystitis with 2-mercaptoethane sodium sulfonate: A histologic study. *J Urol* 1984; 132:580.

40. Freiha FS, Faysal MH, Stamey TA: The surgical treatment of intractable interstitial cystitis. *J Urol* 1980; 123:632.

41. Freiha FS, Stamey TA: Cystolysis: A procedure for the selective denervation of the bladder. *J Urol* 1980; 123:360.

42. Fuloria HK, Jaiswal MSD, Singh RV: Primary hydatid cyst of bladder. *Br J Urol* 1975; 47:192.

43. Gallagher DJ, Montgomerie JZ, North JD: Acute infections of the urinary tract and the urethral syndrome in general practice. *Br Med J* 1965; 1:662.

44. Geist RW, Antolak SJ: Interstitial cystitis in children. *J Urol* 1970; 104:922.

45. Ghazizadeh M, Kagawa S, Yoneda F, et al: Further studies on specificity of red cell adherence test: Nonmalignant bladder lesions. *Urology* 1985; 25:85.

46. Gillenwater JY: Detection of urinary leukocytes by chemistrip-L. *J Urol* 1981; 125:383.

47. Gillespie L: The diaphragm: An accomplice in recurrent urinary tract infections. *Urology* 1984; 24:25.

48. Golin AL, Benson RC: Cyclophosphamide hemorrhagic cystitis requiring urinary diversion. *J Urol* 1977; 118:110.

49. Goodhart GL: In vivo vs in vitro susceptibility of enterococcus to trimethoprim-sulfamethoxazole: A pitfall. *JAMA* 1984; 252:2748.

50. Gordon HL, Rossen RD, Hersh EM, et al: Immunologic aspects of interstitial cystitis. *J Urol* 1973; 109:228.

51. Gow JG: Genitourinary tuberculosis: A 7 year review. *Br J Urol* 1979; 51:239.

52. Greenberg RN, Sanders CV, Lewis AD, et al: Single-dose cefaclor therapy of urinary tract infection. *Am J Med* 1981; 71:841.

53. Hall PW, Vasiljevic M: Beta$_2$-microglobulin excretion as an index of renal tubular disorders with special reference to endemic Balkan nephropathy. *J Lab Clin Med* 1973; 81:897.

54. Hanash KA, Pool TL: Interstitial and hemorrhagic cystitis: Viral, bacterial and fungal studies. *J Urol* 1970; 104:705.

55. Hansman DJ, Brown JM: Eosinophilic cystitis: A case associated with possible hydatid infection. *Med J Aust* 1974; 2:563.

56. Hashida Y, Gaffney PC, Yunis EJ: Acute hemorrhagic cystitis of childhood and papovavirus-like particles. *J Pediatr* 1976; 89:85.

57. Holmberg L, Boman G, Böttiger LE: Adverse reactions

to nitrofurantoin: Analysis of 921 reports. *Am J Med* 1980; 69:733.

58. Horowitz J, Slavin S, Pfau A: Chronic renal failure due to eosinophilic cystitis. *Ann Allergy* 1972; 30:502.

59. Howard TL: My personal opinion on interstitial cystitis. *J Urol* 1944; 51:526.

60. Hoyt H: Cortisone in urologic conditions with report of a trial in interstitial cystitis. *J Urol* 1952; 66:526.

61. Hunner GL: A rare type of bladder ulcer in women: Report of cases. *Trans South Surg Gynecol Assoc* 1914; 27:247.

62. Huovinen P: Trimethoprim-resistant *Escherichia coli* in a geriatric hospital. *J Infect* 1984; 8:145.

63. Jensen J, Uehling DT, Kim K, et al: Enhanced immune response in the urinary tract of the rat following vaginal immunization. *J Urol* 1984; 132:164.

64. Jodal U, Lindberg U, Lincoln K: Level of diagnosis of symptomatic urinary tract infections in childhood. *Acta Paediatr Scand* 1975; 64:201.

65. Jokinen EJ, Alfthan OS, Oravisto KJ: Antitissue antibodies in interstitial cystitis. *Clin Exp Immunol* 1972; 11:333.

66. Juchau SV, Nauschuetz WF: Evaluation of a leukocyte esterase and nitrite test strip for detection of bacteriuria. *Curr Microbiol* 1985; 11:119.

67. Kaijser B, Hanson LA, Jodal U, et al: Frequency of *E. coli* K antigens in urinary-tract infections in children. *Lancet* 1977; 1:663–664.

68. Källenius G, Möllby R, Svenson SB, et al: Occurrence of P-fimbriated Escherichia coli in urinary tract infections. *Lancet* 1981; 2:1369.

69. Källenius G, Winberg J: Bacterial adherence to periurethral epithelial cells in girls prone to urinary-tract infections. *Lancet* 1978; 2:540–543.

70. Kambal A: The relation of urinary bilharziasis to vesical stones in children. *Br J Urol* 1981; 53:315.

71. Kass EH, Finland M: Asymptomatic infections of the urinary tract. *Trans Assoc Am Physicians* 1956; 69:56.

72. Keeton JE: Eosinophilic cystitis. *Soc Pediatr Urol Newslett*, Nov 22, 1978.

73. Kenney M, Loechel AB, Lovelock FJ: Urine culture in tuberculosis. *Am Rev Respir Dis* 1960; 82:564.

74. Kessler WO, Clark PL, Kaplan GW: Eosinophilic cystitis. *Urology* 1975; 6:499.

75. Keys TF, Edson RS: Antimicrobial agents in urinary tract infections. *Mayo Clin Proc* 1983; 58:165.

76. Kraft CA, Platt DJ, Timbury MC: Distribution and transferability of plasmids in trimethoprim-resistant urinary stains of *Escherichia coli*: A comparative study of hospital isolates. *J Med Microbiol* 1984; 18:95.

77. Kramer SA: Genito-urinary infections, in Kelalis PP, King LR, Belman AB (eds): *Clinical Pediatric Urology.* Philadelphia, WB Saunders, 1985, pp 256–274.

78. Krieger JN: Urologic aspects of trichomoniasis. *J Urol* 1981; 18:414.

79. Kumar APM, Wrenn EL, Jayalakshmamma B, et al: Silver nitrate irrigation to control bladder hemorrhage in children receiving cancer therapy. *J Urol* 1976; 116:85.

80. Kunin CM, DeGroot JE, Uehling DT, et al: Detection of urinary tract infections in 3- to 5-year-old girls by mothers using a nitrite indicator strip. *Pediatrics* 1976; 57:829.

81. Kunin CM, McCormack RC: Prevention of catheter-induced urinary-tract infections by sterile closed drainage. *N Engl J Med* 1966; 274:1155.

82. Kuthy E, Ormos J: X-ray microprobe analysis of Michaelis-Gutmann bodies in human and experimental malakoplakia. *Am J Pathol* 1978; 90:411.

83. Lacey RW: Do sulphonamide-trimethoprim combinations select less resistance to trimethoprim than the use of trimethoprim alone? *J Med Microbiol* 1982; 15:403.

84. Larsen S, Thompson SA, Hald T, et al: Mast cells in interstitial cystitis. *Br J Urol* 1982; 54:283.

85. Latham RH, Stamm WE: Role of fimbriated *Escherichia coli* in urinary tract infections in adult women: Correlation with localization studies. *J Infect Dis* 1984; 149:835.

86. Lattimer JK, Wechsler M: Genitourinary tuberculosis, in Harrison JH, Gittes RF, Perlmutter AD, et al (eds): *Campbell's Urology*, ed 4. Philadelphia, WB Saunders Co, 1978, vol 1, pp 557–575.

87. Leach GE, Raz S: Interstitial cystitis, in Raz S (ed): *Urology.* Philadelphia, WB Saunders Co, 1983, pp 351–356.

88. Lebowitz RL: Intravesical chemical cauterization and methemoglobinemia. *Pediatrics* 1980; 65:630.

89. Lecatsas G, Prozesky OW, Van Wyk J, et al: Papovavirus in urine after renal transplantation. *Science* 1973; 241:343.

90. LeDuc A, Cariou G, Jeaneau PL: Case profile: Cystitis emphysematosa. *Urology* 1985; 25:88.

91. Leffler H, Svanborg-Eden C: Glycolipid receptors for uropathogenic *Escherichia coli* on human erythrocytes and uroepithelial cells. *Infect Immun* 1981; 34:920.

92. Lewin KJ, Harell GS, Lee AS, et al: Malacoplakia: An electron-microscopic study: Demonstration of bacilliform organisms in malacoplakic macrophages. *Gastroenterology* 1974; 66:28.

93. Lindberg U, Jodal U, Hanson LA, et al: Asymptomatic bacteriuria in school girls: IV. Difficulties of level diagnosis and the possible relation to the character of infecting bacteria. *Acta Paediatr Scand* 1975; 64:574.

94. Lippert MC, Teates CD, Howards SS: Detection of enteric-urinary fistulas with a noninvasive quantitative method. *J Urol* 1984; 132:1134.

95. Littleton RH, Farah RN, Cerny JC: Eosinophilic cystitis: An uncommon form of cystitis. *J Urol* 1982; 127:132.

96. Lohr JA, Nunley DH, Howards SS, et al: Prevention of recurrent urinary tract infections in girls. *Pediatrics* 1977; 59:562.

97. Lomberg H, Hanson LA, Jacobsson B, et al: Correlation of P blood group, vesicoureteral reflux, and bacterial attachment in patients with recurrent pyelonephritis. *N Engl J Med* 1983; 308:1189.

98. Lorentz WB, Resnick MI: Comparison of urinary lactic dehydrogenase with antibody-coated bacteria in the urine sediment as means of localizing the site of urinary tract infection. *Pediatrics* 1979; 64:672.

99. Maderazo EG, Berlin BB, Morhardt C: Treatment of

malakoplakia with trimethoprim sulfamethoxazole. *Urology* 1979; 13:70.

100. Mahboubi S, Duckett JN, Spackman TJ: Ureteritis cystica after treatment of cyclophosphamide-induced hemorrhagic cystitis. *Urology* 1976; 7:521.

101. Malek RS, Greene LF, Farrow GM: Amyloidosis of the urinary bladder. *Br J Urol* 1971; 43:189.

102. Marberger H, Huber W, Bartsch G, et al: Orgotein: A new antiinflammatory metalloprotein drug evaluation of clinical efficacy and safety in inflammatory conditions of the urinary tract. *J Urol Nephrol* 1974; 6:61.

103. Mattila J: Vascular immunopathology in interstitial cystitis. *Clin Immunol Immunopathol* 1982; 23:648.

104. Matilla J, Pitkänen R, Vaalasti T, et al: Fine-structural evidence for vascular injury in patients with interstitial cystitis. *Virchows Arch* 1983; 398:347.

105. McClurg FV, D'Agostion AND, Martin JH, et al: Ultrastructural demonstration of intracellular bacteria in three cases of malakoplakia of the bladder. *Am J Clin Pathol* 1973; 60:780.

106. McCracken GH, Ginsburg CM, Namasonthi V, et al: Evaluation of short-term antibiotic therapy in children with uncomplicated urinary tract infections. *Pediatrics* 1981; 67:796.

107. McGuire EJ, Lytton B, Corog JL: Interstitial cystitis following colocystoplasty. *Urology* 1973; 2:28.

108. McKay EH: Malacoplakia in ulcerative colitis. *Arch Pathol Lab Med* 1978; 102:140.

109. McKerrow W, Davidson-Lamp N, Jones PF: Urinary tract infection in children. *Br Med J* 1984; 289:299.

110. Meares EM: Bacterial prostatitis vs "prostatosis." *JAMA* 1973; 224(10):1372.

111. Meirowsky AM: The management of chronic interstitial cystitis by differential sacral neurectomy. *J Neurosurg* 1969; 30:604.

112. Messing EM, Freiha FS: Complication of Clorpactin WCS-90 therapy for interstitial cystitis. *Urology* 1979; 13:389.

113. Messing EM, Stamey TA: Interstitial cystitis: Early diagnosis, pathology, and treatment. *Urology* 1978; 12:381.

114. Michaelis L, Gutmann C: Uber Einchlusse in blas ento moren. *Z Klin Med* 1902; 47:208.

115. Michigan S: Genitourinary fungal infection. *J Urol* 1976; 116:390.

116. Mininberg DT, Watson C, Desquitado M: Viral cystitis with transient secondary vesicoureteral reflux. *J Urol* 1982; 127:983.

117. Mulholland SG, Mooreville M, Parson CL: Urinary tract infections and P blood group antigens. *Urology* 1984; 24:232.

118. Mufson MA, Zollar LM, Mankad VN, et al: Adenovirus infection in acute hemorrhagic cystitis: A study in 25 children. *Am J Dis Child* 1971; 121:281.

119. Neu HC: Changing mechanisms of bacterial resistance. *Am J Med* 1984; 77:11.

120. Nicolle LE, Harding GKM, Preiksaitis J, et al: The association of urinary tract infection with sexual intercourse. *J Infect Dis* 1982; 146:579.

121. Nielsen OS, Maigaard S, Frimodt-Moller N, et al: Prophylactic antibiotics in transurethral prostatectomy. *J Urol* 1981; 126:60.

122. Nitze M: *Lehrbuch der Kystoskopie: Ihre Technik und Klinishe Bedeuting*. Berlin, JE Berman, 1907, pp 205–210.

123. O'Conor VJ: Clorpactin WCS-90 in interstitial cystitis. *Q Bull Northwest Med Sch* 1955; 29:392.

124. O'Dowd TC, Ribeiro CD, Munro J, et al: Urethral syndrome: A self limiting illness. *Br Med J* 1984; 288:1349.

125. Oliver JM: Impaired microtubule function correctable by cyclic GMP and cholinergic agonists in the Chediak-Higashi syndrome. *Am J Pathol* 1976; 85:395.

126. Oravisto KJ: Epidemiology of interstitial cystitis. *Ann Chir Gynecol Fen* 1975; 64:75.

127. Oravisto KJ: Interstitial cystitis as an autoimmune disease: A review. *Eur Urol* 1980; 6:10.

128. Oravisto KJ, Alfthan OS: Treatment of interstitial cystitis with immunosuppression and chloroquine derivatives. *Eur Urol* 1976; 2:82.

129. Orskov I, Ferencz A, Orskov F: Tamm-Horsfall protein or uromucoid is the normal urinary slime that traps type 1 fimbriated *Escherichia coli*. *Lancet* 1980; 1:887.

130. Ostroff EB, Chenault OW: Alum irrigation for the control of massive bladder hemorrhage. *J Urol* 1982; 128:929.

131. Padgett BL, Walker DL, Desquitado MM, et al: BK virus and non-haemorrhagic cystitis in a child. *Lancet* 1983; 1:770.

132. Parsons CL: Prevention of urinary tract infection by the exogenous glycosaminoglycan sodium prentosanpolysulfate. *J Urol* 1982; 127:167.

133. Parsons CL, Schmidt JD, Pollen JJ: Successful treatment of interstitial cystitis with sodium pentosanpolysulfate. *J Urol* 1983; 130:51.

134. Pawar HN: Management of vesicouterine fistula following cesarean section. *Urology* 1985; 25:66.

135. Pearl F, Strauss B: Presacral neurectomy and sacral ganglionectomy. *J Urol* 1938; 39:645.

136. Persky L, Austen G: ACTH in radiation cystitis. *J Urol* 1953; 70:724.

137. Persky L, Selman SH: Therapeutic application of epsilon aminocaproic acid in the urology patient. *Contemp Surg* Sept 1980, p 17.

138. Pfau A, Sacks TG, Shapiro A, et al: A randomized comparison of 1-day versus 10-day antibacterial treatment of documented lower urinary tract infection. *J Urol* 1984; 132:931.

139. Phillips FS, Sternberg SS, Cronin AP, et al: Cyclophosphamide and urinary bladder toxicity. *Cancer Res* 1961; 21:1577.

140. Platt R, Murdock B, Polk BF, et al: Reduction of mortality associated with nosocomial urinary tract infection. *Lancet* 1983; 1:893–897.

141. Plotz PH, Klippel JH, Decker JL, et al: Bladder complications in patients receiving cyclophosphamide for systemic lupus erythematosus or rheumatoid arthritis. *Ann Intern Med* 1979; 91:221.

142. Pool TL: Irradiation cystitis: Diagnosis and treatment. *Surg Clin North Am* 1959; 39:947.

143. Principi N, Dalla Villa A, Assael BM, et al: Urinary excretion of *N*-acetyl-β-D-glucosaminidase (NAG) by children with upper or lower urinary tract infections. *Acta Paediatr Scand* 1982; 71:1033.

144. Pugh RCB: Lower urinary tract, in Kissane JM (ed): *Anderson's Pathology.* St Louis, CV Mosby Co, 1985, pp 772–790.

145. Pyeritz RE, Droller MJ, Bender WL, et al: An approach to the control of massive hemorrhage in cyclophosphamide-induced cystitis by intravenous vasopressin: A case report. *J Urol* 1978; 120:253.

146. Redman JF, Seibert JJ: The role of excretory urography in the evaluation of girls with urinary tract infection. *J Urol* 1984; 132:953.

147. Rettig PJ, Nelson JD: Genital tract infections with Chlamydia trachomatis in prepubertal children. *J Pediatr* 1981; 99:206.

148. Rohner TJ, Tuliszewski RM: Fungal cystitis: Awareness, diagnosis, and treatment. *J Urol* 1980; 124:142.

149. Ronald AR: Current concepts in the management of urinary tract infections in adults. *Med Clin North Am* 1984; 68:335.

150. Saxena SR, Laurance BM, Shaw DG: The justification for early radiologic investigations of urinary-tract infection in children. *Lancet* 1975; 2:403–406.

151. Schaeffer AJ, Flynn S, Jones J: Comparison of cinoxacin and trimethoprim-sulfamethoxazole in the treatment of urinary tract infections. *J Urol* 1981; 125:825.

152. Schaeffer AJ, Jones JM, Dunn JK: Association of in vitro *Escherichia coli* adherence to vaginal and buccal epithelial cells with susceptibility of women to recurrent urinary-tract infections. *N Engl J Med* 1981; 304:1062.

153. Schaeffer AJ, Jones JM, Flynn SS: Prophylactic efficacy in cinoxacin in recurrent urinary tract infection: Biological effects on the vaginal and fecal flora. *J Urol* 1982; 127:1128.

154. Schardjin GHC, Statius Van Eps LW, Pauw W, et al: Comparison of reliability of tests to distinguish upper from lower urinary tract infection. *Br Med J* 1984; 289:284.

155. Schultz HJ, McCaffrey LA, Keys TF, et al: Acute cystitis: A prospective study of laboratory tests and duration of therapy. *Mayo Clin Proc* 1984; 59:391.

156. Seddon JM, Best L, Bruce AW: Intestinocystoplasty in treatment of interstitial cystitis. *Urology* 1977; 10:431.

157. Sheehan G, Harding GKM, Ronald AR: Advances in the treatment of urinary tract infection. *Am J Med* 1984; 76:141.

158. Shirley SW, Mirelman S: Experiences with colocystoplasties, cecocystoplasties and ileocystoplasties in urologic surgery: 40 patients. *J Urol* 1978; 120:165.

159. Sidh SM, Smith SP, Silber SB, et al: Eosinophilic cystitis: Advanced disease requiring surgical intervention. *Urology* 1980; 15:23.

160. Silk M: Bladder antibodies in interstitial cystitis. *J Urol* 1970; 103:307.

161. Skudowitz RB, Weintraub CM: Malacoplakia associated with transitional cell carcinoma of bladder. *J Urol* 1977; 118:482.

162. Smellie JM, Katz G, Grüneberg RN: Controlled trial of prophylactic treatment in childhood urinary-tract infection. *Lancet* 1978; 2:175–178.

163. Smey P: Personal communication.

164. Smith CR, Lipsky JJ, Laskin OL, et al: Double-blind comparison of the nephrotoxicity and auditory toxicity of gentamicin and tobramycin. *N Engl J Med* 1980; 302:1106.

165. Smith RB, Van Cangh P, Skinner DG, et al: Augmentation enterocystoplasty: A critical review. *J Urol* 1977; 118:35.

166. Solomkin JS, Flohr AM, Simmons RL: Indications for therapy for fungemia in postoperative patients. *Arch Surg* 1982; 117:1272.

167. Srinivas V, Orihuela E, Lloyd KO, et al: Estimation of ABO(H) isoantigen expression in bladder tumors. *J Urol* 1985; 133:25.

168. Stamey TA: *Pathogenesis and Treatment of Urinary Tract Infections.* Baltimore, Williams & Wilkins Co, 1980.

169. Stamey TA, Condy M, Mihara G: Prophylactic efficacy of nitrofurantoin macrocrystals and trimethoprim-sulfamethoxazole in urinary infections: Biological effects on the vaginal and rectal flora. *N Engl J Med* 1977; 296:780.

170. Stamey TA, Fair WR, Timothy MM, et al: Serum versus urinary antimicrobial concentrations in cure of urinary-tract infections. *N Engl J Med* 1974; 291:1159.

171. Stamm WE, Counts GW, Running KR, et al: Diagnosis of coliform infection in acutely dysuric women. *N Engl J Med* 1982; 307:463.

172. Stanton MJ, Maxted W: Malacoplakia: A study of the literature and current concepts of pathogenesis, diagnosis and treatment. *J Urol* 1981; 125:139.

173. Stark RP, Maki DG: Bacteriuria in the catheterized patient: What quantitative level of bacteriuria is relevant? *N Engl J Med* 1984; 311:560.

174. Stewart BH, Persky L, Kiser WS: The use of dimethyl sulfoxide (DMSO) in the treatment of interstitial cystitis. *J Urol* 1967; 98:671.

175. Streem SB: Genitourinary malacoplakia in renal transplant patients: Pathogenesis, prognostic and therapeutic considerations. *J Urol* 1984; 132:10.

176. Sutphin M, Middleton AW: Eosinophilic cystitis in children: A self-limited process. *J Urol* 1984; 132:117.

177. Svanborg-Eden C, Hanson LA, Jodal U, et al: Variable adherence to normal human urinary-tract epithelial cells of *Escherichia coli* strains associated with various forms of urinary-tract infection. *Lancet* 1976; 2:490–492.

178. Svenson SB, Källenius G, Möllby R, et al: Rapid identification of P-fimbriated *Escherichia coli* by a receptor-specific particle agglutination test. *Infection* 1982; 10:209.

179. Taylor-Robinson P, McCormack WM: The genital mycoplasmas. *N Engl J Med* 1980; 302:1003.

180. Thomas V, Shelokov A, Forland M: Antibody-coated bacteria in the urine and the site of urinary-tract infection. *N Engl J Med* 1974; 290:588.

181. Thompson RL, Haley CE, Searcy MA, et al: Catheter-associated bacteriuria: Failure to reduce attack rates using periodic instillations of a disinfectant into urinary drainage systems. *JAMA* 1984; 251:747.

182. Tolkoff-Rubin NE, Wilson ME, Zurombskis P, et al: Single-dose amoxicillin therapy of acute uncomplicated urinary tract infections in women. *Antimicrob Agents Chemother* 1984; 25:626.

183. Tolley A: The effect of N-acetyl cysteine on cyclophosphamide cystitis. *Br J Urol* 1977; 49:659.

184. Turck M, Browder AA, Lindemeyer RI, et al: Failure of prolonged treatment of chronic urinary-tract infections with antibiotics. *N Engl J Med* 1962; 267:999.

185. Turck M, Goffe B, Petersdorf RG: The urethral catheter and urinary tract infection. *J Urol* 1962; 88:834.

186. Uehling DT: Effect of vesicoureteral reflux on concentrating ability. *J Urol* 1971; 106:947.

187. Uehling DT, Balish E: Antibacterial activity of urinary immunoglobulins. *Birth Defects: Original Article Series.* 1977; 13:421–424.

188. Uehling DT, Jensen J, Balish E: Vaginal immunization against urinary tract infection. *J Urol* 1982; 128:1382.

189. Uehling DT, King LR: Secretory immunoglobulin-A excretion in cystitis cystica. *Urology* 1973; 1:305.

190. Uehling DT, Steihm RE: Elevated urinary secretory IgA in children with urinary tract infection. *Pediatrics* 1971; 47:40.

191. Utz DC, Zincke H: The masquerade of bladder cancer in situ as interstitial cystitis. *J Urol* 1974; 111:160.

192. Utz JP: Chemotherapy for the systemic mycoses: The prelude to ketoconazole. *Rev Infect Dis* 1980; 2:625.

193. Vitenson JH, Grabstald H, Whitmore WF: Ten thousand or more rads supervoltage irradiation to the bladder: Efficacy, sequelae and management. *J Urol* 1972; 107:973.

194. Vosti KL: Recurrent urinary tract infections: Prevention by prophylactic antibiotics after sexual intercourse. *JAMA* 1975; 231:934.

195. Wall RL, Clausen K: Carcinoma of the urinary bladder in patients receiving cyclophosphamide. *N Engl J Med* 1975; 293:271.

196. Walsh A: Interstitial cystitis, in Harrison JH, Gittes RF, Perlmutter AD, et al (eds): *Campbell's Urology* ed 4. Philadelphia, WB Saunders Co, 1978, pp 693–707.

197. Walther PC, Kaplan GW: Cystoscopy in children: Indications for its use in common urologic problems. *J Urol* 1979; 122:717.

198. Watson RA: Ifosfamide: Chemotherapy with new promise and new problems for the urologist. *Urology* 1984; 24:465.

199. Wear JB: Correlation of pyuria, stained urine smear, urine culture, and the uroscreen test. *J Urol* 1966; 96:808.

200. Wechsler H: Update on chemotherapy of renal tuberculosis. *J Urol* 1980; 124:319.

201. Winberg J, Bergstrom T, Lincoln K, et al: Treatment trials in urinary tract infection with special reference to the effect of antimicrobials on the fecal and peri-urethral flora. *Clin Nephrol* 1973; 1:142.

202. Wise GJ, Goldberg PE, Kozinn PJ: Do the imidazoles have a role in the management of genitourinary fungal infections? *J Urol* 1985; 133:61.

203. Wise GJ, Kozinn PJ, Goldberg P: Amphotericin B as a urologic irrigant in the management of noninvasive candiduria. 1982; 128:82.

204. Wise GJ, Kozinn PJ, Goldberg P: Flucytosine in the management of genitourinary candidiasis: 5 years of experience. *J Urol* 1980; 124:70.

205. Witherington R, Branan WJ, Wray BB, et al: Malacoplakia associated with vesicoureteral reflux and selective immunoglobulin A deficiency. *J Urol* 1984; 132:975.

206. Worth PHL: The treatment of interstitial cystitis by cystolysis with observations on cystoplasty: A review after 7 years. *Br J Urol* 1980; 52:232.

207. Worth PHL, Turner-Warwick R: The treatment of interstitial cystitis by cystolysis with observations on cystoplasty. *Br J Urol* 1973; 45:65.

208. Zornow DH, Landes RR, Morganstern SL, et al: Malacoplakia of the bladder: Efficacy of bethanechol chloride therapy. *J Urol* 1979; 122:703.

Chapter 30

Urinary Fistulas

ANUP SINGH, M.D.
FRAY F. MARSHALL, M.D.

A urinary fistula often produces severe disabling symptoms including incontinence, infection, and pain. If basic surgical principles are followed, a successful repair can usually be achieved. Careful investigation is critical to obtain an accurate diagnosis. The etiology of the fistula is important and may determine subsequent management. The principles of closure include adequate urinary and fecal diversion. There should be no tension on the suture lines, and well-vascularized healthy tissue should be utilized for repair. There should be nonapposition of the suture lines. Any infection should be treated adequately. If possible, additional tissue such as omentum is placed between the suture lines to provide an additional buttress of well-vascularized tissue. The operative area should be adequately drained after anatomical reconstruction has been performed. These principles will be stressed in the management of urinary fistulas described in this chapter.

VESICOVAGINAL FISTULA

ETIOLOGY

In medically underdeveloped countries obstetric trauma remains the leading cause of vesicovaginal fistulas (Table 30–1) (Tahzib, 1983). Because of the higher standard of obstetric care in the United States, the incidence of vesicovaginal fistulas after childbirth in all patients with fistula has declined from 39% in Judd's series (1920) to fewer than 6% as reported by Massie in 1964 from the same clinic (Massie, 1964). On the other hand, some recent series have reported an increased incidence of fistula primarily related to the increased frequency of gynecologic operative procedures (Keettel et al., 1978; Counseller and Haigler, 1956). A fistula may follow an uncomplicated operation as a result of multiple sutures that create ischemia or a pelvic hematoma that ruptures into the bladder during the postoperative period. Preoperative radiation, pelvic inflammatory disease, previous surgery, or endometriosis may make surgery more difficult and increase the risk of fistula formation (Lawson, 1972).

Fistulas following aggressive transurethral resection of the bladder neck or transurethral resection of posteriorly based bladder tumors fortunately are now much rarer. Radiation-induced fistulas are commonly associated with carcinoma of the cervix. A few have been reported following radiotherapy of other pelvic malignancies. Fistulas may appear both during the course of radiotherapy or after treatment is completed. Those occurring during therapy are probably secondary to necrosis of tumor in the wall of the vagina or bladder. Later, they are likely secondary to excessive radiation, which leads to endarteritis obliterans and eventual necrosis of the vesical or vaginal wall. These fistulas usually appear during the first two years of treatment and have become less common with improved radiation therapy techniques.

TABLE 30–1.
Etiology of Vesicovaginal Fistulas

Congenital (rare)
Inflammatory
Infection, such as tuberculosis and schistosomiasis
Foreign body in bladder or vagina
Prolonged indwelling urethral catheter
Endometriosis
Trauma
Pelvic surgery
Abdominal hysterectomy
Vaginal surgery
Urologic procedures
Obstetric trauma
Prolonged labor
High forceps delivery
Cesarean section
Direct injury
Neoplastic
Carcinoma of cervix, bladder
Radiation induced

Other less common causes are the destruction of tissue by other malignant tumors, ulceration due to a foreign body, direct trauma (especially automobile accidents associated with pelvic fractures), tuberculosis, schistosomiasis, calculi, and endometriosis.

CLINICAL FEATURES

The most common presenting symptom is continuous leakage of urine from the vagina. Incontinence developing from operative trauma usually occurs 5–14 days postoperatively (Keettel et al., 1978). Leakage of urine is directly related to the size and position of the fistula. With small fistulas there may be slight drainage, more noticeable in some positions. These patients may also void reasonable amounts of urine. Patients with large fistulas have true incontinence of urine and may not void through the urethra at all. Urinary drainage causes irritation of the vagina, vulva, and perineum and usually produces an unpleasant, ammoniacal odor. In neglected cases phosphatic encrustations may be noted. The posthysterectomy fistula is typically located in the vault of the vagina (Tancer, 1980). An obstetric fistula is usually located at a lower level and may be associated with a urethral injury as well.

Graham (1967) described the painful syndrome of postradiation urinary vaginal fistula. He noted that 40% of patients with radiation-induced fistulas developed pain, usually associated with an alkaline urine and deposition of triple phosphate crystals, which further irritate compromised tissue. Urinary leakage can make the patient a social recluse, disrupt sexual activity, and lead to significant mental depression, insomnia, and low self-esteem (Harrison, 1983; Keettel et al., 1978).

DIAGNOSIS

In most patients the diagnosis is obvious (Table 30-.2). However, a complete urologic investigation is mandatory, especially to rule out an associated ureterovaginal fistula. This investigation should include a urinalysis, urine culture, and IV urogram. Other tests, such as a retrograde bulb-tipped ureteropyelogram, can be performed as necessary. In one series 12% of patients

with vesicovaginal fistulas had associated ureterovaginal fistulas (Goodwin and Scardino, 1980).

All patients with fistulas should undergo cystoscopy and vaginoscopy. Every attempt should be made to determine the exact location (relationship to ureteral orifices), size, and the underlying cause of the fistula. Successful repair will depend on these factors. A diligent search should be made for multiple less obvious, additional communications, since many failures have occurred because less obvious fistulas were overlooked. It is also worthwhile to determine the status of the bladder neck and note whether there is any associated loss of urethral tissue. In addition, the patient should be carefully examined using anesthesia and any suspicious lesions biopsied to rule out malignancy. In equivocal cases a simple double-dye test can be performed at the bedside (Raghavaiah, 1974). The vagina is packed with four sterile, wet gauze pads—one in the left and one in the right vaginal fornix, one at the midvaginal level, and the fourth at the vaginal outlet. The bladder is filled with 1% carmine solution (red), and five minutes later 5 ml of indigo carmine (blue) is injected IV. Ten minutes after the injection, the swabs are removed. A red stain (carmine) in the midvaginal pack indicates a vesicovaginal fistula; a blue stain on the upper swabs placed in the vaginal fornices indicates a ureterovaginal fistula; a red stain in the swab at the vaginal outlet indicates leakage through the urethra. Sometimes the findings can be misleading. The amount of indigo carmine excreted depends on renal function, and occasionally dye can reflux up a ureter and gain entry into the vagina via ureterovaginal fistula, providing a false positive result for a vesicovaginal fistula. There is no substitute for a diligent search on the part of the clinician to establish the site or sites of fistulas (Table 30-3).

PREOPERATIVE CARE

Medical and psychological support are necessary in these afflicted women. Cystitis, vaginitis, and perineal dermatitis should be treated with appropriate antibiotics. Catheter drainage may be indicated for small fistulas, and a few will heal spontaneously on this regimen. A small fistula will sometimes close after electrocoagulation and catheter drainage (O'Conor, 1980). The denuding of the fistula tract has also resulted in successful

TABLE 30–2.

Investigations for Urinary-Vaginal Fistulas

Urinalysis
Urine culture
High vaginal swab
Intravenous urogram
Dye test
Retrograde ureteropyelogram
Cystoscopy and vaginoscopy
CT scan

TABLE 30–3.

Differential Diagnosis of Urinary-Vaginal Fistulas

Vesicovaginal fistula
Ureterovaginal fistula
Severe stress or urge incontinence
Vaginitis or enterovaginal fistula

healing (Aycinema, 1977). Local care is very important, with meticulous cleansing and frequent pad changes to minimize inflammatory edema and vulval irritation. Protective creams or ointments are occasionally helpful in reducing vulvovaginitis. Keeping the area as dry as possible, at times by innovative use of collection or drainage devices, is often helpful.

TIMING OF REPAIR

Controversy still remains over the timing of repair. Most surgeons agree that there should be a waiting period of at least 3–4 months before surgical repair is performed (Keettel et al., 1978; Counseller and Haigler, 1956; Moir, 1956; Wein et al., 1980). Collins et al. (1971) recommended a ten-day course of cortisone to decrease inflammation, followed by early repair of the vesicovaginal fistula. Steroids may adversely affect healing, so that this approach appears more hazardous. In their reported series of 38 patients who had repair four weeks after diagnosis, a 28% failure rate was seen. Recently there have been other advocates for early repair (Persky et al., 1979; Raz, 1983), but this early aggressive treatment has not received wide acceptance. Wein et al. (1980) attributed two of their failures to inadequate delay between the initial surgical procedure and a subsequent attempt at fistula repair.

Every patient should be individualized and examined at regular intervals. The optimal time of repair is when all necrotic tissue has disappeared, the inflammatory edema has subsided, the fistula size has stabilized, and all acute inflammatory changes have resolved. The surgeon should not allow his objective judgment to be swayed by a patient's emotions and the referring physician's consternation. The failure rate for repair of the fistula will be higher if the tissue is acutely inflamed, edematous, ischemic, or necrotic. Some fistulas recognized in the immediate postoperative period or following trauma with minimal inflammatory changes may lend themselves to earlier repair. In cases of radiation-induced fistulas, the waiting period should probably be longer.

OPERATIVE MANAGEMENT

A vesicovaginal fistula can be repaired through a vaginal, abdominal, or combined approach. The vaginal approach is used frequently, and a success rate close to 90% is achieved with this technique (Counseller and Haigler, 1956; Massie et al., 1964; Moir, 1956). Compared with the abdominal approach, the vaginal repair of a fistula is a less extensive procedure and usually results in minimal discomfort and disability to the patient (Keettel, 1978; Moir, 1956). However, the abdominal approach often provides better access to the fistula. Improved exposure decreases the likelihood of ureteral in-

jury. If necessary, an omental pedicle graft can be interposed between the bladder and vaginal wall (Wein et al., 1980; Turner-Warwick et al., 1967). Complex fistulas, radiation-induced fistulas, and fistulas close to the ureteral orifices are best approached through the lower abdomen. A combined vaginal and abdominal approach can also be helpful in some instances (Weyrauch and Rous, 1966).

Whether the approach to an individual fistula is vaginal or abdominal depends on the findings and the surgeon, but the principles of repair remain the same. Successs is determined by a watertight, multilayered closure without tension and without overlapping suture lines. Maintenance of blood supply, removal of all necrotic tissue, obliteration of dead space, good bladder drainage, control of infection, and interposition of healthy tissue when appropriate are important technical considerations (Moir, 1956; Persky, 1979). The best chance at closure of the fistula is the first chance.

VAGINAL REPAIR

The inverted lithotomy position provides excellent exposure of the ventral vaginal wall. The bladder is then in a dependent position, and the fistula is more easily dissected. Many surgeons, however, prefer the conventional lithotomy position (Moir, 1956; Keettel et al., 1978; Persky, 1979). Repair of the fistula with the patient in a lithotomy position will be described (Fig 30–1).

After careful evaluation, the ureteral orifices may be catheterized cystoscopically if there is any concern about the ureters. The labia minora are sewn laterally, and a weighted posterior vaginal retractor is placed. Stay sutures are placed in healthy tissue on each side of the fistula, because they are very helpful in identifying the fistula's opening at a later stage in the operation when bleeding and excised tissue may make identification more difficult. Additional Babcock or Allis clamps may bring the fistula track closer to the surgeon. A Foley or Fogarty catheter can be placed through the fistula track as well. This maneuver can greatly aid dissection of a fistula, especially high in the vagina. If the vagina is small, an incision can be made through the dorsal lateral vaginal wall, similar to an episiotomy incision (Graham, 1965), to provide better access. Then a circumferential incision is made around the defect, and all scar tissue is excised. Adequate separation of bladder and vaginal wall is attempted to obtain a tension-free closure. In some instances the separation may be difficult and lead to unnecessary bleeding and devascularization of tissue (Moir, 1956; Patil et al., 1980). On the other hand, it must be remembered that if basic surgical principles are even slightly compromised in these cases, the failure rate is much higher. Caution is also

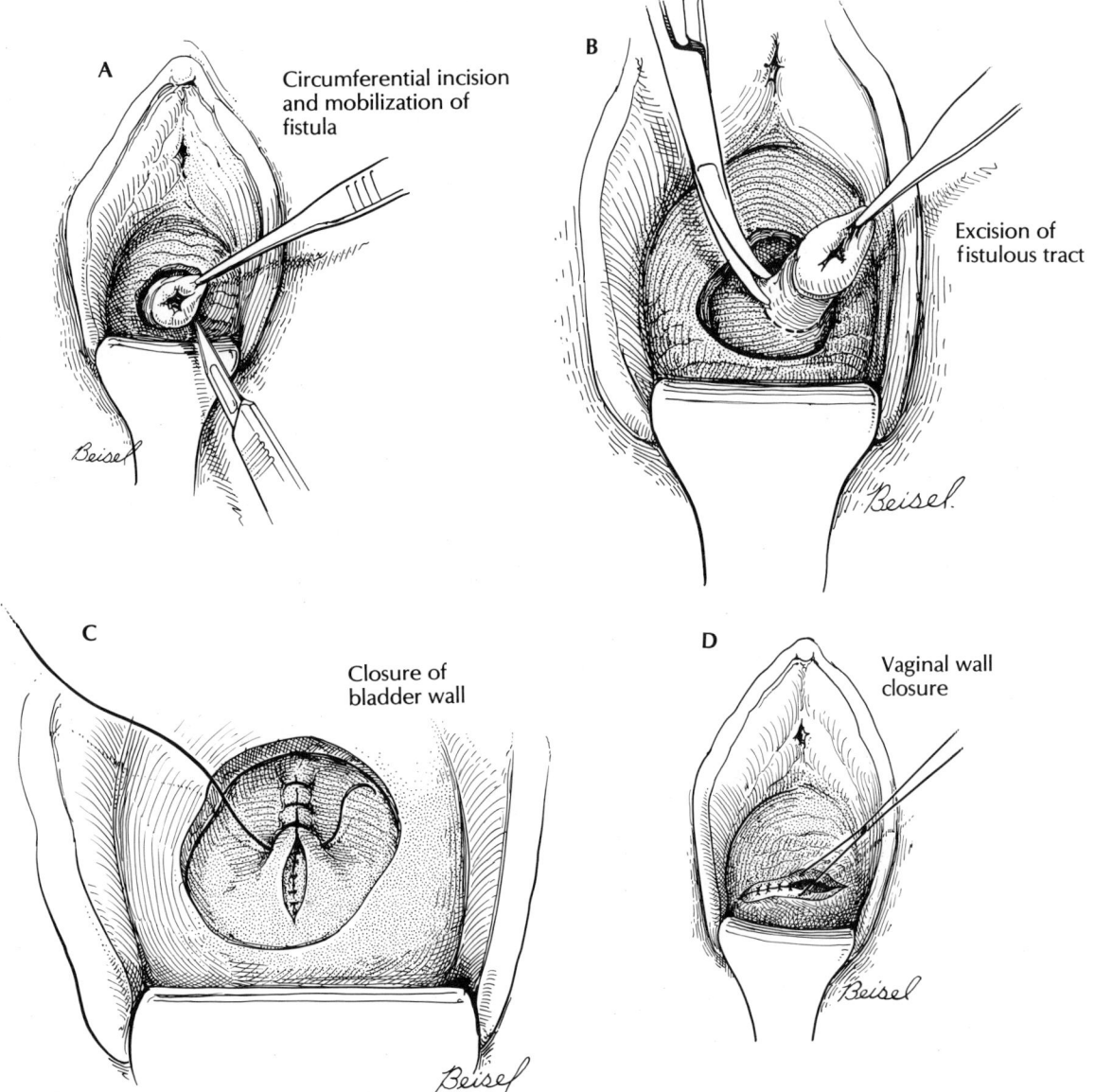

A, Circumferential incision and mobilization of fistula

B, Excision of fistulous tract

C, Closure of bladder wall

D, Vaginal wall closure

Beisel

FIG 30–1.
Schematic representation of a vaginal repair of a vesicovaginal fistula. **A,** circumferential incision around fistula tract with mobilization of the fistula. **B,** fistula is excised. **C,** bladder is closed and then the vagina is closed in multiple separate layers with avoidance of overlying suture lines. **D,** closure of vaginal wall.

exercised in the posthysterectomy fistula to avoid opening the pouch of Douglas and inadvertently creating a bowel fistula, as small bowel may be closely adherent. In high vaginal fistulas there is also the possibility of inadvertent injury to the ureters. After excision of the fistula tract with adequate separation of the bladder and vaginal wall, a three-layer closure is performed (O'Conor et al., 1973; Keettel et al., 1978) with absorbable sutures on an atraumatic needle. Others have even recommended a five-layered closure (Barnes and Bergman, 1964). If there is tension on the anastomosis after the closure, relaxing incisions have been recommended, but in that case the operation has probably been performed inadequately. A vaginal flap advancement or some other maneuver to create a tension-free anastomosis would be a better alternative. Occasionally bleeding may be profuse, but generally vaginal packing will stop bleeding.

One can advance a vaginal flap rather than totally excising the tract. In patients with simple small fistulas such an approach might work, but in more difficult cases the Latzko technique of partial colpocleisis can be performed (Latzko, 1942; Marshall, 1979). This method may cause loss of vaginal length and interfere with sexual function, but has been particularly effective in patients who have had previous radiation (Marshall, 1979). Additional interposition of omentum in patients having radiation is another good adjunct.

Various flaps have been increasingly utilized in recent years (Martius, 1928; Ingelman-Sundberg, 1978; Heckler et al., 1980). In patients with a large defect a pedicle flap will enhance the cure rate. With a low fistula a pedicle graft of labial fibro-fatty tissue (Martius, 1928) can be easily performed. On the other hand, a higher fistula with a large defect may require a buttock flap (Hendren, 1980) or a gracilis muscle flap in patients who have undergone radiation (Ingelman-Sundberg, 1978; Patil et al., 1980; Graham, 1967). For fistulas that involve the trigone or lie close to the ureteral orifices,

a bladder rotation flap or bladder advancement flap with interposition of omentum will often suffice (Kiruta and Goldstein, 1972; Coolsaet, 1984; Quartey, 1972; Su, 1969).

At the end of the operation the bladder is irrigated, and all clots are removed. A Foley catheter is left in place and the vagina packed with gauze.

ABDOMINAL APPROACH

The abdominal approach, exposing the bladder through a lower midline abdominal incision, has become increasingly popular (Fig 30–2). The advantage of this incision is that it can be readily extended upward, and a pedicle flap of omentum can be mobilized. Simple fistulas can be repaired transvesically, but in diffi-

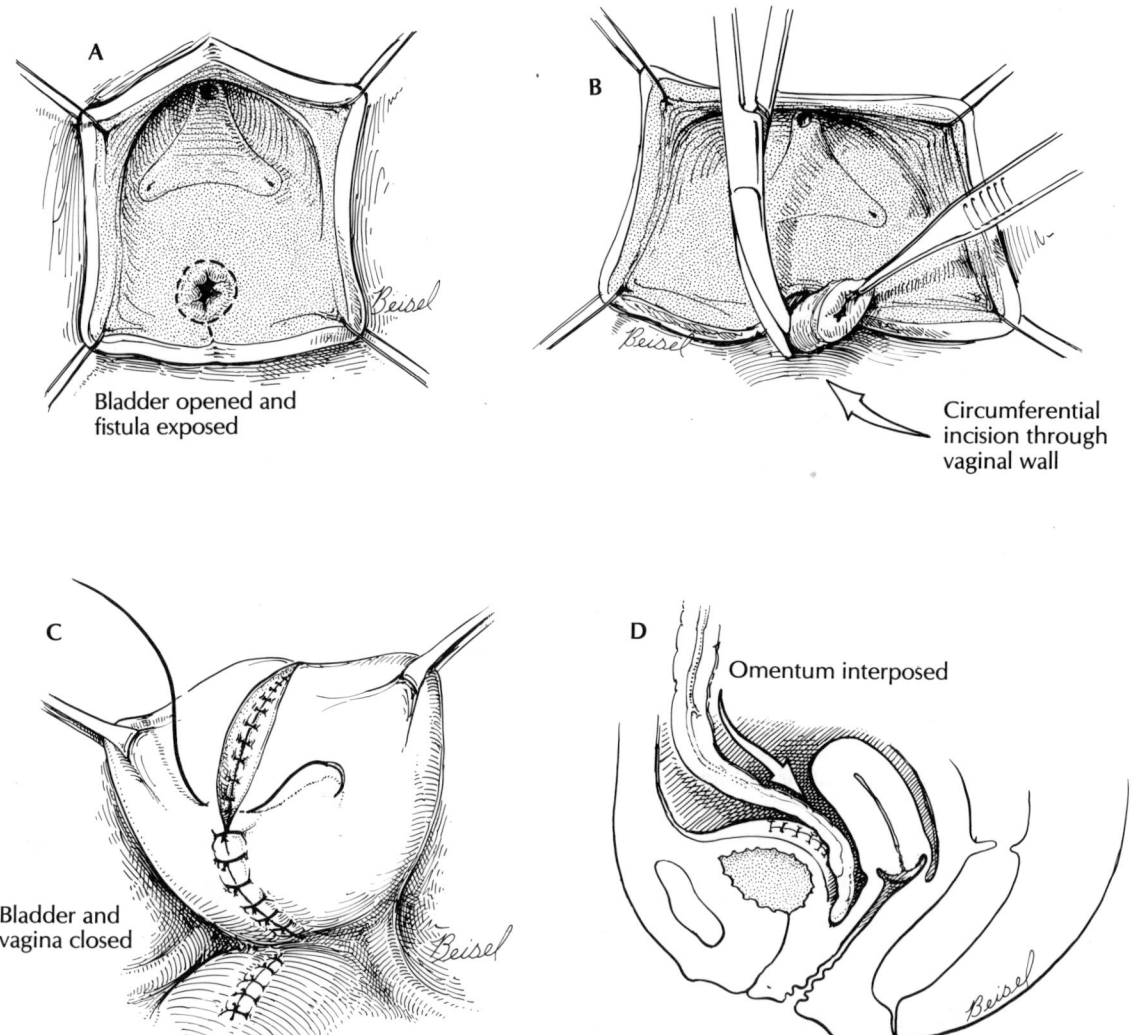

FIG 30–2.
Abdominal repair of vesicovaginal fistula. **A,** transabdominal, transperitoneal exposure of fistula tract reveals fistula above trigone after the bladder is opened. **B,** bladder wall is divided *(dotted line)* and the fistula, including the vaginal wall, is excised transvesically. **C,** bladder and vagina are closed in multiple separate layers. **D,** omentum is interposed between bladder and vagina to provide a vascularized pedicle to prevent further fistula formation.

cult complicated cases a transperitoneal approach is preferable.

The principles of repair are the same. The fistula is exposed through a cystotomy incision, the fistula tract is excised, and a meticulous closure in multiple layers is performed. Closures can be reinforced with a peritoneal flap mobilized in the lateral parietal peritoneum (Eisen et al., 1979). In large defects a pedicle flap of omentum is interposed between at least two suture lines (Turner-Warwick et al., 1967). A urethral and occasionally a suprapubic catheter ensures continuous drainage.

POSTOPERATIVE CARE

The vaginal pack is removed after 24–36 hours. The patient is encouraged to be ambulatory as early as possible to minimize thromboembolic complications, which have been reported in most large series (Counseller and Haigler, 1956). Antibiotics are used intraoperatively and frequently postoperatively until all tubes are removed. Unobstructed vesical drainage is mandatory for success. Probanthine and sedation frequently will reduce bladder spasms. Seven to ten days postoperatively the urethral catheter is removed after verification of no extravasation on a cystogram. It is recommended that the patient abstain from sexual intercourse for at least six weeks postoperatively.

URETHROVAGINAL FISTULA

ETIOLOGY

Urethrovaginal fistula is an uncommon condition that usually results as a complication of such operative procedures as urethral diverticulectomy, anterior colporrhaphy, transurethral resection of the bladder neck, or trauma. In medically deprived countries, obstetric trauma is by far the most important cause of urethral destruction. A fistula may result after trauma, especially with pelvic fractures and urethrovaginal lacerations. In addition, vaginal and urethral neoplasms treated with radiotherapy may undergo necrosis and create a fistula. Pressure necrosis can also occur with a prolonged indwelling urethral catheter. Transsexual surgery and drainage of a periurethral abscess are other rare causes of fistulas. Last, urethrovaginal fistulas can be seen on a congenital basis (Marshall, 1979).

CLINICAL FEATURES

Fistulas involving the bladder neck and proximal 2 cm of the urethra may produce continuous incontinence, while a fistula distal to the external sphincter may be entirely asymptomatic. Usually bothersome urine will drain per vagina, especially when the patient stands. Twenty to forty percent of patients will have associated stress incontinence. Generally the defect is obvious on examination but occasionally may be hidden by the rough, irregular vaginal surface.

DIAGNOSIS

The differential diagnosis includes vesicovaginal fistula, ureterovaginal fistula, severe stress or urgency incontinence, or a serous vaginal discharge. A complete urologic investigation with cystourethroscopy and vaginoscopy will reveal the type and size of the defect. In selected patients, urodynamic studies, urethral pressure profile, voiding cystourethrogram, and the Bonney test will provide additional useful information.

MANAGEMENT

An efficient bladder neck mechanism is critical in females to maintain continence. Studies have shown that the proximal 2 cm of bladder neck and urethra is the continence zone (Lapides, 1958).

The principles of management are similar to those of vesicovaginal fistula. Treatment depends on the presence or absence of symptoms, the etiology, size and location of the fistula, and other local factors. Many techniques have been recommended.

In cases of bladder neck destruction, the Young-Dees-Leadbetter (Leadbetter, 1964) repair may be useful in reconstruction. The transpubic approach as advocated by Waterhouse et al. (1972) may provide additional exposure, but complication rates are higher. Labial fat pad repair (Martius, 1928), gracilis muscle flap interposition (Ingelman-Sundberg, 1978; Patil et al., 1980), pedicle perineal flap (Hendren, 1980), and full-thickness skin graft have been successfully used in reconstruction of the female urethra.

In small fistulas the Martius procedure utilizing the fibro fatty labial tissue may suffice. This tissue is mobilized on a pedicle, preserving its vascular supply on its posterior aspect, and is passed through a subcutaneous tunnel into the vaginal lumen where it is sutured over the repaired urethra. The pedicle buttock flap as described by Hendren (1980), however, provides better vascularized tissue and ensures a greater likelihood of success, especially in larger fistulas (Fig 30–3, A and B).

The bladder is then drained suprapubically for several weeks. In certain patients vesicourethral suspension may be employed at the same time as well.

URETEROVAGINAL FISTULA

ETIOLOGY

The ureter is especially susceptible to injury during vascular, gynecologic, urologic, and colonic operations. Gynecologic surgery remains the most common cause

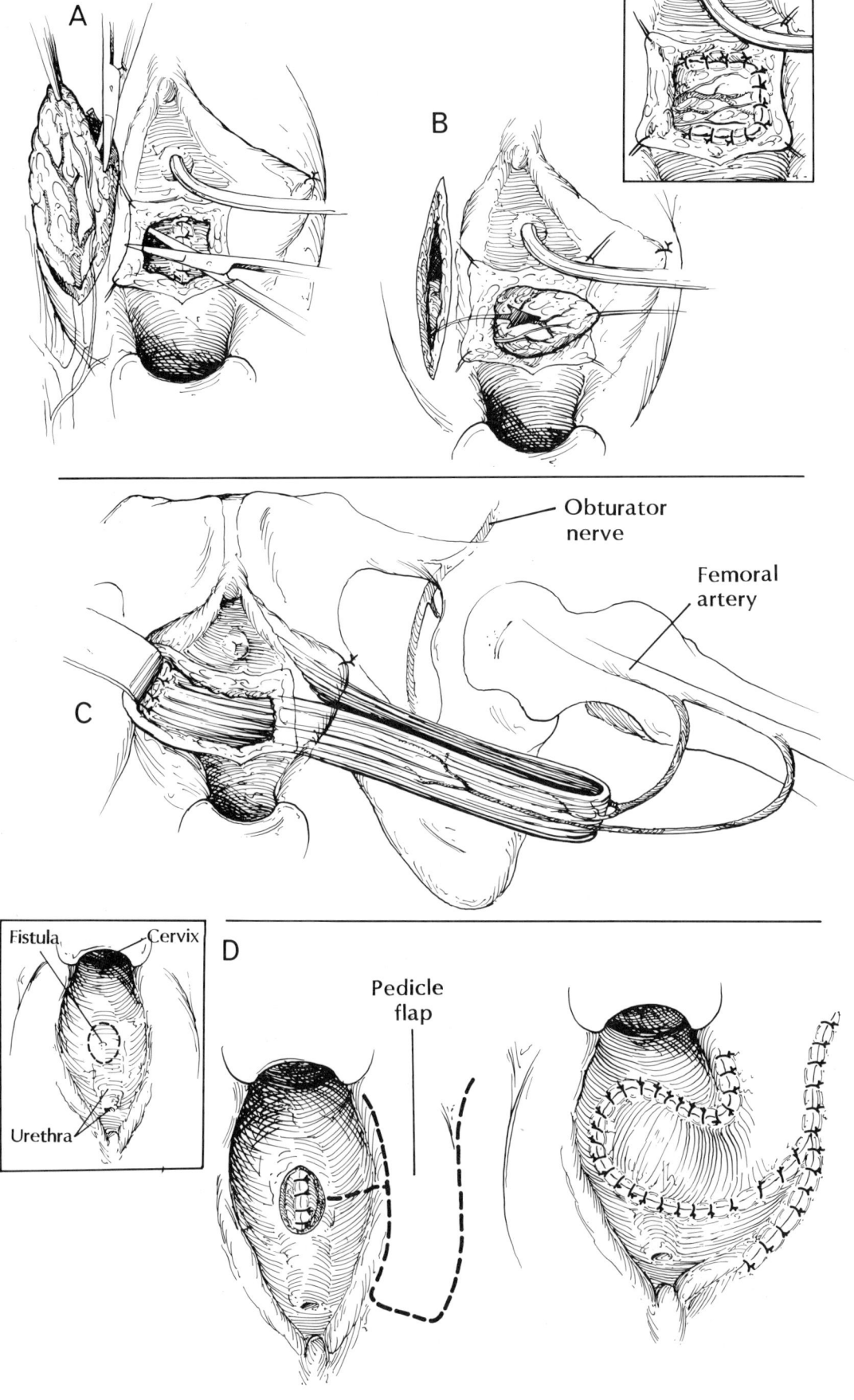

Obturator nerve

Femoral artery

Fistula   Cervix

Urethra

Pedicle flap

of ureterovaginal or ureterocutaneous fistula. Total abdominal hysterectomy is the most common operation responsible for ureteral injury (Wallace, 1972; Murphy, et al., 1982). Injuries are more common if the patient has received preoperative radiation. Postoperative radiotherapy does not appear to be associated with an increased incidence of stricture or fistula formation.

CLINICAL FEATURES

In the postoperative period a ureteral injury is suggested by the presence of unexplained abdominal pain, flank tenderness, abdominal mass, or fever. Subsequent urinary drainage is often discovered by copius drainage through a postoperative drain site. There is usually normal bladder voiding but also continuous urinary leakage from the vagina as might be seen in a small vesicovaginal fistula.

DIAGNOSIS

An IV urogram will demonstrate some degree of ureteral obstruction in most cases (Murphy, 1982; Turner-Warwick et al., 1967). Retrograde pyelography will usually delineate the nature and extent of the injury. Ureteral catheterization will usually reveal obstruction. In equivocal cases, a dye test can be helpful and is quite sensitive (Murphy et al., 1982).

Occasionally an abnormal communication exists between an ecotpic ureter and the vagina but may not be demonstrated by the usual tests (Leary et al., 1979). The upper pole of the kidney may not be visualized on the IV pyelogram because its function is impaired. Presence of a duplication may be suspected from the configuration of the lower pole pelvis. Meticulous cystourethroscopy is mandatory as well as careful inspection of the vagina, but an ectopic orifice can still be missed. Occasionally a mass can be seen in the region of the trigone (Brannan and Henry, 1973). If a mass is appreciated along the wall of the vagina, it can be examined by needle and instilled with contrast; often the ectopic ureter will be demonstrated.

MANAGEMENT

Primary goals include renal preservation and preservation of urosepsis. Years ago a primary nephrectomy was frequently performed. Goodwin and Scardino (1980) reported a reduced nephrectomy rate of 5%. If a ureteral injury is suspected and diagnosed, the patient should rarely require a nephrectomy for ureterovaginal fistula. There is also controversy in the management of ureterovaginal fistula. Early repair has been advocated to prevent irreversible damage to the kidney. On the other hand, a retrograde catheter can be placed to establish drainage or a percutaneous nephrostomy can also be placed with a stent (Lang, 1981). The percutaneous nephrostomy is more easily tolerated than a retrograde ureteral catheter and provides better drainage. Occasionally proximal diversion with a nephrostomy and a stent has been therapeutic in our experience. We have had two patients managed in this way with a follow-up of two years, and there is no evidence of later stricture.

If urinary extravasation or obstruction persists after removal of the ureteral stent, a formal repair can be performed as soon as the inflammation has subsided and infection is under control. A psoas hitch ureteral reimplant has been very successful for us in these cases (Fig 30–4). With the fingers placed intravesically, the mobilized bladder can be displaced superiorly and anchored in position, so that no dissection is necessary in the scarred pelvis. Good results have generally been obtained.

PROSTATIC FISTULA

ETIOLOGY

A rectourethral or prostatic fistula may be congenital or acquired (Table 30–4). The congenital fistula is uncommon and usually occurs in boys in association with congenital anorectal anomalies (Culp and Calhoun, 1964). With a high- or intermediate-level imperforate anus, a rectourethral communication is present in 80% of patients. Treatment is usually in association with anorectal pull-through operations.

The acquired fistula is usually a postprostatectomy complication. In the past, perineal prostatectomy was the most common cause. The majority of patients developed a fistula after a simple prostatectomy or occasionally following an open prostatic biopsy (Weyrauch, 1951; Vose, 1949; Young and Stone, 1917; Kilpatrick and Thompson, 1962; Culp and Calhoun, 1964; Dahl et al., 1974; Olsson et al., 1976). It may also occur as a

---

FIG 30–3.
Alternative flap repair for urethrovaginal/vesicovaginal fistula. **A,** mobilization of the fibro-fatty pedicle from the labia. **B,** tunnel created for interposition. **C,** gracilis muscle flap interposition. **D,** after fistula closure, a posteriorly based buttock pedicle flap is rotated inward and placed in the vaginal wall with the right labium temporarily displaced and returned. (**A–C,** redrawn with permission from Patil U, Water-house K, Laungani G: Management of 18 difficult vesicovaginal and urethrovaginal fistulas with modified Ingelman-Sundberg and Martius operation. *J Urol* 1980; 123:653. **D,** redrawn with permission from Hendren WH: Construction of female urethra from vaginal wall and a perineal flap. *J Urol* 1980; 123:657.)

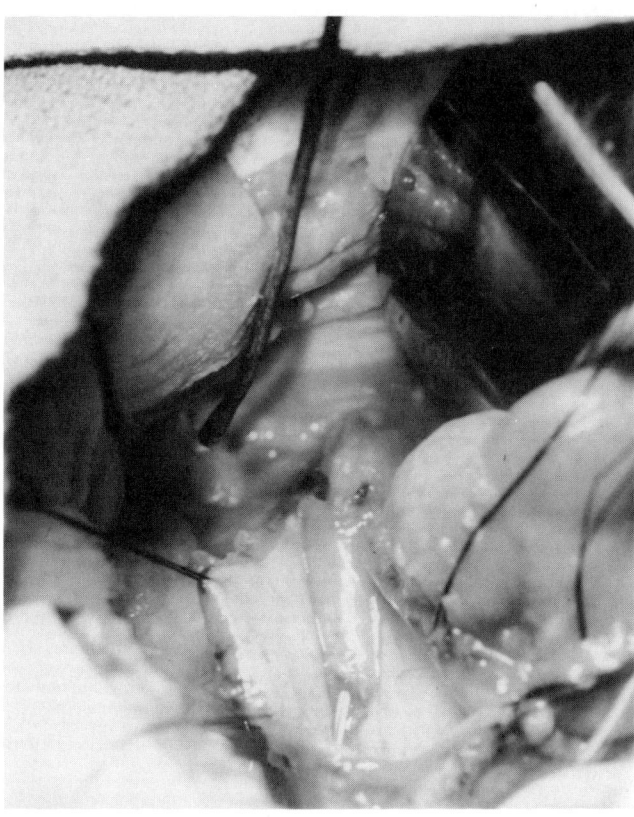

FIG 30–4.
Long ureteral tunnel is seen with a psoas hitch ureteral reimplantation.

complication of suprapubic prostatectomy, retropubic prostatectomy, radical retropubic prostatectomy, or overzealous transurethral resection of the prostate. Rarely a prostatic abscess may rupture into the rectum. A rectourethral fistula has also been described in association with Crohn's disease or inflammatory disease of the colon as well as carcinoma of the rectum or prostatic carcinoma (Buck and Chisholm, 1979). Direct trauma may also result in rectourethral fistula, especially in association with a fractured pelvis (Tiptaft et al., 1983). Rarely, stricture disease or tuberculosis (Hashmonai et al., 1982) may also create a fistula.

CLINICAL FEATURES AND DIAGNOSIS

A rectourethral fistula may be strongly suspected from the patient's history. The patient may present with pneumaturia or fecaluria with leakage of urine from the rectum during micturition. Urinary drainage per rectum may lead to diarrhea. Urinary tract infections and epididymitis are common (Culp and Calhoun, 1964), as are urethral stricture and perineal fistula (Culp and Calhoun, 1964; Tiptaft et al., 1983).

The diagnosis can usually be made by careful inspection of the rectum. Occasionally the opening can be palpated or visualized by direct proctoscopy. A biopsy is performed when necessary to rule out carcinoma. The fistula is also usually visible on careful cystourethroscopy. Radiographic studies of the bowel, as well as additional colonoscopy, can be performed to verify that there is no additional colonic disease.

MANAGEMENT

The variety of methods recommended for the treatment of rectourethral fistulas bears ample testimony that there is no single accepted method of treatment for this uncommon lesion (Table 30–5).

Injury Recognized at Time of Surgery

This is a serious complication and, if not recognized at the time of surgery, may result in severe pelvic infection and urosepsis. The prime objective of early treatment is to divert urine and feces from the site of injury. A suprapubic cystostomy and a diverting colostomy are recommended. After meticulous control of bleeding, the rectum is closed in two layers (the outer

---

TABLE 30–4.

Etiology of Rectourethral Fistula

Congenital
Iatrogenic
   Prostatectomy
   Transurethral resection of prostate
   Urethral dilatation
   Anorectal surgery
Direct trauma, especially associated
   with fractured pelvis
Inflammatory
   Prostatic abscess
   Urethral stricture
   Tuberculosis
   Crohn's disease of colon
Neoplastic
   Prostate
   Urethra
   Rectum

TABLE 30–5.

Principles of Fistula Closure

Adequate urinary and fecal diversion
Maintenance of infection-free
   environment
Adequate drainage
Adequate exposure of the operative
   field
No tension on the suture liness
Transfer of healthy vascular tissue to the
   site of repair in appropriate cases
Nonapposition of suture lines
Closure of nonedematous, healthy tissue

layer closed with silk suture). A Penrose drain is left in situ for 5–10 days to drain the perirectal tissue. If a proximal colostomy is not elected, the risk remains distinctly higher. Allen (1968) successfully managed two cases by primary closure supplemented by rectal tube decompression and neomycin irrigation of the bowel. Manual dilation of the anal sphincter may reduce rectal pressures by temporarily defunctionalizing the sphincter.

Occasionally, a fistula may occur weeks or months after the original operation. Successful treatment depends on the etiology and the general condition of the patient. Controversy remains regarding single-staged and multistaged repairs. For patients in good general condition, a single-staged repair may be effective, but it still carries higher risk than a multistaged repair. For poor-risk patients or patients with major fecal drainage, perineal infection, and inflammation, a multistaged procedure is mandatory. Whatever repair is performed, it is imperative that basic surgical principles be rigorously followed (see Table 30–5).

Adequate bowel preparation is essential for a successful outcome. Intraoperative and postoperative antibiotics are important. Intravenous fluids and an elemental diet are administered postoperatively. On the 14th postoperative day a cystogram is performed, and if there is no leakage of contrast, the cystostomy catheter is removed.

In the multistage procedure, a suprapubic cystostomy and a diverting colostomy (Kilpatrick and Mason, 1969; Mason and Kilpatrick, 1973; Dahl et al., 1974) are done as the initial procedures. With this therapy alone, approximately one third of patients may undergo spontaneous resolution of their fistulas (Olsson et al., 1976). Once infection is controlled and the inflammation has subsided, the defect is closed. A cystogram is performed prior to removal of the urethral catheter. At a later date, the colostomy is closed.

SURGICAL TECHNIQUES

Basically, a urethrorectal fistula may be closed by a transrectal, perineal, posterior, or abdominal approach (Table 30–6). Urethral stricture is a common postoperative complication (Tang, 1978), and this complication can be avoided by suturing the urethral defect transversely or suturing scrotal skin to the edges of the urethral fistulous opening, as in the Johannson technique (Tang, 1978).

Transrectal Repair

In 1949 Vose described a simple technique for the closure of urethrorectal fistulas using the transrectal approach (Vose, 1949). The patient is placed in the lithotomy position. After dilating the anal sphincter, a bi-

TABLE 30–6.

Surgical Techniques

Transrectal approach
Vose procedure
Parks procedure
Perineal approach
Posterior approach
York-Mason procedure
Kraske procedure
Abdominal approach

valve anal speculum is placed to expose the defect. The fistulous tract is excised and the defect closed in layers. This procedure is suitable for small fistulas lying close to the anal verge.

A variation of the transrectal approach is the rectal advancement flap popularized by Parks (Tiptaft et al., 1983), which is simple and suitable for low fistulas. A bivalve anal speculum or a self-retaining anal retractor is inserted, and 1:300,000 epinephrine-saline is injected beneath the mucosa distal to the fistula. An ellipse of rectal mucosa including the fistula is removed. A full-thickness, inverted U-shaped rectal flap is raised above the fistula, brought down over the fistula, and sutured in two layers to the rectal wall (Tiptaft et al., 1983).

The transrectal repair is the only method that can be employed in the early postoperative period. Its disadvantage is that it provides limited exposure of the fistula and therefore is reserved for low-lying, small fistulas.

The Perineal Repair

The patient is placed in the lithotomy position. The fistula is exposed by an inverted U-shaped incision. The fistula is excised and the dissection is carried well above the fistulous site to obtain good exposure (Weyrauch, 1951; Culp and Calhoun, 1964). The urethra is closed with a single layer of 4-0 absorbable suture. A urethral catheter is used as a stent for 10–14 days. The rectal defect is closed in two layers. (The outer layer is closed with silk.) Normal vascularized tissue is interposed between urethra and rectum. The levator ani muscle is readily available and can be approximated in the midline.

For radiation-induced fistulas, the gracilis muscle can be rotated through a subcutaneous tunnel to the perineum and interposed between the urethra and rectum (Ryan et al., 1979). Some surgeons recommend adequate dissection of the rectum and rotation of the rectal fistula site away from the midline or to bring the rectal repair to a level lower than the opening in the prostate, but this dissection is difficult owing to the extensive scarring present around the fistula.

The Young-Stone technique is a variant of the perineal method (Young and Stone, 1917). The prostatic-rectal fistula is divided, and the entire lower portion of the rectum is mobilized and drawn through the anal sphincter; the redundant rectal wall including the fistulous area is excised. The resultant cuff of healthy rectal mucosa is anastomosed to the skin edges after the urethral defect is closed in layers. This operation is technically difficult to perform (Weyrauch, 1951), and fecal soiling has been a troublesome complication (Weyrauch, 1951; Lewis, 1947).

Posterior Approach

In 1962 Kilpatrick and Thompson (1962) described six patients in whom a successful rectourethral repair was carried out utilizing the posterior Kraske approach. York-Mason successfully repaired fistulas in four patients using the posterior transsphincteric approach with no residual incontinence. Beneventi and Cassebaum (1971) have modified this technique using a rectal flap, thus avoiding neurologic and vascular structures.

York-Mason Procedure

The patient lies in the prone jackknife position (Mason and Kilpatrick, 1973). The incision is begun at the level of the anal margin to the left of the midline and extended upward and laterally to the level of the mid-sacrum (Fig 30–5). The mucocutaneous junction is

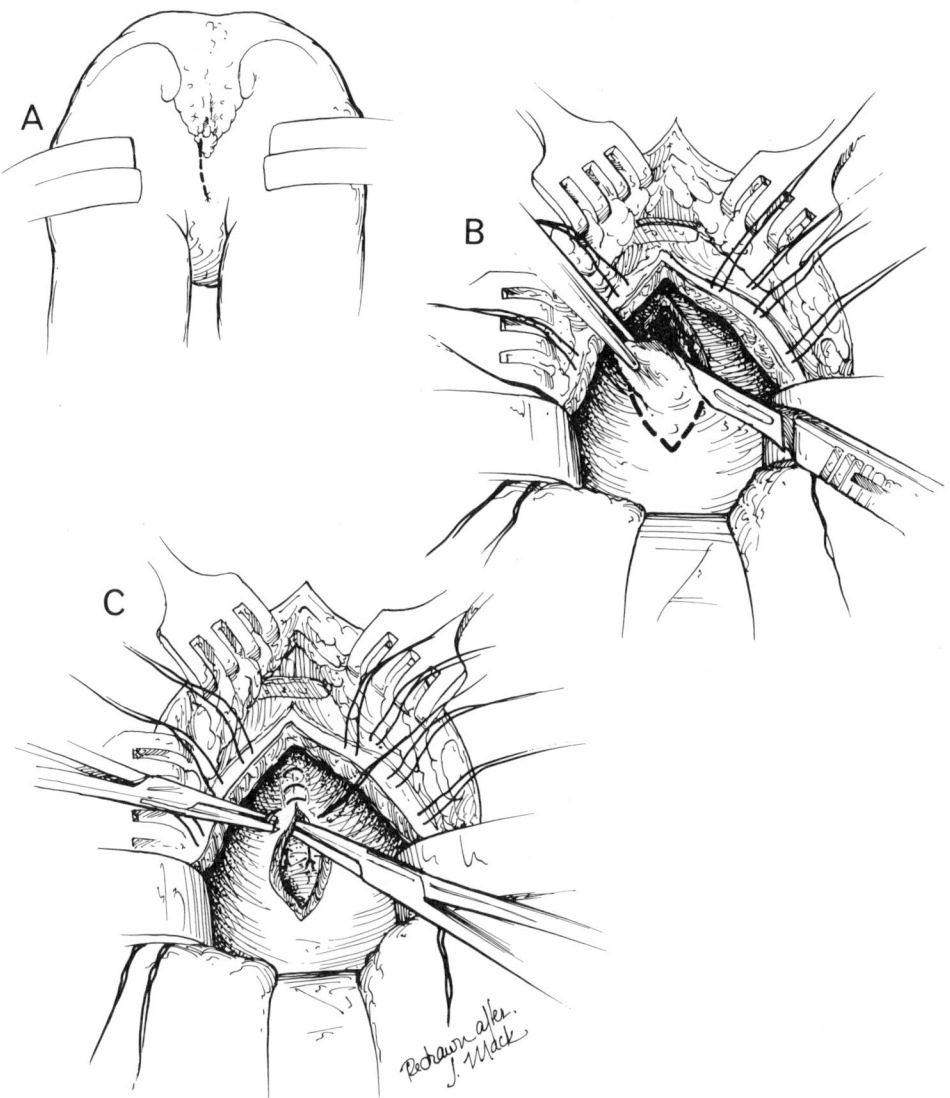

FIG 30–5.
York-Mason procedure. **A,** incision to the left of midline. **B,** transrectal excision of fistula. **C,** closure of posterior rectal wall after closure of urethra and ventral rectum at fistula site. (Redrawn with permission from Dahl DS, Howard PM, Middleton RG: The surgical management of recto-urinary fistulas resulting from a prostatic operation: A report of 5 cases. *J Urol* 1974; 111:514.)

marked with stay sutures. The sphincters are marked with sutures so that they can be easily identified at the end of the operation. The incision is deepened, and the rectal mucosa is divided to expose the defect. The fistulous tract is excised. The urethral opening is closed transversely, and the rectal wall is closed in layers. The sphincter muscles are identified and carefully reapproximated. This approach provides excellent exposure, and careful reconstruction of the sphincters has produced complete anal continence (Kilpatrick and Mason, 1969; Dahl et al., 1974; Prasad et al., 1983).

The Kraske approach involves excision of the coccyx. This approach has been elegantly described (Wiseman and Decter, 1982).

Abdominal Approach

Interposition of an omental pedicle graft was described by Turner-Warwick et al. (1967). This method is indicated for large fistulas and radiation-induced fistulas. The omentum is freed from the gastric margin, and the highly vascularized omental pedicle is interposed between the rectal and urethral wall; the rectal defect is closed in a single layer, but no attempt is made to close the urethral defect. Neoepithelialization occurs on the interposed omentum.

VESICOENTERIC FISTULA

The four categories of vesicoenteric fistula, colovesical, rectovesical, ileovesical, and appendicovesical, are grouped together because of the similarities in clinical features and management.

ETIOLOGY

An abnormal communication between the bowel and bladder was first described in the second century A.D. by Rufus (Moisey and Williams, 1972), but the etiology of acquired vesicoenteric fistulas has changed from the common diseases of the past such as typhoid and tuberculosis to the current common causes such as diverticulitis, malignancy, Crohn's disease, and trauma (Slade and Graches, 1972; Vargas et al., 1974; West, 1973) (Table 30–7). Less commonly, ovarian abscess (Scardino and Lippett, 1968) and foreign bodies in the bowel (Herbst and Miller, 1936) cause fistulas. Congenital fistulas are rare and are often associated with an imperforate anus.

A colovesical fistula is the most common form of vesicointestinal fistula and is most often associated with diverticular disease of the colon but can result from colorectal carcinoma (Karamchandari and West, 1984; Shatila and Ackerman, 1976). It is more common in males, with a ratio of 3:1 (Shatila and Ackerman, 1976; Karamchandari and West, 1984). The lower incidence

TABLE 30–7.
Etiology of Vesicoenteric Fistula

Congenital (rare)
Inflammatory
   Diverticular disease
   Crohn's disease
   Tuberculous ileitis
   Fungal and parasitic disease of colon
Neoplastic
   Adenocarcinoma of colon and rectum
   Carcinoma of cervix
   Carcinoma of bladder
   Carcinoma of prostate
   Radiation enteritis
   Lymphoma
Trauma
   Gunshot wounds
   Pelvic fractures with bony spicules
   Penetrating injuries
   Iatrogenic
   Foreign bodies in bowel, such as fish and
     chicken bones

in women is thought to be due to the interposition of the uterus and adnexa between the urinary bladder and colon (West, 1973; Carson et al., 1978; Talamini et al., 1982; King et al., 1982). Diverticular disease accounts for approximately 50 to 70% of vesicoenteric fistulas (Karamchandari and West, 1984). Malignancy and Crohn's disease account for another 20% and 10%, respectively. Colorectal carcinoma is the most common malignancy associated with a vesicoenteric fistula. Occasionally carcinoma of the bladder, cervix, and prostate are implicated (Karamchandari and West, 1984; King et al., 1982).

Crohn's disease is the most frequent cause of an ileovesical fistula (Talamini et al., 1982; Greenstein et al., 1984; Karamchandari and West, 1984). The mean duration of Crohn's disease at the time of first symptoms of enterovesical fistula formation is 10 years (Karamchandari and West, 1984; Greenstein et al., 1984) and the average age of the patient is 30 years (Karamchandari and West, 1984). Ileovesical fistula develops in 10% of patients with regional ilietis (Karamchandari and West, 1984). Appendicovesical fistula is occasionally reported (Haas et al., 1984; Dalessandri and Swafford, 1983).

CLINICAL FEATURES

Urinary symptoms are the most common presenting symptoms of the disease (King et al., 1982; Talamini et al., 1982; Karamchandari and West, 1984; Greenstein et al., 1984; Shatila and Ackerman, 1976). Suprapubic pain and complaints associated with chronic urinary tract infection are frequent. Chills, fever, and diarrhea are less common. Patients may notice passage of urine through the rectum or passage of mucus in urine. *E. coli* is the most common offending organism, and nearly one third of patients have a mixed-organism infection (Talamini et al., 1982).

Pneumaturia may be intermittent and must be carefully sought in the history. Pneumaturia occurs in approximately 60% of patients but is not pathognomonic of the disease, since it can occur in infections with gas-forming organisms such as clostridia, fermentation of diabetic urine, or urinary tract instrumentation. Fecaluria is less common but is pathognomonic of a fistula.

Symptoms of the underlying disease causing the fistula may be present, but in about one third of patients no symptoms referable to the underlying bowel disease are noted at the time of diagnosis of the fistula (Shatila and Ackerman, 1976).

Greenstein et al. (1984) noted an abdominal mass and abscess more frequently among fistula patients than in a control group of patients with Crohn's disease without a fistula.

DIAGNOSIS

A high index of suspicion is essential for making a diagnosis of vesicoenteric fistula. In most series, patients have been treated for recurrent urinary tract infections extending 4–12 months before a diagnosis of fistula is made.

Cystoscopy remains the most reliable diagnostic test. The presence of a localized area of edema and congestion is a typical finding in the early stage of a fistula. As it matures, it is surrounded by bullous edema and mucosal papillomatous hyperplasia. Fecal material or mucus may be observed in the bladder. Charcoal, oral contrast or ^{51}chromium have been administered by mouth and identified in the urine (Lippert et al., 1984). An attempt can be made to catheterize the suspected tract with a ureteral catheter. A retrograde injection of contrast material can confirm the presence of a fistula, and often a cystogram will demonstrate dye outside the bladder. In diverticular disease and Crohn's colitis the lesion is most commonly on the left posterior wall or the left dome of the bladder, whereas in Crohn's ileitis and in an appendicovesical fistula the lesion is found in the right posterior wall or right dome of the bladder. A biopsy of the lesion is important to rule out malignancy.

Urinalysis may reveal undigested intestinal food residue. A urine culture may disclose an *E. coli* or a mixed-organism infection. Sigmoidoscopy and colonoscopy are useful in assessing any underlying bowel disease.

Cystography is the most useful radiologic examination. The "herald" sign is seen best in oblique views and is a crescentic defect on the upper margin of the bladder and represents a perivesical abscess. A "beehive on the bladder" radiographic configuration may be noted (Kaisary and Grant, 1981).

A barium enema is helpful for evaluating the condition of the bowel, and radiography of the urinary sediment after a nondiagnostic barium enema may enhance the yield of the test (Amendola et al., 1984). Barium detected in the urine sediment can confirm the presence of a fistula.

MANAGEMENT

A thorough and accurate preoperative evaluation helps in planning the treatment. Both single-stage as well as multistage procedures have been utilized in the management of these patients, depending on the underlying disease process and general condition of the patient. A single-stage procedure may be inappropriate in the presence of extensive inflammation, abscess, multiple organ involvement, postradiation changes, or poor-risk patients (King et al., 1982).

In the presence of inflammatory bowel disease, a proximal colostomy does not prevent the development of a fistula (Shatila and Ackerman, 1976). Unlike prostatic-rectal fistula, spontaneous closure of vesicoenteric fistula is rare. To prevent recurrence of fistula, excision of the diseased segment of the intestine is essential (Amin, 1984; King et al., 1982).

After adequate bowel preparation, a single-stage procedure is recommended for patients in a good nutritional state in the absence of abscess or severe inflammation and in the absence of multiple-organ involvement (Shatila and Ackerman, 1976; King et al., 1982). Otherwise, a multistage procedure would be appropriate.

In the absence of malignancy, simple closure of bladder is usually adequate. Whenever possible, omentum is interposed between intestinal and bladder suture lines. A urethral catheter is left in situ for two weeks. A cystogram is usually performed to verify that the bladder is intact before the catheter is removed. In poor-risk patients and in patients with cancer or complex fistulas, a diverting colostomy affords some palliation. A partial cystectomy may be necessary if a colonic carcinoma is present. If a fistula has developed in this setting, an attempt may be made to resect all carcinoma, but the prognosis is often poor.

Recently there has been an interest in managing some cases conservatively. Heiskell et al. (1975) showed in animal experiments that a colovesical fistula can be well tolerated as long as there was no distal urinary or bowel obstruction because they caused an increase in sepsis. Amin (1984) described six patients with colovesical fistulas due to diverticulitis who had successful management with long-term antibacterial therapy alone. Although surgery is the treatment of choice in most patients with vesicoenteric fistulas, conservative therapy has been utilized in a few patients. The longer-term risk of conservative treatment appears to be significant so that restoration of normal anatomical continuity of the urinary and alimentary tract should be the surgeon's goal.

RECTOVAGINAL FISTULAS

ETIOLOGY

A rectovaginal fistula is uncommon in this country. It occurs most often after obstetric trauma. Occasionally, it may follow surgical trauma or direct injury to the perineum. Inflammatory fistulas are principally caused by diverticular disease and Crohn's disease. Rarely, ulcerative colitis, lymphogranuloma venereum, tuberculosis, and actinomycosis may result in a fistula. Malignant disease and irradiation injury also are rare causes of rectovaginal fistulas.

DIAGNOSIS

Even though the presence of a fecal fistula may be obvious, an evaluation and an examination using anesthesia should be carried out in all cases. Barium enema, sigmoidoscopy, and colonoscopy are helpful in evaluating the underlying bowel disease and in determining whether there is stenosis at or below the fistula site. A fistulogram may determine whether a fistula is a simple communication between the vagina and rectum or demonstrate whether there is an associated abscess cavity. All suspicious lesions should be biopsied to exclude malignancy.

MANAGEMENT

The principles of repair are the same as discussed earlier for repair of urinary vaginal fistulas. They include accurate diagnosis, fecal and urinary diversion, excision of the fistula, multilayered closure, maintenance of blood supply, interpretation of such tissues as omentum, and anatomical reconstruction. High obstetric fistulas are often associated with a urinary fistula. The urinary fistula should be treated first, and the cure rate will be enhanced if fecal diversion is performed prior to closure of urinary fistula. A preliminary colostomy is recommended but is at least performed during the repair of a straightforward rectovaginal fistula. There is conflicting opinion regarding early vs. late repair. There are several approaches through which a fistula can be repaired including abdominal, vaginal, anal, and transsphincteric. The choice of approach depends on the experience of the surgeon, the site of the fistula, and the condition of the patient.

REFERENCES

1. Allen TD: Management of inadvertent rectal injury: Report of a technique used in two cases. J Urol 1968; 99:69.
2. Amendola MA, Agha FP, Dent TL, et al: Detection of occult colovesical fistula by the Bourne test. AJR 1984; 142:715.
3. Amin M: Treatment options in vesicoenteric fistulas [editorial]. Am J Surg 1984; 147:705.
4. Aycinema J: Small vesico-vaginal fistula. Urology 1977; 9:543.
5. Barnes RW, Bergman RT: Repair of vesico-vaginal fistula. Am Surg 1964; 30:493.
6. Beneventi FA, Cassebaum WH: Rectal flap repair of prostato rectal fistula. Surg Gynecol Obstet 1971; 133:489.
7. Brannan W, Henry HH II: Ureteral ectopia: Report of 39 cases. J Urol 1973; 109:192.
8. Buck AC, Chisholm GD: Rectovesical fistula secondary to prostatic carcinoma. J Urol 1979; 121:831.
9. Carson CC, Malek RS, Remine WH: Urologic aspects of vesicoenteric fistulas. J Urol 1978; 119:744.
10. Collins CG, Colllins JH, Harrison BR, et al: Early repair of vesico-vaginal fistula. Am J Obstet Gynecol 1971; 111:524.
11. Coolsaet BL: Ventrodorsal vesical repair of complicated vesico-vaginal and vesico-rectal fistulas. J Urol 1984; 131:116.
12. Counseller VS, Haigler FH Jr: Management of urinary vaginal fistula in 253 cases. Am J Obstet Gynecol 1956; 72:367.
13. Culp OS, Calhoun HW: A variety of recto-urethral fistulas: Experiences with 20 cases. J Urol 1964; 91:560.
14. Dahl DS, Howard PM, Middleton RG: The surgical management of recto-urinary fistulas resulting from a prostatic operation: A report of 5 cases. J Urol 1974; 111:514.
15. Dalessandri KM, Swafford GR: Appendico-vesico-colonic fistula. J Urol 1983; 130:777.
16. Eisen M, Jurkovic K, Altwein JE, et al: Management of vesico-vaginal fistulas with peritoneal flap interposition. J Urol 1979; 112:195.
17. Elem B: Urinary calculus in Zambia: Its incidence and relationship to Schistosoma haematobium infection and vesico-vaginal fistula. Br J Urol 1984; 56:44.
18. Goodwin WE, Scardino PT: Vesico-vaginal and uretero-vaginal fistulas: A summary of 25 years of experience. J Urol 1980; 123:370.
19. Graham, JB: Painful syndrome of post-radiation urinary vaginal fistula. Surg Gynecol Obstet 1967; 124:1260.
20. Graham JB: Vaginal fistulas following radiotherapy. Surg Gynecol Obstet 1965; 120:1019.
21. Greenstein AJ, Sachar DB, Tzakis A, et al: Course of enterovesical fistulas in Crohn's disease. Am J Surg 1984; 147:788.
22. Haas GP, Shumaker BP, Haas PA: Appendicovesical fistula. Urology 1984; 24:604.
23. Harrison KA: Obstetric fistula: One social calamity too many. Br J Obstet Gynecol 1983; 90:385.
24. Hashmonai M, Bolkier M, Schramek A: Tuberculous

recto-vesico-cutaneous fistula. *Br J Urol* 1982; 54:324.

25. Hecker WC, Holschneider AM, Kraeft H: Surgical closing of rectovaginal, rectourethral, urethrovaginal and vesicocutaneous fistulas by means of interposition of the gracilis muscle. *Chirurg* 1980; 51:43.

26. Heckler FR, Aldridge JE Jr, Songcharoen S, et al: Muscle flaps and musculocutaneous flaps in the repair of urinary fistulas. *Plast Reconstr Surg* 1980; 66:94.

27. Heiskell CA, Wycki GT, Beal JM: A study of experimental colo-vesical fistula. *Am J Surg* 1975; 129:316.

28. Hendren WH: Construction of female urethra from vaginal wall and a perineal flap. *J Urol* 1980; 123:657.

29. Herbst RH, Miller EM: Vesico-intestinal fistulas, caused by foreign bodies in bowel. *JAMA* 1936; 106:2125.

30. Hudson CN: Rectovaginal fistula, in Rob and Smith's *Operative Surgery*, ed 4. London, Butterworth's, 1983, p 563.

31. Ingelman-Sundberg A: Surgical treatment of urinary fistulae. *Zentralbl Gynakol* 1978; 100:1281.

32. Judd ES: Operative treatment of vesico-vaginal fistulae. *Surg Gynecol Obstet* 1920; 30:447.

33. Kaisary AV, Grant RW: Beehive on the bladder: A sign of colovesical fistula. *Ann R Coll Surg Engl* 1981; 63:195.

34. Karamchandari MC, West CF Jr: Vesicoenteric fistulas. *Am J Surg* 1984; 147:681.

35. Keeltel WC Sehring FG, deProsse CA, et al: Surgical management of urethrovaginal and vesicovaginal fistulas. *Am J Obstet Gynecol* 1978; 131:425.

36. Kilpatrick FR, Mason AY: Postoperative recto-prostatic fistula. *Br J Urol* 1969; 41:649.

37. Kilpatrick FR, Thompson HR: Postoperative recto-prostatic fistula and closure by Kraske's approach. *Br J Urol* 1962; 34:470.

38. King RM, Beart RW Jr, Mellrath DC: Colovesical and recto-vesical fistulas. *Arch Surg* 1982; 117:680.

39. Kiruta I, Goldstein AMB: The repair of extensive vesicovaginal fistulas with pedicled omentum: A review of 27 cases. *J Urol* 1972; 108:724.

40. Lang EK: Diagnosis and management of ureteral fistulas by percutaneous nephrostomy and antegrade stent catheter. *Radiology* 1981; 138:311.

41. Lapides J: Structures and function of the internal vesical sphincter. *J Urol* 1958; 80:341.

42. Latzko W: Postoperative vesicovaginal fistulas: Genesis and therapy. *Am J Surg* 1942; 58:211.

43. Lawson J: Vesical fistulae into the vaginal vault. *Br J Urol* 1972; 44:623.

44. Leadbetter GW Jr: Surgical correction of total urinary incontinence. *J Urol* 1964; 91:261.

45. Leary FJ, Bass RB Jr, Symmonds RE: Watery vaginal discharge in a young woman. *J Urol* 1979; 122:226.

46. Lewis LG: Repair of recto-urethral fistulas. *J Urol* 1947; 57:1173.

47. Lippert MC, Teates CD, Howards SS: Detection of enteric urinary fistulas with a non-invasive quantitative method. *J Urol* 1984; 132:1134.

48. Marshall FF, Jeffs RD, Sarafyan WK: Urogenital sinus abnormalities in the female patient. *J Urol* 1979; 122:508.

49. Marshall VF: Vesicovaginal fistulas on one urological service. *J Urol* 1979; 121:25.

50. Martius H: The repair of vesicovaginal fistulae with interposition pedicle graft of labial tissue. *Zentralbl Gynakol* 1928; 52:480.

51. Mason AY, Kilpatrick FR: Rectoprostatic and recto-urethral fistulae. *Proc R Soc Med* 1973; 66:245.

52. Massie JS, Welch JS, Pratt JH, et al: Management of urinary vaginal fistula: Ten year survey. *JAMA* 1964; 190:902.

53. Moir JC: Personal experiences in the treatment of vesicovaginal fistulas. *Am J Obstet Gynecol* 1956; 71:476.

54. Moisey CU, Williams JL: Vesico-intestinal fistulae. *Br J Urol* 1972; 44:664.

55. Murphy DM, Grace PA, O'Flynn JD: Ureterovaginal fistula: A report of 12 cases and review of the literature. *J Urol* 1982; 128:924.

56. Nesolowski S, Bulinski W: Vesico-intestinal fistulae and recto-urethral fistulae. *Br J Urol* 1973; 45:34.

57. O'Conor VJ Jr: Review of experience with vesicovaginal fistula repair. *J Urol* 1980; 123:367.

58. O'Conor VJ Jr, Sokol JK, Bulkley GJ, et al: Suprapubic closure of vesicovaginal fistula. *J Urol* 1973; 109:51.

59. Olsson CA, Willscher MK, Krane RJ, et al: Management of prostatic fistulas. *Urol Surv* 1976; 25:135.

60. Patil U, Waterhouse K, Laungani G: Management of 18 difficult vesicovaginal and urethrovaginal fistulas with modified Ingelman-Sundberg and Martius operations. *J Urol* 1980; 123:653.

61. Pers M: The closure of urinary vaginal fistulas. *Scand J Plast Reconstr Surg* 1977; 11:147.

62. Persky L, Herman G, Guerrier K: Non-delay in vesicovaginal fistula repair. *Urology* 1979; 13:273.

63. Prasad ML, Nelson R, Hambrick E, et al: York Mason procedure for repair of post operative rectoprostatic urethral fistula. *Dis Colon Rectum* 1983; 26:716.

64. Quartey JKM: Bladder rotation flap for repair of difficult vesicovaginal fistulas. *J Urol* 1972; 107:60.

65. Raz S: *Female Urology.* Philadelphia, WB Saunders Co, 1983, pp 372–377.

66. Raghavaiah NV: Double dye test to diagnose various types of vaginal fistulas. *J Urol* 1974; 112:811.

67. Ryan JA Jr, Beebe HG, Gibbons RP: Gracilis muscle flap for closure of rectourethral fistula. *J Urol* 1979; 122:124.

68. Scardino PL, Lippett WH: Vesical intestinal fistula. *J Urol* 1968; 99:752.

69. Shatila AH, Ackerman NB: Diagnosis and management of colovesical fistulas. *Surg Gynecol Obstet* 1976; 143:71.

70. Slade N, Graches C: Vesicointestinal fistulae. *Br J Surg* 1972; 59:593.

71. Smith AM, Veenema RJ: Management of rectal injury and rectoureteral fistulas following radical retropubic prostatectomy. *J Urol* 1942; 108:778.

72. Su CT: A flap technique for repair of vesicovaginal fistula. *J Urol* 1969; 102:56.

73. Tahzib AU: Epidemiological determinants of vesicovaginal fistulas. *Br J Obstet Gynecol* 1983; 90:389.

74. Talamini MA, Broe PJ, Cameron JL: Urinary fistulas in Crohn's disease. *Surg Gynecol Obstet* 1982; 154:553.

75. Tancer ML: The post total hysterectomy (vault) vesico-vaginal fistula. *J Urol* 1980; 123:839.

76. Tang N: A new surgical approach to traumatic rectoure-thral fistulas. *J Urol* 1978; 119:693.

77. Tiptaft RC, Motson RW, Costello AJ, et al: Fistulae in-volving rectum and urethra: The place of Parks operation. *Br J Urol* 1983; 55:711.

78. Turner-Warwick RT, Wynne EJC, Handley Ashken M: The use of the omental pedicle graft in the repair and reconstruction of the urinary tracct. *Br J Surg* 1967; 54:849.

79. Vargas AD, Quattlebaum RB, Scardino PL: Vesicoenteric fistula. *Urology* 1974; 3:200.

80. Vose SN: A technique for the repair of rectourethral fis-tula *J Urol* 1949; 61:790.

81. Wallace DM: Uretero-vaginal fistula. *Br J Urol* 1972; 44:617.

82. Waterhouse K, et al: The transpubic approach to the lower urinary tract. *Trans Am Assoc Genitourinary Surg* 1972; 64:18.

83. Wein AJ, Malloy TR, Carpiniello VL, et al: Repair of ves-icovaginal fistula by a suprapubic transvesical approach. *Surg Gynecol Obstet* 1980; 150:57.

84. West CF Jr: Vesicoenteric fistulas. *Surg Clin North Am* 1973; 53:565.

85. Weyrauch, HM: A critical study of surgical principles used in repair of urethro-rectal fistula. *Stanford Med Bull* 1951; 9:2.

86. Weyrauch HM, Rous SN: Transvaginal transvesical ap-proach for surgical repair of vesicovaginal fistula. *Surg Gynecol Obstet* 1966; 123:121.

87. Wiseman NE, Decter A: The Kraske approach to the re-pair of recurrent rectourethral fistula. *J Pediatr Surg* 1982; 17:342.

88. Young HH, Stone HB: The operative treatment of ure-throrectal fistula: Presentation of a method of radical cure. *J Urol* 1917; 1:289.

Chapter 31

Bladder Cancer

WILLIAM J. CATALONA, M.D.

EPIDEMIOLOGY AND ETIOLOGY

EPIDEMIOLOGY

Incidence and Mortality Rates

RACE AND SEX.—Bladder cancer is the fifth most common cause of cancer deaths among American men over the age of 75. It is 3–4 times more common among men than women and accounts for 9% of all cancers among men and 4% among women. The incidence (cases per 100,000 person-years) is approximately twice as high among white men as black men (27.0 vs. 13.6), but this discrepancy is much less striking among women (7.1 white vs. 5.5 black) (Young et al., 1981).

The mortality rates from bladder cancer (deaths per 100,000 person-years) are similar for blacks and whites (white men 5.9, black men 5.0; white women 1.8, black women 2.5), suggesting that some of the differences in incidence rates may be due to variable diagnosis or reporting. The mortality rates are generally less than one half of the incidence rates for both men and women (Devesa and Silverman, 1978). In 1985, it is estimated that there will be 40,000 new cases of bladder cancer diagnosed in the United States and that bladder cancer will account for 10,800 cancer deaths (Cancer Statistics, 1985).

REGIONAL DIFFERENCES.—There are some regional differences in the incidence of bladder cancer within the United States, incidence rates being 30%–50% higher in the northern region than in the southern region (Cutler and Young, 1975).

NATIONAL DIFFERENCES.—There are national differences in the incidence of bladder cancer, the United States and England having high incidence rates and Japan and Finland having low incidence rates. In Hawaii the incidence is more than twice as high among Caucasians than Japanese (Waterhouse et al., 1982). There is also evidence that Jewish people have a higher incidence of bladder cancer than non-Jewish people (Sullivan, 1982).

AGE.—Bladder cancer is generally a disease of patients over the age of 50, with incidence rates rising sharply with age (Fig 31–1). Bladder cancer occurs much less frequently in younger patients in whom it tends to express a well-differentiated histology and behave in a more indolent fashion (Benson et al., 1983b).

ETIOLOGY

Several factors have been reported to be causally related to bladder cancer including occupation exposure to chemicals, cigarette smoking, coffee drinking, analgesics, artificial sweeteners, bacterial and parasitic infections, bladder calculi, pelvic irradiation, and cytotoxic drugs. There are considerable data suggesting that bladder cancer is often a carcinogen-induced tumor. Experimental studies suggest that carcinogens initiate the process of malignant transformation by altering the genome. The cancer does not appear until the altered genome is expressed. Oncogenes have been demonstrated in several human bladder cancer cell lines (Fujita, 1984). Promoters are substances that stimulate expression of the altered genome. The process of carcinogenesis usually requires decades. Carcinogenesis is discussed in detail in Chapter 8.

Occupational Exposure

It was reported before the turn of the century (Rehn, 1895) that aniline dyes were urothelial carcinogens. Much later, it was appreciated that the active carcinogens were metabolites of the aniline dyes. In the ensuing years, a considerable body of experimental evidence based on animal studies has accumulated identifying some of the specific chemical carcinogens. These include 2-naphthylamine, 4-aminobiphenyl (xenylamine), 4-nitrobiphenyl, 4,4-diaminobiphenyl (benzidine), and 2-amino-1-naphthol (Morrison and Cole, 1976).

It has been speculated that occupational exposure may account for one fourth to one third of all cases of bladder cancer in the United States (Cole et al., 1972).

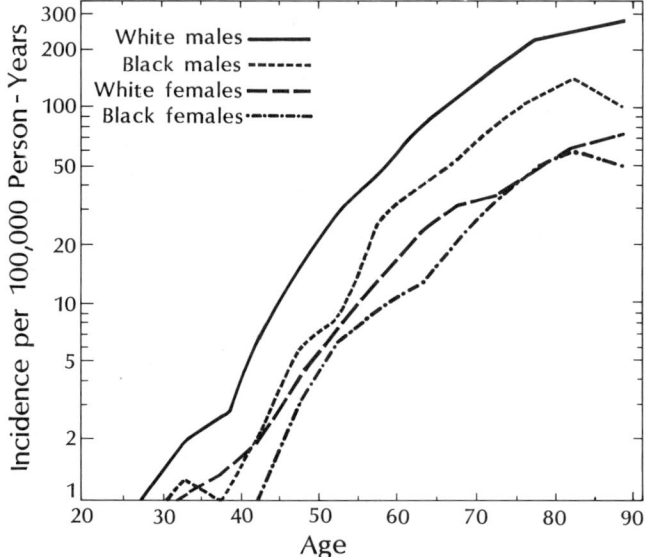

FIG 31–1.
Incidence rates of bladder cancer. (From Young JL Jr, Percy CL, Asire AJ (eds): Surveillence, epidemiology, and end results: Incidence and mortality data, 1973–1977. *Natl Cancer Inst Monogr* 57, 1981.)

Epidemiologic studies have revealed that the latent period may be as long as 40–50 years, although the latent period may be diminished in patients who have longer exposure to the carcinogen (Case et al., 1954).

Many bladder carcinogens are aromatic amines. In addition, there is evidence that dietary nitrites and nitrates can be acted on by bacterial flora to produce carcinogenic metabolites (Chapman et al., 1981). Thus, exposure to endogenous as well as exogenous aromatic amines may be implicated in the etiology of bladder cancer. The potentially carcinogenic endogenous aromatic amines are discussed in the section on tryptophan metabolites.

In addition to the dyestuff industry, several other industries have been associated with an increased risk of bladder cancer. These include leather work, metal machining, and work with organic chemicals (Morrison, 1984).

Cigarette Smoking

There is compelling epidemiologic evidence suggesting that cigarette smokers have up to a fourfold higher incidence of bladder cancer than nonsmokers (Morrison, 1984). The increased risk for bladder cancer correlates with the number of cigarettes smoked as well as the duration of smoking and degree of inhalation. The increased risk of bladder cancer with smoking has been observed in both sexes. Epidemiologic studies examining other forms of tobacco use have not clearly demonstrated an increased risk for bladder cancer. It has been speculated that more than one third of bladder cancer cases may be related to cigarette smoking (Howe et al., 1980).

The carcinogenic substance in cigarette smoke has not been clearly identified. It is known that cigarette smoke contains nitrosamines as well as 2-naphthylamine. It also has been reported that trytophan metabolites may be present in increased concentrations in the urine of cigarette smokers (Hoffman et al., 1976).

Coffee Drinking

A causal relationship has been reported between coffee drinking and bladder cancer; however, not all studies have confirmed this relationship (Cole, 1971; Morrison, 1984). The potential association between coffee drinking and bladder cancer is complicated because coffee drinking is extremely widespread. Furthermore, the use of coffee, artificial sweeteners, and cigarette smoking are often associated.

Analgesics

It has been reported that patients who consumed large quantities of the analgesic phenacetin (5–15 kg over a ten-year period) had an increased risk of transitional cell carcinoma of the renal pelvis. Most of these patients were women. Subsequent studies demonstrated that there is also a two- to four-fold increase in the risk for transitional cell carcinoma of the bladder as well. It has been postulated that the latency period for bladder tumors may be longer than that for upper tract urinary tumors. Phenacetin has a chemical structure similar to that of aniline dyes. Studies of an increased risk for bladder cancer with other analgesics have not revealed a clear correlation (McCredie et al., 1983; Wahlqvist, 1980).

Artificial Sweeteners

Experimental animal studies have revealed that artificial sweeteners in large doses, including saccharin and cyclamates, are bladder carcinogens in rodents. Considerable controversy has arisen from these studies, because the quantities of sweeteners given to the experimental animals were extremely high, and carcinogenesis occurred only in animals who were exposed either in utero or in the neonatal period (Sontag, 1980; Morrison and Buring, 1980). Case control epidemiologic studies have revealed little significant evidence to suggest that the use of artificial sweeteners is associated with an increased risk of developing bladder cancer, although some studies have suggested that certain subsets of patients, such as nonsmoking females or males who are heavy smokers, may have some increased risk (Hoover and Strasser, 1980).

Bladder Infection

Chronic bladder infection, particularly when associated with urinary calculi or indwelling catheters, is associated with an increased risk of the development of squamous cell carcinoma of the bladder (Kunter et al., 1984). Chronic infection with *Schistosoma haematobium* (Lucas, 1982) is associated with a significantly higher incidence of squamous cell carcinoma of the bladder. In countries such as Egypt, where a substantial proportion of the population suffers from chronic schistosomiasis, squamous cell carcinoma of the bladder is the most common malignancy (bilharzial bladder cancer). The development of squamous cell carcinoma usually occurs in patients who have had severe and long-standing infestation. The precise details of carcinogenesis are not clearly understood.

Bilharzial bladder cancer is discussed in more detail in the section on squamous cell carcinoma of the bladder.

Pelvic Irradiation

It has been reported that women who have received ionizing irradiation for carcinoma of the uterine cervix have a two- to four-fold increased risk of developing transitional cell carcinoma of the bladder. This increased risk has not been clearly established in case-controlled epidemiologic studies (Duncan et al., 1977; Palmer and Sprate, 1956).

Cyclophosphamide

Several studies have suggested that patients treated with cyclophosphamide have up to a ninefold increased risk of bladder carcinoma after a relatively short latency period of 6–13 years (Fairchild et al., 1979; Pearson and Soloway, 1978). In the study reported by Durkee and Benson (1980), the vast majority of patients developed muscle-infiltrating bladder carcinomas. The causal relationship between cyclophosphamide and bladder cancer has not yet been formally proved in case-controlled epidemiologic studies.

Endogenous Tryptophan Metabolites

It has been observed that a substantial proportion of bladder cancer patients have increased urinary levels of tryptophan metabolites including kynurenine, acetylkynurenine, kynurenic acid, and 3-hydroxykynurenine (Brown et al., 1969; Wolf, 1973). Brown and associates (1969) and Teulings et al. (1978) reported that the urinary tryptophan metabolite levels correlated with tumor recurrence rates. Leklem and Brown (1976) demonstrated that the administration of pyridoxine resulted in the normalization of the excretion of tryptophan metabolites in some patients. Moreover, a controlled clinical trial demonstrated that pyridoxine administration significantly reduced early tumor recurrence rates in patients with superficial bladder cancer (Byar and Blackhard, 1977); however, in this trial urinary tryptophan metabolite levels were not measured.

Experimental animal studies have failed to induce bladder cancer using tryptophan metabolites as carcinogens. However, the combination of tryptophan plus 2-acetylaminofluorine or administration of tryptophan along with a pyridoxine-deficient diet has induced bladder tumors in an experimental animal model (Bryan, 1971, 1977; Wolf, 1973).

Substantial regional differences have been reported in the incidence of patients having abnormal tryptophan metabolites. For instance, in Boston only 17% of patients had abnormal tryptophan metabolites in the urine, while in Wisconsin 47% had abnormally increased urinary levels (Brown et al., 1969). Interestingly, it has been demonstrated that patients having occupational bladder cancers do not have increased tryptophan metabolite levels. Moreover, increased tryptophan metabolite levels have been reported in patients with cancer other than bladder cancer (Wolf, 1973).

Heredity

It has been reported that bladder cancer may occur in familial clusters (Fraumeni and Thomas, 1967; Aherne, 1974; Smith, 1974; McCullough et al., 1975). Arce and associates (1978) reported an increased incidence of HLA9-B5-CW4 in bladder cancer patients. An alternative explanation for the familial bladder cancers may be similar exposure of family members to the same environmental factors. There is little evidence for hereditary cause of most cases of bladder cancer.

PATHOLOGY

NORMAL BLADDER UROTHELIUM

The normal bladder urothelium is composed of 3–7 layers of transitional cells with large umbrella cells overlapping the cells of the subjacent intermediate cell layers, which in turn rest on the basal cell layer. Characteristically, the nuclei of the intermediate cells are oval shaped and oriented with their long axis perpendicular to the basement membrane, giving the urothelium its normal appearance of cellular polarity (Koss et al., 1974).

EPITHELIAL HYPERPLASIA

Epithelial hyperplasia is characterized by an increase in the number of cell layers without nuclear or architectural abnormalities.

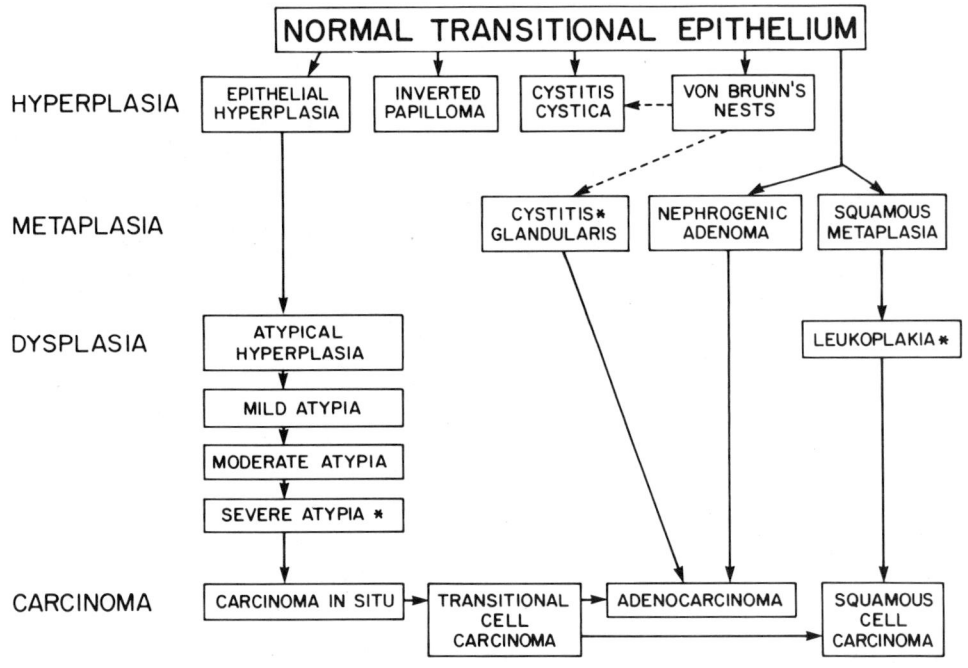

FIG 31–2.
Interrelationships among hyperplastic, metaplastic, dysplastic, and neoplastic bladder lesions. *Asterisks* indicate lesions believed to be premalignant.

UROTHELIAL DYSPLASIA

Preneoplastic Proliferative Alterations of Urothelium

A spectrum of preneoplastic urothelial changes that are postulated to be proliferative and/or metaplastic responses to inflammation, irritation, or carcinogenic influences occur in the urothelium (Fig 31–2).

Atypical epithelial hyperplasia is characterized by epithelial hyperplasia with nuclear abnormalities and partial derangement of the superficial umbrella cell layer (Koss et al., 1974).

Von Brunn's nests are nests of benign-appearing transitional cells situated in the submucosa that are believed to result from inward bud-like proliferation of the basal cells of the urothelium (Fig 31–3) (Patch and Rhea, 1935).

Cystitis cystica is a descriptive term for Von Brunn's nests that have undergone central liquefaction either by cellular degeneration or active secretion (Kunze et al., 1983). Cystitis cystica appears as submucosal nests of peripherally situated transitional cells surrounding a central region of eosinophilic liquefaction (Fig 31–4).

Cystitis glandularis is similar to cystitis cystica except that the transitional cell lining of the microcysts has undergone glandular metaplasia. Cystitis glandularis appears as nests of columnar epithelial cells surrounding a central region of cellular degeneration (Fig 31–5) (Mostofi, 1954). Cystitis glandularis may have a proliferative appearance that looks papillary on cystoscopy; it is believed to be a precursor of adenocarcinoma (Edwards et al., 1972).

In contrast to cystitis cystica and cystitis glandularis, *cystitis follicularis* is a nonneoplastic response to chronic bacterial infection characterized by submucosal lymphoid follicles. Grossly, cystitis follicularis appears as punctate yellow submucosal nodules.

Other Dysplastic Urothelial Changes

Dysplastic urothelium exhibits morphologic changes that are intermediate between normal urothelium and carcinoma in situ. Urothelial dysplasia commonly is categorized as being mild, moderate, or severe; however, it has been claimed that lesions at the severe end of the

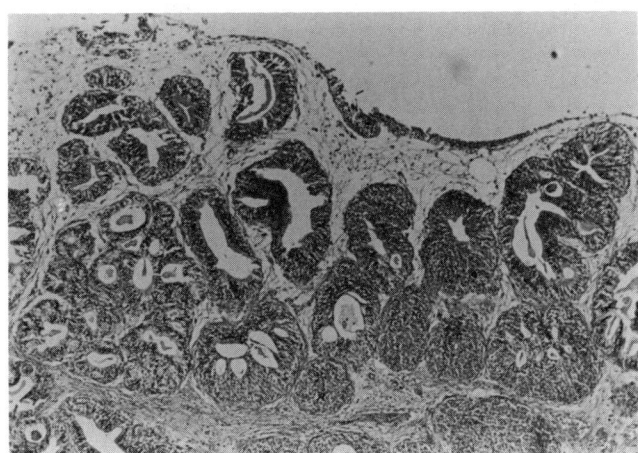

FIG 31–3.
von Brunn's nests. (From Mostofi FK, Sobin LH, Torloni H: *Histological Typing of Urinary Bladder Tumors,* no. 10, International Histological Classification of Tumors. Geneva, World Health Organization, 1973.)

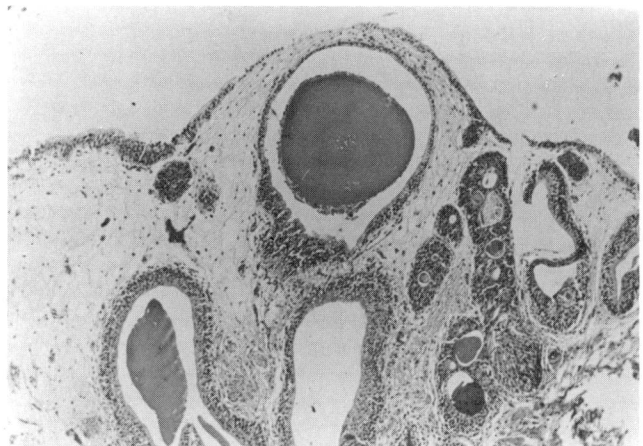

FIG 31–4.
Cystitis cystica. (From Mostofi FK, Sobin LH, Torloni H: *Histological Typing of Urinary Bladder Tumors,* no. 10, International Histological Classification of Tumors. Geneva, World Health Organization, 1973.)

spectrum are truly neoplastic (carcinoma in situ) rather than dysplastic (Murphy and Soloway, 1982). Morphologically, urothelial dysplasia is characterized by epithelial cells having nuclei that are larger than normal, basally situated, notched, and spherical with loss of normal epithelial cellular polarity. Increased numbers of cell layers or mitoses are not consistent findings (Murphy and Soloway, 1982).

Inverted Papilloma

Inverted papillomas are rare, benign proliferative lesions that are believed to be caused by chronic infection and/or bladder outlet obstruction. They occur predominantly in older men with prostatism and frequently

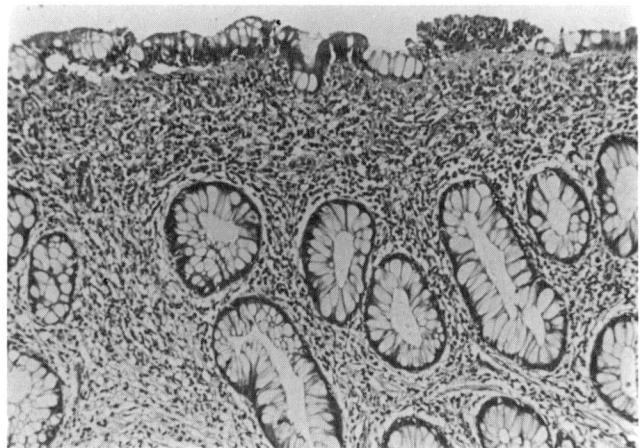

FIG 31–5.
Cystitis glandularis. (From Mostofi FK, Sobin LH, Torloni H: *Histological Typing of Urinary Bladder Tumors,* no. 10, International Histological Classification of Tumors. Geneva, World Health Organization, 1973.)

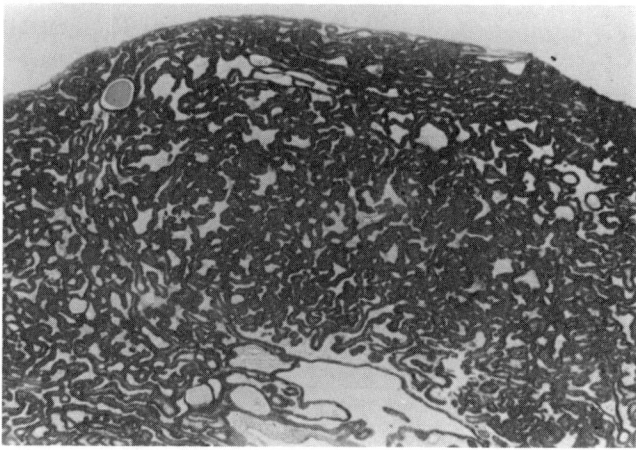

FIG 31–6.
Inverted papilloma. (From Mostofi FK, Sobin LH, Torloni H: *Histological Typing of Urinary Bladder Tumors,* no. 10, International Histological Classification of Tumors. Geneva, World Health Organization, 1973.)

arise in the trigone and bladder neck area (DeMeester et al., 1975).

Histologically, these lesions show papillary fronds projecting away from the bladder lumen into the fibrovascular stroma of the bladder wall, with the entire lesion being covered by normal urothelium (Fig 31–6). Often inverted papillomas contain areas of cystitis cystica and/or squamous metaplasia, but the papillary epithelium has benign cytologic characteristics. There are two basic types of inverted papillomas—trabecular and glandular. The trabecular type arises from proliferation of the basal cells of the urothelium. The glandular type is considered to be a form of cystitis glandularis and as such is considered to be preneoplastic (Kunze et al., 1983) (see Fig 31–2).

The treatment for inverted papilloma is the same as for low-grade transitional cell carcinoma, i.e., transurethral resection and follow-up periodic cystoscopy (DeMeester et al., 1975). Recurrence of inverted papillomas is rare; however, malignant transformation has been reported to occur (Lazarevic and Garret, 1978).

Nephrogenic Adenoma

Nephrogenic adenoma is a rare, adenomatous-appearing tumor of the bladder that derives its name from its histologic similarity to primitive renal collecting tubules. This lesion is believed to arise from metaplastic transformation of normal urothelium in response to trauma, infection, or ionizing radiation. There is generally little nuclear atypia or mitotic activity, but edema and inflammatory cell infiltration are common (Fig 31–7) (Navarre et al., 1982). Nephrogenic adenoma is more common in men and produces symptoms of dysuria and urinary frequency. The treatment is also the same as

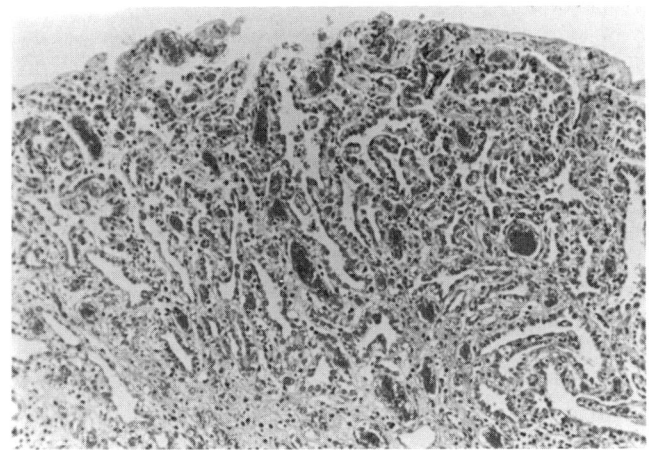

FIG 31–7.
Nephrogenic adenoma. (From Mostofi FK, Sobin LH, Torloni H: *Histological Typing of Urinary Bladder Tumors,* no. 10, International Histological Classification of Tumors. Geneva, World Health Organization, 1973.)

for low-grade superficial transitional cell carcinoma (Navarre et al., 1982). Recurrences have been documented in approximately one half of patients (Schultz et al., 1984).

A malignant counterpart of nephrogenic adenoma called mesonephric adenocarcinoma has been reported (Schultz et al., 1984). Histologically, this lesion is well differentiated and resembles nephrogenic adenoma; however, there is infiltration beyond the lamina propria (Fig 31–8). Mesonephric adenocarcinoma is frequently located in the trigone and bladder neck area. Radical excision is indicated if there is tumor invasion into bladder muscle (Schultz et al., 1984).

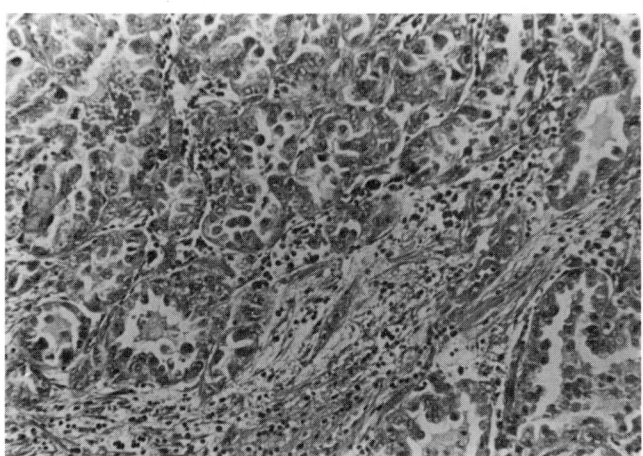

FIG 31–8.
Mesonephric adenocarcinoma. (From Mostofi FK, Sobin LH, Torloni H: *Histological Typing of Urinary Bladder Tumors,* no. 10, International Histological Classification of Tumors. Geneva, World Health Organization, 1973.)

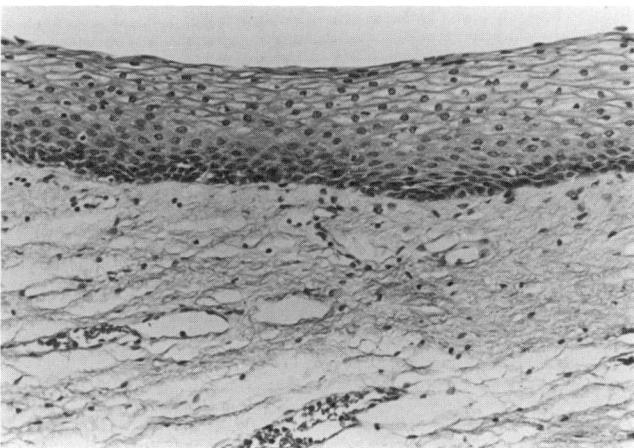

FIG 31–9.
Squamous metaplasia. (From Mostofi FK, Sobin LH, Torloni H: *Histological Typing of Urinary Bladder Tumors,* no. 10, International Histological Classification of Tumors. Geneva, World Health Organization, 1973.)

Squamous Metaplasia

Squamous metaplasia is a proliferative lesion in which the normal transitional cell epithelium is replaced by a mature, nonkeratinizing squamous epithelium. Squamous metaplasia occurs most commonly in the bladder neck and trigone area (Fig 31–9). Although some authors have reported that squamous metaplasia is a precancerous lesion (Mostofi, 1954), others have reported that squamous metaplasia of the vaginal type on the trigone of women is a normal variant that occurs under hormonal influence (Tyler, 1962). Autopsy studies have revealed that squamous metaplasia occurs in the bladder of nearly one half of adult women of all ages but in fewer than 10% of men (Wiener et al., 1979). These studies suggest that squamous metaplasia in the absence of cellular atypia or marked keratinization is a benign condition in either sex.

Vesical Leukoplakia

Leukoplakia is defined as cornification of a normally noncornifying membrane. The histopathologic criteria for vesical leukoplakia include squamous metaplasia with marked keratinization, downgrowth of the rete pegs (acanthosis), cellular atypia, and dysplasia (Benson et al., 1984) (Fig 31–10). Vesical leukoplakia is believed to be a response of the normal urothelium to a variety of noxious stimuli and generally is considered to be a premalignant lesion or one that heralds the presence of frank malignant disease (Benson et al., 1984). There are reports that vesical leukoplakia may progress to squamous cell carcinoma in up to 20% of patients (DeKock et al., 1981; Benson et al., 1984). Vesical leukoplakia is

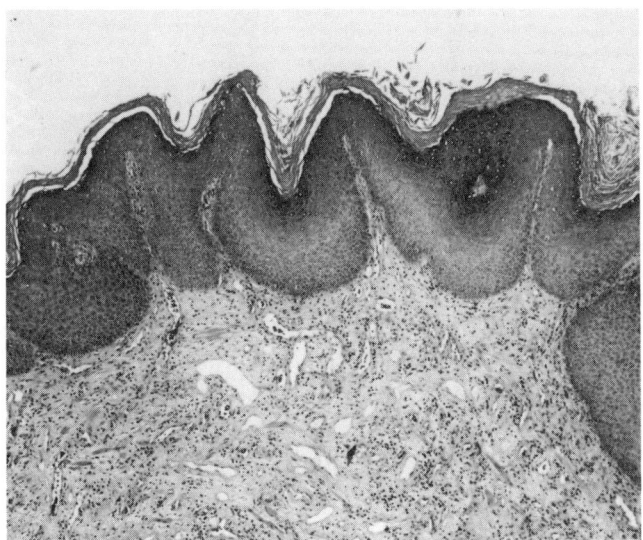

FIG 31–10.
Vesical leukoplakia (see text). (From Farrow GM: Relationship of leukoplakia to urothelial malignancy. *J Urol* 1985; 131:507.)

frequently found in patients who have chronic bladder infection, bladder calculi, long-term indwelling catheters, or schistosomiasis.

TRANSITIONAL CELL CARCINOMA

Carcinoma in situ

Carcinoma in situ is characterized by poorly differentiated transitional cell carcinoma confined to the urothelium. Histologically, there is loss of cellular cohesiveness, widening of intracellular spaces and separation of the cells from the basement membrane as well as loss of the superficial umbrella cell layer (Fig 31–11).

Initially, carcinoma in situ was discovered in associa-

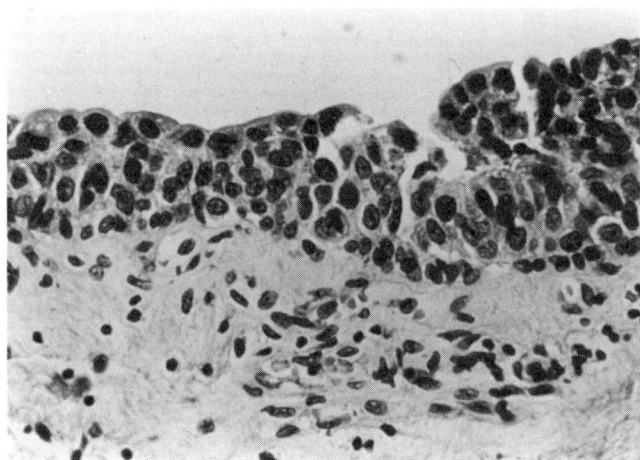

FIG 31–11.
Carcinoma in situ.

tion with invasive bladder cancers, being more common with high-grade and high-stage tumors (Melicow, 1952); however, subsequently it also was found to occur focally or diffusely in bladders containing noninvasive, well-differentiated, papillary transitional cell carcinomas. In these patients, there is a significantly higher incidence of subsequent development of invasive cancer. Carcinoma in situ also may occur in the absence of any visible bladder tumor (Althausen et al., 1976). It has been reported that patients having carcinoma in situ in the presence of a coexistent tumor have an 80% likelihood of cancer progression. Carcinoma in situ in a patient with a history of a prior bladder tumor has a 40% risk of progression (Weinstein et al., 1980).

Grossly, carcinoma in situ appears as a velvety patch of erythematous mucosa. Early in its clinical evolution, carcinoma in situ may produce no symptoms, but later it characteristically produces symptoms of urinary frequency, urgency, and dysuria (Utz et al., 1970). Because of poor cohesiveness of the carcinoma in situ cells, urinary cytopathology sutides are positive in the great majority of patients. Carcinoma in situ is more common among men than women, and often its symptoms are mistaken for symptoms of prostatism, urinary tract infection, or neurogenic bladder dysfunction.

The natural history of carcinoma in situ is unpredictable. Many patients have a protracted clinical course without developing invasive bladder cancer (Riddle et al., 1976; Weinstein et al., 1979), while others progress rapidly to invasive cancer that has a poor prognosis despite definitive therapy (Utz et al., 1970). In the early 1970s, reports from the Mayo Clinic suggested that the malignant potential of carcinoma in situ is substantial (Utz et al., 1970). In this series, many patients developed invasive bladder cancer within a relatively short time, and most ultimately succumbed to cancer despite aggressive therapy (Utz and Farrow, 1984). Other studies have suggested that some patients with carcinoma in situ, particularly those who are relatively asymptomatic, have a longer interval to the development of invasive disease (Riddle et al., 1976). Weinstein et al. (1980) suggested that carcinoma in situ may represent a malignancy that expresses the morphological features of an anaplastic tumor, but has a limited capacity to invade and metastasize.

Patients with carcinoma in situ having marked urinary symptoms generally have a shorter interval to the development of invasive bladder cancer. Approximately 20% of patients who undergo cystectomy for diffuse carcinoma in situ are found to have microinvasive carcinoma (Farrow et al., 1976).

In recent years, intravesical chemotherapy with triethylenethiophosphoramide (thiotepa), etoglucid (Epodyl), mitomycin C, and doxorubicin (Adriamycin) has

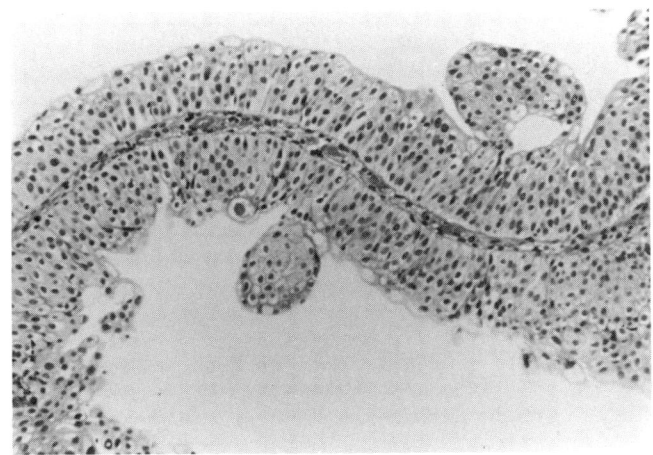

FIG 31–12.
Transitional cell papilloma. (From Mostofi FK, Sobin LH, Torloni H: *Histological Typing of Urinary Bladder Tumors,* no. 10, International Histological Classification of Tumors. Geneva, World Health Organization, 1973.)

been effective in eradicating carcinoma in situ in approximately one third of patients (Soloway, 1984). Intravesical bacille Calmette-Guérin (BCG) therapy currently appears to be the most effective means of conservative therapy for carcinoma in situ, producing complete regression in approximately two third of patients (Kelley et al., 1984). In contrast, radiation therapy has not proved to be effective in eradicating carcinoma in situ (Whitmore et al., 1977).

Papillary or Solid Transitional Cell Carcinoma

More than 99% of bladder tumors are carcinomas, and more than 90% of these exhibit a transitional cell histology (Figs 31–12 to 31–18). The histologic criteria

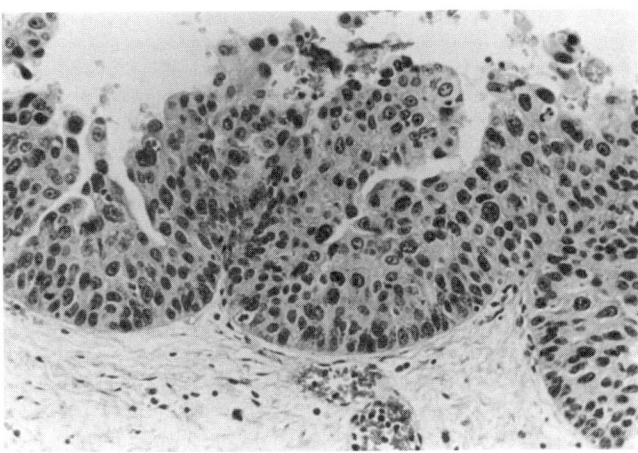

FIG 31–14.
Nonpapillary transitional cell carcinoma. (From Mostofi FK, Sobin LH, Torloni H: *Histological Typing of Urinary Bladder Tumors,* no. 10, International Histological Classification of Tumors. Geneva, World Health Organization, 1973.)

for making the diagnosis of transitional cell carcinoma include abnormalities in epithelial cytologic and architectural features. Transitional cell carcinoma may differ from normal urothelium by having an increased number of epithelial cell layers, papillary folding of the mucosa, abnormalities of nuclear morphology, loss of cellular polarity, abnormalities of the normal cellular maturation from the basal to the superficial layers, the presence of giant cells, nuclear crowding, an increased nuclear to cytoplasmic ratio, prominence of nucleoli, clumping of chromatin, and an increased number of mitoses (Koss, 1975). The most significant of these abnormalities are the presence of cells with large nucleoli and abnormal chromatin and increased numbers of cell lay-

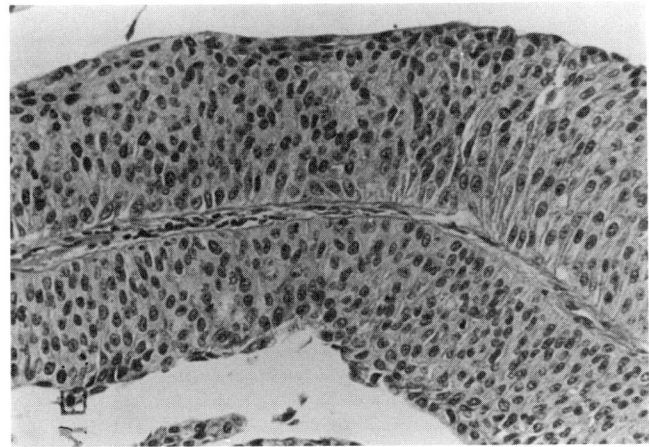

FIG 31–13.
Transitional cell carcinoma, grade 1. (From Mostofi FK, Sobin LH, Torloni H: *Histological Typing of Urinary Bladder Tumors,* no. 10, International Histological Classification of Tumors. Geneva, World Health Organization, 1973.)

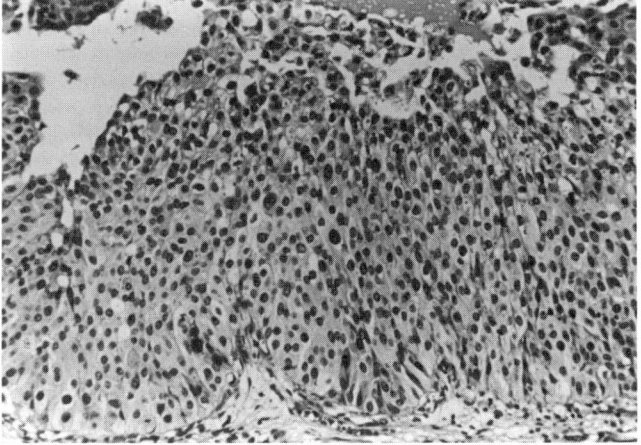

FIG 31–15.
Transitional cell carcinoma, grade 2. (From Mostofi FK, Sobin LH, Torloni H: *Histological Typing of Urinary Bladder Tumors,* no. 10, International Histological Classification of Tumors. Geneva, World Health Organization, 1973.)

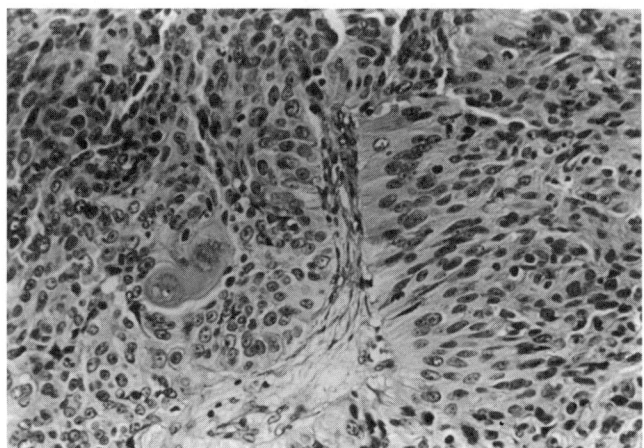

FIG 31–16.
Transitional cell carcinoma with squamous metaplasia.
(From Mostofi FK, Sobin LH, Torloni H: *Histological Typing
of Urinary Bladder Tumors,* no. 10, International Histological
Classification of Tumors. Geneva, World Health Organiza-
tion, 1973.)

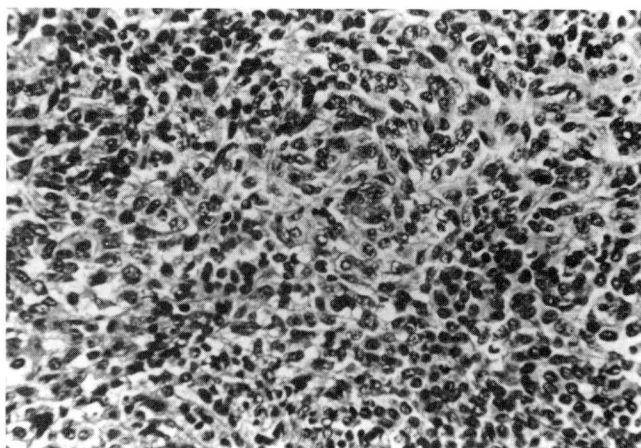

FIG 31–18.
Undifferentiated carcinoma. (From Mostofi FK, Sobin LH,
Torloni H: *Histological Typing of Urinary Bladder Tumors,*
no. 10, International Histological Classification of Tumors.
Geneva, World Health Organization, 1973.)

ers with loss of normal polarity (Murphy and Irving,
1981; Melamed et al., 1960). Some of these changes
also occur in inflammatory, reactive, or regenerative
conditions (Mostofi, 1954). The grading of transitional
cell carcinomas is discussed below.

Transitional cell carcinomas vary in their malignant
potential from papillomas (see Fig 31–12), which are
virtually benign tumors characterized by fibromuscular
stroma covered by normal urothelium, to highly anaplastic carcinomas (see Fig 31–18). Transitional cell carcinomas also manifest diverse patterns of tumor growth
including papillary, sessile infiltrating, nodular, mixed,

and flat intraepithelial growth. It is difficult to demonstrate lamina propria invasion in bladder cancer because a distinct basement membrane is not readily demonstrable in normal bladder mucosa. Moreover,
invaginations of the urothelium into the submucosa, as
occurs with von Brunn's nests, are difficult to distinguish from invasion (Mostofi, 1954). The diagnosis of
cancer is made on the basis of anaplastic changes of the
urothelium, even in the absence of invasion.

Because of the marked metaplastic potential of transitional cell epithelium, transitional cell carcinomas may
contain spindle cell, squamous cell, or adenocarcino-
matous elements (see Figs 31–16 and 31–17). Such elements are present in approximately one third of transitional cell carcinomas. Some tumors may exhibit several
different elements. The World Health Organization
Histological Classification of bladder tumors is shown in
Table 31–1.

Transitional cell carcinomas arise most commonly on
the trigone, bladder base, and lateral bladder walls;
however, they may arise anywhere within the bladder.
Approximately 70% of bladder tumors are papillary,
10% are nodular, and 20% are mixed.

Tumor Grading

Bladder tumors were first graded by Broders (1922),
based on the proportion of anaplastic cells in the tumor.
There is no uniformly accepted grading system for bladder cancer. Most systems are based primarily on the
degree of anaplasia of the tumor cells. The American
Joint Committee (AJC) system includes four grades; the
Union Internationale Contre le Cancer (UICC) system
includes grades 0–3; and the World Health Organiza-

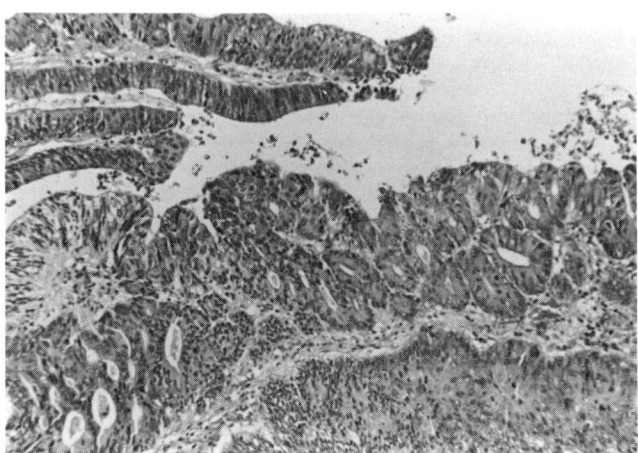

FIG 31–17.
Transitional cell carcinoma with glandular metaplasia. (From
Mostofi FK, Sobin LH, Torloni H: *Histological Typing of Uri-
nary Bladder Tumors,* no. 10, International Histological
Classification of Tumors. Geneva, World Health Organiza·
tion, 1973.)

TABLE 31–1.

World Health Organization Histologic
Classification of Bladder Tumors

Epithelial tumors
  Transitional cell papilloma
  Transitional cell papilloma, inverted type
  Squamous cell papilloma
  Transitional cell carcinoma
    With squamous metaplasia
    With glandular metaplasia
    With squamous and glandular
      metaplasia
  Squamous cell carcinoma
  Adenocarcinoma
  Undifferentiated carcinoma
Nonepithelial tumors
  Benign
  Malignant
    Rhabdomyosarcoma
    Other
Miscellaneous tumors
  Pheochromocytoma
  Lymphomas
  Carcinosarcomas
  Malignant melanoma
  Others
Metastatic tumors and secondary extensions
Unclassified tumors
Epithelial abnormalities
  Papillary (polypoid) "cystitis"
  von Brunn's nests
  "Cystitis" cystica
  Glandular metaplasia
  "Nephrogenic adenoma"
  Squamous metaplasia
Tumor-like lesions
  Follicular cystitis
  Malakoplakia
  Amyloidosis
  Fibrous (fibroepithelial) polyp
  Endometriosis
  Hamartomas
  Cysts

tion (WHO) system includes three grades. All grading systems are somewhat arbitrary.

There is a striking correlation between tumor grade and tumor stage (Jewett, 1946). Most well-differentiated tumors are superficial, and most poorly differentiated tumors are invasive. Stage for stage, there is a significant correlation between tumor grade and prognosis. There is an even closer correlation between tumor stage and prognosis.

In most grading systems well-differentiated (grade 1) tumors (see Fig 31–13) have a thin fibrovascular stalk with a thickened (more than seven cell layers) urothelium exhibiting only slight anaplasia. There may be slight cellular pleomorphism, an increased nuclear-cytoplasmic ratio, and prominence of the nuclear membrane. There may be a mild disturbance of base-to-surface cellular maturation and only rare mitotic figures. Moderately differentiated (grade 2) tumors (see Fig 31–15) have a wide fibrovascular core, and greater disturbance of base-to-surface maturation with loss of polarity. The nuclear-cytoplasmic ratio is higher with nuclear pleomorphism and prominent nuclei. Mitotic figures are frequent. Poorly differentiated tumors (see Fig 31–18) contain cells that do not differentiate as they progress from the basement membrane to the surface. There is marked nuclear pleomorphism with a high nuclear-cytoplasmic ratio. Mitotic figures may be frequent. Most poorly differentiated tumors invade deep into the muscle layers of the bladder wall (Friedell et al., 1980).

Metaplastic Elements

Combinations of different tumor types frequently coexist in the same bladder. All primary epithelial tumors have a common ancestry in the transitional epithelium. The most frequent combination is papillary transitional cell carcinoma with flat carcinoma in situ. Elements of squamous carcinoma also are commonly associated with invasive transitional cell carcinoma (see Fig 31–16). Adenocarcinoma also may occur in association with invasive transitional cell carcinoma (see Fig 31–17) (Koss, 1975). The presence of squamous or glandular metaplasia or both in a transitional cell carcinoma does not change the principal classification of the tumor as a transitional cell carcinoma.

Squamous Cell Carcinoma

There is marked variability in the prevalence of squamous cell carcinoma of the bladder in different parts of the world. Squamous cell cancer accounts for only about 1% of bladder cancers in England (Costello et al., 1984), 7% in the United States (Koss, 1975), and more than 75% in Egypt (El-Bolkainey et al., 1981). The male preponderance is less striking in patients having squamous cell carcinoma (1.3:1). Approximately 80% of squamous cell carcinomas in Egypt are associated with chronic infection with *Schistosoma haematobium* (Fig 31–19). These so-called bilharzial cancers occur in patients who are on the average 10–20 years younger than transitional cell carcinoma patients. Bilharzial cancers are locally exophytic, nodular, fungating lesions that generally are histologically well differentiated and have a relatively lower incidence of lymph node and distant metastases. This low incidence of metastases is believed to be due to the fact that the tumors generally are low grade (El-Bolkainey et al., 1981) rather than because of capillary lymphatic fibrosis resulting from chronic schistosomal infection (Ghoneim and Awad, 1980).

Histologically, squamous cell carcinomas are composed of keratinized cells and usually contain concentric

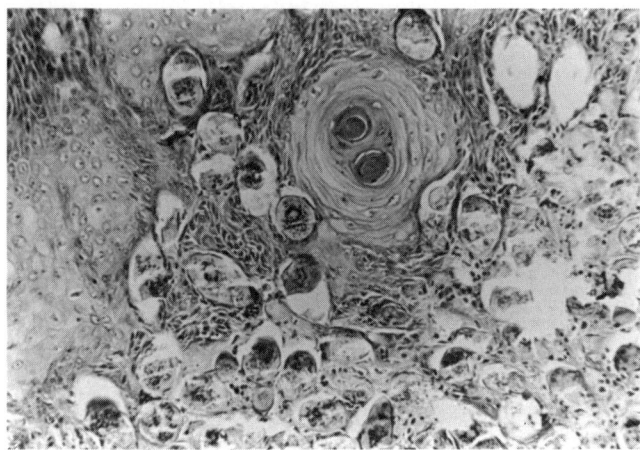

FIG 31–19.
Squamous cell carcinoma associated with schistosomiasis. (From Mostofi FK, Sobin LH, Torloni H: *Histological Typing of Urinary Bladder Tumors,* no 10, International Histological Classification of Tumors. Geneva, World Health Organization, 1973.)

aggregates of cells called squamous pearls (see Fig 31–19). Squamous cell carcinomas may show varying degrees of histologic differentiation and therefore also can be graded (Koss, 1975). Squamous cell cancers shed keratinized cells into the urine that can be detected cytologically in most patients. Squamous cell carcinomas are not frequently associated with carcinoma in situ but often are associated with coexistent squamous metaplasia.

Nonbilharzial squamous cell carcinomas are frequently associated with chronic irritation of the bladder produced by stones, chronic indwelling catheters or vesical diverticula. Up to 80% of paraplegics have squamous changes in the bladder, and as many as 5% develop squamous cell carcinoma (Maruf et al., 1982; Broecher et al., 1981; Kaufman et al., 1977).

It has been suggested that patients with squamous cell carcinoma of the bladder have a less favorable prognosis than their counterparts with transitional cell carcinoma. This impression is created by the fact that a greater proportion of squamous cell carcinoma patients have advanced disease at the time of diagnosis. Several reports suggest that stage for stage, the prognosis of patients with squamous cell carcinoma is comparable to that of those with transitional carcinoma (Richie et al., 1976; Faysal, 1981; Johnson et al., 1976).

Conservative therapy including transurethral resection, partial cystectomy, and radiation therapy for squamous cell carcinoma has not proved to be successful (Newman et al., 1968), although definitive radiation therapy with salvage cystectomy is recommended by some authors (Costello et al., 1984). The best survival results have been achieved with radical cystectomy

with or without preoperative radiation therapy (Ghoneim and Awad, 1980). Although the necessity of preoperative radiation therapy has not been established, Ghoneim and Awad (1980) have recommended short-course preoperative radiation therapy followed by radical cystectomy with pelvic lymphadenectomy as the preferred treatment for squamous cell carcinoma of the bladder. Chemotherapy appears to be less effective for bilharzial squamous cell carcinoma than for transitional cell carcinoma (Maruf et al., 1982).

Adenocarcinoma

Adenocarcinoma of the bladder is rare, accounting for fewer than 2% of all primary bladder cancers. Adenocarcinomas are classified as primary vesical, urachal, and metastatic.

PRIMARY VESICAL ADENOCARCINOMA.—Primary vesical adenocarcinoma arises most commonly on the trigone, lateral walls, and dome of the bladder from metaplasia of transitional epithelium but can occur anywhere within the bladder (Fig 31–20). The vast majority of carcinomas arising in exstrophic bladders are adenocarcinomas, which are believed to occur in response to chronic inflammation (Bennett et al., 1984; Nielson and Nielsen, 1983). Adenocarcinoma also can occur in association with schistosomiasis (Anderstrom et al., 1983).

Histologically, any variant of enteric adenocarcinoma may be encountered in the bladder including signet ring and colloid carcinoma. Tumors may be papillary or solid; most are mucin producing (Koss, 1975). Signet ring adenocarcinoma may produce linitis plastica of the bladder (Choi et al., 1984; Sheldon et al., 1984). Ad-

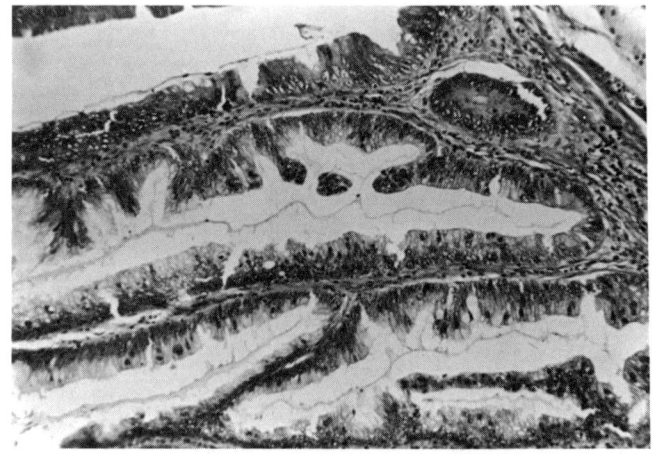

FIG 31–20.
Adenocarcinoma of the bladder. (From Mostofi FK, Sobin LH, Torloni H: *Histological Typing of Urinary Bladder Tumors,* no 10, International Histological Classification of Tumors. Geneva, World Health Organization, 1973.)

enocarcinomas are usually poorly differentiated and invasive. They are not commonly associated with carcinoma in situ but frequently are associated with cystitis glandularis. Adenocarcinomas are poorly responsive to radiation therapy. Radical cystectomy with pelvic lymphadenectomy offers the best chance for cure. Because adenocarcinomas are usually diagnosed at a more advanced stage than transitional cell carcinomas, they generally have a poorer prognosis; however, there is no evidence to suggest that stage for stage the prognosis is different from that of transitional cell carcinoma. Adenocarcinomas exhibit the same pattern of metastases as transitional cell carcinomas. They are also poorly responsive to cytotoxic chemotherapy (Anderstrom et al., 1983).

URACHAL ADENOCARCINOMA.—Technically, urachal carcinomas arise outside the bladder. Histologically, they may be adenocarcinomas, transitional cell carcinomas, squamous cell carcinomas, or sarcomas. Adenocarcinoma is the most common type. Mostofi and associates (1955) established the criteria for the diagnosis of urachal carcinoma. There must be a sharp demarcation between the tumor and the adjacent vesical mucosa, and the tumor must be located within the bladder wall beneath normal urothelium. Urachal tumors may invade through the urothelium and extend into the bladder. This makes the diagnosis of primary urachal cancer difficult to establish (Koss, 1975; Kakizoe et al., 1983).

Most urachal carcinomas are mucin producing. These tumors may extend into the space of Retzius and produce a bloody or mucoid discharge from the umbilicus or present as a palpable mass. Many of these tumors have stippled calcifications on x-ray. Tumors that have invaded into the bladder lumen may produce mucus in the urine.

The prognosis of urachal carcinomas is worse than for primary bladder adenocarcinoma (Mostofi et al., 1955). Many urachal carcinomas have been treated with partial cystectomy, but the local recurrence rate has been 15%–50% (Magri, 1962). Moreover, histologic examination of surgical specimens has revealed unexpectedly wide and deep infiltration of the tumor. These observations suggest that partial cystectomy is inadequate in the majority of cases (Kakizoe et al., 1983; Sheldon et al., 1984). Accordingly, the recommended treatment for all but the smallest, well-differentiated tumors is radical cystectomy and bilateral pelvic lymphadenectomy with en bloc excision of the urachus. Radiation therapy has not been effective in treating urachal carcinoma (Sheldon et al., 1984).

Metastases from urachal carcinoma occur in iliac and inguinal lymph nodes, omentum, liver, lung, and bone (Sheldon et al., 1984). Chemotherapy has not been effective in the treatment of metastatic urachal carcinoma.

METASTATIC ADENOCARCINOMA.—Metastatic cancer from other sites is the most common form of adenocarcinoma of the bladder (Choi et al., 1984). These tumors may derive from primary adenocarcinomas arising in the rectum, stomach, endometrium, breast, prostate, or ovary (Wheeler and Hill, 1954; Klinger, 1951; Nocks et al., 1983). Klinger (1951) reported that only 0.26% of 5,000 autopsy cases had deposits of metastatic adenocarcinoma in the bladder only. Patients with bladder adenocarcinoma should be carefully evaluated for other primary adenocarcinomas before proceeding with definitive therapy.

Carcinosarcoma

Carcinosarcomas are highly malignant tumors that contain both malignant mesenchymal and epithelial elements (Fig 31–21). Mesenchymal elements are usually chondrosarcoma or osteosarcoma (Koss, 1975). Epithelial elements may be transitional cell carcinoma, squamous cell carcinoma or adenocarcinoma. These rare tumors usually occur in middle-aged men, producing gross, painless hematuria. The prognosis is poor despite aggressive therapy with cystoprostatectomy. In the Mayo Clinic series, the five-year survival rate was only 20% (Sen et al., 1985). Carcinosarcomas appear to be resistant to radiation therapy and to chemotherapy (Schoborg et al., 1980; Uyama and Moriwaki, 1981). In fact, it has been postulated that carcinosarcomas may be induced by radiation therapy (Koss, 1975).

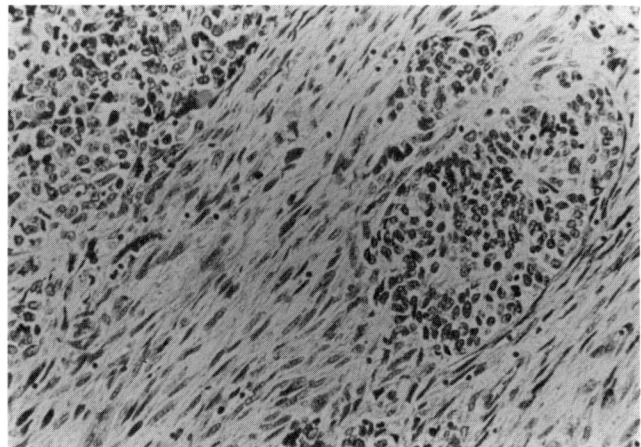

FIG 31–21.
Carcinosarcoma. (From Mostofi FK, Sobin LH, Torloni H: *Histological Typing of Urinary Bladder Tumors,* no. 10, International Histological Classification of Tumors. Geneva, World Health Organization, 1973.)

Metastatic Carcinoma

The bladder may be secondarily involved by malignant tumors from virtually any other primary site. As described above, the most common primary tumors directly invading or metastasizing to the bladder are from the prostate, ovaries, uterus, colon, lung, breast, stomach, or melanoma, lymphoma, or leukemia (Koss, 1975).

NONEPITHELIAL BLADDER TUMORS

Approximately 1%–5% of bladder tumors are nonepithelial in origin. Nonepithelial tumors may be categorized into three groups: (1) primitive connective tissue tumors (leiomyosarcoma, rhabdomyosarcoma, chondrosarcoma, osteosarcoma, liposarcoma, granular cell myoblastoma), (2) tumors of nonconnective tissue origin (angiosarcoma, neurosarcoma, neurofibroma, pheochromocytoma, melanoma, etc.), and (3) secondary nonepithelial tumors (lymphoma, leukemia, plasmacytoma, melanoma) (Rosi et al., 1983).

The malignant connective tissue tumors that arise from normal bladder tissues include leiomyosarcoma, neurofibroma, pheochromocytoma and angiosarcoma.

Leiomyosarcoma

This is the most common malignant mesenchymal tumor of the bladder occurring in adults. Leiomyosarcoma occurs twice as frequently in men as in women. It may appear as a submucosal nodule or a large, ulcerating mass. Histologically, leiomyosarcomas are characterized by spindle cells arranged in parallel bundles (Fig 31–22). Nuclear abnormalities distinguish a malignant leiomyosarcoma from its benign counterpart, leio-

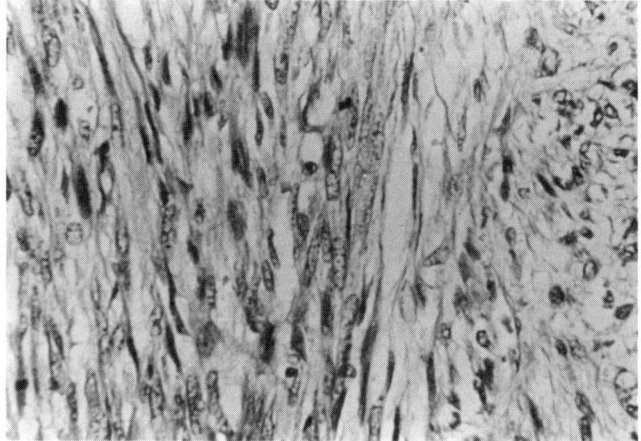

FIG 31–22.
Leiomyosarcoma. (From Mostofi FK, Sobin LH, Torloni H: *Histological Typing of Urinary Bladder Tumors,* no. 10, International Histological Classification of Tumors. Geneva, World Health Organization, 1973.)

myoma, which is the most common benign muscle tumor of the bladder. Although leiomyosarcomas may seem to be ideally suited for treatment with partial cystectomy, survival results generally have been poor with conservative operations for all but the most localized tumors, presumably due to underestimation of the extent of the tumor. Total cystectomy has yielded a five-year survival of 65% (Tsukamoto and Lieber, in press). In contrast to leiomyosarcomas, benign leiomyomas can be treated adequately with simple enucleation.

Neurofibroma of the Bladder

A neurofibroma is a benign nerve sheath tumor that results from overgrowth of Schwann's cells. The tendency for their development is often inherited as an autosomal dominant trait with variable penetrance (Torres and Bennett, 1966). Neurofibromas of the bladder are rare tumors that arise from ganglia in the bladder wall and may occur as either solitary or plexiform lesions. Often vesical neurofibromatosis becomes clinically manifest in childhood with symptoms of urinary tract obstruction, incontinence, vesical irritability, or a pelvic mass (Clark et al., 1977). Conservative management is recommended unless there is urinary tract obstruction or incapacitating symptoms. Malignant degeneration of bladder neurofibromas is rare (Clark et al., 1977).

Pheochromocytoma

Pheochromocytoma is a rare bladder tumor accounting for fewer than 1% of all bladder tumors and fewer than 1% of all pheochromocytomas (Albores-Saavedra et al., 1969). These tumors arise from paraganglionic tissue situated within the bladder wall, usually in the region of the trigone (Koss, 1975). The peak incidence is in the second through the fourth decades. There is no sex predilection. Approximately 10% of bladder pheochromocytomas are malignant, having the capacity to metastasize to regional lymph nodes or distant sites. Malignancy is determined by the clinical behavior rather than the histologic features of the tumor. Most bladder pheochromocytomas are metabolically active, with two thirds of patients having paroxysmal attacks of hypertension on filling or emptying of the bladder. More than one half of patients also have hematuria. Cystoscopy usually reveals a submucosal tumor covered by intact urothelium. Histologically, the tumors are composed of nests of polyhedral cells with eosinophilic cytoplasm (Fig 31–23). The histologic features of a malignant pheochromocytoma cannot be distinguished from those of its benign counterpart (Koss, 1975). Partial cystectomy is adequate therapy for patients having benign pheochromocytomas. The regional lymph nodes should be evaluated preoperatively by CT scan and

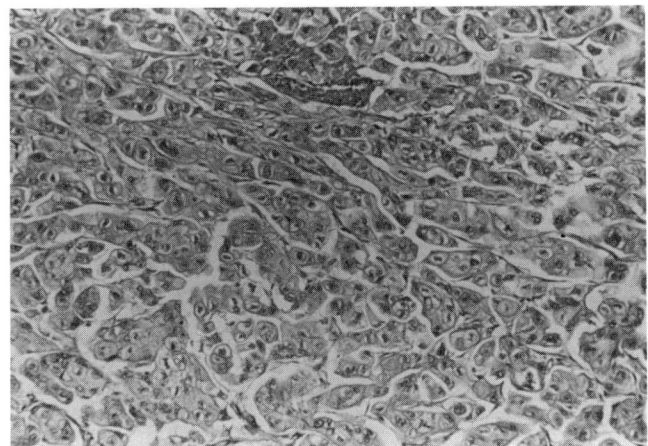

FIG 31–23.
Pheochromocytoma of the bladder. (From Mostofi FK, Sobin LH, Torloni H: *Histological Typing of Urinary Bladder Tumors,* no. 10, International Histological Classification of Tumors. Geneva, World Health Organization, 1973.)

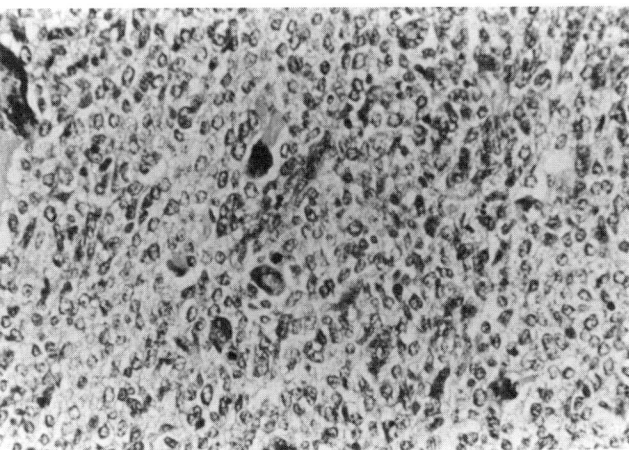

FIG 31–24.
Rhabdomyosarcoma of the bladder. (From Mostofi FK, Sobin LH, Torloni H: *Histological Typing of Urinary Bladder Tumors,* no. 10, International Histological Classification of Tumors. Geneva, World Health Organization, 1973.)

gross inspection at the time of operation. If lymph node metastases are suspected, arteriography may reveal hypervascular metastases. If lymph node metastases are present, pelvic lymphadenectomy should be performed. Patients should be carefully followed up for life, because late metastases may occur. The development of metastases may be heralded by the return of the endocrine manifestations of the tumor (DeKlerk et al., 1975).

Angiosarcomas

These are extremely rare tumors that may arise within the bladder wall. Histologically, they contain dilated vascular channels with prominent papillary endothelial proliferation (Koss, 1975). Cystectomy is the treatment of choice if the tumor has not metastasized.

Malignant connective tissue tumors containing cell types that are not normally present in the bladder include rhabdomyosarcoma, liposarcoma, chondrosarcoma, and osteosarcoma. Sarcomas of the bladder account for fewer than 1% of malignant bladder tumors. It is believed that these tumors arise from pleuripotential mesenchymal tissue of the bladder wall. *Rhabdomyosarcoma* occurs most commonly in young children, but also occasionally occurs in adult life. Embryonal rhabdomyosarcoma in children characteristically produces polypoid lesions in the base of the bladder giving rise to the descriptive term *sarcoma botryoides.* Histologically, the tumor is composed of primitive muscle cells. The characteristic rhabdomyoblast has eosinophilic cytoplasm and an eccentric nucleus (Fig 31–24). Another characteristic histologic feature of embryonal rhabdomyosarcoma is the presence of cytoplasmic cross striations. Rhabdomyosarcomas are biologically aggres-

sive tumors that may metastasize to lymph nodes in 20%–40% of patients or spread hematogenously to other organs. Embryonal rhabdomyosarcomas in children are responsive to multimodality treatment regimens, including chemotherapy with cyclophosphamide, actinomycin-D, vincristine, and Adriamycin; radiation therapy; and excision. The current treatment philosophy is to attempt to preserve bladder function if possible by treating first with conservative excision, chemotherapy, and radiation therapy and to reserve radical excision only for tumors not controlled by conservative means (see Chapter 60).

The adult rhabdomyosarcomas include the spindle cell, alveolar cell, and giant cell types. These tumors behave aggressively and do not respond well to radiation or chemotherapy. In general, the prognosis is poor (Koss, 1975; Tsukamoto and Lieber, in press). Rhabdomyosarcoma also has a rare benign counterpart called rhabdomyoma, which can be treated with conservative excision.

Liposarcoma, chondrosarcoma, and osteosarcoma (Fig 31–25) are very rare tumors of the bladder that may be carcinosarcomas in which the malignant epithelial elements have not been identified. The most effective treatment for these sarcomas is total cystectomy although wide segmental resection may be effective for very localized lesions (Wilson, 1979; Mackenzie et al., 1971; Rosi et al., 1983). In general the prognosis of patients with bladder sarcomas is poor regardless of treatment.

Primary lymphoma of the bladder (Fig 31–26) is believed to arise in the submucosal lymphoid follicles (Koss, 1975). It is the second most common type of nonepithelial bladder tumor. Most patients are in the 40-

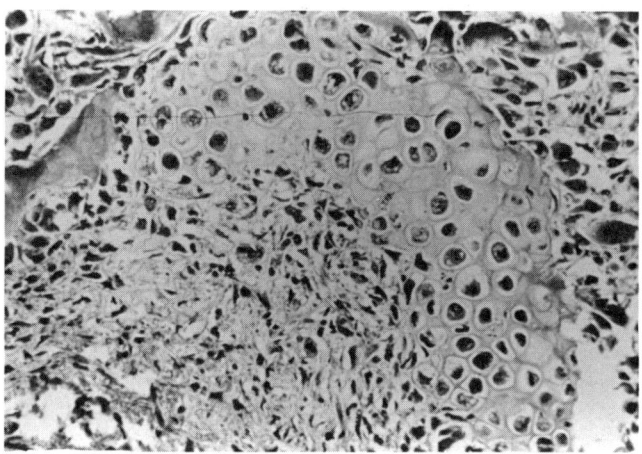

FIG 31–25.
Osteosarcoma of the bladder. (From Mostofi FK, Sobin LH, Torloni H: *Histological Typing of Urinary Bladder Tumors,* no. 10, International Histological Classification of Tumors. Geneva, World Health Organization, 1973.)

to 60-year-old age group, and women are more often affected than men. Any histologic type of malignant lymphoma can be observed. The preferred treatment of localized primary vesical lymphoma is radiation therapy. The prognosis is relatively good, with approximately 50% of patients surviving five years (Koss, 1975).

Other rare primary tumors of the bladder include plasmacytoma, granular cell myoblastoma, malignant melanoma, choriocarcinoma, and yolk sac tumor. These tumors exhibit the same characteristics as their counterparts in other sites of the body and are managed ac cordingly.

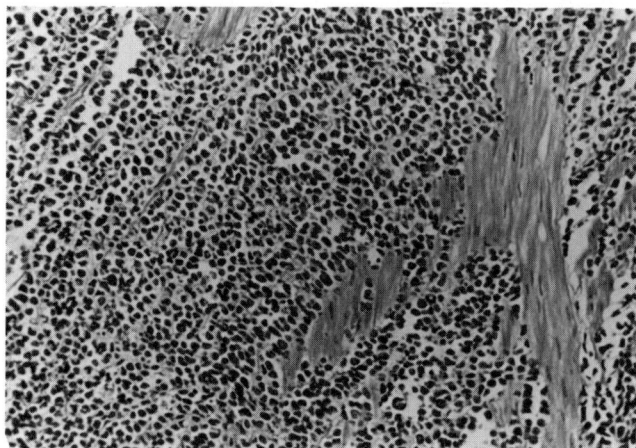

FIG 31–26.
Lymphoma of the bladder. (From Mostofi FK, Sobin LH, Torloni H: *Histological Typing of Urinary Bladder Tumors,* no. 10, International Histological Classification of Tumors. Geneva, World Health Organization, 1973.)

PATTERNS OF DISSEMINATION

MULTICENTRIC ORIGIN

Most experimental evidence suggests that, except for hereditary tumors, epithelial tumors arise from a single cell (monoclonal origin) that has undergone neoplastic transformation. Evidence based on clinical and urinary tract mapping studies suggests that transitional cell carcinoma of the urinary tract often may be a field change disease in which apparently distinct tumors arise at different sites and different points in time. These studies suggest a possible polyclonal etiology of bladder cancer; however, it is not possible to exclude the possibility that such tumors are derived from a single cell clone that has disseminated to other sites in the urinary tract by implantation or lymphatic or vascular dissemination. Studies of chromosomal markers have demonstrated the appearance of the same marker in multiple recurrences of a tumor in the bladder as well as genetically identical cells in a primary bladder tumor and cells from a metastatic site (Summers et al., 1983). Flow cytometric analysis of cells from normal-appearing mucosa in bladders of patients having bladder tumors has demonstrated aneuploid stem cells, suggesting that diffuse premalignant or malignant changes may explain the polychronotopicity (multiple tumors at different sites and times) of bladder cancer (Farsund et al., 1983). Further work is needed to clarify the clonal origin of transitional cell carcinoma.

DIRECT EXTENSION

Three mechanisms of local invasion of bladder cancer have been identified (Jewett and Eversole, 1960). The most common is en bloc spread, which occurs in about 60% of tumors and is characterized by the tumor cells invading in a broad front directly beneath the primary mucosal lesion. The next most common type is tentacular invasion, which occurs in about 25% of tumors. Lateral spread, characterized by tumor cells growing under normal-appearing mucosa, occurs in approximately 10% of tumors.

In general, bladder cancer spreads first by invading through the lamina propria into the submucosa and muscularis of the bladder wall, where it gains access to blood vessels and lymphatics, and then by metastasizing to regional lymph nodes and distant sites. There is a well-established correlation between the presence of muscle invasion and the occurrence of distant metastases (Jewett and Strong, 1946). Bladder tumors also may spread locally to invade adjacent organs including the prostate, uterus, vagina, ureters, rectum, and intestine. Bladder cancers arising in vesical diverticula pose special problems, because they invade directly from the mucosa into the perivesical tissues. Often these tumors are treated with simple diverticulectomy or partial cystectomy, but in most series the survival results have been poor (Faysal, 1981).

Vascular or lymphatic spread in patients having non-invasive or superficially invasive bladder tumors occurs in about 5% of patients with superficial papillary tumors and approximately 20% with high-grade carcinoma in situ. These metastases occur presumably from invasion into superficial lymphatic and vascular channels just beneath the lamina propria.

It is believed that lymphatic metastases may occur earlier and independent of hematogenous metastases in some patients. This is supported by clinical data showing that some patients having limited lymph node metastases apparently can be cured with radical cystectomy and pelvic lymphadenectomy (Skinner and Lieskovsky, 1984). Moreover, autopsy studies have revealed that approximately one fourth to one third of patients dying of diffuse metastatic bladder cancer do not have demonstrable pelvic lymph node metastases (Babaian et al., 1980). The pelvic lymph nodes are the most common sites of metastases occurring in bladder cancer patients (78%). The respective incidences of regional lymph node involvement are paravesical (16%), obturator (74%), and external iliac (65%). The juxtaregional common iliac lymph nodes are involved in approximately 20% of patients (Smith and Whitmore, 1981a).

The respective incidences of distant metastases from bladder cancer are liver, 38%; lung, 36%; bone, 27%; adrenal, 21%; and intestine, 13%. As with most solid tumors, virtually any organ can be involved (Babaian et al., 1980). Prostatic involvement has been reported to occur in approximately 20% of men undergoing cystectomy for invasive bladder cancer (Babaian et al., 1980). Prostatic invasion is generally considered to be a dire prognostic sign.

Bladder cancers also can spread by wound implantation (Weldon and Soloway, 1975). Implantation has been reported to occur in abdominal wound scars as well as in the resected prostatic fossa and urethra. Implantation occurs more commonly from high-grade tumors (van der Werf-Messing, 1985). Clinical studies have demonstrated that implantation of tumor cells into wound scars can be prevented by giving approximately 1,000 rad of preoperative radiation therapy. This is recommended routinely before the performance of partial cystectomy or open interstitial radiation therapy (van der Werf-Messing, 1969). Tumor cell implantation into the resected prostatic fossa is infrequent. Accordingly, some authors have stated that a bladder tumor can be

safely resected at the time of transurethral resection of the prostate without a significantly higher risk of tumor cell implantation into the prostatic fossa or urethra (Green and Yalowitz, 1972).

NATURAL HISTORY

Transitional cell carcinoma of the bladder includes a heterogeneous group of tumors that exhibit a broad spectrum of biologic potentials ranging from superficial, well-differentiated papillomas that behave in an indolent fashion to poorly differentiated invasive cancers that behave in a highly malignant fashion.

In general, tumors that are well differentiated and superficial at the time of diagnosis remain so throughout the life of the patient. Only 10%–15% of these patients subsequently develop invasive or metastatic cancer (Althausen et al., 1976; Heney et al., 1983; Green et al., 1973). The major problem caused by superficial tumors is repeated hospitalizations for the treatment of superficial recurrences. Approximately 70% of patients having superficial transitional cell carcinoma of the bladder have one or more tumor recurrences if treated with endoscopic resection alone (Althausen et al., 1976; Green et al., 1973; Gilbert, 1978) and 25% recur with higher-grade tumors (Gilbert, 1978; Green et al., 1973). It is believed that most recurrences are actually new tumors that arise from other areas of dysplastic urothelium, but some may be true recurrences resulting from inadequate treatment or tumor cell implantation (Page et al., 1978).

The principal challenge of caring for patients with superficial bladder cancer is to minimize tumor recurrences and remain vigilant to identify the minority of the patients having tumors that will progress to invasion or metastases.

Most tumors that have a high malignant potential are found to be invasive or metastatic at the time of diagnosis. In several reported series of patients with invasive bladder cancer, approximately 80%–90% have no prior history of a superficial bladder tumor (Kaye and Lange, 1982; Hopkins et al., 1983). Approximately 50% of patients with invasive bladder cancer have occult distant metastases. This substantially limits the efficacy of local or regional forms of curative therapy. Most of these patients develop overt clinical evidence of distant metastases within one year (Prout et al., 1979; Babaian et al., 1980).

Diffuse carcinoma in situ of the bladder occupies a controversial position in the spectrum of bladder cancers. One view of carcinoma in situ is that it is a diffuse highly malignant neoplasm that is on the brink of becoming invasive or metastasizing. According to this view, unless intravesical therapy induces prompt regression of carcinoma in situ, cystectomy should be performed. In support of this view is the observation that many patients having carcinoma in situ progress to invasive cancer in a relatively short time, and the results of radical surgery after tumor progression are poor (Utz et al., 1970; Koss, 1969).

An alternative view of carcinoma in situ is that it is a peculiar neoplasm composed of tumor cells that express severe morphological anaplasia but a limited biological capacity to invade through the lamina propria or metastasize. This view is supported by the observation that many patients having carcinoma in situ have been followed up for 5–10 years without developing invasive bladder cancer (Barlebo et al., 1972; Riddle et al., 1976). Most clinical evidence suggests that the prognosis of patients with carcinoma in situ who have irritative bladder symptoms is worse than that of asymptomatic patients (Riddle et al., 1976). These data suggest that carcinoma in situ may run a relatively protracted course, with an early asymptomatic phase in which the only clinical manifestation of the cancer may be microscopic hematuria, which may be followed by another phase of irritative voiding symptoms heralding the invasive phase of the disease.

The great majority of patients with metastatic bladder cancer succumb to their disease within two years (Babaian et al., 1980); however, it has been stated that approximately 5% of patients with established metastatic disease have "freak" cancers that run a more indolent clinical course, lasting five years or more (Marshall and McCarron, 1977). Between 10% and 35% of patients having limited regional lymph node metastases survive five years or more without evidence of metastases following radical cystectomy and pelvic lymphadenectomy (Smith and Whitmore, 1981; LaPlante and Brice, 1973; Skinner and Lieskovsky, 1984). To what extent these patients represent surgical cures of regionally metastatic bladder cancer or "freak" cancers is uncertain. These findings suggest that, at worst, patients having only limited nodal metastases generally have a more protracted course than those having visceral or osseous metastases, and, at best, that some patients having limited nodal metastases may be cured by radical excision. In patients having extensive lymph node metastases, the prospects for cure are virtually nil (Smith and Whitmore, 1981). Together, these observations form the basis of the controversial policy of performing cystectomy with very limited prospects for cure in patients having extensive gross nodal metastases (approximately 10% of patients coming to laparotomy), but not performing cystectomy in those having osseous or visceral metastases. The controversy exists because many patients with incurable bladder cancer either may not develop severe local symptoms from the primary tumor or may have them controlled with conservative means such as radiation therapy. Many of these patients conceivably could retain their bladder, even though they may ultimately succumb to bladder cancer.

Because of the uncertainties about the natural history of bladder cancer in individual patients, several clinical and laboratory tests have been examined as potential means of prognosticating the clinical course in individual patients. A detailed discussion is beyond the scope of this chapter; the interested reader is referred to several reviews (Huben, 1984; Javadpour, 1984; Gibas and Sandberg, 1984). In patients having superficial bladder cancer, the most clinically useful prognostic parameters for tumor recurrence and subsequent cancer progression are tumor grade, the depth of tumor penetration in the bladder wall, tumor invasion into lymphatic spaces, tumor size, the presence of associated urothelial dysplasia or carcinoma in situ, papillary or solid tumor configuration, the number and frequency of tumor recurrences, and patient age. Other factors that have been shown to be of prognostic significance are the presence or absence of antigen expression on tumor cells including ABO blood group antigens (Coon et al., 1982; Huben, 1984; Javadpour, 1984), T-antigen (Coon et al., 1982; Javadpour, 1984), and carcinoembryonic antigen (Huben, 1984; Javadpour, 1984).

The presence of chromosomal abnormalities including increased numbers of chromosomes, marker chromosomes, and chromosomes of abnormal size or abnormal position of the centromere also have been shown to correlate with an increased likelihood of tumor recurrence and cancer progression (Gustafson et al., 1982; Falor, 1978; Gibas and Sandberg, 1984). Other tests that have been evaluated for prognostic significance in bladder cancer patients include serum and/or urinary measurements of fibrin degradation products, carcinoembryonic antigen, rheumatoid factor, polyamines, tryptophan metabolites, acute phase reactants, and a variety of enzymes including lactic dehydrogenase, alkaline phosphatase, muraminidase, β-glucuronidase, and creatine phosphokinase (Huben, 1984). None of these tests has proved to be sufficiently sensitive, specific, or practically applicable to be generally adopted into clinical use.

There is compelling evidence suggesting that tumor recurrence and cancer progression rates are higher in patients having high-grade tumors (Heney et al., 1983), tumors that have penetrated the lamina propria of the bladder wall (Dalesio et al., 1983), tumors that have invaded into lymphatic spaces (Anderstrom et al., 1980), tumors that grow with a nonpapillary configuration; in patients having frequent tumor recurrences or recurrences with multiple tumors, large tumors, solid tumors, and tumors associated with urothelial dysplasia or carcinoma in situ (Althausen et al., 1976); and in patients over the age of 40 years old (Benson et al., 1983). Tumor recurrences and cancer progression also are more common among patients who have tumors that do not express blood group antigens (Huben, 1984; Javadpour, 1984), tumors that have marker chromosomes (Gibas and Sandberg, 1984), and tumors that have a large proportion of aneuploid tumor cell clones (Gibas and Sandberg, 1984).

In patients with diffuse carcinoma in situ, an important prognostic factor is the presence of irritative voiding symptoms. Riddle and associates (1976) reported that only one of 13 patients who were asymptomatic died of invasive cancer, while 15 of 23 who had irritative voiding symptoms died of invasive cancer despite treatment with definitive radiation therapy or cystectomy. Another factor that may have prognostic significance in patients with carcinoma in situ is the extent of involvement of the urothelium. Patients having only focal involvement generally have a more indolent course than those with diffuse involvement (Althausen et al., 1976).

The most important factors in predicting the likelihood of lymph node metastases in patients with invasive bladder cancer are tumor grade and depth of tumor invasion (Kern, 1984). Nodal metastases are more common in patients having anaplastic tumors with deep muscle invasion or infiltration into the perivesical fat (Prout et al., 1979; Kern, 1984). In this regard, there is a well-documented direct correlation between tumor stage and tumor grade (Kern, 1984) and a correlation between both tumor grade and stage and the likelihood of distant metastases. There is also a correlation between lymph node metastases and the presence of distant metastases (Prout et al., 1979; Kern, 1984); however, distant metastases can occur in the apparent absence of regional lymph node involvement (Babaian et al., 1980). Paraneoplastic syndromes including hypercalcemia and leukemoid reaction occur in patients with metastatic bladder cancer (Block and Whitmore, 1973; Michel et al., 1984), and generally are associated with a dire prognosis.

DIAGNOSIS

SIGNS AND SYMPTOMS

Painless hematuria is the most common presenting symptom of bladder cancer, occurring in 85% of patients (Varkarakis et al., 1974). Patients with hematuria should be evaluated to rule out bladder cancer, even when other plausible explanations of the hematuria are apparent. The next most common symptom is vesical irritability with urinary frequency, urgency, and dysuria. The latter symptom complex frequently is associated with invasive bladder cancer or diffuse carcinoma in situ. Ureteral obstruction may cause flank pain or azotemia. Other presenting signs and symptoms include lower extremity edema or pelvic mass. Patients rarely present with symptoms of metastatic disease such as weight loss or abdominal or bone pain.

URINARY CYTOLOGY

The diagnosis of transitional cell carcinoma of the bladder is suggested when malignant transitional cells are observed on cytologic examination of the sediment of urine or bladder washings. Malignant transitional cells have enlarged nuclei with irregular, coarsely textured chromatin (Fig 31–27). Conventional microscopic cytology is of limited sensitivity in detecting bladder cancer, since cells from well-differentiated tumors are more cohesive and are not readily shed into the urine. In addition, individual tumor cells from well-differentiated tumors are cytologically normal-appearing. Accordingly, urinary cytology is more sensitive in patients having high-grade tumors; however, even in patients with high-grade tumors, urinary cytology may be falsely negative in 20%. False positive cytologic findings have been reported to occur in 1%–12% of patients and are usually due to severe atypia, inflammation, or changes caused by radiation therapy. Most studies have shown that cytologic screening is not a cost-effective means of detecting bladder cancer unless high-risk populations are studied (Gamarra and Zein, 1984).

Flow cytometry to detect abnormalities of cellular DNA content has been evaluated as another cytologic

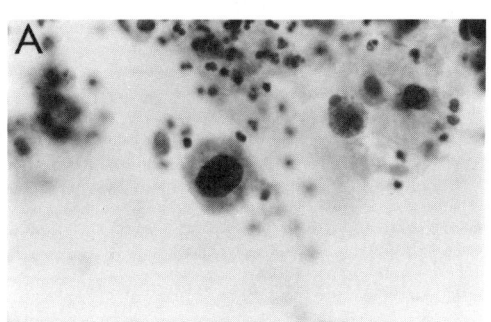
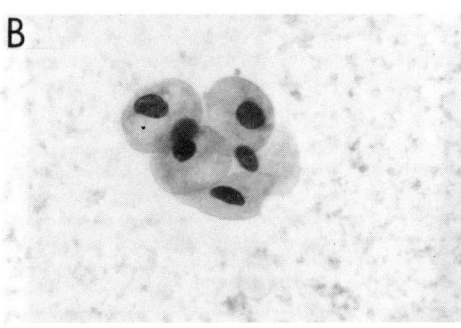

FIG 31–27.
Urinary cytologic preparations. **A,** normal transitional epithelial cells. **B,** malignant transitional cell carcinoma cells. (Courtesy of Michael J. Becich, M.D.)

means of detecting bladder cancer (Frankfurt and Huben, 1984). Bladder wash specimens are required for satisfactory results. Arbitrary limits are used to define normalcy. For instance, if more than 15% of cells are aneuploid, the cytologic findings are considered positive (Melamed, 1984). Most large-scale clinical studies comparing flow cytometry with conventional cytology, with few exceptions (Denovic et al., 1982), have not revealed a significant advantage for flow cytometry. Most superficial, low-grade tumors are diploid, also leading to false negative results. Inflammatory cells in the urine can produce false positives, because many of these cells have rapid DNA turnover. Aneuploidy is common in high-grade tumors. Flow cytometry also is most accurate in patients having carcinoma in situ, in whom more than 90% are correctly identified (Melamed, 1984). The major disadvantage of flow cytometry is its high cost.

It has been reported that saline bladder washings are generally more accurate than voided urine for detecting bladder cancer because the mechanical action of barbotage encourages tumor cell shedding and provides better preserved cells for examination (Trott and Edwards, 1973). Cellular degeneration occurs in urine that remains in the bladder for prolonged periods; therefore, first voided morning specimens should not be used. Other causes for artifactual changes are the presence of urinary tract infection, indwelling catheters, stones, or bladder instrumentation (Gamarra and Zein, 1984). These may cause sheets of transitional cells to be sheared off. These cell fragments resemble papillary fragments of transitional cell carcinomas. The osmotic changes induced in normal cells by contrast media also may cause minor difficulties in interpreting urinary cytologic preparations. Similarly, radiation therapy and intravesical chemotherapy can induce confusing abnormalities in urothelial cells.

Because of the inherent inaccuracies of cytology, many clinicians elect not to send urine samples for cytology when patients have an obvious tumor on excretory urography or cystoscopy; however, it is not uncommon for a patient with a low-grade papillary tumor to have occult (high-grade) carcinoma in situ in the bladder or elsewhere in the urinary tract. In some cases, the presence of carcinoma in situ may be suggested only by the appearance of high-grade tumor cells in the urinary sediment.

EXCRETORY UROGRAPHY

Patients having signs or symptoms suggesting the possibility of a bladder tumor should be further evaluated with excretory urography and cystoscopy. Excretory urography is not a sensitive means of detecting bladder tumors, particularly small ones, but is useful for screening the upper urinary tracts for associated urothelial tumors. Larger bladder tumors may be visualized as filling defects within the bladder or irregularities of the bladder contour on the cystogram phase of the urogram. Ureteral obstruction caused by a bladder tumor usually is indicative of a muscle-infiltrating tumor.

At the time of cystoscopy, the bladder should be carefully inspected, and bimanual examination should be performed before performing a biopsy or transurethral resection. If the upper urinary tracts are not adequately visualized on excretory urography or if there is a question about the presence of an upper urinary tract lesion, retrograde urography should be performed. If indicated, brush biopsy or ureterorenoscopy also should be performed.

RESECTION OF BLADDER TUMORS

There are conflicting opinions about the proper transurethral management of bladder tumors. The traditional teaching has been first to resect the superficial portion of the tumor, sending it as a separate specimen, and then to resect the deep portion along with the underlying bladder muscle, sending it as a separate "deep" specimen. After the tumor has been completely resected, the base of the resected area is fulgurated. This procedure provides histologic documentation of tumor grade and depth of infiltration. Alternatively, it has been suggested that resecting down to muscle for superficial bladder tumors may be unwise, because any recurrence resulting from tumor cell implantation would appear to be invading into muscle (Soloway, 1983), which would mandate aggressive therapy.

Some authors recommend also performing selected-site cold cup mucosal biopsies from areas adjacent to the tumor as well as the opposite bladder wall, dome, trigone, and prostatic urethra. The rationale for obtaining selected-site biopsies is that they provide important prognostic information about the likelihood of tumor recurrence. Between 30% and 70% of invasive bladder tumors are associated with carcinoma in situ. Other authors believe that selected-site mucosal biopsies are unnecessary—even potentially hazardous—because they denude the urothelium and may create areas for tumor cell implantation. Recent studies have shown that the incidence of significant abnormalities is less than 15% in normal-appearing mucosa.

Straining of the bladder mucosa with tetracycline, hematoporphyrins, acridine orange, or methylene blue has been advocated as a means of detecting subclinical areas of bladder tumor. However, these methods have not been adopted into general use because of uncertain sensitivity and specificity of staining (Fukui et al., 1983; Benson et al., 1982).

There is also disagreement about whether it is always necessary to perform a formal resection of superficial, low-grade papillary tumors. Some authors advocate simple fulguration as an outpatient procedure. The disadvantage of this approach is that it does not provide tissue for histologic documentation. This is also a disadvantage of laser therapy for small, superficial bladder tumors.

In patients having extensive, broad-based sessile tumors, particularly in locations that are difficult to reach with the resectocope, it is not always desirable to attempt to resect the tumor completely if it is obvious that the patient is going to require a cystectomy. Attempts at complete resection may result in bladder perforation with dissemination of tumor cells. In such cases, it may be prudent to resect only enough tissue, beginning at the margin of the tumor and progressing toward its center, to establish the tumor grade and presence of muscle invasion. This approach is best accomplished with frozen-section documentation of the adequacy of the biopsy specimens.

In patients having tumors encroaching on the ureteral orifices, a ureteral catheter may be passed up the ureter and the tumor resected around the catheter. However, a ureteral catheter can get in the way and prevent the adequate resection of the tumor. The catheter may be left in place for several days to prevent obstruction of the ureteral orifice by scar tissue. Cauterization around the ureteral orifice should be avoided or minimized.

Resection of tumors on the lateral bladder walls may induce stimulation of the obturator nerve, resulting in contractions of the adductor muscle of the leg. Resection in this area should be performed with the patient under general anesthesia with simultaneous IV administration of pancuronium. This will minimize the risk of inadvertent bladder perforation.

Tumors arising in vesical diverticula should be biopsied only. Patients having these tumors are best treated definitively with partial or total cystectomy and should not be treated with transurethral resection.

STAGING

Accurate staging of bladder cancer is necessary because tumor stage is an important determinant of therapy. Considerable attention has been paid to staging errors in patients with bladder cancer. In general, understaging of the tumor occurs most frequently in patients having high-grade and intermediate-stage tumors, of whom approximately one third are understaged and 10% are overstaged (Wijkstrom et al., 1984).

The first goal of staging is to determine whether the patient has a superficial tumor that can be managed conservatively or whether more aggressive therapy is indicated. Staging techniques other than transurethral resection and bimanual examination are usually reserved for patients having invasive bladder cancer. Detailed imaging studies are performed to determine the local extent of the primary tumor as well as the presence of regional lymph node or distant metastases. Bimanual examination provides limited information about whether there is infiltration of the bladder wall. In general, tumors that are palpable on bimanual examination are found to be infiltrating into the bladder muscle or perivesical tissues.

The second goal of staging is to identify patients who may benefit from curative therapy. CT scanning, ultrasound, and MRI recently have been used to evaluate the local extent of bladder tumors (Fig 31–28). These modalities have replaced triple-contrast cystography and pelvic arteriography (Winterburger et al., 1978). All of these studies provide useful information but are of limited accuracy in determining the presence or absence of microscopic muscle invasion or minimal extravesical tumor spread (Koss et al., 1981). Postoperative changes produced by transurethral resection of the tumor and postradiation therapy fibrosis also may cause difficulties in interpreting CT scans.

CT scanning also provides information about pelvic and para-aortic lymphadenopathy as well as the presence of liver metastases. The accuracy of CT scanning is limited by the fact that it can detect only lymph nodes that are enlarged and only liver metastases that

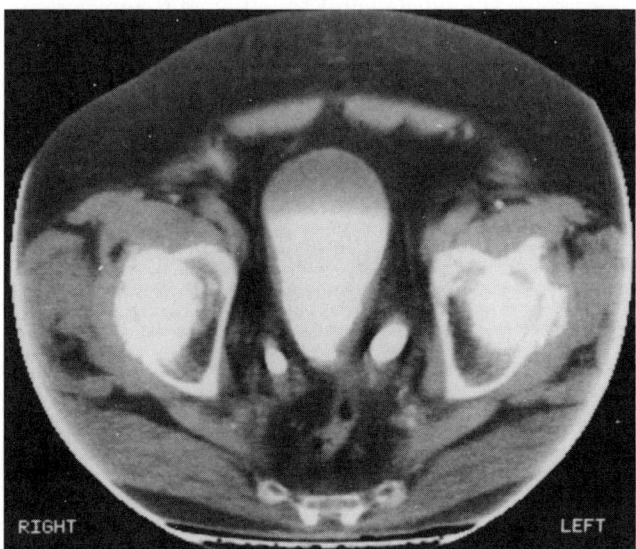

FIG 31–28.
Computed tomographic scan of the pelvis showing an invasive bladder tumor infiltrating the posterior bladder wall. (Courtesy of Dennis Balfe, M.D.)

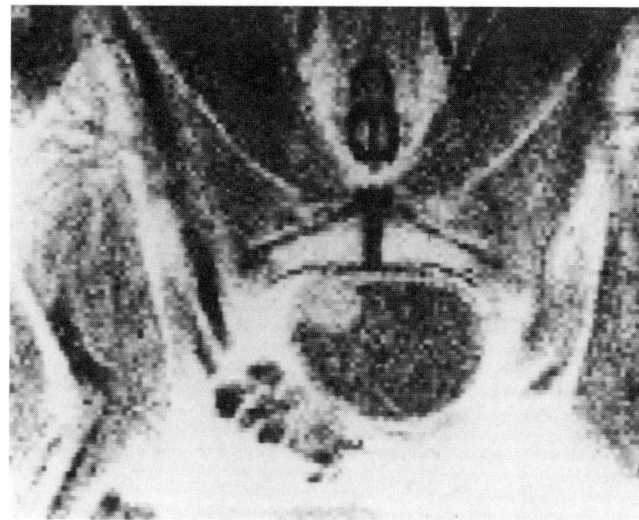

FIG 31–29.
Magnetic resonance imaging scan of the pelvis demonstrating a bladder tumor on the right side of the bladder. (Courtesy of Dennis Balfe, M.D.)

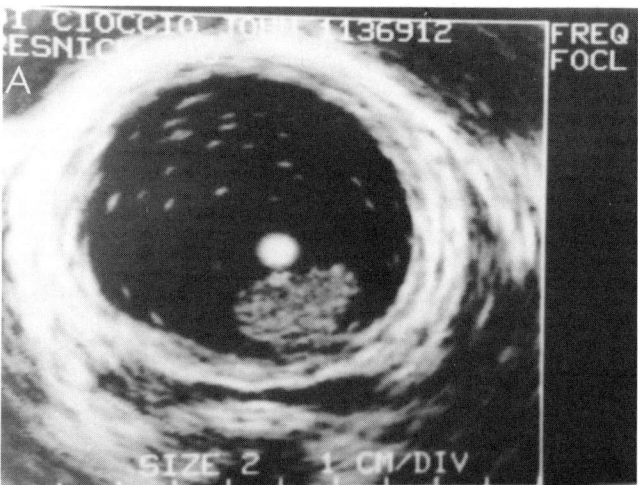

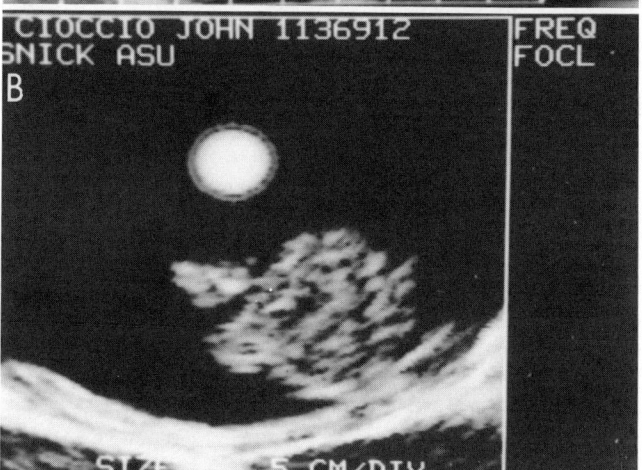

FIG 31–30.
A, transurethral ultrasound scan of superficial bladder tumor. **B,** magnified view showing no evidence of muscle invasion. (Courtesy of Martin I. Resnick, M.D.)

are greater than 2 cm in diameter. Enlarged nodes do not always indicate the presence of metastases, and hepatic lesions such as cavernous hemangiomas may be confused with metastases unless dynamic scanning is performed. CT scans have failed to detect lymph node metastases in 40%–75% of patients (Lantz, 1984).

MRI has been found to be somewhat less useful than CT scanning because its resolution of the pelvic and abdominal anatomy is not as good (Fig 31–29). Theoretically, MRI has the capacity to yield information about different biochemical environments or states of metabolic activity of tissues, but this advantage has not yet been realized. In general, both CT scanning and MRI are more helpful in advanced tumors than early tumors (Vock et al., 1982).

Both abdominal and transurethral ultrasonography have been used to evaluate the local extent of the tumor. Recent studies using a transurethral radial scanner have provided a more accurate assessment of depth of tumor invasion and completeness of tumor resection (Figs 31–30 and 31–31) (Nakamura and Nijima, 1980).

Bipedal lymphangiography provides information about the internal architecture of the external iliac lymph nodes and therefore has the capacity to detect metastases in lymph nodes that are not grossly enlarged (Lantz, 1984; Gibod et al., 1984). However, with few exceptions (Jing et al., 1982), most studies suggest that pelvic lymphangiography is not superior to CT scanning in detecting regional lymph node metastases (Farah and Cerney, 1978). Lymphangiography is associated with more discomfort and morbidity than CT. Lymphangiography can identify only about 50%–60% of patients

with lymph node metastases. False negative studies are more common than false positive studies. Accordingly, at most centers, lymphangiography is seldom performed as a routine staging procedure in patients with invasive bladder cancer. Both lymphangiography and CT scanning can be used in conjunction with fine-needle aspiration biopsy of pelvic lymph nodes to establish, but not exclude, the presence of pelvic lymph node metastases (Gibod et al., 1984).

The most accurate means of staging the pelvic lymph nodes is pelvic lymphadenectomy. There is considerable evidence to suggest that in patients having only limited nodal metastases below the bifurcation of the common iliac arteries and without invasion of the adjacent organs, lymphadenectomy may be of therapeutic benefit as well. The primary fields of lymphatic drainage of the bladder are the perivesical, hypogastric, obturator, external iliac, and presacral nodes. The com-

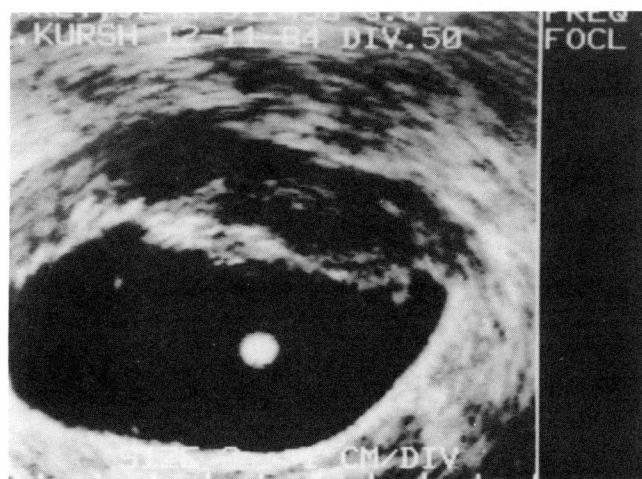

FIG 31–31.
Transurethral ultrasound scan of broad-based invasive bladder tumor. (Courtesy of Martin I. Resnick, M.D.)

mon iliac, inguinal, and para-aortic nodes are the juxtaregional nodes.

The standard pelvic lymphadenectomy for bladder cancer includes the nodes from the iliac bifurcation to the femoral canal and from the genitofemoral nerve to the bladder pedicle. Some authors recommend beginning the dissection at the aortic bifurcation. In contrast, some surgeons do not perform pelvic lymphadenectomy. The incidence of pelvic lymph node involvement correlates directly with the stage and grade of the tumor, ranging from 10% in noninvasive tumors to 40% in high-grade, deeply infiltrating tumors. A review of the literature indicates that of the 10%–40% of patients who have lymph node metastases, 35%–70% have only limited metastases (one or two nodes below the iliac bifurcation), and 10%–35% of these patients may be cured with radical excision. Therefore, the routine performance of pelvic lymphadenectomy would be expected to increase the surgical cure rate by only 1%–10%. Accordingly, in elderly or other high-risk patients, lymphadenectomy may be omitted without substantially altering the ultimate prospects for cure.

Before proceeding with lymphadenectomy, it is desirable to perform a metastatic evaluation to rule out distant metastases. Chest CT scanning is the most sensitive means of detecting pulmonary metastases; however, CT scans frequently detect small, noncalcified pulmonary lesions, most of which are granulomas. There is a direct correlation between the size of a pulmonary lesion and the likelihood of its being metastatic cancer. Most noncalcified lesions 1 cm or larger are metastases. Because routine chest x-rays and chest tomograms do not have sufficient resolution to demonstrate these tiny granulomas but can detect lesions larger than

1 cm in diameter, routine chest films or tomography usually are relied on to rule out pulmonary metastases.

It has been reported that bone scans and liver-spleen scans seldom reveal metastatic disease in patients having normal liver function tests, especially if the alkaline phosphatase level is normal (the bone isoenzyme is heat labile; the liver isoenzyme is heat stable) (Berger et al., 1981). Moreover, CT scans are more accurate than liver-spleen scans in detecting hepatic metastases. Accordingly, liver-spleen scans are not a necessary part of the routine metastatic evaluation of bladder cancer patients, and bone scans are not necessary in patients having a normal alkaline phosphatase level. It is useful, however, to obtain a bone scan as a baseline for future comparison.

Accordingly, the recommended metastatic evaluation for patients having invasive bladder cancer includes a chest film, excretory urogram, CT scan of the pelvis and abdomen, bone scan, and liver function tests. If any of these studies suggests the presence of metastases, histopathologic confirmation should be sought by the least invasive means, usually core needle biopsy.

STAGING SYSTEMS

There are two principal staging systems for bladder cancer. In the United States, urologists generally use the Jewett-Strong (1946) system as modified by Marshall (Marshall, 1952). The other principal staging system is a TNM system developed by the UICC (Wallace et al., 1975), which is generally used by radiation oncologists in the United States. These staging systems are compared in Table 31–2.

The Jewett-Strong-Marshall system considers papillary noninvasive carcinoma and carcinoma in situ both as stage 0. The UICC system classifies them as T_a and T_{is}, respectively. The Jewett-Strong-Marshall system classifies tumors that invade into the submucosa as stage A, while the UICC system classifies them as T_1. The Jewett-Strong-Marshall system subclassifies muscle-invading tumors into B1 and B2, depending on whether there is superficial or deep invasion. The UICC system classifies superficially invasive tumors as T_2 and deeply invasive tumors as T_{3a}. Tumors that have infiltrated into the perivesical fat are classified as stage C in the Jewett-Strong-Marshall system and as T_{3B} in the UICC system. Tumors that invade adjacent pelvic viscera or involve the regional lymph nodes are classified as stage D1 in the Jewett-Strong-Marshall classification. The UICC system designates tumors that invade adjacent viscera as T_4. The UICC classification subclassifies nodal involvement as N_1 for a single homolateral node, N_2 for bilateral regional or contralateral nodes, N_3 for fixed regional nodes, and N_4 for widespread adenopathy involving juxtaregional nodes. The Jewett-

TABLE 31–2.
Staging Systems for Bladder Cancer

FINDING	JEWETT-STRONG-MARSHALL STAGE	UICC CLINICAL	UICC PATHOLOGIC
No tumor in the specimen	0	T_O	P_O
Carcinoma in situ	0	T_{is}	P_{is}
Noninvasive papillary tumor	0	T_A	P_A
Lamina propria invasion	A	T_1	P_1
Superficial muscle invasion	B1	T_2	P_2
Deep muscle invasion	B2	T_{3A}	P_3
Invasion of perivesical fat	C	T_{3B}	P_3
Invasion of contiguous organ	D1	T_4	P_4
Regional lymph node metastases	D1	—	N_{1-3}
Juxtaregional lymph node metastases	D2	—	N_4
Distant metastases	D2	M_1	M_1

Strong-Marshall classification considers patients having juxtaregional nodal metastases or distant metastases as stage D2. The UICC classification considers patients having distant metastases as M_1. Pathologic confirmation of tumor stage is separately designated in the TNM system by the prefix P as $pT_{(1-4)}$ or simply $P_{(1-4)}$.

MANAGEMENT OF SUPERFICIAL BLADDER CANCER (STAGES O AND A)

Approximately 70%–80% of bladder cancer patients have low-grade superficial tumors at the time of clinical presentation. The vast majority of these patients can be treated adequately with transurethral resection or fulguration. The overall survival rates of patients with superficial bladder cancer treated with transurethral resection are excellent (approximately 70% five-year survival) (Nichols and Marshall, 1956; Barnes et al., 1967). Only a small proportion (10%–15%) of patients ultimately develop invasive bladder cancer and require more aggressive therapy such as partial cystectomy, cystectomy, full-course radiation therapy, or integrated radiation therapy and cystectomy. In patients having low-grade, muscle-infiltrating tumors, transurethral resection has yield 30%–40% five-year survival rates (Barnes et al., 1967).

In recent years, one of the major thrusts of laboratory and clinical investigations in the area of superficial bladder cancer has been the development of treatments to eradicate unresectable superficial tumors and to prevent superficial tumor recurrences as well as the development of invasive bladder cancer. To this end, a prodigious investigational effort has been made in studies of intravesical and systemic chemotherapy; intravesical immunotherapy with interferon, BCG, and other agents (Pizza et al., 1984); photoradiation therapy; laser therapy; and vitamin therapy. Other treatments such as hyperthermia and hydrostatic pressure therapy also have been tested but not adopted into general use.

External beam radiation therapy has not proved to be effective in controlling superficial bladder cancer. Radiation therapy does not prevent the occurrence of new tumors (Goffinet et al., 1975) and is associated with a high incidence of morbidity, especially the development of radiation cystitis. It has been reported that interstitial radiation therapy preceded by low-dose preoperative external beam radiation therapy to prevent wound implantation may be more effective than transurethral resection in controlling small superficial bladder tumors (van der Werf-Messing, 1985). Radon seeds,[198] Au seeds, and tantalum wire have been utilized for interstitial radiation therapy. It also has been reported that intracavitary radiation therapy using a radium capsule in a catheter (Hewett et al., 1981) and intraoperative electron beam therapy combined with conventional fractionated external beam radiation therapy (Matsumoto et al., 1981) are effective in controlling superficial bladder cancer, but these treatments have not been widely used in the United States.

CYSTECTOMY

Total or partial cystectomy is rarely required for patients with superficial bladder cancer, with the notable exception of those having symptomatic, diffuse, unresectable papillary carcinoma or carcinoma in situ that does not respond to intravesical therapy. Bracken and associates (1981) reported that the survival rates in patients treated with cystectomy for T_A or T_1 bladder cancer were comparable to those in the age-matched normal population. Most studies show that preoperative radiation therapy does not enhance survival in patients with superficial bladder cancer, although some authors advocate its use because of the possibility of clinical understaging of the tumor.

INDICATIONS FOR ADJUNCTIVE INTRAVESICAL THERAPY

In general, intravesical chemotherapy or immunotherapy should be reserved for patients who are at a high risk for tumor recurrence; i.e., those who have high-grade tumors, recurrent tumors, multiple tumors, tumors associated with urothelial atypia, or carcinoma in situ. Patients who have subsequent recurrences following intravesical therapy with one agent may then be treated with a different agent with reasonable prospects for a favorable response. Of the available agents, thiotepa and BCG are least expensive, and Adriamycin and mitomycin C the most expensive. Intravesical BCG currently appears to be the most effective agent available; however, BCG has not been approved for intravesical use by the FDA, and the optimal strain, dose schedule, and route(s) of administration have not been determined. Intravesical BCG has been used commonly by community urologists, however, because it has been approved by the FDA for use in humans for vaccination against tuberculosis.

Intravesical Chemotherapy

Adjunctive intravesical chemotherapy or immunotherapy is most appropriate for patients having recurrent tumors, multiple tumors, or diffuse carcinoma in situ.

Triethylenethiophosphoramide (Thiotepa)

Early studies using intravesical agents including silver nitrate, trichloracetic acid, and podophyllin for the treatment of superficial bladder cancer met with very limited success. The modern era of intravesical chemotherapy was ushered in with the use of thiotepa in the 1960s (Jones and Swinney, 1961). Thiotepa is an alkylating agent that acts by causing cross linking of nucleic acids and proteins. It is the most commonly used agent for intravesical chemotherapy. Used in doses of 30 mg in 30 ml of saline or 60 mg in 60 ml of saline (1 mg/1 ml) instilled directly into the bladder and retained for two hours weekly for 6–8 treatments followed by monthly treatments, thiotepa has induced complete tumor remissions in approximately 35% of patients and partial remissions in approximately 25% (Koontz et al., 1981). Thiotepa also has been used in varying dose schedules as prophylaxis against tumor recurrences following complete transurethral resection of all visible bladder cancer.

The National Bladder Cancer Collaborative Group A demonstrated that thiotepa in doses of 30 mg or 60 mg given monthly following endoscopic resection of a bladder tumor reduced tumor recurrence rates to 47% compared with a 73% recurrence rate at two years in pa-

tients not receiving thiotepa prophylaxis. Patients with grade 1 tumors benefited significantly from thiotepa prophylaxis; however, those having higher-grade tumors did not appear to derive significant benefit (Prout et al., 1983). In this study, 16% of patients receiving thiotepa prophylaxis had tumor progression, 8% developed muscle invasion, and another 3% developed metastases without evidence of muscle invasion. Green and associates (1984) reported that patients treated with thiotepa prophylaxis had a lower (albeit not statistically significant) incidence of bladder cancer deaths than patients not receiving intravesical chemotherapy. Thiotepa also has been used in the treatment of patients with carcinoma in situ with limited success (Koontz et al., 1981), but a prospective, randomized clinical trial revealed that thiotepa was less effective than BCG in the treatment of superficial bladder cancer (Brosman, 1982).

Because of its low molecular weight (198 daltons), thiotepa is readily absorbed through the urothelium and causes myelosuppression in 15%–20% of patients treated. Accordingly, a WBC and platelet count should be obtained before each treatment. Thiotepa has proved to be most efficacious when administered immediately after tumor resection (Soloway, 1983). A major advantage of thiotepa is that it is relatively inexpensive compared with mitomycin C and Adriamycin.

Etoglucid (Epodyl)

Epodyl is a medium molecular weight (262 daltons) alkylating agent that has been used in Europe in a manner similar to thiotepa. Because it is a larger molecule, it is not as readily absorbed through the urothelium and causes less myelosuppression. It is administered in a 1% solution weekly for 12 weeks and then monthly thereafter. The disadvantage of Epodyl is that it causes more severe chemical cystitis than thiotepa. Complete response rates reported using Epodyl are approximately 45% and partial responses occur in approximately 35% of patients (Lamm, 1983; Robinson et al., 1977).

In a randomized clinical trial, Epodyl was shown to be more efficacious than transurethral resection alone or Adriamycin in preventing tumor recurrences in patients having primary tumors but not in patients with recurrent superficial bladder cancer (Kurth et al., 1984).

Mitomycin C

Mitomycin C is a high molecular weight (334 daltons) antitumor antibiotic that exerts its action principally by inhibition of DNA synthesis. Because of its high molecular weight, transurothelial absorption of mitomycin C is minimal, and myelosuppression is rare. Mitomycin C has been shown to be effective as primary treatment and in many patients who have failed prior thiotepa therapy (Issell et al., 1984; Prout et al., 1982; Soloway, 1985).

The optimal dose schedule for mitomycin C is 40 mg weekly for eight weeks followed by monthly maintenance therapy for one year. In most studies, complete tumor responses occur in approximately 40% of patients, and partial responses also occur in up to 40% (Soloway, 1985; Lamm, 1983; Issell et al., 1984; Prout et al., 1982). Unlike thiotepa, mitomycin C has been reported to yield a better response rate in patients with high-grade tumors (Soloway, 1985). In Soloway's series (1985), 8% of complete responders, 23% of partial responders, and 19% of nonresponders developed muscle invasive cancer, and overall 7% died of metastatic bladder cancer. The main side effects of mitomycin C are chemical cystitis (10%–15%) and genital skin rashes (5%–15%).

Huland and associates (1984) reported that mitomycin C given every two weeks for one year and then monthly for another year was effective in preventing tumor recurrences and the development of invasive bladder cancer; however, all patients in that study had complete tumor excision and negative cytologies before mitomycin C treatment (this excluded patients with multifocal disease), 50% were treated following a resection of a solitary primary tumor, and only 29% had high-grade tumors. Other studies (Lockhart et al., 1983; Denovic, 1983) of mitomycin C prophylaxis in patients with multiple, recurrent, and high-grade tumors have not confirmed the low recurrence rate reported by Huland and associates (1984).

Doxorubicin (Adriamycin)

Adriamycin also is a high molecular weight (580 daltons) antitumor antibiotic that is minimally absorbed through the urothelium. The studies of intravesical Adriamycin therapy have varied considerably in terms of dose schedules and response rates. The available data suggest that at least a 50-mg dose should be used for intravesical therapy. The treatment schedules have ranged from three times per week to monthly. Complete responses occur overall in less than 50% of patients, and partial responses occur in approximately 35% of patients. No significant difference in the response rates of patients with low-grade and high-grade tumors has been reported (Lundbeck et al., 1983).

Adriamycin also has been used for prophylaxis against tumor recurrence in doses of 60–90 mg given at intervals from every three weeks to every three months. Garnick and associates (1984) reported that 47% had recurrences within 18 months, and 16% subsequently developed muscle invasion. Zincke and associates (1983) conducted a clinical trial comparing Adriamycin, thiotepa, and placebo. The recurrence rate in placebo-treated patients was significantly higher (71%) than in thiotepa- (30%) or Adriamycin- (31%) treated patients.

Nijima and associates (1983) compared Adriamycin and mitomycin C as prophylactic agents and found that by day 540 of the trial, approximately 55% of Adriamycin-treated patients were tumor free compared with 42% of mitomycin C-treated patients and 38% of controls. Kurth and associates (1984) found that Adriamycin was not effective in reducing recurrence rates in patients with primary bladder tumors but was effective in patients treated for recurrent tumors. In an unpublished ongoing trial comparing Adriamycin with intravesical BCG, the BCG has proved to yield significantly superior response rates (Lamm et al., 1985). The adverse side effects of Adriamycin include relatively marked chemical cystitis in many patients, in some of whom it has progressed to permanent bladder contractures. Adriamycin is also considerably more expensive than thiotepa or BCG.

In general, similar results have been achieved with all of the agents used for intravesical chemotherapy. When used for the treatment of residual papillary tumors, the complete reponse rates range from 33% to 57%; when used for the treatment of carcinoma in situ, complete regression occurs in 55%–66%; and when used for prophylaxis, tumor recurrence rates are reduced to 30%–44% compared with 70% in controls. The evidence suggesting that intravesical chemotherapy reduced the incidence of subsequent invasive disease or ultimate cancer death rates is inconclusive.

Systemic Chemotherapy for Superficial Bladder Cancer

Systemic chemotherapy has been evaluated as a potential means of controlling superficial bladder cancer. Limited studies have been performed using methotrexate (Hall and Heath, 1981), cyclophosphamide (England et al., 1981), and cisplatin (Needles et al., 1982). Although each of these agents has demonstrated some activity, because of the attendant side effects systemic chemotherapy has not been widely used in patients with superficial bladder cancer.

Vitamin Therapy

Vitamins including pyridoxine (Byar and Blackhard, 1977; Studer et al., 1984), vitamin C (Schlegel et al., 1969), and vitamin A analogue (retinoids) (Alfthan et al., 1983; Studer, 1984) have been evaluated as prophylactic agents against bladder cancer recurrences. The therapeutic efficacy of these vitamins has been of marginal statistical significance. Moreover, retinoids have the undesirable side effects of causing severe dryness of mucous membranes.

Intravesical BCG Therapy

Intravesical BCG therapy was introduced by Morales et al. (1976). In that study, the Pasteur stain of BCG was administered both intravesically (120 mg in 50 ml of saline) and intradermally (5 mg with a Heaf gun) weekly for six weeks.

Several prospective, randomized clinical trials have been conducted comparing the efficacy of intravesical BCG to transurethral resection alone or transurethral resection plus intravesical chemotherapy. Uniformly, the data have supported superior efficacy of BCG therapy (Brosman, 1982, 1984; Camacho et al., 1980; DeKernion et al., 1984; Herr et al., 1984; Lamm et al., 1982; Morales, 1979, 1980, 1983; Morales et al., 1981; Netto and Lemos, 1983; Schellhammer et al., 1984). Most studies have reported the results of BCG treatment for three categories of patients: prophylaxis (tumor-free patients), residual tumor (patients with tumors other than carcinoma in situ), and carcinoma in situ.

Several different strains of BCG have been used including Pasteur, Tice, Connaught, and Moreau. The viability and density of bacilli per milligram of vaccine may vary with the strain used and may vary from lot to lot of the same strain (Kelley et al., 1985). Various routes of administration of BCG also have been used, including combined intravesical and intradermal, intravesical alone, and oral. All have been reported to be successful, but it is unknown which is optimal. Direct intralesional injection of BCG has been evaluated, but was abandoned owing to severe toxic side effects (Martinez-Piniero and Muntanola, 1977).

Several prospective, randomized trials have evaluated BCG for prophylaxis of recurrent tumors (Camacho et al., 1980; Lamm et al., 1980; Pinsky et al., 1982). Lamm and associates (1980) reported that BCG reduced the tumor recurrence rate to 17% as compared with 42% in concurrent controls treated with transurethral resection alone (mean follow-up, 15 months). Many patients who had failed prior intravesical thiotepa were successfully treated with BCG (Brosman, 1982, 1984; Netto and Lemos, 1983; Schellhammer et al., 1984).

Two prospective, randomized trials compared BCG with thiotepa for prophylaxis (Brosman, 1982; Netto and Lemos, 1983). In these studies, the recurrence rates were 0% (mean follow-up, 18–21 months) and 6% (mean follow-up, 36–39 months) with BCG compared with 40% and 43%, respectively, with thiotepa. Moreover, Pinsky and associates (1982) reported data suggesting that BCG may also decrease the rate of progression to muscle invasive disease from 36% in control patients to 9% in BCG-treated patients. However, the Pinsky group population included some patients with initial tumors that had invaded into superficial muscle. Taken together, the various studies have shown that intravesical BCG therapy for prophylaxis has yielded recurrence rates of 0%–41% (most around 20%), while control treatments have yielded recurrence rates of 40%–80% (Morales, 1979; Lamm et al., 1980; Camacho et al., 1980; Brosman, 1982; Netto and Lemos, 1983; DeKernion et al., 1984; Dresner et al., 1984).

Several investigators also have reported the results of BCG therapy for residual tumor (Morales, 1979; Brosman, 1984; DeKernion et al., 1984; Schellhammer et al., 1984; Dresner et al., 1984). Collectively, these series document a complete response rate of 58%. Morales and associates (1979) reported that response rates were higher among patients with low-grade tumors. Morales advocated initiating BCG therapy within ten days of tumor resection to take advantage of increased bacillary adherence to the disrupted bladder mucosa.

Intravesical BCG also has been extensively evaluated in patients with carcinoma in situ (Morales, 1979; Lamm et al., 1982; Brosman, 1984; Herr et al., 1983, 1984; DeKernion et al., 1984; Dresner et al., 1984). Collectively, these studies reveal a complete response rate in 72% of 104 patients treated. Complete responses are usually associated with resolution of urinary irritative symptoms.

The principal side effect of intravesical BCG therapy is vesical irritability, which usually is well tolerated by most patients. The most common symptoms are dysuria (91%), urinary frequency (90%), hematuria (46%), fever (24%), malaise (18%), nausea (8%), chills (8%), arthralgias (2%), and pruritus (1%). Approximately 6% of patients have symptoms severe enough to require treatment with isoniazid (Lamm et al., 1984).

Toxicity is related to the intensity of BCG therapy, which also correlates with its therapeutic success. Brosman (1982) reported greater toxicity with his intensive protocol than that reported by Morales (1979) using a milder regimen. Toxicity also is related to the route of administration. Severe anaphylactic reactions have been reported with intralesional injections, while intravesical and oral administration have not produced severe complications.

The mechanism of action of BCG is unknown. The available clinical data suggest that the development of an immune response to BCG may be involved in the antitumor effect. Several investigators have demonstrated that a delayed cutaneous hypersensitivity response to PPD correlates with a favorable response to BCG therapy (Lamm et al., 1982; Dresner et al., 1984; Brosman, 1982). Intravesical BCG also has been shown to induce a chronic granulomatous response in the bladder of many patients (Morales, 1976; Brosman, 1984; Schellhammer et al., 1984; Lamm et al., 1980).

Most patients who develop granulomas have favorable responses to BCG therapy, while most patiens who do not develop granulomas do not have favorable responses. The correlations between conversion of purified protein derivative (PPD) skin tests and granulomatous response in the bladder and a favorable response to BCG therapy are not perfect. Some patients who have converted skin tests and develop granulomas in the bladder do not respond favorably to BCG therapy and some who do not have converted skin tests or develop granulomas have favorable responses. Moreover, many patients who fail to respond to an initial course of BCG therapy may respond to more intensive treatment. The reasons for this are unclear but may be related to differences in patient immune competence, differences in vaccine potency, or both (Kelley et al., 1985).

Intravesical BCG therapy appears to be more efficacious than intravesical chemotherapy for prophylaxis or treatment of residual tumors and carcinoma in situ. There has been considerable variability in therapeutic regimens used, and the optimal regimen is not known. The cost of BCG is comparable to that of thiotepa, which is the least expensive of the chemotherapeutic agents.

Intravesical Interferon

Interferons (IFNs) are proteins that have diverse biologic properties including the inhibition of tumor cell proliferation. Recombinant DNA technology has produced purified human interferons. One of these, IFN-α_2, has been used in clinical trials in the treatment of superficial bladder cancer (Shortliffe et al., 1984) including carcinoma in situ. The overall complete response rate was 44%. The toxicity was negligible. Further studies will be necessary to compare intravesical interferon with other established treatments for superficial bladder cancer. Other forms of immunotherapy for bladder cancer are discussed under the section on immunotherapy.

Hematoporphyrin Derivative Phototherapy

Hematoporphyrin derivative (HpD) is a complex mixture of porphyrins that appear to be preferentially concentrated in neoplastic or dysplastic tissues. When these tissues are irradiated with light of the proper wavelength, death is induced in sensitized cells. It has been shown that HpD activated by white light could be used to localize, and subsequently to cause necrosis of, human bladder cancer cells.

Recent clinical trials of HpD therapy with subsequent illumination of the bladder with a krypton-ion laser light source have revealed limited success in patients having small superficial tumors or carcinoma in situ but not in those having larger tumors (Benson et al., 1982, 1983a; Hisazumi et al., 1983). HpD produces generalized cutaneous photosensitivity, which limits its applicability.

Laser Therapy

Smith and Dixon (1984) reported on the use of argon laser therapy in patients with small, superficial bladder tumors or carcinoma in situ. The laser beam energy is selectively absorbed by vascular tissues such as transitional cell carcinomas. The argon laser provides a depth of penetration of only 1 mm; therefore, it is safe but cannot be used for large tumors. The neodymium-YAG (yttrium-aluminum-garnet) laser has a greater depth of penetration (4–15 mm) but is somewhat less safe (Hofstetter et al., 1981). Carbon dioxide laser energy is absorbed by water and produces minimal tissue penetration; therefore, it is of limited clinical utility in the treatment of bladder tumors.

The theoretical advantages of laser therapy are that it can be performed through a small cystoscope using local anesthesia with no bleeding or obturator nerve stimulation. The main disadvantage is that not all of the tumor tissue is obtained for histologic examination. Further studies will be necessary to determine the role of lasers in the treatment of patients with superficial bladder cancer.

Hydrostatic Pressure Treatment

Helmstein (1972) reported a technique of treating superficial bladder cancer by filling the bladder under pressure with saline or inserting a balloon in the bladder and inflating it using epidural anesthesia to a pressure above the diastolic blood pressure for 5–7 hours. This resulted in selective necrosis of bladder tumors. A significant complication of hydrostatic pressure treatment is bladder perforation. This method has not been adopted into general use in the United States.

Mucosal Denudation

Complete stripping of the bladder mucosa for diffuse papillomatosis has been attempted with mixed results and a high incidence of morbidity (Hansen et al., 1976). In general, this procedure has been abandoned and supplanted by intravesical chemotherapy or BCG therapy.

MANAGEMENT OF INVASIVE BLADDER CANCER (STAGES B, C, AND D1)

TRANSURETHRAL ENDOSCOPIC RESECTION FOR INVASIVE BLADDER CANCER

Nichols and Marshall (1956) demonstrated that transurethral resection was adequate therapy for the vast majority of patients having superficial, noninfiltrating bladder cancers; however, survival results were poor in patients treated with transurethral resection alone for muscle-infiltrating bladder cancers. The most favorable results of endoscopic management of muscle-invasive bladder cancer are those of Barnes and associates (1967), who reported a 40% five-year survival rate in patients having tumors that infiltrated into but not through the bladder muscle.

In patients having relatively low-grade tumors with only superficial muscle invasion, transurethral resection is a legitimate treatment option. Whether such patients also should receive further therapy is usually decided on an individual basis. Clinical experience has demonstrated that in patients who are poor risks for a major operation having well-differentiated tumors, transurethral resection may be acceptable treatment.

DEFINITIVE RADIATION THERAPY AND SALVAGE CYSTECTOMY

Radiation therapy has been used in its various forms in the treatment of bladder cancer for decades; however, only 20%–36% of patients having invasive bladder cancer can be cured by conventional external beam radiation therapy alone. There is no significant correlation between the histologic grade of the tumor and the response to radiation therapy (Miller and Johnson, 1973; Goffinet et al., 1975; Wallace and Bloom, 1976; Droller, 1982).

In many radiation therapy series, patients have been staged primarily with bimanual examination (stage B2 if there is thickening without a mass, stage C if there is a palpable mass, stage D1 if there is palpable evidence of fixation or extension to adjacent structures) (Miller and Johnson, 1973). This has resulted in considerable inaccuracy in staging.

Only 8%–15% of patients who fail definitive therapy are appropriate candidates for salvage cystectomy (Goffinet et al., 1975; Goodman et al., 1981; Blandy et al., 1980). In these patients, the overall five-year survival rate is approximately 38%, depending on the histopathological findings in the cystectomy specimen (approximately 70% if no residual tumor is found, 50% if superficial residual tumor is present, and 25% if deeply infiltrating cancer is present) (Blandy et al., 1980; Crawford and Skinner, 1980; Smith and Whitmore, 1981b). The operative morbidity and mortality associated with salvage cystectomy is only slightly higher than that of primary cystectomy.

Conventional fractionation for definitive external beam radiation therapy is a tumor dose of 7,000 rad in seven weeks (35 fractions), with 5,000 rad to the pelvis. There is no conclusive evidence that pelvic irradiation can control nodal metastases. Definitive radiation therapy fails to control the primary tumor in 50% of patients (Goffinet et al., 1975). Pallative cystectomy may be required for hematuria, dysuria, and intolerable urinary frequency caused by residual uncontolled tumor. Even in patients whose primary tumors are controlled or eradicated by radiation therapy, palliative therapy including intravesical formalin instillation or cystectomy may be required for refractory radiation cystitis. In addition, approximately one third of patients who have the primary tumor controlled by radiation therapy die of distant metastases.

With Miller and Johnson's (1973) publication of a prospective randomized clinical trial demonstrating that preoperative radiation therapy followed by planned cystectomy (integrated radiation therapy/cystectomy) produced five-year survival results significantly superior to definitive radiation therapy alone, conventional definitive radiation therapy ceased to be considered as the preferred treatment for invasive bladder cancer at most centers in the United States; it generally was reserved for patients who refused cystectomy, were not considered suitable candidates for a major operation, or had unresectable tumor. In Britain and Europe, radiation therapy continues to be the mainstay of definitive treatment for invasive bladder cancer, cystectomy being reserved for those who fail radiation therapy locally. Blandy and associates (1980) reported a 38%, five-year survival rate in stage T_3 patients treated primarily with definitive radiation therapy and salvage cystectomy, when necessary. Goffinet and associates (1975) reported actuarial five-year survival rates following definitive radiation therapy of 35% for stage A, 42% for stage B1, 35% for stage B2, 22% for stage C, and 7% for stage D.

Postoperative radiation therapy has yielded five-year survival rates of 41% for patients having superficial tumors with a 23% incidence of complications and a 6% mortality rate. The survival rate in patients having deeply invasive tumors is dismal (Miller and Johnson, 1973). Therefore, the risk-benefit ratio of postoperative radiation therapy does not seem to justify its routine use in patients with deeply invasive cancer. Combined pre- and post-operative radiation therapy has been reported to produce good three-year actuarial survival results in a trial sponsored by the Jefferson Medical School/Radiation Therapy Oncology Group (Mohuiddin et al., 1981), but these results have not yet been confirmed in other studies.

Advocates of radiation therapy emphasize the advantages of retaining bladder function, avoiding the need for an external urinary appliance, and retention of sexual potency in men. These theoretical advantages may become less compelling with the development and popularization of the potency-sparing technique for radical cystectomy (Walsh et al., 1983), Kock pouch urinary diversion (Skinner et al., 1984), and the Camey enterourethrostomy (Lillien and Camey, 1984).

Newer techniques of primary radiation therapy for

bladder cancer have been developed for the treatment of invasive bladder cancer. van der Werf-Messing and associates (1983) have treated patients with invasive bladder cancers less than 5 cm in diameter with 4,000 rad to the pelvis in four weeks followed by open implantation of the tumor with radium needles, and 3,600 rad of external beam radiation therapy postoperatively to the bladder in four days. This regimen results in the delivery of 7,600 rad to the tumor. Crude three-year survival rates were reported to be 67% for patients with stage T_2 tumors and 43% for those with stage T_3 tumors. The potential for severe radiation cystitis is increased with the use of these high radiation doses.

CHEMICAL RADIATION SENSITIZERS

Chemical radiation sensitizers also have been used in an effort to increase the radiation responsiveness of bladder cancers. Misonidazole is the radiation sensitizer that has been most thoroughly investigated. Albratt and associates (1983) reported on a regimen in which, after 4,000 rad was delivered to the pelvis, patients were treated with two 600-rad fractions delivered at weekly intervals with misonidazole given either systemically or both systemically and intravesically. Complete tumor responses were reported in 73% of patients compared with 43% in historical controls.

COMPLICATIONS OF RADIATION THERAPY

Acute, self-limited complications including dysuria, urinary frequency, and diarrhea occur in approximately 70% of patients. Severe, persistent complications occur in fewer than 10% of patients.

BLADDER SALVAGE PROTOCOLS

With future development of more effective regimens of chemotherapy and radiation therapy and means to identify which tumors are likely to respond favorably, it is anticipated that bladder salvage protocols will apply to more patients. These protocols will most likely combine chemotherapy and radiation therapy. Animal studies have suggested that cisplatin is a radiation sensitizer. Recent studies of cisplatin followed by definitive radiation therapy have yielded local complete tumor response rates of 21%–70% in patients with stages T_2 and T_3 tumors (Shipley et al., 1984).

Hyperthermia also has been combined with radiation therapy with or without chemotherapy as a potential means of achieving improved tumor response rates in patients with advanced bladder cancer. This approach is based on the increased susceptibility of malignant tumors, relative to normal tissues, to the toxic effects of hyperthermia (Kubota et al., 1984; Hall et al., 1974). Complete local responses have been reported in up to 42% of patients. In general, the toxic side effects are greater than observed in patients treated with conventional radiation therapy.

INTRAOPERATIVE RADIATION THERAPY

Intraoperative radiation therapy with electron beams combined with external beam therapy was evaluated by Matsumoto and associates (1981). In this protocol 2,500–3,000 rad was given as a single intraoperative dose followed by 3,000–4,000 rad as fractionated external beam therapy. Good results were achieved in patiens with T_{is}, T_1, and T_2 tumors but not in those with T_3 or T_4 tumors.

INTEGRATED PREOPERATIVE RADIATION THERAPY/CYSTECTOMY AND CYSTECTOMY ALONE

Preoperative radiation therapy for bladder cancer was initiated by Whitmore and associates at the Memorial Sloan-Kettering Cancer Center in 1959. The rationale for preoperative radiation therapy was based on clonogenic assays that demonstrated radiation therapy could destroy 90% of tumor cells and that well-oxygenated cells were more susceptible to the effects of radiation than anoxic cells. It was hoped that preoperative radiation therapy would sterilize peripheral microextensions of the tumor as well as regional lymph node micrometastases. It was speculated that preoperative radiation may damage tumor cells and thus preclude the metastatic potential of tumor cells that may be manipulated into the circulation at the time of cystectomy as well as prevent tumor implantation if the bladder were inadvertently opened during cystectomy. In the initial experience at Memorial Sloan-Kettering Cancer Center, 4,000 rad was delivered in 200-rad fractions over four weeks. However, in 1966, a short-course dose schedule of 2,000 rad in 500-rad fractions over one week was adopted.

Whitmore and Batata (1985) compared the survival figures of patients treated with the long-course, high-dose preoperative radiation and those treated with short-course, low-dose preoperative radiation therapy with a group of patients treated with cystectomy alone at Memorial Sloan-Kettering Cancer Center between 1949 and 1959. They stated that there was a marginally better disease-free survival in patients with deeply invasive tumors who received either course of preoperative radiation therapy as well as fewer deaths from tumor recurrence in the preoperatively irradiated patients.

Miller and Johnson (1973) compared the results of preoperative radiation therapy followed by cystectomy with the results of definitive radiation therapy alone in a prospective study and demonstrated a significant survival advantage for the patients treated with preoperative radiation and cystectomy. In a similar prospective, randomized study, Bloom and associates (1982) reported a small survival advantage for patients receiving preoperative radiation therapy and cystectomy over those receiving definitive radiation therapy alone, but the difference was not statistically significant when all patients were considered. The survival difference was, however, significant in patients who were under 60 years of age and in men.

Two prospective, randomized studies examined whether preoperative radiation therapy plus cystectomy offered a survival advantage over cystectomy alone. One study conducted by the Veterans Administration Cooperative Urologic Research Group (Blackard et al., 1972) failed to show any survival advantage for preoperative radiation therapy over cystectomy alone but also failed to show an advantage of cystectomy over radiation therapy alone. The other study conducted by the National Cooperative Study Group (Prout, 1976) had flaws that produced a bias in favor of integrated radiation therapy and cystectomy. Despite the bias, however, the five-year survival rate of the preoperatively irradiated patients who completed the protocol (36%) was not significantly different from that in patients treated with cystectomy alone (29%). If all patients were considered, the five-year survival rates were not significantly different for patients treated with preoperative radiation therapy (23%) compared with those treated with cystectomy alone (20%).

In 1973, van der Werf-Messing reported that the beneficial results of preoperative irradiation were largely limited to patients who exhibited surgical downstaging. In van der Werf-Messing's series, approximately two thirds of patients exhibited surgical downstaging. Prout (1976) reanalyzed the results of the National Cooperative Study, and also found that patients who exhibited surgical downstaging (35%) had more favorable survival results. The only statistically significant difference between groups in the National Cooperative Study was the comparison of patients treated with preoperative radiation therapy who had complete tumor downstaging to stage P_0 and those who did not have preoperative radiation therapy who had residual tumor in the cystectomy specimen. The group having the highest survival in the National Cooperative Study were patients who did not receive preoperative radiation therapy who had no residual tumor in the bladder. Despite the limitations of the National Cooperative Study, it has been widely cited as supporting the claim that preoperative radiation therapy enhances patient survival, although the data contained therein do not fully support this conclusion.

In a further analysis of the data from the National Cooperative Study, Slack and Prout (1980) reported that surgical downstaging occurred more frequently among patients who had papillary tumors and those who did not have lymphatic invasion. Subsequent studies by Hall and Heath (1981) and by Shipley et al. (1982) failed to demonstrate a correlation between downstaging or tumor histology, tumor grade, papillary configuration or the presence or absence of lymphatic invasion. Whitmore and Batata (1985) reported that in the Memorial Sloan-Kettering Hospital experience, downstaging occurred most frequently among patients having low-grade tumors, low T-category tumors, tumors of transitional cell type, and tumors unassociated with lymph node metastases.

Since the early 1970s to the 1980s, there has been controversy about whether long-course, high-dose or short-course, low-dose preoperative radiation therapy should be employed in patients with muscle-invasive bladder cancer. Whitmore (1980) reported survival results in patients treated with short-course preoperative radiation therapy comparable to those in patients treated with long-course therapy, although surgical downstaging was uncommon in patients treated with short-course therapy. It was speculated that with the short-course preoperative radiation therapy, not enough time elapsed to allow downstaging to be detected. Whitmore also reported that one of the major benefits of preoperative radiation therapy was that it lowered the tumor recurrence rate in the pelvis. This observation was not confirmed in other studies (Prout, 1976; Skinner and Lieskovsky, 1984). In general, integrated preoperative radiation therapy and cystectomy has yielded five-year survival rates of 35%–40% in patients with stages B2 and C (T_3) tumors (reviewed by Droller, 1982).

In the late 1970s and early 1980s, Radwin (1980) and others (Catalona, 1980; Clark, 1978; Daughtry et al., 1977; Mathur et al., 1981; Montie et al., 1984; Vinnicombe and Abercrombie, 1978) decried the uncritical acceptance of the underlying assumptions of preoperative radiation therapy. Radwin stated that data do not exist to support the basic concept that preoperative radiation therapy is beneficial to patients with invasive bladder cancer. Preoperative radiation therapy has been widely accepted because patients treated with preoperative radiation therapy had higher survival rates than historical controls treated with cystectomy alone or radiation therapy alone. However, comparison with historical controls is not necessarily valid. The argument for decreased implantability of tumor cells based on van der Werf-Messing's (1978) data for patients treated with radium implants for superficial bladder tumors also is not necessarily valid, since the bladder is not intentionally opened during cystectomy as it is during radium implantation.

Radwin also pointed out that if (as reported in some studies) two thirds of patients were downstaged by preoperative radiation therapy and these patients had a significant survival advantage, patients treated with preoperative radiation therapy should have a substantially better survival as a group than patients not preoperatively irradiated, but in fact they did not.

Whitmore and Batata (1985) suggested that the effects of preoperative radiation therapy on patients who had surgical downstaging may not be a cause-and-effect relationship, but rather downstaging by radiation therapy may merely identify a subpopulation of patients having biologically favorable tumors who are predetermined to do well (even without preoperative radiation therapy). The same patients treated with cystectomy alone may have fared just as well. Among patients treated with preoperative radiation therapy, certain subgroups seem to have a survival advantage, but others may be disadvantaged. The entire group of preoperatively irradiated patients have a survival that is comparable to that of those treated with cystectomy alone.

Preoperative radiation therapy as given in some protocols may even be harmful to some patients. In patients who received the long-course preoperative radiation there is a substantial delay in the performance of cystectomy (Prout, 1976). In patients whose tumors are not highly radiation responsive, this delay may allow dissemination of the tumor to occur while early cystectomy might prove curative. There are other reported adverse effects of preoperative radiation therapy including an increased incidence of wound infections (Prout, 1976) and bowel complications (Richie et al., 1976), particularly among elderly patients who receive the high-dose preoperative radiation therapy (Whitmore and Batata, 1985).

In recent years, several authors have published data failing to support the survival advantage of preoperative radiation therapy (Vinnicombe and Abercrombie, 1978; Clark, 1978; Daughtry et al., 1972; Montie et al., 1984; Mathur et al., 1981; Skinner and Lieskovsky, 1984). Recently both Mathur et al. (1984) and Skinner and Lieskovsky (1984) have presented modern series of patients treated with cystectomy alone in whom the survival results are comparable to those among patients treated with preoperative radiation therapy and cystectomy.

Skinner's (1984) recent series compared 100 patients who had received short-course preoperative radiation therapy with 97 patients who had been treated with cystectomy alone and found no significant survival advantage for preoperative radiation therapy. The five-year disease-free survival for the patients treated with cystectomy alone for stage P_2 and P_{3A} tumors was 75%, for patients with stage P_{3A} and P_{3B} tumors it was 40%,

and for patients with stage P_4 tumors it was 36%. Skinner observed no difference in the tumor recurrence rate in the pelvis, which was only 9% for patients receiving preoperative radiation therapy but 7% among patients who received cystectomy alone. Moreover, Montie and associates (1984) reported that the decade in which a patient with any given clinical stage of cancer was treated was more important to survival than whether the patient received preoperative radiation therapy. Whitmore and Batata (1985) has pointed out that the comparable survival results in the modern cystectomy series also may be owing to case selection, as it clearly is, for example, in series of patients treated with partial cystectomy alone.

SUMMARY OF PREOPERATIVE RADIATION THERAPY

Critical analysis of the available data suggests that preoperative radiation therapy is not uniformly an effective adjunct when all patients with invasive bladder cancer are considered. There is a subgroup of patients with invasive tumors who have downstaging following radiation therapy who seem to fare better, but as a group, patients treated with preoperative radiation therapy have survival rates comparable to those of patients treated with cystectomy alone. It appears that the response to preoperative radiation therapy may identify a subset of patients whose tumors are biologically favorable, but there is not conclusive evidence that preoperative radiation therapy is responsible for the favorable prognosis. Because of the expense, delay, and potential morbidity associated with preoperative radiation therapy for bladder cancer, there is an increasing tendency among urologic surgeons to omit preoperative radiation therapy prior to radical cystectomy for invasive bladder cancer.

There is marginal evidence based on a randomized clinical study to suggest that preoperative radiation therapy may be beneficial in the treatment of patients having stages P_3 or P_4 bilharzial bladder cancer (Ghoneim and Awad, 1980).

TECHNIQUE OF RADICAL CYSTECTOMY

A midline abdominal incision is made from above the umbilicus to the pubic symphysis, and the peritoneal cavity is explored by palpation and inspection.

The peritoneum is incised on each side of the bladder to the region of the iliac artery bifurcation (Fig 31–32). The ureters are identifed and mobilized distally to the point at which they pass posterior to the vascular pedicles of the bladder. The ureters are divided and the proximal margins sent for frozen section examination (Fig 31–33).

The pelvic lymph node–bearing areas are exposed.

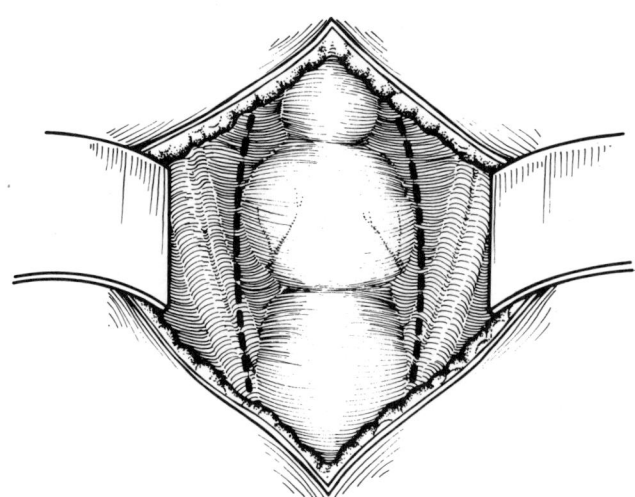

FIG 31–32.
Radical cystectomy. Peritoneum is incised lateral to the bladder to the bifurcation of the common iliac arteries.

This requires ligation and division of the vasa deferentia in men and the round ligaments in women. A pelvic lymphadenectomy is performed. The margins of dissection are the genitofemoral nerve laterally, the hypogastric artery medially, the bifurcation of the iliac artery proximally, and the circumflex iliac vein distally (Fig 31–34).

After completing the pelvic lymphadenectomy, the hypogastric arteries are skeletonized. The first anterior branch, the common trunk of the umbilical and superior vesical arteries, is ligated and divided. The "anterior" bladder pedicles are ligated and divided distally to the level of the urogenital diaphragm. In performing a potency-sparing cystectomy, the vascular pedicles of

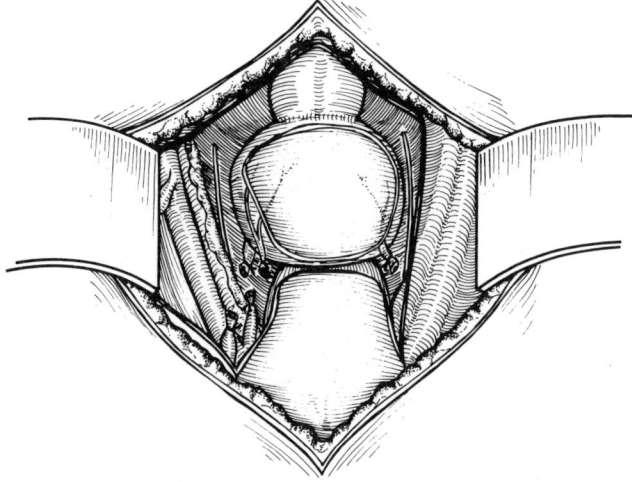

FIG 31–34.
Radical cystectomy. Peritoneum is incised in the pouch of Douglas.

the bladder are divided immediately adjacent to the seminal vesicles and ureters (Fig 31–35).

The peritoneal incisions are joined in the rectovesical pouch of Douglas. The plane posterior to Denonvilliers' fascia between the posterior bladder wall and the anterior rectal wall is developed distally to the prostate in men and to the uterine cervix in women (Fig 31–36).

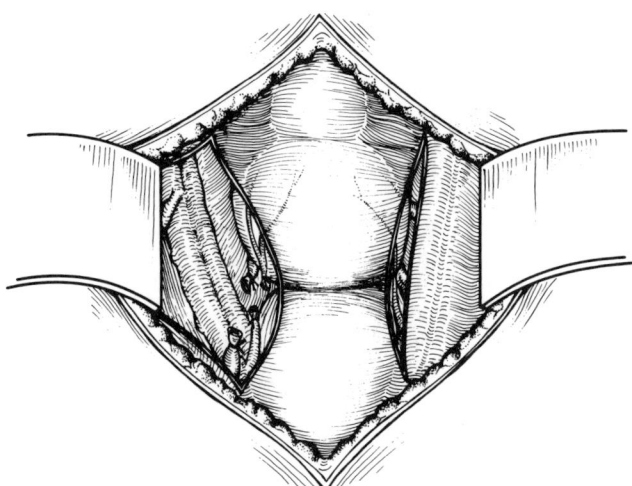

FIG 31–33.
Radical cystectomy. Ureter and superior vesical artery are divided. Note exposed pelvic lymph node-bearing area.

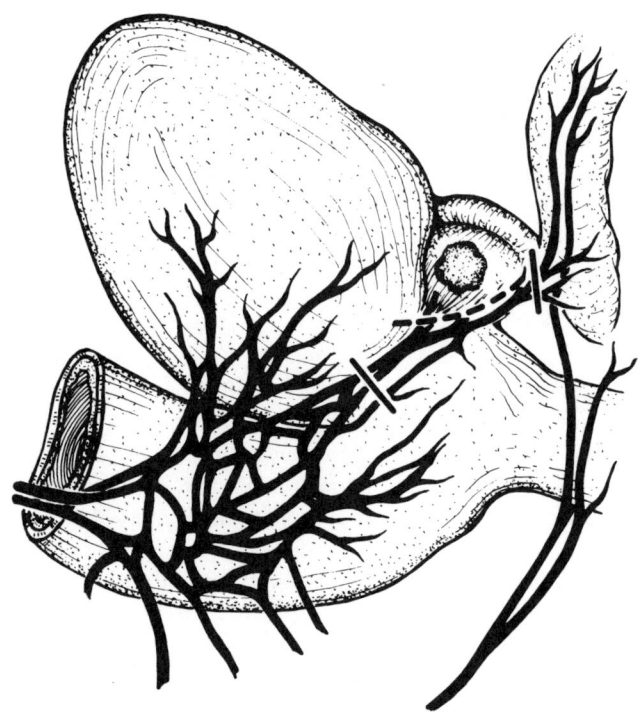

FIG 31–35.
Nerve-sparing cystectomy. *Dotted line* indicates point of incision of endopelvic fascia and prostatic pedicles.

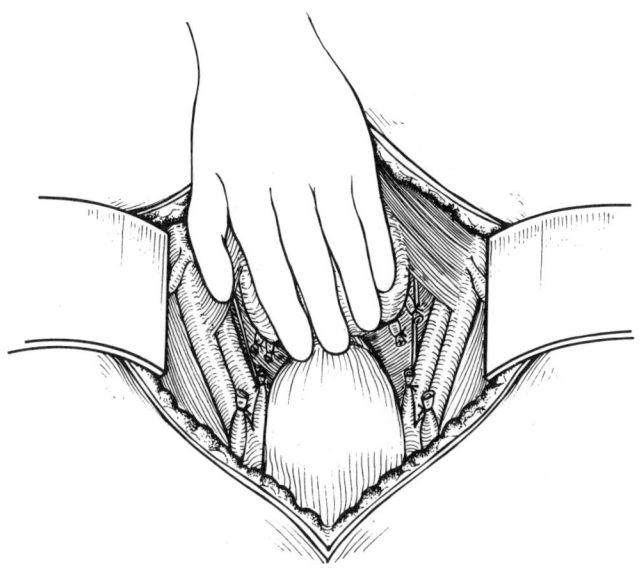

FIG 31–36.
Radical cystectomy. The bladder base is mobilized off the anterior rectal wall and the bladder pedicles are ligated and divided.

This maneuver delineates the "posterior" vascular pedicles of the bladder which are ligated and divided distally to the urogenital diaphragm in men. In women, the vagina is incised posterior to the uterine cervix and the anterior vaginal wall is resected along with the bladder.

The endopelvic fascia is incised on either side of the prostate (bladder neck in women). The puboprostatic ligaments (pubovesical ligaments in women) are incised at their attachments to the undersurface of the pubic symphysis. The venous complex of Santorini is ligated and divided and the urethra is dissected out (Fig 31–37).

In women, total urethrectomy is routinely performed. An inverted U-shaped incision is made anterior and lateral to the urethra and deepened by sharp dissection through the urogenital diaphragm and anterior vaginal wall. This allows for en bloc removal of the bladder, urethra, and anterior vaginal wall. The vagina is then reconstructed with absorbable sutures.

In men, total urethrectomy need not be performed routinely, unless there is histologic evidence of tumor involvement of the bladder neck or prostatic urethra. If urethrectomy is not performed, the membranous urethra is dissected out of the urogenital diaphgram, the urethral catheter is removed, and the urethra is doubly ligated and divided (Fig 31–38). The prostatic apex and attached urethral stump are sent for frozen section examination.

Total urethrectomy is performed by making a midline or inverted Y-shaped perineal incision and dissecting

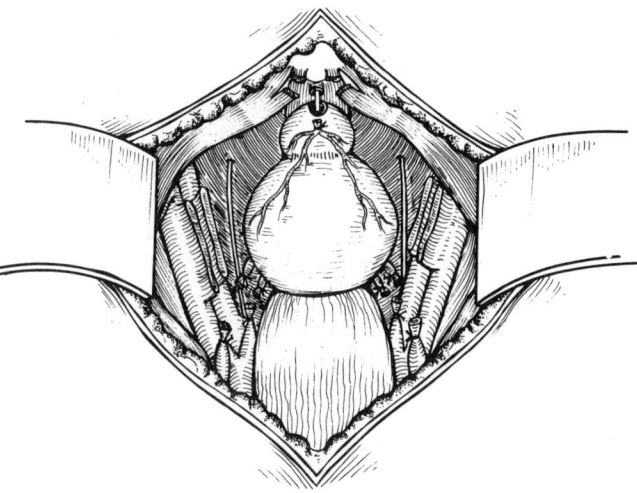

FIG 31–37.
Radical cystectomy. The urethra is dissected out and divided and the surgical specimen is removed.

out the urethra and its surrounding corpus spongiosum. The urethra is dissected free from the corpora cavernosa to the level of the fossa navicularis. A ventral urethral meatotomy is performed, and the fossa navicularis is dissected from the surrounding glans. The bulbar urethra is dissected free from the surrounding tissue and the dissection is deepened to the urogenital diaphragm. If the membranous urethra has been dissected out below the urogenital diaphragm from above, the urethra may be delivered en bloc with the prostate and bladder. Theoretically en bloc removal is desirable, but no data available validate this classic principle of cancer

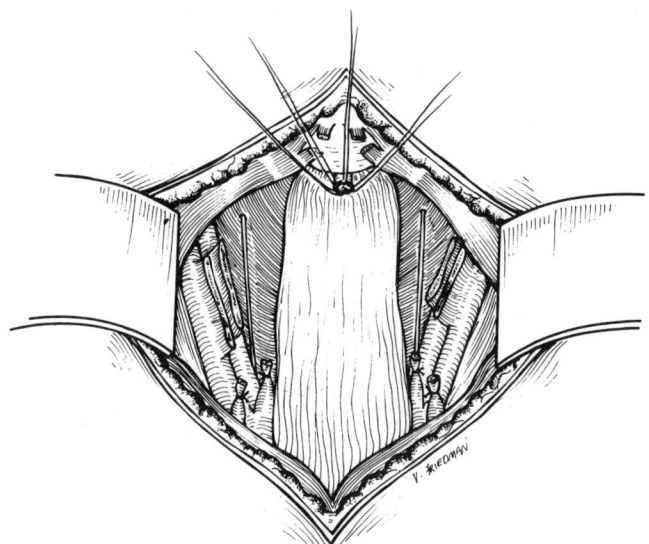

FIG 31–38.
Radical cystectomy. Specimen has been removed. Urethral stump is shown for anastomosis if enterourethrostomy is to be performed.

surgery. The perineum may be drained through the urethral bed with a small drain.

Pelvic lymphadenectomy may control minimal regional lymph node metastases from bladder cancer in about one third of patients (LaPlante and Brice, 1973; Skinner and Lieskovsky, 1984). Approximately 20% of patients having diffuse carcinoma in situ and 30%–40% with deeply invasive tumors have lymph node metastases. Fewer than half of these patients have only minimal nodal metastases. Accordingly, routine pelvic lymphadenectomy is recommended but would be expected to enhance overall cure rates by up to 10%. The scope of the lymphadenectomy should include at least the nodes below the bifurcation of the iliac arteries and from the genitofemoral nerve to the bladder pedicles. There is no evidence that an en bloc node dissection is superior to removing the lymph nodes region by region.

In patients in whom urethrectomy is not performed, urethral recurrences have been reported to occur in approximately 7% (Schellhammer and Whitmore, 1976). In these patients, frozen sections of the prostatic apex should be evaluated at the time of operation and the retained urethra should be evaluated postoperatively with saline washes for cytopathologic examination.

A nerve-sparing modification of radical cystectomy that preserves erectile function in men has been developed by Walsh et al. (1983). Although the parasympathetic neurovascular bundles also can be preserved in patients undergoing total urethrectomy, potency is more readily preserved in patients in whom the urethra is not removed. In the nerve-sparing modification of radical cystectomy, after the superior vesical arteries have been ligated and divided and the bladder has been mobilized in the midline of the anterior surface of the rectum, the endopelvic fascia is incised lateral to the prostate, the puboprostatic ligaments are divided, and the dorsal venous complex is ligated and divided. The urethra is then transected, a purse-string suture is placed around the urethral orifice of the prostatic apex, and the prostate and seminal vesicles are mobilized in a retrograde fashion medial to the neurovascular bundles. The bladder pedicles also are divided medial to the neurovascular bundles. A detailed, illustrated description of the nerve-sparing steps of this operation has been published (Catalona, 1985).

In female patients with invasive bladder cancer, it is customary to perform a modified anterior pelvic exenteration, removing the bladder, uterus, fallopian tubes, ovaries, anterior wall of the vagina, and urethra.

URINARY DIVERSIONS

A variety of urinary diversions has been used in patients undergoing cystectomy for invasive bladder can-

cer. A detailed discussion is beyond the scope of this chapter (see Chapter 32). Ureterosigmoidostomy has been largely supplanted by other forms of diversion because of the potential problems of pyelonephritis, urinary stone disease, hyperchloremic acidosis, and late occurrence of adenocarcinoma of the colon at the site of the ureterocolonic anastomosis. The ileal conduit is the time-honored standard urinary diversion in patients undergoing cystectomy. Peristomal inflammation, peristomal hernia, stomal stenosis, urinary tract stone disease, ureteroileal strictures, and later upper tract deterioration are the principal complications of ileal conduit diversion (Sullivan et al., 1980; Bracken et al., 1981a).

In patients who have sustained radiation-induced or other injury to the ileum, jejunal or transverse colon conduits have been employed. Jejunal conduits are frquently associated with metabolic derangements caused by losses of sodium and chloride from the conduit, which requires careful metabolic monitoring and management. Colon conduits generally are more suitable for the performance of nonrefluxing ureterointestinal anastomosis; however, there may be an increased incidence of anastomotic strictures with nonrefluxing ureterointestinal anastomoses.

Continent urinary diversion including the ileocecal reservoir (Benchekroun, 1982) and the Kock pouch (Skinner et al., 1984) obviate wearing an external appliance but require intermittent self-catheterization instead. Complications of the ileocecal reservoir include leakage of urine through the anti-incontinence valves and infection. Complications of the Kock pouch include leakage, failure of the antirefluxing nipples, stone formation, erosion of the Marlex mesh, and difficulty with catheterization.

The Camey enterourethrostomy (Lillien and Camey, 1984) has been popularized as a means of preserving the normal body habitus and urinary continence while having the patient continue to void through the urethra. With this diversion, a V-shaped loop of ileum is anastomosed directly to the membranous urethra. Complications of the Camey diversion include nocturnal incontinence, daytime incontinence, failure of the antireflux mechanism, and tumor recurrence at the ileal-urethral anastomosis.

With any form of urinary diversion, it is important to obtain frozen-section evaluation of the distal ureters (Zincke et al., 1984). If there is evidence of carcinoma in situ, attempts should be made to obtain a tumor-free ureteral margin. Tumor recurrence rates in the upper urinary tracts in patients having positive ureteral margin are higher (Zincke et al., 1984). In rare circumstances there may be diffuse changes of severe atypia or carcinoma in situ involving the entire length of one or both upper urinary tracts. In such patients, it may not be possible to obtain tumor-free margins, and it may be necessary to proceed with the ureterointestinal anastomosis despite the questionable margins. The overall low tumor recurrence rates in these patients suggest that most patients will fare well (Linker and Whitmore, 1975).

COMPLICATIONS OF RADICAL CYSTECTOMY AND URINARY DIVERSION

The specific complications associated with the various forms of urinary diversion have been discussed above. In general, the overall complication rate for patients undergoing radical cystectomy and urinary diversion is approximately 25%–35%. The most common complications are superficial and deep wound infections (10%), intestinal obstruction (10%) that resolves spontaneously in most patients but requires reexploration in some, hemorrhage, and cardiopulmonary complications. Rectal injury occurs in approximately 4% of patients. If the injury is small and fecal contamination is minimal in patients who have not been irradiated, the injury may be closed primarily. In other instances, a temporary diverting colostomy also should be performed (Flechner and Spaulding, 1982). In recent decades, the mortality rate associated with radical cystectomy has decreased from more than 10% to around 1%–2% (Bracken et al., 1981a).

PARTIAL CYSTECTOMY

In highly selected patients having solitary invasive bladder cancers that cannot be managed safely with transurethral resection in whom there is no prior history of a bladder tumor, and random biopsies reveal no evidence of atypia or carcinoma in situ elsewhere in the bladder or prostatic urethra, and in whom an adequate 2-cm margin around the tumor can be achieved, partial cystectomy is appropriate. Only about 10%–15% of patients having invasive bladder cancer are suitable candidates for partial cystectomy. Numerous studies have demonstrated survival results comparable to those obtainable with more radical excisions when partial cystectomy is performed in properly selected patients (Utz et al., 1973; Resnick and O'Connor, 1973). It is prudent to administer preoperative radiation therapy in doses of at least 1,000–1,200 rad to prevent wound implantation that may result in 10%–20% of patients from opening the bladder (Magri, 1962). The bladder should be closed primarily and drained with a urethral catheter. Ureteral reimplantation may be appropriate when the tumor is located above or lateral to the ureteral orifice but not when it is located between the orifice and the bladder neck. Pelvic lymphadenectomy also may be appropriate in these patients, although data supporting its

efficacy are not available. The obvious advantage of partial cystectomy is that it preserves normal bladder function. Recent studies (Faysal and Freiha, 1979; Schoborg et al., 1979) have reported alarmingly high (70%) tumor recurrence rates, particularly in patients with high-grade tumors. Many of these patients require subsequent total cystectomy (Faysal and Freiha, 1979).

ADJUNCTIVE CHEMOTHERAPY

In recent years, cytotoxic chemotherapy has been used in a small number of trials both preoperatively and postoperatively as an adjunct to radical cystectomy or radiation therapy. The results of these early trials suggest that overall patient survival is not improved and patient compliance is poor (Soloway, 1985). With the development of newer, more effective therapy, it is envisioned that in the future chemotherapy may be a useful adjunct to cystectomy or radiation therapy (Herr et al., 1983b; Herr, 1985; Hill et al., 1985; Soloway, 1985).

MANAGEMENT OF METASTATIC DISEASE (STAGE D2)

CYTOTOXIC CHEMOTHERAPY

Approximately 50% of patients with invasive bladder cancer have occult or overt metastases at the time of diagnosis. In these patients effective systemic therapy is urgently needed.

In recent years, several chemotherapeutic agents including cisplatin, methotrexate, vinblastine, and Adriamycin have been shown to have significant activity against metastatic transitional cell carcinoma. When used as single agents, these drugs induce objective tumor regression in 20%–30% of patients. Most of these responses are only partial, and the vast majority have a duration of only about six months (Yagoda, 1983; Soloway, 1985). Combining these agents in three-drug (CMV) (Harker et al., 1984) or four-drug (M-VAC) combinations (Sternberg et al., 1985), has yielded objective response rates of 57%–70% and, more important, complete responses have been observed in 30%–50% of patients with a longer median duration of response. These regimens are highly toxic, producing severe myelosuppression.

PALLIATIVE RADIATION THERAPY

External beam radiation therapy is often effective in doses of 3,000–3,500 rad given in ten fractions to relieve pain from skeletal metastases. Usually, pain relief is prompt. Prophylactic irradiation is advisable in minimally symptomatic metastases involving weight-bearing bones. Internal fixation may be required for the prevention or treatment of pathologic fractures.

Palliative radiation therapy in doses of 4,000–5,000 rad may be effective in controlling local symptoms from the primary tumor, but sometimes radiation therapy can aggravate the local symptoms.

INTRAVESICAL FORMALIN OR ALUM INSTILLATION

Formalin is a 37% solution of formaldehyde gas dissolved in water. Instillation of 1%–10% formalin solution (0.37%–3.7% formaldehyde gas) in the bladder has been demonstrated to be effective in controlling hemorrhage from advanced tumor or radiation cystitis in many patients (Brown, 1970). Formalin is exceedingly irritating to the bladder, and thus intervesical instillation requires general or regional anesthesia. Because 10% formalin solution has been found to be associated with a high degree of toxicity (obstruction of ureteral orifices), formalin instillation should begin with a 1% formalin solution and repeated with a 4% formalin solution if necessary. The formalin solution is retained in the bladder for 30 minutes. A cystogram should be performed before formalin instillation. If vesicoureteral reflux is present, Fogarty catheters should be passed up both ureters, and the patient should be tilted in a head-up position to protect the upper urinary tracts from the toxic effects of formalin (Fall and Pettersson, 1979).

A 1% alum solution also may be used to treat hemorrhage from radiation cystitis (Ostroff and Chenault, 1982). Alum solution may be administered through continuous bladder irrigation without the need for anesthesia, but at times may require discontinuation because of symptoms of vesical pain and irritability.

PALLIATIVE HYPOGASTRIC ARTERY EMBOLIZATION AND/OR CYSTECTOMY

In rare circumstances, life-threatening hemorrhage may occur from hemorrhagic cystitis or uncontrolled tumor. In these instances it may be necessary to perform transfemoral percutaneous hypogastric artery embolization (Carmignani et al., 1980). If hypogastric embolization or ligation fails to control the hemorrhage, palliative cystectomy may be required.

IMMUNOTHERAPY

The immunobiology of bladder cancer has been extensively studied both in experimental animals and humans (Catalona et al., 1982; Droller, 1984; Haaff et al., 1985); however, it is beyond the scope of this chapter to discuss these studies in detail.

Intravesical BCG therapy for superficial bladder cancer has been discussed above. Although intravesical BCG also has been used in selected patients with invasive bladder cancer (Lamm et al., 1982), convincing evidence of its efficacy in the treatment of invasive disease is lacking. There is some evidence that streptococcal OK-432 may produce beneficial effects similar to those of BCG, although the data are very limited (Kagawa et al., 1979). In contrast, studies using *Corynebacterium parvum* in patients having more advanced disease have not demonstrated therapeutic efficacy (Purves et al., 1979). Recent studies suggest that interleukin-2 also may be effective for superficial tumors (Pizza et al., 1984).

Herr and associates (1978) reported studies suggesting that parenteral administration of the interferon-inducer poly(I:C) may have an effect on reducing early tumor recurrences and may confer long-term benefits in terms of survival of patients with carcinoma in situ, but these studies need further verification (Herr et al., 1978; Kemeny et al., 1981).

Studies of specific passive immunotherapy with immune pig lymphocytes have suggested the possibility of a beneficial effect but also need to be confirmed by other studies (Symes et al., 1978).

REFERENCES

1. Aherne G: Retinoblastoma associated with other primary malignant tumors. *Trans Ophthalmol Soc UK* 1974; 94:938–944.
2. Albratt RP, Sealy R, Tucker RD, et al: Radical irradiation and misonidazole in the treatment of grade III and T₃ bladder cancer. *Int J Radiat Oncol Biol Phys* 1983; 9:629–631.
3. Albores-Saavedra J, Maldonado ME, Ibarra J, et al: Pheochromocytoma of the urinary bladder. *Cancer* 1969; 23:1110–1118.
4. Alfthan O, Tarkkanen J, Grohn P, et al: Tigason (etretinate) in prevention of recurrence of superficial bladder tumors: A double-blind clinical trial. *Eur Urol* 1983; 9:6–9.
5. Althausen AF, Prout GR Jr, Daly JJ: Noninvasive papillary carcinoma of the bladder associated with carcinoma *in situ*. *J Urol* 1976; 116:575–580.
6. Anderstrom C, Johansson S, Nilsson S: The significance of lamina propria invasion on the prognosis of patients with bladder tumors. *J Urol* 1980; 124:23–26.
7. Anderstrom C, Johansson SL, von Schultz L: Primary adenocarcinoma of the urinary bladder: A clinicopathologic and prognostic study. *Cancer* 1983; 52:1273–1280.
8. American Cancer Society Statistics. *Cancer* 1984; 34:7–23.
9. Arce S, Lopez R, Almaguer M, et al: HLA-antigens and transitional cell carcinoma. *Mater Med Pol* 1978; 10:98.
10. Babaian RJ, Johnson DE, Llamas L, et al: Metastases from transitional cell carcinoma of the urinary bladder. *Urology* 1980; 16:142–144.
11. Barlebo H, Sorensen BL, Ohlsen AS: Carcinoma *in situ* of the urinary bladder: Flat intraepithelial neoplasia. *Scand J Urol Nephrol* 1972; 6:213–223.
12. Barnes RW, Bergman RT, Hadley HT, et al: Control of

bladder tumors by endoscopic surgery. *J Urol* 1967; 97:864–868.

13. Beck AD, Gaudin HJ, Bonham DG: Carcinoma of the urachus. *Br J Urol* 1970; 42:555–562.

14. Benchekroun A: Continent cecal bladder. *Br J Urol* 1982; 54:505–506.

15. Bennett JK, Wheatley JK, Walton KN: 10-Year experience with adenocarcinoma of the bladder. *J Urol* 1984; 131:262–263.

16. Benson RC, Farrow GM, Kinsey JH, et al: Detection and localization of in situ carcinoma of the bladder with hematoporphyrin derivative. *Mayo Clin Proc* 1982; 57:548–555.

17. Benson RC Jr, Kinsey JH, Cortese DA, et al: Treatment of transitional cell carcinoma of the bladder with hematoporphyrin derivative phototherapy. *J Urol* 1983a; 130:1090–1095.

18. Benson RC Jr, Swanson SK, Farrow GM: Relationship of leukoplakia to urothelial malignancy. *J Urol* 1984; 131:507–511.

19. Benson RC Jr, Tomera KM, Kelalis PP: Transitional cell carcinoma of the bladder in children and adolescents. *J Urol* 1983b; 130:54–55.

20. Berger GL, Sadlowsky RW, Sharp JR, et al: Lack of value of routine postoperative bone and liver scans in cystectomy candidates. *J Urol* 1981; 125:637–639.

21. Blackard CE, Byar DP, VACURG: Results of a clinical trial of surgery and radiation in stages II and III carcinoma of the bladder. *J Urol* 1972; 108:875–878.

22. Blandy JP, England HR, Evans SJW, et al: T$_3$ bladder cancer: The case for salvage cystectomy. *Br J Urol* 1980; 52:506–510.

23. Block NL, Whitmore WF Jr: Leukemoid reaction, thrombocytosis, and hypercalcemia associated with bladder cancer. *J Urol* 1973; 110:660–663.

24. Bloom HJG, Hendry WF, Wallace DM, et al: Treatment of T$_3$ bladder cancer: Controlled trial of preoperative radiotherapy and radical cystectomy versus radical radiotherapy. *Br J Urol* 1982; 54:136–151.

25. Bracken RB, McDonald M, Johnson DE: Complications of single-stage radical cystectomy and ileal conduit. *Urology* 1981a; 17:141–146.

26. Bracken RB, McDonald MW, Johnson DE: Cystectomy for superficial bladder cancer. *Urology* 1981b; 28:459–463.

27. Broders AC: Epithelioma of the genitourinary organs. *Ann Surg* 1922; 75:574–604.

28. Broecher BH, Klein FA, Hackler RH: Cancer of the bladder in spinal cord injury patients. *J Urol* 1981; 125:196–197.

29. Brosman SA: Experience with bacillus Calmette-Guerin in patients with superficial bladder carcinoma. *J Urol* 1982; 128:27–30.

30. Brosman SA: BCG in the management of superficial bladder cancer. *Urology* 1984; 23(suppl):82–87.

31. Brown RB: Experiences with intravesical formalin administration in advanced carcinoma of the bladder. *Br J Urol* 1970; 42:738–739.

32. Brown RR, Price JM, Friedell GH, et al: Tryptophan metabolism in patients with bladder cancer: Geographic differences. *JNCI* 1969; 43:259–301.

33. Bryan GT: The role of urinary tryptophan metabolites in the etiology of bladder cancer. *Am J Clin Nutr* 1971; 24:841.

34. Bryan GT: The pathogenesis of experimental bladder cancer. *Cancer Res* 1977; 37:2813.

35. Byar D, Blackhard C: Comparison of placebo, pyridoxine, and topical Thiotepa in preventing recurrence of stage I bladder cancer. *Urology* 1977; 10:556–561.

36. Camacho F, Pinsky C, Kerr D, et al: Treatment of superficial bladder cancer with intravesical BCG. *Proc Am Soc Clin Oncol* 1980; 21:359.

37. Carmignani G, Belgrano E, Puppo P, et al: Transcatheter embolization of the hypogastric arteries in cases of bladder hemorrhage from advanced pelvic cancers: Followup in 9 cases. *J Urol* 1980; 124:196–200.

38. Case RAM, Hosker ME, McDonald DB, et al: Tumors of the urinary bladder in workmen engaged in the manufacture and use of certain dystuff intermediates in the British chemical industry. *Br J Ind Med* 1954; 11:75–104.

39. Catalona WJ: Bladder carcinoma. *J Urol* 1980; 123:35–136.

40. Catalona WJ: Nerve-sparing radical retropubic prostatectomy. *Urol Clin North Am* 1985; 12:187–199.

41. Catalona WJ, Ratliff TL, McCool RE: Immunology of genitourinary tumors, in Paulson DF (ed): *Genitourinary Cancer 1.* Boston, Martinus Nijhoff, 1982, pp 169–214.

42. Chapman JW, Connolly JG, Rosenbaum L: Occupational bladder cancer: A case control study, in Connolly JG (ed): *Carcinoma of the Bladder.* New York, Raven Press, 1981, p 45.

43. Choi H, Lamb S, Pintar K, et al: Primary signet-ring cell carcinoma of the urinary bladder. *Cancer* 1984; 53:1985–1990.

44. Clark PB: Radical cystectomy for carcinoma of the bladder. *Br J Urol* 1978; 50:492–495.

45. Clark SS, Marlett MM, Prudencio RF, et al: Neurofibromatosis of the bladder in children: Case report and literature review. *J Urol* 1977; 118:654–656.

46. Cole P: Coffee drinking and cancer of the lower urinary tract. *Lancet* 1971; 1:1335–1337.

47. Cole P, Hoover R, Friedell GH: Occupation and cancer of the lower urinary tract. *Cancer* 1972; 29:1250.

48. Coon JS, Weinstein RS, Summers JL: Blood group precursor T-antigen expression in human urinary bladder carcinoma. *Am J Clin Pathol* 1982; 77:692–699.

49. Costello AJ, Tiptaft RC, England HR, et al: Squamous cell carcinoma of the bladder. *Urology* 1984; 23:234–236.

50. Crawford ED, Skinner DG: Salvage cystectomy after irradiation failure. *J Urol* 1980; 123:32–34.

51. Cummings KB: Carcinoma of the bladder: Predictors. *Cancer* 1980; 45:1849–1855.

52. Cutler SH, Young JL Jr (eds): Third National Cancer Survey: Incidence Data. *Natl Cancer Inst Monogr* 1975; 41:1–454.

53. Dalesio O, Schulman CC, Sylvester R, et al: Prognostic

factors in superficial bladder tumors: A study of the European Organization for Research on the Treatment of Cancer: Genitourinary Tract Cancer Cooperative Group. *J Urol* 1983; 129:730–733.

54. Daughtry JD, Susan LP, Stewart BH et al: Ileal conduit and cystectomy: A 10-year retrospective study of ileal conduits performed in conjunction with cystectomy and with a minimum 5-year followup. *J Urol* 1977; 118:556–557.

55. Devesa SS, Silverman DT: Cancer incidence and mortality trends in the United States: 1935–74. *JNCI* 1978; 60:545–571.

56. DeKernion JB, Huang M, Lindner A, et al: Management of superficial bladder tumors and carcinoma in situ with intravesical bacille Calmette-Guerin (BCG). *J Urol* 1985; 133:598–601.

57. DeKlerk DP, Catalona WJ, Nime FA, et al: Malignant pheochromocytoma of the bladder: The late development of renal cell carcinoma. *J Urol* 1975; 113:864–867.

58. DeKock MLS, Anderson CK, Clark PB: Vesical leukoplakia progressing to squamous cell carcinoma in women. *Br J Urol* 1981; 53:316–317.

59. DeMeester LJ, Farrow GM, Utz DC: Inverted papillomas of the urinary bladder. *Cancer* 1975; 36:505–513.

60. Denovic M, Darzynkiewicz A, Kostryrka-Claps ML, et al: Flow cytometry of low stage bladder tumors. *Cancer* 1982; 48:109–118.

61. Denovic M, Bovier R, Sarkissian J, et al: Intravesical instillation of Mitomycin C in the prophylactic treatment of recurring superficial transitional cell carcinoma of the bladder. *Br J Urol* 1983; 55:382–385.

62. Dische S: The hyperbaric oxygen chamber in the radiotherapy of carcinoma of the bladder. *Br J Radiol* 1973; 46:13–17.

63. Dresner SM, Haaff EO, Ratliff TL, et al: Bacille Calmette-Guerin intravesical therapy for superficial bladder cancer. *Urology Grand Rounds*, 1984, pp 1–7.

64. Droller MJ: Bladder cancer. *Monographs in Urology*. Princeton, NJ, Burroughs Wellcome Co, 1982, pp 131–154.

65. Droller MJ: Immunotherapy of genitourinary neoplasia. *Urol Clin North Am* 1984; 11:643–657.

66. Duncan RE, Bennett DW, Evans AT, et al: Radiation-induced bladder tumors. *J Urol* 1977; 118:43–45.

67. Durkee C, Benson R Jr: Bladder cancer following administration of cyclophosphamide. *Urology* 1980; 16:145.

68. Edwards PD, Hurm RA, Jaeschke WH: Conversion of cystitis glandularis to adenocarcinoma. *J Urol* 1972; 108:568–750.

69. El-Bolkainy MN, Mokhtar NM, Ghoneim MA, et al: The impact of schistosomiasis on the pathology of bladder carcinoma. *Cancer* 1981; 48:2643–2648.

70. England HR, Molland EA, Oliver RTD, et al: Systemic cyclophosphamide in flat carcinoma in situ of the bladder, in Hendry WF, Bloom HJG (eds): *Bladder Cancer: Principles of Combination of Therapy*. London, Butterworths, 1981.

71. Fairchild WV, Spence CR, Solomon HD, et al: The incidence of bladder cancer after cyclophosphamide therapy. *J Urol* 1979; 122:163.

72. Fall M, Pettersson S: Ureteral complications after intravesical formalin instillation. *J Urol* 1979; 122:160–162.

72a. Falor WH, Ward RM: Prognosis in early carcinoma of the bladder based on chromosomal analysis. *J Urol* 1978; 119:44–48.

72b. Farah RN, Cerney JC: Lymphangiography in staging patients with carcinoma of the bladder. *J Urol* 1978; 49:40–41.

73. Farrow GM, Utz DC, Rife CC: Morphological and clinical observations of patients with early bladder cancer treated with total cystectomy. *Cancer Res* 1976; 36:2495–2501.

74. Farsund T, Laerum OD, Hostmark KJ: Ploidy disturbance of normal-appearing bladder mucosa in patients with urothelial cancer: Relationship to morphology. *J Urol* 1983; 130:1076–1081.

75. Faysal MH: Squamous cell carcinoma of the bladder. *J Urol* 1981; 126:598–599.

76. Faysal MH, Freiha FS: Evaluation of partial cystectomy for carcinoma of the bladder. *Urology* 1979; 14:352–356.

77. Faysal MH, Freiha FS: Primary neoplasm in vesical diverticula. *Br J Urol* 1981; 53:141–143.

78. Flechner SM, Spaulding JT: Management of rectal injury during cystectomy. *Urology* 1982; 19:143–147.

79. Fokkens W: Phenacetin abuse related to bladder cancer. *Environ Res* 1979; 20:192–193.

80. Frankfurt OS, Huben RP: Clinical application of DNA flow cytometry for bladder tumors. *Urology* 1984; 23:29–36.

81. Fraumeni JF Jr, Thomas LB: Malignant bladder tumors in a man and his three sons. *JAMA* 1967; 201:507–509.

82. Friedell GH, Parija GC, Nagy GK, et al: The pathology of human bladder cancer. *Cancer* 1980; 45:1823–1831.

83. Fukui T, Yokokawa M, Mitani G, et al: In vivo staining test with methylene blue for bladder cancer. *J Urol* 1983; 130:252–255.

83a. Fujita J, Yoshida O, Yuasa Y, et al: Ha-ras oncogenes are activated by somatic alterations in human urinary tract tumors. *Nature* 1984; 309:464–466.

84. Gamarra MC, Zein T: Cytologic spectrum of bladder cancer. *Urology* 1984; 23:23–26.

85. Garnick MB, Schade D, Israel M, et al: Intravesical doxorubicin for prophylaxis in the management of recurrent superficial bladder carcinoma. *J Urol* 1984; 131:43–46.

86. Ghoneim MA, Awad HK: Results of treatment in carcinoma of the bilharzial bladder. *J Urol* 1980; 123:850–852.

87. Ghione M, et al (eds): *International Symposium on Adriamycin*. Berlin, Springer-Verlag, 1972.

88. Gibas Z, Sandberg AA: Chromosomal rearrangements in bladder cancer. *Urology* 1984; 23:3–9.

89. Gibod LB, Katz M, Cochand B, et al: Lymphography and percutaneous fine needle node aspiration biopsy in

the staging of bladder carcinoma. *J Urol* 1984; 132:24–26.

89a. Gilbert HA, Logan JL, Kagan AR, et al: The natural history of papillary transitional cell carcinoma of the bladder and its treatment in an unselected population on the basis of histological grading. *J Urol* 1978; 119:488–492.

90. Goffinet DR, Schneider MJ, Glatstein EJ, et al: Bladder cancer: Results of radiation therapy in 384 patients. *Radiology* 1975; 117:149–152.

91. Goodman GB, Hislop TG, Elwood JM, et al: Conservation of bladder function in patients with invasive bladder cancer treated by definitive irradiation and selective cystectomy. *Int J Radiat Oncol Biol Phys* 1981; 7:569–573.

92. Green DF, Robinson MRG, Glashan R, et al: Does intravesical chemotherapy prevent invasive bladder cancer? *J Urol* 1984; 131:33–35.

93. Green LF, Hanash KA, Farrow GM: Benign papilloma or papillary carcinoma of the bladder? *J Urol* 1973; 110:205–207.

94. Green LF, Yalowitz PA: The advisability of concomitant transurethral excision of vesical neoplasm and prostatic hyperplasia. *J Urol* 1972; 107:445–449.

95. Gustafson H, Tribukait B, Esposti PL: The prognostic value of DNA analysis in primary carcinoma in situ of the urinary bladder. *Scand J Urol Nephrol* 1982; 16:141–146.

96. Haaff EO, Dresner SM, Kelley DR, et al: Role of immunotherapy in the prevention of recurrence and invasion of urothelial bladder tumors: A review. *World J Urol* 1985; 3:76.

97. Hall RR, Heath AB: Radiotherapy and cystectomy for T3 bladder carcinoma. *Br J Urol* 1981; 53:598–601.

98. Hall RR, Herring DW, McGill AC, et al: Oral methotrexate therapy for multiple superficial bladder carcinomata. *Cancer Treat Rep* 1981; 65:175–178.

99. Hall RR, Schade ROK, Swinney J: Effect of hyperthermia on bladder cancer. *Br Med J* 1974; 2:593–594.

100. Hansen RI, Nerstrom B, Djurhuus JC, et al: Late results from mucosal denudation of urinary bladder papillomatosis. *Acta Chir Scand* 1976; 472:73–76.

101. Harker WG, Freiha FS, Shortliffe L, et al: Cisplatin, methotrexate, and vinblastin (CMV) for metastatic transitional cell carcinoma of the urinary tract (TCC): Chemotherapy evaluation of complete response by site. *Proc Am Soc Clin Oncol* 1984; abstract C-6773:160.

102. Helmstein K: Treatment of bladder carcinoma by a hydrostatic pressure technique. *Br J Urol* 1972; 44:434–450.

103. Heney NM, Ahmed S, Flanagan M, et al: Superficial bladder cancer: Progression and recurrence. *J Urol* 1983; 130:1083–1086.

104. Herr HW: Preoperative irradiation with and without chemotherapy as adjunct to radical cystectomy. *Urology* 1985; 25:127–134.

105. Herr HW, Kemergy N, Yagoda A, et al: Poly(I:C) immunotherapy in patients with papillomas or superficial carcinomas of the bladder. *Natl Cancer Inst Monogr* 1978; 49:325.

106. Herr HW, Pinsky CM, Melamed MR, et al: Long term effect of intravesical BCG on flat carcinoma in situ (CIS) of the bladder. *J Urol* 1986; 135:265–269.

107. Herr HW, Pinsky CM, Whitmore WF Jr, et al: Effect of intravesical bacillus Calmette-Guerin (BCG) on carcinoma in situ of the bladder. *Cancer* 1983a; 51:1323–1326.

108. Herr HW, Yagoda A, Batata M, et al: Planned preoperative cisplatin and radiation therapy for locally advanced bladder cancer. *Cancer* 1983; 52:2705–2706.

109. Hewett CB, Babiszewski JF, Antunez AR: Update on intracavitary radiation in the treatment of bladder tumors. *J Urol* 1981; 126:323–325.

110. Hill DE, Ford KS, Soloway MS: Radical cystectomy and adjuvant chemotherapy. *Urology* 1985; 25:151–154.

111. Hisazumi H, Misahi T, Myoshi N: Photoradiation therapy of bladder tumors. *J Urol* 1983; 130:685–687.

112. Hoffman D, Masuda Y, Wynder EL: Alpha-naphthylamine and beta-naphthylamine in cigarette smoke. *Nature* 1969; 221:254.

113. Hofstetter A, Frank F, Keditsch E, et al: Endoscopic neodymium-YAG laser application for destroying bladder tumors. *Eur Urol* 1981; 7:278–282.

114. Hopkins SC, Ford KS, Soloway MS: Invasive bladder cancer: Support for screening. *J Urol* 1983; 130:61–64.

115. Hoover R, Strasser PH: Artificial sweeteners and human bladder cancer. *Lancet* 1980; 1:837–840.

116. Howe GR, Burch JD, Miller AB, et al: Tobacco use, occupation, coffee, various nutrients, and bladder cancer. *JNCI* 1980; 64:70.

117. Huben RP: Tumors markers in bladder cancer. *Urology* 1984; 23:10–14.

118. Huland H, Otto U, Droese M, et al: Long-term Mitomycin C instillation after transurethral resection of superficial bladder carcinoma: Influence on recurrence, progression, and survival. *J Urol* 1984; 132:27–29.

119. Issell BF, Prout GR Jr, Soloway MS, et al: Mitomycin C intravesical therapy in noninvasive bladder cancer after failure on thiotepa. *Cancer* 1984; 53:1025–1028.

120. Jaske G: Intravesical chemotherapy for carcinoma in situ of the urinary bladder: Five years later. Unpublished data, 1985.

121. Javadpour N: Multiple cell markers in bladder cancer: Principles and clinical practice. *Urol Clin North Am* 1984; 11:609–616.

122. Jewett HJ, Eversole SL: Carcinoma of the bladder: Characteristic modes of local invasion. *J Urol* 1960; 83:383–389.

123. Jewett HJ Strong GH: Infiltrating carcinoma of the bladder: Relation of depth of peneration of the bladder wall to incidence of local extension and metastases. *J Urol* 1946; 55:336–372.

124. Jing B, Wallace S, Zormoza J: Metastases to retroperitoneal and pelvic lymph nodes: Computed tomography and lymphangiography. *Radiol Clin North Am* 1982; 20:511–530.

125. Johnson DE, Shoenwald MB, Ayala AG, et al: Squamous cell carcinoma of the bladder. *J Urol* 1976; 115:542–544.

126. Jones HC, Swinney J: Thiotepa in the treatment of tumors of the bladder. *Lancet* 1961; 2:615–618.

127. Kagawa S, Ogura K, Kurokawa K, et al: Immunological evaluation of a streptococcal preparation (OK-432) in treatment of bladder carcinoma. *J Urol* 1979; 122:467–470.

128. Kakizoe T, Matsumoto K, Audoh M, et al: Adenocarcinoma of urachus: Report of seven cases and review of the literature. *Urology* 1983; 21:360–366.

129. Kaufman JM, Fam B, Jacobs SC, et al: Bladder cancer and squamous metaplasia in spinal cord injury patient. *J Urol* 1977; 118:967–971.

130. Kaye KW, Lange PH: Mode of presentation of invasive bladder cancer: Reassessment of the problem. *J Urol* 1982; 128:31–33.

131. Kelley DR, Ratliff TR, Catalona WJ et al: Intravesical BCG therapy for superficial bladder cancer: Effect of BCG viability on treatment results. *J Urol* 1985; 134:48.

132. Kemeny N, Yagoda A, Wang Y, et al: Randomized trial of standard therapy with or without poly(I:C) in patients with superficial bladder cancer. *Cancer* 1981; 48:2154–2157.

132a. Kern WH: the grade and pathologic stage of bladder cancer. *Cancer* 1984; 53:1185–1189.

133. Klinger ME: Secondary tumors of the genitourinary tract. *J Urol* 1951; 65:144–153.

134. Koontz WW, Prout GR Jr, Smith W, et al: The use of intravesical thio-tepa in the management of non-invasive carcinoma of the bladder. *J Urol* 1981; 125:307–312.

134a. Koss JC, Arger PH, Coleman BG, et al: Ct staging of bladder carcinoma. *AJR* 1981; 137:359–362.

135. Koss LG: *Tumors of the Urinary Bladder*, fascicle 11, *Atlas of Tumor Pathology*. Washington, DC, Armed Forces Institute of Pathology, 1975, pp 1–120.

136. Koss LG, Esperanza MT, Robbins MA: Mapping cancerous and precancerous bladder changes: A study of the urothelium in ten surgically removed bladders. *JAMA* 1974; 227:281–286.

137. Kubota Y, Shiun T, Miura T, et al: Treatment of bladder cancer with a combination of hyperthermia, radiation and bleomycin. *Cancer* 1984; 53:199–202.

138. Kunter AF, Hartge P, Hoover RN, et al: Urinary tract infection and risk of bladder cancer. *Am J Epidemiol* 1984; 119:510–515.

139. Kunze E, Schauer A, Schmitt M: Histology and histogenesis of two different types of inverted urothelial papillomas. *Cancer* 1983; 51:348–358.

140. Kurth KH, Schroder FH, Tunn U, et al: Adjuvant chemotherapy of superficial transitional cell bladder carcinoma: Preliminary results of the European Organization for Research on Treatment of Cancer randomized trial comparing doxorubicin hydrochloride, ethoglucid and transurethral resection alone. *J Urol* 1984; 132:258–262.

141. Lamm DL: Unpublished data, 1985.

142. Lamm DL: Intravesical therapy of superficial bladder cancer. *UAU Update*, series 2. 1983, pp 2–7.

143. Lamm DL, Crawford ED, Montie JE, et al: BCG vs.

Adriamycin in the treatment of transitional cell carcinoma in situ: A Southwest Oncology Group Study, abstract 283. American Urological Association Meeting, 1985.

144. Lamm DL, Stogdill VD, Stogdill BJ: Complications of BCG immunotherapy in patients with bladder cancer. *Proc Am Urol Assoc* 1984, p 140A.

145. Lamm DL, Thor DE, Harris SC, et al: Bacillus Calmette-Guerin immunotherapy of superficial bladder cancer. *J Urol* 1980; 124:38–42.

146. Lamm DL, Thor DE, Stogdill VD, et al: Bladder cancer immunotherapy. *J Urol* 1982; 128:931–935.

146a. Lantz EJ, Hattery RR: Diagnostic imaging of urothelial cancer. *Urol Clin North Am* 1984; 11:576–583.

147. LaPlante M, Brice M II: The upper limits of hopeful application of radical cystectomy for vesical carcinoma: Does nodal metastasis always indicate incurability? *J Urol* 1973; 109:261–264.

148. Lazarevic B, Garret R: Inverted papilloma and papillary transitional cell carcinoma of urinary bladder. *Cancer* 1978; 42:1904–1911.

149. Leklem JE, Brown RR: Abnormal tryptophan metabolism in a family with a history of bladder cancer. *JNCI* 1976; 56:1101–1104.

150. Lillien OM, Camey M: 25-year experience with replacement of human bladder (Camey procedure). *J Urol* 1984; 132:886–891.

151. Linker DG, Whitmore WF: Ureteral carcinoma in situ. *J Urol* 1975; 113:777–780.

152. Lockhart JL, Chaikin L Bondhus MJ, et al: Prostatic recurrences in the management of superficial bladder tumors. *J Urol* 1983; 130:256–257.

153. Lucas SB: Squamous cell carcinoma of the bladder and schistosomiasis. *E Afr Med J* 1982; 59:345–352.

154. Lundbeck F, Mogensen P, Jeppersen N: Intravesical therapy of noninvasive bladder tumors with doxorubin and urokinase. *J Urol* 1983; 130:1087–1089.

155. Lutzeger W, Rubben H, Dahm H: Prognostic parameters in superficial bladder cancer: An analysis of 315 cases. *J Urol* 1982; 127:250–252.

156. Mackenzie AR, Sharma TC, Whitmore WF, Jr, et al: Nonextirpative treatment of myosarcomas of the bladder and prostate. *Cancer* 1971; 28:329–334.

157. Magri J: Partial cystectomy: Review of 104 cases. *Br J Urol* 1962; 34:74–86.

158. Marshall VF: The relation of the preoperative estimate to the pathologic demonstration of the extent of vesical neoplasms. *J Urol* 1952; 68:714–723.

159. Marshall VF, McCarron JP Jr: The curability of vesical cancer: Greater now or then? *Cancer Res* 1977; 37:2753–2755.

160. Martinez-Pineiro JA, Muntanola P: Nonspecific immunotherapy with BCG vaccine in bladder tumors: A preliminary report. *Eur Urol* 1977; 3:11–22.

161. Maruf NJ, Godec CJ, Strom RL, et al: Unusual therapeutic response of massive squamous cell carcinoma of the bladder to aggressive radiation and surgery. *J Urol* 1982; 128:1313–1315.

162. Mathur VK, Krahn HP, Ramse ER: Total cystectomy for bladder cancer. *J Urol* 1981; 125:784–786.

163. Matsumoto K, Kakizoe T, Mikuriza S, et al: Clinical evaluation of intraoperative radiotherapy for carcinoma of the urinary bladder. *Cancer* 1981; 47:509–513.

164. McCredie M, Stewart JH, Ford JM, et al: Phenacetin-containing analgesics and cancer of the bladder or renal pelvis in women. *Br J Urol* 1983; 55:220–224.

165. McCullough DL, Lamm DL, McLaughlin AP III, et al: Familial transitional cell carcinoma of the bladder. *J Urol* 1975; 113:629–635.

166. Melamed MR: Flow cytometry of the urinary bladder. *Urol Clin North Am* 1984; 11:599–607.

167. Melamed MR, Koss LG, Ricci A, et al: Cytohistological observations on developing carcinoma of the urinary bladder in man. *Cancer* 1960; 13:67–74.

168. Melicow MM: Histological study of vesical urothelium intervening between gross neoplasms in total cystectomy. *J Urol* 1952; 68:261–278.

169. Michel F, Gattegno B, Meyrier A, et al: Paraneoplastic hypercalcemia associated with bladder carcinoma: Report of 2 cases. *J Urol* 1984; 131:753–755.

170. Miller LS, Johnson DE: Megavoltage irradiation for bladder cancer: Alone, postoperative or preoperative, in *Proceedings of the Seventh National Cancer Conference*. Philadelphia, JB Lippincott Co, 1973, pp 771–782.

171. Mohuiddin M, Kramer S, Newall J, et al: Combined pre- and postoperative adjuvant radiotherapy for bladder cancer: Result of RTOG/Jefferson Study. *Cancer* 1981; 47:2840–2843.

172. Montie JE, Straffon RA, Stewart BH: Radical cystectomy without radiation therapy for carcinoma of the bladder. *J Urol* 1984; 131:477–482.

173. Morales A: Long term results and complications of intravesical BCG therapy for cancer. *Proc Am Urol Assoc* 1983, p 177.

174. Morales A: Treatment of carcinoma in situ of the bladder with BCG: A phase II trial *Cancer Immunol Immunother* 1980; 9:69–72.

175. Morales A, Eidinger D, Bruce AW: Intracavitary bacillus Calmette-Guerin in the treatment of superficial bladder tumors. *J Urol* 1976; 116:180–183.

176. Morales A, Ersil A: Prophylaxis of recurrent bladder cancer with bacillus Calmette-Guerin, in Johnson DE, Samuels ML (eds): *Cancer of the Genitourinary Tract*. New York, Raven Press, 1979, pp 121–132.

177. Morales A, Ottenhof P, Emerson L: Treatment of residual, non-infiltrating bladder cancer with bacillus Calmette-Guerin. *J Urol* 1981; 125:649–651.

178. Morrison AS: Advances in the etiology of urothelial cancer. *Urol Clin North Am* 1984; 11:557–566.

179. Morrison AS, Buring JE: Artificial sweeteners and cancer of the lower urinary tract. *N Engl J Med* 1980; 302:537.

180. Morrison AS, Buring JE, Verhock WG, et al: An international study of smoking and bladder cancer. *J Urol* 1984; 131:650–654.

181. Morrison AS, Cole P: Epidemiology of bladder cancer. *Urol Clin North Am* 1976; 3:13.

182. Mostofi FK: Potentialities of bladder epithelium. *J Urol* 1954; 71:705–714.

183. Mostofi FK, Sobin LH, Torloni H: Histological typing of urinary bladder tumors, in *International Histological Classification of Tumors*, no. 10. Geneva, World Health Organization, 1973.

184. Mostofi FK, Thomson RV, Dean AL Jr: Mucous adenocarcinoma of the urinary bladder. *Cancer* 1955; 8:741–758.

185. Murphy WM, Irving CC: The cellular features of developing carcinoma in murine urinary bladder. *Cancer* 1981; 47:514–522.

186. Murphy WM, Soloway MS: Urothelial dysplasia. *J Urol* 1982; 127:849–854.

187. Nakamura S, Nijima T: Staging of bladder cancer by ultrasonography: A new technique by transurethral intravesical scanning. *J Urol* 1980; 124:341–344.

188. Narayna AS, Loening SS, Slymen DJ, et al: Bladder cancer: Factors affecting survival. *J Urol* 1983; 130:56–60.

189. Navarre RJ Jr, Loening SA, Platz C, et al: Nephrogenic adenoma: A report of nine cases and review of the literature. *J Urol* 1982; 127:775–779.

190. Needles B, Yagoda A, Sogani P, et al: Intravenous cisplatin for superficial bladder tumor. *Cancer* 1982; 50:1722–1723.

191. Netto NR Jr, Lemos GC: A comparison of treatment methods for the prophylaxis of recurrent superficial bladder tumors. *J Urol* 1983; 129:33–34.

192. Newman DM, Brown JR, Jay AC, et al: Squamous cell carcinoma of the bladder. *J Urol* 1968; 100:470–473.

193. Nichols JA, Marshall VF: The treatment of bladder carcinoma by local excision and fulguration. *Cancer* 1956; 9:559–565.

194. Nielsen K, Nielsen KK: Adenocarcinoma in exstrophy of the bladder—the last case in Scandinavia? A case report and review of the literature. *J Urol* 1983; 130:1180–1182.

195. Nijima T: Intravesical therapy with Adriamycin and new trends in the diagnosis and therapy of superficial bladder tumors, in *WHO, Diagnostics and Treatment of Superficial Urinary Bladder Tumors*. Collaborating Center for Research and Treatment of Urinary Bladder Cancer, 1978.

196. Nijima T, Koiso K, Akaza H, and The Japanese Urological Cancer Research Group for Adriamycin: Randomized clinical trial on chemoprophylaxis of recurrence in cases of superficial bladder cancer. *Cancer Chemother Pharmacol* 1983; 11(suppl):79–82.

197. Nocks BM, Henez NM, Daly JJ: Primary adenocarcinoma of the urinary bladder. *Urology* 1983; 21:26–29.

198. Ostroff EB, Chenault OW Jr: Alum irrigation for the control of massive bladder hemorrhage. *J Urol* 1982; 128:929–930.

199. Page BH, Levison VB, Curwen MP: The site of recurrence of non-infiltrating bladder tumors. *Br J Urol* 1978; 50:237–242.

200. Palmer JP, Sprate DW: Pelvic carcinoma following irradiation for benign gynecological diseases. *Am J Obstet Gynecol* 1956; 72:497.

201. Patch FS, Rhea LJ: The genesis and development of Brunn's nests and their relationship to cystitis cystica

glandularis and primary adenocarcinoma of the bladder. *Can Med Assoc J* 1935; 33:597–606.

202. Pavone-Macaluso M, Caramia G: Adriamycin and daunomycin in the treatment of vesical and prostatic neoplasms: Preliminary results, in Carter SK, DiMarco M, Ghione M, et al (eds): *International Symposium on Adriamycin*. Berlin, Springer-Verlag, 1972.

203. Pearson RM, Soloway MS: Does cyclophosphamide induce bladder cancer? *Urology* 1978; 11:437.

204. Pinsky CM, Camacho FJ, Kerr D, et al: Treatment of superficial bladder cancer with intravesical BCG, in Terry WK, Rosenberg SA (eds): *Immunotherapy of Human Cancer*. New York, Elsevier/North-Holland Inc, 1982, pp 309–313.

205. Pizza G, Severini G, Menniti D, et al: Tumour regression after intralesional injection of interleukin 2(IL-2) in bladder cancer: Preliminary report. *Int J Cancer* 1984; 34:359–367.

206. Price JM: Bladder cancer, in *Canadian Cancer Conference: Proceedings of the Sixth Cancer Research Conference*. New York, Pergamon Press, 1966.

207. Prout GR Jr: The surgical management of bladder carcinoma. *Urol Clin North Am* 1976; 3:149–175.

208. Prout GR Jr, Griffin PP, Nocks BN, et al: Intravesical therapy of low stage bladder carcinoma with mitomycin C: Comparison of results in untreated and previously treated patients. *J Urol* 1982; 127:1096–1098.

209. Prout GR Jr, Griffin PP, Shipley WV: Bladder carcinoma as a systemic disease. *Cancer* 1979; 42:2532–2539.

210. Prout GR Jr, Koontz WW Jr, Coombs J et al: Long-term fate of 90 patients with superficial bladder cancer randomly assigned to receive or not to receive thiotepa. *J Urol* 1983; 130:677–680.

211. Purves EC, Snell M, Cope WA, et al: Subcutaneous *Corynebacterium parvum* in bladder cancer. *Br J Urol* 1979; 51:278–282.

212. Radwin HM: Invasive transitional cell carcinoma of the bladder: Is there a place for preoperative radiotherapy? *Urol Clin North Am* 1980; 7:551–557.

213. Raghaven D, Pearson B, Duval P, et al: Initial intravenous cis-platinum therapy: Improved management for invasive high risk bladder cancer? *J Urol* 1985; 133:399–402.

214. Rehn L: Ueber blasentumoren bei fuchsinarbeitern. *Arch Kind Chir* 1895; 50:588.

215. Resnick MI, O'Connor VJ Jr: Segmental resection for carcinoma of the bladder: Review of 102 patients. *J Urol* 1973; 109:1007–1010.

216. Richie JP, Waisman J, Skinner DG, et al: Squamous cell carcinoma of the bladder: Treatment by radical cystectomy. *J Urol* 1976; 115:670–672.

217. Riddle PR, Chisholm GD, Trott PA, et al: Flat carcinoma in situ of bladder. *Br J Urol* 1976; 47:829–833.

218. Robinson MRG, Sheltz MB, Richards B, et al: Intravesical epodyl in the management of bladder tumors: Combined experience of the Yorkshire Urological Cancer Research Group. *J Urol* 1977; 118:972–973.

219. Rosi P, Selli C, Carini M, et al: Myxoid liposarcoma of the bladder. *J Urol* 1983; 130:560–561.

220. Schellhammer PF, Bean MA, Whitmore WF Jr: Prostatic involvement by transitional cell carcinoma: Pathogenesis, patterns and prognosis. *J Urol* 1977; 118:399–403.

221. Schellhammer PF, Ladaga LE, Fillion MB: Bacillus Calmette-Guerin (BCG) for superficial transitional cell carcinoma (TCC) of the bladder. *J Urol* 1986; 135:261–264.

222. Schellhammer PF, Whitmore WF Jr: Transitional cell carcinoma in the urethra of men having cystectomy for bladder cancer. *J Urol* 1976; 115:56–60.

223. Schoborg TW, Saffos RO, Rodriguez AP, et al: Carcinosarcoma of the bladder. *J Urol* 1980; 124:724–727.

224. Schoborg TW, Sapolsky JL, Lewis CW Jr: Carcinoma of the bladder treated by segmental resection. *J Urol* 1979; 122:473–475.

225. Schultz RE, Bloch MJ, Tomaszewski JE, et al: Mesonephric adenocarcinoma of the bladder. *J Urol* 1984; 132:263–265.

226. Schuhrke J, Barr JW: Intractable bladder hemorrhage: Therapeutic embolization of the hypogastric arteries. *J Urol* 1976; 116:523–525.

227. Sen SE, Malek RS, Farrow GM, et al: Sarcoma and carcinosarcoma of the bladder in adults. *J Urol* 1985; 133:29–30.

228. Sheldon CA, Clayman RV, Gonzalez R, et al: Malignant urachal lesions. *J Urol* 1984; 131:1–8.

229. Shipley WU, Coombs LJ, Einstein AB, et al: Cisplatin and full dose irradiation for patients with invasive bladder carcinoma: A preliminary report of tolerance and local response. *J Urol* 1984; 132:899–903.

230. Shipley WU, Cummings KB, Coombs LJ: 4,000 Rad pre-op radiation followed by prompt radical cystectomy for invasive bladder cancer: A prospective study of patient tolerance and pathologic downstaging. *J Urol* 1982; 127:48–51.

231. Schlegel JU, Pipkin GE, Nishimura R, et al: The role of ascorbic acid in the prevention of bladder tumor formation. *Trans Am Assoc Genitourinary Surg* 1969; 61:85–89.

232. Shortliffe L, Freiha F, Higgins M, et al: Intravesical alpha 2 interferon therapy for superficial bladder cancer, abstract 203. Meeting of the European Association of Urologists, Copenhagen, 1984, p 703.

233. Skinner DG: Management of invasive bladder cancer: A meticulous pelvic node dissection can make a difference. *J Urol* 1982; 128:34–36.

234. Skinner DG, Crawford ED, Kaufman JJ: Complications of radical cystectomy for carcinoma of the bladder. *J Urol* 1980; 123:640–643.

235. Skinner DG, Lieskovsky G: Contemporary cystectomy with pelvic node dissection compared to preoperative radiation therapy plus cystectomy in management of invasive bladder cancer. *J Urol* 1984; 131:1069–1072.

236. Skinner DG, Lieskovsky G, Boyd SD: Technique of creation of a continent internal ileal reservoir (Kock pouch) for urinary diversion. *Urol Clin North Am* 1984; 11:741–749.

237. Slack NH, Prout GR Jr: The heterogeneity of invasive

bladder carcinoma and different responses to treatment. *J Urol* 1980; 123:644–652.

238. Smith JL: Histology and spontaneous regression of retinoblastoma. *Trans Ophthalmol Soc UK* 1974; 94:953–967.

239. Smith JA Jr, Dixon JA: Argon laser phototherapy of superficial transitional cell carcinoma of the bladder. *J Urol* 1984; 131:655–656.

240. Smith JA Jr, Whitmore WF Jr: Regional lymph node metastases from bladder cancer. *J Urol* 1981a; 126:591–593.

241. Smith JA Jr, Whitmore WF Jr: Salvage cystectomy for bladder cancer after failure of definitive irradiation. *J Urol* 1981b; 125:643–645.

242. Soloway MS: Bladder cancer: Management of an increasingly common tumor. *Postgrad Med* 1983; 73:139–151.

243. Soloway MS: Intravesical and systemic chemotherapy in the management of superficial bladder cancer. *Urol Clin North Am* 1984; 11:623–635.

244. Soloway MS: Learning to integrate systemic chemotherapy into a treatment plan for patients with advanced bladder cancer. *J Urol* 1985; 133:440–441.

245. Soloway MS: The management of superficial bladder cancer, in Javadpour N (ed): *Principles and Management of Urologic Cancer*. Baltimore, Williams & Wilkins Co, 1983, pp 446–466.

246. Soloway MS, Einstein A, Corder MP, et al: A comparison of cisplatin and the combination of cisplatin and cyclophosphamide in advanced urothelial cancer: A National Bladder Cancer Collaborative Group A Study. *Cancer* 1983; 52:767–772.

247. Sontag JM: Experimental identification of genitourinary carcinogens. *Urol Clin North Am* 1980; 7:803.

248. Sternberg CN, Yagoda A, Scher HI, et al: Preliminary results of M-VAC (methotrexate, vinblastine, doxorubicin and cisplatin) for transitional cell carcinoma of the urothelium. *J Urol* 1985; 133:403–407.

249. Studer UE, Biedermann C, Chollet D, et al: Prevention of recurrent superficial bladder tumors by oral etretinate: Preliminary results of a randomized double-blind, multicenter trial in Switzerland. *J Urol* 1984; 131:47–49.

250. Sullivan JW: Epidemiologic survey of bladder cancer in greater New Orleans. *J Urol* 1982; 128:281–283.

251. Sullivan JW, Gravstald H, Whitmore WF Jr: Complications of ureteroileal conduit with radical cystectomy: review of 336 cases. *J Urol* 1980; 124:797–801.

252. Summers JL, Falor WH, Ward RM, et al: Identical genetic profiles in primary and metastatic bladder tumors. *J Urol* 1983; 129:827–828.

253. Symes MO, Eckert H, Feneley RC, et al: Adoptive immunotherapy and radiotherapy in the treatment of urinary bladder cancer. *Br J Urol* 1978; 50:328–331.

254. Teulings FAG, Peters HA, Hop WCJ et al: A new aspect of the urinary excretion of tryptophan metabolites in patients with cancer of the bladder. *Int J Cancer* 1978; 21:140–146.

255. Torres H, Bennett MJ: Neurofibromatosis of the bladder: Case report and review of the literature. *J Urol* 1966; 96:910–912.

256. Torti F, Shortliffe L, Williams RD, et al: Superficial bladder cancers are responsive to alpha-2 interferon administered intravesically. ASCO, 1984, Toronto.

256a. Trott PA, Edwards L: Comparison of bladder washings and urine cytology in the diagnosis of bladder cancer. *J Urol* 1973; 110:664–666.

257. Tsukamoto T, Lieber MM: Sarcomas of the kidney, urinary bladder, prostate, spermatic cord paratestis and testis in adults, in Raaf JH (ed): *Management of Soft Tissue Sarcomas*. Chicago, Year Book Medical Publishers. In press.

258. Tyler DE: Stratified squamous epithelium in the vesical trigone and urethra: Findings correlated with menstrual cycle and age. *Am J Anat* 1962; 111:319–335.

259. Utz DC, Farrow GM: Carcinoma in situ of the urinary tract. *Urol Clin North Am* 1984; 11:735–740.

260. Utz DC, Hanash KA, Farrow GM: The plight of the patient with carcinoma in situ of the bladder. *J Urol* 1970; 103:160–164.

261. Utz DC, Schmitz SE, Fugelso PD, et al: A clinicopathologic evaluation of partial cystectomy for carcinoma of the urinary bladder. *Cancer* 1973; 32:1075.

262. Uyama T, Moriwaki S: Carcinosarcoma of urinary bladder. *Urology* 1981; 18:191–194.

263. van der Werf-Messing B: Carcinoma of the bladder treated by suprapubic radium implants. The value of additional external irradiation. *Eur J Urol* 1969; 5:277.

264. van der Werf-Messing BHP: Carcinoma of the urinary bladder treated by interstitial radiotherapy. *Urol Clin North Am* 1985; 11:659–670.

265. van der Werf-Messing B: Carcinoma of the bladder treated by preoperative radiation followed by cystectomy. *Cancer* 1973; 32:1084–1088.

266. van der Werf-Messing B: Cancer of the urinary bladder treated by interstitial radium implant. *Int J Radiat Oncol Biol Phys* 1978; 4:373–378.

267. van der Werf-Messing B, Menon RS, Hopp WCJ: Cancer of the urinary bladder category T_2, T_3 (NxMo) treated by interstitial radium implant: Second report. *Int J Radiat Oncol Biol Phys* 1983; 9:481–485.

268. Varkarakis MJ, Gaeta J, Moore RH, et al: Superficial bladder tumor: Aspects of clinical progression. *Urology* 1974; 4:414.

269. Veenema RJ, Harisiadis L, Chang C, et al: Bladder carcinoma: Preliminary external radiotherapy used as a means for selecting complete treatment, in Connolly JG (ed): *Carcinoma of the Bladder*. New York, Raven Press, 1981, pp 183–191.

270. Vinnicombe J, Abercrombie GF: Total cystectomy—a review. *Br J Urol* 1978; 50:488–491.

271. Vock P, Haertel M, Fuchs WA, et al: Computed tomography in staging of carcinoma of the urinary bladder. *Br J Urol* 1982; 54:158–163.

272. Wahlqvist L: Chemical carcinogenesis—a review and personal observations with special reference to the role of tobacco and phenacetin in the production of urothelial tumors, in Pavone-Maculoso M, et al (eds): *Bladder*

Tumors and Other Topics in Urological Oncology. New York, Plenum Press, 1980, p 47.

273. Wallace DM, Bloom HJG: The management of deeply infiltrating (t₃) bladder carcinoma: Controlled trial of radical radiotherapy and radical cystectomy (first report). *Br J Urol* 1976; 48:587–594.

274. Wallace DM, Chisholm GD, Henry WF: TNM classification for urological tumors (UICC)—1974. *Br J Urol* 1975; 47:1–12.

275. Walsh PC, Lepor H, Eggleston JC: Radical prostatectomy with preservation of sexual function: Anatomical and pathological considerations. *Prostate* 1983; 4:473–485.

276. Waterhouse J, Muir C, Shanmugartanam K, et al: *Cancer Incidence in Five Continents.* Lyon, International Agency for Research on Cancer, 1982, vol 4.

277. Weinstein RS, Alroy J, Farrow GM, et al: Blood group isoantigen deletion in carcinoma in situ of the urinary bladder. *Cancer* 1979; 43:661–668.

278. Weinstein RS, Miller AW III, Pauli BV: Carcinoma in situ: Comment on the pathobiology of a paradox. *Urol Clin North Am* 1980; 7:523–531.

279. Weldon TE, Soloway MS: Susceptibility of urothelium to neoplastic cellular implantation. *Urology* 1975; 5:824–827.

280. Wheeler JD, Hill WT: Adenocarcinoma involving the urinary bladder. *Cancer* 1954; 7:119–135.

281. Whitmore WF Jr: Integrated irradiation and cystectomy for bladder cancer. *Br J Urol* 1980; 52:1–9.

282. Whitmore WF Jr, Batata M: Status of integrated irradiation and cystectomy for bladder cancer. *Urol Clin North Am* 1985; 11:681–691.

283. Whitmore WF Jr, Batata MA, Hilaris BS, et al: A comparative study of two preoperative radiation regimens with cystectomy for bladder cancer. *Cancer* 1977; 40:1077–1086.

284. Wiener DP, Koss LG, Sabley B et al: The prevalence and significance of Brunn's nests, cystitis cystica and squamous metaplasia in normal bladders. *J Urol* 1979; 122:317–321.

285. Wijkstrom H, Gustafson H, Tribukait B: Deoxyribonucleic acid analysis of the evaluation of transitional cell carcinoma before cystectomy. *J Urol* 1984; 132:894–898.

286. Wiley EL, Mendelsohn G, Droller M, et al: Immunoperoxidase detection of carcinoembryonic antigen and blood group substances in papillary transitional cell carcinoma of the bladder. *J Urol* 1982; 128:276–280.

287. Winterberger AR, Wajsman Z, Merrin C, et al: Eight years of experience with preoperative angiographic and lymphographic staging of bladder cancer. *J Urol* 1978; 119:208–212.

288. Wilson TM, Fauver HE, Weigel JW: Leiomyosarcoma of urinary bladder. *Urology* 1979; 13:565–567.

289. Wolf H: Studies on the role of tryptophan metabolites in the genesis of bladder cancer. *Acta Chir Scand* [*Suppl*] 1973; 433:154.

290. Yagoda A: Chemotherapy for advanced urothelial cancer. *Semin Urol* 1983; 1:60–74.

291. Young JL Jr, Asine AJ, Pollack ES: *Cancer Incidence and Mortality in the United States, 1973–1979.* Bethesda, MD, DHEW publication (NIH) 78–1837, 1978.

292. Zincke H, Garbeff PJ, Beahrs JR: Upper urinary tract transitional cell cancer after radical cystectomy for bladder cancer. *J Urol* 1984; 131:50–52.

293. Zincke H, Utz DC, Taylor WF et al: Influence of thiotepa and doxorubicin instillation at time of transurethral surgical treatment of bladder cancer on tumor recurrence: A prospective, randomized, double-blind, controlled trial. *J Urol* 1983; 129:505–509.

Chapter 32

Urinary Diversion and Continent Reservoir

JEAN B. deKERNION, M.D.
ELIAHU MUKAMEL, M.D.

Radical cystectomy remains the most effective method of treatment for high-grade or invasive bladder cancer and despite advances in chemotherapy and radiotherapy, no study has yet demonstrated their equivalence to surgery with respect to survival or local and regional control. Thus, requirement for some form of urinary diversion still persists as it has since cystectomy for bladder cancer was first practiced in the early 1900s. The era of antibiotics and modern anesthetic techniques has markedly reduced morbidity from the cystectomy itself as well as from the urinary diversion. Today, inasmuch as no major therapeutic advances have been made in the management of bladder cancer, attention has turned to improved methods of urinary diversion. This chapter briefly describes these methods and their complications subsequent to radical cystectomy.

URETEROSIGMOIDOSTOMY

Implanting the urinary tract into the colon as popularized by Coffey (1911), though frequently modified, remained the primary method of urinary diversion until the 1950s. Most complications of the procedure stemmed from reflux of infected urine into the upper urinary tract with resulting pyelonephritis and loss of renal function. Incorporation of an antireflux ureteral reimplantation by Leadbetter (1950) and later by Goodwin et al. (1953) largely abrogated the major problems associated with the procedure. A better understanding of the fluid and electrolyte abnormalities such as hypokalemia, hyperchloremia, and hyperchloremic acidosis and their prevention also contributed to renewed confidence in the operation. However, Bricker's description (1950) of the ureteroileal urinary diversion diminished interest in ureterosigmoidostomy and the procedure was abandoned in most institutions.

The chief advantages of ureterosigmoidostomy are the simplicity of the technique and absence of an external stoma. With the use of a long antireflux tunnel, colonic-ureteral reflux is unusual and, indeed, this approach has reduced the incidence of upper tract complications. However, the operation is contraindicated in patients with colonic disease or abnormal upper urinary tract, or who have undergone radiotherapy to the pelvis or in whom postoperative radiotherapy is contemplated.

LEADBETTER TECHNIQUE

Leadbetter (1950) described the formation of a submucosal tunnel in the sigmoid segment. Figure 32–1 compares the tunnel method of reimplantation to earlier techniques that failed to prevent reflux. With this procedure, the left ureter is brought through the sigmoid mesentery and each ureter is implanted in the taenia anteriorly (Fig 32–2). Each taenia is opened to expose the underlying mucosa and the ureters are placed in this submucosal trough measuring 2.5–3 cm. A small opening is then made in the mucosa to which each spatulated ureter is carefully anastomosed with fine 4-0 chromic sutures. The muscularis of the taenia is then closed over the ureter with interrupted silk sutures, which form the tunnel. Both ureters should be placed at least 4 cm apart and as far distally in the sigmoid as possible, to minimize the risk of kinking and possible electrolyte problems. Although stents are not essential, many surgeons pass polyethylene stents from the upper urinary tract through the anastomosis and through the rectum. Decompression of the rectum is then carried out with a rectal tube for at least one week.

GOODWIN TECHNIQUE

The transcolonic technique described by Goodwin et al. (1953) is similar to the reimplantation of the ureter into the bladder described by Leadbetter and Politano. The prepared sigmoid colon is opened anteriorly. Each ureter is brought through a separate submucosal tunnel and sutured to the mucosa from inside the bowel lumen. Ureters can be sutured individually to the mucosa or can be conjoined and anastomosed to the mucosa as a single lumen (Fig 32–3). The anastomoses are stented

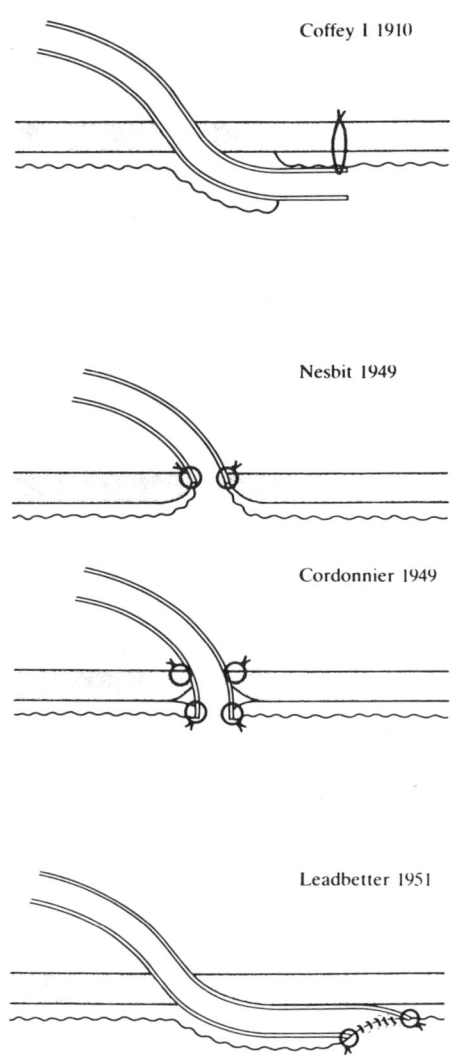

FIG 32–1.
Techniques for ureterosigmoidostomy. (From Walsh A: Urinary diversion in malignant disease, in Ashken MJ (ed): *Clinical Practice in Urology—Urinary Diversion.* Berlin, Springer-Verlag Co, 1982, p 79. Used with permission.)

with fine tubes brought out through the rectum. An alternative method of preventing reflux by an open colonic anastomosis was described by Mathisen (1953), who fashioned an antirefluxing nipple at the end of the implanted ureter.

COMPLICATIONS

Early complications after ureterosigmoidostomy are encountered in about two thirds of patients undergoing this procedure. Wound infection, wound dehiscence, and prolonged ileus are frequent (Zincke and Segura, 1975), but the most serious complication is urinary leakage from the ureterocolic anastomosis or from the colotomy incision. The incidence of this complication can be minimized by watertight ureterocolic anastomo-

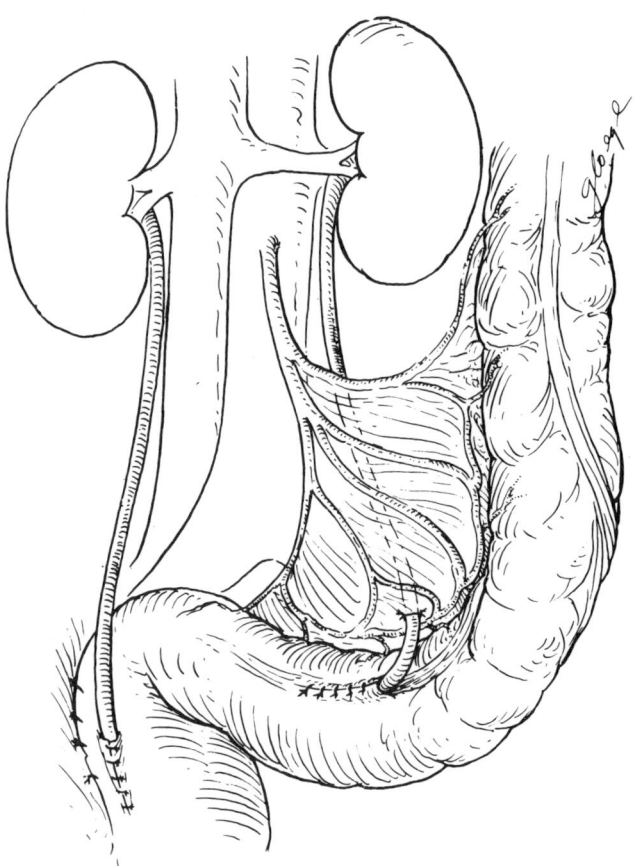

FIG 32–2.
Leadbetter's combined technique of ureterosigmoidostomy. Note that the left ureter is brought through the sigmoid mesentery and implanted on the anterior taenia. The rectosigmoid is fixed to the lateral pelvic wall or psoas muscle in the region of the right ureterocolonic anastomosis. (From Richie JP, Skinner DG: Ureterointestinal diversion, in Walsh PC, et al (eds): *Campbell's Urology.* Philadelphia, WB Saunders Co, 1986, p 2602. Used with permission.)

sis, the use of ureteral stents, and continuous drainage of the rectal content. Proper drainage can be achieved by using two rectal tubes that have multiple perforations. Stents and tubes should be irrigated every four hours to ensure patency. Although minor leaks usually seal spontaneously without further treatment, major leaks involve a high incidence of complications such as prolonged ileus, severe acidosis, and electrolyte imbalance. In patients in whom major leaks develop, immediate reoperation with reanastomosis or cutaneous diversion should be performed and in life-threatening conditions, nephrectomy is indicated.

The combined (nonrefluxing) technique for ureterocolonic anastomosis has improved the long-term results for ureterosigmoidostomy by decreasing the incidence of ureteral reflux and ureteral obstruction (Williams et al., 1969; Wear and Barquin, 1973; Zincke and Segura, 1975). Comparison between the refluxing and nonre-

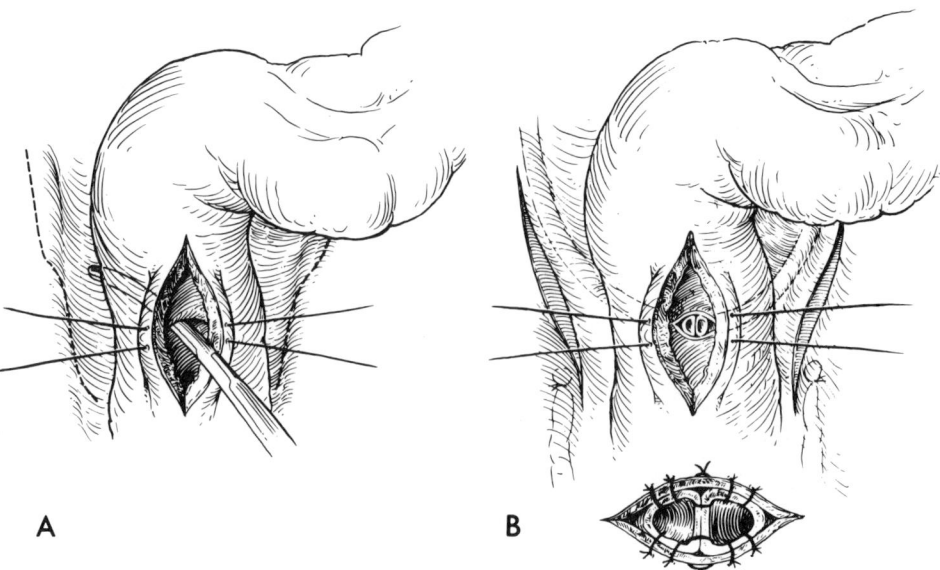

FIG 32–3.
Goodwin's transcolonic technique of ureterosigmoidostomy. Note that both ureters are brought through individual submucosal tunnels in the posterior rectal wall **(A).** The ureters are then sewn together medially before they are secured to the colonic mucosa circumferentially **(B).** (From Richie JP, Skinner DG: Ureterointestinal diversion, in Walsh PC, et al (eds): *Campbell's Urology.* Philadelphia, WB Saunders Co, 1986, p 2603. Used with permission.)

fluxing techniques reported by Wear and Barquin (1973) in fact showed fewer episodes of pyelonephritis and renal damage with the latter technique, although the incidence of acute pyelonephritis and renal deterioration remained high, ranging from 20% to 57% and from 20% to 35%, respectively (Hoffman and Spence, 1965; Williams et al., 1969; Zincke and Segura, 1975) (Table 32–1). Electrolyte disturbance and acidosis, encountered in as many as 47% of patients, are caused by reabsorption of urine from the bowel. Reabsorption takes place not only in the sigmoid colon but also throughout the entire colonic mucosa (Walsh, 1982). Characteristic laboratory findings in the serum subsequent to ureterosigmoidostomy include high urea levels with normal creatinine, high ammonium and chloride levels, low potassium levels, and metabolic acidosis. To improve acidosis and electrolyte imbalance, it is imperative to avoid prolonged contact between urine and the colonic mucosa. All patients should be instructed to evacuate the rectum every 2–3 hours during the day and at least once or twice at night, and a lifelong low-chloride diet and supplemental solution of 10% potassium citrate should be prescribed for most patients.

The association between ureterosigmoidostomy and colonic tumors has been confirmed by numerous studies since the first report by Hammer (1929). In reviewing the literature, Leadbetter et al. (1979) found 45 patients with tumors at the site of the ureterocolonic anastomosis, of which 31 were adenocarcinoma, four were transitional cell carcinoma, and four were benign colonic tumors. It has been estimated that the incidence of colonic carcinoma associated with ureterosigmoidostomy is 500 times higher than in the normal population. The mean lag period for development of carcinoma of the colon for patients following ureterosigmoidostomy after age 40 years was 8.7 years (range, 5 to 14 years), while for those who were younger than 40 years the lag period was 21.4 years (range, 14 to 50 years). The etiology of these tumors is obscure. Mechanical irritation by the fecal stream at the site of the ureterocolic anastomosis (Rivard et al., 1975; Leadbetter et al., 1979), the carcinogenic effects of the mixture of urine and feces (Crissy et al., 1979), and the presence of large amounts of carcinogenic nitroso compounds in the rectal urine (Stewart et al., 1981) were suggested as possible etiologic factors.

Whatever the etiology of the colonic cancer, all patients with ureterosigmoidostomy should be followed up by stool examination for occult blood every three months and undergo a yearly intravenous pyelogram and colonoscopy, starting five years postoperatively. If

TABLE 32–1.
Complications of Ureterosigmoidostomy

SOURCE (YR)	NO. OF PATIENTS	PYELONEPHRITIS, %	ACIDOSIS, %	DETERIORATION ON IVP, %	CALCULI, %
Williams et al. (1960)	57	45	32	35	5
Wear and Barquin (1973)	45	57	47	32	
Zincke and Segura (1975)	173	26	15	20	4

colonic tumor is found, segmental resection of the bowel and cutaneous urinary diversion should be performed.

PERSPECTIVES

The technical simplicity of ureterosigmoidostomy as well as reduction in complications utilizing the antireflux ureteral implantation argue against relegating the procedure to one of historical interest only. When considering quality of life, freedom from external stomas or appliances is always a major concern; moreover, preserving the body imaging by utilizing the ureterosigmoidostomy is one of its paramount assets. Although the ileal conduit and new methods of continent diversion have largely replaced ureterosigmoidostomy, some urologists continue to consider it the treatment of choice for urinary diversion following cystectomy. Reports in recent years, such as that by Zincke and Segura (1975), still recommend this procedure in selected patients. Even though newer methods of direct anastomosis of bowel urinary reservoirs to the urethra may prove to be successful, some patients will still be candidates for ureterosigmoidostomy, which remains the simplest, fastest operative technique for urinary diversion.

URETEROILEAL CUTANEOUS DIVERSION

The complications of a direct ureterocolonic anastomosis prompted the search for a more effective bladder substitute. In 1950, Bricker first reported his method of ureteral anastomosis to an isolated segment of ileum. Freed from some of the major problems of the colonic anastomoses occurring in most medical centers, ileal conduit rapidly became the method of choice for urinary diversion after cystectomy. As noted below, however, the procedure incurred its own immediate and long-term complications. The advances made in stomal appliances and the expansion of the field of enterostomal therapy markedly improved the procedure's feasibility and improved the quality of life. Later recognition of the importance of the everted stoma further minimized the problems with appliances fitting and also reduced the incidence of stomal stenosis. A number of refinements in Bricker's original technique have also contributed to improved results. In an attempt to minimize the incidence of ureteroileal anastomotic stenosis—a common long-term complication—Barzilay (1960) and Barzilay and Goodwin (1968) described the anastomosis of the conjoined ureters to the side of the ileal segment. Wallace (1966) described anastomosis of the conjoined ureters to the proximal end of the isolated ileal segment; this procedure was modified and

popularized in the United States. Until recently, these methods of ileal conduit diversion have been the mainstay of urinary diversion after radical cystectomy and, although associated with some long-term complications, they remain the major methods utilized in the United States today.

BRICKER'S PROCEDURE

The success of Bricker's operation depends on reserving the vascular supply to the ileal segment and the ureters. Most major short-term complications can be traced to problems of vascularity, either due to technical errors or to previous radiotherapy. A short segment of ileum should be used, usually about 12 cm in length. Adequate bowel preparation is essential and the authors have never seen abdominal abscess formation after this operation in a patient who has had adequate antibiotic and mechanical bowel cleansing. In the classic Bricker

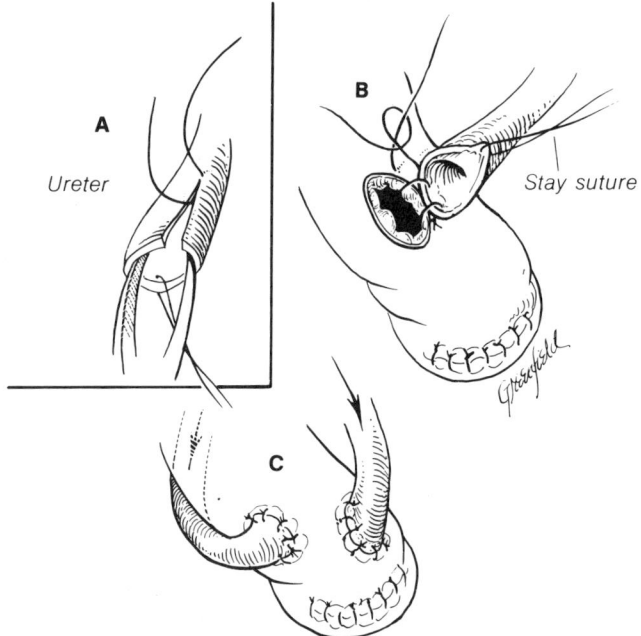

FIG 32–4.
Bricker's procedure for ureteroileal anastomosis. **A,** elliptical spatulation of ureter is performed and apical suture placed. Note use of forceps to spread ureter and prevent handling of mucosa, and stay suture placed at 12 o'clock position to aid in completion of anastomosis. **B,** apical suture of ureteroileal anastomosis has been completed, and additional sutures are being placed in staggered fashion. The stay suture at the 12 o'clock position of the ureter aids in manipulation of ureter from side to side for better visualization. **C,** both ureteral anastomoses have been completed. Right ureteroileal anastomosis is 1 cm distal to left anastomosis, preventing ischemia between the two anastomoses. (From Richie JB: Techniques of ureterointestinal anastomoses and conduit construction, in Crawford ED, Borden TA (eds): *Genitourinary Cancer Surgery.* Philadelphia, Lea & Febiger, 1982, p 233. Used with permission.)

procedure, the left ureter is brought through the sigmoid mesocolon to a point close to the right ureter. The proximal end of the ileal segment is sutured with absorbable sutures and fixed either to the sacral promontory or to the abdominal muscles at the right of the vena cava. Each ureter is spatulated and sutured to a small opening in the lateral aspect of the ileal conduit with nonabsorbable sutures (Fig 32–4). Although stents are usually not required, the authors always prefer to stent the ureteroileal anastomoses for at least 5–7 days.

Construction of the stoma is highly important. The site of the stoma should be tailored to the patient's body habitus; it should be identified and marked preoperatively. A circular opening about 2.5 cm in diameter is made in the skin and the anterior fascia is opened by a cruciate incision. The rectus muscle is separated and the posterior fascia is likewise opened in a cruciate incision sufficient to allow free passage of two fingers. It is imperative that the conduit pass through the rectus muscle rather than lateral to it to prevent herniation. Eversion of the stoma for about 1 cm is perhaps the most critical part of the stoma construction. The everted stoma allows careful and accurate fitting of the stomal appliance, and prevents leaks as well as severe skin changes, which often lead to stomal stenosis.

WALLACE'S PROCEDURE

The Wallace procedure differs from Bricker's operation in that the spatulated ureters are sutured to the proximal end of the isolated ileal segment (Fig 32–5). The theoretical advantage of the procedure is that the conjoined ureters afford a larger anastomosis to the ileum, thereby decreasing risks of ureteral anastomotic stenosis. This is perhaps more important in children and less important in adults undergoing radical cystectomy. Another potential advantage is the ability to visualize the ureteral orifice at the end of the isolated segment, making passage of catheters and contrast studies more feasible. The left ureter is brought over to lie adjacent to the right ureter. Both are trimmed to a convenient point so that the anastomosis is out of the pelvis in the event that radiotherapy becomes necessary at a later date. Moreover, shortening the ureters will prevent subsequent kinking. Both ureters are spatulated on the anterior surface for about 1.5 cm, with care taken not to interfere with the blood supply. The ureters are then joined together with interrupted 4-0 chromic sutures, after which the conjoined ureters are sutured to the proximal end of the isolated segment. The authors prefer to perform the anastomosis with polyethylene stents in place and to bring these stents through the conduit, removing them on the fifth to seventh postoperative day.

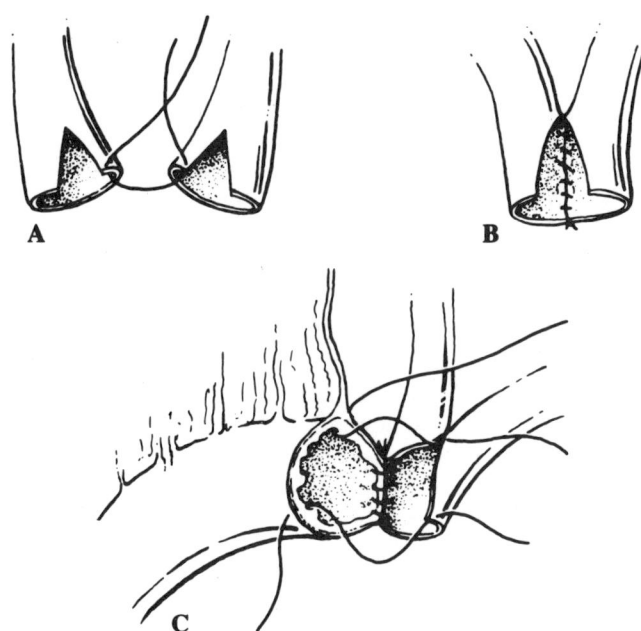

FIG 32–5.
A–C, ureteroileal anastomosis demonstrating the original Wallace technique. (From Roberts MS: Commentary: Ileal conduit diversion in adults, in Whitehead ED, Leiter E (eds): *Current Operative Urology,* ed 2. New York, Harper & Row Publishers, 1984, p 658. Used with permission.)

COMPLICATIONS

Since its first description by Bricker in 1950, ureteroileal diversion has become the prevailing technique for cutaneous diversion, replacing the ureterosigmoidostomy and its recognized incidence of complications. However, as might be expected, experience and time have documented a significant incidence of both late and early complications with this procedure as well. Among the early complications, wound infection and wound dehiscence are noted in 13.8%–25.3% of patients (Johnson et al., 1970; Daughtry et al., 1977; Johnson and Lamy, 1975; Sullivan et al., 1980) and intestinal obstruction in 6.1%–23.6% of cases (Laskowski et al., 1968; Jaffe et al., 1968; Sullivan et al., 1980). Most obstructions respond to conservative treatment, only 3% requiring surgery. The incidence of urinary leakage from the ureteroileal anastomosis may be as high as 11.3% (Malgieri and Persky, 1978), but most series report an incidence of 2.5%–4.4% (Cohen and Persky, 1967; Johnson and Lamy, 1970; Sullivan et al., 1980). The incidence of urinary leakage has been reduced by routine use of ureteral stents. As in ureterosigmoidostomy, small leaks seal spontaneously but major leaks should be treated immediately by inserting a percutaneous nephrostomy tube and facilitating drainage from the loop by inserting a catheter with multiple perforations. If leakage fails to diminish within 24 hours, the ureter should be reimplanted. The rate of early complications is higher in patients who have been irradiated. A high incidence of wound infection, drainage from the anastomosis, and prolonged ileus was reported in patients who received irradiation therapy prior to surgery (Cohen and Persky, 1967; Nichols et al., 1972; Swan and Rutledge, 1974; Malgieri and Persky, 1978). The mortality rate subsequent to ureteroileal diversion was about 20% in the early series (Burnham and Farrer, 1960; Glantz, 1966), decreasing to 2%–3.3% in recent series (Malgieri and Persky, 1978; Daughtry et al., 1977; Brannen et al., 1981). Among the recognized causes of death are pulmonary embolism, myocardial infarction, and infections.

Late complications are encountered in both children and adults (Table 32–2). Insight into delayed effects of urinary diversion on the kidneys, which may appear many years after surgery, has been derived mainly from studies on children followed up for ten years or longer (Schwarz and Jeffs, 1975; Middleton and Hendren, 1976; Shapiro et al., 1975; Smith, 1978; Pitts and Muecke, 1979). These studies show that renal deterioration visible on pyelography occurs in 18%–56% of patients. The incidence of renal damage is directly related to the status of the upper tract at time of diversion and to the length of follow-up. The longer the time interval, the higher the incidence of renal damage. Further renal deterioration in abnormal systems is more frequent than for normal kidneys (Smith, 1978; Pitts and Muecke, 1979; Orr et al., 1981). Smith (1978) showed that subsequent to diverting a normal upper tract, 7% of the renal units at five years, 10% at ten years, and 23% at more than ten years showed deterioration. In the initially abnormal systems, additional deterioration was documented in 16%, 20%, and 45%, respectively, at the same time intervals. Further renal injury to already damaged kidneys may cause serious or fatal complications (Pitts and Muecke, 1979; Richie, 1974). For these reasons, some authors believe that ur-

eteroileal diversion should not be performed when severe renal insufficiency is present but rather, other modes of urinary diversion without interposed bowel should be used (Creevy, 1960; Rickham, 1964; Richie, 1974).

Obstructions of the stoma, the conduit or the ureters are major factors leading to renal damage. The incidence of stomal obstruction in adults is about 5% (Johnson et al., 1970; Johnson and Lamy, 1975; Sullivan et al., 1980) and in children it ranges between 12% and 52% (Rickham, 1964; Smith, 1978). Stomal obstruction may cause back pressure affecting the upper urinary tract, with impairment of urine flow from the loop. In these cases the loop is elongated, not acting as a conduit but as a reservoir containing varying amounts of residual urine. Early diagnosis and revision of stomal stenosis are mandatory to prevent severe, irreversible renal damage. Obstruction may involve not only the stoma, but any part of the loop and, on rare occasions, the whole loop may become fibrotic and stenotic. This late complication was reported in 2%–6.3% of children. Its etiology is not clear, but various factors—including ischemia and infection—have been suggested as possible causes. Ureteral obstruction occurring chiefly at the ureteroileal anastomosis was noted among 7.7%–17.7% adults and 2%–22.3% children, typically on the left side, where the left ureter is brought through the sigmoid mesentery. Intravenous pyelography and antegrade pyelography help to localize the stricture. In adult patients who have had diversion for bladder cancer, tumor recurrence at the site of anastomosis or along the ureter is a possible cause for obstruction. Urine cytology and brush histology of the ureter may establish the diagnosis. Any obstruction subsequent to ileal conduit diversion incurs greater risks of stone formation and urinary infection. Stones develop in about 4% of adult patients and in about 10% of children, and acute episodes of pyelonephritis, in children as well as adults, occur in 10%–20%.

In about 15% of patients with renal deterioration, no

TABLE 32–2.

Complications Affecting the Kidneys in Patients With Ureteroileal Urinary Diversion

SOURCE (YR)	NO. OF PATIENTS	FOLLOW-UP	DETERIORATION ON IVP, %	STOMAL STENOSIS, %	URETERAL OBSTRUCTION, %	PYELONEPHRITIS, %	CALCULI, %
Schwarz and Jeffs (1975)	96 (Children)	2–16 yr	56	32.3	5.2	—	12.5
Middleton and Hendren (1976)	90 (Children)	1–18 yr	41	42	10	20	9
Shapiro et al. (1975)	90 (Children)	10–16 yr	18	38	22.3	16.7	8.9
Johnson et al. (1970)	181 (Adults)	—	29	4.5	7.7	3.9	3.9
Johnson and Lamy (1977)	214 (Adults)	>6 mo	—	5.1	17.7	13.3	2.5
Sullivan et al. (1980)	366 (Adults)	5–15 yr	—	5.1	14.7	19.2	4.0

obstruction is detected, and deterioration in such patients is attributed to infected refluxing urine (Middleton and Hendren, 1976). Experimental studies by Richie et al. (1975) have shown that infected refluxing urine in dogs with urinary diversion produces pyelonephritic changes in most of the kidneys examined. Owing to the increased risk of renal deterioration, all patients with ileal conduits should be followed up for the remainder of their lives by urine cultures, serum creatinine assay, and intravenous pyelography (or renal scan) every six months for the first two years and annually thereafter.

The cause for pyelonephritis, hydronephrosis, and stone formation is usually correctly attributed to obstruction, either at the ureteroileal anastomosis or at the site of the stoma. However, a number of patients develop recurrent pyelonephritis and stones without evidence of obstruction at any point in the urinary diversion. This has called into question the significance of reflux of contaminated urine from the conduit to the upper urinary tracts, and has prompted the recommendation for antireflux implantation into the ileum. Although this is technically feasible, the thin-walled ileum does not adequately support the antireflux tunnel; therefore, this procedure has not gained wide acceptance. Further, subsequent to ileal conduit most adults have excellent preservation of the upper urinary tract with a low rate of pyelonephritis and stone formation.

PERSPECTIVES

While the ileal conduit may not indeed be the ideal bladder substitute, it has stood the test of time and is well suited to the bladder cancer patient after cystectomy. With the use of bowel staplers, the ileal segment can be rapidly isolated and ureteral anastomoses accomplished easily and accurately. Most immediate postoperative complications can be prevented by meticulous surgical technique, adequate preoperative bowel preparation, use of perioperative prophylactic antibiotics, construction of a well-vascularized everted stoma, and use of delicate nontraumatic ureteral stents. Use of a short ileal segment exiting directly through the abdominal wall as well as meticulous anastomosis of the ureters should prevent most major complications. We further believe that routine use of small polyethylene stents for postoperative drainage will minimize the risks due to temporary partial disruption or leak at the anastomotic sites. Long-term complications remain a problem but are less difficult in adult patients with bladder cancer than in the pediatric group. It is important to note that severe short-term complications requiring reoperation are rare in the hands of the experienced surgeon, and patients can easily be fitted with a reliable appliance that requires changing only every 5–7 days.

Even though such patients must tolerate the wearing of an ostomy appliance, they do not need to empty a reservoir and usually resume almost all normal activities. For these reasons, the ileal conduit remains the most popular and frequently utilized method of urinary diversion subsequent to cystectomy for cancer.

COLONIC CONDUITS

For several significant reasons, the colon has been used as a cutaneous urinary conduit. First, the large lumen provides an excellent stoma, which rarely undergoes stomal stenosis. Second, a segment of the colon can be utilized that is well outside previous pelvic radiation fields. Third, the ureters can be implanted into the taenia of the colon in an antirefluxing manner. In rare cases in which viable ureteral length is insufficient to reimplant into an ileal segment, a segment of transverse colon can be directly implanted into each renal pelvis for effective drainage. In children, colonic conduit diversion is superior to the ileal conduit inasmuch as preventing reflux seems more important in children, and the large everted stoma minimizes the risks of hyperkeratosis and stomal stenosis, both major problems in childhood urinary diversion. However, the true significance of these factors in adults with bladder cancer is open to question, and selection of colonic segment urinary diversion must be based on its application in specific incidences.

TRANSVERSE COLON CONDUIT

Transverse colon is well out of the field of any pelvic and lower abdominal radiation, and is therefore suited to patients with significant radiation changes in the ileum. The ends of the transverse colon segment approximate the position of the renal pelves and can therefore be used for very high urinary diversion in patients who have compromised lower ureters. Schmidt et al. (1975) described the technique of transverse colon conduit initially utilizing a direct ureterocolonic anastomosis without attempting a nonrefluxing procedure. The transverse colon is easily isolated with its blood supply from the right colic artery (Fig 32–6,A). Bowel continuity is easily restored by using, as we prefer, a single full-thickness 3-0 silk anastomosis or, alternatively, the GIA bowel staples. When ureteral length allows, the left ureter can be brought through the mesentery to approximate the right ureter. Alternatively, using a longer segment, the left ureter can be anastomosed to the proximal end or side of the conduit. The base of the conduit must be fixed to the retroperitoneum. The ureters are then anastomosed in either a direct fashion or by an antirefluxing method (Fig 32–6,B). The conduit must be constructed so that the distal end (left) emerges as the stoma, in isoperstaltic fashion.

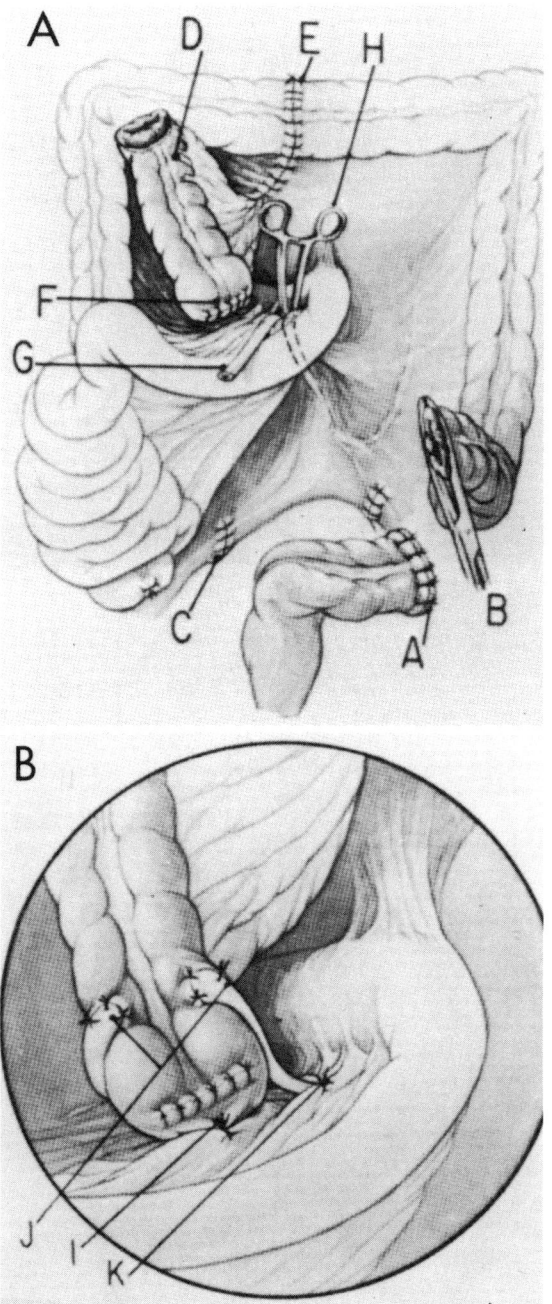

FIG 32–6.
Technique of transverse colon conduit diversion: *E,* bowel continuity restored. **A,** *A* and *B,* sigmoidal resection and colostomy are optional. *C,* peritoneal closure. *D* and *F,* segment isolated. *G* and *H,* ureters dissected. **B,** *I,* fixation of proximal segment. *J,* ureterocolic anastomosis. *K,* retroperitonealization. (From Schmidt JD, et al: *J Urol* 1975; 113:308. Used with permission.)

The method of ureteral implantation remains unresolved. In their initial series, Schmidt et al. (1975) retrospectively concluded that an antireflux anastomosis might have been a better choice. While prevention of reflux may be a laudable goal, it is clear that the risk of stenosis is increased. In most extreme need for urinary diversion, e.g., the patient with complete destruction of the ureters or severe fibrosis of both ureters, use of the transverse colon is the only effective method of diversion. The colon can be directly anastomosed end-to-end to the right renal pelvis and end-to-side to the left renal pelvis before exiting to the skin. In extreme cases, we have even anastomosed each end of the segment to each renal pelvis and fashioned a stoma from the side of the segment as close to the left kidney as possible.

SIGMOID COLON CONDUIT

Supravesical diversion with colonic segments was popularized by Mogg (1965), who emphasized the need for an antireflux ureteral implantation. He utilized a small nipple at the end of the implanted ureter, similar to the principle employed by others to prevent reflux. In fact, popularity of this and other colonic diversion methods was based primarily on the putative importance of preventing reflux. However, Elder et al. (1979) found no significant difference in the rate of upper tract deterioration with colon conduits in children compared to ileal conduits. Nonetheless, the most important application of the nonrefluxing sigmoid conduit is in children, and in patients for whom some type of reconstruction is possible or contemplated. For example, in a patient undergoing radical cystectomy for bladder cancer, the probability of long-term suvival is not high. Furthermore, the importance of preventing reflux, even though widely described in children, has not been established in adults. Clearly, however, if one wishes to prevent reflux, the use of the sigmoid segment is the most efficient method. An important aspect of this and other colonic diversion is the need to demonstrate a normal colon by barium enema, and perhaps, by colonoscopy.

A 15–20 in. segment of sigmoid colon is isolated with its blood supply (Fig 32–7,A and B). To gain mobility of the distal end, the mesocolon and its vessels must be incised all the way through, to include the superior hemorrhoidal artery. A major advantage of the colon conduit is the excellent blood supply to the descending colon; vascular compromise almost never occurs. The segment can then be placed lateral to the reconstituted descending colon, as illustrated in Figure 32–7,C, or medial to it. When the right ureter is shortened or compromised, the conduit is best left medial to the colon because the need for extensive mobilization of the right ureter is thereby minimized. The ureters are then

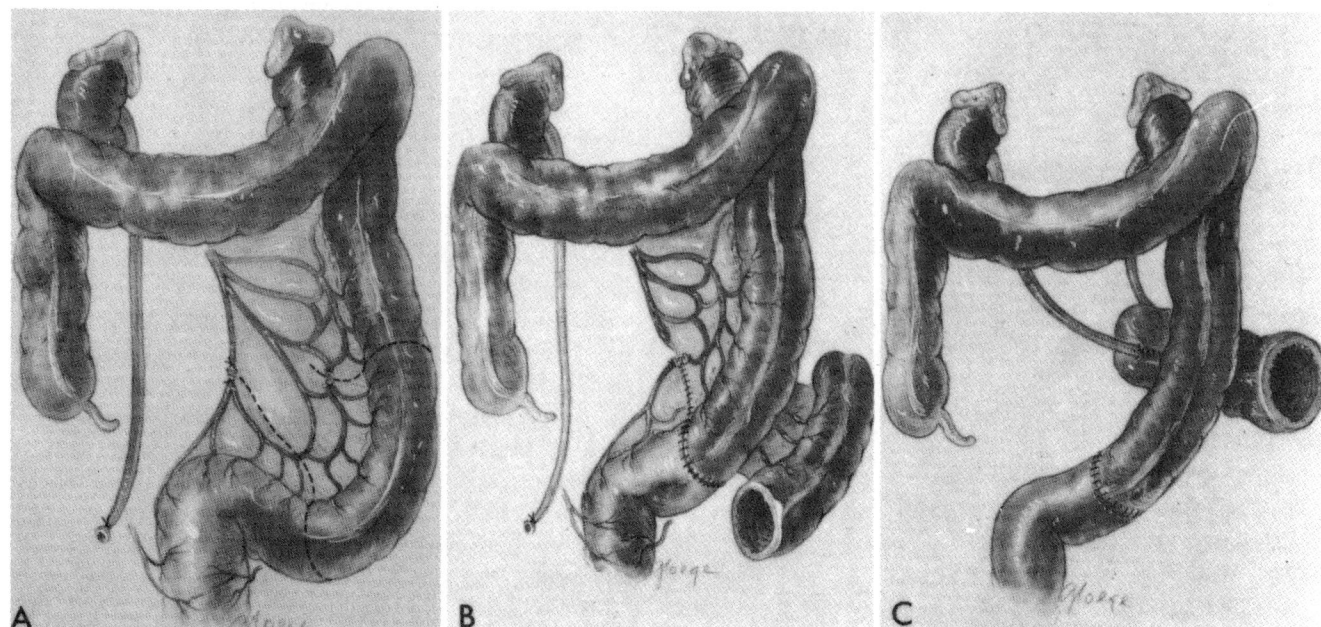

FIG 32–7.
Isolation of a suitable sigmoid segment for sigmoid urinary diversion. Note that the distal sigmoid mesentery is divided all the way to the sacral promontory, including division of the superior hemorrhoidal artery **(A).** This provides maximal mobility of the loop and allows it to function in an isoperistaltic manner. Division of the proximal mesentery should be quite short, and care should be taken to avoid injury to venous drainage from segment. The proximal end of isolated segment is closed, and standard bowel anastomosis is performed medial to the isolated segment **(B).** The right ureter is brought under the sigmoid mesentery **(C).** (From Richie JP, Skinner DG: Ureterointestinal diversion, in Walsh PC, et al (eds): *Campbell's Urology.* Philadelphia, WB Saunders Co, 1986, p 2612. Used with permission.)

implanted into the taenia in a manner similar to that described by Leadbetter for ureterosigmoidostomy (see Fig 32–2). The stoma is fashioned with an everted nipple, an important aspect of all intestinal cutaneous conduits. In Turnbull's stoma (Turnbull and Fazio, 1975) the bowel segment is pulled through the abdominal wall as a loop. Multiple myotomies are then made through the bowel, after which it is opened transversely, creating an effective stoma with minor chance of stenosis. However, the proximal segment occasionally protrudes, leading to difficulties in bag fitting and, in some cases, ureteral obstruction. We have not needed this method except in rare cases of an extremely obese patient in whom compromise of the mesenteric blood supply is a legitimate concern.

COMPLICATIONS

The deleterious effect of ileal conduit diversion on the kidneys has been attributed to the high rate of stomal stenosis and to free reflux of urine from the loop to the kidneys. The significance of this reflux has been demonstrated in experimental studies in dogs (Richie and Skinner, 1975). In these experiments the refluxing ileal loops were compared to nonrefluxing colon conduits. Of the kidneys connected to the ileal loops, 83%

had histological evidence of pyelonephritis compared to 7% of kidneys connected to colon conduits. Significant bacterial growth was present both in the ileal and colon loops, whereas the ureters connected to colonic conduits yielded fewer positive cultures than those connected to ileal conduits. The colon segment was introduced to achieve nonrefluxing ureterocolonic anastomosis and to avoid stomal complications. Early reports on colon conduits were encouraging, showing a low incidence of stomal complications (Morales and Golimbu, 1975; Althausen et al., 1978) (Table 32–3). Stomal stenosis occurred in 0%–2.8% of patients, pyelonephritis in 7.1%–17%, and renal deterioration in 8.6%–22.4% (see Table 32–3). However, the majority of patients included in these studies were followed up for a short period of time. Longer follow-up on patients in other series revealed a higher incidence of complications. Elder et al. (1979) noted that in a group of 41 children who have been followed from nine to 20 years (average, 13.2 years), a 61.5% incidence of stomal stenosis, 22% of ureterocolonic stenosis, and 48.4% of renal deterioration occurred (see Table 32–3). These data suggest that late stomal and renal complications are not prevented by using a colon conduit. A transverse colon segment used for urinary diversion was sug-

TABLE 32–3.

Complications of Colon Conduit Diversion

SOURCE (YR)	NO. OF PATIENTS	FOLLOW-UP, YR	STOMAL STENOSIS, %	URETEROCOLONIC STENOSIS, %	PYELONEPHRITIS, %	STONES %	RENAL DETERIORATION, %
Althausen et al. (1978)	70 (Children, adults)	2–8	2.8	8.6	7.1	4.3	8.6
Morales and Golimbu (1975)	46 (Children, adults)	>1–11	—	13	17	4.3	22.4
Elder et al. (1979)	41 (Children)	9–20	61.5	22	—	16	48.4

gested recently for patients who had undergone irradiation to the pelvis for pelvic malignancies (Schmidt et al., 1975; Beckley et al., 1982). The use of a loop of colon that is less likely to be injured by previous irradiation was associated with a low incidence of ureterocolonic leak (3.3%) and stomal complications (6.6%) (Beckley et al., 1982).

PERSPECTIVE

The use of colon conduits as a urinary diversion in children provides a different set of criteria than does their use in adults with bladder cancer, as discussed in this chapter. In the adult, the problem of stomal stenosis is not major, and can usually be avoided by creating a well-vascularized everted stoma. Prevention of reflux therefore remains the major attribute of the colon conduit. As noted above, the value of the antirefluxing ureteral reimplantation is unclear. The paucity of long-term results makes it possible to discern an actual advantage of antirefluxing procedures vs. standard ureteral implantation, even though a theoretical advantage may exist. Ureteral stenosis increases after antireflux reimplantation, although the impact of this problem in adults may not be significant. The undisputed indica-

tion for the colonic conduit occurs in the patient who has had extensive pelvic irradiation, with radiation damage to the distal ileum. The sigmoid or the transverse colon is preferred for urinary diversion in this situation.

KOCK POUCH RESERVOIRS

Cutaneous urinary diversions, including numerous modifications, still burden the patient with an external urinary collection device. The psychological impact and the altered body image, while not of paramount concern in many adult patients, may impair the life-style of some. In the past several decades, continent urinary reservoirs of various types have been utilized, especially continent ileocecal substitutions, but most of these have not become popular in the United States. In the past few years, encouraged by success with the continent intestinal reservoir, Kock and associates (1978) demonstrated in the laboratory the efficacy of a continent ileal pouch as a urinary reservoir. In their first clinical report, Kock et al. (1982) noted a high complication rate and need for reoperation. Since that time the procedure has been revised numerous times by several authors (Gerber, 1983; Skinner et al., 1984;

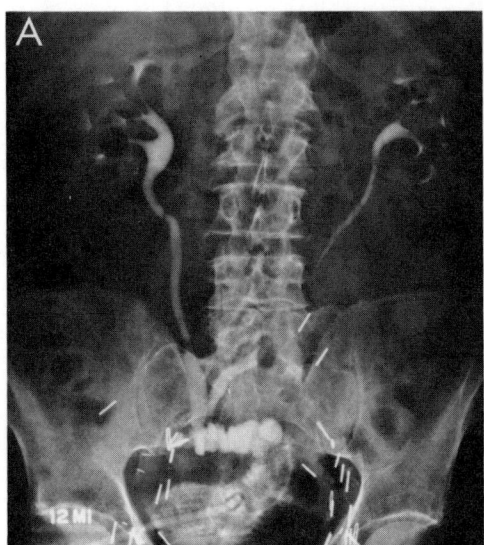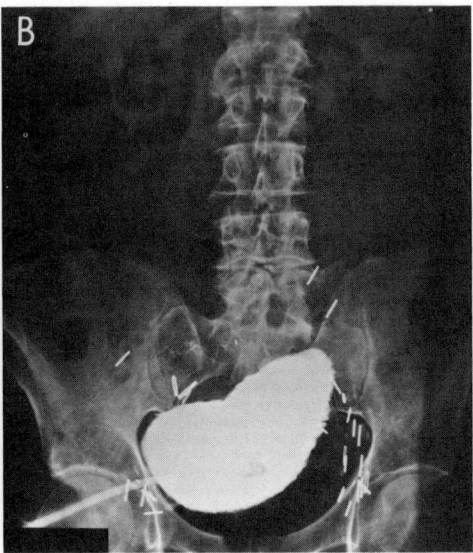

FIG 32–8.
Intravenous pyelography two years after Kock pouch urinary diversion showing a normal upper urinary tract **(A)**. No ureteral reflux is seen when the Kock pouch is filled with contrast material **(B)**.

deKernion et al., 1985). The procedure has already been employed by several centers with reports of good short-term results, offering an alternative to standard cutaneous diversion (Fig 32–8).

TECHNIQUE

Kock initially described the formation of a U-shaped pouch by approximating the open edges of an ileal segment. We have preferred to use an S-shaped segment 78 cm in length. The reservoir is formed by three ileal segments, each 13 cm long (Fig 32–9). After opening the ileum on the antimesenteric border, the edges are sutured together with 2-0 polyglycolic acid sutures in a single layer. The proximal and distal valves are then created by intussuscepting 12 cm of ileum at each end, thus making each valve approximately 6–7 cm long (Fig 32–10). Intussusception is usually facilitated by defatting the mesentery of the segment to be intussuscepted. However, in the case of the fatty mesentery, the mesentery can actually be divided flush with the serosa of the ileum as described by Hendren (1980). The valves are then further fixed with three rows of 4-mm GIA staples. A further method of preventing loss of the intussusception—a critical element in this operation—is the use of a strip of Marlex mesh 2 cm in diameter passed around the ileal segment through a window in the mesentery and then through a similar window in the mesentery of the reservoir (Fig 32–11). After suturing the ends of the mesh together, the mesh is sutured above and below to the seromuscular layers of the bowel.

Having established the proximal and distal valves, the ureters are anastomosed to the proximal segment and stented with no. 8 feeding tubes brought out

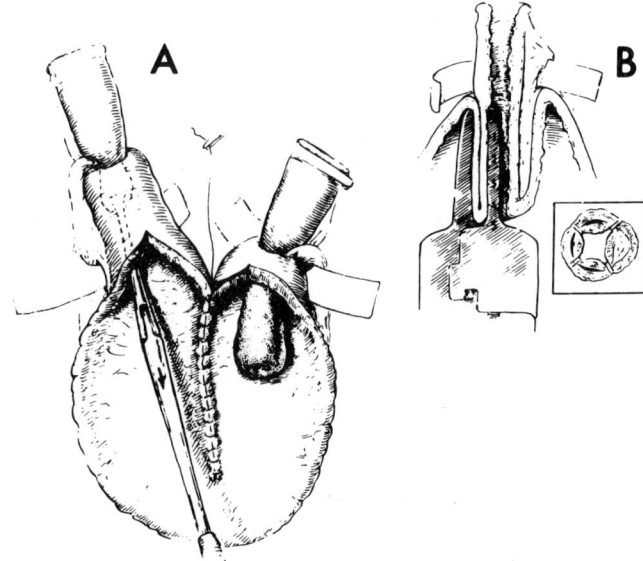

FIG 32–10.
A, after either defatting the mesentery or dividing it, proximal and distal valves are created by intussusception of ileal segment. **B,** fixation is secured with three or four rows of GIA staples. (From Kock NG, et al: Urinary diversion via a continent ileal reservoir. *J Urol* 1982; 128:469. Used with permission.)

through the pouch. This is important to prevent early postoperative urine leaks. The reservoir is then closed with a single running layer of polyglycolic suture. Prior to completing closure, a large catheter is passed into the reservoir and carefully sutured in place. Construction of the distal segment and its anastomosis to the abdominal wall are critical. If the segment is too long or if the Marlex mesh is sutured to the peritoneum, a tight band is created that causes the catheter to curl up, making reservoir catheterization difficult or impossible (Fig 32–12). We currently secure the Marlex mesh to the rectus fascia and use an extremely short (2–3 cm) segment flush with the skin. The opening in the abdominal wall can be created at any convenient point, and is often best placed very low in the right lower quadrant, well below the belt line. During the postoperative period, patency of the drainage catheter is critical and the catheter and stents are left in place for approximately three weeks. They must be irrigated frequently to prevent obstruction by mucus. After removing the catheters, patients are taught to catheterize the reservoir and instructed to do so initially every four hours. Later, many patients catheterize only 4–6 times a day. Many small changes and modifications of the procedure have been devised by every surgeon who performs the operation, most of which have been designed to (1) prevent disintussusception of the distal segment and (2) to improve the ability of the patient to catheterize the reservoir.

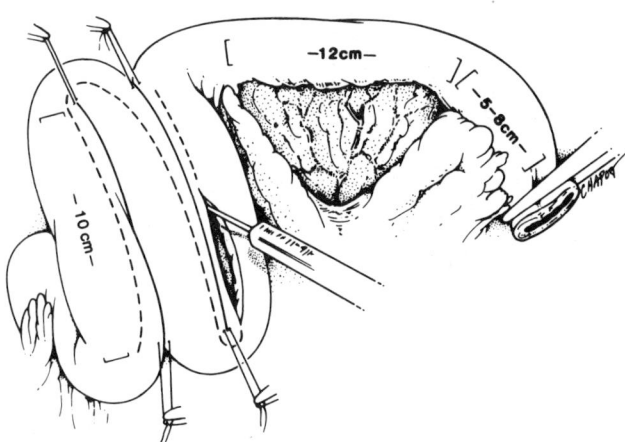

FIG 32–9.
A total of 78 cm of ileum is isolated. The reservoir is formed from an S-shaped configuration of the ileum, utilizing a 13-cm length in each arm, rather than 10 cm. The segment to be intussuscepted for the valve is stripped of mesenteric fat before intussusception. (From *Surg Gynecol Obstet* 1983; 156. Used with permission.)

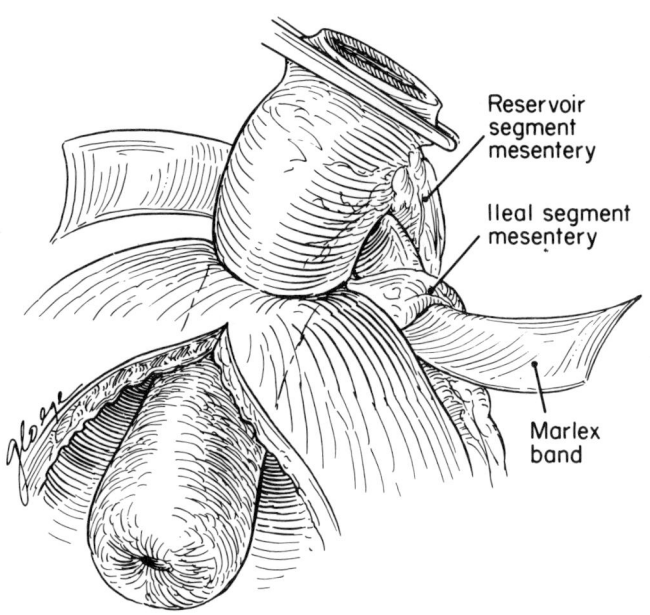

FIG 32–11.
The Marlex strip is passed through a small window in the mesentery of the reservoir and then through a similar window in the mesentery of the adjacent ileal segment. This traps the valve and helps prevent disintussusception. (From deKernion JB, et al: The Kock pouch as a urinary reservoir. *Am J Surg* 1985; 150:84. Used with permission.)

COMPLICATIONS

The Kock operation, while not requiring extensive skill and dissection, is tedious and demanding. In our hands, operative time is increased by 1.5–2 hours over the standard ileal conduit. Hospitalization is also increased by at least several days. In our first 21 patients (deKernion et al., 1985), we encountered the usual problems associated with the operation (Table 32–4), many of which have been resolved by modifying the procedure. In this early series one patient required reoperation two months postoperatively for ureteral obstruction caused by displacement of the pouch and resulting angulation of the ureter as it passed through the mesocolon. This patient represented our only postoperative death, which was due to massive pulmonary emboli ten days after the reoperation. Another patient had bilateral hydronephrosis due to ureteral fibrosis

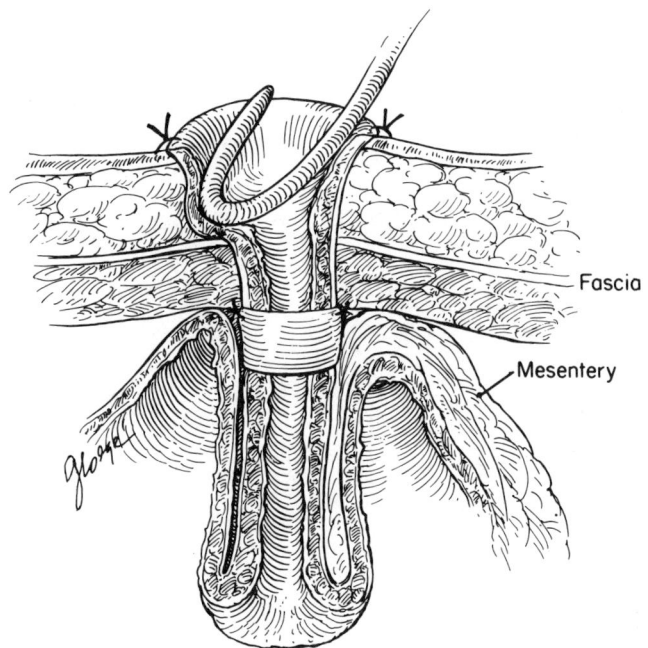

FIG 32–12.
Improper fixation of terminal segment makes catheterization difficult and catheter does not easily enter valve. A short segment, fixation of Marlex to external fascia, and fixation of reservoir to abdominal wall minimize difficulties with catheterization. (From deKernion JB, et al: The Kock pouch as a urinary reservoir. *Am J Surg* 1985; 150:86. Used with permission.)

caused by urinary extravasation and underwent conversion to an ileal conduit. Four patients required subsequent revision of the distal segment, one of whom underwent an emergency operation after failure to catheterize the pouch even while using endoscopic procedures. The pouch was ruptured at the time of exploration and the distal valve was revised. Two patients had revision of the distal segment, usually involving shortening of the segment, and refixation of the distal segment after loss of the intussusception. One of these two patients returned one year after surgery with erosion of the Marlex into the valve. The others have done well and remain continent. Two patients have developed stones in the pouch, one with large stones that could not be retrieved endoscopically (Fig 32–13). Al-

TABLE 32–4.

Complications of Kock Pouch Diversion

SOURCE (YR)	NO. OF PATIENTS	REVISION OF DISTAL VALVE, %	URINARY LEAK, %	REFLUX, %	HYDRONEPHROSIS, %	PYELONEPHRITIS, %	BACTERIURIA, %
Kock et al. (1982)	12	58.3	8.3	25	—	8.3	41.7
deKernion et al. (1985)	21	19.0	4.8	—	4.8	4.8	—
Skinner et al. (1984)	51	9.8	15.7	2	—	5.9	43.1

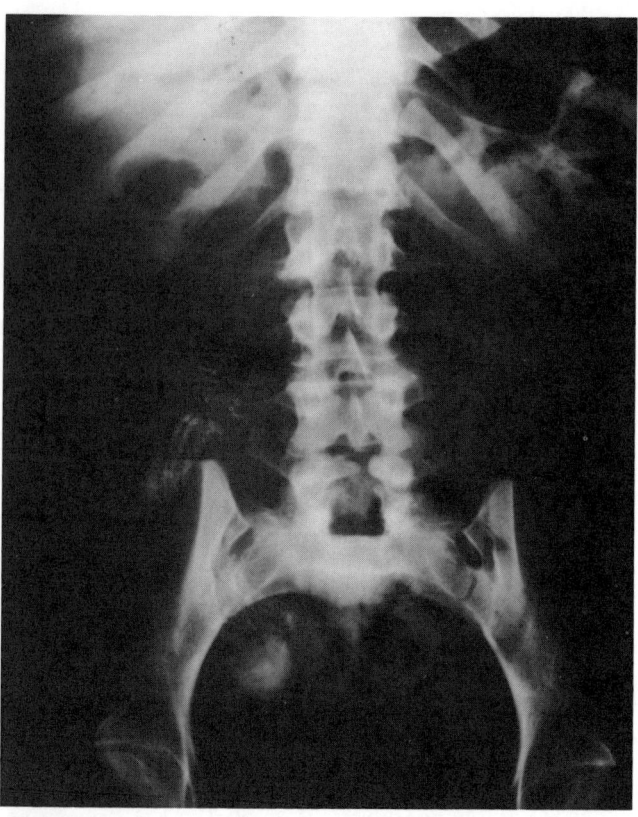

FIG 32–13.
Plain x-ray film of the abdomen six months after Kock pouch urinary diversion showing two stones in the pouch overlying the staples.

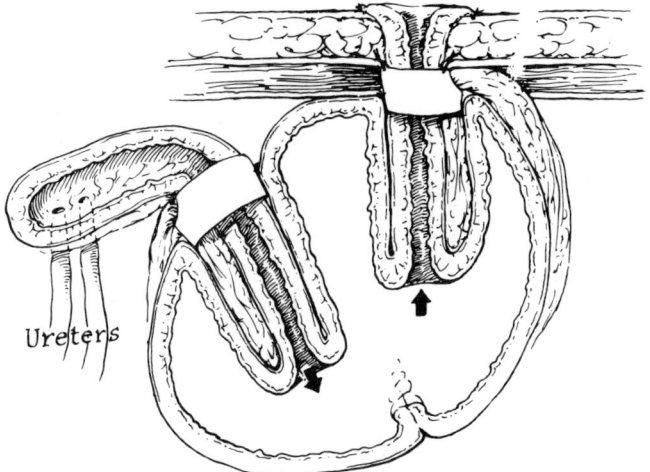

FIG 32–14.
The distal segment is made as short as possible and secured flush with abdominal wall. Then Marlex mesh must be pulled into abdominal wall and sutured to external fascia. The reservoir is then fixed to abdominal wall with several circumferentially placed sutures. (From deKernion JB, et al: The Kock pouch as a urinary reservoir. *Am J Surg* 1985; 150:85. Used with permission.)

though continence of the distal valve is usually satisfactory, some patients have some leakage requiring that a small pad be worn over the opening.

Except for the two patients with ureteral obstruction, the major problems following the Kock procedure have been related to the distal valve and the terminal segment, mainly owing to inability to catheterize the segment. This is often due to a long distal segment and/or loss of the intussusception. A very short terminal segment within the abdominal wall is critical. The stoma must be no more than several centimeters long, just enough to be sutured flush with the skin. Proper passage of the Marlex band through the mesentery also retards loss of the intussusception. The Marlex must be passed through the mesentery of both the reservoir and the distal spout before wrapping it around the ileum (see Fig 32–11), trapping the terminal segment at the junction with the reservoir, thus preventing disintussusception. The Marlex mesh must be pulled well into the rectus muscle and sutured to the anterior rectus fascia, as initially described by Kock (Fig 32–14). If the segment is too long or the Marlex is not fixed properly, the catheter tends to curl out of the segment at the fascial opening or the rigid ring formed by the Marlex

mesh, leading to further edema and swelling, thus making catheterization extremely painful or impossible (see Fig 32–12). The proximal segment and the antireflux valve appear to work well. We have had no cases of hydronephrosis except for the patient cited above. The use of ureteral stents consisting of no. 8 polyethylene infant feeding tubes for a minimum of three weeks promotes healing and bypasses much of the urine from the reservoir, further assuring fixation of the valve.

The major short-term complications have diminished with further experience and nuances of technique. Even in larger series, major reoperation is required in as many as 10% of patients, considerably greater than the early reoperation rate for the standard ileal conduit. Furthermore, the increased length of surgery is demanding, both on the surgeon and the patient. Finally, the long-term complications are as yet unknown. As noted above, stone formation is not uncommon and may prove to be an increasingly difficult problem. In the case of the ileal conduit, major long-term problems did not surface until 5–10 years after surgery and the true risks of infection, hydronephrosis, and severe pyelonephritis due to such factors as stasis have not been identified subsequent to the Kock pouch operation.

PERSPECTIVES

The limitations and shortcomings of the Kock pouch have been detailed above. Even after the serious difficulties are overcome with experience, a significant reoperation rate still exists. Most important, the long-term clinical success rate is completely unknown. Although the Kock pouch incorporates an antireflux mechanism, as noted in the previous discussion, this has not been shown to be a definite deterrent to long-term complications. Therefore, at present, the only known advantage of the Kock pouch is that it frees the patient from wearing an external urinary appliance, and thereby may improve the quality of life. Faced with the prospect of having to wear such a device, many cancer patients will choose a continent reservoir unless the options are placed in balanced perspective. Modern enterostomal therapy provides excellent external devices that obviate all concerns except for the need to change the appliance every 5–7 days. The patient with continent internal reservoir trades this option for the need to catheterize the segment at regular intervals, along with valve incontinence and the possibility of sudden difficulty catheterizing the pouch.

We currently offer our patients the Kock pouch as an option. Most patients, faced with a deadly cancer, however, choose the standard and direct method of diversion, the ileal conduit. The Kock pouch probably has its paramount role in young patients or in those concerned about the impact of an external device on sexual function. Over a four-year period, we have had only one patient with an ileal conduit express any degree of interest in having it converted to a Kock pouch, and most patients are satisfied with the simple external collection device. Clearly, after an initial surge of enthusiasm by the general public, the Kock pouch is finding its appropriate place as a method of urinary diversion. It is appropriate in a selected number of patients who, for psychological reasons, do not wish to have an external device; whether it will seriously compete with the ileal conduit in the cancer patient is doubtful.

CAMEY PROCEDURE

Whether or not a cutaneous urinary diversion is continent, free-flowing, and antireflux, patients still must contend with the altered body image associated with an external stoma and the subsequent complications following such diversions. The ideal bladder replacement is one in which the reservoir is anastomosed directly to the urethra. The reservoir should be adequate to contain a reasonable urine volume, should be antirefluxing (since intraluminal pressure will develop secondary to competent sphincter), and continence must be continued by preserving the external sphincter mechanism. In 1979, Camey and Le Duc reported their experience with 90 patients who underwent construction of a functional ileal bladder after radical cystectomy for bladder cancer. The results were excellent and have been duplicated by other centers in France. The procedure was introduced into the American literature by Lilien and

Camey (1984) and has been performed in a limited number of patients in several centers in this country.

TECHNIQUE

A major component of the Camey procedure is the careful dissection of the urethra from the apex of the prostate gland. This requires meticulous dissection of the urethra with thorough hemostasis, since hematomas can disrupt the delicate anastomosis of the urethra with the bowel. A 35–40 cm segment of distal ileum is then isolated with its blood supply. An opening is created on the antimesenteric border and this 1-cm opening is then sutured to the end of the transected urethra (Fig 32–15) with 6–10 absorbable sutures. On rare occasions, the ileal segment does not reach the urethra, making the procedure impossible.

An interesting adaptation of antireflux surgery was described by Camey and Le Duc (1979). The ureter is brought into the lumen of each end of the bowel; no attempt is made to create an antireflux tunnel. Instead, the mucosa is simply incised and a 3-cm segment of ureter is laid in the trough. The mucosa is then tacked to each side of the ureters and, presumably, eventually grows over the ureteral segment (see Fig 32–15). Each ureter is then stented with a polyethylene stent, which we prefer to bring through a small stab wound in the ileum and then out through the skin. The drainage of the bowel segment is critical inasmuch as a consider-

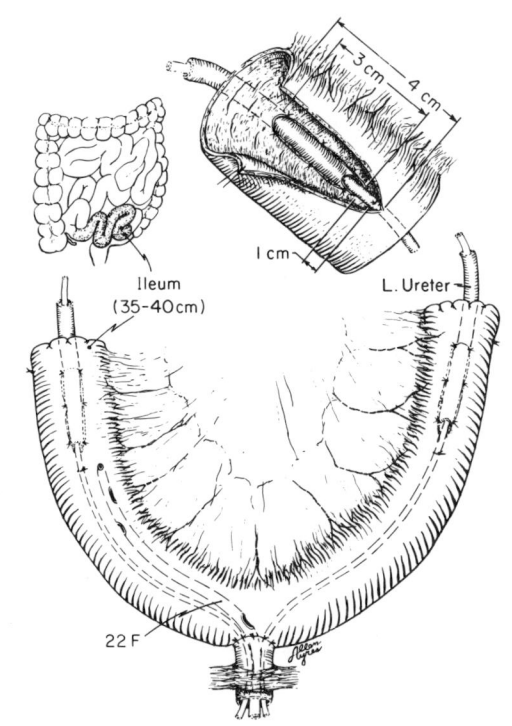

FIG 32–15.
Surgical technique. (From Lilien BM, Camey M: 25 years' experience with replacement of the human bladder (Camey procedure). *J Urol* 1984; 132:886. Used with permission.)

able amount of mucus forms that can obstruct the flow of urine with subsequent disruption of the urethroileal anastomosis. Each limb of the segment is then fixed to the psoas muscle and the pelvis is thoroughly drained.

COMPLICATIONS

In his early experience, Camey preserved the apex of the prostate gland to further ensure continence and facilitate the urethroileal anastomosis. However, this poses the risks of prostate carcinoma or recurrence of urethral tumor for the patient and has been abandoned. Careful division of the urethra at the prostate is sufficient to ensure continence without leaving prostatic tissue. The early major complications usually involve pelvic hemorrhage with disruption of the anastomosis, or urinary leak from other causes with abscess formation. With sufficient experience and attention to detail, such complications can usually be prevented. Pyelonephritis can occur during the postoperative period, but is usually avoided by leaving the indwelling stents for at least three weeks during the healing of the ureteroileal anastomosis.

Camey and Le Duc's method of antireflux implantation of the ureters has proved to be extremely effective. We have recently performed three such operations and no patient has had reflux. Indeed, the upper urinary tracts usually appear pristine after the operation. Camey and his co-workers are to be commended for the cautious way in which they documented the results of their operation, accumulating a large series of patients with long-term follow-up data prior to reporting the results. From their experience, the long-term protection of the upper urinary tracts is excellent (Lilien and Camey, 1984). They also have carefully and accurately reported the major shortcoming of the procedure—nighttime incontinence. In their series, incontinence in some degree occurred in 28 patients who had long-term follow-up, and perhaps in some others in whom long-term documentation was not available. While this remains a major difficulty, it can be minimized by having the patient empty the reservoir every three hours at night. Other patients are quite satisfied to wear an external collection device at night as an acceptable trade-off for the need to have a permanent external urinary bag. Daytime incontinence occurred rarely in Camey's series and more than 90% achieved satisfactory control during the daytime hours. This was facilitated by having the patient empty the reservoir every 2–3 hours and, in some cases, to use a double voiding technique. However, daytime incontinence has been reported in as high as 20% of patients. We have also seen this problem, as have Benson et al. in their series of patients at the Mayo Clinic (personal communication, 1985).

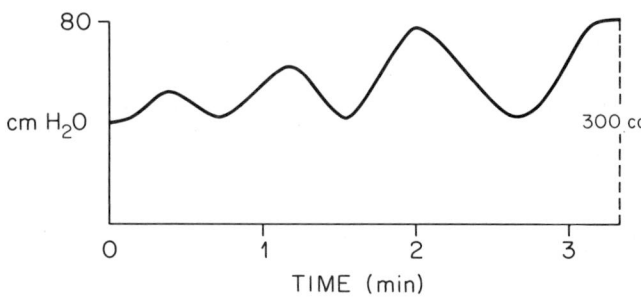

FIG 32–16.
Schematic drawing of the true intraluminal pressure (reservoir pressure minus abdominal pressure) of the Camey loop. During the filling phase with a rate of 100 cc/min of saline several rises of pressure were recorded. The maximum pressure was 80 cm H_2O at a volume of 300 cc of saline.

Whether this is due to the necessary "learning curve" or inadequate preservation of the urethra is unclear. Undoubtedly, the intraluminal pressure in the ileal segment can be very high in some patients and difficult to control by the remaining sphincter mechanism. In one of our patients who has significant daytime incontinence, the measured intraluminal pressure was 80 cm H_2O (Fig 32–16). Vigorous peristaltic waves could be seen through the ileal segment at all times (Fig 32–17). As a solution to this problem Raz (unpublished data, 1986) has recommended a method of incision of the circular fibers of the bowel by side-to-side anastomosis of part of the ileal segment. This and other methods may prove to be more effective than Camey's original U-shaped ileal segment, although the benefits are still theoretical.

PERSPECTIVE

The concept of urethrointestinal anastomosis is now new, and a number of bowel segments have been utilized. Hradec (1965) compared various bowel segments as ileal replacements and concluded that the sigmoid colon or the ileocolic segment was superior to the small bowel because of the larger capacity and purported lesser degrees of nocturnal incontinence. Anastomosis of sigmoid segments to the retained bladder neck was reported by Khafagy et al. (1965) in patients undergoing cystectomy for squamous cell carcinoma induced by schistosomiasis. Continence was preserved and excellent results were achieved but this procedure would not be appropriate for patients with bladder cancer. A number of authors have recently utilized large bowel segments with satifactory results (Kerr et al., 1960; Turner-Warwick and Ashken, 1967), though anastomosis directly to the urethra is still associated with considerable incidence of nighttime incontinence and some incidence of daytime incontinence.

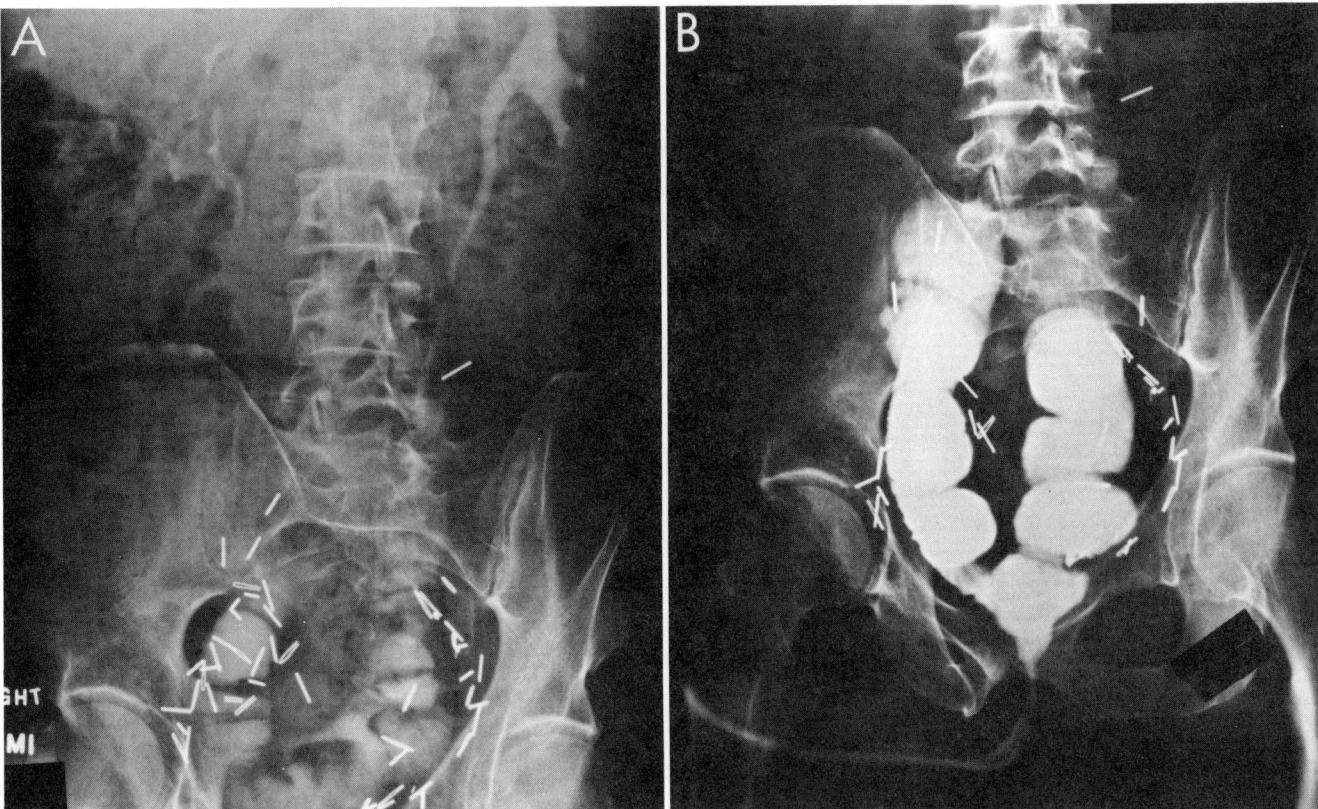

FIG 32–17.
Intravenous pyelography three months after Camey urinary diversion showing a normal upper tract **(A)**. Peristaltic waves are clearly demonstrated on the retrograde loopography **(B)**.

Despite the problems associated with urethrointestinal anastomosis, this approach remains the theoretical best method of postcystectomy urinary diversion. When combined with sparing of the corporeal nerves to the penis, the best possible result can be achieved, with complete restoration of the patient's body image and retention of sexual function. However, the operation is inappropriate in patients with carcinoma in situ and tumors near the bladder neck. This excludes a large population of bladder cancer patients. Moreover, unless the bladder neck is preserved (an inappropriate deviation from standard surgical procedure in transitional cell carcinoma patients), the procedure is unacceptable in females, since complete urinary incontinence will result. Finally, the high incidence of nighttime incontinence is unacceptable to many patients and may represent the major deterrent to the wide adoption of the procedure. Whether this can be overcome by modifying the operation is uncertain. Judicious modifications of the procedure, in some instances combined with use of an external artificial sphincter, may improve acceptance of the procedure, representing a major advance in achieving quality of life for the bladder cancer patient. At present, we carefully select patients for this procedure, but faced with the reality of its shortcomings, many patients still prefer a standard method of external diversion.

REFERENCES

1. Althausen AF, Hagen-Cook K, Hendren WH III: Nonrefluxing colon conduit: Experience with 70 cases. *J Urol* 1978; 120:35.
2. Barzilay B: Experimental study of the technique of uretero-ileal anastomosis. *J Urol* 1960; 83:612.
3. Barzilay B, Goodwin WE: Clinical application of an experimental study of uretero-ileal anastomosis. *J Urol* 1968; 99:35.
4. Beckley S, Wajsman Z, Pontes JE, et al: Transverse colon conduit: A method of urinary diversion after pelvic irradiation. *J Urol* 1982; 128:464.
5. Brannen W, Fuselier HA Jr, Ochsner M, et al: Critical evaluation of one-stage cystectomy-reducing morbidity and mortality. *J Urol* 1981; 125:640.
6. Bricker EM: Bladder substitution after pelvic evisceration. *Surg Gynecol Obstet* 1950; 30:1511.
7. Burnham JP, Farrer J: A group experience with uretero-ileal-cutaneous anastomosis for urinary diversion: results and complications of the isolated conduit (Bricker procedure) in 96 patients. *J Urol* 1960; 83:622.

8. Camey M, Le Duc A: L'enterocystoplastie avec cysto-prostatectomie totale pour cancer de la vessie. *Ann Urol* 1979; 13:114.

9. Coffey RC: Physiologic implantation of the severed ureter or common bile duct into the intestine. *JAMA* 1911; 56:397.

10. Cohen SM, Persky L: A ten-year experience with uretero-ileostomy. *Arch Surg* 1967; 95:278.

11. Creevy CD: Renal complications after ileal diversion of the urine in non-neoplastic disorders. *J Urol* 1960; 83:394.

12. Crissy MM, Steele GD Jr, Gittes RF: Carcinoma in colonic urinary diversion in rats. *Surg Forum* 1979; 30:554.

13. Daughtry JD, Susan LP, Stewart BH, et al: Ileal conduit and cystectomy: A 10 year retrospective study of ileal conduits performed in conjunction with cystectomy and with a minimum 5-year follow-up. *J Urol* 1977; 118:55.

14. deKernion JB, DenBesten L, Kaufman JJ, et al: The Kock pouch as a urinary reservoir: Pitfalls and perspectives. *Am J Surg* 1985; 150:83.

15. Elder DD, Moisey CU, Rees RWM: A long-term follow-up of the colonic conduit operation in children. *Br J Urol* 1979; 51:462.

16. Gerber A: The Kock continent ileal reservoir for supravesical urinary diversion. *Am J Surg* 1983; 146:15.

17. Glantz GM: Cystectomy and urinary diversion. *J Urol* 1966; 96:714.

18. Goodwin WE, Harris AP, Kaufman JJ, et al: Open, transcolonic ureterointestinal anastomosis: New approach. *Surg Gynecol Obstet* 1953; 97:295.

19. Hammer E: Cancer du colon sigmoide dix ans après implantation des uretères d'une vessie exstrophiée. *J Urol Nephrol* 1929; 28:260.

20. Hendren WH: Reoperative ureteral reimplantation: management of the difficult case. *J Pediatr Surg* 1980; 15:770.

21. Hoffman WW, Spence HM: Management of exstrophy of the bladder. *South Med J* 1965; 58:436.

22. Hradec EA: Bladder substitution: Indications and results in 114 operations. *J Urol* 1965; 94:406.

23. Jaffe BM, Bricker EM, Butcher HR Jr: Surgical complications of ileal segment urinary diversion. *Ann Surg* 1968; 167:367.

24. Johnson DE, Jackson L, Guinn GA: Ileal conduit diversion for carcinoma of the bladder. *South Med J* 1970; 63:1115.

25. Johnson DE, Lamy SM: Complications of a single stage radical cystectomy and ileal conduit diversion: Review of 241 cases. *J Urol* 1975; 117:171.

26. Kelalis PP: Urinary diversion in children by the sigmoid conduit: Its advantages and limitations. *J Urol* 1974; 112:666.

27. Kerr WK, Keresteci AG, Kyle VN: Ileocystoplasty: A clinical review of 18 cases. *Can J Surg* 1960; 3:134.

28. Kock NG, Nilson AE, Nilson LO, et al: Urinary diversion via a continent ileal reservoir: Clinical results in 12 patients. *J Urol* 1982; 128:469.

29. Kock NG, Nilson AE, Norlen L, et al: Changes in renal parenchyma and the upper urinary tracts following urinary diversion via a continent ileum reservoir: An experimental study in dogs. *Scand J Urol Nephrol* 1978; 49(suppl):11

30. Khafagy M, El-Bolkainy MN, Borsoum RS, et al: The ileocecal bladder: A new method for urinary diversion after radical cystectomy (a preliminary report). *J Urol* 1975; 113:314.

31. Laskowski TZ, Scott R Jr, Hudgins PT: Combined therapy: Radiation and surgery in the treatment of bladder cancer. *J Urol* 1968; 99:733.

32. Leadbetter WF: Consideration of problems incident to performance of ureteroenterostomy: Report of a technique. *Trans Am Assoc Genitourinary Surg* 1950; 42:39.

33. Leadbetter GW Jr, Zickerman P, Pierce E: Ureterosigmoidostomy and carcinoma of the colon. *J Urol* 1979; 121:732.

34. Lillien OM, Camey M: 25-year experience with replacement of the human bladder (Camey procedure). *J Urol* 1984; 132:886.

35. Malgieri JJ, Persky L: Ileal loop in the treatment of radiation-treated pelvic malignancies: A comparative review. *J Urol* 1978; 120:32.

36. Mogg RA: The treatment of neurogenic urinary incontinence using the colonic conduit. *Br J Urol* 1965; 37:681.

37. Mathiesen W: New method for uretero-intestinal anastomosis. *Surg Gynecol Obstet* 1953; 96:255.

38. Middleton AN Jr, Hendren WH: Ileal conduit in children at the Massachusetts General Hospital from 1955 to 1970. *J Urol* 1976; 115:591.

39. Morales P, Golimbu M: Colonic urinary diversion: 10 years of experience. *J Urol* 1975; 113:302.

40. Nichols WK, Krause AH, Donegan WL: Urinary fistulas after ureteral diversion. *Am J Surg* 1972; 124:311.

41. Orr JD, Shand JEG, Watter DAK, et al: Ileal conduit urinary diversion in children: An assessment of the long-term results. *Br J Urol* 1981; 53:424.

42. Pitts WR Jr, Muecke EC: A 20 year experience with ileal conduits: the fate of the kidneys. *J Urol* 1979; 122:154.

43. Richie JR: Intestinal loop urinary diversion in children. *J Urol* 1974; 111:687.

44. Richie JP, Skinner DG: Urinary diversion: The physiological rationale for non-refluxing colonic conduits. *Br J Urol* 1975; 47:269.

45. Rickham PP: Permanent urinary diversion in childhood. *Ann R Coll Surg* 1964; 35:84.

46. Rivard JY, Bedard A, Dionne L: Colonic neoplasms following ureterosigmoidostomy. *J Urol* 1975; 113:781.

47. Schmidt JD, Hawtrey CE, Buchsbaum HJ: Transverse colon conduit: A preferred method of urinary diversion for radiation-treated pelvic malignancies. *J Urol* 1975; 113:308.

48. Schwarz GR, Jeff RD: Ileal conduit urinary diversion in children: Computer analysis of follow-up from 2 to 16 years. *J Urol* 1975; 114:295.

49. Shapiro SR, Lebowitz R, Colodny AH: Fate of 90 children with ileal conduit urinary diversion a decade later: Analysis of complications, pyelography, renal function and bacteriology. *J Urol* 1975; 114:289.

50. Skinner DG, Boyd SD, Lieskovsky G: Clinical experience with the Kock continent ileal reservoir for urinary diversion. *J Urol* 1984; 132:1101.

51. Smith D: The long-term renal outlook following ileal conduit diversion. *Br J Urol* 1978; 50:69.

52. Stewart M, Hill MJ, Pugh RCB, et al: The role of *N*-nitrosamine in carcinogenesis at the uretero-colic anastomosis. *Br J Urol* 1981; 53:115.

53. Sullivan JW, Grabstald H, Whitmore WF Jr: Complications of ureteroileal conduit with radical cystectomy: Review of 36 cases. *J Urol* 1980; 124:797.

54. Swan RW, Rutledge FN: Urinary conduit in pelvic cancer patients. *Am J Obstet Gynecol* 1974; 119:6.

55. Turnbull RB Jr, Fazio V: Advances in the surgical technique and ulcerative colitis surgery, in Nyhis L (ed): *Surgery Annual*. New York, Appleton-Century-Crofts, 1975, p 315.

56. Turner-Warwick RT, Ashken MH: The functional results of partial, subtotal and total cystoplasty with special reference to ureterocaecocystoplasty, selective sphincterotomy and cystocystoplasty. *Br J Urol* 1967; 39:3.

57. Wallace DM: Ureteric diversion using a conduit: A simplified technique. *Br J Urol* 1966; 38:522.

58. Walsh A: Urinary diversion in malignant disease, in Ashken MJ (ed): *Urinary Diversion*. New York, Springer-Verlag, 1982, pp 75–100.

59. Wear JB Jr, Barquin OP: Ureterosigmoidostomy. *Urology* 1973; 1:192.

60. Williams DF, Burkholder GV, Goodwin WE: Ureterosigmoidostomy: A 15 year experience. *J Urol* 1969; 101:168.

61. Zincke H, Segura JW: Ureterosigmoidostomy: Critical review of 173 cases. *J Urol* 1975; 113:324.

Index